WHO'S WHO IN ENGINEERING

Editor: Gordon Davis
Publisher: Daniel De Simone

SEVENTH EDITION

Published by

AMERICAN ASSOCIATION OF ENGINEERING SOCIETIES, INC.

415 Second Street, Suite 200, Washington, DC 20002

As it is the objective of the American Association of Engineering Societies, Inc. publications to be a forum for the free expression and interchange of ideas, the opinions, positions, and research stated in published works are those of the original researchers and authors and not by the fact of publication necessarily those of the American Association of Engineering Societies, Inc., or related bodies. Contributing researchers and authors are authors are requested and expected to disclose any financial, economic, or professional interests or affiliations that may have influenced positions taken or advocated in these materials.

International Standard Book Number: 0-87615-015-6

The material herein was furnished directly by the individuals. Naturally, only a fraction of their capabilities and backgrounds is included.

Criteria for inclusion in Who's Who in Engineering

An applicant for *Who's Who in Engineering* must conform to one of the following:

- A baccalaureate or higher degree in engineering
- A registered engineer in one or more states
- Achieve more than the first professional membership grade of an engineering society
- Be living and meet the requirements or definitions in one of more of the following:

Member of the National Academy of Engineering

National Engineering and Related Scientific Societies

All current and former officers, directors and executive directors. Includes the senior officer of state or regional chapters.

Honorary Members. This term varies, but applies to grades where the member cannot apply.

All winners of national engineering society prizes, medals and other similar awards, achieved in national competition. *Does not* include students.

Educators

Presidents of engineering colleges
Deans of engineering colleges
Heads of engineering departments in colleges and universities
Tenured Professors

Engineers in Industry

President, vice president or chief engineer of company
Chief engineer or head of major engineering division which has at least 20 professional level engineers or physical scientists.

Consultants

Principals of major consulting engineering firms
Heads of major divisions of large consulting engineering firms which have at least 20 professional level engineers or physical scientists.

Engineers in Government

Civilian or military. Must be in charge of a division of government service with at least 20 professional level engineers or physical scientists.

Other Eminent Engineers

Such other persons who, in the judgment of the American Association of Engineering Societies, have made a substantial contribution to the engineering community and thereby warrant their inclusion in *Who's Who in Engineering*.

PREFACE

The growing importance of engineering in national affairs was a major factor in the formation of the American Association of Engineering Societies, whose purpose is to advance the knowledge and practice of engineering in the public interest. The engineering societies that comprise AAES represent over one-half million engineers in government, industry and the universities.

This volume lists members of the engineering profession who have distinguished themselves in the practice of engineering, the development of engineering resources, and the promotion of technological innovation in the public interest. They are men and women who have extended mankind's capabilities and, in their different ways, contributed to social, economic, and technological progress. Too often these achievements are described as feats of science rather than engineering, and this is due, primarily, to a lack of public understanding and appreciation of the nature and essential role of engineering in society. Yet engineering is indispensable to the quest for greater national productivity and a better quality of life. It is our hope that this volume will help to promote greater public awareness of the importance of engineering and the people who perform this essential role.

We acknowledge and appreciate the many forms of cooperation and assistance from individual engineers, corporations, societies, government agencies, and universities that made this Sixth Edition of *Who's Who In Engineering* possible. And finally, our thanks to Mr. Gordon Davis, Director of Publications at AAES, who edited the volume and shepherded it from start to finish, with abiding care and dedication.

Daniel De Simone
Executive Director
American Association of Engineering Societies

CONTENTS

SOCIETY INDEX

Awards Index

Awards Index

Awards Index

Awards Index

Awards Index

ACCREDITATION BOARD FOR ENGINEERING AND TECHNOLOGY (ABET)

345 East 47th Street, New York, New York 10017
David R. Reyes-Guerra, Executive Director
PRESIDENTS ... Russel C. Jones 1987-88; Gordon H. Geiger 1986-87; Gene M. Nordby 1985-86; Gordon H. Millar 1983-85; Leland J. Walker 1981-83; Richard G. Cunningham 1980-78; P. F. Allmendinger 1978-77; Robert B. Beckmann 1976-75; R. A. Forberg 1974-73.

AWARDS

LINTON E. GRINTER DISTINGUISHED SERVICE AWARD ... the award, instituted in 1972, acknowledges outstanding contributions to the engineering education through the work of individuals in ABET related activities. The award may be conferred annually.
Recipients: Paul F. Allmendinger 1986; Leland J. Walker 1985; George Burnet 1984; Richard G. Cunningham 1983; Gene M. Nordby; William H. Corcoran 1982; Robert B. Beckman 1980; Ernst Weber 1978; William P. Kimball 1976; Harold L. Hazen 1975; William L. Everitt 1973; Linton E. Grinter 1972.

ACOUSTICAL SOCIETY OF AMERICA (ASA)

500 Sunnyside Blvd., Woodbury, New York 11797
Murray Strasberg, Secretary
PRESIDENTS ... Chester M. McKinney, 1987; Ira Dyer 1986; Floyd Dunn 1985; Daniel W. Martin 1984; Frederick H. Fisher 1983; David T. Blackstock 1982; David M. Green 1981; Tony F. W. Embleton 1980; Henning E. von Gierke 1979; James L. Flanagan 1978; John C. Snowdon 1977; Kenneth N. Stevens 1976; Robert S. Gales 1975; Murray Strasberg 1974; Edgar A. G. Shaw 1973.

AWARDS

R. BRUCE LINDSAY AWARD ... presented in the Spring to a member of the Society who is under 35 years of age on 1 January of the year of the Award and who, during a period of two or more years immediately preceding the award, has been active in the affairs of the Society and has contributed substantially, through published papers, to the advancement of theoretical or applied acoustics, or both.
Recipients: Ilene J. Busch-Vishniac, 1987; William E. Cooper, 1986; Peter Mikhalevsky 1984; Ralph N. Baer 1982; Peter H. Rogers 1980; Henry E. Bass 1978; Robert Apfel 1976; Lawrence R. Rabiner 1974; Robert D. Finch 1972; Logan E. Hargrove 1970; Emmanuel Papadakis 1968.

GOLD MEDAL ... presented in the Spring to a member of the Society, without age limitation, for contributions to acoustics.
Recipients: Cyril M. Harris, 1987; James L. Flanagan, 1986; Laurence Batchelder 1985; Robert T. Beyer 1984; Martin Greenspan 1983; Isadore Rudnick 1982; Harry F. Olson 1981; Richard H. Bolt 1979; R. W. B. Stephens 1977; Leo L. Beranek 1975; Philip M. Morse 1973; Warren P. Mason 1971; Frederick V. Hunt 1969.

PIONEERS OF UNDERWATER ACOUSTICS ... presented to an individual, irrespective of nationality, age, or society affiliation, who has made an outstanding contribution to the science of underwater acoustics, as evidenced by publication of research results in professional journals or by other accomplishments in the field.
Recipients: Fred N. Spiess 1985; Arthur O. Williams, Jr. 1982; Claude W. Horton, Sr. 1980; Carl Eckart 1973; Harold L. Saxton 1970.

TRENT-CREDE AWARD ... presented to an individual, irrespective of nationality, age, or society affiliation, who has made an outstanding contribution to the science of mechanical vibration and shock, as evidenced by publication of research results in professional journals or by other accomplishments in the field. The award was established in 1969.

Recipients: Eric E. Ungar 1983; John C. Snowdon 1980; Stephen H. Crandall 1978; J. P. Den Hartog 1975; Elias Klein 1973; Raymond D. Mindlin 1971; Carl Vigness 1969.

WALLACE CLEMENT SABINE AWARD ... presented to an individual of any nationality who has furthered the knowledge of architectual acoustics, as evidenced by contributions to professional journals and periodicals or by other accomplishments in the field of architectural acoustics.
Recipients: Thomas D. Northwood 1982; Cyril M. Harris 1979; Lothar Cremer 1973; Hale J. Sabine 1968.

VON BEKESY MEDAL ... presented to individuals, irrespective of nationality, age or society affiliation, who have made outstanding contributions to the area of psychological or physiological acoustics, as evidenced by publication of research results in professional journals or by other accomplishments in the field.
Recipients: Jozef J. Zwislocki 1985.

THE SILVER MEDAL ... presented to individuals, without age limitation for contributions to the advancement of science, engineering, or human welfare through the application of acoustic principles or through research accomplishments in acoustics.
Recipients: Tony F.W. Embleton in Noise, John G. Backus in Musical Acoustics, Albert G. Bodine in Engineering Acoustics 1986; David T. Blackstock in Physical Acoustics 1985; Arthur H. Benade in Musical Acoustics, William W. Lang in Noise, Vincent Salmon in Engineering Acoustics 1984; Kenneth N. Stevens in Speech Communication, Eugen J. Skudrzyk in Theoretical and Applied Acoustics 1983; Per V. Bruel in Engineering Acoustics 1982; Carleen M. Hutchins in Musical Acoustics, Henning E. von Gierke in Noise, E. Glen Wever in Psychological and Physiological Acoustics 1981; Gunnar Fant in Speech Communication 1980; Herbert J. McSkimin in Physical Acoustics 1979; Benjamin B. Bauer in Engineering Acoustics, Harvey H. Hubbard in Noise 1978; Martin Greenspan in Physical Acoustics, Lloyd A. Jeffress in Psychological and Physiological Acoustics 1977; Hugh S. Knowles in Engineering Acoustics, Theodore J. Schultz in Architectural Acoustics 1976; Franklin S. Cooper in Speech Communication, Isadore Rudnick in Physical Acoustics 1975; Harry F. Olson in Engineering Acoustics 1974.

THE DISTINGUISHED SERVICE CITATION ... awarded to a present or former Member or Fellow of the Society in recognition of outstanding service to the Society.
Recipients: Stanley L. Ehrlich, Samuel F. Lybarger, 1986; William S. Cramer 1984; R. Bruce Lindsay 1981; Robert T. Beyer, Henning E. von Gierke 1978; Gerald J. Franz 1974; Robert W. Young, Betty H. Goodfriend 1973; Laurence Batchelder 1972.

CONGRESSIONAL SCIENCE FELLOWSHIP ... The ASA sometimes sponsors Congressional Science Fellows under the AAAS Program. The major purposes of the program are to make direct public service contributions to the more effective use of scientific knowledge in government and to broaden the perspective and understanding of both the scientific and governmental communities with regard to the process of developing public policy on scientific matters.
Recipients: Charles E. Schmid 1985-86; Joseph W. Dickey 1983-84; James E. Atkinson 1979-80; Barry Leshowitz 1977-78.

AIR POLLUTION CONTROL ASSOCIATION (APCA)

P.O. Box 2861, Pittsburgh, Pennsylvania 15230
PRESIDENTS ... Milton Feldstein 1987-1988: Richard Scherr 1986-1987; Joseph Soporowski 1985-1986; Joseph Padgect 1984-1985; John G. Olin 1983-1984; Donald F. Adams 1982-1983; G. Steve Hart 1981; Robert L. Eisenbach 1980; Frank P. Partee 1979; Richard Boubel 1978; Joseph Palomba 1977; David Standley 1976; Arthur C. Stern 1975; Austin H. Phelps 1974; Alexander Rihm, Jr. 1973; David M. Benforado 1972.

AWARDS

THE FRANK A. CHAMBERS AWARD ... for outstanding achievement in the science and art of air pollution control.

Recipients: Senichi Masuda 1987; Jack G. Calvert 1986; Dr. Julian P. Heicklen 1985; Dr. Robert E. Munn 1984; Bernard Weinstock 1983; James N. Pitts, Jr. 1982; Walker W. Heck 1981; Hiromi Niki 1980; Gaylord Penney 1978; Irving A. Singer 1977; Richard B. Engdahl 1976; E. R. Hendrickson 1975; James P. Lodge 1974; W. L. Faith 1973; Harry J. White 1971; A. Paul Altshuller 1970; W. C. L. Hemeon 1969; Sir Oliver Graham Sutton 1968.

LYMAN A. RIPPERTON AWARD ... this award is for distinguished achievement as an educator in the field of Air Pollution Control. It shall be awarded to an individual who by precept and example shall have inspired students to achieve excellence in all their professional and social endeavors. It recognizes the abilities that only a few in the education profession possess to be able to teach with rigor, with humor, with humility, and with pride. The recipients of this award are representative of the educators we would have chosen for all of our teachers, had we had a choice. They are known by the accomplishments of their students.
Recipients: Melvin W. First 1987; Howard E. Hesketh 1986; Eileen G. Crennan 1985; Harold Cota 1984; Raymond Manganelli 1983.

RICHARD BEATTY MELLON AWARD ... for an individual whose contributions of a civic nature, whether administrative, legislative, or judicial, have aided substantially in the abatement of air pollution.
Recipients: Bruce Walker 1987; William C. Ruckelshaus 1985; John A. Fraser 1983; G. Arthur Webb 1982; Harold J. Paulus 1980; Bernard J. Steigerwald 1977; Maurice F. Strong 1975; W. Brad Drowley 1974; John T. Middleton 1972; Allen D. Brandt 1971; Arthur C. Stern 1970; Vernon C. Mackenzie 1969.

S. SMITH GRISWOLD AWARD ... for outstanding accomplishment in the prevention and control of air pollution by a governmental agency staff member.
Recipients: Jean J. Schueneman 1987; Robert L. Stockman 1986; Dr. Danford G. Kelley 1985; Victor Sussman 1984; Milton Feldstein 1983; Harry H. Hovey 1982; Alexander Rihm 1980; Charles Barden 1979; Don R. Goodwin 1978; Morton Sterling 1977; Charles M. Copley, Jr. 1976; William A. Munroe 1975; John A. Maga 1974; William H. Megonnell 1973; Robert W. Chass 1972.

ALPHA PI MU

Department of Industrial Engineering and Operations Research, Virginia Polytechnic Institute and State University, Blacksburg, Virginia 24061 Dr. Robert D. Dryden, PE, Executive Director,
PRESIDENTS ... Dr. G. T. Stevens, Jr. 1984-1988; Dr. Robert D. Dryden 1982-1984; Dr. Wilbur L. Meier, Jr. 1980-1982; Dr. David D. Bedworth 1978-80; Professor H.J. MacKenzie 1976-1978; Dr. Marvin H. Agee 1974-75; Professor A. Burgess 1973.

AWARDS

DAVID F. BAKER MEMORIAL AWARD ... established in 1973 by the Executive Council to recognize an Alpha Pi Mu Member who makes an "Outstanding Contribution to Industrial Engineering Education." The award is only made in those years when the Executive Council believes that a Significant Contribution has been made. An engraved plaque commemorating the Member's achievement permanently displayed in the Department of Industrial & Systems Engineering of the Ohio State University where Dr. Baker served as a faculty member and Chairman for several years.
Recipients: Dr. Robert D. Dryden 1987, Virginia Polytechnic Institute; Ohio State Chapter 1975; Ohio State Chapter 1974; Ohio State Chapter 1973.

OUTSTANDING CHAPTER AWARD ... presented annually by the National Executive Council to the Chapter which has made the greatest contribution to the purposes of Alpha Pi Mu. Plaques with appropriate engraving are presented to regional winning chapters and an engraved cup is presented to the National winner.
Recipients: Virginia Polytechnic Institute 1983-86; West Virginia University 1982; Virginia Polytechnic Institute 1981; Virginia Polytechnic Institute & State University 1980; Perdue Universi-

ty 1979; Oklahoma State University 1978; Virginia Tech University 1977; Virginia Polytechnic Institute & State University 1976; Virginia Polytechnic Institute & State University 1975; Texas Tech University 1974; Virginia Polytechnic Institute & State University 1973; Texas A&M 1972; Michigan Chapter 1971; Northeastern Chapter 1970; Arkansas Chapter 1969; Arkansas Chapter 1968.

AMERICAN ACADEMY OF ENVIRONMENTAL ENGINEERS (AAEE)

132 Holiday Court, Suite 206, Annapolis, MD 21041
William C. Anderson PE, Executive Director
PRESIDENTS ... Paul L. Busch, 1988; Louis L. Guy, Jr., 1987; Leo Weaver, 1986; Walter E. Garrison 1985; James J. Corbalis Jr. 1984; Ernest F. Gloyna 1983; George P. Hannad, Jr. 1982; M. D. R. Riddell 1981; William J. Carroll 1980; Frank A. Butrico 1979; James B. Coulter 1978; John D. Parkhurst 1977; Richard S. Green 1976; Glen J. Hopkins 1975; Roy F. Weston 1974; Frank R. Bowerman 1973.

AWARDS

EDWARD J. CLEARY AWARD ... this honor is awarded biannually to a person judged by AAEE as an outstanding performer in the management of environmental protection enterprises conducted under either public (local, state, regional, federal, international) or private auspices.
Recipients: E. R. Heiberg, III, 1987; William D. Ruckelshaus 1985; William J. Carroll 1983; Joseph T. Ling 1981; John D. Parkhurst 1979; Paul Haney 1977; Martin Lang 1975; Samuel S. Baxter 1973.

GORDON MASKEW FAIR AWARD ... this honor is awarded annually to a Diplomate of AAEE who is judged to have contributed substantially to the status of the Engineering profession and to the Academy.
Recipients: Robert A. Canham 1987; Joseph F. Lagnese, Jr., 1986; Vinton W. Bacon 1985; Joseph F. Malina Jr. 1984; Arthur C. Stern 1983; Ernest F. Gloyna 1982; Gerald A. Rohlich 1981; William R. Gibbs 1980; Joseph C. Lawler 1979; Stanley E. Kappe 1978; Roy Weston 1977; Ernest Boyce 1976; Wesley E. Gilbertson 1975; Frank A. Butrico 1974; John H. Ludwig 1973; Daniel A. Okun 1972; James B. Coulter 1971.

STANLEY E. KAPPE AWARD ... this honor may be awarded annually to a Diplomate of the American Academy of Environmental Engineers who is judged to have performed extraordinary and outstanding service contributory to significant advancement of public awareness to the betterment of the total environment and other objectives of the Academy.
Recipients: Ralph C. Graber, 1987; James B. Coulter, 1986; Donald L. Snow 1985; Charles F. Niles, Jr. 1984; Ralph C. Palange 1983.

AMERICAN ASSOCIATION FOR THE ADVANCEMENT OF SCIENCE (AAAS)

1333 H St., N.W., Washington, D.C. 20005
Alvin W. Tirvelpiece, Executive Officer
PRESIDENTS ... Sheila E. Widnall 1987-88; Lawrence Bogorad 1986-87; Gerard Piel, 1985-86; David A. Hamburg, 1984-85; Anna J. Harrison, 1983-84; E. Margaret Burbidge, 1982-83; D. Allan Bromley, 1981-82.

AWARDS

AAAS-NEWCOMB CLEVELAND PRIZE ... the Association's oldest award, was established in 1923 with funds donated by Newcomb Cleveland of New York City. Consisting of $5000 and a bronze medal, it is awarded annually to an outstanding research paper published in the Articles or Reports sections of Science magazine during the contest year, January through December.
Recipients: 1986 No Prize Given; James M. Hogle, Marie Chow, David J. Filman 1985; Sally M. Rigden, Thomas J. Ahrens, and Edward M. Stolper 1984; Allan C. Spradling and Gerald M. Ru-

bin 1983; Dennis G. Kleid, Douglas M. Moore, and 8 coauthors 1982; Robert Axelrod and William D. Hamilton 1981.

AAAS SCIENTIFIC FREEDOM & RESPONSIBILITY AWARD ... The AAAS Scientific Freedom and Responsibility Award was established in 1981 to honor scientists and engineers whose exemplary actions, often taken at significant personal cost, have served to foster scientific freedom and responsibility. The annual award consists of $1,000 and a plaque. Eligible for nomination are scientists and engineers who have acted to protect the public's health, safety, or welfare; or focused public attention on potentially serious impacts of science and technology on society by their responsible participation in public policy debates; or established important new precedents in carrying out the social responsibility or in defending the professional freedom of scientists and engineers.
Recipients: Francisco J. Ayala, Norman D. Newell, Stanley L. Weinberg 1987; Victor Paschkis, Colegio Medico de Chile 1986; Werner A. Baum 1985; Jose Westernkamp and Anatoli Koryagin 1983; Morris B. Baslow, Paul Berg, Maxine Singer, and Norton D. Zinder 1982.

AAAS SOCIO-PSYCHOLOGICAL PRIZE ... established in 1952 with funds donated by Arthur F. Bentley, the AAAS Socio-Psychological Prize of $1000 is offered annually for a meritorious paper that furthers understanding of the psychological-social-cultural behavior of human beings. The prize is intended to encourage in social inquiry the development and application of the kind of methodology that has proved so fruitful in the natural sciences.
Recipients: Bob Altemeyer 1986; Dane Archer, Rosemary Gartner 1985; Colin Martindale 1984; David P. Phillips 1983; Richard J. Shweder and Edmund J. Bourne 1982; Gary Wayne Strong, 1981.

AAAS-WESTINGHOUSE SCIENCE JOURNALISM AWARDS ... with support from the Westinghouse Educational Foundation, AAAS makes five annual awards of $1000 each in recognition of outstanding reporting on the natural sciences and their engineering and technological applications, excluding the field of medicine. Awards are for reporting in (1) newspapers of over 100,000 daily circulation, (2) newspapers of under 100,000 circulation, (3) general circulation magazines, (4) television, and (5) radio.
Recipients: Philip M. Boffey, David E. Sanger, William J. Broad, Keay Davidson, Freda Yarbrough, Arthur Fisher, Daniel Zwerdling, Howard Berkes, Paula S. Apsell, Jon Palfreman, and Don Herbert 1986; Boyce Rensberger, Robert W. Andrews, Mark Harden, Andrew C. Revkin, Rick McCourt, Timothy Ferris, Larry F. Botto, and Geoff Haines-Stiles 1985. Paul G. Hayes, James Ehmann, Richard Wolkomir, Lili Francklyn and Mark Heistad, and Linda Harrar and Gary Hochman 1984; Bill Williams, Byron Spice, James S. Trefil, Ira Flatow and Bruce Gellerman, and James Steinbach 1983.

AMERICAN ASSOCIATION OF COST ENGINEERS (AACE)

308 Monongahela Building, Morgantown, West Virginia 26505
Kenneth K. Humphreys, Executive Director
PRESIDENTS ... Jack F. Enrico 1987-1988; Newton F. True, 1986-1987; Joseph M. Sexton 1985-86; Casimir A. DeCwikiel 1984-85; Daniel P. Elliott 1983-84; Lawrence J. Bloch 1982-83; John A. Foushi 1981-82; Brian D. Dunfield 1980-81; L. W. Hedrick 1979-80; Julian A. Piekarski 1978-79; Frank J. Kelly, Jr. 1977-78; Peter G. Zona 1976-77; Dr. James H. Black 1975-76; K. K. Kridner 1974-75; Arthur C. Slater 1973-74.

AWARDS

AACE AWARD OF MERIT ... to recognize outstanding service in Cost Engineering. The award consists of an appropriate framed certificate and plaque.
Recipients: James M. Neil 1987; Harold E. Marshall 1986; Emilio M. Zorilla-Vasquez 1985; Julian A. Piekarski 1984; Hans J. Lang 1983; Jan Korevaar 1982; Frank J. Kelly, Jr. 1981; Arthur C. Slater 1980; Dr. James H. Black 1979; John J. O'Driscoll, 1978; Robert E. Templeton 1977; Henry C. Thorne 1976; Wil-

liam G. Clark 1975; Charles R. Hirt 1974; Sidney Katell 1973; Hollis M. Dole 1972; Thomas C. Ponder 1971; Frederick C. Jelen 1970; Max S. Peters 1969; Wesley J. Dodge 1968.

D.T. ZIMMERMAN FOUNDER'S AWARD (FORMERLY THE AWARD OF RECOGNITION) ... to recognize outstanding contributions by an individual to the aims and purposes of AACE-award consists of an appropriate framed certificate.
Recipients: Edward C. Goodier 1987; Earl K. Clark 1986; Brian D. Dunfield 1985; Francis H. Schiffer 1984; Cletus C. Rivard 1983; Casimir A. DeCwikiel 1982; Paul D. Chamberlin 1981; Robert E. Templeton 1980; Kenneth K. Humphreys, 1979; Thomas C. Ponder 1978; Arthur C. Slater 1977; Julian A. Piekarski 1976; Leslie C. Jenckes 1975; Frank M. Russell 1974; R. Adam Leuchter 1973; K. K. Kridner 1972.

INDUSTRIAL RECOGNITION AWARD ... to recognize outstanding support of the engineering profession in general and cost engineering in particular by major corporations- award consists of an appropriate framed certificate.
Recipients: Sargent & Lundy 1987; E.I. du Pont de Nemours and Company 1986; Camp Dresser and McKee 1985; Stone & Webster Engineering Corporation 1984; Hoffman LaRoche, Inc. 1983; Public Service Electric and Gas Company 1982; American Appraisal Co. 1981; Proctor & Gamble Co. 1980; Stone, Marraccini, and Patterson, Inc. 1979; Dow Chemical, U.S.A. 1978; Stone & Webster Engineering Corporation 1977; Shell Oil Company 1976; Fluor Engineers & Constructors, Inc. 1975; M. W. Kellogg Company 1974; Monsanto Company 1973; Bechtel Corporation 1972.

AMERICAN ASSOCIATION OF ENGINEERING SOCIETIES INC. (AAES)

415 Second Street NE, Washington, D.C. 20002
Daniel De Simone, Executive Director
CHAIRMEN OF THE BOARD OF GOVERNORS ... Richard J. Gowen 1988; Richard W. Karn 1987; John A. White 1986; James Y. Oldshue 1985; Nicholas J. Radell 1984; Robert M. Saunders 1983; Irvan F. Mendenhall 1982; H. Arthur Nedom 1981; Kenneth A. Roe 1980.

AWARDS

THE NATIONAL ENGINEERING AWARD ... established by AAES as an annual award by the engineering profession to an engineer, or group of engineers, in special recognition of outstanding contributions to the benefit of mankind. The award consists of an inscribed scroll and a $5,000 honorarium presented in the recipient's name to an organization of his/her choice active in engineering education.
Recipients: Robert N. Noyce 1987; Erich Bloch 1986; Martin Goland 1985; William H. Pickering 1984; George M. Low 1983; W. Kenneth Davis 1982; Donald Burnham 1981; Harold Brown 1980; Neil A. Armstrong 1979.

THE CHAIRMAN'S AWARD ... established in 1980 to recognize prominent Americans who have demonstrated significant applications of engineering to the uses and needs of mankind. The Award consists of an inscribed scroll.
Recipients: E. R. Heiberg III 1987; Betsy Ancker-Johnson 1986; Donald E. Procknow 1984; Stephen D. Bechtel, Jr. 1983; George A. Keyworth, II 1982; Robert A Frosch 1981; John D. Bulkeley 1980.

KENNETH ANDREW ROE AWARD ... established in 1983 by the Board of Directors of Burns and Roe, Inc. to recognize a member of the engineering profession who has been effective in promoting cooperation, understanding and unity among the engineering societies, thus furthering the good of the engineering community as a whole. The Award consists of an inscribed scroll and an honorarium.
Recipients: James Y. Oldshue, 1987; Nicholas J. Radell, 1986.

THE JOAN HODGES QUENEAU AWARD (PALLADIUM MEDAL) ... given jointly by AAES and the National Audubon Society to honor an outstanding engineering achievement in environmental conservation. The Award consists of an inscribed scroll, the medal cast in palladium, and a bronze replica.

Recipients: Thomas K. MacVicar, 1987; Kenneth R. Daniel, 1986; William A. Jester, 1985; Barbara Ann Gamboa-Lewis 1984; Roy W. Hann, Jr. 1983.

THE AMERICAN CERAMIC SOCIETY, INC. (ACerS)

757 Brooksedge Plaza Dr, Westerville OH 43081
W. Paul Holbrook, Executive Director

PRESIDENTS ... Dale Niesz 1987-88; Joseph L. Pentecost 1986-87; Edwin Ruh 1985-86; Richard M. Spriggs 1984-85; J. Lambert Bates 1983-84; Robert J. Beals 1982-83; James I. Mueller 1981-82; William R. Prindle 1980-1981; Malcolm G. McLaren 1979-80; John B. Wachtman, Jr. 1978-79; Lyle A. Holmes 1977-78; Stephen D. Stoddard 1976-77; Ralston Russell Jr. 1975-76; Joseph E. Burke 1974-75; James R. Johnson 1973-74.

AWARDS

EDWARD ORTON, JR., MEMORIAL LECTURE ... lecturer is to be noted for scholarly attainments in the ceramic or related fields.
Recipients: Robert E. Newnham 1987; Hiroaki Yanagida 1986; Rustum Roy 1985; Gene H. Haertling 1984; Fred M. Ernsberger 1983; Hermann Schmalzvied 1982; John B. Wachtman, Jr. 1981; W. David Kingery 1980; Joseph A. Pask 1979; Julius J. Harwood 1978; Hans Thurnauer 1977; John F. McMahon 1976; Emilio Q. Daddario 1975; James Boyd 1974; George W. Brindley 1973; Henry Eyring 1972; Hobart M. Kraner 1971; Elburt F. Osborn 1970; Edward Wenk, Jr. 1969; W. T. Pecora 1968.

GREAVES-WALKER AWARD ... to recognize and honor members of NICE who have rendered outstanding service to the ceramic engineering profession.
Recipients: Winston Duckworth 1987; James R. Tinklepaugh 1986; James R. Johnson 1985; Stephen D. Stoddard 1984; James I. Mueller 1983; Gilbert C. Robinson 1982; William O. Brandt 1981; W.W. Coffeen 1980; Arthur J. Metzger 1979; Loran S. O'Bannon 1978; John McMahon 1977; Edward C. Henry 1976; Wayne A. Deringer 1975; Karl Schwartzwalder 1974; Ralph L. Cook 1973; Rolland R. Roup 1972; James Edward Hansen 1971; Ralston Russell, Jr. 1970; Emily C. Van Schoick 1969; Raymond E. Birch 1968.

JOHN JEPPSON MEDAL ... to recognize distinguished service within the realms of ceramics. There are no limitations as to nationality, sex, or society membership.
Recipients: George C. Kuczynski 1987; Bernard Schwartz 1986; Harold G. Sowman 1985; Leslie E. Cross 1984; Girard W. Phelps 1983; Della M. Roy 1982; Joseph E. Burke 1981; Gilbert C. Robinson 1980; James I. Mueller 1979; Arnulf Muan 1978; Raymond E. Birch 1977; J. Earl Frazier 1976; D. P. H. Hasselman 1975; Walter H. Gitzen 1974; Elburt F. Osborn 1973; Willis G. Lawrence 1972; Flemmon P. Hall 1971; George J. Bair 1970; Norbert J. Kreidl 1969; Frederick H. Norton 1968.

THE KARL SCHWARTZWALDER PROFESSIONAL ACHIEVEMENT IN CERAMIC ENGINEERING (PACE) AWARD ... a joint ACerS-NICE award to recognize outstanding young ceramic engineers whose achievements have been significant to the profession and to the general welfare of the American people. A nominee must be between 21 and 40 years.
Recipients: Lisa C. Klein 1987; Cullen L. Hackler 1986; John J. Mecholsky 1985; Willard E. Hauth III 1984; W. H. Johnson 1983; Richard E. Tressler 1982; Edmund S. Wright 1981; J. Richard Schorr 1980; H. Kent Bowen 1979; William H. Payne 1978; James C. Waldal 1977; Arthur H. Heuer 1976; W. Richard Ott 1975; Daniel R. Stewart 1974; Gene H. Haertling 1973; Delbert E. Day, Larry L. Hench 1972; Donald L. Kummer 1971; Willard H. Sutton 1970; Robert S. DuFresne 1969; Joseph L. Pentecost 1968.

ROSS COFFIN PURDY AWARD ... to recognize the most valuable contribution to ceramic literature during the previous year. The article must be complete and readily available to the public.
Recipients: Terry A. Michalske & Edwin R. Fuller, Jr. 1987; George W. Scherer 1986; Yao Xi, H. McKinstry and Leslie E. Cross 1985; P.J. Lemaire & H. Kent Bowen 1984; William H. Rhodes 1983; David R. Clarke and Fred F. Lange 1982; Arthur

H. Heuer 1981; Man F. Yan and David W. Johnson, Jr. 1980; Subhash H. Risbud and Joseph A. Pask 1979; Charles D. Greskovich and Joseph H. Rosolowski 1978; Marcus P. Borom, Robert H. Doremus, Anna M. Turkalo 1977; David E. Carlson 1976; A. G. Evans, M. Linzer 1975; Stephen C. Carniglia 1974; O. S. Narayanaswamy 1973; Robert L. Coble 1972; Sheldon Wiederhorn 1971; Robert E. Jech and Dennis W. Readey 1970; Norman M. Tallan-Walter C. Tripp and Robert W. Vest 1969; Orson L. Anderson and Naohiro Soga 1968.

SOSMAN MEMORIAL LECTURE ... a lecture to scientists given by a distinguished scientist in the field of ceramics.
Recipients: Frederick F. Lange 1987; Arthur H. Heuer 1986; Sheldon M. Wiederhorn 1985; H. Kent Bowen 1984; Donald R. Uhlmann 1983; Brian R. Lawn 1982; Anthony G. Evans 1981; James E. White 1980; Robert L. Coble 1979; Richard J. Charles 1978; A. L. Stuijts 1977; Robert J. Stokes 1976; Rustum Roy 1975; John B. Wachtman, Jr. 1974; W. David Kingery 1973.

AMERICAN CHEMICAL SOCIETY (ACS)

1155 16th Street N. W., Washington, D.C. 20036
John K. Crum, Executive Director

PRESIDENTS ... Gordon L. Nelson 1988; Mary L. Good 1987; George C. Pimentel 1986; Ellis K. Fields 1985; Warren D. Niederhauser 1984; Fred Basolo 1983; Robert W. Parry 1982; Albert C. Zettlemoyer 1981; James D. D'Ianni 1980; Gardner W. Stacy 1979; Anna J. Harrison 1978; Henry A. Hill 1977; Glenn T. Seaborg 1976; William J. Bailey 1975; Bernard S. Friedman 1974; Alan C. Nixon 1973.

AWARDS

ACS AWARD IN ANALYTICAL CHEMISTRY SPONSORED BY FISHER SCIENTIFIC COMPANY ... to recognize and encourage outstanding contributions to the science of analytical chemistry, pure or applied, carried out in the United States or Canada. The award consists of $3,000 and an etching. The traveling expenses of the recipient incidental to the conferring of the award are paid.
Recipients: Fred E. Lytle 1988; Gary M. Hieftge 1987; David M. Hercules 1986; James S. Fritz 1985; Allen J. Bard 1984; Thomas L. Isenhour 1983; Ralph N. Adams 1982; Fred W. McLafferty 1981; J. Calvin Giddings 1980; Velmer A. Fassel 1979; Henry Freiser 1978; George G. Guilbault 1977; Howard V. Malmstadt 1976; Sidney Siggia 1975; Philip W. West 1974; James D. Winefordner 1973; W. Wayne Meinke 1972; George H. Morrison 1971; Charles V. Banks 1970; Roger G. Bates 1969; Lockhart B. Rogers 1968.

ACS AWARD IN CHEMICAL EDUCATION SPONSORED BY UNION CARBIDE CORPORATION ... to recognize outstanding contributions to chemical education. The award consists of $5,000 and a suitably inscribed certificate. An allowance is provided to cover travel expense to the meeting at which the award is made.
Recipients: Marjorie H. Gardner 1988; Linus Pauling 1987; Bassam Z. Shakhashiri 1986; Glenn A. Crosby 1985; Arthur W. Adamson 1984; Mitchell J. Sienko 1983; Anna J. Harrison 1982; Derek A. Davenport 1981; Henry A. Bent 1980; Gilbert P. Haight, Jr. 1979; Lloyd N. Ferguson 1978; Robert W. Parry 1977; Leallyn B. Clapp 1976; William T. Lippincott 1975; George S. Hammond 1974; Robert C. Brasted 1973; J. Arthur Campbell 1972; Laurence E. Strong 1971; Hubert N. Alyea 1970; L. Carroll King 1969; William F. Kieffer 1968.

ACS AWARD IN APPLIED POLYMER SCIENCE SPONSORED BY PHILLIPS PETROLEUM COMPANY ... to recognize and encourage outstanding achievements in the science or technology of plastics, coatings, polymer composites, adhesives, and related fields. The award consists of $3,000 and a medallion. An allowance of up to $500 is provided to assist with traveling expenses to the meeting at which the award is presented.
Recipients: David S. Breslow 1988; O. A. Battista 1987; William J. Bailey 1986; James Economy 1985; Donald R. Paul 1984; Frank A. Bovey 1983; Eric Baer 1981; John W. Vanderhoff 1980; Roger S. Porter 1979; John K. Gillham 1978; William A. Zisman 1977; Herman F. Mark 1976; Maurice L. Huggins 1975; Vivian T. Stannett 1974; Carl S. Marvel 1973; Richard S. Stein 1972;

Raymond R. Myers 1971; Raymond F. Boyer 1970; Sylvan O. Greenlee 1969; Harry Burrell 1968.

ACS AWARD IN CHROMATOGRAPHY SPONSORED BY SUPELCO, INC
... to recognize outstanding contributions to the fields of chromatography. The award consists of $5,000 and a certificate. An allowance is provided for covering travel expense to the meeting at which the award is presented.

Recipients: Milton L. Lee 1988; Charles H. Lochmueller 1987; Milos V. Novotny 1986; Leslie S. Ettre 1985; Lloyd R. Snyder 1984; Csaba G. Horvath 1983; Barry L. Karger 1982; Marcel J. E. Golay 1981; J. E. Lovelock 1980; Evan C. Horning 1979; A. J. P. Martin 1978; Raymond P. W. Scott 1977; James S. Fritz 1976; Egon Stahl 1975; Lockhart B. Rogers 1974; Albert Zlatkis 1973; J. J. Kirkland 1972; Julian F. Johnson 1970; Morton Beroza 1969; Lewis G. Longsworth 1968.

ACS AWARD IN COLLOID OR SURFACE CHEMISTRY SPONSORED BY THE KENDALL COMPANY
... to recognize and encourage outstanding scientific contributions to colloid or surface chemistry in the United States or Canada. The award consists of $3,000 and a certificate. An allowance of not more than $500 is provided for traveling expenses to the meeting at which the award will be presented.

Recipients: Howard Brauner 1988; John T. Yates 1987; Eli Ruckenstein 1986; Stig E. Friberg 1985; Brian E. Conway 1984; Janos H. Fendler 1983; Gert Ehrlich 1982; Gabor A. Somorjai 1981; Howard Reiss 1980; Arthur W. Adamson 1979; Harold A. Scheraga 1978; Michel Boudart 1977; Robert J. Good 1976; Robert Gomer 1975; W. Keith Hall 1974; Robert L. Burwell, Jr. 1973; Egon Matijevic 1972; Milton Kerker 1971; Jerome Vinograd 1970; Terrell L. Hill 1969; Albert C. Zettlemoyer 1968.

ACS AWARD FOR CREATIVE ADVANCES IN ENVIRONMENTAL SCIENCE AND TECHNOLOGY SPONSORED BY AIR PRODUCTS AND CHEMICALS, INC.
... to encourage creativity in research and technology or methods of analysis to provide a scientific basis for informed environmental control decision making processes, or to provide practical technologies which will reduce health risk factors. The award consists of $3,000, a certificate of recognition, and an allowance of up to $500 for travel expenses incidental to the conferral of the award.

Recipients: A. Welford Castleman, Jr. 1988; Joseph C. Arcos 1987; Eugene E. Kenaga 1986; Arthur Fontijn 1985; Julian Heicklen 1984; F. Sherwood Rowland 1983; Jack G. Calvert 1982; Philip W. West 1981; James J. Morgan 1980.

ACS AWARD FOR CREATIVE INVENTION SPONSORED BY THE CORPORATION ASSOCIATES
... to recognize individual inventors for successful applications of research in chemistry and/or chemical engineering which contribute to the material prosperity and happiness of people. The award consists of $4,000, and a fine silver medal.

Recipients: Samuel Smith 1988; Robert B. Morin 1987; Alfred Marzocchi 1986; Ralph Milkovich 1985; Edwin P. Plueddemann 1984; O. A. Battista 1983; William S. Knowles 1982; Roy L. Pruett 1981; Stephanie L. Kwolek 1980; Leo H. Sternbach 1979; LeGrand G. Van Uitert 1978; Herman A. Bruson 1977; Manuel M. Baizer 1976; James D. Idol, Jr. 1975; Charles C. Price 1974; Carl Djerassi 1973; H. Tracy Hall 1972; S. Donald Stookey 1971; Gordon K. Teal 1970; J. Paul Hogan 1969; William G. Pfann 1968.

ACS AWARD FOR CREATIVE WORK IN SYNTHETIC ORGANIC CHEMISTRY SPONSORED BY ALDRICH CHEMICAL COMPANY, INC.
... to recognize and encourage creative work in synthetic organic chemistry. The award consists of $3,000, a certificate, and an allowance of not more than $500 for travel expenses incidental to the conferral of the award. The recipient's award address will be reprinted in *Aldrichimica Acta*.

Recipients: Robert E. Ireland 1988; Harry Wasserman 1987; Samuel Danishevsky 1986; Albert I. Meyers 1985; Leo A. Paquette 1984; K. Barry Sharpless 1983; David A. Evans 1982; Barry M. Trost 1981; Yoshito Kishi 1980; George A. Olah 1979; Satoru Masamune 1978; Franz Sondheimer 1976; Herbert O. House 1975; Edward C. Taylor 1974; George Buchi 1973; Bruce Merrifield 1972; Elias J. Corey 1971; Eugene E. van Tamelen 1970; H. Gobind Khorana 1969; Theodore L. Cairns 1968.

ACS AWARD FOR DISTINGUISHED SERVICE IN THE ADVANCEMENT OF INORGANIC CHEMISTRY SPONSORED BY MALLINCKRODT, INC.
... to recognize distinguished services to the advancement of inorganic chemistry by a member of the American Chemical Society. The award consists of $5,000, an appropriate certificate, and an allowance of not more than $1000 for traveling expenses to the meeting at which the award will be presented.

Recipients: M. Frederick Hawthorne 1988; Deward F. Shriver 1987; Jack Lewis 1986; Jack Halpern 1985; Harry B. Gray 1984; Norman Sutin 1983; Arthur W. Adamson 1982; Dietmar Seyferth 1981; Arthur E. Martell 1980; Earl L. Muetterties 1979; Harry J. Emeleus 1978; James L. Hoard 1977; Daryle H. Busch 1976; Fred Basolo 1975; F. Albert Cotton 1974; Ronald J. Gillespie 1973; John C. Bailar, Jr. 1972; Joseph Chatt 1971; Ralph G. Pearson 1970; Anton B. Burg 1969; William N. Lipscomb, Jr. 1968.

ACS AWARD IN INORGANIC CHEMISTRY SPONSORED BY MONSANTO COMPANY
... to recognize and encourage fundamental research in the field of inorganic chemistry. The award consists of $3,000 and a certificate. An allowance of not more than $500 is provided for traveling expenses to the meeting at which the award will be presented.

Recipients: Mark S. Wrighton 1988; Stephen J. Lipperd 1987; John D. Corbett 1986; F. G. A. Stone 1985; M. L. H. Green 1984; George W. Parshall 1983; Ronald Hoffmann 1982; Henry Taube 1981; Alan M. Sargeson 1980; James A. Ibers 1979; Harry B. Gray 1978; Richard H. Holm 1976; James P. Collman 1975; Lawrence F. Dahl 1974; M. F. Hawthorne 1973; Theodore L. Brown 1972; Jack Lewis 1971; Neil Bartlett 1970; Russell S. Drago 1969; Jack Halpern 1968.

ACS AWARD FOR NUCLEAR CHEMISTRY SPONSORED BY AMERSHAM CORPORATION
... to recognize and encourage research in nuclear and radio-chemistry or their application. The award consists of $3,000 and a certificate. Traveling expenses to the meeting at which the award will be presented will be paid.

Recipients: Guenther Herrmann 1988; Ellis P. Steinberg 1987; Victor E. Viola 1986; Gregory R. Choppin 1985; Joseph Cerny 1984; Darleane C. Hoffman 1983; Leo Yaffe 1982; Robert Vandenbosch 1981; Arthur M. Poskanzer 1980; Raymond Davis, Jr. 1979; Paul K. Kuroda 1978; Glen E. Gordon 1977; John O. Rasmussen 1976; John R. Huizenga 1975; Lawrence E. Glendenin 1974; Albert Ghiorso 1973; Anthony Turkevich 1972; Alfred P. Wolf 1971; Paul R. Fields 1970; George E. Boyd 1969; Richard L. Wolfgang 1968.

ACS AWARD IN PETROLEUM CHEMISTRY SPONSORED BY THE AMOCO FOUNDATION
... to recognize, encourage, and stimulate outstanding research achievements in the field of petroleum chemistry in the United States or Canada. The award consists of $5,000 and a certificate. An allowance of not more than $1,500 is provided for traveling expenses to the meeting at which the award will be presented.

Recipients: Werner O. Haag 1988; W. Keith Hall 1987; Frederick G. Bordwell 1986; Edward M. Arnett 1985; Cheves Walling 1984; Robert L. Burwell, Jr. 1983; Irving Wender 1982; Herman Pines 1981; William A. Pryor 1980; Robert L. Banks 1979; Ellis K. Fields 1978; Sidney W. Benson 1977; John H. Sinfelt 1976; Joe W. Hightower 1973; Paul G. Gassman 1972; Gerasimos J. Karabatsos 1971; Lloyd R. Snyder 1970; Alan Schriesheim 1969; Keith U. Ingold 1968.

ACS AWARD IN POLYMER CHEMISTRY SPONSORED BY MOBIL CHEMICAL COMPANY
... to recognize contributions to polymer chemistry. The award consists of $3000 and a certificate, also an allowance of not more than $500 is provided for traveling expenses to the meeting at which the award is made.

Recipients: Pierre G. DeGennes 1988; Vivian T. Stannett 1987; Herbert Morawetz 1986; Joseph P. Kennedy 1985; Harry R. Allcock 1984; Richard S. Stein 1983; John K. Stille 1982; E. J. Vandenberg 1981; George B. Butler 1980; Henri Benoit 1979; Junji Furukawa 1978; William J. Bailey 1977; Paul W. Morgan 1976; Leo Mandelkern 1975; John D. Ferry 1974; Turner Alfrey, Jr. 1973; Arthur V. Tobolsky 1972; Georges J. Smets 1971; Michael M. Szwarc 1970; Frank A. Bovey 1969; Charles G. Overberger 1968.

ACS AWARD IN PURE CHEMISTRY SPONSORED BY ALPHA CHI SIGMA FRATERNITY ... to recognize and encourage fundamental research in pure chemistry carried out in North America by young men and women who have not passed their 36th birthday on April 30 of the year in which the award is presented. The award consists of $3000 and a certificate setting forth the reasons for the award. The traveling expenses to the meeting at which the award will be presented are paid to a maximum of $1000.
Recipients: Jacqueline K. Barton 1988; George Milendon 1987; Peter J. Wolynes 1986; Ben S. Freiser, 1985; Eric Oldfield 1984; Michael J. Berry 1983; Stephen R. Leone 1982; Mark S. Wrighton 1981; John E. Bercaw 1980; Henry F. Schaefer III 1979; Jesse L. Beauchamp 1978; Barry M. Trost 1977; Karl F. Freed 1976; George M. Whitesides 1975; Nicholas J. Turro 1974; John I. Brauman 1973; Roy G. Gordon 1972; R. Bruce King 1971; Harry B. Gray 1970; Roald Hoffmann 1969; Orville L. Chapman 1968.

ALFRED BURGER AWARD IN MEDICINAL CHEMISTRY SPONSORED BY SMITHKLINE BECKMAN CORPORATION ... to recognize outstanding contributions to research in medicinal chemistry. The award consists of an honorarium of $3000 and a plaque commemorating the award event. The traveling expenses of the recipient incidental to the conferring of the award are paid. The award will be presented biennially in even-numbered years and the recipient shall present an award address at the spring meeting of the Division of Medicinal Chemistry.
Recipients: Roland K. Robins 1988; John A. Montgomery 1986; George H. Hitchings 1984; David W. Cushman and Miguel A. Ondetti 1982; T. Y. Shen 1980.

ARTHUR C. COPE AWARD ... to recognize outstanding achievement in the field of organic chemistry, the significance of which has become apparent within the five years preceding the year in which the award will be considered. The award consists of a gold medal, a bronze replica of the medal, and $15,000. Traveling expenses incidental to the conferring of the award will be paid. In addition, an unrestricted grant-in-aid of $25,000 for research in organic chemistry, under the direction of the recipient, designated as an Arthur C. Cope Fund Grant, will be made to any university or non-profit institution selected by the recipient. A recipient may choose to assign the Arthur C. Cope Fund Grant to an institution for use by others than the recipient for research or education in organic chemistry.
Recipients: Kenneth B. Wyberg 1988; Ronald Breslow 1987; Dulio Arigoni 1986; Albert Eschenmoser 1984; Frank H. Westheimer 1982; Gilbert Stork 1980; Orville L. Chapman 1978; Elias J. Corey 1976; Donald J. Cram 1974; Robert B. Woodward and Roald Hoffmann 1973.

CHARLES LATHROP PARSONS AWARD ... to recognize outstanding public service by a member of the American Chemical Society. The award consists of $3,000 and a scroll. An allowance not to exceed $500 is provided to reimburse the awardee for expenses incurred in traveling to the meeting at which the award is presented. The award normally shall be given not more often than once every two years. However, the Board of Directors may at its discretion reduce the interval to one year for a candidate of its choice, if in its judgement circumstances in a given year warrant such action.
Recipients: Norman Hackerman 1987; Franklin A. Long 1985; James G. Martin 1983; Charles G. Overberger 1978; William O. Baker 1976; Russell W. Peterson 1974; Charles C. Price 1973; W. Albert Noyes, Jr. 1970.

CLAUDE S. HUDSON AWARD IN CARBOHYDRATE CHEMISTRY SPONSORED BY MERCK & CO., ... to recognize outstanding contributions to carbohydrate chemistry, whether in education, research, or applications. The award consists of $3,000 and a certificate. An allowance of up to $1500 is provided for traveling expenses to the meeting at which the award is presented.
Recipients: Leslie Hough 1988; Stephen J Amgyal 1987; Gerald O. Aspinall 1986; Hans Paulsen 1985; Laurens Anderson 1984; Bengt Lindberg 1983; Stephen Hanessian 1982; Clinton E. Ballou 1981; George A. Jeffrey 1980; Arthur S. Perlin 1979; Michael Heidelberger 1978; Jack J. Fox 1977; Sidney M. Cantor 1976; Hans H, Baer 1975; Wendell W.Binkley 1974; Roger W. Jeanloz 1973; Derek Horton 1972; Robert S. Tipson 1971; Norman F.

Kennedy 1970; John K. Netherton Jones 1969; Hewitt G. Fletcher, Jr. 1968.

E. V. MURPHREE AWARD IN INDUSTRIAL AND ENGINEERING CHEMISTRY SPONSORED BY EXXON RESEARCH AND ENGINEERING COMPANY ... to stimulate fundamental research in industrial and engineering chemistry, the development of chemical engineering principles and their application to industrial processes. The award consists of $5,000 and a certificate. An allowance of not more than $1000 is provided for traveling expenses to the meeting at which the award will be presented.
Recipients: Jule A. Rabo 1988; Wolfgang M. H. Sachtler 1987; John S. Sinfelt 1986; Michel Boudart 1985; Robert K. Grasselli 1984; Herman Pines 1983; Sol W. Weller 1982; G. Alex Mills 1981; Milton Orchin 1980; John M. Prausnitz 1979; Donald F. Othmer 1978; Alexis Voorhies, Jr. 1977; James F. Roth 1976; Donald L. Katz 1975; Herman S. Bloch 1974; Thomas K. Sherwood 1973; Paul B. Weisz 1972; Heinz Heinemann 1971; Peter V. Danckwerts 1970; Alex G. Oblad 1969; Melvin A. Cook 1968.

THE ERNEST GUENTHER AWARD IN THE CHEMISTRY OF ESSENTIAL OILS AND RELATED PRODUCTS SPONSORED BY FRITZSCHE DODGE & OLCOTT INC. ... to recognize and encourage outstanding achievements in analysis, structure elucidation, chemical synthesis of essential oils, isolates, flavors and related substances. The award consists of $3,000 and a certificate. An allowance of $500 is provided for traveling expenses to the meeting at which the award will be presented.
Recipients: Paul A. Wencler 1988; Wolfgang Oppolzer 1987; Clayton H. Heathcock 1986; David E. Cane 1985; Jerrold Meinwald 1984; Karel Wiesner 1983; Paul A. Grieco 1982; Samuel Danishefsky 1981; Sukh Dev 1980; James A. Marshall 1979; Koji Nakanishi 1978; Robert E. Ireland 1977; Alastair I. Scott 1976; S. Morris Kupchan 1975; Gunther Ohloff 1974; William G. Dauben 1973; Guy Ourisson 1972; Ernest Wenkert 1971; Duilio Arigoni 1970; John W. Cornforth 1969; Elias J. Corey 1968.

FREDERIC STANLEY KIPPING AWARD IN ORGANOSILICON CHEMISTRY SPONSORED BY DOW CORNING CORPORATION ... to recognize distinguished achievement in research in organosilicon chemistry and, by such example, to stimulate the creativity of others toward further advancement of this field of chemistry. The award consists of $3,000 and a certificate. An allowance will be provided to cover travel expenses incidental to conferment of the award. The award will be presented biennially in even-numbered years starting in 1978.
Recipients: Raymond Calas 1988; Peter P. Gaspar 1986; Robert J. P. Corriu 1984; Thomas J. Barton 1982; E. A. V. Ebsworth 1980; Hideki Sakurai 1978; Michael F. Lappert 1976; Hans Bock 1975; Hubert Schmidbaur 1974; Adrian G. Brook 1973; Dietmar Seyferth 1972; Alan G. MacDiarmid 1971; Robert West 1970; Robert A. Benkeser 1969; Ulrich Wannagat 1968.

GARVAN MEDAL SPONSORED BY OLIN CORPORATION ... to recognize distinguished service to chemistry by women chemists, citizens of the United States. The award consists of $3,000, a suitably inscribed gold medal, and a bronze replica of the medal. An allowance of $500 is provided for traveling expenses to the meeting at which the award will be presented.
Recipients: Marye Anne Fox 1988; Janet G. Osteryoung 1987; Jeanette G. Graselli 1986; Catherine C. Fenselau 1985; Martha L. Ludwig 1984; Ines Mandl 1983; Sara Jane Rhoads 1982; Elizabeth K. Weisburger 1981; Helen M. Free 1980; Jenny P. Glusker 1979; Madeleine M. Joullie 1978; Majorie G. Horning 1977; Isabella L. Karle 1976; Marjorie C. Caserio 1975; Joyce J. Kaufman 1974; Mary L. Good 1973; Jean'ne M. Shreeve 1972; Mary Fieser 1971; Ruth R. Benerito 1970; Sofia Simmonds 1969; Gertrude B. Elion 1968.

IPATIEFF PRIZE ... to recognize outstanding chemical experimental work in the field of catalysis of high pressure, carried out by men or women of any nationality who have not passed their 40th birthday on April 30 of the year in which the award is presented. The award will consist of the income from a trust fund and a diploma setting forth the reasons for the award. The financial value of the prize may vary, but it is expected that it will be approximately $3,000 and that it will be awarded every three years.

An allowance will be provided to cover travel expenses incidental to conferment of the award.

Recipients: Robert M. Hazan 1986; D. Wayne Goodman 1983; Denis Forster 1980; Charles A. Eckert 1977; George A. Samara 1974; Paul B. Venuto 1971; Charles R. Adams 1968; Robert H. Wentorf, Jr. 1965; Charles Kemball 1962; Cedomir M. Sliepcevich 1959; Harry G. Drickamer 1956; Robert B. Anderson 1953; Herman E. Ries 1950; Louis Schmerling 1947.

IRVING LANGMUIR AWARD IN CHEMICAL PHYSICS SPONSORED BY THE GENERAL ELECTRIC FOUNDATION ... to recognize and encourage outstanding interdisciplinary research in chemistry and physics, in the spirit of Irving Langmuir. The award consists of $10,000 and a scroll and is presented in even-numbered years. (Selection and presentation is made by the Division of Chemical Physics of the American Physical Society in odd-numbered years). An allowance is provided for traveling expenses to the meeting at which the award is presented.

Recipients: Richard B. Bernstein 1988; Martin Karplus 1987; Sidney W. Benson 1986; Richard N. Zare 1985; Robert Zwanzig 1984; Dudley R. Herschbach 1983; Benjamin Widom 1982; Willis H. Flygare 1981; William Klemperer 1980; Donald S. McClure 1979; Rudolph A. Marcus 1978; Aneesur Rahman 1977; John S. Waugh 1976; Robert H. Cole 1975; Harry G. Drickamer 1974; Peter M. Rentzepis 1973; Harden M. McConnell 1972; Michael E. Fisher 1971; John A. Pople 1970; Charles P. Slichter 1969; Henry Eyring 1968.

JAMES BRYANT CONANT AWARD IN HIGH SCHOOL CHEMISTRY TEACHING SPONSORED BY ETHYL CORPORATION ... to recognize, encourage, and stimulate outstanding teachers of high school chemistry in the United States, its possessions or territories, at both the regional and national levels. The national award consists of $3,000 and a certificate. Expenses incidental to traveling to the meeting at which the award will be presented will be paid.

Recipients: Edmund J. Escudero 1988; Mary C. Johnson 1987; Ronald I. Perkins 1986; Douglas D. Smith 1985; Douglas A. Halsted 1984; Janet A. Harris 1983; Robert Roe, Jr. 1982; Floyd F. Sturtevant 1981; Evelyn R. Bank 1980; Shirley E. Richardson 1979; Samuel H. Perlmutter 1978; Sidney P. Harris 1977; Dorothea H. Hoffman 1976; George W. Stapleton 1975; Wallace J. Gleekman 1974; Melvin Greenstadt 1973.

THE JAMES FLACK NORRIS AWARD IN PHYSICAL ORGANIC CHEMISTRY SPONSORED BY THE NORTHEASTERN SECTION, ACS ... to encourage and reward outstanding contributions to physical organic chemistry. The award consists of $3,000 and a suitably engraved certificate. An allowance of not more than $1,000 is provided for traveling expenses to the meeting at which the award will be presented.

Recipients: Nicholas J. Turro 1988; Paul von R. Schleyer 1987; John I. Brauman 1986; Paul G. Gassman 1985; M. J. S. Dewar 1984; Glen A. Russell 1983; Andrew Streitwieser, Jr. 1982; Jay K. Kochi 1981; Ronald Breslow 1980; John D. Roberts 1979; Jerome A. Berson 1978; Edward M. Arnett 1977; Howard E. Zimmerman 1976; Kurt M. Mislow 1975; Gerhard L. Closs 1974; Kenneth B. Wiberg 1973; Stanley J. Cristol 1972; Cheves Walling 1971; Frank H. Westheimer 1970; Paul D. Bartlett 1969; George S. Hammond 1968.

JAMES T. GRADY-JAMES H. STAEK AWARD FOR INTERPRETING CHEMISTRY FOR THE PUBLIC ... to recognize, encourage and stimulate outstanding reporting directly to the public, which materially increases the public's knowledge and understanding of chemistry, chemical engineering, and related fields. The award consists of $3,000, a gold medal, and a bronze replica of the medal. An allowance of up to $1000 is provided for traveling expenses to the meeting at which the award will be presented.

Recipients: Arthur Fisher 1988; Al Rossiter, Jr. 1987; Ben Patrusky 1986; Joseph Alper 1985; Cristine Russell 1984; Matt Clark 1983; Albert Rosenfeld 1982; Robert W. Cooke 1981; Edward Edelson 1980; Peter Gwynne 1979; Michael Woods 1978; Patrick Young 1977; Gene Bylinsky 1976; Jon Franklin 1975; Ronald Kotulak 1974; O. A. Battista 1973; Dan Q. Posin 1972; Victor Cohn 1971; Robert C. Cowen 1970; Walter Sullivan 1969; Raymond A. Bruner 1968.

JOEL HENRY HILDEBRAND AWARD IN THE THEORETICAL AND EXPERIMENTAL CHEMISTRY OF LIQUIDS SPONSORED BY SHELL COMPANIES FOUNDATION INCORPORATED ... to recognize distinguished contributions to the understanding of the chemistry and physics of liquids. The award consists of $3,000, a certificate, and an allowance of up to $1,000 for travel expenses incidental to conferral of the award.

Recipients: Hans C. Andersen 1988; Stuart A. Rice 1987; Frank H. Stilinger 1986; Berni J. Alder 1985; Robert L. Scott 1984; Jiri Jonas 1983; Joel H. Hildebrand 1981.

NOBEL LAUREATE SIGNATURE AWARD FOR GRADUATE EDUCATION IN CHEMISTRY SPONSORED BY J. T. BAKER CHEMICAL CO. ... to recognize an outstanding graduate student and his or her preceptor, in the field of chemistry, as broadly defined. The graduate student will receive $3,000 and a plaque containing the signatures of Nobel Laureates. The student's preceptor will receive $3,000 and a plaque for permanent display in the institution's Chemistry Department. Traveling expenses of both recipients incidental to the conferring of the award will be paid.

Recipients: David L. Clark and Malcolm Chisholm, Indiana University 1988; Mark D. Hollingsworth and J. Michael McBride, Yale University 1987; Robert L. Whetten, Gregory S. Exra and Edward R. Grant, Cornell University 1986; Peter G. Schultz and Peter B. Dervan, Cal Tech 1985; Christopher S. Gudeman and R. Claude Woods, University of Wisconsin, Madison 1984; David J. Nesbitt, James T. Hynes, and Stephen R. Leone, University of Colorado 1983; Warren S. Warren and Alexander Pines, UCal, Berkeley 1982; James C. Weisshaar, UCal, Berkeley 1981; Wayne L. Gladfelter, Penn State University 1980.

THE PETER DEBYE AWARD IN PHYSICAL CHEMISTRY SPONSORED BY E. I. DU PONT DE NEMOURS & COMPANY ... to encourage and reward outstanding research in physical chemistry. The award consists of $5,000 and a certificate. Traveling expenses to the meeting at which the award will be presented will be paid.

Recipients: Rudolph A. Marcus 1988; Harry G. Drickamer 1987; Yuan T. Lee 1986; Stuart A. Rice 1985; B. S. Rabinovitch 1984; George C. Pimentel 1983; Peter M. Rentzepis 1982; Richard B. Bernstein 1981; Robert W. Zwanzig 1976; H. S. Gutowsky 1975; Walter H. Stockmayer 1974; William N. Lipscomb, Jr. 1973; Clyde A. Hutchison, Jr. 1972; Norman Davidson 1971; Oscar K. Rice 1970.

PRIESTLEY MEDAL ... to recognize distinguished services to chemistry. The award consists of a gold medal designed to commemorate the work of Joseph Priestley, and a bronze replica of the medal. It may not be awarded more than once to the same individual. The traveling expenses incidental to the conferring of the medal are paid.

Recipients: Frank H. Westheimer 1988; John D. Roberts 1987; Karl A. Folkers 1986; Henry Taube 1985; Linus Pauling 1984; Robert S. Mulliken 1983; Bryce Crawford, Jr. 1982; Herbert C. Brown 1981; Milton Harris 1980; Glenn T. Seaborg 1979; Melvin Calvin 1978; Henry Gilman 1977; George S. Hammond 1976; Henry Eyring 1975; Paul J. Flory 1974; Harold C. Urey 1973; George B. Kistiakowsky 1972; Frederick D. Rossini 1971; Max Tishler 1970; Kenneth S. Pitzer 1969; William G. Young 1968.

ROGER ADAMS AWARD IN ORGANIC CHEMISTRY SPONSORED BY ORGANIC REACTIONS, INC. AND ORGANIC SYNTHESES, INC ... to recognize and encourage outstanding contributions to research in organic chemistry defined in its broadest sense. The award consists of a gold medal, a sterling silver replica of the medal, and $15,000. The award will be presented biennially. The recipient shall deliver a lecture at the Biennial National Organic Chemistry Symposium of the American Chemical Society at which time the award will be presented. The travel expenses to the Symposium will be paid.

Recipients: Jerome A. Berson 1987; Donald J. Cram 1985; A. R. Battersby 1983; Nelson J. Leonard 1981; Melvin S. Newman 1979; William S. Johnson 1977; Rudalf Huisgen 1975; Georg Wittig 1973; Herbert C. Brown 1971; Vladimir Prelog 1969.

ACS AWARD IN ORGANOMETALLIC CHEMISTRY SPONSORED BY THE DOW CHEMICAL COMPANY FOUNDATION ... to recognize a recent advancement that is having major impact on research in organometallic chemistry. The award consists of $3000 and a certificate. An allowance of $500 is provided for traveling expenses.
Recipients: Robert H. Grubbs 1988; K. Peter C. Vollhardt 1987; Robert G. Bergman 1986; Richard R. Schrock 1985.

ACS AWARD IN SEPARATIONS SCIENCE AND TECHNOLOGY SPONSORED BY ROHM AND HAAS COMPANY ... to recognize outstanding accomplishments in fundamental or applied research directed to separations science and technology. The award consists of $3,000 and a plaque. Traveling expenses incidental to conferral of the award will be paid.
Recipients: Norman N. Li 1988; Friedrich G. Helfferich 1987; J. Calvin Giddings 1986; Alan S. Michaels 1985; D. B. Broughton 1984.

EARLE B. BARNES AWARD FOR LEADERSHIP IN CHEMICAL RESEARCH MANAGEMENT SPONSORED BY THE DOW CHEMICAL COMPANY ... to recognize outstanding achievements in chemical research management. The award consists of $5,000 and a certificate. Traveling expenses incidental to conferral of the award will be paid.
Recipients: William P. Slichter 1988; Malcolm E. Pruitt 1987; Robert M. Adams 1986; H. W. Coover 1985; James F. Mathis 1984.

FRANK H. FIELD AND JOE L. FRANKLIN AWARD FOR OUTSTANDING ACHIEVEMENT IN MASS SPECTROMETRY SPONSORED BY EXTREL CORPORATION ... recognizes outstanding achievement in the development or application of mass spectrometry. In odd numbered years the award recognizes advances in techniques or fundamental processes; in even numbered years the award recognizes development of the applications of mass spectrometry. The award consists of $3,000 and a certificate. An allowance of up to $1000 is provided for traveling expenses incidental to conferral of the award.
Recipients: Frank H. Field 1988; John Beinon 1987; Klaus Biemann 1986; A. O. C. Nier 1985

AMERICAN CONCRETE INSTITUTE (ACI)
22400 West Seven Mile Road, Detroit, Michigan 48219
George F. Leyh, Executive Vice President

PRESIDENTS ... T. E. Northup 1987; Walter E. Kunze 1986; Emery Farkas 1985; Ignacio Martin 1984; Norman L. Scott 1983; Peter Smith 1982; T. Z. Chastain 1981; Charles J. Pankow 1980; John F. McLaughlin 1979; John L. Goetz 1978; Richard C. Mielenz 1977; Russell S. Fling 1976; Joseph H. Walker 1975; Chester P. Siess 1974; Robert E. Philleo 1973; Edward Cohen 1972; Walter McCoy 1971; Samuel Burks 1970.

AWARDS
ACI CONSTRUCTION PRACTICE AWARD ... presented for a paper of outstanding merit on concrete construction practice. The award is intended to honor the construction man - the man whose resourcefulness comes in between the paper conception and the solid fact of a completed structure.
Recipients: Donald F. Meinheit and Jack F. Monson 1986; Gerry Weiler and Cameron Kemp 1985; Richard C. Meininger 1984; Raymond C. Heun 1983; Mary K. Hurd 1982; John Bickley 1981; H. W. Chung 1980; David Brown 1979; Arvid Grant 1977; Raymond C. Heun, Jesse Schwartz, Kenneth D. Wheeler, William Rohde, Charles Shimel, Joseph R. Wolf 1976; Kenneth G. Jessop 1975; Charles Novacek 1974; John J. White 1971; David J. Fitz Gerald 1970; N. P. Angeles, Lin Y. Huang, Jack L. Korb, Howard R. May and Keith C. Thornton 1969; James M. Keith and Everett W. Osgood 1968.

ALFRED E. LINDAU AWARD ... for outstanding contributions to reinforced concrete design practice.
Recipients: James O. Jirsa 1986; John A. Martin 1985; Robert Englekirk 1984; Richard M. Gensert 1983; Paul F. Rice 1982; Armano Gustaterro 1981; Maurice P. Van Buren 1979; Mark Fintel 1978; Eivind Hognestad 1977; Hubert Rusch 1976; Roger Nicolet 1975; Maz Zar 1974; Fazlur R. Khan 1973; Phil M. Ferguson

1972; Clyde E. Kesler 1971; Alan H. Mattock 1970; E. E. Rippstein and J. F. Seifried 1969; Arthur R. Anderson 1968.

ARTHUR R. ANDERSON AWARD ... presented for outstanding contributions to the advancement of knowledge of concrete as a construction material.
Recipients: Boris Bresler 1986; William L. Dolch 1985; Robert E. Tobin 1984; Floyd O. Slate 1983; Katharine Mather 1982; Robert E. Philleo 1981; Paul Klieger 1980; Joseph J. Waddell 1979; Paul Kliegen 1978; Levi S. (Doc) Brown 1977; T. C. Powers 1976; Milos Polivka 1975; George J. Verbeck 1974; Adam M. Neville 1973.

CEDRIC WILLSON AWARD ... given to an individual who, through his active participation in the American Concrete Institute, has contributed his knowledge to the Institute and the users of lightweight aggregates, lightweight concrete, and lightweight masonry.
Recipients: John W. Roberts 1986; Thomas A. Holm 1984; Joseph J. Shidler 1982; Truman Jones 1980; Robert Tobin 1979; Frank Erskine 1978; J. A. (Bob) Hanson 1977.

CHARLES S. WHITNEY MEDAL ... awarded for noteworthy engineering development work in concrete design or construction. This award may be extended to a firm or agency, or to an individual.
Recipients: Sargent & Lundy 1986; United States Naval Civil Engineering Lab 1984; Figg and Muller Engineers, Inc. 1983; T. Y. Lin International 1981; The Phil Ferguson Structural Engineering Laboratory of the University of Texas at Austin 1980; The Center for Building Technology, National Bureau of Standards 1979; Wiss, Janney, Elstner and Associates, Inc., 1977; Research and Development Division of the Cement and Concrete Association, England 1976; Concrete Technology Corporation, Tacoma, Washington 1975; Concrete Laboratory, Waterways Experiment Station 1974; U. S. Bureau of Reclamation-Engineering Laboratories 1973; Research and Development Laboratories-Portland Cement Association 1972; Engineering Materials Laboratory, University of California, Berkeley 1971; Ammann and Whitney 1970; Eric L. Erikson 1969; Joseph Lacy-Kevin Roche-John Dinkeloo 1968.

HENRY C. TURNER MEDAL ... awarded for notable achievements in or service to the concrete industry.
Recipients: Robert E. Englekirk 1986; Material Service Corporation 1985; Robert E. Philleo 1984; V. Mohan Malhotra 1983; Edward A. Abdun-Nur 1982; Walter Price 1981; William M. Avery 1980; Solomon Cady Hollister 1979; Arthur Anderson 1978; Raymond C. Reese 1977; Phil M. Ferguson 1976; William A. Maples 1975; Ben C. Gerwick 1974; Bryant Mather 1973; George Winter 1972; George E. Warren 1971; Harry B. Zachrison, Sr. 1970; Milo S. Ketchum 1968.

HENRY L. KENNEDY AWARD ... presented for outstanding technical or administrative service to the institute.
Recipients: Clarkson W. Pinkham 1986; Loring A. Wyllie, Jr. 1985; Gerald B. Neville 1984; George C. Hoff 1983; James R. Libby 1982; Emery Farkas 1981; Samuel J. Henry 1980; Francis Principe 1979; Bertold E. Weinberg 1977; Thomas J. Reading 1976; John Arvid (Bob) Hanson 1975; Emil Schmid 1974; Richard C. Mielenz 1973; Edward G. Nawy 1972; Eivind Hogestad 1971; W. J. McCoy 1970; Bruce E. Foster 1969; Samuel Hobbs 1968.

RAYMOND C. REESE STRUCTURAL RESEARCH AWARD ... presented to the author or authors of a paper published by ACI that describes a notable achievement in research related to structural engineering and which indicates how the research can be used.
Recipients: Kwok-Nam Shiu, Dr. J. Dario Aristizabal-Ochoa, W. Gene Corley and Ralph G. Oesterle 1986; Frank J. Heger and Mehdi S. Zarghamee 1985; B. D. Scott, R. Park, and M. J. N. Priestley 1984; Thomas Paulay and Roy G. Taylor 1983; Thomas Paulay 1982; Neil Hawkins & Denis Mitchell 1981; Anders Losberg 1980; John Breen 1979; Neil Hawkins 1978; J. O. Jirsa and John Minor 1977; Denis Mitchell and Michael P. Collins 1976; B. Vijaya Rangan and A. S. Hall 1975; Phil M. Ferguson and K. S. Rajagopalan 1974; Alexander Placas and Paul E. Regan 1973; James G. MacGregor, John E. Breen and Edward O. Pfrang 1972.

ROGER H. CORBETTA CONCRETE CONSTRUCTOR AWARD ... presented to an individual who, as a constructor, has made significant contributions to progress in new and innovative methods of concrete construction.
Recipients: E. Robert Jean 1986; R. E. "Jeff" Kasler 1985; Cecil V. Wellborn 1984; Ontario Hydro-Electric Power Company 1983; Ben C. Gerwick, Jr. 1981; Gerald Miller 1980; Charles J. Prokop 1978; Michael A. Lombard 1977; Phillip R. Jackson 1976; Knut Tovshus 1975; Charles J. Pankow 1974; Arthur R. Anderson 1973.

WASON MEDAL FOR MATERIALS RESEARCH ... awarded to the member or members of the Institute reporting in a paper noteworthy original research work or a discovery relating to materials.
Recipients: Dr. Nicholas J. Carino 1986; George C. Hoff and Alan D. Buck 1985; Y. Yamamoto and M. Kobayashi 1984; Tony C. Liu 1983; James F. Best and Ralph O. Lane 1982; L. M. Meyer & W. F. Perenchio 1981; H. S. Lew and T. W. Reichard 1980; Phillip B. Bamforth & Roger Browne 1979; Kurt F. Wendt and George W. Washa 1977; Ronald A. Coleman and Colin D. Johnston 1976; Norbert K. Becker and Cameron MacInnis 1975; Tony C. Y. Liu and Arthur H. Nilson and Floyd O. Slate 1974; K. W. Nasser-R. P. Lohtia 1973; G. L. Kalousek-E. J. Benton 1972; J. E. Backstrom-L. E. Kukacka-M. Steinberg-J. T. Dikeou 1971; William Gene Corley-Neil M. Hawkins 1970; Larry E. Farmer- Phil M. Ferguson-Egur Ersoy 1969; Alan D. Buck-W. D. Dolch 1968.

WASON MEDAL FOR "THE MOST MERITORIOUS PAPER" ... awarded each year to the author or authors of the most meritorious paper published by ACI.
Recipients: Floyd O. Slate, Arthur H. Nilson, and Salvador Martinez 1986; Robert Brandt Johnson 1985; Ernest K. Schrader 1984; J. S. Ford, D. C. Chang, and John E. Breen 1983; Rudolph C. Valore, Jr. 1982; Russell Fling 1981; Robert E. Tobin 1980; James Shilstone 1979; James Libby 1978; J. O. Jirsa and Jose L. G. Marques 1977; Milton H. Wills, Jr. 1976; Frank A. Randall, Jr. 1975; Y. C. Yang 1974; M. Jack Snyder, F. F. Fondriest, David R. Lankard and Donald L. Birkimer 1973; John E. Breen, James G. MacGregor and Edward O. Pfang 1972; Mark Fintel and Fazlur R. Khan 1971; Hubert Woods 1970; G. N. J. Kani 1969; Leif Ericksson and Torben C. Hansen 1968.

JOE W. KELLY AWARD ... presented for outstanding contributions to education in the broad field of concrete.
Recipients: James G. MacGregor 1986; Paul Zia 1985; Noel J. Everard 1984; Charles G. Salmon 1983; Cutberto Diaz-Gomez 1982; John E. Breen 1981; Edward L. Kawala 1980; George Winter 1979; Boris Bresler 1978; Phil M. Ferguson 1977; Mete A. Sozen 1976; William A. Cordon 1975.

DELMAR L. BLOEM DISTINGUISHED SERVICE AWARD ... presented to a current or recent chairman of a technical committee or to deserving individuals in recognition of outstanding performance.
Recipients: David Darwin, Charles H. Henager, and Henry G. Russell 1986; Timothy J. Fowler, David W. Fowler, Ralph L. Duncan, and Mete Sozen 1985; Stewart C. Watson, Theodore R. Crom, Harry Stavrides, and Charles G. Salmon 1984; V. Mohan Malhotra 1983; William C. Black 1982; Peter Gergely, Donald L. Houghton, and Byron I. Zolin 1981; James T. Dikeou, J. R. Dise, and Eivind Hognestad 1980; Peter D. Courtois, I. Leon Glassgold, and Eric C. Smith 1979; W. Gene Corley and Eugene P. Holland 1978; Richard C. Mielenz and Howard Newlon, Jr. 1977; Joseph A. Dobrowolski, John M. Hanson, and T. E. Northup 1976; Fritz Kramrisch and James G. MacGregor 1975; Calvin C. Oleson and W. Gordon Plewes 1974; Edward Cohen, George F. Leyh, and M. Daniel Vanderbilt 1973; Hector I. King, Albyn Mackintosh, Robert Price, and Peter Smith 1972; Noel J. Everard, Ernest C. Fortier, and Shu-t'ien Li 1971; Russell S. Fling, J. A. Hanson, Paul Kliger, W. H. Kuenning, George H. Nelson, Joseph R. Proctor, Jr., Thomas J. Reading, and Lewis H. Tuthill 1970.

AMERICAN CONGRESS ON SURVEYING AND MAPPING (ACSM)

210 Little Falls Street, Falls Church, Virginia 22046
Richard F. Dorman, CAE, Executive Director
PRESIDENTS ... Alberta Auringer Wood 1987-88; Donald E. Bender 1986-87; John D. Bossler 1985-86

AMERICAN CONSULTING ENGINEERS COUNCIL (ACEC)

1015 15th Street N. W., Suite 802, Washington, D. C. 20005
Howard M. Messner, Executive Vice President
PRESIDENTS ... Lester H. Smith, Jr. 1987-88; Lester H. Poggemeyer 1986-87; Arnold L. Windman 1985-86; Clifford E. Evanson 1984-85; Shelby K. Willis 1983-84; Russell L. Smith, Jr. 1982-83; William R. Ratliff 1981-82; Everett S. Thompson 1980-81; George W. Barnes 1979-80; R. Duane Monical 1978-79; William A. Clevenger 1977-78; Richard H. Stanley 1976-77; Billy T. Sumner 1975-76; Malcolm Meurer 1974-75; William N. Holway 1973-74.

AWARDS

ACEC ENGINEERING EXCELLENCE AWARDS ... to recognize those engineering achievements demonstrating the highest degree of skill and ingenuity and providing a signficant benefit to the public welfare and to the private practice of engineering. Top award chosen as "Grand Conceptor".
Recipients: Howard Needles Tammen & Bergendoff 1987; Sverdrup/Parsons Brinckerhoff 1986; Sverdrup & Parcel 1985; Greiner Engineering Sciences 1984; Sverdrup & Parcel 1983; Williams and Works/Environmental Data, Inc. 1982; McClelland Engineers/CBM Engineers, Inc. 1981; Williams and Works 1980; URS-Madigan-Praeger, Inc. 1979; Ch2m Hill 1978; Kramer, Chin and Mayo, Inc. 1977; Tippetts-Abbett-McCarthy-Stratton 1976; Howard, Needles, Tammen and Bergendoff 1975; Greenleaf-Telesca-Kellerman & Dragnet, Inc. (joint venture) 1974; Ketchum-Konkel-Barrett-Nickel-Austin 1973; Midwestern Consulting, Inc. 1972; Sandwell International, Inc. 1971; International Engineering Co. 1970; Ryckman, Edgerley, Tomlinson, & Associates 1969; Anderson-Bjornstad- Kane 1968.

AMERICAN DEFENSE PREPAREDNESS ASSOCIATION (ADPA)

Rosslyn Center, Suite 900, 1700 No. Moore, Arlington, Virginia 22209
PRESIDENTS ... Lt. General Lawrence F. Skibbie, US Army (Ret.); Gen. Henry A. Miley, Jr. 1977-87.

AWARDS

THE CROZIER PRIZE ... the Crozier Prize is related primarily to scientific research that leads toward the development of weapons systems.
Recipients: Dr. Janet Sue Fender, Dept. of Air Force 1987; Dr. David W. Kerr, Airborne Radar Branch, Naval Research Lab 1986; Dr. Robert E. Weigle, U.S. Army Research Office, Research Triangle Park, North Carolina 1985; Dr. Robert J. Eichelberger, U.S. Army Ballistic Research Laboratory 1984; Dr. Mary C. Henry, U.S. Army Medical Bioengineering Research and Development Laboratory 1983; Carroll Key, Jr., Applied Research Laboratory, Pennsylvania State 1982; Dr. J. Richard Airey, Department of Defense 1981; Leonard W. Price, Hughes Helicopters 1980; Robert G. S. Sewell, Naval Weapons Ctr, China Lake 1979; Theodore M. Benziger, Los Alamos Laboratory 1978; Mark K. Miller 1977.

THE KNOWLES AWARD ... The Knowles Award, established by his family in memory of Harvey C. Knowles, the founder of the American Defense Preparedness Assoc.'s Technical Divisions, shall be presented each year to the American citizen who has been adjudged by the association to have made a major technical contribution to our national armament progress.

Recipients: Weldon Word, Texas Instrument 1982; Frank L. Jarrett, Food Machinery Corp. 1981; William V. Goodwin, RCA Corporation 1980; Abraham Flatau, US Army ARRADCOM 1979; Norman R. Augustine, Martin Marietta 1978; Earl H. Buchanan, US Army ARRADCOM 1977.

SIMON AWARD
Recipients: Willis J. Willoughby, Jr., Dept. of Navy 1987; Victor Linder, US Army Armament Research, Dev. & Engg. Center, and Dr. William C. McCorkle, US Army Missile Command 1986; R Adm. Stephen J. Hostettler, Director Joint Cruise Missiles Project 1985; Richard C. Chapman, Naval Underwater Systems Center 1984; Dr. John D. Weisz, U.S. Army Human Engineering Laboratory 1983.

AMERICAN FOUNDRYMEN'S SOCIETY (AFS)

Golf & Wolf Roads, Des Plaines, Illinois 60016
Charles Jones, Executive Vice President

PRESIDENTS ... Anton Dorfmueller, Jr. 1987-88; John L. Kelly 1986-87; William M. O'Neill 1985-1986; George N. Booth 1984-1985; Hugh M. Sims, Jr. 1983-1984; Eugene E. Paul 1982-1983; Lawrence S. Krueger 1981-82; Charles E. Drury 1980-81; J. R. Bodine 1979-80; John A. Wagner, Jr. 1978-79; Roy F. Nosek 1977-78; Thomas R. Wiltse Dec. 1975-Apr. 1977; Frank S. Ryan May-Dec 1975; Charles E. Fausel 1974-75; S. C. Clow 1973-74.

AWARDS

AWARD OF SCIENTIFIC MERIT ... intended exclusively as a technical citation, to recognize (for example) outstanding technical papers, meritorious contributions or services, and development of a new casting process, production method or engineering advancement having "future possibilities".
Recipients: William B. Huelsen, William D. Scott 1987; Louis J. Pedicini, Victor T. Popp, Robert F. Schmidt 1986; Bradley H. Booth, Richard A. Green and Norris B. Luther 1985; Richard H. Toeniskoetter, and William M. Ball III 1984; Frederic W. Jacobs, Seymour Katz and Donald G. Schmidt 1983; Raymond M. Leliaert, Sr. 1982; Rodney L. Naro, Kenneth J. Pol, William E. Willis, Jr. 1981; Thomas W. Seaton, Arthur F. Spengler 1980; Frank B. Hall, Robert D. Mitchell, Harrison Weaver, James T. Williams 1979; Matt J. Granlund, Robert S. Lee, Alan P. Volkmar 1978; William S. Hackett, Woodrow W. Holden and Paul J. Mikelonis 1977; G. M. Etherington and C. Hibbard Savery 1976; David W. Boyd, Norman P. Lillybeck and William P. Shulhof 1975; Alex L. Graham and Raymond C. Howell 1974; Albert H. Rauch and Frank S. Brewster 1973; Thaddeus Giszczak-Vernon H. Patterson-Charles W. Ward, Jr. 1972; Robert B. Fischer-George J. Vingas-George DiSylvestro 1971; Robert A. Colton-Wayne H. Buell and C. E. Wenninger 1970; Larson E. Wile-Price B. Burgess 1969; Kenneth E. Nelson 1968.

JOHN A. PENTON GOLD MEDAL ... instituted to honor outstanding achievement in the metal casting field.
Recipients: Ralph A. Carlson, American Cast Iron Pipe Co., Birmingham, Alabama 1985; Alex L. Graham, Harry W. Dietert Co., Detroit, Michigan 1982; John O'Meara 1980; Thomas T. Lloyd 1977; David Matter 1975; Carl R. Loper 1972.

JOHN H. WHITING GOLD MEDAL ... instituted to honor outstanding achievements in the metal casting field.
Recipients: Lyle R. Jenkins, Wagner Castings Co., Ductile Iron Society, Decatur, Illinois 1984; Vernon H. Patterson 1977; Fred J. Webbere 1974; Harry Kessler 1971; Samuel Lipson 1968.

JOSEPH S. SEAMAN GOLD MEDAL ... instituted to honor outstanding achievements in the metal casting field.
Recipients: Robert B. Fischer, Ingersoll Rand Co., Easton, Pennsylvania 1983; William J. Willmot 1981; George E. Tubich 1979; Robert L. Doelman 1976; Edwin Walcher, Jr. 1975; Harvey E. Henderson 1973; Anton Dorfmueller, Jr. 1970.

PETER L. SIMPSON GOLD MEDAL ... instituted to honor outstanding achievements in the metal casting field.
Recipients: Charles E. Drury 1987; Ashley B. Sinnett 1986; Thomas R. Wiltse, Central Foundry Division, General Motors Corp., Saginaw, Michigan 1985; Burleigh E. Jacobs, Jr., Crede

Foundries, Inc., Milwaukee, Wisconsin 1983; Glen Stahl, Stahl Specialty Co., Kingsville, Missouri 1982; Charles F. Seelbach, Jr 1979; Bernard N. Ames 1978; Karl L. Landgrebe 1977; Collins L. Carter 1974; John A. Wagner, Sr. 1973; Aldo Dacco 1972; Elmer E. Braun 1969.

THOMAS W. PANGBORN GOLD MEDAL ... instituted to honor outstanding achievements in the metal casting field.
Recipients: John E. Miller 1986; Arthur A. Avedisian 1981; Paul K Trojan 1976; Howard L. Womochel 1976; Kenneth R. Daniel 1974; Sol L. Gertsman 1973; George A. Colligan 1970; Harold W. Ruf 1968.

W. H. McFADDEN GOLD MEDAL ... instituted to honor outstanding achievement in the metal casting field.
Recipients: Richard A. Green 1987; Charles E. Fausel, Lester B. Knight and Associates, Chicago, Illinois 1984; John O. Edwards, Consultant, Ottawa, Ontario 1982; George J. Vingas 1980; Jack Goudzwaard 1978; Bruce L. Simpson 1976; Joseph S. Schumacher 1974; Lester B. Knight 1972; E. C. Troy 1969.

AMERICAN INDIAN SCIENCE & ENGINEERING SOCIETY (AISES)

1085 14th Street Suite 1506, Boulder, Colorado 80302-7309
CHAIRMEN ... Robert K. Whitman 1986-1987; Donald Ridley 1983-1985; George Thomas 1981-82; Alfred Qoyawayma 1979-80; A. T. Anderson 1977-78.

AWARDS

ELY S. PARKER AWARD ... "intended to recognize Indian scientists or engineers who most clearly represent the purpose and goals of AISES."
Recipients: Al Qoyawayma 1986; Mary Ross 1985; Phil Stevens 1984; A. T. Anderson 1983.

AMERICAN INSTITUTE OF AERONAUTICS AND ASTRONAUTICS, INC. (AIAA)

370 L'Enfant Promenade, SW, Washington, D.C. 20024-2518
James J. Harford, Executive Director Nelson W. Friedman, Corporate Secretary

PRESIDENTS ... William F. Ballhaus, Jr. 1988; Larry Adams 1987; Allen E. Fuhs 1986; Alan M. Lovelace 1985; John L. McLucas 1984; Norman R. Augustine 1983; Michael I. Yarymorych 1982; Joseph G. Gavin, Jr. 1981; Artur Mager 1980; George E. Mueller 1979; F. A. Cleveland 1978; Rene H. Miller 1977; Edgar M. Cortright 1976; Grant L. Hansen 1975; Daniel J. Fink 1974; Holt Ashley 1973.

AWARDS

AERO-ACOUSTICS AWARD ... "for an outstanding technical scientific achievement by an individual's contribution to the field of aircraft community noise reduction."
Recipients: David Crighton 1986; Geoffrey Lilley 1984; Marvin Goldstein 1983; Thomas G. Sofrin 1981; Alan Powell 1980; Harvey H. Hubbard 1979; John E. Ffowcs Williams 1977; Herbert Ribner 1976; Michael J. Lighthill 1975.

AERODYNAMIC DECELERATOR AND BALLOON TECHNOLOGY AWARD ... "to recognize significant contribution to the effectiveness and/or safety of aeronautical or aerospace systems through development or application of the art and science of aerodynamic decelerator technology."
Recipients: Domina Jalbert and William B. Pepper 1986; Herman Engel, Jr. 1984; Theodore W. Knacke 1981; (First award presented in 1979).

AEROSPACE COMMUNICATIONS AWARD ... for an outstanding contribution in the field of aerospace communications.
Recipients: C. Louis Cuccia 1986; Sidney Metzger 1984; Walter Morgan 1982; Irwin M. Jacobs and Andrew J. Viterbi 1980; Leonard Jaffey 1978; Robert Garbarini 1976; Arthur C. Clarke 1974; Wilbur Pritchard 1972; Siegfried H. Reiger 1971; Edmund

J. Habib 1970; Eberhardt Rechtin 1969; Dr. Harold A. Rosen 1968.

AEROSPACE CONTRIBUTION TO SOCIETY AWARD ... "to recognize a notable contribution to society through the application of aerospace technology to societal needs."
Recipients: E. Philip Muntz 1987; Engineering Sciences Division, US Air Force Academy 1986; Kenneth Taylor 1985; Ralph R. Nash 1984; Richard Davies 1983; William L. Smith 1982; Fred McFee 1981; Melvin Hartmann 1980; Richard S. Johnston 1979; Elmer P. Wheaton 1978.

AIR BREATHING PROPULSION AWARD ... "for meritorious accomplishment in the science or art of air breathing propulsion, including turbo-machinery or any other technical approach dependent upon atmospheric air to develop thrust or other aerodynamic forces for propulsion or other purposes for aircraft or other vehicles in the atmosphere or on land or sea."
Recipients: Gordon C. Oates 1987; Frank E. Pickering 1986; Robert Hawkins 1985; Bernard Koff 1984; Donald W. Bahr 1983; William Sens 1982; Fred McFee 1981; Melvin Hartmann 1980; Arthur J. Wennerstrom 1979; William J. Blatz 1978; Edward Woll 1977; Frederick Rall, Jr. 1976

AIRCRAFT DESIGN AWARD ... "for the conception, definition or development of an original concept leading to a significant advancement in aircraft design technology."
Recipients: Lawrence A. Smith 1987; James Allburn, Norris Krone and Glenn Spacht 1986; Harry Hillaker 1985; Philip Conslit, Kenneth Holtby and Everett Webb 1984; Kenneth G. Wernicke 1983; Paul B. MacCready, Jr. 1982; Frank W. Davis, Adolph Burstein and Ralph Shick 1981; Albert Rutan 1980; Harold Raiklen 1979; John K. Wimpress 1978; Howard A. Evans 1977; Kendall Perkins 1976; Walter E. Fellers 1975; Richard T. Whitcomb 1974; Herman D. Barkey 1973; Ben R. Rich 1972; Joseph F. Sutter 1971; Harrison A. Storms 1970; Harold W. Adams 1969.

JOHN LELAND ATWOOD AWARD (FORMERLY ASEE AEROSPACE DIVISON-AIAA EDUCATIONAL ACHIEVEMENT AWARD) ... "for improvements of lasting influence to Aerospace Engineering education through the introduction of experimental courses, experimental instruction methods, technical text books, technical articles on education, introduction of new-type laboratory or teaching equipment, demonstration of recent success as a teacher and/or advisor, ability to inspire students to high levels of accomplishment, etc., or significant recent achievements in furthering progress in Aero or Astro Engineering education and its administration by outstanding participation in pertinent educational and professional societies."
Recipients: Jan Roskam 1987; Conrad F. Newbury 1986; W. T. Fowler 1985; Robert L. Young 1984; Leroy Fletcher 1983; Ira Jacobson 1982; Thomas J. Mueller 1981; Donnell W. Dutton 1980; Robert F. Brodsky 1979; Stanley H. Lowy 1978; Barnes W. McCormick, Jr. 1976.

CERTIFICATE OF MERIT AWARD ... "for the purpose of promoting technical and scientific excellence" certificates of merit may be given in connection with a nationally sponsored AIAA activity. It is, however, of utmost importance that such recognition will not in any way duplicate, or interfere with the AIAA Technical Award System.
Recipients: Solid Rockets-T.H. Rytting, T.C. Bruce 1984; Solid Rockets-J.T. Lamberty 1982; Structures Design Lecture-Wolfgang H. Steurer 1981; Materials-Bryan R. Noton 1981; Liquid Propulsion- Gordon A. Dressler, Richard E. Morningstar, Robert Kelso, Robert L. Sackheim, Donald E. Fritz 1981; Liquid Propulsion-Larry B. Bassham, Harry W. Valler 1980; Air Breathing Propulsion-Paul L. Russell, Garry Brant, Richard Ernst, Francis N. Underwood, Jr. 1979; C. M. Willard, F. J. Capone, M. Komarski, H. L. Stevens 1977; Structures Design Lecture-Richard N. Hadcock 1979. Atmospheric Flight-Mechanics-Kenneth W. Illif, Richard E. Maine 1979.

CHANUTE FLIGHT AWARD ... "honoring Octave Chanute, pioneer aeronautical investigator, for outstanding contribution made by a pilot or test personnel to the advancement of the art, science and technology of aeronautics."
Recipients: George R. Jansen 1986; S. Lewis Wallick, Jr. 1983; Raymond L. McPherson 1981; Austin Bailey 1979; T. D. Benefield 1977; Thomas Stafford 1976; Alan L. Bear, Jack R. Lousma, Owen K. Garriott 1975; Charles A. Sewell 1974; Cecil W. Powell 1973; Donald R. Segner 1972; William M. Magruder 1971; Jerauld P. Gentry 1970; William C. Park 1969; William J. Knight 1968.

DE FLOREZ TRAINING AWARD ... "named in honor of the late Admiral de Florez and presented for an outstanding improvement in aerospace training in either aeronautics or astronautics."
Recipients: Edward A. Stark 1987; Waldemar O. Breuhaus 1986; John Sinacori 1985; Walter S. Chambers 1984; Robert M. Howe 1983; Richard J. Heintzman 1982; Carl B. Shelley 1980; James Burke 1979; William Hagin 1978; John E. Duberg 1977; John C. Dusterberry 1975; Hugh Harrison Hurt, Jr. 1974; Carroll H. Woodling 1973; James W. Campbell 1972; Capt. Walter P. Moran 1971; Harold G. Miller 1970; Gifford Bull 1969; Joseph LaRussa 1968.

DISTINGUISHED SERVICE AWARD ... "for an individual member of AIAA who has made great contributions over a period of years by his service to the Institute."
Recipients: Norman C. Baullinger 1987; Tom Gagnier 1986; William R. Williams 1985; Norman Chaffee 1984; George S. Mills 1982; Herbert O. Bair, George J. Frankel, R. Gilbert Moore, William Simmons, Edith L. Woodward 1981; N. A. Tony Armstrong 1980; Charles W. Eyres 1979; Ann Dickson 1978; Kenneth Randle 1977; Warren Curry 1975; Charles Appleman 1974; H. Norman Abramson 1973; William F. Chana 1972; Frederick H. Roever 1971; H. Dana Moran 1970; Peter C. Johnson 1969; Harvey C. Cook, Jr. 1968.

DRYDEN LECTURESHIP IN RESEARCH ... "intended to emphasize the great importance of basic research to programs in aeronautics and astronautics and to be a salute to research scientists and engineers, and named in honor of Dr. Hugh L. Dryden, renowned leader in aerospace research programs."
Recipients: Kenneth W. Iliff 1987; Irvine Glass 1986; Donald Coles 1985; Arthur E. Bryson, Jr. 1984; Edward C. Stone, Jr. 1983; Laurence R. Young 1982; Herbert S. Ribner 1981; Jack Kerrobrock 1980; Dean R. Chapman 1979; Gerald A. Soffen 1978; Abraham Hertzberg 1977; Anatol Roshko 1976; Antonio Ferri 1975; Herbert Hardrath 1974; Herbert Friedman 1973; John C. Houbolt 1972; Coleman Donaldson 1971; Bernard Budiansky 1970; Gerard P. Kuiper 1969; Hans W. Liepmann 1968.

ENERGY SYSTEMS AWARD ... for a significant contribution in the broad field of energy systems, specifically as related to the application of engineering sciences and systems engineering to the production, storage, distribution, and conservation of energy.
Recipients: L.S. 'Skip' Fletcher 1984; Stanford S. Penner 1983; Ken Touryan 1981.

FLUID AND PLASMADYNAMICS AWARD ... "for outstanding contribution to the understanding of the behavior of liquids and gases in motion and of the physical properties and dynamical behavior of matter in the plasma state as related to needs in aeronautics and astronautics."
Recipients: Steven A. Orszag 1986; Edward T. Gerry 1985; Tuncer Cebeci 1984; Seymour M. Bogdonoff 1983; John L. Lumley 1982; Arthur R. Kantrowitz 1981; Eli Reshotko 1980; Charles H. Kruger 1979; Charles E. Treanor 1978; Harvard Lomax 1977; Mark V. Morkovin 1976.

GENERAL AVIATION AWARD ... "for outstanding recent technical excellence leading to improvements in safety, productivity or environmental acceptability of general aviation."
Recipients: Donald G. Bigler 1987; Jan Roskam 1986; David R. Ellis 1985; Joseph Stickle 1984; Gerald Gregorek 1983; Stanley J. Green 1982; William H. Wentz 1981; Donald J. Grommesh 1980; Roger Winblade 1979.

GODDARD ASTRONAUTICS AWARD ... for the most notable achievement in the entire field of astronautics, honoring Robert H. Goddard, rocket visionary, pioneer bold experimentalist, and superb engineer whose early rocket launches opened up the world of astronautics.

Recipients: John McLucas 1987; George E. Solomon 1986; Frederic Oder 1985; Krafft Ehricke 1984; George E. Mueller 1983; John F. Yardley 1982; Peter T. Burr, Kenneth J. Frosh 1981; Robert J. Parks 1980; Maxime A. Faget 1979; Joseph V. Charyk 1978; James S. Martin, Jr. 1977; Edward Price 1976; George Rosen and Gordon Holbrook 1975; Paul D. Castenholz, Richard Mulready, John L. Sloop 1974; Edward S. Taylor, 1973; Howard E. Schumacher, Brian Brimelow, Gary A. Plourde 1972; Gerhard Neumann 1970; Perry W. Pratt 1969; Donald C. Berkey, Ernest C. Simpson, James E. Worsham 1968.

HALEY SPACE FLIGHT AWARD ... "for outstanding contribution by an astronaut or flight test personnel to the advancement of the art, science or technology of astronautics named in honor of Andrew G. Haley, a founder of the American Rocket Society."
Recipients NOAA 8 Recovery Team; Thomas Karras, Michael Cummings, Gay Hilton, Roger Hogan, David Coolidge, Kenneth Ward, Barbara Scott 1985; Byron Lichtenberg, Ulf Merbold, Owen Garriott, Robert Parker, Brewster Shaw and John Young 1984; John W. Young and Robert L. Crippen 1982; C. Gorden Fullerton, Fred Haise, Joe H. Engle, Richard H. Truly 1980; Vance D. Brand, Donald K. Slayton, Thomas P. Stafford 1978; William Dana 1976; Gerald Carr, Edward G. Gibson, William Pogue 1975; Charles Conrad, Jr., Joseph P. Kerwin, Paul J. Weitz 1974; John W. Young, Charles M. Duke, Jr., Thomas K. Mattingly, II 1973; David Worden, David Scott, James Irwin 1972; John Swigert, Fred W. Haise, Jr., James A. Lovell, Jr. 1971; William Anders, James Lovell, Jr., Frank Borman 1970; Donn F. Eisele, R. Walter Cuningham, Walter M. Schirra, Jr. 1969; Lt. Col. Virgil Grissom 1968.

HISTORY MANUSCRIPT AWARD ... "for the best historical manuscript dealing with the science, technology, and/or impact of aeronautics and astronautics on society."
Recipients: James R. Hansen and Fred E. Weick 1987; Scott Pace 1984; E.J.W. Gregory 1982; Richard Hirsh 1980; Roger Bilstein 1979; Edward C. Ezell, Linda N. Ezell 1978; Thomas Crouch 1977; Richard P. Hallion 1975; William M. Leary, Jr. 1973; Richard K. Smith 1972; Dr. Richard C. Lukas 1971; Milton Lomask, Constance McLaughlin Green 1969.

INFORMATION SYSTEMS AWARD ... "for technical and/or management contributions in space and aeronautics computer and sensing aspects of information technology and science."
Recipients: Winston W. Royce 1985; Lynwood C. Dunseith 1983; William A. Whitaker 1981; Barry W. Boehm, Algirdas Avizienis 1979; Albert L. Hopkins, Jr. 1977.

JEFFRIES MEDICAL RESEARCH AWARD ... "named after the American physician who made the earliest recorded scientific observations from the air, for outstanding contributions to the advancement of aerospace medical research."
Recipients: John Patrick Meehan, Jr. 1987; Charles E. Billings 1986; William E. Thornton 1985; Paul Buchanan 1983; Arnauld E. Nicogossian 1982; Sam L. Pool 1981; Stephen L. Kimzey 1980; William L. Smith 1979; Heinz Fuchs 1978; Harold von Beckh 1977; Lawrence F. Dietlein 1975; Malcolm Clayton Lancaster 1974; Karl H. Houghton 1973; Roger G. Ireland 1972; Richard S. Johnston 1971; Walton L. Jones 1970; Frank B. Voris 1969; Loren D. Carlson 1968.

LAWRENCE SPERRY AWARD ... "honoring Lawrence B. Sperry, pioneer aviator and inventor, for a notable contribution made by a young person to the advancement of aeronautics or astronautics."
Recipients: James L. Thomas 1987; Parviz Moin 1986; Vijaya Shankar 1985; Sally K. Ride 1984; Luat Nguyen 1983; Jeffrey N. Cuzzi 1982; Charles W. Boppe 1981; William F. Ballhaus 1980; David A. Caughey 1979; Paul Kutler 1978; Joseph L. Weingarten 1977; David Bushnell 1975; Jan Rusby Tulinius 1974; Dino A. Lorenzini 1973; Sheila E. Widnall 1972; Ronald L. Berry 1971; Glenn Lunney 1970; Edgar C. Lineberry, Jr. 1969; Roy V. Harris, Jr. 1968.

LOSEY ATMOSPHERIC SCIENCES AWARD ... "in recognition of outstanding contributions to the science of meteorology as applied to aeronautics and named after Captain Robert M. Losey, meteorological officer who was the first U. S. officer to die in World War II."

Recipients: John McCarthy 1987; John Speridon Theon 1986; Dennis Camp 1985; John H. Enders 1984; Walter Frost 1983; T. Theodore Fujita 1982; Allan B. Bailey 1981; William W. Vaughan 1980; Allan B. Bailey 1979; Robert A. McClatchey 1978; Robert Knollenberg 1977; Paul W. Kadlec 1975; Norman Sissenwine 1974; George H. Fichtl 1973; David Q. Wark 1972; Verner E. Suomi 1971; Newton A. Lieurance 1970; Robert Fletcher 1969.

MECHANICS AND CONTROL OF FLIGHT AWARD ... "for an outstanding recent technical or scientific contribution by an individual in the mechanics, guidance, or control of flight in space or the atmosphere."
Recipients: Leonard Meirovitch 1987; Heinz Erzberger 1986; Jason Speyer 1985; George Leitmann 1984; John L. Junkins 1983; Angelo Miele 1982; Robert W. Farquhar 1981; Arthur E. Bryson, Jr. 1980; Morris Ostgaard 1979; Richard H. Battin 1978; Joseph R. Chambers, William P. Gilbert 1977; Charles Murphy 1976; Bernard Etkin 1975; Harold Roy Vaughn 1974; Henry J. Kelley 1973; John V. Breakwell 1972; George W. Cherry-Kenneth J. Cox-William S. Widnall 1971; Duane T. McRuer-Irving L. Ashkenas 1970; John P. Mayer 1969; Robert V. Knox 1968.

MISSILE SYSTEMS AWARD ... presented to an investigator or program leader who was instrumental through his technical efforts, systems design skills, and/or system management skills in developing innovative design solutions and/or managing the system development and development of advanced tactical or strategic missile systems.
Recipients: J. Michael Gorman 1987; Daniel M. Tellep 1986; Vahey S. Kupelian 1984; Wayne Meyer 1983; John B. Buescher 1982.

ONR-AIAA RESEARCH AWARD IN STRUCTURAL MECHANICS ... "as a means of fostering excellence in structural mechanics research and to afford the opportunity for close participation in naval structure mechanics activities, enabling the recipient to conduct advanced and original research which will be relevant to naval structural mechanics problems."
Recipients: Edward L. Wilson 1981; William N. Findley 1979; Jack R. Vinson 1977; Edward Stanton 1975; Dave Bushnell 1974; Robert M. Jones 1973; R. E. Nickell 1972; Lawrence H. N. Lee 1971; Stanley B. Dong 1970.

PENDRAY AEROSPACE LITERATURE AWARD ... "named for Dr. G. Edward Pendray, a founder and past president of the American Rocket Society, and presented for an outstanding contribution or contributions to aeronautical/astronautical literature in the relatively recent past."
Recipients: Richard H. Battin 1987; Wayne Johnson 1986; Warren C. Strahle 1985; Leonard Meirovitch 1984; Marvin Goldstein 1983; Angelo Miele 1982; Robert B. Hotz 1981; Fred Culick 1980; Henry J. Kelley 1979; Arnold M. Kuethe 1978; George Leitmann 1977; Stanford S. Penner 1976; William R. Sears 1975; Fred I. Ordway 1974; Marcus F. Heidmann, Richard Priem 1973; Edward W. Price 1972; Nicholas J. Hoff 1971; Wilmont N. Hess 1970; Arthur E. Bryson, Jr. 1968.

REED AERONAUTICS AWARD ... "to honor the most notable achievement in the field of aeronautic science and engineering, in memory of Dr. S. A. Reed, aeronautical engineer, designer, and founder member of the former Institute of Aeronautical Sciences."
Recipients: R. Richard Heppe 1987; Robert J. Patton 1986; Thomas V. Jones 1985; Frederick T. Rall, Jr. 1984; Robert Widmer 1983; John L. McLucas 1982; William R. Sears 1981; Donald Malvern 1980; Paul B. MacCready 1979; James T. Stewart 1978; William Dietz 1977; George Spangenberg 1976; Antonio Ferri 1975; Willis M. Hawkins 1974; I. E. Garrick 1973; Max Munk 1972; Ira Grant Hedrick 1971; Richard T. Whitcomb 1970; Rene Miller 1969; William H. Cook 1968.

GROUND TESTING AWARD ... for outstanding achievement in the development or effective utilization of technology, procedures, facilities, or modeling techniques for flight simulation, space simulation, propulsion testing, aerodynamic testing, or other ground testing associated with aeronautics and astronautics.

Recipients: Robert E. Smith, Jr. 1987; Kazimierz Orlik-Ruckemann 1986; Billy Joe Griffith 1985; James C. Young 1984; James Guy Mitchell 1983; Gerald L. Winchenbach 1982; Glenn D. Norfleet 1981; Frank L. Wattendorf 1980; Jack D. Whitfield 1979; Arthur B. Doty 1978; Bernhard H. Goethert 1976.

SPACE SCIENCE AWARD . . . "for distinguished achievements in studies of the physics of atmospheres of celestial bodies, or of the matter, fields, and dynamic and energy transfer processes occurring in space, or experienced by space vehicles."
Recipients: Frederick L. Scarf 1987; Richard Johnson 1986; Gerry Neugebauer 1985; Edward C. Stone 1984; Charles Barth 1983; James A. Van Allen 1982; Peter M. Bank 1981; Donald M. Hunten 1980; James B. Pollack 1979; Laurence E. Peterson 1978; Bruce Murray 1977; Riccardo Murray Dryer 1975; John H. Wolfe 1974; Paul W. Gast 1973; Norman F. Ness 1972; William Ian Axford 1971; Carl McLlwain 1970; Charles P. Sonett 1969; Kinsey A. Anderson 1968.

SPACE SYSTEMS AWARD . . . "to recognize outstanding achievements in the field of systems analysis, design, and implementation as applied to spacecraft and launch vehicle technology."
Recipients: Robert F. Thompson 1986; Richard H. Kohrs 1984; Seymour Z. Rubenstein 1983; Angelo Guastaferro 1982; Krafft Ehricke 1981; Charles F. Hall 1980; John R. Casani 1979; Wernher von Braun (posthumously) 1978; Walter O. Lowrie 1977; Caldwell C. Johnson, Jr. 1975; Harold Lassen 1974; Harold A. Rosen 1973; Thomas J. Kelly 1972; Anthony J. Iorillo 1971; Maxine A. Faget 1970; Otto E. Bartoe, Jr. 1969.

STRUCTURES, STRUCTURAL DYNAMICS AND MATERIALS AWARD . . . "for outstanding recent technical or scientific contribution in aerospace structures, structural dynamics or materials."
Recipients: James W. Mar 1987; Bryan R. Noton 1986; Chuntsun Hwang 1985; Eric Reissner 1984; Leonard Meirovitch 1983; John H. Wykes 1982; Richard R. Heldenfels 1981; Earl H. Dowell 1980; Lucien A. Schmit 1979; Warren A. Staufer 1978; Walter J. Mykytow 1977; Charles Tiffany 1976; Theodore H. H. Pian 1975; William D. Cowie 1974; Robert T. Schwartz-George P. Peterson 1973; M. Jonathan Turner 1972; Nicholas J. Hoff 1971; Joseph D. Van Dyke, Jr. 1970; Holt Ashley 1969; John C. Houbolt 1968.

SUPPORT SYSTEMS AWARD . . . "for significant contribution to the overall effectiveness of aerospace systems through the development of improved support systems technology."
Recipients: John E. Hart, Jr. 1986; Oscar W. Sepp 1985; Harry Seaman 1984; Richard P. Adam 1983; George E. Marron 1982; John W. Kiker 1981; Joseph J. O'Rourke 1980; Joseph J. O'Rourke 1979; Thomas A. Ellison 1977; Gene Petry 1976.

SYSTEM EFFECTIVENESS AND SAFETY AWARD . . . "for outstanding contributions to the field of system effectiveness and safety or its related disciplines."
Recipients: Charles O. Miller 1987; Robert Peercy, Jr. 1986; Gerald Katt 1985; Paul Dick 1984; John De S. Coutinho 1983; Thomas L. House 1982; George Hirschberger 1981; I. Irving Pinkel 1980; Willis Willoughby, Jr. 1979; F. Stanley Nowlan-Thomas D. Mattson 1977.

THERMOPHYSICS AWARD . . . "for outstanding recent technical or scientific contribution by an individual in thermophysics, specifically as related to the study and application of the properties and mechanisms involved in thermal energy transfer within and between solids, and between an object and its environment, particularly by radiation, and the study of environmental effects on such properties and mechanisms."
Recipients: Alfred L. Crosbie 1987; John T. Howe 1986; Milan Yovanovich 1985; Tom Love 1984; Sanford S. Penner 1983; Richard P. Bobco 1982; Walter B. Olstad 1981; George W. Sutton, Walter B. Olstad 1980; Raymond Viskanta 1979; Allie M. Smith 1978; Chang L. Tien 1977; Donald K. Edwards 1976.

VON KARMAN LECTURESHIP IN ASTRONAUTICS . . . "to honor an individual who has performed notably and distinguished himself technically in the field of astronautics, named for Theodore von Karman, world famous authority on aerospace sciences."

Recipients: Lew Allen 1987; Albert Wheelon 1986; Eberhardt Rechtin 1985; Aaron Cohen and Robert Thompson 1984; George Jeffs 1983; Willis Hawkins 1982; Bruce C. Murray 1981; Daniel J. Fink 1980; Christopher C. Kraft, Jr. 1979; Robert Fuhrman 1978; Joseph Charyk 1977; I. E. Garrick 1975; Harrison Schurmeier 1974; Alan M. Lovelace 1973; Eugene Love 1972; Dr. Irmgard Flugg-Lotz 1971; Erik-L. Mollo- Christensen 1970; Courtland D. Perkins 1969; William R. Sears 1968.

WRIGHT BROTHERS LECTURESHIP IN AERONAUTICS . . . commemorating the first powered flights made by Orville and Wilbur Wright at Kitty Hawk in 1903, and intended to emphasize significant advances in aeronautics by recognizing major leaders and contributors thereto.
Recipients: John M. Swihart 1987; Dwain A. Deets and Lewis E. Brown 1986; John R. Utterstrom 1985; Robert P. Harper, Jr. and George Cooper 1984; Edward Polhamus 1983; John Steiner 1982; Holt Ashley 1981; Bernard Etkin 1980; Nielsen 1979; George B. Litchford 1978; Gero Madelung 1977; J. L. Atwood 1976; Henri Ziegler 1975; A. M. O. Smith 1974; Hermann Schlichting 1973; Franklin Kolk 1972; Robert Lickeley 1971; F. A. Cleveland 1970; Pierre Satre 1969; Charles W. Harper 1968.

WYLD PROPULSION AWARD . . . "for outstanding achievement in the development or application of rocket propulsion systems."
Recipients: Lt. Gen. A. G. Casey 1987; Dominick Sanchini 1986; Rolf Bruenner and Adolf Oberth 1985; Rudi Beichel 1984; Jerry Makepeace 1983; Lloyd F. Kohrs 1982; John Kincaid 1981; Howard Douglas 1980; Derald A. Stuart 1979; William C. Rice 1978; Martin Summerfield 1977; Roderick W. Spence-Howard Seifert 1976; Clarence W. Schnare-James Lazar 1975; Norman C. Reuel 1974; Gerard W. Elverum, Jr. 1973; Karl Klager 1972; Luigi M. Crocco 1971; Hans G. Paul-Joseph Guy Thibodaux, Jr. 1970; Harold R. Kaufman 1969; Harold B. Finger 1968.

DURAND LECTURESHIP FOR PUBLIC SERVICE . . . "for notable achievements by a scientific or technical leader whose contributions have led directly to the understanding and application of the science and technology of aeronautics and astronautics for the betterment of mankind."
Recipients: Robert C. Seamans 1986; Simon Ramo 1984.

AERODYNAMICS AWARD . . . "for meritorious achievement in the field of applied aerodynamics recognizing notable contributions in the development, application, and evaluation of aerodynamic concepts and methods."
Recipients: Robert J. Liebeck 1987; Bruce J. Holmes 1986; Forrester T. Johnson 1985; Lars Eric Ericsson 1984; John E. Lamar 1983.

AEROSPACE MAINTENANCE AWARD . . . presented to an individual who has made a major contribution to aerospace maintenance, specifically in aviation, missiles and space, resulting in a significant improvement in operational and cost effectiveness.
Recipients: James D. A. Van Hoften 1987.

AEROSPACE POWER SYSTEMS AWARD . . . presented for a significant contribution in the broad field of Aerospace Power Systems, specifically as related to the application of engineering sciences and systems engineering to the production, storage, distribution, and processing of Aerospace Power.
Recipients: Gerrit van Ommering 1987; Gary F. Turner 1986; William G. Dunbar 1985; Stephen J. Gaston 1984; Hans S. Rauschenbach 1983; Theodore Ebersole 1982.

PUBLIC SERVICE AWARD . . . presented to "honor an individual outside the aerospace community who has shown consistent and visible support for national aviation and space goals."
Recipients: T. Wendell Butler 1987; Gene Roddenberry 1986.

AMERICAN INSTITUTE OF CHEMICAL ENGINEERS (AIChE)

345 East 47th Street, New York, New York 10017
James F. Mathis, Temporary Executive Director

PRESIDENTS . . . Stanley I. Proctor 1987; H. S. Kemp 1986; J. P. Sachs 1985; J. H. Sanders 1984; R. H. Marshall 1983; R. R. Hughes 1982; W. K. Davis 1981; J. G. Knudsen 1980; J. Y. Oldshue 1979; W. H. Corcoran 1978; A. S. West 1977; K. D. Timmerhaus 1976; K. E. Coulter 1975; Irving Leibson 1974; T. Weaver 1973.

AWARDS

ALLAN P. COLBURN AWARD ... given to encourage excellence in publications on the part of younger members of the Institute, sponsored by E. I. DuPont de Nemours & Company.
Recipients: Klaus F. Jensen 1987; Robert A. Brown 1986; Matthew V. Tirrell 1985; Manfred Morari 1984; James A. Dumesic 1983; George Stephanopoulos 1982; W. Henry Weinberg 1981; Thomas F. Edgar 1980; J. E. Bailey 1979; L. Gary Leal 1978; Clark Kenneth Colton 1977; John H. Seinfeld 1976; Edward L. Cussler 1975; Wm. J. Ward III 1974; C. A. Eckert 1973; Dan Luss 1972; Dale F. Rudd 1971; Ronald B. Root-Roger A. Schmitz 1970; Edward F. Leonard 1969; John B. Butt 1968.

ALPHA CHI SIGMA AWARD ... given to recognize outstanding recent accomplishments by an individual in fundamental or applied research in the field of chemical engineering. Sponsored by Alpha Chi Sigma.
Recipients: Doraiswami Ramkrishna 1987; Keith E. Gubbins 1986; Ronald E. Rosensweig 1985; John C. Berg 1984; Charles W. Tobias 1983; Edward Wilson Merrill 1982; Warren E. Stewart 1981; Roy Jackson 1980; Reuel Shinnar 1979; John A. Quinn 1978; Eli Ruckenstein 1977; Howard Brenner 1976; Arnold A. Bondi 1975; Sheldon K. Friedlander 1974; C. J. Pings 1973; Charles D. Prater 1972; John H. Sinfelt 1971; A. E. Dukler 1970; Rutherford Aris 1969; Klaus D. Timmerhaus 1968.

AWARD IN CHEMICAL ENGINEERING PRACTICE ... given to recognize outstanding contributions by a chemical engineer in the industrial practice of the profession. Sponsored by Bechtel, Incorporated.
Recipients: T. W. Fraser Russell 1987; Raphael Katzen 1986; Frederick A. Zenz 1985; J. Frank Valle-Riestra 1984; William Robert Epperly 1983; Robert B. MacMullin 1982; W. G. Schlinger 1981; David K. Beavon 1980; John Longwell 1979; John W. Scott 1978; David Brown 1977; Jacob M. Geist 1976; James R. Fair, Jr. 1975; Robert G. Heitz 1974.

AWARD FOR SERVICE TO SOCIETY ... given to recognize outstanding contributions by a chemical engineer to community service and the solution of socially oriented problems. Sponsored by Fluor Corporation.
Recipients: William W. Shuster 1987; Richard K. Flitcraft, II 1986; Thomas Harrington Pigford 1985; Robert L. Bates 1983; Gary F. Bennett 1982; Ralph H. Kummler 1981; Chris B. Earl 1980; Gerald L. Decker 1979; Don E. Cox 1978; Frank W. Dittman 1977; Robert T. Jaske 1976; John J. McKetta, Jr. 1975; Gerald A. Lessells 1974; L. K. Cecil 1973.

F. J. VAN ANTWERPEN AWARD FOR SERVICE TO THE INSTITUTE ... given to recognize outstanding contributions by a chemical engineer through service to the Institute. Sponsored by Dow Chemical U.S.A.
Recipients: Gerald A. Lessells 1987; Hugh D. Guthrie 1986; Dr. John J. McKetta, Jr. 1985; Bryce I. MacDonald 1984; A. Sumner West 1983; W. Robert Marshall 1982; Marx Isaacs 1981; George E. Holbrook 1980; J. Henry Rushton 1979; F. J. Van Antwerpen 1978.

FOUNDERS AWARD ... given to recognize outstanding contributions to the field of chemical engineering. Sponsored by Amoco Foundation.
Recipients: John Happel, Frank M. Tiller 1987; Robert C. Reid, John H. Sanders, Charles R. White 1986; Robert H. Marshall and Neal R. Amundson 1985; James Donovan, David Malcolm Mason and James W. Westwater 1984; W. Kenneth Davis, Robert Burns MacMullin, Margaret Hutchinson Rousseau and John P. Sachs 1983; H. W. Flood and Ralph Landau 1982; G. Burnet, Jr., Joseph W. Davison, Jack C. Dart and James Y. Oldshue 1981; Stuart W. Churchill, Richard R. Hughes and Sheldon Isakoff 1980; A. Sumner West 1979; Ernest B. Christiansen and Klaus D. Timmerhaus 1978; Kenneth E. Coulter, Harold S. Kemp, James G. Knudsen and F. J. Van Antwerpen 1977; A. L. Conn, I. Leibson, L. K. Cecil and J. R. Fair 1976; T. W. Tomkowit and T. Weaver 1975; W. H. Corcoran, H. D. Guthrie and M. S. Peters 1974; T. A. Burtis, W. R. Marshall, J. J. Martin and R. L. Pigford 1973; Joseph C. Elgin and Donald A. Dahlstrom 1972; John J.

McKetta, Jr. and Paul D. V. Manning 1971; Jerry McAfee, William A. Cunningham, Walter E. Lobo and Donald B. Keyes 1970; William B. Franklin and William N. Lacey 1968; William R. Collings, Hoyt C. Hottel and Chalmer G. Kirkbride 1967; Manson Benedict, Ernest W. Thiele and Edward R. Weidlein 1966; Francis C. Frary and Carl F. Prutton 1965.

PROFESSIONAL PROGRESS AWARD IN CHEMICAL ENGINEERING ... given to recognize outstanding progress in the field of chemical engineering. Sponsored by the Celanese Chemical Company.
Recipients: James E. Bailey 1987; M. Albert Vannice 1986; Richard C. Alkire 1985; Stanley I. Sandler 1984; Alexis T. Bell 1983; W. Harmon Ray 1982; K. A. Smith 1981; L. Louis Hegedus 1980; Dan Luss 1979; John B. Butt 1978; Morton M. Denn 1977; K. B. Bischoff 1976; John H. Sinfelt 1975; Julien Szekely 1974; P. L. T. Brian 1973; Arthur E. Humphrey 1972; John C. Bonner 1971; James Wei 1970; Cornelius J. Pings 1969; Andreas Acrivas 1968; Thomas J. Hanratty 1967; Leon Lapidus 1966; Robert Byron Bird 1965.

R. H. WILHELM AWARD IN CHEMICAL REACTION ENGINEERING ... given to recognize significant and new contributions in chemical reaction engineering. Sponsored by Mobil Research and Development Corporation.
Recipients: Kenneth B. Bischoff 1987; Dan Luss 1986; Eugene E. Petersen 1985; Gilbert F. Froment 1984; George R. Gavalas 1983; Vern W. Weekman, Jr. 1982; R. A. Schmitz 1981; Charles N. Satterfield 1980; Octave Levenspiel 1979; Paul B. Weisz 1978; Joe Mauk Smith 1977; James J. Carberry 1976; Rutherford Aris 1975; Michel Boudart 1974; Neal R. Amundson 1973.

WARREN K. LEWIS AWARD ... given to recognize distinguished and continuing contributions to chemical engineering education. Sponsored by Exxon Research & Engineering Company.
Recipients: Klaus D. Timmerhaus 1987; Harry G. Drickamer 1986; James Wei 1985; Andreas Acrivos 1984; Joe Mauk Smith 1983; William H. Corcoran 1982; Rutherford Aris 1981; Carroll O. Bennett 1980; M. S. Peters 1979; Stuart W. Churchill 1978; Arthur B. Metzner 1977; Robert C. Reid 1976; Joseph Elgin 1975; R. Byron Bird 1974; W. L. McCabe 1973; Thomas K. Sherwood 1972; Neal R. Amundson 1971; Robert L. Pigford 1970; John T. McKetta 1969; Joseph H. Koffolt 1968; Donald L. Katz 1967; Richard H. Wilhelm 1966; Edwin F. Gilliland 1965.

WILLIAM H. WALKER AWARD ... to encourage excellence in contributions to the chemical engineering literature. This award is sponsored by Monsanto Company.
Recipients: Arthur W. Westerberg, John H. Seinfeld 1987; Howard Brenner 1985; Morton M. Denn 1984; Bruce A. Finlayson 1983; William R. Schowalter 1982; J. L. Duda and J. S. Vrentas 1981; James Wei 1980; Sheldon K. Friedlander 1979; Cedomir Sliepcevich 1978; L. E. Scriven 1977; C. Judson King 1976; Edwin N. Lightfoot, Jr. 1975; Leon Lapidus 1974; J. R. Fair 1973; Harry G. Drickamer 1972; Theodore Vermeulen 1971; Arthur B. Metzner 1970; Stuart W. Churchill 1969; Donald L. Katz 1968; John M. Prausnitz 1967; James W. Westwater 1966; Charles R. Wilke 1965.

MARKETING DIVISION HALL OF FAME AWARD ... given to an AIChE member to recognize outstanding service to marketing and to the Marketing Division of AIChE. Sponsored by United States Steel Corporation.
Recipients: Calvin B. Cobb 1987; Harry Tecklenburg 1986; John H. Sanders 1984.

MATERIALS ENGINEERING AND SCIENCES DIVISION AWARD ... given to recognize outstanding contributions by members of AIChE to the scientific, technological, educational or service areas of materials engineering and science. Its objective is to encourage the continuation of such contributions and enhance the visibility, recognition and value of the Division within AIChE.
Recipients: Stuart L. Cooper 1987; Sheldon E. Isakoff 1986; Donald R. Paul 1985; Nicholas A. Peppas 1984; Alan S. Michaels 1982; John L. Kardos 1981; Curry E. Ford 1979.

ROBERT L. JACKS MEMORIAL AWARD ... given to recognize outstanding individuals who have made substantial contributions to the management of engineers involved in the CPI, or to

management techniques and procedures utilized in those industries. Sponsored by Union Carbide Corporation.
Recipient: Stanley I. Proctor 1986; James F. Mathis 1984.

LAWRENCE K. CECIL AWARD IN ENVIRONMENTAL CHEMICAL ENGINEERING ... given to recognize outstanding chemical engineering contributions and achievements toward the preservation or improvement of man's natural environment (air, water, land). Sponsored by the Standard Oil Company.
Recipients: Peter B. Lederman 1987; Dibakar Bhattachargya 1986; Dr. Richard David Siegel 1985; Robert A. Baker 1984; Andrew Benedek 1983; Berton B. Crocker 1982; David G. Stephan 1981; A. Roy Price 1980; Stacy L. Daniels 1979; Thomas H. Goodgame 1978; Donald A. Dahlstrom 1977; Milton R. Beychok 1976; Gary Bennett 1975; Rienaldo Caban and T. W. Chapman 1974; A. W. Busch 1973; L. K. Cecil 1972.

FOREST PRODUCTS DIVISION AWARD ... given to recognize outstanding chemical engineering contributions and achievements in the forest products and related industries.
Recipients: Joseph M. Genco 1987; George R. Lightsey 1986; Thomas Michael Grace 1984; Louie Yung-Tsu Liu 1983; Delmar R. Raymond 1982; William T. McKean 1981; R. W. Rousseau 1980; Andrew Chase 1979; Steven Prahacs 1978; Louis Edwards 1977; Howard Rapson 1976.

COMPUTING IN CHEMICAL ENGINEERING AWARD ... given to recognize outstanding contributions in the application of computing and systems technology to chemical engineering. Sponsored by Intergraph Corp. & Simulation Sciences.
Recipients: James M. Douglas 1987; David M. Himmelblau 1986; Warren E. Stewart 1985; Gintaras V. Reklaitis 1984; Arthur W. Westerberg 1983; Lawrence B. Evans 1982; Richard S. H. Mah 1981; Brice Carnahan 1980; Richard R. Hughes 1979.

FOOD, PHARMACEUTICAL AND BIOENGINEERING DIVISION AWARD IN CHEMICAL ENGINEERING ... given to recognize outstanding chemical engineering contributions and achievements by an individual in those industries involved in food, pharmaceutical and bio-engineering activities. Sponsored by Kraft, Inc.
Recipients: Henry C. Lim 1987; Robert S. Langer 1986; Robert K. Finn 1985; Jerome Samson Schultz 1984; Charles L. Cooney 1983; Kenneth B. Bischoff 1982; Daniel I. C. Wang 1981; Kenneth H. Keller 1980; Edwin L. Lightfoot 1979; Hans A. C. Thijssen 1978; Allan S. Michaels 1977; Malcolm D. Lilly 1976; C. Judson King 1975; Robert L. Dedrick 1974; Arthur E. Humphrey 1973; Richard Laster 1972; Arthur I. Morgan, Jr. 1971; Elmer L. Gaden 1970.

FUELS AND PETROCHEMICAL DIVISION ANNUAL AWARD ... given to recognize individuals who have made substantial contributions to the technology and advancement of the fuels and petrochemical industries. The basis of the award may be a combination of technical achievement, management skills, business acumen, academic leadership and general service to the profession.
Recipients: John R. Hall 1987; Edwin C. Holmes 1986; John F. Bookout 1985; J. Peter Grace 1984; John J. McKetta, Jr. 1983; Earle B. Barnes 1982; Maurice Granville 1980; Theodore A. Burtis 1979; Jerry McAfee 1978; J. E. Swearingen 1977; C. G. Kirkbride 1976; Maynard P. Venema 1975; Otto N. Miller 1974; H. D. Doan 1973; Ralph Landau 1972; Fred L. Hartley 1971; M. J. Rathbone 1969.

ROBERT E. WILSON AWARD IN NUCLEAR CHEMICAL ENGINEERING ... given to recognize outstanding chemical engineering contributions and achievements in the nuclear industry by a member of AIChE and to promote the objectives of the Nuclear Engineering Division of AIChE.
Recipients: Raymond G. Wymer 1987; Alfred Schneider 1986; Robert Merriman 1985; Leslie Burris 1984; Charles F. Bonilla 1983; Richard C. Vogel 1982; Walton Alexander Rodger 1981; Thomas H. Pigford 1980; William P. Bebbington 1979; Edward A. Mason 1978; John R. Hoffman 1977; Miles C. Leverett 1976; Milton Levenson 1975; James A. Buckham 1974; Lombard Squires 1973; F. C. Culler, Jr. 1972; Stephen Lawroski 1971; Robert B. Richards 1970; W. Kenneth Davis 1969; Mason Benedict 1968; Crawford H. Greenwald 1967.

DONALD Q. KERN AWARD IN HEAT TRANSFER AND ENERGY CONVERSION ... recognizes individuals who have demonstrated outstanding expertise in some field of heat transfer or energy conversion. Sponsored by Hemisphere Publishing.
Recipients: David Butterworth 1987; Stanley J. Green 1986; Dimitri Gidaspow 1984; Duncan Chisholm 1983; James G. Knudsen 1982; L. S. Tong 1981; Geoffrey F. Hewitt 1980; Edwin H. Young 1979; Kenneth J. Bell 1978; Charles H. Gilmour 1977; Jerry Taborek 1976; Alfred C. Muller 1975; Charles F. Bonilla 1974.

T. J. HAMILTON MEMORIAL AWARD FOR GOVERNMENT SERVICE ... presented annually to the Institute member who best exemplifies AIChE government activities during a given year. Sponsored by Siegel-Houston and Associates, Inc.
Recipients: John P. Sachs 1987; Stanley V. Margolin 1986; Frederic W. Hammesfahr 1985; Robert H. Marshall 1984; Daniel S. Maisel 1983; A. Sumner West 1982.

AMERICAN INSTITUTE OF MINING, METALLURGICAL & PETROLEUM ENGINEERS (AIME)

345 East 47th Street, New York, New York 10017
Robert H. Marcrum, Executive Director

PRESIDENTS ... Alan Lawley 1987; Arlen L. Edgar 1986; Norman T. Mills 1985. Robert H. Merrill 1981; M. Scott Kraemer 1980; William H. Wise 1979; Wayne L. Dowdey 1978; H. Arthur Nedom 1977; Julius J. Harwood 1976; James D. Reilly 1975; Wayne E. Glenn 1974; James B. Austin 1973.

AWARDS

ANTHONY F. LUCAS GOLD MEDAL ... recognizes distinguished achievement in improving the technique and practice of finding or producing petroleum.
Recipients: Eugene R. Brownscombe 1987; Henry J. Welge 1986; N. Van Wingen 1985; Joseph E. Warren 1984; Henry J. Ramey, Jr. 1983; Paul B. Crawford 1982; Claud R. Hocott 1981; R. C. Earlougher 1980; Donald L. Katz 1979; Gustave E. Archie 1978; Marshall B. Standing 1977; J. Clarence Karcher 1976; Michael T. Halbouty 1975.

BENJAMIN F. FAIRLESS AWARD ... recognizes distinguished achievement in iron and steel production and ferrous metallurgy.
Recipients: Robert E. Boni 1987; Frank H. Sherman 1986; Frank W. Luersson 1985; John Macnamara 1984; Gerald R. Heffernan 1983; George A. Stinson 1982; Howard O. Beaver, Jr. 1981; J. Peter Gordon 1980; C. William Verity, Jr. 1979; Dennis J. Carney 1978; Edgar B. Speer 1977; David S. Holbrook 1976; Michael Tenenbaum 1975.

CHARLES F. RAND AWARD ... recognizes distinguished achievement in mining administration (the term "mining" regarded in its broad sense to include metallurgy and petroleum).
Recipients: Wayne H. Burt 1986; Ralph E. Bailey 1985; John E. Frost 1984; Ralph E. Hennebach 1983; Edward S. Frohling 1982; Paul C. Henshaw 1981; John E. Swearingen 1980; Ian M. McLennan 1979; Ian MacGregor 1978; Albert P. Gagnebin 1977; Theodore W. Nelson 1976; Robert O. Anderson 1975.

ENVIRONMENTAL CONSERVATION DISTINGUISHED SERVICE AWARD ... recognizes significant contributions to environmental conservation by addition to knowledge; by the design or invention of useful equipment or procedure; or by outstanding service to governmental or private organizations devoted to any field of environmental conservation.
Recipients: Elmore C. Grim 1987; Edward R. Bingham 1986; Seth J. Abbott, Jr. 1985; John S. Lagarias 1984; W. J. Coppoc 1983; Edward W. Mertens 1982; Fred A. Glover 1981; David R. Maneval 1980; Beatrice E. Willard 1979; Kenneth W. Nelson 1978; Max J. Spendlove 1977; H. Beecher Charmbury 1976; Claude E. ZoBell 1975.

ERSKINE RAMSAY MEDAL ... recognizes distinguished achievement in coal mining, including both bituminous coal and anthracite.

Recipients: John Turyn, Jr. 1987; Louis Kuchinic, Jr. 1986; Robert H. Quenon 1985; E. Minor Pace 1984; Raymond E. Zimmerman 1983; Woods G. Talman 1982; William N. Poundstone 1981; Clayton G. Ball 1980; Edwin R. Phelps 1979; Nicholas T. Camicia 1978; James D. Reilly 1977; David A Zegeer 1976; Charles J. Potter 1975.

HAL WILLIAMS HARDINGE AWARD ... recognizes outstanding achievement which has benefited the field of industrial minerals.

Recipients: Richard H. Olson 1987; Frederic L. Kadey, Jr. 1986; James C. Bradbury 1985; Nelson Severinghaus, Jr. 1984; Stanley J. Lefond 1983; J. F. Havard 1982; Sam H. Patterson 1981; Walter D. Keller 1979; Robert L. Bates 1978; Thomas D. Murphy 1977; Haydn H. Murray 1976; R. Gill Montgomery 1975.

HENRY KRUMB LECTURE ... outstanding speakers selected to deliver lectures at AIME Member Society Section meetings.

Recipients: Jagdish C. Agarwal, Maurice C. Fuerstenau, Alfred Petrick, Jr. 1986; Charles F. Barber, Willard Lacy and Alan Lawley 1985; Frederick B. Henderson, III, Ronald Latanision and Paul E. Queneau 1984; John A. Herbst, Bernard H. Kear and Malcolm J. McPherson 1983; Nathaniel Arbiter, Herbert H. Kellogg and Thomas J. O'Neil 1982; Eugene N. Cameron, Raymond L. Smith and Simon D. Strauss 1981; Donald A. Dahlstrom, Nicholas J. Grand and Alfred Weiss 1980; John K. Hammes, John F. Harvard and Milton E. Wadsworth 1979; Herbert D. Drechsler, Thomas V. Falkie and Laurence H. Lattman 1978; Arthur A. Brant, Carl Rampacek and Norman L. Weiss 1977; Paul A. Bailly, Thomas E. Howard and Eugene P. Pfleider 1976; James Boyd, Alvin Kaufman and Donald O. Rausch 1975.

JAMES DOUGLAS GOLD MEDAL ... recognizes distinguished achievement in nonferrous metallurgy.

Recipients: Allen S. Russell 1987; Ernest Peters 1986; Merton C. Flemings 1985; Thomas A. Henrie 1984; Robert I. Jaffee 1983; Morris E. Fine 1982; George E. Atwood 1981; Joe B. Rosenbaum 1980; Albert W. Schlechten 1979; Milton E. Wadworth 1978; Carleton C. Long 1977; John F. Elliott 1976; Petri B. Bryk 1975.

MINERAL ECONOMICS AWARD ... recognizes distinguished contributions to the advancement of mineral economics.

Recipients: Lindsay D. Norman 1987; Richard T. Newcomb 1986; John E. Tilton, 1985; Theodore Robinson Eck, 1984; Walter R. Hibbard, Jr. 1983; Simon D. Strauss 1982; Richard L. Gordon 1981; Hans H. Landsberg 1980; M. A. Adelman 1979; William A. Vogely 1978; John J. Schanz, Jr. 1977; Hubert E. Risser 1975.

MINERAL INDUSTRY EDUCATION AWARD ... recognizes distinguished contributions to the advancement of mineral industry education.

Recipients: George R. St. Pierre 1987; Laurence H. Lattman 1986; William R. Holden 1985; William O. Philbrook 1984; Douglas W. Fuerstenau 1983; Oscar K. Kimbler 1982; Milton E. Wadsworth 1981; Stefan H. Boshkov 1980; Ben H. Caudle 1979; Albert W. Schlechten 1978; Robert Stefanko 1977; Murray F. Hawkins, Jr. 1976; Reinhardt Schuhmann, Jr. 1975.

ROBERT EARLL McCONNELL AWARD ... recognizes beneficial service to mankind by engineers through significant contributions which tend to advance the nation's standard of living or replenish its natural resource base.

Recipients: Charles E. O'Neill 1987; William A. Kriusky 1986; Robert W. Bartlett 1985; Nickolas J. Themelis 1984; F. Kenneth Iverson 1983; Richard G. Miller 1982; Franklyn K. Levin 1981; Vernon E. Scheid 1980; Jim Douglas, Jr.-Donald W. Peaceman-Henry H. Rachford, Jr. 1979; Carl Rampacek 1978; Geoffrey G. Hunkin 1977; Robert H. McLemore 1976; Norwood B. Melcher 1975.

ROBERT H. RICHARDS AWARD ... recognizes achievement in any form which unmistakably furthers the art of mineral beneficiation in any of its branches.

Recipients: Ponisseril Somasundaran 1987; Iwao Iwasaki 1986; Alban J. Lynch 1985; Roshan B. Bhappu 1984; Maurice C. Fuerstenau 1983; Arthur R. MacPherson 1982; William A. Griffith 1981; James E. Lawver 1980; Strathmore R. B. Cooke 1979; Frank F. Aplan 1978; Denis F. Kelsall 1977; Donald A. Dahlstrom 1976; Douglas H. Fuerstenau 1975.

ROSSITER W. RAYMOND AWARD ... recognizes the best paper published by an AIME Member Society and written by a member under 33 years of age.

Recipients: Joseph A. Hrabik 1987; David R. Hanson 1986; Yanis C. Yortsos 1985; William F. Yellig, Jr. 1984; Robert H. Wagoner 1983; Tommy M. Warren 1982; Robert H. Wagoner 1981; Clyde L. Briant 1980; Harold R. Warner, Jr. 1979; Edward J. Novotny, Jr. 1978; John E. Killough 1977; S. N. Singh 1976; Jacob Hagoort 1975.

WILLIAM LAWRENCE SAUNDERS GOLD MEDAL ... recognizes distinguished achievement in mining other than coal.

Recipients: Charles L. Pillar 1987; Richard L. Brittain 1986; John C. Kinnear, Jr. 1984; Frank E. Espie 1983; John Towers 1982; James S. Westwater 1980; Henry T. Mudd 1979; Russell H. Bennett 1978; Frank Coolbaugh 1977; Charles D. Clarke 1975.

AMERICAN INSTITUTE OF PHYSICS (AIP)

345 East 45th Street, New York, New York 10017
Kenneth W. Ford, Director
Chairman of the Governing Board ... Hans Frauenfelder 1986; Norman F. Ramsey 1981-85; Philip M. Morse 1977.

AWARDS

AIP PRIZE FOR INDUSTRIAL APPLICATIONS OF PHYSICS ... to recognize outstanding contributions by an individual or individuals to the industrial applications of physics. This award was established in 1977, and is presented every two years.
Recipients: E. Scott Kirkpatrick and C. Daniel Gelatt 1987; John J. Croat and Robert W. Lee 1985; Joseph A. Killpatrick and Frederick Aronowitz 1983; Alec N. Broers 1981; A. H. Bobeck 1979; Robert D. Maurer 1977.

AIP - SCIENCE WRITER AWARD IN PHYSICS & ASTRONOMY ... given annually, to encourage excellence in science writing in physics and astronomy for the nonspecialist.
Recipients: Shannon Brownlee, Allen Chen and Clifford M. Will 1987; Arthur Fisher and Donald Goldsmith 1986; Edwin C. Krupp and Ben Patrusky 1985; John Tierney and George Greenstein 1984; Martin Gardner and Abraham Pais 1983; Marcia F. Bartusiak and Heinz R. Pagels 1982; Leo Janos and Eric Chaisson 1981; Dennis Overbye and William J. Kaufmann, III 1980; Robert C. Cowen and Dr. Hans von Baeyer 1979; Timothy Ferris and Dr. Edwin C. Krupp 1978; William D. Metz and Dr. Steven Weinberg 1977; Federic Golden and Dr. Jeremy Bernstein 1976; Tom Alexander and Dr. Robert H. March 1975; Patrick Young and Dr. Robert D. Chapman 1974; Edward Edelson and Dr. Banesh Hoffman 1973; Jerry E. Bishop and Dr. Dietrich Schroeer 1972; Kenneth Weaver and Dr. Robert H. March 1971; D.P. Gilmore and Dr. Jeremy Bernstein 1970; Walter Sullivan and Dr. Kip S. Thorne 1969; William J. Perkinson 1968.

DANNIE HEINEMAN PRIZE FOR ASTROPHYSICS (given jointly with the American Astronomical Society) ... to recognize outstanding work in the field of astrophysics. This award was established in 1979; given annually.
Recipients: David L. Lambert 1987; Hyron Spinrad 1986; Sandra M. Faber 1985; Martin J. Rees 1984; Irwin I. Shapiro 1983; P. J. E. Peebles 1982; Riccardo Giaccon 1981; Joseph H. Taylor, Jr. 1980.

DANNIE HEINEMAN PRIZE FOR MATHEMATICAL PHYSICS (given jointly with The American Physical Society) ... to recognize outstanding work in the field of mathematical physics. This award was established in 1959; given annually.
Recipients: Rodney J. Baxter 1987; A. M. Polyakov 1986; David P. Ruelle 1985; Robert B. Griffiths 1984; Martin D. Kruskal 1983; John C. Ward 1982; Jeffrey Goldstone 1981; J. G. Glimm and A. M. Jaffe 1980; Gerard 't Hooft 1979; Elliot H. Lieb 1978; Steven Weinberg 1977; S. W. Hawking 1976; L. D. Faddeev 1975; Subrahmanyan Chandrasekhar 1974; Kenneth G. Wilson 1973; James D. Bjorken 1972; Roger Penrose 1971; Yoichiro Nambu 1970; Arthur Strong Wightman 1969; Sergio Fubini 1968.

THE JOHN TORRENCE TATE INTERNATIONAL MEDAL FOR DISTINGUISHED SERVICE TO THE PROFESSION OF PHYSICS ... to recognize distinguished service to physics on an international level. The Award is intended primarily for foreign nationals. Services that further international understanding and exchange are considered to be of primary importance.
Recipients: Pierre Aigrain 1981; Abdus Salam 1978; Gilberto Bernardini 1972; H. W. Thompson 1966; Paul Rosbaud 1961.

THE KARL TAYLOR COMPTON MEDAL FOR DISTINGUISHED STATESMANSHIP IN SCIENCE ... to recognize the outstanding statesmanship in science of distinguished physicists. The_award, established in 1957, is intended primarily for U. S. nationals.
Recipients: Norman F. Ramsey 1985; Melba Phillips 1981; Samuel A. Goudsmit 1974; Ralph A. Sawyer 1971; Frederick Seitz 1970; Alan T. Waterman 1967; Henry A. Barton 1964; Karl K. Darrow 1960; George B. Pegram 1957.

AMERICAN INSTITUTE OF PLANT ENGINEERS (AIPE)

3975 Erie Avenue, Cincinnati, Ohio 45208
Michael J. Tillar, Executive Director

PRESIDENTS ... Robert R. Ruhlin, PE/CPE 1987; John R. Erickson PE/CPE 1986; Charles M. Lemon 1985; Richard P. Doctor, PE/CPE 1984; Stanley K. Smith CPE 1983; Geoffrey R. Dickinson, PE/CPE 1982; Chester A. Regler, CPE 1981; Don E. Burchard, CPE 1980; John J. Talbert, PE/CPE 1979; Morley Zagury, CPE 1978; Philip L. Pearce, CPE 1977; Clayton C. Verlo, PE/CPE 1976; Louis Gustad, PE/CPE 1975; Frank Holladay, PE/CPE 1974; Joseph Johnstone 1973.

AWARDS

FACILITIES MANAGEMENT EXCELLENCE AWARDS (FAME) ... sponsored by AIPE foundation, the FAME Awards are designed to recognize and promote outstanding accomplishments in plant engineering and facilities management. One top Award of Excellence and several Awards of Merit are presented.
Recipients: Award of Excellence: Bryant Paper Mill, Allied Paper, Inc., 1986; Awards of Merit: Hughes Aircraft Carslbad Plant, Martin Marietta Energy Systems (Oakridge), Beacon Mfg. Co. Swannanoa, Avco Lycoming Textron (Stratford), Matsushita Industrial Co. Franklin Park 1986; Award of Excellence, University of Nebraska Medical Center 1985; Awards of Merit, Allied Corporation, Buffalo Research Laboratory-GTE Service Corporation-IPC Dennison Company-U.S. Postal Service, Denver Bulk Mail Center-University of Richmond 1985; Grand Prize, Bill Hollis, Hewlett-Packard Santa Rosa 1981; 1st Place Large Plant--Randall A. Tassin, CPE, Martin-Marietta Aerospace 1981; 1st Place Small Plant--Charles J. Kolar, CPE, Portland Cement Association 1981.

AMERICAN NATIONAL STANDARDS INSTITUTE INC. (ANSI)

1430 Broadway, New York, New York 10018
William H. Rockwell, Executive Vice President

CHAIRMAN OF THE BOARD ... George S. Wham 1986-88; L. John Rankine 1983-85; William H. Gatenby 1980-82; Frank J. Feely, Jr. 1978-79; John W. Landis 1975-77; Roy P. Trowbridge 1971-1974.

AWARDS

ASTIN-POLK INTERNATIONAL STANDARDS MEDAL ... for distinguished service in promoting trade and understanding among nations through personal participation in the advancement, development, and administration of international standardization, measurement, and certification.
Recipients: William E. Vannah, 1985; William B. Kelly, Jr. and Edward Lohse 1982; Olle Sturen and L. John Rankine 1980; Karl S. Geiges 1978; Francis L. LaQue-Nikolaus Ludwig 1976; Allen V. Astin-Louis F. Polk, Sr. 1974.

THE HOWARD COONLEY MEDAL ... established in 1950 to recognize outstanding appreciation and support of voluntary standardization as a tool of management. It is awarded each year to an executive who has rendered great service through practice and example, and by writing and speeches, in advancing the national economy through voluntary standards.
Recipients: George S. Wham, 1985; Frank J. Feeley, Jr. 1982; Allen F. Rhodes 1980; Baron Whitaker 1978; Roy P. Trowbridge 1976; Francis K. McCune 1969; Miles N. Clair 1968.

STANDARDS MEDAL ... established in 1951 in recognition of outstanding contributions to the philosophy of the ANSI, as well as the constructive development and application of standards and standardization of company, industry, and national levels. It is awarded to an individual who has shown exceptional leadership in the actual development and application of voluntary national standards.
Recipients: Howard P. Michener, 1985; Walter L. Harding and Edward C. Sommer 1982; Carl E. Rawlins 1980; A. Allan Bates 1978; Melvin Myerson 1976; Allen V. Astin 1969; William H. Gourlie 1968.

AMERICAN NUCLEAR SOCIETY (ANS)

555 North Kensington Avenue, La Grange Park, Illinois 60525
Octave J. DuTemple, Executive Director

PRESIDENTS ... Ronald C. Stinson, Jr. 1987-88; Bertram Wolfe 1986-87; E. Linn Draper, Jr. 1985-1986; Joseph M. Hendrie 1984-1985; Milton Levinson 1983-1984; L. Manning-Muntzing 1982-1983; Corwin L. Rickard 1981; Harry Lawroski 1980; Edward J. Hennelly 1979; William R. Kimel 1978; Joseph R. Dietrich 1977; Vincent S. Boyer 1976; Melvin J. Feldman 1975; J. Ernest Wilkins, Jr. 1974; John W. Simpson 1973.

AWARDS

AMERICAN NUCLEAR SOCIETY SPECIAL AWARD ... presented for outstanding work in areas of technical information, biological effects of radiation, novel applications of nuclear energy, reactor materials and reactor physics.
Recipients: John O'Sullivan 1987; Glen J. Schoessow, Alfred C. Tollison, Jr. and Jimmy D. Vandergrift 1986; (co-winners) Milton Levenson, William R. Stratton and Edward A. Warman 1985; (team) Thomas F. Budinger, Stephen E. Derenzo, Ronald H. Huesman and Michael E. Phelps 1984; (co-winners) John Amario and Jack E. Vessely 1983; Allison M. Platt, (team) Comisariat Energie Atomique, R. Bonniaud, N. Jacquet-Francillon, A. Jouan, F. Laude and C. Sombert 1982; Two groups as co-winners, (group) D. O. Campbell, A. P. Malinauskas, W. R. Stratton and (group) Loft/Semiscale Project 1981; Cyril Comar 1980; Yankee Atomic Electric 1979; EBR-II Project 1978; (ONRL Group) J. Blomeke, F. Culler, D. Ferguson, R. Wymer 1977; Norman C. Rasmussen 1976; Edward G. Struxness 1975; Walter Meyer 1974; G. Robert Keepin 1973; Lendell E. Steele 1972; Ely M. Gelbard 1970; Arthur G. Ward 1969; W. Wayne Meinke 1968.

ARTHUR HOLLY COMPTON AWARD ... presented for outstanding contribution to nuclear science and engineering education. This award is in recognition of the late Dr. Arthur Holly Compton, scientist, teacher, national leader, and Nobel Laureate.
Recipients: Alexander Sesonske 1987; Gene L. Woodruff 1986; James J. Duderstadt 1985; Arthur B. Chilton 1984; Dieter Smidt 1983; Nunzio J. Palladino 1982; Merril Eisenbud 1981; John R. Lamarsh 1980; Mohammed M. El Wakil 1979; Kent F. Hansen 1978; Terry Kammash 1977; Charles F. Bonilla 1976; David J. Rose 1975; William Kerr 1974; Harold F. Forsen 1973; Irving Kaplan 1972; Thomas H. Pigford 1971; Raymond L. Murray 1970; Manson Benedict 1969; Samuel Glasstone 1968.

DISTINGUISHED SERVICE AWARD ... outstanding service to the Society or to the Profession for which Fellowship is not appropriate (non-technical).
Recipients: A. Dixon Cullihan 1983; James Schumar 1980; Chih H. Wang 1978; O. J. Du Temple 1978; Ralph G. Chalker 1972.

AMERICAN NUCLEAR SOCIETY SEABORG MEDAL ... in recognition of outstanding scientific or engineering research achievements associated with the development of peaceful uses of nuclear energy.
Recipients: Robert Avery 1986; Manson Benedict 1985; Glenn T. Seaborg 1984.

RADIATION INDUSTRY AWARD ... presented to scientists or engineers whose work has made significant contributions to the application of radiation in industry. The work must have been published in an established scientific journal.
Recipients: John Hubbell 1985; Robin E. Gardner 1984; Harold E. Johns 1983; Russell E. Heath 1982; Howard O. Minlowe 1981; William S. Lyon, Jr. 1980; John W. Cleland 1979; Charles Artandi 1978; Enzo Ricci-Richard L. Hahn 1977; Godfrey N. Hounsfield 1976; Joseph Silverman 1975; Harold Berger 1974; Arthur Rupp 1973; Powell Richards 1972; Bernard Manowitz 1971; Raymond C. Goertz 1969; Robert F. Nystrom 1968.

AMERICAN SOCIETY FOR ENGINEERING EDUCATION (ASEE)

Eleven Dupont Circle, Suite 200
Washington, D. C. 20036
F. Karl Willenbrock, Executive Director

PRESIDENTS ... Robert H. Page 1987; Edmund T. Cranch 1986; Robert Mills 1985; J. C. Hancock 1984; Joseph C. Hogan 1983; Daniel C. Drucker 1982; Vincent S. Haneman, Jr. 1981; Charles E. Schaffner 1980; Joseph J. Martin 1979; Otis E. Lancaster 1978; George Burnet 1977; Lee Harrisberger 1976; Cornelius Wandmacher 1975; John C. Calhoun, Jr. 1974; Joseph M. Pettit 1973.

AWARDS

ALVAH K. BORMAN AWARD, COOPERATIVE EDUCATION DIVISION ... a framed award certificate is presented each year to division members, past or present, who have made significant and sustained contributions to the promotion of the philosophy and practice of cooperative education in engineering and technology.
Recipients: Jack M. Managham and David R. Opperman 1987; Donald R. Borrowbridge 1986; Harold Ratcliff 1985; R. W. Bogener 1984; Malcolm P. Clark and Robert Heyborne 1983; Jack LeMay 1982; Donald W. Lyon and Frank Vandergrift 1981; John L. Campbell 1980; Alvah K. Borman (awarded posthumously), Stewart B. Collins and Henry G. Hutton 1979.

AWARD FOR EXCELLENCE IN LABORATORY INSTRUCTION, DIVISION FOR EXPERIMENTATION AND LABORATORY-ORIENTED STUDIES ... award given to recognize outstanding contributions of an individual who provides and promotes excellence in experimentation and laboratory instruction. Award consists of an award certificate and a stipend of $1,000 provided by the Tektronix Foundation.
Recipients: Ernest O. Doebelin 1987; Robert White 1986; Richard Johnson 1985; Theodore F. Bogart 1983; John M. Radonch 1982; Alexander B. Bereskin 1981.

BENJAMIN J. DASHER BEST PAPER AWARD, EDUCATIONAL RESEARCH AND METHODS DIVISION-IEEE ... plaque and certificate awarded for the best paper presented at the annual "Frontiers in Education" conference cosponsored by ASEE and the Institute of Electrical and Electronics Engineers.
Recipients: Billy V. Koen 1987; Davood Tashayyod 1985; A. L. Riemenschneider and L. D. Feisel 1984; Martha Montgomery 1983; Marilla D. Svincki 1982; Albert J. Morris 1980.

CHEMICAL ENGINEERING DIVISION LECTURESHIP AWARD ... bestowed upon a distinguished engineering educator to recognize and encourage outstanding achievement in an important field of fundamental chemical engineering theory and practice. The award, consisting of $1,000, travel expense reimbursement for four lectures, an honorarium of $500 and certificate, is sponsored by the 3M Corporation.
Recipients: James Christensen 1987; Billy V. Koen 1986; Dan Luss 1985; T. W. Fraser Russell 1984; Warren E. Stewart 1983; Lowell B. Koppel 1982; Arthur W. Westerberg 1981; Klaus D.

Tiommerhaus 1980; Daniel D. Perlmutter 1979; Theodore Vermeulen 1978; Robert C. Reid 1977; Abraham E. Dukler 1976; John M. Prausnitz 1975; Elmer L. Gaden, Jr. 1974; Rutherford Aris 1973; Dale . E. Rudd 1972; William R. Schowalter 1971; Joe M. Smith 1970; Cornelius J. Pings 1969; L. Edward Scriven 1968.

CHESTER F. CARLSON AWARD ... presented annually to an outstanding educator, who by motivation and ability to extend beyond accepted tradition, has made a significant contribution to engineering education. The recipient will have demonstrated an ability to recognize the influence of a changing sociological and technological environment on academic customs and will have responded by applying his creative talents in the design and implementation of a new industrial technique, methodology, or concept. The award is sponsored by Xerox Corporation and consists of $1,000 and a plaque.
Recipients: Lionel V. Baldwin 1987; Lee Harrisberger 1986; Barry Hyman 1985; James Stice 1984; Charles Wales 1983; Dwight B. Aumann 1982; Donald L. Bitzer 1981; Billy V. Koen 1980; William R. Grogan 1979; Robert P. Morgan 1978; Hugh E. McCallick 1977; James E. Shamblin 1976; Robert W. Dunlap and Gordon H. Lewis 1975; Henry O. Fuchs and Robert Steidel, Jr. 1974; Kenneth McCollom and George C. Beakley 1973.

CLEMENT J. FREUND AWARD ... established in 1979 in honor of the 50th anniversary of the Cooperative Education Division, the award honors an individual in business, government or education who has made a positive impact on cooperative education programs in engineering and engineering technology. The award consists of $2,000, reimbursement of travel expenses to attend the Annual Conference, a commemorative plaque, and a certificate and is sponsored through an endowment from the following companies: Caterpillar Tractor Company, Danly Machine Corporation, John Deere, Diamond Shamrock Corporation, Dow Chemical U.S.A., Sundstrand Corporation and Union Carbide Corporation.
Recipients: David Russell Opperman 1987; Robert L. Heyborne 1986; Roy L. Woolridge; Donald C. Hunt 1984; Frank Vandergrift 1983; Stewart B. Collins 1982; James G. Wohlford 1981; Cornelius Wandmacher 1980; Clement J. Freund 1979.

CURTIS W. MCGRAW RESEARCH AWARD ... established to recognize outstanding early achievements by engineering college research workers and to encourage the continuance of such productivity in the future. The award, consisting of $1,000 and a certificate, was established through the efforts of the McGraw-Hill Book Company and the Engineering Research Council.
Recipients: Robert O. Ritchie 1987; George Stephanopoulos 1986; Magdy Iskander 1985; Chung Law 1984; James Bailey 1983; Robert Conn 1982; Alexis T. Bell 1981; Clark K. Colton 1980; Thomas K. Gaylord 1979; Gerald L. Kulcinski 1978; Dan Luss 1977; John N. Seinfeld 1976; Jan D. Achenbach 1975; Julian Szekely 1974; Stephen E. Harris 1973; Jose B. Cruz, Jr. 1972; George S. Ansell 1971; George I. Haddad 1970; Robert G. Jahn 1969; Eric Baer 1968.

DISTINGUISHED EDUCATOR AWARD, MECHANICS DIVISION ... award, consisting of a plaque and certificate, conferred annually for outstanding service to the division with particular emphasis on contributions to the teaching of engineering mechanics.
Recipients: Eugene H. Ripperger 1987; Daniel Frederick 1986; Daniel C. Drucker 1985; C. E. Work 1984; Milton E. Ravine 1983; William B. Stiles 1982; Ferdinand L. Singer 1981; Ferdinand P. Beer 1980; Egor P. Popov 1979; James L. Meriam 1978; Archie Higdon 1977.

DISTINGUISHED SERVICE AWARD, CONTINUING PROFESSIONAL DEVELOPMENT DIVISION ... award, consisting of a plaque and citation, presented by the division to acknowledge outstanding leadership and service in the field of continuing professional development for engineers. It also recognizes noteworthy service within the division itself.
Recipients: George J. Maler 1987; L. W. Ledgerwood, Jr. 1985; Joe G. Gisley 1982; Morris E. Nicholson 1981; Lindon E. Saline 1980; Lionel V. Baldwin 1979; Donald B. Miller and Howard R. Shelton 1978; Monroe W. Kriegel 1977; John P. Klus and Joseph M. Biedenbach 1976.

DISTINGUISHED SERVICE AWARD, ENGINEERING DESIGN GRAPHICS DIVISION ... reflects the deep appreciation of the division and acknowledges by means of a plaque and citation the many distinguished services rendered by a teacher of engineering graphics.

Recipients: Clarence E. Hall 1987; Claude Z. Westfall 1986; Gary Bertoline 1985; Steve M. Slaby 1983; James H. Earle 1982; Robert D. LaRue 1981; Mary F. Blade 1980; William B. Rodgers 1979; Charles G. Sanders 1978; Percy H. Hill 1977; Eugene G. Pare 1976; Robert H. Hammond 1975; Irwin Wladaver 1974; Edward W. Jacunski 1973; Paul M. Reinhard 1972; Matthew McNeary 1971; J. Howard Porsch 1970; Edward M. Griswold 1969; B. Leighton Wellman 1968.

DISTINGUISHED SERVICE CITATION ... granted to ASEE members by the Board of Directors in recognition of long, continuous and distinguished service to education for engineering and engineering technology or related fields through active participation in the work of ASEE. Before 1981 the award was given to organizations, as well as to individuals, to acknowledge significant service. The citation consists of a certificate.

Recipients: Robert Page 1984; Otis Lancaster and William Corcoran 1982; Lawrence P. Grayson, Thomas F. Jones and Ross J. Martin 1981; International Association for the Exchange of Students for Technical Experience/U.S. Inc. 1980; Archie Higdon 1979; Student Competitions on Relevant Engineering (SCORE) 1978; David R. Reyes-Guerra 1978; Joseph M. Biedenbach and Lawrence P. Grayson, Junior Engineering Technical Society 1977; Western Electric Company 1976; Frank D. Hansing 1975.

DONALD E. MARLOWE AWARD ... this award for distinguished education administration is bestowed upon an individual administrator who, by leadership and example beyond accepted tradition, has made significant ongoing contributions to education for engineering and engineering technology. Established in 1981, this award consists of a commemorative plaque and travel expense reimbursement to enable the awardee and spouse to attend the Annual Conference where the award is conferred.

Recipients: Donald D. Glower 1987; L. S. Fletcher 1986; Donald E. Marlowe 1981.

DOW OUTSTANDING YOUNG FACULTY AWARD ... funds contributed annually by Dow Chemical U.S.A. enable the Society to send to the Annual Conference one young faculty member from each of 12 ASEE geographic sections. Each award recipient is provided round-trip fare plus expenses, in addition to an award certificate. Dow also hosts an Annual Conference breakfast to honor the awardees.

Recipients: A.K.M Uddin, Robert J. Bernhard, Kevin J. Scoles, Robert C. Voigt, Robert H. Caverly, Ali Feliachi, Kim A. Stelson, Daniel W. Walsh, David L. Whitman, Marc Hoit, Brian W. Surgenor 1987; A.L. Bunge, R.C. Cox, K.C. Howell, M. Wecharantana, S.R. Eckhoff, R.H. Henry, G.R. Carmichael, W. Whiting, R.T. Gill, N.B. Goldstein, W.P. Wepfer 1986; John Priest, Michael Loui, Marc Donohue, David Delker, Donald Fisher, M. Asghar Bhatti, Jack R. Lohman, James Petersen, Allan Kirkpatrick, Mark E. Davis 1985; Heinz Luegenbiehl, Joe Martin, Muthuraj Vaithianathan, Gerald Jakubowski, Jon Jensen, Larry James, Jeffrey Peirce, Charles Natalie 1984; Gary Viviani, Brian Huggins, Robert Egbert, Clifford Fitzgerald, B. Lee Tuttle, Ian C. Gowter, Ted Eschenbach, Stephen Abt, Fred Gunnerson, Frank Sciremammand 1983; Norman Marsolan, Terrance Akai, Alexander Nauda, Scott Sink, Hermant Pendse, Russell Dean, Max Anderson, Thomas Burton, Marvin Young, Michael Mushak, Frank Croff, Mohamed Rezk 1982; T. J. Anderson, S. M. Batill, J. V. Bukowski, A. Co, J. M. Duff, C. E. Goodson, R. D. Noble, M. W. Riley, R. E. Rinker, P. K. Varshney, T. H. Wenzel 1981; J. Collura, R. C. Eck, T. B. Edil, R. D. Flack, J. W. Haslett, V. K. Kinra, T. W. Lester, C. D. Pegden, H. B. Puttgen, S. W. Stafford, R. S. Subramanian, P. C. Wankat 1980; R. N. Andrews, D. Apelian, T. J. Boehm, H. J. Freeman, M. F. Hein, M. Jolles, I. Kaneko, M. E. Ryan, R. J. Scranton, E. D. Sloan, G. H. Sutherland, L. N. Walker 1979; C. J. Waugaman, R. O. Buckius, R. J. Craig, J. E. Fagan, M. Kupferman, R. D. Gilson, J. P. Sadler, P. P. Fasang, P. F. Pfaelzer, M. Criswell, R. Garcia-Pacheco, D. G. Ullman 1978; W. S. Reed, D. L. Cohn, J. A. Kirk, J. A. Alic, J. W. Cipolla, Jr., J. T. Cain, R. J. Niederjohn, R. Y. Itani, M. M. Cirovic, K. P. Chong, C. E. Halford, R. W. Mayne 1977; G. T.

Craig, T. P. Cullinane, W. T. Darby, L. J. Griffiths, R. B. Hayter, R. G. Hoelzeman, L. L. Northrup, M. E. Parten, H. S. Peavy, W. W. Recker, W. T. Rhodes, J. M. Samuels, Jr. 1976; W. D. Carroll, D. S. Frederick, R. M. Haralick, H. Y. Ko, R. B. Landis, G. W. Lowery, J. F. McDonough, R. W. Mortimer, S. L. Rice, M. A. Sloan, D. R. Utela, D. N. Weiman 1975; W. R. Adrion, R. M. Anderson, Jr., J. S. Boland, III, A. R. Eide, R. C. Pare, O. A. Peku, W. R. Ramirez, R. P. Rhoten, C. A. Rosselli, T. E. Shoup, W. C. Van Buskirk, W. S. Venable 1974; F. L. Bennett, W. L. Carson, W. L. Dickson, J. T. Emanuel, L. L. Hench, J. S. Hirschhorn, T. H. Houlihan, R. G. Jeffers, C. M. Lovas, G. W. May, Lee Rosenthal, R. A. Shaw 1973; R. A. Peura, L. S. L. Bennett, W. L. Dickson 1973; R. A. Peura, L. S. Fletcher, R. D. Findlay, W. G. Wyatt, H. A. Sebesta, T. C. Owens, G. W. Neudeck, H. S. Folger, O. N. Garcia, C. E. Cartmill, K. D. Linstedt, T. G. Stoebe 1972; V. K. Feiste, W. M. Phillips, J. W. Willhide, J. T. Sears, J. Counts, N. C. Kerr, J. R. Hopper, L. D. Goss, W. Petsch, A. A. Torvi, R. E. Lave, Jr., W. B. Krantz 1971; J. M. Able, E. M. Bailey, M. O. Breitmeyer, F. E. Burris, A. M. Despain, R. A. Ellson, J. Goodling, C. A. Kilgore, M. H. Richman, C. H. Sprague, G. T. Taoka, P. K. Turnquist 1970; T. L. Anderson, T. Bonnema, G. A. Cushman, T. A. Haliburton, J. L. Herrington, D. E. Kirk, J. P. Klus, J. T. Pfeffer, W. D. Seider, A. B. Tschantz, G. Watters, R. M. Zimmerman 1969.

JOHN LELAND ATWOOD AWARD, AEROSPACE DIVISION-AIAA ... bestowed upon a distinguished contributor to aerospace engineering education to recognize a recent outstanding educational achievement and to encourage original, innovative improvements in aerospace engineering. The award consists of $2,000 and is sponsored by Rockwell International Endowment. In addition, the American Institute of Aeronautics and Astronautics (AIAA) provides a suitably engraved medal and certificate.

Recipients: Jan Raskam 1987; Conrad F. Newbury 1986; W. T. Fowler 1984; Robert L. Young 1983; L. S. Fletcher 1982; Ira D. Jacobson 1981; Thomas J. Mueller 1980; Donnell W. Dutton 1979; Robert F. Brodsky 1978; Stanley H. Lowy 1977; Barnes W. McCormick 1976; Jack E. Fairchild 1975.

EUGENE L. GRANT AWARD, ENGINEERING ECONOMY DIVISION ... an annual award of $250 to the author of the best paper published in the "Engineering Economist", quarterly publication of the division.

Recipients: T.O. Boucher, L. Abdel-Malek, W. E. Stuckey 1986; William Sullivan, Thomas Ward and Kyung Lee 1985; R. H. Bernhard 1984; Thomas O. Boucher 1983; S. A. Bhimiee, R. Oakford and A. Slazar 1982; C. R. Rao, Kenneth Nowotny, Paul McDevitt 1981; John Freidenfeld, Michael D. Kennedy 1980; James R. Buck, Glenn H. Sullivan, Phillip E. Nelson 1979; Peter G. Sassone 1978; Gerald L. Thuesen 1977; Nabil A. El-Ramly, Richard E. Petersen, K. K. Seo 1976; J. Morley English 1975; Lynn E. Bussey, G. T. Stevens, Jr. 1974; Richard J. Tersine, William Rudko, Jr. 1973; Raymond P. Lutz, Harold A. Cowles 1972; James C. T. Mao, John F. Brewster 1971; Robert F. Kleusner 1970; Gerald W. Smith 1969; Laurence C. Rosenberg 1968.

FRANK OPPENHEIMER AWARD, ENGINEERING DESIGN GRAPHICS DIVISION ... through the generosity of Frank Oppenheimer of West Germany, longtime member of the division, an honorarium of $100 is given by the division to the author(s) of the best paper presented at the midyear conference of the division.

Recipients: Edward Knoblock 1987; Larry Genalo and John F. Freeman, Jr., 1986; Gary Bertoline 1985; Rolan Jenison 1984; Rolan Jenison 1983; Abram Rotenberg 1982; Larry D. Goss 1981; Arvid E. Eide 1980; John T. Demel, Alan D. Kent 1979.

FRED MERRYFIELD DESIGN AWARD ... recognizes an engineering educator for excellence in the teaching of engineering design as well as acknowledging other significant contributions related to engineering design teaching. The award recipient receives $2,000, a $500 stipend for travel to the ASEE Annual Conference and a commemorative plaque. In addition, the awardee's institutional department receives an award of $500.

Recipients: James F. Thorpe 1987; Donald H. Thomas 1986; Max S. Peters 1985; Lee Harrisberger 1984; Geza Kardos 1983; Percy H. Hill 1982; Charles O. Smith 1981.

FREDERICK EMMONS TERMAN AWARD, ELECTRICAL ENGINEERING DIVISION . . . bestowed upon an outstanding young electrical engineering educator in recognition of his contributions to the profession, the award is sponsored by the Hewlett-Packard Company and consists of $2,000, a gold medal and bronze replica, a presentation scroll, and travel expense reimbursement to attend the Annual Conference.
Recipients: Kenneth Short 1987; Lance E. Glasser 1986; Peter Maybeck 1985; M. Vidyasagar 1984; Daniel Siewiorek 1983; Toby Berger 1982; Ben G. Streetman 1981; V. Thomas Rhyne 1980; Martha E. Sloan 1979; Ronald A. Rohrer 1978; J. Leon Shohet 1977; Stephen W. Director 1976; Michael L. Dertouzos 1975; Leon O. Chua 1974; Sanjit Kumar Mitra 1973; Taylor L. Booth 1972; Joseph W. Goodman 1971; Andrew P. Sage 1970; Michael Athans 1969.

GEORGE WESTINGHOUSE AWARD . . . established by the Westinghouse Educational Foundation in 1946 as an annual award to young engineering teachers of outstanding ability to recognize and encourage their contributions to the improvement of teaching methods for engineering students. The award consists of $1,000, a $500 grant for travel expenses to the Annual Conference and a certificate. In addition, the awardee's academic department receives a cash award of $500.
Recipients: John H. Seinfield 1987; Gerald W. Clough 1986; Sunder Advani 1985; Phillip Wankat 1984; Frank Incropera 1983; Alan Willson, Jr 1982; L. S. "Skip" Fletcher 1981; William B. Krantz 1980; J. Michael Duncan 1979; C. Judson King 1978; Roger A. Schmitz 1977; Jerome B. Cohen 1976; Donald G. Childers 1975; Joseph Bordogna 1974; Martin D. Bradshaw 1973; Jack P. Holman 1972; Charles E. Wales 1971; Ali A. Seireg 1970; Arthur E. Bryson, Jr. 1969; Klaus D. Timmerhaus 1968.

GLENN MURPHY AWARD, NUCLEAR ENGINEERING DIVISION . . . bestowed upon a distinguished engineering educator in recognition of notable professional contributions to the teaching aspects of nuclear engineering. The award, consisting of an honorarium and a certificate, is sponsored by former students and friends of the late Glenn Murphy.
Recipients: Kuruvilla Verghese 1987; A. Ziya Akasu 1986; Richard T. Lakey, Jr. 1985; G. H. Sanquist 1984; Virgil E. Schrock 1983; Martin Becker 1982; J. Kenneth Shultis 1981; Allen F. Henry 1980; Glenn F. Knoll 1979; Thomas W. Kerlin, Jr. 1978; Edward H. Klevans 1977; Raymond L. Murray 1976.

JAMES H. McGRAW AWARD . . . awarded each year to an outstanding contributor to engineering technology education, recognizing notable achievement in the areas of teaching, guidance, publications, administration, and/or other related activities. Established in 1950 by the McGraw-Hill Book Company, the award consists of $1,000 and a certificate.
Recipients: Anthony L. Timans and Lawrence Wolf 1987; James Todd 1985; Stephen Cheshire 1984; Walter E. Thomas 1983; Michael Mazzola 1982; Ernest R. Weidhass 1981; Lyman L. Francis 1980; Robert J. Wear 1979; Joseph J. Gershon 1978; Donald C. Metz 1977; Eugene Wood Smith 1976; Louis J. Dunham, Jr. 1975; Merritt A. Williamson 1974; G. Ross Henninger 1973; Richard J. Ungrodt 1972; Melvin R. Lohmann 1971; Hugh E. McCallick 1970; Winston D. Purvine 1969; William N. Fenninger 1968.

JAMES S. RISING AWARD, ENGINEERING DESIGN GRAPHICS DIVISION . . . an honorarium is granted to the outstanding freshman design that has been selected to receive first place in the Creative Engineering Design Display held during the ASEE Annual Conference.
Recipients: Maureen Kerrigan, Inge Oschida, Gail Rupnick 1983; Carol Carfrae, Lauren Healy, Brian Hill, Greg Marsh, John Stillmark, Jeff Wescott 1982; Denise Gustafson, Mark Hermanson, Sue Engel, Eric Mortenson, Kim Kischer, Mohamed Al-Aidy 1981; Denise Gustafson, Mark Hermanson, Sue Engel, Eric Mortenson, Kim Kischer, Mohammad Al-Aidy, Project Advisor: Robert J. Bernhard 1980.

JOHN A. CURTIS LECTURE AWARD, COMPUTERS IN EDUCATION DIVISION . . . a plaque awarded to the person who has presented the outstanding paper in the division's program at the ASEE Annual Conference.

Recipients: George Y. Jumper, Jr., John J. Titus 1986; Robert L. Norton 1985; A. L. Riemenschneider, M. J. Batchelder 1984; Richard C. Harden, Fred O. Simons, Jr. 1983; Trevor N. Mudge 1982; John W. Spadman, Raymond G. Jacquot, Mark N. Hempworth 1981; Dean K. Frederick, Gary L. Waag 1980; C. Norman Kerr 1979.

LAMME AWARD . . . bestowed annually upon a distinguished engineering educator for excellence in teaching and contributions to research and technical literature, achievements which contribute to the advancement of the profession, and engineering college administration. It consists of a gold medal, and a certificate.
Recipients: Arthur E. Bergles 1987; Paul Z. Zia 1986; Richard Engelbrecht 1985; Beitse Chao 1984; H. Bolton Seed 1983; George Burnel 1982; Ernest S. Kuh 1981; John C. Hancock 1980; William H. Corcoran 1979; Mac Elwyn Van Valkenburg 1978; Ascher H. Shapiro 1977; John J. McKetta 1976; John R. Whinnery 1975; George W. Hawkins 1974; Max S. Peters 1973; Glenn Murphy 1972; Richard G. Folsom 1971; Jacob D. Den Hartog 1970; Joseph C. Elgin 1969; Morrough P. O'Brien 1968.

OUTSTANDING BIOMEDICAL ENGINEERING EDUCATOR AWARD, BIOMEDICAL ENGINEERING DIVISION . . . conferred for significant contributions not only to the Society and division but also to the biomedical engineering and education professions as evidenced by the development of successful graduate level programs, curricula, textbooks and professional papers and by membership and activity in other biomedical engineering organizations. The award consists of $300 and a commemorative plaque.
Recipients: Herman R. Weed 1987; Albert H. Cook 1985; T. C. Pilkington 1984; Herman P. Schwan 1983; Glenn V. Gdmonson 1982; Francis M. Long 1981.

RALPH COATS ROE AWARD, MECHANICAL ENGINEERING DIVISION . . . bestowed upon an outstanding teacher of mechanical engineering who has made notable contributions to the engineering profession. Financed from endowment of $40,000 established by Kenneth A. Roe in honor of his father, Ralph Coats Roe, the award consists of $2,000, a plaque and reimbursement of travel expenses to attend the ASEE Annual Conference.
Recipients: John R. Howell 1987; Werner Soedel 1986; George Sandor 1985; L. Glassman 1984; L. S. "Skip" Fletcher 1983; Frank P. Incropera 1982; Jerold L. Swedlow 1981; Wilbert F. Stoecker 1980; Robert H. Page 1979; Ephraim M. Sparrow 1978; Stothe P. Kezios 1977; John R. Dixon 1976.

SENIOR RESEARCH AWARD . . . established in 1979 by the Engineering Research Council of ASEE under the sponsorship of the General Electric Company. Its purpose is to recognize and honor a member of the staff or faculty of a college of engineering who has made significant contributions to engineering research by pushing forward the frontiers of knowledge, by perfecting and applying the latest scientific advances to engineering problems, or by providing administrative leadership to important engineering research programs. This award consists of a gold medal, a certificate, and reimbursement of the recipient's travel expenses to the Annual Conference.
Recipients: Jack E. Cermak 1987; Ray W. Clough 1986; Robert Brodkey 1985; Raymond Viskanta 1984; Alfredo H. S. Ang 1983; Herbert Kroemer 1982; King-sun Fu 1981; August J. Durelli 1980; Thomas J. Hanratty 1979.

W. LEIGHTON COLLINS AWARD FOR DISTINGUISHED AND UNUSUAL SERVICE . . . the highest ASEE award for extraordinary service to education for engineering and engineering technology and allied fields. The award consists of an enameled bronze medallion, which is a representation of the ASEE logo.
Recipients: Anthony Giordano 1987; William Everitt 1980; Joseph M. Pettit 1979; Melvin R. Lohmann 1978; Leslie B. Williams 1974; Linton E. Grinter 1972; W. Leighton Collins 1971.

VINCENT BENDIX MINORITIES IN ENGINEERING AWARD . . . established in 1979 to honor an engineering educator for exceptional achievement in increasing participation and retention of minority and/or women students in engineering curricula. This award is sponsored by The Allied Corporation. The awardee receives $1,500, a certificate and a grant of $500 for reimbursement of ASEE Annual Conference travel expenses.

Recipients: Jane Z. Daniels 1987; William T. Brazelton 1986; Howard S. Kimmel 1985; David Bedworth 1984; Alma Lantz 1983; Raymond Landis 1982; Zbigniew W. Dybczak 1981; Howard L. Wakeland 1980; Frederick William Schutz, Jr. 1979.

AT&T FOUNDATION AWARDS (FORMERLY WESTERN ELECTRIC FUND AWARD FOR EXCELLENCE IN INSTRUCTION OF ENGINEERING STUDENTS) ... awarded to outstanding teachers of engineering students to honor the recipient and to serve as an incentive to make further significant contributions to teaching. The award consists of $1,500 and a certificate and is presented to 16 awardees in 12 geographic sections.

Recipients: Ben G. Streetman, William L. Seibert, Louis Theodore, Louis F. Geschwindner, Howard W. Smith, Charles K. Taft, Roy E. Eckart, Lawrence A. Kennedy, Karl A. Smith, Anil Chopra, Craig K. Rushforth, Jack C. McCormac, Pandeli Durbetaki, Louis J. Galbati, Jr. 1987; J. F. Walkup, R. A. Howland, Jr, F. G. Helfferich, H. Goldstein, C. E. Barb, M. N. Horenstein, D. R. Riley, I. Minkarah, T. P. Herbert, R. Y. Itani, K. C. Crandall, R. D. Nobel, J. L. Hill, R. M. Felder, F. E. Griggs 1985; B. Das, A. Pytel, C. Dym, Y. A. Liu, J. Pommersheim, M. G. Britton, D. J. Bushrem, M. Mikanoff, T. G. Shoup, Y. A. Liu, A. Budak, P. C. Wankat, V. Ramachandran 1984; W. T. Fowler, M. H. Peck, R. L. Kabel, D. E. Malzahn, W. F. Schmidt, G. R. Bell, A. B. Bereskin, R. A. Danofsky, L. Edwards, G. C. Temes, E. D. Sloan, J. L. Gainer, D. T. Kao, A. H. Johannes 1983; W. M. Deen, C. H. Durney, R. A. Heller, B. P. Henderson, A. M. Kanuny, A. Katz, R. J. Kroll, E. C. Lindly, L. D. Luck, R. J. Niederjohn, J. P. O'Connell, W. E. Red, C. J. Savant, M. D. Smyth, B. J. Strait 1982; M. H. Agee, R. O. Buckius, J. T. Cain, G. E. Ford, R. R. Gallagher, A. B. Giordano, D. J. Hamilton, J. P. Hartman, F. S. Hill, Jr., G. E. Klinzing, E. Mikol, E. E. Mitchell, R. H. Nelson, L. D. Pye, R. J. Schmitz, J. E. Stice 1980-81; M. M. Abbott R. G. Levan, J. K. Mitchell, N. A. Peppas, W. M. Portnoy, W. F. Ramirez, Jr., C. G. Salmon, D. S. Shupe, R. K. Toner, B. A. Tschantz, D. T. Tuma 1979-80; S. Ahmed, O. A. Arnas, T. Au, R. M. Brach, M. M. Douglass, W. C. Flowers, R. J. Mattauch, W. F. Phillips, F. N. Rad, L. C. Redekopp, R. P. Santoro, C. L. Sayre, Jr., A. C. Scordelis, J. K. Shultis, J. G. Webster, J. J. Wert 1978-79; D. L. Vines, R. G. Squires, R. Bhattacharyya, B. R. Maxwell, W. C. Turner, W. R. Bennett, Jr., E. O. Doebelin, G. T. Hankins, V. Pavelic, L. T. Bruton, R. S. Elliott, D. F. Tuttle, Jr., J. K. C. Cheng, W. F. Babcock, M. A. Littlejohn, G. R. Youngquist 1977-78; Mo-Shing Chen, G. H. Miley, B. J. Pelan, H. H. West, N. Dean Eckhoff, R. R. Hagglund, J. K. Roberge, J. F. McDonough, D. T. Worrell, J. D. Stevens, T. G. Stoebe, G. C. Beakley, Jr., E. P. Popov, F. M. Long, J. S. Boland III, H. F. Keedy, H. J. McQueen 1976-77; G. E. Anner, A. J. Brainard, L. S. Caretto, M. E. Council, J. Estrin, C. T. A. Johnk, J. M. Kendall, J. R. Kittrell, R. E. Peck, A. J. Perna, R. A. Peura, B. C. Ringo, R. H. Seacat, Jr., F. H. Shair, F. G. Stremler, G. E. Whitehouse, D. J. Wood, P. Zia 1975-76; C. M. Bacon, L.D. Feisel, H. S. Haas, J. R. Hauser, J. F. Lestingi, H. R. Martens, J. L. Massey, W. G. McLean, G. W. Neudeck, E. R. Parker, H. L. Plants, R. Seagrave, J. R. Smith, W. A. Sowa, J. E. Stephens, F. F. Videon, C. R. Viswanathan, F. L. Worley, Jr. 1974-75; F. P. Beer, G. Boothroyd, J. Bordogna, C. M. Butler, A. B. Carlson, V. T. Chow, H. D'Angelo, G. H. Flammer, N. N. Gunaji, H. A. Hassan, W. E. Kastenberg, A. P. Marino, L. Padulo, G. W. Powell, J.T. Sears, B. L. Smith, B. S. Swanson, T. T. Williams 1973-74; D. E. Alexander, B. T. Chao, E. R. Chenette, V. E. Denny, D. L. Dietmeyer, E. T. B. Gross, L. D. Harris, W. H. Hayt, Jr., J. L. Melsa, A. W. Pense, E. L. Saxer, K. E. Scott, J. E. Shamblin, R. F. Steidel, Jr., R. I. Vachon, A. S. Weinstein, P.F. Wiggins, J. W. Willhide 1972-73; V. L. Pass, J. C. Whitwell, E. T. Misiaszek, H. A. Haus, I. B. Thomas, J. W. Bayne, D. T. Wasan, H.B. Kendall, D. G. Childers, M.N. Ozisik, J. Coates, G. H. Duffey, K. N. Reid, Jr., A. S. Levens, J. F. Gibbons, R. C. Smith, D. A. Firmage 1971-72; E. H. Gaylord, E. J. Freise, H. A. Estrin, B. Johnson, F. M. White, R. E. Stickney, E. M. Williams, R. W. Little, P. D. Cribbins, S. F. Adams, G. P. Francis, C. R. Nichols, C. H. Sprague, M. A. Larson, W. R. Tovey, N. J. Castellan, R. D. Strum, W. L. Fletcher 1970-71; J. J. Blum, W. H. Corcoran, R. F. Crank, R. W. Day, A. E. Durling, M. M. El-Wakil, W. C. Hahn, Jr., M. Liu, J. A. M. Lyon, J. R. Melcher, H. Menand, Jr., Z. A. Munir, W. F. Stoecker, H. C. Van Ness, D. H. Vliet, C. E. Wales, E. R. Whitehead, J. C. Williams, III 1969-70; M. D. Bradshaw, A. B. Butler, M. G. Fontana, V. D. Frechette, W. H. Giedt, R. D. Guyton-B. Hazeltine, J. J. Jonsson, E. R. Johnston, Jr., J. C. Liebman, P. W. Likins, A. B. Macnee, M. I. Mantell, L. J. McGeady, G. K. Mesmer, H. A. Moench, R. E. Raven, C. E. Work 1968-69.

WILLIAM ELGIN WICKENDEN AWARD ... The Society Publications Committee, with the assistance of the General Motors Corporation, established this award in 1978 to encourage excellence in writing by recognizing the author of the best paper published in *Engineering Education,* the journal of the Society, during the preceding publication year. The award recipient receives $1,000, a commemorative plaque and reimbursement of Annual Conference travel expenses.

Recipients: Frederick W. Garry 1987; Billy V. Koen 1986; Cynthia Selfe, Freydoon Arbabi 1985; David Bedworth, Hewitt H. Young 1984; Alma Lantz 1983; Lawrence C. Kravitz 1982; Eugene Chesson, Jr. 1981; William R. Grogan 1980; Lawrence P. Grayson 1979; Ali E. Engin, Ann W. Engin 1978.

MERIAM/WILEY DISTINGUISHED AUTHOR AWARD ... to recognize authorship of an outstanding new engineering textbook published after 1984 which embodies technical excellence, clarity of presentation and strong relevance to engineering practice. The award carries with it a stipend of $2,000 and reimbursement of transportation costs to the ASEE Conference.

Recipients: Steven C. Chapra and Raymond P. Canale 1987; William Z. Black and James G. Hartley 1986.

WILEY AWARD FOR EXCELLENCE IN ENGINEERING TECHNOLOGY EDUCATION ... presented to the department with a degree program that has exhibited outstanding leadership in presentation, documentation, innovation and administration in engineering technology education. Funded by John Wiley & Sons, Inc., the award consists of $1,500 and a certificate.

Recipients: Indiana University-Purdue University at Fort Wayne, Electronics Engineering Technology Program 1987.

NEW ENGINEERING EDUCATOR EXCELLENCE AWARDS ... honor up to eight non-tenured engineering educators in the first six years of appointment. The Awards consist of a $5,000 grant and a memento to each recipient, plus transportation expenses to attend the ASEE Annual Conference, donated by the General Electric Foundation.

Recipients: Kevin C. Craig, C. Stewart Slater, Kathleen C. Howell, William E. Lear, Jr., A.K.M. Uddin, Thomas V. Edgar, Richard T. Gill, James F. Thorpe 1987; John W. Zondlo, Danny D. Reible and George Tsiatas 1986.

OUTSTANDING ZONE CAMPUS REPRESENTATIVE AWARD ... honors an Outstanding Zone Campus Representative. A plaque is presented to the ASEE Campus Representative and a certificate to his supporting dean.

Recipients: John N. Clausen and Gerald S. Jakubowski 1987.

MERL MILLER AWARD ... annual prize for the outstanding CoED Journal paper on teaching/instructional methods and is to commemorate the contributions and dedication of Merl K. Miller to the Computers in Education Division.

Recipients: Robert L. Norton 1987.

HARDEN-SIMONS AWARD ... annual award for the outstanding CoED Journal paper on computational methods and is to commemorate the unselfish contributions of Dr. Richard C. Harden and Dr. Fred O. Simons, Jr. to the Computers in Education Division.

Recipients: Raymond C. Jacquot and John Steadman 1987.

RAY H. SPIESS AWARD ... annual prize for the outstanding CoED Journal paper on hybrid analog/digital computers/computation and is to commemorate the contributions of Ray H. Spiess and Comdyna, Inc. to the development of the Computers in Education Division.

Recipients: Samuel J. Lynch 1987.

HELEN L. PLANTS AWARD ... given annually for the most outstanding non-traditional session at the Frontiers in Education Conference of the Educational Research and Methods Division.

Recipients: Billy V. Koen 1987.

RONALD J. SCHMITZ AWARD ... given annually for outstanding contributions to the Frontiers in Education Conference of the Educational Research and Methods Division.

Recipients: John C. Lindenlaub 1987.

BEST CIEC WORKSHOP ... awarded by the Continuing Professional Development Division to the best workshop at the annual College Industry Education Conference.
Recipients: Charles S. Elliott 1987.

OUTSTANDING PAPER ... awarded by the Continuing Professional Development Division to the best paper presented at the annual College Industry Education Conference.
Recipients: Robert M. Anderson 1987.

CERTIFICATE OF MERIT ... awarded for outstanding contributions to the Continuing Professional Development Division.
Recipients: Frederic Rex, John F. Wilhelm and Linda Maynard Hall 1987.

ASM INTERNATIONAL

Metals Park, Ohio 44073
Edward L. Langer, Managing Director

PRESIDENTS ... Raymond F. Ducker 1987; John W. Pridgeon 1986; M. Brian Ives 1985; Donald J. Blickwede 1984; George H. Bodeen 1983; David Krashes 1982; John B. Giacobbe 1981; Raymond L. Smith 1980; Elihu F. Bradley 1979; Nicholas P. Milano 1978; Abraham Hurlich 1977; Robert H. Shoemaker 1976; Dean K. Hanink 1975; Joseph F. Libsch 1974; William D. Manly 1973.

AWARDS

THE ALBERT EASTON WHITE DISTINGUISHED TEACHER AWARD ... this award was established in 1960, in memory of an outstanding teacher and research engineer, who was a founder member and president of ASM International in 1921, to recognize unusually long and devoted service in the teaching of metallurgy, materials science and/or materials engineering, characterized by ability to inspire and impart enthusiasm to students as well as by metallurgical/materials accomplishments.
Recipients: Morris Cohen 1987; George E. Dieter 1986; Lawrence H. Van Vlack 1985; Robert W. Bohl 1984; Ernest F. Nippes 1983; Alan Lawley 1982; William W. Austin 1981; William O. Philbrook 1980; Albert W. Schlechten 1979; G. Marshall Pound 1978; Robert W. Lindsay 1977; E. Eugene Stansbury 1976; Alfred Bornemann 1975; Robert B. Pond 1974; Earl J. Eckel 1973; Robert D. Stout 1972; John F. Elliott 1971; Frederick N. Rhines 1970; Howard L. Womochel 1969; Maxwell Gensamer 1968.

THE ALBERT SAUVEUR ACHIEVEMENT AWARD ... this award, established in 1934 in honor of a distinguished teacher, metallographer and metallurgist, recognizes pioneering metallurgical achievements which have stimulated organized work along similar lines to such an extent that a marked basic advance has been made in metallurgical knowledge.
Recipients: Hubert I. Aaronson 1987; Walter W. Smeltzer 1986; Edward J. Dulis 1985; Robert J. Gray 1984; Frank H. Spedding 1982; Charles J. McMahon, Jr. 1981; Louis F. Coffin 1980; Frederick N. Rhines 1979; Merton C. Flemings 1978; Morris Cohen 1977; Paul A. Beck 1976; Andre Guinier 1975; George R. Irwin 1974; Pol Duwez 1973; William S. Pellini 1972; Victor F. Zackay 1971; Clarence Bieber 1970; Sir Alan Cottrell 1969; Alexander R. Troiano 1968.

ASM AND TMS DISTINGUISHED LECTURESHIP IN MATERIALS AND SOCIETY ... established in 1971 to clarify the role of materials engineering in technology and in society in its broadest sense; to present an evaluation of progress made in developing new technology for the ever changing needs of technology and society; to define new frontiers for materials engineering.
Recipients: James S. Kane 1987; Arden L. Bement Jr. 1986; Robert I. Jaffee 1985; Nathan E. Promisel 1984; Raymond L. Smith 1983; Morris Cohen 1982; The Honorable Dixy Lee Ray 1981; Charles Crussard 1980; Glenn T. Seaborg 1979; Herbert H. Kellogg 1978; Sir H. Montague Finniston, FRS 1977; William O. Baker 1976; Michael Tenenbaum 1975; Cyril Stanley Smith 1974; James Boyd 1973; Sir Alan Cottrell 1972; Harvey Brooks 1971.

THE BRADLEY STOUGHTON AWARD FOR YOUNG TEACHERS OF METALLURGY ... this award was established in 1952, in memory of an outstanding teacher of metallurgy and dean of engineering who was president of the American Society for Metals in 1942, to encourage young teachers of metallurgy by rewarding them for their ability to impart knowledge and enthusiasm to students. The recipient must be under 35 years of age on May 15 of the year in which the selection is made.
Recipients: Jeffery Clarke Gibeling 1987; David L. Bourell 1986; Ronald Gronsky 1985; George M. Pharr 1984; Stephen W. Stafford 1983; J. Barry Andrews 1982; Bruce R. Palmer 1981; Diran Apelian 1980; David K. Matlock 1979; John H. Perepezko, 1978; John W. Morris, Jr. 1977; David L. Olson 1976; John K. Tien 1975; William M. Boorstein 1974; H. R. Piehler 1973; Gordon H. Geiger 1972; Henk I. Dawson 1971; William D. Nix 1970; Richard W. Heckel 1969; George S. Ansell 1968.

CANADIAN COUNCIL LECTURESHIP ... established in 1971 to identify a distinguished lecturer who will present a technical talk to a regular monthly meeting of each of those Canadian ASM chapters who elect to participate.
Recipients: Larry E. Seeley 1987; J. Keith Brimacombe 1986; J. Peter McGeer 1985; A. Brian Rothwell 1984; George B. Craig 1983; T. Raymond Meadowcroft 1982; Bruce M. Hamilton 1981; William C. Winegard 1980; M. Brian Ives 1979; John Convey 1978; W. M. Williams 1977; Gerald R. Heffernan 1976; Carl R. Whittemore 1975; William M. Armstrong 1974; A. D. Fisher 1973; J. S. Kirkaldy 1972; Ralph D. Hindson 1971.

THE EDWARD DEMILLE CAMPBELL MEMORIAL LECTURE ... established in 1926 in memory of and in recognition of the outstanding scientific contributions to the metallurgical profession by a distinguished educator who was blind for all but two years of his professional life. It recognizes demonstrated ability in metallurgical and materials science and engineering.
Recipients: Albert R. C. Westwood 1987; Herbert H. Johnson 1986; Raymond F. Decker 1985; Robert F. Hehemann 1984; Robert A. Rapp 1983; John W. Christian 1982; George T. Hahn 1981; David Turnbull 1980; Morris E. Fine 1979; Harold W. Paxton 1978; Robert I. Jaffee 1977; Jack H. Westbrook 1976; Morris Tanenbaum 1975; Donald J. McPherson 1974; W. A. Backofen 1973; John P. Hirth 1972; William C. Leslie 1971; Mars G. Fontana 1970; Walter R. Hibbard, Jr. 1969; Donald J. Blickwede 1968.

ENGINEERING MATERIALS ACHIEVEMENT AWARD ... established in 1969 to recognize an outstanding achievement in materials or materials systems relating to the application of knowledge of materials to an engineering structure or to design and manufacture of a product.
Recipients: General Electric Company, Metglas Products Division of Allied Signals, Electric Power Research Institute, Empire State Electric Energy Research Corporation 1987; Pratt & Whitney Engineering Division of United Technologies Corporation 1986; General Motors Corporation 1985; International Business Machine Corp. 1984; Corning Glass Works, Western Electric Company/Bell Telephone Labs, Inc. 1983; Sikorsky Aircraft Division, United Technologies Corp. 1982; General Electric Company Lighting Business Group 1981; Engelhard Industries Division, Engelhard Minerals & Chemicals Corp., Corning Glass Works, Technical Ceramic Products Div. 3/M 1980; Battelle Memorial Inst 1979; E. I. du Pont de Nemours & Company Inc. 1978; Joslyn Stainless Steels Division, Joslyn Mfg. & Supply Co. Linde Division, Union Carbide Corporation 1977; Diamond Shamrock Corp, Ford Motor Co., Inland Steel Co., Wyandotte Paint Products Co., 1976; Pratt & Whitney Aircraft, Div of United Technologies Corp. 1975; General Motors Corp., Chevrolet Motor Div., Reynolds Metals Co. 1974; General Electric Co., 1973; Westinghouse Bettis Atomic Power Lab 1972; Armco Steel Corp., Bethlehem Steel Corp., Cleveland Cliffs Iron Co., Ford Motor Co., University of Minnesota, Oglebay Norton Co., Pickands Mather & Co., Republic Steel Corp. 1971; Lockheed Aircraft Corp 1970.

THE GOLD MEDAL ... established in 1943 to recognize outstanding knowledge and great versatility in the application of science to the entire materials industry, as well as exceptional ability in the diagnosis and solution of diversified materials problems.
Recipients: George R. St. Pierre 1987; Morris E. Fine 1986; Oleg D. Sherby 1985; Lawrence H. Van Vlack 1984; Harold W. Paxton 1983; Allen S. Russell 1982; Raymond F. Decker 1981; Sir Alan H. Cottrell 1980; Mars G. Fontana 1979; John R. Low, Jr. 1978; George A. Roberts 1977; Charles S. Barrett 1976; F. Denys

Richardson 1975; Clarence M. Zener 1974; Carl Wilhelm Wagner 1973; Earl R. Parker 1972; Lawrence S. Darken 1971; Kent R. Van Horn 1970; Morris Cohen 1968.

THE HENRY MARION HOWE MEDAL . . . awarded in memory of a distinguished teacher, writer, metallurgist, and consultant to honor the author (or authors) whose paper has been selected as the best of those published in a specific volume of the ASM Transactions, now Metallurgical Transactions.
Recipients: Brent L. Adams 1987; Richard P. Gangloff 1986; Eiichi Takeuchi and J. Keith Brimacombe 1985; G. R. Speice, A. J. Schwoeble and J. P. Huffman 1984; Robert Mehrabian and C. G. Levi 1983; G. A. Irons and R. I. L. Guthrie 1982; H. Dolle & J. B. Cohen 1981; J. K. Brimacombe, F. Weinberg and E. B. Hawbolt 1980; G. W. Simmons, P. S. Pao, R. P. Wei 1979; E. D. Hondros, Martin P. Seah 1978; Paul G. Shewmon 1977; H. Henein, R.I.L. Guthrie, R. Clift 1976; Joseph R. Rellick, C. J. McMahon, Jr. 1975; R. A . Rapp, A. Ezis, G. J. Yurek 1974; M. D. Rinaldi, R. M. Sharp, M. C. Flemings 1973; R. G. Davies-C. L. Magee 1972; John S. Benjamin 1971; B. H. Kear, G. R. Leverant, J. M. Oblak 1970; J. S. Kirkaldy, R. D. Townsend 1969; Kanji Ono-Masahiro Meshii 1968.

JACQUET-LUCAS AWARD FOR EXCELLENCE IN METALLOGRAPHY . . . the ASM/IMS Metallographic Award was established in 1946 for the best entry in the annual ASM International metallographic competition. In 1958 it became known as Francis F. Lucas Metallographic Award, and has been endowed since that date by Adolph I. Buehler. In 1972 ASM joined with the International Metallographic Society in sponsoring the Pierre Jacquet Gold Medal and the Francis F. Lucas Award for Excellence in Metallography. Beginning in 1976 this award is endowed by Buehler Ltd.
Recipients: Stan A. David, John M. Vitek, C. Paul Haltom, Allison G. Barcomb 1987; N. T. Saenz, C. A. Lavender, M. J. Smith, D. H. Parks, G. M. Salazar 1986; Ulrike Taffner and Rainer Telle 1985; Ray H. Beauchamp, Natalio T. Saenz and John T. Prater 1984; Veronika Carle and Eberhard Schmid 1983; J. L. Young 1982; Fumio Kurosawa, Isamu Taguchi, Hirowo G. Susuki 1981; R. H. Beauchamp, K. Fredriksson 1980; M. J. Bridges-S. J. Dekanich 1979; Chris Bagnall, Robert E. Witkowski 1978; Ray H. Beauchamp, Derald H. Parks, Nate T. Saenz, Kenneth R. Wheeler 1977; Lars E. Soderqvist 1976; William C. Coons 1975; G. V. McIlhargie, J. E. Bennett, M. R. Pinnel, D. E. Heath 1974; M. S. Grewal, B. H. Alexander, S. A. Sastri 1973; C. J. Echer, S. L. Digiallonardo 1972; R. J. Gray 1971; Donald R. Betner, Wayne D. Hepfer 1970; R. H. Beauchamp, R. P. Nelson 1969; R. M. N. Pelloux, Mrs. H. Wallner 1968.

THE MARCUS A. GROSSMANN YOUNG AUTHOR AWARD . . . established in 1960 in memory of an eminent metallurgist, research director, and author, who was President of ASM in 1944, to honor the author (or authors) under 35 years of age whose paper has been selected as the best of those published in a scientific volume of the "ASM Transactions", now "Metallurgical Transactions."
Recipients: D. M. Kundrat 1987; Kwai S. Chan 1986; Martin R. Bridge and Gary D. Rogers 1985; G. M. Michal 1982; Thomas M. Devine, Jr. 1981; (not awarded in 1980); Ronald M. Horn-Robert O. Ritchie 1979; Michel Guttmann 1978; A. Grill, K. Sorimachi, J. K. Brimacombe 1977; Siegfried S. Hecker, Amit K. Ghosh 1976; Gregory O. Garmong 1975; L. W. Beckstead, J. D. Miller 1974; M. Y. Solar, R. I. L. Guthrie 1973; M. E. Glicksman, R. J. Schaefer 1971; F. D. Lemkey, E. R. Thompson 1970; T. H. Alden 1969; H. W. Hayden, R. C. Gibson, H. F. Merrick, J. H. Brophy 1968.

THE MEDAL FOR THE ADVANCEMENT OF RESEARCH . . . established to honor an executive of an organization active in the production, fabrication, or use of metals, who has consistently sponsored metallurgical research or development over a period of years, and who has, by his foresight and actions, helped substantially to advance the arts and sciences related to metals and other materials.
Recipients: William D. Manly 1987; Charles W. Parry 1986; Gordon E. Moore 1985; Pierre L. Gousseland 1984; Robert A. Charpie 1983; Martin J. Caserio 1982; William J. De Lancey 1981; Howard O. Beaver, Jr. 1980; Michael Tenenbaum 1979; Arthur M. Bueche 1978; Sherwood L. Fawcett-Lee A. Iacocca 1977; No

Recipient in 1976; Edward N. Cole 1975; James B. Fisk 1974; John W. Simpson 1973; W. P. Gwinn 1972; Frederick J. Close 1971; E. J. Hanley 1970; Harold M. Griffith 1969; Joel Hunter 1968.

THE WILLIAM HUNT EISENMAN AWARD . . . established in 1960 in memory of a founder member of ASM and its first and only Secretary for forty years, to recognize unusual achievements in industry in the practical application of metallurgy to the production of metals or their engineering use.
Recipients: Robert B. Herchenroeder 1987; Norman O. Kates 1986; Terrence G. Bradbury 1985; Michael Korehynsky 1984; G. Bruce Kiner 1983; Francis M. Richmond 1982; John D. Graham 1981; Albert R. Fairchild, Jr. 1980; Clyde A. Furgason 1979; Benjamin Lustman 1978; Donald J. Blickwede 1977; Chester T. Sims 1976; Paul G. Nelson 1975; Muir L. Frey 1974; Max W. Lightner 1973; George Harrison 1972; Kenneth T. Norris 1971; Harold N. Bogart 1970; Joseph V. Emmons 1969; Francis John McMulkin 1968.

DISTINGUISHED LIFE MEMBERSHIP . . . conferred on those leaders who have devoted their time, knowledge and ability to the advancement of materials and the materials industry.
Recipients: Lloyd Reuss and Rev. William T. Hogan, S. J. 1987; F. Kenneth Iverson, George P. Peterson 1986; Richard J. Coar, Robert E. Kirby, Richard P. Simmons, Frank W. Luerssen, Adolph J. Lena, William C. Winegard 1985; Sherwood L. Fawcett, Thomas O. Mathues, Cornell C. Maier 1984; W.H. Krome George, Yoshihiro Inai, Donald L. Ritter 1983; J. Peter Gordon, Frederick C. Langenberg, Robert O. Wilder, Pierre Gousseland, Gerald R. Heffernan 1982; James H. Doolittle, Charles H. Smith, Jr., Frank H. Sherman 1981; Earle M. Jorgensen 1980; Dennis J. Carney, Harry J. Gray, Shintaro Tabata 1979; Theodore Operhall, Henry E. Singleton, Charles B. Baker 1978; Sir H. Montague Finniston, John F. Magee, William L. Neumann 1977; George H. Bodeen, Michael Tenenbaum 1974; Thomas J. Ready, Ian K. MacGregor, Horace A. Shepard 1973; Joseph R. Carter, James C. Hodge, Soichiro Honda, W.F. Rockwell, Jr. 1972; Donald C. Burnham, George G. Zipf 1971; William Blackie, Edwin H. Gott, John Moxon, Buckminster Fuller 1970; E.N. Cole, John D. Harper 1969; James H. Binger, C. William Verity, Jr. 1968; L.C. Mallet, John P. Roche, Bertram D. Thomas 1967; Adolph I. Buehler, Frank R. Milliken 1966; H. George DeYoung 1965.

INTERNATIONAL STUDENT PAPER CONTEST . . . established on the recommendation of ASM's Young Members Advisory committee in 1979. The award recognizes the best student technical paper in the fields of metallurgy, materials science or engineering. It also provides for broader student participation in the Society.
Recipients: Daniel R. Gee, University of Nebraska 1985; Stephen J. Benner, Worcester Polytechnic University 1983; Kenneth R. Hayes, California Polytechnic State University 1982; Lynette Angers, University of Wisconsin 1980.

FELLOW . . . established in 1969 to provide recognition to members for distinguished contributions in the field of metals and materials, and to develop a broadly based forum for technical and professional leaders to serve as advisors to the Society.

HONORARY MEMBERSHIP . . . awarded for distinguished service to the metallurgical profession, to the ASM, and to the progress of mankind.
Recipients: William C. Leslie 1986; Joseph F. Libsch 1985; Raymond L. Smith 1984; Nathan E. Promisel 1983; Francis L. LaQue 1982; William D. Manly 1981; Earl R. Parker 1980; John Convey 1979; Alexander R. Troiano 1978; Donald J. McPherson 1977; Robert I. Jaffee 1976; James B. Austin, Adolph O. Schaefer 1975; Walter E. Jominy 1974; W.G. Burgers, George A. Roberts 1973; Sir Alan H. Cottrell 1972; Samuel L. Hoyt 1971; Francis G. Tatnall 1970; Mars G. Fontana, Donald A. Oliver 1969; Morris Cohen 1968; William J. Kroll 1967; Robert F. Mehl, Albert J. Phillips, Carl E. Swartz 1966; Cyril Stanley Smith 1965.

THE AMERICAN SOCIETY FOR NONDESTRUCTIVE TESTING (ASNT)

4153 Arlingate Plaza, Caller #28518, Columbus, Ohio 43228
Desmond D. Dewey, Managing Director

PRESIDENTS ... Don Earl Edwards 1987; Robert R. Hardison 1986; William Widner 1985; George Wheeler 1984; William Kitson 1983; John Aman 1982; Don L. Conn 1981; Robert Zong 1980; Francis C. Berry 1979; Robert H. Zong 1978; Charles J. Hellier 1977; C. E. Lautzenheiser 1976; Carl B. Shaw 1975; John P. Battema 1973. Zong 1980; Francis C. Berry 1979; Robert H. Zong 1978; Charles J. Hellier 1977; C. E. Lautzenheiser 1976; Carl B. Shaw 1975; John P. Battema 1973.

AWARDS

THE ASNT ACHIEVEMENT AWARD ... presented for a manuscript published in any official journal of the Society which is an outstanding contribution to the advancement of nondestructive testing.
Recipients: Stephen D. Brown 1986; Hermann Wuestenberg, Ulrich Haufe and Anton Erhard 1985; Nathan Ida, R. Palanisimy and William Lord 1984; Richard S. Williams and Philip E. Zwicke 1983; Steven Serabian 1982; Kanji Ono 1981; Davis M. Egle and Don E. Bray 1980; Alex Vary 1979; James C. Crowe 1978; K. Kawashima and R. W. McClung 1977; C. V. Dodd-J. H. Smith 1976; George A. Alers and L. J. Graham 1975; Bernard Stiefeld 1974; Joseph L. Rose, Robert W. Mortimer, Pei Chi Chou 1973 and Donald J. Hagemaier 1972; Justin G. Schneeman 1971; Ralph M. Grant and Gordon M. Brown 1970; Ron J. Botsco 1969; John E. Jacobs 1968.

THE ASNT GOLD MEDAL ... in recognition of a person who has made outstanding contributions to the field of nondestructive testing and its advancement, or who has rendered meritorious service to the Society.
Recipients: Howard E. Van Valkenburg 1986; Gerald J. Posakony 1985; Carl B. Shaw 1987; Ronald J. Botsco 1983; Harold Berger 1982; Donald C. Erdman 1981; Carl E. Betz 1980; Robert J. Roehrs 1979; Paul Dick 1978; Robert C. McMaster 1977; William E. Havercroft 1976; Friedrich Foerster 1975; E. O. Lomerson, Jr. 1974; Royal G. Tobey 1973; W. C. Hitt 1972; E. W. McKelvey 1971; E. L. Criscuolo 1970; C. H. Hastings 1969; B. E. Justice 1968.

THE LESTER HONOR LECTURE ... presented to an outstanding person who is invited to lecture on a subject with direct bearing on the use of nondestructive testing and its application to materials evaluation.
Recipients: Ronald H. Selner 1986; Donald J. Hagemaier 1984; Howard E. Van Valkenburg 1982; Donald D. Dodge 1980; Julian R. Frederick 1978; Gerald J. Posakony 1976; Robert W. McClung 1974; Dr. John T. Norton 1972; Justin G. Schneeman 1970; Samuel A. Wenk 1968.

THE MEHL HONOR LECTURE ... presented to an outstanding lecturer in the field of nondestructive testing. This award is in honor of Dr. Robert F. Mehl for his outstanding research in the field of gamma radiography and other contributions.
Recipients: Charles J. Hellier III 1985; Lutz W. Bahlke 1983; Spencer H. Bush 1981; Emmanuel P. Papadakis 1979; Richard H. Lambert 1977; Harold Berger 1975; Douglas W. Ballard 1973; Robert F. Mehl 1971; Harold Hovland 1969.

TUTORIAL CITATION ... the Tutorial Citation has been established to give recognition to outstanding contributors to the field of nondestructive testing education. Recipients of the citation are selected for their accomplishments in educational activities designed to increase scientific, engineering, and technical knowledge in the field of NDT.
Recipients: Joseph L. Rose 1986; Frank A. Iddings 1985; John L. Summers 1984; Frederick W. Rohde 1983; Charles J. Hellier III 1982; W. Ray Garrett, 1981; Julian R. Frederick 1980; Donald D. Dodge 1979; B. R. "Johnny" Johnson 1978; Harold S. Kean 1977; Vernon L. Stokes 1975; George Pherigo 1974; Robert C. McMaster 1973; Louis B. Strader 1972; Paul Dick 1971.

AMERICAN SOCIETY FOR QUALITY CONTROL (ASQC)

230 West Wells Street, Suite 700, Milwaukee, Wisconsin 53203
Edward A. Buckley, Executive Director

PRESIDENTS ... J. Douglas Ekings 1987-1988; Dana Cound 1986-1987; H. James Harrington 1985-1986; John L. Hansel 1984-1985; Eddie F. Thomas 1983-1984; Robert R. Maas 1982-1983; John D. Hromi 1981-82; Jay W. Leek 1980-81; Philip B. Crosby 1979-80; Robert N. Reece 1978-79; Walter L. Hurd, Jr. 1977-78; Orde R. Weaver 1976-77; Charles H. Brokaw 1975-76; Dr. Howard L. Stier 1974-75; Harry J. Lessig 1973-74.

AWARDS

BRUMBAUGH AWARD ... presented to the author of a paper published either in the "Journal of Quality Technology" or "Quality Progress" which the Committee judges has made the greatest contributions to the industrial applications of quality control.
Recipients: J. Stuart Hunter 1985; John Mandel 1984; Ronald D. Smee 1983; Gerald J. Halm 1982; Edward G. Schilling and Dan J. Sommers 1981; Gerald J. Hahn & Josef Schmee 1980; Edward G. Schilling 1979; John H. Sheesley and Edward G. Schilling 1978; J. Edward Jackson 1977; James M. Lucas 1976; Edward M. Baker 1975; Gerald J. Hahn 1974; Edward E. Schilling 1973; John de S. Coutinho 1972; Ronald D. Snee 1971; Harrison M. Wadsworth 1970; Wayne B. Nelson 1969; Ellis R. Ott 1968.

EDWARDS MEDAL ... presented to an individual who has demonstrated the most outstanding leadership in the application of modern quality control methods, especially through the organization and administration of such work.
Recipients: Raymond Wachniak 1986; Walter L. Hurd, Jr. 1985; John E. Condon 1984; William E. Conway 1983; Thomas C. McDermott, Jr. 1982; Philip B. Crosby 1981; Naomi J. McAfee 1980; Robert W. Peach 1979; Rocco L. Fiaschetti 1977; Edward M. Schrock 1976; William A. Golomski 1975; Walter E. Masing 1974; Arthur Bender, Jr. 1973; E. Jack Lancaster 1972; John J. Riordan 1971; John W. Young 1970; Darian Shainin 1969; J. Y. McClure 1968.

THE EUGENE L. GRANT AWARD ... presented to an individual who has demonstrated outstanding leadership in the development and presentation of a meritorious educational program in quality control.
Recipients: Kenneth S. Stephens 1986; William G. Gage 1985; Lennart Sandholm 1984; Mae-Goodwin Tarver 1983; John J. Heldt 1982; Dorian Shainin 1981; Mason E. Wescott 1980; Max Astrachan 1979; Frank M. Gryna, Jr. 1978; John A. Henry 1977; Yoshio Kondo 1976; W. Grant Ireson 1975; Walter L. Hurd, Jr. 1974; Leonard A. Seder 1973; Harold F. Dodge 1972; Kaaru Ishikawa 1971; David S. Chambers 1970; Gayle W. McElrath 1969; Ellis R. Ott 1968.

SHEWHART MEDAL ... presented to an individual who has made an outstanding contribution to the science and techniques of quality control or who has demonstrated outstanding leadership in the field of modern quality control.
Recipients: Donald W. Marquardt 1986; Ronald D. Shee 1985; Norman L. Johnson 1984; Edward G. Schilling 1983; Jaoru Ishikawa 1982; Richard A. Freund 1981; John Mandel 1980; Hugo C. Hamaker 1979; Lloyd S. Nelson 1978; Albert H. Bowker 1977; John W. Tukey 1976; William R. Pabst, Jr. 1975; Benjamin Epstein 1974; Sebastian B. Littauer 1973; Gerald J. Lieberman 1972; Frank E. Grubbs 1971; J. Stuart Hunter 1970; W. J. Youden 1969; G. E. P. Box 1968.

E. JACK LANCASTER AWARD ... presented to demostrated exceptional leadership in the international scene in promoting the quality profession.
Recipients: Umberto Turelo 1986; Richard A. Freund 1985; Walter E. Masing 1984; Walter L. Hurd, Jr. 1983; Paul C. Clifford 1982; A.V. Feigenbaum 1981.

AMERICAN SOCIETY FOR TESTING AND MATERIALS (ASTM)

1916 Race Street, Philadelphia, Pennsylvania 19103
Joseph G. O'Grady, President

CHAIRMEN OF THE BOARD . . . Robert Baboian 1987; Gladys Berchtold 1986; Lendell E. Steele 1985; Robert G. Redelfs 1984; Richard H. Goodemote 1983; William A. Goodwin 1982.

AWARDS

ADHESIVES AWARD (sponsored by committee D-14) . . . to recognize outstanding work in the science of adhesion or the technology of adhesives. The award consists of a certificate and an honorarium.
Recipients: Girard S. Haviland 1987; Arthur Hockman 1986; Leonard-Hart Smith 1985; James D. Minford 1982: Alan N. Gent 1979; Robert Patrick 1978; Max L. Williams, Jr. 1975; Nicholas J. DeLollis 1973; M. L. Selbo 1972; Henry L. Lee, Jr. 1970.

ARNOLD H. SCOTT AWARD (sponsored by Committee D-9) . . . presented by the ASTM Committee on Electrical Materials for outstanding achievement in the field of electrical insulation technology. The award consists of a plaque.
Recipients: Edward J. McGowan 1986; Ray Barknikas 1985; Carl F. Ackerman, 1984; Paul F. Ast 1983; Phil E. Alexander 1982; Landis E. Feather 1981; Murray L. Singer 1980; Celine M. Paul 1979; William P. Harris 1978; Eugene J. McMahon 1977; Edmund H. Povey 1976; Howard A. Davis 1975; Lester J. Timm 1974; Joseph R. Perkins 1973; Edward B. Curdts 1972; Kenneth N. Mathes 1971; Herbert G. Steffens 1970.

B. F. SCRIBNER AWARD (sponsored by committee E-2) . . . the award may be bestowed on an individual who has rendered outstanding service to the committee for one particular accomplishment (emission spectroscopy). Certificate.
Recipients: Lawrence E. Zeeb 1986; G. Robert Brammer 1985; James V. Derby, 1984; B.F. Scribner 1983; Rudolf Dyck 1982; Monte L. Dawkins 1981; Norma L. Bottone 1980; Ted R. Linde 1979; Donald C. Spindler 1978; G. Leslie Mason 1976.

C. A. HOGENTOGLER AWARD (sponsored by Committee D-18) . . . presented to the author or authors of a paper of outstanding merit on soil for engineering purposes. Walnut-bronze plaque.
Recipients: David Dixon, William Cox, Benton Murphy, Arvid O'Landva, Pierre LaRochell 1986; Toralv Berre 1984; Reginald H. Hardy 1983; Donald H. Shields and Jean-Louis Briaud 1981; Melvin I. Esrig and Robert C. Kriby 1980; V. P. Drnevitch, B. O. Hardin and D. J. Shippy 1979; Rey S. Decker and L. P. Dunnigan 1978; F. C. Townsend and P. A. Gilbert 1977; T. D. O'Rourke and E. J. Cording 1976; Bezalel C. Haimson 1975; H. M. Horn and Yves LaCroix 1974; G. H. Keller, V. J. McDonald, R. E. Olson and A. F. Richards 1973; Peter W. Rowe 1972; D. R. McCreath, R. P. Benson and D. K. Murphy 1971.

CHARLES A. JOHNSON AWARD for insulating liquids and gases (presented by committee D-27) . . .
Recipients: Edwin S. Packer 1982; Paul F. Ast 1981;

CHARLES B. DUDLEY AWARD . . . presented to the author or authors of a paper published by the Society of outstanding merit constituting an original contribution on research in engineering materials. The award consists of a walnut-bronze plaque.
Recipients: Ernest Selig 1987; Alexander Robertson & Tibor Z. Harnathy 1986; Ray Barknikas 1985; Richard A. Conway and B. Charles Malloy-Jones 1984; Kenneth Boldt and B. R. Hall 1980; Robert P. Benedict 1979; Fred A. Smidt, Jr. 1978; John Cairns, Jr. and Kenneth L. Dickson 1977; Eli Freedman, Dale N. Treweek and William H. Seaton 1976; L. F. Coffin, Jr. 1975; JoDean Morrow 1973; Lendell E. Steel and Russell Hawthorne 1972; James R. Rice 1969.

CHARLES W. BRIGGS AWARD (presented by committee E-7) . . . non-destructive testing.
Recipients: Thomas F. Drumwright 1987; Kermit A. Skeie & Jerry J. Haskins 1986; Carlton E. Burley & Joseph Marble 1985; William Dundore, Calvin McKee 1984; John W. Fenton and Patrick C. McEleny 1983; John E. Bobin and Theodore F. Luga 1981.

COPPER CLUB AWARD (presented by Committee B-5) . . . for copper alloys.
Recipients: Arthur Cohen 1983; J.H. Mendenhall 1982;

DANIEL H. GREEN AWARD . . . is presented not more frequently than once a year to a member of Committee D-15 on Engine Coolants of at least three years standing, who has performed outstanding work in the committee. Certificate.
Recipients: Donald L. Wood 1985; Robert L. Chance 1984; William A. Ailor, Jr. 1982; Jonathon B. Craig 1981; Robert Krenowicz 1980; Vincent Richard Graytok 1979; Norman R. Cooper 1976; Matthew A. Boehmer 1975; Leonard C. Rowe 1974; Eric Deynon 1973.

EDITORIAL EXCELLENCE AWARD (presented by Publications Committee) . . .
Recipients: T.T. Chiao 1984; Abel M. Dominguez and Ernest T. Selig 1983.

ELMER THURBER AWARD (presented by Committee F-1) . . . for publication in electronics.
Recipients: W. Murray Bullis and James R. Ehrstein 1984.

FRANK W. REINHARDT AWARD (presented by Committee E-38) . . . terminology on resource recovery.
Recipients: Howard Pincus 1987; Wayne Ellis 1986; Ralph Huntley 1985; Herbert I. Hollander 1984; Arnold I. Johnson 1983; Herbert T. Pratt 1982;

GEORGE IRWIN MEDAL AWARD (sponsored by Committee E-24) . . . in fracture mechanics.
Recipients: Choon Fong Shih 1986; Robert Oliver Ritchie 1985; John Graham Merkle 1984; John W. Hutchinson and James R. Rice 1982; James C. Newman, Jr. 1981.

HARLAN J. ANDERSON AWARD (sponsored by Committee C-26) . . . on Nuclear Fuel Cycle.
Recipients: Merrill Hume 1987; Wayne Delvin 1986; Michael G. Bale 1985; Gerald R. Kilp, Westinghouse Electric Company 1984; Robert L. Carpenter, manager, Rockwell International 1983.

H. V. CHURCHILL AWARD . . . an award to a member or a past-member of Committee E-2 on Emission Spectroscopy for meritorious service to the committee. Certificate.
Recipients: Larry E. Creasy 1986; James Anderson 1985; Norma L. Bottone 1984; John P. Kapetan 1983; John A. Norris 1982; Reed I. Quigley 1981; G. Leslie Mason 1980; Richard F. Jarrell 1979; William R. Kennedy 1978; William H. Tingle 1977; Ted R. Linde 1976; Mrs. Sarah H. Degenkolb 1975; Joseph F. Woodruff 1974; Don C. Spindler 1973; Edwin S. Hodge 1972; P. B. Adams 1971.

H. W. KUMMER AWARD (presented by Committee E-17) . . . in skid resistance.
Recipients: Ralph Haas 1986; Eldridge A. Whitehurst 1984; Rolands L. Rizenberg 1981.

HAROLD DEWITT SMITH MEMORIAL AWARD (sponsored by Committee D-13) . . . for outstanding achievement in the field of textile fiber utilization. The award consists of a bronze-walnut plaque.
Recipients: George Burr & Harold Hindman 1986; Jackson Bauer 1985; Braham Norwick 1984; Fred Fortess 1983; Dame S. Hamby 1982; Ernest Kaswell 1981; Joseph H. Dusenbury 1980; Peter R. Lord 1979; Richard Steele 1978; Bernard Miller 1977; B. Sheldon Sprague 1976; D. C. Prevorsek 1975; Ludwig Rebenfeld 1974; Henry A. Rutherford 1973; Jack Compton 1972; Arnold M. Sookne 1971.

HELEN L. CASEY AWARD (presented by Committee E-41) . . . in laboratory apparatus.
Recipients: William H. Cummines 1987; Robert J. Flately 1986; Robert Burke 1985; Robert C. Burke 1984; Glenn E. Rothwell 1983; Allen R. Fuller 1982.

HENRY A. GARDNER AWARD (presented by Committee D-1) . . . in paint.
Recipients: Joseph Behrle 1986; Carl Fuller 1985; Harry A. Wray 1984; Hiroshi Fujimoto and Eugene A. Praschan 1983; Joy Turner Luke 1981.

JOHN L. HAGUE AWARD (presented by Committee E-16) ... in metal-bearing ores.
Recipients: Robert Sutarno 1986; Jack H. Ornsbee 1985; John D. Selvaggio 1984; John S. Wakeman 1983.

L. J. MARKWARDT AWARD ... an annual award to the author or authors of a technical paper, published by the Society of outstanding merit in the field of wood, wood fibers, and wood base materials, including improvements and developments in the methods of testing. Bronze-walnut plaque.
Recipients: Erwin J. Schaffer 1986; Patrick J. Pellicane 1985; James Dobbin McNatt 1984; Billy Bohannon 1983; Thomas G. Williamson 1982; Herbert W. Eickner 1981; William L. Galligan 1980; Thomas E. Brassel 1979; Russel C. Moody 1978; Ricardo O. Foshi-J. D. Barrett 1977; Robert L. Ethington 1976-James W. Johnson 1975; James P. Goodman 1974; Dr. George Marra 1973; Stanley K. Suddarth 1972.

L. L. WYMAN AWARD (presented by Committee E-4) ... in Metallography.
Recipients: Roger Koch 1986; Ernest C. Pearson 1984; William D. Forgeng 1982.

LOWENHEIM AWARD (presented by Committee B-8) ... in Nonferrous Metals.
Recipients: R. J. Clauss 1984; Fielding Ogburn 1982.

LUNDELL-BRIGHT AWARD ... to a member of Committee E-3 on Chemical Analysis of Metals of at least three years standing who has performed outstanding work in the committee. Certificate.
Recipients: Ford A. Blair 1987; Thomas R. Dulski 1986; Om P. Bharagava 1985; Robert N. Smith 1984; Rudolph B. Fricioni 1983; G. Donald Haines 1982; Robert E. Kohn 1981; Luther C. Ikenberry 1980; James Kanzelmeyer 1979; William R. Bandi 1978; Miss Margaret McMahon 1977; Robert R. Klinge 1976; Joseph J. Aldrich 1975; Francis P. Byrne 1974; Edna F. Jackson 1973; Robert J. Bendure 1972; James I. Shultz 1971; Silve Kallmann 1970.

MARGARET DANA AWARD (presented by Committee F-15) ... in Consumer Products.
Recipients: William McMillan, III 1986; John W. Locke 1985; Robert S. Shane 1984; Malcolm W. Jensen 1983; David A. Swankin 1982; Donald R. Mackay 1981.

MARY R. NORTON MEMORIAL SCHOLARSHIP AWARD FOR WOMEN ... to encourage women college seniors or first-year graduate students to pursue the study of Physical Metallurgy or Materials Science, with emphasis on relationship of microstructure and properties. A certificate and honorarium.
Recipients: Carol A. Baer 1986; Judith A. Hill 1985; Judith Glazer 1984; Michelle W. Gabriel 1982; Phyllis Anne Klein 1981; Kristine M. Stanecki 1980; Loranne M. Brydges 1979; Elma B. Snipes 1977; Margaret D. Weeks 1976; Elane C. Sanderson 1975.

MAX HECHT AWARD (sponsored by Committee D-19) ... presented to a member of this committee (D-19, Water) who has performed outstanding work in the field of water described by the study of water as an engineering material, and standardization of methods of test and specifications in the field. The award consists of a certificate and name enscribed on wall plaque.
Recipients: Ellen Gonter 1986; Lyman H. Howe III 1985; Robert C. Kroner 1984; Charles E. Hamilton 1983; Verity C. Smith 1982; James J. Lichtenberg 1981; Robert J. Baker 1980; Leslie E. Lancey 1979; Russell W. Lane 1978; Charles C. Wright 1977; Donald L. Reid 1976; Sallie A. Fisher 1975; Michael Dannis 1974; James K. Rice 1973; Lester L. Louden 1972; J. F. Wilkes 1971; Earl E. Coulter 1970.

MOYER D. THOMAS AWARD ... is conferred preferably, but not necessarily, on a present or past member of Committee D-22 in recognition of outstanding achievement in the field of the sampling and analysis of atmosphere, with emphasis on contributions leading to the development and use of consensus standards. Gold medal and certificate.
Recipients: James A. Bard 1984; Dr. Morris Katz 1981; Robert D. Keenan 1977; Dr. John B. Clements 1976.

PREVOST HUBBARD AWARD ... established in 1972 by the committee in honor of Prevost Hubbard, and given in recognition of outstanding service to the Committee and work in the field of bituminous road and paving materials. Certificate.
Recipients: James M. Rice 1983; Ward K. Parr 1982; William H. Goetz 1981; J. York Welborn 1980; Bernard F. Kallas 1979; Norman W. McLeod 1978; John M. Griffith 1976; Lloyd D. Rader 1974.

R. A. GLENN AWARD ... is made annually by Committee D-5 to an individual who has made outstanding contributions in the development of ASTM standards relating to coal and coke.
Recipients: David Mason 1987; Louis Verno & Charles L. Wagoner 1986; Richard Mullins 1985; James Cameron and George A. Linton 1984; Ronald J. Morlock 1983; Martial P. Corriveau 1982; Gladys B. Berchtold 1981; William C. Banks 1979; William J. Montgomery 1978; Forrest E. Walker and J. Howard Gwynne 1977; Neil F. Shimp and Robert P. Hensel 1976; M. Perch, J. S. Galbraith and Gregory Gould 1975.

ROBERT J. PAINTER AWARD (Administered by Standards Engineers Society) ... in Standards.
Recipients: Albert L. Batik 1986; Genevieve M. Smith 1985; John A. Blair 1984; Henry J. Stremba 1983; Caryl Twitchell 1982.

ROBERT D. THOMPSON AWARD ... presented not more frequently than once a year to a member of committee E-20 on Temperature Measurement of at least three years standing, who has performed outstanding work in the committee. Certificate.
Recipients: Richard M. Park 1986; Lloyd J. Pickering 1985; James A. Bard 1984; George J. Champagne 1983; N. Ralph Corallo 1981; Jacquelyn Ann Wise 1980; Roy F. Abrahamsen 1979; Joseph D. Sines 1978; Edwin L. Lewis 1977; Donald I. Finch 1976.

SAM TOUR AWARD ... presented to the author or authors of a technical paper, published by the Society, of outstanding merit in the field of improvements and evaluation of corrosion testing methods. The award is a walnut-bronze plaque.
Recipients: P. E. Francis & Antony Mercer 1987; A. Toosooji, R. E. Einziger, A. K. Miller 1986; Ken W. Fertig, Sam L. JeanJaquet, Florian B. Mansfield, M. Eugene Meyer, Shu Tsai 1984; Hugh S. Isaacs and Brijesh Vyas 1983; Antony Donald Mercer 1981; J. H. Payer, R. N. Parker and W. E. Berry 1980; Michael A. Streicher 1979; Ellis D. Verink, Jr., C. M. Chen and M. H. Froning 1978; Fred H. Haynie 1977; J. T. Ryder and J. P. Gallagher 1975; Donald O. Sprowls 1974; H. Lee Craig, Jr. 1973.

SANFORD E. THOMPSON AWARD (sponsored by Committee C-9) ... presented to the author or authors of a paper published by the Society of outstanding merit in the field of concrete and concrete aggregates as this field is defined in the committee scope. The award is a plaque and certificate.
Recipients: Richard Gaynor, Richard Meininger & Tarek S. Khan 1987; Stella L. Marusin 1985; Richard Helmuth, Mohan Malhatra 1984; Surendra P. Shan, Wimal Suaris 1983; Kenneth W. Hall-Richard C. Meininger-Frank T. Wagner 1979; A. M. Barnett and C. F. Orchard 1973; Fred Moavenzadeh and Roberto Kuguel 1971; Zvonimir T. Jugovic, James L. Gillam 1970, Milton H. Wills, Jr. 1969; James J. Malloy, Thomas D. Larson and Philip D. Cady 1968.

WALLACE WATERFALL AWARD ... established by ASTM Committee E-33 on Environmental Acoustics for outstanding contributions to standardization in Acoustics. Walnut-bronze plaque.
Recipients: Ralph Huntley 1983; Charles W. Rodman 1981; Michael J. Kodaras 1977; Hale J. Sabine 1975.

WALTER C. VOSS AWARD ... presented to an engineer or scientist who has contributed notably to the knowledge in the field of building technology, with emphasis upon materials used which constitute significant advances or innovations. The award is a walnut-bronze plaque.
Recipients: Werner H. Gumpertz 1987; Albert W. Isberner, Jr. 1985; J. Ivan Davison 1982; Robert S. Boynton 1980; Rudard A. Jones 1979; William C. Cullen 1978; Austin J. Paddock 1977; Wayne F. Koppes 1976; Carl A. Menzel 1975; Albert G. H. Dietz 1974; William G. Kirkland 1972; Everett C. Shuman 1971.

WILLIAM T. PEARCE AWARD ... presented to a member of Committee D-1 on Paint, Varnish, Lacquer, and Related Products who has performed outstanding work and has had published within the last 5 years a technical paper of merit relating to the science of testing paints and paint materials. Plaque and certificate.

Recipients: Clifford K. Schoff 1987; Harry Wray 1986; Peter Kamarchik, Jr. 1984; Harry Ashton 1983; George W. Grossman 1981; Fred B. Stieg 1976; Melvin H. Swann 1974; Parket B. Mitton 1971.

AMERICAN SOCIETY OF AGRICULTURAL ENGINEERS (ASAE)

2950 Niles Road, P.O. Box 410, St. Joseph, Michigan 49085
Roger R. Castenson, Executive Vice President

PRESIDENTS ... William H. Johnson 1987; Benson J. Lamp 1986; James H. Anderson 1985; Bill L. Harriott 1984; Gerald W. Isaacs 1983; Robert H. Tweedy 1982; Robert H. Brown 1981; Albert M. Best 1980; William E. Splinter 1979; William G. Moore 1978; Frank B. Lanham 1977; S. S. DeForest 1976; Carl W. Hall 1975; L. H. Hodges 1974; C. F. Kelly 1973.

AWARDS

A. W. FARRALL YOUNG EDUCATOR AWARD ... in recognition of outstanding success in motivating others to apply engineering principles to the problems of the agricultural engineering profession.

Recipients: R. Paul Singh 1986; Albert M. Jarrett 1985; Robert J. Gustafson 1984; Gary W. Krutz 1983; Denny C. Davis 1982; Dennis D. Schulte 1981; Dennis E. Buffington 1980; Thomas L. Thompson 1979; David B. McWhorter 1978; Larry J. Segerlind 1976; Bobby L. Clary 1975; E. Paul Taiganides 1974; Donald M. Edwards 1973; Paul K. Turnquist 1972.

AEROVENT YOUNG EXTENSION WORKER AWARD ... in recognition of outstanding success in motivating people to acquire knowledge, skills, and understanding to improve agricultural operations.

Recipients: Elbert C. Dickey 1986; William E. Field 1985; James C. Barker 1984; Loren E. Bode 1983; Gerald R. Bodman 1982; Don D. Jones 1981; Allen R. Rider 1980; Donald R. Price 1979; George A. Duncan 1978; William Mayfield 1977; Peter D. Bloome 1976; L. Bynum Driggers 1975; Myron D. Paine 1974; Byron H. Nolte 1973.

CYRUS HALL McCORMICK MEDAL ... for exceptional and meritorious engineering achievement in agriculture. Now is "Cyrus Hall McCormick Jerome Increase Case Medal."

Recipients: Francis J. Hassler 1987; Joseph K. Jones 1986; Lawrence H. Hodges 1985; Carl W. Hall 1984; Robert E. Stewart 1983; Theodore E. Stivers 1982; Coby Lorenzen 1981; Norval H. Curry 1980; Glen E. Vanden Berg 1979; Russell R. Poynor 1978; Clarence B. Richey 1977; Lester F. Larsen 1976; Orval C. French 1975; Karl H. Norris 1974; Jerome W. Sorenson, Jr. 1973; Stanley A. Witzel 1972; Henry J. Barre 1971; Carlton L. Zink 1970; Edgar L. Barger 1969; Howard F. McColly 1968.

DFISA-ASAE FOOD ENGINEERING AWARD ... to honor those who have made original contributions in research, development, or design, or in the management of food processing equipment or techniques of significant economic value to the food industry and the consumer.

Recipients: Daryl B. Lund 1987; Henry G. Schwartzberg 1985; Judson M. Harper 1983; Dennis R. Heldman 1981; Marcus Karel 1978; Walter M. Urbain 1976; Robert P. Graham 1974; Arthur W. Farrall 1972.

DOERFER ENGINEERING CONCEPT OF THE YEAR AWARD ... to honor an engineer or an engineering team that makes a unique contribution to the development or advancement of a new engineering concept.

Recipients: Michael B. Timmons 1985; Richard L. Ledebuhr 1984; Carl W. VanGilst 1983; Awatif E. Hassan 1982; Paul Jacobson 1981; Hubert Geisthoff 1980; Lambert H. Wilkes 1979; Dale E. Marshall 1978; Donald L. Peterson 1977; Robert B. Fridley 1976; John G. Alphin 1975; Herbert N. Stapleton 1974.

FIEI YOUNG RESEARCHER AWARD ... in recognition of dedicated use of scientific methodology to seek out facts or principles significant to the agricultural engineering profession.

Recipients: David T. Hill 1985; Michael R. Overcash 1984; David R. Thompson 1983; David W. Smith 1982; Richard W. Skaggs 1981; David B. Smith 1980; Louis D. Albright 1979; Billy J. Barfield 1978; Edward A. Hiler 1977; C. Gene Haugh 1976; Charles T. Haan 1975; Dennis R. Heldman 1974; Roscoe L. Pershing 1973; Robert B. Fridley 1972.

FMC CORPORATION YOUNG DESIGNER AWARD ... honors the creator or developer of a technical plan for a machine, building, component, system, subsystem, or agricultural engineering routine. The plan must show promise of significantly influencing agricultural engineering progress.

Recipients: Robert E. Fox 1984; Terrill W. Woods 1983; John A. Collier 1982; Roger P. Rohrbach 1981; David C. Moffitt 1980; Gustaaf M. L. Persoons 1979; John C. Knobloch 1978; Richard W. Hook 1977; Roger W. Curry 1976; Ronald T. Noyes 1975; John E. Morrison, Jr. 1974; Douglas L. Bosworth 1973; James L. Fouss 1972.

G. B. GUNLOGSON COUNTRYSIDE ENGINEERING AWARD ... to honor outstanding engineering contributions to the healthy climate of the American countryside and to a viable economy for its small towns.

Recipients: Melville L. Palmer 1986; James A. Moore 1985; James C. Converse 1984; Roland A. G. Oliver 1983; Donald G. Jedele 1982; Robert H. Brown 1981; Richard K. White 1980; E. A. Olson 1979; Frank J. Humenik 1978; Robert C. Ward 1976; Junius L. Kendrick 1975.

THE GEORGE W. KABLE ELECTRIFICATION AWARD ... for outstanding personal and professional contributions in applying electrical energy to the advancement of agriculture through agricultural engineering.

Recipients: LaVerne E. Stetson 1987; Harold A. Cloud 1986; Lynndon A. Brooks 1985; Landy B. Altman 1984; Lowell J. Endahl 1983; William J. Ridout, Jr. 1982; Troy L. Ingram 1981; Hugh J. Hansen 1980; Richard L. Witz 1979; Dean Searls 1978; Clesson N. Turner 1977; Olin W. Ginn 1976; Frank W. Andrews 1975; Kenneth L. McFate 1974; William E. McCune 1973; Nolan Mitchell 1972; Morris H. Lloyd 1971; Homer S. Pringle 1970; Everett C. Easter 1969.

HANCOR SOIL AND WATER ENGINEERING AWARD ... for noteworthy contributions to the advancement of soil and water conservation engineering in teaching, research, planning, design, construction, management, methods, or materials.

Recipients: John M. Laflen 1987; R. Wayne Skaggs 1986; L. Donald Meyer 1985; Curtis L. Larson 1984; Paul E. Fischbach 1983; Ray J. Winger 1981; A. R. Robinson 1980; Melville L. Palmer 1979; Howard P. Johnson 1978; Benjamin A. Jones, Jr. 1977; Neil P. Woodruff 1976; Elmer W. Gain 1975; Marvin E. Jensen 1974; E. Paul Jacobson 1973; Ernest H. Kidder 1972; Robert P. Beasley 1971; John R. Carreker 1970; James N. Luthin 1969; Glenn O. Schwab 1968.

THE JOHN DEERE MEDAL ... for distinguished achievement in the application of science and art to the soil.

Recipients: Glenn O. Schwab 1987; C. Glenn E. Downing 1986; Neil F. Bogner 1985; Henry D. Bowen 1984; Carl A. Reaves 1983; Marvin E. Jensen 1982; Claude H. Pair 1981; Lambert H. Wilkes 1980; Fiepko Coolman 1979; Lloyd L. Harrold 1978; Jan van Schilfgaarde 1977; John C. Stephens 1976; John H. Zich 1975; Lawrence H. Skromme 1974; J. W. Borden 1973; John Phillips, Jr. 1972; G. Wallace Giles 1971; Wayne D. Criddle 1970; Arthur W. Cooper 1969; Dwight D. Smith 1968.

KISHIDA INTERNATIONAL AWARD ... in recognition of outstanding contributions to engineering-mechanization-technological related programs of education, research, development, consultation, or technology transfer that have resulted in significant improvement in food production and living conditions outside the United States.

Recipients: Lyle G. Reeser 1987; Bill A. Stout 1986; Harry D. Henderson 1985; William J. Chancellor 1984; Byron L. Bondurant 1983; Merle L. Esmay 1982; Emmanuel U. Odigboh 1981; G. Wallace Giles 1980; Bruce H. Anderson 1979; Ralph C. Hay 1978.

THE MASSEY FERGUSON MEDAL ... to honor those whose dedication to the spirit of learning and teaching in the field of agricultural engineering has advanced our agricultural knowledge and practice, and whose efforts serve as an inspiration to others.
Recipients: Mark E. Singley 1987; Gordon L. Nelson 1986; Gustave E. Fairbanks 1982; Francis J. Hassler 1981; Donald B. Brooker 1980; William J. Promersberger 1979; William E. Splinter 1978; S. Milton Henderson 1977; Carl W. Hall 1976; Frank W. Peikert 1975; Frank B. Lanham 1974; Dennis L. Moe 1973; G. E. Henderson 1972; Arthur W. Farrall 1971; Ervin W. Schroeder 1969; Frederick C. Fenton 1968.

MAYFIELD COTTON ENGINEERING AWARD ... to encourage and recognize outstanding engineering contributions to the cotton industry. First presented in December, 1985.
Recipients: Robert M. Fachini 1986; Lambert H. Wilkes 1985.

METAL BUILDING MANUFACTURERS ASSOCIATION AWARD ... for distinguished work in advancing the knowledge and science of farm buildings.
Recipients: Mylo A. Hellickson 1986; Neil F. Meador 1985; Loren W. Neubauer 1984; Hajime Ota 1983; Charles K. Spillman 1982; L. Bynum Driggers 1981; John H. Pedersen 1980; James A. DeShazer 1979; George L. Pratt 1978; Frank Wiersma 1977; G. LeRoy Hahn 1976; Warren L. Roller 1975; John N. Walker 1974; Bruce A. McKenzie 1973; Robert R. Rowe 1972; Thomas E. Hazen 1971; Howard K. Johnson 1970; Landis L. Boyd 1969; Donald W. Richter 1968.

PACKER ENGINEERING SAFETY AWARD ... to encourage and to recognize outstanding contributions to the advancement of agricultural safety engineering in research, design, education, or promotion. These contributions shall have been either in the form of contributions or published literature, notable performance, product innovation, and/or special actions which have served to advance agricultural safety.
Recipients: Rollin D. Schnieder 1986.

AMERICAN SOCIETY OF CIVIL ENGINEERS (ASCE)

345 East 47th Street, New York, New York 10017
Edward O. Pfrang, Executive Director
PRESIDENTS ... Daniel B. Barge, Jr 1987; Robert D. Bay 1986; Richard W. Karn 1985; S. Russell Stearns 1984; John H. Wiedeman 1983; James R. Sims 1982; Irvan F. Mendenhall 1981; Joseph S. Ward 1980; Walter Emanuel Blessey 1979; William Read Gibbs 1978; Leland Jasper Walker 1977; Arthur Joseph Fox, Jr. 1976; William McCoy Sangster 1975; Charles William Yoder 1974; John E. Rinne 1973.

AWARDS

ALFRED M. FREUDENTHAL MEDAL ... is awarded to an individual in recognition of distinguished achievement in safety and reliability studies applicable to any branch of civil engineering.
Recipients: Joseph Penzien 1986; Y. K. Lin 1984; Alfredo H-S. Ang 1982; Jack R. Benjamin 1980; Masanoubu Shinozuka 1978; Emilio Rosenblueth 1976.

THE ALFRED NOBLE PRIZE ... for a technical paper of exceptional merit accepted for publication in any of the technical publications of the following societies: ASCE, AIME, ASME, IEEE, or the Western Society of Engineers. The author must be not more than 31 years of age at the time of the document's publication.
Recipients: David McDowell 1986; William R. Brownlie 1984; George Gazetas 1982; Bharat Bhushan 1981; C. L. Briant 1980; Alan S. Willsky 1979; Maria Comminou 1978; J. E. Killough 1977; S. N. Singh 1976; William L. Smith 1975; V. K. Gupta 1974; Deiter D. Pfaffinger 1973; C. L. Magee 1972; Ben G. Burke 1971; Peter W. Marshall 1970; Ronald Gibala 1969; Richard Holland 1968.

ARTHUR J. BOASE AWARD ... to recognize and honor a person, persons, or organizations for outstanding activities and achievements in the reinforced concrete field, the Arthur J. Boase Award was established by the Reinforced Concrete Research Council.
Recipients: Paul F. Rice 1984; Phil M. Ferguson 1982; Eivind Hognestad 1981; Alfred L. Parme 1980; Arthur R. Anderson 1979; Anton Tedesko 1978; Maurice P. Van Buren 1977; Dugald J. Cameron 1976; Chester P. Siess 1975; Ernst Gruenwald 1974; Douglas McHenry 1973; Eric L. Erickson 1972; Leo H. Corning 1971.

THE ARTHUR M. WELLINGTON PRIZE ... awarded for papers on transportation, on land, on the water, or in the air; and foundations and closely related subjects. The prize, instituted and endowed by "Engineering News-Record" in 1921, consists of a wall plaque and a certificate.
Recipients: Michel Ghosn and Fred Moses 1987; Robert Scanlan 1986; K. Nyman 1985; Michael W. O'Neill 1984; Port Sines Investigating Panel of the Coastal Engineering Research Council, Billy L. Edge (Chairman) 1983; James P. Burket and Hung Jen Kuo 1982; Joseph McDermott 1981; A. James Birkmyer 1980; Charles G. Schilling and Karl H. Klippstein 1979; Srikanth Rao, Thomas D. Larson and Theodore H. Poister 1978; Eugene L. Marquis and Graeme D. Weaver 1977; Rodney E. Engelen 1976; Eli Robinsky, Keith E. Bespflug and Joil P. Leisch 1975; William K. Mackay 1974; James L. Sherard, Rey S. Decker and Norman L. Ryker 1973; Daniel Dicker 1972; Jean J. Janin and Guy F. LeSciellour 1971; Edwin W. Eden 1970; Melvin L. Baron, Hans H. Bleich and Joseph P. Wright 1969; H. Bolton Seed and Stanley D. Wilson 1968.

ASCE NEWS CORRESPONDENT AWARD ... to recognize the efforts of volunteers who have helped to make the ASCE News a success through their special contributions. The prize consists of an inscribed desk-pen set.
Recipients: Jon D. Fricker 1986; Katherine Popko 1985; J. Rodger Adams, Hydraulics Division 1984; Richard D. Woods, Geotechnical Engineering Division 1983; Clifford A. Merritt, Environmental Engineering Division 1982; Stewart Johnson 1981; Ronald Boenau 1980; Thomas Stringfield 1979; John T. Christian 1978; Frank E. Stratton 1977.

ASCE PRESIDENT'S AWARD ... is to be made annually to an ASCE member who has given distinguished services to his or her country.
Recipients: William R. Gianelli 1987; Ralph B. Peck 1986; Stephen D. Bechtle, Jr. 1985; Joseph F. Friedkin 1984; John A. Volpe 1983; John R. Kiely 1982; Ellis Armstrong 1981; Lyman D. Wilbur 1980; Herbert D. Vogel 1979; George R. Brown 1978; Ben Moreell 1977; Lucius D. Clay 1976.

ASCE STATE-OF-THE-ART OF CIVIL ENGINEERING AWARD ... to encourage the most gifted practitioners of civil engineering to review and interpret the state of the art for the benefit of their colleagues. A direct benefit of this award will be the scholarly review, evaluation and documentation of the scientific and technical information needed by the profession.
Recipients: Subcommittee on Vibration Problems Associated with Flexural Members on Transit Systems, Committee on Flexural Members of the Committee on Metals of the Structural Division 1987; James S. Gidley and William A. Sack 1986; James M. Witowski 1985; Committe on Fatigue and Fracture Reliability of the Structural Division's Technical Committee on Structural Safety and Reliability, Paul H. Wirsching (Chairman) 1983; Orville R. Russell, Dennis T. Stanczuk, and John R. Everett 1982; Task Committee on Air Supported Structures, Charles Birnsteil, Chairman 1981; Task Committee on Nuclear Effects, Paul Kruger, Chairman 1980; John A. Focht, Jr. and Leland M. Kraft, Jr. 1979; Carl L. Monismith and Fred N. Finn 1978; John H. Schmertmann 1977; Richard Field and John A. Lager 1976; Committee on Atmospheric Pollution of the Environmental Engineering Division 1975; Joint ASCE-ACI Taks Committee 426 on Shear and Diagonal Tension of the Committee on Masonry and Reinforced Concrete of the Str. Div. 1974; Task Committee on Structural Safety of the Committee on Analysis and Design of the Structural Division 1973; Elory J. Tamanini 1972; Bramlette McClelland, John A. Focht, Fr. and William J. Emrich 1971; Daniel W. Moore and William S. Pollard, Jr. 1970; Ali Sabzevari and Robert H. Scanlon 1969; Boris Bresler and James G. MacGregor 1968.

CAN-AM CIVIL ENGINEERING AMITY AWARD ... is made annually to a member of ASCE for either a specific instance that has had continuing benefit in understanding and good will, or a career of exemplary professional activity that has contributed to the amity of the United States of America and Canada.
Recipients: Peter Smith 1987; Robert F. Legget 1986; Hugh McGrory 1985; W. H. Paterson 1984; Keith A. Henry 1983; Camille Daegenais 1981; James G. MacGregor 1979; Charles B. Molineaux 1978; Alan G. Davenport 1977; John R. Kiely 1976; John B. Stirling 1975; Eugene W. Weber 1974; L. Austin Wright 1973.

CIVIL ENGINEERING HISTORY AND HERITAGE AWARD ... to recognize those persons who through their writing, research or other efforts have made outstanding contributions toward a better knowledge of, or appreciation for, the history and heritage of civil engineering. The award was established in 1966.
Recipients: Robert F. Legget 1987; David P. Billington 1986; Walter H. Cates 1985; Herbert R. Hands 1984; John A. Focht, Sr. 1983; Robert S. Mayo 1982; Emory L. Kemp 1981; Hunter Rouse 1980; Robert M. Vogel 1979; David McCullough 1978; Joseph J. Rady 1977; Neal FitzSimons 1976; Clifford A. Betts 1975; Sara Ruth Watson 1973; Charles J. Merdinger 1972; Carl W. Condit 1971; Stanley B. Hamilton 1970; Gail A. Hathaway 1969; Ulysses S. Grant, III 1968.

CIVIL GOVERNMENT AWARD ... to recognize those persons who have contributed substantially to the status of the engineering profession by meritorious public service in elective or appointive positions in civil government. Primary consideration shall be given to public service which does not require the qualifications of an engineer, but nominees must be registered professional engineers.
Recipients: Douglas T. Wright 1986; George Schoepfer 1985; John S. Lindsay 1984; William A. Gissler 1983; Michael R. Pender 1982; Rex K. Rainer 1981; John C. Kohl 1979; Leroy F. Greene 1978; Martin Lang 1977; Floyd D. Peterson 1976; Raymons J. Smit 1975; John W. Frazier 1974; Milton Pikarsky 1973; Kenneth A. Gibson 1972; Ben E. Nutter 1971; W. Scott McDonald 1970; William J. Hedley 1969.

THE COLLINGWOOD PRIZE ... instituted and endowed in 1894, the competition for this prize is restricted to members of the Society who are 32 years of age or less at the time the paper is submitted in essentially its final form. It is awarded to the author, or authors, of a paper describing an engineering work with which he or they have been directly connected, or recording investigations contributing to engineering knowledge to which he or they have contributed some essential part, and containing a rational digest of results.
Recipients: Colin R. Thorne and Lyle Zevenbergen 1986; R. Kerry Rowe 1985; John M. Mason, Jr. 1984; Thomas D. O'Rourke 1983; Richard H. Dilouie, Jr. 1982; Kirke E. Larson 1981; Philip J. W. Roberts 1980; Jean M. E. Audibert and Kenneth Nyman 1979; Nolan L. Johnson, Jr. 1978; Ronald S. Steiner 1977; John T. Noval and Gail E. Montgomery 1976; Yuet Tsu and G. Wayne Clough 1975; James C. Anderson and Raj P. Gupta 1974; Thomas M. Lee 1973; G. Wayne Clough and James M. Duncan 1972; P. Aarne Vesilind 1971; C. K. Shen, Kenneth L. Lee 1970 and Richard D. Woods 1969; Maurice L. Sharp 1968.

CONSTRUCTION MANAGEMENT AWARD ... is made annually to a member of the Society who has made definite contributions in the field of construction management in general and, more particularly, in the application of the theoretical aspects of engineering economics, statistics, probability theory, operations research and related mathematically oriented disciplines to problems of construction management, estimating, cost, accounting, planning, scheduling and financing, these contributions being made either in the form of written presentation or notable performance.
Recipients: Frank Muller 1987; Norman Nadel 1986; Paul M. Teicholz 1985; Boyd C. Paulson, Jr. 1984; Robert B. Harris 1983; Fred C. Kreitzberg 1982; Alfred C. Maevis 1981; Rolland Wilkening 1980; Robert L. Peurifoy 1979; Louis R. Shaffer 1978; John W. Fondahl 1977; Glenn R. Kosapp 1976; Allen W. Hatheway 1975; Joseph C. Kellogg 1974.

THE DANIEL W. MEAD PRIZES ... awarded to associate members on the basis of papers on professional ethics. The prizes were established and endowed in 1939 by Daniel W. Mead, past president and honorary member of ASCE.
Recipients: Robin Kemper 1987; Karen Irion 1986; Gary Ostroff 1985; Barry A. Rogers 1984; Kenneth L. Carper 1983; Bryant Wing Chew Wong 1982; Daniel Yake 1981; Thomas McCrate 1980; David David T. Biggs 1979; John S. Bruce 1978; Keith Strickland 1977; Glenn R. Koepp 1976; Allen W. Hatheway 1975; John E. Spitiko, Jr. 1974; Charles R. Schrader 1972; David Mann-Charles G. Sudduth 1971; Chester Lee Allen-Hadley Johnson 1970; Theodore Fellinger-George L. Reed 1969; Frances A. Paul-Charles R. Schrader 1968.

EDMUND FRIEDMAN PROFESSIONAL RECOGNITION AWARD ... to recognize the importance of professional attainment in the advancement of the science and profession of engineering by exemplary professional conduct in a specific outstanding instance; an established reputation for professional service; objective and lasting achievement in improving the conditions under which professional engineers serve in public and private practice; significant contribution toward improvement of employment conditions among civil engineers; significant contribution toward improving the professional aspects of civil engineering education; professional guidance of qualified young men who would seek civil engineering as a career; or, other evidence of merit.
Recipients: Celestino Pennoni 1987; David A. VanHorn 1986; James W. Poirot 1985; Erwin R. Breihan 1984; George J. Kral, Sr. 1983; Gene M. Nordby 1982; Russel C. Jones 1981; Gerard Fox 1980; Myron D. Calkins 1979; Willa W. Mylroie 1978; Herbert A. Goetsch 1977; Jack E. McKee 1976; John W. Frazier 1975; Billy T. Sumner 1974; Fred J. Benson 1973; Leo W. Ruth, Jr. 1972; Elmer K. Timby 1971; Orley O. Phillips 1970; Alfred C. Ingersoll 1969; Walter E. Jessup 1968.

ERNEST E. HOWARD AWARD ... to recognize an individual who has made a definite contribution to the advancement of structural engineering either in research, planning, design or construction, including methods and materials, these contributions being made either in the form of papers or through notable performance or specific actions.
Recipients: Mete A. Sozen 1987; Bruno Thurlimann 1986; Paul Weidlinger 1985; William S. Hall 1984; Edward Cohen 1983; Jackson Leland Durkee 1982; George Winter 1981; Gerard Fox 1980; John M. Fisher 1979; Arsham Amirikian 1978; Fazlur R. Khan 1977; Egor P. Popov 1976; George E. Brandow 1975; Bruce G. Johnston 1974; C. Martin Duke 1973; George S. Richardson 1972; Stephenson B. Barnes 1971; Ray W. Clough 1970; Thomas C. Kavanagh 1969; Chester P. Siess 1968.

FRANK M. MASTERS TRANSPORTATION ENGINEERING AWARD ... will be given to a member of the Society on the basis of the best example of innovative or noteworthy planning, design, or construction of transportation facilities. The example shall have been described in published form available to the entire engineering community.
Recipients: C. Michael Walton 1987; Kumares C. Sinha 1986; John Kozak 1984; John J. Fruin 1983; Walter H. Kraft 1982; John Coil 1981; Joseph McDermott 1980; David R. Miller 1979; Richard C. Gallagher 1978; Edward S. Olcott 1977; S. Starr Walbridge 1976.

HARLAND BARTHOLOMEW AWARD ... to recognize contributions to the enhancement of the role of the civil engineer in urban planning and development. The contribution may be in the form of a paper published in a technical journal or may be personal effort and achievement toward that goal.
Recipients: C. Michael Walton 1987; Robert B. Teska 1986; Thomas M. Blalock 1985; Kevin E. Hegnue 1984; Marshall F. Reed, Jr. 1983; George D. Barnes 1982; Albert Grant 1981; Lawrence E. Miller 1980; Edgardo Contini 1979; Floyd D. Peterson 1978; Harold E. Nelson 1977; George C. Bestor 1975; Ladislas Segee 1974; Jack R. Newville 1973; Roger H. Gilman 1972; William H. Claire 1971; Kurt W. Bauer 1970; Alan M. Voorhees 1969.

THE HOOVER MEDAL . . . instituted to commemorate the civic and humanitarian achievements of Herbert Hoover. Inscribed on the medal is the legend "Awarded by Engineers to a Fellow Engineer for Distinguished Public Service." The Hoover Medal Board of Award consists of representatives of four societies: ASCE, AIME, ASME, and IEEE.

Recipients: Lawrence P. Grayson 1986; Robert C. West 1985; Kenneth A. Roe 1984; Joseph J. Jacobs 1983; Michael T. Halbouty 1982; Arnold O. Beckman 1981; Stephen D. Bechtel 1980; Charles M. Brinckerhoff 1979; Donald C. Burnham 1978; Peter Goldmark 1977; James B. Fisk 1976; James Boyd 1975; David Packard 1974; William J. Hedley 1973; Frederick R. Kappel 1972; Louis Ferre 1971; John Erik Jonsson 1970; Edger F. Kaiser 1969; Sir Harold Hartley 1968.

THE J. C. STEVENS AWARD . . . presented to an individual adjudged to have submitted the best discussion in the field of hydraulics (including fluid mechanics and hydrology) in "ASCE Proceedings" and "Civil Engineering". Normally the prize consists of books costing not more than $50.

Recipients: Steven J. Wright 1986; Jacob Bear and P.C. Menier 1985; Nikolaos D. Katopodes 1984; Allen T. Hjelmfelt, Jr. 1983; Gary Parker 1982; George Hebaus 1981; Charles C. S. Song-Chia Ted Yang 1980; George D. Ashton 1979; Brent D. Taylor and Vito A. Vanoni 1978; C. Samuel Martin and David C. Wiggert 1977; Danny L. Fread 1976; A. R. Thomas 1975; Charles R. Neill 1974; Keith D. Stolzenbach and Donald R. F. Harleman 1973; S. David Graber and Thomas R. Camp 1972; Helmut Kobus 1971; William W. Sayre 1970; Fred W. Blaisdell 1969; Ronald W. Jeppson 1968.

THE J. JAMES CROES MEDAL . . . awarded to the author, or authors, of such paper as may be judged worthy of the award and be next in order of merit to the paper to which the Norman Medal is awarded; or, if the Norman Medal is not awarded, then to the author, or authors, of a paper for its merit as a contribution to engineering science. The award was established in 1912.

Recipients: Mohsen M. Baligh 1987; Robert L. Vecellio and Thomas H. Culpepper 1986; George Gazetas and Richardo Dobry 1985; Roy E. Olson and David E. Daniel 1984; Gene F. Parkin and Richard E. Speece 1983; Egor P. Popov 1982; Samuel Aroni and G. Louis Fletcher 1981; Robert Bea, Jean M. E. Audibert and M. Radwan Akky 1980; Charles W. Roeder and Egor P. Popov 1979; Gregory N. Richardson, Daniel L. Feger, Arthur Found and Kenneth L. Lee 1978; John A. Replogle 1977; Archie J. McDonnell and Frank H. Pearson 1976; G. E. Blight 1975; Subrata K. Chakrabarti 1974; David J. D'Appolonia, Harry G. Poulos and Charles C. Ladd 1973; I. M. Idriss and H. Bolton Seed 1972; John H. Nath and Erich J. Plate 1971; John M. Henderson 1970; Hugo B. Fischer 1969; William R. Hudson 1968.

THE J. WALDO SMITH HYDRAULIC FELLOWSHIP . . . offered every third year, and provides $2,000 plus as much as $1,000 in physical equipment connected with the research in the field of experimental hydraulics as distinguished from that of purely "theoretical hydraulics."

Recipients: Lloyd G. Byrd 1985; Leroy G. Holub 1983; Lawrence J. Pratt 1976; Qais N. Fattah 1973; George D. Ashton 1970.

THE JAMES LAURIE PRIZE . . . presented to an individual who has made a definite contribution to the advancement of transportation engineering either in research, planning, design, or construction, whether in the form of papers or through notable performance or specific actions.

Recipients: Oscar T. Lyon 1987; David Saunders Gedney 1986; Jack E. Leisch 1985; Peter G. Koltnow 1984; Salvatore Bellomo 1983; James O. Granum 1982; Harold L. Michael 1981; Ronald W. Pulling 1980; Raymond J. Hodge 1979; Willard F. Babcock 1978; Milton Pikarsky 1977; John P. Veerling 1976; Thomas N. Sullivan 1975; William A. Bugge 1974; Robert W. Brannon 1973; Donald S. Berry 1972; Frances C. Turner 1971; Charles E. Shumate 1970; Walter S. Douglas 1969; Robert Horonjeff 1968.

THE JOHN FRITZ MEDAL . . . established by the professional associates and friends of the late John Fritz, Hon. M. ASCE, of Bethelehem, Pa., on August 21, 1902, his eightieth birthday, to perpetuate the memory of his achievements in industrial progress. At the dinner given in Mr. Fritz's honor, he was presented with a replica in bronze of the artist's original model of the medal as a memento of the occasion.

Recipients: Ralph Landau 1987; Simon Ramo 1986; Daniel C. Drucker 1985; Kenneth A. Roe 1984; Claude Elwood Shannon 1983; David Packard 1982; T. L. Austin, Jr. 1981; Nathan M. Newmark 1979; Robert G. Heitz 1978; George R. Brown 1977; Thomas O. Paine 1976; Mason Benedict 1975; H. I. Romnes 1974; Lyman D. Wilbur 1973; William Webster 1972; Patrick E. Haggerty 1971; Glenn B. Warren 1970; Michael L. Haider 1969; Igor I. Sikorsky 1968.

JOHN I. PARCEL - LEIF J. SVERDRUP CIVIL ENGINEERING MANAGEMENT AWARD . . . to encourage effective leadership and management skills in the civil engineering profession, the award is made annually to a member of ASCE who has made a definite contribution in the field of civil engineering management either in a written presentation or notable performance.

Recipients: Albert Dorman 1987; Fujio Matsuda 1986; John A. Lambie 1984; James C. Howland 1983; Thomas D. Larson 1982; Joseph Lawler 1981; Frank J. Moolin, Jr. 1980; Stephen D. Bechtel, Jr. 1979.

JOHN G. MOFFAT - FRANK E. NICHOL HARBOR AND COASTAL ENGINEERING AWARD . . . is designed to recognize new ideas and concepts that can be efficiently implemented to expand the engineering or construction techniques available for harbor and coastal projects. The award is made annually to a member of ASCE who has made a definite contribution in the field of civil engineering management either in the form of written presentations or notable performance. Development of concepts can occur in any of the academic, design, or construction disciplines.

Recipients: Robert G. Dean 1987; Eugene H. Harlow 1986; John H. Nath 1985; Charles C. Calhoun, Jr. 1984; Basil W. Wilson 1983; James W. Dunham 1982; Omar J. Lillievang 1981; Thorndike Saville, Jr. 1979; Robert L. Wiegel 1978.

JULIAN HINDS AWARD . . . is to be made annually to the author, or authors, of that paper which is judged to be the most meritorious contribution to the field of water resources development such as multi-purpose water projects for irrigation, flood control, municipal and industrial water, hydroelectric power or any combination thereof. The Award can also be made to an individual for notable performance, long years of distinguished service, or specific actions that have served to advance engineering in the field of planning, development, and management of water resources.

Recipients: Gerald T. Orlob 1987; Daniel P. Loucks 1986; Joseph S. Cragwall 1985; Vincent George Terenzio 1984;Munson W. Dowd 1983; William Whipple, Jr. 1982; Leo Beard 1981; Dean Peterson 1980; Carl E. Kindsvater 1979; Ray K. Linsley 1978; Eugene W. Weber 1977; Harvey O. Banks 1976; Victor A. Koelzer 1975.

THE KARL EMIL HILGARD HYDRAULIC PRIZE . . . given to the author or authors of that paper of superior merit dealing with a problem of flowing water, either in theory or practice. Papers published in "ASCE Proceedings" and "Civil Engineering" are eligible. The prize consists of a wall plaque and a certificate.

Recipients: Iehisa Nezu and Wolfgang Rodi 1987; Michael B. Abbott, Andrew D. McCowan and Ian R. Warren 1986; Ben J. Alfrink and Leo C. Van Riun 1985; Syunsuke Ikeda and Takaski Asaeda 1984; Gary Parker, Peter C. Klingeman and David G. Mclean 1983; Dan Zaslavsky and Gideon Sinai 1982; Wolfgang Rodi and Michael Leschizer 1981; Y. H. Tsai and E. R. Nolley 1980; Victor M. Ponce and Daryl B. Simmons 1979; Mehmet S. Uzuner and John F. Kennedy 1978; Poothrikka P. Daily and Enzo O. Macagno 1977; Vito A. Vanoni, Donald C. Bondurant, Jack E. McKee, Paul C. Benedict and Robert F. Piest 1976; Eduard Naudascher and Frederick A. Locher 1975; George D. Ashton and John F. Kennedy 1974; Wayne C. Huber, Donald R. F. Harleman and Patrick J. Ryan 1973; John A. Hoopes, Gerard A. Rohlich and Robert W. Zeller 1972; Hugo B. Fischer, Donald R. F. Harleman and Edward R. Holley 1971; Robert C. Y. Koh and Norman H. Brooks 1970; Donald Van Sickle 1969; Samuel O. Russell and James W. Ball 1968.

KARL TERZAGHI AWARD . . . given to the author of outstanding contributions to knowledge in the fields of soil mechanics, subsurface and earthwork engineering, and subsurface and earthwork construction. The award consists of a plaque and an honorarium of $750.

Recipients: Robert V. Whitman 1987; James K. Mitchell 1985; Lymon C. Reese 1983; Alec Skempton 1981; F. E. Richart, Jr. 1980; Gregory P. Tschebotarioff 1979; Stanley D. Wilson 1977; T. William Lambe 1975; H. Bolton Seed 1973; Lauritis Bjerrum 1971; Ralph B. Pack 1969; William J. Turnbull 1968.

KARL TERZAGHI LECTURE . . . established in 1960 by the Soil Mechanics and Foundations Division of ASCE. A distinguished engineer is invited to deliver a lecture at an appropriate meeting of the society. The lecturer is tendered a certificate and an honorarium of $300.

Recipients: Leonardo Zeevaert 1987; Charles C. Ladd 1986; Jory O. Osterneg 1985; James K. Mitchell 1984; Ronald F. Scott 1983; J. Barry Cooke 1982; Robert V. Whitman 1981; Robert Whitman 1981; G. A. Leonards 1980; George Sowers 1979; Nathan M. Newmark 1978; Robert F. Legget 1977; Lyman C. Reese 1976; G. G. Meyerhof 1975; F. E. Richart, Jr. 1974; Bramlette McClelland 1972; John Lowe, III 1971; T. William Lambe 1970; Stanley D. Wilson 1969; Philip C. Rutledge 1968.

MARTIN S. KAPP FOUNDATION ENGINEERING AWARD . . . is awarded to an individual on the basis of the best example of innovative or outstanding design or construction of foundations, earthworks, retaining structures, or underground construction. Emphasis shall be placed on constructed works where serious difficulties were overcome or where substantial economies were achieved. The example shall have been described in published form available to entire engineering community.

Recipients: George J. Tamaro 1987; Georg Kjerbol 1986; Harry Schnabel, Jr. 1985; Charles L. Guild 1983; Leroy Crandall 1982; Wallace Baker 1981; Henri Vidal 1978; Robert E. White 1977; Ben C. Gerwick, Jr. 1975; Anthony J. Tozzoli 1974.

INTERNATIONAL COASTAL ENGINEERING AWARD . . . to provide international recognition for outstanding leadership and development in the field of coastal engineering. Is made annually to an individual who has made a significant contribution to the advancement of coastal engineering in the manner of engineering design, teaching, professional leadership, construction, research, planning, or a combination thereof.

Recipients: Joe W. Johnson 1987; Eco Wiebe Bijker 1986; Dr. Robert Weigel 1985; Michael S. Longuet-Higgins 1984; Robert G. Dean 1983; Morrough P. O'Brien 1982; Kiyoshi Horikawa 1981; William J. Herron, Jr. 1980; Bernard LeMehaute 1979.

MOISSEIFF AWARD . . . presented to the author or authors of an important paper in "ASCE Proceedings" and "Civil Engineering" dealing with the broad field of structural design, including applied mechanics as well as the theoretical analysis or constructive improvement of engineering sturctures such as bridges and frames.

Recipients: Ahmet E. Aktan and Vitelmo V. Bertero 1987; Howard T. Pearce and Yi-Kwei Wen 1986; Kurt H. Gerstle and Michael H. Ackroyd 1984; David J. Malcolm and Peter G. Glockner 1983; Baidar Bakht and Paul F. Csagoly 1982; S. Van Zyl and S. Scordelis 1981; Arthur Hucklebridge and Ray Clough 1980; Steven E. Ramberg and Owen M. Griffin 1979; Robert Park and Shafiqual Islam 1978; C. Allin Cornell and Hans A. Merz 1977; Cheng-Shung-Lin and Alexander C. Scordelis 1976; Paul Weidlinger 1975; Thomas G. Harmon, Robert J. Hansen, Jose M. Roesset, R. Elangwe, F. Alfred Picardi, Kanu S. Patel and Robert D. Logcher 1974; Leroy Z. Emkin and William A. Little 1973; N. Norby Neilsen, Mele A. Sozen and Toshikozu Takedo 1972; Fred Moses and John D. Stevenson 1971; Harry H. West and Arthur R. Robinson 1970; John A. Blume 1969; Maxwell G. Lay and Theodore V. Galambos 1968.

NATHAN M. NEWMARK MEDAL . . . is made to a member of the Society who, through contributions in structural mechanics, has helped substantially to strengthen the scientific base of structural engineering, these contributions having been made in the form of papers or other written presentations. The field of structural mechanics should be interpreted broadly and should include contributions in dynamics and continuum mechanics related to structural and geotechnical engineering.

Recipients: Emilio Rosenblueth 1987; Robert H. Scanlan 1986; Masanobu Shinozuku 1985; Egor Popov 1981; Olgierd C. Zienkiewicz 1980; Ray W. Clough 1979; Anestis S. Veletsos 1978; Melvin Baron 1977; John E. Goldberg 1976.

THE NORMAN MEDAL . . . instituted and endowed in 1872, this medal is awarded to the author or authors of a paper which shall be judged worthy of special commendation for its merit as a contribution to engineering science.

Recipients: Egor P. Popov, Stephen A. Mahin, Ray W. Clough 1987; James L. Sherard 1986; James L. Sherard, Lorn P. Dunningan and James R. Talbot 1985; Sudipta S. Bandyopadhyay 1984; Theodore V. Galambos, Bruce Ellingwood, James G. MacGregor and C. Allin Cornell 1983; Abdulaziz I. Mana and G. Wayne Clough 1982; John L. Cleasby and James C. Lorence 1980; Anil K. Chopra 1979; Richard D. Barksdale 1978; H. Bolton Seed, Kenneth L. Lee, Izzat M. Idriss and Faiz I. Makdisi 1977; Charles C. Ladd and Roger Foott 1976; Roy E. Olson, David E. Daniel, Jr. and Thomas K. Liu 1975; James R. Cofer and Bobby O. Hardin and Vincent P. Drnevich 1973; Nicholas C. Costes, W. David Carrier III, James K. Mitchell and Ronald F. Scott 1972; John H. Shmertmann 1971; Cyril J. Galvin, Jr. 1970; Basil W. Wilson 1969; H. Bolton Seed and Kenneth L. Lee 1968.

THE O. H. AMMANN RESEARCH FELLOWSHIP IN STRUCTURAL ENGINEERING . . . for the purpose of encouraging the creation of new knowledge in the field of structural design and construction. This fellowship, with a stipend of $2,000, was endowed in 1963 by O. H. Ammann, honorary member of ASCE.

Recipients: W. M. Kim Roddis 1986; Ajay Prakash 1984; Allen Rhett Whitlock 1982; Samuel T. Burguieres, Jr. 1978; Rick Evans 1973; Robin K. McGuire 1972; Eberhard Haug 1970; Mubadda Suidan 1969; John L. Baumgartner 1968.

RAYMOND C. REESE RESEARCH PRIZE . . . to recognize outstanding contributions to the application of structural engineering research, Raymond C. Reese, Hon. M. ASCE, contributed funds to suport a structural engineering research prize.

Recipients: Cheng-Tzu Thomas Hsu 1987; J. O. Malley and Egor P. Popov 1986; Xi La Liu and Wai Fah Chen 1985; Charles W. Roeder and Mahyar Assadi 1984; Avtar S. Pall and Cedric Marsh 1983; Ahmed M. Abdel-Ghaffar 1982; Karl Frank and John W. Fisher 1981; Lawrence F. Kahn and Robert D. Hanson 1980; James G. MacGregor and Sven E. Hage 1979; Hotten A. Elleby, Wallace W. Sanders, Jr., F. Wayne Klaiber and M. Douglas Reeves 1978; Ingvar H. E. Nilsson and Anders Losberg 1977; Joint ASCE-ACI Task Committee 426 on Shear and Diagonal Tension of the Committee on Masonry & Reinforced Concrete of the Structural Division 1976; R. J. Hansen, J. W. Reed and E. H. Vanmarcke 1975; Peter W. Chen and Leslie E. Robertson 1974; Richard A. Parmelee 1973; Munther J. Haddadin, Alan H. Mattock and Sheu-Tien Hong 1972; Otto E. Curth and John F. Wiss 1971; James O. Jirsa, Chester P. Siess and Mete A. Sozen 1970.

RESEARCH FELLOWSHIP . . . for the purpose of aiding in the creation of new knowledge for the benefit and advancement of the science and profession of civil engineering. The $5,000 grant is made on the basis of transcripts of scholastic records, ability to conceive and explore original ideas, descriptions of proposed research and objectives.

Recipients: David C. Goodrich and John A. Mundell 1986; Paul N. Hopping 1985; Jerry E. Stephens 1984; Donald E. Matzzie 1966; David J. Ayres 1965; Joseph F. Lestingi 1964.

RICKEY MEDAL . . . presented to an author or authors of a paper published in "ASCE Proceedings" and "Civil Engineering" in the general field of hydroelectric engineering, including any of its branches, deemed to constitute a worthy contribution to the art.

Recipients: John Parmakian 1987; Arvids Zagars 1986; Stanley D. Wilson 1985; Earl J. Beck 1984; Merlin D. Copen 1983; Klaus R. Ludwig and Richard T. Olive 1982 and Phillip R. Hoffman 1981; Wendell E. Johnson 1980; E. Montford Fucik 1979; George P. Palo 1978; Edward Loane and Franklyn Rogers 1977; Edward S. Loane and Franklyn C. Rogers 1976; Wallace L. Chadwick 1974; Committee on Hydro Power Project Planning of the Power Division 1972; John W. O'Hara, J. George Thon and Clarence H. Whalin 1971; Paul H. Gilbert 1970; Alfred L. Parme 1969; George R. Rich 1968.

ROBERT HORONJEFF AWARD ... to recognize and honor a person(s) or organization(s) for outstanding achievements and contributions to the advancement of the field of air transport engineering.
Recipients: Owen Miyamoto 1986; Nai C. Yang 1984; Joseph Blatt 1981; Herbert H. Howell 1979.

ROYCE J. TIPTON AWARD ... to recognize an individual who has made a definite contribution to the advancement of irrigation and drainage engineering either in teaching, research, planning, design, construction, or management. The award consists of a plaque and certificate.
Recipients: William Oregon Pruitt 1987; Jan Van Schilfgaarde 1986; Finley Burnap Lavery 1985; Herman Bouwer 1984; Ray J. Winger 1983; Marvin E. Jensen 1982; Joseph F. Friedkin 1981; Clyde Houston 1980; John L. Merriam 1979; Elmer W. Gain 1978; Arthur F. Pillsbury 1977; Jerald E. Christiansen 1976; Charles R. Maierhofer 1975; Arthur D. Soderberg 1974; William R. Gianelli 1973; George D. Clyde 1972; William F. Donnan 1971; Ellis L. Armstrong 1970; Harvey O. Banks 1969; Dean F. Peterson, Jr. 1968.

THE RUDOLPH HERING MEDAL ... presented to the author or authors of a paper published in "ASCE Proceedings" and "Civil Engineering" which contains the most valuable contribution to the increase of knowledge in, and to the advancement of, the sanitary branch of the engineering profession. The prize consists of a bronze medal and certificate.
Recipients: Jerry D. Lowry and Jeffrey E. Brandow 1987; George Lee Christensen and Richard I. Dick 1986; Appiah Amirtharajah 1985; Donald J. O'Connor 1984; Linvil G. Rich 1983; Luiz Di Bernardo and John L. Cleasby 1982; Angelos Findikakis and James O. Leckie 1981; John C. Crittenden and Walter J. Weber 1980; John T. Novak, Harry R. Becker, Jr. and Andrew Zurrow 1979; David DiGregorio and Gerald L. Shell 1978; William F. Garber, George T. Ohara and Sagar K. Raksit 1977; Carmen F. Guarino, Michael D. Nelson and Allan B. Edwards 1976; W. C. Boyle, P. M. Berthouex and T. C. Rooney 1975; Kenneth S. Price, Richard A. Conway and Albert H. Cheely 1974; John D. Parkhurst and Richard D. Pomeroy 1973; Harvey F. Collins, Robert E. Selleck and George C. White 1972; Kou-Ying Hsiung and John L. Cleasby 1970; Robert W. Agnew and Raymond C. Loehr 1969; Robert L. Johnson and John L. Cleasby 1968.

SAMUEL ARNOLD GREELEY AWARD ... to recognize the author or authors of the paper published in "ASCE Proceedings" and "Civil Engineering" which contains the most valuable contribution to the sanitary engineering profession.
Recipients: Ronald W. Crites 1987; David Scott Morgan, Jose Ignacio Novoa, and Albert Henry Halff 1986; John P. Hartigan 1985; Joseph A. Salvato 1984; Walter J. O'Brien 1983; Joseph M. Colonell 1982; Alberto Gutierrez and J. Harry Rowell 1981; Otto Brody 1980; David L. Eisenhauer, Ronald B. Sieger and Denny S. Parker 1977; Ralph Stone 1976; James A. Mueller, Thomas J. Mulligan and Dominic M. Di Toro 1974; Ralph G. Berk 1973; Robert C. Moore 1971; Roy E. Ramseier 1970; S. David Graber and Thomas R. Camp 1969.

SIMON W. FREESE ENVIRONMENTAL ENGINEERING AWARD AND LECTURE ... at about yearly intervals upon recommendation of the Executive Committee of Environmental Engineering Division, the Executive Director will invite a distinguished person to prepare for publication and deliver a lecture at an appropriate meeting of the Society. The lecturer shall be tendered a certificate and an honorarium of $1,000.
Recipients: Denny S. Parker 1987; Earnest F. Gloyna 1986; Charles O'Melia 1985; J. O'Connor 1981; Carmen Guarino 1980; Perry L. McCarty 1979; Russel E. Train 1978; Daniel A. Okun 1977.

STEPHEN D. BECHTEL PIPELINE ENGINEERING AWARD ... to recognize outstanding achievements in pipeline engineering, the Bechtel Foundation donated funds to support this Award in honor of the contributions made by Stephen D. Bechtel, F. ASCE, to the engineering profession and for his noteworthy advancements in the design and construction of pipelines throughout the world.

Recipients: Sol Koplowitz 1987; Andrew P. Rollins, Jr. 1986; Robert J. Brown 1985; Walter J. Weber, Jr. 1984; John L. Cleasby 1983; Edward A. Bryant 1982; Frederick E. Culvern 1979; William A. Hunt 1978; Robert V. Phillips 1977; Joe R. Thompson 1975; Maynard M. Anderson 1974; Eldon V. Hunt 1973; Nathaniel Clapp 1972; Joseph B. Spangler 1971.

SURVEYING AND MAPPING AWARD ... made annually to a member of the ASCE who has made a definite contribution during the year to the advancement of surveying and mapping either in teaching, writing, research, planning, design, construction, or management, these contributions being made in the form of either papers or other written presentations, or in some instances through notable performance, long years of service, or specific actions which have served to advance surveying and mapping.
Recipients: William T. Pryor 1987; Harmer A. Weeden 1986; Arthur J. McNair 1985; Alfred Q. Quinn 1984; Gurdon H. Wattles 1983; Gunther H. Greulich 1982; Dean C. Merchant 1981; Ira Alexander 1980; Eldon C. Wagner 1979; Robert H. Lyddan 1978; Morris M. Thompson 1977; Kenneth S. Curtis 1976; William A. White 1975; George D. Whitmore 1974; Philip Kissam 1973; William A. Radlinski 1972; Curtis M. Brown 1971; B. Austin Barry 1970.

T. Y. LIN AWARD ... The ASCE Prestressed Concrete Award was endowed in 1968 by T. Y. Lin, F. ASCE, to encourage the preparation of meaningful papers in the designated field of endeavor.
Recipients: Michael P. Collins and Denis Mitchell 1987; Walter Podolny, Jr. 1986; William C. Stone and Jone W. Breen 1985; Noel D. Nathan 1984; John B. Kennedy and A. M. El-Laithy 1983; Michael P. Collins and Denis Mitchell 1982; Robert Park and Kevin Thompson 1981; Lawrence K. Kahn and Robert D. Hanson 1980; George B. Barney, W. Gene Corley, John M. Hanson and Richard A. Parmelee 1979; Saad E. Moustafa 1978; Zdenek P. Bazant, Domingo J. Carreira and Adolf Walser 1977; Maher K. Tadros, Amin Ghali and Walter H. Dilger 1976; Paul Zia and W. D. McGee 1975; Raoug Sinno and Howard L. Furr 1974; Robert F. Mast and Charles W. Dolan 1973; Stanley L. Paul 1972; Arthur R. Anderson and Soad E. Moustafa 1971; Richard J. Kosiba and Howard W. Wahl 1970; Robert F. Mast 1969.

THE THEODORE VON KARMAN MEDAL ... awarded to an individual in recognition of distinguished achievement in engineering mechanics, applicable to any branch of civil engineering.
Recipients: Richard Skalak 1987; Stanley Corrsin 1986; Philip G. Hodge 1985; Stephen H. Crandall 1984; Albert E. Green 1983; Bernard Budiansky 1982; Chia-Shun Yih 1981; George Hermann 1980; Henry L. Langhaar 1979; Rodney Hill 1978; George Francis Carrier 1977; Yuan-Cheng B. Fung 1976; John H. Argyris 1975; George W. Housner 1974; Hans H. Bleich 1973; Nicholas J. Hoff 1972; Alfred M. Freudenthal 1971; Wilhelm Flugge 1970; Sir Geoffrey Ingram Taylor 1969; Lloyd H. Donnell 1968.

THE THOMAS A. MIDDLEBROOKS AWARD ... presented to the author or authors of a paper published in "ASCE Proceedings" and "Civil Engineering" judged worthy of special commendation for their merit as a contribution to soil mechanics or foundation engineering. The award consists of a cash sum to be determined annually.
Recipients: James M. Duncan and Raymond B. Seed 1987; Richard L. Handy 1986; James K. Mitchell 1985; W. David Carrier III; Leslie G. Bromwell and Frank Somoayi 1984; Max D. Sorota 1983; Edward B. Kinner and Mark X. Haley 1983; Ahmed Abdel-Ghaffar and Ronald F. Scott 1982; John H. Schmertmann and Alejandre Palacios 1981; James M. Duncan 1980; Bobby O. Hardin 1979; James D. Parsons 1978; George F. Sowers 1977; Kenneth L. Lee and John A. Focht, Jr. 1976; Kenneth L. Lee, Bobby Dean Adams and Jean-Marie J. Vagneron 1975; Aleksandar S. Vesic 1974; James K. Mitchell and William S. Gardner 1973; Clyde N. Baker, Jr. and Fazlur Kahn 1972; I. M. Idriss, Kenneth L. Lee and H. Bolton Seed 1971; Richard Campanella, James K. Mitchell and Awtar Singh 1970; Richard F. Brissette, David D'Appolonia and Ekio D'Appolonia 1969; Paul J. Marsal and Luis Ramirez de Arellano 1968.

THE THOMAS FITCH ROWLAND PRIZE ... presented to the author or authors of a paper published in "ASCE Proceedings" describing in detail accomplished works of construction, their cost, and errors in design and execution. The prize consists of a wall plaque and a certificate.

Recipients: James T. O'Connor and Richard L. Tucker 1987; James L. Lammie and Dhirajlal P. Shah 1986; Samir G. Mattar 1985; Yuzo Takeuchi and Akira Kitamura 1984; James E. Olyniec 1983; Alfred M. Petrofsky 1982; C. V. Knudsen 1981; Laurent Hamel and David Nixon 1980; Marcello H. Soto 1979; Charles H. Thornton and Paul A. Gossen 1978; Thor L. Anderson and Melvin C. Williams 1977; Daniel J. Smith 1976; Russell C. Borden and Carl E. Selander 1974; John V. Bartlett, Tadeusz M. Noskiewicz and James A. Ramsay 1973; Hans Sacrison 1972; Richard E. Whitaker 1971; Norman L. Liver 1970; James Douglas 1969; C. Y. Li 1968.

WALTER L. HUBER CIVIL ENGINEERING RESEARCH PRIZES ... awarded to members of the Society in any grade for notable achievements in research related to civil engineering. Preference shall be given to younger members (under 40) who can be expected to continue fruitful careers in research. Each award consists of $100 and a suitable certificate.

Recipients: Achintya Haldar, Yannis Dafalias, Joseph Sherrard, Jean-Louis Briaud, Earl Downey Brill, Jr. 1987; William F. Maloney, Yi-Kwei Wen, Enrique Luco, Michael O'Neill, and Keith W. Bedford 1986; David Darwin, Louis F. Cohn, Stergios A. Debdrou, Jerald Lee Schnoor, Thomas M. Keinath 1985; Patrick J. Ryan, James K. Edzwald, Charles Linwood Vincent, Erik H. Vanmarcke and Ross B. Corotis 1984; J. N. Reddy, Kenneth H. Stokoe, Gerhard H. Jirka, Stephen A. Mahin and Gary S. Logsdon 1983; Ajaya K. Gupta, Raymond E. Levitt, Fred H. Kulhawy and Adib K. Kanafani 1982; Douglas Haith, Gary Hart-Klaus-Jurgen Bathe, Joseph L. Hammack and William F. Marcuson III 1981; Robert M. Clark, Vincent P. Drnevich, Clive Dym, Bruce R. Ellingwood and Boyd C. Paulson, Jr. 1980; Daniel W. Halpin, Ju-Chang, Huang-Hon-Yimko, Ruh-Ming Li, Keith D. Stolzenbach 1979; John D. Borcherding, Thomas J. R. Hughes, Tin-Kan Hung, James O. Jirsa, Philip La and Fan Liu 1978; Ted B. Belytschko, G. Wayne Clough, David H. Marks, James P. Tullis and Harry E. Wenzel 1977; Zdenek B. Bazant, Danny L. Fread, Lester A. Hoel, Paul H. King and John Lysmer 1976; Anil K. Chopra, Izzat M. Idriss, Ignacio Rodriguez Iturbe, Larry A. Roesner and Don James Wood 1975; Asit K. Biswas, Hugo B. Fischer, Alfred J. Hendron, Thomas T. C. Hsu and Paul C. Jennings 1974; C. G. Culver, J. M. Duncan, L. E. King, J. T. Oden and C. T. . Yang 1973; W B. Ledbetter, R. E. Olson, M. Shinozuka, R. L. Street and J. A. Hoopes 1972; C. A. Cornell, M. Gates, R. J. Krizek, K. W. Wong and N. R. Morgenstern 1971; D. P. Louchs, J. Van Schilfgaarde, M. R. Thompson, K. L. Lee and L. A. Schmit 1970; J. W. Fisher, C. J. Galvin, C. C. Ladd, A. R. Robinson and R. F. Scott 1969; A. H. S. Ang, B. O. Hardin, M. E. Jensen, E. Naudascher and J. E. Quan 1968.

THE WASHINGTON AWARD ... awarded annually as an honor conferred upon a brother engineer or by his fellow engineers because of accomplishments which preeminently promote the happiness, comfort, and well-being of humanity. The Washington Award Commission consists of nine members, from the Western Society of Engineers and two members each from the following societies: ASCE, AIME, and IEEE.

Recipients: Rear Admiral Grace M. Hopper 1987; Mark Shepherd 1986; Stephen D. Bechtel, Jr. 1985; Robert W. Galvin 1984; John Bardeen 1983; Manson Benedict 1982; John W. Swearington 1981; Marvin Camras 1980; Dixy Lee Ray 1978; Michael Tenenbaum 1977; Ralph B. Peak 1976; David Packard 1975; John Dunlany deButts 1974; John A. Volpe 1973; Thomas O. Paine 1972; William L. Everitt 1971; Hyman G. Rickover 1970; Nathan M. Newmark 1969; James B. Fisk 1968.

WESLEY W. HORNER AWARD ... presented to the author or authors of a paper published in "ASCE Proceedings" and "Civil Engineering" that makes the most valuable contribution to the sanitary engineering profession (hydrology, urban drainage, and sewerage).

Recipients: T. P. Halappa Gowda and John D. Lock 1987; T. P. Halappa Gowda 1986; James W. Mercer, Lyle R. Silka and Charles R. Faust 1985; Karl E. Kienow, Kenneth K. Kienow and Richard D. Pomeroy 1984; Abu M. Z. Alam, Donald R. F. Harle-

man and Joseph M. Colonell 1983; James V. Olson, Ronald W. Crites and Paul E. Levine 1982; Joseph Cermola, Sergio Decarli, Dev R. Sachdev and Hassan El-Baroudi 1981; Dominic M. Di Toro and Mitchell J. Small 1980; Thomas K. Jewell, Thomas J. Nunno and Donald D. Adrian 1979; Lars-Eric Janson, Stein Bendixen and Anders Harlaut 1978; Naresh K. Rohatgi and Kenneth Y. Chen 1977; Mark E. Noonan, William R. Giessner, Robert J. Cockburn and Frank H. Moss, Jr. 1976; C. G. Gunnerson 1975; Charles V. Gibbs, Stuart M. Alexander and Curtis P. Leiser 1974; Raymond F. Brenner 1971; Joseph A. Cotteral, Jr. and Dan P. Norris 1970; David M. Greer and Douglas C. Moorhouse 1969.

AMERICAN SOCIETY OF GAS ENGINEERS (ASGE)
P.O. Box 936 Tinley Park, Illinois 60477
Charles R. Kendall, Executive Director

PRESIDENTS ... Robert J. Kolodgy 1987-88; Joseph Eckert 1986-87; Richard P. Adams 1985-86; Robert E. White 1984-85; Robert D. Downin 1983-84; Richard E. Coleman 1982-83; John F. Marley 1981-82; Leo Pfister 1980-81; Erwin H. Strassemeyer 1979-80; Robert G. Bush 1978-79; L. J. Swift 1977-78; Daniel J. Kesel 1976-77; Philip J. More 1975-76.

AMERICAN SOCIETY OF HEATING, REFRIGERATING & AIR-CONDITIONING ENGINEERS (ASHRAE)
1791 Tullie Circle, N.E., Atlanta, Georgia 30329
Frank M. Coda, Executive Director/Secretary

PRESIDENTS ... H. E. Burroughs 1987-88; Frederick H. Kohloss 1986-87; Donald R. Bahnfleth 1985-86; Robert O. McDonald 1984-85; Richard P. Perry 1983-84; Clinton W. Phillips 1982-83; Jack B. Chaddock 1981-82; Charles F. Sepsy 1980-81; Hugh D. McMillan, Jr. 1979-80; Morris Backer 1978-79; Bruno P. Morabito 1977-78; William P. Chapman 1976-77; William J. Collins, Jr. 1975-76; David Rickelton 1974-75; Roderick R. Kirkwood 1973-74.

AWARDS

ASHRAE-ALCO MEDAL FOR DISTINGUISHED PUBLIC SERVICE ... recognizes a member of the society who distinguished himself conspicuously, for continued and outstanding participation in public affairs. This award was initiated in 1965, is sponsored by the Alco Valve Company, and consists of a medal and an inscribed scroll.

Recipients: Howard O. Ward 1987; Joseph Kearney, Sr. 1986; Otto A. Tennant 1985; Realto E. Cherne 1984; William W. Gay, Sr. 1983; Warren G. Moses 1982; Howell E. Adams, Jr. 1981; Henry J. Campbell, Jr. 1979; Cullen E. Parmelee 1978; Alwin B. Newton 1976; Earle K. Wagner 1975; George B. Hightower 1974; Bruce L. Evans 1973; Robert W. Flanagan 1972; Harry E. Bovay, Jr. 1971; Carlyle M. Ashley 1970; George A. Linskie 1969; Wilson L. Davis 1968.

ASHRAE-HOMER ADDAMS AWARD ... initiated in 1956 to perpetuate the memory of Homer Addams, charter member and past president of ASHRAE, this award consists of $600 and a certificate presented annually to a graduate student working on an ASHRAE Research Project "to advance his engineering education and knowledge gained through research."

Recipients: James J. Bastian 1987; Polyvious Eleftheriou, Ph.D. 1986; Steven M. Harris 1985; Michael S. Sherber 1984; Mean Shang Chen 1983; Shimano Ni 1982; Charles F. Rhein 1981; Joseph P. Olivieri 1980; Gary L. Reynolds 1979; Sriram Somasundaram 1978; Douglas R. Bryant 1977.

CROSBY FIELD AWARD ... recognizes the Best Paper published by ASHRAE. Award consists of $750 and a certificate.

Recipients: W. K. Brown, Jr. and Charles A. Lynn 1987; William J. Coab 1986; D. W. DeWerth, S. M. Connelly, S. G. Talbert & D. D. Paul 1985; Brian P. Leaderer and William S. Cain 1984; Warren E. Blazier, Jr. 1982; Lawrence G. Spielvogel 1981; George T. Tamura 1980; Istvan L. Ver 1979; Prof. William Rudoy & L. M. Robins 1978; Prof. Clayton A. Morrison, James M. Wheeler, Jr., & Dr. Erich A. Farber 1977.

F. PAUL ANDERSON AWARD ... established in 1930, this award is presented annually for outstanding work or service in any field of the society. The award consists of a medal and a certificate.
Recipients: Everett Palmatier 1987; A. Pharo Gagge 1986; Walter F. Spiegel 1985; William P. Chapman 1983; John Engalitcheff, Jr. 1982; Burgess H. Jennings 1981; Herbert L. Laube 1980; Daniel D. Wile 1979; Dr. Wilbert F. Stoecker 1978; Paul Reece Achenbah 1976; Neil B. Hutcheon 1975; William L. McGrath 1974; Seichi Konzo 1973; Robert H. Tull 1972; Donald K. Tressler 1971; Glenn Muffly 1970; Willis R. Woolrich 1969; Crosby Field 1968.

ASHRAE JOURNAL AWARD ... presented annually for the best paper published in the ASHRAE Journal. The award consists of $250 and a certificate.
Recipients: Carl H. Jordan and Makio Goto 1986; R. L. Webb, Ph.D. and A. Villacres 1985; Francis J. McCabe 1984; Bruce J. Novell 1983; John A. Fox 1982; William B. May, Jr. and Larry Spielvogel 1981.

LINCOLN BOUILLON MEMBERSHIP AWARD ... established in 1967 to commemorate the late past president Lincoln Bouillon's effort in recruiting new members, this award is presented annually to the individual who performs the most outstanding work in increasing the membership of the Society.
Recipients: William C. Bissmeyer 1987; Ross D. Montgomery 1986; Craig M. Goodenow and Kevin Y. F. Chong 1985; Thomas A. Gorman III 1984; Lynn G. Bellenger 1983; G. Allan Gaboric 1982; Frederick E. Stone 1981; Gerald F. Morrow 1980; Lewis E. Seagraves 1979; Oliver K. Lewis 1978; Edward M. Naretto 1977.

LOUISE AND BILL HOLLADAY DISTINGUISHED FELLOW AWARD ... established in 1978, this award is presented annually to a Fellow of the Society, recognizing continuing preeminence in engineering or research work relating to the sciences of heating, refrigeration, air conditioning or ventilation or the allied arts and sciences. Award consists of medal and a special citation.
Recipients: Tamami Kusuda 1987; John H. Fox 1986; Alwin B. Newton 1985; Frank H. Faust 1984; Daniel D. Wile 1983; P. Ole Fanger 1982; A. Pharo Gagge 1981; John I. Yellott 1980; Carlyle M. Ashley 1979.

RALPH G. NEVINS PHYSIOLOGY AND HUMAN ENVIRONMENT AWARD ... presented to a promising investigator, under 40 years of age, for significant accomplishment in the general area of man's response to the environment. Accomplishments are represented by significant papers, published by ASHRAE or by journals of an ASHRAE associate over the five-year period previous to the award, which consists of $200 and a certificate. Award established in 1976.
Recipients: Richard B. Hayter 1987; Lars Molhave 1986; Elizabeth McCullough 1985; Deanna M. Munson 1984; Richard R. Gonzalez 1983; Bjarne W. Olesen 1982; Donald A. McIntyre 1981; Dr. Yasunobi Nishi 1980; Larry G. Berglund 1979; Dr. James E. Woods 1978.

WILLIS H. CARRIER AWARD ... presented for the best technical paper presented at a society meeting by a member of any grade who is 32 years of age or less. The award consists of $500 and a certificate. It is sponsored by the Carrier Corporation and was initiated in 1960.
Recipients: Robert W. Rasmussen 1987; Joseph E. Lavina 1986; Frank E. Jakob 1985; Nancy J. MacDonald Banks 1984; Van D. Baxter 1983; Francesco Pompei 1982; Amir L. Ecker 1981; Stephen E. Selkowitz 1980; Sudhir K. Sastry 1979; A. F. M. Ali 1978; Stephen M. Sessler 1977; Merle F. McBride 1976; Douglas M. Burch 1975; Stanley A. Mumma 1973; Douglas E. McDuffie, Jr 1971; Yosunobu Nishi 1970; Richard E. Barrett 1969; George D. Huffman 1968.

DISTINGUISHED SERVICE AWARD ... conferred to a member of any grade who has served ASHRAE faithfully and with distinction as a member of committees. Award consists of a certificate.

WILLIAM J. COLLINS, JR. RESEARCH PROMOTION AWARD ... established in 1985 and presented to a Chapter Research Promotion Chairman who excels in raising funds for ASHRAE research. Award consists of engraved medal and certificate.
Recipients: Herbert L. Bregman 1987; Bob A. McAlister 1986; Ralph M. Mattison, 1985.

INTERNATIONAL ACTIVITIES AWARD ... conferred to a member who has done the most to enhance the society's international presence or posture. Award consists of a certificate.
Recipients: Richard H. Rooley 1987; Dr. Raymond Cohen 1986; Wilbert F. Stoecker 1985; P. Ole Fanger 1984.

A. T. BOGGS SERVICE AWARD ... initiated in 1986 to honor a Distinguished Service Award recipient of the Society in recognition of particularly notable service continued beyond that which earned the Distinguished Service Award.
Recipients: Frederick J. Reed 1986.

AMERICAN SOCIETY OF LUBRICATION ENGINEERS (ASLE)

838 Busse Highway, Park Ridge, Illinois 60068
Mrs. Maxine E. Hensley, Executive Director

PRESIDENTS ... Lowell C. Horwedel 1987-1988; George Kitchen 1986-1987; Dr. Carleton Rowe 1985-1986; Wayne Coursey 1984-1985; James Steenbeergen 1983-84; Robert Benzing 1982-1983; James R. Dickey 1981-82; Ken A. Wilcox 1980-81; Bernard H. Shelley 1979-80; Victor E. Joll 1978-79; W. K. Stair 1977-78; Donald G. Flom 1976-77; Harry Tankus 1975-76; W. H. Mann 1974-75; L. Robinette 1973-74; C. H. West 1972-73.

AWARDS

ASLE NATIONAL AWARD ... is the Society's highest honor and bestows lifetime membership on recipient. In recognition of outstanding contribution to the field of engineering, not necessarily the field of lubrication. The recipient need not be a member of ASLE.
Recipients: Dr. Herbert Cheng 1987; Toshio Sakurai 1986; Donald H. Buckley 1985; Lowell C. Horwedel 1984; K. L. Johnson and Maxine E. Hensley 1983; Alen Beerbower 1982; F. T. Barwell 1981; Donald G. Flom 1980; Harmen Blok 1979; Vance E. Vorhees 1978; F. F. Ling 1977; E. E. Klaus 1976; A. A. Manteuffel 1975; D. Dowson 1974; P. M. Ku 1973; E. A. Saibel 1972; R. L. Johnson 1971; C. A. Bailey 1970; A. Cameron 1969; A. A. Raimondi 1968.

CAPTAIN ALFRED E. HUNT AWARD ... "ASLE Medal" for award to an "ASLE member for the best paper presented during the year at any Sectional or National ASLE meeting, or published in the Society's official journal, Lubrication Engineering.
Recipients: Dr. Bernard A. Baldwin 1987; Michael N. Gardos 1986; Lowrie B. Sargent, Jr. 1984; Carl N. Rowe 1983; Alen Beerbower 1982; Dr. L. D. Wedeven and Dr. C. Cusano 1981; Dr. E. Erwin Klaus and Dr. Stephen M. Hsu 1980; Harold E. Sliney 1979; Neal D. Rebuck, Alfeo A. Conte, Jr. and Leon Stallings 1978; A. Cameron and A. Jackson 1977; Y. P. Chiu 1975; C. Dayson 1974; A. Beerbower 1972; M. Antler 1971; R. P. Steijn 1970; W. A. Wright 1969; D. H. Buckley 1968.

P. M. KU MERITORIOUS AWARD ... this award, presented for the first time at the ASLE 1979 Annual Meeting, is given to the ASLE member who most typifies the dedicated spirit and hardworking attitude of P. M. Ku, who worked tirelessly in the background for the benefit of the Society, devoting long hours in the performance of many thankless tasks necessary to promote and advance the cause of ASLE. In his honor, the P. M. Ku Meritorious Award has been established to recognize outstanding and selfless achievement. This prestigious award will not be given on an annual basis. It will be presented only when the ASLE Board feels it has been justly earned. To qualify for this honor, the recipient must have been a member of the Society for at least fifteen consecutive years and performed extensive, active, dedicated service for the Society.

Recipients: Robert L. Johnson 1987; Harry Tankus 1986; Douglas Godfrey 1985; James R. Dickey 1984; E. Richard Booser 1983; Donald G. Flom 1982; William K. Stair 1981; Edmond E. Bisson 1980; Robert J. Torrens 1979.

WALTER D. HODSON JUNIOR AWARD ... presented to a member thirty-five years of age or less for the best paper presented or published within the Society on lubrication or an allied subject during the preceding year.
Recipients: Dr. Robert W. Bruce 1987; Leonardo DeChisfre 1986; Robert A. Pallini 1985; Thomas P. Will, Jr. 1984; William D. Marscher 1983; Andrew Jackson 1982; Ralph P. Gabriel 1981; William H. Miller and Dr. R. Gordon Kirk 1980; Robert L. Fusaro 1979; William R. Jones, Jr. 1978; L. D. Wedeven and B. A. Baldwin 1977; M. N. Gardos 1976; I. L. Goldblatt 1975; L. A. Horve 1973; L. C. Lipp 1969; T. A. Harris 1968.

THE WILBUR DEUTSCH MEMORIAL AWARD ... this award is given in recognition of the best paper published on the practical aspects of lubrication during the preceding year.
Recipients: Dr. Shirley E. Schwartz 1987; Mark S. Vukasovich 1986; Andrew G. Papay 1985; Hooshang Heshmat, W. Shapiro and Donald F. Wilcock 1983; Heinz P. Bloch 1982; Jospeh R. Gordon 1981; Larry C. Lipp 1980; Richard E. Cantley 1979; J. E. Snowden, Jr; Jack Conway, Sr. and J. P. Westerheid 1978; D. C. Root 1977; E. R. Booser 1976; E. O. Bennett 1975; R. Bryer 1974; W. Hartmann 1973; C. A. Sluhan 1972; S. Rodman and L. Dromgold 1971; D. Domonoske 1970; J. Gansheimer 1969; P. D. Metzger 1968.

THE AMERICAN SOCIETY OF MECHANICAL ENGINEERS (ASME)

345 East 47th Street, New York, New York 10017
David Belden, Executive Director

PRESIDENTS ... Ernest L. Daman 1988; Richard Rosenberg 1987; Nancy D. Fitzroy 1986; Leroy S. Fletcher 1985; George Kotnick 1984; Frank M. Scott 1983; Serge Gratch 1982; Robert B. Gaither 1981; Charles E. Jones 1980; Donald N. Zwiep 1979; O. L. Lewis 1978; Stothe P. Kezios 1977; Earle C. Miller 1976; Charles L. Tutt, Jr. 1975; Richard B. Robertson 1974; Daniel C. Drucker 1973.

AWARDS

THE AMERICAN SOCIETY OF MECHANICAL ENGINEERS MEDAL ... awarded annually for "eminently distinguished engineering achievement." This award was established in 1920.
Recipients: Phillip G. Hodge, Jr. 1987; Orlan W. Boston 1986; Milton Shaw 1985; Aaron Cohen 1984; Jack N. Binns, Sr. 1983; Robert S. Hahn 1981; Jacob P. Den Hartog 1979; Robert W. Mann 1977; Raymond D. Mindlin 1976; Maxime A. Faget 1975; Nicholas J. Hoff 1974; Christopher C. Kraft, Jr. 1973; E. H. W. Weibull 1972; Horace Smart Beattie 1971; Robert Rowe Gilruth 1970; Lloyd H. Donnell 1969; Samuel C. Collins 1968.

ASME BURT L. NEWKIRK AWARD ... awarded to an individual who is under the age of thirty-five at the time of the nomination and who has made a notable contribution to tribology in research or development as established by papers on tribology accepted for publication in the ASME Journal on Lubrication. The award was established in 1975.
Recipients: Pawan Kumar Goenka 1987; Itzhak Green 1986; Hooshang Heshmat 1984; Bharat Bhushan 1983; Dennis F. Li 1982; Stuart H. Loewenthal 1980; Thomas A. Dow 1979; Pradeep Gupta 1978; Steve M. Rohde 1977; Francis E. Kennedy, Jr. 1976.

ASME CODES AND STANDARDS MEDAL ... awarded in recognition of outstanding contributions to the development of documents, objects or devices used in any part of the national or international ASME programs of technical codification, standardization and certification. The award was established in 1976.
Recipients: Jack B. Levy 1987; William E. Cooper 1986; Paul M. Brister 1985; Melvin R. Green 1984; Walter L. Harding 1983; James W. Murdock 1982; Roy P. Trowbridge 1981; George F. Habach 1980; Joseph F. Sebald 1979; Leonard P. Zick 1978; William G. McLean 1977.

ASME EDWIN F. CHURCH MEDAL ... established in 1972 by the bequest of Edwin F. Church, Jr., to recognize the individual who has rendered eminent service in increasing the value, importance and attractiveness of mechanical engrg. education. The award consists of a medal, a certificate and a $1,000 honorarium.
Recipients: Garland H. Duncan 1987; Emil L. Martinec 1985; Milo Price 1984; Clinton H. Britt 1982; Neal P. Jeffries 1981; Dennis K. Bushnell 1980; Kenneth A. Roe 1979; Frank W. Von Flue 1976; Harry Conn 1975; Hobart A. Weaver 1974; Wilbur Richard Leopold 1973.

ASME GEORGE WESTINGHOUSE GOLD MEDAL ... bestowed for eminent achievement or distinguished service in the power field of mechanical engineering, including contributions of utilization, application, design, development, research and the organization of such activities in the power field. Established in 1952, the award consists of a gold medal and engrossed certificate.
Recipients: Henry O. Pohl 1987; Richard J. Coar 1986; Eugene A. Saltarelli 1985; Joseph R. Szydlowski 1984; Eugene P. Wilkinson 1983; William T. Reid 1982; Earle C. Miller 1981; Fred J. Moody 1980; William R. Gould 1979; Peter Fortescue 1978; John W. Simpson 1976; Charles W. Elston 1974; Bernard F. Langer 1973; William States Lee 1972; Wilfred McGregor Hall 1971; Charles A. Meyer-Robert C. Spencer, Jr. 1970; Ralph C. Roe 1969; Roland A. Budenholzer 1968.

ASME GEORGE WESTINGHOUSE SILVER MEDAL ... established in 1971 for an engineer who has not passed his forty-first birthday on June 30 of the year in which the presentation ceremony is conducted, is awarded for eminent achievement for distinguished service in the power field of mechanical engineering.
Recipients: Albert D. LaRue 1987; Joseph A. Barsin 1986; William J. Bryan 1984; Remco P. Waszink 1983; Leslie D. Kramer 1982; Ronald Pigott 1981; Robert L. Gamble 1980; Edward W. Stenby 1979; Romano Salvatori 1978; James C. Corman 1977; Richard V. Shanklin, III 1976; Shelby L. Owens 1974; Michael A. Ambrose 1973; William Eugene Rice 1972.

ASME HEAT TRANSFER MEMORIAL AWARD ... bestowed on individuals who have made outstanding contributions to the field of heat transfer through teaching, research, design or publications. The award is based on papers in the science, art or general subject of heat transfer.
Recipients: Ralph L. Webb and M. Necati Ozisik 1987; Richard C. Chu and Arcot Ramachandran 1986; Ralph Greif and Virgil E.Schrock 1985; Weng-Jei Wang 1984; Roger Eichhorn 1983; Hasuo Mori 1982; Ared Cezairliyan, Kwang-Tzu Yang & Ivan Catton 1981; John H. Lienhard 1980; Arthur E. Bergles & Yih-Yun Hsu 1979; John A. Clark & Richard J. Goldstein 1978; Robert D. Cess and Rolf H. Sabersky 1977; Warren H. Giedt & Raymond Viskanta 1976; Peter Griffith and Simon Ostrach 1975.

ASME NADAI AWARD ... established in 1975 to recognize distinctive contributions to the field of engineering materials. The award consists of an aluminum plaque.
Recipients: Erhard Krempl 1987; William F. Brown, Jr. 1986; Sumio Yukawa 1985; Thomas J. Dolan 1984; Arthur J. McEvily, Jr. 1983; Iain Finnie 1982; S. Stanford Manson 1981; Michael J. Manjoine 1980; Louis F. Coffin, Jr. 1979; Frank A. McClintock 1978; George R. Irwin 1977; Evan Albert Davis 1976; George M. Sinclair 1975.

ASME RAIL TRANSPORTATION AWARD ... in recognition of an outstanding original paper on a railroad mechanical engineering subject. The paper must be current and contribute to engineering literature in the railroad field. It was established in 1964 and consists of an engrossed certificate.
Recipients: Arthur J. Opinsky, Michael W. Joerms, Daniel H. Stone and Milton R. Johnson 1987; William F. Drish, Jr. and Som P. Singh 1986; Swamidas K. Punwani, Milton R. Johnson, Richard P. Joyce and Charles Mancillas 1985; James C. Paul, Alvin E. Holmes, Paul K. Hichey and Robert L. Gielow 1984; Kenji Hirakawa and Haruo Sakamoto 1982; Pierre P. Marcotte, K. J. Roderick Mathewson and W. Nelson Caldwell 1981; Bruce W. Shute, Eric C. Wright, Charles K. Taft and William N. Banister 1980; Robert W. Radford 1978; M. R. Johnson, R. E. Welch and K. S. Yeung 1976; W. Terry Hawthorn 1975; J. N. Siddall, M. A.

Dokainish and W. Elmaraghy 1974; G. E. Novak and B. J. Eck 1973; W. H. Freeman, L. A. Peterson and J. M. Wandrisco 1972; R. D. Ahlbeck and H. C. Meacham 1970; James A. Bain, Richard T. Gray, Samuel Levy and Estelle J. Playdon 1969; Thomas Schur 1968.

ASME RALPH COATS ROE MEDAL ... established in 1972 to recognize the individual selected by the Society as having contributed most effectively to a better understanding and appreciation of the Engineer's worth to contemporary society. The award consists of a medal, a certificate, an honorarium of $1,000 and a travel expense supplement.
Recipients: T. Lindsay Baker 1987; David Dooling, Jr. 1985; Lee A. Iacocca 1984; Tracy Kidder 1983; Samuel C. Florman 1982; Carl Sagan 1981; Melvin Kranzberg 1980; William D. Carey 1979; David Perlman 1978; Robert C. Seamans, Jr. 1977; Walter Sullivan 1975; Emilio Q. Daddario 1974.

BERNARD F. LANGER NUCLEAR CODES & STANDARDS AWARD ... recognition of an individual(s) who has contributed to the nuclear power plant industry through the development and promotion of ASME Nuclear Codes and Standards or the ASME Nuclear Certification Program. The award, established in 1977, consists of a bronze plaque and an engrossed certificate.
Recipients: Robert L. Dick 1987; Edwin J. Hemzy 1986; Howard F. Dobel 1985; Floyd N. Moschini 1984; Spencer H. Bush 1983; Guy A. Arlotto, Robert J. Bosnak, Robert B. Minogue and G. Wayne Reinmuth 1982; Lawrence J. Chockie 1981; Wendell P. Johnson 1980; William R. Smith, Sr. 1979; William E. Cooper 1977.

BLACKALL MACHINE TOOL AND GAGE AWARD ... awarded for the best paper or papers clearly concerned with or related to the design or application of machine tools, gages or dimensional measuring instruments. The award established in 1954, consists of a bronze plaque and an honorarium of $100 to each author.
Recipients: David A. Dornfeld and Elijah Kannatey-Asibu, Jr. 1986; Paul K. Wright 1985; C. Richard Liu and Moshe M. Barash 1984; Richard E. DeVor, William A. Kline and Igbal A. Shareef 1983; Nam P. Sug and Bruce M. Kramer 1982; Ranga Komanduri and Robert H. Brown 1981; Robert A. Thompson, Subbiah Ramalingam and John D. Watson 1980; Sindre Holoyen and Clayton D. Mote, Jr. 1975; S. P. Loutrel and N. H. Cook 1974; Louis J. Fritz and William P. Koster 1971; Kuo-King Wang, Shien-Ming Wu and Kazuaki Iwata 1968.

CHARLES RUSS RICHARDS MEMORIAL AWARD ... established in 1944 by Pi Tau Sigma as a suitable means of recognizing an engineering graduate who has demonstrated outstanding achievement in mechanical engineering twenty years or more following graduation. The award consists of an honorarium of $1,000, an engrossed certificate, and a travel expense supplement.
Recipients: Allen F. Rhodes 1987; E. Kent Springer 1986; Ephraim M. Sparrow 1985; Ferdinand Freudenstein 1984; Peter A. Engel 1983; Leroy S. Fletcher 1982; Shien-Ming Wu 1981; Albert I. King 1980; John H. Lienhard 1979; John C. Chato 1978; Hassan A. Hassan 1977; Ali Suphi Argon 1976; Carl F. Zorowski 1975; Richard J. Grosh 1974; Ali A. Seireg 1973; Charles E. Jones 1972; Howard L. Harrison 1971; Ralph G. Nevins 1970; Robert E. Uhrig 1969; Bernard W. Shaffer 1968.

INTERNAL COMBUSTION ENGINE AWARD (FORMERLY DIESEL AND GAS ENGINE POWER AWARD) ... in recognition of eminent achievement or distinguished contributions over a substantial period of time, which may result from research, innovation or education in advancing the art of engineering in the field of internal-combustion engines. The award was established in 1966 and consists of a bronze plaque.
Recipients: Hugh A. Williams, Jr. 1986; John M. Bailey 1985; Samuel S. Lestz 1984; James H. Garrett 1983; David B. Field 1982; Phillip S. Myers 1981; Helmuth G. Braendel 1979; William Speicher 1975; Warren J. Severin 1974; Warren A. Rhoades 1973; R. Rex Robinson 1972; Melvin J. Helmich 1971; Leo T. Brinson 1969.

THE ELMER J. SPERRY AWARD ... awarded for a distinguished engineering contribution which, through application, proved in actual service, has advanced the art of transportation

whether by land, sea, or air. This award is jointly administered by ASME, IEEE, SNAME, SAE, and AIAA.
Recipients: Harry R. Wetenkamp 1987; George W. Jeffs, William R. Lucas, George E. Mueller, George F. Page, Robert F. Thompson, John F. Yardley 1986; Richard K. Quinn, Carlton E. Tripp, George H. Plude 1985; Frederick Aronowitz, Joseph E. Killpatrick, Warren M. Macek, Theodore J. Podgorski 1984; Sir George Edwards, General Henri Ziegler, Sir Stanley Hooker, Sir Archibald Russell, M. Andre Turchat 1983; Jorg Brenneisen, Ehrhard Futterlieb, Joachim Korber, Edmund Muller, G. Reiner Nill, Manfred Schulz, Herbert Stemmler, Werner Teich 1982; Edward J. Wasp 1981; William M. Allen, Malcolm T. Stamper, Joseph F. Sutter and Everette L. Webb 1980; Leslie J. Clark 1979; Robert Puiseux 1978; Clifford L. Eastburg and Harley J. Urbach 1977; Jerome L. Goldman, Frank A. Nemec-James J. Henry 1975; Leonard S. Hobbs, Perry W. Pratt and Engineers of Pratt & Whitney Aircraft 1972; George W. Baughman and Sedwick N. Wight 1971; Charles S. Draper 1970; Douglas C. Macmillan, M. Nielsen, Edward L. Teale, Jr. 1969; Sir Christopher S. Cockerell and Richard Stanton-Jones 1968.

FLUIDS ENGINEERING AWARD ... bestowed for outstanding contributions, over a period of years, to the engineering profession and especially to the field of fluids engineering through research, practice and/or teaching. Established in 1978, the award consists of engrossed certificate.
Recipients: Mark V. Markovin 1987; Milton S. Plesset 1986; Apollo M. O. Smith 1985; Hans W. Liepmann 1984; George Rudinger 1983; Ascher H. Shapiro 1981; Robert C. Dean, Jr. 1979.

FREEMAN SCHOLAR AWARD ... awarded to a person of wide experience in fluids engineering. The scholar prepares a review of a coherent topic in his specialty, including a comprehensive state of the art and suggestions for key future research needs. The results are presented in a lecture and published in the ASME Journal of Fluids Engineering. Supported by a fund established in 1926, this biennial award consists of $7,500 honorarium and a travel allowance.
Recipients: Turgut Sarpkaya 1987; John B. Heywood 1986; A. K. M. Fazie Hussain 1984; Simon Ostrach 1982; Edward M. Greitzer 1980; Benjamin Gebhart 1978; William J. McCroskey 1976; Jack E. Cermak 1974; Jack W. Hoyt and Ronald F. Probstein 1971.

GAS TURBINE AWARD ... in recognition of outstanding individual or multiple-author contribution to the literature of combustion gas turbines or gas turbines thermally combined with nuclear or steam power plants. The award was established in 1963 and consists of a bronze plaque.
Recipients: Denis J. Doorley and Martin L. G. Oldfield 1987; Howard P. Hodson 1984; Chunill Hah 1983; Arthur H. Lefebvre and K. V. L. Rao 1982; G. Gordon Adkins, Jr. and Leroy H. Smith, Jr. 1981; Mark S. Darlow, Anthony J. Smalley and Alexander G. Parkinson 1980; Arthur Schaffler 1979; Frank J. Wiesner 1978; I. J. Day, E. M. Greitzer and R. M. Cumpsty 1977; J. P. Gostelow 1976; Edward M. Greitzer 1975; G. L. Commerford and Lynn E. Snyder 1974; John Moore 1973; F. B. Metzger and D. B. Hanson 1972; H. A. Harmon, A. A. Mikolajczak and D. Marchant 1971; Carlyle Reid 1970; O. E. Balje 1969; Arthur D. Bernstein, William H. Heiser and Charles M. Hevenor 1968.

GUSTUS L. LARSON MEMORIAL AWARD ... conferred upon an engineering graduate who has demonstrated outstanding achievement in engineering within ten to twenty years following graduation. This award, established in 1974 by ASME and Pi Tau Sigma, consists of $1,000 honorarium, a certificate and an expense supplement.
Recipients: Van C. Mow 1987; Bharat Bhushan 1986; Klaus-Jurgen Bathe 1985; Robert A. Altenkirch 1984; R. Byron Pipes 1983; Melvyn C. Branch 1982; Terry E. Shoup 1981; Arthur G. Erdman 1980; Gerald R. Seemann 1979; Philip H. Francis 1978; Nam P. Suh 1977; John G. Bollinger 1976; Chang-Lin Tien 1975.

HENRY HESS AWARD ... established in 1914 to honor an original technical paper by a member of ASME who was not yet 31 at the time the paper was submitted for presentation and publication. The award consists of an honorarium of $250, an engrossed certificate and a travel expense supplement.

Recipients: Steven W. Shaw 1986; Richard C. Benson 1984; Bharat Bhushan 1980; Krishna C. Gupta 1979; Maria Comninou 1978; Robert J. Hannemann 1977; Gopal Das Gupta 1976; Lambert B. Freund 1974; Hazem A. Ezzat and Steve M. Rohde 1973; D. C. Gakenheimer 1972; T. L. Geers 1970; James R. Rice 1969.

HENRY R. WORTHINGTON MEDAL . . . bestowed for eminent achievement in the field of pumping machinery. Such achievment may be, for example, in the areas of research, development, design, innovation, management, education or literature. The award, established in 1980, consists of a bronze medal, certificate and $1,000 honorarium.
Recipients: John E. Miller 1987; Warren H. Fraser 1986; Samuel L. Collier 1985; Harold H. Anderson 1984; Calvin A. Gongwer 1983; Allan J. Acosta 1982; Warren G. Whippen 1981; Igor J. Karassik 1980.

HOLLEY MEDAL . . . established in 1924 by George I. Rockwood, Hon. M. ASME, to be bestowed for some great and unique act of an engineering nature that has accomplished a great and timely public benefit. May be bestowed on more than one individual. The award consists of a gold medal, engrossed certificate and lapel button.
Recipients: Robert J. Moffat 1987; Wilson Greatbatch 1986; John V. Atanasoff 1985; Jack St. Clair Kilby 1982; Soichiro Honda 1980; Bruce G. Collipp and Douwe deVries 1979; J. David Margerum 1977; Emmett N. Leith and Juris Upatnieks 1976; George M. Grover 1975; Harold E. Edgerton and Kenneth J. Germeshausen 1973; Willis J. Whitfield 1969; Chester F. Carlson 1968.

HOOVER MEDAL . . . established in 1929 in recognition and appreciation of those principles and ideals of civic obligation and public service exemplified by the life and work of Herbert Hoover. The medal is conferred upon an engineer for great, unselfish, non-technical services to his fellow man. The award is jointly administered by ASCE, AIME, ASME and IEEE.
Recipients: Lawrence P. Grayson 1986; Robert C. West 1985; Kenneth A. Roe 1984; Joseph J. Jacobs 1983; Michael A. Halbouty 1982; Arnold O. Beckman 1981; Stephen D. Bechtel, Jr. 1980; Charles M. Brinckerhoff 1979; Donald C. Burnham 1978; Peter C. Goldmark 1977; James B. Fisk 1976; James Boyd 1975; David Packard 1974; William J. Hedley 1973; Frederick R. Kappel 1972; Luis A. Ferre 1971; John Erik Jonsson 1970; Edgar F. Kaiser 1969; Sir Harold Hartley 1968.

J. HALL TAYLOR MEDAL . . . for distinguished service or eminent achievement in the field of Codes and Standards pertaining to the broad fields of piping and pressure vessels. The scope shall include contributions to technical advancement and administration. The award was established in 1965 and consists of a gold medal and certificate.
Recipients: Stephen A. Bergman 1987; James R. Farr 1986; Robert C. Griffin 1985; William D'Orville Doty 1984; Lowell L. Elder 1983; George V. Smith 1982; Robert J. Cepluch 1981; Paul M. Brister 1980; John F. Harvey 1979; Adolph O. Schaefer 1978; James S. Clark and Raymond R. Maccary 1977; Walter H. Davidson, Frederic A. Hough, Joe J. King, Burton T. Mast and Andrew J. Shoup 1975; Jean E. Lattan 1974; John Dalton Mattimore 1973; William Rolfe Gall 1972; James M. Guy 1971; Bernard F. Langer 1970; Everett O. Waters 1969; Max B. Higgins 1968.

JAMES HARRY POTTER GOLD MEDAL . . . awarded in recognition of eminent achievement or distinguished service in the appreciation of the science of thermodynamics in mechanical engineering through teaching, appreciation or utilization of thermodynamic principles in research, development and design in mechanical engineering. Established in 1980, the award consists of a medal, certificate and $1,000 honorarium.
Recipients: Jack P. Holman 1986; Warren G. Giedt 1985; Robert H. Page 1984; Kenneth C. Cotton 1983; Paul Leung 1982; Joseph Kestin 1981; Alexander Louis London 1980.

JAMES N. LANDIS MEDAL . . . awarded for outstanding personal performance related to designing, constructing or managing the operation of major steam-powered electric stations using nuclear or fossil fuels, coupled with personal leadership in some humanitarian pursuit. The Award, established in 1977, consists of a bronze medal, certificate, $1,000 honorarium and a travel expense supplement.
Recipients: Warren H. Owen 1987; John W. Turk, Jr. 1986; Mendall H. Long 1984; Byron Lee, Jr. 1983; Huberto R. Platz 1982; Vincent S. Boyer 1981; Harvey F. Brush 1980; William E. Hopkins 1978; James N. Landis 1977.

MACHINE DESIGN AWARD . . . in recognition of eminent achievement or distinguished service in application, research, development, or teaching of machine design. The award was established in 1958 and consists of a bronze plaque and engrossed certificate.
Recipients: Gerald G. Lowen 1987; Atmaram H. Soni 1986; Joseph E. Shigley 1985; Bernard Roth 1984; Edward J. Wellauer 1983; Delbert Tesar 1982; Henry O. Fuchs 1981; Merhyle F. Spotts 1980; Robert R. Slaymaker 1979; Ali A. Seireg 1978; Matthew M. Kuts 1977; Charles W. Radcliffe 1976; George N. Sandor 1975; Allan S. Hall, Jr. 1974; Ferdinand Freudenstein 1972; Walter L. Starkey 1971; Reynold B. Johnson 1970; Eugene I. Radzimovsky 1969; C. Walton Musser 1968.

THE MAX JAKOB MEMORIAL AWARD . . . awarded for eminent achievement or distinguished service in the area of heat transfer. This award is jointly administered by ASME and AIChE.
Recipients: Raymond Viskanta 1986; Frank Kreith 1985; Alexander Louis London 1984; Bei Tse Chao 1983; Simon Ostrach 1982; Ralph A. Seban 1981; Stuart W. Churchill 1980; Niichi Nishiwaki 1979; D. Brian Spalding 1978; Ephriam M. Sparrow 1977; Robert G. Deissler 1976; Peter Grassmann 1975; Ulrich Grigull 1974; Karl A. Gardner 1973; James W. Westwater 1972; Warren M. Rohsenow 1971; S. S. Kutaleladz 1970; Shiro Nukiyama 1969.

MAYO D. HERSEY AWARD . . . awarded to an individual in recognition of outstanding and continued contributions to the field of lubrication science and engineering. The award was established in 1965 and consists of a bronze plaque.
Recipients: Marshall B. Peterson 1987; Ward O. Winer 1986; Ernest Rabinowicz 1985; Frederick F. Ling 1984; Donald F. Hays 1983; E. Erwin Klaus 1982; Edmond E. Bisson 1981; Nicolae Tipei 1980; Duncan Dowson 1979; Edward A. Saibel 1978; Robert L. Johnson 1977; John Boyd 1976; Arthur F. Underwood 1975; David Tabor 1974; Donald F. Wilcock 1973; Sydney J. Needs 1972; Dudley Dean Fuller 1971; Merrell Robert Fenske 1970; William A. Zisman 1969; Ragner Holm 1968.

MELVILLE MEDAL . . . best current original paper (not published elsewhere) which has been presented before ASME, or approved for publication by ASME during the two calendar years preceding the year of award. First awarded in 1927, this honor consists of a gold medal, and honorarium and engrossed certificate.
Recipients: Dennis L. Siebers, Robert J. Moffat and Richard G. Schwind 1987; Robert W. Bjorge and Peter Griffith 1986; Lung-Wen Tsai and Alexander P. Morgan 1985; Michael F. Blair 1984; Albert M. C. Chan and Sanjoy Banerjee 1983; Van C. Mow, Steve C. Kuei, W. Michael Lai and Cecil G. Armstrong 1982; Kyung-Suk Kim and Rodney J. Clifton 1981; Raji Chandran, John C. Chen and Fred W. Staub 1980; Thomas J. R. Hughes and W. K. Liu 1979; Donald E. Negrelli. John R. Lloyd and Jerome L. Novotry 1978; Eugene F. Fichter and Kenneth H. Hunt 1977; Bernard J. Hamrock and Duncan Dowson 1976; David M. Sanborn, A. V. Turchina and Ward O. Winer 1975; V. H. Arakeri and Allan J. Acosta 1974; H. W. O'Connor and A. S. Weinsten 1972; Thomas Slot 1971; J. William Holl and A. L. Kornhauser 1970; Leon R. Glicksman 1969; Yian-Nian Chen 1968.

PERCY NICHOLLS AWARD . . . for notable scientific or industrial achievement on the field of solid fuels. The award established in 1942, is jointly administered by the Fuels Division of ASME and the AIME Coal Division.
Recipients: Gordon H. Gronhovd 1986; David A. Zeeger 1985; George K. Lee 1984; E. Minor Pace 1983; James R. Jones 1982; Jack A. Simone 1981; George W. Land 1980; William N. Poundstone 1979; Albert F. Duzy 1978; H. Beecher Charmbury 1977; Richard B. Engdahl 1976; George P. Cooper 1974; Charles H. Sawyer 1972; George Ernst Keller 1971; Richard C. Carey 1970; David R. Mitchell 1969; W. T. Reid 1968.

PI TAU SIGMA GOLD MEDAL . . . established in 1938 as a suitable means of recognizing the accomplishments of outstanding young mechanical engineers. Achievement may be all or in part in any field, including industrial, educational, political, research, civic or artistic. It is awarded within ten years after graduation. The award consists of a gold medal, an honorarium of $1,000, a certificate and a travel expense supplement.

Recipients: David L. McDowell 1987; Dimos Poulikakos 1986; Wing Kam Liu 1985; Michael R. Muller 1984; Polychronis-Thomas Demetriou Spanos 1982; Doyle D. Knight 1980; David A. Peters 1978; Richard E. Lovejoy 1977; John S. Walker 1976; Ted B. Belytschko 1975; Jace W. Nunziato 1974; Christian Ernst and Georg Przirembel 1973; John F. Stephens, III 1972; James R. Rice 1971; Richard Elwood Barrett 1970; Henry K. Newhall 1969; Randall F. Barron 1968.

PRESSURE VESSEL AND PIPING AWARD . . . bestowed for outstanding contributions in the field of pressure vessel and piping technology including, but not limited to, research, development, teaching and significant advancements of the state-of-the-art. The award, established in 1980, consists of a bronze medal and certificate.

Recipients: David H. Pai 1987; Everett C. Rodabaugh 1986; John F. Harvey 1985; Adolph O. Schaeffer 1984; William E. Cooper 1983; Irwin Berman 1982; Gunther P. Eschenbrenner 1981; Dana Young 1980.

PRIME MOVERS COMMITTEE AWARD . . . recognizes outstanding contribution to the literature of thermal electric station practice or equipment. It was established in 1954 and consists of an engrossed certificate.

Recipients: Peter Schofield and David A. Lantzy 1987; William J. Sumner, James H. Vogan and Robert J. Lindinger 1986; David H. Cooke 1985; Kenneth C. Cotton, Harris S. Shaffer, Thomas H. McCloskey and Robert M. Boettcher 1984; Paul G. Albert and William J. Sumner 1983; Erich Raask 1982; Bezalel Bornstein and Kenneth C. Cotton 1981; Bezalel Bornstein and Kenneth C. Cotton 1981; Heinz E. Termuehlen 1980; Henry E. Lokay, D. G. Ramey and W. R. Brose 1979; M. Araoka, J. D. Fox, H. Haneda, K. Setoguchi and W. F. Siddall 1977; Hans-Gunter Haddenhorst, Wolfgang Mattick, Z. Stanley Stys and Otto Weber 1976; Karl A. Gulbrand and Paul Leung 1975; B. Bornstein and Paul Leung 1974; D. W. Rahoi, R. C. Scarberry, J. R. Crum and P. E. Morris 1973; Paul Leung and G. S. Liao 1972; Paul Leung and Raymond E. Moore 1971; Paul Leung and Raymond E. Moore 1970; Charles L. Burton and Paul Goldstein 1969; G. N. Stone and A. J. Clarke 1968.

R. TOM SAWYER AWARD . . . bestowed on an individual who has made important contributions to the advancement of the purpose of the gas turbine industry and to the Gas Turbine Division of ASME. The award was established in 1972.

Recipients: Leroy H. Smith, Jr. 1987; Elvie L. Smith 1986; Anselm Franz 1985; Arthur H. Lefebvre 1984; Sven-Olof Kronogard 1983; R. Noel Penny 1982; Thomas E. Stott, Jr. 1981; Ralph L. Boyer 1980; Sam B. Williams 1979; Sir Frank Whittle 1978; Alexander L. London 1977; Curt Keller 1976; Bruce O. Buckland 1975; Waheeb Rizk 1974; John W. Sawyer 3 1973; R. Tom Sawyer 1972.

RUFUS OLDENBURGER MEDAL . . . established in 1968 by the ASME Automatic Control Division (now the Dynamic Systems and Control Division) to honor Rufus Oldenburger for his distinctive achievements in the field of automatic control and his service to the Society and the Division. The medal is bestowed with a certificate to recognize outstanding contributions to the field of automatic control.

Recipients: Walter R. Evans 1987; Eliahu I. Jury 1986; Karl Astrom 1985; Herbert H. Richardson 1984; J. Lowen Shearer 1983; Bernard Friedland 1982; Shih-Ying Lee 1981; Arthur E. Bryson, Jr. 1980; Henry M. Paynter 1979; Yasundo Takahashi 1978; Gordon S. Brown and Harold L. Hazen 1977; Rudolf Emil Kalman 1976; Hendrik W. Bode and Harry Nyquist 1975; Herbert W. Ziebolz 1974; Clesson E. Mason 1973; Albert Williams, Jr. 1972; Charles S. Draper 1971; John R. Ragazzini 1970; Nathaniel B. Nichols 1969; Rufus Oldenburger 1968.

SPIRIT OF ST. LOUIS MEDAL . . . to recognize meritorious service in the advancement of aeronautics and astronautics. Established in 1929, this award consists of a gold medal and engrossed certificate.

Recipients: Elbert L. Rutan 1987; Bruce McCandless II 1986; Kurt H. Hohenemser 1985; Charles Stark Draper 1984; John W. Young 1983; Frank N. Piasecki 1982; Edgar M. Cortright, Jr. 1981; Michael Collins 1980; Sir Freddie Laker 1979; Paul B. Mac Cready, Jr. 1978; George D. McLean 1977; Abe Silverstein 1974; John F. Yardley 1973; Neil A. Armstrong 1972; Ralph L. Creel 1971; Clarence L. Johnson 1970; G. Merritt Preston 1969; George S. Moore 1968.

TIMOSHENKO MEDAL . . . in recognition of distinguished contributions to applied mechanics. It consists of a bronze medal and engrossed certificate, and was established in 1957.

Recipients: Ronald S. Rivlin 1987; George R. Irwin 1986; Eli Sternberg 1985; Joseph B. Keller 1984; Daniel C. Drucker 1983; John W. Miles 1982; John H. Argyris 1981; Paul M. Naghdi 1980; Jerald L. Ericksen 1979; George F. Carrier 1978; John D. Eshelby 1977; Erastus Henry Lee 1976; Chia-Chiao Lin 1975; Albert E. Green 1974; Eric Reissner 1973; Jacob Pieter Den Hartog 1972; Howard Wilson Emmons 1971; James Johnston Stoker 1970; Jakob Ackeret 1969; Warner T. Koitzer 1968.

WORCESTER REED WARNER MEDAL . . . awarded for outstanding contribution to the permanent literature of engineering. Established in 1930, the award consists of a gold medal, and honorarium of $1,000 and engrossed certificate.

Recipients: Jack P. Holman 1987; Ephriam M. Sparrow 1986; Richard H. Gallagher 1985; Yuan-Cheng Fung 1984; Allan D. Kraus 1983; Herbert Kolsky 1982; Frank Kreith 1981; Olgierd C. Lienkiewicz 1980; Darle W. Dudley 1979; James H. Potter 1978; Joseph E. Shigley 1977; Dennis G. Shepherd 1976; Philip G. Hodge, Jr. 1975; Victor L. Streeter 1974; Max Mark Frocht 1973; Burgess H. Jennings 1972; Stephen H. Crandall 1971; Wilhelm Flugge 1970; Hans W. Liepmann 1969; Merhyle F. Spotts 1968.

SOICHIRO HONDA MEDAL . . . recognizes an individual for an outstanding achievement or a series of significant engineering contributions in developing improvements in the field of personal transportation.

Recipients: Felix Wankel 1987; Lloyd L. Withrow 1986; Shoichi Furuhama 1985; John P. Stapp 1984.

"OLD GUARD" PRIZES . . . awarded annually for the best three presentations of technical papers at the National Contest for Student Members at the ASME Winter Annual Meeting. Supported by contributions from ASME dues exempt members who have reached the age of 65 and have retired. The Prize was established in 1956 and expanded to include second- and third-place winners in 1981.

Recipients: Christopher Della Corte, Thomas Cavallaro and Daniel M. Browning 1986; Ed Rissberger, Michael T. Nelson and Brian D. Berthold 1985; Jeffrey McAlister, John DiMarco and Daniel B. Grandmont 1984; Jonathan R. Wiley, Richard F. Beaufort and Joseph R. Olivier 1983; Gary F. St. Onge, Johnathon I. Macy and Douglas R. Watson 1982; Dan J. Schmitt 1981; John J. Marsal 1980; Zoe D. Kececioglu 1979; Jan D. Dozier 1978; Pauline B. Cramer 1977; Paul E. Hollis 1976; Steven R. Bussolari 1975; Gary L. Smith 1974; Steven H. Blossom and E. J. Strande 1973; Stanley W. Blossom 1972; J. L. Lee 1971; Joseph R. Titone 1970; Walter H. Peters, III 1969; Maurice H. Bunn 1968.

PERFORMANCE TEST CODES MEDAL . . . awarded for outstanding contributions to the development and promotion of ASME Performance Test Codes, including the Supplements on Instruments and Apparatus. The award was established in 1981.

Recipients: P.H. "Pete" Knowlton, Jr. 1987; John H. Fernandes 1986; James W. Murdock 1985; William G. McLean and Kenneth C. Cotton 1984.

ARTHUR L. WILLISTON MEDAL . . . given for the best paper submitted in the Williston Award Contest. The medal was established in 1954.

Recipients: Thomas C. Davis 1987; Stephen J. Schoonmaker 1986; Henry M. Quillian III 1985; Eddie E. Ferrer 1984; Max R. Casada 1983; John H. Pilarski 1982; Charles S. Macaulay 1980;

Steven E. Stephens 1979; Jitendra S. Goela 1978; Harry W. Groot 1977; Enud David Laska 1976; James J. Callas 1974; Frank H. Roubleau, Jr. 1973; Dennis L. Sandberg 1972; James A. Willms 1971; Steven H. Carlson 1970; Arlo Fossum 1969; Frank A. Ralbovsky 1968.

CHARLES T. MAIN STUDENT SECTION AWARD ... recognizes at the national level, Student Members whose leadership and service qualities have contributed to the program and operation of a Student Section of the Society. The Award was established in 1919.
Recipients: Keith G. Benedict and Mark V. Martin 1987; Mark A. Meili and Andre L. Boehman 1986; Anne Bazan and Scott E. Cooper, Jr. 1985; Linda-Marie Hubbard and Douglas L. Wahl 1984; Stephen A. Hight 1983; Brenda B. Elarbee 1982; Scott H. Buehrer 1981; Russell S. Colvin 1980; Richard A. Ferraro 1979; Emily Earle 1978; Charles S. Tamarin 1977; Scott Elliot Baker 1976; Adrian P. Villa 1974; Gary Patrick Pezall 1973; Harold Chapin Lowe 1972; James M. Singleton 1971; Steve H. Woodard 1970; Terry Dean Schmidt 1968.

H. R. LISSNER AWARD ... established in honor of H.R. Lissner for his pioneering contributions in biomechanical research. This award is bestowed for outstanding accomplishments in the area of bioengineering. The award consists of a $1000 honorarium, a certificate and travel expense.

PER BRUEL GOLD MEDAL FOR NOISE CONTROL AND ACOUSTICS ... The award is given in recognition of eminent achievement and extraordinary merit in the field of Noise Control and Acoustics. The achievement must include useful applications of the principles of Noise Control and Acoustics to the art and science of mechanical engineering. The award consists of a $1000 honorarium, a certificate and travel expense.

SAFETY CODES AND STANDARDS MEDAL ... This award is to be presented to one or more individuals who have contributed to the enhancement of public safety through the development and promotion of ASME Codes and Standards or the ASME Safety Accreditation Activity. The award shall consist of a $1000 honorarium, medal and certificate.

AMERICAN SOCIETY OF NAVAL ENGINEERS, INC. (ASNE)
1452 Duke St., Alexandria, Virginia 22314
Captain James L. McVoy, USN (Ret), Executive Director
PRESIDENTS ... Dr. Alfred Skolnick 1985-1989; RAdm. James K. Nunnelley USN 1983-1985; Richard C. Fay 1981-83; VAdm. C. R. Bryan, USN 1979-81; Ivan Monk 1978-79; RAdm. Kenneth E. Wilson, Jr., USN 1977-78; John J. Nachtsheim 1976-77; RAdm. Randolph W. King, USN 1975-76; Mr. Robert Taggart 1974-75; RAdm. David H. Jackson, USN 1973-74; VAdm. Robert C. Gooding, USN 1972-73.

AWARDS
HAROLD E. SAUNDERS AWARD ... "Given annually to that United States citizen who has *demonstrated* productivity, growth, and outstanding accomplishment in the Field of Naval Engineering over the years, with ultimate wide recognition by peers as a leader in the field and of such prestige as to merit acclamation by the Naval Engineering Community."
Recipients: Dr. Reuven Leopold 1986; Capt. Willard F. Searle, Jr. USN (Ret.) 1985; RAdm. Wayne E. Meyer USN 1984; Robert Targgart 1983; RAdm Nathan Sonenshein USN 1982; VAdm Ellis L. Perry USCG 1981; RAdm. John D. Bulkeley, USN (Ret) 1980; Capt. Harry A. Jackson, USN (Ret) 1979; John C. Niedermair 1978; Almon Archie Johnson 1977.

GOLD MEDAL AWARD ... awarded annually to recognize the individual having made the most significant contribution to Naval Engineering during or culminating in the five year period ending in the year under consideration. If the effort has not been consummated, it must have progressed to the stage where significant accomplishment can be shown.
Recipients: Capt. Millard S. Firebaugh, USN 1986; Larry J. Argiro 1985; Peter M. Palermo 1984; Capt. Charles H. Piersall, Jr. 1983; Capt. Brice D. Inman USN 1982; Capt. James W. Kehoe Jr. USN 1981; Capt. Alfred Skolnick, USN 1980; Capt. Henry Cox, USN 1979; Capt. O. B. Nelson, USN Ret 1978; RAdm. Elmer T. Westfall, USN. Ret. 1977; RAdm. Wayne E. Meyer 1976; Mr. James L. Mills, Jr. 1975; Mr. Donald L. Ream 1974; Mr. William M. Ellsworth 1973; Capt. Robert H. Wertheim, USN 1972; CDR Walter J. Eager, (CEC), USN 1971; Mr. Robert A. Meyers 1970; CDR Roderick M. White, USCG 1969; Mr. Lynwood A. Cosby 1968.

SOLBERG ... awarded annually to recognize the United States citizen having made the most signifcant contribution to Naval Engineering through personal research carried out during or culminating in the three year period under consideration. Evidence of personal involvement in the research and a specific assessment of the significance of nominee's direct contribution must be shown.
Recipients: Louis D. Chirillo, Jr. 1985; Jeffrey E. Beach 1984; Howard O. Stevens 1983; A. Erich Baitis 1982; Richard C. Swenson 1981; John Mittleman 1980; Gerald M. Mayer 1979; Roderick D. Turnage 1977; Herbert V. Hitney 1976; Allan G. Ford 1975; Dr. Henry F. Taylor 1974; John W. Henry, IV 1972; Donald L. Folds 1971; George Sorkin 1970; Walter L. Clearwaters 1969; Thaddeus G. Bell 1968.

JIMMIE HAMILTON ... presented annually for the best original paper published in the Naval Engineers Journal.
Recipients: Sigurdur Ingvason, Donald N. McCallum & Capt. Gilbert L. Kraine, USCG (Ret.) 1986; Lcdr. Gregory J. White & Dr. Bilal M. Ayyub 1985; John D. Adams and Walter F. Beverly III 1984; Capt. James Kehoe Jr. USN, Kenneth S. Brower and Edward N. Comstock 1983; J. V. Jolliff and D. L. Greene 1982; R. J. Biondi and E. Kruger 1981; Carl T. Zovko 1980; Capt. James L. McVoy, USN 1979; Lt. Cdr. Stephen R. Olson 1978; George F. Wilhelmi and Henry W. Schab 1977; Jack W. Abbott and Charles M. Atchison 1976; Mr. Peter A. Gale 1975; LCdr W. Lawrence Fulton II, USN 1974; LCdr Clark Graham, USN 1973; Mr. Reuven Leopold, Capt. E. C. Svendsen, USN (Ret) and Mr. Harvey Kloehn 1972; Marcel J. E. Golay 1971; John J. Nachtsheim and Cdr L. Dennis Ballou, USN 1970; LCdr Jack W. Lewis, USCG 1969; Lt. Eugene A. Silva, USNR 1968.

THE FRANK G. LAW AWARD ... awarded, when appropriate, to a member in recognition of outstanding achievements and long term significant contributions resulting in one of the following: (1) Betterment of Society operations (2) Advancement of the objectives of the Society (3) Increased professional stature of the Society. Evidence of personal involvement in the management, administration and financial affairs of the Society measured by achievements while serving as a member, officer, committee chairperson or member. The nominee must have demonstrated leadership, selfless dedication, entrepreneurship, and personal commitment in support of Society interests and goals over a number of years.
Recipients: RAdm. Frank C. Jones, USN (Ret.) 1986; Robert J. Scott 1985; Edward M. MacCutcheon 1984; Daniel J. Weiler 1983; Frank J. Smollon 1982; Dr. James V. Jolliff 1981; Capt. Frank G. Law, USN (Ret) 1980.

AMERICAN SOCIETY FOR PHOTOGRAMMETRY AND REMOTE SENSING (ASPRS)
210 Little Falls St., Falls Church, Virginia 22046
William D. French, Executive Director
PRESIDENTS ... John J. Graham 1987; Alan R. Stevens 1986; Tamsin G. Barnes 1985; Roy A. Welch 1984; William G. Hemple 1983; Alan C. Bock 1982; George J. M. Zarzycki 1981; Rex R. McHail 1980; Francis H. Moffitt 1979; Clifford J. Crandall 1978; Vern W. Cartwright 1977; Hugh B. Loving 1976; John W. Wickham 1975; Joe E. Steakley 1974; Marshall S. Wright, Jr. 1973.

AWARDS
ALAN GORDON MEMORIAL AWARD ... is to encourage and commend the individual(s) who contribute to significant achievements in Remote Sensing and Photo Interpretation.

Recipients: Thomas E. Avery 1986; Joseph K. Berry 1985; Frederick J. Doyle 1984; Robert N. Colwell 1983; Philip N. Slater 1982; Floyd F. Sabins, Jr. 1981; Dr. James R. Anderson 1980; Dr. Roger M. Hoffner 1979; Dr. Richard S. Williams, Jr.-William D. Carter 1978; Dr. Robert B. McEwen 1977.

THE AUTOMETRIC AWARD . . . to recognize achievement in the field of photographic interpretation through special acknowledgement of superior publications on the various aspects of imagery interpretation.
Recipients: William A. Befort 1987; C. J. Tucker, C. L. VanPraet, M. J. Sharman, G. Van lttersum 1986; Eric P. Crist and Richard Cicone 1984; John F. McCauley, G. G. Schaber, C. S. Breed, M. J. Grolier, B. Issawi, C. Elachi, R. Blom and C. V. Haynes 1983; Alexander F. H. Goetz and Lawrence C. Rowan 1982; Charles Elachi 1981; John Welsted 1980; Hugh R. Balkwill 1979; Nicholas M. Short, Paul D. Lowman, Jr., Stanley C. Freden and William A. Finch, Jr. 1978; Nenad Spoljaric, R. R. Jordan and R. E. Sheridan 1977; Robert G. Reeves 1976; Kalman N. Visy 1975; Norman L. Fritz, Malcolm R. Specht and D. Needler 1974; Kenneth R. Piech and John E. Walker 1973; Edward F. Yost and Sondra Wenderoth 1972; Robert L. Mairs 1971; Frank B. Silvestro 1970; H. E. Holt 1969; Norman L. Fritz 1968.

THE TALBERT ABRAMS AWARD . . . to encourage the authorship and recording of current, historical, engineering and scientific developments in photogrammetry.
Recipients: Sabry El-Hakim 1986; Armin W. Gruen 1985; John C. Trinder 1984; Clive S. Fraser 1983; P. Douglas Carman 1982; Salem E. Masry 1981; Dr. Urho A. Rauhala 1980; Vladimir Kratky 1979; Milosh Benesh 1978; Raymond J. Helmering 1977; David F. Maune 1976; Vladimir Kratky 1975; Dean C. Merchant 1974; Maurice E. Lafferty 1973; Eugene E. Derenyi 1972; Roy A. Welch 1971; Kam W. Wong 1970; Gomer T. McNeil 1969; Duane Brown 1968.

LUIS STRUCK AWARD . . . to stimulate Pan American understanding in the field of photogrammetry.
Recipients: Mauricio Aroya Figueroa 1986; Hernan Rivera H. 1985; Bulmaro Cabrera Ruiz 1984; Luis E. Miranda V 1983; Rupert B. Southard 1982; Dr. Placidino Fagundes 1981; Daniel Gut 1980; Jack E. Staples 1979; Robert H. Lyddan 1978; J. Alberto Villasana 1977; Dr. Samuel G. Gamble 1976; Jose A. Saenz 1975; Charles H. Andregg 1974; Teodor J. Blachut 1973; Dr. Arch C. Gerlach 1972; David Landen 1971; Ing. Pablo Arnoldo Guzman 1970; Frederick O. Diercks 1969; George D. Whitmore 1968.

THE PHOTOGRAMMETRIC AWARD (FAIRCHILD) . . . to stimulate the development of the art of aerial photogrammetry in the United States.
Recipients: Lawrence W. Fritz 1987; William C. Mahoney 1986; Zarko Jaksic 1985; Chester C. Slama 1984; Vladimir Kratky 1983; Roy A. Welch 1982; Duane C. Brown 1981; Clifford W. Greve 1980; Michael A. Crombie 1979; Atef A. Elassal 1978; Everett L. Merritt 1977; Edward M. Mikhail 1976; H. Dell Foster 1975; Housam M. Karara 1974; Teodor J. Blachut 1973; Gomer T. McNeil 1972; Duane Brown 1971; Sidney Bertram 1970; Frederick J. Doyle 1969; Gerhardus H. Schut 1968.

PRESIDENT'S AWARD FOR PRACTICAL PAPERS . . . to encourage and commend the individual(s) who publish papers of practical or applied value in the Journal of the American Society of Photogrammetry, *Photogrammetric Engineering and Remote Sensing.*
Recipients: Harold D. Moore & Alan F. Gregory 1987; J. C. Eidenshink 1986; Stephen A. Mundy 1985; B. E. Frazier 1984; Herman H. Arp 1983; Donald Karns 1982; Milton L. Keene 1981; William P. MacConnell 1980; Jon S. Beazley 1979.

THE BAUSCH AND LOMB PHOTOGRAMMETRIC AWARD . . . to stimulate an interest in photogrammetry in college students who display outstanding ability and interest in photogrammetry.
Recipients: Byung-Guk Kim and Daniel C. Oimoen 1987; James Lee and Rudolph J. Meijer 1985; Albert K. Chong and Barry Hill 1984; Charles D. Ghilani and Hal B. Lane III 1983; Bon A. DeWitt and Michael A. Heck 1982; Michael A. Martin 1981; James C. Storey and David C. Goodrich 1980.

THE WILD HEERBRUGG PHOTOGRAMMETRIC FELLOWSHIP . . . to encourage qualified candidates to pursue graduate education in photogrammetry and to promote the development of photogrammetric science.
Recipients: Jeffrey Kretsch 1987; Charles D. Ghilani 1986; Horst A. Beyer 1985; Fidel Calismo Paderes, Jr. 1984; Bon A. DeWitt 1983; Sean Curry 1982; Michael A. Chapman 1981; Jolyon Douglas Thurgood 1980.

AMERICAN SOCIETY OF SAFETY ENGINEERS (ASSE)

1800 E. Oakton St. Des Plaines, IL 60018-2187
Judy T. Neel, CAE, Executive Director

PRESIDENTS . . . Charles R. Dancer 1987-88; William T. Nebraska 1986-87; Delmar E. Tally 1985-86; Thomas J. Reilly 1984-85; Harry A. Partlow 1983-84; James D. Hoag 1982-83; Donald J. Eckenfelder 1981-82; John L. Russell 1980-81; Russell DeReamer 1979-80; Marion L. Jones 1978-79; Dr. William E. Tarrants 1977-78; Theron T. Pinder 1976-77; David V. MacCollum 1975-76; Eugene L. Newman 1974-75; Michael Krikorian 1973-74; B. Gawain Bonner 1972-73.

AWARDS

EDGAR MONSANTO QUEENY SAFETY PROFESSIONAL OF THE YEAR AWARD . . . to achieve public and industry awareness and recognition for the American Society of Safety Engineers and pay tribute to the outstanding accomplishments of Society professionals engaged in the broad field of occupational safety and health. The winner will receive a $1,000 check from Monsanto and an impressive large engraved statuette for permanent display in home or office. These will be presented annually at the Society's Professional Development Conference in June, following judging in the spring.
Recipients: James A. Brodrick, 1987; Jerry L. Williams, 1986; Kim Anderson 1985; James W. Smirles 1984; Frank Lightfoot 1983; John E. Russell 1982; Charles R. Dancer 1981; B. Gawain Bonner 1980.

PROFESSIONAL PAPER AWARDS . . . to extend recognition to the authors of outstanding professional papers published in PROFESSIONAL SAFETY, to stimulate the acquistion of knowledge and its interchange among members of the safety profession. This program is co-sponsored by Veterans of Safety, a national organization of veteran safety professionals, and American Society of Safety Engineers, a professional individual member society dedicated to the advancement of the occupational safety and health profession and the well being and professional development of its members. Three awards are presented annually: First prize: $500 and an impressive wall plaque; Second prize: $400 and a plaque. Third prize: $300 and a plaque.
Recipients: 1st Kenneth Sawyer, 2nd Michael W. Mitchell, 3rd Jerry D. Ramsey 1987; 1st George A. Peters; 2nd James Capps, John D. Pearson and 3rd Henry T. Miller 1981; 1st Charles F. Dalziel, 2nd Charles V. Culbertson and 3rd Jack B. ReVelle 1979.

SCRIVENER AWARD . . . given to the authors of outstanding articles published in non-ASSE but safety or health-related journals or magazines. Grand award: $500.
Recipients: Dr. Ted Ferry 1987; Margaret Carroll 1986; Roman Diekemper 1985; Dr. Ted Ferry 1984.

TENNECO SCHOLARSHIP AWARD . . . $1,000 and handsome plaque given to outstanding student section as judged by Society Awards and Honors Committee -- Funds given to section for subsequent awarding of scholarship to safety/health students deserving of the university's choosing in the name of the section. Honorarium provided by Tenneco Inc.
Recipients: Keene State College (N.H.) Student Section 1987; Texas A&M Univ, 1986; Murray State U., Kentucky, 1985; Indiana Univ. of PA 1984, 1983.

JOHN E. ANDERSON STUDENT OF THE YEAR AWARD . . . sponsored by Maryland Casualty Company . . . recognizes the outstanding ASSE Student Member of the Year on the basis of academic performance, extra-curricular campus and community activities, participation in Society activities, etc.

Recipients: Daniel Bullock, National U., San Diego, 1987; Henry R. Moore, Jr., Oklahoma State University 1985; Hector Teran, Northern Illinois University 1984; Daryl Allegree, Central Missouri State University 1983.

MARSH & MCLENNAN STUDENT PAPER AWARDS ... recognizes outstanding editorial efforts of safety students annually; three prizes given ($500 for first place, $400 for second, $300 for third) by Marsh & McLennan, Inc.

Recipients: Mark Patterson, Indiana U. of Pa., first place, Nicholas Kacbur, Indiana U. of Pa., second place, Keri Holmes, Iowa State U, third place, 1987; Kim Johnson, first place, Auburn University, Alabama; Timothy Kitchen, second place, Iowa State University; William Mitzel, third place, Central Washington College, Ellensburg 1985.

AMERICAN WATER WORKS ASSOCIATION (AWWA)

6666 West Quincy Avenue, Denver, Colorado 80235
Paul A. Scaulte, Executive Director

PRESIDENTS ... John H. Robinson 1987; Robert T. Chuck 1986; Richard Miller 1985; William H. Richardson 1984; Dr. William O. Lynch 1983; John H. Stacha 1982; Kenneth J. Miller 1981; J. B. Gilbert 1980; Donald K. Shine 1979; Curtis H. Stanton 1978; Robert R. Peters 1977; Chester A. Ring III 1976; Walter K. Morris 1975; Robert B. Hilbert 1974; George E. Symons 1973.

AWARDS

THE ALVIN PERCY BLACK RESEARCH AWARD ... established to give recognition of outstanding service by an individual for research work over an appreciable period of time resulting in notable contributions to water science and water works practice.

Recipients: Russell F. Christman 1987; Robert A. Baker 1985; Edwin E. Geldreich 1984; J. Edward Singley 1983; J. L. Cleasby 1982; J. M. Symons 1981; Dr. Richard L. Woodward 1977; Herbert E. Hudson, Jr. 1976; Kenneth E. Shull 1975; Thurston E. Larson 1972; Gordon G. Robeck 1970.

DISTINGUISHED PUBLIC SERVICE ... established in honor of Harry E. Jordan, secretary of the Association from 1936 to 1959, and is presented in recognition of distinguished public service outside the line of duty by an AWWA member.

Recipients: J. Douglas Kline 1987; David L. Crowson 1986; Robert P. Van Dyke 1985; Fredrick H. Elwell 1984; William H. Miller 1983; C. H. Stanton 1982; George C. Sopp 1980; Henry J. Graeser, Jr. 1974; Samuel S. Baxter 1972; Wendell R. LaDue 1969; N. T. Veatch 1968.

PUBLICATIONS AWARD ... the AWWA Publications Award is made annually to the member whose paper published in JOURNAL AWWA represents the most notable contribution to the science or practice of water utility development.

Recipients: Appiah Amirtharajah 1987; Paul V. Roberts & James A. Levy 1986; Brian A. Dempsey 1985; O. Thomas Love, Jr. 1984; Stephen J. Randtke 1983; A. A. Guerrera 1982; J. L. Patton & M. B. Horsley 1981; Dr. Avner Adin 1980; Aaron A. Rosen 1979; Dr. E. Robert Baumann 1978; Dr. Abel Wolman 1977; Parviz Amirhor 1976; William Hutchison 1975; Robert V. Phillips 1974; Dr. Albel Wolman 1973; James O. Jackson 1972; James M. Symons, J. Keith Carswell and Gordon G. Robeck 1971; Paul D. Haney 1970; Richard E. Morris Jr. 1969; Louis Koenig 1968.

WATER UTILITY HALL OF FAME ... established to honor the memory of those men, deceased for at least five years, who have made such significant contributions to community water supply, either through inventions, practices instituted, or technical articles and books, that their influence upon the industry is still felt. (Established in 1970.)

Recipients: Charles R. Cox and Herbert E. Hudson 1987; Louis R. Howson and Thurston E. Larson 1986; Wilfred F. Langelier and Abel Wolmann 1985; Fred Merryfield and Clifford H. Fore 1984; Morrison B. Cunningham and Robert Spurr Weston 1984; Arthur Moses Buswell and Gardner Stewart Williams 1982; Hardy Cross and N. T. Veatch 1981; Harold Eaton Babbit, James Joseph Doland, Nicholas Snowden Hill, John Huey Murdoch and William John Orchard 1980; C. T. Buttermeld, Victor M. Ehlers, Rudolph Hering, Daniel D. Jackson and George R. Spalding 1979; Thomas R. Camp, William Mulholland, Malcolm Pirnie, Marsden C. Smith and W. Victor Weir 1978; Charles H. Spaulding, John Watson Alvord, Lynn H. Enslow, Attmore E. Griffin and John Wesley Hyatt 1977; William H. Brush, John F. Dye, Gordon M. Fair, James H. Fuertes and William Pitt Mason 1976; George G. Earl, Charles P. Hoover, Harry E. Jordan, Samuel B. Morris and Charles F. Wallace 1975; Moses N. Baker, Edward S. Cole, James E. Gibson, Omar H. Jewell and John L. Leal 1974; John M. Diven, Norman J. Howard, Alexander Houston, William C. Stripe and George C. Whipple 1973; Edward Bartow, John R. Baylis, George Warren Fuller, Allen Hazen and Clemens Herschel 1971.

ABEL WOLMANN AWARD OF EXCELLENCE ...
Recipients: Herbert O. Hartung 1986; Gordon G. Robeck 1985.

MEDAL OF OUTSTANDING SERVICE TO AMERICAN WATER WORKS ASSOCIATION ...
Recipients: Thomas J. Blair III 1987; David B. Preston 1986; Donald K. Shine 1985; Kenneth J. Miller 1984; Joseph F. O'Grady 1983; G. A. Colton 1982; E. F. Johnson 1981.

AMERICAN WELDING SOCIETY (AWS)

550 LeJeune Rd., Miami, Florida 33126
Perry J. Rieppel, Acting Executive Director

PRESIDENTS ... J. M. Gerken 1987-88; J.H. Walker 1986-87; H.F. Prah 1985-86; D. C. Bertossa 1984-85; M.D. Randall 1983-84; J.C. Thompson 1982-83; W. T. DeLong 1981-82; H. B. Cary 1980-81; G. K. Willecke 1979-80; A. Lesnewich 1978-79; H. A. Sosnin 1977-78; R. H. Foxall 1976-77; P. W. Ramsey 1975-76; J. W. Moeller 1974-75; J. Edward Dato 1973-74.

AWARDS

COMFORT A. ADAMS LECTURE ... presented annually in honor of AWS's founder and first president to an outstanding scientist or engineer for a lecture presenting some new and distinctive development in the field of welding. The award was created in 1943.

Recipients: G.E. Cook 1987; M.M. Schwartz and J.M. Cameron 1986; D. L. Olson 1984; D. A. Canonico 1983; H. Thielsch 1982; C. D. Lundin 1981; A. W. Pense 1980; G. M. Slaughter 1979; R. K. Sager 1978; P. T. Houldcroft 1977; P. Patriarca 1976; R. W. Nichols 1975; W. T. DeLong 1974; W. A. Owczarski 1973; J. Heuschkel 1972; W. S. Pelini 1971; A. Lesnewich 1970; A. M. Weinberg 1969; J. H. Gross 1968.

AIRCO WELDING AWARD ... established to encourage originality and innovation - through the use of welding techniques - in new products. The award, sponsored by Airco Welding Products Division, Air Reduction Co., Inc., is presented to those whose achievements in the joining or severing of metals improves and benefits mankind and furthers the welding industry.

Recipients: G.M. Goodwin and J.D. Hudson 1986; R.W. Richardson 1985; M. J. Grycko, Jr. 1984; F. M. Thompson 1983; R. M. Gage 1982; P. T. Raty and J. Ross 1981; R. W. Schneider, F. H. Sasse and C. N. Porco 1980; G. F. Nielson 1979; C. E. Oleksiak 1978; R. A. Bishel 1977; W. R. Foley 1976; J. G. Landkrohn 1975; R. P. Robelotto 1974; G. Kazlauskas 1973; F. D. Duffey 1972; C. L. Dooley 1971; A. Amirikian 1970; Richard W. Reynolds and Eugene P. Vilkas 1969.

A. F. DAVIS SILVER MEDAL AWARDS ... established to recognize achievements in welded design in the fields of machine design, maintenance and hard surfacing, and structural design. Sponsored and endowed by the late A. F. Davis, formerly vice president and secretary of the Lincoln Electric Company.

Recipients: S.E. Barhorst, J.C. Majetich 1986; T.A. Aanstoss, D.K. Aidun, C.R. Jordan, R.P. Krumpen, Jr., W.F. Savage, J.M. Weldon 1985; G. E. Sheward, F. W. Schoch, H. Luckow, K. F. Kussmaul, K. I. Johnson, D. A. Edson, R. Crafer and M. Baron 1984; R. A. Morris, W. E. Lukens and O. W. Blodgett 1983; A. J. Turner and R. W. Messler, Jr. 1982; A. J. Moorehead, R. W. Reed, O. Miller and J. J. O'Connor 1981; H. Heusler, W. Packheiser and O. W. Blodgett 1980; J. S. Clark, B. M. Patchett and R. L. Queen 1979; J. F. Kiefner, J. A. Miller and M. Prager 1978;

E. P. Cox and F. V. Lawrence 1977; J. W. Lee and R. M. Losee 1976; P. W. Marshall and A. A. Toprac 1975; O. W. Blodgett and W. B. Root 1974; L. J. Malcolm 1973; J. D. Eyestone, J. S. Huang, D. J. Fielding, F. E. Garriott and B. A. Kalvelage 1972; G. A. Alpsten, J. F. Quaas, R. A. Chihoski and L. Tall 1971; E. C. Garrabrandt, R. S. Zuchowski and H. S. Reemsnyder 1970; F. E. Garriott, J. T. Biskup, J. C. Collins, R. E. Key and H. I. McHenry 1969; T. Kumose, A. Luthy, H. Onove and K. Yamada 1968.

W. H. HOBART MEMORIAL MEDAL . . . awarded annually to each author of the paper published in "Welding Journal" and selected by the AWS Committee on Awards as the best contribution to the progress of welding on the subject of pipe welding, welded piping, or similar applications, including the structural use of pipe, but excluding the manufacturing of pipe.
Recipients: James C. Baker, Peter Howe and James M. Sawhill, Jr. 1987; D.A. Canonico, D.P. Edmonds, G.M. Goodwin, T.L. Hebble, R.K. Nanstad, 1986; W.H. Black, J. Mathew, E.F. Nippes, G.A. Ratz, 1985; M. B. Kasen 1984; D. L. Turner, Jr., B. E. Paton, V. K. Lebedeve, S. Kutchuk-Jatcenko 1983; R. Vasudevan, R. D. Stout and A. W. Pense 1982; D. W. Hood, D. D. Keiser, J. E. Key and P. W. Turner 1981; P. T. Delaune and J. D. Weber 1980; K. P. Havik, R. L. Jones and D. K. Kiltau 1979; J. C. Wormelli 1978; H. W. Ebert 1977; J. M. Sawhill 1976; E. D. Brandon 1975; E. G. Shifrin and M. I. Rich 1974; J. H. O'Connor 1973; H. R. Heap and C. C. Riley 1972; D. R. Stoner 1971; R. P. Lynch and F. J. Pilia 1970; C. M. Cockrell 1969.

JAMES F. LINCOLN GOLD MEDAL AWARD . . . medal and certificate awarded each year for the best paper with one author, which in the judgment of the Committee on Awards is the greatest original contribution to advancement and use of welding published in "Welding Journal". Sponsored and endowed by the late J. F. Lincoln, formerly chairman of the board, The Lincoln Electric Company.
Recipients: D.J. Kotecki 1987; S.G. Forsberg 1986; W.J. Mills 1985; J. C. Lippold 1984; D. G. Howden 1983; S. A. David 1982; C. E. Witherell 1981; D. A. Canonico 1980; D. J. Kotecki 1979; R. P. Simpson 1978; D. J. Widgery 1977; S. J. Matthews 1976; J. C. M. Farrar 1975; F. C. Hull 1974; N. Bailey 1973; N. Kenyon 1972; R. A. Chihoski 1971; L. J. Chin 1970; K. E. Dorschu 1969.

FRED L. PLUMMER EDUCATIONAL LECTURE AWARD . . . this award has been established by the American Welding Society to recognize outstanding contributions to the National Education Lectures at the Annual Meeting of the Society. The lecturer is selected by the Educational Activities Committee of the American Welding Society, and is given a certificate of appreciation. Recognizing Fred L. Plummer's service to the Society as President from 1952-54 and Executive Director from 1957-71, the Educational Lecture was renamed Fred L. Plummer Educational Lecture in 1980.
Recipients: R.D. Thomas, Jr. 1987; W.F. Savage 1986; J.E. Dato 1985; J. C. Papritan, II 1984; J. T. Heile 1983; C. Smallbone 1982; R. Moll 1981; E. Hornberger 1980; G. A. Nikolaev 1979; P. W. Ramsey 1978; C. E. Harbower 1977; A. C. Nunes 1976; B. M. Krantz 1975; A. F. Manz 1974; J. D. Graham 1973; M. D. Thomas 1972.

SAMUEL WYLIE MILLER MEMORIAL MEDAL AWARD . . . donated in 1927 by a former AWS president and executive of Union Carbide and Carbon Company. The award is presented annually by the Committee on Awards for any meritorious achievements which have contributed conspicuously to the advancement of the art of welding and cutting.
Recipients: T.B. Jefferson 1987; K.F. Graff 1986; J.H. Gross 1985; R. A. Dunn 1984; K. H. Koopman 1983; R. H. Foxall 1982; R. J. Cookling 1981; P. W. Ramsey 1980; P. E. Masters 1979; R. B. McCauley 1978; W. F. Savage 1977; I. A. Oehler 1976; L. C. Bibber 1975; R. C. Becker 1974; G. E. Linnert 1973; H. B. Cary 1972.

WILLIAM SPARAGEN AWARD . . . presented annually for the best research paper printed in the Supplement to the "Welding Journal". It consists of a modest stipend (approximately $250) and a certificate, and was first presented in 1961.

Recipients: T. Ogawa 1987; M.J. Cieslak, W.F. Savage 1986; S.T. Furr and J.M. Sawhill, Jr. 1985; N. H. Madsen, J. S. Goodling and B. S. Chin 1984; J. R. Roper and C. Heiple 1983; K. C. Wu 1982; M. J. Cieslak and W. F. Savage 1981; J. C. Lippold and W. F. Savage 1980; N. Bailey and S. B. Jones 1979; C. D. Lundin and W. J. Ruprecht, Jr. 1978; E. F. Nippes, W. F. Savage and E. S. Szekeres 1977; R. F. Heile and D. C. Hill 1976; G. M. Goodwin, R. T. King and J. O. Stiegler 1975; W. A. Petersen 1974; D. S. Duvall, W. H. King, W. A. Owczarski and D. F. Paulonis 1973; T. P. Rich and R. Roberts 1972; G. E. Grotke, R. Stickler and B. Weiss 1971; D. S. Duvall and W. A. Owczarski 1970; W. W. Gerberich, C. E. Hartbower and P. P. Crimmins 1969.

RENE D. WASSERMAN AWARD . . . awarded annually for the best paper published in "Welding Journal" as the greatest original contribution to the progress and/or advancement of the use of brazing or braze welding. A certificate is awarded to each author and an honorarium of $1,000 is divided equally among the authors.
Recipients: R.L. Workman and R.D. Thomas, Jr. 1987; F.M. Hosking, 1986; I. Okamoto, T. Takemoto 1985; A. Sakamoto 1984; H. Ohmura and T. Yoshida 1983; R. Johnson 1982; T. J. Ramos 1981; D. Wielage, B. Z. Weiss, H. D. Steffens and A. H. Engelhardt 1980; W. E. Cooke, J. A. Hirschfield and T. E. Wright 1979; D. A. Canonico, N. C. Cole and G. M. Slaughter 1978; R. R. Wells 1977; D. L. Klarstrom, J. F. Rasmussen and S. Weiss 1976; J. Christensen and K. Rorbo 1975; Roy E. Beal 1973; C. N. Cochran, W. E. Haupin, J. J. Stokes, Jr. and J. R. Terrill 1972; T. F. Berry and G. S. Hoppin, II 1971; R. N. Stenerson 1970; Forbes M. Miller and Nikolajs Bredzs 1969; G. K. Hicken and W. B. Sampte, Jr. 1968.

ADAMS MEMORIAL MEMBERSHIP AWARD . . . This award, established by the American Welding Society, is given as a means of recognizing educators whose teaching activities are considered to have advanced the knowledge of welding of the undergraduate or postgraduate students in their respective engineering institutions.
Recipients: J.H. Devletian 1987; S. Kou, R.G. Thompson 1986; W.A. Baeslack III 1985; J. E. Jones, R. H. Frost and C. Albright 1984; C. L. Tsai, J. M. Samuel, R. W. Richardson and R. V. Kisielewski 1983; R. Roberts and G. R. Edwards 1982; J. J. McCarthy and K. K. Graff 1981; J. C. Williams, D. G. Howard and T. W. Eager 1980; C. Smallbone and D. L. Olson 1979; T. Kobayashi and F. Eichhorn 1978; K. K. Wang, W. Soete, F. V. Lawrence, Jr., B. Jakobsson and G. H. Geerlings 1977; A. E. Miller 1976; S. T. Rolfe, G. C. Lee, J. W. Fisher, J. G. Bouwkamp and C. R. Barret 1975; H. C. Rogers, K. Masubuchi and T. H. Courtney 1974; J. Zotos, V. Weiss, R. A. Moll, R. H. Lambert and T. D. Kay 1973; W. W. Sanders, R. E. Long, H. W. Long and R. G. Gilliland 1972; N. F. Fiore, D. Milner, W. H. Kielhorn, T. H. Hazlett, G. C. Driscoll, A. Choquet and J. G. Bollinger 1971; J. J. Wert, W. F. Savage, R. Nielsen, C. E. Jackson and O. W. Albritton 1970; L. J. McGeady, C. D. Lundin, R. L. Ketter and J. Duboc 1969; R. C. Wiley, A. A. Toprac, E. Pfender and W. J. Hall 1968.

R. D. THOMAS MEMORIAL AWARD . . . This award was established to honor a charter member of the American Welding Society and the AWS representative to the first organization meeting of the International Institute of Welding. It is intended to foster interest in United States participation in the work of the international organization.
Recipients: W.A. Black 1987; C.B. Shaw 1985; R. E. Monroe 1984; D. J. Kotecki 1982; G. W. Oyler 1981; R. D. Thomas, Jr. 1980; B. L. Alia 1979; M. Schwartz 1978; H. D. Link 1977; K. Masubuchi 1976; C. E. Jackson 1975; W. T. DeLong 1974; K. H. Koopman 1973; R. D. Stout 1972; R. B. McCauley 1971; I. A. Oehler 1970; A. Amirikian 1969; J. H. Humberstone 1968; F. L. Plummer 1967; H. Biers 1966; W. Spraragen 1965.

SILVER QUILL EDITORIAL ACHIEVEMENT AWARD . . . The Silver Quill Editorial Achievement Award administered by the Communications Committee is given annually by the American Welding Society to the editors and publishers of the magazine, outside the welding field, which publishes the best original editorial coverage on the subject of welding.

Recipients: Iron Age, R.R. Irving 1986 and 1987; Robotics Today, Robert N. Stauffer 1985; American Machinist, Joseph Jablonowski 1984.

NATTCO NATIONAL WELDING AND CUTTING AWARD ...

This award is sponsored by the National Torch Tip Company and is awarded annually to encourage originality and innovation through the use of new techniques with fuel gas cutting and welding equipment.

Recipients: J.R. Stitt 1987; J. J. Crowe 1984; A. H. Krieg 1983; E. K. Long 1982; J. Pfeifer 1981; L. J. Privoznik and S. C. Light 1980; D. E. Fillipi 1979.

NATIONAL MERITORIOUS AWARD ...

This award, established by the American Welding Society, is given in recognition of the candidate's counsel, loyalty and devotion to the affairs of the Society, his assistance in promoting cordial relations with industry and other organizations, and for his contributions of time and effort on behalf of the Society.

Recipients: D.E. Hamilton, G.E. Linnert 1986; R.J. Cristoffel, W.T. DeLong, G.M. Nally 1985; S. L. Ritter and W. J. Erichsen 1984; A. Lesnewich and K. E. Dorschu 1983; A. L. Collin 1982; W. H. Rice 1981; W. R. Smith, Sr. and R. L. Peaslee 1980; F. C. Saacke 1979; P. J. Rieppel and E. G. Shifrin 1978; F. L. Plummer and J. W. Moeller 1977; J. C. Thompson and R. H. Foxall 1976; R. S. Parrott and L. DeFreitas 1975; H. E. Adkins 1974; H. C. Campbell and R. C. Becker 1973; I. S. Goodman and R. J. Conkling 1972; P. W. Ramsey and H. B. Cary 1971; G. K. Willecke and P. Patriarca 1970; J. J. Chyle and J. J. MacKinney 1969; D. C. Smith 1968; H. E. Rockefeller and J. H. Humberstone 1967; I. A. Oehler and J. E. Norcross 1966; R. D. Stout and J. W. Kehoe 1965.

HONORARY MEMBERSHIP AWARD ...

An Honorary Member shall be a person of acknowledged eminence in the welding profession, or who may be accredited with exceptional accomplishments in the development of the welding art upon whom the American Welding Society may see fit to confer an honorary distinction.

Recipients: R.L. Peaslee, D.L. Sprow 1987; W.T. DeLong, A.W. Pense 1986; E.F. Nippes, A. Pollock 1985; W. C. Rudd and P. P. Puzak 1984; W. R. Smith, Sr. and D. F. Helm 1983; R. Townsend and M. Stepath 1982; E. A. Hobart 1981; J. Holt and A. A. Bernard 1980; F. C. Saacke and H. E. Adkins 1979; G. E. Fratcher and P. J. Rieppel 1978; P. Patricia and P. E. Masters 1977; R. Weck 1976; G. R. Pease and W. D. Doty 1975; G. K. Willecke and A. G. P. Leroy 1974; P. C. Arnold 1973; O. W. Blodgett 1972; W. F. Savage 1971; T. E. Jones 1970; R. D. Thomas, Jr. and A. F. Chouinard 1969; J. H. Humberstone and G. O. Hoglund 1968; La Motte Grover and A. Amirikian 1967; C. I. MacGuffie 1966; V. R. P. Saxe 1965.

McKAY-HELM AWARD ...

for the best contribution to the advancement of knowledge of low alloy steel, stainless or surfacing weld metals involving the use, development or testing of these materials as represented in articles published in the Welding Journal during a given calendar year.

Recipients: P. Burgardt, C. Heiple 1987; C. Dallam, S. Liu, D.L. Olson 1986; D.K. Matlock, D.L. Olson, J.A. Self 1985; V. P. Kujanpaa 1984; D. H. Kah and D. W. Dickinson 1982; C. J. Sullivan, C. D. Lundin and C. D. Chou 1981; Dr. G. M. Goodwin, Dr. S. A. David and D. N. Braski 1980; J. O. Stiegler, G. M. Slaughter, R. T. King, R. J. Gray, G. Goodwin, N. C. Cole and R. G. Berggren 1979; J. Honeycombe and T. G. Gooch 1978; R. A Swift and H. C. Rogers 1977; S. Forsberg, M. Areskoug, A. Backman and A. V. Bernstein 1976.

CHARLES H. JENNINGS MEMORIAL MEDAL AWARD ...

The award will be given to a student or faculty representative of a college or university who is considered to be the source of the most valuable contribution to welding literature which has been published during a calendar year in one of the monthly issues of the Welding Journal.

Recipients: C.E. Albright, W.H. Jones and L.A. Weeter 1987; L. Adler, S.I. Rokhlin, 1986; R.A. Anderson, D.F. Farson, D. Gutow, R.W. Richardson, 1985; J. Szekely, G. M. Oreper and T. W. Eagar 1984; G. L. Leone and H. W. Kerr 1983; E. F. Nippes, W. J. Gestal, Jr. and D. J. Ball 1982; W. F. Savage and J. C. Lippold 1981; T. H. North, E. Koukabi, I. Craig and H. H. Bell 1980; A.

Nowicki, T. H. North, I. Craig and H. B. Bell 1979; W. F. Savage, E. F. Nippes and J. C. Lippold 1978; W. F. Savage, E. F. Nippes and H. Homma 1977; R. W. Heine and J. H. Devletian 1976; P. E. Pence, T. P. O'Brien, R. C. McMaster and E. R. Funk 1975; R. D. Stout 1974.

HOWARD E. ADKINS INSTRUCTOR MEMBERSHIP AWARD ...

is given as a means of recognizing high school, trade school, technical institute or junior college instructors whose teaching activities are considered to have advanced the knowledge of welding of students in their respective schools.

Recipients: P.J. Boes and J.R. Johnson 1987; R.E. Dahl, T.L. O'Brien, 1986; R.D. Graff, J.L. Westfall 1985; F. Wagner and R. Murray 1984; T. E. Reed, Jr., G. J. Klemencic, D. R. Grubbs and F. L. Ellicott 1983; J. H. Pennington, R. F. Smith, R. G. Culbert and J. A. Ciaramitaro 1982; B. B. Moffitt, Jr., T. J. Hudacko, W. W. Culbert and C. Blesh 1981; E. F. Wiltse, J. A. Udy, E. F. Butler and P. A. Ahrens 1980; G. Perez, L. Koellhoffer, V. Dominick and R. L. Collins 1979; R. L. Weisend, W. J. Hammond, K. Frederickson and P. Edmiston 1978; L. F. Wood, R. J. Sacha, C. Martin and M. A. Godley 1977; K. Neely, E. W. Havener, H. B. Cary and J. Bernstein 1976; L. J. Sparrowhawk, Jr., T. Hesse, R. A. Hardin, L. DeFreitas 1975; H. Taylor, H. F. Szczepanski, M. L. Starley and V. J. Cimino 1974; J. E. Pierson III, D. C. Nelson, E. Lopez, Sr. and F. Arning 1973; J. Shotwell, J. W. Price, R. Farthing and E. Ederer, Jr. 1972; R. L. Sysum, L. Koleman, R. H. Johnson and B. W. Carl 1971; H. L. Spellman, J. Orr, P. K. McPearson and G. Hinkley 1970; L. F. Riggins, J. H. Pilszak, W. G. Geyer and R. M. Gates 1969; G. E. Water, J. R. Pedro, J. W. Markwell and J. C. Kasabula 1968; B. B. Rufo, R. F. Kennedy, J. C. Kaluzny and J. E. Gross 1967; R. E. Theiler, R. F. Haynes, I. H. Griffin and J. H. Forrer 1966; R. G. Thomas, T. M. Flack, N. Eberhardt and H. M. Anderson 1965.

WILLIAM IRRGANG AWARD ...

awarded each year to the individual who has done the most to enhance the American Welding Society's goal of advancing "the science, technology and application of welding" over the last five years. The award is sponsored by The Lincoln Electric Company.

Recipients: D.J. Kotecki 1987; R.J. Christoffel 1986.

WARREN F. SAVAGE AWARD ...

recognizes original and innovative research resulting in a better understanding of the metallurgical principles related to welding. It is presented for the best paper on this topic published during the previous calendar year in the Research Supplement of the Welding Journal.

Recipients: B. Damkroger, G.R. Edwards, D. Hayduk and D.L. Olson 1987; S. Lathabai and R.D. Stout 1986.

ELIHU THOMSON RESISTANCE WELDING AWARD ...

presented annually to a living individual who has made an outstanding contribution to the technology and application of resistance welding, as evidenced by: 1) invention or production of new resistance welding equipment innovations; 2) a unique application of resistance welding in a production environment; 3) a technical paper published in the Welding Journal or other similarly prestigious publication; or 4) other contribution as the AWS Awards Committee shall deem worthy of recognition.

Recipients: H.A. Nied 1987; T.E. Jones 1986.

ASSOCIATION FOR COMPUTING MACHINERY, INC. (ACM)

11 West 42nd St., New York, New York 10036
Richard Hespos, Executive Director

PRESIDENTS ... Paul Abrahams 1986- ; Adele Goldberg 1984-86; David H. Brandin 1982-84; Peter J. Denning 1980-82; Daniel D. McCracken 1978-80; Herbert R. J. Grosch 1976-78; Jean E. Sammet 1974-76; Anthony Ralston 1972-74.

AWARDS

DISTINGUISHED SERVICE AWARD ... presented annually to the individual selected on the basis of the value and degree of his services to the computing community.

Recipients: Clair Maple 1986; Donn R. Parker 1985; Saul Rosen 1984; Grace Murray Hopper 1983; Anthony Ralston 1982; Aaron Finerman 1981; Bernard A. Galler 1980; Carl Hammer 1979; Eric Weiss 1978; Thomas B. Steel, Jr. 1975; Richard G. Canning

1976; John W. Carr, III 1975; Saul Gorn 1974; William Atchison 1973; George Forsythe (posthumously) 1972; Don Madden 1971; Franz L. Alt 1970.

DOCTORAL DISSERTATION AWARD ... The Doctoral Dissertation Award is made for the doctoral thesis in computer science and engineering that is judged to be the all-around best of those submitted to ACM. The award is presented at ACM's annual Computer Science Conference. $1,000, certificate, and royalties from sale of publication.
Recipients: John R. Ellis 1986; John R. White 1985; Manolis G.H. Katevenis 1984; Charles E. Leiserson 1983.

ECKERT-MAUCHLY AWARD ... The Eckert-Mauchly award is a joint ACM-IEEE Computer Society award presented to an individual selected for technical contributions to computer and digital systems architecture. The field of computer architecture is considered at present to encompass the combined hardware/software design and analysis of computing and of digital systems. This award has been presented since 1979 at the Symposium for Computer Architecture. $1,000 and certificate.
Recipients: Harvey G. Cragon 1986; Ming T. Liu and John Cocke 1985; Jack B. Dennis 1984; Tom Kilburn 1983; C. Gordon Bell 1982; Wesley A. Clark 1981; Maurice V. Wilks 1980; Robert S. Barton 1979.

GRACE MURRAY HOPPER AWARD ... to be conferred upon the outstanding young computer professional of the year. The winner will be selected on the basis of a single recent major technical or service contribution, clearly outstanding in its own right. Candidates must have been 30 years or less of age at the time the qualifying contribution was made.
Recipients: William N. Joy 1986; Larry Druffel 1985; Daniel H.H. Ingalls, Jr. 1984; Brian K. Reid 1982; Daniel S. Bricklin 1981; Robert M. Metcalfe 1980; Stephen Wozniak 1979; Raymond Kurzweil 1978; Edward A. Shortliffe 1976; Allen L. Scherr 1975; George N. Baird 1974; Lawarence M. Breed, Richard H. Lathwell and Roger D. Moore 1973; Paul H. Cress and Paul H. Dirksen 1972; Donald E. Knuth 1971.

OUTSTANDING CONTRIBUTION AWARD ... established February 1976. Presented annually at the ACM annual conference to individuals selected on the basis of significant service to ACM.
Recipients: Herbert Maisel 1986; Aaron Finerman 1985; Orrin E. Taulbee 1984; Seymour J. Wolfson and Richard Austing 1983; Fred H. Harris 1982; J. A. N. Lee 1981; M. Stuart Lynn 1979; Kathleen Wagner 1978; Bruce W. Van Atta-Smith Dorsey 1976.

SOFTWARE SYSTEM AWARD ... The Software System Award is for a software system that has had a lasting influence, reflected in contributions to concepts or in commercial practice or in both. The award is made either to an institution or to individuals responsible for developing and introducing the software system. Engraved plaque and certificate.
Recipients: Donald E. Knuth 1986; Andries Van Dam 1985; Charles P. Thacker, Robert W. Taylor and Butler W. Lampson 1984; Dennis M. Ritchie and Ken Thompson 1983.

A.M. TURING AWARD ... presented at the ACM Annual Meeting to an individual selected for his contributions of a technical nature made to the computing community.
Recipients: John Hopcroft and Robert Tarjan 1986; A.V. Aho 1985; Nicklaus Wirth 1984; Dennis M. Ritchie and Ken Thompson 1983; Stephen A. Cook 1982; Edgar F. Codd 1981; Charles Richard Antony Hoare 1980; Kenneth Iverson 1979; Robert W. Floyd 1978; John Backus 1977; Michael O. Rabin and Dana S. Scott 1976; Allen Newell and Herbert A. Simon 1975; Donald Knuth 1974; Charles W. Bachman 1973; Edsger W. Dijkstra 1972; John McCarthy 1971; J. H. Wilkinson 1970; Marvin Minsky 1969; R. W. Hamming 1968.

ASSOCIATION OF ENERGY ENGINEERS (AEE)

4025 Pleasantdale Road, Suite 340, Atlanta, Georgia 30340
Albert Thumann, Executive Director

PRESIDENT ... Konstantin Lobodovsky, 1988; Victor Ottaviano 1987; Jerry Taylor 1986; Jon Haviland 1985; Kermit Harmon 1984; Robert Lahey 1983; Raymond Kendall 1982; Harvey Morris 1981; 1981; Gerald Lameiro 1980; Albert Thumann 1979; Albert Thumann 1978.

AWARDS

CORPORATE ENERGY MANAGEMENT AWARD ... outstanding accomplishments in developing, managing and implementing the corporate energy management program.
Recipients: Southwire Company 1987; Abbott Laboratories 1986; Johnson and Johnson 1985; Public Service Electric and Gas Company of New Jersey 1984; American Standard 1983; Western Electric 1982; Pacific Gas & Electric 1981; The City of Houston 1980; United Technologies 1979; Marriott Corporation 1978.

DISTINGUISHED SERVICE AWARD ... outstanding service to the society and energy community.
Recipients: Jerry Taylor 1987; Jon Haviland 1986; Robert Lahey 1985; Kermit Harmon 1984; Raymond Kendall 1983; Gerard Lameiro 1981; Albert Thumann 1980.

ENERGY ENGINEER OF THE YEAR ... outstanding accomplishments in promoting the practices and principles and procedures of energy engineering.
Recipients: John Suptic 1987; Richard Cooper 1986; John R. Sosoka 1985; Verle Williams 1984; Allen R. Margreither 1983; Paul Gilson 1981; Victor Ottaviano 1980; Richard Aspenson 1979; Harvey Morris 1978.

ENERGY EXECUTIVE OF THE YEAR ... outstanding accomplishments in developing, managing and implementing the Corporate Energy Management Program.
Recipients: Robert Rosenberg 1987; Richard Redpath 1986; Edward L. Addison 1985; Gerald Decker 1981; Fred O'Green 1980.

ENERGY MANAGER OF THE YEAR ... outstanding accomplishments in promoting the practices, principles and ideals of energy management.
Recipients: Stephen Roosa 1987; James Warnock 1986; Viktor Boed 1985; Frank Kisicki 1984; William T. Stewart 1983; John Honeycomb 1981.

ENERGY PROFESSIONAL DEVELOPMENT ... outstanding accomplishments in training and development of energy engineers and managers.
Recipients: Hal Guven 1987; John Flynn 1986; John F. Hyfantis 1985; Bruce Colburn 1984; Wayne C. Turner 1983; Donald Pirro 1981; Herbert Reem 1980; Charles Dorgan 1979; Karel Klima 1978.

UTILITY OF THE YEAR AWARD ... outstanding achievements in promoting practices and principles of cogeneration.
Recipient: Northeast Utilities 1987.

COGENERATION COMPANY OF THE YEAR AWARD ... outstanding achievements in promoting practices and principles of cogeneration.
Recipient: Tecogen Incorporated 1987.

COGENERATION PROFESSIONAL OF THE YEAR AWARD ... outstanding achievements in promoting practices, principles and procedures in congeneration.
Recipient: Paul Rivet 1987.

COGENERATION PROJECT OF THE YEAR AWARD ... outstanding achievements in conceiving and implementing an innovative cogeneration project.
Recipients: Sister Phyllis Ann Gerold, RSM and Mercy Hospital Toledo, Ohio 1987.

ASSOCIATION OF ENGINEERING GEOLOGISTS (AEG)

62 King Philip Rd, Sudbury MA 01776
E. A. Blackey, Executive Director

PRESIDENTS ... Theodore Maynard 1987; Norman R. Tilford 1986; Allen W. Hatheway 1985; Robert M. Valentine 1984; Richard W. Galster 1983; William C. Paris, Jr. 1982; Albert J.

Depman 1981; John B. Ivey 1980; Richard J. Proctor 1979; Howard A. Spellman 1978; Noel M. Ravneberg 1977; Jasper L. Holland 1976; Raymond T. Throckmorton, Jr. 1975; Samuel C. Sargent 1974; Frank W. Wilson 1973.

AWARDS

AEG PUBLICATION AWARD . . . established by AEG in 1968 for the person presenting the most outstanding paper published in any AEG publication during the fiscal year.
Recipients: Nikola P. Prokopovich 1986; Roy J. Shlemon 1985; E. L. Krinitzsky and W. F. Marcuson III 1984; David H. Randell, J. B. Reardon, J. A. Hileman, T. Matuschka, G. C. Liang, A. I. Khan and J. A. LaViolette 1983; John E. Costa and Sally W. Bilodeau 1982; Linda Zall and Richard Michael 1981; Alice S. Allen 1979; M. King Hubbert 1976; Robert Strazer, Keith Bestwick and Stanley Wilson 1975; David L. Royster 1974; William H. Easton 1973; Murray T. Dougherty and Nino J. Barsotti 1972; Donald H. Gray 1971; Howard A. Coombs 1970; Richard H. Godson and Joel S. Watkins 1969.

CLAIRE P. HOLDREDGE AWARD . . . awarded to a member of AEG who has published at some time during the five previous years a paper adjudged to be an outstanding contribution to the Engineering Geology Profession.
Recipients: Roy E. Hunt 1984; David L. Royster 1983; C. M. Mathewson 1981; A. B. A. Brink 1980; Douglas R. Piteau 1979; David L. Royster 1978; Shailer S. Philbrick 1977; Frank Netterberg 1975; Robert F. Legget 1974; George M. Hughes (co-authors R. A. Landon and R. M. Farvolden) 1972; Manuel G. Boniila 1971; Joseph F. Poland 1970; Lloyd B. Underwood 1968.

SPECIAL AWARDS . . . a certificate of appreciation to an individual or firm who has performed an outstanding contribution to AEG or to the Profession of Engineering Geology.
Recipients: William D. Bingham, Kenneth E. Kolm, Stephen L. Garrison, George A. Kiersch, Frank W. Wilson, Susan G. Steel, Hugh S. Robertson, Joseph L. Ehasz, Stefi Weisburd 1986; W.W. Peak, T.R. Maynard, N.R. Tilford, R.G. Sherman, R.H. Fickies, C.C. Mathewson, R.C. Kent, M.D. Kent, T.J. McClain, T.R. West, R.G. Thomas, J.B. Ivey, H.P. Oshel, E.L. Krinitzsky and W.W. Hays 1985; N.M. Ravneberg, R.L. Schuster, R.J. Farina, D.L. Hannan, M.D. Kent, R.E. Smith, N.R. Tilford, C.C. Mathewson, T.R. Maynard, J.H. Peck and J.W. Williams 1984; N.R. P. Baczynski, J.E. Bogner, Luc Boyer, J.A. Caggiano, J.W. Cobarrubias, G.L. Hempen, J.B. Ivey, R.C. Kent, C.C. Mathewson, T.R. Maynard, J.H. Peck, R.M. Valentine and J.W. Williams 1983; A.B. Arnold, A.P. Avel, L.A. Brown, C.W. Daugherty, A.J. Depman, R.J. Farina, J.R. Keaton, E.L. Krinitzsky, T.R. Maynard, H.P. Oshel, J.H. Peck, P.G. Smith, R.M. Valentine and J.W. Williams 1982; Gordon E. Cordes, John B. Ivey, Richard W. Galster, Gregory C. Hempen, Mary S. Ray, Richard C. Kent, Christopher C. Mathewson and William C. Paris, Jr. 1981; Albert J. Depman, Richard W. Galster, Christopher C. Mathewson, William C. Paris, Jr., Richard J. Proctor, Norman Tilford 1980; Alice S. Allen, Albert J. Depman, Abraham Dolgoff, Christopher C. Mathewson, Harold W. Stuart, James Hadley Williams 1979; Lynn A. Brown, Axel M. Fritz, Jasper L. Holland, John B. Ivey, Henry W. Maxwell, William C. Paris, Jr., Robert M. Valantine and Charles F. Withington 1978; William J. Berk, Albert J. Depman, Jasper L. Holland, Mary S. Ray and Raymond T. Throckmorton 1977; Noel M. Ravneberg, Howard A. Spellman, Jr., Robert M. Valentine, Richard J. Proctor, J. Hadley Williams, John T. Miller, Abraham Dolgoff, Gerald S. Grainger, Fitzhugh T. Lee, Raymond T. Throckmorton, Jr., Frank W. Wilson, Aubrey D. Henley, Richard W. Lemke, Samuel C. Sargent, Herbert G. Schlicker, Donald C. Rose, Roy A. Hoffman, Gordon B. Oakeshott, James W. Skehan, S. J., Thomas Lenahan, Jr. and W. C. Paris, Jr. 1976; M. J. Spencer, Rep. Joseph A. Battle, Edwin E. Lutzen, William I. Gardner, Bruce M. Hall, Charles E. Hall, Lawrene B. James, John C. Manning, Joseph S. Poland, Robert W. Reynolds, Ted L. Sommers, Harmon R. Tabor and George F. Worts, Jr. 1975; Rummel, Klepper & Kahl, Richard W. Lemke, Robert M. Lindvall, Howard A. Spellman, Jr., Floyd I. Johnston and Frank T. Wheby 1974; Hughes Tool Co. and Harza Engineering Co. 1973; Gordon E. Cordes, Phyllis Cordes, Catherine C. Campbell, Raymond C. Mason, Donald R. Nichols, John M. Weaver, Frank T. Welby, Josephine LeFeber and Douglas C. Moorehouse 1971.

ASSOCIATION OF ENVIRONMENTAL ENGINEERING PROFESSORS (AEEP)

Desmond Lawler, Sec., Univ. of Texas, Dept. of Civil Engineering, EcJ8.6, Austin, TX 78712.

PRESIDENTS . . . E. Robert Baumann 1987; John F. Andrews 1986; F. Michael Saunders 1985; Vernon L. Snoeyink 1984; James K. Edzwald 1983; Francis A. DiGiano 1982; Benjamin C. Dysart III 1981; Roger A. Minear 1980; Charles R. O'Melia 1979; N. Bruce Hanes 1978; George P. Hanna 1977; Paul H. King 1976; Wesley O. Pipes 1975; E. J. Middlebrooks 1974; R. I. Dick 1973.

AWARDS

AEEP ENGINEERING-SCIENCE . . . this award is given for exceptional doctoral research in environmental engineering and science. First awarded in 1969.
Recipients: William W. Clarkson and William J. Jewell 1986; Arthur L. Baehr and M. Yavuz Corapcioglu 1985; Bruce E. Jones and Walter J. Weber 1984; Edward J. Bouwer and Perry L. McCarty 1983; C. J. Yang and Richard Speece 1982; Antonio O. Lau and David Jenkins 1981; Antonio O. Lau and David Jenkins 1980; Bruce E. Rittman and Perry McCarty 1979; Mesut Sezgin and David Jenkins 1978; William Batchelor and A. W. Lawrence 1977; Michael K. Stenstrom and John F. Andrews 1976.

AEEP OUTSTANDING PUBLICATION . . . recognizes authors who have made a lasting contribution to environmental engineering education and research through published work. First awarded in 1983.
Recipients: Richard I. Dick 1986; A.W. Lawrence and P.L. McCarty 1985; W. Stumm and Charles R. O'Melia 1984; J. M. Morgan and W. Stumm 1983.

ASSOCIATION OF SCIENTISTS AND ENGINEERS OF THE NAVAL SEA SYSTEMS COMMAND (ASE)

c-o G. D. Hagedorn, NAVSEA 55X2, Washington, D.C. 20362

PRESIDENTS . . . Matthew Reynolds 1986-87; Gregg D. Hagedorn 1985-86; Donald R. Cebulski 1983-84; R. Keane, Jr. 1982; P. Law 1981; E. Kinney 1980, 1979; C. Geiger 1978, 1977; John R. Buck 1976; Robert Calogero 1975; Lloyd N. Nilsen 1974; Richard S. Carleton 1973.

AWARDS

ASE PROFESSIONAL ACHIEVEMENT AWARD . . . in recognition of those individuals who have accomplished a tangible, substantial and significant achievement of outstanding caliber which contributes to the task of carrying out the mission of the Naval Sea Systems Command, and furnishes inspiration to, enhances the prestige of or stimulates the espirit de corps of the professional staffs of the Naval Sea Systems Command.
Recipients: Jesse Atkins 1987; Richard Steward 1986; Herbert O. Farrington 1985; Matthew T. Reynolds 1984; Stewart M. Williams 1983; Richard S. Carlton 1982; Robert Kean Jr. 1981; Dominic S. Toffolo 1980; William Marlin 1979; Dr. Robert Snuggs 1978; Michael Resner 1977; Floyd Ryan 1976; Daniel J. Weiler 1975; Gerald M. Boatwright 1974; Theodore Sarchin 1973; Anthony Ruffini 1972; Eugene P. Weinert 1971; Ira E. Williams 1970; Stephen M. Blazek 1969; Joseph T. Janik 1968.

THE JOHN C. NIEDERMAIR AWARD . . . presented to the author or authors of the best paper presented to the Annual Technical Symposium of ASE.
Recipients: Arthur Romano, Paul Lain, Timothy Lavalle, Richard Carey 1987; Siu Fung 1986; Jan P. Hope, Vernon Stortz and Michael Fitzgerald 1985; Jeffrey Hough-Chi Chun 1984; Joseph Dettola and Jeffrey Fleming 1983; Alexander Malakhoff and Sidney Davis 1982; Alexander Malakhoff 1981; H. Douglas Marron 1980; Robert Johnson, Nicholas Caracostas and Edward Comstock 1979; Charles Todd and Franz Frisch 1978; Juliani Gatzoulis and Robert Keane 1977; Jan Paul Hope 1976; John W. Fairbanks 1975; Robert G. Tucker and Lawrence Goldberg 1974; John W. Fairbanks 1973; Kenneth C. Morisseau 1972; Harry

Belukjian 1971; Lt. Clark Graham and Walter F. Aerni 1970; Jon R. Buch and Henry W. Stoll, Jr. 1969; Randall G. Thome 1968.

SILVER MEDAL . . . for continuous outstanding contributions to Naval Engineering for a period of not less than seven years.
Recipients: Richard Frey 1987; Preston E. Law, Jr. 1986; Edward T. Kinney 1985; Daniel J. Weiler 1984; James H. Mills, Jr. 1983; Robert T. Hill 1982; Howard J. Friez 1981; Paul R. Sacilotto and Vice Admiral Clarence Russel Bryan USN 1980; Peter Palermo 1978; William Louis 1977; Bertram Remer 1976; Maurice Hauschildt 1975; Laskar Wechsler 1974; Charles P. Roane 1971.

AUDIO ENGINEERING SOCIETY, INC. (AES)

60 East 42nd Street, New York, New York 10017
Donald J. Plunkett, Executive Director
PRESIDENTS . . . Barry Blesser 1981-82; Ray M. Dolby 1980-81; Peter K. Burkowitz 1979-80; John G. McKnight 1978-79; Emil Torick 1977-78; Warren Rex Isom 1976-77; Dr. Duane H. Cooper 1975-76; John M. Eargle 1974-75; John Bubbers 1973-74; Hugh S. Allen, Jr. 1972-73.

AWARDS

THE AUDIO ENGINEERING SOCIETY MEDAL AWARD . . . given annually to a person who has helped significantly in the advancement of the Society. It was established by the endowment of the late C. J. LeBel, one of the founders and the first President of the Society. The first bronze medal was given in 1972. Formerly the Audio Engineering Society Award was a laminated plaque.
Recipients: Derek Tilsley 1981; Raymond E. Cooke 1980; Peter K. Burkowitz 1978; Robert O. Fehr 1977; Hugh S. Allen, Jr. 1976; Arthur E. Gruber 1975; John J. Bubbers 1974; Donald W. Powers 1973; Johan L. Ooms 1972; John G. McKnight 1971; John D. Colvin 1970; Daniel W. Martin 1969.

BOARD OF GOVERNORS AWARD . . . given to a member in recognition of services or accomplishments above and beyond the effective and efficient performance of Committee and Officer duties.
Recipients: Timothy A. Cole and Martin Polon 1981; Almon H. Clegg, Lawrence R. Fincham, Geoffrey M. Langdon, Erik A. Porterfield, Robert Rypinski, Max Cohen and Joerg Sennheiser 1980; Masuo Hayahi Jacques, Robert Dewevre, John M. Bowsher, Jean Bonzon and Fritz P. Sippl 1979; Alan D. Blumien (posthumous), Peter K. Burkowitz, Ethel R. Morris and Dewitt F. Morris 1978; Victor Brociner (posthumous); John J. Bubbers 1977; Jacqueline Harvey 1976; John T. Mullin 1975; George W. Bartlett 1974; Irving L. Joel 1972; Donald W. Powers 1971; John D. Colvin 1970; Alex McKee 1969; Marvin R. Headrick and Ralph Schlegal 1968.

GOLD MEDAL . . . (formerly the John H. Potts Memorial Award) was established by the Society in 1971 and is given in recognition of outstanding achievements, sustained over a period of years, in the field of audio engineering. It was formerly known as the John H. Potts Memorial Award, established by his widow in his memory in 1949.
Recipients: Arthur C. Keller 1981; Hugh S. Knowles and Daniel R. vonRecklinghausen 1978; Georg Neumann and John G. Frayne 1976; Floyd K. Harvey 1974; H. E. Roys 1973; Manfred R. Schroeder 1972; Leo L. Geranek 1971; Rudy T. Bozak 1970; Marvin Camras 1969; Arnold P. G. Peterson 1968.

THE SILVER MEDAL AWARD . . . (formerly known as The Emile Berliner Award) was established by the Society in 1971 in honor of the audio pioneers, Alexander Graham Bell, Emile Berliner and Thomas A. Edison, and is given in recognition of an outstanding development or achievement in the field of audio engineering. It was formerly known as the Emile Berliner Award, established by the Berliner Family in 1953.
Recipients: Barry Blesser 1981; Robert A. Moog 1980; Paul W. Klipsch 1978; Toshiya Inoue and Horst Redlich 1977; Willi Studer 1976; Dr. Takeo Shiga 1974; Erik Rorbaek Madsen 1973; Edgar Villchur 1972; Ray M. Dolby 1971; Arthur C. Haddy 1970; Charles F. Wiebusch 1969; Duane H. Cooper 1968.

THE PUBLICATION AWARD . . . is given each year, alternately, to an author 35 years of age or younger and to an author regardless of age, for the outstanding paper published in the Journal of the Audio Engineering Society during the previous two years.
Recipients: J. M. Berman and Lawrence R. Fincham 1980; Barry A. Blesser 1979.

CANADIAN AERONAUTICS AND SPACE INSTITUTE (CASI)

222 Somerset Street West, Suite 601, Ottawa, Canada K2P 0J1
A. J. S. Timmins, Executive Director
PRESIDENTS . . . Dr. B.D. MacIsaac, 1987-88; Mr. C.B. Wrong, 1986-87; M. C. W. Davy 1985-1986; W. J. Rainbird 1984-1985; H. Halton 1983-1984; I. A. Gray 1982-1983; R. E. Morris 1981-1982; J. D. MacNaughton 1980-81; H. I. H. Saravanamuttoo 1979-80; L. W. Morley 1978-79; K. R. Greenaway 1977-78; J. P. Beauregard 1976-77; W. M. McLeish 1975-76; Ian S. Macdonald 1974-75; F. R. Thurston 1973-74.

AWARDS

C. D. HOWE AWARD . . . for achievement in the field of policy, planning, and overall leadership in Canadian aeronautics and space activities. Consists of a silver wall plaque.
Recipients: Dr. P.A. Lapp, 1987; Mr. A. Morrison, 1986; R. Lefrancois 1985; I. A. Gray 1984; E. L. Smith 1983; L. S. Clarke 1982; F. R. Kearns 1981; W. M. McLeish 1980; R. D. Richmond 1979; M. W. Auld 1977; Lt.Gen. W. K. Carr 1976; E. H. Moncrieff 1975; J. R. Baldwin 1974; H. R. Footit 1973; G. N. Patterson 1972; D. A. Golden 1970; A. Bandi 1969; T. E. Stephenson 1968.

F. W. (CASEY) BALDWIN AWARD . . . for the best paper published in the CASI Journal in the calendar year. Consists of a silver medal and stand.
Recipients: J. DeLaurier, B. Gagnon, J. Wong, R. Williams and C. Hayball, 1987; D.G.D. Watt, K.E. Money, R.L. Bondar, R.B. Thirsk, M. Garneau and P. Scully-Power, 1986; C. L. Cook 1985; P. C. Hughes 1984; G. A. Terwissen and L. K. John 1983; W. Wiebe and R. V. Dainty 1982; R. E. Jones 1981; R. E. Barrington 1980; D. C. Emmerson 1979; B. Eggleston and D. J. Jones 1978; L. H. Ohman 1977; W. S. Hindson and S. R. M. Sinclair 1976; J. J. Gottlieb 1975; F. Mavriplis 1974; C. H. Glenn 1973; F. Mavriplis 1972; A. B. Bauer, J. L. Hess and A. M. O. Smith 1971; L. Brown 1970; D. C. Whittley 1969; M. C. Eames 1968.

McCURDY AWARD . . . for outstanding achievement in the art, science and engineering relating to aeronautics and space research. Consists of a trophy and silver medal.
Recipients: Mr. G. Caiger, 1987; Mr. J. Templin, 1986; C. Wrong 1985; W. R. Franks 1984; J. D. F. MacNaughton 1983; J. Uffen 1982; D. C. MacPhail 1981; H. Kowalik 1980; K. Hawkshaw 1979; F. R. Thurston 1976; M. S. Kuhring 1975; L. W. Morley 1974; D. C. Whittley 1973; E. L. Smith 1972; F. H. Buller 1971; F. C. Phillips 1970; B. Etkin 1969; G. V. Bull 1968.

ROMEO VACHON AWARD . . . for a recent and outstanding display of initiative, ingenuity and his own personal skills in the solution of a particular problem associated with Canadian aerospace technology.
Recipients: Mr. P. Bootsma 1987; Mr. K. Dooley, 1986; Sgt. P. A. McAuley 1985; F. H. Mee 1984; D. Kramer 1983; N. Chan 1981; J. C. L. Richard 1980; R. Cox 1978; Major J. C. Eggenberger 1976; M. A. Laviolette 1974; J. Ratte 1973; E. J. McLaren 1972; J. F. Fairchild 1971; H. E. Rasmussen 1970; S. N. Green 1969.

TRANS-CANADA (McKEE) TROPHY . . . the Trans-Canada Trophy, generally known as the McKee Trophy, is presented annually, except when no qualified recipient is nominated, for outstanding achievement in the field of air operations.
Recipients: Mr. S.W. Grossmith, 1987; Mr. T. Watt, 1986; P. Davoud 1985; Col. G. N. Henderson (posthumously) 1984; D. H. Rogers 1983; E. N. Ronaasen (posthumously) 1982; F. D. Adkins 1981; Col. R. D. Schultz 1978; A-V-M.A.E. Godfrey (Ret) 1977; D. Fairbanks 1976; J. A. M. Austin 1975; R. H. Fowler 1974; M. W. Ward 1973.

W. RUPERT TURNBULL LECTURE . . . an annual prize lecture presented at the Annual General Meeting by an eminent and carefully selected lecturer. The award consists of a certificate and honorarium.
Recipients: Dr. P. Lissaman, 1987; Dr. P. Mandl, 1986; J. L. Orr 1985; R. D. Hiscocks 1984; G. M. Lindberg 1983; J. S. MacDonald 1982; K. E. Money 1981; A. Roshko 1980; G. V. Parkinson 1979; H. Masursky 1978; I. S. MacDonald 1977; E. P. Cockshutt 1976; R. T. Jones 1975; A. W. Fia 1974; J. E. Steiner 1973; J. H. deLeeuw 1972; J. P. Beauregard 1971; R. J. Templin 1970; B. G. Newman 1969; H. S. Ribner 1968.

CANADIAN COUNCIL OF PROFESSIONAL ENGINEERS (CCPE)

116 Albert Street, Suite 401, Ottawa, Ontario, Canada K1P 5G3
Donald Laplante, Executive Director

PRESIDENTS . . . P.A. Lapp 1987-88; Y.C. Dupuis 1986-87; N. A. Johnson 1985-86; R. A. Hemstock 1984-85; J. G. Evans 1983-84; J. E. Benson 1982-83; G. Perron 1981-82; P. Van Vliet 1980-81; F. C. Turner 1979-80; D. W. Zaikoff 1978-79; R. N. McManus 1977-78; C. J. Moull 1977; Eric C. Garland 1976; N. Gilles Tanguay 1975; Russ Hood 1974; D. C. Lambert 1973.

AWARDS

CANADIAN ENGINEERS' GOLD MEDAL AWARD . . . is presented as the highest mark of distinction and exceptional achievement to a Canadian Engineer who has shown himself to be outstanding in his chosen field, or who, by his writings or other endeavours, has made a highly significant contribution to the advancement of standards of excellence and ethical conduct in the field of engineering in Canada.
Recipients: Lloyd R. McGinnis 1986; Bernard Lamarre 1985; Camille Dagenais 1984; John Anderson-Thomson 1983; Larkin Kerwin 1982; John Leslie Charles 1981; J. B. Angel 1980; Robert F. Shaw 1979; George Ford 1978; George B. Langford 1977; George W. Govier 1976; J. A. Ouimet 1975; W. M. Armstrong 1974; R. M. Hardy 1973; R. F. Legget 1972.

MERITORIOUS SERVICE AWARDS . . . to recognize outstanding service and dedication to the Canadian engineering profession through Canadian professional, consulting or technical associations and societies or through voluntary participation in community work and to enhance the role of the associations and societies in the career of the professional engineer.
Recipients: David E. Mitchell, (Community Service) and Eric C. Garland, (Professional Service) 1986; Gilles Perron 1985; Richard M. Dillon 1984; Keith A. Henry 1983; Ralph N. McManus 1982; John H. Dinsmore 1981.

CANADIAN SOCIETY FOR CHEMICAL ENGINEERING (CSChE)

1785 Alta Vista Drive, Suite 300, Ottawa Ontario K1G 3Y6
Patricia M. King, Executive Secretary

PRESIDENTS . . . K. Thompson 1987-88; F.D. Otto 1986-87; A. J. Newton 1985-86; H. K. Rae 1984-85; G. H. Thomson 1983-84; W. Petryschuck 1981-82; J. M. Stewart 1980-81; N. Epstein 1979-80; R. J. Waugh 1978-79; P. J. Carreau 1977-78; E. H. Nenninger 1976-77; E. B. Tinker 1975-76; C. W. Bowman 1974-75; W. J. M. Douglas 1973-74.

AWARDS

ERCO AWARD . . . awarded to a resident of Canada who has made a distinguished contribution to the field of chemical engineering, and who shall not have reached the age of 40 years by Jan 1 of the year in which the nomination becomes effective.
Recipients: Daniel DeKee 1987; Axel Meisen 1986; James F. Kelly 1985; Bruce M. Sankey 1984; J. R. Grace 1983; C. R. Phillips 1982; Martin Ternam 1981; R. Luus 1980; A. P. Watkinson 1979; E. Rhodes 1978; B. B. Pruden 1977; M. E. Charles 1976; C. E. Capes 1975; A. E. Hamielec 1974; M. Moo-Young 1973; N. J. Themelis 1972; I. S. Pasternak 1971; T. W. Hoffman 1970.

R. S. JANE MEMORIAL LECTURE AWARD . . . presented to a person who has made an exceptional achievement in the field of chemical engineering or industrial chemistry.
Recipients: Roger Butler 1987; C.E. Capes, 1986; Leslie W. Shemilt 1985; H. Veltman 1984; E. J. Buckler 1983; H. H. Holton 1982; S. Sourirajan 1981; D. B. Robinson 1980; Douglas Sargent Montgomery 1979; H. K. Rae 1978; J. B. Hyne 1977; R. H. Wright 1976; I. E. Puddington 1975; A. Cholette 1974; O. J. C. Allenby 1973; N. S. Grace 1972; H. Freeman 1971; N. I. Battista 1970; Morris Katz 1969; L. S. Renzoni 1968.

AWARD IN INDUSTRIAL PRACTICE SPONSORED BY ESSO PETROLEUM CANADA . . . presented to a Canadian resident or group for a distinguished application of chemical engineering or industrial chemistry to the industrial sphere.
Recipients: Eric L. Tollefson, 1987; Jasper Mardon 1985; G. E. Courtnage 1984; Technology Group, Sheritt Gordon Mines 1983; H. C. Prime 1982; R. S. Dudley 1981; E. N. Banks 1980; K. Pugi 1979; R. G. Routledge 1978.

ENGINEERING FOUNDATION (EF)

345 East 47th Street, New York, New York 10017
Harold A. Comerer, Director

CHAIRMEN . . . F.F. Aplan 1985-87; I. M. Viest 1983-85; V. A. Haddad 1982-83; Raymond P. Genereaux 1980-82; M. A. Cordovi 1979-78; A. B. Giordano 1978-77 R. W. Sears 1977-76; P. E. Queneau 1975- 74; C. L. Tutt 1973-72.

ENGINEERING INFORMATION (EI)

345 East 47th St., New York, New York 10017

PRESIDENTS . . . Herbert B. Landau 1982-present; Martha E. Williams 1980-82; Walter Grattidge 1978-80; Jay Hilary Kelley 1976-78; Paul E. Irick 1973-76.

THE ENGINEERING INSTITUTE OF CANADA (EIC)

700 E. I. C. Building, 2050 Mansfield Street, Montreal, Quebec H3C 1Z2
T. Chris Arnold, General Mgr

PRESIDENTS . . . W.A.H. Filer 1987-88; Remy Dussault 1986-87; W. B. Rice 1985-1986; SH. L. Macklin 1984-1985; Jack Priestman 1981; V. D. Thierman 1980; Colin D. diCenzo 1979; Russell Hood 1978; A. E. Steeves 1976-77; Robert F. Shaw 1975-76; D. L. Mordell 1974-75; Ian A. Gray 1973-74; William P. Harland 1972-73.

AWARDS

THE JULIAN C. SMITH MEDAL . . . awarded for "Achievement in the Development of Canada".
Recipients: Dr. J.S. Foster, D.A. Chisholm, John Ivor Clark, James Mardon 1987; Bernard Lamarre 1985; D. T. Wright and J. W. MacLaren 1984; W. F. Light 1983; Donald L. Allen and Carson F. Morrison 1980; Camille A. Dagenais and Ian A. Gray 1979; C. A. Dagenais and Ian A. Gray 1978; R. M. Hardy, Colin D. diCenzo, H. L. Macklin and Josef Kates 1977; L. Austin Wright and James W. Kerr 1976; Hon. Jack Davis and Guy Saint-Pierre 1975; George E. Humphries and James D. Dineen 1974; Elsie Gregory MacGill 1973; G. Ross Lord and Alex G. Lester 1972; Arthur R. Harrington and Louis Philippe Bonneau 1971; Francois Rousseau and Robert F. Legget 1970; Robert H. Winters and Irving R. Tait 1969; J. L. Charles and Robert F. Shaw 1968.

THE LEONARD MEDAL . . . awarded for the best paper on mining subjects.
Recipients: M. A. Pfeifer and D. M. Kovachevich 1985; G. S. Zimmer 1979; D. R. Piteau and D. C. Martin 1978; E. T. Robinsky 1976; R. S. Cleland and K. H. Singh 1974; E. J. Klohn 1973; P. N. Calder 1971; G. Grassmuck 1970; J. C. White 1969; H. G. Ewanchuk 1968.

THE PLUMMER MEDAL . . . is awarded for the best paper on chemical and metallurgical subjects.

Recipients: G. M. Swinkels, R. M. G. Berezowsky, P. Kawulka, C. R. Kirby, G. L. Bolton, C. E. D. Maschmeyer, E. F. G. Milner and B. M. Parekh 1979; G. A. Irons and R. I. L. Guthrie 1977; L. J. Pertuck, J. W. Spyker and W. W. Husband 1972; R. F. Feldman and P. J. Sereda 1971.

SIR JOHN KENNEDY MEDAL . . . is awarded by the Council of the Institutes as a recognition of outstanding merit in the profession or of noteworthy contribution to the science of engineering or to the benefit of the Institute.
Recipients: George W. Govier 1987; Camille Dagenais 1985; James M. Ham 1983; Ian A. Gray 1982; J. H. Dinsmore-R. F. Legget-L. M. Nadeau 1977; Jean P. Carriere 1975; J. Georges Chenevert 1973; Joseph Mervyn Hambley 1971; J. Alphonse Ouimet 1969.

ENGINEERING SOCIETIES LIBRARY (ESL)

345 East 47th Street, New York, New York 10017
S. K. Cabeen, Director

CHAIRMEN . . . Theodore Selover 1985-87; Gerald Fox 1983-85; Donald Nesslage 1981-83; F. Lincoln Vogel 1979-81; Reuben Pfennig 1977-79; Reinhard Hellman 1976-77.

ETA KAPPA NU

EE Department, University of Illinois, Urbana, Illinois 61801
Paul K. Hudson, Executive Secretary

PRESIDENTS . . . Joanne Waite 1985; Russell Lueg 1981; Sydney R. Parker 1980; Alan R. Stoudinger 1979; Albert Hauser 1978; Marcus D. Dodson.

AWARDS

OUTSTANDING E. E. STUDENT AWARD . . . The outstanding student is chosen by a jury from the nominations of the chapters. Also several honorable mentions are awarded. An award dinner is held in their honor in August in California with all expenses paid, plus a gift of $1,000.00 to the winner.
Recipients: Yvonne M. Utzig 1985.

OUTSTANDING E. E. PROF. AWARD . . . Chapters are asked to send in their nomination for outstanding Prof in their dept. The jury's selection is presented with a certificate and a $500 gift by Alumni Chapter and the International Board at the school of the winner.
Recipients: No Report

OUTSTANDING YOUNG E. ENGINEER AWARD . . . The jury selects the outstanding young E. E. from the nominations. The candidate must be less than 35 years old and have received their B. S. degree not more than ten years ago. An Award Dinner is held in their honor in N.Y.C., plus a gift of $500.00.
Recipients: William E. Moerner 1985.

EMINENT MEMBER AWARD . . . The Eminent Member Award is made on technical contributions. One award a year.
Recipients: Donald Christiansen 1985.

OUTSTANDING JUNIOR STUDENT AWARD . . . Nominations from Chapters. Several Honorable Mentions. Award Dinner at the school of the winner. Gift of $500.00.
Recipients: Roger A. Davidson 1985.

OUTSTANDING STUDENT AWARDS . . . at City University London, Ecole Superieure d'Electricite Paris and The University of Manitoba Canada. Winners are selected by the Faculty. Each winner at each school receives a gift of $200.00.
Recipients: Clive Croome, London; Dorine Szij, Paris and Sidney W. Allman, Manitoba 1985.

FEDERATION OF MATERIALS SOCIETIES (FMS)

1707 L Street, N.W., Suite 333, Washington, D.C. 20036
Betsy Houston, Executive Director

PRESIDENTS . . . Gilbert M. Ugiansky 1987; James A. Ford, 1986; James A. Ford 1985; Lendell E. Steele 1984; Emanuel Horowitz 1983; Clifford H. Wells 1982; Stanley V. Margolin 1981; Abraham Hurlich 1980; Carl Rampacek 1979; Jerome Kruger 1978.

AWARDS

NATIONAL MATERIALS ADVANCEMENT AWARD.
Recipients: John B. Wachtman, Jr. 1986; Paul C. Maxwell 1985.

FLUID POWER SOCIETY (FPS)

2900 N. 117th Street, Milwaukee, Wisconsin 53222

PRESIDENTS . . . J. Otto Byers, 1987-88; Raymond F. Hanley 1986-87; Norman E. Schmidt 1985-86; Harry R. Holsen 1984-85; Carroll Grigsby 1983-84; Alan Tiedman 1982-83; Donald R. Harter 1981-82; Robert W. Hanpeter 1980-81; Anton H. Henh 1979-80; Richard Read 1977-78; Allen Tucker 1976-77; Robert Wolff 1975-76; Ward Sievenpiper 1974-75; Paul K. Schacht 1973-74; Jack C. McPherson 1972-73.

AWARDS

OUTSTANDING INDUSTRY LEADER AWARD . . . the Society's most prestigious will be awarded to an individual whose efforts have resulted in a broad based impact on the Fluid Power Industry. Consideration is given to dedication, accomplishment and contributions to the Fluid Power Industry; efforts in behalf of one's employer and support given to industry and professional activities such as FPS, NFPA, and ISO, etc.
Recipients: Theodore N. Duncan 1986; William G. Paul, Jr. 1985; Z. J. "Zeke" Lansky 1981; Edward Briggs 1980; Lynn Charlson 1979; R. C. (Bob) Womack 1978; Walter Ernst 1976; Fred Ranney 1975; R. F. O'Keefe 1974; Robert Bellman 1973.

OUTSTANDING FLUID POWER EDUCATOR AWARD . . . this award will be presented to an individual with a minimum of 10 years experience in teaching fluid power in industry or academia. A 10 point list of criteria for selecting this individual has been established by the Educators Awards Committee. Entry forms for the category can be obtained from the FPS Headquarters.
Recipients: Dr. Arthur Akers 1987; James A. Sullivan 1986; Donald Pauken 1985; Ronald Gore 1981; Dr. Jack Slater 1980; Norman Dearth 1979.

DISTINGUISHED ACHIEVEMENT AWARD-INDUSTRIAL
Recipients: Theodore N. Duncan 1986; W.G. Paul, Jr. 1985; G.R. Keller 1984; O. Mahal 1983; E.M. Greer 1982; Z.J. Lansky 1981; E. Biggs 1980; L. Charlson 1979; R. Womack 1978; G. Altland 1977; W. Ernst 1976; F. Ranney 1975; D. Wiberg 1974.

DISTINGUISHED ACHIEVEMENT AWARD-EDUCATIONAL
Recipients: Dr. Arthur Akers 1987; James A. Sullivan 1986; D. Pauken 1985; D. Olson 1984; E.C. Fitch 1983; W. L. Amminger 1982; R.R. Gore 1981; J. Slater 1980; N. Dearth 1979.

THE GEOLOGICAL SOCIETY OF AMERICA, INC. (GSA)

3300 Penrose Place, P. O. Box 9140, Boulder, Colorado 80301
F. Michael Wahl, Executive Director

PRESIDENTS . . . Jack E. Oliver 1987; W. Gary Ernst 1986; Brian J. Skinner 1985; M. Gordon Wolman 1984; Paul A. Bailly 1983; Digby J. McLaren 1982; Howard R. Gould 1981; Laurence L. Sloss 1980; Leon T. Silver 1979; Peter T. Flawn 1978; Charles L. Drake 1977; Robert E. Folinsbee 1976; Julian R. Goldsmith 1975; Clarence R. Allen 1974; John C. Maxwell 1973.

AWARDS

EDWARD B. BURWELL, JR., MEMORIAL AWARD . . . presented to the author or authors of a paper selected for its excellence in advancing knowledge, principles or practice of engineering geology.
Recipients: James F. Quinlan, Ralph O. Ewers 1986; Larry D. Dyke 1985; Roy E. Hunt 1984; Peter W. Lipman and Donal R. Mullineaux 1983; F. Lionel Peckover and Douglas R. Piteau

1982; Allen W. Hatheway and Cole R. McClure, Jr. 1981; Kerry E. Sieh 1980; John W. Bray and Evert Hoek 1979; Nicholas R. Barton 1978; Richard E. Goodman 1977; David J. Varnes 1976; Erhard M. Winkler 1975; Robert F. Legget 1974; J. E. Hackett and Murray R. McComas 1973; Richard J. Proctor 1972; Edwin B. Eckel 1971; Glenn R. Scott and David J. Varnes 1970; Lloyd B. Underwood 1969.

ILLUMINATING ENGINEERING SOCIETY OF NORTH AMERICA (IES)

345 East 47th Street, New York, New York 10017
Jack Richards, Chief Staff Officer

PRESIDENTS ... Richard C. LeVere 1987-88; Robert V. Day 1986-87; Rita M. Harrold 1985-86; James E. Jewell 1984-85; Howard M. Brandston 1983-84; Lewis S. Sternberg 1982-83; Stephen S. Squillace 1981-82; Donald R. Marcus 1980-81; John E. Flynn 1979-80; Will S. Fisher 1978-79; David Patterson 1977-78; Carl Long 1976-77; Kurt Franck 1975-76; George W. Clark 1974-75.

AWARDS

IES MEDAL ... to recognize meritorious technical achievement that has conspicuously furthered the profession, art or knowledge of illuminating engineering. This award was first presented in 1944.
Recipients: John M. Waldram 1986; Stephen S. Squillace 1985; Russell Putnam 1984; Charles Douglas 1983; Stanley McCandless 1982; George W. Clark 1981; Robert T. Dorsey 1980; Richard Kelly 1979; A. H. McKeag 1977; Clarence C. Keller 1976; Charles L. Amick 1975; Domina E. Spencer and Perry Moon 1974; Willem Elenbaas 1973; H. Richard Blackwell 1972; George J. Taylor 1971; Dorothy Nickerson 1970.

LOUIS B. MARKS AWARD ... presented to a member of the Society in recognition of exceptional service to the Society of a non-technical nature.
Recipients: William K.Y. Tao 1986; Francis Clark 1985.

INDUSTRIAL RESEARCH INSTITUTE (IRI)

100 Park Avenue, Suite 3600, New York, New York 10017
Charles F. Larson, Executive Director

PRESIDENTS ... K.W. McHenry, 1989; S.A. Heininger, 1988; R.W. Schmitt, 1987; W. R. Stumpe 1986; S. W. Tinsley 1985; V. A. Russo 1984; Geoffrey Place 1983; H. W. Coover 1982; Jules Blake 1981; J. M. M. Salsbury 1980; E. C. Galloway 1979; F. H. Healey 1978; D. J. Blickwede 1977; A. M. Bueche 1976; N. B. Hannay 1975; H. I. Fusfeld 1974; W. L. Abel 1973.

AWARDS

I. R. I. ACHIEVEMENT AWARD ... is presented to honor outstanding accomplishment in individual innovation and creativity, which contributes broadly to the development of industry and to the benefit of society. Each year, the actual award is an original work of art, symbolizing the creative nature of the achievement.
Recipients: R.D. Maurer, 1986; J.E. Franz, 1985; A. S. Hay 1984; J. W. Backus 1983; H. W. Boyer 1982; R. N. Noyce 1981; A. H. Bobeck 1980; S. D. Stookey 1979; F. B. Colton 1978; R. H. Wentorf, Jr. 1977; L. G. Van Uitert 1976; M. R. Hilleman 1975; W. G. Pfann 1974; W. F. Gresham 1973.

INDUSTRIAL RESEARCH INSTITUTE MEDAL ... is presented to recognize and honor outstanding accomplishment in leadership or management of industrial research, which contributes broadly to the development of industry and to the benefit of society.
Recipients: I.M. Ross 1987; G.E. Pake 1986; R. E. Gomory 1985; H. W. Coover 1984; E. E. David, Jr. 1983; N. B. Hannay 1982; W. H. Armistead 1981; L. H. Sarett 1980; A. M. Bueche 1979; M. E. Pruitt 1978; J. J. Burns 1977; H. B. G. Casimir 1976; James Hillier 1975; R. W. Cairns 1974; W. E. Shoupp 1973; Peter Goldmark 1972; Henri Busignies 1971; W. O. Baker 1970; P. E. Haggerty 1969; J. H. Dessauer 1968.

MAURICE HOLLAND AWARD ... is presented in recognition of the best paper published in RESEARCH MANAGEMENT the previous year based on pertinence to the field of research management, significance to the field of research management, and originality of new management concepts or of research on which the paper was based.
Recipients: Richard Foster, Lawrence Linden, Alan Kanstron, and R.L. Whitely, 1986; Robert Frosch 1985; G. E. Manners, Jr., J. A. Steger, and T. W. Zimmerer 1984; W. R. Stumpe 1983; D. B. Merrifield 1982.

INSTITUTE OF ELECTRICAL AND ELECTRONICS ENGINEERS (IEEE)

345 East 47th Street, New York, New York 10017
Eric Herz, General Manager-Executive Director

PRESIDENTS ... Henry L. Bachman 1987; Brian O. Weinschel 1986; Charles A. Eldon 1985; Richard J. Gowen 1984; James B. Owens 1983; Robert E. Larson 1982; Richard W. Damon 1981; Leo J. Young 1980; Jerome J. Suran 1979; Ivan A. Getting 1978; Robert M. Saunders 1977; Joseph K. Dillard 1976; Arthur P. Stern 1975; John J. Guarrera 1974; Dr. Harold Chestnut 1973.

AWARDS

ALEXANDER GRAHAM BELL MEDAL ... given annually for exceptional contributions to the advancement of telecommunications. It was established in 1976 and is sponsored by the American Telephone and Telegraph Co. It consists of a Gold Medal, Certificate, and $10,000.
Recipients: Joel S. Engel, Richard H. Frenkiel, William C. Jakes Jr. 1987; Bernard Widrow 1986; Charles K. Kao 1985; Andrew J. Viterbi 1984; Stephen O. Rice 1983; Harold A. Rosen 1982; David Slepian 1981; Richard R. Hough 1980; A. Christian Jacobaeus 1979; M. Robert Aaron, John S. Mayo and Eric E. Summer 1978; Eberhardt Rechtin 1977; Amos E. Joel, Jr., William Keister and Raymond W. Ketchledge 1976.

BROWDER J. THOMPSON MEMORIAL PRIZE AWARD ... given annually for the most outstanding paper in any IEEE publication by an author or joint authors under the age of 30. It was established in 1945 and consists of a certificate and $1,000.
Recipients: Daniel Welt, Jon Webb 1987; John K. Ousterhout 1986; Douglas MacGregor, David S. Mothersole 1985; John C. Curlander 1984; Daniel S. Kimes and Julie A. Kirchner 1983; Stig Skelboe 1982; Lawrence H. Goldstein 1981; Alan S. Willsky 1980; Marvin B. Lieberman 1979; David A. Hounshell 1978; Michael R. Portnoff 1977; Dan E. Dudgeon, Russell M. Mersereau 1976; Nuggehally S. Javant 1975: Jorn Justesen 1974; Jerry Mar 1973; G. David Forney, Jr. 1972; L. J. Griffiths 1971; J. David Rhodes 1970.

CLEDO BRUNETTI AWARD ... given annually for outstanding contributions in the field of miniaturization in the electronic arts. It was established in 1976 and consists of a Certificate and $1,000.
Recipients: Michael C. Hatzakis 1987; Richard M. White 1986; Alec N. Broers 1985; Harry W. Rubenstein 1984; Abe Offner 1983; Robert H. Dennard 1982; Donald R. Herriott 1981; Marcian E. Hoff, Jr. 1980; Geoffrey W. A. Dummer and Philip J. Franklin 1979; Jack S. Kilby and Robert N. Noyce 1978.

DAVID SARNOFF AWARD ... given annually for outstanding contributions in the field of electronics. It was established in 1959 and is sponsored by the RCA Corporation. It consists of a Gold Medal, Bronze Replica, certificate, and $1,000.
Recipients: Alan B. Fowler, Frank F. Fang 1987; Yasuharu Suematsu 1986; Henry Kressel 1985; Alan D. White and Jameson D. Rigden 1984; Hermann K. Gummel 1983; Nobutoshi Kihara 1982; Cyril Hilsum 1981; Marshall I. Nathan 1980; A. Gardner Fox and Tingye Li 1979; Stephen E. Harris 1978; Jack M. Manley and Harrison E. Rowe 1977; George H. Heilmeier 1976; Bernard C. DeLoach, Jr. 1975; F. L. J. Sangster 1974; Max Vernon Mathews 1973; Edward G. Ramberg 1972; Alan L. McWhorter 1971; John B. Johnson 1970.

DONALD G. FINK AWARD ... given annually for the outstanding survey, review or tutorial paper in any of the IEEE transactions, journals, magazines, or proceedings. It was established in 1979 and consists of a certificate and $1,000.
Recipients: Shahid U.H. Quereshi 1987; Thomas H. Johnson 1986; Arnold Reisman 1985; Robert A. Scholtz and Enders A. Robinson 1984; Anil K. Jain 1983; Arun N. Netravali and John O. Cimb 1982; Whitfield Diffie and Martin E. Hellman 1981.

EDISON MEDAL ... awarded annually for a career of meritorious achievement in electrical science or electrical engineering or the electrical arts. It was established in 1904 and consists of a Gold Medal, small gold replica, a certificate, and $10,000.
Recipients: Robert A. Henle 1987; James L. Flanagan 1986; John D. Kraus 1985; Eugene I. Gordon 1984; Herman Paul Schwan 1983; Nathan Cohn 1982; C. Chapin Cutler 1981; Robert Adler 1980; Albert Rose 1979; Daniel E. Noble 1978; Henri Busignies 1977; Murray Joslin 1976; Sidney Darlington 1975; Jan A. Raichman 1974; B. D. H. Tellergren 1973; William H. Pickering 1972; John W. Simpson 1971; Howard A. Aiken 1970.

EDUCATION MEDAL ... awarded annually to recognize outstanding contributions to engineering education. It was established in 1956 and consists of a Gold Medal, Bronze Replica, and certificate, and $10,000.
Recipients: Joseph W. Goodman 1987; Richard B. Adler 1986; James Franklin Gibbons 1985; Athanasios Papoulis 1984; Mischa Schwartz 1983; King-Sun Fu 1982; Ernest S. Kuh 1981; Aldert Van Der Ziel 1980; John R. Ragazzini 1979; Harold A. Peterson 1978; Robert M. Fano 1977; John G. Linvill 1976; Charles A. Desoer 1975; John G. Truxal 1974; Lofti A. Zadeh 1973; M. E. Van Valkenburg 1972; Franz Ollendorff 1971; Jacob Millman 1970.

EMANUEL R. PIORE AWARD ... given annually for outstanding achievement in the field of information processing, in relation with computer science, to an individual or team of two individuals. It was established in 1976 and is sponsored by International Business Machines Corporation. It consists of a Gold Plated Bronze Medal, Certificate, $2,000 and $2,500 international travel grant.
Recipients: David J. Kuck 1987; David C. Evans, Ivan E. Sutherland 1986; Azriel Rosenfeld 1985; Harvey G. Cragon 1984; Niklaus Wirth 1983; Kenneth L. Thompson and Dennis M. Ritchie 1982; Lawrence R. Rabiner and Ronald W. Schafer 1980; Richard W. Hamming 1979; J. Presper Eckert and John W. Mauchly 1978; George R. Stibitz 1977.

FOUNDERS MEDAL ... given annually for major contributions in the leadership, planning, and administration of affairs of great value to the electrical and electronics engineering profession. It was established in 1952 and consists of a Gold Medal, Bronze Replica, certificate, and $10,000.
Recipients: James B. Owens 1987; George H. Heilmeier 1986; William C. Norris 1985; Koji Kobayashi 1984; Joseph M. Pettit 1983; Shigeru Yonezawa 1982; James Hillier 1981; Simon Ramo 1980; Hanzo Omi 1979; Donald G. Fink 1978; Jerome B. Wiesner 1977; Edward W. Herold 1976; John G. Brainerd 1975; Lawrence A. Hyland 1974; William R. Hewlett and David Packard 1973; Masaru Ibuka 1972; Ernst Weber 1971; Morris D. Hooven 1970.

FREDERIK PHILIPS AWARD ... for outstanding accomplishments in the management of research and development resulting in effective innovation in the electrical and electronics industry. It was established in 1971 and is sponsored by N. V. Phillips' Gloeilampenfabrieken. It consists of a Gold Medal, certificate and $2,000.
Recipients: Solomon J. Buchsbaum 1987; William Hittinger 1986; George Heilmeier 1985; John K. Galt 1984; Allen E. Puckett 1983; Werner J. Kleen 1981; Dean A. Watkins 1981; William M. Webster 1980; Gordon E. Moore 1979; William E. Shoupp 1978; Koji Kobayashi 1976; C. Lester Hogan 1975; Chauncey Guy Suits 1974; John H. Dessauer 1973; William O. Baker 1972; Frederik J. Phillips 1971.

HARRY DIAMOND MEMORIAL AWARD ... given annually for outstanding technical contributions in the field of government service in any country, as evidenced by publication in professional society journals. It was established in 1949 and consists of a certificate and $2,000.
Recipients: Carl E. Baum 1987; Thomas G. Giallorenzi 1986; Howard S. Jones, Jr. 1985; Sydney R. Parker 1984; Merrill I. Skolnik 1983; Jules Aarons 1982; George Abraham 1981; Martin Greenspan 1980; Henry P. Kalmus 1979; David M. Kerns 1978; Jacob Rabinow 1977; Maxime A. Faget 1976; Louis Costrell 1975; Chester H. Page 1974; Harold Jacobs 1973; William B. McLean 1972; Arthur H. Guenther 1971; Allen V. Astin 1970.

IEEE AWARD IN INTERNATIONAL COMMUNICATION (in honor of Hernand and Sosthenes Behn) ... given annually for outstanding contribution in the field of international communication. It was established in 1966 and is sponsored by International Telephone and Telegraph Corp. It consists of a plaque, certificate, and $2,000.
Recipients: Ken-Ichi Miya 1987; Richard B. Nichols 1986; Frederick T. Andrews, Jr. 1985; Lynn W. Ellis 1983; Hiroshi Inose 1982; Richard C. Kirby 1981; Armig B. Kandoian 1980; A. Nejat Ince 1979; F. Louis H. M. Stumpers 1978; Sidney Metzger 1976; John A. Puente 1975; Leslie H. Bedford 1974; Vladimir A. Kotelnikov 1973; Frank de Jager and Johannes A. Greefkes 1972; Eugene F. O'Neill 1971; Herre Rinia 1970.

JACK A. MORTON AWARD ... one award given annually to an individual or group for outstanding contributions in development of solid-state devices. It consists of a Bronze Medal, Certificate and $2,000. Sponsored by 20 semiconductor organizations of the United States, Europe and Japan. Date established: 1974.
Recipients: Dennis D. Buss, Richard A. Chapman, Michael A. Kinch 1987; Herbert Kroemer 1986; Robert D. Burnham, Donald R. Scifres and William Streifer 1985; Hans S. Rupprecht and Jerry M. Woodall 1984; Nick Holonyak, Jr. 1981; James F. Gibbons 1980; Martin P. Lepselter 1979; Juri Matisoo 1978; Morgan Sparks 1977; Robert N. Hall 1976.

LAMME MEDAL ... awarded annually for meritorious achievement in the development of electrical or electronic apparatus or systems. It was established in 1928 and is supported by the Westinghouse Foundation. It consists of a Gold Medal, Bronze Replica, certificate, and $10,000.
Recipients: Ingolf Birger Johnson 1986; Loren Frank Stringer 1985; William C. MacMurray 1984; Marion E. Hines 1983; Marvin Chodorow 1982; George B. Litchford 1981; Eugene C. Starr 1980; James M. Lafferty 1979; Harry W. Mergler 1978; Bernard M. Oliver 1977; C. Kumar N. Patel 1976; Harold B. Law 1975; Seymour B. Cohn 1974; Charles Stark Draper 1973; Yu H. Ku and Robert H. Park 1972; Winthrop M. Leeds 1971; Harry F. Olson 1970.

MEDAL OF HONOR ... given for a particular contribution that forms a clearly exceptional addition to the science and technology of concern to the institute. It was established in 1917 and consists of a Gold Medal, Bronze Replica, certificate and $20,000.
Recipients: Paul C. Lauterbar 1987; Jack St. Clair Kilby 1986; John R. Whinnery 1985; Norman F. Ramsey 1984; Nicolaas Bloembergen 1983; John W. Tukey 1982; Sidney Darlington 1981; Willima Shockley 1980; Richard Bellman 1979; Robert N. Noyce 1978; H. Earle Vaughan 1977; John R. Pierce 1975; Rudolf E. Kalman 1974; Rudolph Kompfner 1973; Jay W. Forrester 1972; John Bardeen 1971; Dennis Gabor 1970.

MORRIS E. LEEDS AWARD ... given annually for outstanding contribution to the field of electrical measurements. It was established in 1958 and is sponsored by Leeds and Northrup Foundation. It consists of an illuminated certificate and $1,000.
Recipients: William J. Moore 1987; Ian Kay Harvey 1986; Donald C. Cox 1985; Leonard Samuel Cutler 1984; Erich P. Ippen and Charles V. Shank 1983; Lothar Rohde 1982; Frank C. Creed 1981; Wallace H. Coulter 1980; Robert D. Cutkosky 1979; Thomas M. Dauphinee 1978; Arthur M. Thompson 1977; Francis L. Mermach 1976; Norbert L. Kusters 1974; Charles Howard Vollum 1973; Forest K. Harris 1972; Martin E. Packard 1971; Harold I. Ewen 1970.

MORRIS N. LIEBMANN MEMORIAL AWARD ... given annually for important contribution to emerging technologies recognized during preceding three calendar years. It was established in 1919 and consists of a certificate and $2,000.

Recipients: Bishnu S. Atal, Fumitada Itakura 1986; Russell D. Dupuis and Harold M. Manasevit 1985; David E. Carlson and Christopher R. Wronski 1984; Robert W. Brodersen, Paul R. Gray and David A. Hodges 1983; John R. Arthur, Jr. and Alfred Y. Cho 1982; Calvin F. Quate 1981; A.J. Demaria 1980; Ping King Tien 1979; Kuen C. Kao, Robert D. Maurer and John B. MacChesney 1978; Horst H. Berger and Siegfried K. Wiedmann 1977; Herbert J. Shaw 1976; A. H. Bobeck, P C. Michaelis and H. E. D. Scovil 1975; Willard S. Boyle and George E. Smith 1974; Nick Holonyak, Jr. 1973; Stewart E. Miller 1972; Martin Ryle 1971; John A. Copeland 1970.

NIKOLA TESLA AWARD ... one award given annually to an individual or group for outstanding contributions in the field of generation and utilization of electric power. It consists of a plaque and $1,000. Sponsored by IEEE Power Engineering Society. Date established: 1975.
Recipients: J Coleman White 1987; Eric R. Laithwaite 1986; Eugene C. Whitney 1985; Herbert H. Woodson 1984; Sakae Yamamura 1982; Dean B. Harrington 1981; Philip H. Trickey 1980; John W. Batchelor 1979; Charles H. Holley 1978; Cyril G. Veinott 1977; Leon T. Rosenberg 1976.

W. R. G. BAKER PRIZE AWARD ... given annually for the most outstanding paper reporting original work in any of the IEEE transactions, journals, magazines or proceedings. It consists of a certificate and $1,000.
Recipients: James L. Massey, Peter Mathys 1987; Adi Shamir 1986; John W. Adams and Alan N. Willson, Jr. 1985; Yannis Tsividis 1984; Ryszard Malewski, Kurt Feser, Chinh T. Nguyen and Nils Hylten-Cavallius 1983; Carl O. Bozler and Gary D. Alley 1982; Timothy C. May and Murray H. Woods 1981; Gordon M. Jacobs, David J. Allstot, Robert W. Broderson and Paul R. Gray 1980; Stephen W. Director and Gary D. Hachtel 1979; Eugene C. Sakshaug, James S. Kresge and Stanley A. Miske, Jr. 1978; Manfred R. Schroeder 1977; Robert W. Keyes 1976; Stewart E. Miller, Enrique A. J. Marcatili and Tingye Li 1975; David B. Large, Lawrence Ball and Arnold J. Farstad 1974; Leon O. Chua 1973; Dirk J. Kuizenga and Anthony E. Siegman 1972; Andrew H. Bobeck, Robert F. Fischer, Anthony J. Perneski, J. P. Remeika and L. G. VanUiteri 1971; George J. Friedman and Cornelius T. Leondes 1970.

HERMAN HALPERIN ELECTRIC TRANSMISSION AWARD (Formerly, William M. Habirshaw Award) ... given for outstanding contribution in the field of transmission and distribution of electric power. It was established in 1958 and is sponsored by Phelps Dodge Foundation. It consists of a Bronze Medal, certificate, and $1,000.
Recipients: Robert F. Lawrence 1987; Harold N. Scherer 1986; Harry M. Ellis 1985; Ralph S. Gens 1984; Andrew F. Corry 1983; Peter L. Bellaschi 1982; William R. Johnson 1981; Edward W. Kimbark 1980; Howard C. Barnes and Theodore J. Nagel 1979; Martin H. McGrath 1978; Francis J. Lane 1976; Everett J. Harrington 1975; Herbert R. Stewart 1974; Eugene W. Boehne 1973; Lionel Cahill and Jean-Jacques Archambault 1972; Gunnar Jancke 1971; Fred J. Vogel 1970.

INSTITUTE OF INDUSTRIAL ENGINEERS, INC (IIE)

25 Technology Park-Atlanta, Norcross, Georgia 30092
Gregory Balestrero, Acting Executive Director

PRESIDENTS ... Curtis J. Tompkins 1988-89; F. Stan Settles 1987-88; Kenneth E. Case 1986-87; Wilbur L. Meier Jr. 1985-1986; Roger P. Denney, Jr. 1984-1985; John A. White 1983-1984; Barry M. Mundt 1982-1983; Joe H. Mize 1981-82; Leroy O. Gillette 1980-81; Marvin E. Mundel 1979-80; Sidney G. Gilbreath, III 1978-79; Joseph H. Kehlbeck 1977-78; John F. Sweers 1976-77; William A. Smith, Jr. 1975-76; John P. Houston 1974-75; Roy L. Allen 1973-74.

AWARDS
AWARD FOR EXCELLENCE IN PRODUCTIVITY IMPROVEMENT ... recognizes individual(s) or organization who through diligent and innovative means has accomplished significant measurable and extensive achievements which increased productivi-

ty, eliminated human drudgery or improved the quality of working life.
Recipients: Allen-Bradley Co. and Baltimore Gas & Electric Co. 1987; Richard Dauch 1986; Atlantic Steel Company and An-Heuser-Busch, Inc. 1985; Boeing Commercial Airplane Co. and Diablo Systems, Inc. 1984; Black and Decker Corporate Management 1983; A. Joseph Tulkoff 1982; Robert E. Fowler 1981; San Antonio Air Logistics Center 1980.

DAVID F. BAKER DISTINGUISHED RESEARCH AWARD ... given for distinguished research in industrial engineering activity.
Recipients: Mahmoud Ayoub 1987; Jerry L. Purswell 1985; J.J. Solberg 1982; M.M. Ayoub 1981; Richard L. Francis 1977; T. H. Rockwell 1976; H. E. Smalley 1974; E. R. Tichauer and R. L. Disney 1972; L. J. Arp 1971; S. E. Elmaghraby 1970.

FRED C. CRANE DISTINGUISHED SERVICE AWARD ... recognizes an individual's contributions to the Institute. Normally, the individual's contribution must have been rendered at the Institute level and have consisted of long and arduous service.
Recipients: Raymond P. Lutz 1987; Terrence M. Peake 1986; William A. Smith, Jr. 1985; W. H. Reynolds 1983; W. C. Smith 1982; F. J. Johnson 1981; S. T. Wolfberg 1978; A. W. Cywar 1977.

IIE TRANSACTIONS DEVELOPMENT AND APPLICATIONS AWARD ... recognizes the paper judged outstanding in description of innovative and significant developments in industrial engineering which was published in IIE TRANSACTIONS during period of competition.
Recipients: Robert A. Walas and Ronald G. Askin 1985; Bezalel Gavish 1982; A. Dale Flowers 1981; J. J. Marsh, III and Ralph W. Swain 1978.

FRANK AND LILLIAN GILBRETH INDUSTRIAL ENGINEERING AWARD ... bestowed to those who have distinguished themselves by contributions to the welfare of mankind in the field of industrial engineering.
Recipients: Robert N. Lehrer 1987; Donald G. Malcolm 1986; Allan H. Mogensen 1984; Marvin E. Mundel 1982; J. M. Juran 1981; William Grant Ireson 1980; Harold E. Smalley 1979; James M. Apple 1978; B. W. Niebel 1976; W. K. Hodson 1973; Phil Carroll 1970; R. M. Barnes 1969; H. B. Maynard 1968.

JOINT PUBLISHERS BOOK-OF-THE-YEAR AWARD ... a group of publishers sponsors an award of at least $500 plus living and travel expenses to the Institute conference, to the author(s) of an outstanding book on Industrial Engineering.
Recipients: Harrison M. Wadsworth, Kenneth S. Stephens, A. Blanton Godfrey 1987; James A. Tompkins and John A. White, 1986; Arnoldo C. Hax and Dan Candea 1985; David I. Cleland and William R. King 1984; Gerald Nadler 1983; Frank M. Gryna, Jr. 1982; Paul A. Jensen and J. Wesley Barnes 1981; William T. Mooris 1980; Wayne C. Turner, Joe H. Mize, Kenneth E. Case 1979; R. H. Roy 1978; D. T. Phillips, A. Ravindran and J. J. Solberg 1977.

OUTSTANDING ACHIEVEMENT IN MANAGEMENT AWARD ... to recognize executives who have contributed to the industrial profession by creating a climate in their organizations which permits Industrial Engineering techniques to be perfected and used with outstanding results.
Recipients: Melvin R. Goodes 1987; Martin Kuilman 1986; Peter J. Ueberroth 1985; Paul J. McIntire 1984; Robert E. Reiss 1983; James W. McLernon, I. Andrew Rader 1981; Sterling O. Swanger 1980; Marshall McDonald 1979; F. J. Mancheski 1978; L. S. Crane 1976; T. D. Morris 1975; J. Z. DeLorean 1972; W. F. Rockwell, Jr. 1971; A. E. Perlman 1970; G. E. Keck 1969; R. DeYoung 1968.

AWARD FOR TECHNICAL INNOVATION IN INDUSTRIAL ENGINEERING ... recognizes significant innovative technical contributions to the industrial engineering profession as evidenced by (1) significantly expanding the body of knowledge in an industrial engineering function (2) meaningfully establishing yet another functional area of the profession, or (3) providing exceptional technical leadership in a major interdisciplinary project.
Recipient: Kjell B. Zandin, 1986.

IIE DISTINGUISHED EDUCATOR AWARD ... recognizes outstanding educators who have contributed significantly to the industrial engineering profession through teaching, research and publication, extension, teaching/learning innovation, and/or administration in an academic environment.
Recipients: Marvin H. Agee 1987; Ralph L. Disney 1986.

OUTSTANDING IIE PUBLICATION AWARD ... is the highest IIE honor for the outstanding current original publication which has appeared in any IIE sponsored or cosponsored medium. The award is made to the author or authors of that published work which is judged to be a meritorious contribution to the profession of industrial engineering.
Recipients: Randall P. Sadowski 1987; Gary E. Whitehouse 1986.

INSTITUTE OF NOISE CONTROL ENGINEERING (INCE)

P. O. Box 3206, Arlington Branch, Poughkeepsie, New York 12603
PRESIDENTS ... Jiri Tichy 1987; Robert D. Bruce 1986; Eric E. Ungar 1985; James G. Seebold 1984; Lewis S. Goodfriend 1983; Warren R. Kundert 1982; Malcolm J. Crocker 1981; John C. Johnson 1980; Harvey H. Hubbard 1979; Kenneth M. Eldred 1978; Peter K. Baade 1977; Kenneth M. Eldred 1976; George C. Maling, Jr. 1975; Vincent Salmon 1974; K. Uno Ingard 1973; Leo L. Beranek 1972.

INSTITUTE OF TRANSPORTATION ENGINEERS, INC. (ITE)

525 School Street, S.W., Suite 410, Washington, D.C. 20024
Thomas W. Brahms, Executive Director
PRESIDENTS ... Walter H. Kraft 1987; James H. Kell 1986; John D. Edwards, Jr. 1985; Melvin B. Meyer 1984; James L. Foley, Jr. 1983; Leo E. Laviolette 1982; Neilon J. Rowan 1981; Harvey Shebesta 1980; Elmer N. Burns 1979; William Marconi 1978; Edward A. Mueller 1977; D. W. Gwynn 1976; Harold L. Michael 1975; Laurence A. Dondanville 1974; William R. McGrath 1973.

AWARDS
BURTON W. MARSH DISTINGUISHED SERVICE AWARD ... the award is presented to a person who has made outstanding contributions to the advancement of the Institute of Transportation Engineers over a period of several years. It consists of a plaque-mounted medallion and a citation.
Recipients: Paul C. Box 1987; Alan T. Gonseth 1986; David W. Gwynn 1985; Harold L. Michael 1984; Daniel J. Hanson, Sr. 1983; Wilbur S. Smith 1982; Laurence A. Dondanville 1981; William R. McGrath 1980; John C. Landen 1979; David M. Baldwin 1978; Roy W. Jorgensen 1977; Donald M. McNeil 1976; Carlton C. Robinson 1975; Robert S. Holmes 1974; Fred W. Hurd 1973; Guy Kelcey 1972; Edward E. Wetzel 1971; Burton W. Marsh 1970.

THEODORE M. MATSON MEMORIAL AWARD ... Theodore M. Matson contributed greatly to the welfare of his fellow man through the advancement of the techniques of traffic engineering and through training of the men in the profession. This award has been established in his honor for outstanding contributions in the field of traffic engineering, including practical application of traffic engineering techniques or principles, valuable contributions through research, successful adaptation of research findings to a practical traffic situation, or the advancement of the profession through training or administration.
Recipients: Edward Hall, 1987; Herman J. Hoose 1986; Carlton C. Robinson 1985; Robert E. Conner 1984; Woodrow W. Rankin 1983; John F. Exnicios 1982; James H. Kell 1981; Robert S. Holmes 1980; Harold L. Michael 1979; Theodore W. Forbes 1978; H. Robert Burton 1977; Bruce D. Greenshields 1976; Louis E. Bender 1975; Harmer E. Davis 1974; David M. Baldwin 1973; Donald S. Berry 1972; Edmund R. Ricker 1971; Edward H. Holmes 1970; Fred W. Hurd 1969; Henry A. Barnes 1968; Alger F. Malo 1967; Guy Kelcey 1966; Wilbur S. Smith 1965;

George M. Webb 1964; Nathan Cherniack 1963; Charles M. Noble 1962; John L. Barker 1961; Burton W. Marsh 1960; Charles W. Prisk 1959; D. Grant Mickle 1958; O.K. Normann 1957.

PAST PRESIDENTS' AWARD FOR MERIT IN TRANSPORTATION ENGINEERING ... is intended to encourage the conduct and reporting of independent and original research and to provide a means for recognizing outstanding accomplishments. It is an annual award, with the recipient being determined by a committee of judges selected from ITE Past Presidents. Judging is based on written papers submitted in accordance with detailed rules.
Recipients: Juan M. Morales 1987; Chi Y. Lee 1986; Daniel B. Rathbone 1985; Timothy R. Newman 1984; Gary W. Euler 1983; Samuel I. Schwartz 1982; Chia-Juch Chang 1981; Armando Balloffet and George L. Hovey 1980; Robert E. Heightchew 1979; William Kunzman 1978; John E. Fisher and Robert E. Camou 1977; Ronald M. Cameron 1976; Peter N. Scifres and Fred R. Hanscom 1975; Charles F. Sterling 1974; Kenneth G. Courage 1973; Gary D. Long 1972; Peter A. Fausch 1971; Karl Lennart Bang 1970; Charles E. Dare 1969; Robert L. Bleyl 1968.

SECTION TECHNICAL AWARD ... the purpose of the "Section Technical Award" is to encourage active and timely participation by local Section Members in Section Technical Projects by conducting and reporting study, research and investigation of traffic and transportation subjects and to provide a means for the Institute to recognize outstanding accomplishment in such activities.
Recipients: Australia Section Committee on Traffic Signals, Management and Operation of Traffic Signals in Melbourne 1987; An Evaluation of Left-Turn Analysis Procedures, Texas Section, Randy Machemehl, Committee Chairman 1986; Canadian Capacity Guide for Signalized Intersections, Northern Alberta Section, Stan Teply, Editor 1985; Restoring Mobility in Houston, Texas, Texas Section, Dennis L. Christiansen, Coordinating Author 1984; Child Safety on Michigan's Highways-A Solution, Michigan Section, Richard F. Beaubien, Section President 1983; Left-Turn Phase Design, Florida Section, Bruce E. Friedman, Committee Chairman 1982; Washington-Baltimore Section, Robert E. Heightchew, Chairman 1981; No award 1980; San Francisco Bay Area Section Committee on Evaluation of Unwarranted Stop Signs, Charles D. Allen, Chairman 1979.

STUDENT PAPER COMPETITION ... the purpose of the Student Paper Competititon is to encourage Student Members of the Institute of Transportation Engineers to conduct and report on independent and original research and investigation of traffic or transportation subjects and to provide a means for recognizing outstanding accomplishment in this area. Only Student Members in good standing are eligible to enter the competition. The award consists of an appropriately inscribed certificate to be read and presented to the winner by the Chairman of the ITE Technical Council. The paper will be published in one or more appropriate publications of the Institute. The winner will be asked to present a digest of the paper at the Annual Meeting and will receive a $400 cash award from the Institute.
Recipients: Cesar J. Molina, Jr. 1987; Peter S. Lindquist 1986; Hossein Takallou 1985; Yean-Jye Lu 1984; Mitsuro Saito 1983; Warren Hughes 1982; Peter Alle 1981; Anne Taylor-Harris 1980; Michael J. Cynecki 1979; Shih-Miao Chin 1978; Peter George Hart 1977; Panos G. Michalopoulos 1976.

TECHNICAL COUNCIL AWARD FOR OUTSTANDING TECHNICAL COMMITTEE ... is presented annually and recipients are selected through nominations by individual Council Department Heads; by recommendation of a specially appointed Review Committee of the Council; and by vote of the Council during its spring meeting. Selection is based primarily on the significance of the committee's contribution to the traffic engineering profession, the extent to which it met the objective of its project, and the value and usefulness of the form in which its work is reported.
Recipients: Committee 6F-27, Transportation Energy Contingency Planning 1987; A Standard for Pedestrian Signal Indications, Committee 4S-11, Barry W. Fairfax, Chairperson, 1986; Effective Communications for Transportation Professionals, Committee 2-22, William G. van Gelder, Chairman 1985; Traffic and Parking Control for Snow Emergencies, Committee 4A-11, Richard F. Klatt, Chairman 1984; Transportation Energy

Forecasts to the Year 2000, Committee 6F-21, Richard B. Robertson, Chairman 1982; Committee 6F-8 Pedestrian and Transit Malls, Herbert S. Levinson, Chairman 1981; Guidelines for Urban Major Street Design, Paul C. Box, Chariman 1979; Ramifications of the 55 MPH Speed Limit, Committee 4M-2, Charles P. Sweet, Chairman 1978; Trip Generation Rates, Committee 6A6, Carl H. Buttke, Chairman 1977; "Airport User Traffic Characteristics for Ground Transportation Planning" Committee 6F-4, Richard I. Strickland, Chairman 1976; "Guidelines for Planning and Designing Access Systems for Shopping Centers" Committee 5-DD, Richard C. Gern, Chairman; and "Traffic Consideration for Special Events", Committee 6A5, Frederick G. Wegmann, Chairman 1975; "Transportation Planning for Colleges and Universities", Committee 6J Everett C. Carter, Chairman 1974; "Guidelines for Driveway Design and Location", Committee 5N-S, Paul C. Box, Chairman 1972; "Trip Generation of Selected Commercial and Residential Developments" Illinois Section Committee, Neil S. Kenig, Chairman 1971.

DISTRICT/SECTION NEWSLETTER AWARD ... the purpose of this award is to recognize the district or section which, during a specific period of time, produces the best series of newsletters.
Recipients: Australia Section Newsletter, Matthew L. James, Editor; MICHIGANITE, Weldon Borton, Editor; District 6, WESTERNITE, Patricia Van Wagoner, Editor 1987; HoosierIte, Jon D. Fricker, Editor; Washington, D.C. Section Newsletter, W. Scott Wainwright, Editor; District 6, WESTERNITE, Patricia Van Wagoner, Editor 1986; District 6, WESTERNITE, Patricia Van Wagoner, Editor 1985; District 6, WESTERNITE, Patricia Van Wagoner, Editor 1984; District 6, WESTERNITE, Editors: Wayne and Patricia Van Wagoner 1983.

TRANSPORTATION ACHIEVEMENT AWARD ... this award recognizes significant and outstanding transportation achievements by other entities concerned with transportation, such as governmental agencies, legislative bodies, consulting firms, industry, or other private sector organizations. Awards will be presented for outstanding transportation achievements in the categories of operations and facilities.
Recipients: Illinois Department of Transportation/Chicago Area Freeway Traffic Management Program 1987; Glenwood Canyon (CO) I-70 Traffic Control Program and Santa Ana (CA) Regional Transportation Center 1986; Project to Continue Rail Service to the Virginia Eastern Shore - Virginia Department of Highways and Transportation 1985.

TRANSPORTATION ENERGY CONSERVATION AWARD IN MEMORY OF FREDERICK A. WAGNER ... this award recognizes the most notable "activity" conducted or implemented by an agency, firm, association or individual for which the objective is to promote energy conservation in the transportation field.
Recipients: North Carolina DOT and Institute for Transportation Research and Education/North Carolina's Traffic Signal Management Program for Energy Conservation 1987; Australian Road Research Board 1986; California Energy Commission and California Department of Transportation 1985; Municipality of Metropolitan Toronto, Leonard Rach, Director of Traffic 1984.

ANNUAL MEETING AWARD ... the purpose of this award is to recognize the best paper prepared for and presented as part of the technical program at the ITE Annual Meeting.
Recipients: Gregory G. Henk, E-470: Colorado's Public/Private Toll Road 1987; Raymond S. Niedowski, A Proceure for Allocating Road Improvement Costs to Private Developers 1986; Samuel I. Schwartz, Intelligent Traffic 1985; Louis G. Neudorff, Gap-Based Criteria for Traffic Signal Warrants 1984; Robert C. Scales, Transit Malls: Building Flexibility for the Future into Today's Design, and Jonathan E. Upchurch, Guidelines for the Use of Sign Control at Intersections to Reduce Energy Consumption, (co-winners 1982).

INSTRUMENT SOCIETY OF AMERICA (ISA)

67 Alexander Dr., P.O. Box 12277, Research Triangle Park, North Carolina 27709
Glenn F. Harvey, Executive Director

PRESIDENTS ... V. S. Weiss, 1987; T. J. Harrison 1986; M. L. Stanley 1985; M. J. Kopp 1984; L. M. Zoss 1983; Louis G. Good 1982; Darrell R. Harting 1981; J. Robert Middleton 1980; Norman E. Huston 1979; Hugh S. Wilson 1978; John R. Mahoney 1977; Albert Naumann 1976; Walter A. Bajek 1975; W. Spencer Bloor 1974.

AWARDS

ALBERT F. SPERRY MEDAL AWARD ... in recognition of an outstanding technical, educational, or philosophical contribution to the science and technology of instrumentation.
Recipients: Page S. Buckley 1985; George B. Foster 1984; Vincent J. Cushing 1983; Charles C. Waugh 1982; Walter A. Bajek 1981; Walter P. Kistler 1980; Peter J. Herzl 1979; William C. Baumann, Hamisn Small and Timothy Stevens 1978; Arthur C. Ruge 1977; Maxime A. Faget 1976; J. E. Starr 1975; N. B. Nichols and J. G. Ziegler 1974; S. P. Jackson 1973; Henry Roy Chope 1972; Anthony R. Barringer 1971; Henry P. Kalmus 1970; Abe M. Zarem 1969; Nathan Cohn 1968.

ARNOLD O. BECKMAN AWARD ... in recognition of a significant technological contribution to the conception and implementation of a new principle of instrument design, development, or application.
Recipients: Richard E. Self 1985; Charles W. Mundy 1984; Theodore E. Miller, Jr. 1983; Alan E. Rodely 1982; Victor J. Caldecourt, Richard H. Solem, Edward E. Timm and Donald I. Townsend 1979; Ronald S. Newbower 1978; William W. Wilcock 1977; J. F. Holland and J. D. McLean 1975; R. A. Kamper 1974; A. E. O'Keefe 1973; John C. Telinde 1972; Darrell R. Harting 1971; James E. Faller 1970; Kurt S. Lion 1969; Walter F. Gerdes 1968.

DONALD P. ECKMAN EDUCATION AWARD ... in recognition of an outstanding educational contribution to the science and technology of instrumentation.
Recipients: Donald F. Dellenbach, 1986; Chun H. Cho 1985; Ralph L. Moore 1984; F. G. Shinskey 1983; E. Ross Forman 1982; Peter K. Stein 1980; Franklyn W. Kirk 1979; Lowell E. McCaw 1978; R. J. Moffatt 1977; Paul W. Murrill 1976; W. L. Luyben 1975; J. G. Webster 1974; John T. Anagnost 1973; Bernet S. Swanson 1972; Frank J. Ziol 1971; Howard V. Malmstadt 1970; A. D. Moore 1969; Leslie M. Zoss 1968.

EXCELLENCE IN DOCUMENTATION AWARD ... in recognition of the most outstanding article, paper, or other document published under the auspices of the ISA.
Recipients: Donald Ginesi and Gary L. Grebe, 1986; William V. Dailey, Robert H. Sonner, and David E. Wright 1985; M. T. Herrewig and R. B. Michaelson 1984; Gordon C. Ortman 1983; Albert Naumann, Jr. 1982; Harry A. Fertik and Russell G. Sharpe 1981; Donald L. May, B. Nelson Norden, Carl C. Andreasen and Chun H. Cho 1980; Page S. Buckley 1979; Michael J. Flanagan 1978; Renzo Dallimonti 1977; Robert M. Rello 1976; H. C. Carter, D. W. Ferrel, D. R. Nelson, T. C. Sciarrotta and E. A. Standora 1975; R. C. Meyer 1974; M. L. Lyon and R. J. Murphy, Jr. 1973; Richard Villalobos and Robert L. Chapman 1972; Ronald E. Rinehart and Howard R. Kratz 1971; James H. McLaughlin and Walter A. Bajek 1970; Clayton J. Bossart 1969; Donald C. Howey 1968.

ISA STANDARDS AND PRACTICES AWARD ... in recognition of an outstanding contribution to the Society's standards and practices program.
Recipients: Robert S. Crowder, Jr., 1986; Leslie R. Driskell 1985; Leonard F. Griffith 1984; Robert L. Nickens 1983; Albert P. McCauley 1982; Howard T. Hubbard 1981; Boyd A. Christensen 1980; Philip Bliss 1979; Whitney B. Miller 1978; William E. Vannah 1977; Orval P. Lovett, Jr. 1976; D. C. Hughes 1975; E. C. Magison 1974; John R. Mahoney 1973; Ralph E. Clarridge 1972; Frederick L. Maltby 1971.

KERMIT FISCHER ENVIRONMENTAL AWARD ... (certificate, plaque and $500 honorarium) for outstanding achievement in the conception, design or implementation of an application of instrumentation or automatic control in the field of environmental science.
Recipients: Lawrence B. Marsh, 1986; Bipin B. Mishra 1985; Charles E. Everest 1984; Joseph F. Roesler 1983; Thomas L. Stewart 1982; Robert M. Arthur 1981; John A. Almo 1980.

PHILIP T. SPRAGUE ACHIEVEMENT AWARD ... in recognition of an outstanding achievement in the conception, design, or implementation of power plant instrumentation.
Recipients: Henry N. Bowes 1985; William M. Herring 1984; W. J. Gabriel 1983; R. H. Morse 1982; Carroll J. Ryskamp 1981; Gideon Ben- Yaacor 1980; Robert F. Hyland 1979; R. L. Egger, E. J. Nelson and C. F. Sikorra 1978; F. G. Shinskey 1977; Lyle F. Martz 1976; J. C. Pemberton 1975; E. J. Drost 1974; P. S. Buckley 1973; Stanley W. Lovejoy 1972; Warren H. Owen 1970; A. Stirling Grimes 1969; Walter F. Gerdes 1968.

SI FLUOR TECHNOLOGY ACHIEVEMENT AWARD ... offered in recognition of an outstanding achievement in the conception, design, or implementation of instrumentation and/or control in an area of activity covered by the scope of the Society's Technology Department.
Recipients: Vincent V. Tivy 1983; L. J. Griffey 1982; Richard Villalobas 1981; Jerome M. Paros 1980; Irving Cherniack 1979; A. D. Kurtz 1978; Roger L. Frick 1977; Ellsworth R. Fenske 1976; H. S. Harman 1975; E. D. Mannherz and R. F. Schmoock 1974; F. K. Harris 1973; J. Richard Dahm 1971; Joel O'Hougen and Norman E. Brockmeier 1970; Page S. Buckley and William L. Luyben 1969; Leslie R. Driskell 1968.

E. G. BAILEY AWARD ... in recognition of excellence in the design, development or application of instrumentation & control systems in the utilities or process control industries.
Recipient: Ada I. Pressman 1985.

DOUGLAS ANNIN AWARD ... in recognition of an outstanding achievement in the design or development of components of an automatic control system from the input measurement through the final control element.
Recipient: Gordon F. Stiles, 1986; John E. Hewson 1985.

INTERNATIONAL MATERIAL MANAGEMENT SOCIETY (IMMS)

8720 Red Oak Blvd., Charlotte, NC 28217
William Clayburg, Executive Director

PRESIDENTS ... William Shirk 1987-88; William Ransom 1986-87; James A. Tompkins 1984-86; James J. Way 1983-84; Carl H. Franzen 1982-83; James M. Cahill 1981-82.

PRESIDENTS ... David L. Schaefer 1981; Richard A. Schmidt 1980; Theodore A. Krawczyk 1979; John R. Huffman 1978; Richard A. Evans 1977; Theodore A. Allegri 1976; David C. Stewart 1975; Donald W. Francis 1974; Glenn D. Pusch 1973.

AWARDS

MATERIAL HANDLING MANAGER OF THE YEAR ... awarded to a person selected from nominations received from our membership.
Recipients: Thomas V. Garvey 1984; Carleton Lord 1981; Harold W. Weaver 1980; Lavern N. Thompson 1979.

MATERIAL MANAGER OF THE YEAR ... awarded to a person selected as the best material manager of the year from nominations received from our membership.
Recipients: Cecil Voils 1984; E. J. Dempsey 1981; Irving M. Appel 1980; Richard J. Vogt 1979.

IRON AND STEEL SOCIETY INC. (ISS-AIME)

410 Commonwealth Drive, Warrendale, Pennsylvania 15086

PRESIDENTS ... Gerald J. Roe 1987; G. Hugh Walker 1986; Alexander McLean 1985; Ralph T. Brower 1984; H. N. Hubbard, Jr. 1983; J. S. Anslow 1982; Norman T. Mills 1981; Francis D.

Nelson 1980; Edmond I. Whittenberger 1979; Arthur Marcantonio 1978; Edward J. Ostrowski 1977; Harry L. Bishop 1976; William H. Wise 1975; Melvin E. Nickel 1974.

AWARDS

CHARLES H. HERTY, JR. AWARD ... for the best paper presented at the annual Steelmaking Conference and published in the Conference proceedings. This award was established on September 24, 1960 by the National Open Hearth Steel Committee, with the first winner recognized in April, 1962 for his paper presented at the 1961 Conference. The award honors Charles H. Herty, Jr., distinguished physical chemist working in the field of iron and steel technology.
Recipients: J. K. Brimacombe, I. V. Samarasekera, R. Bommaraju 1987; M. Byrne, A. W. Cramb, T. W. Fenicle 1986; A. Delhalle, J. F. Marioton, J. P. Birat, J. Foussal, M. Larrecq and G. Tourscher 1985; K. Sasaki, H. Nakashima, M. Nose, Y. Takasaki and H. Okumura 1984; C. A. Reid, K. W. Heyer and L. E. Hambly 1983; H. Ohzu, H. Ohmori and J. Yamazaki 1982; Guy Denier, Jean-Claude Grsojean, Marcel Lemaire, Francis Schleimer, Romain Henrion and Ferdinand Goedert 1981; N. J. Huettich, D. J. Hurtuk and K. Grieshaber 1980; J. T. von Konijnenburg and E. P. Westerdijk 1979; S. K. Sharma, J. W. Hlinka and D. W. Kern 1978; P. L. Jackson 1977; J. Wendorff, E. Spetzler 1976; H. W. Grenfell and D. J. Bowen 1975; J. E. Lait, J. K. Brimacombe, F. Weinberg and F. C. Muttitt 1974; Paul E. Nilles and Roland A. Holper 1973; D. W. Kern, R. J. Fradeneck and P. D. Stelts 1972; Paul Remy Gosselin and Laszlo Backer 1971; Darwyn I. Brown and George Harry, Jr. 1970; Ronald S. Mulhauser and Dennis G. Hargreaves 1969; J. A. Glasgow and W. F. Porter 1968.

CHARLES W. BRIGGS AWARD (formerly the Electric Furnace Conference Award) ... for the best paper presented at the Annual Electric Furnace Conference and published in the Conference Proceedings. Established in 1961 by the Electric Furnace Committee, with the first winners honored in December, 1962, for the paper they presented at the 1961 Conference. It is not mandatory that this award be made every year.
Recipients: R. A. Heard, N.D.G. Mountford, A. McLean, M. Haissig 1986; S. Argyropoulos 1985; S. Argyropoulos 1984; G. Yuasa, S. Sugiura, M. Fujine and N. Demukai 1983; A. Schei and K. Larsen 1982; J. J. Butler, Jr. and P. J. Grandy 1981; W. L. Wilbern and C. T. Ray 1980; R. A. Swift and W. W. Scott, Jr. 1979; R. I. L. Guthrie 1978; A. McLean and W. J. Maddevez, R. Nikolic and R. S. Segsworth 1977; S. C. Ghorpade, R. G. Gilliland and R. W. Heine 1976; Philip R. Grabfield 1975; R. A. Strimple, C. R. Beechan and J. F. Muhlhauser 1974; A. G. Arnesen and B. Asphaug 1973; John F. Elliott 1972; J. W. Farrell, P. J. Bilek and D. C. Hilty 1971; Rolph A. Person 1970; Kenneth W. Hansen and Robert F. Nale 1969; Malachi P. Kenney 1968.

STEELMAKING CONFERENCE AWARD (runner-up to McKune Award) ... established by the National Open Hearth and Basic Oxygen Steel Committee in 1944, this award is the runner-up to the Frank B. McKune Award, which is given for the best paper not exceeding 5,000 words on some phase of open hearth practice (including design, maintenance, operations, quality, or metallurgy).
Recipients: T. Wada, M. Suzuki, T. Mori 1987; A. W. Cramb, M. Byrne 1986; N. McPherson 1985; J. Hasunuma, A. Ueda, Y. Habu, K. Suzuki, M. Kodama, M. Ohnishi 1984; M. R. Ozgu 1983; E. M. O'Donnell 1982; Daniel Rellis and Clifford C. Smith 1981; D. Kruse and R. O'Neil 1980; A. M. Smillie and R. A. Huber 1979; J. Stravinskas and M. R. Deamer 1978; Dennis M. Balla and Charles R. Beechan 1977; Jan Kristiansen and John E. Hartmann 1976; Donald R. Gillum 1975; S. N. Singh and K. E. Blazek 1974; Ram K. Iyengar and Frank C. Petrilli 1973; Norman A. Robins and J. Marshall Rounsevell 1972; Barry A. Strathdee and Ulrich H. Pause 1971; Frederick W. Irwin and Robert J. Milbourne 1970; Ronald T. Orie 1969; W. S. Suhay 1968.

FRANK B. McKUNE AWARD ... established by the National Open Hearth and Basic Oxygen Steel Committee in 1940, the award is given for the best paper not exceeding 5,000 words on some phase of steelmaking (including design, maintenance, operations, quality, or metallurgy) by an author not more than 40 years of age.

Recipients: N. A. McPherson, S. L. McIntosh 1987; T. Bieniosek 1986; A. Tsuneoka, W. Ohashi, S. Ishitobi, T. Kataoka and M. Tenma 1985; A. W. Cramb and M. Byrne 1984; F. J. Nahlik and E. C. Knorr 1983; D. J. Harris and J. D. Young 1982; Eric D. Anderson and Chris Zimmerman 1981; L. E. Hambly and W. Heyer 1980; S. N. Singh 1979; K. G. Kemlo 1978; Donald J. Muscatello and Gerald J. Grasley 1977; J. G. Dalal and E. H. Wozniak 1976; K. W. Heyer and D. J. Harris 1975; H. W. Grenfell and D. J. Bowen 1974; Ronald L. W. Holmes and John G. Harhai 1973; John W. Onuscheck and Ronald L. W. Holmes 1972; Robert J. Milbourne and T. Raymond Meadowcroft 1971; Donald E. Dorney, Charles E. Houck and Timothy W. Miller 1970; George Moncilovich 1969; J. J. McCarthy and K. R. Bock 1968.

JOSEF S. KAPITAN AWARD (formerly the Ironmaking Conference Award) ... for the best paper presented at the Ironmaking Conference each year, and published in the Conference Proceedings, this award was established in 1953 by the Blast Furnace, Coke Oven, and Raw Materials Committee. It was known as the "Journal of Metals Award" until 1960, when the rules were changed. The first "Ironmaking Conference Award" was presented in 1962 (at the Ironmaking Conference) for the best paper delivered at the 1961 Conference. The first Josef S. Kapitan Award was presented in 1982 at the Ironmaking Conference for the best paper delivered at the 1981 conference.
Recipients: D. D. Kaegi 1987; H. Hasegawa, M. Gono, Y. Ishikawa, S. Okubo, Y. Shimomura, T. Sugiyama, N. Kawamura 1986; G. M. Latshaw, H. R. McCollum and R. W. Stanley 1985; Y. Abe, T. Nishi, Y. Ishikawa, M. Kase, K. Ono and M. Sugata 1984; S. M. Sorensen, Jr., P. A. Rupp and A. A. Arsenault 1983; S. Yabe, I. Kurashige, M. Kojima, T. Miyazaki, Y. Shoji, Y. Kamei and T. Iba 1982; John D. Ashton and Richard R. Schat 1981; A. Ponghis, R. Vidal and A. Poos 1980; P. K. Strangway and M. O. Holowaty 1979; S. Hashimoto, O. Suzuki and H. Yoshimoto 1978; M. Hatano and M. Fukuda 1977; M. Higuchi, T. Shibuya and Y. Niwa 1976; A. Charles Naso and James W. Schroeder 1975; J. D. Ashton, C. V. Gladysz, J. Holditch and G. H. Walker 1974; K. S. Nanavati and G. M. Cohen 1973; T. Kobayashi, T. Miyashita and G. Suzuki 1972; Hans vom Ende, Klaus Grebe and K. G. Speith 1971; R. O. Wanger and V. A. Neubaum 1970; K. Sugasawa 1969.

J. E. JOHNSON, JR. AWARD ... awarded to encourage young people in creative work in the metallurgy or manufacture of pig iron. Recipient must be under 40 when he completes the work that merits recognition.
Recipients: Marvin H. Bayewitz 1987; John A. Ricketts 1986; Richard R. Schat 1985; John H. Schael 1984; Bruce Schiller 1983; William L. Ziegert 1982; Joseph J. Poveromo 1981; Larry N. Fletcher 1980; J. E. R. Holditch 1979; D. B. White 1978; Allen K. Garbee 1977; Thomas R. Meadowcroft 1976; Joseph A. Laslo 1975; Robert W. Bouman 1974.

JOHN CHIPMAN AWARD ... recognizes excellence and originality of an author or authors of a paper on process technology or the application thereof, as it relates to the production and processing of iron and steel.
Recipients: H. M. Pielet, D. Bhattacharya 1987; J. K. Brimacombe, E. Takeuchi 1986; S. Tabuchi and N. Sano 1985; J. W. Hlinka and J. J. Poveromo 1984; R. I. L. Guthrie and G. A. Irons 1983; Richard J. Fruehan, Gerrit J. W. Kor and E. T. Turkdogan 1982; K. Upadhya, I. D. Sommerville, P. Grieveson and J. Taylor 1981; S. Kou, M. C. Flemings and D. R. Poirier 1980; W. Venkateswaran and J. K. Brimacombe 1979; E. T. Turkdogan 1978; Harry P. Leckie and Yong-Wu Kim 1977; Shri Nath Singh 1976; R. I. L. Guthrie, H. Henein and L. Gourtsoyannis 1975; J. W. Farrell, D. C. Hilty, John F. Elliott and Gordon E. Forward 1972.

JOSEPH BECKER AWARD ... recognizes distinguished achievement in coal carbonization.
Recipients: Dietrich Wagener 1987; Ralph J. Gray 1986; J. E. Ludberg 1985; Edward J. Helm 1984; Takashi Miyazu 1983; Michael Perch 1982; Michael O. Holowaty 1981; Andrew H. Brisse 1980; J. P. Graham 1979; J. H. Walsh 1978; K. G. Beck 1977; William Spackman, Jr. 1976; Karl Friedrich Paul Still 1975; Lawrence D. Schmidt 1974; Frederick Denig 1973; George W. Lee 1972; Frans Wethly 1971; Homer H. Lowry 1970; Harold W. Jackman 1969; Alfred Richard Powell 1968.

MECHANICAL WORKING AND STEEL PROCESSING CONFERENCE MERITORIOUS AWARD ... established in 1971 by the Mechanical Working and Steel Processing Committee of the Iron and Steel Division. This award is a "runner-up" to the Michael Tenenbaum Award and is given for the papers chosen as "runners-up" to the best paper presented at the Conference held the previous year, and published in the proceedings.
Recipients: R. M. Hudson, S. Yue, J. J. Jonas, P. J. Hunt, M.J.A. Walker, K. F. Davies, C. Galvani, A. Crawley, R. Holt, E. Cousineau, G. L'Esperance, M. Maeda, T. Ito, N. Tsukiji, S. Umeda 1987; M. Uemura, Y. Maeda, S. Motoyashiki, Y. Ichida, D. N. Thompson, E. L. Klaphake, W. H. Rackoff, A. J. Perella 1986; J. Iyer, J. K. Brimacombe, E. B. Hawbolt, T. Masuda, T. Terada, T. Kawade, H. Ohtsubo, F. Togashi, T. Kurisu, C. Gaspard, P. Cosse, A. Magnee 1985; J. L. Tranchant, P. Boussard, R. Szezesny, C. Chretien, A. Gueussier, H. Abrams, W. J. Jarae, J. T. Corrigan, W. R. Emery, J. R. Cook, T. R. Dishun, D. F. Ellerbrock 1984; W. Tait, H. Abrams, L. A. Troman, J. C. Howard, M. C. Labanow, D. J. Diederich, B. Kawasaki, K. Kaneko, K. Matsuzuka 1983; T. Goda, T. Kimura, H. Naoi, K. Nakajima, S. Yoshiwara, M. Kawaharada, D. Diesburg, T. Cameron, T. Masui, T. Nunokawa, E. Yasui, T. Takagawa, H. Ono, J. Yamada 1982; J. B. Lajournade, E. Wohlfahrt, G. Schanne, R. P. Krupitzer, F. Reis, R. E. Mintus, J. E. Franklin, W. Tait 1981; B. W. Keist, J. Toth, Jr. and J. C. Lucia 1980; E. R. SuVeges, J. A. Straatmann and Y. E. Smith 1979; C. D. Hopkins and R. J. Bassett 1978; T. Taira, T. Yamaguchi, N. Twasaki, K. Tsukada, F. Koshiga, J. Tanaka and T. Osuka 1977; R. Craford and J. Bell 1976; Dewey G. Younger 1975; Gary Van Asperen 1974; T. V. Simpkinson, R. P. Krupitzer, R. E. Mintus and W. F. Barclay 1973; James A. Elias and Rollin E. Hook 1972.

MICHAEL TENENBAUM AWARD ... recognizes best paper given at previous year's Mechanical Working and Steel Processing Conference.
Recipients: R. L. Bodnar, B. L. Bramfitt 1987; H. Abrams 1986; D. L. Paton and S. Critchley 1985; W. Davies 1984; Y. Funyu, T. Okumura and F. Togashi 1983; S. Hansen 1982; D. McCutcheon, H. Wade, B. Armitage and K. Mullins 1981; D. C. Ronemus, R. S. Hostetter and H. Kranenburg 1980; J. W. Morrow, G. Tither and A. P. Coldren 1979; S. K. Kyriakides and J. M. Hambright 1978; E. C. Oren 1977; N. Iwasaki, T. Yamaguchi and T. Taira 1976; A. L. Hillegasse, G. J. Dobrian and P. P. Pincura 1975; Bernard S. Levy and T. E. Fine 1974; J. F. Held and E. J. Paliwoda 1973; Ian G. Thompson 1972.

ROBERT W. HUNT AWARD ... the partners and employees of Robert W. Hunt, desiring to commemorate his great contributions to the steel industry, established a fund to provide the award of a medal and certificate. The fund was presented to the Institute on May 27, 1920, the occasion of Captain Hunt's eightieth birthday. The award is made each year to the author or authors of the best paper for excellence, originality, and potentiality relevant to the technology (applied or fundamental) related to the iron and steel industry. When, in the opinion of the Committee of Award, none of the papers is worthy of the medal and certificate, no award is made. The papers to be considered are those published by the Society within the period since the previous presentation of the Award.
Recipients: M. Byrne, A. W. Cramb, T. W. Fenicle 1987; D. L. Paton, S. Critchley 1986; M. R. Ozgu 1985; J. D. Young and D. J. Harris 1984; I. V. SamaraseKera and J. K. Brimacombe 1983; E. T. Turkdogan, G. J. Kor and R. J. Fruehan 1982; Richard S. Hostetter, Helmut Kranenberg and David C. Ronemus 1981; J. K. Brimacombe, F. Weinberg and E. B. Hawbolt 1980; S. K. Sharma, J. W. Hlinka and D. W. Kern 1979; D. J. Wilson, J. J. DeBarbadillo, E. Snape 1978; J. Wendorff and E. Spetzler 1977; J. Farrell Donald C. Hilty 1976; F. C. Petrilli and R. K. Iyengar 1975; J. W. Onuscheck and R. L. W. Holmes 1974; T. R. Meadowcroft and R. J. Milbourne 1973; L. Backer and P. R. Gosselin 1972; Paul Kozakevitch 1971; G. R. Belton and R. J. Fruehan 1970; W. O. Philbrook, R. H. Spitzer and R. S. Manning 1969; Lawrence S. Darken 1968.

THOMAS L. JOSEPH AWARD ... recognizes distinguished contributions to ironmaking operations which have significantly increased iron production or decreased the cost of doing so.

Recipients: Wei-Kao Lu 1987; G. Hugh Walker 1986; Harry Holiday 1985; Masaaki Higuchi 1984; William D. Millar 1983; Edward J. Ostrowski 1982; Richard J. Wilson 1981; G. L. French 1980; R. L. Stephenson 1979; C. M. Squarcy 1978; John A. Peart 1977; Kenneth M. Haley 1976; Richard H. White 1975; R. V. Flint 1974; John R. Barnes 1973; William A. Haven 1972; Edward W. Davis 1971; Walter R. Trognitz 1970; Hjalmar W. Johnson 1969; D. P. (Pat) Cromwell 1968.

THE METALLURGICAL SOCIETY (TMS)

420 Commonwealth Drive, Warrendale, Pennsylvania 15086
Alexander R. Scott, Executive Director

PRESIDENTS . . . Frank V. Nolfi, Jr. 1988; Alan D. Zunkel 1987; George S. Ansell 1986; Peter Tarassoff 1985; Alan Lawley 1982; Kenneth J. Richards 1981; Dale F. Stein 1980; G. Robert Couch 1979; Robert I. Jaffee 1978; Robert E. Lund 1977; Harold W. Paxton 1976; Thomas A. Henrie 1975; James C. Fulton 1974; Julius J. Harwood 1973.

AWARDS

CHAMPION MATHEWSON MEDAL . . . bestowed on an author of a paper or series of closely related papers, considered the most notable contribution to metallurgical sciences during the period under review.
Recipients: Robert H. Wagoner 1988; Thomas W. Eagar 1987; R. O. Ritchie and S. Suresh 1985; James W. Evans 1984; Paul G. Shewmon 1983; John P. Hirth 1982; Morris E. Fine 1981; J. Keith Brimacombe 1980; William D. Nix 1979; C. M. Wayman 1978; Johannes Weertman 1977; Charles J. McMahon, Jr. 1976; E. Turkdogan 1975; Gilbert Chin 1974; Julian Szekely 1973; James C. M. Li 1972; Bernard H. Kear 1971; Rodney P. Smith 1970.

EXTRACTIVE METALLURGY LECTURER . . . an outstanding man in the field of non-ferrous metallurgy is selected annually to present a comprehensive lecture.
Recipients: T. R. A. Davey 1988; Julian Szekely 1987; Noel Jarrett 1986; Leslie R. Verney 1985; Peter Tarassoff 1984; William G. Davenport 1983; Walter R. Hibbard, Jr. 1982; Allen S. Russell 1981; Derek A. Temple 1980; A. Yazawa 1979; Terkel N. Rosenqvist 1978; Paul E. Queneau 1977; Ernest Peters 1976; John F. Elliott 1975.

EXTRACTIVE METALLURGY SCIENCE AWARD . . . recognizes notable contributions to the scientific understanding of extractive metallurgy.
Recipients: E. Ozberk and R. I. L. Guthrie 1988; J. Keith Brimacombe, Gerald W. Toop and Gregory G. Richards 1987; Howard M. Pielet and Debanshu Bhattacharya 1986; Lawrence M. Cathles III, L. Edmond Eary 1985; R. I. L. Guthrie, Y. Sahai 1984; J. W. Evans, D. Sharma, Y. Zundelevich 1983; Robert G. Robins 1982; Gary H. Kaiura, J. M. Toguri 1981; E. W. Dewing 1980; R. G. Barton and K. K. Brimacombe 1979; J. P. Pemsler and C. Wagner 1978; R. Schuhmann, Jr., Pei-Cheh Chen, P. Palanisamy and D. H. R. Sarma 1977.

EXTRACTIVE METALLURGY TECHNOLOGY AWARD . . . recognizes notable contributions to the advancement of the technology of extractive metallurgy.
Recipients: Roy A. Christini and Marlyn D. Ballain 1988; C. R. Edwards 1987; Mahesh C. Jha, Bruce J. Sabacky and Gustavo A. Meyer 1986; V. P. Keran 1985; D. McKay, E. Parker, H. Salomon-De-Friedberg 1984; R. J. Anderson, R. R. Beck, A. J. Weddick 1983; J. Keith Brimacombe, E. O. Hoefele 1982; John T. Chao, David R. Morris, Frank R. Steward 1981; Derek J. Fray, Robert Gee 1980; J. A. Thornton, H. F. Graff 1979; T. Nagano and T. Suzuke 1978; A. R. Gordn and R. W. Pickering 1977.

INSTITUTE OF METALS LECTURER AND ROBERT FRANKLIN MEHL AWARD . . . recognizes an outstanding scientific leader to present a lecture on a technical subject.
Recipients: William D. Nix 1988; Sir Charles Frank 1987; Howard K. Birnbaum 1986; Peter Haasen 1985; Guy Marshall Pound 1984; Michael F. Ashby 1983; Robert W. Balluffi 1982; Jack W. Christian 1981; John P. Hirth 1980; Erhard Hornbogen 1979; James Chen-Min Li 1978; Earl R. Parker 1977.

LIGHT METALS AWARD . . . for paper which exemplifies the application of science to the solution of a practical problem, presented at Annual Meeting and published in preceding year's *Light Metals*.
Recipients: Ch. Vives, J. P. Riquet, B. Forest and J. L. Meyer 1987; Ho Yu and Daniel K. Ai 1986; Werner K. Fischer and R. Perruchoud 1985; Warren Haupin 1984.

ROBERT L. HARDY MEDAL . . . recognizes outstanding promise for a successful career in the broad field of metallurgy by a metallurgist under the age of 30.
Recipients: Robert B. Beyers 1987; Edward Goo 1986; Francoise K. LeGoues 1985; James J. Komadina 1984; William C. Johnson 1983; Subramanian Suresh 1982; Robert H. Wagoner 1981; Deborah Deun Ling Chung 1980; Sheree Hsaio-Ru Wen 1979; Ronald Gronsky 1978; Clyde Briant 1977; Robert Sinclair 1976; David K. Matlock 1975; Takeshi Egami 1974.

THE WILLIAM HUME-ROTHERY AWARD . . . awarded annually to an outstanding scientific leader in recognition of outstanding scholarly contributions to the science of alloys.
Recipients: Michael F. Ashby 1988; Andre Jean Guinier 1987; David Turnbull 1986; R. E. Watson 1985; Henry Ehrenreich 1984; Leo Brewer 1983; Pol Duwez 1982; A. R. Miedema 1981; Thaddeus B. Massalski 1980; Lawrence S. Darken 1979; Karl A. Gschneidner, Jr. 1978; Jacques Friedel 1977; Charles Barrett 1976; William B. Pearson 1975; Paul A. Beck 1974.

LEADERSHIP AWARD . . . presented annually to an individual who has demonstrated outstanding leadership in some aspect of the fields of metallurgy and materials as a representative of an industrial, academic, governmental or technical organization.
Recipients: Arden L. Bement, Jr. 1988; Morris Cohen 1987; Praveen Chaudhari 1986.

APPLICATION TO PRACTICE AWARD . . . presented annually to an individual who has demonstrated outstanding achievement in transferring research results or findings in some aspect of the field of metallurgy and materials into commercial production and practical use as a representative of an industrial, academic, governmental, or technical organization.
Recipients: King-Ning Tu 1988; G. Dixon Chandley 1987; John J. Gilman 1986.

EDUCATOR AWARD . . . presented annually to an individual who has made outstanding contributions to education in metallurgical engineering and/or materials science and engineering. The award shall not be limited to classroom teachers but shall also recognize contributions through writing textbooks, building of strong academic programs, outreach to high school students, or innovative ways of educating the general populace.
Recipients: Ellis D. Verink, Jr. 1988; John F. Elliott 1987; Mars Guy Fontana 1986.

NATIONAL ACADEMY OF ENGINEERING (NAE)

2101 Constitution Avenue N.W., Washington, D.C. 20418
William C. Salmon, Executive Officer

PRESIDENTS . . . Dr. Robert M. White 1983- ; Courtland D. Perkins 1975-1982; Robert C. Seamans, Jr. 1973-1974.

AWARDS

FOUNDERS AWARD . . . The Founders Award was established in 1965 to honor outstanding engineering accomplishments by an engineer over a long period of time and of benefit to the people of the United States.
Recipients: John R. Whinnery 1986; Willis M. Hawkins, Jr. 1985; John Barttn 1984; Harold E. Edgerton 1983; Kenneth H. Olsen 1982; Jacob P. Den Hartog 1981; Hoyt Hottel 1980; David Packard 1979; George M. Low 1978; John R. Pierce 1977; Manson Benedict 1976; James B. Fisk 1975; J. Erik Jonsson 1974; Warren K. Lewis 1973; Edwin H. Land 1972; Clarence L. Johnson 1971; Charles S. Draper 1970; Harry L. Nyquist 1969; Vladimir K. Zworykin 1968.

ARTHUR M. BUECHE AWARD . . . established in 1983 to honor demonstrated statesmanship in the field of technology, active engagement in the determination of science and technology public policy, active participation on the behalf of technology and active contributions to industry, government and university relationships.
Recipients: W. O. Baker 1986; Jerome B. Wiesner 1985; Edward E. David, Jr. 1984; Simon Ramo 1983.

NATIONAL ACADEMY OF SCIENCES (NAS)

2101 Constitution Avenue, Washington, D.C. 20418
Philip M. Smith, Executive Officer
PRESIDENTS . . . Frank Press 1981- ; Philip Handler 1970 - 1981.

NATIONAL ASSOCIATION OF CORROSION ENGINEERS (NACE)

P. O. Box 218340, Houston, Texas 77218
T. S. Lee, Executive Director
PRESIDENTS . . . T. G. Rankin 1987-88; J. H. Payer 1986-87; W. B. Poff 1985-86; W. J. Neill, Jr. 1984-85; H. H. Lawson 1983-84; W. K. Boyd 1982-83; D. A. Dutton 1981-82; M. D. Orton 1980-81; C. J. Steel 1979-80; E. B. Backensto 1978; L. C. Rowe 1977; A. C. Flory 1976; H. A. Webster 1975; D. D. Byerley 1974; G. M. Jeffares 1973.

AWARDS

A. B. CAMPBELL YOUNG AUTHOR AWARD . . . presented annually to an author not over 30 years old whose paper published in "Corrosion" or "Materials Performance" during the year is judged the most outstanding of those eligible for consideration.
Recipients: P. C. Searson 1987; A. McMinn 1986; John R. Scully 1985; R.C. Newman 1984; C.J. Houghton 1983; R.W. Schutz 1982; Helena T. Rowland 1981; P.E. Manning 1980; Petsuo Shoji 1979; R. D. Kane 1978; R. B. Diegle 1977; K. D. Efird 1976; R. L. Prowse 1975; A. A. Seys 1974; P. E. Morris 1973; R. M. Latanision 1972; W. D. France 1971; J. A. S. Green 1970; E. Snape 1969; T. J. Smith 1968.

FRANK NEWMAN SPELLER AWARD . . . given in recognition of public contributions to the science of corrosion engineering. A contribution to engineering is defined as the development or improvement of a method, apparatus or material by which the control of corrosion is facilitated or made less costly.
Recipients: E. Mattsson 1987; R.N. Miller 1986; Stanley L. Lopata 1985; W. E. Berry 1984; E. W. Haycock 1983; R. F. Stratfull 1982; John H. Morgan 1981; Heinz Spahn 1980; M. C. Miller 1979; Joseph B. Cotton 1978; W. K. Boyd 1977; E. H. Phelps 1976; Bernard Husock 1975; K. N. Barnard 1974; F. M. Reinhart 1973; A. Dravnieks 1971; A. W. Peabody 1970; C. G. Munger 1969; L. P. Sudrabin 1968.

R. A. BRANNON FOUNDERS AWARD . . . in public recognition of a person who has constructively aided in the development and improvement of NACE. It recognizes outstanding work in any area of activity beyond required performance on the part of a member. It is available for organization or promotional, rather than technical performance.
Recipients: D. D. Downing 1987; Frank E. Rizzo 1986; R.I. Lindberg 1985; H.R. Hanson 1984; J.F. Williams 1983; H.G. Byars 1982; Thomas F. Degnan 1981; H. VanDroffelaar 1980; W. J. Neill, Jr. 1979; Darrell D. Byerley 1978; F. W. Hewes 1977; L. C. Rowe 1976; Jack P. Barrett 1975; Ivy M. Parker 1974; R. A. Brannon 1973.

WILLIS RODNEY WHITNEY AWARD . . . given in recognition of public contributions to the science of corrosion. A contribution to science is defined as the development of a more satisfactory theory which contributes to a more fundamental understanding of corrosion phenomena.
Recipients: R. A. Oriani 1987; R.A. Rapp 1986; H.W. Pickering 1985; E.N. Pugh 1984; H. Leidheiser, Jr. 1983; E.D. Verink, Jr. 1982; I.L. Rosenfeld 1981; R. Staehle 1980; R. N. Parkins 1979; Walter A. Mueller 1978; Jerome Kruger 1977; N. A. Nielsen

1976; D. A. Vermilyea 1975; B. F. Brown 1974; M. A. Streicher 1973; G. A. Marsh 1972; W. J. Schwerdtfeger 1971; F. N. Rhines 1970; M. Pourbaix 1969; M. E. Straumanis 1968.

T. J. HULL AWARD . . . given in recognition for an outstanding contribution to NACE in the field of publication.
Recipients: T. J. Hull 1987.

NATIONAL COUNCIL OF ENGINEERING EXAMINERS (NCEE)

P. O. Box 1686, Clemson, South Carolina 29631
PRESIDENTS . . . Dennis F. Meyer 1987-88; J. Harry Parker 1986-87; Edward L. Pine 1985; Sam H. Wainwright 1984; Paul R. Munger 1983; Albert T. Kersich 1982; Eugene N. Bechamps 1981; Alfred H. Samborn 1980; Frederick H. Rogers, Sr. 1979; William J. Hanna 1978; T. E. Stivers 1977; Herman A. Moench 1976; Morton S. Fine 1975; Orland C. Mayer 1974; Roy T. Sessums 1973.

AWARDS

DISTINGUISHED SERVICE CERTIFICATE AND DISTINGUISHED SERVICE CERTIFICATE WITH SPECIAL COMMENDATION . . . acknowledgement of faithful and unselfish service and many contributions to the advancment of registration and the profession.
Recipients: Chappell N. Noble, John W. Pearson 1986; Jack W. Anderson, Marjorie L. Carpenter, Charles L. Kimberling, Edward A. Lobnitz, Paul R. Munger, Paul D. Betz, A. J. Brouillette, Mark Hilton, John T. Merrifield, J. Fred Sauls 1985; Waldemar S. Nelson (with special commendation), George J. Cook, Fred C. Culpepper Jr., Victor C. Fender, R. Duane Monical, Walter K. O'Loughlin, J. Harry Parker, Marion L. Smith and Roy J. Thompson 1984; Eugene N. Bechamps, Michael J. Deutch, Melvin E. Luke (posthumously), Wilbur H. Mechwart, Forrest H. North and George R. Russell 1983; Alfred W. Lane, Edward T. Misiaszek, G. Leigh Morrow, Ferrell J. Prosser, Rex A. Tynes and Sam H. Wainwright 1982; Frederick H. Rogres, Sr. /Special Commendation/ 1981; William E. Carew, Jr., Larry D. Nixon, Alfred H. Samborn, Herman E. Smith, Goodwin G. Thomas, Carlton T. Wise 1981; C. Dale Bratcher, Kenneth R. Daniel, Rudolf A. Jimenez, Fred H. Lyon, Ambrose E. McCaskey, John J. McMahon, Henry S. Steinbrugge and T. E. Stivers 1980; Amos E. Kent/Special Commendation-Posthumously, Herman A. Moench/Special Commendation/, Roger H. Brown, Elmer F. Emrich, Ernest B. Gardow, William J. Hanna, Thomas J. McClellan and Earl C. Radding 1979.

NATIONAL INSTITUTE OF CERAMIC ENGINEERS (NICE)

757 Brooksedge Plaza Dr., Westerville, OH 43081
W. Paul Holbrook, Executive Director
PRESIDENTS . . . William C. Mohr 1987-88; Thomas D. McGee 1986-87; You Song Kim 1985-1986; Norman H. Harris 1984-1985; Willard E. Hauth III 1983-1984; Dale E. Niesz 1982-1983; Herbert W. Larisch 1981-1982; Willis E. Moody 1981; William H. Payne 1980; Thomas S. Shevlin 1979; Everett A. Thomas 1978; Joseph L. Pentecost 1977.

AWARDS

GREAVES-WALKER AWARD . . . this annual award is made to an NICE member who has "rendered outstanding service to the ceramic engineering profession and who, by his or her life and career, has exemplified the aims, ideals, and purpose of the National Institute of Çeramic Engineers".
Recipients: Winston Duckworth 1987; James R. Tinklepaugh 1986; James R. Johnson 1985; Stephen D. Stoddard 1984; James I. Muller 1983; Gilbert C. Robinson 1982; William O. Brandt 1981; William W. Coffeen 1980; Arthur J. Metzger 1979; Loran S. O'Bannon 1978; John F. McMahon 1977; Edward C. Henry 1976; Wayne A. Deringer 1975; Karl Schwartzwalder 1974; Ralph L. Cook 1973; Rolland R. Roup 1972; James E. Hansen 1971; Ralston Russell, Jr. 1970.

KARL SCHWARTZWALDER-PACE AWARD ... this annual award is a joint American Ceramic Society-National Inst of Ceramic Engineers award which honors a young ceramic engineer "whose achievements have been significant to his/her profession and to the general welfare of the American people." The Karl Schwartzwalder-Professional Achievement in Ceramic Engineering (PACE) Award focuses public attention on the outstanding achievements of young persons in ceramic engineering and illustrates the opportunities available in the ceramic engineering profession.

Recipients: Lisa C. Klein 1987; Cullen L. Hackler 1986; John J. Mecholsky Jr. 1985; Willard E. Hauth III 1984; Wallace H. Johnson 1983; Richard E. Tressler 1982; Edmund S. Wright 1981; J. Richard Schorr 1980; H. Kent Bowen 1979; William H. Payne 1978; James C. Waldal 1977; Arthur H. Heuer 1976; W. Richard Ott 1975; Daniel R. Stewart 1974; Gene H. Haertling 1973; Delbert E. Day and Larry L. Hench 1972; Donald L. Kummer 1971; Willard H. Sutton 1970.

NATIONAL RESEARCH COUNCIL (NRC)
(operating arm of NAS-NAE)

2101 Constitution Avenue, Washington, D.C. 20418
Philip M. Smith, Executive Officer

CHAIRMAN ... Frank Press 1981-1993; Philip Handler 1969-1981.

NATIONAL SOCIETY OF PROFESSIONAL ENGINEERS (NSPE)

1420 King Street, Alexandria, VA 22314
Donald G. Weinert, Executive Director

PRESIDENTS ... Charles H. Samson 1987-88; Joseph H. Kuranz 1986-87; Paul E. Pritzker 1985-86; Herbert G. Koogle 1984-85; Louis A. Bacon 1983-84; Marvin M. Specter, 1982-83; Otto A. Tennant 1981-82; William A. Cox 1980-81; Sammie F. Lee 1979-80; Robert L. Nichols 1978-79; Delbert A. Schmand 1977-78; Edward E. Slowter 1976-77; Harry E. Bovay, Jr. 1975-76; Leslie C. Gates 1974-75; Robert L. Reitinger 1973-74; James C. Shivler, Jr. 1972-73.

AWARDS

NSPE-PEI DISTINGUISHED SERVICE AWARD ... for recognition of an individual who has made an outstanding contribution of national scope to advance the causes of the individual engineer in industry or the Professional Engineers in Industry Division of NSPE.

Recipients: Ian Ross 1987; Louis Schindler 1986; Sherwood L. Fawcett 1985; Wilson Greatbatch 1984; David A. Roberson 1983; Alfred W. Andress 1982; Robert L. Nichols 1981; LeRoy Culbertson 1980; Laurence L. Dresser 1979; Horace Lehneis 1978; R. Carleton Spears 1977; James Janssen 1976; L. D. Chipman 1975; Pierce Ellis 1974; Robert Stedfeld 1973.

NSPE-PEPP AWARD FOR OUTSTANDING SERVICE ... for recognition of a professional engineer who has made an outstanding contribution to the private practice segment of the engineering profession.

Recipients: George H. Barton 1987; Robert E. Vansant (posthumous) 1986; James N. De Serio 1985; William L. Carpenter 1984; Robert L. Nichols 1983; David F. Ludovici 1982; Rear Admiral Donald G. Iselin USN (Ret.) 1981; Paul E. Pritzker 1980; William R. Gibbs 1979; James A. Romano 1978; Joseph C. Lawler 1977; Louis A. Bacon 1976; Maxwell Stanley 1975; James F. Shivler, Jr. 1974; Billy T. Sumner 1973; Harry E. Bovay, Jr. 1972; Elmer K. Timby 1971; Edgar B. Boynton 1970; Richard H. Tatlow, III 1969; L. M. Van Doren 1968.

NSPE YOUNG ENGINEER OF THE YEAR AWARD ... for recognition of outstanding scholastic achievements, professional and technical society activities, engineering experience and accomplishments, and civic and humanitarian activities of an engineer 35 years of age or under.

Recipients: H. Lowry Tribble, Jr. 1986; Dr. Ned M. Cleland 1985; Marian Elizabeth Poindexter 1984; Patricia J. Bishop 1983; Ann H. Hansen 1982; Robert T. Berry 1981; Dennis Hirschbrunner 1980; Richard Allen 1979; Max L. Porter 1978; James E. Beavers 1977; Lloyd Piper II 1976; Lowry A. Harper 1975; Lawrence E. Jones 1974; Louis L. Guy, Jr. 1973; John A. Westman 1972; James A. Hackney 1971; Irvin S. Perry (posthumously) 1970.

NSPE DISTINGUISHED SERVICE AWARD ... Recognizes eminence attained by technical contributions to the field of engineering or by exceptional contributions to the engineering profession.

Recipients: Guy H. Cheek (posthumous), Dan H. Pletta, J. Kent Roberts 1987; Paul H. Doll 1986; Leslie C. Gates, William R. Gibbs 1985; George B. Hightower, Robert L. Nichols, James F. Shivler, Jr. 1984.

THE NATIONAL SOCIETY OF PROFESSIONAL ENGINEERS AWARD ... for meritorious service to the engineering profession - restricted to engineers whose sustained and unusual contributions have been to public welfare, the advancement of his profession, and/or service to mankind.

Recipients: Harry E. Bovay, Jr., 1987; Marvin M. Specter, 1986; Wilbur S. Smith 1985; Paul H. Robbins 1984; Luis A. Ferre 1983; Alfred H. Samborn 1982; James Shivier and Kenneth Roe 1981; William S. Lee 1980; Robert C. Seamans, Jr. 1979; Frederick J. Clarke 1978; Edwin H. Young 1977; Samuel S. Baxter 1976; Lyman Dwight Wilbur 1975; Cedomir M. Sliepcevich 1974; Roy F. Weston 1973; Ralph B. Peck 1972; M. R. Lohmann 1971; John W. Beretta 1970; William L. Everitt 1969; Harvey F. Pierce 1968.

OUTSTANDING ENGINEERING ACHIEVEMENT AWARDS ... in an effort to increase public recognition and understanding of engineering NSPE annually selects in nationwide competition up to 10 outstanding engineering achievements. Nominations must come through NSPE Chapters or State Societies. National recognition received by winning projects.

Recipients: 1986 ranged from 4,000-Mile Fiber Optic and Digital Microwave Network in Western USA to Restoration of the Statue of Liberty; 1985 ranged from B-1B Aircraft Weapon System to the Fort McHenry Tunnel in Baltimore, Maryland; Ranged from the world-famous Spacelab to a North Carolina bridge 1984; Ranged from a modernized weather forecasting system to a device which helps paralyzed people walk 1983; Ranged from the technology showcases of the 1982 Knoxville World's Fair and Disney's EPCOT Center to revolutionary new aircraft and waste treatment systems 1982; Ranged from Water Management Program to New Airport 1981; Ranged from Talking Typewriter to Mini-OTEC 1980.

OPERATIONS RESEARCH SOCIETY OF AMERICA (ORSA)

Mount Royal And Guilford Avenues, Baltimore, Maryland 21202
Patricia H. Morris, Executive Director

PRESIDENTS ... Judith S. Liebman 1987; Stephen M. Pollock 1986; Hugh E. Bradley 1985; Michael E. Thomas 1984; David A. Schrady 1983; William P. Pierskalla 1982; George L. Nemhauser 1981; Robert Herman 1980; John D. C. Little 1979; Seth Bonder 1978; Alfred Blumstein 1977; Saul I. Gass 1976; Jack Borsting 1975; David B. Hertz 1974; Robert M. Oliver 1973.

AWARDS

GEORGE E. KIMBALL MEDAL ... for recognition of distinguished service to the Society and to the profession of operations research. Given to an individual. Medal and certificate. Awarded annually. First presented in 1974.

Recipients: Arthur A. Brown and Thornton Leigh Page 1986; Alfred Blumstein 1985; Charles D. Flagle 1984; Merrill M. Flood 1983; Jack R. Borsting 1982; David B. Hertz 1981; Bernard O. Koopman 1980; Charles J. Hitch 1979; John F. Magee 1978; Martin Ernst and William Horvath 1977; Robert Herman 1976; Hugh Miser and Russell Ackoff 1975; Philip Morse, Thomas E. Caywood and George Shortley 1974.

GEORGE E. NICHOLSON, JR. MEMORIAL AWARD ... presented to the winners of a student paper competition. Co-sponsored by the Education and Student Affairs Committees of ORSA. Certificates and (1st place) $500, (2nd place) $250. Top six winners are invited to present their papers at a joint national meeting of ORSA/TIMS.
Recipients: Karl Sigman and Michael A. Trick 1986; Wallace J. Hopp and Paul Tseng 1985; Robin Roundy and Michael Schneider 1984; David D. W. Yao and Patrick T. Harker 1983; K. Sridhar Moorthy and Uday Apte 1982; H. L. Ong and Mostafa H. Ammar 1981; Robert C. Leachman and Gur Huberman 1980; T. S. Wee 1979; James Orlin, John Bartholdi and Paul Dixon 1978; Paul Zipkin and Osman Oguz 1977; P. Orkenyi 1976; Ron S. Dembo 1975; Jerold May 1974; V. Balachandran 1973; Sten E. Wandel 1972.

JOHN VON NEUMANN THEORY PRIZE ... for recognition of the scholar who has made fundamental theoretical contributions to operations research and management science. Engraved medal and $3,000. Given to an individual. Awarded annually. Established in 1974 by the Councils of the Operations Research Society of America and The Insitute of Management Sciences.
Recipients: Samuel Karlin 1987; Kenneth J. Arrow 1986; Jack Edmonds 1985; Ralph E. Gomory 1984; Herbert E. Scarf 1983; Abraham Charnes, William W. Cooper and Richard J. Duffin 1982; Lloyd S. Shapley 1981; David Gale, Harold W. Kuhn and Albert W. Tucker 1980; David Blackwell 1979; John F. Nash, Jr. and Carlton E. Lemke 1978; Felix Pollaczek 1977; Richard Bellman 1976; George Dantzig 1975.

FREDERICK W. LANCHESTER PRIZE ... for the best English language paper in operations research. Presented to the author(s). Engraved medal and $3,000. Awarded annually. Established in 1954 at the Johns Hopkins University.
Recipients: Michael D. Maltz 1985; Narendra Karmarkar 1984; Martin Shubik, Harlan Crowder, Ellis L. Johnson and Manfred M. Padberg 1983; Karl-Heinz Borgwardt 1982; William F. Massy and David S. P. Hopkins 1981; David M. Eddy 1980; Michael R. Gary and David S. Johnson 1979; Richard M. Karp, Gerald Cornuejols, Marshall L. Fisher and George L. Nemhauser 1977; Ralph L. Keeney, Howard Raiffa and Leonard Kleinrock 1976; Laurence D. Stone 1975; Peter Kolesar and Warren E. Walker 1974; Philip Morse 1973; Richard C. Larson 1972; E. E. David, E. J. Piel and J. G. Truxal 1971; Harvey M. Wagner 1969; Anthony V. Fiacco, Garth P. McCormick and Philip M. Morse 1968.

OPTICAL SOCIETY OF AMERICA (OSA)

1816 Jefferson Place, N.W., Washington, D.C. 20036
Jarus W. Quinn, Executive Director

PRESIDENTS ... Robert G. Greenler 1987; Jean M. Bennett 1986; Robert R. Shannon 1985; Donald R. Herriott 1984; Kenneth Baird 1983; Robert P. Madden 1982; Anthony J. DeMaria 1981; Warren Smith 1980; Dudley Williams 1979; Emil Wolf 1978; Peter A. Franken 1977; Boris P. Stoicheff 1976; Arthur L. Schawlow 1975; F. Dow Smith 1974; Robert E. Hopkins 1973.

AWARDS

MAX BORN AWARD ... this award honors the memory of Max Born who made distinguished contributions to physics in general and to optics in particular. It was established in 1982, the centenary of his birth, and is endowed by the United Technologies Research Center. The award consists of a citation, silver medal, and a cash prize of $1,000. It is presented annually to a person who has made outstanding contributions to physical optics, theoretical or experimental.
Recipients: Emil Wolf 1987; Herch M. Nussenzveig 1986; Roy J. Glauber 1985; Adolf W. Lohmann 1984; Joseph W. Goodman 1983; Leonard Mandel 1982.

JOSEPH FRAUNHOFER AWARD ... this award consists of a silver medal, a citation, and $1,000. The first presentation of this award was made in 1982. It is endowed by the Baird Corp. to recognize significant accomplishment in the field of Optical Engineering. The award was named after Joseph Fraunhofer who throughout his career epitomized the finest traditions in creative optical engineering. The work for which this award is given may be characterized as leading to significant practical tools, products or systems. Consideration is given to such criteria as novelty or originality, difficulty, complexity, practical significance, importance to the user, relation to societal needs, or impact on future development.
Recipients: Jerzy A. Dobrowolski 1987; David S. Grey 1986; Peter K. Runge 1985; Donald R. Herriott 1984; Robert E. Hopkins 1983; Robert M. Burley 1982.

FREDERIC IVES MEDAL ... this award was endowed by Herbert E. Ives in memory of his father, Frederic Ives, who is remembered for his pioneering contributions to color photography, photoengraving, three-color process printing, and other branches of applied optics. The award, which consists of a silver medal and a citation and $2,000, is the highest award of the Society for overall distinction in optics. It was presented biennially until 1951, when the donor, in recognition of the rapid growth of optics, increased the endowment so that the medal could be bestowed on an annual basis.
Recipients: Anthony E. Siegman 1987; Amnon Yariv 1986; Emmett N. Leith 1985; Herwig Kogelnik 1984; Boris P. Stoicheff 1983; Lorrin A. Riggs 1982; Georg H. Hass 1981; Aden B. Meinel 1980; Nicolaas Bloembergen 1979; Harold H. Hopkins 1978; Emil Wolf 1977; Arthur L. Schawlow 1976; Ali Javan 1975; David L. MacAdam 1974; Rudolf Kingslake 1973; R. Clark Jones 1972; A. Francis Turner 1971; Robert E. Hopkins 1970; David H. Rank 1969; Edward U. Condon 1968.

ADOLPH LOMB MEDAL ... Adolph Lomb was the Treasurer of the Optical Society from its foundation until his death in 1932. In recognition of his devotion to the interests of the Society and the advancement of optics, The Adolph Lomb Memorial Fund was established by the Society in 1934 with the following directive: "The income of this fund is to be expended, in the discretion of the Board of Directors, for purposes in keeping with the objects of the Society and appropriate as memorials, such as the establishment of a memorial lectureship, awards for distinguished service to the Optical Society of America, awards for outstanding contributions to optical science, and the like." The Adolph Lomb Medal was established in 1940, and it is presented biennially to a person who has made a noteworthy contribution to optics published or accepted for publication before he or she has reached the age of 30 and who has not reached the age of 32 on January first of the year in which the award is presented. The award consists of a medal, a citation, and $1,000.
Recipients: David A. B. Miller 1986; Edward H. Adelson 1984; Won T. Tsang 1982; David M. Bloom 1980; Eli Yablonovitch 1978; Marc D. Levenson 1976; James Forsyth 1974; Robert L. Byer 1972; Marlan O. Scully 1970; Douglas Sinclair 1968.

ELLIS R. LIPPINCOTT AWARD ... this award was established in 1975 jointly by the Optical Society of America, the Coblentz Society, and the Society for Applied Spectroscopy. It honors the unique contributions of Ellis R. Lippincott to the field of vibrational spectroscopy. The award is presented annually to an individual who has made significant contributions to vibrational spectroscopy as judged by their influence on other scientists. Because innovation was a hallmark of the work of Dr. Lippincott, this quality must also be demonstrated by candidates for the award. A recipient will be selected each year by a committee consisting of representatives of each of the sponsoring societies. The award will be presented at the national meeting of one of the sponsoring societies. The award consists of a plaque and $1,000.
Recipients: Wolfgang Kaiser 1986; Ira W. Levin 1985; Jon T. Hougen 1984; John Overend 1983; Michel Delhaye 1982; Ian M. Mills 1981; George C. Pimentel 1980; E. Bright Wilson 1979; Bryce L. Crawford, Jr. 1978; Lionel Bellamy 1977; Richard C. Lord 1976.

C. E. K. MEES MEDAL ... this award was established in 1961 in memory of C. E. K. Mees, who contributed preeminently to the development of scientific photography. The award was endowed by the family of Dr. Mees and consists of a silver medal and a citation and $1,000. It is presented biennially to a recipient who exemplifies the thought that "optics transcends all boundaries"- interdisciplinary and international alike. The recipient is selected by the Board of Directors upon recommendation by a committee of five optical scientists, at least one of whom must have his or her usual residence and place of professional activity elsewhere in in North America.

Recipients: Adolf W. Lohmann 1987; W. Howard Steel 1985; Robert Ditchburn 1983; Maurice Francon 1981; Koichi Shimoda 1979; Andre Marechal 1977; William David Wright 1975; Erik P. Inglestam 1973; Giuliano Toraldo di Francia 1971; Charles H. Townes 1968.

WILLIAM F. MEGGERS AWARD ... this award, which was endowed by the family of William Meggers, several individuals, and a number of optical manufacturers, honors Dr. Meggers for his notable contributions to the field of spectroscopy and metrology. It is awarded annually for outstanding work in spectroscopy. The award consists of a silver medal and a citation and $1,000.
Recipients: Hans R. Griem 1987; Alexander Dalgarno 1986; Theodor W. Hansch 1985; Robert D. Cowan 1984; William C. Martin 1983; George W. Series 1982; Boris P. Stoicheff 1981; John G. Conway 1980; Robert P. Madden 1978; Mark S. Fred, and Frank S. Tomkins 1977; W. R. S. Garton 1976; Jean Blaise 1975; Harry L. Welsh 1974; Curtis J. Humphreys 1973; Charlotte Moore Sitterly 1972; Allen G. Shenstone 1971; George R. Harrison 1970.

DAVID RICHARDSON MEDAL ... in setting up the Richardson medal, the Directors of the Optical Society sought to recognize the unique contribution to applied optics and spectroscopy made by David Richardson. The award which was endowed by Howard Cary, is presented for a distinguished contribution to applied optics. It is intended to recognize individuals who have made significant contributions primarily to technical optics, but not necessarily in a manner manifested by an extensive published record or traditional academic reputation. The award consists of a silver medal, a citation and $1,000. David Richardson received the first award in 1966 for his distinctive contributions to the ruling and replicating of gratings.
Recipients: John W. Evans 1987; John L. Plummer 1986; Norman J. Brown 1985; Erwin G. Loewen 1984; Harold Osterberg 1983; Charles A. Burrus, Jr. 1982; Abe Offner 1981; William T. Plummer and Richard F. Weeks 1980; William P. Ewald 1979; Thomas J. Johnson 1978; Walter P. Siegmund 1977; John H. McLeod 1976; Karl Lambrecht 1975; Roderic M. Scott 1974; William George Fastie 1972; Frank Cooke 1971; Richard S. Hunter 1970; H. Howard Cary 1969; Harold E. Edgerton 1968.

EDGAR D. TILLYER AWARD ... at a meeting of the Society in March 1953, President Brian O'Brien accepted from the President of the American Optical Company the dies of a new medal to be known as the Edgar D. Tillyer Medal, together with an endowment sufficient to permit the striking of the medal biennially. The criteria for selection of recipients of this medal were established by the Board of Directors: The award is to be presented not oftener than once every two years to a person who has "performed distinguished work in the field of vision, including (but not limited to) the optics, physiology, anatomy, or psychology of the visual system." The award consists of a medal, a citation, and $1,000. Subsequently, the Board voted unanimously to make the first award to Dr. Tillyer himself, and the bestowal ceremony took place at the March 1954 meeting.
Recipients: Donald H. Kelly 1986; Mathew Alpern 1984; Leo M. Hurvich and Dorothea Jameson 1982; Fergus W. Campbell 1980; Gerald Westheimer 1978; Floyd Ratliff 1976; Yves Le Grand 1974; Robert M. Boynton 1972; Louise L. Sloan 1971; Lorrin A. Riggs 1969.

CHARLES HARD TOWNES AWARD ... the Townes Award, consisting of a silver medal and a cash prize of $1,000, was established in 1980 to honor Charles Hard Townes, whose pioneering contributions to masers and lasers have led to the development of the field of quantum electronics. The award is presented annually to an individual or a group of individuals for outstanding experimental or theoretical work, discovery, or invention in the field of quantum electronics. The award is made possible by a donation and annual gift from the AT&T Bell Laboratories.
Recipients: Hermann A. Haus 1987; Yuen-Ron Shen 1986; S. E. Harris 1985; Veniamin P. Chebotaev and John L. Hall 1984; Robert W. Hellwarth 1983; C. Kumar Patel 1982; James P. Gordon and Herbert J. Zeiger 1981.

JOHN TYNDALL AWARD ... this award, cosponsored by the Lasers and Electro-Optics Society of the Institute of Electrical and Electronics Engineers and the Optical Society of America, consists of a specially commissioned Steuben crystal sculpture, a scroll, and $1,000. It is endowed by the Corning Glass Works in memory of John Tyndall (1820-1893) who made distinguished contributions to physics. The award is presented to a single individual who has made outstanding contributions in any area of fiber-optics technology including optical fibers themselves, the optical components employed in fiber systems, as well as electro-optic transmission systems employing fibers. The first presentation was made in 1987.
Recipients: Robert D. Maurer 1987.

R. W. WOOD PRIZE ... the Wood Prize was established by the Board of Directors of the Optical Society in 1975 in honor of the many contributions made to optics by R. W. Wood. The prize consists of a cash award of $1,000, a scroll, and a silver medal. It is intended to recognize an outstanding discovery, scientific or technological achievement, or invention in the field of optics. The accomplishment for which the prize is given is measured chiefly by its impact of the field of optics generally, and therefore the contribution is one that opens a new area of research or significantly expands an established one. A five-year grant has been received from the Xerox Corporation in support of the prize. The initial presentation of the Wood Prize was made in 1975.
Recipients: David E. Aspnes 1987; Joseph A. Giordmaine and Robert C. Miller 1986; David H. Auston 1985; Otto Wichterle 1984; Sven R. Hartmann 1983; Linn F. Mollenauer 1982; Erich P. Ippen and Charles V. Shank 1981; Anthony E. Siegman 1980; Peter Franken 1979; Peter P. Soroki 1978; Peter Fellgett 1977; Dr. Theodore H. Maiman 1976; Emmett N. Leith and Juris Upatnieks 1975.

PHI LAMBDA UPSILON (PLU)

Dr. Jack Graybeal, Department of Chemistry, VPI&S.U., Blacksburg, VA 24061

PRESIDENTS ... Dr. Sheldon H. Cohen 1987-1990; Dr. David W. Brooks 1984-1987; Dr. Joseph L. Greene, Jr. 1981-1984; Dr. Thomas E. Daubert 1978-1981.

AWARDS

THE PHI LAMBDA UPSILON FRESENIUS AWARD ... the Fresenius Award consisting of a plaque and suitable honorarium is made each year to a person who has not reached his or her 35th birthday at the time of nomination and who has made notable contributions in chemical research, teaching or administration in any field of pure or applied chemical endeavor.
Recipients: Jacqueline K. Barton 1986; Ben S. Freiser 1985; Mark S. Wrighton 1984; George C. Schutz 1983; Michael J. Berry 1982; Richard P. Van Duyne 1981; John R. Shapley 1980; Tobin J. Marks 1979; Patrick S. Mariano 1978; William P. Reinhardt 1977; Joseph B. Lambert 1976: Robert Vaughn 1975; Richard Zare 1974; Nicholas Turro 1973; Charles Cantor 1972; Willis Flygare 1971; Harry Gray 1970; Rouald Hoffman 1969; John Baldeschweiler 1968.

PI TAU SIGMA

C/O Dr. Edwin Griggs, P.O. Box 5213, Tennessee Technological University, Cookeville, Tennessee 38505
Edwin I. Griggs, National Secretary-Treasurer

PRESIDENTS ... Dr. William B. Cottingham 1986- ; Dr. Hudy C. Hewitt, Jr. 1983-1986; Alexander R. Peters 1980-1983; Suresh Chandra 1978-80, Edward W. Jerger 1975-1977; James W. Bayre 1973-74.

AWARDS

GOLD MEDAL AWARD ... given for outstanding achievement within 0-10 years after graduation.
Recipients: Dimos Poulikakos 1986; Wing Kam Liu 1985; Michael R. Muller 1984; P.D. Spanos 1982; Doyle D. Knight 1980; David A. Peters 1978; Richard E. Lovejoy 1977; John S. Walker 1976; Ted B. Belytschko 1975; Jace W. Nunziato 1974; Christian E. Prziremberl 1973; John F. Stephens, III 1972; James R. Rice 1971; Richard E. Barrett 1970; Henry K. Newhall 1969; Randall F. Barron 1968.

GUSTUS L. LARSON AWARD . . . given for outstanding achievement within 10-20 years after graduation.
Recipients: Bharat Bhushan 1986; Klaus-Jurgen Bathe 1985; Robert A. Altenkirch 1984; R. Byron Pipes 1983; Melvin C. Branch 1982; Jerry E. Shoup 1981; Arthur G. Erdman 1980; Gerald R. Seemann 1979; Philip H. Francis 1978; Nam P. Suh 1977; John G. Bollinger 1976; Chang-Lin Tien 1975.

RICHARDS MEMORIAL AWARD . . . given for outstanding achievement more than 20 years after graduation.
Recipients: E. Kent Springer 1986; Ephraim M. Sparrow 1985; Ferdinand Freudenstein 1984; Peter A. Engel 1983; L.S. Fletcher 1982; Shien-Ming Wu 1981; Albert I. King 1980; John Henry Lienhard 1979; John C. Chato 1978; Hassan A. Hassan 1977; Ali S. Argon 1976; Carl F. Zorowski 1975; Richard J. Grosh 1974; Ali A. Siereg 1973; Charles E. Jones 1972; Howard L. Harrison 1971; Ralph G. Nevins 1970; Robert E. Uhrig 1969; Bernard W. Shaffer 1968.

SIGMA XI, THE SCIENTIFIC RESEARCH SOCIETY

345 Whitney Avenue, New Haven, Connecticut 06511
Deborah Levinson, Executive Director

PRESIDENTS . . . F. Kenneth Hare 1986-87; Lewis M. Branscomb 1985-86; N. Patricia Faber 1984-85; John W. Prados 1983-84; V. Elving Anderson 1982-83; William A. Nierenberg 1981-82; Herbert E. Longnecker 1980-81; Melvin Kranzbert 1979-80; Dean E. McFeron 1978; Harold G. Cassidy 1977; Lawrence M. Kushner 1976; Frederick E. Terman 1975; Linton E. Grinter 1974; Harvey A. Neville 1973.

AWARDS

THE WILLIAM PROCTER PRIZE FOR SCIENTIFIC ACHIEVEMENT . . . awarded annually to a scientist or engineer in recognition of notable achievement in scientific research or in the administration of such research. The Prize consists of a certificate of award and a check for $2000 to the awardee and another $2000 for a Grant-in-Aid of Research to a younger scientist designated by the awardee.
Recipients: Thomas Eisner 1986; George C. Pimentel 1985; Victor F. Veisskopf 1984; Winona and John Vernberg 1983; Joshua Lederberg 1982; George W. Beadle 1981; Herbert A. Simon 1980; Saunders MacLane 1979; Russell W. Peterson 1978; William Nierenberg 1977; Morris Cohen 1976; Dixy Lee Ray 1975; Percy Lavon Julian 1974; William O. Baker 1973; Lewis M. Branscomb 1972; Jacob E. Goldman 1971; Lloyd M. Cooke 1970; Margaret Mead 1969; Athelstan Spilhaus 1968.

THE MONIE A. FERST AWARD . . . for those who have made notable contributions to motivation and encouragement of research through education.
Recipients: Julian Schwinger 1986; Norman F. Ramsey 1985; Richard J. Duffin 1984; John C. Bailer, Jr. 1983; Nickolas J. Hoff 1982; John A. Wheeler 1981; Earnest C. Pollard 1980; Richard L. Solomon 1979; Hans W. Liepmann 1978; E. Bright Wilson 1977.

SOCIETY FOR THE ADVANCEMENT OF MATERIAL & PROCESS ENGINEERING (SAMPE)

843 W. Glentana Street, P.O. Box 2459, Covina, California 91722
Mrs. Marge Smith, Association Manager

PRESIDENTS . . . Charles J. Hurley 1987; Charles J. Weizenecker 1986; N.H. Kordsmeier, Jr. 1985; John T. Hoggatt 1984; Luke C. May 1983; George F. Schmitt, Jr. 1982; Jack M. Dyer 1981; William P. Koster 1980; Raymond W. Fenn, Jr. 1979; Earl Newell 1978; Peggy Moore 1977; Sam Jacobs 1976; D. "Cam" Perry 1975; Mort Kushner 1974; Jack Burroughs 1973.

AWARDS

SAMPE FELLOW PROGRAM . . . the honor of Fellow of the Society was established to provide recognition of members for distinguished contributions in the fields of materials and processes and to develop a broadly based forum for technical and professional leaders to serve as advisors to the Society.
Recipients: Jack Burroughs, John Van Hamersveld 1986; George Gregory, John Hoggatt and Ray Wegman 1985; Stuart M. Lee, Donald Mazenko and Allen P. Penton III 1984; Eugene R. Frye, Samuel M. Jacobs, Clayton A. May, Bryan R. Noton, Nathan E. Promisel and George F. Schmitt Jr. 1983; Gail A. Clark, John Delmonte, Raymond W. Fenn, William P. Koster, Martin Kushner, George Lubin, Andrew C. Marshall, L. E. Meade, C. Cameron Perry, Frank Robinson, Robert J. Schwinghamer and Bernard Silverman 1982.

MERITORIOUS AWARD . . . granted only to a SAMPE Member who has notably enhanced either the stature of the Society or the attainment of its objectives. The bestowal of this award shall be required to be justified by a specific and significant effort, achievement, action or combination thereof by the recipient.
Recipients: James Bell 1987; Eugene Crilly 1986; William Heimerdinger 1984; Jerome Bauer and William Long 1983; Stuart M. Lee 1982; Robert Spira 1981; S. Westerbach 1978.

SAMPE/GEORGE LUBIN MEMORIAL AWARD . . . this award may be granted to any person who has meritoriously fostered the advancement of Material and Process technology and/or has enhanced the attainment of the objectives of the Society and who merits entitlement to the rights and privileges of permanent tenure as an Honorary Life Member of the Society.
Recipients: Stephen Tsai 1987; Mel Schwartz 1986; George Peterson 1982; Robert Schwinghamer 1980; D. Cameron Perry 1979; George Lubin 1978; Morton Kushner 1978; A. C. Marshall 1970.

DELMONTE AWARD . . . this award is granted to a SAMPE member who has notably achieved a technology breakthrough during the current year in the field of Material and Process Engineering.
Recipients: Roscoe Pike 1987; Clyde H. Sheppard 1986; Dave Diveche 1985; John Tipton 1984; Sid Street 1983.

SOCIETY FOR COMPUTER SIMULATION (SCS)

P.O. Box 17990, San Deigo, California 92117
Charles A. Pratt, Executive Director

PRESIDENTS . . . Ralph Huntsinger, 1987; Norbert Pobanz 1985; Walter J. Karplus 1983-84; Stewart I. Schlesinger 1982; Donald C. Martin 1977-79; Per A. Holst 1977; Paul W. Berthiaume 1976; Robert D. Brennan 1974-75; George A. Rahe 1973.

SOCIETY FOR EXPERIMENTAL MECHANICS (SEM)

7 School St. Bethel, CT 06801
K. A. Galione, Managing Director & Publisher

PRESIDENTS . . . Clarence A. Calder 1988; Ian Allison 1987; R. J. Rinn 1986; W. N. Sharpe, Jr. 1985; J. B. Ligon 1984; S. E. Swartz 1983; D. L. Willis 1982; M. E. Fourney 1981; R. L. Johnson 1980; H. F. Brinson 1979; D. R. Harting 1978; E. I. Riegner 1977; J. W. Dallay 1976; E. E. Day 1975; R. C. A. Thurston 1974; C. E. Work 1973.

AWARDS

B. J. LAZAN AWARD . . . given for original contribution to the technology of experimental mechanics.
Recipients: D. L. Brown 1987; A. J. Durelli 1986; A. Lagarde 1985; I. M. Daniel 1984; Karl A. Stetson 1983; S. J. Green 1982; A. S. Kobayashi 1981; James E. Starr 1980; Milton M. Leven 1979; W. T. Bean 1978; J. C. Telinde 1976; M. E. Fourney 1975; J. F. Bell 1974; S. S. Manson 1973; D. Post 1972; D. R. Harting 1971; F. Zandman 1970; A. C. Ruge 1969; Greer Ellis 1968.

F. G. TATNALL AWARD . . . given to recognize individuals who have made outstanding contributions to the Society in service through committee work.

Recipients: B. E. Rossi 1987; D. L. Willis 1986; H. F. Brinson 1985; R. L. Johnson 1984; C. E. Taylor 1983; C. A. Calder 1982; E. I. Riegner 1981; P. H. Adams 1980; R. H. Homewood 1979; C. E. Work 1978; A. E. Johnson 1977; C. C. Perry 1976; L. J. Weymouth 1975; F. C. Bailey 1974; A. S. Kobayashi 1973; M. M. Leven 1972; J. H. Meier 1971; F. B. Stern 1970; C. R. Smith 1969; F. G. Tatnall 1968.

THE HETENYI AWARD . . . SESA's paper-of-the-year award.
Recipients: E. Vogt, J. Geldhacher, B. Dier, M. Krietlow 1986; S. Hashimoto, K. Kawata and N. Takeda 1985; A. S. Voloshin and C. P. Berger 1984; C. A. Sciammarella, P. K. Rastogi, P. Jacquot and R. Narayan 1983; D. Bar-Tikva, A. F. Grandt, Jr. and A. N. Palazotto 1982; M. F. Duggan, J. Lankford and D. L. Davidson 1981; K. A. Stetson 1979; J. G. Blaul, J. Beinert and M. Wenk 1978; R. L. Johnson 1977; I. M. Daniel and R. E. Rowlands 1975; W. J. McAfee and H. Pih 1974; R. J. Sanford and V. J. Parks 1973; Pierre Boone 1972; T. D. Dudderar and R. O'Regan 1971; C. E. Taylor, R. E. Rowlands and I. M. Daniel 1970; E. P. Popov and M. S. Lin 1969; Warren J. Rhines 1968.

M. M. FROCHT AWARD . . . given for achievements in education in experimental mechanics.
Recipients: R. E. Rowlands 1987; W. F. Swinson 1986; K. G. McConnell 1985; C. P. Berger 1984; C. W. Smith 1983; T. Kunio 1982; V. J. Parks 1981; C. A. Sciammarella 1980; E. E. Day 1979; J. P. Pindera 1978; W. F. Riley 1977; J. W. Dally 1976; D. K. Wright 1975; R. D. Mindlin 1974; P. K. Stein 1973; E. O. Stitz 1972; D. C. Drucker 1971; W. M. Murray 1970; C. E. Taylor 1969; M. M. Frocht 1968.

MURRAY LECTURE . . . award to leaders in various fields of technology related to Society interests for outstanding professional achievements.
Recipients: D. R. Harting 1987; E. P. Popov 1986; Erwin Somer 1985; C. C. Perry 1984; A. S. Kobayashi 1983; R. Mark 1982; L. J. Broutman 1981; A. Bray 1980; J. W. Dalley 1979; R. K. Muller 1978; W. E. Cooper 1977; J. E. Starr 1976; N. M. Newmark 1975; C. E. Taylor 1974; G. R. Irwin 1973; M. M. Leven 1972; D. Post 1971; B. F. Langer 1970; T. J. Dolan 1969; C. Lipson 1968.

R. E. PETERSON AWARD . . . for outstanding application paper published in "Experimental Mechanics."
Recipients: C. M. Vickery, J. K. Good, R. L. Lowrey 1986; T. F. Leahy 1985; W. F. Swinson, J. L. Turner, N. H. Madsen, J. L. Milton and J. E. Stone 1984; A. S. Kobayashi and A. Komine 1983; M. E. Duggan 1981; James Dorsey 1979; R. L. Whipple, J. B. Ligon; C. P. Burger; M. S. Coffman 1977; L. W. Hornby and B. E. Noltingk 1975; C. P. Burger 1973; J. C. Telinde 1971.

SOCIETY FOR THE HISTORY OF TECHNOLOGY (SHOT)

School of Engineering, Duke University Durham, N. C. 27706
Aarne Vesilind, Director
PRESIDENTS . . . Bruce Sinclair 1987-1988; Edwin T. Layton, Jr. 1985-86; Melvin Kranzberg 1983-84; Brooke Hindle 1981-82; Thomas P. Hughes 1979-80; Eugene S. Ferguson 1977-78; John C. Brainerd 1975-76; John B. Rae 1973-74.

AWARDS

THE JOAN CAHALIN ROBINSON PRIZE . . . for the best paper presented at the SHOT meeting by a historian under thirty years of age or by a person past thirty who is an accredited graduate student or candidate for a higher degree and who is presenting his or her first paper at a SHOT annual meeting.
Recipients: James H. Capshew 1986; Susan Smulyan 1984; Larry Owens 1983; Mona Spangler Phillips 1982; Christopher Hamlin 1981; J. Lauritz Larson 1980.

THE ABBOTT PAYSON USHER PRIZE . . . given to the author of the best work published by the Society in the preceding three years. It consists of $100 and a certificate.
Recipients: Donald MacKenzie 1986; Eda Fowlks Krauakis 1985; W. G. Vincenti 1984; George Wise 1983; Harold Dorn 1982; T. P. Hughes 1981; S. W . Leslie 1980; L. Bryant 1979; O. Mayr 1978; W. H. TeBrake 1977; Russell Fries 1976; Paul Uselding 1975; Carl Mitcham and Robert Mackey 1974; R. L. Hills

and A. J. Pacey 1973; Cyril Stanley Smith 1972; James E. Brittain 1971; James E. Packer 1970; Eugene S. Ferguson 1969; Carl W. Condit 1968.

THE DEXTER PRIZE . . . awarded to the author of an outstanding book in the history of technology published during any of the three years preceding the award. It consists of $1,000 and a plaque.
Recipients: Walter A. McDougall 1986; Thomas P. Hughes 1985; Ruth Schwartz Cowan 1984; Clayton R. Koppes 1983; Edward W. Constant II 1982; O. Jeremy 1981; L. Hunter 1980; O. P. Billington 1979; R. Jenkins 1978; R. W. Bulliet 1977; Hugh J. G. Aitken 1976; Bruce Sinclair 1975; Daniel J. Boorstin and Donald R. Hill 1974; Donald S. L. Cardwell 1973; Thomas Park Hughes 1972; Edwin T. Layton, Jr. 1971; Lynn White, Jr. 1970; Gotz Quang 1969; Hans Eberhard Wulff 1968.

THE LEONARDO DA VINCI MEDAL . . . the Society's highest award, for outstanding contribution to the history of technology by research, teaching, publication, or otherwise. Consists of a plaque and certificate.
Recipients: Hugh G. J. Aitken 1986; Thomas P. Hughes 1985; Brooke Hindle 1984; Louis C. Hunter 1983; D. S. L. Cardwell 1981; J. B. Rae 1980; J. U. Nef 1979; T. Althin 1978; E. Ferguson 1977; Derek J. daSola Price 1976; Frederick Klemm 1975; Bern Dibner 1974; Carl W. Condit 1973; Ladislao Reti 1972; A. G. Drachman 1971; Bertrand Gille 1970; Lewis Mumford 1969; Joseph Needham 1968.

SOCIETY OF ALLIED WEIGHT ENGINEERS, INC. (SAWE)

344 East J Street, Chulla Vista, California 92010
Fred Wetmore, Executive Secretary
PRESIDENTS . . . Angelo S. Karadimos 1986-88; Frank Fong 1985-86; William F. Young 1983-84; James T. Werner 1981-82; J. S. Wisniewski 1980; J. McLaughlin 1978-79; B. B. Coker 1976-77; R. Crossen 1974-75; James C. Mitchell 1973.

AWARDS

EDGAR L. PAYNE AWARD . . . awarded to the young engineer (age less than 35) who has contributed significantly to the Society of Allied Weight Engineers or the weight engineering profession. The society may choose to not select an award recipient in any given year.
Recipients: Cathy B. Griesinger and Gregory J. Burch 1987; Fred L. Burkhart and Larry J. Linner 1985; Gary W. Balthrop 1981; Tom Oole 1979; Roger S. St. John 1972; Robert J. Taylor 1971.

GEC CUP (replacing Revere Cup) . . . awarded to the author of the most outstanding technical paper presented on the SAWE Annual International Conference.
Recipients: Jerry Pierson 1987; Larry J. Linner 1985; Russell G. Maguire 1984; Russell G. Maguire 1983; John R. Atkinson, Roy N. Staton 1982; John D. Brown and Charles J. Thomas 1981; Roger S. St. John and Paul F. Piscopo 1980; John M. Hedgepath, Lawrence A. Finley and Karl K. Knapp 1979; Robert R. Sorrals and Andy J. Holten 1976; Larry E. Lewis and Roger S. St. John 1975; Wolfgang Schneider 1974; David Simpson 1973; Paul W. Scott 1972; Joseph Lotto and Joseph Lagana 1971; T. Lamb 1970; Karl C. Sanders 1969; Henry B. Ruble and Roger P. Johnson 1968.

THE SOCIETY OF AMERICAN MILITARY ENGINEERS (SAME)

607 Prince St., P. O. Box 21289, Alexandria, Virginia 22320-2289
Brig. Gen. Walter O. Bachus, Executive Director, USA (Ret)
PRESIDENTS . . . Lt. Gen. E. R. Heiberg III 1987; Rear Adm. John Paul Jones, Jr. 1986; Rear Adm. Kenneth G. Wiman 1985; Maj. Gen. Clifton D. Wright, Jr. 1984; Rear Adm. William M. Zobel 1983; Lt. Gen. Joseph K. Bratton 1982; Louis W. Riggs 1981; Maj. Gen. William D. Gilbert, USAF 1980; R. Adm. Donald G. Iselin, USN 1979; Lt. Gen. John W. Morris 1978; Mr. Seymour S. Greenfield 1977; MG. Robert C. Thompson 1976;

RAdm. A. R. Marshall 1975; Lt. Gen. William C. Gribble, Jr. 1974; Brig. Gen. Hubert O. Johnson, Jr. 1973.

AWARDS

BLISS MEDAL . . . offered as an annual award to the engineering professional or instructor at a college or university at which a unit of the Reserve Officers Training Corps of the Armed Forces is established for the most outstanding contribution to military engineering education, or serving to promote recognition of the importance of technical leadership in the National Defenses establishment.
Recipients: Dr. Wilbur L. Meier, Jr. 1985; J. David Irwin 1984; Arthur Akers 1983; Col. James D. Strong 1982; William R. Kimel 1981; John D. Haltiwanger 1980; John W. Knapp 1979; R. C. Edward 1978; Adrian R. Chamberlain 1977; B. R. Teare 1976; Dean Arthur A. Burr 1975; Howard V. Smith 1974; Fred J. Benson 1973; Lionel V. Baldwin 1972; Fred N. Peebles 1971; Lauress L. Wise 1970; Charles H. Weaver 1969; T. H. Kuhn 1968.

COLBERT MEDAL . . . offered as an annual award to a member of the United States Coast and Geodetic Survey, officer or civilian, active or retired, in recognition of the most outstanding contribution to military engineering through achievment in design, construction, administration, research or development.
Recipients: Charles D. Kearse 1986; Comdr. A. Nicholas Bodnar 1985; Richard C. Patchen 1984; Bobby J. Taylor 1984; LCdr. Gerald B. Mills 1983; LCdr. Dan E. Tracy 1982; Lawrence W. Fritz 1981; LCdr. Donnie M. Spillman 1980; Melvin S. Asato 1979; D. E. Northrup 1978; Raymond H. Carstens 1977; Carroll I. Thurlow 1976; LCdr. Ludvik Pfeifer 1975; Raymond W. Tomlinson 1974; Charles D. Kearse 1973; Col. Frederick O. Diercks 1972; Capt. John O'Boyer 1971; Lloyd L. Clay 1970; Cdr. Clinton D. Upham 1969; George B. Lesley 1968.

GODDARD MEDAL . . . the Goddard Medal is named for MG Guy H. Goddard, Past President of the Society and former Director of Civil Engineering, USAF. The Medal is offered as an annual award to an active duty enlisted member of the U. S. Air Force to recognize outstanding contributions to military troop construction and/or base maintenance by demonstrated technical and leadership ability. The award is presented to an individual within one of the following Air Force career fields: pavement maintenance, construction equipment operator, carpenter mason, protective coating specialist (painter), site development specialist, electrical exterior (power), electrical interior, electrical power production, refrigeration and air conditioning, liquid fuels, heating systems, metal processing (welder), sheet metal base maintenance equipment repair, special vehicle maintenance, general purpose repairman, and vehicle body repairman.
Recipients: Senior Master Sergeant Richard P. Driscoll 1986; Senior Master Sergeant James R. Keefe 1985; Senior Master Sergeant James T. Merek 1984; Senior Master Sergeant Bob G. Buckles 1983; Chief Master Sergeant Woodrow F. Giddens 1982; Senior Master Sergeant Dennis J. McAvoy 1981; Senior Master Sergeant Arthur J. Hanrahan 1980.

GEORGE W. GOETHALS MEDAL . . . offered annually to an engineer in civil or military practice for the most eminent and notable contribution in the field of engineering, particularly in design, construction, and methods.
Recipients: William L. Stevens 1986; Edward Cohen 1985; Alexander G. Tarics 1984; George J. Zeiler 1983; Euclid C. Moore 1982; Michael A. Kolessar 1981; Robert J. Taylor 1980; E. Montford Fucik 1979; A. Casagrande 1978; Garland L. Watts 1977; Henry W. Holliday 1976; James Polk Stafford, Jr. 1975; Col. John E. Catlin, Jr. 1974; Robert H. Hayes 1973; Michael Yachnis 1972; Robert Y. Hudson 1971; Arsham Amirikian 1970; Wendell W. Ralphe 1969; Col. Jerome O. Ackerman 1968.

MOREELL MEDAL . . . offered annually as an award to a member of the Civil Engineer Corps of the Navy, regular, reserve, or civilian, on active or inactive duty or retired, in recognition of the most outstanding contribution to military engineering through achievement in design, construction administration, research, or development connected with military engineering.
Recipients: Robert E. Hammond 1986; Rear Adm. Thomas S. Maddock 1985; Capt. William J. O'Donnell 1984; Comdr. George W. Holland 1983; Milon Essoglou 1982; LCdr. Harrison B. Whittaker, II 1981; Capt. Eric R. Wilson, Jr. 1980; LCdr.

George W. Yankoupe 1979; W. F. Burke 1978; LCdr. Ole L. Olsen 1977; Comdr Lawrence K. Donovan 1976; Cdr. Tracy C. Tucker 1975; Cdr. Gordon W. Tinker 1974; Cdr. Jack B. Moger 1973; Cdr. Malcolm J. MacDonald 1972; RAdm. George Reider 1971; Glenn W. Zimmer 1970; Capt. Blake W. Van Leer 1969; Cdr. Malcolm T. Mooney 1968.

NEWMAN MEDAL . . . offered as an annual award to a member of the Directorate of Civil Engineering of the Air Force, regular, reserve, National Guard, or civilian, on active or inactive duty or retired, in recognition of the most outstanding contribution to military engineering through achievment in design, construction, administration, research or development.
Recipients: Col. John S. Donovan 1986; Col. Ray D. Schwartz 1985; Brig. Gen. Joseph A. Ahearn 1984; Col. Roy M. Goodwin 1983; Brig. Gen. Sheldon J. Lustig 1982; Brig Gen. Paul T. Hartung 1981; Col George E. Ellis 1980; Col M. Gary Alkire 1979; J. P. Thomas 1978; Col Elo Mussetto 1977; Lt. Col William R. Sims 1976; Col. Charles W. Lomb 1975; Maj. Arthur Y. Kishiyama 1974; Col. Louis F. Dominguez 1973; Col. Warren E. Campbell 1972; BGen Archie S. Mayers 1971; Capt. James L. Baushke 1970; Maj. James F. Hagen 1969; Col. John B. Rose 1968.

OREN MEDAL . . . offered for the first time in 1968 as an annual award to a member of the Coast Guard, officer or civilian, active or retired.
Recipients: Lt. Thomas H. Biggs 1986; Bob D. Lamm 1985; LCdr. James M. Seagraves 1984; James M. Sherman 1983; David Lebofsky 1982; LCdr. Gregory H. Magee 1981; Cdr. Harry W. Tiffany 1980; Martin J. Boivin 1979; R. C. Boggs 1978; Raquel McAllister 1977; Hisashi Imanaka 1976; Cdr. John R. Wells, Jr. 1975; Cdr. Walter E. Peterson 1974; Mr. Robert G. Stafford 1973; Mr. Byron J. Clark 1972; LCdr. K. L. Reichelt 1971; Gerald B. O'Hara 1970; Jonah Mack 1969; Glen E. Logan 1968.

SARGENT MEDAL . . . the Sargent Medal is named for Vice Adm. Thomas R. Sargent, III Former Chief of Civil Engineering and Vice Commandant of the Coast Guard from 1970-1974. This annual award is given in recognition of outstanding contribution to Coast Guard Civil and/or Facilities Engineering by an active duty Warrant Officer or Noncommisioned Officer of the Coast Guard or a Coast Guard civilian employee of equivalent grade. The achievement must have occurred during the calendar year preceding the nomination or may be based on the completion, during that year, of a multi-year activity.
Recipients: CWO2 Paul D. Dangreau 1986; CWO2 Daniel Sathre 1985; Harold Spraggins 1984; CWO2 Douglas M. Carroll 1983; CWO4 Roger Lewis 1982; MICS Charles R. Shepherd 1981; CWO3 Earle J. Madden 1980.

SHIELDS MEDAL . . . the Shields Medal is named in memory of Petty Officer Marvin G. Shields, who was posthumously awarded the Congressional Medal of Honor for Valor in combat while serving in Vietnam. The medal is offered as an annual award to an active duty enlisted member of the U. S. Navy to recognize outstanding contributions to military troop construction and/or base maintenance by demonstrated technical and leadership ability. The award is presented to an individual within one of the following Navy career ratings: engineering aid, equipment operator, construction mechanic, builder, steelworker, construction electrician, and utilitisman.
Recipients: Steelworker First Class George M. Havash 1986; BU1 Rickie D. Deems 1985; CM1 Daniel J. Brunger 1984; BU1 Theodore T. Posuniak 1983; Claud V. VanGurp 1982; Jack Benny Feagins 1981; Utilitiesman First Class Robert E. Zulick 1980.

STURGIS MEDAL . . . the Sturgis Medal is named for LTG Samuel D. Sturgis, Jr. former Chief of Engineers of the U. S. Army. The Medal is offered as an annual award to an active duty enlisted member of the U. S. Army to recognize outstanding contributions to military troop construction and/or base maintenance by demonstrated technical and leadership ability. The award is presented to an individual within one of the following army career fields: combat engineering, construction and utilities, power production, heavy equipment and maintenance, drafting, surveying, printing, special electrical devices, instrument repairman, and reproduction repairman.

Recipients: Sergeant First Class George R. St. Onge 1986; Sergeant First Class Dennis J. Ansay 1985; Staff Sergeant James A. Kochara 1984; Seargent First Class Eugene Middleton 1983; Seargeant First Class (P) Michael L. McGuiggan 1982; Staff Seargent Jackie L. Thomas 1981; Staff Seargent Billy D. Miller 1980.

SVERDRUP MEDAL . . . the Sverdrup Medal was established by The Society in memory of the distinguished military engineer, Maj. Gen. Leif J. Sverdrup, U. S. Army Ret. The medal is awarded annually to an active duty military engineer member of The Society who has not passed his/her 36th Birthday in the year for which nominated for the most outstanding engineering contribution, or similar achievement of significance to The Society, the military service, or the Nation.
Recipients: Lt. Michael D. Kiehnau 1986; Maj. Michael H. Schmidt 1985; Maj. Patrick M. Coullahan 1984; Maj. Peter A. Topp 1983; LCdr. Charles D. Wurster 1982; Capt. M. Stephen Rhoades 1981; LTC Eugene A. Lupia 1980.

TOULMIN MEDAL . . . awarded annually to the author of the article judged to be the best published in "The Military Engineer" during the year.
Recipients: Dr. Paul E. Torgersen *The Health of Engineering Education* 1986; George B. Korte, Jr. *CADD's Impact on A/E Firms* 1985; Lt. Col. Andrews M. Perkins, *Operation URGENT FURY: An Engineer's View* 1984; Jacques S. Gansler, *Industrial Preparedness: National Security in the Nuclear Age* 1983; Thomas H. Spencer, *The A/E's Role in Mobilization* 1982; Hungdah Chiu, *A Rapprochement Between the Two Chinas* 1981; James G. Stanley 1980; Leland R. Johnson 1979; R. P. Howell 1978; Harry E. Bovay 1977; Leland R. Johnson 1976; BGen. William Whipple, Jr. 1975; BGen. George A. Lincoln 1974; Leland R. Johnson 1973; Maj. Gen. Richard H. Groves 1972; Maj. William B. Willard, Jr. 1971; LCdr. Robert D. Smart 1970; Walter J. Tudor 1969; Lt. Col Maurice K. Kutz, Jr. 1968.

TUDOR MEDAL . . . offered annually as an award to a civilian member of SAME (under 36) for the most outstanding contribution to military engineering design. The award was made for the first time in 1966.
Recipients: Alan E. Murphy 1986; Rod E. Markuten 1985; Dennis R. Duke 1984; Thomas Edison 1983; Russel E. Milnes 1982; Jean-Yves Perez 1981; Ashraf N. Wahba 1980; Ernest K. Schrader 1979; R. F. Lovejoy 1978, Jonathon M. Nash 1977; James R. Van Orman 1976; Gary D. Vest 1975; Carter J. Ward 1974; John W. Rushing 1973; Alton A. Bradford 1972; Michael A. Boyd 1971; Stanley Y. Yasumoto 1970; Phillip E. Lammi 1969; Robert N. Bradley 1968.

WHEELER MEDAL . . . offered annually as an award to a member of the Corps of Engineers of the Army, regular, reserve, National Guard or civilian, on active or inactive duty or retired, in recognition of the most outstanding contribution to military engineering through achievement in design, construction, administration, research or development.
Recipients: Col. John I. Coats 1986; Col. Steven G. West 1985; Maj. Stewart H. Bornhoft 1984; John J. Blake 1983; Maj. Gen. John F. Wall, Jr. 1982; Joe G. Higgs 1981; Maj Wendell L. Barnes 1980; Col Cranston R. Rogers 1979; LTC R. H. Gates 1978; Benjamin G. DeCooke 1977; William J. Flathau 1976; Capt. Anthony V. Nida 1975; Mr. Lloyd A. Duscha 1974; Mr. Ronald H. Gelnett 1973; Col. John L. Lillibridge 1972; MGen. Robert P. Young 1971; col. Robert O. Brugge 1970; Lee Stanford Garrett 1969; Eugene Groden 1968.

SOCIETY OF AUTOMOTIVE ENGINEERS, INC. (SAE)

400 Commonwealth Drive, Warrendale, Pennsylvania 15096
Max E. Rumbaugh, Jr., Executive Vice President
PRESIDENTS . . . William S. Coleman 1987; Frank Walter 1986; Elliott A. Green 1985; Gordon H. Millar 1984; Charles C. Colyer 1983; N. John Beck 1982; Philip J. Mazziotti 1981; Harold C. MacDonald 1980; Lewis E. Fleuelling 1979; Leo A. McReynolds 1978; Dr. Gordon L. Scofield 1977; Rodger F. Ringham 1976; George J. Huebner, Jr. 1975; Wilson A. Gebhardt 1974; J. C. Ellis 1973.

AWARDS

ARCH T. COLWELL COOPERATIVE ENGINEERING MEDAL . . . to be presented to an individual for a unique and outstanding contribution over a period of time to the Society's technical committees engaged in the development of SAE standards, specifications, technical reports, and data development through cooperative research.
Recipients: George P. Townsend, Jr. 1984; Ray V. Riggs 1983; Ralph H. K. Cramer 1982; Lawrence H. Hodges 1981; Roy P. Trowbridge 1980; G. E. Burks 1979; Francis L. LaQue 1978; William F. Sherman 1977; Arch T. Colwell 1976.

ARCH T. COLWELL MERIT AWARD . . . awarded annually to the authors of papers of outstanding technical or professional merit which shall have been presented at a meeting of the Society or any of its Sections during the calendar year.
Recipients: K. H. Muhr, K. H. Virnich, B. Chehroudi, S. H. Chen, F. V. Bracco, Y. Onuma, M. E. G. Sweeney, R. G. Kenny, B. P. Selberg, Y. Kimbara, F. Sugasawa, Y. Shiraishi, G. B. G. Swann, G. P. Blair, N. D. Brinkman, R. F. Stebar, B. W. Schumacher, J. C. Cooper, W. Dilay, S. C. Y. Tung, K. Rokhsaz, K. Shinoda, H. Kobayashi, Y. Tateishi, J. L. Linden, S. Kamiya, H. Kawamoto, M. Murachi, S. Kato, S. Kawakami, Y. Suzuki, M. R. Hoeprich, N. Kobayaski, H. Koide, T. Kakimoto, L. D. Willey 1985; P. F. Flynn, K. L. Hoag, M. M. Kamel, R. J. Primus, Masanori Miyachika, Tetsuaki Hirota, Kenji Kashiyama, C. Thomas, F. Hartemann, J. Y. Foret-Bruno, C. Tarriere, C. Chillon, C. Got, A. Patel, G. Hufschmitt, Nathaniel R. Baker, Franklin E. Lynch, Lito C. Mejia, Lars G. Olavson, Gino Sovran, T. M. Liou, M. Hall, D. A. Santaicca, F. V. Bracco, Shigeru Onishi, Souk Hong Jo, Pan Do Jo, Satoshi Kato, Donald C. Siegla, Charles A. Amann, G. R. Edgar, D. A. Gore, Gordon H. Holze, C. Paul Pulver, Yoseph Gebre-Giorgis, Joseph T. Skladany, F. A. Costello, Tadakuni Hayashi, Masahiro Taki, Shinji Kojima, Teruaki Kondo, D. E. Steere, J. C. Wall, S. K. Hoekman, Robert L. Seaman, Charles E. Johnson and Ray F. Hamilton 1984; John C. Bierlein, Arnold O. DeHart, Wallace R. Wade, Jerry E. White, James J. Florek, Harry A. Cikanek, Otto A. Ludecke, David L. Dimick, Randall C. Davis, David L. Hagen, Huel C. Scherrer, David B. Kittelson, Erwin D. Lowell, Allyn G. Tidball, P. A. Lakshminarayan, John C. Dent, Chiaki Tsumuki, Katsuhiko Ueda, Hitoshi Nakamura, Katsumi Kondo, Tetsuya Suganuma, Walter Mortara, C. Canta, James C. Ulezelski, David G. Evans, Raymond J. Haka, John D. Malloy, Daniel C. Garvey, James D. Cronkhite, Klaus-Dieter Johnke, Herbert Fehrecke, Kazuo Kontani, Shinishi Gotoh, R. C. Yu, T. W. Kuo, S. M. Shahed, T. W. Chang, L. C. Lindgren, P. W. Heitman, S. R. Thrasher, David L. Harrington, R. R. Toepel, J. E. Bennethum and R. E. Heruth 1983; Ather A. Quader, Jack D. Whitfield, J. L. Jacocks, William E. Dietz, Jr., Samuel R. Pate, Lowell A. Reams, Timo A. Wiemero, Michael B. Levin, Wallace R. Wade, Richard Gibbs, Benjamin Hill, Robert Whitby, Robert Johnson, Shirish Ambadas Shimpi, Robert A. Jorgensen, John D. Horsch, Kent R. Petersen, David C. Viano, John M. Beardmore, Douglas R. Hamburg, David Klick, Charles M. Oman, Kenneth T. Menzies, Kevin J. Beltis, Rose E. Fasano, Larry D. Mitchell, W. W. Daws, Kouzo Nakamura, Kenzi Mihara, Yasutada Kibayashi, Takeshi Naito, J. Sacreste, F. Brun-Cassan, A. Fayon, Claude Got, C. Tarriere, A. J. Patel, James C. Clerc and John H. Johnson 1982; Timothy R. Erdman, John N. Johnson, Howard L. Benford, Maurice Leising, Wallace R. Wade, Jerry E. White, James J. Florek, Shoichi Furuhama, Masaaki Takiguchi, Kenji Tomizawa, Steven Martin Japar, Ann Cuneo Szkarlat, Hikmat F. Mahmood and Antoni Paluszny 1981; John B. Cole, Martin D. Swords, Medhi Namazian, Steven P. Hansen, Edward Lyford-Pike, J. Sanchez-Barsse, John B. Heywood, Joseph M. Rife, Voigt R. Hodgson, L. Murray Thomas, Homi J. Tata, Edward R. Driscoll, John J. Kary, Hikaru Kuniyoshi, Hideaki Tanabe, G. Takeshi Sato and Hajime Fujimoto 1980; Robert A. Ayres, Earl G. Brewer, Steven W. Holland, Shoichi Furuhama, Masaaki Takiguchi, Leonard B. Graiff, James A. Bennett, Mark F. Nelson, Alec H. Seilly, T. Michal Dyer, William A. Moore, Andrew M. Skow, Gary E. Erickson and Dale J. Lorincz 1979; Hideo Sakai, Henry Leidheiser, Jr., Dong K. Kim, David S. Howarth, Ralph V. Wilhelm, Jr., Samuel D. Hires, Rodney J. Tabacqynski, James M. Novak, Armin Felske, Gerhard Hoppe, Horst Matthai, Donald L. Stivender, Harold A. Brownfield, Darrell O. Rogers, Duane L. Abata, Phillip S.

Myers, Otto A. Uyehara, Leonard Walitt, Chung-Yen Liu and James L. Harp, Jr. 1978; John F. Cassidy, Jr., David C. Chang, Kuang-Huei Lin, James A. Bennett, Mark F. Nelson, John A. Verrant, David A. Kittelson, W. C. Schnell, R. L. Grossman and G. E. Hoff 1977; David G. Adams, Subimal Dinda, Robert A. George, Russell W. Karry, Arthur S. Kasper, John Pogorel, Jr., Willard E. Swenson, Jr., William L. Weeks, William A. Ashe, Ather A. Quader, L. E. Craig and K. H. Hummel 1976; Baird Wallace, Ralph S. Shoberg, Gerald W. Nyquist, Clarence J. Murton and M. R. Mitchell 1975; Trevor O. Jones, Jean A. Tennant, Narendra J. Sheth, S. L. Bussa, N. M. Mercer, William L. Brown, James E. Bernard, J. T. Tiekin, P. S. Fancher, R. E. Wild, W. H. Hucho, H. J. Emmelmann, Robert J. Donohue, Bernard W. Joseph, D. B. Wimmer, R. C. Lee, Frank Tao, Walter Waddey, Don Hollinger, Arvin Mueller, Martin Kramer and Stanley A. Shatsky 1974; Leonard H. Caveny, Martin Summerfield, Don B. Chaffin, Rodney K. Schutz, Richard G. Synder, Albert S. Paul, Rodney J. Tabaczynski, John B. Heywood, James C. Keck, Charles A. Amann, Wallace R. Wade, Mason K. Yu, William J. Galloway, Glen Jones, Keith B. Termaat, Kenneth A. Freeman, Teodors Priede, P. E. Waters, Larry R. Oliver, Dewey D. Henderson, Willard A. Sanscrainte, Melvin L. Chazen, Dominique Cesari, Robert Quincy and Yves Derrien 1973; Gorman L. Fisher, Charles D. Lennon, Robert F. Wheaton, Eugen Hanke, D. R. Liimatta, R. R. Hurt, W. L. Hull, R. W. Deller, Walter Bergman, Harold R. Clemett, Narendra J. Sheth, Harold J. Mertz and Lawrence M. Patrick 1972; Walter Bergman, Harold R. Clemett, G. L. Fischer, R. F. Hurt, C. D. Lennon, D. R. Liiamatta, H. J. Mertz, L. M. Patrick, N. J. Sheth-R. W. Deller 1971; T. Asanuma, W. J. Brown, L. S. Cessna, J. V. Foa, S. A. Gendernalik, A. Kaplan, T. K. Hasselman, H. Hayakawa, H. W. Heinzman, C. M. Jones, R. V. Kerley, K. J. Lloyd, F. J. Marsee, A. P. Penton, J. L. Perry, S. A. Short, F. L. Ver Synder, W. R. Wade, R. F. Wheaton-H. Yamakawa, Walter Cornelius and Wayne A. Daniel 1970; G. P. Blair, E. C. Fitch, Jr., J. R. Goulburn, L. Harned, Frances E. Heffner, C. R. Hilpert, L. E. Johnston, L. N. Mattavi, J. S. Meurer, A. A. Miklos-E. J. Miller, G. Schafpf and A. D. Urlamb 1969; J. P. Barrack, Irving N. Bishop, M. S. Bonura, J. K. Jackson-Kaye, J. V. Kirk, Laurence Maggitti, Jr., W. A. McGowan, D. E. Muehlberger, L. J. Nestor, J. D. Parisen, D. F. Putnam, P. E. Rubbert, G. R. Saaris and Aladar Simko 1968.

DANIEL GUGGENHEIM MEDAL ... to honor persons who make notable achievements in the advancement of aeronautics.
Recipients: Thornton Arnold Wilson 1985; Thomas H. Davis 1984; Nicholas J. Hoff 1983; David F. Lewis 1982; Edward C. Wells 1980; Gerhard Neumann 1979; Edward H. Heinemann 1978; Cyrus R. Smith 1977; Marcel Dassault 1976; Dwane L. Wallace 1975; Floyd L. Thompson 1974; William McPherson Allen 1973; William C. Mentzer 1972; Sir Archibald E. Russell 1971; Jakob Ackeret 1970.

EDWARD N. COLE AWARD FOR AUTOMOTIVE ENGINEERING INNOVATION ... this award honors the memory of Edward N. Cole and the inspiration he provided to other engineers by his search and drive for product innovation. It also recognizes Mr. Cole's own achievements in the field of automotive engineering. To perpetuate recognition of Mr. Cole's achievements and dedication as an automotive engineer, the Society of Automotive Engineers will administer an annual award for outstanding innovation and achievement in the engineering development of automobiles, their components, systems and accessories. The Cole Award is made possible through a fund originally established by contributions from General Motors Corporation SAE members and other employees and retirees who were associated with Mr. Cole. The award, consisting of a certificate and an appropriate commemorative gift, shall be awarded annually to an SAE member whose innovative design is described in an SAE paper, or whose lifetime of accomplishment is judged to be a significant achievement in automotive engineering.
Recipients: Vernon D. Roosa 1986; William F. Milliken, Jr. 1985; Konrad F. Eckert 1984; Frank J. Winchell 1983; Alan G. Loofbourrow 1982; George H. Muller 1981; Achilles C. Sampietro 1980; Oliver K. Kelley 1979.

ELMER A. SPERRY AWARD ... is given in recognition of a distinguished engineering contribution which, through application, and proved in actual service, has advanced the art of transportation whether by land, sea or air.

Recipients: Harry R. Wetenkamp 1987; George W. Jeffs, William R. Lucas, George E. Mueller, George S. Page, Robert F. Thompson, John F. Yardley 1986; Carlton E. Tripp, George H. Plude, Richard K. Quinn 1985; Frederick Aronowitz, Joseph E. Killpatrick, Warren M. Macek, Theodore J. Podgorski, Carlton E. Tripp and George H. Plude 1984; Sir George Edwards, Gen. Henri Ziegler, Sir Stanley Hooker, Sir Archibald Russell and M. Andre Turcat 1983; George Brenneisen, Ehard Frutterlieb, Joachim Kober, Edmund Muller, G. Reiner Nill, Manfred Schulz, Herbert Stemmler and Warner Teich 1982; Edward J. Wasp 1981; William M. Allen, Malcolm T. Stamper, Joseph F. Sutter and Everette M. Webb 1980; Robert Puiseux 1979; Clifford L. Eastburg and Harley J. Urbach 1977; Jerome F. Goldman, J. J. Henry and Frank Nemec 1975; Leon S. Hobbs and Perry W. Pratt 1972; Sedtwick N. Wight and George W. Baughman 1971; Dr. Charles S. Draper 1970.

THE HORNING MEMORIAL AWARD ... recognizes accomplishment and acknowledges contributions through research and development of the better mutual adaptation of fuels and internal combustion engines. Established in 1938 this award is in honor of Harry L. Horning, past-president of SAE.
Recipients: M. J. Hall and F. V. Bracco 1986; Jay K. Martin and Claus Borgnakke 1985; J. C. Wall and S. K. Hoekman 1984; Anthony J. Giovanetti, Jack A. Ekchian, John B. Heywood and Edward F. Fort 1983; Bruce D. Peters 1982; Robert Ching-Fang Yu, Syed M. Shahed, Tanvir Ahmad, Steven L. Plee and James P. Myers 1981; Edward G. Groff and Frederic A. Matekunas 1980; T. Michal Dyer 1979; Gary L. Borman and I. A. Voiculescu 1978; W. T. Lyn 1977; Harold Gibson 1976; Milton K. McLeod 1975; Ather A. Quader 1974; Lamont Eltinge 1973; John M. Pierrard and Richard A. Crane 1972; J. A. Bolt, S. J. Derezinski and David L. Harrington 1971; M. M. Roensch 1970; David K. Trumpy 1969; C. L. Cummins 1968.

L. RAY BUCKENDALE LECTURES ... sponsored by the Rockwell-Standard Corporation, these lectures are directed to the needs of young engineers and students and deal with automotive ground vehicles for either on or off-road operation, either commercial or military service. The lecturer receives an honorarium of $250 and a certificate of honor.
Recipients: Richard A. Drollinger 1987; Trevor O. Jones 1986; Tom Gillespie 1985; Stan Chocholek 1984; H. J. Bajaria 1983; Ray W. Murphy 1982; John C. Walter 1981; John C. Walter 1980; John R. Kinstler 1979; Martin J. Hermanns 1978; Kenneth W. Cuffe 1977; Raymond E. Heller 1976; Ernest R. Sternberg 1975; Phillip S. Myers 1974; Richard L. Staodt 1973; John W. Durstine 1972; P. R. Kyropoulos 1971; V. M. Perkins 1970; Gary Smith 1969; Jack A. Davisson and J. J. Hartz 1968.

FRANKLIN W. KOLK, AIR TRANSPORTATION PROGRESS AWARD ...
Recipients: Arthur P. Adamson and Richard J. Coar 1985; Donald P. Hetterman 1984; Elliott A. Green 1983; Herbert H. Slaughter, Jr. 1982; Martin C. Hemsworth 1981; Joseph F. Sutter 1980; John G. Borger 1979.

THE MANLY MEMORIAL MEDAL ... awarded annually to the author of the best paper relating to theory or practice in the design or construction of, or research on, aerospace engines, their parts, components, or accessories, which shall have been presented at a meeting of the Society or its sections during the year.
Recipients: Dana Morris Dunham, William L. Sellers III and Joe W. Elliot 1985; Michael J. Harris and James J. Nichols, Jr. 1984; D. J. Dusa, D. W. Speir and D. K. Dunbar 1983; B. Hugh Little and B. L. Hinson 1982; Joseph A. Bluish and John Kniat 1980; Bernard J. Rezy, Kenneth J. Stuckas, J. Ronald Tucker and Jay E. Meyers 1979; Richard R. Wysong 1978; J. E. Chirivella, W. A. Menard and L. A. Duke 1977; Howard G. Lueders and D. A. Nealy 1976; Robert A. Howlett 1975; William D. McNally 1973; R. M. Schimer 1970; Rose Worobel and Frederick V. Metzger 1973; R. M. Schirmer 1970; Joseph S. Alford and Brian Brimelow 1969; Ivan H. Bush, G. K. Richey and D. J. Stava 1968.

RALPH H. ISBRANDT AUTOMOTIVE SAFETY ENGINEERING AWARD ... presented to author or authors of best paper or best combination of papers on automotive safety engineering or to an individual for distinguished accomplishment in automotive safety engineering.

Recipients: John D. Horsch, Ian V. Lau, David C. Viano and Dennis V. Andrezejak 1984; Sam Saint 1983; A. J. Patel, J. Sacreste, F. Brun-Cassan, A. Fayon, C. Tarriere and Claude Got 1982; Charles K. Kroell, Mary E. Pope, David C. Viano, Charles Y. Warner and Stanley D. Allen 1981; V. R. Hodgson and L. Murray Thomas 1980; Donald F. Huelke 1979; J. Y. Forret-Bruno, F. Hartemann, C. Thomas, A. Fayon, C. Tarriere, C. Got, A. Patel 1978; Roy Haeusler 1977; R. T. Bundorf and R. L. Leffert 1976; John Versace and Roger J. Berton 1975; Lawrence M. Patrick, Nils Bohlin and Ake Andersson 1974; T. N. Louckes, R. J. Slifka, T. C. Powell and S. G. Dunford 1973; Ralph H. Isbrandt 1972.

THE RUSSELL S. SPRINGER AWARD ... presented annually to youngest author of paper published in SAE Transactions.
Recipients: Kenneth J. Stoss and Marvin D. Beaseley 1985; Michael A. Savoia 1983; Phillip G. Fuerst 1982; Edmund J. O'Keefe 1981; Kim H. Wilson 1980; Willie J. Culpepper 1979; Dilworth D. Lymann 1978; Kolar Seshasai 1977; Larry E. Albright 1976; Lary L. William 1975; J. L. Rau 1974; Thomas W. Hosler 1973; Joseph C. Selby 1972; William Giles 1971; F. W. Ratliff 1970; Kermit O. Allard 1969; Donald I. Manor 1968; D. H. Iacovani 1967.

VINCENT BENDIX AUTOMOTIVE ELECTRONICS ENGINEERING AWARD ... the purpose of the Vincent Bendix Automotive Electronics Engineering Award is to help focus the attention of SAE members in the field of automotive electronics engineering by recognizing outstanding contributions to this field. The Award, consisting of a medal and a certificate, will be presented annually to the author, or authors, who deliver the most outstanding paper at Society or Section meetings addressed to the subject of automotive electronics engineering.
Recipients: A. I. Cohen, K. W. Randall, C. D. Tether, K. L. VanVoorhies and J. A. Tennant 1984; Melvin J. Johnson 1983; Lowell A. Reams, Timo A. Wiemero, Michael B. Levin and Wallace R. Wade 1982; Thomas A. Spoto and Marcello Veneziano 1981; William J. Fleming 1980; Alex Seilly 1979; David S. Howarth and Ralph V. Wilhelm, Jr. 1978; John F. Cassidy, Jr. 1977; S. C. Hadden, L. R. Hulls and E. M. Sutphin 1976; T. O. Jones, T. R. Schlax and R. L. Colling 1975.

WILLIAM LITTLEWOOD MEMORIAL LECTURE ... this lecture is developed on an annual basis with the objective to advance civil air transport engineering and to recognize those who make personal contributions to the field.
Recipients: John F. McDonald 1986; Paul M. Johnstone 1985; Robert W. Rummell 1984; John C. Brizendine 1983; Paul Besson 1982; Norman Parmet 1979; Willis M. Hawkins 1978; Frank Kolk 1977; Raymond Kelly 1976; Gerhard Neuman 1975; Richard Jackson 1974; Edward Wells 1973; John Borger 1972; Peter Masefield 1971.

THE WRIGHT BROTHERS MEDAL ... awarded annually to the author of the best paper on aerodynamics or structural theory, or research, or construction, or operation of airplanes or spacecraft which shall have been presented at a meeting of the Society or its sections during the year.
Recipients: Charles W. Boppe 1985; Joseph C. Eppel, Michael D. Shovlin, Robert J. Englar, James H. Nichols, Jr. and Michael J. Harris 1984; Carol A. Simpson 1983; Jean-Jacques Speyer 1982; Raymond M. Hicks 1981; Walter S. Cremens 1980; William A. Moore, Gary E. Erickson, Dale J. Lorincz and Andrew M. Skow 1979; Raymond A. Hicks and Garrett N. Vanderplaats 1977; J. A. Alic and H. Arcang 1975; M. J. Wendl 1974; Richard E. Hayden 1973; D. H. Bennett and R. P. Johannes 1972; J. Hong 1970; W. N. Redisch, A. E. Sabroff, P. C. Wheeler and J. G. Zaremba 1969; J. R. MacGregory and L. J. Nestor 1968.

CLIFF GARRETT TURBOMACHINERY ENGINEERING AWARD ... To promote engineering developments and the presentation of SAE papers on turbomachinery engineering. This Award honors Cliff Garrett and the inspiration he provided to engineers by his example, support, encouragement and many contributions as an aerospace pioneer. To perpetuate recognition of Mr. Garrett's achievements and dedication as an aerospace pioneer, SAE administers an annual lecture by a distinguished authority in the engineering of turbomachinery for on-highway, off-highway, and/or spacecraft and aircraft uses.

Recipients: Arthur J. Wennerstrom 1986; Frank E. Pickering 1985; David P. Kenny 1984.

SOCIETY OF FIRE PROTECTION ENGINEERS (SFPE)
60 Batterymarch Street, Boston, Massachusetts 02110
D. Peter Lund, CAE, Executive Director
PRESIDENTS ... Rolf H. Jensen, P.E. 1988; Ralph E. Collins, III, P. E. 1986-87; Harold E. Nelson, P. E. 1984-85; Maurice R. Boulais, P. E. 1982-83; Jack Bono, P. E. 1980-81; Harry C. Bigglestone, P.E. 1978-79; William H. McClarran, P.E. 1976-77; James W. Nolan, P.E. 1974-75; Martin M. Brown, P.E. 1973.

AWARDS

ARTHUR B. GUISE MEDAL ... created to ensure that the engineering and scientific contributions of Arthur B. Guise will not only be remembered but will live on as a continuing inspiration and encouragement to those who follow in the pursuit of scientific achievement. It is bestowed upon an individual for eminent achievement in the advancement of the science and technology of fire protection engineering. Established in 1982.
Recipients: Oliver W. Johnson 1987; Howard W. Emmons 1986; Gunnar Heskestad 1985; Arthur B. Guise (posthumously) 1983.

FIRE PROTECTION MAN-WOMAN OF THE YEAR ... designed to recognize significant personal achievement in Fire Protection from areas beyond SFPE. Criteria: 1.) Contribution must have been made in recent years 2.) Contribution must be broad in scope at least of national importance 3.) Contribution must have apparent lasting effects. Established in 1973.
Recipients: Richard A. Waters 1987; Congressman Doug Walgren (D-PA) 1985; Gunnar Heskestad 1984; Dougal Drysdale 1983; Howard W. Emmons 1982; Jackson Weaver 1981; Duane D. Pearsall 1980; John L. Bryan 1977; Dick Van Dyke 1976; Anne W. Phillip, M.D. 1975; Richard E. Bland 1974; Arthur Sampson 1973.

HAROLD E. NELSON SERVICE AWARD ... for dedicated and inspired service to the advancement of the ideals and goals of the Society of Fire Protection Engineers.
Recipient: Harold E. Nelson 1987.

SOCIETY OF MANUFACTURING ENGINEERS (SME)
One SME Dr., P.O. Box 930, Dearborn, Michigan 48121
Kenneth Thorpe, Executive Vice President & General Manager
PRESIDENTS ... John E. Mayer 1987-88; Donald G. Zook 1986-87; Marvin F. DeVries 1985-86; Forrest D. Brummett 1984-85; Reginald W. Barrett 1983-84; Charles F. Carter, Jr. 1982-83; Robert L. Vaughn 1981; Robert A. Dougherty 1980; Joseph C. Hall 1979; Robert W. Militzer 1978; Edward S. Roth 1977; M. Eugene Merchant 1976; William B. Johnson 1975; William J. Huot 1974; Nicholas J. Radell 1973.

AWARDS

EDUCATION AWARD ... in recognition of an individual's development of dynamic curricula, fostering sound training methods, or inspiring students to enter the profession of tool and manufacturing engineering.
Recipients: George E. Dieter 1987; Dr. Inyoung Ham 1986; Gustav J. Olling 1985; Eugen A. Matthias 1984; Martyn Farley 1983; Laboratory for Machine Tools & Industrial Organization - Technical Univ. of Aachen 1982; Merrill L. Ebner 1981; Milton C. Shaw 1980; Geoffrey Boothroyd 1979; Keiji Okushima 1978; Nathan H. Cook 1977; Oyvind Bjorke 1976; Dr. Dell K. Allen 1975; Dr. Shien-Ming Wu 1974; Roger L. Geer 1971; Dr. Arthur E. Keating 1970; Arthur E. Thompson 1969; Harold P. Rodes 1968.

ELI WHITNEY MEMORIAL AWARD ... to honor an individual, either an SME member or non-member, who has distinguished himself within the broad concept of mass-production manufacture.

Recipients: Richard E. Dauch 1987; Gordon G. Fee 1986; Reuben R. Jensen 1985; James F. Barcus, Jr. 1984; Quentin C. McKenna 1983; James C. Bakken 1982; Armando Fiorelli 1981; James C. Danly 1980; Robert W. Truxell 1979; Ritsu Tohmine 1978; Martin J. Caserio 1977; Henry D. Sharp Jr. 1976; Robert Anderson 1975; Alonzo G. Decker, Jr. 1974; William D. Innes 1973; L. V. Whistler 1971; U. A. Whitaker 1970.

GOLD MEDAL ... in recognition of outstanding service through published literature, technical writings, or papers dealing with tool and manufacturing engineering subjects.
Recipients: Norman R. Augustine 1987; Dr. Kurt H. Lang 1986; Taylan Altan 1985; Toshio Sata 1984; Shiro Kobayashi 1983; William B. Rice 1982; Hidehiko Takeyama 1981; Daniel B. Dallas 1980; Jiri Tlusty 1979; William P. Koster 1978; Lester V. Colwell 1976; Harry Conn 1975; Dr. John A. Schey 1974; Roy L. Williams 1973; Clyde A. Sluhan 1972; E. J. Weller 1970; Robert L. Vaughn 1969; Michael Field 1968.

FREDERICK W. TAYLOR RESEARCH MEDAL ... in recognition of significant published research leading to a better understanding of materials, facilities, principles, and operations, and their application to improve manufacturing processes.
Recipients: Kuo-King Wang 1987; Dr. Nam P. Sah 1986; Robert Hocken 1985; Erwin G. Loewen 1983; P. A. McKeown 1982; Janez Peklenik 1981; Erich G. Thomsen 1980; Ehud Lenz 1979; John G. Bollinger 1978; James B. Bryan 1977; Branimir F. Von Turkovich 1976; Father Jaques Peters 1975; Dr. John F. Kahles 1973; Kenjiro Okamura 1971; Richard E. Reason 1969; Eugene Merchan 1968.

JOSEPH A. SIEGEL MEMORIAL AWARD ... this award is conferred only upon SME members. Recipients are members who have made contributions through leadership, voluntary support, or other timely acts which benefit the Society.
Recipients: Marcus B. Crotts 1987; James Throop 1986; Clarence W. Doty 1985; William B. Johnson 1984; Myron M. Martin 1983; R. William Taylor 1982; William M. Spurgeon 1981; Emery P. Miller 1980; Bernard J. Wallis 1979; Philip R. Marsilius 1978; Guy Bellows 1977; Clyde A. Sluhan 1976; Francis J. Sehn 1975; Joseph L. Petz 1974; Andrew W. Williams 1973; Harry B. Osborn 1972; Burt F. Raynes 1971; J. J. Demuth 1970; Harold E. Collins 1969; H. Dale Long 1968.

ALBERT M. SEARGENT PROGRESS AWARD ... in recognition of accomplishments in the field of manufacturing processes, methods, or management.
Recipients: Richard P. Lindsay 1987; David McMurtry 1986; Michael W. Davis 1985; LaRoux K. Gillespie 1984; Hans J. Warnecke 1983; Robert H. Brown 1982; Leo J. Lapwing 1981; Michael Humenik, Jr. 1980; Joseph F. Engelberger 1979; Carl J. Oxford, Jr. 1978; Frank R. L. Daley, Jr. 1977; Robert C. Wilson 1976; Albert F. Welch 1975; William E. Brandt 1974; William L. Naumann 1973; John R. Conrad 1972; Frederick W. Braun 1971; Semon E. Knudsen 1970; Richard K. Wilson 1969; Ralph H. Ruud 1968.

DONALD C. BURNHAM MANUFACTURING MANAGEMENT AWARD ... awarded for achieving significant gains in organizational competitiveness through innovation, recognition or leadership in the effective use of human, technical or financial resources.
Recipients: Stanley E. Huffman 1987; Thomas J. Murrin 1986; James A. Baker 1985; Robert A. Fuhrman 1983.

OUTSTANDING YOUNG MANUFACTURING ENGINEER AWARD ... Conferred in recognition of significant achievement and leadership in Manufacturing Engineering. Award nominees must be age 35 or younger.
Recipients: Lisa D. Andrus, Bryan V. Baker, Gary F. Benedict, Ming-Yung Chern, Taewoo Han, Dr. K. J. Kim, Jack E. Lohmann, H. Lee Martin, Miguel R. Martinez, Randy H. Moss, Mitchell R. Nichols, Barbara A. Plonski, Bahram Ravanti 1987; Jane C. Ammons, Terrance BeCraft, Valerie C. Bolhouse, Ian D. Cannon, Kam C. Lau, Jyotirmoy Mazumber, Jimmy Lee Mills, Ezat T. Sanii, George Chryssolouris, Timothy Greene, Matraj C. Iyer, A. Galip Ulsoy, James K. West 1986; David P. Aldrich, William S. Bathgate, Dennis Bray, Paul Richard Dawson, Steven D. Hoffer, Vincent W. Howell, Sr., David Liu, Scott Mackelprang,

George F. Reinerth, Michael D. Sedler, John Sosoka, Charles L. Tucker III, Muthuraj Vaithianathan and Chi-haur Wu 1985.

SOCIETY OF MINING ENGINEERS OF AIME (SME-AIME)

8307 Shaffer Parkway, P.O. Box 625002, Littleton, Colorado 80162-5002
Claude L. Crowley, Executive Director

PRESIDENTS ... B. A. Kennedy, 1987; A. T. Yu 1986; T. V. Falkie 1985; F. L. Kadey, Jr. 1984; L. Kuchinic, Jr. 1983; Maurice C. Fuerstenau 1982; Alfred Weiss 1981; Nelson Severinghaus, Jr. 1980; Robert Stefanko 1979; Robert S. Shoemaker 1978; Donald O. Rausch 1977; John F. Havard 1976; Robert L. Llewellyn 1975; Donald A. Dahlstrom 1974; Robert H. Merrill 1973.

AWARDS

ANTOINE M. GAUDIN AWARD ... for scientific or engineering contributions that further understanding of the technology of mineral processing.
Recipients: G. A. Parks 1985; M. E. Wadsworth 1984; L. G. Austin 1983; P. Somasundaran 1982; I. Iwasaki 1981; J. E. House 1980; N. Arbiter 1979; M. C. Fuerstenau 1978; D. W. Fuerstenau 1977; W. A. Griffith 1976; D. W. Frommer 1975.

ARTHUR F. TAGGART AWARD ... for a paper or series of closely related papers considered the most notable contribution in the area of mineral processing during the period under review. Only those papers appearing in AIME publications are considered for this award.
Recipients: K. P. Ananthapadmanabhan and P. Somasundaran 1986; Jan Hupka and Jan D. Miller 1985; F. F. Aplan 1984; M. D. Flavel 1983; D. A. Dahlstrom 1982; I. Iwasaki 1981; N. Arbiter and E. K. C. Williams 1980; R. L. Wiegel 1978; M. C. Fuerstenau and Bruce R. Palmer 1977; J. B. Hiskey and M. E. Wadsworth 1977; N. W. Stump and A. N. Roberts 1976; D. B. Brimhall and M. E. Wadsworth 1975; L. S. Diaz and P. M. Musgrove, Jr. 1974.

D. C. JACKLING AWARD AND LECTURER ... recognizes significant contributions to technical progress in the fields of mining, geology, or geophysics.
Recipients: C. O. Branner 1986; Anthony R. Barringer 1985; R. J. Robbins 1984; R. D. Mollison 1983; W. H. Callahan 1982; R. W. Ballmer 1981; H. E. Hawkes 1980; Paul A. Bailly 1979; A. Weiss 1978; F. K. Wilson 1977; R. L. Akre 1976; T. E. Howard 1975; S. R. Wallace 1974; L. H. Hart 1973; J. B. Knaebel 1972; J. K. Gustafson 1971; J. David Lowell 1970; C. D. Michaelson 1969; F. Cameron 1968.

GEM AWARD ... to recognize outstanding contributions to the SME Government, Energy and Minerals (GEM) program.
Recipients: Patricia C. Petty 1985; Alfred Ransome 1980; Page L. Edwards 1979; Alfred G. Hoyl 1977.

HOWARD N. EAVENSON AWARD ... for distinguished contributions to the advancement of coal mining.
Recipients: A. W. Calder 1986; R. Eugene Samples 1985; W. N. Poundstone 1984; R. E. Murray 1983; L. C. Gates 1982; D. F. Crickmer 1981; Irvin C. Spotte 1980; Thomas L. Garwood 1979; R. G. Heers 1978; Otes Bennett, Jr., 1977; W. F. Diamond 1976; C. J. Potter 1975; J. A. McCormick 1974; S. M. Cassidy 1973; C. T. Holland 1972; E. R. Phelps 1971; W. G. Talman 1970; R. Stefanko 1969; J. W. Leonard 1968.

ROCK MECHANICS AWARD ... recognizes distinguished contributions to the advancement of the field of rock mechanics.
Recipients: Frederick D. Wright 1986; H. Douglas Dahl 1985; Z. T. Bieniawski 1984; M. D. G. Salamon 1983; James J. Scott 1982; Jack Parker 1981; Charles T. Holland 1980; Thomas A. Lang 1979; Donald F. Coates 1978; Walter Wittke 1977; Richard E. Goodman 1976; Manuel R. Rocha 1975; William F. Brace 1974; Charles Fairhurst 1973; Leopold Muller 1972; John W. Handin 1971; Neville G. W. Cook 1970; Wilbur I. Duvall 1969; Leopold Obert 1968.

ROBERT PEELE MEMORIAL AWARD ... for the most outstanding paper on a mining, geology or geophysics subject published in the "Transactions," of which one or more authors are AIME members not over 35 years of age at the time the paper was submitted.
Recipients: Frederick D. Fox 1986; Dirk J. A. van Zyl 1985; J. G. Clevenger 1984; S. M. Miller 1983; D. E. Nicholas 1982; J. R. Kyle 1981; Glenn L. Krum 1980; E. R. Mueller 1979; M. W. Roper 1978; P. J. Goossens 1977; A. K. Burton 1976; G. C. Barrientos 1975; A. E. Lewis and R. L. Braun 1974; J. B. Seller, G. R. Haworth and P. G. Zambas 1973; M. B. Kahle and A. D. Pernichele 1972; J. A. Bratt, B. A. Fahm, B. A. Kennedy and K. E. Niermeyer 1971; Wilson Blake 1970; C. B. Manula and T. J. O'Neil 1969; M. J. Coolbaugh 1968.

YOUNG ENGINEER AWARD ... brings recognition of engineering professionalism to young people working in the coal industry.
Recipients: H. Erik Sherer 1986; Bruce P. DeMarcus 1985; J. W. Parkinson 1984; J. R. Ackerman 1983; R. W. Cope 1982; W. D. Hake 1981; P. L. Longenecker 1979; T. J. Sawarynski 1978; J. W. Francis 1977; M. P. Miano 1976.

SOCIETY OF MOTION PICTURE AND TELEVISION ENGINEERS (SMPTE)

595 W. Hartsdale Avenue, White Plains, NY 10607
Lynette Robinson, Executive Director

PRESIDENTS ... M. Carlos Kennedy 1987-88; Harold Eady 1985-1986; Leonard F. Coleman 1983-1984; Charles E. Anderson 1981-82; Robert M. Smith 1979-80; William D. Hedden 1977-78; Kenneth M. Mason 1975-76; Bryson S. Roudabush 1973-74.

AWARDS

THE AGFA-GEVAERT GOLD MEDAL ... to honor the recipient by recognizing the individual's outstanding leadership, inventiveness and/or other achievement in the research, development, or engineering of new techniques and/or equipment which result in a significant improvement to the interface between motion picture film and television imaging system, whereby the combined advantages of both contribute to the further development of visual communications systems.
Recipients: Roland J. Zavada 1986; Leroy DeMarsh 1985; Bengt Modin 1984; Karel G. M. Staes 1983; C. B. B. Wood 1982; Heinrich Zahn 1981; Kenneth G. Lisk 1980; John D. Millward 1979; Donald B. Milliken 1978; John D. Lowry 1977; Daan Zwick 1976; Rodger J. Ross 1975.

DAVID SARNOFF GOLD MEDAL AWARD ... to honor recipient by recognizing outstanding contributions in the development of new techniques or equipment which have contributed to the improvement of the engineering phases of television, including theatre television.
Recipients: Michael O. Felix 1986; Richard J. Taylor 1985; Richard S. O'Brien 1984; Frank Davidoff 1983; Takashi Fujio 1981; Maurice Lemoine 1980; Masamiko Morizono 1978; Renville H. McMann 1977; Adrian B. Ettlinger 1976; John L. E. Baldwin 1975; Joseph A. Flaherty 1974; Arch C. Luther, Jr. 1973; Peter Rainger 1972; Walter Bruch 1971; Charles H. Coleman 1970; Peter C. Goldmark 1969.

EASTMAN KODAK GOLD MEDAL ... to honor the recipient by recognizing outstanding contributions which lead to new or unique educational programs utilizing motion pictures, television, high-speed and instrumentation photography or other photographic sciences.
Recipients: Arthur Knight 1986; Allan Curtis 1985; Jay Leyda 1984; Harold E. Edgerton 1983; Erik Barnouw 1982; Robert W. Wagner 1981; Irwin A. Moon 1980; Roman Vishniac 1979; Reid H. Ray 1978; James Card 1977; Herbert E. Farmer 1976; Richard B. Hull 1975; Charles F. Hoban, Jr. 1974; John Flory 1973; Eric M. Berndt 1972; John A. Maurer 1971; Jamison Handy 1970; Walter I. Kisner 1969; Edgar Dale 1968; Samuel N. Postlethwait 1967.

HERBERT T. KALMUS MEMORIAL AWARD ... to do honor to the recipient by recognizing outstanding contributions in the development of color films, processing, techniques or equipment

useful in making color motion pictures for theatre or television use.
Recipients: John L. Baptista 1986; Keith E. Whitmore 1985; Christoph Geyer 1984; Harry R. Beilfuss 1983; Hirozo Ueda 1982; Manfred G. Michelson 1981; Alan M. Gundelfinger 1980; Judith A. Schwan 1979; Roderick T. Ryan 1978; Roland G. L. Verbrugghe 1977; Bernard Happe 1976; Hans-Christoph Wohlrab 1975; Frank P. Brackett, Jr. 1974; Charles J. Hirsch 1973; Dean M. Zwick 1972; Linwood G. Dunn 1971; Willi G. Engel 1970; Howard W. Vogt 1969; Walter A. Fallon 1968; John Monroe Waner 1967; Vernon J. Duke 1966; Henry N. Kazonowski 1965.

JOHN GRIERSON INTERNATIONAL GOLD MEDAL AWARD ... to honor the recipient by recognizing significant technical achievements related to the production of documentary motion-pictures films.
Recipients: Leonard A. Green 1986; William C. Shaw 1985; Kenneth Richter 1984; Stefan Kudelski 1983; Jean-Pierre Beauviala 1982; Peter Parks 1981; Chester E. Beachell 1980; Jacques-Yves Cousteau 1979; Forrest O. Calvin 1978; William H. Offenhauser, Jr. 1977; Gerald G. Graham 1976; Lloyd Thompson 1975; Eric M. Berndt 1974; George W. Colburn 1973.

JOURNAL AWARD ... to recognize the outstanding paper originally published in the Journal of the Society during the previous calendar year.
Recipients: W. Tuckerman Biays and John L. E. Baldwin 1986; D. Q. Humphreys, R. C. Sehlin, J. Mutter and William F. Schreiber 1985; Wilfried Liekens and John L. E. Baldwin 1984; Ralph C. Brainard, Arun N. Netravali, D. E. Pearson, Glenn L. Kennell, Frank Reinking, Stephen W. Spakowsky, Richard C. Sehlen and Geoffrey L. Whittier 1983; C. Bradley Hunt and Haruo Sakata 1982; Toyohiko Hatada, Haruo Sakata and Hideo Kusaka 1981; Andrew Oliphant and Martin Weston 1980; David K. Fibush 1979; Eugene Leonard 1978; Margaret W. Beach 1977; K. Kano, T. Imai, M. Inaba, A. Sugimoto, Y Itah, Y. Inoue and T. Shimizu 1976; A. D. Berg, R. J. Cormier, J. S. Courtney-Pratt 1975; W. J. Hannan, R. E. Flory, M. Lurie and R. J. Ryan 1974; LeRoy E. DeMarsh 1973; Rebville H. McMann, Jr., Leo Beiser, Wendell Lavender and Robert Walker 1972; Peter C. Goldmark 1971; C. B. B. Wood 1970; C. J. Bartleson 1969; J. H. Altman 1968; Walter C. Snyder 1967; Harold Wright 1966; Otto H. Schade, Sr. 1965.

PHOTO-SONICS ACHIEVEMENT GOLD MEDAL AWARD ... to do honor to the recipient by recognizing outstanding contributions to the development of new techniques or equipment which have contributed to the improvement of the engineering phases of instrumentation and/or high-speed photography.
Recipients: Donal L. Clayton 1983; Alexander E. Huston 1982; Gong Zutong 1981; Rudi Schall 1980; Hallock F. Swift 1979; Tsuneyoshi Uyemura 1978; A. Earl Quinn 1977; A. S. Dubovic 1976; W. P. Dyke 1975; Roy Edwards 1974; Robert D. Shoberg 1973; R. J. North 1972; Ernest M. Whitley 1971; C. David Miller 1970; William G. Hyzer 1969; Frank Frungel 1968.

PROGRESS MEDAL AWARD ... to do honor to the individual by recognizing outstanding technical contributions to the progress of engineer-phases of the motion-picture and/or television industries.
Recipients: Masahiko Morizono 1986; Roland J. Zavada 1985; Joseph A. Flaherty 1984; Ray M. Dolby 1983; Frank Davidoff 1982; Daan Zwick 1981; Dr. August Arnold 1980; Donald G. Fink 1979; Robert E. Gottschalk 1978; E. Carlton Winckler 1977; Edward H. Reichard 1976; William T. Wintringham 1975; Sidney P. Solow 1974; Wilton R. Holm 1973; Norwood L. Simmons 1972; Rodger J. Ross 1971; Peter C. Goldmark 1970; J. S. Courtney-Pratt 1969; Charles R. Fordyce 1968.

SAMUEL L. WARNER MEMORIAL AWARD ... to do honor to the individual by recognizing outstanding contributions in the design and development of new and improved methods and-or apparatus for sound-on-film motion pictures, including any step in the process.
Recipients: Richard J. Stumpf 1986; Ioan Allen 1985; Ronald E. Uhlig 1984; Terry Beard 1983; Erik Rasmussen 1982; John O. Aalberg 1981; Arthur C. Blaney 1980; Norman Prisament 1979; Ray M. Dolby 1978; Albert P. Green 1977; Alfred W. Lumkin

1976; Joseph D. Kelly 1975; Waldon O. Watson 1974; Loren L. Ryder 1973; George R. Groves 1972; James R. Corcoran 1971; Gordon E. Sawyer 1970; Fred G. Albin 1969; Stefan Kudelski 1967; Fred Hynes 1966.

THE ALEXANDER M. PONIATOFF GOLD MEDAL FOR TECHNICAL EXCELLENCE... It is the purpose of this award to honor the recipient by recognizing outstanding technical excellence of contributions in the research or development of new techniques and/or equipment that have contributed significantly to the advancement of audio or television magnetic recording and reproduction.
Recipients: Charles E. Anderson 1986; Thomas G. Stockman, Jr. 1985; Allen J. Trost 1984; Michael O. Felix 1983; Ray M. Dolby 1982.

THE PRESIDENTIAL PROCLAMATION ... It is the purpose of this award to recognize individuals of established and outstanding status and reputation in the motion-picture and television industries worldwide.
Recipients: Herbert E. Farmer, Guy Gougeon, Daniel E. Slusser 1986; Harold Greenberg, William H. Smith, Joseph D. Kelly 1985; Edward H. Reichard and David W. Samuelson 1984; Norman R. Grover 1982.

THE CITATION FOR OUTSTANDING SERVICE TO THE SOCIETY ... The purpose of this citation is to recognize individuals for dedicated service to the Society (at Section or national levels) over a sustained period, and for whom other recognition (Fellow Membership, etc.) is not justified at the time.
Recipients: Vernon L. Kipping 1986; John Barry, Richard Cornell, John F. Donovan, Yvon Jean, Thomas J. McCormick 1985; Grant Dearnaley, Charles D. Kircher 1984; Ivan Barclay, Ted H. Horn 1984; Robert B. Desrosiers, W. Russell McCown and Richard B. Glickman 1983; Paul F. Brown, John W. Caluger and Frederick R. Nobbs 1982; Arthur E. Florack and Alvin J. Siegler, 1981; William R. Ahern, Eugene R. Myler and Paul Yang 1980; Douglas V. Dove, Irving Rosenberg, Joseph A. Semmelmayer and William E. Youngs 1979; John Corso, Jr., Frederick M. Remley, Paul F. Wittlig and Kurt Wulliman 1978; Edward J. Blasko, Mathias J. Herman, Julian D. Hopkinson, Andrew Kufluk and Burton Stone 1977; Frank J. Eberhardt, Maurice L. French, Donald V. Kloepfel, M. Warren Strang and John R. Sullivan 1976; Harry Teitelbaum 1975; C. Carroll Adams 1974.

THE SOCIETY OF NAVAL ARCHITECTS AND MARINE ENGINEERS (SNAME)

601 Pavonia Avenue, Suite 400, Jersey City, New Jersey 07306
Robert G. Mende, Secretary and Executive Director
PRESIDENTS ... Edward J. Campbell 1987-88; Perry W. Nelson 1985-86; C. L. French 1983-84; John J. Nachtsheim 1981-82; Lester Rosenblatt 1980-79; Robert T. Young 1977-78; L. V. Honsinger 1976-75; Phillip Eisenberg 1974-73.

AWARDS

THE BLAKELY SMITH MEDAL ... for outstanding accomplishment in ocean engineering.
Recipients: John A. Mercier 1985; Joe W. Key 1983; Ben C. Gerwick, Jr. 1981; Blakely Smith 1979.

THE CAPTAIN JOSEPH H. LINNARD PRIZE ... for the best paper contributed to the Proceedings of SNAME.
Recipients: Peter Blume, Alfred M. Kracht 1986; Dwight K. Koops, Robert X. Caldwell, Maurice Gordon 1985; Richard J. Baumler, Toshio Watanabe and Hiroshi Huzimura 1984; John P. Breslin, Robert J. Van Houtin, Justin E. Kerwin and Carl-Anders Johnsson 1983; William H. Garzke, Jr. and George D. Kerr 1982; Helge Johannessen and Knut T. Skaar 1981; Stanley G. Stiansen, Hsien Y. Jan and Donald Liu 1980; Justin Kerwin and Chang-Sup Lee 1979; Goeffrey G. Cox and Adrian R. Lloyd 1978; E. Scott Dillon, Ludwig C. Hoffmann and Donald P. Roseman 1977; Theodore A. Loukakis and Chryssostomos Chryssostomidis 1976; Jacques B. Hadler, Choung M. Lee, J. T. Birmingham and H. D. Jones 1975; Vito R. Milano 1974; Willem B. van Berlekom and Thomas A. Goddard 1973; Fred C. Bailey, Edward V. Lewis and Robert S. Little 1972; Roderick Y. Edwards, Jack W. Lewis and Vincent W. Ridley 1971; Frank M. Lewis

1970; Warren I. Signell 1969; Samuel R. Heller, Arthur R. Lytle, Richard Nielsen Jr. and John Vasta 1968.

THE DAVID W. TAYLOR MEDAL ... for notable achievement in naval architecture and marine engineering.
Recipients: Robert N. Herbert 1986; J. Randolph Paulling, Jr. 1985; Jan Dirk Van Manen 1984; Jens T. Holm 1983; Jacques B. Hadler 1982; Erwin Carl Rohde 1981; Peter M. Palermo 1980; Philip F. Spaulding 1979; John J. Nachtsheim 1978; James J. Henry 1977; Harry Benford 1976; James B. Robertson, Jr. 1975; Roger E. M. Brard 1974; Jerome L. Goldman 1973; John R. Kane 1972; Phillip Eisenberg 1971; Ludwig C. Hoffman 1970; Douglas C. MacMillan 1969; Matthew G. Forrest 1968.

THE DAVIDSON MEDAL ... for outstanding scientific accomplishment in ship research.
Recipients: William B. Morgan 1986; John V. Wehausen 1984; John P. Breslin 1982; Manley St. Denis 1980; Louis Landweber 1978; J. Harvey Evans 1976; William E. Cummins 1974; Georg P. Weinblum 1972; B. V. Karvin-Kroubovsky 1970.

THE VICE ADMIRAL E. L. COCHRANE AWARD ... for the best paper delivered before a section of SNAME during the year.
Recipients: Vlodek Laskow, Paul A. Spencer, Ian M. Bayly 1986; Michael G. Parsons, Richard W. Harkins 1985; Archibald C. Churcher, Alex Kolomojcer and Geoff Hubbard 1984; Watt D. Burton, Jr. 1983; Robert D. Tagg 1982; Roy L. Harrington 1981; Walter C. Cowles 1980; Thomas N. Sanderlin, Stuart M. Williams and Robert D. Jamieson 1979; John R. Kane 1978; Frank S. Chou 1977; Alexander C. Landsburg and John M. Cruiskshank 1976; Guy C. Volcy 1975; William H. Garzke, Jr., Ralph E. Johnson and Alexander C. Landsburg 1974; John E. Ancarrow and Roy L. Harrington 1973; Frederic A. Thomas 1972; Scott Dillon 1971; John P. Breslin 1970; Thoma Lamb 1969; D. C. Tolefson and Leonard Brand 1968.

THE VICE ADMIRAL "JERRY" LAND MEDAL ... for outstanding accomplishment in the marine field.
Recipients: C. L. French 1986; Thomas B. Crowley 1985; Lester Rosenblatt 1984; Edwin Murray Hood 1983; Robert I. Price 1982; Ellsworth L. Peterson 1981; P. Takis Veliotis 1980; A. Dudley Haff 1979; William M. Benkert 1978; E. Scott Dillon 1977; Robert J. Blackwell 1976; John V. Banks, 1975; Helen Delich Bentley 1974; James F. Goodrich 1973; Andrew E. Gibson 1972; Daniel D. Strohmeier 1971; Andrew Neilson 1970; Ralph K. James 1969; Donald A. Holden 1968.

THE WILLIAM M. KENNEDY AWARD ... for outstanding service and contribution in the development of systems and planning applied to shipbuilding and ship repair.
Recipients: Ralph Anselmi 1987; Thomas C. Rosebraugh 1986; John J. Garvey 1985; Peter E. Jaquith 1984; Charles J. Starkenburg 1983; Daniel M. Mack-Forlist 1982; Louis D. Chirillo 1981.

THE SOCIETY OF PACKAGING PROFESSIONALS (SPHE)

Reston Intl Ctr., Reston, Virginia 22091
William C. Pflaum, Executive Director
PRESIDENTS ... Thomas J. Lowery 1987-88; George F. Long 1986-87; William E. Loy 1984-85; Gerald Zimmer 1982-83; Roger Stewart 1980-81; A. McKinley 1978-79; P. Dixon Majesky 1977; Joseph J. Burrowes 1976; Richard M. Reutlinger 1975; Peter Henningsen 1974; Clarence A. Moore 1973.

AWARDS

PACKAGING COMPETITION ... users and suppliers of packages and material-handling devices are eligible to enter SPHE's Annual International Packaging Week Competition.

BEST OF SHOW ... the "Best of Show" Award is presented the most outstanding exhibitor at SPHE's Annual Packaging Competition.
Recipients: Julius Kupersmit, Container Corp 1986; Gary Gallagher, IBM 1984; Walter C. Vincent 1980; William C. Kropf and Patrick E. Wright 1978; Joseph F. Malleck 1977; Leo Kalmerton 1976; Richard J. Schwartz 1975; J. Aubrey Olson 1974; John Beaumont 1973; David Collier 1972; Kenneth B. Kraft-David L.

Root 1971; Victor L. Scott 1970; T. D. Kornegor 1969; J. C. Mabe 1968.

SOCIETY OF PETROLEUM ENGINEERS INC (SPE)

P. O. Box 833836, Richardson, Texas 75083-3836
Dan K. Adamson, Executive Director

PRESIDENTS ... R. Lyn Arscott 1988; Noel D. Rietman 1987; Dennis E. Gregg 1986; Kenneth W. Robbins 1985; James R. Jorden 1984; T. Don Stacy 1983; W. C. Barton, Jr. 1982; Arlen L. Edgar 1981; Marvin L. Katz 1980; Charles L. Bare 1979; Hugh B. Barton 1978; F. F. Craig, Jr. 1977; C. A. Hutchinson, Jr. 1976; E. E. Runyan 1975; D. G. Russell 1974; Gerald E. Sherrod 1973.

AWARDS

ANTHONY F. LUCAS GOLD MEDAL ... awarded from time to time to recognize distinguished achievement in improving the technique and practice of finding and producing petroleum engineering. It was the first AIME award specifically created for the petroleum members of the Institute.
Recipients: Eugene R. Brownscombe 1987; Henry J. Welge 1986; N. Van Wingen 1985; Joseph E. Warren 1984; Henry J. Ramey, Jr. 1983; Paul B. Crawford 1982; Claude R. Hocott 1981; Robert C. Earlougher 1980; Donald L. Katz 1979; Gustave E. Archie 1978; Marshall B. Standing 1977; J. Clarence Karcher 1976; Michael T. Halbouty 1975; William L. Horner 1974; Albert G. Loomis 1972; M. K. Hubert 1971; H. G. Doll 1970; C. J. Coberly 1969; A. F. van Everdingen 1968.

CEDRIC K. FERGUSON MEDAL ... gives recognition for significant contribution to the permanent technical literature of the profession by a younger member of the Society. It is awarded annually for an outstanding technical paper in any field of petroleum engineering written by a member of the Society under the age of 33.
Recipients: Woon Fong Leung 1987; Jeffrey G. Southwick 1986; Tommy M. Warren 1985; Steven G. Shryock 1984; Ronald L. Sparks 1983; Walter B. Fair, Jr. 1982; Robert E. Jones and Geoffrey Thorp 1981; Charles J. Glover, Maura C. Puerto, John M. Maerker and Erik I. Sandvik 1980; Gurmeet S. Brar 1979; D. J. Hammerlindl 1978; Medhat M. Kamal 1977; Robert H. Rossen 1976; Thomas L. Gould 1975; Bruce H. Walker 1974; M. R. Todd 1973; Roy T. McLamore 1972; A. R. Sinclair 1971; Thomas A. Edmondson, M. J. Jeffries-Harris and Charles A. Jordan 1970; Mark A. Childers 1969; D. A. T. Donohue and Harvey S. Price 1968.

DEGOLYER DISTINGUISHED SERVICE MEDAL ... Established in 1965 to recognize distinguished and outstanding service to the Society of Petroleum Engineers, to the profession of engineering and geology and to the petroleum industry.
Recipients: Donald G. Russell 1987; Charles W. Arnold 1986; L. B. Curtis 1985; Edmund C. Babson 1984; Henry J. Gruy 1983; John C. Calhoun, Jr. 1982; H. A. Nedom 1981; Claude R. Hocott 1980; Jack J. Reynolds 1979; Joseph G. Richardson 1978; Charles R. Dodson 1977; Robert H. McLemore 1976; Robert B. Gilmore 1975; John S. Bell 1974; Thomas C. Frick 1973; Wayne E. Glenn 1972; Lincoln F. Elkins-M. T. Halbouty 1971; Jan Jacob Arps 1970; Charles V. Millikan 1969; John R. Suman 1968; John M. Lovejoy 1967; Everette L. DeGolyer (posthumously) 1966.

JOHN FRANKLIN CARLL AWARD ... established to recognize distinguished achievement in or contributions to petroleum engineering. It is awarded from time to time on recommendation by the Award Committee. It is presented for either technical or non-technical achievements that further petroleum engineering through advancement of the profession or the application of engineering principles to petroleum development and recovery.
Recipients: Hossein Kazemi 1987; Kenneth A. Blenkarn 1986; L. W. Holm 1985; Aziz S. Odeh 1984; Kermit E. Brown 1983; Floyd L. Scott 1982; Leo W. Fagg 1981; Donald G. Russell 1980; Murray F. Hawkin 1979; John M. Campbell 1978; Forrest F. Craig, Jr. 1977; Wilbur F. Cloud 1976; H. J. Ramey, Jr. 1975; Ted M. Geffen 1974; Paul D. Torrey 1973; Harold Vance 1972; Fred H.

Poettmann 1971; Daniel N. Dietz 1970; H. G. Bostet 1969; George H. Fancher 1968.

LESTER C. UREN AWARD ... created by the Society in 1963 to recognize distinguished achievement in the technology of petroleum engineering by a member of the profession who made his contribution before age 45.
Recipients: W. Barney Gogarty 1987; G. Paul Willhite 1986; Fred I. Stalkup 1985; Keith H. Coats 1984; Harley Y. Jennings, Jr. 1983; William C. Goins, Jr. 1982; Julius S. Aronofsky 1981; Murray F. Hawkins, Jr. 1980; Robert C. Earlougher, Jr. 1979; Thomas K. Perkins 1978; Joseph G. Richardson 1977; Harold W. Winkler 1976; Charles S. Matthews 1975; Michael Prats 1974; H. J. Ramey, Jr. 1973; J. J. Arps 1972; Herman Dykstra-Richard L. Parsons 1971; A. F. van Everdingen 1970; Morris Muskat 1969; C. J. Coberly 1968.

SPE DISTINGUISHED SERVICE AWARD ... The SPE Distinguished Service Award, established in 1947, recognizes contributions to the Society of Petroleum Engineers that exhibit such exceptional devotion of time, effort, thought and action as to set them apart from other contributions and thereby create an added indebtedness of the Society to the contributor.
Recipients: Chapman Cronquist and Wesley W. Eckles 1987; Howard B. Bradley and T. Don Stacy 1986; W. Clyde Barton 1985; Lyman L. Handy 1984; Arlen L. Edgar 1984; Edward E. Runyan and James A. Klotz 1983; Dwayne E. Godsey, Roland F. Krueger, Marvin L. Katz and James G. Litherland 1982; Charles L. Bare and Jan Geertsma 1981; David L. Riley, J. Don Clark and Hugh B. Barton 1980; Robert L. Whiting, Richard A. Morris and Granville Dutton 1979; H. Arthur Nedom and Charles W. Arnold 1978; John P. Hammond 1977; Kenneth W. Robbins and Lawrence B. Curtis 1976; Gerald E. Sherrod and John S. Bell 1975; M. Scott Kraemer and John M. C. Gaffron 1974; A. J. Horn and Robert C. Earlougher 1973; Claude R. Hocott 1972; John R. McMillan, H. Mark Krause, Jr. and Rupert C. Craze 1971; Jack M. Moore, Hal M. Stanier and Ben H. Caudle 1970; A. L. Vitter, Jr. 1969; Floyd E. Schoonover and Howard C. Pyle 1968.

SPE YOUNG MEMBER OUTSTANDING SERVICE AWARD ... presented in recognition of contributions and leadership by members of the Society under the age of 36 to the public and the community, as well as to the Society or the petroleum industry or profession.
Recipients: DeAnn Craig, Lucy B. King, Andrew A. Young 1987; Mark A. Klins 1986; Philip C. Crouse and Claudia N. Roberts 1984

SPE PUBLIC SERVICE AWARD ... presented in recognition of distinguished service to the country, state, community or public through excellence in leadership, service or humanitarianism.
Recipients: Wm. P. Clements, Jr. 1987; Robert L. Parker, Sr. 1985; Dean A. McGee 1984

SPE DRILLING ENGINEERING AWARD ... recognizes outstanding achievements in or contributions to the advancement of drilling engineering.
Recipients: Frank J. Schuh 1986; Leon H. Robinson, Jr 1985; Keith K. Milheim 1984.

SPE PRODUCTION ENGINEERING AWARD ... recognizes outstanding achievements in or contributions to the advancement of production engineering.
Recipients: A. Buford Neely 1986; Robert H. Gault 1985; Fred W. Gipson 1984.

SPE RESERVOIR ENGINEERING AWARD ... recognizes outstanding achievements in or contributions to the advancement of reservoir engineering.
Recipients: Khalid Aziz 1987; W. John Lee 1986; Donald W. Peaceman 1985; Henry J. Welge 1984.

THE SOCIETY FOR IMAGING SCIENCE AND TECHNOLOGY (SPSE)

7003 Kilworth Lane, Springfield, Virginia 22151
Calva A. Lotridge, Executive Director

PRESIDENTS ... William Towns 1985-1989; Vivian K. Walworth 1981-85; Allie C. Peed 1977-81.

SOCIETY OF PHOTO-OPTICAL INSTRUMENTATION ENGINEERS (SPIE)

PO Box 10, 1022-19th Street, Bellingham, Washington 98225-0010

Joseph Yaver, Executive Director

PRESIDENTS ... R. Barry Johnson 1987; James C. Wyant 1986; Lewis Larmore 1985; Robert E. Fischer 1984; Warren J. Smith 1983; Richard J. Wollensak 1982; Andrew G. Tescher 1981; Robert R. Shannon 1979-80; Joseph B. Houston, Jr. 1977-78; Brian J. Thompson 1976-75; Charles N. DeMund 1973-74.

AWARDS

ALBERT M. PEZZUTO AWARD ... a commemorative plaque, may be given to a currently serving or former National Officer or Governor who has served at least one complete term as an Officer or Governor, in recognition of exceptional services rendered to the Society. It shall not be awarded more often than once every five years. Honorarium: $1,000. Donor: Mrs. Albert Pezzuto.
Recipients: Robert E. Fischer 1986; Charles N. DeMund and Brian J. Thompson 1978; J. E. Durrenberger 1971.

GEORGE W. GODDARD AWARD ... a commemorative plaque, may be given in recognition of exceptional work in the field of aerospace photo-optical instrumentation technology. Honorarium: $500. Donor: Itek Corporation.
Recipients: Morris Birnbaum 1987; Louis J. Cutrona 1986; Leon van Speybroeck 1985; Aden Baker Meinel 1984; Milton Chang and John Matthews 1983; Joseph B. Houston, Jr. 1982; Robert R. Shannon 1981; F. Dow Smith 1980; Carl E. Duckett 1979; General Lew Allen, Jr. 1978; Roderic M. Scott 1977; Walter J. Levison 1976; Earle Brown 1975; Lewis Larmore 1974; Leon Kosofsky 1971; Donald B. Milliken 1970; Harry Davis 1969; Richard W. Philbrick 1968.

RUDOLF KINGSLAKE MEDAL AND PRIZE ... a silver-plated bronze medal may be awarded annually in recognition of the most noteworthy original paper to appear in the Society's official journal, *Optical Engineering*, on the theoretical or experimental aspects of optical engineering. All papers published in the journal are automatically eligible for consideration for this award. Honorarium: $1,000.
Recipients: Arthur D. Fisher, Lai-Chang Ling, John N. Lee, Robert C. Fukuda 1986; Armand R. Tanguay, Jr. 1985; Gene R. Gindi and Arthur F. Gimitro 1984; James R. Palmer 1983; David M. Pepper 1982; Robert A. Sprague and William D. Turner 1981; G. Ferrano and G. Hausler 1980; J.R. Fienup 1979; Norman J. Brown 1978; David B. Kay and Brian J. Thompson 1977; Richard E. Swing 1976; James M. Burch and Colin Forno 1975; Irving R. Abel and Bruce R. Reynolds 1974.

SPIE GOVERNORS' AWARD ... this is a service award certificate which may be given to an individual who, in the opinion of the Board of Governors, has rendered a significant service of outstanding benefit to the Society.
Recipients: Roger H. Schneider 1987; Henry John Caulfield 1984; Irving J. Spiro 1981; Clyde W. Tombaugh 1980; James C. Wyant, Andrew G. Tescher and Theodore T. Saito 1979; Edward L. Caplan (posthumous) 1978; Yale H. Katz 1977; J. E. Durrenberger 1970; Revere G. Sanders 1969; Clyde T. Holliday 1968; John Kiel 1967; Carl N. Brewster 1966; Eugene B. Turner and Edward E. Rich 1965.

SPIE PRESIDENT'S AWARD ... a discretionary award plaque, it may be given to an individual who, in the opinion of the President and the Board of Governors, has rendered a unique and meritorious service of outstanding benefit to the Society.
Recipients: Hermann Walter, Irving J. Spiro, Hans J. Frankena and Jumpei Tsujiuchi 1987; National Research Council Canada and Roderic M. Scott 1986; Gerd Herziger and Theodore Maiman 1985; Andrew G. Tescher and Richard J. Wollensak 1984; Adolf Lohmann, Andre Marechal, Jean Ebbeni and Andre Monfils 1983; Bernard G. Ponseggi 1982; John Kiel 1981; Allan H. Gott 1980; Joseph B. Houston, Jr. and Joseph Yaver 1979; Rob-

ert Lewis and Peter Poulsen 1978; H. John Caulfield, John B. DeVelis and David A. Treffs 1977; John Kiel, Andrew E. Trolio and John H. Waddell (posthumous) 1976; Charles N. DeMund 1975; Russell C. Bunting 1970; Duncan MacDonald 1969; George C. Higgins 1968; Brian J. Thompson 1967; Helen F. Gustafson 1966.

TECHNOLOGY ACHIEVEMENT AWARD ... a medal may be awarded annually in recognition of outstanding national or international accomplishment in the conception or reduction to practice of optical engineering technology. The technology should have broad application in contributing to the advancement of optical engineering or optical instrumentation engineering. The award shall reflect strongly the diversity of interests and accomplishments of SPIE in the fields of optics, and electro-optics. The medal may be awarded to an individual or to members of a team.
Recipients: Balzers, AG, Basic Research Laboratory 1987; The Perkin-Elmer Corporation 1986; Eastman Kodak Company 1985; Darryl E. Gustafson and Thomas I. Harris 1984; Harold E. Bennett and Jean M. Bennett 1983; Optical Coating Laboratory, Inc., accepted by Rolf Illsley 1982; Tatsuo Izawa, Donald B. Keck, John B. MacChesney and Peter C. Schultz 1981; James B. Bryan, Philip Steger and Theodore T. Saito 1980; Charles DeVoe and Clarence Babcock 1979.

THE GOLD MEDAL OF THE SOCIETY ... the Gold Medal, the principal award of the Society may be awarded annually in recognition of outstanding national or international accomplishment in the discipline of photo-optical instrumentation engineering. The award shall strongly reflect the diversity of interests and accomplishments of SPIE in the fields of photography, optics and electro-optics. The medal may be awarded to an individual or to two or more members of a team.
Recipients: H. Angus Macleod 1987; Brian J. Thompson 1986; Warren J. Smith 1985; Frank Cooke 1984; Robert E. Hopkins 1983; Harold H. Hopkins 1982; Harold E. Edgerton 1981; Rudolf Kingslake 1980; Edwin H. Land 1979; James G. Baker 1978; John Strong 1977.

DENNIS GABOR AWARD ... may be presented annually in recognition of outstanding inventive accomplishment in the field of optics in the broadest sense. The award shall reflect the interests of the Society in the reduction to practice of new technology, and may be presented to an individual or to members of a team.
Recipients: Joseph W. Goodman 1987; A. Jacques Beaulieu 1986; Robert A. Jones 1985; Abe Offner 1984; Emmett N. Leith and Yuri N. Denisyuk 1983.

SOCIETY OF PLASTICS ENGINEERS, INC. (SPE)

14 Fairfield Drive, Brookfield Center, Connecticut 06805-0403

Robert D. Forger, Executive Director

PRESIDENTS ... John R. Kretzschmar 1987; Greg R. Thom 1986; Richard G. Johnson 1985; William C. Kuhlke 1984; Robert J. Schaffhauser 1983; Thomas W. Haas 1982; James E. Chinners, Jr. 1981; George P. Schmitt 1980; Joseph Magliolo, Jr. 1979; Jack L. Isaacs 1978; Lawrence J. Broutman 1977; George E. Pickering 1976; Harold A. Holz 1975; John H. Myers 1974; Theodore S. Stoughton 1973; Ronald J. Cleveringa 1972; Irwin L. Podell 1971; Ralph A. Noble 1970.

AWARDS

SPE INTERNATIONAL AWARD IN PLASTICS ... presented to an individual who has made outstanding and unique contributions to plastics engineering and science. Consists of a gold medal, $3,000 honorarium, and travel allowance.
Recipients: Dr. Wolfgang Mack 1987; Nathaniel C. Wyeth 1986; Daniel W. Fox 1985; W. Lincoln Hawkins 1984; George Menges 1983; George Lubin 1982; Roger S. Porter 1981; Eric Baer 1980; Charles G. Overberger 1979; Vivian T. Stannett 1978; William Willert 1977; Bryce Maxwell 1976; Allan S. Hay 1975; Alfred C. Webber 1974; J. Lawrence Amos 1973; Michael Szwarc 1972; Albert G. H. Dietz 1971; Michael Szwarc 1970; Richard S. Stein 1969; Raymond T. Boyer 1968.

SPE AWARD IN PLASTICS EDUCATION ... consists of bronze plaque and $1,000 honorarium.

Recipients: Dr. James L. White 1987; Kurt C. Frisch 1986; Rudolph D. Deanin 1985; Musa Kamal 1984; Charles C. Winding 1983; Raymond B. Seymour 1982; Paul S. Bruins 1981; Roger S. Porter 1980; James M. McKelvey 1979; Motowo Takayanagi 1978; Russell W. Ehlers 1976; Louis R. Rahm 1968.

SPE AWARD IN PLASTICS RESEARCH . . . consists of bronze plaque and $1,000 honorarium.
Recipients: James E. McGrath 1987; Christopher W. Macosko 1986; Frank E. Karasz 1985; Martin G. Broadhurst 1984; Thor L. Smith 1983; Donald R. Paul 1982; Lawrence E. Nielsen 1981; Alan N. Gent 1980.

SPE AWARD IN PLASTICS ENGINEERING AND TECHNOLOGY . . . consists of bronze plaque and $1,000 honorarium.
Recipients: Walter Schrenk 1987; Lee L. Beyler, Jr. 1986; Stephanie L. Kwolek 1985; Otis D. Black 1984; Zehev Tadmor 1983; Bruce H. Maddock 1982; Nathaniel C. Wyeth 1981; Shiro Matsuoka 1980.

SPE AWARD IN PLASTICS BUSINESS MANAGEMENT . . . consists of bronze plaque and $1,000 honorarium.
Recipients: John C. Reib 1987; Frank H. Wheaton, Jr. 1986; Ross H. Dean 1985; Robert A. Hoffer 1984; William F. Mericle 1983; Samuel N. Shapiro 1982; Wolfgang A. Mack 1981; David S. Weil 1980.

SPE AWARD . . . John W. Hyatt (Service to Mankind) . . . consists of a bronze plaque and $1,000 honorarium.
Recipients: Bernarr J. Hall 1987; W. Brandt Goldsworthy 1986; Henry E. Griffith 1985; Wilfred Lynch 1984; Robert K. Jarvik 1983.

SOCIETY OF WOMEN ENGINEERS
(SWE)

345 East 47th Street, New York, New York 10017
B.J. Harrod, Managing Director

PRESIDENTS . . . F. Suzanne Jennoches 1988-89; Kathleen Harer 1987-88; B. K. Krenzer 1986-87; Susan K. Whatley 1985-86; Sharon Lindquist-Skelley 1984-85; Barbara Wollmershauser 1983-84; Helen Grenga 1981-82; Sharon V. Loeffler 1980-81; Ada I. Pressman 1979-80; Paula L. Loring 1978-79; Arminta J. Harness 1976-78; Carolyn F. Phillips 1974-76; Naomi J. McAfee 1972-74.

AWARDS

SWE ACHIEVEMENT AWARD . . . conferred each year upon a woman who has made significant contribution to engineering. This honor is given to a woman in the field of engineering practice, research, education, or administration. The award consists of membership for life in SWE, an award pin, and a certificate.
Recipients: Nance K. Dicciani 1987; Yvonne C. Brill 1986; Y.C. L. Susan Wu 1985; Geraldine V. Cox 1984; Joan B. Berkowitz 1983; Harriett B. Rigas 1982; Thelma Estrin 1981; Carolyn M. Preece 1980; Jessie G. Cambra 1979; Giuliana C. Tesoro 1978; Mildred S. Dresselhaus 1977; Ada I. Pressman 1976; Dr. Sheila E. Widnall 1975; Barbara C. Johnson 1974; Dr. Irene C. Peden 1973; Nancy D. Fitzroy 1972; Dr. Alva Matthews 1971; Dr. Irmgrad Fluge-Lotz 1970; Alice Stall 1969; Isabella L. Karle 1968.

DISTINGUISHED ENGINEERING EDUCATOR AWARD . . . established 1985. Presented to a woman, who as a full time or Emerita engineering educator, has demonstrated excellence in teaching and the ability to inspire students to high levels of accomplishments, has shown evidence of scholarship through contributions to research and technical literature, and who, through active involvement in professional engineering societies, has made significant contribution to the engineering profession. The award consists of an inscribed momento and plaque.
Recipient: Dr. Rosalia N. Andrews 1986.

STANDARDS ENGINEERING SOCIETY
(SES)

1709 N. West Ave, Suite 122, Jackson, MI 49202
Donald A. Elinski, Executive Director

PRESIDENTS . . . Lorraine Tressel 1987-88; Donald A. Elinski 1985-86; Roland B. Pinkston 1983-84; Thomas R. Kelly 1981-82; Richard L. M. Rice 1979-80; Roscoe M. Brown 1977-78; John A. Fedorochko 1975-76; Raymond E. Monahan 1972-74.

AWARDS

LEO B. MOORE MEDAL . . . to an individual for highest achievement, extraordinary contribution and distinguished service in the field of standardization, and its advancement through original research and writing, creative application and development or service - any individual is eligible.
Recipients: Raymond E. Monahan 1986; Thomas R. Kelly 1985; Jan A. V. Ollner 1984; Donald R. Mackay 1983; Robert B. Toth 1982; Dr. Howard I. Forman 1981; Dr. S. David Hoffman 1980; Jesse W. Caum 1979; Kenneth W. Tuchn 1978; William A. McAdams 1977; William T. Cavanaugh 1976; Dr. Francis L. LaQue 1975; Dr. Robert F. Leggett 1974; Lowell W. Foster 1973; Adm. George F. Hussey, Jr. 1972; H. A. Roy Binney 1971; Samuel H. Watson 1970; Roy P. Trowbridge 1969; Harry E. Cheseborough 1968.

SES-ASTM ROBERT J. PAINTER MEMORIAL AWARD . . . to an individual for special service in the field of standardization through company program, managerial support or education research-any individual is eligible.
Recipients: Albert L. Batik 1986; Genevieve A. Smith 1985; John A. Blair 1984; Henry J. Stremba 1983; Caryl E. Twitchell 1982; John F. Corey 1981; Wayne P. Ellis 1980; William Pferd 1979; Dr. Howard I. Forman 1978; S. David Hoffman 1977; Creighton J. Hale 1976; Lester Fox 1975; Margaret Dana 1974; Dr. Lal C. Verman 1973; Jean A. Caffiaux 1971; Leroy C. Gilpert 1970; H. Baron Whitaker 1969; George H. Hamden 1968.

SES AWARD FOR OUTSTANDING PAPER . . . to the author of the best article, book or other publication dealing with the subject of standardization (articles in other fields, which might contribute to standardization but would first have to be judged on their technical merits are not eligible.) Award is to an author or publication. Any individual or publication is eligible.
Recipients: Norman A. Hagan 1986; Robert B. Toth 1985; Lester V. Ottinger and Robert N. Stauffer 1984; Joan A. Koenig 1983; Lester Sumner 1982; Pierre Ailleret 1981; Charles E. Gastineau and Donald L. Kear 1980; Riddell H. Ford 1979; R. Philip Preston 1978; Robert D. Gidel 1976; Jeffrey J. Hamm 1974; Robert B. Toth 1973; Leon V. Whipple 1972; Shirleigh Silverman 1970; Dr. Allen V. Astin 1969; J. Herbert Hollomon 1968.

THE TAU BETA PI ASSOCIATION, INC.

P.O. Box 8840, University Station, Knoxville, Tennessee 37996
James D. Froula, Secretary-Treasurer

PRESIDENTS . . . Martha S. Martin 1986-90; Paul H. Robbins 1982-86; Donald L. Bender 1978-82; Edward T. Misiaszek 1976-78; Thomas M. Linville 1974-76; George P. Palo 1970-74.

TECHNICAL ASSOCIATION OF THE PULP AND PAPER INDUSTRY (TAPPI)

Technology Park, P.O. Box 105113, Atlanta, Georgia 30348
William L. Cullison, Executive Director

PRESIDENTS . . . Dr. Vincent A. Russo 1987-1988; William H. Griggs 1985-1986; Terry O. Norris 1983-1984; Sherwood G. Holt 1981-82; William O. Kroeschell 1979-80; Peter E. Wrist 1977-78; Harris O. Ware 1975-76; M. J. Osborne 1973-74.

AWARDS

THE TAPPI ENGINEERING DIVISION AWARD . . . granted in recognition of an individual's outstanding technical contributions in the area of interest of the TAPPI Engineering Division.

Recipients: Edward F. DeCrosta 1987; N. L. Heberer 1986; W.B. A. Sharp 1985; L. H. Busker 1984; A. H. Tuthill 1983; P. B. Wahlstrom 1982; J. M. Bentley 1981; E. J. Justus 1980; W. K. Metcalfe 1979; R. P. Derrick 1978; W. B. Wilson 1977; H. M. Canavan 1975; C. A. Lee 1974; E. W. Hopper 1973; J. M MacBrayne 1972; H. O. Teeple 1971; W. C. Bloomquist 1970; P. E. Wrist 1969; G. E. Shaad 1968.

ENGINEERING LEADERSHIP AND SERVICE AWARD ... in recognition of outstanding contributions to the division.
Recipients: Robert A. Bareiss 1987; Earl W. Ramsdell 1986; R.G. Spangler 1985.

THETA TAU FRATERNITY
9974 Old Olive Street Road, St. Louis, Missouri 63141
Robert E. Pope, Executive Director

GRAND REGENT ... Randall J. Scheetz 1986; A. Thomas Brown 1982-1986.

AWARDS

ALUMNI HALL OF FAME ... Awarded to members who have excelled in their profession or the fraternity, selected from nominations by the local chapters.
Recipients: Charles W. Bripzius, C. Raymond Hanes, Joseph W. Howe, Paul L. Mercer, Robert E. Pope 1987. Erich J. Schreder, Elwin L. Vinal, William M. Lewis, Isaac B. Hanks 1986.

TRIANGLE FRATERNITY (TF)
51674 U.S. 33, South Bend, In 46637
H. Jack Sargent, Executive Director

PRESIDENTS ... Kenneth W. Quayle 1987; David C. Jordon, PhD 1986; Gerald S. Jakubowski, PhD 1985; Rufus A. Brown 1984; Randy G. Kerns 1983; Gerald S. Jakubowski, PhD 1982; William R. McGovern 1981; James R. Marshall 1980; Robert J. Mosborg 1979; Richard Beaubien 1978; Robert N. Hettrick 1977; Randall E. Drew 1976; Richard H. Sudheimer 1975; Edward A. Bischoff 1974; Robert J. Mosborg 1973.

AWARDS

TRIANGLE CITATION ... certificate presented occasionally of a member of Triangle in acknowledgement of his having received outstanding honorable recognition in such fields as engineering, architecture, science, education, industry, government, and the like, and by virtue of this having enhanced the prestige of Triangle. (Not an annual award.)
Recipients: Frederick R. Kappel, Maynard P. Venema, Daniel V. Terrell, Stothe B. Kazios, Jay S. Hammond, Ellison S. Onizuka.

UNITED ENGINEERING TRUSTEES (UET)
345 East 47th Street, New York, New York 10017
Alexander D. Korwek, Secretary & General Manager

PRESIDENTS ... A. S. West 1986-87; C. E. Ford 1984-85; G. F. Habach 1982-83; O. S. Bray 1981; R. W. Sears 1979-80; H. D. Guthrie 1977-78; R. H. Tatlow 1976; L. N. Rowley, Jr. 1974-75; W. K. MacAdam 1973.

AWARDS

DANIEL GUGGENHEIM MEDAL ... awarded annually for notable achievements in the advancement of aeronautics.
Recipients: Paul B. MacCready 1987; Hans Wolfgang Liepmann 1986; Thornton Arnold Wilson 1985; Thomas H. Davis 1984; Nicholas J. Hoff 1983; David S. Lewis 1982; Clarence L. (Kelly) Johnson 1981; Edward C. Wells 1980; Gerhard Neumann 1979; Edward H. Heinemann 1978; Cyrus R. Smith 1977; Marcel Dassault 1976; Dwane L. Wallade 1975; Floyd L. Thompson 1974; William McPherson Allen 1973; William C. Mentzer 1972; Sir Archibald E. Russell 1971; Jakob Ackeret 1970; H. Julian Allen 1969; H. M. Horner 1968.

JOHN FRITZ MEDAL ... awarded annually for notable scientific or industrial achievement in any field of pure or applied science. The award is made by a board of 15 appointed in equal number from the membership of ASCE, AIME, ASME, AIChE, and IEEE.
Recipients: Ralph Landau 1987; Simon Ramo 1986; Daniel C. Drucker 1985; Kenneth A. Roe 1984; Claude E. Shannon 1983; David Packard 1982; Ian MacGregor 1981; T. Louis Austin, Jr. 1980; Nathan M. Newmark 1979; Robert G. Heitz 1978; George R. Brown 1977; Thomas O. Paine 1976; Manson Benedict 1975; H. I. Romnes 1974; Lyman Dwight Wilbur 1973; William Webster 1972; Patrick E. Haggerty 1971; Glenn B. Warren 1970; Michael L. Haider 1969; Igor Ivan Sikorsky 1968.

VOLUNTEERS IN TECHNICAL ASSISTANCE (VITA)
1815 N. Lynn, Suite 200, Arlington, Virginia 22209
Henry R. Norman, Executive Director

CHAIRMAN ... Ambassador Jean Wilkowski

WATER POLLUTION CONTROL FEDERATION (WPCF)
601 Wythe Street, Alexandria, VA 22314-1994
Dr. Quincalee Brown, Executive Director

PRESIDENTS ... Beth Turner 1988; Carl V. Huber 1987; H. Gerard Schwartz, Jr. 1986; Harry A. Tow 1985; Earnest F. Gloyna 1984; Charles H. Jones 1983; E. J. Newbould 1982; Carman F. Guarino 1981; Geoffrey T. G. Scott 1980; Martin Lang 1979; Richard S. Englebrecht 1978; Horace L. Smith 1977; Victor G. Wagner 1976; Sam L. Warrington 1975; John D. Parkhurst 1974; J. F. Byrd 1973.

AWARDS

CHARLES ALVIN EMERSON MEDAL ... for outstanding service in the collection and treatment of wastewater, as related particularly to the problems and activities of the Water Pollution Control Federation in such terms as the stimulation of membership, improving standards of operational accomplishments, fostering fundamental research, etc.
Recipients: Franklin D. Dryden 1986; Joseph B. Hanlon 1985; Charles F. Niles 1984; Frederick G. Pohland 1983; J. Edward Meers 1982; Elmer E. Ross 1981; J. Floyd Byrd 1980; John D. Parkhurst 1979; Arthur O. Caster 1978; Earl R. Howard 1977; Kenneth S. Watson 1976; Paul D. Haney 1975; Carl C. Larson 1974; Ralph E. Fuhrman 1973; William H. Wisely 1972; Clair N. Sawyer 1971; Gordon E. McCallum 1970.

COLLECTION SYSTEM AWARD ... the nominee must have contributed, by original concept and outstanding practical application, to the advancement of the state of the art of wastewater collection, and the nominee's service must have been distinguished in planning, operation, and maintenance, facility design, education, training, or research.
Recipients: Richard W. Arbour 1986; Charles Button 1985; Gerald M. Cadwell 1984; Thomas E. McMahon 1983; Stein Bendixen 1982; Robert H. Hinkson 1981; Richard D. Pomeroy 1980; Robert L. Reed 1979; C. E. G. Bland 1978; Kenneth D. Kerri 1977; Gerald D. Underwood 1976; Horace L. Smith 1975; Alvin A. Appel and Sidney Preen 1974.

GEORGE BRADLEY GASCOIGNE MEDAL ... for an outstanding contribution to the art of wastewater treatment plant operation through the successful solution of important and complicated operational problems.
Recipients: Michael C. Mulbarger, Kenneth L. Zacharias, Farooq Nazir, and Dale Patrick 1986; Frederic J. Winter and Dale F. Ogle 1985; Glen T. Daigger, Glenn A. Richter, Jerry R. Collins and John W. Smith 1984; Dan P. Norris, Denny S. Parker, Marvin L. Daniels and Eben L. Owens 1983; John C. Murk, Jerry L. Frieling, Liberato D. Tortorici and Clyde C. Dietrich 1981; James D. Pastika and Glenn J. Lindroos 1980; Robert A. Paschke, Y. S. Hwang and Douglas W. Johnson 1978; Warren T. Uhte 1977; Manmohan N. Bhatla 1976; Gary L. Nickerson, C. Mi-

chael Robson, Roncie D. Morrison and Robert C. Clinger 1975; Richard E. Finger 1974; Bernard W. Dahl, John W. Zelinski and Owen W. Taylor 1973; William F. Milbury, Donald McCauley and Charles H. Hawthorne 1972; V. J. Jordan, Jr. and Clarence H. Scherer 1971; Raymond D. Leary, Lawrence A. Ernest and William J. Katz 1970.

GORDON MASKEW FAIR MEDAL . . . for outstanding contributions to the field in the area of engineering education, particularly teaching excellence.
Recipients: John Charles Geyer 1986; George Tchobanoglous 1985; Jack A. Borchardt 1984; Hovhaness Heukelekian 1983; Abel Wolman 1982; Dr. Albert E. Berry 1981; Gerard A. Rohlich 1980; Ernest F. Gloyna 1979; Daniel A. Okun 1978; Clair N. Sawyer 1977; George J. Schroepfer 1976; Harold Benedict Gotaas 1974; Don E. Bloodgood 1971.

HARRISON PRESCOTT EDDY MEDAL . . . for outstanding research contributing in an important degree to existing knowledge of fundamental principles or processes of wastewater treatment.
Recipients: T. Sekizawa, K. Fujie, H. Kubota, T. Kasakura, and A. Mizuno 1986; Antonio O. Lau, Peter F. Strom, and Davis Jenkins 1985; Brian J. Eis, John F. Ferguson and Mark M. Benjamin 1984; Bill Batchelor 1983; Peter O. Nelson, Anne K. Chung and Mary C. Hudson 1982; Robert G. Kunz, Thomas C. Hess, Alan F. Yen and Arthur A. Arseneaux 1981; Richard G. Luthy, Samuel G. Bruce, Jr., Richard W. Walters and David V. Nakles 1980; Jack Famularo, James A. Mueller and Thomas Mulligan 1979; John F. Ferguson and Thomas A. King 1978; Kenneth J. Williamson and Perry L. McCarty 1977; Douglas T. Merrill and Roger M. Jorden 1976; Stephen P. Graef and John F. Andrews 1975; John F. Ferguson, David Jenkins and John Eastman 1974; Joseph L. Pavoni, Mark W. Tenney and Wayne F. Echelberger 1973; Harry E. Wild, Jr., Clair N. Sawyer and Thomas C. McMahon 1972; Paul M. Berthouex and Lawrence B. Polkowski 1971; Donald E. Evenson, Gerald T. Orlob and John R. Monster 1970.

HARRY E. SCHLENZ MEDAL . . . for distinguished service in promoting public awareness, understanding, and action in water pollution control and may go only to someone whose principal employment is outside the technical field.
Recipients: William O. Taylor 1985; Frances H. Flanigan 1984; William G. Reinhardt 1983; James T. Detjen 1982; Marion R. Stoddart 1981; Tom Davey 1980; Betty Klaric 1979; Charles W. Heckroth 1977; Murray Stein 1976; Clem L. Rastatter 1975; Stuart Finley 1972; Gladwin Hill 1971.

PHILIP F. MORGAN MEDAL . . . for originality, significance, comprehensiveness, effort, and most importantly, verification of an idea by the application of in-plant studies of operational processes.
Recipients: W. Fred Cebalt 1986; Terry Mudder and Philip R. Karr III 1985; Ralph E. Pfister and Edward Goscicki 1984; Robert B. Klausegger and Richard R. Dague 1983; Paul L. Brunner 1982; Carmen F. Guarino and Clarence R. Shade 1980; Charles D. Malone and Daryle R. Smith 1979; George A. Brinsko 1977; DeFro Tossey 1976; John L. Puntenney, David B. Cohen and William C. Hebard 1975; Robert E. Derrington, David H. Steven and James D. Phillips 1974; David A. Long, Ralph D. Black and William C. Anderson 1973; Berard W. Dahl 1972; Herman R. Zablatzky and George T. Baer, Jr. 1971; B. R. Brown and L. B. Wood 1970.

THOMAS R. CAMP MEDAL . . . for demonstration by design or the development of a wastewater collection or treatment system, the unique application of basic research or fundamental principles.
Recipients: Wilhelm von der Emde 1986; David Jenkins 1984; David G. Argo 1983; Ross McKinney 1982; W. Wesley EcKenfelder, Jr. 1981; Russell L. Culp 1980; John S. Jeris 1979; Anton A. Kalinske 1977; Perry L. McCarty 1975; Richard L. Woodward 1973; Martin Lang 1972; Edwin F. Barth 1971; George J. Schroepfer 1970.

WILLEM RUDOLFS MEDAL . . . for outstanding contribution by an industrial employee on any aspect of industrial waste control.

Recipients: B. F. Severin and R. A. Poduska 1986; J. Floyd Byrd, Maxine D. Ehrke and Jan I. Whitfield 1985; Robert T. Stoller, Rono Pellissier and Paul Studebaker 1984; Michael Alsop and Richard A. Conway 1983; David W. Averill, Peter M. Huck, Garabet H. Kassakhian, Duncan Moffett and Robert Webber 1982; Paul V. Knopp and Tipton L. Randall 1981; John H. Robertson, Judy Y. Longfield and Vincent S. Wroniewicz 1980; W. A. Eberhardt, J. L. Lewis, R. A. Scharp and C. A. Barton 1979; Kenneth S. Watson, Arthur E. Peterson and Richard D. Powell 1978; Hugh J. Campbell, Jr. and Robert F. Rocheleau 1977; William C. Hutton and Sebastian A. LaRocca 1976; William L. Rose and Ronald E. Gorringe 1975; Joseph C. Hovious, Richard A. Conway and Charles W. Ganze 1974; Robert N. Simonsen, R. F. Peoples and P. Krishnan 1973; Jack T. Garrett 1972; Bruce A. Wing and William M. Steinfeldt 1971; Lloyd M. Grames and Ray W. Kuenman 1970.

WILLIAM J. ORCHARD MEDAL . . . for distinguished service to the Water Pollution Control Federation.
Recipients: Elmer E. Ross 1985; Paul D. Haney 1979; George J. Schroepfer 1977; Frank H. Miller 1974; Harris F. Seidel 1971.

WHO'S WHO
IN ENGINEERING

Aaron, Howard B
Business: 466 Stephenson Hgwy, Troy, MI 48084
Position: VP & Gen Mgr, Transmission Products Division *Employer:* DAB Industries Inc. *Education:* PhD/Met Engr/Univ of IL; MBA/Bus Ad/Univ of MI; MS/Met Engr/Univ of IL; BMetE/Met Engg/Cornell Univ. *Born:* 12/2/39. in B'klyn, NY. Served to Capt ORDC USAR 1967-69. Ford Motor Co 1967-78 (Scientific Res, Quality Assurance, Auto Assy Div, to hd of corp mtls testing lab, gen sruss) VP ENG RES DAB Ind Inc 1978-81-. Resp for R & D Mtls Dev, QC, and New Prod & Proc Design & Release. VP & Genl Mgr Transmission Prods Div- P & L responsibility for Div incl R & D, Eng, Mfg, Mktng, sales, etc. Part-time teaching in Engg & Bus AD Schools at Wayne State Univ, Detroit. Bd of Dir Epilepsy Ctr of MI, Mbr Advisory Bd Protection & Advocacy Committee, State of MI, Det Chamber of Commerce Comm. Married Judith E Kirsch (1962), daughter Nadine Wendy, born 1969. Enjoy theatre, racketball. *Society Aff:* ASM, TMS of AIME, SAE, ASTM, AMA.

Aaron, M Robert
Business: ATT. Bell Labs, Holmdel, NJ 07733
Position: Head, Digital Communications Dept. *Employer:* Bell Labs *Education:* MS/EE/Univ of PA; BS/EE/Univ of PA *Born:* 8/21/22 With Bell Labs since 1951. Early in his career he made fundamental contributions to computer aided design and applied these techniques to the development of the first repeatered transatlantic cable system. Since 1956 he has been involved in the development of digital systems. He was a key contributor to the design of the first degital system used in the exchange telephone plant. He presently is responsible for planning activities in digital communications. Fellow IEEE (1968); co-recipient IEEE Alexander Graham Bell Medal (1978); elected to NAE (1979). Active in IEEE for more than thirty years, member of several committees, former pres IEEE Circuits and Systems Society. During his spare time he may be found on the golf course. *Society Aff:* IEEE, NAE

Aaronson, Hubert I
Business: Dept of Met Engrg & Mat Sci, Pittsburgh, PA 49931
Position: R.F. Mechl Prof. *Employer:* Carneige-Mellon Univ. *Education:* PhD/Met Eng/Carnegie, Mellon Univ; MS/Met Eng/Carnegie-Mellon Univ; BS/Met Engrg/Carnegie-Mellon Univ. *Born:* July 1924. BS, MS, PhD from Carnegie-Mellon Univ. Native of NY City. USA Air Corps 1943-46. Staff Mbr, Metals Res Lab Carnegie-Mellon Univ 1943-47. Staff Mbr, Met Dept Sci Res Lab, Ford Motor Co 1958-72. Sr Visiting Fellow Univ of Manchester, UK spring 1972. Prof of Met Engrg MI Tech Univ 1972-1979. R.F. Mechl Prof of Met Engrg and Met Sci at Carnegie-Mellon Univ 1979- . Principle res area: mechanisms of diffusional phase transformations. C H Mathewson Gold Medal TMS/AIME 1968. E.C. Bain Award, Pittsburgh Chapter ASM 1980. Elected Fellow of ASM, 1980; elected Fellow TMS-AIME 1983. Past Chmn TMS Heat Treatment & Ferrous Met Cttes Detroit Sect AIME, TMS R.L. Hardy Gold Medal Ctte, TMSAME Inst of Metals/RF. Mechl Gold Medal Ctte. Past Chrmn ASM Technical Div Bd, Materials Science Div, Phase Transformation Ctte. *Society Aff:* TMS-AIME, ASM, IMS, MS, ΣΞ.

Abata, Duane L
Business: College of Engrg, Houghton, MI 49931
Position: Assoc Prof/Assoc. Dean *Employer:* MI Technological Univ. *Education:* PhD/ME/Univ of WI-Madison; MS/ME/Univ of WI-Madison; BS/ME/Univ of WI-Madison. *Born:* 6/9/49. Dr Duane Abata received a PhD in 1977 from the Univ of WI-Madison and is currently a mbr of the faculty of MI Technological Univ of Houghton, MI where he is teaching various mechanical engg courses on the undergraduate and graduate levels. In 1983 Dr Abata was awarded the Distinguished Teaching Award from the faculty and student body of Michigan Tech Univ. He has subsequently served as Dean and Assoc Dean in the Coll of Engrg at Michigan Tech. Dr Abata has done research in several areas of combustion which include combustion kinetics, internal combustion engines and pollutant formation from the detonation of explosives. In 1978, Dr. Abata was awarded the Outstanding Educator Award from the Society of Automotive Engrs and in 1979 he was awarded the Arch T Cowell Merit Award, also from the Society of Automotive Engrs, for his research work on gasoline engines. In 1983 Dr Abata was awarded the Distinguished Teaching Award from the faculty and student body of Michigan Tech Univ. Dr Abata is a mbr of SAE, ASEE, the Combustion Inst, the leadership society of Phi Kappa Phi, the leadership society of Omicron Kappa Delta and the Hon Res Society of Sigma Xi and is a Reg PE in the State of WI. *Society Aff:* SAE, CI, ASEE, ΣΞ.

Abbott, Henry H
Business: 53 Hope Rd, Tinton Falls, NJ 07724
Position: Telecommunications Consult (Self-emp) *Education:* BEE/EE/OH State Univ *Born:* 5/27/01. From 1922 to 1960, I was a mbr of Technical Staff, Supv, and Dept hd of Bell Tel Labs (BTL) and AT&T Co. From 1960 to 1966 I was Dir, Customer Tel Systems Lab of BTL. From 1966 to 1973, I was Mbr of Div Technical Staff of MITRE Corp and subsequently a consultant to MITRE on Military Telecommunications. I am a Fellow of IEEE, a Reg PE of NY State, and I have 37 US patents. *Society Aff:* IEEE, AFCEA, ΣΞ, HKN, TRIANGLE

Abbott, Michael M
Business: RPI, Troy, NY 12181
Position: Assoc Prof *Employer:* RPI *Education:* PhD/ChE/RPI; BS/ChE/RPI *Born:* 10/3/38 in Niagara Falls, NY. Married Mary Les Veaux in 1962; have one child (Mark). Worked for Exxon R&E Co, 1965-1969. With RPI since 1969, as Research Assoc (1969-1974) and as Assoc Prof of Chem Engrg (1974-present). Author/coauthor of 20 technical papers and three books. Major teaching interests are in the thermal sciences and in chem process design. In 1979, received an ASEE Western Electric Award for excellence in instruction of engrg students. Have outside interests in music and history. *Society Aff:* AIChE

Abbott, Wayne H, Jr
Business: PO Box 8647, 100 N Fifth Ave, Ann Arbor, MI 48107
Position: Utilities Dir. *Employer:* City of Ann Arbor. *Education:* BS/Chem Eng/MI State Univ *Born:* 6/9/25. With City of Ann Arbor 25 yrs as Utilities Dir. Responsible for dev & operations including technical, economic, financial planning, forecasts & planned expansions of water, wastewater storm water & Hydropower system serving over 100,000 persons. Natl Dir AWWA, MI Sec 1980-83 & Chrmn Mgt Div AWWA, member water utility council AWWA. Served Army Air Corp 1943-45 pilot. Previously worked for State of MI Water Resources Commission & Cities of E Lansing & Lansing, MI. Profl Reg Civ Engr, MI. Mbr Kiwanis Club & active with United Way. Active sports spectator; enjoy fishing & skiing. *Society Aff:* WPCF, AWWA, AAEE.

Abdel-Khalik, Said I
Business: Nuclear Engrg Dept, 1500 Johnson Dr, Madison, WI 53706
Position: Prof *Employer:* Univ of WI *Education:* PhD/ME/Univ of WI-Madison; MS/ME/Univ of WI-Madison; BSc/ME/Univ of Alexandria-Egypt *Born:* 8/9/48 Native of Alexandria, Egypt. Taught Mechanical Engrg at the Univ of Alexandria (1967-69); postdoctoral work at the Univ of WI Rheology Research Center (1973-74) and Solar Energy Lab (1974-75); Senior Engr at the Nuclear Power Generation Division of Babock and Wilcox (1975-76). Joined the Nuclear Engrg Dept of the UW-Madison in 1976; assumed current position as Prof in 1982. Authored or co-authored over 100 papers and publications in heat transfer, reactor safety, and fusion technology. Enjoy sailing, photography and chess. *Society Aff:* ANS, PROS

Abderhalden, Ross
Business: 2439 Royale, Beauport, Beauport Quebec, Canada G1C 1R8
Position: VP *Employer:* Extincto Inc *Education:* Engrg/ME/Tech Coll Herisau, Switzerland *Born:* 2/3/36 Native of Switzerland, came to Canada in 1958 after completion of education in native country. With Canadian Underwriters Assoc as fire inspector from 1961 to 1968. Loss prevention conslt with Reed Stenhouse from 1968 to 1972. Involved with Major Fire Protection Equipment installer from 1972 to

Abderhalden, Ross (Continued)
1979. VP, responsible for all fire protection Engrg Design Work for Extincto since 1980. Pres of SFPE Montreal-Ottawa Chapter 1980 to 1982, previously VP for 2 yrs. Enjoys all outdoor activities, especially skiing, active pilot. *Society Aff:* SFPE

Abdul-Rahman, Yahia A
Business: 515 S Flower St - Rm AP 3113, Los Angeles, CA 90071
Position: Mgr, Natl Oil Co Contracts. *Employer:* ARCO Petrol Products. *Education:* PhD/CE/Univ of WI; MS/CE/Univ of WI; BS/CE/Cairo Univ; MA/Intl Finance/Univ of TX. *Born:* 7/9/44. 7/1965 BS with honors Cairo Univ - taught till 2/1968. MS & PhD CE Univ WI, Madison. Developed for first time engg data for foam spray drying of milk (MS) & production of dry ammonium phosphate in spray reactor (PhD). Both works published. Joined ARCO in 1971 as sr engr. Then, principal engr, administrative assoc, & sr strategic planning consultant. Now, Mgr, Natl Oil Co Contracts responsible for $4 billion dollar crude oil purchasing contracts for ARCO. At ARCO developed new shale oil process. Holds eight patents on shale oil production & processing. Developed for first time engg package for in-situ uranium leaching. Developed long term energy supply/demand methodology & strategy options. PE, TX, No 36708 since 1974. Married to Dr (Eng) Magda Abaul- Rahman (with C F Braun) and have two daughters Maie (9) and Marwa (5). *Society Aff:* AIChE.

Abegg, Martin G
Business: 1501 W Bradley Ave, Peoria, IL 61625
Position: Pres. *Employer:* Bradley Univ. *Education:* PhD/Civil Engg/Rensselaer Poly Inst; MS/Civil Engg/Univ of CO; BS/Gen Engg/Bradley Univ. *Born:* 10/3/25. Engg instructor at Bradley Univ, 1947-50. Served as asst prof, assoc prof, and prof, 1950-60. Civil Engg Dept head, 1960-63. Dean of the Coll of Engg and Tech, 1963-70. Univ Pres, 1971 to present. Engg aide with IL Div of Hgwys, 1946. Civil engr in Peoria, IL, 1948. Park Dist Engr, 1953-55. Construction Engr with Norman Porter & Assoc, NYC 1956-57, 59. Served to Lt (jg) USNR, 1943-46. Recipient of Putnam Award for Excellence in Teaching at Bradley Univ, 1961, Recipient - Honorary Degree from Illinois College, 1982; Recipient - Distinguished Engineering Alumnus Award, University of Colorado, 1986. Reg PE and Land Surveyor in IL. Rotarian, past pres. Honorary Societies: Sigmi Xi, Sigma Tau, Phi Kappa Phi, Omicron Delta Kappa, Tau Beta Pi, Chi Epsilon. *Society Aff:* ASEE, ASCE, NSPE.

Abel, Carl R
Business: 3919 Madison Ave, Greensboro, NC 27410
Position: Consulting Struct Engr (Self-employed) *Born:* 6/12/14. Born Wilson, NY 1914. Reared Sou Pines, of AL. During WWII Structural Design of Army & Navy ordnance facilities with M H Connell of Miami. Struct design Brookhaven Lab 1947-1948 with H K Ferguson Co. Later with J N Pease & Co, Charlotte, NC. Structural design commercial industrial & Inst Bldgs. Chief Engr Brich Assoc of NC 1952 to 1977 res & dev structural appraisals & investigations - Reg NC Past Pres NC Section ASCE. Author two papers published in Intl proceedings of two symposiums. Presently in private practice, consulting structural engrg self-employed. *Society Aff:* ASCE.

Abel, Irving R
Home: 22 Partridge Rd, Lexington, MA 02173
Position: Mgr Optical Design & Engg. *Employer:* Honeywell Radiation Center. *Born:* Feb 1923 Rochester NY. MS Syracuse Univ, BS Univ of Rochester. US Navy 1945-46. Optical Engr Bausch & Lomb Optical Co, Farrand Optical Co, & Norden Div of United Aircraft; Mgr Optical Engrg Baird-Atomic Inc 1964-68; Honeywell Radiation Ctr 1968- . 6 pats incl Multispectral Scanner Optical Sys used on Skylab. H W Sweatt Engr-Scientist Award Honeywell 1970; NASA Cert of Recog for creative dev of tech 1974; 1st Kingslake Medal & Prize 1975. Enjoys tennis, hiking, & classical music.

Abel, John F, Jr
Home: 310 Lookout View Ct, Golden, CO 80401
Position: Prof. *Employer:* CO Sch Mines. *Education:* DSC/Mining/CO Sch Mines; MSC/Mining/CO Sch Mines; EM/Mining/CO Sch Mines. *Born:* 11/24/31. Prof Mining Engg, CO Sch of Mines, 1974 to present. Assoc Prof Univ of AZ, 1966-74. Res Mining Engr (WAE) US Geol Survey 1965-75. Rock Mechanics Engr Terrametrics, Inc 1962-64. Formerly Coal Mine Engr, Proj Leader Ice Mining. Visiting Prof Univ of Newcastle U/TYNE 1972-73, New South Wales 1978, Western Australia Sch of Mines 1978. Consultant to industry on rm and pillar, caving, subsidence, shaft lining and pit slope design. *Society Aff:* AIME, AEG.

Abernathy, Bobby F
Business: 800 Oil & Gas Bldg, Ft Worth, TX 76102
Position: Owner *Employer:* Abernathy Exploration Co *Education:* BS/Petrol Engg/Univ of TX; AA/-/Henderson County Jr College; Adv Mgt Course/-/Univ of Western Ontario. *Born:* 6/25/33. in Athens, TX. Reg engr in TX & Alberta. Won student AIME paper contest in 1954. Recipient of Cedrick K Ferguson medal in 1963. Have written a number of papers on waterflooding & waterflood analysis for AIME & SPE presentations. Recipient of Henderson County Jr College Distinguished Alumnus Award for 1980; University of Texas VIP Alumni. Past Distinguished Lecturer of SPE. Served in various engg positions in US & Canada formerly Senior V.P. Explor & production, Champlin Petrol Co. Elected Dist Mbr of SPE-AIME in 1983. Formed own Oil and Gas Explor and Prod firm in April 1981. Enjoy golf, hunting & fishing. *Society Aff:* SPE OF AIME, ΣΓΕ, ΤΒΠ, ΦΘΚ, CIM.

Abernathy, George H
Business: Agri Engg Dept, PO Box 3268, Las Cruces, NM 88003
Position: Hd of Dept. *Employer:* NM State Univ. *Education:* PhD/Agri Engr/OK State Univ; ME/Agri Engr/Univ of CA; BS/Agri Engr/NM State Univ. *Born:* 11/9/29. Native of NM. Served in US Army Corps of Engrs during Korean War. Accepted position at NM State Univ in 1957 where res included machine design & application to cotton production, tillage machine power requirement, & energy conservation. Received PE license in 1964. Was regional chrmn of ASAE in 1964 and am currently Regional Dir. Have been active in NMSPE having been NM State Pres in 1978-79 and named New Mexico engineer of The Year, 1981.Am currently Hd of Agricultural Engg Dept. *Society Aff:* ASAE, NSPE, NMSEA, SIGMA, BETTA ΣΤ, ΦΒΤ.

Abonyi, Erwin
Home: 2620 High Ridge Rd, Stamford, CT 06903
Position: Sr VP *Employer:* PRC Harris, A Div of PRC Engrg Inc *Education:* M/CE/NYU; B/CE/NYU *Born:* 8/8/22 Affiliated with Frederick R Harris, Inc, Consulting Engrs, now PRC Harris, A Division of PRC Engrg Inc since 1950 as design engr through Project Mgr and as an officer principal since 1964. Extensive periods overseas in the U.K. Spain, Holland and Iran as Project Mgr. General Mgr of Holland subsidiary and Deputy Managing Dir of Iran partnership. Principal activities involved port and harbor development work, highway engrg, petroleum and marine terminal facilities, transportation and urban planning, and construction management. Registered PE in NY and NJ. Colonel, CE USAR (Retired). *Society Aff:* ASCE, PIANC, SAME, NSPE

Abraham, George
Home: 3107 Westover Dr SE, Washington, DC 20020
Position: Research Physicist & Consultant. *Employer:* US Naval Research Lab. *Education:* PhD/Physics/Univ Of MD; SM/Communication Engg/Harvard Univ; ScB/Elec Engrg/Brown Univ *Born:* July 15, 1918. Head, Systems Applications, Office of the Dir of Res and Tech Applications. US Naval Res Lab. Prev Hd of Exper Devices & Microelectronics Sects & Cons Electronics Div/NRL. As Training Officer estab Grad Sci Educ Prog at Lab Res Assn, lect in elec engrg & physics Univ of MD & lect in elec engrg G W Univ. Lecturer at American Univ; Industrial Cons. Founder, Pres, Chmn & Dir Intercollegiate Broadcasting System Inc. Dir Harvard Series, World Wide Broadcasting Foun (Sta WRUL & WRUW) Boston

Abraham, George (Continued)
MA. Radio Engr RCA. Engrg duty officer US Navy WWII. Capt US Navy Res. Fellow Award IEEE, AAAS, WAS, NYAS; 1981 Harry Diamond Award IEEE; IEEE Centenial Award, 1984; 1981 Washington Society of Engrs Award; Distinguished Serv Award IEEE; District of Columbia Science Citation; Edison Award & FBM Award US Navy. Navy Research Publication Award 1974, 84; Government Microcircuit Applications Conference, Founders Award, 1986;Outstanding Service Award, D.C. Soc Prof Eng, 1986; -National Capital Engineer of the Year Award, Dist of Columbia Council of Engrg & Architectural Societies 1987. Bd Mbr & Reg Dir IEEE 1971-72; Bd of Mgrs & Pres WAS; Bd Mbr & Pres WSE; Mbr, Bd of Registration for Prof Engrs, District of Columbia. *Society Aff:* IEEE, APS, ASNE, AAAS, ΣΞ, ΤΒΠ, HKN.

Abrahamson, Royal T
Business: PO Box 14025, Suite 536 Petroleum Bldg, Amarillo, TX 79101
Position: Owner. *Employer:* Abrahamson & Assoc, Consulting Engrs. & Surveyors *Education:* BSCE/Traffic Engg/TX Tech Univ; BBA/Mgt & Govt/W TX State Univ; Asso Sci/Pre-Engg/Amarillo College. *Born:* 10/21/30. Chief X-Ray Technician, Yokohama, Japan in Army Medical Corp, 1951-53. Traffic Engr, City of Amarillo, 1957-72. Pres Amarillo City Fed Credit Union, 1962. Pres Dale Carnegie Class, 1968. Pres Panhandle Chapter TSPE/NSPE, 1969. Chrmn Accident Review Bd, City of Amarillo, 1965 & 1966. Listed in "Who's Who in the South and Southwest–, 1971. Rotary Intl Mbr, 1972-. Bd Mbr, St Paul Methodist Church, 1973-. Bd Mbr Girl Scout Council. Mbr Grand Jury, Randall County, 1968. Boy Scouts 47 yrs, Eagle Scout, Woodbadge Course Director, Silver Beaver, Scoutmaster, Cubmaster, Explorer Advisor. Bd Mbr - Eighty Scouts, Inc. Mbr Bldg Bd of Appeals. Classical music. Registered Professional Engineer (TX); Registered Public Land Surveyor (TX). *Society Aff:* ASCE, NSPE, ITE, CSI, ASSE, NFC, NAARE, AAFS, NAFS, NAFE

Abrams, John W
Home: 120 Glen Rd, Toronto Ont Canada M4W 2W2
Position: Prof. *Employer:* Univ of Toronto. *Education:* PhD/Astrophys/Univ CA; AB/Astronomy/Univ CA. *Born:* 12/15/18. Native of San Francisco, CA. Served as OR officer in Royal Canadian AF & in Royal AF (1940-45). Taught Univ Manitoba & Wesleyan Univ, CT (1945-49). Defence Res Bd Canada in Ottawa, London, (Eng), Paris (France) as OR officer. Chief of OR (1961-62). Prof Ind Engg, Univ Toronto (1963-79). Dir Inst History and Philosophy Sci & Tech (Univ Toronto 1966-72). Pres CDRS 1962, Pres Can Soc History & Phil of Sci 1971-74 Secy Gen, Intl History of Tech Committee 1974-. Coronation Medal, Centennial Medal. *Society Aff:* ORSA, CORS, ORS, SHOT, HSS.

Abramson, H Norman
Business: PO Drawer 28510, 6220 Culebra Rd, San Antonio, TX 78284
Position: Executive VP. *Employer:* Southwest Res Inst. *Education:* BS/Mech Engrg/Stanford Univ; MS/Engrg Mechs/Stanford Univ; PhD/Engrg Mechs/U Texas-Austin. *Born:* 3/4/26. Prin technical areas include fluid and structural dynamics, especially interaction problems. Assoc Prof of Aero Engg at TX A&M, 1952-56 and affiliated with Southwest Res Inst since 1956. Past VP-Basic Engg and mbr of the Bd of Governors of ASME, Technical Dir of AIAA, Chrmn of Engg Sec of AAAS, Mbr of the Coun of the US Natl Acad of Engrg, and active in various other profl affairs and advisory committees. Author of numerous publications. *Society Aff:* ASME, AIAA, SNAME, NAE, AAAS.

Abramson, Harold I
Home: 107 Fieldboro Dr, Lawrenceville, NJ 08648
Position: Staff Dir, Educational Services. *Employer:* Amer Inst of Chem Engrs. *Education:* BChE/Chem Engg/Cooper Union for the Advancement of Sci; MSChE/Chem Engg/Drexel Univ. *Born:* 10/7/37. Worked in product dev and res for FMC Corp 1959-62; Tech Ed of Chem Engg. progress pub by AIChE; 1962-66; Asst Exec Secy, Hd of Continuing Education Dept & Ed of AIChE Continuing Education Series, AIChE Modular Instruction Series & AIChE Workshop Series 1966- . Dir, Educational Services 1980-Present. Mbr ASEE, ASEE Bd of Dirs 1983-1985, ASEE Continuing Prof Dev Div Chrmn-Elect 1984-85; Chrmn 1985-86, Council of Engg & Sci Society Execs.Community Activities: Boy Scouts & Little League Baseball. *Society Aff:* AIChE, ASEE, AAAS.

Abramson, Norman
Business: The Aloha System, Honolulu, HI 96822
Position: Prof. *Employer:* Univ of HI. *Education:* PhD Stanford Univ; MS UCLA; AB Harvard. *Born:* 4/01/32 Taught communication theory Stanford 1958-64; Berkeley 1965; Harvard 1965- 66. MIT 1981 Appt Prof of Elec Engg Univ of HI 1967; Prof of Comp Sci 1968; Chmn Comp Sci 1968-71. Dir of The Aloha Sys, a univ res proj concerned with new forms of computer communication networks; Cons to indust labs US govt agencies & the UN consultant, international telecommunications union, Geneva. Chairman of the Board of Director, Technology Education Associates, Sydney Australia, 1980- ; member Board of Directors, Public Service Satellite Consortium, Washington; Member Board of Directors Pacific Telecommunications Council, Honolulu. Fellow IEEE, P Chmn Admin Ctte of IEEE Grp on Info Theory. IEEE 6th Region Achieve Award 1972.

Abramson, Paul B
Home: 280 Beechmont Dr, New Rochelle, NY 10804
Position: Chief Engr *Employer:* Intl Tech Services, Inc *Education:* PhD/Phys/Univ of CO; JD/(Law)/Loyola Univ of Chicago; MS/Phys/CA State Univ; BS/Engr Mech/Lehigh Univ *Born:* 12/19/39 1980-present, Chief Engr ITS (consltg engrs); 1974-1984 Mgr, Light Water Reactor Safety Section, Argonne Natl Lab; 1968-1974 Chrmn, Phys, MSC Denver CO; 1965-1967 Conslt, Martin Marietta; 1962-1965. Res Engr, Atomics Intl. Awarded Citation US Nuclear Regulatory Comm, 1983, outstanding tech mgmt; Mbr TMI-2 (NSAC) investigatory team 1979-1983; performed and directed nuclear power safety analyses 1962-present; keynote speaker and lecturer, Intl Atomic Energy Agency; Author 'Guidebook To Light Water Reactor Safety Analysis; Hemisphere 1985, approx 100 articles. Mayor - Wheat Ridge, CO 1974. Reg PE (Calif-Nuclear); Admitted NY and IL Bar. *Society Aff:* ANS, ABA

Aburdene, Maurice F
Business: Dept of Elec Engrg, Lewisburg, PA 17837
Position: Prof *Employer:* Bucknell Univ *Education:* PhD/EE/Univ of CT; MS/EE/Univ of CT; BS/EE/Univ of CT *Born:* 5/22/46 Assoc Prof of Electrical Engrg, Bucknell Univ. Taught at Swarthmore Coll, The St Univ of NY at Oswego, and The Univ of CT. Visiting Research Scientist at the Lab for Information and Decision Systems at M.I.T. (1978-79); NASA-ASEE Fellow at Stanford Univ-Ames Research Center (1977-78); DOE-ASEE Fellow at Jet Propulsion Lab (1981); Navy-ASEE Fellow at the Nav Res Lab (1982); Project Mgr at ACCO-Bristol, Waterbury, CT (1968-70). IEEE Control Systems Chapter Pres (Philadelphia) (1980-81); IEEE Student Activities Chrmn, Region 2 (1981-1984). Consultant for intl companies in the selection and implementation of computer systems for management, business, industrial applications, and CAD. *Society Aff:* IEEE, ΣΞ, ACM, ASEE.

Achenbach, Jan D
Business: Dept of Civ Engg, Northwestern U, Evanston, IL 60201
Position: Prof. *Employer:* Northwestern Univ. *Education:* PhD/Aero & Astro/Stanford Univ; BS/Aero Engr/Delft. *Born:* 8/20/35. Born in the Netherlands, naturalized US citizen. Walter P. Murphy Prof of Civ Engg & App Math at Northwestern Univ. Visiting Appointments at Columbia Univ, Univ of CA, San Diego Technical Univ of Delft. Author and Editor of 5 books and more than two hundred technical papers. Mbr US Natl Acad Engg. Fellow ASME & Am Acad of Mechanics. Recipient of ASEE Curtis W McGraw Res Award (1975). Editor of Wave Motion, an Intl Journal reporting res on Wave Phenomena, & Mbr Editorial Bds of 3 other journals. Pres assn of Chrmn of Depts of Mechanics (1971-73) & Pres Am Acad of Mechanics (1978-79). Mbr at large US Natl Committee T & AM (1972-78). Mbr

Achenbach, Jan D (Continued)
Exec Committee App Mechanics Div ASME (1979-85). *Society Aff:* ASME, ASA, AGU, AAM.

Acheson, Allen M
Business: 1500 Meadow Lake Pkwy, Kansas City, MO 64114
Position: Pres *Employer:* Black & Veatch International *Education:* BS/ME/IA State Univ; LHD/Honorary/Tarkio College *Born:* 6/12/26 Pres of Black & Veatch Intl Engrs-Architects and Exec Partner of Black & Veatch, the parent organization. Since joining the firm in 1963, has directed overall mgmt of engrg assignments for the firm in the design, const, and commissioning of thermal electric generating plants; gas turbine and combined cycle generating plants; electric and water transmission and distribution systems; electric substations; water and wastewater systems; storage and distribution of grains and other foods; studies and reports related to utility infrastructure. Proj work under his direction has taken place in 39 countries outside the US. Extensive travel for 20 years. Prior to 1963 he was Mgmt and Operating Adv to the Yanhee Electricity Authority, Bangkok, Thailand; Gen Superintendent of City Power & Light, Independenc, MO; and Supt of the Carroll Power Station for Iowa Public Service Co. BS degree from Iowa State Univ 1950 and the Professional Achievement Citation from Iowa State Univ 1976. Fellow of ACEC. Mbr of Missouri Soc of PE. Active in the Intl Fed of Consulting Engrs (FIDIC) through the ACEC and also has been Chrmn, Intl Div, ACEC. Reg PE in IA, MO, KS, MI, and AK and Queensland, Aus. Lic pilot with instrument and multi-engine ratings. Served on a U.S. Navy destroyer in the Pacific 1944-1946. Former Chrmn of the Bd of Tarkio Coll; Dir of the Kansas City Foreign Trade Zone. A Dir of AMBRIC (a joint venture of Amer and British firms carrying out the Greater Cairo Sew undertaking). *Society Aff:* ASME, MSPE, NSPE, SAME

Achtarides, Theodoros A
Home: Apt 117 D Torbay Estates, St John's, Newfoundland, Canada A1A 4B8
Position: Asst Prof of Naval Arch-Conslt *Employer:* Memorial Univ *Education:* PhD/Naval Arch/Univ of Newcastle Upon Tyne, England; OE/Ocean Engrg/MIT; SM/Naval Arch & Marine Engrg/MIT; BA/Math/Augustana Coll, SD. *Born:* 7/30/48 in Thessaloniki, Greece. Valedictorian, BA Summa Cum Laude, S.M., O.E., Ph. D. Teaching and Research Assistant MIT 1968-73, Research Engr R&D American Bureau of Shipping 1973, Demonstrator and "Ridley" Research Fellow Univ of Newcastle Upon Tyne 1974-78. Greek Army Signal Corps 1980. At Stevens Inst of Technology since 1981-83 assinged to expand the teaching and Research Program in Ship and Ocean Engrg Structures. Fulltime conslt at the level of Assoc, C.R. Cushing and Co. June-Dec. 1983, Ship structural analysis and design. Jan, 1984- present at Memorial Univ of Newfoundland, Dept of Shipbuilding, in charge of ship structures. "Naval Survey–, Technical Journal of the Greek Navy, annual national paper competition first prize 1975; SNAME Graduate paper honor prize 1973. Admitted to the US as a "Scientist of Exceptional ability and intl recognition–, 1981. *Society Aff:* ΣΞ, SNAME, MTS, RINA, IMARE, SUT, NECIES, IESS, HIMT, ATMA, UNDERWATER ASSOC.

Ackerman, Roy A
Home: 1610-C Quail Run, PO Box 5072, Charlottesville, VA 22905
Position: Technical Dir. *Employer:* ASTRE. *Education:* PhD/ChE/-; SM/Chem-Biochem Engg/MIT; BS/ChE/Poly Inst Brooklyn. *Born:* 9/9/51. ChE for Tri-Flo Res Labs, Inc 1972-1974. Sr Proj Engr for Thetford Corp 1975. With ASTRE since 1976. Became Technical Dir in 1978. Responible for corporate educational group, consulting div, & contract R&D group. Tau Beta Pi, Sigma Xi, Samuel Ruben Scholar in ChE. Reviewer & Editor for several journals. Holds 10plus patents. Authored 20plus technical publications & 2 monographs. Enjoy swimming, bicycling, & tennis. *Society Aff:* AIChE, ACS, SIM, ASAIO, WPCF, ASM, AIC

Ackermann, William C
Business: 2521 Hydrosystems Lab, 208 North Romine, Urbana, IL 61801
Position: Emeritus Prof. *Employer:* Univ of IL. *Education:* DSc/Water Sci/Northwestern Univ; BS/Civ Engg/Univ of WI. *Born:* 10/7/13. Profl employment has been with the TN Valley Authority (1935-1954), Agri Res Service (1954-1956), Chief of IL State Water Survey (1956-1979), Exec Office of the Pres (1963-1964), and Adjunct or Emeritus Prof of Civ Engg at the Univ of IL (1980-). Maj profl interests have been in water resources engg including res, planning, design, and policy. Related activities have included mbrship on numerous Natl Res Council Commissions, Bds, and Committees. Active consultant in US and abroad. *Society Aff:* ASCE, AWRA, AWWA, AGU.

Ackley, John W
Home: 2415 - 18 St –C–, Moline, IL 61265
Position: Retired Agri. Engr. Consultant *Employer:* Self *Education:* BSAE/Agri Engg/OH State Univ; Capt. Pilot, Multi Engine/Pilot Training/US Army Air Corps. *Born:* 5/8/22. Piloted P-38's in WW11. BS Agri Engr, OH State Univ 1951. Joined Deere & Co Jan 1952. 12 yrs Product Design & Dev. Developed liquid fertilizer and field sprayer equipment line. Deere & Co European office 1964-70, Mgr, Product Engg Coordination. Sr Staff Engr, Deere Technical Center, 1970-84. Reg PE. Nine farm equipment pats. ASAE Fellow, Mbr. SAE, Sigma Xi, Council for Agri Science & Tech & Rotary Intl. Past Chrmn ASAE PM Div, 2nd V-Chrmn of Section 111 of CIGR., 1977 Chrmn Industrial Profl Advisory Council, PA State Univ Sch of Engg, Dir Power & Machinery Div of ASAE, Chrmn Applications Systems Task Force of ASTM E- 35.22, Resource Person Conservation Tillage Info Ctr. Regional VP of ASAE. *Society Aff:* ASAE, SAE, ΣΞ, AFA, ROTARY INTNL

Ackmann, Lowell E
55 E Monroe St, Chicago, IL 60063
Position: Senior Partner *Employer:* Sargent & Lundy *Education:* BS/EE/Univ of IL *Born:* 7/2/23 Native of Elgin, IL. Married, 3 children. Recipient of Distinguished Alumnus Award, U of I Electrical Engrg Alumni, 1979. U. S. Navy, 1943-1946. Sales Engr for Allis-Chalmers (Chicago, Peoria, Milwaukee) 1946-1954. Roland Construction Co, Dallas, TX (family construction business) 1954-1956. Joined Sargent & Lundy, a power engrg consulting firm principally serving the electric utility industry, in 1956. Admitted to Partnership 1966, Appointed Manager Elect Dep 1968, appointed Dir of Services in 1977. Elected Sr. Partner 1984. Has presented numerous technical papers and has had articles published in trade literature. Registered PE in 20 sts. Member of BD of Dir of the Western Society of Engrs. Enjoys golfing. *Society Aff:* NSPE, IEEE, AIF, ANS, ISPE, WSE

Ackoff, Albert K
Home: 250 S Danbury Cir, Rochester, NY 14618
Position: Retired *Education:* MSChE/CE/MA Inst of Tech; BSChE/CE/Univ of PA. *Born:* 12/7/14. Phila Pa. BSChE Univ of PA; MSChE MA Inst of Tech. Engr Walbeck Foods; Chemist DuPont Co; Dev Engr Westvaco Chlorine Prods; with Eastman Kodak Co 1942- 1977 as ind engr, chem plants proj mgr and Engg training coordinator. Active in profl dev and continuing education of engrs. On natl committees of NACE, AIChE, ASEE. Fellow AIChE-1971, secy-treas of ASEE Continuing Engg Studies Div 1971-73, Pioneer Award ASEE 1977. Chem Engg advisory Committee Univ of Rochester 1970-77. Retired 1977. Hobbies include foreign travel, camping, the arts, cooking and wine. *Society Aff:* ASEE, AIChE, ACS.

Acrivos, Andreas
Business: Dept of Chemical Engineering, Stanford, CA 94305-5025
Position: Prof *Employer:* Stanford Univ *Education:* PhD/ChE/Univ of MN; MS/ChE/Univ of MN; B/ChE/Syracuse Univ *Born:* 06/13/28 Born in Greece where he received his early education. Following his arrival in the US in 1947, he enrolled at Syracuse Univ and then at the Univ of MN from which he received a PhD in Chem Engrg in 1954. He taught at the Univ of CA in Berkeley (1954-1962) and has been a Prof of Chem Engrg at Stanford Univ since 1962. Among his honors are: Guggenheim Fellowships in 1960 and in 1977; the Profl Progress & W.K. Lewis Awards of the AIChE in 1968 & 1984, respectively; and election to the Natl Academy of Engrg

Acrivos, Andreas (Continued)
in 1977. He is a member of the US National Ctte on Theoretical & Applied Mechanics & since 1982, he has been the editor for Fluid Dynamics of the Physics of Fluids. *Society Aff:* AIChE, APS, ACS, NAE

Adair, Jack W
Home: 3829 Wren, Ft Worth, TX 76133 *Employer:* USDA-SCS-Retired. *Education:* BS/Agri Engr/TX A&M. *Born:* 9/1/16. Fellow in ASAE 1978. SCS - 1941-1978 worked in OK field offices and Asst State Dir until 1965. Then became Hd of Engg and Watershed Planning for 12 states giving guidance and approval to planning, design and construction of multi-purpose dams for flood prevention, water supply, fish & wildlife and recreation. 1978-present self employed consultant for engg and inspection of Water Resource Dams. *Society Aff:* ASAE, SCSA.

Adam Kozma
Business: c/o Syracuse Research Corp, Merrill Lane, Syracuse, N.Y 13210
Position: VP-Dir Defense Electronics Engrg Division *Employer:* Syracuse Research Corp. *Education:* PhD/Elec Eng/Imperial College, Univ of London; MSE/Instrumentation/Univ of MI; MS/Eng Mech/Wayne State Univ; BSE/Mech Eng/Univ of MI. *Born:* 2/2/28. Native of Cleveland, OH. Res Engr & group leader spec in optical processing of radar data & holography Willow Run Labs Univ of MI 1958-69; Genl Mgr of Electro-Optics Ctr, Harris Corp dev holographic & data processing equip 1969-73; with Radar Div ERIM 1973-86, 1st as Mgr of Tech & Application Progs, Mgr Electromagnetics & Electronics, Dir Tech Depts, VP-Dir Radar Div 77-87. VP-Dir Defense Electronics Eng Div, Syracuse Res Corp 1987-; Fellow OSA, Chmn Tech Group on Info Processing & Holography 1973-75, Assoc Editor, Optics Letters, 1977-79. Fellow IEEE, Mbr Rad Sys Panel, Aero & Elec. Sys Soc, Mbr Sigma Xi, SPIE, Amer Defense Preparedness Assoc, Mbr Avion Sec Steer Comm. *Society Aff:* OSA, IEEE, ADPA, ΣΞ, SPIE

Adam, Paul J
Business: P. O. Box 8405, Kansas City, MO 64114
Position: Exec Partner *Employer:* Black & Veatch *Education:* BS/ME/Univ of KS *Born:* 10/26/34 Joined Black & Veatch Consltg Engrs, Kansas City, MO in 1956. Served as First Lt USAF 1956 to 1959. Rejoined Black & Veatch in 1959 as Mech Engr. Subsequently promoted to Project Engr, then Project Mgr of major power projects. Named a partner in 1975 and became Exec Partner and Head of the Power Div in 1979. Responsible for all of the firm's power station and transmission and distribution design and construction mgmt. Led in development of the Power Div proprietary computer aided engg system, POWRTRAC. Registered in 16 states. *Society Aff:* ASME, NSPE, ANS.

Adam, Stephen F
Business: 350 West Trimble Rd, San Jose, CA 95131
Position: R & D Mgr *Employer:* Hewlett-Packard Co *Education:* PhD/EE/Polytechnic Inst of Budapest; MS/EE&ME/Polytechnic Inst of Budapest; BS/EE&ME/Polytechnic Inst of Budapest *Born:* 02/28/29 Born, raised and educated in Budapest, Hungary. 1952-1956: Mbr of the technical staff, Research Inst for Telecommunications. 1957-present, Hewlett- Packard Co, Palo Alto/San Jose, CA. Mbr of the Technical Staff, Proj Mgr, R & D Sect Mgr. Holds California Life Teaching Credentials. Author of the textbook: "Microwave Theory and Applications," Prentice Hall 1969. Wrote many articles in trade journals. Holds several patents in the field of microwave engrg. Fellow of IEEE, past pres of IEEE Microwave Theory and Techniques Soc, Admin Committee Mbr. Past secretariat of the IEC-TC/66 dealing with microwave measurement standardization. *Society Aff:* IEEE

Adamczak, Robert L
Home: 7556 Beldale Ave, Dayton, OH 45424
Position: Asst for Plans. *Employer:* USAF. *Education:* PhD/Chemistry/Univ of Buffalo; MA/Chemistry/Univ of Buffalo; BA/Chemistry/Univ of Buffalo. *Born:* 8/16/27. As a res analyst (Olin Mathieson Corp 1950-1952) developed analytical techniques for high energy fuels for military applications; at Esso Res and Engr Co (1956-1958) did combustion res on compression and spark ignition engines; entered civ service (USAF) to work on advanced fluids, lubricants and fuels (1959-72); as staff scientist developed and defined future AF Mtls Res & Dev (1972-1975); Promoted to Asst for Plans of Wright Aeronautical Labs (1975 to 1979), Promoted to Chief of Plans of Wright Aeronautical Labs (1979 to present). *Society Aff:* ASME, ASLE, ACS, AAAS, ΣΞ.

Adams, Arlon T
Home: 102 Kensington Rd, Syracuse, NY 13210
Position: Prof. *Employer:* Syracuse Univ. *Education:* PhD/EE/Univ of MI; MS/EE/Univ of MI; BA/Applied Sci/Harvard. *Born:* 4/26/31. Prof Adams has been active in the area of numerical methods for electromagnetic problems. He has published two books (Electromagnetics for Engineers, John Wiley, 1972, and Intersystems Electromagnetic Compatibility (coauthor) Holt, Rinehart, and Winston, 1972) and over seventy papers in the area of electromagnetics. He has received five awards (three natl, two local) from the IEEE. He is listed in Who's Who in the East and is a Prof of Elec and Comp Engg at Syracuse Univ. *Society Aff:* HKN, ΦΚΦ, IEEE.

Adams, Charles A
Home: 18103 Spellbrook Dr, Houston, TX 77084
Position: Mgr Bus Dev. *Employer:* Litwin Engrs & Constructors, Inc. *Education:* BS/ChE/Tri-State University *Born:* 6/24/24. Native of Xalapa, Mexico. Served with USN 1942-45. Held various engg & supervisory positions with Monsanto Co, Chemstrand Div to 1958. Joined the Fluor Corp, as a process engr & held positions as proj engr, construction & sales engr, assisted in the formation of Fluor's subsidiaries in Venezuela and marine engg & construction firm. Joined the Lummus Co as mgr of Latin Am Bus Dev. Independent consultant to Venezuela & Peruvian Govts in the refinery and petrochem activities. Joined Litwin in 1976 as Mgr of Latin Am Bus Dev. Since 1983 self-employed as domestic and international bus dev consultant/advisor, also involved in the search & placement of mgmt engrg profs. Enjoys music, ballet, & golf. *Society Aff:* AIChE.

Adams, Joe J
Business: 1400 E Tremont St, Hillsboro, IL 62049
Position: Pres. *Employer:* Hurst-Rosche Engrs, Inc. *Education:* BS/Civ Engg/Univ of IL. *Born:* 2/1/23. Served with 24th Inf Div, USA 1943-45. Construction Engr - Triangle Consturctin Co, Kankakee, IL 1951-54. Gen Mgr - Hurst-Rosche, Inc, Consulting Engrs Hillsboro, IL 1954-65. Pres & Chrmn of Bd of Dirs, Hurst-Rosche Engrs, Inc with offices in IL, IN, KY, MD and IA. Firm provides complete civ engg services to private industry and fed, state and local governmental agencies. *Society Aff:* NSPE.

Adams, John L
Home: 674 Clermont St, Denver, CO 80220
Position: Consulting Engr (Self-employed) *Education:* BS/CE/Univ of MD. *Born:* 10/23/10. Bureau of Relamation, 8 yr Concrete Dams, Corp of Engr Flood Control, 1 yr Corp of Engrs, 1 yr Army Base Construction. Stearns-Roger, 14 yrs Mining Ore Benefication Plants, Private practice Civ Engr and Land Surveying 20 yrs. Reg PE and Land Suveyor CO No 566. Dept of Interior Bureau of Land Mgt Commission as US Deputy Mineral Surveyor. Mining Claim Retracement and Public Land Surveys to Patent Mining Claim. *Society Aff:* NSPE-PEC, PLSC, TBΠ

Adams, Ludwig
Home: 205 Thompson Dr, Pittsburgh, PA 15229
Position: Partner *Employer:* Ahala Consultants *Education:* BSCE/Civ Engg/NYU; MCE/Civ Engg/Rensselaer Poly Inst; -/-/Grad studies at Pitt, Penn State, Carnegie Mellon Univ. *Born:* 8/12/16. Pittsburgh-Des Moines Steel; Natl Advisory Committee Aeronautics (Cleveland); Mellon Inst Industrial Res; Univ Pittsburgh (ACP); Steel Structures Painting Council; Blaw-Knox (D&R Staff & Mattoon); VA Erection (VP); returned PDM (1961) Asst Chief Engr. Consolidated Natural Gas (1982-85); Robert W. Hunt Company (1987). Author-Lecturer wind tunnels, cryogenics, thermal insulations, welding, corrosion. NYU Centennial Engr (1954). Tau Beta Pi, Iota Alpha, Sigma Xi, Sigma Pi Sigma, Sigma Beta Sigma, Sigma Beta Alpha. Reg PE. PM PER Fellow ASHRAE & ASCE; Mbr ASTM, AWS. Wife Alberta Howard (Wellesley, Bryn Mawr, Pitt); Children Alberta Anne (Penn State '74), Ludwig Howard (Penn '76 & 77, Columbia Law '79, Dept Justice). Now private. *Society Aff:* ASCE, ASTM, AWS, ASHRAE, Past Chrmn of PRS of ASTM, AWS, ASHRAE

Adams, Robert R
Business: PO Box 428, Corvallis, OR 97339
Position: VP. *Employer:* CH2M Hill. *Education:* MEngr/Structural Engg/Yale Univ; BS/Civ Engg/OR State Univ; BS/Mech/Univ of CO. *Born:* 9/25/24. Active duty USNR WWII 1943-46, American and Pacific Theatre. With CH2M Hill since 1949. Partner 1961; VP 1966; Dir of Civ Engg 1972. Past Pres, OR Sec ASCE and Structural Engrs Assn of OR. Past Chrmn, Benton County Planning Commission and OR Judicial Fitness Commission. Natl Examination Committee for ICET. Mbr Lions Intl, Sigma Phi Epsilon fraternity. Reg PE 5 states. Hobbies music, golf, travel. Wife Mary Heumann Adams, daughters Kathryn Adams Johnson and Susan Adams Stutzman. ASCE OR Civ Engr of the Yr 1978. Member, Oregon State Board of Higher Education; Oregon Council for Economic Education. *Society Aff:* ASCE, NSPE, ACEC, APWA, PIANC.

Adams, William E
Home: 228 View Dr, Las Vegas, NV 89107
Position: City Mgr. *Employer:* City of Las Vegas. *Born:* Aug 30, 1924. BSCE Univ of KS 1951. Mbr Nev St Bd of Reg PE; Western Reg V Chmn Prof Engrs in Govt; Natl Dir (Nev) to natl Soc of Prof Engrs; PPres NV Soc of Prof Engrs. Mbr: ASCE, Soc of Amer Military engrs. Reg Engr in NV and MO. Mbr: Amer Pub Works Assn, Internatl City Mgrs Assn, Adv Ctte on Career Educ Clark County Sch Dist. St Rep Natl Lib of Sci & Communication Comm; VP Las Vegas Fed Credit Union; Life Mbr KS Alumni Assn.

Adams, William J, Jr
Home: 774 Bellerose Dr, San Jose, CA 95128
Position: Conslt-New Prod & Bus Dev *Employer:* Self employed *Education:* BSME/Mech Engrg/Santa Clara University *Born:* 2/9/17. in Riverdale, CA on Feb 9, 1917. BSME Santa Clara University 1937. Engr GE Co 1937-45, 6 yrs designing first remote-controlled aircraft gun turrets. FMC Corp 1946-1980. 7 yrs Ch Engr Outdoor Power Equi Div. 18 yrs Asst Gen Mgr Central Engg Labs on advanced dev incl vehicle traction and mobility and agri mechanization. 5 yrs Engg Consultant and Dir Planning & Ventures on forestry vehicles and tech exploration. 4 yrs Mgr New Bus Dev. ASAE Fellow, Dir (1968- 72). Region IX History & Heritage Chrmn & Public Affairs Chrmn ASME; SAE; Tau Beta Pi; Pres Santa Clara Valley Engrs Council (1956, 1970, 1971). Prof Engr CA ME 7672 & AG 234. Charter Memb. Tau Beta Pi Zeta Chptr; Regent Santa Clara University. Patentee in US, Canada and numerous foreign countries. Other interests: boy scouts, skiing, fishing, hunting. *Society Aff:* ASME, SAE, ASAE, TBΠ.

Adams, William M
Home: 900 62 St, Downers Grove, IL 60515
Position: Ret Chief Engr *Education:* BS/Agri Engr/IA State Univ. *Born:* 7/27/19. 17 yrs experience in Grain combine design & dev. Attained rank of Dept Asst Chief Engr. 7 yrs experience in tillage equip design & dev, six as chief engr for west coast plant operations, 10 yrs as chief engr of value analysis, for Agri equip Div FH. have worked 35 yrs for present employer, served term as tech VP, ASAE. *Society Aff:* ASAE.

Adamson, Dan K
Business: 6200 N Central Expressway, Dallas, TX 75206
Position: Exec Dir. *Employer:* Soc of Petrol Engrs. *Education:* BA/English-History/Southwestern Univ. *Born:* 10/12/39. Native of Vernon, TX. Taught secondary sch in Denver, CO, area 1962-1964. Editor of Journal of Petrol Tech, 1967-1971. Gen Mgr Soc of Petrol Engrs, 1972- 1979. Prog Mgr, 1969-1979 Offshore Tech Conf. Exec Dir SPE & Exec Mgr OTC. *Society Aff:* SPE, CESSE.

Adare, James R
Business: 490 Broadway, Buffalo, NY 14204
Position: Pres *Employer:* Buffalo Forge Co *Education:* BASC/Eng. Physics/University of Toronto *Born:* 11/23/27 Native of Toronto, Ontario, Graduate of Toronto Univ 1950. Started with the sales office of Canadian Blower/Canada Pumps Limited which is the Canadian subsidiary of Buffalo Forge Co. Moved to Kitchener, Ontario manufacturing facility in 1952. Worked 3 years in design, 10 years in sales, 12 years in manufacturing ending as VP of Manufacturing, then 2 years as Exec VP and finally 5 years as Pres. Moved to the position of Pres of Buffalo Forge Co in Buffalo, NY in 1982. The companies make fans, air conditioning equip, pumps and machine tools at 8 plants in the U.S.A., Canada and Mexico. Mgr, Assoc, Prof Engrs, Ontario.

Adcock, Willis A
Home: 3414 Mt. Bonnell Dr, Austin, TX 78731
Position: Professor *Employer:* Univ of Texas *Education:* PhD/Physical Chemistry/Brown Univ; BS/Chemistry-Math/Hobart College *Born:* Nov 1922. PhD Brown Univ; BS Hobart. Mbr tech staff Stanolind Oil & Gas Co 1948-53; Semiconductor R&D TX Instruments 1953-64; Tech Dir Sperry Semiconductor 1964-65; With TX Instruments 1965-1986; Univ of Texas at Austin 1986- . Fellow IEEE 1965; Mbr NAE 1974. *Society Aff:* IEEE, ACS, AAAS

Addae, Andrews K
Business: School of Engineering, Washington, DC 20059
Position: Assoc Prof *Employer:* Howard Univ *Education:* ScD/Nuclear Engrg/MIT; SM/Nuclear Engrg/MIT; BSc/ME/Univ of Sci and Tech, Ghana *Born:* 09/27/39 Native of Ghana. Married with two children. Responsible for the Reactor Physics Design of the Massachusetts Inst of Tech (MIT) Research Reactor II 1968- 70. Served as Research Assoc at MIT 1970-71. Taught Engrg at Howard Univ 1971- 74. Served Ghana Atomic Energy Commission as Principal Scientific Ofcr and Head Dept of Reactor Tech 1974-77. Returned to teach Engrg at Howard Univ 1977-81. Served as Conslt to utility companies in the area of Nuclear Power Safety. Currently on leave of absence from Howard Univ and serving the government of Ghana as Technical Dir, Ministry of Fuel and Power. Enjoy modern music and tennis. *Society Aff:* ANS, GIE

Addoms, Hallett B
Home: 8903 Griffon Ave, Niagara Falls, NY 14304
Position: Director Design Engineering *Employer:* Hooker Chem Co. *Education:* BE/ChE/Yale. *Born:* 8/1/18. in Kew Gardens, NY. BE from Yale. With Hercules Powder Co 1940-45 in res at Wilmington, production supervision at Volunteer & Sunflower Ordnance Works. Then with Niagara Alkali Co in res, production, & engrg. Chief Engr 1953 until 1955 Hooker Merger. With Hooker Chem Co 1955 to 1982 as Proj Mgr and then Dir Design Engrg, responsible for Civ, Electrical, & Mech Engrg, Drafting, & Engrg Studs. Presently an independent conslt to Chem Ind. PE in NY & formerly British Columbia. Mbr & Past chrmn Western NY Sec AIChE, received its Profl Achievement Award 1973. Former mbr & chrmn CMA Engrg Advisory Ctte. Former cruiser owner, and LaSalle Yacht Club mbr. Pres of Niagara Frontier Cross Country Ski Club, mbr The Niagara Club, mbr & past Commander Niagara Power Squadron. Mbr ACS and elected Fellow AIChE Jan 1981. *Society Aff:* AIChE, ACS.

Addy, Alva L
Business: 140 MEB, 1206 W. Green St, Urbana, IL 61801
Position: Prof of ME & Head *Employer:* Univ of IL. *Education:* PhD/ME/Univ of IL; MS/TAM/Univ of Cincinnati; BS/ME/SD Sch of Mines & Tech. *Born:* 3/29/36.

Addy, Alva L (Continued)
Native of Dallas, SD. Engr & Honors Prog participant for GE Co 1958-60. At the Univ of IL: Grad College Fellow 1960-61; Grad Asst 1961-63; Faculty Mbr 1963-present; Associate Head for Research and Graduate Study 1980-87; Head, Department of Mechanical & Industrial Engg. Consultant in fluid dynamics to US Army, NATO- AGARD, various industrial firms. Fellow ASME; Associate Fellow AIAA. Cooperative Res Fellow, US Army Res Office, Summer 1971. Invited NATO-AGARD Lectr, von Karman Inst for Fluid Dynamics, Brussels, Belgium 1968, 1975, 1976. Works primarily in base flow, high energy lasers, & ejector pumping systems. *Society Aff:* ASME, AIAA, ASEE, ΣΞ

Ade, C William
1842 Newburyport Rd, Chesterfield, MO 63017
Position: Pres *Employer:* Gateway Resources *Education:* D Sci/Chem Engg/WA Univ; MGT/Gas Tech/IL Inst of Tech; MS/Chem Engg/Univ of MO; BS/Chem Engg/Univ of MO. *Born:* 2/10/27. in Moline, IL and raised in Washington, Missouri, Currently Self Employed as a consult in fields of energy & natural gas marketing Joined Mississippi River Fuel Corp as Chem Dev Engg in 1951. Served this co and its successor in various respon VP 1972-83 Respon for gas control operation, gas supply, and computer areas. Employed by Northern Natural Gas 1970-72. Served with the Chemical Corps, US Army 1953-55. Consultant in energy, desalination, cryogenics, and computer software. Enjoy hunting, fishing, outdoor life, photography. *Society Aff:* AAAS, AIChE, ACS, SPE

Adelberg, Kenneth
PO Box 424, Woodstock, VA 22664
Position: Pres *Employer:* Patken Corp *Education:* BS/CE/Columbia Univ *Born:* 12/15/17 Native, Tampa, FL; Experience construction Middle East Rd, Hgwy, Airport Design and construction, Housing and Commercial bldgs, site development. New towns and smaller projects. Preliminary studies double decking George Washington Bridge and Crosstown Expressway, NY. Executive Engr, Levitt and Sons, Inc. Civil consultant Rouse Co. New Town Columbia, MD. Chief Engr New Town, Crofton, MD. Founded Kenneth Adelberg Assocs 1967 Consulting Firm, site and surveys, Pres, Chief Engr. Furnished consulting services and court testimony to U.S. Dept of Interior, U.S. Department of Justice. Qualified expert Federal and St Courts. Home and Bldg inspections. Licensed PE, NY, NJ, etc. Forensic engrg since 1970. Founded Patken Corp 1983. *Society Aff:* ASCE, NSPE, VSPE, NAFS.

Adkins, Alonzo F
Business: P. O. Box 447, Amarillo, TX 79178
Position: Chrmn & Prof *Employer:* Amarillo Coll *Education:* PhD/EE/TX Tech Univ; MS/EE/TX Tech Univ; BS/EE/TX Tech Univ *Born:* 06/09/31 Native of Fort Worth, TX. Taught at Texas Tech Univ from 1963 to 1970. Came to Amarillo Coll in 1970. Promoted to Dept Chrmn in 1972. Appointed Adjunct Prof in the Dept of Electrical Engrg in May 1979. *Society Aff:* NSPE, IEEE, ASEE

Adkins, Howard E
Home: 424 N Ashland, Park Ridge, IL 60068
Position: Pres. *Employer:* Howard E Adkins & Assoc. *Education:* MS/-/Univ of WI; BS/-/West St Univ KY. *Born:* 10/4/12. Teacher - Univ WI - Assoc Prof - MEngr 1947-57. Mgr Weld Engr - Field Services - Kaises Al 1957-78. Pres - Howard E Adkins & Assoc Inc 1979-. Dir AWS 1970-1973 & 1975-78. Chrmn, Natl AWS Awards Comm 1973. Chrmn Natl aWS Nominating Comm 1969. Lincoln Gold Metal Award - AWS. Honorary Mbr AWS 1979. AWS Handbook Comm Chapters 69, 72, 97. Chrmn Alum Assoc Welding Comm 1969-1971. Many articles published on welding & aluminum. AWS Natl Meritorious Award - 1974. Kaiser Alum Tech Services Engg Award of Merit, 1969. Methodist, Mason Married - Wilma. Four children, Howard E Jr, Michael R, Wilma Carol, Martha Lee. *Society Aff:* AWS, ASM.

Adler, Irwin L
Home: 643 Primrose Lane, River Vale, NJ 07675
Position: Pres. *Employer:* FFI, Corp *Education:* D. Eng Sc/ChE/NYU; MS/ChE/NYU; BChE/ChE/New Coll of Eng. *Born:* 7/30/28. Faculty of Engg Depts at Newark College of Engg and NYU. Engg research Supr positions at General Foods, 1961-68. Specialized in new process dev with implementation to mfg facilities. Mgr technology centers at Beechnut, 1968-71. Responsible for basic and applied research in food engg and biological sciences. Pres Dumont Foods, 1971-73. Consultant on product and process research and dev, 1973-present. Patents on various food processes including extraction, roasting, dehydration, carbonation, agglomeration. Various papers. *Society Aff:* AIChE, ACS, IFT.

Adler, Lawrence
Business: C.O.M.E.R, White Hall, Morgantown, WV 26505
Position: Prof *Employer:* West VA Univ *Education:* PhD/Mining Engrg/Univ of IL; MS/Mining Engrg/Univ of UT; BS/Mining Engrg/Columbia Univ; AB/Math/NYU *Born:* 06/06/23 New York, NY. Served with Army Air Force 1943-45. In 8th A.F. awarded Air Medal, Purple Heart (with cluster). Worked underground summers as mine helper. Employed as junior engr to asst cvl engr in tunnel projs, City of Los Angeles, CA (1951-1955). Asst Prof, Mining Engrg at Univ Missouri-Rolla, Lehigh Univ, Michigan Tech Univ (1955-1961). Assoc Prof to Prof at Virginia Polytechnic Inst (1963-1978). Prof at West Virginia Univ (1979-Current). Registered PE, Conslt, Faculty Senates (V.P.I. & W.V.U.). Contributing editor and contributor to SME handbook, wrote two text books, and over 50 journal articles dealing with ground control, equipment analysis, and mine design, dist mbr AIME. *Society Aff:* AIME, ASCE

Adler, Richard B
Business: 77 Mass Ave, Rm 38-409, Cambridge, MA 02139
Position: Prof and Assoc Dept Hd of Elec Engg *Employer:* MIT. *Education:* ScD/-/MIT; SB/-/MIT *Born:* May 9, 1922 NY City. Harvard 1939-41; SB MIT 1943, ScD 1949. Mbr fac MIT 1949- ; Assoc. Hd Elec. Engg 1978- . Prof Elec Engg 1959- ; Cecil and Ido Green prof. 1974-76, Res Asst Res Lab of Electronics 1946-56; Gr Ldr Solid State & Transistor Grp, Lincoln Labs 1951-53. USNR 1944-46. Rec Sloan Teaching Award MIT 1955 & 56; Prem Paper Award Royal Aero Soc 1955. Fellow Amer Acad Arts & Sci, IEEE; Mbr Sigma Xi, Eta Kappa Nu, Tau Beta Pi. Flds of inter: solid-state phys electronics; circuit theory; electromagnetic theory; biomed engrg. Auth or co- auth 8 coll texts, 16 educ films, 10 tech and 2 educ papers. *Society Aff:* ΣΞ, HKN, TBΠ

Adler, Robert
Business: 3005 MacArthur Blvd, Northbrook, IL 60062
Position: VP Research. *Employer:* Extel Corp. *Education:* PhD/Physics/Univ of Vienna (Austria). *Born:* 12/4/13. Res Group, Zenith Radio Corp from 1941; Dir of Res 1963-77. Since Jan 1978, Dir of Research of Extel Corp, manufacturers of teleprinters. Also, Adjunct Prof of Electrical Engg, Univ of IL, Urbana. Work includes phasitron modulator used in early FM transmitters, receiving tubes for FM detection and color demodulation, transverse-field traveling wave tubes, electron beam parametric amplifiers; also, first electromechanical IF filter, dev of ultrasonic remote control devices for television receivers; acousto-optical interaction (light deflection and modulation) and acoustic surface waves (filters and amplifiers). Now involved in research on teleprinter systems and devices. *Society Aff:* NAE, IEEE, AAAS.

Adler, Stanley B
Home: 2428 Bellefontaine, Houston, TX 77030
Position: Manager of Training and Special Projects in Process Department *Employer:* M. W. Kellogg Co. *Education:* BSChE/Chem Engg/Univ of PA. *Born:* 1/11/17. With M. W. Kellogg since 1947. Specializing in (1) applications of thermodynamics to chemical processing, and (2) the evaluation and correlation of data for the design of chemical, petrochemical, and petroleum plants. Formerly mgr of Technical Data Services Div which publishes an engg data book for Co engrs and which recommends thermodynamic data and correlations for use in engg calculations on

Adler, Stanley B (Continued)
or off the computer. Active in many phases of engg society work, particularly within the Amer Inst of chemical Engrs. Enjoys tennis, ham radio. *Society Aff:* AIChE, API.

Adomat, Emil Alfred
Business: 9250 W Flagler St, Miami, FL 33152
Position: Exec VP. *Employer:* FL Power & Light Co. *Education:* BS/EE/GA Inst of Tech; AdvMgtProg/Business/Harvard. *Born:* 9/29/24. Since joining FPL in 1950 he has held a variety of technical and mgt positions at the dist, div and co-wide level. His present responsibilities include operations, engg, construction, proj mgt, system planning, fuels, corporate dev, licensing and environmental planning, quality assurance, and res and dev. He is a mbr of the Bd of Dir of the Atomic Ind Forum, the Bd of Dirs of the Associated Industries of FL, the Natl Advisory Bd of the GA Tech Alumni Assn, and a past pres of the Harvard Business Sch Club of Southern FL. *Society Aff:* AIF, IEEE.

Adrian, Donald D
Business: Louisiana State Univ, 2401 A CEBA Bldg, Baton Rouge, LA 70803
Position: Prof of Civil Engg. *Employer:* Louisiana State Univ. *Education:* PhD/Civil Engg/Stanford Univ; MS/Hydraulic & Sanitary Engg/Univ of CA, Berkeley; BS/Civil Engg/Notre Dame Univ; BA/Liberal Arts/Notre Dame Univ. *Born:* 7/29/35. Grew up near White River and attended high school at Winner, SD. Married Joan Hogan in 1961, children are Nancy E (b 1962), Sandra L (b 1963), Bryan G (b1964) and Edward D (b1965). Employed by California Health Dept (1959), Water Resources Engg, (1964), Vanderbilt Univ (1964-67), Univ of Massachusetts (1967-82). Colorado State Univ 1982-86), and Louisiana State Univ. 1986-present). Sabbatical appointment at Princeton and US Environmental Protection Agency (1973-74) & Colorado State (1980-81). Dir Environmental Engg Graduate Program (1974-79). Dir of Louisiana Water Resources Institute 1986-. Reg PE in NJ and TN. *Society Aff:* ASCE, WPCF, AWRA, AWWA.

Adrian, Ronald J
Business: Univ. of Illinois Dept. of Theoretical & Applied Mechanics, 216 Talbot Lab, 104 So. Wright Street, Urbana, IL 61801
Position: Prof *Employer:* Univ of IL. *Education:* PhD/Phys/Cambridge Univ; MS/ME/Univ of MN; BSME/ME/Univ of MN. *Born:* 6/16/45. Native of Minneapolis, MN. Engr for Boeing Co. NSF Fellow and Churchill Scholar. With the Univ of IL, Dept of Theoretical and Applied Mech since 1972. Conslt in Fluid Mech and instrumentation. *Society Aff:* AIAA, APS, OSA.

Advani, Sunder H
Business: 209 Boyd Lab, 155 W. Woodruff, Columbus, OH 43220
Position: Prof & Chrmn *Employer:* Ohio State Univ *Education:* PhD/Eng/Stanford Univ; MS/ME/Stanford Univ; BS/ME/Bombay Univ *Born:* 08/10/41 Grad Teaching and Research Asst at Stanford Univ, 1962-64; Senior Research Engr at Northrop Corp 1965-67; Prof & Assoc Chrmn at West Virginia Univ 1978; Prof & Chrmn of Engrg Mech at Ohio State Univ 1978-present, Assoc Dean 1984- present. Author of numerous papers and textbook chapters, conslt to several industries; Recipient of Outstanding Teacher Award (1972-73). *Society Aff:* ASME, ASEE, AAM

Agarwal, Jagdish C
Business: 200 Clarendon St, Boston, MA 02116
Position: VP/Engg. *Employer:* Charles River Assoc Inc. *Education:* DChE/Poly Inst of NY; MChE/Poly Inst of NY; BSc/Banaras Hindu Univ. *Born:* 9/8/26. Charles River Assoc Inc, VP/Engg. Responsible for strategic planning in the minerals & metals industry, Amax Specialty Metals Corp, VP, Tech, 1982-1984, responsible for new product and business dev Kennecott Copper Corp, 1969-1979, Dir of Tech. Responsible for all process dev activities in nonferrous & coal industries. US Steel Corp, 1954-1969. Responsible for providing engg & economic analysis for new devs in natural resources. Dravo Corp, 1951-1954, Sr Process Engr. Certified Cost Engr, PE's License, twenty two patents *Society Aff:* AIChE, TMS, SME, ACS, ISS.

Agarwal, Paul D
Business: 773 Kirts, Troy, MI 48084
Position: Head-Electrical Engg Dept. *Employer:* General Motors Research Labs. *Education:* PhD/EE/Polytechnic Inst, Brooklyn, NY; MS/EE/IL Inst of Tech; BS Hons/EE & ME/Benares Hindu Univ, Benares, India. *Born:* 1/21/24. Born in India 1924. In US since 1947. Instructor and Asst Prof of Electrical Engg, Polytechnic Inst of Brooklyn 1951-57; Prof of Electrical Engg, Univ of MA 1957-61. Consultant to General Electric Co on electromagnetic and electromechanical problems 1954-61; Head, Electric Power and Propulsion at General Motors Defense Research Labs, Santa Barbara, CA 1961-67; Head, Electrical Engg Dept at GM Res Labs, 1967-85; Exec Engr, GM Adv. Engg Staff, Warren, MI 1985-present. Directs research in energy conversion. electromagnetic fields, automotive controls, and electric propulsion systems. Received several patents and AIEE and Sigma Xi awards. Fellow IEEE. *Society Aff:* IEEE, SAE.

Agarwal, Ram G
Home: 7025 East 70th St, Tulsa, OK 74133
Position: Research Assoc. *Employer:* Amoco Production Co. *Education:* PhD/Petroleum Engr/Texas A&M Univ; DIC/Petroleum Reservoir Engr/Imperial College of Science, London; BS/Petroleum Engr/Indian School of Mines, Dhanbad (India). *Born:* 7/6/38. Native of Calcutta (India), came to U.S. in 1963. After obtaining a PhD degree in petroleum engineering from TX A&M, joined the Research Dept of Amoco Production Co in Tulsa, OK in 1967. Currently, I am a Res Supvr in the production res division of Amoco. Am responsible for developing pressure transient techniques for evaluating performance of oil and gas wells. Also involved in simulation, evaluation and performance predictions of low permeability gas reservoirs. During 1978-79 was chrmn of the well testing committee of the Soc of Petroleum Engg. Has published a number of papers. *Society Aff:* SPE.

Agbabian, Mihran S
Business: Department of Civil Engineering, Los Angeles, CA 90089-0242
Position: Chrmn and Fred Champion Prof of Engrg *Employer:* Univ of Southern California *Education:* PhD/Struct Engr/UC Berkeley; MS/CE/Cal Tech; BS/CE/Amer Univ of Beirut; BA/Physics/Amer Univ of Beirut *Born:* 12/9/23 Cyprus, PhD U C Berkeley; MS CalTech; BS & BA Amer Univ of Beirut. Licensed cvl, struct, & mech engr; CA & licensed engr; HI, FL, OR. Assumed current position in 1984. Former positions include Pres of Agbabian Assoc an engrg & cons firm, since founding of company, Dec 1962 (now Chrmn of the Bd), and engrg positions at Bechtel San Francisco Parsons Los Angeles, & John K Minasian Pasadena. Also lectr at U C Berkeley & UCLA various yrs. Current tech activities are in fields of earthquake hazard mitigation and in design of structures against dynamic forces. Pres of Earthquake Engrg Res Inst (1983 - 1985); elected mbr of Natl Acad of Engrg in 1982. Fellow ASCE. *Society Aff:* ASCE, AIAA, ACI, ASTM, EERI, SSA, NYSS, ΣΞ.

Agee, Marvin H
Home: Rt 3, Box 45-H, Floyd, VA 24091
Position: Assoc Prof. *Employer:* VPI & State Univ. *Education:* PhD/Ind Eng/OH State; MS/Ind Eng/VPI & Su; BS/Ind Eng/VPI & SU. *Born:* Floyd VA. PhD from the OH St Univ; BS & MS from VPI & SU. Faculty mbr of the Dept of Indus Engrg at VPI & SU from 1957-64 & from 1968-present; ser as Dept Hd from 1970-73. Fac mbr at OH State from 1964-68. Natl Offices held in Alpha Pi Mu, Inc: Exec Sec, VP, Pres, Adv Bd (1969-). IIE Natl Dir of Continuing Educ 1972-74, Dir of Facilities Planning and Design Div (1980-82); Regio III Award for Excellence (1979); Fellow Awd (1984). Outstanding Teaching Awds 1961, 1964, 1976, 1979, 1980, 1982. Mbr of IIE and SME. Co-author of three books: *Quantitative Analysis for Management Decisions* 1976, *Principles of Engrg Economic Analysis* (1977, 1984), *Study Guide for Professional Engrs' Examination in Indus Engrg* (1982). *Society Aff:* IIE, SME

Aggarwal, J K
Business: Dept of Elec and Computer Engg, Austin, TX 78712
Position: Prof. *Employer:* Univ of TX. *Education:* PhD/Elec Eng/Univ of IL; MS/Elec Eng/Univ of IL; BE/Elec Eng/Univ of Liverpool; BS/Math/Univ of Bombay *Born:* Nov 1936. PhD & MS Univ of IL Urbana. Came to Univ of TX 1964, became Prof 1972. Fellow IEEE 1976. Ed of Newsletter of Circuits & Sys Soc 1972-73; Assoc Bd of Transactions of the Circuits & Sys Soc 1973-75. Book entitled "Notes on Nonlinear Systems" publ by Van Nostrand Reinhold; publ numerous papers on digital filters, image processing & pattern recognition. Hobby: competitive sailing. Additional Books: Computer Methods in Image Analysis, IEEE Press 1977; Digital Signal Processing, Western Periodical Company, 1979; Nonlinear Systems: Stability Analysis, Dowden Hutchinson Ross, 1977. Deconvolution of Seismic Data 1982. Additional Offices Held: General Chairman, IEEE Computer Society Conference on Pattern Recognition and Image processing, 1981, Dallas; Program Chairman, Workshop on Computer Analysis of Time Varying Imagery Sponsored by IEEE Computer Society, Philadelphia, 1979. Associate Editor: Pattern Recognition, Image and Vision computing, computer vision graphics and image processing, & IEEE Expert. *Society Aff:* IEEE, HKN, PRS, SPIE, AAAI

Agnew, Harold M
Business: 322 Punta Baja, Solana Beach, CA 92075
Position: Retired *Education:* PhD/Physics/Univ of Chicago; MS/Physics/Univ of Chicago; AB/Chemistry/Univ of Denver. *Born:* 3/28/21. Los Alamos Sci Lab 1943-46, Alt Div Ldr 1949-61, Ldr Weapons Div 1964-79, Dir, 1970-79. Pres, GA Technologies 1979-84. Sci Advisor Supreme Allied Comdr in Europe 1961-64, Chmn Army Sci Adv Panel, 1965-70; USAF Sci Adv Bd 1957-68; Consultant, Joint Com on Atomic Energy 1968; USAF Minuteman Planning Com 1961; Chmn US Army Sci Adv Panel 1964-70; Chmn US Army Combat Dev Com Sci Adv Grp 1965; Pres's Sci Panel 1968-74; Sr Fellow, Woodrow Wilson Natl Fellowship Fdth 1973- . Gen Adv Comm, US Arms Control and Disarmament Agency 1974-80, Chrmn 1974-76; Council on Foreign Rel, 1975- White House Science Council 1982- ; DOE, Ennico Fernie Award 1978, AEC, E. O. Lawrence Award 1979. Phi Beta Kappa, Sigma Xi, Omicron Delta Kappa.Served two terms NM State Senate 1955-61;Los Alamos School Bd 1950-55; Governor's (NM) Radiation Advisory Council 1959-61; NM Health and Social Serv Bd, 1971-73. *Society Aff:* NAS, NAE, APS, AAAS.

Agnew, Robert W
Business: PO Box 2022, Milwaukee, WI 53201
Position: Div Mgr. *Employer:* Envirex Inc-Environ Sci Div. *Born:* Apr 18, 1942. BSCE Marquette Univ 1964; MS & PhD Environ health engrg Univ of KS 1967, 69. Joined Envirex (formerly Rex Chainbelt) 1969 as proj engr & is now Div Mgr of Environ Sci Div with 50 employees & approx $2 million/yr in bus. Mbr: WPCF, Natl & Wis Soc of Prof Engrs, Sigma Xi, Chi Epsilon & ASCE where he served on numerous ASCE cttes. Received ASCE Rudloph Hering Medal 1969 & ASCE Young Civil Engr of the Year Award 1975.

Agnew, William G
Business: Gen Motors Res Lab, Warren, MI 48090
Position: Tech Dir. *Employer:* Gen Motors Res Labs. *Education:* PhD/ME/Purdue Univ; MS/ME/Purdue Univ; BS/ME/Purdue Univ. *Born:* 1/12/26. Born in Oak Park, IL, 1926. Purdue Univ: BSME 1948, MSME 1950, PhD Mech Engg 1952. Served in US Army Manhattan Dist Corps of Engrs 1944-46. With Gen Motors Res Labs since 1952. In 1971 appointed Technical Dir of five depts which conduct res in areas of automotive emissions, powerplants for automotive vehicles, vehicle structures, vehicle noise, fluid mechanics, aerodynamics, fuels and lubricants, and combustion. Received SAE Horning Meml Award 1960. Mbr of Natl Academy of Engg, Combustion Inst (a dir 1960-1976), SAE, ASME, AAAS, Sigma Xi, Engg Soc of Detroit. Served on working committees in CRC, SAE, and ASTM. *Society Aff:* AAAS, ASME, SAE, ΣΞ

Agosta, Vito
Home: 42 Cherry La, Huntington, NY 11743
Position: Prof Emeritus *Employer:* Polytechnic Univ *Education:* PhD/ME/Columbia Univ; MSME/ME/Univ of MI; BME/ME/Polytechnic of Brooklyn; ME/CCNY *Born:* 07/26/23 NYC. PE NY State, Fulbright Prof, England. Research on combustion instability on liquid and solid propellant rocket motors for Air Force and NASA. Analysis and design of supersonic combustors for Navy. Past Mbr of JANNAF on liquid and solid propellant rocket motors and on air breathing engines. Patents on mixing of liquid fuels and on emulsification of non-miscible liquids. Design and mfg of fuel processing sys for alternate fuel utilization in boilers, engines and turbines. Design and mft of aerothermochemical gas blast electric power switch for high energy transfer. P Pres of Propulsion Sciences Inc, Pres of Fuels Sys Design Corp., Pres of Propulsion Dynamics. *Society Aff:* AIAA, ASME, CI

Agrawal, Ashok K
Business: Bldg 130, Upton, NY 11973
Position: Nuclear Engr *Employer:* Brookhaven Natl Lab *Education:* ScD/Nuclear Engrg/Mass Inst of Tech; MSc/Phys/Banaras Hundu Univ, India; BSc/Phys, Math, Chem/Banaras Hindu Univ, India *Born:* 10/30/42 Born in India. Has worked in nuclear breeder reactor dev at Argonne Natl Lab, Westinghouse Elec Corp, and now for past eleven years with Brookhaven Natl Lab. Was posted, on assignment from Brookhaven, with the UKAEA Culham Lab for a year. Author and co-author of numerous pubs and reports. Editor of two books (Decay Heat Removal and Natural Convection in LMFBRs; and Modeling and Simulation of Energy Sys). Mbr, Exec Ctte, ANS Nuclear Reactor Safety Div. Actively involved in organizing tech meetings on nuclear reactor safety. *Society Aff:* ANS, ASME

Agrawal, Dharma P
Business: Box 7911, Electrical & Computer Engineering, Raleigh, NC 27695-7911
Position: Professor, E&C Engineering *Employer:* N.C. State University *Education:* D.SC. Tech./Computer Engrg/Swiss Federal Inst of Tech, Lausanne, Switzerland; M.E. (Hons.)/Electr. & Comm/Univ. of Roorkee, Roorkee, India; B.E./Electr. Engrg. /Ravishankar Univ. Raipur, India *Born:* 04/12/45 Native of Balod, India; Faculty member at Indian universities, Switzerland, Southern Methodist (Dallas), Wayne State (Detroit), before joining N.C. State Univ. Author of more than 130 papers. Group leader of 24-node loosely coupled B-HIVE Multicomputer Project. Tutorial text on Advanced Computer Architecture. Program Ch of 1984 International Symposium on Computer Architecture and member of Program Cttees for International Conferences. Fellow of IEEE. Was a member of Computer Society Publications and Education Activity Boards. Current Interest: Hardware & Software for Parallel and Distributed Processing, Computer Architecture, Reliability & Fault-tolerance. Editor of IEEE Computer and Journal of Parallel and Distributed Computing. *Society Aff:* IEEE, ACM, SIAM

Aguirre, Vukoslav E
Business: 14 Inverness Dr E, Unit 8H, Englewood, CO 80112
Position: President *Employer:* Aguirre Engineers, Inc. *Education:* MS/Soil and Rock Mech/Univ of IL, Urbana Ill; BS/Civil Engr/USMA West Point, NY *Born:* 11/02/41 Santiago, Chile; came to US, 1960, naturalized, 1971; married Emma Jeannette Bendana, November 15, 1970; children Sergio Eneas, Tonka Lily. Proj engr Ackenheil Assoc, Pittsburgh, 1965-69; soil specialist Harza Intl Co, Chicago, 1969-70; proj mgr Law Engrg Testing Co, Washington, 1971-74; pres dir Colorado Testing Lab, Denver, 1974-75; pres Geotek, Inc, Denver, 1975-77; pres dir Aguirre Engrs, Inc. Englewood, Colorado, 1977--. Served with US Army, 1960- 64. Reg professional Engr, CO, DC, IL, MD, MI, PA, UT, VA, WY, Republic of Chile. Mbr Colorado Consulting Engrs Soc Soil Mech and Foundation Engrg. Roman Catholic. Author: (with Earl G. Planty) *A Survey of Personal Practices of Firms* Located in Central IL. *Society Aff:* ASCE, ISRMFE, CEC

Agusta, Benjamin J
Home: 92 Oakcrest Dr, Burlington, VT 05401
Position: Exec Engr. *Employer:* IBM. *Education:* PhD/EE/Syracuse Univ; MS/EE/MIT; BS/EE/MIT. *Born:* 7/1/31. Brooklyn NY. Attended Brooklyn Tech HS; Mbr: Tau Beta Pi, Eta Kappa Nu, Sigma Xi Academic Honorary Socs. Licensed PE NY. Fellow IEEE for contributions to Monolithic Semiconductor Memory Devices & Processes. 9 patents; sev IBM outstanding invention awards; publ numerous tech papers. Mbr Ed Bd of IBM Jrnl of R&D; Engg Mgr of IBM Advnaced Tech Lab Burlington VT. *Society Aff:* IEEE, ΤΒΠ, HKN, ΣΞ, PES.

Ahern, John J
Business: 3044 W Grand Blvd, Detroit, MI 48202
Position: Consultant (Retired) *Born:* Sept 20, 1913 Chicago. BS Fire Protection & Safety Engrg IIT 1935. Dir Dept of Fire Protection & Safety Engrg IIT 1945-59; Cons 1945-59; Cons 1945-59; Cvl Serv Comm Chicago 1953-59; Pres Greater Chicago Safety Council 1952-55; 1st Pres Soc of Fire Protection Engrs 1950-53; Chmn & Mbr Occupational Safety and Health Subctte of Motor Vehicle Mfgrs Assn 1971-73. Mbr GM-UAW Natl Safety & Health Ctte; Chmn GM Occupational Safety & Health Ctte; PPres Natl Fire Prot Assn; Chmn Electronics Computer Sys Ctte; Mbr NSPA Stds Council; PPres, Chmn Bd, Amer Soc for Indus Security; Mbr US Delegation, NATO's Indus Security Conf, Paris France, April 1967.

Ahlen, John W
Business: 100 Main St, Little Rock, AR 72201
Position: Pres *Employer:* Arkansas Science & Technology Authority *Education:* Ph.D./Physiology/University of Illinois at the Medical Center; B.S./Bioengineering/University of Illinois at Chicago Circle *Born:* 01/04/47 A native of Chicago. 10 years science adviser of Illinois Legislative Council: energy, environmental, natural resources, medical issues. 4 years chief executive officer of quasi-public science and engrg organization, Arkansas Science & Technology Authority . VP-Public Information, ASME 1981-85; Dedicated Service Award, ASME 1986; Senior VP-Public Affairs, ASME 1987-89. *Society Aff:* ASME, IEEE, ΣΞ

Ahlert, Robert C
Home: PO 27, Buttzville, NJ 07829
Position: Prof. II *Employer:* Rutgers Univ. *Education:* PhD/Chem Engr/Lehigh; BChE/Chem Engr/Poly Inst of Brooklyn; MSEng/Engr/UCLA. *Born:* 1/22/32. Native of NY. Worked in a chem process industry for several yrs and then served as a res engg and res admin in liquid propellants for the Rocketdyne Div, North American Aviation. Joined Rutgers Univ as Dir of Engg Res and Assoc Prof of Chem Engg in 1964 after completion of doctoral studies at Lehigh Univ. Since 1964 has been concerned with grad teaching, res, res dev and res mgt. Became Prof of Chem & Biochem Engg in 1970 and Prof II in 1979. Primary technical interests include water quality planning, water quality mgt and modeling, solid and hazardous waste management and the application of novel chem processes to water and waste treatment. Licensed profl engr in NJ. Active in a number of engg and scientific societies. Public service activities include mbrship in local bd of education and regional health planning council. Consultant in various areas of water resources and water resources res. *Society Aff:* AIChE, AIC, ASEE, NSPE, AWRA, ΣΞ.

Ahmed, Nasir
Business: Dept. of Elect. Engg, Manhattan, KS 66506
Position: Prof *Employer:* Kansas State University *Education:* Ph.D/EE/Univ of NM; MS/EE/Univ of NM; BS/EE/Univ of Mysore-India *Born:* 8/11/40 Born in India and have lived in the U.S. since 1961. From 1966 to 1968 I worked as Principal Res Engr in the area of information processing at the Systems and Res Cntr of Honeywell, Inc., St. Paul, Minn. Since 1968 I have been at KS State Univ, where I am currently a Prof of Elect Engrg. I am the leading author of *Orthogonal Transforms for Digital Signal Processing*, (Springer- Verlag, 1975), and *Computer Science* Fundamentals, (Merill, 1979), with E.A. Unger. *Society Aff:* IEEE, ASEE, ΣΞ

Ahmed, S Basheer
Business: 900 St Rd, Princeton, NJ 08540
Position: Pres *Employer:* Princeton Economic Research *Education:* PhD/Economics/Texas A&M Univ; MS/Economics/Texas A&M Univ; MS/Economics/Osmania Univ; BA/Economics/Madras Univ *Born:* 1/1/34 in India and came to U.S. in 1961. Received M.S. and Ph.D. from TX A&M Univ in 1963. Became U.S. citizen in 1972. Taught at TN Tech Univ, OH Univ, and Western KY Univ, 1966-1980. Visiting Fellow, Princeton Univ, 1977. Consultant to Oak Ridge Natl Lab, 1969-1977. Current Pres of a consulting and research firm, Princeton Economic Research Inc. and a past Pres of IEEE-SMC Society and a Fellow of AAAS. Recip of 1984 IEEE Centennial Medal and Certificate. Also Prof of Mgmt Sci, Pace Univ *Society Aff:* IEEE-SMCS

Ahrens, Frederick W
Business: Dept. of Mechanical & Aerospace Engrg, Columbia, MO 65211
Position: Prof *Employer:* The Univ of Missouri *Education:* Ph.D./ME/Univ of WI-Madison; MS/ME/Univ of ID; BS/ME/Case Inst of Tech *Born:* 9/9/40 Born in Two Rivers, Wisc Engaged in nuclear reactor safety res at Phillips Petroleum Co. 1962-65. Investigated MHD phenomena in electric arcs at U. Wisc, 1965-69. With Whirlpool Corp. 1969-76. Developed computer simulation programs for design and analysis of air conditioners, dryers, refrigerators; res on appliance energy consumption reduction ideas. At Argonne Nat Lab 1976-80. Res on: compressed air energy storage processes and system optimization; pulse combustion burners. With Inst of Paper Chem 1980-84. Res on paper drying processes; teaching heat transfer. At U Missouri-Columbia since Jan 1985. Associated with Design optimization Lab. Author or co-author of over 30 reports and papers. Two patents. Elected to Sigma Xi. *Society Aff:* ASME, TAPPI.

Aigrain, Pierre RR
Business: 173 Bouldevard Haussmann, Paris, France 75008
Position: Genl Technical Mgr. *Employer:* Thomson *Born:* Sept 1924 French Naval Acad 1942, DSc Carnegie Inst of Tech Pittsburgh USA 1948: Dr of Sci Paris 1950. Engr Navy AEC 1948-52; Asst prof Lille 1952-54. Parie 1954- 58; prof 1954-65: Sci Dir DR. ME. Ministry of Defense 1961-63; Genl Dir Higher Educ 1965-68; Genl Delegate R&D French Govt 1968-73; prof MIT 1973-74; Genl Tech Mgr Thomson Group 1974. Fellow Academy Arts & Sci Boston. Acad of Engg Sci Stockholm Engrg Sci Washington. Several distinctions-Legion of Honour France. Order of Merit France. F R Germany. Polar Star Sweden.

Aiken, George E
Business: PO Box 23210, 1800 Harrison St, Oakland, CA 94623
Position: Chief Mining Engineer *Employer:* Raymond-Kaiser Engrs Inc *Education:* MS/Mining Engrg/Univ of MN; BS/Geological Engrg/Univ of AK. *Born:* 8/1/21. Instrumental in promoting the use of bucketwheel excavators in current N Am mining by publications in profl journals & books "Surface Mining" & "Mining Engineers' Handbook--. Assisted in editing of book on bucketwheels by German author. Superintendent of large mines in S Am & responsible for design contributions for maj ferrous & non-ferrous mines with consulting firms of Bechtel Corp & Raymond-Raymond-Kaiser Engrs. Proj mgr & Chief Mining Engineer for Kaiser Engrs, Oakland, CA since 1970. Reg PE in CA & published extensively on mining related subjects including maintenance, transportation, processing, and mining methods. Former chrmn of SME surface mining-coal div & mbr of other society committees. *Society Aff:* AIME, SME.

Aillet, Robert R
Business: 1055 Louisiana Ave, Shreveport, LA 71101
Position: Secretary. *Employer:* Aillet, Fenner, Jolly & McClelland, Inc. *Education:* BS/Civil Engg/LA Tech Univ. *Born:* 1/28/28. Native of Youngsville, LA. Reg Prof Civil Engg in the state of LA, TX, and AR. Civil Engg experience includes employment with the LA Dept of Highways, engg dept of United Gas Pipline Co, and E M Freemany & Assocs, Consulting Engrs. In 1960 became partner in firm of Aillet,

Aillet, Robert R (Continued)
Fenner, Jolly & McClelland, Inc, Consulting Civil and Structural Engrs. Serves as Corporate Secretary and Managing Partner. Pres ASCE Chapter and Pres ACEC local Chapter. *Society Aff:* ASCE, NSPE, LES, SAME, WPCF, ACEC, AWWA.

Ailor, William H, Jr
Business: Metallurgical Res Div, 4th & Canal Sts, Richmond, VA 23261
Position: Res Engr. *Employer:* Reynolds Metals Co. *Education:* BChE/Chem Engr/NC State Univ; BS/History/Univ of Tampa, FL. *Born:* 7/15/17. Ailor, William H, Jr, corrosion engr; b Knoxville, TN, July 15, 1917; BS Univ of Tampa 1939; BChE NC State Univ 1948; m Clara Louise Horne, May 1, 1942; children-William H III, James R, David C Research chem ACLRR 1948-53; Res engr. NC State Univ 1953-54; Math Instructor. VA Commonwealth Univ 1959-79. Reg PE Engr 1975- . Bd dirs. Richmond Council Boy Scouts of Amer, Chrmn of Sioux and Capitol Districts. LCDR USNR 1941-46, 52-53. Recipient Silver Beaver Award Boy Scouts Amer. Fellow ASTM (award of merit 1970); Fellow AAAS; Mbr Natl Assoc Corrosion Engrs, US Naval Ist, ASM, Amer Soc Naval Engrs. Editor; Metal Corrosion in the Atmosphere, 1977; Handbook on Corrosion Testing and Evaluation, 1971; Engine Coolant Testing, 1980. *Society Aff:* 0AAAS, ASTM, ASM, SAE, ASNE, USNI

Aitala, Roger
Home: 5402 Imogene, Houston, TX 77096
Position: Pres. *Employer:* Isonex Chem Engg, Inc. *Education:* BS/ChE/Syracuse Univ. *Born:* 4/18/37. Grad from Syracuse Univ in 1957. Worked in industry with large chem manufacturers and engg design companies intil 1974. Founded ISONEX CHEM ENGRS and CONSULTANTS in 1974, which incorp in 1978, to provide chem engg and chem marketing services to process industry. Firm is especially well known for its services to the worldwide soda ash industry. *Society Aff:* AIChE, CMRA.

Ajamil, Luis
Home: 8600 NW 36 S, Miami, FL 33155
Position: Dir, Intern Div *Employer:* Post, Buckley, Schuhi & Jernigan *Education:* BS/CE/Unif of FL *Born:* 09/17/50 Cuban born, arrived at age of 10 in the US. Studied engrg at the Univ of FL, and began upon graduation as a design engr for Public Works projs. Heads up Intl Div of a major consltg engr firm with responsibility for mgmt and marketing throughout Latin America and the Caribbean. Works under his mgmt include major highway, seaport, airport, and planning projs with values over $500 million. *Society Aff:* NSPE, ASCE, FES, SAME

Akers, Sheldon B
Home: 110 Treeland Cir, Syracuse, NY 13219
Position: Staff Computer Scientist. *Employer:* G E Co Syracuse NY. *Education:* BS/EE/U of MD; MA/Math/U of MD. *Born:* Oct 1926 Garrett Park MD. USNR Radar Tech Prog WWII. Empl in Wash DC area as Electronic Engr with Natl Bureau of Stds, US Coast Guard, ACF Indust 1948-56; with G E Electronics Lab and Mil Elec Syst Oper 1956- . Author of over 35 papers in areas of switching cir theory oper res, design automation; co-auth text on design automation. Active in sev IEEE prof groups; Mbr: MAA, Sigma Xi. Adj Prof at Syracuse U. Computer Soc Dist Visitor. Fellow IEEE. *Society Aff:* MAA, IEEE.

Akers, William W
Business: P.O. Box 1892, Houston, TX 77251
Position: VP for Administration *Employer:* Rice Univ. *Education:* PhD/ChE/Univ of MI; MS/ChE/Univ of TX; BS/Che/TX Tech College. *Born:* 12/31/22. Lifelong resident of TX. Began profl career with Atlantic Refining Co in 1947. Mbr of the Rice Univ faculty since 1947. Chrmn of the Dept of Chem Engg, 1955-65. Dir of the Bio-Medical Engg Lab, 1964-71. VP for External Affairs 1975- 1980. Assumed present position as VP for Admin in July 1980. *Society Aff:* AIChE, AAAS, ACS, ΣΞ, ΤΒΠ.

Akesson, N B
Business: Agr. Engi. Dpt, Univ. of Calif, Davis, CA 95616
Position: Prof Emeritus of Ag Engr. *Employer:* Univ of CA. *Education:* BSAE/Agr Engr/ND State Univ; MSAE/Agr Engr/Univ of ID. *Born:* 6/12/14. USN Torpedo res 1942-47. Univ of CA 1947 onward, Full Prof 1962. Dir ASAE 1972-74; Pres Pacific Region ASAE 1966; Pres CA Weed Soc 1965. Consultant to FAO pesticide application 1973-74 (sabbatical leave), Consultant Who vector control 1972 & 1976, Consultant to United Fruit 1959, State of Israel (1968) & Japan 1973. Elected Fellow ASAE 1975, Faculty Res England 1967, Fulbright Fellow 1957- 58 England & Tanzania. Over 200 papers and 3 manuals on pesticides application by air & ground. CA reg PE, Assoc editor ASAE PM publications. *Society Aff:* ASAE, ESA, WSSA, ASTM

Akhurst, Denys O
Home: P. O. Box 481, Boca Raton, FL 33432
Position: Dean, Coll of Engrg *Employer:* Florida Atlantic Univ *Education:* PhD/EE/Univ of Nottingham; BSc/EE/Univ of Nottingham *Born:* 03/02/27 Ferranti Research Scholar of the Inst of Elecl Engrs, 1951-53. Fulbright Travelling Scholar for Lecturing and Research at MIT, 1955-57 Coll Apprentice, Metropolitan-Vickers Elecl Co, 1953-55 and 1957. Lecturer, then Asst Prof, Dept of Elecl Engrg, MIT, 1955-57, 1957-59. Prof and Chrmn, Dept of Elecl Engrg, Univ of Miami, 1959-61. Prof and Head, Dept of Elecl Engrg, Univ of AR, 1961-70. Dean of Coll of Engrg, FL Atlantic Univ, 1970-. Assoc Dir, Engrg & Indus Experiment Station, Univ of FL 1977-. NSF Postdoctoral Research Award, Stanford Univ, summer 1962-63. C Eng, Chartered Engr, Council of Engrg Institutions(London), 1965. PE, Reg PE, AR, FL. FIEE, Fellow, IEE, 1967. Engr of the Year, Broward County, 1973. Pres, FES, Broward County Chapt, 1976. Appointed to Fl Metric Council by Governor, 1976. Certificate, Inst for Educ Mgmt, Harvard Univ, 1977. Paul Harris Fellow, Rotary Intl, 1979. Pres, Boca Raton Rotary Club, 1970-80. Fellow, FL Engrg Soc, 1981. Publications in professional journals. *Society Aff:* ASEE, NSPE, IEEE, IEE, FES, ΤΒΠ, HKN, BLUE KEY, ΣΞ, ΠΜΕ, ΦΔΚ, ΦΚΦ, AND ΘΤ

Akin, Lee S
Home: 48 Cypress Way, Rolling Hills Estates, CA 90274
Position: Tech Dir - Engrg Sci also Adjust Prof Cal St. Univ *Employer:* Western Gear Corp, ATD. *Education:* PhD/ME/W Coast Univ; MS/Appl Phys/W Coast Univ; BS/Gen Engg/W Coast Univ; AA/Machine Design/Los Angeles Tech Trade College; CB/Bus Mgt/UCLA. *Born:* 8/5/27. Dr L S Akin is personally responsible for much of today's advanced gearing tech over the past 30 years. He developed an approach for evaluating the scoring failure phenomenon in gears as a function of the operating lubrication regime based on a ratio (dimensionless film thickness) and contact temperature. In cooperation with co-workers he developed the math relationships needed to evaluate the impingement depth and cooling effect of jet oil sprays on gear teeth. He has done much work on the topography of manufactured tooth surfaces, especially in terms of math characterization. Recently he has devised an advanced technique to analyze and predict noise level in some kinds of large geared drive systems. *Society Aff:* ASME.

Akins, Richard G
Business: Dept of Chem Engrg, Manhattan, KS 66506
Position: Prof *Employer:* Kansas State University *Education:* PhD/ChE/Northwestern Univ.; MS/ChE/Univ of Louisville; BS/ChE/Univ of Louisville *Born:* 12/14/34 Born and raised in Louisville, Ky. Attended Louisville public schools and the University of Louisville. Employed by Argonne National Lab. in 1962-63. Has taught in the dept. of Chem. Engrg, Kansas State University since 1963. Married, three children. Hobbies: photography, fishing. *Society Aff:* AIChE, ASEE, ASME.

Albach, Carl R
Home: PO Box 4235, Santa Fe, NM 87502-4235
Position: Semi-retired Electrical Engineer *Employer:* Lewis Poe & Assoc. Part time *Education:* MA/Psychology/Univ of Buffalo, NY; EE/Elec. Eng./Rensselaer Poly Inst. *Born:* 2/21/07. AT&T Co 10 yrs, Western Elec Co 3 yrs, Corps of Engr 2 yrs Arch firm 3 yrs Consult Elec Engg 23 yrs, NM PE Reg Bd 2 yrs, NM Elec Code Comm 9 yrs UNM Elec Consult 15 yrs, UNM Lecturer on Lighting 11 yrs, 42 articles in trade and prof magazines. Founder mbr & 2nd pres. NM IES, Fellow IES, 25 year Service Award, Founder & Natl Dir CEC/US, now ACEC CEC/NM 20 & 30 yr Service Award as founder & 1st pres. Life Fellow ACEC. Listed in Who's Who in Engg, in the West & in NM since '57. Who's Who In America '86. Engr of Distinction '73. Exec Dir for Consltg Engrs Coun of NM 1967-85 and Elec Engr Lewis Poe & Assoc. *Society Aff:* IEEE, ACEC, CEC/NM.

Alban, Lester E
Business: US 52 South, Lafayette, IN 47902
Position: Asst Chf metallurgist. *Employer:* Fairfield Mfg Co, Inc. *Education:* BS/Met Eng/Purdue Univ :Toastmaster Internatl. Authored *Systematic Analysis of Gear Failures*, 1985, American Society for Metals. *Society Aff:* ASM, SAE, IMS

Albaugh, Frederic W
Home: 2534 Harris Ave, Richland, WA 99352
Position: Retired *Education:* PhD/Phys Chemistry/Univ of MI; MS/Phys Chemistry/Univ of MI; AB/Chemistry/UCLA. *Born:* 4/17/13. Chem analyst and R&D engr, petrol industry, 1935-37, 1941-43, 1945-47; licensed Petrol Engr (CA). Plutonium res, Manhattan Proj, 1943-45; res mgr, GE- Hanford, Battelle-Northwest, 1947-70; Dir, Battelle-Northwest Labs, 1967-70; corporate technical dir, Battelle Meml Inst, 1970-76. Independent consultant, 1976-87. Prog Leader, Hanford Plutonium Utilization Prog, 1956-65; conceived and directed Hanford Fast Flux Test Facility (FFTF) Prog, 1963-70; Dir, Battelle Energy Prog (coal), 1973. Fellow, ANS, 1963; GE Co rep, AEC-GE Task Force on Hanford Diversification, 1963-64; WA State Human Rights Commission, 1966-68; elected to NAE, 1978. Nuclear tech, energy policy planning and analysis, management, storage and disposal of nuclear wastes; nuclear plant licensing and regulation. *Society Aff:* ANS, ACS, AAAS, NAE.

Alberigi, Alessandro
Business: Phys Dept, Via Campi/213 A, Modena, Italy 41100
Position: Prof. *Employer:* Univ of Modena. *Education:* Laurea in Fisica/Cum Laude/Univ of Rome, Italy. *Born:* 6/15/27. Native of Reggio E (Italy). Scholar (1948-52) and Assoc Prof (1952-60) in Univ of Rome. Assoc Prof in Univ of Bologna (Italy) (1960-66). Full Prof in Univ of Modena, Italy. *Society Aff:* IEEE, APS, SIF, AEI.

Albert, Edward V
Home: 4486 Monahan Rd, La Mesa, CA 92041
Position: Conslt Engr, Aircraft Propulsion *Employer:* Self-E.V. Albert, PE *Education:* BS/ME/Univ of KY; Cert/Space Tech/Univ of CA; Cert/Macro Systems/Univ of CA; Cert/Bus Law/Univ of Cincinnati. *Born:* 04/02/20 Lynch, KY. Attended public schs FL and KY. Employed as Mech Engr, Texaco Inc, NY, 1941-52, for engine testing and product application Aviation Fuels and Lubricants. Joined Gen Elec Co, Evendale, OH, as Installation Engr 1952, progressed to Mgr Product Planning, Jet Engine Dept 1955-56. Transferred to Field Programs, San Diego, CA, 1956. Managed district office for all aerospace and defense products, 1958-75, and Aircraft Engine Field programs 1976-82. Retired 1982 to form consltg engrg serv for all aspects aircraft propulsion. Reg PE (mech); KY-1949, OH-1953, CA-1960. SAE Natl EAB and Aerospace Council 1967-69. Elected to Bd of Dir 1981-83, Exec Committee and Publications Committee. *Society Aff:* SAE, AIAA, ANA, AFA

Alberts, James R
Home: 45 Birchwood Rd, Glen Rock, NJ 07452
Position: VP. Retired *Employer:* Mobil Chem. Retired *Education:* BS/Ind Engg/Univ of WA; BS/CE/Univ of WA. *Born:* 12/10/24. C F Braun & Co for 8 yrs. Participated in process, mech design, startup of chem plants & petrol refineries. Aramco 2 yrs Section Hd in Engg Dept doing crude distribution, refinery design & refinery operation. Joined Chem Group of Dart Industries in 1961. Held positions of VP Engg & Construction, VP Production & VP Marketing for Rexene Polymers & Pres, Rexene Styrenics Co. Spent 5 years with Mobil Chemical as VP & Gen Mgr of Styrenics. Presently Pres & Owner of Intercontinental Engrg Inc. *Society Aff:* AIChE.

Alberts, Richard D
Home: 7616 Rustic Dr, Tampa, FL 33614
Position: VP *Employer:* Greiner Inc *Education:* BS/CE/Univ of MD *Born:* 12/24/43 As a VP in charge of Greiner's Planning Division, has been responsible for management of environmental studies for projects worth in excess of ten billion dollars. Diversity of complex projects have included major hub airports, new communities, hgwys, high level bridges and downtown redevelopments. Managed preparation of environmental guidebook and training program for FAA Academy. Instructor of environmental study management at Academy for past ten years. Presented technical papers to FHWA, IBTTA, and ASCE. Strong emphasis on innovative resolutions of environmental issues. *Society Aff:* ASCE.

Alberts, Warren E
Home: 1140 Old Mill Rd, Unit 403F, Hinsdale, IL 60521
Position: Consultant *Education:* BS/ME/Univ of WI; EDP/Business/Stanford Univ. *Born:* 9/20/16. Military Pilot USAF 1940-45, Rank - Col. Joined United Air Lines 1946. Appointed Asst to VP - Flight Operations 1947. Dir of Ind Engg 1953. Elected VP - Ind Engg 1958 & VP & Asst to Pres 1960. Responsible for planning United/Capital Merger. Elected VP - Mgt Services and Controls 1961 & VP - System Operations Services 1971. Corporate responsibility for Facilities Engg, Airport Planning, Operations Res, Ind Engg and Purchasing. Retired from United Air Lines - Sept., 1980. Consultant. Pres EJC 1975-76, Dir EJC 1969-76, Fellow IIE, Mbr AIAA. *Society Aff:* AIAA, IIE.

Albrecht, E Daniel
Business: 41 Waukegan Rd, P.O. Box 1, Lake Buff, IL 60044
Position: Chmn & Pres., President and CEO *Employer:* Buehler Ltd., Buehler International Inc. *Education:* Metallurgical Engg/Hon/Univ of AZ; PhD/Phys Metallurgy/Univ of AZ; MS/Phys Metallurgy/Univ of AZ; BSMetE/Metallurgical Engg/Univ of AZ. *Born:* 2/11/37. Native of Kewanee, IL. Staff metallurgist Los Alamos Sce Lab 1959-61. Phys and Program Mgr Lawrence Livermore Lab 1964-71. Founder and Pres Metallurgical Innovations Inc 1969-72. With Buehler Ltd since 1971, Chmn & Pres 1976- . Pres & CEO Buehler International Inc. 1985- . Dir. John Mowlem & Co., PLC 1985- . Fellow of ASM. Member National Advisory Board, The Heard Museum of Anthropology, Phoenix, AZ 1980- . Mbr, Bd of Trustees, Millicent Rogers Museum, Taos, 1982-. Mbr, Bd. of Directors Coal Gasification Inc., Des Plaines, IL 1986- and Sauk Southwest Steel Corp, Inc, Tulsa, OK 1987- . Mbr Bd of Trustees Lake Forest Acadamy-Ferry Hall College Prep School 1977-81. Hobbies golf, gardening, antique collecting and collecting American Indian Art. *Society Aff:* ASM, IMS, TMS, RMS, DGM.

Albright, Louis D
Business: 206 Riley Robb Hall, Ithaca, NY 14853
Position: Prof. *Employer:* Cornell Univ. *Education:* PhD/Agri Engg/Cornell Univ; MS/Food Sci/Cornell Univ; BSAE/Agri Engg/Cornell Univ. *Born:* 12/31/40. Native of Newfield, NY. Served with US Army Ordnance; discharged with rank of Capt. On faculty of the Univ of CA at Davis in Agri Engg, 1972-1974. With Cornell Univ in Agri Engg since 1974. Interests relate to environmental control and energy mgt in plant and animal structures. ASAE 1979 Engg Achievement Young Researcher Award. Hobbies include bonsai and woodworking. *Society Aff:* ASAE, ASHRAE, AS/ISES, ISHS.

Albright, Lyle F
Business: Sch of Chem Engg, W Lafayette, IN 47907
Position: Prof of Chem Engr. *Employer:* Purdue Univ. *Education:* PhD/ChE/Univ of

Albright, Lyle F (Continued)
MI; MS/ChE/Univ of MI; BSE/ChE/Univ of MI. *Born:* 5/3/21. Ind experience with Dow Chem Co (1939-1941), E I duPont deNemours on Manhattan Atomic Proj (1944-1946), Colgate-Palmolive (1950-1951), and with over 20 chem & petrol cos as consultant (1951-present). Chem Engg staff, Univ of OK (1951-1955), Univ of TX (1952), Texas A and M Univ (1985), & Purdue Univ (since 1955). Mbr AIChE (Fellow), Dir AIChE, 1982-1984; ACS. Editor for Chem Processing & Engg book series, Marcel Dekker, and editorial bd of Modern Engg Tech. Author or editor of seven books on plastics & their monomers, nitration, pyrolysis (2 books), alkylation, coke formation on metals, and high temperature measurements respectively. Res mainly in petrochem, petrol refining, energy conversion, & process dev. Hobby is magic. *Society Aff:* AIChE, ACS.

Albright, Robert J
Business: 5000 N Willamette Blvd, Portland, OR 97203
Position: Prof *Employer:* Univ of Portland. *Education:* PhD/Elec Engg/Univ of WA; MS/Elec Engg/OR State Univ; BS/Elec Engg/OR State Univ. *Born:* 3/6/41. Native of Gervais, OR. Prof of elec engg at the Univ of Portland. Served as Chrmn of elec engg, 1971-86, and as Acting Dean of engg, 1976-77. Researcher and consultant in the fields of biomedical instrumentation, energy conversion, and control systems. Particular areas of interest and endeavor deal with engg education, random processes, energy economics, and systems stability. Author of several professional papers. Reg engr in OR. Recipient of the Univ of Portland Culligan Award (1977) for outstanding service and teaching. Much of free time spent renovating an old farm-style home. Married Libbi Gerber. Daughters Mary Ann and Anna Marie. *Society Aff:* IEEE, NSPE, ASEE, AAUP, AAAS.

Alcabes, Joseph
Home: 483 Steven Ave, West Hempstead, NY 11552
Position: V Pres Engg *Employer:* St Lukes/Roosevelt Hospital Center *Education:* MBA/Bus Admin/Bernard M Baruch Sch of Bus & Publ Adm.; BCE/ Engg/City Univ. *Born:* 3/16/31. Engr, Bechtel Corp, NYC, 1956-60; mgr estimating, cost and scheduling div Stone & Webster, Garden City, NY, 1960-69; dir engg planning LI Jewish Med Center, New Hyde Park, NY, 1968-74; dir planning Lincoln Med Center, Bronx, NY, 1974-77, dir admin services, 1977-78; vp, planning and operations, Bronx-Lebanon Hospital Center, 1978-1987; V Pres Engg-St Lukes Roosevelt Hospital Center, 1987- ; lectr in field, Tech advisor Franklin Sq Citizens Assn, 1969-70, Reg PE NY, SC, Mem Greater NY Hosp Assn (chmn engg adv com 1976-77), Am Assn Cost Engrs, (certified cost engg, dir 1973-75), Am Coll Hosp Adminstrs, Nat Soc Prof Engrs, Am Hosp Assn Systems Soc, Hosp Exec Engrs Assn Greater NY, Am Soc Hosp Engrs, Alumni Assn City Coll NY, Contbr articles to profl publs, V Pres Society for Advancement of Management 1981-82, Who's Who in the East. Am Bar Assoc Register of Expert Witnesses in the Construction Ind. *Society Aff:* AACE, NSPE, ACHA.

Alden, John D
Business: 345 E 47th St, New York City, NY 10017
Position: Accreditation Consultant *Employer:* Accred Bd for Engg & Tech (formerly ECPD) *Education:* BS/Elec Engg/MIT; AB/Chem/Cornell Univ. *Born:* 10/23/21. Commander, US Navy (Retired). Dir of Manpower Activities, Engrs Joint Council 1965-1978. Accred. Dir. ABET 1978-1986. Author: *Flush Decks and Four Pipes*; *American Steel Navy*; *The Fleet Submarine in the U.S.N.*. Co-author: *Careers and Opportunities in Engineering*. *Society Aff:* ASEE, ASNE.

Alden, Raymond M
Home: 469 Ena Rd #3604, Honolulu, HI 96815
Position: Retired *Education:* AB/EE/Stanford Univ. *Born:* 11/17/21. Following graduation from Stanford Univ in 1944, served in the Naval Reserve for 2 yrs in communications assignments. Field Engr with Western Union for 4 yrs, primarily working with telegraph carrier equipment and in operating circuit design. With Hawaiian Telephone Co from 1951 until 1964, mostly in engg aassisgnments, including Chf Engr and VP - Operations. Joined United Telecommunications, Inc, in 1964 as Exec VP. Became a Dir in 1968, and Pres in 1973. Served as Chf Operating Officer. Became Vice Chairman 1/1/81, in charge of strategic planning. Retired as V.Ch. 10/15/85, and as Director 4/14/87. Reg for the practice of Engg in HI and KS. *Society Aff:* IEEE, NSPE.

Aldinger, Paul B
Home: 115 Laurel Ave, Providence, RI 02906
Position: Pr Geot Engr Asst, Dept Mgr *Employer:* Maguire Group *Education:* PhD/Geot Engr/Univ of RI; MS/Civ Engr/Univ of VT; BS/Civ Engr/Univ of VT. *Born:* 1/22/47. Born in NY City, raised in VT and served in the USN Civil Engr Corps in CONUS and Vietnam 1969-71. Worked part-time for Knight Consulting Engineers and teaching asst while in grad sch 1971-73. With Maguire Group since 1973 as Geotechnical Engr assuming respon as Mgr of Dept in 1978. Previously served as Pres of RI Section of ASCE, and is V P of RI Chapter NSPE. Completed PhD degree at Univ of RI in 1983. Am married with 3 children and enjoy spectator and participatory sporting activities, travel and reading as well as puttering around the house. *Society Aff:* ASCE, NSPE, NWWA, AGU, EERI.

Aldrich, Robert A
Home: 295 Wormwood Hill Rd, Mansfield Ctr, CT 06250
Position: Prof & Dept Hd. *Employer:* Univ of CT. *Education:* PhD/Agri Engr/MI State Univ; MS/Agri Engr/WA State Univ; BS/Agri Engr/WA State Univ. *Born:* 4/25/24. Native of NY State. Served in US Army Corps of Engrs, 1942-1946. Joined faculty of WA State Univ in 1951. Served on faculties of Univ of KY, 1958, at MI State Univ, 1959-62. PA State Univ, 1962-1979. Did res and teaching in agri struct at all inst. Res in greenhouse engg and energy conservation. Married Roberta Bowlby, 8/27/46. Four daughters. Hobbies include golf, skiing, canoeing, woodworking. *Society Aff:* ASAE, NSPE, ASHRAE.

Aldrich, Robert G
Business: 300 Erie Blvd West, Syracuse, NY 13202
Position: Dir R & D Projs. *Employer:* Niagara Mohawk Power Corp. *Born:* July 2, 1940. PhD Syracuse Univ; B Met E RPI. With Niagara Mohawk Power Corp 1974- ; curr Dir R & D Projs. Adjunct Prof Syracuse Univ & RPI. Presently Indus Cons; Cons Scientist, Oak Ridge Natl Lab. Empire St Elec Energy res Corp; Chmn Clean Fuels Task Force; mbr Energy Ctre Elec Pwr Res Inst; mbr Fossil Fuel Task Force ASM; V Chmn Syracuse Chap Amer Inst of Mining; Tech club of Syracuse; Bd of Gov NY Acad of Sci. Internatl Scholars Directory; Who's Who in the East; Amer Men of Sci.

Aldridge, Melvin Dayne
Business: 108 Ramsay Hall, Auburn, AL 36849
Position: Asst Dean for Res, Dir Engrg Exp Station and Prof Elec Engrg *Employer:* Auburn Univ *Education:* DSc/EE/Univ of VA; MEE/EE/Univ of VA; BS/EE/W VA Univ *Born:* 7/20/41 Resp for dev and maintence of res activities in College of Engrg at Auburn Univ since 1984. Dir of a Univ Energy Res Program with emphasis on coal 1978 to 1984. Asst to Full Prof of Elect Engrg, West VA Univ 1968 to present. Conducted res on application of electronic monitoring and communications to coal mine safety and productivity problems 1969-present. Electronic engineer with NASA, 1963-1968. *Society Aff:* IEEE, IAS, ASEE

Aldworth, George A
Business: 33 Yonge St, Toronto Ontario M5E 1E7 Canada
Position: VP Mech Design. *Employer:* MacLaren Engineers Inc/Lavalin *Education:* BASc/Mech Engg/Univ of Toronto; RMC/Mech Engg/Royal Military College. *Born:* 7/13/33. Natl pres of Canadian Soc for Mech Engg 1976; Fellow of the Engg Inst of Canada in 1978. From Hamilton, Ontario. Served with Royal Canadian AF. Plant engg work for Dow Chem in 1956. Mech design work from 1957 to present for MacLaren Engrs Inc/Lavalin, consulting engrs in environmental pollution control, energy conservation, & industrial engineering. As VP, Mech Design, responsible for

Aldworth, George A (Continued)
process/facility dev, estimating, & services for water, wastewater, & solid- waste treatment facilities. Special interest in pumps, drivers, & waterhammer protection; R&D on cold region shallow-buried water systems; design of heat- recovery incinerator facilities. *Society Aff:* CSME/EIC, ASME, AWWA, CIE.

Alers, George A
Home: 13108 Sandstone Pl NE, Albuquerque, NM 87111
Position: Senior Research Scientist *Employer:* New Mexico Engineering Research Institute *Education:* PhD/Phys/State Univ of IA; MS/Phys/State Univ of IA; BA/Phys/Rice Univ. *Born:* 11/22/28. Trained in solid state phys with a specialty in ultrasonics. Used ultrasonics for the study of metals under extreme conditions of stress, temperature, pressure & magnetic fields. Author of over sixty technical papers in the reviewed literature. Employed in basic res at Westinghouse, Ford Motor Co & Rockwell Intl. Currently developing new techniques for nondestructive testing. Pres, IEEE Sonics & Ultrasonics Group 1979-80. Technical Prog Chrmn IEEE Ultrasonics Symposium 1975 & 1979. Session organier &/or Chrmn for annual meetings of IEEE, ASNT, ASA & AIME. ASNT Achievement Award 1976. *Society Aff:* AAAS, IEEE, ASNT, ASA.

Alexander, Donald C
Home: Box 59, Chichester, NY 12416
Position: Technical Dir Cable Tech *Employer:* ITT *Education:* BS/EE/Worcester Poytechnic Inst *Born:* 09/01/21 *Society Aff:* IEEE, CIGRE

Alexander, John A
Home: 568 Miller Rd, Wyckoff, NJ 07481
Position: Dir of Technology *Employer:* Arwood Corp. *Education:* BS/Met Eng/Carnegie Inst of Technology; Grad Study/Met Eng/Carnegie Inst of Technology; Grad Study/Met Eng/MA Inst of Technology; -/Res Mgmt/George Washington Univ. *Born:* 2/3/31. Native of Swissvale, PA. Started as QC Engr for Inland Steel Co. Ordnance officer in Korea, 1954-56. Jr Engr for Westinghouse followed by Res Metallurgist position at Carnegie Tech. Seven yrs as Res Metallurgist studying dilute alloys and thin films at AMMRC, Watertown, MA. Joined venture capital R&D firm, General Technologies Corp, for metal matrix composites work, 1965-69. Progressed at TRW Inc from Mgr, Metal Matrix Composites Applications to Mgr Casting Technology to Dept Mgr for Matls Res, 1969-1980. Had respon for technical liaison with TRW foreign investment casting licensees. Joined PCC in 1980 to direct R&D, Proc Engr & Proc. Control Depts for 3 plant large superalloy & titanium investment casting operation. Assumed Director of Technology responsibility in 1985 for the seven plant investment casting company, Arwood Corp. Active in community affairs. Enjoy music, golf, tennis and bowling. *Society Aff:* AIME, ASM, ICI.

Alexander, Suraj M
Business: Dept of Indust Engrg, Louisville, KY 40292
Position: Prof *Employer:* Univ of Louisville *Education:* PhD/Indus Engrg & Oper Res (IE & OR)/VA Poly Inst & State Univ (VPI & SU); MS/IE & OR/VPI & SU; BS/Mech Engrg/Indian Inst of Tech (Madras, India) *Born:* 01/26/48 Born in Quilon, Kerala State, India. Naturalized as US Citizen in 1980. Employed as an Equip Engr, Corning Glass works, in the Sci Prods and the Consumer Prods Divs 1973-75. Instructor, Dept of IE & OR, VPI & SU, 1975-76. Asst Prof, Dept of Indust Engrg, Univ of Louisville, 1976-81. Assoc Prof 1981-87, Professor 1987-present. Specialist in the areas of mfg, planning and ctrl, facilities planning, quality control and reliability. Serv as an instructor, researcher and conslt in the above areas. Mbr of the Louisville Jaycees 1976-81. Mbr of honor socs Alpha Pi Mu, Phi Kappa Phi and Sigma Xi. Initiated as an eminent engr in Tau Beta Pi in 1980. Enjoy tennis and the outdoors. *Society Aff:* NSPE, IIE, SME, ASQC, AAAI, TIMS

Alexander, William D
Business: P. O. Box 770, Pawleys Island, SC 29585
Position: Consultant (Self-employed) *Education:* CE/Civ Engr/NC St Univ; BS/Chem/VMI *Born:* 06/20/11 Colonel USAF retired 1962. Awarded Legion of Merit, Air Force Commendation Medal. VP, Partner, Pres, Seelye Stevenson Value & Knecht, NY; Asst General Mgr, Metropolitan Atlanta Rapid Transit Authority. Responsible for design, construction and startup of new rail transit system. Reg Engr in eleven states. Distinguished Engrg Alumnus, NC State Univ 1976. Phi Kappa Phi, Tau Beta Pi. The Moles. *Society Aff:* NAE, ASCE, SAME

Alexander, William R
Business: 1205 W Columbia St, Springfield, OH 45504
Position: Document Control Administrator *Employer:* SPECO Div Kelsey-Hayes Co. *Born:* Aug 1917 Columbus, OH. BME & MS OH St Univ. Design Engr Gleason Works Rochester NY 1939-48; Inst Dept of Mech Engrg OH State Univ 1948-51; Asst Dir of Engg SPECO Div 1951; VP-Engrg 1965, specializing in des of precision tracking missile antenna mounts, helicopter and maring gear transmission sys & aircraft actuation sys. Reg PE OH; Assumed current responsibility May, 1978.

Alexandridis, George G
Business: One Aerial Way, Syosset, NY 11791
Position: Pres *Employer:* Lockwood, Kessler & Bartlett, Inc. *Education:* B.S./C.E./MIT; Grad Studies/City Planning/Columbia Univ; Transp/Brooklyn Polytech/Inst *Born:* 10/11/34 Joined DeLeuw Cather 1956, as a transportation engr. Joined LKB, Consulting Engrs 1961, Senior Engr. Dept Head 1972. VP/Engrg Design for solid waste, civil, hgwy, site development projects 1975-78. Became Pres/Chrmn in 1978, and Pres/Chief Operating Officer of parent co, Viatech, Inc. 1981. Registered PE; NY, NJ, MA, MD, DE. Past Pres Long Island Branch ASCE 1979. Stimulated continued development of LKB's aerial mapping, field survey & construction administration depts. Expanded LKB's involvement in solid waste, resource recovery and environmental consulting. Contributing author Hgwy Engrg section, McGraw Hill's 1972 Yearbook of Sciences and Technology. Numerous technical papers including solid waste treatment in European resource recovery plants, landfill technology, leachate generation. *Society Aff:* ACEC, ASCE, AWWA, ITE, NSPE

Alexeff, Igor
Business: Ferris Hall., Univ. of Tennessee, Knoxville, TN 37996-2100
Position: Prof EE. *Employer:* Univ of TN. *Education:* BA/Phys/Harvard; MS/Phys/WI; PhD/Phys/WI *Born:* 1/5/31. Published over 100 refereed papers on plasma and controlled thermonuclear fusion, consultant (worked in Switzerland, Japan, India, S Africa, Brazil), worked as a group leader in Oak Ridge Natl Lab & as a res engr at Westinghouse. Holds several patents. Reg PE (TN) Fellow, American Physical Society Fellow, Institute of Electical and Electronics Engineers, Has held numerous local and national offices in IEEE. IEEE Centennial Medal 1984, Univ. of Tenn. Chancellors Scholar 1984, IEEE Outstanding Engineer Award for the Southeast, 1987. *Society Aff:* IEEE, APS, ΣΞ, MENSA.

Alexiou, Arthur G
Business: 6010 Executive Blvd, Rockville, MD 20852
Position: Assoc Dir. *Employer:* Office of Sea Grant, NOAA. *Education:* MS/EE/Univ of NH; BS/EE/Univ of NH; Grad Sch/Metrorology/MIT; Grad Sch/Public Admin/American Univ. *Born:* 2/26/30. Native of Manchester, NH. 1951-53, Communications Construction Engg & Meteodrologist, US Air Force; 1953-65, Deputy Dir, Oceanographic Instrumentation Dev, US Naval Oceanographic office; Mgr. Spacecraft Oceangraphy Project, US Navy/NASA; 1967-70, Dir, Inst Support Program, National Sci Foundation; 1970- present, Assoc Dir Office of Sea Grants, NOAA. Responsible for dev of a network of Sea Grant Colleges via a grant program in education, research and advisory services. Specialty: Remote Sensing. DOC Silver Medal/ 1973. *Society Aff:* MTS, WSE

Alexis, Robert W
Business: 240 Michigan St, Lockport, NY 14094
Position: VP. *Employer:* Chem Design, Inc. *Education:* BS/CE/Syracuse Univ. *Born:* 3/19/36. Native of Falconer, NY. Did R & D on their sales of Molecular Sieves for Union Carbide Corp. Joined Chem Design, Inc as Proj Engr in 1964. Became VP and co-owner in 1972. Long time mbr of AIChE and Tau Beta Pi. Hobbies include boating, skiing and fishing. *Society Aff:* AIChE.

Alford, Cecil O
Business: School of Elec Engineer, Atlanta, GA 30332
Position: Prof *Employer:* GA Inst of Tech *Education:* PhD/EE/MS State Univ; MS/EE/GA Tech; B/EE/GA Tech *Born:* 09/28/33 Industrial experience with Radiation, Inc, (now Harris Corp) Melbourne, FL and Martin Marietta Corp, Orlando, FL. Teaching experience with Mississippi State Univ, Starkeville, MA, and Tennessee Tech Univ, Cookeville, TN prior to current position. Currently, Prof of Elecl Engrg at Georgia Tech teaching graduate and undergraduate courses in computer architecture, microprogramming and computer algorithms. Direction of the Computer Architecture Laboratory and the Robotics Laboratory at Georgia Tech. Research in Special Purpose Computer Architectures and the coordinated control of robots. *Society Aff:* IEEE

Alford, Jack L
Business: Harvey Mudd College, Claremont, CA 91711
Position: Prof of Engg. *Employer:* Harvey Mudd College. *Education:* PhD/Mech Engg/CA Inst of Tech; MS/Mech Engg/CA Inst of Tech; BS/Mech Engg/CA Inst of Tech. *Born:* 11/19/20. Served USNR, 1942-46, Lt; Engg, Northrop Aircraft, CA Inst of Tech, US Naval Ordnance Test Station, Technicolor Corp, Jet Propulsion Lab; with Harvey Mudd College 1959-; Kindelberger Prof, 1965-; Interest: mechanics, engg seismology, experiential education. Hobby: sailing. *Society Aff:* ASME, NSPE, EERI, ASEE, ΣΞ, ΤΒΠ.

Algazi, Vidal R
Business: Dept of Elec and Comp Engr, Univ of California, Davis, CA 95616
Position: Prof *Employer:* Univ of CA. *Education:* PhD/Elec Engr/MIT; MS/Elec Engr/MIT; BS/Radio Engr/Ecole Superievre *Born:* 2/25/30. Asst Prof of Elec Engg, MIT, 1963-1965; Univ of CA at Davis since 1965, Asst Prof 1965-1969, Assoc Prof 1969-1974. Prof 1974-present, Chrmn 1975-86; Dev Engr, 1954-1959 consultant: Lawrence Livermore Lab, Signal Processing, 1966 to present. Fulbright Travel Grant 1954. Ford Post Doctoral Fellow 1963-65, Specialization; Signal & Image Processing. *Society Aff:* IEEE, AAAS, ASP, ΣΞ, ΤΒΠ.

Alic, John A
Home: 1425 4th St SW, Apt A412, Washington, DC 20024
Position: Senior Associate *Employer:* Office of Tech Assessment *Education:* PhD/Engrg Materials/Univ of MD; MS/ME/Stanford Univ; B/ME/Cornell Univ *Born:* 11/24/41 Case Writer, Stanford Univ, 1965-66. Insructor, mechanical engrg, Univ of MD, 1966-72. Asst and assoc prof, mechanical engrg, Wichita State Univ, 1972-79. Proj Dir, Office of Tech Assessment, 1979-present. Profl interests: tech assessment; engrg design; fatigue and fracture of materials. Responsibilities at OTA have centered on studies of the intl competitiveness of US industries. Author or co-author of more than 65 tech publications. Wright Brothers Medal, Soc of Automotive Engrs, 1975. Dow Outstanding Young Faculty Award, American Soc for Engrg Education, 1977. *Society Aff:* SAE, ASM, AAAS, ASEE.

Alig, Frank S
Home: 8080 North Pennsylvania St, Indianapolis, IN 46240
Position: President. *Employer:* Frank S. Alig, Inc. *Education:* BS/Civil Eng/Purdue Univ. *Born:* 10/10/21. *Society Aff:* ASCE, NSPE, PCI.

Alizadeh, Mike M
Business: 2258 Meldon Pkwy, St Louis, MO 63146
Position: Pres *Employer:* Geotechnology, Inc. *Education:* BS/Civil Engrg/Tehran Univ; MS/Civil Engrg/Univ of MI *Born:* 8/24/32. Born in Iran in 1932, migrated to the US in 1958. Following graduation from the Univ of MI worked for Fruco and Assoc in St Louis for 10 yrs. Left Fruco in 1969 to assist in opening a branch office for McClelland Engrs in St Louis, MO. In 1975 opened a branch office for Shannon & Wilson, Inc in St Louis. Established the firm of Geotechnology, Inc in 1984 and assumed present position. *Society Aff:* ASCE, ACEC, NSPE, DFI, SAME, ISSMFE.

Alkire, Richard C
Business: 1209 West California St, Urbana, IL 61801
Position: Alumni of Chem Engrg Head of Chem. Engrg. Dept. *Employer:* Univ of IL *Education:* PhD/ChE/Univ of CA at Berkeley; MS/ChE/Univ of CA at Berkeley; BS/ChE/Lafayette Coll (Easton) *Born:* 04/19/41 Alkire joined the Univ of IL in 1969 where he established a research prog in electrochemical engrg. Potential and current distribution principles were applied to practical configurations involving bipolar processes, porous and resistive electrodes, electrodeposition & etching, electro-organic synthesis, plasma etching and localized corrosion. Carl Wagner Award (ECS, 1985); Professional Program Award (AIChE, 1985); Mbr, National Materials Advisory Board (1985-88); Pres, The Electrochemical Society (1985-86). *Society Aff:* AIChE, ECS, AAAS, NACE, ISE, ACS.

Allaire, Robert E
Home: 146 Boy St, Bristol, CT 06010
Position: Dir Corp Facilities Eng. *Employer:* Stanley Works-New Britain Conn; Retired *Born:* 1927 Bristol CT. Cert Plant Engr P.E. Reg State of Conn. No.10692. New Departure Div C M Plant Engg, Supervisor of Engrg 1952-58; The Stanley Works Ch Plan engr organized facilities engrg funct 1958-62; Mgr Engrg Maint incl facilities engrg, new plant plan & environmental engr maint of bldgs & equip approx 2.5 mil SF 1962-72; Mgr Plant Engrg admin energy environ security, maint, facilities engrg at main plant & corp hqtrs New Britain CT 1979-1981 Dir Plant Eng & Maint. 1981-Dir Corp Facilities Eng. - Resp. New Plant Planning, facilities Eng Corp Energy Communications 1972-. Offices held: Certification Ctte Div & Reg VP Am Inst of PE, P Pres Cap Region AIPE, AIPE Plant Eng. of the Year 1979 RH Morris Medal. Chmn YWCA Bldg Advisory Ctte, Conn. D.E.P. Hazardous Waste Mgt. Task Force. Dir Conn Land Trust Inc. Hobbies: hunting, fishing, gardening.

Allan, Donald S
Home: Wiswall Rd RFD, Newmarket, NH 03857
Position: Independent Conslt *Employer:* Self Employed *Education:* BS/Chem Engg/Northeastern Univ. *Born:* 7/25/21. Served as Communications Officer, Army Air Corps, 1943-46. Employed by Arthur D Little, Inc 1948-81. Res engr 1948-60. Unit Mgr Combustion Technology 1961-68, Safety and Fire Technology 1968-76. Section Mgr Safety and Engg Technology 1976-81. VP 1978. Conslt Safety and Engrg Tech 1981-present. Respon for evaluating technological risks and failures, including fires and explosions. *Society Aff:* AIChE, SSS, NFPA.

Allard, Kermit O
Business: PO Box 8000, Waterloo, IA 50704
Position: Division Engr. *Employer:* John Deere Product Engr. *Education:* Baccalaureate/Agri Engg/Univ of ME. *Born:* 4/14/36. Currently I am Manager, OEM Engineering for John Deere Tractor Works. The group I am responsible for designs and maintains non traditional products. I am a registered professional engineer in Iowa *6063. Society Aff:* SAE, ASAE.

Alldredge, Glenn E
Home: 5041 S Irvington Pl, Tulsa, OK 74135
Position: Exec VP. *Employer:* The Rainey Corp. *Born:* Sept 1926 Tulsa OK. Navy enlisted manaboard aircraft carrier WWII. BS Chem Engrg Univ of OK 1949. Entire prof career with heat transfer equipment mfg co's: Joseph A Coy Co; Yuba Heat Transfer; Disler Engrg. Jobs incl Estimating Sales Engr, Design Engr, Sales Mgr. Joined Rainey Corp 1966; assumed respon as Exec VP & Ch Operating Officer 1971. Chmn of Air Cooled Exchanger Sec 9th Ed Engrg Data Book of Gas Processors Suppliers Assn; Chmn of Air Cooled Manufac's Ctte API. Active in Baptist Church. Main leisure activity: golf.

Allen, C Michael
Business: Computer Science Dept, Charlotte, NC 28223
Position: Prof *Employer:* Univ of NC at Charlotte *Education:* PhD/Systems Engr/State Univ of NY at Buffalo; MS/Elec Engr/Carnegie Inst of Tech; BS/Elec Engr/Carnegie Inst of Tech. *Born:* 9/1/42. in New Castle, PA. Taught at SUNY/Buffalo from 1968-74. Engg Asst - CIA, Engr - Sylvania, Chief Engr - Len Lab from 1965-74. Joined UNCC in 1974 and developed the Digital Systems/Computer Engg Prog. Res interests are digital systems, multiple-valued logic, and medical/forensic imaging systems. Co-founder & Pres of DataSpan, Inc., 1984. Co-founder of the Intl Symposium on Multiple-Valued Logic; co-founder of the Charlotte Microcomputer Soc; Mbr of Bd of Dir Charlotte Sec, IEEE. Recipient of several res grants and hold numerous publications. *Society Aff:* IEEE, ACM, ASEE.

Allen, Charles W
Home: 1691 Filbert Ave, Chico, CA 95926
Position: Prof. *Employer:* CA State Univ. *Education:* PhD/Mech Engr/Univ of CA, Davis; MS/Mech Engr/Case Inst of Tech; BS/Mech Engr/Univ of London. *Born:* 7/24/32. Native of Newbury, England. Came to US in 1957. Engr, Lear Siegler Cleveland, OH 1957-62, specializing in rotating electromechanical machinery. Res Asst, Case Inst of Tech 1961, designed splint strictive for orthotic arm aid. Group Leader, bearing dev Aerojet-Gen Corp, Sacramento, CA, 1962-63. Assoc Prof, CA State Univ, Chico 1966-71, Prof 1971 to present. Head, Dept of Mech Engg 1976-79, 1982-84. Lubrication res for NASA 1968-74. Visiting Fellow. Univ of Leicester, England 1974. NSF URP Grant 1975. Publications in areas of engg education and bearing res. Research & Lecturing Univ of Guadalajara, Mexico 1986. *Society Aff:* ASME, ΣΞ

Allen, Charles W
Business: Materials Science Division, Argonne National Lab, Argonne, IL 60439
Position: Instrument Scientist, HVEM-Tandem Facility; Prof. Emeritus (Univ. of Notre Dame) *Employer:* Argonne National Laboratory *Education:* PhD/Metallurgy/Univ of Notre Dame; MS/Metallurgy/Univ of Notre Dame; BS/Physics/Univ of Notre Dame *Born:* 01/07/32 A univ engrg faculty mbr 1959-86, active in teaching and research in areas of materials sci and engrg including phase transformations, radiation damage, defects in magnetic materials and photovoltaics, and the electron microscopy of materials. Fulbright Research Scholar in 1970/71. Active in Boy Scouts of America. Summer research participant at Argonne Natl Laboratory in 1972, 74 and 81; principal scientist for high voltage electron microscopy at Argonne National Laboratory 1985-present. Scientist-in-Residence in 1982/83. *Society Aff:* AIME, ASM Intl, EMSA, MRS.

Allen, Chester L
Business: PO Box 1963, Harrisburg, PA 17105
Position: Partner, Project Mgr. *Employer:* Gannet Fleming Transportation Engrs. *Education:* BS/Civil Engr/WVU; MS/Civil Engr/WVU. *Born:* Aug 16, 1948 Buckhannon W VA. BS & MS W VA Univ. Empl Gannett Fleming Transportation Engrs, Inc July 1972- . Project Mgr for numerous projects in design and rehabilitation of bridges and buildings including structural integrity inspections. Has been engaged in final des computation and const drawings for num types bridges, buildings & underground subway & tunnel sys. Daniel W. Meade Prize for Students ASCE 1970 for paper: The Ethics of Teacher Evaluation. Enjoys sports, bible study and Christian Witnessing. *Society Aff:* ASCE.

Allen, Clarence R
Business: Seismological Lab, Pasadena, CA 91125
Position: Prof of Geology & Geophysics. *Employer:* CA Inst of Tech. *Born:* Feb 15, 1925. BS Physics Reed College 1949; PhD Geol & Geophys CA Inst of Tech 1954. USAAF 1943-46; B-29 Nav Far East. Prof of Geol & Geophys Caltech 1955-. Advisor to CA Dept Water Res, Natl Sci Fdn, US Geological Survey, Bureau Reclamation. Pres Geol Soc Am 1973-74; Pres Semiological Soc Am 1975-76; Chan CA Min & Geol Bd 1975. Mbr: Am Acad Arts & Sol 1974, Natl Acad Engg 1976. Natl Acad Sci 1976. Res interests: seimotectonics, active faults, seismicity, geological hazards, earthquake prediction.

Allen, David C
Business: 510 Cottage Grove Rd, Bloomfield, CT 06002
Position: Pres. *Employer:* Koton, Allen & Assoc, Inc. *Education:* BSME/Mech Engg/Univ of ME. *Born:* 4/22/29. Application Engg Baker Ice Mach Co; Appl Engg Worthington Corp; 1952-1974 VP, Chem Engr The Horton Co, Hartford, CT Mech Contractors, Pres Hartford Mech Contr Assn 1970-71; 1974-76 Prof Admin Hoen Constr Inc, Hartford CT, Gen Contr Mgrs. Since 1977 Pres Kofon, Allen & Assoc, Inc, Bloomfield, CT. Consulting Engrs doing Commercial, Institutional, industrial & Multiple Residential projs. Licensed PE in several states. Mbr adjunct faculty Univ of Hartford. Maj Profl policy is to seek practical rather than theoretical solution to engg questions. *Society Aff:* NSPE, ASHRAE.

Allen, Dell K
Business: 435 Clyde Engg Bldg, Provo, UT 84602
Position: Prof. *Employer:* Brigham Young Univ. *Education:* EdD/Ind Education/UT State Univ; MS/Ind Education/Brigham Young Univ; BS/Tool Engr/UT State Univ; Grad Work/Met Enrg/IL Inst of Tech. *Born:* 12/1/31. Prof of Mfg Tech; Dir of Comp-Aided Mfg Lab; active in Info Systems dev, Mfg Education, and Stds dev; mbr of Committee on Comp-Aided Mfg of the Natl Acad of Engrg; former mbr of the ASEE Bd of Dirs, SME Natl Education Committee; Chrmn CAM-I Stds Committee; Reg PE, Certified Mfg Engr; Recipient of SME Intl Education Award 1975. Educator, author, researcher, inventor, consultant; married to the former Wanda Israelsen; children: Curtis, Mark, Melody, Heidi, Paul, Don, Romay, and Robert. *Society Aff:* ASEE, SME, ASM.

Allen, Douglas L
Business: 11501 Northlake Dr, Cincinnati, OH 45242
Position: Exec. VP NDCC, President-NDCC Chemical Group. *Employer:* National Distillers & Chemical Corp. *Education:* BS/Chem Engg/Univ of Toronto. *Born:* 6/3/24. Toronto, Ontario, Canada. RCAF 1943-45. Univ of Toronto 1950. Shell Oil of Canada to 1952. US citizen 1957. With US Industrial Chem Co 1952-date, in IL, TX and NYC & Ohio. Contributed to Encyclopedia of Chemistry, Encyclopedia of Physics - Reinhold. Licensed PE in IL & TX. VP Manufacturing US Industrial Chemicals Co., Gen Mgr, Emery Chemicals VP National Hydrocarbon Inc. Dir, Quimic SA de CV, Mexico Dir National Hydrocarbon, Inc. Dir, National Helium Corp. Currently Executive Vice President, National Distillers & Chemical Corp., and President - NDCC Chemical Group, comprising Emery Chemicals and USI Chemicals. *Society Aff:* AIChE.

Allen, Edmund W
Business: 16 Exchange Place, Salt Lake City, UT 84111
Position: President *Employer:* E.W. Allen & Associates *Education:* BS/CE/Univ of UT *Born:* 11/18/27 Founded structural engrg firm in 1960. Structural engr for many projects in Utah & surrounding states including the Metropolitan Hall of Justice Bldg., SLC; the Skaggs Pharmany Bldg., The Behaviorial Scis Bldg., Art & Architecture Bldg., & 1981 Medical Cntr Bldg. Addition at the U of U; the Chase Fine Arts Cntr & Spectrum at Utah State Univ.; the Valley Fair Mall, SLC; the J.C. Penney Office Bldg., SLC; Snowbird Lift Facilities, Commercial Plaza, Hotel & Mountain Restaurant near Salt Lake City; The Eston-Kenway tower, SLC; Governor's Plaza Condominiums, SLC; & The Salt Palace Exhibition Hall Addition, Salt Lake City. Pres, Consulting Engrs Council of Utah 1979-1980. Pres, Structural Engrs Assoc of Utah, 1981-1982, Natl Dir, Consulting Engrs Council of Utha, 1984-85. Awarded "Outstanding Utah Engr of 1985" by the Utah Engineers Council. Awarded a merit

Allen, Edmund W (Continued)
award for a bridge over the Green River So by ACIS in 1984. *Society Aff:* ASCE, ACI, AISC, ACEC, TMS, EERI

Allen, Gerald B
Home: 18312 Springtime Lane, Huntington Beach, CA 92646
Position: VP Chemicals. *Employer:* Wilson & Geo Meyer & Co. *Education:* BS/ChE/MO Sch of Mines & Met. *Born:* 10/31/37. Spent youth in St Louis, MO attended MO sch of Mines, Rolla. Served in the US Army Corp of Engr as a First Lt. Have been employed as a Service Engr for Dow Industrial Service, Chemical Engr for C K Williams Co, Process Engineer for Hercules, Inc, and vairous mktg and mgmt positions with Wilson & Geo Meyer & Co. Active as a Boy Scout Leader in Huntington Beach, CA and a mbr of the Huntington Beach First Christian Church. *Society Aff:* AIChE, LASCT, SCPCA, SCIFT, ΣΤΤ

Allen, Harry S
Business: 15 Stelton Rd, Piscataway, NJ 08854
Position: Exec V P/Consulting Engr. *Employer:* Kupper Associates *Education:* MS/CE/Newark College of Engg; BS/EE/Lehigh Univ; BA/Am Hist/Amherst College. *Born:* 1/1/33. Lancaster, PA., USAF, 1954-1956. Directs the operations of Kupper Associates and the design of water and wastewater treatment plants, water distribution and sewage collection systems, pumping stations and control and instrumentation systems. Mbr of IEEE, AWWA, WPCF, ASCE; Diplomate - Am Acad of Environmental Engrs; Licensed PE: NY, NJ, DE, PA, MD, LA, CT, TX; Licensed Water and Wastewater Treatment Plant Operator: NJ; Chrmn, Legislative Committee, Consulting Engrs Council of NJ. Mbr of Tau Beta Pi - Engg Honorary Fraternity, Mbr of Eta Kappa Nu - Elec Engr Honorary Fraternity, Honorary Mbr of Chi Epsilon Fraternity. *Society Aff:* WPCF, ASCE, AWWA, IEEE, ACEC, ISA, NSPE, NSSPE, AAA.

Allen, Harvey G
Home: 3190 Pinecrest Dr, Murrysville, PA 15668
Position: Advisory Engr. *Employer:* Westinghouse Electric Corp. *Born:* May 1922. BSME Penn State Univ 1943; addl studies Carnegie-Mellon Univ grad sch. Engg officer USN 1943-46. Centrifugal pump hydraulic engr Ingersoll- Rand Co 1947-52; Westinghouse Electric Corp specializing in hydraulic des & dev of centrifugal pumps 1952-. Principal hydraulic designer of main coolant pumps used in Westinghouse navy & commercial nuc power plants, & of boiler circulating pumps for fossil fuel service. Respon for hydraulic des and dev of Liquid Metal Fast Breeder Reactor pumps used in the FFTF. 6 patents in hydraulic mach. Fellow ASME 1976.

Allen, Herbert
Business: PO Box 1212, Houston, TX 77251
Position: Consultant. *Employer:* Cameron Iron Works, Inc. *Education:* BS/ME/Rice Univ. *Born:* 5/2/07. Ratcliff, TX. Engaged in res 1929-31. With Cameron Iron Works, Inc, mfg of oilfield equip and ball valves and the melting, forging and extrusion of high- strength metals, since 1931, serving as chief engr, vp and gen mgr, pres and chrmn, retiring in 1977. Continues to serve as dir and consultant. Also advisory dir of TX Commerce Bank. Retired from Bd of Big Three Indus. Former chrmn of bd of trustees, Rice Univ. Elected to Natl Academy of Engg 1979; elevated from Fellow to Honorary Mbr ASME 1979. Tau Beta Pi. Holder of 150 patents in field. Episcopalian. *Society Aff:* AIME, ASME, NAE, API.

Allen, James
Business: 5700 Old Seward Hgwy, Suite 206, Anchorage, AK 99502
Position: Principal *Employer:* Franklin & Allen *Education:* BS/CE/Univ of IL *Born:* 04/18/40 Alton, IL. Postgraduate work Univ of Alaska. Design engr with VanDoren Hazard Stalling Svhnacke consltg firm, specialized in structures. Moved to Anchorage, AK, 1970. Established own consltg engrg firm 1974, providing consltg engrg services structures, Arctic structures, and seismic design. Pres, Anchorage Branch ASCE 1979, Pres Elect Alaska Soc of PE, 1981. Secy Alaska Soc of PE, 1980. Enjoys hunting, classical music, skiing and reading. *Society Aff:* ASCE, NSPE, EERI, ACI, AITC

Allen, James B
Home: Rd 7 Box 24, Wash, PA 15301
Position: Pres. Retired. *Education:* BS/ChE/MIT *Born:* 4/15/14. Early education in Kansas City, MO, employed by Dupont, Manhattan Proj, Koppers, Pitts Coke and Chem from 1948 to 1961. Pres of Abhem Inc and Custom Chemicals Inc 1961-86. Retired 1986. *Society Aff:* ACS, AIChE, AAAS.

Allen, James D
Home: 1904 Bromley Estates, Pine Hill, NJ 08021
Position: Mgr *Employer:* The FPE Group *Education:* BSME/Aero Engr/Syracuse Univ. *Born:* 08/18/41 Native of Syracuse NY. Field Engr and dist Supervising Engr for Indust Risk Insurers (formerly FIA) Syracuse NY, Richmond VA 1963-74; Consltg Engr Johnson & Higgins Insurance brokers Richmond VA, Charlotte NC 1974-1978; Dir Fire Protection Engrg Insurance Co of North Amer Phila, PA 1978-1982 - specialize in HPR indust fire protection power plants; 1982-present mgr the FPE Group Phil office. Reg PE VA, PA, NJ, MD. Past Pres Phila chap SFPE. Dir SFPE Intl; NSPE Mbr. Chrmn SFPE - PE Registration Ctte, Chrmn PEAC ctte of NCEE. NCEE Coun Record holder. Enjoys sailing, soaring, shotgunning, canoeing, tennis, biking, intl travel. *Society Aff:* SFPE

Allen, James L
Home: 10502 Lacera Dr, Tampa, FL 33618
Position: Prof. *Employer:* Univ of S FL. *Education:* PhD/EE/GA Inst of Tech; MSEE/EE/GA Inst of Tech; BEE/EE/GA Inst of Tech. *Born:* 9/25/36. in Graceville, FL. Present position - Full Prof, Elec Engg, Univ of S FL (Tampa). Career history includes positions with GA Tech Engg Experiment Station, CO State Univ, and Sperry Microwave Electronics. Current technical areas of interest include microwave GAsFET Amplifiers, microcomputer applications & computer-aided circuit design. Fellow of IEEE (1977). Chief Editor IEEE-MTT Transactions (1977-79). IEEE-MTT Natl Administrative Committee (1974-present), Secy (1973). Technical chrmn, 1979 Intl Microwave symposium. FL West Coast "Engineer of Year" (1979) (Sponsored by 18 Engg Socs). FL West Coast Sigma Xi "Outstanding Research Scholar (1981)–. *Society Aff:* IEEE, ISHM.

Allen, John L
Business: Ste 1120, 1901 N. Ft. Myer Dr, Arlington, VA 22209
Position: Pres. *Employer:* John L. Allen Assoc., Ltd. *Education:* PhD/Communications Biophysics/MIT; SM/EE/MIT; BS/Engg Sci/Penn State. *Born:* 6/13/31. HRB Singer, Inc; Jr engr, 1954-1958, radar and infrared systems. MIT Lincoln Lab, 1958-1971; antennas and radar systems res. Res on invertebrate vision at MIT, 1966-1968. Navl Res Lab, Assoc Dir of Res for Electronics, 1971- 1974. Dept of Defense, Deputy Dir of Defense Res and Engg (Res and Advanced Tech), 1974-1977. Gen Res Corp; VP, 1977-1978, Pres, 1978-1981: private consultant, 1981 to present. Fellow, IEEE, past IEEE Natl Lecturer on Microwave Theory and Techniques, mbr Tau Beta Pi. Awarded the Navy Distinguished Civilian Service Award for his work at NRL. *Society Aff:* IEEE.

Allen, Keith J
Home: 347 Wilshire Dr N, Salem, OR 97303
Position: Civ Engr. Retired. *Employer:* Soil Conservation Service. *Education:* BS/Agri Engr/OR State Univ. *Born:* 12/16/25. I am a native Oregonian. I started my engr career with USDA, Soil Conservation Service, in 1954 working in field locations throughout OR until 1967, work involved irrigation water mgmt, stream channel improvement, and drainage. In 1967 I became a part of the state SCS staff. I was field engr for the Willamette River Basin Survey, 1967-69. I was engr for Oregon River Basins, 1969-74. I was planning engr for Oregon Watershed Planning Staff, 1974-81. As engr, 1967-81, I was responsible for river basin studies and watershed project planning throughout OR. Irrigation facilities were as large as several

Allen, Keith J (Continued)
thousand acres, multipurpose dams as high as 150 feet with storage capacities to 25,000 acre-ft. Retired from SCS Nov 1981, available as a consultant. I am a mbr Am Soc of Agri Engrs. I was OR Sec chrmn for 1976. Im a reg Engr and Land Surveyor in OR. Retired, Nov. 1981. *Society Aff:* ASAE.

Allen, Matthew A
Business: Stanford University (P. O. Box 4349), Stanford, CA 94305
Position: Department Head *Employer:* Stanford Linear Accelerator Center *Education:* Ph.D./Physics/Stanford University; B.Sc./Physics/University of Edinburgh *Born:* 04/27/30 Native of Edinburgh, Scotland. U.S. Citizen, 1961. Graduate student/research asst, Microwave Lab, Stanford Univ. Developed theory of coupled cavities for high-power traveling-wave tubes, 1955-59. Mgr of Tube Research, Microwave Assoc, 1961-65. At SLAC since 1965. Responsible for designing & building RF systems for all storage rings at SLAC. Head of Accelerator Physics Dept., 1982-84. Since 1984, Head of Klystron/Microwave Engrg Dept. Mbr of Administrative Cttee and Chairman of Particle Accelerator Science & Technology Technical Cttee, Nuclear and Plasma Sciences Society, IEEE, 1978-86. Elected Fellow of the IEEE, 1987 Active in local politics as Mayor and City Councilman. *Society Aff:* IEEE, APS, ΣΞ

Allen, Merton
Home: 2307 Dean St, Schenectady, NY 12309
Position: Pres. *Employer:* Merton Allen Associates. *Education:* M/ChE/Villanova Univ; B/ChE/NYU. *Born:* 9/30/29. Served as 1st Lt with Army Corps of Engrs, 1951-1953. Chem Dev Engr with IRC, Inc, 1953-1956. Chem Engr and Proj Mgr with GE R&D Ctr, 1956-1969. Founding charter mbr and div mgr and corporate mgr of mtls with Environment One Corp, 1969-1975. President of Alprax Enterprises Limited 1975-1980. Manager of reports and data management for Mechanical Technology Incorporated 1977-1981. Inventor of phytarium controlled environment plant growth chamber. Total of 10 US patents, numerous foreign patents and author of over 50 published papers and presentations. Abstractor for chem abstracts services of ACS. Recipient GE Silver Award to Inventors. President of Merton Allen Associates, a technical information consulting firm. Also president of The InfoTeam, publisher of "Innovator's Digest" news service. *Society Aff:* AIChE, ACS, AAAS.

Allen, Robert L
Home: 3005 Dove St, Rolling Meadows, IL 60008 *Employer:* Now Retired *Education:* BSEE/Power Maj/Univ of ND. *Born:* 6/16/23. Zone Sales Activities are to assist sales personnel on product tech, new product needs and lighting applications. Worked in power line construction 1949- 1951; for a consulting engg co 1951-1953 and 1953 to present with the Westinghouse Elec Corp. Past Chrmn of Chicago Lighting Club, Mbr of IES Chicago Roadway Lighting Forum, Past Chrmn of IES Roadway Lighting Forum Committee (Natl); Past Regional VP & Natl Dir of IES as well as Past Chicago Sec IES Chrmn. Recipient of Chicago Sec IES Merit Award & Chicago (City) Industry "Leadership" Award. Retired Lt Col USAFR. Currently active in the IES and on the IES Roadway Lighting Cttee. *Society Aff:* IES, IEEE.

Allen, Robert S
Business: 9505 W Central, Suite 101, Wichita, KS 67212
Position: Owner. *Employer:* Robert S. Allen Assoc, Inc. *Education:* BS/CE/Worcester Poly Inst. *Born:* 11/12/31. Warwick R 1. Worked for Am Cyanamid, Dewey & Almy Div of W R Grace, Thompson Chems, & Neutron Prods Inc; incl work on SBR, PVC & radiation polymerization of vinyl acetate; Mgr New Projs for Conoco Plastics responsible for erection of 1st large reactor PVC plant in US 1968; VP Chem Proj Worldwide Const Services Inc 1972; started own co for design of refinery chem, & plastics facilities 1976. Area of speciality incl PVC Resin, compounds, plasticizers, & assoc areas. *Society Aff:* AIChE, ISA.

Allen, Roy L
Business: 4117 Boca Bay Dr, Dallas, TX 75244
Position: Independent Contractor *Employer:* Computer Information Svcs. *Education:* BS/Ind Engg/OK State Univ. *Born:* 1/8/26. Partner, investor, developer of computer info sys for indus and govt; Bd of Dir, Computer Co; writer and lecturer on robotics, factory and office automation, and decision support sys. Ind experience includes twenty yrs in Ind Engg & Data Processing mgt. Prog Mgr, Exec Advisor - Business Systems, Asst to the VP & Mgr - Mgt Systems with Rockwell Intl. Served 12 years with City of Dallas as Dir, Dept of Data Services. Past Natl Pres, IIE. Fellow, IIE. Mbr, Tau Beta Pi. Reg PE (TX). Mbr, Am Mgt Assn eight years. Mbr, Ind Engg Advisory Committee, OK State Univ. A contributing author, third edition "Handbook of Industrial Engineering–. Served as Chrmn and Lectr at Am Mgt Assn seminars & courses. Mbr of the Editoral Advisory Board of the international journal, "Computers and Industrial Engineering–. *Society Aff:* IIE.

Allen, William H, Sr
Business: P.O. Box 368, 650 Park Ave, King of Prussia, PA 19406
Position: Sr. V.P. *Employer:* Gannett Fleming Affiliates. *Born:* in Mobile AL. Civ Engrg Univ of SC 1941; refresher courses Carnegie Inst of Tech. Proj Engr on airports, US Engrg Dept SC, MD, WA DC; with Gannett Fleming Affiliates Engrs 1944-: Sr. VP & Eastern Regional Mgr. Phila & NJ, Exec Offices Responsible for admin & supervision of field & office personnel engaged in design of large hgwy & rail trans projs in principally urban areas of Eastern & Southern USA & W Africa Officer & on cttes of 10 engrg & civic orgs. Reg Engr in PA, NY, CT, NJ. Natl Council State Bds of Engrg Exam. Reg. Landscape Arch. *Society Aff:* ASCE, NSPE, ASHE, ITE, ARTBA.

Allen, William S
Home: 521 Moran St, Bryan, TX 77801
Position: Owner. *Employer:* W S Allen & Assocs. *Education:* MS/Agricultural Engg/TX A&M Univ; BS/Agricultural Engg/Univ of TN; Grad- /IND College of Armed Forces. *Born:* 11/23/19. Native Wilson County, TN. Served in India-China during WWII as infantry and logistics officer. Became Extension Engr for TX A&M Univ in 1947; responsible for housing structures, storage and processing programs. Served in Korean action 1950-52. Returned to TX A&M Extension Service as Extension Agricultural Engr; became project leader with five engrs in sec. Was author of over 60 publications for TX A&M. Retired 1975. Pres, mfg firm, for automatic feed processing equip; owner consulting firm (12 million $ under construction currently.) Retired colonel, Engr Corps, USAR. Former officer grain storage, construction, and land development firms. Mbr: ROA, Masonic bodies, CSI, SBCCI, Post Tensioning Inst, ACI. *Society Aff:* ASAE, NSPE, SAME, IIE, TSA.

Allentuch, Arnold
Home: 323 Martin Luther King, Jr. Blvd, Newark, NJ 07102
Position: Assoc VP for Academic Affairs. *Employer:* NJ Inst of Tech. *Education:* PhD/Appl Mechanics/Poly Inst of Brooklyn; MS/Appl Mechanics/Cornell Univ; BS/ME/Worcester Poly Inst. *Born:* 12/4/30. Born and raised in Worcester, MA. Served in US Navy. Res Assoc at Poly Inst of Brooklyn 1962/63. Preceptor, Columbia Univ 1963/64. Asst Prof of Civ Engg, the Cooper Union, 1964-1966. Prof and Assoc. VP for Academic Affairs, NJ Inst of Tech, 1966- present. Res for Office of Naval Res on stiffened shells, automated design, 1955-1973. Member of the Board of Trustees, NJ Marine Science Consortium. Play the violin. Widely read and knowledgeable about music and the arts. Member of the Board of Overseers, Foundation at New Jersey Institute of Technology (NJIT). *Society Aff:* ASME, ASEE, ΣΞ, AAAS.

Alley, E Roberts
Business: Peach Ct, PO Box 224, Brentwood, TN 37027
Position: Pres. *Employer:* E Roberts Alley & Assoc. *Education:* MS/Env Engg/Vanderbilt Univ; BE/CE/Vanderbilt Univ. *Born:* 9/17/38. in Greenville, MS; US Army 1960-61; Oman Construction Co, Proj Engr, Memphis, TN, 1961-62; J R Wauford & Co, Inc, Sr Engr, Donelson, TN, 1962-1968; Alley & Brown, Inc, Pres,

Alley, E Roberts (Continued)
Nashville, TN, 1968-1971; Hart-Freeland-Roberts, Inc, VP, Nashville, TN, (merged with Alley & Brown), 1971-1974; E Roberts Alley & Assoc, Inc, Pres, Brentwood, TN; 1974-present; Deacon, First Presbyterian Church; Scoutmaster; Past Pres, Brentwood Rotary Club; Mbr, Nashville Chamber of Commerce; Reg Engr, TN, KY, AL, MS, GA, FL, SC, NC, VA, MO, AR, TX, LA. *Society Aff:* NSPE, ASCE, ACEC, AWWA, WPCF.

Allgoever, George L
Home: 4224 Bower Ln, Winston-Salem, NC 27104
Position: Engr. *Employer:* Southern Bell T & T Co. *Born:* May 20, 1946 W Reading PA. BE Elec Engg Villanova Univ; working towards MS at NC A&T Univ. Mbr Eta Kappa Nu at Villanova. Reg PE NC. Plan engr in defense activities div Western Elec Co 1968-73; transferred to Southern Bell 1973 as outside plant engr spec in engg local telecommunications facilities.

Alligood, Bob
Business: 4019 Boulevard Ctr Dr, Jacksonville, FL 32207
Position: Sr VP. *Employer:* Reynolds Smith & Hills Inc. *Education:* BIE/Univ of Fl Coll of Engg *Born:* Dec 5 1932 Mitchell County GA. Grad Moultrie HS; Mbr IIE, P Pres FL Chap; FL Engrg Soc Legis Ctte; mbr Univ of FL Alumni Assn, FL Blue Key Hon Leadership Frat, FL Council of 100. State Rep in State Legis from Orange Cty for 3 1/2 yrs. Selected Outstanding Young Men of Amer 1966; selected Engr of Yr 1966 by FES Central FL Chap; Life Mbr Phi Kappa Phi Hon Soc & Tau Beta Pi.

Allison, James M
Business: Driftmier Engineering Ctr, Athens, GA 30602
Position: Assoc Prof *Employer:* Univ of GA *Education:* PhD/Agric Engrg/Cornell; MS/Agric Engrg/Clemson; BS/Agri Engrg/Clemson *Born:* 06/06/40 Native of Brevard, NC. Prof career at Univ of Georgia. One year as Extension Engr with remainder of time in teaching and research. Teaching responsibilities in the area of Electrical Circuit Analysis and Electronics; Research effort devoted to control systems for use on solar energy application and microprocessor controls in Agriculture. Author or Co-author of some 35 research publications and reports. Past Chairman Agricultural Engg Div of ASEE. Committee mbr of four ASAE committees. Held various offices in the GA Section ASAE. Enjoys gardening, wood working, fishing and working with Boy Scouts. *Society Aff:* ASAE, ASEE, ΣΞ, ΓΣΔ

Allison, William W
Business: PO Box 1498 Lancaster Ave, Reading, PA 19603
Position: Loss Control. *Employer:* Gilbert Commonwealth Engrs. *Education:* -/Ind Engg/Univ Pittsburgh. *Born:* 4/30/14. Fellow ASSE. Forty published papers include safety res on zirconium, pressure systems, and unique profitable risk control concept of the vital few high potential loss risks. Now loss control consultant with Gilbert Commonwealth Engrs after 13 yrs at Sandia Labs. Developed and managed pioneering loss control at Westinghouse Atomic Power following work as chem div safety and health dir. Serve on Utilities, NFPA and ANSI Committees. Lecture at univs and confs. Alumnus Pitt and Bethany. PE and CSP with special interest in advances available in health and safety through the significant few high potential risk control methodology. *Society Aff:* AIHA, AIChE, ASSE, SSS.

Allmendinger, Paul F
Business: 345 East 47 St, New York, NY 10017
Position: Exec Dir, retired *Employer:* Amer Soc of Mech Engrs. *Education:* BS/Engg/US Naval Acad. *Born:* 03/02/22. A career predominately engrg mgmt in industry with significant interest and activity in the profession of engrg. A native of Moline, IL, with an engrg degree from the US Naval Acad, 1944. Increasing administrative responsibilities in assoc mgmt, leading to Exec Dir, in 1982, of ASME, intl prof engrg soc., ret. 1987. Active in engrg socs, serving on major bds and cttes. Held leadership positions in natl standards programs through ANSI; involved in engrg accreditation by serving on Engrg Accreditation Ctte, leading to Natl Pres of Engrs' Council for Prof Dev in 1977-78. *Society Aff:* ASME, ASEE, SAE, SME, IMechE.

Alluzzo, Gasper
1148 Nugent Avenue, Bayshore, NY 11706
Position: Project Engr *Employer:* Cosentini Assocs *Education:* MS/Physics Educ/CW Post Ctr of LIU; BS/CE/Polytechnic Inst of Brooklyn *Born:* 05/23/47 Native of Brooklyn, NY. Graduated Engrg Sch in 1970. Worked for NYS Dept of Transportation as Field Inspector with NYC Bd of Water Supply. Worked on the planning and construction of a water distribution system that would deliver one billion gallons of water a day. With Bowe Walsh & Assocs was Proj Engr and promoted to Construction Proj Supvr. Responsible for over 86 million dollars of construction. With Richard Browne & Assocs, Headed NYC Office Construction Management Dept. Presently democratic candidate for Suffolk County Legislator in the 6th District. *Society Aff:* NSPE

Almenas, Kazys K
Home: 8806 49th Ave, College Park, MD 20740
Position: Assoc Prof. *Employer:* Univ of MD. *Education:* PhD/Nucl Engg/Warsaw Univ; BS/CE/Univ of NB. *Born:* 4/11/35. Resident in MD. Staff mbr at ANL Reactor Engg Div up to 1965. Joined Univ of MD faculty in 1969. Early publications in reactor core neutronics, current res interest in light water reactor safety field. Primarily containment design. Consultant to Bechtel Corp, ANL, ACDA, NRC and IVIC (Venezuela). *Society Aff:* ANS, ASME.

Almgren, Louis E
Business: Sierra Consultants International, 145 Natoma St, San Francisco, CA 94105
Position: Pres. *Employer:* Sierra Consultants International *Education:* MS/Civ Engg/Stanford Univ; BA/Gen Engg/San Diego State College. *Born:* 4/2/26. Grad from San Diego State College in 1947. Grad from Stanford Univ in 1949. Accepted for mbrship in SFPE in 1950. Became reg Civ Engr in CA in 1966. Became principal & vp of Gage-Babcock & Assoc in 1967. Served as pres of the N CA Chapter of SFPE from 1971-72. Founded & became Pres of the FPE Group in 1974. Served on the Bd of Dirs of SFPE 1973-1979. Became reg Fire Protection Engr in CA in 1975. Founded and became President of Sierra Consultants International in 1980. *Society Aff:* SFPE, ASCE, NFPA, API.

Almond, John H
Home: 882 Timberlain, Jackson, MS 39211
Position: Mgr *Employer:* Allen & Hoshall, Inc. *Education:* Bach of Science/Civil Engg/MS State Univ. *Born:* 1/5/47. Native of Robinsonville, MS. 1st Lt, US Army Ordinance Corps, honorable discharge. Design engr with Allen & Hoshall, Inc 1972-74. Promoted to proj mgr in 1974. Presently in charge of full service office established for Allen & Hoshall, Inc in Jackson, MS in 1976. Reg PE in State of MS, AL, AR, LA and TN, Okla. Control Association VP 1978, Pres 1979. Enjoy classicals music and hunting. *Society Aff:* ASCE, SAME, WPCF, NSPE.

Almquist, Wallace E
Business: 1700 MacCorkle Ave, Charleston, WV 25314
Position: VP. *Employer:* Columbia Gas Transmission Corp. *Education:* BS/CE/PA State Univ. *Born:* 12/23/22. Since grad from Penn State, I have worked in res, plant operations, process engg and design. The majority of my career has been devoted to the gas utility industry, where I have performed in all engg and operating functions, with maj responsibility in each area. I serve on the American Gas Assn Managing Committee, Pipeline Res Committee and the Interstate Natural Gas Assn Construction and Operations Committee. In recognition of work on Stress Corrosion Cracking of Pipelines, I spoke on this subj at the 1979 World Gas Conf. *Society Aff:* AGA.

Alpern, Milton
Business: 2635 Pettit Ave, Bellmore, NY 11710
Position: Principal. *Employer:* Alpern & Soifer Cons Engrs. *Education:* MSCE/Struc/Columbia Univ; BCE/CE/The Cooper Union *Born:* 6/25/25 MSCE Columbia Univ; BCE Cooper Union. Served Army Corps of Engrs 1945-47; Engg empl: Edo Aircraft Co 1945, Frederick Snare Corp 1947-50, others part-time while teaching: Instr, Asst Prof Civ Engg Cooper Union NY 1948-60; Prin Milton Alpern, PE, Cons Engr 1960-71; Ptnr Alpern & Soifer Cons Engrs 1971-. P Pres: NYSSPE, L I Chapt ACEC, Nassau Chapt NYSSPE, Met Sec ASCE Jr Branch; Past Dir: NSPE, L I Chap CSI, ABET; Mbr many prof & tech soc's and Cttes; PE lic in 13 states & Natl Cert. Welding Awards 1961, 1969. Engr of Yr PE & ASCE Nassau County NY. Married, 2 grown children. Interested in tech antiques, history, etc; Member, (Chair 1984-86) N.Y.S. Board for Engg. and Land Surveying 1978-; Adjunct Prof Structures, N.Y.I.T., 1978-1979 & Pratt Inst. 1986-86; Member Engg Manpower Comm. 1978-1980; Hon Mbr: Chi Epsilon, Tau Beta Pi; Author: Various Tech and Prof articles. *Society Aff:* SAME, NCEE, AISC, CSI, ASCE, NSPE, ACEC, ACI, ASTM, AWS, AAFS, IABSE.

Alpert, Sumner
Home: 15833 Castlewoods Dr, Sherman Oaks, CA 91403
Position: Director, Strategic Planning *Employer:* HR Textron *Education:* MS/ME/Case Western Reserve; BS/ME/Northeastern Univ; MBA/Mgt/San Diego State College. *Born:* 9/24/22. Native of Boston, MA. Aero Res Scientist, Lewis Lab NASA 1944-51. Chief Technical Engr and Mgr Aerospace and Ind Products for Solar Turbines Intl, 1951- 1969. VP Marketing & VP engr for Aeronca, Inc, 1969-1972. VP Turbo Products, the Marquardt Co, 1973-1976. VP, Rotary Compressor Div for Chicago Pneumatic Tool Co, 1977-1980. Fellow ASME. Project Management Professional. *Society Aff:* ASME, AIAA, SAE.

Al-Saadoon, Faleh T
Home: 906 W Park Ave, Morgantown, WV 26505
Position: Assoc Prof *Employer:* WV Univ *Education:* Ph.D./ChE/Univ of Pittsburgh; M.S./Petro Engr/WV Univ; B.S./Petro Engr/WV Univ; GCE/Sci/S. Herts Coll of Further Edu, Eng *Born:* 1/11/43 Faleh T. Al-Saadoon is an Assoc Prof of Petroleum Engrg at WV Univ. Prior to his present position, he was an assistant prof at the Coll of Petroleum and Mining Engrg, Al-Fateh Univ, Tripoli, Libya. He also worked for four years as the Head of the Petroleum Engrg Dept, Coll of Engrg, Univ of Baghdad, Baghdad, Iraq. He was on the Bd of Dirs of the Iraqi Society of Engrs for two consecutive years. Dr. Al-Saadoon holds a Ph.D. degree from the Univ of Pittsburgh. He is a Registered Professional Petroleum Engr in the st of WV. *Society Aff:* SPE of AIME, SPWLA, AIChE, AACE

Alsberg, Dietrich A
Business: 123 Pine St, Lincroft, NJ 07738
Position: Retired *Employer:* Bell Labs. *Education:* BS/EE/Stuttgart Tech Univ. *Born:* 6/5/17. Native of Kassel, Germany; dev engr with several cos in OH, 1940-1942; US Army, Ordnance Corps, 1942-1945; Bell Labs, 1945-82. Consultant 1982- Precision high- frequency & microwave transmission measurements, early transistor characterization, Thor & Titan ballistic missile radioinertial guidance systems for military & space missions, Nike-X, Sentinel, & Safeguard phased-array radars, nuclear hardening, Electromagnetic effects of nuclear weapons, millimeter waveguide transmission, microwave radio remote surveillance systems and satellite communications systems. Life-Fellow IEEE. Nineteen patents & numerous publications contribution to books. Past mbr Bd of Ed, Chrmn Environmental Commission. *Society Aff:* IEEE.

Alsing, Allen A
Business: 20 E Main St, Ashland, OR 97520
Position: Dir of Public Works. *Employer:* City of Ashland. *Education:* BS/CE/OR State Univ; AA/CE/College of Marin, Kentfield, CA. *Born:* 8/13/27. Has been with the City of Ashland as Water Supt, City Engr, and/or Dir of Public Works for 36 yrs. Areas of responsibility include managing the public Works Dept including Divs of Water, Sewer, Street, Engg, Equipment Pool & Airport. Past Chrmn, Pacific Northwest Section, Am Water Works Assn; Past Pres. OR Chapter, Am. Public Works Assn; Top Ten Public Works Leader, Am. Public Works Assn. 1985; Past Pres. Rogue Valley Chapter PE of OR. Fuller Awardee, Am. Water Works Assn, 1976; International Director, Am. Water Works Assn, 1980-1983; Engr. of the Year, Rogue Valley Chapter, PE of OR, 1971; PE, State of OR. Profl. Land Surveyor, State of OR. Diplomate, Am. Academy of Environmental Engineers. *Society Aff:* NSPE, PEO, AWWA, APWA, AAEE.

Alspaugh, Dale W
Business: Westville, IN 46391
Position: Chancellor *Employer:* Purdue Univ North Central Campus *Education:* PhD/Engg Sci/Purdue Univ; MS/Engg Sci/Purdue Univ; ME/Mech Engg/ Univ of Cincinnati. *Born:* 5/25/32. in Dayton OH. Employed as a proj engg res and future products engg, Frigidaire Div GMC 1955-56. Received General Motors Fellowship for graduate study 1956-57. Military svc, 1st Lt US Army - Interior Ballistics Lab, Aberdeen Proving Ground MD, 1958-59. Instructor, Purdue Univ 1959-64. Became Asst Prof of Aeronautics & Astronautics in 1964, Assoc Prof in 1968, Prof in 1981. Consultant to various military and industrial laboratories. Areas of specialization: Optimization, systems engg, flight and orbit mech, design. Active in community affairs, mbr West Lafayette School Bd (1976-81, Pres 1978-79), and Park Bd (1976-81), Bd of Dir. Michigan City (IN) Memorial Hosp. (1984-), Industrial Task Force Valparaiso Chamber of Comm (1984-).

Altan, Halil N
Home: 317 Park Road, Ambler, PA 19002
Position: Sr Proj Engr *Employer:* Aydin Monitor Systems *Education:* MSEE/Telecomm Elect/Tech U of Istanbul *Born:* 5/25/46 Native of Turkey. Graduated from ITU in 1971. Started to work at Badger Irayteon. Served at NATO ENF., Turkey for 1.5 yrs. Joined Makyal Inc. and worked as VPres and then as Pres of Maktel, the Electronics division of Makyal for 5 yrs. Invited to AMS, U.S.A. to work as proj engr. Became U.S. resident and has been with AMS since 1978. Enjoys soccer, tennis and music. Between Oct 1982 and Feb 84 worked at ICI Americas as Sr Elect Engr in dev dept. In Feb 84 returned to Aydin Monitor Systems as Sr Elec Engr. Currently a bd mbr of the Aydin's joint venture in Turkey.

Altan, Taylan
Business: Ohio State Univ, Baker Rm 216, 1971 Neil Ave, Columbus, OH 43210
Position: Prof *Employer:* Ohio State Univ *Education:* PhD/ME/Univ of CA; MS/ME/Univ of CA; Diploma Engr/ME/Tech Univ Hannover, W- Germany *Born:* 2/12/38. Native of Trabzon, Turkey. Worked as Res Scientist at Dupont 1966-68. Battelle Columbus Labs, 1968-86. Assumed present responsibility as Prof at OSU in 1986. Conducts and leads R&D in metal forming, extrusion, forging, rolling, sheet metal forming and computer-aided design and mfg of dies and molds. Past chrmn of ASME's production engg div. Since 1978, Chairman of ASM's Forging Activity. Co-authored two books and more than 150 technical papers. *Society Aff:* ASME, SME, ASM, FIERF

Altenhaus, Julian L R
Home: 49 Claremont Dr, Maplewood, NJ 07040
Position: Prin. *Employer:* Julian LeRoi Altenhaus, PhD, PE. *Education:* PhD/IE/ECU; IE/IE/NYU; MS/IE/Stevens; BSAE/Aero E/Purdue; BSME/Mech Engg/Purdue. *Born:* 6/9/25. Native NYC; USAFR, Capt; Ind Engr, Wright Aeronautical; Mgr of Consulting Services, Service Bureau Corp, Div of IBM; Private Practice 1955 as Gen Mgt Consultant serving most industries in matters of organization, systems, data and word processing, business evaluation, cost and profit optimization, mergers, aquisitions, exec search, market res, sales progs and incentives, and related areas. Married Dr Corrinne Batlin Altenhaus, Clinical Psychologist, mbr Alpha Pi Mu; holder of Silver Medal, AMA; yachtsman, private pilot, ham radio

Altenhaus, Julian L R (Continued)
operator; mbr Masonic order and Shrine. *Society Aff:* ASME, IIE, NSPE, PEPP, NJSPE.

Altenhofen, Robert E
Home: 1371 Bobolink Cir, Sunnyvale, CA 94087
Position: Chrmn Bd of Photogram, West Map Ctr. *Employer:* Div of Topography USGS. *Born:* BS Civ Engg Marquette Univ, Milwaukee WI June 1934. Apptd Jr Topographic Engr with Topographic Div 1936; presently Chrmn Branch of Photogrammetry, Western Mapping Ctr US Geological Survey. 4 yrs WWII officer Corps of Engrs Army Map Service, terminal rank of Maj. Mbr ASCE, Amer Soc of Photogram & several foreign societies in mapping field. Auth num tech papers for Photogram Engg & chapt on rectification in 2nd & 3rd ed of Manual of Photogrammetry. Pres of Amer Soc of Photogram 1972. Granted Dist Service Award Dept of Interior.

Alter, Amos J
Home: Box 020304, Juneau, AK 99802
Position: Principal. *Employer:* Amos J Alter Civ Engr. *Education:* CE/Civ Cold Regions Engg/Purdue Univ; MPH/Pub Hlth Engg/Univ of MI; BSCE/Sanitary-Engrg/Purdue Univ. *Born:* 8/4/16. Past experience includes work in studies, reports and designs relating to water resources, water supply, waste water, air quality, solid waste disposal, occupational health, water and waste treatment as well as construction mgt and teaching and research on the above topics. *Society Aff:* ASCE, NSPE, WPCF, AWWA, IPA, APHA, AINA, AAEE.

Altiero, Nicholas J
Business: Dept of Met, Mech, Mtls Sci, E Lansing, MI 48824
Position: Prof. *Employer:* MI State Univ. *Education:* PhD/Aero Engg/Univ of MI; MA/Mathematics/Univ of MI; MSE/Aero Engg/Univ of MI; BS/Aero Engg/Univ of Notre Dame. *Born:* 9/22/47. Native of Niles, OH. Completed PhD in 1974. Post-doctoral Scholar and Lecturer at Univ of MI 1974-1975. Consultant to US Bureau of Mines 1974-80. Joined faculty at MI State Univ in 1975. Has published technical papers in areas of Boundary-Integral Methods, Geomechanics, Fracture Mechanics. Fulbright Scholar, 1981, Milan, Italy; Von Humboldt Scholar, 1982, Aachen, W. Germany. *Society Aff:* AAM, ASME, ISCME, SES, AIAA, ASEE, ΣΞ, ΤΒΠ.

Altman, Peter
Home: 23300 Providence Dr Apt 605, Southfield, MI 48075
Position: Engg Consultant. *Employer:* Self employed. *Education:* BAeE/Aero Engg/Univ of Detroit. *Born:* 2/24/02. Montreal, Canada. Active in education, res, airplane & engine design for over fifty yrs. Co-builder Powell Racer, winner all Light Airplane events in 1925 Intl Air Races. 1925-40 - Instr, prof & dir - Univ of Detroit Aero Dept. 1940-43 - Dir, Mfg Res - Vultee Aircraft. Developed composite structures and mfg processes. 1943-48 - Dir Aeronca Bd & consultant Continental Motors. 1948-64 VP- Continental Motors. 1964-75 - Exec Consultant, Teledyne Continental Motors. 1975 to date - Consultig Engr. Elected Engg Soc of Detroit Fellow in 1974 & Soc of Automotive Engrs in 1976. *Society Aff:* AIAA, SAE, ESD.

Altschuler, Helmut M
Home: 22550 Bluebell Ave, Boulder, CO 80302
Position: Sr Res Scientist. *Employer:* Conslt *Education:* PhD/Electrophysics/Polytech Inst of Brooklyn; MEE/EE/Polytech Inst of Brooklyn; BEE/EE/Polytech Inst of Brooklyn *Born:* Feb 6, 1922. BEE, MEE, PhD Electrophys Polytech Inst of Brooklyn. Res staff Microwave Res Inst, Facuty Electrophys Dept PIB to 1964. Specialized microwave network theory & measurements. Natl Bur of Stds 1964-; Ch Radio Stds Engg Div; Acting Ch Rdio Stds Lab; Sr Res Sci Electromagnetics Div (concern with stand of phys measurements & electromagnetic measurement techniques). Sr Res Sci Inst for Basic Stds (concern with intl involvement) Sr Stds Specialist Ofc of Domestic and Intl Measurement Standards (concerned with US role in Intl Organization of Legal Metrology) Chrmn G-MIT of IEEE 1964-65; Chrmn CPEM Conf & Exec Cttes 1966-70; Chrmn Comm I USNI of URSI 1970-73; Chrmn Comm A URSI 1975- 78; Fellow IEEE 1967. *Society Aff:* IEEE, USNC/URSI

Altstetter, Carl J
Business: 302 MMB, 1304 W. Green St, Urbana, IL 61801
Position: Prof. *Employer:* Univ of IL. *Education:* ScD/Met/MIT; MS/Met/IL Inst Tech; MetE/Met/Univ of Cincinnati. *Born:* 10/26/30. Lima, OH. Married, two sons. Asst Prof Univ of IL 1958-63, Assoc Prof, Univ of IL 1963-69, Prof 1969-. Acting Hd, Dept of Met, 1974-75. Max Planck Inst Metallforschung, Stuttgart Germany, 1968. Max Planck Inst Plasma Physics Garching, Germany, 1977. NATO Sr Sci Fellow 1977. CSIR Fellow, South Africa, 1985. Res areas: fatigue, hydrogen embrittlement, stress corrosion cracking, thermodynamics of alloys, diffusion. Product liability consulting. *Society Aff:* ASM, AIME, MRS.

Altwicker, Elmar R
Business: Dept. of Chem. Eng. & Env. Eng, Troy, NY 12181
Position: Prof *Employer:* Rensselaer Poly. Inst. *Education:* PhD/Chem/OH St Univ; BS/Chem/Univ. of Dayton *Born:* 4/4/30 Born in Germany, U.S. Citizen 1959, Post-doctoral at the Univ of Vermont; Industrial exprience at American Cyanamid, Princeton Chem Res, and UOP (Univ Oil Prod), Groupleader; Res and Dev. 1968 Joined RPI as Assoc Prof in the Bio- Environ Mental Div. Since 1975 Prof of Env. Eng. in the dept of chem eng & envir engrg. *Society Aff:* ACS, APCA

Alvarez, Ronald J
Business: Engg Dept, Hofstra University, Hempstead, NY 11550
Position: Prof of Engrg *Employer:* Hofstra Univ *Education:* PhD/Civ Engg/NYU; MS/CE/Univ of WA; B/CE/Manhattan Coll. *Born:* 02/17/35 Twenty-seven yrs of varied experience in cvl engrg specializing in structures and solid waste mgmt. Junior engr 1957-58, Assoc engr 1958-60, Engrg designer, 1960-61 The Boeing Co, Seattle, WA; Senior research engr, Republic Aviation Corp, Farmingdale, NY 1961-62; Asst Prof of Engrg 1962-68, Assoc Prof of Engrg 1968-74, Prof of Engrg 1974-, Dir of Continuing Engrg Educ 1970-, Hofstra Univ; Consltg Engr, Glen Cove, NY 1968-; Reg PE in NJ, NY and WA; Certificate of Qualifications from Natl Council of Engrg Examiners; author of numerous technical papers on Solid Waste Mgmt; 1981 Engr of Year, Long Island Branch, Met Section, ASCE. *Society Aff:* NSPE, ASCE, ASME.

Alvine, Raymond G
Business: 360 Aquila Ct, Omaha, NB 68102
Position: Pres *Employer:* Raymond G Alvine & Assoc. Inc *Education:* BS/ME/IA State Univ. *Born:* 8/22/26 Burlington, IA. Wife Dorothy Ann Lux. Fairmont Foods, Columbus, OH, 1949- 50; Carl A Goth. PE, Omaha. 1910-51; Silas Mason Co. Burlington, IA 1951-53; Leo A Daly Co, Omaha, 1953-61; Pres. Raymond G Alvine & Assoc, Inc, Consulting Engrs, Omaha. 1961 to date. Chrmn. Omaha Bldg Bd of Review, 1973 to date. Chrmn, Omaha Steamfitters Bd Examiners, 1970 to date. Chrmn. City of Omaha High Rise Bldg Ordinance, 1973 to date. Reg PE IA, NE, SD, NV, AZ, NCEE, MN, KS, CA. State Building Energy Standards Board 1980-86, State Fire Marshal's High Rise Code Committee 1981; Chairman, ASHRAE Standard 100.2; Chairman, ASHRAE Standard 90- Section 4; Pi Tau Sigma, ASHRAE-Fellow, APEC-Fellow; ASHRAE Standards Ctte; Tau Sigma Delta; Past Pres, Conslt Engrs Council of NE; Pres 1975 & 1976, Automated Procedures for Engrg Conslts; Contributor, Architectural Graphic Standards. Hobby - Sailing. *Society Aff:* ASHRAE, ACEC, NSPE, ASME, SAME, APEC.

Aly, Adel A
Business: Sch of Ind Engg, Univ of Oklahoma, Norman, OK 73019
Position: Prof. *Employer:* Univ of OK. *Education:* PhD/Ind Engg & Operations Research/VPI & SU; MS/Ind Engg/NC State Univ BSc/Production Engg/Univ of Cairo. *Born:* 10/30/44. Native of Cairo, Egypt. Worked with the Natl Inst of Mgt Dev as a jr consultant, 1966-1967. The Univ of Cairo 1967-1969 as instr in the production engg dept. With the Univ of OK since 1/75. Associate Editor of IIE Transactions. Reg PE in OK & Egypt. Past-Pres of OK City Chapter of IIE. Res interests

Aly, Adel A (Continued)
are in facilities location & planning, routing, distribution systems, and system analysis of public sector. *Society Aff:* IIE, SME, ORSA, ASEE, TIMS, NSPE.

Amann, Willard L
Home: 410 Hollomon Rd, Kerrville, TX 78028
Position: Vice Pres Retired *Employer:* Houston Natural Gas Corp *Education:* BS/Natural Gas Engrg/TX A & I Univ; BS/Chem/TX A & I Univ *Born:* 09/01/26 Native of San Antonio, TX. Spent entire business career with Houston Natural Gas Corp, joining upon graduation from TX A & I Univ, Kingsville, TX in 1947. Advocate of codes and standards in gas industry, spent over twenty-five yrs in developing gas and mech codes. Received "Distinguished Service Award" in 1976 from Southern Gas Association for code activities. Past Chrmn and Mbr of ANSI Z223.1 and NFPA-54. Responsible for feasibility of central cooling and heating plants as a commercial venture for HNG, resulting in four plants totaling in excess of 46,000 tons of air conditioning and 900,000 pounds of steam. Retired as VP of HNG in 1984, and is presently doing consulting activities. *Society Aff:* NSPE, ASHRAE

Amber, Wayne L
Business: 401 N Bendix Dr, South Bend, IN 46620
Position: Chief Met. *Employer:* Bendix Corp. *Education:* BSE/Met/Purdue Univ. *Born:* 6/4/34. A lifetime resident of South Bend, IN, working on various met engg capacities at the Bendix Corp, since grad from Purdue Univ. Past Chrmn of the Notre Dame Chapt "ASM–, served on the Natl Advisory Awareness Council of "ASM–, committee mbr on "ASM" wear failures handbook. Won the *Non-ferrous Materials Award* of ASM (1966) & *The Die Casting of the Year Award* in 1968 from Die Casters Assn. Presently active as mbr of *American Deep Drawing Research* Group. *Society Aff:* ASM, SAE.

Amberg, Arthur A
Home: 13 Maple Ct, Plainfield, IL 60544
Position: Product Engr *Employer:* Intl Harvester *Education:* MS/Agric Engrg/Univ of IL; BS/Agric Engrg/Cornell Univ *Born:* 08/24/45 Entered engrg profession as design engr with Intl Harvester in agricultural tractor product design, 1969. Progressed to proj engr tractor cab design, 1972. Promoted to mgmt position of product engr tractor cab design, 1979. Appointed to current position of product engr in charge of human factors and operator comfort for tractor and harvester cab designs, 1981. Reg PE. FIEI Roll Over Protective Structure Committee Vice-Chrmn, FIEI Human Factors Committee mbr. US delegate to ISO Human Factors Sub-Committee in Copenhagen, Denmark, 1980. ASAE Section offices held: Secy-Treas, Prog Chrmn, Chrmn, Past Chrmn, 1980 ASAE Natl Winter Meeting Host Committee Chrmn. *Society Aff:* ASAE, SAE, FIEI

Ambrose, Tommy W
Business: 505 King Ave, Columbus, OH 43201
Position: VP BMI & Dir Multi-Compoment Operation *Employer:* Battelle Memorial Inst. *Education:* BS (Ch.E) U of I, 1950, MS (ChE) U of I 1951; PhD (Ch.E.) Oregon State *Born:* 10/14/26 USNR 1944-46; Supr Mgr res & engg GE Co 1951-65; Douglas United Nucl Co 1965-69; Exec Dir Battelle Seattle Res Ctr 1969-75; Dir Res, Dir-VP Battelle Memorial Inst Pac NW Div 1975-1979. VP BMI & Dir Multi Component Oper 1979-. Chrmn Bd of Trustees Columbia Basin Coll 1967-69; VP Bd-Trustees Pacific Sci Ctr Fdn 1975; Mbr: Coll of Engg Univ of WA 1974-, Adv Ctte Coll of Engg Univ of ID 1974-. PE WA & OH; Bd of Trustees/Exec Ctte, Columbus Symphony Orchestra 1980-; Chrmn, Communications Ctte, Mgmt Div. AIChE 1981-; Industry/Academia Relations Cttee, Ohio Acad of Science 1982-; Bd of Dirs/Chrmn, Planning Ctte, Jefferson Academy, Columbus 1984-. *Society Aff:* AIChE, ANS, ΣΞ, Φ LANDU E

Amdahl, Gene M
Business: 1250 E Arques Ave, Sunnyvale, CA 94086
Position: Chrmn of the Bd of Dirs. *Employer:* Amdahl Corp. *Born:* Nov 1922 Flandreau SD. BSEE SD St Univ 1948; PhD Theor Phys Univ of WI 1952. USN 1942-44. IBM Fellow 1965-70, Ch Architect of System/360; Chrmn/Bd Amdahl Corp mfg of lrg-scale IBM-compatable computer sys 1970-. Mbr NAE; Fellow IEEE; NATO Summer Lectr; DPMA Computer Sci Man of the Yr Award 1976. Enjoys golfing.

Amelio, Gilbert F
Business: 101 Bernal Rd, San Jose, CA 95111
Position: VP & Gen Mgr. *Employer:* Fairchild Semiconductor. *Education:* PhD/Physics/GA Tech; MS/Physics/GA Tech; BS/Physics/GA Tech. *Born:* 3/1/43. New York City; raised and educated in Miami, FL; degreed in Physics from GA Institute of Tech with PhD 1968. Prin in Information Sciences, Inc 1962-65. Consultant for SIR, Inc and Emory Anesthesiology Dept 1967-68. MTS, Bell Labs 1968-71. Pioneer in CCD with first experimental results and publication. Fairchild 1971 to present. In R&D and Dir of CCD until 1976. In MOS as Dir of Engg until 1977. 1977-78 MOS Div VP and Gen Mgr. In 1978 to Present VP and Gen Mgr MOS Products Group. Fellow IEEE 1978. In 1981 named General Manager of Microprocessor Products Division. 16 patents in MOS and CCD granted or pending. *Society Aff:* IEEE, ΣΞ.

Amerman, Carroll R
Business: PO Box 478, Coshocton, OH 43812
Position: Res Hydraulic Engr. *Employer:* US Dept of Agri, Sea/Agri Res *Born:* July 1930 Danville, IL. US Army 1949-52. BS, MS & PhD Purdue Univ. Began work with USDA-ARS in 1958 Coshocton OH, transferred to Madison WI 1967, to Columbia MO 1973, to Coshocton, OH, 1980. Performs res on agri hydrology, concentrating on infiltration and subsurface aspects. Chrmn Amer Soc Agri Engrs Soil & Water Div 1975-76; mbr: Am Geophys Union, Soil Conservation Soc of Am, Sigma Xi, Tau Beta Pi.

Amick, Charles L
Home: 344 Halcyon Dr, Glendale, MO 63122
Position: Lighting Consultant *Employer:* Day-Brite Lighting Div Emerson Elec Co. *Education:* Elec Engr/Thesis was on Spectral Distribution of Light Sources/IA State Univ; BS/EE/IA State Univ. *Born:* 11/21/16. GE student engg course - 1937-38 and various engg positions in Lamp Div at Nela Park, Cleveland OH 1938-1955. With Day-Brite Lighting Div, Emerson Elec Co from 1955 through 1979, now a Lighting Consultant 1980-present. Reg PE OH & MO. Author of Fluorescent Lighting Manual published by McGraw-Hill Book Co in 1942, 1947 and 1961. Many papers and articles on lamps and illumination. Chrmn or mbr of many Illuminating Engg Soc committees, was Natl IES Pres 1964-65, and received their Gold Medal 1975. Also MSPE Distinguished Services Award, ISU Profl Achievement Citation, and St Louis Engrs' Club Award of Merit and Honorary Membership. *Society Aff:* IES, NSPE, MSPE.

Amico, G Vincent
Home: 831 Longhaven Dr, Maitland, FL 32751
Position: Dir of Res. *Employer:* Naval Training Equip Ctr. *Education:* MS/Systems Engg/FL Tech Univ; MBA/Ind Mgt/Hofstra; BAE/Aero Engg/NYU. *Born:* 10/20/20. Dir of Res, Naval Training Equip Ctr, Stress Engr with Curtis Wright Corp 1941-45; Army Air Corps 1945-47; joined Special Proj Group of Republic Aviation 1947-48 as struct design engr on adv aircraft & missile; joined Naval Train Equip Ctr as pro engr 1948; Dir/Engr 1969, respon for engg dev & acquisition of military train equip & simulators. 1979 Dir of Res, conducts res in simlation tech & training methodology applicable to military training systems. Dept of the Navy Sr Exec; Prof Engr OH. Mbr: AIAA, AFCEA, SESA, SAME, Tau Beta Pi, Alpha Pi Mu. *Society Aff:* AIA, AFCEA, SESA, SAME.

Amirikian, Arsham
Business: 35 Wisconsin Cir, Chevy Chase, MD 20815
Position: Pres. *Employer:* Amirikian Engg Co. *Education:* DTechSc/Struct Engg/Tech

Amirikian, Arsham (Continued)
Inst/Vienna, Austria; CE/Struct Engg/Cornell Univ. *Born:* 5/17/99. Native of Armenia. Came to US in 1919 Entered Govt service in Bureau of Yards and Docks (now Naval Facilities Engg Command) in 1928 and served through grades to Chief Designing Engr and Chief Engg Consultant. Pioneered in welded design and precast concrete construction and dev many framing innovations some of which were patented. Retiring in 1971 established own consulting firm for private practice in his specialized field. Received over a score of awards, including Distinguished Service Award of Navy (1966) and Dept of Defense (1969), AWS Airco Award (1969), SAME Goethals Medal (1971), and ASCE Howard Award (1978). Enjoys classical music and literature. *Society Aff:* NAE, ASCE, AWS, ACI, SNAME, SAME.

Amrine, Harold T
Home: 107 Sunset Ln, W Lafayette, IN 47906
Position: Ret. (Prof Emeritus of Indus Engrg, Purdue Univ) *Education:* MS/Mech Engg/Univ of IA; BS/Mech Eng/Univ of IA. *Born:* 6/9/16. Began engg career as an Ind Engr with Swift and Co and W A Sheaffer Pen Co 1939-40. Taught at OH State Univ 1940-41 and 1945-46. Served in Army Ordnance Dept 1941-45. Mbr of Ind Engg Faculty at Purdue Univ 1946 to 1981. Hd 1955-69. Assoc Dir, Div of Sponsored Progs, Purdue Res Fdn, 1970-73. Hd, Dept of Freshman Engg, 1973 to 1981. VP, Pritsker and Assocs, 1981-82. Fulbright Lecturer, Univ of Melbourne, 1959. Visiting Prof of Engg, AZ State Univ, 1969-70. Co-author, "Manufacturing Organization and Management." Honorary Dr of Aeronautical Engg, Embry-Riddle Aeronautical Univ, 1972. Fellow in IIE. Past Officer IIE. *Society Aff:* IIE, ASME, ASEE.

Anand, Dave K
Home: 8824 Burningtree Road, Bethesda, MD 20817
Position: Professor *Employer:* University of Maryland *Education:* Ph.D./Mech. Engr. /The George Washington University; M. S./Mech. Engr./The George Washington University; B.S./Mech. Engr./The George Washington University *Born:* 04/04/39 Simulated the dynamics and control of the first gravity gradient and drag-free satellite; conducted theoretical and experimental work on the first heat pipe used in space for thermal control for which a patent was awarded; organized and chaired the first International Heat Pipe Technology Symposium; Chaired the first national meeting on the use of Energy considerations for solar energy systems. Received superior and outstanding performance award from the National Science Foundation for a new research program; Outstanding Alumnus award (1986) from GWU; Outstanding Performance (1985) from Univ of Maryland; Published 100 papers and three books. PE in Maryland. *Society Aff:* ASME, TBPI, ΣTAU, ΣXI, ΠΤΣ

Anand, Subhash C
Business: Dept of Civ Engg, Clemson, SC 29631
Position: Prof. *Employer:* Clemson Univ. *Education:* PhD/Struct Engr & Struct Mechanics/Northwestern Univ; MS/Struct Engr/Northwestern Univ; BSc/CE/Banaras Hindu Univ. *Born:* 7/27/33. in Lyallpur (India) and finished high sch and college education with a BSc in Civ Engg in India. Worked with Govt and private agencies from 1955-1958. Went to W Germany in 1958 and worked there with private firms as structural design engr. Came to US in 1964 and pursued grad studies at Northwestern earning MS and PhD degrees. Taught at CA State Univ at Sacramento from 1968-1970 as Asst Prof, at IL Inst of Tech, Chicago, from 1970-1972 as Assoc Prof. Joined Dept of Civ Engg at Clemson Univ as Assoc Prof in 1972. Was promoted to prof in 1976. *Society Aff:* ASCE, ASEE, AAM, TMS, Int. Assoc. Comp. Mech.

Anastasion, Steven N
Business: 3300 Whitehaven St. NW, Washington, DC 20235
Position: Exec Dir *Employer:* NACOA *Education:* MS/EE/MIT; BS/-/US Naval Acad *Born:* 04/09/21 A 1942 graduate of the US Naval Acad. He also attended MIT, the Naval War Coll and the Industrial Coll of the Armed Forces. Anastasion served in the US Navy from 1942 to 1972. His assignments included command of the guided missile cruiser USS LEAHY, Technical Asst to the Asst Secy of the Navy for Research and Development, and Commander of the Naval Weapons Laboratory in Dalhgren, VA. He twice received the Legion of Merit for his accomplishments in the US Navy. In 1972, Anastasion joined the Natl Oceanic and Atmospheric Admin (NOAA) as Exec Secy of the Interagency Committee on Marine Sci and Engrg. In 1973, he was appointed Chief, Plans and Prog Coordination Office in NOAA's Marine Resources Office, and became Dir of NOAA's Office of Ocean Engrg in 1976. He was appointed to the NACOA position in January 1980. *Society Aff:* SNAME, MTS

Ancker, Clinton J, Jr
Home: 23908 Malibu Knolls Rd, Malibu, CA 90265
Position: Prof Emeritus. *Employer:* Univ of Southern CA. *Education:* PhD/Engr Mechs/Stanford; Mech Engr/Engr Mech/Univ at CA Berkeley of CA; MS/ME/Univ of CA; BSME/ME/Purdue. *Born:* 6/21/19. Chrmn, ISE Dept USC - 1968-84. Prof Emeritus USC-1984-present. Dir, Natl Hgwy Safety Inst, DOT 1967- 68. Hd, Math & OR Prog, System Dev Corp 1959-67. Mgr, Analco Services Co 1958- 59. Also with Booz Allen Appl Res, Operations Res Office, U C Bekeley, Purdue Univ and Detroit Edison Co. Earlier, US Army WWII 1941-46. Reg ME CA. Mbr several Honor Socs. Author numerous papers on engg and Operations Res topics. Natl Council Mbr of ORSA 1970-1973. Natl Pres of Omega Rho 1976-1978. *Society Aff:* ORSA.

Ancker-Johnson, Betsy
Business: 14th & Constitution NW, Washington, DC 20230
Position: Asst Secy for Sci & Tech. *Employer:* US Dept of Commerce. *Born:* Apr 27, 1929 St Louis MO. BA Wellesley Coll 1949; PhD Phys Tuebingen Univ Germany 1953. AAUW Fel 1950-51; Horton-Hallowell Fel 1951-52; Jr res physicist & lectr phys U C Berkeley 1953-54; mbr staff Inter-varsity Christian Fel 1954-56; sr res physicist microwave phys lab Sylvania Elec Prods Inc 1956- 58; mbr tech staff David Sarnoff Res Ctr RCA 1958-61; Visiting Sci Bell Tech Labs 1968; res specialist Boeing Sci Res Lab 1961-70. Supr solid state & plasma electronics 1970-71; Mgr Adv Energy Sys Boeing 1971-73; Affl Prof Elec Engg Univ of WA 1967-. Assoc Prof 1964-67. A/S of Commerce for Sci & Tech 1973-; Assoc Dir Res Non-Nucl Res Argonne Natl Lab 1977-. NAE, Fellow APS, Councillor-at-lrg 1971-76; Fellow IEEE, AAAS.

Andersen, James C
Home: 1433 Comet Rd, Sioux Falls, SD 57103
Position: Reg Engr. *Employer:* SD Dept of Water & Natural Resources. *Education:* Masters/Public Health/Univ of MN; BS/Civil Engg/SD State Univ *Born:* 4/20/24. SD State Univ. Pres Club & Guest Lecturer; Prof Engr (Sanitary) SD; Mbr NSPE SD - State Dir SD Engg Soc 1978-80; WPCF SD Assn: VP 1971, Pres 1972, Pres & Mbr Exec Bd 1973, Sec-Treas 1978-81, National Dir 1974-77, Arthur Sidney Bedell Award 1979, WPCF Federation Nominating Committee 1977, Life Mbr 1987, Select Soc of Sanitary Sludge Shovelers 1985, AWWA: SD Chmn 1971 - George Warren Fuller Award 1971; P Chmn & Mbr of Exec Bd 1972, Life Mbr 1982; Commissioned Officers Assn of US Public Health Service: Inactive Commission as Sanitary Engg Dir; SD Water and Wastewater Association: Board of Dirs 1961. Davenport Award 1970. VP 1972, Pres 1984, Honarary Mbr 1981; Sioux Falls Masonic Bodies. El Riad Shrine, Philatelist. Retired 1987. *Society Aff:* WPCF, AWWA.

Andersland, Orlando B
Home: 901 Woodingham Dr, E Lansing, MI 48823
Position: Prof. *Employer:* MI State Univ. *Education:* PhD/Civil Engg/Purdue Univ; MSCE/Civil Engg/Purdue Univ; BCE/Civil Engg/Univ of MN. *Born:* 8/15/29. Prof of Civil Engg at MI State Univ. Received the univ's Distinguished Faculty Award in 1979. Formerly res engr, Purdue Univ; staff engr for the Natl Academy of Sci AASHO Rd Test; and officer, Army Corps of Engrs 1952-54. Directed res grants from the Natl Sci Fdn on the mech behavior of frozen ground and res on mech properties of fibrous organic soils supported by the Environmental Protection Agency and the Natl Council of the Paper Industry. Co-editor and contributor to

Andersland, Orlando B (Continued)
GEOTECHNICAL ENGINEERING FOR COLD REGIONS, co-author of *GEOTEK-LAB GEOTECHNICAL SOFTWARE FOR THE IBM PC*, and author of numerous technical publications. *Society Aff:* ASCE, ASTM, ASEE, ISSMFE.

Anderson, A Eugene
Home: 647-E Sawmill Rd Haycock, Quakertown, PA 18951
Position: Retired. *Education:* MSEE/EE/OH State Univ; BS/EE/OH State Univ. *Born:* 4/22/16. in Lima, OH. Married Helen Roll in 1942, five children. Sr Exec, MIT. Bell Telephone Labs, 1939-57 - Electron tube and transister R&D. Dir of Electron Device Dev 1955. Signal Corps Engg Labs 1942-47. Electron tube dev and standardization. Merit citation. Western Elec 1957-75, Dir of Engg. 19 patents, many papers and talks. Many tech and advisory committees. Fellow IEEE, AAAS. Tau Beta Pi, Sigma Xi, Phi Eta Sigma, Sigma Pi Sigma. Reg Engr PA Dist Engr Lehigh Valley PSPE. Distinguished Alumnus OH State Univ. *Society Aff:* IEEE, AAAS.

Anderson, Arthur G
Business: 100 Westchester Ave, White Plains, NY 10604
Position: IBM VP; & Group Exec-D P Product Group. *Employer:* IBM Corp. *Education:* PhD NYU: BS Univ of San Fran; MS Northwestern Univ. *Born:* Nov 1926. USN 1944-46; with IBM 1951-, holding num engg & mgr positions incl Dir of Res. elect IBM VP 1969; Pres Gen Prod Div 1972-1979. Group Exec 1979-. Rec IBM Invention Achievement Award. Patents in field of spin echo storage techniques & digital circuitry. Fellow Am Phys Soc; Fellow IEEE. *Society Aff:* ASEE, ASQC, AAS, NAE, AIP, NYAS, IEEE

Anderson, Arthur R
Business: 1127 Port of Tacoma Rd, Tacoma, WA 98421
Position: Bd Chrmn. *Employer:* ABAM Engrs Inc. *Education:* Dr of Sci/Civ Engg/MIT; MS/Civ Engg/MIT; BS/-/Univ of WA. *Born:* 3/11/10. Staff MIT 36-38, 39-41, 41-46. Hd, Technical Dept Cramp Shipyard, US Buships Cons Engr. Stamford CT 46-51. Co-founder & principal Concretetech Corp & ABAM Engrs 1951 to date. Visiting Cttee, Univ of MA. WA State Council for Post Secondary Education. Pres Puget Sound (Wash) Sci Fair, 1954-58; mbr Pub Utility Bd, City of Tacoma, Wash, 1954-69, Chrmn 1968-69; mbr Pacific Lutheran U Collegium, 1976-; mbr Vis U Wash; mbr Wash State Council for Post-Secondary Edn, 1977-. Reg PE, Wash, Conn, B.C., Can. *Society Aff:* ASCE, ASTM, ACI, PCI, SNAME, NAE, NSPE, FIP.

Anderson, Arvid E
Home: 448-C Cheshire Ct, Lakewood, NJ 08701
Position: Retired. *Education:* DEngg/Honorary/Worcester Poly Inst; EE/Profl/ Worcester Poly Inst; BS/EE/Worcester Poly Inst. *Born:* 5/26/97. BSEE;EE; DEngg Worcester Poly Inst. Native Falmouth, MA. Entire career with GE. Received 72 US patents and made 95 foreign assignments of inventions in Power Switchgear directed mainly to improved continuity of electric power and reduced operating costs. Mbr numerous technical and standardizing committees 1931-1961 within GE; also AIEE, NEMA, etc, at Headquarters or Natl level. Mbr AIEE Bd of Examiners 1950-1961. Author 21 published technical papers and articles. Received Coffin Award in 1942. Reg PE in PA. Mbr TBP and SX. Life Fellow IEEE and Fellow AAAS. Hobbies: Reading and music. *Society Aff:* AAAS, IEEE.

Anderson, Brian D O
Business: Australian National Univ, Canberra ACT 2601, Australia,
Position: Prof of EE. *Employer:* Univ of Newcastle. *Born:* Jan 1941 Sydney Australia. PhD Stanford Univ; BE, BSc Sydney Univ. Fac mbr Stanford Univ, then Elec Engg Newcastle Univ Australis - Dept Hd 1967-75. Fellow Aust Acad of Sci, Aust Acad of Technal Sci, IEEE, Inst of Engrs Aust. Auth of many papers & several books.

Anderson, Bruce H
Home: 1486 Maple Dr, Logan, UT 84321
Position: Ret. (Exec Dir CID) *Education:* Doc of Engr/Irrigation Engr/Univ of CA-Davis; MS/Irrigation Engr/UT State Univ Logan; BS/Irrigation Engr/UT State Univ Logan *Born:* 8/30/17. Dr Anderson has an intl reputation for his work, res, prog dev and admin of agriculture-related projs all over the world. 1950-Alberta, Canada - water and drainage studies. 1951-53 - Chief, Agricultural Div TCA, Shiraz, Iran. 1954-60 - Adivsor, Co-Dir, Dept Agri, Ministry of Agri, Tehran, Iran. 1963-64 - Extension Water Resource Specialist, UT State Univ. 1964-69 - Dir, Inter-American Ctr for Integral Dev of Land & Water Resources Merida, Venezuela, to develop a personnel training prog for all South and Central American countries which is ongoing today. 1973-1980 Exec Dir, Consortium for Intl Dev (CID) Comprised of 11 Univs emphasizing improvements in food production, living conditions, and educational progs. 1980-Retired from the Univ of (Prof Emeritus). 1980-1984 - Many consltg assignments on Water Resource Dev in Peru, Dominican Republic, Egypt, and Saudi Arabia. *Society Aff:* ASAE

Anderson, C Henry
Business: SE Tower-Prudential Ctr, Boston, MA 02199
Position: VP Dir of Engg Pulp/Paper Div. *Employer:* Charles T Main Inc. *Born:* Aug 1928 MA. ABS Northwestern Univ. Joined Chas T Main Inc 1950, made Assoc Mbr of Firm 1962, Asst VP 1969, VP 1971. Pres respon, Main's Pulp & Paper Div, engg design activ incl estab the design stds for the Div. Proj Mgr of Honor award-winning design Sulphate Pulp Mill, Nova Scotia 1968: presented by Natl Cons Engrs Council. Reg PE many states.

Anderson, Charles E
Home: 2138 Hillcrest Rd, Redwood City, CA 94062
Position: Product Planner. *Employer:* Ampex Corp. *Born:* BSEE Case Sch of Applied Sci 1948. Worked as elec engr for several cos in fields of ultrasonic testing & data recording; joined Ampex 1954, was mbr of team which developed 1st video tape recorder; 18 yrs as engr, video tape recording - hd dept before moved to adv prod plan 1972.

Anderson, Charles J
Home: 1006 Lawndale Ln, Wash, IL 61571
Position: Heat Treat Supt. *Employer:* Caterpillar Tractor Co. *Education:* BS/Met Engr/Univ of IL. *Born:* 1/17/39. Chicago, IL area native. Attended Univ of IL from 1960-64. Worked as dev foundry net for Duriron Co Dayton, OH after grad. With Caterpillar since 1965. Varied experience up to third level mgt. Including castigs, heat treatment, inspection, powde met, welding & cutting, laser tech and application. Holder of (1) US Patents, several other filed. *Society Aff:* ASM, ASQC, AWS.

Anderson, Charles R
Business: 2 Corporate Place S, Piscataway, NJ 08854
Position: Home Off Dir Loss Control-Asst Security *Employer:* Continental Insur Cos - Ctck (Conti at Tech Ser) *Born:* Nov 19, 1920. BCE CCNY 1942. Served US Army 1942-46 Infantry Unit Cdr Capt. Supt of Const with Gen Contractors Lexington KY & NY City 1946-51. With Continental Ins Co 1951-. Serves as engg cons to co mgt & over 400 field engrs throughout US & foreign countries; prepares manuals, bulletins on fire protection & allied property fields which are distrib primarily to co personnel worldwide. Served as Pres NY Chapt SFPE, presently on Bd of Dir SFPE; on fire protection cttes of both SFPE & Amer Ins Assn. Hold PE Lic in NY State 1952, TN 1965, CA 1978.

Anderson, Clayton O
Home: 121 Washington Ave S, Minneapolis, MN 55401
Position: Pres *Employer:* Steel Structures Inc. *Education:* BCE/Civ Engrg/U of MN *Born:* 3/16/31. I was formerly VP & Dir of Design for Rauenhorst Corp, former Pres of Mpls Sec ASCE, former Pres of MN Chapter of MSPE. Formerly VP Sales & Engg for Phoenix Steel Inc of Eau Claire, WI a steel fabricating co of 45M sales in bridge, bldg, & reinforcing steel. I am married, have two children. Have served as YMCA Dir, & several United Fund positions. Active in youth athletics & serve as

Anderson, Clayton O (Continued)
church usher. Enjoy skiing, cross-country and downhill, also racquetball, hunting, fishing, golf & tennis. Hold Registration in MN, WI, SD, MT and MI. Currently Pres of Steel Structures Inc, a gen construction and construction mgmt firm located in Minneapolis. *Society Aff:* ASCE, MSPE

Anderson, Danny R
Business: 2525 E Yandell Dr, El Paso, TX 79903
Position: President *Employer:* Danny R Anderson, Consultants, Inc. *Education:* MS/Civ Engg/Univ of TX; BS/Civ Engg/Univ of TX. *Born:* 2/9/43. Resident of El Paso, TX for eighteen yrs. Employed by TX Hgwy Dept as bridge designer and mtls technician from 1967 to 1971. Principal geotechnical engr for ABDW, Inc 1971 to 1978. Guest lecturer in soil mechanics at the Univ of TX at El Paso, 1974 to 1976. President and chief geotechnical engr for Danny R Anderson, Consultants Inc since 1978. Past Pres, TSPE El Paso Chapter; Dir-at- Large, ASCE TX Sec; Pres ASCE, El Paso Branch, 1974; Outstanding Young Engr award, TSPE El Paso Chapter, 1975; Mbr El Paso County Historical Commission, 1978; Mbr El Paso Historic Landmark Commission, 1981. Reg PE, TX and NM. *Society Aff:* ASCE, NSPE, ASTM, ACI.

Anderson, Darryl W
Home: PO Box 248, Rock Creek, MN 55067
Position: Consultant. *Employer:* Darryl Anderson PE. *Education:* MBA/Bus Admin/ Golden Gate Univ; BSChE/Chem Engg/Univ of ND. *Born:* 9/11/47. Responsible for engg, design and construction of several different chem production facilities for clients in the chem mfg industry. Served in a variety of mgt policy and proceeduremaking positions. Performed process res and dev studies leading to production of new chem products. Currently pres of own private consulting firm. Born in Pine City, MN. Have had interest and activity in local, state, natl politics and in community affairs. Fish, play tennis, run. Influenced gatherings of world wide acclaimed dignitaries through pertinent correspondence. Leading intentionally to increased amicasility between nations. *Society Aff:* AIChE, NSPE.

Anderson, David Melvin
Business: 7059J Commerce Cir, Pleasanton, CA 94566
Position: Pres. *Employer:* Anderson Engg. *Education:* Doctor of Engg/Mech Engg/ Univ of CA; MS/Mech Engg/Univ of CA; BS/Mech Engr/Univ of CA. *Born:* 7/2/44. Worked for Hexcel Corp in 1967. At the UC Bio-Mechanics Lab, designed and patented a powered wheelchair that adjusts in length, width, and seat height in addition to reclining and climbing curbs. Worked for MB Assoc on Govt contracts in 1974. At Filper Corp (A DiGiorgio Co), designed and patented a new generation food processing machine (peach pitter) as a replacement for the 3000 existing machines in canneries world-wide. The design goals of profitability and sanitation resulted in linkage-driven, non-lubricated, all stainless machines. Industry reaction encouraged the formation in 1977 of an independent Consulting service with an emphasis on cost-effective design. In 1979, formed Anderson Engg to specialize in the design and construction of automation machinery and specialized Robot grippers and interface hardware. *Society Aff:* ASME, SME, ASM, SAE.

Anderson, Dean B
Home: 8521 Colima Rd, Whittier, CA 90605
Position: Consultant-Optics & Microwaves *Employer:* D B Anderson, Conslt *Education:* BSEE/Elec Engrg/MT State Univ *Born:* 10/22/21 Radio Propagation Engr, Signal Corp (1943-46). Supervising Engr, Hazeltine Electronics Corp Little Neck, NY (1946-54). Various Mgt & Staff positions in Rockwell Intl Electronics Res Ctr & its predecessors, NAR & NAA at Anaheim, Thousand Oaks & Dowey, CA (1954-1980). Adjunct Professor, Electrical & Computer Engineering, University of Cincinnati 1980-, Consultant, Optics & Microwaves 1980- Pioneered interest in optical waveguide in 1964 developing integrated optics tech using (GaAl) As and SiO2 on Si. Elected Fellow, IEEE "for contributions to optical waveguide and optical parametic amplication" (1972). Pres (1979); Treasurer (1977-78); IEEE/Quantum Electronics and Applications Soc. Mbr QEAS AdCom (1976-79), MTTS AdCom (1974-76); Chrmn IEEE 1963 MTT International Symposium. IEEE Centennial Medal 1984. *Society Aff:* IEEE, OSA, LEOS, MTTS

Anderson, Donald K
Business: Dept of Chem Engg, Michigan State Univ, E Lansing, MI 48824-1226
Position: Chrmn & Prof. *Employer:* MI State Univ. *Education:* PhD/Chem Engg/ Univ of WA; MS/Chem Engg/Univ of WA; BS/Chem Engg/Univ of IL. *Born:* 7/15/31. in Iron Mountain, MI. Employed with: Hanson Chem Equip Co, S Beloit, IL, 1950-51. Woodword Governor Co, Rockford, IL, 1951-52. Univ of WA, 1956-60. MI State Univ (Asst Prof 1960-1964; Assoc Prof 1964-69; Prof 1969-77; Joint appointment, Dept of Physiology, 1970-; Chrmn, Chem Engg 1977-). Distinguished Faculty Award, MI State Univ 1973. Reg PE (MI). *Society Aff:* AIChE, ACS, ASEE.

Anderson, Eric D
Home: 20179 Wellesley Dr, Riverview, MI 48192
Position: Chief Metallurgist-Primary *Employer:* Republic Steel Corp. *Education:* BS/Met Engg/Univ of Cincinnati. *Born:* 4/2/48. Joined Republic Steel Corp in 1971 and had quality assurance responsibilities in the finishing and primary operations prior to current assignment as chief metallurgist - primary operations at the Warren, OH Plant. Has been active in promoting the field of met and steel tech through involvement with ASM, Warren Chapt Chrmn 1977-78, Natl Mbrship Committee 1978-81, Natl Young Member Committee 1981-84, as well as AIME, Chrmn PA-OH Sec 1980-81. Also promoted the steel industry by serving as an Am Iron & Steel Inst Steel Fellow 1976-79. Received the Frank B McKune award in 1981 as coauthor of technical papaer "Bath Oxygen Analysis and Utilization." *Society Aff:* AIME, ASM.

Anderson, Frank A
Home: 410 S 11th St, Oxford, MS 38655
Position: Assoc Dean & Prof Emeritus. *Employer:* Univ of MS. *Education:* PhD/ChE/LA State Univ; MS/ChE/Univ of ME; BS/ChE/Univ of S CA. *Born:* 6/22/14. Birthplace: Bridgeport, CT. Married Mary Alla Courtney. Two children: Frank, Phyllis. Worked for Shell Oil, Gaylord Container, Chemstrand, Oak Ridge Natl Lab. Chrmn, Chem Engg, Univ MS, 28 yrs; last position: Assoc Dean Engr, Prof Chem Engr. Served as Chrmn, Memphis Sec, ACS. Distinguished Engrg Service Award by Mississippi Engrg Society; Pres, MS Acad Sci; Council Chrmn, Oak Ridge Assoc Univs; Chrmn, Chem Engr Educ Projs Comm, AIChE; Pres, New Sci Advocates. Awards: Outstanding Alumni Citation, Univ Bridgeport; Outstanding Teacher, Univ MS; Achievement Award, Memphis Sec, AIChE Honors: Fellow, AIChE; Who's Who; American Men Science; Phi Kappa Phi; Tau Beta Pi. Interests: Impact of Tech, Ethics, Music, Vikings. Distinguished Engineering Service Award, MS Engineering Society, 1981. Engineer of Distinction Award, Univ. Miss. Engrg Alumni Chapter, 1983. Community-Government Service Award, Oxford and Lafayette County, 1983. *Society Aff:* AIChE, ACS, NSPE, ISUS.

Anderson, Gery F
Business: 1828 Tribute Rd, Suite H, Sacramento, CA 95815
Position: Pres. *Employer:* Anderson Geotechnical Consultants, Inc. *Education:* MS/Geological Engg/San Jose State Univ; BS/Geological Engg/San Jose State Univ. *Born:* 3/30/42. BS (1964), MS (1968), Geological Engg, San Jose State Univ. Owner-Founder & Pres of Anderson Geotechnical Consultants, Inc, Consulting Geotechnical Engrs & Geologists. Started co in 1971 & have been active in the investigation of wastewater treatment & mgt as related to soil & geologic conditions. Forerunner in the dev of land cpability studies to determine best land use utilizing geological & soil restraints. Santa Cruz County office mgt for Gribaldo, Jones and Assoc (1969-1971). Reg Civ, CA & NV; Reg Geologist & Certified Engg Geologist, CA. Mbr Consulting Engrs Assn of CA, Assn of Engg Geologists. *Society Aff:* ACEC, ASFE, AEG, AEP.

Anderson, Harold H
Home: 14 Glen Road, Bridge of Allan, Scotland, UK FK 9 4 PL
Position: Design Eng Consultant Semi Retired *Employer:* Weir Pumps plc *Education:* Hons BSc Mech Eng/Univ of Leeds, England; BSc/Math & Elec Eng/ Univ of Leeds, England *Born:* 07/21/06 Four years workshop experience 1921-24 and Univ vacations 1927 Pump Design Eng, Mather & Platt. 1938 Chief Design Eng Harland Eng Co, Pumps, Turbines up to 100 MW. 1963 Design Consultant Weir Pumps plc 1983, retired, still active. Consultant, Watson Eng Cons 1975 to date. 1939-45, Royal Elec & Mech Eng Ist Class, Lt Colonel British Army in France, Belgium and W Desert Books, Centrifugal Pumps 3rd Ed, 1981, Submersible Pumps, 1986, Score of papers to ASME, IME, IEE and lectures world wide, twelve visits to USA 1951-85. Several patents. Chairman Pump Committees ISO and BSI up to 1981. Member Tech Committees Inst Mech Eng Steam Group and Fluid Group, British Hydromechanics Research Assoc, Brit Pump Mfrs Assoc Europump. Consultant Editor Trade & Tech Press Ltd. Inst Mech Engineers Prize, 1947, IME James Clayton Prizes 1955 and 1961 ASME Worthington Prize 1984, Calvin W Rice Award 1985. *Society Aff:* InstCE, IMechE, CEng, ASME

Anderson, J Edward
Business: 111 Church St SE, Minneapolis, MN 55455
Position: Prof. *Employer:* Univ of MN. *Education:* PhD/Aeronautics/MA Inst of Tech; MSME/Mech Engg/Univ of MN; BSME/Mech Engg/IA State Univ. *Born:* 5/15/27. NACA Langley Field, 1949-51; Honeywell, Inc, 1951-63; Univ of MN Dept of Mech Engg, 1963-. Presently Prof and Dir, Ind Engg Div. Teach courses on tech- soc, tech assessment, ethics in engg, transit systems analysis and design, design. Res and consulting in automated transit, former natl dir of Am Inst of Aero & Astro, founder and first president of Advanced Transit Assn, Chrmn of Intl Conf on Personal Rapid Transit and editor of its proceedings, author of textbook *Transit Systems Theory*, overseas speakers bureua for US Info Agency in 1973 and 1977, spent ten months in USSR as exchange prof sponsored by Natl Academy of Sciences in 1967-68. *Society Aff:* IIE.

Anderson, J Hilbert
Business: 2422 S Queen St, York, PA 17402
Position: Pres. *Employer:* Sea Solar Power, Inc. *Education:* MS/Mech Engg/Penn State; BS/Mech Engg/Penn State. *Born:* 10/20/08. Lewisburg, PA. After Penn State degrees one yr at sea as marine engr. 17 yrs till 1952 with Ingersoll Rand as turbo-machinery designer and Res Mgr developing axial, centrifugal, and reciprocating compressors. Also Diesel engines, turbines, rock drills, and heat exchangers. 1952 to 1963 at York Corp in charge of refrig compressor design. 1963 to present consulting engr. Founded Coupling Corp of America to mfg flexible couplings. Also Sea Solar Power, Inc to develop power, fresh water, and chemicals from tropical waters. Also designed country's first hot water geothermal power plant in Imperial Valley, CA. *Society Aff:* ASME, ASHRAE, NSPE.

Anderson, Jack W
Business: 165 Wildwood Ave, White Bear Lake, MN 55110
Position: Pres. *Employer:* Jack Anderson Associates. *Born:* Oct 30, 1925 St Paul MN. Undergrad Univ of MN Engg; Grad San Fran St Coll Sci & Math 1951; Cert Yale Bureau of Hgwy Traffic 1966. P Pres N Central Sect Inst of Transp Engrs; Fellow Inst of Transp Engrs. PE MN & WI. Prev empl CA DOT & MN Hgwy Dept specializing in hgwy geometric design, traffic opers and traffic safety improvements. Curr applying traffic opers expertise on cons basis to small cities, counties & priv land developers. Expert witness in traffic accident reconstruction.

Anderson, James D
Business: 4751 N 15th St, Phoenix, AZ 85014
Position: VP. *Employer:* Adam, Hamlyn, Anderson Consulting Engg, Inc. *Education:* BS/Civil Engg/Colorado State Univ. *Born:* 2/15/28. Native of Colorado, moved to AZ 1960. Early engg experience was in highways with US Bureau of Public Roads and Alaska Road Commission. Since 1956, has been with Western consulting engg firms, prin since 1965. PPres AZ Sect of ASCE and Fellow in that organization. Served on ASCE National Committees on Surveying and Mapping, Highway Geometrics and Highway Res Bd Committee, AZ, A02. Enjoys golf and antique cars. *Society Aff:* ASCE, NSPE, APWA, SAME, ACEA,.

Anderson, James D
Business: 251 S Lake Ave, Pasadena, CA 91101
Position: VP Business Dev *Employer:* Jacobs Engg Group Inc. *Education:* BSChE/Chem Engg/Univ of TX. *Born:* 7/18/35. Native of Little Rock, Arkansas. High Sch in El Dorado, Arkansas. Grad Bus Mgmt Courses at UCLA. Started up US First Nickel-Cobalt Hydrometallurgical Plant in 1959 producing nickel & cobalt from sulfide deposits mined at Moa Bay, Cuba by Freeport Nickel Proj/Process Engr for Diamond Shamrock in optimizing operation of a Montecatini Acetylene Plant in 1962-64, specializing in reactor design & yield improvement & carbon removal from product gases. Proj Engr for Diamond Shamrock's Agri Chemicals Div in 1964-67, specializing in producing DDT; Arsonates, 2,4,5-T & 2,4-D. Sales engr for Fluor Corp & later for Jacobs Engg. When Jacobs Engrg recently centralized all business dev and mktg efforts nationwide, Jim Anderson was appointed Group VP, Domestic Business Dev. Later he took a new position which he is responsible for all Jacobs' govt facilities sales efforts. *Society Aff:* AIChE, AIME, SME, SAME.

Anderson, James H
Business: 104 Agri Hall, E Lansing, MI 48824
Position: Dean, and V Provost College of Agri & Natural Resources. *Employer:* MI State Univ. *Education:* PhD/Agri Engr/IA State Univ; MS/Agri Engr/NC State; BS/ Agri Engr/Univ of GA. *Born:* 1/11/26. Native of Odom, GA. Served in WWII, Assoc Prof Clemson Univ 1957-60, Prof and Hd of Agri Engr Dept Univ of TN 1960-61, Prof & Hd of Agri Engr Dept MS Ste Univ 1961-67, Dean of Resident Instruction MS State Univ 1967-68, Dir of Agri Expt Station MS State Univ 1969-77, Dean College of Agri & Natural Resources MI State Univ 1977 to present. V Provost 1984 to present, Mich State Univ. Presently responsible for planning, coordinating & administrating progs in res, teaching & ext in College of Agri and Natural Resources at MI State Univ. Elected Fellow mbr of ASAE in 1979. Pres of ASAE, 1984-85. Enjoy flying, gardening & camping. *Society Aff:* ASAE, AAAS.

Anderson, James J
Business: 2304 University Ave, St Paul, MN 55114
Position: Chairman *Employer:* Watermation, Inc. *Education:* Bof CE/Sanitary/Univ MN; MSCE/Sanitary/Univ of MN. *Born:* 6/21/29. Project Engr, Anaerobic Digestion of Meat Packing Wastes 1952-54. WPCF Harrison Prescott Eddy Medal for project - 1956. Detail design of first full scale anaerobic plant at Albert Lea, MN. Project engr TKDA, St Paul, MN 1954-60. Pioneered computer control and modeling for combined sewer overflow control 1966-69. Founded Watermation, Inc in 1969, specializing in modeling and computer control of sewer systems and treatment works in major US cities. Community activities include Jaycees, United Fund, Red Cross, YMCA, Rotary International. Hobbies are antiques airplaces, amateur radio. fishing, hunting. *Society Aff:* ASCE, ISA, WPCF, AWWA.

Anderson, James T
Home: 2170 Royal Dr, Reno, NV 89503
Position: Emeritus VP Academic Affairs and Professor Mechanical Engrg. *Employer:* Univ of NV *Education:* PhD/Heat Transfer/Univ of London; DIC/Imperial Coll; MS/ME/MI State Univ; BS/ME/MI State Univ *Born:* 06/18/21 From instructor to assoc prof, mech engrg, Michigan State Univ 1946-59; prof, 1950-60; visiting prof Tulane, 1960; Chrmn, Dept of Mech Engrg, West Virginia Univ, 1960-63; Dean of Engrg, Univ of Nevada-Reno, 1963-71, VP Academic Affairs, 1970-; Acting Pres, 1973-75; Const Jet Engine Dept, Aircraft Nuclear Propulsion Dept, Gen Elec Co, Cincinnati, 1957-61; Alleghany Ballistic Missile Laboratory, 1961-63. Served to first lieutenant, Signal Corps, OSS, 1943-46. Reg PE, MI and NV. Consultant to Nevada Dept. of Rehabilitation 1984 to Present. *Society Aff:* ASME, NSPE, ТВП, ΣΞ, ΦΚΦ

Anderson, John E
Business: 2083 E. Center Circle, Minneapolis, MN 55369
Position: VPres/Chief Engr *Employer:* GME Consultants, Inc. *Education:* PhD/Structural Eng/Northwestern Univ; MS/Geotech Eng/Northwestern Univ; BS/CE/Northwestern Univ *Born:* 5/25/47 Native of Chicago, IL. Geotech engr specializing in soil-structure interaction problems. Consulting has included nuclear power plant desing, dams, offshore structures, heavy industrial facilities and commerical high rise structures in Houston, Chicago and Minneapolis. His doctoral work was concentrated in geotech and structural analysis for soil-structure systems. He has had extensive experience consulting for soil dynamics and earthquake engrg in the U.S. and overseas. He is registered in 9 states and the District of Columbia. He has authored publications on soil-structure interaction and the use of the pressuremeter. *Society Aff:* ASCE, NSPE, EERI

Anderson, John G
Business: 100 Woodlawn Ave, Pittsfield, MA 01201
Position: Mgr-High Voltage Progs. *Employer:* GE Co. *Education:* BS/EE/VPI. *Born:* 8/21/22. in Dante, VA. Radar and communications officer, S Pacific 13th AF, 1943-45; test engr (GE) 1946; Lightning and High Voltage Res Engr (GE) 1947-1957; Res Engr-System Insulation (GE) 1957-64; Technical Dir, Proj EHV (GE) 1964-67; Mgr, Proj UHV (Ultra High Voltage Transmission) 1967-72; Mgr, AC Transmission Studies (GE) 1972-74; Mgr, GE High Voltage Lab, 1974-79; Mgr, High Voltage Progs, 1979- present. Former Chrmn, IEEE Power Engg Society Transmission and Distribution Committee. Presently Chrmn, Natl Public Affairs Subcommittee. *Society Aff:* IEEE, AAAS, CIGRE.

Anderson, John P
Business: Univ Station, Birmingham, AL 35294
Position: Dean & Prof. *Employer:* Univ of AL in B'ham. *Education:* PhED/Engg Mechanics/GA Inst of Tech; MS/Engg Mechanics/GA Inst of Tech; MS/App Mathematics/GA Inst of Tech; MBA/Business Admin/Univ of AL in B'ham; BS/App Mathematics/GA Inst of Tech. *Born:* 3/27/39. While working on my undergrad degree, I was employed in the Cooperative Education Prog working as an engr. I was appointed Asst Prof at GA Tech upon completion of my second Master's degree. From there, I moved to an Asst Professorship at the US Air Force Academy. After acquiring the rank of Assoc Prof at the Academy, I moved to a position at The Univ of AL in Birmingham as Assoc Prof of Engg. i was appointed to my first admin position, that of Asst to the VP for Univ College in 1971. I served for two yrs as Asst to the Pres for Planning. I now hold the admin position of Dean of UAB Special Studies and the academic positions of Prof of Engg and Prof of Mathematics at The Univ of AL in Birmingham. I continue to teach engg and math courses and sponsor the Society of Black Engrs at UAB. *Society Aff:* ASEE, ASME, ASCE, ΣΞ, ΦΚΦ, ΒΓΣ.

Anderson, Keith E
Business: 161 Mallard Drive, Boise, ID 83706
Position: VP *Employer:* James M. Montgomery, Consulting Engineers, Inc. *Education:* MS/Geology/Univ of Rochester; BS/Geology/CIT. *Born:* 2/27/20. Native of Los Angeles, CA. Ground-water geologist and engr for US Geol Survey and States of Iowa and Missouri, 1941-50. US Army (Europe) 1944-46. Regional drainage and groundwater engr for Pacific Northwest, US Bur Reclamation 1950-57. Consultant in ground water, irrigation, and drainage 1957-date with frequent foreign assignments for UN, World Bank, and Foreign Governments in Latin America, Asia, and Pacific. Editor, Water Well Handbook. PPres Consulting Engrs of ID. PPMbr, Boise, ID City Council. PTrustee, Boise Independent School District. PMbr Idaho Bd. Registr. Prof. Geologists. *Society Aff:* ASCE, NSPE, ICID, AGU, NWWA, AWWA, ISPE, IAPG.

Anderson, Lawrence K
Business: 555 Union Blvd, Allenton, PA 18103
Position: Lab Dir *Employer:* AT&T Bell Labs *Education:* PhD/EE/Stanford Univ; MS/EE/Stanford Univ; BEngg/Eng Physics/McGill Univ. *Born:* 10/2/35. Native of Canada. Grad from McGill Univ, Montreal, Canada as Ernest E Brown Gold Medalist, 1957. Joined Bell Labs in 1961 and worked in optoelectronics, device lithography, integrated circuit packaging, and lighwave component and subsystems. Currently Dir, Electronic Components and Subsystems Laboratory. Fellow of IEEE (1973) for "contributions to holographic optical memories." Pres, IEEE Electron Devices Soc, 1976-77; mbr, IEEE Board of Dirs, 1979-80. *Society Aff:* IEEE.

Anderson, Melvin W
Business: Dept of Civil Eng, Univ S FL, Eng 118, Tampa, FL 33620
Position: Chrmn/Prof. *Employer:* Univ South FL *Education:* PhD/Civ Engg/Carnegie-Mellon Univ; MS/Civ Engg/Carnegie-Mellon Univ; BS/Civ Engg/VA Military Inst. *Born:* 1/9/37. Native of Baltimore, MD. Second Lt US Army Corps at Engr 1959-60. Asst Prof/Assoc Prof LA State Univ 1963-69. Assoc Prof/Prof Univ of S FL 1969- present. Dir of Prog of Distinction in Urban Water Resources 1976-78 USF, Chrmn of Dept of Civil Engrg and Mechs 1979-present. USF, Natl Bd of Dir AWRA 1978-81, VP AWRA 81, Pres Elec AWRA 82, Pres AWRA 83, Pres FL Sec AWRA 1972-73, ASCE Natl Water Resources Systems Comm 1975-present. VP FL Engr Soc 84-85. Pres FL Engr Soc 86-87. Editor Journal of the Hydraulics Div ASCE 1980-82. A mbr of Sigma Xi, Tau Beta Pi, Chi Epsilon, Omicron Delta Kappa, Phi Kappa Phi, and Order of the Engr. Reg PE in LA & FL. *Society Aff:* ASCE, ASEE, AWRA, NSPE.

Anderson, Paul M
Home: 13335 Roxton Circle, San Diego, CA 92130
Position: Conslt *Employer:* Power Math Assoc, Inc *Education:* PhD/EE/IA St Univ; MS/EE/IA St Univ; BS/EE/IA St Univ *Born:* 1/22/26 Native of IA. Recevied BS, MS, and PhD degrees from IA St Univ. IA Public Service Co transmission and distribution engr for six years. Taught and performed research at IA St Univ from 1955-1975. Program mgr in Power Sys Planning and Operations at the Electric Power Research Inst, Palo Alto, 1975- 1978. Founded Power Math Assocs, analytical conslt firm, in 1978. Currently President and Principal Engineer of firm. Electric Power Sys Prof at AZ St Univ, 1980-1982. Fellow IEEE. Enjoys classical music and piloting own aircraft. Author of Analysis of Faulted Power Sys and Co-author of Power Sys Control and Stability. *Society Aff:* IEEE, ASME, CIGRE, ΣΞ.

Anderson, Philip P, Jr
Home: 9933 Crestview Pl, Newburgh, IN 47630
Position: Engg Consultant Retired. *Education:* PhD/Phys Chem/Univ of TX; MS/CE/Univ of TX; BS/CE/Univ of TX. *Born:* 9/20/10. in Calvert, TX. Refrigeration engr 1936-46, Res Supervisor 1946-56, Servel, Inc. Dir of Dev for Absorption 1956-58, Carrier Corp Chief Engr 1959-60, VP Engg Res & Dev 1960-75, Arkla Air Conditioning Co Engg Consultant for Arkla Industries Inc 1975-1980. Recipient of 1975 Technical Achievement Award of Tri- State Council for Sci & Engg. 1979 Fellow ASHRAE. Enjoy gardening & hiking. *Society Aff:* ASHRAE, ACS, ΣΞ, AAAS. ΤΒΠ.

Anderson, Raymond L
Home: 601 Courtland St, Greensboro, NC 27401
Position: Ch Ind Engr. *Employer:* Cone Mills Corp. *Education:* BS/Ind Engg/GA Inst of Tech. *Born:* 12/12/25. Memphis, TN. BIE GA Tech. Served in Army Air Corps, 1944-45. Has worked as Indus Engr in metal forming & fabrication; served as plant Ind Engr, div Oper Res Analyst, Corp Ind Engr for textile mfg: Milliken & Co & Cone Mills Corp. Assumed Chief Ind Engr of Denim Div, Cone Mills Corp, Greensboro, NC 1971. Has served in IIE offices: Chapt Pres 2 different chaps; Reg III Chap Dev Chrmn; IIE Dir of Fac Plan & Design. *Society Aff:* IIE, ATIE.

Anderson, Richard L
Business: Votey, Elec. Engrg. Dept, University of Vermont, Burlington, VT 05405
Position: Prof. *Employer:* Univ of VT. *Education:* PhD/Solid State Elect/Syracuse Univ; MS/Elec Engg/Univ of MN; BS/Elec Engg/Univ of MN. *Born:* 2/4/27. Res Engr with IBM 1952-61. Prof of Elec Engg, Syracuse Univ 1961-79, Univ of VT

Anderson, Richard L (Continued)
since 1979. Visiting Prof Univ of Madrid 1960-61, Univ of Sao Paulo 1967- 69. Maj interests: Electronic materials and devices, solar energy conversion, Low-Temp Elec. Lectured extensively in Latin America. Developed concept of heterojunction and first phototype. Originator of heterojunction solar cell. Published extensively in scientific literature. Fellow IEEE, received honorary doctorate from Univ of Sao Paulo 1970. Received first Brazilian Prize in Microelectronics 1980. *Society Aff:* IEEE, APS, SBF, ECS.

Anderson, Robert
Home: 7315 Glengate Ct, Dayton, OH 45424
Position: Aerospace Engr. *Employer:* USAF. *Education:* BS/Aero Engr/Tri-State College. *Born:* 3/8/32. Native of Kaleva, MI. Served USN 1952-1956. Employed by Rockwell Intl, Vought Corp, Chrysler & Boeing as Weight Engr & McDonnell as Wind Tunnel Test Engr. Employed by USAF at WPAFB, OH since 1967. Responsible for mass properties methodology dev, estimation & control for advanced design tech studies & contractor evaluation conducted at AFFDL. Have received sustained superior and letter of commendation. Activities in local SAWE chapter include pres, vp & treas. Active in church & scouting. Enjoy outdoor sports. *Society Aff:* SAWE.

Anderson, Robert C
Business: 919 FM Rd 1959, Houston, TX 77034
Position: Chrmn of Bd *Employer:* Anderson & Assocs, Inc. *Education:* BS/Met Engr/Univ of IL. *Born:* 7/18/26. Employed Sheffield Steel, div Armco Steel. Employed as Chf Metallurgist WKM div ACF Industries. Employed as Pres Anderson & Assocs, Inc 1957-present. Anderson & Assocs has an extensive lab devoted solely to the analysis of metal failures. About 50% of bus is related to litigation. Have pioneered in the use of metallurgical techniques to solve product liability controversys. Have qualified as an expert in 26 states in both State & Fedl Court. *Society Aff:* ASM, NACE, AWS, ASME, SME, AIMME, ASTM, SNT, NSPE, ACIL, IMS.

Anderson, Robert E
Home: 21 Theodore St, Wharton, NJ 07885
Position: Prof. Emeritus *Employer:* NJIT. *Education:* MS/EE/Univ of NH; BS/EE/Newark College of Engg. *Born:* 2/22/19. Engg field of interest - Logic and Digital Circuit Design. Engg Consulting- Fed Shipbldg and Dry Dock Co, Convair (Astronautics), Boeing, Douglas Airplane Co, TVA, Brookhaven Natl Labs, AT&T, NYT, NJBT, ITT, IBM, Westinghouse, Bryne Assocs. LCDR-USNRET. PE License NJ & NH. Councilman and Mayor, Wharton, NJ 1954- 60. Mbr, Bd of Dirs Microlab/FXR 1969-76. Chrmn, IEEE NY Sec Communications Group 1974-75. Pres, ECSPE 1975-76. Faculty mbr, UNH 1947-49. Faculty mbr, NJIT, 1949-83. Acting Chrmn, EE Dept, NJIT 1975-76. Acting Assoc. Dean of Engineering 1981-83. Professor-Emeritus 1983-. Hobbies-organic gardening, traveling, history (Panama), and poetry. *Society Aff:* IEEE, NSPE, ASEE.

Anderson, Robert H
Business: 91 Roseland Ave, Caldwell, NJ 07006
Position: Vice President *Employer:* Purcell Assoc *Education:* M/BA/Bus Admin/Univ of New Haven; BS/CE/PA State Univ *Born:* 9/3/39 Native of Rahway, NJ. Employed by a major consulting engrg firm in Western PA till 1967. City Engrs Office in CT community handling engrg at airport and capital projects within the City until 1972. General Mgr of a Housing Construction Co in CT, bldg approximately 1,000 units annually, until 1976. Assumed position as Principal Engr/General Mgr of this office in 1976 with responsibility of the technical and administrative duties of performing consulting engrg services. This includes domestic and intl assignments. Promoted to current position of VP in 1983. *Society Aff:* ASCE, NSPE, ACEC

Anderson, Robert L
Business: 3707 Cody Rd, Columbus, GA 31907
Position: Self Employed *Education:* BS/CE/MS St Univ; LLB/Contract Law/Southeastern Univ *Born:* 2/22/98 Native of Jackson, MS. Civilian employee with Bureau Docks, Navy Dept, Washington, DC. I was in responsible charge of Structural Design of framing and fdns for many large and important structures. My particular interest was promoting the use of precast and prestressed concrete for water front structures. Since retirement I have been active as consultant to architects, contractors, attorneys and Insurance Cos for vehicle accidents reconstruction and structural problems. Numerous appearances in court as expert witness or consultant for causes of accidents at train crossings, vehicle accident reconstruction and structural or fdn failures. Reg PE in five states. *Society Aff:* ASCE, SAME, ACI, PCI, NSPE, GSPE.

Anderson, Robert M
Home: 128 Hurd Rd, Trumbull, CT 06611
Position: Mgr-Tech Ed Oper *Employer:* GE Co-260SW *Education:* PhD/EE-Solid State/Univ of MI; MS/Phys/Univ of MI; MSE/EE/Univ of MI; BSE/EE/Univ of MI; Assoc/Pre-Engineering/Flint Jr College. *Born:* 2/15/39. Began as Asst Prof of Elec Engg at Purdue Univ 1967. Specialized in solid- state mtls and devices. Assoc Prof of EE 1971-79. Dir of Continuing Engg Education 1973-79. Ball Brothers Prof of Engg 1976-79. Prof of EE 1979. Joined GE in 1979 as Mgr of Engg Education and Training for Corporte Consulting Services. In 1982 named mgr of Tech Ed Oper for Corp Engrg and Mfg. Active in ASEE holding Chapter, Section and Natl offices. Active in American Vacuum Soc holding Chapter, Region and Natl offices. Sr mbr of IEEE. 1974 Outstanding Young Faculty IL-IN Sec ASEE. 1986 Distinguished Service Award, Continuing Prof Dev. Division, ASEE. Listed in Who's Who in America. PE in IN. Married, father of two, Christian. *Society Aff:* IEEE, ASEE

Anderson, Robert T
Business: 2255 McCoy Rd, Columbus, OH 43220
Position: Pres. *Employer:* Robert T Anderson Assoc. *Education:* BS/Ind Engg/OH State Univ. *Born:* 3/26/31. Section Mgr, Nondestructive Test Engg, Aerojet-General Corp, Sacramento, CA 1960-66. Hd, Nondestructive Testing Research Engg, General Dynamics Convair, San Diego, CA 1966-75. Tech Dir, ASNT, 1975-80. Faculty mbr, San Diego Mesa College, 1968-75. EJC Engg of Distinction 1973. ASNT National Secretary 1974-75; Dir 1971-74. Lecturer and author of more than 40 tech papers. Consultant in nondestructive testing and reg PE. *Society Aff:* ASNT, ASTM, ASM, ASQC, AWS, OSPE.

Anderson, Roy E
Business: 65 Valley Stream Parkway, Suite 110, Malvern, PA 19355
Position: Vice President *Employer:* Mobile Satellite Corp *Education:* MSEE/EE/Union College; BA/Phys/Augustana College. *Born:* 10/30/18. 2 yrs Navy WWII incl radar sch at Bowdoin & MIT; 2 yrs phys instr Augustana Coll. With GE 1947-1983; R & D in indust electronics, doppler radio dir finders, satellites for small terminal & mobile applications, ionospheric propagation. Originated & dev tech for vehicle pos surveiliance by ranging from satellites. Co-founder of Mobile Satellite Corp. Participant in govt studies & prof soc cttes. 100 tech publs; 18 patents. Fellow IEEE; mbr: AIAA, ION, Fellow AAAS, Coolidge Fellow of GE Corp R & D. Fellow, Radio Club of America. *Society Aff:* IEEE, AIAA, ION, AAAS, NASAR, RCA.

Anderson, Stanley W
Business: PO Box 767, System Planning Dept, Chicago, IL 60690
Position: Staff Asst. *Employer:* Commonwealth Edison. *Education:* PhD/Electric Power/IL Inst of Tech; MS/Electric Power/IL Inst of Tech; BS/Electrical Engg/IL Inst of Tech. *Born:* 8/2/21. Fellow of IEEE. Reg PE. US Army Signal Corps, 1942-46. Power transformer design and dev at GE, 1950-52. Engg consultant and project leader, Middle West Services, 1952-68. Here he applied OR, math and computer techniques. From 1968, Commonwealth Edison, as consultant to System Planning. Used OR, engg and manth tech. Taught company courses. From 1970 taught grad courses at IIT additionally. Sponsored many successful PhD theses and many MS theses. Chmn electrical div of the American Power Conference. Author of thirty-five

Anderson, Stanley W (Continued)
tech papers including 1973 IEEE-PES prize paper. *Society Aff:* IEEE, ORSA, ASEE, ΣΞ

Anderson, Thomas P
Home: 1218 Sherwood Rd, Columbia, SC 29204
Position: Retired. *Education:* MSSE/Sanitary Engg/Univ of NC; BSCE/CE/Clemson Univ. *Born:* 2/20/08. Retired Chief, Bureau of Water Hygiene & Special Services, SC Dept of Health & Environmental Control. Secy-Treas SC Bd of Certification of Environmentl Systems Operators. Tau Beta Pi. Fuller Award AWWA. Wiedeman Awad SE Sec AWWA Life Mbr SC Public Health Assoc. Past Secy-Treas & Chrmn SE Sec AWWA. Past Pres Assoc of Bds of certification Past Pres SC Water & Sewage Works Assoc. Past Chrmn Interstate Sanitation Seminar. Elder Presbyterian Church. Kiwanis Club (Golden K). Retired Lt Col Sanitary Corps US Army. Reg PE (SC). *Society Aff:* AWWA.

Anderson, Thor L
Business: 417 Montgomery St, San Francisco, CA 94104
Position: Sr VP *Employer:* Sverdrup & Parcel and Assoc, Inc. *Education:* MSCE/Structures/Lehigh Univ; BSCE/Struct Design/WA Univ. *Born:* 3/16/32. Native of St. Louis, MO; Officer in USCG 1953-55; Jointed Sverdrup in 1955; became mgr of Sverdrup's Phoenix office in 1973. Became mgr of Sverdrup's San Francisco office 1975, currently Sverdrup's Western Regional Mgr. Significant consulting engg assignments include: Proj Mgr for several cement plant expansions; Proj Engr for the design of the LA Superdome; Proj Structural Engg for the HH Humphrey Bldg in Washington, DC. Engg Awards include James F Lincoln Arc Welding Fdn 1975; ASCE Thomas Fitch Rowland Prize 1977. PPres of the St. Louis Sect ASCE, Pres of Golden Gate Branch ASCE. *Society Aff:* ASCE, SAME.

Anderson, Wallace L
Business: 4800 Calhoun, Houston, TX 77004
Position: Prof. *Employer:* Univ of Houston. *Education:* ScD/Elec Engg/Univ of NM; MA/Phys/Rice Univ; BS/Elec Engg/Univ of ND. *Born:* 9/2/22. Employment: 1948-56, party mgr, res engr, McCollum Exploration Co; 1957-61, res assoc, teaching assoc, lecturer, Univ NM; 1961-64, Assoc Prof EE, NYU; 1964- 68 Sr Res Engr and 1968-69, Staff Scientist, Southwest Res Inst; 1965-69, Assoc Prof Physics (part-time), Trinity Univ; 1969-72, Prof, EE, 1972-77, Chrmn of Dept and 1977-present, Prof EE, Univ of Houston. Res grantee Natl Inst of Health and Natl Sci Fdn during most of past ten yrs. Approximately 60 publications in the profl literature. Sr mbr IEEE, mbr numerous honorary societies. *Society Aff:* AAAS, IEEE, APS, OSA, SPIE.

Anderson, Walter T
Business: Sch of Tech, Houghton, MI 49931
Position: Prof & Dir *Employer:* MI Tech Univ. *Education:* MSEE/Power/MI Tech Univ; BSEE/Electronics/MI Tech Univ. *Born:* 3/30/23. Test Engr GE 1943-44; Process Engr TN Eastman Oak Ridge 1944-45; EE Instr MTU 1946-51; Proj Engr Union Carbide Nuclear Oak Ridge 1952-54; EE Prof MTU 1954- Admin Asst 1970-81; Asst Dept Head, 81-84; Dir, Sch of Tech, 84-. Cons Engr 1956-. Prof Engr: MI, WI. Mbr: MI Bd Regis for PE 1966-81; for Land Surveyors 1970-81. VP & Dir NCEE 1973-75; Sr Mbr IEEE; Mbr: ASEE, NSPE, IES, MI Assn of Professions; Affiliate Mbr MI Soc of Reg Land Surveyors. Award of Merit Central Zone NCEE 1973. Distinguished Service Certificate NCEE 1978. *Society Aff:* NSPE, IEEE, ASEE, IES.

Anderson, Warren R
Home: 573 Jeffrey Dr, San Luis Obispo, CA 93401
Position: Emeritus Head and Professor EL/EE Dept *Employer:* CA Polytech State Univ. *Education:* BS/Elec Engr/LA State Univ; BS/Agri Sci/Univ of MN; AA/Science/Bethel College. *Born:* 7/31/14. Native of Houston, MN. Taught high school agriculture, Windom, MN prior to World War II. Served in US Army Signal Corps, 1942-45; detached to Plant Engg Agency, Phila. Switching circuitry engr, Automatic Electric Co, Chicago. With CA Polytechnic State Univ since 1946; Prof since 1959; Head, Electronic & Elec Engr Dept since 1975-80. Part-time and consulting engg assignments with Gen Elec Co, Northrup Aircraft Co, Western Gear Corp, IBM. Charter mbr, CA Society of Professional Engrs. Numerous comm chairmanships and assignments with the Los Angeles Chapter of AIEE and the Los Angeles Sec of IEEE. Enjoys gardening, reading and politics. *Society Aff:* IEEE, ASEE.

Anderson, William C
Business: 69 South St, Auburn, NY 13021
Position: Managing Partner. *Employer:* Pickard and Anderson. *Education:* BS/Civ Engg/IA State Univ. *Born:* 9/24/43. IA and NY State Health Depts 1963-67; US Navy 1967-68; Dir of Environmental Health, Cayuga County, NY, 1968-73; Formed Pickard and Anderson in 1973; Managing Partner 1978-; PE in NY, NJ, PA and IA. WPCF Philip F Morgan Medal 1973; NYWPCA Lewis Van Carpenter Award 1978; Diplomate Am Acad Environ Engrs; Mbr of Chi Epsilon; ASCE Solid Waste Mgt Committee, Chrmn 1974-76; ASCE Environ Engg div, Chrmn 1979-80; ACEC Environmental Committee; 1977-1981 ACEC Solid Waste Task Force, Chrmn 1977-80. Expert Witness US Congress 1975, 1976, 1978, 1980. Adjunct Assoc Prof Civ Engr, Syracuse Univ 1978-. Consultant to Masters of Engrg Program Cornell Univ 1979-80. ASCE Management Group D 1983- ; Hon Conceptor Award-MI CEC; Trustee Amer Acad of Environ Engrs 1982- ; Author of 14 prof papers and articles. *Society Aff:* CSI, ASCE, NSPE, WPCF, ASTM.

Anderson, William M
Business: 6551 S. Revere Pkwy, Englewood, CO 80111
Position: Pres *Employer:* Norsaire Corp *Education:* MS/ME/IL Inst of Tech; BS/ME/Marquette Univ *Born:* 4/29/43 Graduated Marquette Univ, BSME 1965. Awarded National Science Foundation Fellowship; IL Inst of Tech, MSME, 1967. Areas of specialization include fluid dynamics, heat and mass transfer, HUAC. Hold 6 patents in areas of solid waste incineration and air pollution control. Employed by CE-Air Preheater, Wellsville, NY, 1967-1977 in various technical and management positions. In 1977 joined Cargocaire Engrg Corp. Served as VP of Operations, responsible for all engrg and manufacturing activities. Recently joined Norsaire Corp as Pres. Norsaire manufacturers and sells low energy air conditioning equipment for commercial and industrial applications. Married. Two children. *Society Aff:* ASME, ASHRAE

Andrade, Joseph D
Business: Coll of Engrg, Salt Lake City, UT 84112
Position: Prof and Dean *Employer:* Univ of UT *Education:* PhD/Matls Sci/Univ of Denver; BS/Mater Sci/San Jose State Univ *Born:* 7/13/41 Professor of Bioengineering, materials science, and pharmaceutics, Univ of UT. Dean, Coll of Engrg since 7/83. Research activities: protein interactions with solid surfaces, biochemical sensors, fiber optics, polymer surface chemistry, biomedical materials. *Society Aff:* ACS, APS, BMES, ASEE, OSA.

Andresen, T Richard
Business: 414 Nicollet Mall, Minneapolis, MN 55401-1993
Position: Supt-Plant Engineering *Employer:* Northern States Power Co *Education:* BME/Mech Engg/Univ of MN *Born:* 6/27/41 Native of Crystal, MN. Erie Mining Co-1965-1967. Joined Northern States Power Co in 1967 at Riverside Plant. 1972 High Bridge Superintendent - Plant Engrg. Past Pres NSP Engrs Assoc. Organized SETAC (Scientists and Engrs Technological Assessment Council). Taught advanced steam engrg - Dunwoody Industrial Inst. Past Pres, MN Society of PE. Toastmasters. Past Pres of MN-Prof Engrs in Industry. Natl NSPE Dir. NSPE Regional Vice Chairman - Professional Engineers in Industry. Past Pres of TwinLakes Chapter of MSPE. Engineer-in-Charge, one of Minnesota's Seven Wonders of Engrg. Registered Prof in MN & SD. St. Paul River Corridor Planning Task Force. Past Pres of Minnesota Prof Engrs Foundation. *Society Aff:* NSPE, ASME, ASQC.

Andrew, James D, Jr
Home: 5599 Shadowlawn Dr, Sarasota, FL 33581
Position: Exec Engr. *Employer:* Babcock & Wilcox. *Education:* BS/IE/Yale-Sheffield. *Born:* 3/24/04. Brooklyn NY. With United Engrs and Constructors 1926-30; Babcock & Wilcox 1930-69; Uniform Boiler & Pressure Vesels Laws Soc Chrmn 1969-73. Life Fellow ASME; Boiler & Pressure Vessel Code Ctte 1958-69; ISO/Tech Comm Boilers Plenary Chrmn Sweden 1970, Power Div Chrmn 1966. B & W: Mgr Codes & Stds; Proj Engr-1st supercritical pressure once-through boiler in USA 1957. Hon Mbr: ASME, B&PV Code Ctte; Natl Bd of Boiler & Pressure Vessel Inspector; Am Boiler Mfg Assn.

Andrews, Al L
Business: Division of Technology, Tempe, AZ 85287
Position: Asst Prof. *Employer:* AZ St Univ. *Education:* EdD/Technical Education/Univ of MO; MS in Ed/Trade & Industrial/Southern IL Univ; BS/Mfg Processes/Southern IL Univ; Assoc of Tech/Tool & Die/Vocational Tech Ins; Assoc of Pre Engr/Mech Pre Engr/Centralia Jr College. *Born:* 9/15/42. Employed by Rock Valley College since 1967 and presently is chairperson of the Div of Tech. A Certified Mfg Engg and past chairperson of the Rockford Chapter American Soc of Metals. During the past three yrs a mbr of the Education Committee natl ASM and Outstanding Young Mbr for 1978. Has worked as Plant Mgr; Fox Valley Packaging Co; Asst R&D Engr; Ipsen Industries; Lubrication Engr; LSP Industries. A training Specialist for a number of Rockford cos and Mech Technical Engg educator for Rock Valley College. Bd of Dir and Chrmn of Ed Ctte of newly formed Comp Automated Sys Assoc Rockford Chapter. *Society Aff:* SME, ASM, CASA, ATEA, NAIT, FES, ASEE.

Andrews, Arlan R
Business: First and Sycamore St, Coldwater, OH 45828
Position: Product Mgr *Employer:* AVCO New Idea *Education:* BS/Agric Engr/IA State Univ *Born:* 3/13/41 Native of Central IA. Design engr for AVCO New Idea 1964-1970 specializing in design of powered harvesting and processing equipment. Product engr with Huffy Corp 1970-1975 specializing in design and production of powered equipment for lawn and garden care. Returned to AVCO New Idea in 1975 as Product Mgr. Current responsibilities relate to overall product management with primary activities associated with identification of new products, product design, and marketing. Chrmn OH Section ASAE 1981. Vice Chrmn OH Section ASAE 1980. Secretary-Treas OH Section ASAE 1979. *Society Aff:* ASAE

Andrews, Harold J
Business: 449 N Cox Rd, Gastonia, NC 28054
Position: Pres. *Employer:* Lithium Corp of Am. *Education:* BChE/CE/Univ of MN. *Born:* 7/1/28. Faribault, MN 1928. Educated MN. Primary HS & Univ. Joined Lithium Corp of Am 1951. Started Great Salt Lake Minerals & Chem in UT as JV with LCA 1965-1973. Served as Pres 3 yrs during construction and start up. VP Planning Gulf Resources & Chem Houston, TX 1973-1976. Gulf Resources Parent Co of GSL and LLCA. Pres Lithium Corp of Am (LCA) 1/8/77 thru present. *Society Aff:* AIChE.

Andrews, John F
Business: Civ Engg Dept, Houston, TX 77004
Position: Prof. *Employer:* Univ of Houston. *Education:* PhD/Sanitary Engg/Univ of CA; MS/Civ Engg/Univ of AR; BS/Civ Engg/Univ of AR. *Born:* 7/10/30. Native to AR. Asst to Assoc Prof at Univ of AR from 1953-60. Assoc Prof to Dept Hd at Clemson Univ from 1963 to 1974. Prof of Civ & Environmental Engrg at Univ of Houston 1975-1982. Prof of Environ Engrg at Rice Univ in 1982. Engaged in teaching and res on the design and operation of wastewater treatment plants. Consultant on design and operation of treatment plants to several large cities and industries. Has organized several intl workshops to bring researchers and practioners together. Awarded Harrison Prescott Eddy Medal by Water Pollution Control Fed in 1975 for outstanding res. Received Assn of Environmental Engrg. Prof Award in 1976 for guidance of outstanding PhD dissertation in Environmental Engrg. Reg P E in AR, SC, TX. *Society Aff:* AAEE, ASCE, AIChE, ASEE, AEEP.

Andrews, Raynal W, Jr
Home: 150 Guyasuta Rd, Pittsburgh, PA 15215
Position: Consultant (Self-employed) *Education:* ME/ME/Rensselaer Poly Inst . *Born:* 6/23/06. Instr in ME at Rensselaer and Engg Ext for PA State. 41 yrs with Alcoa including 5 yrs as Chief Mech Engr and 4 yrs as Chief Environmental Engr. 2 yrs as consultant to EPA. 30 yrs as ASME rep on Dust Explosion Hazards Committee. Do limited engg consulting. Expertise in Dust Explosion Hazards, Light Metal Fabrication, Heavy Ind Hydraulics, Gen Mech Engg & Env Engg. Currently listed in International Who's Who in Engineering, First Edition, International Biographical Center, Cambridge, England. *Society Aff:* ASME.

Andrews, Robert V
Home: 114 Medford Dr, San Antonio, TX 78209
Position: Prof. *Employer:* Trinity Univ. *Education:* BS/Chem Engr/OR State Univ; MS/Chem Engr/TX A&M Univ; PhD/Chem Engr/TX A&M Univ *Born:* 7/4/16. Portland, OR. Chem Engr for Tidewater Assoc Oil Co 1938-39. Served as Airborne Radar Officer for US Air Force, WWII. Taught Chem Engg, TX A & M Univ 1940-52 with res work for TX A & M Res Fdn. Dean of Engg and Dir of Res Ctr, Lamar Univ until 1961. Dean of Engg, Trinity Univ, until 1976. Prof of Engg Sci, Trinity Univ, 1976-82. Visiting prof, Monsanto Chem Co, 1948-49; Gen Elec Co, Hanford Works, 1951-52; Dow Chem Co, 1954. Mbr Res Advisory Comm to coordinating Bd, TX College and Univ System, 1966-67. Trustee, Southwest Res Inst. Author of articles in field. Prof Emeritus Trinity Univ 1982- . Reg PE TX & OH. *Society Aff:* ΣΞ, TBΠ, NSPE, TSPE, AIChE, ASEE.

Ang, Alfredo H S
Business: 3129 Newmark C.E. Lab, 208 N. Romine, Urbana, IL 61801
Position: Prof. *Employer:* Univ of IL. *Education:* PhD/Structural Engg/Univ of IL; MS/Structural Engg/Univ of IL; BSCE/Civ Engg/Mapua Inst of Tech. *Born:* 7/4/30. Came to US 1955, naturalized 1968. Struct design 1954-55 Manila; Asst Prof 1959-62; Prof 1965 Univ of IL, Mbr NAE; ASCE res prize 1968, State-of-Art award 1973, AM Freudenthal Medal 1982, ASEE Sr Res Award. Res Struct Engr IL. Assoc Fellow AIAA; Mbr: ASCE, ASME, IABSE, SSA, EERI, Sigma Xi, Phi Kappa Phi; Chrmn ASCE STD Tech Ctte on Struct Safety & Reliability. ASCE STD Exec Committee 1985-89. Res & cons in struct reliability probability-based design, risk & decision analyses. Auth over 120 tech papers & articles, two-volume text on "Probability Concepts in Engrg Planning and Design–. *Society Aff:* ASCE, ASME, AIAA, ASEE, SSA, EERI.

Angel, Edward S
Business: Dept Computer Science, Univ. New Mexico, Albuquerque, NM 87131
Position: Prof & Chmn, Comp Sci & Prof, Elec Comp Eng. *Employer:* Univ NM *Education:* PhD/EE/Univ So CA; MS/EE/Univ So CA; BS/Engrg/CA Inst of Tech *Born:* 1/6/44 Received PhD 1968. Academic positions at Univ of CA at Berkeley, Univ of Southern CA, Univ of Rochester and Univ of NM. Author or Coauthor of three books and over 50 papers. Senior Fulbright Lectureship 1981-82, Indian Inst of Science Bangalore. Research in image processing, systems, biomedical engrg, computer graphics, robotics, course author and lecturer for integrated computer sys, inc. Chmn, Comp Sci Dept, Univ of New Mexico. *Society Aff:* IEEE, COMPUTER SOC, ΣΞ, HKN, ACM.

Angelakos, Diogenes J
Business: Cory Hall, Berkeley, CA 94720
Position: Prof of EE, Dir of Elec Res Lab. *Employer:* Univ of CA. *Education:* PhD/Engg Sci Applied Physics/Harvard Univ; MS/Engg Sci Appl Physics/Harvard Univ; BS/Elec Engg/Univ of Notre Dame. *Born:* 7/3/19. Prof of Elec Engg (since 1951) Univ of CA, Berkeley. Married Helen Hatzilambrou (1946)(Dec 1982); two children: Erica (dance Instructor) and doctoral student (UCB) Demetri (dec 1979

Angelakos, Diogenes J (Continued)
age 25). Liaison scientist ONR London (1946); Dir. of the Electronics Res. Lab (1964-85); Guggenheim fellow (1957); Fellow IEEE; Sigma Xi; EKN; TBP; Coauthor (Everhart) Microwave Communications, 1968, numerous technical articles. *Society Aff:* AAAS, IEEE.

Anghaie, Samim
Business: 202 Nuclear Sciences Center, University of Florida, Gainesville, FL 32611
Position: Assoc Prof of Nuclear Engrg *Employer:* Univ Florida, Dept Nucl Engrg Sci *Education:* PhD/Nuclear Engrg/PA State Univ; MS/Phys (Engrn, Phys)/Pahlavi (Shiraz) Univ; BS/Phys (Engrn, Phys)/Pahlavi (Shiraz) Univ *Born:* 09/18/49 Faculty member at Pahlavi (Shiraz) Univ 1974-75; Faculty mmbr at Univ of Baluchistan, 1975-1978; Asst Prof of Nuclear Engrg at Univ of Florida 1982-84; Asst prof of Nuclear Engrg at Oregon State Univ 1984-86; Assoc Prof of Nuclear Engrg at Univ of Florida 1986-present. Con-current position: Senior Research Engr, Florida Nuclear Associates, 1984-present. Sci. Res.: Power reactor design, space reactor design, reactor thermal hydraulics, mass, momentum, energy and radiation transport phenomena, two-phase flow and heat-transfer, advanced computational methods, development of large computer codes. *Society Aff:* ANS, ASME, ΣΞ, ΦΚΦ, ISA

Angiola, Alfred J
Business: One Aerial Way, Syosset, NY 11791
Position: Pres *Employer:* Lockwood, Kessler & Bartlett, Inc *Education:* MS/Engrg Mech/Columbia Univ; BS/Civ Engrg/Columbia Univ *Born:* 02/20/47 An Alumnus of Stuyvesant High School in NY City, Mr Angiola joined the Grumman Aerospace Corp upon graduating from Columbia Univ in 1968. While one of Grumman's first Masters Fellows, he participated in the Lunar Module proj as a struct design engr. He joined Parsons Brinckerhoff Quade & Douglas, Inc in 1973, and five years later was placed in charge of that Company's Advanced Tech Div. After a stint as Eastern Regional Dir of a natl environ consltg firm, Mr Angiola joined Lockwood, Kessler & Bartlett, Inc as Environ Prog Mgr in 1982. In 1983, he was named Gen Mgr of LKB's parent co, Viatech, Inc, and in 1984 returned to LKB as Pres. A PE, Mr Angiola has written and testified extensively in the field of environ analysis. *Society Aff:* ASCE, APCA, NSPE, SAME, WPCF

Angus, Donald L
Business: 1127 Leslie St, Don Mills, Ontario, Canada M3C 2J6
Position: Pres. *Employer:* H H Angus & Assoc Ltd. *Education:* BASc/Engg/Univ of Toronto. *Born:* 1917 Toronto. BASc Univ of Toronto 1941. Can Army 1941-45 in Canada & overseas in NW Europe; Ret Capt. Pres H H Angus & Assocs Ltd, 150 man firm of cons engrs specializing in mech & elec engg. P Pres of PE of Ontario. Fellow: ASHRAE, EIC, RSH. Married; 2 children. *Society Aff:* ASHRAE, EIC.

Angus, John C
Business: Dept of Chem Engr, Case Western Reserve Univ, Cleveland, OH 44106
Position: Prof of Engr. *Employer:* Case Western Reserve Univ. *Education:* PhD/CE/Univ of MI; MS/CE/Univ of MI; BS/CE/Univ of MI. *Born:* 2/22/34. in Grand Haven, MI. Res Engr with 3M Co, St Paul 1960-63. At Case Inst of Tech since 1963 as Asst Prof, Assoc Prof & Prof. Chrmn of Chem Engg Dept 1974- 80. Interim Dean of Engg, 1986-87. Sr NATO Fellow & Visiting Lectr, Univ of Edinburgh, Scotland, 1972-73. Visiting Prof. Northwestern University, 1980-81. Principal res interests are in crystal growing, thermodynamics, electrochemistry, coal utilization, sulfur chemistry, & towed trolling devices. Active since 1965 in various aspects of metastable diamond growth & the morphology of diamond surfaces. Also have worked extensively in advanced electrochemical plating techniques and in the thermodynamics of complex redox systems. *Society Aff:* AIChE, ACS, ECS.

Annestrand, Stig A
Business: PO Box 3621, Portland, OR 97208
Position: Manager Research and Development *Employer:* Bonneville Power Administration. *Education:* MSEE/EE/Royal Inst of Tech; BSEE/EE/Royal Inst of Tech. *Born:* 9/18/33. in Husby, Sweden. Naturalized 1972. MSEE, Royal Inst of Tech, Stockholm Sweden. ASEA's high voltage lab, Sweden: res engr 1958-61, mgr of res 1962-67. Bonneville Power Admin since 1967: hd, high voltage unit 1967-74; hd, elec investigations section 1974-77; chief, branch of labs 1977-1981; manager, research and development 1981-85. 1986-87 Congressional Fellow, US Congress. 1987- manager, research and development, Bonneville Power Administration. Responsible for extensive elec, mech, & chem test progs & res & dev projs associated with EHV and UHV power transmission. Coauthor std handbook for elec engrs. Fellow IEEE. Reg PE, OR. Active in scouting. *Society Aff:* IEEE, CIGRE, SER, CF.

Annicelli, John P
Home: Troy Ln, Bedford, NY 10506 *Employer:* Prudential Bache *Education:* BCE/Civil Engg/Manhattan College; MS/Env Engg/RPI; Juris Dr/Law/St John's Univ. *Born:* 1/20/36. Resident of Bedford Village, NY. Investment Banker, Former Partner Loeb, Rhoades & Co 1974-1978. Served with the NY State Environmental Facilities Corp and Chief Engr Metrolpolitan Region 1970-1972. Former Pres of NY State Soc of PE *Society Aff:* NYSSPE.

Ansel, Gerhard
Home: 317 Linden Ln, Lake Jackson, TX 77566
Position: Sr Res Specialist. *Employer:* Dow Chem Co. *Born:* June 1912. MS Phys Met Carnegie Inst of Tech; SB MIT. X-ray diffraction studies at Metals Res Lab CIT 1935-37. Co-recip AIMME Henry Marion Howe Award for work at MRL. With Dow Chem Co Midland MI, Madison IL, Freeport TX 1937-. Work has covered res on magnesium, magnesium alloys & fabrication processes. Responsibilities have covered engg, production & mgt functions in plant casting, extruding & rolling alluminum & magnesium alloys. 6 patents on fabrication of magnesium & 7 tech publs.

Ansell, George S
Home: 1722 Illinois St, Golden, CO 80401
Position: Pres *Employer:* CO School of Mines *Education:* PhD/Met Engr/RPI; MMetE/Met Engr/RPI; BMetE/Met Engr/RPI. *Born:* 4/1/34. Served with US Navy 1955-58. Metal Physics Consultant Staff, US Naval Res Lab 1957-58. Faculty mbr RPI 1960-1984. Robert Hunt Prof 1967-84, Chrmn Materials Div 1969-74. Dean of Engg 1974-1984. Pres CO School of Mines 1984- date. Chrmn Inst of Metals Div AIME 1975. Exec Comm, 1974-78, and Treas, 1979-84, Pres 1986 the Metallurgical Soc of AIME, Advisory Committees for various govt agencies and natl labs, industrial consultant, Hardy Gold Medal AIME 1961. Bradley Stoughton Award ASM 1968. Fellow ASM 1972. Fellow The Metallurgical Society of AIME 1980. Distinguished Faculty Award RPI 1973. *Society Aff:* AIME, ASM, NSPE, ASEE.

Anthony, Rayford G
Home: 1102 Sul Ross Dr, Bryan, TX 77802
Position: Prof *Employer:* TX A&M Univ *Education:* PhD/ChE/Univ of TX (Austin); MS/ChE/TX A&M Univ; BS/ChE/TX A&M Univ *Born:* 12/26/35 Co-author of the textbook, Fundamentals of Chem Reaction Engrg. Published more than 60 manuscripts on kinetics, thermodynamics and Chem Reaction Engrg. General Chrmn of Instrumentation Symposium for the Process Industries 1966 - present. Employed by TX A&M Univ 1966 - present. Industrial consultant 1966 - present. *Society Aff:* AIChE, ACS

Anthony, William S
Business: PO Box 256, Stoneville, MS 38776
Position: Agri Engr. *Employer:* US Cotton Ginning Lab. *Education:* MS/Agri Engr/MS State Univ; BS/Agri Engr/MS State Univ. *Born:* 7/18/45. Native of Leland, MS. Agri Engr at the US Cotton Ginning Res Lab in Stoneville, MS since 1971 with emphasis on mtls handling and noise abatement. Author of 40 publications. Received Army Commandation Medal in 1977 for meritorious service as Engr Co Commander. Presently serves as a Construction Engr for the US Army Reserve and holds the rank of Maj. Served as Secy-Treas of the MS Sec of the ASAE in 1973-74 and as

Anthony, William S (Continued)
Chrmn in 1976-77. Received Young Engr of the Yr Award for the MS Sec of the ASAE in 1978. *Society Aff:* ASAE, SAME, ΓΣΔ.

Anton, Nicholas G
Home: 2501 Antigua Terrace A3, Coconut Creek, FL 33066
Position: Consultant (Self-employed) *Education:* -/Physics/Columbia Univ; -/Engg/Tech Inst Leonards da Vinci. *Born:* 12/14/06. Consultant, Physicist, Engr; Born Trieste, Austria 1906. Graduate Technical Ist Leonard Da Vinci 1926. Student Columbia Univ 1926-28. Various Position Engr Duovac Radio Tube Corp 1928-31. Bd Chrmn, Pres, Dir Engg-Res & Dev-Electronic Labs Inc 1931-32-Amperex Electronic Products Inc 1932-48-Anton Electronic labs Inc 1948-61-Eon Corp 1961-78. Bd Chrmn, Pres-Anton IMCO Corp 1959-61-Dosimeter Corp 1963-75 Fellow-Inst Elec Electronic Engrs. Fellow-Amer Physical Soc. Fellow-NY-Acad Scis Fellow-Amer Assoc Advances Scie. Assoc Fellow-NY Acad Medicine. Mbr Amer Mathematical Soc. Mbr-Soc Mechanical Soc. Mbr-Society Mech Engr, etc. Published numerous papers. 28 pats in fields. *Society Aff:* IEEE, APS, NYAS, NYAM, AAAS.

Anton, Walter F
Home: 1959 Wrenn St, Oakland, CA 94602
Position: VP *Employer:* Tudor Engrg Co *Education:* MS/CE/Univ of CA (Berkeley); BS/CE/Univ of CA (Berkeley) *Born:* 1/24/36 Two years in USNC Civil Engrs Corps and two years with Kaiser Engrs. Seven years with Harza Engrg Co, with increasing responsibilities in the design of large dams, underseepage control systems, large cooling reservoirs for two steam power plants, and other hydraulic structures. Ten years in charge of the Engrg Dept of East Bay Municipal Utility District involving distribution system planning, design and field engrg activities of a major San Francisco area water agency -- most recently Assistant General Mgr and Chief Engr. Currently VP in charge of engrg operations of the San Francisco regional office of Tudor Engrg Co. *Society Aff:* ASCE, AWWA, AAEE, USCOLD, ΤΒΠ

Antoni, Charles M
Home: Bldg III, Apt 500, 211 Lafayette Rd, Syracuse, NY 13205
Position: Prof Emeritus *Employer:* Syracuse Univ. *Education:* SB/Civ Engrg/MIT; MS/Civ Engrg/Lehigh. *Born:* 12/7/16. Born New York, NY. Educated in grammar and high school-Ridgefield, CT. Taught for 2 yrs (Civil Engg) at PMC Construction Superintendent four yrs. Applied physics Lab, JHU, two yrs. Two yrs teaching mechanics and structures at Cornell. Thirty-five yrs at Syracuse Univ. Four times acting chrmn of CE Dept. Have worked on a number of classified projs, consulting on structures problems, consulting on accident reconstructions. Worked during several summers with NYSDOT and Bureau of Budget. Mbr Chi Epsilon MIT. Retired from Syracse Univ 1983. Retain Consltg Reg PE in NY and PA. *Society Aff:* ΣΞ, AAUP, ASCE.

Antrobus, Thomas R
Business: PO Box 778, Doylestown, PA 18901
Position: Owner. *Employer:* Thomas Antrobus & Assoc. *Education:* BS/ME/Drexel Univ. *Born:* 7/14/40. Proprietor of Thomas R Antrobus & Assoc Consulting Engrs & Surveyors. A consulting engg firm primarily dedicated to providing plant engg consulting services to ind clients, including new facility design & construction mgt, facility evaluation & various studies & renovation design. Previously employed by Union Carbide as the Chief Proj Mgr & employed by Am Electronics Labs, Inc as a Facilities Mgr. Past Natl VP of Technical Services, of Am Int of Plant Engrs (AIPE.). Member Am Soc of Mech Engrs, and Nat Soc of PE Certified Plant Engr by AIPE, Reg PE in several states and reg surveyor. *Society Aff:* AIPE, ASME, NSPE.

Aoki, Masanao
Business: 4731 Boelter Hall, Los Angeles, CA 90024
Position: Prof. *Employer:* Univ of CA. *Education:* PhD/Engg/Univ of CA; DSc/Engg/Tokyo Inst of Tech; MS/Physics/Univ of Tokyo; BS/Physics/Univ of Tokyo. *Born:* 5/14/31. Born in Japan. Has taught at Univ of CA, LA, UC Berkeley, Univ of IL, Univ of Cambridge (England) and Osaka Univ. Elected Fellow of IEEE Control Systems Soc for contribution to theory and applications of control systems. Elected Fellow of the Econometric Soc. Currently Prof of Dept of Computer Sci, Univ of CA, LA. *Society Aff:* IEEE, AEA, SEDC.

Aplan, Frank F
Business: 5 Mineral Science Bldg, Univ Park, PA 16802
Position: Prof of Met & Mineral Processing. *Employer:* PA State Univ. *Education:* ScD/Met Engrg/MIT; MS/Minl Proc/MT Coll Min Sci & Tech; BS/Met Engrg/SD Sch of Mines & Tech. *Born:* 8/11/23. Prof of Met & Mineral Processing, PA State Univ, 1968 to date; Hd, Dept of Mineral Preparation 1968-71; Chrmn, Mineral Processing Sec 1971-78; Chrmn, Met Sec 1973-75. Union Carbide Corp (Mgr, Mineral Engrg R&D) 1957-67; Kennecott Copper Corp (Sr Scientist) 1957; Univ of WA (Asst Prof) 1951-53; Climax Molybdenum Co (Mill Engr) 1950-51. Robert H Richards Award, AIME, 1978, Distinguished Mbr, Soc of Mining Engrs (SME 1978), A.F. Taggart Award, 1985 Chrmn, Mineral Processing Div, 1972-73; mbr Bd of Dirs, Engrg Fdn, 1977-, Chrmn, 1985-87. Wilson Distinguished Teaching Award, Penn State 1977. Profl interests: ore & coal preparation, flotation, gravity concentration, hydrometallurgy. *Society Aff:* AIME, AIChE, ASM, ACS.

Apperson, James L
Home: E61 Orchard Lane, Shelton, WA 98584
Position: Managing Partner *Employer:* Associated Engineers III, Inc *Education:* BS/IE/OR State Univ; BS/CE/OR State Univ *Born:* 6/29/27. Mr Apperson is Managing Partner of Associated Engineers III, Inc from 1978 to present. He was formerly VP and Gen Mgr of the midwest operations of STRAAM Engrs, Inc and Pres of Anderson Apperson Inc. His experience includes five yrs as city engg of the City of Portland, OR, where he managed seven bureaus. Respon for streets, bridges, solid waste and wastewater utilities and treatment. In his 12 yrs with the Boeing Co, Seattle, WA, he was assigned to a wide variety of management roles on numerous aerospace projects. He was bridge and structural engr for the City of Portland from 1951-57. *Society Aff:* ASCE, NSPE, WPCF, AWWA, APWA

Appleby, Loran V
Home: 25 Joques Farm Rd, Lake George, NY 12845
Position: Pres. *Employer:* L V Appleby, Inc. *Education:* BME/ME/RPI. *Born:* 4/7/27. Born in Schenectady, NY attended RPI grad in 1950 after serving in USN; worked as sales engr & consulting engr prior to forming above engg rep co in 1959. Active in ASHRAE entire working career, elected Dir at Large 1977; distinguished service Award in 1972. *Society Aff:* ASHRAE.

Appleman, W Ross
376A Avenida Castilla, Laguna Hills, CA 92653
Position: Retired. *Education:* EE/EE/Univ of IL; BS/EE/Univ of IL. *Born:* 4/25/06. Married Alberta Kohlenbach (deceased) 1930. 3 children, 7 grandchildren. Married Mildred Cutler 1980. Design and Dev Engr, Mgr R&D, Chief Elec Engr, Chief Engr, Factory Mgr elec motor and generator mfg including Emerson Elec, Marathon Elec & Reliance Elec, 35 yrs. Asst Prof CA Polytechnical Univ Pomona. Formerly mbr following committees: NEMA Gen Engg, AIEE Rotating Machinery, Single Phase & FHP, Single Phase Test Code, Carbon Brush. Chrmn: Dayton Section AIEE, Area 2 Prize Paper Comm, Secret Clearance. Formerly Reg OH & CA. Life Fellow IEEE. 7 patents, 13 technical articles. Mutual Dir Leisure World Laguna Hills 5 yrs. Enjoy dancing, travel, bridge, music & grandchildren. Designed first motor for garbage disposal ,hospital bed and Jacuzzi. *Society Aff:* IEEE.

Appleton, Joseph H
Business: Univ Sta, Birmingham, AL 35294
Position: Prof of CE. *Employer:* Univ of Al in Birmingham *Education:* PhD/Civ Engr/Univ of IL; MS/Civ Engr/Univ of IL; BCE/Civ Engr/Auburn Univ. *Born:* 8/5/27. in Collinsville, AL. Res Asst Univ of IL, 1947-49 & 1951-54. Structural Res Engr, US Bureau Public Rds, 1949-50. Instr, NC State Univ 1950-51. Structural

Appleton, Joseph H (Continued)
Engr, AL Cement Tile Co, 1954-59. Prof of Civ Engg, Univ of AL at Birmingham, 1959-. Liaison rep, Chrmn, Dean, Sco of Engg 1963-1978. Fellow ASCE and ACI. Structural Consultant. Author numerous articles. Mbr and Chrmn numerous ASCE and ACI committees. *Society Aff:* ASCE, NSPE, ACI, ASEE, ΣΞ.

Arams, Frank R
Business: 180 Marcus Blvd, Hauppauge, NY 11787
Position: VP. *Employer:* LNR Communications, Inc. *Education:* DEE/Electrophysics/Poly Inst of NY; MS/Appl Phys/Harvard Univ; MS/Business/Stevens; BSEE/-/Univ of MI. *Born:* 10/18/25. VP of LNR Communications, Inc (Satellite Communications Equip) 1971- present; Hd, Infrared & Electroptics Dept, AIL Div of Cutler-Hammer 1965-71. Consultant AIL 1956-65; Sr Staff Mbr, RCA Microwave Div, Harrison, NJ 1948-56. Fellow IEEE for contrib to microwave quantum electronics. Author book "Infrared to Millimeter Wave Detectors," Artech House. Author of 50 papers & 5 patents in Microwave Communications and Laser Electronics. *Society Aff:* IEEE, OSA.

Arata, Sil L Sr
Business: Hewlett-Packard Co, 1000 NE Circle Blvd, Corvallis, OR 97330
Position: Packaging Engg. *Employer:* Hewlett-Packard Co.. *Education:* BS/IE/Oregon State Univ. *Born:* 3/22/30. Concord, CA, Served USNR 1949-52. Former Mgr Packaging Engg Tektronix, Inc; (1960-78). Former Packaging Engg for John Fluke Mfg Co, Inc (1979). Now employed by Hewlett-Packard, Corvallis Div. Western Regional VP Soc of Packaging & Handling Engrs (SPHE) 1969-70, 1977-78; Pres Columbia River Chap SPHE 1972-74. Fellow Mbr SPHE 1975. Ch, Recertification Cttee; SPME. 1985-87. Enjoy collecting railroad memorabilia, boating and travel. *Society Aff:* SPHE.

Arbiter, Nathaniel
Home: X9 Ranch, Vail, AZ 85641
Position: Prof Emeritus. *Employer:* Krumb School of Mines. *Born:* 1/2/11. in Yonkers, NY; Res Asst - Columbia Sch of Mines '37-'43; Res Metallurgist- Battelle Meml Inst Columbus, OH '43-'44; Res Metallurgist-Phelps Dodge Corp-'44- '51; Assoc Prof, Prof, Krumb Sch of Mines, Columbia Univ 1951-69; Prof Emeritus '77-; Dir Res, Group Consulting Metallurgist, the Anaconda Co 1969-77; numerous offices AIME; RIchards Award, 1961; Mineral Industry Education Award, 1971; Hon mbr 1976; Gaudin Award-1980 Taggart Award, 1981; 1982 Henry Krumb Lecturer Adj Prof Univ UT 1983- Natl Acad of Engg-1977. Numerous technical papers, patents, editorships; active in consulting. *Society Aff:* AIME, SME.

Arbogast, Ronald G
Home: 16205 W 13 Mile Rd, Southfield, MI 48076
Position: Project Superintendent *Employer:* Turner Construction Co. *Education:* PhD/CE/Wayne State Univ; MS/CE/WV Univ; M/Urban Planning/Wayne State Univ; BS/CE/WV Univ *Born:* 2/10/49 Native of Weston, WV. Employment experiences have entailed the academic, public and private sectors. Conducted various research projects and taught several classes in academia prior to 1978. Structural Engr from 1972-1973 in WV. Fifteen years of diversified construction experience on commercial, industrial, civil, and environmental projects. Several publications in national journals. Registered PE in MI, PA, NJ, & Va. *Society Aff:* ASCE, NSPE

Archang, Homayoun
Business: Machine Design Dept, Tehran Narmak, Iran
Position: Asst Prof. *Employer:* Iran College of Sci & Tech. *Born:* Jun 1946 Isfahan Iran. BS & MS Wichita St Univ, Wichita KS. Resided in US 1964-75. Mbr of faculty of Mech Engg Dept Iran Coll of Sci & Tech Feb 1975-. Mbr ASME. SAE 1975 Wright Bros Medal. Enjoys Jazz, painting & sports.

Archer, Robert R
Business: Civ Engg Dept, Univ. of Mass, Amherst, MA 01003
Position: Prof. *Employer:* Univ of MA. *Education:* PhD/Math/MIT; SB/Math/MIT *Born:* 9/8/28. I have taught engineering at MIT, Case-WR, and Univ of MA over the last 32 yrs. My res has been published in several Engg journals, Math journals, and Biological journals. I have served as a consultant to a number of engrg firms and foreign national laboratories. I have served as Visiting Prof at 6 foreign Universities. *Society Aff:* ASME, ΣΞ

Ardell, Alan J
Business: Materials Dept, 6412-A BH, Los Angeles, CA 90024
Position: Prof. *Employer:* UCLA. *Education:* PhD/Materials Sci/Stanford Univ; MS/Materials Sci/Stanford Univ; BS/Metallurgy/MIT. *Born:* 3/24/39. Born in Brooklyn, reared in Queens and Roslyn, NY. Spent two yrs as NSF Post-Doctoral Fellow at Univ of Cambridge, England learing tech of transmission electron microscopy. Was Asst Prof at CA Inst of Tech 1966-68. Joined Materials Div at UCLA as Assoc Prof in Sept 1968; promoted to Prof in 1974. Spent 1974-75 on sabbatical as Fulbright-Hays Research Fellow at CEA-CEN, Saclay, France, working on Niplus-ion irradiation of Ni-base alloys; Asst Dean, Instruc Resources, Sch Engrg & Appl Sci, 1979-82. Current research interests: Precipitation hardening of alloys; Irradiation effects; Creep of single crystals. Mbr of AIME, ASM, EMSA. Author and co-author of approx 70 publs, tech lit. Enjoys tennis, racquetball, classical music. *Society Aff:* AIME, ASM, EMSA

Arden, Richard W
Business: 950 Industrial Way, Sparks, NV 89431
Position: Pres *Employer:* SEA Engrs/Planners *Education:* MS/CE/Univ of NV-Reno; BS/CE/Univ of NV-Reno *Born:* 4/3/35 As Pres of SEA, Mr Arden administers the engrg, planning, surveying and geotechnical organization-consisting of 140 Professional and Technical Employees, with corp headquarters in Sparks, NV and branch office in Las Vegas, NV. He coordinates many of the firm's projects with the governmental agencies and oversees their completion during the design and construction phase. He specializes in Water Rights and appears as an expert witness. Mr Arden has been active in many professional organizations and has served as Pres along with being selected as "Engr of the Year" for the State of NV. *Society Aff:* ASCE, ASTM, NSPE, ACI.

Ardis, Colby V
Business: Dept of Civ Engg, Toledo, OH 43606
Position: Chrmn & Prof. *Employer:* Univ of Toledo. *Education:* PhD/Civ & Environmental Engr/Univ of WI; MS/Meteorology/Univ of WI; BS/CE/Univ of WI. *Born:* 5/21/35. My profl career began in education at the Univ of WI as an instr in engg mechanics. USAF weather officer 1958-1960. While a sr engr with WI MI Power Co I taught engg part-time in the evenings. Became Area Coordinator for Univ of WI in 1966, then Inst Dir in the Univ Ext Prog (Engg) in 1967. Progressively Civ Engr, Data Mgt Supv, then Prog Analyst on the corporate staff from 1967 until 1979 with the TN Valley Authority. Past pres of 7 technical or profl dev organizations. Numerous honors and awards. Author of 40 papers and three chapters in two textbooks. PE in WI & OH. Professor and Chairman of Civil Engrg Department, The Univ of Toledo, Toledo, OH 1979 to date. *Society Aff:* ASCE, ASEE, AGU, NSPE.

Areghini, David G
Home: 2526 Angelcrest Dr, Hacienda Heights, CA 91745
Position: Generating Station Mgr *Employer:* Southern CA Edison. *Education:* BS/Civil Engg/Univ of AZ; MBA/Mgmt/UCLA. *Born:* 10/30/42. Born in Cottonwood, AZ. Attended HS in Phoenix, AZ. Spent two years in the US Army Corps of Engrs as a Lt. Planning and Scheduling Engr for Fluor Corp for approx three years (1969-72). Served as a cost engr and cost engr supervisory for approx four years for Sothern CA Edison. One yr as a construction superintendent and one year as a Proj Mgr for Southern CA Edison. Served as Chief Steam Engr for Four years and Overhaul Mgr for one year for Southern CA Edison. Assumed present position as Generating Station Mgr in June 1984. Respon for Managing all activities on Edison's larg-

Areghini, David G (Continued)
est fossil fired generating station. Reg PE in CA. Enjoys: jogging, reading and working with children activities. *Society Aff:* AACE, IAE.

Arenz, R James
Home: Jesuit House, Gonzaga University, Spokane, WA 99258
Position: Prof of Mechanical Engrg *Employer:* Gonzaga Univ *Education:* PhD/Aero/CA Inst of Technology; MS/ME/St Louis Univ; MST/Theology/Santa Clara Univ; BS/ME/OR State Coll. *Born:* 8/20/24. After education through college in OR, was stress analyst and aerodynamicist for five yrs with Douglas Aircraft Co, Santa Monica, CA. Joined Jesuit religious order in 1950, ordained to Catholic priesthood in 1965. Res Fellow in Aeronautics, Caltech, 1963-66; Visiting Res Scientist, Ernst-Mach- Inst, Germany 1966. Mechanical Engg Prof, Loyola Marymount Univ, Los Angeles, 1967-78; Consulting Res Scientist, NASA Jet Propulsion Lab, 1969-78. NASA-ASEE Summer Faculty Res Fellow, Johnson Space Center, 1968. Dean of the Faculty, Parks Coll of St Louis Univ, 1978-80. Dean of Arts and Sciences, 1980-83, and Prof of Mech Engg, Gonzaga Univ, 1985. (on leave of absence as Sr Analyst, U.S. Congress, Office of Tech Assessment, 1983-85). Founder Mbr, Amer Acad of Mechs; Mbr, board of reviewers, Applied Mechs Reviews; 27 res publications in dynamic viscoelasticity and behavior of aerospace matls. *Society Aff:* AIAA, ASME, SEM, ASEE.

Arffa, Gerald L
Business: 799 W Michigan St, Indianapolis, IN 46202
Position: Chrmn-Dept of Supv. *Employer:* Purdue Sch of Engg & Tech at Indianapolis. *Education:* PhD/Admin & Engg Systems/Union College; MBA/Production Mgt/Syracuse Univ; BS/CE/Clarkson. *Born:* 9/25/28. in Syracuse, NY. Educated in Binghamton, Syracuse, Potsdam & Schenectady, all in NY. Design engr, process engr & Gen Foreman at Solvay Div, Allied Chem, 1955-60, Process engr, production supervisor, shop mgr, mgr mft admin ar GE, semiconductor products, Syracuse, NY, 1960-1971. Mgr Mfg Studies Progs in GE Corporate Education 1971-1978, in Schenectady, NY. Maj career shift in 1979, teacher-admin in the Hoosier Capital. *Society Aff:* ASEE.

Argon, Ali S
Business: 77 Massachusetts Ave Rm 1-306, Cambridge, MA 02139
Position: Prof. *Employer:* MA Inst of Tech. *Education:* ScD/Materials/MIT; SM/Mech Eng/MIT; BS/Mech Eng/Purdue Univ *Born:* 12/19/30. Native of Turkey. Started engg career at High Voltage Engg Corp as structures and mtls expert. Taught at Middle East Technical Univ Ankara, Turkey 1958-1960. Joined MIT faculty in Mech Engg Dept Asst Prof 1960 - Prof 1968. Quentin Berg Prof 1982. Teaching and res in area of mechanics and phys of mech behavior of engg mtls. Consultant to govt and industrial res labs. Published 4 books, over 170 technical papers. ASME Charles Russ Richards Award, 1976, Fellow, American Physical Society. Visiting Prof of Polymer Physics Univ of Leeds, UK 1972. Collector of random antiques. *Society Aff:* APS, AIME-TMS, AAAS

Arias, Javier S
Business: Baja CA 210-403, Mexico DF 7
Position: Pres. *Employer:* Arias y Asociados, SC. *Education:* BS/Mech & Elec Engg/Iberoamericana Univ. *Born:* 4/14/38. Prof of IE - Natl Univ of Mexico. Prof of IE - LaSalle Univ. Past Hd of IE Dept - Iberoamericana Univ (1968-1969). Product Engr - GE (1960-65). Consultant - Intl Bus Cons De Mexico, SA (1965-68). Pres - Arias y Asociados, SC - IN Cons 1968 up to date. VP Region Mexico IIE - 1975-79. Honorific mention on thesis work. Enjoys: music, history, reading, golf, swimming, frontennis. *Society Aff:* IIE, CIME.

Aris, Rutherford
Business: 151 Chem Engg, Minneapolis, MN 55455
Position: Regents' Prof. Chem Engg-Mtls Sci Dept. *Employer:* Univ of MN. *Education:* DSc/Undesignated/Univ of London; PhD/Math & Chem Engng/Univ of London; BSc/Math/Univ of London. *Born:* Sept 15, 1929 Bournemouth Eng. BSc 1948; PhD 1960; DSc 1964 Univ London. Empl by Imperial Chem Indust 1946-48 & 1950-55; Lectr in Tech Math Edinburgh Univ 1956-58; Dept of Chem Engg Univ of MN 1958-, Prof 1963. Alpha Chi Sigma Award for Chem Engg Res AIChE 1969; Chem Engg Lectureship Award ASEE 1973; NAE 1975; Wilhelm Award AIChE 1975. Lewis Award AIChE 1981. Auth of the Math Theory of Diffusion & Reaction in Permeable Catalysts, 2 vols 1975 & sev other books & papers. Mbr of Soc for Natural Philosophy; Fellow Inst of Math & its Applications; Soc for Math Biol; Medieval Acad of Amer, AIChE; ACS; Soc. Sci and Illuminators (Lay), etc. *Society Aff:* SSI.

Arman, Ara
Home: 1148 Verdun Dr, Baton Rouge, LA 70810
Position: Director & Prof *Employer:* LA State Univ. *Education:* MS/Civ Engg/Univ of TX; BS/Civ Engg/Robert College. *Born:* 9/12/30. Native of Istanbul, Turkey. Immigrated to US in 1955. Worked at the LA Dept of Hgwys as dist lab engr & soil design engr. Joined LA State Uiv as Asst Prof in 1963. Asst Dir LSU Div of Engg Res 1965, Assoc Prof 1967, Prof of Civ Engg 1970, Chrmn of Dept of Civ Engg 1976-1980. Assoc. Dean, College of Engineering 1980-87. VPShelltech Inc, 1963-68; VP Sado Engrs, Dir ARTBA, 1966-67; VP ARTBA, 1969-72; Pres, Ed Div ARTBA, 1972-73; Chrmn, ASCE Comm on Soil Improvement and Placement, 1978-; 1986- Director, LA. Transportation Research Center. 1987-La. Board of Registration for Engineers and Land Surveyors-Board Member. Chrmn, TRB Comm on Cement Stabilization, 1978-. Member ASTM - committee D-18 on Soils & Rock, Mbr, Exec Council Intl Geotextiles Soc. Enjoys fine art & fishing. *Society Aff:* ASCE, ASTM, TRB, IGS, ISSMFE.

Armand, Jean-Louis
Business: Dept of Mech and Environ Engrg, Santa Barbara, CA 93106
Position: Prof and Chrmn *Employer:* Univ of CA *Education:* PhD/Aeronautics and Astronautics/Stanford Univ; MS/Aeronautics and Astronautics/Stanford Univ; Ingenieur Diplome/-/Ecole Polytechnique, Paris *Born:* 04/17/44 Born in Annecy (France). Joined Univ of CA, Santa Barbara as Prof in the dept of Mech and Environ Engrg in 1983. Dept Chrmn since 1984. Was prior to this res engr (1971-73) with Chantiers de l'Atlantique shipyard in Saint-Nazaire, then Scientific Dir (1973-1983) of French Ship Res Inst. Taught Appl Mechs and Ocean Engrg at various French Univs, among them the Ecole Polytechnique in Paris, with visiting appointments at the Univ of Ca, Berkeley. Royal Institution of Naval Architects medal for his work on dynamics of marine structures. Mbr, Editorial Bd, 3 journals. Author or editor of 4 books. Author of more than 40 tech papers in dynamics of struct, stochastic dynamics, structural optimization and reliability, with emphasis on marine vehicles and structures. *Society Aff:* ASME, SNAME

Armeniades, Constantine D
Business: Chemical Engrg Dept, Houston, TX 77001
Position: Prof *Employer:* Rice University *Education:* Ph.D/Polymers/Case Western Reserve Univ; MS/Eng/Case Inst of Tech; BS/ChE/Northeastern Univ *Born:* 5/29/36 Native of Thessaloniki, Greece. Naturalized U.S. citzen in 1962. Joined the Res and Dev group on plastics and coating at Arthur D. Little, Inc. in 1961. Joined the Rice Univ faculty in 1969; current position: Prof of Chem Engrg and Master of Will Rice College. Res in structure-property relations of synthetic polymers and biomaterials. Consultant to the chem industry on process design and polymer properties. Enjoy classical music and travel.

Armentrout, Daryl R
Home: 1117 Burning Tree Lane, Knoxville, TN 37923
Position: Chief, Procurement Quality Assurance. *Employer:* TVA Office of Nuclear Power. *Education:* PhD/Civil Engrg/Univ of TN; MS/Structural Engg/VPI and State Univ; BS/Civil Engg/ Univ of TN. *Born:* 4/6/42. Native of Telford, Tn. Worked with Humble Oil Co in New Orleans 1965-66 as production engg. Joined TVA in 1968 as Civil engg doing structural steel design for fossil and nuclear plant design.

Armentrout, Daryl R (Continued)
In 1974 was named Chief, TVA's Engg Services Staff. In 1977-85 served in position as Asst to the Mgr, Office of Engg Design and Construction. In 1986 assumed current position as Chief, TVA's Procurement Quality Assurance in the Office of Nuclear Power. Has directed all TVA activities since 1979 under the US-China Hydroelectric Protocol including seminars both in the US and China. Served as Pres, TN Valley Sect, ASCE, 1979. TN State Chrmn, NSPE Scholarship Chrmn in Tennessee 1980-81, served as Pres, Tennessee Society of Professional Engineers 1986-87. *Society Aff:* ASCE, NSPE, NMA, ANS.

Armiger, William B
Business: 66 Great Valley Pkwy, Malvern, PA 19355
Position: Pres. *Employer:* BioChem Technology, Inc. *Education:* PhD/Chem-Biochem/Univ of PA; BSE/ChE/Princeton Univ. *Born:* 7/17/49 Native of Baltimore, MD. Founded BioChem Tech, Inc in 1977. BioChem Technology sells systems for monitoring and controlling fermentation processes. Proprietary products include sensors for measuring NADH and F420, computer interface hardware, computer controlled peristaltic pumps, and software for on-line calculation of cell mass and process control. Published or presented over 25 technical papers. Editor: *Computer Applications in Fermentation Technology, Biotech & Bioengg Sym Series No 9*, John Wiley & Sons, NY, 1979. *Society Aff:* AAAS, AIChE, ACS, ASM, ISA, SIM.

Armour, Charles R
Home: 2699 Johnson Rd, Atlanta, GA 30345
Position: Chrmn of the Bd *Employer:* Armour Cape & Pond, Inc. *Education:* BA/Arch Engr/GA Inst Tech; BS/Arch Engr/GA Inst Tech *Born:* 3/17/29 Native of Atlanta, GA. served as air intelligence and photo intelligence officer with the US Navy aboard the USS Yorktown during the Korean War 1952- 1955. Assoc with Harry G Hunter Assocs as structural engr from 1955 to 1965. Founder of Armour Cape & Pond, Inc 1965. Served as State Chrmn of PEPP 1983/1984. PEPP Eng of Year 1985. Served as State Chrmn MATHCOUNTS 1984/1985/1986 NSPE. Pres Atlanta Chapter ACI 1986 - Pres GSPE 1987. Serving on Bd of Dirs of the ACI and the Ga Concrete Adv Bd. Hobbies include radio control model airplanes, rebuilding old sports cars, and micro-computers. *Society Aff:* ASTM, GSPE, ACEC, ACI, PCI.

Armour, Thomas S, Jr
Home: 2166 Shady Ln, Columbia, SC 29206
Position: Pres. *Employer:* Elec Design Consultants. *Education:* MS/EE/LSU; B/EE/Clemson. *Born:* 3/24/23. Military service AUS June '43--Oct '46, 1st Lt, Reg PE SC, NC, GA & TX; Past Pres CESC, IES, Columbia Chap SCSPE, Augusta CSI; Natl Dir ACEC(SC), Bd Dir SCSPE; First Pres Sr Citizens Council Augusta/Richmond County (GA), Mbr Greater Columbia Chamber of Commerce, Comm of 100, St Martins Episcopal Church, Woodlands Country Club; Self Employed 1970, Inc Oct, 1974, Design Consultants, Inc, d/b/a/ Elec Design Consultants, Who's Who in the South and Southeast. *Society Aff:* NSPE, SCSPE, PEPP, ACEC, CESC, IES, SAME, AEE.

Armstrong, Charles H
Business: 660 Bannock, Denver, CO 80204
Position: Partner *Employer:* R W Beck & Assoc. *Education:* BME/Mech Engg/Marquette Univ. *Born:* 10/16/41. Awarded Scholastic Honors at Marquette Univ for Outstanding Achievement in the Thermal Power Option. Powr plant design engr with Burns & McDonnell Engg Co, 1964 to 1972. Joined R W Beck & Assoc in 1972, elected to Assoc status in 1975. Selected as partner, April, 1980. Currently assigned as supervisor of Project Management for the Central Design Office of R W Beck. Recreational pursuits include skiing, sailboat racing, mtn climbing, & classical music. *Society Aff:* NSPE, ASME.

Armstrong, Ellis L
Home: 3709 Brockbank Dr, Salt Lake City, UT 84124
Position: Pres. *Employer:* Ellis L. Armstrong Assocs, Engrs & Cons Inc. *Education:* Dr.Engr/Civil Engr/Newark Coll of Engr; Dr. Sci/Engrg/So. Utah State Coll; BS/Civil Engrg/Utah State Univ. *Born:* 5/30/14. Honoris Causa (Dr Engrg) Newark Coll of Engrg served with USBR 1934-53. Proj Engr for cons managing US portion of St Lawrence Power & Seaway Proj 1953- 56; Dir of Hgwys UT State Hwy Comm 1957; Comm of US Public Roads 1958-61; US Commissioner of Reclamation 1969-73. Supervised over 500 proj as principal in Engrg firm; Recipient of numerous awards & National honor mbr in ASCE, AWWA, APWA, Chi Epsilon and Public Works Historical Society. Author and editor, *History of Public Works in the US 1776-1976*. Chm, US Committee of the World Energy Conference-1972-75. Adj Prof, Civil Engineering, Utah U, Utah State, BYU 1978-present. *Society Aff:* ASCE, APWA, AWWA, USNC/WEC, IWRA, ARTB.

Armstrong, John C
Business: 2 Park Ave, New York, NY 10016
Position: Pres. *Employer:* Geo S Armstrong & Co, Inc. *Education:* AB/Chemistry/Williams College; BS/Chem Engg/NY Univ. *Born:* 11/25/18. Worked 1942-46 E I duPont Electrochemicals Dept in Res, mfg dev, and industrial engg dept, with alkali cyanides, peroxides, and misc chemicals. Worked 1946-date with Geo S Armstrong & Co in capacity as mgmt consultant to industry in USA & abroad. Reg Prof Engr NY State Emphasis on economic evaluation for capital programs, merger analysis, market res in forest products, pulp and paper, steel, chemicals, oil, shipping, cement, aluminum, and plastics, as well as corporation reorganizations. *Society Aff:* AIChE, ACS, NYSSA.

Armstrong, Neil A
Business: Arrow Space Engg, Location 70, Cincinnati, OH 4522' Univ Prof of Engg, Univ of Cincinnati, Cincinnati, OH 45221. B. Aug 1930. MS Univ of SCa. BS Purdue Univ. Born in Wapakoneta, OH. Prof Armstrong is presently involved in teaching & research. Naval aviator during Korean action. In capacity as aeronautical res pilot for NACA & NASA, Neil Armstrong performed as an X-15 project pilot, flying that aircraft to over 200,000 ft. and approx 4, 000 miles per hr. Prof Armstrong was selected as an astronaut by NASA in Sept 1962. He was command pilot for the Gemini 8 mission, launched on March 16, 1966. Prof Armstrong performed the first successful docking of 2 vehicles in space. As spacecraft commander for Apollo 11, the first manned lunar landing mission, Armstrong gained the distinction of being the first man to land a craft on the moon and the first to step on its surface. He has been decorated by 17 countries and is the recipient of many special honors. He is married and has 2 sons.

Armstrong, Ronald W
Business: Dept of Mech Engg, College Park, MD 20742
Position: Prof. *Employer:* Univ MD. *Education:* PhD/Met Engg/Carnegie-Mellon; MSc/Met Engg/Carnegie-Mellon; BES/Engg Sci/Johns Hopkins. *Born:* 5/4/34. Employed at Westinghouse Res Labs, Pittsburgh, 1955, 1958, 1959-64. Taught at Brown Univ prior to joining MD as Prof in 1968. Spent study periods: in the US at US Steel Res, Oak Ridge Natl Lab, Lawrence Livermore Lab, Inst for Defense Analyses, United Tech, Natl Sci Fdn, Naval Surface Weapons Ctr, Natl Bur of Standards, Naval Res Lab, and Naval Ship Res and Dev Ctr; and, abroad, at Leeds Univ, U.K., CSIRO Div of Tribophysics (Univ of Melbourne), Australia, and DSIR Phys and Engrg Lab, New Zealand (as Sr Fulbright -Hays Fellow). Was 1979 NATO Advanced Study Inst Lecturer on X-ray (Reflection) Topography, 1985 NATO ASI Lecturer on Deformation of Energetic Crystals, and 1980 and 1983 US-France Coop Sci Prog Lecturer on X-ray Topography. Appointed Scientific Liaison Officer, European Scientific Office, Office of Naval Research, London, 1982-84. Visiting Prof, Dept of Metallurgy, Univ of Strathclyde, Glasgow, UK, 1982-85. Visiting Fellow, Clare Hall, Univ of Cambridge, UK, 1984. Mainly involved in res on strength of mtls via dislocation processes. AIME Robert Lansing Hardy Gold Medal for 1962, and Fellow of ASM, 1985. *Society Aff:* ASM, AIME.

Armstrong, William D
Business: 3111 E Frank Phillips Blvd, Bartlesville, OK 74003
Position: VP. *Employer:* Greenawalt-Armstrong, Inc. *Education:* BS/Arch Engr/OK Univ *Born:* 12/12/36. in Willow Springs, MO. My wife's name is Pat and we have three sons. After grad from the Univ of OK, I worked for the MO Hgwy Dept, a consulting firm in Joplin, an architectural firm in Bartlesville, and the City of Bartlesville as City Engr and Dir of Public Works. In 1973 I joined Greenawalt Engg, now Greenewalt-Armstrong, Inc, of which I am VP. Reg in OK, MO and KS as PE. Reg in OK as Land Surveyor. *Society Aff:* NSPE

Armstrong, William M
Home: 2194 W 57th Ave, Vancouver BC Canada V6P 1V4
Position: Retired (Exec Dir) *Education:* BASc/CE/Univ of Toronto; DSc/CE/Univ of BC. *Born:* 10/27/15. 1937-43 Steel Co of Canada, Hamilton, Supervisor Met Lab. 1943-46 Ontario Res Fdn, Res Met. 1946-54 Assoc Prof of Met, Univ of BC. 1954-64 Prof of Met, Univ of BC. 1964-65 Prof & Hd of Dept of Met, Univ of BC. 1964 Pres Assoc of PE of BC. 1966 Dean, Faculty of Appl Sci - Univ of BC. 1966-1973 Mbr of Sci Council of Canada. 1966-67 Pres Canadian Council of PE. 1969 Fellow, Royal Soc of Canada. 1967-74 Deputy Pres UBC & Secy to Bd of Governors. 1974 Fellow, Am Soc for Metals. 1974-78 chrmn of Univs Council for BC. 1974 Gold Medal Canadian Council of PE. 1978 Exec Dir, Secretariat on Sci, Res & Dev. Retired - Oct 31, 1980. *Society Aff:* FASM, FRSC, AIME, CIME, APE.

Arnas, Ozer A
Business: Mech Engrg Dept, Sacramento, CA 95819
Position: Professor *Employer:* CSU-Sacramento *Education:* Ph.D./ME/North Carolina State Univ.-Raleigh; MS/ME/Duke University; BS/ME/Robert College, Istanbul *Born:* 09/01/36 Professor Emeritus, Mech Engrg. LA State Univ., Jan '86-Present; Prof-Assoc Prof-Asst Prof, Engrg, LA State Univ., 1962-'86; Distinguished Faculty Fellow, LA State Univ Foundation, 1984-85; Consulting Lecturer, the Catholic Univ of America, 1967-68; Research Analyst, Wright-Patterson Air Force Base, Materials Laboratory 1967-68; Visiting Prof & Research Fellow, Univ of Liege, Belgium, 1972-73 and 1979-80. Visiting Research Professor, Eindhoven Univ of Technology, the Netherlands, 1979-80. May 1967: Excellence in Teaching Award of $1500-, Halliburton Educational Foundation, Inc; 1974 Ralph E. Teetor Award winner of SAE; 1979 Western Electric Award winner of ASEE; 1987 Meritorius Performance award of Calif State Univ; Author of over 80 publications; Editor of two Journals and a series in Mechanica Engrg. *Society Aff:* ASME, ASEE, ΣΞ, ΠΤΣ, ΤΒΠ

Arndt, Roger E A
Business: St Anthony Falls Hydraulic Lab, Mississippi River at 3rd Ave SE, Minneapolis, MN 55414
Position: Dir & Prof. *Employer:* Univ of MN. *Education:* PhD/Civ Engg/MA Inst of Tech; SM/Civ Engg/MA Inst of Tech; BCE/Civ Engg/CCNY. *Born:* 5/25/35. Native of NYC, NY. Res & preliminary design of under-water rockets & high speed marine vehicles at Allegany Ballistics Lab and subsequently at Lockheed-CA Co. Until 1977, Assoc Prof of Aerospace Engg at the PA State Univ, teaching and res in low speed aerodynamics, aeroacoustics & cavitation. Presently, Dir & Prof of Hydromechanics at the St Anthony Falls Hydraulic Lab, Univ of MN; responsible for an extensive prog of fundamental and applied res in hydraulic engg. Consultant to numerous cos & governmental agencies; Lorenz G Straub Award, 1968; First Theodor Ranov Lecturer, 1978. Avid interest in private aviation. *Society Aff:* ASA, AIAA, ASCE, ASME, IAHR, ΣΞ.

Arnette, Harold L
Business: 8337 Martinsburg Rd, Mount Vernon, OH 43050
Position: Pres *Employer:* Information Control Corp *Education:* M.S./Comp Sci/OH St Univ; B.S./Math/OH St Univ *Born:* 2/21/41 Employed by TRW Inc., electronic design. OH St Univ, ultrasonics and x-ray television research. Rockwell Intl, intelligence systems development for US Navy. Bell Telephone Labs, translation table compilation. Assoc Prof of Computer Science developing the major and computer facilities. Since 1970, consultant to telecommunication and NC Industries. Formed Information Control Corp. in 1978 providing consulting and research services and CAD/CAM systems development for the flat stock metal working industry. Currently designing Flexible Manufacturing System (FMS) for flat stock applicants. *Society Aff:* ACM, SME, SCCP

Arno, Norman L
Business: P.O. Box 2711, 300 North Los Angeles St, Los Angeles, CA 90053
Position: Chief, Engrg Division *Employer:* US Army Corps of Engrs *Education:* MS/Coastal Engrg/Delft, Netherlands; BS/CE/Univ of WA, Seattle *Born:* 3/10/32 Native of Seattle, WA. With US Army Corps of Engrs since 1954. Active on Bd of Dirs, ASCE Div of Waterway, Port, Coastal & Ocean and Committee on Tidal Hydraulics CoE. Currently in Los Angeles, Chief, Engrg Division for water resource projects in five southwestern states. Previously Chicago District, engrg management, water resource projects for four states; North Central Division, engrg management, water resource projects in 5 districts; Office, Chief of Engrs, Washington, DC, as Chief, Coastal and Estuaries Sections; AK District, Chief Coastal Engrg and Hydraulic Design, project engr for earthquake damaged waterfront facilities; and Seattle District, project planner and hydraulic designer. *Society Aff:* ASCE, ASBPA, USCOLD

Arnold, Charles W
Business: PO Box 2189, Houston, TX 77252-2189
Position: Sr Res Advisor. *Employer:* Exxon Production Res Co. *Education:* PhD/ChE/Univ of TX; MS/ChE/Univ of TX; BS/ChE/Univ of OK. *Born:* 6/13/23. Native of Goree, TX. Served with USN 1943-45. Taught in Goree public schools 1945-46. With Exxon Corp (formerly Std Oil of NJ) since 1954. In charge of res on reservoir engg & enhanced oil recovery methods 1964-1975. Assumed present position as Sr Res Advisor specializing in enhanced oil recovery methods 1975. Dir, Soc of Petrol Engrs 1974-76; SAME: Distinguished Lectr SPE 1974-75; Distinguished Service Award SPE 1978. Dir, Am Inst of Mining, Met, and Petrol Engrs 1980-83; SAME: VP 1982-83. *Society Aff:* SPE of AIME, ACS.

Arnold, David W
Home: 2026 Country Club Dr, Yazoo City, MS 39194
Position: Senior VP *Employer:* Mississippi Chem Corp. *Education:* PhD/ChE/IA State Univ; MS/ChE/IA State Univ; BS/ChE/Univ of MS. *Born:* 12/6/36. Native of Dundee, MS. Parents Mr and Mrs J B Arnold. Served in US Navy 1958-63. Junior Engr for Ames Lab, IA State Univ 1963-66. With MS Chem since 1966. Held positions of Process Study Engrg, Research Project Mgr, Chief Process Engr. Mgr of Process Engrg, Dir of Engrg, VP Engrg & Sr VP- Engrg prior to becoming Sr VP - Research and Engrg in 1987. Married one child. Mbr of Trinity Episcopal Church. Registered Prof Engr, Pres of Mississippi Engrg Soc, Chmn Univ of MS Engrg Adv Bd, Mbr Yazoo City Public Service Commission. Mbr Tau Beta Pi. Past activities include Pres of Univ of MS Engrg Alumni Chapter, Pres of Yazoo County Chamber of Commerce, Mbr Governor's Engrg Adv Council *Society Aff:* AIChE, NSPE.

Arnold, Gerald E
Home: 837 S Windsor Blvd, Los Angeles, CA 90005
Position: Cons, Engr (Self-employed) *Education:* BS/CE/CO State Univ. *Born:* 11/23/01. Hydraulic and sanitary consulting engr. Formerly hd of water depts, San Diego CA and Phila PA. Consultant to many foreign govts on water supply, sewerage and environmental projs. Organized and administered agencies for operation and maintenance of utility and environmental installations. Mbr of technical and profl societies, held local and natl offices. Recipient of several awards and citations. Formerly chrmn of Engg Manpower Commission of Engrs Joint Council. *Society Aff:* ASCE, AWWA, AAEE WPCF.

Arnold, Lynn E
Home: 5154 Montgomery Rd, Cincinnati, OH 45212
Position: Research Engr *Employer:* Xtek Inc. *Education:* MSME/ME/Univ of IL; MetE/Met/Univ of Cincinnati. *Born:* 11/17/34. Native of Cincinnati, OH. With Xtek Inc since 1959. Currently Research Engr. Past pres Engrs & Scientists of Cincinnati, Cincinnati Technical Council, Cincinnati OSPE, Cincinnati ASM, Cincinnati ASME, and OH Council ASME. VP NSPE 1978-79. Pres Univ of Cincinnati Engg Alumni 1979-81. Chairman Cinti SME 1981-82; Chairman AGMA Metallurgy and Materials Comittee 1981-present. Co-Chrmn AIME-ISS Roll Tech Ctte, 1986-88. Pres Elect Cincinnati Soc for Advancement of Mgmt 1987-88. Fellow ASM. Fellow AAAS. Past Pres Cincinnati Editors Assn, Cincinnati SAR, Audubon Soc of OH. OH Audubon Council, Cincinnati Mgt Assn, Tool Steel Mgnt Club, Polygons, SYA Intergroup Council. Honors incl Univ of Cinti. Engg Distinguished Alumni Award, Eisenman Meml Award from Cinti ASM, Community Service Award from Cinti Technical Council. Trustee of Ctr for Mfg Tech. *Society Aff:* NSPE, ASM, ASME, AIME, AISE.

Arnold, Orville E
Business: 815 Forward Dr, Madison, WI 53711
Position: Pres *Employer:* Arnold and O'Sheridan, Inc *Education:* BS/CE/Univ of WI *Born:* 9/30/33 After employment in a steel co, Corps of Engrs and an architectural/engrg firm, Mr Arnold was a founder of Arnold and O'Sheridan, Inc, consulting engrs, Madison, WI. The firm offers structural, mechanical, electrical, civil engrg services and land surveying. Mr Arnold served for eight years as chrmn of the Structural Chapter 53 Code Committee which rewrote the Structural Requirements of the WI State Building Code. Mr Arnold served as a member of the WI Examining Bd of Architects, Engrs, Designers and Land Surveyors for three years. General Chrmn--National Structural Engrg Conference, August, 1976. Committee on Professional Registration, 1977-1981. Pres WI Sect ASCE 1972. Received Dist Service Citation from Univ of WI Coll Engrg 1982. *Society Aff:* ASCE, NSPE, ACEC, AIA.

Arnold, Thaddeus R
Business: P.O. Box 1305, North Pearman Dairy Rd, Anderson, SC 29622
Position: VP *Employer:* Russell & Axon Engrs-Planners-Architects, Inc *Education:* BS/Arch Engrg/Clemson Univ *Born:* 4/13/30 Native of Anderson, SC. Served with US Army Ordnance Corps in Korea 1952- 1954. Associated with Robinson Engrg Service 1956-1974 attaining the level of Senior Engr. Joined Russell and Axon Engrs-Planners-Architects, Inc. Am presently VP and Regional Office Mgr and also serve as Corp Dir of communications Engrg. Registered professional in SC, NC, GA, FL, AL and TN. Elder in the Presbyterian Church. Enjoy hunting, fishing and golf. *Society Aff:* ASCE, NSPE, WPCF, ITPA, ACE

Arnold, William Howard
Business: PO Box 1970, Richland, WA 99352
Position: V.P. Engrg & Development *Employer:* W Hanford Co. *Education:* PhD/Phys/Princeton Univ; MA/Phys/Princeton Univ; BA/Chem & Phys/Cornell Univ. *Born:* 5/13/31. After PhD in Physics in 1955 (Princeton), joined Westinghouse Atomic Power Div. Became Sr Engr and then reactor Phys Design Mgr 1961-62 Dir Nuclear Fuel Mgt, NUS Corp 1962-8, W Nerva Prog as Test Mgr and Prog Mgr 1968-70, Mgr Weapons Dept, Underseas, W Defense 1970-79, Engg Mgr and Genl Mgr, W Pressurized Water Reactor Systems Div 1979-81, Pres W Nuclear Intl div 1974 elected to Natl Acad Engg. 1952 married Josephine Routheau - 5 children. 1981-87, Gen. Mgr., W Advanced Reactors Div. 1987- , V.P. Engineering and Development, W Hanford Co. *Society Aff:* ANS, NAE, APS, AAAS, ΣΞ.

Arnwine, William C
Business: Rockwell International, 3370 Miraloma Ave, M, S FC14, Anaheim, CA 92803
Position: Mbr Tech Staff VI. *Employer:* Rockwell Intl. *Education:* PhD/Engr Valuation/IA State Univ; MS/IE/IA State Univ; BS/IE/OK State Univ. *Born:* 7/28/29. Native of McCurtain, OK. Officer in the Army Corps of Engrs, 1954-55. Methods Engr with E I duPont. On faculty at IA State Univ & Univ of AR. Prof & Hd of IE, Univ of NB (Omaha) & NM State Univ. Since 1972, Levels V & VI Technical Staff position at Rockwell Intl, mfg of aerospace & marine electronic guidance systems. Natl Dir for the IIE Aeropsace Div, 1978. Engr of the yr, Electronic Systems GP, Rockwell Intl, 1977, & Outstanding Engg Merit Award, Orange County Engg Council, 1979, & for MTM Asso: Fellow, 1984, VP, 1984-85, Chm of Research Comm, 1982-84, on Board of Directors, 1980-83 & 1986-pres. Reg PE in OK & CA. *Society Aff:* IIE, ORSA, ΣΞ, NMA, MTM ASSO.

Aronson, David
Home: 9 Riverview Dr W, Upper Montclair, NJ 07043
Position: Consulting Engr. *Employer:* Self-employed. *Education:* BS/Chem Engr/Cooper Union Night Sch; BS/Chem Engr/Polytechnic, Brooklyn. *Born:* 9/24/12. Brooklyn NY. BSChE Cooper Union. Prior to WWII worked with Zola G Deutsch cons to heavy chem indust; for Sanderson & Porter 1942-43 in chg of chem equip installation & startup at CWS Arsenal Pine Bluff AR; on Manhattan Proj 1943-45 dev methods for controlling barrier quality; Elliott Co 1945-51 in chg of heat transfer & heavy oil combustion equip; at Worthigton Corp 1951-70 served headquarters & operating divs as cons on energy conversion & process design. Since 1970 consultant on energy conservation and conversions to industry and government. Rec Worthington Award World-wide Engr 1964. Reviewer AIChE Journal, AMR. Participates in community affairs. Interests: philosophy, economics, history. *Society Aff:* ASME, AIChE, ACS, AAAS.

Arrigoni, David M
Business: 231 O'Connor Dr, San Jose, CA 95128
Position: Pres. *Employer:* David M Arrigoni & Assoc, Inc. *Born:* 5/16/33. San Francisco. Reg PE CA E7019. In private practice 1960-; responsible for elec design of over $500 million commercial, ind, institutional and religious bldgs, work includes many multi-million-dollar high rise structures, campus type schools, single level bldgs. Dir design of all elec sys; lighting, pwr, signal, etc; prov elec engg calculations; supervised all field installation; collaborated with client in "teamlike" effort; prepared const cost istimates & budgets; investigated new concepts in Elec & Illuminating Engg using an IBM/370 computer. Developed an economical "CAD" computer assisted design system for Architects & Engrs using micro-computers. 3 times 1st place winner for the best lighting design spons by the Illum Engg Soc. *Society Aff:* IES, IEEE, CEA, ACEC.

Arthur, Harold G
Home: No 52 Cherry Hills III, 2800 South University Blvd, Denver, CO 80210
Position: Consulting Engg (Self-employed) *Education:* CE/Civil Engg/SD School of Mines & Tech; BS/Civil Engg/SD School of Mines & Tech. *Born:* 8/23/14. Starting in 1935 as a rodman on a survey party, rose through the ranks of the Bureau of Reclamation to become Dir of Design and Construction in 1972, with respon for all Bureau construction in the 17 Western States. Retired in 1977 after 40 yrs service and now an individual consultant on dams and powerplants serving on international consulting boards. Among other awards received Dept of Interior's Distinguished Service Award and the "Beaver" Engg Award for Outstanding Achievement in Heavy Engg Construction. Expert Witness on Heavy Engrg Const, construction contract admin, and geotechnical engrg. *Society Aff:* ASCE, USCOLD, USICID.

Arthur, Paul D
Business: Mech Engg, Irvine, CA 92717
Position: Prof Mech Engr *Employer:* Univ of CA, Irvine. *Education:* PhD/Aero/Caltech; MS/Mech Engr/Univ of MD; BS/Mech Engr/Univ of MD. *Born:* 2/23/25. *Society Aff:* ASME.

Arthur, Robert M
Business: 548 Prairie Rd, Fond du Lac, WI 54935
Position: Pres. *Employer:* Arthur Tech. *Education:* PhD/Envir Health Engg/Univ of IA; MS/Envir Health Engg/Harvard Univ; BS/Civ Engg/Northwestern Univ; BA/

Arthur, Robert M (Continued)
Math/Ripon College. *Born:* 3/21/24. Fond du Lac, WI US Army 43-46, BA Math Ripon College 1949, BSCE Northwestern Univ 1953; MS in Sanitary Engg Harvard Univ 1956, PhD in Sanitary Engg Univ of IA 1963. Prof & Chrmn, Biological Engg Dept Rose-Hulman Inst of Tech until 1973. Pres, Arthur and Assoc Consulting Engrs 1957 to 1977 - Pres, Arthur Tech 1965: Pres Tech-Line Instruments 1984- . Patents and Publications in Field. Past-Chrmn Instrumentation Subcommittee WPCF. Mbr WPCF *Standard Methods* Committee, Mbr ASTM Committee D- 19, Tau Beta Pi PE WI. *Society Aff:* WPCF, ASTM, ISA, ASEE, TAPPI.

Arthur, Wallace
Home: 22 Raymond St, Harrington Park, NJ 07640
Position: Dean/College of Sci & Engg. *Employer:* Fairleigh Dickinson Univ. *Education:* PhD/Phys/NYU; BS/EE/NYU; BS/Engg Sci/NYU. *Born:* 11/22/32. Served USA Signal Corps 1952-54. Asst Res Sci NYU 1957-61; Lectr Rutgers Univ 1961-62; Asst Prof to Prof Fairleigh Dickinson Univ 1962-73; Dean Coll of Sci & Engg 1973-. Cons on atmospheric ionization, cosmic radiation, health physics. Respon for sch of 65 full-time 100 part-time faculty; 1700 students. Enjoys sports, classical music. *Society Aff:* ASEE, AIP, NJAS.

Artley, John L
Business: School of Engg, Durham, NC 27706
Position: Prof, Research *Employer:* Duke Univ. *Education:* DrEngg/EE/Johns Hopkins Univ; MS/EE/Univ of MI; BS/EE/Univ of MI. *Born:* 9/27/23. Have taught Elec Engg at Univ of MI, Univ of AR, US Naval Academy, Johns Hopkins Univ and Duke Univ. Engr in Training, Ford Motor Co 1941-43. Engg Officer in USNR, WWII. Design Engg, elec motors and generators, 1948-49. Author of Generalized Fields Book for Engrs, Prin Investigator in Res on: Thin Superconducting Films (AEC), Ferromagnetic Materials (US Army), Spin Waves (AROD), Relativistic Models (US Army). Engineered Self Control present main Interest. *Society Aff:* AAAS, ASEE, IEEE, ΣΞ, TBΠ.

Arzbaecher, Robert C
Business: Il Inst. Tech, Chicago, IL 60616
Position: Dir, Pritzker Inst of Medical Engrg *Employer:* IL Inst of Tech *Education:* PhD/EE/Univ of IL; MS/EE/Univ of IL; BS/EE/Fourier Inst of Tech. *Born:* 10/28/31. Prior to grad study, worked in nuclear reactor engg at Argonne Natl Lab and Army Reactors Branch, AEC. Arthur J Schmitt scholar during PhD work at Univ of IL. Following receipt of PhD, taught at Christian Brothers Coll, Memphis, and Univ of IL. Chrmn of Dept of Electrical Engrg and Prof of Internal Medicine at Univ of Iowa for 5 yrs, before assuming present position. Res interests and publications are in cardiac electrophysiology. Inventor of pill-electrode for esophageal electrocardiography. Interests: skiing, sailing, wife, five children, electrocardiology. *Society Aff:* IEEE.

Asbury, Carl E
Home: 4418 N Bay Cir, N Ft Myers, FL 33903
Position: Retired. *Education:* BSEE/Power/Univ of IL. *Born:* 6/7/12. NY Power Pool, Schnectady, NY; Planning Mgr for System Generation & Maj Transmission (1970-73). Southern Services Inc, Birmingham, AL; Asst to the Pres (1968-70) Special studies such as interconnected area reliability & bulk power supplies "System Budget Engr" (1964-68) Review & advise on all system construction budget to the pres. Commonwealth Assoc Inc, Jackson, MI; VP & Mgr of Elec Engg Section (1962-64). Proj Engr for application & design of ten 500 kv substations. VP & Chief Elec Engr (1960-62). In charge of all Elec Application & Design. Published several technical articles. Chrmn of many profl technical committees. I enjoy the privileges of an "Advanced" License in Amateur Radio. *Society Aff:* IEEE.

Aseff, George V
Home: 1015 Avondale Ave, SE, Atlanta, GA 30312
Position: Mgr, Engg Failure Analyses, Corporate Materials Consultant *Employer:* Law Engg Co, Inc. *Education:* MS/Met/GA Inst of Tech; MS/Chemistry/GA Inst of Tech; BS/Ind Chemistry/Case Inst of Tech. *Born:* 5/26/21. Home: Atlanta, GA. Lab instr ind chemistry prior to WWII. Army Signal Corps 1943-45, Pacific theatre. Consultant, team leader & sr engr, Lockheed GA Co. Staff consultant, prof lectr, Lockheed Nuclear Products, 1958-67. Scientist principal investigator, technical marketing, Lockheed R&D Labs, 1958-1970. Administrator, GA State Univ, 1970-71. Mgr of engg failure analyses, Corporate Mtls Consultant, 1971-current, Law Engg Co. Chrmn SE Dist, ASTM, 1976-1979, chmn Atlanta chapter ASNT 1977-78, chrmn Atlanta chapter ASM 1976-77. Staff abstractor for *Chemical Abstracts* Journal, 25 yrs. Reg PE 1951-current. Photographic techniques and equip as a hobby. Reg PE GA 1953- . *Society Aff:* ASM, ASTM, ASNT, GA Acad Sci, NACE, AAA

Asfahl, C R
Business: Ind Engg, Fayetteville, AR 72701
Position: Prof. *Employer:* Univ of AR. *Education:* PhD/Ind Engg/AZ State Univ; MS/Ind Engr/Stanford Univ; BS/Ind Engg/OK State Univ *Born:* 8/31/38. Joined Univ of AR faculty in 1969. Promoted to Prof of Ind Engg in 1972. Former faculty at OH Univ & Long Island Univ. Res consultant & author on the subj of Occupational Safety & Health. Author of books on robotics, automation, and safety engineering. Consultant on safe design of automatic manufacturing control systems. Research in computers, expert systems, and artificial intelligence. *Society Aff:* IIE, ASEE, AIHA, ASSE, SME.

Ash, Alvin G
Home: 8417 - 24th Ave NW, Gig Harbor, WA 98335
Position: Pres. *Employer:* Olympic Chem Corp. *Education:* BS/Chem/Univ of CA, Berkeley. *Born:* 2/26/23. Reg PE (Chem Engg) State of WA. Served in WWII in 8th Air Force. Founded Northwest Petrochemical in 1959 and was its pres until merge in 1963. Joined Hooker Chem as Dev Mgr, was Managing Dir of the Magnesium Project, a joint venture with Hooker and NL Industries at Rowley, Utah. After Hooker merged with Occidental Petroleum, became VP of Occidental Research. Founded Olympic Chem in 1976, a manufacturer of sodium sulfite and sodium bisulfite. Chief tech field is process design. Interested in hunting, fishing, hiking and skiing. *Society Aff:* AIChE, ACS.

Ash, Eric A
Business: Imperial College, London, SW7 2AZ, England
Position: Rector Imperial College (Prof of Electronics) *Employer:* Univ of London. *Education:* DSc/EE/Univ of London; PhD/EE/Univ of London; BSc/EE/Univ of London. *Born:* 1/31/28. Obtained BSc 1948 and PhD 1952 in Elec Engg at the Imperial College of Sci, London Univ. Following 2 yrs at Stanford Univ and 8 yrs at Stanford Telecommunication Labs Ltd, joined Univ College, London Main profl interests are in physical electronics, specifically acoustic imaging, signal processing, and integrated optics. Since 1985, Rector of Imperial College. *Society Aff:* FRS, FIEEE, FIEE.

Ash, Robert L
Home: 1503 Powhatan Ct, Norfolk, VA 23508
Position: Eminent Prof and Chrm, Mech Engrg and Mechanics Dept *Employer:* Old Dominion Univ. *Education:* PhD/Mech Engg/Tulane Univ; MS/Mech Engg/Tulane Univ; BS/Mech Engg/KS State Univ. *Born:* 12/27/41. Native of Wetmore, KS. Taught at Old Dominion Univ since 1967. 1983-84 Acting Dean of Engrg. Actively involved in a variety of aerospace res projs. Prin investigator on numerous contracts for NASA and Navy. Sr Resident Res Assoc at Jet Propulsion Lab, CA Inst of Tech for two years. Responsibilities there included dev of in situ propellant production systems for planetary sample return missions. Has been eminent prof since 1982. Author of more than forty refereed publications. Enjoy jogging and outdoor sports. *Society Aff:* ASME, ASEE, AIAA, ΣΞ

Ashby, C Edward, Jr
Business: 900 East Eighth Ave, Suite 200, King of Prussia, PA 19406
Position: Pres *Employer:* Envirosafe Services, Inc. *Education:* BS/ChE/WVA Univ. *Born:* 10/16/42. Numerous Dev assignments led to position as Asst Operations Mgr for Shell Chemical Co's Polymers Div in Houston TX prior to joining Rollins Environmental Services as Plant Mgr in 1973. Served as Technical Dir and Operational Mgr of this Industrial Waste Mgmt Firm prior to promotion to VP in 1977. Joined CSI in 1979 to launch program to enter new market. Promoted to Sr. VP in 1981. New div, Envirosafe Services, developed into multilocation service company employing over 200. Appointed Pres, 1985, and successfully took company public in 1987. Now employ approx 250 and generate $40plus mm in revenues. *Society Aff:* AIChE, NSPE.

Ashley, Carlyle M
Home: 7320 Barberry Ln, Manlius, NY 13104
Position: Ret. (Self-employed) *Education:* ME/-/Cornell Univ. *Born:* 8/17/99. Employed by Carrier Corp 43 yrs as R&D Engr, Dir of R&D, Chf Staff Engr. Contributed 67 US pats, over 30 papers and articles. Specialized in thermodynamics, heat and mass transfer, fluid and mechanics, acoustics, air conditioning and refrigerating, mechanisms. Presidential Mbr and Fellow ASHRAE, F. Paul Anderson, Louise and Bill Holladay. ASHRAE-Alco and Dist 50 Year Mbr Awards: Modern Pioneers in Industry 1966 of NAM; former Mbr Acoustical Soc of America, EJC, IIR. Currently consulting engr in energy use and conservation. *Society Aff:* ASHRAE.

Ashley, Holt
Business: Dept of Aero/Astro, Durand 369, Stanford, CA 94305
Position: Prof. *Employer:* Stanford Univ. *Education:* ScD/Aero Eng/MIT; MS/Aero Eng/MIT; BS/Phys Scis/ Univ of Chi. *Born:* 9/10/23. Prof, Depts of Aero & Astro (1967-present) and Mech Engg (1975-present), Stanford Univ Dir, Advanced Tech Applications Div, Res Applications, Nat'l Sci Fdn 1973-1974. Prof, MIT (1962-1966). Assoc Prof, MIT (1954-1961). Asst Prof, MIT 1948-1953). MIT Goodwin Medal, 1952; Fellow, Amer Acad of Arts and Sciences; 50th Anniversary Medal, Amer Meteorological Soc; Structures, Structural Dynamics and Mtls Award, AIAA; Mbr, Nat'l Acad of Engg; AF Exceptional Civilian Service Award (twice); Honorary Fellow, AIAA; Mbr (elected 1975) Intl Acad of Astro. Past-Chrmn, Assn for Cooperation in Engg. Mbr, Exec Ctte, Aerospace Div, ASME, 1976-1982. Pres, AIAA 1973-74. Mbr, NRC Aeronautics & Space Engrg Bd (1978- 1982). Author of 4 books and over 100 res papers and reports. Wright Brothers Lecturer, AIAA, 1981. *Society Aff:* AIAA, ASME, AAAS.

Ashley, J Robert
Home: 18 Skyline Dr, Nashua, NH 03062
Position: Sr. Prin. Elec. Eng. *Employer:* Sanders Associates, Inc. *Education:* Ph.D./EE/Univ of Florida; MSEE/EE/Univ. of Kansas; BSEE/EE/Univ. of Kansas *Born:* 10/22/27 J. Robert Ashley, Fellow IEEE, received BSEE, 1952, and MSEE, 1956, from Univ. of Kansas and joined Sperry Electronic Tube Division, Gainesville, FL. Univ. of Florida granted his Ph.D., 1967. 1967 to 1981, Dr. Ashley taught Elect Engrg at Univ of Colorado and then joined Sperry in Clearwater, FL. In 1987, he joined Sanders Associates, Nashua, NH. Dr. Ashley holds 12 US patents, has written one book and 21 major papers, and has given over 75 convention presentations. His awards include IEEE Centennial medal, IEEE Region 3 1985 Outstanding Engineer, IEEE MTT-S ARFTG Automated Measurement Technology Award, American Men and Women of Science, and Who's Who in Technology Today. *Society Aff:* IEEE

Ashley, Ramon L
Business: 12400 E Imperial Hwy, Norwalk, CA 90650
Position: Prin Engr *Employer:* Bechtel Power Corp *Education:* BS/Phys/Tufts Univ; -/Chem Engr/Clarkson Coll of Tech *Born:* 07/22/28 Born in Springfield, MA; raised in Holyoke, MA. Initially employed by Brookhaven Natl Lab as res asst. Joined North Amer Aviation (subsequently, Atomics Intl) and became Nuclear Analysis Supervisor; thereafter, directed an internal nuclear safety review team for plant designs and operations (1951- 1970). Joined Bechtel Power Corp and became Chief Nuclear Engr, served as Exec Asst to VP, Planning and Special Studies. Currently perform special licensing related tasks. Active with ANS; elected to several offices and served on numerous cttes; currently serving on Exec Ctte, Bd of Dirs and as Finance Ctte Chrmn. Enjoy jogging, biking, tennis and reading. *Society Aff:* ANS

Ashton, George D
Home: 86 Bank St, Lebanon, NH 03766
Position: Research Physical Scientist *Employer:* US Army Cold Regions Res & Energy Lab. *Education:* PhD/Mechanics & Hydraulics/Univ of IA; MS/Civil Engrg/Univ of AZ; BSCE/Civil Engrg/Univ of IA *Born:* 12/7/39. Native of Davenport, IA. Served with Army Corps of Engrs 1962-64. Civ engr for Bechtel, Inc, San Francisco as structural engr, 1964-67. Res Assoc and Instr, Univ of IA, 1968-1971. With USA CRREL since 1971 as Hydrologist, 1971- 1975. cf, Snow Ice Branch 1975-1981. Resch Phys Scientist 1981-83. cf, Geophysical Science Branch 1983-85. Assumed pres position 1985. NSF Natl Merit Scholarship 1957-1961. J Waldo Smith Fellow (ASCE) 1970-1971. Lorenz G Straub Award (Univ MN) 1971. Hilgard Prize (ASCE) 1974. J C Stevens Award (ASCE), 1979. Army R&D Achievement Award 1982. Collect antique duck decoys. *Society Aff:* ASCE, IAHR, IGS, ΣΞ

Ashton, Jackson, A, Jr
Business: 11863 Market Place Ave, Baton Rouge, LA 70816
Position: Pres. *Employer:* Ashton & Assoc, Inc. *Education:* BS/Elec Engg/LA State Univ. *Born:* 7/9/51. Native of Jeanerette, LA -- now residing in Baton Rouge, LA. Worked as an elec engr in industry and a consulting engg firm prior to co-founding own engg firm in 1975. Became Pres of the firm changing the firm's name to Ashton & Assoc, Inc in 1977. Ashton & Assoc, Inc is a consulting engg firm of forty engrs, designers and draftsmen providing elec and instrumentation engg services to industrial, commercial, institutional and petro-chem clients. Reg Elec Engr in the State of LA. Married with two children. Enjoy traveling, attending sportin events and saltwater fishing. *Society Aff:* NSPE, LES, IEEE, ISA, CEC/L.

Ashwood, Loren F
Box 346 Route 1, New Franken, WI 54229
Position: Mgmt Consultant *Employer:* self employed *Education:* BS/ME/Univ of IL *Born:* 2/27/17 and educated in Moline, IL. Various engrg and management positions in farm implement, aircraft, paper converting, and machinery manufacturing industries since 1939. Management consultant to corrugated paper converting companies and their suppliers worldwide since 1977. Listed: "Who's Who in the Midwest–. Officer of several TAPPI committees. Chrmn-Corrugated Division 1967-8. Hobbies: amateur radio (W9DH), travel. *Society Aff:* TAPPI, ARRL

Asmus, John R
Business: P.O. Box 3965, San Francisco, CA 94119
Position: VP *Employer:* Bechtel Civil & Minerals, Inc *Education:* BS/EE/Univ of TX *Born:* 9/30/25 Asmus' early career was with Pacific Gas & Electric Co, and Intl Engrg Co. In 1960 he joined Bechtel on thermal power work and in 1963 was named Chief Electrical Engr of the Bechtel joint venture responsible for design and construction of the San Francisco Bay Area Rapid Transit system. He held increasingly responsible positions on this project until job closeout in 1974. Asmus returned to Bechtel as Mgr of Transportation Projects. He was elected VP and Mgr of Division Operations in 1977. In this capacity he is responsible for all transportation, environmental and telecommunications work. *Society Aff:* IEEE.

Aspenson, Richard L
Business: 42-7E-04, P.O. Box 33331, St Paul, MN 55133
Position: Dir, Energy Mgmt *Employer:* 3M Co *Education:* BS/ME/Univ of MN *Born:* 10/7/25 Native of MN. Served US Army Air Force 1943-1945. Chief Engr Wilson and Co 1950-1955. 3M Co 1955-Present, starting as Design Engr, progressing to Mgr of Mechanical and Utilities. Presently is Dir of Energy Management. Bd

Aspenson, Richard L (Continued)
of Dirs Council of Industrial Boiler Owners. Chrmn Energy Conservation Committee National Association of Manufacturers. Steering Committee Process Gas Consumers Group. National Energy Resources Organization. Scientists Inst for Public Information. MN Business Partnership. 3M Engrg Achievement Award, 1976. Energy Saver's Award of Excellence, 1979. Engr of the Year Award, 1979, NERO Merit Award for Conservation, 1979. Enjoys golf, gardening, hunting and fishing. *Society Aff:* ASHRAE, ASME

Asquith, David J
Business: 5 Hawthorne Rd, Sturbridge, MA 01566
Position: Consulting Metallurgist. *Employer:* Sorby Lab. *Education:* BA/Chem/Brown Univ. *Born:* 2/25/18. Lincoln, RI. Reg Eng - MA. Employed Moore Drop Forging Co, Chem- metallurgist, 1940-57; Chief Metallurgist, 1957-70; Dir Quality Control, 1970- 74. Active in dev induction heating applications, automated forging and powder metal forging. Represented co on National Titanium Fabrication Committee, 1952- 54; ASTM Committees A-1, E-4. Contributor ASM Metals Handbook, Vols 2, 7. Retired 1974, formed own company, Sorby Lab. Collect old publications on metallography and enjoy rowing. *Society Aff:* TMS of AIME, ASM Intl, ASTM, AES, IMS.

Astill, Kenneth N
Business: Anderson Hall, Medford, MA 02155
Position: Associate Dean of Engineering *Employer:* Tufts Univ. *Education:* PhD/ME/MIT; MS/ME/Harvard Univ; MAE/Auto Engg/Chrysler Inst of Engg; BS/ME/Univ of RI. *Born:* 7/16/23. Lab engr with Chrysler Corp before joining faculty of Tufts Univ, 1947. Consultant in heat transfer & fluid mechanics for industry including AF Cambridge Res Ctr, Kaye Instruments, US Army Natick Labs & C S Draper Labs. Author of numerous papers on secondary flows & annular heat transfer. Author of "Numerical Algorithms" with B Arden, Addison-Wesley Publishing Co; NSF Faculty- Fellow, 1968. Visiting Fellow at Univ of Sussex, 1968 & Leeds Univ, 1976, Visiting Prof, Univ of Sussex, 1983. Held following offices in ASME: mbr, Fluid Mechanics Committee; Region I College Relations Chrmn, 1977-78; Boston Sec Exec Committee, 1979; Co-chrmn Symposium in Biomedical Fluid Mechanics 1965-66, Chrmn Boston Section, 1981; ESNE Exec Ctte 1984-. *Society Aff:* ASME, ASEE, ESNE.

Astin, Allen V
Home: 5008 Battery Ln, Bethesda, MD 20014
Position: Dir Emeritus. *Employer:* Natl Bureau of Stds. *Education:* PhD/Phys/NYU; MS/Phys/NYU; AB/Phys/Univ of UT. *Born:* 6/12/04. Native of UT. PhD in Phys from NYU. Postdoctoral Fellowship at Johns Hopkins. Joined Natl Bur of Stds staff in 1932 for work in Dielectrics, radio telemetry and, in WWII, proximity fuzes. Appointed chief electronics ordnance Div in 1948 and NBS dir in 1952. Retired as dir emeritus in 1969. Consultant state dept 1969-77. Home secy NAS 1971-75. His Majesty's Medal (UK) for service 1946. Rockefeller Public Service Award 1963, Harry Diamond Award (IEEE) 1970, Stds Medal (ANSI) 1969, French Legion of Honor 1977; Honorary Mbr AIAA, ASHRAE, ISA, SES, ASTM, ANSI. *Society Aff:* NAS, APS, IEEE, AAAS, AAAS, APS.

Astley, Wayne C
Business: 2301 Market St, Philadelphia, PA 19101
Position: VP. *Employer:* Phila Electric Co. *Education:* Masters/ME/Univ of PA; BS/ME/Univ of PA. *Born:* 12/17/'6. Joined Phila Elec Co in 1939. Dir of Res (1962-1965); Mgr of Corporate Planning (1966-1973) Present position - VP of Gen Admin since April 1973. Mbr of the Franklin Inst & Sigma Tau. Has served on various committees in the Am Soc of ME (Fellow) & the Edison Electric Inst. A mbr of the Advisory Council of the Dept of ME & Applied Mechanics, Univ of PA. Dir of: World Affairs Council of Phila; Contact - Phila; Susquehanna Electric Co; Susquehanna Power Co; Phila Electric Power Co. *Society Aff:* ANS, ASME, AIF, EEI.

Astrahan, Morton M
Business: 5600 Cottle Rd, San Jose, CA 95193
Position: Res Staff Mbr. *Employer:* IBM Res Div. *Born:* 1924. PhD Northwestern Univ 1949. IBM Res Staff Mbr 1970-, dev relational data base sys; joined IBM 1949; participated in design of IBM 701 computer and managed sys planning & part of dev of AN/FSQ-7, SAGE air defense sys. Fellow IEEE 1969. Organizer & 1st Chrmn of IRE Prof Group on Electronic Computers now IEEE Computer Soc. Involved in mgt of Joint Computer Conferences 1952-. Now serv on AFIPS Natl Computer Conf Ctte. AFIPS Dist Serv Award 1975; 2 IBM Corp invention awards, 1 for drum associative memory & 1 for input/output program interrupt.

Atchison, Joseph E
Home: 84 Edgewood Ave, Larchmont, NY 10538
Position: Pres. *Employer:* Joseph E Atchison Consultants Inc. *Education:* B.S.-LSU, MS & PhD.Inst of Paper Chemistry *Born:* Dec 1914. Ser with Amy Chem Corps in Pacific Theatre 1942-46; 1st Lt to Lt Col & Commanding Officer of Chem Battalion; devtech for firing 4.2 in ch mortars from landing craft. Tech Dir John Strange Paper Co 1946-48; Ch Pulp & Paper Branch for ECA (Marshall Plan) 1948-52; Resident Mgr Portorican Paper Prods 1952-53; with Prsons & Whittemore Inc 1953-67 in chg of Pulp & Paper Proj Div, becoming VP 1954 & Sr VP 1964; during the period 1953-67, 41 turn-key projs carried out in 24 countries on 5 continents; Founder & Pres Joseph E Atchison Cons Inc 1967-. Pres Am Arab Assn for Commerce & Indus 1964-68. Recipient TAPPI Pulp Mfg's Div Award 1974. Elected as TAPPI fellow, 1978. Chrmn TAPPI Non-wood Plant Fibers Ctte 1970-. Auth over 90 pub papers, reports & chapts of textbooks. Enjoys tennis & bowling.

Atchley, Bill L
Business: PO Box 992, Clemson, SC 29631
Position: Pres. *Employer:* Clemson Univ. *Education:* PhD/Civ Eng/TX A&M Univ; MS/Civ Eng/Univ of MO-Rolla; BS/Civ Eng/Univ of MO-Rolla. *Born:* 2/16/32. Athlete, educator, admin - native of Cape Girardeau, MO. Minor league pitcher NY Giants baseball organization. Taught engg mechanics Univ of MO-Rolla 1957-75; developed Doctor of Engg prog; designed "Profl Dev Degree–; Chrmn, MO Energy Council; Sci Advisor to Governor, MO; Assoc Dean, Engg, 1970-75. Dean, Engg, Prof, Mech Engg and Mechanics 1975-79, WV Univ; Elected Pres, Clemson Univ, 1979. Chrmn, Council of Pres for the state-supported colleges and univs in SC 1980-81; Chrmn, Res and Education Ctte of the Governor's Advisory Council on Alcohol Fuels 1980; Ctte on Energy and Environment, Natl Assn of State Univs and Land Grant Colleges 1981; Bd of Dirs, SC Energy Research Inst 1981; Fossil Energy Advisory Ctte of US Dept of Energy 1981; Awards Ctte, NSPE 1981. Bd of Trustees, SC Res Authority; mbr, D.O.E. Adv Ctte on Fed Asst for Alternative Fuels Demonstration Facilities, 1981-84. *Society Aff:* ASEE, ASCE, NASULGC, NSPE

Athans, Michael
Business: 77 Mass Ave, Rm 35-308, Cambridge, MA 02139
Position: Prof. *Employer:* MIT. *Education:* PhD/EE/Univ CA; MS/EE/Univ CA; BS/EE/Univ CA. *Born:* 5/3/37. Staff mbr MIT Lincoln Lab from 1961-1964. Joined MIT faculty in 1964; current rank prof of Systems Sci and Engg. Also, Dir MIT Lab for Info and Decision Systems. Co-author of three books and over two hundred technical papers. Recipient of Eckman Award (1964),Terman Award (1969) and AACC Education Award (1980). Fellow of IEEE and AAAS. *Society Aff:* IEEE, AAAS.

Atkin, George, Jr
Business: RD 1, PO Box Q, Tidioute, PA 16351
Position: Pres. *Employer:* Northwest Engg, Inc. *Education:* Bach of Sci/Math & Phys/Grove City College. *Born:* 2/4/25. Native of Tidioute, PA, served with US Army in Europe 1943-1946. Attended Grove City College - 1946-1950, received BS degree with majors in math and Phys. VP of McMinns Ind, Inc 1952-1959, in private practice as individual practitioner and pres of Northwest Engrg, Inc 1959 to present. Pres of PA Soc of PE 1979-80. Natl Dir of Natl Soc of PE 1974-1977 and

Atkin, George, Jr (Continued)
1979-1981. Trustee NSPE PAC and Chrmn PA soc Pac 1979-84. Past Pres of Warren County Chamber of Commerce. Enjoy fishing. *Society Aff:* ASCE, ASTM, ASHE, SAME, NSPE.

Attinger, Ernst O
Business: Dept of Biomedical Engg, Box 377 Medical Center, Charlottesville, VA 22932
Position: Chm Biomedical Engg. *Employer:* Univ of VA. *Education:* PhD/BME/Moore School of EE Univ of PA; MS/Med Instr/Dres Inst of Tech; MD/-/Zurich Univ Medical School; BA-/Winterthur Cantonal School. *Born:* 12/27/22. Postdoctoral Fellowships: Cardiopulmonary Lab, National Jewish Hospital, Denver Colorado, 1943-54; Lung Station, Bost City Hospital, Tufts Univ, 1954-59 and Resident Intership Presbyterian Hospital, Phila, PA, 1959-62. Prof Experience: Chief Resident, Diseases of the Chest, Heilstraette Du Midi, Davos, Switzerland, 1950-52; asst resident, Internal Medicine, Lincoln Hospital, New York, 1952-53; asst prof of medicine, Medical School, Tufts Univ, 1956-59; from asst prof to assoc prof of physiology, School of Veterinary Medicine, Univ of Pennsylvaniz, 1961-67; Research dir Res Inst, Presbyterian Hospital, 1962-67; Prof of Physiology and Chmn of Biomedical Engg, Univ of VA, 1967- . Forsyth Prof. of Applied Science 1984- . Consultant; WHO, NIH, PHS, HEW, NSF, NAS, and NRC. Research: Analysis of biological systems, systems analysis with particular emphasis on control hierarchy in biological & social systems, Technology assessment. *Society Aff:* IEEE, BMES, APS, AAAS, ISTAHC, ΣΞ, BT.

Atwood, Donald J
Business: 3044 West Grand Blvd, Detroit, MI 48202
Position: Vice Chrmn *Employer:* Gen Motors Corp *Education:* MS/EE/MIT; BS/EE/MIT *Born:* 5/25/24 Served on technical staff of MIT Instrumentation Lab associated with the research which pioneered development of inertial guidance systems. Served in US Army 1943-46. Joined Gen Motors in 1959; in 1970 named mgr of Detroit Diesel Allison Division, Indianapolis Operations; in 1974, general mgr of Delco Electronics Division, Kokomo, IN; in 1978, VP and general mgr of Detroit Diesel Allison Division; in 1981, VP and Grp. Exec. of Truck and Bus Group; in 1984, elected exec. vice pres. and member of Bd. of Dirs. of Gen Motors; in 1985, also made pres. of GM Hughes Electronics Corporation; and in 1987, elected Vice Chairman of Gen. Motors, his present position. Member, Bd. of Dirs. of Aviation Hall of Fame, Inc; member, Corp & Exec Cttee MIT; member, Bd. of Dir of Charles Stark Draper lab, Inc; mbr, Motor Vehicle Manufactures Assn Policy Cttee. *Society Aff:* NAE, AIAA, AFA, SAE

Atwood, Glenn A
Home: 725 Ridgecrest Rd, Akron, OH 44303
Position: Prof & Assoc Dean *Employer:* Univ of Akron. *Education:* PhD/Chem Engr/Univ of WA; MS/Chem Engr/IA State Univ; BS/Chem Engr/IA State Univ. *Born:* 10/24/35. Native Le Mars, IA. Worked for Esso Res and Engg as Process Engr 62-65. Joined Univ of Akron in 1965 to help start the Dept of Chem Engg. Prof since 1977. Asst Dean 1983-85. Assoc Dean 1985- . Was Dir of Environmental Studies for 2 yrs, Danforth Assoc. Have consulted for Gen Tire, B F Goodrich, Diamond Foundry, various local govt agencies and community groups. Active in community affairs, received Akron Council of Engr & Scientific Societies Humanitarian Award in 1977. Married & four children, enjoy canoeing and woodworking. *Society Aff:* AIChE, ASEE, ΣΞ, ΤΒΠ, NSPE.

Atwood, James D
Business: 3315 N Oak Trafficway, PO Box 7305, Kansas City, MO 64116
Position: VP. *Employer:* Farmland Industries, Inc. *Education:* BS/Chem Engg/TX Tech. *Born:* 1/1/31. Attended Pan American College in his hometown of Edinburg, TX. Grad from Tx Tech Univ at Lubbock in 1952 with a BS in Chem Engg. Prior to joining Farmland Industries, spent 20 yrs with Celanese Chem Co at various Positions, including process engg, process design, and operations management. Was employed by Farmland in Jan, 1973 as project mgr for the construction of the first Enid, Oklahoma ammonia plant. Subsequently, promoted to plant mgr and was assigned the respon for the construction of the second ammonia plant at Enid. In March, 1976, was promoted to VP, Nitrogen Manufacturing, for Farmland in Kansas City, MO. In his present position, he is respon for all nitrogen fertilizer manufacturing, including plants at Fort Dodge, Iowa; Hastings, Nebraska; Dodge City and Lawrence, Kansas; Enid, Oklahoma; and Pollock. Louisiana, and is also responsible for ammonia distribution facilities & ammonia terminals. *Society Aff:* AIChE.

Atwood, Kenneth W
Home: 140 West 8600 South, Midvale, UT 84047
Position: Prof *Employer:* Univ of UT *Education:* PhD/EE/Univ of UT; MS/EE/Univ of UT; BS/EE/Univ of UT *Born:* 12/21/22 in Cedarview, UT. Married Ruth Johnson. Children Marylou, Kenneth L, Judy Ruth. Served in US Air Corps in World War II. Worked as Radio Engr for stations KSL and KSL-TV, Salt Lake City 1949-55. Senior Engr at General Electric Co, Richland, WA 1962-63; Instructor in Electrical Engrg Dept, Univ of UT 1956-58; Assistant Prof 1958-65; Assoc Prof 1965-76; Prof 1976-present. Author; Electronic Engrg 1962; 2nd Edition 1966; 3rd Edition 1973. Semiconductor Devices and Circuits, 1971. *Society Aff:* IEEE, ASEE

Atwood, Theodore
Home: 6 Oak Pkwy, Sparta, NJ 07871
Position: Refrig Specialist. *Employer:* Allied Chem Corp. *Education:* BS/Marine Transp & Engg/MA Inst of Tech. *Born:* 11/7/25. Native of Swampscott, MA. Worked in Marine Dept of United Fruit Co. Served as Staff Engr and in Plant Mgt for Merchants Refrig Co, NYC from 1953 til 1960. Since 1960 has been associated with Allied Chem Corp. Active in ASHRAE. Held all offices in North Jersey Chapter. Served as regional Chrmn and Dir of ASHRAE from 1973 til 1976. Received ASHRAE's Best Technical Paper Award, 1971. *Society Aff:* ASHRAE.

Aubell, Gregory G
Business: 4405 Talmadge Rd, Toledo, OH 43623
Position: Partner *Employer:* Finkbeiner, Pettis & Strout, Ltd *Education:* BS/CE/Univ of Toledo *Born:* 1/11/34 Native of Toledo, OH. B.S.C.E., Univ Toledo, 1957. Received Reserve Officer's Commission in 1957. Became Field engr for Finkbeiner, Pettis & Strout in 1958. Was promoted Office Engr in 1962. Became registered engr in OH in 1963 and registered Land Surveyor in OH in 1968. Also registered in MI, NC & CA as engr. Acted as Project Engr and Assoc on numerous water and wastewater projects from 1962-1980, with same firm. Was named a partner in firm in 1980-main responsibility as Chief Production Engr. Past Pres OH Council- ASCE Current Secy of OSPE, Past Pres TSPE & Natl Dir to NSPE from Ohio, Member of OACE Bd of Dir. Member Township Zoning Commission. Secy Engg Alumni of UT. Enjoy music and sports. *Society Aff:* NSPE, ASCE, AWWA, OACE.

Aubin, Bruce R
Home: 11 Merton Crescent, Hampstead, Quebec H3X 1L5. Canada
Position: Vice President-Facilities & Supply and Senior Corporate Technical Advisor *Employer:* Air Canada. *Education:* BAe/Aero Engg/St Louis Univ; MBA/Bus Adm/McGill Univ. *Born:* 3/1/31. Native of Hamilton, ON. Grad of St. Louis Univ in Aeronautical Engrg. in 1951 and McGill Univ MBA in 1976. Employed with Canadian Car & Foundry as Grp Leader, Stress Analysis & Structures. Joined Air Canada in 1953 and successively promoted to Proj Engr, DC8; Prog Mgr, DC9/L1011; Dir, Maintenance Planning in 1971, Dir, Maintenance Engrg. in 1974. Gen Mgr, Engrg with responsibility for overall airworthiness, spec dev and perf of Air Canada aircraft fleet, 1976. Appointed V P - Purchasing & Sup with responsibility for all corp technical and material mgmt 1980. 1985 Vice President - Facilities & Supply with overall responsibility for purchasing, properties, facilities, and fuel acquisition. Currently Senior Corporate Technical Advisor and Vice President - Facil-

Aubin, Bruce R (Continued)
ities & Supply. A Natl Dir of the Soc of Auto Engrs, Chmn of Engrg & Mtce Comm. of ATAC, Chmn, Materials Mgmt Comm - ATA and a mbr of other int'l indus cttee incl. IATA, IAAAC. Has pub. a no. of tech. papers and made presentations to SAE, IATA, AIAA, the Armed Forces and other engg and scientific meetings. *Society Aff:* RAS, AIAA, CASI, SAE, CORS, OEQ.

Audibert, Jean M E
Home: 18039 Bambrook Lane, Houston, TX 77090
Position: Mgr of Engrg and Associate *Employer:* ERTEC (The Earth Technology Corporation) *Education:* PhD/Civil Engg/Duke Univ; Diplome d Ingenieur/Engg/Ecole Nationale Superieure; Arts et Metiers/-/Paris, France. *Born:* 8/27/46. Received educ in France and graduated from Duke Univ in 1972. Res Assoc at Laval Univ, Quebec, 1972-73. Asst Project Mgr for Geotechnical Engrs, Inc, 1976, Sr Proj Engr. Mgr of Engrg and Associate with Ocean Services Div of Woodward- Clyde Consultants, Houston, TX 1977-1981. Presently Mgr for Engrg and Associate for ERTEC, Houston, TX where he directs offshore site investigations, platform foundation and pipeline routing projs, both for site-specific designs and general State of the Art technology developments. He undertook res into soil- pipeline interaction since 1975 and applied results to design of pipelines subjected to seafloor instabilities and faulting. ASCE Collingwood Price 1979 for this work. Also, ASCE J. James Croes Medal 1980 for paper on Response of Offshore Structures to Earthquakes. *Society Aff:* ASCE, ISSMFE, XE.

Auerbach, Isaac L
Business: 455 Righters Mill Rd, Narberth, PA 19072
Position: Pres. *Employer:* Auerbach Conslts. *Education:* MS/Applied Phys/Harvard Univ; BS/EE/Drexel Univ. *Born:* 10/9/21. Mr Auerbach was a mbr of the original design team for Univac I from 1947 to 49 with Eckert Mauchly. From 1949-57 was with Burroughs where he directed their entire defense & space res & dev efforts. In 1957 started his own group of cos; Auerbach Corp for Sci & Tech, Auerbach Assoc Inc (now Calculon Corp), Auerbach Publishers Inc & Auerbach Consultants, providing consulting services in computer tech, mgmt info sys, telcomms and strategic business planning. Mr Auerbach was the founder, first pres & first honorary life mbr of the Intl Federation for Info Processing. In addition to mbrship in the Natl Acad of Engg, Mr Auerbach has been elected a Distinguished Fellow of the British Comp Soc, a fellow of IEEE & AAAS, & an Honorary Mbr of the Info Soc of Japan. *Society Aff:* IEEE, NAE, AAAS, ACM.

Aufiero, Frederick G
Home: 6 Volunteer Way, Lexington, MA 02173
Position: Principal *Employer:* SEA Consultants Inc *Education:* MS/Civil Engg/Northeastern Univ; BS/CE/Northeastern Univ *Born:* 1/7/45 Native of Boston, MA. Currently living in Lexington, MA. Wife and five children. Working in civil and environmental engrg since 1967. Employed by SEA- Consultants Inc since 1971. Began as staff engr and progressed to position as principal and Executive in 1979. Broad experience in all aspects of engrg projects and management. Currently holds position as chief operating officer responsible for all SEA manpower and production control as well as other administrative duties. Extensive project involvement as well, particularly in areas of client responsibility and marketing. Active in professional societies with emphasis in legislative affairs and their effect on engrg community. *Society Aff:* ASCE, WPCF, ACEC, APWA

Aulenbach, Donald B
Business: Dept of Chem & Envir Engg, Troy, NY 12181
Position: Prof. *Employer:* Rensselaer Polytechnic Inst. *Education:* PhD/Sanitation/Rutgers Univ; MS/Sanitation/Rutgers Univ; BS/Chem/Franklin and Marshall College. *Born:* 3/7/28. in Berwick, Pa. All elementary education in Allentown, Pa public schools. Worked 6 yrs with Delaware Water Pollution Commission in Dover as chemist- bacteriologist in charge of all water pollution control labs. Teaching environmental engg at RPI since 1960. Res on nutrient inputs to lakes and land application of wastewater. Private consulting in environmental engg. Active in NY Water Pollution Control Assoc. Elder, Christ Community Reformed Church, Clifton Park. Play violin, bass viol, sing in church choir. *Society Aff:* WPCF, AWWA, NSPE, ACS, NWWA, HPS, AEEP, ASLO, AAEE.

Ault, Eugene Stanley
Home: 822 Sixth St, Traverse City, MI 49684
Position: Prof Emeritus. *Employer:* Purdue Univ. *Education:* MME/Mach Des/Cornell Univ; ME/Mach Des/Cornell Univ; BME/-/Johns Hopkins Univ. *Born:* 6/9/98. Native of Baltimore. Taught graphics, machine design, and mgt at Cornell, Rice, Lehigh, Case & Purdue, where I went as Prof in Charge of Machine Design. Ind work with Kingsbury Ordnance Plant, also as Asst Bridge Engr, City of Houston, designed vertical lift bridge and machinery. Chrmn Cleveland Engr Soc, first chrmn of Machine Design comm, ASME, co-founder of Mechanicsm Conferenced, and Automation Conferences, Co-author of "Fundamentals of Machine Design–, and contributor to technical magazines. Hobbies: music, art, travel in US, Canada & Mexico. *Society Aff:* ASME, ASEE, MEC, ΣΞ, ΠΤΣ.

Ault, G Mervin
Home: 25920 Myrtle Ave, Olmsted Falls, OH 44138
Position: Pres *Employer:* Gateway Technology Associates, Inc. *Education:* BS/Mech Engg/Iowa State Univ. *Born:* 8/23/21. Consultant, was Dir of Energy Programs at NASA-Lewis Res Center; managed major portions of DOEs terrestial energy R&D for alternative auto engines, phosphoric acid fuel cells, wind energy, utilization of synthetic fuels from coal, and NASA's space power, materials, and structures technology conducted by four divisions of about 400 professionals. Recipient of NASA Distinquished Service Medal, NASA Exceptional Scientific Achievement Medal, NASA Exceptional Service Medal, Greater Cleveland Career Service Award for Exceptional Performance in Federal Service, Presented a Horace Gilette Memorial Lecture for ASTM. Served on National Advisory Board of National Academy of Sciences. Served in City elective office. *Society Aff:* AIME, ASM, AAAS.

Aured, Charles F
Business: 720 Green Crest Dr, Westerville, OH 43081
Position: Pres. *Employer:* Panels, Inc. *Education:* BME/Mech Engg/OH State Univ. *Born:* 9/30/34. Native of Columbus, OH. Field Engr for Sohio after graduation 1957-61. Co- founder of Panels, Inc 1961, designers and builders of industrial elec and pneumatic control panesl and instrumentation sys. PPres of Columbus Chapter ISA. Currently serving on SP60 "Control Centers" Committee ISA. Hobbies: golf a photography. *Society Aff:* ISA.

Austin, Donald S
Business: Austin, Tsutsumi & Associates, Inc, 501 Scmner St. Suite 521, Honolulu, HI 96817-5031
Position: Pres *Employer:* Austin, Tsutsumi & Assoc, Inc. *Education:* BS/CE/Univ of HI *Born:* 12/2/22 Born in Honolulu, HI; attended Stanford Univ 1941-43 and Univ of HI 1948- 51, BSCE; Married Ruth Woolley of Hondlulu 1947; Children Donald S. Jr., Margaret, Herbert, Allan; Jr. Engr., Hawaian Airlines Ltd, Honolulu 1944-47; Asst. Civil Engr, H.A.R. Austin & Assoc, Ltd, Hondlulu, 1951-57, V.P. & Treas. 1957- 59; Pres, Austin, Smith & Assoc, Inc., Hondlulu, 1959-75; Pres/Chrmn of Bd, Austin, Tsutsumi & Assoc, Inc, Hondlulu, 1975 to present; Instructor CE. Dept. Univ of Hawaii 1952; Fellow, American consulting engrs council 1967 to present; Fellow, ASCE; Pres, Hawaii Section 1956-57, Natl. Dir 1976-79, Natl V.P. 1979-80; Mbr, NSPE; Mbr, Engrg Assn of Hawaii 1951 to present; Mbr, SAME; Chamber of Commerce of Hawaii; Mbr: The Pacific Club; Prof, Registration: Hawaii, Guam; Mormon. *Society Aff:* ASCE, NSPE, SAME, WPCF, AWWA, ACEC, XE, ACI

Austin, James B
Home: 114 Buckingham Rd, Pittsburgh, PA 15215
Position: Retired. *Education:* PhD/Physical Chemistry/Yale Univ; BS/Chem Engg/Lehigh Univ. *Born:* 7/16/04. Wash Grove, MD, a suburb of Wash. After grad study

Austin, James B (Continued)
at Yale, joined the newly formed Res Lab of US Steel at Kearny, NJ. Became Dir of Lab 1946. Moved to Pittsburgh with the Lab in 1956. Admin VP, Res and Tech for US Steel, 198. Retired 1968. AIME, Pres 1973, Fairless Award 1969, Honorary Mbr, 1979. ASM Pres, 1954, Fellow 1969, Honorary Mbr 1975. Metals Soc, London, Honorary Mbr, Japanese Iron and Steel Inst Honorary Mbr, 1965, Tawara Gold Medal 1979. Collect Japanese prints. Advisor on Japanese prints, Museum of Art Carnegie Inst, Pittsburgh. *Society Aff:* AIME, ASM, Metals Society, JISI.

Austin, John H
Business: ESE-401 Rhodes Ctr, Clemson, SC 29631
Position: Prof & Hd Environ Sys Engg. *Employer:* Clemson Univ. *Born:* Feb 22, 1929 WA DC. BS Syracuse Univ 1951; MS MIT 1953; PhD Univ of CA, Berkeley 1963. Fulbright Scholar Netherlands 1953-54. Sanitary Engr USPHS Cincinnati OH & Saigon Viet Nam 1955-59; Asst & Assoc Prof of Sanitary Engg Univ of IL 1963-69; Prof Clemson 1969-. Pres Assn of Environ Engg Prof's. Received Fulbright Scholarship; Tau Beta Pi, Sigma Xi, Chi Epsilon.

Austin, Walter J
Business: Civil Engrg Dept, PO Box 1892, Rice Univ, Houston, TX 77251
Position: Prof & Chrmn *Employer:* Rice Univ *Education:* PhD/CE/Univ of IL; MS/CE/Univ of IL; BS/CE/Rice Inst *Born:* 2/6/20 Native of Chicago, IL. Structural Designer, Chicago Bridge and Iron Co 1942-46. Research Assoc, Assistant Prof and Assoc Prof, Univ of IL 1948-60. Prof at Rice Univ since 1960; Chrmn of Civil Engrg Dept 1963-64 and 1977-present. Visiting Prof, Univ of CO, summers 1964, 1965 and 1968; Visiting Prof, Univ of TX, summer 1977. Fellow, American Society of Civil Engrs; Moisseiff Award 1958; Life Member, Structural Stability Research Council; Executive Committee 1976- present. Consulting in structural analysis and design. *Society Aff:* ASCE, SSRC

Autio, Marvin E
Home: 10509 Karen Ave, NE, Albuquerque, NM 87111
Position: Dir, Facilities & Construction Mgr Div. *Employer:* US Dept of Energy. *Education:* BS/Civ Engr/SD Sch of Mines & Tech. *Born:* 5/23/19. Native of Black Hills of SD. Yr of interesting & challenging experience in 1942 on CANOL Proj, Yukon Territory, Surveyor for 550-mile traverse under severe conditions; snow-shoes, tents, dog teams. Served with Combat Engrs, WWII in European Theatre. Completing thirty yrs Civilian Govt Service in Engg- Construction, proj mgt, for maj nuclear facilities; including, fuel production plants, gaseous diffusion plant, test reactors for Admiral Rickover's fuel dev progs, & nuclear weapons dev & production facilities. Majority of projs have been unique to the Nuclear Industry; state-of-art, one-of-a-kind, to satisfy stringent criteria.

Avanessians, Alexander A
Business: 60 Broad St, New York, NY 10004
Position: Dir Services Engg. *Employer:* RCA Global Communications Inc. *Born:* Feb 1923; MSEE Sch of Engg Columbia Univ (with grad citation) June 1961; BS & MS Tehran Univ June 1946; BS Iran Military Acad June 1945 (top of the class). Avion Electronics Div ACF Indus 1959; Design Engr RCA Global Communications Inc 1959-62, Group Leader Automation & Terminal Sys 1962-65, Mgr Adv Sys Engg 1965-66, Mgr Sys Engg 1962-72, Mgr Computer Engg Proj 1972-73, Dir Computer Proj 1973-75; assumed current respon, Dir Services Engg for RCA Global Communications 1975. David Sarnoff Fellowship Award 1963-64. Mbr IEEE.

Avedesian, Michael M
Business: 240 Hymus Blvd, Pointe Claire, Quebec Canada H9R 1G5
Position: Project Mgr-Engg. *Employer:* Noranda Res Centre. *Education:* PhD/Chem Engg/Univ of Cambridge (UK); BEngg(Hon)/Chem Engg/McGill Univ (Canada). *Born:* 2/13/45. Native of Cambridge, Ontario. Early schooling in St Catharines, Ontario. Awarded Ernest Brown Gold Medal, CIC medal and Soc of Chem Industry Merit Award for undergraduate work at McGill Univ, Montreal. Awarded an Athlone Fellowship to study for PhD degree at Univ of Cambridge in England. Joined Noranda Research Centre in 1972 as Res Engr. Left Noranda in 1977 to initiate and launch new Coal Conversion Section in the Canadian governments Energy Res Labs. Joint SNC Inc in 1978 as consultant process engg. Rejoined Noranda Res Centre in 1979 as Project Mgr - Engg. Presently Secretary of the CSChE. Enjoys hockey and soccer. *Society Aff:* CSChE, AIChE, IChemE, OEQ, APEO.

Averbach, Benjamin L
Home: 165 Somerset St, Belmont, MA 02178
Position: Prof of Mateials Sci. *Employer:* MA Inst of Tech. *Education:* ScD/Metallurgy/MA Inst of Tech; MS/Metallurgy/Rensselaer; BS/Metallurgy/Rensselaer. *Born:* 8/12/19. Was Chief Matallurgist for US Radiator Corp and a metallurgist with the General Electric Company. Appointed Asst Prof at MIT in 1947, Assoc Prof in 1952 and Prof in 1958. Principal res activities in fracture of high strength steels, structure of amorphous materials, X-ray and neutron diffraction, magnetic materials. Was pres of Alloyd Res Corp. PPres of International Conference on Fracture. Approximately 200 publications. *Society Aff:* ASM, APS, TIM, TMS

Avila, Charles F
Home: 272 Atlantic Ave, Swampscott, MA 01907
Position: Retired. *Education:* BS/EE & Bus Admin/Harvard Univ. *Born:* 9/17/06. Early work as engr, maintenance of lines dept of Boston Edison Co. Studied failures and redesigned or specified new mtls. Specialties were electrolysis and corrosion mitigation and high voltage cable. Worked through grades of div hd, section hd, to vp of engg and construction, 1956; vp and asst gen mgr, 1957; corporate vp, 1959; pres and CEO, 1960; and chrmn, 1967. Have been a dir of 19 corps including Boston Edison, Raytheon, Liberty Mutual, John Hancock, Shawmut Bank, and Chas T Main. During WWII designed high altitude military cameras. Retired, 1971; some directorships continued to 1984. *Society Aff:* IEEE, NAE, HES.

Avins, Jack
Home: 178 Herrontown Rd, Princeton, NJ 08540
Position: Consultant. *Employer:* Self-employed *Education:* MEE/EE/Poly Inst of Brooklyn; AB/Math, Phys/Columbia Univ. *Born:* 3/18/11. From 1941 to 1946 served in the US Army Signal Corps attaining the rank of Maj. In 1947 joined RCA. Did res & dev in field of FM & color TV receivers. Elected a fellow of the IEEE in 1956 "For Contributions to the Development of Television Receivers." Received the RCA David Sarnoff Award in 1971 & the IEEE Consumer Electronics Award in 1976 for the dev of integrated circuits for color TV receivers. Hold 54 US patents. Retired from RCA in 1976. *Society Aff:* IEEE.

Avril, Arthur C
Business: PO Box 17087-St Bernard, Fischer Ave & B & O RR, Cincinnati, OH 45217
Position: Pres. *Employer:* Sakrete Inc. *Education:* BEM/Mining/OH State Univ. *Born:* 1/8/01. Pres of Sakrete Inc, Sakrete Sales Inc, Sakrete of Indiana, Inc, A & T Dev Corp, and Partner Sakret Trockenbaustoffe & Co KG Inventor, holder of 18 patents for processing equip. Reg Mining and Civil Engg in OH and ID. Mbr of Rotarty Club, Delta Tau Delta, Queen City Club, Cincinnati Country Club, Browns Run Country Club, Cincinnati Chamber of Commerce. Listed in Whos Who in America, Whs Who in Finance & Industry, Community Leaders & Noteworthy Americans, Men of Achievement, International Biography, Royal Blue Book. Served on following committees: ASTM C-9 amd G-1, Engg Soc of Cincinnati, O Transportation, Childrens Hospital Bldg, and Chamber of Commerce Environmental. *Society Aff:* ASTM, ACI, NAM.

Avula, Xavier J R
Home: PO Box 1488, Rolla, MO 65401
Position: Prof. *Employer:* Univ of MO. *Education:* PhD/Engg Mech/IA State Univ; MS/APpl Mech & A E/MI State Univ; BTech/Aeronautical & Mech Engr/Indian

Avula, Xavier J R (Continued)
Inst of Tech; BSc/Math/Andhra Univ *Born:* 1/8/36. in Kavali, Andhra Pradesh, IN. Section Officer- Mechanical, Neyveli Lignite Corp, Madras, IN 1960-61. Guest Scientist, Inst for Fundamental Res, Braunschweig, West Germany 1961-62. Grad Res Asst at MI State, 1962-64 and at IA State, 1964-67. Asst Prof of Engg Mechanics, Univ of MO-Rolla 1967-73. Assoc Prof 1973-77. Natl Academy of Sciences - National Acad of Engg Sen Post Doct, Resident Res, Assoc, Aerospace Med Res, Lab Wright-Patterson Air Force Base, OH 1974-76. Prof of Engg Mech, Univ of MO-Rolla, 1977-present. Initiated the International Conferences on Mathematical Modelling. Editor J of Math and Computer. Modelling published by Pergamon Press. Organized Int Assoc Math Modelling. *Society Aff:* ASME, SES, IAMM.

Axford, Roy A
Home: 2017 S Cottage Grove Ave, Urbana, IL 61801
Position: Prof of Nuclear Engr. *Employer:* Univ of IL. *Education:* ScD/Nuclear Engg/MIT; SM/Nuclear Engg/MIT; SB/Sen Sci/MIT; AB/Phys/Williams College. *Born:* 8/26/28. Native Detroiter. High School valedictorian. Completed five-yr combined plan between Williams and MIT Received in Feb, 1958 first doctorate awarded by MIT in nuclear engg. Supervisor, Reactor Analysis Group, Atomics Intl, 1958- 1960. Assoc Prof & Prof of Nuclear Engg at TX A & M Univ, 1960-1963, Northwestern Univ, 1963-1966, Univ of IL, 1966 to present. Consultant to industry, Argonne Natl Lab, Los Alamos Scientific Lab for civilian and military applications of nuclear energy. Responsible for formulating ECPD accredited baccalaureate curricula and grad res progs in nuclear engg. Received recognition certificate for excellence in undergrad teaching, 1979 and 1981. Tau Beta Pi; Sigma Xi. Numerous res publications. *Society Aff:* AIAA, ANS, ASME.

Ayars, James E
Business: 5544 E Air Terminal Dr, Fresno, CA 93727
Position: Agricultural Engineer *Employer:* USDA-SEA-AR *Education:* PhD/Agri Engg/CO State Univ; MS/Agri Engg/CO State Univ; BAgriEngr/Agri Engg/Cornell Univ. *Born:* 1/28/43. and raised in Penns Grove, NJ. Served with USAF in a Titan II missile wing 1965-1969. Hydraulic Engr with NY State Dept of Environmental Conservation 1969- 1971. Researcher with Dept of Agri Engg CO State Univ 1971-1976. Asst Prof in Dept of Agri Engg at Univ of MD 1976-1980. Taught and did research in the areas of environmental quality and soil and water resource mgt. Hobbies include cooking, woodworking and backpacking. 1980-Present research Scientist with Water Management Research Unit, USDA-SEA-AR in the areas of groundwater recharge and irrigation and drainage water management on irrigated lands in arid and semi- arid areas. *Society Aff:* ASAE, ΣΞ, ASA, SSSA.

Aycock, Kenneth G
Home: 887 Cherokee Road, Auburn, AL 36830
Position: Civil Engineer *Employer:* USDA-Soil Conservation Service *Education:* BS/Agri Eng/Auburn Univ *Born:* 10/20/46 Native of Tuscumbia, Ala. Began working with the Soil Conservation Service (SCS) in the state design section. Transferred to project office working on watershed planning staff as a hydrologist. In 1976 was assigned duties as a civil engr on watershed planning staff to design flood control structures. Transferred in 1980 to SCS Ala. River Basin Staff. As civil engr on the staff, duties include determination of location and feasibility of structures for soil and water conservation in Alabama. *Society Aff:* ASAE

Ayer, William E
Business: 3000 Sand Hill Rd, Menlo Park, CA 94025
Position: Business Cons & Private Investor (Self-emp) *Education:* BA/EE/Stanford Univ; MA/EE/Stanford Univ; PhD/EE/Stanford Univ. *Born:* 11/12/21. PhD Stanford Univ 1951; EE & BA Stanford Univ. Officer USNR WWII. Dev Engr Mackay Radio & Telegraph 1946-51; Res Assoc Applied Electronics Lab Stanford Univ 1951-57; VP Granger Assoc 1957-59; Founder & Pres Applied Tech Inc 1959-67; VP Itek Corp 1967-69; Presently Dir of a number of corps including BR Comm of Sunnyvale, CA, Spectra-Physics Inc of Mtn View CA, CA Water Serv Co of San Jose, ARCO Sys, Inc of Sunnyvale, CA plus several others. Trustee of Stanford Univ & Fellow AIEE; mbr: Tau Beta Pi, Sigma Xi. Reg Engr CA. *Society Aff:* AIEE, ΤΒΠ, ΣΞ.

Aynsley, Eric
Business: 607C Country Club Dr, Bensenville, IL 60106
Position: Pres. *Employer:* Almega Corp. *Education:* PhD/CE/Univ of Birmingham; BSc/CE/Univ of Birmingham. *Born:* 9/8/40. Obtained PhD on ind sponsored air pollution control proj in Dept of Chem Engg, Univ of Birmingham, England in 1965. Worked in British Steel Ind & lectr in Chem Engg Dept, Univ of Newcastle, England. Res Engr at IIT Res Inst, Chicago on air pollution control projs. Lectr on air pollution & control in Environmental Engg Dept of ITT in Chicago. Presently Pres of the Almega Corp in Bensenville (Chicago) IL providing in plant OSHA & air sampling & analysis and stack emission & air pollution control equip testing & evaluation for ind clients. *Society Aff:* AIChE, Inst Chem E, APCA.

Ayoub, M M
Business: Dept of IE, Box 4130 TX Tech, Lubbock, TX 79409
Position: Horn Prof *Employer:* TX Tech Univ *Education:* PhD/Ind Engrg/State Univ of IA; MS/Ind Engrg/State Univ of IA; BS/ME/Univ of Cairo, Egypt *Born:* 2/17/31 Dr Ayoub is a PE who devoted most of his professional time to the area of occupational ergonomics. He is internationally known for his work in occupational Biomechanics and Work Physiology. He is currently Horn Prof of Industrial Engrg at TX Tech Univ. He also serves as the Dir of the Inst for Ergonomics Res at TX Tech. Dr Ayoub is a fellow of the Human Factors Society, Ergonomic Res Soc, Inst of Industrial Engrs, American Society of Biomechanics, and the American Industrial Hygiene Association. In 1975 he was awarded the Paul M Fitts award for outstanding contributions to Human Factors education. In 1980 the Inst of Industrial Engrg awarded him the Ergonomics division award for his contributions to the field. In 1981 he was awarded the David Baker Distinguished Research Award by the Inst of Industrial Engrs for his research work in Ergonomics. In 1982 he was awarded the Faculty distinguished research award by the Texas Tech Dads Association.. *Society Aff:* IIE, HFS, ERS, ASB, AIHA

Ayoub, Mahmoud A
Business: Dept of IE, PO Box 5511, Raleigh, NC 27650
Position: Prof *Employer:* NC State Univ. *Education:* BS/Civil Engg/Cairo Univ; MS/Ind Engg/TX Tech Univ; PhD/Ind Engg/TX Tech Univ. *Born:* 1/1/42 Mahmoud A Ayoub is Prof of Industrial Engg at NC State Univ, Raleigh, NC. He is also Adjunct Prof of Environmental Sci and Engg, School of Public Health, the Univ of NC at Chapel Hill. Dr Ayoub teaches biomechanics, ergonomics and safety. Dr Ayoub received a BSCE from the Univ of Cairo and holds MS and PhD degrees from TX Tech Univ. he is a senior mbr of IIE, Alpha Pi Mu, Ergonomics, Human Factors and American Soc of Safety Engrs. He has been published widely. He is also a frequent speaker at national and international conferences and symporia. Dr Ayoub is a consultant to industry and government in matters concerning ergonomics, safety and health in the workplace. *Society Aff:* IIE, HFS, ERS, ΑΠΜ.

Ayres, James Marx
Business: 1180 So Beverly Dr, Ste 600, Los Angeles, CA 90035
Position: Pres *Employer:* Ayres Ezer Lau, Inc. *Education:* MS/ME/Purdue Univ; BS/ME/Univ of CA Berkeley *Born:* 11/28/22 Pomona, CA. Officer, US Submarine Serv WW II. Teaching Assist, 1 yr Univ Berkeley. Chief mechanical engr, various arch/engrg firms, 1949-56. Entered private practice, in Los Angeles 1956. Pres Ayres & Hayakawa, 1959-74. Pres Ayres & Hayakawa Energy Management, 1974-77 Pres. Ayres Assoc., 1977-84, Pres. Ayres Ezer Lau, 1985--- Author numerous technical papers and articles on Air Conditioning and building energy systems, computer calculations. Off-peak cooling, solar design & earthquake non-structural damage. Active in ASHRAE and other various technical societies. Reg Engr 7 staes & Nat

Ayres, James Marx (Continued)
Council Engr. Examiner, member State of CA. Seismic Safety Com, 1975-79. *Society Aff:* ASHRAE, ACEC, APEC, NCEE, EERI.

Azar, Jamal J
Business: 600 S College, Tulsa, OK 74104
Position: Dir & Prof. *Employer:* Univ of Tulsa. *Education:* PhD/Mechanical/OK Univ; MS/Aerospace/OK Univ; BS/Aerospace/OK Univ. *Born:* 9/19/37. Prof of Petroleum Engg and Dir of Drilling Res at the Univ of Tulsa. World wide lecturer in Drilling Engg and Structural Analysis and Design. Author of several technical publications including text books in Finite Element Analysis, Drilling in Petroleum Engg and Marine Structures. Served as a consultant for numerous private and govt agencies. Reg PE. Enjoys hunting and tennis. *Society Aff:* AIME/SPE.

Azer, Naim Z
Business: Dept of Mech Engrg, Manhattan, KS 66506
Position: Prof *Employer:* Kansas State Univ *Education:* PhD/ME/Univ of IL; MS/ME/Univ of Alexandria-Egypt; BS/ME/Univ of Alexandria-Egypt *Born:* 7/20/27 From 1959-1964, was on the Faculty of the College of Engrg, Univ of Alexandria, Egypt. Since 1964, has been on the faculty of the Dept of Mech Engrg at Kansas State Univ. He teaches courses in the thermal scis area at the undergraduate and graduate levels. He does res in the areas of heat transfer, fluid mech and environmental engrg. He is a Res Assoc at the Inst for Environmental Res at Kansas State Univ. Has numerous technical publications. Received a Distinguished Service Award from A.S.M.E. for serving over ten yrs as reviewer for Applied Mech Reviews. Mbr of Sigma Xi, Phi Kappa Phi and Pi Tau Sigma *Society Aff:* ASME, ASHRAE, ASEE

Aziz, Khalid
Business: 112 Peter Coutts Circle, Stanford, CA 94305
Position: Prof & Chmn Petroleum Engineering *Employer:* Stanford University *Education:* PhD/ChE/Rice Univ; MSc/Petro Engrg/Univ of Alberta; BSc/Petro Engrg/Univ of Alberta; BSE/ME/Univ of MI. *Born:* 9/29/36 Born in Pakistan, now a permanent resident of the US. Currently Prof & Chmn of Petroleum Engrg at Stanford Univ Previous positions held include Prof of Chem and Petroleum Engrg at the Univ of Calgary and Mgr of the Computer Modelling Grp. Author or co-author of over 120 technical papers and three books. Current res interests include multiphase flow in pipes, computer modelling of enhanced oil recovery processes, fluid properties and natural gas engrg. *Society Aff:* AIChE, SPE, SIAM, APEGGA.

Azzam, Rasheed MA
Business: Electrical Engrg Dept, New Orleans, LA 70148
Position: Dist Prof *Employer:* Univ of New Orleans *Education:* Ph.D./EE/Univ of NB; M.Sc./EE/Univ of NB; B.Sc. /EE/Cairo Univ *Born:* 3/9/45 Postdoctoral Fellow, Univ of NB, 1972-1974. Assoc Prof of Research in Engrg and Medicine, Univ of NB, 1974-1979. Fulbright Senior Research Scholar, France, 1985-86. Presently Dist Prof of Electrical Engrg, Univ of New Orleans. Author and coauthor of over 100 research papers in 20 refereed national and intl journals. Author of the book "Ellipsometry and Polarized Light," (North-Holland, 1977), translated into Russian by Mir, Moscow. Coorganizer and coeditor of Proceedings of the 1975 and 1979 Intl Conferences on Ellipsometry. Cited for outstanding contribution to SPIE in 1977 and 1978 for chairing two of the Society's national seminars. Fellow of OSA. Patentee in the field. Recipient of NSF grants. *Society Aff:* OSA, SPIE, ΣΞ

Baade, Peter K
Business: Principal Consultant, Noise and Vibration Control, Inc, 171 Brookside Lane, Fayetteville, NY 13066
Position: Prin Conslt *Employer:* Noise and Vibration Control, Inc *Education:* Dr-Ing/Acoustics/Tech Univ Berlin; Dipl-Ing/ME/Tech Univ Karlsruhe; BS/ME/Robert College. *Born:* 7/29/21. Goettingen, Germany. BSME Robert College, Istanbul Turkey 1942; Dipl Ing Technical Univ of Karlsruhe, Germany 1947; Dr Ing Environ Engrg Tech Univ of Berlin, Germany 1971. Conslt for noise and vibration control. Visiting Prof at Sch of Mech Engrg, Purdue Univ. Past Pres Inst of Noise Control Engrg; num other posts in prof soc's in the heating, air-cond, acoustics, & standards fields. Fellow Acoustical Soc of Amer, Fellow Amer Soc of Heating Refrig & Air-Cond Engrs. Enjoys gardening & teaching *Society Aff:* ASA, ASHRAE, INCE.

Babb, Albert L
Home: 3237 Lakewood Ave S, Seattle, WA 98144-9998
Position: Prof Nuc Engg, Chem Engg; Adjunct Prof Bioengg *Employer:* Univ of Washington. *Education:* PhD/ChE/Univ of IL; MS/ChE/Univ of IL; BAppSci/Che/Univ of BC *Born:* 11/07/25 Native of Vancouver, Brit Col. U S citizen 1954. Univ of Wash fac mbr since 1952; first chmn of Nuclear Engrg Dept; Prof of Chem Engrg. Elected to Natl Acad of Engrg 1972. Elected to Institute of Medicine, 1981. Cited by State Joint Ctte on Nuclear Energy 1967. 'Engr of Yr' 1968; 'Outstanding Educator of Amer' 1971. Co-inventor of first central dialysate delivery sys for ctr hemodialysis of patients & first home artificial kidney machine for overnight treatment. Patents on sys for stabilizing structures & oil pipelines in permafrost and biomedical devices. Inventor of extracorporeal sys for the continuous treatment of blood of patients with Sickle Cell Anemia and Auto Imune Diseases. Co-inventor of minature wearable insulin pump for diabetics; Co- inventor of sodium bicarbonate proportioning system for kidney machines; P Dir of ANS; Mbr Exec Ctte and chmn of Environ Sci Div. Memberships in: ASAIO, AIChE, ANS ASN and ISAO; Mbrship Ctte Natl Acad of Engrg; Institute of Medicine; Tr Pacific Sci Ctr Foun. Reg P E Chem & Nuclear Engrg. Natl Kidney Found Pioneer Award, 1982; Fellow, Amer Inst Chemists; Fellow Amer Inst Chemical Engineers Sigma X Award, 1982; College of Engineering Outstanding Teacher Award, 1987. Enjoys classical music & horticulture. *Society Aff:* ANS, AIChE, ASAIO, NAE, IOM, ISAO, ΣΞ, ΤΒΠ, ASN

Babcock, Daniel L
Home: 1347 California Dr, Rolla, MO 65401
Position: Prof of Engg Mgmt. *Employer:* Univ of MO-Rolla. *Education:* PhD/Engg/UCLA; SM/Chem Engg/MIT; BS/Chem Engg/Penn State Univ. *Born:* 11/25/30. in PA. Beginning 1953 served 3 yrs each as Dev Test Officer in USAF (now Lt Col retired), chemist/technical writer for Dow Corning Corp, tech editor for Chemical Propulsion Information Agency, Johns Hopkins Univ; Supervisor and Systems Engr in Apollo rocket propulsion, Space Div, Rockwell International Corp 1963-69. Founding Exec Dir ASEM, past Chmn Engg Mgmt Div ASEE, Cert Quality Engr, ASQC. Author numerous articles on engg management education and other topics. Reg PE (MO). *Society Aff:* ASEE, ASEM, ASQC, SAME, SSS.

Babcock, Henry A
Business: P.O. Box 463, Golden, CO 80401
Position: Retired Professor and Head. *Employer:* Colorado School of Mines. *Born:* Feb 23, 1917. PhD, MS & BS Univ of Colorado at Boulder. Reg PE & LS in Colorado (1084). Worked for U S Engrs prior to WW II, & U S Bureau of Reclamation following discharge from US Navy. With Colorado Sch of Mines since Sept 1946, advancing from Instructor to Hd of Basic Engrg Dept. Cons work in design of concrete grain elevators, design of lrg subdivisions & hydraulic transportation of solids. Enjoy playing bassoon in amateur symphony orchestras.

Babcock, James H
Home: 6714 Melrose Dr, McLean, VA 22101
Position: VP, Mgr *Employer:* IRT Corp, Washington Office *Education:* PhD/EE/Stanford; MS/EE/MIT; BS/EE/Univ of IA. *Born:* 8/22/36. Native of Wash, DC. 17 yrs with CIA. Office of Secy of Defense 1975-1981. 1975-76 Staff Specialist, Communications Satellites; 1976-77, Dir, Sys, ODTACCS; 1977-79 Dir, Intelligence Sys, Office of Asst Secy Def Communications Command and Control and Intelligence; 1979-81 Deputy Asst Secy Defense (Intelligence), OASD (C3I). 1981 Asst Dept Under Secy (Intelligence). 1981-83 VP, R&D, Planning Res Corp Govt Infor Sys, McLean, VA.; 1983-present VP Mgr Washington Office, IRT Corp. Sr Mbr IEEE. Enjoy photography, music, travel. *Society Aff:* IEEE, AIAA, AFCEA, ΣΞ.

Babcock, John A
Home: Coffee Run RR4 B3G, Hockessin, DE 19707
Position: Design Proj Mgr - Retired *Employer:* DuPont *Education:* BS/Petro & Nat Gas Engrg/PA St Univ *Born:* 1/7/17 Employed Sinclair Refining 1940-41, Dupont 1941-77 in process dev, consultation, and chemical plant design. Mbr AIChE since 1951 including service as Natl Dir, Chrmn Govt Programs Steering Cttee, and Chrmn Wilmington and Delaware Vly Local Sections. Co-Chrmn EJC-Joint Societies Ctee 1978 Second Edition "Guidelines for Employment of Professional Engrs and Scientists–. Author magazine articles on plant design and technical mgmt. Registered Professional Engr including serv with DE Assoc of Prof Engrs (state registration bd) as member of Council 1977-85 in various offices, including Pres 1983-85, and member of Examining Cttee since 1977. Chrmn since 1984 Bldg Ctte, Catholic Diocese of Wilmington. *Society Aff:* AIChE

Babcock, Robert E
Business: Rm 223, Ozark Hall, Fayetteville, AR 72701
Position: Professor, Chem. Engr. *Employer:* Univ of Arkansas. *Education:* PhD/Ch Eng/Univ of OK; MSch/Ch Engg/Univ of OK; BS/Petr Engg/Univ of OK. *Born:* Oct 10, 1937. Sr Res Engr 1964-65 at ESSO Production Res Co, Houston Texas. Asst Prof 1965, Assoc Prof 1968, Prof 1974 Dept of Chem Engrg, Asst Dean of Grad School 1970-76, Dir of Ark Water Resources Res Ctr 1970-present, all at Univ of Arkansas. Pres of Ark Sect of Amer Water Resources Assn 1972-73. Chmn AIChE-SPE Joint Symposium Ctte 1973-75. Ark Governor's Coord Ctte on Water Resources 1970-71. State of Ark State Stream Preservation Ctte 1972-82. Outstanding Educators of Amer 1970 & 1974, Arkansa Conservation Achievement Award, 1980. *Society Aff:* AICHE, SPE

Babcock, Russell H
Home: 165 Clapboardtree St, Westwood, MA 02090
Position: Consulting Engg. *Employer:* Private Practice. *Education:* MS/Sanitary Engg/Harvard Univ; BS/Civil Engg/Tufts College. *Born:* 4/1/24. Author of over 50 papers & 2 books. P E in MA, ME, VT, NH, RI, CT, NY, FL, NJ. Diplomate Amer Acad of Environ Engrs. Formerly Dir AWWA 1966-70 & 1974-77, dir of Lrg WPCF 1965-68. Director NEWWA 1980-1983. VP NEWWA 1983-1984, Pres NEWWA 1984-1985. Recipient of Dexter Brackett Mem Medal NEWWA 1963. Recipient P Presidents Award NEWWA 1976 and 1984. Mbr Tau Beta Pi. *Society Aff:* ASCE, AWWA, IWES, WPCF

Baber, Michael F
Business: 8326 Olive St Rd, St Louis, MO 63132
Position: Pres. *Employer:* Baber Supply & Engg. *Education:* M/ChE/Univ of DE; BS/ChE/VPI & SU *Born:* 5/22/37. Native of Hopewell, VA. Corp Planner for Reynolds Metals Co 1963-68. Mgr Market Res at Monsanto Co 1968-72. Prof Chem Engr Numerous publications. Active in AIChE, Natl Speakers Assoc, Toastmasters and Dale Carnegie. Mgmt Conslt. Presents seminars and training programs. Develops cleaning engrg programs. In 1972 he started Baber Supply & Engrg, A cleaning supply & engrg corp. In 1978 Baber Seminars; and in 1980 Baber Profl Maintenance, Inc. These companies specialize in profl cleaning engrg products and services plus mgmt consltg and training, nationwide. *Society Aff:* AIChE.

Babson, Edmund C
Business: 15605 Carmenita Aord, Ste 107, Santa Fe Springs, CA 90670
Position: Petroleum Engr *Employer:* Babson and Sheppard *Education:* Metallurgical Engr//Stanford Univ; AB/Engrg/Stanford Univ *Born:* 4/27/08 Union Oil Co: 1934-1936 Research Chemist, 1936-1944 Production Engr, 1944- 1946 Production Foreman, 1949-1951 Asst Chief Petroleum Engr, 1951-1959 Mgr of Operations, Candian Division, 1959-1963 Mgr of Foreign Operations, 1965-1978 Babson and Burns, Petroleum Engrs, 1978 to date Babson and Sheppard, Petroleum Engrs. Other Business and Professional Activities: 1950 Chrmn, Pacific Coast District, API; 1951 Acting Chief, Reserves Section, Production Division, Petroleum Administration for Defense; 1954-1960 Dir, AIME; 1957-1960 VP, AIME; 1959 Pres, Canadian Petroleum Association; 1966-1967 Distinguished Lecturer, Society of Petroleum Engrs, AIME; 1967-1972 CA State Bd of Registration for PE, Member, Pres 1970; 1967-1974 Trustee for Jade Oil & Gas Co in Chapter X proceedings; 1972-1973 Chrmn, Los Angeles Basin Section, Society of Petroleum Engrs of AIME; 1976-1978 Dir, Society of Petroleum Engrs of AIME. Registered Petroleum Engr CA & OR. *Society Aff:* SPE/AIME, API

Babu, Addagatla J
Business: IMSE Department, University of South Florida, Tampa, FL 33620
Position: Assist Prof *Employer:* Univ of South Florida *Education:* PhD/Operations Res/Southern Medhodist Univ; MS/Operations Res/Southern Methodist Univ; MA/Appl Math/Univ of Alabama; MSc/Math/Osmania Univ, India; BSc(Hons)/Math, Phys, Chem/Osmania Univ, India *Born:* 07/25/49 Native of India. Res Fellow in Regional Res Labs, Hyderabad, 73-75. Asst Prof of Indust Engrg, State Univ of NY at Buffalo, 1979-83. Visiting faculty of Indust Engrg, Univ of Central FL, 83-85. Prin Investigator for Computerized authoring sys for Naval Training Ctr, 80-82. Computer sys analyst involved in automation at Naval Training Equip Ctr, summer 1985. Asst Prof of Indust Engrg, Univ of South Florida, since Fall 85. Assoc Prog Chrmn of the Intl Conference on Robotics and Factories of the Future, 1984. Reg PE (FL) 1984. Outstanding Young Man of America, 1984. Terman awd for outstanding graduate student, 1979. Reddy award for Elocution, 1971. Author of several published articles and recipient of several literary awards. *Society Aff:* IIE, HFS

Bacci, Guy J, II
Business: 401 No Michigan Ave, Chicago, IL 60611
Position: Mgr. Advanced Mfg. Technololgy, Truck Group *Employer:* Internatl Harvester Co. *Education:* MBA/Bus/Univ of Chicago; BS/IE/IL Inst of Tech. *Born:* Dec 18, 1925 Chgo, Ill; s. Guy Joseph & Lydia M (Bagnatori) B; m. Angeline Florda, Oct 26, 1946; c. Guy Joseph III, Geoffrey J. Truck Mfg co exec. Indus Engr Internatl Harvester, Melrose Pk Ill 1946-53; supr wage & salary 1953- 59; Ch Indus Engr 1960-61; Corp Indus Engr Mfg Res 1962-66; Genl Supr Indus Engrg Mfg Serv 1967-72; Corp Mgr 1972-1978 Mgr. Indus Engr Trk Group 1978-81. Reg P E Calif. Instructor Ill Inst of Tech 1967-1978. Adviser, Jr Achievement 1958-60, judge natl awards 1965-67; mbr bus adv council Sch Bus Admin Loyola Univ 1972- . USNR 1944-46. Fellow Award Amer Inst Indus. Engrs, 1980, Outstanding Alumnus Award IIT 1973, award Amer Inst Indus Engrs 1965. Mbr Natl Adv Council Indus Engrg 1967- ; Methods, Time, Measurement Assn Dir 1970-1978; Amer Inst Indus Engrs (Tres 1971-72). Club: Glenview Men's Golf. Home: 1039 Meadowlark Ln, Glenview Ill 60025. *Society Aff:* IIE.

Bacher, Alfred A
Home: 4609 Merivale Rd, Chevy Chase, MD 20015
Position: President. *Employer:* Baco & Bacher Assocs. *Education:* PhD/Chem Engg/Tech Univ Vienna; Postgraduate/-/Oxford Univ; Postgraduate/- /Univ of Munich; Postgraduate/Pastenr Inst, Paris. *Born:* Feb 1910. Res Assoc Oxford Univ; Prof, Hd Inst for Fermentation Indus, Tech Univ Vienna; Res Assoc Natl Univ Dublin. Sr Engr Narragansett Brewing Co; Tech Dir Bradley & Baker; Hd Chem Processes, Polaris Proj, Navy Dept; Ch Engrg Eval, Office of Saline Water, Interior Dept; Ch, Prog Review & Eval EPA. 1975 Pres Baco Corp & Bacher Assocs, Engrs - Cons, chem indus, energy, environ. Reg P E. AIChE Fellow, ACS, Water Pollution Control Fed, Internatl Assn on Water Pollution Control. *Society Aff:* AIIChE, ACS, WPCF, IAWPC.

Bachman, Charles W
Home: 11 Percy Rd, Lexington, MA 02173
Position: Pres *Employer:* Bachman Information Systems, Inc. *Education:* MS/Mech Engg/Univ of PA; BC/Mech engg/MI State Univ. *Born:* Dec 11, 1924 Manhattan, Kansas. Founder & Pres of new co 4/83, that designs & manufacturers computer software. Current prods focus on new paradigms for intelligent business info sys. Bd mbr of the Computer Museum, Boston MA. The Dow Chemical Co 1950-60; Genl

Bachman, Charles W (Continued)
Electric 1961-70 in assignments relating to mfg Control Sys & Database Mgmt Sys; Honeywell Info Sys 1970-1981 in assignments relating to dev software for new (computerbased) info sys. VP of Res, Cullinane Database Sys, Inc. Awarded the A M Turing Award by the Assn of Computing Machinery 1973 for pioneering work in database sys tech (inventor of the Integrated Data Store). Outside interests: family (wife, Constance & children Margaret, Thomas, Sara & Jonathan) and orchid growing. Elected Distinguished Fellow of British Computer Soc in 1977, Inventor of Data Structure Diagrams now commonly called Bachman Diagrams-V Chrmn ANSI Study Group-Database Mgmt (1972-75) Chrmn ANSI Study Group-Distributed Sys (1977-80) Chrmn ISO/TC97/SC16-Open sys Interconnection (1978-80). Founder-Mbr of CODASYL Database Task (1965-68). Holds 13 patents data processing/database mgmt. *Society Aff:* ACM, ΤΒΠ

Bachman, Henry L
Home: 5 Brandy Road, Huntington, NY 11740
Position: VP *Employer:* Hazeltine Corp *Education:* M/EE/Polytechnic University; B/EE/Polytechnic University; Advanced Management Program/-/Harvard Business School *Born:* 4/29/30 1971 to present - Joined Hazeltine Corp when a subsidiary co, Wheeler Labs, was merged into the parent co in 1970. Presently, VP, Engg responsible for the development and implementation of engineering design policies, practice, & development & analysis tools. Previously served as VP Operations, with responsibility for purchasing, mfg, and quality. Prior to that, served as VP, Customer Serv & Quality, of the Indus Prods Div, Computer Terminal Equipment Prod Line, and prior to that, as VP, Prod Assurance, with respon for quality, reliability, maintainability, field engrg and logistics support for Govt Prods. 1951-1970 - Started career at Wheeler Labs as a Dev Engr working on antennas and microwave components. Held various tech and managerial positions, including Assist Chief engr and VP, until becoming Pres in 1968, responsible for directing all phases of co operations. *Society Aff:* IEEE, ASQC, NSIA, EIA

Bachman, Walter C
Home: 21 Wayside, Short Hills, NJ 07078
Position: V Pres & Chief Engr-retired *Employer:* Gibbs & Cox Inc, N Y. *Education:* MS/Indus Eng/Lehigh Univ; BS/Indus Eng/Lehigh Univ. *Born:* Dec 1911. Lic P E NY. Taught mech engrg at Lehigh Univ 1935-36. Marine Engr at Fed Shipbldg & DryDock Co 1936. Gibbs & Cox Inc 1936-70; Ch Engr 1958; V Pres & Ch Engr 1963-70. Assoc with the design of a wide variety of merchant & naval ships & other marine engrg projects. Fellow ASME; Mbr SNAME, ASNE, N Y Acad of Sci. Mbr Natl Acad of Engrg since 1967. Active on var tech & adv cttes. *Society Aff:* ASME, NAE, SNAME, ASNE.

Bachus, Walter O
Home: 3808 Great Neck Ct, Alexandria, VA 22309
Position: Executive Director *Employer:* The Society of American Military Engineers *Education:* BSCE/Texas A&M 1950; MS/NY Univ 1957; AMP/Harvard 1973 *Born:* October,1926 Various army assignments through Exec to Army Asst Vice Chief of Staff 1971. Seattle District Engr 1972-1973; promoted to B Gen & asgd Div Engr, No Cntl Div, Chicago 1973-1975. Became Army's Director, Fac. Engr, Off Ch. of Engr 1975-1978. Retired Sept 1978, becoming SAME Ex Dir, 1978-present. Awards: Dist Service Medal; 2 Legions of Merit, 3 Br. Stars, SAME Gold Medal 1975. Wife: Helen, 2 children. Reg Prof Engr: Texas and D.C.

Bachynski, Morrel P
Home: 78 Thurlow Rd, Montreal H3X 3G9 Canada
Position: Pres. *Employer:* MPB Tech Inc. *Education:* PhD/Physics/McGill Univ; MSc/Physics/Univ of Saskatchewan; B Eng/Eng Physics/Univ of Saskatchewan *Born:* 7/19/30. Since 1976, Pres & founder of MPB Tech Inc. Previously VP, Res & Dev for RCA Limited. Has over 80 publications in scientific & engg journals. Co-author of "The Particle Kinetics of Plasmas–. Fellow of IEEE, Am Physical Soc, Royal Soc of Canada, Canadian Aero & Space Inst. Winner of David Sarnoff Outstanding Achievement Award in Engg (1963), Prix Scientifique du Quebec (1973), Canada Enterprise Award (1977), Queens 25th Anniversary Silver Medal (1977). Canadian Assoc of Physicists Medal for Achievement in Physics (1984). *Society Aff:* IEEE, APS, AIAA, AGU

Bachynski, Nicholas
Business: 101 Huntington Ave, Boston, MA 02199
Position: V Pres & Treasurer. *Employer:* Chas T Main Inc. *Education:* MBA/Bus Admin/Harvard Univ; BSEE/Elec Engg/Univ of Manitoba. *Born:* Feb 1925. Native of Ontario Canada. Process control engr for Abitibi Paper Co, spec in newsprint mfg. Elec engr for Canadian Comstock Ltd, responn for frequency conversion of elevators. Mgr of market res for Reliance Elec Co. Joined Chas T Main Inc 1959. Assumed current pos of V P & Treas in 1974. Respon incl data processing, accounting, & budgeting functions. Elected Dir 1976. 1979-80. Pres Dir, Cons Engrs Council of New England Inc; 1971-72 Chmn Engrg Econ & Mgmt Ctte TAPPI. Reg P E sev states. *Society Aff:* IEEE, NSPE, CECNE, TAPPI.

Backensto, Elwood B
Business: Engg Dept, Paulsboro, NJ 08066 *Employer:* Mobil Res & Dev Corp. *Education:* BS/Chem Eng/Lehigh Univ *Born:* 12/-/21. Native of Emmaus, PA. Joined Mobil Res upon grad in 1943 where he was Group Leader in charge of light products refining, reforming catalyst dev, and corrosion & met. Assigned several US Patents; co-authored 18 papers on corrosion. Served on API Sub-ctte on corrosion (1955-69). Mbr NACE since 1955; active at section, region & natl levels. Historian for Family Assn & prepared a book on its genealogy. Enjoys music, jazz, fishing, bridge, bowling & golf. Citation of Recognition 1970 NACE; Dir 1972-5, 1977-80, 1984, 1985 NACE; Exec Comm 1973 and 1977-80 NACE. Pres NACE 1978-9, Retired 1982.

Backer, Morris
Business: PO Box 8098, Houston, TX 77004
Position: Sr Vice President. *Employer:* Bovay Engrs Inc. *Education:* BS/Mech Engg/TX A&M Univ. *Born:* Dec 1927. US Navy 1946-48, Jr Engr Corps of Engrs 1951-52. Since 1952 with Bovay Engrs Inc. Progressed with Bovay from Design Engr spec in air-cond & plumbing design to present respon of Dir & Sr V Pres Bovay & Corp Mgr-Operation- Reg. Offices. Special interests incl mech design & coord of major institutional & commercial bldgs. Dir ASHRAE 1969-74, Treas ASHRAE 1974-75, V Pres ASHRAE 1975-77, Pres-elect ASHRAE 1977-78, Pres ASHRAE 1978-79. 1975 awarded Fellow grade mbr ASHRAE. *Society Aff:* ASHRAE, ASPE, NSPE, TSPE, ACEC, CECT.

Backlund, Brandon H
Business: 10535 Pacific St, Omaha, NB 68114
Position: Pres *Employer:* Backlund Engineering Co, Inc *Education:* BS/CE/IA St Univ *Born:* 06/13/18 Amarillo, TX. Wife Emily is artist, teacher, realtor, appraiser. Sons Dr Mark H. at Mount Vernon, WA, Gregory, industrial engr at Minneapolis, daughter Nancy Freeman, interior decorator, Omaha. Served in US Army WWII, Pacific and Europe, 2d Lt to Lt Col. Deputy Commander, NB State Guard. Reg PE and land surveyor in 8 midwest states and CA. Consltg engr with own firm since '51 in midwest US, Caribbean, and Southeast Asia. Past pres of NB Soc of PE, Natl Soc of PE, and American Council of Consltg Engrs of NB. Mason, Shriner, Rotarian. Ambassador of NB Diplomats. Private Pilot, hobbies golf, sailing, hunting, fishing. *Society Aff:* ASCE, ACEC, NSPE, SAME, ACSM

Bacon, Charles M
Home: 1023 Graham Dr, Stillwater, OK 74075
Position: Prof, Sch of Elec and Computer Engrg *Employer:* Oklahoma State Univ. *Education:* BS/Agri/OK State Univ; BS/Elec Engr/OK State Univ; MS/Elec Engr/OK State Univ; PhD/Elec Engr/MI State Univ. *Born:* 12/20/33. BS & MS OK State Univ; PhD MI State Univ 1964. Bendix Fellow, NSF Fellow & Asst Prof of Elec Engg at MSU. Returned to OK State Univ in 1966. Prof & Hd of Elec Engg 1978-

Bacon, Charles M (Continued)
1984. Teaching & res interests are computer engrg, control theory & computer simulation of large-scales sys. Awarded Okla Soc of Prof Engrs Outstanding Engg Achievement 1971; Midwest Sect ASEE - Western Elec Award for Excellence in Teaching 1975; named Halliburton Prof of Engg 1975. *Society Aff:* IEEE, ASEE, ΦΚΦ, ΣΞ, SME

Bacon, David W
Business: Queen's Univ, Kingston, Ontario, Canada K7L 3N6
Position: Dean of Appl Sci *Employer:* Queen's Univ *Education:* PhD/Statistics/Univ of WI; MS/Statistics/Univ of WI; BASc/Engrg Phys/Univ of Toronto *Born:* 09/12/35 Dr Bacon is a Reg PE in the Province of Ontario with past employment exper with Canadian Gen Elect Co and Du Pont Canada Inc. Since 1968 he has been a prof in the Dept of Chem Engrg at Queen's Univ and Dean of the Faculty of Appl Sci at that institution since 1980. His fields of interest in teaching and res include statistical methodology for process modelling, process dynamics and control, optimization and simulation. He is an active conslt to a number of companies in Canada and USA. *Society Aff:* CSChE, ASA, RSS, SSC

Bacon, Louis A
Home: 5240 W Kingston Ct NE, Atlanta, GA 30342
Position: VP & Dir. Federal Marketing *Employer:* Heery International, inc. *Education:* BSCE/Structural Engg/Univ of IL. *Born:* Apr 1921 Champaign, Ill. Wing designer Douglas Aircraft Co 1943-44; US Navy as Radar Technician 1944-46; 20 yrs with Shaw Metz & Assocs (Arch-Engrs) Chgo, maj as Ch Struct Engr & Assoc Ptnr. Joined P & W Engrs Inc in 1966 & became Pres in 1969. Joined Stanley Cons Inc in 1974 as V Pres & Hd Atlanta Div. In 1976 joined arch-engrg firm Heery & Heery as VP & Dir Engg Div. Pres Ill Soc of Prof Engrs 1964-65; Natl Dir NSPE 1966-69; V Pres NSPE 1969-71; Chmn NSPE/PEPP 1971-72. Pres NSPE 1983-84. 'Illinois Award' from ISPE 1968; NSPE/PEPP 'Chmn Award' 1973 NSPE/PEPP 'PEPP Award' 1976; Honor Mbr Chi Epsilon, Member National Society of Professional Engineers; Member Georgia Society of Professional Engineers; Honorary Member Illinois Society of Professional Engineers; Fellow Member American Society of Civil Engineers. Outstanding Achievement Award, Engineers of Metro Atlanta 1980; Engr of The Year, Engr of Metro Atlanta 1982; Engr of The Year, GSPE 1984. LoyaltY Award, U. of Ill. Alumni Assoc. 1985, Distinguished Alumnus Award, U. of Ill. Civil Engineering Alumni Assoc. 1985. *Society Aff:* ASCE, NSPE, SAME, ASAE, SEAI

Bacon, Merle D
Home: 6265 Altman Dr, Colorado Springs, CO 80907
Position: Commander (Colonel USAF). *Employer:* USAF-Frank J Seiler Res Lab. *Education:* BS/Aero Engr/Wichita State Univ; MS/Aero Engr/Univ of OK; PhD/Engr Mech/Univ of OK. *Born:* Nov 1931. PhD & MS Univ of Oklahoma; BS Wichita State Univ. Native of McPherson, Kansas. Liaison & struct engr for Beech Aircraft 1951-54. USAF active duty 1954- , with duties as interceptor pilot, Assoc Prof of Engrg Mech at the USAF Acad, & res & dev of high energy laser sys. Assumed current assignment as Commander, Frank J Seiler Res Lab located at the USAF Acad in 1975. Respon for planning & execution of basic res in chem, applied math, & aero-mech; also respon for encouraging & supporting res by the USAF Acad fac & students.

Bacon, Vinton W
Home: 2616 143rd Place S.E, Mill Creek, WA 98012
Position: Professor. *Employer:* Univ of Wisc-Dept of Civil Engrg. *Born:* Dec 1916. Grad from Univ of Calif, valedictorian of class in civil engrg, with sanitary engrg in 1940. Captain, U S Public Health Serv 1943-46. Exec Secy Northwest Pulp & Paper Assn, Seattle Wash, handled res programs on paper & pulp processes, waste treatment methods, & related fields 1956-62. Exec Officer Calif State Water Pollution Control Bd, dev policy, coord regional water pollution control bds, liaison with State Legislative & Congressional Cttes 1950-56. Spec int in wastewater treatment. Bd Chmn EEIB 1967-69; Officer AAEE 1967-69; WPCF Bd of Cont 1969; NAE; ASCE Civil Gov Award 1967: Construction's Man of Yr 1967 Engrg News-Record; Ill Sect ASCE 'Engr of Yr' 1965.

Baddour, Raymond F
Business: MIT Chem Engrg Dept, 66-440, Cambridge, MA 02139
Position: Lammot duPont Prof of Chem Engrg. *Employer:* MIT. *Education:* ScD/Chem Engg/MIT; MS/Chem Engg/MIT; BS/Chem Engg/Univ of Notre Dame. *Born:* 1/11/25. Native of Laurinburg, N C. Fac at MIT since 1951. Hd of Dept 1969-76. Dir Amgen, 1980-, Cofounder, Breh, Inc. 1983. Cofounder, Energy Resources Co, Inc., 1974; Cofounder, Erco, AG (Switzerland) 1980; Cofounder, Abcon, Inc. 1963. Dir Raychem Corp 1972-80. Dir Lam Research, 1980-82. Genl Motors Sci Adv Ctte 1971- 82. Cons Mobil Chem Co 1962-84. Freeport Minerals Co., 1977-82. Mbr of Corp, Museum of Sci 1964- . Reg P E Mass. Fellow Amer Inst of Chem Engrs; Fellow Amer Inst of Chemists; Fellow Amer Acad of Arts & Sci; Amer Chem Soc; Amer Assn for the Advancement of Sci; N Y Acad of Sci. Res in high temp chem, chem reactions in gaseous discharges, heterogeneous catalysis, suface diffusion, enzyme catalysis, membrane separation processes, adsorption, & ion exchange *Society Aff:* AIChE, ACS, AAAS, AAA&S, AIC, NYAS

Badertscher, Robert F
Business: 505 King Ave, Columbus, OH 43201
Position: Projects Mgr. *Employer:* Battelle Columbus Labs. *Education:* MSc/Aero Engg/OH State Univ; B Aero Engg/OH State Univ. *Born:* 6/4/22. With Battelle's Columbus Lab since 1952. Currently mgr, combat vehicle systems projects. Res engr from 1952-61. Since 1961 have served in a variety of tech mgt positions including Chief, Aeronautics Div; Mgr, Tech Assessment Programs; and Assoc Dir, Battelle's Advanced Systems Lab (an off-site operation directly supporting the Army's Tank-Automotive Res & Dev Command). Broad experience in performing and managing technological assessments and evaluations of the performance and effectiveness of military systems. Aerodynamicist, Aeroproducts Div, GM Corp, 1950-52. Instr, Franklin Univ, 1947-50. Army Air Corps, 1942-45. *Society Aff:* AIAA.

Badler, Norman I
Business: Computer & Information Sci, University of Pennsylvania, Philadelphia, PA 19104-6389
Position: Professor *Employer:* Univ. of Pennsylvania *Education:* PhD/Comp Sci/Univ. of Toronto; MSc/Math/Univ. of Tornoto; BA/Creative Studies-Math/Univ of Santa Barbara *Born:* 5/3/48 Inventor of Computer Graphic Display Software for 3D Biomechanical Models of Body, Hands, and Face. Interested in studying human motion through computer simulation, analysis, notations, and language. Generally interested in computer graphics, interactive systems, and computer vision; especially animation, motion, 3D shape representation, archaeological site reconstruction. Dissertation first on computer analysis and segmentation of 3D object motion. Active in special interest group on computer graphics (ACM-Siggraph), elected VChrmn 1979-82. General Chrmn 1979 IEEE workshop in computer analysis of time- varying imagery; Program Chrmn 1983 SIGGRAPH workshop on motion representation and perception. Assoc editor of *IEEE Computer Graphics and Applications*, and Sr editor of *Computer Vision, Graphics and Image Processing*. Received res grants from NSF, ONR, NASA, and Army Res. *Society Aff:* IEEE, ACM, ACL, AAAI, CSS

Baeder, Donald L
Business: 10889 Wilshire Blvd, Los Angeles, CA 90024
Position: Exec VP - Sci & Tech *Employer:* Occidental Petroleum Corp *Education:* BA/Chem/Baldwin-Wallace College; BS/CE/Carnegie Inst Tech. *Born:* 8/23/25. Presently Exec VP - Sci and Tech of Occidental Petroleum Corp. 1977-81 Pres & Chief Operating Officer of Hooker Chem Co, the chem div of Occidental Petrol Corp, hdquartered in Houston, TX. From Nov 1975 to Jan 1977, served as exec vp of Occi-

Baeder, Donald L (Continued)
dental Petrol Corp responsible for its entire res & dev activities. Prior to joining Occidental, was associated with Exxon Corp since 1951, serving in various capacities including mgr, Plastics Div, Exxon Chemicals and, more recently, as vp in charge of corporate & govt res. *Society Aff:* AIChE, ACS.

Baer, Alva D
Business: Dept of Chemical Engg, Salt Lake City, UT 82112
Position: Prof *Employer:* Univ of UT. *Education:* PhD/Chem Engg/Univ of UT; MS/Chem Engg/Univ of UT; BS/Chem Eng/Univ of UT. *Born:* 11/17/27. Native of Salt Lake City, UT. Married with three children. Industrial experience with FMC Corp in high temperature processes and with Atomccs Intl Div of now Rockwell Inter. Prof at Univ of UT since 1959 and dept chrmn 1978 to 1981. Hobbies are skiing, backpacking and hiking, and Boy Scout leader. *Society Aff:* AIChE, ISA.

Baer, Eric
Home: 2 Mornington Ln, Cleveland Heights, OH 44106
Position: Leonard Case Prof of Macromolecular Science *Employer:* Case Western Reserve Univ. *Education:* Doct Eng/Chem Eng/Johns Hopkins Univ; MA/Chem/Johns Hopkins Univ *Born:* 7/18/32. Upon grad, joined EE DuPont as res engr at their Exper Station to work on struct property relationships in polymeric mtls. In 1960, was appointed asst prof of chem engg at the Univ of IL, & 2 yrs later joined the faculty of Case Western Reserve Univ to dev an educational & res prog in polymer sci and engg. This prog grew into the Dept of Macromolecular Sci which he headed until his appointment in 1978 as Dean of Case Inst of Tech of Case Western Reserve Univ. In 1983, returned to full teaching and res as the Leonard Case Prof of Macromolecular Sci. In 1968 rec'd Curtis W McGraw Awd from the Am Soc of Engg Educ which is given annually to an engr under the age of 35 for outstanding res contributions. Served on numerous natl & intl prof cttes and serves as a conslt to industry. In 1980, received Intl Awd of the Soc of Plastics Engrs and in 1981, received the Bordon Awd of the Amer Chem Soc. *Society Aff:* ACS, APS, SPE, AAAS, ASM, ASME.

Baer, Sheldon
Business: 19699 Progress Dr, Strongsville, OH 44130
Position: VP Marketing. *Employer:* Anderson IBEC. *Born:* 6/4/29. Married, 3 children. OH State Univ Engrg Sch 1947-48; BS Chem E Penn College Cleveland 1952; grad courses Chem Engrg Case Inst of Tech, Cleveland; Medical Repair St Louis Medical Depot, US Army, grad 1955; Prof Mgmt, Assoc Degree John Carroll Univ, Cleveland 1975. B F Goodrich Chem Co, Avon Lake OH: Dev Engr Pilot Plant Oper, Supr Process Dev 1952-56; Anderson IBEC 1956- : Pilot Plant & Lab Mgr reporting to VP of Sales & Res June 1956; Applied Engr with cont respon for Pilot Plant & Lab activities & added respon of sales & field dev of newly dev products & processes - reporting to VP of Sales & Res oct 1960; Tech Dir respon for R&D incl sales, engrg & field activities of newly dev products reporting to Pres - Mbr of Admin Ctte Jan 1964; Tech Dir respon for R&D & Engrg, reporting to Pres Jan 1966;l VP of Engrg & Prof, 1969. Genl Mgr Process Plant group 1976-78; VP Mktg 1978. *Society Aff:* AIChE.

Baerwald, John
Business: John E. Baerwald, P.C, 1006 Calle Catalina, Santa Fe, NM 87501
Position: Pres *Employer:* John E. Baerwald, P.C. *Education:* PhD/CE/Purdue Univ; MSCE/CE/Purdue Univ; BSCE/CE/Purdue Univ. *Born:* 11/2/25. BSCE, MSCE, PhD Purdue Univ. Univ of IL acad staff 1955-1983, Prof of Transportation & Traffic Engg; Dir of Hwy Traffic Safety Ctr; Fellow ASCE; Governor's Official Traffic Safety Coord Ctte 1962-69; Fellow Inst of Transportation Engrs (Internatl Pres 1970); Tr Champaign-Urbana Mass Transit Dist 1973-1983, Chmn 1974-1983; Natl Safety Council (Bd of Dir 1975-80); Transportation Res Bd; Amer Rd & Trans Bldrs Assn, Pres Educ Div 1979-80; Bd. of Dir. 1979-83; Sigma Xi; Chi Epsilon; Ed & Contributor, articles in prof journals; Reg PE IN, IL, & CA; cons traffic and transportation engr. Technical Advisory Comm, IL Transportation Study Commission, 1977-81. Pres John E. Baerwald, P.C. *Society Aff:* ITE, ASCE, ARTBA, TRB.

Bagci, Cemil
Business: Dept of Mech Engg, Box 5014, Cookeville, TN 38505
Position: Prof. *Employer:* TN Tech Univ. *Education:* PhD/ME/OK State Univ; MS/ME/OK State Univ; MS/Tech Educ/OK State Univ; BS/Tech Educ/OK State Univ; Diploma/Mech Design/Tech Teachers College Ankara, Turkey *Born:* 12/1/32. Citizen of Turkey, permanent resident of USA. Has been teaching and involved with res in mech design: Mechanisms, dynamics of machines, lubrication and bearing design, robotics, and finite element applications in machine design, in the Dept of Mech Engg at TTU since the Fall of 1967. Have published over 150 tech papers in these areas, primarily in elastodynamics of mechanisms, balancing, finite element dev, robotics, critical speeds and stress analysis and fatigue design. Had AID scholarship for two yrs, 1961-62. Had Sigma Xi res awards 1971, 1973; IBM Mechanism Design Award, 1977. Proctor and Gamble certificate of merit for Mechanical Design 1979, 1981 merit award for mechanism design. The first Coplanor Research Award of TTU, 1984. Has been reviewer for several tech journals. Authored a text book. Presently completing three engineering textbooks. *Society Aff:* AAM, ASME, SAE, USCIFTOM, ASLE.

Bahar, Leon Y
Business: Dept Mech Engr and Mechanics, Philadelphia, PA 19104
Position: Prof *Employer:* Drexel Univ *Education:* PhD/ME/Lehigh Univ; MS/ME/Lehigh Univ; BS/ME/Robert Coll *Born:* 04/04/28 In engrg teaching and research since 1957. Instructor in Mechanics 1957- 1963 at Lehigh Univ. Asst Prof 1963-1966 City Univ of NY. Assoc and full Prof 1966 to date at Drexel Univ. Research in the fields of dynamics, vibrations, earthquake engrg, applied mathematics and solid mechanics leading to 80 journal articles. Consltg for IBM, United Engrs and Constructors, Gulf and Western, INA Corp, etc. *Society Aff:* ASME, AAM, SES, NYAS

Bahder, George
Home: 24 Highpoint Dr, Edison, NJ 08817
Position: VP. *Employer:* Genl Cable Intl. *Education:* PhD/Elec Engg/Warsaw Politechn, Poland; MS/Elec Engg/Warsaw Politechn, Poland; BS/Elec Engg/Waublerg & Rotward Eng Sch, Poland. *Born:* 1/17/25. Native of Poland 1948-62 Asst and later Assoc Prof in Elec Inst, Warsaw, Poland. 1962-70 Genl Cable Corp 1970-73 VP, Res and Dev, Phelps Dodge Cable and Wire Co. 1973-present VP Res and Dev, Genl Cable Corp and recently Genl able Intl. Special field of activity - elec insulation, high voltage technique, electric cables. *Society Aff:* IEEE, CIGRE.

Bahnfleth, Donald R
Business: ZBA, Inc, 23 E Seventh St, Cincinnati, OH 45202
Position: Pres & Prin *Employer:* ZBA, Inc. *Education:* MS/ME/Univ of IL; BS/ME/Univ of IL. *Born:* July 1927. Born W Chicago Ill. Res Asst Prof, Univ of Ill 1952-58; Editor, Heating, Piping & Air Cond 1962-71; current affiliation since 1972. Chmn ASHRAE Res & Tech Ctte 1974-75 & Fellow ASHRAE. Treas of St Paul Lutheran Village Inc, a non-profit corp org to provide housing for the elderly. Dir-at-large, ASHRAE 1978-81; Treasurer ASHRAE 1981-82; VP ASHRAE 1982-84; Pres-Elect ASHRAE 1984-85; Pres ASHRAE 1985-86; Bd of Dirs Univ of IL. Mech & Indus Eng Alumni Assoc, Rotary Intl. Chairman, St. Paul Evangelical Lutheran Church of Madisonville, 1987-89; Cincinnati Distinguished Engineer 1987. *Society Aff:* ASHRAE, APCA, NSPE, ΣΞ, ACEC

Bailey, Burton P
Home: c/o James River Corp, Richmond, VA 23217
Position: Mkt Dev Mgr. *Employer:* James River Corp. *Education:* BSCHE/Chem Eng/Univ of Cincinnati. *Born:* Native of Springfield, OH. Capt in AF. Employed 3 yr at duPont in film dept in Product & Process Dev. Worked nine yrs in R&D. Mead Corp then Product- Sale Dev in Paper field. Eventually employed by James

Bailey, Burton P (Continued)
River Paper Corp to manage Adams Div. Transferred to Market Dev. Mgr, James River Corp, Richmond Va. Married with three children. *Society Aff:* NSPE, TAPPI.

Bailey, Cecil D
Business: Dept of Aero/Astro Engineers, 2036 Neil Ave Mall, Columbus, OH 43210
Position: Prof *Employer:* The Ohio State Univ *Education:* PhD/Aeronautical Engr/Purdue Univ; MS/Aeronautical Engr/Purdue Univ; BS/Aeronautical Engr/MS State Univ *Born:* 10/25/21 Zama, MS. Parents, James D. and Matha (Roberts) Bailey. Married, Myrtis Taylor, Sept 8, 1942. Two daughters, Marilyn and Beverly. Commissioned 2nd Lt, US Army Air Corps, 1944. Advanced through grades to Lt Col, USAF, 1965. Air Corps and USAF pilot, 1944-56; Senior Pilot, 1957-60; Command Pilot, 1961-67. Retired from USAF, 1967. Asst Prof, USAF Inst of Tech, 1954-58; Assoc Prof, USAFIT, 1965-67; Assoc Prof, The OH State Univ, 1967-69; Prof, OSU, 1970-. Awarded Air Force Commendation Medal by the USAF Research and Development Command, 1962, and by the USAF Air Univ Command, 1967. Received the McQuigg Distinguished Teaching Award, The OH State Univ, 1976. Directed the USAF-ASEE Summer Faculty Research Prog, Wright-Patterson AFB, OH, 1976-78. Contributor of many articles and discussions to the recent literature on the motion of matter and/or energy in time and space. *Society Aff:* ASEE, SEM

Bailey, Eugene C
Business: RR6 Box 252, Columbus, IN 47201
Position: Conslt *Employer:* Self Employed *Education:* BSME/Power/Purdue; MSME/Power/Purdue. *Born:* Apr 7, 1910. Native of La Grange, Ill. Tr & Pres, Lyons Township High School 1955-69; Tr & V Pres, Coll of Du Page 1970-77. Captain, Ordnance Dept 1942-45: respon for Prod Engrg of self-propelled artillery. Commonwealth Edison Co 1933-75: supervisor of Mech Constr 1948-52; Asst to Mgr of Production 1952- 54; Sys Mech & Bldg Engr 1954-62; Admin Engr 1962-75. V.P. Bus. Dev, John Dolio & Assocs (now Dolio & Metz Ltd,) Chicago, 1975-82. Mbr ASME Boiler Code Special Ctte on Nuclear Power; Inservice Inspection of Nuclear Units Subctte 1955-75; ASME Fellow 1961 & Life Mbr 1975. *Society Aff:* ANS, ASME, AWS, NSPE, WSE.

Bailey, Fred C
Home: 48 Coolidge Ave, Lexington, MA 02173
Position: Group Executive *Employer:* Teledyne Inc. *Education:* SM/Mech Engg/MIT; SB/Mech Engg & Matls/MIT. *Born:* Oct 5, 1925. Attended local schools in Claremont, N H. US Navy 1944-46. Res Engr Fatigue & Stress Analysis, Caterpillar Tractor Co 1949-52. Natl Acad of Sci-Natl Res Council 1952-55; Admin of Ship Fracture Res. Teledyne Engg Services (Div of Teledyne Inc), formerly Lessells & Assocs Inc since 1955; Engrg Mech, Stress Analysis, Materials Engrg. Lexington Bd of Selectmen 1969-78. SESA: Fellow 1976; Tatnall Award 1974; Pres 1967-69; Treas 1961-66. SNAME: Linnard Prize 1971.. *Society Aff:* SNAME, ASME, SESA, AWS.

Bailey, Herbert R
Business: 5500 Wabash Ave, Terre Haute, IN 47803
Position: Prof of Mathematics. *Employer:* Rose-Hulman Inst of Tech. *Education:* PhD/Math/Purdue Univ; MS/EE/Univ of IL; BS/ChE/Rose-Hulman; BS/EE/Rose-Hulman. *Born:* Nov 2, 1925. Applied mathematician with Naval Ordinance Plant, Marathon Oil Co, & Martin. Teaching pos at Genl Motors Inst, Colorado State Univ & Rose- Hulman Inst of Tech. Res & cons in biomathematics, heat transfer & oper res. Activities incl racquet sports & church. *Society Aff:* MAA.

Bailey, James E
Business: Box 2901, Tuscaloosa, AL 35487
Position: Prof. *Employer:* Univ of AL. *Education:* MS/Aerospace Engg/MS State Univ; BS/Aerospace Engg/MS State Univ. *Born:* 4/16/33. Sr Engr Gen Dynamics, Ft Worth, TX 1958-1962; Assoc Prof Aerospace Engg, the Univ of AL, 1963-1979; Natl Sci Fdn Faculty Fellow, MIT Dept of Aero, 1966- 1969; Prof of Aerospace Engg, the Univ of AL, 1979-present; Dir of Flight Dynamics Lab, the Univ of AL 1977-present; Res interests are flight simulation, flight control, & flight mechanics. *Society Aff:* AIAA, SAE.

Bailey, James E
Home: 2032 N. Gentry, Mesa, AZ 85203
Position: Assoc Prof *Employer:* AZ State Univ *Education:* PhD/IE/Wayne St Univ; MS/IE/Wayne St Univ; BS/IE/Wayne St Univ *Born:* 01/16/42 Born and raised in Detroit, MI. Educated at Wayne State Univ. Worked at Gen Motors, Natl Bank of Detroit, Wayne State and AZ State Univs. Past Chief Exec Ofcr of Kurta Corp. Active mbr of American Inst of Industrial Engrs. Regional VP, Chapter Pres and Computer Div Dir of IIE. Published several books and numerous technical articles. Mbr of Alpha Pi Mu and MacKenzie Honor Socs. Interested primarily in production control, information sys and engrg mgmt. Married to Petra K. with two children, Michael and Heidi. *Society Aff:* IIE, TIMS, SIM.

Bailey, John A
Business: 2100 Phila Natl Bank Bldg, Phila, PA 19107
Position: Partner. *Employer:* Louis T Klauder & Assoc. *Education:* PhD/Pol Sci/Univ of PA; MGA/Govt Admin/Univ of PA; BS/Mech Engg/TX A&M Univ. *Born:* 7/6/18. Served WWII in US Army Ordnance Dept, 2nd Lt to Capt. City Mgr from 1949- 1953. Was Deputy St Commissioner, St Comm and Deputy Mng. Dir, City of Phila 1953-1961; Exec Dir (PSIC) and Deputy Gen Mgr, SEPTA, 1961-1967. Dir of the Transportation Ctr of Northwestern Univ, 1967-1975, followed by VP, Murphy Engg Co, Chicago, 1975-1976. Became partner Louis T Klauder & Assoc 1976, handling Planning and Reports activities. Primary interests are: transportation policy, mgt of engg, education and public transportation activities. Was Mbr of Bd APWA (1971-1974); Bd of Visitors, USA Transportation School (1969-1972), Pres, Metro Housing and Planning Council of Chicago (1972-1974). *Society Aff:* NSPE, APWA, ICMA, APSA, ASPA.

Bailey, Raymond V
Business: School of Engrg, New Orleans, LA 70118
Position: Assoc Dean of Engrg. *Employer:* Tulane Univ. *Born:* Nov 1923. PhD & MS Louisiana State Univ; BS from Louisiana Tech. Chemist for Cities Service Refining Corp 1945. US Army 1945-46. Assoc Prof of Chem Engrg at Univ of Miss 1948-51. Hd of Chem Engrg Dept at Tulane Univ 1951-73. Asst Dean at Tulane Univ 1967-73. Assoc Dean at Tulane Univ 1973- . 28 yrs cons on fluid dynamics, heat transmission, reaction kinetics, simulation & optimization, res mgmt, & corp mgmt. Reg P E La. Current duties are admin of grad programs, res programs & financial matters for the Sch of Engrg.

Bailey, William A
Business: Marbil Engineering Asso, 6910 Furman Pkwy, Riverdale, MD 20737
Position: Consulting Agri Engr & Sole Proprietor *Employer:* Marbil Engg Assoc *Education:* BSC/Agri Engg/Univ of NB. *Born:* Dec 31, 1921 Palmyra, Nebr. ScB Univ of Nebraska 1950. US Marine Corps 1940-46 & during 1950. Jr Engr USDA Cotton Ginning Lab 1950-53. Proj Engr FMC 1953-55 & USDA at Storrs Conn, College Park Md, Beltsville Md 1955-72. Sr Engr USDA, Beltsville Md 1972-85 (Retired), Consulting 1985- . Mbr ASAE, ASHRAE, Amer & Internatl Soc's of Horticultural Sci Cttes concerning environ control for plant growth, handling & transportation of farm produce & animals. Chmn Wash D C-Md Sect ASAE, V Chmn for Americas of ISHS Comm on Horticultural Engrg. Publ 73 papers & received 4 pats. *Society Aff:* ASAE, ASHRAE, AATA.

Bailey, William J
Business: Dept of Chemistry, University of Maryland, College Park, MD 20742
Position: Research Prof of Chemistry. *Employer:* Univ of Maryland. *Education:* BChem/Chemistry/Univ of MN; PhD/Organic Chem/Univ of IL. *Born:* Aug 11, 1921. Native of Walker, Minn. Taught at Wayne State Univ, Detroit Mich before coming to the Univ of Md in 1951. Res area is in Organic Chem of Polymers. Pres

Bailey, William J (Continued)
of the Amer Chem Soc 1975. Chmn of Bd of Dirs of the Amer Chem Soc 1979-81 Chmn of the Div of Polymer Chem 1968. Chmn of the Natl Res Council Ctte on Macromolecules 1968-78 . Received the First Fatty Acid Producers Res Award 1955. Honor Scroll Dist of Columbia Inst of Chemists 1975. Outstanding Achievement Award of Univ of Minn 1976. Amer Chem Soc Award in Polymer Chem 1977. Annual Achievement Award of Univ of Md Sigma Xi Chapter 1979. Amer Chem Soc Award in Applied Polymer Chem 1986. Hillebrand Prize Chem Soc Washington 1985. Abbott Lecture Award Univ MD, 1987. *Society Aff:* ACS, AAAS, ΣΞ, AXΣ.

Bailie, George E
Business: 6733 W 65th St, Chicago, IL 60638
Position: President. *Employer:* Union Carbide, Films-Pkging Div. *Education:* BS/Chem Eng/Northwestern Univ; MBA/Chem Eng/Univ of Chicago. *Born:* Mar 1923. MBA Univ of Chicago; BS Chem Engrg Northwestern Univ. Army Air Corps WW II. Became associated with Union Carbide Corp in 1956 with The Visking Corp (a predecessor of the corp's Films - Pkging Div) in Chgo. Assumed current respon of Pres for this div in 1971. Mbr Amer Inst of Chem Engrs. Served as Chmn of the Chgo sect of AIChE 1958-59; Chmn 1962 AIChE Natl Convention. Mbr Chgo Mgmt Assn. *Society Aff:* AIChE, AMA, SPI.

Baillif, Ernest A
Business: Monte Rd, Benton Harbor, MI 49022
Position: VP, R&E. *Employer:* Whirlpool Corp. *Education:* PhD/ME/Univ of Minn; MS/ME/Univ of Minn; BM/ME/Univ of Minn; BA/EE/Univ of Minn *Born:* 12/8/25. in Pine County MN. Joined Seeger, Whirlpool's predecessor, in 1953. Received assignments of increasing responsibility in engg until 1968 when elected VP, corporate engg. Became VP, res and engg, in 1970. Has written numerous articles, holds seven patents, and is a mbr of various honor societies, including Tau Beta Pi. A mbr of Heil-Quaker's bd of dirs (Whirlpool subsidiary), serves on mech engg advisory committee, Univ of MN; industrial advisory committee, Purdue; bd of trustees, GA Tech Res Inst; and bd of trustees at local hospital. He and his wife reside in St Joseph, MI, also Bd of Visitors, Memphis State Univ. *Society Aff:* ASEE, IEEE, ASHRAE

Baily, William E
Home: 6598 Bose Lane, San Jose, CA 95120
Position: Mgr, Fuel Mechanical Design. *Employer:* G E Co, Nuclear Energy Sys Div. *Born:* Oct 1932. BS, MS Univ of Pittsburgh, Met Engrg. Engr with Westinghouse Atomic Power Dept 1957-62 spec in UO(2) fabrication & inpile testing. Assigned by Westinghouse for 18-month tour at Hanford Wash to study plutonium. Joined Nuclear Energy Div of G E in 1963. The 1963-73 period spent in Fast Breeder fuel dev activities. Wide range of mgmt assignments from Proj Engr, R&D on fuel, & plutonium facil oper. Since 1973, assumed respon for fuel testing & facil oper for water reactor group. Current pos is Mgr of Fuel Mech Design, Boiling Water Reactor Oper. Over 45 pubs. Active in local & natl ANS. Enjoy mountain climbing & skiing. Fellow Award Natl ANS.

Bain, James A
Business: 2901 E Lake Rd, Bldg 14, Erie, PA 16531
Position: Senior Mechanical Engineer *Employer:* General Electric Co. *Born:* Oct 1935. MS & BSME Lehigh Univ. USAF 1958-61 at Air Force Cambridge Res Labs. Employed by Genl Elec Co, Ordnance Dept 1961-62; Corp Res & Dev Ctr 1962-71; Transportation Sys Bus Div 1971- . Current respon incl analysis & measurement of stress, vibration, acoustics, & fatigue on transportation products; also incl investigations to improve Div tech in those areas. Received ASME's Rail Transportation Award 1969 for analytical work on pantograph-catenary sys. *Society Aff:* ASME

Bainbridge, Danny L
Home: Rt 9 Box 65-A, Fairmont, WV 26554
Position: Proj Admin/Consltg Engr *Employer:* Freelance Technical Services.
Education: BS/Engrg/WVA Univ. *Born:* 2/12/51. Native of Fairmont, WV, grad in 1973 from the College of Engg at WVU. Masters Degree Candidate in Mining Eng. Construction engr for 2 yrs for small firm. Field Engr for Coleman Assoc specializing in sub-div dev. Environmental Engg, 1975-1984, for Eastern Coal specializing in pollution control & reclamation. Responsible for design & installation of all water treatment plants for Eastern. Pres ASAE 1978-79. Presently gen mgr for consltg firm in charge of Construction Projects and General Engrg Consltg Serv to a wide variety of clients. Enjoy hunting & golf. *Society Aff:* ASAE.

Bainer, Roy
Home: 623 Miller Dr, Davis, CA 95616
Position: Dean Emer, College of Engrg. *Employer:* Univ of Calif at Davis.
Education: BS/Agri Engr/KS State Univ; MS/Agri Engr/KS State Univ; LLD/Honorary/Univ of CA; Doc Eng/Honorary/KSU. *Born:* Mar 7, 1902. Res staff KSU 1926-29. 1929-69, advanced from Asst Prof to Prof Univ of Calif-Davis. Chmn Agri Engrg 1947-61. Dean of Engrg 1961-69. Cons in England, Chile, Peru, Brazil, Japan, Laos, Thailand, Guatemala, Spain, Iran & Mexico and Malawi. McCormick Medal 1948 ASAE, Bendix Medal 1962 ASEE, Distinguished Service Award 1960 KSU & 1962 Univ of Mo. Pres ASAE 1956-57. Elected to Natl Acad of Engrg 1965. Co-author 2 textbooks in field. Publ over 100 papers. Enjoy golf & organ playing. *Society Aff:* ASAE, AAAS, ΣΞ, ASEE.

Bainum, Peter M
Business: Dept of Mech Engrg, Washington, DC 20059
Position: Prof of Aero Engrg *Employer:* Howard Univ. *Education:* BS/Aerospace Engr/TX A&M; SM/Aeronautics and Astronautics/MIT; PhD/Aerospace Engr/The Catholic Univ of Amer. *Born:* 2/4/38. With Howard since 1969. Spec: spacecraft attitude control, orbital mech, stability theory. Appl Physics Lab Johns Hopkins Univ 1965-69; 1969-72, Cons. IBM Fed Sys Div 1962-65. Martin-Marietta Corp 1960-62. AAS: Fellow and VP, International. AIAA: Assoc Fellow. Fellow British Interplanetary Soc. NASA/ASEE Summer Fac Fellowship 1970-71. SAE Teetor Award for Engrg Educators 1971. Phi Eta Sigma, Tau Beta Pi; Sigma Gamma Tau, Phi Kappa Phi, Sigma Xi. VP Res, Cons WHF & Assoc Inc, 1977-86. Howard Univ Oustanding Res Award, 1981. Vice Chrmn, Intl Astronautical Federation Astrodynamics Tech Ctte, 1984- . Intl Academy of Astronautics, 1985-. Member NASA(Italian) PSN Tether Applications Simulation Working Group, 1987-. *Society Aff:* AAS, AIAA, BIS, DGLR.

Baird, Avorald L
Home: 3044 W Shangri-La Rd, Phoenix, AZ 85029
Position: Deputy Dir Streets & Traffic *Employer:* City of Phoenix. *Education:* MSCE/Civil & Trans/Calif State Univ LB; BSCE/Civil/NMSU. *Born:* 1/8/39. Employed by L A Cty Rd Dept, CA 1962-66 as Civil Engr. City of Long Beach, CA 1966-70 as Asst Traffic Engr with City of Phoenix since 1970. Assumed current respon as Deputy Director in 1985. Desicgon, Planning & Safety Div Hd, 10 yrs previous. Previously was proj dir of a federally founded proj to estab Traffic Safety Div. Conduct accident analysis & dev a computerized records sys. To date, projs are continuing to reduce total accs by 50%, inj accs by 56%, fatals by 75%, & are saving more than 1/2 million $ per yr. Reg PE MBr Inst of Transportation Engrs. PPres 1973-74 of the AZ Chap of ITE, Secy-Treas 1972-73. Mbr Phi Kappa Phi. Cert instructor of a variety of engineering courses at Phoenix College. *Society Aff:* ITE, ΦΚΦ.

Baird, Hugh A
Business: 1000 S Fremont Ave, Alhambra, CA 91802
Position: President. *Employer:* C F Braun & Co. *Education:* MS/Chem Eng/Caltech; BS/Chem Eng/Caltech; AA/Eng & Science/Pasadena Jr College. *Born:* Feb 1921. Born in Scotland, reared in Southern Calif. Headed Test Dept for Rocket Res & Dev at Caltech on a Navy contract admin by the Office of Sci Res & Dev during WW II. Joined C F Braun & Co 1946 as Process Engr. Moved to Chem Engrg 1948,

Baird, Hugh A (Continued)
then to Project Mgmt 1956, Mgr Chem Engrg Dept 1968, Asst Mgr Engrg Div 1970, V P 1971, Exec V P 1974. Pres 1979, Fellow AIChE 1976. Fellow IAE. Hobbies: hiking, backpacking, color photography. *Society Aff:* AIChE, IAE, EJC, MPC, ACS.

Baird, Jack A
Business: 295 N Maple Ave, Basking Ridge, NJ 07920
Position: Vice President. *Employer:* AT&T. *Education:* PhD/Elec Engg/TX A&M Univ; MS/Elec Engg/Stevens Inst of Tech; BS/Elec Engg/TX A&M Univ. *Born:* 5/27/21. in Omaha, TX. US Navy 1943-45. With Bell Tele Labs 1946-73. VP Bell labs 1967-73. VP Engg AT&T 1973-74. VP Engg & Network Services AT&T 1974-76. Became VP Customer Services AT&T 1976-78. VP Engg & Network Services AT&T July 1978. Became VP Network Planning & Design AT&T Nov 1978. Fellow IEEE. Mbr Natl Acad of Engg. *Society Aff:* NAE, IEEE.

Baird, James A
Business: 307 Wynn Dr, Huntsville, AL 35807
Position: SR. VP & Dir, Systems Technologies Div *Employer:* General Research Corp. *Education:* PhD/Engr Mech/Univ of AL; MS/Engg Mech/Univ of AL; BS/Mech Engg/Univ of AL. *Born:* 8/10/35. Native of Alabama. US Air Force 1958-61, Res engr in aerospace studies with Teledyne-Brown and Site Defense Project. Assumed current pos as Dir, Systems Technologies Div for Genl Res Corp 1982. Enjoy outdoor sports & flying *Society Aff:* AIAA   ΤΒΠ, ΣΞ.

Baird, James L
Business: 73 Pond St, Waltham, MA 02154
Position: Sr Vice President. *Employer:* Artisan Industries Inc. *Education:* BS/ChE/MIT. *Born:* Nov 9, 1917 Richmond, N H. Have been with Artisan Indus Inc since grad. V P Chem Engrg 1953, Dir 1953, Sr V P 1973. Former Dir of Kontro Co. Treas, V Chmn, Chmn Local Sect AIChE. Fellow AIChE, Genl Chmn AIChE Natl Meeting. Founding Mbr AACE. *Society Aff:* AIChE, AACE, ADCS, AOCS, ISA, IES.

Baisch, Stephen J
Business: 809 Hyland Ave, Kaukauna, WI 54130
Position: Bd. Chairman *Employer:* S J Baisch Assoc Inc. *Education:* BS/ME/Univ of WI. *Born:* 10/28/17. Ironwood, MI. Univ of WI, 1942. Taught Mech Dwg Univ of WI. US Army 1942- 1946 2nd Lt to Maj. N Africa, Sicily & Italian campaigns. With Thilmany Paper Co 1946-1958. Pres S J Baisch Assoc 1958-84. Dir Petratex Paper, Peoria, IL. Partner Potsdam Power Co. Patentee in field. Enjoy glee club, boating, fishing, woodworking. *Society Aff:* ASME, TAPPI, WSPE, NSPE.

Bajcsy, Ruzena
Business: Dept of Computer and Information Sci, Philadelphia, PA 19104
Position: Prof and Chairman *Employer:* Univ of Pennsylvania *Education:* PhD/Comp Sci/Stanford Univ; PhD/EE/Slovak Tech Univ, Czech *Born:* 05/28/33 Dr. Bajcsy is a native of Bratislava, Czechoslovakia where she was an assst prof in the Electrical Engrg Dept until 1968. Then she came to Stanford Univ where she worked at the artificial intelligence lab for 4 1/2 yrs. In 1972 she joined the Computer Sci Dept at the Univ of PA. Her research is in Robotics and Computer Vision. She was the editor of the Intl Pattern Recognition Newsletter, Assoc Editor of IEEE Trans on PAMI, and other editorial bds. She is the Dir of a Lab called 'GRASP' an acronym for General Robotics Active Sensory Processing, Lab, which encompasses an interdisciplinary research group of Mechanical, Electrical and Computer Engrs and Scientists. *Society Aff:* IEEE, ACM

Bajura, Richard A
Business: Dept of Mech & Aerospace Engrg, West Virginia University, Morgantown, WV 26506-6101
Position: Prof & Dir of Energy Research Center *Employer:* WV Univ. *Education:* PhD/Mech Engr/Univ of Notre Dame; MSME/Mech Engr/Univ of Notre Dame; BSME/Mech Engr/Univ of Notre Dame. *Born:* 2/2/41. Native of Duquesne, PA. Research Engr at Alliance res Ctr of Babcock & Wilcox Co for 1 yr following completion of doctoral program in 1967. Postdoctoral study at Johns Hopkins Univ for 1 yr before joining WV Univ in 1969. Promoted to Prof and Assoc Chrmn for Grad Studies in 1978. Named Dir of WV Univ Energy Research Center in Sep, 1984. Assoc Editor of ASME Journal of Fluids Engg 1978-80 and 83-85; mbr of ASME Fluid Meters and Fluids Engg Div ctte. Enjoy jogging, camping and sports activities. *Society Aff:* ASME, APS, AIAA, ASEE.

Bak, Eugene
Business: Two Mile Run Rd, Franklin, PA 16323
Position: VP, Operations. *Employer:* Mooney Chem, Inc. *Education:* MBA/Mgt/Seton Hall Univ; BChE/Chem Engr/OH State Univ. *Born:* 7/18/33. Joined Mooney Chem in 1970 as Plant Mgr. Became VP Operations in 1974 with the responsibilities to manage mfg. In 1969-1970, Partner in Radom chem Co. Previous to Radom Chem, employed by Diamond Shamrock for 12 yrs. Held a number of positions with Diamond in the field of engg, process dev and mfg. Grad of OH State with BChE and Seton Hall with MBA. PE in the States of NJ, OH & PA. Attended mgt seminars at Harvad, Syracuse Univ & MI Univ. Married, 3 children, live in Mineral Township, PA. *Society Aff:* AIChE, ACS.

Bakeman, Charles T
Home: 2315 E Sound View-Bx 487, Langley, WA 98260
Position: Semi-retired. *Employer:* Self. *Education:* BS/Elec Engg/Univ of WA. *Born:* May 23, 1903. Fellow Mbr Emer, Illuminating Engrg Soc. Regional V P Illuminating Engrg Soc. Cons: Western Defense Command, Regional Civilian Defense Agency, & State of Washington WW II. Cons Engr License Calif 1945-60. Served on Natl Ctte establishing codes & tech standards 1930 to 1960. Served as Indus Mgr Puget Sound Power & Light Co, Seattle to 1945. Self-employed State of Calif 1945 to 1957. Space & Aircraft engrg The Boeing Co 1957 to 1969. Self-employed 1969 to date. *Society Aff:* IES.

Baker, Arthur, III
507 Casazza Dr, Suite B, Reno, NV 89502
Position: Consltg Mining Geologist *Employer:* Self-employed *Education:* PhD/Geology/Stanford; MS/Geology/Stanford; BA/Geology/Wesleyan. *Born:* Apr 29, 1923 Waterbury Conn. BS Wesleyan Univ 1947; MS, PhD Geology Stanford Univ 1953. Married Dorothy M Stamps 1945, 3 children. AUS 194446, Corporal. 1949-54 mine geologist Coronado Copper & Zinc Co; 1955-62 geologist - ch geologist Callahan Mining Corp; 1963 Mgr Huntley Indus Minerals Inc; 1963-67 cons Bishop, Calif; 1967-72 Assoc Dir Nevada Bureau of Mines; 1972 Dean, Mackay Sch of Mines, Univ of Nevada, Reno. 1981 Consltg mining geologist. *Society Aff:* SEG, GSA, AIME.

Baker, Bernard S
Business: 3 Great Pasture Rd, Danbury, CT 06810
Position: Pres. *Employer:* Energy Research Corp. *Education:* PhD/ChE/IIT; MS. ChE/Univ of Penna; BS/ChE/Univ of Penna. *Born:* 6/26/36. Educated in Chemical Engg my career began as a Fulbright Fellow at the Lab of Electrochemistry, Univ of Amsterdam in 1959. I have pursued a career in Electrochemical Engg involving energy conversion and energy storage R&D and system engg. I am married with four children. *Society Aff:* ACS, ECS, AIChE.

Baker, Broughton Leonard
Business: College of Engrg, Columbia, SC 29208
Position: Distinguished Prof Emeritus. *Employer:* Univ of South Carolina. *Education:* BS/ChE/Univ of SC; PhD/ChE/NC State Univ. *Born:* 7/26/12. PhD NC State Univ; BS Univ of SC. Native of SC. Indus exper: Allied Chemical, Chem Warfare Serv, Naylee Chem, Ellicott Labs (pt owner). Acad exper: organized & hd Dept Chem Engg Univ of SC 1946-70, Prof 1956-77, presently Distinguished Prof Emeritus. Mbr SC State Bd of Engg Examiners since 1963. Mbr Tau Beta Pi, Sigma Xi, AIChE (Chmn Savannah River Sect AIChE 1964), ASEE (Pres Southeastern

Baker, Broughton Leonard (Continued)
Sect 1962-63, Chmn Chem Engg Div 1972-74). Distinguished Service Award, Natl Council engg Examiners 1974, Fellow AIChE 1975. *Society Aff:* AIChE, NSPE, ASEE.

Baker, Dale B
Business: PO Box 3012, Columbus, OH 43210
Position: Deputy Exec Dir *Employer:* Chemical Abstracts Service. *Education:* BChE/Chem Engr/OH State; MSc/Chem/OH State. *Born:* Sept 19, 1920. Internatl Council of Sci Unions Abstracts Bd; Amer CChem Soc; AAAS; Amer Soc for Info Sci; Amer Inst of Chem Engrg Natl Fed of Abstracting & Indexing Services; P Pres OSU Alumni Assn; Author num pubs; Distinguished Alumnus A OSU Coll of Engrg March 1970; Amer Men of Sci; Who's Who in Sci; Sigma Xi; Columbus Tech Cou Man-of-the-Yr Award 1968; Boss-of-the- YrInternatl; North Broadway United Medthodist Church. 1979 Recipient ACS Patterson-Crane Award. 1983 Recipient of the ASIS Award of Merit. *Society Aff:* ACS, ASIS, ILSU AB, NFAIS, AAAS.

Baker, George S
Home: 8065 N. Mohawk, Milwaukee, WI 53217
Position: Prof *Employer:* Univ of Wisconsin-Milw *Education:* PhD/Physics/Univ of IL; MS/Physics/Univ of IL; BS/Physics/Purdue Univ *Born:* 08/02/27 US Army 1945-47. Asst Prof Univ of Utah 1957-62-Physics Dept; Tech Spec Aerojet General 1962-67; Prof Univ of Wisconsin-Milw 1967-; Asst Dean for Graduate Prog 1972-76; Materials Dept Chrmn, 1978-. Main interests are in mechanical forming; mfg, failure analysis and product liability. *Society Aff:* ASM

Baker, Glenn A
Business: 7410 E Helm Dr, Scottsdale, AZ 85260
Position: President. *Employer:* Environmental Research Labs. *Education:* BSEE/EE/Univ of AZ. *Born:* 8/6/26. BSEE Univ of AZ 1950. Native of AZ. US Navy, CEC WWII. Instructor Univ of AZ 1949-51. C & A Engr Westinghouse Elec Copp 1951-55. Power Engr, Salt River Project 1955-57. Pres Baker, Moody & Fredrickson, Cons Engrs Phoenix, AZ 1957- 72. Pres Environ Res Labs, Scottsdale AZ 1972- . IES Regl VP 1967-69; IES Natl Dir 1969-70; IES Natl Treas 1970-71. IES VP-Res & Technical Activities 1979-81. AZ Electrical Indus Man of the Yr 1963. AZ Soc Prof Engrs - Prof Engr of the Yr 1964. Fellow IES. Mbr IEEE. Mbr - ASHRAE. Reg PE AZ, NM, NV &CA. *Society Aff:* IES, IEEE, ASHRAE.

Baker, Hugh
Business: Metals Park, OH 44073
Position: Program Manager *Employer:* American Society for Metals. *Education:* BS/Metallurgy/Univ of Notre Dame; BS/Mech Engg/Univ of Notre Dame. *Born:* May 1926 Muskegon, Mich. Worked in met lab of Campbell, Wyant & Cannon Foundry 1946-48. From 1951-70, res & dev engr at Dow Chem, spec in properties, fabrication & uses of magnesium & aluminum. From 1970-79, on publications staff of Amer Soc for Metals , where respon for tech content of the Metals Handbook. Since 1979, ASM Manager of joint ASM/NBS Data Program for Alloy Phase Diagram and Managing Editor of the Bulletin of Alloy Phase Diagrams.. *Society Aff:* ASM, AIME/TMS

Baker, James L
Business: 303 East Locust, Angleton, TX 77515
Position: Partner. *Employer:* Baker & Lawson. *Education:* BS/Civil Engg/Univ of MO. *Born:* 4/16/10. Native of Brownwood, TX. Attended Howard Payne College. Transfered to MO Univ Sophomore year-Employed by TX Hwy. Dept upon graduation as Materials Inspector. Advanced through various Engg positions to Dist Designing Engr, Houston Dist in 1955. Responsible for all Hwy Design including Houston Freeway System, outside the Loop, until retirement in 1965. Established Consulting Engg Office, maintained to present, performing all types of Civil Engg related work for Cities, Counties and Industry in the middle Gulf Coast Area of TX. Married Bethe Ethridge 1933-Two Daughters Patsy Lawson and Judi Stradinger-Active in Church and Community Activities including Schoolboard Pres. *Society Aff:* NSPE, TSPE.

Baker, John R
Home: 12335 Queensbury Ave, Baton Rouge, LA 70815
Position: Ch Structural Engr. *Employer:* Howard Needles Tammen & Bergendoff. *Born:* Aug 5, 1921. BCE Ohio State Univ. Native of Polo, Mo. Served with Army Corps of Engrs in Canada & Europe during WW II. Has been employed by HNTB since grad in 1948, engaged in all phases of design & plan preparation for fixed & movable hwy & railroad bridges. Among these assignments was that of Ch Bridge Engr, Joint Venture of Barnard & Burk & HNTB for 150 miles of La Interstate Hwys which included the award winning I-10 crossing of the Atchafalaya Floodway. P Pres Baton Rouge Branch of ASCE, present Dir La Sect ASCE.

Baker, Merl
Business: Box x, ORNL, Oak Ridge, TN 37835
Position: Mgmt. *Employer:* Oak Ridge Natl Lab, Union Carbide Corp. *Education:* PhD/Mech Engg/Purdue Univ; MS/Mech Engg/Purdue Univ; BS/Mech Engg/Univ of KY. *Born:* July 1924. Native of Cadiz, Ken. Prof of mech engrg at Univ of Kentucky until 1963. First chancellor Univ of Missouri-Rolla 1964-73. Spec Asst to the Pres, Univ of Missouri sys, since 1973. Prepared respon for UM compliance with fed gov regulations in sev areas, incl affirmative action. Distinguished Serv Award ASHRAE 1971; Univ of Kentucky Hall of Distinguished Alumni 1965; Distinguished Engrg Alumnus Award Purdue Univ 1965; Award of Merit for outstanding teaching ASHRAE 1959. Pi Tau Spigma Gold Medal Award, 1950. 1977- present, Energy Mgmt Outreach, Oak Ridge Natl Lab, Union Carbide Corp. Enjoy gardening & refinishing furniture. *Society Aff:* ASME, ASEE, NSPE, ASHRAE, ASEM.

Baker, Raymond G
Home: 2206-144th S E, Bellevue, WA 98007
Position: Mgr, Test Planning & Management *Employer:* Boeing Aerospace Co. *Education:* BSEE/Electronics/UT State Univ. *Born:* 12/18/36. Joined Boeing Aerospace Co 1958; currently Mgr of Environmental Test Laboratories Test Planning & Management Grp, respon for test sys planning design, & conduct supporting all Boeing Aerospace Co test progs. Active in Instrument Soc of Amer serving as Genl Chmn of 21st Intl instrumentation symposium, Phila May 1975. P Pres ISA Seattle Sect & P Dir of Test measurement Div ISA. & VP-Elect Technology Dept of ISA. *Society Aff:* ISA, IES

Baker, Robert F
Business: 9422 Holland Ave, Bethesda, MD 20014
Position: President. *Employer:* Robert F Baker & Assocs. *Education:* MSCE/Soil Mechs/Purdue; BCe/Structures/OH State. *Born:* Sept 1917. Reg Engr in Ohio, Kentucky, W Va & Md. Native of Ohio. Airfield constr, Army CE 1942-46. Soils engr W Va Rd Commission 1949-54. Prof & Res Dir, Cvl Engrg , Ohio State Univ 1954-62. Dir of R&D U S Bureau of Public Rds 1962-67. Cons & part-time prof since 1967. Roy W Crum Distinguished Service Award Hwy Res Bd 1967. Distinguished Alumnus Award Purdue Univ 1965, Ohio State Univ 1970. Author of 'Highway Risk Problem', Wiley 1971. Co-author 'Public Policy Dev', Wiley 1975. Ed of 'Handbook of Hwy Engrg', Van Nostrand Reinhold 1975. *Society Aff:* ASCE, AAPT, SAME, TRB

Baker, Thomas C, Jr
Business: 500 E 8th - Rm 994A, Kansas City, MO 64106
Position: Div Supr-Ckt Prov. *Employer:* Southwestern Bell Tele Co. *Education:* BS/EE/St Louis Univ; MBA/Bus/St Louis Univ. *Born:* Dec 27, 1928. US Marine Corps 1946-48. Employed by Southwestern Bell Telephone Co since 1952 in var engrg mgmt pos. Natl Dir NSPE 1975-1979; Sr Mbr IEEE; P Pres MO Soc of Prof Engrs; P Pres MSPE PEI Div, VP-NSPE. *Society Aff:* IEEE, NSPE, MSPE.

Baker, Walter L
Home: 22 Oliver Hazard Perry Rd, Portsmouth, RI 02871
Position: President *Employer:* Baker Engineering Co. *Education:* MS/EE/PA State Univ; BEE/EE/Clarkson College. *Born:* Aug 7, 1924; Tech Supervisor Tenn Eastman Corp 1944-45; Sr Engr Philco Corp 1945-49; Res Assoc to Prof Engrg, Pa State Univ 1949-82; now Prof. Emeritus; Sr Mbr Grad Fac 1965-82; served as principal scientist & Program Mgr Applied Res Lab, Pa State Univ; Tech Program Mgr Torpedo Mk 48 1967-71; Pres, Baker Engrg Co, Portsmouth, RI, 1981-; Cons indus & U S Gov agencies 1952- ; Tech Dir Bellefonte Lab, John I Thompson & Co 1964-67; Reg P E Pa; Sr Mbr Inst Elec & Electronics Engrg. Mbr N Y Acad Sci, Acoustical Soc of Amer, Sigma Xi. U S Navy Dept Meritorious Public Serv Citation 1975. Contributor of articles to professional journals. *Society Aff:* IEEE, ASA, ΣΞ, NYAS.

Baker, Warren J
Business: San Luis Obispo, CA 93407
Position: President. *Employer:* CA Polytechnic State Univ. *Education:* PhD/Civ Engr/Univ of NM; MS/Civ Engr/U of Notre Dame; BS/Civ Engr/U of Notre Dame. *Born:* 9/5/38. Native of Westminister, MA. Res Assoc & Lecturer E H Wang Civil Engrg Res Facil, NM 1962-66. Assoc Prof of Civil Engrg Univ of Detroit 1966-71. Visiting Engr & NSF Fac Fellow MIT 1971-72. Prof of Civil Engrg Univ of Detroit 1972-79. Dean of Engrg 1973-78. Academic VP 1976-79. Chmn ASCE Geotech Div Ctte on Reliability 1973-76, Mbr Geotech Div Publ Ctte 1974-82. Pres Detroit Chap NSPE 1976-77. City of Detroit Mayor's Mgmt Adv Ctte 1975-78. 1979 Pres Cal Poly. Fellow, Engineering Society of Detroit. Mbr AASCU Bd Dirs; Mbr Cmmittee on Science and Technology. Presidential appointments: Brd for Intl Food and Agricultural Dev 1983-86; Natl Sci Brd 1986- . *Society Aff:* ASCE, NSPE, ASEE.

Baker, Wilfred E
Business: Wilfred Baker Engineering, 8103 Broadway, Suite 102, P.O. Box 6477, San Antonio, TX 78200
Position: Pres *Employer:* Wilfred Baker Engineering *Education:* Dr Eng/Mech Eng/Johns Hopkins Univ; MSE/Mech Eng/Johns Hopkins Univ; BE/Mech Eng/Johns Hopkins Univ. *Born:* 1/3/24. Native of Baltimore Md. US Navy 1943-46. Instructor in mech engrg Johns Hopkins Univ 1947-49. Res Engr for Ballistic Res Labs 1949-61. Pricipal Dev Engr for AAI Corp 1961-64. Southwest Res Inst, 1964-84 Inst Scientist, 1972-84, with Wilfred Baker Engineering since 1984. Pres, 1984-. Cons to many gov labs & private firms on dynamics, scale modeling, explosion effects, & hazards analysis. P Chmn Shock & Vibration Ctte ASME. Assoc Fellow AIAA 1979, Sig Xi. Fellow ASME 1971. Enjoy sailing, handball, classical music *Society Aff:* ASME, AIAA, ΣΞ, SAME

Baker, William A
Home: P. O. Box 122, Hingham, MA 02043
Position: Naval Architect *Employer:* Private Practice *Education:* SB/Naval Arch & Marine Engrg/MIT *Born:* 10/21/11 Native of New Britain, CT. Employed as a naval architect with Bethlehem Steel Co, Shipbldg Div, 1934-1964. Curator, Hart Nautical Museum, MIT, 1963- 1981. Mbr Com for Revision of 1948 Safety of Life at Sea Conv, 1957-1960; SOLAS Working Com on Stability and Subdiv, 1961-1981; Com on Revision of 1930 Load Line Conv. Plans for historic vessels CJOA, MAYFLOWER II, MARYLAND DOVE and others. Author: The New Mayflower: Her Design and Construction, 1958; Colonial Vessels, 1962; The Engine-Powered Vessel, 1965; A Maritime History of Bath, Main, 1973; plus misc technical papers, articles and chapters. Other activities include music, photography and boating. *Society Aff:* SNAME

Baker, William O
Home: Spring Valley Rd, Morristown, NJ 07960
Position: Chrmn, The Rockefeller University *Education:* PhD/Phys Chemistry/Princeton Univ; BS/Phys Chemistry/WA College. *Born:* 7/15/15. Native of Chestertown, MD. Joined Bell Labs 1939; named Chmn 1979. Early sci work on matls, solid state properties of dielectrics & polymers cables and waveguide strucs, pioneering work leading to dev of ablative heat shields. Following earlier mbrship on Pres's Sci Adv Com, Pres's Foreign Intelligence Adv Bd, & V Chmn of Pres's Com on Sci & Tech, and natl Cancer Adv Bd, Consultant of Office of Sci & Tech Policy & other Coms of the US Senate, Currently Mbr, Council Federal Emergency Management Agency (FEMA) Advisory Bd, Nat'l, Commission on Excellence in Ed., Trustee Health Effects Inst., Mellon Fnd. (Chmn), Princeton and Carnegie-Mellon Univ.; Mbr of natl Acad of Sci, natl Acad of Engg, Inst of Medicine, Amer Acad of Arts & Sci, Amer Philosophical Soc; Hon Mbr Chemists' Club NY. Pres. Nat'l Sci, Medal, 1982. Vannevar Bush Award, 1981, Sarnoff Prize (AFCEA) 1981, Jefferson Medal NJ Patent Law Am 1981, Madison Marshall Award (ACS) 1980, von Hippel Award 1978; Willard Gibbs Medal 1978; Fahrney Medal 1977; Parsons Awd 1976; Awd for Distinguished contributions to Res Admin 1976. AIC Gold Medal 1975; Mellon Inst Awd 1975; James Madison Medal, Princeton Univ 1975; Proctor Prize 1973; Frederick Philips Awd 1972; Indus Res Inst Medal 1970; Edgar Marburg Awd 1967; Priestly Medal 1966; Per Kin Medal 1963. *Society Aff:* APS, SLA, ACS, AAAS, MRS.

Bakken, Earl E
Business: 7000 Central Ave NE, Minneapolis, MN 55432
Position: Founder and Senior Chairman of the Board *Employer:* Medtronic Inc. *Education:* Grad Work/EE/Univ of MN; Bach/EE/Univ of MN. *Born:* 1924 Minneapolis, Minnesota. Bach of EE 1948, followed by grad work at Univ of Minnesota. Cofounder in 1949 of Medtronic, Inc., currently the world leader in cardiac pacemaker systems. CEO until March 1976; Chairman of the Board, 1976-1985; Senior Chairman of the Board, 1985 - present. Personally financed and chairs Museum of Electricity in Life in Minneapolis. Developed first wearable external, battery-powered pacemaker in 1958 for Dr. C. Walton Lillehei. Chairs: Minnesota Medical Alley, Archeaeus Project; Board of Directors: Children's Heart Fund. Awards: College of St. Thomas Centennial Medal; Med-Tech Outstanding Achievement Award; NASPE's Distinguished Service Award; University of Minnesota Outstanding Achievement Award. *Society Aff:* IEEE, ISA, AAMI, NASPE, AAS

Balabanian, Norman
Business: Elec & Comp Engg, Syracuse University, Syracuse, NY 13244
Position: Prof & Chrmn *Employer:* Syracuse Univ. *Education:* PhD/Elec Engr/Syracuse Univ; MSEE/Elec Engr/Syracuse Univ; BSEE/Elec Engr/Syracuse Univ; Dipl/Sci/Aleppo College. *Born:* 8/13/22. Served in USAAF 1943-46. Author of books (10), technical articles (30), articles on education (10), encyclopedia articles (7). Editor of books on education (2), IEEE Trans on Circuit Theory (1963-65), IEEE Trans on Communications (1976), IEEE Tech and Soc (1975-76, 1979-), IEEE Spectrum Editorial Bd (1973-1975). Chrmn EE Div ASEE (1966-67). UNESCO field staff in Mexico (1969-70); Visiting Prof, Univ CA Berkeley, (1965-66); Fulbright Fellow Jugoslavia (1974-75); Academic Advisor for Engg, Natl Electronics Inst, Algeria (1977-78). Fellow of IEEE and AAAS. Deeply concerned over intrusions on civ liberties. *Society Aff:* AAAS, IEEE, ASEE, AAUP.

Balanis, Constantine A
Business: Dept of Elec and Computer Engrg, Arizona State University, Tempe, AZ 85287
Position: Prof of EE. *Employer:* AZ State Univ. *Education:* PhD/Elec Engg/OH State Univ; MEE/Elec Engg/Univ of VA; BSEE/Elec Engg/VPI. *Born:* 10/29/38. Born in 1938, Trikala, Greece. From 1964 to 1970 was with NASA Langley Res Ctr, Hampton, VA. From 1968-70 held part-time appointment as Asst Professorial Lecturer with George Washington Univ grad extension at Langley. From 1970 to 1983 was with Dept. of Elec. Engr., WV Univ., Morgantown, WV. Since 1983 with the Dept of Elec and Computer Engrg, AZ St Univ where he is now a Prof. Teaches undergrad and grad courses in electromagnetic theory, microwave circuits and devices, and antennas. Res interests in application of the Geometrical Theory of Diffraction

Balanis, Constantine A (Continued)
(GTD) and Moment Method (MM), Radar Cross Section Predictions & Measurements, multipath interference, electromagnetic wave propagation in microwave integrated lines, and reconstructive geophysical tomography. Served as Assoc Editor of *IEEE Trans Antennas and Propag* (1974- 77), and Delegate-at-Large, Sec-Treas, V-Chrmn and Chrmn of Upper Monongahela IEEE Subsection. Presently Assoc Editor of *IEEE Trans Geoscience and Remote Sensing* (1982-84). Also served as Editor of Newsletter for IEEE Geoscience and Remote Sensing Soc (1982-83). Fellow of IEEE and author of *Antenna Theory: Analysis and Design*, Harper & Row Publishers, Inc. (1982). *Society Aff:* ΣΞ, HKN, TBΠ, ΦΚΦ, IEEE.

Balazs, Bill A
Home: 7500 Woodlake Dr, Walton Hills, OH 44146
Position: Chief Engr *Employer:* GE-Reuter-Stokes *Education:* BS/Indus Eng/Univ of Budapest; APM/Mgmt/John Carrol U, Cleveland *Born:* 06/13/33 PE. Born in Hungary. Educated in Hungary and USA. Taught in ACME Tech Inst in the field of Engrg Tech. Prod Engr for Morrison Product Inc, specializing in Heating Equip Design. with Reuter-Stokes, Inc, Subsidiary of Gen Elect, since 1965. Assumed current respon as Engrg Mgr for manufacture of nuclear detectors for reactor instrumentation in 1973. Respon includes design and implementation of nuclear sensors for PWR plant application, traversing in-core probes, source and intermediate range, bottom-entry sensors for neutron or gamma radiation monitoring in BWR. Certified Engr in the field of Mfg, Plant and Cost Engrg. *Society Aff:* ANS, ASME, NSPE, IIE, SME

Balch, Clyde W
Business: 2801 W Bancroft St, Toledo, OH 43606
Position: Emeritus Dean Cont Ed & Prof of Chem Engg. *Employer:* Univ of Toledo. *Education:* MSES/Engr Sc/Univ of Toledo; MS/Chem Engr/Univ of MD; BS/Chem Engr/Univ of MD. *Born:* June 11, 1917 Winterset Iowa; m. Mary Jo Mitchell 1940; c. Charles M, Thomas S, John R. Phys Chemist U S Naval Res Labs 1938-39; chem eng E I duPont De Nemours & Co 1939-46; V P Maumee Chem Co 1946-65; Prof Chem Engrg, Chmn of Dept Univ o f Toledo 1964-68; Dean Cont Education 1967-77; cons in field 1960- . Bd of Dir Toledo Goodwill Indus 1967- . Toledo Engr of Yr 1962. Fellow Amer Inst Chem Engrs. reg P E OH & AL. *Society Aff:* AIChE.

Baldwin, Allen J
Home: 3518 St. Paul Ave, Minneapolis, MN 55416
Position: Engg Supervisor *Employer:* Honeywell *Education:* MBA/Exec Mgmt/Univ of MN; BS/ME/Univ of MN *Born:* 6/20/35 Graduated from the Univ of MN with a degree in Mech Engrg in 1963. At that time he joined Honeywell Inc and has progressed to the position of Engg Supervisor in Quality Assurance. He has acted as an interface between Honeywell and the society in areas of prof dev. While at Honeywell he earned his MBA degree in 1970 working evenings and the Univ of MN. As pres of the MN Society of Prof Engrs, he spearheaded a mbrshp drive which resulted in a 30% increase in the society mbrshp. He served for 2 yrs as chrmn of natl ASME energy standards committee and chaired the Minnesota ASME section. Baldwin was Pres of the Minnesota Federation of Engg Societies in 1985-86. He is a Reg Professional Engg in 3 states in the disciplines of mech & elec engg. *Society Aff:* NSPE, ASME, ASQC.

Baldwin, Clarence J
Business: Westinghouse Electric Corp, 777 Penn Center Blvd, Pittsburgh, PA 15235
Position: Mgr, Advanced Sys Tech. *Employer:* Westinghouse Electric Corp. *Education:* EE/EE/MIT; MS/EE/Univ of TX, Austin; BS/EE/Univ of TX, Austin. *Born:* Aug 1929. Fortescue Fellowship. PMD Harvard Bus School. Native of San Antonio, Texas. Joined Westinghouse Electric Corp, Electric Util Engrg Dept, 1952. Mgr Generation, Advanced Dev 1962-1969, Mgr Res & Dev 1969-1980, Mgr Advanced Sys Tech Div since 1980 respon for advanced res and dev, application software, energy mgt systems, & cons serv for power sys. Authored over 75 papers. Served on Engrg Accred Comm, US Technical Committee of CIGRE, US Member Committee of World Energy Conference. Eta Kappa Nu Outstanding Young EE Award 1961; Distinguished Engrg Grad Univ of Texas 1967; Fellow IEEE; Reg P E. *Society Aff:* IEEE, ASME, ASEE, NSPE.

Baldwin, David M
Home: 5101 River Rd.- Apt. 816, Bethesda, MD 20816
Position: Retired (Consultant) *Education:* MS/CE/Univ of IL; BS/CE/Univ of IL. *Born:* 7/31/12. Urbana, Ill. City traffic engr Evanston, Ill. Safety Engr Va State Police. Traffic Engr Natl Safety Council 13 yrs. Exec Secy Inst of Traffic (now Transportation) Engrs 1956-62. Various titles with Bureau of Public Rds & FHWA 1962-79. Pres ITE 1968. Matson Mem Award 1973. B W Marsh Award, 1978. Chmn Natl Ctte on Uniform Traffic Laws & Ordinances 1973-79, Chmn Ctte on Hwy Oper, Pan Amer Hwy Congress 1975-80. Special int in traffic accident records. Chmn ANSI D- 20.4, developing roadway environ data base 1974-79. Fellow ITE. Tau Beta Pi. Chi Epsilon. Sigma Xi (Assoc.) *Society Aff:* ITE.

Baldwin, James W, Jr
Business: Dept of Civil Engg, Columbia, MO 65211
Position: Prof. *Employer:* Univ of MO. *Education:* PhD/Thro&App Mech/Univ of IL; MS/Thro&App Mech/Univ of IL; BS/Gen Engg/Univ of IL. *Born:* 6/29/29. Native of White Hall IL. Officer US Army Corps Engrs 1951-54. Assoc Prof CE Univ of MO 1960-66. Sr Design Engr Rust Engg Pittsburgh 1965-66. Prof CE Univ of MO 1967-present. Civil Engr Dept Chmn 1967-72. Visiting Prof CE Warrick Univ Coventry, England 1972. Extensive res in Civil Engr materials and structures. Failure analysis. Field testing of full scale structures. PE. *Society Aff:* ASCE, ACI, ASEE, Chi Epsilon, Tau Beta Pi, Sigma Xi.

Baldwin, Lionel V
Home: 1900 Sequoia, Fort Collins, CO 80524
Position: Dean *Employer:* CO State Univ *Education:* PhD/ChE/Case Inst of Tech; SM/ChE/MIT; BS/ChE/Univ of Notre Dame *Born:* 05/30/32 Dr Baldwin has served as Dean of the Coll of Engrg at CO State Univ since 1964. Prior to joining CSU in 1961 as an Assoc Prof of Cvl Engrg, he worked for NASA in Cleveland, OH for six yrs. He has published over 50 technical papers in turbulent transport in the atmosphere and educational tech. He served as first Chrmn, Bd of Dirs, Association for Media-based Continuing Education for Engrs (AMCEE); Mbr, Gen Motors Inst Visiting Committee; and Mbr, NASULGC Commission on Education for the Engrg Profession; Chrmn, Engrg Dean's Council, ASEE. He is reg engr in the State of CO. *Society Aff:* ASEE, NSPE, ΣΞ.

Baldwin, Miles S
Home: 507 Wimer Cir, Pittsburgh, PA 15237
Position: Actg Mgr, Pwr Gen Syst Engrg *Employer:* Westinghouse Elec Corp *Education:* BS/EE/Wash Univ (St Louis) *Born:* 12/20/21 Native of St Louis, MO. Served in US Air Corps 1943-45. Joined Westinghouse in 1949, where experience includes application of electrical equipment on elec utility and industrial power systems, solution of system problems, power system design, establishing equipment design requirements, and protection of systems and equipment. Named IEEE Fellow; active on IEEE Power Generation Committee- Station Design Subcommittee, Auxiliaries Systems Working Group. Reg PE in PA and OH. Enjoys gardening and photography. Prior to present position, was well known in the industry as a Generation Conslt (also with Westinghouse). *Society Aff:* IEEE

Baldwin, Robert D
Business: Suite 306, 10905 Fort Washington Rd, Fort Washington, MD 20022
Position: VP. *Employer:* Booker Assocs, Inc. *Education:* Bach of Science/Civil Engg/ Univ of MD. *Born:* 11/10/20. Native of Washington, DC. Graduate of Univ of MD. Bach of Sci degree, Civil Engg. Served 4 yrs in USAF as an Engg Officer, flight Engr and Dir of Engg Sch. From 1946-65 held various positions on the former PA Railroad in Maintenance, Operations, Engg and Mgmt. Subsequent to this, was Dir of Transporation, Planning & Engg for a "Top 35" East Coast Engg Firm. Then served

Baldwin, Robert D (Continued)
a tour as Chief Engr, Facilities, Assessment & Planning, US Railway Assoc, after which joined Booker Assocs, Inc and is presently VP/Mgr, Eastern Region. *Society Aff:* AREA, SAME, NSPE, IIE, TRB, ARTBA, ACEC.

Balent, Ralph
Business: 8900 DeSoto Ave, Canoga Park, CA 91304
Position: VP & Genl Manager, Atomics Intl Div. *Employer:* Rockwell Intl Corp, Energy Sys Group. *Education:* BS/Mech Eng/Univ of Denver; MS/Melch Eng/Univ of CA. *Born:* 1/23/23. in Denver, CO. BS Mech Engg Univ of Denver 1948; MS Mech Engg Univ of CA 1949. Instructor Melch Engg Univ of Devner 1949-50. With Atomics Intl Div, Rockwell Intl Corp since 1950. Exec Dir Compact sys Div in 1960 - respon for successful flight test of space nuclear reactor sys. From 1969-78, VP Liquid Metal Fast Breeder Reactor Programs. Currently VP & Genl Manager, Atomics Intl Div. San Fernando Valley Engr of Yr Award 1964. Outstanding engr alumnus award Univ of Denver. Reg P E (Nu & M E CA). Amer Nuclear Soc, Amer Soc of Mech Engrs. *Society Aff:* ANS, ASME.

Balestrero, Gregory
Business: 25 Technology Park/Atlanta, Norcross, GA 30092
Position: Acting Exec Dir *Employer:* Institute of Industrial Engineers *Education:* Bachelor of Science/Industrial Engineering/Georgia Institute of Technology *Born:* 07/16/47 Acting Exec Dir of the Institute of Industrial Engineers, a membership assoc for the industrial engrg profession. He is responsible for maintaining the headquarters office and staff of 85. Greg also holds the position of Group Mgr - Publications and Technical Services where he is responsible for support of IIE's 21 divisions, technical information services, the Industrial Engrg and Management Press, and the *Industrial Engrg* magazine. He holds a BIE degree from Georgia Institute of Technology and has served on the IIE staff since 1981. *Society Aff:* CESSE, GSAE, ASAE

Balk, Kenneth
Business: 14500 South Outer Forty Rd., PO 1038, Chesterfield, MO 63017
Position: President. *Employer:* Kenneth Balk & Assocs Inc. *Education:* BS/Civil Eng/WA Univ. *Born:* Jan 16, 1934. Attended Washington Univ's Grad Sch of Bus & Inst of Cont Education. Estab KBA in 1959 as a struct engrg firm & has guided the firm's growth into a multi-disciplined arch/engrg firm. Reg P E, Struct Engr, holds the cert granted by the Natl Council of Engrg Examiners. Mbr Natl Assn of Home Bldrs, Prestressed Concrete Inst & Fellow of Amer Soc of Civil Engrs. Reg PE in CO, FL, IL, IA, KS, MO, NY, TX. *Society Aff:* ASCE, ACEC, NSPE, MSPE, NHBA, SAME, PCI.

Ball, Ben C, Jr
Business: P O Box 1166, Pittsburgh, PA 15230
Position: Vice President. *Employer:* Gulf Oil Corp. *Born:* June 1, 1928. BS & MS MIT 1948 & 1949 in Chem Engg; Grad Harvard Bus School's Advanced Mgmt Program 1976. With Gulf Oil Corp since 1949; V P & Hd of Planning Res Dept since 1975. Visiting Sr Lecr at MIT's Sloan Sch of Mgmt, assigned to MIT Energy Lab since 1976. Mbr AIChE. Reg P E Texas. Former Pres of Texas Chap of N Amer Soc for Corp Planning. Mbr Bd of Consultants, Inst of Cultural Affairs. Hobbies: ham radio (WA3VTX), duplicate bridge (Natl Master), woodcarving.

Ball, Charles E
Business: 5501 West I-40, Amarillo, TX 79106
Position: Executive Vice President. *Employer:* Texas Cattle Feeders Assn. *Education:* M.S./Iowa State University 1948; B.S./Texas A&M University 1947 *Born:* Feb 1924 Lamar County Texas. For 20 yrs Regional Ed for Farm Journal; Author of horse book 'Saddle Up' which became 'best seller' in 2 mos; over 125, 000 copies sold. Oct 1972- , Exec V Pres Texas Cattle Feeders Assn Amarillo Texas. Honors incl: Man of the Yr, Texas County Agri Agents Assn 1973; Outstanding Agriculturist Texas Tech Univ 1971; Outstanding Farm Magazine Writer, American Seed Trade Assn 1959. Enjoys: horse training, youth work, photography. *Society Aff:* ASAE Am. Soc. Agricultural Engineers

Ball, Harry R
Home: 6072 Horizon Dr, East Lansing, MI 48823
Position: Exec Dir *Employer:* MI Society of PE *Education:* BS/AE/Agric Equip/Univ of MO *Born:* 10/29/15 Currently PE Soc Exec. Native of Columbia, MO. Served to Lt Commander US Navy 1941-46. Test and development engr Intl Harvestor 1946-1956 - resident supervisor during construction of lab; specialty refrigerators. With Whirlpool Corp 1955-1976 established and directed functions of engrg services and div planning, Marion (Ohio) Div (1955-59), Laundry Group Planning (1959-64) and Corporate Planning (1964-1973) St Joseph, MI. Mbr bd Builders Resources Corp 1972-1974. Mbr bd Chicago Theological Seminary 1974-1976. Pres MI Soc of PE 1964-65 and MI Association of the Professions 1975-76. Reg PE Indiana (5066) and Ohio (21894). *Society Aff:* NSPE, ASAE, ASEE, AWWA, AAAS, SAME

Ball, Richard T
Business: 901 N. Pitt St, Alexandria, VA 22314
Position: Senior VP *Employer:* VVKR Incorporated *Education:* BSCE/Structural/Purdue Univ. *Born:* Mar 1922 Lafayette Ind. Tau Beta Pi, Chi Epsilon, Sigma Delta Chi, Scabbard & Blade. Served in field artillery USA 194346; ret from Army Reserve as Col 1974. Struct Engr 1946- . Reg Engr Ind 1949; currently reg: Ind, Va, Md, DC & W VA. N.Y. Minn, Texas, N.J. Senior Vice President VVKR Incorporated 1968- . Fellow & Life ASCE, P Chapt Pres NSPE, Reg VP NSPE; P Pres Kiwanis Club of Arlington VA, Lt Gov. 20th Div Dist Kiwanis, State Coordinator Mathcounts. *Society Aff:* NSPE, ASCE, AITC, SAVE, ACI

Ballantyne, Joseph M
Business: 312 Day Hall, Ithaca, NY 14853
Position: VP Res & Advanced Studies *Employer:* Cornell Univ *Education:* PhD/EE/MIT; SM/EE/MIT; BS/EE/Univ of UT; BS/Math/Univ of UT *Born:* 12/16/34 Dr Ballantyne was instrumental in establishing the Natl Research and Resource Facility for Submicron Structures at Cornell; served as Acting Dir during its first yr and sits on its prog committee and policy bd. During 1970- 71, he held an NSF Senior Fellowship at Stanford Univ and during 1978-79 was a Visiting Scientist at the IBM Watson Research Labs. His work has focused on electromagnetic and optical properties of semiconductors, including epitaxial growth of III-V compounds, injection laser, high speed photodetectors, and submicron lithography for integrated optics. He has presented over 120 research papers, and holds three patents. *Society Aff:* IEEE, TMS, AVS, HKN, TBΠ, ΦΚΦ, ΣΞ

Ballard, Douglas W
Home: 2204 Dietz Farm Rd NW, Albuquerque, NM 87107
Position: Conslt-Nondestructive Testing, Quality Assurance & Contamination Control *Employer:* Retired from Sandia National Labs, Albuquerque, 1983 *Education:* MS/Mech Engr/Univ of NM; BS/Mech Engr/Univ of NM. *Born:* 10/4/22. Alabama. Served USA Corps of Engrs with Manhattan Proj 1943-46 Oak Ridge & Los Alamos. With Sandia Labs 1950-1983, Supr of Nondestructive Testing Div, Quality Assurance, Product Evaluation & Mfg Res Div. BS & MS in Mech Engrg 1950, from UNM 1952. National Pres ASNT 1963, NMAB panels for NDT & Materials Shortages; Chmn of conservation in materials utilization study NCMP; represents ASNT in Federation of Materials Soc's. Fellow ASNT 1973, Mehl Honor Lectr 1974, Honorary Mbr ASNT 1985, Registered PE. Awards in SW watercolor competition *Society Aff:* ASNT, ASM.

Ballard, Richard L
Home: 9316 Coronado Terrace, Fairfax, VA 22031
Position: Chief, Aero Tech *Employer:* Dept of Army. *Education:* BS/Aero/VPIESU; MS/Aero/NYU. *Born:* 3/7/29. in Lindside, WVA. Aero Engg Grumman 1950-54. US Army BRL 1954-56. Staff Aero Engg Martin Co 1956-60. Aero Engg Army R &

Ballard, Richard L (Continued)
D Staff 1960-62. Staff Responbility for Aero Tech for Army 1962-71. Chief Aero Tech Army Staff 1974- present. *Society Aff:* AIAA, AHS, AAA, AUSA.

Ballard, Stanley S
Home: 4320 NW 17th Place, Gainesville, FL 32605
Position: Distinguished Service Prof of Physics, Emeritus *Employer:* Univ of FL *Education:* PhD/Physics/Univ of CA-Berkeley; MA/Physics/Univ of CA-Berkeley; AB/Physics/Pomona Coll *Born:* 10/01/08 Born in CA and educated there. Teacher of physics in several univs. On active Navy duty during World War II, in charge of research and development of optical instruments, Bureau of Ordnance. Chief of radiometry for Bikini atom bomb tests, 1946. Conslt in infrared instrumentation to several aerospace companies, including The RAND Corp. Collects, produces, and publishes data on new infrared-transmitting optical materials. Coauthor of two books and many scientific articles and technical reports. Pres, Intl Commission for Optics, 1956-59. Distinguished Service Award, OSA, 1977. Oersted Medal, AAPT, 1985. *Society Aff:* SPIE, OSA, APS, AAPT, IOP (London)

Ballhaus, William F
Business: 2500 Harbor Blvd, Fullerton, CA 92634
Position: Pres. *Employer:* Beckman Instruments, Inc. *Education:* PhD/Aerodynamics and mathematics/CA Inst of Tech; ME:AERO/Mech Engg Aero/Stanford Univ; AB/Gen Engg/Stanford Univ. *Born:* 8/15/18. Native of San Francisco, CA. Reg civil and mech engr. Douglas Aircraft: aerodynamicist, structures engr, design specialist, and proj engr, 1942-50. Developed the Aircraft Growth Factor. Convair: chief, preliminary design, 1950- 53. Northrop Corp: Norair Div: Asst Chief engr, chief, engr, VP and chief engr, 1953-57. Established Nortronics Div: VP and gen mgr, 1957-61. Exec VP and dir of Northrop Corp, 1961-65. Beckman Instruments, Inc: Pres and dir since 1965. Enjoy golf and training cutting horses. *Society Aff:* AIAA, NAE, ISA.

Balli, Macedonio, Jr
Business: 954 E Madison St, Brownsville, TX 78520
Position: Architects, Engineers, Planners. *Employer:* Balli & Associates Inc. *Education:* BSCE/Civil Engr/Univ of TX. *Born:* Apr 11, 1931 Pharr Texas. AA Texas Southmost Coll; Proj Engr Settles & Claunch; Asst City Engr City of Brownsville; taught Vocational Drafting Brownsville HS 1968; Ptnr Fernandez & Balli; Owner Balli & Assocs. Navy Airman 1951-53 Korean Conflict. Immediate P Pres Soc of Prof Engrs; Cons Engrs Council. Boys Club of Brownsville; Rotary Internatl. No hobbies. Enjoys assisting children's HS with campus dev. Southmost Coll Tech Board. *Society Aff:* NSPE.

Ballmer, Ray W
Business: P. O. Box 3299, 7000 S. Yosemite St, Englewood, CO 80155
Position: Exec VP *Employer:* Amoco Minerals Co *Education:* MS/Ind Mgmt/MIT; BS/Mining Engrg/NM Inst of Mining Tech *Born:* 05/06/26 Santa Rita, NM. Active duty US Navy 1943-46. Sloan Fellow 1959/60. 1949- 1969 Kennecott Copper Corp, Ray Mines, responsibilities included mine operation- smelter operations. Developed and installed truck haulage at Bingham Canyon. Mine superintendent at Bingham, later Acting General Superintendent of Mining & Concentrating at UT Div. 1969-1975 Bougainville Copper Ltd. Constructed and operated large mining complex in Papua New Guinea as Managing Dir. 1975-present Amoco minerals Co-currently Exec VP. Married, two children. Resident of Denver, CO. 1981 recipient of the Daniel Jackling Award. *Society Aff:* AIME, AIMM

Balloffet, Armando F
Business: 3955 East Exposition Ave, Denver, CO 80209
Position: Vice Pres *Employer:* URS Engineers *Education:* ME/Environ Engr/Univ of CO; BS/CE/Columbia Univ; BA/Math/Columbia Coll *Born:* 10/10/45 Born in Buenos Aires, Argentina. Served in Peru with Peace Corps 1969, 70. Systems Analyst for Tippetts-Abbett-McCarthy-Stratton, 1971-73 (TAMS). Joined URS in 1973 as Project Mgr, Environmental Projects, especially air, noise and water pollution modeling. Systems Analyst for TAMS, 1976-78, during assignment in Nairobi, Kenya on Kenya Natl Master Water Plan. Back to URS in 1978. Managed CO State Rail Plan, EPA Needs Survey, and Bureau of Reclamation Salinity Study of CO River among others. VP in charge of Planning Div, 1980. Winner Past Pres Award, Inst of Transportation Engrs, 1980. *Society Aff:* APCA, WPCF

Ballotti, Elmer F
Business: 222 S Riverside Plaza, Chicago, IL 60606
Position: Partner. *Employer:* Greeley and Hansen. *Education:* BSCE/Civil Engg/ Purdue Univ. *Born:* Nov 4, 1929 Calumet City Ill. BSCE Purdue Univ 1952. Test Engr United Aircratf Corp 1952-53; Principal Asst Engr Greeley and Hansen 1953-68, Ptnr 1968- , Bus Mgr 1972-75, Bus & Proj Dev Sponsor 1975- . Proj Mgr for following clients: City of N Y, Sioux Falls S D, Des Plaines Ill, St Petersburg Fla, Highland Park Ill, City of Chicago, N Y State, Philadelphia Pa, Orange County Calif; Oscar Meyer & Co, Abbott Labs. Mbr: ACEC, ASCE, AWWA, WPCF, AAEE; Chmn ASCE EED Awards Ctte 1976. Reg 9 states. *Society Aff:* ACEC, ASCE, AWWA, WPCF, AAEE.

Baloun, Calvin H
Business: Chem Engg, Room 180, Stocker Center, Athens, OH 45701
Position: Prof. *Employer:* OH Univ. *Education:* PhD/Met Engr/Univ of Cincinnati; MS/Chem Engr/Univ of Cincinnati; ChE/Chem Engr/Univ of Cincinnati. *Born:* 4/14/28. Corrosion consultant to industry since 1963. Teaching areas Corrosion, Thermodynamics, Solid State Reactions. Chmn ASTM Subcom G01.09 Corrosion in Natural Waters. *Society Aff:* AIChE, AIME, ASM, ASTM, NACE, ASEE.

Baltes, Robert T
Business: 5151 N. 16th St, D212, Phoenix, AZ 85016
Position: Pres *Employer:* Baltes & Assoc Ltd *Education:* BS/EE/Univ of WI *Born:* 5/10/38 Mr. Baltes is the KEY FACTOR in our company's growth, personally responsible for master planning and scheduling, economic studies, design development and contract negotiations. Bob has a broad background in the consulting field, and under his leadership the firm has built a solid reputation for excellence in electrical and mechanical design. He has received over 500 commissions, involving office bldgs, industrial, institutional, government and commercial projects. Bob holds registrations in AZ, CA, CO, FL, IL, IA, NV, NM, TX and WI. He was also the Consulting Engr with the High Rise Committee, writing the electrical portion of the Phoenix High Rise Code in 1973 and 1977. *Society Aff:* ACEC, NSPE, IEEE, IES

Balzhiser, James K
Business: 860 McKinley, Eugene, OR 97402
Position: Retired *Employer:* Balzhiser/Hubbard & Assoc *Education:* BS/ME/MT State Univ *Born:* 03/30/20. Native of Montana. Worked for Wright Aeronautical during World War II and from 1945 to 1954 was with BF Goodrich in Akron, OH. Started Balzhiser Engrg in 1954 and am currently retired from Balzhiser/Hubbard & Assocs, Consltg Mech and Electrical Engrs. Served 3 yrs on the Natl Bd of Dirs for CSI and have served continuously since 1960 on various Mech and Appeals Bds. Recently I have specialized in the Utilization of Geothermal Heat for heating and industrial processing, and the development of Alternate Energy Resources. Balzhiser has been a mbr of ASHRAE since 1955 and was elected to Fellowship status in 1984. Balzhiser has been a member of NSPE since 1956 and was recently (1987) awarded Life Membership status. *Society Aff:* NSPE, ASHRAE, CSI, PEO.

Bamford, Waldron L
Business: PO Box 628, 6800 Montrose Rd, Niagara Falls, Ontario L2E 6V5 Canada
Position: Pres. *Employer:* Can-Eng Manufacturing Ltd. *Education:* BSc/Mech Engg/ Univ of Detroit. *Born:* Aug 1932. Engrg Officer Royal Canadian Navy 1955-58. Sales & Proj Engr - industrial furnaces & alloy castings Walker Metal Prods Ltd 1958-59; Mgr Furnace & Fabrication Div Welmet Indust 1959-64; V Pres Can-Eng Mfg 1964- , respon for Engrg Mgr & Sales Industrial Furnaces, and Robots; Pres of

Bamford, Waldron L (Continued)
subsidiary commerical heat treating co, Can-Eng Metal Treating Ltd. Dir & Chmn Niagara Dist Health Unit. Dir of minor hockey. Trustee M.T.I. *Society Aff:* APEO, ASM, AMPI, MTI.

Ban, Thomas E
Home: 1156 Berwick Lane, South Euclid, OH 44121
Position: Pres *Employer:* Ban Prod Inc *Education:* MS/Met Engg/Univ of MN; BChE/Chem Engg/Univ of MN. *Born:* 3/13/25. Born and raised in Hibbing MN, served U.S. army WWII in Pacific Theatre. Graduate work on Taconite Fellowship Univ. of MN, pioneered in dev of iron ore pelletizing process. Cleveland-Cliffs Iron Co. in 1951 and headed R & D in pyrometallurgy. McDowell-Wellman Eng. Co. Cleveland Construct & Operate Dwight- Lloyd Research Laboratories, Dir. of Res. 1955, V. Pres. R & D McDowell-Wellman 1962. V. Pres. R & D Cleveland Res. Center Dravo Eng & Const., 1980. Sr. Exec. Eng & Mgr Res & Dev. Davy McKee Corp. 1981, Authored plus 45 Tech publications and awarded plus 40 patents in fields of Agglomeration, Metallurgy and Fuels Conversion. *Society Aff:* AIME.

Bance, Edlow S
Home: 450 Old Country Rd, Garden City, NY 11530
Position: Consultant. *Employer:* Retired *Education:* ME/Engrg/Stevens Inst of Tech. *Born:* July 2, 1912. Employed by Jersey Central Power & Light Co in Gas Operation 1934-43. Ordnance Officer USN Reserve, Hingham Shipyard, Boston Navy Yard 1943-46. Employed by Long Island Lighting Co, Gas Sys Dispatcher, Mgr of Gas Sys Oper, Asst to Mgr Gas Prod & Oper 1946-72. Mbr ASME (Chmn Long Island Sect 1960-61); Exec Ctte Metropolitan Sect 1956-58, Fellow 1971, Public Affairs Ctte 1974-77, Admissions Ctte 1976-87, NSPE NJSPE NAPE. Class Secy Stevens Alumni Assn 1951- . Founder Long Island Stevens Club. Article 'Use of Instrumentation in the Operation of a Gas Distribution System' ISA Power Conference 1958; contributor to article 'Engrg in Scandanavia' NJSPE Journal 1970. Instr Appalachian Gas Measurement Short Course 1959. "A Look at History of ASME" Centennial Meeting Stevens Inst of Tech Hoboken NJ Apil 1981, "One Homeowner's Experience With Gas Heat" Limelight Long Island Section Bulletin 1982. *Society Aff:* ASME, NSPE, SAR.

Bandel, John M
Home: 40 Whitehall Blvd, Garden City, NY 11530
Position: Director of Power. *Employer:* Union Carbide Corporation. *Born:* Aug 22, 1911 Baltimore Md. BE Johns Hopkins Univ 1932. Was Design Engr & Special Asst to Engrg V Pres Foster Wheeler Corp; held following positions with Union Carbide Corp Feb 1941- : Res Engr, Plant Mgr, Asst to Engrg V Pres-Metals Div, Engrg V Pres & Budget Dir Metals Div, Dir of Power. Fellow ASME; Mbr IEEE & AMA; Reg Prof Engr.

Bandler, John W
Business: Dept Electrical and Computer Engrg, Hamilton, Ontario Canada L8S 4L7
Position: Prof *Employer:* McMaster Univ. *Education:* DSc/Engrg/Univ of London; PhD/Microwaves/Univ of London; BSc/Elec Engrg/Univ of London *Born:* 11/09/41 Studied at Imperial Coll of Sci & Tech, London, England, from 1960-66. PhD in Microwaves, Univ of London 1967; DSc (Engrg), Univ of London, in Microwaves, Computer-aided design and Optimization of Circuits and Systems in 1976. Joined Mullard Res Labs, Redhill, Surrey, England in 1966. 1967-69, Postdoctorate Fellow & Sessional Lectr, Univ of Manitoba, Winnipeg. Joined the Dept of Elec Engrg, McMaster Univ, Hamilton, Canada, as Asst Prof in 1969. Became Prof in 1974, Chrmn of the Dept in 1978 & Dean of the Faculty of Engrg in 1979. Current position is Prof in the Dept of Elec and Computer Engrg. Directs res, Simulation Optimization Sys Res Lab at McMaster Univ. Pres of Optimization Sys Assocs Inc., Dundas, Ontario, Canada. Over 210 publs in elec circuits, computer-aided design and mathematical optimization and internationally recognized for advanced software tech transfer to indust in CAD/CAM. Fellow, IEEE & Fellow, IEE (Great Britain). Mbr of the Assn of PE of the Provence of Ontario (Canada). Fellow, Royal Society of Canada. *Society Aff:* IEEE, IEE, PEng, RSC

Banerjee, Sanjoy
Business: Dept of Chem and Nuclear Engrg, Santa Barbara, CA 93106
Position: Prof *Employer:* Univ of CA *Education:* Ph.D./ChE/Univ of Waterloo, Canada; B.E./ChE/Indian Inst of Tech, Kharagpur *Born:* 11/13/43 Dr. Banerjee joined the Chem and Nuclear Engrg Dept at UC, Santa Barbara as a prof in 1980. He is currently Chairman of the Department. He previously held the Westinghouse chair in the Dept of Engrg Physics, McMaster Univ, Canada from 1976-1980. Other appointments include: acting dir of Applied Science, Atomic Energy of Canada, Pinawa, Manitoba during 1975-76; and head of the Reactor Analysis Branch at Atomic Energy of Canada from 1972-75. Dr. Banerjee was a founding member of the Canadian Advisory Committee on Nuclear Safety, and represented Canada at numerous IEA and OECO meetings. He has also served on several program and national committees of the ANS. Dr. Banerjee obtained his Ph.D. in 1968. In 1982 he was elected Fellow of the Nuclear Soc and received the ASME Award for Best Heat Transfer Paper. In 1983 he was the recipient of the Melville Medal, ASME. This is the highest ASME research award for current work in mech engrg. In 1985 Dr. Banerjee was named the A.S. Neaman Distinguished Visiting Prof at the Israel Institute of Technology (TECHNION). *Society Aff:* ASME, AIChE, ANS, NAS.

Bang, Karl L
Business: TFK Grev Turegatan 12 A, Stockholm, Sweden 114 46
Position: Managing Dir *Employer:* Transp Res Comm — Swed Roy Acad of E *Education:* BS/CE/Royal Inst of Technology, Stockholm; MSc/CE/OH State Univ; PhD/CE/Lund Univ, Sweden. *Born:* Feb 1941. Traffic signal engr Stockholm Municipal Services until 1969; traffic planner VBB Cons Engrs Malmo Sweden until 1974; Permanent Secy Transport Res Comm 1975-82, Managing Director 83- . P Pres Award Inst of Traffic Engrs 1970. V Pres European Cooperation in field of Scientific & Tech Res (COST) proj 30 Electronic Traffic Aids for Major Roads 1970-78. Responsible for M.Sc. traffic engineering education at Institut Teknologi; Banduag, Indonesia 85-87 (82-program). *Society Aff:* SVR, TTF, ITE

Banicki, John
Business: 1333 Rochester Rd, PO Box 249, Troy, MI 48099
Position: Pres. *Employer:* Testing Engrs & Consultants, Inc. *Education:* BSCE/Civil/Wayne State Univ. *Born:* 6/20/22. Native of Detroit, MI. Served as Air Officer, USN Air Corps, Pacific Area WWII. Civ Engg with Spencer, White & Prentis, NY, NY specializing in fdns. Construction Mgr OW Burke Co, Detroit, & Herbert Realty Co, Chicago on bldgs and engg works. Civ & structural engg with Hazen & Sawyer, Consulting Engrs, NY, NY. Mtls & soils Engg with US Army Corps of Engrs, Detroit Dist. Founder in 1966 of Testing Engrs & Consultants, Inc geotechnical & mtls testing engrs, Detroit MI Soc of PE in Detroit; PPres Construction Specifications Inst, Detroit Chapter; Recipient Engg of the Yr Award, 1979 MI Soc of PE, Detroit; Elected Fellow, Engg Soc of Detroit, 1978. *Society Aff:* ACEC, NSPE, ASRM, ASNT, SAME.

Banker, Robert F
Business: 1500 Meadow Lake Pkwy, Kansas City, MO 64114
Position: Partner *Employer:* Black & Veatch *Education:* BS/CE/Univ of KS *Born:* 01/14/27 Native of Muskogee, OK. Served with Navy 1944-1946. Employed with Black & Veatch, Consltg Engrs of Kansas City, MO continuously since 1949. Became a partner in Black & Veatch in 1977. Currently serve as Project Mgr on public utility financial related projects including water and sewage rates, appraisals, feasibility studies, official statements related to bond sales, and cost manuals. Am Past Chrmn of the Water Rates Committee of the American Water Works Association. Am currently a mbr of Financial Mgmt Cttee and Econ Res Cttee of the AWWA. Enjoy classical music, cards, and fishing. *Society Aff:* AWWA, APWA, WPCF, NSPE, ASCE.

Bankoff, S George
Business: Chemical Engineering Dept, Evanston, IL 60201
Position: Walter P Murphy Professor *Employer:* Northwestern Univ. *Education:* Ph.D./Chem. Eng./Purdue Univ.; M.S./Mineral Dressing/Columbia Univ.; B.S./Mineral Dressing/Columbia Univ. *Born:* 10/07/21 DuPont 1942-48 (explosives, atomic bomb project, and plastics); Rose Polytechnic Inst, Terre Haute, Ind 1948-58, (Chairman, Chemical Engrg Dept); NSF science faculty fellow, Cal Tech 1958-59; Northwestern Univ., Evanston, IL 1959- (Walter P Murphy Professor, Chem Engrg & Mech Engrg). Pres, SGP Assoc, Evanston, IL 1984-. Co-inventor of resistivity probe for void fraction measurement in gas-liquid flow. Inventor of original process for production of stable polytetrafluorofthylene latex. Author of first radial distribution model for bubbly flow, 2nd highest cited paper in technology area (1983). About 200 papers in refereed journals, conference proceedings, 45 Ph.D. Students Distinguished Engrg Alumnus, Purdue; 1987 Max Jakob Memorial Award in Heat Transfer; Fellow ASME, AIChE; Chairman, Heat Transfer & Energy Conversion Div, AIChE. *Society Aff:* AIChE, ASME, ANS

Banks, Harvey O
Business: 710 South Broadway, Walnut Creek, CA 94596
Position: Pres, Water Resources Div. *Employer:* Camp Dresser & McKee Inc. *Education:* BSCE/CE/Syracuse Univ; MS/CE/Stanford Univ. *Born:* Mar 1910 Chaumont N Y. 22 yrs public service. Dir Water Resources, St of Calif 1955-60; initiated $2.8 billion State Water Proj now in oper; completed Calif Water Plan. Officer Corps of Engrs 1942-45. Private cons practice 1961- , serving principally Fed, State & local agencies in resource planning, inst & mgmt assignments; expert witness. Mbr NAE; Honorary Mbr ASCE, Honorary Mbr AWWA; Diplomate AAEE; Mbr WPCF; Mbr AWRA. Royce J Tipton & Julian Hinds Awards ASCE. Ikbo Iben Award AWRA. Arents Pioneer Medal Syracuse Univ. Reg Prof Engr Calif & other states. *Society Aff:* ASCE, AWWA, WPCF, AAEE, NAE, AWRA.

Banks, John V
Home: 6041 Camino de la Costa, La Jolla, CA 92037
Position: Vice Chairman. *Employer:* National Steel & Shipbuilding Co. *Education:* BS/Civil Engg/Univ of ID. *Born:* Jan 1917. Reg Prof Struct Engr: Ore. 1946-61 held various tech & managerial positions with Kaiser Frazer, Willys Motors, & Kaiser Steel Corp. 1962-76 successively Exec V Pres & Genl Mgr, Pres & Ch Exec Officer, & V Chmn of the Board Natl Steel & Shipbuilding Co. Mbr of Council & mbr of T & R Ship Prod Ctte, Soc of Naval Architects & Marine Engrs; Mbr Board of Mgrs Amer Bureau of Shipping, Mbr Board of Dir Shipbuilders Council of Amer. Gold Knight of Mgmt Award 1967, Indus Man of the Year, San Diego 1973, Vice Admiral Land Medal 1975. Hobbies: boating, agriculture. *Society Aff:* SNAME.

Bannister, Ronald L
Home: 1049 Edgewood Chase Dr, Glen Mills, PA 19342
Position: Manager *Employer:* Westinghouse Electric Corp. *Education:* Mast/Mech Engg/Villanova Univ; Bachelor/Mech Engg/GA Inst of Tech. *Born:* 2/13/33. Officer Army Signal Corps 19578-59. Dev Engr for Steam Turbine Div Westinghouse Elec Corp, specializing in noise control of high-speed machinery; in 1970 assumed Div respon for advance instrumentation design & acoustic tech; assumed respon as Proj Mgr to design, install & check out all new engg test facilities 1974; in 1979 assumed respon as program mgr for the conceptual design of a advanced pulvereized coal power plant. Appointed mgr., Project coordination and Advanced Planning for the Steam Turbine Development dept., in 1981. Has written 35 tech papers on vibration, testing, instrumentation & noise control, and turbine & power plant design. INCE Bd/Dir, ASA Tech Committee in Noise 1978-80, ASA Tech Committee Shocks & Vibration 1980-82. *Society Aff:* ASA, INCE, ASME.

Bansal, Ved P
Home: 8476 Thames St, Springfield, VA 22151
Position: Chief Mechanical Engineer. *Employer:* VVKR Partnership. *Born:* Sept 1938. BSME Howard Univ Wash DC. Ch Mech Engr VVKR 1968- , specializing in energy recovery heating & air conditioning designs for hospitals, office buildings & schools. Presently manage staff of 8 engrs & draftsmen. Designed one of 1st energy recovery sys 8 yrs ago for res lab in White Oak MD. Designed many bldgs with water to air heat pumps for energy conserv. Recent involved in solar energy studies NSF. Mbr NSPE, ASHRAE.

Bany, Herman
Home: 47W Stratford Ave, Lansdowne, PA 19050
Position: Consulting Electrical Engineer - ret. *Employer:* G E Co, Schenectady & Philadelphia. *Education:* BS/EE/IA State Univ. *Born:* May 1896 Tripoli Iowa. 2nd Lt Corps of Engrs AEF 1918-19. Student Engr GE 1919, Asst Head Large Motor & Generator Test. Pioneered dev of Automatic Switchgear in Switchgear Div for control & protection of power generation, conversion & utilization apparatus in elec power stations & substations. As Application Engr, Sect Head, Div Staff Asst on special assignments & Cons Engr 1953-61 (retirement) was granted 77 US (& also foreign) patents covering new dev in Switchgear. Tau Beta Pi. Awarded Prof Achievement Citation 1968 & Marston Gold Medal 1973 by ISU. *Society Aff:* IEEE.

Bara, Zygmund J
Home: 13929 Heather Glen Dr, Chesterfield, MO 63017
Position: Consultant *Education:* BS/EE/Univ of Rochester, NY. *Born:* Oct 1924 S Hadley MA. Grad work George Washington Univ. With USN 1943- 46. Communications Field Engr Philco Corp, assigned to var USAF locations 1949- 58 including Ch, Engrg Planning Office 1823d AACS Group (Globecom E & I) 1953- 58; Ch, Telecommunication Sys Engrg HQ GEEIA 1958-69; Dir, Communications Engrg Hq. Air Force Communications Service (AFCS) 1969-78; Chief, Communications Sys Engrg at 1842d Electronics Engrg Group (AFCC). Respon for sys & standards engrg for Air Force, HF, microwave, tropospheric scatter, TV, switching, COMSEC, data, teleprocessing. Sr Mbr IEEE. Reg Prof Engr MA & MO. Pres, K Z Electronics, Inc, 1980-present. Consultant SOTAS, Inc. 1986-present. *Society Aff:* IEEE.

Barany, James W
Business: Indus Engrg, West Lafayette, IN 47907
Position: Professor & Associate Head. *Employer:* Purdue University. *Education:* PhD/Ind Engr/Purdue; MSIE/Ind Engr/Purdue; BSME/Mech Engr/Notre Dame. *Born:* 8/24/30. South Bend Ind. Production Liaison Engr 1953 & 56 Bendix Aviation Corp. Army 3rd Engrg Bn 1954-55. Engrg faculty, mbr Purdue 1961- ; appointed Assoc Head of Sch of Indus Engrg 1970. Cons to Taiwan Productivity Ctr, Western Electric, Gleason Works, Amer Oil, Timken Res Ctr. IIE Dir of Student Conferences & Awards 1977- (Fellows Award 1982); ECPD Student Dev Ctte 1975-77; ABET ad hoc visitor for accreditations 1972-83; Editorial Bd for Human Factors Journal 1980-83; mbr MTM Res Assn, Order of Engrg, Sigma Xi, Tau Beta Pi Alpha Pi Mu. NSF and Easter Seal Found res awards for hemiplegic gait 1961-65. Author Work Measurement sect of Productions Handbook, Roland Press 1972. *Society Aff:* IIE, ASEE, HFS, MTM.

Barbe, David F
Business: Pentagon, Room 5E 683, Washington, DC 20350
Position: Dir, Submarine & ASW Prog *Employer:* Dept of Navy ASN (R, E & S) *Education:* PhD/EE/Johns Hopkins Univ; MSEE/EE/WV Univ; BSEE/EE/WV Univ. *Born:* 5/26/39. Native of Webster Springs, WV. Taught Electrical Engg at West VA Univ from 1962-64. Westinghouse Solid State Fellowship at Johns Hopkins University 1965- 69.Fellow engg at Westinghouse Adv Tech Lab from 1969-71. Section Head and Branch Head at the Naval Res Lab from 1971-79 specializing in electron devices. Has presented numerous invited papers and has organized and chaired numerous conference sessions in the USA and Europe. Has contributed several chapters and had edited two books. Fellow of the IEEE. Asst for Electronics and Physical Sci, Office of Asst. Sec. Navy for Res, Engrg and Sys from 1979-1983. *Society Aff:* IEEE, APS, ΣΞ.

Barber, Dean A
Home: Rt 1 - Box 381, Deer Park, WA 99006
Position: Principal Geologist *Employer:* Shannon & Wilson Inc. *Education:* BA/Geol/Fresno State College; MS/Geol/Univ of ID. *Born:* February 1, 1928 MS Geol Univ of ID. Was employed by Core Labs as petro reserve engr; has been engg geologist for S Calif Edison Co. Was private consulting geologist prior to joining Shannon E Wilson Inc, 1971. Is currently Principal Geologist Shannon & Wilson Inc Geotechnical Cons Spokane Wash. Special interests: engrg problems related to geologic hazards *Society Aff:* AEG.

Barber, John W
Business: 655 W. Thirteen Mile Rd, Madison Heights, MI 48071
Position: Chief Engr *Employer:* Spalding, DeDecker & Assoc *Education:* BS/CE/Detroit Inst of Tech *Born:* 12/24/36 Upon graduation, worked with City of Pontiac, MI as an engr responsible for urban renewal demolition and site improvement contracts. Became a systems analyst for the Oakland County DPW before moving to the City of Warren, MI as an engr responsible for construction. Promoted to Asst City Engr in charge of design office and field personnel. Joined a major Detroit-area underground contractor as engr/estimator. Joined Spalding, DeDecker & Assocs, Inc in 1969. Assumed current responsibility as VP and Chief Engr in 1972. Pres, ASCE Michigan Section 1986-87. Pres, MSPE Oakland Chapter. Received 1986 MSPE Outstanding Engineer Award in Private Practice. *Society Aff:* ASCE, MSPE

Barber, Mark R
Business: AT&T Bell Labs, 1247 South Cedar Crest Blvd, Allentown, PA 18103
Position: Supr Integrated Circuit Test Group. *Employer:* AT&T Bell Labs. *Education:* PhD/Elec Engrg/Univ of Cambridge England; BE/Elec Engrg/Univ of New Zealand; BSc/Math/Univ of New Zealand. *Born:* 7/23/31. While in England studied electron flow in high-power klystrons. Worked on underwater acoustic signal processing New Zealand Naval Res lab 1959-61; joined Bell Telephone Labs Inc NJ 1962 to work on solid-state microwave switches, amplifiers, mixers & Gunn oscillators; involved in bipolar semiconductor memory dev & dev of computer controlled apparatus for testing silicon integrated circuits 1968- . Assoc Ed Journal Solid State Circuits 1968-71; IEEE Solid State Circuits Council 1971-76, Chmn 1976-77; Sponsors Ctte Integrated Solid State Circuits Conf 1972-75; Fellow IEEE 1975; Program committee Semiconduct Test Conference 1978-81. *Society Aff:* IEEE.

Barber, Paul D
Home: 4404 W. 94th St, Prairie Village, KS 66207
Position: Chief, Engrg Div *Employer:* US Army Corps of Engineers *Education:* Juris Doctor/Law/Univ of MO at Kansas City; BS/CE/KS Univ *Born:* 10/04/34 Profl career is with the Kansas City District of US Army Corps of Engrs (24 yrs). Chief of the Engrg Div (300 employees) since March 1971. Directs the planning and design effort for water resources projects in the District and also the engrg and design effort in the District's Military mission. Past pres of the Kansas City Section of ASCE. Have been active in boy scouts, coached baseball and basketball, am active in church activities and am past pres of the Sch Bd for a 1,000 student private high sch. Reg PE and Licensed Attorney in the State of KS. *Society Aff:* ASCE, SAME

Barbera, Salvatore J
Business: 295 N Maple Ave, Basking Ridge, NJ 07920
Position: Assistant VP-Dir of Major Projects *Employer:* A T & T. BS Phys Long Island Univ. Started engrg career as Maintenance Engr with N Y Telephone 1953; spent next 7 yrs in N Y City as Equip Engr, Planning Engr & Budget Engr, reaching level of Div Engrg Supr Const Plans 1960; next 7 yrs spent in a num of plant assignments, reaching pos of Genl Plant Mgr Manhattan 1967. Named Asst V Pres Network Oper 1971; named to position Engrg Dir, Switching at A T & T 1973. In July 1977 assigned to General Manager Corporate Engineering at Western Electric Company. Appointed Assistant Vice President-Director of Major Projects January 1980 at AT & T.

Bar-Cohen, Avram N
Business: 8100 34th Avenue South, Minneapolis, MN 55440-4700
Position: Executive Consultant *Employer:* Control Data Corporation *Education:* PhD/Mech. Eng./Mass Inst of Technology; SM/SB/Mech. Eng./Mass Inst of Technology *Born:* 01/19/46 Native of Brooklyn, NY. Started professional career in 1968 with Raytheon Co. in Mass, performing and directing thermal R&D studies. Joined the Mechanical Engrg faculty at the Ben Gurion Univ, Israel in 1973; Senior Lecturer, tenured, 1975; Assoc Prof 1981. In 1984 assumed corporate responsibility for Packaging & Physical Modeling at Control Data Corp., while holding a visiting appointment in the Mechanical Engrg Dept. at the Univ of MN and while on extended leave from Ben Gurion Univ. Active member of ASME's Heat Transfer Div, past Chrmn of Cttee on Heat Transfer in Electronic Equipment, 1981-85, and CHMT Society of IEEE. Married Annette Pavony in NY on 09/11/66; three children. Played soccer at Lincoln High School in Brooklyn, NY (1960-1963) MIT (1963-1966) and for Boston Tigers of the American League (1966-1968). Authored many papers dealing with thermal phenomena in electronic equipment and co-authored text (with A.D. Kraus) "Thermal Analysis and Control of Electronic Equipment" in 1983. Member of Pi Tau Sigma, Tau Beta Pi, Sigma Xi and NY Academy of Sciences. Fellow of ASME and Senior Member of IEEE. *Society Aff:* ASME, IEEE

Bard, Gerald W
Home: 1215 Arbor Vitae Rd, Deerfield, IL 60015
Position: V Pres *Employer:* Drake Beam Morin *Education:* MBA/-/Iona Coll; BS/Bioehem Eng/Univ of WI *Born:* 10/7/32. 32 yrs progressive respon in Food Engrg Dev with Rexnord, Duncan Foods Co (Div of Coca Cola), Squibb-Beechnut; Genl Prod Mgr, Genl Mgr Frozen Food Div Libby McNeill & Libby. US Army Quartermaster Food & Container Div 1956-57. Prof Mbr: Inst of Food Techs, Chrmn Ctte on Div 1970-72; NSPE; AIChE (Food & Bio Div, Dir 1969-71, V Chrmn 1973); Alpha Chi Sigma; Phi Tau Sigma. PE Wisc. Boy Scouts of Amer, Deerfield Youth Council, Instr: Bus Admin, Coll of Lake County IL, Lake Forest Grad Sch of Mgmt. 1978-80 Dir Process Engrg Kraft Inc. 1981 Facilities Conslt-Abbott Labs. 1984- V Pres Drake Beam Morin. *Society Aff:* AIChE, IFT, ΑΧΣ, ΦΤΣ, WFS, MSPC.

Bard, Richard O
Business: Evansport Rd, Bryan, OH 43506
Position: Pres. *Employer:* Bard Mfg Co. *Education:* BIE/-/OH State Univ. *Born:* 3/27/33. Grad from OH State 1959. Joined the family business, Bard Mfg Co. Promoted to VP of Mfg in 1969 and became Pres of the co in Sept, 1976. Bard Mft manufactures a complete line of oil, gas and electric furnaces, electric central air conditioners and heat pumps for residential and light commercial bldgs. Bard products are sold throughout the United States and Canada and overseas.

Bardeen, John
Business: Dept of Physics, U of ILL, 1110 W. Green St, Urbana, IL 61801
Position: Prof of Elec Engrg & Physics, Emer. *Employer:* University of Illinois. *Education:* BS/EE/Univ of WI; MS/EE/Univ of WI; PhD/Math & Physics/Princeton Univ. *Born:* 5/23/08. Geophysicist, Gulf Oil Co, 1930-33. Jr Fellow, Soc of Fellows, Harvard Univ, 1935-38. Asst Prof Physics Univ of MN 1938-41; Physics Naval Ordnance Lab 1941-45; Res Physicist Bell Tel Lab 1945-51; prof of EE & Physics Univ of IL 1951-75 Emer 1975- . Main interest in solid state & low temp physics. Pres APS 1968-69, VPres 1967-68. Michelson-Morley Award 1968, Natl Medal of Sci 1965, Vincent Bendix Award 1964, Medal of Honor, IEEE 1971, Nobel Prize (Phys) 1956 & 1972. Pres Medal of Freedom, 1977. Founder's Award, NAE, 1984. *Society Aff:* APS, NAS, NAE, Am Phil Soc, AAAS, IEEE.

Barduhn, Allen J
Business: Hinds Hall, Syracuse, NY 13244
Position: Emeritus Prof. *Employer:* Retired *Education:* PhD/Chem Engrg/U of TX 1955; MS/Chem Engrg/Univ of WA 1942; BS/Chem Engrg/U of WA 1940. *Born:*

Barduhn, Allen J (Continued)
8/24/18. Sr Engr Tide Water Assoc Oil Co 1941-49. Prof Chem Engr Syracuse Univ, 1954 to 1986. Emeritus Prof. 1986 retired. Ind Consultant (Freezing & Gas Hydrates). Fellow AIChE, Reg Chem Engr, CA. VP Comm IX Intl Inst Refrig 1967-71. Editorial Bd "Desalination" 1968-85 and Syracuse Univ Press 1967-74. UNESCO Expert, Bucaramanga, Colombia 1969- 70. Consultant Natl Univ, Mexico 1972-74. Visiting Prof, Shiraz, Iran 1977-78. Mbr Tau Beta Pi 1940, Sigma Xi 1940. Res Interests: Desalting Sea Water by Freezing & Hydrating, Ice Growth Rates, Kinetics & Thermodynamics of Gas Hydrates, Waste Water Treatment by Freezing, Inventor of Eutectic Freezing Process for Waste Water Treatment (1961). Author of 53 papers & 20 US Govt reports. *Society Aff:* AIChE, ACS, ASEE, AAAS

Bardwell, AG
Home: 130 Mimosa Lane, Las Cruces, NM 88001
Position: Conslt *Employer:* Self *Education:* BS/Mech Engr/NM State Univ. *Born:* 7/11/21. USAFF 1942-46. With Stanley Cons 1946-83. Reg Prof Engr 13 states. Mbr: NSPE. Married, 2 children. Areas of specialization: proj mgmt, mktg. *Society Aff:* NSPE

Bare, Charles L
Business: 116 Park St, London, England W1Y 4NN
Position: Director and Gen'l Mgr. - Planning and Admin *Employer:* Conoco Inc. *Education:* BS/Pet Engr/Univ of OK *Born:* 4/22/32. Native Oklahoman. Early career with Magnolia Petr Co in various field locations in W TX and Gulf Coast. Joined the Conoco Inc Res Lab in 1961 working on economic modeling. Since 1967 worked in automating Conoco's US Producing Operations, moving into mgt in 1972 with responsibility for headquarters admin, computing, and automation. 1980-83 responsible for Production Operations in FarEast, Latin America and Africa. Currently responsible for partner opers, and staff depts for Conoco UK Limited in London. *Society Aff:* SPE, AIME

Barenberg, Ernest J
Business: Dept of Civ Engg, 208 N. Romine, Urbana, IL 61801
Position: Prof of Civ Engg. *Employer:* Univ of IL. *Education:* PhD/Civ Engg/Univ of IL; MS/Civ Engg/Univ of KS; BS/Civ Engg/KS State Univ. *Born:* 4/9/29. Native of Rawlins County KS. Worked construction prior to Univ education; Aircraft Design, Cessna Aircraft Co (Landing Gears) 1953. Served US Army Corps of Engrs 1953-55; Univ of KS 1955-60, Instr and Asst Prof; Univ of IL 1960- present; from Res Asst to Full Prof in 1971; Teaching and Res in Civ Engrg; held concurrent positions with Univ of IL and US Army Engrs CERL (Acting Branch Cheif) 1971-73. Assoc Hd, Dept of Civil Engrg Univ of IL 1981. Served as advisor and consultant to many govt agencies (Corps of Engrs, Air Force, FHWA, FAA) and industry. Cited by Engrg Council for Teaching Effectiveness (1968), received Everett Award for Teaching Excellence (1973). Author of over 80 technical publications on pavement design and on paving materials. *Society Aff:* ASCE, TRB, NSPE, ASTM.

Barer, Seymour
Business: 145 Ocean Ave, Lakewood, NJ 08701
Position: Pres. *Employer:* Barer Engg. *Education:* MS/Chem Engg/Univ of MN; BS/Chem Engg/Univ of MN. *Born:* 4/10/23. Native of Brooklyn, NY. Res Engr at Univ of MN. Served USN to CDR, expert Naval Air Intelligence & Engg, decorated WWII. Proj Mgr with Allen Porter Lee in Bolivia; Proj Mgr, Mgr Design & Stds, Cost Engg at Toms River Chem Corp. Pres of Barer Engg since 1966, designs & dev ind wastewater treatment plants, specialist in industrial process eng & product liability investigations. Developed/designed food drying plant, automatic hazardous chem system, dental cleansers, etc. Reg PE; reg Fallout Shelter Analyst; Chrmn, Lakewood Ind Commission (1960-64); Mbr- State of NJ Energy Commission; Licensed Indust Wastewater Treatment Operator (NJ); Licensed Bldg. Inspector; Licensed Subcode Official; Licensed Construction Official; Qualified Energy Analyst by NJDOE; Arbitrator, Amer. Arbitration Assoc.; Conciliator, NJ Dept Public Advocate. Qualified ASBESTOS handling and removal; expert hazardous and toxic waste disposal and decontamination; expert Forensic Engr and court testimony; Chem, Mech, Environ Engr. *Society Aff:* CEC, NSPE, WPCF, PEPP, AAEE, NAFE.

Barfield, Billy J
Business: Agri Engg Dept, Lexington, KY 40546
Position: Prof of Agri Engr. *Employer:* Univ of KY. *Education:* PhD/Agri Engr/TX A&M Univ; BS/Civ Engg/TX A&M Univ; BS/Agri Engr/TX A&M Univ; Prof Cert/Met/TX A&M. *Born:* 10/8/38. Native of Logansport, LA. Served in the USAF as a Meteorologist 1961-64. Grad Asst, TX A&M 1964-68; Prof of Agri Engg, Univ of KY, 1968-present. Teaching and res in the area of soil & water engg, with more than 100 technical papers and publications, including one monograph & 3 textbooks. Outstanding Young Researcher Award, ASAE, 1978. *Society Aff:* ASAE.

Barfield, Robert F
Home: 703 Shallow Creek Road, Tuscaloosa, AL 35406
Position: Dean of Engrg *Employer:* Univ of AL. *Education:* PhD/ME/GA Inst of Tech; MSME/ME/GA Inst of Tech; BME/ME/GA Inst of Tech. *Born:* 2/8/33. Native of Thomaston, GA; Preliminary Design Engr, AiRes Corp, Los Angeles, 1957-58; Asst Prof of ME, GA Inst of Tech, 1958-65; Visiting Prof, Kabul Univ, Afghanistan, 1963; Corp Mech Engg, Thomaston Mills Corp 1965-67; Prof of Mech Engg, the Univ of AL, 1967-present, Dean, Coll of Engrg, Univ of AL 1983- present; Visiting Prof and Dept Hd, Univ of Petroleum and Minerals, Saudi Arabia, 1971-73; Sr Advisor, Shiraz Technical Inst, Iran, 1975-77; Reg PE: AL, GA; Dir, Quadtech Corp; enjoy painting, camping, woodwork. USICA consultant to Univ of Jordan, Summer 1981; Yarmouk Univ, Summer 1986, and Birzeit Univ. 1984; Cons. to King Saud Univ, Saudi Arabia, 1983- ; Member of Board, of Directors of Inter. Assoc. for Exchange of Students for Tech Experience; Chrmn Inter. Div of ASEE. *Society Aff:* ASME, ASEE, ΣΞ, TBΠ, ΠΤΣ.

Barge, Daniel B, Jr
Business: 604 Hillwood Blvd, Nashville, TN 37205
Position: Chmn of Bd *Employer:* Barge, Waggoner, Sumner & Cannon. *Education:* BE/Civil Eng/Vandervilt Univ.; Diploma Butler Co High Sch *Born:* Jan 1922 Butler Cty Ala. USA Field Artillery European Theater 1944-46. 10 yrs-NC & StL Railway Nashville on design & const of major improvements to track, terminals, & yards; Principal Barge Waggoner, Sumner & Cannon 1955- , in genl chg of all cvl engrg projs for firm. Engr of Yr Nashville engrg orgs 1971. 1st Pres Vanderbilt Engrg Alumni Assn 1972; Dir Dist 6 ASCE 1967-69; V Pres Zone 2 ASCE 1975-76, President 1987. Active in num civic, religious & prof orgs. Recipient of Vanderbilt Engrg Sch Distinguished Alumnus Award 1981. *Society Aff:* ASCE, NSPE, SAME, ACEC, APWA, ARTBA

Bargellini, Pier L
Business: P.O. Box 256, Clarksburg, MD 20871
Position: Conslt *Employer:* Self-employed *Education:* Dr Engr/EE/Polytechnic of Turin 1937; MS/EE/Cornell U 1949. *Born:* 2/7/14. Consultant 1984-present. Sr Scientist Dir's Office COMSAT Labs 1968-1983. Prior to 1968 on fac of Moore Sch of Engrg, Univ of PA 1950-68. Chrmn of Ed Bd of "COMSAT Tech Review–; author over 60 pubs in US & abroad. Life Mbr and Fellow IEEE & Assoc Fellow AIAA, Emeritus AEI, Mbr HKN, Mbr Sigma Xi. *Society Aff:* IEEE, AIAA, AAAS, ARRL.

Barger, Charles S
Home: 11141 Irwin Ave S, Bloomington, MN 55437
Position: Principal-Sr Proj Mgr *Employer:* Rieke, Carroll, Muller Assocs *Education:* BS/CE/Univ of MN. *Born:* 02/23/28 Minneapolis, MN, served in US Army 1946-48, 50-51; 1954-59 Cvl Engr with Burns and McDonnell Engrg Co Kansas City, MO; 1959-66 Proj Engr Toltz, King, Duvall, Anderson, St Paul Minn; since 1966 principal and sr proj mgr Rieke, Carroll, Muller Assocs Minnetonka MN. 1974-79; 1986-present ACEC representative on Engr Joint Contract Document Committee; 1979-80 President Minneapolis-St Paul Chapter Construction Specification Inst; 76-81 Bd mbr Minneapolis-St Paul Chapter Construction Specification Inst; 1982-

Barger, Charles S (Continued)
83 Pres Consltg Engrs Council of MN; 1977- 84 Bd mbr Consltg Engrs Council of MN; 1979-present Construction Industry arbitrator for American Arbitration Assoc; 1987-1988 Conference Chair Midwest Engrs Conference & Ex position. *Society Aff:* NSPE, ASCE, AWWA, ACEC, CSI.

Bari, Robert A
Business: Bldg 130, Upton, NY 11973
Position: Assoc Dept Chrmn *Employer:* Brookhaven Natl Lab *Education:* PhD/Phys/Brandeis Univ; AB/Phys/Rutgers Univ *Born:* 09/03/43 Native of Brooklyn, NY. Held positions in solid state theory at MIT Lincoln Lab and Brookhaven Natl Lab, 1969-1973. Visiting Asst Prof of Phys at State Univ of NY, Stony Brook, 1973-1974. Since 1974 involved in nuclear reactor safety analysis at Brookhaven. Currently Assoc Chrmn of Dept of Nuclear Energy. Special interests include probabilistic risk assessment and severe accident analysis for light water reactors. Adj Prof of Nuclear Engrg at Polytech Inst of NY and Manhattan Coll. Numerous publs in nuclear reactor safety and solid state phys. Active in tech prog cttes of the Amer Nuclear Soc and mbr of Exec Ctte of Nuclear Reactor Safety Div, 1982-1985. *Society Aff:* ANS, APS

Barkan, Philip
Business: Dept of Mech Engineering, Stanford, CA 94305
Position: Prof *Employer:* Stanford Univ *Education:* PhD/ME/PA State Univ; MS/ME/Univ of MI; BS/ME/Tufts Univ *Born:* 03/29/25 Dr Philip Barkan is presently a Prof of Mech Engrg, Design Div, Stanford Univ. Prior to 1977, he served the Gen Elec Co, Switchgear Business Div in Philadelphia, PA as Mgr, Applied Physics and Mech Engrg and other technical capacities. In 1973, he was the first recipient of the Charles P. Steinmetz Medal awarded by the Gen Elec Co for technical excellence. He holds over 50 issued US Patents and has over 40 publications. He is a member of the Natl Acad of Engrg, and a Fellow of the IEEE. *Society Aff:* ASME, IEEE, ASEE, NAE, ΣΞ, SME.

Barkemeyer, O'Gene W
Home: 3401 Chaparral Dr, Temple, TX 76501
Position: Asst State Conservation Engr. *Employer:* USDA-SCS. *Education:* BS/Agri Engr/TX A&M Univ. *Born:* 1/6/38. Native of Rosebud, TX. Served US Air Force as Transportation Officer and mgmt Engr 1960-63. Served USAF Reserve for 15 yrs. Employed by USDA Soil Conservation Service, full time since 1963. Worked at various locations in TX in engr positions which included area, design, field specialsst, and soil mechanics. Assumed current respon as Asst State Conservation Engr for statewide engg functions in 1977. Outstanding performance award 1979. Pres, TX Section ASAE 1978-79. Interest in Agribusiness and leadership in local Lutheran Church. Reg PE in TX. *Society Aff:* ASAE, SCSA.

Barker, Dee H
Business: 270 CB, Provo, UT 84602
Position: Prof. Emeritus *Employer:* Brigham Young Univ. *Education:* PhD/ChE/Univ of UT; BS/ChE/Univ of UT. *Born:* 3/28/21. Born in Salt Lake City, UT. Educated in Salt Lake City sch system. Completed BS in Chem Engg in 1948, PhD in 1951. This was the first PhD engg degree offered by the Univ of UT. Worked with the E I duPont de Nemours & Co in Wilmington, at the Savanna River plant for a period of 8 yrs at which time I joined the Chem Engg faculty of Brigham Young Univ as an Assoc Prof. I've been hd of the dept for a total of 9 yrs in two different terms and have had extensive work in foreign countries particularly India. Have participated as chrmn of paper sessions in both socs to which I belong & am currently a mbr of the education accreditation committee of the AIChE as well as the hd of the Intl Relations Committee for the AIChE. In addition. I have published about 25-30 papers dealing with education, heat transfer, fluid flow, nuclear reactor safety and trace element analyis of human hair. I served as Fulbright Scholar Aug-Dec 1980 in Korea. Presently Prof Emeritus BYU. *Society Aff:* AIChE, ASEE.

Barker, Eugene M
Home: 6103 Warwood Rd, Lakewood, CA 90713
Position: Dir Quality Improvement *Employer:* Douglas Aircraft Co *Education:* BS/IE/Northwestern Univ *Born:* 06/13/38 Early career in quality assurance with Richardson Co, Rocketdyne, Inland Steel Prods and Allis-Chalmers. Held quality mgmt positions with Mattel, Audio Magnetics and Diamond Reo Trucks. VP Intl Oper Diamond Reo and VP Mktg and VP Engrg John Blue. Since 1985, Dir Total Quality Assurance, Douglas Aircraft. Fellow ASQC and IAE, past Sect and A&D Div. Chrmn ASQC. Instructor at West Coast Univ Coll. Authored over 25 papers and presentations. Enjoy video production, Middle Eastern dance and theatrical makeup. *Society Aff:* ASQC, SAE, IIE

Barker, Richard C
Business: 525 Becton Cntr, Yale University, New Haven, CT 06520
Position: Prof *Employer:* Yale University *Education:* PhD/EE/Yale Univ; M. Eng/EE/Yale Univ; B Eng/EE/Yale Univ *Born:* 3/27/26 Born in Bridgeport, Conn. He has been active in res on magnetic materials and devices, ferromagnetic resonance in metals, electron tunneling in metals and semiconductors, and digital data processing systems for spacecraft. He has received an Humboldt Sr Scientist Award from Germany, a Japan Soc for the Promotion of Sci Visiting Sci Fellowship, and has held visiting faculty appointments at several universities in Europe and Japan. He has been a consultant to several corporations and government labs. He is a Fellow of the IEEE and a mbr of the APS, AAAS, and Sigma Xi. *Society Aff:* IEEE, APS, AAAS, ΣΞ

Barksdale, Richard D
Business: School of Civil Engrg, Atlanta, GA 30332
Position: Prof of Civil Engineering. *Employer:* Georgia Institute of Technology. *Education:* PhD/Civil Engg/Purdue Univ; MSCE/Civil Engg/GA Inst of Technology; BCE/Civil Engg/GA Inst of Technology; AS/Civil Technology/Southern Tech Inst. *Born:* May 2, 1938. Taught Cvl Tech at Southern Tech Inst 1958-60; Asst, Assoc & Full Prof Ga Inst of Tech 1965- , specializing in teaching, res & cons in geotechnical engrg; Ch Cons & V Pres Soil Sys Inc 1972-80. Has conducted extensive res & cons in areas of materials & geotechnical engrg. Has over 40 publs & tech reports. P Pres Ga Sect ASCE 1975; P Chmn Transp Res Bd Ctte, P Chrmn ASCE Ctte, 2 TRB Cttes & 2 natl ASCE ctte. Reg P E 7 states. P Pres GA Tech, Chapter Phi Kappa Phi. *Society Aff:* ASCE.

Barlow, Harold M
Home: 13 Hookfield, Epsom, Surrey England
Position: Professor Emeritus. *Employer:* Univ of London, Univ Coll - ret. *Education:* PhD/Physics/Univ of London; BSc/Elec Engg/Univ of London; Hon DSc/-/Heriot Watt Univ, Edinburgh; Hon DEng/-/Univ of Sheffield. *Born:* Nov 15, 1899 England. Sub Lt RNVR 1916-18. E Surrey Ironworks Ltd 1923; Barlow & Young Ltd 1923-24; Acad Staff Univ College London 1924-67; Supt Radio Dept Royal Aircraft Estab, Farnborough 1939-45; Prof of Elec Engrg Univ Coll London 1945-67. Mbr BBC Sci Adv Ctte. Sr Ptnr Barlow, Leslie & Ptnrs London Cons Engrs. Natl Electronics Council; Council on Environ Sci & Engrg. Fellow Royal Soc London, IEE. Faraday Medal 1967; Dellinger Gold Medal Internatl Union of Radio Sci 1969; Harold Hartley Silver Medal Inst of Measurement & Control 1973; Fellow IEEE, Mervin J Kelly Award; Mbr Polish Academy of Sci 1968; Hon Fellow IERE 1971; Hon Mbr Inst of Electronics & Communications Engrs of Japan 1973; Chmn Brit Natl Com of Internatl Radio Union (URSI) 1967-73; FEng, U K, For Assoc US Natl Acad Eng 1979. *Society Aff:* RS, IEE, IMechE, IEEE.

Barlow, Jewel B
Business: Wind Tunnel Bldg 081, College Park, MD 20742
Position: Dir/Prof. *Employer:* Univ of MD. *Education:* PhD/Aerospace Engg/Univ of Toronto; MS/Aerospace Engg/Auburn Univ; BS/Engg Phys/Auburn Univ. *Born:* 11/9/40. and raised in MS. Worked at Lockheed GA Co 1959-61. Mbr of faculty at

Barlow, Jewel B (Continued)
Auburn Univ 1966-67. Doctoral dissertation on aerodynamics of propellers in turbulent flow. Joined faculty of aerospace engg dept at Univ of MD in 1970. Special interests and res in aerodynamics and flight mechanics of aircraft and helicopters. Dir of Glenn L Martin Wind Tunnel. *Society Aff:* AIAA, SAE, ISA, ACM, AHS, SPIE

Barnard, Richard H
Home: 5001 Frederick Dr, Charleston, WV 25312
Position: Prof of Chem Engg. *Employer:* W Virginia College of Grad Studies. *Education:* PhD/ChE/W VA Univ; MSChE/ChE/W VA Univ; BSChE/ChE/W VA Univ. *Born:* May 1934. BSChE, MSChE, PhD from W Virginia University. Native of Pennsboro, W Va. Res Chemist & Plant Supt for Amer Cyanamid Corp, 1961-67 at Multi-prod plant. Served as Dir of Engrg at the Kanawha Valley Grad Center of W Va Univ from 1967-72 and as Dean of Engrg & Science, W Va Coll of Grad Studies 1972-77. Interim Dean of Facul ty 1977-78. Major assign for past 12 yrs has been dev a new & innovative grad college for employed engrs & scientists pursuing part time grad study. Pres, Charleston Sect AIChE 1973-74. Enjoy golf & Antique collecting. *Society Aff:* ASEE, AIChE, AIR.

Barnes, Bruce F, Jr
Business: 4658 Gravois Ave, St. Louis, MO 63116
Position: Pres *Employer:* Barnes, Henry, Meisenheimer & Gende, Inc. *Education:* BS/ME/Wash Univ-St. Louis *Born:* 11/18/26 Evanston, IL. Served with U.S. Air Force 1944-45. Sales Mgr, Colt Industries-Fairbanks Morse Beloit Division, 1949-1968. "Assoc" - Warren & Van Praag, Inc., Consltg Engrs - Architects, 1969-1972. Pres & General Mgr of Barnes, Henry, Meisenheimer and Gende, Inc., 1972 to present. Registered PE in thirteen (13) sts. Active in Methodist Church, member of Administrative Bd, and in Boy Scouts, New Merit Badge Counselor. Mbr of American Legion. Enjoys camping and leisure sports. *Society Aff:* ASME, NSPP, MSPP, CEC.

Barnes, Casper W
Business: Sch of Engg, Irvine, CA 92717
Position: Prof of EE *Employer:* Univ of CA. *Education:* PhD/EE/Stanford Univ; MSE/EE/Univ of FL; BEE/EE/Univ of FL. *Born:* 11/24/27. From 1954 to 1956 Prof Barnes was a Sr Engg with the Sylvania Microwave Tube Lab, Mtn View, CA, & from 1956 to 1966 Prof Barnes was a Sr Res Engr with Stanford Res Inst, Menlo Park, CA. Since 1966, Prof Barnes has been on the faculty of the Sch of Engg at the Univ of CA, Irvine, where he is currently Prof and Chrmn of EE. *Society Aff:* IEEE, AAAS.

Barnes, Charles Ray
Business: 6640 Sunset Terr, Des Moines, IA 50311
Position: Charles R Barnes, Cons Struct Engr. *Employer:* Owner. *Education:* BS/Civil Engg/IA State Univ, Amer IA. *Born:* Aug 8, 1923. Native of Coeur d'Alene, Idaho. Ser in US Naval Const Battalions in WWII & in Reserve to date retiring as Cdr. Engr with Amer Bridge Co 1946-49, Ch Struct Engr with Wetherell & Harrison Archs 1949-73; formed cons firm 1973- . Ser as Dir in Iowa Sect ASCE; Tres, Iowa Engrg Soc; P Struct Engrgs of Central Iowa. Worked on Hoover Library, Fed Office Bldg, Des Moines; Terrace Hill Mansion, Marshall Cty Courthouse, and renovation of many bldgs significant to Iowa. Fellow ASCE. Director Iowa-MinnesotaChapter of American Concrete Institute. Taught 5 quarters as assist Visiting Professor at Iowa State University. *Society Aff:* ASCE, ACI, NSPE, CEC.

Barnes, Charlie H, Jr
Business: 1001 Chinaberry Blvd, Suite 350, Richmond, VA 23225
Position: Chief Transportation Engineer *Employer:* Harland Bartholomew & Assoc. *Education:* MS/CE/Tranp/WV Univ; BS/CE/VA Polytechnic Inst *Born:* 06/07/38 Principle Transportation Planning Engr for VA Dept of Hgwys and Transportation 1964-1968. Dir of Traffic & Planning for City of Petersburg, VA 1968-1971. Joined Wiley & Wilson in 1971, became Proj Mgr in 1973 and Assoc mbr of firm in 1974. Responsible for preparation of comprehensive plans, water quality mgmt plans, environmental impact statements, land use planning, traffic engrg studies, review of ordinances for various localities involving zoning and subdivision, and expert witness in annexation cases. Past Mbr of Lynchburg Bd of Zoning Appeals (1977-87). Enjoys camping, fishing, swimming and photography. Became Chief of Transportation Planning & Traffic Engineering for Harland Bartholomew & Associates in their Richmond Office in 1987. *Society Aff:* NSPE, ITE, APA, AICP

Barnes, Clair E
Home: 4525 Keever Ave, Long Beach, CA 90807 *Born:* 12/17/17. Reg PE in CA. Born in CO. Served in ETO in the US Corps of Engrs 1940-45. Employed at Douglas Aircraft Co 1952-1981. Coordinated between engg depts. Wrote several Mfg Manuals used in plant and in some colleges including: "Numerical Control Engg Manual–, "Machine Tool Design Stds–, and "Engg Design Considerations for Numerical Control Tooling–. Inventor and designer of numerically controlled mfg equip widely used in the aircraft industry. Also interested in Theology. Wrote "Geophysical Evidence of a Global Deluge" On March 31, 1981 retired from Douglas Aircraft Co., and is now doing occassional consulting on producibility of parts. Also lectures on Engineering , Scientific, and Religious subjects. *Society Aff:* NSPE, SME.

Barnes, Ellis O
Home: 8034 Hunters Grove Rd, Jacksonville, FL 32216
Position: Consultant *Education:* BChE/Chem Engg/Univ of Louisville. *Born:* Jan 1922. Native of Hopkinsville, Ky. Union Camp Corp 1943. 1944-47 Process Engr By-Prods, 1947-52 Group Leader, By Prods Dev, 1952-54 Superintendent By-Prods Dev, 1954-56 Proj Engr-Const, 1956-62 Mgr Chem Prods, 1962-67 Mgr Chem Dev, 1967-73 Dir Chem Dev & Engrg 1973-1980 Gen Mgr Titanium Enterprises, a joint venture of Union Camp Corp and Amer Cyanamid. 1980-86 Consultant, Chemical Group Union Camp Corp. 1981-to date Consultant in private practice. Co-author of 'The Naval Stores and Tall Oil Industries', 'Tall Oil' and 'Production Methods for the Manufacture of Crude Tall Oil'. Co-founder Coastal Georgia Sec AIChE, Elected Fellow of AIChE May, 1981. *Society Aff:* AIChE, ACS.

Barnes, Frank S
Business: Dept Elec & Computer Engrg-Campus Box 425, Boulder, CO 80309-0425
Position: Prof, Dept Elec & Comp Engrg *Employer:* University of Colorado. *Education:* PhD/EE/Stanford; MS/EE/Stanford; Eng/EE/Stanford; BS/EE/Princeton. *Born:* 7/31/32. Fulbright Prof, Baghdad, Iraq 1957-58. Res Engr Colorado Res Corp 1959. Physicist Natl Bureau of Standards 1959. Dept of EE, 1959- . Chmn 1964-1980. Res activ in the field of lasers, flash lamps and the applic of lasers to bio materials. Current activity in the area of microwave damage to biol materials & applic of ultrasound medicine. Edit Bd IEE Letters 1969-75. Editor of IEEE Student Journal 1968-70. Pres of the IEEE Group on Electron Devices 1974. V P for Pubs of IEEE 1974-75. Fellow IEEE 1970. Curtis McGraw Award from ASEE 1965; Stearns Award, University of Colorado, 1980. Fellow AAAS. Bd of Dir BEMS 1981-. Bd of Dir ABET 1975-1982. ANSI C95 Ctte on Microwave Safety Standards. *Society Aff:* IEEE, AAAS, ΣΞ, APS, HKN, ΤΒΠ, AHA, BEMS.

Barnes, Fred E
Business: Aviation Division, 575 Market, San Francisco, CA 94105
Position: Manager, Aviation Operations *Employer:* Chevron U.S.A. *Education:* MS/ME/U of MO; BS/ME/U of MO. *Born:* Clinton, Missouri. MSME & BSME from Univ of Missouri. Joined Shell Oil Co in 1965. Responsibility in Fuels & Lubricants Res including lab projects, field office responsibilities & Head office coordination. Joined Chevron Research Company in 1976, Chevron Chemical Company in 1977 and assumed present position with Chevron U.S.A. in 1981. Responsibilities include product development, product management staff and currently Manager of Aviation Operations in Aviation Division, Chevron U.S.A. Previously on Bd of Dirs, Soc Automotive Engrs. Previously Chmn, Sect BD Younger Mbrs Ctte, Dir Ad Hoc Younger Mbrs Ctte, Sect Bd Exec Ctte; Chairman, No. California Section

Barnes, Fred E (Continued)
and member of Board of Governors for N. California and St. Louis sections. Previously Regional Section Coordinator. Chrmn, Natl Mbrship Comm, 1985. Presently Assistant Secretary, Subcommittee J on Aviation Fuels, ASTM Ctte DO2. *Society Aff:* ASTM, SAE.

Barnes, George C
Home: 3245 Dorsett Ln, York, PA 17402
Position: Senior Research Engineer *Employer:* Alloy Rods Div, Allegheny-International, Inc.. *Education:* BS/Met Engg/VPI. *Born:* Sept 1931. Prior to joining Alloy Rods in 1971, held metallurgical positions with Fansteel, Inland Steel, TeledyneMcKay, & Allis-Chalmers. Assumed present position, in Filler Metal Product Development, in 1980.Lecturer, Evening Div, Purdue Univ Calumet Ctr, 1962-64. Lectured at a number of AWS Sect Mtgs since 1971. Bd of Dirs, York Symphony, 1970-74; Chmn, York-Central Pa Sect AWS 1971-72; Chmn, York Chap ASM 1976-77; Dist 3 Dir AWS 1976-84. Hobbies include classical music, chess, bridge, swimming, & sailing. *Society Aff:* ASM, AWS.

Barnes, George D
Home: 8617 Sycamore Trail, Germantown, TN 38138
Position: VP. *Employer:* Buchart-Horn Inc. *Education:* BS/CE/Univ of FL; Cert/ Traffic Eng/GA Inst of Tech. *Born:* Dec 1934, Linden, NJ. Adv degree work, Memphis State Univ. Employment: Wisc Dept of Transportation - design, resident engr; Boeing Airplane Co - Liaison Engr, BOMARC Missile system; Beiswenger, Hoch & Assocs - Transportation Engr; Reynold s, Smith & Hills - Reports/Airports Engr. With Buchart-Horn since 1968. Mgr Mid-South Div; VP 1974. Current responsibilities: direction of sales & production for firm in southern US; intl liaison. Nat'l Director ASCE 1977-80. Pres, Tenn Div ITE 1974-75 ; Chmn Urban Planning Dev Div, ASCE 1974-75. Reg Prof Engr 15 states. Hobbies: private flying, tennis, history, music. *Society Aff:* ASCE, NSPE, ITE.

Barnes, George W
Business: Box 634, Brunswick, ME 04011
Position: President *Employer:* Wright-Pierce. *Education:* BS/Civil Engg/Univ of ME. *Born:* 11/14/27. With Wright-Pierce, since 1950 in various postions. Genl Mgr since 1968. Pres since 1976. Reg prof Engr ME, NH, VT, MA, NY. Civic Mbrshis: Corporator Brunswick Savings Inst, Trustee Brunswick Reg Hosptial, VP ME Div Amer Trauma Soc. Prof & Bus Socs: Economic Resources Council of ME, Construction Specifications Inst (vaious offices ME Chap), Amer Soc of Civil Engrs, Cons Engrs of ME (Pres 1975-76), ACE Council (Chmn Cocmts Ctte 1975-76, Treas 1976- 78, Pres elect 1978-79, Pres 1979-80) Maine State Chamber Commerce (Chmn Energy Comm.1981). *Society Aff:* ACEC, NSPE, ASCE, CSI.

Barnes, Ralph M
Business: 5242 GSM Bldg, UCLA, Los Angeles, CA 90024
Position: Prof Engrg & Prod Mgmt Emeritus. *Employer:* Retired. *Education:* PhD/Industrial Engg/Cornell Univ; MS/Industrial Engg/Cornel. Univ; BSME/mech Engg/WVA Univ; ME/Mech Engg/WVA Univ. *Born:* Oct 1900. Univ of IA. Asst Prof to Prof of Indus Engrg. UCLA 1949-69, Prof Engrg and Prod Mgmt. Cons USA, England, Norway, Sweden, Spain, Yugoslavia, Uruguay, Peru, Costa Rica, Mexico, Japan. Author of books in field of Indus Engrg with translations in six languages. Rec SAM Gilbreth Medal 1941; Natl Indus Incentive Award 1951. Univ Missouri Award for Distinguished Ser in Engrg 1967. IIE Frank and Lillian Gilbreth Indus Engrg Award 1969. Fellow AAAS, Internatl Academy of Mgmt. Fellow and Life Mbr ASME, IIE, SAM. Honorary Mbr Acad of Indus Engr W VA Univ, 1981. *Society Aff:* ASME, IIE, SAM, AAAS.

Barnes, Robert C
Business: 1400 Sheridan Rd, N Chicago, IL 60064
Position: VP Corp Engg. *Employer:* Abbott Labs. *Education:* MBA/Bus Admin/Univ of Chicago Exec Prog; BS/Chem Engg/Northwestern Univ. *Born:* 3/12/27. Has been Corp VP for Engg at Abbott Labs since Sept 26, 1969. Grad with honors from Northwestern Univ in 1949 with BS degree in chem engg and in June of 1966 was awarded a Master's Degree in Business Admin from the Univ of Chicago's Exec Prog. Coming to Abbott from S C Johnson & Son in 1951, Mr Barnes held a number of managerial positions in production, maintenance, admin and engg leading to election as a VP in 1969. Mbr of Tau Beta Pi, an engg honorary society, and Beta Gamma Sigma, a business honorary society. Mbr of American Inst of Chem Engrs, Chicago Assoc of Commerce and Industry, Pharmaceutical Manufacturer's Assoc, Director, First Federal Savings & Loan Ass'n. of Waukegan, Dir of Murdoch & Coll Inc of Chicago,, Dir of Murdoch & Coll Inc of Chicago, and is a Reg PE in the State of IL. *Society Aff:* AIChE, PMA, ISPE.

Barnes, Stephenson B
Business: 2236 Beverly Blvd, Los Angeles, CA 90057
Position: Chairman of the Board. *Employer:* S B Barnes Assoc. *Education:* BS/Civil Engg/Purdue Univ; Hon Doctorate/-/Purdue Univ. *Born:* Jan 6, 1900, Omaha Neb. South Bend High School 1918. Los Angeles City 1921-24, Struct Designer 1924-31. Private Pract cons Engr 1933- . Reg Calif, Texas, Ariz Washington, Wisconsin. Served 8 yrs on Calif Bd of Reg. Awards: Engr of the Year, Los Angeles 1963, Howard Award ASCE 1971, AISI Special Citation Award 1972, Los Angeles Chamber of Commerce Achievement Award 1957. Engrg Soc: SEAOC, P Pres; EERI, P VP; ASCE,ACI SSA. Hobbies: Tennis, banjo. Now Retired. *Society Aff:* ASCE, CEAC, EERI, ACI, SSA, SAME.

Barnett, Donald O
Business: 312-C Cudworth Hall, Birmingham, AL 35294
Position: M.E. Prog Dir *Employer:* Univ of AL in Birmingham *Education:* PhD/ME(Thermosci)/Auburn Univ; MSE/ME/Univ of AL; BS/ME/Univ of KY *Born:* 09/09/36 Native of Cleveland, OH. Aeronautical research engr at NASA/ Lewis, 1958-60; Mech Engr, USAF (AFSC), 1960-63. Aerospace Technologist for NASA/Marshall from 1963-65 specializing in thermal stratification of cryogenic fluids. Lead engr of Crossed-Beam, Cross-correlation fluid dynamics proj and Mgr, Aerothermodynamics Branch for Northrop/Huntsville, 1966-73. Senior research engr and Head, Laser Velocimetry and Electro-Optical Instrumentation, ARO, Inc; Tullahoma, TN 1973- 78. Currently, Assoc Prof of Mech Engrg and Mech Engrg Prog Dir at the Univ of AL in Birmingham. Responsible for many innovations in LV utilization and system design concepts such as single-beam and universal LV's. *Society Aff:* ASME, ASEE, ΣΞ

Barnett, Irvin S
Business: P.O. Box 1679, Wichita, KS 67201
Position: Sr. V. Pres *Employer:* Martin K. Eby Constr Co Inc *Education:* BS/CE/KS St Univ *Born:* 7/28/23 Born in Morland, KS. Served in U.S. Navy 1944-47. Employed as Field Engr and Supt for Martin K. Eby Constr Co Inc 1948-1960. VPres for same company to present. Corporate Dir of Martin K. Eby Constr Co Inc since 1960. Pres, AGC of Kansas 1985. *Society Aff:* ASCE, NSPE, AGC.

Barnett, Samuel C
Home: 1938 Gotham Way, N.E, Atlanta, GA 30324
Position: Program Manager *Employer:* Institute of Nuclear Power Operations *Education:* Ph.D./Mech. Engrg./Georgia Institute of Technology; M.S./Mech. Engrg./ Georgia Institute of Technology; B.I.E./Indus. Engrg./Georgia Institute of Technology *Born:* 05/10/22 Born in Chatsworth, Georgia. Served with U.S. Army 1942-45. Served in various research, teaching, and administrative positions at Georgia Inst of Tech, including Prof of Mech Engrg, Asst Dir of School of Mech Engrg, and Assoc Dean of Undergraduate Div; awarded title of Prof Emeritus of Mech Engrg after retirement in 1980. Manager of Educational Assistance Prog at Inst of Nuclear Power Operations since 1980. Life Fellow, ASME. *Society Aff:* ASME

Barnett, Stanley M
Business: Chem Engg Dept, Univ of RI, Kingston, RI 02881
Position: Prof *Employer:* Univ of RI. *Education:* PhD/ChE/Univ of PA; MS/ChE/Lehigh Univ; BS/ChE/Columbia Univ; BA/-/Columbia College. *Born:* 4/8/36. Native of NYC. Experience with Gen Dynamics Corp, Esso Res and Engg Co and Shell Chem Co. Developed biochem and food engg prog at Univ of RI. Mbr, Adv Bd to RI Solid Waste Mgt. Corp; Dir, RI Lung Assn and Dir, RI Conservation Law Fdn. Chrmn, AIChE survey of world utilization of food wastes and agri residues. Pres, AIChE Mbrship Committee; Pres, AIChE Food, Pharmaceutical and Bioengg Div; Co- editor, two AIChE Symposium Series issues on alternate sources of food, fuel and chem; AIChE Fellow (1986) AIChE Mbrship Award (1975), Pres, RI AIChE; Biotechnology Progress editorial board; URI Engineering School Research Award (1986) and Honors Faculty Fellow (1987); IFT Dir Food Engrg Division & patent on multiphase reactor/fermenter. *Society Aff:* AIChE, ACS, NAMS, SIM, ASEE.

Barnett, W John
Business: PO Box 24, East Moline, IL 61244
Position: Chief Design Engr. *Employer:* McClure Engrg Assoc Inc. *Born:* Nov 9, 1944. BSCE from Finley Engrg College, Kansas City, Mo. Design Engr for Black & Veatch of Kansas City 1967-71; Design Engr & Const Project Mgr for Warran & Van Praag Decatur, Ill 1971-74; Office Mgr of Tiernan-McClure Engrs of Macomb, Ill 1974-76. Assumed current responsibility of Chief Designer for the East Moline Div of McClure Engrg Assocs May 1976. Also, responsible for Sanitary Engrg Projects for McClure Engrg Assoc. Mbr of ASCE, WPCF, ISPE, CSI.

Barney, Kline P, Jr
Business: 125 West Huntington Dr, Arcadia, CA 91006
Position: Pres *Employer:* Engg-Sci, Inc. *Education:* MPA/Public Adm/San Diego State Univ; BSCE/Civil Engg/Univ of Utah. *Born:* 12/16/34. Served three yrs as a combat engg officer in the US Marine Corps (Capt). Following three yrs as assistant engg of the Fallbrook Public Utility Dist in San Diego County, CA, he joined Engg-Sci, Inc, (ES) in 1963. Progressing from Project Mgr, Office Mgr, Reg Mgr and Chief Engg, he became President of ES in 1981 and is a Prin of the firm and serves as a mbr of the Bd of Dir of ES. Enjoys running, hiking, classical music. *Society Aff:* AAEE, TBP, CE, PES.

Barnhart, Edwin L
Business: 611 Ryan Plaza Dr, Arlington, TX 76011
Position: Pres. *Employer:* Hydroscience Inc. *Education:* -/Civil Engg/Manhattan College; -/Environ Engg/Manhattan College. *Born:* Nov 1936. BC & MSE from Manhattan College. After initial res work at Manhattan College, was instrumental in the formation of Hydroscience and now Pres. Principal area of interest is treatment of indus wastes, in particular the solution of complex problems dealing with biological systems & the effect of organics on the environ. Has served as advisor to a wide variety of indus clients concerning the achievement of water quality goals. Recent work has dealt with the recovery of chemicals from indus wastes. *Society Aff:* ASCE, AIChE, WPCF, AAEE.

Barnhart, James H
Business: PO Box 38209, Dallas, TX 75238
Position: VP. *Employer:* Ford, Bacon & Davis Texas. *Education:* BS/ChE/TX A&M. *Born:* 9/17/24. St Louis, MO. Educated in TX. 3 1/2 yrs US Army in WWII. Sixteen yrs in Petrol refining in Lab, Operations, Maintenance, Plant Engg, R&D, Plant Design, Engg Mgt. Eighteen yrs in engg contracting in process design, plant startup, engg mgt, corporate officer. *Society Aff:* AIChE, API.

Barnhill, A Virgil, Jr
Business: 351 NW 40th Ave, Ft Lauderdale, FL 33317
Position: Owner. *Employer:* Barnhill & Assocs. *Education:* BIE/Engg/Univ of FL. *Born:* May 3, 1927, Tampa, Fla. Served in USNR 1945-46. Graduated from Univ of Florida in 1951 with BIE degree. Engr with Florida Power & Light Co 1951-60. Field engr for Line Matl Indus 1960-67. Organized Barnhill & Assocs, Cons Engrs in 1968. Specialize in all phases of electrical engrg in the power distribution field. Mbr of FES, NSPE, FICE, IEEE, IES. Reg Engr in States of Florida, Georgia, Texas, Mississippi, Alabama, Tenn, N Carolina, S Carolina, & Va. Certified by NCEE. Hobbies, photography & genealogy. *Society Aff:* ACEC.

Barnhill, Walter O
Home: 2991 St Claire Rd, Winston-Salem, NC 27106
Position: Dept Chief-Indus Engrg. *Employer:* Western Electric Co Inc. *Education:* BS/IE/NC State Univ. *Born:* May 1934. BSIE from N C State Univ. Native of Wilson, N C. Served with the US Army in Europe 1954-56. With Western Elec Co since 1959. Has held supervisory pos in Mfg Operations, Tech Publications, Salary Admin & Indus Engrg. V P & Mbr of the Bd of Trustees, American Inst of Indus Engrs 1973-75. Recipient of a regional IIE Award of Excellence 1975. A past mbr of the Jr Chamber of Commerce & active church mbr. Enjoys amateur athletics. Was Awarded a Fellow Mbrship by The American Inst of Industrial Engrs in 1979. *Society Aff:* IIE, STC.

Barnwell, Joseph H
Home: Rt. 2, Box 2659, Ruston, LA 71270
Position: Professor Emeritus *Employer:* LA Tech Univ *Education:* M.S./Mech. Engr. /Texas A & M; B.S./Elec. Engr./Georgia Tech *Born:* 12/19/09 Native of Centerville, Tenn. Partner, machinery development shop, Barnwell, Shreveport, La. 1935-41. Supervisor, War Production Training then Assoc Prof. of Mech Engrg, Louisiana Tech Univ 1941-51. Leave from Tech to General Electric Co., Schenectady, N.Y. 1951-58. Mgr, Submarine Advanced Reactor and Service Engrg 1956-57. Mgr, Automatic Equipment Engrg 1957-58. Returned to Louisiana Tech 1958 as Prof of Mech Engrg. Developed feedback control systems and experimental stress analysis courses. Published and presented papers. Patents: Mechanical and Bioengineering areas. Professor Emeritus of Mech Engrg 1976. Pres, Joseph H. Barnwell and Assoc, Inc. (consulting firm). Hobbies: Telescope making and cabinet making. *Society Aff:* ASME

Barocio, Alberto J
Home: Av de las Fuentes 668, Mexico D F, Mexico 20
Position: Ch Exec. *Employer:* Ministry of Agri and Hidraulic Resources Mex. *Education:* CE Degree/-/Natl Autonomous Univ. *Born:* April 1922 in Puebla, Mexico. Professional Title 1944. Urban-Indus developments 1944-48. Co-founder Civil Engrs Associated (ICA) 1948-64. Genl Supt - Construction Mgr - Chief Engr - Hydraulic Plants, Dams, Power & Irrigation Projects. Const Mgr - Mexican Petroleum (PEMEX) 1965-1970. Fed Government's Ministry of Hydraulic Resources 1972-74. Directorship Admin, 1974-76 Directorship Financial, Tech Studies; actually, Mexico's Valley Water Commission, Ch Exec. Mbr & Past Pres 1944-76 Civil Engrg College, Civil Engrg College Federation, Cost Engrg Soc, Engrs & Architect's Soc, Mexico Sect Amer Soc of Civil Engrs. Lecturer Civil Engrg, Hydraulic Plants, Dams. Co-author book 'Constructions in Rock' 1967. Mexico, D F. *Society Aff:* CICM, SMIA, SMIEC, ASCE.

Baron, Melvin L
Business: 110 East 59th St, New York, NY 10022
Position: Partner & Dir of Res. *Employer:* Weidlinger Assocs, Cons Engrs. *Education:* BCE/Structural Engg/City Univ of NY; MS/Structural Engg/Columbia Univ; PhD/Applied Mechanics/Columbia Univ. *Born:* Feb 27, 1927. Married - Muriel, children: Jaclyn & Susan. Partner & Dir of Res for Weidlingere Assocs, Cons Engrs, NY City. Also Adjunct Prof of Civil Engrg at Columbia Univ. Special interest in applied mechanics, undersea structures, vibrations of structures, ground shock & earthquake analysis. Chmn Exec Ctte, Engrg Mechanics Div, ASCE 1968-69. Mbr, US Natl Ctte on Theoretical & Applied Mechanics (NRC) 1975- . Mbr, Mgmt Group C, Amer Soc of Civil Engrs, 1943-47. Spirit of St Louis Jr Award, 1958, ASME; J Jaes Croes Medal, 1963, ASCE; Walter L Huber Res Prize 1966, ASCE; Arthur M Wellington Prize 1969, ASCE. Nathan M Newmark Medal, 1977,

Baron, Melvin L (Continued)
ASCE. Also mbr, Natl Acad of Engg 1978. Chairman, Committee on Computational Mechanics (NAE-NRC)-1981-. Fellow, ASME, ASCE. *Society Aff:* ASME, ASCE, NYAS.

Baron, S
Business: Brookhaven Nat'l Lab, Upton, NY 11973
Position: Assoc Dir. *Employer:* Ass. Univ. Inc/Brookhaven Nat'l Lab. *Education:* PhD/Chem Engg/Columbia Univ; MS/Science/Johns Hopkins Univ; BS/Science/Johns Hopkins Univ. *Born:* 4/5/23. Presently, Assoc Dir of Brookhaven Natl Lab, responsible for Dept. of Nuclear Energy, Dept of Applied Science, and Tech Transfer. Previously Sr Corp VP-Engg and Tech Dir of Burns and Roe' responsible for advanced power tech such as fast breeder, fusion, MHD, fluidized bed, and the design of many commercial nuclear power projects. Mbr, National Academy of Engineering A Fellow of the AAAS, ANS and the ASME. Mbr of the US Delegation to Russia on the Fast Breeder Exchange. Mbr of the Argonne Natl Laboratory Review Cttees in fusion, nuclear and chem tech. Mbr of the NJ Commission on Radiation Protection. *Society Aff:* ASME, ANS, AAAS, NAE.

Barone, Richard V
Business: PO Box 100 Geo Wash Hwy, Lincoln, RI 02865
Position: Mgr Engg Applications. *Employer:* NIFE Incorp. *Education:* ScD/Metallurgical Engg/MA Inst of Technology; SM/Metallurigcal Engg/MA Inst of Technology; SB/Metallurgical Engg/MA Inst of Technology. *Born:* 1/4/37. Specializes in mgmt & matls oriented methods for profit & effectiveness of client cos. As a cons has worked on key projs incl fatigue problem on Lockheed Electra; matls analysis on ocean liner SS Unites States, TX Towers, Wrightsville Beach Desalination Facility for US Dept of Interior & nuclear aircraft carrier Enterprise for Adm Rickover's Group. Elected to Sigma Xi; Mbr MIT Educ Council & Alumni Council; ASM, AIME-TMS & ASM Sheet Metal Forming Cttes. Has held mbrships in ASTM, NACE, SNT, SESA, AOA, & AFS while active in those fields. ScD from MIT in Met engg is supplemented with a minor at Harvard Bus Sch. He most recently was a prof & Dept Hd of Indust Technology in the Coll of Engg at the Univ of Lowell. His current interest is to develop the Application Engg Function at NIFE, Incorp as a maj program to assit the sale and specification of NIFE, Incorp products in the USA. He has recently developed a seminar covering these areas. This seminar is available for presentation to selected audiences. Technical papers are scheduled for presentation and publication at the International Tele Communications Energy Conf and the offshore Tech Show on the battery use for solar and engine cranking applications. *Society Aff:* ASM, AIME.

Barr, Eugene A
Home: 418 Country Club Rd, Bridgewater, NJ 08807
Position: Retired (Div VP - Corp Dir) *Education:* PhD/Chemistry/Univ of PA; MS/Chemistry/Univ of PA; BS/CE/Villanova Univ. *Born:* 11/20/15. Spent 40 yrs with Union Carbide Corp in capacities ranging from a bench chemist and chemical engr working on the structure of phenolic resins to process work on reinforced laminates to dev work on new phenols, alcohols and related products. Have managed res and dev at all levels from group leader to Sr VP and have pioneered new products dev work on radically new polymers. Was in charge of the export business of chem-plas as Sr VP for 3 yrs. In my last position before retirement on Jan 1, 1981 was ambassador to the academic community in charge of Univ Rel Dept with responsibility for recruiting, aid to education, placement and relations with the leaders of the Academic Community. *Society Aff:* AIChE, ACS, ASEE.

Barr, Frank T
Home: 39951 Pierce Drive, Three Rivers, CA 93271
Position: Retired. *Employer:* Frank T Barr, Energy Cons. *Education:* PhD/ChE/Univ of IL; MS/Phys Chem/WA Univ; BS/ChE/WA Univ. *Born:* July 1910. Grew up in St Louis. Taught Ill Inst Tech 1934-36; with Exxon family 1936 to retirement in 1973; Exxon Res & Engr to 1969, then Exxon Nuclear; Sr Assoc 1952-73. Now Energy Consultant. Areas of activity include oil & gas process development, but for most of career was Exxon's man on other energy sources, & on unconventional energy conversion & utilization, with emphasis on analysis of future needs & supplies, development & promotion, at technico- economic-socio levels. Fellow AIChE. Avocations: music, camping, hiking, geography. *Society Aff:* AIChE, ACS, ANS.

Barr, Harold T
Home: 409 Cornell Ave, Baton Rouge, LA 70808
Position: Prof (Emeritus). *Employer:* Louisiana State Univ. *Education:* BS/Agri Engg/Univ of MO; MS/Agrif Engg/IA State. *Born:* Jan 1899, Fresno, Calif. 1924-29 Teaching Univ of Arkansas, 1929-36 Teaching and Res Louisiana State Univ, 1936-56 Head of A E Res, 1956-69 Head of Teaching & Res Retired. Author of 24 bulletins & 37 technical papers. Past secy, Vice Chmn, & Chmn of both Louisiana & S W Sect of ASAE. Assisted in writing Louisiana Regulations in 1953 governing the sale, storage, transportation & handling of Anhydrous Ammonia. Developed procedures & testing facilities for Anhydrous Ammonia pressure vessels, valves, regulators, & hose. Official for Louisiana, approved by ASME. Designed many city & rural homes, grain storage & drying structures, & misc bldgs. Honorary Fellow of ASAE. Elected to Alpha Zeta, Gamma Sigma Delta, & Phi Kappa Phi. Mbr of Louisiana Engrg Soc. Reg PE in Louisiana. Hobbies, gardening, flowers, bldg furniture. *Society Aff:* ASAE, AZ, ΣΔ, ΦΚΦ.

Barr, Harry F
Home: 25620 Meadowdale, Franklin, MI 48025
Position: Retired VP. *Employer:* GM Corp. *Education:* BME/Automotive/Univ of Detroit. *Born:* 8/22/04. Test engr Cadillac 1929-1930. Design engr 30-44. Staff engr 44-50 Asst chief engr 50-52. Asst chief engr Chevrolet 52-56. Chief Engr 56-63. VP GM Engg Staff 63-69. Honorary doctorate Univ of MO 1966. NAE 1965. Trustee Detroit Inst of Tech 1964-79. Governor Cranbrook Inst of Sci 1968-78. Pres SAE 1970. Detroit Athletic Club (D.A.C.). Bloomfield Hill C C - Detroit Athletic Club. *Society Aff:* SAE, NAE, ESD.

Barranger, Glynn D
Business: 1315 Franklin Rd SW, Roanoke, VA 24016
Position: Sr Partner. *Employer:* Hayes, Seay, Mattern & Mattern A&E. *Born:* July 1920 Roanoke, Va. BS in Civil Engrg from VPI 1943; Field engr in const 1946-48; Town Mgr, Marion, Va 1948-1951; with HSMM as Ch Engr, Assoc & partner since 1951; qualified as municipal expert in Va Courts; Reg Engr Va, Tenn, NC, Ky. Past Pres of Va Sect of ASCE. Currently V C of ASCE Natl Ctte on Local Sects & Dist Councils & on ACEC Natl Ad Hoc Ctte on Government Affairs. On Bds of various local civic orgs & local businesses. Elder Presby Church.

Barre, Henry J
Home: 1683 Arlingate Dr North, Columbus, OH 43220
Position: Prof Emeritus & Cons Engr. *Employer:* Ohio State Univ; Self. *Education:* PhD/Engg & Physics/IA State Univ; MS/Agri Engg/IA State Univ; BS/Agri Engg/KS State Univ. *Born:* April 1905. Native Kansan. Agri Engrg Staff Iowa State Univ 1930-43; in charge Grain Storage Res US Dept Agri 1938-43; Head Agri Engrg Purdue 1943-52; Private engrg cons 1953-62; Prof Agri Engrg Ohio State Univ 1963-73. Ford Foundation short-term consultant, Punjab Agri Univ India 1966, 67,70 & Internatl Rice Res Inst, Sri Lanka, 1974. Fellow ASAE; mbr ASEE Council 1947; Engr Rep, Agri Bd, Natl Res Council, 1956-59; mbr AAAS, Sigma Xi, Phi Kappa Phi, Tau Beta Phi, Natl Soc Prof Engr Reg (Ohio) 1956. Cyrus HallMcCormick Medal ASAE 1971; Golden Plate Award, Amer Acad Achievement 1972; Distin Engrg Service Award, Kansas State Univ 1974. Short term consultant World Bank Grain Storage India 1978, and Burma 1979. *Society Aff:* ASAE, ASEE, ΣΞ, ΤΒΠ, AAAA.

Barrentine, William M
Home: 1233 Hillcrest Dr, San Jose, CA 95120
Position: Mgr-Control Rm Engg. *Employer:* General Electric Co. *Education:*

Barrentine, William M (Continued)
MSEE/EE/Syracuse Univ; BSEE/EE/MS State Univ. *Born:* 5/30/27. In present pos as Mgr of Control Rm Engg, I am resp for the instrumentation and control equipment design; control rm configuration; standard design practices and in leading the efforts of the organ to achieve tech superiority and in meeting co tech and production goals effectively and efficiently. Prior to present position, was tech Mgr on a number of assignments performing and directing Engg activities assoc with: Electronic Desigh, Instrumentation Center Design; Data Retrieval/Handling; and Advanced Engg Dev activities. *Society Aff:* IEEE, PE.

Barrett, Bruce A
Business: Stanley Bldg, Muscatine, IA 52761
Position: VP & Head, Energy Systems Group *Employer:* Stanley Conslts, Inc *Education:* BS/EE/Univ of IA *Born:* 03/20/49 Presently, VP & Head, Energy Systems Planning and Design Group of Stanley Conslts, Inc. Has worked for SCI since 1971. Technical specialist in power sys communications and control as well as substation design. Principal assignments have included design mgr for a large rural electrification prog for Bolivia, 1975-78, and VP and Resident Mgr of SCI's West African subsidiary in Monrovia, Liberia, 1978-80. Chrmn, Student Chapter IEEE of Univ of IA in 1970; Mbr of 1984 delegation to the People's Rep of China for rural electrification; Chrmn, IEEE IA-IL Section 1987-88; Hometown is Muscatine, IA; interests include community activities, English sportscars and landscape gardening. *Society Aff:* IEEE, NSPE, TBП, HKN

Barrett, John H
Business: Prudential Ctr, Boston, MA 02199
Position: Group Vice Pres *Employer:* Chas. T. Main, Inc *Education:* BS/CE/Worcester Polytechnic Inst; Advanced Mgmt Prog/Harvard Business Sch *Born:* 04/22/25 Native of Hudson, MA. Served in Cvl Engr Corps, US Navy, 1945-1954. Proj Engr, Procter & Gamble Co 1954-1956. Joined Chas. T. Main, Inc on St Lawrence Hydro Proj, 1956. Served as Asst Engr Mgr, Asst to Pres, VP, Group VP, and mbr of Bd of Dirs. Former Pres Consltg Engr Council of New England. Natl Dir, ACEC. Vice Chrmn, Intl Engr Com. Delegate for US to Intl Federation of Consltg Engr 1981. *Society Aff:* USCOLD

Barrett, Michael H
Business: 7456 W 5th Ave, Denver, CO 80226
Position: Chairman *Employer:* Ketchum, Konkel, Barrett et al. *Education:* MBA/Bus/Univ of Denver; BS/CE/Univ of CO. *Born:* June 20, 1932 Dove Creek, Colo. Married Barbara Jane Kreutz; children- Robert, Mary, Bonnie, William. Ketchum & Konkel, Denver, 1955-63; Pres Ketchum, Konkel, Barrett, Nickel, Austin, Inc 1969- ; Dir Testing Cons Inc; Teacher U Colo 1963-64, Univ Denver, 1968-69, Dept Def Civil Def, 1962-68. Mbr Planning Comm, Westminster, Colo 1971-72; Chmn Bd Dir Denver Boys Inc; Pres Denver Area Council BSA 1974, Area VP 1976. Served with USNR, 1951-54. Recipient award Lincoln Arc Welding, 1966-68. Award Amer Inst Steel Const, 1969. 1st pl award Cons Engrs Council US, 1973. Reg Prof Engr Colo, Calif, Fla, Wis, NC, NY, Mich, NM, Utah, Alberta Canada. Mbr NSPE, ACI, Fellow ASCE; ACEC, SESA, Pres PEC 1970; Pres Elect CEC/C 1981; SEAC, Harvard Bus Sch Club, Amer Arbitration Soc, Denver C of C, Phi Kappa Tau, Chi Eps Denver Rotary Club.. *Society Aff:* NSPE, ASCE, ACEC, ACI.

Barrett, Richard E
Business: 505 King Ave, Columbus, OH 43201
Position: Proj Mgr. *Employer:* Battelle-Columbus Labs. *Education:* MS/Mech Engr/OH State Univ; BA Mech Engr/Mech Engr/OH State Univ. *Born:* 6/7/38. MS & B Mech Engrg from OH State Univ; native of Columbus OH. Joined Battelle in 1962, conducted & directed analytical & experimental studies in areas of coal-fired boilers combustion, air pollution emissions measurment & control, and space heating; author of 40 papers & 3 book chapters. Chrmn ASME Fuels Divs, 1978-79. Mbr of ASME Fuels Divs Exec Ctte, 1974-79. Various other offices in ASME & ASHRAE. ASME Pi Tau Sigma Gold Medal Award, 1970; ASHRAE Willis H Carrier Award, 1969; ASME Henry Award, 1967; Columbus Technical Council, Technical Person of the Year, 1979; Columbus Junior Chamber of Commerce, Outstanding Young Man, 1967. Currently Mbr: ASME Energy Conversion Group Operating Bd, ASME Res Committee on Ash deposits and Corrosion from combustion gases. Reg in OH E-032088. Mbr ASME Natl Nominating Ctte, 1982-1984 Chrmn Engrg Fdn Conference on Slagging and Fouling, 1984. *Society Aff:* ASME.

Barrett, Ronald E
Business: Stanley Bldg, Muscatine, IA 52761
Position: Exec VP *Employer:* Stanley Consultants Inc. *Education:* ScD/-/Rutgers Univ *Born:* 6/27/23. Dir, Stanley Cons Inc 1966, VP/Caribbean Regional Mgr with Stanley Cons, Inc until 1971. Hd of Intl Div until 1978. Hd Power Div until 1982 when assumed present position. Reg Prof Engr. Fellow ACES, Mbr NSPE, ASME, AMA & Iowa Engrg Soc Dir of Stanley Cons Nigeria Ltd, Dir of Stanley Cons Liberia Ltd. Author of numerous prof papers. Award: Knight Commander, Order of African Redemption, Liberia. Married Elaine N Davis. 4 children: Ronald Jr, Bruce, Bryan, Mary. US Navy veteran. Areas of specialization: feasibility, financial & planning studies, project mgmt, genl mgmt *Society Aff:* ASME, NSPE, AMA, ACEC.

Barrientos, Gonzalo C
Business: P. O. Box 349 COMIBOL, La Paz Bolivia
Position: Asst Mine Mgr (Sub Gerente). *Employer:* Corporacion Minera de Bolivia COMIBO. *Born:* March 26, 1942, Potosi, Bolivia. 1974 Reg Mining Engr, Bolivia. 1968 MS Sc Mich Tech; 1966 BSc Univ Tomas Frias, Bolivia. With COMIBOL, Bolivia: Asst Mine Mgr EMCOROCORO, Hd of The Dept of Mining Projs, Shaft Engr, respon for the design & sinking of a 2000' shaft; Rock Mech Engr. Mine Res Engr, White Pine Copper Co; Instr of Rock Mech, Mich Tech; Mine Supt Huari Huari Bolivia. Awards: 1975 Peele Award, SME of AIME. Socs: AIME, Assoc Mbr. Mich Tech Alumni Assn. Bolivian Engrg Natl Council. Federation of Engrs of COMIBOL.

Barron, Randall F
Business: Dept of Mech Engrg, College of Engineering, Ruston, LA 71272
Position: Prof & Dir, Div of Engg Res. *Employer:* Louisiana Tech Univ. *Education:* PhD/Mech Engr/OH State Univ (1964); MS/Mech Engr/OH State Univ (1961); BS/Mech Engr/LA Tech Univ (1958). *Born:* May 16, 1936. Native of Oak Grove, La. Taught in ME Dept, Ohio State Univ 1958-65. Cons for Herrick L. Johnston, Inc 1959-61 & CVI Corp 1962-65. With La Tech Univ since 1965. Author of *Cryogenic Systems*, McGraw-Hill 1966 Teach & conduct res in heat transfer, cryogenics, & acoustics. Promoted to Prof rank in 1970. Cons for Riley-Beaird, Inc since 1966. R.M. Teetor Award 1966; La Engr Soc Award of Merit 1967; Pi Tau Sigma Gold Metal Award 1968; La Tech Sigma Xi Outstanding Res Award 1971. Louisiana Tech Alumni Professorship Award - 1979, Engineering and Scientific Council Profl Engrg Achievement Award, 1981. Faculty Adv, La Tech Student Chap ASME 1966-69; Treas, Monroe Sub-Section ASME 1975-80. Mbr of Cryogenic Engineering Conference Bd. 1979-83. Apptd Dir of the Div of Engg Res at La Tech Univ in Sept 1977. Apptd Dir Engr Research and Graduate Studies at La Tech Univ in Sept. 1983. Consultant for Olinkraft, Inc, West Monroe, LA, 1977-1983. *Society Aff:* ASME, AAAS, ASEE, CSA

Barros, Sergio T
Business: Lgo. Arouche, 96, Sao Paulo, 01219 Brazil
Position: Pres *Employer:* Then De Barros Ltda. *Education:* Doctor in Engrg/Transp/Univ Sao Paulo; CE/Transp/Univ Sao Paulo *Born:* 06/03/25 Native of Sao Paulo, Brazil. Graduated in Cvl Engrg 1948 and Doctor in Engrg 1973 at Escola Politecnica of the Univ of Sao Paulo. Asst Prof in Transportation Engrg in same Univ. Pres of Then de Barros Ltda, Consltg Engrs in the area of planning and design of transportation facilties, and supervision of transportation projects. Has been doing work

Barros, Sergio T (Continued)
mostly for Government agencies and industry. Enjoy travel and yachting. Married, presently living in Sao Paulo, Brazil. *Society Aff:* ITE

Barrow, Bruce B
Home: 9840 Canal Rd, Gaithersburg, MD 20879
Position: Supr. Electronics Engr *Employer:* Natl Communications Sys *Education:* PhD/EE/Tech Univ at Delft; EE/EE/MIT; MS/EE/Carnegie-Mellon Univ; BS/EE/Carnegie-Mellon Univ. *Born:* 4/12/29. With GTE Labs & GTE Sylvania 1962-73, managing res & dev in telecommunications. Designed & executed Rake experiments to characterize tropospheric-scatter propagation channels. Exec VP, Genl Systems Dev Corp, 1973- 75. Genl Mgr, Weinschel Engrg Co, 1975-76; Sr Staff, Defense Communications Engg Center, 1976-1981, respon for communications sys planning; Sr Staff, Natl Communications Sys, since 1981, natl security emergency preparedness telecomm planning. Dir IEEE 1973; Fellow IEEE; Fellow AAAS; Tau Beta Pi Fellow; Fulbright Scholar. Recipient IEEE Centennial Medal, and IEEE 1987 Charles Proteus Steinmetz Award. *Society Aff:* IEEE, AAAS, ASTM.

Barrow, Robert B
Home: 116 Dublin Hill Rd, Higganum, CT 06441
Position: Retired *Education:* BS/Met Engg/MO Sch of Mines. *Born:* 1/11/26. Buffalo, NY. Served with US Air Corps 1943-1946. Met with Bausch & Lomb Optical, Ritter Co, Consolidated Vacuum Corp. Supervisor Lab Processing Pratt & Whitney Aircraft since 1961. Publications & Patents in Directional Solidification of Vacuum melted superalloys. Editor 1968 Vacuum Met Conf. Chrmn Rochester, NY Chapter ASM 1961 & New Haven, CT ASM 1971. Scouting at Troop & Dist Level for nine yrs. Enjoy swimming, golf, sailing & music. Married Barbara Pearson Nov. 21, 1979. Retired Pratt & Whitney Aircraft 1985. *Society Aff:* ASM, AVS.

Barry, B Austin
Home: 4415 Post Rd, Bronx, NY 10471
Position: Prof. *Employer:* Manhattan College. *Education:* MCE/Structures/NY Univ; BCE/Civil/Rensselaer Polytechnic Inst; BA//Catholic Univ of America. *Born:* 7/23/17. in Newburgh, NY, became a mbr of the Brothers of the Christian Schools (FSC) in 1933. Since 1943 has taught civil engg at Manhattan College, NYC, specializing in surveying, transportation, and economics; chaired the Civil Engg Dept for nine years. In ASCE held Metropolitan Sec offices (including pres), subsequently serving on national Bd of Direction. Served as pres of NE Region of American Society of Photogrammetry. In 1961-2 served as Pres of American Congress on Surveying and Mapping. Within American Society for Engg Education was pres of Mid-Atlantic Sec, later served two years as ASEE VP, Secs East. Received the first ASCE Surveying and Mapping Medal (1972) for contributions to the advancement of education and professionalism in surveying and mapping. *Society Aff:* ASCE, ACSM, ASEE, ASPRS, ΣΞ, XE.

Barry, John E
Business: 6105 Center Hill Rd, Cincinnati, OH 45224
Position: Sect Head - Pulp, Paper & Bevg Cost Engrg *Employer:* Procter & Gamble *Education:* BS/Eco/St. Joseph's Univ *Born:* 09/14/27 I am a Natl Dir - American Association of Cost Engrs - with over 20 yrs active mbrship. My current assignment (Engrg Div - Procter & Gamble) includes responsibilities for Cost Engrg for all Pulp, Paper and Beverage proj worldwide. This includes long range capital planning, feasibility and appropriation estimates, capital cost optimization and profitability analysis. I supervise twenty engrs in this capacity. From 1974 to 1978, I was a guest lecturer at the University of Cincinnati Engrg Coll. Our daughter is a Junior in Chemical Engrg at MI State Univ. Our son is a Freshman in Electrical Engrg at the Univ of Detroit. *Society Aff:* AACE

Barsom, George M
Business: P.O. Box ESE, Gainesville, FL 32602
Position: Pres & CEO. *Employer:* ESE, Inc. *Education:* DSc/Civ Engrg/WA Univ; MSCE/Civ Engrg/Univ So CA; BSCE Civ Engrg/Univ S CA *Born:* 9/18/40. Native of Los Angeles, CA. VP for RETA, Inc., St. Louis, MO 1969-1974. Currently Pres & CEO of ESE, Inc., a large consulting organization. Academic background in engg and chemistry and biology of natural systems. Career in regional resource mgt, quantifying environmental impacts of maj projects (airports, power plants, industries, surface mines) & minimizing them through the design disciplines. As CEO provides leadership and mgt to the organization. Serves as a Dir of ESE, Inc., Reynolds, Smith and Hills, FL Inst of Consulting Engrs; mbr, Advisory Council, Sch of Engg, Univ of FL. *Society Aff:* ASCE, NSPE, AWWA, WPCF, AAEE

Barstow, Robert J
Business: 401 Commonwealth Ave, Boston, MA 02215
Position: President. *Employer:* Barstow Engineering Inc. *Education:* AA/Mgt/Boston Univ. *Born:* 1/6/30. Educated at Franklin Inst 1950, Boston Univ 1964. 9 yrs exper with mfgs of heating & air conditioning equipment, specializing in automatic temperature control systems; 6 yrs with cons engrs in private practice, specializing in heating, ventilating & air cond, design & field representation. Started private practice in 1962, specializing in cons engrg services for architects for heating, ventilation & air conditioning sys. Expanded to incl sanitary engrg services in 1966. Moved to own bldg in 1970 at 401 Commonwealth Ave. Prof Organizations; American Cons Engrs Council of New England; Amer Soc of Heating, Refrg & Air Conditioning Engrs; Construction Specifications Inst; Association of Energy Engineers. *Society Aff:* ACEC/NE, ASHRAE, CSI, AEE.

Barteau, Mark A
Business: Dept of Chem Engrg, University of Delaware, Newark, DE 19716
Position: Assoc Prof *Employer:* Univ of DE *Education:* PhD/Chem Engrg/Stanford Univ, 1981; MS/Chem Engrg/Stanford Univ, 1977; BS/Chem Engrg/WA Univ, 1976 *Born:* 09/08/56 Native of St Louis, MO. Recipient of NSF Grad Fellowship. PhD dissertation recognized by the Victor K LaMer Award of the Div of Colloid and Surface Chem, ACS. Recipient of NSF Post-doctoral Fellowship, post-doctoral res at the Inst of Solid State Phys, Tech Univ, Munich, West Germany. Asst Prof of Chem Engrg and Assoc Dir, Center for Catalytic Sci and Tech, Univ of DE, since 1982. Recipient of NSF Presidential Young Investigator Award, 1985. Res specialities: Surface Sci and Catalysis. *Society Aff:* AIChE, ACS, ΣΞ, TBП, AAAS, MRS, CATALYSIS SOCIETY

Bartell, Howard F
Home: RD 6 - Box 324, Meadville, PA 16335
Position: Pres, Bartell Metallurgical Group, Inc *Education:* -/Met Engr/Lewis Inst; -/Met Engr/Univ of Cincinnati; -/Met Engr/Youngstown Univ; -/Accounting/Sinclair College. *Born:* 6/14/21. Formerly Dir of Met & Quality Control for Channellock, Inc, where I have been employed since 1952. Instr of met at Crawford County Area Vocational- Technical Sch. On Exec Committee of Northwestern PA Chapter of ASM. Past Chrmn & Secy-Treas of the same Chapter. Served on Natl Chapter Advisory Committee of ASM. Served with the Army Air Corps from 1942-1945 as met at Air Matl Command, Dayton, OH. Returned to same position under Civ Service. Worked on Standardization of Hardenability Test for SAE. Did met analysis of enemy mtl & failure analysis. Enjoy racing homing pigeons & playing the trombone. Pres of Bartell Metallurgical Group, Inc, a high-tech consortium for conslt and res and dev in the mech, chem, electrical, and metallurgical engrg fields. *Society Aff:* ASM.

Barthel, Harold O
Business: 101 Transportation Bldg, 104 S. Mathews Ave, Urbana, IL 61801
Position: Assoc Prof. *Employer:* Univ of IL. *Education:* PhD/ME/Univ of IL; MS/ME/Univ of IL; BS/ME/Univ of IL. *Born:* 9/17/25. on farm near Milledgeville, IL. Attended rural grade sch. Served with US Navy 1944-1946. Employed by current employer since 1954. Specialty expertise in unsteady waves in gas dynamics; current focus on second shock formation in spherical shock tubes. Have published several papers on structure of detonations. Served as Trustee for the Wesley Fdn at Univ of

Barthel, Harold O (Continued)
IL. Married Eileen Baumeister, we have two daughters. Enjoy reading, sports & woodworking. *Society Aff:* ΠΤΣ, ΤΒΠ, ΣΞ, ΦΚΦ

Barthold, Lionel O
Home: 1482 Erie Blvd, Schenectady, NY 12305
Position: Chairman. *Employer:* Power Technologies Inc. *Education:* BS/Physics/Northwestern Univ. *Born:* Mar 1926. Apparatus engrg assignments with GE until 1954, then various technical assignments in GE s System Engrg Group in Schenectady, Technical Dir of GE s Project EHV - a 500 kV & 765 kV res project in 1963. Later as Mgr of Transmission Engrg, organized 'Project UHV', a similar program for 1000 to 1500 kV. Founded Power Technologies, Inc', an analytical cons co in 1969, & served as Pesident 1969-1986, now Bd Chmn. Author of over 40 technical papers, chaired various cttes in IEEE, IEC, & Conf Internatle des Grandes Reseaux Electriques a haute tension, CIGRE. 1974 Admin Ctte of IEEE Power Engrg Soc & Chmn of its Fellows Ctte. President of that society 1981-1983. 1976 named internatl Chmn of a new CIGRE Ctte on the future of ele sys. Dir of the Oxygen Enrichment Co, Ltd & Schenectady Trust Co. Elected to the National Academy of Engineering, 1981. *Society Aff:* PES.

Bartholomew, Dale C
Business: PO Box 800, Richland, WA 99352
Position: Dir-Prod Operations. *Employer:* Rockwell Hanford Operations. *Education:* BS/Met Engg/Univ of WI. *Born:* 4/7/30. Eau Claire, WI native. Held various engg, manufacturing and management positions at Fansteel Metallurgical Corp (1954-59), General Electric Co (1959- 62), Westinghouse Electric Co (1962-68), and General Atomic (1968-76). Hired as Manager of Engg and subsequently became manager of operations for Atlantic Richfield Hanford Co in 1976. Became Dir of Production Operations at Rockwell Hanford Operations in 1977. Enjoy Chamber of Commerce work, golf and music of all kinds. *Society Aff:* ASM, ANS.

Bartholomew, Harland
Home: 19 Wydown Terrace, St Louis, MO 63105
Position: Consultant. *Employer:* Harland Bartholomew & Associates. *Born:* 1889. CE & ScD Rutgers College. Civil Engr & City Planner. Prepared comprehensive city plans, zoning ordinances, special plans & studies for numerous Amer cities. Former hd Harland Bartholomew & Assocs. Engr City Plan Comm, St Louis 1916-54. Chmn Natl Capital Planning Comm 1953-60, exProf Civic Design, Univ of Illinois. Hon Mbr ASCE, Amer Inst of Planners, & Amer Inst of Landscape Architects. *Society Aff:* ASCE.

Bartilucci, Nicholas J
Home: 9 Andover Dr, Syosset, NY 11791
Position: Partner. *Employer:* Dvirka & Bartilucci, Consulting Engrs. *Education:* MCE/Sanitary Engg/NYU; BCE/Sanitary Engg/Manhattan College. *Born:* 5/24/31. Served with US Army 1954-1956. Assignment involved application of sanitary engg techniques for the prevention of insect borne tropical diseases. Proj engr 1956-1964, partner 1964to present with consulting engg firms specializing in water supply, wastewater & solid waste projs. Commissioner, Jericho Water Dist, Nassau County, NY. Serving in this position as an elected official since 1968. Pres, NYWPCA 1977; Bd of Control, WPCF, 1971-1974; Diplomate AAEE; NYWPCA, Kenneth Allen Award 1966, Charles Agar Award 1978; WPCF, Bedel Award 1978; Nassau County Chapter NYSSPE, Community Service Award 1978; Dir, Nassau Ctr for the Developmentally Disabled. *Society Aff:* WPCF, NSPE, APWA, AAEE, AWWA

Bartlett, Donald L
Business: 1525 South Sixth St, Springfield, IL 62703
Position: Assoc Partner *Employer:* Hanson Engrs Inc *Education:* M.S./Structural Engrg/Univ of IL; B.S./CE/Univ of IL *Born:* 4/11/47 Assoc Partner in charge of Hanson Engrs structural engrg and computer depts and project mgr on special projects. Registered engr in six sts. Awards: Univ of IL Young Engr Award, 1981; the Structural Engrs Association of IL Award of Merit for innovative design, 1980; and the U of M.I.T. Rural Undergraduate Research Award, 1972. Project Mgr on the design, investigation or evaluation of many bldgs and bridges and on massive underpinning and protective construction structures in Atlanta, Philadelphia and Boston. Design project mgr on 50 million dollar bridges for Saudi Arabia. Published papers on concrete and engineering administrative use of computers. *Society Aff:* ASCE, AISC, NSPE, AWS, AIA

Bartlett, James V
Business: 21 Magnolia Crossing, Savannah, GA 31411
Position: Conslt *Employer:* Self *Education:* MCE/Civil/Rensselaer Poly Inst; BCE/Civil/Rensselaer Poly Inst; BSc/-/U.S. Naval Academy. *Born:* Oct 28, 1917 West Va. Served in US Navy 1942-72, retired as Rear Adm, Civil Engr Corps. Assignments incl duties as Cdr, Third Seabee Brigade in Vietnam & as Deputy Ch of Civil Engrs of the Navy. With Raymond Internatl Inc of Houston, Texas 1972 to 1982 (retired), as Sr VP-Engrg & V Chmn/Bd of Raymond Tech Facilities Inc, cons engrs. Registered Prof Engr. *Society Aff:* NSPE, ASCE, SAME, ΣΞ, ΤΒΠ, ΧΕ

Bartlett, Robert W
Business: PO Box 27007, Tucson, AZ 85726
Position: VP of Res. *Employer:* Anaconda Minerals Co. *Education:* PhD/Met/Univ of UT; BS/Met Engr/Univ of UT. *Born:* 01/08/33 PhD, BS from Univ of UT. Chem & Extractive Metallurgy, VP of Res, Anaconda Copper Co, Tucson; Materials Sci Dept Dir, Stanford Res Inst; Assoc Prof, Stanford Univ; Mgr Hydrometallurgy Res Dept Kennecott Copper Corp; Electrochem Soc Turner Award 1965; TMS Extractive Metallurgy Tech Award 1975; Dir of the Metallurgical Soc of AIME. *Society Aff:* TMS-AIME, SME-AIME.

Bartnikas, Ray
Business: 1800 Descente Ste-Julie, Varennes Quebec Canada J0L 2P0
Position: Maitre de Recherche *Employer:* Hydro-Quebec Inst of Res. *Education:* PhD/EE/McGill Univ; MEng/EE/McGill Univ; BASc/EE/Univ of Toronto. *Born:* 1/25/36. Maitre de Recherche, Hydro-Quebec Inst of Res Education; PhD/EE/McGill Univ; MEng/EE/McGill Univ; BASc/EE/Univ of Toronto. Cable Dev Labs, Northern Elec, Lachine, Que (1958-1963), res engr, Bell-Northern Labs, Ottawa, (1963-1968), mbr of scientific staff, Hydro-Quebec Inst of Res, Varennes, Que, (1968-pres) Dir, Mtls Res & Maitre de Recherche. Dr Bartnikas is a specialist in the field of corona discharges & dielectric losses. He is the author of numerous papers in his field of endeavor has contributed a number of chapters on the subject. Currently, he is the Editor of the ASTM Book Series on Engg Dielectrics; also he is the Editor of a book entitled Elements of Cable Engineering. Dr Bartnikas is a Fellow of the IEEE (1977) & from 1976-78 he was the pres of the IEEE-Elec Insulation Soc; in 1980, he was accorded the T.W. Dakin Distinguished Scientific Achievement Award; in 1984, he received the IEEE Centennial Medal. 1980-86 he was the chrmn of the ASTM Committee on Elec Insulation. He is a fellow of ASTM & in 1985 received the ASTM Award of Merit, the ASTM Dudley Medal, & the ASTM Arnold Scot Award. In 1986 he was presented with the CSA Award of Merit. He is also a mbr of the Canadian Stds Assn (CSA), the Canadian Electrical Assoc (CEA), the Intl Electrotech Commission (IEC) & the Ordre des Ingenieurs du Quebec. From 1970-76 he was the official Canadian Delegate on the Electrical Insulation Committee of CIGRE. (Con Internationale des Grands-Reseaux Electriques). Dr Bartnikas is an adj prof in the Elec Engrg Dept of the Univ of Waterloo, the Elec Engrg Dept. of McGill Univ and in the Engrg Physics Dept of Ecole Polytechnique (Univesite de Montreal) Society Aff: IEEE, ASTM. *Society Aff:* IEEE, ASTM.

Bartolo, Adolph M
Business: PO Box 9, Sugar Land, TX 77487-0009
Position: Exec VP & Chief Operating Office *Employer:* Imperial Sugar Co. *Education:* BS/Chem Engr/LSU. *Born:* Apr 1929. Native of Cairo, Egypt. Chemical Engr to Asst Supt Southdown Sugars, Houma, La 1951-58. With Imperial Sugar Co at Sugar Land, Texas, 1958 Asst Plant Mgr, 1965 Plant Mgr, 1968 VP Refinery Op-

Bartolo, Adolph M (Continued)
erations, 1970 elected to Bd/Dir. 1980 Exec. V.P. Mbr AIChE; past Bd/ Dir of Sugar Proc Res, Inc; past Pres & Bd Mbr of Sugar Industry Technologists, Inc; Technical Task Force Mbr of US Cane Sugar Refiners Assn and Sugar Assn. *Society Aff:* SIT, SPRI, NSDA, AIChE, TSAI, USCSRA.

Barton, Cornelius J
Business: 1875 Thomaston Ave, Waterbury, CT 06714
Position: Genl Mgr. *Employer:* Chase Nuclear-Div Chase Brass/Copper. *Born:* 1936. BS, MS, & PhD from Rensselaer Polytechnic Inst, Troy, NY. With US Steel Res Ctr until 1969, then became Dir of R&D for Chase Brass & Copper Co, Inc, then respon for mfg and engg for the Chase Nuclear, US, Div. In 1976 made Genl Mgr of Chase Nuclear, respon for all activities at the unit's two plants in Waterbury, CT and Arnprior, Ontario. VP and Mbr of the Bd of Dirs of Chase Nuclear (Canada) Ltd. The facilities produce zirconium alloy pressure tubes for use in heavy water nuclear power stations around the world, as well as a broad spectrum of titanium tubular and extruded shape products, primarily for use in the chemical industry. *Society Aff:* AIME, ASM, ASTM, ACA.

Barton, David K
Home: Prospect Hill Rd, R D No 2, Box 78, Harvard, MA 01451
Position: Conslt *Employer:* Self *Education:* AB/Physics/Harvard College. *Born:* Sep 1927, Greenwich, Connecticut. Army Signal Corps 1946-48. Radar engr, White Sands Proving Ground, 1949-53; Signal Corps Engrg Labs 1953-55; RCA 1955- 63 & Raytheon Co 1963- . Recipient of first RCA David Sarnoff Award for outstanding achievement in engrg 1958, for work on AN/FPS-16 precision instrumentation radar; IEEE M Barry Carlton Award, 1962. Author Radar System Analysis 1964; co-author Handbook of Radar Measurement 1969; editor, Artech House Radar Library, 1974- . Fellow IEEE 1971, for contributions to precision tracking radar & radar systems engrg. Chmn IEEE Radar Systems Ctte, 1972-75. Lecturer on radar Univ Pennsylvania 1960-61, GW Univ 1975-present. *Society Aff:* IEEE.

Barton, George H
Business: 415 Norway St, York, PA 17403
Position: Pres. *Employer:* Barton, Butcher & Assoc. *Education:* BEE/-/Cornell Univ. *Born:* 10/4/23. Native of York, PA. BEE 1950, Cornell Univ following service with Army Air Force. Design engr for an elec construction firm 1950-60. Responsible charge of electrical systems for industry, public bldgs, commercial projs & process plants, 1960-1968. Pres of consulting firm since 1968, mech & elec, energy studies, instrumentation & control. ; PastChrmn NSPE/PEPP. *Society Aff:* NSPE, IES, IEEE, SAME.

Barton, Hugh B
Business: 800 Bell, PO Box 2180, Rm 1977, Houston, TX 77001
Position: Regulatory Affairs Mgr. *Employer:* Exxon Co, USA. *Education:* BS/Petroleum Engg/LA State Univ. *Born:* 1/31/21. Native of LA and a petroleum Engg grad of LA State Univ. Joined Exxon Co, USA in 1941. Following approximately four yrs in the Air Force, served in engg and managerial assignments of increasing responsibility in TX, CA, & LA. In 1971, transferred to New Orleans as the Southeastern Production Div Environmental Conservation Mgr. Currently Regulatory Affairs Mgr in the Headquarters Production Dept in Houston. Was Pres of the Soc of Petroleum Engrs of AIME in 1978 and was VP of the American Inst of Mining, Metallurgical, and Petroleum Engrg, Inc in 1979. *Society Aff:* AIME, SPE, NOIA, MOGA, API.

Barton, Sherwin M
Business: 6526 Randolph Dr, Boise, ID 83709
Position: Consultant *Employer:* Self *Education:* BS/Mining Engr/Univ of ID. *Born:* in Moscow, ID. Univ. of Idaho B.S. Mining Engineering, Mining industry to 1942. Navy Photo Interpretation Officer to 1945. Retired USN LCDR. Cons practice to 1970; Pres BSMH Inc, Boise, Denver - transportation, photogrammetry, civil, mining. Merged 1970 Intl Engg Co, Inc, P VP. P Mbr Idaho Nuclear Energy Comm, P Ada County Planning Comm. P Mbr Univ of ID Engg Advisory Bd, P CHmn ID Bd Engg Examineers, ECPD Ad Hoc Visiting Ctte, P Intl Northwest Aviation Council, Treas ID Soc Prof Engrs, NSPE, AIME, ASCE, P DOT Natl Defense Executive Reserve. Disting Service Award NCEE. Retired 1979, serving as consultant to Intl Engr, Co, Inc, Boise, ID to 1980. *Society Aff:* AIME, ASCE, ACEC, NSPE.

Barton, Thomas H
Business: Faculty of Engg, Calgary Canada T2N 1N4
Position: Dean. *Employer:* Univ of Calgary. *Education:* DEng/EE/Univ of Sheffield; PhD/EE/Univ of Sheffield; BEng/EE/Univ of Sheffield. *Born:* 5/28/26. Dr Barton is Dean of the Faculty of Engg at the Univ of Calgary & is a Fellow of IEEE. His main profl interest is in rotating elec machines & their associated electronic equip & he has numerous publications in this area. He is a past chrmn of the Rotating Machinery Committee of the Power Engg Soc of IEEE & is a mbr of the Elec Drive Committee of the Industry Applications Soc of IEEE. *Society Aff:* IEEE, IEE, ΣΞ, ISA.

Barton, William L
Business: 1383 Airport Rd No., Naples, FL 33942-9990 *Employer:* Wilson, Miller, Barton, Soll & Peek Inc *Education:* B/CE/Auburn Univ *Born:* 9/4/39 Alton, IL 1939; Resident FL since 1950; Superintendent, Barton Construction Co. 1961-63; Engr, Smally-Welford-Nalvin, 1963-65; VP Harlan-Barton Engrg Labs 1966-69; VP & Principal Stockholder of Wilson, Miller, Barton, Soll & Peek, Inc. , 1969-present. Married: Patricia Mead, Children: Bret, Tracy, Berne. Positions held and awards: Pres Calusa Chapter FL Engrg Society 1972-73; Pres Naples Kawanis Club 1973-74; Young Engr of the Year - Calusa chapter 1974; Kawanian of the Year-Naples Kawanis Club 1975; Collier County Outstanding Citizen of the Year 1979; Fellow Member FL Engrg Society 1981; Currently FL Inst of Consulting Engrs-VP; Currently Pres National Federation of Parents for Drug Free Youth, office in Washington, DC. *Society Aff:* ASCE, ASTM, NSPE

Baruch, Jordan J
Business: 1200 18th St NW, Washington, DC 20036
Position: Conslt in Mgmt & Tech *Employer:* Jordan Baruch Assoc. *Education:* ScD/Instrumentation/MIT; MS/EE/MIT; BS/EE/MIT. *Born:* 8/21/23. Fellow Mbr: IEEE NY, Acoustical Soc of Amer Wash DC. Mbr NAE Wash DC. Celotex fellowship in Acoustics; Armstrong Cork fellowship in Acoustics; Eta Kappa Nu Award, Outstanding Young Elec Engr *Place of birth:* NYC. Lectr in Bus Admin Harvard Bus Sch; Prof of Engrg Thayer Sch and of Bus Admin Amos Tuck Sch of Bus Admin Dartmouth; Lectr EE MIT. Asst Secy of Commerce for Sci & Tech 1977- 81. Enjoys sailing. *Society Aff:* NAE, IEEE, ASA

Barus, Carl
Business: Swarthmore, PA 19081
Position: Prof of Engg. *Employer:* Swarthmore College. *Education:* MS/EE/MIT(1948); AB/Physics/Brown U (1941) *Born:* 9/21/19. S Orange, NJ. Radar maintenance officer, USNR, WWII. Asst engr with Raytheon Mfg Co, 1945. Staff of Res Lab of Electronics, MIT, 1946-1952. Asst, Assoc, Prof of Engg at Swarthmore College, 1952 to present. Visiting Prof at Ahmadu Bello Univ in Zaria, Nigeria, 1963-64. Returned to Nigeria as Visiting Prof at the Inst of Mgt and Tech, Enugu, for the yr 1976-77. In 1971 served as visiting staff mbr at the Natl Ctr for Energy Mgt and Power, Univ of PA. Married; three sons. *Society Aff:* IEEE, AAAS, ASEE, AAUP, FAS

Barzelay, Martin E
Home: 5070 N. Ocean Drive #6C, Singer Island, FL 33404
Position: Consltg Engr *Employer:* Self-employed *Education:* MS/Mech Engg/Harvard; BS/ME/Northeastern. *Born:* 01/15/18 During and following WWII Chief of structures and aerodynamics, Cluett Aircraft and Pratt, Read Co; design safety engr, Republic Aviation; aeronautical engr, FAA. Prof (now Emeritus) of Mech and

Barzelay, Martin E (Continued)
Aerospace Engrg, Syracuse Univ (31 yrs). Visiting Prof, Harvard, 1963-1964. Research for numerous government agencies and private companies. Published over 30 papers, research reports and multi-volume text "Scientific Automobile Accident Reconstruction." Editor of book on Product Safety Mgmt. Awarded two patents. NY PE License. Invited lectures, seminars and presentations at universities, government agencies and societies. Currently consltg engr specializing in Accident Analysis and Reconstruction. *Society Aff:* ASEE, AIAA, CEC, ΠΤΣ, AAUP, ΣΞ, ΤΒΠ

Basel, Louis
Home: 106 Carriage Dr, Stamford, CT 06902
Position: Pres-CEO. *Employer:* Crawford & Russell Inc. *Education:* MS/CE/MIT; BS/CE/MIT. *Born:* 6/28/26. Native of Stamford, CT. Served in US Navy 1944-46. Dev engr at Oak Ridge Natl Lab 1950-51. Process engr with United Engrs & Constructors 1951-59. With Crawford & Russell since 1959 in position of Chief Process Engr, VP-Sales, Exec VP & since 1975 Pres. *Society Aff:* AIChE, ACS.

Bashe, Robert
Business: 300 Robbins Lane, Syosset, NY 11791
Position: Marketing Manager. *Employer:* Fairchild Camera & Instrument Corp. *Education:* BS/Physics/Queens College; MS/EE/Columbia Univ. *Born:* Feb 20, 1933. BS Physics Queens Coll of NY 1954, MSEE Columbia University 1961. Marketing Mgr Imaging Systems Div of Fairchild Camera & Instrument Corp, Syosset, NY, respon for market dev for advanced systems in govt-related applications. Engaged in extensive liaison with mgmt & tech personnel at govt & prime contractor facilities. Prof societies: Soc of Photo- Optical Instrumentation Engrs (SPIE); Natl Governor 1971- , NY Regional Pres 1974-76 , recipient of Robt Morris Memorial Award. IEEE Mbr. Author of numerous technical papers. *Society Aff:* SPIE, IEEE, ADPA, AAA.

Bashkow, Theodore R
Business: Columbia University, Dept of Computer Science, New York, NY 10027
Position: Prof. *Employer:* Columbia Univ. *Education:* PhD/Elec Engg/Stanford Univ; MS/Elec Engg/Stanford Univ; BS/Mech Engg/Washington Univ St Louis, MO. *Born:* 11/16/21. in St Louis, MO. Served in Army Air Force, 1943-45. Mbr Tech Staff: RCA (Sarnoff) Lab 1950-52 (research on color TV deflection yokes), Bell Telephone Lab 1952-58 (research in circuit theory, transistor circuits). With Columbia Univ since 1958, revising, creating and teaching courses related to digital computer and microcomputer architecture and programming. Do research in computer architecture, data communications. Chmn of Dept of EE and Computer Sci 1968-71. Chmn, Tech Program Committee 1968 Spring Joint Computer Conference (SJCC). Chmn, Scientific Secretaries Committee, 1965 International Federation of Information Processing Congress. *Society Aff:* IEEE, ACM.

Basile, Norman K
Business: 1 World Trade Ctr-3047, New York, NY 10048
Position: Executive Vice President. *Employer:* John J McMullen Associates Inc. *Born:* Feb 1934, Utica, New York. BS in Marine Engrg from New York State Maritime College 1956. MSME from City University of New York 1962. Instructor in Marine Engrg at NY State Maritime College. 6 years as marine engr for design of commercial & naval ship propulsion sys. Joined John J McMullen Associates in 1965 as Supv of Marine Engrg. Became VP of Engrg Div in 1968 & was active in gas turbine propulsion, LNG sys design & advanced marine systems dev. Assumed current respon as Executive VP in 1974 with respn for establishing & implementing corporate mgmt, admin & fiscal policies. Active in ASNE Gas Turbine Cttes & various SNAME Technical Cttes. Mbr of ASNE, SNAME, IEEE & Amer Defense Preparedness Assn.

Baskin, Herbert B
Business: P.O. Box 7156, Berkeley, CA 94707
Position: Pres *Employer:* General Parameters Corp *Education:* MS/EE/Syracuse Univ; BS/EE/City Univ of NY. *Born:* 11/24/33. Joined the Univ of CA in 1969 as Prof of Elec Engg & Comp Sci. Was Prin Investigator of Comp Systems Res Lab, 1970-74. Founded Comp Lock Systems Corp in 1974 and was pres. In 1976, Comp Lock was acquired by Datapoint Corp, at which time, he became VP of Datapoint with responsibility for the Western Dev Ctr. In 1980 and 1981 left Univ of CA and Datapoint to start General Parameters Corp. *Society Aff:* IEEE, ACM, SID, ΣΞ.

Basore, Bennett L
Business: 103 Engg North, Stillwater, OK 74078
Position: Prof & Hd. *Employer:* OK State Univ. *Education:* ScD/EE/MIT; BS/EE/OK State Univ; BS/Math/OK State Univ. *Born:* 8/31/22. in Okla City, OK. Served on active duty in USNR during WWII on submarines. Employed by Lincoln Labs, 1950-52, Sandia Corp, 1952-57, Dikewood Corp 1957-63. With US ACDA as chief of office of Weapons Technology, 1963-67. Joined OK State Univ as Prof in 1967, served as Assoc Dir of Instruction for engg, and now hd of Sch of Gen Engg. Served as chrmn of Albuquerque-Los Alamos Section of IRE, 1957- 58. *Society Aff:* IEEE, ASEE, ΤΒΠ, ΦΚΦ.

Bass, Daniel M
Business: Petro Consultant, 17401 River Road, Channelview, TX 77530 *Employer:* Self *Education:* PhD/Pet Eng/TX A&M; MS/Pet Eng/TX A&M; BS/Pet Eng/LA State Univ. *Born:* 10/7/26. Native of Westlake, Louisiana. Served with US Navy 1943-46, Engr for Mobil Oil Co. Taught at TX A&M 1954-62; with CO Sch of Mines since 1962-79, as Hd of the Petroleum Engrg Dept. 1980- Kerr McGee Prof of Petroleum Engrg. Mbr SPE of AIME, Sigma Xi, Tau Beta Pi, Pi Epsilon Tau; author, co-author or ed of 3 books & 5 co engrg manuals. Retired May 1986 from Colorado School of Mines - a petroleum engineer consultant full time since July 1, 1986. *Society Aff:* SPE, ΣΞ, ΤΒΠ, ΠΕΤ.

Bass, Max S
Business: 1 Woodbridge Center, Woodbridge, NJ 07095
Position: President. *Employer:* M&T Chemical/Subsidiary Elf Aquitaine. *Education:* MBA/Finance/Fairleigh Dickinson Univ; BChE/-/City College of NY. *Born:* Jul 1928. BChE, CCNY; MBA Fairleigh Dickinson University. Prof Engr in several states incl New York & New Jersey; Plant Engr, Allied Chemical Corp, Claymont, Delaware 1950-55; Project Engr, Tenneco Chemical 1955-63. With M&T Chemical since 1963 in various capacities incl Proj Mgr, Dir of Mfg Services, Dir Operations Plating Div. VP of Chemical Div, and Sr VP Operations with total P&L respon. Mbr MCA, SPI, AIChE. Devoted to tennis. *Society Aff:* SCI, AMA, CMA, AIChE.

Bass, Michael
Business: Denney Research Bldg, Univ Park, Los Angeles, CA 90007-0271
Position: Prof E.E. *Employer:* Univ of Southern CA *Education:* Ph.D./Physics/Univ of MI; M.S./Physics/Univ of MI; B.S. /Physics/Carnegie- Mellon Univ *Born:* 10/24/39 Prof Michael Bass is a Prof of Electrical Engrg & Chmn of EE-Electrophysics at U.S.C. His research interests are in the optical properties of matter and laser materials processing. Prof Bass is a Fellow of the IEEE and the OSA. *Society Aff:* IEEE, OSA, AAAS.

Bassingthwaighte, James B
Business: Univ. of Washington WD-12, Seattle, WA 98195
Position: Prof Ctr for Bioengineering *Employer:* U of Washington School of Medicine. *Education:* MD/Medicine/U of Toronto; PhD/Physiology/U of MN. *Born:* 9/10/29. in Canada. Major res interest-cardiovascular bioengg & ion exchanges in the heart. University of Washington: Prof Bioengineering and Biomath since 1975. Director, Center for Bioeng. 1975-1980. Mayo Grad Sch of Medicine, Prof/Medicine 1975; Prof/Physiology 1973-75; Assoc prof/Bioengg 1972-75. Biomedical Engg Soc since 1968, Dir 1971-74, Pres 1976; Recipient Alza Award 1986; Microcirculatory Soc USA since 1973, Council 1975-78, 1979-82; Amer Heart Assn since 1965 Research Review Comm, 1979-82; Biophysical Soc since 1968, Assoc. Editor Journal 1980-83; Amer Physiological Soc since 1965, Circulation Group, Editorial

Bassingthwaighte, James B (Continued)
Bd 1972-76 and 1980-1983. NIH Career Dev Award 1964-74. Chrmn, Biotechnology Resources Advisory Committee, Div Research Resources, Natl Inst of Health 1977-79. Mbr USA Natl Committee of the Intl Union of Physical Scis (IUPS) 1979-86 & Chrmn 1983-86. Chairman, IUPS Commission on Bioengineering in Physiology 1986- . *Society Aff:* APS, BMES, SCS, SMB, AAAS, AHA, MCS

Batch, John M
Business: 505 King Ave, Columbus, OH 43201
Position: VP. *Employer:* Battelle Meml Inst. *Education:* PhD/ME/Purdue Univ; MS/ME/MT State Univ; BS/ME/MT State Univ. *Born:* 11/16/25. Dr Batch has been active in res, res mgt, & education since 1950. In various positions within Battelle he has managed all physical sci res & dev of long-range, interdisciplinary energy progs (Battelle-Northwest); responsible for a staff of 2,500 performing res & dev studies (Battelle-Columbus); & is now responsible for Mgt of dev contracts in specific areas of tech dev including the Office of Nucl Waste Isolation (ONWI) for the US Dept of Energy. He taught as an Instr in mech engg at SD State College, Purdue Univ; & later held an Adjunct Assoc Prof appointment at the Univ of WA State Univ & OR State Univ. *Society Aff:* ANS, ASME, AIF.

Batchelder, David G
Business: Agri Engrg Dept, Stillwater, OK 74074
Position: Prof. *Employer:* Oklahoma State University. *Education:* BS/Agr Engr/KS State Univ; MS/Agr Engr/OK State Univ. *Born:* Apr 30, 1920. Reg PE in OK. P Chmn of SW Region of ASAE 1975-76. Mbr of PM 45 Ctte of ASAE. Received Engrg Design Award, OSPE 1968 & 1972. Listed in Who's Who in South & Southwest & in American Men of Sci. Currently employed by OK State U (since Feb 1955) for res (75%) & teaching (25%) in the Agricultural Engrg Dept. Interim Dept Head of Agr Eng Jan 85-July 85 Mbr & Elder Presbyterian Church. Retired July 1, 1985. *Society Aff:* ASAE.

Batchelder, Howard R
Home: 12253 Sanigo Road W, San Diego, CA 92128
Position: Retired (Consulant Chem Engr) *Born:* Sept 1906 Mass. SM MIT 1930, SB MIT. Employed by Standard Oil Co Ind, United Gas Improvement Co, Penna Indus Chem Co, Inst Gas Tech, U S Bureau of Mines, Office Synthetic Fuels. Battelle Mem Inst - ret 1971. 18 pubs, 11 pats. m. Christina Dickson 1930; 2 children: Ann & Susan; 6 grandchildren. Fellow AIChE. Cons mainly in field of syn liquid & gaseous fuels.

Batchelor, John W
Home: 11709 Joan Dr, Pittsburgh, PA 15235
Position: Retired. *Education:* BSEE/Elec Engg/Purdue Univ; AB/Physics & Math/Butler Univ. *Born:* Nov 1914. Design engr on steam turbinegenerators for Westinghouse Electric 1938. Mgr, Turbine Generator Engrg 1956-74. Asst Mgr Large Generator Dept 1974-1978. Member National Academy of Engineering 1980. Fellow IEEE; Mbr ASME; Registered Engr in PA. Retired July, 1978. *Society Aff:* IEEE, ASME.

Batdorf, Samuel B
Home: 5536 B Via La Mesa, Laguna Hills, CA 92653
Position: Adjunct Prof *Employer:* UCLA *Education:* PhD/Physics/UC Berkeley; MA/Physcis/UC Berkeley; BA/Physics/UC Berkeley. *Born:* 3/31/14. Jung Hsien, China. Asst to Assoc Prof Physics, Univ of Nevada 1938-43. Aeronautical Res Scientist, NACA Langley Lab 1943-51. Mgr Dev, Westinghouse Elec Corp 1951-56. Asst Dir of Res, Electronics, Lockheed Missile & Space Co 1956-58. Inst for Defense Analyses 1958-59. Dir of Res in Physics & Electronics, Aeronutronic Div of Ford 1960-62. Principal Scientist, Aerospace Corp 1962-77. Adj Prof, Engg & Appl Sci, UCLA 1972-85. Author of about 120 papers on elastic stability, plasticity, fracture statistics, fracture mechanics, composites, etc. Fellow AIAA, Fellow APS, Fellow and Past Pres of AAM. Hon Mbr & P Chmn of Applied Mechanics Div ASME. *Society Aff:* AIAA, APS, AAM, ASME.

Bate, Geoffrey
Business: School of Engg, Santa Clara University, Santa Clara, CA 95053
Position: Assoc Dean of Engg and Prof of Elec Engg and Comp Sci *Employer:* Santa Clara Univ *Education:* PhD/Physics/Univ of Sheffield, Eng; BSc (Hons Phys)/Physics/Univ of Sheffield, Eng *Born:* 3/30/29. Educated in England. Four yrs as a Scientific Officer in the Royal Naval Scientific Service, working on infra-red photoconductors. Then, Univ of British Columbia, Vancouver, Canada, Asst Prof. Joined IBM in 1959 to work on magnetic recording and later became Mgr of Recording Phys, IBM Lab in Boulder, Co. 1978, joined Verbatim corp as VP for Advanced Tech. Elected Fellow of IEEE in 1978 and in 1980 was chosen a Distinguished Lecturer of the Magnetic Society. Awarded the Centennial Medal by IEEE in 1984. Assoc Ed of the Journal of Magnetism and Magnetic Materials. Now Assoc Dean of Engg & Professor of E.E. and L.S., Santa Clara Univ. *Society Aff:* APS, IEEE, RESA

Bateman, Barry L
Business: Anthony Hall, Carbondale, IL 62901
Position: Executive Dir for Computing Affairs. *Employer:* Southern Ill Univ-Carbondale. *Education:* PhD/Computer Sci/TX A&M; MS/Computer Sci/TX A&M; BA/Math/TX A&M. *Born:* Sept 15, 1943. BA, MS, PhD from Texas A&M Univ. Native of Pettus, Texas. Acting Dir of Computer Center & Head of Dept of Computer Science, Univ of Southwestern Louisiana 1969-72; Chmn of Computer Science, Texas Tech Univ 1972- 76; assumed current position Jun 1976. Currently on Pres's staff & respon for all computing activities for SIU Carbondale & the SIU Medical School. Mbr of Alpha Pi Mu, Upsilon Pi Epsilon (National Officer 1972-), Sigma Xi, ACM (Ctte Chmn 1974-). Published over 100 articles, 3 books & 2 books in progress, *Who's Who in America*, Fellow in TX Acad of Sci-1976, Natl Judge for Intl Sci & Engg Fair (ISEF) 1977-79.. *Society Aff:* APM, UPE, PME, ΣΞ, ACM.

Bateman, Leonard A
Home: 231 Brock St, Winnepeg, Manitoba R3N 0Y7 Canada
Position: Chairman & Ch Executive Officer. *Employer:* Bateman & Associates Ltd. *Education:* MSc/Math Analysis/Univ of Manitoba; BSc/Elec/Univ of Manitoba. *Born:* 1/14/19. Joined Manitoba Hydro 1956. Held Postions of Dir; Sys Operations, Sys Planning, Asst Ch Engr, Asst Genl Mgr & Ch Engr and Ch Engr & Genl Mgr. Apptd Chrmn & Chief Exec Officer in 1972. Formed Bateman Consulting Services in January 1979. Formed Bateman & Associates in 1980. Held directorships of Banff Sch of Advanced Mgmt, Canadian Elec Assoc, Mbrship on Canadian Natl Ctte World Energy Conf. Apptd Chmn CIGRE Study Ctte on DC links. Serve on exec Canadian Natl Ctte CIGRE. Elected Chairman of CNC/CIGRE 1980. P Pres APEM. Fellow Engg Inst of Canada. Sr Membr IEEE. Enjoy Photography, hunting & classical music. *Society Aff:* EIC, CEA, CNC/CIGRE, CANWEC, IEEE

Bates, Clayton W, Jr
Business: Dept of Materials Sci and Engg, Stanford, CA 94305
Position: Prof. *Employer:* Stanford Univ. *Education:* PhD/Physics/WA Univ; MEE/Elec Engg/Poly Inst of Brooklyn; BEE/Elec Engg/Manhattan College; MS/Applied Physics/Harvard Univ. *Born:* 9/5/32. in NYC and attended public schools there. industrial experience includes work at RCA, AVCO, Syvania and Varian Assocs. Varian Sabbatical Yr Award (1968). Spent as visiting prof at Imperial College of the Univ of London. At Stanford since 1972. Made full prof in 1977. 1978-1979 Sabbatical yr spent as visiting prof at Princeton Univ (EE & Comp Sci Dept) & Resident Visitor at Bell Labs, Murray Hill, NJ. Res Areas: Luminescent phosphors, structure of glasses, X-Ray Image Intensification, photoelectron spectroscopy of Photoelectronically active surfaces, optical and electrical transport properties of inhomogeneous composites. *Society Aff:* APS, OSA, IEEE, AAAS, AAUP, RPS.

Bates, Richard H T
Business: Electrical Engrg Dept, Christchurch 1, New Zealand
Position: Full Prof *Employer:* Univ of Canterbury *Education:* D.Sc./EE/Univ of London-Eng; B.Sc./EE/Univ of London-Eng *Born:* 7/8/29 Industry, Britain - Canada - USA, 1952-66. Univ of Canterbury, New Zealand, since 1967. Personal Chair in Electrical Engrg, 1975. New Zealand official member of U.R.S.I., I.A.U. & I.C.O. Elected Fellow of Royal Soc of New Zealand (RSNZ - 1976) and of IEEE (1980). Pres Australasian Coll of Physical Scientists in Medicine (ACPSM - 1981/3). Cooper Mem prize (RSNZ - 1980). Mechaelis Mem Prize (Univ of Otago - 1980). Over 200 publications on microwaves, antennas, diffraction theory, optics, image processing, biomedical engrg, astronomy, molecular structure, computers in music. *Society Aff:* RSNZ, IEEE, IEE, ACPSM

Bates, Robert L, Jr
Business: PO Box 17216, Jacksonville, FL 32216
Position: Exec VP. *Employer:* Hubbard, Bates, Associates, Inc. *Education:* MS/Sanitary Engg/GA Tech; Bach/Civil Engg/GA Tech. *Born:* Sep 1937. BCE & MS from Georgia Tech. Native of Waycross, Ga. Sanitary engrg designer with Robert & Co Assoc, Atlanta 1961-65. Water quality control engr, Georgia Water Quality Control Bd 1965-68. With Flood & Assocs Inc, Jacksonville, 1968-77, VP 1969-77. Co-owner and Exec VP, Hubbard, Bates & Assoc, Inc, 1978-present. Respon for all environ eng design and pre-design engrg, environ & economic studies & reports for water & wastewater improvements for municipal clients; heavy emphasis on 201 facility planning for EPA grant program. Mbr FES, NSPE, WPCF. Registered Prof Engr in Florida & Georgia. Enjoy fishing & reading. *Society Aff:* NSPE, WPCF, PEPP, ACEC.

Bates, Robert T
Home: 13652 West 54th Ave, Arvada, CO 80002
Position: Principal *Employer:* Meurer & Assocs, Inc *Education:* BS/CE/Univ of WI-Platteville *Born:* 09/21/47 Principal with Meurer & Assocs, Inc. Previous experience with MSM Conslts of Denver, and the Wisconsin Dept of Transportation. Reg Engr in WI, WY, CO & NM. Mbr of the American Concrete Inst (ACI) and ACI Committee 344 Circular Prestressed Concrete Structures, and Post Tensioning Inst (PTI) and the PTI Committee on Tendon Stressed Concrete Tanks and Silos. Has performed research on the behavior of post tensioned concrete tanks with the PTI and Univ of CO. Responsible for Meurer & Assocs, Inc structural engrg. Enjoys music, reading and athletics. *Society Aff:* ASCE

Bathen, Karl H
Business: Manoa Campus-(Honolulu), Keller Zoid, Honolulu, HI 96822
Position: Prof/Researcher *Employer:* Univ of HI *Education:* PhD/Phys Oceanography/Univ of HI; MS/Oceanography/Univ of HI; BS/ME/Univ of CT *Born:* 11/28/34 He worked in the field of engrg for 7 yrs as proj and R&D Engr on aerial and underwater photographic equipment, data processing and display techniques, electronics, and oceanographic research equipment. The past 11 yrs he has conducted oceanographic research in the areas of coastal, physical and chem oceanography, heat storage and the mixed-layer depth in the North Pacific Ocean, oceanographic instrumentation, disposal of wastes in the ocean, environmental impact studies, and numerical modeling techniques for coastal circulation. He is the author of 41 publications on these subjects. Dr. Bathen is presently a prof and researcher at the Univ of HI, researching in OTEC and environmental studies, and teaches courses in oceanography, marine instrumentation, nearshore survey techniques and numerical modeling. Dr. Bathen's interests at present are focused on OTEC, alternate energy sources, and nearshore physical oceanography. *Society Aff:* ΣΞ, AGU, MTS, CCH, ASME

Batra, Romesh C
Business: Dept of Engg Mech, Rolla, MO 65401
Position: Prof *Employer:* Univ of MO-Rolla. *Education:* PhD/Engg Mech/Johns Hopkins Univ; MASc/Mech Engg/Univ of Waterloo, Canada; BSc/Mech Engg/Panjabi Univ, Patiala India *Born:* Dherowal, Panjab, India, Aug 16, 1947; s. Amir Chand & Dewki Bai (Dhamija) B.; MASc, Univ Waterloo (Can), 1969; PhD, Johns Hopkins Univ, 1972; Res assoc McMaster Univ, Hamilton, Ont, Can, 1973-74; asst prof Engr Mech Univ of MO-Rolla, 1974-76, assoc prof, 1977-Prof 1981-; asst prof Univ AL, Univ, 1976-77; vis res scientist Johns Hopkins Univ, Baltimore, summer 1977, Visiting Prof, Univ Pisa, Italy, July 1980. Visiting Scientist, Sandia National Laboratories, Albuquerque, Summer 1983, Visiting Senior Research Fellow, Ballistic Res. Lab., Aberdeen, 1983-84. Visiting Prof. Univ. of Minnesota, Jan-June 1985. Visiting Prof. Technical Univ. of Berlin - June 1987. Johns Hopkins Faculty Arts & Sci fellowship, 1969-72, Whitehead tuition fellowship, 1969-72; recipient teaching excellence award Univ AL, 1977, School of Engineering- Halliburton Excellence Award 1986. Res grants Univ MO, summer 1976, 78, 79-81, NSF 81-83, ONR 1982-83, ARO 81-82, 1985-88, NSF 1986-87; Contributes articles to profl journals and Natl/Intl Conferences. *Society Aff:* ASME, AAM, SES, SNP, SR, ASEE.

Batterman, Steven C
Business: Dept. of Bioengineering, Univ. of Pennsylvania, Philadelphia, PA 19104-6392
Position: Prof. *Employer:* Univ of PA. *Education:* PhD/Engg/Brown Univ; ScM/Engg/Brown Univ; BCE/Civ Engg/Cooper Union. *Born:* 8/15/37. Native of Brooklyn, NY. With Univ of PA since 1964 having held appointments in the Sch of Engg and Appl Sci (Mech Engg and Appl Mechanics, Bioengg), Sch of Medicine (Orthopaedic Surgery) and Sch of Veterinary Medicine. Maj interests in biomechanics, structural mechanics and stress analysis. Consultant to govt, industry, insurance cos and attorneys, and has served as a forensic expert in hundreds of cases involving accident reconstruction, structural failures, products liability, crashworthiness and mech of human injury. Chrmn, Mechanics Div, ASEE, 1971-72. Associate Editor, Journal of the Engg Mechanics Div, ASCE, 1972-74. Married (Judith), has 3 children and resides in Cherry Hill, NJ. Chrmn, Eng'g Section, AAFS, 1986-87, Bd of Dir, AAFS, 1987-90. *Society Aff:* ASCE, ASME, SESA, AAM, ASEE, BMES, ASSE, SAE, AAFS, AAAM.

Battin, Richard H
Business: 555 Tech Square MS4A, Cambridge, MA 02139
Position: Associate Department Head. *Employer:* Charles Stark Draper Laboratory Inc. *Education:* PhD/App Math/MIT; SB/Elec Eng/MIT. *Born:* 3/3/25. Dir of Mission Dev for MIT Apollo program providing navigation, guidance & control software for each of Apollo flights. Author of book *Astronautical Guidance* 1964 & co-author with J H Laning, Jr of *Random Processes in Automatic Control* 1956. MIT Adj Prof of Aero & Astro. Mbr NAE 1974; IAA 1971; Fellow AIAA 1970. Fellow AAS 1984. Recipient of AIAA Louis W Hill Space Transportation Award 1972 with D G Hoag, AIAA Mechs and Control of Flight Award, 1978. AIAA Pendray Aerospace Literature Award 1987. The Institute of Navigation Superior Achievement Award for Continuing Outstanding Contribution to the Advancement of Navigation for 1980 and the MIT Dept. of Aeronautics and Astronautics Teaching Award for 1980-1981. Mbr Air Force Studies Bd of the NAE, Mbr Bd of Dirs, Hon & Awards Committee of the AIAA, Chrmn of the AIAA Student Activities Committee Mbe Aerospace Safety Adv Panel & Special Advisor to Div Advisory Group of USAF Aeronautical Sys Div. *Society Aff:* NAE, AIAA, AAS, IAA.

Baudat, Ned P
Business: 5400 Mitchelldale, Suite A-3, Houston, TX 77092
Position: Pres. *Employer:* B & C Assoc, Inc. *Education:* BS/Chem Engg/Rice Univ. *Born:* 11/27/37. Native of Houston, TX. Process engg with Brown and Root from 1960-1966. Dev vapor-liquid equilibrium correlation for light hydrocarbons. Process engg, project manager and Manager of Process Engg for Hudson Engg Corp from 1966-76. With B & C Assoc since 1977. Dev power recovery system to utilize waste heat to produce mech horsepower. Motive fluid used is propane or butane. *Society Aff:* AIChE.

Baudry, Rene A
Home: Rue de L'Yvette, Paris, 75016 France
Position: Consultant-Westinghouse Elec Corp. *Employer:* Westinghouse Elec Corp-ret. *Born:* May 1899, France. Grad Engr from Ecole Nationale dArts et Metiers Aix-en-Provence. Retired in 1964 as Mgr, Dev Engrg Dept, Lge Rotating Apparatus Div, Westinghouse Electric Corp. First Paper Prize AIEE Power Div, 1951 Fellow Awards ASME & IEEE. Special patent awards from Westinghouse Elec Corp in 1952 & 1961. IEEE Lamme Medal 1966 for significant contribs to the design of lge generators. Certificate of Commendation United States Navy - Bureau of Ships - for outstanding service to the US Navy during WWII. Life Fellow ASME & IEEE. *Society Aff:* ASME, IEEE, SEE, ICF.

Bauer, Jerome L
Home: 5134 Finehill, LaCrescenta, CA 91214
Position: Director of Research *Employer:* Furane Products Div. M&T Chemicals Inc. *Education:* MS/ChE/PA State Univ; BS/ChE/Univ of Dayton. *Born:* 10/12/38. Native of Pittsburgh, PA. Course work to PhD, OH State Univ. Employment '63-'67 Univ of Dayton, Asst Prof Chem Engr '67 OH State Univ. Instr Chem Engg '68-'73 Ferro Corp, Mgr Adv Comp Dept '73-'74 Lockheed-Sr Mtl Engr '74-'77 Convair/Gen Dynamics, Design Specialist, '77-80 JPL member of the Technical Staff '81- present Furane Div, M&T Chemicals, Director of Research. Chapter Chrmn SAMPE (Los Angeles & San Diego) current Natl Treasurer. Chapter Chrmn AIChE (Dayton, OH) Wife former Karen J Dowling, children Lori (17) Trish (15) Jeff (13) Keith (11) Brian (9). *Society Aff:* SAMPE, AIChE, ACS, ASTM.

Bauer, Kurt W
Business: P.O. Box 1607, Waukesha, Wisconsin 53187-1607
Position: Executive Director. *Employer:* SE Wisc Regional Planning Comm. *Education:* PhD/Civil Eng/Univ of WI; MS/Civil Eng/Univ of WI; BS/Civil Eng/Marquette Univ. *Born:* 8/25/29. Prof work has incl cnduct of lrg-scale control survey & topographic mapping proj; preparation of comprehensive urban & regional dev plans & plan implementation devices; design of sys & facility plans for arterial street, mass transit, airport, sanitary sewerage, storm water drainage, water supply facilities; preparation of plans for lrg-scale land dev projects. Dir of SE Wisc Regional Planning Comm. Recipient 1978, Univ of WI-Madison College of Eng Distinguished Service Citation; 1977, US Dept of Transportation, Federal Highway Administration Public Service Award, June 1974, Award, Great Lakes Comm; April 1973, Disting Prof Achievement Award, Marquette Univ; Oct 1970 Harland Bartholomew Award ASCE. Editor, Manual of Practice on Engineering Surveying, American Society of Civil Engineers, 1985. Author of 40 prof papers. *Society Aff:* ASCE, ACSM, TRB, SCSA, AICP.

Bauer, Robert F
Business: 7624 So. Painter Ave. Ste. D, Whittier, CA 90602
Position: Retired *Employer:* Owner Bauer Land & Cattle Co. *Education:* BS/Eng/USC. *Born:* Jan 1918. Joined Union Oil Co of Calif in 1942 as Res Engr, then Foreman, Production Superintendent & Field Engr. Became Mgr of Offshore Operations for the 'CUSS Grp', a joint venture of Continental, Union, Shell & Superior Oil Cos, in 1953. Formed Global Marine Inc in 1959 & was Pres of GMI until 1966 when became Ch/Bd. Present Mbr of NAE & received the AIME 1971 Engrg Achievement Award. *Society Aff:* USC.

Baughman, George W
Home: 1255 Sara Ct, Winter Park, FL 32789
Position: Vice President - Retired. *Employer:* Westinghouse Air Brake Company. *Education:* EE/OH State Univ; BSEE/OH State Univ.; Dr. Eng/OH State Univ./Dec 13, 1985 *Born:* 2/11/00 - Gilboa Native of Gilboa, OH. With AT&T for 3 yrs. Went with Union Switch & Signal in 1923 where main contributions were in Automatic Train Control & Centralized Traffic Control. Advanced through various positions from Dev Engr to VP Engrg. Then 5 yrs at Hdquarters of Westinghouse Air Brake Co, followed by 1 yr as Cons to the US Dept of Transportation. 111 patents in his name. Fellow IEEE & Mbr Ext Awards Ctte 1972-73. Elmer A Sperry Award 1971. Benjamin G Lamme Gold Medal from OH State Univ 1978 for achievements in Engg. Fellow Inst of Railway Signaling Great Britain. Main hobby is golf. Mbr Eta Kappa Nu, Tau Beta Pi and Sigma Xi. *Society Aff:* IEEE - Fellow and Life Member, IRS-Fellow.

Baughman, Richard A
Home: 21 Beech Ln, Lock Haven, PA 17745
Position: Dir Mfg Engr. *Employer:* Piper Aircraft Co. *Education:* BS/Met Engr/CO Sch of Mines. *Born:* 9/21/27. Native of Grand Junction, CO. 15 yrs in jet engine met at GE, Cincinnati, specializing in titanium and high temperature rolling contact bearing mtls. Publications and patents. Skilled in mgt of mft engg and processes including welding and brazing. Reg PE in OH & CO. Piper Aircraft since Dec 1977. Enjoys politics and all sporting activities. Qualified aircraft pilot and basketball referee. *Society Aff:* ASM.

Bauld, Nelson R, Jr
Business: Mech Engrg Dept, Clemson, SC 29631
Position: Prof *Employer:* Clemson Univ *Education:* PhD/Theoretical and Applied Mech/Univ of IL; MS/Mech/WV Univ; BS/ME/WV Univ *Born:* 5/18/31 Native of Clarksburg, WV. Taught engrg mechanics at WV Univ, VA Polytechnic Inst and State Univ, Univ of IL, and Clemson Univ. Author of text book Mechanics of Materials (Brooks/Cole Publishing Co 1982). Post doctoral leave to Division of Engrg and Applied Physics, Harvard Univ (1970). Specialty: stability of isotropic and composite shell type structures. *Society Aff:* ASME, SCSPE

Baum, Carl E
Home: 5116 Eastern SE, Unit D, Albuquerque, NM 87108
Position: Senior Scientist *Employer:* U.S. Air Force Weapons Lab *Education:* Ph.D./EE/California Inst of Tech; M.S./EE/Calif Inst of Tech; B.S./EE/Calif Inst of Tech *Born:* 02/06/40 Sr Scientist for Electromagnetics, Air Force Weapons Lab, Albuquerque, NM. Air Force (Capt.) 1962-71. Honeywell Award for best undergraduate engineer 1962. Air Force R&D Award 1970. Pres. Electromagnetics Society 1983-85. Founder, Pres., SUMMA Foundation. IEEE EMC Soc Richard R. Stoddart Award 1984. Fellow IEEE (1984), "for pioneering the singularity expansion method and electromagnetic topology in electromagnetic theory, and for dev of EMP simulation and electromagnetic sensors." IEEE Harry Diamond Memorial Award (1987), "for outstanding contributions to the knowledge of transient phenomena in electromagnetics." Numerous books and technical articles (various awards). Choir director and composer. *Society Aff:* IEEE, EMS, USNC/URSI

Baum, David M
Business: 1313 Dolley Madison Blvd, McLean, VA 22101
Position: Asst VP *Employer:* Dynalectron Corp *Education:* ME/Mgt Engrg/RPI; BS/Mgt Engrg/RPI *Born:* 10/19/50 Author of test and courses in Security and Safety Systems. Taught Engrg at RPI. Served also on staff of VP-Finance and Admin. With Dynalectron Corp since 1976 having served as Systems Engr, Proj Engr, Proj Engrg Mgr, Mgr of Special Projects, and Operations Dir. Elected Asst VP in 1979. Current responsibilities include corporate wide direction in communications and advanced technology systems. Recognized by A.S.I.S. as Certified Protection Profl. Enjoy classical music and volleyball. *Society Aff:* IIE, ASEE, IEEE

Baum, Richard T
Home: 9 Ivy Hill Road, Chappaqua, NY 10514
Position: Consultant. *Employer:* Jaros, Baum & Bolles. *Education:* MS/ME/Sch of Engg, Columbia University; BS/ME/Sch of Engg, Columbia University; BA/Liberal Arts/Columbia University *Born:* 10/3/19. Mech Engr, Elec Boat Co, Groton, CT, 1941-1943. Served as 1st Lt US Army AF, 1943-46. With Jaros, Baum & Bolles, Consulting Engrs, NY, NY, 1946- ; Partner, 1958- ; Partner Emeritus, Consultant Firm, 1986 to present. Mbr, Advisory Council, Faculty of Engg and Appl Sci, Columbia Univ, 1972. Reg PE in NY and 21 states and the Natl Bureau of Engg Reg-

Baum, Richard T (Continued)
istration. Mbr Natl Acad of Engrg. Fellow, American Consulting Engrs Council, American Soc of Heating, Refrig and Air-Conditioning Engrs, American Soc of Mech Engrs. Mbr, Natl Soc of PE, Member, Board of Building Research Board of National Research Council, American Arbitration Assn (Panel of Arbitrators, 1973), NY Assn of Consulting Engrs, Natl Soc of Energy Engrs. Mbr of Steering Group and Chrmn of Mechanical, Electrical and Vertical Transportation Cttees of Council on Tall Buildings and Urban Habitat (Vice Chairman North America). *Society Aff:* ASHRAE, ASME, NSPE, NAE.

Baum, Sanford
Business: Dept of Mech and Industrial Engrg, Salt Lake, UT 84117
Position: Assoc Prof *Employer:* Univ of UT *Education:* PhD/IE/Stanford; MS/Stat/Stanford; BS/Chemist/Univ of CA-Berkeley *Born:* 10/22/24 Dr Sanford Baum is an Assoc Prof in the Mechanical and Industrial Engrg Dept of the Univ of UT. His research interests include studies of inflation in engrg economy, multicriterion decision analysis, and the use and development of random search methods in mathematical programming. He has worked at Stanford Research Inst, Martin-Marietta Aerospace and the US Naval Radiological Defense Lab. *Society Aff:* IIE, TIMS

Baum, Sidney J
Home: 17001 Mooncrest Dr, Encino, CA 91436
Position: Ret, Consultant *Education:* PhD/Chem Engg/Columbia Univl; MS/Chem Engg/MIT; BS/Chem Engg/NJIT. *Born:* Jan 1916, Newark, New Jersey. Chem Engr duPont 1936-38; Celanese 1941- 42; Pres Amer Polymer Corp 1943-53; Mgr Polyco Div Borden 1953-56; VP Foster Grant Co 1956-61; Pres Baum Chem Corp 1961-76. Scientific Investigator US Dept Commerce 1947. Major exper in polymerization, plastic processing, light polarizers. 8 US patents. Conslt Leisure activities: golf, furniture making, opera. Retired 1978. *Society Aff:* AIChE, ACS, SPE.

Baum, Willard U
Home: 2737 Fairview St, Allentown, PA 18104
Position: Retired VP, Sys Power & Engrg. *Education:* BS/EE/Drexel Inst of Technology. *Born:* 9/21/06. With Pennsylvania Power & Light Co 40 yrs in Operating, System Planning & Engrg. Dev concept of using automatically switched capacitors for voltage control & reactive supply. Mgr System Planning, Ch Electrical-Mech Engr, & 1967- 70, VP System Power & Engrg. Joint participant in planning & construction of Keystone & Connemaugh 1800 mw steam-electric stations & associated 500 kv tramsmission. Respon for conception, planning, design & construction of 5-800 mw PP&L steam-electric units. Named Fellow IEEE for contributions to the planning of lge utility systems. Retired in 1970. Interested in music; choirmaster for 31 yrs. Volunter Actg. Bus. Man., Christ Lutheran Church, Allentown PA since 1970. *Society Aff:* IEEE, NSPE, PSPE.

Bauman, Richard D
Home: 2110 Vassar Dr, Boulder, CO 80303
Position: President. *Employer:* Trans Plan Associates, Inc. *Education:* PhD/Transportation/AZ State Univ; MS/Civil Engr/Univ of UT; BS/Civil Engr/IA State Univ. *Born:* 4/26/38. Active in transportation planning and engineering projects since 1960. Worked for Harnischfeger Corp and the UT Highway Dept. Taught as an Assistant prof and later as a tenured Assoc Prof at the Universities of CO, OK & HI, Served as City Planner for Scottsdale, AZ and then became a consultant in transportation planning and traffic engg. Founded Trans Plan Associates, Inc in 1978. Pres CO/WY Section of ITE 1978-79. *Society Aff:* ITE.

Baumann, Edward Robert
Home: 1627 Crestwood Cir, Ames, IA 50010
Position: Anson Marston Disting Prof of Engrg. *Employer:* Dept CE, Engrg Res Inst, Iowa St U. *Education:* BSE/Civil Engr/Univ of MI; BS/Sanitary Engr/Univ of IL; MS/Sanitary Engr/Univ of IL; PhD/Sanitary Eng/Univ of IL. *Born:* May 12, 1921, Rochester, NY. Grad Rome Free Acad, Rome, NY; Postdoc appointment at Univ of Durham, King s College England in 1959-60. Faculty leave at 9 INCOFILT (Internatl Consortium of Filtration Res Grps) Univs 1972-73. Grad Asst, Instructor, Res Assoc at Univ of Illinois 1946-53. Assoc Prof 1953, Prof 1957, & Disting Prof 1972 Iowa State Univ 1953- . Res interests: solids/fluids separation tech with emphasis on deep-bed granular media & diatomite filtration of water & wastewater. Over 120 tech articles in referred journals. Mbrship in numerous societies. Registered Prof Engr. *Society Aff:* AWWA, WPCF, NSPE, ASCE, AAUP, ASEE, AEEP, AIChE, AAEE.

Baumann, George P
Business: PO Box 101, Florham Park, NJ 07932
Position: Engg Advisor. *Employer:* Exxon Res & Engg Co. *Education:* PhD/CE/Cornell Univ; BS/CE/IIT. *Born:* 1/10/19. A native of Chicago, joined a local E I Dupont plant for a year before coming east for a grad degree & working for the Rubber Reserve. Now have 36 yrs of experience with Exxon Res & Engg Co. This includes four yrs in refinery planning, seven yrs in process dev & 25 yrs in process design, plant startups, and technical services. His experience has included all petroleum processing, and in particular, fluid solids processing, chems and fertilizer processing, and such specialities as flue gas desulfurization, coal gasification and liquefaction. Often used as a company consultant and committee mbr for reviews such as uranium enrichment, nuclear spent fuel reprocessing, fluidized iron ore reduction, and offshore crude stripping. Presently responsible for the process engg output of 20 engg professionals. *Society Aff:* ACS, AIChE, API, ΣΞ, ΤΒΦ.

Baumann, Hans D
Home: 32 Pine Street, Rye, NH 03870
Position: Pres *Employer:* H D Baumann Assoc, Ltd *Education:* PhD/Mech Engr/Columbia Pacific Univ; MS/Mech Engr/Columbia Pacific Univ; BS/Indust Engrg/Metallfachschule Bielefeld/Germany *Born:* 10/12/30 Native of Germany. Chief Dev Engr, Worthington S/A France 1966-1969. VP Tech - Masoneilan Int Inc until 1974. Intl Valve Conslt and currently Pres of H D Baumann Assoc, Ltd. Fifty (50) publs and co-author of handbooks plus over 80 patents. Well known for practice and theory of control valves including throttling noise. Fellow mbr ISA and recipient of ISA Chet Beard Award. Active on US (ANSI) and intl standards. Chrmn Industry Liaison Ctte Fluid Controls Inst. Reg PE in 4 states. Served as technical expert in patent litigations. Member Admission Ctte IVA, Honorary Member Spanish Society of Industr. Engineers. *Society Aff:* ASME, ISA, SIIE

Baumeister, Philip
Business: 2789 North Point Pkwy, Santa Rosa, CA 95407
Position: Chief Scientist. *Employer:* OCLI. *Education:* PhD/Physics/UC Berkeley; BS/Physics/Stanford. *Born:* 3/17/29. Full-time faculty mbr at Univ of Rochester (Rochester, NY) 1959-1978. Engr at Aeroject Electrosystems 1978-1979. From 1979-present, Chief Scientist at OCLI. Specialties: Optical interference coatings, thin films. *Society Aff:* OSA, AVS, SPIE.

Baus, Bernard Villars
Business: 17 Emajagua St, Santurce, PR 00913
Position: President. *Employer:* Industrial Chemicals Corporation. *Education:* PhD/Chem Engg/Cornell Univ; BEChE/Chem Engg/Tulane Univ. *Born:* Sep 1925, Gramercy, Lousiana. Various positions with DuPont in Texas, West Virgnia & Delaware involving process dev, plastics res, & internatl scene; joined Commonwealth Oil Refining Co in San Juan, PR in 1962, soon becoming VP for chemicals; played a leading role in the dev of 400 million dollar petrochemical complex in Puerto Rico; business cons 1971- , clients incl Govt of Puerto Rico, & new refinery constructed by ECOL, Ltd; since 1974, Pres of Industrial Chemicals Corp, new co for sulfuric acid manufacture and related inorganics. *Society Aff:* ACS, AIChE, AAAS, ΣΞ.

Baxter, Meriwether L Jr
Business: 17 Braintree Cres, Penfield, NY 14526
Position: Ret. (Gear Consultant Self-employed) *Education:* BSME/Yale Univ. *Born:* Dec 7, 1914, Nashville, Tenn. Spent over 40 yrs with Gleason Works, Rochester, NY as a specialist in bevel & hypoid gears, incl supervision & in- plant teaching. Titles incl Ch, Mathematical Res; Ch Engr, R&D; Dir of Engrg. Numerous patents relating to gears & gear-making machinery, US & foreign. Numerous published papers on cam & gear theory. Author of Chapter 1, Gear Theory, in McGraw Hill Gear Handbook. Served on numerous ASME & AGMA cttes. Retired from Gleason Works Dec 1975. Presently self-employed gear cons. Fellow ASME. *Society Aff:* ASME, IMS.

Baxter, Samuel S
Business: 7048 Castor Ave, Philadelphia, PA 19149
Position: Consulting Engineer. *Education:* Dipl/Municipal Engg/Drexel Univ. *Born:* Feb 6, 1905, native of Philadelphia. Diploma, Drexel Inst of Technology 1925. D Engrg, hon, Drexel University 1967. Served various positions 1923-72 with engrg & public works depts of City of Philadelphia, incl Asst Dir Public Works 1940-42, Ch Engr 1950-51, Commissioner & Ch Engr-Water Dept 1952-72. With Manhattan Proj, C of E 1942-46, rank Major. Career incl work on airports, highways, bridges, water & sewage projects. Natl Pres ASCE 1971; Natl Pres AWWA 1965, & Hon Mbr; Natl Pres APWA 1947 & Hon Mbr; Mbr Natl Acad of Engrg; Diplomate, Amer Acad of Environmental Engrs; Mbr NSPE, SAME; Chmn APWA Res Foundation 1959-77. Registered Prof Engr Pennsylvania; Trustee Drexel Univ. NSPE Award 1976, APWA Award-Top Ten 1960. Chmn/Bd East Girard Savings Assn Philadelphia; Pres Phila Council Boy Scouts of Amer 1968-69; Mbr Chi Epsilon, Union League, Philadelphia. M. *Society Aff:* ASCe, AWWA, APWA, NAE, AAEE, SAME, WPCF.

Bay, James B
Business: GMI 1700 W 3rd Ave, Flint, MI 48502
Position: Asst. Dept. Head, Mechanical Engrg Dept. *Employer:* General Motors Institute. *Born:* Jun 19, 1932. BSME from Mich Technological University 1954. Native of Ontonagon, Michigan. Registered Prof Engr Michigan. Employed by Detroit Edison Co 1954, 1956-57. Served as officer in Army Corps of Engrs 1954-56. Joined faculty at GMI in 1957. Served as Dept Chmn from 1963-78. Serves on SAE Mid- Michigan Section Governing Bd and named Engineer of the year 1981. Mbr of MSPE & served as Pres in 1973-74. Chairman, MSPE Scholarship Trust. Served as Natl Dir of NSPE. Active in church, civic & Explorer Post affairs. Mbr of Blue Key Fraternity, Theta Tau Fraternity. Enjoys all outdoor activities.

Bay, Robert D
Business: 1500 Meadow Lake Parkway, Kansas City, MO 64114
Position: Engineering Manager *Employer:* Black & Veatch, Engrs-Arch *Education:* BSCE/Structures/University of Missouri-Rolla *Born:* 09/15/26 St. Louis native. Engrg Mgr, Black & Veatch, Engrs-Architects that design water and wastewater facilities worldwide. Previously served as Asst. Chief Engr, Water Dept, St. Louis, and Dir Tech Sers, LaClede Steel. Known for research-development of deformed wire; load transfer devices for highway pavement; designed patented composite floor system, World Trade Center Bldgs, New York; expert, industrial paint systems. Military service 1944-83, Retired Reserve Major General, awarded Distinguished Service Medal; Honorary PhD Missouri U. Rolla 1986; Infantry OCS Hall of Fame 1982; Merit Award, Engrs Club, 1977; Academy of Civil Engrs MUR 1976; Young Engr of the Year-Missouri, 1952; National Pres ASCE, 1986. *Society Aff:* ASCE, NSPE, SAME, ACI, AWWA, WPCF

Bayazitoglu, Yildiz
Business: Mech Engrg and Material Science Dept, Houston, TX 77001
Position: Assoc Prof *Employer:* Rice Univ *Education:* PhD/ME/Univ of MI; MS/ME/Univ of MI; BS/ME/Metu-Turkey *Born:* 7/27/45 Taught Mechanical Engrg Cources in Univ of M, Metu, Univ of H and Rice Univ. Served in technical committees of AIAA, ISES/AS and ASME. Registered PE in TX. Presently Coordinator of Graduate Studies in MEMS Dept at Rice Univ. Research interest is in the field of thermal fluid sciences, including related subjects in physics and mathematics, e.g. Natural Convection, Radiation, Phase Change Heat Transfer and Solar Engrg. Enjoys the growth of her sons, Ozgur, Mert and Kunt. *Society Aff:* ASME, AIAA

Bayer, David M
Business: Dept of Civil Engineering UNCC, Charlotte, NC 28223
Position: Prof Civil Engrg *Employer:* Univ of NC at Charlotte *Education:* PhD/Structural Mech/Vanderbilt Univ; MS/Structural Mech/Vanderbilt Univ; BS/CE/GA Inst of Tech *Born:* 5/2/41 Native of Nashville, TN. Served as Captain in the US Army Corps of Engrs at the Waterways Experiment Station in Vicksburg, MS. With Univ of NC at Charlotte since 1970. Promoted to Prof of Structural Engrg in 1984. Recently appointed as Coordinator of Cooperative Education for the Coll of Engrg. Currently District 6 Director , District 6 ASCE & Member of Bd of Dir. Registered PE in NC and MS. *Society Aff:* ASCE, NSPE, ASEE, PENC.

Bayles, Allison L
Business: 6112 Alder St, Pittsburgh, PA 15206
Position: Principal. *Employer:* Allison L Bayles & Assocs-Cons Engrs. *Education:* ME/Lehigh Univ. *Born:* 6/11/04 Field Engr and Asst Supt, Liberty Bridge River Piers, Pittsburgh, Dravo Corp. Plant Sys Engr, Engineering Works Dept Neville Island; Corp Reorganization Work 1932-39, New York. VP Hill Diesel Engine Co. Lansing, MI 1940-41;Dir Res, Rogers Diesel & Aircraft Corp. NYC 1941-42; VP Operations, American Engrg Co, Power Plant-Ship Machinery. Phila 1942-44; Cons Engr, Mgr, Fine Particle Processes Dept, C H Wheeler Manufacturing Co Phila 1945-50; VP Engrg Scaife Co, Pressure Vessels Pittsburgh 1950-52; Cons engr, private practice Allison L. Bayle & Assoc. 1952- , US, India, Turkey, Venezuela. Life Fellow ASME; Mbr Phi Beta Kappa, Vice ChrmnTau Beta Pi, Omicron Delta Kappa, Vice Chrmn Pittsburgh District Export Council 1977-81, Vice Chrmn Western Pa Dist Export Council 1981- ; Reg Prof Engr, NY, NJ, PA, NC. Avocation: Eastern philosophies. Author.

Bayne, James W
Business: 152 M.E. Bldg, Urbana, IL 61801
Position: Prof & Assoc Hd M.E. *Employer:* Univ of IL *Education:* MS/ME/Univ of IL; BS/ME/Univ of IL *Born:* 4/2/25 Served in Navy 1943-46. Prof of Mechanical Engrg, Univ of IL. Assoc Head of Dept, Acting Head 1974-75. National Secretary Treasurer of PiTauSig 1959- 71, National Pres, 1971-74. Pres, Association of Coll Honor Societies, 1974-75. Executive Committee, ME Division of American Society of Engrg Education. Selected Danforth Fellow, 1978. Recipient, Western-Electric ASEE Outstanding Teacher Award; IL-IN Section, 1971. Selected "Outstanding Educators of America–, 1970. Recipient of Coll Everitt Award and Piece Award. Co-author of "Opportunities in Mechanical Engrg–. *Society Aff:* ASME, ASEE, ΤΒΠ, ΠΤΣ

Bays, Neil E
Business: Bays, Inc, 200 George Street - Suite 4, Beckley, WV 25801 *Education:* BSEM/Mining Engg/W VA Univ. *Born:* 4/28/35. Native of WVA. Worked as Design Engr 1961-62 with Lively Mft & Equip Co; Preparation Mgr Slab Fork Coal Co 1962-63; Ch Engr with Lively 1963-67; VP of Engg 1967-70; Exec VP with Lively Mfg 1970-1984. Active in NSPE, P Chap Pres, Natl Dir from NSPE, P Chmn PEI WVSPE, PV Chmn NSPE-PEI Central Region. PPres of WVA Soc of Profl Engrs. Chrmn-Shady Spring Public Service District. Mbr of Natl Fire Protection Assoc Sub-Committee on Fire Protection Assoc Sub-Committee on Fire Protection for Coal Preparation Plants. Enjoy golf tennis & fishing. Past President - College of Mineral and Energy Resources Alumni Chapter West Virginia University. Past VP Central Region NSPE. VP Engineering Industrial resources - Carnegie, PA. *Society Aff:* NSPE, AIME.

Bazant, Zdenek P
Business: Dept of Civil Engrg, Evanston, IL 60201
Position: Professor of Civil Engrg. *Employer:* Northwestern University. *Education:* PhD/Engrg Mech/Czech Acad of Sci. *Born:* 12/10/37. in Prague. Postgrad dipl in Phys 1966 Charles Univ. Bridge engr, researcher, lectr (Docent) in Prague to 1967. At Northwestern Univ, Evanston, IL since 1969; Prof 1973, Dir, Ctr for Concrete and Geomaterials, 1981-present. Coordinator Struct Engrg Prog 1974-78. IL Reg Struct Engr. Cons Argonne Natl Lab, Ontario Hydro, Babcok & Wilcox, OakRidge Natl Lab, Sargent & Lundy Engrs, Sandia Lab, W R Grace & Co. Published book-creep of concrete; over 230 papers on creep, inelastic behavior, fracture, concrete, stability. 5 patents. Editorial Bds of Cement & Concrete Res, J Engrg Mech Div ASCE; Solid mechanics Achives; Mtls and Struct (RILEM); Num & Analytical Methods in Geomechs. Chrmn ASCE-EMD Comm on Mtls, Chmn RILEM inferm Ctte on Creep, Chmn ASMIRT Div (Concrete Struct). Cttes in ASCE, ACI, ASME, CEB. Dir of grants & contracts from NSF, NSF- RANN, ERDA, DOE, DNA, AFOSR, ARPA, LASL, WES, EPRI, LANL. French Govt Fellowship 1966; Election Fellow Amer Acad of Mechs 1978, of ACI 78, ASCE 79, Ford Sci Fellowship 68; RILEM Medal 75; Huber Res Prize ASCE 76, T.Y. Lin Prestrened Concrete Award ASCE 77; Guggenheim Fellowship 78-79; IR-100 Award, Indust Res & Dev, 82. Enjoy music, tennis, skiing. *Society Aff:* ASCE, ACI, ASME, ACS, SEAOI.

Bazzini, Robert J
Home: 1 Carolina Place, Ridgewood, NJ 07450
Position: Marketing Mgr. *Employer:* Imo Delaval *Education:* ME/Mech Engg/ Stevens Inst of Technology. *Born:* Sep 1926, Hoboken, New Jersey. Degree in Mech Engrg from Stevens Inst of Technology. Registered Prof Engr in New York, New Jersey & Pennsylvania & holds mbrships in AWWA, WPCF, New Jersey Soc of Prof Engrs, NSPE, Hudson County Prof Engrg Soc, as well as full mbrships in both the ASME & the Soc of Naval Architects & Marine Engrs. Marketing Manager of Imo Delaval Inc. headquartered in Ridgewood, NJ. *Society Aff:* AWWA, ASME, NSPE, NJSPE,WPCF, SNAME, IME.

Bea, Robert G
Business: 7330 Westview Dr, Houston, TX 77055-5199
Position: VP *Employer:* Woodward-Clyde Consultants *Education:* MSE/CE/Univ of FL; B/CE/Univ of FL; AA/Engrg/Jacksonville Univ *Born:* 1/14/37 VP, Sr Consultant, Ocean Services Divn, Woodward-Clyde Consultants, 1977- 81. Sr Staff Civil Engr, Head Office Design Group, Shell Oil Co, 1976-79. Engr- in-Training, US Army Corps of Engrs, 1975-78. Intl experience in design, siting and construction of offshore platforms and pipelines for oil and gas industry. Received ASCE James R Croes Medal 1980. Married to Joan Panning. Two sons, Robert and Joel. Hobbies include sailing and skin diving. *Society Aff:* ASCE, ACE, MTS, SPE, SSA, EERI

Beachem, Cedric D
Business: Naval Res Lab, Washington, DC 20375
Position: Supervisory Metallurgist. *Employer:* Naval Research Laboratory. *Education:* BS/Met Engr/NC State U. *Born:* 02/05/32 Beaufort, NC. BS Met Engr NC State Univ 1957. USAF 1953-55. Employed at the Naval Res Lab since 1957. Sabbatical, Cambridge Univ 1967-68. Res activities ctr around microscopic crack tip fracture mechanisms & dev of failure analysis tools. Washington DC Chap ASM burgess Award 1970. ASTM Sam Tour Award with B F Brown 1969. Fellow Washingt Acad of Scis. Fellow Amer Society for Metals. Formed & chaired the ASTM E24.02 Subctte on Fractography & Related Microstructures. Metallurgical Transactions Bd/Review 1971-74. ASM Handbook Ctte 1976-79. AGARI Fractography Cons 1967-68. Mbr ASM, AIME, ASTM, NACE. *Society Aff:* ASM, AIME, NACE, ASTM.

Beadle, Charles W
Home: 420 E 12th St, Davis, CA 95616
Position: Professor *Employer:* Univ of CA. *Education:* PhD/Engr Mechanics/Cornell Univ; MSE/Engr Mechanics/Univ of MI; BS/Mech Engr/Tufts Univ. *Born:* 1/24/30. Res engr, Gen Motors Corp, Radio Corp of America, Ford Motor Co specializing in advanced instrumentation and testing methods. Mbr of faculty of Dept of Mech Engg, Univ of CA at Davis since 1962. Chrmn of Dept 1978-1986. Teaching, res and consulting in mech design, solid mechanics, structural dynamics. *Society Aff:* ASME, ASEE

Beadling, David R
Business: 0112 SW Pennoyer St, Portland, OR 97201
Position: Principal. *Employer:* Pease & Beadling Engrs, Inc. *Education:* MS/Struct Engg/Univ of CA, Berkely; BS/Civil Engg/OR State Univ. *Born:* 6/28/35. Native of Portland, OR. Lt with Army Corps of Engrs 1957-59. Project engr with Portland General Electric Co 1960-64, C V Burgstahler & Assoc 1965-71, Moffat, Nichol & Bonney Consulting Engrs 1972-73; Principal of Pease & Beadling Engrs, Inc since 1974. Pres Oregon Section ASCE 1972; Chmn Pacific Northwest Council ASCE 1978; Pres SEAO 1978. Reg PE in OR, WA, CA, ME. *Society Aff:* ASCE, SEAO, FPRS, ACEC.

Beakes, John H
Business: 10650 Hickory Ridge Rd, Columbia, MD 21044
Position: Exec VP *Employer:* General Phys Corp *Education:* MS/Environ Eng/The Johns Hopkins Univ; BS/-/US Naval Acad *Born:* 02/24/43 1966-1974 Engr Officer, US Naval Nuclear Power Program. 1966-1968 Nuclear Power Training. 1968-72 USS Nautilus. 1972-74 Chief Engrg Officer, USS George Bancroft. 1974-present General Phys Corp. 1974-78 Sr Engr. 1978-79 Principal Engr/Dir Engrg Services. 1979-82 VP, Engrg. 1982-87, Sr VP, Power Tech. 1984-86 Pres, Gen Tech Services (subsidiary). 1987-present Executive Vice President and Chief Operating Officer. *Society Aff:* ASME, ANS, SME

Beakley, George C
Business: Coll of Engrg and Applied, Sciences, Arizona State University, Tempe, AZ 85287
Position: Assoc Dean of Engrg *Employer:* AZ State Univ *Education:* PhD/ME/OK State Univ; MS/ME/Univ of TX; BS/ME/TX Tech Univ *Born:* 2/3/22 Native of Levelland, TX. Combat infantry officer WWII. Design engr for Bell Helicopter Corp. Prof of Engrg and Assoc Dean at ASU since 1956. Author or co- author of 29 coll engrg textbooks and recipient of ASEE's first Chester F Carlson Award for creative endeavor in the teaching of engrg design. Also named recipient of ASEE's Western Electric Award for excellence in engrg teaching and ASU's Distinguished Faculty Achievement Award. Awarded Presidential Medallion & named Distinguished Engineer by Texas Tech Univ. State pres of AZ Society of PE and national VP of NSPE. Registered engr AZ, TX, OK. Fellow of ASEE and ASME. Enjoy classical music, tennis, and fishing. *Society Aff:* ASME, IIE, ASEE, ASHRAE, NSPE.

Beale, D Anthony
Business: 6207 Old Keene Mill Ct, Springfield, VA 22152-2385
Position: Pres *Employer:* Advance Engrs Ltd *Education:* M/Engrg Admin/George Washington Univ; B/CE/George Washington Univ *Born:* 10/13/37 Principal designer of major projects with full responsibility for all work executed by the firm. Types of projects include warehouses, shopping centers, office bldgs, churches, recreational bldgs, fire stations, schools and townhouses. Native of Alexandria, VA. Surveyor for Wright Contracting Co 1960- 1963. Chief Engr with Southern Iron Works 1963 to 1970. VP Profl Engrs Inc 1970 to 1975. *Society Aff:* NSPE, ACEC, CEC/VA, CECMW, VSPE.

Beall, James H
Home: 9836 Stephenson Dr, New Port Richey, FL 34655
Position: Engrg Consultant *Employer:* Self *Education:* B.S.EE./(Power Option)/ Clemson University *Born:* 02/06/30 Born and raised in Chicago, Ill, he is currently an Engrg Consultant in Florida. He retired from the AT&T Teletype Corp in 1984. At the time of his retirement, he was a Sr Engr in Plant Engrg with project responsibility for circuit bd electrochemical installations and research MOS facility installations involving exotic new technologies. He also had engrg responsibility for all plant electrical operation, maintenance and expansion at the 1,300,000 square-foot Skokie location. Other work included a two-year period in which he worked on electromagnetic compatibility problems. He has been very active in the IEEE. *Society Aff:* IEEE

Beall, Samuel E, Jr
Home: RFD 14, Houser Rd, Knoxville, TN 37919
Position: Consultant. *Education:* BS/Chem Eng/Univ of TN. *Born:* 5/16/19. Grad, Univ TN 1942; 1943-1948: Performed res in explosives and nuclear energy (E.I. DuPont and Monsanto); 1948-1962: Responsible for construction and operation of first enriched plate nuclear reactor, and of first liquid fuel reactor to generate electricity; 1962-1974: Dir Reactor Div, Oak Ridge Natl Lab; 1974-1975: Organized and became first dir of Energy Div, ORNL, for res on renewable energy sources and environmental effects related to energy. (Union Carbide). Societies: Fellow Amer Nuclear Soc, former Chmn of Power Div and Mbr Bd Dir; Tau Beta Pi; Amer Assoc Adv Sc; Other: Reg PE, TN; Special Advisor (on Nuclear Safety) to Governor TN; Former Bd Mbr, TN Energy Authority. *Society Aff:* AAAS, TBII.

Beam, Benjamin H
Business: 732 N Pastoria Ave, Sunnyvale, CA 94086
Position: Pres. *Employer:* Beam Engg, Inc. *Education:* BS/Elec Engg/OR State Univ. *Born:* 4/6/23. Naval Aviator during World War II. Served with Natl Aeronautics and Space Admin (and its predecessor agency the Natl Advisory Comm for Aeronautics) from 1948 to 1974 in various capacities including wind tunnel testing, instrumentation design and development, and facilities design. Founded Beam Engineering, Inc in 1974 as consulting elec and mech engrs. The firm now provides consulting engg services, and gen engg contracting, in CA and HI. Am currently pres of the co. Reg PE in OR and CA. Five patents. Twenty technical papers. Enjoy golf. *Society Aff:* ASHRAE, ASME, ASPE, APS, NSPE

Beam, Walter R
Home: 3824 Fort Worth Ave, Alexandria, VA 22304
Position: Vice President, Research & Development *Employer:* Sperry Division, Sperry Corp. *Education:* PhD/EE/U of MD; MS/EE/U of MD; BS/EE/U of MD. *Born:* 8/27/28. in Richmond, VA. BS, Mbr Tech Staff RCA Labs princeton, NJ 1952-56; Mgr Microwave Adv tech 1956-59; Prof of Elec Engg Rensselaer Poly Inst 1959-64, Dept Hd 1959-62; Mgr Exploratory Memory, IBM Res Ctr, Yorktown Hts, NY 1964-67; Dir of Engg Tech IBM Sys Dev Div 1967-69; Engg Cons, private practive 1970-74; Deputy for Adv Tech, Office of Asst Secy, US Air Force 1974-81. VP Research and Development, Sperry Div., Sperry Corp, 1981-83. Inpd Conslt 1983- . Registered Prof Engr NY. Eta Kappa Nu Award, Hon Mention 1957; Fellow Award IEEE 1968; US Air Force Decorations for Exceptional Civilian Service, 1979 and 1981; Mbr US Air Force Scientific Adv Bd, 1982- . *Society Aff:* IEEE, AIAA, ACM, ADPA, AOC.

Beamer, Scott
Home: 36 King Ave, Piedmont, CA 94611
Position: Owner. *Employer:* Scott Beamer & Assocs. *Education:* BS/EE/Univ of CA, Berkeley. *Born:* Apr 1914. Mbr Faculty Univ of Calif 1948-63. Part-time teaching regular classes in Architecture & Engrg in Berkeley, & Extension Div in Oakland & San Francisco. Lighting Res at Univ of Calif for State School System. Lighting cons to Bay Area Rapid Transit District Joint Ventures. Served as Major, AUS, Office of Ch of Ordnance 1942-46, in Washington, DC, France & Germany, on Radio Proximity Fuse Proj. Army Commendation Medal. Bronze Star Medal. Disting Service Wreath & Star. Fellow IES; Life Mbr IEEE. Diplomate NAFE; Mbr NSPE; Mbr CSPE. 40 tech articles for prof journals, etc. Patentee. Registered in 3 states. *Society Aff:* IEEE, IES, NAFE, NSPE, CSPE.

Beard, Arthur H, Jr
Business: 6900 Wisconsin Ave, Chevy Chase, MD 20015
Position: President. *Employer:* Arthur Beard Engineers Inc. *Education:* CE/Sanitary Engr/Purdue Univ; BSCE/Civil Engg/Purdue Uni. *Born:* Nov 1917, Chicago, Illinois. After coll grad in 1940, worked with several leading sanitary engrg cons firms in Indiana, Arizona & Florida. Taught water & wastewater courses at the Univ of Arizona. In 1958, formed his own cons firm based in Tucson & Pheonix, Arizona. In 1972 reorganized co as Arthur Beard Engrs, Inc, with additional offices in Maryland & Virginia. Exper in all aspects of sanitary engrg incl water resources & advanced wastewater treatment & reclamation technology. AWWA George Warren Fuller Award. Hobby is collecting stamps. *Society Aff:* ASCE, AWWA, AAEE, WPCF.

Beard, Charles I
Home: 4309 Ann Fitz Hugh Dr, Annandale, VA 22003
Position: Physicist. *Employer:* Retired Aug 1982 *Education:* PhD/Physics/MA Inst of Technology; BS/EE/Carnegie Inst of Technology. *Born:* 11/30/16. Teaching Fellow in physics MIT 1939-41. Radar officer Army Signal Corps 1941-46. In res continuously thereafter at Field Res Labs of Magnolia Petro Co, Applied Physics Lab of Johns Hopkins Univ, Sylvania Electronic Defense Lab, Boeing Scientific Res Lab, & Naval Res Lab. Fellow IEEE. Bolljahn Memorial Award of IEEE G-AP 1962. Chmn of US Comm II of URSI 1970-72. Assoc Editor of Radio Science 1973-75 *Society Aff:* IEEE, APS, URSI, AAAS.

Beard, Henry O
Home: 1310 Huber Lane, Glenview, IL 60025
Position: Vice President & Genl Mgr. *Employer:* Merrill Machinery Div/Pennwalt Corp. *Education:* BS/Chem Engg/MI Tech Univ. *Born:* Jan 14, 1922, Detroit, Mich. Grad from Michigan Technological University 1944 with honors, BS degree in Chem Engrg. Who's Who in Amer Colleges & Universities. Class President. Lettered in Track. With Pennwalt Corp since 1944. Positions held: Supr Production Engrg, Safety Engr, Material Handling Supr, Planning Engr, Mgr of Services Dept. Assumed current pos of VP & Genl Mgr of Merrill Machinery Div for manufacture of pharmaceutical counting equip. Respon for dir, policy & earnings of Div. Mbr of AIChE Natl & Chicago Chap. Active in PMMI, Kiwanis, Rotary, Junior Achievement, scouting, church activities. Enjoy all sports, bridge & gardening. *Society Aff:* AIChE, PMMI, NMA.

Beard, Leo R
Home: 606 Laurel Valley Rd, Austin, TX 78746
Position: Senior Staff Engineer *Employer:* Espey, Huston & Assoc., Inc. *Education:* BS/CE/Caltech. *Born:* Apr 6, 1917, West Baden, Indiana. Hydraulic Engr, Corps of Engrs, Los Angeles, Washington & Sacramento. Retired as Dir of the Corps Hydrologic Engrg Ctr in 1972. Married Marian Wagar 1939, 3 children. Wife deceased. Married Marjorie Pierce Wood 1974. Specialize in water resources mgmt & statistical & mathematical modeling. Emeritus Prof of Civil Engrg Univ. of Texas and Staff Engineer with Espey, Huston & Assoc; Registered Civil Engr in California & Texas; Mbr NAE; Honorary Member & P Chmn of Water Resources Div ASCE; Pres, AGU Section of Hydrology; Honorary Member of AWRA; Mbr. Executive Bd, IWRA; Mbr Water Resources Systems Com, International Asso of Hydrological Sciences; Editor Journal of Hydrology. *Society Aff:* ASCE, AGU, AWRA, NSPE.

Beard, M Craig
Home: 11012 Raccoon Ridge Ct, Reston, CA 22091
Position: Dir of Airworthiness. *Employer:* Federal Aviation Administration. *Education:* BS/Aero Eng/Wichita Univ. *Born:* 6/9/33. BS from Wichita Univ. Native Wichita, KS. Served with USN, Korean War. Aircraft structures engr at Cessna Aircraft Co & LTV. Ch Engr at Wren Aircraft Corp. With FAA since 1965 involved at progressive levels of respon in design & airworthiness cert of civ aeronautical products; holding positions at: Fort Worth, TX; Brussels, Belgium; Wash, DC & LA, CA. Current position as Dir of Airworthiness since Aug 1979; last prior position as Ch,

Beard, M Craig (Continued)
Aircraft Engg Div, FAA Western Region, LA, CA. TX PE 23927 & private pilot. Amateur photographer.

Beard, Samuel J
Home: 364 Westmoreland Dr, Richland, WA 99352
Position: Vice President *Employer:* Exxon Nuclear Co, Inc. *Education:* MS/Mgmnt/MIT; BSChE/Chem Engg/Univ of WA; BA/Chem/Univ of MN. *Born:* 5/3/30. in Elk River, MN. Joined Gen Elec Co in 1954 as a chem engr to investigate chem tech for reprocessing irradiated nuclear fuel. Subsequently held engr and mfg dept mgr positions at Atlantic Richfield when this co replaced Gen Elec as prime contractor to the Atomic Energy Comm. Joined Exxon Nuclear Co in 1971 to head the engg effort for entering the commercial spent fuel reprocessing business. Held position as Gen Mgr of the Laser Enrichment Dept from 1978 to 1981. In 1981 assumed current position as Vice President of Engg and Technology. Mbr of the board of directors of Exxon Nuclear Co., Exxon Nuclear Idaho Co. and Jersey Nuclear Avco Isotopes. Inc. Nominated an MIT Alfred P Sloan Fellow in 1968. *Society Aff:* AIChE, ANS.

Bearden, Larry J
Business: Dept of Biomedical Engg, Birmingham, AL 35294
Position: Asst Prof. *Employer:* Univ of AL. *Education:* PhD/Bio-Engg/Clemson Univ; MS/Biochemistry/Clemson Univ; BA/Chemistry/Berry College. *Born:* 10/27/46. Native of Cartersville, GA. Married, one child. Received PhD in Bioengg, 1976. Selected to "Outstanding Young Men in America," 1977. Presently with Neurosciences Prog and Dept of Biomedical Engg, Univ of AL in Birmingham. Res and teaching interests in (a) the behavioral correlates of the electrical activity of the brain, behavioral neurochemistry, (b) the automation of clinical lab instrumentation and (c) neur stimulation systems. *Society Aff:* AAAS, SN, ΣΞ, IEEE

Bearss, William S
Business: 81-83 E Market St, Buffalo, NY 14204
Position: President. *Employer:* Buffalo X-Ray Company. *Education:* BS/Bus Adm/ Univ of Buffalo. *Born:* Feb 1930. BS from University of Buffalo 1952. Native of Buffalo, NY. Started Buffalo X-Ray Co in 1952 specializing in nondestructive testing & sales of nondestructive equip. Chmn of Western NY Sect for nondestructive Testing 1965-66. Elected Fellow of Amer Soc for Nondestructive Testing 1974. *Society Aff:* ASM.

Beaton, Roy H
Home: 21315 Lumbertown Lane, PO Box 1018, Saratoga, CA 95071
Position: Ret. 1981 *Education:* DSc (hon)/-/Northeastern Univ; D Engg/Chem. Eng./ Yale Univ; BSChE/Chem. Eng./Northeastern Univ. *Born:* 9/1/16. Served Manhattan Project (DuPont Co) 1943-45, Chicago, Oak Ridge, Hanford. With GE 1946-81; Genl Mgr Design & Const Hanford Works 1956-57; Genl Mgr Neutron Devices Dept, Milwaukee 1957-63; Genl Mgr Spacecraft Dept, Philadelphia 1963-64; Genl Mgr Apollo Sys Dept, Daytona Beach 1964-67; VP & Genl Mgr Electronic Sys Div, Syracuse 1968-74; VP & Genl Mgr Energy Sys & Technology Div 1974-75. VP & Genl Mgr-Nuclear Energy Sys Division 1975; VP & Group Executive- Nuclear Energy Group 1977; Sr VP & Group Executive-Nuclear Energy Group 1979; Retired Sept. 1981. *Society Aff:* NAE, AIC, ANS, AIChE, IEEE, AIAA, AAS, NSPE

Beatty, Millard F
Business: Dept of Engng Mechanics, Lexington, KY 40506
Position: Prof. *Employer:* Univ KY. *Education:* PhD/Mechanics/Johns Hopkins Univ; BES/Mech Engrg/Johns Hopkins Univ. *Born:* 11/13/30. Native of Baltimore, MD. Taught at Johns Hopkins (Instr) and Univ DE (Assoc Prof). Joined Univ KY Dept Engg Mechanics in 1967. Named Prof 1976. Author of numerous technical papers which have appeared in Archive for Rational Mechanics and Analysis, Intl Journal of Solids and Structures, Intl Journal of Engg Sci, Acta Mech, Am Journal of Phys, Quarterly Journal of Mechanics and Appl Math, Journal of Appl Mechanics, Journal of Elasticity, Intl Journal of Nonlinear Mechanics, Quarterly of Applied Math and other Soc proceedings. *Society Aff:* SES, AAM, AAPT, MAA, SNP, ASME.

Beatty, Robert W
P. O. Box 591, Boulder City, NV 89005
Position: Consulting Electronics Engineer. *Employer:* Self. *Education:* PhD/Engg/Unif of Tokyo; SM/Elec Communications/MIT; BS/Elec Eng/Geo Washington Univ. *Born:* 5/31/17. Consulting Radio Engr, Washington, DC 1938-39 Jansky & Bailey, 1946-48 Lohnes & Culver; NRL, Anacostia DC 1940-41, Underwater Sound 1941-42, Radio Direction Finding 1942-43, Radar Fire Control 1944, Lt (jg.) U.S. Naval Reserve 1944-46, NBS Washington, DC 1946-54, UHF Stds, Boulder CO 1954-74, microwave stds res. IEEE Fellow 1967. Mbr MTT AdCom 1961-75, Ed IEEE Trans on MIT 1963-65, Chmn TPC & MIT Symposia 1962 & 1973. Chmn URSI Comm I (USA) 1957-60. Mbr URSI 1954- . Author of numerous res articles & monographs. IEEE ARITG Automated Measurements Career Award 1983, IEEE Microwave Theory and Techniques Soc 1975 MIT National Lecturer, IEEE Microwave Theory and Techniques Soc 1987 Microwave Career Award. *Society Aff:* IEEE, URSI.

Beauchamp, James M
Business: 1020 Engg Bldg, Columbia, MO 65211
Position: Dir, Prof. *Employer:* Univ of MO. *Education:* MS/Ind Engg/Lehigh Univ; BS/Ind Engg/Lehigh Univ. *Born:* 11/5/19. Reg PE in MO. VP, NSPE and Chrmn, NSPE PEE Div (1975-1976) NSPE Bd of Dir (1969-1976). Pres MSPE (1971-1972), Trustee MSPE Educational Fdn (1978-1981). Lt USNR - Engg Duty (1942-1946). Instr and Asst Prof of Ind Engg, Lehigh Univ (1946-1951), Asst Plant Engr Olin Mathieson Chem Corp (1951-1955). Sr Service Engr, E I duPont de Nemours and Co (1955-59). Last employer: Univ of MO - Columbia Coll of Engg - Assoc Prof, Ind Engg (1959-1965), Dir, Engg Ext (1965-), Prof (1965-), Retired as Emeritus Prof, Univ of MO-Columbia, 1982. Pres, Columbia MO Northwest Rotary Club (1970-1971). *Society Aff:* NSPE, NYSPE, ASEE

Beauchamp, Jeffery O
Business: 2600 Southwest Freeway, Suite 455, Houston, TX 77098
Position: Pres *Employer:* INTERMAT Intl Materials Management Engrs *Education:* M.S./ME/Univ of Houston; B.S./ME/Univ of Houston *Born:* 1/19/43 in Alice, TX. Married Toni Nobler September 7, 1963. Mechanical Designer at Great Lakes Petroleum Service, Houston, 1963-64; Mechanical engr at Elliott Co., Houston, 1964-68; Research Asst. and Teaching Fellow at the Univ of Houston, 1968-70; Machine Design Chief Engr at Mallay Corp., Houston, 1970-74; Project Mgr at Fluor Engrs & Constructors, Houston, 1974-79; Founder and Pres of INTERMAT Intl Materials Management Engrs, Houston, 1979 to present. Speaker, lecturer, author. Registered PE in TX; member of the National and TX Societies of PE; Outstanding Young Engr of 1974, Past Pres of the Sam Houston Chapter, TSPE; Past Pres, of the Engrs Council of Houston; member of the American and TX Consulting Engrs Council, ASME; Houston Museum of Fine Arts, Contemp Arts Museum, Smithsonian Assoc, Sigma Xi, Phi Kappa Phi, Pi Tau Sigma, and the Amer Arbit Assoc; Past Pres Science & Engrg Fair of Houston, Mbr of Leadership Houston, Dir Westheimer Natl Bank. *Society Aff:* NSPE, ASME, ACEC, ΣΞ, ΦΚΦ, ΠΤΣ, AAA, NACLO.

Beauchemin, Roger O
Business: 1134 St Catherine West, Montreal, Quebec H3B 1H4 Canada
Position: Sr Partner BEAUCHEMIN, Beaton Lapointe, Inc *Employer:* Arrowby Consultants Inc. *Education:* BASc/Engg/Ecole Polytechnique de Montreal/Univ de Montreal *Born:* 5/20/23. BA Sc Ecole Polytech de Montreal. Res & Lec on Concrete, Canada Cement Co 1950-55; 1955- , cons. Major works involve transp planning, 1967 World Exhibition Montreal, several hwys & bridges in Canada, Africa & Mid-East. Dir of CANSULT Ltd, Proj Mgr of Mirabel, Montreal's new internatl airport; Chmn of Montreal Port Authority, C.E.O. Port of Montreal P Pres Inst of Transp Engrs, Canadian Sect; P Pres Montreal Port Coun; was Chmn of Proj Ctte for pub-

Beauchemin, Roger O (Continued)
lishing Manual of Geometric Design for Canadian Rds & Sts'; Dir. Natl Westminster Bank of Canada, Mont Tremblant Lodge Inc; Canadian Marconi Co, Chrmn of the Bd United Provinces Insurance Co, Economic Coun of CD & other numerous Canadian Co's; mbr of Quebec Order of Engr; mbr The Canadian Soc for Civ Engrg, Inst of Transp Engrs & other internatl socs, Recipient of: Canadian Des of Merit Citation in 1966; Awd for services rendered to soc as an engr by the Ecole Polytech Grad Soc 1978. *Society Aff:* CSCE, ITE, ACEC.

Beaufait, Fred W
Business: College of Engrg, Wayne State University, Detroit, MI 48202
Position: Dean & Prof *Employer:* Wayne State Univ. *Education:* PhD/Structural Engg/VPI & SU; MSc/Structural Engg/Univ of KY; BSc/Civil Engg/MS State Univ. *Born:* 11/28/36. Mbr of the faculty of the Civil Engg Dept at Vanderbilt Univ in Nashville, TN for 14 yrs prior to moving to WV. Assumed the duties of Chrmn of Civil Engg at WV Univ in August of 1979; became Assoc Dean for Academic Affairs & Res for the College of Engrg July 1, 1983. Became Dean of the College of Engrg at Wayne State Univ. Sept, 1986. Have held visiting faculty positions at Univ of Liverpool, England, and Univ of Wales - Cardiff, Wales. Authored two textbooks in the area of structural analysis and wrote numerous tech papers. Have been active in ASCE, ASEE and ACI. *Society Aff:* ASCE, ASEE, ACI, MSPE.

Beaver, Howard O, Jr
Home: 320 Hain Ave, Reading, PA 19605
Position: Chrmn & CEO. *Employer:* Carpenter Technology Corp. *Education:* BS/Metallurgy/PA State Univ. *Born:* 5/18/25. *Society Aff:* AIME, ASM.

Beaver, Roy L
Business: 2901 E Lake Rd, Erie, PA 16531
Position: General Manager. *Employer:* General Electric Co. *Education:* BSEE/EE/VPI. *Born:* 8/1/31. Joined GE Co in 1954. Served in Army Signal Corps 1954-56. Exper incl work with power transformers, fractional horsepower motors, industrial control equip, missile propellant systems, medium AC motors, electronic countermeasures, motor drives, automatic test equip, numerical control equip for machine tools & transportation equip. Grad of GE's Advanced Engg Prog & supervised the A Course of the program. Presently General Mgr Transportation Equipment Products Dept. *Society Aff:* IEEE.

Beavon, David K
Home: 1280 Afton St, Pasadena, CA 91103
Position: VP (Retired) *Employer:* Ralph M. Parsons Co *Education:* BS(hon)/Applied Chem/CA Inst of Tech *Born:* 08/02/17. Employed Texaco, Inc, 1938-65 in petroleum refinery technical operations and mgmt; R M Parsons 1965-82. Holder of 39 US pats in fields of sulfur tech, catalytic cracking, alkylation, deasphalting. Developed Beavon Sulfur Removal and BSR/Selectox Processes for control of Claus Sulfur Plant Emissions. Fellow of AIChE, mbr Tau Beta Pi. Publications in sulfur and petroleum refining tech. Various awards for technical achievement, including 1980 Chemical Engrg Practice Award of AIChE and Outstanding Personal Achievement Award of Chemical Engrg. Enjoy woodworking, stained glass, golf. *Society Aff:* AIChE

Beazley, Jon S
Home: 330 Ponce St, Tallahassee, FL 32303
Position: State Topographic Engr. *Employer:* Florida Department of Transportation. *Born:* Retired Head, State Topographic Office since 1964, respon for aerial photography, remote sensing, surveying, photogrammetry & mapping needs of Florida State Govt. Mbr Secy Cabinet appointed Mean High Water & other state advisory cttes. Registered Prof Engr, Civil. Fellow Florida Engrg Soc; Sr VP 1972, Dir 1963- 65; received Outstanding Technical Achievement Award 1974; Dir, Mbr ASP; Mbr Ex Comm 1974- ; Life Mbr Amer Cong Sur & Mp, Dir 1962-65; Hon Mbr Fla Soc Prof Land Surveyors; Honorary Mbr ASCE, April, 1987. Chmn S&M Div 1969; Amer Geo Univ; Amer Assn State Hwy Office & Natl Soc Prof Engrs. Pres No Florida Georgia Tech Alumni Club 1966.

Bebbington, William P
Home: 905 Whitney Dr, Aiken, SC 29801 *Employer:* Retired. *Education:* Phd/Chem Engg/Cornell Univ; B.Chem/Cornell Univ *Born:* 9/10/15. Born in Painted Post, NY. Entire professional career with DuPont: high- pressure synthesis & technology, economic studies & planning. From 1950 until retirement at end of 1974 in Atomic Energy Div at Savannah River Plant, heavy water tech & chem separations. Last 12 yrs in charge of direct tech asst & tech services to nuclear production cycle. Now amateur botanist & plant ecologist. Cons or request in chem engg & environmentl protection. Fellow AIChE, P Chmn Nuclear Engg Dir. P Chmn Nuc Subdiv, ACS. Mbr Southern Appalachian Botanical Club. Recipient: Robert E Wilson Award of the Nuclear Engg Div of AIChE, 1979. *Society Aff:* AIChE

Bechamps, Eugene N
Business: 495 Biltmore Way Suite 301, Coral Gables, FL 33134
Position: President *Employer:* Bechamps Aylward Associates, Inc *Education:* BS/Civ Engr/Univ of Miami. *Born:* 10/7/29. 1953-54 Structural Engr - North Am Aviation. 1954-56 US Army Corps of Engrs, Pusan, Korea. 1956-63 Bridge Design Engr, County Public Works, Miami, FL. 1965-69 Chief, Hgwy Engr, County Public Works, Miami, FL. 1968-70 Mbr, FL Bd of Engg Examiners. 1970-71 State Pres, FL Engg Soc Adjunct Facility - Univ of Miami Sch of Engg. 1969-80 Sr VP & Principal, Carr Smith & Associates Mgr for large Civ, Structural, Hgwy, Bridge & Transit Projs. 1980-1981 President, Bechamps Aylward and Associates, Inc. 1973-79 Mbr, FL Bd of PE & Land Surveyors. 1975-78 Mbr & Chrmn, Uniform Examinations Committee, 1979-80 Pres-Elect, 1980-81, President NCEE. *Society Aff:* FSBPE, NCEE, NSPE.

Becher, William D
Business: E.R.I.M., 3300 Plymouth Rd, Ann Arbor, MI 48105
Position: Owner *Employer:* Widbec Engr *Education:* PhD/Elec Engg/Univ of MI; MSE/Elec Engg/Univ of MI; BS/Radio Engg/Tri-State Univ. *Born:* 11/26/29. Married Helen Hager 1950, children Eric and Patricia. Bogue Elec, Proj Engr, 1950-53; Goodyear Aircraft, Sr Dev Engr, 1953-58; Beckman Instruments, Sr Systems Engr, 1957-58; Bendix Systems, Staff Engr and Sec Supervisor, 1958-63; Univ of MI, Res Engr Willow Run Labs, 1963-68, Prof and Chrmn, Dearborn, 1968- 78, Adjunct Prof Ann Arbor, 1978-79 & 1981- ; NJ Inst of Tech, Dean of Engrg, 1979-81; ERIM, Dept Hd, 1977-79, Assoc Dir 1981-?; WIDBEC Engr, Consultant, 1976-; MI Computers and Instr, Pres, 1983-87; US Army, 1956-55; Interest areas: Computer Programming, Image Processing, digital and analog computer system design, switching theory, radar systems, electronic circuits and systems, digital instrumentation systems, computer-aided design. *Society Aff:* IEEE, ASEE.

Bechtel, Stephen D
Business: 155 Sansome St, San Francisco, CA 94104
Position: Senior Director. *Employer:* The Bechtel Group. *Born:* 1900, Aurora, Indiana. Attended University of California, Berkeley. WWI service in France 20th Engrs, AEF. Sr Dir, Bechtel Group of Co's which he has served for 50 yrs, 25 as Pres, 5 as Chmn. Elected Sr Dir 1965. Recipient first Berkeley Citation, Univ of Calif College of Engrg 1975. Same yr named one of Top Ten Construction Men of past half century by ASCE. Mbr Calif Inst of Tech Assocs, former Trustee Stanford Univ, Dir Emeritus Stanford Res Inst. Hon LLD Univ of Calif Berkeley, Loyola & Golden Gate Univ. Hon Dr Public Serv, Univ San Francisco, Hon LLD Carroll Coll, Moles Award, Golden Beaver Award, John Fritz Medal & Certificate, Medal of Honor, Govt of Indonesia, Conslt Constructors Council of Amer, Amer Petroleum Inst. Hon Dr of Engrg, Univ of Pacific & Washington Univ. *Society Aff:* ASCE.

Bechtel, Stephen D, Jr
Business: 50 Beale St, San Francisco, CA 94105
Position: Chairman *Employer:* Bechtel Group, Inc *Education:* -/Civil Engrg/Student,

Bechtel, Stephen D, Jr (Continued)
Univ of CO 1943-44; BS/Civil Engrg/Purdue Univ (Hon Dr Engrg 1972); MBA/Bus Admin/Stanford 1948 *Born:* 05/10/25 American Engr and Business Exec. Chrmn of the Bechtel Group, Inc 1973- present. Engrg and mgmt positions with Bechtel since 1941. Dir IBM Corp, Chm of Business Council. Former Chrmn Nat'l Acad of Engrg; Chrmn Industry Advisory Brd of Nat'l Acad of Engrg. Life Councillor and past Chrmn Conf Bd. Fellow ASCE. Mbr: Natl Action Council on Minorities in Engrg; Presidents Council Purdue Fdn-Purdue Univ; American Inst of Mining, Metallurgical & Petroleum Engrs. Trustee CA Inst of Tech. Engrg News-Record Man of the Yr 1974. Hon Fellow Institution of Chemical Engrs - U.K. Foreign Member, Fellowship of Engrg - U.K. Mbr, Chi Epsilon, Tau Beta Pi. Moles Award for Outstanding Achievement in Constrn. 1977; Engrg Mgmt Award ASCE 1979; Herbert Hoover Medal 1980; Chairman's Award AAES 1982; President's Award ASCE 1985; Washington Award, Western Soc Engrgs 1985. Reg PE in NY, MI, AK, CA, MD, HI, OH, DC, IL and VA. *Society Aff:* ASCE, AIME, ASEE

Bechtold, James H
Home: 3513 Ridgewood Dr, Pittsburgh, PA 15235
Position: Div Mgr-Materials Sci Div. *Employer:* Westinghouse Research Laboratories. *Education:* BS/Metallurgy/Univ of IL; MS/Metallurgy/Univ of IL. *Born:* May 1922, Boonville, MO. Instructor Met Engrg Univ of Illinois 1948-49; Westinghouse Res Labs 1949- : Res Engr Met Dept 1949-55, Advisory Metallurgist Met Dept 1955, Mgr Met Dept 1955-62, Assoc Dir Met & Ceramic Tech 1962-64, Dir Materials Cons 1964-69, Dir Materials & Processing 1969-74, Div Mgr Materials Science Div 1974- . Head div respon for all R&D on metals & ceramics at Westinghouse Central Res Lab. Mbr ASM, AIME. Bd/Dir Metal Properties Council- Exec & Finance Comms, Mbr Materials Advisory Panel, Mbr Penna Governor's Sci Advisory Ctte; Fellow ASM 1970. Rotary mbr Forest Hills, Pa Chap. Holder 6 patents-granted. *Society Aff:* ASM.

Beck, Arnold H
Business: Univ Engrg Dept, Trumpington St, Cambridge CB2 1PZ England
Position: Prof, Head of Elec Engrg Div. *Employer:* University of Cambridge, England. *Education:* MA/-/Cambridge; BSc/Elec Comm/London. *Born:* 8/7/16. Professor of Engrg 1966-83, Emeritus Professor since 1983, Head of Electrical Div 1971-1981, Univ of Cambridge; Fellow of Corpus Christi College, Cambridge since 1962. Fellow pf Univ Coll, London, 1979. Education: Gresham s School, Holt; Univ College, London. Res engr Henry Hughes & Sons 1937-41; seconded to Admty Signal Estab 1941-45; Standard Telephones & Cables 1947-58. Lec Cambridge Univ 1958-64; Reader in Elec Engrg 1964-66. FIEEE 1959. Publications: Velocity Modulated Thermionic Tubes 1948; Thermionic Valves 1953; Spacecharge Waves 1958; Words & Waves 1967, with H Ahmed; Introduction to Physical Electronics 1968; Statistical Mechanics, Fluctuations and Noise, 1976. Handbook of Vacuum Physics, Vol 2, Parts 5 & 6 1968; Papers in JI.IEE, Inst Radio Engrs etc. Home address: 18 Earl St, Cambridge England. *Society Aff:* IEEE, IEE.

Beck, Charles F
Home: 1508 E Campbell, Arlington Heights, IL 60004
Position: Assoc & Dir of Purchasing *Employer:* Sargent & Lundy Engineers. *Education:* BS/Civil Eng/Ill Inst of Tech. *Born:* 08/07/26. Served with USNR 1944-46. With Sargent & Lundy since 1948 engaged in civil & structural engrg of elec generating stations. Since 1957 implemented computer techniques in solution of complex engrg design problems & in info systems. Appointed Assoc 1967. Hd, Computer Svcs Div since its inception, responsible for software dev and admin of computer facilities. Since 1979 conslt on Engrg & Computer Mgmt. Licensed SE & PE IL. Fellow ASCE. Mbr ACM, Western Soc of Engrs, & ASCE Tech Council on Computer Practices. P Pres, IL Section ASCE. P Chrmn ASCE Power Div. Authored numerous publications. Enjoys classical music, literature, & the outdoors. *Society Aff:* ASCE, ACM.

Beck, Earl J
Business: 150 South Wacker Dr, Chicago, IL 60606
Position: Exec VPres *Employer:* Harza Engrg Co *Education:* MS/CE/Univ of WI; BS/CE/Univ of WI *Born* 1/19/24 Native of Plymouth, WI; worked for United States Bureau of Reclamation - Yuma. Arizona 1946-1947; With Harza Engrg Co since 1947. Exec VPres, 1979 to date; Chief Engr, 1971-79; Elec to Bd of Dir, 1971; V Pres since 1967; mainly engaged in design of large dams and hydroelectric plants on such rivers as the Columbia (USA); Tigris (Iraq); Angelina (USA); Suriname (Suriname); Angat (Philippines). Reg Prof Engr in 11 states. Chrmn, Exec Ctte of Energy Div (ASCE) 1979; Chrmn, Committee on Hydraulics of Dams (USCOLD) since 1975. *Society Aff:* ASCE, USCOLD, NSPE, ACEC

Beck, Edwin L
Business: PO Box 1293, Alburquerque, NM 87103
Position: Director of Transportation. *Employer:* City of Albuquerque. *Born:* May 1928, Texarkana, Texas. Attended University of New Mexico & Tulane University. Served in US Army Corp of Engrs 1951-53. Asst Traffic Engr, City of Tulsa, Oklahoma 1953-57; Palmer & Baker Engrs, New Orleans La 1957-59; Asst Traffic Engr, City of Albuquerque 1959-62; Traffic Engr 1962-73; Highway Program & Planning Engr 1973-74; Dir of Transportation 1975- . In charge of Traffic Engrg, Transit, Highway Programming, Street Maintenance Parking. Pres New Mexico Section ITE 1963, 69, & 73. *Society Aff:* NSPE, ITE, APWA.

Beck, Franklin H
Business: 116 W 19th Ave, Columbus, OH 43210
Position: Prof. *Employer:* OH State Univ. *Education:* PhD/ME/OH State Univ; MS/ME/OH State Univ; BS/ME/PA State Univ. *Born:* 4/5/20. Native of Bethlehem, PA. Metallurgical Engr with E I duPont at Experimental Station, Wilmington, DE; Grasselli Labs, Cleveland, OH; and Manhattan Proj, Richand, WA for period 1943-46. With the OH State Univ from 1946-present-Res Prof, Supervisor and Dir of Metals Res at the OSU Engr Expt Station (1946-70); Res Assoc, Assist Prof and Prof in Dept of Metallurgical Engr (1946-present). Active teaching of Corrosion, Materials Sciences and Mechanical Met. Res in metallic corrosion and alloy development. *Society Aff:* ASM, NACE, ΣΞ.

Beck, James V
Business: Engineering Building, Dept of Mech Engg, E Lansing, MI 48824
Position: Prof. *Employer:* MI State Univ. *Education:* PhD/Mech Engg/MI State Univ; SM/Mech Engg/MA Inst of Tech; BS/Mech Engg/Tufts Univ. *Born:* 5/18/30. Married Barbara Louise Rettie 1960; children Sharon Louise and Douglas James. BSME Tufts Univ 1956; SMME MIT 1957; PhD ME MI State Univ 1964. ASME, Tau Beta Pi, Pi Tau Sigma, Sigma Xi. Worked at AVCO RAD from 1957-1962, MSU from 1962-present. Prof of Mech Engg at MI State Univ since 1971. Published res in areas of heat transfer inverse conduction problems, parameter estimation. Co-author (with K J Arnold) of "Parameter Estimation in Engg and Sci," Joh Wiley 1977. Co-author with B. Blackwell and C. R. St. Clair, Jr., of "Inverse Heat Conduction: Ill-posed Problems," John Wiley 1985. Reg PE Consultant. Prin fields: heat transfer and parameter estimation. *Society Aff:* ASME, AIAA, ΣΞ.

Beck, Paul A
Business: Department of Materials Science & Eng, 1304 W Green Street, University of Illinois, Urbana, IL 61801
Position: Prof Emeritus *Employer:* Univ of IL. *Education:* MS/Metallurgy/MI Tech Univ; Mech Eng/Mech Engg/Budapest Tech Univ; Dr MIn(hc)/-/Austria Tech Univ. *Born:* 2/5/08. Born in Budapest. I came to the US at age 20, became a citizen and worked here until retirement for the Univ of IL, Urbana, after 25 yrs of service. Over 200 tech publications resulting from res in physical metallurgy (recrystallization, grain growth, annealing textures, alloy phase stability) and in the physics of metals (electronic specific heat of the alloys of transition metals, magnetism in alloys). Member National Academy of Engineering. *Society Aff:* TMS-AIME, ASM, APS.

Beck, Paul J, Jr
Business: Indus Engrg B-56, KP, Rochester, NY 14650
Position: Technical Advisor. *Employer:* Eastman Kodak Co. *Born:* Jun 4, 1925. BS Indus Engrg Penn State 1953. US Navy WWII. Continuous employment since 1953, Industrial Engrg Div, Kodak Park, Eastman Kodak Co. Have worked in all phases of industrial engrg. Currently respon for Materials Handling & Facilities Planning Dev, training & cons. VP Education IMMS 1976-77. Certified Prof in Material Handling & Material Mgmt. Chrm. International Fork Truck Rally 1981. Member IIE. Material Handling Rep. for U.S. Commerce Dept. 1978 Leipzig, E. Ger.

Beck, Steven R
Business: Dept. of Chem. Eng, Texas Tech University, Lubbock, TX 79409
Position: Chrmn of Chem Engr *Employer:* TX Tech Univ *Education:* Ph.D./ChE/Univ of TX; B.S./ChE/KS St Univ *Born:* 1/26/47 Native of Manhattan, KS. Five years with Atlantic Richfield Co as Sr. Research Engr working on R&D of oil shale recovery and coal conversion. Officer of Dallas section of AIChE. Joined TX Tech as Asst Prof of Chem Engr in 1977. Named Chrmn in 1981. Research interests include biomass gasification, hydrolysis of lignocellulosic residues, and fermentation of biomass hydrolyzates. *Society Aff:* AIChE, ACS, ΣΞ, ASEE.

Beck, Theodore R
Home: 10035 31st Ave NE, Seattle, WA 98125
Position: Pres. *Employer:* Electrochem Tech Corp. *Education:* PhD/ChE/Univ of WA; MS/ChE/Univ of WA; BS/ChE/Univ of WA. *Born:* 4/11/26. Res Engr, E I du Pont, Deepwater, NJ, 1952-54. Group Leader, Kaiser Aluminum and Chem Corp, Permanente, CA, 1954-59. Res Section Hd, Am Potash & Chem Corp, Henderson, NV, 1959-61. Tech Staff Boeing Aerospace 1961-65 Sr Basic Res Sci. Boeing Scientific Res Labs, Seattle, WA, 1965-72. Res Scientist, Flow Res, Inc, Kent, WA, 1972-75. Pres, Electrochem Tech Corp, Seattle, WA, 1975-. VP, ECS, 1972-75; Pres, 1975-76. Res & Affiliate Prof, Univ of WA, Dept of CE, 1972-80. Served as Consultant to govt (DOE, NSF), 1973-. Reg PE, WA State. *Society Aff:* AAAS, ACS, AIChE, ECS, AIME.

Becken, Eugene D
Home: 52 Rutland Rd, Glen Rock, NJ 07452
Position: Formerly Chmn Executive Ctte Bd/Dir. *Employer:* RCA Global Communications Inc. *Education:* MS/Industrial Mgmt/MIT; MSEE/Communications/Univ of MN; BSEE/Communications/Univ of ND. *Born:* Apr 29, 1911. Employed 41 years until mandatory retirement age with RCA Global Communications Inc, having served in many engrg, operating & executive capacities incl VP & Ch Engr, Executive VP Operations, Pres & Chmn Executive Ctte Bd/Dirs. Mbr Bd./Governors Communication Soc IEEE. Genl Chmn MIT Alumni Ctr of New York. Mbr of Amer Assn for Advancement of Science, NY Acad of Sci, Sigma Tau & Sigma Xi. Fellow IEEE, Radio Club of America & a Sloan Fellow of MIT. Received the de Forest Aud ion Award from the Veteran Wireless Operator Assn & the Disting Alumni Sioux Award from Univ of North Dakota; Disting Service Award MIT Alumni Assoc; Prof Engr lenses in the States of New York & New Jersey. Active in photography, golfing & travelling, Mayor Glen Rock, NJ. *Society Aff:* IEEE, AAAS, NYAS, RCA, VWOA, AFCEA.

Becker, Edward P
Home: R D 7 Seidersville Rd, Bethlehem, PA 18015
Position: Chief Engineer. *Employer:* Lehigh Structural Steel Co. *Education:* MS/Civil Engg/Lehigh Univ; BSCE/Civil Engg/Lehigh Univ. *Born:* 10/20/29. Native of Bethlehem, PA. Grad with hon Lehigh Univ, 1951 BSCE; MS Lehigh Univ 1954. Army Corps of Engrs Korean War; Design engr with Berger Assoc. & Gannett, Fleming, Corrdry & Carpenter, Inc. Harrisburg, PA. spec in steel structures 1954-61. Joined Lehigh Structural Steel Co 1961, structural engr 1963, since 1968 Ch Engr in charge of design & welding. Fellow ASCE; Registered Prof Engr PA & SC; Lincoln Arc Welding Awards 1964, 73, 81 & 86; Pres PSPE 1977-78; Natl Dir NSPE 1974-79. VP PSPE 1974-76; Pres PSPE (LV) 1971; Chmn AWS (LV) 1970; Pres ASCE (LV) 1968 ASCE NATL Comm on Registration of Engrs. 1978-82, Sec 1979- 80 V.C. 1980-81, Chairman 1981-82; Engr of the Yr-PSPE (LV) Chapter 1979; PA. Engr of the year 1980 - PSPE; *Who's Who in East*, 17th & 18th Editions; *Men of Achievement* 7th Ed by Inter. Bio. Center, Cambridge England; *Who's Who Finance & Industry*, 22nd Edition; NSPE Environmental Comm 1981-82; AISC Bldg Task Force (1983-84); ASCE Natl Task Ctte on Design Resp. Chrmn (1983-85); Technical lecturers before ASCE, AISC & AWS sections, ASCE Natl Conv (1981) AICS Natl Conv (1985) and Struct Stability Res Council (1979); Pub of Tech Papers in "Struct Stab Res Council Proceeding 1979" –Modern Arc Welded Structures" Vol IV, 1980, & ASCE proceedings - "Journal of Prof. Issues in Engr', Apr 1986. *Society Aff:* NSPE, ASCE, AWS, PSPE, AISC.

Becker, Gerhard W M
Business: Bundesanstalt fur Materialforschung und-prufung BAM Unter den Eichen 87 Berlin 45, Germany D-1000
Position: President *Education:* Dr. rer. nat./Physics/Techn. Univ. Braunschweig; Dipl. -Phys./Physics/Techn. Univ. Braunschweig *Born:* 08/13/27 *Society Aff:* ASME

Becker, Norbert K
Business: 875 Ovellelk Ave, Windsor, Ontario Canada
Position: President. *Employer:* N K Becker & Assocs Ltd. *Education:* PhD/Civil Engg/Univ of Windsor; BASc/Civil Engg/Univ of Windsor. *Born:* Oct 1944. PhD from University of Windsor in Civil Engrg. Deputy Mgr & Assoc M M Dillon Ltd, Cons Engrs & Planners until 1976, specializing in both structural & municipal engrg. Also taught Civil Engrg at Univ of Windsor 1970-71 & has been Res Assoc at the Univ since 1971. Has authored several technical papers in appl mechanics of concrete & received the Wason Medal for Materials Res from the ACI in 1975. 1976-78 Respon for the construction & engrg operations of the Collavino Brothers Construction Co, a major Canadian contractor active in Canada, the USA & abroad. Since 1978, pres of N K Becker & Assocs Ltd, a co engaged in Civil Engg & property dev. Active in Canada, the USA & Egypt. Also Exec VP of Briarwood Properties a Land Dev Co active in MI. *Society Aff:* ACI, ASCE, CSCE, APEO.

Becker, Peter E
Business: 3900 S Clinton Ave, South Plainfield, NJ 07080
Position: Executive Vice President. *Employer:* Metz Metallurgical Corporation. *Education:* BS.Ch.E./Chemical Engineering/NCE/Newark Institute of Technology *Born:* Oct 1939. BSChE Newark College of Engrg 1964. US Army Intellegence 1956-59 Far East. Chemical & Mechanical Process Engr for Metz Refining Co 1964-67. VP Metz Metallurgical Corp 1967. 1968 Elected to Board of Directors, assumed manufacturing & dev respon of precious metals in wrought, powder, & chemical forms for electrical, electronic & aero-space applications. 1971 Executive Vice President. 1986 Division Manager Metallurgical Products; member Edward Weston Society, New Jersey Institute of Technology. Mbr of ACS, AIChE, Internatl Soc Hybrid Microelectronics, IEEE, IMME. *Society Aff:* ACS, IEEE, ISHM, IPMI, IMME

Becker, Roger J
Business: P.O. Box 351, Mt. Laurel, NJ 08054
Position: Mgr Engrg Services *Employer:* Oakite Prods Inc. *Education:* BIE/-/OH State Univ. *Born:* Oct 1920. BS Indus Engrg Ohio State University, Tau Beta Pi Hon. University Chicago Grad School of Bus. Native of Toledo, Ohio. Served USAAF, 2nd Lt 1944. Hoover Co, indus engr, Assembly Supt 1946; Tech Service Rep Oakite Products, Inc 1949, Mgr Petroleum & Chem Div 1965. Formed Oakley Serv Co Div 1968. Accredited Corrosion Specialist, Natl Assn Corrosion Engrs. Registered Prof Engr Illinois. Assoc Mbr AIChE; VP Barbie Productions Inc, former Dir Northeastern Maintenance Services Inc 1973. Hobbies incl wood carving, canoeing, bicycling, darkroom photography. *Society Aff:* AIChE, NACE, CTI.

Beckett, George A
Home: PO Box 97, St Louisville, OH 43071
Position: Conslt Plant Mgr *Employer:* Self *Education:* BSc/ChE/Northeastern Univ;

Beckett, George A (Continued)
Adv Mgt/Banff School of Adv. Mgt *Born:* 11/12/19. Army Corps Engrs (Manhattan Proj Oak Ridge, Los Alamos) 1944-46. With Georgia-Pacific Corp & predecessors since 1947, Vancouver, B.C.; Seattle Washington; Newark, Ohio. Technical, sales, production, engrg, mgmt. Present position Nov. '85 to Present-Conslt-Plant Mgr. Saranda Corp, Newark, Ohio. Internatl Pres IMMS 1971-72. Mbr Chamber/Commerce; Rotary Club; Score, mbr-chrmn numerous community/business ctte. Enjoys gardening & sports. *Society Aff:* IMMS, SCORE

Beckett, John C
Home: 260 Coleridge Ave, Palo Alto, CA 94301
Position: Conslt *Employer:* self *Education:* AB/Engineer/Stanford Univ; EE/Elec Engr/Stanford Univ *Born:* 8/6/17 Goldfield, Nevada. Grad Stanford Univ, AB Engrg 1938, EE 1941. Phi Beta Kappa, Tau Beta Pi. Fellow IEEE; Section Chrmn, AIEE & IEEE; Chrmn AIEE Ctte Elec Tech in Medicine & Biology. Named Bay Area Elec Engr of Yr 1956, IEEE Centennial Award 1984, San Francisco Chap Engrs Joint Council. US Navy WWII, Cdr USNR, Ret. Retired Govt Relations Dir HP Co, Palo Alto, CA. Formerly Pres & Genl Mgr Palo Alto Engr Co, Engr Co, subsidiary HP; Ch Engr Wesix Elec Heater Co, San Francisco 1945-60. Public Serv: Chrmn Metropolitan Trans Comm, San Francisco Bay Area 1973-76; Founding Mbr BART Bd & Comm. Author of tech papers on air quality in living spaces & urban mass transit in the auto age. Mbr Bd of Dirs ANSI, 1981-83. Currently conslt Hewlett Packard Co TRW Defense Electronics Group, Defense Science Board (DOD) & special limited partner of Alpha Partners Venture Capitalists, Menlo Park, CA. Bd of Dirs, Cochlea Corp, San Jose, and Microbot Co, Sunnyvale, CA. Senior Advisor to the President's Blue Ribbon Commission on Defense Management 1985-86. *Society Aff:* IEEE

Beckjord, Eric S
Business: US ERDA RDD, Washington, DC 20545
Position: US Coordinator, Intl Nuclear Fuel Cycle Evaulation. *Employer:* US Dept of Energy. *Education:* AB/Physics/Harvard Univ; MSEE/Elec Engg/MIT. *Born:* Feb 1929. MSEE MIT; AB cum laude Harvard. LT, jg, USNR 1951-54. General Electric Co 1956-63: Dev Engr, Boiling Water Reactors 1956-60; Proj Engr, Nuclear Reactor Thermionic Conversion 1960-63. Westinghouse Electric Corp Nuclear Energy Systems 1963-75: Mgr, Plant Dev 1963-66; Mgr Systems Engrg 1966- 68; Engrg Mgr, Nuclear Fuel Div 1968-70; Engrg Mgr, Pressurized Water Reactors 1970; VP & Tech Dir, Westinghouse Nuclear Europe 1970-73; Dir, Uranium Enrichment Operations 1973-74; Mgr Strategic Planning, Water Reactor Div Bus Unit 1975. Federal Energy Admin, Assoc Dir for Policy Analysis, Office of Nuclear Affairs 1975. Energy R&D Admin, Dir of Div of Reactor Dev & Demonstration 1976-77; respon for the Liquid Metal Fast Breeder Reactor Program Dir of Div of Nuclear Power Dev, US Dept of Energy 1977-78; Respon for Light Water Reactor Dev and Safety; Nuclear Fuel Cycle; Gas-Cooled Reactors. US Coordinator for Intl Nuclear Fuel Cycle Evaulation, US Dept of Energy 1978-79. Lectr at Norwegian Inst for Atomic Energy 1962. Lectr at MIT in nuclear reactor safety 1966-67. Chmn ANS Ctte for Pre. *Society Aff:* IEEE, ANS.

Beckman, Arnold O
Business: PO Box C-19600, Irvine, CA 92713
Position: Chairman of the Board. *Employer:* Beckman Instruments Inc. *Born:* Native of Cullom, Ill. BS in Chem Engrg 1922, MS Physical Chem 1923 at University of Illinois. Received Doctorate in Photo-chemistry in 1928 at California Institute of Technology. Served on the Caltech chemistry faculty until 1940 when he left the teaching profession to devote full time to the dev & manufacture of scientific instruments. Founder of the Instrument Soc of Amer, a natl prof assn; Pres ISA 1952. Hon Life Mbr ISA 1959. Mbr NAE, ACS, the Newcomen Soc, & an Hon Mbr of Amer Inst of Chemists; Fellow of the Amer Assn for the Advancement of Sci, & a Benjamin Franklin fellow of Great Britians Royal Soc of Arts. Many yrs leading role in campaign against air pollution. In 1970 was named by Pres Nixon to the Fed Air Quality Advisory Bd.

Beckmann, Petr
Business: Elec Engrg Dept, Boulder, CO 80309
Position: Professor. *Employer:* University of Colorado. *Education:* Dr Sci/EE/Czeschoslovakia Acad of Sciences; PhD/EE/Prague Tech Univ; MSc/EE/Prague Technical Univ. *Born:* 1924, Prague, Czechoslovakia. Dept Hd, Inst of Radio Engrg, Czechoslovak Acad of Scis until 1963 when invinted to Univ of CO; defected to US and stayed as Prof of Elec Engrg. Author of 60 scientific papers & 10 books, incl "Depolarization of EM Waves–, "Scattering of EM Waves from Rough Surfaces–, "Probability in Communication Engrg–, "Orthogonal Polynomials for Engrs & Physicists–, "The Structure of Language–. Fellow IEEE; Reg PE CO. Mbr Amer Nuclear Soc Health Phys Soc Sigma Xi. Appointed by Ronald Reagan to Energy Task Force in Sept 1980. Editor & publisher of monthly *Access to Energy. Society Aff:* IEEE.

Beckmann, Robert B
Business: Chem Engg, College Park, MD 20742
Position: Prof. *Employer:* Univ of MD. *Education:* PhD/ChE/Univ of WI; BS/ChE/Univ of IL; -/ChE/Univ of OK. *Born:* 9/15/18. Native of St Louis, MO and Buena Vista, CO. Res Chem Engr for Humble Oil & Refining Co, Baytown TX (1944-46); mbr of Chem Engg faculty, Carnegie-Mellon Univ (1946-61); chrmn of Chem Engg (1961-66) and Dean of the College of Engg (1966-77) at Univ of MD; currently Prof. Chrmn, Ind & Engg Div of ACS, 1966; chrmn of Engg sec AAAS, 1978; Pres of Engrs Council for Profl Dev, 1975, 1976. Currently mbr of the Bd of Dirs of Versar, Inc, Oak Ridge Associated Univs (ORAU) and council on Postsecondary Accreditation(COPA). *Society Aff:* AIChE, AIC, ACS, ASEE, AAAS.

Beckwith, Robert W
Business: 11811 62nd St N, Largo, FL 33540
Position: Chairman. *Employer:* Beckwith Electric Co Inc. *Born:* Jul 1919. BS in EE Case Western Reserve, MEE Syracuse University. Native of Kent, Ohio. Spent 15 yrs in Power Line Carrier Engrg, GE & 6 yrs Mgr Computers Electronics Lab GE with Gulton Industries 3 yrs. Formed Beckwith Electric Co May, 1967. Fellow IEEE for pioneer work in FSK telemetering 1944. *Society Aff:* IEEE, FES, NAPE.

Beckwith, Sterling
Home: 1824 Doris Drive, Menlo Park, CA 94025
Position: Consulting Engineer. *Education:* PhD/Elect Engr/Caltech; MS/Elect Engr/Univ of Pittsburgh; AB/Elec Engrg/Stanford. *Born:* 10/29/05. Elec design Westinghouse 1927-32, MWD 1933-35, Allis-Chalmers 1935-52. Cons Engr since 1952. Design of large generators and motors. Cyrogenic tank design/const for pioneering LNG transporation. Dev & design of first air curtain cabinet fpr frozen foods. Lamme Medalist IEEE 1958. *Society Aff:* IEEE.

Bedrosian, Samuel D
Business: Moore Sch 200 S 33rd St, Philadelphia, PA 19104-6930
Position: Prof Electrical Engrg. *Employer:* University of Pennsylvania. *Education:* PhD/EE/Univ of PA; MEE/EE/Poly Inst of NY; AB/Math Scis/SUNY at Albany. *Born:* Mar 1921. Served in US Army Signal Corps SW Pacific 1943-46. SC Engrg Labs, Ft Monmouth, dev of circuits & multichannel systems 1946-55. Systems Engr at Burroughs Res Ctr, Paoli, data processing problems in lrg systems. With Univ of Pennsylvania since 1960; teaching & res computer-communications, fractal-graphs for lrg scale networks, fuzzy sets and image processing, graph theory & automatic fault diagnosis. Chairman of SE Dept 1975-80. Director M.B.A./M.S.E. Dual Degree Program 1977-82, 1980-81 NAVELEX Research chair Prof. at Naval Postgraduate School, Montery, CA. Assoc Editor and Mbr of Committee on Sci & the Arts J Franklin Inst; Fellow IEEE. General Chairman IEEE Int Symp. on Circuits & Systems May 1987 (in Phila). Cons. *Society Aff:* IEEE, HKN, ΣΞ.

Bedworth, David D
Business: Arizona State Univ, Tempe, AZ 85287
Position: Prof of Engrg *Employer:* Arizona State University. *Education:* PhD/Industrial Eng/Purdue; MSIE/Industrial Eng/Purdue; BSIE/Industrial Eng/Lamar Univ. *Born:* 11/29/32. Manchester, England. Worked as indus engr with Boeing, Pure Oil & General Electric. Served with US Navy 1955-57. With Arizona State Univ since 1963. Appointed Full Professor 1969 & Chair 1976-1980. Won IIEs Book of Yr Award (for Industrial Systems: Planning, Analysis & Control) 1974; Co-Receipient of ASEE's W.E. Wickenden Award for Best Paperin Engineering Education-1984. Natl Regional Dir IIEs Production Control Div 1976; editorial Bd IIE Transactions 1969-74. Mbr Natl Executive Council Alpha Pi Mu 1965-88, Natl Pres 1978-80. Registered Arizona Prof Engr. Enjoy classical music, gardening, mysteries, jogging & tennis. *Society Aff:* ASEE, IIE, SME.

Beech, Gary D
Home: 1701 West Glen Oaks Lane, Mequon, WI 53092
Position: Deputy Program Director *Employer:* CH2M Hill *Education:* MS/Civil Engr/Univ of IL; BS/Mil Engr/US Military Academy *Born:* 04/15/37 Native of Parsons, KS. Serv with Army Corps of Engrs 1959-85. Served with engr troop units in US, Germany and Korea. Commanded engr constr battalion. In contract constr mgmt since 1971 as resident engr, area engr and 1982-85 as the St. Louis District Engr. Respon for navigation and flood control proj in MO and IL. Joined CH2M Hill in 1985. Currently Deputy Program Director for CH2M Hill on $1.8B Milwaukee Water Pollution Abatement Program. *Society Aff:* ASCE, SAME, PMI

Beedle, Lynn S
Business: Lehigh University, Bldg. #13, Bethlehem, PA 18015
Position: Director, High-Rise Institute *Employer:* Lehigh University. *Education:* BS/CE/Univ of CA, Berkeley; MS/CE/Lehigh Univ; PhD/CE/Lehigh Univ. *Born:* 12/7/17. in Orland, CA. Attended grade schs in Hollywood & Glendale. Short period with Todd-CA, shipbuilding corp; US Navy 1941-47. Taught in Post-grad Sch at US Naval Acad & served as Officer-in-Chg of Underwater Explosions Res at Norfolk Naval Shipyard. Participated in tests of atomic bombs at Bikini. Lehigh Univ as res engr 1947. Now Univ Disting Prof of Civil Engrg & Dir of High-Rise Inst; res at Lehigh: studies in plastic design in steel structs, influence of residual stresses, high-strength bolts, behavior of welded plate girders & planning & design of tall bldgs. Served in work of AREA, AWS, SSRC, WRC, AISC & ASCE. Currently Struc Stability, Dir of Res Council & Dir of ASCE-IABSE-AIA-UIA-AIP- IFHP Council on Tall Bldgs and Urban Habitat. Board of Dir., Lehigh Valley Sect. ASCE; Past Int'l Contact Dir. ASCE; Mbr NAE. Const Indus Award ENR 1966 & 1972. Silver Medal Am Wdd Soc., 1958 Regional Techncial meeting award of An Ivon & Steel Inst.; 1973 TR Higgins Award, AM Inst Steel Construction; 1977 Engr of Year Award, Lehigh Valley Sect, Natl Soc Profl Engrs. Av Plastic Design of Steel Frames. Editor in Chief, Mgr of Planning & Design of Tall Bldgs. Hon Mbr IABSE. Life Mbr SSRC. Hon Mbr ASCE. *Society Aff:* ASCE, NAE, SSRC, AWS, CTBUH, ULI.

Beehler, James E
Home: 256 Melbourne Pl, Worthington, OH 43085
Position: Consult Elec Engr. *Employer:* Self *Education:* MS/EE/Purdue; BS/EE/Purdue *Born:* March 1923 Indiana. BS, MS Purdue Univ. Electronics Technician US Navy WWII. Reg Prof Engr Indiana. 1948, Ohio 1982; relaying work with AEP System's I&M Electric Co; transferred to AEP Serv Corp NY 1957 as Sr Engr; 1959 transferred to Switchgear; 1972 promoted to Cons Elec Engr. IEEE Fellow; P Chmn Switchgear Ctte; Mbr PCB Subctte & working groups; P Mbr IEEE Standards Bd, P Chmn PES Standards Coordinator's Ctte; Chmn PES Tech Operations Dept; Mbr AEIC Ctte on Elec Power Apparatus; V Chmn ANSI C37 Ctte on HV Circuit Breakers. Chmn EL & P C37 Ctte, IEEE Awards: Switchgear Ctte, 1980; Standard Boards Medallion, 1984; Centennial Medal, 1984. *Society Aff:* IEEE, CIGRE.

Beeman, Ogden
Business: 319 S.W. Washington St, 614, Portland, OR 97204
Position: Principal *Employer:* Ogden Beeman & Assocs *Education:* Dipl Engr/Hydraulics/Tech Univ of Delft; B.S./CE/Stanford Univ *Born:* 3/13/35 Served as ships officer U.S. Coast & Geodetic Survey, 1956-58. Civil Engr in charge of dredging for Corps of Engrs, Portland District 1961-67. Served as consultant to U.S. Army and U.S. St Dept in S.E. Asia 1965-67. Marine Dir, Port of Portland, in charge of marine terminal planning, development and operations 1967-74. Project Mgr, Korea Ports Study 1974-76. Founded Ogden Beeman & Assocs, consulting engineers 1976. Consulting in development of Ports, Waterways and Marine Facilities in U.S., Canada, Mexico, Africa and East Asia. Author OR Ports Study (1980) and numerous professional papers. Active in civic affairs including member and past pres, City Club of Portland. *Society Aff:* ASCE, WEDA

Beer, Craig E
Business: Agri Engrg Dept, Ames, IA 50011
Position: Professor. *Employer:* Iowa State University. *Education:* PhD/Civil Eng-Agri Eng/IA State Univ; MS/Agri Eng/IA State Univ; BS/Agri Engr/IA State Univ. *Born:* Feb 1927. PhD/MS/BS Iowa St Univ. Native of Keosauqua Iowa. Employed by USDA, SCS part time during grad work; present position half res, half teaching in Soil & Water Engrg. Res activity in erosion, drainage, irrigations; current proj leader in multidisciplinary proj in On Site Sewage Disposal. Tech Dir Soil & Water Div, ASAE & Mbr Bd of Dir ASAE. Hobbies: hunting, fishing, woodworking. Reg Engr. *Society Aff:* ASAE, SCSA, ASEE.

Beer, Janos M
Business: Dept of Chem Engg, 25 Ames St, Cambridge, MA 02139
Position: Prof. *Employer:* MIT. *Education:* DSc/ChE/Univ of Sheffield; PhD/ChE/Univ of Sheffield; BSc/ME/Budapest Univ. *Born:* 2/27/23. Native of Budapest, Hungary. Taught combustion engg at the Tech Univ of Budapest 1952-56; Res engr Babcock & Wilcox, Renfrew Scotland 1957; CEGB Res Bursar, Univ of Sheffield, England 1958-60; Hd, Res Station, Intern Flame Res Fdn, Ijmuiden, Holland 1960-63; Prof of Fuel Science, Penn State Univ. USA; Prof of Fuel Tech & Chem Eng Univ of Sheffield, England 1965-76; Dean of Engg Univ Sheffield 1972-75; Prof of Chem and Fuel Engg, Dept of Chem Engg MIT Cambridge, MA. Sc. Director MIT Combustion Research Facilities; ASME Moody Award 1964; Australian Commonwealth Visiting Fellowship Award 1972; elected to the Fellowship of Engg (UK) 1979; Fellow of ASME 1975. Melchett Medal, Inst. Energy; 1985; Alfred Egerton Gold Medal Combustion Institute 1986; Coal Science Gold Medal BCURA 1986. *Society Aff:* ASME, AIChE, InstF., F. Eng.

Beers, George L
Home: 3622 N. Ocean Shore Blvd, Ormond Beach, FL 32074
Position: Retired. *Education:* BS/EE/Gettysburg College; ScD/Gettysburg College. *Born:* May 1899. Held various admin positions in res & engrg at RCA; retired 1964. N A M Modern Pioneer Award 1940; Fellow IEEE, SMPTE; Mbr Natl Television Systems Ctte that proposed the standards for color television; lic PE in NJ; over 80 US Pats. *Society Aff:* IEEE, SMPTE.

Bees, John H
Business: Old Ridgebury Rd, Danbury, CT 06817
Position: Division Pres *Employer:* Union Carbide Corp *Education:* MS/ChE/WV Univ; BS/ChE/WV Univ *Born:* 2/25/27 in Huntington, WV. Mr Bees joined Union Carbide in 1952 as a production supvr at the corp's chems facility in Inst, WV. He moved to the Central Purchasing Dept in the Corp's NY office in 1964. He became asst mgr for agricultural chems in 1966, operations mgr for agricultural chems and amines in 1967, and product dir of agricultural chems in 1972. In 1973 he was appointed a VP of Chems and Plastics, and gen mgr of the Agricultural Chems Dept. In 1975, he became gen mgr of the Industrial Chems Dept, and in 1977 was named gen mgr of the ethylene oxide/glycol and derivatives dept and VP of the former Industrial Chems Div. He was appointed pres of the Ethylene Oxide/Glycol Div in Dec 1979. He lives in West Redding, CT. He is married to the former Bonita Jean

Bees, John H (Continued)
Nielsen and has three sons, John H. Jr., Terrance J.; and David L. He served in US Army from 1944-1947. *Society Aff:* AIChE, ΤΒΠ, ΣΞ, ΦΛΥ, ΣΓΕ

Begell, William
Home: 46 East 91st St, New York, NY 10028
Position: President. *Employer:* Hemisphere Publishing Corp. *Education:* MChE/Chem Engg/Polytechnic Inst-Brooklyn; BchE/Chem Engg/CCNY; D. Sci/Heat Transfer/Ac Sci BSSR *Born:* 5/18/28. Native of Poland. Interned WWII in German concentration camp, escaped & finished high school in Munich. Arrived US 1947, naturalized citizen. 1953 joined res staff & faculty, Dept of Chem Engrg Columbia Univ; headed Heat Transfer Res Facility. 1961 founded & became Exec VP of Scripta Technica; 1974 founded & became Pres of Hemisphere Publishing Corp. Founded, 1975, Washington Publishers Group/V Chrmn, Exec Council, Profl & Scholarly Publishing Div, AAP; Chrmn, Govt Relation Comm, AAP; Mbr, Center for the Book, Library of Congress; Rep of ASME to AAES Publishing & Services Overview Comm; Chm, Budget Comm, ASME Policy Board of Communications; Mbr at Large, Communications Board, ASME; Mbr, Bd of Dir, AAP; Mbr, Exec Council of Intl. Centre for Heat & Mass Transfer and recipient of ICHMT Award of Merit; recipient B. Gomez Award for distinguished achievement in publishing; Lecturer on publishing, George Washington Univ, and NY Univ, Conslt, Heat Transfer Lab, Columbia Univ, Coeditor, forthcoming Handbook of Journal Publishing; Editor of 7 books; contributor of numerous articles to professional journals in the field of heat transfer; patentee in the field. *Society Aff:* AIChE, ASME, ASEE, AAP, ICHMT, AAAS, ANS, SSP

Beggs, H Dale
Home: 108 E 131 St, Jenks, OK 74037
Position: Petroleum Consultant *Employer:* Self *Education:* PhD/Pet Engr/Tulsa Univ; BS/Physics/Tulsa Univ; BS/Math/Tulsa Univ *Born:* 12/12/35 Native of Tyler, TX; 7 years as consulting Petroleum Engr; Asst Prof Petr Engr at LA Tech Univ, 1971-73; Assoc Prof Petr Engr at Tulsa Univ, 1973-1981. Author of numerous technical papers; co-author of books Two-Phase Flow in Pipes and Technology of Artificial Lift Methods, and author of book Gas Production Operations. SPE Distinguished Lecturer, 1987-1988. Currently teaching short courses and consulting in production engineering. Registered PE in OK and LA; petroleum consultant. *Society Aff:* SPE of AIME, ASEE

Begun, Semi J
Business: 3109 Mayfield Road, Cleveland Hts, OH 44118
Position: President. *Employer:* Auctor Assocs Inc. *Education:* Phd/Engrg/Berlin Inst of Tech; Master/Engrg/Berlin Inst of Tech; Bachelor/Engrg/Berlin Inst of Tech *Born:* Dec 2, 1905. Undergrad & grad studies Berlin Inst of Tech, Doctor of Engrg 1933. Contrib to sound & data recording- mech & by way of magnetic recording; author of Magnetic Recording 1949, reprinted 5 times in US; also printed in UK. Pres Citation of Merit for mbrship of Natl Defense Res Ctte; Emil Berliner Award Audio Engrg Soc. Fellow Acoustical Soc, IEEE. VP Technology Brush Dev Co 1938-52, Clevite Corp 1952-69, Gould Inc 1969-70. Auctor 1970-present. Life fellowship IEEE. Director Emeritus-Bally Mfg, Pyromatics, Inc. The IEEE Centennial Medal 1984 John H. Potts Memorial Awd for Outstanding Achievement in the Field of Audio Engrg - Audio Engrg Soc 1960. *Society Aff:* IEEE, ASA

Behal, Victor G
Business: PO Box 460, Hamilton, Ontario Canada L8N 3J5
Position: Engg Matls Specialist. *Employer:* DOFASCO Inc. *Education:* Cert/Metallurgy/McMaster Univ. *Born:* 3/2/22. Prague Czechoslovakia. Studied Chem, Physics & Metallurgy; McMaster Univ. With Metallurgical Dept at Dofasco 1942, responsible for Heat Treating & Nondestructive Testing Depts; later all steel castings specifications & Engrg Dept metallurgical specifications. ASM: Ontario Chapter Chmn 1960-61, Canadian Council Chmn 1964-66; Bd of Trs 1969-71; Tech Div Bd 1972-74; Fellow 1976. ASNT: Ontario Sect Chmn 1956-57; Canadian Soc for NDT: Natl Pres 1967-70; CSNDT Foundation: Chmn of Bd of Trs 1976-80 ; Can Inst Mining & Met Chmn Hamilton Branch 1960-61, Natl Councillor 1961-64. Can Govt Specifications Bd Certificate of Merit 1975; Steel Founder's Soc of Amer: Spec Ctte Chmn 1974-. Active in ASTM, CSA, WIC, ISO. ASNT Fellow 1978. W E Havercroft Award from CSNDT-1978; Charles N Briggs Meml gold Medal from Steel Founders' Society of America-1979. Award of Merit and Fellow ASTM 1981. DOFASCO Materials Engineering Award from the Comochiam Institute of Mining and Metallurgy - Mettallurgical Society - Materials Engineering Section - Winner 1981. Hon Life Mbrship, CSNDT. *Society Aff:* ASM, ASTM, ASNT, CGSB, CIMM, CSA, CSNDT, CSNDTFM, SFSA, ISO, WIC

Behnke, Wallace B, Jr
Business: PO Box 767, Chicago, IL 60690
Position: Vice Chrm. *Employer:* Commonwealth Edison Co. *Education:* BS/EE/Northwestern Univ. *Born:* 2/5/26. Native of Chicago, IL. Served in US Naval Reserve 1943-54. With Commonwealth Edison Co since 1947, serving in various engg and admin capacities. Elected VP in 1969, Exec VP in 1973, with respons for engg, operations, Dir in 1979 and VChrmn in 1980; Chmn and Dir of Proj Mgmt Corp since 1972, involved in liquid metal fast breeder reactor dev and demonstration. Pres WSE 1977, and Hon Mbr, Chrm AIF 1982-83, Fellow IEEE 1978, VP Power Engr. Society Also, Dir of Lake View Trust and Savings Bank, Duff & Phelps Selected Utilities, Inc., Calumet Industries Inc., Tuthill Pump Co. Atomic Industrial Forum, Il Inst of Tech, Research Inst., Materials Properties Council, Northwestern Mem Hosp, United Way of Chicago, and Protestant Fnd. Bd of Gov. Argonne National Laboratory. Reg Eng, State of IL. NAE 1980. *Society Aff:* IEEE, WSE, ANS.

Behrend, William L
Home: 479 Carnegie Drive, Pittsburgh, PA 15243
Position: Retired *Employer:* RCA *Education:* MS/EE/Univ of WI; BS/EE/Univ of WI. *Born:* 1/11/23. Native of Wisconsin Rapids Wisc. Served US Navy 1944-46. Res Engr 1947-64 for David Sarnoff Res Center spec in television res; with RCA Broadcast Systems 1964-1984. Responsible for Advanced Development & systems analy in Broadcast Transmitters. Independent Conslt 1984- . Sig Xi 1950; IEEE Scott Helt Memorial Award 1971; IEEE Fellow 1975; RCA Corp Awards 1950/56/59/63/75. Enjoy ice skating, skiing & swimming. *Society Aff:* AAAS, IEEE, ΣΞ, NYAS.

Beimborn, Edward A
Business: Center for Urban Transp Studies, Milwaukee, WI 53201
Position: Prof *Employer:* Univ of WI-Milw *Education:* PhD/CE/Northwestern Univ; MS/CE/Northwestern Univ; BS/CE/Univ of WI-Mad *Born:* 1/18/42 Prof Beimborn is the Dir of the Center for Urban Transportation Studies and Prof of Civil Engrg at the Univ of WI-Milwaukee. He has been involved in a wide range of research, planning and teaching activities which have focused on: mass transit systems, transportation impacts, evaluation procedures, transportation planning, research utilization, and freight transportation. He has experience with the City of Milwaukee, States of WI and AK and the National Bureau of Standards. He also has been a visiting Fellow at the Transport Studies Unit of Oxford Univ. He is a Registered PE in the State of WI. *Society Aff:* ASCE, ITE, TRB, TRF

Beisel, Robert O
Business: PO Box 266, Houston, TX 77001
Position: Exec Vice President. *Employer:* REX International Corp. *Born:* BS 1950 Mississippi St Univ, Chem Engrg; grad work heat transfer Univ of Houston. Enlisted in Cadets Army Air Corps, Radar Training. Formerly with Occidental Petroleum Corp. Proj Mgr Chem Complex in Saudi Arabia & proposed sulphur plant at Aramco. Since 1971 Partner & Exec VP Rex Internatl. Engaged in cons engrg for ammonia & gas plants & procurement mgmt for oil & petrochem indus. P Pres of

Beisel, Robert O (Continued)
Mississippi St Univ Alumni Houston Texas Chap. Mbr Sigma Alpha Epsilon. Enjoy sailing, woodworking & hunting.

Beitler, Samuel R
Home: 71 W Beaumont Rd, Columbus, OH 43214
Position: Emeritus Prof of Mech Engrg. *Employer:* Ohio State University. *Education:* Mech Engg/ME/OH State Univ; BME/ME/OH State Univ. *Born:* March 1899 Carey, Ohio. Mbr of Mech Engrg Faculty at Ohio St 1921- ; Progressed from Instr to Prof of Mech Engrg; Budget Dir & Admin Asst to Pres of Univ 1954-59. Res cons & Dir Res ASME 1963-72; Cons Engr on Flow Measurement 1929- ; Exec Dir & Secretary Internatl Assn for the Properties of Steam 1963-74. Fellow ASME; Chmn of many Res & Standardization Cttes both natl & internatl; Chmn & Mbr of Bds of Trs of United Methodist Church at local, district & conference levels. *Society Aff:* ASME.

Beitscher, Stanley
Business: Rocky Flats Plant, North American Space Operations, Rockwell International Corporation, P.O. Box 464 Golden, CO 80402-0464
Position: Assoc Scientist. *Employer:* Rockwell Intl. *Education:* PhD/Met Engg/CO Sch of Mines; MS/Met Engg/RPI; Met E/Met Engg/CO Sch of Mines. *Born:* 11/19/35. in Brooklyn, NY. Attended public schs in Brooklyn, Atlanta, and Denver. Genl Elec Engg Program 1957-58 Cleveland, Phila and Lynn, MA. Mech Metallurgist at Knolls Atomic Power Lab 1958-63. Rocky flats Plant Dow Chemical & Rockwell since 1963. Lecturer Mechanical Engg at CO Univ - Denver and Boulder 1976- present. Current position Assoc Scientist - Physical Metallurgy. Rocky Flats Engr of Year - 1978. ASM - Rocky Mountain Chapter, Chrmn 1980-81; MEI Natl Committee Secy, Rocky Mountain SAAB Club Pres 1980. Colorado Stepfamily Assn. Enjoys sports, music, mechanics, history. *Society Aff:* ASM, ΣΞ, ASME, CSMAA, RPIAA.

Bejczy, Antal K
Home: 1455 Riviera Dr, Pasadena, CA 91107
Position: Senior Research Scientist *Employer:* JPL-Caltech *Education:* Ph.D./Physics/University of Oslo, Norway; B.S./El. Eng./Techn. Univ. of Budapest, Hungary *Born:* 1/16/30 Native of Hungary. Lecturer at Univ of Oslo, Norway, in Applied Physics. Sr Research Fellow with NATO-Fulbright fellowship at Caltech. Sr Research Scientist and Technical Mgr; at JPL since 1969. Author or coauthor of over 90 papers on Robotics and Advanced Teleoperation, has two U.S. patents in robot sensing and over twenty NASA Tech Brief awards. Serves as editorial bd mbr for several technical journals on robotics. IEEE Fellow since 1986. Was General Chairman of IEEE International Conference on Robotics and Automation. President of IEEE Council on Robotics and Automation in 1987 *Society Aff:* IEEE

Bekey, George A
Business: Dept of Elec Engg, Los Angeles, CA 90089-0782
Position: Prof of Elec Engrg, Biomedic Engrg, & Comp Science *Employer:* University of Southern California. *Education:* PhD/Engrg/UCLA; MS/Engrg/UCLA; BS/Engrg/UC Berkeley *Born:* 6/19/28. Res Engr UCLA 1950-54. Service US Army Signal Corps 1954-56. Dir Computer Center for Beckman Instruments 1956-58; TRW Systems 1958-62 as Hd of Analy & Simulation Sect & Sr Staff Engr; since 1962 on faculty Univ of Southern Calif. Chrmn of Elec Engrg Sys Dept 1970-72 and 1978-82. Currently Chairman of Computer Science Dept. Pub 2 books & over 100 tech papers in system identification, biological systems & robotics. Fellow IEEE 1972. Enjoy problem solving, philosophy & backpackg. *Society Aff:* IEEE, AAAS, BMES, SCS, ACM, AAAI, SME

Bekker, Mieczyslaw G
Home: 224 E Islay, Santa Barbara, CA 93101
Position: Retired (Self-employed) *Education:* DrIng/ME/Munich Tech Univ; DrEng/ME/Carleton Univ; MSc/ME/Warsaw Inst of Tech. *Born:* 5/25/05. Native of Strzyzow (Poland). Polish MN of Ntl Defense: 1931-1939; French MN of Armament; 1940; Served with Canadian Army 1943-1956; Chief Land Locomotion Lab, US Army Tank Automotive Command: 1954-1961; also Res Prof, Stevens Inst of Tech: 1950-1952, & staff mbr Operations Res Office. Johns Hopkins Univ: 1952- 1954; Lectr, Univ of MI, 1956-1957; Hd Ground Mobility Res, GM Corp, 1961-1970. Responsible for dev of systems analysis & off-road locomotion res for Planetary & Terrestrial locomotion 1970-present, Conslt Engr, Lecturer, author and expert court witness. *Society Aff:* SAE, AAAS.

Bekooy, Rodger G
Business: 3320 S.W. First Ave, Portland, OR 97201
Position: Principal *Employer:* Carson, Bekooy, Gulick & Assocs, Inc *Education:* BS/ME/OR St Univ *Born:* 1/10/42 Native of Portland, OR. US Navy 1965-1970. Staff Engr with Morrison, Funatake & Assocs 1970-1977. One of three founding principals of Carson, Bekooy, Gulick & Assocs, Consulting Engrs. Firm founded in 1977. Specialize in the design of HVAC systems and plumbing systems for bldgs and in energy conservation studies. Authored HVAC computer program used by engrg firms in US and Canada. Pres, OR Chapter ASHRAE, 1981-1982. *Society Aff:* ASHRAE, ASPE, ACEC, NSPE

Belcher, Donald W
Home: 550 Prospect St, Westfield, NJ 07090
Position: Pres *Employer:* Belcher Engineering Inc. *Education:* BChE/Chem Eng/Yale Univ. *Born:* Jul 13, 1922. Attended Loomis Inst 1936-40. Served USNR 1943-46; Engr & Exec Officer LSM-95 & 232; Lt jg. Chem Supr & Asst Dept Mgr E I du Pont 1946-48; employed by Bowen Engrg 1948-1982, made Ch Engr 1956, VP 1957, Tech Dir 1973, Exec VP & Ch Oper Officer 1976-78, 80-82 & Pres 1978-80. PE NJ since 1953, PE LA 1979, & Mbr AIChE since 1960. Var publications & pats relating to spray drying. Native & resident of Westfield NJ; active in civic affairs; married, 5 children, 5 grandchildren. Hobbies: music, boating, swimming; left Bowen Engrg, Inc 1982 formed Belcher Engrg, Inc. 1982 as Pres, and additionally formed Drytec Coffee Inc 1983 as Pres. Mbr NSPE since 1953, Mbr ACC&CE since 1982. Mbr APCA since 1980. *Society Aff:* AIChE, APCA, ACC&CE, NSPE

Belden, David L
Business: 345 East 47th Street, New York, NY 10017
Position: Exec Dir. *Employer:* Amer Soc of Mech Engrs *Education:* PhD/-/Stanford Univ; MSIE/-/Stanford Univ; BGE/-/Univ of Omaha. *Born:* 1/9/35. Native of Red Wing, MN. Served 22 yrs in the US Air Force retiring in 1976 as a Colonel. Assignments in fields of procurement and production. Also served on faculty of the Air Force Inst of Tech, Final assignment on Staff of Secretary of the Air Force. Distinguished Grad of Industrial College of the Armed Forces. Bd of Governors of AAES since 1979, Bd of Dirs of the Council of Engrg and Scientific Soc Exec, 1980-86, Pres 84-85. *Society Aff:* IIE, ASEE, ASME.

Belden, Reed H
Business: Allied Corp, 6 Eastmans Road, Parsippany, NJ 07054
Position: VP & Gen Mgr Metglas Prod Dep *Employer:* Allied Corp *Education:* BSC/Chemistry/Univ of NB. *Born:* 2/18/32. Native of Brainerd, MN. Joined Allied Chem in 1953 after receiving a BSC degree from the Univ of NB. Most recent assignments with Allied Chem have been Assoc Dir-Eng Tech Ctr, 1977-78; Dir-Process Tech, 1978-79; 1979-80 Dir-METGLASR Products Dept; and presently Vice President & General Manager-Metglas Products Department.. *Society Aff:* AAAS, AIC, ACS, SPE.

Belding, Harry J
Home: 315 Papermill Road, Oreland, PA 19075
Position: Supervisor Quality Assurance. *Employer:* Alan Wood Steel Co. *Born:* Jan 21, 1921. BS Phil Coll of Textiles & Sci. Native of Erie Pa. Served US Navy 1942-52. With Alan Wood Steel Co since 1948; held positions of Customer Serv Engr &

Belding, Harry J (Continued)
Metallurgical Statistician; assumed current responsibility as Supr Quality Assurance 1972. Dir ASQC 1971-74; VP ASQC 1975-76. Enjoy model building & carpentry.

Belding, John A
Home: 9828 Dellcastle Rd, Gaithersburg, MD 20879
Position: Pres *Employer:* Res & Mgmt Technicians, Inc *Education:* PhD/Mech Engg/Univ of AL; MSA/Engg Admin/George Washington Univ; BSME/Mech Engg/Univ of AL. *Born:* 1/12/43. Hartsville SC. Teaches part-time GWU; Design Engr Saturn V Boeing Co 1965- 66; Prog Mgr airbreathing missile propulsion, Naval Air Sys Command 1967-74; Prog Mgr Automotive Combustion, Conversion Equipment & Shale Oil, Natl Sci Foundation 1974-75; Dir Conservation Res & Tech ERDA 1975-77, respon for planning & implementation of progs totaling $50 million. Dir Fossil Fuel util, Doe 1977-78, coal & alt fuel util, value approx $175 million; Asst VP & Mgr Wash Operations, Energy, Technology, & Engg, Sci Applications, Inc, 1978-79; VP & Mgr Energy Technology & Engg, Sci Application, Inc 1979-1982, energy engineering, comp mgt consulting. Pres Res & Mgmt Technicians, Inc and Comprep, Inc 1982- present, microcomputer, engrg & mgmt conslt. Outstanding Sr Award 1965, Outstanding Performance Award 1969 & 72; Navy Nominee, Pres's Personnel Interchange Prog 1974. Mbr Pi Tau Sigma & Sigma Xi published many energy, mgt & microcomputer papers. *Society Aff:* ASME, ΣΞ, ΠΤΣ.

Belefant, Arthur
Business: 305 Oak Street, Melbourne Beach, FL 32951
Position: President. *Employer:* Belefant Assoc Inc. *Education:* MS/Mgmt/FL Inst of Technology; BEE/Elec Engg/City College of NY; BA/Mgmt/Univ of MD. *Born:* Jun 17, 1927. Sr Mbr FES, IEEE; Past Chmn of Canaveral Council of Tech Socs & Brevard Arts Council; recognized Engr-of-the-Year 1971 for Brevard County. Reg in 5 states. Owner of cons firm in east central Florida. Previously was Proj Mgr & Ch Elec Engr for DMJM in Tokyo & Asst Proj Mgr for Frank E Basil Co Athens Greece. *Society Aff:* IEEE, IES, ASME, NSPE, ASME, ASHRAE, SAME, FICE, PES, CCTS.

Belfort, Georges
Business: Dept of Chem Engrg, Troy, NY 12180-3590
Position: Prof *Employer:* Rensselaer Polytechnic Inst *Education:* PhD/Environ Engr/Univ of CA, Irvine; MS/Environ Engr/Univ of CA, Irvine; BS/ChE/Univ of Cape Town, So Africa *Born:* 5/8/40 Dr Belfort teaches and conducts research in the separation and recovery area. Various research projects have in common liquid-solid interactions. In membrane separation processes, studies include polarizing fields, retention of organics, hollow fiber ultrafiltration, hollow fiber cell (hybridoma) culture reactors, pervaporation, membrane fouling phenomena and fundamental studies such as particle dynamics in porous ducts and solute-membrane interactions. Dr Belfort and his group are actively involved in the theoretical (thermodynamics) and experimental (isotherm and mini-column studies) aspects of aqueous phase adsorption. Magnetic Resonance Imaging is also used to study the fluid flows and mass transfer characteristics in bioreactors. Dr Belfort consults extensively with industry, both in the USA and overseas. *Society Aff:* AIChE, ACS, AAUP

Belfry, William G
Business: 1210 Sheppard Ave E, Toronto, Ontario, Canada M2K 2S8
Position: A/Mgr Natl Sales (Retired) *Employer:* Texaco Canada Inc *Education:* MS/Mech Engr/U of TX; BASc/ChE/U of Toronto *Born:* 7/14/19 Born Ottawa Canada. Worked Texaco Refinery Toronto after graduation. Became Chief Chemist. Served 3 yrs Canadian Navy. Naval delegate in Govt. purchasing stds and natl res council cttes. Took M.S. and instructed in thermodyanics after demobilization. Became atuomotive engr with Texaco. Did fuel and lubricant dev with car and truck mfg. Progressed through tech service and marketing to a/mgr national sales. Helped establish first 2 Canadian ASLE sections. Was chrmn Hamilton 1952. Toronto 1962. Also chrmn SAE Toronto 1973, general chrmn SAE intl fuel and lubricants 1978 and natl dir SAE 1980 to present. Hobbies golf, philately and wildlife conservation. *Society Aff:* SAE, ASLE, APEO

Bell, Burton J
Home: 208 Marina Drive, Eufaula, AL 38027
Position: Retired. *Born:* Jul 10, 1899 Warren Pa. Educ: public schools, Grove City Coll, Case Inst Tech. Married, 1 daughter, 3 grandchildren, 1 great-grandchild. Paving Engr Corps of Engrs, Wash Natl Airport; Tech Liaison Officer, C of E, US Army Atlanta 26 yrs, ret 1965; Exec Secy Chattahoochee River Basin Comm 1965-74. Organizer, P Chmn Coosa Valley Area Planning & Dev Comm, P.Chmn Coosa Valley Employment & Training Council; County Historian, P.Pres Gordon County Hist Soc, Editor '1976 Gordon Cty History'. Fellow, Life Mbr, P Pres Ga Sect ASCE; Fellow Hon Mbr, P Pres Ga Architectural & Engrg Soc; Editor Georgia Engr 18 yrs; Fellow, Life Mbr, P Pres Atlanta Post SAME; Res VP SAME 1956-66, Natl Dir SAME 1946-72.

Bell, Carlos G, Jr
Home: Rt 10, Box 369C, Charlotte, NC 28213
Position: Prof. *Employer:* Univ of NC at Charlotte *Education:* SD/Appl Sci/Harvard; SM/Appl Sci/Harvard; ORSORT/Nuclear Engg/Oak Ridge S O Reactor Technology;BS (CE)/Civ Engg/TX A&M Tech. *Born:* 12/18/22. Combat navigator (Pacific) in WWII. Distinguished student at TX A&M. Won Clements Herschel Prize in Hydraulics at Harvard. Worked for TIAA, a US State Dept Corp in the Amazon Valley. Harvard doctoral dissertation on nuclear weapons effects. Worked on reactor safety for twelve years at Oak Ridge. Presently Celanese Prof of Civil Engg at UNCC. Mbr of Union of Concerned Scientists. *Society Aff:* ASCE.

Bell, Charles V
Business: Academic Affairs, Angwin, CA 94508
Position: Academic VP *Employer:* Pacific Union Coll *Education:* PhD/EE/Stanford Univ; MS/EE/Stanford Univ; BS/EE/MS State Univ. *Born:* Mar 19, 1934 Starkville, Miss. Cons to Bendix Res Labs on parametric amps 1959-62; Visiting Asst Prof of Engrg at Harvey Mudd Coll 1965-66; Head Microwave Electronics Sect Hughes Aircraft Ground Systems Div 1962-72, primarily responsible for design & dev of low-noise microwave amplifiers; Assoc Prof of Physics Walla Walla College 1972-74; Dean of Walla Walla College School of Engrg 1974-1984; became VP for Academic Affairs at Pacific Union Coll 1984, Mbr ASEE, IEEE (Sr Mbr), Sigma Xi, Internatl Double Reed Soc, Rotary International. Enjoy music, family & church activities. *Society Aff:* IEEE, ASEE, ΣΞ, IDRS, AAHE.

Bell, Chester Gordon
Business: 15 Walnut St, Wellesley Hills, MA 02181
Position: Chief Tech Officer *Employer:* Encore Computer Corp *Education:* MS/EE/MIT; BS/EE/MIT *Born:* 8/19/34. Co-op student GE & American Elec Power; Fulbright Scholar to Univ of New S Wales Australia 1958. Computer Speech analysis & synthesis res MIT 1959-60. Manager of computer Design for DEC from 1960-1966. During that time he was responsible for DEC's PDP-4,-5, and -6 computers. He consulted for DEC from 1966-1972 while at CMU working on various computers and products including the PDP-11. From 1972 to 1983 he was VP, Eng. DEC, responsible for the company's research, design and development activities in computer hardware, software, and systems. Joined Encore Computer Corp in 1983 as Chief Tech Officer respon for overall product strategy. He has worked in the computer field on computer architecture, modularity of design, multiprocessors, and applications. He is dedicated to industry, standard compatible products. He is the co-author of four books, and has written over 50 papers and holder of 6 patents. Mbr: Eta Kappa Nu, ACM, AAAS, NAE, Fellow IEEE, The Computer Museum BOD; Awards: IEEE *Society Aff:* NAE, IEEE-Fellow, AAAS-Fellow, ACM

Bell, Frank R
Business: 3340 Arctic Blvd, Suite 201, Anchorage, AK 99503
Position: Partner. *Employer:* Bell, Herring and Assoc. *Education:* BS/Civ Engg/WA State Univ; Masters/Engg Mgt/Univ of AK. *Born:* 3/24/43. Began career with city of

Bell, Frank R (Continued)
LA in 1966 as Civ Engg Asst I. Went on active duty as 2nd Lt USAR in 1967. Spent one yr as Construction Platoon Leader Ft Belvoir VA. One yr as operations officer 497th engg co Vietnam. Moved to AK 1969. 5 yrs as design engr with Tryck Nyman and Hayes in Anchorage then 2 yrs with ARCO as construction engr at Prudhoe Bay on the Arctic Coast. Partner of Bell, Herring and Assoc since 1974. Bell, Herring and Associates was listed as 348 in ENR's top 500 design firms last yr. I enjoy hunting and fishing. *Society Aff:* ASCE, NSPE, PEPP.

Bell, James G
Business: 117 Knollwood Drive P.O. Box 7197, Rapid City, SD 57709
Position: President. *Employer:* Bell, Galyardt & Assoc Inc. *Education:* BSCE/Civil Engg/Univ of NB. *Born:* 1/1/26. Worked for US Bureau of Reclamation 1948-51; 1951-68 at Kirkham, Michael & Assocs as Engr-in-Charge & Principal of SD Corp; formed Bell Galyardt & Assoc Inc 1968, served as Ch Exec Officer. P CHmn Urban Ctte, Past Mbr Public Affairs/Legislative Action Ctte & Current Tr of Eads Genl Trust, PPres & PNatl Dir Cons Engrs Council of SD. Amer Water Works Assn. Water Pollution Control Federation. ASCE, PPres SD Sect. Natl Soc of Prof Engrs. SD Engg Soc; of SD Water Congress; PPres Rapid City Ch of Commerce; VP Bd of Dir, Rapid City Airport; Bd or Dir, Water & Natural Resources; Past Chrmn, Bd Dir SD Ch of Commerce; Bd of Dir Rapid City Dev Fdn, Past Mbr. *Society Aff:* ASCE, NSPE, ACEC.

Bell, James P
Business: Chem Engg U-139, Storrs, CT 06268
Position: Prof. *Employer:* Univ of CT. *Education:* ScD/ChE/MIT; BS/ChE/Lehigh Univ. *Born:* 11/4/34. Sr. Res Engr, DuPont Plastics Dept 1956-62, Monsanto Chemstrand Res Ctr 1966-69. Assoc Prof in Chem Engg Dept, Univ CT, 1969-75. Prof 1975-current Res interests: Structure-Property relationships in polymers, particularly epoxy resins and synthetic textiles, and cement dev for orthopedic surgery. Sr Fullbright Lecturer, Freiburg W Germany, 1975-76. German Academic Exchange Service fellowship, 1975. Guest of fdn for Res Conf in Chemistry, Osaka, Japan 1977. Visiting Prof CHE Dept, Univ of Naples 1983. Layd Davis fellowship, The Technion, Haifa, Israel 1983. Visiting Prof CHE Dept, Univ of Naples 1983. Lady Davis fellowship, The Technion, Haifa, Israel 1983. *Society Aff:* AIChE, APS, ACS.

Bell, John F
Home: P.O. Box 19127, Phoenix, AZ 85005 *Education:* BSEE/Purdue *Born:* Jan 1914. Teaching & res 1938-39 school year. Entered Zenith Radio Television Dept 1939; Dept Head Advanced TV tuner dev 1947-71; also engrg computer facility 1966-71; responsible for early UHF tuner & UHF channel strip dev & engrg. Chmn Broadcast & TV IEEE Group 1962-63; Chmn of R-4.4 TV Measurements Stds Ctte 1971-76. Fellow IEEE 1966; Chmn Chicago Sect Fellow Awards Ctte 1972-76. 1978-present Bd Chrmn, Bell Asso. Refiner.

Bell, John M
Business: Sch of Civ Engg, W Lafayette, IN 47907
Position: Assoc Prof. *Employer:* Purdue Univ. *Education:* PhD/Environ Engg/Purdue Univ; MSCE/Environ Engg/Purdue Univ; BSCE/Civ Engg/Clarkson College. *Born:* 10/8/35. Native of Sparkill, NY. Asst Prof of Environmental Engg, Purdue Univ, W Lafayette, IN 1962-1967. Sr Engr, H B Steeg & Assoc, Indianapolis, IN 1968. Assoc Prof of Environmental Engg, Purdue Univ, 1969 to date. Teaching & res in the areas of: a) municipal & agri solid wastes; b) math modelling of lakes and rivers; c) stream biology & pollution; & d) stormwater runoff. Editor of: a) Environmental Engineering News (Monthly publication); & b) Proceedings, Purdue Industrial Waste Conference. Honor socs: Chi Epsilon, Tau Beta Pi, Sigma Xi. Recipient of "Sagamore of the Wabash" Award. Enjoy tennis & fishing. *Society Aff:* ASCE, AWWA, ASEE, APWA, WPCF, AEEP.

Bell, John S
Home: 8237 Columbia Drive, Tyler, TX 75703
Position: Petroleum Exec retired. *Employer:* Exxon Co USA. *Education:* B.S./Petro Engrg/Oklahoma Univ *Born:* Jul 12, 1907 Ft Worth, Tx. BS in Petro Engrg Okla Univ 1930. Employed by Humble Oil & Refining Co (now Exxon Co USA) 1933-72. Various engrg & mgmt assignments. Final-Asst to VP-Prod. Mbr Tau Beta Pi, Sigma Tau & Sigma Gamma Epsilon Hon Fraternities; mbr of Ctte to write Emergency Petroleum & Gas Admin procedure manual; Mbr AAPG & AIME; Pres SPE of AIME 1959; Pres AIME 1971; received SPE's DeGolyer Medal 1974 & SPE's Disting Serv Award 1975; Hon Mbr AIME 1976; AIME rep to EJC 1959-62. Active with Boy Scouts, holding their Silver Beaver & Silver Antelope Awards. Elder First Christian Church, Tyler. Mason & Shriner. m. Alma Dunbar, 1 son. Activities: Scouts, church, golf, SPE of AIME & 'Honey-do'. *Society Aff:* AIME.

Bell, Kenneth J
Business: Sch of Chem Engg, Okla. State Univ, Stillwater, OK 74078
Position: Regents Prof of Chem Engg. *Employer:* OK State Univ. *Education:* PhD/ChE/Univ of DE (1955); MChE/ChE/Univ of DE (1953); BS/ChE/Case Inst of Tech (1951) *Born:* 03/01/30 Born in Cleveland, OH. Engr, Gen Elec, Hanford Operations, Richland, WA, 1955-56. Asst Prof, Chem Engg, Case Inst, 1956-1961. Visiting Prof, Oak Ridge Sch Reactor Tech, 1958. OK State Univ, School of Chem Engg: Assoc Prof, 1961-67; Prof, 1967-; Regents Prof, 1977-; Acting Head, 1977-78. Senior Chem Engr, Argonne National Laboratory, 1985-present. Sr Res Engr, HTRI, 1968-69, AIChE Heat Transfer/Energy Conversion Div, Chrmn, 1966; Natl Heat Transfer Conf Coordinating Comm 1967-75; Delegate, Assembly Intl Heat Transfer Conferences, 1971-75, AIChE/HTEC Donald Q Kern Award, 1978. OSPE Outstanding Engineer, 1980 Founding Editor, *Heat Transfer Engg.* Consultant, heat exchanger design, dev, application. Reg PE, OK. *Society Aff:* AIChE, ASEE, NSPE

Bell, Robert A
Home: 11309 Broad Green Dr, Potomac, MD 20854
Position: Dir Div of Mgmt Systems & Policy. *Employer:* US Food & Drug Admin. *Education:* BSIE/Indus Engg/TX Tech Univ. *Born:* Nov 26, 1937. m. Harriet Ann Fleck 1965; children Cynthia, Robert. Joined FDA 1968 & assumed current respons as Sr Mgmt Analysis Officer Jan 1974; past Dir Government Div of Amer Inst of Indus Engrs (IIE); previously selected to all local chapter offices including Pres & Bd of Dir. Received awards of appreciation from IIE for participating in revision of CSC classification standards for indus engrs & as Government rep to Amer Natl Standards Inst Ctte to standardize indus engrg terminology. Received sev government awards including FDA s Commendable Serv Award for Superior Performance & Outstanding Leadership in current position, and FDA's highest award the Award of Merit, for sustained superior contribution toward management and administrative activities of the FDA. Reg Prof Engr, Mbr Washington Soc of Engrs, Mbr Alpha Pi Mu, IE Hon Soc & Sr Mbr of IIE. *Society Aff:* IIE, WSE.

Bell, Ronald L
Business: 611 Hansen Way, Palo Alto, CA 94303
Position: Dir Solid State Lab. *Employer:* Varian Associates. *Born:* Sept 1924. BS/PhD Durham Univ England. British sci civil serv (SERL Baldock) 1950-57; worked on microwave tubes & gas discharge physics; joined Varian Microwave Tube Div 1957, moved to Corp Res 1960; worked on microwave applications of III-V compound semiconductors; directed negative affinity photoemitter dev, GaAs solar cell dev; Dir Corp Solid State Lab 1972. Fellow IEEE 1971.

Bell, Thaddeus G
Home: 9 Colonial Drive, Waterford, CT 06385
Position: Physicist *Employer:* Naval Underwater Sys Ctr-New London. *Education:* BS/Physics/Yale. *Born:* Feb 6, 1923. Native of Leominster Mass. Commissioned Ensign USNR 1945. Served 1 year active duty Navy; with Naval Underwater Systems Center as physicist 1947-85; Consultant 1986- ; Married Charlotte Ann Groh 1950, 2 children. Fellow Acoustical Soc of Amer 1959 for contrib to Naval sonar performance prediction tech & sonar system design. Solberg Award 1968 from Amer Soc of Naval Engrs for basic design of AN/SQS-26 sonar. Navy Superior Ci-

Bell, Thaddeus G (Continued)
vilian Service Award 1985. David Bushnell Award 1986 from the Amer Defense Preparedness Assn for contributions to the science and practice of underater acoustics. *Society Aff:* ASA, ASNE.

Bellaschi, Peter L
Home: 7417 S E Yamhill St, Portland, OR 97215
Position: Cons Engr. *Employer:* Self-Employed. *Born:* Feb 13, 1903 Piedmont Italy. BS, MS MIT; DSc Washington & Jefferson College. With Westinghouse Electric until 1947 & self-employed thereafter. Award of Merit Westinghouse; Eta Kappa Nu Oustanding Engr Award. World-wide cons activities in electric power transmission & transformation at extra-high- voltages, esp in USA, Canada, South America & Europe. Author of more than 150 tech papers in Transactions of Elec & Electronics Engrs USA; Internatl Conf of Large High-Voltage Sys & elsewhere. Assigned 6 pats USA, including lightning- stroke generator. Enjoyed traveling & combined natl & internatl activities. Linguist, fluent in 3 languages; W. N. Habnshaw Award of Inst of Elec & Electronics Engrs for contributions in the field of transmission and distribution of electric power and to the development of extra-high voltage apparatus and systems. Awarded IEEE Centennial Medal 1984.

Bello, Gastone
Business: 556 Morris Avenue, Summit, NJ 07901
Position: Sr. V.P. Technical Operations *Employer:* CIBA-GEIGY Corp Pharm Div. *Education:* PhD/Chem/Univ of Trieste, Italy *Born:* 1/15/31. Resident of Summit NJ. Employed as R/D chemist, Grp Ldr & Mgr for Montecatini, Milan Italy; with GEIGY Chem Corp 1966-70 in Res & Prod; with CIBA- GEIGY Corp 1970-74 as Dir, Chem Dev Serv & Dir, Corp Prod & Engrg Serv; 1974 appointed VP Prod & Engrg in Pharmaceuticals Div of CIBA-GEIGY Corp. 1980 SR VP Tech Oper. Mbr ACS, AIChE, Soc of Plastic Engrs. Member, New Jersey State Right-To-Know Advisory Council. *Society Aff:* PMA, ACS, AIChE, SPE

Bello, Phillip A
Business: 220 Reservoir St, Needham Heights, MA 02194
Position: President. *Employer:* CNR Inc. *Education:* ScD/EE/MIT; SM/EE/MIT; BS/EE/Northeastern Univ. *Born:* Oct 1929. ScD/SM MIT, BS Northeastern Univ, all in Elec Engrg. Asst Prof of Communications at Northeastern Univ 1955-57; Engrg Spec Sylvania Appl Res Lab 1958-61; Sr Scientist Adcom 1961-65; VP Signatron Inc 196572; since 1972 Pres of CNR Inc Needham Mass, a co devoted to R/D in communications & allied fields; works includes both dev & analy of tech for high-speed digital communications, nonlinear & linear communications channel simulation & communications channel characterization for wide variety of communications media. Pub extensively in transactions of IEEE. Elected Fellow IEEE 1970. *Society Aff:* IEEE.

Bellows, Glen L
Home: 210 Foster Dr, Normal, IL 61761
Position: Pres *Employer:* Buchanan, Bellows & Assoc Ltd *Education:* BS/ME/Univ of IL *Born:* 1/9/37 Native of Spencer, IA. BSME Univ of IL, 1959. Mechanical engr (heating, ventilating, air conditioning system design) for Brown, Manthei, Davis & Mullins, Champaign, IL - 1959-1965. Founded Buchanan, Bellows and Assoc., Consulting Engrs, Bloomington, IL - 1966. VP - 1966-1981. Pres - 1981- . Registered PE - IL. Charter Member Central IL Chapter of CSI. Program Chrmn - 1971/1972. Honors Committee Chrmn, IL Society of PE - 1971/1972. Occupational Development Center, Bloomington, IL (sheltered workshop) - Bd Member 1969-1978, Secretary 1973-1975, Treasurer 1972/1973, 1975-1978. Chrmn, Town of Normal Heating and Cooling Bd 1980-present. Bishop, Church of Jesus Christ of Latter-day Saints 1977-1984. *Society Aff:* ASHRAE, NSPE, CSI, NFPA, AEE.

Bellows, Guy
Home: 1776 Larch Ave., #307, Cincinnati, OH 45224
Position: Consultant *Employer:* Retired *Education:* BSEE/Elec Engr/Purdue Univ. *Born:* June 1914. Native Erie Pa. Transportation Control Designer & Supr 1936-47; Asst Super & Superintendent Control Manufacturing, Motor & Generator Mfg & DC Motor & Gen Mfg 1947-59; Mgr Mfg Engrg & Mfg-Jet Engine Dev Prog 1959-62; Mgr Mfg Engrg Res Lab for Jet Engines 1962-65; Sr Mfg Engr Advanced Mtls & Processes 1965-74; currently Consultant, Nontraditional Machining Processes & Surface Integrity. 7 pats, over 50 tech papers. Mbr SME Bd 1972-76; Mbr ASM, SAMPE, Cincinnati Union Bethel (Neighborhood Servs) Bd, P Mbr Community Chest Allocation Bd, 1974 Engrg Merit Award-San Fernando Valley Engrg Council, 1977 Joseph A Siegel Mem Award-SME. Prof Engr Ohio & Pa. Fellow of SME in 1986. Philatelist, clock collector & camera enthusiast. *Society Aff:* SME.

Bellport, Bernard P
Home: 855 Terra Calif Dr 4, Walnut Creek, CA 94595
Position: Engrg Cons. *Employer:* Self. *Education:* BS/Engg/Poly College of Engg. *Born:* May 1907. Native of La Crosse Kansas. Prior to 1935 engr with Montana Highway, Phoenix Utility Co & mining companies in west; 1936-72 US Bur of Recl-engr, const engr, Reg Dir, Ch Engr, Dir Design & Const. 1972- , cons. Reg PE Colorado. Mbr NAE, Fellow ASCE, USCOLD, ICID & FC, AAA, IWRA. Honors: Dept of Int Gold Medal, Golden Beaver for Engrg, 1 of APWA Top 10 Public Works Men 1970; Hon Mbr Chi Epsilon. *Society Aff:* NAE, ASCE, ICID & FC, AAA, IWRA, USCOLD

Belohoubek, Erwin F
Home: 37 Kingsley Road, Kendall Park, NJ 08824
Position: Head Microwave Circuits Research. *Employer:* David Sarnoff Research Center *Education:* PhD/EE/Techn Univ of Vienna, Austria; Dipl Ing/EE/Techn Univ of Vienna, Austria *Born:* May 1929 Vienna Austria. Taught as Asst at same Univ until 1955; joined RCA in Harrison NJ, 1957 transferred to RCA Lab in Princeton NJ as mbr of Tech Staff; worked on magnetrons, traveling wave tubes, delay tubes & electrostatically focused microwave tubes; since 1969 responsible for group working in microwave integrated circuits, computer aided measurement & design, linear & saturated solid state amplifiers & small subsystems. April 1987, RCA Laboratories became David Sarnoff Research Center, a subsidiary of SRI. Elected Fellow IEEE 1975. Married, 2 boys. Active in sports, enjoys music. *Society Aff:* IEEE.

Belt, Robert M
Home: 2542 Olopua Street, Honolulu, HI 96822
Position: Retired *Education:* BS/Civil Engg/OR State Univ. *Born:* Jun 5, 1907. 1926-30 Ore St Highway Dept; 1931 USBPR, Hawaii, location engr; 1932-41 District Engr Territorial Highway Dept Kauai Hawaii; 1941-46 Civil Engr Corps US Navy; 1947-52 Territorial Highway Engr, Superintendent of Public Works, Chmn Harbor Bd Territory of Hawaii; 1953- co-founder & Bd Chmn Belt, Collins & Assocs Ltd cons engrs 1953-80. ASCE Fellow & Hon Mbr 1975; Natl Dir 1964-66; Pub Ctte Chmn 1966; Prof Practice Ctte Chmn 1975; ACEC Natl Dir 1972. Mbr Chi Epsilon. Hawaii Engr of the Year 1966. *Society Aff:* ASCE, ACEC, XE

Belter, Walter G
Home: 5117 White Flint Dr, Kensington, MD 20895
Position: Sr. Exec Consultant. *Employer:* NUS Corp. *Education:* MS/Civil Engg/Univ of WI; BS/Civil Engg/Univ of WI. *Born:* 6/16/23. Sanitary engr with IN State Bd of Health and US Air Force, specializing in industrial waste treatment 1950-57. Responsible for planning, dev and tech dir of R&D programs for the treatment and disposal of radioactive wastes and associated environmental studies with the Atomic Energy Commission, Energy Research and Dev Admin, and the Dept of Energy 1957-78. Joined NUS Corp in 1978 in consulting engg for assessment of environmental impacts of siting, construction, and operation of utility and industrial facilities. Pres, FWQA 1969; Arthur Sidney Bedell Award, 1970. Conference of Federal Environmental Engrs, first Federal Environ Engr-of-the-Yr award, 1974. *Society Aff:* ASCE, ANS, WPCF, FWQA, CFEE.

Beltz, Fred W, Jr
Business: Turbine & Compressor Div, PO Box 8788, Trenton, NJ 08638
Position: VP & Gen Mgr. *Employer:* Transamerica Delaval Inc. *Education:* BSE/Naval Arch & Marine Engr/Univ of MI. *Born:* 12/29/22. Served US Merchant Marine as Engg Officer rising to Chief Engr during & following WWII. Joined Transamerica Delaval, then Delaval Steam Turbine Co, Tenton, NJ in 1949 as Field Service Engr, served various engg capacities in plant & field. 1965 appointed Mgr of Service & Repair Dept, Turbine Div. 1970 appointed Gen Mgr of Condenser & Filter Div, Florence, NJ. 1978 to present VP & Gen Mgr, Turbine & Compressor Div, Trenton, NJ. Pres, Heat Exchange Inst 1977. Chrmn Phila Sec SNAME 1977-78. Reg PE NJ & PA. Active Episcopal Church, Rotary, & several Sailing Clubs. *Society Aff:* ASME, ASNE, IME, SNAME.

Belytschko, Ted B
Business: Dept of Civil Engg, Evanston, IL 60201
Position: Prof of Civil Engg. *Employer:* Northwestern Univ. *Education:* PhD/Mechanics/IIt; BS/Mech/IIT. *Born:* 1/13/43. Prof of Civil & Nuclear Engg at Northwestern since 1977, Univ of IL at Chicago 1968-77. Res in computer methods of struct analy, particularly for nonlinear problems, such as reactor safety analysis, weapons effects & vehicular crashworthiness; Cons to IIT Res Inst, Argonne Natl Lab, & other firms. Editor, Nuclear Engineering and Design & Engineering with Computers. Phi Tau Sigma Gold Medal, ASME 1975, Walter Huber Research Prize, ASCE, 1977. Thomas Jaeger Research Prize, 1983; Chrmn, Engrg Mechanics Div., ASCE, 1981-82; Exec Cttee, Applied Mechanics Div, ASME, 1985-1990. *Society Aff:* ASME, ASCE.

Belz, Paul D
Home: 2401 Wood Stream Court, Ellicott City, MD 21043
Position: Consulting Engineer *Employer:* Self *Education:* MSE/Applied Mechs/Johns Hopkins Univ; BSME/Mech Eng/VPI & SU. *Born:* Dec 31, 1920. Served with Coast Artillery Corps (Anti-Aircraft), US Army in Pacific as Radar Repair & Maintenance Officer 1943-46. Design engr 1946 & fellow engr 1952-84 for Westinghouse Electric Corp responsible for all phases of mech design of military electronic equip for aircraft, ground stations, surface water vessels & underwater craft; conceived & devised a unitary packaging philosophy for a class of airborne fire control systems, which is widely used in indus. Fellow Mbr ASME. Reg PE Md. Mbr Maryland St Bd of Registration for PE 1975-86, chairman 1983-86. *Society Aff:* ASME, ADPA, ΤΒΠ, ΦΚΦ, ΠΤΣ.

Bement, Arden L, Jr
Home: 509 Zorn Lane, Mayfield Village, OH 44143
Position: VP Tech Resources *Employer:* TRW *Education:* PhD/Met Engr/Univ of MI; MS/Met Engr/Univ of ID; EMet/Metallurgy/CO Sch of Mines. *Born:* 5/22/32. Native of New Castle, PA. Employed as Sr Res Engr with GE Co, Hanford Labs 1954-1965 and as Sr R&D Mgr with Battelle-Pacific Northwest Labs 1965-1970. Last position was Mgr, Fuels and Mtls Dept. Appointed Prof of Nuclear Mtsl in the Dept of Nuclear Engg and Dept of mtls Sci and Engg at the MA Inst of Tech in 1970 and remained until 1976, at which time appointed Dir, Office of Mtls Sci, Defense Advanced Res Projs Agency, Office of the Secy of Defense. Deputy Undersecretary of Defense, Res & Advanced Tech, 1979-1980 job at TRW begun Dec 1980. Fellow of American Nuclear Soc, Am Soc of Metals, and Am Inst of Chemists. Outstanding Achievement Award, CO Engrs Council. Lt Col, US Army Corps of Engrs (Ret). Editor or Co-editor 4 books, 90 publications. Reserve Medal U.S. Army 1979, Outstanding Perf Award, Dept of Defense 1977, Dist Civil Serv Medal, Dept of Defense 1980, Award for Outstanding Contrib for Adv of Black Amer in DoD 1981, Mbr of Natl Acad of Engrs 1983, Dist Achievement Award, CO Sch of Mines 1984, Engineering Citation Award, UCLA, 1987. *Society Aff:* TMS-AIME, ANS, ASM, ASTM.

Bendelius, Arthur G
Home: 1220 Witham Dr, Dunwoody, GA 30338
Position: Sr VP *Employer:* Parsons Brinckerhoff Quade & Douglas, Inc *Education:* MMS/Mgmt/Stevens Inst of Technology; BE/Engg/Stevens Inst of Technology. *Born:* 5/21/36. Engr with several Consulting & Mfg firms. Joined Parsons Brinckerhoff Quade & Douglas, Inc in 1960 Assumed current respons as Sr VP in 1978. Licensed Profl Engr in AL, AR, FL, GA, KY, LA, MD, MN, MS, NJ, NY, NC, OK,TN, TX, SC, UT, OH & NCEE. Pres-Atlanta Post-SAME 1978; Dir GA Soc of Profl Engrs 1976-78; GA Engg Foundation Pres 1982-83, Life Member 1983; Technical Committee Chrmn - ASHRAE - 1974-79; Metro-Atlanta Engr of Year-Private Practice-1978; SIT Alumni Achievement Award 1978; Who's Who in South and Southwest 1976-87; Licensed Pilot - Photographer. Mbr Tau Beta Pi; Who's Who in Tech Today 1982-83. Who's Who in America 43, 44 & 45 Editions; Member Bd of Dirs Transportation Association of South Carolina; Mbr SAME National Bd of Dirs 1983-86; Fellow SAME. *Society Aff:* ASME, SAME, NSPE, ASHRAE, TRB, AOPA.

Bender, Donald L
Home: E1532 Pinecrest, Spokane, WA 99203
Position: Prof-Civil Engrg *Employer:* Gonzaga Univ *Education:* PhD/Civil Engg/Univ of WI; MS/Irrigation Engg/CO State Univ; BS/Civil Engg/Univ of WY. *Born:* 9/7/31. Native of WY. Joined Civil Engg faculty at WA State Univ in 1956. Tech speciality is hydrology with emphasis on hydrometerology and floods. Held admin positions in College of Engg from 1972 to 1983 as Asst, Assoc, and Acting Dean. Returned to teaching at Gonzaga Univ in 1983. Participated 1970-71 in the Am Council on Educ Fellows Program in Academic Admin at the Univ of CA at Irvine. Dir, Civil Engg Div of the Am Society for Engg Educ 1973-76 and chrm 1978-79. Outstanding Educators of Am Award 1974. Natl Pres of Tau Beta Pi 1978-82. VChrm, Western, PE in Educ-NSPE 1981-82. Ex Board, PEE-NSPE 1983-85. *Society Aff:* ASCE, ASEE, NSPE.

Bender, Frederick G
Home: 310 E Hickory St, Sutton, NB 68979
Position: VP *Employer:* Jacob Bender & Son *Education:* BSC/Agr Engr/Univ of NB *Born:* 10/13/22 Engaged in retail farm equipment business as part owner & operator for 37 yrs. VP Jacob Bender & Son Inc. Vice Chrmn NB Section ASAE 1981-82. 10 yrs membership on local public school advisory councils. Avocation: Liscenced as KOFBP amatuer radio operator for 28 yrs. Other organizations: Master Mason, Scottish Rite, Shrine. Served in WWII with 1327th Engr Regt Ledo Road Project, Burma. *Society Aff:* ASAE

Bender, Victor W
Home: 44127 No Halcom, Lancaster, CA 93534
Position: Unit Ch, Flight Test Engr. *Employer:* McDonnell Aircraft Co. *Born:* Jan 1932. MS in Engrg Mgmt Univ of Missouri at Rolla; BS in Aeronautical Engrg Wichita Univ. Native of Cornlea Nebraska. Taught public school prior to entering USAF 1952; received pilot wings & comm in USAF; currently hold commercial pilot license; engr in charge of F4 & F15 aircraft testing for McDonnell Aircraft Co, Edwards AFB Calif. Chmn Antelope Valley AIAA, P V Chmn & Secretary. Enjoy flying, sports & outdoor activities.

Bendixen, Stein L
Business: Box 5038, S-102 41 Stockholm Sweden
Position: Consultant engr. *Employer:* VBB. *Education:* BS/Civ Engg/Stockholms Tekniska Inst. *Born:* 11/3/28. Norwegian nationality. With VBB, Sweden since 1958. Participated in projs in Am, Asia, Africa & Europe. Specialized in urban hydrology. Currently with VBB's dept of Water, Soil & Environment Dev, engaged in problems connected with the renewal of technical systems in densely built-up areas. Wesley W Horner's Award 1978. *Society Aff:* SVR, SKIF.

Benedetti, Aldo J
Business: Highway Licences Bldg, Mail Stop B, Olympia, WA 98504
Position: Commissioner. *Employer:* WA Utilities & Transportation Commission. *Education:* BE/Math-Science/Univ of Puget Sound; BS/Genl Engr/USNA, Annapo-

Benedetti, Aldo J (Continued)
lis. *Born:* 1/30/22. Native Tacoma WA. US NAvy 1945-47; US Navy 1951-55; present rank Capt USNR. Civil Engr Water Div, Dept of Public Utilities 1947-51, 1953-55; Asst Supt Water 1955-57; Supt Water Div 1957-62; Asst Dir Utilities 1962-72; Dir Utilities 1972- 79. Mbr Amer Water Works Assn, Amer Pub Pwr Assn, Tacoma Engrs Club; 1971 Disting Serv Award Tacoma Chamber of Commerce; 1969 Water Utility Man of the Year Pacific NW Sect AWWA, P Chmn Distrib Div AWWA, P Chmn Pacific NW Sect AWWA; P Dir Amer Pub Pwr Assn; P Secretary-Treasurer Western Systems Coordinating Council; P Pres Columbia Storage Pwr Enchange. 2 articles pub in AWWA Journal. P Ch Admin Officer Dept Pub Utilities-municipal elec, water & railroad switching operations. 1979-Commissioner, WA Utilities & Transportation Commission. Enjoys golf & fishing.

Benedict, Robert P
Business: STDE Box 9175 Bldg N206, Philadelphia, PA 19113
Position: Fellow Engr. *Employer:* Westinghouse Electric Corp. *Education:* MS/Mech Engrg/Cornell Univ; BME/Mech Engrg/Rensselaer Poly Inst *Born:* May 1924. Army Air Corps Pilot P-40, B-25 WW II. Pi Tau Sigma, Tau Beta Pi, Sigma Xi; Instrumentation Dev Engr Westinghouse Electric Corp, Aviation Gas Turbine & Steam Turbine Divs since graduation; has been Mgr of the sect. Adjunct Prof Drexel Univ Evening College teaching Instrumentation, Thermodynamics, Gas Dynamics, Heat Transfer, Fluid Mech. V Chmn ASTM E-20 Temperature Measurement; Chmn ASME PTC 19 Instruments & Apparatus; Mbr ASME Fluid Meters Res Ctte; P Chmn ASME Temperature Res Ctte; Fellow ASTM, Fellow ASME. Author numerous papers & 11 books. *Society Aff:* ASME, ASTM

Benes, Peter
Business: 2937 7th Ave S, Suite 212, Birmingham, AL 35233
Position: Proj Mgr. *Employer:* Stanley Consultants. *Education:* BS/CE/Case Inst of Tech. *Born:* 11/19/28. Native of Cleveland, OH. Grad of Case Tech after serving three yrs in the Marine Corp. Worked as an ind water treatment consultant specializing in problems associated with boiler and cooling systems. Also been involved with sales mgt positions with a maj pollution control equip mfg. Has served as pres of the AL Soc of PE. *Society Aff:* NSPE.

Benfey, Rudolf L
Home: 231 Blackberry Rd, Liverpool, NY 13090
Position: Mgr Over-the-Horizon Radar Proposal. *Employer:* General Electric Co. *Education:* BE/EE/Yale Univ. *Born:* 12/4/27. Following grad from Yale in 1947, I worked at GEs Pittsfield, Schenectady and Syracuse plants on program assignments, then became a design engg in radar displays. From 1951-55, I was in the US Air Force as Tech Intelligence Officer, analyzing foreign developments in electronics and missiles. Returning to GE, I have worked in systems engg, developing space surveillance and tracking radars. I have held mgt positions since 1966, radar systems engg, test, including tactical, air defense and missile/space defense radars, and for major radar proposals. *Society Aff:* IEEE.

Benforado, David M,
Business: 900 Bush Ave, Bldg 21-2W, St Paul, MN 55144
Position: Sr Environmental Specialist *Employer:* 3M *Education:* BS/Chem Engg/ Columbia Univ. *Born:* Nov 1925. Reg PE in NYS. Natl Pres APCA1972-73. Mbr Natl Acad Sciences Comm for 1979 Report Odors from Stationary & Mobil Sources - 1979 Pres Woodbury Lions Club 1978-79. varied indus exper, incl cons with a specialty in air pollution control, incineration & odor measurement. Chrmn APCA. TT-4 Odor Comm. Board of Trustees of Am Acad Environmental Engrs, AIChE, Environmental Div & ASME Spec: pollution control of indus wastes & assessment of environmental impact of prods; presently responsible for corporate environmental Legislative activities. Bd of Dir 1968-74 APCA Awarded Honorary Membership APCA in 1981. *Society Aff:* APCA, AIChE, AAEE.

Benford, Harry B
Home: 6 Westbury Court, Ann Arbor, MI 48105
Position: Prof Emeritus Naval Arch & Marine Engrg. *Employer:* University of Michigan. *Education:* BSE/NA & ME/Univ of MI. *Born:* 8/7/17. in Schenectady, NY. Practiced engg at Newport News Shipbldg Co (design, prod, cost estimating) until 1948. Taught naval architecture at Univ of MI since 1948-82, except 1 yr serv with staff of Natl Res Council 1959-60; Chmn of Dept of Naval Arch & Mar E 1967-72, Ret 1983. Author of numerous papers on engg economics & ship design. Fellow & Hon Mbr of SNAME & active on serv cttes; currently serving as VP. Hobby: Gilbert & Sullivan; also music, theater & sailing. *Society Aff:* SNAME, MTS, RINA, AACE, NEC, ASEE, TAU BETA P, ΦΚΦ

Benham, David B
Business: 9400 N. Broadway, PO Box 20400, Oklahoma City, OK 73156-0400
Position: Chmn of Bd/Ch Exec Officer. *Employer:* The Benham Group *Education:* Cert/Naval Arch/US Naval Acad P G Sch; BS/Engg/US Naval Acad. *Born:* 11/11/18. WW II US Navy aircraft carrier program. SecNav Citation & Commendation Ribbon for outstanding performance. Chmn of Bd/CEO The Benham Group; Past Director Academy Computing Corp. & Oklahomans for Energy & Jobs, Director Navy League of US (Okla. Council). Reg PE 24 states & Dist of Columbia. Engrg Hall of Fame Okla St. Univ; member, Coll of Engrg Bd of Visitors Univ of Okla; Fellow, American Cons Engrs Council; fellow, Soc of Amer Military Engrs (Outstanding Service Award); Diplomat Academy of Environmental Engrs; life member, Amer Water Works Assoc, member, Intl Bridge Tunnel & Tnpk Asso. Past Natl Pres US Navel Acad Alumni Assn (1975-77). Numerous business & professional magazine contributions and awards. *Society Aff:* ACEC, ASCE, SAME, NSPE, AAEE, AWWA, IBTTA.

Benioff, Ben
Home: 371-D Avenida Castilla, Laguna Hills, CA 92653
Position: P Pres (Ret) *Employer:* Self *Education:* BS/Civil Engg/CA Inst of Technology. *Born:* Colonel, Corps of Engrs US Army; Bronze Star US Army; Military Honor Republic of China; Dir of Transportation Republic of China. Advisory Council ASTM, AAA, Forest Prod Res Soc; Past Secretary Amer Inst of Timber Const; Past Pres Struct Engrs Assn of So Calif; Hon Mbr Struct Engrs Assn of So Calif; P Pres King-Benioff-Steinmann-King Cons Engrs. R&D Engr 1936-56 Summerbell Roof Structs L A Calif. *Society Aff:* ASTM, SAME, FPRS, TROA, ACEC, SEA, SEAOC, ACSC.

Benjamin, Jack R
Business: 260 Sheridan Ave, Ste 205, Palo Alto, CA 94306
Position: Chrmn *Employer:* Jack R Benjamin & Assoc, Inc *Education:* ScD/CE/MIT; MS/CE/Univ of WA; BS/CE/Univ of WA *Born:* 7/24/17 *Society Aff:* ACI, ASCE, EERI

Benjamin, John S
Business: Sterling Forest, Suffern, NY 10901
Position: Res Mgr Materials Res. *Employer:* The Internatl Nickel Co Inc. *Born:* May 1939 Beverly Mass. BS, MS, DSc MIT. Res Metallurgist at Paul D Merica Res Lab of Internatl Nickel Co Inc 1965-67; dev mech alloying process & dispersion strengthened superalloys combining an oxide dispersion with precipitation hardening; Supr of Composites Sect 1967-72 & Res Mgr Matls Res 1972- . Author of 14 tech papers & 24 US pats. Awarded Howe Medal of ASM 1971. Hobbies incl hiking, winemaking & painting.

Benjamin, Richard W
Home: 3415 W Hurley Pond Rd, Wall, NJ 07719
Position: Prof of Electronic Engrg and Special Assist to the Provost *Employer:* Monmouth College. *Education:* MS/Engg Sci/NJ Inst of Tech; BS/Electronic Engg/ Monmouth College. *Born:* 6/20/30. Native of Spring Lake, NJ. Has been a mbr of the Faculty of Monmouth College (NJ) since 1959. Presently Prof of Electronic Engrg and Assist to the Provost. Dean (1981-86) of Science and Professional

Benjamin, Richard W (Continued)
Studies. Chrmn (1971-73) of the Electron Devices/Microwave Theory and Techniques Chapter/NJ Coast Sec IEEE. Consultant in the area of microwave devices and systems. Enjoys tennis and gardening. *Society Aff:* IEEE, ASEE, SPS, HKN.

Benjes, Henry H
Home: 10230 Lamar, Overland Pk, KS 66207
Position: Consultant *Employer:* Kapco, Inc. Black & Veatch Engrs-Architects *Education:* BS/Civil Engrg/KS Univ. *Born:* March 1914. Native of Kansas City Mo. With Black & Veatch since 1938; partner 1956-80; 1951-80Engrg Mgr of Civil Environmental Div having genl dir of design & opers in home office of Div. Mbr ASCE, AAEE, NSPE, AWWA; Mbr of a number of natl engrg org cttes; P Chmn of AWWA Standards Council, AWWA Standards Ctte on Butterfly Valves & AWWA Ad Hoc Ctte on Evaluation of Pipe Design Criteria & Safety Factors. Author of numerous tech papers. Reg Prof Engr. Hobby: horseback riding & golf. *Society Aff:* ASCE, NSPE, AWWA, AAEE.

Benn, Donald M
Home: 5066 San Julio Ave, Santa Barbara, CA 93111
Position: President *Employer:* Omnitek Engineering, Inc. *Education:* MS/System Management/Univ So CA; BAeA/Aeronautical Engr/Univ Detroit; Grad/Exec Management/The Pres Fed Exec Inst *Born:* 6/8/28 Native of Detroit, MI. Naval pilot aircraft and blimps 1953-1957. Flight tested aircraft and missiles 1949-1983. Left Navy in 1965 for Air Force Western Test Range as Chief, Flight Analysis, Range Safety. Advanced to Chief Engr, Deputy Dir Safety, WSMC, 1978. Retired 1983 and began Omnitek. Define missile impact, penetration & overpressure environment for offshore oil units endangered by Vandenberg launches as basis for design of shelters. Respon for defining all planning and engineering tasks. PE-Safety-CA. *Society Aff:* AIAA, ITEA, THA, ANA, ES

Bennett, A Wayne
Business: ECE Dept, Clemson University, Clemson, SC 29634-0915
Position: Head of Electrical and Computer Engineering *Employer:* Clemson University *Education:* PhD/EE/Univ of Florida; MS/EE/VPI; BS/EE/VPI. *Born:* 5/5/37. Native of Rocky Mount, VA. Worked as systems engr for Ind Control Dept of GE. Prof of EE VPI, Prop and Head ECE at Clemson U.Outstanding Educator of Am 1973 & 1975. Chrmn Computers Educ Div (CoED) of ASEE. Chrmn Profl Interest Council IV of ASEE. VP of Intl Assn for Math & Computers in Simulation. *Society Aff:* IEEE, ASEE, IMACS, ACM.

Bennett, Alan J
Business: VP - Research, Varian Associates, 611 Hansen Way, Palo Alto, CA 94303
Position: VP, Research & Dev. *Employer:* Varian Associates. *Education:* B.A./Physics/U of Pennsylvania; M.S./Physics/U of Chicago; Ph.D/Physics/ U of Chicago *Born:* Phila., June 13, 1941; s. Leon Martin and Reba (Perry) B.; m. Frances Kitey, June 16, 1963; children: Sarah, Rachel, Daniel. BA, U. Pa, 1962; MS, U Chgo., 1963, PhD, 1965. Physicist Gen. Electric Research and Devel. Ctr., Schenectady, N.Y., 1966-74, br. mgr., 1975-79; dir. electronics lab. Gould Inc., Rolling Meadows, Ill., 1979-84; v.p. research Varian Assn., Palo Alto, Calif., 1984- Contbr. articles of profl. jours. Fellow NSF 1963-65, 66. Mem. AAAS, IEEE (sr.), Am. Phys. Soc., Phi Beta Kappa, Sigma Xi. Avocations: linguistics, amateur radio. *Society Aff:* NAE, APS, IEEE

Bennett, Carroll O
Business: Dept of Chem Engg, U-139, Storrs, CT 06268
Position: Prof. *Employer:* Univ of CT. *Education:* DEng/Chem Engr/Yale Univ; BS/ Chem Engr/Worcester Poly Inst. *Born:* 4/1/21. Taught at Purdue Univ 1949-1959 and did res on diffusion and high-pressure thermodynamics. Mgr process R & D at the Lummus Co, 1959-1964. Since then at the Univ of CT, doing res on heterogeneous catalysis. On various leaves taught in France, Algeria, and Chile. Co-author of *Momentum, Heat and Mass Transfer.* 1980, Warren K Lewis Award from the AIChE. *Society Aff:* AIChE, ACS.

Bennett, Charles H
Business: Box 4000, Green Bay, WI 54303
Position: President. *Employer:* F Hurlbut Co. *Education:* BS/Civil Engg/ Northwestern Univ. *Born:* June 1930. Native of La Grange Ill. Served in US Army 1954-56. Employed by F Hurlbut Co since 1956, being elected Pres & Ch Exec Officer 1974; also Pres of 4 affiliated companies; dir of sev mfg co's in northeastern Wisc. P State Pres of Trade Assns; P Pres Green Bay Area Chamber of Commerce 1972; P Chmn of Precast Systems Inc - 1973; Pres of Wisc Soc of Prof Engrs 1975-76; Engr of the year Wisc Soc of Prof Engrs 1972. Sr. Natl Dir for WI Soc Professional Engrg 1977-present. *Society Aff:* ASCE, NSPE.

Bennett, Dwight H
Home: 6461 Shire Way, Long Beach, CA 90815
Position: F-18Program Mgr. (Northrop) *Employer:* McDonnell Aircraft Co. *Education:* BSME/Aero Option/CA Inst of Tech. *Born:* Nov 1917 Oklahoma City. Married Katherine Mason, 3 children. 23 yrs Convair San Diego various engrg managerial positions; Chmn San Diego Sect Inst of Aero Sci 1949-50; Natl Council IAS 1951-52. 1 yr Aerocommander, V P & Asst to Genl Mgr; 17 yrs McDonnell Aircraft, various Prog Engrg Positions; currently F- 18 Progr. Mgr (Northrop). SAE Wright Bros Medal 1972. Active Flight Instructor aircraft & instruments. Mbr Aircraft Operations Tech Ctte, AIAA. *Society Aff:* AIAA.

Bennett, F Lawrence
Business: School of Engineering, 539 Duckering Bldg, Fairbanks, AK 99775
Position: Prof and Head, Engg Mgt *Employer:* University of Alaska. *Education:* PhD/Civil Engg/Cornell Univ; MS/Civil Eng/Cornell Univ; BCE/Civil Eng/ Rensselaer Polytechnic Inst *Born:* 4/4/39. Troy NY. Empl by United Engrs & Const Philadelphia 1965-68; at Univ of AK 1968, Dept Hd 1968-1980, 1983-present. Asst to Chancellor 1977-79 Vice Chancellor for Acad Affairs 1979-82. Pres AK Sect ASCE 1975-76; Mbr ASEE, ASEM, NSPE. Dow Awd for Outstanding Young Faculty Mbr Pacific NW 1973. Engrg & mgmt cons. Author of 25 papers & reports, and 1 book. Former lay Leader AK Methodist Conf. Married Margaret Ann Musgrave 1962, Children: Matthew Lawrence, Andrew Lee. Hobbies: gardening, hiking, skiing, camping. *Society Aff:* ASCE, NSPE, ASEE, ASEM

Bennett, Frederick R
Business: Sch of Engg, College Pl, WA 99324
Position: Prof of Engrg *Employer:* Walla Walla Coll. *Education:* PhD/Engg Science/ WA State Univ; MSCE/Civil Engg/WA State Univ; BSME/Mechanical Engg/Walla Walla College. *Born:* 2/11/30. Born in Boston, MA. worked in New England construction industry and engg aid. Transferred out west to finish undergraduate sch. Field and design engr for Hoover Construction Co, 1955. US Navy, Naval Arch (Structural) at Puget Sound Naval Shipyard; special assignment with US Navy Bureau of Ships. Wash DC, preliminary design branch ('56-'61). Took advanced degrees. Teaching civil engg (structural) at Walla Walla Coll ('61-present). Pres, Columbia Section ASCE ('78-'79). Chrmn: Transporation Committee, Walla Walla Chamber of Commerce, Port of Walla Walla Commissioner, Fire and Ambulance Volunteer, Enthusiast of sailing and good music. *Society Aff:* ASCE, ASEE, NSPE, SNAME.

Bennett, G Bryce
Home: 2680 Skyfarm Dr, Hillsborough, CA 94010
Position: Consultant *Employer:* Self employed *Education:* MS/Civ Engg/Univ of WA; BS/Civ Engg/Univ of ID. *Born:* 4/9/21. Native of Shelley ID. Officer UNR WWII. 1946-51 Asst Prof Seattle Univ & Univ of ID; 2 yrs Mtls Engr ID Dept of Hwys; Div Asphalt Mgr of Shell Oil Co 2 yrs; 1955 returned to ID Dept of Hwys as Asst St Hwy Engr & 1956-65 served as St Hwy Engr; since 1964 has moved through various positions with IECO & since mid- 1968 VP & Gen Mgr or Exec VP and

Bennett, G Bryce (Continued)
Dir; transferred to Morrison-Knudsen Co. Cerrejon Coal Project 1981; since January, 1983, active as engineering management consultant. *Society Aff:* ASCE

Bennett, G Kemble
Business: Coll of Engrg, Industrial Engrg, Texas A&M Univ, College Station, TX 77843
Position: Chrmn of Industrial Engrg *Employer:* Univ of South FL *Education:* PhD/IE/TX Tech Univ; MS/Engr Math/San Jose St Univ; BS/Math/FL St Univ *Born:* 4/2/40 Native of FL. Worked as a systems engr with Martin Co and Lockheed Missiles and Space Division, 1962-66. Assistant Dir of TX Tech Univ Computer Center, 1966-69. Visiting Scientist, NASA Spacecraft Center, 1969-70. Received Ph.D. degree 1970 and became a member of faculty at VA Tech. Joined faculty of Univ of South FL in 1974. Became chrmn of Industrial Engrg 1979-86. Became Prof and Hd of Ind Eng at Texas A&M in 1986. Published widely in numerous national and intl journals and books. Mng Ed of The Logistics Spectrum, and is a mbr of the Bd of Referees for The Annals of the Soc of Logistics Engrs. Registered PE in FL & TX. Serves as a consultant on logistic engineering and telecommunications to several industries. *Society Aff:* IIE, ASEE, SOLE

Bennett, Gary F
Business: 2801 W Bancroft, Toledo, OH 43606
Position: Prof of Biochem Engg. *Employer:* Univ of Toledo. *Education:* PhD/ChE/Univ of MI; MSE/ChE/Univ of MI; BSc/ChE/Queen's Univ. *Born:* 7/22/35. Native of Windsor, Ontario. Was a teaching fellow in the Dept of Chemistry and Dept of Chem Engg from 1958-1963 (at the Univ of MI). Asst Prof of Biochem Engg, Univ of Toledo, 1963-1966. Assoc Prof of Biochem Engg, Univ of Toledo, 1966-1972. Prof of Biochem Engg, Univ of Toledo, 1972-. Am environmental consultant for several cos. Have 25 books & conference proceedings and written more than 40 articles & reports. Am also a speaker on the AIChE Natl Tour. Have received 30 grants in pollution control with a total dollar amount of $800,000. Am editor of the Journal of Hazardous Mtls and Environmental Progress. *Society Aff:* AIChE, AAEE, WPCF, APEO.

Bennett, Marlin J
Home: 15922 NE 153rd, Woodinville, WA 98072
Position: Pres. *Employer:* Freezing Systems Inc. *Education:* MS/Chem Engg/Oregon State Univ; BS/Chem Engg/Stanford. *Born:* 7/20/39. Completed BS in 1961, MS in 1962. R & D Engg for Stauffer Chem 1962-66. Dev of chlorinated hydrocarbon processes. License eng for Stauffer 1966-69. Design & startup of VCM and other plants throughout world. Mgr process dev for Basic vegetable products 1969-75. New food processes dev & started. Lewis Refrigeration 1975-1981. VP Engg. Started new company - Freezing Systems Inc. 1981-present. Pres & owner. Enjoy outdoors, Hunting & Fishing. *Society Aff:* AIChE.

Bennett, Ralph W
Business: 1107 Fairfield Ave, Eugene, OR 97402
Position: Pres/CEO. *Employer:* Bonney, Bennett & Peters, Inc. *Education:* BS/Mathematics/Univ of OR. *Born:* 12/8/36. As Pres and Chief Exec Officer of Bonney, Bennett & Peters, Inc, He is responsible for developing and managing a leading consulting engg firm for the wood products industry involved in the design, engg, and construction mgt responsibilities for new plant construction and/or retrofit existing plants. The co has enjoyed substantial growth over the last few yrs and is currently one of the most prominent consultants in the wood products industry. The corporate headquarters are in Eugene, OR with branch offices in Atlanta, GA and Vancouver, British Columbia, Canada. *Society Aff:* PEO, NSPE, ACEC, AIC, PSBMA.

Bennett, Richard T
Business: 7 S 600 County Line Rd, Hinsdale, IL 60521
Position: Sr Project Engr. *Employer:* J.I. Case, A Tenneco Co. *Education:* MS/Agri Engg/Univ of IL; BS/Agri Engg/Univ of MO. *Born:* 6/10/36. in St Louis, MO. Employed by Intl Harvester after grad from Univ of IL; worked in test dev of agri & indus tractor & outdoor power equip, especially in areas of acoustics, vibration & exper stress analysis; worked with photographic & electronic instrumentation; 1974-79. Ch Engr in charge of Outdoor Power Equip design. 1979-83 Mgr Product Dev Agri Tractors & Equipment. Currently Sr Proj Engr, JI Case Co. Mbr SAE; SR Mbr ASAE; active in local sects & natl cttes SAE & ASAE; Dir of ASAE 1972-74. Married, 2 daughters. *Society Aff:* SAE, ASAE, NSPE.

Bennett, Robert A
Business: 175 East Old Country Rd, Hicksville, NY 11801
Position: Mgr Gas Supply. *Employer:* Long Island Lighting Co LILCO. *Education:* MME/Thermodynamics/Polytech Inst of Brooklyn; BSME/Thermodynamics/Cornell Univ. *Born:* 11/25/34. Native of New York. With Navy 1943-47 incl Pacific & China areas; retired Naval Reserve 1973; with LILCO since 1948; electric prod plants; System Engrg & Planning Service, both elec & gas functions; Mgr Gas Supply in 1974 with responsibility for all gas supply activities of Co., added gas planning in 1984; ASME VP Region II 1971-75; ASME rep EJC Bd of Dir 1975-77, presently ASME Secretary and Treasurer and serving on other ASME natl cttes. Hobbies: Golfing, gardening & photography. *Society Aff:* NSPE, ASME.

Bennett, Robert W
Home: 1503 Ashford Ave 14C, Condado, PR 00911
Position: Dir Tech Services. *Employer:* Roche Prod Inc Puerto Rico. *Education:* BS/Ceramic Eng/Alfred Univ. *Born:* June 1918. Native of New York City. Served with US Navy 1942-45. Design Engr James P ODonnell Engrs 1946-50; Blaw-Knox Chemical Plants Inc 1950-69; Design Engr, Proj Engr, Proj Mgr, Mgr of Projs; Hoffmann-La Roche Inc 1969 Capital Proj Coordinator, Mgr Tech Service Texas Div. Registered PE in NY and PA. Staff lecturer in Proj. Mgnt Univ Pgh Grad. Schl. San Juan Chbr. of Commerce, Puerto Rico Wing CAP (pilot). *Society Aff:* AIChE.

Bennett, W Burr, Jr
Business: 20 N Wacker Dr 2608, Chicago, IL 60606
Position: Pres *Employer:* W Burr Bennett Ltd *Education:* B/CE/Syracuse Univ *Born:* 10/11/20 Honesdale, PA, US Army Air Force 1941-1945. Air Medal Distinguished Flying Cross. Syracuse Univ 1980 Bachelor Civil Engrg, Magna Cum Laude, Member Tau Beta Pi, Sigma Pi Sigma. Plant engr designed early prestressed Concrete plant 1954 for Frontier Dolomik Concrete Products, Lockport, NY. Portland Cement Assoc 1956-1968 - structural engr to dir engrg services - then Executive VP Prestressed Conc Inst 1968-1979. Started consulting firm to serve conc industry Jan 2, 1979 - currently Dir Am Conc Inst, formerly chrmn ACI Prestressed Concrete Comm, and Bldg Code Requirements Committee - ACI Raymond E Davis lecture Quebec, Sept 1981. *Society Aff:* ASCE, PCI, ACI

Bennett, William R
Home: 29 Tulip Lane, Colts Neck, NJ 07722
Position: Charles Batchelor Prof of Elec Engr. *Employer:* Columbia Univ New York. *Education:* PhD/Physics/Columbia Univ; AM/Physics/Columbia Univ; BS/Elec Engg/OR State Univ. *Born:* June 1904. Born in Des Moines Iowa. Mbr Tech Staff Bell Tel Lab 1925-65; retired as Head of Data Theory Dept 1965; Prof Elec Engrg Columbia Univ 1965-72; apptd Charles Batchelor Prof Emer of Elec Engrg 1972; early contributor to elec circuit theory, signal transmission theory & use of pulse code modulation. Publ papers & books on elec noise, multiplex telephony, modulation methods & data transmission. Editor IEEE Trans on Circuit Theory 1957-60. Life Fellow IEEE, Fellow Amer Phys Soc; Kelly Award in Telecommunications 1968 IEEE. Interested in psychical res. *Society Aff:* IEEE, APS, ASPR, SPR.

Bennett, William R, Jr
Business: 102 Dunham Lab, New Haven, CT 06520
Position: Prof. *Employer:* Yale Univ. *Education:* PhD/Physics/Columbia Univ; BA/Physics/Princeton Univ; MHA/(honorary)/Yale Univ; DSc/(honorary)/Univ of New Haven. *Born:* 1/30/30. Instructor in Physics, Yale Univ 1957-1959. Mbr of Technical Staff, Bell Labs, 1959-1962. Assoc Prof of Physics and Applied Science, Yale Univ 1962-1965. Prof of Physics and Applied Science, Yale Univ 1965-. C B Sawyer Prof of Engg and Applied Science and Physics 1972-. Master of Silliman College, Yale University 1981-87. Co-inventor of the first gas laser (1960). Morris Liebmann Award of the IEEE (1965) for res on lasers. Western Elec Fund Award of the ASEE (1977) for teaching engg. Outstanding Patent Award of the Res and Dev Council of NJ for the Invention of the Helium-Neon Laser (1977). A P Sloan Fdn Fellow 1963- 1964; J S Guggenheim Fdn Fellow, 1967. Hobbies: playing chamber music and scuba diving; color photography. *Society Aff:* IEEE, APS, OSA.

Bennion, Douglas N
Business: Chem Engg Dept, 350 CB, Provo, UT 84602
Position: Prof of ChE. & Dept Chairman *Employer:* Brigham Young Univ. *Education:* BS/ChE/OR State Univ; PhD/ChE/Univ of CA. *Born:* 3/10/35. Specialize in electrochem reactor design, electrochem energy storage, ion transport with applications to batteries, fuel cells, membrane transport, and electrochem separations. *Society Aff:* AIChE, ACS, ECS.

Bennon, Saul
Home: 2701 W Twickingham Dr, Muncie, IN 47304
Position: Engg Consultant. *Education:* MS/EE/Univ of PA; BSEE/EE/Univ of PA. *Born:* 8/9/14. b in Phila. Phi Beta Kappa, Sigma Xi, Eta Kappa Nu. Joined Westinghouse Elec Corp as Transformer Design Engr at Sharon Penna 1937; proj Engr in Ordnance 1943; received Westinghouse Order of Merit 1945; Sec Mgr of Distrib. Transformer Design & then Sect Mgr of Power Transformers Design 1949-62; Mgr of Power Transformer Dev 1962-66; 1966-78 Mgr of Engrg Large Power Transformers, Muncie, Ind. Presently Engg Consultant. Fellow IEEE, P Chrmn Sharon Section IEEE P Chrmn IEEE Transformers Committee. Reg Engr in PA. Enjoy amateur radio, photography & tennis. *Society Aff:* IEEE, NSPE.

Beno, Paul S
Home: 621 W Laramie Lane, Milwaukee, WI 53217
Position: SALES Mgr. *Education:* BM/Engr/Marquette Univ *Born:* June 1921. Pi Tau Sigma (Hon Frat). Native of Milwaukee Wisc. Experimental Engr & Inspection for LeRoi engines & compressors to 1950; held positions as Supt, Plant Mgr. Mfg Dir & Genl Mgr for Simplicity Mfg Garden Equip 1950-74; assumed responsibilities as V P - Genl Mgr for manufac & sale of custom gears & matl handling equip 1975-77 Joined Apex Industrial as Sales Mgr 1978-82. Assoc with Capitol Stampings (sprocket & idler pulley mfg) 1959-79. Became Sales Mgr for Kapco, Inc (Metal Fabrication & Stampings) in 1982. Held local & regional offices in Soc of Mfg Engrs; elected internatl Dir 1976-80. Received "Award of Merit" in 1984, and Olin Simpson Award in 1986, and selected as "SME Fellow" in 1987. Certified Mfg (S M E) & Wisc Prof Engr. Mbr Internatl Rotary & Paul Harris Fellow. *Society Aff:* SME

Benoit, Richard C, Jr
Home: 10 Windsor Terrace, Utica, NY 13501
Position: Communications Processing & Distribution Sect. *Employer:* US Air Force Rome Air Dev Center. *Education:* D Engr/Eng Adm/Clayton Univ; D Int Aftrs/Hon/Clayton Univ. *Born:* May 16, 1917 E Orange NJ. BSEE & D Engr Clayton Univ St Louis Mo. D. Intl Affairs (Hon) Fellow IEEE, Fellow AAAS; Dir IEEE 1974-75. Prof assoc with fields of communications & electronics 1940- , spec in telecommunications R&D mgmt, sys engrg & internatl progs; presently Ch Communication Processing & Dist Sect of USAF Rome Air Dev Center; served as tech advisor to USAF Europe & Spanish Air Force & rep USAF at various NATO telecommunications confs. Author 30 pub tech papers in fields of radio navigation & telecommunications. *Society Aff:* IEEE, AAAS, D. Int'l Affairs (Hon).

Benson, Frank A
Business: Univ of Sheffield, Mappin St, Sheffield S1 3JD UK
Position: Prof of Electronic & Elec Engrg. *Employer:* University of Sheffield. *Education:* DEng/-/Univ of Sheffield; PhD/-/Univ of Sheffield; MEng/-/Univ of Liverpool; BEng/-/Univ of Liverpool. *Born:* Nov 21 1921. Fellow IEEE, IEE. Offices Held: 1943-46 Res Officer Admiralty Signal Establishment Witley; 1946-49 Asst Lecturer in Elec Engrg Univ of Liverpool; 1949-67 Lecturer, Sr Lecturer, Reader in Electronics, Univ of Sheffield; 1967- Prof & Head of Dept of Electron & Elec Engrg Univ of Sheffield. 1972-76 Pro-VChancellor Univ of Sheffield. *Society Aff:* IEEE, IEE.

Benson, Fred J
Business: Faculty Exchange P.O. Box 9624, College Station, TX 77840
Position: President *Employer:* Engitech, Inc *Born:* Sept 1914. MS Texas A&M Univ; BS Kansas St Univ. Served with US Navy 1943-46; USNR Officer 1943-74. Worked for Kansas Hwy Comm. Internatl Boundary Comm, Cities of Wichita Falls & College Station Texas, Purdue Univ. Texas A&M Res Fdn. Currently President, Engitech Inc Retired, Professor, Dean of Engr., Dir Engrg Research Vice-President and Vice Chancellor, Texas A&M University System. Bd of Dir Univ Natl Bank, Bryan Building & Loan Assn, H B Zachry Co, Gen Security Life Insurance Co. Edmund Friedman Award ASCE; Wilbur S. Smith Award, ASCE; Tasker H Bliss Award SAME: Honorary member ASCE; Honorary Doctor of Engineering Degree, Kansas St. Univ. Hon mbr, APWA, mbr Cosmos Club. *Society Aff:* ASCE, ASEE, NSPE, ΣΞ, ΦΚΦ, ET, APWA, ARTBA

Benson, Homer E
Business: 615 Washington Road, Pittsburgh, PA 15228
Position: President. *Employer:* The Benfield Corp. *Education:* BS/Chem Engg/Case Inst of Tech. *Born:* Feb 25, 1918 Cleveland Oh. R&D in synthetic fuels from coal 1942-58; Ch Gas Synthesis Sect US Bur of Mines; Asst Dir Res Consol Natural Gas Co 1958-66 working primarily on coal gasification; Pres The Benfield Corp since 1966 spec in purification of natural gas, synthesis gases, hydrogen & gas derived from coal. *Society Aff:* AIChE, ACS, AACE.

Benson, Melvin L
Home: 1756 Tamworth Ct, Dunwoody, GA 30338
Position: Pres. *Employer:* M. L. Benson & Assoc., Inc. *Education:* BS/Chem Eng/Univ of WI *Born:* 1/8/28. Asst Sales Mgr, Eclipse Fuel Engrg Co, Rockford, IL (1955-60), Chattanooga (1960-65), Regional Sales Mgr APC equip, Western Precip Div Joy Mfg Co, Chicago (1965-70), Atlanta (1970-72). Mfg Rep APC equip, Applied Engrg Co (1972-75). VP Carotek, Inc, Mfg Rep, Atlanta, GA, (1975-80). In Chattanooga served as Gen Chrmn Engrg Week (1963), Secy Chattanooga Chapter TSPE (1964-65). Co-author of ABC's of Dowtherm Engrg. Football official high sch & college (1955-75). Swimming official (1965-75). Mbr Dunwoody Country Club VP & Pres DHS-PTA (1973- 75), Pres DHS Team Booster Club (1974), VChrmn DCC Swim-Tennis Cttee (1972-73). (1980-) Pres, M L Benson & Assoc, Inc. Reg PE, Mbr of AIChE, TAPPI, APCA, MANA. *Society Aff:* AIChE, TAPPI, APCA, MANA

Benson, Walter
Business: Dept of Applied Science and Engrg, New London, CT 06320
Position: Prof *Employer:* US Coast Guard Academy *Education:* PhD/EE/PA State Univ; MS/EE/PA State Univ; BS/EE/PA State Univ *Born:* 9/11/40 Native of Mascon, PA. Since 1970, has been a Patent Examiner at the US Patent Office; Research Assistant in Electrical Engrg, Instructor and Assistant Prof of Electrical Engrg at the PA State Univ. He had two patent disclosures and has presented papers at the Transportation Research Bd Annual Meeting and the 1980 SAE Congress and Exposition on the contents of these disclosures. Assumed current position of Prof of Electrical Engrg in 1979. Summer research last year resulted in ATE methods for remote microcomputer systems in the Coast Guard. Treasurer, IEEE 1980. *Society Aff:* IEEE, NSPE

Benson, Willard R
8 Grover Road, Dover, NJ 07801
Position: Pres *Employer:* Beacon Tech Inc *Education:* MS/Applied Mech/Univ of VA; MS/Statistics/Stanford; ME/ME/Stevens Inst *Born:* MS Statistics Stanford; MS Applied Mech Univ of Vir; ME Stevens Inst. 35 yrs of Government serv starting with serv in Army during WW II & followed by 3 yrs with Natl Adv Ctte for Aeronautics; 1952-69 at Picatinny Arsenal in Ordnance R/D with primary emphasis on conception & evaluation of new munitions & weapon systems; later Ch of the Fuze Div at Munitions Command and ARMCOM; finally Ch of the Applied Sci Div of ARRADCOM. Since 1981 Pres of a conslt firm concentrating on the design and analysis of munition sys. *Society Aff:* ADPA, AUSA, ASME

Benstein, Eli H
Home: 5224 Coldstream, Toledo, OH 43623
Position: Chief Scientist *Employer:* Teledyne CAE. *Education:* MAeE/Aero Engg/Cornell Univ; BASc/Aero Engg/Univ of Toronto. *Born:* Aug 1929 Windsor Canada. Indentured to Teledyne CAE 1953- , originally in Performance Dev; today Generalist in R/D Mgmt; concept, design, dev and prog. mgt of all forms of military gas turbines; J69, J100, XLJ95, J402, F106 & Advanced Tech of Turbomachinery; Dir Design Engrg 1970; Dir of Engrg 1975; Dir Adv Dev Prog 1978. Chief Scientist 1987. Active in SAE, NASA Advisory Comm. Awarded 'Most Outstanding Air Breathing Propulsion Paper of 1974 by AIAA/SAE. Classical music buff; also hooked on reading. *Society Aff:* SAE.

Bentley, Lawrence H
Home: 100 Fort Hill Road, Groton, CT 06340
Position: Pres. *Employer:* DiCesare, Bentley-Engineers Inc. *Education:* BS/Civ/Univ of CT. *Born:* 9/8/33. Principal in firm of DiCesare-Bentley Engrs, Inc, since 1960. Mbr ACSM, ASCE, ACEC & NSPE. Past State Pres CT Soc of PE; past Natl Dir NSPE; past mbr CT Bd of Reg for PE and Land Surveyors; Mbr CT Assn of Land Surveyors; Chrmn of Groton Bldg Code Bd of Appeals. Enjoys boating and fishing. *Society Aff:* NSPE, ACSM, ASCE, ACEC.

Benza, Michael
Home: 3798 Brecksville Rd, Richfield, OH 44286 *Education:* BS/CE/OH State Univ *Born:* 1/20/40 Mr Benza graduated from OH State Univ in 1963 with a BS in Civil Engrg. He has been employed with several Consulting Engrg firms specializing in Environmental Engrg projects. Since starting his own firm in 1972 he has broadened his scope to include all facets of Civil Engrg. He is the active 1981- 82 Pres of The Cleveland Section of American Society of Civil Engrs. He was Pres of the Northeast Section of the OH Water Pollution Control Conference in 1980- 81. He is still very active in both organizations. *Society Aff:* ASCE, CEC, AWWA, WPCF

Benziger, Charles P
Home: 11116 Farr Dr, Knoxville, TN 37922
Position: Conslt *Employer:* Self-employed *Education:* AB/Geol/Univ of TN, Knoxville *Born:* 9/6/20 Native of Knoxville, TN. Served Army Air Force, 1941-1945-TVA, 1948-1956, Bechtel Corp 1956-1957, Proj Geologist, Niagara Power Project 1957-1963, Chief Geologist, Chas T Main Inc; 1963-83. In responsible charge of all geological-geotechnical work of Main Hydro-projects in US, Central & South America, Africa, Mid-East and Far-East. Includes some 100 Hydro projects. Assoc of Main since 1971. Retired from Main in 1983, have been in private consulting since, involved in foundations for hydro generation plants. *Society Aff:* AEG, AIPG, USCOLD, IAEG

Benzing, Robert J
Home: Box 3755, RD 3, Montpelier, VT 05602
Position: Retired *Education:* BS/Chem Engg/Purdue Univ; MS/Chem Engg/Purdue Univ. *Born:* April 1930. Assoc with Air Force Materials Lab 1953-85; major area of exp is in Aerospace fluid & lubricant materials R/D. Natl Dir of Amer Soc of Lubrication Engrs 1971-78; Natl Treas 1978-80; Fellow of Amer Soc of Lubrication Engrs 1976 Natl Secretary 1980-81 Vice President at Large (President Elect) 1981-1982. *Society Aff:* ACS, ASLE, ΣΞ, RESA.

Beranek, Leo L
Home: Suite 804, 975 Memorial Drive, Cambridge, MA 02138
Position: Conslt and author *Employer:* Self *Education:* DSc/Communication Engg/Harvard Univ; MSc/Communication engg/Harvard Univ; BA/Math & Physics/Cornell (IA). *Born:* 9/15/14. Bd of Overseers Harvard Univ 1984- ; Chr Bost Sym Orch 1983-87; Trustee, 1977-87; VP 1980-1983. Hon D. Sc. Cornell (IA) , Hon D. Commerical Sc. Suffolk Univ.Asst Prof Harvard; Assoc Prof MIT; Hon D. Laws, Emerson Coll, Boston; Hon D. Pub Service Northeastern Univ. Pres Bolt Beranek & Newman 1953-69. Pres, Boston Bcasters, Inc, 1963-79, Chr, 1980-83. Author 6 books on noise control & acoustics. Fellow: Am Acad Arts & Sci, Ac Soc Am Pres 1954-55, APS, AAAS, IEEE, AES Pres 1967-68; Mbr: NAE Council 1970-71, INCE Pres 1971-73. AIAA, MA Broadcasters Assn Pres 1978-79, World Affairs Council Pres 1975-78. Honors: Gold Medals from Ac Soc Am & AES, Abe Lincoln Television Award, Hawksley Lecture London, USA Merit, ASA Sabine, Silver Medal French Acous Soc, Phi Beta Kappa, Sigma Xi, Eta Kappa Nu. Reg PE MA *Society Aff:* ASA, APS, AES, IEEE, INCE, NAE, AAAS

Berbeco, George R
Home: 170 Dartmouth Street, West Newton, MA 02165
Position: President. *Employer:* Charleswater Products, Inc. *Education:* SM/Chem Engg/MIT; SB/Chem Engg/MIT. *Born:* 3/17/44. Natl Merit Scholar at MIT. Student engr for Inland Steel Co; Mgr of Radiation Res Lab at High Voltage Engg Corp; assumed VP position at Charleswater Assoc 1975 with genl mgmt, mkting respon; assumed crrent position as Pres Charleswater Prod Inc in 1977 with operational duties for Static Protection Packaging. Sigma Xi, Guest lecturer at Harvard Bus Sch. Boston Univ & lecturer at Boston State College. Who's Who in MA, 16 publs, 2 pats. Who's Who in the East. Enjoy Country, western & classical music, play squash & classical piano. Who's Who in Business and Finance. *Society Aff:* ACS, AIChE.

Bereman, John S
Home: P.O. Box 2107, Cody, WY 82414
Position: Principal (Self-employed) *Education:* BS/CE/Univ of CO *Born:* 1/1/26 Since graduation in 1947 have been engaged in the private practice of civil engrg and land surveying. Established my own firm in Cody, WY in 1952 and have practiced for clients in and out of the state of WY since. Special interest in water resource development and related diciplines. *Society Aff:* ASCE, ACEC, WES, WACES

Bereskin, Alexander B
Business: 898 Rhodes Hall 30, Cincinnati, OH 45221
Position: Prof Emeritus of Elec Engrg. *Employer:* University of Cincinnati. *Education:* MSc/EE/univ of Cincinnati; BSc/EE/Univ of Cincinnati *Born:* Nov 15, 1912 San Fran Calif. ASEE Mbr; Engrg Soc of Cincinnati Mbr, Secy 1969-70; Eta Kappa Nu Mbr; IEEE (IRE-AIEE) Fellow; Natl Offices in IRE: Dir Region 4 1961-62, Chmn Prof Group on Audio 1959-60, Editor Trans on Audio 1955- 58; Natl Cttes in IRE: Sects 1952-53, Prof Groups 1959-60, Adm Comm Prof Group on Audio 1956-59, Educ 1955-61; Natl Cttes in IEEE: Mbr & Transfer 1965-68, Review 1964-65, Fellow 1967-69, Internatl Hospitality 1964-70. Sigma Xi Mbr; Tau Beta Pi Mbr; Hon & Awards: Fellow IRE 1958, IRE-PGA Achievement Award 1959, Disting Cincinnati Engr of the Year 1976, ASEE (DELOS) Excellence in Laboratory Instruction Award 1981; Western Electric Fund Award, ASEE 1983; Award of Merit, Cincinnati Chapter Eta Kappa Nu, 1983. *Society Aff:* IEEE, ΣΞ, ΤΒΠ, HKN

Berg, Daniel
Home: 1623 Branning Rd, Pittsburgh, PA 15235
Position: Provost Carnegie-Me *Employer:* Carnegie-Mellon Univ. *Education:*

Berg, Daniel (Continued)
BS/Physics & Chem/City College of NY; MS/Physical Chem/Yale Univ; PhD/Physical Chem/Yale Univ. *Born:* 6/1/29. Provost at CMU. Respons include: the activities of Engineering at CMU. An industrial scientist (BS, CCNY, 1950; MS, PhD, Yale Univ, 1951-53, Physical Chemistry) & technical exec with Westinghouse for more than 20 yrs, Dr Berg has been mgr of Energy Sys Res & Technical Dir of Uranium Resources. He has contributed extensively to the technical & mgmt literature. He is a Mbr of the Natl Acad of Engg and a Fellow of the Inst of Electronics & Electrical Engrs, the Amer Assoc for the Advancement of Sci, and the Amer Inst of Chemists. As a mgmt consultant, he has worked extensively for the govmt & industry. He serves on the bd of dirs of EIA, Inc and Argonne. He also serves boards as Science Advisor. He has taught courses in mgmt of technological innovation at the Grad Sch of Ind. Adm. *Society Aff:* ACS, IEEE, AAAS, AIC, APS.

Berg, Eugene P
Business: Automatic Spring Coiling Co, 4045 W Thorndale Ave, Chicago, IL 60646
Position: Chmn of Bd & Pres. *Employer:* Automatic Spring Coiling Co. *Education:* BS/Mech Eng/Purdue Univ; MBA/-/Univ of Chicago. *Born:* May 25, 1913 Chicago Ill. For past 16 yrs assoc with Bucyrus-Erie Co in South Milwaukee Wisc & serves as Chmn of Bd & Pres; responsible for conceptual engrg of prods serving useful purpose in competitive marketplace incl excavating, hoisting & drilling equip incorporating advanced mech & hydraulic design features to give superior performance characteristics with low operating & maintenance costs; established an advance design div to incorporate latest dev in basic & applied scis to insure prod superiority during yrs to come. Retired from Bucyrus-Erie Co 5/31/78. Currently associated with Automatic Spring Coiling Co as Chrmn. *Society Aff:* ASME.

Berg, Kenneth E
Business: 649 Mission St, San Francisco, CA 94105
Position: President *Employer:* FPE Group. *Education:* BS/ME/IL Inst of Tech. *Born:* 4/27/39. Engr for Gage-Babcock & Assoc, Inc, 1962-1968. As the corporate fire protection engr for Kaiser Aluminum & Chem Corp, 1968-1971, was responsible for entire corporate loss prevention prog for 150 maj facilities. Returned to Gage- Babcock as Supervising Engr, 1971-1973. Mgr, fire protection div of San Francisco office of the Sierra Group, 1973-1975. In 1975, cofounded the FPE Group, an independent fire protection engr consulting corp; Sr VP til 1979 and Pres since then. Recognized internationaly for work in loss control engg. Active mbr of NFPA technical committes. PE: Mech, IL; both Mech & Fire Protection, CA. *Society Aff:* SFPE, NFPA, ASSE, NSPE.

Berg, Lloyd
Business: Dept of Chem Engg, Bozeman, MT 59717
Position: Prof of Chem Engr. *Employer:* MT State Univ. *Education:* BS/ChE/Lehigh Univ; PhD/ChE/Purdue Univ. *Born:* 8/8/14. Taught chem engg at Univ of Pittsburgh, KS Univ & MT State Univ 1943- present. Hd of Chem Engg Dept at MT State 1946-79. Dir, ASEE Summer Schools 1962 & 1967. Developed Diesel oil desulfurization process for Husky Oil Co in 1952, currently Dept of Energy principal investigator for up-grading solvent refined coal to transportation quality fuels. MT State was leading producer of female chem engrs 1960-1979 during tenure as Dept Hd. Completed 26plus mile marathons in 1978 & 1979 at age 63 & 64. Currently leading inventor and publisher in separations by extractive distillation. *Society Aff:* AIChE, ACS.

Berg, Paul O
Home: 910 Illsley Dr, Fort Wayne, IN 46807
Position: Cons Engr. *Employer:* Paul Berg Engrg. *Education:* BSME/Mech/IA State Univ. *Born:* Jan 1907. Native Burlington Iowa. Asst Ch Engr John Morrell & Co Ottumwa Iowa; Ch Engr Central Soya Co Fort Wayne Ind; established Cons business 1947 spec plant design for feed mills, flour mills, grain elevators & fertilizer plants. Reg engr New York, Pennsylvania, Ohio, Indiana, Illinois, Minnesota & Tennessee. Hold 10 USA pats. Mbr ASME, State Secy Indiana Soc of Prof Engrs; Chmn Indiana's Admin Building Council. ACEC Honor Award winner 1969 for Engrg Excell. Selected by feed indus to design feed prod school for Kansas St College Manhattan Kansas. Also qualified for reg engr MO, KY, LA, WI, MD, GA and prov of Saskatchewan. Held cert Natl Council of Engrg Examiners. *Society Aff:* NSPE, ASME, ISPE.

Berg, Philip J
Business: One Oliver Plaza, Pittsburgh, PA 15222
Position: Exec VP, Operations. *Employer:* Dravo Corp. *Education:* BS/Mech Engg/Lehigh Univ. *Born:* 7/27/23. One of two Exec VP, Mr Berg is respon for all of the firm's diverse operations and for Purchasing and Traffic. He joined Dravo as a proj engr in 1946 and served in various engg and sales capacities until being named mgr of genl construction sales, Engg and Construction Dept in 1956. He later served as mgr of the sales, dev and operations functions of the Engg Construction Div before being named genl mgr of the division in 1966. He was elected a VP in 1967. Mr Berg was elected a group VP in 1970, in which post he headed the firm's engg construction operations, and was elected Sr VP, operations in 1974. He was elected to his present position in 1978 and has been a Mbr of Dravo's Bd of Dirs since 1977. *Society Aff:* AIMPE.

Berg, William D
Home: 1102 Chapel Hill Rd, Madison, WI 53711
Position: Prof *Employer:* Univ of WI *Education:* PhD/CE/Univ of IL; MS/CE/Purdue Univ; BS/CE/Purdue Univ *Born:* 9/26/42 Member of the Civil Engrg faculty at the Univ of WI since 1970. Responsible for teaching and research in traffic engrg, transportation planning, and civil engrg systems analysis. Served as Highway Research engr with the Federal Highway Administration from 1974 to 1976. Since 1971, have served as a consultant to federal, state, and local governmental agencies, and to numerous railroads, consulting engrg firms, and attorneys. Author of over 40 technical reports and publications. *Society Aff:* ASCE, ITE.

Bergen, William B
Home: Box 747, St Michaels, MD 21663
Position: Aerospace Cons. *Employer:* self. *Education:* BS/Aero Eng/MIT. *Born:* March 1915. Rockwell Internatl Corp: 1974-1975 Corporate VP Aerospace, 1974 Pres Space Operations, 1970-74 Pres Aerospace Group, 1967-1970 Pres Space Div; Martin Marietta Corp 1961-67 Corporate VP & Pres Aerospace Div; The Glenn L Martin Co: 1959-61 Pres, 1955-59 Exec VP, 1953-55 VP Operations, 1951-53 VP Engrg. Awards: 1943 Lawrence Sperry Award, 1966 NASA Pub Serv Award Gemini, 1969 NASA Pub Serv Award Apollo VIII, 1969 Pres Award Apollo XI. *Society Aff:* NAE, AIAA, AAS, NSI.

Berger, Bernard B
Home: 344 East Hadley Rd, Amherst, MA 01002
Position: Consultant *Employer:* Univ of MA/Amherst *Education:* MS/Sanitary Engrg/Harvard; BS/Pub Hlth Engrg/MIT *Born:* 8/21/12 Member USPHS Commissioned Corps 1941-1966. Dir water supply and water pollution control research, Robert A. Taft Sanitary Engrg Center 1954-1963. Dir, Univ of MA Water Resources Research Center and Prof of Civil Engrg 1966-1978. Water Resource Specialist, Office of Science & Technology, Executive Office of the Pres 1968-1969. Founder Member, Secretary (1962-1968), and VP (1970-1976) Intl Association on Water Pollution Research. ASCE Rudolph Hering Medal 1971. Honorary Fellow Inst of Water Pollution Control (UK). Hon Mbr WPCF 1982, Honorary Sc.D. Univ of MA, 1979. Elected member National Academy of Engrg 1979. *Society Aff:* ASCE, WPCF, AWWA, AAEE

Berger, Bruce S
Business: Dept of Mech Engg, College Park, MD 20742
Position: Prof. *Employer:* Univ of MD. *Education:* PhD/Engg Mechs/Univ of PA; MSc/Engg Mechs/Univ of PA; BSc/Engg/Univ of PA. *Born:* 5/23/32. Employed by Gen Elec. Co, Math Analyst, 1956-1957, Asst Instr & Res Asst Univ of PA, Engg

Who's Who in Engineering

Berger, Bruce S (Continued)
Mechs, 1957-1961, Asst Prof, Dept of Appl Math, Vanderbilt Univ 1962- 1963, Asst, Assoc & Prof, Dept of Mech Engg, Univ of MD, 1964-present. Fields of Res & publication are acoustic fluid-solid dynamics & vibrations, numerical inversion of integral transforms, applications of diff geometry & tensor analysis to numerical solutions in continuum mechanics & chaos. Publications in ASME Jour of Appl Mechs, Math of Comp, Quart Applied Math, Jour of Engg Mechs, ASCE, Jour of Am Acoust Soc, etc. *Society Aff:* ASME, SIAM.

Berger, Harold
Business: Industrial Quality, Inc, P.O. Box 2397, Gaithersburg, MD 20879-0397
Position: Pres *Employer:* Industrial Quality, Inc. *Education:* BS/Physics/Syracuse Univ; MS/Physics/Syracuse Univ. *Born:* 10/7/26. Exper NDE imaging sys using X-rays, neutrons, ultrasound, light, infrared; X-ray & light detection exper G E 1950-59; semiconductor devices Battelle 1959- 60; NDE methods Argonne Natl Lab (as Sr Physicist & Group Leader 1960-73); NDE measurements & stds. Natl Bureau of Standards, Chief, Office of NDE, 1973-1981; NDE Consulting and R&D, Industrial Quality, Inc., Pres and founder, 1981; Holds 9 U S pats; has over 100 tech publs & 3 books. IR-100 Award 1965, ASNT Achievement Award 1967, ANS Radiation Indus Award 1974. Dept of Commerce Silver Medal, 1979 ASNT Gold Medal 1982.ed NDT Soc Great Britian Pres Hon Lecture 1971 & ASNT Mehl Hon Lecture 1975. Tech Chrmn, 11th world conference on NDT, Las Vegas, Nov. 1985. Fellow ASNT & ANS Mbr the Brit Inst NDT. Tech Editor Materials Evaluation ASNT 1969-1987, ASNT Honorary Member Award, 1987. *Society Aff:* ANS, APS, ASNT, ASTM, ASM, AWS, ΣΞ.

Berger, J Walter
Business: Midland Bldg-505 C.B, 101 Prospect Ave, Cleveland, OH 44115
Position: Mgr-Cost Engr *Employer:* The Standard Oil Co (OH) *Education:* MBA/Managerial Control/Lynchburg Coll; BS/IE/VA Polytechnic Inst *Born:* 7/1/47 Native of Lynchburg, VA. Graduated 1969 and began employment with Babcock and Wilcox as an estimator. After several internal transfers, became Mgr, Estimating in 1974. Shifted careers from Manufacturing to Engrg and Construction in 1976 with employment as Supvr, Cost Engrg with Arthur G McKee, concentrating on petroleum and iron and steel projects. In 1978, joined Sohio as Mgr, Cost Control responsible for all capital expansion projects other than exploration/production. In 1980, promoted to Mgr, Cost Engrg, assuming responsibility for Estimating. Active in church and civic activities. Enjoy bridge, tennis, and golf. *Society Aff:* AACE

Berger, Louis
Business: 100 Halsted St, East Orange, NJ 07019
Position: Chrmn. *Employer:* Louis Berger International Inc *Education:* PhD/-/Northwestern Univ; MS/Geology/MA Inst of Tech; BS/Civil Engg/Tufts Univ, Doctor of Sci/Hon/Tuft Univ. *Born:* 4/12/14. Since 1950 pres of cons group bearing his name. Author of many papers on engr soil mech & foundations & mgmt problems; prior to 1950 Prof of C E Penn St Univ & taught soil mech & highway engrg; also lecturer at Northwestern Univ; Officer in USCG WW II; Hd of Foundations & Soils Sect of St Louis Dist Corp Engrs. Hobbies: golf, photography & sport-scars. Sci interests are in the field of non-destructive evaluation of pavements. Dir 1972 IRF. 1987 Pres. of IRF Educ Found. *Society Aff:* IRF, ACEC, IECIC, ASCE.

Berger, Richard L
Business: 3211 Newmark Civil Engg Lab, Urbana, IL 61801
Position: Prof of Civ & Material Science *Employer:* Univ of IL. *Education:* PhD/Geol/Univ of IL; MS/Geol/Univ of IL; BS/Geol/Univ of IL. *Born:* 7/31/35. Native of Belleville, IL. Res scientist for American Cement Corp 1962-70. With Civ and Ceramic Engg Depts, Univ of IL from 1971. Full Prof from 1975. Res interest in developing the relationship between microstructure and engg performance of cementitious construction mtls. ASTM Bates Award 1974. Chrmn Am Ceramic Soc Cements Div 1972. Fellow Am Ceramic Society 1982. *Society Aff:* ACerS, ΣΞ, ACI, MRS.

Berger, Robert C
Home: 6212 Raymond Ct, Erie, PA 16505 *Education:* BEE/EE/OH State Univ. *Born:* 3/31/18. Born in Columbus, OH. Emloyed by Genl Elec Co since graduation in June 1941. Assignments in Cleveland, OH, Schenectady, NY and Erie, PA. Chrmn Cleveland Section AIEE - 1956-57. Reg PE in State of OH, No 10335-1945. Mbr of Technical Assoc of the Pulp & Paper Industry 1946-66. Published 26 technical articles in the Trade Press; Profl Soc Publications and 4 Genl Elec Technical Courses and Application Manuals. Retired Dec 31, 1979. *Society Aff:* IEEE.

Berger, Stanley A
Business: Dept of Mech Engg, University of California, Berkeley, CA 94720
Position: Prof of Engg Sci. *Employer:* Univ of CA Berkeley. *Education:* PhD/Appl Math/Brown Univ; BS/Math & Physics/Brooklyn College *Born:* in Brooklyn, NY. Res Assoc at Princeton Univ 1959-60. With Dept of Mech Engg, Univ of CA, Berkeley since 1961, rising from Lecturer to full Prof. Gen area of interest is fluid mechanics. In particular, have or am working in following areas: explosion theory, wake flows, compressible and viscous flows, MHD. Since 1968 have worked extensively in bioengg, in addition to more traditional areas of fluid mechanics. Have consulted for, among others, The RAND Corp. Lockheed Missiles and Space Co, Todd Shipyards, Sci Applications Inc (SAI), IBM. *Society Aff:* ASME, AIAA, AAAS, SIAM, APS, SES

Berger, Toby
Business: 308 Philips Hall, Ithaca, NY 14853
Position: Prof. *Employer:* Cornell Univ. *Education:* PhD/Appl Math/Harvard Univ; MS/Appl Math/Harvard Univ; BE/EE/Yale Univ. *Born:* 9/4/40. Attended Yale Univ under scholarships from Grumman Aircraft and Boeing Aircraft. Grad studies at Harvard under Raytheon Fellowship Prog. Sr Scientist at Raytheon Co, Wayland, MA, through Aug 1968. Joined Cornell Univ faculty in Sept 1968; currently Prof & Acting Director of School of Elec Engg. Guggenheim Fellowship for studies in info theory and ergodic theory, 1975-76. Fellow of the IEEE. Co-Chrmn of 1977 IEEE Intl Symposium on Info Theory. Pres of the IEEE Info Theory Group, 1979. Fellow of the Japan Society for the Promotion of Sci, 1980-81. Fellow of the Ministry of Educ of the People's Republic of China, 1981. Author of "Rate Distortion Theory–, Prentice-Hall EE Series, 1971. Author of many technical papers on info theory and coherent signal processing. Recipient of 1982 Frederick Emmons Terman Award of the Amer Soc of Engrg Ed awarded annually to a young elec engrg educator for outstanding contrib to teaching and res. Editor-in-chief, IEEE Trans, on Information Theory, 1987- . *Society Aff:* IEEE, AAAS, ASEE, ΤΒΠ.

Bergeron, Clifton G
Business: Dept of Ceramic Engg, Urbana, IL 61801
Position: Prof Dept Ceramic Engg. *Employer:* Univ of IL. *Education:* PhD/Ceramic Engg/Univ of IL; MS/Ceramic Engg/Univ of IL; BS/Ceramic Engg/Univ of IL. *Born:* 1/5/25. Native of Los Angeles, CA. Attended Univ of KS prior to WWII. US Army 1943- 46, ETO. Sr ceramic engr for A O Smith Corp 1950-55. Staff engr Whirlpool Corp 1955-57. Faculty ceramic engg, Univ of IL 1957-date. Taught courses in high temperature kinetics, phase equilibria, glass science & tech, properties of mtls. Res in high temperature coating, glass, crystallization. Served as hd of dept Jan 5, 1979, to Aug 1986. Currently full time teaching and research. *Society Aff:* ACS, AAAS, ASEE, NICE.

Bergey, Karl H, Jr
Business: AMNE, Univ of Oklahoma, Norman, OK 73019
Position: Prof *Employer:* Univ of OK *Education:* MS/Aeronautical Engrg/MIT; BS/Aeronautical Engrg/PA St Univ *Born:* 12/25/22 Served with US Navy, 1943-1946, as Aerial Navigator. Project Engr, North American Aviation; Assistant Chief Engr, Piper Aircraft Corp; VP, Research & Engrg, General Aviation Division, Rockwell Intl. Presently, Prof of Aerospace & Mechanical Engrg, Univ of OK, and Chairman, Bergey Windpower Co, Inc (2001 Priestley Ave, Norman, OK 73069), manufacturer

Bergey, Karl H, Jr (Continued)
of advanced windpower generators. Designer of a number of general aviation aircraft, including the Piper Cherokee series. Major interests continue to be aircraft design and windpower technology. *Society Aff:* AIAA, SAE, AAAS, ΠΤΕ.

Bergh, Arpad A
Business: 600 Mountain Ave, Murray Hill, NJ 07974
Position: Dept Head *Employer:* Bell Labs *Education:* PhD/Physical Chem/Univ of PA; MS/Chem/Univ Szeged-Hungary *Born:* 4/26/30 From 1959-1966 he was with Bell Labs, Allentown, PA, where he worked on the development of semiconductor devices and on various aspects of MOS technology. In 1966 he was transferred to Bell Labs, Murray Hill, NJ, to supervise the development of GaP electroluminescent devices. He is currently Head of the Lightwave Devices Dept developing LED based optical interfaces with emphasis on long wavelength LEDs and photodetectors. With P. J. Dean (of Royal Signals and Radar Establishment) as a coauthor, he has written a book on Light Emitting Diodes which was published by Oxford Press in 1976. He is a Fellow of the IEEE and a member of the SID. *Society Aff:* IEEE, SID

Bergh, Donald A
Home: 1396 Westerrace, Flint, MI 48504
Position: Assoc Prof. *Employer:* GM Inst. *Education:* MS/Met/MI State; BS/Met/MI State. *Born:* 2/20/21. Native of Flint, MI. Alfred Noyles Scholarship, mbr of Phi Lambda Tau MSU '43. Started teaching career in Army 1944-46. Instr Foundry Metals & Alloys MSU 1947-52, Instr Foundry Muskegon Public Schools 1953-55. Assoc Prof of Mech Engg, Foundry & Processes 1955-present. Active in ASEE, VP Young Engr Teachers 1950, Chrmn of Saginaw Valley ASM 1970, Chrmn of Saginaw Valley AFS 1975. Numerous natl committees, Chrmn SME Casting & Molding Council 1977-1981. Several papers on Foundry Processing, Gray Iron met & handbook chapters. Western MI Advisory Committee 1975-76. SME Certified Life Mfg Engr in Casting and Molding. Patent application. *Society Aff:* ASM, AFS, SME.

Bergles, Arthur E
Business: Mechanical Engineering, Rensselaer Polytechnic Institute, Troy, NY 12180-3590
Position: Prof *Employer:* Rensselaer Polytechnic Institute *Education:* PhD/ME/MA Inst of Tech; SM/ME/MIT; SB/ME/MIT. *Born:* 8/9/35. Natl Magnet Lab staff and Mech Engg faculty, MIT, 1962-69. Prof of Mech Engg, GA Tech, 1970-72. Prof & Chrmn of Mech Engg, 1972-83 IA State Univ Prof and Dir Heat Trans Lab, 1983-86. Clark and Crossan Prof of Engg, RPI, 1986-. Published technical papers, co-authored or edited books, prepared reports, invited lectures, and presented short courses. Consultant to numerous corporations, ASME Board of Governors. Previously: ASME VP, Chrmn ASME Heat Transfer and numerous committees. Recipient of: Fulbright Award, Fellow ASME, ASME Heat Trnsfer Meml Award, Alexander von Humboldt Sr Scientist Award of Fed Republic of Germany, Anson Marston Distinguished Professor of Engrg IA State Univ, ASME Dedicated Service Award, Fellow ASEE, IA State Univ Faculty Achievement Award in Research, SAE Ralph R. Teetor Award, and ASEE Lamme Award. *Society Aff:* ASME, AIChE, ASHRAE, ASEE, AIAA, SAE, AAAS

Bergman Donald J
Home: 55 S. Greeley Apt. 210, Palatine, ILL 60067
Position: Retired. *Education:* PhD/EE/Santa Clara Univ –17–; BS/EE/Univ of CA Berkeley, 15. *Born:* Nov 3, 1895. Asst Const Engr Shaw-Batcher Shipyd 1917; taught ME & EE at SCU 1919; Engr Draftsman C F Braun 1919; Engr Draftsman UOP 1920; Ch Engr 1935; Cons 1958; retired 1962. 14 pats. Chmn Chi Sect ASME 1951; Chmn Auto Control Div 1958; VP Gt Lakes Region 1960-63. ANSB31. Piping Code Mech Design Com 1957; ANSI31. 3 Refinery Piping Code 1951-63. Hon Pi Tau Sigma Northwestern Univ 1951, Merit Award Chicago Tech Soc Council 1957 ASME Centennial Award 1980; Fellow ASME 1963, Fellow ISA 1963; ACS, Catgut Accoustical Soc, Sheridan Shore Y C, Michigan Shores Club, Amer Canoe Assn, Cal Alpine Club. *Society Aff:* ASME, ISA, ACS.

Bergman, Richard I
Home: 134 Leabrook Ln, Princeton, NJ 08540
Position: President *Employer:* Savant Associates Inc *Education:* SB/ChE/MIT; SM/ChE Practice/MIT *Born:* 1/18/34 Born in NY City. SB MA Inst Tech 1955; SM 1956. Executive Dev Engr Esso Res & Engrg Co 1956-60; dir engrg dev Princeton Chem Res Inc NJ 1960-67; VP Systemedics Inc Princeton Inc, 1971-79, Special Projs Equifax Inc 1978-1980; Pres Savant Assoc 1980- . Exec Dir White House Task Force on Workplace Safety & Health 1977-78, also Dir; Dir Response Analy Corp 1975-77. Adj Prof CE Newark Coll Engrg 1957-58. Mbr Copr Vis Comm Med Dept MIT 1973-83, 1986- ; Whitaker Coll MIT 1979-1985; Mbr Amer Inst Chem Engrs (Chrmn NJ Sect 1965-66, Ethics Ctte 1972-1977); NY Acad Scis, AAAS, ACS, Amer Pub Health Assn; Sigma Xi. Contrib articles to prof journals. Patentee in field. *Society Aff:* AIChE, ACS, APHA, AAAS

Bergougnou, Maurice A
Home: 24 Foxchapel Rd, London, Ontario N6G 1Z2 Canada
Position: Prof of Chem Engg. *Employer:* University of Western Ontario. *Education:* PhD/Chem Eng/Univ of MN; Masters/Chem Eng/ENSIE-Univ of France; BESc/-/Nancy, France. *Born:* Sept 7, 1928 France. PhD Chem E Univ of Minnesota USA. Petroleum process dev & chem engrg res (fluidized bed reactors) for Exxon Res & Engrg 1958-67 USA; Prof of Chem Engrg Univ of Western Ontario Canada 1967- . Mbr Exec Can Soc Chem Engrs 1972-75; VP Engrg & Co-Founder Elstat Ltd. Fellow Chem Inst of Canada. Res & indus cons on design & scale-up of commercial fluidized bed reactors, on electrostatic separation of solids & minerals while in fluidized state, on energy conservation. *Society Aff:* ACS, AIChE, CIC, CSChE, AAAS.

Bergrun, Norman R
Business: 26865 St Francis Rd, Los Altos Hills, CA 94022
Position: CEO. *Employer:* Bergrun Res & Engrg. *Education:* LLB/CA Law, Property/LaSalle Univ; BSME/Aeronautics/Cornell Univ.; DSC (Hon.)/Science, Humanities/World Univ. *Born:* Aug 1921. Native of Marion Ohio. Aeronautical Res Sci spec in icing flight investigations & airplane stability & control Ames Res Ctr NACA (now NASA) to 1957; Mgr Analysis & Planning of flight tests for Polaris Proj Lockheed Missile & Space Co to 1970; Dir Information Mgmt Systems responsible for computerized mgmt aides Nielsen Engrg & Res Inc to 1972; subsequently entered private practice. AIAA cerificate for 40 yrs of sustained contrib to aeronautics & astronautics 1975; Calif St Dir to NSPE 1975-76; & 1979-84; Assoc Fellow AIAA. Books: "Tomorrow's Technology Today" L C No 73-76119 1972 and "Ringmakers of Saturn–, ISBN No. 0946270-33-3, 1986. Enjoy music, skiing & nature walks. Calif Reg P E. Ch Bd, Calif Prof Engrg Ctr, 1985-87; Public Policy Dir, San Francisco Section AIAA 1986-87; First VP, CSPE, 1987-88. *Society Aff:* NSPE, AIAA, AAAS, EASAH.

Berkeley, Frederick D
Home: 50 Old Mill Rd, Rochester, NY 14618
Position: Pres. *Employer:* Graham Mfg Co, Inc. *Education:* BS/ME/RPI. *Born:* 8/7/28. Entire Business career with Graham Mfg Co, Inc. Started working for Graham in Sept 1950 in the Engg Dept in Batavia, NY. Worked in Corp's NYC office as a Sales Engr in training. Transferred to Corp's Chicago sales office in 1955 as a Sales Engr. Returned to Batavia as Dir of Res in 1956 & subsequently promoted to Asst to the VP. In 1962 elected Pres & Chrmn of the Bd succeeding Frederick D Berkeley, Jr. Profl activities include Pres of Tubular Exchager Mfg Assn, 1975-1976, Dir of The Business Council of NY State, Trustee of Williston Northampton Sch, Easthampton, MA, Past Dir & Chrmn of the Bd, Federal Reserve Bank of NY-Buffalo Branch 1978-80, Dir of Voplex Corp, Rochester, NY. Advisory cttee mmbr for Roberts Wesleyan College; Vice Chrmn & Dir of Heat Transfer Research, Inc.; Mmbr of the Young President's Organization 1965-79; Mmbr of the World Business Council. *Society Aff:* ASME, AIChE, PE.

Berkley, James E
Business: 3700 Buffalo Speedway, Houston, TX 77052
Position: Group Vice Pres Engrg & Const. *Employer:* The Pace Companies. *Born:* Nov 23, 1933. BSME Purdue Univ 1955. Native of Crete Illinois. Began career 1955 as Mech Engr with Esso Standard Oil Co; joined Standard Oil of Ohio 1959 as Proj Engr with their central engrg group; 1962 joined The Pace Companies as Cons Engr in area of refining, petrochem & chem; served in sev mgmt capacities with Hydrocarbon Const Co, a Pace Co subsidiary performing engrg & const services; elevated to Pres of Hydrocarbon 1975; recently promoted to Group VP of The Pace Companies responsible for engrg & const activities of Pace subsidiaries.

Berkness, I Russell
Home: PO Box 5224, Richmond, VA 23220
Position: President. *Employer:* Berkness Control & Equip Corp. *Education:* BSME/-/Purdue. *Born:* June 1910. Employed by Fairbanks Morse, Raymond Impact Pulverizer, Thermal Engrg, Virginia Engrg (now Basic Const); Naval Officer 1943-46; cons engr 1946-47; started manufac agency 1947 - Berkness Control & Equip Corp spec in steam power plant equip. Mbr ASME; named Engr of the Year by Virginia Soc of Prof Engrs 1966; appointed Chmn of Tech Adv Ctte to St Air Pollution Control Bd 1967-86; elected Natl Dir of Natl Soc of Prof Engrs 1973 & 1976. Elected as Dipl in The Amer Acad of Environmental Engrs. Elected Life Fellow of ASME June 1985. Elected to membership in the Newcomen Society July 1986. *Society Aff:* NSPE, ASME, AAEE.

Berlad, Abraham L
Home: 1 Harmon Ct, Stony Brook, NY 11790
Position: Prof of Engrg *Employer:* State Univ of NY *Education:* PhD/Physics/OH State Univ; BA/Physics/Brooklyn Coll *Born:* 9/20/21 Prof of Engrg and Dir, Energy Technology Lab, State Univ of NY at Stony Brook. Chief, Combustion Fundamentals Section NASA Lewis Research Center, 1951- 56. Senior Staff Scientist, Convair Scientific Research Lab, 1956-1964. General Research Corp, 1964-1966. Chrmn, ME Dept, SUNY at Stony Brook, 1966-1969. Member, Bd of Dirs, The Combustion Inst (Intl). Visiting Professorships held at The Univ of CA and The Hebrew Univ (Jerusalem). Consultant to NASA, USDOE, USNRC, Brookhaven National Lab, and others. *Society Aff:* Comb Inst (Intl), APS

Berlekamp, Elwyn R
Home: 1836 Thousand Oaks Blvd, Berkeley, CA 94704
Position: Pres. *Employer:* Cyclotomics Inc. *Education:* PhD/Elec Engg/MIT; MS/Elec Engr/MIT; BS/Elec Engr/MIT. *Born:* 9/6/40. 1964-67 Asst Prof of Elec Engg at Berkeley; 1967-71 mbr of Mathematics Res Ctr at Bell Telephone Labs; since 1971 Prof of Mathematics & Elec Engg & Comp Sci at Berkeley; Chrmn of Comp Sci 1974-75. Now Pres of Cyclotomics, Inc., which became a Kodak subsidiary in 1985. 1973 Pres of IEEE info Theory Soc; Fellow IEEE, 1971 Outstanding Young Elec Engr selected by HKN, elected to NAE in 1977, Council of American Math Soc 1978-81. *Society Aff:* IEEE, NAE, AMS.

Berman, Abraham S
Business: Dept Aerospace Engg & Mechanics, Minneapolis, MN 55455
Position: Prof. *Employer:* Univ of MN. *Education:* PhD/Physical Chem/OH State Univ; BChEngg/ChE/CCNY. *Born:* 5/20/21. in NYC. Engr PA Ord Wks 1942-43; Res Engr SAM Labs, Columbia Univ, 1943- 1946; Hd Flow Res Dept, Union Carbide Nuclear, Oak Ridge, TN, 1950-1966; Prof Fluid Mechanics, Dept Aerospace Engg & Mech, Univ of MN, 1966-. Numerous publications in technical journals. IT Student Award for Outstanding Teaching, 1977; Morse-Amoco Award for Outstanding Contributions to Undergrad Education, 1978. Consultant to industry. *Society Aff:* ACS, APS.

Berman, Baruch
Home: 28739 Trailriders Dr, Rancho Palos Verdes, CA 90274
Position: Senior Engrg Specialist *Employer:* Rockwell Intl *Education:* MS/EE/Columbia Univ; Dipl in Ind Tech/EE/Technion, Israel Inst of Tech; BS/EE/Technion, Israel Inst of Tech. *Born:* 11/10/25. Baruch Berman was born in Tel Aviv, Israel. His many yrs of experience include control of power from microwatts to megawatts where both high & low-voltage applications were implemented & associated frequencies ranged from dc to gigacycles. Before joining Rockwell Intl, Seal Beach, CA, he held positions such as VP of Engrg, Chief Engr, Mgr of Engrg & Sec Hd with aerospace and ind firms. His activities included the evolution of concepts, analysis, trade-offs, design & hardware dev. At present he is applying his extensive experience to the fields of satelite power sys fusion energy, solar elec, magneto-hydro dynamic, fuel cell, battery charging, and energy conservation. Mr Berman holds US pats for transistorized regulators, triggers for SCR's (thyristor) and thyristor light, heat and motor control. He has received awards for his work. He is a Mbr of the Natl Society of Profl Engr, Reg in CA, A Fellow of the Inst for the Advancement of Engrg and A Fellow of the IEEE. Married, Rose S. Berman-children, Orrie A. Berman, Sharon J. Berman. *Society Aff:* IEEE.

Berman, Donald
Business: 504 Beaver St, Sewickley, PA 15143
Position: Exec VP. *Employer:* Green Intl, Inc. *Education:* MS/Met/Carnegie Mellon Univ; BS/Met/Carnegie Mellon Univ. *Born:* 4/15/25. Native of Pittsburgh, PA. Grad of Carnegie Mellon Univ (BS & MS). Served with US Army 1942-44. Over 20 yrs experience with Green Intl, Inc, Consuting Engrs. Design & proj mgt experience in hgwys, bridges, structures, water pollution control facilities, waste disposal. Assumed current responsibiity as Exec VP in 1977. Served as Dir of Allegheny County, PA Dept of Waste Systems Mgt 1971-73 & Dept of Works 1974-77. Adjunct Assoc Prof, Grad Sch of Public Health & Dept of Civ Engg at Univ of Pittsburgh. Pres, Pittsburgh Sec ASCE 1979-80. *Society Aff:* ASCE, SAME, WPCF, AAEE, APWA.

Berman, Irwin
Business: 12 Peach Tree Hill Rd, Livingston, NJ 07039
Position: Chrmn, Technical Directorate. *Employer:* Foster Wheeler Dev Corp. *Education:* PhD/Applied Mechs/Poly Inst of NY; MS/Applied Math/Stevens Inst of Technology; BS/Mech Engg/City Univ of NY. *Born:* 10/16/25. Infantry Scout in ETO WW II, Purple Heart. Head Struct Analy Sect Curtiss Wright Corp 1948-54; Res Assoc Poly Inst of N Y 1954-56; Adjunct Prof of M E NYU 1960-71; Foster Wheeler Energy Corp Res Div 1956- ; Positions held: 1956-58 Cons to Div & Corp, 1958-76 Mgr, Solid Mech Dept incl Analysis, Computer Tech, Dev, Testing, Instrumentation, Advanced Fabrication Techniques etc; 1976-78 Tech Dir Engrg Sci & Technology Res., Chrmn Technical Directorate, 1978- . ASME Fellow Grade, Sr VP Chrmn Council on Engrg; formerly VP Communications, Chmn & mbr of code cttes, tech cttes & natl & internatl mtgs. Tech Ed of Journal of Pressure Vessel Technology; SESA Papers Ctte; PVRC of Amer Welding Soc Exec & Main Ctte. Ed of 3 tech compilations in computer technology; 2 compilations in explosive fabrication Author & Co-author of 60 papers on struct analy, experimentation, reliability etc. Holder of 16 US pats. UN expert for IAEA in Argentina. *Society Aff:* ASME, SESA.

Bernard J Yokelson
Business: AT&T Bell Laboratories, Naperville-Wheaton Rd, Naperville, IL 60566
Position: Dir Toll Digital Switching Laboratory *Employer:* Bell Telephone Laboratories, Inc. *Education:* MEE/Elec Engg/Polytechnic Inst of New York; BS/Elec Engg/Columbia Univ. *Born:* 9/14/24. 1943-46 served with Army Signal Corps, since 1948 with Bell Labs, currently respon for dev of toll digital electronic switching sys, upon joining Bell Labs was concerned with a variety of developments in defense projs, transmission sys & multifrequency signaling, 1954-66 worked in various capacities on dev of No. 1 Electronic Switching Sys, 1966-74 Dir Oper Sys Lab respon for sys automating operator functions, 1974-76 Dir Electronic Power sys Labs, 1977-1980 Dir. Local Digital Switching Systems Laboratory respon for development of No 5ESS Electronic Switching Sys. 1980-present responsible for dev of Toll Digital

Bernard J Yokelson (Continued)
Switching Systems, Fellow IEEE; Mbr Sigma Xi & Tau Beta Pi. *Society Aff:* IEEE, T,BETAΠ, ΣΞ.

Bernard, James E
Business: Dept of Mech Engg, East Lansing, MI 48824
Position: Assoc Prof & Dir for Case Center. *Employer:* MI State Univ. *Education:* PhD/Engg Mech/Univ of MI; MS/Engg Mech/Univ of MI; BS/Engg Mech/Univ of MI. *Born:* 12/1/43. At Genl Motors, he developed a hybrid simulation and graphic display of vehicle ride (1966, 1968). His work at the Highway Safety Research Inst (1971- 76) included dev of computer-based mathematical models for the study of commercial vehicle braking and handling. In his current position as Dir of The Case Center for Computer-Aided Design at MI State Univ he is engaged in teaching and research in the analysis of the design of mechanical sys using interactive computer graphics. SAE - Arch T Colwell Award, 1973; Ralph Teetor Award, 1978;; MSU - Teacher Scholar Award, 1978. Eminent Engr, MSU chapter of Tau Beta Pi, 1979. *Society Aff:* ASME, SAE, SCS.

Bernstein, I M
Business: MEMS Department, Carnegie Mellon Univ, Pittsburgh, PA 15213
Position: Prof and Hd of Dept of Metallurgical Engrg and Materials Sci *Employer:* Carnegie-Mellon University. *Education:* PhD/Metallurgy/Columbia Univ; MS/Metallurgy/Columbia Univ; BS/Metallurgy/Columbia Univ. *Born:* 10/14/38. NYC. Post Doctoral Fellow Central Elec Bd England 1966-67; Scientist E C Bain Lab for Fundamental Res US Steel Corp 1967-72; at Carnegie-Mellon Univ 1972-87. Provost and Academic VP, Illinois Inst. of Technology 1987- . Liason Scientist ONR London 1977-8. Co-organizer & co-editor of proceedings of 3 internatl confs on hydrogen embrittlement held 1973 1975 and 1980; assoc editor of Handbook of Stainless Steel. Active in CMU univ affairs (V Chmn, Fac Senate 1975 Asso Dean of Engrg 1978-82) & local politics (democratic ctteman City of Pgh 1970-74). Fellow ASM, 1985 Outstanding Young Mbr award 1974, Pittsburgh Night Lecturer, 1982; Golden Triangle Chapter ASM, Charles Barnett Silver medal-Rocky Mtn Chapter ASM 1983. *Society Aff:* AAAS, ASM, AIME, ASEE.

Bernstein, Jerome
Business: PO Box 2226, Baton Rouge, LA 70815
Position: Mgr-Exxon R&D Labs. *Employer:* Exxon Co. *Education:* BS/Chem Engg/City College of the City Univ of NY. *Born:* 6/19/34. Began career at the Exxon Res and Engg Co in Florham Park, NJ. In 1957 where he was actively involved in the process dev and design of new octane processing ad refinery desulfurization facilities. He spent two yrs in Europe providing tech service to Exxon affiliates in a wide range of refining processes. Has held managerial positions in Exxon Engg Petrolelum Dept at Florham Park and at Esso Eastern, Inc in Houston, TX. Currently Manager of the Exxon Res and Dev Lab in Baton Rouge. A native of NY City, he is married and has three daughters. *Society Aff:* AIChE, API.

Bernstein, Paul R
Home: 2 Balter Rd, New City, NY 10956
Position: Prog Mgr *Employer:* Seelye, Stevenson, Value and Knecht, Inc. *Education:* PhD/Engr Sci/NJ Inst of Tech (in progress); MSCE/Civil Engr/NJIT; BSCE/Civil Engr/CCNY *Born:* 08/22/39 Born in Bronx, NY. Finished CCNY after a tour in the Army with the Corps of Engrs. Attended NJ Inst of Tech receiving a Masters in Civil Engr and included in the "Who's Who in American Colleges and Universities–. Currently in the Doctoral Program at NJIT. Worked for several Consult Engrg firms, proceeding from Designer to Program Manager. Currently heading company's office at Co-op City in charge of Engrg Dept. Active in Veterans affairs, presently County Commander of JWV. Professional Engr in N.Y. and N.J. Professional Planner in N. J. *Society Aff:* NSPE, ITE, AAA, TBP

Berrett, Paul O
Business: Department of Electrical & Computer Engrg, 459 CB, Brigham Young Univ, Provo, Utah 84602
Position: Prof *Employer:* Dept. of Electrical & Computer Engrg *Education:* PhD/EE/Univ of UT; MS/EE/Univ of So CA; BS/EE/Univ of UT *Born:* 3/1/28 Riverton, UT. Hughes Master of Science Fellow 1953-55. Research positions at Hughes Aircraft Co, Sperry UT Labs, Upper Air Research (Univ of UT), Battelle Northwest, Goddard Space Flight Center (NASA), Naval Weapons Center, Eyring Research Institute Inc, (Current consulting in a research science role). Faculty positions at Univ of UT (1957-1964) and Brigham Young Univ. Full Prof since Sept 1971. Student Activities Chrmn Region Six IEEE 1965-68, Advisory Committee 1967-71. Secretary Engrg Coll Council ASEE 1971-73. Chrmn Metrication Coordinating Committee ASEE 1977-78. Chrmn Rocky Mountain Section ASEE 1978-79. *Society Aff:* ASEE, IEEE, HKN, ΣΞ, TBΠ.

Berrin, Elliott R
Home: PO Box 255, Freeport, NY 11520
Position: Pres. *Employer:* Elliott R Berrin, PE, PC. *Education:* BME/Mech Engg/Poly Inst of Brooklyn. *Born:* 4/9/39. Reg PE in 5 states. Mbr EJC Committee on Employment Guidelines. Advisor on Tech Reports to SFPE. Author of several papers in various engg and business fields. Native of NY. Field engr with Factory Insurance Assn 1961-68; VP of Schiff Terhune, Inc 1968-75; Natl Mgr Technical Services, GAB Business Services 1975-78; Faculty of the College of Insurance 1978-1981; owns and operates a consulting engg firm specializing in loss investigation (fire, arson, structural collapse, machinery damage, etc), 1978-present. Hobbies: Applications of photography to solution of engg problems, and writing of technical papers. *Society Aff:* SFPE, IAAI, ASME, EPIC, NFPA.

Berrio, Mark M
Home: RR 2 Box 314, Greencastle, IN 46135
Position: Prof *Employer:* Rose-Hulman Inst of Tech *Education:* Ph.D./Structures/MI St Univ; M.S./Structures/ Univ of MI; C.E. /Structures/Univ of Guatemala; BA/Math/Univ of Salvador. *Born:* 7/19/33 I was born and raised in Spain. I became a Civil Engr in Guatemala, where I practiced engrg as a structural designer and as a general contractor. In 1966 I immigrated to the U.S. At first I worked as a structural designer for Roof Structures, Inc., in St. Louis, MO, which in those days could boast having built the two largest steel domes in the world: Astrodome in Houston, TX, and Hopodromo in San Sebastian, Spain. I designed the one for Spain. Later I decided to further my formal education, went back to school and obtained a doctoral degree in engrg mechanics. Upon graduation I decided to remain in academia. In 1972 I joined the faculty of Rose-Hulman Inst of Technology, in Terre Haute, IN, where I have been teaching civil engrg and doing consulting. In 1971 I became a full member of the ASCE. This is the ninth year in a row that I serve as the Campus Activity Chrmn of the American Society for Engrg Education at Rose- Hulman. After serving as Treas for eight consecutive years, I was elected President of our SigXi Chapter for the 1981-82 period. I enjoy classical music, photography and swimming. *Society Aff:* ASEE, ASCE, ΣΞ.

Berry, Bill E
Business: PO Box 70, Boise, ID 83707
Position: Agri Engr. *Employer:* ID Power Co. *Education:* MSAE/Engg/Univ of ID; BSAE/Engg/Univ of ID. *Born:* 5/18/25. Homestate-ID. WWII Veteran-Korean Conflict Veteran. 23 yrs Irrigation Pumping Specialist-ID Power Co. Base at Twin Falls, ID. 7 yrs customer service/energy mgt staff mbr. Irrigation pumping specialist. The Twin Falls area added over 450,000 hp of pumps during the 23 yrs at Twin Falls. The ID Power Co now serves, over 1,200,000 hp of irrigation pumps-about 6500 irrigation customers-& I am the specialist for the co. *Society Aff:* ASAE, NSPE, ISPE.

Berry, Carl M
Business: PO Box 66459, Seattle, WA 98166
Position: Cons Engr (Self-employed) *Born:* March 1904. Univ Wash: forestry, logging, C E; Univ Oregon C E. Field & Office Engr Weyerhaeuser Timber Co 1920-32; Field Engr US Coast & Geodetic Survey 1932; Ch Triangulation US Bur Recla-

Berry, Carl M (Continued)
mation 1932-38; Engr US Bonneville Power Admin 1938-42, 1946-48; served with USAAF 1942-46; cons civil, geodetic, photgrammetric & forest inventory engr Seattle 1948- . Author Combining Geodetic Survey Methods with Cadastral Surveys Vol 105, Transactions ASCE 1940; tech journal contribs. Fellow Amer Soc C E (Natl Chmn Exec Comm Surveying & Mapping Div 1958, Natl Chmn Comm on Geodetic Surveying 1957-65, P Pres Seattle Sect). Fellow Amer Congress on Surveying & Mapping. Mbr Amer Cons Engrs Council; Charter Mbr Amer Soc Photogrammetry (P Pres Puget Sound Region); St Adv Bd Surveys & Maps 1956-1982; Chmn 1960-75, 77 Mbr Intl Right of Way Assoc. *Society Aff:* ASCE, ACEC, ACSM, ASP&RS, IRWA.

Berry, Daniel M
Business: Computer Sci Dept, 3532 Boelter Hall, Los Angeles, CA 90024
Position: Assoc Prof. *Employer:* UCLA. *Education:* PhD/Applied Math/Brown Univ; BS/Math/Rensselaer Poly Inst. *Born:* 7/28/48. in Cleveland Heights, OH. He received his BS in Math from Rensselaer Polytechnic Inst in 1969. He went to grad school at Brown Univ, during which time he worked at General Electric R&D Center in Schenectady and taught at the Hebrew Univ in Jerusalem. He joined the UCLA faculty as an Acting Asst Prof in Sept 1972. After some delay, he completed his PhD thesis in Sept 1973 for his degree in Computer Science from Brown. Shortly thereafter he was promoted to Asst Prof. He was promoted to Assoc Prof in 1977. In 1978 he spent a sabbatical at the Hebrew Univ and the Weizmann Inst in Israel. *Society Aff:* IEEE, ACM.

Berry, Donald S
Business: Dept of Civil Engrg, Evanston, IL 60201
Position: Murphy Prof of Civil Engrg, Emeritus *Employer:* Northwestern Univ. *Education:* BSCE/Civil Eng/SD Sch of Mines; MS/Highway Engg/Iowa State College; PhD/Civil Engg/Univ of Michigan *Born:* 1/1/11. Traffic Engr & Dir Traffic & Transp Div of Natl Safety Council 1936-48; Asst Dir Inst of Transp & Traffic Engr U of Calif 1948-56; Prof Engrg Purdue Univ 1956-57; since 1957 Prof of C E Northwestern Univ, in charge of grad prog in transp engrg. Chmn Dept of C E 1962-68; Chmn Highway Res Bd 1965; Chmn ASCE Comm on Natl Transp Policy 1972; Mbr Comm on Transp & Comm on Public Engrg Pol of Natl Acad of Engrs. Mbr, Natl Acad Engrg, since 1966. T M Matson Mem Award ITE 1972; James Laurie Prize ASCE 1972; Becoming Murphy Prof Emeritus (Sept 1, 1979 by retirement). *Society Aff:* ITE, ASCE, NAE

Berry, Francis C
Business: 1500 N 50th St, Birmingham, AL 35201
Position: R&D and Special Proj Engr *Employer:* Chicago Bridge & Iron Co. *Education:* BS/Elec Engg/Univ of AL. *Born:* 9/1/24. Native Alabamian. BS in EE 1950 Univ of Alabama. 34 yrs with Chicago Bridge & Iron Co; presently R&D and Special Proj Engr Houston, TX. Has held post incl: Staff Asst (NDE Mgr-B'ham), mgr of Chmn PVRC Subctte on Nondestructive Examination of Material for Pressure Components; Mbr ASME, AWS, ASNT; served as Natl Bd of Dirs of ASNT; Chmn of ASNT Standing Ctte for Level III Certification. 1979-80 - Natl past Pres & Chrmn of the Bd of Amer Society Nondestructive Testing. Nondestructive Testing Cons. Reg PE Sts of Al & Cal. Temporary Assignment - Scottish Marine Operations (CBI), Hunterston, Scotland. *Society Aff:* ASNT, AWS, ASME, Tau Beta Pi.

Berry, J Douglass
Business: P.O. Box 1963, Harrisburg, PA 17105
Position: Pres Senior VP & Treasurer *Employer:* Gannett Fleming, Inc & GANCOM, Inc. *Education:* BS/Elecrical Engr/Univ of PA. *Born:* 7/26/26. in West Reading, PA. Served with Army 1944-47; Army Res 1947-69. Telephone Engr with Bell Telephone Co of PA 1947; with Gannette Fleming Inc since 1952 as utility valuation engr, dir of data processing & other serv functions within corp; Pres GANCOM (Div Gannett et al) serving Public in data processing/reprographics fields. Reg in PA as Elec Engr. Director and Sr VP & Treasurer of the consulting engrg firm of Gannett Fleming Inc since 1976. Director of local bank and large health foundation. *Society Aff:* IEEE, NSPE, ACEC, PEPP, PSPE.

Berry, James R
Business: P.O. Box C-50, Little Rock, AR 72116
Position: VP *Employer:* Garver & Garver, Inc *Education:* BS/CE/Univ of AR; Post Grad Studies/Water Resource Engrg/Univ of AR Grad Inst of Tech *Born:* 11/10/33 Graduated from high school at Springdale, AR. Spent four years in US Navy during Korean War. After graduating from the Univ of AR School of Civil Engrg, worked for the AR Highway Dept nine months and have been in consulting engrg the past 21 years. Currently VP and partner in Garver & Garver, Inc, Little Rock, AR. Responsible for all public works projects, solid and hazardous waste work and other special engrg projects. Have extensive experience on projects with the US Army Corps of Engrs, AR Highway Dept, and many cities. *Society Aff:* ASCE, NSPE, SAME, APWA

Berry, John T
Business: School of Mechanical Engrg, The GA Inst of Tech, Atlanta, GA 30332
Position: Prof *Employer:* The GA Inst of Tech *Education:* PhD/Metallurgy/Univ of Birmingham; BSc/Metallurgy/Univ of Birmingham *Born:* 5/16/31 in Leicester England. Educated at Gateway School.and Polytechnic there. Received scholarship to Birmingham Univ (BSc Hons, PhD). Served in Royal Naval Scientific Service 1954-57. Career divided between education, industry & research including NGTE, NCRE, PERA & SKF in UK, Climax Molybdenum, IITRI, the Univ of Bridgeport (CT) & VT and (since 1974) The GA Inst of Technology in USA. Avocations include local history, music (horn) & antebellum home. *Society Aff:* ASME, AFS, ΣΞ

Berry, Kay L
Business: 2851 So Parker Rd, Aurora, CO 80014
Position: Mgr-Research. *Employer:* Rio Blanco Oil Shale Co. *Education:* BS/Chem Eng/Univ of CO. *Born:* 7/3/22. Native of SD. Moved to Colorado in 1938. Served in US Navy 1943-46. From 1947 to 1953 with US Bureau of Mines Oil Shale Demonstration Plant in Rifle, CO. From 1953-56 with Petroleo Brasileiro (Petrobras) in Brazil as mgr oil shale pilot plant. After eight yrs with Cameron and Jones, consulting oil shale engrs in Denver, joined Standard Oil Co (Indiana) in 1964 and was assigned to the Rio Blanco Oil Shale Co in 1974. *Society Aff:* AIChE.

Berry, Richard C
Home: 111 Kate, Longview, TX 75605
Position: Dean of Engrg *Employer:* Le Tourneau College. *Education:* MS/Engg Sci/Univ of AR; BD/THEology/New Orleans Baptist Theological Seminary; BS/Che/LA State Univ. *Born:* 9/3/27. in Wichta Falls, TX. Reared in Lake Charles, LA. Served in US Navy 1945-46. Engr with Coastwise Petrol Co and Johnson Service Co specializing in control systems with both firms. Pastor of churches in LA and MS. Taught at Bluefield College, Bluefield, VA, before joining faculty at Le Tourneau College in 1968. Taught Mech Engg, primarily thermal sciences. Registered PE, State of Texas. Dean of Engrg at Le Tourneau College since 1986. Enjoy tennis and sci fiction. *Society Aff:* ASEE, ASME.

Berry, Richard C
Business: One Technology Drive, Rogers, CT 06263
Position: Vice Pres, Technology. *Employer:* Rogers Corp. *Education:* BS/Chem Eng/MIT; MS/Chem Eng/MIT. *Born:* July 1928. R/D work in fiber/polymer composition for elec insulating & mech applications prior to R/D mgmt duties. Former V Chmn ASTM Ctte F-3 on gaskets; former Chmn of IEEE Ctte on Dielec Materials; former Chmn & 1 of the founders of Eastern Conn Chem Engrs Club; Mbr Indus Adv Council U. of Conn. Inst. of Polymer Science; Mbr. Adv Board U of Conn College of Engineering; Former Delegate, Industrial Research Institute. Mbr Conn St Bd of Tr for Regional Comm Coll; CT St Board of Higher Education; and New England Bd of Higher Educ; Hon Mbr Pi Tau Sigma Mech Engrg Hon Soc. Numerous patents in non-metallic materials field. *Society Aff:* ACS, AIChE, SPE, IEEE, IEPS.

Berry, Richard N
Home: 465 Congress St, Portland, ME 04101
Position: Cons Engr. *Employer:* Self Employed *Education:* BS/Genl Engg/Univ of ME. *Born:* Nov 4, 1915 Malden Mass s. Edward & Maude Berry. Educ Univ of Maine 1937 BS Gen Engrg w Dist; Beta Theta Pi; m. Sylvia Snow Jan 2, 1975. Pac/ETO US Army Engrs 1941-47; cvl engrg cons 1951- . Mbr City Council 1940, Fire Comm 1947-50 Malden Mass; Mbr Maine St Legislature 1960-78; Senate Majority Floor Leader 1970-74; Col CE USAR (Ret); Pres, Stratton Water Co, Winter Harbor Water Co, Woodland Water & Elec Co, Bristol Shellfish Farms Inc., Roberts Office Supply Co; Dir, Malden (MA) Trust Co; Mbr Maine St Museum Comm. Tau Beta Pi. Reg P E; Reg Geologist; Reg Land Surveyor. Fellow ASCE. Clubs: Cumberland (Portland), Portland Country, The Algonquin Club (Boston) Portland Yacht, Appalachian Mtn (Life Mbr). *Society Aff:* ASCE, SME.

Berry, Virgil J, Jr
Business: 600 Hunter Dr, Oak Brook, IL 60521
Position: Pres. *Employer:* Chicago Extruded Metals Co. *Education:* PhD/Chem Engg; Phys/CA Inst Tech; MS/Chem Engg/CA Inst Tech; BE/Chem Engg/Vanderbilt Univ. *Born:* 12/20/28. Res group supervisor, Pan American Petrol Corp., Tulsa, 1951-60; Dir petrol production res, Sinclair Res, Inc., Tulsa 1960-63; Mgr systems and computing, Sinclair Oil & Gas Co, Tulsa, 1963-66; Dir planning & coordination & Secy Mgt Committee, Sinclair Oil Corp, NYC, 1966-69; Mgr intl finance, control & planning, Atlantic Richfield Co, NYC, 1969-70; Sr vp operations, Joseph Schlitz Brewing Co, Milwaukee, 1970-78, Mgt consultant 1978-79. Pres, Chicago Extruded Metals Co, Oak Brook, 1979-. Lectr math, Univ Tulsa, 1952-66. Over 40 articles on transfer processes, oil recovery tech and applied math. *Society Aff:* AIChE.

Berry, William A
Business: 2630 Galiano Street, Coral Gables, FL 33134
Position: President. *Employer:* Berry & Ellis Assocs Inc. *Education:* BEE/Math & EE/Youngstown Univ. *Born:* Nov 1925. Native of Mineral Ridge Ohio. Aviation Machinist Mate US Navy 1943-46. Entered Private Practice 1956 as Cons Engr; Inc 1968. Former Tr of Miami Chapter of Florida Engrg Soc; chaired St Ctte that wrote 1st Interprof Manual of Practice for Cons Engrs of Florida Practice Sect of FES. Presently Bd of Tr Chmn for Florida Inst of Cons Engrs Insurance Prog; Elec Cons to Dade County Aviation Dept, Miami Jackson Memorial Hospital Maintenance Dept & Good Samaritan Hospital in West Palm Beach Florida. Fellow Mbr Florida Engrg Soc. *Society Aff:* NSPE.

Berry, William L
Business: 19 Forest Glen, Iowa City, IA 52242
Position: Senior Associate Dean and C. Maxwell Stanley Professor of Production Management *Employer:* University of Iowa *Education:* DBA/Production Mgt/Harvard Univ; MS/Industrial Engg/VPI & SU; BS/Economics/Purdue Univ. *Born:* 12/24/35. Grad of the GE Mfg Training Prog 1960 Mfg supervisor with the Ind Control Dept of GE 1960-1964. Asst and Assoc Prof of Ind Mgt at the Krannert Grad Sch of Ind Admin 1968-1976. Prof of Operations and Systems Mgt, Grad Sch of Bus, IN Univ (1976-82), C. Maxwell Stanley Professor of Production Management and Senior Associate Dean, College of Business Administration, University of Iowa (since 1982). President-elect and President, Decision Sciences Institute 1987-88; President, Operations Management Association 1986-87; Associate Editor, Decision Sciences Journal, Member Editorial Board of the Journal of Industrial Management & the Journal of Manufacturing and Operations Management; Visiting Professor at the London Business School, the Cranfield School of Management, the University of Warwick (England). Author of four books and numerous res articles in the field of Producion Planning and Control Systems. VP of Systems Engg and Technical Divs, IIE 1979-1980; Mbr *IIE Transactions* Editorial Bd; Dir Conf Planning and Coordination 1979; Dir and Dir-Elect Production Planning and Control Div IIE, 1971-1973. Received IIE Production Planning and Control Div Award 1979. Fellow, Decision Sciences Institute 1979. *Society Aff:* IIE, ORSA, TIMS, DSI, OMA, APICS.

Bers, Eric L
Home: 3829 Palmetto Ct, Ellicott City, MD 21043
Position: Policy Analyst *Employer:* U.S.Dept of Trspn *Education:* MS/CE/GA Inst. of Tech.; MS/Systems Eng/GA Inst of Tech.; BS/CE/Wash Univ (MO) *Born:* 3/13/50 Native of Baltimore, MD. and High School graduate of Baltimore City College. Obtained Professional Engrg Assoc (PEA) lic from Maryland and is a reg Professional Engr (PE) in the Dist of Columbia. Since 1975, has published over 20 private papers and technical articles in natl and international journals on virtually all modes of passenger and freight transportation. Testified formally on numerous occasions before the Interstate Commerce Commission on passenger train discontinuancs, natl energy issues, mergers, consolidations, and acquisitions. Elected President of the Columbia Commuter Bus Corp. Currently reviews papers for four engrg publications. Enjoys numismatics and household construction. *Society Aff:* ASCE, ITE, TRB, ASEE.

Bershad, Neil J
Business: School of Engineering, Irvine, CA 92717
Position: Professor *Employer:* Univ. of Calif., Irvine *Education:* Ph.D./Elect Engr./Rensselaer Poly Inst; MSEE/Elec Engr. /Univ of Southern Calif; B.EE./Elec Engr/Rensselaer Polytechnic Inst *Born:* 10/20/37 Native of Brooklyn, NY. Engineer for Hughes Aircraft Co.-Radar Signal Processing 1958-1965. 1st Lieutenant USAF, Cambridge Research Laboratories- research in communication theory and part time lecturer, Northeastern University-1962-1965. Joined Engineering Faculty at UC Irvine as Assistant Professor of Electrical Engineering 1966, promoted to Associate Professor 1969, Professor 1976. ViceChair EE Dept (1986-) Research interest include communication theory, stochastic signal processing and acoustic array processing. Also provides consulting services in underwater acoustic signal processing. Associate Editor- Communication Theory, IEEE Trans. on Communications (1974-1980). Assoc Editor-Adaptive Filtering-IEEE Trans on Acoustics, Speech and Signal Processing (1987-). Special interest-communication theory as applied to high fidelity sound reproduction and stochastic analysis applied to adaptive filters. *Society Aff:* IEEE, ΣΞ, HKN, TBΠ

Bert, Charles W
Business: 2516 Butler Dr, Norman, OK 73069
Position: Benjamin H Perkinson Chair Prof of Engg. *Employer:* University of Oklahoma, Norman. *Education:* BS/ME/Penn State Univ; MS/ME/Penn State Univ; PhD/Engg Mechs/OH State Univ. *Born:* Nov 1929. Lt USAF Eglin AFB 1952-54. Design Engr Fairchild Acft 1954-56; Principal ME through prog dir, solid-struct mech Battelle Columbus Labs 1956-63; Instructor mech Ohio St 1959-61; Assoc Prof 1963-66, Prof 1966-, Benjamin H Perkinson Professor of Engg 1978-86, George L Cross Research Professor 1981-; Benjamin H. Perkinson Chair Professor of Engg 1986-. Univ of OK; dir, Sch of Aerospace, Mech & Nuc Engrg 1972-77. Reg P E Ok & Pa. Founder Mbr, Bd of Dir, Fellow Amer Acad of Mech; Fellow AAAS, ASME, SEM; Assoc Fellow AIAA (on Nat Struct Tech Comm 1969-72, Deputy Director-Tech., Region); Mbr: ASEE, NSPE, SES. Cons on composite-material mech to Sandia Labs, Inst for Defense Analy etc. Author of over 240 tech papers. *Society Aff:* AAM, AAAS, ASEE, AIAA, ASME, NSPE, SES, SEM.

Bertch, Thomas M
Home: Box 399, Lexington, TX 68947-399
Position: Pres *Employer:* Thomas M Bertch & Assoc Inc *Education:* BS/Pet Engr/Univ of TX *Born:* 10/14/17 Spent 17 yrs with a major oil co and 14 yrs with an independent oil co and 16 yrs as a consultant. I am a registered PE in the states of TX, LA, and OK. Served twice as secry-treas of SPEE and once as Pres (1974). *Society Aff:* AIME-SPE, NSPE, SPEE, SPWLA, HGS

Berthouex, Paul M
Home: 533 N Blackhawk AVe, Madison, WI 53705
Position: Prof Civil & Environ Eng. *Employer:* Univ of Wisconsin, Madison. *Education:* PhD/Civil Engr/Univ of Wisc; MS/Civil Engr/Univ of Iowa; BS/Civil Engr/Univ of Iowa *Born:* 8/15/40. 2 yrs as Ch Res Engr GKW Cons Engrs Mannheim West Germany, spec in water supply & water pollution control in Europe, Asia & Africa; 2 yrs as Asst Prof Univ of Conn-Storrs; with Univ of Wisconsin-Madison, Dept of Civil & Environ Engrg since 1971 & Chmn of Water Resources Engrg Div 1974-76. Chmn, Env. Engrg. Div. 1980-present. Mbr ASCE, WPCF, AAAS, IAWPR, Chi Epsilon, Tau Beta Pi, Sigma Xi. WPCF Harrison Prescot Eddy Medal 1971; ASCE Rudolph Hering Medal 1975; WPCF Awards Ctte 1972-74; ASCE Awards & Publ Ctte 1975-76. Co-Author of 'Strategy of Pollution Control 1977 & more than 80 articles in prof journals. Natl Acad of Sci Committee on Natl Statistics, 1975-76 Consultant to TN Valley Authority, World Health Organization & Asian Development Bank. Intl conslt experience in W. Samoa, India, Singapore, Korea, Indonesia, Nigeria *Society Aff:* ASCE, WPCF, IAWPRC

Bertoni, Henry L
Business: 333 Jay St, Brooklyn, NY 11201
Position: Prof *Employer:* Polytechnic Inst of NY *Education:* PhD/Electrophysics/Polytechnic Inst of Bklyn; MS/EE/Polytechnic Inst of Bklyn; BS/EE/Northwestern *Born:* 11/15/38 Joined the faculty of the Polytechnic Inst of Brooklyn (now of NY) in 1966 and currently holds the title of Prof of Electrophysics. His studies in ultrasonics have included the analysis of waveguides, transducers and Raleigh angle reflection of acoustic beams. mode converting transducers and resonators via equivalent network techniques. He has also studied electromagnetic wave propagation in anisotropic and lossy media, and grating and prism transducers for integrated optics. Currently, his interests include the application of ultrasonics to nondestructive testing and to the acoustic microscope. *Society Aff:* IEEE, ASA, URSI/USNC

Bertossa, Donald C
Home: 19161 Bellwood Dr, Saratoga, CA 95070
Position: Prin Engr *Employer:* Gen Elec Co-Nuclear *Education:* BS/Metallurgical Engr/Univ of IL-Urbana *Born:* 11/23/22 PE-State of CA. Past Pres. American Welding Society. Past Chrm, Golden Gate Welding and Metals Conference. Member; Welding Research Council High Alloys Committee, ASME Section III (Nuclear Power) Code, ASME Section XI (In Service Inspection). U.S. Delegate, International Institute of Welding. Industrial Experience: Aluminum R&D, Aircraft Forgings, Pressure Vessel Plant, Field Construction, Welding R&D, Corrosion, Nuclear Power, Space System and Vacuum Systems. *Society Aff:* AWS, ASM, ASME, IIW

Bertram, Robert L
Business: 1100 Milan Bldg, Houston, TX 77002
Position: VP-Operations. *Employer:* Aramco Service Co. *Education:* MS/ChE/Carnegie Inst of Techn; BS/ChE/Northeastern. *Born:* 5/26/40. Native of MA; extensive oil co overseas experience with CAL-TX Oil Co and Arabian Amer Oil Co. Process Design & Proj Mgmt experience with Crawford and Russell. Currently respon for all Engg, Electronic Data processing, Petroleum Engg and Communications for Aramco Services Co in Houston, TX. Attended Columbia's Advanced mgmt Program in 1978. Enjoy long-distance running, sailing and tennis. *Society Aff:* AIChE, API.

Bertram, Sidney
Home: 1210 Oceanaire Dr, San Luis Obispo, CA 93401
Position: Volunteer Lecturer (Retired) *Employer:* California Polytechnic State University *Education:* PhD/Physics/OH State Univ; MS/Comm Engr/OH State Univ; BS/Elec Engg/CA Inst of Tech. *Born:* July 1913 Canada. Asst Prof Ohio St; Bureau of Ships commendation for contrib to sonar dev WWII; responsible for design of automatic equip for measuring terrain altitudes from aerial photographs; currently involved in teaching electronic engineering & in fundamental studies relating to nature of electromagnetic fields. Fellow IEEE; recipient Photogrammetric Award Amer Soc of Photogrammetry. Publs in electromagnetic theory, photogrammetry, atmospheric refraction, circuits & digital filtering. *Society Aff:* IEEE, ΣΞ.

Bertran, Enrique
Business: Humacao Industrial Park, Box 2521, Humacao, PR 00661
Position: VP. *Employer:* PCR Puerto Rico, Inc/SCM Corp. *Education:* BS/Chem Engg/Univ of Puerto Rico. *Born:* 6/6/49. Native of Havana, Cuba. Immigrated to USA in 1961, Became US Citizen in 1972. Process Engr with E I DuPont de Nemours, specializing in new products start-ups. With PCR since 1976, named VP of Puerto Rico Operation following yr. Enjoy working in the pharmaceutical industry. Hobbies include tennis and sailing. *Society Aff:* AIChE.

Bertrand, S Peter
Business: 2186 Dallas Drive, Baton Rouge, LA 70806
Position: President. *Employer:* S P Bertrand & Assocs Inc. *Education:* BS/ME/LA State Univ. *Born:* Jan 1936. BS Louisiana St Univ. Native of Eunice Louisiana. Engr with Mobil Oil Co 1957-60; Proj Mgr, Asst VP with Walk, Haydel & Assocs New Orleans 1960-70; founded S P Bertrand & Assocs 1970; Cons Engrs to Chem & Petrochem Mfg Indus; spec in Proj Mgmt & Design of Mech Facilities. P Chmn Baton Rouge Sect ASME; Mbr ACEC, NSPE. Tennis Enthusiast. *Society Aff:* ASME, ACEC, NSPE, ASHRAE, LES.

Berty, Jozsef M
Home: 310 West Grandview Blvd, Erie, PA 16508
Position: Prof of Chem Engrg *Employer:* The Univ of Akron *Education:* DSc/Chem Engg/Technical Univ of Budapest, Hungary; Dipl/Chem Eng/Technical Univ of Budapest, Hungary. *Born:* 10/25/22. Native of Hungary, worked in industry during and after WWII. 1950 joined Hungarian Oil & Gas Res. Inst to build a petrochemcial industry in Hungary. Taught in Veszprem coll of Chemical Engg. 1956-57 visiting prof in Leuna- Merseburg, East-Germany. 1957-76 Union Carbide Corp in USA, technology manager for over 1 billion lbs of ethylene oxide production and all catalyst manufacture. 1973-74 Sr Fulbright Lecturer at Techn Univ of Munich, Germany. Since 1976 independent consultant, propietor of Berty Reaction Engrs, Ltd. 31 papers, 18 pats. Since 1981 prof at Univ of Akron, O. Inventor of catalytic reactor named after him. *Society Aff:* AIChE, ACS, NYAS.

Beshers, Daniel N
Business: Columbia University, 1105 Mudd Bldg, New York, NY 10027
Position: Prof. *Employer:* Columbia Univ. *Education:* PhD/Phys/Univ of IL; MS/Phys/Univ of IL; BA/Math/Swarthmore College *Born:* 8/13/28. Met Div, US Naval Res Lab, 55-57; Henry Krumb Sch of Mines, Columbia Univ, Asst Prof of Met 57-60, Assoc Prof 60-69, Prof 69; Summer & special leaves: Lawrence Livermore Lab, 61, 64; Brookhaven Natl Lab 63; Univ of Poitiers, France 79-80; Res interests: Mech properties of mtls including elastic constants, internal friction, ultrasonics, magneto-mechanical effects, dislocation motion, plastic deformation; Diffusion, interstitial elements; Physical Met; Nondestructive testing; Acoustic emission. *Society Aff:* TMS-AIME, ASM, APS, ΣΞ.

Beskos, Dimitrios E
Business: Department of Civil Engrg, Univ of Patras, Patras 26110, Greece
Position: Professor *Employer:* Univ of Patras, Greece. *Education:* PhD/Struct Mechanics/Cornell Univ; MS/Solid Mechanics/Cornell Univ; Dipl/CE/Natl Tech Univ of Athens. *Born:* 1/26/46. in Athens, Greece in 1946. He came to the USA in 1969 for grad studies. He obtained a PhD from Cornell in 1973 and served as Instr there 1973-74. He joined the Civ Engg Dept of the Univ of MN as Asst Prof in 1974 and was promoted to Assoc Prof in 1979. In 1983 he left Minnesota to become Professor of Civil Engrg at the Univ of Patras, Greece. He specializes in struct engg and appl mechanics. He likes history- archaeology and soccer. *Society Aff:* ASCE, ASME, AAM.

Besse, C Paul
Home: 44 Gull Street, New Orleans, LA 70124
Position: Retired. *Education:* BS/Civil Engg/TX A&M Univ. *Born:* Jan 1912. Varied Engrg positions 1934-38; employed by Standard Oil Co of Californias Chevron Oil Co 1938- ; assumed position of Asst Ch Engr Const 1960; significant contrib to ocean engrg, offshore technology & res have been made in areas of wave force measurement & analy, soil mech, platform innovations, oil well drilling & prod systems through Standards world-wide offshore oil activities. Mbr SPE of AIME, NACE & Tau Beta Pi. Recipient of AIME Robert Earl McConnell Award 1974. Reg P E Louisiana. *Society Aff:* SPE OF AIME, NACE, TBΠ.

Bessen, Irwin I
Business: Mining Products Dept, Box 7428, Houston, TX 77008
Position: Mgr-Engg. *Employer:* Genl Elec Co. *Education:* PhD/Metallurgy/Columbia Univ; AM/Physics/Columbia Univ; BS/Physics/Rensselaer Poly Inst. *Born:* 6/30/26. Native of New Rochelle, NY. Physicist for North American Philips Co in development of x-ray and electron-optical methods for study of matls and biological tissues (from 1950-58). Joined Jones & Laughlin Steel Corp as physical metallurgist and was Asst Dir of Res with product development respons (until) 1966. Became Technical Dir of Specialloy, Inc before joining Genl Elctric's Aircraft Engine Grp in 1968. Heat managerial assignments in welding, coatings and ceramics until transfer to Mining Products Dept in 1979. Winner of GE's Badger Award for matls achievements. Enjoy sailing, hiking and music. *Society Aff:* ASM.

Besson, Pierre A
Home: 5 Avenue Constant-Coquelin, Paris France 75007
Position: Ing Genl des Ponts et Chaussees. *Employer:* French Government. *Born:* 12/5/01. in Privas (Ardeche) France. Ecole Polytechnique Paris; Ecole Nationale des Ponts et Chaussees; Ecole Superieure D'Electricite; Universite de Paris. Ingenieur au Service des Phares et Balises 1926-34; au Service Central de l'Electricite 1934-39; Directeur des Telecommunications et de la Signalisation au French Air Ministry - Cvl & Military 1939-42; Ch du Service Central de l'Electricte au Ministere de l'Industrie 1944-49; Directeur de l'Ecole Superieure d'Electricite 1949-58 Paris; Ingenieur Genl a la Direction de l'Electricite 195867; Pres du Comite Technique de l'Electricite 1958-67; Pres du Comite des Radiocommunications (TC 12) de la Commission Electrotechnique Internationale (IEC) 1951-69. Mbr du Comite d'Action Scientifique de la Defense Nationale; Pres, Societe des Radioelectriciens 1950; Pres, Societe Francaise des Electriciens 1965; Pres du Comite de la Rechercha scientifique et technique de Uinistere de l Equipment; Fellow (and Life Mbr) de "The Institute of Electrical and Electonics Engineers–. *Society Aff:* IEEE, SEE, ISF.

Best, Albert M
Home: 5 Kutz Avenue, New Holland, PA 17557
Position: Engrg Res Director. *Employer:* Retired 1982 *Education:* BS/Agri Eng/PA State Univ. *Born:* 9/20/19. Knox Pennsylvania. Grad BS in agri engrg Pensylvania St Univ 1942. Retired as Dir of Engrg Res for Sperry New Holland Div of Sperry Rand. P Chmn of Pennsylvania Sect & V Chmn of the North Atlantic Region ASAE; served as Chmn of Educ & Res Div, Dir of Power & Machinery Div & Admin V Pres; ASAE President 1979-80. Represented ASAE on the AAES Board and Member of AAES Executive Committee for 1981. Reg P E, Hon Mbr of Pennsylvania Chapter Alpha Epsilon; listed in Who's Who in America & Engrs of Distinction Pres ASAE 1979-80 - now Life Fellow ASAE *Society Aff:* ASAE.

Best, Cecil H
Business: Civil Engrg, Seaton Hall, KSU, Manhattan, KS 66506
Position: Prof *Employer:* Kansas State University *Education:* PhD/Eng. Sci/Univ. of CA-Berkeley; MS/Eng. Sci/Univ. of CA-Berkeley; BS/Eng Physics/Univ. of CA-Berkeley *Born:* 10/3/28 Native of Canada. Naturalized in 1954 following active duty in United States Army. Royal Norwegian Council for Scientific and Industrial Research Post-Doctoral Fellow, Technical University of Norway, Trondheim, 1960-61. With Kansas State University since; Assoc Prof of Applied Mechanics, 1961; Prof, 1964; Assoc Dean of Engrg, 1968; Prof of Civil Engrg, 1974. Lic Prof Engr, Kansas, 1962. Enjoy hunting, fishing, birding, 3-rail billiards, handcrafts, and most music. *Society Aff:* ACI, ASCE, ASTM, NSPE

Bethea, Robert M
Business: Dept of Chem Engrg, Texas Tech Univ, P.O. Box 4679, Lubbock, TX 79409
Position: Prof of Chem Engr *Employer:* TX Tech Univ *Education:* Ph.D./ChE/IA St Univ; M.S./ChE/IA St Univ; B.S./ChE/VA Polytech Inst *Born:* 12/12/35 NASA-Langley (1964-66) evaluating contaiminant control systems. Joined TX Tech 1966, prof 1975. Sabbatical (1972) in engrg analysis of air pollution control systems at EPA. Consultant numerous companies for air pollution and occupational environment control. Author of: "Statistical Methods for Engrs and Scientists," (2nd ed, 1985), "Air Pollution Control Technology," (1978) and over 110 journal articles and presentations. Registered PE (TX). Research areas: air pollution and occupational dust control, instrumental methods development, environ effects on solar concentrator mirrors. Teaching activities include developing accelerated MSChE program; co-developer interdisciplinary MS program in occupational safety and health. Member editorial review bd AIChE Journal. Monsanto and Ford Foundation Fellowships, Sigma Xi Research Award at ISU; Diplomate, Amer Acad Env Engrs, certified in air pollution control; numerous awards at TX Tech for excellence in teaching and res. *Society Aff:* AIChE, APCA, ACS, AIHA, ΣΞ, TBΠ

Bethea, Thomas Jesse
Business: 249 Johnston St Bx 10786, Rock Hill, SC 29730
Position: President. *Employer:* Bethea Engrg Co Inc. *Education:* MS/ME/Clemson Univ; BS/ME/Clemson Univ. *Born:* May 12, 1933. BS 1958 & MS 1965 Clemson Univ. Engr with Martin Marietta Corp & Chrysler Corp prior to 1963; Sr Dev Engr & Group Supr with Celanese Corp of Amer to 1972; Cons in Mech & Elec Engrg, Pres Bethea Engrg Co Inc. Mbr: Tau Beta Pi, Natl Soc of Prof Engrs, ASHRAE; P Chmn Scholarship Ctte S C Soc of Prof Engrs 1975-76; Mbr Elec Bd City of Rock Hill S C; P Chmn Bd of Deacons, 1st Presbyterian Church of Rock Hill S C 1975-76, Mbr Amer Consulting Engineers Council, Mbr ACFC Education Committee. *Society Aff:* NSPE, ACEC, ASHRAE, TBΠ.

Bethke, Donald G
Home: 34380 Sunset Drive, Oconomowoc, WI 53066
Position: Director of Engrg. *Employer:* HUSCO - Div of Koehring Co. *Education:* BSME/IN Inst-Tech. *Born:* Feb 1938. BSME Indiana Inst of Tech. Native of Waukesha Wisc. Design Engr Hydraulic Unit Specialties Co (Husco) 1961-63 spec in the design of hydraulic control valves (sectional) & high performance hydraulic relief valves; experimental test engr for Pratt & Whitney Aircraft 1963-64 spec in main hydraulic control sys for jet engines; Proj Engr for Husco Div spec in excavator hydraulics 1964-69; Dir of Engrg Husco Div 1969- . Mbr of SAE & NFPA. Enjoy outdoor activities & photography. *Society Aff:* SAE, NFPA.

Betner, Donald R
Home: 430 Kindig Rd, Indianapolis, IN 46217
Position: Sr Experimental Met. *Employer:* Detroit Diesel Allison - GM. *Education:* AB/Chemistry/Franklin College. *Born:* 1/11/26. Born at Greenwood, IN, Jan 11, 1926. Attended Ctr Grove grade and high sch grad in 1944. Served two yrs & three months during WWII in Hospital Corp, USNR in the S Pacific theatre, being honorable discharged in 1946. Attended Franklin College, Franklin, IN, receiving an AB degree in Chemistry in 1950. Married in 1949 & father of two children. Worked in the field of X-Ray until joining Detroit Diesel Allison Div of GM Mtls Labs in 1955. Fields of expertise include Met Facture Analysis & Electron Microscopy. Co-recipient of Francis F Lucas Award in 1970. Articles published on Transformations in High Temperature Alloys & Beryllium. Mbr ASM Handbook Committee for six yrs. *Society Aff:* ASM, EMSA, IMS.

Bettigole, Neal H
Business: 601 Bergen Mall, Paramus, NJ 07652
Position: Principal. *Employer:* N.H. Bettigole, P.C. *Education:* BECE/Civ Engg/Yale Univ. *Born:* 10/22/29. Mr Bettigole, who established the N H Bettigole Co in 1966, has been involved in the engr profession for more than 30 yrs. Prior to forming his own Co, he worked for the firm of Hardesty & Hanover, rising from Designer to Assoc Engr. Since his organization was founded 21 yrs ago, it has served many maj public agencies, numerous town & counties in NY, NJ, and Conn., and varied commercial & ind concerns. The firm has designed a maj movable bridge over the Hudson River, a fixed bridge over the Flushing River, the rehabilitation of a maj movable railroad bridge, widening the Tappan Zee Bridge, the rehabilitation of the 11 miles of the Henry Hudson Pkwy, emergency reconnection of NY thruway at Schoharie Creek, inspection of over 3,000 bridges of all types, & many other projs. Mr Bettigole is active in many profl organizations, Pres of the Assn for Bridge Construction & Des, Chrmn of the Construction Council of NY State. He holds US and Canadian patents on exodermic bridge deck, a major innovation in bridge deck design and construction. *Society Aff:* ASCE, NSPE, IBTTA, IABSE, ASTM

Bettis, John R
Home: 179 Third Ave, Charleston, SC 29403
Position: Chief Oper Officer Mgr & Engg. Retired. Consult *Employer:* Commissioners of Public Works. *Education:* BS/Mech Engg/Clemson Univ. *Born:* 8/20/17. in Greenville, SC. Joined Comm of Public Works as student engg upon grad from Clemson July 1940. 4 yrs Army - WWII- Europe/ Mgr & Engg (top position) Comm of Public Works since 1954. Retired July 1986 - now Consult to CPW. Dir WPCF, 1968-70. Fuller Award, American Water Works Assoc, 1961. Dir American Water Works Assoc, 1976-79. Herman F Wiedeman Award, 1965. PChmn Charleston Mech Engg Club. PPres Charleston Rotary. PChmn Salvation Army Advisory Bd. Contributed to prof journals. Outstanding Achievement Award - S C Water Resources Comm 1979. Pres. Charleston St. Andrews Society, Founded 1729. *Society Aff:* AWWA, WPCF.

Beu, Eric R
Home: 339 Reeves Ave, Dover, OH 44622
Position: Exec VP. *Employer:* Dover Chem Corp. *Education:* BS/CE/Univ of MI. *Born:* 5/24/24. Native of Buffalo, NY. Process engr for Birds Eye-Snider Labs 1946-50, designing & operating frozen juice concentrate plants. Process engr for Genessee Res Corp 1950-54, designing constructing & operating specialty organic chem mfg facility. With Dover Chem Corp since 1954. Assumed current Exec VP and Chief Operating Officer position in 1978. Am responsible for bus results for this mfg of chlorinated paraffin, Commercial & reagent hydrochloric acid, sodium hypochlorite & paradichlorobenzene. Pres TVSPE 1971. Play classical piano. Enjoy swimming, tennis. *Society Aff:* AIChE, ACS, NSPE.

Beumer, Richard E
Business: Sverdrup Corporation, b801 N. Eleventh, St. Louis, MO 63101
Position: Sr. Vice Pres. *Employer:* Sverdrup Corporation *Education:* BS/EE/Valparaiso Univ. *Born:* Feb 1938. Reg P E in 8 states. Native of St Louis Mo. Joined Sverdrup & Parcel 1959. Between 1959 & 1969 had design & proj mgmt responsib on variety of indus process & governmental res & defense facility projs; 1969 became Asst V Pres & Mgr of S&P s Phoenix Az office; returned to S&P s St Louis office 1973 as AVP & Asst Ch Engr responsib for all energy conversion related projs; became V Pres & Ch Engr Indus Div 1975; became VP & Mgr Indus Div 1977 & Sr VP & Central Region Mgr in 1979; elected Pres in 1982. Currently Sr VP and GM-Central Group for Sverdrup Corporation. Active on professional, civic, charitable org & church cttes. *Society Aff:* ACEC, ANS, NSPE

Beutler, Frederick J
Business: EECS Department, EECS Building, The University of Michigan, Ann Arbor, MI 48109
Position: Prof and Chrmn. Electrical Engineering: Systems Graduate Program *Employer:* Univ of MI. *Education:* PhD/Engr Sci/Caltech; SM/Engr Sci/MIT; SB/Econ and Engr/MIT. *Born:* 10/3/26. Res engr (AC Spark Plug Div of GMC, Autonetics Div of North Am Aviation, Ramo-Wooldridge Corp), 1949-1957. Engr faculty at Univ of MI since 1957; Prof of Info and Control Engr, 1963-. Visiting Prof of Elec Engr, Caltech, 1967-68. Chrmn, Computer Info and Control Engr Program at Univ of MI, 1977-. Managing Editor, SIAM J Appl Math, 1970-75. Elected mbr, Council of SIAM (governing body), 1968-1974. Listed Who's Who in America, etc. Chrm, Elec. Engr.: Systems Graduate Program, 1985- ; Elected Fellow, Institute of Electrical & Electronic Engrs. (IEEE), 1980; Elected Member, Tau Beta Pi (as Eminent Engineer), 1981. *Society Aff:* IEEE, ORSA, SIAM, IMS, AMS.

Bevacqua, C Joseph
Home: 133 Glenwood Ave, Leonia, NJ 07605
Position: VP. *Employer:* Wyssmont Co Inc. *Education:* MS/ChE/Stevens Inst of Tech; BS/ChE/Columbia Univ; BS/Biology/Fordham Univ *Born:* 8/27/32. After receiving a BS degree in Biology, served in the US Army Medical Nutrition Lab conducting research with irradiated foods. After receipt of BS degree in Chem Engg accepted position with General Foods ad Process Dev Engg (1959-63). Joined Wyssmont Co in 1963. Later promoted to Sales Mgr and in 1971 promoted to VP, in charge of Sales Engg Dept. Mbr & Pres - Leonia NJ Bd of Ed 1971-1977 Mbr & Chrmn-Leonia Planning Bd 1977-present. *Society Aff:* AIChE

Bever, Michael B
Business: Rm 13-5026, Cambridge, MA 02139
Position: Prof of Matls Sci & Engg, Emeritus/Sr Lecturer *Employer:* MIT *Education:* ScD/Metallurgy/MIT; SM/Metallurgy/MIT; MBA/-/Harvard Univ; Dr. Jru. utr. /Law & Admin./Univ of Heidelberg. *Born:* Aug 7, 1911. Dept of Metallurgy at MIT: Asst 1940, Prof of Metallurgy 1956-75, Prof of Materials Sci & Engrg 1975-77; Prof of Matls Sci & Engg, Emeritus; Sr Lecturer. Mbr ARPA Materials Res Council 1966-74 . Reg P E Mass. Fellow Amer Acad of Arts & Scis. Corresponding Mbr, Berliner Wissenschaftliche Gesellschaft. Mathewson Gold Medal AIME 1965; Recycling Award Natl Assn of Secondary Materials Indus 1972. Genl field: Metallurgy, Materials Sci & Engrg. Special interests: thermodynamics, calorimetry, gas-metal reactions, alloy theory, surface hardening of steel, deformation of metals, characterization of structs, prod & recycling of materials, economics of materials. Mbr Editorial Bd McGraw-Hill Series in Materials Sci & Engrg. Co-Editor *Conservation & Recycling;* Consulting Editor *Environmental Impact Assessment Review*; Editor- in-Chief, *Encyclopedia of Materials Sci and Engrg* (Pergamon Press and MIT Press). *Society Aff:* AIME/TMS, ASM, MRS, ΣΞ, AAAS, INSTITUTE OF METALS (LONDON)

Beverage, Harold H
Home: 23 Quaker Path-Box BX, Stony Brook, NY 11790
Position: Retired part time Cons. *Employer:* RCA Labs. *Education:* BS/EE/Univ of ME. *Born:* Oct 1893. D Alexanderson Radio Lab G E Co 1916; transferred to RCA 1920; RCA Communications, Ch Res Engr 1929, V Pres R/D 1941 & also Dir Radio Res RCA Labs 1942. Issued 40 pats incl Traveling Wave Antenna & Diversity Reception. Pres Inst of Radio Engrs 1937. Part time Expert Cons Office of Secy of War 1942- 46. Received many honors incl Liebmann Prize & Medal of Honor Inst of Radio Engrs, Signal Corps Certificate of Appreciation, Presidential Certificate of Merit, AIEE Lamme Medal, Eminent Mbr Eta Kappa Nu, Honorary Mbr Tau Beta Pi, Marconi Gold Medal from VWOA. Fellow IEEE & AAAS. Listed in Whos Who in America & other reference publs. *Society Aff:* IEEE, AAAS.

Bevis, Herbert A
Business: 312 Weil Hall, Gainesville, FL 32611
Position: Associate Dean for Academic Affairs *Employer:* College of Engineering Univ of FL. *Education:* PhD/Env Engg/Univ of FL; MS/Nucl Phys/US Naval Postgrad Sch; MSE/Sanitary Engg/Univ of FL; BCE/Civ Engg/Univ of FL. *Born:* 9/28/29. Grew up in Gainesville, FL. After grad from the Univ of FL served as a Commissioned officer in the US Public Health Service for 9 yrs which included assignments in the Region II Office (NYC), Region III (Wash DC) Radiological Health Hdquarters Staff (Wash, DC), Taft Sanitary Engg Ctr (Cincinnati, OH), Nuclear Weapons Test Ctrs (NV & Eniwetok) and with the TX State Dept of Health (Austin, TX). Returned to the Univ of FL in 1961. Progressed through the ranks to full prof including a tour as Asst Dean. Appointed Associate Dean of the College of Engg in January 1981. A principal and consultant for Water & Air Res, Inc. Primary hobby is woodworking. *Society Aff:* NSPE, ASCE, AAEE, HPS, CFSE, AEEP, ΤΒΠ, ΦΚΦ, ΟΔΚ.

Bezier, Pierre E
Home: 12 Avenue Gourgaud, Paris, France 75017
Position: Ret Engr *Education:* D.SC./Math/Universite Paris VI; Ingr Mechanics/ Mechanics/Ecole Nationale Arts et Metiers; Ingr Electricity/Elect/Ecole Superieure D'Electricite *Born:* 09/01/10 1933-75 Engineer Renault Automobile Co., Dir. Tool and Machine-Tool Drawing Office; Dir Manufacturing Engrg. Responsible for CAD/CAM "Unisurf" System *Society Aff:* ASME

Bezzone, Albert P
Home: 829 Senior Way, Sacramento, CA 95831
Position: Chief, Structure Construction. *Employer:* California Dept of Transportation. *Born:* June 1931. Attended Calif St Univ Sacramento. Fellow ASCE. P E Calif. Joined Calif Dept of Transportation as Bridge Engr 1953; had charge of Bridge Design Sect 1963-74; natl involvement in dev of prestressed concrete bridges; responsib for design of 1st cast-in-place segmental bridge in US; Ch of Opers for state owned toll bridges 1974-75; Exec Asst to Chief Engr 1975-77. Author of number of articles & papers on design of hwy bridges. ASCE Greensfelder Prize 1976. Prof Engrs in Calif Government's 'Man of the Year' 1974; Engrg News-Record Men Who Made Marks 1972. *Society Aff:* ASCE.

Bhada, Ron K
Business: Babcock & Wilcox Co, R&D Div, P.O. Box 835, Alliance, OH 44601
Position: Mgr Advanced Products *Employer:* The Babcock & Wilcox Co, R&D Div *Education:* PhD/Chem Eng/Univ of Mich; MS/Chem Eng/Univ of Mich; MBA/ Management/Univ of Akron; BS/Chem Eng/Univ of Mich *Born:* 1935. Res Asst Univ of Michigan 1953-59; joined Babcock & Wilcox 1959 as Res Engr on advanced power generation & combustion & gasification of coal; promoted 1972 to Mgr of Chem Engrg with respon for about 30 profs involved in R/D in combustion, fuels handling, pollution control & related process dev; promoted 1977 to Manager of Planning, R&D Div, respon for all R&D Planning for the Co. Promoted 1981 to Mgr of Planning & Adv Tech with added respon for planning and implementation of all Strategic Tech Prog of CO. Mbr of OH Acad of Sci; Mbr Natl Res Ctte AIChE. Authored 9 tech publs. Special Faculty, Youngstown State Univ. *Society Aff:* AIChE, OAS, PLU, BGS

Bhagat, Pramode K
Business: Wenner Gren Research Lab, Univ of KY, Lexington, KY 40506
Position: Assoc Prof *Employer:* Univ of KY *Education:* PhD/EE/OH State Univ; MS/EE/Univ of Cinn; B Tech/EE/Indian Inst of Tech, Madras, India *Born:* 10/7/44 Ranchi, India Worked as project engr for Systems Research Labs, Dayton (1966-67). Have been working for Univ of KY since 1973. Assumed current responsibility as Assoc Prof in 1980. Current areas of interest are biomedical engrg, digital signal processing and control systems. Present research focus has been in biomedical applications of ultrasound, non destructive evaluation and defect characterization. *Society Aff:* IEEE.

Bhagat, S K
Business: Dept. of Civil & Env. Engineering, Washington State University, Pullman, WA 99164-2910
Position: Chrmn/Prof *Employer:* WA State Univ *Education:* PhD/Env Engrg/Univ of TX (Austin); MS/Sanitary Engrg/Univ of TX (Austin); BS/CE/Univ of TX (Austin); BSc/Sci/Panjab Univ (India) *Born:* 5/14/35 Have over 25 years experience in research regarding water quality engrg, municipal and industrial wastes, toxic and radioactive materials, eutrophication and lake quality restoration, environmental impact assessment, and teaching of graduate/undergraduate courses in environmental engrg. Author of 50 publications dealing with various aspects of water pollution assessment and control. Recipient of 1980 Arthur Sidney Bedell Award from Water Pollution Control Federation for extraordinary personnel service in water pollution control field. Have served as consultant to federal, state, local and private organizations dealing with environmental problems; served on numerous univ and professional society committees; have approximately 10 years experience in univ administration. *Society Aff:* ASCE, ASEE, WPCF, IAWPR, UCOWR

Bhappu, Roshan B
Home: 4370 S. Fremont, Tucson, AZ 85714
Position: Vice Pres & General Mgr. *Employer:* Mt States Mineral Enterprises Inc. *Born:* June 1953. D Sc Colorado School of Mines; Met E & MS from C S M. Native of Karachi Pakistan; naturalized citizen of USA 1957. Proj Engr for C S M Res Foundation 1952-54; Resident Metallurgist for Miami Copper Co 1954-59; Sr Metallurgist, Res Prof & V Pres for N M Tech Res Foundation, Chmn of Metallurgical & Environmental Engrg N M Inst of Mining & Tech 1959-72; V Pres & Genl Mgr of Mountain Sts R/D Div of Mountain Sts Mineral Enterprises Inc since 1972. Mbr: AIME, ASM, ACS, AAAS, Sigma Xi; Chmn Mineral Processing Div SME, AIME 1976. Van Diest Gold Medal of C S M 1968. Special Cons to UN.

Bharathan, Desikan
Home: 9786 W 22nd Pl, Apt 103, Lakewood, CO 80215
Position: Sr Engr *Employer:* Solar Energy Research Inst *Education:* PhD/Aerospace/Univ of VA; MS/Aerospace/Univ of VA; BTech/Aeronautical/ Indian Inst of Tech. *Born:* 1/11/49. in India. Educator and Researcher. Have worked in areas of low-speed aerodynamics, supersonic and hypersonic aerodynamics, rarefied gas dynamics, turbulence, two-phase flows, automatic control sys, and applied mathematics. Topics of past res: aerodynamic forces on cones in rarefied flows, magnetic model suspension sys for wind tunnels, aerodynamic stability of missiles, and glow discharge flow visualization, and safety aspects of nuclear reactors. Currently working on studies of Flash Evaporators and direct-contact condensers for use in open cycle ocean thermal energy conversion systems. *Society Aff:* ASME, AIAA, ΣΞ.

Bhatt, Suresh K
Home: 954 Twinleaf Ct, Bethel Park, PA 15102
Position: Chief Engr *Employer:* Gates Engrg Co *Education:* MS/ME/VA Polytechnic Inst & SU; BS/ME/Indian School of Mines *Born:* 9/24/46 Native of India. Worked with Indian Iron and Steel Co as asst mine mgr. Started with US Steel in 1970, worked five years in various engrg positions in coal division. Joined Gates Engrg in 1975. Assumed current position in 1980. Have taught several mining courses to engrg students, BS, level. Served as chrmn of Professional Ethics Committee and member of Scholarship Committee for Appalachian, Chapter NSPE. As chief engr responsible for all activities, engrs, project management, for the dept and engrs of other offices depending project size. Responsible for projects specializing coal mining, planning, design, economic evaluations. *Society Aff:* AIME, NSPE, AMC

Bhowmik, Nani G
Business: State Water Survey, 2204 Griffith Drive, Champaign, IL 61820
Position: Principal Scientist *Employer:* IL Dept of Energy and Nat Res, Water Survey Div *Education:* PhD/Civ Engg/CO State Univ; MS/Civ Engg/CO State Univ, BS/Civ Engg/Dacca Univ. *Born:* 1/6/40. in Bangladesh, taught for three yrs at the Bangladesh Engg Univ. Received BS degree in Civil Engineering from Dacca Univ. Received MS and PhD from CO State Univ. Has been working for the IL State Water Survey for the last 19 yrs. Have done extensive res work in open channel flow problems, lab experimentation, circulation patterns in lakes, lake shore protec-

Bhowmik, Nani G (Continued)
tion, hydraulic geometries of streams and floodplains, bank erosion, sediment transport, physical impacts of navigation, soil erosion, river mech and ecology of large rivers, secondary circulation, and others. Have published and presented technical papers extensively. 1977 Freeman Fellow of ASCE, and has travelled around the world presenting seminars and invited papers at various univs, tech conferences and Water Resources Centers. Enjoys fishing, gardening, and camping. *Society Aff:* ASCE AGU, IWRA, IAS.

Bhushan, Bharat
Home: 14527 Eastview Dr, Los Gatos, CA 95030
Position: Scientist Mgr *Employer:* IBM Corp *Education:* Ph.D./Mechanical Eng./Univ of Colorado, Boulder, CO; M. B.A./Management/Rensselaer Polytechnic Inst., Troy NY; M.S. /Mechanics/Univ. of Colorado, Boulder, CO; M.S./Mechanical Eng./Massachusetts Institute of Technology, Cambridge *Born:* 09/30/49 Native of India. Worked as a research staff mbr at Mass. Inst. of Tech. 1971-72. Was a consulting engr for Automotive Specialists, Denver 1973-76. Was an instructor at Univ. of Colo. 1975-76. Was a Sr Eng. Scientist/Program Mgr, R & D Div of Mechanical Technology Inc. Latham, NY. 1976-80. Was a Research Scientist at Technology Services Div. of SKF Industries Inc., King-of-Prussia, PA. 1980-81. Was Advisory Engr/Mgr, General Products Div., IBM Corp., Tucson AZ 1981-86. Presently, Sr Engr/Mgr, IBM Almaden Research Center, San Jose, CA, 1986. Responsible for research activities in thin film disks for magnetic storage systems. Authored more than 100 technical papers and hold several patents in the field. Recipient of ASME Henry Hess Award, ASME/ASCE/IEEE/WSE/AIME Alfred Noble Prize, ASME Burt L. Newkirk Award, ASME Gustus L. Larson Award, Univ. of Colo. Regents Distinguished Service Award, Univ. of Colo. Gold Medal, IBM Invention Achievement and Special Recognition Awards. *Society Aff:* ASME, IEEE, STLE, NSPE, ΣΧΙ, ΤΒΠ

Bianchi, Leonard M
Home: 240 Donner Ave, Monessen, PA
Position: Consultant (Self-employed) *Education:* MS/Met Engrg/RPI; BS/Met Engrg/Carnegie Mellon *Born:* 1/10/30 Native of Monessen, PA. Studied sculpturing and basketball. Technical Dir for Refractomet Div. Universal Cyclops 62-65. Marketing Mgr for TAPCO-TRW 65-70. Prof at WVa Univ in School of Mineral & Energy Resources 72-76. VP & Gen Mgr of Evaporative Products for Airco Temesal. Currently Pres of Metallurgy Inc., a consulting firm specializing in vacuum metallurgy & electron beam coatings. Over 25 technical publications and patents in Vacuum Metallurgy, Electron Beam coatings, & INFAB processing. Former A.V.S. Dir. Current mbr V.M.D. Exec Committee. Reg Engr PA-WV. *Society Aff:* AVS, ASM, AIME

Bianchi, William C
Home: 4375 San Simeon Cr Rd, Cambria, CA 93428
Position: Consultant-Ground Water Mgmt (Self-employed) *Education:* PhD/Soil Sci/UC-Davis; BS/Irrigation Sci/UC-Davis. *Born:* June 1929 L A Calif. Asst Prof Univ of Nev-Reno 1956-59; Res Scientist USDA-ARS 1959; Res Leader Fresno Water Mgmt Ground-Water Recharge 1966-79. Currently Consultant in Ground Water Mgmt. Author or co-author of 40 sci publs relating to agri drainage, evapotranspiration, soil-water movement instrumentation, geophysical measurements, artificial ground-water recharge phenomena & facility design, ground-water quality & ground-water hydrology. Mbr Sigma Xi. FAO Cons, received AWWA Best PaResources Div 1975. Skier, hunter & fisher. *Society Aff:* SSSA-ASA, AAAS, SCSA.

Bias, Frank J
Home: 5 Incognito Lane, Ossining, NY 10562
Position: Consultant *Employer:* Viacom Intl Inc. *Education:* BSEE/-/IA State College. *Born:* Oct 1, 1919 Des Moines Iowa. 1941joined General Electric Co; during WWII designed radar & radar countermeasures equip; after the war designed television transmitters & held var engrg design & mgmt positions, becoming Mgr-Engrg Visual Communication Prods Dept 1963; joined the predecessor to TeleVue Systems Inc 1969; became V Pres Engrg 1974;. VP Science & Technology, Viacom Intl Inc 1979 respon technical, film syndication, pay-TV Programming Cable Television & Broadcasting. June 1986 Semi-Retired. Consultant to Viacom. Fellow, P Chmn Syracuse Sect & Broadcasting Group IEEE; Mbr Tau Beta Pi, Eta Kappa Nu, Phi Kappa Phi, Pi Mu Epsilon; active in EIA, IEC & NCTA. Reg P E in NY & Calif. *Society Aff:* IEEE, SMPTE, SCTE, SBE, NSPE

Biberman, Lucien M
Home: 5904 Lenox Road, Bethesda, MD 20817
Position: Staff Member. *Employer:* Institute for Defense Analyses. *Education:* BS/Chem/R.P.I. *Born:* May 31, 1919 Phila Pa. Sci and Tech Div Applied physicist - turned to countermine warfare when the submarine threat materialized early in WW II; later, consecutively, became engaged in exterior ballistics & rocketry, missile guidance, IR & UV detection & low-light-level TV & IR imaging. Recent texts for which served as writer & editor incl the monograph Reticles in Electrooptical Devices (Pergamon), the 2-volume Photoelectronic Imaging Devices' (Plenum) & 'Perception of Displayed Information' (Plenum). Fellow Optical Soc of Amer, Soc for Info Display, Inst of Elec & Electronic Engrs, Soc of Photo-optical Instrumentation Engrs. *Society Aff:* OSA, IEEE

Bickart, Theodore A
Business: Engr Deans Office, 209 Link Hall, Syracuse University, Syracuse, NY 13244-1240
Position: Prof & Dean. *Employer:* Syracuse Univ. *Education:* DEng/EE/Johns Hopkins Univ; MSE/EE/Johns Hopkins Univ; BES/EE/Johns Hopkins Univ. *Born:* 8/25/35. As a grad student and until 1961, taught as an Instr in the Dept of Elec Engg at the Johns Hopkins Univ. As a 1st Lt in the US Army during the subsequent two yrs, worked as a Systems Engr with the Satellite Communications Agency at Ft Monmouth, NJ. Affiliated with Syracuse Univ since 1963, now as a Prof in the Dept of Elec & Comp Engrg & Dean of Enrgr. Was Visiting Scholar during 1977 in the Dept of EE and Comp Sciences at the Univ of CA in Berkeley. Author of more that 80 res papers & coauthor with N Balabanian of the books *Electrical Network Theory* and *Linear Electrical Networks: Analysis, Properties, Design, and Synthesis.* Is a Fellow of the IEEE, been an Assoc Editor of the IEEE Transactions on Circuits & Systems, and served as Technical Prog Chrmn of the 1980 IEEE Intl Symposium on Circuits & Systems. Been an IEEE Delegate to the Popov Soc (USSR) and IEEE Circuits and System Soc. representative to the Chinese Inst of Electronics. Mbr of Eta Kappa Nu, Tau Beta Pi, and Sigma Xi. Active in research on mathematical software for the analysis of stiff systems. *Society Aff:* IEEE, ACM, SIAM, AMS, AAAS.

Bickford, Thomas E
Business: 5085 Reed Rd, Columbus, OH 43220
Position: Director, Human Resources *Employer:* Burgess & Niple, Limited *Education:* BS/CE/Northeastern Univ *Born:* 5/22/29 Born Newton, Mass, Edward Milton/Mabel Etta (Eldridge) B. Married Carol Shay 5-24-81 Children: Douglas, Linda. Employment: Civil Engr Columbia Gas System, 1957-60;, Scioto Conservancy Dist., 1962-64; Burgess & Niple, Limited, Civil Engr 1961 & 62 & 64-70, Asst. Personnel Dir 66-70, Personnel Dir 1971- present. Senior Professional in Human Resources 1978. Civic activities incl. PTA, United Way, Civic Assoc. Reg PE Ohio, Florida. Who's Who in Midwest. Mbr ASCE, Personnel Association of Central Ohio (V.P. 1972, Treas 1977, Dir 1981), Amer Soc Personnel Administration. *Society Aff:* ASCE, PACO, ASPA

Biddison, Cydnor M, Jr
Business: 1730 W Olympic Blvd, Los Angeles, CA 90015
Position: President. *Employer:* Hillman, Biddison & Loevenguth. *Education:* BSCE/Civil Engr/CA Inst of Tech. *Born:* 1918 Oakland Calif. Engr Bd Ft Belvou Va 1941-43; Lt USNR Pacific Ocean 1944-46. With Hillman etal since 1952; became Partner & V P 1965. Reg Civil & Struct Engr Calif; Reg P E Nevada & MT, UT & OR. Pres Struct Engrg Assn of So Calif; Pres Struct Engrs Assn of Calif; Secy-Treas

Biddison, Cydnor M, Jr (Continued)
Western Sts Conf of Struct Engrs; Treas Cons Engrs Assn of Calif. Fellow Inst for Advancement of Engrs; Treas Los Angeles Community Design Center; Dir Const Specification Inst; Dir Los Angeles Council Engrg Soc; Dir CA Inst of Tech Alumni. *Society Aff:* CEC, ACI, CSI, ASTM.

Biddle, Walter A
Home: 1819 E Coolidge St, Phoenix, AZ 85016
Position: VP, Engg. *Employer:* Triandria Design Group *Education:* BS/ME/Univ of NM *Born:* 3/2/15. Native of McKeesport PA. Taught in engg college & flight training school of NM during WWII. Sales engg AZ 1947-58. Principal in cons firms 1958-76. Biddle & Young Cons Engrs Inc; respon for air-conditioning plumbing & related mech design. VP Engg in Dorsett Industries, Inc 1976-1979, Industr engr consultant. Fellow Mbr ASHRAE & Life Mbr ASME. VP Engg in Triandria Design Group 1980- present. Air-cond. consultant. *Society Aff:* ASHRAE, ASME, NSPE

Bidstrup, Wayne W
Home: PO Box 263, Remington, VA 22734
Position: Vice President. *Employer:* PCL Packaging Inc. *Education:* BS/Elec Engg & Chem Engg/MTU; SM/Chem Eng/MIT. *Born:* April 9, 1931. Taught at Kansas St in ChE Dept 1955. Worked in plastics converting indus for 15 yrs with Mobil Plastics and finally as V Pres of Polymer International. Became V Pres of O'Brien Homes 1974. VP of PCL in 1978. Enjoy wood working as hobby. *Society Aff:* AIChE, IEEE, ΣΞ.

Bieber, Herman
Business: 143 Dorset Drive, Kenilworth, NJ 07033
Position: Owner *Employer:* Bieber Enterprises *Education:* D Eng Sci/Chem Eng/Columbia Univ; MS/Chem Eng/Columbia Univ; BS/Chem Eng/Columbia Univ. *Born:* 1/13/30. Dr H Bieber is currently owner of Bieber Enterprises, specializing in information research and creativity management. He was with Exxon Res and Engrg Co. for 31 years & also taught at Columbia, CUNY and Stevens Tech. He holds 21 US Patents and has co-authored a textbook and several journal articles. He is active in the AIChE Education & Accreditation Committee & is now AIChE representative on the Boards of the Engg Foundation and the Engrg Soc Library. He is also past chmn of the Engg Foundation Conferences Committee, a trustee of Midwest Research Inst, and on the Bd of The Society of Columbia Graduates. *Society Aff:* AIChE, ΤΒΠ, ΣΞ, ΦΛΥ, ΘΤ.

Biedenbach, Joseph M
Home: 168 Dorset Dr, Columbia, SC 29210
Position: Dir Cont Ed for Engr. *Employer:* Univ of SC. *Education:* PhD/Higher Ed/Univ MI State; MBA/Finance/Southern IL Univ; MS/Physics/Univ of MI; MS/Education/Univ of IL; BS/Engr/Univ of IL. *Born:* 1/29/27. He has served on the faculties of Gen Motors Inst, FL Atlantic Univ and Purdue Univ; as Associate Dean, Purdue Univ, Indianapolis; Dir of Education Resources, Hershey, PA Medical Ctr, and is currently Dir of Continuing Engg Education and Health Sciences at Univ of SC. Industrial experience includes res physicist at A C Sparkplug Div of Gen Motors; Dir of Continuing Education, Corp Staff, RCA. Professional Soc Activities include: Pres, Education Group, IEEE; Pres, CPD, ASEE. *Society Aff:* IEEE, ASEE, ASTD, NUCEA, MPI.

Biehl, Francis W
Business: N66 W12659 Ravine Dr, Menomonee Falls, WI 53051
Position: Pres. *Employer:* Biehl Engineering Inc. *Education:* MSME/Mech Engg/Wayne Univ; BSME/Mech Engg/Marquette Univ. *Born:* 11/20/28. Certification by Written Examination. Reg PE - WI & IL. Natl Cert - Natl Council Engg Examiners. Reg Struct Engg - IL. Certified Safety Professional. Reg Sanitarian - WI. Certified Soil Tester - WI. Reg Architect - WI. Cert Dwelling Inspector - WI. Commercial Bldg Code Inspector - WI; Private Detective - WI; Certified In Plumbing Engrg - ASPE; Snowmobile Safety Instructor - WI DNR. Cert. Fire Investigator IAAI. Cert. Electrical Inspector WI. Cert Gen Mechanic - Natl Inst Automotive Excellence. Mbr: WI Amusement Device Code Committee, WI Bldg Code Advisory Committee, ASTM D21 Polishes & Slip Resist. NIBS Consultants Committee Nat. Exec. Service Corps, Amer. Arbitration Assn. Boy Scouts of American-Merit Badge Counsel. Occupation: Consulting Eng 33 yrs in broad range of disciplines; *Society Aff:* NSPE, ASPE, PEPP, ASHRAE, ASSE, SAE, IES, NAFE, IAAI, ASTM, NFPA, IAEI, ASM

Bieniek, Maciej P
Business: Broadway & 116th St, New York, NY 10027
Position: Prof & Dept Chrmn. *Employer:* Columbia Univ. *Education:* DSc/Appl Mech/Gdansk Inst of Tech; MS/CE/Gdansk Inst of Tech. *Born:* 1/5/27. Native of Vilno, Poland. From 1955-1958 Assoc Prof of Struct Mechanics at Gdansk Inst of Tech. Assoc Prof at Columbia Univ from 1959 to 1963; 1963-1969, Prof at Univ of S CA; 1969-present, Prof of Civ Engg and currently Dept Chrmn at Columbia Univ. Teaching and res in appl mechanics, and struct analysis and design. Reg PE. *Society Aff:* ASCE, AIAA, IABSE.

Bienstock, Daniel
Business: P.O. Box 10940, Pittsburgh, PA 15236
Position: Deputy Div Dir, Proj & Tech Information Mgmt *Employer:* US Dept of Energy. *Education:* MS/Chem Eng/Carnegie Mellon; MS/Chemistry/NYU; BA/Chemistry/Brooklyn College. *Born:* 12/1/17. Deputy Div Dir, Pittsubrg Energy Tech Ctr. US Dept of Energy. Active in the fields of combustion, magnetohydrodynamics, air pollution & synthetic fuels. With 130 publications & patents in these areas. Received the Meritorous Service Awards from the US Dept of Interior & the US Energy Res & Dev Admin & the Vermeil Award (Medal) from the Societe D Encouragement Pour La Rocherche Et L Invention (Paris France), the McAfee Award, Amer Inst of Chem Engrs and the Hornfleck Award, Pennsylvania Soc of Professional Engrs. *Society Aff:* AIChE, NSPE, ASME, APCA, ACS, AAEE

Bierbach, Edward R
Business: 78 Fairway Lane, Littleton, CO 80123
Position: President. *Employer:* Bierbach Cons Engrs Inc. *Education:* BSCE/-/Univ of Denver. *Born:* Aug 1922. BSCE Univ of Denver 1950. Native of Colorado. Struct Engr Phillips Chemical Co 1950-53; Milo S Ketchum 1953-58. Formed own practice of Struct Engrg 1958, performing struct engrg on buildings of all types. Recent awards: Rocky Mountain ACI Regional Award for concrete building design 1975, Colorado Masonry Assn for reinforced brick high rise building design 1974. V Pres CEC/Colo & State Commissioner for CCE/Colo. Enjoy sports.

Bierbaum, J A
Business: 7787 S Perry Park Blvd, Larkspur, CO 80118
Position: Consultant (Self-employed) *Education:* BS/ChE/Northwestern Univ; MS/ChE/Northwestern Univ *Born:* 6/29/24. Native of Golf, IL. Served in US Navy 1942-45. Proj Engr for Am Oil Co 1948-53. Sales Engr for Universal Oil Products Co 1953-56. Technical Dir for Natl Cooperative Refinery Assn 1956-58. VP & Treasurer for Gen Carbon & Chem Corp 1958-61. Consultant 1962-64. VP of Mfg for Midland Cooperatives, Inc 1964- 72. Sr Vp for Tosco Corp 1972-77. Pres & Chief Exec Officer for Gary Energy Corp 1977-79. Pres & Chief Exec. Officer for US Ethanol Corp 1980-83. Currently Consultant in Petrol Industry & Met Carbon. Enjoy golf & sports. *Society Aff:* AIChE

Bierlein, John C
Home: 5481 Vincent Trail, Washington, MI 48094
Position: Staff Research Engr. *Employer:* Res Labs - General Motors Corp. *Education:* PhD/Engg/Univ of WI; MBA/Finance, Economics & Mgmt/Univ of Detroit; MS/Engg/Univ of WI; BS/Engg/MI State Univ. *Born:* 11/13/36. 1959-65. Engrg Instr at U of WI and Res Ast conducting res on $Na(2)SiO(3)$ sys, & fatigue mechanisms; since 1965 conducted res at GM Res Labs on material res for bearings & elastomeric seals, failure analysis, scoring & embedability, and friction of materials in sliding contact under marginal lubrication conditions for numerous engine

Bierlein, John C (Continued)
components. Sev pats and pubs. Cover article in Sci American on 'The Journal Bearing', First Place Award for Outstanding Tech Paper in 1979 Triple Engrg Show & Conf, Outstanding Sr Metallurgical Engr. Award, SAE Excellence in Oral Presentation Award, SAE Arch T. Colwell Merit Award, Reg P E in MI. Married with 2 children. Enjoys model railroading and dollhouses & folk and square dancing. Currently mbr: ASM, TMS of AIME, SAE, Sigma Xi, Alpha Sigma Mu, Phi Lambda Tau, Beta Gamma Sigma. Presently Coun Mbr of Mech Working and Forming Tech Div of ASM, SAE Non-ferrous Metals Comm. Soc Aff: ASM, AIME, SAE, Sigma Xi, Alpha Sigma Mu, Beta Gamma Sigma. *Society Aff:* ASM, AIME/TMS, SAE.

Biesenberger, Joseph A
Home: 14 King Ct, Wayne, NJ 07470
Position: Prof *Employer:* Stevens Inst of Tech. *Education:* PhD/Chem Engg/Princeton Univ; MSE/Polymers/Princeton Univ; BS/Chem Engg/NJ Inst of Tech. *Born:* 11/23/35. With Stevens Inst of Tech since 1963; Asst Prof, 1963-68; Assoc Prof, 1968- 71; Prof of Chem Engrg, 1971-present; Head of Dept of Chemistry and Chem Engrg, 1971-1979; Mbr, Editorial Advisory Bd of *Polymer Engineering and* Science journal; Editor of Polymer Process Engrg journal; co-author of book, *Principles of Polymerization Engrg*; Editor of book, *Devolatilization of Polymers*; author or co-author of more than 100 research papers; conslt to indus on polymer process engrg. Hobby is sailing. *Society Aff:* AIChE, SPE.

Bigda, Richard J
Home: 6732 So Columbia Avenue, Tulsa, OK 74136
Position: President, Owner. *Employer:* Richard J Bigda & Assocs. *Education:* BS/Ch.E/Wayne State Univ; MS/Ch.E./Univ of MI; MS/Indi Mgt/Univ of MI *Born:* May 1930 Detroit Mich. Formerly res engr with Shell Chemical Corp & a refinery engr wth Exxon; 2 yrs with the Army Chem Corps in Alaska; Tech Mgr Wyandotte Chemicals Internatl; Asst to Pres Chemicals, Helmerich & Payne Inc; Chemical Market & Corp Dev Cons since 1967. Mbr: AIChE, Amer Chem Soc. AAAS, OSPE, NSPE, PE in TX, OK. Republican St Ctteman 1973-77. Speaker for Natl AIChE & Natl Assn of Dev Orgs, Amer Soc of Military Engrs & other engrg socs. Served on Governor's study & internatl trade cttes. Chairman, Tulsa Chapter (Downtown) OSPE; Emiment Engineer, Tau BETA PI; Represenative for Technology Exchange. U.S. -Czechoslovakia; CEO Technotreat Corp. Tulsa; Author *Used Oil Re-Refining: Collection-Technology- Economics*. Copyrighted 1980; Author of 29 technical publications; Advisory member of Indian Dept Committe of Philbrook Art Museum. Tulsa; President of Philbrook Friends of Amer Indian Art. Author: The Technobyte Microcomputer Prog Compendium for Chem Engrg & Chemistry. Author: Speciality Chem Opportunities copyright 1983. *Society Aff:* AIChE, ACS, NSPE, TBII, OSPE

Bigelow, Robert O
Business: 25 Research Dr, Westborough, MA 01582
Position: VP. *Employer:* New England Electric System. *Education:* MS/Elec Engg/MIT; BS/Elec Engg/MIT. *Born:* March 1926. BS & MS 1950. Joined New England Electric System 1950 in field of power sys planning & computer applications; appointed Ch Elec Engr 1969, Asst Ch Engr 1970 & Dir of Planning & Power Supply 1973 with respon for planning of power supply & generation & transmission facilities. VP 1977. Mbr of Planning Ctte New England Power Pool since its inception 1968, served as V Chmn 1971 & 1972 & Chmn 1973 & 1974. Fellow IEEE. Mbr of Sys Planning Ctte- Edison Elec Inst & Chrmn 1985-87. *Society Aff:* IEEE.

Bigelow, Wilbur C
Business: Materials Engr, Univ. of Michigan, Ann Arbor, MI 48109-2136
Position: Prof *Employer:* University of Michigan *Education:* Ph.D./Physical Chem/Univ. of Michigan; M.S./Chem/Univ. of Michigan; B.S. /Biological Chem/Pennsylvania State Univ *Born:* 3/18/23 Born in Pennsylvania. Served at U.S. Naval Res. Labs. during WWII. Res. Assoc at U. of Mich Until joining the faculty in 1955. Prof of Materials Engrg, 1962. Presentrly Undergraduate Program Advisor for Materials and Metallurgical Engrg and Dir of the Electron Microscopy and Electron Mircobeam Analysis Labs. *Society Aff:* EMSA, MAS, ACS

Biggs, David T
Business: 291 River St, Troy, NY 12180
Position: VP. *Employer:* Ryan-Biggs Associates, PC *Education:* Master of Engg/Civil Engg/ Rensselaer Polytec Inst; BS/Civil Engg/Rensselear Polytech Inst. *Born:* 3/28/50. Mr Biggs is a structural engr specializing in masonry design, comm and industrial const, and bridge design. Lic as an engr in PA, NY, NH, MA, MI and VT, he is a mbr of Amer Concrete Inst, Amer Soc of Civil Engg, NYS Society of Prof Engg, and Consulting Engg Council of NY. Mr Biggs grad from RPI in 1972 with a BS and in 1973 with a ME in Civil Engg. He received the 1979 Daniel W Mead Prize, & also the 1983 Edmund Friedman Young Engr Award from ASCE. *Society Aff:* ASCE, ACI, NSPE, TMS, ACEC.

Bigley, Paul R
Business: 99 N Front St, Columbus, OH 43215
Position: Sr VP. *Employer:* Columbia Gas of OH (Columbia Gas Distribution Companies) *Education:* BS/Chem E/OH State Univ. *Born:* 12/11/29. Native of Marion, OH. Served with Weather Service, USAF 1948-52. With Battelle Meml Inst, specializing in titanium research. With Columbia Gas since 1958. Became VP & Chf Engr in 1970, respon for facility planning. In 1974 became Sr VP for all operations, engg, mkting and gas supply of Distribution Cos. *Society Aff:* NSPE, OSPE.

Bilbo, Bill C
Business: 402 Rowell Bldg, Fresno, CA 93721
Position: President. *Employer:* Sortor Engrg Inc dba Resources Intl. *Education:* BS/Ag Engrg/UC - Berkeley/Davis *Born:* 12/5/27. , TX US Navy. Prof background includes ranch and shop supvr, personnel training and supv, safety educ, equip evaluation and selection--Anderson-Clayton Vista del Llano Farms; Farm Adv, Kern Co., CA; private farming; Proj Dir USDA- CSUF fresh fruit mech harvesting res; joined Sortor Engrg as agri engr cons, 1969; assumed presidency, 1972; domestic and foreign conslt activities include investigative and feasibility studies, design, implementation and mgmt, agri dev and reclamation projs for private indus, public utilities, govt entities, including over 20 World Bank Missions-Middle East, India, Far East, and S Amer. ASAE Pacific Region Profl Reg Chrmn; instrumental in helping to obtain prof status for agri Engrs in CA, 1975; Engr of the Yr, Pacific Region--ASAE, 1976. Favorite sport/fishing. *Society Aff:* ASAE, ASFMRA, NSPE, IFA

Bilello, John C
Home: 32 Gaul Road South, Setauket, NY 11733
Position: Prof *Employer:* State Univ of NY *Education:* PhD/Met Engr/Univ of IL (Urbana); MS/Mat Sci/NYU; B Met E/Met Engr/NYU *Born:* 10/15/38 Visiting NATO Fellow, Politecnico di Milano, Center for Nuclear Engrg "E. Fermi, 1973, Visiting Prof, same 1974, Dir Synchrotron Topography Project, National Synchrotron Light Source, Brookhaven National Labs, 1980-. Research interests, Synchrotron X-ray diffraction topography, defects in solids, mechanical behavior, fracture mechanics, defects in electronic materials, Dean, Coll of Engrg and Applied Sciences, SUNY Stony Brook 1977-81. *Society Aff:* ASM, APS, AIME

Biles, William E
Business: Dept of Industrial Engrg, Baton Rouge, LA 70803
Position: Prof & Chairman *Employer:* Louisiana State Univ *Education:* PhD/Ind Engg/VPI & SU; MSE/Ind Engg/Univ of AL; BSChE/Chem. Engg/Auburn Univ. *Born:* 6/28/38. A native of Mobile, AL. Served as a Lt in the US Army Signal Corps in 1961- 1962. From 1962 to 1966, a Product & Process Dev Engr with Union Carbide Corp's Advanced Mtls Lab in Lawrenceburg, TN. From 1966 to 1969, a Process Engr for Thiokol Chem Corp in Huntsville, AL. From 1971 through 1978, on the faculty of Aerospace & Mech Engg at the Univ of Notre Dame. From 1979 to 1981, Professor and Head of Industrial Engg, PA State Univ. Dir of the Operations Res Div of IIE from 1978-1980. Dir of Res for IIE from 1983-86, mbr of the Engrg

Biles, William E (Continued)
Accreditation Commission of ABET from 1983 to present. Specialities are simulation, optimization, response surface methodology, mtl handling design, and mftg automation. *Society Aff:* IIE, SME, TIMS.

Biles, William R
Business: 6161 Savoy, Houston, TX 77036
Position: Pres. *Employer:* Biles & Assoc. *Education:* PhD/ChE/Penn State; MS/ChE/Penn State; BS/Che/Univ Cincinnati. *Born:* 12/20/23. William R Biles was born in Cincinnati, OH. His exper includes position as Chem Prod Suprv, American Cyanamid Company; Supervisor Engg Res, Supervisor Math & Computing, Shell Oil; Exchange Scientist, Royal Dutch/Shell; Sr Process Engg, Mgr Process Computer Control, Shell; VP, Davis Computer Systems. Since 1970 Pres and founder of Biles & Assoc, engrs and contractors specializing in process computer control and factory automation. He has worked in the filds of Distillation, Catalytic Cracking, Molten Salt Processes, Simulation and Optimization, Process Control and Project Management of Computer Control Projects. Currently he consults with sr management on computer control policy and organization. *Society Aff:* AIChE, IEEE, ISA, ACS, SME.

Billet, Arthur B
Home: 2322 Via Siena, La Jolla, CA 92037
Position: Pres *Employer:* A.B. Billet and Associates *Education:* BS/Aero Engr/Univ of MI. *Born:* 5/20/20. Sperry-Vickers 1940-67 Exec Engr dev aircraft hydraulic components; the Boeing Co 1968-70 Res Engr for 747 hydraulic sys; Teledyne Ryan Aeronautical 3 yrs in Sys Engrg of remotely piloted vehicles; Rohr Industries Ind, 1973-81 Sr Engr in turbine engine, rail & bus vehicle dev. Supervisor all fluid power systems 3000 ton Surface Effect Ship. HYDRO Products Principal Engr. Pres A.B. Billet and Associates. Advanced Marine Vehicles Soc of Automotive Engrs 34 year mbr & mbr of Natl Bd of Dir. Elected Fellow Member 1983. Headed expeditions to Greenland, North Africa & Panama for SAE/USAF for aircraft equip retrieval in environmental studies. Amer Inst for Aeronautics & Astronautics Elected Fellow Mbr 1984 & 23 year mbr. Inst of Environmental Scis Natl Pres 1961-62, 29 year mbr & 1st Fellow Mbr. Pres Amer Defense Preparedness Assn San Diego Sect. Author of 84 tech papers given throughout the world. Cons to Us Dept of Commerce for world test equip export needs. Licensed pilot since 1936. *Society Aff:* SAE, AIAA, IES, ADPA, MTS.

Billett, Robert L
Home: 5510 Laurette Street, Torrance, CA 90503
Position: Section Hd. *Employer:* TRW Defense & Space Systems Group. *Education:* Bs/Packaging/Michigan State Univ. *Born:* June 1939. BS in Packaging Michigan St Univ 1962. Res Engr in Packaging, Material Handling & Transportability Engrg for Space & Autonetics Divs of Rockwell Internatl; with TRW since 1968, presently Sect Head respon for Packaging Handling & Transportability res & design activities for the Defense & Space Sys Group. Active in Soc of Packaging & Handling Engrs, holding many offices, incl Pres of Los Angeles Chap. Mbr of Sigma Pi Epsilon, honorary fraternity of SPHE. Selected as Mbr-of-the-Year for Los Angeles Chap 1973. *Society Aff:* SPHE

Billig, Frederick S
Business: Johns Hopkins Road, Laurel, MD 20707
Position: Chief Scientist, Assoc Supervisor Aeronautics Div. *Employer:* Johns Hopkins Univ Appl Physics Lab. *Education:* BE/Mech Eng/Johns Hopkins Univ; MS/Mech Eng/Univ of MD; PhD/Mech Engg/Univ of MD. *Born:* Feb 28, 1933. Mbr of Principal Prof Staff Chief Scientist, Assoc. Supr. Aeronautics Dept. APL/JHU & lecturer in Aerospace Engrg Univ of Maryland.Fellow Vice President Member Services 1981-83 & Dir of Amer Inst of Aeronautics & Astronautics. Received Disting Young Sci Award from Md Acad of Sci 1966 and the Silver Medal from the Combustion Inst 1970. *Society Aff:* AIAA, Comb Inst

Billings, Bruce H
Business: 7303 North Marina Pacifica Dr, Long Beach, CA 90803
Position: Dir of Research *Employer:* Intl Tech Assocs, Inc *Education:* PhD/Physics/Johns Hopkins Univ; MA/Physics/Harvard Univ; BA/Physics/Harvard Univ *Born:* 07/06/15 Dr. Billings was pres, Thagard Research Corp; VP, Aerospace Corp; Special Asst (Sci and Tech) to US Ambassador, Taiwan; VP and Gen Mgr, Baird Corp; Asst Dir, Defense Research and Engrg, US Dept of Defense; Physicist, Polaroid Corp. Profl activities include pres of Optical Soc of Amer, VP of Intl Commission for Optics, UN Sci Advisory Committee Mbr, Secy and Fellow of the American Acad of Arts and Scis, and mbr of Air Force Advisory Bd. He holds seven patents and has written numerous papers. *Society Aff:* OSA, APS, AIP, SPIE.

Billingsley, David L
Business: 2000 Fairfield Ave, Shreveport, LA 71104
Position: Pres *Employer:* Billingsley Engr Co. *Education:* MS/Petrol Engr/Univ of TX; BS/Petrol Engr/Univ of TX. *Born:* 11/8/35. Native of Houston, TX. Formed Billingsley Engr Co in June, 1963. Reg PE in states of AR, LA, OK, & TX. Mbr of Tau Beta Phi and Sigma Lamma Epsilon, honorary fraternities. *Society Aff:* AIME, NSPE, LES, SPEE.

Billington, David P
Business: Dept Civil Engineering, Princeton Univ, Princeton, NJ 08544
Position: Prof *Employer:* Princeton Univ *Education:* BS/E/Princeton Univ *Born:* 06/01/27 Fulbright Fellowship, Belgium, in Structural Engrg of Univ of Louvain 1950- 51 and at Univ of Ghent 1951-52. Structural designer Roberts and Schaefer Co, New York City, 1952-60; proj engr for piers, hangars, utility structures, launch facilities. At Princeton Univ as Visiting Lecturer 1958-60, Assoc Prof 1960-64, Prof 1964-present; research in thin shell concrete structures, structural dynamics, thin plates, and in the history and aesthetics of structures. Co-dir of the prog Humanistic Studies in Engrg. Teaching courses in concrete structures, structural analysis and thin shells for engrs, in structural design for architects and in the history and aesthetics of structures for liberal arts students. Author of Thin Shell Concrete Structures, 2nd Ed, 1982 and Robert Maillart's Bridges: The Art of Engrg, 1979 (winner of the Dexter Prize in the History of Tech). The Tower & The Bridge, 1983, mbr of the NAE. Married to Phyllis Bergquist, six children. Episcopalian and Republican. *Society Aff:* ASCE, ACI, PCI, IABSE, IASS, ASEE, SHOT

Billington, Ted F
Business: PO Box 422, 1203 Johnson Ave, Murray, KY 42071
Position: Reg PE/Owner (Self-employed) *Education:* BS/Civil Eng/ Univ of KY. *Born:* 11/22/38. Grad of Univ of KY, August, 1961; Asst Res Eng, KY Dept of Trans, Sept 1961 to Oct 1962; Chief Structural Eng for Lee Potter Smith & Assocs, Arch, Paducah, KY, 1962-67; Established private practice consltg firm in Murray, KY, June, 1967. Have served as prin and owner of firm since its establishment; Specialty field of practice, architecture, structural-earthquake eng, land surveying, Reg PE-KY, TN, ID, IL; Reg Land Surveyor-KY, TN; State Chrmn NSPE/KSPE/PEPP 1978, planning. *Society Aff:* NSPE, ASCE, ACI, KSPE.

Billinton, Roy
Business: Electrical Engg Dept, Saskatoon, Saskatchewan S7N 0W0 Canada
Position: Associate Dean, Graduate Studies Research & Extension. College of Engg. *Employer:* Univ of Saskatchewan. *Education:* DSc/-/Univ of Saskatchewan; PhD/Elect Engg/Univ of Saskatchewan; MSc/Elect Engg/Univ of Manitoba; BSc/Elect Engg/Univ of Manitoba. *Born:* 9/14/35. Born in Leeds, England - Immigrated to Canada in 1952. Worked in Operations and Planning Depts of Manitoba Hydro prior to joining the Univ of Saskatchewan. Author and co-author of five books on Reliability and Power System Reliability Evaluation. Author and co-author of over 280 papers on the subject of Power System Analysis, Operations and Reliability Evaluation. Pres of the Independent Consulting Co, Power comp Assoc. Fellow of IEEE, RSC, & EIC. PE, Province of Saskatchewan. *Society Aff:* IEEE, CEA, EIC, CSEE, RSC.

Bills, Glenn W
Home: 1854 Crone Ave, Anaheim, CA 92804
Position: Mbr of Technical Staff *Employer:* Hollands Assocs *Education:* MSE/Elec Engg/UCLA; BSEE/Elec Engg/Univ of WA. *Born:* 8/15/15. Born in Tacoma, WA. Taught at Univ of WA in 1938-39. Was Electrical Engr at BPA from 1939-55. Was at North American Aviation from 1955-71. Set up consulting practice in 1971 and consulted with Boeing Computer Services, General Atomic Co, Rockwell Interntl and Systems Dev Corp. Am now teaching at Univ of MO at Rolla, MO. Have one patent and 20 published papers. Am life sr mbr of IEEE, Tau Beta Pi Mbr and a Reg PE in State of CA. I perform research in application of computers and Control Theory to Electric Power Systems and provide consltg services to electric utilities. *Society Aff:* IEEE.

Binder, Sol
Home: 7652 Wyndale Ave, Philadelphia, PA 19151
Position: Chief-Dynamic Sys & Structures Design *Employer:* Boeing Helicopter Co. *Education:* BME/Mech Engg/City College of NY. *Born:* March 25, 1928. Mbr Pi Tau Sigma, Natl Honorary Mech Engrg Fraternity. Grad work at Drexel Univ. Joined the Piasecki Helicopter Co (subsequently the Boeing Helicopter Co) 1951 as a stress engr; appointed Ch Transmission Stress Engr 1960; CH-47 Asst Proj Engr 1968; assumed current pos as Chief - Dynamic System Design 1976; also respon for providing support for the Boeing Co on wind energy & aircraft & hydrofoil drive sys. Have co-authored SAE articles on Drive Sys Design & a chapter in the McGraw-Hill Gear Design Handbook. *Society Aff:* AHS, ΠΤΣ, AAAA.

Bindseil, Edwin R
Home: 333 Fernwood Lane, Erie, PA 16505
Position: Sr. Vice President. *Employer:* American Sterilizer Co. *Education:* MBA/Bus/Harvard Univ; BChE/Chem Engr/Univ of Detroit. *Born:* 10/5/30. Served with USAF 1954-56. 1953-58 Chem Engr Esso Standard Oil; 1958- American Sterilizer Co in following capacities: Marketing Engr, Prod Mgr, Market Planning Dir, VP R&D, Genl Mgr, Industrial Div, VP Marketing, Group VP, Sr VP Corp Dev & Planning, Sr VP. Res & Dev. *Society Aff:* AIChE, ACS, AAMI.

Bingeman, Jonas B
Business: PO Box 1090, Hightstown, NJ 08520
Position: Dir Corp Engg. *Employer:* NL Industries, Inc. *Education:* PhD/CE/LSU; MS/CE/MN; MS/Phys Chem/Detroit; BSc/CE/Queen's. *Born:* 2/21/25. S River, Ont, Canada 1925. Came to US 1946. Naturalized 1955. Undergrad ChE Queen's Univ, Canada MS Phys Chem Univ of Detroit. MS ChE Univ of MN. PhD ChE LA State. Married Agnes Kathleen Macdonald 1948. 4 children: Grant William, Leslie Kathleen, John Macdonald, Claire Eileen. Ethyl Corp 1950-1960, process engr, economic analyst, spvsr process design. Rexall Chem proj mgr, asst plant mgr, vp technical Fibers Div, asst to corp pres, dir corp R&D. NL Industries 1974-, Dir of Corp Engg. Mbr Advisory Council VA Tech Services; mbr natl Defense Exec Reserve, US Dept Commerce; served with Canadian Navy 1944-46; Ind rep Engr Joint Council; Am Inst Chem Engrs: chrmn Tidewater Sec, chrmn tech symposia Los Angeles, 1964, Tidewater 1968; VA Chamber of Commerce: chrmn education committee. Clubs: Mendham, (NJ) golf & tennis; Chemists; Canadian Club of NY; Metropolitan (NY). *Society Aff:* AIChE, EJC.

Binger, Wilson V
Business: 655 Third Ave, New York, NY 10017
Position: Consulting Engr. *Employer:* Self. *Education:* MS/Civil Engg/Harvard Univ; AB/Engg Sci/Harvard Univ. *Born:* 2/28/17. Reside Chappaqua NY. Prior to WWII soils & foundation engr on Chesapeak & Delaware Canal & on Third Locks Proj of the Panama Canal; during war with Army Corps of Engrs: 2nd Lt & Instr at Robert College Istanbul Turkey. After war, soils & foundation engr on dams in Venezuela & Argentina - on Sea Level Canal Proj of the Panama Canal, on Missouri River dams for the Corps of Engrs; const contractor. Joined TAMS 1952, Partner 1962, Chmn 1975, Retired 1985. Natl Acad of Engg. Exec Comm 1965-69 USCOLD, Secy 1961-79 USCOLD; VP 1978-81 ICOLD; Pres 1973 AICE, V Pres 1973-75 ACEC; Exec Comm & Treas 1976-80 FIDIC, VP 1980-81 FIDIC, Pres 1980-82 FIDIC; Dir IRF 1975-1983, Business Council for International Understanding - Director 1982-85; Regional Plan Association (N.Y.) - Director 1983-; Trustee 1970-Present, Robert College of Istanbul. Foreign Member, Fellowship of Engineering (U.K.). Received Steinmetz Award of Consulting Engineer Magazine 1985. *Society Aff:* ASCE, NSPE, ACEC, ICE, NAE.

Bing-Wo, Reginald
Business: 2220 12th Ave - Suite 220, Regina, Saskatchewan S4P OM8 Canada
Position: Registrar & Secretary-Treasurer. *Employer:* Assn of Prof Engrs of Saskatchewan. *Education:* BSc/Civil Engg/Univ of Saskatchewan. *Born:* Feb 1921 Lethbridge Alberta Canada. Design Engr with Canada Dept of Agri, P F R A 1943-67; since 1968 Registrar & Secretary-Treasurer Assn of Prof Engrs of Saskatchewan. Mbr of Assns Council 1949-51. Genl Manager & Chief Exec Officer, Assoc of Consulting Engrs of Saskatchewon. Prof affiliations: Made a Fellow Jan 28, 1980 The Engrg Inst of Canada, The Canadian Soc for C E, The Military Engrs Assn of Canada. Also now an International Member, The Society of American Military Engineers (SAME) Military: LCdr Royal Canadian Naval Reserve, awarded the Canadian Forces Decoration. Fraternal: Masonic & Shriners. *Society Aff:* FEIC, MCSCE, MEAC.

Binney, Stephen E
Business: Department of Nuclear Engineering, Corvallis, OR 97331
Position: Associate Professor *Employer:* Oregon State University *Education:* PhD/Nuclear Engrg/U of CA, Berkeley; MS/Nuclear Engrg/U of CA, Berkeley; BS/Engrg Phys/Oregon State U *Born:* 11/19/41 Native of Klamath Falls, OR. Nuclear Engr at NASA-Lewis Res Ctr. Sr Scientist at EG&G, Inc., Las Vegas, involved in aerial radiological measurements. At Oregon State Univ since 1973, currently as an Assoc Prof. Conslt to several prof organizations. Chrmn, ANS Radiation Protection and Shielding Division, 1984-85. Fields of specialization in applications of nuclear instrumentation and techniques, gamma ray and neutron spectrometry, environmental radiation monitoring, and radiation shielding. *Society Aff:* ANS, HPS

Binning, Robert C
Home: 3473 Tall Timber Trail, Dayton, OH 45409
Position: Director Environmental R/D. *Employer:* Monsanto Res Corp. *Education:* PhD/Phys Chem/LA State Univ; MS/Chem/LA State Univ; BS/Chem/LA State Univ. *Born:* Feb 1921 Baton Rouge; s. Francis Henry & Mary Lillian (Romero) Binning. Married Lucille Annette Giese 1944, children: Robert, Janet (Mrs Donald Briedenbach), John, Adele. R/D petrochem & separations Amer Oil Co; with Monsanto Co since 1958; Dir Environ R&D Monsanto Res Corp. Served with USAAF 1940-45, decorated DFC, Air Medal with 2 Oak Leaf Clusters. Pres Dad's Assn Butler Univ 1973; listed in Whos Who in the Midwest 1974; named Fellow AIChE 1975. Contributor of articles to prof pubs. Patentee in field (33 US, 43 ex-US issued). Office: 1515 Nicholas Rd, Dayton, Ohio 45407. *Society Aff:* AIChE, AIC, ADPA, WFS.

Biot, Maurice A
Home: Avenue Paul Hymans 117/34, B-1200, Brussels Belgium
Position: Private Cons. (Self-employed) *Education:* PhD/Aeros/Caltech; D Sc/Math-Phys/Louvain; -/Elect Eng/Louvain; -/Min Eng/Louvain; -/Philosophy/Louvain *Born:* May 1905. Graduate from Louvain Univ, Philosophy 1927. Mining Engrg 1929, Elec Engrg 1930, Doctor of Sci 1931; received PhD in Aeronautics from Caltech 1932. Held professorships at Louvain in Applied Sci, Columbia in Physics, Brown in Applied Math. LCd USNR 1943-46. Cons Shell Dev Co 1946-66 & Cornell Aeronautical Lab Inc 1946-69; Cons Mobil Res, private contractor USAF Office of Sci Res. Author numerous sci papers & 3 books on applied math & mech. Mbr Natl Acad of Engrg, Amer Acad of Arts & Scis & Royal Acad of Belgium. Award Royal Acad (Belgium 1934); Timoshenko Medal 1962; Von Karman Medal 1967; Honoris

Biot, Maurice A (Continued)
causa Liege 1967; Medal of Liege Univ 1969. Hon Fellow AcSOC 1983. *Society Aff:* ASME, AMS, APS, AIAA, SSA, AGU

Birch, John R
Business: 3525 N 124th St, Brookfield, WI 53005
Position: Prin. *Employer:* Birch-Grisa-Phillips, Inc. *Education:* BS/Civ Engg/IA State Univ. *Born:* 11/25/22. Surveyor and dynamiter, NOB, Bermuda, 1942. Airplane Commander, B-29, WWII, Capt. Engr, Jensen & Johnson, 1948-51. Dir of Public Works, Whitewater and Beaver Dam, WI, 1951-58. Dir of Engg, Megal Dev Corp, Contractor and Developer, 1958-61. Prin, Birch-Grisa-Phillips, Inc, Architects-Engrs-Planners, 1961 to present. Engaged in bldg design in WI and other states. Principal planner for maj industrial parks. Prin, BGP/CM, construction mgrs. Partner, real estate investment cos. Past chrmn, League of WI Municipalities Engg & Public Works Sec. Past pres, Lions Club. Mbr, State bldg code committees. Registered engr, architect and surveyor. Registered engr, architect and surveyor. Commercial pilot. Commercial pilot. *Society Aff:* NSPE, AIA.

Bird, George T
Business: PO 37, Vienna, VA 22180
Position: Pres. *Employer:* Bird Engg - Res Assoc Inc. *Education:* BS/EE/TX A&M. *Born:* 4/6/16. Native of TX. Formerly Assoc Prof EE Univ of Houston, served with USAF in WWII (Lt Col, Reserve), staff Engr ARINC Res Corp 52-60. Founded-Pres Bird Engg- Res Assoc Inc 1960 to current. Author of several Engg Handbooks on System Reliability and Maintainability published by GPO for US Navy, mbr of NSPE, IEEE, and others. BSEE TX A&M (1940); Reg PE in VA, MD, DC, TX. Amateur Radio (W4DRQ) as hobby. Married to Katie M McClure, two daughters, Evelyn and Carolyn, and one son George T Jr. *Society Aff:* NSPE, IEEE.

Bird, R Byron
Business: Chem Engr Dept., Univ of Wisc, 1415 Johnson Dr, Madison, WI 53706-1691
Position: Vilas Res Prof. *Employer:* Univ of WI. *Education:* PhD/Chemistry/Univ of WI; BS/ChE/Univ of IL *Born:* 2/5/24. Co-author of books: *Molecular Theory of Gases and Liquids* (Hirschfelder, Curtiss, Bird), *Transport Phenomena* (Bird, Stewart, Lightfoot), *Een Goed Begin* (Bird, Shetter), *Comprehending Technical Japanese* (Daub, Bird, Inoue), *Dynamics of Polymeric Liquids*, Vol. 1, Fluid Mechanics (Bird, Armstrong, Hassager), *Dynamics of Polymeric Liquids*, Vol. 2, Kinetic Theory (Bird, Curtiss, Armstrong, Hassager), *Reading Dutch* (Shetter, Bird). Fulbright fellow, Univ. of Amsterdam, The Netherlands (1950-51); Fulbirght professor & Guggenheim fellow, Tech. Univ. of Delft (1958); Fulbright professor, Kyoto and Nagoya Universities (1962-1963). Honorary doctorates: Lehigh U., Washington U., Tech. Univ. of Delft, Clarkson College, Colorado School of Mines. *Society Aff:* ASR, AIChE, ACS, APS, KIVI, NNV, NAE, Amer Acad. of Arts and Sci, AANS, KNAW, Wisc. Academy of Sciences, Arts, & Letters.

Birdsall, Blair
Business: c/o Steinman, 45 Broadway Atrium, New York, NY 10006
Position: Partner. *Employer:* Steinman Boynton Gronquist & Birdsall. *Education:* BSCE/CE/Princeton Univ; CE/CE/Princeton Univ. *Born:* May 1907 Newark N J. BSCE 1929, CE 1930 Princeton Univ; Master's Thesis: First Struct Model of Suspension Bridge. Roebling Bridge & Tramway Div CF&I Steel Corp 1934-65, Ch Engr & Genl Mgr 1952-65; Design, manufacture & const, wire supported struct. Charter mbr & Passenger Tramway Standards-Drafter, ANSI Ctte B77 Passenger Tramway Safety Standard; Assoc 1965-68, Partner 1968- of present firm. Personal specialties incl wire, structs using wire & mech-struct sys, incl suspension bridges, suspension sys, suspended roofs, freight & passenger tramways, heavy lifting devices & bulk material handling equip. Fellow & Life Mbr ASCE; Founding Chmn Ctte on Contract Admin; Fellow ASME; licensed P E in 16 states. *Society Aff:* ASCE, ASME, ASTM, AISC, IBTTA, IABSE, ASCE, ACEC, NSPE

Birdsall, Charles K
Business: EECS Dept, Univ of Calif, Berkeley, CA 94720
Position: Prof *Employer:* Univ of California *Education:* PhD/EE/Stanford Univ; MS/EE/Univ of MI; BSE/EE/Univ of MI; BSE/Engr Math/Univ of MI *Born:* 11/19/25 From 1951-1959 he worked in industry on microwave amplifiers using electron beams in general admittance walls or slow-wave structures, with 27 patents. In 1959 he joined the Univ of CA with research on high temperature plasmas, studying instabilities, heating and transport, recently adding plasma-assisted materials processing, all using computer simulation with many particles. He has been conslt to LLNL Livermore since 1960, was Miller Research Prof (1963-64), and at Osaka Univ (1966). He was founding chrmn of the Energy and Resources Prog (1972-74). He was at Reading Univ (1976), Nagoya Univ (1981-82) and at CA Inst of Tech (1982). He was founding coordinator for the Physical Electronics and Bioelectronics faculty in his dept. Author of *Election Dynamics of Diode Regions* (with W.B. Bridges, 1966) and *Plasma Physics via Computer Simulation* (with A. B. Langdon, 1985). *Society Aff:* IEEE, APS, AAAS.

Birdsall, J Calvin
Business: Box 218789, Houston, TX 77218
Position: Pres *Employer:* Galvanizing Technologies, Inc. *Education:* BS/ChE/Univ of Texas *Born:* 8/23/39. Native of Corpus Christi TX. Reg P E TX. Assoc with Gulf Oil Chems 1962-70 working with manufacture & end use dev of polymers & new business ventures; Pres of ASKCO Engr 1970-75 spec in mfg of control sys for extrustion & heating processes & design consultation; Pres of All Containers Corp 1975; respon for dev of new tech, design & const of form polystyrene extrustion & forming facil. Leisure interests incl aviation, amateur radio, golf & sailing. Now responsible for design & construction of equip for the galvanizing indus. *Society Aff:* AIChE

Biringer, Paul P
Business: Dept of EE, U of Toronto, 35 St George St, Toronto Ont M5S 1A4 Canada
Position: Professor. *Employer:* Univ of Toronto. *Education:* PhD/Elec Engg/Univ of Toronto; MASc/Elec Engg/Royal Inst of Technology, Stockholm; Dipl Eng/Elec Engg/Techn Univ of Budapest. *Born:* 10/1/24. Hungary. Res Asst & Res Assoc with Royal Inst of Tech Stockholm 1947-52; with Univ of Toronto since 1952, Full Prof since 1965; P Pres, Dir Elec Engrg Consociates Ltd Engrg Cons. Inventor of magnetic & solid st conversion devices. Over 100 pubs & 35 pats. Fellow Ctte IEEE, Same Ad Com Magnetics Soc 1972-86, Same Educ Activities BD 1968-70, Same Chmn Toronto Sect 1964-65, Same Fellow 1970. Centennial Medal (IEEE) 1984. Bd of Specialization Assn of Prof Engrs of Ontario 1973-76, Same Bd of Examiners 1960-68, Same Sons of Martha Medal 1968. V Chmn Bd of Governors of the George Brown College of Applied Arts & Tech 1975-81, Chmn Can Natl Ctte of TC27 (& TC68) of Inter Elec Commission 1971-. Centennial Medal (IEEE) 1984. *Society Aff:* APEO, CIGRE, ASEE, ICER.

Birkeland, Halvard W
Business: 500 So. 336th St, Federal Way, WA 98003
Position: Consulting Engr. *Employer:* ABAM Engrs Inc. *Education:* CE/Structures/Univ of WA; MSCE/Structures/Univ of WA; BSCE/Genl Civil Engg/Univ of WA. *Born:* 1907 Oslo Norway. Reg P E. Exp: Bur Reclam Denver - design of dams, canals 1934-39 & 1946-48; Panama Canal struct design 1939-41; Corps Engrs fortifications design 1942-43; USN Lt, ship repair ashore, damage control afloat 1943-46; Internatl Engrg Co, Ch Design Engr 1948-55; City of Tacoma, Hd Engr 1955-57; ABAM Engrs Inc Tacoma, Cons Engr 1957- . Affiliate Prof Univ of Wash 1963-80. Author of numerous tech articles. Mbr Panama Canal Zone Civic Council 1940-41; Tacoma Bd of Buildings Appeals 1964- ; Wash Bd of Registr 1970-80. *Society Aff:* ASCE, ACI.

Birkimer, Donald L
Home: 16 Bob White Trail, Gales Ferry, CT 06335
Position: Tech Dir. *Employer:* Coast Guard Res & Dev Center. *Education:* PMD/Bus Adm/Harvard Grad Sch of Bus Adm; PhD/Structures/Appl Mech/Univ of Cincinnati; MS/Structures/Appl Mech/Univ of Cincinnati; BSCE/Civil Engg/OH Univ. *Born:* 9/6/41. in New Lexington, OH. Civil Engg with Wright Patterson Air Force Base. Res civil engg with US Army Corps of Engrs. Res Structural Engg with Battelle Mem Inst. Acting Branch Chief, Construction Materials Branch, US Army construction engg res lag. Advanced from branch chief, materials branch, to Asst Dept Hd, Advanced Weapons Dept, Naval Surface Weapons Center. Became first tech dir of Coast Guard Res & Dev Center in 1975. Winner of ACI WASON medal, elected to Connecticut Academy of Science and Engineering. *Society Aff:* ASTM, AAAS, TIMS, AMA, ASCE.

Birks, Neil
Business: 848 Benedum Hall, Pittsburgh, PA 15267
Position: Prof *Employer:* Univ. of Pittsburgh *Education:* PhD/Met/Univ of Sheffield; BMet (Hon)/Met/Univ of Sheffield. *Born:* 10/16/35 Native of Sheffield England, NATO Res Fellow at Max Planck Inst for Physical Chem Gottingen W. Germany 1960-62. Res Investigator with the United Steel Companies Rotherham UK. 1962-64. Senior Lecturer in Metallurgy Sheffield University 1964-78. Professer of Mat. Sci & Eng University of Pittsburgh 1978- . Consultant in Materials. Special interests High Temp Deterioration of Metals, physical chem of process metallurgy. *Society Aff:* ASM, ECS, IOM, SMEA.

Birnbaum, Howard K
Business: Materials Research Lab, 104 S Goodwin Ave, University of Illinois, Urbana, IL 61801
Position: Prof *Employer:* Univ of Illinois *Education:* PhD/Metallurgy/Univ of IL; MS/Metallurgy/Columbia Univ; BS/Metallurgy/Columbia Univ *Born:* 10/18/32 Born and raised in Brooklyn, NY and educated in the public sch sys. Attended Brooklyn Technical High Sch followed by Univ educ at Columbia Univ. and Univ. of Illinois. He has held positions at the Univ of Chicago (1958-1961) and at the University of IL (1961-present). Director of the Materials Res. Lab. Research interests include lattice defect behavior, H in solids, hydrogen embrittlement, hydride properties, diffusion and application of acoustics to the study of the properties of solids. Conslt for several corps and government agencies. *Society Aff:* AIME, ASM, APS, AAAS, ΣΞ

Birnbaum, John D
Home: 2305 Aspen Lane N E, Cedar Rapids, IA 52402
Position: Manager. *Employer:* Collins Radio Grp, Rockwell Intl. *Born:* Aug 1926. Reg P E in Michigan. Educ: Michigan St Univ ME; Univ of Michigan Indus Engrg. Served with USAF during WWII. With Collins Radio Group Cedar Rapids Iowa, div of Rockwell Internatl since 1961; presently Mgr Internatl Service Ctr: Mgr of Indus Engrg, Mgr of Indirect Measurement Prog. Has been pioneer in the field of Indirect Work Measurement. Natl Region XI V Pres IIE 1969-71; Engr of the Year Award Region XI IIE 1973; Fellow Award IIE 1976. Active in community affairs.

Birnstiel, Charles
Business: 108-18 Queens Blvd, Forest Hills, NY 11375
Position: Consultant *Employer:* Self (Prin) *Education:* Eng. ScD./Strutures/NYU; M. C.E./Structures/NYU; B./C.E. /NYU *Born:* 12/6/79 Since 1974 consulting structural engr. Previously Prof of Civil Engrg at Polytechnic Inst of New York, 1973-74; Instructor to Prof of Civil Engrg at NY Univ, 1954-1973. Offer structural engrg services for bldgs and bridges, including movable structures, and structural mechanics research. *Society Aff:* ASCE, ASEE, IABSE, ASTM

Biro, Joseph E
Business: 60 W. Broad St, Bethlehem, PA 18018
Position: Pres *Employer:* J. E. Biro & Assoc, Inc *Education:* BS/AE/Lafayette Coll *Born:* 06/04/25 Biro has had 38 yrs experience in consltg work: 34 as principal with his co. He is a PE with natl registration since 1962; registered in PA since 1953, and is also registered in NJ, NY, GA, MA, MS, AL, FL, DE, MD, and VA. In addition to his business activities, Biro has been adjunct prof of mech-electrical engrg at Northampton County Area Community Coll and a conslt to the PSAE section of the American Inst of Architects in Washington, DC. Biro combines his interest in golf with the love of travel and has played in Scotland, Portugal, Hawaii, as well as throughout the US. Charter Member of the Veterans of the Battle of the Bulge (WWII), Director of the 94th Infantry Division Assoc. *Society Aff:* ASHRAE, ASPE, IFMA, HEA

Biron, George R
Business: 1822 Drew St, Clearwater, FL 33575
Position: Vice Pres *Employer:* McFarland-Johnson Engineers, Inc *Education:* B/CE/Catholic Univ of AM *Born:* 03/26/25 Born in Hartford, CT. Served with US Army in Europe 1944-48. Finished Coll 1948. Various highway and bridge design jobs with Gannett Fleming, Blauvelt Engr and McFarland Johnson 1948-84. Since 1971 Reg Mgr, Southeastern Region, for McFarland-Johnson Engrs. Also respon for construction monitoring group of MJE. Married with 9 children, 2 grandchildren. *Society Aff:* ASCE, SAME, NSPE, CEC

Bischoff, Alfred F H
Home: 79 Cyupress in the Wood, Daytona Beach, FL 32019
Position: Cons Engr. (Self-employed) *Education:* BSEE/Elec Engg/Union College. *Born:* Nov 1912. With General Electric Co as Advanced Dev Engr & Mgr for 34 yrs; Design, dev & prod of Loran Navigation Sys 1940; Mgr in the Cons Lab; respon for integrated sys design & field troubleshooting; Mgr of Coolidge Lab 1955 respon for design & quality of medical diagnostic & radiation therapy equip, dev of the initiator for atomic weapons & in 1959 prod engrg of same; 1962-70 Mgr Apollo Spacecraft Reliability & Safety, providing tech guidance to NASA OMSF. Boy Scout Silver Beaver AWD. Fellow AIAA, IEEE. *Society Aff:* ANS, IEEE, AIAA.

Bischoff, Kenneth B
Business: Dept of Chem Engrg, Newark, DE 19711
Position: Unidel Prof of Biomed & Chem Engrg. *Employer:* University of Delaware. *Education:* PhD/Chem Eng/IL Inst Tech;BS/ibid/ibid. *Born:* Feb 1936. Native of Chicago Ill. Fellow Rijksuniversiteit Gent Belgium 1960-61. Mbr Faculty Univ Texas Austin 1961-67, Univ Maryland 1967-70, Walter R Read Prof Engrg, Cornell Univ 1970-76 (Dir School Chem Engrg 1970-75); Unidel Prof Biomedical & Chem Engrg Univ Delaware 1976- chrmn Dept Chem Engg, 1978-82. Cons to Natl Inst Health, Exxon Res Eng, General Foods, WR Grace, Envt Protection Agency (past). Mbr of Council (Dir) AIChE 1972-74, Chmn cttes; Bd of Dirs ECPD 1972-78. Chmn, Council for Chem Res 1985, Co-author or co-editor of 4 books, numerous articles. Editorial Bds: J Pharmacokinetics & Biopharmaceutics, ACS Advances in Chem Series, Received Amer Pharm Assn Ebert Prize 1972, AIChE Prof Progress Award 1976–, AIChE Thirty Fourth Annual Inst Lecturer, 1982; AIChE Food Pharmaceutical and Bioengg Div. Award 1982; AIChE R.H. Wilhelm Award 1987; AIChE Fellow 1987. Fellow AAAS 1980- Registered Engr. (P.E.) Texas. *Society Aff:* AIChE, ACS, AAAS, ASAIO, CS.

Bishop, Albert B
Business: 1971 Neil Avenue, Columbus, OH 43210
Position: Prof, Dept of Indus & Systems Engr. *Employer:* The Ohio State Univ. *Education:* PhD/Ind Engg/OH State Univ; MS/Ind Engg/OH State Univ; BEE/Electronics/Cornell Univ. *Born:* 4/7/29. Native of Philadelphia, PA. USAF, 1951-53; Bell Telephone Labs, 1954. Dept of Industrial & Systems Engrg, OH State Univ, 1954-present, Chrmn 1974-82; teaching and res in facility design, optimization, and systems. Conslt to industrial and government organizations. Reg PE, OH. Fellow of IIE and ASQC. Chrmn, Council of Industrial Engrg Academic Dept Heads, 1980-81. Dir of Academic Affairs, 1981-82, Dir. for Research 1985-present, IIE. Engrg Accreditation Commission, ABET, 1981-87. Tau Beta Pi, Natl Exec Council, 1966-

Bishop, Albert B (Continued)
70. Author: book on *Discrete-Time Control Systems.* Columbus Tech Council, Tech Person of the Year, 1983. *Society Aff:* IIE, IEEE, ASQC, ASEE, SME.

Bishop, Asa O, Jr
Business: 210 SMC, Knoxville, TN 37996-0520
Position: Assoc Dir/Research *Employer:* Univ of Tennessee *Education:* PhD/EE/Clemson Univ; MS/EE/Clemson Univ; BS/EE/VA Military Inst *Born:* 06/30/38 Joined the faculty of the Univ of TN Electrical Engrg Dept in 1970. Proceeded on a tenure track to Prof in 1978. Author of a dozen papers in the past 15 yrs in the area of medical data analysis and data acquisition. Selected as a Presidential Exchange Exec in 1978-1979. Received Meritorious Civilian Service Award for exceptional service in the position of Special Asst for ADP Planning to the Asst Secy of the Navy for Financial Mgmt for developing an authoritative departmental statement of major policies and goals for the next 20 yrs and institutionalizing formal Dept of the Navy APD planning. Currently working in computer aided engrg and integrating video disk tech and personal computers. *Society Aff:* IEEE, ASEE, NCGA

Bishop, Eugene H
Business: College of Engg, Clemson Univ, Clemson, SC 29634-0921
Position: Prof of M.E. *Employer:* Clemson Univ *Education:* PhD/ME/Univ of TX; BS/ME/MS State Univ *Born:* 10/29/33 Native of Hattiesburg, MS. Served as a commissioned ofcr in the Cvl Engrg Corp, US Navy from 1955-1961. Assoc Prof of Mech Engrg, MS State Univ; Prof and Head Mech Engrg Dept, MT State Univ; Prof and Head, Mech Engrg Dept, Clemson Univ; Assoc. Dean of Engrg, Clemson Univ; Prof of Mech Engrg, Clemson Univ. Teach undergraduate and graduate courses in Heat Transfer and do research in Basic and Applied Heat Transfer. Enjoy Golf. *Society Aff:* ASME

Bishop, Floyd A
Business: 500 E 18th St, P.O. Box 53, Cheyenne, WY 82001
Position: VP *Employer:* Banner Assocs, Inc. *Education:* B.S./CE/Univ of WY *Born:* 8/10/20 Ten years last past, VP of Banner Assocs, Inc, Conslt Engrs and Architects, with total staff of 100 employees. Personal specialization in water rights, hydrology, water supply and engrg aspects of water law. Prior to current assignment was WY St Engr for 12 years with broad responsibility for supervision of waters of the St including issuing, adjudication, and regulation of water rights, and extensive interstate water responsibilities. Past assignments include 13 yrs, partner in small consltg firm, and 4 years, irrigation engr. USDA, Soil Conservation Ser. *Society Aff:* ASCE, NSPE, ACEC.

Bishop, George A, III
Home: 116 Bellridge Rd, Glastonbury, CT 06033
Position: Lighting Consultant *Employer:* Spectrolume Inc *Education:* BS/Elec Engg/Southeastern MA Univ. *Born:* 9/12/38. Native of Fall River, MA. Joined Holophane Co 1960 as sales engr specialist in lighting design and application. Reg PE in CT. Past chrmn IES CT Chap; Chrmn Natl Bd of Examiners of IES 1968-69. Author of four IES lighting data sheets published in IES Journal. Author *Energy Saving Industrial Lighting for the 1980's* presented at IEEE 1979 annual industry applications society meeting and "Engrs Guidelines for Competitive Analysis of HID Industrial Luminaires" at IEEE's 1981 meeting. Mbr IEEE production and application of light ctte. Active in Sailing and Motorboating. *Society Aff:* IES, IEEE, AIPE.

Bishop, Harold F
Business: 12 Lakeside Lane, Denver, CO 80212
Position: Pres *Employer:* Bishop, Brogden, Rumph Inc *Education:* B.S./CE/Univ of UT *Born:* 2/13/35 Long Beach, CA. Graduated Univ of UT 1958. Served as naval officer 3 years, followed by 2 years with Forest Service. Joined Denver consulting firm Tipton & Kalmbach 1962. Between 1963 & 1971 worked for the firm on the Indus basin project in Pakistan. From 1971 to 1978 was VP and an owner of Leonard Rice Consulting Water Engrs, followed by 2 years as VP of Tipton & Kalmbach. In 1980 founded Bishop Assocs, Inc., and is currently pres. Was Pres of consulting engrs council/CO 1980-81 and is active in other civic and professional organizations. *Society Aff:* ACEC, ASCE, ICID.

Bishop, Harry L, Jr
Business: AL Tech Speciality Steel, PO Box 91, Watervliet, NY 12189
Position: Asst VP-Technology *Employer:* Al Tech Specialty Steel Corp *Education:* ScD/Process Metallurgy/MIT; Met E/Metallurgy/Univ of Cincinnati. *Born:* 12/3/27. Program for Execs 1965, CMU; IRI program for Res. Mgr 1974; Native of Youngstown Ohio. Student Engr with with Youngstown Steel & Tube Co 1951-52; Res Asst Metallurgical Dept M I T 1952-55; Dev Engr-Process Metallurgy for Jones & Laughlin Steel Corp 1955-56; with Allegheny Ludlum Steel Corp from 1956-1982 and was Mgr-Process R/D 1973 to 1982; assumed current position of Asst VP-Technology, Al Tech Specialty Steel Corp in 1982; Author of six technical papers and eleven patents; Pres of ISS-AIME, Dir AIME; Mbr AISE, AVS, ASM & the Metals Soc (London). Allegheny Ludlum Lounsberry Award 1958, Univ of Cincinnati Disting Engr Alumnus Award 1976 Distinguished Member ISS-AIME 1977 and ASM Fellow 1982. *Society Aff:* ISS-AIME, ASM, AISE, AVS, THE METALS SOCIETY (LONDON).

Bishop, Stephen L
Business: Metcalf & Eddy Inc, 10 Harvard Mill Square, Wakefield, MA 01880
Position: V President. *Employer:* Metcalf & Eddy Inc. *Education:* MS/CE/Tufts Univ; BS/CE/Northeaster Univ. *Born:* Feb 14, 1931. Native of Wakefield Mass. With Metcalf & Eddy since 1953; V Pres Metcalf & Eddy Inc 1973; V P Metcalf & Eddy Internatl 1976; spec areas of exper involve filtration of water & wastewater, water treatment, water transmission & distrib sys planning & design in both natl & internatl fields. Author & co-author of tech papers regarding waterworks facilities dealing with resources dev, water plant sludge treatment & other waterworks related subjects. Received NEWWA Dexter Brackett Award 1971, AWWA Resources Div Award 1972 & Metcalf & Eddy-H L Kinsel Award 1973. NEWWA Dexter Brackett Award 1981, NEWWA Past Presidents Award 1985, and the Institute of Water Engineers and Scientists (Great Britain) Diploma 1987. Reg. P E. *Society Aff:* AWWA, ASCE, NEWWA, ΣΞ, MWUA

Bishop, Thomas R
Home: 2202 Viking Dr, Houston, TX 77018
Position: Chief Engr R&D Retired *Employer:* Bowen Tools Inc *Education:* BS/ME/Univ of Houston *Born:* 10/26/25 Engr of the Year Award in Cost Reduction, Apollo Prog 1966; AISI Award- Design in Steel 1975; Patentee-oil well equip. Publication of Articles in magazines. Related to oil field equipment Marquis Who's Who in South & Southwest since 1967; Who in Houston 1980 publication. Reg PE TX, LA, and AL. Mbr-Masonic order. USMCR 1944-1946. . *Society Aff:* TSPE

Bishop, William H IV
Home: 1235 Walker Ave, Baltimore, MD 21239
Position: Pres *Employer:* Bishop Engrg Inc *Education:* MES/Comp Sci/Loyola; BS/ME/Johns Hopkins Univ *Born:* 6/10/39 1971-Left Lloyd E Mitchell, Mechanical Lontractor as chief engr. 1975-Left McNeill & Baldwin, Consulting Engr as project engr. Present-Pres of Bishop Engrg. Consulting Engrs. Primary Disciplines - HVAC, Energy Conservation. Testing and Balancing, fire protection, plumbing, sanitary, pollution control. Teaching-Plumbing Design, Basic & Intermediate & Advanced Air Conditioning, Testing and Balancing of mechanical building systems. Interests-computer programming, problem solving, family, people. Currently - writing Engrg Programs for General use in conslts offices. *Society Aff:* NSPE, ASHRAE, ASPE, ESB.

Bisio, Attilio L
Business: Chemical Engineering; Progress, 345 E 47th St, New York, NY 10017
Position: Editor-Chemical Engineering Progress *Employer:* American Institute of

Bisio, Attilio L (Continued)
Chemical Engineers *Education:* MS/CHE/Columbia School of Engg; BS/CHE/ Columbia School of Engg; BA/Chem/Columbia College. *Born:* 8/21/30. Was Engineer Advisor and manager Elastomers and Plastics Div, Exxon Engg and Dir of Engg, Fiber FMC. Has taught at the school of Engg, Columbia Univ and and Kean College. Executive Director, Materials Technology Institute. *Society Aff:* ACS, AIChE, AEA, CDA, NACE.

Bisplinghoff, Raymond L
Business: Exeter, NH 03833
Position: Co Dir and V.P. for Research *Employer:* Tyco Labs. *Education:* PhD/Physics/Swiss Fedl Inst of Techn; SM/Physics/Univ of Cincinnati; SB/Aero Engg/Univ of Cincinnati. *Born:* 2/7/17. Served as Instructor at the Univ of Cincinnati; Engr at Wright Field. Naval Officer 1943-46; Apptd Asst Prof at MIT in 1946; advanced through grades to Prof & hd of Dept of Aeros & Astronautics 1966; Dean Sch of Engg 1968-70; on leave for 4 yrs during 1962-66 as Assoc Admin NASA. Deputy Dir NSF 1970-74; Chancellor, Univ of MO-Rolla 1974-77, Mbr Bd of Dirs & VP Tyco Labs 1977- . Pres AIAA 1965-66; Dir AIAA 1965-67; Dir EJC 1967; Dir ECPD 1967; Natl Acad of Engg 1965; Natl Acad of Scis 1967; Von Karman Lecturer 1965; Godfrey L Cabot Award 1972; Disting Service medal NASA, 1967; Disting Service Award NSF, 1973; Mbr Natl Sci Bd, Chrmn USAF Scientific Advisory Bd, Mbr Defense Sci Bd. *Society Aff:* AIAA.

Bissell, Roger R
Business: 700 Clearwater La, Boise, ID 83707
Position: Assistant Discipline Director *Employer:* CH2M Hill *Education:* BS/ME/Univ of ID; AA/Engrg/Boise St Univ *Born:* 11/11/39 21 yrs of experience in alternative energy systems including energy recovery, solar, geothermal, liquified natural gas, and biomass conversion. He has a degree in mech engrg from the Univ of ID, and is a reg PE. He is a mbr of the ID Soc of PE, and received their "Outstanding Young Engr" award in 1974 and is presently serving as Natl Dir of NSPE. He served on the Bd of Dirs of Chem Hill from 1980 to 1982. *Society Aff:* ASME, NSPE, AEE.

Bisson, Edmond E
Home: 20786 Eastwood Ave, Fairview Park, OH 44126
Position: Cons Engr (Self-employed) *Education:* ME/Mech Engg/Univ of FL; BS/ Mech Engg/Univ of FL. *Born:* July 1916. Native East Barre Vermont. Retired 1973 as Assoc Ch Fluid Sys Components Div NASA Lewis Res Center. Res on lubrication, friction & wear NACA- NASA 1939-73; now Cons Engr in Tribology (sci of lubrication). Instructor short courses (1961-) UCLA, Univ of Tenn, VPI, Case-Western Res. Natl Pres Amer Soc Lubrication Engrs 1963. Recipient P. M. Ku Medal ASLE 1980; Mayo D. Hersey Medal, ASME 1981; NASA Medal for Exceptional Scientific Achievement 1968; Natl Award ASLE 1967; Jacques de Vaucanson Medal French Soc Groupement pour l Avancement de la Mecanique Industrielle 1966; Alfred E Hunt Medal ASLE 1954. Reg P E Ohio. Author (with Anderson) Advanced Bearing Technology 1964. Author of over 70 papers on lubrication, friction wear; published var journals, NACA & NASA. Editor in chief, Amer Soc Lubrication Engineers. (1976-). Fellow ASME & ALSE 1981-Adjunct Prof, RPI. *Society Aff:* ASLE, ASME, NYAS

Biswas, Asit K
Home: 28 Elvaston Avenue, Nepean, Ont., Canada K2G 3T4
Position: Director. *Employer:* Biswas & Associates *Education:* Ph.D./Civil Engineering/University of Strathclyde Glasgow; M.Tech./ Civil Engineering/Indian Institute of Tech., Kharaglour, India; B.Tech./ Civil Engineering/Indian Institute of Tech., Kharaglour, India *Born:* Feb 25, 1939. PhD Univ of Strathclyde UK. D.Sc. (Honorary), University of Lund, Sweden, Lecturer Univ of Strathclyde 1963-67; Prof of CE Queen s Univ Canada 1967-68; 1st Rockefeller Foundation Fellow in Internatl Relations 1974-75; with Dept of Environment 1968-76; Dir, Environ Sys Branch 1973-76. Director, Biswas & Associates 1976-present. Sr Cons to United Nations Sys Branch since 1973. Sr Cons to 16 governments, UN Sys & OECD. Vice-Pres, Internatl Water Resources Assn 1978-81. President, International Society for Ecological Modelling 1976-present. ASCE Walter L Huber Award 1974. Author of 31 books & well over 350 tech papers. Enjoys classical music & literature. *Society Aff:* ASCE

Bittel, Lester R
Home: 106 Breezewood Terr, Bridgewater, VA 22812
Position: Professor *Employer:* James Madison Univ. *Education:* MBA/-/James Madison University; B.S. Ind Eng'g/Lehigh University *Born:* 12/09/18 Author/editor of numerous textbooks, including Encyclopedia of Professional Management, Management By Exception, What Every Supervisor Should Know. Former editor/publisher of Factory magazine. Honors from ASME: Fellow, Henry Robinson Towne Lecturer, Frederick W. Taylor Award, Contennial Medal. Recipient of Wilbur McFeeley Award for Outstanding Contribution to Management Education from International Management Council. Named a Virginia Eminent Scholar in 1986. *Society Aff:* ASME

Bittenbinder, Nicholas A
Home: 9433 Katherine Dr, Allison Park, PA 15101
Position: VP. *Employer:* Pullman Swindell. *Education:* PhD/Engg Sci/Univ of Vienna; MScCE/Structural Engg/Univ of Vienna; BScCE/Structural Engg/Univ of Vienna. *Born:* 4/10/15. Throughout 45 yrs of profl experience in engg, was associated with many govt agencies & publicly-owned corps specializing in structures & transportation, as: Bridge Engr for German Autobahn, Berlin, since 1941. Chief Engr Bridges & Structures, Dept of Transportation, Upper Austria, since 1946. Consulting Engr with Preload Co, NYC, NY, since 1956. Proj Mgr with raymond Intl, NYC, NY, since 1957. Dist Engr, Bridges & Structures, PA Dept of Transportation, since 1962. Group VP, Civ Works Group, Pullman Swindell, Pittsburgh, PA, since 1972. Enjoy classical music, fishing, tennis, swimming. *Society Aff:* AAAS, ASCE, NSPE, SAME, NY Academy of Sciences.

Bitzer, Donald L
Business: 252 ERL, 103 S. Mathews Ave, Urbana, IL 61801
Position: Dir-Computer-based Educ Res Lab. *Employer:* University of Illinois. *Education:* PhD/EE/Univ of IL; MS/EE/Univ of IL; BS/EE/Univ of IL. *Born:* Jan 1, 1934 E St Louis Ill. Faculty Univ of Ill Urbana; Assoc Prof 1963-67; Prof Elec Engrg, 1967- Dir Computer-based Educ Res Lab 1967- . Data Processing Mgmt Assn Computer Sci Man of Yr Award 1975; Natl Acad Engrg (Vladimir K Zworykin Award), 1973, ASEE Chester F.Carlson Award 1981, Amer Soc Engrg Educ; Laureate, Lincoln Academy of Illinois, 1982. Co-inventor of plasma display panel which received Indus Res 100 Award 1966. Pioneer PLATO- Large computer-based educ sys, electronic displays, phone line communication devices. *Society Aff:* NAE, ASEE, IEEE.

Bixby, William Herbert
Home: 2308 SE 21st St, Ft. Lauderdale, FL 33316
Position: V P Applied Res - Retired. *Employer:* ITT Power Systems Div *Education:* PhD/EE/Univ of MI; MSEE/EE/Univ of MI; BS/EE/Univ of MI; BS/Math/Univ of MI; MME/ME/Chrysler Inst of Engg. *Born:* Dec 28, 1906; m. Dorothy Bancroft Tibbits Jan 17, 1963 BSE EE BS Math 1930, MSEE 1931, PhD EE 1933 Univ of Mich; MME 1935 Chrysler Inst of Engrg; Special Problems Engr Chrysler 1933-36; Instr to Prof EE Wayne Univ 1936-56; Cons Engr Power Equip Co Detroit 1937-56, V Pres for Applied Res, Power Equip Div, North Electric Co, Galion Ohio 1956. Mbr: Sigma Xi, Tau Beta Pi, Engrg Soc of Detroit; Fellow AAAS; Life Fellow IEEE. Spcl achievements, numerous pats issued on electronically controlled rectifiers, line voltage regulators, generator exciter regulators & assoc devices. *Society Aff:* IEEE, AAAS, ESD, ΣΞ, ΤΒΠ

Bjorke, Oyvind
Business: Production Engrg Lab NTH-SINTEF, Richard Birkelands vei 2B, Norway 7034 NTH-Thj
Position: Professor. *Employer:* Univ of Trondheim. *Education:* Dr Ing/-/Univ of Trondheim; MS/-/Univ of Trondheim. *Born:* April 1933. Skilled mech from machine tool indus 1955. Native of Trondheim Norway. Served in Army 1952-53 rank second lieutenant. Prof at Univ of Trondheim from 1966 & head of Production Engg Lab NTH-SINTEF at the Univ of Trondheim from 1969; current respon as Dean of Mech Engr Dept Retired 3 yrs ago of Univ of Trondheim. P of Sci Tech Ctte Optimization of CIRP. Retired 2 yrs ago, Mbr of Bd of Dirs in CAM-I. SINTEF Award for sci contrib 1975. SME Educational Award 1976. Enjoy cross country orienting as sport & hiking in the mountains. VP of CIRP. *Society Aff:* DKVS, NTVA, IVA.

Bjorklund, David S Jr
Business: PO Box 3240, Monroe, CT 06468-0324
Position: Pres *Employer:* Spath-Bjorklund Assoc., Inc. *Education:* MS/Environ Engr/ Northeastern Univ; BS/CE/New England Coll *Born:* 7/8/51 I graduated June, 1973, B.S.C.E. from New England Coll in Henniker, NH and received a Masters Degree from Northeastern Univ, Boston, MA, June 1975. From 1975 to 1978, I was involved in various projects which examined the results of various types of land use on water quality. In 1978, with a partner, formed the firm of Spath-Bjorklund Assocs, Inc., in Monroe, CT. I presently serve as firm pres. I direct a staff in the design of wastewater treatment facilities, land use planning, and evaluation of the impacts of cultural activities on aquatic systems. I am registered in CT and NH. *Society Aff:* NSPE, WPCF

Blaauw, Gerrit A
Home: 124 Mozartlaan, 7522 HP Enschede Netherlands
Position: Prof of Elec Engrg. *Employer:* Tech Hogeschool Twente Enschede NL. *Education:* PhD/Appl Sci/Harvard Univ; BS/EE/Lafayette College. *Born:* July 17, 1924 The Hague Netherlands. Worked at the Harvard Computation Lab on the design of the Mark IV Calculator; 1952-55 at the Mathematical Center Amsterdam Netherlands; 1955-65 with IBM at the Poughkeepsie N Y Dev Lab. Participated in the design of the Stretch computer 1960 & Sys/360 1964; from 1965 Prof of EE & Chmn of the subdept of Digital Tech at the Technische Hogeschool Twente at Enschede Netherlands. 1969-73 Chmn Dept of EE. Mbr Sigma Xi, Fellow IEEE, Mbr Royal Dutch Academy of Sci. *Society Aff:* ACM, IEEE.

Blachman, Martin M
Business: 820 Davis Street, Evanston, IL 60201
Position: President. *Employer:* Barton-Aschman Assocs Inc. *Education:* BS/Civil/Univ of IL. *Born:* Aug 1926. BSCE Univ of Illinois. Native of Chicago Ill. Served with Army Air Force in WWII. With Barton-Aschman Assocs since 1951; directed transportation studies & projs in US, Canada, Belgium, France, Spain, Mexico, Puerto Rico, Venezuela, Peru & Columbia; became Pres 1972, assuming respon for exec admin of all of the firm's operations in the transportation & urban planning fields. Lecturer in Traffic Engrg short courses Univ of Illinois, Univ of Wisconsin, Northwestern Univ, Internatl Council of Shopping Centers. P Pres Mid-west Sect & P Mbr Internatl Bd of Dir, Inst of Transportation Engrs. Mbr APA, APWA & ITE. *Society Aff:* ITE, APWA, APA.

Blachman, Nelson M
Business: Bldg IV, Box 188, Mountain View, CA 94042
Position: Sr Scientist, Off of the Chief Engr *Employer:* GTE Sylvania Sys Group. *Education:* PhD/Engg Sciences & Appl Physics/Harvard Univ; AM/Physics/Harvard Univ; BS/Physics/Case Sch of Appl Science *Born:* 10/27/23. Cleveland Ohio. John Tyndall Scholar at Harvard 1946, Gordon McKay Scholar 1946-47, AM & PhD (engrg sci & appl physics) 1947. Sp Res Assoc Underwater Sound Lab Harvard 1943-45; Res Assoc Cruft Lab 1945-46; Assoc Scientist, Brookhaven Natl Lab 1947-51; Physicist, computer br, Office of Naval Res 1951-53, math br 1953-54; Sr Scientist Western Div, Sylvania Sys Grp, GTE Inc 1954- . Lectr, Univ of Md 1951-52; Liaison Scientist, London Br, Office Naval Res 1958-60, 1976-78; Lectr, Engrg Ext, Univ Calif 1961-63; Fulbright Sr Lectr, Univ Madrid & Sch Telecommun Engrg, Spain 1964-65; Lectr, Stanford Univ 1967. Fellow AAAS, IEEE, IEE; Mbr Acous Soc Amer, Amer Statist Assn, Inst Math Statist, Math Assn Amer, Soc Indus & Appl Math, Internatl Sci Radio Union, Sigma Xi. Statistical communication theory. *Society Aff:* ASC, OSC, IEE, IEEE, URSI, IMS, ASS

Blachut, Teodor J
Home: 61 Rothwell Dr, Ottawa, Ontario K1J 7G7 Canada
Position: Guest Scientist, Nat. Res. Council of Canada *Education:* DrSc/Photogrammetry/Tech Univ Zurich; MEng/Geodesy/Tech Univ of Lwow. *Born:* 2/10/15. in Poland. MSc in geodesy & photogrammetry Lwow Tech Univ. Dr Sc ETH Zurich. WWII with Polish Army in France. 1941-45 Asst & Lecturer ETH Zurich UniRes Council of Canada as the founder & hd of Photogrammetric Res Sect. Holder of numerous pats, author of over 100 publs in many languages on geodetic, photogrammetric & cartographic subjs. Fellow of the Royal Soc of Canada. Received num hon distinctions & sci awards. 1974 hon doctor degree from Mining & Metallurgy Acad Cracow Poland. Respon for dev of num novel photogrammetric concepts & instruments. Initiated and promoted establishemnt of mfg photogrammetric instruments in Canada. *Society Aff:* RSC, CIS, ASPRS SGP.

Black, Charles A
Business: 7201 NW 11 Pl-Bx 1647, Gainesville, FL 32602
Position: Sr Vice President. *Employer:* Black, Crow & Eidsness Inc. *Education:* BSCE/Civil Pub Health/Univ of FL. *Born:* July 1920 Gainesville Fla. US Army WWII ETO. Holds Sanitary Engr Res Commission US Public Health Serv. Mbr AWWA since 1947, serving on all levels from Chmn Fla Sect through succession of natl offices from V Pres-Elect through P Pres 1968-73; Mbr Exec Ctte, Chmn Genl Policy Council, numerous other cttes. Water Utility Man of Year 1961. Co-founder, Sr Principal Black, Crow & Eidsness Inc Cons Engrs, Dir Business Dev. Spec: Water Treatment. Hobbies: hunting, fishing, rock collecting. Mbr: ASCE, APHA, WPCF, NSPE, CEC, ASME, Royal Soc of Health (Great Britain), Amer Acad of Environ Engrs (diplomate). Listed: Whos Who in America, Engrs of Distinction, Amer Men of Sci Sr Consultant, CH2M-Hill, Engrs, 1976-1981. *Society Aff:* AAEE, AWWA, WPCA.

Black, David L
Home: 1889 F. St., N.W, Washington, DC 20006
Position: Deputy Director, Science & Technology *Employer:* Organization of American States *Education:* BA/Philosophy/Baylor Univ; Postgrad/Mgmt/Univ of TX; Postgrad/Urban Res/Trinity Univ. *Born:* 4/3/34. Plainview Texas. AB degree in Liberal Arts; post-grad study Univ of Texas in mgmt. Postgrad study Trinity Univ-Urban Res. Chmn Latin Amer Relations Ctte ASM, Mbr AAAS. Position at Southwest Research Institute, San Antonio Texas - in charge of internatl progs relating to const, housing, grain storage, water supply tech in dev countries. Cons for 7 years to UNIDO, UNESCO & UNEP in areas of materials tech, res dev for universities in dev countries, low-cost housing progs for dev countries. Mbr of spec ctte for NAS on food storage & food losses in dev countries. Author of papers on Latin American economics & tech transfer. Deputy Dir Sci And Tech at OAS since 1980. Specific responsibility implementing S/T progrms Caribbean mbr states as well as Cen/So Amer mbr countries. Responsible for program design, budgeting and implementation of programs in tech, basic sci, applied sci and S/T policy and mgt (Listed in Who's Who in South). *Society Aff:* ASM, AAES.

Black, Harold S
Home: 120 Winchip Road, Summit, NJ 07901
Position: Private Cons (Self-employed) *Born:* April 14, 1898 D Eng & BS Worcester Polytechnic Inst. Positions held: Western Electric Systems Engr, Bell Labs Res Engr, General Precision Principal Res Sci. Inventor negative feedback amplifier 1927.

Black, Harold S (Continued)
Author: Feedback Amplifiers, Modulat ion Theory, 33 encyclopedia articles. Contrib 19 scientific papers to prof journals. Granted 333 US and Foreign pats. Literary critic, teacher, lecturer. Certificate of appreciation US War Dept 1946. Received numerous prizes, 10 medals, 11 Fellowships, 19 Mbrships, 9 awards & numerous hons. First to produce pulse position modulation and pulse code modulation multichannel moicrowave radio srelay sys for military & domestic applications. Devised new feedback control sys applicable to communications, consumer electronics, weaponry, computers, servos, biomech, medicine & many other diverse fields. Inducted into the Natl Inventors Hall of Fame, Feb. 8, 1981 for his invention of the negative feedback amplifier (U.S. Patent No. 2,102,671) thereby recognizing an inventor whose invention and scientific contributions have enriched our natl welfare. On June 6, 1981 Worcester Poly Inst, Ima mater, was awarded the Institutes highest honor, the Robert H. Goddard award for 1981 in Recognition of Outstanding Professional Achievement.

Black, J Temple
Business: 308 Dunstan Hall, Auburn, AL 36849
Position: Prof of IE and Dir of Adv'd Mfg'g Tech Ctr *Employer:* Auburn Univ. *Education:* BSIE/IE/Lehigh Univ '60; MSIE/ME Mfg'g/WVa Univ '63; PhD/M&IE/U of IL Urbana '69 *Born:* 6/10/37. Raised on Delmarva Peninsula. Taught mfg engrg ar WV Univ, Univ of VT, Univ of RI, OH State, Chrmn of I & SE at UAH 1981-84. Joined AU in Sept 84. Author of over 50 tech papers in metal cutting, electron microscopy, continuous extrusion, mgfg systems engrg, cellular manufacturing systems and Group Tech. Author (along with E Paul DeGarno & L Kohser) of "Mtls and Processes in Mfg--7th Ed-MacMillan. Former Mfg-Systems Div Dir, IIE and Current Chrmn of Production Engrg Div of ASME. Conslt to E I duPont, Am Safety Razor, Schick, IBM, and numerous other cos. Outside interests include music writing and tennis. Married to Carol Strom Black with three children. *Society Aff:* ASME, IIE, SME, NAMRI

Black, James H
Business: Dept Chem & Met E-Box G, University, AL 35486
Position: Prof of Chem Engrg *Employer:* Univ of AL *Education:* PhD/Chem Engg/Univ of Pittsburgh; MS/Chem Engg/Univ of Pittsburgh; BS/Chem Engg/Univ of Pittsburgh; AB/Chemistry/Cornell Univ. *Born:* Aug 14, 1921. Res Chemist - Koppers Co; Grad Teaching Asst & Instr Chem E Univ of Pittsburgh; Fellow (Anthracite Fellowship) Mellon Inst of Indus Res; Asst Proj Engr Standard Oil Indiana; Sr Tech & Supervising Tech US Steel Corp; Prof of Chem Engrg Univ of Alabama. Fellow AIChE, AACE & Amer Inst of Chemists. Mbr ACS, AAUP. Various local sect offices in AIChE, AACE & AAUP. Exec Secy, Natl Dir, Admin V Pres, Pres & Past Pres of Amer Assn of Cost Engrs. Pub numerous articles in chem & cost engrg. Author of two books in Cost Engrg. *Society Aff:* AIChE, AACE, AIC, ACS, AAUP.

Black, Joseph E
Business: Dept of Tech, Bellingham, WA 98225
Position: Prof. *Employer:* Western WA Univ. *Education:* PhD/Met Engg/Lehigh Univ; MME/Mech Engg/NYU; BS/Met Engg/Lehigh Univ; BSME/Mech Engr/Cooper Union. *Born:* 10/21/17. Univ Prof specializing in metallurgical engg, production and computer aided mfg. US Army Officer 1943-66; Deputy at the Ballistics Res Labs, Aberdeen, MD, 1959-63; Commanding Officer of US Army Materials Res Agency, Watertown, MA, 1962-66. Assumed current position at Western WA Univ in 1967. Licensed PE in WA and CA. Industry consultant in mfg metallurgy. Leisure interests: music and birds. *Society Aff:* SME, ASM, ISA, ΣΞ.

Black, Robert O
Home: 9604 Todd Mill Rd, Huntsville, AL 35803
Position: Dir, Test & Eval Directorate. *Employer:* US Army Missile Command. *Education:* MS/Mgmt/MIT; MS/Ind Engr/UAH; BS/Ind Mgmt Eng/OU. *Born:* 3/11/33. MS in Mgmt MIT; MS in Indus Engg UAH; BS in Indus Mgmt Engg OU. Born & raised in OK City. Worked as prod engr for MN Honeywell; joined Govt serv 1958 as reliability engr; 1970 organized & became 1st Dir for prod Asurance at US Army Missile Command; currently Dir, Test & Evaluation in the Army Missile Engg Lab. Dept of the Army Exceptional Civilian Serv Award, Dept of The Army Meritorious Civilian Serv Award (2 occasions). Tech grad courses in mgmt part-time. Married, 3 sons. Hobbies are sailing & woodworking. *Society Aff:* AUSA.

Black, William Z
Business: School of Mech Engineering, Atlanta, GA 30332
Position: Prof *Employer:* Georgia Inst of Tech *Education:* PhD/ME/Purdue Univ; MS/ME/Univ of IL; BS/ME/Univ of IL *Born:* 10/09/40 in Champaign, IL. Moved to Atlanta after receiving PhD from Purdue in 1967. Currently Prof of ME with research primarily in Heat Transfer. Emphasis in research since 1975 has been Heat Transfer from Electrical Equip. Has directed research work funded by USAF, USN, NASA, EPRI, AND NSF. *Society Aff:* ASME, IEEE, ASTM, NSPE

Blackburn, Douglas B
Business: 2 Broadway, New York, NY 10004
Position: Chrmn & Pres. *Employer:* Ford, Bacon & Davis Inc. *Education:* BS/Engg/Cornell Univ. *Born:* 11/20/18. Resident of Sparta NJ. 1942 commissioned ensign in US Navy & released 1945 a Lt. 1945-51 Cuban Sugar Mill; 1951 joined Ford, Bacon & Davis as Sr Engr; elected Dir of Ford, Bacon & Davis 1970. Pres 1972 & Chrmn & CEO 1973. Dir Trident Engg Assocs. Mbr ASME, NSPE, PE reg in 19 states. Univ Club, City Midday Club, Tau Beta Pi; Cornell Soc of Engrs; LA Soc of Engrs. Dir of Stratford/Graham Engg Corp (Kansas City, MO). *Society Aff:* ASME, NSPE.

Blackburn, J Lewis
Home: 21816 8th Place West, Bothell, WA 98021
Position: Retired. *Employer:* Westinghouse Electric Corp. *Education:* BS/Elec Engg/Univ of IL. *Born:* 10/2/13. Native of Kansas City, MO. Mbr Phi Kappa Phi, Tau Beta Pi, Eta Kappa Nu, Sigma Xi. Westinghouse from 1936-78 now retired part-time; taught Power Sys Analysis, Relaying, Symmetrical Components, Westinghouse Grad Progs, as Adjunct Prof Brooklyn Polytech Inst, Newark College of Engg & for IEEE. Editor & Author of Applied Protective Relaying book; author, Protective Relaying, Principals & Applications, Mareel Dekker, NY 1987. Reg PE. Life Fellow IEEE; Westinghouse Order of Merit; 1978 Distinguished Service Award of IEEE Power Sys Relaying Committee; 1979 outstanding service award of the IEEE Educational Activities Board; 1980 Westinghouse Auditorium-classroom in Coral Spring FL named "Lewis Blackburn Room–, 1984 IEEE-Power Engineering Soc Centennial Medal for "leadership in relay engineering, education aud service to the Society–. Attwood Associate Award, CIGRE US National Cmttee 1986. Member - P Chmn IEEE PES Power Sys Relaying Ctte; P Chmn Publs Dept; Past Secy of IEEE Power Engg Soc; Past Treas-Dir now President of the China Stamp Soc, & Treas Dir of the Amer Soc of Polar Philatelists. *Society Aff:* IEEE, CIGRE.

Blackledge, James P
Business: Univ Park Campus, Denver, CO 80208
Position: Assoc Director. *Employer:* Denver Res Inst Univ of Denver. *Born:* April 1920. MSME 1949, BSCE 1948 Univ of Utah; completed course work toward DSc 1957 Colo Sch of Mines. Assoc Dir DRI, DU 1949- . Prof Met, C of Engrg DU; Dir Office of Intl Progs DRI, DU 1973- . Distin Faculty Award 1970 C of Engrg DU. Mbr ASEE, ASM, Sigma Xi. Reg P E Colo. Mbr Adv Panel, USAID Com on S&T in Latin Amer. Mbr NAS Panel on mgmt. Training Progs for Tech Insts in LDCs. Mbr NAS Ad Hoc Panel for feasibility of an Internatl Indus Inst. Mbr Bd of Dirs VITA Inc. Consul for Guatemala.

Blackwell, Elwood T
Home: 933 W Sycamore Street, Chase City, VA 23924
Position: Retired *Education:* BS/Agri Engg/NC State. *Born:* Dec 1920. Native of Granville County NC. 2 yrs Campbell Coll Buies Creek NC. 38 mos US Army Air Force, Elec Specialists on B-24, B-27 & B-29. Degree in Articultural Engrg NC

Blackwell, Elwood T (Continued)
State; completed Basic Modern Mgmt course & other spec courses in Public relations, Mbr Educ Prog, Annual Work Plans & Budget. Dir Servs Dept 1956- ; 1948-56 Electrification Advisor; Chrmn VA Sect ASAE 1971-72; Pres Mecklenburg Egg Corp; chmn of sev cttes for VA Farm & Home Eletrification Counc. Presented 3 papers to ASAE Convention. Awards: VA State Hon Farmer, Life Mbr in P T A, 4-H Alumni Award. Listed in Who's Who in the South & Southwest. *Society Aff:* ASAE.

Blackwell, H Richard
Business: 1314 Kinnear Road, Columbus, OH 43212
Position: Director Inst for Res in Vision. *Employer:* Ohio State University. *Born:* 1921. BS Haverford Col 1941; AM Brown Univ 1942; PhD Univ Michigan 1947. Res scientist in vision, visibility, physiological optics & light & vision; 1943-45 Res Scientist Polaroid Corp & Tiffany Foundation MIT; 1945-58 Faculty Univ Michigan; since 1958 Faculty Ohio State Univ. 1945-55 Exec Secretary Armed Forces-Natl Res Council Vision Ctte. Author of over 300 tech articles. Fellow: IES, OSA, Amer Acad Optometry. OSA Lomb Medal 1952, IES Gold Medal 1972. Mbr OAS Bd of Dirs 1973-77; Mbr US Natl Ctte, Commission Internationale dEclairage. Internatl Chrmn CIE Tech Ctte on Visual Performance, 1963-79. *Society Aff:* IES, OSA, AAO, ARVO.

Blackwell, O Mortenson
Business: 1314 Kinnear Road, Columbus, OH 43212
Position: SR Res Assoc Inst for Res in Vision. *Employer:* Ohio State University. *Education:* Grad Studies/Psychology/Univ of MI; AB/Pyschology/Univ of MI. *Born:* 7/16/23. Res scientist in vision & illumination from 1954; Inst for Res in Vision OSU from 1958. 33 publs. Research sponsored by Illuminating Engg Res Inst spec in visibility under hwy lighting & visual performance as a function of illumination for observers of different ages from 1960. Mbr: Res Subctte & committee on Tunnel Lighting, IES Roadway Lighting Ctte, Optical Soc of Amer, Assn for Res in Vision & Ophthalmology, Internal Res group on Color Vision Deficiencies, US Natl Ctte, Internal Commission for Optics (1975-78), US Natl Ctte, Commission Internationale d'Eclairage. Fellow: IES, Amer Acad Optometry & Optical Society of America, Intl Secretary, CIE Tech Ctte on Visual Performance (1963-79), Consultant (1979-). *Society Aff:* OSA, IES, ARVO, IRGCVD, AAO.

Blahut, Richard E
Business: Owego, NY 13827
Position: IBM Fellow *Employer:* IBM *Education:* PhD/EE/Cornell Univ; MS/Physics/Stevens Inst of Tech; BS/EE/MIT *Born:* 6/9/37 Named a Fellow of the IBM Corp in 1980. Has been with IBM since 1964 with general responsibility for analysis and design of caterent signal processing systems, digital communication systems, and statistical information processing systems. He is a courtesy prof of electrical engrg at Cornell Univ. Where he has taught since 1973. He is a Fellow of IEEE and VP of the Information Theory Group. He is author of the forthcoming book "Theory and Practice of Error Control Codes" to be published by Addlson-Weslpy. *Society Aff:* IEEE, AAAS

Blaisdell, Fred W
Home: 4540 30th Avenue South, Minneapolis, MN 55406
Position: Res Hydr Engr; Collaborator *Employer:* US Dept of Agri, Agri Res. Ser. *Education:* CE/Civil/Univ of NH; MS/Hydraulics/Mass Inst Tech; BS/Civil/Univ of NH. *Born:* 7/21/11. With US Dept of Agri since 1935; Soil Conservation Ser res on res instrumentation at Natl Bureau of Standards DC 1936-40, hydraulic R/D of generalized designs for soil & water conservation structs at St Anthony Falls Hydraulic Lab MN 1940-83; activities transferred to Agri Res Ser 1954 & to Water Conservation Structures Lab. 1983-86; collaborator 1986- major contribs are SAF stilling basin & closed conduit spillway hydraulics. J C Stevens Award ASCE 1969. Fellow ASCE & ASAE. *Society Aff:* ASCE, ASAE, IAHR, SCSA.

Blake, Alexander
Home: 550 Escondido Circle, Livermore, CA 94550
Position: Engineer at Large Nuclear Test Engg Div. *Employer:* Lawrence Livermore Natl Lab *Education:* MSc/Mech/London Univ; Dipl Ing/Mech Eng/ Univ College, London. *Born:* 9/24/20. Poland. Served with Allied Armed Forces in Middle East & Europe 1939-45. Held sr posts with the major British & US Indus. Gained natl recognition for extensive pub work incl several grad level textbooks for engrs. Editor in chief of new engrg handbook. Joined Lawrence Livermore Lab 1967 in support of res for natl defense. Testified before the US Senate Subctte on earthquake legislation. Currently Chrmn of several cttes and design review panels at the Lab. Became active as ASME Sect Exec in regional & natl affairs incl design tech transfer & applied mech reviews. Honors incl chartered Mech Engr of Great Britain & the rank of Fellow of ASME. Interested in music, bridge & chess. *Society Aff:* ASME.

Blake, Jules
Business: 909 River Road, Piscataway, NJ 08854
Position: V Pres R/D. *Employer:* Colgate-Palmolive Co. *Education:* PhD/Organic Chem/Univ of PA; MS/Chem/Univ of PA; BS/Chem/Univ of PA. *Born:* July 1924 NY N Y. PhD in Organic Chem 1954, MS Chem 1951, BS 1949 Univ of Pennsylvania (Phila). E I du Pont de Nemours Co 1954-66 Wilmington Delaware; Dir of R&D Chem Group Mallinckrodt Chemical Works 1966-71 St Louis Mo; V Pres R/D Kendall Co 1971-72 Boston Mass; V Pres R/D Colgate-Palmolive Co 1973- Piscataway N J. Mbr: Amer Chem Soc, Sigma Xi, Chemists Club, Univ Club, Past Pres of Indus Res Inst. PPres Assoc of Res Dir (NY). Advisory Bd to Science Facilities-Univ Pennsylvania. Board of Overseers School of Dental Medicine-Univ Pennsylvania. *Society Aff:* ACS, ΣΞ, SCI.

Blake, Lamont V
Home: 800 Copley Lane, Silver Spring, MD 20904
Position: Electronics Consultant *Employer:* Self *Education:* Mast of Sci/Physics/Univ of MD; BS/Physical Scis/MA State Coll. *Born:* 11/7/13. Native of MA. Employed AR Pwr & Light Co, 1937-40, radio interference investigator. Res staff of Naval Research Lab, WA, DC 1940-72; Hd of Radar Geophysics Branch at retirement. Sr Scientist, Technology Service Corp. 1972-1985. Now self-employed Electronics Consultant. Author *Antennas* (Wiley, 1966) (republished by Artech House, Inc, 1984), *Transmission Lines and Waveguides* (Wiley, 1969), Chapter 2 of *Radar Handbook* (McGraw-Hill, 1970; M. Skolnik, ed), *Radar Range-Performance Analysis* (DC Heath, & Co., 1980) (republished by Artech House, Inc., 1986). Chrmn WA Chapter IEEE Grp Antennas and Propagation, 1965-66. Recipient RESA (NRL Chapter) Applied Sci Award, 1963, Navy Superior Civilian Service Award, 1972. Elected Fellow IEEE, 1979. Hobbies: photography, gardening, writing, computer programming. *Society Aff:* IEEE, APS, ΣΞ, RESA

Blake, Wilson
Home: PO Box 928, Hayden Lake, ID 83835
Position: Cons Mining Engr (Self-employed) *Education:* PhD/Mining Engg/CO Sch of Mines; MS/Engg Sci/UC Berkeley; BA/Geology/UC Berkeley. *Born:* Aug 1934 San Francisco Calif. 5 children. 1/Lt US Army 1958-60. Sr Scientific Data Analyst at UC LRL 1961-65; Supr Res Civil Engr USBM Denver Mining Res Ctr 1965-72; Adj Asst Prof Colorado School of Mines 1971-72; Dir Mining Res for GEC-AMINES Lubumbashi Zaire 1972-74; Cons Mining Engr Hayden Lake Idaho 1974- . Specialist in application of rock mech to design & stability of underground openings particularly deep mining with rock burst problems. Design installation and application of micro seismic monitoring systems to detect and delineate high stress areas in deep mines. SME Peele Award 1970, Interdisciplinary Award. US Natl Ctte for Rock Mech 1972. *Society Aff:* AIME, NWMA, ISRM, CIM.

Blake, Winchester G
Home: 484 Ridgeway, White Plains, NY 10605
Position: Application Engr-Turbines -ret. *Employer:* General Electric Co (Retired). *Education:* SB/Mech Engg/MIT *Born:* April 14, 1899 Fitchburg Mass. Fellow ASME.

Blake, Winchester G (Continued)
P E New York. With Genl Electric 1923 - their New York Office 1926; increasing power plant desgn work in that area for const throughout the country, made it advantageous to have someone in the city to aid cons in integrating turbines in their plant design to the best advantage. Starting when turbines were small, pressures & temperatures low, in growing up with the art I became sufficiently saturated with turbine lore to allow me to perform that duty. Interests: Physical Sciences, Astronomy, History, Haydn symphonies; once climbed mountains, wish I still could. *Society Aff:* ASME.

Blakely, Jack M
Business: Bard Hall, Ithaca, NY 14853
Position: Prof. *Employer:* Cornell Univ. *Education:* PhD/Phys/Glasgow Univ; BSc/Phys/Glasgow Univ. *Born:* 4/8/36. Ayrshire, Scotland. Res Fellow at Harvard Univ 1961-63, Div of Engg and Appl Phys. Cornell Univ 1963-present, Prof, 1971. Guggenheim Fellow, Cavendish Lab, Cambidge 1970-71. NSF Fellow, Berkeley 1976. Science & Engr Res Council (UK) Fellow, York Univ. 1984. Res activities mainly in area of surface sci; author of "Introductin to Properties of Crystal Surfaces" (1973) and editor "Surface Physics of Materials," VI, II (1975), "Interfacial Segregation–, (1978). Grad & undergrad courses taught in areas of statistical thermodynamics, phase transitions, elec & magnetic properties of mtls, surface sci. *Society Aff:* AIME, APS, AVS, Inst P. (UK), MRS.

Blakely, Thomas A
Home: 52 Argyle Ave, Blackwood, NJ 08012
Position: Dir of Plant Operation *Employer:* Kennedy Memorial Hosp. *Education:* BS Ind Eng'r (88)/Ind. Engr/Trenton State College; A.A.S./Mech. Engr/Camden County College *Born:* 12/21/45 Native of Baton Rouge, LA., now residing in Blackwood, NJ served 8 years at sea with the Navy and Merchant Marine. I hold a Red Seal Engrs License in New Jersey. In nine years I have risen from mbr status to National VP in the AIPE. I am in charge of all facility operation at Kennedy Memorial Hospital, Stratford, NJ. Married with three sons. *Society Aff:* AIPE, ASHE

Blakey, Lewis H
Home: 5606 Massachusetts Ave, Bethesda, MD 20816
Position: Dir, Engg and Housing Support Ctr *Employer:* US Army Corps of Engrs. *Education:* BSCE/Civ Engg/Univ of Notre Dame; MSE/Engg Sci/Geo Wash Univ; PhD/Civ Engg/Catholic Univ of Am; MBA/Bus Admin/Univ of Chicago. *Born:* 11/28/33. Burlington, NC, Prof Exp: Asst prof civ engg, 65-67, Dep Chief, Spec Engg Div, Engr Studies Ctr, 68-70, Chief, Engg Div, N Cent Div, 71-76, Dep Dir, Facil Engg, 76-78, Chief, Office of Policy, Civ Works, US Army Corps Engrs, 78-80, Chief, Planning Division, Civil Works, 80-86, Dir, Engg & Hsg Supt Ctr, 87-present. Honors & Awards: PACE Award, Secy Army, 65; Meritorious Srv Award, US Army Corps Engrs, 76, 78; Pub Works Leader of Yr, Am Pub works Asn, 78, Meritorious Exec in the Sr Exec Serv, 1982. Mbr: Am Soc Civ Engrs; Soc Am Mil Engrs; Am Assn of Univ Profs, Tau Beta Pi. *Society Aff:* ASCE, SAME, AAUP, TBΠ.

Blanco, Jorge L
Home: 3632 Inwood Ave, New Orleans, LA 70114
Position: VP Operations. *Employer:* Amax Nickel Div. *Education:* BS/CHE/GA Inst of Tech. *Born:* 3/13/28. Born in Havana, Cuba. Grad high school from Worcester Academy, Worcester, MA. Worked for Freeport Minerals (10 yrs) and Kaiser Aluminum (8 yrs) before joining Amax in 1971. Assumed current position of VP-Operation for the Amax Nickel Div July 1979. Prior to it I was VP and Gen Mgr of Amax Nickel Refining Co Inc Specializes on hydrometallurgy of nickel, cobalt, copper. *Society Aff:* AIChE, LES, AIME.

Blank, William F
Business: 2623 East Pershing Rd, Decatur, IL 62526
Position: Pres & Gen Mgr *Employer:* Blank, Wesselink, Cook & Assocs, Inc *Education:* BS/CE/Sanitary & Structural/NYU *Born:* 06/23/20 In 1965, I founded William F. Blank & Assocs, Engrs-Conslts, in Decatur, IL. I became Senior Partner of Blank and Wesselink & Assocs in 1966. On January 1, 1981, this company was succeeded by Blank, Wesselink, Cook & Assocs, Inc. The company has its principal central office in Decatur IL and principal office in St. Louis MO. I am Pres of Blank, Wesselink, Cook & Assocs, Inc. (BWC). I am a Member & past pres of the Metro. Decatur Chamber of Commerce, Pres of Decatur Macon County County Economic Development Foundation, Vice Chairman of Board of Directors of Decatur Memorial Hospital, Pres of Continental Engg Corp, & Member of Rotary International. *Society Aff:* AAEE, ASCE, ASME, NSPE, APCA, ACI, WPCF, ISPE, CWSPCA, ACEC, CECI, AAA (Arbitrator)

Blankenburg, R Carter
Home: 712 N Curtis Ave, Alhambra, CA 91801
Position: Consultant (Self-employed) *Education:* BSc/EE/Cal Tech. *Born:* Feb 7, 1905 San Diego Calif. Reg P E Calif. Engr S C E Co 1927-1970 incl 15 yrs in charge of underground power sys specs, design & const; Dir Underground R/D 5 yrs; active duty WWII 5 yrs Army Corps of Engrs; 2 yrs ETO incl restoration of utilities Le Mans France & vicinity. Col USAR (ret). Initiated IEEE URD Confs 1963; Gen Chmn Anaheim 1969. Respon for refinement & application of many concepts resulting in substantial cost reductions for underground power sys. Sponsored Underground Power Sect meetings PCEA 1949-54. Charter mbr & sponsor of Western Underground Comm, Chmn 1954; Fellow IEEE 1966. *Society Aff:* IEEE, NSPE, PES.

Blasland, Warren V, Jr
Business: 1304 Buckley Road, Syracuse, NY 13221
Position: VP. *Employer:* O'Brien & Gere Engrs, Inc. *Education:* MSSE/Santiary/Syracuse Univ; BCE/Civil/Manhattan College. *Born:* 1/16/45. New York City. Project Engg - Havens Emerson consulting Engrs - 1966. Subsequently joined O'Brien & Gere - 1967. Presently VP Process Engg Div with offices in Syracuse, New York City, White Plains, and Poughkeepsie, NY. Lecturer - Syracuse Univ Engg Grad School - 1976. PPres - New York State Soc of Prof Engrs (NYSPE) 1975-76-Dir 1976-78. Present Dir - New York State Chapter American Public Works Assoc (APWA). *Society Aff:* NSPE, APWA, WPCF, AWWA.

Blattner, Meera M
Business: Livermore, CA 94550
Position: Assoc Prof *Employer:* Univ of CA, Davis and Lawrence Livermore Natl Lab *Education:* PhD/Engrg/UCLA; BS/Math/Univ of So CA; BA/Lib Arts/Univ of Chicago *Born:* 8/14/30 in Chicago, IL. Mother of three sons, Douglas, Robert and William. Taught mathematics and computer science at Harbor Coll in Los Angeles, CA. Research fellow in Engrg and Applied Science at Harvard, after receiving PhD in 1973. Assistant Prof Rice Univ, Rice Univ 1974-79. Adjunct, TX Medical Center, 1975- 81. Visiting Prof, Univ of Southern CA and Univ of Paris. Program Dir, Theoretical Computer Science, National Science Foundation, 1979-80. Accepted current position in 1980. *Society Aff:* IEEE, SWE, ACM

Blayden, Lee C
Home: 247 White Oak Drive, New Kensington, PA 15068
Position: Manager, Environ Labs. *Employer:* Alcoa. *Education:* MS/Metallurgical Engg/Univ of WI; BS/Metallurgical Engg/MI State Univ; Assoc/Eng/Grand Rapids, J C. *Born:* Aug 1941 Opportunity Wash. MS Met Engrg Univ of Wisc; BS Cum Laude Michigan St Univ. Joined Alcoa Labs 1965; major activities incl dev & implementation of non-polluting molten metal fluxing processes, energy saving melting sys & sys to recover aluminum from municipal refuse. Principal inventor of Alcoa 469 Process licensed throughout the world to control air pollution. Presently directs corporate environmental R&D and manages environmental sampling throughout Alcoa. Served ASM locally & nationally; Chmn Pgh Chapter 1975-76. Mbr Tau Beta Pi & Sigmi Xi. Listed in Outstanding Young Men in Amer 1972 and American

Blayden, Lee C (Continued)
Men & Women of Science. Married, father of 3. Enjoy skiing, golfing & fishing. *Society Aff:* ASM, AIME.

Blaylock, Albert J
Business: 1909 McKee St, San Diego, CA 92109
Position: Pres. *Employer:* Blaylock-Willis & Assoc. *Education:* MS/Civ Engg/Stanford Univ; BS/Civ Engg/San Deigo State College. *Born:* 10/21/18. in Phila, PA. Attended grad & high sch in Binghamton, NY. Naval aviator in WWII through 1948. College through 1953. Civ Engg Reg 1957, Structural Engg 1960. Opened consulting office 1960. Presently pres Blaylock-Willis & Assoc, structural engrs. Past pres Structural Engrs Assn of CA (1977). Presently pres of San Diego Sec ASCE (1980), Mbr of Bd of Dirs of CEAC (1979). Designer & Structural Engr of record of many highrise, institutional, governmental & waterfront structures in US, Mexico, & Central Pacific. *Society Aff:* ASCE, SEAOC, CEAC, ACI, ASME.

Blazey, Leland W
Home: 1218 Fairy Hill Rd, Jenkintown, PA 19046
Position: VP Mfg & Engg. *Employer:* Retired. *Education:* MS/CE/OH State Univ; BA/Chem-Biol/Hobart College. *Born:* 10/6/13. Macedon, NY. Taught in Shortsville, NY public high sch 1936-39. Joined Merck & Co, Inc 1941 in pilot plant. Served in several engg & managerial responsibilities at Merck & Co, Inc. Assumed position of Dir of Engg McNeil Labs, Inc in 1958 & VP of Mfg & Engg in 1960 and later to Bd of Dirs. Mbr Bd of Dirs Philarea Easter Seals Soc. Enjoy cabinet making, bridge, golf, hunting and fishing. *Society Aff:* AIChE (Fellow).

Blecher, Franklin H
Business: Whippany Road Rm 3F-326, Whippany, NJ 07981
Position: Dir Mobile Communication Lab. *Employer:* Bell Labs. *Education:* Doctorate/Elec Ent/Poly Inst of Brooklyn; Masters/Elec Eng/Poly Inst of Brooklyn; Bachelor/Elec Eng/Polytechnic Inst of Brooklyn. *Born:* 2/24/29. Since Aug 1952 has been with the Bell Telephone Labs & has been involved in a wide range of activities incl design of transistor circuits, dev of solid- state short-haul carrier sys, design of millimeter wave networks, dev of the L4 coaxial c arrier sys & the dev of SIC mask-making equip; is presently Dir of the Mobile Communications Lab & respon for all Bell Sys dev work in this field. Recieved the Browder J Thompson Memorial Award 1959 for his work on transistor feedback amplifiers. Has p articipated in many IEEE activities incl serv as Chmn of the Circuit Theory Group, Chmn of the Solid-State Circuits Conf & Council, Genl Chmn of EREM & Chemn of the NEREM Bd of Dirs & Chmn of the INTERCON Prog Ctte. VP for Technical Activities & Mbr of the Bd of Dirs-1977. Elected Mbr of the Natl Acad of Engrg in 1979. *Society Aff:* HKN, TBΠ, IEEE, AAAS, NAE.

Bleich, Hans H
Business: Columbia Univ, New York, NY 10027
Position: Prof - Dir Inst of Flight Structs. *Employer:* Columbia Univ. *Born:* Educ in Vienna Austria: Eng Diploma 1933, DSc 1934. Designer (bridges & buildings) in Austria, Britain & US 1933-50; since 1950 Prof of CE Columbia Univ; since 1954 Dir of Inst of Flight Structs. Author Design of Suspension Bridges, Springer 1935; Editor Buckling Strength of Metal Structs, McGraw Hill 1951. Author of many papers in applied mechs. Awards by ASCE: Laurie Prize 1951, Croes Medal 1963, Wellington Prize 1969, Karman Medal 1973. Cons to Government & indus structs, buildings & dynamics.

Blenkarn, Kenneth A
Home: 9115 E 37 Court South, Tulsa, OK 74145
Position: Research Dir (Retired) *Employer:* Amoco Production Co *Education:* PhD/ME/Rice Univ; MS/ME/Rice Univ; BS/ME/Rice Univ; BA/ME/Rice Univ *Born:* 05/17/29 In charge of Amoco Offshore Tech Research 1972-86 pursuing improvements of offshore structure design or construction practice and development of new petroleum production systems. Publications (15) cover petroleum drilling and production, ice-structure interaction, ocean environmental forces, as well as structure design and reliability. Participation in technical committee work includes offshore structure engineering practices and rules for organization in Norway, United Kingdom and the US. Served as Distinguished Lecturer for the Soc of Petroleum Engrs. Mbr Marine Bd of Acad of Engrg. Mbr Acad of Engrg. Retired 1986. *Society Aff:* SPE

Blesser, William B
Business: 333 Jay St, Brooklyn, NY 11201
Position: Prof, Co-Dir of Bio E *Employer:* Polytechnic Univ *Education:* M/EE/Polytechnic Inst of Brooklyn; B/ME/Rensselaer Polytech Inst; ScD/Bio E/IN N Univ *Born:* 2/19/24 Engrg experience includes: Plant Engr with Beaunit Mills Inc; Chief Mechanical Engr at Anton Electronic Labs; Design Engr at Bulova Research and Development Labs. Joined Mechanical Engrg teaching staff at the Polytechnic in 1955. Later switched to teaching in The Electrical Engrg Dept. Started the Bioengrg Dept in 1961. Am presently Co-Dir of program. Presently developing courses in Robotics and Simulation for Mech Eng Dept. Past consultant activities have included AORD Tech Advisor (recipient of outstanding civilian service award), development of special courses in Bioengrg and computers at Downstate Medical Center and Mt Sinai Medical Center and Rehab Engrg Res with the VA. Present conslt activities include devel and supervision of comp facilities at a Holter Cardiac Testing Lab and the Cardiac Rehab Exercise Lab at Montefiore Hosp and Medical Ctr. Hold PE License in NY. *Society Aff:* BMES, ASEE, SME/RI, SCS.

Blessey, Walter E
Home: 5546 Dayna Ct, New Orleans, LA 70124
Position: Prof Emeritus Tulane U & Consulting Eng *Employer:* Self-employed *Education:* CE/Civ Engg/Tulane Univ; BS/Civ Engg/Tulane Univ. *Born:* 10/2/19. Natl Pres, ASCE 1979, Hd of Tulane Univ, New Orleans Civ Engg Dept 1959- 1984 & consulting engr in the fields of structural & fdn engg. Authored numerous books and publications in these fields. Held many responsible ASCE leadership positions since first ASCE service as Pres of the Tulane Student Chapter. Active in civic, social & religious activities of New Orleans. Presented lectures in Central & South America, China, Australia, England and to various American groups. Ardent golfer and outdoor enthusiast. Dir First Financial Bank of New Orleans 1986-. Named Outstanding Alumnus of College of Engineering at Tulane, 1979. *Society Aff:* ASCE.

Blewett, John P
Business: 310 W 106th St, New York, NY 10025
Position: Sr Physicist. *Employer:* Brookhaven Natl Lab. (Retired) *Education:* BA/Physics and Mathematics/Univ of Toronto; MA/Physics/Univ of Toronto; PhD/Physics/Princeton. *Born:* April 1910. Roy Soc Canada Fellow Cambridge 1936-37. Physicist in Res Lab Schenectady 1937-46; worked at GE on thermionics, semiconductors, radar, radar countermeasures, particle accelerators; Sr Physicist Brookhaven Lab 1947-1978. Co-designer 3-GeV Cosmotron, 33 GeV Alternating Gradient Synchrotron and Natl Synchrotron Light Source. Co-author with Livingston of 'Particle Accelerators'. Author of numerous papers on accelerators & other topics. Supervised Brookahavens energy prog 1973-75. Fellow of Amer Phys Soc, IEEE AAAS & N Y Acad of Scis. Retired. *Society Aff:* APS, IEEE, NYAS, AAAS.

Bleyl,Robert L
Home: 7816 Northridge Ave, NE, Albuquerque, NM 87109
Position: Traffic & Accident Resconstruction Engr (Self-employed) *Education:* PhD/Transportation Engg/PA State Univ; CHT/Highway Traffic/Yale Univ; MS/Traffic Engg/Univ of UT; BS/Civil Engg/Univ of UT. *Born:* 3/25/36. Native of Salt Lake City, UT; Deputy State Traffic Engr, UT State Dept of Highways; Taught Traffic Engrg, Bureau of Highway Traffic, Yale Univ & PA State Univ; developed & taught Transportation Engrg Courses, Univ of NM; Currently Highway Traffic & Accident Reconstruction Engr; Pres, ITE NM Section, 1978-79; PPres's Award, ITE, 1968. *Society Aff:* ITE, NSPE, TRB, SAE, NAFE.

Blickwede, Donald J
Business: Suite 310, 437 Main St, Bethlehem, PA 18018
Position: Conslt *Education:* BS/Chem Eng/Wayne State; ScD/Metallurgy/MIT; Bus AD/-/Harvard Bus School *Born:* 7/20/20. Native of Detroit. Has spent most of his career with Bethlehem Steel Corp which he joined 1950 as a res engr; was elected a V Pres 1964. Retired in 1982. Received BSChE from Wayne St Univ 1942 & DSc in metallurgy from M I T 1948; attended the Advanced Mgmt Course at Harvard School of Business Admin 1969. Mbr of ASM which he is presently serving as Natl Pres, AIME, & the Indus Res Inst of which he is a P Pres. *Society Aff:* ASM, AIME, IRI, ISIJ

Blight, Geoffrey E
Business: U of Witwatersrand CE Dept, Jan Smuts Ave, Johannesburg South Africa
Position: Prof of Construction Materials. *Employer:* Univ of the Witwatersrand. *Education:* DSc/Mech of Soils & Matls/London Univ; PhD/Mech of Soils/London Univ; MSc/Mech of Soils/Witwatersrand Univ; DSC/Mech of Soils & Matls/ Witwatersrand Univ. *Born:* 7/30/34. in Transvaal SA. Educ Univ of the Witwatersrand BSC (Engg) 1955, MSc (Engg) 1958; 2 yrs with cons engr on design const of earth dams, then studied at Imperial Coll London; received PhD from Univ of London 1961; lectured at Witwatersrand & res into properties of unsaturated soils until 1964 when joined SA Natl Bldg Res Inst 1964, here concentrating on indus & mining waste disposal Witwatersrand 1969. Current res interests incl Mechs of unsatureated soils, Mine waste disposal, Pressures in silos. Awards DSc (Engg) by Univ of London in field of Mech of Soils & Engrg Matls in 1975 same degree awarded by Univ. of Witwatersrand in 1985. Many papers publ by ASCE. Awarded J James R Croes Medal by ASCE 1975. *Society Aff:* ASCE, AAPT, SAICE.

Bliss, E John
Business: Assn of Prof Engrs, New Brunswick, 123 York Street, Fredericton, N.B., Canada E3B 3N6
Position: Reg of Prof Engrs *Employer:* City of Fredericton *Education:* B.Sc./CE/Univ of New Brunswick *Born:* 1933 Native of Montreal, Quebec. 1979-present, City Engineer, City of Fredericton. 1980 Pres of Assoc of Prof Engrs of New Brunswick. Interest: Sailing, Flying (holds private pilots lic). *Society Aff:* APENB, CSCE, CTAA, AWWA, APWA

Blizard, John
Home: 3 Farm Rd-Waveny Ctr, New Canaan, CT 06840
Position: Dir of Res Emeritus & Cons. *Employer:* Foster Wheeler Energy Corp. *Education:* MS/Durham Univ. *Born:* Aug 1882 in England. Prof of Engrg at McGill Univ 1906-11; Canadian Dept of Mines engaged in solid fuels res-combustion & gasification 1911-20; came to US 1920 & served as Ch of Fuels Div at the US Bureau of Mines in Pittsburgh; joined Foster Wheeler (then Power Specialty Co) 1923; was appointed Dir of Res at Foster Wheeler Corp 1927. Percy Nichols Award for achievement in the field of solid fuels 1957. Elected Honorary Mbr of ASME 1958. Retired from active service 1959 - assumed pos for Foster Wheeler at that time I was elected a Fellow of the ASME. *Society Aff:* ASME.

Bloch, Erich
Business: NSF, 1800 G Street, Washington, DC 20550
Position: Director *Employer:* NSF *Education:* BS/EE/Univ of Buffalo *Born:* 1/9/25 Studied at Swiss Polytechnic Inst, Zuerich. IBM 1952-1984. Various assignments: Engrg Mgr Stretch Computer System; Mgr Solid Logic Tech Program; Dir Poughkeepsie Lab; VP Operations, Components Division; Gen Mgr & VP, East Fishkill. Assistant Group Executive-Tech. Associations: Fellow of IEEE; Member of Computer Society; Member National Academy of Engrg; 1984- ; Director, National Science Foundation. *Society Aff:* NAE, IEEE, SME

Bloch, Lawrence J
Home: 9828 Gene St, Cypress, CA 90630
Position: Principal Engr *Employer:* Fluor Engr & Constr, Inc *Education:* Cert of Completion/Indust Mgmt/No. Orange County Coll; Cert of Completion/ME-Des/No Orange County Coll; Cert of Completion/Metalurgy/Fullerton Coll *Born:* 1/8/32 Bloch is a Certified Cost Engr and a Principal Engr with Fluor Engrs and Constructors. He supervises estimate preparation on a multi-billion dollar project. Presently Administrative VP of the American Association of Cost Engrs and previously Technical VP, Dir-Project Management, Dir-Cost Management, Member-Certification Bd and Chrmn-Capital Cost Estimating Committee. Bloch has contributed to the development of the cost engrg science through his development of various computer aided estimating and cost control programs. These include "FLAME–, "EXPONE–. "SCORE" and "ESP–. He has presented various papers on estimating, some have been reprinted all over the world. *Society Aff:* AACE

Block, Irving G
Business: 1000 Western Ave, Mail Zone 27729, Lynn, MA 01910
Position: Mgr Lynn Utilities Operation. *Employer:* General Electric Co. *Education:* BS/EE/Univ of ME. *Born:* 1/15/29. BSEE Univ of ME. Childhood in Portland ME & presently residing in Marblehead MA. Served with US Army Signal Corps 1951-53. Joined Genl Electric Co 1950 as Elec Engr in pwr distrib; numerous positions in Plant Engrg, Facilities & Utilities field until apptd to present position in 1976. Num articles in trade magaiznes. Instructed apprentices in elec engrg. Enjoy skiing, traveling, photography & classical music.

Block, Robert C
Business: Rensselaer Polytechnic Institute, Troy, NY 12180-3590
Position: Professor *Employer:* Rensselaer Polytechnic Institute *Education:* PhD/Nuclear Phys/Duke Univ; MS/Phys/Columbia Univ; BS/Elec Engrg/Newark Coll of Engrg (now NJ Institute of Tech) *Born:* 02/11/29 Robert C Block is currently Prof of Nuclear Engrg and Engrg Physics and Dir of the Gaerttner LINAC Lab at the Dept of Nuclear Engrg and Engrg Physics, RPI. His res interests include experimental neutron phys, nuclear data measurements, multiphase flow measurements, industrial applications of radiation, applications of radiography for nondestructive testing and radiation effects in microelectronics. He was formerly at the Oak Ridge National Laboratory and has had research sabbaticals at the UKAEA Harwell Laboratory in Great Britain, at the Kyoto Univ Res Reactor Lab in Japan and Sandia National Lab in NM. He is a consult to industry and government and has authored numerous technical publications. *Society Aff:* ANS, APS, AAAS, IEEE

Block, Robert J
Home: P.O. Box 2211, Norman, OK 73070
Position: Prof. *Employer:* Univ of OK. *Education:* PhD/Met/Univ of IL; MS/Met/ Columbia Univ; SB/Met/MIT. *Born:* 2/18/35. Brooklyn, NY. Attended Midwood HS. Married, W Nancy, three children, Deborah, Jennifer, John. Maj interests, plastic deformation of metals, failure analysis, products liability, consulting to the legal profession. *Society Aff:* ASM.

Blodgett, Omer W
Home: 2013 Aldersgate Dr, Lyndhurst, OH 44124
Position: Sr. Design Consultant *Employer:* The Lincoln Electric Co. *Education:* Bach. Metallurgical Engr. (w. distinction)/University of Minnesota; Mech. Engineer/ University of Minnesota *Born:* 11/27/17 In 1962, 73, 80, 83 received the A.F. Davis Silver Metal for best paper on Structural Welding. In 1968 was the lecturer for the Educational Lecture Series of the AWS Annual Meeting. In 1983 received the T. R. Higgins Lectureship Award by AISC. Authored numerous technical papers and chapters in handbooks. Authored the 460 page text, "Design of Weldments," and the 830 page text, "Design of Welded Structures." Member of the AWS Structural Welding Code and AISC Specifications Cttee. A licensed PE listed in "Who's Who in America" and "Eminent Engineers." Conducted seminars in Australia (1971, 75, 78) in South Africa (1981) and Republic of China (1987). *Society Aff:* ASCE, ASME, AWS

Blok, Harmen
Home: Univ of Technology, Dept. Mech. Engrg, Delft, Holland
Position: Emeritus Prof of Mech Engrg; Consultant in Mech Engg, including Tribology. *Employer:* Self *Education:* MS/Mech Engg/Univ Tech, Delft, The Netherlands. *Born:* 9/8/10. Amsterdam Holland. 1932 graduate Mech Engr Tech Univ Delft Holland. 1933-51 Res Engr Royal Dutch/Shell Lab Delft Holland; 1951-1981 Prof of Mech Engrg Mgr Lab for Machine Elements & Tribology Tech Univ Delft Holland. 1952-57 Initiator & Chmn Wear Div Dutch Assn for Materials Sci. 1966 Mayo D Hersey Award ASME. 1973 Tribology Gold Medal Inst Mech Engrs London. 1979 Natl Award ASLE; 1982 Medal of Merit, Ass. Materials Sci. (Netherlands); V Pres Internatl Tribology Council. Since 1936 authored about 60 papers on machine elements, incl. design, friction, lubrication & wear. *Society Aff:* ASME, ASLE, KIVI, BMK.

Blomberg, Charles R
Business: 800 The Calif Co Bldg, New Orleans, LA 70114
Position: Chief Petroleum *Employer:* Chevron Oil Co. *Born:* Oct 1917 Denver Colo. Petroleum Engr Colo School of Mines 1939. Phillips Petroleum Co 1940-42; US Navy 1942-46; Inspector Petroleum Prods Saudi Arabia Chevron Oil Co 1947- ; Reservoir Engrg Sr Staff Resv Engr 1967-72; Asst Ch Engr Petr 1972- ; respon for all Prod & Res Engrg; functional supervision over 100 engrs. Chief Engineer Chervon Petroleum (OK) Ltd, London 1976-1978. General Manager Operations Chevron Petroleum (OK) Ltd Aberdeen Scotland 1978- 1980 Chief Engr. Chevron Petroleum USA - Eastern Region 1981-Bd of Dirs Soc of Petroleum Engrs of AIME 1975-77. Hobbies: golf, reading, gardening.

Bloom, Martin H
Business: Long Island Center, Rt 110, Farmingdale, NY 11735
Position: Inst Prof *Employer:* Polytechnic Inst of NY *Education:* PhD/Applied Mech/Polytechnic Inst of Brooklyn; MS/Applied Mech/Polytechnic Inst of Brooklyn; BS/ME/Polytechnic Inst of Brooklyn *Born:* 05/19/21 Industrial experience: Celanese Plastics Corp, Squier Signal Corps Lab, General Applied Sci Labs. Joined faculty of Polytechnic Inst of Brooklyn, Aeronautical Engrg and Applied Mech 1951, Dept Head 1964-1966, Dir of Aerodynamics Labs 1964-1980, Dean of Engrg 1966-1974, Inst Prof 1974-, Dir Polytechnic's Long Island Center 1978-1979. Author over 70 papers and patents. Reg PE in NY. Conslt for industry and government. Founding editor Journal of Computers and Fluids (Pergamon), Bd of Dirs Accreditation Bd for Engrg and Tech (ABET, formerly ECPD). US Army Outstanding Civilian Service Medal. NY Acad of Scis award for research. *Society Aff:* AIAA, ASME

Bloom, Ray A
Business: 1835 Dueber Ave S W, Canton, OH 44706
Position: Mgr Process Res. *Employer:* The Timken Co. *Education:* MS/Met E/ Carnegie Inst TRech; Dipl/Ind Chem/Rochester Inst Tech. *Born:* June 21, 1921. Mbr of Tau Beta Pi & Sigma Xi. Joined The Timken Co 1950 as Res Metallurgist in the Steel & Tube Div working primarily in res concerning steelmaking & processing; was appointed Sect Ch Process Metallurgy Res, Res Div; 1975 became Mgr Process Res; respon for corporate res in manufacture of alloy steel, tubing, tapered roller bearings & rock bits. Mbr of ASM, BISI, AVS & ACS 1977 Asst Genl Mgr Res; 1979 Genl Mgr Avanced Process-Steel. *Society Aff:* ASM, BISI.

Bloome, Peter D
Business: 116 Mumford Hall, 1301 West Gregory Dr, Urbana, IL 61801
Position: Asst Dir, Agri, Natural Resources & CRD. *Employer:* Univ of IL *Education:* PhD/Agr Engrg/Univ of IL; MS/Agr Engrg/Univ of IL; BS/Agr Engrg/ Univ of IL. *Born:* 1/7/43. Native of Carlinville, IL. Served as Area Advisor with Ext Service in Northern IL. Was Extension Agricultural Engr at OK State Univ from 1970 to 1984. Areas of interest: grain drying, handling & storage, machinery management, & feed processing. Aerovent Young Extension Man Award, 1976 from ASAE. Sabbatical to Queensland Dept of Primary Industries, Australia, 1978. Distinguished Young Agri Engr Award, 1980 from SW Region ASAE. USDA Superior Service Award in 1986. *Society Aff:* ASAE.

Bloor, W Spencer
Business: 1904 Jody Road, Meadowbrook, PA 19046
Position: Consulting Engineer *Employer:* Self *Education:* BS/Elec Engg/Lafayette College. *Born:* 10/16/18. Trenton N J. 1981-Present Assoc Network Systems Dev Assoc and S.T. Hudson Intl. Also Conslt Staff Univ of Penn, Moore School of Elect Engrg and Editor ISA Transactions. Employed Leeds & Northrup Co as Field Engr 1940-55; sev mgmt posts 1955-69 & Manager Steam & Nuclear Power Systems 1969-81. Officer US Navy 1943-46. ISA Pres 1974 & V Pres Publs 1968-70. Currently Vice-Chrmn, Franklin Institute Committee on Science and Arts, Mbr, IFAC Working Group on Electric Power Plants and Chairman ISA History Ctte. Fellow AAAS, IEEE & ISA. Mbr Natl Acad of Engg, Phi Beta Kappa, Tau Beta Pi & Eta Kappa Nu. Engineer of Year Delaware Valley 1980 Honorary doctor of engg Lafayette College 1981. Reg P E PA. *Society Aff:* AAAS, IEEE, ISA, NAE.

Blossom, John S
Business: 35 East Seventh St, Cincinnati, OH 45202 *Employer:* Retired *Education:* /Mech Engg/Lawrence Inst of Tech (2 yrs). *Born:* 8/10/17. Reg PE by examination and reciprocity in 10 states and NCSBEE Cert of Qualification. Principal in conslt engrg firm since 1957. Chrmn, Mich/ Socy Prof Engrs Bldg Code Comm - rewrite of Bldg, Heating and Refrigeration Codes (1953- 56). Co-author paper "Pressurizating High Temperature Water Systems" awarded ASHRAE Wolverine Diamond Key Award, 1959. On Natl Panel of Arbritrators since 1968. On two ad hoc Natl Academy of Sciences Bldg Res Advisory Bd Comm: Chrmn "Methods of Refuse Handling in Low Rise Multi-Family Structures" (1966-71); mbr - "Collection, Reduction and Disposal of Solid Wastes" (1967-74). Presented technical papers on "High Temperature Water Systems" to Natl Coal Conf (Purdue Univ, 1963; Univ Ky, 1963); Amer Power Conf, Chicago (1964); Inst Power Plant Chief Engrs (Univ Ill, 1964). Presented opening paper on 11 energy conservation seminars (1974-75). Mbr and chrmn Handbook, Publishing, Research and Technical Comm's (1973-1983), mbr, Technology Council (1984), ASHRAE, ASHRAE Fellow, 1978; ASHRAE Dist Service Award, 1981. Retired 31 Oct., 1986. *Society Aff:* ASHRAE, NSPE.

Blue, E Morse
Business: 2642 Saklan Indian Dr, Walnut Creek, CA 94595
Position: President *Employer:* E.M. Blue & Associates, Inc. *Education:* SM/ChE/MIT; BS/Chemistry/Univ of CA. *Born:* 6/27/12. Native of Spokane, WA. Joined Chevron Res Co in 1938 as process design engr, specializing in catalytic conversion processes. In 1956 became Engg Consultant to Std Oil Co of CA refineries on catalytic processes. In 1964 became Mgr, Invention Dev, Chevron Res Co responsible for licensing processes & tech developed by the Chevron Cos. Retired from Chevron in 1977 & became independent Consulting Chem Engrs. Formed E.M. Blue & Associates, Inc. in 1979. Served in USN, 1941-1946. Present status, Capt, USN (Ret). Lectr in Chem Engg, Univ of CA, Berkeley, from 1959 to date. Enjoy golf & 10 grandchildren. *Society Aff:* AIChE, ACS, LES.

Blum, George H
Business: 600 Hempstead Tpke, West Hempstead, NY 11552
Position: Partner *Employer:* Reiffman & Blum *Education:* BCE/Civil/CCNY *Born:* 5/11/22 Native of New York City. Air Force Engrg and Intelligence Officer, World War II and Korean War. Partner in present firm for past 16 years specializing in commerical industrial and institutional bldgs. Active in Queens chapter NY SSPE for 25 years. *Society Aff:* NSPE

Blum, Joseph J
Business: Commander, 1843 EI Group, Wheeler AFB, HI 96854
Position: Commander *Employer:* US Air Force. *Education:* MS/EE/AF Inst of Tech; BS/EE/IA State Univ. *Born:* 2/26/34. in Earling, IA. Active duty USAF 1956- ; promoted to Colonel 1977. Assoc Prof of Ele Engg USAF Acad 1966-72; Chmn Weapons & Sys Engg Dept US Naval Acad 1971-77; presently Commander, 1843 Engi-

Blum, Joseph J (Continued)
neering Installation Group, AF Communications Command, Wheeler AFB Hawaii; Reg PE Mass; Winner 1070 ASEE Western Electric Fund Award; Author college textbook: "Introduction to Analog Computation–, Harcourt Brace Jovanovich, Inc.; Military crewmember and parachutist; Hobbies: backpacking and philately. *Society Aff:* ASEE.

Blum, Michael E
Home: 1055 Duchess Ave, W Vancouver, V7T 1G8 Canada
Position: Mgr, Q A. *Employer:* Chemetics Intl. *Education:* MS/Corrossion Engg/Univ of London, England; BS/Metallurgy Engg/Univ of Witwatersrand, S Africa, Africa; BA/Sociology Psychology/Univ of South Africa; Dipl/Bus Admin/Ipade, Mexico, DF. *Born:* 10/26/42. Raised in Johannesburg, S Africa, where he initially studied and then worked as a metallurgist in the chemical industry. After further study in England, joined INCO in London as Corrosion Engr. In 1972 moved to Mexico and as Technical Dir helped to establish Services Metallurgicos, SA; a co providing metallurgical consulting services to Mexican Industry. Chrmn of Education Committee, Mexico, DF Chapter of ASM. Became quality Assurance Mgr of Babcock & Wilcox, Mexico in 1975. Presently Mgr of Quality Assurance and Matls Eng, Chemetics Intl, Vancouver, Canada. Married, two children. Enjoys Reading and Outdoors. *Society Aff:* NACE, ASME, ASM, AWS, ASTM.

Blum, Seymour L
Home: 39 Pilgrims Path, Sudbury, MA 01776
Position: President. *Employer:* SLB Associates. *Education:* ScD/Ceramics/M.I.T.; B. S/Ceramic Eng./Alfred Univ. *Born:* ScD in Matls Sci at MIT 1954, BS Alfred Univ 1948. Bowdoin Coll Army Air Force Meteorology 1943. Process & Plant Engr Commercial Decal 1948-51; Raytheon 1954-63 basic matls res & Mgr of Adv matls prod; IITRI Dir of Ceramics Res 1963- 68, Grp VP 1968-71; MITRE Dir of Energy & Resource Planning 1971-78; Northern Energy Corp VP 1978-83. Charles River Assoc. VP 1984-86; SLB Associates, President 1986-. Chmn of Natl Mat Adv Bd of NAS Commission on Eng. & Tech. Systems, NAS. Trustee Electronics Div, Chmn of Publications, Orton Award, VP Am Ceramic Soc. Cons to US Bureau of Mines and US Congress Office of Tech Assess. Visiting Committee MIT & IIT. Am Soc of Metals, Public Service Comm. Fed of Mat Soc. Chamber of Commerce (Gov/Reg Delivery Systems). Comm of Societoechical Systems (NAS). Fellow Am Ist of Chem. Fellow Am Cer Soc, Distinguised Ceramist (NE 1975). *Society Aff:* AmCerSoc.

Blumberg, Marvin E
Home: 22 Robert Drive, Hyde Park, NY 12538
Position: President. *Employer:* Carlgen Inc. *Education:* BSChE/ChE/Drexel Univ. *Born:* Oct 1, 1926 Phila Pa. Served US Army 1944-46 in Pacific Theater Operations. Married, 2 children. Atlantic Richfield Co 1953-61 as Engr; Sharples Corp Phila Pa 1961-65 Mgr Engrg; DeLaval Separator Co Poughkeepsie N Y 1965-69 Mgr Engrg; Drew Chemical Co 1969-71 Boonton N J Dir of Engrg; 1971-73 founded Chemec Process Systems Consortium & served as Pres; 1973-76 employed by General Signal Corp as Pres B I F Carlgen Inc; in 1976 purchased Carlgen Inc from General Signal Co & serves as Pres & Chmn of Bd. Reg P E New Jersey, also Reg PE in: MA, VT, CT, and New Hampshire. Mbr AIChE, AWWA WPCF. Holds Process pats in field. *Society Aff:* AIChE, AWWA, WPCF.

Blume, John A
Business: Sheraton-Palace Hotel, San Francisco, CA 94105
Position: Chrmn of the Board *Employer:* URS/John A Blume & Assoc, Engrs. *Education:* PhD/Civil Engg/Stanford Univ; -/Civil Engg/Stanford Univ; AB/Civil Engg/Stanford Univ. *Born:* 4/8/09. Has pioneered in the dev & application of new design concepts & procedures for analysis of bldgs & structs in response to earthquakes & other forms of energy release. Author of over 100 comprehensive writings on engg sujs. PPres San Francisco Sec ASCE, CEAC & SEAOC; Founding Mbr of EERI, Hon Mbr Pres, 1978- 81; Fellow ACEC, AAAS, SAME; Hon Life Mbr NYAS, ASCE, IAEE, SEAONL, & ACI. Elected to natl Acad of Engg 1969. The John A Blume Earthquake Engrg Center at Stanford is named for him. Consulting Prof of Civil Engg, Stanford Univ.

Bly, James H
Home: 56 Fieldstone Dr, Syosset, NY 11791
Position: Accelerator Applications Engr; Prod Mgr. *Employer:* Radiation Dynamics Inc. *Education:* -/Physics/Univ of Chicago. *Born:* 1917. Supr Radiographic Testing Pratt & Whitney Aircraft 1940-53; Res Dir & Div Mgr X-ray Inc 1953-56; High Voltage Eng Corp 1956-65; final pos Dir Sci Applications & Acting G M ARCO Div; Radiographic Prod Mgr Appl Radiation Corp 1965-70; Cons Radiation & Accelerators 1969-72; Accelerator Applications Engr Radiation Dynamics 1971-85 (Prod Mgr 1980-85). Long E B Processing 1985- . Mbr Amer Phys Soc, Amer Soc Nondestructive Testing, Amer Soc Testing & Materials. Coolidge Award ASNT 1964; Fellow 1974. Chmn ASTM Comm Nond estr Testing 1948-62 & 1965-70. Fellow & Award of Merit ASTM 1975. Hon Mbr ASTM 1978. Enjoy classical music. *Society Aff:* APS, ASTM, ASNT.

Blythe, Ardven L
Home: 7814 Haven St, Huntsville, AL 35802
Position: Sr Engr *Employer:* Triad Microsystems, Inc *Education:* MS/IE & OR/OH State; BS/IE/TX A&M. *Born:* 7/25/34. Native of Pittsburgh, PA. Served as pilot & dev engr in USAF, received numerous decorations & awards. Assumed position with Sperry Univac Defense Systems Div engaged in ind engg tasks on indirect operations. Appointed Dir of Operations Res projs; received inaugural Univac Award of Excellence. Currently engaged in the dev & application of math models in the fields of financial mgt & life cycle cost analysis. Mbr of Adjunct Faculty, FL Inst of Tech grad prog. Active in USAF Ready Reserve as an aero systems program mgr. Regional VP & Mbr of Bd of Trustees, IIE, 1977-1979. *Society Aff:* IIE, ΤΒΠ, ΑΠΜ.

Blythe, David K
Home: 975 Edgewater Drive, Lexington, KY 40502
Position: Assoc Dean. *Employer:* Coll of Engrg Univ of KY *Education:* MCE/Transportation/Cornell Univ; CE/CE Matls/Univ of KY; BS/CE/Univ of KY. *Born:* May 18, 1917 Georgetown Kentucky. Jr Civil Engr US Forest Serv 1940-41; Lt to Major US Air Force - Maintenance Engrg Officer 1941-46; Matl Engrs 1946-47; Instructor to Prof College of Engrg Univ of Kentucky 1947-57; Chmn Civil Engrg 1957- 69; Assoc Dean 1969- . Reg P E & Land Surveyor s Fellow ASCE. Mbr ACSM, Mbr ASEE Continuing Professional Dev Div P Pres Kentucky Sect ASCE; P Chmn District 9 Council ASCE. Mbr: Tau Beta Pi, Omicron Delta Kappa, Triangle Fraternity. Chmn Mining Engr Dept, 86-87. Retired June 30, 87. *Society Aff:* ASCE, ASEE, ACSM, NSPE.

Blythe, Michael E
Business: P. O. Box 1000, Carrollton, GA 30119
Position: Chief Mech Engr *Employer:* Southwire Co *Education:* B/ME/GA Inst of Tech *Born:* 05/23/48 Native of Cedartown, GA. Served in GA Army Natl Guard 11 yrs Tank Commander and Platoon Sergeant, Co-op student with Southwire Co and Coca-Cola USA. With Southwire since graduating from GA Tech, BME, 1972. Assumed current position of Chief Mech Engr, Corporate Design Engrg in 1981. Responsible for industrial plant design, office design, and equipment installation. Treasurer GSPE 1979-80, '80-81. Reg PE in GA, AL, NH, & UT. Mbr Ga. State Heating and Air-Conditioning code Advisory Comm. Chairman Carrolton Planning & Zoning Commission 1984. *Society Aff:* NSPE, ASHRAE, GSPE.

Boardman, Bruce E
Business: 3300 River Dr, Moline, IL 61265
Position: Staff Engr. *Employer:* Deere & Co. *Education:* BS/Metallurgy/Purdue Univ. *Born:* 8/26/42. IL resident since grad in 1965. Specialize at Deere & Co in product liability, mtl properties and correlation of metallurgical factors affecting fatigue. Teach metallurgy at Blackhawk College and industry sponsored ASM Metals

Boardman, Bruce E (Continued)
Eng Inst courses. Author of papers on failure analysis and fatigue; most recent include SEM/79 paper "Failure Analysis-How to Choose the Right Tools" and ASM Metals Handbook, Ninth Edition, article "Fatigue Resistance of Steels." Chrmn Tri-City Chapter ASM 1973/74 and elected Tri-City Chapter's Outstanding Young Mbr 1976. Active ASTM Committee on Statistical Aspects of Fatigue and SAE Fatigue Design and Evaluation Committee. *Society Aff:* ASM, SAE, ASTM.

Boaz, Virgil L
Home: 601 Jefferson Dr, Lake Charles, LA 70605
Position: Prof. *Employer:* McNeese State Univ. *Education:* PhD/Elec Engr/MS State Univ; MS/Elec Engr/TX A&M Univ; BS/Elec Engr/TX A&M Univ. *Born:* 12/3/37. Native of Whitesboro, TX. Served 2 yrs in Army at the Signal Corps Res & Dev Lab, Ft Monmouth, NJ 1961-1963. Hd of Avionics Sec, Engr Exp Station, MS State Univ, 1963-1967, while taking post-grad courses. Sr Design Engr at Westinghouse Large Power Transformer Div, Muncie, IN, 1967-1974. Prof of Elec Engr at McNeese State Univ, Lake Charles, LA, since 1974. Have received 5 transformer related patents. *Society Aff:* IEEE.

Bobart, George F
Home: 1320 Twin Lakes Rd, Athens, Ga 30606 *Education:* BS/EE/Univ of MD *Born:* 1/18/30 Native of MD. Ground Electronics Officer (Capt) USAF 1952-1954. Joined Westinghouse Electric Corp 1954 as design engr in induction heating. Held positions of project engr, supvr engr & engr mgr 1958-1973. Product line mgr of Induction Heating 73 to 78. Mgr Induction Heating & Ultrasonic Cleaning Dept 78 to 1986. Energy chrmn - Forging Industry Assoc 78 to 86. Active IEEE in many areas since 1963. Hold positions of office since 1976 & including pres of Industry Applications Society of IEEE. Currently consultant in electrotechnologies & factory automation for the Electric Power Research Inst & ASM Intl. *Society Aff:* IAS/IEEE, ASM

Bobeck, Andrew H
Business: AT&T Bell Laboratories, 600 Mountain Ave, Murray Hill, N.J 07974
Position: Supervisor. *Employer:* Bell Labs. *Education:* MS/EE/Purdue Univ; BS/EE/Purdue Univ. *Born:* Tower Hill, Pa. Honored by Purdue as Distinguished Engrg Alumnus 1968, & in 1972 that univ awarded him an hon Dr of Engrg degree. Received Stuart Ballantine medal from Franklin Inst 1973 & in 1975 elected to Natl Acad of Engg/ In 1976 the Danish Acad of Technical Scis awarded the Valdemalr Poulsen Gold Medal for his pioneering work on magnetic bubble memories. He has been granted more than 125 patents. *Society Aff:* NAE.

Bobeczko, Michael S
Home: 2838 Jennifer Dr, Castro Valley, CA 94546
Position: Mgr Corporate Noise Control Engrg *Employer:* Kaiser Aluminum & Chemical Corp *Education:* MS/ME/Carnegie-Mellon Univ; BS/ME/Purdue Univ *Born:* 08/06/46 in Cleveland, OH. Internationally recognized Engrg Conslt in Accoustics, Noise Control and CAD/CAM. Research Engr - Westinghouse Acoustics Lab, 1968. Senior Acoustical Conslt - Bolt Beranek and Newman Inc 1970. Since 1973 have been with Kaiser Aluminum & Chemical Corp as Mgr of Corporate Noise Control Engrg and CAD/CAM Dept. Developed State-Of-The-Art tech for reducing industrial noise particulary in can plants, saws and blowers. On the Bd of Dirs of the Inst of Noise Control Engrg; Chrmn 1977-1979 of the Aluminum Association - Noise Control Committee; Chrmn of the Noise Control Codes and Standards Committee for the American Soc of Mech Engrs - Pi Tau Sigma and Tau Beta Pi; contributed articles to professional journals and patentee in field. Enjoy construction projects, swimming and travelling. *Society Aff:* INCE, ASME, ASA, NCGA, CASA/SME

Bobrowski, Jan J
Business: Grosvenor House, Grosvenor Rd, Twickenham Mddx, TWI 4AA England
Position: Sr Partner. *Employer:* Jan Bobrowski & Ptnrs, Cons Engrs. *Education:* PhD/-/Univ of Surrey BSC/-/London Univ; ACT/Battersea College of Advanced Technology. *Born:* 3/31/25. BSc (Eng)(Hons) London, Ph.D. Surrey, F.Eng., PEng (Alberta) & British Columbia, FICE, FIStructE, MCSCE. Lwow Grammar School before WW II. Served with Polish 2nd Corps (1941-46). Exper with steel suppliers, cons engrs & contractors. 1962 founded firm of cons engrs in England, 1969 in Canada. Dev & applied new tech ranging from prestressed concrete to stainless steel & plastics. Spec in prefabrication & industrialised construction. VP of Federation Internationale de la Precontrainte. VP (85/86) of Inst of Structural Engrs & Concrete Soc. Pres (86/87). Mbr of the CEB Advisory Ctte of Res and Application, Visiting Prof of Civil Engrg at Imperial Coll of Sci & Tech, London Univ. Mbr of the CEB, IABSE, IASS, FIP, ACI, PCI *Society Aff:* FICE, FISE, ACI, PCI, F Eng.

Bochinski, Julius H
Home: 26 Redding Ridge Dr, Gaithersburg, MD 20760
Position: VP. *Employer:* Enviro Control, Inc. *Education:* PhD/Chem Engg/IA State Univ; BS/Chem Engg/Univ of Detroit. *Born:* 6/25/22. Native of Detroit, MI. Served with US Navy 1942-46. Process design engg for Fluor Corp 1954-60. Res engg for Bell & Howell Co 1960-62. Engg manager for process instruments with Beckman Instruments, Inc 1962-68. Tech consultnat on production problems to Rockwell International, 1968-70. Manager of engg and manufacturing for Instrument Div, Enviro Control, Inc. Assumed respon at Enviro Control Inc, in 1975 for activities in occupational health, tech feasibilty and economic impact of complying with proposed government regulations. *Society Aff:* AIChE, ACS, AIHA.

Bock, John E
Home: 417 Lindsay Ave, Scotia, NY 12302
Position: Retired. *Employer:* Genl Elec Co - ret. *Education:* EE/Eng/Norrkoping Technical Sch Sweden. *Born:* Aug 1888. Grad Electrical-Mechanical line in Sweden. 1912 Illum Engrg Lab, later Lighting Div Gen Elec Co. Prin work photometry, street, highway, baseball, football field lighting, calculations & testing. Searchlight testing for Army, Navy in WW I & II. Coord spectacular lighting features at Washington Arms Conf Buffalo Centennial, Detroit. Conventions, elec expositions, state fairs & Niagara Falls. Author, co-author tech papers. Fellow Emer IES, Hon Mbr Testing Procedure Ctte. Stamp collecting, gardening.

Bockhop, Clarence W
Home: 1419 McKinley, Ames, IA 50010
Position: Professor & Hd-Agri Engrg Dept. Retired. *Employer:* International Rice Research Institute. *Education:* PhD/Agr Engr & T&AM/Iowa State Univ; MS/Agr Engrg/Iowa State Univ; BS/Agr Engrg/Iowa State Univ *Born:* Mar 28, 1921 Paullina, Iowa. Prof & Hd, Dept of Agri Engrg Iowa St Univ 1962 to 1980. Ed & Serv Mgr The Stewart Co, Dallas, Texas 1948-53. US Army 1943- 48, Capt Artillery & Air Officer, Philippines 1945-46. Hd Dept of Agri Engrg Univ of Tenn 1957-60, Prof Iowa St Univ 1960, Visiting Prof Univ of Ghana 1969- 70. Fellow ASAE, Dir 1973-75, chmn of sev cttes; Mbr ASEE; Mbr Iowa Engrg Soc; Chmn of Prof Engrs in Ed Chap 1975; Reg PE Iowa. Mbr Gamma Sigma Delta Sigma Xi, Phi Kappa Phi, Phi Mu Alpha, Alpha Epsilon. Author num papers on tillage, mach mgmt, mech in dev countries. Head, Dept of Agricultural Engineering, International Rice Research Institute, Manlia, Philippines, 1980-1986. Retired. *Society Aff:* ASAE, ΣΞ, ΦΚΦ

Bode, Loren E
Business: Agric Engr. Dept, 1304 W. Penn. Ave, Urbana, IL 61801
Position: Prof and Assoc Head *Employer:* Univ of IL. *Education:* PhD/Agri Engr/Univ of MO; MS/Agri Engr/Univ of MO; BS/Agri Engr/Univ of MO. *Born:* 7/9/43. Raised on grain farm near Hannibal, MO. Conducted res for USDA, ARS from 1965-1973 on the dev of equip for applying agricultural chemicals. With Univ of IL Agri Engg Dept since 1973. Responsible for teaching, res and ext progs regarding application of chemicals. Natl Pres Alpha Epsilon Honor Soc 1975, Chrmn ASAE PM-41 Committee 1974, VChrmn ASTM E29.04 Committee, 1978. Enjoy all sports. *Society Aff:* ASAE, ASTM, WSSA.

Bodeen, George H
Business: 8600W. Bryn Mawr Ave, Suite 800 North, Chicago, IL 60631
Position: Chairman, President and CEO *Employer:* Lindberg Corp. *Education:* BS/Civil Engg/Northwestern Univ; MS/Matls Sci/Northwestern Univ. *Born:* Mar 1924. Native of Chgo, Ill. Prior to WW II, worked in R&D Dept of Teletype Corp. Served with Army Air Corps 1943-46. Sales Engr for Neenah Foundry Co. With Lindberg Corp since 1952, elected Pres in 1965. Headed Lindberg Corp team for heat treating solid propellant rocket motor cases. Worked as cons to SNECMA of Paris, France 1963. Distinguished Life Mbr ASM 1975. Northwestern Univ Alumni Merit Award in Engrg 1975. Mbr of the Advisory Council of Northwestern University Technological Inst. Elected a Fellow of ASM in 1979. Enjoy fishing, golf, jogging, skiing & photography. Mbr of the Bd of Dir of Benefit Trust Life Insurance Co, DICKEY-john Corp, Imperial Clevite, Inc., Pepper Const Co and Lindberg Corp. Past Pres and Trustee, Amer Soc for Metals. Mbr, The Metallurgical Soc for AIME. *Society Aff:* ASM, TMS/AIME.

Bodenheimer, Vernon B
Home: 4929 Cambridge Dr, Dunwoody, GA 30338
Position: Asst Mgr Pulp & Paper Div. *Employer:* Patchen, Mingledorf & Assocs. *Born:* Dec 1922 High Point N C. BS (hon) Chem Engrg from N C State College. Served 1st Lt US Air Force 1942-45 (air medals, Distinguished Flying Cross). Chem Engrg Champion Paper & Fibre Co Canton, N C 1948-51. Asst Pulp Mill Supt Riegel Corp, Acme NC 1951-52. Southern Mfg Stebbins Engrg Co Pensacola, Fla 1952-56. Tech Dir Continental Can Co Augusta Ga 1966-69. V P Oper, Prince Albert Pulp Co, Saskatchewan, Can 1969-72. V P Paper Indus Engrs Atlanta Ga 1972-75. Asst Mgr Pulp & Paper Div Patchen, Mingledorf & Assocs 1975- . Author papers in pulp & paper field. Mbr: AIChE, TAPPI, Tau Beta Pi, Phi Kappa Phi, Baptist, Mason, Shriner. Listed in Who's Who in Indus & Commerce. Int: reading, golf, bridge.

Bodinus, William S
Home: 6600 Keating Ave, Lincolnwood, IL 60646
Position: Mgr & VP *Employer:* Carrier Corp. *Education:* BS/Engrg/IL Inst of Tech. *Born:* Feb 24, 1909. BS Ill Inst of Tech (Lewis). Native of Chgo, Ill. Reg PE Ill. Fellow & Life Mbr ASHRAE. Instructor air cond 8 yrs Grad School Armour Inst & N W Univ. Western Zone Mgr & V P Carrier Contracting (engrg) Carrier Corp 45 yrs - ret. Author approx 90 articles on air cond & refrig equip & applications. Natl Chmn for sev yrs air cond ctte MCA. Chmn bldg ctte Medina Shrine Hosp & Scottish Rite Bodies Chgo. Also mbr Bldg Cttee IL Masonic Med Ctr & Shrine Hospital. Enjoy fishing & hunting. *Society Aff:* ASHRAE.

Bodman, Gerald R
Business: 217 Agrt Engg Bldg, Lincoln, NB 68583-0771
Position: Assoc Prof/Extension Agricultural Engineer *Employer:* Univ of NB. *Education:* MS/Agri Engr/PA State Univ; BS/Agri Engr/PA State Univ. *Born:* 5/22/44. Raised near Catawissa, PA; Reg PE in 29 states; Licensed Surveyor; Co-owner, Space Preceptors Assoc, part-time consulting firm; areas of specialization include farmstead engg, structural design, livestock environmental control systems, manure mgt, grain-storage, mastitis control, farmstead electrification, milking sys design/analysis, extraneous (stray) voltage in livestock environment, and use of solar energy in agri structures; experience includes work as design engr/construction supervisor--New England Pole Builders, Inc, Ludlow, MA; irrigation system design and sales; Extension Agricultural Engînerer - Penna State Univ; Extension Agri Engr (livestock systems)--Univ of NB; author of over 250 articles, bulletins and papers; recipient of 17 ASAE Educational Aids Competition blue ribbons; mbr four ASAE natl committees. Mbr Sigma Xi; NSPE; PEPP; ASHRAE; AAAS; Natl Mastitis Council; recipient: "Young Extension Man of the Year" award, ASAE 1982; recipient: "Excellence in Extension Programming" award, Epsilon Sigma Phi - 1982 and Univ of NE 1982 & 1983. Recipient--Walnut Grove Livestock Service award, 1986. Licensed private pilot. *Society Aff:* ASAE, NSPE, ASHRAE, NMC, ADSA

Boe, Rollin O
Business: P O Box 389, Ogden, UT 84402
Position: Pres & Chmn of Bd/Dir. *Employer:* R I Corporation. *Education:* BS/Physics/San Diego State College. *Born:* May 1927. Native of South Dakota. Reg PE Utah. V P R&D wth Kittell-Lacy Inc 1954-64. Sr Mbr of Advanced Tech Staff with Marquardt Co 1964-69. Pres & owner R I Corp (engrg & cons firm) since 1969. Num inventions incl supersonic combustion nozzle. Mbr: Cons Engrs Council of Utah, Acoustical Soc of Amer, Amer Inst of Aeronautics & Astronautics, Institute of Noise Control Engineering. Enjoy classical music, flying & Golf. *Society Aff:* CECU, AIAA, INCE, ASA.

Boeckel, John H
Home: 9207 Davidson St, College Park, MD 20740
Position: Dir of Engrg. *Employer:* NASA/Goddard Space Flight Ctr. *Education:* MS/Mech Engg/Univ of Rochester; BS/Mech Engg/Duke Univ. *Born:* Mar 13, 1927. Employed by Eastman Kodak Co at Camera Wks, taught at Univ of Syracuse. For nearly a decade was involved in environ testing of ordnance at US Naval Ordnance Lab (now Naval Surface Weapons Ctr). Joined NASA (Greenbelt, MD) in 1959 & worked progressively in satellite testing, mech design of satellites, project mgmt of Landsat/Nimbus, use of satellite tech for practical applications. Became Dir of Sys Reliability in 1975 & Dir of Engrg 1976. Int: sports cars. *Society Aff:* AIAA, Fellow AAS

Boehm, Arnold H
Business: 1 Complexe Desjardins, Montreal, Canada H5B 1CB
Position: Consultant to Steel Industry. *Employer:* The SNC Group. *Born:* Oct 1906. M Eng Univ of Prague. In Canada since 1939. With The Steel Co of Canada Ltd, Montreal 1940-62: appointed Ch Engr 1946. With the SNC Group 1962 to date: Cons to Steel Indus & Mgr Internatl Dev. 1969 UNIDO adviser to the Turkish gov re natl expansion of Turkish steel indus. Presently Dir of Quebec Steel project for Gov of Quebec. The Engrg Inst of Canada: Regional V P 1973-75. The Amer Soc for Metals, Metals Park, Ohio: Mbr Natl Adv Tech Awareness Council 1973-75. Enjoys nature on his farm, studies the impact of the engr's work on the economy.

Boehm, Elmer L
Home: 24 Brookwood Acres, St Louis, MO 63131
Position: Major Project Mgr. *Employer:* Monsanto Co. *Education:* ChE/Univ of Cincinnati. *Born:* 1/3/23. Cincinnati, Ohio. Chem Engrg Univ of Cincinnati 1941-3 & 1946-9. Civil Engr Harvard 1943-44; Shrivenham Univ (England) 1946. Asst Plant Engr Davison Chem Co Cincinnati 1949-52. Tech cons to Jos Crosfield & Sons, Warrington, England 1953. Plant Engr W R Grace, Lake Charles La 1953-59. Proj-ect Mgr Monsanto Co, St. Louis, Mo 1959-64. Gen Supt Monsanto's Chocolate Bayou & Krummerich Plants 1964- 67. Mfg Mgr St Louis 1968. Pres Monsanto Biodize Sys, N Y 1968-70. Monsanto Corporate Engrg Mgr 1970-83 Monsanto Mgr of Environmental Control 1984- Sect Chmn AIChE, Alpha Chi Sigma, Phi Delta Theta. Served under Patton WW II. Decorated with Purple Heart & several battle stars. Asst Scoutmaster Boy Scouts; Bd of Dir Camp Fire Girls; AIChE Fellow; Director AIChE; *Society Aff:* ΑΧΣ, ΦΔΘ, AIChE.

Boehm, Robert F
Business: Mech & Industrial Engineering Dept, Univ of Utah, Salt Lake City, UT 84112
Position: Prof *Employer:* Univ of Utah *Education:* PhD/ME/Univ of CA Berkeley; MS/ME/WA St Univ; BS/ME/WA St Univ *Born:* 01/16/40 in Portland, OR. Industrial positions with: Westinghouse, Boeing, Lawrence Livermore Lab, Gen Elec Atomic Power Equip Dept, Jet Propulsion Lab and Sandia Laboratories. Joined Univ of UT in 1968; 1975-1976, 1981-84, Chrmn of Mech & Industrial Engrg. Worked in heat transfer problems in solar and geothermal energy technologies and in biotech. ASME Section Chrmn, ASME Regional Secy and Chrmn of ASME Committee on Heat Transfer in Energy Systems. Mbr of UT State Energy Conservation

Boehm, Robert F (Continued)
and Development Council and Chrmn of UT State Solar Advisory Committee. Fellow of ASME. Author of two texts, numerous papers & chapters. *Society Aff:* ASME, ASEE, ISES, ΣΞ.

Boehringer, Ludwig C, Jr
Home: 58 Sunrise Cresent, Rochester, NY 14622
Position: Assistant Dir Mgmt Services Div *Employer:* Eastman Kodak Co *Education:* B.S./ME/Univ of Rochester; B.A./Econ/Univ of Rochester *Born:* 11/6/24 Native of NY City. Served in the U.S. Navy as an ensign during WWII. Joined the Eastman Kodak Co in 1947 as an industrial engr; where, after many assignments in all of the co's worldwide divisions, became assistant dir of its Industrial Engrg Division in 1976. In 1977, assumed current responsibility as assistant dir of the Management Services Division, which provides corporate-wide industrial engrg and information systems consulting services. Region V VP, IIE, 1981-1982. Listed in Marquis 1981-1982 editions of "Who's Who In The East" and "Who's Who In Finance And Industry." *Society Aff:* IIE.

Boenau, Ronald E
Business: 400 Seventh Street, S.W, Washington, DC 20590
Position: Transportation Engineer *Employer:* U.S. DOT, Urban Mass Transp. *Education:* MSCE/transportation/University of Maryland Grad. School; Certificate/traffic engr./Northwestern Univ. Grad. School; BCE/civil engr./University of Florida; AS/engr. tech./Manatee College; AA/gen. ed./Manatee College *Born:* 08/08/43 Developed and coordinated legislative strategies for a major U.S. Senate Bill resulting in $10 billion cost saving, U.S. House of Representatives' initiatives to develop joint public/private transportation projects, the implementation of transportation program plans in highway and mass transit safety. Recipient of U.S. Government's Congressional Fellowship; ASCE News National Award; Young Engineer of the Year (NSPE Society); Young Engineer of the Year (DC Engineers/Architects Council); Special Citations from UMTA, U.S. DOT and FHWA. Selected for training as U.S. representative to Nepal by the Peace Corps. Author/co-author of twelve transportation reports and publications. Special Assistant to U.S. Senator Paula Hawkins; U.S. Congressman Harold Hollenbeck; UMTA Administrator Arthur Teele; transportation engr with DeLeuw, Cather; Andrews & Clark; Florida DOT; FHWA and UMTA. Interested in Biblical counseling. *Society Aff:* ASCE, NSPE, ITE, TRB, AAES

Boerger, Philip T
Home: 4660 Elmherst Dr, Beaumont, TX 77706
Position: Vice President, Fossil Projs *Employer:* Gulf States Utilities Co. *Born:* May 1924. MSCE Univ of Minnesota; MS Internatl Affairs from George Washington Univ; BS from USMA. Ret as Brigadier General in US Army Corps of Engrs. Served as Engr, UNC/USFK/Eighth Army; Engr US Army, Pacific; & Div Engr, USA Engr Div, Missouri River. Joined Giffels in 1975 to dev & monitor projects outside the USA. Joined GSU in 1978 to supervise construction of fossil fueled power plants. NSPE; Fellow SAME; formerly pres Tehran & Omaha Posts, SAME; Far Eastern Regional V P SAME.

Boesch, Francis T
Business: Castle Point Station, Hoboken, NJ 07030
Position: Hd EECS Dept. *Employer:* Stevens Inst of Tech. *Education:* PhD/EE/Poly Inst of NY; MS/EE/Poly Inst of NY; BS/EE/Poly Inst of NY. *Born:* 9/28/36. in NY. On faculty of Poly Inst of NY, and Univ of CA at Berkeley. Mbr Res Staff Bell Tel Labs. White House Consultant to Exec Office of Pres of US. Founding editor of *Networks* journal; on editorial bd of *Journal of Graph Theory*. Elected Fellow of IEEE, and New York Acad. Sciences, and to mbrship in Eta Kappa Mu, Sigma Xi, and NY Acad of Sciences. Currently Charles Batchelor Prof & Hd of Elec Engg and Comp Sci Dept at Stevens Inst. Dist Commissioner Boy Scouts of Am. *Society Aff:* IEEE, ACM, NYAS, SIAM.

Boesch, Henry J, Jr
Business: 213 Tomahawk Ct, Bolingbrook, IL 60439
Position: Pres *Employer:* Boesch Consulting Engrs Inc *Education:* BS/CE/Tech Inst of Northwestern Univ *Born:* 11/28/27 Native of IL, Chicago Metropolitan Area. Broad background in construction and engrg for state, county, and heavy construction contractors. VP of Advance Consulting Engrs, Inc 1968-1976. Pres of Boesch Consulting Engrs, Inc 1976 to present. Firm is engaged in civil engrg for industrial parks & residential subdivisions, storm water management, land planning, & other related services. Enjoy classical & not-so-classical music & handguns. *Society Aff:* ASCE, NSPE, ACI, ITE, AFA, AREA, ACEC, AWWA

Boettcher, Harold P
Home: 19285 Lothmoor Dr Lower, Brookfield, WI 53005
Position: Prof. *Employer:* Univ of WI. *Education:* PhD/Mechanics & EE/Univ of WI; MS/EE/Univ of WI; BS/EE/Univ of WI. *Born:* 7/24/23. Eagle, WI. Served with US Navy 1944-46. Instr in Mechanics 1946-54 at Univ of WI-Madison. Dir Elec Motor Res Lab A O Smith Corp 1954-61. Assoc Prof (1961- 66) & Prof (1966-present), Dept of Elec Engg & Comp Sci, Univ of WI-Milwaukee. Served as Dept Chrmn 1965-70 & 1974-76. Interests in the areas of control systems, electromagnetomech energy conversion, and applications of computers & computing techniques to problem solving. Registered Professional Engineer in Wisconsin. *Society Aff:* IEEE, ASEE, ΣΞ.

Bogart, Harold N
Home: 2988 Ishpening Trail, Traverse City, MI 49684
Position: Dir-Mfg Engg & Sys Office; Private Conslt *Employer:* Ford Motor Co - Retired 12/31/78. *Education:* SM/Indstrial Mgmt/MA Inst of Technology; BS/Chem Engg/MI State Univ. *Born:* Sept 1916. BS Mich St Univ; MS Indus Mgmt (Sloan Fellow) MIT. Reg Prof Engr chem, met engrg. Joined Ford 1937. Respon for dev & introduction of new mfg processes in foundry, mech working & machining plants, met opers, & painting & plating activities. P Pres Amer Soc for Testing & Materials. Hon Fellow Amer Soc for Metals; received ASM William Hunt Eisenman Award. Mbr SAE, AFS, ASM, Amer Chem Soc, Amer Inst of Mining & Met Engrs. *Society Aff:* SAE, ASM, AFS, SME, AIME, ACS.

Bogdanoff, John L
Business: 120 A&ES Bldg-Purdue, West Lafayette, IN 47907
Position: Professor. *Employer:* Purdue Univ. *Born:* May 25, 1916. BME Syracuse Univ, MSE Harvard Univ, PhD Columbia Univ. In 1950 became Assoc Prof in School of Civil Engrg at Purdue Univ attached to the Mech Dept. Became Prof in Div of Engrg Sci in 1953. From 1967-72 was Assoc Hd & Hd in Sch of Aeronautical & Astronautical Engrg. Fellow of Am. Assoc. Adv. Science Mbr Tau Beta Pi, Sigma Psi, Amer Soc of Mech Engrs, Amer Inst of Aeronautics & Astronautics, Amer Phys Soc; Fellow Amer Soc of Mech Engrs & Mbr Natl Acad of Engrg. Fellow of Am. Assoc. Adv. Sci. Int: dynamics, vibration, fatigue, crack growth & wear & stochastic processes applied to engrg. Has num pubs.

Boge, Walter E
Home: 2004 Kenley Ct, Alexandria, VA 22308
Position: Chief Systems Studies Div *Employer:* US Army Engr Topographic Labs. *Education:* MS/CE/Purdue Univ *Born:* 05/01/37 With the Army Corps of Engrs since 1960, involved in R&D of advanced map data reduction, & image exploitation sys. Attended grad sch at Syracuse Univ 1962-63 & Purdue Univ 1972-73 where he received MS in Civil Engrg. Presently Chief of Systems Studies Div Computer Scis Lab. Mbr Chi Epsilon, Phi Kappa Phi, and Amer Soc of Photogrammetry. Likes to play tennis & bridge. *Society Aff:* ASP

Bogen, Alfred T, Jr
Home: 6068 Pear Orchard, Jackson, MS 39211
Position: V P, Engrg & Dev. *Employer:* First Mississippi Corp. *Education:* BS/ChE/MS State Univ; Postgrad/ChE/LA State Univ. *Born:* Jan 14, 1919 Ariz. BS

Bogen, Alfred T, Jr (Continued)
ChE Mississippi State Univ 1941. Major U S Army 1941-45. Process Engr Esso Standard Oil Co 1946-52. Tech Adv to Process Supt STANVAC - Sumatra 1952-55. Mgr Process & Eval Chemstrand Corp 1955-64. Genl Mgr Monsanto Internatl Engrg - Latin Amer 1964-67. Dir Engrg First Mississippi Corp 1967-72. Elected V P 1972. Respon for corporate engrg, dev, real estate, oil & gas explor, & aviation activities. Reg Engr Fla & Miss. Hobbies: flying, ham radio, work shop. *Society Aff:* AIChE, ACS, ASM.

Bogen, Samuel Adams
Home: 20 Reid Ave, Port Washington, NY 11050
Position: Pres, Treas, Dir. *Employer:* Bogen Jenal Engrs P C. *Education:* EE/-/Columbia; BSE/-/Columbia; BS/-/Columbia. *Born:* Mar 1913. Pres Bogen Jenal Engrs PC, 20 Reid Ave, Port Washington, NY. Born NYC, ed NYC public schools, Columbia 1928-34. Var employers 1934-47. Formed S A Bogen Engrs 1947, succeeded by Bogen Jenal Engrs PC 1971. Vice-Pres. & Dir. Bogen, Johnston, Lau & Jenal P.C., 1976-1981.Adj Prof of Elec Engrg, Polytechnic Inst of Brooklyn 1950-51. Secy CEC/US 1963-64, Treas 1964-65, Pres Elect 1966- 67, Pres 1967-68. Dir & Chmn of Bd, Design Professionals Insurance Co 1971; Pres 1972-76. Port Washington School Bd 1968-73. Honor Award 1969 CEC/US. Fellow ACEC. *Society Aff:* IEEE, IES, ACEC.

Bogert, Ivan L
Business: 2125 Center Avenue, Ft. Lee, NJ 07024
Position: Managing Partner. *Employer:* Clinton Bogert Assoc. *Education:* BS/CE/Cornell Univ *Born:* 3/24/18. Mr Bogert, managing partner of Clinton Bogut Associates, has over 47 years experience in civil and sanitary engrg. Mr. Bogert joined Clinton Bogert Associates (predecessor firms) in 1942, and became managing partner in 1949. He has been directly involved with project design/design direction since joining the firm in such project areas as water pollution control, water supply/treatment/distribution, solid waste management, drainage/flood control. He has been responsible for the development of numerous design innovations that reduced project operating and capital costs. He is a member of seventeen major professional organizations and holds P.E. Licenses in fourteen states. *Society Aff:* SAME, ASCE, IAASE, NSPE, AAA, AGU, ACEC, APHA, APWA, ASPO, AWWA, IAWPRC Intl, IWRA Intl, ISWPCA Intl, IWSA Intl, RSH, WPCF.

Boggess, Jerry R
Home: 3280 Heatherfield Dr, Hacienda Heights, CA 91745
Position: Corp VP *Employer:* Grinnell Co. *Education:* B/ME/Univ of UT *Born:* 2/1/44 Native of Midvale, UT. Designed the Fire Protection System in the U of Engrg Bldg. Active in the dev and acceptance of hydraulically calculated sprinkler systems. Tenth person in the State of CA to become a PE. Pres of the Soc of Fire Protection Engr, Chrmn of the Joint Apprentice Committee for the training of pipe fitters in L.A. on Bd of Regents of the Glendale Community College. Enjoys backpacking, snorkeling, and jogging. Active mbr of the Mormon Church. *Society Aff:* ASME, SFPE, NASA

Boggs, Andrew T, III
Home: Box 1127 Hemlock Farms, Hawley, PA 18428
Position: Exec VP Emeritus *Employer:* ASHRAE. (Retired) *Education:* BS/Chemistry/Norwich Univ. *Born:* Aug 1920. Chemistry degree from Norwich Univ 1947. Armed Forces ETO 1943-46 (Belgian Military Cross); USAR 1946-65; retired Major. Asst to Asst Dir- Secy Edison Electric Inst 1947-55 developing specialized indus studies; secy of indus heat pump ctte. Joined ASHRAE as Tech Secy 1955; Assoc Secy 1961; Exec Secy/Corp Secy 1967; Exec VP 1978; Exec VP Emeritus 1985. *Society Aff:* ASHRAE.

Boggs, Robert G
Home: 119 Colony Rd, Groton, CT 06340
Position: Prof and Chmn, CE *Employer:* US Coast Guard Academy. *Education:* PhD/Applied Mech/Univ of CT; MS/Mech Engrg/Univ of CA-Berkeley; BS/Mech Engrg/Univ of Co; BS/Bus/Univ of CO. *Born:* 3/6/30. Design engr at US Naval Ordnance Test Station 1953-54. Military service in Coast Guard as Exec officer of USCGC Hornbeam and USCGC Ewing. Systems Engr, Gen Dynamics/Elec Boat Div 1959-62. Faculty mbr at US Coast Guard Academy since 1962. Developed Coast Guard Academy Civil Engg Prog, staff and faculty from its inception to full ABET accreditiation. Two terms as Academy Faculty Senate Pres. ASEE Mechanics Div Chrmn 1979-80; Professional Interest Council Chairman, 1985-87; Board of Directors 1985-87, Vice President 1986-87. SAME Oren Medal 1978. Western Electric Fund Award, 1979. Author *Elementary Structural Analysis,* Holt, Rinehart, Winston 1984. *Society Aff:* ASEE, ASCE, SAME.

Bogina, August, Jr
Business: 9020 Rosehill Rd, P.O. Box 14515, Lenexa, KS 66215
Position: Pres. *Employer:* Bogina Consulting Engr. *Education:* BS/Engr/KS State U *Born:* 9/13/27. Native Kansan. US Army 1945-46 - Field Engr, US Geol Survey - Civ engr, Corps of engrs - Chief of Civ Engr Dept, Bayles & Assoc - Construction Engr, Campbell & Assoc - Owner, Bogina & Assoc - Pres, Bogina Consltg Engrs, Inc - Partner-Bogina Petroleum Engrs - City engr (Retainer Basis) Lenexa and Fairway, KS and Harrisonville, MO - Pres, Lenexa Area Chamber of Commerce — Chrmn Lenexa Planning Commission - Chrmn Lenexa Bd of Zoning Appeals - Chrmn, Lenexa Bd of Code Appeals - KS House of Rep (30th Dist) (three terms) Ks Senate (10th Dist) Second Term) Chrmn, House and Senate Joint Committee on Bldg Construction. Chrmn, Senate Ways and Means Committee. *Society Aff:* NSPE, KES, KCE, MSPE.

Bogner, Neil F
Home: 6903 Gillings Rd, Springfield, VA 22152
Position: Assoc Deputy Chief, Tech Dev & Application *Employer:* USDA Soil Cons Ser. *Education:* BS/Engg/Univ of IL; BS/Agri/Univ of IL. *Born:* 5/13/26. Born and raised on a farm near Henry, Ill. Graduated from the Univ of IL in 1950 with degrees in engrg and in agriculture. Joined Soil Conservation Service as a student trainee. Following graduation spent next 11 yrs in various field engrg positions. In 1962 became a regional construction engr. Served in this capacity in the Midwest and in the Northeast. In 1972 after 5 yrs as hd of Northeast engrg office moved to Natl Office. In 1974 named as Dir, Engrg Div for Soil Conservation Ser. In Feb 1981 became Assoc Deputy Chief for Tech Dev & Application. In this position shares in the leadership and direction of engr economics, social, science, ecological scis and intl programs. Mbr of numerous profl societies-ASAE, ASCE, SCSA, ICID, USCOLD; Active in natl activities of ASAE. ASAE Fellow 1980. Also elected to position of Tech Dir SW term to start June 1982. Mbr of Interior Review Grp that investigated failure of Teton Dam. *Society Aff:* ASAE, ASCE, SCSA, ICID, USCOLD.

Bogue, Donald C
Business: Dept Chem, Met & Polymer Engg, Knoxville, TN 37916
Position: Prof. *Employer:* Univ TN. *Education:* PhD/ChE/Univ DE; BS/ChE/GA Tech. *Born:* 1/9/32. Recipient 1967 Allan P Colburn Award (AIChE) employment Esso Std Oil, Baton Rouge; since 1960 Univ of TN; now active in Polymer Engg; spent three yrs at Kyoto Univ, Japan. *Society Aff:* AIChE, SPE, ACS, SR, SR(J).

Bogue, Stuart H
Business: 17000 Twelve Mile Rd, Southfield, MI 48076
Position: President. *Employer:* Pate, Hirn & Bogue Inc. *Education:* MS/Civil Engr/Univ of MI; BS/Civil Engr/MI State Univ. *Born:* Dec 1923 Wabash Cty, Ind. Entered US Army in 1943 & served as tank gunner in the European theater. Received a BSCE from Michigan State Univ 1949 & MSCE from Univ of Michigan in 1951. Design Engr for Pate & Hirn for sev yrs before becoming Secy & then Pres of Pate, Hirn & Bogue Inc. Principal designer of var water & waste water projects in southeastern Michigan. Mich Sect Chmn of Amer Water Works Assn 1965 & Natl Dir in 1971-74. Received AWWA's Fuller Award in 1973. Mbr Amer Acad of Environ Engrs. *Society Aff:* AWWA, WPCF, AAEE, NSPE, ASCE.

Bohl, Robert W
Business: 206 MMB, 1304 W. Green St, Urbana, IL 61801
Position: Prof. *Employer:* Univ of IL. *Education:* PhD/Met Engr/Univ of IL; MS/Met Engr/Univ of IL; BS/Met Engr/Univ of IL. *Born:* 9/29/25. Native of Peoria, IL. On staff of Univ of IL at Urbana since 1946; currently Prof of Met & Nuclear Engg, Emeritus Special interests are nuclear mtls, physical metallurgy, and failure analysis. Consultant to several cos and various legal firms. Fellow of Am Soc for Metals. Publications in several technical journals. Distinguished Teacher Award-Urbana campus-1979, A.E. White Award (ASM) 1980. *Society Aff:* ASM, ANS, ASEE.

Boice, Calvin W
Home: PO Box 81, Richland, WA 99352
Position: President *Employer:* Braun Hanford Co *Education:* BS/EE/IA State Univ. *Born:* 7/5/25. Native of CA. Served in USN 1943-1946. With C F Braun & Co since 1951, involved in design of large industrial facilities. Became Mgr of EE in 1963. Became VP & Mgr of Power Div in 1977. In 1981 became Pres of Braun Hanford Co, a wholly-owned subsidiary of C F Braun & Co. Braun Hanford Co provides architect engrg service for US Dept of Energy, Richland Operations Office. Reg engr in 10 states. *Society Aff:* IEEE, ASME.

Boily, Michel S
Business: 361 Speedvale W, Guelph Ontario N1H 6M3 Canada
Position: Technical Marketing Mgr. *Employer:* Foseco Canada Inc. *Education:* MBA/Marketing/McGill Univ; BEng/Met/McGill Univ. *Born:* 12/18/48. Born in Montreal, Quebec, 1948. Studied engg at McGill Univ (Met Dept) from which grad in 1972. For first yrs after grad, worked in refractory sales then in process tech (heat treating) of gear components for aircraft engines. Opportunities led to the steel industry first as QC Engr (physical met of rolling) then as consultant and supplier to primary steel fabricators. With Foseco Canada since early 1976 & presently steel mills div technical marketing mgr. Obtained my MBA in Dec 78 after 4 1/2 yrs of part time studies at McGill. Currently responsible for product policy & dev as well as acting as asst gen mgr for the div. Enjoy cross country skiing, music & camping. *Society Aff:* ASM, AIME, OEQ.

Boland, Joseph S, III
Business: Elec Engrg, Auburn, AL 36830
Position: Prof & Asst Dean of Engrg *Employer:* Auburn Univ. *Education:* PhD/Elec Engg/GA Inst of Tech; MSEE/Elec Engg/Auburn Univ; BEE/Elec Engg/Auburn Univ. *Born:* Sep 23, 1939, Montgomery, Alabama. Army Signal Corps 1962-64. Served on Auburn Univ fac since 1968. Reg PE Ala. Principal investigator & project leader on sev res contracts. Supervised num PhD & Master's theses. Author num papers pub in refereed journals & tech reports on control of space vehicles. Presented num papers at tech conferences. Cons 1973- on terminal homing missile guidance sys. Mbr Sigma Xi, Eta Kappa Nu, Tau Beta Pi, Phi Kappa Phi, ASEE. Sr Mbr IEEE. ASEE Dow Outstanding Young Fac Award. ASEE Western Electric Fund Award. *Society Aff:* IEEE, ASEE

Bolduc, Oliver J
Home: 719 N Douglas, Arlington Heights, IL 60004
Position: Operations Manager *Employer:* Comstock Engineering, Inc. *Education:* BS/Chem Engr/Univ of IL. *Born:* 7/19/24. Native of Chicago, IL. Served with Army Air Corps 1943-46, 1st Lt Navigator. Chem Process Engr, Armour & Co 1950-54. Proj Engr Blaw-Knox Co 1954- 58. Chief Process Engr, Davidson-Kennedy Assoc 1958-61. Asst Div Mgr, The Austin Co 1961-71. VP of Engg, Mid-America Engg 1971-77. VP, Midwest Region, Jacobs Engg Group 1977-1980. General Mgr, Intensa Inc Consltg Engrs. 1980-1982, Mgr Engrg Services, Comstock Engrg Inc. Twenty five yrs in Chem, Pharmaceutical and Industrial Plant Design & Construction. Reg PE in 13 states. Interests include golf, sailing and travel. *Society Aff:* AIChE, AACE, AISE.

Boley, Bruno A
Business: Dept. of Civil Engrg & Eng. Mechanics, Columbia Univ, New York, NY 10027
Position: Prof of Civil Engrg. *Employer:* Columbia Univ *Education:* ScD/Aero Engg/Polytechnic Inst of Brooklyn; MAeE/Aero Engg/Polytechnic Inst of Brooklyn; BCE/Civil Engg/College of the City of NY. *Born:* 1924 Italy. Asst Dir of Struct Res at PIB; Goodyear Aircraft Corp 1948- 50; Assoc Prof Aero Engrg Ohio State Univ 1950-52; Prof of Civil Engrg Columbia Univ 1952-68; J P Ripley Prof & Chmn, Mech Dept Cornell Univ 1968-72; Dean & W P Murphy Prof, Northwestern Univ 1973-87. Prof of Civil Engineering, Columbia University 1987- . Mbr Natl Acad of Engrg; ASME, (Hon. Mbr) Amer Acad Mech Soc Eng Sci; AIAA Fellow; NSF Fellow; NATO Fellow; AAAS Fellow. Pres: Amer Acad Mech; Soc of Engrg Sci & Assoc of Chmn of Depts of Mech. Chmn: U S Natl Ctte on Theoretical & Applied Mech; Applied Mech Div ASME; Advisor Genl, Struct Mech in Reactor Tech. Ed-in-Chief: 'Mech Res Communications'. Author: *Theory of Thermal Stresses* (Wiley 1960); *Thermoinelasticity* (Spring 1970); *High Temp Struct & Mtls.* (Pergamon 1964) *Crossfire in Prof Educ,* (Pergamon 1977); about 100 tech papers. Secy, Intl Union of Theoretical & Appl Mechs Congress Exec Committee; Chrmn, Engrg Education Study Panel of the Natl Acad of Engrg; (Chrmn, IL Council for Res Dev & Demonstration; Chrmn, Midwest Program for Minorities on Engrg). Sc.D. (Hon.) CCNY 1982; Mbr of Bd of Gov, Argonne Natl Lab 1983-86; Mbr of Bd of Gov, ASME, 1984-86. *Society Aff:* ASME, AIAA, ASEE, SES, WSE, AAM, AAAS

Bolgiano, Ralph, Jr
Business: Phillips Hall, Ithaca, NY 14853
Position: Prof. *Employer:* Cornell Univ. *Education:* PhD/Radiophys/Cornell Univ; MEE/EE/Cornell Univ; BEE/EE/Cornell Univ; BSEE/EE/Cornell Univ. *Born:* 4/1/22. Mr Bolgiano, born in Baltimore, MD, earned the BEE from Cornell Univ in 1947, following four yrs with the Signal Corps in hf communications. On completion of the MEE in 1949 he spent five yrs in uhf communications systems dev with GE Co. In 1958 he was awarded the PhD, by Cornell, in radiophysics. Since that time he has been a mbr of the EE Faculty at Cornell. In addition to his teaching he has worked primarily in the field of radiowave propagation & its relation to atmospheric structure, the aerodynamics of that structure & remote exploration thereof by radio techniques. *Society Aff:* AMS, AGU, IEEE.

Boll, Harry J
Business: 600 Mountain Ave, Murray Hill, NJ 07974
Position: Supvr *Employer:* Bell Labs *Education:* PhD/EE/Univ of MN; MS/EE/Univ of MN; BS/EE/Univ of MN *Born:* 03/05/30 After completing graduate work in secondary electron emission, he joined Bell Labs, Murray Hill, NJ where he worked in the general areas of semiconductor devices and integrated circuits, including solid state transducers, random access memories thyristor switching arrays and custom integrated circuit logic design. He is presently a group supvr in the Advanced Development Lab responsible for fine line MOS process monitoring and forthe design and testing of very high speed MOS circuits. In 1980 he was elected to the fellow grade of the IEEE "For the development of novel semiconductor devices and for leadership in large scale integrated circuit design." *Society Aff:* IEEE

Bollard, John R H
Business: Dept of Aero & Astro, Seattle, WA 98195
Position: Prof. *Employer:* Univ of WA (205 Gugg Hall). *Education:* PhD/Aeronautics/Purdue; MS/Struct Mech/Univ of New Zealand; BS/CE/Univ of New Zealand. *Born:* Jul 13, 1927 New Zealand. BS & MS Univ of New Zealand in Struct Mech. PhD Purdue Univ 1954 in Aeronautics. Lect at Univ of New Zealand & cons engr to New Zealand aircraft indus 1954-. Prof & dir aerospace lab, Purdue Univ 1956-60. Visiting Prof Caltech 1960. Prof & Chmn Dept of Aeronautics & Astronautics, Univ of Wash 1961-76, Prof 1976- . Founder, bd mbr, & cons Mathematical Sci Northwest Inc. Assoc Fellow AIAA, P Mbr Structures Ctte, Ed Ctte. Mbr ASEE, P Chmn Aerospace Div. Mbr JANNAF, P Chmn Struct Integrity Ctte,

Bollard, John R H (Continued)
Tau Beta Pi, Sigma Xi. Visiting Ctte Jet Propulsion Lab, Materials Ctte NAS. Cons expert NASA. ACE Fellow 1967 in residence Univ of Calif Berkeley. Some 60 pubs in field. *Society Aff:* AIAA

Bolle, Donald M
Business: 308 Packard Lab 19, Bethlehem, PA 18015
Position: Dean, College of Engineering and Physical Sciences *Employer:* Lehigh University *Education:* PhD/EE/Purdue Univ; BSc/EE/Durham Univ. *Born:* 3/30/33. Born in Amsterdam, the Netherlands, where he received his primary education. Secondary and college education were pursued in the UK. Emigrated to the US in 1955, naturalized 1961. Married Barbara J Girton, 1957, and has four sons. Res Engr, EMI, Hayes, UK, 1954-55. Held various positions at Purdue Univ, 1956-62. NSF Postdoctoral fellowship at Cambridge Univ 1962-63. Brown Univ in 1963-1980. Joined Lehigh University - Chairman, Department of Electrical and Computer Engineering in August 1980, Dean, College of Engineering and Physical Sciencies, from July 1, 1981. Has consulted widely with industrial and fed labs. Enjoys travel, classical guitar, art, squash and golf. *Society Aff:* IEEE, ASEE, AAAS.

Bollinger, John G
Business: 1513 Univ Ave, Madison, WI 53706
Position: Dean, Coll of Engrg, Prof of Mech Engrg *Employer:* Univ of Wisc-Madison. *Education:* PhD/Mech Engr/Univ Wisc-Madison; MS/Mech Engr/Cornell Univ; BS/Mech Engr/Univ of Wisc-Madison *Born:* 05/28/35 Grand Forks, ND. Raised in NY, graduating from Manhasset H S 1953. BS ME from Univ WI in 1957, MS ME from Cornell 1958 continuing for PhD at Univ WI in 1961. 1973 Bascom Prof of ME, 1974-79 Chrmn Dept of ME, Univ WI-Madison. Author over 80 tech pubs, num pats & textbook in field of automatic controls. Received num awards, incl Fulbright Postdoc Fellow 1962, Donald P Eckman Award 1965, Pi Tau Sigma Gold Medal 1965, Adams Memorial Mbrship Award of AWS in 1970, Pi Tau Sigma Gustus L. Larson Memorial Awd, ASME, 1976; Res Medal, SME, 1978; ASME Centennial Awd 1980; Fulbright Scholar 1980, Visiting Prof at Cranfield Inst of Tech England, and Univ of Stuttgart, Germany. Teaching, cons & res in automatic control, computer control & noise control. Mbr, Natl Acad of Engrg, 1983. Reg PE WI. 1981 Dean Coll of Engrg, Univ of WI-Madison. *Society Aff:* ASME, ASEE, SME, AWS, CIRP, ΣΞ

Bolt, Bruce A
Business: Seismographic Sta, Earth Sci Bldg, Berkeley, CA 94720
Position: Prof *Employer:* Univ of CA *Education:* DSc/Geophysics/Univ of Sydney; PhD/Applied Math/Univ of Sydney; BSc/Math/Univ of Sydney *Born:* 2/15/30 Native of Australia. Appointed Prof of Seismology and Dir of Seismographic Stations, Univ of CA, Berkeley, 1963. Naturalized US 1976. Pres Seismological Society of America 1974-75. Pres Intl Assoc of Seismology and Physics of the Earth's Interior, 1979-83. Earthquake and Wind Forces Committee, Veterans Administration, 1971-75. Chmn, Electric Power Res Inst Seismic Center Advisory Bd, 1984- . Commissioner, Seismic Safety Commission, CA 1978 to present, chrm 1984-. Trustee, CA Acad Sciences, 1981; Pres 1982-. Elected member National Academy of Engrg, 1978. Consultant on strong ground motion assessment. Author of texts on applied math, seismology, and hazards. Enjoys swimming and sailing. *Society Aff:* EERI, AGU, SSA.

Bolt, Jay A
Home: 2300 Melrose, Ann Arbor, MI 48104
Position: Prof of Mech Engrg. *Employer:* Univ of Michigan. *Education:* MS/Mech Eng/Purdue Univ; MME/Automotive/Chrysler; BSXME/Mech Eng/MI State. *Born:* Oct 1911. Native of Mich. Taught at Univ of Mich, Univ of Ill, Univ of Notre Dame before WW II. Dir of aircraft carburetor res for Bendix Corp 1941-47. Assoc Prof of Mech Engrg at Univ of Mich 1947-53. Prof 1953- . Dir Walter E Lay Automa tive Lab, & respon for much of teaching & res concerning internal combustion engines at Univ of Mich. SAE Horning Mem Award 1972. Cons to indus & gov on automotive engrg, engines, fuel controls, & exhaust emissions. Adv to legal profession on products liability & patents. *Society Aff:* IME, SAE, ASME, ASEE, NSPE.

Bolt, Richard H
Home: Tabor Hill Rd, Lincoln, MA 01773
Position: Chmn of the Bd, Emeritus. *Employer:* Bolt Beranek & Newman Inc. *Born:* Apr 22, 1911 Peking, China. PhD Physics Univ of Calif 1939. MIT: Tech Dir, Underwater Sound Lab, during WW II; Dir, Acoustics Lab 1945-57; Prof of Acoustics, Dept of Elec Engrg; Lecturer, Political Sci. Assoc Dir, Natl Sci Foundation 1960-63. Chmn of Bd, Bolt Beranek & Newman Inc until retirement 1976 - now cons. Former Pres Acoustical Soc of Amer; received its Biennial Award. Pres Internatl Comm on Acoustics 1951-57. Fellow IEEE, Acoustical Soc of Amer, Amer Phys Soc, Amer Acad of Arts & Sci. Pubs in architectural acoustics, noise control, sonics, sci & tech in public policy.

Bolte, Charles C
Home: 809 Philadelpha Ave, Chambersburg, PA 17201
Position: Exec VP *Employer:* American Business Consultants *Education:* MBA/Gen Mgr/Colgate Darden, UVA; BIE/Ind Engrg/GA Tech *Born:* 1/8/45. ANAK, ODK, KOSEME, Who's Who while at GA Tech. Served as Commissioned Officer in Southeast Asia with Army Ordnance Battalion. Marketing & Commercial Product Dev for W R Grace & Co on "Letterflex" photopolymar relief printing plate 1972-1974. VP for Wm Byrd Press from 1974 to 1984 in charge of all new product dev and tech implementation. On faculty of PIA's Exec Dev Prog since 1975. Dir, Graphic Arts Div, AIIE, 1978-1980. Chairman of the Techno-Economic Forecasting Committee & on the Exec Bd of the Graphic Arts Technical Foundation Serve as Elder on the Session of Presbyterian Church of the Falling Springs. Serve on the Bd of Dir of the Chambersburg Rotary Club. Published in numerous Graphic Arts trade journals. Has co- chaired a dozen tech conferences for printers. *Society Aff:* ODK, ANAK, KOSEME, IIE.

Bolz, Harold A
Home: 3097 Herrick Rd, Columbus, OH 43221
Position: Dean Emeritus-Coll of Engrg. *Employer:* Ohio State Univ. *Education:* MSME/Mech Engg/Case Western Reserve Univ; BSME/ME/Case Western Reserse Univ *Born:* May 27, 1911. Dev engr Weatherhead Co 1935-38. Asst & Assoc Prof mech engrg 1938-46 & headed Genl Engrg Dept 1946-54 at Purdue. At Ohio State Univ Assoc Dean & Prof of Mech Engrg 1954-58, Dean 1958-76, Dean Emeritus 1976- . Interim Pres, OH Northern Univ 1979- . NSPE & OSPE, Fellow ASME, Pres ASEE 1971- 72, Sigma Xi, Tau Beta Pi, Pi Tau Sigma, Theta Tau, Trustee Goodwill Indus Central Ohio. BSME 1933, MSME 1935, Case Western Reserve Univ.; D.Eng. (Hon. 64) Purdue Univ.; D.Eng. (Hon. 73) Tri-State Univ.; D.Hum. Let. (Hon 80) Ohio Northern Univ.. *Society Aff:* ASME, NSPE, ASEE.

Bolz, Ray E
Home: 21 Lantern Ln, Worcester, MA 01609
Position: V P & Dean of Faculty (Retired) *Employer:* Ret'd. *Education:* D Eng/ME/Yale; MS/ME/Yale; BS/ME/Case Inst of Tech *Born:* 1918. In 1942 joined Langley Lab of NACA (now NASA), transferred to Lewis Lab in Cleveland, Ohio, & became Hd, Jet Engine Combustion Sect 1945. In 1946 returned to Yale for D Engrg work (received in 1949) & then taught at RPI as Asst Prof of Aeronautics 1947-50. In 1950 accepted the post of Assoc Prof of ME at Case Inst of Tech & successively held the pos of Hd ME Dept 1956, Leonard Case Prof & Hd of the Engrg Div 1962, & Dean of School of Engrg of CWRU (after federation of Case & Western Reserve Univ) in 1967. In 1973 he became VP & Dean of Faculty at WPI in Worcester, MA. Has been involved in many activities in ASME & ASEE incl mbrship on ASME Bd of Ed (1966-70), Chmn of Bd & V P for Ed of ASME (1968-70), & Chmn of the Grad Studies Div of ASEE (1966-67). Has been active as an advisor to sev activities of NSF, mbr of the Ed & Accreditation Ctte for ECPD (1962-67). He retired 6/30/84. *Society Aff:* ASME, ASEE.

Bombach, Otto F
Home: 4023 Exultant Drive, Rancho Palos Verdes, CA 90274
Position: Hd Plant Engrg. *Employer:* Hughes Aircraft Co. *Education:* BS/ME/IA State Univ. *Born:* Jun 24, 1922. Reg PE Ill & Calif. Native of Blairstown, Iowa. Served with Army Corps of Engrs 1942-45 & 1951-53. Plant Engr with All-Steel Equip Co for 19 yrs. Designed material handling sys & special automated manufac equip. With Hughes Aircraft Co since 1968. Respon for facils, maintenance, construction, office services, environ impact. Plant Engr of Yr for Los Angeles chap of AIPE 1973 & for Western Region of U S in 1974. Presently Chrmn Western Region AIP Conference Committee. Fellow in Amer Inst of Plant Engrs. Active in church work. *Society Aff:* AIPE.

Bomberger, Howard B
Home: 385 Bradford Dr, Canfield, OH 44406
Position: Conslt, Titanium *Employer:* Self Employed *Education:* PhD/Met Engg/OH State Univ; MS/Met Engg/OH State Univ; BS/Met/PA State Univ. *Born:* 2/8/22. The candidate has made important contributions to the full range of titanium tech over the past 35 yrs and has also conducted basic studies on other metal systems. From this effort he authored over 65 technical papers and received 31 US patents and more than 60 foreign patents. This work included directing the activities of more than 170 technical personnel and it resulted in the dev of several new titanium alloys and a number of lower-cost production processes. Responsible positions were held in indus and in local, natl & intl technical organizations. A series of grad-level met courses are being taught at the Youngstown State Univ. He consults for the US Air Force and indus. He served in the US Army in WWII. Elected Fellow-ASM. Awards for Professional Achievement include: D.F. McFarland, Penna State U.; U.S. Congress, Apollo *Society Aff:* ASM, AIME, TMS, NACE, ECS, ΣΞ.

Bonar, Ronald L
Business: 616 South Harrison St, Ft Wayne, IN 46802
Position: Pres. *Employer:* Bonar & Assoc, Inc. *Education:* BSCE/Civ Engg/IN Inst of Tech. *Born:* 11/8/37. Native of Farmdale, OH. Employed by City of Ft Wayne, IN for 15 yrs working under 5 different mayors. Held positions of Chief WPC Engr, City Engr, Mbr of Bd of Public Works, Mbr of City Plan Commission; VChrmn & Chrmn of City-County Bldg & Minimum Housing Dept. Directed all engg dept with over 120 personnel. IN 1976 began own consulting engg firm which now has over 30 employees. Specializes in municipal/sanitary engg. Past Pres and current Dir of In Water Poll Control Assn, Past Pres and Dir of the IN Chapter of APWa, Diplomate of the American Academy of Environmental Engrs. *Society Aff:* IWPCA, APWA, ACEC, AWWA, ISPE, AAEE.

Bonasso, Samuel G
Business: PO Box 1200, Morgantown, WV 26505
Position: President. *Employer:* Alpha Assocs Inc. *Education:* MS/Civil Engrg/West VA Univ; BS/Civil Engrg/Univ of Miami *Born:* Dec 1939 Wyatt, W Vir. Attended public schools in W Vir. Taught at W Vir Univ Civil Engrg Dept 1962-64. Fortune Engrg Associates, Alexandria, Vir 1964- 65. Founded Ski Lift Internatl, aerial tramway design-manufacturing construction firm 1965. Founded Apha Associates Inc cons engrs and architects 1969. Adjunct Professor, Civil Engineering, West Virginia University. Past Secretary, A.N.S.I. B77 Committee on Aerial Tramways. Listed in Engrs. of Distinction, 2nd edition. Registered in W Vir, Colorado, Pennsylvania, North Carolina, Ohio, MD, DE, VA, and LA. Int: golf, skiing, the Creative Process. *Society Aff:* IABSE, ASCE, NSPE, NAFE

Bond, Frederick E
Business: PO Box 92957, Los Angeles, CA 90009
Position: Sr Engr - Sat Sys Div. *Employer:* Aerospace Corp. *Education:* DEE/Communication/Polytech Inst of Brooklyn; MSEE/Communications/Rutgers Univ; BSEE/Communications/Drexel Univ. *Born:* Jan 1920. Army Signal Corps in Europe 1942-46. In communications R&D at Signal Corps Labs, Ft Monmouth 1946-57; TRW Inc 1957-60; Ryan Communications Inc 1960-61; & Westinghouse Astroelectronics Labs 1961-64. Satellite Communications Sys Engrg at Aerospace Corp 1964-73 & 1977-present. Deputy Dir DCA, with respon for sys arch for all DOD Satellite Communications 1973-76. Fellow IEEE. Mbr Sigma Xi, Tau Beta Pi, Eta Kappa Nu. Lt Col US Army Ret Reserve. *Society Aff:* IEEE.

Bond, Horatio L
Home: 30 First Ave, PO Box 393, Hyannis Port, MA 02647
Position: Consulting Engr (retired) *Education:* S.B./Eng.Adm. M & E/M.I.T. *Born:* 11/30/00 Native of Barnstable, MA. Chief Engr, National Fire Protection Association, 1938-1968. Provided service to industries, institutions, municipalities and other agencies. WW II consultant to NDRC, 1942-1945. Member Committee on Fire Research, NAS-NRC, 1956-1965. Authored, edited and contributed to books, reports and periodicals which are standard references to fire. Since 1968, until present retirement, continued similar work in private practice. Pres, M.I.T. Alumni Association, 1953-1954, Corp of M.I.T. Member, 1953-1959. From Engrg Societies of New England, "New England Award 1982–. *Society Aff:* SFPE, NFPA, AWWA, IFIREE (UK)

Bond, Theodore E
Business: Bldg 005, BARC-West, Beltsville, MD 20705
Position: Supervisory Agri Engr. *Employer:* U S Dept of Agri. *Education:* DrEngg/-/Univ of CA, Davis; MS/Civil Engg/IA State Univ, Ames, IA; BS/Agri Engg/Univ of CA, Davis. *Born:* 10/31/17. Dr Engg & BS Univ of CA, MS IA State Univ. Engg sect Army Corps Hdqtrs 1943-45. With ARS-USDA since 1948. From 1948-70, conductred res & led a natl team on problems of defining & providing optimum growing environments for livestock. From 1970-73, conducted res on control of livestock-produced pollution at the US Meat Animal Res Ctr, Clay Ctr, NE. 1973-79 res leader on housing problems for low income rural families at the ARS Rural Housing Res Unit, Clemson, SC. ASAE Metal Bldg Manufacturer's Award 1959; ASAE Fellow 1968. Since 1979 Program Priorities Analyst, Sci & Educ Adm USDA, Beltsville, MD. *Society Aff:* ASAE, ISES.

Bonder, Seth
Business: P O Box 1506, Ann Arbor, MI 48106
Position: President. *Employer:* Vector Research Inc. *Education:* BS/Mech Engg/Univ of MD; PhD/Industrial Engg/OH State Univ. *Born:* Jul 14, 1932. PhD Indus Engrg (Operations Res) Ohio State Univ 1965. Currently Adj Prof at Dept of Indus & Oper Engrg at Univ of Michigan. Council Mbr & PPres of Oper Res Soc of Amer & P Pres of Military Oper Res Soc. Former mbr of Army Scientific Adv Panel. Also mbr of Amer Assn for Advancement of Sci, Internatl Inst for Strategic Studies, The Institute of Management Sciences, Tau Beta Pi, & Omicron Delta Kappa. Principal field of res int has been creation & application of sys, pol & oper analysis in both pub & private sectors, with primary exper in the defense community. *Society Aff:* ORSA, MORS, AAAS, IISS, TIMS.

Bondurant, Byron L
Business: 2073 Neil Ave, Columbus, OH 43210
Position: Prof & Asst Chrm/Agri Engrg *Employer:* Ohio State Univ *Education:* BAE/Ohio State University; MS/(Civil)/Univ of Connecticut *Born:* Nov 11, 1925 Attended Case, RPI & Purdue. Engrg cons, res & teaching in Somalia, USA, Kenya, India, Sierra Leone. Adv to Dean & Dean, Coll of Agri Engrg, Punjab India. Hd, Dept of Agri Engrg, Univ of Maine. Life Mbr ASAE, Sigma Xi, NYAS, SID. Fellow AAAS, ASAE, ISAE & IIE. Mbr ASCE, ASEE, AAUP, MAE, OAS, & NSPE. Held offices of Secy, Conn Valley, Chmn Acadia & V Chmn North Atlantic Sect, Chmn of Ohio Sect, Chmn of Tri-state Sect, Secy-Treas of Ohio & Tri-state Sects, ASAE; Chmn MSPE & Dir MAE. P E, Director International Dept. ASAE 1980-82, Fulbright Professor in Kenya 1979-80, Mgr for UNDP/FAO/MUCIA (Midwest Consortium for International Activities) Project in Somalia 1976-78, Land Surveyor. Recipient of 1983 Kishida Intl Award (ASAE) Vice Chmn & Chmn Elect Intl Div ASEE *Society Aff:* ASAE, ASEE, ASCE, NSPE.

Bondurant, Donald C
Home: 1316 Lakeshore Dr, Heber Springs, AR 72543
Position: Consulting Services, Retired *Employer:* Corps of Engrs, US Army-ret. *Education:* BSCE/Civil Engg/Univ of MO. *Born:* Jul 1908 Charleston, Mo. Civilian employee, Corps of Engrs, US Army 1931-72. Hydraulic design & supervision sedimentation investigations. C of E rep on var interagency bds; cons on stream channel & fluvial sediment problems. Cons, AID auspices, Gov Egypt on Nile River. Co-author ASCE M & R 'Sedimentation Engrg'. Life Mbr ASCE. Co-recipient ASCE 1976 Karl Emil Hilgard Hydraulic Prize. *Society Aff:* ASCE.

Bondurant, James A
Business: Rte 1, Box 186, Kimberly, ID 83341
Position: Agri Engr *Employer:* U.S. Dept of Agri, Agric Res Service *Education:* MSc/Agri Engg/Univ of NB; BSc/Agri Engg/KS State Univ. *Born:* Oct 1926. Ness City, Kansas. USNR 1944-46. Agri Engr with USDA-ARS 1958- . Res in surface irrigation practices, runoff & sediment control, trash and weed seed removal from irrigation water, & use of food processing waste water for irrigation. Mbr ASAE, Bd of Dir 1973-75, Chmn ASAE-SW Stds Ctte 1968-75, Chmn PNW Region ASAE 1970, Mbr NSPE, Idaho SPE-PEG Chmn 1973. PE-AE. Pubs in tech journals. Patentee on Automatic Irrigation Gates. *Society Aff:* ASAE.

Bondy, Hugo A
Business: P O Box 4207, Altanta, GA 30302
Position: Chief Engr. *Employer:* WAGA-TV. *Born:* Dec 1908. Attended Columbia Univ extension, grad Capitol Radio Engrg Inst. Marine radio operator, studio engr/supervisor WNEW Radio prior WW II. Sr field engr USAFIME & AFHQ 1942-45 North Africa & Italy. Built radio stations WIKK, WZIP, WHHH. Staff engr Storer Broadcasting Co 1951. Since 1952 Chief Engr WAGA-TV. Life Mbr IEEE; Mbr AES, SOBE. Fellow SMPTE. Served var SMPTE cttes. Recipient 1976 Univ of Georgia, Sch of Journalism Pi Kappa Gamma Pioneer Broadcaster Award.

Bonell, J Frank
Home: 1585 Monaco Ave, Salt Lake City, UT 84121
Position: President. *Employer:* Engrg Assocs Inc. *Education:* BA/Industrial Technology/San Jose State College; BS/Civil Engg/San Jose State College; MS/Civil Engg/Univ of UT. *Born:* May 1927. BA & BS San Jose State Coll. MS Univ of Utah. US Navy 194546 Northwest Pacific, USAF 1951-53 Korea. Cons Civil Engr 1954-67. Formed Engrg Assocs Inc 1967 & have been a corporate officer since that date. Have been in charge of all civilstruct projects & have extensive exper on projects in Mex, Brazil, Australia & other foreign assignments involving salt & saline minerals, their feasibility, dev & production as related to the engrg aspects of each project. Mbr ACEC, ACEC-Utah & ASCE. Enjoy hunting, fishing & skiing. *Society Aff:* ASCE, ACEC, ACEC.

Bonell, John Arthur
Home: 1125 Williams Ave, Reno, NV 89503
Position: Prof Emeritus of Civil Engg. *Employer:* Univ of Nevada. *Education:* CE/Civil Engg/SD State Univ; MS/Civil Engg/CA Inst of Tech; BS/Civil Engg/SD State Univ. *Born:* Dec 25, 1913 S Dak. Asst City Engr, Huntington Beach, Calif 1938-41. Served with US Army Corps of Engrs 1941-46 in US, Africa & Europe. Discharged with rank of Major. Struct Engr with cons firms, Los Angeles 1946-49. Joined staff of the Civil Engrg Dept, Univ of Nevada in 1949. Dept Chmn 1957-72. Chmn Exec Ctte, Ed Div ASCE 1973-75. V Pres NSPE, Prof Engrs in Ed 1974-75, Dir, ASCE, 1977-80. Dir, NSPE, 1958-59 & 1962-64. V Pres, NSPE, 1964-67. *Society Aff:* ASCE, NSPE, ASEE, ACI.

Bonilla, Charles F
Business: 520 W 120th St, New York, NY 10027
Position: Prof Emeritus of Chem Engg & of Nucl Engg. *Employer:* Columbia Univ. *Education:* AB/Liberal Arts & Sci/Cuenca Inst, Spain; AB/Pre Engg/Columbia Univ, NY, NY; BS/EE/Columbia Univ, NY, NY; Ch Eng/Ch Engg/Columbia Univ, NY, NY; PhD/Electrochemistry/Columbia Univ, NY, NY. *Born:* Jul 1909 Albany, N Y. Specialize in analysis, testing & design of unit oper & nuclear equip & sponsored res for grad student thesis support. Tutor CCNY 1932-37; to Prof & Chmn John Hopkins 1937-49; Prof to Chmn, Senate, etc. Columbia 1948- . Major ChE cons: BEW-FEA, Rubber Reserve, Phillips Chem, Belfort Instrument, NSF, duPont, Chlormetals. Thermal-hydraulic, nuclear: Navy EES, BNL, ANL, Knolls & GE Space, Bettis & Westinghouse, APDA, Thermowheel, National Lead. Some 200 articles, patents, HBK chapters, gov missions, etc. AIChE: Fellow, Kern Award 1974; ANS: Founder, Compton Award 1976; Mbr, etc: ACS, ASEE, K-7, 13 (ASME); Ed Nuclear Engrg & Design (Internatl Journal). PE N Y, Md. *Society Aff:* AIChE, ACS, ANS, ASEE.

Bonini, William E
Business: I-58 Guyot Hall, Dept Geol & Geophy Sci, Princeton, NJ 08544
Position: Prof *Employer:* Princeton Univ *Education:* PhD/Geol & Geophysics/Univ of WI-Madison; MSE/Geol Engrg/Princeton Univ; BSE/Geol Engrg/Princeton Univ *Born:* 8/23/26 Native of Washington, DC. On faculty of Princeton Univ since 1953. Currently on joint appointment between Dept of Civil Engrg and Dept of Geological and Geophysical Sciences. 1970-present George J Magee Prof of Geophysics and Geological Engrg. 1971-74 Chrmn, Water Resources Program; 1973- present, Chrmn, Geological Engrg Program; 1971-1973 Pres, Yellowstone-Bighorn Research Assoc, 1971-73-; Chrmn NY-Phila Section, Assoc of Engrg Geologists. Teaches Engrg Geology, and Applied Geophysics. *Society Aff:* GSA, SEG, AEG, EAEG, AAPG, AGU

Bonkosky, Brian
Business: Attn: DRSTA-VR, Warren, MI 48090
Position: Sr Project Engr. *Employer:* US Army Tank Automotive Command. *Education:* MBA/Univ of Detroit; BS/Aerospace Eng/Northrop Univ; Grad Studies/Indus Eng/Wayne State Univ. *Born:* 9/15/45. Grad studies in Indus engg at Wayne State Univ. Respon for combat vehicle survivability. Dev wideley used finite element tech & ballistic design computer codes for combat vehicle design. Vice Chrmn MI Sect AIAA; Chief Judge Detroit Metro Sci Fair; Mbr ADPA, AUSA. *Society Aff:* AIAA, ADPA, AUSA.

Bonnell, Ronald D
Business: Center for Machine Intelligence, University of South Carolina, Columbia, SC 29208
Position: Prof & Dir. *Employer:* Univ of SC. *Education:* MSME/ME/Univ of KY; BSME/ME/Univ of KY; Grad Work/Control/Univ of TN; NSF Traineeship/Computers/GA Inst of Tech. *Born:* 3/4/34. Ronald D Bonnell received the BS and MS degrees in mech engg from the Univ of KY and has completed additional grad work in computers and control at the Univ of TN and GA Inst of Tech. Prior to joining the faculty at the Univ of SC in 1970 he held engg positions for thirteen yrs with IBM, the Univ of TN, and the Univ of KY. He is currently the Director of the Center for Machine Intelligence at the Univ of SC. His res activities include computer control, expert systems, machine learning, and database engrg. *Society Aff:* IEEE, ACM, AAAI.

Bonnell, William S
Business: P O Drawer 2038, Pittsburgh, PA 15230
Position: Gen Mgr, Admin Serv Div. *Employer:* Gulf Sci & Tech Co. *Born:* Dec 1916. BS Chem Engrg Univ of Ill. Native of Mich. Employee of Gulf Oil Corp. Worked at Gulf Res & Dev Co Process Dev 1939-43. Foreman, Supt & Dir of Refinery & Petrochem Oper at Port Arthur Refinery 1944-60. Mgr Oper at Phila Refinery 1961. GOC Petrochem Production & Special Proj Mgr 1962-64. GOC Petrochem Mgr 1965-67. Proj Mgr (Japan) Gulf Residual Hydrodesulfurization 1968- 71. Europoort Mgr Chemicals (Holland) 1972. Gulf Res Dir Licensing (Japan) 1973-74. Proj Mgr GOC Middle East 1975. Admin Serv Genl Mgr, Gulf Sci & Tech Co at Hammarville Res Labs 1976.

Bonner, Andrew V
Business: Star Rt. 2, Box 746, Yucca Valley, CA 92284
Position: Lighting Consultant. *Employer:* A.V. Bonner and Associates *Education:* AA/Bus Admin/Chaffey College. *Born:* 12/24/21. in Minneapolis, MN. Lived in CA since 1923. Served with the Quartermaster Corps, US Army 1944-47. Employed by The Southern CA Edison Co, from 1947 to 1980 as Lighting Cons, & Prin Lighting Cons from 1970. Presently Heads the Firm A.V. Bonner and Associates, Lighting Consultants. Active member of the Southern California Sect, Illum Engg Soc (IES). Served as Regional VP of the South Pacific Coast Region, RVP, Dir & VP Operational Activities. Mbr IES Distinguished Service Award Ctte; the Southern CA Lamplighters the Designers Lighting Forum; & Fellow Inst for the Advancement of Engg *Society Aff:* IES.

Bonner, Z David
Business: 8700 Tesoro Drive, San Antonio, TX 78286
Position: Vice Chairman of the Board *Employer:* Tesoro Petroleum Corporation *Education:* BSChE/Chem Engg/Univ of TX. *Born:* Feb 23, 1919 San Antonio, Texas. Joined Gulf after serving in the Navy in WW II. Has held respon pos with Gulf in Pittsburgh, Wash D C, Port Arthur, Tokyo, London, Cleveland & Houston. In 1968 received Distinguished Grad Award, Coll of Engrg, Univ of Texas. Dir Gulf Oil Corp 1974 to 3/31/79; Vice Chairman of the Board, Tesoro Petroleum Corporation since 9/1/80. *Society Aff:* ΤΒΠ

Bonnet, William E
Home: 12 So Bryn Mawr Pl, Media, PA 19063
Position: Retired *Employer:* Sun Co Inc. *Education:* PhD/Chem Eng/Univ of DE; MChE/Chem Eng/Univ of DE; BS/Chem Eng/Drexel Univ. *Born:* 1/26/27. Employed by Sun Co since 1950. R&D 1950-60; Commercial Dev 1960-63; Mgmt Sci, 1963-70; New Bus Dev 1970-75; Corp Tech Staff 1975-76; Environmental Assessment 1977-1983. Retired.

Bonnett, G M
Business: 4513 Croatan Highway, P.O. Box 1066, Kitty Hawk, NC 27949
Position: President *Employer:* Gambits, Ltd. *Education:* BChE/ChE/CCNY. *Born:* Sep 1918. Mech Engr at Oxweld Acetylene Co prior to WWII. Navigator & radar operator US Air Force 1943-46. Artillery USA in Korea 1950-52; awarded Silver Star. With Newport News Shipbldg & Dry Dock Co 1946-76. Involved in all facets of ship design & construction as Nuclear Engr, Non-Destructive Testing Engr, Mgr of Steel Hull Div, Production Control Mgr. With Deepsea Ventures Inc developing equip & procedures for deep ocean mining oper 1976-79. Currently working as a consultant. ASNT Tech Council Chmn 1967, V Pres 1969, Fellow 1974. Enjoy golf & traveling. *Society Aff:* ASNT, SNAME.

Bonnichsen, Vance O
Business: PO Box 2385, Pensacola, FL 32503
Position: President. *Employer:* Chem-Quip Inc. *Education:* BS/ChE/IA State Univ; MS/ChE/TX A&M Univ. *Born:* Apr 1918. Draftsman & Engr TVA. Chem Engr Northern Regional Res Lab USDA. Chem Engr Proj Leader Corn Prods Refining Co. Sr Supr Engr & Ch Design Engr Chemstrand Corp. Founded Chem-Quip Co to engage in providing process equipment sales & service as manufacturers' reps. *Society Aff:* AIChE.

Bono, Jack A
Business: 333 Pfingsten Rd, Northbrook, IL 60062
Position: Pres. *Employer:* Underwriters Labs Inc. *Education:* BS/Gen Engr/Northwestern Univ. *Born:* 5/10/25. Grad of Northwestern Univ. Majoring in Gen Engg, Reg PE in six states. Joined Underwriters Labs in 1946. Managing Engr of Fire Protection and Heating, Air-Conditioning, and Refrig Depts. Elected VP - External Affairs (1977) and Pres (1978). Past Pres of Soc of Fire Protection Engrs. Award of Merit and elected Fellow of ASTM. *Society Aff:* NSPE, SFPE.

Bonourant, Leo H
Home: 100 Glenborough Dr. #950, Houston, TX 77067
Position: President *Employer:* FW Management Operations, Ltd (USA) *Education:* BS/Chem Engrg/LA Tech Univ *Born:* Sept 1935 Wisner, La. BS Louisiana Tech Univ 1957. Worked for Universal Oil Prod 1957-69 in areas of process dev, oper, & process licensing. Genl Mgr of Quaker State Congo, W Vir refinery during construction, startup & oper betwwen 1969-74. VP & Gen. Mgr of Refining Sunland Refining 1974-1977. Area Sales Mgr Foster Wheeler Energy Corp 1977-1979. VP, FW Management Operations, Ltd (USA) 1980 President FW Management Operations, Ltd. (USA). 1981 Member AIChE, NPRA, API. Registered P.E., Texas, LA other Interests: Motorcycle trail riding, tennis, travel. *Society Aff:* AIChE.

Bontadelli, James A
Home: 11517 N. Monticello Dr, Knoxville, TN 37922
Position: Director, Industrial Engineering *Employer:* Tennessee Valley Authority *Education:* Ph.D/Ind. Engr./The Ohio State University; MS/Ind. Engr./Stanford University; BS/Ind. Engr./Oregon State University *Born:* 09/27/29 Native of St. John, WA. Served with U.S. Army Corps of Engrs, 1951-62. Engr & mgr with Battelle Columbus Labs, specializing in indust engrg & systems analysis in private industry and government operations. Assumed current pos as Dir of Indust Engg at TVA in 1974. Faculty (Part-time), Dept of Indust Engrg, Univ of TN. VP, IIE, 1985-87. Distinguished Alumnus Award, College of Engrg, Ohio State Univ, 1984. *Society Aff:* IIE

Bonthron, Robert J
Business: 3300 S. Federal St, Chicago, IL 60616
Position: Assoc Prof *Employer:* IL Inst of Tech *Education:* PhD/Mech/IL Inst of Tech; MS/Mech/IL Inst of Tech; BS/ME/IL Inst of Tech *Born:* 11/12/22 USNR, Naval Research Lab World War II Instructor, Mechanics 1947-54; Asst Prof, M.E. 1954-61; Assoc Prof since 1961; Asst to Dean of Engr 1952-59; Dean of Students 1968-75; all at IIT. *Society Aff:* ΠΤΣ, ΤΒΠ, ΣΞ

Book, Wayne J
Business: School of Mech Engrg, Atlanta, GA 30332
Position: Prof *Employer:* GA Inst of Tech *Education:* PhD/ME/MIT; Engr/ME/MIT; SM/ME/MIT; BS/ME/Univ of TX *Born:* 11/28/46 in San Angelo TX. After completing his formal education at MIT he began a teaching and research career at the GA Inst of Tech. Research and publications have focused on design and control of mechanical arms for teleoperators and robots. He was coinventor of a computer controlled exercise machine and cofounder of the co developing the prototype machine. He was recently Assoc Editor for the ASME Journal of Dynamic Systems Measurement and Control for the IEEE Trans. Automatic Controls and Visiting Scientist at The Robotics Inst, Carnegie-Mellon Univ. From 1983-87 he was Dir of the Computer Integrated Mfg Sys Prog at the Georgia Inst of Tech. *Society Aff:* ASME, AAAS, IEEE, ΣΞ, ΤΒΠ, SME.

Booker, Henry G
Home: 8696 Dunaway Dr, La Jolla, CA 92037
Position: Prof of Applied Physics. *Employer:* Univ of Calif, San Diego. *Education:* PhD/Mathematics/Univ of Cambridge; BA/Mathematics/Univ of Cambridge. *Born:* 1910 England. Naturalized Amer Citizen 1952. Ed at Cambridge Univ where received bach degree 1933, & doctor's degree 1936. During WW II was in charge of theoretical res at the radar res estab of the Royal Air Force, & subsequently became Prof of Elec Engrg at Cornell Univ. From 1959-63 was Dir of Sch of Elec Engrg at Cornell & then IBM Prof of Engrg & Applied Math. In 1965 became Prof of Applied Physics at the Univ of Calif, San Diego. Fellow IEEE; Mbr Natl Acad of Sci. *Society Aff:* IEEE, AGU, APS, NAS.

Bookman, Robert E
Home: 8962 E. Indian Canyon Rd, Tucson, AZ 85715
Position: President *Employer:* Command and Control System Engineering Co. *Education:* BS/EE/Univ of AZ. *Born:* 4/16/20. Mgr, Combat Info Ctr design, Bureau

Bookman, Robert E (Continued)
of Ships, during WWII. Engr Mgr for Navy command & control automation, Univ of MI. Res Prof, Tufts Univ, concerned with interdisciplinary res on & teaching of system analysis & synthesis. Dir of Engg for medium size communications-electronics mfg. Dept Mgr, System Dev Corp, involved in design & implementation of several large automated progs. Dir of Engg for Natl Military Command System at Defense Communications Agency. System Engg of the Worldwide Military Command and Control System, a $9B complex of communications, sensors, facilities and ADP, deployed worldwide. Currently manager of a small engg consulting firm emphasizing top-down design, and evolutionary development and implementation of military command and control systems. *Society Aff:* IEEE.

Boone, Enoch Milton
Home: 2579 Berwyn Rd, Columbus, OH 43221
Position: Prof Emeritus-Elec Engrg. *Employer:* Ohio St Univ, 2015 Neil Ave. *Education:* MS/Elec. Eng./Michigan University *Born:* Millersburg Ky 1903. BA, MS (physics). Instr engrg math, physics, Univ of Colorado 1926-30; res physicist, Genl Elec Co 1930-32; Head, Engrg Dept, Amarillo College 1932-37; M.S. Degree E.E. Mich. Univ. 1937; Prof EE Ohio State Univ 1937-73; mbr tech staff, Bell Tele Labs 1944-45; OSU Dir Electron Device Lab 1947-65, Bd of Dir Res Foundation 1963-66; Fellow IEEE (Dir Region 4, IRE 1955-56, mbr natl cttes 1946-66); author, tech papers-reports; text, Circuit Theory of Electron Devices, John Wiley & Sons Inc 1953; Mbr Eta Kappa Nu, Tau Beta Pi, Phi Beta Kappa; Sigma Xi; ASEE.

Boorstein, William M
Business: P O Box 2430, Grand Rapids, MI 49501
Position: President (1978). *Employer:* Franklin Metal Trading Corp. *Education:* BSE/Met Engr/Univ of MI; MSE/Met Engr/Univ of MI; PhD/Met Engr/Univ of MI. *Born:* May 28, 1938. 1967-70 Asst Prof of Met Engrg at Ohio State Univ, Columbus; 1970-73 Assoc Prof of Met Engrg at Ohio St Univ; 1973-74 Foundry Cons for The Viking Corp, Hastings, Mich; 1974-1978 V Pres of Franklin Metal Trading Corp, Grand Rapids, Mich; 1974 to present cons in metallurgy & metallurgical engrg. 12 pubs & 1 pat. ASM Bradley Staughton Outstanding Young Teacher Award for 1974. m. Donna; c. Jeffrey, Stephen, Michelle. *Society Aff:* ASM, AFS.

Booth, Murray A
Business: P O Box 3707, Seattle, WA 98124
Position: Dir of Tech-New Commercial Prog's. *Employer:* Boeing Commercial Airplane Co. *Born:* Aug 15, 1937 Vancouver, B C. BA Sc Engrg Physics, Univ of British Columbia 1960. Joined Boeing Commercial Airplane Co 1960, assigned to Aerodynamics Staff spec in aircraft stability & control analysis & design. Ch of Tech Staff for 747 Product Dev - 1970, Dir of Tech for New Commercial Programs - 1976. Enjoy classical music & most sporting activities.

Booth, Taylor L
Home: 451 Wormwood Hill Rd, Mansfield Ctr, CT 06250
Position: Prof of Elec Engrg & Computer Sci. *Employer:* Univ of Connecticut. *Education:* BS/EE/Univ of CT; MS/EE/Univ of CT; PhD/EE/Univ of CT. *Born:* 9/22/33. Native of Manchester, Conn Chrmn Comp Sci and Enrg 1982- . Engr Westinghouse Elec Corp 1956-59. Fac Univ of Conn since 1959. Chmn Comp Sci Program, EECS Dept 1969-78. Chrmn Comp Sci & Engr 1982-. Fortesque Fellow 1955, Frederick Emmons Terman Award 1972, Fellow IEEE 1975, IEEE Centenial Medal 1984, Assoc Ed, IEEE Trans Computers 1972-75. Editor in Chief 1978-1981 Governing Board IEEE Comp Society 1981-1983 Secy IEEE Comp Society 1981 First VP 1982, VP Education 1983 Author *Sequential Machines & Automata Theory* 1967, Digital Networks & Comp Systems 1971 Second edition 1978; (with Y T Chien) Computing: Fundamentals & Applications 1974, Introduction to Comp Engr 1984 *Society Aff:* IEEE, ACM, ΣΞ, ASEE.

Booth, Weldon S
Business: One South Franklin St, Nyack, NY 10960
Position: Prin Pres *Employer:* WS Booth & Co, Coakley & Booth Inc *Education:* BS/Civil Engrg/Columbia Univ Sch of Engrg; AB/Liberal Arts/Columbia Coll *Born:* 06/10/16 Fellow and Life Mbr ASCE. Founded Coakley & Booth Inc and WS Booth & Co; Ensign USNR CEC Corp 1944-5; Life Mbr Moles; Inventor and holder of 4 patents: 1) "Underground *Sheeting* Methods" (Contact Sheeting); 2) "Contact Sheeting Fastening Devices–; 3) "Contact *Foundation* Method" (Simultaneous Construction of Substructure and Superstructure); 4) "Method of Making A High-Capacity Earthbound Structural Reference–. Pioneered Constr and Engrg Tech i.e., "Tie-Backs in Soil" (See Civil Engrg Sept. 1966); First use of OCM Catalytic Exhaust in Tunnelling (See Construction Methods and Equipment Jan 1953). Former Pres of Columbia Engrg Sch Alumni Assoc. Recipient of Columbia Univ Alumni Medal 1964. Numerous Cttes ASCE Met Section. *Society Aff:* ASCE

Boothe, Allen P
Business: 9455 Ridgehaven Ct, San Diego, CA 92123
Position: VP and Reg Mgr *Employer:* George S. Nolte & Assocs *Education:* MS/Mgmt/Engrg/US Naval Postgrad Sch; BS/CE/RPI; BS/Engrg/US Naval Acad *Born:* 05/27/37 A native Californian; Progressed through increasingly responsible positions in construction, contract admin and planning during 20-yr career in Navy Cvl Engr Corps. With Nolte since 1979 as Regional Mgr for Southern CA, having full opeational and mgr responsibility. Reg in CA and AZ; Taught undergrad and grad level courses in Cvl Engrg at San Diego State Univ; Pres of San Diego Section of ASCE, 1980-81; SAME Post Dir, 1974-76; Dir, San Diego Association for Retarded Citizens, 1979-81; Charter Mbr, San Diegans for Water; Member of San Diego Bd of Zoning Appeals 1981-83; Married to the former Shirley Roberts of Charleston, WV; Three children Douglas 1961; Steven 1964 and David 1971. *Society Aff:* ASCE, SAME, APWA, ABC

Boothroyd, Albert R
Business: Dept of Electronics, Carleton Univ, Ottawa, Ont K1S 5B6 Canada
Position: Prof *Employer:* Carleton Univ, Ottawa, Ont, Canada. *Education:* PhD/Elec Eng/Imperial College; BSc/Elec Eng/Imperial College. *Born:* 1925 England. In 1951, joined Imperial College as Lectr in Elec Engrg; in 1959 appointed Reader in Electronics. Since 1952, engaged in res on solid- state devices & circuit electronics. In 1964 appointed Prof Electronics & Hd of Dept of Elec Engrg at Queen's Univ, Belfast. From 1968 Prof of engg in Dept of Electronics, Carleton Univ, Ottawa, Ont, Canada. Chmn of Dept 1968 to 1985. Current res int: theory, modeling & application of semiconductor devices, & computer-aided design. *Society Aff:* IEEE, ΣΞ.

Boothroyd, Geoffrey
Business: Amherst, MA 01003
Position: Prof of Mech Engrg. *Employer:* Univ of Massachusetts. *Education:* DSc/Eng/Univ of London; PhD/Eng/Univ of Londen; BSc/Eng/Univ of London. *Born:* Nov 18, 1932. After serving an apprenticeship with Mather & Platt Ltd, Manchester, Eng, joined the Atomic Power Div of the English Elec Co, Leicester, Eng, as a design engr. In 1958 joined the fac of the Univ of Salford, Eng, & was Visiting Assoc Prof at Georgia Inst of Tech during 1964-65. Since 1967 has been Prof of Mech Engrg at Univ of Mass. *Society Aff:* SME, ΣΞ, ASEE, CIRP

Bootle, Benny T
Business: PO Box 8187A, Greenville, SC 29604
Position: Pres. *Employer:* Bootle Equip Sales & Service, Inc. *Education:* BSME/ME/Univ of SC. *Born:* 11/2/31. & educated in SC. Worked way through sch (100%) Dean's list last two yrs. Attended six months grad course in air conditioning & refrig in WI. Reg engr in SC by written examination. Pres of four corps involved in the distribution, sales & service of heating, refrig & air conditioning products. Past Pres Piedmont Chapt SCSPE. Past Pres Greenville Chapt ASHRAE. First recipient ASHRAE Region IV Regional Award of Merit. Past ASHRAE Regional Chrmn & mbr BOD. 1981 recipient of ASHRAE Distinguished Service Award. Currently MBR. of ASHRAE A&A Comm . Sports & fishing. Motorcycling (Natl Champion 1961). Antique automobiles. *Society Aff:* ASHRAE, NSPE.

Booy, Emmy
Home: Rt 4, 43 Karlann Dr, Golden, CO 80401
Position: Asst Prof. *Employer:* CO Sch of Mines. *Education:* PhD/Geology/Columbia Univ; MS/Geology/Columbia Univ; BS/Ceramic Engg/Alfred Univ. *Born:* 11/17/38. Locust Valley, NY. Married, no children. Taught geology at Queens College, NY, 1964-1968, Bemidji State College, MN, 1968-1969; geological engg at MI Tech Univ 1969-1977, CO Sch of Mines, 1977-present. Visiting Prof at McGill Univ, 1976. Reg PE in CO, MI, & VT. Specialties - mineralogical controls of soil & rock behavior, environmental impact of mining, industrial rocks and minerals, areas 33, 34, 210, 158, and 12. *Society Aff:* AEG, SEPM, AIPG, SEASSM.

Booy, Max L
Home: 2636 W-Robino Dr, Wilmington, DE 19808
Position: Principal Consultant. *Employer:* E I duPont de Nemours. *Education:* Ingenieur/Mech Engr/TH Delft; Doctor TW/Flow/TH Delft *Born:* 5/13/16. After 1 yr NLL joined Netherlands Navy Yard, Java, Indies 1939-48. Ch Engg Denver Ept Co, Johannesburg 1948-49, Res Engg Denver 1950-52, Designer Dorr Co, Stamford 1953-55. With DuPont's engg since 1955, Principal consultant since 1967. Fellow ASME. Retired from Dupont 1980. Consulting Engineer in Polymer Processing Eqt. Adj Prof Drexel Univ *Society Aff:* ASME, SPE

Borcherding, John B
Business: Arch & Civil Engg, ECJ 5.200, Austin, TX 78712
Position: Assoc Prof. *Employer:* Univ of TX. *Education:* PhD/Civil Engg/Stanford Univ; MS/Civil Engg/Stanford Univ; BS/Civil Engg/Missouri Univ. *Born:* 12/29/43. in Houston, TX; raised in St Louis Missouri; married to former Constance R Hardi of Coupland, TX; cost & Scheduling Engg with Rust Engg Co and Fruin-Colnon Contr Co. Served as commissioned officer in US Public Health Svc. Currently, Assoc Prof in Arch & Civil Engg at Univ of TX at Austin. Specialist in Motivation, Productivity and Organizational Behavior in Construction. Consultant to AGC, NECA, CE Lummus, Bechtel, Ebasco, Kirkwood Electric, Owens Corning Fiberglass, Riley Stoker, Brown & Root, Procter & Gamble, Daniel, Voss. Author of over 40 articles and reports and two books. Chairman of three National Society committees. Recipient of ASCE Walter L Huber Res Prize. *Society Aff:* ASCE, PMI, ASEE.

Bordogna, Joseph
Business: School of Engineering and Applied Sci, Philadelphia, PA 19104
Position: Dean, School of Engrg and Applied Sci *Employer:* Univ of Pennsylvania. *Education:* PhD/EE/Univ of PA; SM/EE/MIT; BSEE/EE/Univ of PA. *Born:* Mar 22, 1933 Scranton, Pa. Line Officer US Navy 1955-58. Res & dev engr RCA Corp prior to prof career. Presently Alfred Filter Moore Prof of Elec Engg, Dir Moore Sch of Elec Engrg, Dean, School of Engrg & Applied Sci, Univ of Penn. Prof ints in field of electro-optics & ed innovation. Fellow IEEE & AAAS, mbr ASEE, NSPE, Franklin Inst, Tau Beta Pi, Eta Kappa Nu, Sigma Xi. Co-author var of papers & 2 bks. Recipient of Lindback & United Engrs teaching awards & the Western Elec & Geo Westinghouse ed awards of ASEE. Engineer of the Year in Philadelphia 1984. *Society Aff:* IEEE, ASEE, AAAS, NSPE.

Borel, Robert J
Business: 1127 Dearborn Dr, Columbus, OH 43085
Position: VP Engg. *Employer:* Worthington Industries, Inc. *Education:* MSc/EE/OH State Univ; MBA/Finance/Univ of Rochester; BSEE/EE/OH State Univ. *Born:* 3/9/43. Native of Columbus, OH. Held engg design & mgt positions at Eastman Kodak Co from 1966 through 1973. Joined Worthington Industries, Inc in 1974. Elected WII VP, Engg in 1977, & mbr of Bd of Dirs in 1979. Responsible for corporate mgt of plant, facilities, & equip engg, Group VP, 1981. *Society Aff:* OSPE, HKN, TBΠ.

Boresi, Arthur P
Home: 3310 Willett Dr, Laramie, WY, IL 82071
Position: Prof & Head-Civil & Architectural Engg. *Employer:* Univ of WY, Laramie. *Education:* PhD/Theo & Applied Mechs/Univ of IL; MS/Theo & Applied Mechs/Univ of IL; BS/Elec Engg/Univ of IL. *Born:* 8/27/24. Res Engr N Amer Aviation 1950; Matls Engg Natl Bureau of Stds 1951 Mbr fac Univ of IL at Urbana 1953-79; Distinguished Visiting Prof Clarkson College Tech, Potsdam, NY 1968-69; Distinguished Visiting Prof, Naval Postgraduate School, Monterey, CA 1979; Cons in field, USAAF, AUS 1943-46. Found Mbr & Treas Amer Acad mech; Fellow Amer Soc ME, Mbr Amer Soc CE, Amer Soc Engg Ed, Soc. Exper. Stress Analysis Sigma Xi. Author: Engg Mech 1959; Elasticity in engg Mech, 2nd Edit 1974; Advanced Melchs of Matls 3rd Edition 1978; also articles. *Society Aff:* ASME, ASCE, ASEE.

Borg, Sidney F
Business: Hoboken, NJ 07030
Position: Prof Civ Engg. *Employer:* Stevens Inst of Tech. *Education:* DrEng/Aeronautics/Johns Hopkins Univ; MCE/Civ/Poly Inst Brooklyn; BSCE/Civ/Cooper Union. *Born:* 10/3/16. Author of seven textbooks and about 100 res papers. Fulbright Prof to Royal Danish Tech Univ, State Dept. Visiting Scientist to Polish Academy of Sci Visiting Prof to Technische Hochschule Stuttgart, Technion (Israel Institute of Tech), AF Inst of Tech. Distinguished Alumnus Award, Cooper Union and Brooklyn Poly Inst. Recipient of Natl Sci Fdn Res and Teaching contracts. Postdoctoral Fellow in Physiology at Woods Hole. Mbr Tau Beta Pi, Sigma Xi, Chi Epsilon, Reg PE. Hon Master of Engg Degree, Stevens Inst of Tech. *Society Aff:* ASCE, AAAS, ASEE, NYAS, EERI.

Borgiotti, Giorgio V
Business: Hartwell Rd, Bedford, MA 01730
Position: Technical Staff. *Employer:* Raytheon, Missle System Div. *Education:* Eng Dr/EE/Univ of Roma, Italy. *Born:* 11/23/32. Native of Roma, Italy. Worked for Selenia, Italy, From 1958 to 1965 as a senior engineer and manager of the Microwave Antenna R&D section. Joined Raytheon, Missle Systems Div, in 1967 as prin engr and mgr of the Electromagnetic Analysis and design of phased array antennas. From 1974 to 1975 he was a visiting scientist at the Air Force Cambridge Res Ctr, Bedford, MA doing res in the area of limited scan antennas. From 1976-1977, he was associated with Jet Propulsion Lab, dev near field/far field reconstruction algorithms. Since rejoining Raytheon in 1977, as a mbr of the Tech Staff of the Radar Lab, he has been involved in R&D in the area of radar tech. *Society Aff:* IEEE, URSI, ΣΞ.

Borgnis, Fritz E
Home: Bergstrasse 99, Ch Zurich-8032, Switzerland
Position: Prof EE *Employer:* Federal Inst of Tech, Zurich. *Education:* PhD/Elec Engg/Tech Univ, Munich (W. Germany). *Born:* Dec 24, 1906 Mannheim, Ger. Munich Inst Tech 1932-38; Telefunken Co Labs 1938-40; U Graz 1940-47; Fed Inst Tech 1947-50; Institut fur Hochfrequenztechnik 1960- ; Caltech 1951-54; Harvard 1954-57; Dit Philips Res Labs Germany 1957-60. Fellow IEEE, APS, AAS; mbr Swiss soc, Swiss Soc for Natural Sci, Swiss Soc Elec Engrs. Author books & articles. Res on ultrasonics, waves in crystals, electromagnetic theory, plasma & microwave physics, med electronics. m. Gerda M Maschinda Mar 10, 1964. *Society Aff:* Swiss Soc. of Elec Engrs

Borie, Edward T L
Business: 600 Carondelet St, New Orleans, LA 70130
Position: Senior VP *Employer:* Walk, Haydel, & Assocs, Inc. *Education:* BS/ChE/Univ of Southwestern LA *Born:* 9/5/27 Native - New Orleans, experience with American Oil, Cities Service, U.S. Army Chem Corps; Walk, Haydel & Assocs since 1960 - Responsible for staff of over 50 chem and environmental engrs providing process and environmental consulting and design for process industries. Past Vice-Chrmn, Chrmn Baton Rouge Section AIChE. Member, Executive Committee, Fellow AIChE; recipient 1965 Award for Outstanding Civic and Community Service by Baton Rouge Council Engrg and Scientific Societies. Recip 1984 Coates Award of AIChE and ACS for Ser to the Profession, Soc, and Commun. Baton Rouge Chamber of Commerce. Mbr, LA Commission on Ethics for Public Employ-

Borie, Edward T L (Continued)
ees. Former Senior Warden, St. Luke's Episcopal Church. Political activities include Delegate to 1964, 1968 and 1976 Republican National Conventions, Congressional Campaign Mgrs, Presidential and Gobernatorial Candidates. *Society Aff:* AIChE, NSPE, ACEC, APCA, PEPP, LES.

Borjesson, Bryan F
Business: 514 Steel Creek Rd, Fairbanks, AK 99712
Position: Consult Eng *Employer:* Self *Education:* M/CE/Univ of AK; BS/CE/Univ of AK *Born:* 10/7/37 Camas, WA, resident of Fairbanks, AK for past 31 years, married and two children. Consulting Engr, Estimator. Arctic design in soils and structures, Design and Construct in all areas of Cold Regions structures and thermodynamics, Forensic Engg. Taught in Community Coll. Owner, VP and Chief Engr of King Steel, Inc. Former Pres local chapter ASPE, former treasurer of St Chapter ASPE. *Society Aff:* ASCE, ASPE, ASTM, ACI, ICBO.

Borland, Whitney M
Home: 8412 B N. Central Ave, Phoenix, AZ 85020
Position: Self-employed *Education:* MS/ME/Hydraulics/Univ of CA; MS/CE/Structures & Geol/Univ of CO; BS/ME/Univ of NB; Post Grad/Univ of Cornell *Born:* 11/06/05 Hydraulic Engr with Bureau of Reclamation - 1930-36 Testing Hydraulic Models; 1936-42 Spillway Designs; 1946-72 Sediment Section. Lt Co, US Army-10th Mountain Div, Asst Div Engr and Operation Ofcr to various engrg groups. 1972- Consltg Engr to various firms - Thailand, Columbia S.A., Puerto Rico, Central America-Reservoir Sedimentation. River Morphology Stable Waterway 1972-1987 conslt to Bureau of Indian Affairs river boundary definition and river hydraulics. *Society Aff:* ASCE, USCOLD

Borman, Gary L
Business: 1513 Univ Ave, Madison, WI 53706
Position: Prof of Mech Engrg. *Employer:* Univ of Wisconsin-Madison. *Education:* PhD/Mech Engg/Madison; MS/Math/Madison; MS/Mech Engg/Madison; BS/Math/Madison. *Born:* Mar 1932. Worked for Genl Electric, Evandale 1957-60 & fac mbr Madison since 1964. Director, Engine Research Center at U.W. since October 1986. Tech ints are combustion & engine simulation. Active in SAE at both local & natl level & past mbr of SAE Bd of Dir. Awards incl Horning Colwell Award & Ralph R Teetor Ed Award. In 1970 spent semester in Yugoslavia under Fulbright Grant. Author of more than 50 tech papers. Hobbies: travel, hiking, photography. *Society Aff:* SAE, Combustion Institute

Born, Harold J
Business: P O Box 102, Tulsa, OK 74101
Position: Chrmn and C.E.O *Employer:* Born Inc. *Education:* BA/Pre Law/Univ of Tulsa; BS/Petroleum Refining/Univ of Tulsa. *Born:* 6/13/18. Assumed current respon as Pres for Direct Fire Heaters of Born Inc in 1949. Dir of Born Heaters Ltd, Brighton Sussex Eng, Societe des Fours. Born Paris, France. Reg PE OK, LA, TX. *Society Aff:* AIChE, OSPE, NSPE, AIME.

Born, Robert H
Home: 15 Georgetown, Irvine, CA 92715
Position: Pres. *Employer:* Robert H. Born Consulting Engrs, Inc. *Education:* MSCE/Civ Engr/Univ So Calif; BE/Civ Engr/Univ So Calif *Born:* Nov 1925. Native of Los Angeles, Calif. Hydraulic Engr with Calif Dept Water Resources 1949-58. Chf. Engr San Luis Obispo Cty Flood Control & Wtr Consrv Dist, 1958-70. Joined Camp Dresser & McKee Inc, Environ Engrs in 1970 V P, 1970-78. Born, Barrett & Assocs VP 1978-86, Organized Robert H. Born Consulting Engrs, Inc, in 1986. President 1986-Present. Spec water resources mgmt & environ control. Award from AWWA for dir of Lopez Proj as Outstanding Public Works Proj in San Luis Obispo, Ventura, & Santa Barbara Counties, Calif in 1969. P Chmn Hydraulics Tech Group, LA Sect ASCE. Mbr & P. Chmn, Surface Water Committee, & Sec, Ex. Com Irrig & Drain Div, ASCE. P Mbr, Exec Comm and P Chrmn, Water Quality & Resources Policy Comm, CA-NV Section, AWWA. Mbr AAEE. *Society Aff:* ASCE, USCOLD, AWWA, AAEE

Bornemann, Alfred
Home: 63 Cliff Rd, Nantucket, MA 02554
Position: Prof Emeritus. *Employer:* Stevens Inst of Tech. *Education:* ME/Engrg/Stevens Inst of Tech. *Born:* 4/6/06. During grad study in Germany (1927-30) I married Lieselotte Just of Dresden (now deceased, 1981). We had had three children, Peter, Barbara (Mrs David Stainton) and Toni (Mrs Martin McKerrow). Starting at Stevens as Asst Prof (1930) I became Prof and Hd of Met, a dept I created, in 1948. After a full life of teaching, res, consulting and admin I was named Prof Emeritus in 1974. Alongside my academic career I've been entrusted with several honorary positions, carrying considerable responsibility. I was Pres of the Nantucket Maria Mitchell Assn (1974-81) which does educational work and res in Astronomy and Natl Sci and currently am a Trustee or Director of the Atheneum (a public library), the Coffin School, the Nantucket Chamber Music Center, the Nantucket Arts Council and the Land Council, all on Nantucket. *Society Aff:* ACS, ASM, TMS, ASTM.

Borrego, Jose M
Business: Rensselaer Polytechnic Institute, 110 8th Street, Troy, NY 12181
Position: Prof. *Employer:* RPI. *Education:* ScD/EE/MA Inst of Tech; EE/EE/MA Inst of Tech; SM/EE/MA Inst of Tech; IME/EE/Ins Tec y de Est Sup de Monterrey. *Born:* 3/21/31. Jose M Borrego received the BS degree in mech & elec engg from the Instituto Tecnologico y de Estudios Superiores de Monterrey, Mexico in 1955. During 1955-1956 he was a Fulbright student at the MA Inst of Tech, Cambridge, where he received the SM degree in 1957 & ScD degree in elec engg in 1961. Since 1965 he has been on the faculty at RPI in Troy, NY, where he is currently a Prof in the Elec & Systems Engg Dept. His res interests are on solar cells, radiation effects on semiconductor devices & solid state microwave devices. He is a Professional Engineer registed in New York State. *Society Aff:* IEEE, ΤΒΠ, ΣΞ, HKN.

Borst, David W
Home: 2104 Chelsea Rd, Palos Verdes Estates, CA 90274
Position: Consultant *Employer:* International Rectifier Corp. *Education:* ScB/Elec Engg/Brown Univ *Born:* 8/5/18. 1940-61: Employed by General Electric Co as design engr and application engr first for mercury arc and later for seimconductor power rectifier equipment. 1961-1986: Employed by International Rectifier as application engr on semiconductor rectifier devices. 1986-present, Consultant to International Rectifier. 1977-80: Chmn, Industrial Power conversion Systems Dept, Industry Applications Soc, IEEE, 1974-present: Chmn, JEDEC Committee JC-22 of Electronic Industries Assoc. Reg PE, MA. Co-founder WBRU, Brown Univ and IBS, Intercollegiate Braodcasting System. Past Tech Mgr, Operations Mgr, Pres and Chmn of IBS; 1977-present: VChrmn, West and Secretary. Active in community theatre, singing, world travel. *Society Aff:* ΣΞ, IEEE

Borsting, Jack R
Business: Room 3E836, The Pentagon, Department of Defense, Washington, DC 20301
Position: Asst Secty of Defense (Comptroller) *Employer:* Dept of Defense *Education:* PhD/Math Statistics/Univ of OR; MA/Math/Univ of OR; BA/Math/OR State College. *Born:* 1/31/29. Since Aug 1980 Dr. Borsting has been Ass't Sec. of Defense (Comptroller) Since 1959, Dr Borsting has held various positions at the Naval Postgrad School. This includes Prof and Chrmn, Dept of Operations Analysis, 1964-71; Chrmn, Dept of Operations Res and Admin Sciences, 1971-74; and Provost and Academic Dean 1974-1980. Other academic positions held include Visiting Prof, Univ of CO, Boulder, 1967, 1969, and 1971; Visiting Distinguished Prof, OR State Univ, 1968. He has acted as consultant for numerous cos and has been a mbr of numerous Bds and Panels. He is Past Pres, ORSA; Past Pres, Military Operations Res Soc; Fellow of AAAS; and is treas, Intl Federation of Operations Res Societies. *Society Aff:* ORSA, TIMS, ΣΞ, ASA, IMS, SIAM, AAAS.

Bosakowski, Paul F
Business: P.O. Box 11248, Salt Lake City, UT 84147 *Employer:* Kennecott Minerals Co., Engrg Ctr *Education:* BS/CE/NJ Inst of Tech. *Born:* 4/8/43. Served in US Navy Civ Engr Corps, including service with the Seabees in Vietnam. Dist Engr for NJ Public Service Elec and Gas Co, 1969-70. Proj Mgr, Construction, Ole Hansen and Sons, Inc 1970-73. With Fluor Corp since 1973. Assumed current position as Mgr of Cost and Scheduling Engg at Fluor Mining & Metals in 1978. Technical Prog Chrmn, 1978 AACE Annual Meeting. Elected Dir, AACE, 1979. Active in church and civic affairs. Enjoy camping and travel. *Society Aff:* AACE, PMI.

Bosch, John A
Business: P O Box 5000, Binghamton, NY 13902
Position: Manager-Engrg. *Employer:* General Electric Co. *Education:* BS/Engg/Penn State Univ. *Born:* 3/14/29. USAF Maintenance Officer 1951-53. With Genl Electric since 1953. Assumed current respon as Genl Manager-Aerospace Electrical Sys Programs. Mbr SAE, IEEE, AIAA, NSIA & NSPE. Co-authored papers *Reconfigurable Redundancy Mgmt for Aircraft Flight Control, Solftware Development for Fly-By-Wire flight Control Sys*, VSCF Aircraft Electrical Pwr. Private pilot. *Society Aff:* SAE, IEEE, ACAA, NSTA, NSPE.

Boschuk, John, Jr
Business: 101 West Mall Plaza, Carnegie, PA 15106
Position: Principal/Mgr *Employer:* Orbital Engrg, Inc. *Education:* PhD/Geotech Engrg/Century Univ; MS/CE/Univ of CA & Berkeley; BS/CE/Lehigh Univ; AS/CE/Trenton Jr. Coll *Born:* 6/9/45 Native of Trenton, NJ. Received direct commission Lt. Corps of Engrs (1969) Promoted to Capt. (1970), Served as geotechnical engr: HI, Okinawa & Korea. Prepared status/safety documents for all impoundments in Ryuku Islands for reversion to Japan. Employed: Dames & Moore 1966-1969; Woodward-Clyde (1972- 1979); Orbital Engrg Inc.-Present: Principal, Mgr, Research Dir: Geotechnical/Earth Science Division; Chairs Prof. Devel. and Training Program. Specialist in Acoustic Emissions. PE: PA, TX, KY, TN, AR. Expertise: Dams, Surface Mining, Foundations, Testing, Groundwater. Enjoys Fishing *Society Aff:* ASCE, ASTM, USCOLD, ISSM&FE, NSPE, PSPE, ΤΒΠ, ΧΕ

Bose, Nirmal K
Business: Penn State Univ, 121 Elec. Engr. East, University Park, PA 16802
Position: Singer Prof *Employer:* The Pennsylvania State Univ *Education:* PhD/EE/Syracuse Univ; MS/EE/Cornell Univ; B Tech/Electronics & Tele- Comm/IIT, Kharagpur *Born:* 08/19/40 Currently I am the Singer Prof of Signal Processing & Prof of Electrical Engg at The Pennsylvania State Univ. I was elected to be a fellow of IEEE in 1981 for my contributions to multidimensional systems theory & circuits and systems education. I served as a visiting faculty at the Univ. of CA, Berkeley, Univ. of MD. and the American Univ. of Beirut at Lebanon. I received CNRS-IRIA Grant to do research in France and DAAD Grant to conduct research in West Germany. *Society Aff:* ΣΞ, IEEE, AMS, AAAS, NYAS.

Boshkov, Stefan
Business: H Krumb Sch of Mines, New York, NY 10027
Position: Chrmn & Prof. *Employer:* Columbia Univ. *Education:* EM/Mining/Columbia Univ; BS/Mining/Columbia Univ. *Born:* 9/29/18. in Bulgaria, entered USA 1938. Service in US Army, Military Intelligence, CBI, 1943-1946. Consulting engr 1950-present. Distinguished Prof (Fulbright) Yugoslavia, 1969. Guest lectr: Taiwan, 1972, 1976; OAS Chile, 1972; USSR, 1974; Poland, 1976; Bulgaria, 1976; Bolivia, 1977. China 1980, 1981. Mbr Intl Organizing Committee of World Mining Congress 1962-present. Pres Benedict Fdn 1973-present. Dir SME of AIME (4 yrs). *Society Aff:* AIME, AAA, ΣΞ.

Bossart, Clayton J
Home: 409 Woodhaven Dr, Monroeville, PA 15146
Position: Project Supervisor. *Employer:* Mine Safety Appliances Co. *Born:* Dec 4, 1930. BS Ch E Univ of Wisconsin. With MSA since 1956; primary work in dev & design of analysis instrumentation for process indus & environ monitoring. Mbr ACS, ISA. Presently Director of Analytical instrumentation Div of ISA. Lect at ISA short courses. Recipient of 1969 ISA 'Excellence in Documentation' award. Pats & pubs in instrumentation field. Extra-curricular activities incl: Treas of co Credit Union & TreasStatistician of Annual Fund Drive for retarded children.

Boston, Orlan W
Home: 121 Miner Circle, V.O.C. Country Club West, Sedona, AZ 86336
Position: Prof-Emer Mech & Production Engrg. *Employer:* Univ of Michigan-ret. *Education:* ME/Mech Engg/Univ of MI; MSE/Mech Engg/Univ of MI; BME/Mech Engg/Univ of MI. *Born:* Nashville, Mich 1891. m. Stella R Roth (dec'd) 1916, m. Marion Miller Langford 1968. Prof & Chmn Production Engrg Univ of Mich 1921-56. Cons Natl Acad of Sci & Indus, Amer Stds Assn, Indus. Pub sev books on materials, metal cutting, metal processing, production engrg, & 300 papers on res, publ by var tech societies. Wrote sections of Mark's Mech Engrg Handbook, ASM, ASTE (now SME), Encyclopaedia Britannica & McGraw-Hill Encyc of Sci & Indus. Mbr sev hon soc. Hon Mbr ASME, Fellow Engrg Soc of Detroit. Life Mbr ASEE, ASM, SME, Natl Defense Preparedness Assn. Listed in sev Who's Whos. *Society Aff:* ASME, ASM, SME, ESD, ASEE, NDPA.

Bostwick, Willard D
Business: School of Engg and Tech, 799 W Michigan St, Indianapolis, IN 46202
Position: Assoc Dean/Prof. *Employer:* Purdue Univ. *Education:* PhD/Education/Univ of KY; MA/Education/Roosevelt Univ; BS/Math/Northern IL Univ. *Born:* 12/04/37. Native of Fayette County, IL. Served with US Navy, 1955-58. IL Div of Hwys, 1960-62. Research Analyst, IL Dept of Labor, 1962-65. Univ of Louisville, Speed Scientific School, 1965-76. Presently Prof & Assoc Dean, Purdue Univ School of Engg & Tech at Indianapolis. Speciality Co-op Educ 1976-77. Chmn, Cooperative Educ Div 1978-79; Chmn, ILL-IND Section 1987-88; of Amer Soc for Engg Educ 1978-79; Pres, Inpls Sci & Engrg Fdn 1983-85. mbr IIE, ASEE Co-author var of papers. *Society Aff:* IIE, ASEE.

Boswell, Howard L
Business: 78 Mt Vernon St, Ridgefield Pk, NJ 07660
Position: President. *Employer:* Boswell Engrg Inc. *Born:* Apr 1921. BSCE Purdue Univ. Prof courses from Cooper Union. Native of Ridgefield Pk, N J. Served as Ensign US Navy WW II. Proj Engr Boswell Engrg Co 1948-55. Partner Boswell Engrg Co 1955-76. Pres Boswell Engrg Inc 1961- . Cons Civil Engr to Central Amer countries, also N Y Dept of Transportation, N J DOT. Fellow ASCE. Enjoys boating & golf. Engr of the Year 1979 Award by NJ Soc of Municipal Engrs for Hirsohfeld Brook Proj, Dumont, NJ. Gov Committee Restoration of Monuments, NJ. Expert Witness before NJ Supreme Ct on Tideland cases. *Society Aff:* ACEC, ASCE, NSPE, NJSME.

Bosworth, Cyrus M
Business: Cyrus M. Bosworth & Assoc, RR #3 Box 267, Waldoboro, ME 04572
Position: Environ Control Consultants *Employer:* Self-employed *Education:* MS/Chem Engg/Syracuse Univ; BS/Chem/Antioch College. *Born:* Oct 1917 Cleveland, Ohio. P E New York, Maine & Ill. Employed by Carrier Corp 1945-72: as Materials, Res, Dev, Cons., Engr, & Dev FreezeWash Separation Process for Water Desalinization & for Odor Control Processes with 4 patents. At Teepak 1972-83 designing & installing equipment to control internal & external air contamination. Consulting Engineer 1983-present. Active in tech soc: Mbr AICHE, Former Chmn Syracuse Sect & Central Ill Sect; Fellow ASHRAE, chmn tech cttes; Mbr APCA & tech ctte; ACS. Awarded Wolverine-ASHRAE Diamond Key 1965 & Distinguished Serv 1972. Active in Volunteer Organizations Chmn citizens ctte on Air Pollution 1966. *Society Aff:* AIChE, ASHRAE, APCA, ACS

Bosworth, Douglas L
Home: 4432 37th Ave, Rock Island, IL 61201
Position: Mgr, Manufacturing *Employer:* John Deere Harvester Wks *Education:*

Bosworth, Douglas L (Continued)
MS/Agri Engr/Univ of IL; BS/Agri Engr/IA State Univ *Born:* Oct 1939. Native of Goldfield, Iowa. Reg P E, with Deere & Co since 1959 with exper in Reliability, Test Engrg & Design Engrg, and Manufacturing Engrg . Assumed current job in August 1985 with respon for Manufacturing, Materials Management and Manufacturing Engineering for Farm Machinery Manufacturing Factory. Mbr ASAE, ASAE Bd of Dir 1974-76 and 1979-82. ASAE Young Design Engr of the Yr 1973. Chmn Ill Wisc Region ASAE 1973. VP ASAE 1979-82. VP Quad-Cities Engr & Science Council 1984-85, Rotary, ABET Engineering Accreditation Commission 1985-1990. *Society Aff:* ASAE.

Botsaris, Gregory D
Business: Pearson A-102, Tufts University, Medford, MA 02155
Position: Prof. *Employer:* Tufts Univ. *Education:* PhD/Chem Engg/MA Inst of Techn; Chem E/Chem Engg/MA Inst of Techn; MS/Chem Engg/MA Inst of Technology; Dipl/Chemistry/Univ of Athens (GREECE). *Born:* 1/6/30. Taught at the Univ of Athens 1952-54. With Dennison Mfg Co, Framingham, MA 1960-61. With Tufts Univ since 1965. Asst Prof 1965, Assoc Prof 1969, Prof 1976. Consultant to Industry in the fields of Crystallization and Surface and Colloid Science. *Society Aff:* AIChE, ACS, AAGS.

Botsco, Ronald Joseph
Business: NDT Instruments, Inc, 15622 Graham St, Huntington Beach, CA 92649
Position: President/Chrmn of the Bd. *Employer:* NDT Instruments Inc. *Education:* BS/Met Engrg/Case Inst Tech *Born:* Sep 4, 1937. Native of Youngstown, Ohio. Sr Res/Project Engr N Amer Aviation (Rockwell Internatl) 1961-67; Genl Mgr Microwave Instruments Co 1967- 70; Pres of NDT Instruments since 1970; Reg P E Calif (two fields);; ASNT Natl Achieve Award 1969. ASNT Gold Medal Awd (Highest Awd of ASNT), 1983. Mbr ASNT & ASM; Hon Fellow Status ASNT, listed in Marquis Who's Who, Noteworthy Americans, Directory of Internatl Biography & recog as Prominent Engr by Library of Congress. Author/Lectr of 50 papers/reports on nondestructive testing; patent holder; Married: Margaret Enjoy snow skiing, boating and other sports. *Society Aff:* ASNT, ASM

Botset, Holbrook G
Home: W Waldheim Rd, Pittsburgh, PA 15215
Position: Prof Emeritus, Petro Engrg. *Employer:* Univ of Pittsburgh. *Education:* BSChE/-/Purdue Univ. *Born:* Oct 23, 1900 Plymouth, Ind; Ch Chem Radium Dial Co 1926-28; Res Engr Gulf Res & Dev Co 1928-46; Prof & Chmn Petro Engrg Dept, Univ of Pittsburgh 1946-66; Prof Emer 1966- . Author sev papers on petro engrg & related areas; Tech Res Adv Penna Grade Crude Oil Assn 1949-55; Mbrengrg ctte Interstate Oil Compact Ctte 1955-65; given John Franklin Carll Award by S P E of AIME 1969. Res Prof Engr in Penna since 1946. *Society Aff:* AIME, ACS, ΣΞ.

Bott, Walter T
Business: 100 N E Adams St, Peoria, IL 61629
Position: Plant Metallurgist. *Employer:* Caterpillar Tractor Co. *Education:* BS/Met Engrg/Univ of IL. *Born:* Jun 24, 1931. Met Engrg. Officer USAF Res. Started Caterpillar 1956 Heat Treat Engrg, Met Engrg, Met Lab Oper & Foundry. In Brazil 4 yrs-Plant Metallurgist. Assumed current respon of Plant Metallurgist at Caterpillar's Basic Engine Plant 1972. Mbr Amer Soc for Metals, Amer Foundryman's Soc, Soc of Automotive Engrs, Iron & Steel Soc of AIME. P Chmn Amer Soc for Metals 1973-74. Chmn cttes such as arrangements, program, publicity, mbrship, etc. *Society Aff:* ASM, AFS, AIME, SAE.

Bottaccini, Manfred R
Business: AME Dept, Tucson, AZ 85721
Position: Prof. *Employer:* Univ of AZ. *Education:* PhD/Fluid Mech/State Univ of IA. *Born:* 1/20/22. Prof Bottaccini is a recognized authority in the Application of Fluid Mechanics to Medicine. He is the author of one book, co-author of five other book and has published more than 100 articles in nationally recognized journals. *Society Aff:* AAAS, ASME, AUS.

Boubel, Richard W
Home: P.O. Box 3630, Sunriver, OR 97707
Position: Prof Emeritus of Mech Engrg. *Employer:* Oregon State Univ. *Education:* PhD/Environ Engg/Univ of NC; MS/Mech Engg/OR State Univ; BS/Mech Engg/OR State Univ. *Born:* 8/1/27. Fac Dept of Mech Engg at OR State Univ since 1954. Cons to indus & gov on air pollution control. Bd of Dir, VP Pres (1978-79), Pres (1979-80) Air Pollution Control Assn. Tech Program Chmn for 1976 Intl Mtg, Air Pollution Control Assn. Diplomate Amer Acad of Environ Engrs. Author *Fundamentals of Air Pollution* with Stern, Turner & Fox. Over 30 reports & articles in conf proceedings & tech journals. 3 patents. *Society Aff:* APCA, ASME, AAEE, ΣΞ, ΠΤΣ, ΤΒΠ

Bouchard, Carl E
Home: 1548 Wickerton Dr, West Chester, PA 19380
Position: Water Res Planning Specialist. *Employer:* Soil Conservation Service USDA. *Education:* BS/AE/Univ of ME. *Born:* Mar 19, 1941, Fort Kent, Maine. Reg PE Maine. USAF 1959-63. Asst to Engr, Walter F Carpenter Assoc, Orono, ME 1964-66; Agri Engr, New Brunswick Dept of Agri, Fredericton N B Canada 1966-67; Civil Engr (Hydrology & Planning), S C S, Orono, ME 1967-76, respon for Water Resource Planning. 1976-date, Water Resource Planning Specialist, Northeast Technical Service Center, SCS, Broomall, PA. ASAE Acadia Sect, Secy 1970-71, V Chmn 1971-72, Chmn 1972-73; SCSA, Pine Tree Chap, V P 1974, Pres 1975-William Penn Chapter VP 1981; NSPE, Maine Soc, 1st V P 1976-77 2nd V P 1975-76, E Me Chap, Secy 1972-73, V P 1973-74, Pres 1974-75. PA Soc, Chester Cty Chap, VP 1978-79. Pres 1979-80. S C S Cert of Merit 1974, 1976 & 1980. NSPE, Maine Young Engr of Yr 1975. Acctg Warden 1978-81, St Francis-In-The-Fields Church, Malvern; Boy Scouts of America, Webelos Den Leader, 1980-81. *Society Aff:* NSPE, ASAE, SCSA.

Bouchard, Richard J
Business: 3250 Wilshire Blvd, Los Angeles, CA 90010
Position: VPres, Dir of Corp Programs *Employer:* Daniel, Mann, Johnson & Mendenhall *Education:* M/CE/Univ of NC; B/CE/RPI *Born:* 12/21/36 1958-1964, Bureau of Public Rds, Hwy and Transit engr; 1964-1968, State of Rhode Island, Dir of Planning and Dev; 1968-1975, U.S. Dept. of Transportation, Dir of Transportation Planning Assistance; 1975-present Daniel, Mann, Johnson & Mendenhall, Dir of Corp Programs. Recipient of U.S. Dept of Transportation Secretarial Awards for Meritorious Service. Mbr of the Natl Academy of Engrg BART Impact Advisory Bd. Author of over 40 professional papers. *Society Aff:* ASCE, ITE, TRP, APTA, ARA

Bouck, David W
Business: PO Drawer 14024, 520 N Semoran Blvd, Orlando, FL 32857
Position: Chief Envir. Engr *Employer:* Dawkins & Assocs, Inc. *Education:* MS/Environ Engrg/Univ of Central FL; BS/CE/Univ of Central FL *Born:* 9/26/49 Corning, NY in 1949. Attended Univ of TN, Knoxville. BS and MS from Univ of Central FL, Orlando. Project Mgr and Civil/Environmental Engr with Dawkins & Assocs, Inc., Consulting Engrs, Orlando since 1971. Responsible for process engrg design/specification and construction management for design conformance on water, wastewater and solid waste projects. Project emphasis is advanced technology pollution control techniques, economic feasibility and environmental compatibility. Registered PE in FL and GA. Biographee in Marquis' Who's Who in the South and Southwest and Who's Who in Technology. *Society Aff:* NSPE, WPCA, ACEC, ASCE

Boudart, Michel
Business: Dept of Chem Engrg, Stanford, CA 94305
Position: Prof of Chem Engrg. *Employer:* Stanford Univ. *Education:* PhD/Chem/-. *Born:* Jun 18, 1924 Brussels, Belgium; Amer Citizenship 1957. BS 1944, MS 1947

Boudart, Michel (Continued)
Univ of Louvain; PhD 1950 Princeton Univ (Chemistry). Res work at Princeton to 1954, on fac to 1961 of Princeton's Dept of Chem Engrg; UC Berkeley 1961-64 as Prof of Chem Engrg; Stanford 1964 to present, Prof of Chem & Chem Engrg; Visiting Prof Univ of Louvain 1969, Rio de Janeiro 1973, Tokyo 1974 & Paris 1980. Co-founder & Dir of Catalytica Assoc Inc. Mbr Amer Chem Soc, Amer Inst of Chem Engrg, Chem Soc of London, as well as mbr Natl Acad of Sci. Natl Acad of Engrg. Belg Amer Ed Fd Fellow, Procter Fellow, Curtis-McGraw Res Award, R H Wilhelm Award in Chem Reaction Engrg, Kendall Award and MURPHREE Award of Am Chem Soc. *Society Aff:* AIChE, ACS, AAAS.

Boulais, Maurice R
Home: 925 Greenville Ave, Greenville, RI 02828
Position: Asst. Secy. *Employer:* Allendale Ins *Education:* BS/Mech Engg/Univ of CT. *Born:* 6/6/25. Married: Nancy L. Hamilton Nov. 27, 1947. Children - Nancy M., Suzanne, and Donald H. Served in USN during WW11 as aviation mechanic. Entire post graduate career with Allendale Mutual and predecessor companies. Presently Maximum Foreseeable Loss Evaluator for the Underwriting Staff. Member of UL's Fire Engineering Council. Past Chairman of fire Safety Com. Greater Providence Chamber of Commerce. Past President of New England Chapter SFPE. Assumed Presidency of the Society of Fire Protection Engineers in May 1981. Elected to the grade of Fellow in 1986. *Society Aff:* SFPE.

Bouldin, Donald W
Business: 420 Ferris Hall, Dept of Elec Engr, Knoxville, TN 37996
Position: Prof. *Employer:* Univ of TN. *Education:* PhD/EE/Vanderbilt Univ; MSEE/EE/GA Tech; BE/EE/Vanderbilt Univ. *Born:* 12/1/45. Dr Bouldin served as Electronics Officer in the US Navy for four yrs. He joined the faculty of the Univ of TN in 1975 and is now Prof of Elec Engg. He is a reg PE in TN and serves as a consultant to Oak Ridge Natl Lab. His interests include computer-aided design of VLSI systems for parallel processing. *Society Aff:* IEEE, ΣΞ, ΤΒΠ.

Boulet, Lionel
Home: 11285 Pasteur, Montreal PQ H3M 2N8, Canada XX
Position: Chaiman of the Board. *Employer:* Groupe LGL *Education:* Dr/EE/Univ of IL; MSc/EE/Univ of IL; BSc/EE/Laval Univ; BA/-/Laval Univ *Born:* 7/29/19. Born & educated in Quebec City (Que). Jr Engr with RCA Victor (1943-45). Champaign Res Assoc at Univ of IL (1946-47). Lecturer (1948). Asst Prof (1950). Prof (1953). Hd (1954-64) of EE Dept and Res Labs of Laval Univ. Specialized in hyperfrequencies, antennas and corona. Consultant for Hydro-Quebec in 1964 on the establishment of res facilities. Dir of Inst de recherche d'Hydro-Quebec (IREQ) since 1967. Pres of IERE (1976). Officer of Order of Canada (1975). Honoris Causa Dr from five canadian universities. Sevl other honors 1982 - Exec VP - Hydro-Quebec 1984 - 1985 - Foundation for Dev of Sci and Tech. 1986 Groupe LGL, Chairman of the Board. *Society Aff:* IEEE, NRCC, SCC.

Boulger, Francis W
Home: 1817 Harwitch Rd, Columbus, OH 43221
Position: Sr Tech Advisor, retired *Employer:* Battelle's Columbus Labs. *Education:* MS/Metallugy/OH State Univ; Met Eng/Metallurgy/Univ of MN. *Born:* Jun 1913 Minneapolis, Minn; m. 1940. Conducted & managed res & dev programs in met, mech & production engrg at Battelle since 1937. Author more than 140 articles & co-author 7 bks in those specialties. Mbr AIME, Fellow SME, Fellow ASM, Fellow ASME; V P 1973-74, Pres 1975-76 Hon Mbr, 1979 Internatl Inst of Prod Engrg Res (CIRP). Recipient of Hunt Medal 1955 AIME, Gold Medal SME 1967, 1975 Achieve Award Amer Machinist. Org & led Metal Forming Div SME 1968-70, Chmn Res Grants Ctte 1971-73. Active on adv cttes for OECD & U S Gov agencies, invited lectr in 11 foreign countries. Mbr, Natl Acad of Engg, 1978 Founder Mbr, Intl Cold Forging Group 1967 and of N Amer Manufacturing Res Inst of SME, 1981; Named "Distinguished Alumnus–, by Ohio State Univ, 1984. *Society Aff:* SME, NAE, ASME, AIME, ASM, CIRP.

Bouman, Robert W
Business: Bethlehem Steel Corp, Research Dept, Bethlehem, PA 18016
Position: Mgr-Process Research. *Employer:* Bethleham Steel Corp. *Education:* MS/Mining Engr/OH State Univ; BS/Mining Engr/OH State Univ; -/Met Engr/Lehigh Univ. *Born:* 12/20/37. Native of Cleveland, OH. Worked as a dev engr for Youngstown Sheet and Tube Co during 1961-64. With Bethlehem Steel since Sept 1964. Maj activities have been in primary steel mgt operations including ore preparation, sintering, the ironmaking blast furnace steelmaking and continuous casting. Assumed present duties in 1985. Respon for new continuous casting process developments and all outside funded programs in primary steelmaking. Enjoy golf and hunting. *Society Aff:* AIME.

Boundy, Ray H
Home: 600 South Ocean Blvd, Boca Raton, IL 33432
Position: Consultant. *Employer:* Self. *Education:* DSc/-/Croye City College; MS/ChE/Case Western Res Univ; BS/ChE/Case Western Res; BS/ChE/Grove City College. *Born:* Jan 1903. Early ed in Grove City, Pa. Worked in oil fields in W Va & Pa during college vacations. With Dow Chem Co 1926-68; First in R&D 1926-39. Then Asst to Pres & finally as Mgr of R&D 1952-68. Retired 1968 as V P for R&D. Mbr of Bd of Dir & Exec Comm. After retirement, self-employed cons in R&D mgmt as well as working with the Internatl Exec Comm in Taiwan & US Info Agency in E Europe. Gold Medal 1965 IRI, Modern Pioneers of Creative Indus 1965 NAM. *Society Aff:* NAE, AIChE, ACS.

Bouquard, Joseph P
Business: 117 E 7th St, Chattanooga, TN 37402
Position: President. *Employer:* Bouquard Engrg Co Inc. *Education:* BS/Struct Engg/Cleveland State Univ Cleveland, OH. *Born:* Feb 1917. Tau Beta Pi. Native of Cleveland, Ohio. Co-op student with TVA 1937-40. Struct Engr with TVA 1940-43. US Combat Engrs 1943-46 with 16 mos in Europe. Schmidt Engrg Co, Chattanooga, V P 1946-64. Bouquard Engrg Co, Chattanooga, Pres 1964- . Cons Engr primarily in hwy & sanitary engrg fields with services in all phases of civil engrg. Reg P E Ala, Ga, Ohio & Tenn. Mbr ASCE, ACEC, AWWA, NSPE, WPCF, Kiwanis Club, Chamber of Commerce & Chattanooga Engrs Club. *Society Aff:* ASCE, ACEC, NSPE, AWWA.

Bourbaki, Socrates Peter
Home: 4950 Chicago Beach Dr, Chicago, IL 60615
Position: Supervising Civil Engr. *Employer:* Metro Sanitary Dist of Greater Chgo. *Education:* BSCE/Tunnels/Univ of KY; Mgmt/Supervision/Univ of Chicago. *Born:* Oct 27, 1916. Univ of Kentucky Award. Outstanding Engrg Alumnus 1968. Fellow Royal Soc of Health-London, Eng. Subways & Superhwys-City of Chgo. Spec Engrg Div, Panama Canal. US Army Corp of Engrs Overseas Duty-Iran. Supt of Maintenance & Util TWA Air Base-Abadan, Iran. Began working for Metro Sanitary Dist of Greater Chgo 1946-Dir of Labs, Engr of Sewer Control, Dir of Fulton Cty Land Reclam Proj, Proj Mgr Construction Div, Supervising Civil Engr for Elec & Fulton Cty Projects, Soil & Rock Mech Sect. Ill P E License 1947-Ill Sanitarian License Dec 1966. Mbr Natl Soc of Prof Engrs, Ill Prof Engrs Soc, Water Pollution Control Fed, Chgo Public Health Engrs Club, P Commander 2nd Dist Council of Dept of Ill, Amer Legion. Cook Cty Clean Streams Ctte, Hellenic Prof Soc of Ill. *Society Aff:* FRSH, NSPE, ISPE, WPCF.

Bourdon, Joseph H, III
Home: 2830 Glen Elyn Way, Baldwin, MD 21013
Position: Senior VP *Employer:* Kidde Conslts, Inc *Education:* BS/CE/Univ of MD *Born:* 10/28/30 Native of Baltimore, MD. Served as Communications Ofcr in the USAF (1952- 1954). Assoc Engr for Whitman Requardt & Assocs (1954-1980). With Kidde Conslts, Inc since 1980. Currently Chief of KCI's Environmental Div responsible for planning and design of primarily water and wastewater facilities. Active in various youth progs. *Society Aff:* ASCE, NSPE, WPCF, AWWA.

Bourgault, Roy F
Home: 9 Einhorn Rd, Worcester, MA 01609
Position: Prof of ME. *Employer:* Worcester Poly Inst. *Education:* MS/Met/Stevens Inst of Tech; BS/ME/Worcester Poly Inst. *Born:* 2/25/20. Mbr of faculty of WPI since 1955; five yrs as its elected Secy. Prior to academic career, was met for US Steel & for Warner-Hudnut Corp. Served in USAAF 1943-46. Presently consultant to mfg, ins cos, legal firms & individuals as expert in failure analysis. Past chrmn, Worcester Chapter, ASM (1967-68). Chrmn, Mtls Div, ASEE (1979-80). Significant avocational expertise as woodcarver & as photographer. Well-known in New England as volunteer leader, Boy Scouts of Am. *Society Aff:* ASEE, ASM.

Bourgoyne, Adam T
Business: Petroleum Engineering Dept, Baton Rouge, LA 70808
Position: Chrmn & Prof *Employer:* Louisiana State Univ *Education:* PhD/Petrol Engrg/Univ of TX at Austin; MS/Petrol Engrg/LA St Univ; BS/Petrol Engrg/LA St Univ *Born:* 07/01/44 Native of Baton Rouge, LA. Is married to Kathryn Daspit also of Baton Rouge and has six children. Has worked for several yrs in the oil industry in various petroleum engrg assignments. Started teaching at LA State Univ in 1970 and assumed the responsibility of chrmn of the Petroleum Engrg Dept in 1977. Has several works published including a textbook. Also has two inventions on the market. Has recently received SPE distinguished faculty award. *Society Aff:* TBΠ, ΠΕΤ, SPE/AIME, API

Bourne, Henry C, Jr
Business: Atlanta, GA 30332
Position: Acting Pres. *Employer:* Georgia Institute of Technology *Education:* ScD/EE/MIT; SM/EE/MIT; SB/EE/MIT. *Born:* 12/31/21. Served an Asst Prof at MIT, from 1952-1954. He was an Asst and then Assoc Prof, Univ of CA, Berkeley, 1954-1963; Prof 1963-1979, chrmn, Electrical Engg, 1963-1974, Rice Univ; Div Dir, Engg, 1977-1979, Deputy Asst Dir, Directorate for Engg and Applied Science, 1979-1981, National Science Foundation; and VP for Academic Affairs, Georgia Inst of Tech 1981-1986. Appointed Acting Pres of Georgia Institute of Technology 1986-present. *Society Aff:* IEEE, AAAS, ASEE, APS.

Bourne, William H
Home: 510 Vittorio Ave, Coral Gables, FL 33146
Position: VP & Director *Employer:* Post, Buckley, Schuh, & Jernigan. *Education:* BS/Civil Engg/MO Sch of Mines & Metallurgy. *Born:* Sep 1920; Resident of Florida 41 yrs; Capt CEC USNR-ret, on active duty 1942-46 & 1951-53. V P Joseph G Moretti Inc, Genl Contractor 11 years; Dept Hd, Contract Admin & Specifications; VP, Director; Post, Buckley, Schuh, & Jernigan Inc, Cons Engrs 19 yrs. Pres S Fla Sect ASCE 1973; Pres Miami Chap, Fla Engrg Soc 1960; Director, Fla. Engr. Soc. 1961-1964; Pres Greater Miami Chap CSI 1974. Hobby: sailing. *Society Aff:* ASCE, NSPE.

Bourquard, Everett H
Business: 1400 Randolph St, Harrisburg, PA 17104-3497
Position: Chrmn of Bd *Employer:* E. H. Bourquard Assocs, Inc. *Education:* BS/Engrg/MS St Univ. *Born:* 1/10/14 Born and raised in Vicksburg, MS. Planning and hydraulic design of flood control projects in Yazoo and Susquehanna River Basins, US Army Corps of Engrs, 1936-41. Ens to LCD in US Navy Civil Engr Corps on Construction of Air Stations in Pacific and Atlantic, 1941-45. In charge navigation and traffic studies for sea level canal at Panama, The Panama Canal, 1946-48. Supervised design and construction of State flood control projects as Chief Flood Control.....Engr for PA, 1948-55. Pres and Chief Engr of engrg firm specializing in water resources engrg, 1955 to date. One major project was $88 M Neshaminy Program which included 10 reservoirs, 2 county parks, 95 MGD river pumping facilities, 46 MGD booster pumping facilities, 10.5 miles trans. mains, and 20 MGD water plant with 17.2 miles of trans mains. *Society Aff:* NSPE, AWWA, ASCE, ASTM

Bours, William A, III
Home: The Devon, 2401 Pennsylvania Avenue, Wilmington, DE 19806
Position: retired *Employer:* E I du Pont de Nemours - Wilmington. *Education:* MS/Engg/Columbia Univ; BA/Chemistry/Princeton Univ; BS/Engg/Princeton Univ. *Born:* 7/20/18. Was on Exec Ctte & Bd of Dir of Natl Paint & Coatings Assn, was Pres. Mbrship in Amer Chem Soc; Amer Inst of Chem Engrs; AM section of Society of Chem Indust; Phi Lambda Upsilon hon chem frat; Tau Beta Pi hon engrg soc, & Sigma Xi hon sci res soc. Clubs: Princeton Club of N Y; Wilington Club; Pine Valley Golf Club; Wilmington Country Club; Greenville C.C. Mr. Bours was pres of the Bd of Trustees of the Univ of Delaware Res Fdn. He is on the Chemistry Department's Adv Counc of Princeton Univ. He is a trustee of Philadelphia's Frankin Institute. Awarded Heckel Award and Indust Statesman Award by Nat'l Paint & Coatings Assn. *Society Aff:* AIChE, ACS.

Bousha, Frank N
Home: 2513 S 87th St, Omaha, NE 68124
Position: Dep Dir, Missile Facil, Engrg & Serv. *Employer:* SAC, Offutt AF Base, Nebr. *Born:* 1926. BS Univ of Nebraska-Lincoln. 30 sem hrs for MBA. Native of Omaha, Nebr. Hydraulic struct design engr for Army Corps of Engrs 194953. Civil & struct engrg in Maintenance, Engrg, & Programs Directorates, Strategic Air Command Hdqtrs 1953-68. Deputy Dir, Missile Facilities, SAC Hdqtrs 1968- . Mbr SAME, NSPE, AFA, BPOE. P E Nebr 1960. Meritorious Civilian Serv Medal 1975. Enjoy all sports.

Boutwell, Harvey B
Home: 30 Bates Dr, Cheshire, CT 06410
Position: Traffic Engineer. *Employer:* Hwy Traffic Cons. *Education:* BS/Eng/Yale. *Born:* May 16, 1924. Cert Bureau of Hwy Traffic, Yale Univ 1961. Cert Traffic Safety Mgmt N Y Univ 1962. Native Concord, N H. US Naval Res WW II. N H Dept P W & Hwys (Traffic Res Engr) 1954-64. Res Assoc Yale 1964-68. Traffic Engr Kaehrle Traffic Assocs 1968-71. Owner Hwy Traffic Cons 1971- . P E in N H, Mass, Ct, Calif. Fellow Inst of Transportation Engrs. Int Dir ITE 1974-76. Pres New Eng Sect ITE 1970. Disting Service Award, New Eng Sect ITE 1975. Mbr T R B. Exchange Club of Cheshire, Ct. Andrew Award, National Exchange Club, Feb. 1980. Distinguished Service Award, District One, ITE, May 1981. *Society Aff:* ITE, ACEC.

Bouwer, Herman
Business: 4331 E Broadway, Phoenix, AZ 85040
Position: Director. *Employer:* U S Water Conserv Lab, USDA-ARS. *Education:* PhD/Agri Engg/Cornell Univ; MS/Agri Engg/Natl Agri Univ, Wageningen, The Netherlands; BS/AGri Engg/Natl Agri Univ, Wageningen, The Netherlands. *Born:* Jul 1927 The Netherlands. BS & MS Dutch Agri Univ at Wageningen; PhD Cornell Univ N Y 1955. Res & grad teaching Auburn Univ, Ala 1955-59. Res Hydr Engr, U S Water Conservation Lab, Phoenix, Ariz 1959-72, Dir 1972- . Res in groundw recharge with sewage effl, irrigation, & drainage. Adj Prof groundw hydrol, Ariz State U. Cons to gov & local agencies. Reg engr & mbr var prof soc's. Author more than 190 tech pubs & textbook in groundwater hydrology. Rec OECD Sci Fellowship 1964, ASCE W Huber Res Prize 1965, ASCE R. J. Tipton Award 1984, Best Paper Awards, Amer Soc Agri Eng 1964, Soil Cons Soc Amer 1969, Nat Water Well Assoc 1978, USDA Superior Serv Awards 1963 & 1973, & USDA Scientist of the Year Award 1985. *Society Aff:* ASCE, ASAE, NWWA, AGU, WPCF, IAWPRC

Bovay, H E Jr
Business: 3355 W. Alabama, Suite 1140, Houston, TX 77098
Position: Pres & Ch Exec Officer. *Employer:* Mid-South Telecommunications Company *Education:* BS/Civil Engg/Cornell Univ. *Born:* Sep 1914 Big Rapids Mich. Engr Humble Oil & Ref'g, later proj mgr for first plant to make Toluene synthetically from petro, & for first plant making alkylate blending agent for aviation gasoline. Formed engrg firm 1946, incorporated 1962 Bovay Engrs Inc Pres then 1973 Bd Chmn & Ch Exec Officer. Retired 1984. Currently Pres Mid-South Telecommunica-

Bovay, H E Jr (Continued)
tions Co. Prof engrg license 17 states. Pres NSPE 1976, V P 1970-72; Pres TSPE 1968; Fellow and Life Member ASCE; council AICE 1970-73; recipient NSPE-PEPP Award 1972, TSPE Outstanding Engrg Award 1974, NSPE Award 1987, ASHRAE/ALCO Medal Award 1971; Fellow & Life Member, ASHRAE 1981, Distinguished Engineer Texas Engineering Foundation 1980. Enjoy hunting, fishing & golf. Active in Boy Scouts of America, President of San Houston Area Council 1964-1966, President of South Central Region 1981, Mbr Natl Exec Bd 1982-1987, Awarded Silver Beaver, Silver Antelope, and Silver Buffalo for Service to Scouting. *Society Aff:* NAE, NSPE, ASCE, ASHRAE, TBΠ, APPA.

Bowden, Bertram V
Home: 'Pine Croft' - 5 Stanhope Rd, Bowdon near Altrincham, Cheshire, England
Position: Retired (Academic Admin-ret-Member House of Lords) *Born:* Jan 1 1910 Chesterfield Derbyshire. MA, PhD, HonDS, HonLLD, MSc Tech. Fellow IEE, IEEE, ICE (cr Life Peer 1963). Ed Chesterfield Grammar Sch; Emmanuel College, Cambridge. Worked with late Lord Rutherford on problems of nuclear structure 1931-34; PhD 1934. Became Physics Master at Liverpool Collegiate School 1935-37; Chief Physics Master at Oundle School 1937-40; radar res in England 1940-43; radar res in USA 1943-46 - took 20 Englishmen to join 300 Americans in Combined Res Group at the Naval Res Lab in Washington to make new, combined identification & beacon sys. Work was stopped at end of war & simplified version of Sys now used universally in all civil aircraft. Sir Robert Watson-Watt & Partners 1947-50; Ferranti Ltd, Manchester 1950-53 - took charge of the exploitation of digital computers. Came to Manchester College of Tech in 1953 as Principal & remained as Principal of what became known as UMIST Univ of Manchester Inst of Sci and Tech until Sept 1976, except for 1 yr, 1964-65, when became Minister of State, respon for sci in this country. (1960-64 was Chrmn, Electronics Res Council, Ministry of Aviation). Hon DS Rensselaer Polytech 1974.

Bowden, Warren W
Home: 519 S. 5th St, Terre Haute, IN 47807
Position: Prof Chem Engr *Employer:* Rose-Hulman Inst of Tech *Education:* PhD/ChE/Purdue Univ; MS/ChE/Rose-Hulman Inst of Tech; BS/Chem/Univ of ME *Born:* 11/28/25 S. Penobscot, ME. Served in Merchant Marine in World War II, 1944-1946; Worked for Commerical Solvents Corp, 1949-1954 as research engr in pilot plants; served with US Army, 1954-1956; Progressed from instructor to professor of Chem Engrg, Rose-Hulman Inst of Tech, 1956 to date. *Society Aff:* ACS, AIChE, NYAS, IAS

Bowditch, Frederick W
Home: 2777 Orchard Trail, Troy, MI 48098
Position: VP *Employer:* MVMA *Education:* PhD/ME/Purdue Univ; MSME/ME/Purdue Univ; BS/ME/Univ of IL. *Born:* 11/17/21. Currently Vice President of the Technical Affairs Division of MVMA and therefore responsible for all technical activities of the Association. Was Exec asst to the vp of GM Environ Act Staff. Previously dir of automotive emission control for GM and represented GM on matters involving automotive air pollution control, assuring that GM metal automotive emission regulations, and was responsible for coordinating GM emission control act, the analysis of current and proposed leg and assessment of future control requirements. *Society Aff:* SAE, APCA, ESD.

Bowen, Carl H
Home: 115 W. Central Rd, P.O. Box 834, Arlington Heights, IL 60006
Position: Consultant *Employer:* PRC Consoer Townsend, Inc. *Education:* AB/Economics/Univ of KS *Born:* 12/5/10 1937-1941 operated sales engrg firm designing heating and air conditioning systems. 1941 employed as mech design engr - Consoer, Townsend & Assoc, consulting engrs. 1946 made chief mech following WWII service - US Navy. Elected Ptnr 1950 and continued until 1976 - became VP of PRC Consoer Townsend, Inc., the succeeding firm. Retired as Consultant to the firm 1980. As solid waste mgt Ptnr designed the first processing system using shredded refuse to fire steam boilers at Hamilton, Ont., and the first co-disposal system at Duluth, MN utilizing shredded refuse to fuel fluidized bed reactors incinerating sewage sludge filter *Society Aff:* NSPE, ASME, ASHRAE, WSE, ACEC

Bowen, H Kent
Business: Room 12-009, Cambridge, MA 02139
Position: Prof of Ceramic Engrg *Employer:* MIT *Education:* PhD/Ceramics/MIT; BS/Ceramic Engrg/Univ of UT *Born:* 11/21/41 A native of Magna, UT. Spent all of my profl career as a prof at MA Inst of Tech. Authored or coauthored a textbook and over fifty published papers. Research on processing of materials and the relationship of processing to physical properties. Conslt to over 20 corps. Assoc Dir of MIT Materials Processing Center. Fellow of the American Ceramic Soc. Schwartzwalder Award from the Natl Inst of Ceramic Engrs (1974), F. H. Norton Award (1980), Richard M. Fulrath Award (1981). *Society Aff:* ACS, APS, Electrochem Soc, Amer Chem Soc

Bowen, J Ray
Business: FH-10, Seattle, WA 98195
Position: Dean *Employer:* Univ of WA *Education:* PhD/ChE/Univ of CA; SM/ChE/MA Inst of Tech; SB/ChE/MA Inst of Tech. *Born:* 1/9/34. Married Priscilla J Spooner 1956 Children J Ray II, Sandra L, & Susan E ages 31, 27, 24 respectively. At Univ Washington Seattle since 1981. Served as Dean of Engrg to present. At UW Madison 1962-1981. Served as departmental chrmn 1971-1973 & 1978-1981 & as assoc vchancellor (1972-1976). NATO-NSF Postdoctoral fellow in sci at Cambridge Univ (1962-1963). ASEE-NASA Faculty Fellow at Stanford Univ 1965. NATO-NSF Sr Postdoctoral Fellow at Imperial College 1968. Richard Merton Visiting Prof of Deutsche Forschungsgemeinschaftt at Univ of Karlsruhe 1976-1977. Res interests: combustion physics, detonations, and chem lasers publications in scientific journals. Editor of Proceedings of 7th, 8th, 9th and 10th Intl Colloquium on Dynamics of Explosions & Reactive Systems. *Society Aff:* AIChE, APS, AAAS, ASEE, TBΠ, ΣΞ.

Bowen, Paul H
Home: PO Box 122, Brigeton, NC 28519
Position: Engr *Employer:* Conslt *Education:* B of ME/-/OH State Univ. *Born:* Oct 1918. Native of Londonville, Ohio. Army Corp of Engrs 1940-45. Proj Engr Wright Patterson AF Base, spec in jet engine dev 1946-52. With Westinghouse Res Labs 1953-1984, spec in lubrication, wear, & friction of rolling & sliding surfaces in air & hostile environments. Enjoy golfing & sailing. *Society Aff:* ASLE.

Bowen, Ray M
Business: 177 Anderson Hall, Lexington, KY 40506-0046
Position: Dean, Coll of Engrg *Employer:* Univ of KY *Education:* PhD/ME/TX A & M Univ; MS/ME/CA Inst of Tech; BS/ME/TX A & M Univ *Born:* 03/30/36 Ft. Worth, TX. Active duty USAF (1961-64) at the Air Force Inst of Tech. Post Doctoral Fellow in Mechanics (1964-65) at Johns Hopkins Univ. Assoc Prof Engrg Mechanics (1965-67) at LA State Univ, Faculty mbr at Rice Univ since 1967. Prof of Mech Engrg since 1973. Chmn Mech Engrg Dept 1972-77. Div Dir, Natl Sci Fdn, 1982-83. Dean, Coll of Engrg, Univ of KY since Sept 1983. Mbr of the Soc of Scholars of Johns Hopkins Univ. Mbr of the Houston Philosophical Soc. Author of approximately fifty technical articles and of two textbooks. Awarded Erskine Fellowship at the Univ of Canterbury, Christchurch, NZ 1974. *Society Aff:* ASME, AAM, ASEE, ΣΞ, ΦΚΦ, TBΠ

Bowering, Richard E
Business: Stelco Inc, Utilities Dept, Hilton Works, Hamilton Ontario L8N 3T1 Canada
Position: Ass't Supt, By Products. Hilton Wks. *Employer:* Stelco Inc *Born:* Jan 21, 1942. Grad B Sc Met Engrg 1963. Sev pos in qual control at the Steel Co of Canada until present appointment as Supt Canada Works East Mill in 1974. P Chmn of Ontario Chap ASM; ASM Young Mbr's Award 1972; President's Pin 1971. Sev natl ASM cttes, 1979 Ass't Supt-Utilities Hilton Wks - Stelco Inc..

Bowerman, Frank R
Business: 125 W Huntington Dr, Aradia, CA 91006
Position: Sr VP. *Employer:* Engg-Science Inc. *Born:* Jul 1922. BS, MS Caltech. Army Air Corps WW II. Los Angeles Cty Sanitation Dists 18 yrs, 8 yrs Asst Ch Engr. Asst to V P - Dev Aerojet-Genl Corp. Group V P - Zurn IndusProf, Dir Environ Engrg, Univ of S Calif, Chmn Civil Engrg 1970-75, now Adj Prof. Pres Amer Acad Environ Engrs 1973, Diplomate. Pres Inst Solid Wastes APWA 1966. Natl Dir APWA 1974-77. Natl Dir WPCF 1965-68. V P Los Angeles Sect ASCE 1975, Fellow. Rudolph Hering Medal ASCE 1962. Nichols Award APWA 1965. Tau Beta Pi, Sigma Xi, Hon Mbr Chi Epsilon.

Bowers, David F
Business: 1043 E. S. River St, Appleton, WI 54911
Position: Section Leader *Employer:* The Inst of Paper Chem *Education:* Ph.D/Metallurgical Eng/OH St Univ; M.S./Metallurgical Eng/Univ of WI; B.S. /Metallurgical Eng/Univ of AL *Born:* 5/8/37 Reared in Birmingham, Alabama. Worked as a Plant Metallurgist for Ross Meehan and Dayton Malleable Iron in Chattanooga, TN and Columbus, Ohio respectively, 5 yrs each company between academic enrollments. Also served in U. S. Army Reserves C of E. for 7 1/2 yrs. Assumed current position at the Inst of Paper Chem in 1976 - teaching and dir corrosion res for the pulp and paper ind Recipient of Fondry Educ Fellowship, 1966 and International Nickel Company Fellow, 1972-1975. Enjoy "free time" with my family. *Society Aff:* ASM, NACE, TAPPI

Bowers, James C
Business: Fowler Ave, Tampa, FL 33620
Position: Full Prof. *Employer:* Univ of S FL. *Education:* ScD/Controls/WA Univ; MS/Controls/Univ of TN; BE/EE/Vanderbilt. *Born:* 2/13/34. Dr James C Bowers is a Univ Prof with a background of over 20 yrs of ind experience & 12 yrs of teaching experience. His field of study includes Control System Analysis & Solid State Design, in addition to his extensive pioneering work in Comp Aided Design & Analysis & Solid State Modeling. Prof Bowers originated & headed the team which developed the popular Super Sceptre computer prog used by over 200 cos & agencies around the world. He has given numerous Seminars on a variety of comp oriented topics. He has co-authored over 25 technical articles & four textbooks. He is presently a Sr Consultant for the Army, Navy & AF. *Society Aff:* IEEE.

Bowers, Klaus D
Business: 600 Mountain Ave, Murray Hill, NJ 07974
Position: Vice Pres. *Employer:* AT&T Bell Labs *Education:* PhD/Physics/Oxford, England; MA/Physics/Oxford, England; BA/Physics/Oxford, England. *Born:* 1929 Stettin, Germany. Ed in England. BA 1950, MA & PhD 1953 Res Lecturer at Christ Church, Oxford 1952-56. With Bell Labs since 1956, first in res & then in var mgmt pos in electronics dev, except for leave of absence to Sandia Labs as Managing Dir & V P 1971-75. Presently VP, Electronics Technology. Fellow Inst of Electrical & Electronics Engrs., member National Academy of Engineering. *Society Aff:* IEEE, NAE

Bowers, Wendell
Home: 1010 Osage Dr, Stillwater, OK 74075
Position: Ass't dir, Agri Programs *Employer:* Oklahoma State Univ. *Education:* BS/Agri Engr/Univ of IL; MS/Agri Engr/Univ of IL. *Born:* Sept 1926. Native of Potomac, Illinois. Extension , Univ of Ill 1951-67. Ass't Dir. Agri: Programs, Cooperative Extension Service, Oklahoma State Univ 1967- . USDA Superior Serv Award 1971. ASAE Fellow 1978. Since 1951 served as cons to farmers, public serv & indus in genl area of farm power & machinery. Internatl cons on farm machinery mgmt. Author 3 text books on Agri Engrg & Farm Machinery Mgmt. *Society Aff:* ASAE.

Bowhill, Sidney A
Business: Dept of Elec Engrg, Urbana, IL 61801
Position: Dir, Aeronomy Lab. *Employer:* University of Illinois. *Education:* PhD/Physics/Cambridge;BA, MA/Cambridge Univ. *Born:* Aug 8, 1927. Native of Dover, England. Came to USA 1955, naturalized 1962. Res engr with Marconi's Wireless Telegraph Co Ltd, Chelmsford, Eng 1953- 55, working on long-distance radio propagation. Prof of Elec Engrg & Dir of Aeronomy Lab, Univ of Ill Urbana, since 1962. Pres of Aeronomy Corp, Champaign, Ill from 1969. Engaged in teaching & res in theory & applications of the upper atmosphere & ionosphere, incl rocket & satellite tech & radio remote sensing. Fellow IEEE *Society Aff:* NAE, IEEE, AGU, AAS, AAAS, ASEE.

Bowie, Robert M
Home: 2404 Hamilton Dr, Ames, IA 50010-8248
Position: Tech Mgmt Cons (Self-employed) *Education:* PhD/Physics/IA State Univ.; MS/Physics/IA State College; BS/Chemistry IA State College *Born:* Aug 24, 1906 Table Rock, Nebr; c. 2. Physicist Sylvania Elec Prods 1933- 34, Dir Phys Lab 1935-49, Dir Engrg 1949-55, Dir Res 1955-58, V Pres Sylvania Res Labs 1958-60, V Pres. Gen Tel & Electronics Labs Inc 1960-64; Tech Mgmt Consult 1964- . Mbr Natl TV Sys Cmt 1951-54; NY State ADv Coun Adv Indus Res & Dev; Fellow Phys Soc; Fellow IEEE. Thermionic emission from nickel; cathode ray tube res & design such as negative ion traps, metal bodied cathode ray tubes; microwave tube res & design; microelectronic circuits; Member and Trustee of Palisades Inst. for Research Services 1970-1984. *Society Aff:* APS, IEEE.

Bowles, C Quinton
Business: 600 West Mechanic, Truman Engrg Labs, Independence, MO 64050
Position: Mech & Aerospace Engr *Employer:* Univ of MO-Columb *Education:* PhD/Mat Sci & Engrg/Delft Univ of Tech, The Netherlands; MS/Physics/Univ of MO-KC; BS/Physics/Univ of MO-KC *Born:* 4/22/35 Dr Bowles has been Coordinator for Mech Engrg in Kansas City and is responsible for the materials science portion of the Engrg curriculum. He is past chrmn of the Educational Ctte, past pres and currently mbr of the Exec Bd of the Kansas City Chapter of Amer Soc for Metals. He is presently vice Chairman elect of Committee E-9 (Fatigue) and is also active in committee E-24 (Fracture) of the Amer Soc for Testing and Materials. Dr Bowles was the recipient of a Natl Sc Fdn Grant and has reviewed proposals for the Natl Sci Fdn. During 1976-1977 he was a recipient of a two year Fellowship to the Aeronautical Engrg Dept of the Delft Univ of Tech, Delft The Netherlands. In 1984 he was elected a Univ of MO-Kansas City Trustees Faculty Fellow. Dr Bowles is the author of numerous tech papers, reports, etc and has presented numerous seminars on the subject of Fatigue and Fracture. *Society Aff:* ASM, ASTM, ΣΞ

Bowles, Joseph E
Home: 6810 N Hi-Wood Ct, Peoria, IL 61614
Position: Owner *Employer:* Engrg. Computer Software *Education:* MSCE/Structures-Geotechnical/GA Inst of Tech; BSCE/-/Univ of AL. *Born:* 7/12/29. Resident of IL. Served Korean war, Taught Southern Technical Inst, Univ of WI and Bradley Univ. With Bradley Univ 1963-1980 with specialties in Soil Mechanics/Fdn Engg and Structural Engg. Author of six textbooks: Analytical and Comp Methods in Fdn Engg, Engg Properties of Soils and their Measurement (3rd Ed), Fdn Analysis and Design (4th Ed), Physical and Geotechnical Properties of Soils (2nd Ed), Structural Steel Design and Structural Steel Design Data Manual, all with McGraw-Hill Book Co. Texts represent leading Civ Engg to SI in US. Owner of Engg Computer Timesharing/Software service to consulting engrs since 1979. *Society Aff:* ASCE, ACI, ASTM, ASEE.

Bowman, Craig T
Business: Dept of Mech Engg, Stanford University, Stanford, CA 94305
Position: Prof. *Employer:* Stanford Univ. *Education:* PhD/Aero Engg/Princeton Univ; MA/Aero Engg/Princeton Univ; BS/ME/Carnegie Inst of Tech. *Born:* 9/19/39. Sr Res Engr with United Technologies Res Ctr, E Hartford, CT 1965-1976, specializing in combustion of hydrocarbon fuels, combustion-generated air pollution and air-breathing propulsion systems. At Stanford since 1976. *Society Aff:* CI, ASME

Bowman, Donald R
Home: 1605 Walthour Rd, Savannah, GA 31410
Position: Pres *Employer:* Bowman Engrg, Inc *Education:* BS/ME/KS State Univ; // Aeronautical Engr/Univ of VA; //Aeronautical Engr/Wichita State Univ; //Sanitary Engr/IL Inst of Tech *Born:* 1/27/27 Native of Kansas. Served in Army Air Corps in World War II, 1944-46. Worked in Aircraft Industry 1950-70 specializing in Aircraft Stability and Control and external stores. Process Engr for Metro San Dist of Chicago. Water Pollution Control Administrator for City of Savannah, GA. Now Engrg Consultant in Private Practice in Water and Wastewater and subdivision development. Pres Savannah Chapter, GSPE, 1979-80, VP GSPE 1980-82. *Society Aff:* NSPE/GSPE, WPCF, GW&PCA, APWA.

Bowman, E Dexter, Jr
Business: 4715 E Fort Lowell Rd, Tucson, AZ 85712
Position: Sr Vice Pres-Engrg *Employer:* Duval Corp - Tucson, Ariz. *Education:* BS/ChE/Univ of CO *Born:* May 1922. BS ChE Colorado Univ 1947. Native of Denver Colo. US Naval Forces 1942-46, Engrg Officer aboard LST, Discharged as Lt. Employed with natl engrg-constr firm 1948-63 as purchasing agent, field engr, project engr & project mgr. Employed by Duval Corp since 1963 in processing & plant dev areas. Elected V Pres in 1969, Sr Vice Pres in 1982. Respon charge for design of all potash, sulphur, metals & hydrometallurgical plant installations since 1965. Mbr AIChE & AIME. Reg P E Colo & Ariz. Enjoy tennis, classical music & needlepoint. *Society Aff:* AIChE, AIME

Bowman, Louis
Business: 233 N Michigan Ave, Chicago, IL 60601
Position: Pres *Employer:* Alfred Benesch & Co *Education:* BS/CE/Univ of IL, Urbana *Born:* 10/18/29 1951 Construction Foreman for Vermillion Construction Co. 1951-53 Military Service, Office of Chief of Engrs. 1953-54 Zonolite Contractors - Estimating Engr. 1954-Pres Alfred Benesch & Co: Soils Engr, Project Engr, Project Mgr, Project Design Engr for interstate highway projects, Chief Transportation Engr in charge of preparation of major route location studies and contract documents for highway, railroad and other related civil engrg projects; VP responsible for general administration of transportation projects, new business development and coordination of all multidisciplinary projects. Presently Pres responsible for daily operation of office and general administration of all projects. *Society Aff:* ASCE, ARTBA, ITE, NSPE.

Bowman, Mark M
Home: 101 Forrest Park Rd, Bartlesville, OK 74003
Position: Market Specialist *Employer:* Phillips Chem Co. *Education:* BS/Engr/Univ of Houston; MS/Engr/Univ of Houston. *Born:* Johnson City, Tenn, Feb 22, 1923. Instr Materials Engrg, Univ of Houston 1950-51; Physicist, Fracture Dynamics at Naval Res Lab 1951-55; Metallurgist, Test Div. Phillips Petro Co 1955-60; Advanced Plastics Applications, Phillips R&D Div 1960-64; Market Res, Phillips Chem Co 1964. Prof Engr Okla. P Chmn Tulsa Chap ASM. Mbr PPI. Contrib made in polyolefin processing, product dev, & marketing. Special field: plastics applications. *Society Aff:* ASM, SPE, CMRA.

Bowman, Rush A
Business: Rush A Bowman & Assoc, 3723 Oakley Ave, Memphis, TN 3811
Position: Owner *Employer:* Rush A Bowman & Assoc *Education:* BS/Hydraulics/Univ of MI. *Born:* 7/-/15. in Harrisburgh, PA. WWIIBB1940-46 Chief Gage Section Pffn Ord Diwt; Ret Lt Col. Mgr Cost Control Walworth Co, mfg valves & fittings. Works Mgr Berry Div Oliver Iron & Steel, Corinth MS, mfg hydraulic pumps & power sys. Since 1956 cons Indust Engr. Spec Work Measurement, Cost Control Sys & spec mech cost reduction equip. Dev sys for constructing Std Data. Late 40's early 50's, Extension Insltructor I E subjs Univ of Pitts & PA State Coll. Contrib McGraw- "Indus Engg Handbook" & "Encyclopaedia Sci & Tech–. Reg Engr PA, MS, TN & AR. Class II cons mbr MTM Assn. Hobby: mfg of working frames for needlework-hold pats. *Society Aff:* IIE, CEC, PEPP, NSPE.

Bowman, Thomas E
Business: Coll. of Science &, Eng, 150 W. University Blvd, Melbourne, FL 32901
Position: Dean, Coll of Science & Engrg *Employer:* Florida Institute of Tech *Education:* Ph.D/ME & Astro Sci/Northwestern Univ; M.S./ME/ CA Inst of Tech; B.S./ME/CA Inst of Tech *Born:* 8/3/38 s. Nina H Bowman & the late Robert A Bowman (8th-10th Eds, W W in Eng) Propulsion Res Sci, Martin-Marietta Denver Div, 1963-69, & Visiting Lect Appl Math, Univ Reading England, 67-68. At FIT since 69 as Assoc Prof to 75, Prof since 75, ME Dept Head 78-86. Dean, Grad. Sch 1982-86; Dean Sci & Engg since 1986. Chmn, Canaveral Section ASME, 66-67. Past Chmn, Natl ME Dept Hds Comm & Memb Bd on Engrg Education, ASME. Past Chmn Policy Adv Bd, Florida Solar Energy Cntr. Res interests include low-g fluid mech/heat xfer & solar/thermal energy conversion. Primary avocations antique car restoration & small sailboat racing. *Society Aff:* ASME, ASEE

Bowman, Waldo G
Home: 268 Ridge Rd, Douglaston, NY 11363
Position: Retired. *Education:* BS/Civil Engg/Univ of KS. *Born:* Sep 1900 Lawrence, Kansas. Grad study bus admin Harvard. McGraw-Hill Pub Co 1925-65. Ed-in-chief Engrg NewsRecord 1940-63. Publisher Engrg News-Record & Constr Methods & Equip 1963-65. Eastern Mgr Black & Veatch cons engrs 1967-77. Fellow Amer Soc of Civil Engrs; Pres 1964, V Pres 1958-59, Dir 1948-51. V Pres Amer Inst of Cons Engrs 1972. Mbr Internatl Comm on Large Dams & its U S Ctte 1950- . Dir Engrs Joint Council 1950 and 1970. Travelled throughout U S & to all parts of the world on engrg & constr writing & study assignments for both Engrg News-Record & Black & Veatch. Lic P E NY & NJ. *Society Aff:* ASCE, ACEC.

Bowne, Sidney B, Jr
Business: 161 Willis Ave, Mineola, NY 11501
Position: Partner. *Employer:* Sidney B Bowne & Sons. *Education:* BS/CE/Lehigh Univ. *Born:* 1920 Glen Cove, N Y. BS CE Lehigh Univ. Army Corps of Engrs 1942-46. With present firm (founded 1922) since 1947, partner since 1950. Spent considerable time in Near East, Africa & Southeast Asia as Sr Partner of Litchfield, Whiting, Panero, Severud Assoc & Litchfield, Whiting, Bowne & Assoc (AE firms) in 1950's & early 60's. Principally concerned with ground water dev, supply, storage, treatment & distrib. Mbr var tech, prof & civic orgs & bds.

Boyar, Julius W
Business: PO Box 15394, 4324 S Sherwood Forest Blvd, Baton Rouge, LA 70895
Position: VP & Genl Mgr *Employer:* Catalytic, Inc. *Education:* BS/Civil/Univ of Cincinnati. *Born:* 1/1/23. Born in Cincinnati, OH, served as Capt C O E, South Pacific, 1942-46. Awarded Bronze Star. Graduated as Civil Engr 1950-59. Reg PE OH, PA, LA. With Catalytic, Inc since 1951 in various proj engg and proj mgmt assignments. Have been in respon charge of engg and construction on over 50 major projs for the chemical, pharmaceutical and nuclear industries. Assigned as Genl Mgr of Gulf Coast Opers in June 1978, with respon to establish a full service engg/ construction office in Baton Rouge to serve the chemical, petrochemical and refining industries. Promoted to VP in 1980. *Society Aff:* AIChE.

Boyce, Earnest
Home: 1601 Granger Ave, Ann Arbor, MI 48104
Position: Prof of Civil Engrg-Emeritus. *Employer:* Univ of Michigan-ret. *Education:* MSC/Sanitary Engg/Harvard; CE/Civil/IA State Univ; BSCE/Sanitary/IA State Univ. *Born:* Jul 11, 1892 Winterset, Iowa. Civil & Sanitary Engr. m. Elsie J Green 1919; One son James. Mbr staff of School of Engrg Univ of Kansas, Lawrence Kansas 1924-41. Also Chief Engr & Dir of Sanitation, Kansas St Bd of Health 1942-41. Sr San E ngr U S Public Health Serv 1941-44. Prof of Municipal & San Engrg 1941-62 Univ of Mich. Chmn CE Dept Univ of Mich 1947-61. San Engr Cons Pan Amer Health Org & World Health Org. Life Mbr ASCE, Hon Mbr 1975. Water Pollution Control Fed P Pres & Hon Mbr. Engrg Soc of Detroit P Pres & Fellow. Amer Water Works Assn, Amer Pub Health Assn. Diplomate Amer Acad of

Boyce, Earnest (Continued)
Environ Engrg & Gordon Fair Award. Sigma Xi, Tau Beta Pi, Rotarian Paul Harris Fellow. Amer Pub Wks Assn. *Society Aff:* ASCE, AWWA, APHA, WPCF.

Boyd, Dwight A
Business: 7999 Knue Rd-Suite 406, Indianapolis, IN 46250
Position: President. *Employer:* Boyd/Sobieray Assocs Inc. *Education:* BSME/Mech Engg/Purdue Univ; Assoc-Engr/Engg/Benton Harbor Jr College; Dip/mathematics/Benton Harbor HS. *Born:* May 28, 1930 Benton Harbor, Mich. BS ME Purdue Univ 1955. Bldg elec sys designer/administrator for local cons engrg firm 1955-67. Dir of Engrg for local Audio/Video/RF Sys Contractor 1967-69. Owned & oper Cons M/E firm 1969-75. Formed & oper Cons A/E firm 1975. Reg P E Ind, Ill, Mich, Ohio, Kentucky, Fla & Georgia. Mbr Acoustical Soc of Amer, Audio Engrg Soc, Amer Cons Engrs Council, Amer Artibration Assn & ASHRAE & Illuminating Engg Soc. *Society Aff:* ASA, AES, ACEC, AAA, ASHRAE, IES.

Boyd, Gary D
Home: 56 E River Rd, Rumson, NJ 07760
Position: Mbr of Tech Staff *Employer:* Bell Labs Holmdel N J. *Education:* PhD/EE/CA Inst Tech 1959; MS/EE/CA Inst Tech 1955; BS/EE/CA Inst Tech 1954 *Born:* 9/14/32. Thesis in plasma phys. With Bell Tel Labs 1959- ; Lectr in Applied Phys Harvard 1966-67. Chmn 1972 Gordon Conf on Nonlinear Optics. Fellow IEEE 1976; Mbr Amer Physical Soc. Has contributed papers & patents in res of laser resonator theory, solid state and semiconductor lasers, nonlinear optics, ultrasonics & liquid crystal display devices. *Society Aff:* IEEE, APS, SID.

Boyd, J Huntly, Jr
Business: Bremerton, WA 98314
Position: Captain, USN. *Employer:* US Navy-Puget Sound Naval Shipyard. *Education:* NavE/Naval Engg/MIT; SM/Naval Arch & Marine Engg/MIT; BS/Engg/US Naval Acad. *Born:* 12/12/31. NavE & SM from MIT, BS from US Naval Acad. Native of Trenton, MI. Commissioned Officer US Navy since 1953 in ship engg specialty. Served at sea, on Fleet staffs, & in tech/engg activities ashore. Addl subspecialty of diving & salvage. Fleet Salvage Office O in C Navy Exper Diving urit 1969-71. ir of Ocean Engg Supervisor of Salvage & Diving, Naval Sea Sys Command 1973-76. Overall respon for clearing all (ten) block ships from Suez Canal 1974. Planning Office Norkfork Naval Shipyard 1976-79. Presently Commander, Puget Sound Naval Shipyard. Mbr SNAME, Hon Life Mbr ASNE. ASNE Gold Medal 1966. *Society Aff:* SNAME, ASNE.

Boyd, James
Business: 228 Del Mesa Carmel, Carmel, CA 93921
Position: Consultant *Employer:* Retired *Education:* BSc/Engr & Ecco/CA Inst of Tech; MSc/Geophysis/CA Sch of Mines; DSc/Geology/CA Sch of Mines. *Born:* Dec 1904. Native of Australia, naturalized U S citizen. Field Engr Radiore Co 1927-29, geophysics; Instr thru Assoc Prof C S M 1929-41; Dean 1946- 47; US Army Materiel Command 1941-44; Dir of Indus, Military Gov, Germany 1945- 46; Dir U S Bureau of Mines 1947-51; Defense Minerals Administrator 1949-51; Exploration Mgr Kennecott Copper Corp 1951-55; V P of Exploration 1955-60; Pres of Copper Range 1960-70, Chmn 1971; Exec Dir Natl Comm on Matls Pol 1971-73; Pres Matls Assocs 1973-77; Pres Mining & Metallurgical Soc 1960-63; AIME Pres 1969. Rand Medal 1963; Jackling Lecturer 1967; Hon Mbr 1973; Mbr Natl Acad of Engrg 1967-Life; Chmn Materials Adv Ctte, Office of Tech Assess 1973-79; Hoover Medalist 1975; Chmn Comon Surface Mining & Reclamation 1978-79; Mbr Natl Matls Advisory Bd (NRC) 1973-76; Mineral Resource Bd 1975-78; Bd on Mineral and Energy Resources 1978-81; Member Tech. Advisory Com. Office of Nuclear Waste Isolation, 1979-83, Engr Grp 1983- . *Society Aff:* AIME, GSA, MMSA, NAE

Boyd, James A
Business: PO Box 3, Houston, TX 77001
Position: VP. *Employer:* Brown & Root, Inc. *Education:* Dr of Jurisprudence/Law/S TX College of Law; BSChe/Chem Engg/Univ of Tx, Austin. *Born:* 1/31/39. in Austin, TX. Process Engg with Soughwestern Oil & Refining Co Corpus Christi 1963-65. Process/Project Engg Brown & Root, Inc Houston, TX 1965-68; Project Mgr 1968-73; Chief Engg Brown & Root (UK) Ltd London 1974-78 (VP); VP Brown & Root, Inc Houston 1978-present Petroleum & Chem International Group - Worldwide. Enjoy hunting, sports, working with young people. *Society Aff:* AIChE, NSPE, TSPE, BIP, ABA, TBA.

Boyd, James S
Home: 172 Orchard, E Lansing, MI 48823
Position: Ret. *Employer:* MI State Univ. *Education:* BS/Agri Engg/SD State Univ; MS/Agri Engg/MI State Univ; PhD/Agri Engg & CE/IA State Univ. *Born:* 3/8/17. Engg officer US Navy for 4 yrs. Engr for soil conservation service for 2 yrs. On faculty at MI State Univ for 33 yrs doing teaching, res & extension work in farm bldgs. Res was concerning dairy bldg design, solar energy, animal wast mgt. Spent 2 yrs in Nigeria starting agri engg dept. Lectured in Norway, Sweden, Finland and England. Received Metal Bldg Award from ASAE. Received Blue Ribbon Awards for publications from ASAE. *Society Aff:* ASAE.

Boyd, John A, Jr
Business: 800 West 47th St - Suite 600, Kansas City, MO 64112
Position: President. *Employer:* Boyd, Brown, Stude & Cambern. *Education:* MS/Civil Engg/Univ of KS; BS/Civil Engg/Univ of KS. *Born:* Dec 20, 1930 Kansas City, Mo; Reg P E MO & KS & CO. Served in US Navy as Naval Aviator aboard Aircraft Carrier, Korean Conflict. Instr of Civil Engrg Univ of Kansas; Design Engr, Howard, Needles, Tammen & Bergendoff; Partner & Pres of Boyd, Brown, Stude & Cambern, Cons Engrs since 1966. Natl Dir of Soc of Amer Military Engrs; Pres Greater Kansas City Post SAME; Pres Midwest Concrete Indus Bd; Pres Santa Fe Chap Naval Res Assn; Chmn Kansas City Prof Engrs in Private Practice; Chmn Missouri PEPP; Pres, Engrs Club of KS City; Pres, KS City Section, Amer Soc of Civil Engrs; Pres, Consulting Engrs Council of MO, Natl Dir ACEC; Exec Bd, NSPE-PEPP Bd of Governors; Natl Dir, ASCE. *Society Aff:* SAME, ASCE, NSPE, ACEC.

Boyd, John H
Business: 1900 E Pleasant, Dekalb, IL 60115
Position: Mgr Eng. *Employer:* GE. *Education:* Bach/Elec Engg/Univ of MN. *Born:* 1/17/39. Current position of Mgr Engg for Appliance Motors (1977) in Genl Elec has respon for advanced engg, product design, and engg labs. Previous assignments include: Mgr Advance Engg for heating and Conditioning Fan Motors (1974), Mgr Quality Control and Reliability Engg for Hermetic Motors and Small Industrial Motors (1970), and Mgr of Hermetic Motor Product Design (1968). Joined GE after graduation from the Univ of MN with a BEE degree in 1965 as a design engr specializing in computer aided design techniques. *Society Aff:* IEEE.

Boyd, John T
Business: 400 Oliver Bldg, 535 Smithfield St, Pittsburgh, PA 15222
Position: Chrmn *Employer:* John T Boyd Co. *Education:* BS & EM/Mining Engg/OH State Univ. *Born:* 1/25/13. BS & EM, OH State Univ, 1935. Native of OH. Reg PE in 6 states & 1 province. Pioneered in mine mechanization. Consultant to the mining industry for 44 yrs. Traveled extensively throughout US, Canada, Mexico, Europe, Greece, United Kingdom, S America, Africa, Japan & Australia. Mbr AIME, NSPE, AMC. Married, two sons. Dist Alumni Award, OH State Univ, 1983; Distinguished Member, Society of Mining Engineers, 1987. *Society Aff:* AIME, NSPE, AMC.

Boyd, Joseph A
Home: 1406 S Riverside Dr, Indialantic, FL 32903
Position: Chrmn. *Employer:* Harris Corp. *Education:* PhD/EE/Univ of MI; MS/EE/Univ of KY; BS/EE/Univ of KY. *Born:* 3/25/21. Since 1978, has been chrmn and chief exec of Harris Corp, a billion-dollar- per-year producer of communication and

Boyd, Joseph A (Continued)
info processing equipment. Had been pres since 1972. Joined co in 1962 after 13 yrs at the Univ of MI as prof of electrical engg and the dir of UM's Inst of Sci and Tech. Fellow of the Inst of Electrical and Electronics Engrs, and dir of Machinery and Allied Products Inst. Served two terms as pres of the Armed Forces Communications and Electronics Assoc Ind. Mbr of the Univ of FL Business Advisory Council, dir of the FL Council of 100. *Society Aff:* IEEE, AFCEA, AUSA, ΣΞ.

Boyd, Landis L
Home: 1725 Concord Drive, Ft. Collins, CO 80526
Position: Director-at-Large, WAAESD. *Employer:* WA State Univ. under contract to WAAESD. *Education:* PhD/Agri Engr-Engr Mech/IA State Univ; MS/Agri Engr/IA State Univ; BSAE/Agri Engr/IA State Univ; Post Doc/Higher Ed-Ed Admin/Univ of MI-Fed Exec Inst. *Born:* 12/1/23. Asst Prof to Prof & Coordinator of Grad Instruction in Agri Engg, Cornell Univ, 9/48-8/64; Hd of Dept & Prof of Agri Engg (8/64-10/72) & Asst Dir, Agri Experiment Station (10/72-5/78), Univ of MN; Dir and Assoc Dean (effective 7/1/79), College of Agri and Home Economics (1983) Res Ctr, WA State Univ, 4/78-4/85. Director-at-Large, Western Association of Agricultural Experiment Station Directors (WAAESD) 4/85-present. Reg engr in NY & MN. Scientific capabilities in structural analysis, crop storage, animal housing, human housing, automatic control systems, systems modeling, computer utilization & waste heat utilization. Active in profl soc, ASAE, serving as VP-Regions, 1970-73, & a mbr and/or chrmn of over 20 ASAE committees. Natl Assoc of Univ and Land Grant Coll (NASULEC): Chair, Exp Station Section, 1980-81; Experiment Station Ctte on Policy (ESCOP), 1981-84 and ex officio, 1985-present; also Executive Vice Chair, 1987. Western Assoc of Agricultural Experiment Station Dirs (WAASED); Exec Com, 1980-81, 1983-84 ex officio, 1985-present; Chair, Res Implementation Ctte (RIC), 1983-84; Ctte of Nine (C-9), 1984-86. *Society Aff:* ASAE, ASEE.

Boyd, Marden B
Home: 304 Montaign Dr, Vicksburg, MS 39180
Position: Chief Hydraulic Analysis Div. *Employer:* US Army Engr Waterways Exper Sta. *Education:* MS/Hyd. Engg/CO State Univ; BS/Civil Engg/MS State Univ. *Born:* Dec 1934. Continuous assn with Waterways Exper Station (Vicksburg) since grad in 1956, except for serv with Navy 1957. Worked at var assignments within 3 diff divs of Hydraulics Lab before assuming current assignment in June 1978. Respon for planning & dir studies of complex hydraulic engrg & sys problems. Reg P E Miss & mbr ASCE & NSPE. 1965 Hilgard Hydraulic Prize from ASCE. Meritorious Civilian Serv Award 1973. Active in church & prof orgs. Married to former June Newcomb. *Society Aff:* ASCE, NSPE.

Boyd, Noel F
Business: 1 Oliver Plaza, Pittsburgh, PA 15222
Position: Sr VP-Dravo Engrs and Constructors *Employer:* Dravo Corp. *Born:* BS Chem Engrg Univ of Pittsburgh 1942. Native of Pittsburgh, PA. Joined Blaw-Knox Chem Plants (acquired by Dravo Corp in 1973) 1946 as Process Engr. Also served as Dir of Engrg, VP - Engrg, VP - Exec Mgr & VP - Mktg. Asst Genl Mgr and VP - Genl Mgr. Present position Sr VP-Dravo Engrs and Constructors since February 1981.

Boyd, Robert Lee
Home: 413 Brewers Creek Ln, Carrollton, VA 23314-9641
Position: Pres. *Employer:* Mature Enterprises Inc *Education:* AB/Phys/Kenyon College. *Born:* Nov 1914. Since 1936, exper in air cond, heating & ventilating, incl contracting, servicing, trouble-shooting, design & application engrg. Formerly active in trade & prof org cttes & pubs. Inventor, author & exponent (before public acceptance) of heat pumps, electric comfort heating, high level thermal bldg insulation, unitary air conditioning sys, electric demand controllers, applying heaters to combat specific comfort disturbers, & simplified calculation of heating-cooling sys actual energy consumption & maximum demands. Since 1982, pres, Mature Enterprises, Inc., advocate for the elderly. ASHRAE: Fellow mbrship 1967, Life Mbrship 1979. *Society Aff:* ASHRAE.

Boyd, Robert N
Home: 1457 Woodbine Rd - R R 3, Kingston, Ontario K7L 4V2. Canada
Position: Cons Eng'r *Employer:* Self Emplyd *Education:* BSc/Mech Eng/U of Toronto *Born:* Feb 1918 Toronto. 1939-41 Generating station maintenance & construction, Ontario Hydro. 1941-69 & 1974 to 1982 duPont of Canada as Proj Engr, Wks Engr, & Mgr, Specialist Engrg concerned with design, construction & maintenance of synthetic yarn & petrochem plants. 1969-72 Dir of Design, Inco. 1972-74 Dir of Engrg Branch, Canada Post Office. 1982 to Present Cons. Engr in field of hydrocarbon safety. Fellow EIC, ASME & CSME. Reg Ont & Que. Awarded EIC Plummer Medal. Author of many papers on design, construction & computer applications. Oper amateur radio station VE3SV, play golf. *Society Aff:* EIC, ASME, CSME

Boyd, Spencer W
Business: One Northside 75, Atlanta, GA 30318
Position: Partner. *Employer:* Self. *Education:* BS/EE/GA Tech. *Born:* 10/14/04. Grad GA Tech 1926. 2 yrs GE Co 1928-43. Newcomb & Boyd, 1943-45. Active duty USNR, Lt Commander. 1945, Newcomb & Boyd. Mbr Atl Bldg Code Comm 1945. Mbr ASHRAE Council 1949-51. Fellow ASHRAE, Life Mbr IEEE. *Society Aff:* ASHRAE, IEEE.

Boyd, Thomas A
Home: 1016 Harvard Rd, Grosse Pte Park, MI 48230
Position: Genl Motors Res Labs-ret. *Employer:* Retired. *Education:* ChE/BchE/OH State Univ. *Born:* Oct 10, 1888. Entire prof career as staff mbr GM Res Labs, chiefly doing & supervising pioneering res on fuels & combustion. Pres Engrg Soc of Detroit 1943-44; Pres ASTM 1947-48. Horning Mem Award SAE 1950. Hon degrees: D Engrg Univ of Detroit 1952; D Sc Ohio State Univ 1953, & Wayne State Univ 1955. Pub many tech papers & 4 books: 'GASOLINE, What Everyone Should Know About It'; 'RESEARCH, The Pathfinder of Sci & Indus'; 'PROFESSIONAL AMATEUR, The Biography of Charles Franklin Kettering'; 'PROPHET OF PROGRESS, Selections from the Speeches of Charles F Kettering'.

Boyd, Walter K
Home: 7891 Green Lawn, Houston, TX 77088
Position: Sr Res Leader. *Employer:* Battelle Meml Inst-Houston Operations. *Education:* Bach/Mech Engg/OH State Univ. *Born:* 10/1/20. Native of Sherman , NY, Previously Mgr Corrosion Sect at Battelle's Columbus Labs since 1959, Now Sr Res Leader, Battelle-Houston spec in hightemp alloys & spec chem resistant alloy sys which have been patented. Extensive exper in corrosion problems assoc with var types of water, incl marien environ & the high-temp, high-purity water used in fossil fuel & nuclear power plants. Received citation of recog for outstanding contrib to the Natl Assn of Corrosion Engrs in May 1973. Received 1976 Frank Newman Speller Award for NACE for outstanding contributions to corrosion engg. Designate distinguished alumnus College of Engg OH State Univ. Reg PE OH, TX & CA. Reg corrosion specialist, Enjoys golf. *Society Aff:* NACE, ASM, ICORS&T

Boyer, Kenneth L
Home: 2601 Hawthorn Rd, Cuyahoga Falls, OH 44221
Position: President. *Employer:* Harry Fisher Assocs Inc. *Education:* BEE/Elec Engg/OH State Univ. *Born:* 8/19/16. Native of OH. Engr, Allis Chalmers 1940. Served in Army Signal Corp 1940- 45. Elec Equipment Buyer, B F Goodrich 1946. Harry Fisher Assocs, Manufacturers Representatives, 1947-55 as Sales and Application Engr for elec apparatus. VP and Dir of Harry Fisher Assoc 1955-68. 1968-present time, Pres, Treasurer and Dir of Harry Fisher Assoc involved in Sales and Marketing Mgmt. Served in all offices of IEEE Section organization and Conference organization. Active in church, college engg alumni organization. Enjoy sports, reading and travel. *Society Aff:* IEEE, NSPE, OSPE, AISE, HKN.

Boyer, LeRoy T
Home: 1 Concord, Urbana, IL 61801
Position: Prof. *Employer:* Univ of IL. *Education:* PhD/Civ Engr/Univ of MN; BS/Civ Engr/Univ of MN. *Born:* 3/10/37. Native of Ada, MN. Roadway & bridge design with MN Dept of Transportation; hydraulic model studies at St Anthony Falls Hydraulic Lab; taught at Univ of MN; lectr at numerous colleges & univs & consultant to contruction ind & compute mfg in areas of construction & cost. *Society Aff:* ASCE.

Boyer, Robert O
Business: 100 Erieview Plaza, Cleveland, OH 44114
Position: VP. *Employer:* OH Bell Tel Co. *Education:* BS/Industrial Engg/OH State Univ. *Born:* 8/6/20. Native of Columbus, OH. Educated in the Columbus Schools and Attended OH State Univ. On graduation enter the US Army and served in the Corps of Engrs (1942-46) Attained the Rank of Maj. With the OH Bell Tel co since 1946 in various managerial assignments-became VP-Personnel in 1967. Active in numerous civic organizations in the Cleveland area. *Society Aff:* NSPE.

Boyer, Vincent S
Home: 1322 Grenox Rd, Wynnewood, PA 19096
Position: Retired. *Employer:* Consultant *Education:* BS/Mech Engg/Swarthmore College; MS/Mech Engg/Univ of PA *Born:* Apr 1918. Joined Phila Elec Co 1939 as engr of plant tests in Elec Oper. Served as engrg officer of a destroyer 1944-46 US Navy. Returned to Phila Elec following discharge & served in var supervisory pos in power stations. Became first superintendent of the Peach Bottom Atomic Power Sta 1960, & in 1965 named Genl Supt of Sta Oper Dept. 1967 apptd Mgr of Elec Oper Dept, V P Engrg & Res Dept 1968. Past Dir & Fellow Amer Nuclear Soc & P/Pres ; also served as Chmn of the Reactor Oper Div. Fellow Amer Soc of Mech Engrs & P Chmn Phila Sect. Serve on many indus cttes, among which are those involved with the EPRI. Enjoy golf Elected Engr of Year - 1979 by DEL Valley Tech Soc. Received Hon Dr of Engg Sci from Spring Garden College in 1979 Elected to National Academy of Engineering 1980. Retired 3/1/87. *Society Aff:* ANS, ASME, SAME, NSPE.

Boyette, John V, Jr
Business: 420 Park Ave, Box 1717, Greenville, SC 29602
Position: VP *Employer:* Piedmont Engr, Arch & Planners *Education:* BS/CE/Clemson Univ *Born:* 12/31/43 Native of Lyman, SC. 1965 - 1968 Nuclear Process Engr, Charleston Naval Shipyard. Currently VP and member of Bd of Dirs of Piedmont Engrs, Architects & Planners responsible for total business development and marketing activities. Sales volume has increased over 200% since taking over total business development activities. Previous positions with firm have included sales, design project manager and principal-in-charge on major industrial projects. Served as State Education Committee SCSPE 1978-1979 Piedmont Chapter SCSPE Engrs Week, Committee Chrmn 1979, Treasurer Piedmont Chapter SCSPE 1980, Secretary Piedmont SCSPE 1981. Enjoys running and sailing. *Society Aff:* NSPE, SCSPE

Boykin, William H, Jr
Home: 1030 NW 39th Street, Gainsville, FL 32605
Position: Pres *Employer:* System Dynamics, Inc (SDI), Adj Prof Univ of FL *Education:* PhD/Engg/Stanford Univ; MS/ME/Auburn Univ; BS/Engr Phys/Auburn Univ. *Born:* 7/22/38 Native of Birmingham, AL. Instr (full time) of Mech Engg, Auburn Univ while pursuing master's degree, 1961-64. Res Scientist with Outstanding Termination Awards, Hayes Intl Corp 1961 & 1964. Natl Sci Fdn Fellow & Res Asst, Stanford Univ, 1964-67. Consultant to US Army, Nuclear Defense Agency and maj defense and energy cos. Outstanding Service Award, Univ of FL, Engg, 1975. Assoc Dir, Ctr for Intelligent Machines & Robotics, 1977-1981 Pres System Dynamics Inc, 1979- present. System Dynamics does sys engrg, design and test of defense and space sys with offices in Gainesville, FL; Orlando, FL; Huntsville, AL, and St. Louis, MO, 1979-present. Served on various Army blue ribbon panels, red teams and boards 1969-present. *Society Aff:* ΣΞ, ΤΒΠ, ΦΚΦ, ΠΜΕ.

Boylan, Bernard R
Business: 4333 Trans World Rd-100, Schiller Park, IL 60176
Position: Regional Engr. *Employer:* G E Co - Lamp Marketing Dept. *Education:* BS(Eng)/Eng/US Naval Academy *Born:* 7/11/27. BS U S Naval Academy. Native of East Gary, Ind. US Navy 1950-54. Engrg & Sales assignments with Genl Elec in N Y, Indianapolis, Chicago 1954- . Frequent lecturer at Midwest univs & indus grps. Instructor Coll of DuPage, Univ of Notre Dame. Mbr sev indus cttes, Regional V Pres IES 1976-78. Author - *The Lighting Primer* - an elementary textbook. Fishing, golf, racquetball. *Society Aff:* IESNA

Boylan, David R
Business: Coll of Engrg, Ames, IA 50011
Position: Dean, College of Engrg. *Employer:* Iowa State Univ. *Education:* PhD/Chem Engg/IA State Univ; BS/Chem Engg/Univ of KS. *Born:* Jul 1922. Native of Belleville Kansas. Instr Univ of Kansas 1942-43; Proj Engr Genl Chem Co; Sr Engr Amer Cyanamid Co; Plant Mgr Arlin Chem Co 1943- 48; joined ISU fac as instr in chem engrg 1948; became Assoc Dir Engrg Res Inst 1959; Dir Engrg Res Inst 1966; Dean Coll of Engrg 1970. Fellow AIChE 1973 ; Fellow, AAAS,1981; Exec Reservist Natl Def Exec Res 1975; Governor's Sci Adv Comm 1968; Natl Adv Council On Res & Energy Conserv 1974; NSPE. Field of expertise: fluid flow, filtration, proc design, fertilizer tech, Heavy Inorganic Processes. Mbr Tau Beta Pi, Phi Kappa Phi, Sigma Tau, Phi Lambda Upsilon, Omega Chi Epsilon. Reg P E Iowa. Vice President of the National Society of Professional Engineers, Chairman of the Division of Professional Engineers in Education (1985-86). *Society Aff:* NSPE, ASEE, ACS, AIChE, ΤΒΠ, ΦΚΦ, ΦΛΥ, ΩΧΕ, ΣΤ, ΣΞ

Boyle, Willard S
Home: Wallace, Nova Scotia, Canada
Position: Consulant. *Employer:* self *Education:* BSc/Physics/McGill; MSc/Physics/McGill; PhD/Physics/McGill. *Born:* 1924 Amherst Nova Scotia. Joined Bell Labs 1953; engaged in various studies relating to solidstate spectroscopy res. Instrumental in dev 1st continuously operating ruby laser. Co-inventor Charge-Coupled Devices. Joint recipient of 1973 Franklin Institute Stuart Ballantine Medal; 1974 IEEE Morris N Liebmann Award. Retired Exec Dir Res, Communications Sci Div in Holmdel NJ. Res activities of Dev incl quantum electronics, lightwave communications, signal processing & communication networks. Fellow of IEEE, Amer Phys Soc; Mbr of NAE. Hon. LLD 1984 Dalhousie University. *Society Aff:* APS, IEEE, Nat Ac Eng.

Boyle, William R
Business: Manpower Ed, Res & Training Div (MERT), PO Box 117, Oak Ridge, TN 37831-0117
Position: Chrmn, MERT Div. *Employer:* Oak Ridge Assoc Univs. *Education:* PhD/Chem Engr/W VA Univ; MS/Nuclear Engr/W VA Univ; BS/Chem Engr/Newark College of Eng. *Born:* 5/27/32. Native of Paterson, NJ. Served as proj officer, Special Weapons & Nuclear APplications USAF 1953-55. Fac mbr W VA Univ 1961-77. Dir Nuclear Engg Program 1967-68; Acting Chemn Chem Engg 1969; Asst Dir engg Exper Stat 1969-77; Prof Chem Engg 1972-77. Chemn, Manpower Educ, Res & Training Div, Oak Ridge Associated Univs, 1977- . 15 pubs, incl *Manpower Reorientation Alternatives for Increasing Univ Participation for Training New Agency Staff in the Coal/Energy Res Area.* Pres W VA Soc of Prof Engrs 1976; Natl Dir NSPE 1977-79 Registered Prof. Engr. (P. E.). *Society Aff:* NSPE, ANS, ASEE, AICHE.

Boylston, John W
Home: P O Box 311, Solomon's, MD 77001
Position: Genl Manager - Marine. *Employer:* El Paso LNG Co. *Born:* Jun 1939. Degrees from USMMA & Univ of Mich - BS Marine Transportation & BS Naval Arch & Marine Engrg. Sr Naval Architect - Sea Land Service 1965-75. Respon for construction & conversion of over 50 vessels incl the world's largest, fastest container ship, the SL-7. Currently respon for mgmt of El Paso's nine ship LNG fleet. SNAME Council Mbr. SNAME Grad Paper Award & Linnard Prize. Natl Acad of Sci panels on Natl Def & Human Error in Navigation. SNAME panels on maneuvering, powering, coatings, structures & nuclear propulsion. Panel on U S metrification. Enjoy sailing & swimming.

Boynton, Edgar B
Home: 1600 Westbrook Ave, Apt. 427, Richmond, VA 23227
Position: Wiley & Wilson Inc-ret. *Employer:* Retired. *Education:* MS/Mech Engr/Univ of IL; BS/Mech Engr/VA Poly Inst *Born:* Apr 18, 1899 Illinois. Army reserves 1921 to 1958, retired as Lt Col; Active duty with Army Engrs 4 yrs. 48 1/2 yrs with Wiley & Wilson, Cons Engrg, 23 yrs as Partner. Fellow ASME. Mbr & past V Pres NSPE. Vir Engr of Yr 1961. Received NSPE-PEPP Natl Outstanding Serv Award 1970. Mbr & past Chmn of State Air Pollution Control Bd. Mbr Adv Res Ctte (on School Bldgs) to State Dept of Ed 1950-72. Mbr Bd of Dirs Retreat Hospital, Chrmn Mch '82 to Mch '84. *Society Aff:* ASME, NSPE

Bozatli, Ali N
Home: 1609 Furman St, Fayetteville, AR 72701
Position: Asst Prof. *Employer:* Univ of AR. *Education:* PhD/Engg Mechanics/VPI & SU; MS/Mech Engg/CO State Univ; BS/Mech Engg/Bogazici Univ. *Born:* 12/1/52. Native of Turkey. Worked in Germany for Peddinghaus in the summers of 1972 and 1973. Extensive res and publications in heat transfer, boundary layer control, applied mathematics, and solar energy. Asst prof of Engg Sci at the Univ of AR since 1978. Rotary Intl Grad Fellowship recipient. Enjoy folk dancing. *Society Aff:* ASME, ASEE.

Bozzacco, Sil C
Home: 2513 S 18th St, Philadelphia, PA 19145
Position: Elevator Consulting Engr. *Employer:* Private Practice. *Education:* BSEE/Elec Engg/Villanova Univ. *Born:* 2/7/22. Reg PE: PA, NJ, DE, DC, FL, MA. 1968-present: Elevator Consulting Engr. 1950-1968: 14 yrs application engr with maj intl elevator co; 4 yrs as Sales Mgr with maj independent elevaotr co; instr in mechanics of elevators (1950-1963); teacher of Industrial Electricity at Spring Garden Technical Inst (1956-1960).

Braaksma, John P
Home: P.O. Box 795, Manotick, Ontario Canada KOA 2NO
Position: Prof *Employer:* Carleton Univ *Education:* Ph.D./Trans Eng/Univ of Waterloo; MASc./Trans. Eng/Univ of Waterloo; BASc. /CE/Univ of Waterloo *Born:* 12/05/38 Prof of Civil Engrg since 1973. Teaching courses in Transportation Planning, Traffic Engrg, Transportation Terminal Design, and Airport planning. Conducting res in the design of transportation terminals, pedestrian traffic flow and safety. Providing consulting services to government agencies and private consulting firms. Published over 50 papers on transportation matters. *Society Aff:* ASCE, ITE, TRB, CSCE, APEO

Braatelien, Edwin H, Jr
Business: 10101 Linn Station Rd, Louisville, KY 40223
Position: Branch Office Mgr. *Employer:* Camp Dress & McKee Inc. *Education:* MSE/Civil Engg/AZ State Univ; BA/Chemistry/AZ State Univ. *Born:* 2/7/34. in Phoenix, AZ. BA & MSE AZ State Univ. USMC 1953-56. Sr Environ Engr with Camp Dresser & McKee since June 1978. With city of Phoenix 1961-78, Asst Water & Sewers Dir 1971-78. Fac Assoc Environ Engg, AZ State Univ 1968, 70, 75. Reg PE AZ, PE WI, Mbr ASCE. Pres AZ Sects of AWWA & WPCF 1973. Chmn WPCF Personnel Advancement Ctte 1973-78. Dir WPCF 1976-78, WPCF Exec Comm 1978, WPCF Hatfield Award 1966; WPCF Bedell Award 1971; AWWA Fuller Award 1976. *Society Aff:* ASCe, AWWA, WPCF.

Braccia, Anthony A
Business: 730 Montgomery St, San Francisco, CA 94111
Position: President. *Employer:* Braccia & Assocs - Arch/Engg/Planning. *Education:* BME/Mech Engg/Pratt Inst. *Born:* Apr 8, 1920 Brooklyn, N Y. BME Pratt Inst 1942. Submarine serv WW II. Rear Admiral USNR (ret). Arch Engrg & construction of significant commercial, indus, institutional type projects. Was tech adv to Calif Olympic Comm for feasability study, planning, design, & constr of facil at Squaw Valley for the VIIIth Olympic Winter Games 1960. Serves on Pratt Inst's Bd of Visitors, and Cogswell College CHMN Bd of Trustees, is Chmn of the Fleet Admiral Nimitz Scholarship Fund for Oceanography. Pres Navy League, San Fran. Served as Genl Chmn of San Fran Bay Area Engrs Week, Pres of San Fran Post SAME, Pres of Engrs Club of San Fran. Legion of Merit Award 1976, Secy of Navy Pub Serv Award 1975, Silver Beaver 1975, Gold Medal SAME 1971 and 1979. Enjoy sports incl football, baseball, swimming & jogging. *Society Aff:* SAME, NSPE, CSPE.

Brach, Philip L
Home: 3010 Legation St. N.W, Washington, DC 20015
Position: Dean *Employer:* Univ of DC *Education:* PhD/Transportation/Catholic Univ; MS/Soil Mechanics/Lehigh Univ; BCE/Civil Structures/Manhattan Coll *Born:* 10/29/36 Originally from NYC, a registered Prof Engr in NY State. Involved in virtually every aspect of civil engrg, from surveying, design and constr, both in practice and educ. Active in prof assocs; Past Pres of the Dist of Columbia Soc of Prof Engrs. Currently Dean of the Coll of Phys Sci, Engrg and Tech at the Univ of the Dist of Columbia. Consult on engrg and related tech educ, an expert witness on special or unique engrg problems. Avocation in presenting tech concepts to general interest groups. *Society Aff:* ASCE, NSPE

Brackett, Frank P
Home: 11035 Landale Street, No. Hollywood, CA 91602
Position: Consulting Engr. *Employer:* Technicolor Motion Picture Corp-ret;. *Education:* PhD/Physical Chem/Harvard Univ; MA/Physical Chem/Harvard Univ; BA/Chem & Math/Pomona College. *Born:* Oct 9, 1906; Claremont Calif home 1906-28; Teaching Asst both schools; m. Davida Wark 1929; c. Patricia & Alison. Started Motion Picture Processing at Paramount Lab 1933. With Technicolor 1935 to retirement 1972; pos incl Foreman, Superintendent, Tech Supervisor, Tech Dir, & Res Dir. The work covered the many chem & phys opers unique to the Technicolor imbibition processes as well as the later intro of the Eastman Kodak Color systems. Now cons with State cert as a chem engr. Life Fellow Soc of Motion Picture & T V Engrs; recipient of the SMPTE Herbert T Kalmus Gold Medal Award; acted on the Bd of Governors & Chmn of the Color Ctte for the SMPTE. Pres of Hollywood Optimists 1962-63. Phi Beta Kappa, Alpha Chi Sigma, Sigma Xi. Contrib articles to prof Journals. *Society Aff:* SMPTE, ΣΞ, ΑΧΣ, ΦΒΚ.

Bradbury, James C
Business: IL State Geological Survey, 615 E. Peabody Dr, Champaign, IL 61820
Position: Prin Geologist & Hd, Mineral Resources Grp Emeritus *Employer:* Illinois State Geological Survey. *Education:* PhD/Geology/Harvard Univ; MA/Geology/Harvard; BA/Geology/Univ of IL. *Born:* 7/7/18. Oregon IL. Army service, Corps of Engrs 1941-45. IL State Geol Survey 1949- 84. Hd of Mineral Resources Grp since 1981, respon for Survey's program in mineral deposits, incl coal, oil & gas, indust minerals, metals. Non-prof int: performing arts & outdoor activities (fishing, skiing, et al). Bd of Dir 1969-72 SME. *Society Aff:* SME, SEG, GSA, AAAS

Bradford, Michael L
Home: 11924 Fairhaven Dr, Baton Rouge, LA 70815
Position: Pres. *Employer:* M L Bradford Engrs, Inc. *Education:* MS/CE/Northwestern Univ; BS/CE/OH Univ. *Born:* 7/18/42. Native of OH. Moved to Baton Rouge, LA, in 1965. Employed by Ethyl Corp from 1965 to 1974 in technical services & process design. Worked for G R Stucker & Assoc, Inc, a consulting firm, from 1974 to 1977. Ending position was Chief Process Engr. Founded M L

Bradford, Michael L (Continued)
Bradford Engrs, Inc in 1977, a consulting firm which offers process design services to the chem industry. Technical Mgr. Jacobs Engrg Grp Inc., 1982-present. *Society Aff:* AIChE.

Bradford, Ralph J, Jr
Business: Harbor Dr & 28th St, PO Box 80278, San Diego, CA 92138
Position: Chief Marine Engr. *Employer:* Natl Steel & Shipbldg Co. *Education:* Dipl/Marine Engg/US Merchant Marine Acad. *Born:* 5/19/26. Native of Upper Darby PA. Served as Engg Officer in Merchant Marine & US Navy 1947-1952. Engg Supervisor & Marine Engg Mgr at Elec Boat Div & Quincy Shipbldg Div, Gen Dynamics, responsible for submarine & surface ship machinery design 1952-1973. Chief Marine Engr at Natl Steel & Shipbldg Co in San Diego, CA, responsible for surface ship machinery design since 1973. Chrmn San Diego Sec SNAME 1977-1978. Co-author of two SNAME papers. Chrmn of Marine Committee Gas Turbine Div ASME 1976-1978 Reg PE in CA, CT & MA. US Coast Guard License as Chief Engr of steam vessels any horsepower. *Society Aff:* SNAME, ASME, ASTM.

Bradish, John P
Business: 1515 W Wisconsin Ave, Milwaukee, WI 53233
Position: Asst Dean-Engr *Employer:* Marquette Univ *Education:* PhD/ME/UW; MS/ME/UW; BS/ME/Marquette Univ *Born:* 11/12/23 Lawrence, MA. Served in US Navy Reserves, WWII, 1942-46. Married, 7 children. Employment: Marquette Univ, instructor, 1946-51; Asst Prof, 1951-63; Assoc Prof, 1963- in Mechanical Engrg; Dept Chrmn, Mechanical Engrg, 1958-61; Asst Dean 1971-Present. Consultant to Argonne National Lab, Center for Educational Affairs, 1958-present. Current responsibilities include administrative responsibility for large optional cooperative education program in engrg, evening division program for adult students, and relations with industry. Have also been active for 30 years consulting in areas of Mechanical Engrg, especially Thermodynamics, Heat Transfer, Nuclear Engrg and related fields. *Society Aff:* ASME, ASEE, CEA, ΣΞ

Bradley, Elihu F
Home: 71 Crestwood Rd, West Hartford, CT 06107
Position: Chief Materials Engineer - Retired. *Employer:* Pratt & Whitney Aircraft. *Education:* BE/Metallurgy/Yale. *Born:* Started career with Seymour Mfg Co Seymour Conn as chemist; with Pratt & Whitney Aircraft, United Technologies Corp 1941- . Assumed Ch Materials Engrg pos 1961 with respon for Pratt & Whitney Aircraft engine matls; directed dev of titanium for compressor parts & precision cast nickel superalloys for turbine blades. Delivered Keynote Speech 1964 Golden Gate Metals Conference; was guest lectr on Aircraft Special Materials for Natl Comm on Materials Policy. Holds 10 patents on high temp materials. Fellow ASM; Assoc Fellow AIAA; Natl Trustee ASM 1974-77; 1975 ASM Engrg Materials Achievement Award; Pres 1978-79 Amer Soc for Metals. Prof Engr/Conslt. *Society Aff:* ASM, TMS, AWS, ASCE, ASAE

Bradley, Eugene B
Business: Anderson Hall, Lexington, KY 40506
Position: Prof EE. *Employer:* Univ of KY. *Education:* PhD/Physics/Vanderbilt Univ; MSEE/EE/Univ of KY; BS/Physics-Math/Georgetown College. *Born:* 11/4/32. Native of Georgetown, KY. Served in Army Signal Corps 1957-60. Instructor in Physics, Electronics, Vanderbilt Univ 1962-64. Asst Prof EE 1964-68; Assoc Prof EE 1968-75; Prof EE and Physics 1975-. Dir of Grad studies 1977-1981. Areas of Res-Quantum Electronics, Thin films on metallic surfaces. Enjoy classical music, amateur radio, fishing. *Society Aff:* APS.

Bradley, Frank L Jr
Business: One Penn Plaza 250 West 34th St, NY, NY 10119
Position: Ch of the Bd. *Employer:* Stone & Webster Mgmt Constr *Education:* BME/Mechanical/Cornell; MS/Mgmt/Stevens Inst. of Tech *Born:* 04/26/24 Married (Kathleen) 10 children, 12 grandchildren. Worked for Public Service Electric & Gas Co. for 7 1/2 years in the electric generation dept. Worked for Stone & Webster since 1957 in areas of generation consulting, organization studies, financed analysis and forecasting, supply & demand forecasts. was a Director of Upper Pennsylvania Power Co. and South Jersey Industries. *Society Aff:* ASME

Bradley, Howard B
Home: 215 Fresh Meadow Dr, Roanoke, TX 76262
Position: Professional Technical Training Conslt *Employer:* Self Employed *Education:* MS/Petrol Engrg/Univ of TX; BS/Petrol Engrg/Univ of TX *Born:* 3/24/20 Chicago, IL. Assistant Prof Petroleum Engrg, Univ of TX and Univ of AL, 1950-53. Senior Research Technologist, Magnolia Petroleum Co., Dallas, 1953-57. As Technical Training Advisor established training activity for professionals, Mobil Oil Corp, Field Research Lab, Dallas, 1958-1968. Served as Corporate Mgr of Technical Training responsible for Mobil's worldwide technical training program for engrs, geologist and geophysicists, NY City, 1968-80 and in Dallas 1981-82. Ex-Officio mbr SPE's (AIME) Symbols and Metrication Committee and SPE's representative on American Petroleum Institute's Metric Transition Committee, Subcommittee on Units. SPE Distinguished Mbr & Distinguished Service Award both in 1986. Enjoys golf and bowling. *Society Aff:* SPE.

Bradley, Hugh E
Business: Stanford Industrial Park, Palo Alto, CA 94304
Position: Dir, Infor Sys. *Employer:* Syntex Corp. *Education:* PhD/OR-Stat/Johns Hopkins; MS/EE/MIT; BS/EE/MIT. *Born:* 11/4/34. Native of Olean, NY. Formerly Res Engr Sperry Gyroscope Co, Asst Prof Univ of MI, Adj Assoc Prof Western MI Univ, & Adj Prof Grand Valley State Colleges. Assumed current pos of dir, Info Sys, Syntex Corp 1979. Respon for corp infor sys, computer services. Ed Intl Abstract in Oper Res. Pres. Michiana Chap TIMS 1973; Councilman ORSA 1975-78; V Chmu TIMS College on Infor Sys 1976-78; VPTIMS 1977-80; Chmn ORSA Publications Committe 1978-79; Dir PMSA 1977-80; Treas. ORSA 1979-81. Formerly Mgr, Mgmt Infor Serv, Upjohn co & Group Mgr, Computer Sys & Services, Lab Procedures, Inc (Div of Upjohn Co) 1967-1979. *Society Aff:* ORSA, TIMS, PMSA.

Bradley, Thomas W
Home: 2407 Sagamore Hills Dr, Decatur, GA 30033
Position: Director of Research & Development *Employer:* Atlanta Gas Lt Co, Box 4569 Atlanta. *Education:* PhD/Business/CO State Christ. Coll; MS/Indus Mgt/GA Inst. of Tech; BCS/Statistics/Univ of GA; BBA/Business/GA State; BS/ME/Ga Inst. of Tech. *Born:* May 1923. Engrg grad (BS) Georgia Tech: PhD, MS, BCS, BBA business. Native of Atlanta. Reg P E Ga. State Lic HVAC Contractor, Certified Energy Mgr (LIFE) (AEE),. Naval Officer WW II & Korean: Asia, Africa, Europe theaters. Admiral's Staff, Fleet Radio Officer Mediterranean. Retired Commander. Atlanta Gas Light Co since 1951, Dir R & D. Spearheaded gas industry's stricter, 'Back to the Source' energy conservation concept. Indus Adv, St Energy off. AGA's natl 'Hall of Flame' Award 1975. Pres Ga Soc Prof Engrs 1976-77. Num Engineer Of The Year awards NSPE, AIPE. Dept of Def Merit Award 1965. Founder & Hd Navy-Marine Corps MARS in Ga 1962-67. 'Extra Class' Amateur Radio (N4TB). Pioneer of Ham Radioteletype & world's first proven 2-way television to all continents via ionospheric propagation. Dist TV Coord & Net Control of Navy's exper transcontinental television MARS network. Numerous citations for Voluntary Disaster and Welfare Communications Worldwide. *Society Aff:* NSPE, ASHRAE

Bradley, William A
Home: 1919 W Kalamazoo St, Lansing, MI 48915
Position: Prof Civil Engr. *Employer:* MI State Univ. *Education:* PhD/Civil Engr/Univ of MI; MS/Civil Engr/Univ of IL; BS/Civil Engr/MI State Univ. *Born:* 11/11/21. Lansing, MI. Engr, Douglas Aircraft, 1943-4; bridge engr, G M Foster Co, 1944-6. MI State Univ faculty, 1947-; Prof, Civil Engg, Mech, 1961-75; Prof, Civil Engr, 1975- . MI State Univ Distinguished Faculty Award, 1963; Western Elec Fund instructional excellence award, 1966. Reg PE - MI. *Society Aff:* ASCE, AAAS, ACI, ASEE, IASS.

Bradner, Hugh A
Business: University of California, La Jolla, CA 92093
Position: Prof of Engg Phys & Geophysics, Emeritus *Employer:* Univ of CA. *Education:* PhD/Phys/CA Tech; AB/Phys, Math/Miami. *Born:* 11/5/15 Tonopah, NV, Nov 5, 1915. Married (Marjorie Hall) 1943. PhD CA Tech 1941. DSC Miami 1960. Res in electron microscope, mine warfare, nuclear weapons, high energy physics, neutrino astrophysics (DUMAND), accelerators and detectors, seismology, underwater operations, diving, controls instrumentation. Res employment with Champion Paper and Fibre Co, US Naval Ordnance Lab, Los Alamos, Univ of CA, Berkeley Rad'n Lab, Univ CA, San Diego. Consulting with Gen Atomics, Natl Res Council, Gen. Phys. Corp. & others. *Society Aff:* AIAA, APS.

Bradshaw, John P, Jr
Home: 3101 Stoneridge Rd, Roanoke, VA 24014
Position: Partner. *Employer:* Hayes Seay Mattern & Mattern. *Education:* BS in CE/Civil/MA Inst of Tech.;MS in CE/Soil Mechanics/MA Inst of Tech. *Born:* 1/7/33. A Professional Engr in VA and a number of other Southern states. Involved as a mbr of a consulting firm in Civil/Structural projs, in particular those associated with Geotechnical, Fdn Engg and Port facilities. Lead Design engr on two Secs of Washington DC's Metro as well as other office & industrial facilities. Active in Professional and Community Actions including Past Pres of the VA Sec ASCE & Past Chrmn of the City of Roanoke Planning Commission. *Society Aff:* ASCE, NSPE, VSPE.

Bradshaw, Martin D
Business: EECE Dept, Albuquerque, NM 87131
Position: Prof. *Employer:* Univ of New Mexico. *Education:* PhD/Elec Engg/Carnegie Inst of Tech; MSEE/Elec Engg/Univ of Wichita; BSEE/Elec Engg/Univ of Wichita. *Born:* 6/24/36. Native of KS. Engg Teacher since 1958; Univ of Wichita (1958-61), Carnegie Tech (1961-63), Univ of New Mexico (1963-present). Assoc Prof (1967), Full Prof (1972), Asst Dean, College of Engg (1974-76). Western Electric Fund Award (ASEE), Gulf-Southwest Section, 1969, George Westinghouse Award (ASEE), 1973, Board of Directors, 1975-77. Spent yr on Sabbatical leave with State Electricity Commission of Victoria, Melbourne, Australia. Sr Mbr IEEE. Specific interests: electric power systems, electromagnetic fields, teaching techniques. Hobbies: RC model airplanes, amateur radio (WA5VAL). *Society Aff:* ASEE, IEEE.

Brady, Frank B
Home: 111C Severn AVe, Annapolis, MD 21403
Position: Exec Dir *Employer:* Inst of Navigation *Education:* -/Radio & Elec Engrg/Univ of Cincinnati *Born:* Jun 1914. Radio Engrg Univ of Cincinnati Evening Engrg College 193339. Engrg Asst, Crosley Radio 1933-39. Radio Engr Army Air Corps 1939-46. Flight Proj Engr Air Transport Assn 1946-55. Partner A&E Firm 1955-57. Sr Staff Cons Singer Co 1957-74. Independent Cons 1974-76. Dir, Natl Airspace Sys Engg, Air Transport Association, 1976-79. Exec Dir, Inst of Navigation 1979- . Reg P E DC. Staff Cons President's Aviation Adv Comm 1970-73. Chmn U S Delegation to NATO Planning Landing Sys 1970-73. Trustee Cosmos Club Foun 1974- . Bd Mbr Mohawk Air Serv Inc 1974-1977. War Dept Medal of Freedom Award 1947. IEEE Fellow 1969. Radio Tech Comm for Aeronautics Citation 1963. 2 pats. Author book 'A Singular View'. Participating author book 'Avionics Navigation Sys'. Author 30 tech papers & articles. Guest Editor Inst of Radio Engrs Transactions June 1959 Special Issue on Instrument Approach and Landing. Guest Editor Oct 1983 IEEE Proceedings on "Global Navigation–. *Society Aff:* IEEE, ION, AFCEA.

Brahms, Thomas W
Home: 1810 N Shore Ct, Reston, VA 22090
Position: Exec Director. *Employer:* Inst of Transportation Engrs. *Education:* BS/Civil Engg/Northeastern Univ. *Born:* Mar 12, 1945. Transportation Planner Boston Redev Auth, Boston Mass 1965-69; Sr Traffic Engrg Aide, Town of Brookline, Mass 1970-71; Proj Engr Tippetts-Abbett-McCarthy-Stratton, Brookline, Mass 1971-73; Dir Tech Affairs, Inst of Transportation Engrs (Formerly Inst of Traffic Engrs), Arlington, Va 1973-76; Exec Dir Inst of Transportation Engrs, Arlington, Va April 1976- . Mbr Inst of Transportation Engrs, Mbr Council of Engrg & Sci Soc Execs. Mbr American Society of Association Executives. *Society Aff:* ITE, CESSE

Brainard, Alan J
Business: 255 Benedum Hall, Pittsburgh, PA 15261
Position: Assoc Prof. *Employer:* Univ of Pittsburgh. *Born:* Jun 1936. PhD & MS Univ of Michigan; BS Fenn College. Native of Cleveland, Ohio. Process Engr with Esso Res Labs 1964-66. With Univ of Pittsburgh since 1967. Assoc Prof Chem Engrg since 1971. Dir Freshman Engrg Program 1973-77. Res ints incl heat transfer, kinetics, thermodynamics, and coal conversion. Active mbr ACS, AIChE, ASEE. Western Electric Fund award for teaching excellence 1975. Enjoy classical music, golf, turtle figurine collecting and people.

Brainerd, John G
Business: Moore Sch of Elec Engrg, University of Penn, Philadelphia, PA 19104
Position: Univ Prof EE & Hist of Tech. (em.) *Employer:* Univ of Pennsylvania. *Education:* ScD/EE/Moore Sch Univ of PA; BS/EE/Moore Sch, Univ of PA. *Born:* Aug 7, 1904. Dir Moore Sch of Elec Engrg 1954-70; Proj Supervisor of ENIAC (world's first electronic large-scale genl purpose digital computer) 1943- 46; Fellow & Life Mbr IEEE; former mbr Bd of Dir IRE & IEEE; chmn of num AIEE, IRE & IEEE cttes; Mbr & former Secy of IEC ctte; Sr Engr & Asst St Dir PWA 1935- 36; Fellow of Royal Soc of Arts; Pres of Soc for the Hist of Tech 1975-76. Founder's Medal IEEE; Honeywell Medal for Engrg & Sci. IEE Commorative Medal. *Society Aff:* IEEE, AAAS, ΣΞ, TBΠ, HKN, FRANKLIN INST, SOCIETY FOR THE HISTORY OF TECHNOLOGY.

Braisted, Paul W
Business: Dean' Office, 1010 Engineering, College of Engineering, University of Missouri-Columbia, Columbia, Missouri 65211
Position: Prof-Mech & Aerospace Engrg *Employer:* Univ of Missouri-Columbia. *Education:* PhD/Mech Engg/Stanford Univ; MSME/Mech Engg/Syracuse Univ; ScB/Mech Engg/Brown Univ. *Born:* Apr 1928. Dev engrg IBM 1948-56; Instr, Assoc Prof, Syracuse 1956-60, 1963-66; Assoc Prof, Prof Univ of Missouri-Columbia since 1966; Chmn Dept of Mech & Aero Engrg Univ of Missouri-Columbia 1967-1984; Asst Dean, Engr Placement; Dir & Dir Minorities Engrg Program since 1985; Active in Amer Soc of Mech Engrs: Chmn Mech Engrg Dept Hds Ctte 1973-74; mbr Policy Bd-Ed 1973-79; V P, Region VII, mbr ASME Council & Ctte on Regional Affairs 1976-80; Chmn Ctte on Regional Affairs 1980- 81; mbr Ctte on Staff since 1982; mbr Ctte of Planning and Organization since 1982; Chmn since 1986. Also mbr Amer Soc of Engrg Ed, NSPE, Missouri Soc. Prof Engrs. Reg P E Mo. *Society Aff:* ASME, ASEE, NSPE, MSPE.

Bralye, George C
Home: 2700 All View Way, Belmont, CA 94002
Position: Sr. Constr. Consultant *Employer:* Jacobs Assocs *Education:* BS/CE/Univ of the Pacific *Born:* 9/16/11 Presently active as consultant after 42 years of heavy construction engrg and management. 18 years with UT Intl retiring as mgr, operations engrg. Registered PE, CA. Past pres Shasta Branch, ASCE, member of ASCE and USCOLD ad- hoc committees. Co-Author, Handbook of Heavy Construction. US Army Platoon Leader Combat Engrs. Project engrg: Shasta, ML Morris and Littleton Dams and powerhouses. Ops. Engr Mgr., domestic and o'seas, missile bases, command center (NORAD), pumped storage project, underground power house, shafts and tunnels. Metallurgical process plant. Constr. Mgr. Tankage, oil storage (Iran), ore shiploading facilities (AUS) and Mgr of Construction Control. Industrial complex (Saudi Arabia). *Society Aff:* ASCE, USCOLD

Bramer, Henry
Home: P.O. Box 10369, Pittsburgh, PA 15234
Position: Pres. *Employer:* Datagraphics, Inc. *Education:* Phd/Economics/Univ of

Bramer, Henry (Continued)
Pittsburgh; MS/Chem Engg/Univ of Pittsburgh; BSChE/Chem Engg/Univ of Pittsburgh. *Born:* 9/22/21. Native of Pittsburgh, Pa. Served with Army Air Corps (1942-45). Mellon Inst Senior Fellow (1950-66) on American Iron and Steel Inst Fellowship; res and dev of industrial water pollution control methods. Senior VP Cyros W Rice Div NUS Corp 1966-69, VP Synectics Corp 1969-71. Pres of Datagraphics, Inc since 1971. VChrmn and Tech Dir of PENYSIS, Inc (a Datagraphics affiliated co) since 1975. Since 1950, have specialized in industrial water pollution control. Have additionally worked flue gas and automotive exhaust control systems since 1974 (2 patents). *Society Aff:* AAAS, AIChE, WPCF, ACS, AAEE.

Bramfitt, Bruce L
Home: 16 Pleasant Dr-RD 7, Bethlehem, PA 18015
Position: Research Fellow *Employer:* Bethlehem Steel Corp. *Education:* PhD/Met Engg/Univ of MO-Rolla; MS/Met Engg/Univ of MO-Rolla; BS/Met Engg/Uni of MO-Rolla. *Born:* 2/4/38. in Troy, NY. Taught at the Univ of MO-Rolla 1960-66. Metallurgist at Watervliet Arsenal summers 1959-64. With Bethelehem Steel Corp since 1966. Assumed position of Supervisor Sheet Steel Metallurgy 1979, Supvr Physical Metallurgy and Metallography 1982, Sr Scientist- Metallurgy 1984, and present position of Research Fellow in 1987. 10 pats and numerous technical papers. Exec Committee Lehigh Valley Chapter ASM 1976-78 and 1986-87. ASTM Joseph R. Vilella Award 1974, AIME/ISS Eastern Section C D Moore Award 1977 & 1979 AIME/ISS Michael Tanenbaum Award 1986. Sigma Xi, Alpha Sigma Gamma Epsilon Hon Societies. Who's Who in the East 1977, Men of Achievement 1979, Men and Women of Distinction 1979. Enjoy classical music, backpacing, fishing and sailing. *Society Aff:* AIME, ASM, IMS.

Bramley, Jenny Rosenthal
Home: 7124 Strathmore St, Falls Church, VA 22042
Position: Staff Scientist. *Employer:* Night Vision Lab, Dept of Army. *Education:* PhD/Physics/NY Univ; ScM/Physics/NY Univ; ScB/Math/Univ of Paris. *Born:* 7/31/09. Natl Res Fellow, John Hopkins. Grad School Instructor in physics, Brooklyn College & NYU prior to 1942. Physicist Signal Corps 1942-50. Between 1953 & 1966 proj engr Du Mont, cons to major electronics firms, sr scientist Melpar. Since 1967, Staff Scientist US Army Night Vision Lab. Fellow IEEE, Amer Phys Soc, Washington Acad Sci. Mbr OSA, SID, AMS, Sigma Xi, AAUW. 53 papers in recog journals, sixteen pats. P Chmn Joint IRE-AIEE Ctte on Optoelectronic Devices, P Chmn Wash IEEE Chap ED P Chmn IEEE N Va Sect. *Society Aff:* IEEE, APS, OSA, AMS, SID

Brammer, Forest E
Home: 1312 Devonshire Rd, Grosse Pointe Park, MI 48230
Position: Prof Emeritus, Undergraduate Advisor. *Employer:* Wayne State Univ *Education:* PhD/EE/Case Inst of Tech; BS/EE/NC St Univ; AB/Physics/Concord Coll; Grad Study//Univ NC & Johns Hopkins Univ *Born:* 7/21/13 1933-36-high school teacher, Beaver, WV; 1937-42-geophys engr for Schlumberger Well Surveying Corp, in TX, IL, and MI; 1946-1948-research engr & unit supervisor at Johns Hopkins Applied Physics Lab; 1948-1960--faculty of Case Inst of Technology, Cleveland, OH; 1960-1970--prof & chrmn of electrical engrg dept at Wayne State Univ, Detroit, MI. A consultant for Goodyear Aircraft Corp 1954-1958; for Republic Steel Corp 1958-1966. Served as Captain in the Signal Corps, AUS, 1942-1946. *Society Aff:* IEEE, ASEE, ΤΒΠ, HKN, ΣΞ

Brancato, Emanuel L
Home: 7370 Hallmark Rd, Clarksville, MD 21029
Position: Consultant *Employer:* Elec Power Res Inst (EPRI) *Education:* MS/EE/Columbia Univ; BS/EE/Columbia; BA/Pre-Engr/Columbia Univ. *Born:* 11/3/14. Native of NYC. Served the Navy since 1938; Elec Engr, Sci Sect N Y Naval Shipyard in diagnostics of shipboard elec plants. With the Naval Res Lab since 1946 in instrumentation dev, environ effects on dielectric & semiconductor materials & devices spec in aging effects on elec insulation. 1946 Head Insulation Sect, 1954 Head Solid State Applications Branch, 1970 Cons to Dir of Res, 1974 Head Cons Staff in Spec Projects Div. Ret. 1979, Consultant to EPRI 1979. USA rep on Internatl Electrotech Comm, on aging phenomena. IEEE Fellow, & Washington Acad of Sci Fellow for contrib to res in elec insulation. Hobbies: sound reproduction & music. *Society Aff:* IEEE, ΣΞ, WAS, ASTM, IEC.

Brancato, Leo J
Home: 137 Chambers Rd, Danbury, CT 06811
Position: President. *Employer:* Mite Corporation. *Education:* MS/Mech Engg/Columbia Univ, NYC; BME/Mech Engg/Cooper Union, NYC. *Born:* Oct 27, 1922 NYC. Reg P E Conn. Design Engr ERMOLD CO 1946-51; with Heli-Coil Corp 1952-70, Exec V P 1963-70, Pres 1970; V P, Dir MITE Corp (Heli- Coil merged into MITE) 1970-74; Pres MITE Corp 1974- . Incorporator of Union Savings Bank, Danbury, Ct; Commissioner of Conservation Danbury Ct, 1974-79; Trustee Danbury Hosp 1961- . Served as First Lieutenant, Corps of Engrs, AUS 1943-46, Pacific Theater. Mbr N Y Acad of Sci; Fellow ASME; Tau Beta Pi. Patentee in field of fastener tech. Hon Citation Award by Cooper Union 1976. *Society Aff:* ASME, ASM, NYAS

Branch, Robert E
Business: P O Box 161223, Memphis, TN 38186
Position: Pres *Employer:* Robert Branch Contract Services Inc. *Education:* MChE/Chem Engr/Univ of DE; BChE/Chem Engr/Univ of IL. *Born:* Jun 20, 1925. Served as Chmn & on Exec Ctte of Memphis AIChE. Naval Air Force 1943-45. duPont 1950-55 as Dev Engr; FMC 1955-58 as Plant Mgr; Chapman Chem Co 1958-70 as Exec V P & V P of Mfg; Robert Branch Contract Services Inc since 1970 functioning as Manufacturer's Agent for a num of companies offering custom processing & packaging services. In 1973 acquired Chapman Chem Co's Contract Manufacturing activity which is now being operated as PRO-SERVE INC. PE TN 1960. Hobbies: handball, guitar, flying. *Society Aff:* ASQC, AIChE.

Brand, John R
Business: 16 Pelham Rd, Welshire, Wilmington, DE 19803
Position: Retired. *Employer:* E I du Pont de Nemours & Co Inc. *Education:* BSME/Mech Engg/Worcester Polytechnic Inst. *Born:* May 1915. BSME Worcester Poly Inst. Employed by du Pont since 1936 with assignments in mfg, engrg, & R&D. On Manhattan Proj (Oak Ridge, Tenn) during WW II. At Engrg Dev Lab since 1953, Dir-Engrg Dev since 1969, respon for dev of advanced mfg processes & equip. Retired Feb 1978. Fellow ASME, P Chmn Design Engrg Div. Have been active in civic affairs incl School Bd Mbr, officer of Civic Assn, Community Concerts, & church. *Society Aff:* ASME.

Brandenburg, George P
Home: 10396 Southwind Dr, Cincinnati, OH 45242
Position: Sr Engr. *Employer:* GE Co. *Education:* JD/Law/Northern KY State Univ; MBA/Mgt/Xavier Univ; BS/Met Engr/Univ of Cincinnati. *Born:* 5/2/44. Married with three children. Chem Technician Battelle Meml Inst, Columbus, OH, 1965-66; Physical Met, Nucl Systems Progs, GE, Cincinnati, OH 1966-72; Res Met, Mtl & Process Tech Labs, GE, 1972-74; Sr Mtls Application Engr, M&PTL, GE, 1978-present. Work in dev & application of high temperature mtls in space jet engine application. Significant work in technical soc activities. Admitted to OH & Fed Bars in 1978, Sr partner in McTigue, Farrish & Brandenburg, 602 Gwynne Bldg, 6th & Main Sts, Cincinnati, OH 45242, Gen Practitioners of Law, Mbr of Cincinnati, OH, Am & Fed Bar Assn, Assn of Trial Lawyers of Am, OH Acad of Trial Lawyers, Phi Alpha Delta. (Reg PE in OH, mbr of local technical socs, technical publications in field). *Society Aff:* AIME, ASM, NSPE.

Brandin, David H
Business: 333 Ravenswood Ave, Menlo Park, CA 94025
Position: Mgr - Information Sys Group. *Employer:* Stanford Research Inst. *Born:* Jan 1939 Chgo Ill. BS Math (Physics) from Illinois Inst of Tech. Res Mathematician

Brandin, David H (Continued)
with IIT Res Inst 1959-68; Mgr/Application Systems with Informatics Inc 1968-69; V P Symbolic Control Inc 1969-72. With SRI since 1972. Taught in undergrad & grad evening schools at IIT from 1965-68. Mbr Council of the Assn for Computing Machinery 1974- , Chmn ACM SIG/SIC Bd 1974- , Organizing Chmn ACM Spec Int Group on Simulation 1967-68, Mbr IFIP Working Group 5.3 (Discrete Manufacturing); pubs in computational analysis, microcomputer applications & simulation.

Brandinger, Jay J
Business: David Sarnoff Research Center, 201 Washington Road, Princeton, NJ 08543-5300
Position: VP Manufacturing & Materials Research *Employer:* David Sarnoff Research Center. A Subsidiary of SRI International *Education:* PhD/EE/Rutgers Univ; MS/EE/Rutgers Univ; BEE/EE/Cooper Union. *Born:* 1/2/27. Native NYC. 1951 joined RCA Labs tech staff spec in radio communications. 1959 headed a res group in Data Communications, 1966 Sys & Display Res, 1971 Hd of Television Sys Res. 1974 Div VP of TV Engg, RCA Consumer Electronics Div. In 1979 Assumed Respon As RCA Div VP, "SelectaVision" Video Dis Operations and made VP & GM in 1981. Hold 4 patents, mbr Sigma Xi, Eta Kappa Nu; Sr Mbr IEEE, Mbr Inst of Mathematical Statistics & Soc of Info Display & NY Acad of Sci. 3 outstanding Achieve Awards from RCA. Enjoy amateur radio & flying. *Society Aff:* IMS, SID, IEEE, NY Acad Sci

Brandon, Robert J
Business: 175 Curtner Ave, San Jose, CA 95124
Position: Mgr, Nuclear Services Engg. *Employer:* GE Nucl Engg Div. *Education:* MS/CE/RPI; BS/CE/CCNY. *Born:* 8/6/30. and raised in NYC, receiving a degree in Chem Engg from CCNY in 1953. He joined GE in 1955 on the Chemet Prog, and received his Masters Degree from RPI in 1957. He worked in a variety of design functions at Knolls Atomic Power Lab, with his last assignment as Mgr of Reactor Engg on the AIG Proj. Between 1972 & 1974 he worked at CR&D on dev of centrifuges for uranium enrichment. Since 1975 he has been Mgr of Nucl Services Engg for Nucl Energy Business Group in San Jose. In this position he has staffed a group of engrs dedicated to the support of both domestic and overseas BWR operating plants. *Society Aff:* ANS.

Brandow, George E
Business: 1660 W Third St, Los Angeles, CA 90017
Position: Chairman *Employer:* Brandow & Johnston Assoc *Education:* BS/CE/Univ of So CA *Born:* 10/27/13 Structural Engr, b Crookston MN. s. Harry William and Laura (Ramstad) Brandow; m. Anita Dunn, July 1, 1938; children: Peter Dunn and Gregg Everett; Structural Engr and L. A. Refinery Chief Engr, Union Oil Co of CA, 1943-45; Consulting practice Brandow & Johnston 1945- ; ASCE Past Pres, Los Angeles Section, National Dir; Pres American Inst of Consulting Engrs, 1969; Trustee Univ of Southern CA, 1969-86; Overseer Huntington Library and Art Gallery; Los Angeles Engr of Year, 1975; ASCE's Ernest Howard Medal, 1975 and Honorary Member, 1980; Los Angeles Chamber of Commerce Construction Industries Achievement Award, 1981; Outstanding Achievement Award Univ of So Calif School of Engrg, 1982. *Society Aff:* ASCE, ACEC, SEAC

Brandt, Kent H
Business: 4640 Executive Dr, Columbus, OH 43220
Position: Exec VP *Employer:* Brubaker/Brandt, Inc - Archs and Planners. *Education:* Bach of Arch/Arch/OH State Univ. *Born:* 10/4/27. Partner & Exec VP of Brubaker/Brandt, Inc. Archs/Planner, Columbus, OH since 1957. PPres, Columbus Chapter Amer Inst of Archs and Bd of Trustees, Archs Society of OH. Lecturer; visiting critic; former Asst Prof of the Sch of Arch, The OH State Univ; Past Chrmn Advisory Committee, Dept of Arch, The OH State Univ. Lecturer, Natl Educational Television. Past Chrmn of Bd of Zoning Adjustment and Mbr of Planning Commission, Upper Arlington, OH. 1964 recipient of The OH State Univ Texnikoi Outstanding Alumnus Award. 1970 recipient of The OH State Univ Coll of Engg Disting Alumnus Award. *Society Aff:* AIA, ASO.

Brandt, Roy G
Home: P O Box 31, Winamac, IN 46996
Position: Product Mgr - Agri Div. *Employer:* Winamac Steel Prod - McIntosh Div. *Born:* Jan 4 1921 Spencer, S Dak. BS Agri Engrg South Dakota State Univ 1944. Proj Engr for Harry Ferguson Inc, Detroit, Mich; Proj Engr for Laver's Engrg Co, Chgo, Ill; Asst Chief Engr Oliver Corp, S Bend, Ind; Ch Engr Sidewinder, FMC, Minden, La; Terr Mgr Oliver Corp, Mason City, Iowa; Prod Mgr Agri Div, Winamac Steel Products. P Pres Oliver Mgmt Club; P Chmn FIEI-ASAE Student Awards Ctte; P Chmn ASAE Ctte PM-42; P Chmn Ind Sect ASAE. Current work consists of Engrg & sales of heat treated steel used in tillage equip (shanks & teeth) to forty diff customers in U S & Can.

Brannock, N Fred
Business: 1400 N Harbor Blvd, Fullerton, CA 92635
Position: VP. *Employer:* Simulation Sciences. *Education:* BS/ChE/Univ of MD. *Born:* 11/27/39. 25 yrs experience in the field of chem process plant simulation. Co-author of a number of papers on simulation and optimization. Experience in the engg contracting field prior to present position. Reg engr in CA. In present position am responsible for development and marketing functions of a co providing a process computing service to industry. Present position is VP. *Society Aff:* AIChE.

Branscomb, Lewis M
Business: 79 J.F. Kennedy St, Cambridge, MA 01742
Position: Prof. *Employer:* Harvard Univ. *Education:* PhD/Physics/Harvard U. *Born:* 08/17/26 Dir, Science, Technology and Public Policy Program, Harvard Univ. 1986- Former Director, National Bureau of Standards 1969-72. Chief Scientist & VP IBM Corp. 1972-86. Ch. National Science Bd 1980-84. Recipient Arthur Bueche Award, National Acad of Engrg 1987. Mbr Natnl Acad of Engrg, Natnl Acad of Sciences, Natnl Acad of Public Admin Natnl Institute of Medicine. *Society Aff:* IEEE, NAE

Branum, William H
Home: 3122 Old Blue Ridge, San Antonio, TX 78230
Position: Exec VP. *Employer:* Professional Service Industries, Inc. *Education:* MS/Civil Engg/Univ of MO-Rolla; BS/Civil Engg/Univ of MO-Rolla. *Born:* 9/8/41. Post grad study at West VA Univ. Reg in ARK, LA, TX, OK, AZ, NM. 1969-70, Asst Prof at Univ of MO-Rolla; 1970-72. Served in US Army, Master Instr at US Army Engr Sch, Ft Belvoir VA, concurrent assignment as Asst Prof Lecturer, Geo Wash Univ; 1970-72, With Law Engg Testing Co-Served as Sr Soils Engr & Raleigh, NC Engg Mgr 1973-74 & Norfolk Branch Mgr & Sr Engr 1974-79. Exec VP Professional Service Industries, Inc 2/79-present. *Society Aff:* ASCE, ISSMFE.

Brashears, Alan D
Business: Rt 3 Box 215, Lubbock, TX 79401
Position: Agri Engg. *Employer:* US Dept of Agri, SEA, AR. *Education:* PhD/Engr/Texas Tech Univ; MS/Agri Engr/TX A&M Univ; BS/Agri Engr/TX A&M Univ. *Born:* 1/22/38. Native of Graham, TX. Worked for the US Dept of Agri in College Station, TX, 1962 on cottonseed storage res; Albany, GA, 1962-1965 peanut cleaning and sizing res; and Lubbock, TX, 1965 to present on cotton production engg systems. Chrmn and secy-treas, Southwest Region, TX Sec ASAE. Secy-treas, TX Sec ASAE. Enjoy hunting, fishing, & woodworking. *Society Aff:* ASAE, ΣΞ.

Brasunas, Anton deSales
Home: 8030 Daytona Dr, St. Louis, MO 63105-2510
Position: Emeritus Prof of Metallurigical Engg. *Employer:* Univ of Missouri-Rolla. *Education:* ScD/Met/MIT; MS/Met/OH State Univ; BS/Chem Engg/Antioch College; AChE/Chem Engg/Newark College Engg. *Born:* 3/11/19. Native of Elizabeth N J. Taught at Univ of Tennessee 1952-54, res engr at Battelle Mem Inst in high temp

Brasunas, Anton deSales (Continued)
alloys & corrosion, res metallurgist at Oak Ridge Natl Lab - liquid metal corrosion & ed dir of ASM Int'l & its Metals Engrg Inst. Assumed University position Aug 1964; also Mid-Central Regional Mgr of the US Metric Assn; Natl Bu Stds speaker. Natl Pres Alpha Sigma Mu 1968-69; Fellow Amer Soc Metals 1975; Citation for Outstanding Contrib to Natl Assn of Corrosion Engrs 1974; enjoys tennis, camping, travel, metrication. *Society Aff:* ASEE, ASM Int'l, NACE, USMA.

Braswell, Robert N
Home: 1563 San Luis Ro, Tallahassee, FL 32304
Position: Professor of Electrical Engg; Director, Computing Center *Employer:* Florida State Univ *Education:* PhD/Engrg/OK State Univ; MS/Engrg/Univ of AL; BS/IE/Univ of AL *Born:* 7/23/32 Born in Boaz AL and attended public schools in Marshall County. 3 1/2 yrs active USAF duty in the Korean War. Engrg employment has been with Rouse Engrg Co (1956-57) as a senior designer; Hughes Aircraft Co (1958) as an electronics design and reliability engr; the Univ of AL (1957-61) as an Asst Prof; Teledyne Brown Engrg Co (1959-64) as Mgr of the Systems Engrg Div; the Univ of FL (1964- 72) as Prof and Chrmn of the Industrial and Systems Engrg Dept; Natl Aeronautics & Space Admin (1967-68) as a Systems Engrg Consult; Dept of the AF (1972-86) as a member of the US Senior Exec Service in Weapons Systems research, dev & systems engrg and Director of Research in Mathematics and Computer Science; Prof of Electrical Engineering & Director, Computer Center, Florida State Univ (1986-present). Member of Tau Beta Pi, Alpha Pi Mu, Pi Mu Epsilon, Omicron Delta Kappa, Chi Alpha Phi, Sigma Tau, Sigma Xi, Alpha Lambda Chi & Blue Key. Reg PE-FL, AL, LA. *Society Aff:* IIE, ASEE, AIAA, ADPA, ORSA, IEEE.

Bratcher, C Dale
Home: 655 Upland Rd, Louisville, KY 40206
Position: Dir PW Engr. *Employer:* Naval Ordnance Sta. *Education:* ME/Civ Engr/Univ of Louisville; BSCE/Civ - Arch Engr/Univ of Louisville. *Born:* 1/10/32. Consulting Engr (1957-62) - Dir PW Engr (1962 to present). Reg Civ, Ind, Sanitary Engr and Land Surveyor. KSPE Louisville Chapter's Distinguished Engr Award (1979); KSPE Resolution of Appreciation for service to the Engr Profession (1979); Director Louisville Chapter KSPE; State Director Kentucky Society of Professional Engineers; Served as Secy and as VP of KSPE Chrmn KY State Bd of Registration for Engrs and Land Surveyors (1971-3) - appointed a bd mbr emertus after 8 yrs of service (1979); Numerous committee appointments and certificates of appreciation from Natl Council of Engrs Examiners; Distinguished Service Award (1980), National Council of Engr Examiners; Published Engr Article in "The Navy Civil Engineer–; Extensive US Navy career (26 yrs) retired with rank of Capt; Inducted into Tau Beta Pi Honor Society as an eminent engr (1979). *Society Aff:* NSPE, ΤΒΠ.

Bratkovich, Nick F
Home: 5015 Knoll Crest Ct, Indianapolis, IN 46208
Position: Welding Engr. *Employer:* Detroit Diesel Allison Div, GM-ret. *Born:* May 1914. Attended Illinois Inst of Tech, Chgo Ill. Most of indus career spent in Chgo, Ill & Indianapolis, Ind. Last pos held: Sect Ch of Materials Joining Processes Lab, Detroit Diesel Allison Div GMC. Spec in metal joining with GMC since 1941. Co-inventor of Lamilloy (pat 3,584,972). ASM & AWS mbr & have participated in handbook cttes for both soc's. Cited in recog of distin serv & tech contribs. Authored & presented sev tech papers in joining tech. Awarded the ASM-MAC award 1964, A F Davis Silver Medal cert 1965, & hon as ASM Fellow 1973. *Society Aff:* AWS, ASM.

Bratt, Richard W
Business: PO Box 977, Syracuse, NY 13201
Position: Chief Met High Speed & Tool Steels. *Employer:* Crucible Specialty Metals Div Colt Ind. *Education:* BS/Met Engg/Syracuse Univ. *Born:* 1/16/31. Have 26 yrs experience in specialty steel industry in various met engg & supervisory capacities - all at Crucible. Primary product specialties are high speed & tool steels. Coordinates mill met practices, quality evaluations, field applications, & customer technical service in high speed & tool steels. Also coordinates dev activities of new products, most notably the Crucible particle Met (CPM) process for high speed & tool steels. Native of Syracuse, NY. Enjoy music, golf, photograhy & travel. *Society Aff:* ASM, ASTM.

Bratton, Joseph K
Business: 20 Mass Avenue, NW, Washington, DC 20314
Position: Chief of Engineers *Employer:* Dept of the Army *Education:* Grad/Command & Gen Staff Coll and Army War Coll; M/Nuclear Eng/MIT; B/Eng/U.S. Military Academy-West Point *Born:* 4/4/1926 Lieutenant General J. K. Bratton became Commander, U.S.Army Corps of Engrs, Oct 1, 1980. Command assignments include: the 24th Engr Battalion, 4th Armored Div, Europe; the 159th Engr Group, Vietnam; the South Atlantic Div of Corps of Engrs. Staff assignments include: Dir of Military Application for Dept of Energy, Washington, D.C.; Chief of Nuclear Activities, Supreme Hdquarters Allied Powers, Europe; Exec to the Supreme Allied Commander, Europe; Secretary to the Joint Chiefs of Staff; and Military Assistant to the Secretary of the Army. Among Bratton's military awards are the Defense Distinguished Service Medal and the Army Distinguished Service Medal. *Society Aff:* SAME, ASCE, ASNE, USCOLD, AUSA

Brauer, Donald G
Home: 6116 Parnell Ave So, Edina, MN 55424
Position: Pres *Employer:* The Brauer Group Inc *Education:* MAPA/Pub Admin/Univ of MN; BSCE/Civ Eng/IA State Univ. *Born:* 11/24/29 Clinton, IA. Special weapons officer, USAF; Pittsburgh-Des Moines Steel Co- Bridge Division: Asst City Mgr, Edina, MN, Partner, Harrison/Brauer/Rippel, Minneapolis office; Founder Brauer and Assocs, Minneapolis and Brauer & Assocs, Rocky Mountain, Denver; part-time faculty Univ of MN; Visiting lecturer MN Univ, IA State Univ, TX A&M Univ, UT State Univ, and Univ of GA. Responsible principal planner for cities in seven states and for state, regional, local and national parks in fifteen states, Canada and Mexico. Special interest in Environmental Mediation and citizen participation process and procedures. Dir, Suburban Natl Bank, Eden Prairie and Sibley Co. State Bank, Henderson, MN. *Society Aff:* ASCE, NSPE

Brauer, Joseph B
Home: 1309 Carroll St, Rome, NY 13440-2607
Position: Ch, Microelectronics Reliability Division *Employer:* USAF, Rome Air Dev Ctr. *Education:* Master/Chem Eng/Syracuse Univ; BS/Metallurgical Eng. /Purdue Univ. *Born:* Feb 1930. Native of Lawrenceburg Ind. Worked in Genl Motors Central Foundry Lab, Danville Ill prior to serv with USAF as R&D Proj Officer 1951-53. As Air Force civilian since 1953, conducted & supervised R&D in electronic materials, miniaturization, solid state physics, microelectronics, reliability & accelerated testing. Initiated Air Force Prog on Physics of Failure in Electronics Became Ch of Solid State Applications in 1962 and Ch of MicroElectronics Reliability Division in 1987 with respon for quality & reliability assurance of monolithic, hybrid & multichip mircocircuits since 1968. U S mbr for solid state devices and microcircuits on NATO Group of Experts AC/301 (SG/A) (WG/1). Fellow & Prof Chem Engr, Accredited in AIC, mbr ACS & ASM. 6 patents for microwave devices. Enjoy hunting, fishing, sailing. *Society Aff:* AIC, ASM.

Braun, J S
Business: P O Box 35108, Minneapolis, MN 55435
Position: President. *Employer:* Braun Engrg Testing Inc. *Education:* MSCE/Soil Mechanics/U of MN; BCE/Soil Mechanics/U of MN *Born:* 1/19/33. Native of St Cloud, Minn. Founder & Pres of Braun Engrg Testing Inc, cons soils & materials engrg firm in Minneapolis, Hibbing, St Cloud, Rochester and St Paul, MN and Williston and Bismark, N.D. and Billings, MT. V P ACEC 1975-76. Sr V P 1976-77 *Society Aff:* MSPE, ACEC, ASCE, ASTM

Braun, Ludwig
Business: Coll of Engrg, Stony Brook, NY 11794
Position: Prof *Employer:* State Univ of NY *Education:* D/EE/Polytech Inst of Bklyn; M/EE/Polytech Inst of Bklyn; B/EE/Polytech Inst of Bklyn *Born:* 5/14/26 Native of Long Island, NY. 1951-55, leader of electronic design team, Anton Electronic Labs, specializing in nuclear instrumentation systems. 1955-1972, engrg faculty at Polytechnic Inst of Brooklyn teaching bioengineering and system engrg. 1972-present, Prof of Engrg, SUNY at Stony Brook, teaching bioengineering, computer literacy, and computers-in-education courses. 1968-74, Dir, Huntington Computer Project, developing high-school science computer courseware. 1975-present, Dir, Lab for Personal Computers in Education at SUNY at Stony Brook. Advisor on computers in education to National Science Foundation, National Inst of Education, and many school districts and teachers groups nationally. *Society Aff:* AAAS, IEEE

Braun, Walter G
Home: 199 Twigs Lane, State College, PA 16801
Position: Assoc Dean for Instr, Coll of Engrg *Employer:* Penn State Univ (Ret) *Education:* BChE/ChE/Cooper Union; MS/ChE/PA State Univ; PhD/ChE/PA State Univ. *Born:* 6/23/17. Res in Raman & infrared spectroscopy, mass transfer, heat transfer, thermodynamic & phys properties. Ed *Tech Data Book-Petro Refining,* Amer Petro Inst. *Society Aff:* AIChE, ACS.

Braun, Warren L
Home: 680 NY Ave, Harrisonburg, VA 22801
Position: CEO-Chairman *Employer:* ComSonics Inc. *Education:* Dr of Science/Honorary/Shenandoah Coll & Conservatory; BSEE/Comm Engg/ Valparaiso Tech Inst; Dipl/Bus/Alexander Hamilton Inst; Dipl/Comm Engg/Capitol Radio Engg Inst. *Born:* 8/11/22. Postville, IA. Taught ESMWT courses, VPI 1942-1945. Constructed & designed radio stations WSVA-AM-FM, WTON, WJMA & WSIR; TV stations WJZ-TV & WSVA-TV, Blue Ridge CATV system. Gen Mgr WSVA-AM-FM-TV 1962-1965. Consulting Engr 1965-; Pres, ComSonics 1972- , CEO & Chairman 1986; VSPE Engr of the Yr, 1965; VSPE Distinguished Service Award 1969. Fellow, AES, 1966. Pres VA Assn of Profession 1972-74; Mbr State of VA Air Pollution Control Bd, 1966-69; VA State Water Control Bd 1974-82; VChrmn 1976- 77, Chrmn 1977-78. ORSANCO Bd 1974-82, Chrmn 1978-1980;; NCTA Engr Committee 1979-; Bd mbr ESOP Council of Am 1979- , Vice Pres 1980-81, Vice Chrmn 1981- , Chairman 1987- ; Executive of the Year - PSI 1983. Businessman of the Year H-RCH of Commerce 1986; Ex Committee & Bd Mbr WVPT-TV; James Madison Univ. Bus School Advisory Committee 1986- ; Member Bd H/R Sewer Authority 1985- ; Bd Mbr US Senate Productivity Committee 1986- . *Society Aff:* IEEE, NPSE, AES, ASA, ASTM, SMPTE, SCTE.

Brauner, Herbert A
Home: 110 Sleepy Hollow Dr, San Anselmo, CA 94960
Position: VP, Concrete Products Division *Employer:* V.H. Pomeroy & Co. Inc. *Education:* BS/Aero Engr/Univ of TX. *Born:* 6/6/24. USNR 1942-46. Field Engr City of San Francisco 1949-53. Since 1953 var constr engrg & prestressed concrete engrg pos with Ben C Gerwick Inc, later Pomeroy-Gerwick, later Santa Fe-Pomeroy Inc, now J H Pomeroy's & Co, Inc 1962-69 Ch Engr prestressed concrete mfg plant, Petaluma CA 1969- , Plant Mgr 1968-81. Mbr 1976-77 Pres Northern CA Chap ACI. Hobbies: boating & leathercraft. Chmn ACI Committee 543 (Piung). 1979 Pres Prestressed Concrete Manufactures Association of CA. Present VP Concrete Products Division, V.H. Pomeroy & Co. Inc. Current Chmn PCI Pile Committee and Chmn ACI Committee 543, Concrete Piling. *Society Aff:* ASCE, ACI, SEAONC

Braus, James M
Business: PO Box 2463, Houston, TX 77252-2463
Position: Gen Mgr Engg *Employer:* Shell Oil Co. *Education:* BS/Chem Engg/Unov of ND. *Born:* 11/24/36. Native of Bismarck, ND. Joined Shell Oil Co in 1958 at the Wood River IL Refinery. Have had assignments in refinery design, planning, operations, and construction in NY, Houston, TX, and Martinez, CA. In the current assignment, am responsible for engineering matters associated with chemical & oil operations in manufacturing, transportation, & marketing. *Society Aff:* AIChE, CII.

Brautigam, Dale P
Business: 58391 Main St, New Haven, MI 48048
Position: President. *Employer:* New Haven Foundry Inc. *Born:* Jackson Mich, Jun 1928; BA Albion College 1952; BS Washington Univ 1953; MBA Univ of New Mexico 1965; Reg P E Mich. AIPE: Fellow 1976, Internatl Plant Engr of yr 1972, Internatl V P 1972-73, Treas 1970-72 Region 2 VP 1968-70; Amer Foundrymen's Soc: Serv Citation 1972, Dir of Detroit Chap ; SME: Cert Mfg Engr- Mfg Mgmt; IIE: Mbr, V P NM Chap 1966; Alpha Pi Mu; Kappa Mu Epsilon; Received Mr Industry Award 1970 IMPO Magazine; Ch Plant Engr Westran Corp 1955-62; Staff Indus Engr; Sandia Corp 1953-55 & 1962-66; Dir of Plant & Indus Engrg CWC Castings/Textron 196675; Pres Geo Fischer Foundry Sys & Sutter Prod Co 1975-76. Written, presented, & pub num of papers in fields of decision making, human dev, proj control, & engrg sys.

Braverman, Nathaniel
Home: 215 E Berkshire Ave, Linwood, NJ 08221
Position: Aviation Sys Conslt (Self-employed) *Education:* MS EE/Elec Eng/Columbia Univ; BEE/Elect Eng/CCNY. *Born:* Jul 4, 1915. Civilian engr Wright Field (Signal Corps-later USAF) 1939- 58. Airways Modernization Bd, then Fed Aviation Admin 1958 until retirement from Gov serv 1973. Positions: R&D engr, mgr, ch of planning, expert cons, ch tech res. Respon early dev many now common air navigation & safety devices, e g DME. Since 1945: analy, planning & testing of navigation, landing, traffic control & communication sys. Mbr many natl cttes & tech orgs. RTCA citation 1950; Fellow IEEE 1964; FAA award for pioneering res into applications of satellites to civil aviation 1967. *Society Aff:* IEEE, ION.

Bray, David W
Business: Elec & Comp Engg Dept, Potsdam, NY 13676
Position: Prof & Chrmn *Employer:* Clarkson Coll of Technology. *Education:* PhD/Sys & Info Sci/Syracuse Univ; MEE/EE/Rensselaer Poly Inst; BEE/EE/Rensselaer Poly Inst. *Born:* 3/23/29. Prior to present position, with the GE Co where work included computer simulation of adaptive signal processing appl to radar, sonar and satellite identification, theoretical dev of formal language concepts, design of computer languages, implementation of compiler-generation software, and digital circuit test generation software. Current res activities at Clarkson College include specialized computer system design and simulation, and fault tolerant computer studies. *Society Aff:* IEEE, ACM, ASEE, SCS.

Bray, Oscar S
Home: 18 Lakeview Dr, Lynnfield, MA 01940
Position: Sr Conslt *Employer:* Camp Dresser & McKee Inc *Education:* BS/Civil Engg/Univ of Cincinnati. *Born:* Dec 1905 Dover N J. s. Oscar S & Bertha (Janner) m. Helen L Shanley Jan 11, 1933; c. Helen M (Mrs W A Jeffers), Mary E (Mrs E P Womack). Surveyman, DL & WRR Hoboken N J 1923-27; Constr Supt Natl Park Service Cold Spring N Y 1934-35; Field Engr Boston & Salem Ma 1936-38; Engr Washington D C 1939-40; Struct Engr, Asst Ch Str Engr, Proj Mgr Jackson & Moreland, Boston 1941-58; Pres & Ch Engr Jackson & Moreland Int 1959-68; Pres & Ch Engr Bray Backenstoss & Co 1969-71; V P CDM Int 1972-76; Consultant CDM Inc 1977- ; Lectr Northeastern Univ; Guest Lectr Univ of Ill; Mbr Lynnfield Water Bd 1943-51; Mbr Lynnfield Planning Bd 1968-69, Chmn 1970; Fellow ASCE (Dir 1964-66, Pres 1971-72); Fellow, ACEC Trustee United Engr Trustees 1975-84. 1981-82; Dir, Engg Fnd 1975-1981, 1984- Pres Tau Beta Pi; Chi Epsilon (Nat Honor Mbr): Sigma Alpha Epsilon; Repub; Presbyterian. *Society Aff:* ASCE, ACEC, Tau Beta Pi, Chi Epsilon, Sigma Alpha Epsilon.

Brazinsky, Irving
Home: 6 Rustic Lane, Matawan, NJ 07747
Position: Process Dev Mgr *Employer:* Foster Wheel Energy Corp *Education:*

Brazinsky, Irving (Continued)
ScD/Chem Engg/MA Inst of Tech; MS/Chem Engg/Lehigh Univ; BChE/Chem Engg/ Cooper Union. *Born:* 10/27/36. and raised in NYC. Received BChE from Cooper Union (1958), MS from Lehigh Univ (1960) and ScD from MIT (1967). Research Engr with National Aeronautics & Space Adm (1958-61) and Polaroid Corp (1966-69). Sr research engr with Celanese Corp (1969-76) and Halcon R&D Corp (1976-1981). Process Dev Mgr Foster Wheeler Energy Corp (1981-present). Adjunct Prof at New Jersey Inst of Tech (1971- present); teach grad courses in chem reactor kinetics & catalysis; A number of contributions to prof journals & patents in the areas of thermodynamics, heat and mass transfer at low temperatures, cryogenics, fiber spinning, & plastics processing. Enjoy theatre, sports and running marathons. *Society Aff:* AIChE.

Brebbia, Carlos A
Business: 52 Hemstead Rd, Southampton, England
Position: Director *Employer:* Computational Mechanics Institute. *Education:* PhD/Civ Engg/Univ of Southampton; Dipl Eng/Civ Engg/Univ of Litoral. *Born:* 12/13/48. Dr Brebbia started his academic career as a res assoc at the Univ of Litoral, Argentina. In 1969 he was appointed Lecturer at Southampton Univ, England where he was promoted to Sr Lecturer in 1976. He was Prof of civ engg at UCI and is now Reader in computational mechanics at Southampton University England and Director of the Computational Mechanics Inst. Dr Brebbia has written ten books on numerical methods in engg and edited 30 others. He has written over 150 scientific papers. *Society Aff:* ASCE, ISCME.

Breeding, Lawrence E
Business: 1343 H St, Anchorage, AK 99501
Position: Owner. *Employer:* Self employed. *Education:* BS/ME/Univ of WA; BA/ ME/Marin Jr College. *Born:* 7/29/22. Education: Marin Jr College, BAME 1941; Univ of WA, BSME 1949; US Navy, Flight Engr, 1941-1946. 1949 Union Oil Co of CA, Seattle, WA. Design of piping and pumping facilities for oil storage and docking installations. 1951 Consulting Engg Firms, Seattle, WA. Design, specification writing, inspection, and project engr for commercial and industrial projects with heating, air conditioning, piping, plumbing, and fire protection systems. Design, specifications and inspection of construction of new pulp mills, waste wood facilities and studies for investment in new mills. 1970 Opened my own office in Anchorage, AK for Mech Engg Consulting Service for Heating, Ventilation, Air Conditioning, Piping, Plumbing, Fire Protection Systems In Schools, Hospitals, Labs, Office Bldgs, Military Installations, Power Plants, Water Treatment Plants, Commercial Installations and Industrial Facilities. Energy Conservation Mgt Programs have been started with the Fed and State Dept of Energy. *Society Aff:* NSPE, PEPP, ASHRAE, NFPA.

Breen, Dale H
Business: 7 S 600 Cty Line Rd, Hinsdale, IL 60521
Position: Mgr,Metallurgy Technaology Center *Employer:* Intl Harvester Co, Science & Technology Laboratory *Education:* MS/Indust Met/Univ of MI; MBA/Bus/Univ of Chicago; BS/Mech Eng/Bradley Univ. *Born:* 3/10/25. Native of Charleston, IL. Univ of MI teaching staff 1952-53. With Intl Harvester co since 1953 as Res Metallurgist, then Matls Engr, then Ch Engr Metallurgical Res then Mgr, Matls Res & engg, & since 1978, Mgr, Met Technology Center Science & Technology Laboratory. Currently serving on ASM Publications Council as Chrmn & Council mbr of Matls Sys & Design Div. Reading, religious studies, AAU Masters swimming, and scuba. Co-author: *the Hardenability of Steels - Concepts, metallurgical Influences, & Industrial Applications* published by ASM *Society Aff:* ASM, AIME, SAE, ASTM, TMS, ISIGB.

Breen, John E
Home: 8603 Azalea Trail, Austin, TX 78759
Position: Prof of Civil Engrg. *Employer:* Univ of Texas, Austin 78712. *Education:* PhD/CE/Univ of TX at Austin; MSCE/CE/Univ of MO; BCE/CE/Marquette Univ. *Born:* May 1932, Buffalo N Y. Struct designer Harnischfeger Corp Milwaukee 1952-53. Lieutenant, Civil Engr Corps USN 1953-56. Asst Prof Civil Engrg Univ of Missouri, Columbia 1957-59; Instr, Asst Prof Civil Engrg 1959-65, Assoc Prof 1965-68, Prof 1968- , Al-Rashid Chair in Civil Engrg, 1984- . Univ of TX at Austin. Reg P E Texas, Missouri. Author num tech papers & reports on behavior & design of reinforced & prestressed concrete structs. Mbr Natl Acad Engrg; Fellow ACI; Fellow ASCE; ACI Wason Medal, Most Meritorious Paper 1972 and 1982; ACI Raymond C Reese Res Medal 1972 and 1979 ACI Joe Kelly Award 1980; ACI Arthur Andersen Award 1987; ASEE T.Y.Lin Award 1986; RCRC A.J. Boase Award 1987. *Society Aff:* ACI, ASCE, NAE.

Brehm, Richard L
Home: 3340 Ross Road, Palo Alto, CA 94303
Position: Conslt *Employer:* Self *Education:* PhD/Engr/UCLA; MSE/Nucl Engr/Univ of MI; BSE/Aero Engr/Univ of MI; -/Post Grad Engr & Phys/Technische Hochshule. *Born:* 8/17/33. Cadillac, MI. Supervised nucl analysis & methods dev at Atomics Intl, 1958- 65. Prof of Nucl Engg at the Univ of AZ 1965-1980. Specialities included neutronic, heat transfer, mtls behavior, dynamics, powerplant engineering and economic analysis. Consultant to AEC, JPL, Sandia, Los Alamos, BBC (Germany), Nucl Services Corp, San Diego Gas & Elec, GCRA. Fulbright Fellow, Technische Hochscule, Aachen, Germany, 1958-59. Sr NATO Fellow, Atomic Res Ctr, Karlsruhe, Germany, 1975. VP of ECE Corp, San Diego, 1979-81; Dep. Mgr, Energy Projects Division, Science Applications, Inc, La Jolla, 1981. Priv Conslt as ECE, 1981-85. VP of Rasor Assoc, Sunnyvale, 1985-86. Priv Conslt 1987-Date. *Society Aff:* ΣΞ, TBΠ.

Breihan, Erwin R
Home: 5200 Oakland Ave, St Louis, MO 63110
Position: President. *Employer:* Horner & Shifrin Inc, Cons Engrs. *Education:* BS/Civil Engrg/WA Univ *Born:* 10/31/18. Native St Louis. Served in Naval Aviation during WW II 1942-46 & as Reserve Officer 1946-71, also in Naval Aviation retiring as a Captain. With Horner & Shifrin since 1946 & was elected Pres in 1973. Pres of Engrs Club of St Louis 1975. Pres Cons Engrs Coun of MO 1976. Pres MO Soc of PE 1982. Chmn Air Field Pavement Ctte ASCE 1968-1970. Chmn Exec Ctte Air Transport Div ASCE 1973. Chmn Ctte for Maintenance of Qualifications for Continuing Engrg Regis 1973.75. Chmn Ctte on Standards of Practices ASCE 1975; Chmn Professional Practices Div ASCE 1979; Chmn Mgmt Grp C ASCE 1978; Chmn Manual 45 Subctte ASCE 1975-1981; Mbr Professional Activities Comm ASCE 1980-1984. Hon Mbr Chi Epsilon Univ of MO- Rolla 1973. Awd of Merit Engrs Club of St Louis 1971 - Hon Mbr Engrs Club 1978, Edmund Friedman ASCE Medal, 1984. *Society Aff:* ASCE, ACEC, Cec MO, APWA, NSPE, MSPE, AWWA, AAAE, SAME, ASTM, TRB, ARTBA.

Breinan, Edward M
Business: Silver Ln, E Hartford, CT 06108
Position: Sr Mtls Scientist. *Employer:* United Technologies Res Ctr-24. *Education:* PhD/Physical Metallurgy/Rensselaer Poly Inst; BS/Metallurgical Engg/Rensselaer Poly Inst. *Born:* 9/27/42. Initial career interest in composite mtls, including directionally- solidified eutectics and metal-matrix fiber composites. Began pioneering res in the field of Laser Mtls Processing in 1972, including use of continuous, high power, CO2 lasers for welding, cutting, heat treating, and surface structural modification of mtls. Discovered Fusion Zone Purification during laser welding of ferrous alloys. Currently interested in application of lasers to novel mtl fabrication processes, especially rapid solidification laser processing. Co- inventor of the Laserglaze TM process for surface modification of mtls and the Layerglaze TM process for sequential build up of bulk, controlled structures using lasers. Also developing techniques for applications of electron beams to materials processing. *Society Aff:* ASM, AIME, AWS, SPIE, JWS.

Breipohl, Arthur M
Business: 1013 Learned Hall, Lawrence, KS 66045
Position: Professor of Elec Engrg *Employer:* Univ of Kansas. *Education:* ScD/Elec Engg/Univ of NM; MS/Elec Engg/Univ of NM; BS/Elec Engg/Univ of MO. *Born:* Nov 14, 1931 Higginsville Mo. Presently Prof of Dept of Elec Engrg Univ of Kansas. From 1964 to 1970 was Asst Prof, Assoc Prof & Prof of Elec Engrg at Oklahoma St Univ. Worked for Sandia Corp 1957-64 & prior to that worked for Westinghouse. Pub num of papers in areas of reliability, communication, sys studies and power systems & authored a book, Probabilistic Sys Studies, which was pub by John Wiley 1970. Mbr Sigma Xi, Tau Beta Pi, Eta Kappa Nu. *Society Aff:* IEEE, ASEE.

Breitenstein, Charles Jay
Home: 8805 Sheridan Dr, Buffalo, NY 14221
Position: Pres. *Employer:* C J Breitenstein Co, Inc. *Education:* BS/Chem Engg/Univ of IL. *Born:* 12/10/21. Nine yrs chem process engr at Linde Div, Union Carbide (1942-51); 6 yrs - VP & Plant Engr at Metal Cladding,Inc - coating applicator (1951-57). Prin in engg manufacturer's representative business (1957 to date). Filtration exchange, valves, coatings, Blasting Abrasives, FRP Vessels are our main products. *Society Aff:* AIChE, NACE.

Breitling, Thomas O
Business: 5296 South 300 West, Murray, UT 84107
Position: Pres *Employer:* Newbury Engrg Corp *Education:* M/Bus Admin/Univ of UT; BS/CE/Univ of AL *Born:* 8/12/23 Native of Theodore, AL. Worked in defense effort prior to World War II. Served with Army Corps of Engrs 1943-46. Worked in engrg with various companies in AL, MS, Republic of Honduras, CO, and UT. Did two short term assignments in Panama and one in Peru. Assumed current responsibility as pres and general mgr of Newbery Engrg Corp in December 1979. Newbery Engrg does plant and facilities designs for mining and other industrial concerns. Chrmn, Mineral Processing Division of AIME 1978; Dir of Society of Mining Engrs of AIME 1979-1981. Hobby: Alfalfa farming. *Society Aff:* AIME, ACEC

Breitmayer, Theodore
Home: 148 Draeger Dr, Moraga, CA 94556
Position: Principal Engr. *Employer:* Bechtel Natl. *Education:* MS/Chem Engg/Univ of WA; BS/Chem Engg/OR State Univ. *Born:* Jun 1922. R & D Engr Crown Zellerbach 1948-51, by-product & pollution abatement dev. Sr Process Design Engr Monsanto Chem 1951-55, petrochemicals. Sr Process Design Engr C F Braun & Co 1955-64, petro & genl chemicals. Chief Process Engr, Bechtel Group Inc 1964- , genl chemicals & nuclear fuel reprocessing, plutonium processing, processing for uranium enrichment. AIChE Fellow; tennis, camping, photography. *Society Aff:* AIChE, ANS.

Breitwieser, Charles J
Home: 2738 Caminito Prado, La Jolla, CA 92037
Position: President. *Employer:* Design, Development Cons. *Education:* DSc/Elec Engg/Univ of ND; MS/Elec Engg/CA Inst of Technology; BS/Elec Engg/Univ of ND. *Born:* Sep 1910. Native of Colorado. Taught math Univ of N Dak 1931-33. Cons engr & assoc of Dr Lee DeForest 1934-39. Staff engr & later hd of Engrg Labs & Electronics for Consolidated Vultee Aircraft Corp 1940-50. Dir of Engrg P R Mallory Corp 1950-54. V P of Lear Inc 1954-57; a co-founder of Cubic Corp 1950 & became active as Exec V P 1960-73. Also Chmn of Bd of United States Elevator Corp, a subsidiary of Cubic. Retired in 1973 to head group of Mgmt & Dev Cons. Mbr Sigma Xi, Sr Mbr & Fellow IRE. Fellow AIEEE. Contrib papers in field. Holder sev patents in aircraft elec sys & missile guidance. *Society Aff:* ΣΞ, IRE, IEEE.

Breitzig, R William
Business: P O Box 1958, Huntington, WV 25720
Position: Product Mgr. *Employer:* Huntington Alloys Inc. *Education:* BS/Metallurgical Eng/Case Inst of Tech *Born:* 9/14/38. BS Met Engrg 1960 Case Inst of Tech. Native of Glenshaw PA. Started with Huntington Alloys Inc 1960. Joined Navy Apr 1962. Grad from OCS Aug 1962. Part of Navy career spent exper with explosive cladding. Returned to Huntington Alloys Sep 1965 as tech serv rep. Areas of respon incl giving advise on machining, annealing, pifkling, drawing, rolling, & fabricating high-nickel alloys. Became mkt coord in 1975 with respon for the thermal processing indus. Duties incl promoting high-nickel alloys to the indus heating mkt. Become Product Mgr respon for Plate, forgings and large rounds. Presently Marketing Servs Admin. Pres W Vir Chap ASM 1976; Asst Chmn Sales & Mkt Activities, ASM Heat Treating Div rounds. P. *Society Aff:* ASM, NRA, ROA

Brennan, James J
Home: 2441 Liberty, Beaumont, TX 77702
Position: Prof *Employer:* Lamar Univ *Education:* PhD/ME/Univ of TX at Austin; MS/IE/Univ of AR-Fayetteville; BS/EE/IA State Univ *Born:* 4/02/24 Native of Hot Springs, AR. US Navy, 1943-46. Research and development in Materials Handling, Navy Dept, 1948-53. Consultant to National Clay Pipe Manufacturers, 1953-54. Consultant Plant Location, AR Industrial Development Commission. Plant Engr, Baldwin Piano Co, Fayetteville, AR. Assistant Prof, Industrial Engrg, Univ of AR, 1960-65. Consultant to plastic moulding firm. Research in materials handling and automation, Univ of AR Medical Center. National Science Foundation Faculty Fellow. Instructor, Mechanical Engrg, Univ of TX. Industrial Engrg Dept, Lamar Univ, Assoc Prof, Prof, Dept Head. Consultant in Industrial Safety and Product liability since 1969; over 700 cases nationwide. *Society Aff:* ASTM, ASQC, ASM, ASSE, ASNT, NSC, NSPE, TSPE, NAFE, ASME, IIE

Brennan, Peter J
Business: PO Box 1345 Gracie Station, New York, NY 10028
Position: President *Employer:* Brennan Partners, Inc. *Education:* BChE/Chem Eng/ Catholic Univ of Amer *Born:* 7/30/31. Dublin, Ireland. Early engg experience in chem and nuclear industries. Took up tech journalism when joined *Chem Engrg* magazine in 1959. Later Editor of EJC *Engineer Magazine*, founding editor *Industria Quimica* and *International Instrumentation*, Editor-in-Chief *Instruments & Control Systems*. Frequent consultant and author on business, tech and energy to United Nations and US Govt Agencies with particular emphasis on int'l industrial dev and trade. Author of numerous articles, books, on engrs and engg, specific industries, tech and intl dev. Jesse H Neal Award 1977. Married.

Brennan, Robert D
Business: 1501 California Ave, Palo Alto, CA 94304
Position: Dev Specialist. *Employer:* IBM. *Born:* Jul 1928. MSEE Stanford; BS & BA Calif State Univ, San Jose. Field Engr, radar navigation & bombing sys; electronics instr; simulation studies of missile guidance sys; intell analysis. With IBM since 1960. Recog as IBM's 'top technically-oriented professional within his specialty'--digital simulation of continuous sys. Pioneered dev of continuous sys simulation languages; proj leader for IBM's Continuous Sys Modeling Programs. Visiting Prof of Computer Sci, Univ of Albuquerque 1974-76. Pubs on concepts & application of continuous sys simulation. Simulation studies in endocrinology. Broad cons in engrg, phys & biol sci, bus & social sci. V P Soc for Computer Simulation 1972; Pres 1973-75. Chmn Summer Computer Simulation Conf Ctte 1969-76.

Brenner, Egon
Business: 500 W 185 St, New York, NY 10033
Position: Exec VP *Employer:* Yeshiva Univ *Education:* DEE/Elec Engg/Polytechn Inst of Brooklyn; MEE/Elec Engg/Polytechn Inst of Brooklyn; BEE/Elec Engg/City College of NY. *Born:* 7/1/25. Taught Elec Engg at CCNY beginning in 1946. Prof since 1966. Assoc Dean 1966, Dean of Grad Studies (Engg) 1967-71. Dean of Sch of Engg 1971-73. Provost City Coll (CCNY) of Cuny 1973-76. Act'g Vice-Chancellor, City Univ of NY 1976- 77. Deputy Chanceller, City Univ of NY 1978-81. Exec VP Yeshiva Univ 1981- Chief Academic Officer responsible for all colleges

Brenner, Egon (Continued)
and sch of the Univ, academic support services and related activities. *Society Aff:* IEEE, AAAS, ASEE

Brenner, Howard
Business: Dept of Chem Engrg, M.I.T, Cambridge, MA 02139
Position: W.H. Dow Prof of Chem Engg *Employer:* M.I.T. *Education:* Eng ScD/Chem Engg/NY Univ; MChE/Chem Engg/NY UNiv; BChE/Chem Engg/Pratt Inst. *Born:* Mar 1929. Native of NYC. Taught at NYU from 1955-66. At CMU from 1966- 77. At UR from 1977-81. At MIT since 1981. Co-authored monograph 'Low Reynolds Number Hydrodynamics'. Bingham Medal, Soc. Rheol. 1980; Fellow AICHE 1980; Fellow Amer. Acad. Mechs. 1984; Fellow Amer. Assoc. Adv. Sci. 1982; National Academy of Engineering 1980. AIChE Alpha Chi Sigma Award 1976; AIChE W.H. Walker Award 1984; Fairchild Distinguished Scholar, Caltech 1975-76; Sr Visiting Fellow, Sci Res Council, Great Britain 1974; 11th Annual Honor Scroll IEC Div ACS 1961. *Society Aff:* AIChE, ACS, SNP, AAAS, AAM, Soc. Rheology, SIAM, NAE, Intern. Assoc. ColDoid & Interface Scientists, APS, Soc. Natl. Phil

Brenner, Mortimer W
Home: 24 Kent Rd, Scarsdale, NY 10583
Position: Pres *Employer:* Brenner Conslts, Inc. *Education:* ScM/Chem/NY Univ; BS/Chem Eng/NY Univ. *Born:* Jul 28, 1912. Chem Engr for Natl Aniline Co, Res Chem Engr for United Fruit Co, Ch Chemist for Schwarz Labs, Process Engr & Assoc Plant Supt for Pabst Brewng Co. Radar Officer US Navy 1944-45. Asst Tech Dir, Ch Engr & V P for Schwarz Labs 1946-58. V P Oper for Jacob Ruppert 1958-60. Exec V P 1960-65 & Pres 1965-75 of Schwarz Services Internatl Ltd. Self-employed cons 1976-81. Pres of Brenner Conslts, Inc. 1982- . P Pres Amer Soc of Breswing Chemists. Cincinnati Award of Master Brewers Assn of Amer 1965. Author or co-author approx 100 tech papers. First recipient of Award of Merit of Masters Brewers Assn of the Americas for outstanding contributions to the brewing industry, 1980. *Society Aff:* AIChE, ACS, AIC, IFT, MBAA, I of B, NYAS.

Brenner, Theodore E
Business: 475 Park Ave S, New York, NY 10016
Position: President. *Employer:* The Soap & Detergent Assn. *Education:* MS/Sanitary Engg/Johns Hopkins Univ; BCE/Civil Engg/Manhattan College. *Born:* Apr 1930. Native of NYC. Reg P E NJ & Pa. Sanitary engr in USAF where served to Capt 1952-59. Presently Lt Col USAFR (ret). Mgr Waste Treatment Dept of the Permutit Co. Principal in Hydroscience Inc, biological waste treatment consultants. With The Soap & Detergent Assn since 1963, a natl trade group, & assumed current pos as Pres in 1972. Mbr Amer Soc of Civil Engrs, Amer Inst of Chem Engrs & Fellow Amer Public Health Assn. Diplomate, Amer Acad of Environ Engrs. Author num tech papers & articles in recog sci & engrg journals. Mbr sev fed & state environ adv groups. Rumson Bd of Ed 1968-74, First VP 1973-74; Rumson-Fair Haven Regional Bd of Ed 1974-77, VP 1976-77. *Society Aff:* ASCE, AIChE, APHA, AAEE.

Bresler, Aaron D
Home: 12 Helene Ave, Merrick, NY 11566
Position: Technical Assistant to the President. *Employer:* AIL Div, Eaton Corp *Education:* DEE/EE/Polytechnic Inst of Brooklyn; MEE/EE/Polytechnic Inst of Brooklyn; BEE/EE/CCNY. *Born:* Jun 1924 NYC. Fellow IEEE. Mbr Tau Beta Pi, Eta Kappa Nu, Sigma Xi. First Lt, US Army Signal Corps 1944-48. Fields of tech spec incl network theory, electromagnetic theory, antenna & microwave sys engrg. Mbr fac CCNY Engrg School 1948-51 & 1953-55. Res staff of PIB Microwave Res Inst 1951-53 & 1955-59. V P & Ch Engr Jasik Labs Inc 1959-68. Independent cons 1968-72. Director, Antenna Sys Div of AIL Div, Eaton Corp 1972-1986. Assumed present position July 1986. *Society Aff:* IEEE, TBΠ, HKN, ΣΞ.

Bresler, Boris
Home: 570 Vistamont Ave, Berkeley, CA 94708
Position: Principal *Employer:* Wiss, Janney, Elstner, and Assoc Inc *Education:* MS/Aero Engr/CA Inst of Tech; BS/CE/Univ of CA-Berkeley *Born:* 10/18/18 Born in Harbin, China. Worked in shipbuilding and aircraft industries in CA, 1941-1945. Taught structural engrg at Univ of CA, Berkeley, 1946-1977. Joined Wiss, Janney, Elstner and Assocs, Inc., Consulting and Research Engrs, Principal and Mgr of CA office in 1977. Member of National Academy of Engrg; Fellow ASCE and ACI; Member Reinforced Concrete Research Council, Intl Association of Bridge and Structural Engr, Structural Engrs of CA. NSF Postdoctoral, Guggenheim fellowships; Wason Medal, State-of-the-Art of Civil Engrg, J. W. Kelly awards. Co-author of Structural Steel Design, Reinforced Concrete Engrg, numerous articles in technical journals. *Society Aff:* ASCE, ACI, IABSE.

Bresler, Sidney A
Business: PO Box 86 Cathedral Station, New York, NY 10025
Position: Pres. *Employer:* Bresler and Assoc, Inc. *Education:* MChE/Chem Engg/Polytechnic Inst of Bklyn; MBA/Bus Adm/Columbia Univ; BChE/Chem Engg/CCNY. *Born:* 8/12/23. Consulting Chem Engg. Founded present co in 1969. Active in the US and the Far East. Previously Project Mgr for both A/E and operating companies. Respon for design and construction of major petrochemical projects. Areas of specialization include fertilizer plant design and operation, coal gasification and liquefaction, methanol production, inorganic chemicals production, brackish water desalting, tech and econ feasibility studies. Chmn of symposia and author of more than thirty articles and patents relating to above subjects. Prof Engr, NY. Chmn, NY Section AIChE, 1975-76. Fellow, AIChE. *Society Aff:* AIChE, ACS, AWWA, AACE.

Bressler, Marcus N
Business: 400 W. Summit Hill Dr, W9A61 C-K, Knoxville, TN 37902
Position: Sr. Engrg Specialist *Employer:* Tennessee Valley Authority *Education:* MSME/Heat Power/Case Institute of Technology; B of ME/Machine Design/Cornell University *Born:* 07/31/29 Born in Havana, Cuba. Came to the US in 1942. Served in US Army 1952-54. Employed by the Babcock & Wilcox, Co. As design draftsman, stress analyst and materials engr, 1955-66; Lenape Forge as chief design engr, 1966-70; Taylor Forge as mgr, product design and devel 1970-71; and the TVA as supervisor, codes, standards and materials and senior engrg specialist, 1971 to present. Mbr ASME Boiler & Pressure Vessel Cttee. Consultant and lecturer on nuclear and fossil codes and standards and material applications. ASME Century Medallion, 1980; elected Fellow, ASME, 1983. *Society Aff:* ASME, ASM, ASTM, ASQC

Breuer, Coy L
Business: P. O. Box 270, Jefferson City, MO 65102
Position: Asst Div Engr-Construction *Employer:* MO Hgwy & Transportation Dept *Education:* BS/CE/Univ of MO-Rolla *Born:* 4/3/24 Native of Phelps County MO. Entered the US Army shortly after high school graduation and served approximately three years (1943-1945) primarily with 843rd Engr Aviation Battalion which earned Battle Participation Bronze Service Stars for the Normandy Campaign, Northern France Campaign, Rhineland Campaign and the Central Europe Campaign. With the MO Highway Dept since 1949. Promoted to present position in 1969 with responsibility for Contract Administration for Highway and related construction. Chrmn of PE in Government Practice Division of NSPE and as NSPE VP 1980-81. Past Pres of Mid-MO Section of ASCE. *Society Aff:* NSPE, ASCE

Breuer, Delmar W
Home: 5539 Oakshire Cir, Dayton, OH 45440
Position: Hd, Dept of Mech & Engrg Sys. *Employer:* Wright-Patterson AF Base, Ohio. *Education:* PhD/Aero Castro/OH State Univ; MS/CE/Univ of MO at Rolla; BS/Aero/IA State Univ. *Born:* 6/21/25. Prof Emeritus, Aerospace Eng, Air Force Inst of Tech. PhD OH State Univ; MS Univ of MO at Rolla; BS IA State Univ. Born near St James, MO, attended rural sch & the St James HS. Navy July 1943-Aug 1946, V-12 student and Ensign. Taught Mech courses at MO Sch of Mines Mar 1947-June 1950. Stress analyst North Amer Aviation July 1950-Sept 1951. Asst

Breuer, Delmar W (Continued)
Prof 1951-55, Assoc Prof & Acting Dept Hd 1955-58, Hd, Mechanics & Engrg Sys Dept 1958- , & Prof since 1961. Oct 1951 to present with the Air Force Inst of Tech. Assoc Fellow AIAA; Mbr ASEE; Mar 1977- Hd, Dept of Aeros & Astronautics, Air Force Inst of Technology. Retired Dec 1980. Self Employed, teacher, consultant. *Society Aff:* AIAA, ASEE, ADPA.

Breuer, Melvin A
Business: SAL 324, Los Angeles, CA 90089-0781
Position: Professor *Employer:* Univ of S Calif-E E Dept. *Education:* PhD/Elect Engr/Univ CA Berkeley; MS/Engr/UCLA; BS/Engr/UCLA *Born:* Feb 1938. Native of Los Angeles. Currently Prof of Elec Engrg & Computer Sci at Univ of Southern Calif, L A. Main interests lie in area of fault tolerant computer design & computer aided design of digital sys. Author over 100 papers as well as ed of 'Design Automation of Digital Sys: Theory & Tech', 'Digital Sys Design Automation: Languages, Simulation & Data Base', & co-author 'Diagnosis & Reliable Design of Digital Sys'. Mbr Tau Beta Pi, Sigma Xi, Eta Kappa Nu, Fellow IEEE. *Society Aff:* TBΠ, ΣΞ, HKN.

Breuning, Siegfried M
Home: 129 School St, Wayland, MA 01778
Position: Prof *Employer:* Southeastern MA Univ *Education:* ScD/Transp/MIT and Harvard Univ; Dipl Ing/CE/Techn Univ-Germany *Born:* 7/25/24 Native and educated in Germany (first degree) and USA (highest degree) Transportation Planner, Montreal Transit Com 1951-52. Teaching Appointments: Techn Univ Stuttgart, Germany, 1951; Univ of Alberta, Canada 1957-59; MI State Univ, 1959-63; MIT 1965-69, Harvard Grad School of Design, 1969. Since 1970 Prof of Transportation Engrg and Interdisciplinary Studies, SMU - 1963-65 Chief, Transportation Systems Division, National Bureau of Standards - Consultant in Transportation Planning, Research, and Techn Assessment to VW, Mercedes-Benz, Cabot Corp, Burma, Taiwan, Nigeria, South Australia, etc - Over 30 Publications - Visiting Prof at Univ of OK, Techn Univ Stuttgart, Karlsruhe, Muenchen, Germany. *Society Aff:* TRB, FG & d SW.

Breyer, Donald E
Business: 3801 West Temple Blvd, Pomona, CA 91768
Position: Prof *Employer:* CA State Polytechnic Univ, Pomona *Education:* ME/Structural/Univ of CA, Berkeley; BS/CE/CA St Polytechnic Univ, Pomona *Born:* 12/20/41 Prof Breyer is a registered civil engr in CA and TX and serves as a consultant on a variety of structural engrg projects. He is a member of the Technical Advisory Committee of the American Inst of Timber Construction and a member of the General Design Committee of the Intl Conference of Bldg officials. Prof Breyer is the author of the timber design book Design of Wood Structures, 1980, published by McGraw-Hill. *Society Aff:* ASCE, AITC, ICBO

Breyer, Norman N
Home: 1615 Robinhood Pl, Highland Park, IL 60035
Position: Prof, Dept of Met & Mat Engrg *Employer:* Illinois Inst of Technology. *Education:* PhD/Met Engr/Ill Inst of Tech; MS/Met/U of Michigan; BS/Met Engr/Mich Tech Univ *Born:* Jun 21, 1921 Detroit, Mich. Indus exper: NASA 1947-48, Detroit Tank Arsenal 1948-52, Natl Roll & Foundry Co 1952-54, Blaw-Knox Co 1955-57, LaSalle Steel Co 1957-64. Prof Ill Inst of Tech 1964- . Author 25 tech papers, 3 books & 4 patents. Area of expertise: ferrous metallurgy, failure analysis, armor & its application to military vehicles. Soc: Tau Beta Pi, Sigma Xi, Alpha Sigma Mu. *Society Aff:* ASM, AIME, ASTM, TMS

Brian, P L Thibaut
Business: Air Products & Chemicals, Inc, Allentown, PA 18195
Position: VP, Engrg. *Employer:* Air Products & Chemicals Inc. *Education:* ScD/Chem Eng/MIT; BS/Chem Eng/LA State Univ. *Born:* Jul 8, 1930 New Orleans, La. Chem Engrg: MIT fac, Dept of Chem Engrg: Asst Prof 1955-62; Assoc Prof 1962-66; Prof 1966-72. Overseas Fellow, Churchill College, Cambridge, England 1969-70. Author of textbook & more than 50 articles on heat & mass transfer, fluid mech, applied chem kinetics, numerical math, & water desalination. Air Products & Chemicals Inc: V Pres Engrg 1972- ; Bd of Dir 1973- . 1973 Prof Progress Award AIChE; Elected mbr of the Natl Acad of Engrg 1975/ LA State Univ Engg Hall of Distinclion Charter Mbrs, 1979. *Society Aff:* AIChE, ACS, NAE.

Briant, Clyde L
Business: Corporate R&D Center, P.O. Box 8 Met 269-K1, Schenectady, NY 12301
Position: Staff Metallurgist *Employer:* Gen Elec Co *Education:* Eng ScD/Mat Sci and Metallurgy/Columbia Univ; MS/Mat Sci and Metallurgy/Columbia Univ; BS/Mat Sci/Columbia Univ; BA/Chem/Hendrix Coll *Born:* 5/31/48 Born in Ashdown, AR. Postdoctoral fellow at Univ of PA before assuming current position in 1976. Current research involves properties of interfaces and their role in metallurgical problems. Author or co-author of over 100 technical publications. Winner of the 1977 Hardy Gold Medal of the Metallurgical Society of AIME for a young metallurgist showing exceptional promise. Winner of the 1980 R.W. Raymond Award for best paper by a young author published by AIME (presented by AIME). Winner of the 1980 Alfred Noble Prize awarded by ASCE for the best paper by a young author published by AIME, ASCE, IEEE, ASME, or the Society of Western Engrs. Chosen by Science Digest as one of the 100 outstanding young scientists in America, 1983. *Society Aff:* AIME, ASM.

Brick, Donald B
Home: 4 Blueberry Lane, Lexington, MA 02173
Position: Conslt *Employer:* Donald B. Brick & Co, Inc *Education:* PhD/Applied Physics/Harvard; SM/Applied Physics/Harvard; AB/Applied Physics/Harvard. *Born:* 10/1/27. AB cum laude Harvard; SM 1951, PhD 1954 (applied physics); Harvard Teaching Fellow 1950-52, Asst 1952-54, Res Fellow 1954-55, GTE Sylvania, Engrg Spec 1954- 55, Sr Engrg Specialist 1957-61, Sr Scientist 1961-64; Sci Dir 1964-65; Founded, Pres & Tech Dir Info Res Assn Inc (now General Terminal Co. Inc) 1965-70, Chmn Bd 1970-71; Treas IRA Sys 1967-68; V P Data Sys Dev & G M, Tech Ctr, Addressograph Multigraph 1971-72. Tech & Mgmt Cons 1954-55, 1965-75 INSTR Northeastern Univ 1962-68; P L 313, Tech Dir, Deputy for Dev Plans, Elect Sys Div, USAF Nov 1975-July 1983, Engrg Conslt 1983. PE, Comm of Mass. Chmn IEEE Prof Group on Sys Sci & Cybernetics, IEEE & AA Cybernetics 1969-70. President, AFCEA Chapter 1980-81. Mbr num Bds of Dirs 1965-75, 1983- . Elected to Sigma Xi, Fellow of N Y Acad of Sci, IEEE, AIAA, & AAAS. Pub over 40 papers; 2 patents. Electromagnetic (microwave & antenna) theory, computer tech; info & communication sci; decision theory; applied math; signal processing, electronics, systems analysis and design *Society Aff:* ΣΞ, AAAS, IEEE, AIAA, AFCEA.

Brick, Robert M
Home: 32552 Sea Island Dr, Laguna Niquel, CA 92677
Position: Metallurgical Cons. *Employer:* Self. *Education:* PhD/Metallurgy/Yale Univ; MetE/Metallurgical Engg/Lehigh Univ. *Born:* Nov 18, 1908. Asst Prof Yale until 1945, then founded & became Dir of Sch of Metallurgical Engrg at Univ of Pennsylvania. From 1955-73 was at Continental Can Co Res Ctr, in var capacities, incl Genl Mgr of Corporate R&D. Author of over 75 pub tech articles & engrg text 'Structure & Properties of Alloys', now in its 4th Edition. Cons for 40 yrs as met generalist, incl continuing in such capacity at the Los Alamos Sci Labs. Recipient of AIME Mathewson Gold Medal for best pub paper 1937, elected Fellow Amer Soc for Metals 1972. *Society Aff:* ASM, AIME.

Brickell, Gerald L
Business: 3706 Ingersoll Ave, Des Moines, IA 50312
Position: VP; Principal Engr *Employer:* Johnson, Brickell, Mulcahy & Assocs. *Education:* BSCE/Civil Engg/KS State Univ. *Born:* Nov 1, 1935; Traffic Engr with Kansas State Hwy Comm & City of Aurora, Colo prior to 1966. With Johnson, Brickell, Mulcahy & Assocs Inc since 1966. Current respon: VP & Principal. P Pres Missouri Valley Sect Inst of Transportation Engrs. Chmn Dist IV Inst of Transpor-

Brickell, Gerald L (Continued)
tation Engrs 1976. Chmn Consultants Council of Institute of Transportation Engineers 1986, 1987. *Society Aff:* ITE, ASCe, NSPE, APWA

Bridgers, Frank H
Business: 213 Truman St NE, Albuquerque, NM 87108
Position: Chrmn of the Bd of Dir and Chief Exec Officer *Employer:* Bridgers & Paxton Cons Engrs Inc. *Education:* MS/Mech Engg/Purdue Univ; BS/Mech Engg/Auburn Univ. *Born:* Jan 13, 1922. Native of Birmingham, Ala, but have lived in Albuquerque, N M for most of prof career. Served in Army Corps of Engrs 1942-45. Asst Proj Engr for Charles S Leopold Engrg Co in Phila. Formed Bridgers & Paxton Cons Engrs May 5, 1951. Codesigner of world's first commercial solar heated bldg. Pres of Amer Soc of Heating, Air Cond & Refrig Engrs 1970-71. Fellow ASHRAE 1972. Have designed a total of 30 solar energy bldg projects that are operational, mostly commercial or inst bldgs, the largest being the 300,000 sq ft Community College of Denver/North Campus. *Society Aff:* ASHRAE, NSPE, ACEC.

Bridges, Jack E
Home: 1937 Fenton Lane, Park Ridge, IL 60068
Position: Sr Engrg Advisor. *Employer:* IIT Research Inst. *Education:* BS/EE/Univ of CO; MS/EE/Univ of Co. *Born:* Jan 1925. US Navy 1943-46. Taught elec engrg at Iowa State & Northwestern Univ. Conducted advanced dev on radio & television receivers & related consumer electronics for Zenith Radio, Magnavox & Warwick. With IIT Res Inst since 1961, spec in electromagnetic sys & effects. Presently Sr Engrg Adv respon for dev & supervision of projects concerned with electromagnetic sys, radio frequency interference, biological effects & nondestructive tests. IEEE (IRE) Browder J Thompson Mem Prize Award for 1956. Cert of Achievement IEEE Group on Electromagnetic Compatibility 1972, Fellow IEEE 1974. 1983 Dist Engrg Alumus Award, Univ of CO; Cert of Appreciation, 1976, IEEE Group on Electromagnetic Compatibility; Prize Paper Award, 1981, Power Engrg Soc of the IEEE. *Society Aff:* HKN, TBII, ΣΞ.

Bridges, William B
Business: 128-95, Pasadena, CA 91125
Position: Carl F Braun Prof of Engg *Employer:* CA Inst of Technology. *Education:* PhD/EE/Univ of CA Berkeley; MS/EE/Univ of CA Berkeley; BS/EE/Univ of CA Berkeley *Born* 11/29/34. Inglewood, CA. Hughes Res labs 1960; Sr Staff 1965; Sr Scientist 1968-1977; Mgr Laser Dept 1969-70. Joined CA Inst of Tech 1977; Prof of Elec Engg & Appl Phys 1977-1983; Carl F Braun Prof of Engrg 1983- ; Exec Officer for Elec Engg, 1978-81. Res on gas lasers, millimeter waves, microwave tubes, optical comm & radar sys. Discovered noble gas ion laser 1964. Eta Kappa Nu Outstanding Young EE (Hon Mention) 1966. A. L. Schawlow Medal of Laser Inst of Amer 1986. Fellow IEEE. Fellow OSA. Mbr, Natl Acad of Engg; Mbr, Natl Acad Sci; Co-Chmn 1971 Conf on Laser Engrg & Appli. Mbr, Optical Soc of Amer Bd of Dirs 1982-84 and 1986- ; VP 1986, Pres Elect 1987. Mbr, U.S. Air Force Sci Advisory Bd 1985- . Sherman Fairchild Distinguished Scholar Caltech 1974-75. Assoc Ed, IEEE J Quantum Electronics 1977-1982; Assoc Ed J Opt Soc America 1977-1982. *Society Aff:* IEEE, OSA, NAE, NAS

Bridwell, John C
Home: 2021 Alexandria Dr, Lexington, KY 40504
Position: Dir, Div of Sys Engrg *Employer:* Kentucky Dept of Info Sys. *Education:* MSCE/Civil Engg/Univ of KY; BSCE/Civil Engg/Univ of KY. *Born:* Jan 1938. Native of Frankfort, Ky. Bridge design & bridge plan review for Kentucky Dept of Hwys 1959-70. Asst to State Hwy Engr for engrg computer applications 1970-72. Director, Technical Computing, Ky DOT, 1972-1981 Dir, Sys Engrg, Ky DIS, 1981-87. Presently respon for designing & implementing all computer applications for Commonwealth of Ky. Ky Pres ASCE 1973; Natl Pres Hwy Engrg Exchange Program 1974; Pres Ky Info Processors Adv Ctte 1975. Licensed Prof Engr in KY. *Society Aff:* ASCE.

Bridwell, John D
Home: 297 Weed St, New Canaan, CT 06840
Position: VP *Employer:* GTE Comm Products *Education:* PHD/EE/Poly. Inst of Brooklyn; MBA/Finance/Babson Coll; MS/EE/MIT; BS/EE/KS St Univ *Born:* 8/3/39 Born in Ottawa, KS, graduate of Will Rogers High, Tulsa, OK. Served in US Navy 1958-62. Joined Bell Labs, which sponsored MS and PhD Programs at MIT and Brooklyn Poly. Joined MIT Lincoln Lab in 1973. Mgmt conslt at Boston consltg group 1977-79, followed by GTE Corporate planning experience led to becoming VP in one of GTE's Divs involved with telecommunications systems. *Society Aff:* IEEE

Briggs, Edward C
Business: 1105 Coleman Ave, Box 1201, San Jose, CA 95108
Position: Senior Staff Engr *Employer:* FMC Corp - Ord Engr Div *Education:* BS/ME/Purdue Univ *Born:* 1/27/29 Native of Chagrin Falls, OH. Attended Purdue Univ becoming AlpTauOme Fraternity member. Served as USAFR aircraft maintenance officer. Developed first lift truck production hydrostatic drive, artic vehicle with common engine, transmission and hydraulic systems. As Senior Staff Engr has division design responsibility for auxilliary automotive systems. FMC hydraulic consultant and educator. PE License CA CS001683. CA Teaching Credential 1976. FPS National Distinguished Achievement Award 1980. FPS National Regional and Bd Dir 1976-. FPS Chapter Pres 1974-75. ANSI B-93 Fluid Power Standards Committee 1979-. *Society Aff:* FPS, ANSI, ADPA

Briggs, William B
Home: 1819 Bradburn Dr, St Louis, MO 63131
Position: Dir Program Dev, Fusion Energy *Employer:* McDonnell Douglas Corp (Astro-St Louis). *Education:* DSc/Physics/Phillips Univ; MS/ME/GA Tech; BA/Phyiscs/Phillips Univ. *Born:* Dec 13, 1922. BA Physics Phillips Univ 1944. USNR 1943-46, Atlantic & West Pacific Theaters. Aeronautical Scientist with NACA, Lewis 1948-52 doing axial flow compressor res. 1952-64 Chance Vought Aircraft/LTV propulsion engrg for missiles & Mgr on Dyna Soar. Wash rep 1959-62. At the McDonnell & McDonnell Douglas Co, Mgr Adv Space & Dir. Energy Programs. AIAA Assoc Fellow, Dir Region V 1974-77, V P Mbr Services 1977-79. Mbr NASA Planetary Quarantine Adv Panel. Chrm., Bd of Dir Christian Bd of Publication. Enjoy horticulture & golf. *Society Aff:* AIAA, ANS.

Briggs, William W
Business: Briggs Engineering, Inc, Suite D-2, 4619 Emerald, Boise, ID 83706
Position: Pres *Employer:* J-U-B Engrs, Inc *Education:* BS/CE/Univ of ID; AA/CE/Boise State Univ *Born* 10/25/28 Native of Boise, ID. Assoc of Briggs and Assocs, Consulting Engrs, from 1950 to 1962. Partner of Johnson, Underkofler & Briggs, Consulting Engrs, from 1962 to 1969. Senior VP of J-U-B Engrs, Inc, from 1969 to 1981. Currently Pres and CEO of J-U-B Engrs, Inc. Past Pres of the Consulting Engrs of ID, past Dir of the ACEC, past Dir and VP of Greater Boise Chamber of Commerce, past Dir of Home Builders Association of Southwestern ID, past member of Ada County Planning Commission and Boise City Fire Code Committee. *Society Aff:* ASCE, NSPE, ULI

Brigham, Elbert Oran
Business: E-Systems ECI Division, Box 12248, St. Petersburg, FL 33707
Position: VP & GM. *Employer:* E-Systems, Inc. *Education:* PhD/EE/Univ of TX; MS/EE/Univ of TX; BS/EE/Univ of TX; MA/Engg Admin/George Washington Univ. *Born:* 9/13/40. Dr. Brigham has been VP and GM of E-Systems ECI Division since 1980. Previously he had served as VP and GM of E-Systems Melpar Division in Falls Church, Va. He joined the Melpar Division in 1976 as VP-Engineering and became GM in 1979. Earlier, he held various senior positions with the Department of Defense, the National Security Agency, and private industry. He was the head of NSA's HFDF Wideband Systems Branch and Director of Reconnaissance and Surveillance for the Office of the Assistant Secretary of Defense (SIGINT). Dr. Brigham has consulted in the academic, government and business communities, has taught at several universities, and has lectured throughout the U.S. and Europe. In addition to numerous articles, he has published a textbook Fast Fourier Transform (Prentice-Hall, 1974), and others are in preparation. He is a Fellow member of the IEEE and a member of Tau Beta Pi, Eta Kappa Nu and Phi Eta Sigma. *Society Aff:* IEEE, AOC, AFCEA.

Brigham, William E
Business: Petrol Engr Dept, Rm 115 Lloyd Noble Bldg, Stanford, CA 94305
Position: Prof. & Assoc. Chairman *Employer:* Stanford Univ. *Education:* PhD/Chem Engr/Univ of OK; MS/Chem Engr/Univ of OK; BS/Chem Engr/IA State Univ. *Born:* 4/1/29. From 1950 to 1954, pilot plant engr for Johnson's Wax; two yrs of this period were spent in USMC. From 1956 to 1958, Instructor, Univ of OK. From 1958 to 1971, held a series of technical and supervisory positions with Continental Oil Production Res Div. From 1971 to present, Prof, Petroleum Engg, Stanford Univ. His maj interests are geothermal and oil and gas reservoir engg. He has published widely and consulted for both govt and industry in these fields. His res interests include geothermal engg, enhanced recovery of heavy oil, transient flow, and other oil and gas recovery phenomena. *Society Aff:* SPE-AIME, GRC.

Bright, Kenneth M
Business: Stanley Bldg, Muscatine, IA 52761
Position: V Pres & Project Mgr. *Employer:* Stanley Consultants Inc. *Education:* BS/Civil Engr/Univ of IA. *Born:* Apr 1922 Iowa City Iowa. Air Force 1942-46. Reg P E 6 states. Joined Stanley Consultants 1948. Report & Design Engr on num civil, water & wastewater projects. Proj Mgr since 1957. V P since 1972. Num papers on water & wastewater, public works financing, & rate sys. Chmn IES-AGC-APWA Joint Ctte on Municipal Constr, Co-editor of Standard Iowa Public Works Specifications. Mbr NSPE, IES (Dir), CEC/Iowa, APWA (Dir of Iowa Chap), AWWA, SAME, WPCF (Natl Dir, Pres-Iowa Assn). Mbr Adv Ctte to Iowa Water Pollution Control Comm. *Society Aff:* NSPE, APWA, AWWA, SAME, WPCF.

Brighton, John A
Business: School of Mech Engrg, Georgia Inst of Tech, Atlanta, GA 30332
Position: Prof and Dir, School of Mechanical Engrg *Employer:* GA Inst of Tech *Education:* PhD/Mech Engg/Purdue Univ; MS/Mech Engg/Purdue Univ; BS/Mech Engg/Purdue Univ. *Born:* 7/9/34. in Gosport, IN. Worked as a design draftsman, Schwitzer Corp, then Instr for Purdue Univ & finally as mbr of tech staff, Aerospace Corp until PhD. Became Asst Prof of Mech Engg at PSU. Chrmn of Dept of Mech Engg at MI State Univ (1977-82). Dir and Prof, School of Mech Engrg, GA Inst of Tech (1982-) Mbr 4 engg soc & 3 soc. Became active in Artificial Heart Res 1971. 37 pubs, 18 presentations,13 res proposals granted. Tech ED *Journal of bioengineering*. Active in prison volunteer work. Hobbies: backpacking & tennis. *Society Aff:* ASME, ASEE, ASAIO.

Briley, George C
Business: P O Box 3-C, San Antonio, TX 78217
Position: Vice President, Marketing *Employer:* Refrigeration Engrg Corp. *Education:* BSEE/Elect Engr/LA Tech Univ *Born:* Nov 1925. US Army 1943-46. Employed York Div B W 1949-60 Regional Mgr, 1960-61 Frick Co. Field Sales Mgr, 1962-75 V P Lewis Refrigeration Co; Reg P E Texas, La, Ms, Ala, Ga. Wisc. Active mbr ASHRAE, AIChE, founding Pres Internatl Inst of Ammonia Refrigeration; listed in Who's Who in Engrg 1959 & 1980, Who's Who in Southwest 1974-75; co-holder patent covering ammonia recovery from textiles and quick freezing of fruits and vegetables & Slush Ice Maker. Present duties involve mgmt in marketing & engineered sales. Enjoy hunting & spectator sports. *Society Aff:* ASHRAE, ASChE, IIAR, IIR, RETA

Brill, Edward L
Business: 3510 Mahoning Ave, Youngstown, OH 44509
Position: Mgr *Employer:* Floyd Browne Assoc *Education:* B/CE/Youngstown State Univ *Born:* 8/19/47 Native of Warren, OH. Officer US Army Corps of Engrs, 1971 to present. Hydraulic Engr for OH Dept of Transportation and Mgr of Hydraulic Engrg for Michael Baker Corp. Assumed current position of Mgr of the Youngstown Office of Floyd Browne Assocs, Limited in 1979. Mgr of Marketing and Professional Engrg Activities of firm in NE OH and Western PA. Pres of the Mahoning Valley Society of PE. *Society Aff:* ASCE, NSPE, SAME, WPCF

Brill, Lawrence F
Business: 2901 Ponce de Leon Blvd, Coral Gables, FL 33134
Position: Pres. *Employer:* Lawrence F Brill, Inc. *Education:* BSCE/Civ Engg/Univ of Miami. *Born:* 3/24/36. & attended grad through secondary sch in NYC. Worked as construction worker while attending Univ of Miami. Upon grad worked for Loyd Frank Vann, Arch, leaving as assoc & chief engr. Joined Pancoast, Ferendino, Grafton & Skeels as chief structural engr. Left that firm in 1964 to form Lawrence F Brill, Inc, Consulting Engrs. Mr Brill designed maj engg projs in S FL and is a recognized expert in the design of vehicle maintenance facilities, having designed over 500 such facilities throughout the US & Canada. Mr Brill is presently reg in 38 states and has a natl reg certificate. *Society Aff:* ACEC, FICE, FES.

Brillhart, S Edward
Home: Oaklands Cott-RD 4, Box 313, Easton, MD 21601
Position: Retired. *Education:* BSME/Carnegie Inst of Tech. *Born:* Jul 21, 1896 Lancaster Pa, s. Samuel & Anna Christina (Sachs) Brillhart. CutlerHammer Mfg Co Jul 1923-Dec 1925; Engr Mfg Dev, Western Elec Co Jan 1926Sep 1928; Asst Ch Engr, Victor Talking Machine Co Oct 1928-Feb 1929; Process Dev Engrg, Asst Engr of Manufacture, Dir of Res & Dev Engrg, Dir of Res Cons, Western Electric Co Mar 1929-Jun 1961. Retired June 1961. 11 patents; ASME Fellow; AAAS, Cosmos Club. *Society Aff:* ASME, AAAS.

Brinckerhoff, Charles M
Home: 784 Park Ave, New York, NY 10021
Position: Consultant. *Employer:* Ret Chmn & CEO The Anaconda Co. *Education:* Met E/Columbia Univ; BA/Columbia Univ. *Born:* 1901 Minneapolis. Phelps Dodge Corp 1925-26 Testing Engr; Inspiration Con Copper Co 1926-35 Mine Foreman; Andes Copper Mining Co 1935-48 Asst Mine Supt to Genl Mgr; Chile Exploration Co 1948-58 Genl Mgr; The Anaconda Co 1958-68 Pres, Chmn CEO; Cons 1969- . Columbia Univ Wendell Medal; Columbia Univ Egleston Medal; AIME Saundes Medal; Hon Mbr AIME; Chile Mining Soc Gold Medal Mining; Chile Mining Soc Gold Medal 50 yr; Holland Soc Gold Medal for Indus; SME/AIME Distinguished Mbr; Sigma Xi; Natl Acad Engrg; Univ of Ariz-Hon Dr of Sci. WW I - 12th Div 212th F S Bn. In Feb. 1980 at Las Vegas was presented with the Hoover Medal by (ASME). Following retirement from Anaconda 1/1/69, did consltg work incl. Sar Cheshmeh, Iran; copper deposit in Panama. *Society Aff:* AIME, ΣΞ, NAE.

Bringhurst, Lynn H
Business: 180 E First S, PO Box 11368, Salt Lake City, UT 84139
Position: Dir, Technical Operations *Employer:* Mtn Fuel Supply Co. *Education:* MEA/Ind Engg/Univ of UT; BS/Mech Engg/Univ of UT. *Born:* 3/14/39. Employed by Hercules, Inc 1962-1966 at Bacchus, UT. Reviewed design changes to solid propellant rocket motors and approved changes for the production dept. Employed by Mtn Fuel Supply Co since 1966. Assignments have been mech system design engr, gas engr, ind engg supervisor, supervising rate engr and Dir. Codes and Standards. Currently Dir, Technical Operations. Responsible for working with Codes and Stds developing organizations, presenting Co testimony, and internal implementation of codes and stds requirements, tooling development and materials evaluation at Mtn Fuel Supply. Coordinate Federal Energy Code changes with State Energy Office and have served on the State Bldg Bd Committee and Salt Lake Energy Advisory Committee to develop codes for energy conservation. *Society Aff:* ASHRAE.

Brinkers, Henry S
Home: 190 W 17th Ave, Columbus, OH 43210
Position: Assoc Prof *Employer:* OH State Univ Coll of Engrg *Education:* MCP/Planning/MIT; M/Arch/Univ of IL; B/Arch/Yale Univ *Born:* 11/21/31 Native of NY City. Taught at the Univ of IL and the Boston Architectural Center before joining the faculty of the School of Architecture at the OH State Univ along with professional positions at Edwin S Voorhis & Son, Engrs, Long Island, NY; Metcalf & Eddy, Engrs, Boston; and the Boston Redevelopment Authority. *Society Aff:* AAAS, ASEE

Brinkmann, Charles E
Business: 101 Madison St, PO Box 780, Jefferson City, MO 65102
Position: VP-Operations. *Employer:* MO Power & Light Co. *Education:* BS/EE/MO Sch of Mines. *Born:* 8/20/27. Native of Washington, MO. Served in US Army 1945-46. Worked for McDonald Aircraft for one year developing tests on experimental aircraft. Employed by MO Power & Light Co, power operations dept, in 1951. In 1963 became dept hd, respon for operation of transmission lines, generating stations & system dispatching. Current position since 1976; respon for operation of co's elec & natural gas distribution systems, mbr of labor negotiations committee and mbr of corporate general exec committee. Former officer & active in local and state profl organizations. Interests include hunting, fishing, camping and muzzle loading primitive weapons. *Society Aff:* NSPE, IEEE.

Brinkmann, Joseph B
Business: 1285 Boston, Bldg 25BW, Bridgeport, CT 06602
Position: Sr Metallurgical Engr. *Employer:* Genl Electric Co. *Education:* MS/Matls Engrg/Worcester Poly Inst 1964; BME/Tool Engrg/Genl Mot Inst 1958. *Born:* 8/30/31. P E Mich, Met Engrg. Journeyman Toolmaker, 1954. Career oriented to materials and manufacturing process dev. Have filled positions from Manufacturers Rep (2 yrs) to College Prof (5 yrs). GE Ferrous Diecasting Proj Engr 1967. GE Appliance Met Dev in Powder Met & all casting processes. GE Mgrs Award 1971. Since 1973 have dev continuous casting of electronic qual copper rod--Dip Forming. Hold (3) Patents for Dipforming. Enjoy children & running. *Society Aff:* SME

Brinsko, George A
Business: Pima County Wastewater Management Dept, 130 West Congress, Tucson, AZ 85701-1317
Position: Dir, Wastewater Management Dept *Employer:* Pima County Government *Education:* BS/Geology/Univ of Pittsburgh *Born:* 8/6/28 Native, Duquesne, PA, US Maine Corps 1950-52. Employed Alcosan 1956, Resident Engr, 7-mile interceptor tunnel construction project. Appointed Gen Supt, Operations, 200 mgd regional treatment facility, 1959. Directed upgrading of facility to secondary treatment; improvements in process application; process control, aeration, diinfection, sludge dewatering, sludge disposal, incineration, odor control. Directed mgmt, control, Hazardous Waste mtls in 215- square-mile area. Pioneered use of hydrogen peroxide to oxidize sulfides. Exec Conslt, NUS Corp 1978-79. Director, Wastewater and Solid Waste Mgmt, Pima County Govt, Nov 1979-present. Directs wastewater treatment activities for Tucson metropolitan area, providing service through 3 major plants, 13 smaller plants, and 50 on-line pump stations, treating 50 mgd. Also directs programs in Industrial Wastewater Control, the disposal of sludge through subsurface injection on agricultural lands. *Society Aff:* AAEE, NSPE, ASCE, WPCF, SAME, AWWA, AMSA, ASPA, PSPE, ASCE of PA, WPCAP, WPCAWP, SAME, AWEP.

Brinson, Halbert F
Business: Engrg Sci & Mech Dept, Blacksburg, VA 24061
Position: Chairman, Center for Adhesion Science and Professor of Engineering Sciences and Mechanics *Employer:* Vir Polytechnic Inst & State Univ. *Education:* DHC/Honorary Doctor of Science/University of Brussels (VUB);1986; Phd/-/Engr Mech Standord Univ; -/Theo & Appl Mech/Northwestern Univ; MS/Civil Engr/NC State Univ; BCE/Civil Engr/NC State Univ. *Born:* 7/30/33. Stress Analyst Lockhead Aircraft Corp 1956-57. US Army Res Officer; 2nd Lt, 1st Lt & Capt 1956-65. Instr N Car State Univ 1957-61. NSF Sci Fac Fellow Northwestern Univ 1961-62 & Ford Foundation Fellow Stanford Univ 1962-65. Asst Pro 1965-69, Assoc Prof 1969-75, Prof 1975- , Chrmn, Ctr for Adhesive Sci, 1980- , at VPI & SU. Res & Grad specialities incl behavior of polymeric adhesives & composite matls, viscoelasticity, fracture, photomechanics. Mbr & chmn num cttes of Soc for Exper Mechanics (SEM) 1965- . VP SESA 1976-78, Pres. 1978-79. Fellow of SEM, 1984. Honorary Doctory of Science from the Free Univ. of Brussels (VUB), 1986. Hobbies: golf & tennis. *Society Aff:* SEM, ASTM, ASME, ΣΞ, AS.

Brinson, Robert J
Home: PO Box 12658, Charleston, SC 29412
Position: Pres *Employer:* The Halcyon Group *Education:* Masters/Bus Admin/Univ of Chicago; BS/Chem Eng/Purdue Univ *Born:* 11/19/36. Native of South Carolina. Served in the Navy 1954-56. Technical Engr for Procter and Gamble 1960-66. Joined Morton Chem Co 1966. Named Mgr of Central Engg 1970. Joined Wallerstein Div of Baxter Travenol in 1974. Named VP of successor co, GB Fermentation Industries, in 1977. Formed a consltg company, Coastal Consultants, in 1982; consltg to the chemical and fermentation industries. Currently Pres of the Halcyon Group, successor to Coastal Consultants. Enjoy sailing and outdooor sports. *Society Aff:* AIChE

Brisbin, Sterling G
Business: 10 Albany St, Cazenovia, NY 13035
Position: Managing Partner. *Employer:* Stearns & Wheler, Engineers & Scientists. *Education:* SM/Sanitary/MA Inst of Tech; SB/Civil/MA Inst of Tech. *Born:* 4/8/29. Native of Gloversville, NY. Served with Corps of Engrs 1952-54 in France. Formerly Sanitary Engr with Chas T Main, Dev and Sales Engr with Dorr Oliver, Inc. Partner of Stearns & Wheler, Civil & Sanitary Engrs since 1962. Currently managing partner. In 1974 was Adjunct Assoc Prof, Cornell Univ, grad school teaching industrial wastes. Avid golfer, woodworker and hunter. *Society Aff:* ASCE, WPCF, AWWA, NSPE.

Briskman, Robert D
Business: Geostar Corporation, 1001 22nd Street NW (8th Floor), Washington, DC 20037
Position: Senior Vice President, Engineering & Operations *Employer:* Geostar Corporation *Education:* MS/Elec Engg/MD Univ; BSE/Elec Engg/Princeton Univ. *Born:* 10/15/32. Engr with IBM & Army Security Agency 1954-59. Employed by NASA 1959-64 respon for ground instrumentation dev on Apollo, Gemini, Mariner & Echo. With COMSAT since 1964 originally involved in satellite command & control & dev/implementation of INTELSAT global transmission sys. Subsequently Asst VP, Space & Info Sys of COMSAT Genl in charge of domestic communications services by satellites incl COMSTAR program serving AT&T/GT&E, then VP, Systems Implementation respon for engrg of satellites, earth stations and communications tech facilities. Senior VP of Engineering and Operations at Geostar Corporation since 1986. Fellow, Former Secr-Treas, VP for Tech Activities, Dir & Stds Bd Chmn IEEE. Recipient Army Commendation Medal, NASA Apollo Achievement Award, EASCON Founder Award, USAB Citation of Honor & IEEE Centennial Medal. *Society Aff:* IEEE, AOC, AIAA, AFCEA, WAS, WSE

Brisley, Chester L
Business: California Polytech State Univ, San Luis Obispo, CA 93407
Position: Prof. *Employer:* California Polytech State Univ. *Education:* PhD/Speech-IE/Wayne State Univ; MS/IE/Wayne State Univ; BS/IE/Youngstown Univ; Diploma/IE/Gen Motors Inst *Born:* 4/3/14 Served on teams sponsored by US State Dept, giving technical assistance in West Berlin, Japan, Indonesia, Korea, and India. Presented the distinguished Service Award, December, 1974, by the Univ of WI-Extension. Positions held: Industrial Engrg Mgr, Wolverine Tube Division of Calumet and Hecla, Inc, Detroit; Staff Assistant to Dir of Production, Chance-Vought Aircraft, Inc, Dallas; Consultant, A.T. Kearney & Co, Chicago; Chief Industrial Engr, Allis- Chalmers, Milwaukee; Mgr in Management Services Touche-Ross, Certified Public Accountants, NY. Prof & Assoc Chrmn, Univ of WI-Extension; Prof CA Polytechnic State Univ. Published many articles relating to Industrial Engrg. IIE, Awarded Fellow Membership 1968; MTM, Fellow 1978-80; VP, IIE - Region XI 1979-81; SAM (Society for Advancement of Management) Fellow 1959 Exec VP, IIE 1983-85. *Society Aff:* IIE, ASME, NSPE/WSPE, MTM Assoc, ASEE

Bristol, Thomas W
Home: 526 E. Meadowbrook Ave, Orange, CA 92665
Position: Dept. Mgr. *Employer:* Hughes Aircraft Co. *Education:* PhD/EE/Syracuse Univ; MS/EE/Univ. of MI; B/ME/Gen Motors Inst. *Born:* 11/17/37 Native of Rochester, N.Y. Instructor at Syracuse Univ 1963-1967. Mbr of Tech Staff at Rockwell International, specializing in Electromagnetic Penetration Studies and Surface Acoustic Wave (SAW) device res 1967-1971. Joined Hughes Aircraft Co in Fullerton, CA in 1971 to Direct SAW Device Res and Design, and assumed current responsibilities as mgr of Microelectronics Tech Dept in 1980. In 1975, Received L.A. Hyland patent award from Hughes. Served as Mbr of ADCOM of IEEE Sonics and Ultrasonics Grp 1978-1980 and Pres of SAME during 1981. *Society Aff:* IEEE

Britner, George F
Business: 300 Park Ave, New York, NY 10022
Position: VPres *Employer:* Phelps Dodge Industries, Inc. *Education:* BS/Electrical Engrg/U.S. Naval Academy *Born:* 04/02/18 Served as an unrestricted line officer in U.S. Navy 1941-1972. Deputy Commander (1967-8) and Commander (1968-9), U.S. Naval Ship Engrg Cntr, Wash, DC. Since 1972, with Phelps Dodge Industries, manufacturer of copper and alloy products. Energy, technical matters and R&D are among responsibilities as VPres since 1979. Studied Naval Eng at the U.S. Naval Postgraduate School.

Brittain, James E
Home: 2440 Nancy Lane, NE, Atlanta, GA 30345
Position: Assoc Prof *Employer:* Georgia Tech *Education:* PhD/History of Science & Tech./Case Western Reserve; M.A./History of Science & Tech/Case Western Reserve; M.S./Electr. Engr./University of Tennessee; B.S./Electr. Engr./Clemson University *Born:* 05/20/31 Native of Mills River, NC. Instructor in USAF Radar & ECM Schools, 1951-54. Taught EE at Clemson Univ, 1959-66. Have taught history of science and technology at Georgia Tech since 1969. Mbr of Editorial Bd of IEEE Press since 1983 and Editorial Bd of IEEE Spectrum since 1986. Member of IEEE History Cttee 1973-81 and 1983-present. Centennial Medal of IEEE, 1984. Fellow of IEEE, 1987. Fellow of Radio Club of America, 1985. Fellow of Royal Society of Arts, 1969. Avocations include industrial archeology and trout fishing. *Society Aff:* IEEE, AAAS, RSA

Brittain, John O
Business: Materials Science & Eng, Northwestern University, Evanston, IL 60201
Position: Professor. *Employer:* Northwestern Univ. *Education:* PhD/Metallurgy/Penn State Univ; BS/Metallurgy/Penn State Univ. *Born:* Feb 15, 1920 Pgh, Pa. Res Asst, instr & INCO Fellow Penna State Univ. Metallurgist, res lab ALCOA, New Kensington, Pa. 1st Lt Infantry-24th Div Army WW II. Battelle Mem Inst 1950-51; Sci Staff Columbia Univ 1951-55; Fac Northwestern Univ 1955- : Chmn Dept of Matls Sci 1968-73, Chmn Matls Res Ctr 1976-79. Res Assoc Argonne Natl Lab summer 1967. Prof socs: Sigma Xi, Tau Beta Pi, Alpha Sigma Mu (Natl Pres 1975-76), Alpha Chi Sigma, AIME (Bd of Review & ECPD accreditation), ASM (Chmn Chgo-Northern Chap 1976-77), Amer Phys Soc, Amer Ceramic Soc, AAUP, ASEE. Cons Johnson Service Co, Amphenal Corp, US Air Force Materials Lab, Universal Oil Products, Speedfam Corp. Enjoy photography, camping & fishing, opera, gardening and orchids. *Society Aff:* AIME, ASM, APS.

Britton, Myron (Ron) G
Business: Univ of Manitoba, Winnipeg, Manitoba R3T 2N2 Canada
Position: Prof *Employer:* Dept of Agricultural Engrg *Education:* PhD/Agric Engr/TX A&M Univ; MSc/Agric Engr/Univ of Manitoba; BE/CE/Univ of Saskatchewan *Born:* 11/21/40 Employed by Shell Oil (Winnipeg) and wood industry (London, England; Winnipeg; Toronto). Academic career as Lecturer, Univ Manitoba; Asst Prof, Texas A&M. Returned Manitoba 1973, served as Asst Dir, School of Agriculture, then through academic ranks to Prof Ag Eng. Responsible for teaching & research in light-frame structures area. Chmn, North Central Region ASAE, 1980-81; VP (Tech) CSAE, 1981-83; Chmn, Canada Expert Ctte on Ag Structures, 1985-87; Pres CSAE, 1987-88. North Midwest Section ASEE Western Electric Teaching Award, 1984; North Central Region ASAE Ag Engr of the Year, 1985; Univ of Manitoba Saunderson Teaching Award, 1986; CSAE-CSSBI Buildings Award, 1987. *Society Aff:* ASAE, CSAE, ASEE.

Britzius, Charles W
Business: 662 Cromwell Ave, St Paul, MN 55114
Position: Employee & shareholder ASTM Representative *Employer:* Twin City Testing Company (Huntington Plc) *Education:* MSCE/CE/Univ of MN; BSCE/CE/Univ of MN. *Born:* 6/26/11. Reg Civil Engr Minn. Hall Testing Labs 1933-36; Univ of Minn as Instr 1936-37; Minn Hwy Dept 1937-38; Twin City Testing & Engrg Lab Inc 1938 to date; Soil Exploration Co 1938 to 1985. Employee of Twin City Testing Co. Mbr Minn & Natl Soc of Prof Engrs: State Pres 1958, Natl Dir 1962-64, V Pres 1965-67; CEC/Minn; P Bd & Ctte mbr; ASCE: P Pres Northwest Sect; P Dist Dir 1959-61. P Grand Regent Theta Tau. Native of Rochester, MN. Resides in Deephaven, MN, with wife. Hobbies are hunting, fishing & sports in genl, especially tennis. *Society Aff:* ASCE, NSPE, SES, CSI, CEC, ASTM.

Broad, Richard
809 Riverside Dr, Newport News, VA 23606
Position: VP, Engg. ret 1987 *Employer:* Newport News Shipbldg. *Education:* MS/Nav Arch & Mar Eng/U of MI; BS/Nav Arch & Mar Eng/U of MI *Born:* 10/1/21. Native of Newport News, VA. Grad Newport News Shipbldg Apprentice Sch. BS & MS degrees in Naval Architecture and Marine Engg, Univ of MI. Attended the Oak Ridge Sch of Reactor Tech and Harvard Advanced Mgt Prog, AMP74. At NNS&DDCo: 1952-1958 Nuclear Aircraft Carrier Enterprise design, Atomic Power Div; 1958-1963 established and directed nuclear quality control; 1963 appointed Hd of Atomic Power Div, 1970 elected VP, and 1978 VP of Nuclear and Non-Nuclear Engg. Engr of the Yr, VA Peninsula by Natl Soc of PE, 1978. Enjoys sailboat cruising. Retired 1987 after 48 yrs. service. *Society Aff:* SNAME, NSL, USNL

Broadman, Gene A
Business: P O Box 1204, Livermore, CA 94550
Position: Dept Head. *Employer:* Lawrence Livermore Lab-U of C. *Education:* BS/ME/Univ of AZ, Tucson; MS/Eng Sci/UC Berkeley. *Born:* Jun 1934. MS Engrg Sci Univ of Calif, Berkeley; BS Univ of Arizona, Tuscon in Mech Engrg. Employed past 29 yrs at Lawrence Livermore Lab, one of Nation's leading Nuclear Res & Dev Labs. Dept Hd, Plant Engg Dept Div Leader for the Nuclear Explosives Engrg Div. Consulting Engineer, President of Safety Analysis Associates. Reg Mech Engr CA & Reg Nuclear Engr CA. Married, 2 children. Enjoy outdoor sports. *Society Aff:* ASME, NSPE

Brock, Harold L
Business: P O Box 8000, Waterloo, IA 50704
Position: Engrg Cons, *Employer:* Yanmar Diesel Engine Co, Ltd. *Born:* 11/23/14. Grad Henry Ford Trad & Apprentice Schs/ Chief Tractor Engr - Ford Motor Co - 1940-58. Asst Dir of Engg Res, John Deer - 1959-61. Mgr of Product Engg - 1962-72. Engg Cons, John Deer - 1973- . 1971 President of Soc of Automotive Engrs. Chmn SAE Finance Ctte, 1973-77. Bd Chmn, Hawkeye Inst of Tech 1965- . Bd Mbr 12 local, state & natl activities. Farmer-President John Deere Circle W Dealership. *Society Aff:* SAE, ASAE.

Brock, Louis M
Business: Dept of Engrg Mech 00461, Lexington, KY 40506
Position: Prof *Employer:* Univ of KY *Education:* PhD/Theor & Appl Mech/Northwestern Univ; MS/Struct Mech/Northwestern Univ; BS/Sci Engrg/Northwestern Univ *Born:* 4/16/43 Davenport, IA. Job experience with Black & Veatch Consulting Engrs, General Dynamics/Convair, Sargeant-Welch Scientific and American Can Co during period 1962-65. At KY since 1971 with sabbatical at Oxford Univ in 1980-81 Academic Year. Currently Prof of Engrg Mechanics and Correspondent ASCE Engrg Mechanics Division. Elastic waves and fracture research is sponsored by National Science Foundation and has resulted in 60 technical publications. Hobbies include hiking, jogging, classical music and tennis. Also does work in human body 3D kinematics and kinetics and serves as Navy/ASEE Summer Research Faculty Member at Naval Research Laboratory. *Society Aff:* ASCE, ASME, ΣΞ, ΤΒΠ, AAUP

Brock, Richard R
Business: Dept of Civil Engrg, Fullerton, CA 92634
Position: Prof of Engg. *Employer:* CA State Univ Fullerton. *Education:* PhD/CE/Caltech; MS/CE/UC Berkeley; BS/CE/UC Berkeley. *Born:* Mar 1938. Taught at Univ of Calif Irvine & Calif State Univ Fullerton since 1967. Chair of Civil Engrg Fac at CSUF 1973-77. Teaching interests in hydraulics & hydrology with res interests in open channel & groundwater flow. ASCE J C Stevens Award 1967. Served on Hydromech Ctte of Hydraulics Div ASCE 1973-77, Chmn 1975-76. Reg C E Calif. *Society Aff:* ASCE, AGU, IAHR.

Brock, Robert H, Jr
Business: Dept of Forest Engrg, Syracuse, NY 13210
Position: Prof and Chrmn, Forest Engrg Dept *Employer:* SUNY-Coll of Environ Sci & Forestry. *Education:* PhD/Geodetic & Photogrammetric Engr/Cornell Univ; MS/Photogrammetry/Syracuse Univ; BS/Forestry/Syracuse Univ. *Born:* Feb 1933 New Rochelle NY. US Army Airborne 1953-55. Taught in Civil Engrg, Syracuse 1959-66. Sr Sys Analyst with CBS 1966-67. Taught Mapping & Photogrammetry at SUNY-CESF at Syracuse 1967- . Conducted Photogrammetric Engrg Res for US Army, US Air Force, NSF, U S Forest Service & Private Indus 1962- . Mbr Cornell Soc of Engrs, N Y Acad of Sci & Sigma Xi. Dir of ASP 1975-78. Exec Ctte of ASP 1976-78. Active mbr AAU & Syracuse Chargers Track Club, Certified Photogrammetrist (ASP), Inventor in area of Remote Sensing Mapping Systems and Calibration procedures, U.S. Pat. No. 4, 199,759, US Pat. No. 4,489,322. *Society Aff:* ASAE, ACSM, ASPRS

Brockenbrough, Thomas W
Home: 5 South Dillwyn Rd, Newark, DE 19711
Position: Prof and Chrmn, CE Dept *Employer:* Univ of DE *Education:* SM/Civil Engrg/Mass Inst of Tech; BS/Civil Engrg/VA Poly Inst *Born:* 07/14/20 Born 1920 in Buena Vista, VA. Educ at Washington & Lee Univ, Virginia Poly Inst and Massachusetts Inst of Tech. Served on faculty at Univ of Delaware since 1953, 10 yrs as Assistant Dean of Coll of Engrg and two terms as chrmn of Civil Eng Dept. Mbr of Coun of Delaware Assoc of Prof Engrs and Exec Com of Delaware Acad of Sci. Past Pres of Delaware section of ASCE and Delaware Coun of Engrg Societies. Elder in Presbyterian Church and Vice Chrmn of Presbytery Cttee on Stewardship of Accumulated Resources. *Society Aff:* ASCE, ACI, ΤΒΠ, ΧΕ, ΣΞ, ASEE

Brockett, Roger W
Business: Pierce Hall, Cambridge, MA 02138
Position: Gordon McKay Prof of Applied Math. *Employer:* Harvard University. *Education:* PhD/Engg/CASE Western reserve; MS/Engg/CASE Western Res; BS/Engg/CASE Western Reserve. *Born:* 10/22/38. From 1963 to 69 taught in Dept of Elec Engrg at MIT. Presently Gordon McKay Prof of Applied Math in Div of Engrg & Applied Physics, Harvard Univ, Cambridge, Mass, where participates in the interdisciplinary Decision & Control Program through teaching & res. Co-editor (with J.C. Willems) of the Journal Systems and Control Letters & author of book ('Finite Dimensional Linear Sys') (N Y: Wiley) & co-editor (with D Q Mayne) of ('Geometric Methods in Sys Theory') (Dordrect, Netherlands: Reidel) 1973. Assoc Editor *Siam Journal on Control & Optimization*; *Siam Journal on Mathematical Analysis & Its Applications*; *Journal on Nonlinear Analysis* and *Mathematical Modeling*. Has held post-doc pos at Warwick Univ, Imperial College, & Univ of Rome, University of Nagoya, University of Granirsgen & a Guggenheim Fellowship for study in math sys theory. Recipient of the Amer Automatic Control Council's Donald P Eckman Award. *Society Aff:* IEEE, SIAM, AMS, ΣΞ.

Brockman, Donald C
Business: 3290 W. Big Beaver Rd. STE 500, Troy, MI 48084
Position: VP *Employer:* Ellis/Naeyaert/Genheimer Assoc, Inc *Education:* B/IE/Gen Motors Inst *Born:* 5/1/42 Native of Western IL. Cooperative student of General Motors Inst from 1960- 1965 at Chevrolet - Janesville, WI. Industrial engr from 1966 thru 1969 at Giffels Assocs. Joined Ellis/Naeyaert/Genheimer in October, 1969, progressing from Industrial Engr to Group Leader to Team Mgr to VP and Bd member in 1980. Reporting to him are all project teams & 2 branch offices. Past chrmn and member for 12 years of ESD Credentials Committee. Past Chrmn & Mbr for 3 years of ESD Civic Affairs Cttee. Mbr ESD Construction Activities Cttee. Enjoy athletics as participant and spectator. *Society Aff:* ESD, ACEC.

Brockway, Charles E
Business: R&E Ctr, Kimberly, ID 83341
Position: Prof. *Employer:* Univ of ID. *Education:* PhD/Water Resources/UT State Univ; MS/Civ Engr/CA Inst of Tech; BS/Civ Engr/Univ of ID. *Born:* 12/4/36. Born and raised in Ketchum, ID. Commissioned US Army; field engr Converse Fdn Engrs. Res engr US Bureau & Reclamation, Denver. Joined Univ of ID in 1965. Currently in charge of water mgt, groundwater and irrigation res at Kimberly, ID Res Station. Outstanding engr for ID 1968. Pres ID Soc of PE 1978, Chrmn NSPE Young Engrs Comm, Registered Comm Chrm ASCE I'D Div O'M Comm 1977-79, Water Quality Committee 1982-84. Chr. Idaho Board of Professional Engineers an Land Surveyors. Married, 3 children, enjoy skiing, fishing. *Society Aff:* ASCE, NSPE, ΣΞ, NCEE.

Brockway, Daniel J
Home: 5904 Holland Rd, Rockville, MD 20851
Position: Consultant, Aerospace *Employer:* Brockway Associates, Inc. *Education:* BAE Catholic Univ; post-grad work at Maryland Univ, UCLA, CUA. *Born:* 1/26/25 Served with First & Third Army Artillery in Europe during WW II. Design engr for Navy Transonic Wind Tunnel 1950. Aerodynamicist for Republic Aircraft 1956. Consultant to Navy in supersonic aerodynamics. Ch Engr, Navy Space Defense 1960-65, joined OSD in 1965. Repson included dev of maneuverble spacecraft, satellite early warning sys, atomic energy detection sys, electro-optics tech, & advanced space sys. Commendations & awards in aircraft design, space defense analyses, & space shuttle concepts. *Society Aff:* Nat'l Space Club; American Defense Preparedness

Brockway, George L
Business: 1626 E Virginia Ave, College Park, GA 30337
Position: Pres. *Employer:* Brockway & Assoc Inc *Education:* BSCE/Civ/Duke Univ. *Born:* 8/14/23. 1946-1952 Draftsman to structural engr designing all types of structures pertaining to US Govt air base. 1952-1965 Structural design engr for heavy industrial projs - 25 million dollar fertilizer plant, 18 million dollar chem plant, 30 million dollar sintering plant. Also included agri-field design as structural engr, proj engr & chief engr as feed mills, grain storage facilities, rendering plants, poultry processing plants. Specialized in "Slip Form Concrete" design of bldg & storage bins. 1965-present Owner & Pres of arch-eng consulting firm. Reg PE in 16 states. Firm also provides design-construct services for industrial projects. *Society Aff:* ASCE, NSPE, ACI, SAME.

Broderick, James J
Home: 5387 High Ct, West Bloomfield, MI 48033
Position: VP-Commercial. *Employer:* Edward C Levy Co. *Education:* MBA/Finance/Rutgers Univ; MS/ChE/NJIT; BS/ChE/MIT. *Born:* Mar 1929. From 1950 to 1971 held pos with Amer Cyanamid in Engrg, Sales & Sales Mgmt, Corp Dev & Internatl Div. Has an extensive background in rubber from Cyanamid's Rubber Chemicals & Polymers Dept in addition to having been Product Mgr for Colored Pigments. Left Cyanamid in 1971 to assuem Presidency of Genl Abrasive Co with plants in Niagara Falls, N Y & Ontario, Canada. Left Genl Abrasive in 1975 to form Micro Chemical Industries Inc, the production & marketing or for OPAL BLACK rub ber filler & MiCRO-CRETE, a new acid-resistant cement. In 1979 elected to be VP-Commercial of Micro chemicals Parent the Edward C Levy Co Slag Recovery, Cement, Concrete and Asphalt are main project lines. *Society Aff:* AIChE, ASTM.

Broderick, James Richard
Business: 6234 Richmond Ave S105, Houston, TX 77057
Position: Principal. *Employer:* Pieratt Broderick Assocs Inc. *Born:* Apr 1935. Resident of Houston, Texas, married, 3 children. BS Arch engrg 1958, Univ of Texas in Austin. With Walter P Moore & Assocs Inc Houston Texas 1958-72, Assoc with major highrise project respon, Officer & Dir of Corporation. Since 1972, Principal in firm of Pieratt Broderick Assocs Inc, Struct Engrs, Houston Texas. Mbr CSI & Prof Affiliate Mbr AIA. Outside interests incl tennis, jogging, genealogy, investments.

Brodersen, Arthur J
Business: Box 1628, Station B, Nashville, TN 37235
Position: Prof of Electrical Engrg *Employer:* Vanderbilt Univ. *Education:* PhD/Elec Engr/Univ of CA Berkeley; MS/Elec Engr/Univ of CA Berkeley; BS/Elec Engr/Univ of CA Berkeley. *Born:* 8/31/39. Faculty mbr, Univ of FL 1966-1974 and Dir of the Microelectronics Lab; Vanderbilt Univ since 1974, Chrmn of Elec and Biomedical Eng 1976-1981 and Associate Dean of Engineering 1979-86; Nashville Sec Chrmn, IEEE 1978-79; Res and teaching interests in microelectronics, computer-aided design of integrated circuits, and application of AI to electronic circuits and training. *Society Aff:* IEEE, AAAI, ASEE.

Brodkey, Robert S
Business: 140 W 19th Ave, Columbus, OH 43210-1180
Position: Prof *Employer:* The Ohio State Univ *Education:* PhD/ChE/Univ of WI; MS/ChE/Univ of CA-Berkeley; BS/Chem/Univ of CA- Berkeley; AA/-/SF City Coll *Born:* 9/14/28 A native of CA. Worked for Standard Oil of NJ and have since been at OH State. Active in teaching and research with over 100 publications of technical papers and books. Research areas cover most of fluid mechanics with an emphasis on imaging techniques in turbulence and non-Newtonian flow. Have had active programs of joint research with Max Planck Inst for Flow Studies in Goettingen, West Germany. *Society Aff:* AIChE, ACS, APS, AAAS, SR.

Brodnax, Marlynn G
Business: 350 Jordan, Shreveport, LA 71101
Position: Partner. *Employer:* Shoemaker Colbert Brodnax. *Education:* BA/Chemistry/Rice Univ. *Born:* 2/12/23 Native Shreveport, LA. Worked for various consulting engrs and corps as a Jr Engr from 1955 through 1963. Partner with Frey-Shoemaker-Colbert-Brodnax, Architects-Engrs, 1964 to present. During the period from 1955 to the present, designed mech and elec systems for commercial, institutional, and some residential and industrial bldgs. Reg PE in LA, AR, and TX. Pres Ark-La-Tex Chapter Illuminating Engg Soc 1971. Hobbies: sailing Thistle one design, physical fitness prog, music and reading. *Society Aff:* ASHRAE, IES, NSPE.

Brodsky, Marc H
Business: T.J. Watson Research Center, P.O. Box 218, Yorktown Heights, NY 10598
Position: Program Director *Employer:* IBM *Education:* Ph.D/Physics/U. of Pennsylvania; MS/Physics/U. of Pennsylvania; BA/Physics/U. of Pennsylvania *Born:* 08/09/38 A solid state physicist, materials and device engineer and R & D manager. Currently Program Dir of IBM's Advanced Gas Technology Lab. Major research has been in the field of amorphous semiconductors. Edited two editions of the book *Amorphous Semiconductors* (Springer, 1979 & 1985). Past member of Exec Ct of the American Physical Society's Div of Condensed Matter Physics. Editorial Bds of Thin Solid Films and Applied Physics. *Society Aff:* IEEE, APS, AAAS

Brokaw, Charles H
Home: 2597 Mercedes Dr, Atlanta, GA 30345
Position: VP, Product Integrity *Employer:* Edwards Baking Company *Education:* BS/Agric. Chemistry/Univ of FL; -/Chemistry & Mgmt/Stevens Inst of Tech. *Born:* 11/17/19. in Orlando, FL. USAF 1946-52; Capt; awarded Bronze Star. Jr Res Chemist Nat Res Corp dev frozen orange juice; Food Technologist Genl Foods Corp; Dir of Qual Control Minute Maid Corp; Dir of Quality Assurance, Dir of Scientific Regulatory Affairs, Cola-Cola USA; Retired 1982. Fellow Mbr ASQC; VPres 1972-74, Pres 1975-76, Chmn/Bd 1976-77. Mbr Inst Food Technologists Chrmn Florida & Dixie Sections; Chmn, Intl. Relations Cttee. Pres Soc of Soft Drink Technologists; Phi Tau Sigma hon food sci frat; Assn of Food & Drug Officials; Regis Prof Quality engr CA; Who's Who in America; Lt Col USAF Res (Ret). *Society Aff:* ASQC, IFT, SSDT, AFDO, SBA, NCWW.

Bromberg, Robert
Business: 1 Space Pk, Bldg E-1, Redondo Beach, CA 90278
Position: V Pres-Res & Engrg. *Employer:* TRW Defense & Space Sys Group. *Born:* Aug 1921. PhD Univ of Calif L A; MS & BS Univ of Calif Berkeley. Mbr tech staff U C Berkeley, res in heat transfer & instrumentation 1943-46. Assoc Prof of Engrg & Asst Dir Inst of Tech Cooperation UCLA. Joined Ramo Wooldridge Corp (later TRW) 1954. Managed R&D in engrg mech, materials & chem engrg, dev of propulsion & energy conversion sys, & dev of software & info sys. V P 1962. Currently concerned with planning & dir of company res & assuring qual of tech activities throughout TRW Sys Group. Respon for liaison on tech matters with TRW Inc & outside orgs. Mbr UC Engrg Adv Council & Trustee of UCLA Foundation. Chmn Corp Mbr Comm 1973 AIAA. Fellow AIAA 1972; Engrg Alumnus of Yr 1959 UCLA.

Bronson, Harold R
Business: 4900 Augusta Ave, Richmond, VA 23230
Position: Mgr Loss Prevention. *Employer:* Utica Natl Insurance Co. *Education:* BS/Eco Mgt/Lycoming College. *Born:* Jul 1925. Cert Safety Prof (143) 1970. Native of Boonville, N Y. Served as Air Force Combat Crewman. Joined Utica Natl Insur Co (Middle Atlantic Regional Office) 1950 assigned Richmond serving industries with total safety & implementation of indus & fleet safety programs. Made Mgr of Field Force 1960. Veterans of Safety 1966, Bd of Dir, Vir Safety Assn 1967. V Pres Region IX (& Natl Bd of Dir 1973) 1000 Safety Professionals in Region 1974- . Bd of Dir Southern States Safety Conf 1976. Pres Central VA Safety Assoc 1977-78. Cert of Award Colonial Vir Safety Engrs 1971. Safety Prof. of Year 1979, Region IX ASSE. Enjoy swimming, golf & tennis. *Society Aff:* ASSE.

Bronzino, Joseph D
Home: 6 Wyngate, Simsbury, CT 06070
Position: Prof of Engg. *Employer:* Trinity College. *Education:* PhD/EE/Worcester Poly Inst; MS/EE/US Naval Postgrad School; BS/EE/WPI. *Born:* 9/29/37. Native of Brooklyn, NY. Worked for NY Telephone Co for two yrs then in 1964 taught at UNH in the elec engg dept. In 1967, received an NSF Sci Faculty Fellowship to study at the Worcester Fdn for Experimental Biology. Joined Trinity College in 1968. Appointed Dir of the joint Trinity College/Hartford Grad Ctr Biomedical engg Prog in 1969. Established greater Hartford Clinical Engg Internship in 1974. Appointed the Vernon Roosa Prof of Appl Sci (an endowed chair at Trinity College) in 1978. Author of texts "Technology For Patient Care" published by C V Mosby in

Bronzino, Joseph D (Continued)
1977 and "Computer Applications for Patient Care" to be published by Addison - Wesley Publishing, Co Jan 1982. "Biomedical Engineering & Instrumentation Basic Concepts and Affluations" PWS Publishers Jan 1986. Dir IEEE - GEMB 1975-1978, Chrmn ASEE - BMED 1979. VP Tech Actives 1984-85 - IEEE/EMBS Pres - IEEE/EMBS 1985-87. Active in town govt. *Society Aff:* IEEE/EMBS, AAMI, SN, AAAS, ASEE.

Brooker, Donald B
Home: 2314 Ridgemont, Columbia, MO 65201
Position: Prof Emeritus of Agri Engrg *Employer:* Univ of Mo-Columbia *Education:* MS/Agri Engg/Univ of MO; BS/Mech Engg/Univ of MO; BS/Agri Engg/Univ of MO. *Born:* Dec 1916. Native of Troy Grove, Ill. Grew up on Ill farm. Clerk supervisor with Missouri St Agri Stabilization & Conservation Serv prior to WW II. Army Air Corps 1942-45. Instr Purdue Univ 1949-51. Teaching & res in area of crop processing. Some 40 tech pubs. Sr author, 'Drying Cereal Grains' (textbook). Winner 2 ASAE Journal Paper Awards. Reg Engr. Honor soc incl Tau Beta Pi, Pi Tau Sigma, Sigma Xi, Phi Kappa Phi, Gamma Sigma Delta. Co-founder of Alpha Epsilon, Natl Agri Engrg Hon Soc. Fellow 1971 ASAE. Dir ASAE 1976-78. Life Fellow 1983 ASAE. *Society Aff:* ASAE, ASEE.

Brookes, Charles E
Business: 7379 Route 32, Columbia, MD 21056
Position: Sr VP/Dir of Spec Res Proj *Employer:* W R Grace & Co. *Education:* BS/CHE/Yale Univ. *Born:* Feb 1925. Native of Bernardsville N J. US Infantry 1942-45. With Sun Oil Co 1949-52. Joined Dewey & Almy Chem Co 1952, later acquired by W R Grace & Co. Var mkting & mfg assignments. Appointed V P Dewey & Almy 1962. Transferred to Davison Chem Div as Pres 1967. Corp V P 1973. Transferred to NY Hdquarters 1977 as Deputy Grp Exec, Industrial Chemicals Grp. Apptd Group Exec, Industrial Chemicals Group July 1, 1978 elected Sr VP 1978. Spec int: control of auto emission pollutants through catalyst tech, resulting in probably world's largest catalyst plant for purpose. Num civic activities with spec int in econ dev. Awards & hon soc: Tau Beta Pi, Sigma Xi, Yale Engrg School Award, AIChE Student Award. Active in sailing & tennis. *Society Aff:* ACS, CDA, AIChE.

Brookner, Eli
Home: 282 Marrett Rd, Lexington, MA 02173
Position: Cons Scientist. *Employer:* Raytheon Co. *Education:* Dr Sc/Elec Engg/Columbia Univ; MS/Elec Engg/Columbia Univ; BEE/Elec Engg/CCNY. *Born:* Apr 1931. Dr Sc & MS (Elec Engrg) 1962 & 1955 Columbia Univ; BEE CCNY 1953. Evaluated Armstrong FM over-the-horizon radar, summer 1952, while jr engr at RADC. Sys analy on Nike Zeus & long range def radars at Columbia Univ Elec Res Lab 1953-57, 1960-62. Signal processing, Fed Sci Corp 1957-60. Raytheon Co 1962-, presently Cons Sci-Waveform analy, propagation, target signatures, characterization of laser & mm communication channels, decoy discrimination, space radars. Designed Wake Measurement Radar. Tech support to Cobra Dane, MSR, Pave Paws, Aegis, Milirad, Sea Sparrow. Organizer & lecturer of radar courses for Raytheon, IEEE & Microwave Journal. Fellow IEEE, Assoc Fellow AIAA, TBP, EKN, URSI Comm B and C, Reg Prof Engr Mass, Franklin Inst Annual Journal Premium Award. *Society Aff:* IEEE, AIAA, URSI, TBΠ, HKN.

Brooks, Carson L
Home: 3029 E Heatherbrae Dr, Phoenix, AZ 85016
Position: Dir Res Tech & Applied Sci. *Employer:* Reynolds Metals Co. *Education:* BSc/Elecochemical Eng/MIT. *Born:* Aug 1913. 40 yrs met practice in the aluminum indus with Reynolds & Alcoa in R&D. Areas of spec incl aluminum melting, skim reclamation, scrap reclamation & recycling, molten metal qual, ingot casting incl electromagnetic, & energy conservation. Author book 1970 'Basic Principles of Aluminum Melting, Metal Preparation & Molten Metal Handling'. These fields, having many unresolved problems & need for major change, represent a challenge to younger indus metallurgists in whom this engr has always had a major interest. Enjoy sports, gardening. & youth activities. *Society Aff:* ASM, AIME.

Brooks, David W
Business: Dept of Chemistry, 227 Hamilton Hall, Lincoln, NB 68588
Position: Prof. *Employer:* Univ of NB. *Education:* PhD/Chemistry/Columbia Univ; MA/Chemistry/Columbia Univ; BA/Chemistry/NYU. *Born:* 2/22/42. David W Brooks is a chemistry teacher & supervises res in the area of chem education. *Society Aff:* ΦΛΥ, ACS, ΦΒΚ, NARST, AAAS.

Brooks, Frederick M
Business: 6942 Titian Ave, Baton Rouge, LA 70806
Position: Exec VP *Employer:* Barbay Engrs, Inc *Education:* BS/EE/LA State Univ *Born:* 11/17/43 Native of New Iberia, LA. Has been a principle of Barbay Engrs since 1968. As Operations Administrator of the firm is responsible for overall project management, cost control and resource allocation. Experienced in project engrg and project management for commercial, institutional, governmental and industrial projects, setting and implementing design procedures and standards. Intl experience includes project work in South and Central America and Africa. Registered PE in the State of LA. Particular expertise in forensic engrg relating to the electrical utility industry including extensive expert witness testimony. *Society Aff:* NSPE, LES, IEEE, PEPP, IAS

Brooks, Frederick P Jr
Business: Dept of Computer Sci, Sitterson Hall 082A, Chapel Hill, NC 27514
Position: Kenan Prof of Computer Sci. *Employer:* Univ of North Carolina-Chapel Hill. *Education:* PhD/Computer Science/Harvard; MS/Computer Science/Harvard; BS/Physics/Duke *Born:* Apr 1931. BA Duke. North Carolina native. Kenan Prof of Computer Sci Dept, Univ of N Car at Chapel Hill. With IBM Corp 1956-65: Mgr Dev of Sys/360, OS/360; an architect for IBM Stretch & Harvest Computers. Bd of Dir: Triangle Univs Computation Ctr; N Car Ed Computing Service. Natl Acad of Engrg. Fellow Amer Acad of Arts & Sci. Fellow IEEE. IEEE McDowell Award 1970. DPMA Computer Sci Man of Yr Award 1970. Guggenheim Fellow. Defense Science Board 1982-1986. National Medal of Technology 1985. National Science Board 1987. Latest work as author: The Mythical Man-Month. *Society Aff:* IEEE, ACM

Brooks, George H
Business: Dept of Indus Engrg, Auburn, AL 36849
Position: Prof Indus Engrg. *Employer:* Auburn Univ. *Education:* PhD/IE/GA Tech; MS/IE/GA Tech; B/IE/FL. *Born:* Apr 1919. 1979 to present, Prof Indus Engrg Auburn Univ; 1966 to 1979, Prof & Hd Indus Engrg Auburn Univ; 1964-66 Prof Indus & Sys Engrg Univ of Florida; 1959-64 Assoc Prof Indus Engrg Purdue; 1951-59 Cons & Cons Supervisor E I duPont de Nemours & Co. Cons: Expert Witness, Several edses IBM, Burlington Indus, US Navy, US Air Force, Ohio Univ, Univ of Missouri, Hospitals. Amer Inst of Indus Engrs Sr Mbr 1959, Fellow of the Inst 1972; V P Region VII, Mbr Bd of Trustees 1971-73. ECPD Curriculum Visitor 1968-78. Chmn, & all offices, Council of I E Academic Dept Hds 1967-71. Chmn 3 Inst Task Forces. Mbr ASEE, Alpha Pi Mu, Tau Beta Pi, Sigma Tau, Phi Kappa Phi, Sigma Xi, Phi Eta Sigma, Order of the Engr. Reg P E Ala. Who's Who in Amer, Amer Men of Sci, Contemporary Authors. NSF Sci Fac Fellow 1961-64. Military Service, 1941-46, 47, 50, Capt US Army. Lic Radio Amateur. *Society Aff:* IIE.

Brooks, Harvey
Home: 46 Brewster St, Cambridge, MA 02138
Position: Prof of Tech & Pub Pol. *Employer:* Harvard Univ 02138. *Education:* PhD/Physics/Harvard; AB/Mathematics/Yale. *Born:* Aug 5, 1915 Cleveland, Ohio. Ed Hawken School, the Hill School, Yale Univ, Cambridge Univ, Harvard Univ. Soc of Fellows, Harvard 1940-41; Harvard Underwater Sound Lab 1941-45; Ordnance Res Lab, Penn St 1945-46; Knolls Atomic Power Lab, Genl Elec 1946-50; Gordon McKay Prof of Applied Physics, Harvard 1950- ; Dean of Engrg & Applied Phys 1957-75. Mbr President's Sci Adv Ctte 1959- 64, Natl Sci Bd 1962-74; Chmn Ctte on Sci & Public Policy, Natl Acad of Sci 1969-74; Chmn Comm on Sociotech Sys NRC 1974-1979 ; Pres Amer Acad of Arts & Sci 1971-76. Mbr Natl Acad of Sci, Natl Acad of Engrg, Inst of Medicine, Amer Philosophical Soc. *Society Aff:* ANS, ASEE, APS, SHOT, AAAS.

Brooks, Henry S
Home: 201 Gibbon St, Alexandria, VA 22314
Position: Dir of Govt Relations *Employer:* VVKR Incorporated *Education:* BS/Engrg/NC State Univ *Born:* 06/28/11 Born Oxford, NC. Married-two children, thirty-nine yrs US Govt Public Works agencies and Public Works Planning Staff. The White House eleven yrs active duty US Army Corps of Engrs and USAF Civil Engrs-Foreign service, China, Burma, India and Europe. Mbr Alexandria Planning Commission, Architectural Bd of Review and Bd of Directors Alexandria Hospital. Chrmn, Bd of Education, Alexandria VA. Chrmn, Joint Committee on Federal Construction. Mbr, Bldg Secton Ctte American Natl Metric Council. *Society Aff:* NSPE

Brooks, M Scott
Business: Maplewood Ave, Bloomfield, CT 06002
Position: Mgr, Met Services. *Employer:* J M Ney Co. *Education:* BS/Met Engrg/VA Poly Inst. *Born:* May 6, 1929. Res Metallurgist NBS 3 yrs on metal fatigue & failure analy of commercial aircraft accidents. Joined J M Ney Co 1958. Performed res on noble metal alloys for dentistry & electromech devices. Author papers for dental & bio-medical applications. Presently Mgr, Met Serv, involved in analytical & mech control of proprietary alloys. Also involves customer & intra-company metallurgical trouble shooting. Mbr tech team that lectures nationwide on noble metals in electromech applications. ASM Natl Chap Adv Ctte, Chmn 1969, 1970; ASM Natl Chap & Mbrship Council, Chmn 1970-71; ASTM Cttes B.02, B.04, E.02, E. 04.05.01; ASM tech lecturer-2 chapters, 1 conf; ASM President's award 1967. *Society Aff:* ASM, ASTM, IADR.

Brooks, Norman H
Business: Keck Labs 138-78, Pasadena, CA 91125
Position: Dir Envir Quality Lab & Prof of Envir & Civ Engg; Executive Officer for Environmental Engrg Science. *Employer:* Caltech. *Education:* PhD/Civil Engg & Physics/Caltech; MS/Civil Engg/Harvard Univ; AB/Mathematics/Harvard College. *Born:* Jul 2, 1928, Worcester Mass; married, 3 children. Caltech fac since 1953, promoted to prof in 1962; organized new Environ Engrg Sci program 1969, & directed it until 1974, and 1985-present; Dir of Environ Qual Lab 1974-present; named James Irvine Prof of Environ & Civil Engg in 1976. Visiting assignments at SEATO Grad Sch of Engrg, Bangkok 1959-60, MIT 1962-63, Scripps Inst of Oceanography Fall 1971 & Swiss Fed Inst of Tech (E.T.H.), Zurich, 1984-85. Res on pollution control in the ocean, hydraulic transport processes, sedimentation, hydraulic engrg, & environ policy. Author over 50 tech papers & reports. Cons on outfall design for sewage effluents & thermal discharges. Has served on num adv cttes & bds dealing with environ issues (e g Environ Studies Bd, NAS/NAE 1973-76; Commission on Engineering and Technical Systems, NAS/NAE 1986-). Elected to NAE 1973. Elected to Natl Acad of Scis, 1981. *Society Aff:* ASCE, AAAS, AGU, WPCF, NAE, NAS.

Broom, Harry P
Home: Kirby Lane, Rye, NY 10580
Position: Independent Consultant. *Education:* BS/Chem Eng/Univ of PA. *Born:* Aug 1911 Camden N J. Diploma, Phillips Exeter Acad. B.S. ChE. Univ of Penna. Mgr Dev Lab, Houdry Process Corp 1936-40. Participated dev first commercial catalytic cracking process which produced high octane aviation gasoline needed WW II. Contract Engr E B Badger & Sons Co 194052. 1952-78, Ralph M Parsons Co (world-wide engrg, const serv in chem-petrochem, gas processing, mining-met, power sys areas). Bus dev, admin respon; charge NY oper. Exec V P 1971-78. Headed 1961 formation, dev Parsons-Jurden Corp (subsidiary) serving mining-metallurgical indus. Pres, Dir this unit . Presently Independent Consultant. *Society Aff:* AIChE, AIME, AAAS.

Broome, Taft H
Business: Dept of Civil Engrg, 2300 6th Str, N.W, Washington, DC 20059
Position: Dir, Large Space Structures Institute *Employer:* Howard University *Education:* Sc.D./CE/George Washington Univ; M.S.E./CE/George Washington Univ; B.S./C. E./Howard Univ; MS/Sci/Tech. Studies/Rensselaer Polytechnic Institute *Born:* in Washington, DC Dr. Broome has been teaching at Howard Univ since 1972 and has practical experience as a consultant in finite element analysis and as a fellow with the Nuclear Regulatory Commission and the Natl Science Foundation. He has research experience with NASA and has received over $15M in research grant support in the structural analysis of large space structures. He is Chrmn of the Ethics Ctte of AAES. *Society Aff:* ASCE, ΣΞ, TBΠ, AAAS, SES

Brophy, James J
Business: 210 Park Bldg, Salt Lake City, UT 84112
Position: VP for Research *Employer:* Univ of Utah *Education:* PhD/Physics/IL Inst of Tech; MS/Physics/IL Inst of Tech; BS/EE/IL Inst of Tech *Born:* 6/6/26 Native of Chicago. Research and research administration at IIT Research Inst, 1951-1967. Research interests are solid state physics and electrical fluctuations. Books include Semiconductor Devices (McGraw-Hill), Basic Electronics for Scientists (McGraw-Hill), now in Fifth Edition, Electronic Processes in Materials (McGraw-Hill), Organic Semiconductors (ed) (Macmillan), and contributor to Fluctuation Phenomena in Solids (Academic Press). Academic VP, IL Inst of Technology, 1967-1976. Senior VP, Inst of Gas Technology, 1976- 1980, responsible for US and intl education programs. VP for Research, Univ of UT since 1980, responsible for total academic research program. Private pilot, skier, scuba diver. *Society Aff:* AAAS, APS.

Brophy, Jere H
Business: One NY Plaza, New York, NY 10004
Position: Director, Adv. Technology Initiation *Employer:* INCO, Ltd. *Education:* PhD/Metallurgy Engg/Univ of MI; MS/Metallurgy Engg/Univ of MI; BS/Metallurgy/Univ of MI; BS/Chem Engg/Univ of MI. *Born:* Mar 11, 1934. Asst Prof MIT 1958-63; 1963- , Inco, Paul D Merica Res Lab - var pos - Sect Supervisor-Nickel Alloy Section; Res Mgr - Non-Ferrous Group; Asst Mgr & Mgr Res Lab. Mbr Bd of Dir IMD Div of AIME 1973-76; Mbr R&D Council AMA 1975- ; Mbr AIME & ASM; AAAS; Elected Fellow ASM 1975, Elected Fellow AAAS 1979. Commodore Nyack Boat Club. Co-author *Thermodynamics of Structure* pub by Wiley. Author 30 tech papers. Director, Advanced Technology Initiation. *Society Aff:* ASM,AIME, AAAS

Broselow, Stanley D
Home: 2643 W 232nd St, Torrance, CA 90505
Position: Procurement Manager *Employer:* Hughes Aircraft Co. *Education:* BS/EE/Drexel Univ. *Born:* 8/3/25. Native of Southern NJ. Construction mgmt engr for Army Corps Engrs for 20 yrs-15 for the Baltimore District on civil works and military construction and 5 for the Ballistic Missile Construction office at Los Angeles and Norton AFB. Served as Renegotiator, Dir of Renegotiating and last 4 yrs chrmn of the Western Renegotiation Bd. Retired from US Govt in March 1979 and now with Hughes Aircraft Co. Respon for acquistion of mechanisms for guidance and control systems of geosynchronous communications satellites. *Society Aff:* IEEE, ASCE, SAME, NCMA.

Brosilow, Rosalie
Business: 614 Superior Ave. W, Cleveland, OH 44113
Position: Editor *Employer:* Welding Design & Fabrication *Education:* B.S./Metallurgical Eng/Univ of MI *Born:* 7/7/39 Editor of Welding Design & Fabrication, a national magazine that covers the metals welding and fabricating industry. Specializes in robotic arc welding; thermal cutting of metals; welding for the power and process industries. Was chairman of 1981 national conference on Robotic Arc Welding, sponsored by AWS. Mbr ASM Metals Joining Council. Mbr AWS Committee on Educational Conferences and Seminars. Previously, research metallurgist with Electric Storage Battery Co., Philadelphia. *Society Aff:* AWS, ASM, ASNT, AAAS

Brosz, Donald J
Business: Univ Station-Box 3354, Laramie, WY 82071
Position: Prof *Employer:* Univ of Wyoming. *Education:* BS/Agri Engg/SD State Univ; MS/Agri Engg/SD State Univ. *Born:* Jan 12, 1933; Reg Engr 642 Wyoming. S Dak St Univ 1955-60, Res & Extension; Kansas State Univ 1960-62, Extension Irrigation Engr; Univ of Wyoming 1963- , Extension Irrigation Engr; cons to USAID Team, Mogadiscio, Somalia, Africa 1965; Secy 1964 & Chmn 1972 Rocky Mountain Sect ASAE; Blue Ribbon Award, Ed Exhibit 1971 ASAE; Outstanding Serv Award, Land Improvement Contractors of Amer 1972; Distinguished Serv Award, Upper Missouri Water Users 1973; Outstanding Extension Serv Award Alpha Zeta 1976; Governor's Water Agency Reorg Study Ctte 1976, 4 states Heaogate Award, 1979; Water Use and Mgmt Natl Task Force, 1978-79; ASAE Regional Council, 1981-82. *Society Aff:* ASAE, IA, ESP, SCSA, NWRA, LICA.

Brothers, James T
Home: 54 Sylvian Way, Los Altos, CA 94022
Position: Retired. *Employer:* Aeronutronic-Ford-ret. *Education:* BS/WA Square College/NY Univ. *Born:* Apr 1908. Undergrad physics WSC-NYU; Grad work Columbia, CCNY, U of Penn, U of Detroit. Employed as engr by Polymet, Pilot, A H Grebe, RCA Patent Dept, Philco 1933-73, Div Engr in charge of Component & Material Div Phila, Staff to Dir of Res (semiconducters) Lansdale, Staff to Dir Product Assurance (aerospace) W D L Palo Alto. Cons to DOD 1939-61, mbr AGREE, Chmn of RDB Panel of EM Devices, Chmn AGEP Panel on Capacitors, mbr Task Group 5 (Darnell Com) U S Natl Delegate to I E C-Tech Comm 12 Radio Communication, active in E I A & I E E E cttes. Life Fellow I E E E, received 1964 Radio Fall Mtg Plaque from E I A. Retired 1973, enjoy Cons electronic devices, financial analy of investments, music & gardening. Propagating. Rhododendron, Azalea, Fuschia. Mbr Amer Fuchsia Sci Mbr Morris Arboretum- See American Soc Metals - Author Permanent Magnet Mat, ASM Manual. *Society Aff:* IEEE.

Brotzen, Franz R
Business: Rice University - MEMS, P.O. Box 1892, Houston, TX 77251
Position: Professor, Materials Sci. *Employer:* Wm Marsh Rice Univ. *Education:* PhD/Physical Metallurgy/Case Inst of Techn; MS/Physical Metallurgy/Case Inst of Techn; BSM/Met Engg/Case Inst of Techn. *Born:* Jul 4, 1915 Berlin; son of George & Lena (Paculiy) B; Materials Scientist; m. Frances B Ridgway Jan 31, 1950; c. Franz Ridgway, Julie Ridgway. With A Quimica Bayer Ltda 1934-40, J Magnus e Cia Ltda 1940-41, A G McKee & Co 1941-42, R G LeTourneau Inc 1947-48, U S Bur Mines 1950-51, Case Inst Tech 1951- 54; fac Dept Mech Engrg Rice Univ Houston 1954- , Prof 1959- , Dean Engrg 1962- 66. Vis Lectr Univ of Brazil 1963, 65. Guggenheim Fellow 1960-61. visiting Scientist Polytechnic Inst Zurich 1966-67; U S Sr Sci Award, Res Fellow Max- Planck Institute Stuttgart 1973-74. Piper Prof 1975. *Society Aff:* TMS-AIME, ASM (Fellow), APS, ASEE

Broughton, Donald B
Home: 1639 Hinman Ave, Evanston, IL 60201
Position: Mgr Separation Processes. *Employer:* Univ Oil Products, Process Div. *Education:* ScD/Chem Eng/MIT; MS/Chem Eng/MIT; BS/Chem Eng/PA State College. *Born:* 1917 Rugby England. Emigrated to USA 1924; naturalized 1934. BS & MS Penna State College. ScD MIT 1943. Asst Prof Chem Engrg MIT 1943-49. Indus cons 1944-49. With Universal Oil Products Co since 1949, in dev of new processes for separation of chemicals (Process Dev Dept). Main areas liquid-liquid extraction & adsorption. Mbr Amer Inst Chem Engrs; Mbr Amer Petro Inst, Amer Chem Soc, Natl Acad of Engrg. Alpha Chi Sigma award in Chem Engrg res from same 1967. Amateur pianist, classical. *Society Aff:* AIChE, ACS, API, NAE.

Broutman, Lawrence J
Business: 3424 S. State, Chicago, IL 60616
Position: Prof. *Employer:* IL Inst of Tech. *Education:* ScD/Mtls Engg/MIT; SM/Mtls Engg/MIT; SB/Civ Engg/MIT. *Born:* 2/9/38. Native of Chicago, IL. Sr Res Engr at IL Inst of Tech Res Inst from 1963 to 1966. Assoc Prof of Mechanics, IL Inst of Tech, 1966 to 1970; Prof of Mtls Engg in Dept of Metallurgical and Mtls Engg and Dir, Mechanics of Mtls Lab at IIT, 1970 to present. Pres, Soc of Plastics Engrs, 1977-1978. Co-editor of Modern Composite Mtls, 1967; Co-editor, Composite Mtls (8 volumes) 1974; Co-author of Analysis and Performance of Fiber Composites, 1980; published 130 technical papers; holds 4 patents. Consultant to several industrial corps and govt agencies a Pres of L J Broutman and Assoc, Ltd. *Society Aff:* SPE, SEM, ASTM.

Broward, Hoyt E
Home: 4270 Hemlock La, Titusville, FL 32780
Position: VP *Employer:* Reynolds, Smith and Hills *Education:* BS/ME/GA Inst of Tech *Born:* 2/4/16 Native of Jacksonville, FL. Engr in Steam Turbine Dept, Allis-Chalmers Mfg Co, Milwaukee, prior to World War II. In US Army Air Corps 1942-45. With Reynolds, Smith and Hills, Architects-Engrs-Planners, Inc since October 1946. Specialized in steam electric generation 1946-1962 as Design Engr and Project Mgr. Chrmn, NE FL Section of ASME in 1958 and 1959. Involved in NASA's Saturn V - Apollo Moon Rocket Program 1962-69, and from 1972-81 in the Space Shuttle Program for the design of ground support facilities. Promoted to VP - Merritt Island Office in 1977. *Society Aff:* ASME

Browder, Cecil L
Home: 16607 SE 14th St, Bellevue, WA 98008
Position: Cons Engineer. *Employer:* Cecil L Browder & Assocs. *Education:* BS/Arch Eng/Univ of Kansas. *Born:* Nov 1922; Lic Civil & Struct Engr Ariz & Wash; NSPE/WSPE Pres 1975-76; NSPE Task Force on Minority Engrg Employment & Participation 1973-74; NSPE/WSPE State Guidance Chmn; EJC Engrg Schools Visitation Program 1975; Northwest Regional Amer Arbitration Assn ; Seattle/Puget Sound Engrg Council Engr of Yr 1972; Visiting Lectr Wash State Univ Engrg Sch 1975; Order of The Engr 1971; Earthquake Engrg Research Inst. 1985; Amer Soc of Civil Engrs-Fellow; Struct Engrs Assn of Wash; Amer Concrete Inst; Phi Beta Epsilon, Beta Chapter Univ of Kansas Nov. 1986. Proj Engr Seattle-Tacoma Internatl Airport Parking Terminal & Tunnels for People Mover; WA. State-Dept Transportation I 90 Hywy 1984-1988. Enjoy golf, fishing, civic activities. *Society Aff:* NSPE, ASCE, SEAW, EERI, ACI, WSPE.

Brower, Ralph T
Business: Mapleton, IL 61547
Position: Retired *Education:* BS/Metallurgy/Purdue Univ *Born:* 12/27/24 Graduate of Purdue Univ in Metallurgical Engrg. Employed by Caterpillar Tractor Co since 1950. Held technical, supervisory, and management positions including Foundry Mgr, Manufacturing Mgr. Have plant responsibility for foundry tooling design and build, quality control, process engrg and control, manufacturing and materials development, purchasing, and metallics purchasing and sales. Have served as chapter chrmn of AFS, trustee of Foundry Education Foundation, chrmn of the Electric Furnance Div of the Iron & Steel Society and dir of ISS. Have had papers published in AFS transactions, Modern Castings and Foundry Magazines, and the Iron & Steelmaking magazine and the Iron & Steel Society transactions. *Society Aff:* AFS, ISS, SAE

Brown, A Thomas
Home: 311 Buckingham Place, Clarksville, TN 37040 *Employer:* US Army. *Education:* BS/CE/Univ of AL. *Born:* 9/3/55. Very active in Theta Tau, a Natl PE Fraternity, both during college & after grad. Represented the local chapter at two natl conventions. Received the Outstanding Delegate Award which is the highest award that a student mbr can achieve, is one of seven alumni mbr elected to the Exec Council of Theta Tau Fraternity. Commissioned into the US Army Corps of Engrs upon grad. Spent 18 months with the AK Dist C of E designing & supervising construction of a $165 million flood control proj near Fairbanks. Presently assigned to the 20th Engr Battalion (Combat) at Ft Campbell, KY. Is the company commander of Headquarters and Headquarters Company. *Society Aff:* SAME, ΘT.

Brown, Ann T
Home: 901 Timberline Dr, Akron, OH 44313
Position: Research Librarian. *Employer:* The Timken Co-Canton Oh 44706. *Education:* MS-LS/Info Sci/Case Western Res Univ; MBA/Mgmt/Akron Univ; BS/Div of Nature Sci/Akron Univ. *Born:* Sep 1928. Also studied at Purdue. Addl studies in Bus Admin at Univ of Akron. Mbr Amer Soc for Metals; Chmn Metals Info Ctte. Mbr Amer Soc for Info Sci; Chmn Northern Ohio Chap. Mbr AAAS & ASTM. Listed in Who's Who in Amer 38 ed. Author sev papers on tech info. Coord & lect for prof dev seminars on info sys. Enjoy golf, tennis, photography & camping. *Society Aff:* ASIS, ASM, ASTM, ASAS.

Brown, Anthony C
Home: 17E Black Horse Pike, Williamstown, NJ 08094
Position: Sr VP. *Employer:* J J Henry Co. *Education:* -/EE/Drexel. *Born:* 6/22/18. Native of NJ. Thirty two years with NY ship, serving as VP of Engg from 1957 to 1967. Currently Sr VP of J J Henry Co. Author of numerous papers. Dir of several cos, & officer & committee mbr of technical socs. Numerous Engrg Awards, Fellow of SNAME. Holder of Prof Engrg License in numerous states. *Society Aff:* SNAME, ASNE, AIEE, AOA, PES.

Brown, Bevan W, Jr
Business: 140 Evans Bldg. c/o TVA, Knoxville, TN 37902
Position: Director of Air and Water Resources *Employer:* Tenn Valley Auth. *Education:* BS/Mech Engg/Clemson Univ; MS/Mech Engr/Clemson Univ; MS/Civil Engg/Stanford Univ. *Born:* in Starr SC. 1950 employed Tenn Valley Auth, Civil Engr in Flood Control Branch. Respon for hydrology needed to plan struct in TN Valley with particular reference to floods. 1974-79, Asst Dir of Water Mgmt. 1979-1982, Asst. Mgr of Natl Resources, 1982-date, Director of Air and Water Resources. 1972-75 served as ASCE Natl Dir of dist 6. 1967 received apptment to Pres's Mid-Career Program in systematic Analysis. Mbr Tau Beta Pi. Reg PE TN. *Society Aff:* ASCE; USCOLD

Brown, Bob Diggs
Business: P O Drawer 1431, Duncan, OK 73533
Position: V Pres-Sales & Advertising. *Employer:* Halliburton Services. *Education:* BS/ME/OK State Univ. *Born:* Jul 1, 1918. Since grad, employed by Halliburton Services. In 1969 became V P of Sales & Advertising. Reg Engr, Mbr SPE of AIME, API, OSPE & NSPE. Served as officer in local sections & natl cttes incl P Natl Dir of SPE. Distinguished lecturer SPE 1981-82. Past V Pres of South Central Chap OSPE and currently mbr of Bd of Trustees OSPE. Elected to hon mbrship to Pi Epsilon Tau by Dept of Petro Engrg, Marietta Coll & recog as Engr of Yr by South Central Section OSPE. Served on Indus Adv Bd to 5 colleges. Mbr of Exec Bd of IPE (Intl Petroleum Exposition). *Society Aff:* API, SPE, AIME, OSPE, NSPE.

Brown, Burton P
Business: 50 Brown St, Baldwinsville, NY 13027
Position: Sys Cons, Electronic Sys Div. *Employer:* General Electric Co. *Education:* MS/EE/Univ of VT; BS/EE/Univ of CO. *Born:* Dec 1917. Native of Denver Colo. Joined Genl Electric 1941. Engaged in dev of radar antennas, waveguide & transmission line circuits. Became mgr of advanced dev in 1952. Current spec: military radar sys & def sys. Asst Prof at Univ of Vermont 1946-47. Asst Dir of Def Res & Engrg, Office of Secy of Defense, respon for DOD res & dev activities pertaining to air & missile defense of Continental U S 1962-63. Since 1965, active on num military panels in DOD & the Exec Office of the Pres. Enjoy fishing, camping, & painting. Fellow IEEE 1967. Mbr NAE 1973. Chmn Electronics Panel 1966-73 Air Force Sci Adv Bd. Mbr 1968-73 Army Sci Adv Panel. *Society Aff:* IEEE, NAE.

Brown, C Alvie
Business: 1285 Boston Ave, Bridgeport, CT 06602
Position: Mgr Systems Engg. *Employer:* Genl Elec Co. *Education:* BS/EE/Purdue. *Born:* 12/2/18. Many different jobs all with The Genl Elec Co. Two yrs of lab work, nine yrs of advance electronic dev four yrs of sales engg, eight yrs mging design engg, three yrs of strategic planning, Fifteen yrs of managing sys engg. Citizen of the whole US, not just one location, having in eleven places during this career. Interests: the profession, the technology of energy, boats, music, working with my hands.

Brown, C Carter
Business: PO Box 711, Baton Rouge, LA 70821
Position: Sr Advisor and Conslt, Cons Engrs firm *Employer:* Brown & Butler, Inc., Cons Engrs *Education:* BS/Civil Engg/LA State Univ *Born:* 10/-/12 Natchitoches, Louisiana. Married Sylvia Ernestine Gass. Children: Barbara Anne, Carter Gass, Sylvia Ellen Brown Swinger. Observer USC & GS 1934-35; hydraulics engineer SCS 1935-41; design engineer, state planning engineer, dir Louisana Dept of Public Works 1941-1948. Lt. (j.g.) to Lt. USNR 1944-46. Partner Brown & Butler 1948-1970, owner 1970-1983. Sr advisor and conslt to successor firm. Pres Louisana Section ASCE 1971. Vice-president ACEC 1967. Director ACEC 1960-61. Committee of Fellows ACEC 1978-79. Fellow ACEC. President LA CEC 1961. Fellow ASCE. Honor member Chi Epsilon. Active in interest of engrg education and scholarship *Society Aff:* ACEC, ASCE, WPCF, WPCF, ASCE

Brown, Charles D
Business: Engrg Dept. P.O. Box 6090, Newark, DE 19317
Position: VP-Engrg *Employer:* E.I. Du Pont De Nemours & CO. *Education:* BS/Civil Engrg./University of Arkansas *Born:* 10/31/27 Native of Mineral Springs, Arkansas. Received bachelor's degree in civil engrg from the Univ of Arkansas and joined the engrg dept of E. I. du Pont de Nemours & Company in 1951. Held numerous assignments in the project environment, including construction, financial, and design organizations of the corporate engrg dept. Assumed current position as VP, Engrg in 1982. Chairman of the task force for The Business Roundtable's Construction Industry Cost Effectiveness Program 1978-81, and of the Construction Cttee, 1981-83. Received Engineering-News Record "Man of the Year" award in 1983 and Construction Industry Institute's "Award of Excellence" in 1986. Avid golfer.

Brown, Charles M
Business: 4625 Royal Ave-Box 579, Niagara Falls, NY 14302
Position: Mgr, Specialty Market Dev. *Employer:* Union Carbide Corp, Metals Div. *Education:* BS/Engg/Univ of MI. *Born:* Nov 23, 1919; public school in Lewiston N Y ASM Howe Award 1950. First specialization was in austenitic stainless steels, & later with metals such as vanadium, titanium, tantalum, columbium & tungsten. Dev alloying agents for these metals & for aluminum, superalloys, low alloy high strength steel & tool steel. Pubs & patents in these areas. *Society Aff:* ASM, AIME, ASTM, ADPA.

Brown, Charles T
Home: 8219 E. Lippizan Trail, Scottsdale, AZ 85258
Position: Retired. *Education:* BS/ChE/TX A&M Univ *Born:* 6/9/24. Native of Shreveport, LA. US Army 1943-46. Various Tech, operating and mgmt positions in petro refining and petrochemical operations of Cities Service Co at Lake Charles Operations 1947-68 including mgr of Tech Ser. Coordinator of Mfg, NY, 1969-71. General Mgr of AZ Copper Mining operations 1972-76. VP metals 1977- 82. Appointed Exec VP Pinto Valley Copper Corp in 1983. Active in tech and industry org including AZ Mining Assn and AMer Mining Congress. Retired 1986. Married to June Ann Roeger, Children: Laura Street, Stephen, Paul. *Society Aff:* AIME.

Brown, Daniel M
Business: Box 1600, Station B, Nashville, TN 37235
Position: Assoc Prof Civil Engrg. *Employer:* Vanderbilt Univ. *Education:* PhD/Civil Engg/Univ of IL; MS/Civil Engg/Univ of IL; BSCE/Civil Engg/Duke Univ. *Born:* 5/26/39. ; J A Jones Scholar (Duke Univ) 1957-61; grad with dept distinction. NDEA Title IV Fellow 1961-64; ASCE O H Ammann Res Fellowship 1964-65. Teaching & Res Asst Univ of IL 1961-65. Asst Prof Civil Engrg Univ of IL 1965-69. Assoc Prof Civil Engrg Vanderbilt Univ 1969- ; Prog Dir 1975-76, respon for

Brown, Daniel M (Continued)
var courses in struct analy & design, sys engg, surveying, urban dev & transportation. Res in struct optimization, struct reliability, guyed tower analy & design, urban dev & transportation. Commissioner, Nashville Metro Transit Auth 1975-76. Mbr Nashville Univ Club; Bd of Dir 1972-75, Pres 1973-74. Mbr Percy Priest Yacht Club; Mbr Phi Eta Sigma, Tau Beta Pi, Chi Epsilon, Sigma Xi. Mbr ASCE. P E Tenn, Minn, Kans. *Society Aff:* ASCE, XE, TBΠ, ΦΗΣ.

Brown, David
Home: 003 Andros Rd, Ocean Reef Club, Key Largo, FL 33037
Position: Retired *Education:* MSc/Chem Eng/MIT; MS/Chem Eng/MIT; BA/Math Chem/Swarthmore College *Born:* 04/25/17 Native of Rochelle, NY. Engr Std Oil Calif 1941, MW Kellogg Co 1942-44, US Navy 1944-46, Shell Dev Co 1946-52; Scientific Design Co/Halcon 1952-81 (Vice Chrmn). Nat'l Acad of Engrg 1978. *Society Aff:* AIChE, ACS

Brown, David R
Home: 1470 Sand Hill Rd., Apt. 309, Palo Alto, CA 94304
Position: Dir, Advanced Computer Sys Dept. *Employer:* SRI International. *Education:* BS/EE/Univ of WA; SM/EE/MA Inst of Tech. *Born:* Oct 1923. Worked on Whirlwind computer at MIT beginning in 1946. At MIT Lincoln Lab, led group that dev the ferrite memory core, & TX-O & TX-2 computers. Was head of SAGE sys office at the MITRE Corp 1958-63. Dir of Info Sci Lab at SRI since 1963. Fellow IEEE. U S Co-chmn of 2nd USA-Japan Computer Conf in 1975. *Society Aff:* IEEE, ACM.

Brown, Delmar L
Home: 3128 SE Woodstock Blvd, Portland, OR 97202
Position: Owner. *Employer:* Northwest Engrg Labs. *Education:* BS/Mech Engg/OR State Univ. *Born:* 3/28/09. Served with Portland Genl Elec Co (utility) as engr, chemist & supt of Testing 1931 to retirement 1974. Avocation interests in sound & photography led to inventions in automatic hi-speed photo registry of fish, & sys of hwy sound tracks, U S Pat 2,851,539. Fellow IEEE. Pres Prof Engrs of Oregon 1944. Pres Valley Telephone Co (Ore) 1964-66. Chmn of Sunnyside Telephone Co (successor) 1966-69. Active as owner of Northwest Engrg Labs, product designers. *Society Aff:* IEEE, NSPE.

Brown, Donald C
Business: 906 Creekdale, Richardson, TX 75080
Position: Pres *Employer:* Don C. Brown and Assoc. *Education:* B.S./Chem. Eng./University of Texas at Austin *Born:* 12/05/24 Worked in Sun Oil engrg and mgmt assignments in exploration & production 1947-84; Area Engr 1950; District Engr 1952; Dir of Field Operations 1967; Mgr Compensation & Benefits 1970; Chief Operations Engr 1974. Retired 1984 and organized consulting firm. Member of SPE of AIME since 1953; among offices held Chmn East Texas Section of SPE 1963; Chmn of SPE Engrg Manpower Cttee 1981, and Ex Officio member since. Commissioner of Engineering Manpower Commission of AAES since 1984; Chmn of EMC Compensation Cttee since 1985; VChmn of EMC 1987. Reg P.E.,TX. *Society Aff:* SPE OF AIME

Brown, Elgar P
Business: 515 Pittsfield Dr, Worthington, OH 43085
Position: Principal/Owner *Employer:* Elgar Brown, Consulting Engrs *Education:* B/CE/OH State Univ *Born:* 10/2/18 Belfast, OH. Graduated from OSU 1945 BCE. Ang Construction Co 1947-1964. Principal, Elgar Brown, Consulting Engrs 1964 to present. *Society Aff:* ASCE, NSPE, ACI, AISC, OSPE

Brown, F Leslie
Home: 4824 Stonewall Ave, Downers Grove, IL 60515
Position: Dir of Engrg. *Employer:* Mechanical Contractors Assn. *Education:* B.S.M.E./Mechanical Engineering/Univ of Illinois *Born:* Mar 16, 1920; BS ME Univ of Illinois; P E Ill, etc; Natl Dir NSPE 1974-77; P V Pres ISPE 3 terms; P Pres Chgo Chap ISPE; P Pres Ill Architects--Engrs Council; P Mbr Ill Gov's Adv Council; P Pres Ill Chap ASHRAE; Mbr Ill School Code Adv Bd; Prof Affiliate Mbr AIA; Recog Bldg & Constr Code Expert; Recipient Distinguished Service Award ASHRAE; Distinguished Service Citation, Ill Governor's Adv Council; Secy Manpower Qualifications Comm 1966-72. Fellow ASHRAE; Mbr. NSPE; Mbr. ISPE; Mbr Chicago Energy Policy Council, 1978- . Mbr. Chicago Code Advisory Comm.; Honorary Life Mbr.-IL Archts-Engrs. Council Mbr. CACI Code Advisory Comm.;P Chm ISPE Bldg. & Constr. Code Comm.; Mbr. Church Building Comm.; Hobbies: Photography, painting, woodcarving, writing. *Society Aff:* NSPE, ASHRAE, ISPE, CACI, AIA.

Brown, Frederick R
Home: 105 Stonewall Rd, Vicksburg, MS 39180
Position: Retired. *Education:* BS/Civil Eng/U of Ilinois *Born:* Feb 15, 1912 Peoria, Ill. Employed by Waterways Exper Station 1934- ; Chief, Hydrodynamics Branch 1945-63; Ch, Nuclear Weapons Effects Div 1963-64; Asst Tech Dir & Ch, Tech Programs & Plans 1964-69; Tech Dir 1969- . Recipient of Dept of Army Meritorious Civilian Service Award 1947 & 1968 & Exceptional Civilian Service Award 1973. Dept of Defense Distinguished Service Award 1979. Acting Pres, Mid-South Sect ASCE 1956-57; Exec Ctte Hydraulics Div ASCE 1963-65; Pres Miss Sect ASCE 1970-71; Natl Dir ASCE 1974-76. Natl VP 1976-78 Hon Mbr ASCE Chmn Vicksburg Planning Comm 1960-78. *Society Aff:* ASCE, NSPE, IAHR, SAME, PIANC

Brown, Garry L
Business: Pasadena, CA 91125
Position: Prof *Employer:* Caltech *Education:* D Phil/Engrg Sci/Oxford; BE/Engrg/Univ of Adelaide *Born:* 5/14/42 An Australian Rhodes Scholar in 1964. Research Fellow in Aeronautics at Caltech 1967-71. Univ of Adelaide, Australia 1971-78. Prof of Aeronautics at Caltech 1978-. Fields of research include turbulence, combustion and noise. Consultant to a wide range of aerospace companies and government research labs. South Australian Engrg Award 1975. *Society Aff:* AIAA, AAAS

Brown, George H
Home: 117 Hunt Dr, Princeton, NJ 08540
Position: Exec V Pres-ret. *Employer:* RCA Corp-ret. *Education:* PhD/EE/U of WIS; MSC/EE/U of WIS; BS/EE/U of WIS *Born:* Oct 1908. Natl Acad of Engrg. Edison Medal, DeForest Audion Award. Eminent Mbr Eta Kappa Nu. Silver Beaver, Silver Antelope, BSA. Modern Pioneer Award, Natl Assn of Manufacturers. Cert of Appreciation, War Dept. Distinguished Service Citation, Univ of Wisc. Citation 4th Interntl Television Symposium, Montreux. Shoenberg Mem Lectr, Royal Inst. Fellow IEEE, AAAS, Royal Television Soc. Bd of Dir, The Trane Co. 1967-79, First Natl Bank of Hamilton Sq 1969-78, Hamilton Hospital 1965-77. Radio Corp of Amer 1933-73. Bd of Dir RCA 1965-73. 80 U S patents, Award for Outstanding Achievement in Radio and Television, University of Arizona, 1980; IEEE Centennial Medal, 1984; Dr of Eng. (hon) Univ of RI, 1968. 1986 Engineering Achievement Award, National Association of Broadcasters. Armstrong Medal, Radio Club of America. Author, autobiography, "and part of which I was–. *Society Aff:* ΣΞ, IEEE, RTV, NAE, RCA

Brown, Glenn R
Home: 32250 Burlwood Dr, Solon, OH 44139
Position: Sr VP *Employer:* Standard Oil. *Education:* PhD/Chem Engg/Case Inst; MS/Chem Engg/Case Inst; BS/Chem Engg/Penn State. *Born:* 6/1/30. Dr. Brown joined Sohio as an engr in 1953 in the Chem & Phy Res Dept. After 10 yrs of res exper, he undertook assignment in managerial roles in operations res, mgmt sys, and chem operations. In 1970, Dr. Brown became mgr of all res activities for Sohio, and on July 1, 1975 was elected VP, Res & Engrg. In early 1978, he assumed direction of Corp. Planning in addition to his res and engrg activities. In December 1979, he was elected Sr. VP. His present duties include resp for tech and strategic planning activities. His title is Sr. VP, Tech and Planning, and he is a mbr of the Sohio Bd of

Brown, Glenn R (Continued)
Dir. He is also a Dir of the Centran Corp. and Central Natl Bank. *Society Aff:* AIChE, ACS, SCI, SAE.

Brown, Gordon M
Home: 3191 Bluett, Ann Arbor, MI 48105
Position: Prin Research Engr. *Employer:* Sci Res Labs, Ford Motor Co. *Education:* MSNE/Nuclear Eng/Univ of MI; BSME/Mech Eng/Gen Mtrs Inst. *Born:* 2/17/34. Resident of Ann Arbor, MI. Fisher Body, GMC, Tooling Proj Enttr. Bendix Aerospace Sys, Proj Engr spec in nuclear & electro-opital instrumentation design. Dir of Engg for GCO, R&D of holographic nondestructive test methods & equipment; invented holographic method & engg equip for testing tires; Dir of Mfg. Prin Res Engr, Res Labs, Ford Motor Co since 1973; design & dev of electro- optical test equip for qual & process control. Mbr OSA, ASNT, & ASNT Achievement Award 1970. Enjoy investments, photography & fishing. *Society Aff:* OSA, ASNT.

Brown, Gordon S
PO Box 272, Grantham, NH 03753
Position: Inst Professor Emeritus. *Employer:* MIT. *Education:* ScD/Elec Engg/MA Inst Tech; SM/Elec Engg/MA Inst Tech; SB/Elec Engg/MA INst Tech; Dr Eng (Hon) /-/Purdue; DSe(hon)/-/Dartmouth; DrEng (hon)/-/Stevens Inst.; Dr Eng (hon)/-/Tech Univ Denmark.; Dr Eng (hon)/-/South Methodist *Born:* Aug 1907 Drummoyne, New South Wales, Australia. Diploma Mech & Elec Eng Royal Melbourne Tech College 1926. Came to MIT as student 1929 & received SB in 1931, SM in 1934 & ScD in 1938. Became mbr of fac in 1939. In 1952 appointed Hd of Dept of Elec Engrg & embarked on major program of ed innovation. In 1959 appointed Dean of Sch of Engrg. Dugald C. Jackson Prof 1968, Institute Professor 1973. Has been a key figure in estab of Ctr for Materials Sci & Engrg, the Ctr for Adv Engrg Study, the Info Processing Services Ctr & Proj Intrex. Received the President's Cert of Merit, the Naval Ordnance Dev Award, the Geo Westinghouse Award & the Lamme Medal, both from ASEE, the Medal in Elec Engrg Ed from AIEE, & the Joseph Marie Jacquard Annual Mem Award from the Numerical Control Soc. Rufus Oldenburger Medal from Automatic Control Div of ASME. Mbr of NAE, ASEE, Sigma Xi, Tau Beta Pi, & Eta Kappa Nu, & is a Fellow of AAAS & IEEE. The Gordon Stanley Brown Building dedicated at MIT. Dec 6, 1985. *Society Aff:* NAE, IEEE, AAAS, ASEE

Brown, Harold
Home: 1015 33rd St. N.W, Washington, DC 20007
Position: Consultant & Corporate Director *Employer:* Self-employed *Education:* PhD/Physics/Columbia Univ; AM/Physics/Columbia Univ; AB/ Physics/Columbia Univ. *Born:* Sep 1927. PhD, AM, AB Columbia Univ. Became Pres of Caltech 1969. Before joining Caltech, was Secy of the Air Force. During 1947-48, was a lectr in physics on the Columbia staff, in 1949-50 at the Stevens Inst of Tech, & in 1951-52 at the Univ of Calif at Berkeley. In 1952 joined the Radiation Lab staff of the Univ of Calif, moving from Group Leader to Dir during the period 1952-61. An appt by Pres Kennedy in 1961 placed Dr Brown in the highest Tech pos in the Dept of Def when he was made Dir of Defense Res & Engrg. As a mbr of the US delegation to the Strategic Arms Limitation Talks, he has participated in SALT I, & SALT II. Award for Exceptionally Merit Serv 1969 DOD. Exceptional Civilian Serv Award 1969 AF. Appt Secretary of Defense Jan 1977. The Harold Brown Award for Outstanding Achievement in R&D 1969 AF. *Society Aff:* NAS, NAE, AAAS, APS.

Brown, Harry L
Home: 1106 Winding Dr, Cherry Hill, NJ 08003
Position: Prof Mech Engr *Employer:* Drexel Univ. *Education:* PhD/Nuclear Engr/Stanford Univ; MS/Engrg Sci/Stanford Univ; BS/EE/Drexel Univ *Born:* 3/27/36 Native of NJ, completing graduate studies at Stanford Univ. Served two years in Corps of Engrg at Atomic Energy Commission working on design of new reactor cores and safety of Army nuclear reactors. Worked additional year and one half of Fast Breeder Reactor Core, pressure vessel and safety on FFTF and commercial systems. Tenured prof at Drexel Univ, Dir of Energy Inst at Drexel since 1970. Initiated major industrial process energy information system for Dept of Energy. Pres of General Energy Assocs which has developed major Industrial Plant energy Profile (IPEP) Data Base for over 300,000 industrial plants in the country. *Society Aff:* ANS, ASME

Brown, Jack H U
Business: 8848 W Commerce St, San Antonio, TX 78284
Position: Professor of Biology *Employer:* Univ of Houston. *Born:* 1918 Nixon Texas. BS Southwest Texas State Univ 1936, PhD Rutgers 1948. Engr for Army AF 1940-43; Dir Biol Res, Mellon Inst 1948-50; Medical School Prof, Emory, Univ of N Carolina, Geo Washington 1950-60; Chmn Emory 1959; NIH & HSMHA, U S Gov 1960-73; Dir NIGMS 1969; Assoc Deputy Admin HSMHA 1972; Coord S W R Consortium 1974- . Assoc Provost, Univ of Houston, Prof of Biology 1978- Mbr Natl Acad of Engrg; Fellow IEEE, AAAS. Pres Biomed Engrg Soc 1966; Natl Secy GEMB 1963 Ad Hoc PSAC Cttes, Sigma Xi Res Award 1958. Gerard Swope Award 1948; NASA Special Award, 1976 Prof, Environ Sci Univ of Texas 1972.

Brown, James Andrew
Home: 3800 Pine Rd, Portsmouth, VA 23703
Position: Rear Admiral-ret. *Employer:* USN. *Education:* BS/EE/US Naval Academy; MS/Naval Construction/MIT Cambridge Mass *Born:* Aug 19, 1914 Columbia Tenn. From Ensign USN 1936 to Rear Admiral 1963. Retired from Navy 1970. Formerly V Pres J L Smith Corp-Genl Contractor. Mbr NAME, Soc N Engrs, U S Naval Inst, CSI; V P Soc Naval Arch & Mar Engrs; Mbr Council, Chmn Chesapeake & Hampden Rds Section; Mbr Council Amer Soc Naval Engrs; Mbr Sigma Xi 1952. Formerly Mgr and Chief Engr Hampton Office, CDI Marine Co. Now Sr Engineer Q.E.D. Systems Inc. Commissioner Tidewater Regional Transit. Chairman Bd Trustees Supplemental Retirement Sys and Fire & Police Retirement Sys, City of Portsmouth, Member Bd Dir. Portsmouth Chamber of Commerce, Chairman Highways and Mass Transit Committee. Mbr Trans Ctte, Hampton Roads Chamber of Commerce, Mbr Navy League, Naval Inst Proceeding. *Society Aff:* SNA&ME, ASNE, N Inst Proc

Brown, James G
Business: 2505 W Holcombe Blvd, Houston, TX 77030
Position: Pres. *Employer:* James G Brown & Assoc Inc. *Education:* BS/Mech Engr/TX A & M. *Born:* 11/16/17. Grad TX A & M 1938, 4 yrs with irrigating co adding increased capacity, one yr with airplane mfg in plant layout dept, 4 yrs in repair of aircraft in US Navy, 7 yrs with eng construction cos in responsible charge of design of paper mills, chem plants and refineries, and 25 yrs in consulting practice. Over the past 5 yrs, the consulting firm has developed and patented a new concept for a combined offshore drilling and production platform. In addition, a sinkable form for sand has been developed as an underwater raised level floor to increase the operating water depth using slotted barges. *Society Aff:* NSPE.

Brown, Jesse E
Home: 631 Drexel Ave, Glencoe, IL 60022
Position: Retired. . *Born:* Sep 1902. Native of E Marion, N Y. Studied Elec Engrg Cornell Univ. Engr with Fed Communications Comm & its predecessor 1924-36. With Zenith Radio Corp television & radio engrg 1936-40 & Ch Engr 1940-50. Asst V P 1950-58. V P Engrg & Res 1958-72. Chmn of var cttes of Radio Tech Planning Bd & Natl Television Sys Ctte. Fellow Mbr of the Inst of Elec & Electronic Engrs. Mbr of Dir Inst of Radio Engrs in 1937 & 1939. Owner & operator Cottage Plantation, St Francisville, La. *Society Aff:* IEEE.

Brown, John Webster
Business: 1135 Terminal Way, Suite 108, Reno, NV 89502
Position: Pres. *Employer:* John Webster Brown-Civil & Structural *Education:* BS/Civil Engr/Univ of NV., Reno. *Born:* 7/17/26. Reno, NV, US Navy Aircrewmen WWII, Bridge Design Experience NV Hgwy Dept, Industrial Bldg Design Experi-

Brown, John Webster (Continued)
ence Kaiser Engrs, Founded his civil and structural engg firm in 1953, past Pres NV Branch ASCE, Past Pres NV Society of Professional Engrs, Past Dir Natl Society of Professional Engrs, NV Society of Professional Engrs 1969 Engr of the yr, NV Chap AGC 1974 SIR Award, past mbr Selective Service Appeal Bd for NV, Authority on Snow Loads on Structures. *Society Aff:* ASCE, NSPE, WPCF.

Brown, Joseph L, Jr
Business: 521 34th Street, Gulfport, MS 39507
Position: Owner *Employer:* Brown Engrs, Inc. *Education:* BS/CE/MS St Coll *Born:* 10/26/25 Native of Hattiesburg, MS. Served with U. S. Air Force 1944-45 and 1951-52. With American Oil Co. 1951-63 as Ms Division Engr and Senior Staff Engr, Chicago, IL. Principal in several consulting businesses on the Mississippi Gulf Coast since 1963. Major projects include Harrison County Sand Beach Restoration Project, Popps Ferry Bridge Project and City of Gulfport Water and Sewer Improvement Program. Pres, MS Engrg Society 1972-73; Dir, NSPE, 1973-75; MES Engr of the Year Award 1976; Editor, MS Engr 1973-80; Pres, MS Consulting Engrs Council 1981. Enjoy boating and fishing. *Society Aff:* NSPE, ASCE

Brown, Kermit E
Business: 600 S College Ave, Tulsa, OK 74104
Position: Prof. *Employer:* Univ of Tulsa-Dept of Petroleum Eng. *Education:* PhD/Petroleum/Univ of Texas; MS/Petroleum/Univ of Texas; BS/Petroleum/Texas M&M Univ; BS/ME/Texas A&M Univ *Born:* Nov 2, 1923. BS Petro Engrg & Mech Engrg from Texas A&M College, & MS & PhD Petro Engrg from Univ of Texas. Taught on Petro Engrg fac at Texas Univ. At Univ of Tulsa has served as Hd of Petro Engrg Dept; V P for Res; & Assoc Dean of College of Engrg & Phys Sci. Was nominated Disting Lectr, SPE of AIME 1969-70; Outstanding Educator in America 1970; Fulbright Lecturer Ecuador 1970; received Teaching Excellence Award for College of Engrg & Phys Sci Univ of Tulsa 1975. Has conducted schools on gas lift in most oil areas of the world. Rec'd John Franklin Caroll Award through SPE 1983; Selected Distinguished Mbr SPE 1983. *Society Aff:* SPE, ASEE

Brown, Leonard N
Home: 108 Munson Dr, Syracuse, NY 13205
Position: Mgr Engg & Sales. *Employer:* Air Custom Systems, Inc. *Education:* BS/ME/Syracuse Univ. *Born:* 10/31/09. Native Syracuse, NY. 36 yrs Carrier Corp in Contract, Application & Dev Engg (1938-1974). Specialist in Low Temperature Refrig & Heat Pump Systems, 10- 450 tons (1949-1958). Minute Man Silo air conditioning (1962-63). Author of technical papers on air conditioning. Retired Carrier 1974 as Mgr of Dev Engg with responsibility for Product Dev. ASHRAE Fellow, Life Mbr, Past Chapter Pres & recipient Distinguished Service Award. Author, Revisor, Editor or Chrmn 34 times for chapters & sections ASHRAE Handbook. Served on several ASHRAE & ARI natl engg committees. Currently Mgr Systems Sales & Engg for Air Custom Systems, Inc, Syracuse, NY. Listed in "Who's Who in Technology Today–. *Society Aff:* ASHRAE.

Brown, Leonard S
Business: 50 Briar Hollow, Suite 200 W, Houston, TX 77027
Position: Pres. *Employer:* Mercer-Brown Engrs, Inc. *Education:* MBA/Mgt/Univ of Houston; BS/Civ Engg/TX A & M Univ. *Born:* 10/11/32. PE with career emphasis on civ engg design and construction projs. Now involved in bus mgt and admin as Pres of Mercer-Brown Engrs, Inc, formed in 1975 to provide engg services for land dev projs. Active in consulting engg in Houston area since 1968. Earlier served 12 yrs with TX Hgwy Dept as Supervising Resident Engr in Buffalo, TX. Outstanding Young Engr, 1967, Brazos Chap. TSPE Lt Col (Retired), US Army Reserve, Corps of Engrs. Reg Public Surveyor. Active in Church of Christ. Interested in outdoors sports: hunting, fishing, scuba diving. *Society Aff:* ASCE, NSPE, ACEC.

Brown, Louis J
Home: 101 Briar Lane, Newark, DE 19711
Position: Specialist Counselor. *Employer:* Univ of Delaware. *Education:* ME/Mech/NY Univ; BS/Civil/NY Univ. *Born:* Aug 1906 Brooklyn N Y. Engr Consolidated Edison Co N Y 1928-42 testing, start-up & improvements to operation. Participated in dev new method for incrementally loading boiler & turbines. With duPont Co 1942-71. As Cons Mgr, provided specialized engrg consultation, company wide for design, installation & oper of power facil. Instrumental in estab energy conservation program made available to all indus. Specialist Counselor Univ of Delaware Tech Services. Chmn ASME Power Div Exec Ctte 1969-70. Mbr ASME Power Dept Policy Bd 1970-71. Chmn ASME-IEEE Joint Power Generation Conf Sponsors Ctte 1969. Reg P E NY & Delaware. Elected Fellow ASME 1971. *Society Aff:* ASME.

Brown, Marshall J
Home: 307 Monahan Drive, Fort Walton Beach, FL 32548
Position: Chief, Range Dev Div. *Employer:* USAF, Eglin AF Base, Florida. *Born:* Aug 18, 1922. BSEE Univ of Maryland; MBA, FSU; Air Force 1942-46 in electronic countermeasures testing. Began Civil Serv career as test engr on AF bomb-nav sys at Eglin AF Base 1950. Since 1956 held var pos in dev of range instrumentation. Currently respon for this program with emphasis on ground facil which simulate threat radar & missile characteristics. Reg in Florida. Current Regional Dir, Assn of Old Crows (Electronic Countermeasures). Enjoy sailing & duplicate bridge.

Brown, Martin M
Business: 40 Midchester Ave, White Plains, NY 10606
Position: Consultant. *Employer:* Self employed. *Education:* BChE/Chem Engr/Cooper Union, NY. *Born:* 1/17/16. Fire protection eng consultant, Investigator and Expert witness with office and residence at 40 Midchester Ave, White Plains, NY. BChE from Cooper Union 1939, PE in NY and CA. Former pres SFPE, and dir of engg for American Int'l (insurance) Group, Inc. Instructor at NY Univ and Mercy College. *Society Aff:* SFPE.

Brown, Norman
Business: 3231 Walnut St, Philadelphia, PA 19104
Position: Prof of Matls Scie & Engg. *Employer:* Univ of Penna - LRSM Bldg. *Education:* PhD/Metallurgy/Univ of CA Berkeley; MS/Metallurgy/Stanford; BS/Metallurgy/MIT. *Born:* 2/7/21. 1942-46 Metallurgist at Naval Torpedo Sta, AF Officer, Infantry, Ordnance 32nd Div. Ballistics Lab, Aberdeen Proving Ground 1946-48. Prof Univ of Penna since 1952. Dept Matls Sci and Engg and Dept of Orthopedee Surgery. Areas of teaching & res, "Mech Behavior of Solids–, "Phase Transformations–, & "Polymer Fracture–, Biomaterials, cartilage and bones. Guggenheim Fellow at Cavendish Lab 1958-59. NIH Fellow & Visiting Prof in Physics at Bristol Univ 1966-67. Fellow AM. Phy. Soc. Engaged in restoration & preservations of an historic site. Wife, Dr Miriam Jones Brown (noted educator). *Society Aff:* ASIM, APS, ACS.

Brown, Paul J
Home: 10611 Glenwild Rd, Silver Spring, MD 20901
Position: Dir-Elect & Hybrid Veh Div. *Employer:* US Dept of Energy. *Education:* MEA/-/WA Univ, St Louis; BSME/-/WA Univ, St Louis. *Born:* 4/18/24. Since 1977 resp for the Dept of Energys program for the res, dev and demonstration of electric and hybrid vehicles. Established the National Lab for Auto Safety and served as it dir from 1967-76 in the National Bureau of Standards and the US Dept of Transportation. Prior to Government Svc, served 16 yrs in private industry as Design Engg, Project Mgr and Production Mgr in the aerospace and defense industry. *Society Aff:* SAE.

Brown, Ralph E
Business: P. O. Box 98008, Atlanta, GA 30359
Position: VP for Engr & Research *Employer:* Law Engrg Testing Co *Education:* Ph.D./CE/Carnegie-Mellon Univ; MS/CE/Carnegie Inst of Tech; BS/CE/Duke Univ *Born:* 8/12/43 Native of Durham, NC. Has worked for Law Engrg Testing Co since 1968. Served as Senior Engr in Birmingham, AL office; Engrg Dept Mgr and Chief Engr in Washington, DC; and is currently Dir of Engrg and Research, a VP and a member of the Bd of Directors in Atlanta, GA corporate headquarters. He is responsible for developing and administering co engrg policies and programs, quality assurance activities and engrg computer programs. A member of Phi Beta Kappa and Tau Beta Phi, his principal interests are boating and photography. *Society Aff:* ASTM, ASCE, ASFE

Brown, Richard L
Business: P O Box 405, Glennville, GA 30427
Position: V Pres & Genl Mgr of Mfg. *Employer:* Duramatic Prod, div of Rotary Mower. *Born:* Jul 1948; BIE (Co-op Program) Georgia Tech; Native of Lyons, Ga. Joined Foster Wheeler Corp as Constr Engr upon graduation. Accepted present pos as Gen Mgr of Mfg in 1972 involved in Capital Equip budgeting & specification, implementation of all manufacturing processes, long range corporate planning. Enjoy camping with wife & 2 children as well as involvement with church-related activities.

Brown, Richard M
Business: Loomis Lab of Physics, Urbana, IL 61801
Position: Prof of EE & Physics. *Employer:* Univ of IL. *Education:* PhD/Physics/Harvard Univ; MS/Physics/Harvard Univ; AB/Electronic Physics/Harvard Univ. *Born:* 5/17/24. *Society Aff:* IEEE, ACM, APS.

Brown, Robert A
Home: 3315 O'Hara Rd, Huntsville, AL 35801
Position: Prof *Employer:* Univ of AL in Huntsville *Education:* PhD/IE/OH State Univ; MSc/IE/OH State Univ; BS/Gen/US Naval Academy *Born:* 9/25/27 Following service in the US Navy during the Korean conflict, began his teaching career while a graduate student at The OH State Univ. Developed a consulting specialty in Systems Engrg and Management. Moved to the new engrg program at The Univ of AL in Huntsville during the development of the Lunar Lander. Serves UAH as Industrial and Systems Engrg Dept Chrmn. Also served as chrmn of the Huntsville Section, IEEE, and on its Bd for a number of years. Most recently a member, Engrg Manpower Committee, ASEE. *Society Aff:* IIE, ASEE, IEEE, ORSA.

Brown, Robert O
Business: 6885 Washington Ave South, Suite 200, Edina, MN 55435
Position: President. *Employer:* Robert O Brown Co. *Education:* BChemE/Chem Engg/Univ of MN. *Born:* Jul 1917. With Pillsbury Co from 1939-56. With Robert O Brown since 1956, assumed Presidency 1972. *Society Aff:* ASHRAE, AICHE, IFT, ACEC, AACC, NYAS.

Brown, Robert W
Business: 45 Manor Rd, Smithtown, NY 11787
Position: Associate *Employer:* Sidney B Bowne & Son. *Born:* 10/7/41. With Sidney B Bowne & Son since 1961 establishing the firm's Eastern Long Island branch office in 1967. Licensed in NY, PA & CA & responsible for proj design & construction for drainage, parking, recreation, and hgwy projs representing local, state, & federal agencies. Assoc partner, Sidney B. Bowne & Son since 1980. Natl Chrmn ASCE Cttee on Engrg Surveying Treasurer of LI Branch of CEC. *Society Aff:* ASCE, NSPE, ASAME, CEC.

Brown, Ronald D
Business: Stanley Consultants, 225 Iowa Ave, Muscatine, IA 52761
Position: VP *Employer:* Stanley Consultants *Education:* BS/EE/Univ of IA *Born:* 12/10/36 Native of Mount Pleasant, IA. Employed by Stanley Consultants for 23 yrs beginning in 1960. Served as Sys Engr for Western IL Power Cooperative 1964-1968. Was appointed Chief Electrical engr for Stanley Conslts in 1975. Since 1977, has held position of VP. Is registered PE in IA and ten other states. Was a mbr of IA State Bd of Engrg Examiners for nine yrs beginning in 1973. Past mbr of Engrg Accreditation Commission of the Accreditation Bd for Engrg and Tech. Is past IEEE Section Chrmn and received Engr of the Year award, Quint Cities Council for Natl Engrs Week in 1980 and IEEE Centennial Medal in 1984. Received NCEE Central Zone Distinguished Service Award in 1984. Past Pres of Muscatine Chapter of Iowa Engrg Soc. Member of Univ of IA College of Engrg Industry Advisory Bd. *Society Aff:* IEEE, NSPE, ASEE

Brown, Roy F
Business: 2500 Harbor Blvd, Fullerton, CA 92634
Position: V P-Mgr, Process Instru & Controls. *Employer:* Beckman Instruments Inc. *Education:* BSMe/Mech Eng/NJ Inst of Tech. *Born:* Sep 1922 East Orange NJ. Employed at Westinghouse prior to WW II. USAF 1942-46. Field engr for Honeywell Indus Div. Joined Beckman Instruments Inc in 1956 as Eastern Regional Mgr, Process Div. Presently Corp V P & Mgr of Process Instruments & Controls Group. Respon incl integrated world-wide oper of company in fields of Process Analyzers, Pollution Monitoring Instrumentation, Automotive Test Instrumentation, & Process Control Instrumentation. 1976-78 Pres, Process Measurement & Control group, Sci Apparatus Makers Assn. Sr Mbr ISA, Mbr ASME, Lic P E NY & CA. *Society Aff:* ASMe, ISA.

Brown, Samuel J
Business: 7500 North Belt Dr, Houston, TX 77396
Position: President/Consultant *Employer:* Quest Engineering Development *Education:* PhD/Civil/Mechanics/Univ. of Akron Ohio; M.S.E./Mechanics/Univ. of Florida; B.S.M.E./Mechanical/Univ. of Southwestern La *Born:* 05/06/41 Consultant/Designer on numerous govt (NASA, Navy, DOE, Sandia, Langley, NRC) and industry (Process, Power, Petrochemical, Manufacturing, Commercial Building, and Equipment) sponsored projects. Author of over 40 papers, 200 reports, and editor/author of 20 technical volumes in the field of fluid-structure dynamics, mechanics, Hazardous Release Protection, and Failure Analysis. ASME fellow, P.E., ASME/PVP technical paper award, and ASME Code & Standards Subcttee Chrmn. A lecturer (visiting Ohio State, Univ Wis, and Akron Univ) and Professional Development Founder of Quest Engrg Devel Corp, (a D & A and R & D applied mechanics consulting firm) and co - founder of Intertech Corp, a full engineering services company. *Society Aff:* ASME, ASCE, ASM, SEM, SME, SIGΞ, PTI

Brown, Samuel P
Home: 32 Cayuga Way, Short Hills, NJ 07078
Position: Dir & Mbr of Exec Ctte of Bd/Dir. (Self-employed) *Education:* BS/Engg Admin/MA Inst of Tech. *Born:* 5/16/13. Native of Maplewood NJ; m. Helen M Cook of Maplewood Oct 1934; remained happily married to her until her death in 1976; 2 children, Donald K & Joan C (Mrs J A Winston); 7 grandchildren and 10 step-grandchildren. Married (2nd) Feb 77 to Harriet Runcie Hartz who died April 1978; married (3rd) to Natalie Earl Humphrey Jan '79. BS 1935 MIT. After 6 yrs (3 jobs) diverse work, more business than engrg, served Oct 1941-Nov 1945 in US Army Air Force at Wright Field. Awarded Legion of Merit. Dec 1945 until retirement May 1975 with COVERDALE & COLPITTS Cons Engrs NYC; became partner 1952, Sr Partner 1969, & Chmn of Coverdale & Colpitts Inc 1970. P Pres 1965 Amer Inst of Cons Engrs. Presently Dir & Mbr of Exec Ctte of Brooklyn Union Gas Co, Fuel Resources Inc & Dir of Kaiser Steel Corp. Sr Mbr of The Conference Bd. Reg PE-NY, NJ, CA & PA. Mbr ASME since 1950. *Society Aff:* ASME, ACEC.

Brown, Seymour W
Home: 866 United Nations Plaza, New York, NY 10017
Position: Pres *Employer:* Seymour W. Brown & Associates *Education:* BME/-/City College of NY. *Born:* 11/1/14. A chief engr Carrier Corp 1945-54. Founded cons engrg firm 1955, merged into Michael Baker Jr Inc 1969, elected V P-Dir, Pres-Dir 1971. Chrmn of the Bd 1980; Pres, Seymour W. Brown & Assoc 1982 Distinguished Service Award 1963; Fellow ASHRAE 1967. Life Mbr ASHRAE 1980; Fellow SAME 1976; Fellow NY Acad of Scis 1976. Author-Assoc Ed ASHRAE Guide &

Brown, Seymour W (Continued)
Data Book Applications. Inventor- pioneer, high velocity air distribution & heat pump cycle. Representative Intl Inst of Refrigeration, NAS. Advisor to Adm Rickover, environ control sys for first nuclear power submarine. Pres Engrg & Arch Alumni 1967-69; Bd Dir, City College Fund; 50th Anniv Award; Career Achievement Award 1980; Sch of Engrg CCNY, Contrib to Indus & Commerce. 1973 Mayor, NYC, presented Medal Commemorating 125 Yrs of Urban Higher Ed. Named Fellow ACEC 1973. Dir & Pres NYC Post SAME 1976. Mbr Newcomen Soc of N A. Engrg Excell: 1967 CEC & 1970-1-2-3-4- 5-8 1980 NYACE; 1974 ACEC Honor Award. NY Chap ASHRAE Outstanding Design Award 1972; PE NY, NJ, PA, CT. Bd/Gov & Chamn. Ad Hoc Cttee on Energy of NY Bldg Congress; Mbr Advisory Council, Natl Energy Fdn; Mbr Dean's Advisory Council, Univ of Miami. Mbr Mayor Koch's Cttee on Cogeneration; Cert Energy Mgr, Assn of Energy Engrs 1982. *Society Aff:* ASHRAE, SAME, ACEC, ASME, NSPE, SNAME, ASNE, ASPE.

Brown, Timothy J
Home: RR1, Box 438, Saunderstown, RI 02874
Position: Project Engr *Employer:* CE Maguire Inc *Education:* BS/CE/Univ of RI *Born:* 1/16/53 Native of Warwick, RI. Have been working for the past 6 years as a project Engr for CE Maguire Inc and now acting Project Mgr for all Wastewater projects within the City of Warwick, RI. Currently Pres of the RI section ASCE and contact member for the student chapter at the Univ of RI. *Society Aff:* ASCE, AWWA, WPCF, ΦΚΦ, TB PI

Brown, Wayne S
Business: Coll of Engrg, Salt Lake City, UT 84112
Position: Prof, Mech Engg & Dir UT Innovation Center. *Employer:* Univ of Utah. *Education:* PhD/Mech Engg/Stanford Univ; MS/Mech Engg/Univ of TN; BS/Mech Engg/Univ of UT. *Born:* 3/19/28. Dev Engr Oak Ridge Natl Lab 1951-53. Asst Prof Mech Engg, Univ of UT 1953- 57, Acting Asst Prof Stanford Univ 1957-59. Asst Genl Mgr & Dir of Engg UT Res & Dev Co 1960-64. Mbr US NAtl Ctte for Rock Mechanics, Natl Acad of Sci 1971-75. Fellow Amer Council on Ed, Academic Admin Intern Program 1968-69. Univ of UT: Prof & Chmn Mech Engg 1964 to present; Co-Dir Biomedical Engg Inst 1970-72; Assoc Dean Coll of Engg 1971-73; Dean College of Engg 1973-78. Cons Terra Tek Inc; Res: thermodynamics; heat transfer; nuclear engg, rock mech, dental engg. Chrmn UT Innovation Center 1978-present; Chrmn of Bd of Dirs: Terra Tek Inc, NPI Fellow Amer Soc of Mech Engrs, Mbr UT Advisory Council for Sci and Tech. *Society Aff:* ASME, SESA.

Brown, William F, Jr
Business: 21000 Brookpark Rd, Cleveland, OH 44135
Position: Distinguished Res Assoc *Employer:* NASA-Lewis Res Ctr. *Education:* MSc/Met Engg/Case Inst of Tech. *Born:* 7/23/22. Res Fellow Case 1945-48. 1948 to present, NASA-Lewis; Distinguished Res Assoc. 90 tech papers & 3 books on fracture. Founder ASTM E-24 Ctte on Fracture Testing of Metals. Regional Editor, *Intl Journal of Fracture*. Chief Tech Ed, Aerospace Struct Metals Handbook. Sigma Xi. Lincoln Welding Fdn Award 1948. ASTM Dudley Medal 1961. NASA Medal for Exceptional Scientific Achievement 1969. Fellow ASTM 1970. Hon Mbr Exec Ctte ASTM E-24 Ctte 1975. Hon Mbr ASTM 1976. NASA Award for Technical Contributions to the Apollo Program 1981. ASTM E-24 Fracture Comm Founders Award 1984. ASME Nadai Award for Fracture Research. *Society Aff:* ASTM, ASME.

Brown, William Fuller, Jr
Business: Dept of Elec Engrg, 123 Church St, SE, Minneapolis, MN 55455
Position: Professor Emeritus. *Employer:* Univ of Minnesota. *Education:* PhD/Physics/Columbia Univ; BA/English/Cornell. *Born:* Sep 1904. Taught at Columbia & Princeton Univ's 1928-41. With Naval Ordnance Lab 1941-45; Sun Oil Co 1946-55; 3M Co 1955-57. Prof of Elec Engrg Univ of Minnesota 1957-73; Prof Emeritus 1973- . Fulbright Scholar 1962. Fellow Amer Phys Soc, AAAS, IEEE, N Y Acad Sci. Winner A Cressy Morrison Award, N Y Acad Sci 1967; Hon Life Mbr Magnetics Soc of IEEE 1974. Res in ferromagnetism & related fields. *Society Aff:* APS, AAAS, IEEE, NYAS.

Brown, William H
Business: PO Box E-LSU, Baton Rouge, LA 70893
Position: Asst Dir & Prof *Employer:* LA Agri Experiment Station *Education:* PhD/Agri Engg/Univ of MO; MS/Agri Engg/Univ of MO; BS/Agri Engg/Univ of MO. *Born:* 5/22/42. Wm H Brown grew up in Sweet Springs, MO and attended local schools. He recieved BS, MS & PhD degrees in Agri Engg at the Univ of MO-Columbia. He served as Asst & Assoc Prof of Agri & Biological Engg at MS State Univ from 1968 to 1976. In 1976 he was appointed Hd of the Agri Engg Dept at LA State Univ where he was responsible for all personnel, fiscal and other adm matters of a major res and teaching prog. In Oct, 1983 he was appointed Asst Dir of the LA Agri Exper Sta. His responsibilities include contract adm and other mgmt aspects of a major agri res and devel prog. *Society Aff:* ASAE, ASEE, NSPE, AIBS.

Browne, David N
Home: 92 Rose Drive, Columbia, SC 29205
Position: Instructor. *Employer:* Midlands Technical Coll. *Education:* BS/Civil Engg/ Univ of SC. *Born:* Aug 1930; s. Lorenzo N & Christine M (Deaton) B; m. Nell Singley 1960; CE with U S Forest Service 1961; CE with City of Columbia 1962-63; CE with B P Barber & Assoc 1964-69; private practice 1970-72; instructor at Midlands Tech College 1972- ; PE & RLS in S C; Pres of S C Sect ASCE 1972; Pres of S C Council of Engrg Soc 1972; ACSM; ASEE; Mbr of S C Soc of Engrs; author of course in hwy design & constr for the S C State Bd for Tech Ed; home 92 Rose Drive, office Midlands Tech College, Columbia, S C. *Society Aff:* ASEE, ACSM.

Browne, John F X
Business: 525 Woodward Ave, Bloomfield Hills, MI 48013
Position: Pres *Employer:* John F. X. Browne & Assocs, P.C. *Education:* BS/Math Physics, E.E./Univ of Detroit *Born:* 12/5/35 Established consulting engineering practice in 1965. Specializing in broadcasting practice before Federal Communications Commission including radio/TV/microwave/CATV/satellite communications system design. Reg PE in 10 states and DC. Active in national society affairs including the Soc of Motion Picture and Television engrs, NSPE, the Assoc of Federal Communications Consulting Engineers and senior member of IEEE. President of AFCCE 1979-1980. Elected fellow of SMPTE 1979. Governor SMPTE 1982-1986. In addition to consulting practice serves as forensic engineer/expert witness/appraiser. Co-author Broadcasting Section Reference Data for Radio Engineers 7th ed. Avocations including flying and Arctic research/navigation (holds airline multiengine pilot license/pilot-in-command of around-the-world record flight 1984, transpolar record flight 1985). *Society Aff:* BKSTS, SMPTE, AFCCE, IEEE, NSPE, ESD, AAA, SBE, BP

Browne, Thomas E, Jr
Home: 5 Woodside Rd, Pittsburgh, PA 15221
Position: Cons Engr (retired) *Employer:* Self *Education:* PhD/EE/CA Inst Tech; MS/EE/Univ of Pittsburgh; BS/EE/NC State Univ. *Born:* Jun 1908 Murfreesboro N C. Employed by Westinghouse Elec Corp 1928-33 & 1936-73, Grad Asst Caltech 1933-36. Westinghouse res engr, sect mgr, & cons engr Power Circuit Breaker Div & Res Labs. Author or co-author 31 tech papers, articles and book on Circuit Interruption (Marcel Dekker, 1984), 21 patents incl as co-inventor basic patent on SF(6) circuit breaker. Reg P E Pa 1946. Elected Fellow AIEE 1958 for 'research contrib to rapid arc quenching for high-voltage circuit interrupters'. *Society Aff:* IEEE.

Browning, David Lee
Home: 6642 Aranda Ave, La Jolla, CA 92037
Position: Manager, Advanced Structures and Design *Employer:* Genl Dynamics, Space Systems Div. *Education:* BS/Civil Struct/San Diego State Univ. *Born:* Feb 1939. San Diego Calif. Grad studies at SDSU & UCSD in struct mech. Employed by GD since 1960 with assignments in: Atlas missile launch site activation; prod design & analysis of liquid propellant rocket tankage & interstage structs; preliminary design/analysis of advanced launch & entry vehicles, aircraft, & propulsive stages for use with the Space Shuttle; Program Management of automated space fabrication systems development; and management of all design & analysis departments involved in R&D of composite materials, applications, advanced structures & superconducting magnets. Mbr Chi Epsilon hon soc. Winner: SAWE Revere Cup 1976. AIAA (San Diego Section) Outstanding Contribution to Aerospace Engrg 1980. *Society Aff:* XE, AIAA.

Browning, Robert C
Business: 510 St Mary's St, Raleigh, NC 27605
Position: Consulting Engr (Self-employed) *Education:* BCE/CE/NC State Univ. *Born:* 10/31/19. Field engr with gen contractor, consulting engr, and USN (sanitary engr.) 1940-42. Officer US Army Corps of Engrs 1942-45. Assigned British Royal Engrs on construction of 12000 bed in hospitals in England. 1946-50 field and office engr WM C Olsen Consulting engr, Raleigh, structural and power plant piping. Opened private practice 1950. Consultant to hospitals, industry and institutions. Pres Prof Engrs of NC 1969-70. Pres, NC Assn of the Professions 1977-78. Pres Triangle Intl Trade Assoc 1972-73, First VP NC World Trade Assoc 1974-75. *Society Aff:* NSPE, IES, IHF, RSH, IABSE, IES, ASHRAE, NCWTA.

Brownlee, William R
Home: 780 Montgomery Dr, Birmingham, AL 35213
Position: Cons Engineer (Self-employed) *Employer:* Self *Education:* EE (Professional) 1954/-/University of Arizona; BSEE 1927/EE/University of Arizona *Born:* Dec 1, 1905 Cleveland, Tenn; son Colin C & Jessie Louisa (Hazelton) Brownlee; BS Univ of Arizona 1927, EE 1954. Married Elizabeth Cecil Marquet June 18, 1932; 1 daughter, Jane (Dr Jane Brownlee Hazelrig). Westinghouse 1927-28; Relay engr Tenn Elec Power Co 1929-39; Planning Engr, Commonwealth & Southern Corp, Jackson, Mich 1939-51; Southern Co Services, Birmingham, Ala 1951-70, Planning Engr, Power Pool Mgr, V P, Exec V P, Pres, Chmn. Dir Southern Company 1968-75. Mgr Southeastern Electric Reliability Council, Birmingham 1971-1986; Director Emeritus 1986- . Birmingham Engr of Yr 1968; Life Fellow IEEE-Outstanding Engr Region III 1970; Tau Beta Pi, Phi Kappa Phi, Eta Kappa Nu. *Society Aff:* IEEE, CIGRE.

Brownstein, Arthur M
Home: 432 George Pl, Wyckoff, NJ 07481
Position: Mgr - New Ventures Tech Div. *Employer:* Exxon Chem Co. *Education:* PhD/Chemistry/OH State Univ; BS/Chemistry/Rutgers Univ. *Born:* 10/30/33. Native of Trenton, NJ. Married Rosalie Silverman, 1958. Two children. Joined Allied Chem as sr res chemist eventually becoming Group Leader for process res. Became Proj Mgr in Commercial Dev for Sun Oil Co subsequently. Joined Chem Systems as Dir-Process Evaluation & Res Planning in 1972. Became VP in 1976. Assumed current responsibility as Mgr of the New Ventures Tech Div for Exxon Chem Co in 1978. Have full technical, administrative & budgetary responsibility for new petro-chem ventures. Adjunct Prof in chem engg at Univ of PA & City Univ of NY. Visiting Prof at Univ of Waterloo. Author of two books. *Society Aff:* AIChE.

Broxmeyer, Charles
Business: 400 7th St SW, Washington, DC 20590
Position: Deputy Assoc Admin for Mgmt and Demonstrations. *Employer:* Urban Mass Transportation Admin, DOT. *Education:* MS/EE/Univ of PA; BS/EE/Drexel Univ. *Born:* 4/17/23. US Army AF 1943-46. US Naval Air Dev Ctr 1949-55. MIT Instrumentation Lab (now Draper Lab) 1955-63. Raytheon Co 1963-64. Instrumentation Lab 1964-70. Design & analy of inertial navigation sys; num tech papers on inertial navigation & guidance; Author: Inertial Navigation Sys (McGraw-Hill 1964). With Urban Mass Transportation Admin since 1970. Respon for res & dev of new automated transportation sys, Policy dev, admin of tech asst program. DOT Meritorious Achievement Award 1972. *Society Aff:* IEEE, AIAA, ΣΞ.

Bruce, Albert W
Home: 1850 Alice St. Apt. 505, Oakland, CA 94612
Position: Supervising Mech Engr-ret. *Employer:* Pacific Gas & Elec Co-ret *Education:* BS/Mech Engg/Univ of IL. *Born:* Jul 1903. Native of Oak Park, Ill. Engrg employee Utilities Power & Light Corp, Chgo 1925-32. Officer Civilian Conservation Corps 1933-36. Engrg Dept Pacific Gas & Electric Co San Fran 1937-68 except for duty during WW as officer in Army Corps of Engrs 1941-45. From 1958-68 supervised the engrg of the geothermal power plants at The Geysers, Calif. Prof Mech Engr Calif. Life Fellow Amer Soc of Mech Engrs. Mbr Soc of Amer Military Engrs, Reserve Officers Assn, Pacific Coast Elec Assn. *Society Aff:* ASME, SAME, ROA, PCEA, TROA.

Bruce, Marvin H
Business: 1085 S Robert St, West St Paul, MN 55118
Position: President. *Employer:* Environmental Engrs Inc. *Education:* BEE/-/Univ of MN. *Born:* Jul 1, 1922. Army Air Forces 1942-45. Elec Engr Architectural-Engrg Firms 1952-64. V P Anderson-Bruce Engrs Inc 1965-66. Pres & Founder of Environ Engrs Inc 1967. *Society Aff:* IEEE, IES, NSPE, CEC.

Bruch, John C, Jr
Business: Dept of Mech & Environmental Engg, Santa Barbara, CA 93106
Position: Prof. *Employer:* Univ of CA. *Education:* PhD/Civ Engg/Stanford Univ; MS/Civ Engg/Stanford Univ; BS/Civ Engg/Univ of Notre Dame. *Born:* 10/11/40. Native of WI. Presently associated with the Dept of Mech and Environmental Engg, Univ of CA at Santa Barbara, Sant Barbara, CA. Started in 1966 and is now serving in the capacity of Prof. Author of numerous technical articles. Principle expertise involves Engg Mathematics, especially numerical analysis and methods of solution. *Society Aff:* ASCE, ΤΒΠ, ΣΞ.

Bruchner, Gordon W
Business: 760 Horizon Dr, Grand Junction, CO 81501
Position: VP *Employer:* ARIX *Education:* Geological Engr/Geol/CO School of Mines *Born:* 5/19/32 Mr Bruchner is a VP of ARIX and heads the management group for the Grand Junction office. He has been with ARIX 10 years and has over 15 years total experience as a Civil Engr. Mr Bruchner has had direct supervisory responsibility for the planning and design of many projects for municipalities in Western CO and WY. These projects include water and waste water treatment as well as streets and drainage. He also serves as City Engr for energy growth impacted communities and has gained an insight into area problems and mitigation measures associated with rapid growth. *Society Aff:* WPCF

Brudner, Harvey J
Home: 812 Abbott St, Highland Park, NJ 08904
Position: Pres. *Employer:* WLC, Incorporated. *Education:* PhD/Physics/NY Univ; MS/Physics/NY Univ; BS/EE/NYU Univ. *Born:* 5/29/31. A native of NY City, Dr Brudner is a Fellow of the IEEE for "leadership in the development and application of computers and electronic audio-visual systems in education and training." He was Dean of Science & Technology at NY Inst ofTechnology, 1962-64; he worked as a Principal Engr of Physicist with Bendix Corp, Emerson Radio, Amer Can Co, Power Authority of the State of New York and the US Naval Ordnance Lab. He was Pres of the Westinghouse Learning Corp (1971- 76). His hobbies are Mathematics and Art. *Society Aff:* IEEE, SMPTE, AAAS.

Bruene, Warren B
Home: 7805 Chattington Dr, Dallas, TX 75248
Position: Staff Engineer *Employer:* Electrospace Systems, Inc. *Education:* BS/EE/IA State Univ. *Born:* 11/1/16. Staff Engr Electrospace Systems Inc. Apr 1984 to present. Sr Engr Collins Radio Co, now part of Rockwell Internatl, Nov 1939-Apr 1984. Major contrib in HF single sideband communications. Conferred grade of Fellow IEEE 1961. Co-authored book 'Single Sideband Principles & Circuits', McGraw Hill

Bruene, Warren B (Continued)
1964. Authored chaps for 5 other handbooks & 14 tech papers. Granted 22 patents. IEEE Regional Dir 1962, Fellow Awards Ctte 1961-62, Awards Bd 1964. Engr of Yr Award 1975 from Preston Trail Chap of TSPE. P E Texas. Mbr ARRL & United Methodist Church *Society Aff:* IEEE, ARRL.

Bruggeman, Warren H
Business: 1 River Rd-Bldg 500-104, Schenectady, NY 12345
Position: Genl Manager. *Employer:* G E Co-Gas Turbine Engrg & Mfg Dept. *Born:* Dec 1925. BS & MS RPI. US Navy 1942-46. Reactor Design & Nuclear Site Mgr with Knolls Atomic Power Lab, Genl Elec Co. Genl Mgr, Mach Apparatus Oper, Genl Elec Co 1966-71. Current pos-Genl Mgr, Gas Turbine Engrg & Mfg Dept, G E Co. Fellow ASME, Fellow Amer Inst of Chemists, Mbr ANS.

Brugger, Robert M
Home: 1012 Bourn, Columbia, MO 65203
Position: Dir, Res Reactor. *Employer:* Univ of MO. *Education:* PhD/Nuclear Physics/Rice Univ; MA/Nuclear Physics/Rice Univ; BA/Physics/CO College. *Born:* 1/13/29. Native of Colorado Springs. Worked 19 yrs at ID Natl Engg Lab for Phillips Petroleum Co, ID Nuclear Co and Aerojet Nuclear Co. In 1974 became dir of the Univ of MO Res Reactor Facility. Fellow of the ANS and APS. *Society Aff:* ANS, APS.

Bruhn, Hjalmar D
Home: 5418 Lake Mendota Dr, Madison, WI 53705
Position: Prof Emeritus of Agri Engrg & Private Conslt in Agricultural Engrg *Employer:* Univ of Wisconsin, Madison. *Education:* Ms/Mech Engr/MA Inst Tech; BS/Mech Engr/Univ of WI; BS/Agri Engr/Univ of WI. *Born:* 8/5/07. Teaching, res Farm Power & Machinery. Dept Chmn 1962-66. Author, co-author 150 tech papers on res & dev of machinery & equip relating to production of protein from plant juices; alfalfa cubes (wafers); aquatic vegetation harvesting, utilization; self-propelled cherry harvesting; forage harvesting, conditioning, drying; crop irrigation & well-jetting procedures; tree planting. Fellow ASAE. Amer Forage & Grassland Council, Amer Assoc for Advancement of Sci, WI Acad of Sci, Arts, and Letters, Reg Prof Engr WI. Mbr Alpha Zeta, Alpha Epsilon, Pi Tau Sigma, Gamma Sigma Delta, Natl Society of Professional Engineers. WI Soc of Professional Engrs, Soc for Green Vegetation Res. *Society Aff:* ASAE, AAAS, NSPE, WASAL, WSPE.

Bruins, Paul F
Home: 708 Harris Ave, Austin, TX 78705
Position: Prof Emer, Chem Engrg. *Employer:* Polytechnic Univ. *Education:* PhD/Chem Eng/IA State Univ; MS/Chem Eng/IA State Univ; BS/Chemistry/Central College. *Born:* Dec 1905. Fellow AIChE. Lic Prof Engr NY. Educator in chem engrg since 1927, spec in plastics tech. Cons to many major corporations Author many tech papers, 30 patents, editor of 11 books on plastics. Supt of Sunday School 25 yrs, officer of Douglaston Community Church 32 yrs. Lic instrument pilot, 2800 hrs. Raised & educated 6 daughters. *Society Aff:* ΦΛΥ, ΤΒΠ, ΣΞ, ΩΧΕ, AIChE, ACS, SPE

Bruley, Duane F
Business: Biomedical Engrg Dept, Ruston, LA 71272
Position: Head, Biomedical Engrg and Dir of Re-Hab Center *Employer:* Louisiana Tech University *Education:* PhD/ChE/Univ of TN; MS/ME/Stanford Univ; ORSORT/NE/Oak Ridge Sch of Reactor Tech; BS/ChE/Univ of WI. *Born:* 8/3/33. Union Carbide Nuclear Co 1956-59. Prof Clemson Univ 1962-73. Prof & Hd Chem Engg, Tulane Univ 1973-77. Visiting Prof Princeton Univ 1972. Visiting Prof (JSPS) Yamagata Univ, Yamagata Japan, 1975. Intl Ctte-OTT, Outstanding Educator of Amer, P E SC, N O Sect-AIChE Treas & exec ctte, NSF grant to Japan, First Place res awd SE Sect ASEE 1967, natl Res Coun Grant to Sweden, NIH res grants; cons - duPont, WESVACO, ENKA Elec Assoc Inc. Mbr USPTA. Enjoy tennis & skiing, Grad Oak Ridge Sch of Reactor Tech. VP For Acad Affairs & Faculty Dean Rose- Hulman Inst Tech, 1977-1981. Hd Biomed Engrg, Dir Rehab Ctr 1981-84. Dean, Sch Engrg & Tech, CA Polyt St Univ, San Luis Obisbo, 1984- , Dir, Natl Inst Handicapped Res Ctr Lic Vehicles 1983- ; LA Tech Univ Awd, Engrg Res 83. Dean, Sch of Engrg & Tech, CA Poly, San Luis Obisbo, CA 1984-present; Dir Natl Inst Handicapped Res Ctr for Pers Licensed Veh 1983-present; LA Tech Univ Awd for Engrg Res, 1983; LA Engrg Soc Chas Kerr Awd, 1983; Consltg w/Exxon, El Paso Polyolefino Co, Masterline Corp, Venture Res. *Society Aff:* AIChE, ASEE, ΣΞ, ISOTT, BME.

Brumbaugh, Philip S
Home: 1359 Mason Rd, St Louis, MO 63131
Position: Consultant *Employer:* Self-employed *Education:* PhD/Production Mgt/WA Univ (St. Louis); MBA/Mgt/WA Univ (St. Louis); AB/Economics/WA Univ (St. Louis). *Born:* 11/14/32. Formerly operations analyst, Humble Oil & Refining Co & ind engr, Falstaff Brewing Corp. Taught ind engg & related subjs for ten yrs at WA Univ (St. Louis) & at Univ of MO-St Louis. President Qualtech Systems, Inc 1974-84. Consultant QC engg since 1984. ASQC - Certified Quality Engr since 1974. Past Dir of IIE (Quality Control & Reliability Engg Div); Past Pres, St Louis Chapter IIE; Past Chrmn, St Louis Sec ASQC. Served with Army Intelligence 1954-56. Mbr Engrs' Club of St Louis. *Society Aff:* IIE, ASQC, ASA

Brummett, Forrest D
Home: 1795 Michigan St, Martinsville, IN 46151
Position: Chief Engr *Employer:* Detroit Diesel Allison GM *Education:* B.S./ME/Purdue Univ *Born:* 6/24/30 Certified Manufacturing Engr -- CMfgE, and Chief Engr, Detroit Diesel Allison GMC. With GM (35) years serving in various engrg and design capacities. Served in the Army Engrs during the Korean Conflict in charge of machining operations. Served as Chrmn of the 800 member SME World Technical Council for three years, SME Bd of Directors since 1972, and currently an International SME VP. SME Award of Merit 1981. Enjoy all spectator sports, water skiing, and downhill snow skiing. Past Pres of the SME, 80,000 members internationally. Pres of "Real World Management Consultants, Inc" Gov Task Force High Technology. *Society Aff:* SME, ISEF, AMA.

Brunelle, Eugene J, Jr
Business: 4006 Eng. Ctr.; R.P.I, Troy, NY 12180-3590
Position: Prof *Employer:* Rensselaer Polytechnic Inst *Education:* Sc.D./Aerothermoelasticity/MIT; M.S.E./Aero Engr/Univ of MI; B.S.E./Aero Engr/ Univ of MI *Born:* 3/17/32 Montpelier, VT, Sc.D., MA Inst Tech., Asst. Prof aerospace and mechanics. Princeton 1960-64, Assoc Prof mech engr, aero engr, and mechanics Rensselaer Polytech Inst 1964-. Cons. U.S. Army Weapons Command, U.S. Navy Bureau Ships and many Pri. companies. Over 60 papers published in leading national and intl journals. Co-author of Chapter 10, Principles of Aeroelasticity. Research in elasticity, non-linear mechanics, plates and shells, aeroelasticity, hydrofoil instability, mechanics of composite materials, super flywheel technology. Have discovered many fundamental ideas about composite materials/structures including the Similarity Rules that interlink all vibration and stability problems of composite structures in an extraordinarily simple fashion. Enjoys classical music and remotely piloted vehicle modeling. *Society Aff:* AIAA, AAM, ASME, SIAM

Brunner, Richard F
Business: P O Box 50280, New Orleans, LA 70150
Position: Exec. V.P. Chief Operating Officer, Bd. of Directors *Employer:* Avondale Industries Inc. *Education:* BS/ME/Tulane. *Born:* 9/9/26. US Naval Air Corp WWII. Ch Engr Service Foundry Div, Avondale Shipyards Inc 1955; Ch Estimator, Avondale Shipyards Inc 1962; VP, Advanced Programs, Avondale Shipyards Inc 1968. VP Engr. 1972, Senior V.P. 1978. Mbr SNAME, Navy League, Intl House, Propeller Club., Bd of Admin Tulane Sch of Engg, ABS Technical Ctte., Reg PE LA. *Society Aff:* ABS, SNAME, ΤΒΠ, NAVY LEAGUE PROPELLER CLUB.

Bruns, Robert F
Business: 10408 Manchester Rd, St Louis, MO 63122
Position: Owner. *Employer:* Self-Employed. *Education:* BS/Mech Engr/MO School of Mines. *Born:* 8/19/21. Native of St Louis, MO - Navy pilot in WWII - Obtained BS in Mech Engg in 1947 - Various employment as design engr from 1947 thru 1957. Formed own co in 1957 as a Consulting Engg firm specializing in mech and elec design of commercial and institutional bldgs. Variety of other types of projs in mech, elec and communications field. Reg in home state of MO and 8 other states. Enjoy golf and other sports. *Society Aff:* ASHRAE, NSPE, PEPP.

Brunski, John B
Business: Dept Biomed. Engg, Room 7040 JEC, Troy, NY 12181
Position: Assoc Prof *Employer:* Rensselaer Poly Inst *Education:* PhD/Met & Mat Sci/Univ of PA; MS/Met & Mat Sci/Stanford Univ; BS/Met & Mat Sci/Univ of PA *Born:* 6/10/49 Native of Phila, PA. Currently Assoc Prof of Biomed Eng at Rensselaer Poly Technic Inst, Troy NY (since 1977). Res in progress on : (1) The Biomechanics of Dental Implants; (2) Biomechanics of Bone Remodeling. Teaching at RPI includes Statics and Dynamics, Modeling and Design, Materials Sci, Biomaterials, and Biomechanics. *Society Aff:* ASME, ASM, SFB, AADR, SESA.

Brush, Harvey F
Business: PO Box 3965, San Francisco, CA 94119
Position: Exec VP & Dir. *Employer:* Bechtel Group, Inc. *Education:* BS/Chem Engg/ Penn State. *Born:* 9/21/20. Native of Templeton, PA. Fellow ASME, Mbr AIChE, Fellow ANS. Sponsoring director of research and engineering; finance and controller operations; internal auditing; information services; personnel; procurement; security; and special management services. Previous positions include sponsoring director of refinery, chemical, petroleum and pieline work for Bechtel world wide; responsible senior officer for Bechtel's power world wide; Deputy General Manager of an operating division; Div Mgr Engr, Mgr of Engr; Mgr Nuclear Engr. Member World Trade Club, San Francisco; Peninsula Golf & Country Club, San Mateo; The Family, San Francisco; Commonwealth Club, San Francisco. Enjoys gardening, craftwork, fishing, boating. *Society Aff:* ANS, ASME, AIChE, AAAS.

Bruton, Robert O
Business: PO Box 582528, Tulsa, OK 74158
Position: Pres. *Employer:* Bruton Knowles & Love Inc. *Education:* BS/Civil Engg/ Univ of OK. *Born:* Jan 1, 1927. Native of Norman, OK. Army Engr 1945-46, OK State Hwy Dept 1949-55. Var cons firms 1955-65. Partner in Fell & Assocs 1966, now Pres of Bruton Knowles & Love Inc. Respon for Civ Engg projs. Pres OK-Sect ASCE 1975, Prog Chrmn ASCE Natl Transp Mtg 1973, Pres CEC-OK 1975-76. Pres Tulsa Post SAME 1983-4 Enjoy golf, fishing, boating. *Society Aff:* ASCE, ACEC, NSPE, SAME.

Brutsaert, Wilfried H
Business: Hollister Hall, Ithaca, NY 14853
Position: Prof of Civil & Environ Engrg. *Employer:* Cornell Univ. *Education:* PhD/Engg/Univ of CA; MS/Irrigation Engg/Univ of CA; Eng/Agri Engg/Univ of Ghent, Belgium. *Born:* May 1934 Ghent, Belgium. Res Asst Univ of Calif Davis 1958-62. Teaching & res at Cornell since 1962. Hydraulic Engr with TAMS, Cons Engrs NY 1965-66 Visiting Scientist Geophys Inst, Tohoku Univ, Japan 1969-70. Dept Civil Engineering, Tohoku Univ, Japan 1983-84. Leaves at Dept Civil Engrg. Tokyo Univ Japan 1973 & Isotope Dept, Weizmann Inst, Israel 1975. Visiting prof hydraulics-hydrology Univ of Wageningen, The Netherlands 1976-77 and Federal Insitute of Technology, Zurich, Switzerland, 1978. Present res mainly in ground water; atmos turbulent transport & evaporation; floods & droughts; hydrological sys analy. Phi Kappa Phi; John W Gilmore Award, Univ of Calif 1962; Freeman Fellowship in Hydraulics, Amer Soc Civil Engrs 1975; Fulbright-Hays Award, Council Internatl Exchange Scholars 1 976. *Society Aff:* ASCE, AGU, AMS.

Bruun, George T
Home: 613 Bridgeman Terrace, Towson, MO 21204
Position: Operations Mgr. *Employer:* AZ Bogert Co Inc. *Education:* BS/Chem Engg/ Univ of Pittsburgh. *Born:* Jan 1922 Pittsburgh Pa. BS Univ of Pittsburgh. Naval Electonic courses Bowdoin College & MIT; served with Motor Torpedo Boat Squadron 10 & mine sweeper command 1944-46. Served in production, process engrg & as Production Control Mgr with Koppers Co Chem Div. Started 1966 with Miracle as Production Mgr. Became V P for Corporate Mfg, R&D, Purchasing, Traffic Functions 1973. Regined from Miracle Adhesives Corp 1/29/79. Began with AZ Bogent Co Inc, 6/1/79, as Operations Mgr. Prof engrg lic PA. Married, 2 daughters. Enjoy golf & bridge. *Society Aff:* AIChE.

Bryan, Billy B
Business: Dept of Agri Engrg, Fayetteville, AR 72701
Position: Professor & Head. *Employer:* Univ of Arkansas. *Education:* MS/Univ of NB; BS/Agri Engr/Univ of NB. *Born:* 8/8/21. Reg Engr, AR. Native of Forrest City, Ark. Served with the Field Artillery 1942-46 with 2 yrs duty in Europe. Continued to serve in the USAR; retired Colonel USAR. Extension Irrigation Engr Kansas State College 1951-53; Agri Engrg Dept, Univ of Arkansas with respon of res & teaching in soil & water engrg area 1953- . Appointed to current respon as Prof & Hd of Agri Engrg Dept 1964. Respon for admin the res & teaching program. Hobbies: photography, sports, golf. *Society Aff:* ASAE, ASEE.

Bryant, Edward A
Home: 20 E 9 St, New York, NY 10003
Position: Vice President. *Employer:* Hazen & Sawyer. *Education:* BCE/Sanitary/RPI; MSE/Sanitary/Univ of MI. *Born:* Feb 1931. P E NY, Mass, NJ, MI, NC. After spending 2 yrs in CEC in the Philippines, joined Hazen & Sawyer in 1956 working on all facets of sanitary engrg. 1963-69 in charge of sanitary engrg dept at Tippetts-Abbett-McCarthy- Stratton & worked on a num of overseas projects. Returned to Hazen & Sawyer 1970 as assoc spec in water resources & water treatment process design. ASCE: Chmn Natl Ctte on Session Programs 1970; Chmn of Natl Ctte on Employment Conditions 1974; Metropolitan Sect Pres 1975, Natl Dir 1977-80. *Society Aff:* ASCE, AWWA, WPCF.

Bryant, Edward K
Home: 4620 N Park Ave, Chevy Chase, MD 20815
Position: Consultant. *Employer:* Tippetts-Abbett-McCarthy-Stratton. *Education:* CE/-/Rensselaer Polytechnic Inst. *Born:* Apr 1902. Born in Norwood, Mass. Prior to WW II associated with cons engrg firms in Rochester N Y & Trenton N J & conducted own cons bus for 10 yrs in Mt Holly N J. US Navy SeaBees 1942-46 ultimately as Officer-in-charge of Eleventh Construction Battalion. From 1946 to retirement in 1974 with Tippetts- Abbett-McCarthy-Stratton becoming a partner in 1956, prin engaged in the internatl field. Reg P E 6 states; Life Mbr ASCE, Fellow Amer Cons Engrs Council; Mbr Soc of Amer Military Engrs, Soc for Internatl Dev, Amer Arbitration Assn. Award from Federation International des Ingenieurs Conseils after 5 yrs as Chmn of the FIDIC/United Nations Dev Program Liaison Ctte for promoting the interests of cons engrs throughout the world. ACEC past Presidents Award for signifigant contributions to the International Consulting Engineering Profession.. *Society Aff:* ASCE, ACEC, SAME, SID, AAA, FIDIC.

Bryant, Howard S
Home: 25 Golfview Rd, Upper Saddle River, NJ 07458
Position: Corp V Pres, Engrg. *Employer:* Witco Chemical Corp. *Education:* ScD/Chem Eng/MA Inst of Tech; SM/Chem Eng/MA Inst of Tech; BChE/Chem Eng/GA Inst of Tech. *Born:* May 1928. Major exper is in engrg & in R&D of the chem, petrochem & petro refining indus. Chem exper incl specialty products, large volume organic intermediates & inorganics, fertilizers, pesticides, polymers, minerals processing, viscose rayon. In current pos since 1974 with respon for design & construction of all new facilities. *Society Aff:* AIChE, ACS.

Bryant, J Franklin
Business: 219 Mariner Square, Ste 200, Tampa, FL 33609
Position: Principal *Employer:* Gahagan and Bryant Assocs *Education:* BS/IE/GA Inst of Tech *Born:* 11/16/29 Graduate of GA Tech with post-graduate work at GA Tech and George Washington Univ. Taught Electrical Engrg at USNA, 1953-55. Field and Office Engr for McWilliams Dredging Co, 1955-57. Chief Civil Engr for Gahagan Dredging Corp in Venezuela, 1957-59. Division Mgr Gahagan Dredging, 1959-69. Division Mgr Atlantic, Gulf & Pacific Co, 1969-1974. Principal of Gahagan and Bryant Assocs 1974-Present. Registered PE in FL, GA, NY and TX. Member of Tau Beta Pi and Alpha Pi Mu. Specializing in studies, design and construction supervision of dredging projects and dredging equipment. *Society Aff:* ASCE, NSPE, PIANC

Bryant, John H
Home: 1505 Sheridan Dr, Ann Arbor, ME 48104 *Education:* PhD/Elec Engg/Univ of IL; BS/Elec Engg/TX A&M Univ. *Born:* Apr 1920. Army Signal Corps 1942-45, Europe. IT&T Labs 1949-55; Bendix Corp Res Labs 1955-62. Pres & Founder of Omni Spectra Inc 1962, Chmn 1974. Consit to Omni Spectra, subsidiary MIA-COM, Inc 1980-85. Adjunct Research Scientist, U Mich, 1985, Pres of the Microwave Theory & Techniques Soc of the IEEE 1970 and Distinguished Microwave Lecturer 1986-7. Fellow IEEE; Univ of Illinois, Electrical Engrg Alumni Assn Distinguished Alumnus Award 1971; Univ of IL College of Engg Alumni Award for Disting Service in Engg; Tau Beta Pi; Eta Kappa Nu; Sigma Xi. *Society Aff:* TBΠ, HKN, ΣΞ, IEEE.

Bryant, Kendred L Jr
Business: Box 691 Annex, Burlington, NC 27215
Position: Division Engrg Mgr *Employer:* Burlington Industries *Education:* BS/Elec Engrg./Duke University *Born:* 06/21/27 Native of Roanoke Rapids N.C. Served with U.S. Navy 1943-47 and 1950-52. Project Engr for E.I. duPont in high temperature metal research. With Burlington Industries since 1967. Assumed div engrg magr duties in 1976. Am responsible for facilities, air & water pollution, hazardous materials, energy mgmt and plant engrg manpower devt. Pres AIPE Ch *23, 1984. Ch Plant Engr of Year 1983. Mbr AIPE National Certification Bd. AEE - Tarheel Ch. Energy Mgr of Year 1987. Hobbies are golf and grandchildren. Society Aff:* AIPE, AEE

Bryce, John B
Home: 157 Alexandra Blvd, Toronto, Ontario Canada M4R 1M3
Position: Formerly Mgr, Hydraulic Studies (Retired 1978) *Employer:* Ontario Hydro. *Education:* MASc/Hydraulics/Univ of Toronto; BASc/Civil/Univ of Toronto. *Born:* Jan 1, 1913 Toronto. With Ontario Hydro 1935-78, except Natl Res Council 1938-40. From 1953, respon for specialized hydraulic design & testing of hydraulic features of Hydro's hydro-elec & thermal stations, incl dir of Hydraulic Model Lab; more recently also phys environ studies. Represented Ontario 1953-77 on Internatl St Lawrence River Bd of Control. 1978-to date- consultant on hydraulic and irrigation projects. Mbr Assn Prof Engrs Ontario; Mbr Engrg Inst of Canada-elected Fellow 1974; Received EIC Angus Medal 1960, Keefer Gold Medal 1968. *Society Aff:* EIC.

Bryers, Richard W
Home: Box 139 Barley Sheaf Rd, Flemington, NJ 08822
Position: Senior Research Associate *Employer:* Foster Wheeler Dev. Corp. *Education:* MS/ME/New Jersey Institute of Technology; BS/ME/Lehigh University *Born:* 10/17/31 Former mbr Engrg Foundation Conference Cttee, Chairman of the ASME Research Cttee on Deposits and Corrosion Due to Impurities in Combustion Gases, mbr of ASME National Nominating Cttee. Organized and chaired three international conferences on various aspects of heat exchanger fouling, initiated a program to assemble and publish a compendium of 5000 abstracts on fireside deposits and corrosion. He is editor of three major publications on fouling and shagging. He project-managed and engineered numerous programs related to Navy heavy oil evaluation; micronized coal; fire side deposits due to impurities in coal oil and refuse; conceptual design of atmospheric and pressurized fluidized bed steam generators, multisolids fast circulating fluidized bed boiler, chemically active fluid bed gassifiers; and R&D of desalination equipment and processes. Holds 9 patents on fluidized bed combustors, Fellow mbr ASME and Institute of Energy. *Society Aff:* ASME

Bryniarski, Americ J
Business: 200 East 8th St., Rm. 210, Mountain Home, AR 72653
Position: Water Resources Eng *Employer:* AR Soil & Water Conservation Commission. *Education:* MS/CE/Univ of AR; BS/CE/Univ of AR *Born:* 6/10/34 Native of Baxter County, AR. Bridge designer with MO Highway Dept 1963- 1967. With AR Soil & Water Conservation Commission since 1969. Am responsible for dam safety and water rights for the entire state. Pres AR Section ASCE 1980; VP 1979. Enjoy classical music, fishing and camping. *Society Aff:* ASCE, ASDSO.

Bryson, Arthur E, Jr
Durand Bldg, Stanford, CA 94305
Position: Prof Dept of Aero & Astro. *Employer:* Stanford University. *Education:* BS/Aero Eng/IA State Univ; MS/Aero Eng/CA Inst of Tech; PhD/Aero Eng/CA Inst of Tech. *Born:* Oct 1925. Aircraft maint officer, US Navy 1944-45. Paper mill engr, Container Corp of Amer 1947. Res Engr United Aircraft Corp 1948. Res Asst, Caltech 1949-51. Res Engr Hughes Aircraft 1951-53. Assoc Prof 1956-61, Prof 1961-68, Harvard. Hunsaker Prof MIT 1965-66. Author 100 tech papers, 1 book & 1 film. At Stanford since 1968. Chmn, Aero & Astro 1971-79, Paul Pigott Professorship 1972-present. Fellow AIAA & Amer Acad of Arts & Sci. Mbr ASEE, Sigma Xi, Tau Beta Pi, NAE & NAS. Honorary Member IEEE. Spec int in guidance & control, flight mech, & fluid mech. Chmn Natl Comm Fluid Mech Films 1964-68; Dir-Tech AIAA 1965-67; Westinghouse Award ASEE 1969; Pendray Award AIAA 1968; Mech & Control of Flight Award AIAA 1980; Oldenburger Medal ASME, 1980; Control Systems Award IEEE, 1984. *Society Aff:* AIAA, NAE, NAS, IEEE, ASEE.

Tecklenburg, Harry
17 Eaton Ave, Norwich, NY 13815
Position: Sr VP & Gen Mgr *Employer:* Norwich Div, The Procter & Gamble Co *Education:* MS/ChE/Univ of WA; BS/ChE/MIT. *Born:* 11/3/27. Native of Seattle, WA. Joined Procter & Gamble Co in 1952, holding positions of increasing responsibility in res and dev activities. Became mgr of mfg and product dev for paper products, & later dir of corporate res div Since 1976, Sr VP with overall responsibility for the co's worldwide scientific res and 1984, Sr VP & Gen Mgr of Norwich Eaton Pharmaceuticals, Inc. Outside activities include: Mbr, Bd of Dirs, Pharmaceutical Mfg Assoc; Mbr, Engrg Soc of Cincinnati; Mbr, Am Inst of Chemical Engrs; Mbr Am Chem Soceity; Dir Emeritus, Goodwill Indust of America; Mbr, Visiting Ctte, Dept of Chem Engrg, Univ of WA. *Society Aff:* AIChE, ACS, TAPPI, AAAS.

Bublick, Alexander V
Home: 100 Cheney Rd, Marlborough, CT 06447
Position: Asst Chief. *Employer:* Pratt & Whitney Aircraft. *Education:* MS/Met/Stevens Inst of Tech; BS/ME/Stevens Inst of Tech. *Born:* 7/30/34. Grad from Northfield Mt Hermon Sch, MA '52. Served USAF 1954-57. Started carrier with PWA in 1961 as Process Planning & Dev Engr. In 1964 joined Mtls Engg & Res Lab as Sr Met responsible for evaluation of jet engine disk & shaft mtls; 1966-70 Sr Mtls Engr, dev of high temperature aerospace sheet alloys; 1970-77 Mtls Design Proj Engr, selection of mtls for advanced jet engines; 1977- present, direct activities of Mfg Support Group in Mtls Control Lab. 1976-77 ASM Hartford Chapter Chrmn. 1979-81 mbr of ASM Handbook Committee. *Society Aff:* ASM, AIME, AWS.

Buchanan, R C
Business: Dept of Ceramic Engineering, 105 So. Goodwin, University of Illinois, Urbana, IL 61801
Position: Prof of Engrg *Employer:* Univ of IL. *Education:* ScD/Ceramic Sci/MIT;

Buchanan, R C (Continued)
BS/Glass Tech/Alfred Univ *Born:* 04/10/36 Prof of Ceramic Engrg, Univ of IL at Urbana-Champaign. Previously employed by IBM Corp as res engr, dev of electronic ceramic materials for component, packaging and display applications. Areas of res included: solder glass and coatings, substrates, ceramic metal composites and thick film tech. Current res interests encompass: PZT and dielectric capacitor materials, struct ceramics and dielectric films for device applications. Emphasis is on powder synthesis, sol- gel techniques, processing and low temp densification studies. Teaching areas include: crystal and solid state chem, crystallography and elec ceramics. Dr Bauchanan is a past Chrmn of the Elec Div and Fellow of the Amer Ceramic Soc, Mbr Natl Inst of Ceramic Engrs, Electrochem Soc, AAAS and the Intl Soc for Hybrid Microelects. Holds several patents and has pub widely in the above res areas. Dr Buchanan received a DSc from MIT in Ceramic Sci with a Minor in Metallurgy and a BS degree in Glass Tech from Alfred Univ. *Society Aff:* ACS, ES, AAAS, NICE, MRS, ISHM.

Buchele, Wesley F
Business: 107 Davidson Hall, Ames, IA 50011
Position: Prof of Agricultural Engrg *Employer:* IA State Univ *Education:* PhD/Ag Engr-Soil Physics/IA State Univ; MS/Agr Engr-Engr Mech/Univ of AR; BS/Agr Engr/KS State Univ *Born:* 3/18/20 Reared on a farm near Cedar Vale, KS. Served in Coast Artillery, Infantry and Ordnance in US Army during WW II (1943-46). Junior Engr, John Deere Waterloo Tractor works 1946-48. Taught at Univ of AR, MI State Univ-currently at IA State Univ. Teach and conduct research on design of agricultural harvesting, planting and tillage machines. Developer of large round bale Ridge Till System of Farming and Axial Flow Threshing Cylinder concepts. Holder of 26 patents. Member of SigXi, GamSigDel, PhiLamTau and AlpEps. Became Fellow in ASAE in 1977. Traveled in more than 55 countries, lectured in 12 countries. Hobbies - photography and golf, travel, appropriate technology applications in third world. *Society Aff:* ASAE, SAE, NSPE, ASA, NIAE

Buchheit, Richard D
Home: 3786 Quail Hollow Dr, Columbus, OH 43228
Position: Res Engr. *Employer:* Battelle Columbus Labs. *Education:* BS/Met Engrg/Columbia Univ. *Born:* Aug 1924. US Naval Res 1942-46. Met Observer for Crucible Steel Co of Amer 1946-48. With Battelle Columbus Labs since 1948 where served as Hd of the Optical Metallographic Labs for 20 yrs. Assoc Section Mgr 1954-76. Respon for conduct of wide var of metallographically oriented contract res studies and failure analyses. Reg P E Ohio. Mbr ASM, & Charter Mbr, Bd of Dir, & Treas of IMS. Received num ASM Metallographic Exhibit Awards incl Best-in-Show 1954. Pub sev tech articles & contrib to ASM Metals Handbook volumes. *Society Aff:* ASM, IMS.

Buchsbaum, Norbert N
Home: 4982 Dumfries, Houston, TX 77096
Position: Eng'g Consultant *Employer:* Self *Education:* MS/ChE/Poly Inst of NY; BE/ChE/Cooper Union. *Born:* 4/15/29. Grad in 1953 with a Bachelor's Degree in Chem Engg from Cooper Union in NY. Has been assoc with two intlly-known engg contractors where he was responsible for the design of numerous plants including several "first-of-its-kind" projs. Later assumed exec positions, first with a pharmaceutical co & then with a maj oil co. 1986-present Engrg Consultant. 1972-86 Gulf Oil Corp: VP Engg Div. 1970-1972 Norchem, Inc: Pres. 1967-1969 Hoffmann-La Roche: Dir of Engg. 1960-1967 the Lummus Co: Mgr of Projs, Mgr Proposals, Mgr Process Design - Chem Plants. 1956-1960 Scientific Design Co: Proj Mgr, Proj/Process Engr. 1953-1956 the Lummus Co: Process Engr. *Society Aff:* AIChE.

Buchsbaum, Solomon J
Business: AT&T Bell Laboratories, Crawfords Corner Rd, Holmdel, NJ 07733
Position: Exec VP - Customer Sys. *Employer:* AT&T Bell Laboratories, Holmdel. *Education:* PhD/Physics/MA Instof Technology; BS/Physics & Math/McGill Univ; MS/Physics/McGill Univ. *Born:* 12/4/29. in Stryj, Poland. Joined Bell Labs 1958 & engaged in exper & theoretical res in gaseous & solid plasmas. In 1961, named hd of Solid State & Plasma Phys Res Dept; in 1965, Dir Electronics Res Labs & in 1968 VP res at the Sandia Corp's Sandia Lab. Returned to Bell Labs as Exec Dir, Res, Communications Sci Div 1971. In 1975, apptd Exec Dir, Transmission Sys Div and in April 1976 VP, Network Planning & Customer. Assumed present post May 1979. Mbr NAE, NAS, Fellow AAAS, APS, IEEE & Amer Acad of Arts & Sci. Chairman, White House Science Council; Senior Consultant, Defense Science Board; Trustee, Rand Corporation; Director, Draper Laboratory; Member Governing Board, Argonne National Laboratory. *Society Aff:* AAAS, AAAS, APS, IEEE, NAE, NAS.

Buck, Alan D
Home: 105 Herrod St, Vicksburg, MS 39180
Position: Supervisory Geologist. *Employer:* US Army Engr Waterways Exper Sta. *Education:* BS/Geology/Univ of IL; MS/CE/Purdue Univ. *Born:* Jan 1925 Waynesville, Ill. BS Univ of Illinois; MS Purdue Univ. WW II in Europe with a US Armored Div. Employed since 1952 by US Army Corps of Engrs as geologist (petrographer). Field, applied concrete res. Corecipient, Amer Concrete Inst Wason Medal for Res 1968. Recipient of Secy of Army Fellowship for R&D & of USAE Waterways Exper Station Dir's Res & Engrg Achievement Award 1973. Mbr Soc Sigma Xi, Amer Concrete Inst. *Society Aff:* ACI, CMS, ΣΞ.

Buck, Frank A M
Business: 2384 San Diego Ave, San Diego, CA 92110
Position: Managing Principal *Employer:* King, Buck & Assocs., Inc. *Education:* PhD/Phys Chem/Purdue Univ; MSc/Chem Engg/Univ of British Columbia; BSc/Chem Engg/Univ of British Columbia. *Born:* 6/26/20. Native of Canada, US citizen 1954. Dev engr, Powell River Co, 1944-45. Instructor in Chem Eng Dept, Univ of WA, 1948-49. With Shell Oil & Shell Chemical Cos in R&D, mfg operations, mktg, and bus planning functions, 1949-79, except for 1968-69 when served as Industrial Specialist in US Office of Oil and Gas, Washington, DC. Since 1979 managing principal of King, Buck & Assocs, Inc. consultants in energy and chemical processing industries. Member, Hearing Board, San Diego County APCD (air pollution control district). Dir AIChE 1978-80, Chrmn AIChE Energy Coordinating Committee 1976-77. Avocation: avocado and macadamia nut orchards. *Society Aff:* AIChE, ACS.

Buck, Richard L
Business: 4 Cermak Blvd, PO Box 128A, St Peters, MO 63376
Position: Plant Mgr. *Employer:* Airwick Ind, Inc. *Education:* MS/Engg Mgt/Univ of MO; BS/ChE/Univ of MO. *Born:* 5/22/40. Born & raised in the central sec of the US. Early career experience were in process - proj engg & product dev. With Airwick Ind, a Ciba Geigy Co since 1977. Assumed current responsibilities as Plant Mgr for Manufacturing of Consumer and Ind products in 1977. Secy Ozark Chapter MSPE 1968. Bd of Dirs mbr of the St Louis Area Chapt Natl Fdn - Mar of Dimes. Enjoy music, sports & outdoor activities. *Society Aff:* AIChE, NSPE, MSPE, AMA.

Buckham, James A
Business: PO Box 847, Barnwell, SC 29812
Position: Exec VP. *Employer:* Allied-Gen Nucl Services. *Education:* PhD/ChE/Univ of WA; MS/ChE/Univ of WA; BS/ChE/Univ of WA. *Born:* 9/15/25. Dr Buckham assumed his present position in 1976, and has worked in the nuclear engg field since 1953, specializing in nucl fuel reprocessing & radioactive waste mgt. Awarded the Robert E Wilson Award by AIChE in 1974 for excellence in this field; served AIChE as Chrmn of its Natl Prog Committee, Nuclear Engg Div, & ID Sec, & is a Fellow and Director of that Inst; also served as chrmn of the Eastern ID Sec of ANS; taught chem engg full time for two yrs at the Univ of WA & part time for the Univ of ID for 15 yrs. *Society Aff:* AIChE, ANS, AIF, ACS, AAAS.

Buckingham, Frank E
Home: 6451 E Skyline Dr, Springfield, MO 65804
Position: Free-Lance Writer & Equip Consultant *Employer:* Self Employed Agri-

Buckingham, Frank E (Continued)
Business Writing *Education:* BS/Agric Engr/IA State Univ *Born:* 12/12/32 Grew up on IA farm. Agricultural Engrg BS included five technical journalism courses. Professional career devoted to communications--primarily translating technical materials for dealers and equipment users. Three years writing for Implement & Tractor Magazine; eight years with Massey-Ferguson product training and advertising; advertising mgr for Clay Equipment; publicity dir, Paul Mueller, Co. Most of last 14 years, free-lance writer and equipment consultant to manufacturers, publications and US Dept of Commerce as Industry Technical Representative to England's Royal Agricultural Show. Active in ASAE 28 years--area and national committees, elected Dir of Publications Dept 1980-82. *Society Aff:* ASAE

Buckley, John D
Business: Langley Res Ctr-219, Hampton, VA 23665
Position: Tech Asst to Ch of Fabrication Div. *Employer:* Natl Aero & Space Admin (NASA). *Education:* PhD/Cer Engg/IA State Univ; MS/Cer Engg/Clemson Univ; BS/Cer Engg/Clemson Univ; BS/Pre Med/St Lawrence Univ. *Born:* Oct 1928. Native of Saranac Lake, N Y. USAF 1952-56. Materials Engr, Proj Engr & Sect Hd with NASA-Langley Res Ctr since 1959. Assumed current respon as Tech Asst to Ch of Fabrication Div 1972. Major engrg tasks at NASA-Heat Shield for Space Shuttle; Hardware for Viking Mars Lander; NASA Safety Survey Team B-1 Bomber. Who's Who in Aeronautics; Fellow Amer Ceramic Soc; Cert Mfg Engr; Material Sys & Design Div Council ASM; Chmn of Ceramic Metal Sys Div ACS 1974. City of Newport News Planning Commission; Mbr Vir St Bd of Health 1972-76. Assoc Prof Geo Washington Univ Ext NASA-Exceptional Service Medal-1977; NASA- Technology Utilization Award-1978; Who's Who in Government Amer Men & Women of Science Lt/Col USAF Res Air Force Sys Command-Nondestructive Evaluation & Thermal Protection Sys (projs). Major Eng. Proj at NASA-National Transonic Facility. *Society Aff:* AIAA, ASM, ACerSoc, SME.

Buckley, John D
Business: 8600 NW 36th St, Miami, FL 33166
Position: Consultant *Employer:* Post, Buckley, Schuh & Jernigan Inc. *Education:* BS/CE/Univ of ME; MS/Sanitary Eng/Harvard Univ. *Born:* Dec 1923 Boston, Mass. As consultant, respon for guiding quality control firm. PBS&J provides cons engrg/planning serv throughout Southeast; headquartered in Miami; 34 regional offices. Engr of Yr Dade Cty 1971; Fellow ASCE & Fla Engrg Soc; frmr Chmn NSPE/PEPP Prof Liability Ctte; frmr Mbr Bd of Dir, Fla Inst of Cons Engrs; frmr Mbr NSPE Nominating Ctte, Former Bd Mbr; Former Chmn ASCE Natl Ctte on Standards of Practice & Former ASCE Dist 10 Chmn, Pres, Fla. Engineering Society 1981, Natl Council of Engrg Examiners. Univ of Miami Fac, Adj Instr 1965-81. Advisory Council, Engg School, Fl. Int Univ; Pubs: 'Pump Control Maintains Uniform Pressure', Public Works Magazine 1-74; 'Reverse Osmosis: Moving from Theory to Practice', Cons Engr 11-75. Reg Engr Fla, Maine, Mass. *Society Aff:* NSPE, ASCE, AWWA, WPCF.

Buckley, Lawrence J
Home: 216 Neck Rd, Madison, CT 06443
Position: V P & New England Mgr. *Employer:* Goodkind & O'Dea Inc, Cons Engrs. *Education:* BCE/CE/Rensselaer Polytech; MCE/CE/Rensselaer Polytech. *Born:* Oct 1924. USNR 1943-46. Field engr on constr projects 1949-54. Employed by Goodkind & O'Dea Consulting Engineers 1954- . Proj Mgr of var design & constr projects in the field of public works in Mass, N Y & Conn unitl 1973; then became V Pres co-owner resp for New England Projs in water supply, sewerage, foundations, roads and bridges. Reg Prof Engr Conn, N Y, Mass. Mbr Natl Socs Prof Engrs, Connecticut Society of Professional Engineers, ASCE, Society of American Military Engineers, Permanent International Association of Navigational Congresses Madison Conn. Inland Wetlands Comm 1974- . Arbitrator Amer Arbitration Assn. Participant in boating activities on Long Island Sound. *Society Aff:* ASCE, NSPE, PIANC, SAME, CSPE.

Buckley, Merrill W Jr
Home: 263 Lewis Rd, Springfield, PA 19064
Position: Administrator Planning *Employer:* RCA Corp - Govt Systems Division *Education:* MS/EE/Univ of PA; BS/EE/Villanova Univ *Born:* 10/11/26 Served in the US Navy as an electronic technician and radar instructor, 1945-46. Early engrg employment included the Naval Air Development Center (test and evaluation), EL-TRONICS (design) and the Frankford Arsenal (radar R&D). Since joining RCA in 1953 has held a variety of positions, E.G., Systems Analyst, Project Dir, Mgr Logistic Support, Mgr Field Planning, Mgr of Planning and Manufacturing Coordination, on such projects as TERRIER, TALOS, NIKE, ATLAS, MINUTEMAN, BMEWS, APOLLO and AEGIS. Frequent seminar speaker in the US and abroad on management and engrg topics. Lectures for several universities and professional societies and is a Naval Reserve Officer (retired). *Society Aff:* IEEE

Buckley, Page S
Business: Engrg Dept, Wilmington, DE 19898
Position: Principal Design Consultant. *Employer:* E I duPont de Nemours & Co. *Education:* BS/ChE/Columbia Univ. *Born:* Jun 1918. Hon Dr of Engrg Lehigh Univ 1975. 1940-49 Monsanto Co: process dev, design, plant trouble-shooting, statistical qual control, instrumentation. Joined duPont 1949: instrumentation, process control res, dev & design. Mbr ISA (Fellow) & AIChE. Book, 'Techniques of Process Control', Wiley 1964. About 35 papers. Mbr Natl Acad of Engrg 1981. *Society Aff:* ISA, AIChE.

Buckley, Robert F
Business: 40 Federal St, Lynn, MA 01910
Position: Mgr-Engineering. *Employer:* Genl Elec Co, Instrument Prod Oper. *Education:* BS/ME/Tufts Univ. *Born:* Salem, Mass 1923. After obtaining BS from Tufts Univ 1944 & serving in US Navy WW II, joined the Meter & Instrument Dept as Tech Engr 1946. In 1951, became Mgr-Aircraft Flowmeter Engrg, followed by assignments as Mgr Reactor Instrument Engrg, Mgr Process Instrumentation Engrg & Mgr Prod Planning of the Instrument Dept. Present pos March 1976.

Buckner, W Don
Business: 4D1 Phillips Bldg, Bartlesville, OK 74004
Position: Professional Training Coordinator *Employer:* Phillips Petroleum Co. *Born:* BS EE Oklahoma State Univ. Working toward MBA from OSU. Began with Phillips'Engrg Dept 1955; worked as Assoc Dir of Recruitment 1965- 71; Worked as Director, East Coast Region, Refined Products Supply from 1971 through 1979; current position includes developing technical training programs for engineers throughout Phillips Petroleum Co. President Bartlesville Chapter Okla Soc of Prof Engrs; State Dir OSPE; Gov, State of Okla Toastmasters; International Dir of Toastmasters International. Past President Okla Soc of Prof Engrs & Natl Dir Soc of Prof Engrs. Outstanding Engr of Yr 1974, OSPE.

Bucy, J Fred
Business: P. O. Box 225474, Dallas, TX 75265
Position: Pres & CEO *Employer:* TX Instruments Inc *Education:* M/Physics/Univ of TX-Austin; B/Physics/TX Tech Univ *Born:* 7/29/28 Native of Tahoka, TX. Joined TX Instruments Inc in 1953-, gen mgr apparatus division, corporate VP 1953-67, corporated group VP components group, 1967-72, executive VP 1972-75, executive VP, chief operating officer 1975-, pres 1976-, CEO, April 84, Chrmn of Bd of Regents of TX Tech Univ and TX Tech Univ Health Science Center 1980-82. Recipient Distinguished Engr award TX Tech 1972, Fellow IEEE 1974, Natl Acad Engrs. Member of SigPiSig and TauBetPi. Clubs: Cosmos in WA and Northwood Country Club, Petroleum, & Univ Club in Dallas. *Society Aff:* NAE, IEEE

Bucy, R S
Business: Aerospace Eng, Univ. of Southern Calif, Los Angeles, CA 90089-1191
Position: Prof of Math & Aerospace Engrg. *Employer:* USC *Education:* SB/Math/MIT; PhD/Math Statistics/UC Berkeley. *Born:* Jul 20, 1935. Mbr Amer

Bucy, R S (Continued)
Math Soc, Fellow IEEE. Major field of int is filtering theory. Author Interscience book 'Filtering for Stochastic Process with Application to Guidance' 2nd Edition, Chelsea, 1987. Recipient Humboldt Prize Govt of W Germany. Visiting Prof France Toulouse, 1973, Berlin 1975, and Nice 1983. Delegate to IEEE-Soviet Academy Workship of Information Theory Moscow 1975, Conslt M.I.T. Lincoln Lab, Aerospace Corp, T.R.W., Thompson C.S.F. Editor "Nonlinear Stochastic Problems, Reidel. Editor "Information Sciences-, journal of North Holland. *Society Aff:* AMS, IEEE.

Budenholzer, Roland A
Business: Ill Inst of Tech, Chicago, IL 60616
Position: Prof Emeritus Mech Engrg, Chairman Amer Power Conf. *Employer:* Illinois Inst of Tech. *Education:* PhD/Mech Engg/CA Inst of Technology; MS/Mech Engg/CA Inst of Technology; BS/Mech Engg/NM State Univ. *Born:* Nov 1912. Res Fellow Caltech 1939-40. Joined IIT as Instr 1940, becoming Prof of Mech Engrg 1947. Cons IIT Res Inst 1946- . Resident Res Assoc Argonne Natl Lab, summer 1961. Dir Midwest Power Conf 1949-51 & Amer Power Conf 1952-78, Chrmn, Amer Power Conf, 1978- . Trustee WSE 1969-72, Chmn Exec Comm, Power Div, ASME 1971-72. Mbr Exec Comm, Bd of Dir, US Natl Comm, World Energy Conf 1972-78. Recipient ASME Geo Westinghouse Gold Medal 1968. Chicago Technical Societies Council Award of Merit, 1975. Honorary Member and Life Fellow ASME etc. Fellow ASME & Mbr ANS, ASEE, Sigma Xi, Tau Beta Pi & WSE. Major int: thermodynamics, power generation & energy. Enjoy travel, hiking & photography. *Society Aff:* ASME, ANS, ASEE, WSE.

Budiansky, Bernard
Business: Pierce Hall, Harvard University, Cambridge, MA 02138
Position: Prof of Structural Mech. *Employer:* Harvard Univ. *Education:* PhD/Appl Math/Brown Univ; ScM/Appl Math/Brown; BCE/Civil Engg/CCNY. *Born:* Mar 8, 1925. BCE CCNY 1944; PhD appl math Brown U 1950. Aero struct res at NACA, Langley Field Va 1944-55. Teaching & res at Harvard, Div of Applied Sciences since 1955. Res int incl plasticity, fracture, shells, buckling, geophys problems. Dryden Res Lectr AIAA 1970; Townsend Harris Medal CCNY 1974; von Karman Medal ASCE 1982 Evinger Medal Soc. Engrg Sci. 1985 Honorary Sc.D. Northwestern Univ. 1986; Mbr Natl Acad of Sci, Natl Acad of Engrg, Amer Acad of Arts & Sci. Foreign Member, Royal Netherlands Academy of Arts & Science. Foreign mbr Danish Ctr for Applied Math & Mech. *Society Aff:* ASCE, ASME, AIAA, AGU.

Budinger, Fred C
Business: N 920 Lake, Spokane, WA 99212
Position: Prin. *Employer:* Budinger & Assoc. *Education:* BA/Math & Phys/AZ State Univ. *Born:* 6/13/36. After grad in 1959, taught Mth & Phys. Joined Warne, Sergent & Hauskins, Phoenix, AZ in 1961; and S & G Labs, Lompoc, CA in 1967. After 10 yrs experience, established branch office and lab in Bakersfield, CA for BSK. In 1976, established own consulting office and testing lab in Spokane, WA. Budinger & Assocs offers full range of geotechnical services, as well as drilling, testing, and inspection. VP of WSPE 1987-88; Pres, Inland Empire sec ASCE 1981-82. Presents geotechnical seminars throughout US, and has published articles in "Civil Engineering" magazine. Enjoy camping, photography & flying. *Society Aff:* NSPE, ASCE, ASFE, ASTM, ISSMFE.

Budinski, Kenneth G
Business: Kodak Park Div B-23, Rochester, NY 14650
Position: Sr Met. *Employer:* Eastman Kodak Co. *Education:* MS/Met Engg/MI Tech Univ; BME/ME/GM Inst. *Born:* 6/29/39. Currently a Sr Consultant on Engrg Mtls for Eastman Kodak specializing in the area of Tribology. Present V Chrmn of ASTM G2 Committee on Erosion & Wear. Author of numerous papers dealing with friction and wear and a textbook on engrg mtls: properties & selection (1983 Reston Inc 2nd Ed). Technical contributor to ASM & ASME handbooks. Past chrmn & dir of the Rochester Chapt of ASM. Educational activities include teaching mtls engrg at Rochester Inst of Tech & Monroe Community College in addition to numerous tutorials for technical socs. Maj contribution to engrg has been fundamental res leading to a better understanding of the mechanisms of wear of tool mtls. Participant in 1983 NSF Tech Mission to China on Wear (Shenyang PRC). *Society Aff:* ASM, ASTM, WRC, SAE.

Budlong, Dudley W
Business: 9017 Reseda Blvd 208, Northridge, CA 91324
Position: President *Employer:* Dudley W. Budlong Consultants *Education:* BSEE/Elec Engg/IL Inst of Technology. *Born:* May 9, 1922. Ch Engr May Engrg Co; Pres & Ch Engr Budlong & Assocs; Exec V P QuintonBudlong, Los Angeles 1954-73; Group VP Boyle Engrg Corp, Northridge Calif 1974-1981 President Dudley W. Budlong Consultants 1981-. Planning Cabinet, Amer Cons Engrs Council/U S 1970-76, Chmn 197576. Reg P E Calif, N J, N Y, Nev, Vir, Fla, Mich, Minn, Alaska, Utah. Fellow Inst for the Advancement of Engrg; Mbr Cons Engrs Assn of Calif (p dir); Cons Elec Engrs of S Calif (p pres & dir); Mech-Elec Cons Engrs Council, Calif (p state chmn). Tau Beta Pi, Eta Kappa Nu. Engrg Prof Adv Council, Calif State Univ Northridge, Calif. *Society Aff:* ACEC, AIPE, AEE.

Budny, Bernard R
Home: 434 East Wilbur Ave, Milwaukee, WI 53207
Position: Prof *Employer:* Milwaukee School of Engrg *Education:* MS/EE/Univ of WI-Mad; BS/EE/Marquette Univ *Born:* 11/13/27 Native of Milwaukee, WI. Aviation Electronics Technician in US Navy 1945- 46. Field Engr with AC Electronics Div of GM 1951-54. Guidance system design with General Dynamics 1954. Project Engr and Dept Head with Allen Bradley 1954- 68. Project Engr with Modern Co 1969. With Milwaukee School of Engrg since 1970. Assumed current status as Prof, EE dept in 1971. Consultant in solid state electrical energy conversion components, circuits, and systems. Past dir WI Society PE. Thomas Mor PTA. Past Chrmn IEEE Magnetics Group (Milwaukee). Interested in flying and art. *Society Aff:* IEEE, HKN.

Bueche, Arthur M
Business: 3135 Easton Turnpike, Fairfield, CT 06431
Position: Sr VP-Corp Technology. *Employer:* General Electric Co. *Education:* BS/Chemistry/Univ of MI; PhD/Physical Chem/Cornell Univ. *Born:* Nov 1920. PhD Cornell Univ; BS Univ of Michigan. Native of Flushing, Mich. WW II Synthetic Rubber Prog. Res Assoc, chem, Cornell Univ 194750. Joined Genl Electric as polymer chemist 1950. Var managerial pos in chem res prior to being named V P-R&D 1965. In 1978 promoted to Sr VP-Corporate Technology. IS GE's prin technical officer & a mbr of the Co's Corp Policy Bd. Mbr Natl Acad of Sci & Natl Acad of Engrg. Var gov adv ctte assignments, hon degrees, & assn offices, incl Pres-Indus Res Inst 1975-76. *Society Aff:* NAS, NAE, AAAS, ACS, AIP, APS, NSPE.

Buehler, Adolph I
Home: 621 Central St, Evanston, IL 60201
Position: Retired. *Education:* Dipl/-/Optical College; Dipl/-/Alexander Hamilton Inst. *Born:* 11/1/93. Native of Lucerne Switzerland. Attended Public schools in Switzerland. Optometry Grad Optical College Berlin Germany 1912-1913. Asst Optical Bus & mgr in Zurich & Basel Switzerland. Emigrated to the USA 1922. Sales Engr with Leitz NY. In 1929 arranged & managed Midwestern Office for Leitz NY. Established my own Bus 1936 Adolph I Buehler & later Buehler Ltd specializing, inventing & mfg equip for microstructural Analysis. Retired as Pres & Chrmn of both 1976. *Society Aff:* ASM, AFS, IMS.

Buehring, Norman L
Business: Los Angeles Dept. Wtr & Pwr, P O Box 111, Los Angeles, CA 90051
Position: Exec Asst to the Gen Mgr *Employer:* Los Angeles Dept of Water & Power. *Education:* BSCE/Civil Engr/CSU-Long Beach; MSCE/Civil Engr/USC; MPA/Public Ad/USC; Certif/Pub Works Mgt/UCLA. *Born:* 7/23/42. Cert in Pub Works Mgmt UCLA. PE CA. Engr for DWP since 1965. Designer on reservoir & open channel projects, res engr on steel tank & earthwork projs, Exec Staff Engr & currently in

Buehring, Norman L (Continued)
charge of water sys Dams Matls & Geology Sect. 1975 Edmund Friedman Young Engr Award for Prof Achievement. IAE Fellow & Pres Elect for College of Fellows. Los Angeles Outstanding Young Engr in ASCE Activity 1973 & 1974. 1974-76 Chmn, Natl ASCE Ctte on Student Services. ECPD Ctte on Student Dev 1975-76. LA Sect ASCE Treas 1975-76, Dir 1974, Pres AMF 1973. Chrmn ASCE Mbr Activities Div Exec Committee, 1978-80 and chairman ASCE Committee on Grants, Fellowships and Bequests 1979-80. *Society Aff:* ASCE, AWWA, IAE.

Buelow, Frederick H
Business: 460 Henry Mall, Madison, WI 53706
Position: Prof of Agri Engrg *Employer:* Univ of Wisconsin-Madison. *Education:* PhD/Agri Engr/MI State Univ; MS/Engg/Purdue Univ; BS/Agri Engr/ND State Univ. *Born:* 3/13/29. Native of Drake, N Dak. Officer US Air Force 1952-54. Fac of Agri Engrg Michigan State Univ 1956-66. Prof of Agri Engrg Univ of Wisconsin-Madison 1966- . Chmn of Agri Engrg Univ of Wisconsin-Madison 1966-1983. Dir Amer Soc of Agri Engrs 1972-74, 1977-79 and 1985-87; Also Chmn Wisc Sect ASAE 1975-76. Member ABET/EAC 1980-85. Fellow in ASAE. *Society Aff:* ASAE, ASEE, Sigma Xi.

Bueltman, Charles G
Home: 12 Ransom Rd, Birmingham, AL 35210
Position: Pres. *Employer:* C B, Inc. *Education:* BCE/Sanitary Engr/RPI; BS/Math/RPI. *Born:* 7/21/26. RPI, BS '47, BCE '48. Currently Pres C B, Inc - Engineers-Surveyors. Experience as mgr of Tech/Profl operations; in charge of process design & engg, res & dev acquisitions, industry - govt relations in prior positions. Served as Pres - Cahaba Water Renovation Systems, Inc.; Pres - TII Ecology Div; Mgr - Wheelbrator Water & Wastewater Systems Div; VP - Teledyne Brown Sci & Engg Div & Pres - Townsley Mullins (A & E subsidiary); Chrmn - Govt/Industry Technical Task Force on Eutrophication; Pres - the Soap & Detergent Assoc; Mgr - Permutit Co Wastewater Treatment Dept and Resident Advisor to Shinko Pfaudler Co, Lt (Japanese Joint Venture). *Society Aff:* AIChE, WPCF, AWWA, NSPE.

Buenz, Peter R
Home: 4730 Gulfway Dr, Baytown, TX 77521
Position: Exec Vice Pres. *Employer:* Chemical Exchange Industries *Born:* Jul 1937. Native of San Antonio Tex. Naval Officer 1960-63. Process Engrg Supt ARCO Chemical, Channelview Tex 1967-70. Assumed current pos as V P Chem Exchange Co Inc 1970. Plant Mgr of Galena Pk, Tex plant & in charge of engrg & oper of both Galena Pk & Baytown Tex plants. Dir & Secy Baytown Area Water Auth. Elder of Baytown Presby Church. Bd mbr Harris-Galveston Coastal Subsidence District.

Buffington, Dennis E
Business: Agricultural Engrg Dept Rogers Hall, Gainesville, FL 32611
Position: Assoc Prof *Employer:* Univ of FL *Education:* Doctor of Philosophy/Agric Engr/Univ of MN; MS/Agric Engr/PA State Univ; BS/Agric Engr/PA State Univ *Born:* 8/11/44 Native of Elizabethville, PA. Assumed current faculty position in Agricultural Engrg Dept at Univ of FL in 1971 with primary roles in teaching and research. Teach graduate and undergraduate courses in structures and environment. Research interests include thermal analysis of structures, environmental modification systems for structures, environmental physiology of plants and animals, and systems analysis. Involved with ASAE activities and continuing education workshops. Recipient of A. W. Farrall Young Educator Award in 1980. *Society Aff:* ASAE, ASHRAE, ASEE, ΣΞ

Bugbee, Percy
Home: 5 Howes Ln, Plymouth, MA 02360
Position: Hon Chrmn of Bd. *Employer:* Natl Fire Protection Assn. *Education:* BS/Chem Engg/MA Inst of Tech. *Born:* 9/5/98. Have been active in fire protection engg work for more than fifty yrs. Gen mgr of Natl Fire Protection Assn for thirty yrs, now Hon Chrmn of the Bd. Helped to found the Soc of Fire Protection Engrs. Charter mbr no. one and honorary life mbr. Author of text book Principles of Fire Protection published in 1978. Listed in Who's Who in America. *Society Aff:* SFPE.

Bugliarello, George
Business: 333 Jay St, Brooklyn, NY 11201
Position: President. *Employer:* Polytechnic Univ. *Education:* ScD/Hydrodynamics/MIT; MS/-/Univ of MN; Dott Ing/-/Univ of Padua. *Born:* 1927. ScD MIT; MSCE Univ Minn; Dott Ing Univ Padua. LL.DD. (Hon) CMU; P E NY. Pres Polytechnic Inst of N Y 1973- . Huber Res Prize 1967; NATO Sr Post-Doc Fellow Technical Univ Berlin 1968; Alza Lect, Biomedical Engrg Soc Aug 1976. P Chmn ASCE Engrg Mech Div; former Dir Biomedical Engrg Soc; Mbr Bd of Sci & Tech for Internatl Dev of the NAS, & Chmn of its Adv Ctte on Tech Innovation. Chrmn, NSF Advisory Ctte for Sci Education. Chmn, Adv Panel, Tech Transfer to Mid-East, Off Tech Assmnt, Many pubs in bioengrg, fluid mech, computer lang & social tech. Books authored or edited: Bioengineering-An Engrg View; Computer Sys & Water Resources; The Impact of Noise: A Socio-Tech Intro; Technology, Community & the Univ; Issues & Public Policies in Ed Tech. Chmn COSEPP, AAAS 86- ; Steer. Comm, Sci for Stability, NATO, 1984- , Member, HAE; Member, Council on Foreign Relations. *Society Aff:* ASCE, ASEE, NYAS, NAE.

Buist, Thomas B
Business: 2270 SW 36th St, Ft Lauderdale, FL 33312
Position: VP. *Employer:* Larry L Wilson, Inc. *Education:* BSCE/Civ/Univ of Miami. *Born:* 4/2/48. Participated in student exchange with Univ of Sheffield, 1971. Proj Engr for Connell Assoc 1971-1974, responsible for regional environmental planning. Designer Engr from 1974-1975 with Black, Crow, Eidsness. Project Mgr for Hazen & Sawyer from 1975-77 in charge of environmental design & construction projs. In 1977, became principal partner in Larry L Wilson Construction, responsible for contract negotiations and administrations. Pres of Am Soc of Civ Engrs - S FL Sec - 1977. Enjoy skiing & sailing. *Society Aff:* ASCE, FES, NSPE.

Bulkeley, Peter Z
Business: MECH-EL INDUSTRIES, 17 Everberg Rd, Woburn, MA 01888
Position: VP Technical Operations *Employer:* Mech-El Industries *Education:* PhD/Engg Mech/Stanford; MS/Mech Engg/MIT; BS/Mech Engg/MIT; BA/Mathematics/Bowdoin College. *Born:* Mar 1934. Native of San Francisco. Taught at Stanford 1957-72; held pos as Dir, Design Div of Mech Engrg, Dean of Students, & Assoc Chmn of Mech Engrg. At Bradley Univ since 1972 as Dean of the College of Engrg & Tech. in Sept. 1980 Position change to V.P. Tech. Ops. at Mech-El. interest in Mech Vibrations, Engrg Design, New Company Start-ups. *Society Aff:* ASEE, ASME, SME, ΣΞ.

Bull, Stanley R
Business: 1617 Cole Blvd, Golden, CO 80401
Position: Senior Scientific Adv *Employer:* Solar Energy Research Inst *Education:* PhD/ME/Stanford Univ; MS/Engr Sci/Stanford Univ; BS/ChE/Univ of MO *Born:* 5/15/41 Native of LaPlata, MO. Held positions of Assistant Prof, Assoc Prof, and Prof in Coll of Engrg at The Univ of MO-Columbia, 1967-80. Fulbright-Hays Prof and Visiting Scientist at the CENG, Grenoble, France in 1973-74. Service on national committees of ANS and AAPM. Chrmn of the Mo-Kan Section of ANS, 1977. Consultant to Argonne National Lab, Ellis Fischel State Cancer Hospital, Univ of Louisville Health Sciences Center, and Boone County Hospital. Senior Engr at Solar Energy Research Inst, 1980-81. Senior Scientific Advisor In Deputy Dirs office, 1981-present. *Society Aff:* ISES-AS, ASEE, APS-AAPM, HPS, ASNT

Bull, Wesley A
Home: 2621-2nd Ave 1905, Seattle, WA 98121
Position: Pres. *Employer:* Wesley Bull & Assoc, Inc. *Education:* BS/EE/Univ of WA. *Born:* 3/25/22. Worked for PT&TC, including Bell Tel Labs (2 yrs), for a total of 16

Bull, Wesley A (Continued)
yrs in the Chief Engrs Dept. Formed my own telecommunications consulting engg firm "Wesley Bull & Assocs" in 1962. The firm was inc in 1978. Wesley Bull is Pres and Chief Exec Officer. Wesley Bull & Assocs, Inc is devoted to performing telecomm conslt and design engg services for telephone util, govt and commercial industry throughout the US. Currently employment is 11, present volume of business is $600,000 per yr. *Society Aff:* NSPE, IEEE, AFCEA, WSPE, ACE.

Bungay, Henry R
Business: 204 Ricketts Bldg, Troy, NY 12180
Position: Prof of Chem & Environ Engrg. *Employer:* Rensselaer Polytechnic Inst. *Education:* PhD/Biochem/Syracuse; BChE/Chem Engrg/Cornell Univ *Born:* Jan 22 1928. Teaching exper at Virginia Polytechnic Inst, Clemson Univ & now at RPI. Served as Program Mgr at NSF for Enzyme Tech & at ERDA for Fuels from Biomass. Spent 7 yrs at Eli Lilly & Co in fermentation dev & 3 yrs at Worthington Biochem Corp as Vice Pres-Tech Dir. Active in Amer Chem Soc as Chmn, Div of Microbial Chem & Tech & Councilor. Fellow of Amer Inst Chem Engrg, Amer Soc Microbiol, & Sigma Xi. Reg P E Vir & N J. Life master in Amer Contract Bridge League, private pilot license. Enjoy tennis. Dean's award for innovations in effective teaching at Clemson Univ, James Van Lanen Service Award Am Chm Soc, Am Assoc of Publishers Award for best technical book for "Energy, The Biomass Options" Wiley (1981). *Society Aff:* ACS, ASM, AIChE, SGM, ΣΞ

Bunker, Frank C
Business: PO Box 8368, Chicago, IL 60680
Position: Dir, Refining and Project Development, Int *Employer:* Amoco Oil Co *Education:* BS/Chem Engr/Rice Univ. *Born:* 7/11/23 Native of Houston, TX. US Navy 1941-46. Served as Deck & Eng. Officer, South Pacific 1943-45. With Std Oil, IN since 1947. In var domestic & intl Res, Dev & Mfg assignments. Currently Dir, Refining and Project Development, Intl for Amoco Oil Company. Registered PE-TX. Dir overseas subsidarys. *Society Aff:* AIChE, API.

Bunn, Joe M
Business: Agricultural Engrg Dept, Clemson Univ, Clemson, SC 29634-0357
Position: Prof *Employer:* Clemson Univ *Education:* PhD/Ag Engrg-Math/IA State Univ; MS/Ag Engrg/NC State Univ; BS/Ag Engrg/NC State Univ *Born:* 1/10/32 Native of Wayne County, NC. Worked in a teaching and research position at the Univ of KY from 1960 to 1978, except for 2 years tour with and AID-Ran D team in Thailand. Responsibilities at Univ of KY were in the area of on-the-farm processing, curing and handling burley tobacco. In 1978 moved to present position of Prof of Agricultural Engrg with teaching and research responsibilities in the area of materials handling and crop processing. Have authored over 100 technical articles, 3 chapters for textbooks, and numerous other educational materials. *Society Aff:* ASAE, ΣΞ, ΓΣΔ, CAST

Bunshah, Rointan F
Business: 6532 Boelter Hall, Univ of CA, Los Angeles, CA 90024
Position: Prof. *Employer:* UCLA. *Education:* DSc/Metallurgy/Carnegie Mellon Univ; MS/Metallurgy/Carnegie Mellon Univ; BSc/Metallurgy/Banaras Hindu Univ. *Born:* 12/18/27. Educator, researcher and consultant in materials synthesis, vapor deposition processes, vacuum metallurgy, structure-property relationships, Tribiological coatings Tribiological coatings solar cells, and biomaterials. Author of 120 publications and Editor of Reference Set *Techniques of Metals Research.* Co-Author of recently published book on "Deposition Techs and Their Applications–. *Society Aff:* ASM, AVS, IIM, IVS.

Bunzl, Rudolph H
Home: 406 Lakeway Ct, Richmond, VA 23229
Position: President. *Employer:* Amer Filtrona Corp, Richmond. *Education:* BS/ChE/GA Tech. *Born:* Jul 1922. Engrg & supervisory pos with Shell Chemical Co in Calif & N Y 1943-54. Served in US Army 1944-46. Reg chem engr Calif. V Pres 1954-59 & Pres of American Filtrona Corp since 1959. *Society Aff:* AIChE.

Buonicore, Anthony J
Business: Commerce Park, 4695 Main St, Bridgeport, CT 06606
Position: President *Employer:* Buonicore-Cashman Associates, Inc. *Education:* MChem Eng/-/Manhattan Coll; BChem Eng/-/Manhattan Coll *Born:* 2/25/50. Native of Bronx, NY. Process Environmental Engr, Stauffer Chem Co, 1970- 1971. USAF Principal Environmental Engr 1972-1975. Proj Mgr, Entoleter, Inc 1975-1978. Gen Mgr, Air Pollution and Solid Waste Services Div, York Res Corp 1978-1979. VP, York Services Corp 1980-81. Reg PE, OH, CT, NY & MA. Chrmn, APCA Tech Council, 1984-. Author more than four dozen publications and seven books. Founder and First Gen Chrmn, Annual AIChE/APCA Conferences on Energy and the Environment. Mbr, Tau Beta Pi, Sigma Xi. AIChE Recognition Award for Outstanding Leadership, 1974. Diplomate, AAEE, APCA. *Society Aff:* AIChE, APCA, AAEE, ISPE.

Burbank, Farnum M
Home: 6034 Forrest Hollow La, Springfield, VA 22152
Position: Chief Equip. Dev. Engr *Employer:* USDA-Forest Service *Education:* B.S./Mech Engrg/IA St Univ *Born:* 6/21/23 Native of Los Angeles, CA. Engrg officer with U.S. Navy, 1943 to 1946. Mechanical design engr for U.S Electrical Motors, 1947 to 1950. Quality control and reliability engr, Rocketdyne Div., North American Aviation (Rockwell intl), 1950 to 1965. With USDA - Forest Service, Equipment Development since 1965, at San Dimas, CA and Missoula, MT, then to headquarters, Washington, DC as Chief Equipment DevelopmEnt engr. Responsible for programming, budgeting, policy development, coordination, and monitoring of entire F.S. Equipment Development Program. National Forest engrg committee, PM-55, ASAE; local section chrmn, ASAE. Registered PE, CA, No. AG00183. Hobby-model boats. Enjoy hiking, fishing, outdoor recreation. *Society Aff:* ASAE, AFA

Burchsted, Clifford A
Business: Oak Ridge N Lab-Box X, Oak Ridge, TN 37830
Position: Engrg Specialist. *Employer:* Union Carbide Corp Nuclear Div. *Education:* MS/Ind Mn/Univ of TN; BS/Mech Engg/Northeastern Univ; -/Engg/Ripon College. *Born:* Aug 1921 Braintree Mass. Reside in Clinton Tenn. Army AF 1942-46. Sample-Durick Co 1948-52. Joined Union Carbide Nuclear Div 1952, with Oak Ridge Natl Lab since 1956. International authority in high efficiency and nuclear air and gas cleaning specialist, sr author 'Nuclear Air Cleaning Handbook', author num papers on nuclear & high efficiency air cleaning. Mbr sev natl standards cttes. Cons on nuclear air cleaning, lectr Harvard Sch of Public Health. Fellow ASME, Royal Soc of Health. ASME Centennial Medal, 1980. IES Seligman Award 1974. Mbr Phi Kappa Phi, Sigma Xi. Married Elizabeth Johns Simmons of Miami, Fla. Presently mbr of US Dept of Energy Nuclear Standards Management Center at Oak Ridge Natl Lab (oper by Unin Carbide), cons on specifications for spec equip. *Society Aff:* ASME, NSPE, RSH, IES, ASTM.

Burdette, Edwin G
Business: Dept of Civ Engr, Knoxville, TN 37996-2010
Position: Prof. *Employer:* Univ of TN. *Education:* PhD/Civ Engr/Univ of IL; MS/Civ Engr/Univ of TN; BS/Ag Engr/Univ of TN. *Born:* 9/8/34. Edwin G Burdette was born in Martin, TN, on Sept 8, 1934. Since 1969 he has been a faculty mbr of the Dept of Civ Engg at the Univ of TN, Knoxville, and has done extensive consulting work in Structural Engg. His primary field of profl interest is the analysis, design, and behavior of reinforced and prestressed concrete structures; he has been heavily involved in testing of anchors in concrete and in the analysis of structures for impulsive and impactive loads. He is married to the former Patsy Hill of Paris, TN, and they have five children. *Society Aff:* ASCE, ASEE, ACI, TRB

Burdette, John C, Jr
Business: Burdette, Koehler, Murphy & Assoc., Inc, 300 Quadrangle, Cross Keys, Baltimore, MD 21210
Position: Partner *Employer:* Burdette, Koehler, Murphy and Assocs *Education:* BE/EE/Johns Hopkins Univ *Born:* 7/1/21 During the 16 years as partner in Burdette, Koehler, Murphy and Assoc, Mr. Burdette has been responsible for more than 500 building power and lighting projects and seven major electrical distribution projects. Included in his design were two community colleges, major st office bldgs, hospitals and research bldgs. Mr. Burdette has held office in IEEE and IES and is recipient of the Founder's Award, presented by the Engrg Society of Baltimore in recognition of his contribution to the engrg field. Mr. Burdette is active in civic and prof organizations, among them Baltimore Rotary, the Building Congress and Exchange, Sec for the Board of Lutheran Hospital and Treas for St. Luke Lutheran Health Care, Inc. *Society Aff:* NSPE, CSI, IEEE, IES, ACEC

Burdick, Glenn A
Home: 1005 Curlew Place, Tarpon Springs, FL 33589
Position: Dean. *Employer:* Univ of S FL. *Education:* PhD/Physics/MIT; MS/Physics/GA Tech; BS/Physics/GA Tech *Born:* 9/9/32. Attended the Intl Inst for Theoretical Physics in 1959. Worked in the Div of Sponsored Res at MIT and served as sr mbr of the res staff of Sperry Rand Corp. Res was selected for publication in the series Frontiers in Physics. Served as Assoc Prof of Engg at Univ of S FL from 1965 to 1968 and has held the rank of full prof of engg from 1968 to the present. Dean of the College on Sept 1, 1979. Science, Engineering & Technology Service to Industry, Vice Chmn of Florida/Task Force- 1980-81. Expert Accident Reconstructionist. Retired as Dean at the University of South Florida in June 1986 and was given the title of Dean Emeritus and Distinguished Professor. *Society Aff:* IEEE, ISHM, ΤΒΠ, ΣΠΣ.

Burger, Christian P
Business: Dept of Eng Sci & Mech, Ames, IA 50011
Position: Professor. *Employer:* Iowa State University. *Education:* BSc(Eng)/Mech Engr/Univ of Stellenbosch South Africa; PhD/Mech Engr/Univ of Cape Town South Africa *Born:* 1929. Engr with BRUSH ABOE Group of Companies in United Kingdom 1952-55 spec in diesel & steam/elec power plants. Tech Engr Mobil Oil Co South Africa 1956-62--cons to heavy industrial & mining companies. Univ of Cape Town 1962-71- -cons to industries; teach in mech engrg; res in thermal stress analysis & combustion. Dept of Engrg Sci & Mech, Iowa State Univ 1971- . Res ints: exper stress & strain analy. Mbr SESA, Mbr ASNT Chmn paper's review ctte for 'SESA', Peterson Award 1971-73; Peterson Award 1975-76, Frocht award 1984, Hetenyi award 1984, Technical Editor for "EXPERIMENTAL MECHANICS-..
Society Aff: SEM, ASNT

Burgers, William Gerard
Business: Laboratorium voor Metaalkunde, Inst of Tech, Delft, The Netherlands
Position: Prof of phys chem, emeritus. *Employer:* State of Netherlands-ret. *Education:* Dr Chem/Univ of Groningem Netherlands *Born:* 1897 Arnhem, The Netherlands. Ramsay Fellow, Davy Faraday Res Lab, London 1925-27; Res worker, Philips Res Lab, Eindhoven 1927-40; Prof Phys Chem, Inst of Tech, Delft 1940-69; Visiting Prof Purdue Univ, Lafayette Ind USA 1949- 50. About 200 papers on recrystallization, phase transformations, x-ray & electron diffraction (Bibliography in J Less Common Metals 28, 1972. Mbr Royal Dutch Acad of Sci; Hon Mbr Societe Francaise de Metallurgie; Amer Society for Metals, Metals Soc-London; Royal Dutch Chem Soc; Hon doctor Universities of Strasbourg, Paris, Louvain, Bruxelles; Acta Metallurgica Gold Medal 1976 & others. *Society Aff:* KNCV.

Burgess, Price B
Home: 208 East Oak, Albion, MI 49224
Position: Consultant *Employer:* Independent *Education:* MBA/Indus Mgmt/Univ of MI; B Met E/MEt Engg/Rensselaer Poly Inst. *Born:* 1923 New Bedford, Mass. Reg Ind, Mich. Revere Copper & Brass 1943-44; Wyman-Gordon 1944-46; instr Univ of Mich 1947-49; Foundry Services 1949-50. Res Assoc Standard Oil (Ind) 1950-55. Ch Metallurgist Muncie Malleable 1955-57. Tech Dir John Sherry Lab 1957-60. 1960-81, Hayes-Albion Corp as Ch Metallurgist, Res Mgr, Tech Dir, Internatl Service Mgr, Controls Mgr since 1981, Independent Consultant. AFS Award of Sci Merit 1969, Res Bd 1970-72, Chmn Malleable Div 1969-71. Secy ASTM Ctte A-4 on Iron Castings 1970-72. ASM Engrg Program & Handbook Cttes 1969-75. Pubs- malleable iron processes, controls, & properties. *Society Aff:* ASTM, ASM, AFS, IBF, SPE, AIME.

Burggraf, Odus R
Business: 328 CA Bldg, 2036 Neil Mall, Columbus, OH 43210
Position: Prof *Employer:* OH State Univ *Education:* PhD/Aeronautics/CA Inst of Tech; MS/Aero Engrg/OH State Univ; BAAE/Aero Engrg/OH State Univ *Born:* 2/27/29 Fort Wayne, IN. Military Service, USAF 1954-56. Aerodynamicist Douglas Aircraft Co 1952; Engrg Specialist, Curtiss-Wright Corp 1956-60; Staff Scientist, Lockheed Missiles and Space Co 1960-64. Asst Prof, USAF Inst Tech 1954-56; Assoc Prof, Ohio State Univ 1964-66; Prof, 1966-Present. Guggenheim Jet Propulsion Fellow, CA Inst Tech 1952-54; Consultant, Astro Research Corp 1959- 69; Honorary Research Fellow, Univ Coll London 1972; Senior Postdoc Fellow, National Center for Atmospheric Research 1977; Special Lecturer in Applied Math, Univ Coll London 1978; AIAA National Technical Committee on Fluid Dynamics 1979- 80. Visiting Res Fellow, Imperial Coll London 1984. Research Interest: Computational Fluid Dynamics. *Society Aff:* AIAA, AAUP

Burghardt, M David
Business: Hempstead, NY 11550
Position: Prof & Chrmn of Engr *Employer:* Hofstra Univ *Education:* PhD/Mech/Univ of CT; MS/Mech/Univ of CT; BS/Marine/US Merchant Marine Acad *Born:* 05/24/42 A native of Groton, CT; sailed in the merchant marine as an engrg officer. Taught thermal engrg 12 yrs at US Merchant Marine Acad before assuming chrmnshp of the engrg prog at Hofstra Univ. Has publ several texts, most noteworthy *Engineering Thermodynamics with Applications*, 2nd Ed, Harper & Row. Consult on preventive maintenance and condition monitoring. Outside activities include pottery and a variety of sports. *Society Aff:* ASME, ASEE, IMARE

Burghart, James H
Business: 1983 E 24 St, Cleveland, OH 44115
Position: Prof of Elec Engg. *Employer:* Cleveland State Univ. *Education:* PhD/Systems Engg/Case Inst of Tech; MSEE/EE/Case Inst of Tech; BSEE/EE/Case Inst of Tech. *Born:* 7/18/38. in Erie, PA. Previously held faculty positions at the AF Inst of Tech and State Univ of NY at Buffalo. Since 1975, Prof of Elec Engg at Cleveland State Univ, Cleveland, OH (Chairman, 1975-85). Res interests in the gen areas of control systems, optimization and artificial intellegence. Has held several offices including chairman (1980-81) with the IEEE Cleveland Sec. Chrmn of IEEE Industry Applications Soc committee on student activities (1977-present), and Western Area Chairman of IEEE Region 2 (1986-present). *Society Aff:* IEEE, ΣΞ, ASEE.

Burghoff, Henry L
Home: 175 Mt Vernon Ave, Waterbury, CT 06708
Position: Metallurgical Consultant. *Employer:* Self. *Education:* DEng/Metallurgy/Yale Univ; MS/Metallurgy/Yale Univ; BS/Elec Eng/Yale Univ. *Born:* June 14, 1907. Native of Yalesville, Conn. With Chase Brass & Copper Co Inc 1928-68. Many patents & tech papers on copper alloys. Spec in phys metallurgy, melting, casting, fabrication, corrosion & powder metallurgy of copper & copper alloys; extrusion of titanium & zirconium; direction & admin of R&D programs in above areas, plus rhenium processing. Assumed respon as Dir of R&D in 1960. Currently Cons to Battelle Columbus Labs & Copper Dev Assn. Mbr Sigma Xi; ASM Fellow 1971. *Society Aff:* ASM.

Burkart, Matthew J
Business: PO Box 219, Southampton, PA 18966
Position: Pres. *Employer:* AEGIS Corp. *Education:* BSCE/Civil Engg/MO School of Mines and Metallurgy. *Born:* 1/10/40. Worked as office engr, heavy construction. Regional engg mgr for maj fdn contractor. Dir of Construction Services and Loss Control maj intl insurance carrier. Formed consulting firm in 1975. Consultant to insurance and construction industry. Experienced in heavy rigging and overland transport involving maj intl heavy equip moves. *Society Aff:* NSPE, ASCE, ASSE, ASTM, ACI.

Burkavage, William J
Business: Abington Executive Park, Clarks Summit, PA 18411
Position: Chairman - CEO *Employer:* Burkavage Design Associates, P.C. *Education:* BS/Civil Engg/Lehigh Univ. *Born:* 11/20/19. Retired Commander, Civil engr Corps, US Navy 1942-1946, 1951-1953 founder, president, NE Engg Co. 1958-1978 partner Von Storch & Burkavage, Architects and Engrs 1950-1978, Burkavage Evans Associates 1978-1984, successor firm Burkavage Design Assoc. Former mbr, secretary PA state registration bd, prof engrs former mbr, Natl Council Engg Examiners mbr Bd of Trustees Marywood Coll Scranton, PA, mbr Corp Advisory Bd Keystone Jr. Coll, La Plume, PA, partner in charge-projects operation moon ralay, transmitter and receiver facilities MD and HI; weather satellite tracking facility, Wallops Island, VA; ionspheric probe, worlds largest radio telescope, Arecibo, PR; prototype hangar - C5A - worlds largest aircraft various locations around the world. *Society Aff:* ASCE, NSPE, ASME, ADPA.

Burke, Jack W
Business: 12687 West Cedar Dr, Lakewood, CO 80228
Position: Mng VP *Employer:* Converse Ward Davis Dixon *Education:* MS/CE/TX A & M Univ; BS/CE/TX A & M Univ *Born:* 8/20/40 Born, raised and formal education in the state of TX. Carried out research studies for the TX Highway Dept at TX A & M Univ. Spent over 7 years as a Geotechnical Engr for Howard, Needles, Tammen and Bergendoff in Kanses City, MO. Geographically, project assignments located in fifteen mid-western states. Started with Converse Ward Davis Dixon in 1971 as a Project Engr in Pasadena, CA. Appointed Managing VP of Pasadena, CA office in 1976 and was in responsible charge of all technical, administrative and management aspects of the office. In 1981, designated as Managing VP of companies' newest office in Denver, CO. Life- time hobby of jogging 100 miles per month. *Society Aff:* ASCE, NSPE, ISSMFE, USCOLD

Burke, Joseph E
Business: 33 Forest Rd, Burnt Hills, NY 12027
Position: Consultant, Matls Sci & Tech (Self-employed) *Education:* BA/Chem/McMaster Univ; PhD/Chem/Cornell Univ *Born:* Sep 1914. Worked for Internatl Nickel Co, Norton Co, Manhattan Dist (Los Alamos) 1943-46, Assoc Prof Metallurgy, Univ of Chicago 1946-49; Mgr Metallurgy G E Knolls Atomic Power Lab 1949-54, Mgr Ceramics G E CRD 1954-72, Mgr Planning & Resources 1972-74; Mgr Special Projs 1974-79, Retired from GE 1979. Consultant, Matls Science & Technology. Mbr Natl Acad of Engrg 1976, Fellow Amer Ceramic Soc, Distinguished Lecturer 1971-73, Pres 1974-75 Jeppson Medal, 1981; Fellow Amer Soc for Metals, Amer Nuclear Soc. Primary tech ints: reaction kinetics & origin of microstructure in metals & ceramics; materials for nuclear energy. Hobbies: photography, antique cameras, woodworking. *Society Aff:* ACS, NAE, ASM, NICE.

Burkert, John Wallace, Sr
Home: 29 Stella Dr, Rd 1, Hockessin, DE 19707
Position: President. *Employer:* Delaware Engrg & Design Corp. *Education:* BS/EE/Penn State Univ. *Born:* 6/29/26. Native of Philadelphia Penna. Power Specialist Engr, E I duPont de Nemours Co Inc, Newark Del 1950-66, Pres Delaware Engrg & Design Corp, Newark Del 1966- . Guest lectr Pratt Inst on Life Support Sys 1973-74; Genl Chmn Engrs Week Banquet for State of Delaware 1972; Mbr Amer Soc Heating, Refrig & Air Cond Engrs. (Pres & organizer of Delaware Chap 1969-70, mbr & P Secy T C 1.8, Natl Res Promotion ctteman 1974-76. Past mbr Bd of Dir & Region III Chmn. Bd of Dir of Sanford School. *Society Aff:* ASHRAE, NSPE, CEC, SAME.

Burkhardt, Donald B
Home: 2 The Horseshoe, Newark, DE 19711
Position: VP *Employer:* Von Bree Inc. *Education:* BS/ChE/Univ of RI. *Born:* Apr 1920 Providence R I. Employed by Monsanto-9 yrs production supervision & 3 yrs process design. Then employed by duPont Engrg Dept-6 yrs process eval & design. Formed Von Bree Inc, Newark 1960 to provide cons in all aspects of sulfuric acid plant design & modification incl catalyst kinetics. Have designed entire new sulfuric acid plants & catalytic converters. From 1969 thru 1983 was engaged in process design cons for Catalytic Inc in projects involving inorganic acids, dyes, pesticides, furfural, pickle liquor recovery, nerve gases, & waste incineration. In 1984, was responsible for process design for two pharmaceutical plant proj with Carew Assoc. Have taken recent courses in microbiol, biochem, and genetic biol at the Univ of DE. Experienced in expert witness testimony. Enjoy gardening, sailing. Reg P E Del. *Society Aff:* ΤΒΠ, ΦΚΦ, AIChE.

Burkhardt, Kenneth J, Jr
Home: Quaker Lane Farm, Box 420, Quakertown, NJ 08868
Position: Exec VP *Employer:* Dialogic Corp *Education:* PhD/Comp Sci/Univ of WA; PhC/Comp Sci/Univ of WA; MS/Comp Sci/Rutgers Univ; AB/Phys/Cornell Univ. *Born:* 3/30/45. Born in Elizabeth, NJ on March 30, 1945. Presently an asst prof of Elec Engg at Rutgers Univ and an independent consultant in computer system design. Previously served as Systems Architect at Burroughs Corp., Ass't Professor at Rutgers University, VP of Systems at Intec Inc, as a Res Assoc in the depts of Psychology and of Physiology and Bio-Physics at the Univ of WA and as a systems programmer at the American Cyanamid Co. Current res interests are the ditital voice store and forward technology, computer architecture and real-time system design. *Society Aff:* IEEE, ACM, ΣΞ

Burnet, George
Home: 4813 Dover Drive, Ames, IA 50010
Position: Department Head. *Employer:* Iowa State Univ. *Education:* PhD/ChE/Iowa State Univ; BS/ChE/Iowa State Univ *Born:* Jan 1924 Fort Dodge, Iowa. US Army 1944-48; USAR until 1969; separated as Lt Col. Process design engr, Commercial Solvents Corp 1951-56. Iowa State Univ since 1956; Hd, Chem Engrg Dept & Ch, Chem E Div, Ames Lab, USERDA 1961-72; Hd & Nuclear Engrg Dept 1972-1983. Coordinator, Engg Education Projs Office since 1977. Anson Marston Distinguished Prof since 1975. NSF/AID cons on sci ed in India 1967. Reg P E. Chmn Div of Fertilizer & Soil Chem ACS; Natl Pres Omega Chi Epsilon; Chmn Engrg Ed & Accreditation Ctte ECPD; Fellow and Pres ASEE; Fellow AIChE. Chrmn, Educ Affairs Council, Am Assoc of Engrg Soc, 1980-82. Bd of Dirs, Accreditation Bd for Engrg and Tech, 1977-83. US representative to World Federation of Engrg Organizations, Ctte on Educ and Training, 1983- . Adv Ctte, NSF Directors for Sci and Engrg Educ, 1984 - 86. Fellow, AAAS. *Society Aff:* AIChE, ASEE, NSPE, ACS, AAAS

Burnett, James R
Business: 1 Space Park-Bldg E115076, Redondo Beach, CA 90278
Position: VP & Genl Mgr *Employer:* TRW Defense Sys Grp *Education:* PhD/EE/Purdue; MSEE/EE/Purdue; BSEE/EE/Purdue. *Born:* Nov 1925. BS, MS & PhD Purdue. On fac Purdue until joining TRW 1956. Now V Pres & Genl Mgr TRW's Defense Sys Group. Winner of Charles A Coffin Fellow Award. Has been mbr of Selection Bd for Natl Sci Foundation; cons for Argonne Natl Lab Dept of Defense & other govt agencies. Also received AF Sys Command Award. Awarded hon degree of Dr of Engrg Purdue 1969; elected to NAE 1975. Active in IEEE, AIAA, Eta Kappa Nu, Tau Beta Pi & Sigma Xi. *Society Aff:* IEEE, AIAA, ΣΞ, ΤΒΠ, NAE.

Burnett, Jerrold J
Home: 13590 W. Colfax Ave, Golden, CO 80401
Position: Prof. Emeritus *Employer:* Retired *Education:* PhD/Engg Sci/Univ of OK; MS/Physics/TX A & I Univ; BA/Math/TX A & M Univ; ASEE/Elec Engg/Univ of TX at Arlington. *Born:* 5/31/31. Equip Engr, Southwestern Bell Telephone Co 1955-57; Instr, Univ of Dallas 1958-59; Asst Res Physicist, NM Tech 1959-61; Assoc Prof and Hd of Physics, Northwestern OK State Univ 1961-64; Prof of Physics, Radiation Safety Officer, CO School of Mines 1966-86, now professor emeritus; Program mgr, Grad Fellowships Prog, Natl Sci Fdn 1977-78. Dir of programs for Superior High School Students, 1967- 1976; Dir of Pre-College Teacher Grad Training Progs, 1962-present. Mbr, Bd of Dir of "The Diversity–, a non-profit education corp. Consultant to several industrial groups, cities and governmental agencies on environmental and educational problems. 1982-1984 Mgr, Radiation, Photometry and Spectrometry Laboratories, Measurement Standards Laboratory, Univ of Petroleum and Minerals, Dahahran, Saudi Arabia. *Society Aff:* NSTA, NSSA, ΣΞ, AAUP, ΣΠΣ.

Burnett, Richard E
Home: 9728 Camino Del Sol NE, Albuquerque, NM 87111
Position: Consulting Engineer. *Employer:* Self. *Education:* BS(EE)/University of Colorado *Born:* 1/8/11. Native of CO. Field Engr on large hyrdroelectric pwr & water supply projs USBR 19 yrs. With MAIN 1955-78 & VP 1971-78 Retired. Design, constr & review of large hydroelectric, water supply & irrigatiön projs worldwide. Life Mbr ASCE, M USCID, M USCOLD. Reg P E 4 states. *Society Aff:* ASCE, USCOLD, USCID

Burnham, Donald C
Home: 615 Osage Rd, Pittsburgh, PA 15243
Position: Retired *Education:* BSME/Automotive/Purdue Univ. *Born:* Jan 28, 1915. Hon Dr Purdue, Drexel, Indiana Inst of Technology, Polytechnic Inst of Brooklyn. Mgr of Mfg Oldsmobile Div of Genl Motors 1948-52. Asst Engrg Mgr Oldsmobile Div 1952-54. V P Mfg Westinghouse Elec Corp 1954-63. Pres Westinghouse Elec Corp 1963-69. Chmn Westinghouse Elec Corp 1969-75. Dir- Officer Westinghouse Elec Corp 1975-80 . ASME Richards Mem Award 1958. Mbr Natl Acad of Engrg 1968. Fellow ASME 1969. ASME Leadership Award 1970. Gantt Mem Medal 1972. SAM Taylor Key Award 1969. Hoover Medal 1978. Nat Engr Award of Am Assoc of Engrg Socs 1981. Hon Mbr SME 1974. Life Mbr ASM. Mbr IEEE, AIEE, SAE, and the Business Council. V Chmn Natl Comm on Productivity & Work Qual 1974-75. Dir, Amer Prod Ctr, 1978-80 . Dir, Soc Mfg Engrs, 1978-80 Dir. Goodwill Industries of America, Dir Logistics Management Institute. Emeritus trustee Carnegie - Mellon University, & The Carnegie. *Society Aff:* ASME, IEEE, SME, ASM, SAE, IIE.

Burns, Elmer N
Business: 25 S Front St, Columbus, OH 43215
Position: Traffic Control Engr. *Employer:* OH Dept of Transp. *Education:* BS/Civ Engg/Univ of Pittsburgh. *Born:* 5/23/32. Native of Mt Lebanon, Pittsburgh, PA. Served as Civ Engr, US Army Signal Corps, 1955-57. Hgwy Design Engr, Richardson, Gordon & Assoc. With OH DOT since 1962. Assumed present. Chairman, Public Relations, Ohio Society of Professional Engineers 1979-81. Technical Member, Martketing Committee, National Committee on Uniform Traffic Control Devices 1981-present. position in 1966, responsible for dev of stds & specifications for traffic control device construction improvement plans coordinated on a statewide basis. Reg PE, OH & PA. Presently Secy, TRB Freeway Operations Committee. Chrmn, ITE Technical Council, 1975-77. Pres, ITE, 1979. Mbr, Sigma Tau, Honorary Engg Fraternity. Mbr, Res Subcommittee, Natl Advisory Committee on Uniform Traffic Control Devices, 1975-77. Alternate Delegate, Natl Committee on Uniform Traffic Laws & Ordinances, 1975-present. Chrmn, Public Relations, OSPE 1979-81. Technical Mbr, Markings Cttee, Natl Cttee on Uniform Traffic Control Devices 1981-present. *Society Aff:* ITE, NSPE, OSPE, TRB.

Burns, Fredrick B
Home: 1768 Poplar Ave, South Milwaukee, WI 53172
Position: VP R&D. *Employer:* E Z Paintr Corp. *Education:* BS/EE/Univ of MO School of Mines. *Born:* 6/6/29. Springfield, MO. US Army 1954-56. Product dev engr for McGraw-Edison Co. Evening div faculty Marquette Univ College of Engg 1954-59. Reg PE since 1961. Joined E Z Paintr Corp 1961. Named VP R&D 1965. Respon for many product dev in painting tools industry. Granted more than a dozen US and foreign patents. Presented five tech papers at natl and local Societies for Paint Technology. Guest lecturer in paint and polymer conferences at Univ of MO-Rolla. Mbr ASTM. Married. 3 sons and a daughter. *Society Aff:* ASTM.

Burns, James F
Business: 3939 San Pedro NE, Suite D, Albuquerque, NM 87110
Position: VP Dir. *Employer:* Boyle Engg Corp. *Education:* BS/Civ Engg/Univ of NM; BFA/-/Dallas Art Ctr. *Born:* 11/18/28. Structural Engr & draftsman for Albuquerque Dist Corps of Engrs from 1948- 1953, Structural Engr & draftsman for R L Rolfe, Consulting Engr, Dallas, TX. Staff Engr for Gordon Herkenhoff & Assoc from 1956 to 1963, Dist Engr for Hydro- Conduit Corp, Albuquerque, NM from 1964 to 1970. Principal Engr for Fred Burns (self-employed) from 1972 to 1975; Principal Engr from 1972-1975 for MacCornack & Burns, Consulting Engrs. Presently Principal Engr, VP & Dir of Boyle Engg Corp, Albuq, NM. Active in profl & community affairs. Mbr of Albuquerque Environmental Planning Commission 1974-present (chrmn 1976-77) Bd of Albuquerque Ctrs Inc, Bd of Neighborhood Housing Services of Albuquerque. Past Pres of ASCE NM. Currently Pres-elect Natl Dir CEC/NM. Awards: Outstanding Sr Civ Engr Student, Univ of NM. Engr of the Yr 1978-NM Soc of PE. Bronze Star Metal for Valor, Korea 1952. *Society Aff:* ASCE, NMSPE, CECNM, APWA.

Burns, Joseph P
Home: 5501 York Ave S, Minneapolis, MN 55410
Position: Civil Engr. Consultant *Education:* BS/Civil Engg/Notre Dame Univ. *Born:* 1914. US Naval Reserve in Civil Engrg Corps 1943-46, incl 2 yrs on waterfront constr in S W Pacific. With U S Geological Survey as Photogrammetric Engr & in supervisory pos 1938-52. Asst Ch Engr Aero Service Corp 1952-53. V P Mark Hurd Aerial Surveys Inc 1954-70. 1970 -76 Civil Engr with U S Geological Survey. 2nd V P Amer Soc of Photogrammetry 1969-70, Dir ASP 1969. Pres ASP 1971- 72. 1976-Pres Private Consultant, Photogrammetry. *Society Aff:* ASPRS, ACSM, ASCE, NSPE.

Burns, Robert C
Business: Suburban Sta Bldg, Philadelphia, PA 19103
Position: President. *Employer:* Robert C Burns Assoc. *Education:* BME/Mech Engr/Catholic Univ of Amer. *Born:* Aug 1921. Born & raised in Pottsville, Pa. Army Engrg in WW II. Worked in aircraft & alum industry before college. After grad 1951, joined a Wash D C cons engrg firm & later became an assoc in the firm. Opened office in Phila 1958 for the genl practice of Mech & Elec engrg with airport lighting & elec distribution a specialty. P Pres CEC, Phila. Mbr ACEC, NSPE, IES, ASHRAE, Amer Arbir Assn, Chmn Mayor's Adv Bd for the Elec Code Phila. Reg Pa & 11 other states. Hobbies: jogging, music & gardening. Engr Excl 1972, CEC. *Society Aff:* ACEC, NSPE, ASHRAE, IES.

Burns, Robert G
Business: 245 Summer Street, P.O. Box 2325, Boston, MA 02110
Position: Quality Assurance Mgr *Employer:* Stone & Webster Engrg Corp *Education:* Assoc Degree/Mech Engrg/Northeastern Univ *Born:* 01/03/38 R G Burns, a native of Quincy, MA, is presently Quality Assurance Mgr for Stone & Webster Engrg Corp., respons for general QA Dept Mgmt covering field procurement and hdqts operations. He is also respon for Dept-wide quality engrg activities. His work at SWEC supports central station design, constr, project sponsorship, and licensing: including consulting for nuclear and non-nuclear power projects. Mr. Burns has been

Burns, Robert G (Continued)
active at the local and natl levels in the ASQC in which he is a Fellow. He is a founding mbr of the Nuclear Power Tech Ctte of the ASQC, which evolved into the Energy Div of which Mr Burns is VChrmn. He is also a member of the Amer Soc of Mech Engrs, a mbr of the Main Ctte on Nuclear Quality Assurance and Chrmn of the Standards Coordinating Subctte. *Society Aff:* ASME, ASQC

Burns, Stephen J
Business: Dept of Mech Engrg, Univ. of Rochester, Rochester, NY 14627
Position: Prof *Employer:* Univ of Rochester *Education:* PhD/Mat Sci & Engr/Cornell Univ; MS/Engrg Physics/Cornell Univ; BS/Engrg Sci/Pratt Inst *Born:* 1/31/39 Post-doctoral Fellow and Assistant Prof in Division of Engrg at Brown Univ 1967-1972. Specialty in dynamics fracture. Joined the Mechanical Engrg faculty at the Univ of Rochester in 1972. Research activities include, fracture mechanics and fracture toughness testing, design of fracture tough materials, microstructural deformation at crack tips in metal, ceramics and polymers, creep cracking and rate dependent fracture. Research techniques include: transmission electron microscopy, scanning electron microscopy, x-ray diffraction, mechanical testing and non-linear fracture mechanics testing. Invented the monolithic x-ray tube. *Society Aff:* ASME, AIME(TMS), ASM

Burr, Arthur A
Business: RPI, Troy, NY 12181
Position: Professor & Dean Emeritus *Employer:* Rensselaer Polytechnic Inst. *Education:* PhD/Physics/Penn State; MA/Physics/Univ of Saskatchewan; BA/Physics/Univ of Saskatchewan. *Born:* Aug 1913. Native of Manor, Saskatchewan. Naturalized Amer citizen. Taught in Saskatchewan public schools 1932-36. Worked during WW II as a res & dev engr for Armstrong Cork Co. Joined the RPI fac 1946. Hd of Dept of Met Engrg 1955-61. Dean of the Sch of Engrg 1962-74. Appointed to Chair of Rensselaer Prof 1974-78; Prof emeritus July 1978. Mbr ASEE, AIME, ASM. Award for outstanding teacher ASM 1952. Lester Mem Lecturer 1966. SAME Bliss Medal 1975. Hobbies incl gardening, fishing, camping. *Society Aff:* ASM, ASEE.

Burr, Arthur H
Business: Dept. of Mechanical Engineering, E.T.C. 5.160, Univ. of Texas at Austin, Austin, TX 78712
Position: Visiting Prof. Mech. Engrg. *Employer:* Univ. Texas/Austin *Education:* PhD/-/Univ of MI; MS/-/Univ of Pittsburgh; BS/WPI *Born:* 5/1908 Westinghouse Res Labs 1929-33. Rice Univ 1933-41, Univ of Missouri 1941-44, Asst Dir Aerial Measurements Lab of Northwestern Univ 1944-47. Cornell Univ: Hd, Dept Machine Design 1947-68, Cons Aero Lab 1951-53, Sibley Prof 1954-73. Universidad de las Americas, Mex, Prof 1973-76 & Hd Mech Engrg 1973-1975. Visiting Professor, Univ. of Texas at Austin 1984-88. Also, at times, visiting prof & cons at universities in Brazil, Columbia, & India. Prof ints & pubs in the analy & design of mechanisms & machines, incl impact, pitting, & lubrication. Fellow ASME. A book published by Elsevier in March 1981, namely Mechanical Analysis and Design. *Society Aff:* ASME, ASEE

Burr, Griffith C, Jr
Business: 139 Scott, Memphis, TN 38112
Position: Pres. *Employer:* Griffith C Burr, Inc. *Education:* BS/Mech Engg/Univ of TN. *Born:* 2/29/32. Attended college (GA Tech and Univ of TN) as co-op student employed by father's engg firm and is now Pres of the firm. Has directed mechanical and electrical design over last twenty-one years. Reg PE in TN, AL, AR, MA, MS, and FL. Fellow mbr ASHRAE 1978; Pres, Consulting Engrs of TN 1973; Chrmn of Energy Comm of the American Consulting Engrs Council for two terms, 1977-1979; Mbr of the Technical Advisory Group serving Fed Govt in its formulation of Bldg Energy Performance Stds (BEPS). Hobbies include tennis, sailing and hunting. Married with three daughters. *Society Aff:* ASHRAE, NSPE, ACEC.

Burr, Richard W
Business: P.O. Box 925009, Houston, TX 77292-5009
Position: President *Employer:* Burr Engineers, Inc *Education:* MBA/Business Admin/Texas A&M Univ; BS/Mech/Engrg/Texas A&M Univ *Born:* 11/08/40 I am a registered prof engr in eleven states. Since 1971, I have had my own consulting engrg practice, Burr Engrs. We now have three offices, in Houston, Austin, and San Antonio, Texas. Respons for mech and elect engrg design for commercial and institutional bldgs, i.e. high-rise office bldgs, hospitals, hotels, etc. 1969 - Design Engr in Mech and Elect Depts for John and Glasco, Inc. in Corpus Christi, Texas. 1970-71 VPres and Hd of Mech Dept, Thomas John Engrs, Houston, Texas. Active mbr of Rotary International. Interested in snow skiing. *Society Aff:* ASME, ASPE, ASHRAE

Burrell, Edward R
Business: 1 New York Plaza, New York, NY 10004
Position: President *Employer:* The Internatl Nickel Co Inc. *Education:* B Met E/Metallurgy/Rensselaer. *Born:* 8/3/32. Native of NYC. Attended RPI Grad Ctr, Windsor, CT. Served with Army Ordnance 1955-57. Met Engr & Supervisor at Genl Motors, New Departure Div 1957- 65. With Intl Nickel since 1965. Var sales & mktg capacities. Assumed current respon as VP-Exec 1979. Respon for Inco mktg efforts in US, Latin Amer, Hobbies: tennis, squash & golf, excelling at none.

Burrell, Montrust Q
Business: P O Box 9356, Baton Rouge, LA 70813
Position: Prof of Engrg. *Employer:* Southern University. *Education:* MS/ME/Tulane Univ; BS/ME/Howard Univ. *Born:* Oct 22, 1927; Native of Houston, Texas. BS Howard Univ, MS Tulane Univ. Reg ME La. 31 yrs Engrg Teaching & Admin. Part-time Cons for 23 yrs in Mech Equip for buildings. Chmn Baton Rouge Sect ASME 1976-77. Mbr Amer Soc for Engrg Ed, Mbr Louisiana Engrg Soc. Past Chmn Bd of Dir of Minority Engrs of Louisiana Inc Cons Engrs. Enjoy bowling & chess. *Society Aff:* ASME, ASEE.

Burrier, Harold I, Jr
Business: 1835 Dueber Ave, SW, Canton, OH 44706
Position: Sr. Research Specialist *Employer:* Timken Co. *Education:* Master's/Met/Case Western Reserve Univ; BS/Met/-. *Born:* 6/23/38. in Canton, OH. Metal powder applications engr, GE Co 1960-62. Met Engr, TRW, Inc 1962-65. Joined the Timken Co in Canton, OH in 1965 as Res Met. Involved in dev of matls & heat treating processes for tapered roller bearings, properties of high-temperature materials. Chrmn of Canton Massilon Chapter, ASM in 1976. Mbr & ASM Constitution & Rules of Govt Committee. Mbr & Secy ASM Chapter & Mbrship Council Hobbies are horses & photography. Married, 3 children. *Society Aff:* ASM.

Burris, Conrad T
Business: 4513 Manhattan Coll, Parkway, New York-Bronx, NY 10471
Position: Professor and Chairman, Chemical Engineering Dept. *Employer:* Manhattan College. *Education:* PhD/Chem/Cath Univ of Amer; MS/ChE/Univ of Alberta; BS/ChE/Univ of Alberta. *Born:* May 1924. Employed by Consolidated Mining & Smelting Co 1946-47 & by Canadian Indus Ltd 1948-50. Assoc with Manhattan Coll since 1955. Served as Chmn, Chem Engrg Dept 1961-71, 1983-present & Dean, Sch of Engrg 1971-80. Res ints: pressure effects on liquid phase reactions; dissolved oxygen analy by polarography; factors influencing oxygen uptake rate in aqueous sys. Served as Dir of Particulate Solid Res Inc, an indus-spon res org located at Manhattan College. Mbr Sigma Xi, & Tau Beta Pi. *Society Aff:* AIChE, ASEE, ACS.

Burris, Frank E
Business: Rt 38, Bldg 204-2, Cherry Hill, NJ 08358
Position: Mgr, Engg Education. *Employer:* RCA Corp. *Education:* PhD/EE/Univ of Cincinnati; MS/EE/Univ of Cincinnati; BSEE/EE/Univ of Cincinnati. *Born:* 12/29/43. Native of Ft Thomas, KY. Faculty mbr in Dept of Elec Engg at Univ of Cincinnati from 1967-1974. Faculty mbr in Dept of Elec Engg & Elec Engg Tech at

Burris, Frank E (Continued)
Bradley Univ from 1974-1978. Mgr, Engg Education, RCA Corp since 1978. Responsible for mgt of RCA's Corp Continuing Engg Education unit which enrolls approximately 1700 RCA technical personnel annually in approximately 150 classes in 30 locations. Principal profl interest: continuing profl dev of technical personnel. VP & mbr, ASEE Exec Committee, 1977-78; Mbr, IEEE Education Soc Ad Com, 1979-81; Mbr, ASEE Continuing Professional Development Division Executive Bd, 1980-83. *Society Aff:* ASEE, IEEE, SME, TBΠ, HKN.

Burris, Harrison R
Business: One Space Park, Redondo Beach, CA 90276
Position: Mgr, Advanced Computer Architecture R&D. *Employer:* TRW Defense & Space Systems Grp. *Education:* MS/Comp Sciences/PA State Univ; MBA/Mgt/Fairleigh Dickinson Univ; BS/EE/PA State Univ. *Born:* 7/13/45. Managed data processing operations of Applied Sci Labs 1968. Chief Technologist, Office of the Proj Mgr for US Army Tactical Data Systems 1970-74. Instituted Army surety prog & microprogramming res culminating in the military comp family. Joined TRW in 1974, became Mgr of Advanced Comp Architecture Res and Dev in 1977. Created the synthetic workload & instrumented architectural level emulation tools for comp design. Also, comp sci lectr at CA State Univ at Fullerton & editor for microprogramming & emulation of Simulation magazine. *Society Aff:* IEEE, ACM, SCS.

Burroughs, Frank S
Business: 1000 S Fremont Ave, Alhambra, CA 91802
Position: Vice President. *Employer:* C F Braun & Co. *Education:* MS/ChE/Univ of So CA; BE/ChE/Univ of So CA. *Born:* Jul 3, 1925 Los Angeles Calif. BE & MS Chem Engrg Univ of Southern California. Cert in Bus Mgmt Program for Tech Personnel from Univ of California at Los Angeles. Joined C F Braun & Co 1948. Respon incl Mgr of Process Engrg Dept & Mgr of Engrg Div. Currently have corporate respon for Engrg, Computer Sys, & Personnel. Mbr AIChE & ACS. Nonprof interests incl civic activities & golf. *Society Aff:* AIChE, ACS.

Burroughs, Jack E
Home: 2005 Milam St, Fort Worth, TX 76112
Position: V P-Res, Dev, Engrg & Mfg. *Employer:* S & B Bio-Medics Inc. *Education:* BSCrE/Cer Engg/Univ of TX; MSE/Nuclear Engg/Southern Methodist Univ; -/Cer Engr & Biomedical/OH State Univ. *Born:* Jan 1926. Grad course work at Ohio State towards PhD. Native of Marion, Ohio. Was associated with Genl Dynamics/FW for 13 yrs, respon for all ceramic R&D. Dresser Minerals for 3 yrs. Proj Mgr for calcined Kaolin dev & production. Assumed current respon as V P of S & B Jan 1974. Developer of many aerospace materials, bio-engrg materials. Pub over 30 articles in var journals. Fellow Amer Ceramic Soc, P Chmn ACS, P Natl Pres SAMPE, started the J of SAMPE, Mbrship Chmn Soc for Biomaterials. Hon Keramos, NICE, & Sigma Xi. Bus: S & B Bio-Medics Inc, 512 S Freeway, Ft Worth TX 76104. VP Res Dev Engr, Mfgs Mkt for Seecor, Inc. VP New Products, Aquisition & Mfg Intl Biomedical, Inc. Seminar Instructor at the Univ of TX at Arlington on Biomatls. *Society Aff:* ACS, SAMPE, SFB, NICE, TSPE.

Burrowes, Joseph J
Home: 7 N Drexel Ave, Havertown, PA 19083
Position: President. *Employer:* PAKOIL CO. *Education:* AB/Political Science/Temple University *Born:* Jan 7, 1919. AB Political Science Temple University, 2 yrs Temple Univ Law School. 1942-46 Pilot US Army Air Corps--reverted to inactive status with rank of Major. Sev yrs of Tech Sales with R M Hollingshead Corp spec in Gov Specification Rust Preventives. June 1952 founded Pakoil Co, which represents sev companies supplying packaging materials. Emphasis is on designing of specific packs for specific sophisticated equip to meet the requirements of either governmental or indus programs. Eastern Regional V P SPHE 1968-69; Natl V P SPHE 1970-71; Exec V P SPHE 1972-73; Pres SPHE 1974-75; Bd Chmn SPHE 1976-77. *Society Aff:* SPHE, ADPA, NIPHLE

Burrows, F G Alden
Home: 921 Fickle Hill Rd, Arcata, CA 95521
Position: Professor. *Employer:* Humboldt State Univ. *Education:* PhD/Civil Engg/Univ of WA; MSc/Civil Engg/Univ of Manitoba; BSc/Civil Engg/Univ of Manitoba. *Born:* Jun 1930. Native of Treherne, Man. With Bridge Sect, Hwy Dept, Province of Manitoba 1952-66. Started as resident engr & became Asst Ch Bridge Engr 1957. Resigned in 1966 to enter grad sch. Finished grad studies 1971 & assumed teaching career at Humboldt St Univ. Cons Engr-Diver in underwater inspection of bridges, dams & docks, rivers & coastal areas throughout U S & Canada. Enjoy skiing & swimming. Freeman Fellowship 1969 ASCE; Canadian Good Rds Fellowship 1966 CGRF. *Society Aff:* ASCE, ARB, BA, APEM.

Burrus, C S
Business: EE Dept. Rice Univ, P.O. Box 1892, Houston, TX 77251
Position: Prof of EE *Employer:* Rice Univ *Education:* PhD/EE/Stanford Univ; MS/EE/Rice Univ; BS/EE/Rice Univ; BA/EE/Rice Univ *Born:* 10/9/34 in Abilene, TX. Married in 1958, two children born in 1959 and 1961. Joined the Rice Univ faculty in 1965. Was Master of Lovett Coll at Rice from 1972-78; chrmn of EE Dept 1984-. Was a Visiting Prof at the Univ of Erlangen in West Germany in 1975-76 and again in 1979-80. Visiting Fellow at Trinity Coll, Cambridge Univ, summer 1984. Publ 2 books and over 100 articles in Digital Signal Processing. Investigator in 13 grants. Registered PE in TX since 1967. Received Rice Univ teaching awards in 1969, 74, 75,76, and 1980. IEEE Society Senior Paper Award in 1974, Alexander von Humboldt Senior Award in 1975, Senior Fulbright Fellow in 1979, and IEEE Fellow in 1981. IEEE Society Technical Achievement Award for 1985. *Society Aff:* IEEE, ΣΞ, TBΠ

Burrus, Charles A, Jr
Business: AT & T Bell Laboratories, Crawford Hill Lab, Holmdel, NJ 07733
Position: Mbr Technical Staff. *Employer:* AT&T Bell Labs *Education:* PhD/Physics/Duke Univ; MS/Physics/Emory Univ; BS/Physics/Davidson College. *Born:* Jul 1927 Shelby N C. Res in microwave spectroscopy at mm- & sub-mm wavelengths, mm-wave semiconductor diodes, electroluminescent diodes, photodetectors, single-crystal optical fibers, optical-fiber devices. Fellow of AAAS, APS, IEEE, and OSA. Rec'd David Richardson Medal, OSA, 1982, Distinguished Technical Staff Award - AT&T Bell Labs, 1982. Active Methodist. Hobby: photography. *Society Aff:* AAAS, APS, IEEE, OSA, ΦBK, ΣΞ

Burte, Harris M
Home: 3006 Hedge Run Court, Dayton, OH 45415
Position: Chief Scientist *Employer:* AFWAL Matls Lab, Wright-Patterson AFB, Ohio 45433B. *Education:* PhD/Chem Eng/Princeton; MS/Chem Eng/MIT; BS/Chem Eng/NYU *Born:* Nov 14, 1927 NYC. Sr res engr Textile Res Inst, Princeton N J 1948-53. AF Officer 1953-56. Since 1956 has held sev sr civilian pos in what is now the AF Materials Lab & as Ch Engr-Materials for the AF Aeronautical Sys Div. Contrib have covered a broad interdisciplinary range from fundamental res to engrg dev & production incl topics such as: non-linear viscoelasticity of protein fibers; hydrogen embrittlement in titanium alloys; design with brittle materials; metals process modeling; adhesive bonding; nondestructive evaluation; composite materials. *Society Aff:* ACS, AIME, ASM, AIAA

Burtis, Theodore A
Business: 100 Matsonford Rd, Radnor, PA 19087
Position: Pres & Chief Operating Officer. *Employer:* Sun Company Inc. *Born:* May 1922. BS Chem E 1942 Carnegie Inst of Technology; MS 1946 Texas A&M Univ; Hon Doctorate of Sci 1972 Ursinus Colllege. Pres Houdry Process & Chemical Co 1956-67. Also served as Dir & V P of Houdry's parent co, Air Products & Chemicals Inc. In 1967, joined Sun Co as Dir of Commercial Dev, subsequently appointed Dir of Res & Engrg, V P of Res & Dev, Pres Sun Res & Dev Co, & V P for Dev Projects. 1972, V P for Mkting, 1973 Pres of Sun Oil Co of Penna, 1975 Exec V P & Dir of parent company. June 1976 elected Pres & Ch Operating Officer of Sun

Burtis, Theodore A (Continued)
Company. Served as Dir, United Engrg Trustees 1963-65. Pres AIChE 1967; Treas AIChE 1970.

Burtness, Roger W
Home: 2113 Zuppke Dr, Urbana, IL 61801
Position: Assoc. Professor *Employer:* University of Illinois *Education:* PhD/EE/Univ of IL; M.S./EE/Univ of IL; B.A./Econ/St OLAF Coll; BS/EE/Univ of IL *Born:* 5/13/25 Native of Chicago. Navy V-12, WWII, commissioned. Test engr, General Elec, Schenectady, Syracuse, Erie, 1947-48. Field engr, Genl Elec, Chicago, 1948-49. Teaching asst and instructor, Univ of IL, 1949-53. Proj engr, then Section Engr, electro-Motive Div, Genl Motors, 1953-56. Proj engr, Grp Hd, and Mgr of Engrg nd Res, Stewart Warner Electronics, 1956-59. Assoc Prof of Electrical Engrg, Univ of IL, 1959 to present, now specializing in linear integrated circuits and solid state drive systems. Also served as consulting engr for a number of companies including Borg Warner Res and Electromotive Div of Genl Motors. Reg Prof Engr, State of IL. *Society Aff:* IEEE

Burton, Alan K
Business: Project Finance Group 5044, P.O. Box 37020, San Francisco, CA 94137
Position: VP *Employer:* Bank of America *Education:* MBA/Bus/Golden Gate Univ; MS/Mineral Econ/Stanford Univ; ACSM/Mining/Camborne School of Mines, U.K. *Born:* 6/11/39 Underground mining in Zambia and open pit and dredge mining in Malaysia and Australia, 1959-68. Assistant Mine Mgr for Osborne & Chappel in Malaysia; Mine Superintendent, Alcoa of Australia, Kwinana, Western Australia, 1968-70; Project Engineer, Senior Mining Engineer, Bechtel Inc., 1971-80; VP/Senior Mining Eng., Bank of America, 1980 to present. Affiliations: Chartered Engr - United Kingdom, Fellow - Inst of Mining and Metallurgy (London), Member - American Inst of Mining, Metallurgical and Petroleum Engrs. *Society Aff:* AIME, MMSA, IMM.

Burton, Bruce E
Home: 5664 Co Rd 35, Ada, OH 45810
Position: Dean College of Engr'g. *Employer:* Ohio Northern Univ. *Education:* PhD/Eng Mech/Univ of CO; MS/Eng Mech/OH State Univ; MAE/Automotive/Chrysler Inst of Engg; BS/ME/OH Univ. *Born:* Oct 19, 1933. P E Ohio. Native of Columbus, Ohio. Engr in chassis design with Chrysler Corp 1955-58. With Ohio Northern Univ since 1958. Assumed respon as hd of the mech engrg dept 1964. Promoted to full professor 1969. Appointed dean, College of Engineering in 1985. Mbrship in ASME, Tau Beta Pi, Sigma Xi ASEE. Name appeared in Outstanding Educators of Amer, 1970 Edition & Who's Who in Ohio 1975 Edition. Enjoy traveling. *Society Aff:* ASME, TBΠ, ΣΞ, ASEE, ΦHΣ, ΦKΦ, O/DELTAK

Burton, Conway C
Business: 3360 Commercial Ave, Northbrook, IL 60062
Position: President. *Employer:* Chicago Testing Lab Inc. *Education:* BChE/Chem Engr/Univ of MN. *Born:* Oct 7, 1920 Kansas City, Kan. Beta Theta Pi Frat. US Navy 2 yrs Electronic Technician. 1946-50 Tech Serv Engr Amer Bitumuls subsidiary of Standard Oil Calif. 1950-54 Product Engr Standard Oil Ind. 1955 V P & elected Pres in 1963 of Chgo Testing Lab Inc, a testing & cons lab on bituminous materials & paper. Authored many tech papers on asphalt tech & contrib editor of 'Bituminous Materials'. Reg engr Ill, Mich & Wisc. 1974-76 Pres of Amer Council of Independent Labs Inc. Hobbies: golf & skeet. *Society Aff:* AIChE, ASTM, AAPT, TAPPI, NSPE

Burton, Ralph A
Home: 1825 Ridge Road, Raleigh, NC 27607
Position: Pres *Employer:* Burton Technologies Inc. *Education:* PhD/Mech Engr/Univ of TX; MS/Mech Engr/Univ of TX; BS/Mech Engr/Univ of AR. *Born:* 10/31/25. Asst prof, MA Inst of Technology 1952-54; Assoc Prof Univ of MO, 1954-58; Section Mgr Southwest Research Inst, 1958-69; Liaison Scientist ONR London (on Leave) 1967-69; Chrmn Mechanical Engg, Northwestern, 1969-71; Prof to 1980 Leave 1978-81 Consultant to office of Naval Res. 1980 to present, Head Mechanical and Aerospace Engineering Dept, NCSU (1981-84). Pres, Burton Technologies Inc., 1984- . Author Vibration and Impact (Addison Wesley, 1958; Dover 1968); Dover 1968); Bearing and Seal Design in Nuclear Power Machinery Amer Soc of Mech Engrs (1967); Thermal Deformation in Frictionally Heated Sys (Elsevier 1980). Mbr Cosmos Club, Past Chrmn Lubrication Div, ASME, Fellow ASME, Past Mbr ASME Natl Nominating Committee Basic Engg Dept of Exec Committee, Best Paper, Award Lub Div ASME 1966 ASME Centennial, Medallion (Chicago Section), 1980. *Society Aff:* ASME, AIAA, AAAS.

Burton, Robert S, III
Business: PO Box 2687, Grand Junction, CO 81501
Position: Oper Mgr. *Employer:* Occidental Petroleum Corp. *Education:* MSChE/ChE/West VA Univ; BSChE/ChE/VPI. *Born:* June 4, 1947. MSChE W Vir Univ; BSChE VPI. 1969-72: Proj Engr on EPA spon dev of dual-fluidized bed to produce high Btu gas from solid waste. Conducted res in 'fast' fluidized bed coal gasification & liquification for City Coll of N Y. 1972- : Oper Mgr for Occidental Oil Shale. Have been instrumental in dev the 'in-situ' process for the extraction of oil from shale. Hold num pats concerning combustion techs & oil shale operational procedures. Mbr AIChE, AIME, 'Chem Engrg' Product Res Panel. Hobbies: flying, reading. *Society Aff:* AIChE, AIME.

Burton, William J
Home: 9451 Lee Highway, #713, Fairfax, VA 22031
Position: Prog Mgr *Employer:* Dept of the Navy *Education:* PhD/Mech. Engrg.-Phys. Oceanog'fy/Texas A&M University; MS/Mech. Engrg.-Fluid Mech./University of S. C.; BS/Mech. Engrg.-Power/University of S. C. *Born:* 3/22/31 Born, Gaffney, SC. US Army, Military Police, 1951-53. Pioneered dev ski-equipped cargo aircraft, Lockheed-Georgia Co., 1957-62. Developed advanced gas turbines, compressors, supersonic diffusers, General Motors Corp., 1964-67. Created/ taught courses energy conversion; directed research: wave-structure interactions, oil-spills, dynamic face seals while on faculties of Texas A&M and Univ Tenn 1970-74. Projects Manager, fluidized-bed combustion, cogeneration, advanced power cycles, emerging technologies, Tennessee Valley Authority, 1974-79. Currently Prog Mgr, mooring/ocean systems dev, Dept of Navy. Fellow, ASME. Reg P.E., TN. 46 publications. *Society Aff:* NSPE, ASME, SNAME, SME, ΣΞ

Burzell, Linden R
Business: 2750 4th Ave, San Diego, CA 92103
Position: Genl Mgr & Ch Engr. *Employer:* San Diego County Water Auth. *Education:* BS/Civil Engr/Caltech. *Born:* 9/16/24. Retired Cdr, Civil Engr Corp, US Naval Res. Fellow ASCE. Past Chrmn San Diego Sect ASCE. P Chmn CA Sect AWWA. Dir AWWA 1976-79; Ambassador 1974. Geo A Elliot Award 1971. George Warren Fuller Award 1976. APWA top Ten Public Works Leader-of-the-Yr 1980. Genl Mgr & Ch Engr, San Diego Cty Water Auth 1964- . This agency supplies imported water to San Diego Cty. Genl Mgr & Ch Engr, Vista Irrigation Dist 1951-63. *Society Aff:* ASCE, AWWA, APWA.

Burzlaff, Arthur A
Business: P O Box 1108, Arvada, CO 80001
Position: President. *Employer:* BF Sales Engrg Inc. *Education:* BS/Chem Engg/IA State Univ. *Born:* 10/26/22. Reg P E Chem Engrg Iowa & Colo. Active Mbr AIChE. Held pos of plant engr, res & dev engr, dept & co mgr for var firms in chem mfg. Started BF Sales Engrg 1965 to provide chem engrg knowledge to the mkting of indus engineered equip. This mkting is through cons engrg firms & corp engrs of indus accounts. Now Pres & major stockholder. Enjoys skiing, jeeping, fishing & genl mountain activities in spare time. *Society Aff:* AIChE.

Busby, Edward O
Business: Univ. of Wisconsin, 1 Univ. Plaza, Platteville, WI 53818
Position: Dean *Employer:* Univ of WI-Platteville *Education:* PhD/Civil Engr/Univ of WI-Madison; MS/Civil Engr/Univ of WI-Madison; BS/Civil Engr/Univ of WI-Madison *Born:* 06/22/26 Native of Madison, WI. Served in US Navy during WWII. Resident engr WI Hgwy Commission 1950-51. Asst City Engr, LaCrosse, WI 1951-53. Sales engr, WI Culvert Co 1953-59. Lecturer in Civil Engrg Univ of WI-Madison 1959-66. Assumed leadership Univ of WI-Platteville's Coll of Engrg as Prof of CE and Dean in 1966. WI Soc of Prof Engrs-Exec Dir (part-time) 1962-66; Pres 1972; Natl Dir 1976-81; Engr of the Yr 1982. NSF Faculty Fellowship 1970-71. Visiting Prof Univ. of Tenn 1984-85. Vice-Chair Registration Bd of Prof. Engrs 1981-84. *Society Aff:* NSPE, ASCE, ASEE

Busby, Michael R
Business: 620 N First St, Nashville, TN 37207
Position: Pres *Employer:* Watt Count Engrg Systems, Inc *Education:* PhD/ME/Univ of TN; MS/Aero & Astro/MIT; BS/Engr Sci/TN Tech Univ *Born:* 1/3/45 Native of Nashville, TN. Research engr with Arnold Research Organization. 1969-1972 specializing in energy transfer in high energy molecular beams. Dept chrmn, Dept of Mechanical Engrg, TN St Univ, Nashville, TN, 1972-1975. Assumed current position of Pres of Watt Count Engrg Systems, Inc. in 1975. Responsible for the implementation and sales of energy conservation systems in residential and commercial structures. Enjoy astronomy and caving. *Society Aff:* ASME

Busch, Arthur W
Business: 6211 W Northwest Hwy, Suite C258, Dallas, TX 75225
Position: Consultant *Employer:* Self *Education:* SM/Envr Engrg/MIT; BS/Civil Engrg/Texas Tech Univ *Born:* 10/09/26 Asst to dir res and devel, Infilco, Tucson, 1952-55; mem faculty Rice U 1955-75, prof environ engrg 1964-75, chmn dept 1967-70; adminstr region VI, US EPA, Dallas, 1972-75; Chmn SW Fed Regional Council, 1973-74; VP for environ affairs, Southwest Res Inst, San Antonio, 1975-76; cons in field, 1955-, pvt practice, 1976-. Cons to Water Quality Office, Mem Pres's Air Quality Adv Bd 1960-71. AUS 1944-46. Harrison Prescott Eddy medal WPCF, 1961, Distinguished Engr, Texas Tech U 1972, Distinguished Alumnus Awd, Texas Tech U 1975, Envir Div Awd, AIChE, 1973, Registered prof Engr Texas, Tau Beta Pi, author book, numerous papers. *Society Aff:* AIChE, ACS, WPCF, AAEE, NFS

Busch, Joseph H
Home: 2341 Valleywood Dr, San Bruno, CA 94066
Position: Genl Mgr, Ch Engr. *Employer:* H J Wickert & Co Inc. *Education:* BS/Mech Engg/Univ of NM. *Born:* Sept 1929. BSME Univ of New Mexico 1951. 1951-54 commissioned service US Navy. 1954-57 Sales Trainee, Application Engr, Sr Res Engr Enterprise Engine & Machinery Co. Since 1957, H J Wickert & Co Inc, manufacturers agent, representing US & foreign manufacturers, marine field, for prod appli, sales & service. Presently respon for oper & mgmt of co. Mbr SNAME, Chmn Northern Calif Sect 1973-74; Mbr ASNE & Assoc Mbr ASME. Mbr Naval Res; 1974-present commanded Naval Res Engrg Co 12-3, 1976 commanded Naval Shipyard Unit 4020. Active leader BSA last 12 yrs. Enjoy bldg & outdoor activities. *Society Aff:* SNAME, ASME, ASNE.

Busek, Robert H
Business: 601 Bergen Mall, Paramus, NJ 07652
Position: Assoc Engr. *Employer:* N H Bettigole Co. *Education:* MSCE/Struct Engg/Purdue Univ; BCE/Civ Engg/CCNY. *Born:* 6/2/47. Prior to joining the Bettigole firm, Mr Busek seved as an engr-in-training for the CA State Hgwy Dept. He was also a grad res asst at Purdue Univ, where he wrote his Master's thesis - and later a hgwy res paper - on optimizing the design of hgwy bridge girders. Mr Busek first joined N H Bettigole Co in 1969, shortly after receiving his BCE from the CCNY. Following a two-yr hiatus to earn his MSCE Degree from Purdue Univ, he returned to full-time employment at the firm. He served as a Proj Engr for five yrs before becoming an Assoc Engr in 1979. *Society Aff:* XE, ΣΞ, ASCE.

Busemann, Adolf
Home: 970 Lincoln Pl, Boulder, CO 80302
Position: Prof Emer of Fluid Mech. *Employer:* Univ of Colorado. *Education:* Dr Ing/-/T H Braumschweig; MS/Mech Eng/T H Braumschweig; BS/Mech Eng/T H Braunschweig. *Born:* Apr 1901. Res Asst Max Planck Inst of Flow Res at Goettingen 1925-31. Asst Prof for Fluid & Thermodynamics at Tech Univ Dresden 1931-35. Prof & Div Hd for Gas-dynamics & Rocketry at Aeronautical Res Inst Braunschweig 1936-46. Sr Scientist in Aerodynamics NASA Langley Res Ctr 1947-64. Prof Emer since 1969. Proposed swept wings for supersonic flight (publicly) 1935, for subsonic flight (confidentially) 1936. Auth book 'Heat & Matter Transfer' & auth num handbook articles. Sylvanus Albert Reed Award 1967 AIAA. *Society Aff:* NAE, AIAA, AAS.

Busey, Harold M
Home: 5007 Benton Ave, Bethesda, MD 20814
Position: Physical Scientist. *Employer:* Dept of Energy, Retired. *Education:* PhD/Chemistry/Tulane Univ; MS/Chemistry/Tulane Univ; BA/Chem & Physics/IL College. *Born:* 1917. Scientist at Los Alamos Scientific Lab 1947-65; plutonium chem, lab design, cryogenics, exper nuclear reactor dev. Sci Advisor to State Governor 1963-65. Asst to Dir, Douglas Labs 1965-69. Sr Res Assoc, Battelle Northwest Lab 1969-70. Environ Scientist Douglas-United Nuclear Corp 1970-71. Scientist Argonne Natl Lab 1972-73. Safety, Environ & Nuclear Waste Mgmt Specialist ERDA 1974-77, Physical Scientist, DOE 1977-1985, Retired. Fellow Amer Nuclear Soc. Publs, pats. Mbr sci expeditions. Int: tech sys design, energy, nuclear safety. *Society Aff:* ANS, AAAS, ACS

Bush, G Frederick
Business: 15705 Piedmont, Detroit, MI 48223
Position: (Ret) *Education:* AB/English/Univ of Detroit; BChE/Metallurgy/Univ of Detroit. *Born:* June 8, 1914 Guelph, Ont, Canada. P E. Fellow ASM. Foundry Metallurgist, Auto Specialties; Ch Metallurgist, McKinnon Indus GMC; Ford Motor Co since 1949 spec in matls & processing such as painting, plating & plastics. Automotive corrosion & its preventions a major interest. Mgr, Matls Engrg, Engrg & Res Staff retired. P Chmn Detroit Chap ASM & Matls Engrg Ctte SAE; Group Dir ISTC, SAE. Hobbies - travel & photography. *Society Aff:* SAE, ASM.

Bush, Spencer H
Home: 630 Cedar Avenue, Richland, WA 99352
Position: Owner/Conslt *Employer:* Self employed *Education:* PhD/Met Engg/Univ of MI; MSE/Met Engg/Univ of MI; BSE/Met Engg/Univ of MI; BSE/Chem Engg/Univ of MI. *Born:* 4/4/20. Manhattan Proj 1944-46 while in Army. USAEC (now DOE) Hanford since 1953 as individual contributor, mgr, staff cons in reactor matls for GE 1953-65 & Battelle Northwest 1966-1983. Extensive work in reactor safety at Hanford & as mbr USNRC Adv Ctte on Reactor Safeguards 1966-77. Adj Prof Jt Ctr for Grad Study. Fellow ANS 1970, ASM 1971, ASME 1981, Mbr NAE 1970- . Reg P E CA (MetE, NucE); ASTM Gillett Lectr 1975; ASNT Mehl 1981; ASME Bernard Langer Codes & Stds Award 1983; ANS T.J. Thompson Award 1987. Chmn ASME Section XI; mbr, ASME Bd Nuclear Codes & Stds and NDE Div Exec Ctte 1983-; mbr, ANS Bd Dir 1984-; Chrmn, Washington State Bd of Boiler Rules 1980-86. Conslt on materials and reactor safety. *Society Aff:* ANS, AIME, ASM, ASME, NAE.

Bushnell, James D
Business: P O Box 101, Florham Pk, NJ 07932
Position: Sr Engrg Advisor. *Employer:* Exxon Res & Engrg Co. *Education:* BChE/Chem Engg/NY Univ. *Born:* Mar 1922 Ilion N Y. Tau Beta Pi, Psi Upsilon. 3 yrs with Standard Oil Ind at Whiting, on 100 Octane Avgas res. Joined Exxon 1947. Career mostly in process engrg. 1958, became hd of lubes & treating processes. Since then have been respon for Exxon's lube & specialties engrg & res guidance,

Bushnell, James D (Continued)
leading to dev & intro of sev major lube process innovations incl Exxon's Integrated Lube Plant, DILCHILL dewaxing, all catalytic white oil hydrogenation & EXOLN extraction processes. Author NPRA, API & World Petroleum Congress papers, 22 U S pats. AIChE, ACS. Hobbies: antique cars, photography, choral music. *Society Aff:* ACS, AIChE.

Busot, J Carlos
Business: Coll of Engrg, Tampa, FL 33620
Position: Chrmn Chem. Eng. Dept. *Employer:* Univ of So Florida. *Education:* PhD/Chem Engg/Univ of FL; MS/Chem Engg/Univ of Gainesville, FL; BS/Chem Engg/Univ of Gainesville, FL. *Born:* Aug 1938. Native of Cuba. PhD, MS & BS Univ of Florida; however, had basically completed undergrad training at the Universidad de Villanueva Cuba where he was pres of the Student Council during the first yr of Fidel Cartro's regime. Res Engr with E I duPont de Nemours 1963-67, spec in continuous polymerization (U S pats 3487049 & 3534082). Chrmn Chem Engrg at the Univ of So Fla, Tampa since its beginning in 1970. Active Prof Engr in State of Fla. Enjoys fishing & chess. *Society Aff:* ΣΞ, AIChE, NSPE, TBΠ, FES.

Bussey, Lynn E
Business: Dept of Indus Engrg, Manhattan, KS 66502
Position: Assoc Prof, Indus Engrg. *Employer:* Kansas State Univ. *Education:* PhD/Ind Engg/OK State Univ; MS/Ind Engg/OK State Univ; BSME/Mech Engg/Cornell Univ. *Born:* 1920 W Palm Beach Fla. PhD 1970 & MS Oklahoma State Univ; BSME Cornell Univ Ithaca N Y 1947. Secondary school, Riverside Mili Acad (honor grad 1938). Exper proj engr & Capt USAF 1942-46. Plant Mgr, later Corp V P & Treas Cox Assoc Co's Springfield Mo 1947-67. Pres Okla Fertilizer & Chem Co 1952-55. Cons engr (structs, mech & elec sys, processing sys) 1950-66. Reg P E Mo, Kan, Okla. Author book & num tech articles on engrg econ analysis. Recipient of ASEE Eugene L Grant Award 1974 for best refereed tech article in 'The Engrg Economist'. Mbr Tau Beta Pi, Phi Kappa Phi, Sigma Xi, Alpha Pi Mu, IIE, ASME & other soc's. Assoc Prof Kansas State Univ 1970- . *Society Aff:* IIE, ASME, ORSA, ASEE.

Bussgang, Julian J
Business: 110 Hartwell Ave, Lexington, MA 02173
Position: President. *Employer:* Signatron Inc. *Education:* PhD/Applied Physics/Harvard Univ; MS/Elec Eng/MIT; BSc/Telecomm/Univ of London, UK. *Born:* Mar 1925. PhD Harvard Univ; MSEE MIT; BS Engrg Univ of London U K. Was Res Asst at MIT Res Lab for Electronics; Mbr Tech Staff at Lincoln Lab 1951-55; first Mgr - radar dev, then Mgr - appl res at RCA 1955-62. Founded SIGNATRON 1962, providing cons & R&D services & instrumentation to Gov agencies & many major corps in communications & radar. Lectr Northeastern Univ 1962-65 & Harvard 1964. Assoc Editor 'Radio Sci' for Communications 1975-78. Fellow IEEE 1973; Reg P E Mass; IEEE Admin Ctte of Prof Group on Info Theory 1966-73; active in Smaller Bus Assn of N Eng & in Town of Lexington Mass. *Society Aff:* IEEE, ΣΞ.

Bussolini, Jacob J
Business: Dept 693, Bethpage, NY 11714
Position: Dir, Bus Operations. *Employer:* Grumman Aerospace Corp. *Education:* BS/EE/Norwich Univ; PMD32/Mgt/Harvard Bus Sch. *Born:* Jan 1936. Native of Avon Ct. 1st Lt US Army Signal Corps. Served as a Sr Res Engr for Hazeltine Corp. Moved to Grumman 1961, eventually achieved the pos of Ch of Reliability. Current respons incl Admin, Resources & Finance Mgmt, reporting to the VP of Admin & Resources. Holder of 1 pat & credited with the delivery of more than 15 tech papers. Lectured in Europe under NATO AGARD sponsorship & a regular lectr for Amer Mgmt Assn. Served as mbr of Naval Air Sys Effectiveness Adv Bd. Chmn Design-to-Cost Ctte of Long Island Assn. Mbr Toastmasters Internatl, AOPA. Lic private pilot. AIAA Finance Committee, Pres of Bd of Dirs of East Plains Mental Health Corp - Mbr of Long Island Early Fyers, Coord of Grumman product Restoration team. Active with Huntinghaton, NY Boys Club. *Society Aff:* AIAA.

Butcher, Nathan B
Business: 631 E Crawford-Bx 1648, Salina, KS 67401
Position: Consultant *Employer:* Wilson & Co, Engrs & Archs. *Education:* TE/Textile Engg/TX Tech Univ. *Born:* Dec 1915 Carlsbad N M. Texas Tech 1936, T E. Early C E exper, U S Bureau of Reclamation & Soil Conservation Serv. With Wilson & Co as civil & struct engr since 1942. 1950-51, served as mgr of Wichita branch office during design & const of major USAF base & plant expansion of Boeing Co. Recalled to home office, Salina, as Genl Mgr 1952. Became co-owner & managing partner 1959. In 1963, partners founded Wilson-Murrow co's for internatl work, mainly hwy engrg in Saudi Arabia. Fellow ASCE & ACEC. Hobby Arabian & Quarter Horses. Partner in BNB & B&R Ranches, Kansas. *Society Aff:* NSPE, ASCE, ACEC, PEPP.

Butler, Blaine R
Business: Aeronautical Engg, Embry-Riddle Aero Univ, Prescott, AZ 86301
Position: Professor & Head *Employer:* Embry-Riddle Aeronautical Engg *Education:* PhD/Aero/Purdue; MSAE/Aero/Purdue; BSME/Mil Engg/USMA. *Born:* 8/11/25. 1984 - present: Professor & head, Department of Aeronautical Engineering; Embry-Riddle Aeronautical University, Prescott, Arizona 86301. National Pres of Sigma Gamma Tau, the Aerospace Honor Soc (1979-1982). 1972-1983: Professor of Engineering, Purdue University, W. Lafayette, Indiana 47907. 1970-1971 - Vice Commander Space Defense Ctr; NORAD; Colorado Springs, CO. 1963-1968 - Prof; Dept of Aeronautics; USAFA, Colorado Springs, CO. 1949-1972 - Military Aviator - 6,000 hours. *Society Aff:* AIAA, CASI, ASEE, BIS.

Butler, Donald J
Home: P.O. Box 405, Cuttingsville, VT 05738
Position: Prof of Civ Engg. *Employer:* Rutgers Univ. *Education:* PhD/Engg Mechanics/Columbia Univ; MS/Civ Engg/Columbia Univ; BS/Civ Engg/Columbia Univ. *Born:* 5/16/25. Native of NYC. Served with Army Corps of Engrs 1943-46. Instructor through Assoc Prof of Civ Engg, Columbia Univ 1949-66. Gulbenkian Res Fellow, Natl Civ Engg Lab, Lisbon, Portugal 1965-66. Prof of Civ Engg, Rutgers Univ, 1966-83. Prof Emeritus since 1983. Served as Civ Engg Dept Grad Prog Dir 1967-75 and as Exec Officer 1967-75, 1979-83. Visiting Scientist, Bldg Res Establishment, Garston, England 1975-76. Res interests: structural stability, probabilistic methods, experimental mechanics. Other interests: music, art, oenology. *Society Aff:* ASCE, AAUP.

Butler, Earle B
Home: 720 Summit Rd, Marion, OH 43302
Position: Consultant, Engrg Mgmt. *Employer:* Self. *Education:* MSCE/Civil/Univ of TNl; BSCE/Civil/Rose-Hulman Inst of Tech. *Born:* May 1912. 1935-41 Engr, TVA Hydraulic Lab. 1941-63 US Army Corps of Engrs. Area Engr for military const airfields & bases in US. Commanding Officer, Engr Sect, Atlanta ASF Depot; Baton Rouge Engr Depot; Engr Maintenance Ctr; 83rd Engr Const Battalion France; 2nd Engr Const Group Korea. Dir, Supply & Procurement, Hanau Engr Depot Germany. Dist Engr Buffalo Dist. Retired as Colonel 1963. 1963-69 Genl Mgr, Port of Cleveland. 1969-78, Floyd G Browne & Assocs Ltd, Cons Engrs - Planners; Principal & Chmn of Bd, 1978- , Consultant, Engg Mgmt. Fellow ASCE. *Society Aff:* ASCE, SAME.

Butler, Gilbert L
Home: 4819 Tabard Pl, Annandale, VA 22003
Position: Civil Engr. *Employer:* Urban Mass Transp Admin. *Born:* June 1934. BSCE Univ of Colo. Reg P E Colo. Struct engr with Bureau of Reclamation spec in rock mech & field testing of underground openings. With UMTA since 1974. Serve as proj mgr for the tunneling tech & track res progs. V Chmn of the Fed Interagency Ctte on Excavation Tech, Mbr ASCE, active in sev cttes of ASCE & TRB - involved in var aspects of underground const tech. Active in civic affairs; enjoy golf.

Butler, James L
Business: Coastal Plain Exp Sta, Tifton, GA 31793
Position: Research Leader *Employer:* US Dept of Agri Agr. Research Service. *Education:* PhD/Agri Engg/MI State Univ; MS/Agri Engg/Univ of TN; BSAE/Agri Engg/Univ of TN. *Born:* 1/8/27. Service in US Army Air Corps 1944-46. Res with Ga Exper Station 1951-56 & 1958-60. With ARS-USDA since 1960. Became Res Investigations Leader, Forage & Oilseeds, Harvesting & Processing 1962. Res Leader & Tech Advisor for Forage, Oilseeds & Sugarcane 1972, & Principal Investigator for Solar Energy Applications to Crop Drying 1975. Became Manager to the USDA SEA Southern Agr. Energy Center in 1980. Research Leader in 1985. Sr Mbr ASAE, Chmnship of sev natl cttes plus chmnship of the Southeast Region & Ga Section & Mbr ASAE Bd of Dirs. President APRES, 1981- 82. Assoc Editor-Engg for Peanut Science (1974-80). Editorial Bd of Energy in Agri, 1981-present. 1960 ASAE Paper Award. 1978 Bailey Award. 1979 Golden Peanut Res Award. 1980 Certificate of Merit, USDA.; 1981 Engr of the Year, Ga. Sec. ASAE; 1982 Certificate of Merit, USDA; 1985 Fellow ASAE; 1986 Certificate of Merit, USDA. *Society Aff:* ASAE, NSPE, APRES, GAS, NYAS, CAST.

Butler, Jerome K
Business: Elec Engg, Dallas, TX 75275
Position: Prof. *Employer:* Southern Methodist Univ. *Education:* BSEE/EE/LA Poly Univ; MSEE/EE/Univ of KS; PhD/EE/Univ of KS. *Born:* 9/27/38. Jerome K Butler received the BS EE from LA Poly Univ and MS EE and PhD degree from the Univ of KS. In 1965, he joined the faculty at Southern Methodist Univ, Dallas, TX where he is now Prof of Elec Engg. His primary interest areas are injection lasers and communication systems. He has held consulting appointments with TX Instruments and Earl Cullum Assocs. In the summers from 1969 to 1979 he was at RCA Labs, Princeton, NJ. Dr Butler is a mbr of Sigma Xi, Tau Beta Pi, Eta Kappa Nu, and a reg PE in the state of TX. *Society Aff:* IEEE.

Butler, Jon T
Business: Evanston, IL 60201
Position: Assoc Prof *Employer:* Northwestern Univ *Education:* PhD/EE/OH State Univ; M/EE/RPI; B/EE/RPI *Born:* 12/26/43 Assoc Prof in the Dept of Electrical Engrg and Computer Science, Northwestern Univ, Evanston, IL since 1974. Area of specialty multiple-valued logic, fault tolerant computing, and array processing. National Research Council Post doctoral Assoc - 1980-81 Naval Postgraduate School, Monterey, CA, - 1973-74 Air Force Avionics Lab, Wright-Patterson AFB, OH. Chrmn, Intl Symposium on Multiple-Valued Logic, 1980, chairman, IEEE Computer Society Technical Committee on Multiple-Valued Logic,1980-1982, Editor, IEEE Trans on Computers, 1982- present. Rensselaer Scholarship 1966-67, IBM Fellowship 1970-71, NASA Traineeship 1971-72. Registered PE - State of OH. *Society Aff:* IEEE, ΣΞ.

Butler, Margaret K
Business: 9700 S Cass Ave, Argonne, IL 60439
Position: Dir, Natl Energy Software Ctr (Sr Computer Scientist). *Employer:* Argonne Natl Lab. *Education:* AB/Math/IN Univ. *Born:* 3/7/24. Grad work USDA Grad Sch Wash DC, Univ of Chgo, Univ of MN - MN. worked as statistician with US Bureau of Labor Statistics Washington 1945, St Paul MN 1949-51, USAF in Europe 1946-48. Employed as mathematician at Argonne Natl Lab briefly 1948-49 & then from 1951 to date. Hd of Reactor Engrg Computing Group, then Computing Services Div, then lab-wide computer applications; from 1960 to date in charge of agency software exchange ctr. In Amer Nuclear Soc, served as Chmn Math & Computation Div 1966-67; Fellow 1972; Bd of Dir 1976- 79; Exec. Committee 1977-78; Chmn Bylaws & Rules Comm 1979-1982. Served as Chair for Pioneer Day National Computer Conf. 1985 and as Chair for Technical Program National Computer Conf. 1987. *Society Aff:* ANS, ACM, AWIS, AAAS, IEEE-CS, AMA.

Butler, Roger M
Business: 2500 University Dr NW, Calgary, Alberta Canada T2N 1N4
Position: Prof Petroleum Engrg *Employer:* Univ of Calgary *Education:* PhD/Chem Engg/Imperial College of Sci & Technology; BSc/Chem Engg/Imperial College of Science & Technology. *Born:* 10/14/27. Asst Prof Chem Engrg Queen's Univ Kingston Ont 1951-55. Imperial Oil Res Dept 195568. Process dev, dewaxing, wax recryst, extraction, hydrofining, reforming, petrochems, Athabasca bitumen recovery, proj dev. Engrg Div 1968-70. Heavy crude upgrading, Sarnia Refinery 1970-71. Implemented supervisory refinery computer control. Exxon Logistics N Y 1971-73. Advisor Marine Tankage & Terminalling. Dev statistical method for tankage requirements. Imperial Engrg Div Sarnia 1973-74, Mgr New Tech - defined energy conservation prog. Sci Advisor New Energy Resources 1974-75. 1976-82, in charge of res on in-situ recovery of heavy crudes. Esso Resources Canada Ltd 1982-83 Dir Tech Prog AOSTRA. Since Nov 1983 Endowed Chair Pet Eng. P Chmn Bd, CIC, FCIC, Fellow I Chem E, Fellow AIChE, Prof Engr, Cert Engr *Society Aff:* AIChE, CSChE, CIM, SPE, ICE.

Butler, Vern W
Home: 8751 Roundtree Ave, Englewood, CO 80111
Position: Water Resources Engr *Employer:* Rio Blanco Oil Shale Co *Education:* BS/CE/KS State Univ *Born:* 4/10/36 Native of Harveyville, KS. Grew up on farm. Five years with KS Water Resources Bd. Moved to Brookings, SD as Mgr-Engr of East Dakota Conservancy Sub- District, 1964-1971. With SD Water Resources Commission 1971-1973. Appointed Secretary, SD Dept of Natural Resources Development, 1973-1978 administering statewide programs in water rights, water resources planning and development, geological survey, weather modification and oil and gas conservation. With CH2M Hill's Denver Regional Office 1979-1981. Water Resources Eng. with Rio Blanco Oil Co 1982-84. Water Resources Eng. with Rio Blanco Oil Shale Co 1982-84. ASCE State Section Pres 1971 and District 7 Council 1971-73. NSPE Chapter Pres 1975. North American Interstate Weather Modification Council Chrmn 1975-1976. Married and 3 children. *Society Aff:* ASCE, NSPE, AWWA.

Butt, Jimmy L
Home: 2572 Stratford, St. Joseph, MI 49085
Position: Retired *Employer:* Retired *Education:* MS/Agri Engg/Auburn Univ; BS/Agri Engg/Auburn Univ. *Born:* 10/13/21. Native of Tippo, MS; grad from high school in Wetumpka, AL. Served as field artillery and military govt officer in Europe during and following WWII. Conducted res in artificial drying and storage of seeds, grain, and hay for nine yrs at Auburn Univ. Served as exec VP, ASAE 1956-86; Pres 1987-88. Active in the Council of Engg and Scientific and Soc Executives (CESSE) and served as pres in 1977-78. Former Bd Mbr, AAES. Presented Chevalier du Merit Agricole, 1979, by French Minister of Agriculture. Member, Research Advisory Council, Auburn University. *Society Aff:* ASAE (fellow), NSPE

Butt, John B
Business: Dept of Chemical Engineering, Evanston, IL 60201
Position: Walter P. Murphy Prof of Chem Engrg *Employer:* Northwestern Univ *Education:* BS/Chem Eng/Clemson Univ; M Eng/Chem Eng/Yale Univ; D Eng/Chem Eng/Yale Univ *Born:* 09/10/35 Born Norfolk, VA. Served on faculty at Yale Univ prior to present position at Northwestern. Currently Walter P. Murphy Prof of Chemical Engrg. Allan P. Colburn Award, AIChE, 1968; Profl Progress Award, AIChE 1978. Chrmn, Gordon Res Conference on Catalysis, 1979. AIChE Natl Dir, 1978-77. Alexander von Humboldt Prize, 1985. *Society Aff:* AIChE, ACS, The Catalysis Society

Buttery, Lewis M
Home: 407 W First St, Lampasas, TX 76550
Position: VP. *Employer:* Bonner & Moore Assoc, Inc. *Education:* BS/ChE/Univ of TX. *Born:* 3/20/24. Native of San Angelo, TX. Served in the USN as a Gunnery Officer in the Pacific, WWII. Engr for Fire Prevention & Engg Bureau of TX, 1947-51. Production engr for Monsanto, 1951-66. Consultant with Bonner & Moore since

Buttery, Lewis M (Continued)
1966. Named VP - Production Mgt Services in 1978 with responsibilities world-wide for systems & services in support of maintenance & operation of offshore production facilities, refineries & petrochem plants. *Society Aff:* AIChE.

Button, Kenneth J
Business: MIT, Box 72 Mit Branch, Cambridge, MA 02139-0901
Position: Sr Scientist. *Employer:* MA Inst Technology. *Education:* DSc/EE/Tokyo Inst Technology; MS/Physics/Univ Rochester; BS/Physics/Univ Rochester *Born:* 10/11/22. Undergrad & grad work Univ of Rochester NY 1946-52. Army 1942-46. Microwave Engr & Solid-State Physicist MIT 1952-62. Co-author book: Microwave Ferrites & Ferri-magnetics. Charter mbr & Group Leader MIT Natl Magnet Lab 1962. Co-founder of Intl Org of Crystal Growth 1965. Demonstrated early applications & founded series of annual intl oonferences on Infrared & Millimeter Waves. Former Mbr OSA Tech Council & Ch Tech Group on Far Infrared. Former Mbr Admin Committee of IEEE-MTT Soc. Former Mbr IEEE Energy Ctte, Quantum Electronics Council, TAB Mtgs Ctte & MA Engg Council. Fellow IEEE. Fellow Amer Physical Soc. Editor of Acad Press book series entitled, *Infrared and Millimeter Waves*; Editor of Plenum Press journal entitled, *Journal of Infrared and Millimeter Waves*. *Society Aff:* IEEE, APS

Butts, Bennie J
Business: P O Box 21207, Greensboro, NC 27420
Position: Area Industrial Engr. *Employer:* Burlington Industries. *Education:* BS/IE/VA Poly Inst & SU. *Born:* Nov 11, 1922. Alpha Pi Mu, Tau Beta Pi, O D K, Phi Kappa Phi. Native of Richmond Virginia. Major Army Air Corps 1942-46. Employed by Burlington Industries 1950- as Plant, Div, Area & Corp IE. Currently IE responsibility for 7 operating divs in Home Furnishings Area, Fellow of IIE, V P Region III 1969- 71, Publs Pol Bd 1975-81; Human Factors Soc & Adv Bd of VPI IEOR Dept. *Society Aff:* HFS, IIE.

Buzzell, Donald A
Home: 4 Roxbury Rd, Norwalk, CT 06855
Position: Asst Exec Dir *Employer:* Amer Society of Civil Engrs. *Education:* BS/Civil Engrg/Geo Washington Univ *Born:* 1/25/27. Englewood NJ. US Navy 1943-45. Struct Engr Army Corps of Engrs 1950-53. Mgr Heavy Const Assoc Genl Contractors 1953-61; Exec Dir Cons Engrs Council 1962-75; Exec VP Amer Cons Engrs Council 1975-77. Amer Soc of Civil Engrs, Asst Exec Dir beginning Apr 1982. Amer Soc of Assn Execs Certified Assn Exec; ASAE Awards: Mgmt, Gold Key & Serv Award for Private Enterprise support. Hobbies: fishing, philately. *Society Aff:* ASCE, ASAE

Byars, Edward F
Home: 4 Riverpoint, Clemson, SC 29631
Position: Consultant *Employer:* Self *Education:* BSME/Mech/Clemson Univ; MSCE/Civ/Clemson Univ; PhD/eng Mehanics/Univ of IL. *Born:* 3/22/25. Educator: b Lincolnton, NC, Mar 22, 1925; s Edward Hayes and Lois N (Ford) B: BS in Mech Engg, Clemson Univ, 1946, MCE, 1950: PhD, Univ IL, 1957; m Betsy Alice Cromer, June 25, 1950; children - Laurie, Betsy, Nan, Guy. Instr Clemson Univ, 1947-53, asst prof, 1953-55, asso prof, 1957-60; instr Univ IL, 1955-57; prof, chrmn dept mech engg and mechanics WV Univ, Morgantown, 1960-1984; Dean, Coll of Engrg, WV Univ, 1979-1980; Exec Asst to the Pres, Clemson Univ, 1980-85. Cons in field. Served with Signal Corps, AUS, 1946-47. Mbr ASME, SESA, ASEE, Soaring Soc Am, Tau Beta Pi, Sigma Xi, Pi Tau Sigma. Author: Engineering Mechanics of Deformable Bodies, 1963. Contbr articles to profl jours. Patentee in field. *Society Aff:* ASEE, ASME, SESA.

Byer, Robert L
Business: Stanford University Bldg 10, Stanford, CA 94305
Position: Prof, Dean of Research *Employer:* Stanford Univ *Education:* PhD/Applied Physics/Stanford; Masters/Applied Physics/Stanford; BS/Physics/Berkeley *Born:* 05/09/42 Ph.D. degree in applied physics from Stanford Univ 1969. His interest has been non-linear interactions including harmonic generation and parametric oscillators. Joined Applied Physics Dept, Stanford. Research in remote sensing using tunable laser sources, leading to the development of the unstable resonator Nd:YAG laser and to high-power tunable infrared generation in LiNBO3 parametric tuners. 1974 he and colleagues initiated research in coherent anti-Stokes Raman spectroscopy (CARS), named the effect, and continued research in high-resolution Raman spectroscopy. 1976 he suggested use of stimulated Raman scattering in hydrogen gas to generate 16 micron radiation from a CO2 laser source. Research at Stanford confirmed the expected simplicity and efficiency of the approach. In 1980 he began research on slab geometry solid-state laser sources, leading to successful theoretical and experimental devel in high-peak and average-power solid-state laser sources. He has worked in industry at Spectra Physics, 1964-65. Chromatix, 1969-74, helped found Quanta Ray Inc., in 1975 and today consults for Newport Inc., Hoya Optics, Spectra Technology, and Lightwave Electronics Corp. Chairman of the Applied Physics Dept at Stanford from 1981 to 1984, and appointed Assoc Dean of Humanities & Sciences in 1985. Sept. 1, apptd. Vice Provost & Dean of Research. Dr. Byer is a member of AAAS, APS, a Fellow of the OSA, and was President of the IEEE Lasers and Electro-Optics Society for 1985. *Society Aff:* IEEE

Byer, Robert Louis
Business: Stanford Univ, Stanford, CA 94305
Position: Prof *Employer:* Stanford University. *Education:* PhD/Quantum Electronics/Stanford Univ; Masters/Quantum Electronics/Stanford Univ; BA/Physics/Univ of CA Berkeley. *Born:* May 1942. Employed at Spectra Phys Inc 1964-65. Mbr OSA, APS, AAAS, IEEE. Awarded Adolf Lomb Medal OSA 1972. Sloan Fellow 1975; Fellow OSA 1977. Cons Westinghouse Res Lab, Allied Chem, Bethlehem Steel Inc, Quanta Ray & US Army. Married 1964, 4 children. *Society Aff:* OSA, APS, AAAS, IEEE.

Byloff, Robert C
Home: 565 Maywood Rd, York, PA 17402
Position: Pres. *Employer:* Incom International. *Education:* BS/Elec Engg/Univ of CA-Los Angeles. *Born:* 4/6/28. Native of Los Angeles, CA. Sr Proj Engr, Garrett AiResearch 1951-1963, rotating machinery and control equipment. VP of Engg, Louis-Allis Co-1963-71, in the industrial large motor bus. Pres of Hydrospace and Wolfe Res Divs of EG&G 1971-74, engaged in oceanographic science, underwater acoustics and computer software. Pres of Fincor Div of Incom Intl 1974-present. Pres of Magnetics Soc of IEEE 1971-72. Fellow IEEE-1978. *Society Aff:* IEEE, ASE.

Byrd, Jack, Jr
Business: Engrg Sciences/Industrial Engr, Morgantown, WV 26506
Position: Prof and Acting Chrmn *Employer:* WV Univ *Education:* PhD/Engrg/WV Univ; MS/IE/WV Univ; BS/IE/WV Univ *Born:* 12/27/44 Native of Fairmont WV. Has taught industrial engrg at WV Univ since 1969. Is the author of two text books in the area of applied decision sciences. An active consultant in the areas of productivity improvement. Currently Dir of the Center for Entrepreneurial studies and Development, a non-profit corp established to aid business development in WV. Has published numerous in professional journals related to industrial engrg. *Society Aff:* IIE, TIMS, NSPE

Byrd, Lloyd G
Business: 2921 Telestar Court, Falls Church, VA 22042
Position: Sr VP, Mgr. *Employer:* Byrd, Tallamy, MacDonald & Lewis *Education:* BS/Civil Engg/OH State Univ. *Born:* 5/6/23. Native of Atlanta, GA. 1952-60, Deputy Project Dir and Maintenance Engg. OH Turnpike Commission; 1960-63, Assoc Editor Public Works Magazine and Editor of Street and Highway Manual. Joined Byrd, Tallamy, MacDonald & Lewis in 1963 as a partner. Partnership merged with Wilbur Smith & Assoc in 1972. Dev intl conslt practice in mgmt sys for trans dept responsibilities in maintenance and operations. Served as special consultant to Transportation Res Bd in developing concepts and report for $150 mil-

Byrd, Lloyd G (Continued)
lion Strategic Transportation Res Study (STRS). PChmn Highway Div Exec Committe, PPres Natl Cap Section, Dir of Dist 5, ASCE. PChmn, Group 3 Council, Transportation Research Board. PPres, Fairfax County Chamber of Commerce. Past Chmn, Human Rights Commission, Fairfax County. In 1978 received the Distinguished Alumnus Award from OH State Univ. *Society Aff:* ASCE, NSPE, APWA, TRB.

Byrer, Thomas G
Business: 505 King Ave, Columbus, OH 43201
Position: Sec Mgr. *Employer:* Battelle Columbus. *Education:* BMetE/Met/OH State Univ. *Born:* 11/25/29. Native of Canton, OH. Employed as met in sheet mill and casting shop at Chase Brass & Copper, Euclid, OH, from 1954-57. Joined Battelle in 1957 working on copper alloy dev and titanium extrusion processes. Became Assoc Mgr in 1970 and Mgr in 1974 of Metalworking Sec in Engg and Mfg Dept. Responsible for all aspects of section working in metalforming dev, solution of mfg problems, CAD/CAM, and application of metal deformation analysis to metalforming processes. He has served on several technical society committees and was recently appointed to the Technical Div Bd of ASM. Hobbies are photography and furniture refinishing. *Society Aff:* ASM.

Byrne, Bernard F
Business: Dept of Civil Engg, Morgantown, WV 26506
Position: Assoc Prof. *Employer:* WV Univ. *Education:* PhD/Civil and Urban Engg/Univ of PA; MSE/Civil Engg/Univ of CA; BSCE/Civil Engg/Carnegie Inst of Tech. *Born:* 6/12/44. in Charleston, WV. Raised near Pittsburgh, PA. Worked for DeLeuw, Cather & Co 1971-73 as Transportation Planning Engr. Started at WV Univ Aug, 1973. Co- Prin Investigator, Bridge Shoulder Width Study; Co-Prin Investigator, Rural Transit Demand Study. Promoted to Assoc Prof Aug, 1978. Mbr, TRB Rural Transit Comm. *Society Aff:* ASCE, ASEE, ITE.

Byrne, J Gerald
Home: 1159 1st Ave, Salt Lake City, UT 84103
Position: Prof & Chairman *Employer:* Dept of Metallurgy, Univ of Utah *Education:* PhD/Metallurgy/Northwestern Univ; MS/Metallurgy/Stevens Inst of Tech; ME/Mechnical Engg/Stevens Inst of Tech. *Born:* 1930. Engr with Crucible Steel Co, Amer Brake Shoe Co, Dow Chemical Co; faculty Stevens Inst 1962-66 & Univ of Utah 1966- . Chmn. Author 1965 MacMillan book Recovery, Recrystallization & Grain Growth & over 90 tech pubs in strengthening, fatigue, chem vapor deposition & positron annihilation applied to deformation & hydrogen embrittlement. Natl Acad of Sci & NMAB ctte serv, ASMI, AIME, ASTM cttes. Elected Fellow of Amer Soc for Metals, 1982. Appointed as 1st Ivor D. Thomas Professor of Physical Metallurgy, 1987. Ski instructor Alta, UT. *Society Aff:* ASMI, AIME, ASTM, PSIAI.

Byrne, Joseph
Business: 1201 W 5th St, Los Angeles, CA 90017
Position: VP Corporate Human Resources *Employer:* Unocal Corp. *Education:* ScD/Chem Engg/MIT; SM/Chem Engg/MIT; BA/Chem Engg/Stanford Univ. *Born:* June 1923 Santa Fe NM. Served with US Navy in WW II. Asst Prof MIT 1950- 53; with Unocal Corp. (formerly Union Oil Co.) since 1953. Held various prositions in: Union's refining dept 1953-62, Unocal's planning dept 1963-68, VP Marketing Western Region 1968- 75, VP Operations Coordination for Unocal's 76 Div 1975-77. VP Natl Mktg & Operations Coordination 1977-81. Involved with API activities concerning cost of removing lead from gasoline, costs of changing volatility of gasoline, wast oil studies. Mbr API, AIChE, Amer Soc Advancement of Sci Feb 1981: VP, Corporate Human Resources. *Society Aff:* AIChE, API, ASAS, ΣΞ, ΦΒΚ, ΑΧΣ, ΦΛΥ.

Byrne, Patrick J
Business: 1000 E Union, Olympia, WA 98501
Position: Pres *Employer:* Byrne-Stevens *Education:* BS/Gen Engrg/Univ of Portland *Born:* 4/30/34 Mr Byrne's engrg career includes 25 years of service to municipal government. He is currently retained by five towns and municipal districts in Western WA where he has been responsible for the planning, design, construction and financial management of water and sewerage systems; also transportation planning, design and construction of urban streets and rural roads. His interest in hydrology and urban drainage has led to the writing of computer programs. Utilizing the data base available for the "Rationale Method–, to synthesize hydrographs for overland flow in urban environments. *Society Aff:* ASCE, CECW, AWWA, APWA.

Byrnes, James J
Home: 5368 Bamboo Court, Orlando, FL 32811
Position: VP. *Employer:* Burns & Roe Inc. *Education:* BSChE/Chem Engr/Columbia. *Born:* July 1922. PE NY, NJ, WA, FL; reg Health Physicist. Mbr AIChE & ISA. Native of NYC. Work for Carbide & Carbon Chemicals Co during WW II on Manhattan Proj at Oak Ridge Tenn; with H K Ferguson Co on design & operation of graphite nuclear res reactor at Brookhaven Natl Lab; worked on early dev of coml nuclear power at Walter Kidde Nuclear Lab & Stone & Webster Engrg Co; served as Proj Dir for Washington Pub Power Supply System Nuclear Proj No 2. Serving as Project Manager on Semole Elect Corporation 2-600 MW coal fired power plants. Responsible for all engrg, prepurchase of equip & preparation of const contracts. *Society Aff:* AIChE, ISA.

Cabble, George M
Home: 295 F Lakemoore Dr., NE, Atlanta, GA 30342
Position: Vice President *Employer:* RRME, Inc. *Education:* PhD/ME/Univ of IL; MS/ME/VPI; BS/ME/VPI; -/Economics/Princeton Univ. *Born:* 9/24/15. NYC 1915. Taught at VPI & Univ of IL 1948-1954. Specialty, Railway Mech Engg. Westinghouse Air Brake Co, res, engg, sales & Sr Technical Specialist in airbrakes and composition brake shoes 1954-1975. Amtrak, chief Wheel & Brake Systems 1975-1980. Amtrak'S rep on Brake and Brake Equip Ctte, and Wheels, Axles, Bearings & Lubrication Ctte of Assn of Am Railroads. Retired. Now consulting as Vice-President, RRME, Inc. Author 18 papers. Fellow, Am Soc of Mech Engrs. Mbr, Air Brake Assn. Reg PE. *Society Aff:* ASME, ABA.

Cabeen, S Kirk
Business: 345 East 47th St, New York, NY 10017
Position: Dir *Employer:* Engrg Societies Library *Education:* MS/Library Sci/Syracuse Univ; BA/Chem/Lafayette Coll *Born:* 1/22/31 After graduation from coll, spent two years in Army Chem Corps. After discharge went to work as Assistant Librarian for American Metal Co. (now Amax). Then became Librarian at Ford Instrument Co., Div. of Sperry-Rand. Came to Engrg Socities Library as Assistant to the Dir in 1964. Became Dir in 1968. *Society Aff:* SLA, ALA, ASIS, ACS, A of C

Cacciamani, Jr, Eugene R
Home: 11409 Rolling House Road, Rockville, MD 20852
Position: Senior VP *Employer:* Hughes Network Systems, Inc. *Education:* Ph.D./Elect. Eng./Catholic University of America; M.E.E./Elect. Eng./Catholic University of America; B.E.E./Elect. Eng./Union College *Born:* 09/15/36 Native of Nyack, New York; has lived in Washington, DC, since 1958. Served as 1st Lt. USAF 1958-62 and RCA Data Systems 1962-65. Since 1965 has been involved in Satellite Communications including: COMSAT Labs 1965-72; American Satellite 1972-81; Hughes Network Systems (formerly M/A-COM Telecommunications) 1981-present. Key achievements include: Co-inventor of the INTELSAT SPADE system; development of high speed TDMA; development of INTELSAT 56kbps data services (all during 1967-72). Developed and implemented first private satellite data networks (1973-80). Implemented first all digital satellite system (1979). Received INTERFACE award for extraordinary achievements in data communications (1984). Elected to Fellow IEEE 1987. *Society Aff:* IEEE

Cachat, John F
Business: 4620 E 71 St, Cuyahoga Heights, OH 44125
Position: VP. *Employer:* Tocco Div Park-OH Ind. *Education:* MBA/Marketing/Case Western Reserve Univ; BS/Elec Engg/Case Western Reserve Univ. *Born:* 9/1/16. Fellow of IEEE since 1975; Sr Mbr since 1946. Served on Ad Com of Industry. Applications Soc of IEEE. Have been VP and Gen Mgr of Tocco Div of Park-OH Industries Inc for 13 yrs. *Society Aff:* IEEE.

Cacioppo, Anthony J
Home: 2871 Stauffer Dr, Xenia, OH 45385
Position: Prof, Human Factors Engg *Employer:* College of Engineering Wright State University *Education:* PhD/Behavioral Sci/Univ of IA; MA/Exp Psych/Kent State Univ; BS/Chemistry/Kent State Univ. *Born:* 10/13/23. Native of Akron, OH. SErved with the Army Air Corps 1942-45. Chem process engr with the Ravenna Ordnance Plant, OH, (1952). With Goodyear Aero Corp from 1953-1963, progressing through the orgizational structure from sr engr to departmental mgr. Entered govt service in 1963 as the chief scientist of the AF's Foreign Tech Div. Upon retirement from AF's Foreign Tech Div. (June 1986), accepted position of Prof. of Human Factors Engrg with Wright State Univ, Dayton, OH. Twice awarded the presidential Ran, of Meritorious Executive in the Senior Executive Service. *Society Aff:* AAAS, OAS, APA, OPA, HFS.

Cacuci, Dan G
Business: Oak Ridge National Lab, Bldg. 6025, P.O. Box X, Oak Ridge, TN 37831-6363
Position: Section Hd/Sr Scientist/Prof *Employer:* Oak Ridge Natl Lab *Education:* PhD/Applied Phys & Nuclear Engrg/Columbia Univ; M.Phil (Master of Philosophy) /Applied Phys/Columbia Univ (NYC); MS/Nuclear Engrg/Columbia Univ *Born:* 05/16/48 Section Head, Systems Analysis and Shielding Section, Oak Ridge Natnl Lab Editor, *Nuclear Science & Engineering*, The Res Journal of the ANS. Prof (Part-time), Dept of Nuclear Engrg and Dept. of Mathematics, Univ of TN. Sec, Fellow, ANS. Natl Planning Ctte (1983-86), ANS. Publ over 90 journal articles & int'l conference papers. Biographical listings in: Who's Who in Frontier Science & Technology, 1st & 2nd Eds; Who's Who in the World, 7th Ed; The International Who's Who of Contemporary Achievement, 1985 Ed. *Society Aff:* ANS, AAAS

Cadden, James M
Business: P.O. Box 42214, Houston, TX 77042
Position: Manager of Research *Employer:* Getty Oil Co. *Education:* BS/TX A & I Univ-Petr & Nat Gas Engr; BS/TX A & I Univ-Chem & Math; worked on MS in PE at Univ of H. *Born:* 12/19/23 A Registered Petroleum Engineer in State of Texas since 1952. Served with USN 1943-46. Employed at Getty Oil Company since 1950, working in various line, staff and Management positions. Was Chief Engineer of Getty and is currently Manager of Research. Member of API and Society of Petroleum Engineer of AIME. Served in various official capacities of AIME. Enjoys grandchildren, gardening, bridge, fishing and golf.

Cadden, Jerry L
Home: 17333 Montero Rd, San Diego, CA 92128
Position: Group Tech Mgr. *Employer:* Union Carbide Corp. *Education:* MS/Met Engr/MO Sch of Mines; BS/Met Engr/MO Sch of Mines. *Born:* 7/21/38. Native of Kirksville, MO. Began career at Y-12 plant in Oak Ridge, TN in 1962. Transferred to NY in 1975 to new bus res. Later transferred to San Diego with Union Carbide Electronics Div. Pres of Oak Ridge Chapt of ASM 1971-1972. Enjoy tennis & skiing. *Society Aff:* ASM, AACG.

Cadoff, Irving B
Business: Polytechnic U, 36 Saw Mill River Rd, Hawthorne, NY 10532
Position: Prof. *Employer:* Polytechnic Univ. *Education:* DEngSci/Metallurgy/NY Univ; MME/Mech Eng/NY Univ; BME/Mech Eng/City Coll of NY. *Born:* 8/7/27. Appointed to NYU Engg Faculty 1949. Advanced through Prof of Materials Sci, 1965 and chmn Dept of Mettallurgy and Materials Sci, 1968. Transferred to Polytechnic Inst of NY, 1973. Currently Prof Materials Sci. Res, pub, consulting in areas of elect materials, materials science, energy conversion. *Society Aff:* AIME, ECS, APS.

Cadogan, William P
Home: 35 Fox Den Rd, W Simsbury, CT 06092
Position: Dir, R&D. *Employer:* Emhart Corp. *Education:* ScD/CE/MIT; SM/CE/MIT; SB/CE/MIT. *Born:* 11/2/19. 1946-48: MIT doctoral res on adsorption and separation methods for light hydrocarbon gases. 1948-51: Amoco Oil Co., process design and economics of petrol processing, and synthetic liquid fuels. 1951-56: Process Res, Inc., VP. Consulting engg and process design for soap & detergent, pharmaceutical, and catalyst mfg. 1956-63: AMF, Inc., Mgr Chem Dev, in fields of food tech, ion membrane processes, and plastics applications in new products. 1963-date: Emhart Corp, Dir, R&D, Res for new products in fields of fire detection, electronic security and building access control; inspection machinery; process control systems; discrete parts mfg systems automation. *Society Aff:* AAAS, AIChE, ASME, IEEE, ACS, RESA.

Cady, Philip D
Business: 212 Sackett Bldg, Univ Park, PA 16802
Position: Prof. *Employer:* Penn State Univ. *Education:* PhD/CE/Penn State Univ; MS/CE/Penn State Univ; BS/CE/Penn State Univ. *Born:* 6/26/33. Native of Mansfield, PA. Worked for affiliates of the Std Oil Co (NJ) in Baltimore, MD and Aruba, Netherlands Antilles 1956-62. Employed by the PA State Univ since 1962. Currently prof of civ engg, specializing in mtls. Over 75 publications and one patent. Mbr of technical committees in the American Soc for Testing and Mtls, the American Concrete Inst, and the Transportation Res Bd. ASTM Sanford E Thompson Award, 1968. *Society Aff:* ASCE, ASTM, ACI, TRB

Caffrey, Ronald J
Business: 507 E Michigan St, Milwaukee, WI 53201
Position: VP Marketing. *Employer:* Johnson Controls, Inc. *Education:* BS/Ind Eng/Yale Univ; BS/Bus Admin/Yale Univ. *Born:* 3/16/28. NYC. Ex-Cpt Field Artill USA. Entire bus career with Johnson Controls Inc NY Branch Mgr, Southwest Regional Mgr-Dallas, Midwest Regional Mgr-Chicago, VP Marketing Milwaukee. Presently responsible for marketing, long range planning, Business Unit Coordination, advertising & training in computerized bldg automation & control systems for energy conservation, fire & security. ASHRAE Dir 1974-7, Chrmn Educ Committee 1973, DSA 1976. Chrmn of the Intelligent Buildings Inst 1986-88. Hobbies: tennis, sailing. *Society Aff:* ASHRAE, AAEE, IBI.

Cagle, A Wayne
Home: 6622 Englewood, Raytown, MO 64133
Position: Sr Staff Engr. *Employer:* AT & T Technology Systems (Ret 12/'85) *Education:* MS/Chem/U of Louisville; BS/Chem/High Point College. *Born:* May 1924 High Pt NC. B.S. High Point College, High Point, NC. MS Univ Louisville, Grad study UNC, VPI. WWII infantry-ETO 3-cpgns; Bronze Star- V/Cluster. Staff-Cone Mills Res-Synthetic/natural blends. AT & T '51-chem, electrochem & organic coatings; solder & salt bath brazing; electroforming; plant design/construction; supv-grad eng education: advanced mfg processes. R&D- Mtls-prod, tooling & fixture design-telephone transmission, and electron components. Consultant-mfg mtl & processes. "Reinforced Plastics Design Award" 1972. 1 pat/3pat Digest. Past Chrmn & Director/KCChapt-ASM/Central NC Sec-ACS. Fellow, & Secy Natl Mbr Com-ASM. SME. Dir, KC Chapt-SPE: Dir ED/Newsletter, councilman, E&E Div, & Tech Vol Committee-SPE. Sigma Xi and Alpha Chi Sigma. AF&AM: 32 degrees Scot; Shrine. "Who's Who/Methodist Church." Retired AT&T, 12/85. Lecturer, Pittsburgh St. Univ. KS, 86-87. *Society Aff:* ACS, ASM, SPE, SME.

Cagley, Leo W
Business: 2015 Grand Ave, Des Moines, IA 50312
Position: Senior VP *Employer:* Green Construction Co *Education:* M.S./CE/Harvard Univ; B.S./CE/IA St Univ *Born:* 9/2/19 Native of IA. Before World War II worked as Civil Design Engr, Panama Canal Co. Served as Navy Civil Engr Corps officer from 1943-46. Returned to Panama Canal serving in progressive positions to Chief, Civil Engrg Branch. From 1955 to 1965 held position of VP-Chief Engr for heavy civil construction co in TX and OK. Since 1965 have been with Green Construction Co as Chief Engr, VP-Engrg, and since 1976 as Senior VP. A member of the Bd of Directors and the executive committee of the Bd. Active in engineering education work. *Society Aff:* SAME, AIC, ICOLD, ICID

Cagnetta, John P
Business: P.O. Box 270, Hartford, CT 06101-0270
Position: Sr. Vice Pres *Employer:* Northeast Utilities *Education:* PhD/Chem Engr/Poly Inst of NY; B/Chem Engr/Poly Inst of NY *Born:* 08/15/32 He is responsible for regulatory planning and relations; business planning and corporate dev; economic and load forecasting; and dev of electric and gas rates. Dr Cagnetta has 30 yrs of professional exper in the electric industry, involving the planning, design, constr, and operation of electric generation facilities. Dr. Cagnetta has testified before state and federal regulatory agencies on Corp integrated demand and supply planning, utility economics, nuclear and environmental matters. He has served on a variety of industry adv grps, including the Electric Power Research Institute and the Edison Electric Institute. He is a mbr of the Connecticut Energy Advisory Board and the Connecticut Academy of Scientists and Engrs. *Society Aff:* ANS

Cahill, William J, Jr
Business: P.O. Box 220, St. Francisville, LA 70775
Position: Senior Vice President *Employer:* Gulf States Utilities Co. *Education:* BME/ME/Poly Inst of Brooklyn. *Born:* 6/13/23. VP, Nuclear Licensing, Quality Assurance & Reliability, Res & Dev, Consolidated Edison Co of NY, Inc. BME, Poly Inst of Brooklyn, 1949. Joined Consolidated Edison in 1949, design of steam power plants. From 1954 to 1956, on loan to Knolls Atomic Power Lab, design of nuclear power plants for submarines Seawolf & Triton. Aided in test operation of prototype nuclear plant at West Milton, NY. During 1957 employed by Nuclear Dev Assoc, in design of BR-3 test reactor at Mol, Belgium & dev prog for reactor proposed for Alaska. From 1957 to 1980, technical & managerial assignments associated with Indian Point No 1, proposed Ravenswood Nuclear Plant, Indian Point No 2 with 3, as well as several conventional steam power plants. Mbr: Natl Soc of PE. PE licensed in NY State, Louisiana, and Texas. Retired from Con Ed July 1980. Since August 1980, Senior Vice President for Gulf States Utilities Company, in charge of design, construction, and operation of the River Bend Nuclear Station at St. Francisville, LA. Recipient in 1986 of Amer Nuclear Soc's Walter H. Zinn award for outstanding contributions to the advancement of nuclear power. *Society Aff:* ASME, ANS, NSPE.

Cahn, David S
Business: CalMat Co, 3200 San Fernando Rd, Los Angeles, CA 90051
Position: VP.-Regulatory Matters *Employer:* CalMat Co. *Education:* DEngg/Mineral Engg/Univ of CA., Berkeley; MS/Mineral Engg/Univ of CA., Berkeley; BS/Mining Engg/Univ of CA., Berkeley. *Born:* 1/12/40. in Los Angeles, CA. Married to Sharon Ann Marting and have three children. Engr, Homer Res Labs, Bethlehem Steel Corp, 1966-68. Amcord, Inc,. Res Engr and Group Leader, 1968-1971. dir, Environmental Matters, 1971-1977. VP 1977-80. Dir, Environmental Matters, 1980-82, VP-Regulatory Matters, 1982-present, California Portland Cement Company and CalMat Co. Currently responsible for corporate-wide environmental, health and safety, energy and governmental programs. NDEA Fellow, 1963-1966; AIME Rossiter W Raymond Meml Award, 1972. *Society Aff:* AIME, AIChE, APCA, ASTM.

Cahoon, John R
Business: Dept of Mech Engg, Winnipeg, Manitoba R3T 2N2 Canada
Position: Prof Dept ME. *Employer:* Univ of Manitoba. *Education:* PhD/Metallurgical Engineering/University of Alberta; MSc./Metallurgical Engineering/University of Alberta; B.Sc./Metallurgical Engineering/University of Alberta *Born:* Aug 1939 Calgary, Alberta. B Sc Met Engg Univ of Alberta Edmonton. Two yrs, 1966 & 68 as Fellow, Mellon Inst, Pittsburgh PA. With the Dept of ME, Univ of Manitoba since 1968. Head of the Dept from 1975-86. Mbr of the Assn of PE of the Province of Manitoba. Chrmn of Manitoba Chapter Am Soc for Metals during 1971-72. Fellow of Am Soc for Metals 1983. Mbr Can. Inst. for Min and Met. *Society Aff:* ASM, CIM, APEM.

Cain, Charles A
Business: 155 Elec Engg Bldg, Urbana, IL 61801
Position: Prof. *Employer:* Univ of IL. *Education:* PhD/EE/Univ of MI; NSEE/EE/MA Inst of Tech; BEE/EE/Univ of FL. *Born:* 3/3/43. in Tampa, FL. He received the BEE degree (with highest honors) from the Univ of FL, Gainesville, in 1965, the MSEE degree from MA Inst of Tech, Cambridge, in 1966, and the PhD degree in elec engg from the Univ of MI, Ann Arbor, in 1972. During 1965-1968 he was a Mbr of the Technical Staff at Bell Telephone Labs, Inc, Naperville, IL, where he worked in the electronic switching systems dev area. Since 1972, he has been with the Dept of Elec Engg, Univ of IL, Urbana-Champaign, where he is currently a full Professor & Chairman of the Bioengineering faculty. He has been involved with res on the biological effects and medical applications of ultrasonic & electromagnetic radiation. *Society Aff:* IEEE, ΣΞ, ΦΚΦ, RRS, NAHG, BS.

Cain, James T
Home: 1058 Blackridge Rd, Pittsburgh, PA 15235
Position: Assoc. Prof *Employer:* Univ of Pittsburgh *Education:* PhD/EE/Univ of Pitt; MS/EE/Univ of Pitt; BS/EE/Univ of Pitt *Born:* 5/2/42 Native of Pittsburgh, PA, presently, Assoc. Prof of Elec Engrg, Univ of Pittsburgh. Has worked for Westinghouse Elec Corp, Bell Telephone Labs, and Bell Telephone of PA. A recipient of the ASEE Dow Outstanding Young Faculty Award, the IEEE Computer Society Special Group Award, and the ASEE Western Elec Fund Award for excellence in Engrg Education. Present IEEE Computer Society. Activities include VP for educational activities; Assoc Editor-in-Chief IEEE Micro; Rep Dir to CSAB (Computing Sciences Accreditation Board); Ad hoc visitor for Computer and Electrical Engrg, Accreditation Bd for Engrg and Tech (ABET); mbr - IEEE Educ Activities Bd. *Society Aff:* IEEE, ASEE, ΣΞ

Cain, John L
Business: Wilmore Hall, Auburn, AL 36830
Position: Dir Pub R&T. *Employer:* Auburn Univ. *Born:* Aug 1924 Cocoa FL. B Chem E GA Tech. Employment: Pres-Dir Public Res and Tech Prog, Auburn Univ; Previous-Dir, Engg Prog Dev, Auburn Univ; Dir, Engg Extension Service, auburn Univ; Dir, Corperative Education, VPI; Assoc Dir, Cooperative Education, GA Tech; Res Assoc, TN Cooperation: Lab Technician, DuPont. Positions held: Chrmn, Cooperative Education Div, Am Soc for Engg Education-ASEE; Chrmn, Relations with industry Div, Southeastern Section, ASEE; VP, ASEE; Natl Commission for Cooperative Education; Prog Advisory Council, Southern Growth Policies Bd; Elder, Presbyterian Church; Tau Beta Pi; Phi Kappa Phi.

Cakmak, Ahmet S
Business: Dept of Civ Engg. & Op. Res, Princeton, NJ 08544
Position: Prof *Employer:* Princeton Univ. *Born:* Aug 1934 Izmir Turkey. PhD, MSE Princeton Univ, BS Robert College Istanbul. Asst in Instruction at Princeton 1958-60. Res Assoc at Columbia Univ 1960-62, Lectr at Princeton in 1962. Promoted to Asst Prof in 1963, Assoc Prof in 1969 and Prof in 1972. Chrmn of the Dept of Civ Engg at Princeton 1971-1980. Visiting Professor, CA Institute of Technology, 1980-81. Currently Assoc Dean, Sch of Engrg and Applied Sci. Author of over fifty technical reports and a contributor to journals in mechanics including: Journal of Applied Mechanics, Acoustical Society of America Journal, and Intl Journal of Solids

Cakmak, Ahmet S (Continued)
& Structures. Editor-in-Chief of the International Journal of Soil Dynamics and Earthquake Enginering. *Society Aff:* ASCE, ASME, ASR

Cal Y Mayor, Rafael
Home: Dr Pallares Portillo 176, Mexico City, D F 04040, Mexico
Position: Dir. *Employer:* Cal Y Mayor Y Asociados, S.C. *Education:* BS/Civil Engg/National Univ Of Mexico; Cert/Traffic Engg/Yale Univ. *Born:* 9/10/23. Made first Origin and Destination study in Mexico in 1951; first urban planning studies for a highway network in 1951; first channelization of urban intersections in 1951. Admin of the first rural expressway, 1952 and first prof of Traffic Engg in Mexico, in 1954. Founder and first head of the Traffic Engg Commission of the Ministry of Public Works, 1965-70; founder and first Dir of the Traffic Engg and Transportation Dept, of Mexico City (1971-77). PPres of the Mex Assoc of Traffic Dir; founder and VP of the Mex Assoc of Traffic and Transp Engg; Fellow and Dir of District 8 of the Institute of Transportation Engg (1977-81). *Society Aff:* ITE, AMC, AMDT, AMITT, AI.

Calabretta, Victor V
Home: East Passage Estates, Jamestown, RI 02835
Position: Asst VP *Employer:* CE Maguire, Inc *Education:* MS/CE/Worcester Polytechnic Inst; BS/CE/Worcester Polytechnic Inst *Born:* 7/29/46 Career has specialized in the field of Marine and Heavy Civil Engrg. Currently responsible for US and worldwide port design operations and the firm's Geotechnical Engrg group. Served as officer Navy Civil Engrs Corps 1969-71 in Vietnam. Author of numerous publications in the Marine field and recipient of Geotechnical Group Award BSCE/ASCE, April, 1979. Member of the North Atlantic Ports Assoc and Propellor Club of the US Active in community service. Elected Jamestown Town Moderator 1981, Dir, Jamestown Lions Club, Member, Knights of Columbus. Currently enrolled in Ph.D. Program, Ocean Engrg, Univ of RI. *Society Aff:* ASCE, NSPE, SAME, SNAME

Calaceto, Ralph R
Home: 248 Palmer Ct, Ridgewood, NJ 07450
Position: Director *Employer:* Hydronics Engg Corp. *Education:* PhD/Chem Engg/BPI; MChE/Chem Engg/BPI; BChE/Chem Engg/Pratt Inst. *Born:* 8/17/21. Pioneer in the Dev of Gas cleaning apparatus and techniques in the fertilizer, miscellanious chemical and steel industry. Holder of some fifteen pats in the US and abroad concerning features of gas cleaning devices using venturi, counterflow towers and wet precipitators. Through Calaceto's Devs, particularly in the FL fertilizer industry and in numerous foreign areas the removals and recovery of valuable gaseous metal effluents were established. Presently active as consultant. Numerous papers covering the fields of Air Pollution Control and Equipment have been published. *Society Aff:* AIChE, AISE.

Calahan, Donald A
Business: Dept of Electrical Engg. & Computer Science, Univ of Michigan, Ann Arbor, MI 48109
Position: Prof. *Employer:* Univ of MI. *Education:* PhD/Elect Engg/Univ of IL; MSEE/Elect Engg/Univ of IL; BSEE/Elect Engg/Univ of Notre Dame. *Born:* 2/23/35. in Cincinnati, OH. On faculties of Univ of IL (1960-65), Univ of CA (1963- 64, visiting), Univ of KY (1965-66), Univ of MI (1966-), Interests have ranged from circuit theory (1958-65), computer-aided circuit design (1963-72), and algorithms for large scale serial and parallel processing (1972-). Author of two books and coauthor of one book. Presently consultant to industrial and government organizations on applications of vector & parallel processors to engineering & scientific simulation. *Society Aff:* IEEE.

Calder, Clarence A
Business: Dept of Mechanical Engrg, Oregon State Univ, Corvallis, OR 97331
Position: Assoc Prof *Employer:* OR State Univ *Education:* Ph.D/ME/Univ of CA-Berkeley, MS/ME/Brigham Young Univ; BS/ME/Oregon State Univ *Born:* 10/30/37 He has been employed at WA State Univ as an Asst Prof, at Lawrence Livermore National Lab as a senior research engr, and is currently an Assoc Prof of Mechanical Engrg at OR State Univ. Research interests include experimental mechanics, laser-material interaction, and computer-aided. He has served on the Executive Board of the Society for Experimental Stress Analysis was Treasurer of the Society from 1981-84, and is President, 1987-88. He is registered in the state of OR. *Society Aff:* SEM, ASME, ASEE, BSSM

Caleca, Vincent
Home: 14 Mohegan Trail, Natick, MA 01760
Position: High Voltage Specialist. *Employer:* Chas T Main, Engrs. *Education:* BEE/Elec Power/Cornell Univ. *Born:* 5/27/19. in NYC. WWII Army service 1943-46. Employed by Am Elec Power Service Corp in 1951 as Relay Engr, in time assuming system-wide responsibilities. In 1964 assigned to industry sponsored AC/DC System Operation Res Proj, becoming Asst Mgr. At AEP in 1967, headed Elec Res Section during final stages of 765kV system design. Staff High Voltage Engr in 1971 with leading role in res for ultra high voltage up to 2000kV. Joined Chas T Main in 1975. IEEE Fellow 1979: "For contributions to the design, protection, and operation of high-voltage ac & dc power transmission systems.–. *Society Aff:* IEEE, AAAS, CIGRE, HKN.

Calhoun, John C, Jr
Business: Petroleum Engineering, 201 Doherty Bldg, College Station, TX 77843
Position: Deputy Chancellor for Engrg Emeritus *Employer:* Retired - Self-employed *Education:* PhD/Petrol Engg/Penn State Univ; MS/Petrol Engg/Penn State Univ; BS/Petrol Engg/Penn State Univ. *Born:* 3/21/17. Director, Crisman Institute for Petroleum Reservoir Management, Tx A&M Univ. 1987-88; Deputy Chancellor for Engineering, 1980-1983; Exec VChancellor for Progs, the TX A&M Univ. System 1977-80; also Distinguished Prof of Petrol Engg and Dir of the TX Petrol Res Committee, TX A&M Univ.; VP for Academic Affairs, TX A&M Univ. 1971-77, asst and sci advisor to the Secy of the Interior and acting dir of Office of Water Resources Res. 1963-65; From 1955-69 served at TX A&M as VP for Progs, Dean of Geosciences, Dir Ctr for Marine Resources, and Dean of Engrg; Sea Grant Award 1984; SPE DeGolyer Medal 1982; 1976 awarded Penn State Alumni Fellow; 1975 award Honorary DSc from Ripon College. Member, National Academy of Engr. Past Pres ASEE, Honorary Mbr ASEE (1978); Past Pres Marine Tech Soc; served on Bd of Dirs of Engrs Joint Council and Bd of the Dirs of ECPD; past Pres Soc of Petrol Engrs; Hon mbr of AIME; Fellow, AAAS. Has served on Naval Studies Bd, and as Chrmn of the Ocean Affairs Bd, Natl Acad of Sci. 1972 presidential appointee to Natl Advisory Com for Oceans and Atmosphere. Served as Chmn of Bd of Trustees of Univ Corp for Atmospheric Res., Mbr of TX Coastal & Marine Council (1971-1983), Chmn of Res Coord Panel of the Gas Res Inst. and chmn of Bd of Dirs of Inst of Nautical Archeology. *Society Aff:* ASEE, SPE/AIME, AAAS, ΦΚΦ, ΣΞ, MTS.

Calhoun, Tom G
Business: 511 N Akard, Suite 1100, Dallas, TX 75201
Position: Pres. *Employer:* Calhoun Engg Inc. *Born:* June 1927. BS TX A&M Univ. Jr engr-Stanolind 1950-52, Reservoir engr-Sun Oil Co 1952-55. Joined De Golyer & MacNaughton in 1955, elected vp in 1961. Calhoun Engg Inc 1965-present. Also officer and/or dir of Petrol Tracers Inc, P- R-O Mgt Inc, Amity Oil Co Inc. Reg PE in TX & LA. Mbr of Soc of Petrol Engrs of AIME, dir 1974-77; mbr of Soc of Petrol Scientists, Natl Soc of PE, TX Soc of PE.

Calhoun, William D
Business: PO Box 532, Chambersburg, PA 17201
Position: Consulting Engr (Self-employed) *Education:* MSCE/Civ-Structural Engg/WV Univ; BSCE/Civ Engg/WV Univ. *Born:* 12/30/41. and raised in Moorefield, WV. Located in Chambersburg, PA, in 1968. Proj Engr for Delta Engr 1965-68, Principal Engr with Nassaux-Hemsley Inc 1968-74, Dir of Engg for Delta Intl 1974-

Calhoun, William D (Continued)
76 and opened Private Practice June 1976. Licensed Engr in eight (8) states and licensed surveyor in two states. Received patent for New Concrete Decking System in 1970. Pres of Franklin Chapter, PSPE, 1976-78. Selected to Who's Who in the East, Seventh Edition. Hobbies include camping, hunting and fishing. *Society Aff:* ASCE, NSPE, PSPE, ACI, PCI.

Calkins, Myron D
Business: 414 East 12th St, Kansas City, MO 64106
Position: Dir of Public Works. *Employer:* City of Kansas City. *Education:* Hon Prof/Mgmt Engr/Univ of MO-Rolla; BS/Civil Engr/WA State Univ. *Born:* 10/1/19. in Tacoma, WA. Wife: Lenore. Children: Sue, Ron and Don. WWII Navy veteran. Former City Engr of Tacoma; currently Dir of Public Works in Kansas City, MO (1964-present). PPres of national APWA and two of its chapters, Tacoma Section ASCE, Kansas City Post SAME, Kansas City Engrs Club, State Soc MSPE, Western Missouri Chap MSPE. Trans Officials Div ARTBA and Highway Engrs Assoc of Mo. Active in exec leadership positions in ASCE, APWA, MSPE, NSPE, ARTBA and Highway Engrs Assoc of MO. Honors: "Top Ten Public Works Men of the Yr" 1973; first Hon Mbr of Inst of Adm Mgmt, APWA, 1978; 1979 Edmund Friedman Prof Recognition Award, ASCE. Mayor,s Commendation Award 1981. Reg PE in MO and WA. *Society Aff:* ASCE, APWA, SAME, NSPE, ARTBA, AAEE.

Callahan, Frank T
Business: 300 E Carpenter Freeway, Suite 1210, Irving, TX 75062
Position: Pres - CEO *Employer:* Greiner Engg Inc *Education:* BSCE/Civ Engg/Univ of Notre Dame. *Born:* 7/22/28. Born in Swedesboro, NJ in 1928. Grad from Notre Dame in 1950. Joined Greiner, Inc in 1950. Later named an Assoc of the firm, promoted to VP in 1969, Exec VP in 1970, and Pres in 1975. In charge of engrg work for Greiner's offices including Tampa Intl Airport, NASA's Space Shuttle Landing Facility, portions of WA Mass Transit Sys, to name a few. Active in local C of C and other civic org, a Fellow in FL Engrg Soc. Mbr Natl Bd of Dirs Univ of Notre Dame Alumni Assn 1978-81. Mbr Bd of Dirs Jr Achievement of Greater Dallas. Recipient of Univ of Notre Dame award for Distinguished Service to the Engrg Profession. *Society Aff:* ARTBA, NSPE, IBTTA, ACEC, ACI.

Callahan, Harry L
Business: 1500 Meadow Lake Pkwy, Kansas City, MO 64114
Position: Exec Partner & Div Mgr. *Employer:* Black & Veatch. *Education:* BSCE, Univ of KS *Born:* 01/11/21 Corps of Engrs & Butler Mfg prior to WWII. Joined Black & Veatch in 1946. Hd of Special Projs Div Civ-Structural design 1954 specializing in dynamic analysis of structures subj to blast loadings. Presented papers and participated in res panels. Proj Engr in 1964 on largest compressed air shock tube in US, Ballistic Res Labs, MD. Proj Mgr, Perimeter Acquisition Radar Power Plant, Safeguard, 1968. Div Mgr Special Projs Div, 1970. Partner 1971, Exec Partner 1979. Dir Black & Veatch Intl 1979. Active in 5 Megawatt Solar Thermal Test Facility. One of NSPE outstanding Engg Achievements 1978. Partner in charge several major hazardous waste designs. Div Mgr Industrial/Special Projects 1981. Bd of Dirs Black & Veatch Intl, Pritchard Co. construction claims & litigation. *Society Aff:* ASCE, ACEC, NSPE, ACI, ANS, WPCF.

Callaway, Samuel R
Business: 9301 W 55th, LaGrange, IL 60525
Position: Chief Mtls Engr. *Employer:* GM. *Born:* Mar 1919 Minneapolis, MN. BS Met E Univ of MN 1940, MBA Univ of Chicago Exec Prog 1960. Entire career with GM Corp. Met at Res Labs Div 1940-52. Mtls Stds Activity-GM Mgt Staff 1953-54. Chief Met, Electro-Motive Div LaGrange, IL 1955 to date. In charge of Mtls Lab, respon for metals, polymers & elec insulation used in locomotive mfg. Received Bachelor's Degree with High Distincton and elected to Tau Beta Pi. Mbr Am Soc for Metals since 1940 and named a Fellow of ASM in 1970.

Callen, James D
Business: Engineering Research Bldg, 1500 Johnson Dr, Madison, WI 53706
Position: Kerst Professor of Nuclear Engineering and Engrg Physics, and Physics *Employer:* Univ of Wisconsin-Madison *Education:* 1968 PhD/Nuclear Engrg/Massachusetts Inst of Tech; 1964 MS/Nuclear Engrg/Kansas State Univ; 1963 Fulbright fellow/Mech Eng and Phys/Technische Hogeschool Te Eindhoven, Eindhoven, Netherlands; 1962 BS/Nuclear Engrg/Kansas State Univ *Born:* 01/31/41 Native of Wichita, KS. Fulbright, AEC, NSF graduate and NSF postdoctoral fellowships 1962-9. In 1968-9 was postdoctoral fellow at Institute for Advanced Study, Princeton, NJ working on fusion plasma theory. From 1969 through 1972 was Asst Prof of Aeronautics and Astronautics at MIT teaching fusion plasma theory courses. In 1972 joined Fusion Energy Div at Oak Ridge Natl Lab, and served as Plasma Theory Section Hd there 1975-9. Have been Prof of Nuclear Engrg & Engrg Phys. and Phys at the Univ of Wisconsin-Madison since 1979, conducting res and teaching predominantly graduate courses on fusion plasma theory. Fellow of ANS, APS; Chairman of Div of Plasma Phys of the APS in 1986. Guggenheim fellow 1986-87 at JET Project, England. *Society Aff:* ANS, APS

Callinan, Joseph P
Business: 7101 W 80th St, Los Angeles, CA 90045
Position: Dean, Science & Engrg *Employer:* Loyola Marymount U *Education:* PhD/Engrg/Univ of CA, LA; MS/Engrg/Univ of CA, LA; BS/Mech Engr/Loyola Univ, LA *Born:* 06/21/34 Employed as a Research Engineer at the Rocketdyne Div of North American Aviation, Inc in 1957 and 1958. A mbr of the faculty of Loyola Marymount Univ since 1958. Present rank is Prof of Mech engrg. Assumed current respons as Dean of the Coll of Sci and Engrg in 1982. Consulted for a variety of firms in the area of industrial heat transfer and thermodynamics. Was NSF Science Faculty Fellow. Registered Professional Mech Engr in California. *Society Aff:* ASME, ASEE, TBΠ, ΣΞ.

Callison, Gerald J
Business: Suite 855, 1250 Poydras Plaza, New Orleans, LA 70112
Position: Area Sales Mngr *Employer:* Dresser Atlas *Education:* BS/Petro & Nat Gas Eng/TX A & I Univ *Born:* 06/18/35 Joined Lane Wells Co (later Dresser Atlas) as a field engr in 1962. Worked as a senior field engr, sales engr, log analyst, sales mgr, asst area mgr, senior sales mgr and account rep, assuming his current position, offshore area sales mgr for Dresser Atlas based in New Orleans, in 1978. Lectures on wireline operations at Tulane Univ and the Univ of Southwestern LA. Has served SPE as membership chrmn for the East TX Section, and as sec, program chrmn and 1979-80 chrmn of the Delta Section. Will serve as SPE dir beginning October, 1981. *Society Aff:* SPE, SPWLA

Calo, Joseph M
Business: Div of Engrg, Box D, Providence, RI 02912
Position: Assoc Prof *Employer:* Brown Univ *Education:* Ph.D/ChE/Princeton Univ; AM/ChE/Princeton Univ; BS/ChE/Newark Coll of Engrg *Born:* 11/9/44 Native of Newark, NJ Graduate School at Princeton Univ, 1966-70. Served with the Air Force (Capt), Air Force Cambridge Research Labs, 1970-74. Research Engr for Exxon, Research & Engrg Co in reactor fundamentals and fluid mechanics, 1974-76. Asst Prof of Chem Engrg at Princeton Univ, 1976-81. With Brown Univ since 1981. Contributing to development of new chem engrg program. Air Force Commendation Medal, 1974. *Society Aff:* AIChE, ACS, IUPAC.

Calton, Lynn B
Home: 1700 E. College, Broken Arrow, OK 74014
Position: Partner. *Employer:* CSP Engineering Co. *Education:* BS/Civil Engg/Univ of MO at Rolla. *Born:* 6/7/42. Native of Lamar, MO. Inactive mbr of MO Nat'l Guard. Specialist Engr from 1970 to 1975. Proj Engr from 1975 to 1977. Utilities Dept Supervisor of 50 plus employees, 1978-84. Owner/Partner of Engineering Company since 1984. PEPP chrmn with OK Society of Professional Engrs- Tulsa Chap. State Chairman- Engr's Week. Bd of Dir, V-P Publicity, Programs, Mbrship-Tulsa Chapter of OSPE. Former mbr of Advisory Comm- Regional Wastewater Fa-

Calton, Lynn B (Continued)
cilities Plan. Former Bd of Dir of local Civic Organization and Little League Baseball. Advisory Bd-Tulsa Jr. College. *Society Aff:* ASCE, NSPE, SAME, PEPP.

Calvert, Seymour
Business: 4901 Morena Blvd, Suite 402, San Diego, CA 92117
Position: Pres. *Employer:* APT, Inc. *Education:* PhD/ChE/Univ of MI; MS/ChE/Univ of MI; BS/ChE/MI Technical Univ. *Born:* 1/9/24. Native of MI. Service in US Army Engg Corps 1943-46. Employed by Battelle Inst, Vitro Corp & Univ of MI. Prof at Case Inst of Tech. Dir of Ctr for Air Environment Studies, PA State Univ. Dean of Engg & Dir of Air Pollution Res Ctr, Univ of CA, Riverside. Founder & Pres of Air Pollution Tech, Inc & Calvert Environmental Equip Co. *Society Aff:* AIChE, ACS, APCA, NSPE, AIHA.

Cambre, Ronald C
Business: PO Box 61520, New Orleans, LA 70161
Position: Executive Vice President *Employer:* Freeport Phosphate Rock Co. *Education:* PMD/-/Harvard Bus Sch; BS/CE/LA State Univ. *Born:* 9/22/38. Native of New Orleans, LA. Proj Engr for Int Paper Co, 1960-64. Joined Freeport in R&D in 1964 & have served in numerous process, design & mgt capacities in the US & abroad including Process Engr Supervisor, Sr Process Engr, Operations Mgr, VP of two divisions, responsible for the corp R&D & Engg functions. Currently responsible for Florida operations of the Chemical and Phosphate rock mining activities of Freeport. Married, 3 children & am an avid tennis player. *Society Aff:* AIChE.

Cameron, Joseph A
Home: 160 Craig Dr, Greensburg, PA 15601
Position: Consul *Employer:* Self *Education:* BS/Met Eng/Carnegie Inst of Tech. *Born:* 11/23/20. BS Carnegie Inst of Tech. Employed by GE Co 1941-45. Employed by Elliot Co Subsidiary of United Technologies Corp. 1945-86. Now a consul in current position since on apparatus for power & process industries- compressors, turbines, and turbocharges. Mbr of group awarded first prize in Lincoln Electric Co arc welding design contest 1967. Mbr Fellow ASM, NACE, ASME, ASTM, Author of compressor mtl section of Encyclopedia of Chem Engg & Processes. *Society Aff:* ASM, NACE, ASME, ASTM.

Camp, Albert T
Home: Rt 1, Box 1278, Welcome, MD 20693
Position: President *Employer:* The Brentland Corporation *Education:* MS/Ind Mgt/MIT; BE/CE/Yale. *Born:* 1/6/20. Grad of Yale Univ 1941 in ChE. Employed Hercules, Inc 1941-1950 as Res Assoc. Naval Ordnance Test Station 1950-59 as Hd Propellants Divs. Lockheed Propulsion Co 1959-1964 Dir Propellant Dev. Naval Ordnance Station 1964-1980, in various positions as GS-16 Supervisory Engr. Retired 1980 as mbr of Sr Exec Service . Holder of over 20 patents on propellants, protective coatings, rocket designs. Consultant in solid propellants & propulsion. President, The Brentland Corporation, which is engaged in human nutrition and national defense activities. *Society Aff:* ACS, ADPA, ΣΞ, RI.

Camp, David T
Home: 3918 Estates Dr, Troy, MI 48084
Position: Prof & Chrmn. *Employer:* Univ of Detroit. *Born:* Nov 1937. BS, MS, PhD Chem Engg Carnegie-Mellon. Fac mbr in Met Engg at alma mater 1962-66. With Univ-Detroit since 1966. As Asst Dean of Engg 1966-70, guided dev of innovative Dr of Engg prog. Dept Chrmn 1972-present. Technical interests focus on process & extractive met as well as energy usage and policy and its relationship to the environment. Bd/Dir TMS-AIME 1974-76. Mbr AIChE, ASEE, Tau Beta Phi. PE MI. Native Toledo, OH.

Campanella, Richard G
Business: Dept of Civ Engg, Vancouver, Canada V6T 1W5
Position: Prof and Head of Civ Engg Dept. *Employer:* Univ of British Columbia. *Born:* May 1936. CCNY, UCLA & UC Berkeley; BS, MS, PhD UC-Berkeley; Have been in Vancouver BC Canada since 1965 when appointed to fac at Univ of British Columbia. Full Prof in 1976; Head of Dept in 1978; Teach Soil Mech & Fdns, direct res on Mechanics of Natural Clays, Slope Stability and Insitu testing & evaluation. Especially interested in adv experimental techniques in lab and field. Dir of Soil Mech Labs & field testing prog. Elected mbr of Bd of Dirs for Canadian Geotechnical Soc CGS 1972-74, 1974-77. Organizing Chrmn 1976, Canadian Geotechnical Conf on Slope Stability in Vancouver. Married, 3 children; enjoy sailing & camping. Native Brooklyn NY.

Campbell, Bobby D
Business: 4338 Belleview, Kansas City, MO 64111
Position: Pres. *Employer:* B D Campbell & Co, Inc. *Education:* MS/Arch Engrg/Univ of TX; BS/Arch Engrg/Univ of TX. *Born:* July 1925 Stroud, OK. Geo Wash & Nettie-Arbaugh-C. m. Lolis Jane Lasater Dec 1945; children Mark Alan, Christopher Craig, Connie Sue. Struct engr. BS Univ of TX 1951, MS 1952. Instr Civ Engr Univ of TX 1952-53; design engg draftsman Wilbur Kent, Lufkin TX 1953-55; res engr Balconies Res Ctr Austin TX 1955; Jr Partner W Clark Craig & Assoc Engrs 1955-56; pri practice struct engg Kansas City MO. Registered 16 States. Served with USNR 1943-45. Mbr ASCE, ACI, CEC, NSPE, SAME Tau Beta Pi, Chi Epsilon. *Society Aff:* ACI, ASCE, NSPE, ACEC, SAME

Campbell, Bonita J
Business: 18111 Nordhoff St, Northridge, CA 91330
Position: Assoc Prof of Engg. *Employer:* CA State Univ (CSUN). *Education:* PhD/Operations Res/Univ of CA, Los Angeles; MBA/Quantitative Methods/Pepperdine Univ, Los Angeles; MS/Engg/Univ of Redlands; BS/Mathematics/CO State Univ. *Born:* 9/15/43. Fifteen yrs experience in tech, mgt and consulting capacities with The Aerospace Corp, GE, Kaiser Steel, and several surveying and consulting firms. Faculty positions held at Pepperdine Univ and CSUN. Natl officer in AWIS (1979- 81); Dir of several NSF grants for Women in Engg projs (1976-present); Author of two books and several articles and reports; recipient of special ASEE Achievement Award (1979); Included in several bibliographic references, most recently *American Men and Women of Science* (1979). *Society Aff:* SWE, AWIS, ASEE, IIE, AWM.

Campbell, Calvin A, Jr
Business: 4834 South Halsted St, Chicago, IL 60609
Position: Pres & CEO *Employer:* Goodman Equipment Corp. *Education:* JD/Law/Univ of MI; BS/Chem Engg/MA Inst Tech; BA/Economics/Williams College. *Born:* 9/1/34. Native Midland, MI. With Exxon co NYC from 1961-69; Chmn Board & Treas John B Adt Co, PA & NY 1969-70. Since 1971 Pres & CEO Goodman Equip Corp, Manufacturers mining & tunneling equipment. Also Chmn Bd Improved Plastics Machinery Corp, wholly owned subsidiary of goodman, manufacturers injection and blow molding machinery. Mbr American & NY Bar Assoc; Newcomen Soc North America; Young President's Organization; Racquet Club Chicago; Glen View Club; PSI Upsilon & Phi Delta Phi Fraternities; Chairman, Bd of Governors & Chmn Manufacturers Div, American Mining Congress; Director & Member Executive Committee. American Mining Congress; Who's Who in World; Bd Dirs IL Manufacturers' Assoc; Member Economic Development Commission, Chicago. Member Board of Dirs, Thermatics, Inc. (Houston). *Society Aff:* AIChE.

Campbell, Charles A
Business: 747 Third Ave, New York, NY 10017
Position: VP. *Employer:* Chem Systems, Inc. *Education:* BS/ChE/RPI. *Born:* 5/20/22. 24 yrs with Tidewater Oil (Getty Oil) and 5 yrs with Commonwealth Oil (Puerto Rico) in R&D, Engg, Manufacturing, Planning, Administrative, and Executive assignments both domestic and international. Consultant to hydrocarbon process and energy industries since 1971 specializing in venture analysis, project feasibility, strategic planning, and project management. *Society Aff:* AIChE, ACS.

Campbell, Delmer J
Business: 3000 Bissonnet, Houston, TX 77005
Position: VP. *Employer:* Trunkline Lng. *Education:* BS/ChE/KS Univ. *Born:* 9/20/37. Native Independence, MO. Grad college 1959. 20 yrs in various assignments with Panhandle Eastern Pipe Line Co. Currently Admin VP Trunkline Lng Co. Importation of Lng from Algeria. *Society Aff:* AIChE.

Campbell, George S
Home: 73 Hillyndale Rd, Storrs, CT 06268
Position: Prof. *Employer:* Univ of CT. *Education:* PhD/Aero/CA Inst of Tech; MS/Aero/CA Inst of Tech; BAE/Aero/RPI; BS/Aero/RPI. *Born:* 11/29/26. in Sauquoit, NY. Served in USNR 1944-46. Aero Res Scientist with Natl Advisory Committee for Aero, 1947-53. Res Engr for Hughes Aircraft, 1954-63. Hd of Aerospace Engg Dept, Univ of CT, 1963-71. Currently Prof, Mech Engg Dept. *Society Aff:* AIAA, SNAME, SCS.

Campbell, George S
Home: 203 East Brow Road, Lookout Mountain, Chattanooga, TN 37350
Position: Chmn Bd, Emeritus *Employer:* Geo S Campbell & Assoc, Inc. *Education:* BS/Mech Engg/Univ of TN. *Born:* 3/29/11. Native: Chattanooga, TN. At Univ of TN: Tau Beta Pi, Phi Kappa Phi, and Sigma Chi, 1935 entered HVAC industry. 1940 started own consulting engg office. 1942-46, Lt, USNR, USS Columbia, CL-56, Pacific Theater. 1946 re-established engg practice. 1950 started Campbell Industries, Inc, manufacturing precision carbon deposited resistors; 1953, sold to Clarostat. Returned to consulting practice. Licensed engr: AL, GA, TN, VA; NCEE Certificate 3301. PPres: (1) TN Chapter of ACEC, (2) TN Valley Chapter of ASHRAE, (3) Chattanooga Civitan Club. Secretary during founding yr of TN Chapter of NSPE. Mbr; ACEC, ASME, ASHRAE, Chattanooga Chamber of Commerce, Chattanooga Engineer of the Year-1981, Rotary Club and Episcopal Church, Wife: Jo. Children: Nancy & Jeff. *Society Aff:* ASHRAE, ASME, ACEC.

Campbell, Harvey A
Business: 5855 Gladys, Beaumont, TX 77706
Position: President *Employer:* Campbell Consultants, Inc. *Education:* BS/ChE/Univ of TX. *Born:* Dec 1930 Tyler TX. Married, 4 children. BSChE Univ of TX. Celanese Corp TX, NJ, WV 1952-57. Ch Process engr & oper mgr TX Butadiene & Chem Corp Houston 1957-62. Oper mgr, plant mgr Sinclair Petrochem Houston 1962-68. Oper plan mgr Sinclair Oil Corp NY 1968-69. Corporate mgr current business analysis Atlantic Richfield NY 1969-71. VP Vulcan Mtls Co Chem Div 1971-1978. Executive VP - Texas Oil & Chemical Terminal, Inc. 1978-1981. President, Campbell Consultants, Inc. 1981- Technical, energy and management consultant. Mbr Am Inst Chem Engrs, Tau Beta Pi, Phi Lambda Upsilon, Omega Chi Epsilon. Reg PE TX.

Campbell, Henry J, Jr
Home: 1300 Lowell Ave, New Hyde Park, NY 11040
Position: Principal. *Employer:* Henry J. Campbell, Jr., P.E. *Education:* ME/ME/Stevens Inst of Tech. *Born:* 1/3/22. Richmond Hill, NY. Married: Lillian E Gallimore. Children: Nancy Elizabeth, Barbara Ann. Merchant Marine and positions in industry until 1953. Formed own consulting firm and been in continuous practice since then. Licensed in 28 states. Certified NCEE. Mbr and past chrmn NY State Bd for Engg and Land Surveying. Officer/dir of several natl, state and chapter engg societies. Received a number of awards for design, profl and charitable contributions including 1979 ASHRAE-ALCO award for Distinguished Public Service. Officer/dir in several natl, local charitable organizations. Author of technical papers. AAA arbitrator. 1980 NYSSPE Engr of the Yr. *Society Aff:* AAEE, ACEC, ASHRAE, IEEE, AWWA, NSPE, ASME.

Campbell, John B
Home: 319 Richmond Rd, Douglaston, NY 11363
Position: Editorial Director *Employer:* Hearst Business Publishing Group *Education:* BS/ChE/Univ of MI; BS/Met E/Univ of MI. *Born:* 2/11/28. Ridley Park, PA. BS Chem & Met Engg Univ of MI 1949. US Army Chem Corps 1951-52. Wrote for Mtls & Methods Mag, now called Mtls Engg 1949-56; Managing Ed 1957-63. Managing Ed Space/Aeronautics 1964; Ed 1965-68. Tech Ed Business Week 1969-71; Assoc Ed in charge of BW Indus Edition 1972-75. Sr Ed overseeing BW indus coverage 1976-79. Editor-in-Ch Chemical Week 1980-83. Currently Editorial Dir, Hearst Business Publishing Group. P Pres Village Light Opera Group, Ltd & Douglaston Community Theater. Married, 2 grown children. Enjoy choral singing, sailing, computer, racquet ball.

Campbell, John M
Business: 1333 N Glengarry Rd, Birmingham, MI 48010
Position: Ret. *Education:* BS/Chem Engg/MA Inst Tech. *Born:* 8/8/03. General Motors Res Lab, 1926-1968: Hd Fuels and Lub Dept., 1947-1954; Technical Dir., 1954-1957: Scientific Dir. , 1957-1967: Spec Asst to VP Res, 1967-1968; Ret 1968-present. Pioneering investigation and discovery of rel between molecular structure of hydrocarbons fuels and their knocking tendency and susceptiblility to anti-knock effect of tetraetnyl lead and hence the efficiency with which they can be burned, in gasoline engines. Participation in development of engine and procedure for measuring knocking tendency of motor gasolines which has become standard for octane number. Dis of effectiveness of org phosphorous add for controlling preignition in gasoline engines. Organ of early investigations to measure and to control harmful emissions from auto engines. *Society Aff:* SAE, ACS, AIC, ASME, AAAS.

Campbell, John M
Home: 1215 Crossroads Blvd, Norman, OK 73072
Position: Bd Chm *Employer:* Campbell Co. *Education:* PhD/ChE/OK Univ; MS/ChE/OK Univ; BS/ChE/IA State Univ *Born:* 3/24/22 Virden, IL. Engr with DuPont 1943-46 on atomic energy. Mgr Process Equip Div BS&B 1950-54. Dept Dir, Distinguished Prof and Dir Petroleum Research Center 1954-68 at OK Univ. Bd Chrmn Campbell Group of companies 1968 to present. Author or co-author of 9 books and 127 technical articles in engrg area. Holder of patents and developer of major gas processing equip. National Chrmn of several AIME Committees. Distinguished lecturer SPE and winner of 1978 John Franklin Carll Award for outstanding contributions to petroleum engrg. Consultant to over 100 companies and governments throughout the world. *Society Aff:* AIME, SPE, SPEE

Campbell, R Neal
Home: 7 Marshall Ct, Greenville, SC 29605
Position: Pres-Owner Retired *Employer:* AE Engineers *Education:* BEE/Elect/Clemson Univ *Born:* 2/18/22. US Navy 1943-46. Pres CEO Consulting Engg Firm, AE Engrs, 1949-1986. Married 1943. 7 children. Presbyterian. Hobbies golf, photography. Registered S.C., N.C. *Society Aff:* ACEC (Fellow), CESC.

Campbell, Regis I
Business: 3101 Euclid Ave, Cleveland, OH 44115
Position: VP. *Employer:* Barber & Hoffman Inc. *Education:* BSSE Cleveland State/Structural Design/Cleve. State Univ.-Fenn Col. *Born:* May 1923. Served in infantry and gen service engrs in Europe in WWII. Part time instr in Civ Engrg Dept Cleveland State Univ, & guest lectr. Cons Engr on fed, state & local govt struct problems, as well as those of sch systems and the private sector. Contact rep & field, spec problems man for Barber & Hoffman Inc. Reg PE 12 states incl OH. P VChrmn Consulting Engrs of OH. Natl Dir Natl Soc PE P. Pres. Ohio Soc. Prof. Engrs, Bd. Member Engrs, Foundation of Ohio. Received OH Soc of PE Citation & Award as Engr of the Yr-1981. Received the Fenn Coll of Engrg, Cleveland State Univ Outstanding Alumnus Award - 1981. Active in Kiwanis, Church and Civic Affairs plus the co-mgt of six children. *Society Aff:* NSPE, OSPE, CSPE, CCEP, OACE, EFO

Campbell, Richard J
Business: 3600 Pammel Creek Rd, la Crosse, WI 54601
Position: VP & Gen Mgr. *Employer:* Trane Co. *Born:* Dec 1929. Welding Engg

Campbell, Richard J (Continued)
Degree, OH State Univ 1955. First Lt USAF 1955- 57. Welding Res work. Welding Engr ACF Industries 1957-59. GE Co Welding Engr 1959-61. Trane Co 1961-63 Welding Engr, 1964-65 Sr Mfg Engr, 1966-68 Mfg Mgr, Epinal France, 1969-71 VP Purchasing 1971-73 VP Marketing, 1973 VP & GM Commercial Air Cond Div. Mbr ASME Boiler Code Welding Section 1963-66, Am Welding Soc, Am Society for Metals, Purchasing Assn, Reg PE WI & OH, Am Soc of Heating & Refrig Engrs, Honorary Recipient James F Lincoln Welding Award 1953 & 61, OH State Univ Outstanding Engg Alumnus 1973.

Campbell, Robert A
Business: 1215 Crossroads Blvd, Suite 200, Norman, OK 73072
Position: Chrmn *Employer:* Petrotech Consultants, Inc. *Education:* MS/Petroleum Engr/Univ of OK; BS/Petroleum Engr/Univ of OK *Born:* 4/26/49 Native of Norman, OK. Worked 71/2 years for Enserch Exploration (formerly Lone Star Producing Co.) with final position of District Engr responsible for Appalachia, Rocky Mountain and Mid Continent Regions. Was responsible for reservoir evaluation of world's two deepest wells (30,000 ft). Last eleven years independent consulting petroleum engr forming and heading Four consulting and producing companies. Member SPE of AIME. Chrmn SPE Symbols & Metrication Committee. Co-author of Mineral Property Economics and instructor for John M. Campbell & Co. Registered PE in OK. Enjoy all sports. *Society Aff:* SPE

Campbell, Stephen J
Business: 4403 N Central Expressway, Suite 300, Dallas, TX 75205
Position: Pres *Employer:* Gunnin-Campbell Consulting Engrs, Inc *Education:* MSKE/CE/So Methodist Univ; BSKE/CE/So Methodist Univ *Born:* 9/4/42 Native of Mission, TX. Lecturer, So Methodist Univ and co-author of professional publications. Structural Analyst, Collins Radio, Dallas, TX, 1967- 72. Senior Structural Engr, Ellisor & Tanner, Dallas, TX 1972-78. VP, Ellisor & Tanner, 1978-79. Pres, Gunnin-Campbell Consulting Engrs, Inc, 1978-present. Licensed Professional Engr in nine states. *Society Aff:* ASCE, ACI, CEC, CSI

Camras, Marvin
Business: 10 W 35th St, Chicago, IL 60616
Position: Sr Scientific Advisor. *Employer:* IIT Research Institute. *Education:* LLD/Hon/IL Inst of Tech; MS/Elec Engg/IL Inst of Tech; BS/Elec Engg/Armour Inst of Tech. *Born:* Marvin Camras is an engr, inventor, and physicist whose contributions to the art of magnetic recording include tape and wire recorders, high frequency bias, recording heads, tape material, magnetic sound for motion pictures, multitrack recorders, stereo sound, and video tape recorders. He is the author of many articles, has been awarded over 500 patents, and is sometimes called "The Father of Modern Tape Recording–. *Society Aff:* AAAS, IEEE, SMPTE, AES, AcSA, WSE, NAE.

Canavaciol, Frank E
Home: 7119 Juno St, Forest Hills, NY 11375
Position: Emeritus Prof. *Employer:* Poly Inst of NY. *Education:* Elec Engg/Elec Engg/Poly Inst of NY; Profl Engr/PE/NY State. *Born:* 1/16/96. Air Service (Now AF) 1st WW. Prof of Elec Engg Poly Inst of NY. PE, State of NY 1438. At time of retirement - Hd of evening undergrad elec engg served as examiner in elec engg for NY State Profl License from inception of the law to 1962 when I retired. Honorary Doctor of Engineering, Poly Inst of NY 5/80. *Society Aff:* IEEE, RCA.

Cancilla, Myron A
Business: 4097 N Temple City Blvd, El Monte, CA 91731
Position: Pres. *Employer:* Sparling Div. *Born:* Nov 1933. MS Chem Engg from USC; BS from Poly Inst of Brooklyn. Native of Brooklyn, NY. Started as Res Engr with the Aerophys Lab of N Am Aviation, Inc in 1955. Spent next eighteen yrs in the Engg Dept of the Rocketdyne Div of N Am (and subsequently Rockwell Intl) during the Saturn/Apollo and Space Shuttle Progs. Gen Mgr & Pres of United Controls Div of Envirotech Corp in 1973. Pres Sparling Div of Envirotech Corp 1975 to present. Mbr ISA, AIChE and AWWA. Enjoy reading & family activities.

Candy, James C
Business: HOH-L137, NJ 07733
Position: MTS. *Employer:* Bell Labs. *Education:* PhD/Engg/Univ of North Wales; BSc/Engg/Univ of North Wales. *Born:* 9/27/29. From 1954 to 1956 was with S Smith & Sons, Guided Weapons Dept, for the next four yrs worked on nuclear instrumentation at the Atomic Energy Res Establishment, Harwell. In 1959 came to the US to take up an appointment as a Res Assoc at the Univ of MN, and a yr later joined Bell Labs, Holmdel, NJ, where he has investigated digital circuits and pulse transmission methods, and is now concerned with methods for processing telephone and video signals. *Society Aff:* IEEE.

Canevari, Gerard P
Business: PO Box 101, Florham Park, NJ 07932
Position: Engg Adv. *Employer:* Exxon Res & Engg Co. *Education:* MS/Mech Engg/Stevens Inst of Tech; BS/Mech Engg/ Stevens Inst of Tech. *Born:* 11/28/24. Joined Exxon Res and Engg Co in August 1953 and worked on both the design and dev of petroleum processing equipment and tech. During the past 15 yrs he has been involved with the area of surface chem-part its application to engg problems such as the separation of immiscible dispersed liquids. Some tech dev during this period are now in comm application, such as the use of a chem demulsifier to recover water-free oil from tanker was water, a hydrocarbon vapor suppressant and a chem dispersant to eliminate oil slicks. Mr Canevari presently holds 24 patents. Recipient of National Inventors Hall of Fame Medallion. *Society Aff:* ASTM.

Canfield, Frank B, Jr
Business: 3000 S Post Oak, Houston, TX 77056
Position: Chrmn of Bd. *Employer:* ChemShare Corp. *Education:* PhD/CE/Rice Univ; BS/CE/Univ of AR. *Born:* 12/6/36. Native of Searcy, AR. Taught chem engg at the Univ of OK 1962-1974. Chrmn of dept at OU 1966-69 and 1973-74. Co-founded ChemShare Corp, an intl supplier of process engg software, in 1969. With ChemShare full-time since 1974. Enjoy opera, sailing, snow skiing and snorkeling. *Society Aff:* AIChE, ACS.

Canham, Robert A
Business: 2626 Pennsylvania Ave, NW, Wash, DC 20037
Position: Exec Dir. *Employer:* Water Pollution Control Federation. *Education:* MSCE/Sanitary Engrg/Purdue Univ; BSCE/Sanitary Engrg/Purdue Univ. *Born:* 1/3/21. Assn Exec; B Virden, IL, Jan 3, 1921; s Howard Ambrose & Rhoda Ann (Haneline) C; BS in Civ Engrg, Purdue Univ, 1942, MS (Natl Polio Fdn fellow), 1947; m Margie Lu Bruton, Dec 21, 1951; children-Patricia Ann, Robert Allen, Jane Claire. Civ engr M W Kellogg Co, 1942-47; san engr Natl Canners assn, 1947- 57; assoc editor Water Pollution Control Fedn, Wash, 1957-64, asst sec, editor, 1964-69, exec dir, 1969-. Adj prof san engrg Howard Univ, Wash, 1967-; ons USPHS, 1962-65, Food canning industry, 1957-; mbr adv group EPA & US Geol Service 1972-. Pres PTA, 1966; treas. Fairfax County Council PTA's 1970-; citizen adviser Fairfax County Sch Bd, 1968-. Served from ensign to lt (jg) USNR, 1944-46. Reg PE, DE, IN. Diplomate Am Acad Environmental Engrs. Honorary Fellow in Inst of Public Health Engrs (UK). Honorary Fellow Institute of Water Pollution Control (UK)Mbr Water Pollution Control Fedn (hon mbr), ASCE, Am Water Works Assn, (hon mbr)VA. Water Pollution Control Assn, Inter-Am Assn San Engrs, Am Acad Environmental Engrg, Sigma Nu. Methodist. Clubs; Cosmos, Engrs, (Balt), Editor: Jour Water Pollution Control Fedn, 1957-69. Contrib articles to Profl journs. *Society Aff:* WPCF, ASCE, AWWA, AAEE.

Cannerelli, Gary D
Home: 7454 Waxwood Cir, North Syracuse, NY 13212
Position: Managing Engineer *Employer:* O'Brien & Gere Engrs, Inc *Education:* BS/Civil & Environ Engrg/Clarkson Coll *Born:* 1/6/51 Graduated Clarkson Coll

Cannerelli, Gary D (Continued)
1972 with a BS in Civil & Environ Engrg. Elected to membership of Tau Beta Pi & Chi Epsilon (National & Civil Honorary Societies). Joined O'Brien & Gere Engrs Inc in 1972 as a Design Engr. Currently, a managing engineer in responsible charge of major utility design projects in the fields of water, sewage & drainage and also specializing in roads, railroads and industrial work. Registered Professional Engr in NY State. Served on the Syracuse Section ASCE Bd of Dir and is a Past Pres of the Syracuse Section. Also served on the NSPE Scholarship Selection Committee, and is currently serving on the Board of Directors of the Central New York Homebuilders Association. *Society Aff:* ASCE, NSPE, AWWA

Cannon, Charles E
Business: 87 Kilby St, Boston, MA 02109
Position: VP *Employer:* Coffin & Richardson, Inc *Education:* MS/Sanitary Engrg/Harvard Univ; BS/CE/Worcester Polytechnic Inst *Born:* 5/9/23 Served CE Corps USN 1944-46. Field Engrg Metcalf & Eddy 1947-50 dam construction. Project and Sr Project Engr 1950 to 1965 for Metcalf & Eddy on sanitary engrg facilities and soil mechanics problems. 1965 to date with Coffin & Richardson. VP in charge of water works projects since 1968. Registered Engr in 12 states. Appeared as expert witness before courts and regulatory agencies. Performed duties on consulting engrg under trust indentures of public authorities financed by revenue bonds. *Society Aff:* AWWA, ACEC, ASCE

Cannon, Charles N
Business: 3333 Michelson Dr, Irvine, CA 92730
Position: Pres *Employer:* Fluor Engrs & Constructors, Inc *Education:* BS/ChE/Univ of MI *Born:* 5/7/22 in Salt Lake City, UT. Married Margie Moore, 3/2/45; two daughters, Carolyn and Martha. Process Engr 1947-51, Consulting Engr 1951-61, Sales Executive-Fluor Corp 1961-68, Managing Dir-Fluor Nederland BV - The Netherlands 1968-71, VP and Sr VP-Marketing-Fluor Engrs and Constructers, Inc 1971-77, Pres-Fluor Engrs and Constructors, Inc 1977-present, Group VP and Dir-Fluor Corp, 1977-present. Awards: Meritorious Achievement Award-Orange County Engrg Council. Memberships: Balboa Bay Club, Big Canyon Country Club-Newport Beach, CA, Eldorado Country Club, Vintage Country Club-Indian Wells, CA. Military Record: First Lieutenant, US Army 1942-1945. Hobbies: Golf, fishing & skiing. *Society Aff:* AICHE, API, NPRA

Cannon, Francis V
Business: 1025 N Milwaukee St, Milwaukee, WI 53201
Position: Provost *Employer:* Milwaukee Sch/Engg. *Education:* Phd/Economics/Univ Wisconsin-Milwaukee; MSEE/Elec Eng/Marquette Univ; BSEE/Elec Eng/MSOE. *Born:* Nov 1935. PhD Univ Wisconsin-Milwaukee; MSEE Marquette Univ; BSEE MSOE. Prof EE Dept, MSOE 1963-. Dean of Engg, 1970-78. Vice President, Academics and Dean of Faculty 1980, Exec VP & Provost 1984-. Regional VChrmn, Engg Tech Ctte, EXCP 1974-. Salgo-Noren Fdn Outstanding Educator Award (1970). ASEE visiting Engr Prog Engg Tech Curriculum Consultant (1972-73). ASEE Engg Manpower Ctte 1976-77 & 1984-. Enjoy all sports, comtemporary music & American history. Mbr ASEE, IEEE, NSPE, Engrs & Scientists of Milwaukee. *Society Aff:* IEEE, ASEE, NSPE

Cannon, Robert H, Jr
Business: Div of Engg, Pasadena, CA 91125
Position: Chrmn Div/Engg. *Employer:* CalTech. *Born:* Fellow of AIAA & a mbr of its Bd of Dirs. Chrmn of NASA Res Advisory Subcommittee on Guidance & Control, & also Chrmn of Res Advisory Group of NASA's Electronics Res Ctr in Cambridge. VChrmn of AF Scientific Advisory Bd. From 1966 to 1968 served as Chief Scientist of USAF. Received BS from Univ of Rochester 1944, and ScD in Mech Engg from MIT 1950. Has been a consultant to aerospace firms since 1959 & was a radar & CIC officer in the UNS in WWII. Was an instr at MIT 1949-50, Visiting Asst Prof at UCLA 1954-57, & Assoc Prof of Mech Engg at MIT 1957-59. Came to Stanford Unv 1959. Is author of num technical papers & articles & a textbook "Dynamics of Physical Systems–. Received the Exceptional Civilian Service Award in 1968. Mbr Natl Acad of Engg.

Canonico, Domenic A
Business: 911 W. Main St, Chattanooga, TN 37402
Position: Director, Metallurgical & Materials Lab *Employer:* Combustion Engineering Inc. *Education:* PhD/Met Eng/Lehigh Univ; MSc/Met Eng/Lehigh Univ; BS/Met Eng/MI. *Born:* 1/18/30. Born and raised in Chicago, IL. Served in the USAF during the Korean conflict. Following service did welding and brazing research at Armour Research Fnd (1953-58). Staff of Lehigh Univ as instructor and researcher, 1958-62, while pursuing grad studies. Employed by Union Carbide Corp-Nuclear Div (Oak Ridge Natl Lab) 1965-81. Fellow of ASM, cited for leadership & mgmt in heavy section steels associated with nuclear reactor research and dev. Employed by Combustion Engineering Inc. Fossil Power Group, since 1981. Fellow of ASM, cited for leadership & mgmt in heavy section steels associated with nuclear reactor research and dev. AWS Adams Lecturer (1983), recipient of AWS Lincoln Gold Medal (1980), Rene D. Wasserman Award (1978). Member Main Committee ASME Boiler and Pressure Vessel Code; Chrmn Subcommittee Properties; Member Subcommittee Power Boilers; Mbr Subgroup Strength Ferrous Alloys. Mbr Sigma Xi. Lecturer-Univ of TN, ASM, Constultant, Adj Prof - Univ of TN. Member Advisory Committees School of Engrg, Univ of Tennessee, Knoxville and Univ of Tennessee, Chattanooga. *Society Aff:* ASM, ASME, AWS, ASTM.

Canter, Larr W
Business: Norman, OK 73069
Position: Dir & Assoc Prof. *Employer:* Univ of OK. *Born:* May 1939. PhD Univ of TX at Austin; MS IL; BE from Vanderbilt. Native of Nashville, TN. Commissioned Officer in US Public Health Service 1962-65. Asst Prof Tulane Univ 1967-69. Technical Prog Chrmn ASCE Natl Conf 1969. Univ of OK 1969-present. Outstanding Educators of Am 1975. Conduct res & teaching in air & water pollution, solid wastes mgt, and environmental impact assessment.

Canter, Walter H Jr
Home: 1449 Fountaine Dr, Columbus, OH 43221
Position: VP Mfg. Retired *Employer:* AccuRay Corp. *Education:* MS/ME/OH State Univ; BS/ME/OH State Univ. *Born:* 10/8/20. Native of Columbus, OH. Served in USAF as officer & pilot 1940 to 1945. Associated with GE Co one yr, worked on jet engine dev proj. Joined AccuRay Corp 1951 as mech engr, 1964-82 VP Mfg. Responsible for operation of Ind Engg, Production, Mtl Mgt, Master Schedule divs. Mbr Pi Eta Sigma, Pi Tau Sigma, Tau Beta Pi honorary fraternities, Newcomen Soc, 1970-1974 OSU Mech Engg Visiting Committee, OSU Pres's Club, OSU Distinguished Alumnus Award-1975, Columbus Technical Council Man of Yr-1976. Thirteen patents, disclosures, written number of articles for PE journals. VP; Corporate Quality & Productivity Management 1982-85; Retired 1985. *Society Aff:* ASME, ASQC.

Cantieri, William F
Home: 310 Hillside Ave, Lancaster, OH 43130
Position: Consultant (Self-employed) *Education:* MS/ME/VPI; BS/ME/VPI. *Born:* 6/16/13. Native of Lynchburg, VA. In Power Engr Fields of Operation, testing, equip design and mgt since grad. Lt (USNR) Ship repair, Oran, Algeria, 1943-45. With Lynchburg Foundry, NY State Elec & Gas & Babcock & Wilcox before war. Chapman Chem and Diamond Power Specialty after war. Later, as VP Engg and new product dev, initiated policy of expansion into foreign markets. Life Fellow ASME. Chrmn ASME Power Div, JPGC and Energy Committees, ASME Rep to Coordinating Committee on Energy. Member-at-Large ASME Policy Board, Power Dept. PE State of OH. Independent Energy Consultant since 1976. *Society Aff:* ASME.

Cantor, Irwin C
Business: 919 Third Ave, New York, NY 10022
Position: Chief Exec Officer *Employer:* Office of Irwin G Cantor. *Education:* BCE/-/CCNY *Born:* 9/17/27 NYC. USAAF 1945-46. Design Engr Tippetts, Abbott McCarthy 1951-54. Sr Engr Jacob Feld 1954-55. Ch Engrg Abrahams & Hertzberg 1955-60. Part Hertzberg & Cantor 1960-71. CEO Office of Irwin G. Cantor 1971- . Respon for design and supervision of construction of hi-rise office bldgs, schools, hospitals, apt & institutional structures in steel, reinforced & prestressed concrete. Lic in NY & 13 other states. Bd Mbr of NYACE; NYCIB-Ctte on Hi-Strength Concrete, Hi-rise Construction, and Bldg Codes; PCI; ACI; Comm on Hi Strength Concrete; K P Mbr NYC Community Planning BD 7; Exec Comm NYASCE Pres of Coop Housing Dev 1955-61; Lectr & Author of numerous technical articles. Received numerous awards from CIB, AISC, NYACE, ASCE for "The Galleria" 1978, Palace Hotel 1981, Nassau Comm College 1979, Trump Tower 1983, and many other major projects. *Society Aff:* NYACE, ACE, PCI, CIB, ASCE, NYSPE

Capacete, Jose L
Business: PO Box 11490, San Juan, PR 00922
Position: Partner. *Employer:* Capacete, Martin Assoc. *Education:* MSCE/Soil Mechanics/Univ of IL; BSCE/CE/Univ of Dayton. *Born:* 7/15/23. Bd of Direction ASCE (1976-78). Tau Beta Pi honorary Engg Soc. Acad of Arts and Sci of PR. Pres Colegio de Ingenieros y Agrimensores de PR (1955-57). Distinguished Civ Engr of PR 1975. Distinguished Civ Engr award from APWA 1975. Certificate of Honor Civ Engrs Inst of PR 1978. Author of many technical papers in Geotechnical Engg. Mbr Bd of Governors Ashford Community Hospital PR and Bd Governors Dorado Beach Golf Club; Pres Bd of Dirs, PR Inst of Culture; Vice- Chrmn PR Acad of Arts and Sci. *Society Aff:* ASCE, EERI, SSA, AGU.

Capants, John M
Business: 575 Mtn Ave, Murray Hill, NJ 07974
Position: VP Production. *Employer:* Airco Ind Gases. *Education:* BSME/ME/Univ of MD. *Born:* 4/2/36. in Riga, Latvia. Educated in Germany and MD. Served in USAF as transportation officer '59-'62. With present employer since '62 - working in various operations jobs. In current position since '78. Amoung hobbies include woodworking, skiing and photography. *Society Aff:* AIChE.

Capehart, Barney L
Business: ISE Dept, 303 Weil Hall, Gainesville, FL 32611
Position: Prof. *Employer:* Univ of FL. *Education:* PhD/Systems Engg/Univ of OK; MS/EE/Univ of OK; BS/EE/Univ of OK. *Born:* 8/20/40. Currently involved in energy efficiency res, & dev of energy policy. Working on microprocessor based energy mgt systems for residential & commercial applications. Author of 30 journal articles, 41 conf papers, "Florida's Electric Future–, publ by the FL Conservation Fndn, 1982; a chapt on electrical metering tech in a book "Innovative Electric Rates" publ by Lexington Press, 1982, & a chapter on coal transportation in a book, "Coal Burning Issues–. Publ by Univ of Florida Press, 1980. Director, Energy Management Division of IIE, 1986-87. Editor, *I. Journal of Energy Systems,* 1985-88. Energy Efficiency Consultant to Florida Legislature, 1984-87. With Univ of FL since 1968. Previous employment with Aerospace Corp. Served two yrs in USAF, Electronic Systems Div. Activist in energy policy, energy conservation, & power plant siting. Born in Galena, KS. Married with 3 children. Wife is attorney specializing in environmental law. Enjoy hiking, canoeing & do-it-yourself projs. Listed in Who's Who in the World, Who's Who in America, and Who's Who in the South and Southwest, Fellow, Amer Assoc for the Advancement of Science. *Society Aff:* IEEE, IIE, SCS, AEE.

Caplan, Aubrey G
Home: 5401 Hobart St, Pittsburgh, PA 15217
Position: Owner *Employer:* Self *Education:* BS/EE/Carnegie-Mellon Univ. *Born:* 10/22/23. Owner of elec engg consulting firm located in Pittsburgh, PA, specializing in lighting, power, fire prevention, communications and utility rate study. Regional VP of Illuminating Engg Soc 1975-77; Pres of Greater Pittsburgh Consulting Engrs Council, 1976 to 1979. Mbr of Masonic Blue Lodge (Master in 1954), Consistery and Shrine. Mbr of City of Pittsburgh Bldg Codes Panel. Hobby is photography. Employed by local power co for fourteen yrs; in private practice 1959-present. *Society Aff:* IES, ACEC, AIEE, CSI.

Caplan, Frank
Home: 8 Kent Court, Smithtown, NY 11787
Position: President *Employer:* Quality Services Inc. *Education:* Bachelor/Mech Engrg (BME)/Cornell Univ *Born:* 10/15/19 US Army Retired Reserve Lt. Col. with service during World War II. Life Mbr, ADPA. Teacher of Industrial Statistics/Quality Management in-plant, in adult education, and in graduate degree programs for numerous schools. Author of "The Quality System: A Sourcebook For Managers and Engineers–; Chilton Book Co; 1980. Present Editor-in-Chief, "Quality Engineering" magazine. Chairman of the Long Island Section, Past Electronics Division Chairman, and Fellow of ASQC. Quality Manager/Director/Internal Consultant/Vice-president for several US Corporations. External Consultant with numerous companies and government agencies in Europe, Africa, and the US. Currently hd of own consulting firm. *Society Aff:* ASQC, ADPA

Caplan John D
Home: 2515 Covington Pl, Birmingham, MI 48010
Position: Exec Dir GM Res Labs *Employer:* General Motors Res Labs. *Education:* MS/ME/Wayne State Univ; BS/CE/OR State Univ. *Born:* 3/5/26. Weiser, ID. MS Wayne St Univ. BS OR St Univ. Ed at OR St Univ was interrupted by WWII-during which he attended Stanford Univ in the Army Specialized Training Reserve Prog & later served in Australia, New Guinea, and the Philippine Islands. Joined the GM Res Labs in 1949 & is presently Exec Dir. Mr Caplan is married, and resides in Birmingham, MI. He is a Mbr of the NAE, ACS, AIChE(F), IRI, SAE, AAAS(F). Phi Kappa Phi, Tau Beta Pi, Sigma Tau & Phi Lambda Upsilon. Crompton-Lanchester Medal 1964 IME. Dir 1971-present, Pres 1976, CRC. *Society Aff:* AAAS, ACS, AIChE, IRI, NAE, SAE.

Caplan, Knowlton J
Business: PO Box 9342, Minneapolis, MN 55440
Position: Pres. *Employer:* Industrial Health Engg Assoc, Inc. *Education:* MS/CE/WA Univ; BS/CE/WA Univ. *Born:* 6/23/19. Ind hygiene & air pollution for MO, MI, long before EPA, OSHA. Wrote first version of ACGIH Ind Ventilation Manual. Organized Health Dept for Uranium Div, Mallinckrodt; designed control for radiation, radioactive dusts, gases, toxic chemicals; later was involved in design of new $55 million plant; later became Div Mgr of Engg. At Kaiser Engrs, designed ventilation aluminum reduction process, BOF furnace. As VP R&D, Carter-Day Co, brought 3 products to market. Assoc Prof, Univ MN, 7 yrs. Pres, Ind Health Engg Assoc, last 10 yrs. Published 42 peer-review articles. Reg PE in 7 states. *Society Aff:* AIHA, AIChE, ASHRAE, AAIH.

Caplinger, Robert M
Business: PO Box 27456, Houston, TX 77027
Position: VP. *Employer:* Raymond Intl Inc. *Education:* PhD/Industrial Psych/Bradford Univ Pasadena CA; MBA/Bus Admin/Pacific Northwestern Univ Seattle WA; MS/Industrial Psych/PA State Univ; BS/Psych/Davis & Elkins College. *Born:* 9/4/28. Native of Elkins, WV. Worked for GE in various activities of Employee Relations for 14 yrs. Mgt consultant with Cresap McCormack & Paget for 2 yrs. Deep involvement in tech recruiting, training, motivational studies of Engrs. Joined Raymond Intl in 1969 in charge of all human resources; recruiting planning, replacement. Overall policy dev. EJC Corp Liaison Council; Adjunct Prof - Univ of FL; Intl Business Comm of Hou C of C VChrmn; Intl Studies Advisory Bd Univ of St Thomas; Veteran US Army/US Navy. *Society Aff:* SAME, ASCE.

Capon, Jack
Home: 6 Saddle Club Rd, Lexington, MA 02173
Position: Staff Mbr. *Employer:* MIT Lincoln Lab. *Born:* b Apr 1932. BEE CCNY, MSEE MIT, PhD Columbia Univ. Staff mbr at Lincoln Lab from 1965-present. Fellow of IEEE, mbr of Am Geophysical Union, Eta Kappa Nu, Tau Beta Pi, Sigma Xi, Fellow of Am Assn for the Advancement of Sci. Assoc Ed IEEE Transactions on Info Theory 1974-present.

Capstaff, Arthur E Jr
Business: PO Box 712, Marblehead, MA 01945
Position: Managing Partner *Employer:* ENDEAVOUR GROUP *Education:* SM/Ocean Engg/MIT; MBA/Mrkt/Boston Coll; BS/Met Eng/NC State *Born:* 11/7/44. Native of Newport News, VA. Industrial experience with INCO & TX Instruments. At MIT, developed first comparative cost analysis of ocean manganese nodule mining operations. In 1978, founded TPA, an eng consulting firm specializing in technological innovation and its subsequent commercialization, in 1987, merged firm into ENDEAVOUR GROUP, a technical mgmt consulting firm. Chapter Chrmn: ASM (1974), MTS (1976). ASM Pres's Award (1976). Enjoy boatbuilding, sailing, and lobstering. *Society Aff:* ASM, MTS, SME, AIME, AACE, BCS

Carastro, Sam
Business: 3636 So Westshore Blvd, Tampa, FL 33629
Position: Pres. *Employer:* Carastro, Aguirre & Assoc, Inc. *Education:* Bach/EE/Ga Inst of Technology. *Born:* 11/27/27. Native of Tampa, FL. Served in USAF 1952-55 and retired at Lt Col in USAF Reserves. Fellow of FL Engg Society. Founder and pres of Carastro, Aguirre and Assocs, Inc, consulting engrs since 1960. Engaged in proj engg, proj dev, and mgmt of firm. Reg PE in AL, CA, FL, IL, IN, IA, KY, LA, MA, NC, OH, PA, SC, TX, WA & VA. NEC Certificate No 3833. *Society Aff:* IES, IEEE, NSPE.

Carberry, James J
Business: Chem Engg Dept, Notre Dame, IN 46556
Position: Prof. *Employer:* Univ of Notre Dame. *Education:* PhD/Chem Engg/Yale Univ; MS/Chem Engg/Univ of Notre Dame; BS/Chem Engg/Univ of Notre Dame. *Born:* 9/13/25. Native of Brooklyn, NY. Served in US Navy 1944-46. Process Engr, duPont Co (1951-53). Teaching Fellow, Yale Univ (1953-55). Yale Garland Fellow (1956-57). Sr Research Engr duPont Experimental Station (1957-61). Faculty at Univ of Notre Dame (1961-date). NSF Fellow, Cambridge Univ (UK) 1965-66. Hays-Fulbright Sr Scholar, Italy (1974). Churchill Fellow, Cambridge Univ (1979 & 1982). Mellon Fellow. Cambridge Univ (1979); Fellow of Royal Society of Arts (London) 1980; Advisory Council, Princeton University (1980-date). Yale Engrg Assoc Award for Adv of Pure & Applied Sci (1967). RH Wilhelm Award in /Chem Reaction Engg (AICHE) 1976. Author of "Chemical & Catalytic Reaction Engrg" (1976) McGraw- Hill; Co-Editor "Catalysis Reviews" Marcel Dekker (NY). Cambridge Philosophical Soc; Yale Club of NY Metropolitan Opera Guild. *Society Aff:* AIChE, ACS, NYAS, ΣΞ.

Carberry, Judith B
Business: Department of Civil Engineering, Newark, DE 19716
Position: Associate Professor *Employer:* Univ of Delaware *Education:* PhD/Envir Eng/Univ of Notre Dame; MS/Env Chem/Univ of Notre Dame; BS/Chem/Cornell Univ *Born:* 03/10/36 Born, raised, and attended public schools in Oil City, Pennsylvania. After graduating from Cornell University, worked as a spectroscopist for duPont in Wilmington, DE. After completing graduate work at Univ. of Notre Dame, completed a post doctoral assignment at University College London. Assumed faculty position at Univ of Delaware in 1973, specializing in water, wastewater, and sludge treatment processes as topics for research. Received numerous awards and funding from Nat Sci Found, Off of Water Res and Tech, Env Protect. Agency, North Atlantic Treaty Org, and Lady Davis Found. *Society Aff:* WPCF, AEEP, SWE

Carbon, Max W
Business: Nuclear Engrg and Engineering Physics Dept, Univ of Wisconsin, 1500 Johnson Dr, Madison, WI 53706
Position: Prof and Chrmn *Employer:* Univ of WI-Madison, 1958 to Present *Education:* PhD/Heat Transfer/Purdue Univ; MS/ME/Purdue Univ; BS/ME/Purdue Univ *Born:* 1/19/22 Currently Prof and Chrmn Dept of Nuclear Engrg and Engrg Physics, Univ of WI- Madison. 1975-87: member of the Congressionally-established Advisory Committee on Reactor Safeguards which advises the US Nuclear Regulatory Commission and the Dept of Energy on nuclear reactor installations. 1955-58: Chief of Thermodynamics Section at Avco Mfg Corp engaged in design of Titan and Minute Man Nose Cones. 1949-55: Engr and Group Leader at Gen Elec Co involved in establishing safe thermal-hydraulic limits for operation of the plutonium-producing piles at the Hanford Engrg Works. 1967-68: Group Leader of Ford Foundation-Coll of Engrg program at Polytechnic Institute, Republic of Singapore. Distinguished Engrg Alumnus, Purdue Univ. Fellow, ANS. *Society Aff:* AAAS, ANS, AAUP, ASEE

Carbonara, Robert S
Business: Battelle, 505 King Ave, Columbus, Ohio 43201
Position: Market Development *Employer:* Battelle Columbus Labs. *Education:* PhD/Mat Sci/Univ of Cincinnati; BS/Physics/Univ of Pittsburgh *Born:* Mar 1937. PhD Univ of Cincinnati, BS Univ of Pittsburgh. Res Asst Scaife Radiation Lab, Univ of Pittsburgh 1957-59; Res Asst-Mellon Inst/Carnegie-Mellon Univ 1960-64; NASA Trainee Univ of Cincinnati 1964-68; Sr Res Scientist- Prog Mgr, Battelle Columbus Labs. Res interests: noble & transition metal alloys, intermetallic & nonstoichiometric compounds, x-ray & surface analysis, surface chem, thermodynamics & statistical mech. Mgt of lrg scale res progs. Rapid solidification of metals. Phi Eta Sigma, Phi Lambda Upsilon, Natl Ed Ctte of Am Soc for Metals, Presidents Award, Am Soc of Metals. VChrmn ASTM Ctte E42, Chrmn Columbus Chapter ASM, Chrmn Acad. for Metals & Materials Committee of ASM, Chrmn Advancement of Res Med Ctte of ASM. Materials Processing Ctte of ASM, Co-Chrmn Intl Conf on Rapidly Solidified Materials-ASM. *Society Aff:* ASM, APMI.

Carbone, Walter E
Home: 25 Hickory Pl, Chatham Twp, NJ 07928
Position: President *Employer:* Wilputte Corp. *Education:* ME/ME/Stevens Inst Tech. *Born:* 6/8/14. Ch Engg Columbia Univ & Brooklyn Poly Inst. Native NY, NY. PE NY & FL. Have been active in the Coke Oven & Chem Processing Industries for over 46 yrs on a world-wide basis. Associated with Allied Chem Corp over 35 yrs; when div was sold and established as Wilputte Corp was appointed exec VP. Authored and presented numerous papers relating to all phases of Coke Oven & Coal Chem Tech. Mbr Tau Beta Pi; Fellow of AIChE Am Iron & Steel Engrs; Mbr ACS, Newcomen Soc and numerous other profl organizations. Enjoy photography, high fidelity, golf and other sports. *Society Aff:* AIChE, AIME, ACS.

Carbonell, Celso A
Home: PO Box 6-194 El Dorado, Panama, Republic of Panama
Position: Pres. *Employer:* Protex Panama, SA. *Education:* Civ Engr/Concrete Const/State Univ of IA; Military Engr/-/Peruvian Military Academy. *Born:* 7/5/22. in Panama. Worked with US Corps of Engrs in Panama. Chief Municipal Engr of the City of Panama 1949-1951. Minister of Public Works 1951-1952 & 1968-1969. Dist Mgr Sika Chem Corp 1952-1957, VP Sika Panama 1957-1959, Pres Sika Panama 1959-1963. Candidate to VP of the Republic of Panama 1964. Pres Protex Panama, SA 1965 to date. Twice Pres of Panamanian Society of Engrs and Architects. Pres local Chap ASCE. VP Pan Am Federation of Engg Assn 1978-83. Honorary Consul of Paraguay in Panama 1976 to date. Main profl activity has always been concrete tech and repair of concrete structures. Considered an expert in these matters. *Society Aff:* ASCE, ASTM, ACI, USCOLD, PCI, ASCC, FESAE, SPIA, SAME

Carbonell, Ruben G
Business: Dept of Chem Engg, Davis, CA 95616
Position: Prof. *Employer:* Univ of CA. *Education:* PhD/Chem Engr/Princeton Univ; MA/Chem Engr/Princeton Univ; BE/Chem Engr/Manhattan College. *Born:* 12/27/47. in Havana, Cuba. Moved to NYC, Dec 1958. Began teaching at the Univ of CA in Fall 1973. Consultant for Lawrence Livermore Lab in areas of in-situ shale oil recovery and underground nuclear waste disposal. Maj res areas include: quantum mechanics, biochem engg, transport processes in porous media. Recreational activities: scuba diving, cross-country skiing, drawing and painting. *Society Aff:* AIChE, ACS, AAAS, ΣΞ, ASEE.

Carcaterra, Thomas
Home: 4210 Isbell St, Silver Spring, MD 20906
Position: Principal *Employer:* Sel-Employed *Education:* MCE/Struct Engg/College of the City of NY; BCE/Civil Engg/College of the City of NY. *Born:* 11/11/22. in NY City. Served with Army Corps of Engrs 1944-46. Structural engg for several consulting engg firms, prior to entering private practice. Partner of Smislova and Carcaterra, 1961-66. Prin of Carcaterra and Assocs, 1966-1976. Partner and Dir of Engg, Chatelain, Samperton and Carcaterra, 1976-78. Pres of E/A Design Group, Chtd, Architects & Engineers, 1978-84. Established individual consult practice in 1984 as Thomas Carcaterra, PE.. *Society Aff:* ASCE, ACI.

Card, Howard C
Business: Dept. of EE, Univ. of Manitoba, Winnipeg, Man, R3T 2N2 Canada
Position: Prof. *Employer:* Univ of Manitoba *Education:* PhD/Solid State Electronics/Univ of Manchester; MSc/Materials Science/Univ of Manitoba; BScEE/Elec Engg/Univ of Manitoba. *Born:* 3/30/47. in Winnipeg, Manitoba, Canada. Studied in Canada then in England. Athlone Fellow 1969-71. Postdoctoral res Univ of Manchester (UMIST) 1972-4. Mullard Res Fellow. Asst Prof, Univ of Waterloo, 1974-5. Assoc Prof, Columbia Univ in Dept of Elec Engg and Mbr of Columbia Radiation Lab 1975-1980. Presently Prof of Electrical & Computer Engineering, University of Manitoba, Winnipeg, Canada. Instructor, Continuing Education Program, Bell Labs and Consultant, IBM Watson Res Ctr. NSERC E.W.R. Steacie Fellow 1984-86. Author or coauthor of more than 140 publications in scientific and technical journals. Rh. Institute Award for Interdisciplinary Research 1983; Stanton Award for Excellence in Teaching 1983; Sigma Xi Senior Scientist Award 1987; Honary Editorial Board, Solid State Electronics; Chairman, Gordon Conference on MIS Systems 1982; Chairman, 1987 Canadian Conference on VLSI. Principal research interests: Solid State Electronics, Computer Engrg. *Society Aff:* IEEE, APS, ECS, OSA, AAAS, ACM, ΣΞ, HKN, NYAS.

Cardello, Ralph A
Business: Baytown R&D, PO Box 4255, Baytown, TX 77522
Position: Gen Mgr *Employer:* Exxon Res & Eng Co *Education:* MS/ChE/Princeton Univ; BS/ChE/Princeton Univ. *Born:* 1/24/27. Joined Std Oil Dev Co - Exxon R&D - 1951 first as Process Engr, Proj Design, then Planning & Economics to Asst Dir, Petrol & Chem Processes. Res Coordinator and Corp Strategic Planning, Humble Oil in Houston, 1965. Mgr, Exxon Synthetic Fuels Res 1967-69. Gen Mgr Exxon Tech Dept 1970-72. Mgr Gas Tech, LNG & Light Hydrocarbons Exxon Corp 1972-78. Gen Mgr Exxon R&D, Petrol Dept 1978-79, Managing Dir, Esso Coal Tech, Rotterdam, Netherlands 1980-83, Gen Mgr Baytown Div, Exxon R&D 1983-. *Society Aff:* API, AIChE, ΦBK.

Carden, Arnold E
Business: PO Box 2908, University, AL 35486
Position: Prof of Engr Mech *Employer:* Univ of AL. *Education:* BS/ME/Auburn Univ; MS/Eng Mech/Univ of AL; PhD/Metallurgy/Univ of CT *Born:* 4/27/30. in Birmingham, AL, Phillips High School, US Army Signal Corp Officer 52-54. Married Patsy Walker 1952; Children; Daniel, Timothy, Evelyn, Rebekah, Benjamin, Elizabeth; Summer employment at ORNL seven summers, LANL seven summers; conslt in failure analysis, mech engg, fatigue, Reg in AL (PE 6595). *Society Aff:* ASME, ASM, NACE

Cardinal, Robert J
Business: 333 Ravenswood Ave, Menlo Park, CA 94025
Position: Dir *Employer:* SRI, Intl *Education:* MS/IE/IIT; BS/Engrg/US Coast Guard Academy *Born:* 12/24/35. As Executive VP of LB Knight & Assocs, Inc, planned and designed over one billion dollars of complex manufacturing and government facilities plus major assignments for ARAMCO in Saudi Arabia, the President's Commission on American Shipbuilding, Dept. of Defense, EEOC and many major international corps. Taught in IIT's graduate school, is a univNow Pres!Served as Pres Trustee, and part Pres of it's Alumni Assoc. Publications include: ASCE, NSPE, and IIE Journals and Consulting Engr and Material Handling Engrg. Now Pres of a 200 person consulting and service firm and a frequent writer and lecturer on health care cost containment. Certified Mgmt Conslt (IMC), Now Dir of Mgmt Consltg Center for SRI Intl - A 3000 person scientific, Engrg and Mgmt research and sultg organization. Registered PE - IL & CA. *Society Aff:* IEEE, IIE, IMC, NSPE

Carelli, Mario D
Business: Advanced Energy Systems Division, P.O. Box 158, Madison, PA 15663
Position: Fellow Engineer *Employer:* Westinghouse Elec Corp *Education:* PhD/Nuclear Engrg/Univ of Pisa, Italy; BS (equivalent)/Mech Eng/Univ of Florence, Italy; Post-Doctorate Fellowship/Nucl Engr/Univ of Pisa, Italy *Born:* 02/06/42 Major area of expertise: Nuclear reactors core design; space thermal mgmt. Has been respon for thermal-hydraulic design of fast breeder cores; dir devel of analytical methods and design criteria. Currently chrmn patent cttee, W-AESD. Adjunct faculty prof Univ of Pittsburgh, PA, Energy Resources Program where he teaches courses in nuclear tech and is mbr of faculty adv ctte. Executive ctte ANS power div; tech program chrmn and international steering cttee chrmn 1987 ANS/ENS international conference on fast breeders. Chrmn, IAHR international working group on liquid metals thermal-hydraulics, Secretary IAHR fluid mechanics in energy prod section. Published over 70 papers and articles. Registered Professional Engr in Pennsylvania and Italy. *Society Aff:* ANS, IAHR, ASME

Caretsky, Marvin
Business: 60 E 42 St, New York, NY 10017
Position: Sr Partner. *Employer:* Caretsky & Assoc. *Born:* May 1927, NY. Sr Partner Caretsky & Assoc Cons Engrs. BME NY Univ. Ser in AF WWII. After grad from college held engg pos with architectural & cons engg firms. Went into private practice in 1959 forming own firm. Reg PE in 13 states. Have specialized in design of heating, ventilating & air cond, plumbing & sanitary & lighting & electric power for bldg projs with spec emphasis on hospitals & health facilities. P Pres, Treas & Dir of NYACE. Fellow ACEC. Mbr NSPE & ASHRAE. Awards for Engr Excellence in Mech/Elec Engg 1972, 1976, NYACE.

Carew, Wm E, Jr
Business: 2005 Concord Pike, Wilmington, DE 19803
Position: Sr Conslt *Employer:* Carew Assoc, Inc. *Education:* BS/Elec/Worcester Poly Inst. *Born:* 5/26/14. Reg PE, grad of Worcester Poly Inst, with BS degree in EE 1937, Mbr of ASHRAE, ASME, NCEE. Past Pres Natl Council of Engrg Examiners; Chrmn Engrg Affairs Council of Amer Assoc of Engrg Socs, Delaware Engineer of 1986 and Mbr of Engrg Accreditation Commission of ABET. Experience: Three yrs with Factory Ins Assoc in Fire Protection Engg; 8 yrs with E I duPont de Nemours & Co, Inc in safety and fire protection operations & design of power, heating and air conditioning systems, and 35 yrs as a consulting engr in DE. Design has included residential, commercial & institutional bldgs and industrial plants. *Society Aff:* ASME, ABET, NCEE, ASHRAE.

Carey, Donald E
Home: 507 St Clair Ave, Spring Lake, NJ 07762
Position: Water Supt. *Employer:* City of New Brunswick. *Education:* BSCE/Civil Engg/Penna Military College. *Born:* 10/28/25. in NYC. Junior Engg NY Water Ser-

Carey, Donald E (Continued)
vice Corp. Distribution Engg Hackensack Water Co. Asst Chief Engg Elizabethtown Water Co. Supt Transmission and Distribution and then supt Planning and Engg Elizabethtown Water Co. Exec Dir South Brunswick Utilities Authority Manager Water and Sewer Leisure Technology. Exect Dir FL Keys Aqueduct Authority. Presently water supt New Brunswick, NJ. *Society Aff:* AWWA, ASCE, AWRA, NJOA, WPCF.

Carey, John J
Home: 3486 Woodland Rd, Ann Arbor, MI 48104
Position: Consulting Engr. *Employer:* Self. *Education:* MSEE/EE/MIT; BSEE/EE/MIT. *Born:* 12/10/11. Native of Boston, MA. Electro-Mech Engr, the Panama Canal, 1934-41. Elec Designer, Jackson & Moreland Engrs, 1941-42. Corps of Engrs, US Army, 1942-46, Post Engr, Ledo Assam, India 1943-45. Prof, Elec Engg, Univ of MI, 1946-72. Co- Dir Power Systems Lab, Univ of MI, 1960-1972. Prof Emeritus, 1972-. Elec Consultant, John J. Carey, PE, 1972-. Reg Engr, MI. Chrmn, SE MI Section, AIEE, 1954-55. Pres, MSPE, 1967-68. Natl Dir, NSPE, 1971-77. Am or have been consultant for Am Metal Products, Consumers Power, Detroit Edison, GE, Westinghouse, State of MI, US Govt, et al. *Society Aff:* IEEE, NSPE.

Carey, Julian D
Home: 1916 Bentley Rd, Schenectady, NY 12309
Position: Conslt Welding Process *Employer:* Self *Education:* BS/Metallurgical Engr/VA Polytech Inst *Born:* 9/17/18 Native of Baltimore, MD. Metallurgist for Wright Aeronautical Corp. (1942- 45). Head of Metallurgical Section, Bridgeport Works Lab, Gen Elec Co (1950-55). Operated process lab for GE Welding Dept, York, PA (1955-58). Welding Engr, Borg Warner Corp., York PA (1958-59). Mgr Welding Development and Welding Engrg at GE Knolls Atomic Power Lab (1959-69). As Mgr Metal Joining Operations, Corporate Consulting Services, GE Co., he was concerned Co-wide with welding and related materials joining activities. Retired from GE Co 1982. District Dir of American Welding Society (AWS), 1981-84 and past chrmn of Northern NY Section (AWS). Enjoys sailing, hunting, travel, and shop projects. *Society Aff:* ASM, ASME, AWS

Carley, Charles T, Jr
Business: Drawer ME, Mississippi State, MS 39762
Position: Prof & Head Mechanical & Nuclear Engr Dept. *Employer:* MS State Univ. *Education:* PhD/Mech Engr/NC State Univ; MS/Mech Engr/VA Poly Inst & State Univ; BS/Mech Engr/MS State Univ. *Born:* 12/27/32. Native of Vicksburg, MS. After receipt of BS worked for Gen Elec Co in Lynn, MA prior to attending Naval OCS. Served three yrs as a Naval officer in the Civ Engg Corps, served as an Instr of Mech Engg at VA Poly Inst from 58 to 60, an Asst Prof of Mech Engg at MS State Univ from 60 to 61. After receipt of PhD from NC State Univ in 1964. He rejoined the Mech Engg faculty at MS State Univ and was subsequently made Hd of the Mech Engg Dept in 1969. Served as VP of ASME - 1972-74, have served as chrmn of the MS Technical Advisory Committee on Boiler and Pressure Vessel Safety since 1974. Fulbright Sr Lecturer at Univ of Buenos Aires, Argentina 1986 & Follow-up Fulbright Award at Univ of Catamarca, Argentina, 1987. Enjoys marathon running. *Society Aff:* ASME, NSPE, MES, ANS, ASEE

Carley, Thomas G
Business: Dept of Engg Sci & Mechanics, Knoxville, TN 37916
Position: Prof. *Employer:* Univ of TN. *Education:* PhD/Theoretical & Appl Mechanics/Univ of IL; MS/Engg Mech/LA State Univ; BS/ME/LA State Univ. *Born:* 7/3/35. Native of Vicksburg, MS. Faculty positions at Univ of KY & Southern Methodist Univ and Univ of TN. Industrial experience with Oak Ridge Nat Lab, Boeing Co, Naval Res Lab, ARO Inc, TN Valley Auth. Principal teaching and res activities and industrial Consulting in areas of structural mechanics, dynamics, vibration and noise. *Society Aff:* ASEE, ASME, AAM.

Carlile, Robert E
Business: 6909 SW Freeway, P.O. Box 36306, Houston, TX 77036
Position: VP *Employer:* Petty-Ray Geophysical, Geosource, Inc. *Education:* PhD/Petro Engrg/TX A&M Univ; MS/Petro Engrg/Tulsa Univ; BS/Petro Engrg/Tulsa Univ *Born:* 10/16/33 18 years Univ teaching in Petro Engrg, Computer Science, 11 years industry; Sunray D X oil Co.; Keener Oil Co.; Kewanee Oil Co.-Petty-Ray Geophysical, Geosource Inc. 9 years consultant; 5 major oil company's, US Bureau Mines, US AID (Vietnam) 1967-1974, Successful negotiations engrg contracts, equipment Vietnam, Brazil, China (8 trips 1977-present), Russia, joint ventures (Romania), author/co-author over 100 articles, three (3) books. Member Pi Epsilon Tau, Tau Beta Phi, Phi Eta Sigma. *Society Aff:* AIME, SPE, SEG

Carlile, Robert N
Business: Dept. of ECE, Tucson, AZ 85721
Position: Prof. *Employer:* Univ of AZ. *Education:* PhD/EE/Univ of CA; Engr/EE/Stanford; MS/EE/Stanford; BA/Phys/Pomona College. *Born:* 4/24/29. Robert N Carlile received the BA degree from Pomona College, Claremont, CA, in 1951, the MS & Engrs' degrees from Stanford Univ in 1953 & 1956, respectively, & the PhD from Univ of CA, Berkeley, CA in 1963. From 1954 to 1959 he was engaged inthe R&D of microwave tubes at Hughes Aircraft Co, where he developed a nonreciprocal travelling-wave tube circuit. In 1963, he joined the faculty of the Univ of AZ, Tucson, where he was appointed full prof in 1969. He held a visiting position at the Plasma Physc Lab, Princeton Univ, from 1969- 1970. His res interests are in experimental plasma engg. Recently, he has been testing some of the models which are currently used to predict EMP generation & propagation. *Society Aff:* IEEE, APS.

Carlile, V Sam
Home: R R 1, Box 172, Anderson, IN 46011
Position: Retired *Born:* Mar 1917 AR. Spec schooling incl home studies, GM Inst & Inhouse Delco- Remy train. Attend Trans Corp during WWII. Marine Off Cadet Sch, later was Exec Officer of Training Vessel. Work exper: Tool estimating; Process Engrg; Supervisor Suggestion Plan; Plant Engrg; Plant Layout; Proj Writing a Methods Analysis. Sup Project Control. Respon for all capital proj activity through coord & follow-up. VP IIE 1972-77; IIE Nominating Ctte 1976; IIE Cert Task Force 1975-76; Regional Community Serv Chrmn 1970-72; IIE Regional Conf Chrmn 1971; IIE Chap Pres 1967. Enjoys speech-making & golfing. Certified Mfg Engr, SME. Activities include work for church and as Chrmn of a citizens organization which concerns itself with environmental problems. Although I still retain active mbrship status in both SME and 11E, I am no longer active in these fields. Retired 1977. *Society Aff:* IIE, SME.

Carlin, Herbert J
Business: Cornell University, Ithaca, NY 14853
Position: J. Preston Levis Prof of Engng. Emeritus. *Employer:* Cornell Univ. *Education:* PhD/EE/Poly Inst of NY; MS/EE/Columbia Univ; BS/EE/Columbia Univ. *Born:* 5/1/17. in NYC. Westinghouse relay engg five yrs. Polytech Inst NY 1945-1966. Dept Hd Fellow IEEE. "Award for Outstanding Achievement" 1960-66. USAF 1965. Chrmn IEEE Prof-Group Circuits 1958. Visiting committees Natl Bureau Stds 1967-70, Lehigh Univ 1967-74, Univ of PA: 1979-. NSF Sr Postdoctoral Fellow Award 1964, Visiting Prof MIT - 1972-73, NATO Res Prof Genoa Italy 1974. Visiting Scientist Natl Ctr for Telecommunications, Issy Les Moulineaux, France 1979-1980. Visiting Prof Tianjin Univ PR of China 1982, Visiting Prof Univ Coll, Dublin 1983. Dir Cornell EE 1966-75. J Preston Levis Prof Cornell 1966-. Over 75 technical papers on networks & microwaves. *Society Aff:* IEEE.

Carlisle, Thomas C, III
Business: P.O. Box 32414, Charlotte, NC 28232
Position: Corp. Safety & Fire Prot *Employer:* Celanese Corp *Education:* B.S./ME/Univ. of NC *Born:* 10/19/47 Native of Charlotte, NC. Served in the Army 1967-69. Spent 6 1/2 yrs with factory mutual as a loss prevention consultant. Currently employed with Celanese Corp as a safety and fire protection engr serving the chemical and specialities plants. I am responsible for systematic engrg evaluations

Carlisle, Thomas C, III (Continued)
of existig processes to identify and correct conditions leading to unreliability spills, fres and explosions, Mbr of SPFE national society and pres at Carolina chapter of SFPE serving North and South Carolina. *Society Aff:* SFPE

Carlson, Dale A
Home: 9235 - 41st Ave NE, Seattle, WA 98115
Position: Dean Emeritus Dir Valle Scandinavian Exchange Prog *Employer:* Univ of WA. *Education:* PhD/Sanitary Engr/Univ of WI; MSCE/Civ Engr/Univ of WA; BSCE/Civ Engr/Univ of WA. *Born:* 1/10/25. Educator, engrg; born Aberdeen, WA; son of Edwin C. G. Carlson & Anna A. (Anderson); married Jean M Stanton, Nov 11, 1948; children -- Dale Ronald, Gail L Manahan, Joan M Lee, Gwen D. Elliott. Water engr, acting city engr, City of Aberdeen, WA, 1951-55; design engr, Grays Harbor Construction Co, 1954-55 (part time); Asst to Full Prof, Dept of Civ Engr, Univ of WA, 1955-71; Prof & Chrmn, 1971-76, Dean, College of Engg 1976-1980, Dir Valle Scandinavian Exchange Program 1980-present, Prof Emeritus 1983, Dean Emeritus 1980- . Chrmn, Civil Engrg, Seattle Univ 1983- present, Tres, Pacific Northwest Synd LCA, Consultant miscellaneous engrg projs. Chrmn, WPCF Publications Committee, 1978-1980; mbr, ECPD Accreditation Visiting Team-ASCE; Am Assn of Environmental Engr; WPCF-USANC Committee 1978-1984; Science Advisor Committee U. of Illinois, EPA Center 1981-1983. Dean's List, Alfred Univ, 1943-44; Pres's Award, Grays Harbor College, 1974; Magplus *Society Aff:* ASCE, AWWA, WPCF, ASEE, IAWPR, SWHA, AAEE.

Carlson, David E
Home: 514 Nancy Rd, Yardley, PA 19067
Position: Gen Manager Solarex Thin Film Div *Employer:* Solarex Corp., Sub of Amoco Corp. *Education:* PhD/Phys/Rutgers Univ; BS/Phys/Rensselaer Poly Inst. *Born:* 3/5/42. in Weymouth, MA. R & D Physicist at US Army Nuclear Effects Lab, Aberdeen Proving Grounds, MD 1968-1969. Capt in US Army, Vietnam (Bronze Star Medal) 1969-1970. RCA 1970-83. RCA Labs Outstanding Achievement Awards, 1973 and 1976. Received the Ross Coffin Purdy Award from the American Ceramic Soc in 1976. Co-receipent of Morris N. Liebmann Award, IEEE, 1984. Walton Clark Medal, Franklin Institute, 1986. Published more than 75 technical papers and issued 22 US Patents. Presently, Gen Manager of Solarex Thin Film Div. 1983-present. Invented Amorphous silicon solar cell. *Society Aff:* IEEE, APS, AVS, ΣΞ.

Carlson, Donald E
Business: 216 Talbot Lab, 104 S. Wright St, Urbana, IL 61801
Position: Prof. *Employer:* Univ of IL. *Education:* PhD/Appl Math/Brown Univ; MS/Theoretical & Appl Mechanics/Univ of IL; BS/Engg Mechanics/Univ of IL. *Born:* 3/8/38. Native of Tampico, IL. Formal education received at the Univ of IL at Urbana-Champaign and Brown Univ. Currently Prof of Theoretical and Appl Mechanics at the Univ of IL at Urbana-Champaign. Present interests include elasticity, thermoelasticity, general continuum thermomechanics, and dimensional analysis. *Society Aff:* SNP, AAM, SES, ASME, AMS, SIAM.

Carlson, Harold C R
Home: 611 Lake Point Dr, Lakewood, NJ 08701
Position: Chief Eng, Consulting Eng (Self-Employed) *Employer:* The Carlson Co. Inc. *Education:* see below *Born:* 7/18/08. Consulting Engr Lic Profl Engr; consulting indus engr, inventor, author, lecturer, machine desr; b Easthampton, MA, July 18, 1908; ed Smith Tech Sch, MA; US Naval Schl, VA; Pratt Inst, Brooklyn Poly. Inst; Columbia Univ Design engr, Otis Elevator Co, NY, 1929-40; chf engr, Lee Spring Co, Brooklyn, 1940-46;mgr and chf engr, Chas Fischer Co, Brooklyn, 1946-48; pres The Carlson Co 1948-70 Consult on indus mgmt, mfg, prod, meth des, etc, to many outstanding firms. Partic in des of elevators for Empire State Bldg, Chrysler Bldg, No 1 Wall St, etc. Holds pats on Spring Coilers, Testing Equip and mechanical prods. Author: many arts, pub tech jls. Hons: Cert of Testimonial of Appreciation for Advancement of the Prof of mechanical engg from ASME (1947); Modern Design Award for Exellence in Genl Mechanical Design from Design News, Mem: NSPE; Life Fellow ASME & ASTM Author Spring Designer's Handbook (1978), Springs: Troubleshooting & Failure Analysis (1980), Spring Mftg Handbook (1982). *Society Aff:* ASME, ASTM, SMI

Carlson, O Norman
Business: 122 Metals Dev Bldg, Ames, IA 50011
Position: Prof Emeritus *Employer:* IA State Univ. *Education:* PhD/Chemistry/IA State Univ; BA/Chem & Math/Yankton College. *Born:* 12/21/20. Married Virginia Jyleen in 1946. Three children, Gregory, Richard & Karen. Joined IA State Univ faculty in 1950. Appointed Chrmn of Met dept & Div Chief of Met, Ames Lab DOE 1962-66. Prof of Mtls Sci & Engg & Sr Met with Ames lab, 1975-1987. Currently emeritus professor: IA State Univ Faculty Citation in 1971. Visiting Scientist with Max Planck Inst fur Metallforschung, Stuttgart 1974-75 and 1983. Mbr of Bd of Regents of Waldorf College 1964-74. Mbr of AIME, ASM, ACS, Phi Kappa Phi, Sigma Xi & Who's Who in Am. Mbr of Lutheran Church & Lions Club. Hobbies are golf, bowling, genealogy, Civil War history, & photography. *Society Aff:* AIME, ACS, ASM.

Carlson, Robert C
Business: Dept of Industrial Engg, Stanford, CA 94305
Position: Professor *Employer:* Stanford Univ. *Education:* PhD/Operations Res/Johns Hopkins Univ; MS/Operations Res/Johns Hopkins Univ; BS/Mech Engg/Cornell Univ. *Born:* 1/17/39. Worked as operations analyst for Bell Labs, 1962-70. At Stanford since 1970. Spent 1978-79 as visiting prof at Amos Tuck Business School, Dartmouth College. Spent part of 1984 as visiting faculty mbr at the Intl Mgmt Inst in Geneva, Switzerland. VP of Palo Alto Assocs, a mgt systems consulting corp, since 1973. Dir of Stanford's special one-week summer program in Manufacturing Strategy since 1986. Mbr of bd of dirs of Palo Alto Chap of APICS since 1975. Has consulted with cos throughout the US and has published numerous articles in leading academic journals. Res specialties include production and inventory control and manufacturing strategy. Hobbies include squash, wine tasting, and salmon fishing. *Society Aff:* ORSA, TIMS, APICS, IIE, ISIR

Carlson, Walter M
Home: 216 W Hilton Dr, Boulder Creek, CA 95006 *Employer:* IBM (Retired) *Education:* MS/Chem Engg/Univ of CO; BS/Chem Engg/Univ of CO. *Born:* 9/18/16. Joined DuPont as Industrial Engr 1939 and worked on process improvement, planning, plant design and engg service assignments. In 1954 installed DuPont's first large computer. Joined US Dept of Defense in 1963 as Dir of Technical Info. Joined IBM in 1967; Corporate Marketing Consultant 1968-1985; Retired 1985. Formed Univac Users' Group 1955. Founded AIChE Machine Computation Committee 1958. AIChE Dir 1964-66. AIChE Fellow. EJC Treasurer 1967-75. Pres Assn for Computing Machinery 1970-72. Advisory panel TO NBS Inst for Computer Sci and Technology 1972-78. Dir & Exe Cttee, Charles Babbage Foundation 1984-87. Dir. Engrg Information Inc., 1985-87. *Society Aff:* AAAS, AIChE, ACM, ACS, ORSA.

Carlson, Walter O
Business: 225 North Ave, NW, Atlanta, GA 30332
Position: Prof *Employer:* GA Inst of Tech *Education:* PhD/ME/Univ of MN; MS/ME/Univ of MN; B/AERO/Univ of MN *Born:* 10/11/21 Native of Minneapolis, MN. Served with US Navy 1943-46. Aircraft stress analyst for McDonnell Aircraft Corp, 1946-47. Instructor in mechanical engrg, Univ of MN, 1947-56. Research specialist in re-entry heat transfer, GE Co, Philadelphia, PA, 1956-59. Research engr and mgr physics section, RCA, Moorestown, NJ, 1959-62. GA Inst of Tech since 1962. Prof of Mechanical Engrg. Areas of research include heat transfer, MHD and plasma physics, 1967-1980 various administrative positions, associate dean of engrg, acting dean of engrg, executive dir of So Tech Inst. Consulted for Minneapolis Honeywell, General Mills and Lockheed GA. *Society Aff:* AAAS, ASME, ASEE

Carlyle, Jack W
Home: 3245 Barry Ave, Los Angeles, CA 90066
Position: Prof. *Employer:* UCLA. *Education:* PhD/EE/Univ of CA; MA/Statistics/Univ of CA; MS/EE/Univ of WA; BA/Math/Univ of WA. *Born:* 2/23/33. Cordova, AK. Teaching experience in EE at Univ of WA; Univ of CA (Berkeley); Princeton; UCLA. Currently Prof of Engg & App Sci at UCLA; also mbr of Comp Sci Dept. Chrmn of Technical Committee on Math Fdns of Computing, IEEE Computer Soc, 1975-78. Recipient of IEEE Comp Soc Honor Roll Award, 1978. *Society Aff:* IEEE, ACM, SIAM, IMS, AMS.

Carmi, S
130 Crosshill Rd, Philadelphia, PA 19151
Position: Prof. & Head Mech. Eng. & Mechanics Dept. *Employer:* Drexel Univ. *Education:* PhD/Aero Engrg/Unif of MN; MS/Aero Engrg/Univ of MN; BS/Mineral Engrg/Univ of the Witwatersrand Johannesburg, SA. *Born:* 7/18/37. Born in Romania and educated in Israel, came to the States in 1963 and subsequently became US citizen. After receiving the PhD from the Univ of MN in 1968, joined Wayne State Univ. where promoted in 1978 to Prof in Mech Engg with WSU until 1986 when accepted current position at Drexel Univ. as Prof and Head, Mech. Eng. & Mechanics Dept. During Sabbatical in 1977/78 held the I Taylor Chair at the Technion, Israel Inst of Tech and during sabbatical in 1985/86 was Congressional Fellow serving as Science Advisor to Senator Carl Levin (Mich.). Res grants from the US Army Res Office and DOE Mbr: Grad Council, Univ Coun, Policy Comm, Tenure and Promotion Committees, ASME Committees, Session Organizer. Chrmn: Sessions, ME Grad Committee, College of Engg Faculty Assembly Univ Res Comm. Refereed papers published, seminars and presentations at profl meetings. One book (Ed). Assoc. Editor, Journal of Fluids Engineering. *Society Aff:* ASME, APS, ΣΞ, ΤΒΠ, ΠΤΣ, SOC. FOR NAT. PHILOSOPHY.

Carnes, Walter R
Business: Drawer DE, Mississippi State, MS 39762
Position: Assoc Dean. *Employer:* MS State Univ. *Education:* PhD/Mechanics/Univ of IL; MS/Aeronautical/GA Inst of Tech; BAE/Aeronautical/GA Inst of Tech. *Born:* 10/7/22. Seven yrs active duty - USAF - pilot ind experience - Lockheed Aircraft Corp. Taught math and aerospace engg - GA Tech. Taught aerospace engg and engg mechanics - MS State. Since 1970 - Assoc Dean - dir of Instruction. *Society Aff:* NSPE, ASEE

Carnes, Wm T
Home: 5505 Namakagan Rd, Bethesda, MD 20816
Position: Consultant & Author. *Education:* BSEE/EE-Comm/Univ of KS. *Born:* 2/17/15. (Retired). Dir of Avionics for Aeronautical Radio, Inc was chairman for 25 years and then elected Chmn. Emeritus of the Airlines Electronic Eng Comm Ctte responsible for dev. all airline avionics. Currently serving as Tech Adv to the Exec Comm of the Radio Tech Commission for Aeronautics. Author of textbook "Effective Meetings for Busy People-Let's Decide It and Go Home---McGraw-Hill Book Co. hard cover 1980, paperback 1983, & republished by IEEE Press 1987. Dir, Div III, IEEE, 1971-73. *Society Aff:* IEEE, IPA.

Carnevale, Dario
Home: 1325 NE 138th St, N Miami, FL 33161
Position: Gen Mgr. *Employer:* UOP Processes Intl Inc. *Education:* PhD/Civ Engg/Univ of Rome. *Born:* 1/11/35. Mr Carnevale, a petrol & petrochem co exec. Native of Paola (Italy) received his Dr of Engg at the Univ of Rome. His business career has been primarily in the oil industries, encompassing marketing of oil refinery technologies, design, construction & overall mgt of oil & petro chem plants throughout the world & on behalf of several govts. Mbr of AIChE is Gen Mgr of UOP, Colombian Branch, at Bogota, Colombia, S Am & has been handling the marketing of tech for oil & petrol, chem plants in the northern part of S Am (Colombia, Venezuela, Panama, Netherland Angilles, Trinidad & Tobago), the overall proj mgt for oil refinery projs in Colombia & in Venezuela which costs are in the excess of billion dollars. Recently assumed positions as Regional Exec for Europe Signal Companies, Sr. Advisor for Europe Allied Signal Int Inc, & the Henley Group Inc. Mr Dario Carnevale is listed in the 1978-79 & 1980-81 edition of Who's Who in Am; Who's Who in the south & Southeast; Personalities of, America, Men and Women of distinction, Men of Achievement, Bk of Hon, Int Who's Who of Intelectuals. *Society Aff:* AIChE.

Carney, Dennis J
Business: PO Box 118, Pittsburgh, PA 15230
Position: Chmn, CEO. *Employer:* Wheeling-Pittsburgh Steel Corp. *Education:* ScD/-/MA Inst of Tech; BS/Met/PA State Univ. *Born:* 3/19/21. Married: Virginia Mae (Horvath), 1943, five children. Served to Lt (JG) USNR, 1943-46. With US Steel Corp, 1942-74: Gereral Sup, 1963-65; VP Long Range Planning, 1965-68; VP Applied Res, 1968-72; VP Res 1972-74. With Wheeling- Pittsburgh Steel Corp, 1974-present; VP Operations, 1974-75; Exec VP, Dir, 1975- 76; Pres 1976-79; Chief Operating Officer 1976-77; Chief Exec Officer 1977--; Chmn 1978--. Dir: International and Amer Iron and Steel Inst, Wheeling Coll, Duquesne Club, United Way of Southwestern PA, Allegheny Trails Council of Boy Scouts of Amer. Mbr: Sigma Xi, Tau Beta Pi, Sigma Nu. Author; (with others) Gases in Metal, 1956. *Society Aff:* ASM, AIME, AISE.

Carney, Terrance M
Business: School of Engineering, Univ of Tenn at Chatta, Chattanooga, TN 37403
Position: Prof. of Engineering, Dir EE & Computer Appl *Employer:* Univ of TN. *Education:* PhD/Engg/Rice; SMAE/Aero Engg/MIT; SBAE/Aero Engg/MIT. *Born:* 12/2/33. Native of Lansing, MI. Guidance and control systems engr for NACA, then NASA from 1958 until 1970. Involved in the Apollo proj and other progs while at Langley Res Ctr, Manned Spacecraft Ctr and the Electronics Res Ctr. Since 1970 an engr faculty mbr at the Univ of TN at Chattanooga, currently Prof of Engg and Dir of Elect Engr and Computer Appli. *Society Aff:* IEEE, NSPE, ASEE, OE.

Carney, Thomas P
Business: 910 Skokie Blvd, Northbrook, IL 60062
Position: Pres. *Employer:* Metatech Corp. *Education:* PhD/Organic Chem/Penn State Univ; MS/Organic Chem/Penn State Univ; BS/Chem Engrg/Univ of Notre Dame. *Born:* 5/27/15. MS & PhD Penn St, BSChE Univ of Notre Dame, 1937. VP R&D & Control Eli Lilly & Co; Exec VP G D Searle & Co; Chrmn Exec Ctte Natl Patent Dev Corp. Formed own Co, Metatech, in 1976. Centennial of Sci Award 1972, Centennial of Engg Award 1974 from Univ of Notre Dame. Disting Serv Award from Assn of Cons Chem & Chem Engrs 1976. Bd of Trustees, Barat College. Chrmn Bd of Trustees, Univ of Notre Dame Hon LLD Univ of Notre Dame, 1961; Ernest Stewart Award of the Council for Adv and Support of Educ, 1982. *Society Aff:* ACS, AAAS, ATC

Carney, William D
Business: PO Box 270, Jefferson City, MO 65102
Position: Div Engr-Bridge. *Employer:* MO Hgwy Dept. *Education:* BS/Civ Engg/Univ of MO. *Born:* 9/23/23. A native of MO. Served in US Army - 1943 to 1946. Was promoted to his current position in 1968. He has served as district engr, sr engr-surveys and plans, construction engr, construction inspector & proj engr with the hgwy dept. He served as pres of the southeast chapter of the MO Soc of PE, pres & secy- treas of the Mid-MO sec of the Am Soc of Civ Engrs and VP of the Hgwy Engrs Assn of MO. He is a mbr of an advisory panel of the Natl Cooperative Hgwy Res Prog. *Society Aff:* ASCE, NSPE, AWS, PCI, ΤΒΠ.

Carnino, Annick
Business: EDF Direction Generale, 32 rue de Monceau, Paris, France 75008
Employer: Electricite De France *Education:* Ingenieur ESE/Ecole Superieure D'Electricite (France) (equivalent to PhD in Electricity and Electronics); Ingenieur en Genie Atomique (equivalent to PhD in Nuclear Engrg) Institut National Des Sciences et Techniques Nucleaires (France) *Born:* 09/11/40 After my deg in Nuclear

Carnino, Annick (Continued)
Engrg, I joined in 1963 the French Atomic Energy Comm. My first areas of job were in the field of neutronics, thermal hydraulics and remote control of safety experiments. In 1972, I headed the group of 20 engrs for the probabilistic safety assessment of nuclear installations. I became in 1980 deputy of the Dir for Safety of the French Atomic Energy Comm. In 1982, I joined ELECTRICITE DE FRANCE as deputy to the Dir Inspector General for the Safety and Security of the Nuclear Installations of EDF. I have publ many papers and articles in intl conf and magazines in the fields of safety, reliability, risk assessment, and human factors and I am considered an intl expert in these fields. *Society Aff:* ANS, ENS, SRS, SFEN, 3SF

Carpenter, Carl H
Home: 1246 E 100 So, Springville, UT 84663
Position: Principal Eng of Dept. *Employer:* Provo City Water & Wastewater Dept. *Education:* BS/CE/UT State Univ; AS/CE/Snow JR Coll *Born:* 2/4/32 Native of Manti, UT. Served 2 years in US Army 1952-54 in Korea and Japan. Design Engr with UT Power & Light Co 1956-57. Hydraulic Engr, US Geological Survey in Southern and Central UT on ground-water investigations, 1957-65. District Engr, Central UT Water Conservancy District in charge of planning and water treatment plants, 1965-73. Principal in consulting engrg firm, Salt Lake City, 1973-81. District Engr, Central UT Water Conservancy District in charge of water treatment plants, 1981-85. Principal Engineer, Provo City Water & Waste Water Dept., 1985 to present. Served as Pres of UT Section ASCE, 1979-80 and Chrmn of Pacific SW Council ASCE, 1979-80. Served as Trustee for Intermountain Section of AWWA, 1970-71. *Society Aff:* ASCE, AWWA, AIH, WPCF.

Carpenter, Edward F
Home: 2117 Saguaro Rd, Grand Junction, CO 81503
Position: Pres. *Employer:* Plateau Engg, Inc. *Education:* BS/Civ Engg/Stanford Univ. *Born:* 1/27/22. Native of CO, served in AK as Ordnance Officer, WWII. Survey Engr, CA Div of Hwys 4 yrs; Chief Designer (Civ), Geo Nolte Engrs, Palo Alto, CA; Entered consulting practice Santa Rosa, CA 1957, as Carpenter & Mitchell to 1968; N Counties Engg Co. Ukiah, CA, to 1974; Plateau Engg, Inc, Grand Junction, CO, since 1975. Consult, design & supervise construction of water & wastewater treatment systems for many municipalities & districts in Western CO. Active in local NSPE & AIME Chapters; Rotary & Presbyterian Church. *Society Aff:* ASCE, NSPE, AWWA, WPCF, AIME.

Carpenter, Joyce E
1697 Old Lost Mtn Rd, Powder Springs, GA 30073
Position: Scientist *Employer:* Lockheed-GA Co. *Education:* MS/Aero Engrg/GA Tech; B/AE/GA Tech. *Born:* 5/11/50. Native of Atlanta, GA. With Lockheed-GA Co since 1975, engrg sys analyst with background in structures and materials. Officer in Atlanta Section SWE, AIAA. Mbr Sigma Gamma Tau. Biography included in "Breakthrough: Women in Aviation–, by Elizabeth Simpson Smith, Walker and Co, NY, 1980. Private Pilot. *Society Aff:* SWE, AIAA, AAS.

Carpenter, William L
Business: PO Box 5229, Greenville, SC 29606
Position: Chm., Pres, CEO *Employer:* Sirrine Environmental Consultants, Inc. *Education:* BS/EE/US Naval Acad; -/Aero/NC State Univ. *Born:* 5/26/26. in Columbia, SC. Grad US Naval Acad. Served in US Navy as Engg Officer and Instr at US Naval Acad. Mbr of NSPE since 1960, having served on the Profl Practices Committee in 1978. Hd of Planning & Dev for SCSPE in 1978, served on the Office Mgr Committee of PEPP in 1977-1978. Twice held office of Pres of CESC (1970 & 1976). On Bd of Dirs of ACEC from 1969 to 1972. Currently Pres and Chief Exec Officer for Sirrine Environmental Consultants, Inc. responsible for mgt and admin of 75-plus employees, Environmental engg firm with offices in Greenville, SC, Raleigh, NC; Charleston, SC. Enjoys golf, skiing, walking & spectator sports. *Society Aff:* ACEC, CESC, NSPE, PEPP, SCSPE, TAPPI.

Carpentier, Michel H
Home: 49 Blvd de France, F91220 Bretigny France
Position: R&D Corp Exec. *Employer:* Thomson. *Education:* -/-/Ecole Polytechnique, Paris; -/-/ENSAE, Paris; -/-/ESE, Paris *Born:* 1/16/31. Native of France. Scientist in French Ministry of Defense 1956-62. International expert in NATO Hdquarters 1962-63. Technical Mgr Cotelec 1963-66. Tech Mgr Thomson-CSF Radar Div 1966-73. Tech Mgr Thomson-CSF Electronic 1974-75. Tech Mgr Thompson-CSF Corpate 1975-1977. R&D Corp Exec Thomson 1978--. Prof; ESE-ENSTA-ENSAE-France. Grand Prix de L Electronique 1969-Framce/ Several books about Servo-mechanisms and Radars in France, USA, Russia, China - Fellow IEEE. *Society Aff:* IEEE, SEE

Carr, Charles D
Business: P.O. Box 211, Ada, MI 49301
Position: Pres. *Employer:* Fishbeck, Thompson, Carr & Huber. *Education:* BS/Civil Engg/MI State Univ. *Born:* 5/25/36. Native of Lansing, MI. Served in US Air Force 1959-62. With FTCH since 1963, becoming a Principal in 1969 and Pres in 1984. Selected as "Young Engineer of the Year" by Grand Valley Chapter of MI Soc of Prof Engrs (MSPE) in 1971. Pres of State Prof Engrs in Private Practice 1971-72. State Pres of MSPE 1976-77. National Dir (NSPE) from MI 1980-81. State Pres. of Consltg Engrs Council/MI, 1984-85. Pres of Lansing Exchange Club in 1973. Trustee-Lansing Exchange Club Youth Foundation 1974-1984. Bd of Directors-MSU Engineering Alumni Assn 1976-80. Married and have three children. Enjoy sports and travel. *Society Aff:* NSPE, WPCF, ACEC.

Carr, Gerald P
Business: 5619 Fannin St, Houston, TX 77004
Position: Sr VP *Employer:* Bovay Engrs, Inc *Education:* MS/Aero Engrg/Princeton; BS/Aero Engrg/US Naval PG Sch; BS/ME/Univ of So CA *Born:* 8/22/32 Native of Santa Ana, CA. Upon graduation from the Univ of So CA, began a military career in the US Marine Corps as a fighter pilot. Served in this capacity until selected in 1966 as a NASA Astronaut. NASA assignments included astronaut support crews and capsule communicator for the Apollo 8 and 12 flights. Also involved in the development and testing of lunar modules 2, 4 and 6 and the lunar roving vehicle. Designated prime crew commander in 1970 for the third manned Skylab mission which flew from 16 Nov 1973 until 8 Feb 1974 and established a world record for time in space which prevailed for 4 years. Holder of many national awards as a result of the mission. Retired from USMC in 1975 and from NASA in 1977. Joined Bovay Engrs, Inc in 1977 and am Mgr of Houston Area Operations. *Society Aff:* NSPE, AAS, SAME, SETP, SUNSAT

Carr, Grover V
Business: 2627 Schaul St, Columbus, GA 31906
Position: Partner. *Employer:* G V Carr & Co. *Education:* BSCE/Struct, San Hgwy, City Planning/GA Inst of Tech. *Born:* 8/15/11. Vidalia, Toombs Co, GA to Wm M and Mary McCellan Carr; married Grace Truman Ellison, 4-27-36 in GA, Also certified as fallout shelter analyst by US Dept of Defense. Have practiced in all areas in Columbus GA since 1940. Have been country surveyor of Muscogee Co since 1941; also pres McClellan Home Furnishings owner of Carmel Farms: Mbr of GA State Bd of Engg Examiners for Engg and Surveyors, 1954 to 1961, (Chrmn 1961) mbr of GA Bd of Engg Revue and GA Bd of Space and Aeronautics, Committee on Surveying in Natl Council of State Bd of Examiners. Charter mbr of GSPE, twice past pres of chapter as state pres received the Ned Mellett Award 1961; named Engr of the Yr 1973 (in private practice) for state of GA. Sta. Life Mbr ASCE (Fellow Grade). Life Mbr NSPE. *Society Aff:* ASCE, NSPE.

Carr, Stephen H
Business: Materials Research Center, Technological Inst, Northwestern Univ, Evanston, IL 60201
Position: Prof. *Employer:* Northwestern Univ. *Education:* PhD/Polymer Sci/Case Western Reserve Univ; MS/Polymer Sci/Case Western Reserve Univ; BS/Chem

Who's Who in Engineering

Carr, Stephen H (Continued)
Eng/Univ of Cincinnati. *Born:* 9/29/42. Native of Dayton, OH. Worked with Inland Div, GMC, as co-op student and engr, involved with manufacture and processing of polymeric matls, 1960-65. Was awarded an NIH Fellowship for his graduate res. Joined faculty of Northwestern Univ in the Depts of Matls Science and Engg and Chemical Engg; promoted to Prof in 1978. Active res is concerned with polymeric matls in the solid state, polymer processing, and selected biopolymers. Consultant to industry on subjs related to polymeric matls selection, processing, and evaluation. Awarded 1980 Ralph Teetor Award from Soc of Automotive Engrs and Fellow of the Amer Physical Soc in 1983. Reg Engr; OH E-39976. *Society Aff:* AAAS, AIChE, ASM, SPE, APS, ACS, SAE, TBΠ, ΣΞ

Carreira, Domingo J
Business: 55 E Monroe St, Chicago, IL 60609
Position: Concrete Technologist. *Employer:* Sargent & Lundy Engrs. *Education:* MS/Civ Engg/IIT; MS/Urban Planning/Escuela Tecnica Superior De Arq; BS/Architecture/Univ of Havana. *Born:* 4/5/33. Began career in the Natl Regulatory Plan, Cuba, 1957. From 1960 to 1968 was Associated Prof at the Univ of Havana, Technical Dir of Industrias Siporex, hd of the Mtl Section at the Bldg Res Station, & Assesor for the Natl Soil and Mtl Testing Lab. 1969 recipient of a research scholarship at the Instituto E Torroja, Madrid, Spain; and 1977 co-recipient of the T Y Lin Award. Has been with Sargent & Lundy, Chicago, IL since 1970. Present chrmn of the ACI Ctte 209, mbr of the ASME subctte on nuclear certification, reg PE, and ASME/ACI level III nuclear inspector. *Society Aff:* ASCE, ACI.

Carreker, John R
Home: 373 Hampton Ct, Athens, GA 30605
Position: Res Leader Emeritus. *Employer:* USDA - Retired. *Education:* BS/Agri Sci/AL Poly Inst; MS/Agri Engg/AL Poly Inst. *Born:* 8/15/08. 1933-34: CCC Camp Superintendent; 1934-38: USDA-SCS Proj Engr; 1938-55: USDA-SCS & ARS Res Agri Engr; 1955-58: USDA ARS Res Experiemtn Station Dir; 1958-61: USDA-ARS & SCS Res Liasion Rep; 1961-73: USDA-ARS Res Agri Engr and Res Leader for Water Mgt Studies in the Southeast. *Society Aff:* ASAE, SCSA.

Carreno, Pablo A
Business: P.O. Box 86, South Bay, FL 33493
Position: VP. *Employer:* Gulf Western Food Prod. *Education:* Agricultural Eng. MS & Chemical Eng. BS. *Born:* Apr 1925. BS Sci Belen (Jesuit Sch) 1933, letters thru 1947 Univ of Havana Agricultural & Chem Engg, BS Post-Grad LSU. 1947-48 Am Cynamid- Res Ion Exchange; 1950-61 Covadonga, Cuba- Res & Production Supr; Lutgarda - Cuba Sugar Mfg Supt Tech Asst, Bd/Dir. Cia Rayonera, spinning area supervisor. Present - Okeelanta Sugar Refinery, G&W Food, Ch Chem, Production Supr, Asst Mgr, Mgr, VP. VP G&W Tech Chinchilla Varona do Brazil Engg Div. C & A Engineers, Officer AIChE Am Soc Sugar Cane Technologists.

Carrier, George F
Business: Harvard University, 311 Pierce Hall, Cambridge, MA 02138
Position: Prof. *Employer:* Harvard Univ. *Education:* PhD/Appl Mech/Cornell Univ; BS/Mech Engr/Cornell Univ. *Born:* 05/04/18. BS Mech Eng, Cornell Univ 1939; PhD Applied Mech, Cornell Univ 1944; Brown Univ-Asst Prof 1946-47, Assoc Prof 1947-48, Prof 1948-52. Harvard Univ- Gordon McKay Prof ME 1952-72; T Jefferson Coolidge Prof of Applied Math 1972-; Act Dean Div Engrg & Applied Phys 1975-76. Assoc Editor, Quarterly of Applied Maths. Mbr NAS, NAE, SIAM. Fellow AAAS. Hon Mbr ASME. Hon Fellow Brit Inst Math & Its Appli. Mbr Cornell Univ, Engrg College Council, Am Philosophical Soc. *Society Aff:* ASME, SIAM, AAAS, NAS, NAE.

Carrier, W David, III
Business: P.O. Box 5467, Lakeland, FL 33807-5467
Position: Pres *Employer:* Bromwell & Carrier, Inc *Education:* ScD/CE/MIT; SM/CE/MIT; SB/CE/MIT *Born:* 12/21/43 Raised in San Diego, CA. Served as lunar soil mechanics specialist at the Johnson Space Center, NASA, Houston 1968-73, during the Apollo program. Responsible for development and performance of experiments to measure the mechanical properties of the lunar surface, including Astronaut-operated devices and lab testing of lunar samples. Asst Chief Soils Engr at Bechtel, San Francisco, 1973-77, specializing in geotechnical analysis and design of fossil and nuclear power plants, earth dams, and industrial plants. Mgr, Solid Waste Systems at Woodward-Clyde Consultants, San Francisco 1977-78. Presently, Pres involved in advanced mineral waste disposal tech. Author of more than 50 papers. Norman Medal and Middlebrooks Award from ASCE and Certificate of Commendation from JSC. Registered PE in CA, FL, etc; Charterd Engr, Great Britain. *Society Aff:* ASCE, NSPE, AIME, CGS, ICE, ISSMFE, BGS.

Carroll, Billy D
Business: CSE Department, P.O. Box 19015, Arlington, TX 76019
Position: Professor and Chrmn *Employer:* University of Texas at Arlington *Education:* PhD/EE/Univ of TX at Austin; MS/EE/Univ of TX at Austin; BS/EE/Univ of TX at Austin. *Born:* 10/28/40. A native of Gatesville, TX. Ind experience includes stays at TX Instruments, Inc & Gen Dynamics Corp. Joined the faculty of Auburn Univ in Sept 1970. Visiting assoc prof, Univ of CA, Berkeley (1979-80) Prof of Computer Science and Engineering, Univ. of Texas at Arlington (1981-present), Dept Chrmn (1982-present). Co-author of *An Introduction to Computer Logic* (Prentice-Hall, Inc 1975). Co-author of "Tutorial: Fault-Tolerant Computing" (Computer Society Press, 1987). Author or co-author of numerous technical papers & reports. Technical areas of interests include computer arch, fault-tolerant computing, computer aided design, distributed computing, & applications of microcomputers. Chapter organizer & first pres-AL Chapt of IEEE Comp Soc (1975-76). Southeastern area chrmn-IEEE Comp Soc (1977-79). ASEE/Dow Award (1975). Co General Chrmn, First Intl Conference on Distributed Computing Sys. Editor-In-Chief, IEEE Computer Soc Press (1983 to 1986). Mbr, IEEE Computer Soc Model Program Task Force (1981- 1983). Mbr Tau Beta Pi, Eta Kappa Nu, Sigma Xi, Upsilon Pi Epsilon, IEEE, IEEE- CS, ACM, ASEE. Chrmn, Assoc for Computer Sci & Computer Engrg Chairs Steering Cttee (1985-87). *Society Aff:* IEEE, ACM, ASEE, TBΠ, ΣΞ, HKN, YΠE, ΩP.

Carroll, Charles F, Jr
Home: 2007 Gatewood Pl, Silver Spring, MD 20903
Position: Sci & Tech Intell Off. *Employer:* Navy Dept WA. *Born:* BSEE Univ of CA 1937, JD Georgetown Univ 1948, Radio Design Engr in RCA, Patent Examiner & Patent Attorney, sci & tech intelligence work for guidance of defense res, dev & system acquisitions. Was proj mgr for missile guidance computers in Burroughs Res. In charge of Naval air defense radars & fighter of interceptor control prog during WWII. Currently Naval Mtl Command- proj future Naval threats to natl security, & provide oper res & systems analyses, evaluating the effectiveness of our planned responses.

Carroll, J Raymond
Business: Office of the Architect, US Capitol, Washington, DC 20515
Position: Dir of Engrg *Employer:* Off of Architect of US Capitol. *Education:* MS/Engrg/Univ of IL; BS/ME/Univ of IL. *Born:* 7/8/22. BS, MS Univ of IL. Army Corps of Engrs 1943-46; Depts Mech Engg & Arch Univ of IL teaching grad & undergrad heat power & mech & elec equip and systems for bldgs 1947-75; Pres Carroll-Henneman & Assoc engr cons 1954-75; Dir Engg Office of the Arch of US Capitol Wash DC 1975-; Order of the Engineer 1965; Reg PE IL & WI; Dist Alumni Award Dept Mech Engg Univ of IL 1968 & 1983; IL Award IL Soc PE 1970; Fellow ASHRAE 1972; Dist Serv Award ASHRAE 1973; IL PE Exam Bd 1969-74; Pres ISPE 1966-67; Bd Dir NSPE 1961-65, 1968-75; VP NSPE 1973-75; Asst Treas NSPE 1975-; NSPE Rep to NCSBCS and AGC Liaison 1976-; Res & Tech Ctte ASHRAE 1971-76; Public Info Comm ASHRAE 1978-80, 1981-; VP Urbana Chamber of Comm 1972-74; Mbr Rockville Energy Comm. 1980-85, Chrmn 1982-84. Loyalty Award, Univ. of IL Alum Ass. 1986. Author; Lecturer; Episcopal Ch; Mbr

Carroll, J Raymond (Continued)
INCE, WSE, ASEE, SIPE, DCSPE, NSPE, ASHRAE, AEE, Pi Tau Sigma, Tau Beta Pi. Lambda Alpha International. *Society Aff:* AEE, WSE, INCE, NSPE, ASHRAE

Carroll, Lee F
Home: 43 Evans St, Gorham, NH 03581
Position: Proprietor. *Employer:* L F Carroll, PE; Elec Consult. *Education:* BS/EE/Northeastern Univ *Born:* 10/14/37 CARROLL, LEE FRANCIS, cons engr; b Berlin, NH, Oct 14, 1937 s Alton Francis and Mary Elizabeth (Cushing) C; BS in Elec Engrg, Northeastern Univ, 1960; m Judith Ann Magoun, Apr 9, 1960; children - Shawn, Pamela, Bruce. Plant elec engr Fraser Paper Ltd, Madawaska ME, 1960-64; maintenance engr Amer Optical Co, Southbridge MA, 1964-65; plant elec engr GA Pacific Group, Lyons Falls NY, 1965- 66; chief elec engr central engrg dept Brown Co, Berlin NH, 1966-70, Wright & Pierce Engrs, Topsham ME, 1970-73; prin L F Carroll, PE elec cons, Gorham NH, 1973- ; pres Gorham Devel Corp Mem Water and Sewer Comm, Gorham 1975- , mem budget com, 1973-82, chrmn, 1976-82; moderator Sch Dist Gorham, 1978- ; Town of Gorham 1987-; mem Area Vocat Ed Adv Com 1978- , chrmn, 1982-. Trustee; Gould Acad, Bethel ME, 1983- . Reg PE, VT, PA, VA, TX, LA, NH, MA, ME, NY, RI. Mem IEEE, Nat Soc Profl Engrs, Nat Fire Protection Assn, Illuminating Engrg Soc, Flying Engineers. Congregationalist. Home: 43 Evans St Gorham NH 03581 Office: 1 Exchange St Gorham NH 03581. *Society Aff:* IEEE, NSPE, IES, NFPA, Flying Engineers.

Carroll, Paul F
Business: 195 Spangler Ave, Elmhurst, IL 60126
Position: Pres. *Employer:* Semiconductor Specialists, Inc. *Education:* BS/Mech Engr/Northwestern Univ; Grad Courses/Elec Engg/Northwestern Univ. *Born:* 7/10/28. Served 2 yrs as an officer in the Navy after graduation from Northwestern Univ in 1951. Profl experience includes Co-op Student at Glen L. Martin Co, sales engg at Northwestern Elec and Bogue Elec & District sales Mgr for Transition Electronics. In 1959 founded Semiconductor Specialist, Inc. Chmn of the Bd of Midwest College of Engg, Lombard, IL in 1974. Dir of the Intl Trade Club of Chicago. Show Dir Midcon 1977, IEEE Regl Dir 1975-76, IEEE VP Regl Activities 1978-79. Treasurer NEDA 1981-85, Director EDS Show, 1981-87, Elected to VIP Electronics Club 1980. *Society Aff:* IEEE.

Carroll, William J
Business: P.O. Box 7009, Pasadena, CA 91109-7009
Position: Chairman of Board. *Employer:* James M Montgomery Consulting Engrs Inc. *Education:* MS/CE/CA Inst of Tech; BS/CE/CA Inst of Tech. *Born:* 12/23/23. Married Louise Judson, 1944. 4 children, Charisse, Gary, Christine and Pamela. IN Waste Engr LA, CA 1949-51. James M Montgomery, 1951 to present. Became Pres and C.E.O. in 1969. Presently Chairman of Board. Co specializes in water and sewage planning and design, both US and Intlly. Pres, Los Angeles Sec. ASCE 1967. Pres of Consulting Engrs Assoc of CA 1972. Pres Alumni Assoc of Cal Tech 1975-76. Pres, Am Acad of Environmental Engrs, 1979-80. Vice-Pres. ASCE 1986-88. *Society Aff:* ASCE, WPCF, AWWA, AAEE, NAE

Carson, Gordon B
Home: 5413 Gardenbrook Dr, Midland, MI 48640
Position: VP *Employer:* Mich Molecular Inst *Education:* DEng/ME/Cas Western Reserve Univ; ME/ME/Yale Univ; MSME/ME/Yale Univ; BSME/ME/Case Western Reserve Univ. *Born:* 8/1/11. Career started as grad asst at Yale Univ. Then instr to asst prof in charge of ind div at Case Inst of Tech, concurrently dir of res, Cleveland Automatic Machine Co. This was followed by a tour of duty as mgr of engg & secy of the corp for Selby Shoe Co. From this point, became dean of engg and dir of the Engg Experiment Station, OH State Univ. Next, vp bus & finance, and vp of the Res Fdn, OH State Univ. Thence to Albion College, as exec vp. Followed by Northwood Inst as dir of finance. Currently VP, MI Molecular Inst. Fellow ASME, IIE, AAAS. Mbr USNI, ASEE, SAMPE, Natl treas, Sigma Xi. Mbr Tau Beta Pi, Phi Eta Sigma, Alpha Pi Mu, Omicron Delta Epsilon. *Society Aff:* IIE, ASME, AAAS, USNI, ASEE, SAMPE

Carson, Larry M
Business: 3220 SW 1st, Portland, OR 97201
Position: Pres *Employer:* Carson, Bekooy, Gulick and Assocs, Inc *Education:* BS/ME/OR State Univ *Born:* 6/27/37 Principal, Carson, Bekooy, Gulick and Assoc, Consulting Engrs. Mechanical engrg consultant specializing in HVAC and energy conservation. Past Pres of OR chapter, ASHRAE, currently ASHRAE national nominating committee alternate. Member city of Portland HVAC code review board. *Society Aff:* ASHRAE, NSPE, ACEC, CSI

Carson, Norman R
Home: 2374 Magnolia Blvd W, Seattle, WA 98199
Position: Consultant to the Managing Partner *Employer:* R W Beck & Assoc. *Education:* BS/Elec Engg/Univ of IL. *Born:* 2/15/16. Commonwealth Edison Co, Chicago - 10 yrs - generating station engr; WWII - US Army Signal Corps, Maj-Bronze Star Medal; Erik Floor and Assoc Consulting Engrs, Chicago - 5 yrs - chief elec engr for 5 US Corps of Engrs hydro plants; R W Beck & Assoc, Engrs and Conslts, since 1955, retired Partnership Chrmn, Exec Partner, Chief Design Engr. Reg PE IL, WA, OR, AK, CO. *Society Aff:* IEEE, CEC, HKN.

Carson, Ralph S
Business: Dept of Electrical Engg. UMR, Rolla, MO 65401
Position: Prof. *Employer:* Univ of MO-Rolla. *Education:* PhD/EE/IL Univ; MS/EE/MI UNiv; BS/EE/IN Tech, Davidson College; -/-/IN Univ (Fort Wayne). *Born:* 12/22/22. in Durham, NC. Attended primary and secondary schools in Mooresville, NC. Taught radio in NYA school during part of WWII. With Farnsworth Radio Corp 1945- 46 specializing in FM equipment. Has taught electrical and electronics engg since 1946. Assumed present position in 1965. Author of "High-Frequency Amplifiers–, Wiley Interscience, 1975 (2nd edition in 1982) and "Principles of Appled Electronics–, McGraw-Hill, 1961, and numerous technical articles. Married Phyllis Creaser, East Tawas, MI, 1949. Two Children Gary and Mary. *Society Aff:* ASEE, IEEE.

Carson, William M
Business: PO Box 1000, Halifax, Nova Scotia B3J 2X4 Canada
Position: Asst Dean of Engg. *Employer:* Nova Scotia Tech College. *Born:* Nov 1934. Grew up on ID cattle ranch. Grad from Univ of ID with BSc Engg 1956. Grad studies in Meteorology at MIT follows by 4 yrs active duty with USAF. Owned & operated irrigated farm in ID until 1968. Completed PhD at Univ of ID 1970 & immigrated to Canada as Hd of the Bio-Resources Engg Dept at the Nova Scotia res progs. Prof involvements incl Chrmn Acadia Sect ASAE 1973, Regional Dir CSAE 1974, Bd Examiners of the Assoc of Prof Engrs of Nova Scotia and Pres of Aquatech Ltd.

Carstens, Marion R
Home: PO Box 803, Dahlonega, GA 30533
Position: Consultant. *Education:* PhD/Hydraulic Engg/State Univ of IA; MS/Hydraulic Engg/State Univ of IA; BS/Agri Engg/WA State Univ. *Born:* 10/1/19. Native of Reardan, WA. Served with US Army Corps of Engrs 1941-46. Taught at WA State Univ, Asian Inst of Tech, and GA Inst of Tech. Prof Emeritus of GA Tech since 1978. Pres, GA Sec of ASCE 1972. J C Stevens Award, ASCE 1954. Designated as one of the "Men who made marks in 1972" by Engg News Record. Holder of twelve US Patents mostly in capsule transport. *Society Aff:* ASCE.

Carstensen, Edwin L
Business: Dept Elec Engrg, University of Rochester, Rochester, NY 14627
Position: Prof of EE and Biophysics *Employer:* Univ of Rochester. *Education:* PhD/Phys/Univ of PA; MS/Phys/Case Inst of App Sci; BS/Phys Sci/NB State Teachers College. *Born:* 12/8/19. Early yrs in Oakdale, NB. During WWII, served with Columbia Univ's (later the US Navy's) Underwater Sound Ref Labs. At the Univ of

Carstensen, Edwin L (Continued)
Rochester, assisted in the dev of one of the first training progs in biomedical eng. Principle res contributions: Mechanisms of the absorption of ultrasound in biological mtls, dielectric properties of biological mtls, biological effects of ultrasonic and low frequency electric fields. Assoc Prof & Prof Elec Engrg, Univ Mentor, 1982- . Book "Biological Effects of Transmission Line Fields," Elsenior 1987. *Society Aff:* IEEE, ASA, AIUM, Biophysical Soc, Biomedical Engineering Soc, NAE.

Cartelli, Vincent R
Home: 14 Murchison Pl, White Plains, NY 10605
Position: Consulting Engr (Self-employed) *Education:* MCE/Adv Struct/Engr Sch NYU; BSCE/Civ Engr/Coll Engr NYU. *Born:* 6/14/11. Italy; USA Citizen. Reg PE - NY, NJ, DE, IL, MN, OH, Reg Struct Engr, IL. Asst Chief Airfields, US Dept Def 1941-1945. Engr Weskopf $ Pickworth 1945-1952. Consultant, Severud Assoc 1952-1975. Chief Engr, Rome Italy, LWPS 1956-1957. Wolchuk & Mayrbaurl 1975-1976. Private Consultant-USA, Egypt Bolivia Italy, Korea 1975-1984. Advisory Commission, Adjunct Prof CUNY 1966-. ACSE-IABSE Comm Tall Bldgs; ASCE-ACI Shear & Diagonal Tension; Welding Res Council Struct Steel Comm; Am Iron & Steel Inst Task Group Earthquake Resistant Structures ASCE State of Art Engg Award 1974. ASCE Raymond C Reese Award 1976. Enjoy classical music, art & theatre. *Society Aff:* ASCE, ACI.

Carter, Archie N
Business: 4820 Minnetonka Blvd, Minneapolis, MN 55416
Position: Vice Pres. *Employer:* Buan, Carter & Assocs Inc. *Born:* Native of Fremont IA. BA Cornell Coll, Mt Vernon IA 1933; BSCE Univ of IA 1935. US Army Engrs Rock Island IL 1935-37; Civ Engr Firestone Tire Plantations, Liberia W Africa 1937-39, bldg hgwys & dev plantations; 1939-48 Editorial staff of Engg News-Record magazine with last 5 yrs Regional Editor in Wash DC; 1948-56 Mgr Natl Hgwy Div of Assoc Gen Contractors Inc Wash DC; since Jan 1 1957 cons in MN. P Pres of MN Sect ASCE & former Dir ASCE natl Bd/Dir.

Carter, Charles F, Jr
Home: 5700 Charteroak Dr, Cincinnati, OH 45236
Position: Executive Director *Employer:* Institute of Advanced Manufacturing Sciences, Inc. *Education:* BS/EE/Northeastern Univ. *Born:* 10/28/27. Executive Director of the Institute of Advanced Manufacturing Sciences located in Cincinnati, Ohio. The Institute conducts projects relating to manufacturing processes, equipment and support systems. Was Technical Director at Cincinnati Milacron. Career in the area of machine tool development, machine elements and computer-aided manufacturing. Most recent efforts have focused on the use of machine tools incorporated in systems. Expert in flexible manufacturing systems, papers and articles on concepts of machine utilization, role of computer-based technologies for improving utilization. Certified Manufacturing Engineer, Professional Engineer--Massachusetts. 1982-83 President of the Society of Manufacturing Engineers. Member of the Institute of Electrical and Electronics Engineers. *Society Aff:* SME, IEEE.

Carter, Charles L, Jr
Home: 1211 Glen Cove Dr, Richardson, TX 75080
Position: VP of Quality Assurance & Human Res; Consulting *Employer:* C.L. Carter Jr. & Assoc. Inc. Management & Personnel Consultants *Education:* PhD/Mgt & Psychol/NCU; MA/Mgt/NCU; Diploma/Industrial Mgt/NCE. *Born:* 5/26/26. Reg PE in Quality; Certified Quality Engr; Certified Reliability Engr; Certified Mfg Engr; Reg PE in Mfg in Canada; Accredited Prof. Mgt. Consultant; Certified Prof. Manager; Certified Counseling Psychologist; Over 38 yrs experience in Quality Assurance, Quality Control, Reliability, Safety, Mfg, Production, Engg, Mtls, Human Resources, Training & Dev of People, Mgt Consulting. Written & Published 11 books, 1 training & dev film, & several mgt seminars as conducted nationwide for colleges and univs, business & industry. Executive Mgt, engg & consulting positions with: VMX, Inc, Rath & Strong, Inc /Plantronics-Action /Carter & Assoc Inc /Collins Radio /Martin-Marietta /EMR /Westinghouse /Western Elec. Currently VP of Quality Assurance & Human Resources Consulting for C.L. Carter, Jr. & Assoc. Inc. of Richardson, Texas. *Society Aff:* ASQC, SME, ASSE, ASTD, NMA, AACD

Carter, Geoffrey W
Home: 14 Church Farm Garth, Leeds, LS17 8HD UK
Position: Retired. *Employer:* Univ of Leeds. *Education:* MA/Math-Mech Sci/Univ of Cambridge. *Born:* 5/21/09. After training I worked for ten yrs as a res engr in ind, engaged on special investigations & measurements arising from a wide area of power engg. For the remainder of my working life I was hd of a univ dept of elec & electronic engg; this time saw a great increase in the scope of the dept's activity and in the number of students. My chief technical interest has been in the engg applications of electromagnetic theory. I have written some papers in this area and also several successful textbooks of univ std. *Society Aff:* IEEE, IEE.

Carter, Hugh C
Business: 10200 Willow Creek Rd, P.O. Box 26759, San Diego, CA 92126
Position: Pres *Employer:* Carter Engrs. *Education:* BS/Mech Engg/Caltech, CA. *Born:* 12/7/25. Mr Carter established his firm in 1957 as well as an innovative engr and designer of all types of mech systems, and has been a pioneer in energy conservation methods and practices. He designed the first solar heated school in CA. A prodigious writer and lecturer, Mr Carter has to his credit over 100 articles on all aspects of mech engg in building design. He contributes monthly to *Specifying Engineer Magazine* a column called "Carter's Cost Index." Books to his credit are "Mechanical Estimating" (used as a text at the USC), "Simplified Mechanical Specifications" and "Carter's Design Details." Mr Carter has long been active in Southern CA civic and community groups, and has served on a number of city and state governmental committees. *Society Aff:* ACEC, AEE, ASHRAE.

Carter, Jimmy
Business: Richard B. Russell Bldg, 75 Spring Street, Atlanta, GA 30303
Position: Private Citizen *Employer:* Office of Jimmy Carter *Born:* 10/1/24. Plains GA. James Earl Carter Jr, the 39th Pres of the US, entered Naval ROTC at GA Inst of Tech, grad US Naval Acad 1947, Post-grad Union Coll, Schenectady NY. Comm Navy, served in nucl submarine prog under Adm Rickover. Resigned comm 1953 to run peanut, fertilizer, seed business. Elected GA Senate 1962, became GA's 76th Governor 1970. Mbr Trilateral Commission; Chrmn Southern Regional Educ Bd. Elected US Pres 1976. Married Rosalynn Smith in July 1946. Has three sons & one daughter. Hobbies include fishing, hunting and bottle collecting.

Carter, Joseph R
Business: 105 Madison St, Worcester, MA 01613
Position: Chmn *Employer:* Wyman-Gordon Co. *Education:* BS/Metallurgy/PA State Univ. *Born:* 10/8/18. Native of Youngstown, OH 1940-Sharon Steel Corp. 1942-46-Officer in US Navy, WWII. 1946-57-VP-Operations, Wheeling-Pittsburgh Steel Corp subsidiary. 1957-Div Operations Mgr, Wyman-Gordon Co; 1960-VP, 1961-Dir, 1964-Exec VP, 1967- Pres, 1972-Chf Exec Officer, and 1978-Chrmn. Directorships: Avco Corp; State Mutual Securities; Chmn. ITTA Business Roundtable. Trusteeships, Worcester Memorial Hospital, Foundation for Economic Res; ASM David Ford McFarland Award; Doctor of Bus Sci Degree-Central New England College. Enjoy sailing and golf. *Society Aff:* ASM, AISI.

Carter, Lee
Business: 622 Belson Ct, Kirkwood, MO 63122
Position: Owner. *Employer:* Lee Carter Reg PE. *Education:* MSEngg/Chem Engg/Cornell; BSEngg/Chem Engg/Purdue. *Born:* 9/8/17. Employed in private practice as Chem Engr, Engrg Economist and Tech advisor in settlement of claims. Previously Sr VP, R W Booker & Assoc, Inc; Mgr Western Canada for Stone & Webster; Asst Sales Mgr London, England for Stone & Webster; Process Engr, Res Engr Oil Refining and Hydrogen Peroxide. Commander, Ordnance (Ret) US Naval Reserve. *Society Aff:* AIChE, IChE, AAAS, AEA, CIC, ACEC.

Carter, Nathan B, Jr
Business: 5823 fifth St, Meridian, MS 39301
Position: Pres. *Employer:* Carter Engg Corp. *Education:* BSCE/Civil Engg/Univ of TN. *Born:* 12/7/36. Mbr Chi Epsilon, Jaycee, Disting Service Award Nominee, Past Mbr Bd of Dirs of the Steel Joist Inst and the Steel Joist Committee for the Amer Inst of Steel Construction, 1960-62 - Chem Corp, Fort Detrict, 1963-66 - Asst Chief Engr at Todd Steel, 1966-69 - Chief Engr at Tucker Steel, 1969-present - Pres of Carter Engg Corp with offices in Jackson & Meridian, Pres Computer Graphics, Inc, Pres Arch Printing & Supply co, MS Engg Soc, Kiwanis, Trustee Bd of Northview Assembly of God Church, Jaycee Football Coach & Reg Engr in MD, TX and MS. *Society Aff:* NSPE, AISC, ACI.

Carter, Robert L
Business: 221 EE Bldg, Columbia, MO 65211
Position: Prof. *Employer:* Univ of MO. *Education:* PhD/Phys/Duke Univ; BS/Engr Phys/Univ of OK. *Born:* 8/22/18. Native of OK. Served with army field artillery as officer during WWII (Southwest Pacific area). Proj engr and engg group leader sodium-cooled nuclear reactor dev, Atomics Intl, 1949-63. Prof of EE and Nucl Engr, Univ of MO since 1962; specialities unconventional energy conversion, nuclear reactor materials. Visiting staff mbr Los Alamos Scientific Lab 1968-69, Argonne Natl Lab, 1967, Oak Ridge Lab (Y-12) 1945-46. Dir, MSPE, 1978-81; pres, Central Chap, MSPE, 1977-78. Trustee, MSPE Ed Foundation, 1982- , Chrmn, Mo Low Level Radioactive Waste Task Force, 1981-84. Hobby: violin and viola playing. *Society Aff:* NSPE, APS, ANS, AAAS.

Carter, Roger V
Home: 20635 - 4 Ave SW, Seattle, WA 98166
Position: Chief Met. *Employer:* Boeing Co. *Education:* Ms/Automotive Engg/Chrysler Inst; BS/Met Engg/MI Tech Univ. *Born:* 8/28/29. Met Engr at Chrysler Central Engg 1951-1955. Joined Boeing Co 1955 as Met and Mtls Engr, became Mtls Engg Mgr in 1964, Chief Met in 1973. Chrmn of 1980 Northwest Metals & Minerals Conf. Chrmn of SAE Aerospace Mtls Specifications Committee. Past-chrmn ASM Chapter. *Society Aff:* ASM.

Carter, Seymour W
Home: 3 Coolidge St, Sherborn, MA 01770
Position: Consultant *Employer:* Stone & Webster Engg Corp. *Born:* Apr 1923. BBA Engg Northeastern Univ. Elected Fellow ASNT 1973. Watertown Arsenal labs Watertown MA 1941-57 progressing from Sci Aide to Electronic Scientist; D S Kennedy Co Cohasset MA 1959-61 Qual Control Mgr; Army Mtl Res Agency Watertown MA 1961-63 Electronics Scientist; AVCO Corp Lowell MA 1963-70 Group Leader NDT; Stone & Webster Engg Corp Boston MA 1971-80, Asst Ch Engr NDT Div, 1980-present, Consultant.

Carter, William R
Business: 101 S Main St, Fayetteville, TN 37334-0753
Position: Chrmn. *Employer:* Carter Ltd. *Education:* BS/Civil Engg/Univ of TN. *Born:* 7/8/21. Married with three children. 1950-1982 - C F W Construction Do, Inc, Fayetteville, TN. Served as Pres & Chrmn of the Bd. 1948-49 - Mack Hamilton Construction Co, Fayetteville, TN. 1946-47 - Univ of TN, Knoxville, TN. Instructor in Civil Engg. 1943-46 - United States Navy. Earned the Bronze Star, serving in the Atlantic and Pacific Campaigns, and is a Lt (jg) retired. 1982- present Carter Ltd-Chrmn. *Society Aff:* ASCE, ASPE.

Carton, Allen M
Home: 8607 Bells Mill Rd, Potomac, MD 20854
Position: Deputy Asst Chief of Engrs *Employer:* U.S. Army Corps of Engineers *Education:* Master of Liberal Arts/Liberal Arts/The Johns Hopkins Univ; Bachelor of Engrg/CE/The Johns Hopkins Univ; Grad/Logistics/Industrial Coll of Armed Services *Born:* 07/07/26 Deputy Asst Chief of Engrs, U.S. Army Corps of Engrs, 1982 to present: supervise installation management activities, the devel & defense of Army Military Construction, Family Housing, and Real Property Operation & Maintenance programs, including necessary authorization and appropriation legislation. Army expert witness before Office, Secretary of Defense; OMB and Congressional Cttees on all programs and cost variations. Began government career in 1948 ad Jr. Highway Engr with Bureau of Public Roads. Served in government since as Highway, Railroads, Hydraulic, Civil, Construction and General Engr. Meritorious Executive, Sr Exec Service. Fellow ASCE. *Society Aff:* NSPE, ASCE, SAME, APWA.

Cartwright, Vern W
Business: Cartwright Aerial Surveys, Inc, 5979 Freeport Blvd, Sacramento, CA 95822
Position: Bd Chrmn *Employer:* Cartwright Aerial Surveys, Inc. *Education:* Undergrad/Remote Sensing/Univ of CA, Berkeley; Undergrad/Remote Sensing/Stanford Univ; Undergrad/Comp Graphics/Harvard *Born:* 3/6/22 Vern is Bd Chrmn of Cartwright Aerial Surveys, Inc., a company he founded in 1946. He is also Pres of Cartwright Res Corp, which manufactures and markets equip worldwide that he has patented. He set up the Intl Remote Sensing Inst in 1966 and serves as its continuing Pres. Past Natl Pres of the ASP and past Natl Pres of the LCP. Vern served on a number of boards for Governor Reagan when he wasgovernor of the state of CA, and serves on Harvard University's Grad School of Design Program Advisory Bd for computer graphics week. Reg PE, CA. *Society Aff:* ASP, LCP, ASCE, SAME, NSPE, APWA

Cary, Howard B
Business: 600 West Main St, Troy, OH 45373
Position: Senior Advisor Welding Tech *Employer:* Hobart Brothers Co. *Education:* BS/Welding Engr/OH State Univ. *Born:* 5/24/20. BS Indus Engrg OH State Univ - specialty: Welding Engr. Native of Columbus, OH. Employed by: Tank Div of Fisher Body Div GMC Grand Blanc MI; US Navy 1944-46 Western Pacific; Battelle Mem Inst Columbus OH-dev Mig Welding process 1946-48; Marion Power Shovel Co Marion OH welding engr, Welding Supr, Plant Mgr, manufacturer of power shovels & cranes 1948-58; Hobart Bros Co Troy-OH-Dir of Tech Ctr respon for application engg & training and currently VP respon for above plus Advanced Welding Technology. Samuel Wyllie Miller Mem Medal 1971 AWS, Natl Merit Award 1970, AWS, Bd/Dir 1973-76 AWS. Author. Reg Prof Engr OH. Natl Pres American Welding Soc. 1980-1981. Author Modern Welding Technology published by Prentice Hall 1979. *Society Aff:* AWS, ASME, ASM, WRC, EWI, AWI.

Cary, Robert A
Business: PO Box 151, Latrobe, PA 15650
Position: Vice President - Asst to the Pres *Employer:* Teledyne Vasco *Education:* Bach of Science/Met Engg/Lehigh Univ. *Born:* 9/2/18. Graduated from Lehigh Univ 1940. Tau Beta Pi. US Army 1941-45. Last 2 yrs Chief Ordnance Officer anti-Aircraft Command South West Pacific Theater, Lt Colonel (Bronze Star Decoration). Joined Vanadium Alloys Steel Co (now Teledyne Vasco) 1946 as metallurgical engr. Successive position as mgr technical service, process metallurgy, quality control. co-inventor patented tool steel. Promoted to VP-Technical Dir 1965. Past Chrmn AISI Tool Steel Technical Committee. CoAuthor ASM Book "Tool Steels–. Secy ASTM Tool Steel Subcommittee 1967-date. Mbr ASM Tool Steel Committee. Co-author ASM Book *Tool Steels.* Vice Chrmn Westmoreland Cty Indus Dev Auth 1967 to date. *Society Aff:* ISS, AIME, AISI, ASM, ASTM, APMI.

Casabona, Anthony M
Home: 28 Carline Dr, Clifton, NJ 07013
Position: Electronics Consultant Retired. *Employer:* ITT Avionics Div-Nutley NJ. *Education:* BEE CCNY 1941; grad courses Columbia Univ 1942 & Brooklyn Polytech 1943. *Born:* June 16, 1920 Joined ITT Nov 1941, completed mgmt courses at AMA 1956. Mbr Tau Beta Pi & Eta Kappa Nu; elected Fellow IEEE 1968. In early yrs at ITT, pioneered in dev of ILS & Tacan. Many pos during 35 yrs svc with ITT incl Mgr Engrg & Flight Opers at Westchester City Airport 1947, Dir Engrg 1959, Mgr Independent R&D 1970. Served sev RTCA Tech Panels on Navigation. Hold

Casabona, Anthony M (Continued)
11 pats & authored num tech papers incl chaps in the Radio Engrg Handbook and the Encyclopedia of Electronics. Retired June 1, 1977 and Served as special consultant to ITT until June 1982. Enjoy fishing, target shooting & traditional jazz.

Casagrande, Leo
Business: 40 Massachusetts Ave, Arlington, MA 02174
Position: Consulting Engr. *Employer:* Self-employed. *Education:* DrTech/CE/Technical Univ; Diplom Ingenieur/CE/Technical Univ. *Born:* 9/17/03. Native of Austria. 1928-30 Design Engr in Germany. 1930-32 Res Asst at MIT. 1932-33 Asst to Prof Terzaghi at Technical Univ in Vienna. 1933-34 Organizing Soil Mechanics Inst at Technical Univ at Berlin. 1934-45 Hd of Soils and Fdn Div of German Hgwy Dept. 1940-45 Honorary Prof of Soil Mechanics at Technical Univ in Braunschweig. 1946-50 Res Engr at Bldg Res Station, Watford, England. 1956-72 Prof of Fdn Engg at Harvard Univ. 1950-to date Consulting Engr. *Society Aff:* ASCE, BSCE, NAE

Casasent, David P
Business: Dept Elec Engr, Pittsburgh, PA 15213
Position: Prof Elec Engg. *Employer:* Carnegie-Mellon Univ. *Education:* PhD/Elec Engg/Univ of IL; MS/Elec Engg/ Univ of IL; BS/Elec Engg/Univ of IL. *Born:* 12/8/42. Tenured Prof of EE at Carnegie-Mellon Univ. Author of 2 textbooks on electronics; editor of 1 book, 7 Conference Proceedings and 6 Journal special issues on optical computing. Res spec; Optical signal and image processing. Author of over 350 journal and conference tech papers. Fellow IEEE, OSA and SPIE, various tech awds. *Society Aff:* IEEE, OSA, SPIE

Casazza, John A
Business: 1000 Connecticut Ave, NW, Washington, DC 20036
Position: President *Employer:* Ransom & Casazza, Inc. *Born:* Jan 1924. BEE Cornell Univ. Elec Officer USN 1944-46. With PSE&G Co 1946-1977 as VP Planning & Res in 1974. Respon for forecasting, dev of expansion plans, economic evals & res & dev. VP Stone and Webster Management Consultants 1977-1979. President Ransom and Casazza, Energy Consultants, 1979 to present. Fellow IEEE, Chairman IEEE Energy Committe, Inc PE. Active in past in EEI, NERC, and CIGRE.

Casciani, Robert W
Business: 400 Orange St, Ashland, OH 44805
Position: VP Engg & Mfg. *Employer:* F E Myers Co. *Born:* July 1920. BChE OH State Univ 1943. Proj engr on Manhattan Proj with Union Carbide 1943-44. Taught engg courses at Ashland College & Ashland High School Adult Educ. With F E Myers Co 1944-. Respon incl Prod Dev Engr, Plant Engr, Mgr Mfg Engg, Dir of Engg, VP Engg. Present respon as VP Engg & Mfg assumed Jan 1976. Reg PE OH. Cert Mfg Engr in SME. P Comdr US Power Squad - Holds "Navigator" rating. Avid cruising yachtsman.

Case, Edward T
Business: 500 Kiesel Bldg, Ogden, UT 84401
Position: Pres. *Employer:* Case Lowe & Hart Inc. *Education:* BS/Mech Engg/OR State Univ. *Born:* 4/18/27. Born in Caldwell, ID. Const engr Bonneville Power Adm 2 yrs; Ship's Engg Officer US Navy - Korean War; Mech Design Engg Lyle Marque & Assocs & Kaiser Aluminum in Spokane WA 4 yrs; Hd Facilities Engg Dept, Thiokol Corp Brigham City, UT 4 yrs. In private practice since 1963. Pres Case, Lowe & Hart, Inc Architects/Engrs. Field of practice is primarily indus facil design. PPres ACEC, UT. PE WA, OR, ID, UT, WO, AZ, TX. *Society Aff:* ASME, CECU, ASHRAE, NSPE.

Case, James B
Home: 1210 Colonial Rd, McLean, VA 22101
Position: Phys Scientist. *Employer:* Defense Mapping Agency. *Education:* PhD/Geod Sci/OH State Univ; MS/Geod Sci/OH State Univ; BS/CE/Stanford Univ. *Born:* 10/26/28. Born Lincoln, IL. Attended schools in CA. BS Civ Engrg Stanford Univ 1950; MS, PhD Geodetic Sciences OH State Univ 1957 & 1959. US Army 1951-53. Geodetic Surveyor Inter-American Geodetic Survey 1950-51 & 1953-55. Sr Photogrammetrist Broadview Res Corp 1960-61. Principal Scientist Autometric Oper of Raytheon Co 1961-71. Phys Scientist Defense Mapping Agency Hydrographic/Topographic Ctr 1971-79. Phys Scientist Hdquarters Defense Mapping Agency 1979-1982. Phys Scientist, Defense Mapping Agency systems center, 1982-. Dir Photogrammetric Surveys Div, 1 of 3 tech divs of ASPRS 1974-76. Ed-in-Chief of *Photogrammeric Engrg & Remote Sensing*, official journal of ASPRS 1975-. *Society Aff:* ASPRS, CIS, BCS, RGS, ACSM

Case, Kenneth E
Business: Sch/Indus Engg, Mgt, Stillwater, OK 74078
Position: Prof and Director of the Graduate Program *Employer:* OK State Univ. *Education:* BSEE/Electrical Engineering/Oklahoma State Univ; MS/Industrial Engineering/Oklahoma State Univ; PhD/Industrial Engineering/Oklahoma State Univ. *Born:* Aug 1944. BSEE, MSIE, PhD OK State Univ. Former Mgt Scientist for GTE Data Services & Assoc Prof of IEOR at VA Tech. Res & publishing efforts primarily in quality control. Fellow IIE, Senior Member ASQC; International President IIE 1986-87, Distinguished Service Award IIE 1984, Award of Excellence 1980, Department Editor IIE Transactions, Chairman Tulsa Section ASQC, Editorial Board Journal of Quality Technology. OSPE Outstanding Engineer in Oklahoma 1987. Received Distinguished Eagle Scout Award 1986. Reviewer and author for num tech journals; co-author of 3 books.Reg PE; ASQC Cert Qual/Reliability Engr. Mbr Sigma Chi, Blue Key, Phi Kappa Phi, num honoraries. Amateur radio station K5KC. *Society Aff:* IIE, ASQC, NSPE, ASEE.

Caselli, Albert V
Business: PO Box 3105, 1100 Milam, Houston, TX 77001
Position: Sr Staff Engr. *Employer:* Shell Oil Co. *Education:* ChE/Chem Engg/Columbia Univ, NY; BS/Chem Engg/Columbia Univ, NY. *Born:* 8/24/15. Native of NY, NY. With Shell since 1937: various assignments in US and overseas (Shell Petroleum Co, Ltd, London and Shell Internationale Chemie Mij, The Hague). Activities include: engg res and dev, process design and engg in petrochemicals, technical economics and group planning. Currently Sr Staff Engr, Shell Oil, Houston: coordination of technical training for engrs. Active in profl affairs, AIChE; Natl Dir, 1975-77; chrmn, Program and Publications committees, etc; Awards committee. Sigma Xi, Phi Lambda Upsilon. Reg PE, CA, TX. Hobbies: travel, classical music, photography, Intl Wine and Food Society. *Society Aff:* AIChE, ACS, AAAS, SCI, NYAS, SWSF, CMRA.

Caseman, Austin B
Business: School of Civil Engrg, GA Tech, Atlanta, GA 30332
Position: Prof of CE *Employer:* GA Inst of Tech *Education:* ScD/Civ Engrg/MIT; MSCE/Civ Engrg/UT State Univ; BSCE/Civ Engrg/UT State Univ *Born:* 02/13/22 Native of Monroe, UT. Married (Louise), daughter (Cathy). 1948-56, Instructor Asst Prof of Civil Engrg, WA State Univ; Since 1956, Associate Prof- Prof of Civil Engrg, GA Tech (specializing in structures) Undergraduate Coordinator. Summer employment as Structural Engr with Corps of Engrs and several aircraft companies. Awards: National Science Foundation Fellowship, ASCE Faculty Award by Student Chapter at GA Tech. Listed: Who's Who in So and SW; American Men of Science. Member: Pi Kappa Alpha Fraternity; Blue Key Honor Society; LDS Church. Hobbies: Gardening and photography. Reg PE; Mbr, ASCE, ACI. *Society Aff:* ASCE, ACI, ΦΚΦ, ΣΞ, ΧΕ, ΣΤ

Casey, H Craig, Jr
Business: Engg 130, Durham, NC 27706
Position: Chrmn & Prof. *Employer:* Duke Univ. *Education:* PhD/EE/Stanford Univ; MS/EE/Stanford Univ; BS/EE/OK State Univ. *Born:* 12/4/34. Worked at Hewlett-Packard in Palo Alto, CA as a dev engr from 1957 to 1962 while attending Stanford Univ. Joined Bell Labs at Murray Hill, NJ in 1964 as a mbr of Technical Staff and

Casey, H Craig, Jr (Continued)
since 1970 was in the Solid-State Electronics Res Lab. Principal area of work has been III-V compound semiconductor mtls and devices. In Aug 1979, became a Prof & Chrmn of the Elec Engg Dept of Duke Univ. *Society Aff:* IEEE, APS, ECS.

Casey, John R
Business: 58 River St, Milford, CT 06460
Position: Principal Partner *Employer:* Milford Engrg Assoc *Education:* BS/CE/Univ of CT *Born:* 12/14/38 Native of Milford, CT; Served with United Sts Army as a Helicopter Pilot from 1960-1963; Construction Engr for C. W. Blakeslee, General Contractors, on major hgwy, bridge, dredging and Sewage Treatment Facility projects, 1963-1968; Project Supvr to Dir of New England Regional Office for Bowe-Walsh and Assocs, Consulting engrs, on Municipal Water Pollution Control and Storm Water Management Projects 1968-1976; Principal Partner of Milford Engrg Assocs, Consulting civil engrs-Land Surveyors, 1976- to date, providing engrg services to private, industrial and Municipal Clients. Also, from 1979- to date, serve as Assistant Secretary to the St of CT Bd of Registration For PE and Land Surveyors. *Society Aff:* ASCE, NSPE, WPCF

Casey, Leslie A
Business: U.S. Dept of Energy, Salt Repository Project Office, 110 North 25 Mile Avenue, Hereford, TX 79045
Position: General Engr *Employer:* US Dept of Energy *Education:* MS/Nuc Engg/Univ of CA, Los Angeles; BS/Chem/State Univ of NY, Brockport; AA/-/Univ of FL. *Born:* 6/13/52. 1974 Ms Casy graduated from SUNY at Brockport, NY, with a BS in chemistry. After attaining a MS in nuclear engg in 1976 from UCLA, Casey entered a two-yr intern prog with the US Nuclear Regulatory Commission. The majority of the intership was conducted as a nucl engg in the Waste Mgmt Div. At the conclusion of her intership Ms Casey accepted a position within NRC as Risk Assessment Engr in the Office of Research. In February 1980, Ms Casey joined the US Dept of Energy (DOE). Casey has responsibilities related to the establishment of repositories for high level nuclear waste as part of the Civilian Radioactive Waste Mgmt program within the DOE. These responsibilities focus on systems engg activities for dev of mined repositories *Society Aff:* SWE, ANS, ACS.

Cashin, Francis J
Home: 219 Gosling Hill, North Hills, NY 11030-4015
Position: Chairman of the Board *Employer:* Cashin Assocs, PC. *Education:* BS/ME/Stevens Inst of Tech. *Born:* 6/3/24. BS Stevens Inst of Tech. Native of Mineola NY. US Navy: WWII Commanding Officer AKL; Korean War Exec Officer Destoyer. Principal DeFlorez Engg Co Inc; Gen Mgr Huck Co Engrs; formed Cashin Assoc P C July 1959 spec in environ & engg sys design. Co-auth NY State "Manual for Incinerator Plant Operators–. NYSSPE Pres 1982-83 ABET Director 1983-89; President-Elect 1987; Community Serv Award 1976; & Pres's Award 1976 Meritorious Service Award 1987. Engr of the year, Nassau County, NY 1983. ABET Pres-Elect 1987-88. Hold sev US pats. Formally Tr Stevens Inst. Enjoy classical music, flying, duplicate bridge. Society Aff: ASCE, ASME, IEEE, NSPE, AAEE, IES, SME. *Society Aff:* ASME, NSPE, IEEE, ASCE, SME, IES, AAEE.

Cashin, Kenneth D
Business: Goessmann Lab, Amherst, MA 01003
Position: Prof *Employer:* Univ of MA *Education:* PhD/CHE/RPI; MS/CHE/WPI; BS/CHE/WPI *Born:* 5/10/21 Served in US Navy 1944-46. Initiated Undergraduate Chem Engrg at Univ of MA in 1948. Was tenured in 1952. Sabbatical Leave in 1954-55. Resulted in PhD from RPI in 1955. Leave of absence 1968-70 to initiate a Dept of Chem Engrg at the Univ of Petro & Minerals, Dhahran, Saudi Arabia. Accepted Assoc Deanship, School of Engrg at Univ of MA for 3 years (1978-81). Am a member of Tau Beta Pi and served (1981-82) as Pres of Sigma Xi local chapter. Enjoy hunting, fishing and sailing. *Society Aff:* AIChE, ACS, ASEE, ΣΞ.

Caskey, Jerry A
Business: 5500 Wabash Ave, Terre Haute, IN 47803
Position: Prof *Employer:* Rose-Hulman Inst of Tech *Education:* PhD/ChE/Clemson Univ; MS/ChE/Clemson Univ; BS/ChE/OH Univ *Born:* 9/8/38 1973-present at Rose-Hulman Inst of Tech; VA Polytechnic Inst 1967-73; Research Engrs Organic Chemicals Development; The Dow Chem Co 1965-67. Education: PhD Clemson Univ 1965; Publications: 16, concerning Water Resources Engrg, Solar Energy, Polymer Chem; Organizations: AIChE, Am. Soc. for Eng. Ed.,; Honors/Awards: Visiting research prof, Israel Inst of Tech 1972; Public Health Service Trainee 1963-65; Principal Expertise: Polyelectrolytes in solid-liquid separations, Solar Energy use for space and water heating; Related Fields: Energy, Environmental. *Society Aff:* AIChE, ASEE

Caso, Ralph J
Business: 2466 N Jerusalem Rd, N Bellmore, NY 11710
Position: Owner/Principal. *Employer:* Self-employed/Consulting. *Born:* Sept 1924. BME NYU. Native of NY. Reg Engr NY, NJ, CT, MA, & NV. Sr Public Health Engr State of NY. USAF 1943-45. Ch Mech Engr for NASA Vehicle Assembly BLDG while with Seelye, Stevenson, Value & Knecht. Private practice since 1968. Many projs for local municipals & the priv sector. Currently involved in solar energy, energy conservation & energy mgt. Mbr Li Chap CEC, ASHRAE, LI BLDG Congress (Bd/Gov), NSPE. Actively engage in athletics, hunting and fishing.

Caspe, Marc S
Business: 1640 Oakwood Dr, San Mateo, CA 94403
Position: President *Employer:* M S Caspe Co. *Education:* MSCE/Structural/Lehigh Univ; BSCE/Civ Engg/Univ of City of NY. *Born:* 1/30/37. As a Chief Engr for Intl Engg Co, Mr Caspe was involved in the dev & implementation of proj mgt systems for both IECO & Morrison-Knudson Co. Mr Caspe is a licensed Civ & Structural Engr in CA and 13 other states, with extensive experience in the the management of projects. His technical proficiency extends beyond the management of projects to a detailed involvement with all disciplines as they apply to power dev, ind & commercial construction, water & waste treatment facilities, transportation systems & environmental design. Currently, Mr Caspe is serving in support of several private & governmental agencies who are seeking to strengthen their proj mgt performance. *Society Aff:* ASCE, PMI, AWWA, SEAOC, ASTM, WPCF, ASME, ACEC, CEAC.

Casperson, Lee W
Business: Department of Electrical Eng, Portland State University, P.O. Box 751, Portland, OR 97207
Position: Prof *Employer:* Portland State University *Education:* PhD/EE & Phys/CA Inst of Tech; MS/EE/CA Inst of Tech; BS/Physics/MA Inst of Tech. *Born:* 10/18/44. Native of Clackamas, Oregon. 1971-85 with the Electrical Engg Dept of the School of Engg and Applied Sci, Univ of CA at Los Angeles. Since 1983, Prof., Electrical Engg the School of Engg and Applied Sci, Portland State Univ. Portland, OR. Prin research interests include quantum electronics, gas lasers, high gain laser media, and optical resonators, and this research has led to over 100 publications. Have dev several new lecture and lab courses and consult for local industries. Awarded the Centennial Medal of the IEEE 1984, and the Faculty Excellence Award of the Oregon Legislature in 1986. *Society Aff:* IEEE, OSA, ΣΞ, HKN, AAAS

Cass, A Carl
Home: 37 Nicholson St, NW, Wash, DC 20011
Position: Intl Engr & Mgt Consultant (Self-employed) *Employer:* Self *Education:* DSc/Honorary Doctorate-Economics/Univ of Nuevo Leon; PhD/Met/Fed Univ State of Rio de Janeiro; BS/Met/Fed Univ State of Rio de Janeiro; CE/RR Engr/Purdue Univ. *Born:* 10/26/06. Prof of Metallurgy & internatl consltg engr. Edu: Dr Engrg, Sch of Engrg, Fed Univ, State of Rio de Janeiro, 1965; Hon DSc, Univ Nuevo Leon, Mexico, 1958; B in Civ Engrg, Purdue Univ; Metalgy & Geodetic Surveying, RPI; Higher Math & Astron, Buffalo NY. Reg PE in DC; ASCE Life Mbr. Hon Mbr ILAFA (Lat Am); JCitation from Italian Govt, 1968, Commendatore dell'Ordine Al Merito della Rep It; author, 11 maj publs, in Spanish and English;

Cass, A Carl (Continued)
Chief, Engrg Div, Export- Import Bank of US 1951-76. Prof & internatl consltg engr since 1977. *Society Aff:* ASCE, ILAFA.

Cass, James R, Jr
Home: 9324 Arlington Blvd, Fairfax, VA 22031
Position: VP & Chief Engr *Employer:* James C. Buckley, Inc. *Education:* MS/CE Structures/Northeastern Univ; SB/CE/MIT *Born:* 1/29/27 Was structural and project engr for consulting firm, port engr with MA Port Authority, quality mgr for prestressed concrete manufacturer, all in Boston, MA area. Since 1973, have specialized in studies, design and construction of intl maritime port and airport facilities with James C. Buckley, Inc and affiliate firm, Airways Engrg Corp. Currently VP and Chief Engr of Intercontinental Consultants, Inc, and subsidiaries, Buckley and Airways. Chrmn, New England Council ASCE - 1967; Sec.-Treas, MA Section, ASCE - 1958-63; Member Ports and Harbors Committee, ASCE - 1975 to date (Chrmn-1983). Registered PE in 9 states, District of Columbia, and Puerto Rico. *Society Aff:* ASCE, PIANC

Cassara, Frank A
Business: Route 110, Farmingdale, NY 11735
Position: Assoc Prof *Employer:* Polytechnic Inst of NY *Education:* PhD/EE/Polytechnic Inst of Brooklyn; MS/EE/Polytechnic Inst of Brooklyn; BS/EE/Rutgers Univ *Born:* 4/19/44 Since 1970 he has been a member of the faculty at the Polytechnic Inst of NY (formerly Polytechnic Inst of Brooklyn) where he has taught undergraduate and graduate courses in electronics and communication systems and has engaged in research in the areas of FM communication systems and surface acoustic wave devices. He is currently Assoc Prof of Electrical Engrg. *Society Aff:* IEEE

Cassaro, Michael A
Home: 9702 Westport Rd, Louisville, KY 40222
Position: Prof. *Employer:* Dept of Civil Engg. *Education:* PhD/Structural Engg/Univ of FL; MS/Struct Mechs/Univ of CA; BCE/Civil Engg/RPI. *Born:* 3/15/31. Native of Brooklyn, NY. Served with US Navy as Photographic interpreter 1954-57. Structural Eng for the Prescon Corp; involved in design and construction of post tensioned prestressed concrete structures 1958-61. Presently Prof of Civil Engg; served with Univ of FL 1961-68, Univ of Louisville 1968-present. PPres, Louisville Chapter of Soc of Sigma Xi for scientific res. Faculty advisory to student Chapter of Chi Epsilon, Civil Engg Honor Soc. Sr structural consultant, Schimpeler-Corradino Assocs Consulting Engrs. Mbr of the Governor's Earthquake Hazards and Safety Technical Advisory Panel, State of KY. Mbr of the Mayor's Economic Development Advisory Ctte, City of Louisville. Mbr Air Pollution Control Bd of Jefferson County, KY. Associate Prog Dir, Urban Studies Center, Univ of Louisville. Reg PE in KY, FL & IN. *Society Aff:* ASCE, ACI, PCI, NSPE, EERI, ASTM.

Cassidy, John J
Business: Hydraulics/Hydrology, Bechtel Civil, Inc, P.O. Box 3965, San Francisco, CA 94119
Position: Engrg Mgr, Hydraulics/Hydrology *Employer:* Bechtel Civil & Minerals, Inc. *Education:* PhD/Hydraulic Engineering/Univ of IA; MS/Civ Engg/MT State Univ; BS/Civ Engg/MT State Univ. *Born:* 6/21/30. Born in Gebo, WY, June 1930. Received BSCE and MSCE degrees from MT State Univ and PhD from Univ of IA. Mgr. Hyd Engr, Bechtel Civil, Inc. 1985-pres. Dir of WA Water Res Ctr and Prof of Civ Engg at WA State Univ 1979- 1981. Chief Hydrologic Engr with Bechtel, Inc 1974-79; 1981-85. Respectively Asst Prof, Assoc Prof, Prof, and Hd of Civ Engg at Univ of MO 1963-74. Instr at MT State Univ 1958-60 and Univ of IA 1960-62. Ford Fdn Fellow 1962. Bechtel Fellow 1986. Co-author of text "Hydrology for Engineers and Planners." Pres ASCE Mid MO Sec 1972-73. Mbr ASCE Hydraulics Div Exec Committee 1976-84, Chrmn 1978-79, Chrmn ASCE Natl Water Policy Ctte 1984-85, Mbr 1982-85. Co-author "Small and Mini Hyd Sys" McGraw-Hill 1984; Co-author: "Spillways for High Dams–, chap 4, *Develop in Hydraulic Engrg-2*, Elsevier Applied Sci Publ, 1984. Reg PE in MT, MO, CA, and WA. *Society Aff:* ASCE, USCOLD, AWRA, AGU, IAHR

Cast, Karl F
Home: 9345 Hill Trace, Baton Rouge, LA 70809
Position: VP Engrg & Data Processing *Employer:* Ethyl Corp. *Education:* MS/CE/MIT; BA/Chem/Lawrence Univ. *Born:* 2/14/17. Native of Appleton, WI. Mbr Phi Beta Kappa, Prof Engg No 2700-WI, previous employment: DuPont, Marathon Paper, & Bechtel Corp. Previous experience with Ethyl in economic evaluation, mfg technical service (gen supt) & gen mgr Operations-Intl Div. Currently responsible for Ethyl's - Central Systems & Data Processing Dept, Corporate Engg Dept, & Land Dept. *Society Aff:* AIChE.

Castagnos, Lee J, Jr
Business: 1211 Lafayette St, Houma, LA 70360
Position: Principal. *Employer:* Castagnos-Goodwin & Assoc *Education:* BSME/Mech/Univ of SWLA. *Born:* Aug 5, 1939. Native of Houma, LA. BSME Univ of Southwestern LA 1960. Civ Engr Corps USN; Dir Physical Facilities at Univ of Southwestern LA 1965-70; 1970 to present, Principal in the firm of Lee J Castagnos Jr & Assoc & Castagnos- Goodwin & Assoc; Mbr LA Engg Soc, NSPE, ASHRAE, Am Cons Engrs Council, Cons Engrs Council, LA; Res Engr LA & MS; Tau Beta Pi. *Society Aff:* LES, NSPE, ACEC, CECL, ASHRAE.

Castenschiold, Rene
Home: Lee's Hill Rd, New Vernon, NJ 07976
Position: Pres *Employer:* LCR Consulting Engineers, P.A. *Education:* BEE/Elec/Pratt Inst. *Born:* 2/7/23. Upon graduation joined Genl Elec Co, worked on the Manhattan Proj and later design engr at Schenectady, NY. In 1951 joined Automatic Switch Co, Florham Park, NJ where he worked on the dev and application of equipment for emergency power systems. Served in various mgmt positions. In 1986 Founded LCR Consulting Engineers, P.A. specializing in Forensic Engineering. From 1967-79 also lecturer at NJ Inst of Technology. Granted nine pats, authored over 30 Technical Publications and contributed to several engg books. In 1979 elected Fellow of the IEEE and in 1983 appointed to IEEE Standards Bd. Reg Prof Engr in New York and New Jersey. Named to Distinguished Alumni Board of Visitors, Pratt Institute, 1979. Recipient of The James H. McGraw Award, 1986. Listed in Who's Who in Technology Today, Amer Men and Women in Science, Who's Who in Finance and Industry, Who's Who in the World, Who's Who in America. Chairman U.S. Technical Advisory Group and U.S. Delegate to International Electrotechnical Commission. *Society Aff:* IEEE, NSPE, ISA, ACEC, NFPA, IAEI

Castleman, Louis S
Business: 333 Jay St, Brooklyn, NY 11201
Position: Prof. *Employer:* Poly Inst of NY. *Education:* ScD/Physical Met/MIT; BS/Met/MIT. *Born:* 11/24/18. Employed as met by Sunbeam Elec Mfg Corp (1939-41), as sr scientist, supervisory scientist, & acting section mgr by Westinghouse Atomic Power Div (1950-54), and as met specialist by Gen Tel & Electronics Corp (1954-64). Prof of physical met at Poly Inst (1964-now). Served as reserve officer in the AUS (1940-78); active duty in US, Corsica, & France (1941-46); Lt Col, AUS-Ret (1978). Chrmn, LI Chapter, ASM (1963-4); Pres, Metal Sci Club of NY (1973-4). Poly Distinguished Teacher Award (1975). Fellow, AAAS (1978). Listed in Who's Who in Am, Outstanding Educators of Am, Dictionary of Intl Biography, etc. *Society Aff:* AIME, APS, ASEE, ASM, AAAS, ΣΞ.

Castorina, Anthony R
Business: 835 Main St, Bridgeport, CT 06609
Position: Asst VP. *Employer:* Bridgeport Hydraulic Co. *Education:* BS/Chemistry/Tufts Univ; Post-Grad/Engg/Univ of Bridgeport. *Born:* 8/26/25. Native of Bridgeport, CT. Served with US Army Signal Corps 1943-46. Chemistry Instructor, Univ of Bridgeport, 1960-75. Began service with Bridgeport Hydraulic Co in 1949 as Chemist and attained Asst VP in 1979. Respon for Lab, Cross Con-

Castorina, Anthony R (Continued)
nection Control, Watershed Control and Water Treatment Processes throughout career. George Fuller Award (AWWA) 1977. Chrmn - Pollution Abatement Committee 1974-77, now Past Chairman, Chmn AWWA Comm - Quality Control in Reservoirs CT Section AWWA. Reg PE CT 1978. Publications - Ann Arbor Press and AWWA. Journal on Watershed Control, Raw Water Quality Improvement, Treatment Plant Equipment Maintenance and Lab Operation. *Society Aff:* AWWA, NEWWA, ACS.

Castro, Walter E
Business: 113 Riggs, Clemson University, Clemson, SC 29634
Position: Prof of Mech Engg. *Employer:* Clemson Univ. *Education:* PhD/Theo & App Mechanics/WV Univ; MSME/Mech Engr/Clemson Univ; BSME/Mech Engr/Indiana Inst of Tech; AAS/Bldg Construction/SUNY at Delhi. *Born:* 12/31/34. and raised in NY state. Attended SUNY at Delhi from Aug 1952 - June 1954, receiving AAS degree in Bldg Construction. Employed in residential and commercial construction from June 1954 - Aug 1956. Pursued career in engg August 1956 - June 1966, specializing in Fluid Mechanics. Entered engg education in 1966 at Clemson Univ, teaching and conducting research in fluid mechanics. Promoted to Prof of Mech Engg and Mechanics in 1974. Served as asst dept head and coordinator of grad studies in the Mech Engg Dept. until 1984. Presently serving as Assistant Dean for Undergraduate Studies in the College of Engineering at Clemson University. *Society Aff:* ASEE, ASME, ΣΞ, TBΠ.

Cataldo, Charles E
Business: PO Box 1900 W Station, USBI/BPC, Huntsville, AL 35807
Position: Staff Scientist *Employer:* United Space Boosters, Inc., Booster Production Co. *Education:* Grad courses/Met & Bus/Univ of AL; BS/Chem/Univ of AL. *Born:* 2/12/27. Cardiff, AL. Resident of Birmingham to 1951. BS Chem Univ of AL 1950. Army Air Corps WWII. Joined Army Ballistic Missile Agency Huntsville, AL in 1951, later transferring with the group to NASA in 1960. Began work in mtls engg related to selection control, res & dev for space launch vehicles, spacecraft & payloads, & held var of pos in lab mgt & proj engg. Presently Chf Sci, United Space Boosters Inc, Matls & Processes Tech Dept. NASA Exceptional Serv Award 1969, Exceptional Engrg Achievement Award in 1982, Yuri-Gagarin Diploma for role in Apollo-Soyuz project in 1977. Author num papers on space vehicle mtls, esp environ effects. Mbr NASA Mtls Adv Ctte. Mbr AIAA, ASM(Fellow), SAMPE, serving on both local & natl cttes. *Society Aff:* AIAA, ASM, SAMPE, ADPA.

Catanzano, Samuel
Business: 600 Grant St, Suite 1400, Pittsburgh, PA 15219
Position: VP *Employer:* M&M Protection Consultants *Education:* BS/ChE/Univ of Pittsburgh *Born:* 10/6/28 Native of Pittsburgh, PA. Served with Army Signal Corps 1946-1947. Research Engr specializing in combustion research for an Atomic energy Commission project. field engr for Factory Mutual engrg Division, specializing in fire protection. With M&M Protection Consultants since 1962. Assumed current responsibility as Mgr, Pittsburgh Office in 1971. Also serving as Dir, United Sts Steel Corp Fire Protection Division since 1971. Charter Pres of Pittsburgh- Three Rivers Chapter SFPE; currently a member of Bd of Directors of SFPE. *Society Aff:* SFPE

Catenacci, Giorgio O
Business: Cesi-Via Rubattino 54, Milan, Italy 20134
Position: President and Managing Director *Employer:* CESI. *Education:* Doctor in Elec Engrg/Poly of Milan *Born:* 03/25/27 Prof Elec Installations. Transmission Line Engr with Edison Co Milan 1950- 56. With CESI since its estab in 1956 as Ch Engr network studies dept & successively nigh power dept. Gen Dir since 1965. Fellow IEEE. Chrmn Tech Ctte 17 "Switchgear" of the Italian Electrotech Ctte-Dir of the technical Journal "L'ENERGIA ELETTRICA."

Cathey, Jimmie J
Business: Dept. of E.E, 453A Anderson Hall, Lexington, KY 40506
Position: Assoc Prof *Employer:* University of Kentucky *Education:* Ph.D/EE/TX A&M Univ; MS/EE/Bradley Univ; BS/EE/TX A&M Univ *Born:* 5/16/41 Born in Whitney, TX. Res Engr with Caterpillar Tractor Co. 1965-68. Served in following capacities with Marathon LeTourneau Co. specializing in design of traction electrical machinery and controls: Proj Engr 1968-69, 1972-74; Chief Electrical Engr 1974-77; Dir Applied Res 1977-81. Presently Assoc Prof of Electrical Engrg at the Univ of Kentucky. Author of 20 technical papers and holder of three U.S. Patents in field. Reg Professional Engr (Texas) and mbr of Eta Kappa Nu, Tau Beta Pi, and Phi Kappa Phi. *Society Aff:* IEEE, HKN, TBΠ, ΦΚΦ.

Cathey, W Thomas, Jr
Home: 228 Alpine Way (PBH), Boulder, CO 80302
Position: Prof *Employer:* Univ of CO *Education:* PhD/EE/Yale Univ; MS/EE/Univ of SC; BS/EE/Univ of SC *Born:* 11/26/37 Autonetics Research Center, Rockwell Int'l doing research and heading small group in holography, laser systems, adaptive arrays, image formation, optical information processing (6 years). Since 1968 at Univ of CO specializing in optical processing, optical communications and electro optics. Consulting in laser systems, electromagnetic compatability, imaging, holography, acoustic tomography, adaptive optical elements. In charge of EE program at Denver Campus (1970-75), Dept Chrmn, 1984. Director, Center for Optoelectronic Computing Systems, 1985-date. Fellowships: NSF (1960-62), Univ of CO Faculty Fellowship (1972-73), Croft Fellowship (1982). Elected OSA Fellow (1979), Author of Optical Information Processing and Holography; (Wiley, 1974). *Society Aff:* IEEE, OSA, SPIE

Caton, Robert L
Business: 101 W Bern St, Reading, PA 19601
Position: Mgr New Prod. Dev. *Employer:* Carpenter Tech Corp. *Education:* BS/Met/PA State Univ. *Born:* 2/28/37. Native of Uniontown, PA served 3 yrs (1959-62) on active duty with USN. Retired from Naval Reserve (1979) as a Commander. Joined Carpenter in 1962 in R&D. Assumed current responsibility in 1983. Responsible for New Product Development function. Holds 5 US Patents & has published/presented several technical papers. *Society Aff:* ASM, SAE.

Caudle, Ben H
Business: CPE 2.502, The University of Texas, Austin, TX 78712
Position: B.J. Lancaster Prof of Petroleum Engg *Employer:* Univ of TX. *Education:* PhD/Petrol Engg/Univ of TX; BS/Chemistry/Univ of TX. *Born:* 4/27/23. Dallas, TX, Jr Chemist, US Bureau of Mines 1943-1947. US Army 1944-1946. Res scientist, Atlantic Refining Co, Production Res Lab 1947-1961. Univ of TX 1961-present. Present position: Prof of Petrol Engg. Bd of Dirs, SPE of AIME 1964-1967, SPE distinguished Service Award 1969, AIME Distinguished Educator Medal - 1979, Gen Dynamics Teaching Award-1974. Hon Mbr A.I.M.E. 1987. Present res interest in Petrol Recovery. *Society Aff:* AIME, SPE.

Caufield, H John
Home: 385 Old Beaverbrook Rd, Village/Nagog Woods, MA 01718
Position: Pres *Employer:* Innovative Optics, Inc *Born:* Mar 25, 1936. PhD IA State Univ, BA Rice Univ Indus Rand at TX Instru, Raytheon, Sperry Rand, & Block Engg. Author of two books num papers, pats, reports, bk chapters, presentations in coherent optics, holograph, & coherent optics. Chrmn Intl Mtgs in optics for Gordon Res Conf, IEEE, & SPIE. Ed bds of & Laser Focus. Sr Mbr IEEE. Fellow Optical Soc of Am and SPIE, Symposia VP, SPIE. Am Men of Sci, Who's Who in East, Intl Dictionary of Biography. Editor, *Optical Engrg.*

Caulton, Martin
P.O. Box 9442, Aspen Hill Branch, Silver Spring, MD 20906
Position: Member of Technical Staff *Employer:* RCA Labs *Education:* PhD/Physics/Rensselaer Polytechnic Inst; MS/Physics/Rensselaer Polytechnic Inst; BS/Physics/Rensselaer Polytechnic Inst *Born:* 8/28/25 Served in US Parachute Infantry during WW-II. Fulbright Scholar at Imperial Coll, London 1954-55 (nuclear

Caulton, Martin (Continued)
physics). Research at Bell Telephone Labs, Murray Hill, NJ, 1955-58, on low noise microwave tubes. Physics Faculty in Union Coll, Schenectady, NY, where co-authored textbook PHYSICAL ELECTRONICS. On RCA Labs Princeton, NJ Technical Staff since 1960, research on: microwave power tubes, electron beams, plasmas and solid-state devices. Directing projects in microwave integrated circuits and devices since 1966, authored numerous papers in the field. He has taught microwaves and modern physics from 1961 as Adjunct Prof of Electrical Engrg, Drexel Univ, Philadelphia and as Visiting Prof at the Technion, Israel Inst of Tech in 1971. Elected Fellow of IEEE in 1979. *Society Aff:* IEEE, APS, ΣΞ

Cavaliere, Alfonso M
Home: 150 Huntington Road, New Haven, CT 06512
Position: Consultant *Employer:* Self *Education:* BS/CE/Northeastern Univ. *Born:* 12/28/14. With Truscon Steel Co, Ct Dept of Public Works, CT State Highway Dept & US Engr Dept before WWII. US Army 1942-46. Staff mbr Photo Interpretation Section of Military Intelligence School Camp Ritchie, MD. Joined Westcott & Mapes Inc in 1946. Elected VP Structural Engg 1963. Also respon for New Business Dept. Pres CT Section ASCE 1970-71. Pres CT Engrs in Private Practice 1970-71. Mbr Zoning Bd of Appeals New Haven CT - Building Commissioner New Haven, CT. Mbr Local Advisory Council of Regional Construction Industry Arbitration Committee. Enjoys music, theatre and travel. *Society Aff:* ASCE, NSPE, ACI.

Cave, Jere S
Home: PO Box 222, Short Hills, NJ 07078
Position: Retired *Education:* BE/EE/Vanderbilt Univ. *Born:* Sept 1914 Youngstown OH. BE, Summa Cum Laude Vanderbilt Univ 1936. Res Short Hills, NJ. Employed by OH Bell 1936-61 & left there as Asst VP in 1961 to become Asst VP AT&T. Elected VP AT&T Const Plans Dept in 1975. Respon for Bell Sys Capital Budget Planning, Depreciation Studies, Inventory Mgt & Purchase of Gen Trade Communications Prod. Maj in US Army Signal Corps S W Pacific WWII. Elected Fellow IEEE 1971. Retired Oct 1, 1979. *Society Aff:* IEEE.

Cavell, George R
Home: 3342 Avenida Hacienda, Escondido, CA 92025-7244
Position: Prog Mgr. *Employer:* US Postal Service. *Education:* MEA/Engrg Adm/Geo Washington Univ; BS/Ind Adm/Yale Univ. *Born:* 8/19/26. Employed as Industrial Engr at Westinghouse and then Postal Service. Managed the dev, design and installation of ten mail-flo tray conveyor material handling systems for the Postal Services costing $10 million. Directed the dev of the concept, design and implementation plan for the Postal Service National Bulk Mail System. A $950 million project with 21 plants throughout the country. Gn Mgr of the NY Bulk and Foreign Mail Center during its activation. Formerly, Genl mgr Engg Div in Western Region USPS. Currently Prog Mgr/Facilities Western Area Maintenance Overhaul and Tech Service Center. Reg in Mech and Indus Engrg. Masters in Engrg Admin. *Society Aff:* IIE.

Cawley, John H
Business: 6010 Exec Blvd, Rockville, MD 20852
Position: Dir, Natl Ocean Service Engrg Staff *Employer:* US Dept of Commerce/NOAA. *Education:* MS/EE/Univ of CA; BS/EE/Notre Dame. *Born:* 2/28/25 Native San Diego, CA. Sr engr Univ of CA, Scripps Inst of Oceanography 1949-1957; responsible for design & evaluation of solid state oceanographic instruments. Res Staff Mbr Gen Dynamics/Gen Atomic 1957-1962; responsible for design & evaluation of solid state nucl monitoring & safety instrumentation; Owner & Mgr of Mission Instruments Corp 1957-1961; developed specialized ocean measurement equip. Sr Staff Mbr Arthur D Little, Inc, 1963-1977; responsible for technical & mgt studies for govt & ocean industries. Dir Advanced Technology Office, Natl Oceanic & Atmospheric Admn 1977 to 1983; responsible for dev of advanced ocean tech & environmental measurement & monitoring systems; from 1983 to present, Dir Natl Ocean Service Engrg Staff, Natl Oceanic & Atmospheric Admin, and serves as the primary engrg, tech dev and advanced sys planning/design office in NOS. *Society Aff:* IEEE, MTS, AAAS.

Cayton, David W
Home: 14166 Reservation Rd, Salinas, CA 93908
Position: VP. *Employer:* Bud Antle Inc. *Education:* Masters/Business Admin/Univ of So CA; Bachelors/Ag Engg/CA State Poly Univ. *Born:* 8/30/36. *Society Aff:* ASAE.

Caywood, James A
Business: 1133 15th St, N.W, Washington, DC 20005-2701
Position: President *Employer:* DeLeuw, Cather & Co. *Education:* BSCE/CE/Univ of Kentucky. *Born:* Jan 28, 1923. BSCE Univ of KY 1944. Served to LTjg USNR 1944-46; Asst Engr & Sr Instrumentman L&N Railroad Co Latonia KY 1946-47; Asst Engr B&O RR Co Cincinnati 1947-50; Asst Div Engr 1950-52; Div Engr 1953-57, Engr Maintenance of Way 1957-60, Asst Ch Engr 1960-61, Ch Engr 1961-64; Gen Mgr Egr Planning for C&O/B&O RRs 1963-64; Pres & Dir Royce Kershaw Co Inc 1964-65; Sr VP & Dir DeLeuw, Cather & Co Washington DC 1965-1978. Reg PE 50 states and DC. Mbr Am Soc of Civ Engrs, Natl Soc of PE, Am Railway Engg Assn, the Moles, ARTBA. Pres De Leuw, Cather & Co Washington DC 1978-present. *Society Aff:* ASCE, ARTBA, AREA, NSPE, ITE, APTA, NCEE, ACEC, RPI, SAME.

Caywood, Thomas E
Business: Graduate School of Business, Chicago, IL 60637
Position: Sr. Lecturer *Employer:* Univ of Chicago *Born:* May 1919. AB Cornell College; MA Northwestern Univ; PhD Harvard Univ, Math & Phys. Univ of Chgo, Inst for Air Weapons Res: Sr Mathematician & Coord of Res 1947-52. Armour Res Fdn: Supr of OR, 1952-53. Caywood-Schiller Assoc: Partner 1953-70 A.T. Kearney: Vice Presidents' 1970-77. Panel on Ordnance, Trans & Supple: Chrmn 1960-64. Defense Sci Bd: Mbr 1960-64. Defense Sci Bd, Task Force for Gun Sys Acquisition: Chrmn 1975. ORSA: Pres 1969; Ed Operations Res 1961-68; Chrmn Guidelines for Prof Practice 1971; Geo E Kimball Medal 1974. Cornell College: Bd of Trustees 1964-; Pres 1970-71. Private Pilot. *Society Aff:* IIE

Cedroni, Ernest
Business: 735 Randolph, Detroit, MI 48226
Position: Gen Mgr. *Employer:* Detroit Metro Water Dept. *Education:* BSEE/Elec Power/Detroit Inst of Tech. *Born:* 3/4/25. in Italy, raised in Detroit, MI. Served in US Army, 1943-1946. Employed by Detroit Water Dept since 1950. Developed control systems for various water treatment operations, designed a central control system to monitor & remotely control numerous pumping stations, designed a sewer monitoring system to monitor levels, storm overflows & rainfall. Proj Engr responsible for the design & construction of the Lake Huron Water Treatment Plant. Assumed position of Gen Mgr of the Dept in 1974. Chrmn of Mgt Div of AWWA, 1965. Chrmn of Stds Council of AWWA, 1978. *Society Aff:* AWWA, WPCF.

Cermak, Jack E
Home: 407 E Prospect St, Ft Collins, CO 80525
Position: Univ. Distinguished Professor *Employer:* CO State Univ. *Education:* PhD/Engrg Mech/Cornell Univ; MS/Hydraulics/CO State Univ; BS/Civ Engrg/Co State Univ. *Born:* Sept 1922. Educator, researcher & consultant in fluid mechanics & wind engg. With CO State Univ since 1949 and serves as Univ Distinguished Professor, 1985-present. NATO Post-doctoral Fellow, Cambridge Univ, 1961. Established Fluid Dynamics & Diffusion Lab in 1959 & served as Lab Dir & Prof-in-Charge, Fluid Mech & Wind Engg Prog, 1959-85. Developed unique meteorological wind tunnel for simulation of the atmospheric boundary layer & significantly advanced physical modeling of wind forces on structures, diffusion over urban areas & behavior of plumes emitted from industrial stacks. Organized Engg Sci Maj at CO State Univ & served as first Chrmn, 1962-73. Officer US Army Field Artillery, 1943-46, retired as Lt Col Ordnance Corps Reserve, 1971. Chrmn, Fluid Dynamics

Cermak, Jack E (Continued)
Comm, Engg Mech Div, ASCE, 1960-61. Chrmn, Mech Div, ASEE, 1966. Chrmn, Rocky Mtn Sec, AIAA, 1969. Chrmn, Comm on Natural Disasters, NAE, 1981-1983. Pres, CO State Univ Res Found, 1968-72. Founding Mbr & Pres, Wind Engrg Res Council, Inc 1970-present. Chmn Fifth Intl Conf Wind Engrg 1979. Conslt to major engrg & arch firms on wind effects on structures & air pollution. Mbr Editorial Adv Bd for Ind Aero Abstracts & Regional Ed, Intl Journal of Wind Engineering ASME Freeman Scholar 1974. Natl Lecturer Sigma Xi 1976-77, Distinguished Lecturer American Society of Mechanical Engineers, 1987-present, Pres, Cermak Peterka Petersen, Inc., 1983-present. *Society Aff:* ASCE, ASME, AIAA, NAE, AGU, AMS, APCA, ASHRAE, ASEE, AAAS, AAM, ACI.

Cermola, Joseph A
Business: 99 Colony St, Meriden, CT 06450
Position: Pres. *Employer:* Cardinal Engrg Assocs Inc *Education:* BSCE/Civil/Ind Inst Tech *Born:* May 9 1930 Derby CT. BSCE IN Inst of Tec 1952. Hgwy engr for State of IL; Resident Engr Madigan-Hyland, Cons Engrs; Partner & Pres Cardinal Engrg Assocs Meriden CT. P Pres CT Engrs in private practice & CT Assn of Land Surveyors. Mbr CT SCE, ACSM, ACEC, NSPE, WPCF, Am Arbitration Assn. Reg PE CT, MA, NY. Chrmn-State of CT Bd of registration for PE and Land Surveyors. *Society Aff:* NSPE.

Cernica, John N
Home: 611 Plymouth Dr, Youngstown, OH 44512
Position: Prof, Civil Engr *Employer:* Youngstown St Univ *Education:* Ph.D./CE/Carnegie Mellon Univ; M.S./CE/Carnegie Mellon Univ; B.S. /CE/Youngstown St Univ *Born:* 5/14/32 Native of Romania. Prof at Youngstown St Univ from 1957 to present. Dept chrmn from 1959 to 1975. Author of several textbooks: "Fundamentals of Reinforced Concrete–, 1964, Addison Wesley Publishing Co.; "Strength of Meterials–, 1967, Holt, Rinehart & Winston; "Strength of Materials" SI Units, 1977, Holt, Rinehart & Winston; "Fundamentals of Geotechnical Engrg–, 1981, Holt, Rinehart & Winston. Consultant to Industry on Geotechnical Problems. Owner of Engrg Firm: "J.N. Cernica Assocs–. Married to Patricia Marinelli. Daughters, Kathy, M. Jude, Alice, Johanna, Tricia & Sarah. Listed in Who's Who in America; Who's Who in Midwest; American Man of Science; Distinguished engr Award, St of OH, 1964. Registered PE, Sts of FL, SC, IA, MI, OH, PA, IN, MD, NY, NC, NJ, VA, WV, KY. *Society Aff:* OSPE, ASCE, ACI, ASEE, NSPE, MVSPE, ΣΞ, ΦΚΦ, ΣΤ

Cerny, Rodney A
Business: 1 Erieview Plaza, Cleveland, OH 44114
Position: VP Engg. *Employer:* H K Ferguson Co. *Born:* June 28, 1925 Cleveland OH. BSCE Cornell Univ; grad studies for MS San Engg Case inst of Tech. US Marine corps 1943-46 & 1950-51. Employed by the H K Ferguson Co since 1947 except for 1953-56 Havens & Emerson Ltd. Elected VP Engg Jan 1974. Formulates & monitors engg policies & activities of co. Testifies as expert witness in spec tax & pollution litigation. Mbr: ASCE, Cleveland Engg Soc, NSPE, WPCF.

Chabries, Douglas M
Business: Dept of Electrical Engrg, Provo, UT 84602
Position: Prof *Employer:* Brigham Young Univ *Education:* PhD/EE/Brown Univ; MS/EE/CA Inst of Tech; BS/EE/Univ of UT *Born:* 2/18/42 Native of Los Angeles, CA. Attended the CA Inst of Tech under a Navy Lab Graduate Academic Program. After graduating from Brown Univ in 1970, returned to the Naval Ocean Systems Ctr, San Diego, CA and concentrated research activities on underwater acoustic scattering and diffraction theory. In 1973 research activities focused on digital signal processing. Served as head of Advanced Methods then the Electronics Branch from 1969 to 1977, then assumed responsibility of Weapons Tech Div at NOSC until joining the faculty of BYU in 1978. Current research activities center around adaptive signal processing, image processing, & speech processing. Serves as chairman of the department of Electrical & Computer Engineering. Enjoy fishing and hunting. *Society Aff:* IEEE, ΣΞ, ΦΚΦ, ΤΒΠ, ΗΚΝ

Chacey, Lloyd Adair
Home: 191 Chacey Ln, Worthington, OH 43085
Position: Retired *Education:* BCSE/Civil Engg/OH Northern Univ. *Born:* 10/10/99. Native of Topeka, KS. Prin Methodist Mission School, Malaysia, 1927-29. Starting 1930, Engg for Dept of Highways OH; 1943 part-time Secy for OH SPE, 1945 full-time; 1964 Dev 1st statewide Engg Foundation with an innovative & creative prog; 1968-79. Exec Dir Engg Foundation of OH - 1971 Engg Citation & Award redip OSPE; successfully promoted the Order of the Engg in OH & the nation. Secy of Natl Bd of Govs of Order of the Engg 1970-1980.

Chaddock, Jack B
Home: Ten Learned Pl, Durham, NC 27705
Position: Prof & Chrmn. *Employer:* Duke Univ. *Education:* ScD/Mech Engr/MIT; SM/Mech Engr/MIT; BS/Mech Engr/WV Univ; BS/Naval Sci/Univ of SC. *Born:* 12/6/24. US Naval Reserve 1943-46, service in S Pacific as Navigator, USS Rotanin. Asst Prof, MIT (Mech Engr), 1953-57; Assoc Prof, Renselaer Poly t (Mech Engr), 1957-59; Prof and Assoc Dir Herrick Labs, Purdue Univ, 1959-66; Sr Consultant, Carrier Corp, 1966; Prof Chrmn, Duke Univ (Mech Engr Mtls Sci) 1966-86; Assoc. Dean for R&D, School of Engr, 1987- . Fulbright Lecturer, Finland Inst of Tech, Helsinki, Fin, 1955-56; Visit Sr Res Assoc, CSIRO Mech Div, Melbourne, Australia 1973; Visiting Prof, Univ of New South Wales. Sydney, Aust, 1973. Visiting Prof, Building Research Establishment, Watford, U.K., 1980. Visiting Sr. Scientist, Lawrence Berkeley Lab, U. of CA, 1986-87. Phi Beta Kappa, Tau Beta Pi, ASHRAE Awards: Diamond Key, 1958 (for Outst Publ), Fellow 1971, Distinguished Service Award 1970, E K Campell Award of Merit, 1972 (for excell in teaching). Profl activities: President, ASHRAE 1981-82, Chrmn, Prof Engrs in Ed, Prof Engrs of NC, Pres of John B Pierce Fnd, New York, NY, General Chrmn, Intl Symposium on Moisture & Humidity, 1985. *Society Aff:* ASHRAE, ASME, NSPE, IIR, ΣΞ.

Chadwick, George F
Home: 133 Pin Oak Drive, Buffalo, NY 14221
Position: Sr. Project Engineer *Employer:* Ethox Corp. *Education:* MS/Chemistry/Pennsylvania State Univ; BA/Chemistry/Univ of Buffalo. *Born:* July 6 1930. MS Penn State, BA Univ of Buffalo. Native of Buffalo NY. Synthetic fluids res at Penn State's Petrol Lab 1951-54. Basic Plastics Res at Durez (Hooker) 1954-57. With Airco 1957-77. Respon for product & process development for electronics div. 1977-85 Process Eng Consultant in chemical & environmental areas. 1986-present, Ethox Corp. Respon for process & product engng studies for medical devices. Fellow PRI (Great Britain 1973); Mbr ACS, AICHE, SME-Rbotics and ASTM. Active churchworker. Enjoy photography & good music. *Society Aff:* ACS, AIChE, SME-Robotics, ASTM.

Chadwick, Wallace L
Business: Suite 239, 250 W First St, Claremont, CA 91711
Position: Independent Consultant. *Employer:* Various. *Education:* D Engg Sci/-/Univ of Redlands. *Born:* 12/4/97. Various engg capacities for S CA Edison Co 1922-31; design and construction CO River Aqueduct, Metropolitan Water Dist S CA 1931-37; Civ engr, Chief Civ Engr, Mgr of Engg, VP of Engg and Construction Southern CA Edison Co, 1937-62. Independent consultant 1962 to date, including, Southern CA Edison, California Department Water Resources, Churchill Falls (Labrador) Corp, James Bay Energy Corp., Bechtel Companies, Harza Engrg, SNC, Hydro Quebec. Chrmn, Bd Consultants James Bay Comples, Jubail Review Bd, Independent Panel Teton Dam Failure and others. Various awards, including ENR 1978. *Society Aff:* ASCE, IEEE, ASME, NAE.

Chae, Yong S
Business: Coll of Engrg, P.O. Box 909, Piscataway, NJ 08854
Position: Prof & Assoc Dean *Employer:* Rutgers Univ *Education:* PhD/CE/Univ of

Chae, Yong S (Continued)
MI; MS/CE/Dartmouth Coll; AB/CE/Dartmouth Coll *Born:* 7/29/30 Taught at Rutgers since 1964. Presently, Prof of Civil Engrg and Assoc Dean of Engrg. Extensive research experience for the US Government agencies and the NJ Dept of Environmental Protection. Published numerous articles in the areas of geotechnical engrg, soil dynamics and earthquake engrg. Consultant to a number of consulting firms and government agencies. Active in the professional society committee activities, NATO Senior Fellow in Science 1975. *Society Aff:* ASCE, ASEE, NSPE, TRB

Chaklader, Asoke C D
Business: Dept of Met Engg, Vancouver BC V6T 1W5
Position: Prof. *Employer:* Univ of British Columbia. *Education:* PhD/Ceramic Engg/Leeds Univ; Grad diploma/Ceramic Engg/College of Ceramic Tech; BSc/Natl Sci/Calcutta Univ. *Born:* 9/1/30. Born in Bamrail, India on Sept 1, 1930. Schooled in several States of India. Obtained BSc degree with Physics, Chemistry and Math from the Calcutta Univ in 1949. Obtained a Grad Diploma in Ceramic Engg from the College of Ceramic Tech (Formerly Bengal Ceramic Inst) Calcutta in 1951 with First Class. Worked in the Natl Met Lab during 1951-54. Joined the Univ of Leeds, Leeds, UK in 1954 and obtained PhD in 1957. Awarded the Sr Studentship of the Royal Commission for the Exhibition of 1851 for the period 1957-59. Since Nov 1, 1959 at the Univ of BC, Vancouver. Published 70 papers and has 3 patents in the USA. *Society Aff:* ASM, ACS, NICE, CCS, BCS.

Chakrabarti, Subrata K
Home: 191 E Weller Dr North, Plainfield, IL 60544
Position: Dir, Marine Res. *Employer:* CBI Research Corp. *Education:* PhD/Engg Mechanics/Univ of CO; MS/ME/Univ of CO; BS/ME/Jadavpur Univ. *Born:* 2/3/41. I received my BS degree in 1963 from the Jadavpur Univ, India and MS & PhD's in 1966 and 1968 respectively from the Univ of CO. My previous experience includes design of power plants & piping for Simon Carves Ltd and Kuljian Corp in India. I taught for one yr at the Univ of CO. Since 1968 I am employed by the Chicago Bridge & Iron Co, Plainfield, IL where I hold the position of Dir of Marine Res Div. Previously, I was the hd of the analytical group in this div. I am involved in the design, dev & testing of offshore structures and have over 100 scientific publications. I am a reviewer for NSF, ASCE, MTS, JPT & ASME. I received the 1974 James Croes medal and 1979 Freeman Fellowship from ASCE; Outstanding New Citizenship Award in 1982. I am a reg PE in IL. I received the 1984 Ralph James Award from ASME. I am Vice Chmn of the OMAE Committee of ASME & the Technical Editor of the Journal of OMAE. My textbook on "Hydrodynamics of Offshore Structures" will be published in 1987 by CML Publications. *Society Aff:* ASCE, ΣΞ, AAAS, ASTM, ASME

Chakrapani, Durgam G
Business: 2233 SW Canyon Rd, Portland, OR 97201
Position: VP *Employer:* MEI-Charlton, Inc *Education:* PhD/Metallurgical Engrg/Univ of IL; M/Foundry Engrg/Indian Inst of Sciences, India; B/ME/Univ of Madras, India *Born:* 11/28/40 Engrg education in India included industrial training in Foundries. Following graduation in India, was awarded post graduate fellowship to under go advanced training at various metallurgical institutions in Germany. After returning from Germany, as a technical mgr for a non Ferrous Foundry, was responsible for developing Foundry and finishing processes. Pursued Doctoral work at the Univ of IL in metallurgical engrg with the thesis emphasis in stress-corrosion cracking. Was mgr of corrosion control and water treatment lab at the Univ. Currently responsible for corrosion and metallurgical projects at MEI-Charlton. Registered PE in the states of OR, CA and WA. Authored or co- authored over 30 publications on stress-corrosion cracking, hydrogen embrittlement and corrosion control methods. *Society Aff:* AIME, ASM, ASTM, NSPE, NACE, ECS, ASCE, TAPPI

Chalabi, A Fattah
Business: Institute Rd, Worcester, MA 01609
Position: Prof *Employer:* Worcester Polytech *Education:* PhD/Structural, Construction/Univ of MI; MSc/CE/Univ of MI; BSc/CE/Univ of Baghdad *Born:* 04/12/24 Civil Engr, educator, Married Beatrice Austin Oct 14, 1956. Civil Engr Ministry of Public Works Iraq 1946-49, Structural Engr (Ayres, Lewis, Norris and May 1951-56, Asst Prof Univ of Baghdad 1956-59, Structural Engr (TAMS 1956-59) Asst Prof of Civil Engrg, Worcester Polytech Inst 1959, Prof of Civil Engrg 1966, George I. Alden Prof of Engrg 1978. Reg PE, MA. *Society Aff:* ASCE, ASEE, ACI, XE, ΣΞ

Chalker, Ralph G
Business: 8900 DeSoto Ave, PO Box 309, Canoga Park, CA 91304
Position: Sr Stds Advisor. *Employer:* Energy Systems Group, Rockwell Intl. *Education:* BS/ME/Univ of UT. *Born:* 7/27/18. Sr Technial Advisor reporting to the VP of Engg. Directed the Engg design and fabrication of many res, dev and power nuclear reactor plants including the SNAP 10 nuclear powered satellite. Prog mgr for first nuclear reactor sold on the commercil market. Developed and implemented std design practices and engg mgt procedures for nuclear systems. Performed analysis and experiments which data resulted in approval to locate nuclear testing and reactor operations in Santa Susana mtns near metropolitan Los Angeles. Extensively involved in co industry, Natl & Intl Nuclear Stds. Chrmn of the Am Natl Stds Inst (ANSI) Nuclear Technical Advisory Group & Exec Committee. *Society Aff:* ASME, ANS.

Chalmet, Luc G
Business: 303 Weil Hall, Gainesville, FL 32611
Position: Asst Prof. *Employer:* Univ of FL. *Education:* Doctorate/Mgt Engg/Cath Univ of Louvain; MS/Mgt Engg/Cath Univ of Louvain; MS/Civ Engg/Rijks Univ Ghent. *Born:* 1/11/49. Citizen of Belgium. Software engr for CIG, Brussels 1972-1973. Teaching asst Louvain 1973-1975. Current position since 1978. Teaching Production Control, Facilities design, Operations Res, Engg Economy, & Markov Processes. Res in multi-objective location & layout, network modeling of bldg evacuation, and vehicle dispatch problems. Officer of IIE North FL Chapter. Enjoy music and photography. *Society Aff:* IIE, ORSA, TIMS.

Chamberlain, James E
Home: 13454 Ronnie Way, Saratoga, CA 95070
Position: Cartographer; Retired. *Employer:* US Geological Survey. *Education:* BS/Forest Engg/Univ of MT. *Born:* 8/14/28. Mr James E Chamberlain received his BS degree from the Univ of Montana in 1951 and has been employed by the Topographic Div, US Geological Survey since that time. He worked in the Pacific Region form 1951-59 on field and photogrammetric surveys. In 1960 he was transfered to the staff in Washington, DC. He served in both the Office of Research and Technical Standards and the Office of Plans and Program Dev. In 1968, Mr Chamberlain returned to the Western Mapping Center hdquarters in Minlo Park, CA, as a Dist Engg in the Banch of Field Surveys. In 1975 he was appointed Spec Asst for Mapping Requirements, and in jan 1978, he was named Chief, Branch of Plans and Production. Mr Chamberlain is a Fellow Mbr of the Amer Congress on Surveying and Mapping and has served on the Board of Dirs. He is also a mbr of the Amer Soc of Photogrammetry & also served on that Bd of Dirs. He served as a Dir of Cartography Div of ACSM & as the prog Chrmn for the 1967 Annual meeting. In 1971 he was the Dir of San Francisco Convention of ASP-ACSM, and in 1978 he was Chrmn of the third Intl Symposium on Comp Assisted Cartography. *Society Aff:* ASCE, ACSM, ASP, CLSA.

Chamberlin, Paul D
Home: 10427 W Arkansas Pl, Lakewood, CO 80226
Position: Staff Met. *Employer:* Occidental Minerals Corp. *Education:* MBA/Bus/AZ State Univ; BS/Met Engg/MI Tech Univ. *Born:* 1/28/37. Saginaw, MI; s Marshall Vernon & Ruth Carrier (Davis) C; BS in Met Engrg with honors, MI Tech Univ, 1959; MBA, AZ State Univ, 1968; m Dixie Ann Coughran, Aug 9, 1958; children - Michael S, Julie A. res engr US Steel Corp, MN, 1959-60; res engr & mill met, Anaconda Co, Butte, MT, 1962-64; res engr & smelter met, Inspiration Consol Copper Co, 1964-67; process engr, Hazen Res, Golden, CO, 1968-73; sr met Continental Oil Co, Denver, 1973-1979; staff met, Occidental Minerals Corp, Denver, 1979-. Served with US Army, 1960-2. Reg PE, CO Mbr AIME, Am Assn Cost Engrs (certified cost engr; assn dir, 1974-76; dir certification bd, 1976-). Republican. Office: Occidental Minerals Corp, 777 S Wadsworth, Bldg 4, Lakewood, CO 80226. *Society Aff:* AIME, AACE.

Chambers, Carl C
Home: 322 Estero Woods Village, Estero, FL 33928
Position: Emeritus Univ Prof. *Employer:* Univ of PA. *Education:* ScD/EE/Univ of PA; BS/Math/Dickinson College. *Born:* 5/8/07. RCA Vacuum Tube Engr 1929-32; Bartol Res Rdn 1932-33 through all teaching ranks from Instr in EE to Univ Prof of Engg and admin Dir of Res, Supv of Res of Moore Sch. Dean of Moore Sch, VP of Univ for Engg and Res, 1933 to date. (Inactive since 1975). *Society Aff:* AAAS, ASEE, IEEE, NSPE, NAE.

Chambers, Carlon C
Business: 526-20 1/4 Road, Grand Junction, CO 81503
Position: Pres. *Employer:* Technology Management, Inc. *Born:* 4/4/37. Early career included fifteen yrs with American Gilsonite Co, a Standard Oil of CA subsidiary, progressing through the ranks to Tech Asst to the Refinery Mgr followed by three yrs as Sn Process Engg with Occidental Oil Shale, Inc. Pres of Tech Management, Inc since 1973 where respon include directing the consulting engg and contract R&D efforts of the co. Reg PE in CO. 1979-80 Pres of Ute Chapter, Prof Engrs of CO, NSPE and 1982 Chairman of CO Plateau Section of the Society of Mining Engg of AIME. Twenty-six patents in a broad range of fields. *Society Aff:* AIChE, NSPE, SME-AIME, SPE-AIME, AAAS, ASTM.

Chambers, David S
4810 Old Hickory Blvd, Hermitage, TN 37076
Position: President *Employer:* Industrial Controls, Inc. *Education:* MBA/Statistics/Univ of TX, Austin; BA/Mathematics/Univ of TX, Austin *Born:* 01/26/17 From 1958-81, Prof of Statistics, Univ of TN, Knoxville. Since 1981, Pres & CEO, Industrial Controls, Inc, Nashville. Executive Secretary, Pres (1970) and Bd Chrmn (1971) of the American Soc for Quality Control. Past Mbr, Bd of Dirs, EJC. Academician, International Academy for Quality Work in Productivity and Quality in the US and Throughout the Free World. *Society Aff:* ASQC, ASA

Chambers, Fred
Home: 13815 Britoak Ln, Houston, TX 77079
Position: Consultant (Self-employed) *Education:* BS/EE/Auburn Univ. *Born:* 2/17/12. GE 1930-32; TN Elec Power 1933-39; TN Valley Auth 1939-71 in var pos in power oper and plan and engg, made respon for power supply planning and for planning, Engg and Const. of transmission sys in 1963, made asst mgr of power in 1970; Bovay Engg 1971-73; NY state PSC 1973-74; Consultant Houston, TX 1974-. Who's Who in America, Who's Who in the World; Fellow IEEE, Chrmn East TN Sec 1959-60; Named Engr of decade by TSPE Chattanooga Chapt 1970; Mbr CIGRE; Auth technical papers and served on var comm of prof socs and natl and intl study groups. Served with USNR 1943-45. *Society Aff:* IEEE, CIGRE, PES.

Champagne, Joseph S
Business: 370 Second St, PO Box 1001, Troy, NY 12180
Position: Prin. *Employer:* Champagne Assoc. *Education:* MSCE/Civil Engg/Rensselaer Poly Inst; BSCE/Civil Engg/Rensselaer Poly Inst; /Traffic Engg/Yale; Bus Adm/NY, Univ, NY. *Born:* 8/27/33. Mr Champagne began his career with the NYSDOT. During this time he was in charge of the traffic control program and worked on various highway design projects. He also served as chief engg for a private firm in charge of planning, design and installation of traffic control devices. He has held the position of Sr Traffic Engg for the Port of NY Authority and was in charge of traffic studies, research and safety section and worked in planning and design of port facilities. As a principal and owner of his own firm during the past 12 yrs, he has been involved in over 600 transportation and traffic engg projects including urban and mass transportation planning, appraisal and design of large traffic generators, shopping centers, convention centers and traffic signal design projects. He has also provided expert testimony in both planning & zoning boards for municipalities, & in court for plaintiffs and defense of county & state govts. *Society Aff:* ASCE, NSPE, ITE.

Chan, Shih H
Business: Dept of Mech Engg, Milwaukee, WI 53201
Position: Prof & Chrmn. *Employer:* Univ of WI. *Education:* PhD/ME/Univ of CA; MS/ME/Univ of NH; Dipl/ME/Taipei Inst of Tech. *Born:* 11/8/43 in Taiwan. Came to the US for grad studies in 1964. Received PhD from Univ of CA, Berkeley, in 1969. Taught at NYU as Asst & Assoc Prof in 1969-1972 and at Poly Inst of NY as Assoc Prof in 1972-74. Res Engr at Argonne Natl Lab in 1974- 75, specializing in heat transfer and breeder reactor safety. Reg PE in 1978. With Univ of WI-Milwaukee since 1975. Prof 1978 & chrmn of ME Dept in 1979. *Society Aff:* ANS.

Chan, Shu-Park
Business: EECS Dept, Santa Clara, CA 95053
Position: Nicholson Family Professor. *Employer:* Univ of Santa Clara. *Education:* PhD/EE/Univ of IL; MS/EE/Univ of IL; BS/EE/VA Military Inst. *Born:* 10/10/29. Canton, China. Is presently Nicholson Family Professor of the Elec Engg & Comp Sci Dept at the Univ of Santa Clara. Is author of "Introductory Topological Analysis Elec Networks–, co-auth of "Analysis of Linear Networks & Systems: a Matrix-Oriented Approach with Computer Applications" and "Introduction to the Applications of the Operational Amplifier–, and ed & major contributor of "Network Topoloty & Its Engineering Applications" 1975. Mbr Tau Beta Pi, Eta Kapp Nu, Pi Mu Epsilon, Phi Kappa Phi, & Sigma Xi hon soc. Pres of Chinese Arts & Culture Inst. Former VP & Trustee, Triton Museum of Art, Santa Clara, CA; former mbr of Exec Bd, deSaisset Art Council, Univ of Santa Clara; recipient of Pres's Special Recognition Award for Outstanding teaching, research, and leadership, 1978-79, University of Santa Clara; Honorary Professor at University of Hong Kong, 1980-81; Visiting Professor, Academia Sinica, Peking, China, Summer 1980; Tamkang Chair, Tamkang University, Taiwan, June 1981. Awarded Nicholson Family Professorship, 1987- . *Society Aff:* IEEE, ASEE.

Chandler, John P
Business: Murray Hill RD, Box 74, Hill, NH 03243
Position: Consulting Engineer *Employer:* Self *Education:* SM/Civil Engrg/Harvard Univ; MS/Math/Rennselaer Polytech Inst; MS/Bus Admin/George Washington Univ; BS/Gen Engrg/US Military Academy *Born:* 5/18/26 Mbr: National Soc of Professional Engineers. Retired from US Army in June 79. VP and Mgr Hydropower, Anderson-Nichols & Co. 1979-80. Lecturer, Plymouth State College 1981- . *Society Aff:* NSPE, SAME

Chandler, R L
Business: 5930 Beverly, Mission, KS 66202
Position: General Manager; Retired *Employer:* Water Dist 1, Johnson Cty KS. *Education:* BS/Civ Engrg/KS St Univ *Born:* BS Civ Engr KS St Univ 1949. Lic PE KS. City Engr Newton, KS. Ch Engr Kansas City Suburban Water Co Inc. Ch Engr water Dist No 1 of Johnson Cty KS. Gen Mgr Water Dist NO 1 of Johnson City KS 1959. Mbr AWWA. P Chrmn KS Sect AWWA. P Mbr Bd of Dir & Exec Council AWWA. Recipient AWWA Geo Warren Fuller Award. Mbr ASCE. Cons to US/AID Philippines 1970-. "Provincial Water Supply-Pre-feasibility Study-Philippines–. Governor's Task Force on Water Resources (Kansas) 1977-78, Pres Kansas River Alliance 1984. First Vice-Pres Mo-Ark Assoc 1984-85. Retired January 1987. *Society Aff:* AWWA, ASCE

Chandra, Suresh
Business: Sch of Engg, Greensboro, NC 27411
Position: Dean-Sch of Engg. *Employer:* NC Agri & Tech St Univ. *Born:* July 1939. PhD CO St Univ; M Chrmn Engg Univ of Louisville; B Sc-Chem E Banaras Hindu Univ India; B Sc Univ of Allshabad India. Asst Prof Mech Engg Univ of Miami 1966-71. Prof & Chrmn Mech Engg 1971-74; Acting Dean Sch of Engg 1974- 76 NC A&T State Univ. Natl VP Pi Tau Sigma, mech engg hon frat, since 1970. Mbr Am Soc for Engg Ed, Am Soc of Mech Engrs, Am Geophysical Union. Num presentations & pubs in heat transfer & fluid mech areas. Mbr US delegation to First Spec Assembly of Intl Assn of Meteorology & Atmospheric Phys Melbourne Australia 1974. Hon Mbr Tau Beta Pi & Pi Tau Sigma. Enjoy music & tennis.

Chang, Anthony T
Business: Castle Point Station, Hoboken, NJ 07030
Position: Prof *Employer:* Stevens Inst of Tech *Education:* Dr-Ing/Applied Mech/ Tech Univ Berlin; Dipl-Ing/ME/Tech Univ Darustadt *Born:* 10/25/37 *Society Aff:* ASME

Chang, Chin-Hao
Business: P.O. Box 2908, Tuscaloosa, AL 35487
Position: Prof. *Employer:* Univ of AL. *Education:* PhD/Engg Mechanics/Univ of MI; MS/Engg Mechanics/VPI & State Univ; BS/Civ Engg/Natl Taiwan Univ. *Born:* 7/2/26. in Hainie, Zejiang, China. Structural engr, Taiwan Power Co, 1953-1955. Asst Prof, Assoc Prof and Prof of Engg Mechanics, the Univ of AL since 1960. Visiting Prof in Dept of Civ Engg, Natl Taiwan Univ, 1966. Specialized in Structural Mechanics. Res was supported by NSF; Marshall Space Flight Ctr, NASA; and US Army Missile Command, Redstone Arsenal, AL. Technical papers were published in profl journals. Reviewer of the Shock and Vibration Digest. Father, Fou-Kon Chang; Mother, Qau-Pau Ton, two sisters, one brother, married to Kathy Chun-Er Yang; Son, Yu Yang. *Society Aff:* ΣΞ, AAM

Chang, George C
Home: 29615 Lincoln Rd, Bay Village, OH 44140
Position: Prof of Aeronautical Engg & Dean of School of Graduate Studies and Research *Employer:* Embry-Riddle Aeronautical Univ *Education:* PhD/CE/Univ of IL; MS/CE/Univ of IL; BS/CE/Natl Cheng Kung Univ-Taiwan *Born:* 08/23/35 During 1966-69, I worked for Boeing Aerospace Co on missile sys structures. Performed res and teaching duties at the US Naval Acad prior to joining US Energy Res & Development Admin 1975. Also made a major consltg effort in analyzing INTELSAT IV spacecraft airframe design for COMSAT during 1969-73. Was responsible for the initiation and admin of a natl prog of energy res valued at $21,000,000. Significant achievements included energy storage technologies applicable to solar photovoltaics, wind sys, and elec utility peaking operation. 1979-87, I worked as Professor and Associate Dean of Engg at Cleveland State Univ. Have been an Assoc Dir of the NASA-Univ Joint Inst for Aerospace Propulsion and Power since July 1982. Assumed current position in August 1987. Enjoy reading and traveling. *Society Aff:* ASEE, AIAA, ASME, ΣΞ, AAM.

Chang, Herbert Y
Business: Naperville, IL 60540
Position: Department Head *Employer:* Bell Labs. *Born:* Nov 1937 Shanghai China. BS, MS & PhD in EE from Univ of IL with Bell Labs since 1964. Past experience incl: Fault-tolerant ESS-Electronic Switching Sys-design & anal, microprogrammed processor dev & dev of software tools and software engineering techniques. Current respon deals with dev of computer aided design tool for aiding hardward design. Outstanding New Citizen Award Chicago 1972; IEEE Fellow. Enjoys music, photography & tennis.

Chang, Hsu
Business: T J Watson Res Ctr, Yorktown Heights, NY 10598
Position: Res Staff Mbr. *Employer:* IBM Corp. *Born:* Feb 3,1932 Yangchow China; Naturalized US citizen 1967. BS Natl Taiwan Univ 1953; MS & PhD Carnegie-Mellon Univ 1957-59. With IBM Res since 1959 spec in magentics memories, VLSI. Served as Mgr & Tech Cons. Many patents & papers on magnetic films & bubble devices, several books including, " Magnetic Bubble Tech" book for IEEE Pres 1975. Current int in VLSI implementation of computer arch. Fellow IEEE, served as Ed-in-Chief IEEE Trans on Mag 1970-72, Tech Prog Chrmn for Intl Magnetics Conf 1970, 71, 74 Prof. Prgg chrmn for Electro 1981. Adjunct prof at Natl Taiwan Univ 1972-74 Carnegie-Mellon Univ 1964-67, 1979-now.

Chang, John C
Home: 575-19 Otay Lakes Rd, Chula Vista, CA 92010
Position: Sr Met. *Employer:* Rohr Ind. *Education:* MS/Met/Univ of CA; BS/Mining & Met/Chiao-Tung Univ. *Born:* 4/19/19. Responsible for res & dev progs in metals res; involving titanium, nickel, aluminum & iron alloys; author of Rohr's Mtls Manual & more than 50 technical papers in diversified subjs in met for various organizations & magazines; presented papers in both Intl Conf of Titanium in London, 1968 and in Boston 1972, instr of profl courses on "Titanium & Titanium Alloys–, "Ferrous Metallurgy" –Alluminum and Aluminum Alloys" & "Engineering Materials" in Southwestern College, Chula Vista, CA; Profl Met Engr in CA; Chrmn of ASM, San Diego, 1971-1972. *Society Aff:* ASM.

Chang, Nai L
Business: 1043 E. Southriver St, Appleton, WI 54912
Position: Assoc Prof of Engrg *Employer:* Inst of Paper Chemistry *Education:* MS/ChE/Columbia Univ; BS/Chem/Chinese Natl SW Assoc Univ-China *Born:* 08/05/22 Naturalized US citizen. Was born in Peking, China. Production mgr of Kaohsiung Ammonium Sulfate Works, Inc, 1953-1957, and Div Head of Planning Div, 1957-59. Was an Asst Prof of Engrg, 1966-70, have been an Assoc Prof of Engrg and Res Assoc since 1970 at the Inst of Paper Chemistry. Courses taught: Mass & Energy Balances, Transport Processes, and Dynamics of Papermaking, and etc.

Chang, Richard K
Business: Yale Univ, Applied Physics and Elec Eng, PO Box 2157 Yale Station, New Haven, CT 06520
Position: Prof of Applied Physics And Elec Eng *Employer:* Yale Univ. *Education:* PhD/App Phys/Harvard; MS/App Phys/Harvard; BS/Elec Engrg/MIT. *Born:* 6/22/40. Res interests: Light scattering from microparticles, optical diagnostics for combustion & turbulence studies, nonlinear optics, solid-state physics, and surface physics/science. *Society Aff:* OSA, APS, IEEE.

Chang, Sheldon S L
Home: 5 Seaside Dr, Pt Jefferson, NY 11777
Position: Prof of Engg. *Employer:* State Univ of NY. *Education:* PhD/EE/Purdue Univ; MS/Phys/Natl Tsinghua Univ; BS/Phys/Natl SW Assoc Univ, China. *Born:* Jan 1920 Peking China. Res & Dev Engr Robins & Myers Inc Springfield OH 1948-52. Taught NYU 1952-63. State Univ of NY 1963-present; Dept Chrmn 1963-69; Acting Dean 1971; Leading Professor, 1983 Visiting MacKay Prof Univ of CA Berkeley 1969-70. Two books: Synthesis of Optimum Control Sys 1961 McGraw-Hill; Energy Conversion 1963 Wiley. Ed in Chief, Fundamentals Handbook of Electrical and Computer Engrg, three volumes, Wiley, 1983. Twelve patents. Over 100 articles in control sys theory, info theory, elec machinery, network theory, computers, electronics, & econ. Cons to Indus. Fellow IEEE. *Society Aff:* IEEE

Chang, Tsong-how (Phil)
Business: Dept of Industrial & Systems Engrg, Milwaukee, WI 53201
Position: Assoc Prof *Employer:* Univ of WI-Milwaukee *Education:* PhD/ME/Univ of WI-Madison; MS/IE/WV Univ; BS/ME/Nat'l Taiwan Univ *Born:* 10/11/29 Quality Control Engr with Taipei Cotton Mills, Taiwan 1953-55; operations research analyst with Honeywell Co. 1960-61. Taught industrial engrg at MS State Univ 1963-66. On faculty of Univ of WI-Milwaukee since 1968. Appointed advisory Prof of

Chang, Tsong-how (Phil) (Continued)
Shanghai Inst of Mechanical Engrg in 1979 for the development of a Master program in industrial systems engrg in China. Areas of interest include statistical modeling and economic analysis of engrg problems. IIE WI Pres 1978, IIE Region XI Dir 1979-81. *Society Aff:* IIE, ORSA, ΣΞ, IIE, IIF

Chang, William S C
Business: C-014, La Jolla, CA 92093
Position: Prof. *Employer:* Univ of CA. *Education:* PhD/Elec Engg/Brown Univ; MSEE/Elec Engg/Univ of MI; BSEE/EE/Univ of MI. *Born:* 4/4/31. Born in China, 1931. Son of Mr and Mrs T W Chang. There are three children, Helen, Hugh, and Hedy. Wife's name is Margaret K Chang. Lecturer and res associate, Stauford Univ, 1957-59. Asst prof, Dept of Elec Engg, The OH State Univ, 1959-62. Assoc prof, the OH State Univ, 1962-65. Prof of Elec Engg, Wash Univ, St Louis, 1965-79. Chrmen of Elec Engg, Wash Univ, 1965-71. Samuel Sachs Pof of EE, 1976-79. Professor, Department of Electrical and Computer Engineering, University of California, San Diego, 1979-present. *Society Aff:* IEEE, APS, OSA.

Chang, Y Austin
Business: Dept of Metal and Min Engrg, 1509 University Avenue, Madison, WI 53706
Position: Prof and Chairman. *Employer:* Univ of Wisconsin-Madison *Education:* PhD/Metallurgy/Univ of CA; MS/Chem Engr/Univ of WA; BS/Chem Engr/Univ of CA. *Born:* 11/12/30 PE, State of WI. Industrial experiences include Stauffer Chem Co, 1956-59, Aerojet-Gen Corp, 1963-67 and Sandia, Summer, 1971. Taught at Univ of WI-Milw 1967-80; Res Specialty in thermodynamics, phase diagrams, and oxidation. Authored or co-authored more than 100 publications in the field including several books. Mbr of TMS-AIME (Bd of Dirs, 1978-81; Chrmn, Physical Chemistry Comm, 1975-76); Comm on Alloy Phases, 1974-76). Fellow and mbr of ASM (Trustee, 1981-84, and Chrmn, Thermodynamics Activity Comm, 1976-78) and mbr of NACE, and the Electrochemical Soc. Mbr of Sigma Xi, TAU Beta Pie, and Phi Tau Pi. Was voted by students in the College of Engr and Appl Sci to be an Outstanding Instructor in 1972, awarded as an Outstanding Educator of America in 1973, and a Community Lenders and Noteworthy American, 1976-77, Pres of Alpha Sigma Mu (Metallurgical and Materials engrg Honor Soc), 1984-85 and the Recipient of the 1984 Byron Bird Award for Excellence in a Res Pub, Univ of Wisconsin-Madison. *Society Aff:* NACE, TMS-AIME, ASM.

Chant, Raymond E
Business: Dept of Mech Engrg, Univ. of Manitoba, Winnipeg Manitoba, R3T 2N2 Canada
Position: Dir, Office of Ind Engg. *Employer:* Univ of Manitoba. *Education:* M Eng/ Thermal Power/McGill; B Eng/Thermal Power/McGill *Born:* 3/25/21. Prof Dept Mech Engg Univ of Manitoba 1953-pres. 1956-73, Hd Dept of Mech Engg Univ of Manitoba. 1973-74 sabbatical leave, Inter-Tech Corp Warrenton VA. 1974-1984 Dir, Office of Ind Res Univ of Manitoba. Also active mbr of Solar Energy Soc of Canada Inc-SESCI Chrmn of Bd 1977/78, ASME, ISES, ASHRAE, APEM, SESCI, Wpg Chamber of Commerce, FCSME & FEIC. In connection with the above, was Pres of Assoc of Prof Engrs of Province of Manitoba 1961-62; VP Engg Inst of Canada 1963-70; Pres Univ of Manitoba Chap of Sigma Xi 1970 and President Canadian Society of Mechanical Engineers 1981/82. *Society Aff:* ASME, ASHRAE, ISES, APEM, CSME.

Chao, Bei T
Home: 704 Brighton Dr, Urbana, IL 61801
Position: Prof of Mech & Ind Engg. *Employer:* Univ of IL, Urbana *Education:* PhD/ME/Univ of Manchester; BS/EE/Chiao-Tung Univ. *Born:* 12/18/18. Naturalized 1962. PhD Engg Univ of Manchester; BS highest honor, Natl Chiao Tung Univ China. Assoc Engr & Supr Tool & Gage Div, Central Machine Works 1939- 45 & assumed current position in 1955, Hd of Dept of Mech & Ind Engg 1975-86. Tech Ed, Journal of Het Transfer 1975-81, Mbr Ed Adv Bd, Int'l J. Heat and Mass Transfer, Int'l Communications in Heat and Mass Transfer, Numerical Heat Transfer. Boxer Indemnity Scholar 1945-48; ASME Blackall Award 1957, Heat Transfer Award 1971, Fellow 1974; ASME/AIChE Max Jakob Mem Award, 1983, ASEE Western Elec Fund Award 1973, Mech Engrs Div Outstanding Teacher Award 1975; Benjamin Garver Lamme Award, 1984, Fellow 1986; AAAS, Fellow 1984; Russell S Springer Prof of Mech Engg, Univ of CA, Berkeley 1973; Mbr US Engg Education Delegation to Visit PRC, 1978; Mbr National Academy of Engrg 1981, Academia Sinica 1986; Mbr Advisory Screening Comm. in Engg, Fulbright-Hayes Awards Program, 1979-81, Chrmn 1980, 1981. *Society Aff:* ASME, ASEE, NAE, AAAS, Academia Sinica.

Chao, Kwangchu
Business: School of Chem Engg, W Lafayette, IN 47907
Position: Prof. *Employer:* Purdue Univ. *Education:* PhD/Chem Engg/Univ of WI; MS/Chem Engg/ Univ of WI; BS/Chem Engg/ Chekiang Univ. *Born:* 6/7/25. in Chongqing, China. Came to the US in 1954. Married to Jiunying Su in 1953. Have three sons; Howard, Albert and Bernard. Since leaving Wisconsin in 1957 have been with Chevron Res Corp (1957-63). Il Inst of Tech (1963-64), and OK State Univ (1964-68). With Purdue since 1968. *Society Aff:* AIChE, ACS.

Chao, Paul C S
Business: 655 Third Ave, New York, NY 10017
Position: Chief Project Engr, Tarbela *Employer:* Tippetts-Abbett-McCarthy-Stratton *Education:* Ph.D./Engrg/Univ of London; D.I.C./Applied Fluid Mech/Imperial Coll, London; B.Sc./Hydraulic Engrg/Nat Central Univ, China *Born:* 7/21/17 Paul Che Shen Chao, PE, Ph.D. (U. of London 1948), Consulting Engr, Forest Hills, NY; concurrently Chief Project engr, Tarbela Dam Project and Staff Consultant, Tippetts-Abbett-McCarthy-Stratton, NYC. Born in Yibin, Sichuan, China. Came to U.S. 1949, naturalized 1954. Has 37 years experience with British and American Univ and Construction and Consulting Engrg firms involving total planning, design and supervision of construction of multi-purpose projects including 7 years on Shihmen Dam in Taiwan and 21 years on Tarbela Dam in Pakistan. Engrg Consultant to Yangtze Valley Planning Office of People's Republic of China. *Society Aff:* ASCE, USCOLD

Chao, Raul E
Home: 412 Henley Dr, Bloomfield Hills, MI 48013
Position: Chrmn Chem Engg Dept. *Employer:* Univ of Detroit. *Education:* PhD/Chem Engg/Johns Hopkins Univ; MS/Chem Engg/Johns Hopkins Univ; BS/ Chem Engg/Univ of PR. *Born:* 12/21/39. in Havana, Cuba. Process Engr with Exxon, 1964-1968. Chrmn, ChE Dept, Univ of PR, 1968-1977. Extensive consulting work with the Economic Dev Admin of PR, NASA, the Dept of Energy, the Water Resources Res Inst, the EPA, Rockwell Intl and others, in the fields of Process Dev, Economics and Mgt strategies. Exxon Inventor Award, 1970; mbr of TBP, Sigma Xi and Phi Kappa Phi. Serval articles, patents and one book published. Married in 1964 to Olga Nodarse, a Sch Psychologist. Two children: Raul Octavio and Maria Isabel. *Society Aff:* AIChE, ASEE, ACS.

Chapin, William F
Business: 923 Emerald Bay, Laguna Beach, CA 92651
Position: Retired *Education:* BS/ChE/Calif Inst of Tech. *Born:* Mar 1920. Good Hope, LA. BS CA Inst of Tech 1941. Alum Assoc Dir 1967- 70. Alum Fund Council 1973-76. Process engr Permanente Metals Corp 1941-44. With Fluor since 1944 involved in engg & const of process plants. Process Engr Los Angeles 1944-48; Ch Process Engr Houston 1948-57; Mgr Projects Los Angeles 1957- 62; VP-Process Engineering & Development 1963-69; VP- Engineering-1970-1972.VP Tech Serv & Subsidiaries 1972-73; VP Proj Mgt 1973-present. Var activities in AIChE incl Exec Comm & Chrmn 1972 of Fuels & Petrochem Div. Elected AIChE Fellow 1976. Hobbies incl hiking, swimming, & Golf, skiing & scuba diving. Retired 1982. *Society Aff:* AIChE.

Chapman, Clabe, Jr
Business: PO Box 1208, Valdosta, GA 31601
Position: Agricultural Engr. *Employer:* GA Power Co. *Born:* Sept 8, 1939 Bordon Springs AL. BSAE 1965 Univ of GA. Agri Engr GA Power Co, all agri marketing activities coordinating activities in S GA. Mbr Am Soc of Agri Engrs. Chrmn Natl ASAE Ctte-Electrical Controls for Farmstead Equip. P Chrmn of GA Sect ASAE & secy-treas of Southeast Region ASAE. Mbr of many civ & agri orgs incl Kiwanis Club, Boys Club of Am, GA Farm Bureau, GA Agribusiness Council, Baldosta/Lowndes Cty Chamber of Commerce, Hon Mbr of GA Young Farmers Assn and Future Farmers of Amer. Mbr First Baptist Church, Valdosta & Fellowship of Christian Athletes.

Chapman, Dean R
Business: NASA-Ames Res Ctr 200-4, Moffett Field, CA 94035
Position: Dir of Astronautics. *Employer:* Natl Aeronautics & Space Admin. *Born:* Mar 8, 1922. PhD CA Inst of Tech 1948; BS & MS CA Inst of Tech 1944. Joined NASA-formerly NACA in 1944 since that time has been active at Ames in theoretical fluid mech as well as in many areas of experimental res. Assumed current pos as Dir of Astronautics in 1973. Received the Lawrence Sperry Award 1952; Rockefeller Public Sev Award 1959; NASA Award for Exceptional Scientific Achievement 1964; H Julian Allen Award 1972. Mbr NAE & Fellow AIAA.

Chapman, Lloyd E
Home: 41 Forest St, Reading, MA 01867
Position: Mgr-Engg. *Employer:* GE Co. *Born:* Grad Univ of NH 1947; BSME Cum Laude. Elected to Tau Beta Pi & Phi Kappa Phi Hon Soc. Reg PE MA. Employed by GE Co Wilmington MA for past 29 yrs. Held Design Engg pos with Home Laundry Prod from 1948-54. Prod Engg respon in Aircraf Acessory Turbine Dept from 1954-60. Joined Direct Energy Conversion in 1960 as Mgr-Fuel Cell Prod Engg & in 1970 became Prof Mgr-Fuel Cell Tech Dev. In Apr 1974 appointed Mgr-Engg & Mgr-Oper for Direct Energy Conversion Progs.

Chapman, Robert L
Business: 2500 Harbor Blvd, Fullerton, CA 92634
Position: Mgr Tech Serv. *Employer:* Beckman Instruments Inc. *Education:* MBA/Mgt/Univ of S CA; BS/Engg Phys/Univ of ME. *Born:* Oct 1921. Physicist for Am Cyanamid Co 1943-52, spec in infrared spectroscopy & allied spectro-analytical techniques. Application engr for Liston Becker Instrument Co on non-dispersive infrared analyzers until purchased by Beckman Instruments. With Beckman since 1954 as application engg supr & principal application engr, spec in air pollution instrumentation. VP. ISA Tech Dept 1975. *Society Aff:* ISA, SAE, ASTM, ACS.

Chapman, Robert R
Business: One Seagate, Toledo, OH 43666
Position: Manager-Corrugated Pkg Design *Employer:* Owens-IL Inc. *Born:* Sept 1936. Native Toledo OH. Attended Univ of Toledo. Served in US Army Forces 1954-57. Employed by Owens-IL since 1959. Mbr of the Glass City Chapter of the Society of packaging and handling engineers. Field of specialization - packaging for Glass Containers, Material Handling, Package Testing, Package Cast Reduction programs and knowledge of shipping and government package regulations and creative corrigated package design.

Chapman, William P
Business: P.O. Box 591, WI 53201
Position: VP. *Employer:* Johnson Controls, Inc. *Education:* MS/ME-Heat Transfer/Purdue; BS/Mechanical Engg/Univ of CA. *Born:* 10/19/19. Served in US Infantry 1940-1945; separated with rank of Capt. Engr with Natl Tube Div of US Steel; joined Johnson Controls, Inc in 1956 as Admin Dir of Res; in 1958 made Dir of Res and in 1964 VP, Operations, position now held. Licensed PE in PA & WI; Pres of ASHRAE 1976-77; elected 1977 "Engr of the Yr" by Engrs & Scientists of Milwaukee, Inc. Dir of St Michael and St Joseph's Hospitals; Univ of WI-Milwaukee Fdn; Milwaukee Urban League; also serves on Mayor's Sci and Tech Council of Milwaukee and its Steering Committee. *Society Aff:* ASHRAE, NSPE.

Chappee, James H
Home: Box 495, Friendswood, TX 77546
Position: Deputy Chief, Safety *Employer:* NASA, Johnson Space Ctr *Education:* BS/GE/Univ of IL *Born:* 12/20/28 Currently Deputy Chief of Safety for Johnson Space Ctr, NASA. Responsibilities include industrial safety, fire prevention and detection systems, and system safety analysis for Shuttle space-craft systems and operational techniques. Consultant to private industry in the petro-chem, heavy metals reclamation and product safety disciplines. Professional Engr, CA (Safety) and TX (Industrial). Nat'l Pres of System Safety Society, 1981-1983. *Society Aff:* SSS

Chappelear, Patsy S
Business: P.O. Box 218218, Houston, TX 77218
Position: Sec Leader Process Engr *Employer:* Hudson Engrg Corp *Education:* BS/ChE/Rice Inst; BA/ChE/Rice Inst *Born:* 10/23/31 Engr in lube oils at Shell Oil Deer Park Refinery 1954-55. Assoc in thermodynamics research in gas processing at Rice Univ, 1955-1976. Author, lecturer and consultant, 1955-1975. Member, Editorial Review Board for GPSA Engrg Data Book, 1975-present. Fellow AIChe. Registered PE (TX). With Hudson since 1975 in Process and Project engrg for onshore and offshore oil/gas processing facilities. Married, four children. *Society Aff:* AIChE ΣΞ

Charles, Michael E
Business: Vice-Dean, Faculty of Ap.Sc. & Engineering, University of Toronto, Toronto, Ont, Canada M5S 1A4
Position: Vice-Dean, Faculty of Appl Science & Engrg *Employer:* Univ of Toronto. *Education:* PhD/Chem Engg/Univ of Alberta; MASc/Chem Engg/Univ of Alberta; BASc(hon)/Chem Engg/Imperial College of Sci and Tech, Univ of London. *Born:* 12/20/35. Leicester, England. Educated at Universities of London and Alberta. Employed by Research Council of Alberta and Imperial Oil Limited before joining Univ of Toronto in 1964. Prof and Chmn, Dept of Chem Engg and Applied Chem 1975-85. Now Vice-Dean, Faculty of Applied Sc. & Engineering. Received 1976 ERCO Award of the Canadian Soc for Chem Engg. Dir of Chem Engg Research Consultants Limited since 1965. Research in field of fluid mech and heat transfer relates directly to resource dev, including design of process equipment and pipeline transportation systems. Conslt on Arctic pipeline tech and multi-phase flow. *Society Aff:* CRMA, CSChE, APEO, SCI.

Charles, Richard J
Business: GE Corporate R&D, Schenectady, NY 12301
Position: Ceramist. *Employer:* GE Co. *Education:* SeD/Metallurgy/MIT; M.A.Sc./Metallurgy/UBC (Canada); B.A.Sc./Mining/UBC (Canada) *Born:* Sept 8, 1925. ScD from MIT, BS & MS from Univ of British Columbia Canada. Asst Prof Dept of Met MIT 1954-56. 1956-GE, Res Mgr in Met & Ceramics 1969-83. Currently R&D ceramist. 1950-Rossiter W Raymond Award AIME. 1972-Geo W Morey Award Am Ceramics Soc. 1974-Coolidge Fellow GE Co. 1976-Adjunct Prof Dept Mtls Sci & Engg MIT. *Society Aff:* Am Cer Soc

Charley, Philip J
Business: 4101 N Figueroa St, Los Angeles, CA 90065
Position: Pres. *Employer:* Truesdail Labs Inc. *Born:* Aug 1921 Melbourne Australia. PhD Univ of S CA, MS Southern CA, BS Univ of WI. Test Engr GE 1943-44, Lt Royal Canadian Elec & Mech Engrs, 1944-46, Instr Univ S CA 1946-47, Engr Std Oil Co of CA 1947-54. VP Truesdail Labs Los Angeles 1954-71, Pres 1971 to date.

Charley, Robert W
Home: 14255 SW Barlow Rd, Beaverton, OR 97005
Position: Director, Water Resources *Employer:* CH2M Hill. *Education:* MSCE/Hydraulics-Water Resource/MT State Univ; BSCE/Water Resource/UT State Univ. *Born:* 7/16/37. Native of Humboldt County, CA. Worked for CA Div of

Charley, Robert W (Continued)
Hgwys, CA Dept of Water Resources and the US Geological Survey. Employed by CH2M Hill 1966. Presently director water resources, northwest District. Became part owner in 1975. Specialties include computer applications, hydrology, hydraulics, dams, irrigation works, and related structures. Experience in planning and design. Now a dir in the OR Water Resouce Congress. Pres of Shasta Branch ASCE, 1973. Pres of OR Sec ASAE 1978. Mbr US Committee on Irrigation Drainage and Flood Control, the Natl Water Resource Assn, and Sigma Tau. Enjoy sports, river rafting, music, and travel. *Society Aff:* ASCE, ASAE, NWRA.

Charpentier, David L
Home: 151 Hickory Cir, Middletown, CT 06457
Position: Forman. *Employer:* Pratt & Whitney. *Education:* -/Struct Engg/North Eastern Univ; Metallurgy/Engg/Univ of CT. *Born:* 3/13/31. TX Instruments-Engg aid 1957 to 1967. Pratt & Whitney Aircraft-NDT engr 1967 to 1971. United Nuclear Corp-Qual engr supv 1971 to 1975. Metals Testing Co-QCMgr to pres 1975 to 1978. Pratt & Whitney Aircraft-Forman 1978 to date. Fellow American Soc for Nondestructive Testing. PE state of CA. Who's Who in Engineering, Who's Who in America, Who's Who in the Community, Men of Achievement Throughout the World; Who's Who of Intellectuals Throughout The World. *Society Aff:* ASM, ASME, ASQC, ASTM, ASNT.

Chartier, Vernon L
Home: 5190 SW Dover La, Portland, OR 97225
Position: Chief High Voltage Engr *Employer:* Bonneville Power Admin. *Education:* B.S./EE/Univ of CO; B.S./Bus/Univ of CO *Born:* 2/14/39 Native of Ft. Morgan, Co. Research Engr for Westinghouse Electric Corp from 1963-1975, specializing in corona and field effects of high voltage transmission lines. With BPA since 1975. Assumed current responsibility as Chief High Voltage Engr for the development of high voltage transmission technology up to 1200 kV in 1977. Chrmn of IEEE Corona & Field Effects Subcommittee, 1978-82. Chrmn IEEE T&D Ctte. Technical Advisor to USNC/IEC on CISPR Subc C. Chrmn of Subcommittee 4 (High Voltage Lines & Apparatus) of ANSI C63. Fellow IEEE. Member of Bd of Directors of IEEE Electromagnetic Compatibility Society 1978-81. Registered engr, PA. *Society Aff:* IEEE, CIGRE, ASA

Charyk, Joseph V
Business: 950 L'Enfant Plaza SW, Washington, DC 20024
Position: Retired. *Education:* Ph.D/Aeronautics/California Institute of Technology; M.S./Aeronautics/California Institute of Technology; B.Sc./Engg Physics/Univ of Alberta. *Born:* in Canada. BSc Engrg-Phys Univ of Alberta; MS & PhD Aero Cal Tech; Hon LLD Univ of Alberta; Hon Dr Ing Univ of Bologna; married Edwina Rhodes, 4 children; Jet Propulsion Labs 1945-46; Asst & Assoc Prof Aero Princeton Univ 1945-55; Dir Aerophys & Chem Lab Missile Sys Div Lockheed Aircraft Corp & Mgr Space Tech Div 1955-59; Asst Secy USAF, R&D 1959, Under Secy 1960-63; Pres Dir Comsat 1963-79. Ch Exec Officer 1979-85; Chrmn 1983-85. Dir Communications Satellite Corp, Dir Abbot Labs; Dir MNC Financial, Inc., Dir American Security Corp, & American Security Bank, N.A., Dir C.S. Draper Labs. Mbr Natl Acad Engrg. Fellow IEEE, Intl Acad Astro; Fellow AIAA. *Society Aff:* AIAA, NAC, IEEE, IAA, NISS, NSC, AFC&EA.

Chase, Arthur P
11204 Old Post Rd, Potomac, MD 20854
Position: Senior Consultant *Employer:* The Nelson Group, Inc. *Education:* BSCE/Struct/Cooper Union Inst of Tech. *Born:* 6/14/20. Native of NYC. Before joining CRS Group Engrs employed by MacLean Grove Co, NY; Third Tube Lincoln Tunnel; W DE Water Supply Tunnel, NY; Rhine River Tunnel, Duesseldorf, Germany; Riverdale Tunnel, NY; Canyon Power Tunnel, CA. Designed tunnels and tunnel shields used in Mexico City; Florence and Milan, Italy; Wash, DC; New York, NY; Buenos Aires, Argentina. Designed medical locks for Compressed Air Medical Ctr, San Francisco. Served on Bd of Engg Consultants on Wash Metro System and San Francisco Bay Area Rapid Transit System. Inventor of precast concrete segmented tunnel lining system. *Society Aff:* ASCE.

Chase, Robert L
Business: Bldg 535, Upton, NY 11973
Position: Sr Electronic Engr. *Employer:* Grookhaven Natl Labs. *Born:* Mar 19, 1926, Brooklyn, s Ephraim & Rose-Roslofsky-C; BS in Elec Engg Columbia 1945; MEE Cornell Univ 1947; PhD Univ Uppsala Sweden 1973; m Ellen J Blackburn July 2, 1970; children-Polly E, William S, Claudia A. Electronic engr Brookhaven Natl Lab Upton NY 1947-72, 74-, Laboratoire de l'Accelerateur Lineaire, Orsay, France 1972-74. Served with USNR 1943-46. Fellow IEEE. Author: Nucl Pulse Spectrometry 1961. Home: 300 Edwards St, Roslyn NY 11577.

Chase, Thomas L
Business: 1000 Rimrock Rd, Columbus, OH 43219
Position: Pres. *Employer:* Elec Heat Treating Co. *Education:* Bach of Met/Engg/OH State Univ. *Born:* 7/26/26. Native of Columbus, OH. Served in US Army WW11 for 2 yrs, 17 months in Europe. Partner, owner and now pres of Elec Heat Treating Co since 1954, providing heat treating services of highest quality to industry utilizing science of metallurgy. Past Chrmn and Active supporter of Local ASM Chapter. Active in Greek Orthodox Church Locally and Nationally and Internationally in Greek-Amer organizations and affairs. Delegate to IFHT Congresses in Warsaw, Shanghai, London, Berlin. Reg Prof Metallurgical Engr (State of Ohio). Arction of Ecumenical Patriarchate, Constantinople. *Society Aff:* NSPE, ASM, AWS.

Chase, William J
Business: Fourth & Vine Bldg, Seattle, WA 98121
Position: Pres. *Employer:* URS Co. *Education:* BS/Civ Engg/Univ of WA. *Born:* 3/16/29. William J Chase received his BS in Civ Engg from the Univ of WA in 1950 and has been with the URS Co since 1950 except in 1952-55 when he served as an officer in the USAF. He is now pres of the URS Co in Seattle. He has served as: Chrmn of the Pacific Northwest Chapter of the Young Presidents' Organization & alumni organization, mbr of the Bd of Dirs and Tres of the Economic Dev Council of Puget Sound, chrmn of the Engg Advisory Bd for WA State Univ, Pres of the WA Society of PE, mbr of the Pacific Univ Bd of Trustees, mbr of the Bd of Dirs of the Western Region Natl Council for Christians and Jews, mbr of the Steering Committee for the Council for Washington's Future and mbr of the Bd of Dirs of Pacific First Fed Savings and Loan Assn. Board of Directors of Washington Council on International Trade, member, Economic Panel, member, World Business Council, member, Chief Exec Forum. He also is a mbr of the Rainier Club and the WA Athletic Club. *Society Aff:* ASCE, APWA, WPCF, WSPE.

Chastain, Theron Z
Home: 2336 Leafmore Dr, Decatur, GA 30033
Position: Chairman of Board *Employer:* Chastain Forensics Corp *Education:* MS/Civil Engg/GA Inst of Tech; BS/Civil Engg/GA Inst of Tech. *Born:* 7/15/20. Native of GA. Naval Architect 1943-45. Served with Navy Construction Battalion 1945-46. Structural Engr with I E Morris and Assoc, Atlanta, 1947-53. Established own firm in 1954, Chastain and Tindel, Inc, in 1959, and Chastain Forensics Corp in 1983. Firm specializes in Forensic Engineering services including consulting and research. VP, ACI, 1979-80; Pres, ACI, 1981. Consulting Engineer of the Year, 1973, by CEC/GA; Engr of the Yr, 1972-73, by GSPE; Honor Award, 1977, By CRSI and Honorary Mbr Rutgers Univ Chapter of Chi Epsilon, 1975. Hobbies include golf and woodworking. *Society Aff:* ACI, ASCE, ACEC, ASTM, NSPE, RCRC, PCI, PTI, EERI, NAFE, AAA, IABSE, AIPE.

Chastain, William Roy
Home: 505 Arlington Ave, Berkeley, CA 94707
Position: Account Exec *Employer:* Occidental Chem Corp *Education:* Masters/Bus Adm/Univ of Akron; BS/Chem Engg/MO Sch of Mines/Assoc in Sci/Chem/Southwest Baptist College. *Born:* 2/7/32. in Clever, MO. Graduated with honors

Chastain, William Roy (Continued)
form MO School of Mines in 1955. Employed by Columbia Southern Chem (PPG) at Barberton, OH--assigned to two new plant startups--titanium tetrachloride and trichlorethylene. General foreman of trichlor and chlorine plants while at Barberton. Assigned to PPG home office in Pittsburgh, Penna for five yrs (1963-68) as nationwide tech service eng on chlorine and the oxidizing agents. Author of PPG film---Chlorine Safe Handling" and articles on "Calcium Hypochlorite-. Transferred to San Francisco in 1968 as marketing representative and solvent specialist. In 1979 joined Continental Chemical Co as VP of this chlorine repackaging firm. Joined Hooker Chemicals & Plastics Corp., 12/15/80 as Account Mgr Recently promoted to Account Exec for Occidental Chem Corp. *Society Aff:* NSPE, AIChE, ΑΧΣ.

Chaston, A Norton
Business: Electrical Engrg Dept, 459 CB, Provo, UT 84602
Position: Assoc Prof *Employer:* Brigham Young Univ *Education:* MS/EE/Brigham Young Univ; BS/EE/Univ of UT; AS/Science/Univ of ID *Born:* 4/7/26 Native of Salt Lake City, Utah. USNR, 1943-46. Missionary to England for The Church of Jesus Christ of Latter-Day Saints, 1946-48. Gen Elec Co, 1951-57. Test Engrg Program, Guided Missile Dept and Atomic Power Equip Dept at Schenectady, NY and San Jose, CA. Faculty member with Elect Engrg Dept, Brigham Young Univ 1957 to present; presently, Supv of Electric Power Option. Sabbatical leaves: 1963-64, So CA Edison, 1972-73, Electric Utilities Engrg, Gen Elec Co. Summer employment 1958-1972: Convair, Montek, Lockheed, Gen Elec, UT Power and Light, LADWP. Private consulting: Architectural Engrg, 1970-present. Author of text *Electric* Machinery, published by Prentice-Hall, 1986. *Society Aff:* ΣΞ, ΤΒΠ, IAEI

Chatel, Bertrand H
Business: 1 United Nations Plaza, Rm DC 1052, New York, NY 10017
Position: Chief, Tech App. *Employer:* United Nations. *Born:* 8/22/20. Born in Ecot-France. Naval Acad-Brest 1939 & Elec-ESE Paris 1949-Engg Dipl Economist-Paris 1954. French Navy 1939-47; Engg in a cons firm & electric manufac co 1949-58; Mgr of a factory manufacturing prefabricated bldgs 1958-62; Ch Engg Tech Facilities, European Space Res Org-ESRO 1962-69; Ch Sci Applications Office for Sci & Tech, United Nations 1969-present. Mbr of AAAS, IEEE, and ANS. *Society Aff:* AAAS, IEEE, ANS.

Chato, John C
Business: 132 Meb, University of Illinois, 1206 W. Green St, Urbana, IL 61801
Position: Prof of Mech Engg and Bioengg. *Employer:* Univ of IL. *Education:* PhD/ME/MIT; MS/ME/Univ of IL; ME/ME/Univ of Cincinnati. *Born:* 12/28/29. Native of Budapest, Hungary, now citizen of USA. Married Elizabeth Owens. Have three children. Asst Prof MIT, 1958-64. NSF Post-Doctoral Fellow at Technical Univ of Aachen, Germany, 1961. Assoc Prof, Univ of IL, 1964-69, Prof since 1969. NASA/ASEE Summer Faculty fellow 1966-67. Chrmn, Bioengg Faculty 1972-78, 1982-83, 1984-85, Univ of Cincinnati Distinguished Engg Alumnus 1972, ASME Fellow 1975, ASME Charles Russ Richards Meml Award 1978, Fogarty Sr Intl Fellow at Inst for Biomedical Engg in Zurich, Switzerland, 1978-79. Visit to Hungary under sponsorship of US and Hungarian Academies of Sci, 1978. Cryogenic Engg Conf Russell B Scott Meml Award, 1979. Honorary visiting Prof., Univ New South Wales Kensington, Australia, 1986. Mbr Urbana Plan Commission 1973-78. Member Urbana Exchange Club, Member Urbana Park District Advisory Committee 1981-84. Elder First Presbyterian Church of Urbana. Other activities are tennis, outdoor hikes, bicycling, birding, color photography, music, and theater. *Society Aff:* ASME, ASHRAE, IEEE, IIR, ASEE, ΣΞ

Chaudhary, Kailash C
Business: 3272 Villa La, Napa, CA 94558
Position: Pres *Employer:* Chaudhary & Assoc *Education:* B.S./CE/Heald Engrg Coll *Born:* 8/21/37 KAILASH C. CHAUDHARY, PE, Pres and Founder of Chaudhary & Assocs, is a Civil Engr with registration in OR, WA and CA. He is also registered as a Professional Land Surveyor in OR. His 25 years of experience in design, inspection and contract administration allows him close contact with the firm's activities and its staff of 14 members engaged in the performance of general civil engrg, land surveying and land planning activities in northern CA. Mr. Chaudhary emigrated from India in 1957 and has settled in Napa Valley, with his wife and 5 children. *Society Aff:* NSPE, ACSM, AWWA, SAME, CSPE, CCCE&LS.

Cheatham, John B, Jr
Home: 4402 Briarbend, Houston, TX 77035
Position: Prof *Employer:* Rice Univ *Education:* PhD/ME/Rice Univ; MS/ME/SMU; BS/ME/SMU *Born:* 6/29/24 Mechanical Engr, GE, Link Belt, Atlantic Refining and Shell (1948-63). Joined Rice faculty (1963), chrmn ME Dept (1968-72). Past Chrmn ASME Petro Div Rock Mechanics and Transactions Papers Committees. Tech. Editor J. Pressure Vessel Tech (1976-78). Founding Tech Editor ASME Trans J. Energy Resources Tech (1979). Elected ASME Fellow (1977). NAS-NRC Committees on Rock Mechanics (1966), Rock Mechanics Research (1980), Consultant Drilling Tech (1975-76). SPE Editorial Board (1976). Pres Techaid Corp (1977-present). Pres Cheatham Engrg Inc (1977-present). Registered PE (TX). ASME Pet Div Ralph James Award (1980). *Society Aff:* ASME, SPE of AIME, ASEE, ISRM

Check, Paul S
Home: 7810 Maple Ridge Rd, Bethesda, MD 20014
Position: Assistant Director for Plant Systems *Employer:* US Nuclear Regulatory Commission. *Education:* MS/Nuclear Engg/Univ of Cincinnati; BS/Physics/Fairfield Univ. *Born:* 5/24/34. Native of Trumbull, CT. With NRC (and AEC) since 1966. Assumed present position in 1980. Previously served as Chf, Reactor Safety Branch, Chief, Core Performance Branch and as Leader, Reactor Physics Section; and before that for several yrs as a proj mgr. Reactor Physicist for Pratt & Whitney Aircraft in Middletown, CT, 1959-66, and Honors Program Engr with Genl Elec Co in Evendale, OH, 1957-59. Enjoys classical music and tennis. *Society Aff:* ANS.

Cheek, Guy H
Home: Route 5, Box 469, Monroe, NC 28110
Position: Dir-Energy Conservation Div. *Employer:* Air Conditioning Corp. *Education:* BME/ME/NC State Univ. *Born:* Jan 25, 1926, Bennett NC. 1951. Naval Aviation Serv 1943-47, South Pacific Theatre. Background: Aircraft power plants-Pratt & Whitney-Refrig, HVAC, Electric Unil, Constr Dev & Mgt, Energy Utilization & Cons. Since 1970-cons-Mech Elec-Energy Utilization & Sys. ASHRAE PChrmn & Member-TC 9.4-Applied Heat Pump; Chrmn-TC9.6-Systems Energy Utilization, P Pres of PE NC; Past V.P. & Dir NSPE & NC Assn of Professions. Mbr NC Air Qual Council. Holds patents for concepts, controls & hardward for Built-Up Heat Pumps. Registered Engineer - NC, SC, VA, TN, GA. *Society Aff:* NSPE, ASHRAE.

Cheek, Robert C
Business: 1-7-/Yurakucho 15-N, Chiyoda-Ku, Tokyo 100, Japan
Position: Vice President Technology *Employer:* Westinghouse Elec (Japan) Tokyo. *Born:* Nov 1917. Charleston SC. BSEE GA Tech, MSEE Univ of Pittsburgh. Joined Westinghouse Elec Corp 1939. Power Sys Engr 10 yrs. Var sales & engrg mgmt pos 5 yrs. Gen Mgr Indus Electronics 5 yrs. Dir Tele-Computer Ctr 8 yrs. Dir Mgt Sys 4 yrs. Pres Westinghouse Tele-Computer Sys Corp 4 yrs. Dir Ebara-Infilco (Joint Venture) 4 yrs. Assumed present pos 1980. Eta Kappa Nu Award 1979. Fellow IEEE 1962. Westinghouse Order of Merit 1963. Westinghouse Patent Award-10 patents. "Computer Sci Man of Yr 1972-, DPMA. Many tech papers, articles, co-author *Transmission & Distribution*-bk. Reg PE, PA. Hobbies: Amateur Radio-W3VT, golf, languages.

Cheeks, John R
Business: 2520 Regency Rd, Ste 106, Lexington, KY 40503
Position: Pres *Employer:* Stokley-Cheeks & Assoc *Education:* MS/CE/Geotech/Univ of KY; BS/CE/Univ of KY *Born:* 10/3/48 Native of Pikeville, KY. After Coll, project engr for Law Engrg Testing Co in Jacksonville, FL 1972-1975 and Sr Engr and

Cheeks, John R (Continued)
Engrg mgr in Nashville, TN 1975- 1978. With Stokley-Cheeks since 1978, became Pres in 1980. The firm provides consulting engrg services in the geosciences. Mr. Cheeks also serves as adjunct instructor at the Univ of KY Dept of Civil Engrg. Serves as Chrmn of Professional Engrs in Private Practice in Kentucky, and is Chrmn of the Professional Quality Ctte of ASPE. *Society Aff:* NSPE, KSPE, ASCE, ASFE.

Cheeseman, Charles E, Jr
Business: 1800 Volusia Ave, Daytona Beach, FL 32015
Position: Gen Mgr, Simulation and Control System Dept *Employer:* Genl Elec. *Education:* PhD/Sys Engg/Univ of PA; MS/Sys Engg/Univ of PA; BS/Engg Sci/USAF Acad. *Born:* 8/3/40. Dr Cheeseman has held a wide range of engg and mgmt positions in the GE Aerospace Group. He has been respon for Nuclear Weapon Effects Testing, Space Environ Simulation, Manned Space Hardware Design and Test, Microwave Remote Sensing Dev, Natl Resources Survey Sys Dev, and Space Instrument Dev. More recently, he has been respon for the genl mgmt of GE's Marine and Industrial Control Sys business at its Daytona Beach operation. In 1980 he was appointed General Mgr at Daytona Beach, where he is in charge of a 1300 employee dept which designs and manufactures electronic control systems and computer - generated - imagery simulators. His major outside interest is flying. *Society Aff:* AIAA, ASNE, SNAME.

Cheh, Huk Y
Business: Dept. of Chem. Engg, Columbia University, New York, NY 10027
Position: Prof. *Employer:* Columbia Univ. *Education:* PhD/ChE/Univ of CA; BASc/ChE/Univ of Ottawa. *Born:* 10/27/39. H Y Cheh was born in Shanghai, China on Oct 27, 1939 and became a US citizen in 1974. He received his BASc from Ottawa Univ in 1962 and his PhD in chem engg from the Univ of CA, Berkeley in 1967. He was with Bell Tel Labs in Murray Hill, NJ, from 1967-1970, as a mbr of technical staff. Dr. Cheh came to Columbia Univ in 1970 and is currently the Ruben-Viele Professor of Electrochemistry. During 1978 to 1979, he served as a Prog Dir in the Div of Chem and Process Engg, Natl Sci Fdn. Dr. Cheh's res area deals generally with the field of chem engg and surface and electrochemistry. He is married and has two daughters. *Society Aff:* AIChE, ELECTRCHEM SOC, AES, ΣΞ, NY ACAD SCI

Chelton, Dudley B
Home: 500 Mohawk Dr. 308, Boulder, CO 80303
Position: Cryogenics Consultant *Employer:* Self *Education:* MS/Mech. Engr./Mass. Institute of Tech.; BSME/Mech. Engr./Ohio State Univ. *Born:* July 17, 1928 Baltimore MD. BSME OH State Univ 1948, MS Mech Engg MIT 1949. Res assoc in cryogenic engg Los Alamos Scientific Lab, NM 1950-51. Joined staff of Natl Bureau of Stan, Cryogenics Div in 1951. From 1963 until 1968 Ch of Cryogenic Sys Sect, 1968-74 Ch of Cryogenics Div of NBS-Inst for Basic Stds, 1974-1977 Sr Engg Cons within the Cryogenics Div, Retired 1977. Private Cryogenic Engineering Consultant 1977-present. Author or co-author of more than 30 pub papers on cryogenic engg subjects. Prin fields of res at NBS-cryogenic refrig & liquefaction tech, cryogenic sys eval, liquid hydrogen bubble bhambers, aerospace vehicle cryogenic propellants, Liquefied Natural Gas (LNG) Tech and safety in the use of cryogenic fluids.

Chen, Chi Hau
Home: 415 Bradford Place, N Dartmouth, MA 02747
Position: Prof & Chrmn *Employer:* Southeastern MA Univ *Education:* PhD/EE/Purdue Univ; MS/EE/Univ of TN; BS/EE/Nat Taiman Univ *Born:* 12/22/37 Received PhD from Purdue Univ in 1965. Has been with the Electrical Engrg Dept of Southeastern MA Univ since 1968. Author of the books *Statistical Pattern Recognition*, Hayden Book Co 1973, *Digital Waveform Processing and* Recognition, CRC Press 1981, and *Nonlinear Maximum Entropy Spectral Analysis Methods for Signal Recognition*, Wiley 1982. Editor of seven books in pattern recognition and seismic signal processing areas. *Society Aff:* IEEE, ASEE, SEG

Chen, Chiou S
Business: 302 E Buchtel Ave, Akron, OH 44325
Position: Prof. *Employer:* Univ of Akron. *Education:* PhD/Elec Engg/Univ of Rochester; MS/Elec Engg/Univ of Rochester; BS/Elec Engg/Natl Taiwan Univ. *Born:* 1/22/38. Worked for Taylor Instruments Co, Rochester, NY, 1964-1966 in process control instrumentation. Did res in Navigation and Control at NASA Ames Res Ctr and Satellite image processing at NASA Goddard space flight ctr. Taught at Univ of Akron since 1968; Elec Engrg Dept hd since Mar 1984; res interest in computer control and digital signal processing. *Society Aff:* IEEE, ΣΞ, ASEE.

Chen, Chuan F
Business: Dept of Aerospace & Mech Engrg, Univ of AZ, Tucson, AZ 85721
Position: Dept Chrmn. *Employer:* Univ of Arizona *Education:* PhD/Aero Engg/Brown Univ; MS/Mech Engg/Univ of IL; BS/Mech Engg/Univ of IL. *Born:* 11/15/32. I was born in China, came to US in 1950, became a naturalized citizen in 1961. Served as Res Scientist, Sr Res Scientist, and Asst to the Chief Engr at Hydronautics, Inc, Laurel, MD 1961-63. Joined the Rutgers faculty as Asst Prof in 1963, Assoc Prof 1966, Prof 1969, Chairman 1976-80. Joined University of Arizona faculty in 1980. Was a Sr Visitor at the Dept of Appl Mathematics and Theoretical Physics, Cambridge Univ, England, 1971-72, a Visiting Scholar at the Res School of Earth Sciences, Australian Natl Univ, Canberra, Australia, summer 1978. Mbr of Geologicl expedition to Skaergaard Intrusion in East Greenland within the Arctic Circle, summer, 1979. Current res interest: double-diffusive convection, hydrodynamic stability. *Society Aff:* ASME, AIAA, APS, ASEE

Chen, David H T
Home: 106 Hickory Spring Road, Wilmington, DE 19807
Position: Exec VP. *Employer:* Helix Assoc, Inc. *Education:* PhD/Chem Engg/Univ of Rochester; MS/Chem Engg/Univ of RI; BS/Chem Engg/Cheng Kung Univ. *Born:* 7/12/35. Native of Chekiang, China. Came to the US for grad studies in 1959. Worked for Combustion Engg, Celanese Research and DuPont in a variety of R&D positions. Assumed current position in 1978, have over all operating respon of a specialty chem manufacturer. Concurrently serving as an adjunct Assoc Prof of Engg at Widener Univ. A reg PE in the State of DE. Enjoy reading, and traveling. *Society Aff:* AIChE, ASEE.

Chen, Di
Business: Advanced Memory Laboratory, Magnetic Periphrals Inc, 2766 Janitell Rd, Colorado Springs, CO 80906
Position: Director *Employer:* Magnetic Periphrals Inc. *Education:* PhD/EE/Stanford Univ; MS/EE/Univ of MN; BS/EE/Natl Taiwan Univ. *Born:* 3/15/29. From 1959 to 1962, Di Chen taught & did res in magnetism at the EE Dept of the Univ of MN as an asst prof. From 1962-1980, he did R & D work at Honeywell Corp Mtl Sci Ctr (CMSC). His main res activity was in the field of laser applications, magneto-optic memory, & optical communication & optoelectronics. He joined the advanced memory lab of the Magnetic Periphrals Inc in 1980 as the dir of the optical systems. He has published over 60 technical papers & was issued thirteen patents. In 1974, he received the Honeywell Sweatt Engrs & Scientist award, & in 1975 he was promoted to the fellow grade of IEEE. *Society Aff:* IEEE, OSA, ΣΞ, HKN.

Chen, Francis F
Business: Boelter Hall, Rm 7731, Los Angeles, CA 90024
Position: Prof. *Employer:* UCLA. *Education:* PhD/Physics/Harvard Univ; MA/Physics/Harvard Univ; AB/Astronomy/Harvard Univ *Born:* 11/18/29. After completing an experimental PhD thesis in high energy phys in 1954, I joined the Princeton Plasma Phys Lab (Proj Matterhorn) to work on nucl fusion. In the 15 yrs at Princeton I worked on the problems of magnetic confinement in toruses & initiated the study of fundamental plasma phys phenomena in small experimental machines. In 1969 I became Prof of Elec Sciences at UCLA where I now teach courses in Applied Plasma Phys, controlled fusion, & basic electrical engrg. conduct a res prog in

Chen, Francis F (Continued)
laser-plasma interactions and laser-driven accelerators. *Society Aff:* APS, IEEE, NYAC, ΣΞ

Chen, Fu-Hua
Business: 96 S Zuni St, Denver, CO 80223
Position: Chairman *Employer:* Chen & Assoc Inc. *Born:* July 1912. MS Univ of IL, BS Univ of MI. Native of China. Ch Engr Burma Rd, China 1941-43. Ch Engr Koknor-Tibet Hgwy, China 1943-45. Prof Natl Fu Tan Univ 1945-49. Lab Dir Public Works Dept Hong Kong 1949-56. Engr Woodward, Clyde, Sherard & Assoc Denver CO 1957-61, Pres Chen & Assoc, Cons Soil Engrs, Denver, CO 1961 to date. Professor CO St Univ Ft Collins, CO. Author "Foundations on Expansive Soils–, Elsevier Pub Co 1975. Pres 1974-75, CEC, VP ACEC. Hon. Doctor degree Colo State Univ. Hon Member ASCE.

Chen, Hui-Chuan
Business: P.O. Box 6316, University, AL 35486
Position: Assoc Prof *Employer:* Univ of AL *Education:* PhD/Operations Res/State Univ of NY, Buffalo; MSIE/Ind Engrg/Univ of AL; BS/Ind Mgt/Taiwan Cheng-Kung Univ *Born:* 4/9/38 Native of Taiwan. Came to U.S. in 1964. Naturalized in 1976. Taught at Univ of AL since 1970. Current - Assumed responsibility as assoc prof of Comp Sci, specializing in comp simulation, data structures, and artificial intelligence. Comp consulting for transportation industry, pharmacy business, and accounting firms. Publications appeared in various journals on computers and operations research. *Society Aff:* IIE, ACM, APM, PME, MSR

Chen, Hung T
Home: 40 Tilton Dr, Freehold, NJ 07728
Position: Prof & Asst Chrmn. *Employer:* NJ Inst of Tech. *Education:* PhD/ChE/Poly Inst of NY; MS/ChE/Poly Inst of NY; BS/ChE/Natl Taiwan Univ. *Born:* 8/23/35. Educator; Native of Taiwan, China; came to US 1960, naturalized, 1972; BS, Natl Taiwan Univ, 1958; MS, Poly Inst of NY, 1962, PhD, 1964; Process Engr, FMC Corp 1964-66; asst Prof, NJ Inst of Tech, 1966-70, Assoc Prof 1970-75, Prof of Chem Engg and Asst Chrmn, 1975-; Consultant, Brookhaven Natl Lab. LI NY 1967-. Recipient awards NSF. Author numerous articles. Licensed PE (NJ). Mbr AIChE, NSPE, Sigma Xi, Omega Chi Epsilon. *Society Aff:* AIChE, NSPE, ΣΞ, ΩXE.

Chen, Juh W
Business: College of Engrg and Technology, Southern Illinois Univ, Carbondale, IL 62901
Position: Prof & Chrmn. *Employer:* Southern IL Univ. *Education:* PhD/Ch.E./Univ of IL; MS/CH.E/Univ of IL; BS/CH.E. /Taiwan College of Engg. *Born:* 11/10/28. Shanghai, China, Naturalized Citizen. Asst Prof, Bucknell Univ, 1959-65. Assoc Prof and Prof of Engg, 1965-71, Chrmn of Thermal and Environmental Engg, 1971-85 , Assoc Dean of Engrg, 1985-present, Cons to Upjohn Co, Staley Mfg Co, Harza Engg Co, Reg PE, Diploma of AAEE. Res interests in kinetics, catalytic desulfurization, coal fines recovery and utilization. Enjoy playing duplicate bridge and tennis. *Society Aff:* AIChE, ASME, ACS, AAEE, NSPE.

Chen, Kan
Business: Ann Arbor, MI 48109
Position: Prof. *Employer:* Univ of MI. *Education:* ScD/EE/MIT; SM/EE/MIT; BEE/EE/Cornell Univ. *Born:* 8/28/28. Native of Hong Kong. Attended schools in China. Mgr of Systems Tech R&D at Westinghouse Elec Corp, 1954-1965. Dir of Urban Dev Prog at Stanford Res Inst, 1966-1970. Prof of Elec & Comp Engg and Dir of Ph.D. Prog in Planning at the Univ of MI, 1971-present. Dir & Chief Scientist of Acumenics Res & Tech, Inc, 1978-present. IEEE Fellow & AAAS Fellow. Authors of 6 books and over 70 articles. *Society Aff:* IEEE, AAAS.

Chen, Kenneth Y
Business: Environmental Engg Prog, Los Angeles, CA 90007
Position: Prof & Dir. *Employer:* Univ of S CA. *Education:* PhD/Environ Sci & Engg/Harvard Univ; MS/Sanitary Engg/Univ of RI; BS/Civ Engg/Natl Taiwan Univ. *Born:* 2/12/41. Native of Taiwan, China. Joined Univ of S CA - Environmental Engg Prog in 1970 as an asst prof, became assoc prof in 1974. Prof & Dir in 1978. Responsibilities include res, teaching & admin. Specializing in environmental effects of energy dev; migration of trace contaminants in the environment; and disposal of hazardous wastes. Consultant to office of Energy-Related Invention, Natl Bureau of Stds; Argonne Natl Lab; other fed, state, & local agencies; as well as ind concerns. Co-recipient of Wesley W Horner Award of Am Soc of Civ Engrs in 1977. *Society Aff:* AAAS, ASCE, AWWA, WPCF

Chen, Kun-Mu
Home: 4608 Tacoma Blvd, Okemos, MI 48864
Position: Prof. *Employer:* MI State Univ. *Education:* PhD/Appl Phys/Harvard Univ; MS/Appl Phys/Harvard Univ; BS/EE/Natl Taiwan Univ. *Born:* 2/3/33. Prof of Elec Engg. MI State Univ. PhD & MS Harvard Univ. BS Natl Taiwan Univ. Born Taipei, Taiwan. Came to US in 1957, citizen 1969. Gordon McKay & C T Loo. Fellows at Harvard Univ. Res Assoc at Univ of MI, 1960-64. Prof of Elec Engg at MI State Univ since 1967. Dir of Grad Prog of Elec Engg, MI State Univ from 1968-73. Published numerous papers in the areas of Electromagnetic Radiation, plasma phys and Interaction of EM fields with Biological Systems. Fellow of Inst of Elec and Electronics Engrs, Fellow of AAAS, Mbr of US Commissiiin ABC of Intl Union of Radio Sci, Socs of Sigma Xi, Phi Kappa Phi and Tau Beta Pi. Recipient of Distinguished Faculty Award from MI State Univ in 1976. Recipient of outstanding Achievement Award in Science and Engineering from Taiwanese Amer Foundation, 1984. *Society Aff:* IEEE, AAAS, URSI, BEMS.

Chen, Leslie H
Home: 203 Yoakum Parkway #709, Alexandria, VA 22304
Position: Retired *Education:* Sc.D./Mech. Engr./Harvard University; M.M.E./Mech. Engr./Univ. of Delaware; B.S./Aero. Engr./Chiaotung University *Born:* 11/22/23 After receiving Sc.D. degree, worked for consulting firms of Jackson & Moreland in Boston and Sverdrup & Parcel in St. Louis as a senior mech engr. In 1956, joined General Dynamics Corp, Electric Boat Div, Groton, Connecticut and 1962-77, became mgr of technology dev supervising over 200 prof engrs and physical scientists for devt of advanced submarine technologies. In 1978, joined the US Coast Guard Office of R & D and in 1981 bcame asst div chief, Marine Technology Div. ASME Distinguished Service Award, ASME Fellow, 1982, and was a founding member of the Connecticut Academy of Science and Engrg. *Society Aff:* ASME

Chen, Michael M
Business: 1206 W. Green St, M E Bldg, Urbana, IL 61801
Position: Prof *Employer:* Univ of IL *Education:* PhD/ME/MIT; SM/ME/MIT; BS/ME/Univ of IL *Born:* 03/10/32 Area of specialization: Fluid mechanics and heat transfer in energy, mfg, and bioengrg. Concurrent positions: Conslt on Heat Transfer and Fluid Flow. *Society Aff:* ASME, APS

Chen, Ming M
Business: 110 Cummington St, Boston, MA 02215
Position: Prof *Employer:* Boston Univ *Education:* B.S./ME/Nat Wu-han Univ-China; M.S./ME/Univ of IL; M.S. /Aero Engr/Univ of WA; Ph.D./Theo & Appl Mech/Univ of IL *Born:* Dr. Ming Chen has experience in managing large multi-disciplined research projects. During his professional career, he has held numerous and academic institutions throughout the United Sts and abroad. He also served as visiting chair prof and subsequently as VP at the National Cheng-Kung Univ where he was involved in the development of the Engrg Science Research Center and worked with the faculty in basic and applied research. Dr. Chen has served as visiting prof at the Dept of Aeronautics and Astronautics at MA Inst of Technology and as special lecturer at the Hua-zhong Univ of Science & Technology & later at the Fuzhou Univ. Dr. Chen negotiated and designed a five year educational cooperation program between Boston Univ and the Hua-zhong University of Science & Technology. Dr.

Chen, Ming M (Continued)
Chen was on the first Bd of Trustees of the Univ of Lowell and served as the Chrmn of the Committee on Faculty and Academic Affairs. *Society Aff:* AIAA, AAUP, SESA, CAETI, ΣΞ, TBΠ, ASEE

Chen, Mo-Shing
Business: 506 Carlisle Hall, Arlington, TX 76019
Position: Prof of Elec Engg Dir, Energy Sys Res Ctr. *Employer:* Univ of TX at Arlington. *Education:* PhD/Elec Engg/Univ of TX; MS/Elec Engg/Univ of TX; BS/Elec Engg/Natl Taiwan Univ. *Born:* 8/20/31. Dr Chen joined the UTA facualty in 1962 and in 1968, founded the Energy Sys Res Ctr, a unique facility where students engage in innovative res under the guidance of experienced faculty. Annually, Dr Chen & ESRC staff offer the short course "Modeling and Analysis of Modern Pwr Sys," the oldest course of its kind. In his career, Dr Chen has consulted with many companies and has published forty-five articles in various journals. Among the awards he has received are the Exxon Award (1974), EEI Engg Educator Award (1976), ASEE Western Elec Fund Award (1977), and USCIE Achievement Award (1979). *Society Aff:* IEEE, ASEE, USCIE.

Chen, Peter W
Home: 12 Willow Dr, Englewood Cliffs, NJ 07632 *Employer:* Self-employed. *Education:* DSc/Structural Engg/MIT; MSc/Structural Engg/MIT; BSc/Structural Engg/MIT. *Born:* 3/19/34. BS, MS & ScD MIT. Tau Beta Pi, Chi Epsilon & Sigma Xi. Struct Engr Special Structures at Ammann & Whitney 1959-64. Ch Struct Engr Trylon Inc 1964-65. With Skilling, Helle, Christiansen, Robertson 1965-77 1965-77. *Society Aff:* ASCE, NYAS.

Chen, Rong-Yaw
Business: 323 High St, Newark, NJ 07102
Position: Prof *Employer:* NJ Inst of Tech. *Education:* PhD/ME/NC State Univ; MS/ME/Toledo Univ; BS/ME/Natl Taiwan Univ. *Born:* 1/6/33. in Taiwan. Taught at Univ of Toledo 1961-63. Atlantic Christian College, Wilson NC 1965-66 and NJ Inst of Tech 1966-. Have been a principal investigator for 3 grants from the US Army to investigate fluidic contamination and two grants from industries to suudy electrical connectors and strainers. *Society Aff:* ASME, ΣΞ, NATPA.

Chen, Shoei-Sheng
Business: 9700 S. Cass Av, Argonne, IL 60439
Position: Sr. Mechanical Engr. *Employer:* Argonne National Lab. *Education:* PhD/Eng. Mech./Princeton University; MA/Civil Eng. /Princeton University; MS/Civil Eng./Princeton University; BS/Civil Eng./National Taiwan University, Taiwan *Born:* 01/26/40 Native of Taiwan. Joined Argonne National Lab 1968. Currently a senior Mech Engr. Conducts extensive studies on flow-induced vibration. Published over 70 papers and a book on this subject. A short-term Technical Expert for IAEA in 1977, 79 and 80. Chairman of Technical Subcttee on Fluid/Structure Interactions, pressure vessel and piping Div of ASME. Co-Editor and Co-Organizer of a series of Symposia on Flow-induced vibration; Co-winner of the Distinguished Performance Award of the Univ of Chicago in 1986. *Society Aff:* ASME, ASA, AAM

Chen, Ta-Shen
Business: Department of Mechanical and Aerospace Engineering, University of Missouri Rolla, Missouri 65401
Position: Professor *Employer:* University of Missouri-Rolla *Education:* Ph.D./Mech. Engrg./University of Minnesota; M.S./Mech. Engrg./Kansas State University; B.S./Mech. Engrg./National Taiwan University *Born:* 02/05/32 Worked as Mech Engr at Taiwan Shipbuilding Corp., 1955-57 and Ingalls-Taiwan Shipbuilding & Dry Dock Co., 1957-59 in Taiwan. Held the position of Mech Engr, 1963-66, part-time and Research Engineer, 1966-67 with the U.S. Bureau of Mines, Twin Cities Mining Research Center. Academic positions held include Asst Prof 1967-69, Assoc Prof, 1969-73, Professor, 1973-present, and Graduate Coordinator, 1986-present at the Dept of Mech & Aerospace Engrg, Univ of Missouri-Rolla. Pioneered studies on thermal fragmentation of rocks and ores at U.S. Bureau of Mines. Active in heat transfer research, particularly in the areas of mixed convection and instability of buoyancy-affected flows, and have published over 120 technical papers. Received U.S. Bureau of Mines Special Service Award, 1966, Univ of Missouri-Rolla Outstanding Teacher Award, 1968, and Univ of Missouri-Rolla Faculty Excellence Award 1987. Active in ASME Heat Transfer Division's Cttee on Theory and Fundamental Research and AIAA's Thermophysics Technical Cttee. A Fellow of ASME, 1986-present and an Assoc Editor of Journal of Thermophysics and Heat Transfer. *Society Aff:* ASME, AIAA

Chen, Tien Y
Business: Dept of Civil Engrg, 2070 Neil Ave, Columbus, OH 43210
Position: Prof *Employer:* OH State Univ *Education:* PhD/CE/Univ of IL; M/CE/Brooklyn Polytechnic Inst; BS/CE/St. John's Univ *Born:* 1/18/23 With the OH State Univ since 1962. Previously was on the faculty of St. John's Univ, Shanghai, Univ of IL at Urbana-Champaign, Dartmouth Coll. Also was structural designer and engr for a number of engrg firms in China, Hong Kong, New York and Chicago. Author of a number of articles and textbook on reinforced concrete design. *Society Aff:* ASCE, ACI

Chen, Tsong M
6704 Maybole Place, Temple Terrace, FL 33617
Position: Prof *Employer:* Univ of South FL *Education:* PhD/EE/Univ of MN; BS/EE/Natl Taiwan Univ *Born:* 11/25/34 Born in Taiwan, China. Asst Prof and then Assoc Prof of Engrg Sci at the FL State Univ, Tallahassee, FL, from 1964 to 1972. Joined the faculty of the Univ of South FL, Tampa, FL in 1972. Prof of Electrical Engrg from 1974 to present. Have been active in electrical noise research for many yrs and have published more than forty-five technical papers in profl journals and conferences. Was selected as the Outstanding Prof of the Yr in 1971 by the Engrg Sci Honorary, Phi Sigma Upsilon and the Outstanding Educator of Amer in 1972. *Society Aff:* IEEE, ISHM, AVS

Chen, Wayne H
Business: College of Engg, Gainesville, FL 32611
Position: Dean. *Employer:* Univ of FL. *Education:* PhD/EE & Math/Univ of WA; MSEE/EE/Univ of WA; BSEE/EE/Natl Chiao Tung Univ. *Born:* 12/13/22. in China, came to US in 1947, and naturalized in 1957. On staff at Univ of WA as Electronic Engr, Cyclotron Proj, 1949-50, and Assoc, Dept of Math 1950-52. With Bell Tel Labs as Mbr of Technical Staff (Summers 1953 & 1954) and Consultant (summer 1955, and March 1956 to April 1960). With Univ of FL College of Engg since 1952. Assumed current responsibility as Dean of Engg, and Dir, Engg & Ind Experiment Station (EIES) in 1973. Visiting Scientist, Natl Acad of Sciences to USSR, 1967. Author of 2 books and patentee in field. Fellow IEEE 1969. *Society Aff:* IEEE, ASEE, FES, NSPE.

Chenault, Woodrow C, Jr
Business: 816 Dennison Dr, Champaign, IL 61820
Position: Chief Structural Engr *Employer:* Daily & Assoc Engrs *Education:* MS/CE/Univ of IL-Urbana; BS/CE/Univ of IL-Urbana *Born:* 8/18/42 Native of Belleville, IL. Served as Battalion Operations Officer and Co Commander in 9th Engr Bn US Army in Germany from 1964 to 1967. Joined Daily and Assocs Engrs in 1967 as Structural Design Engr. Chief Structural Engr since 1973 and currently Corporate Secy. Projects include responsible charge of: structural design of curved girders and structural steel for approx 6000 ft of elevated urban expressways, structural design for numerous municipal wastewater/water treatment facilities, structural design for numerous highway bridges, Dam Safety Inventory for Corps of Engrs in IL and IN, Failure and other special investigations involving expert testimony, Hydraulic studies and Reports. *Society Aff:* ASCE, ACI, PCI, ASTM, ACEC

Chenea, Paul F
Business: Res Labs, GM Tech Ctr, Warren, MI 48090
Position: VP in charge of Res Labs. *Employer:* GM Corp. *Born:* May 1918. PhD Univ of MI, D Sci Rose Poly Inst, D Engg Purdue Univ, D Engg Tri-State College, D Humane Letters Clarkson College, D Eng Drexel Univ. 1940-41 Proj Engr Contractors Pacific Naval Air Bases Alameda CA. 1941-46 Militay Serv. 1946-52 Faculty, Univ of MI. 1952-61 Educator-Administrator Purdue Univ. 1961-67 VP for Academic Affairs Purdue Univ. 1967-present GM Res Labs; appointed VP, GM Corp, in charge of Res Labs 7/1/69. Mbr: EJC; NAE; AAAS; ASEE; ASME; Fellow Am Acad of Arts & Sci.

Cheney, Donald E
Business: Box 147, Eau Claire, WI 54701
Position: Pres. *Employer:* Finley Engg Co, Inc. *Education:* BS/EE/Univ of WI. *Born:* 4/22/25. Grad Univ of WI BS (EE) 1948. Worked as engr for Gen Telephone Co of WI 1948-1950, Gustau Hirsch Organization 1950-1953, Reliable Elec Co 1954-1959, Pres of Finley Engg Co, Inc 1960-present. Specialize in the design of telephone communication and CATU systems. Reg in WI, IL, MN, PR, Guam, MI, UT and other states. *Society Aff:* IEEE.

Cheney, James A
Home: 658 Elmwood Dr, Davis, CA 95616
Position: Prof. *Employer:* Univ of CA. *Education:* PhD/Engg Mech/Stanford Univ; MS/Engg/UCLA; BS/Engg/UCLA. *Born:* 2/2/27. Born & raised in Los Angeles CA. Spent one yr in the US Navy at the end of WWII. Obtained bachelors & masters degree at UCLA majoring in soil mechanics, married Frankyee Jane Jackson, 1951. Worked for L T Evans, Fdn of Engg for two yrs & became reg PE in CA. Work in strength analysis for Lockheed Missile & Space Co for 10 yrs & obtained PhD from Stanford Univ & then joined teaching staff at Univ of CA, Davis in 1962. Wife died 1966, 6 children survive, remarried Barbara Louise Chadwick, 2 children. Currently the Dir of UC Davis NASA - NSF Geotechnical Centrifuge Facility at Moffett Field, CA & Prof of Civ Engg at UC Davis. *Society Aff:* ASCE, ASEE.

Cheney, Lloyd T
Home: 16858 Avon Rd, Detroit, MI 48219
Position: Prof Emeritus *Employer:* Wayne State Univ. *Education:* MSCE/Struct/Lehigh Univ; BCE/Aerial Photo/Syracuse Univ. *Born:* 11/11/17. in Buffalo, NY. Degree in Civ Engg, Magna Cum Laude, Syracuse Univ 1938. Steel column res at Lehigh Univ as Am Inst of Steel Construction Res Fellow 1938-40. Commendation from office of Scientific Res and Dev 1945. Naval Ordnance Dev Award 1946. Taught at Case-Western Reserve and Cornell Univ before joining faculty at Wayne Univ, Sept 1948. Visiting and adjunct appointments at Univ of MI and Univ of Detroit. Employment by architect-engr firms in Detroit area. Private consulting particularly in area of fire damage to structures. Pres, MI Sec ASCE 1957. Natl dir ASCE 1965-67. *Society Aff:* ASCE, ASEE, ACI.

Cheng, Alexander H D
Business: Dept of Civ Engrg, Newark, DE 19716
Position: Assoc Prof of Civ Engrg *Employer:* Univ of DE *Education:* PhD/Civ Engrg/Cornell Univ; MS/Civ Engrg/Univ of Missouri-Columbia; BS/Civ Engrg/Natl Taiwan Univ *Born:* 05/25/52 Taught at Cornell Univ (1981-82) and Columbia Univ (1982-85) before joining Univ of DE as Assoc Prof. Specializing in flow through porous media, hydraulic facturing and numerical methods. *Society Aff:* ASCE, AGU, IAHR, AAM

Cheng, David H
Business: School of Engg, New York, NY 10031
Position: Dean. *Employer:* City Univ NY. *Education:* PhD/Structural Mech/Columbia Univ; MS/Structural Engg/Univ of MN; BS/Appl Mathematics/Franco-Chinese Univ. *Born:* 4/19/20. b I-Shing, China, Apr 19, 1920; Came to US 1945, naturalized, 1956; MS Univ MN, 1947; PhD (Wm Richmond Peters, Jr Fellow), Columbia, 1950; M Lorraine Hui- Lan Yang, Sept 4, 1949; children - Kenneth, Gloria. Instr Rutger Univ 1949-50; lectr to Prof Coll City NY 1955-66; dir grad studies and exec officer PhD prog in engg, 1977-78; dean of engg, 1979-; cons M W Kellogg Co, Inst Def Analyses, NYC Transp Adm. Recipient 125th Anniversary medal Coll City NY, 1973; Am Soc Engg Ed, NASA-Faculty Fellow 1964-65; Sigma Xi, Tau Beta Pi (Outstanding Teacher Award 1972), Chi Epsilon, Phi Tau Phi. Author: *Nuclei of Strain in the Infinite Solid, 1961; Analysis of Piping Flexibility and Components*, 1973; 30 res papers and 2 monographs. *Society Aff:* ASME, ASCE, ASEE.

Cheng, David H S
70 Farmington Ave 1-J, New London, CT 06320
Position: Electronics Eng. *Employer:* Naval Underwater Systems Center. *Education:* PhD/EE/Univ of MO; MS/EE/Univ of MO; BS/EE/Univ of MO; BA/Pol.Sc./St. John's U; MA/Journ/Univ of MO. *Born:* 10/28/22. in Shanghai, China. Became US citizen in 1961. Educated in Shanghai, China & Columbia, MO. Has been teaching elec engg since 1958 at Univ of MO-Columbia. 1967 & 1968 as ASEE-AEC Summer Faculty Fellow worked at Goddard Space Flight Ctr on VLF propagation and time schronization. As visiting scientist 1967-68 worked at Radiation Lab, Univ of MI on multidimensional antennas. Summers 1970-79 worked at Naval Underwater Systems Ctr, New London Lab, New London, CT on submarine antennas and antenna systems. Mbr of Tau Beta Pi, Eta Kappa Nu, & Sigma Xi. *Society Aff:* IEEE, APS, URSI, TSGB, ΣΞ.

Cheng, David K
Home: 4620 N. Park Ave, Apt. 1405E, Chevy Chase, MD 20815
Position: Prof. Emeritus *Employer:* Syracuse Univ. *Education:* D. Eng (Hon)/Electrical Engineering/Natl Chiao Tung Univ.(Taiwan); ScD/Communication Engg/Harvard Univ; SM/Communications Engg/Harvard Univ; BSEE/Electronics/Natl Chiao Tung Univ. *Born:* 1/10/18. Published two books, "Analysis of Linear Systems," in 1959 and "Field and Wave Electromagnetics" in 1983, and more than 120 journal articles on electromagnetic theory, antennas and arrays, communication and signal-processing systems. Won several best paper awards, an annual res award of Sigma Xi, an Achievement Award of the Chinese Inst of Engrs, and a Distinguished Service Award of the Phi Tau Phi Scholastic Honor Soc. Appointed Centennial Prof. 1970; awarded Chancellor's Citation for exceptional academic achievement in 1981, Syracuse Univ. Was a Guggenheim Fellow in 1960-61, a mbr of IEEE Publications Bd for 1968-70, an Natl Academy of Sci Exchange Scientist to Hungary in 1973, to Yugoslavia in 1974, and to Poland and Romania in 1978. Has been a Consulting Editor for Addison-Wesley Publishing Co. and Intext Educational Publishers, and an Engrg Consultant for GE, IBM, Syracuse Research Corp., and TRW. *Society Aff:* IEEE, IEE, AAAS, ΣΞ, HKN, URSI, ΦΤΦ

Cheng, Franklin Y
Business: Civil Engg Dept, Rolla, MO 65401
Position: Prof. *Employer:* Univ of MO-Rolla. *Education:* PhD/Civil Engg/Univ of WI-Madison; MS/Civil Engg/Univ of IL-Urbana; BS/Civil Engg/Taiwan Natl Cheng-Kung Univ. *Born:* 7/1/36. Prof Cheng was born in China and became a citizen on the US by naturalization. He is married to Dr Pi-yu Chang whose doctoral degree in biochemistry was conferred by the Univ of WI-Madison. They have two children, George and Deborah, aged 21 and 16, respectively. Dr Cheng has been very active in research which is demonstrated by an impressive list of publications - more than 140 papers and tech reports and a textbook. He initiated the First Interntl Symposium on Earthquake Structural Engg and edited the two-volume symposium proceedings. Prof Cheng is not only teaching a wide range of undergraduate and grad courses, but also he is active in engg consulting and directing and teaching short courses and seminars for practicing engrs and academic researchers. He has directed and taught more than ten short courses & has participants from every state of the Union & many foreign countries. Dr Cheng always takes the leading step of responding to the changing natl res needs by serving as chairman of several profl

Cheng, Franklin Y (Continued)
committees & currently directing four res proj sponsored by the Natl Sci Fdn. Prof. Cheng received honorary Professorship from the Harbin Civil Engg Institute in China, 1981, & was awarded the Halliburton Excellence Award & Excellence Faculty Award from the Univ of Missouri-Rolla in 1986-87. *Society Aff:* ASCE, ASEE, EERI, SSRC, IASSAR.

Cheng, Henry M
Home: 4106 Landgreen St, Rockville, MD 20853
Position: Asst for Ships & Vehicles Plans. *Employer:* Dept of Navy. *Education:* Marine Mech Engr/Naval Arch/MIT; MS/Naval Arch/MIT; BS/Marine Engg/Univ of MI. *Born:* 7/21/24. Hold BS degree from Univ of MI & MS degree from MA Inst of Tech (MIT) in the fields of naval arch & marine engg. Also hold Profl Marine Mech Engr degree from MIT. Have been a designer and design supv at the Newport News Shipyard; a res naval architect at advisor for ships and vehicles in the Office of Chief of Naval Operations, and presently serve as the Asst for Ships ad Vehicles Plans in that Office. Author of technical reports and papers on ship machinery, hydrodynamics, propulsion systems, advanced ship design, ship systems, and economics. Mbr of the Am Inst of Aero and Astro, and of the Soc of Naval Architects and Marine Engrs (SNAME). Served on SNAME Ship Machinery Committee and Technical and Res Panels. Recipient of Capt Joseph H Linnard prize from SNAME in 1965. *Society Aff:* SNAME, AIAA.

Cheng, Ping
Business: Dept of Mech Engg, 2540 Dole St, Honolulu, HI 96822
Position: Prof. *Employer:* Univ of HI. *Education:* PhD/Aero & Astro/Stanford Univ; MS/Mech Engg/MIT. *Born:* Born in China. Naturalized 1972. Visting Prof at Natl Taiwan Univ 1968- 1970. Joined Univ of HI as Assoc Prof of Mech Engg in 1970; promoted to the rank of Prof since 1974. Visting Prof of Petroleum Engrg, Stanford Univ. (1976-1977), Guest Prof of Tech Univ of Munich, 1984. Served as ctte mbr on Heat Transfer in Energy Sys. ASME Heat Transfer Div since 1975. Served as Session Chrmn for various heat transfer confs since 1977. Published papers on radiative heat transfer, radiative gasdynamics, & heat transfer in porous media & geothermal sys. Listed in "Who's Who in the West–, "Who's Who in the Republic in China–, "Who's Who in Technology Today–, "American Men & Women of Science–, "Int Who's Who in Engineering–, and "Men of Achievement–. Recipient of 1969 Chung-Shan Award for Res Excellence. Natl Sci Fdn & Dept of Energy Grantee. Fellow of ASME 1986. *Society Aff:* ASME, ΣΞ.

Cheng, Shang I
Business: 51 Astor Pl, New York, NY 10003
Position: Prof. *Employer:* Cooper, Union, The. *Education:* PhD/ChE/Univ of FL; MS/ChE/Univ of FL; BS/ChE/Natl Chekiang Univ. *Born:* 6/5/20. in China. After grad from Natl Chekiang Univ with a BS degree in Chem Engg, he worked for Chinese Govt owned RD Organizations for 13 yrs before coming to US for grad studies in 1958. He worked for B F Goodrich, Chem Co as a dev scientist before he joined the Cooper Union faculty in 1965. He is married with five children. He is teaching applied math in chem engg, dynamics and control and chem process design. His current res interests are in the areas of dynamic modeling of chem engg systems, pollution abatement and integrated gasification. *Society Aff:* AIChE.

Cheng, William Jen Pu
Home: 705 Louwen Dr, St Louis, MO 63124
Position: Vice President. *Employer:* Petrolite Corp *Education:* MS/Chem Engg/Washingon Univ, St Louis, MO; BS/Chemistry/Tsing Hua Univ, China. *Born:* 9/26/15. Changsha, China. Served China Vegetable Oil Corp as plant superintendent, 1941-44 Chinese Army 23rd Arsenal as chemical engr, 1944-46. Came to US in 1948, naturalized. married Chuan Huan Wu in 1954. Children: Elizabeth, James, Nancy and Helen. With Petrolite Corp since 1952. Served as Hd, Pilot Plant, 1960-63, Engg Res Mgr, 1963-67, Dir of Engg, 1968-1979 and Vice President, Oil Field and Indust Chem Grp, since 1980. Specializing in chemicals for petroleum production, refining and pollution control with 31 patents in field. Elected Fellow of AIChE. Enjoy photography and classical music. *Society Aff:* AIChE, ACS.

Chenoweth, Darrel L
Home: RR 3, Paoli, IN 47454
Position: Prof. *Employer:* Univ of Louisville. *Education:* PhD/EE/Auburn Univ; MS/EE/Auburn Univ; BEE/EE/GMI. *Born:* 11/6/41. Native of Richmond, IN. Proj engr for Gen Motors Corp in 1963. Employed by Auburn Univ as Res Asst and Instr between 1964 and 1969 while pursuing PhD degree. With LTV Aerospace Corp in 1969, specializing in fly-by-wire control systems. Joined Univ of Louisville in 1970, promoted to Assoc Prof of Elec Engg in 1972, and Prof of Elec Engg in 1978. Responsibilities include teaching and sponsored res in microrompuoters and control systems. Chrmn of Southeast Region of ASEE Electrical Engg Div in 1975. Married in 1963 and have two daughters. *Society Aff:* IEEE, ASEE, ISPE.

Cheo, Peter K
Business: United Tech Research Center, E Hartford, CT 06108
Position: Sr Principal Scientist *Employer:* United Tech Research Ctr *Education:* Honorary Doctor/Science/Aurora Univ; PhD/Physics/OH State Univ; MS/Physics/Inst; BS/Physics/Aurora Univ *Born:* 2/2/30 Taught in Colls for 7 years from 1954-1961. A member of technical staff at Bell Telephone Labs from 1963-1970. A mngr of Laser Applications Research at Aerojet Electro Systems from 1970-1971. A Sr Principal Scientist at United Technologies Research Ctr from 1971 to present. An adjunct professor in Electrical Engineering at the Hartford Graduate Center from 1978-present. Author of a text book on Fiber Optics & an Editor of a series of Handbooks on Lasers and optical Technologies. *Society Aff:* OSA, IEEE

Cherna, John C
Business: 817 Sherbrooke St W, Montreal Quebec H3A 2K6. Canada
Position: Assoc Prof. *Employer:* McGill Univ. *Education:* DiplEng/Mech/Swiss Fed Inst of Tech; CandEng/Mech/Zurich, Switzerland. *Born:* 4/6/21. Native of Budapest, Hungary - Studied at Budapest and Zurich, Switzerland. Ind positions in Switzerland, France & Canada. Since 1961 at McGill Univ, currently teaching courses in Engg Graphics & Design and Ind Engg. Past Chrmn of Montreal Section of SAE, & of Montreal Branch of CSME, Faculty Advisor at McGill for SAE & CSME. Fellow of the Engineering Institute of Canada 1984. Ralph R Teetor award 1979, SAE. McGill Engineering Outstanding Teacher Award 1986. *Society Aff:* SAE, ASM, ΣΞ, CSME.

Cherne, Realto E
Home: 2500 Elmwood Ave, Rochester, NY 14618
Position: Consultant. *Employer:* Harold Bumpus, Jr, Conslt Engr *Education:* ME/ME/Univ of MN; BME/ME/Univ of MN. *Born:* 12/10/07. Native of Duluth. Employed by Carrier Corp following grad in 1929. Between then and entering private practice in 1945: handled a number of engg and mgt assignments of increasing responsibility, NYC and Newark offices; Chief Engr, Sydney, Australia; Branch mgr, Cincinnati; Dir of Sales in Syracuse hdquarters. Consulting Engr for over 1600 projs of various sizes and complexity: schools, colleges, hospitals, high-rise bldgs, athletic facilities, banks, and various commercial and industrial projs. Co-author of "Modern Air Conditioning, Heating and Ventilating" (3 editions); author of over 60 articles and papers; 1970 Rochester Engr of the yr; 1974 Distinguished Service and 1979 Distinguished Fifty-Year Member Awards from ASHRAE; 1975 Outstanding Contribution Award from CEC/NYS; Lighting Design Award of Merit from IES; 1980 "Others" Plaque and Citation from The Salvation Army; served on several local, state, and national volunteer boards. *Society Aff:* ASHRAE, IES, IAEI.

Chernock, Warren P
Business: 1000 Prospect Hill Rd, Windsor, CT 06095
Position: VP-Dev Nuclear Power Systems. *Employer:* Combustion Engg Inc. *Born:* Jan 1926. BS Met Engg Columbia Univ; MS Met NYU. US Army 1943-46. Met Engr Argonne Natl lab & Sylvania Elec Prod spec in met R&D & nucl fuels dev.

Chernock, Warren P (Continued)
Wth Combustion Engg Inc since 1956. Assumed current pos as VP Dev, Nucl Power Sys in 1974. Chrmn Met Dept Hartford Grad Ctr of RPI since 1956. Formerly visiting Assoc Prof of Nucl Engg at MIT. Fellow ANS & ASM. Mbr Tech Adv Ctte of Metal Properties Council. Chrmn MPC Subctte on Nucl Mtls.

Cherrington, B E
Business: Ericsson School of Engineering and Computer Science, Univ. of Texas at Dallas, P.O. Box 830688 Richardson, TX 75083-0688
Position: Dean, School of Engrg & Computer Science *Employer:* Univ of Texas at Dallas *Education:* PhD/EE/Univ of IL; MASc/EE/Univ of Toronto; BASc/EE/Univ of Toronto. *Born:* 3/16/37. Assumed current position as Dean of the School of Engineering and Computer Science and Ericsson Professor of Electrical Engineering, Univ. of Texas at Dallas, in 1986. Previously served Prof & Chrmn of Elec Engg, Univ of FL, Prof of Elec Engrg, Prof of Nucl Engg & Res Prof of the Coordinated Sci Lab at the Univ of IL at Urbana-Champaign. Served as an ACE Fellow in Academic Admin 1977-78. Author of thirty-one articles in profl journals and one book, "Gaseous Electronics and Gas Lasers" published by Pergamon Press. *Society Aff:* IEEE, ASEE, APS, TBΠ, HKN, ΣΞ, SME.

Cherry, C Douglas
Business: 55 S Main St, Phillipsburg, NJ 08865
Position: Pres. *Employer:* C Douglas Cherry & Assoc P A. *Education:* MS/Civ Engr/Newark Coll of Engr; BS/Civ Engr/Lafayette Coll *Born:* 1936. MS Newark Coll of Engg, BS Lafayette Coll. Native New Brunswick, NJ & Abington, PA. Hgwy & Bridge designer, PA Dept of Hgwy 1958-63, Municipal Engr, Phillipsburg NJ 1963-67, Cty Engr Hunterdon Cty 1967-70. Pres C Douglas Cherry & Assoc P A 1970-present. Municipal Engr for 8 NJ Municipalities. Design lrg subdivs & P U D's P Pres Northwest Chap NJ Soc of PE. Young Engr of Yr 1971, Spec award NJ Soc Municipal Engrs 1971 for "Untiring efforts to estab high standards of practice & ethics–. auto restoration. 1983 First Place "Project of the Year Award" NJ Soc of Municipal Engrs. PE NJ & PA, Prof Planner & Land Surveyor NJ. Hobbies-Antique auto restoration. *Society Aff:* NJPE

Chertow, Bernard
Home: 139 Sunnyside Park Rd, Syracuse, NY 13214
Position: Sr VP/Operations *Employer:* Bristol Myers Co, Indus Div *Education:* ScD/Chem Engg/MA Inst of Tech; MS/Chem Engr/IL Inst of Tech; BS/Chem Engr/IL Inst of Tech. *Born:* 12/30/19. To 1949 as Asst Prof in Chem Engg at MIT did and directed res in areas of gas adsorption, emulsion breaking (several publications); Also taught engg courses directed MIT field station at Parlin NJ. Since 1949 to date have been with Bristol Myers Antiobiotics and Anti Cancer Div in Syracuse. Hold patents for production of tetra cycline and mutamycin (anti cancer). After 25 yrs in dev of processes became gen mgr of Syracuse Bulk Operations bringing automation and process control via computer to antibiotic operations. Jan 1980 became VP Manufacturing, worldwide operation; Dec 1980 became VP Oper; Jan 1983 became Sr VP Oper; all since '74 Bristol Myers Company Industrial Division, Dev. & Mfr. of Antibiotics Anticancer and the prescription pharm products. Jan '85 retired Bristol Myers, joined Galson Research Corp, E. Syracuse - Pres 1985-87. Current (1987) Bd Mbr Galson Research Corp and Dir Chemical Eng'g Galson & Galson Consulting Engrs. Galson Research Corp is dedicated to process development for management or destruction of toxic and hazardous materials in the environment. *Society Aff:* AIChE, ACS, AAAS, ISPE, ΣΞ.

Chesnut, Donald R
Home: 1330 Pin Oak Cir, Wichita, KS 67230
Position: Prog Mgr *Employer:* Boeing *Education:* BS/EE/KS State Univ *Born:* 1/2/26 1944-46 served in US Navy. 1950-56 Experimental Flight Testing on B-47, B- 52 in Wichita, KS. 1957-61 in charge of Bomarc Missile Launches. 1961-64 Design & Test Mgr Minuteman II Instrumentation. 1964-67 Design Mgr Lunar Orbiter Spacecraft Power & Communication Systems. 1967-69 Technology Mgr Apollo Fire Investigation. 1969-72 Boeing Base Mgr for Minuteman III R&D launches. 1972-4 Boeing Area Mgr for Apollo Skylab, Huntsville, AL. 1974-9 Pres of Boecon, Boeing Engrg & Construction Subsidiary. 1979-80 Minuteman Prog Mgr. 1980-82 B-52 Prog Mgr, Wichita, KS. 1983-present, VP Avionics & Weapon Sys Integration Prog. Current organizations has over 1600 profl level engrs. *Society Aff:* AIAA

Chesnut, Dwayne A
Business: The Forum at Cherry Creek, 425 S. Cherry Street - Suite 300, Denver, CO 80222
Position: Pres. *Employer:* Critical Resources, Inc *Education:* PhD/Theo Chem/Rice Univ; BS/Chem Engg/Rice Univ. *Born:* 3/8/36. Native Texan. Lab experience in oil & gas industry during summers from high school through grad school; extensive scientific prog, Rice Computer Proj. Joined Shell Dev Co's Exploration and Production Res Div, Jan 1963 as Res Chemist; Consultant in Theoretical Physics, Lawrence Radiation Lab, 1964-5; Sec Leader, Physical Chemistry, Shell Dev, 1965-8; Staff and Sr Staff Reservoir Engr, Shell Oil, Denver, 1968-1974. Founder of Energy Consulting Assoc, 1974 and President, 1974-1981. Founder of Critical Resources, 1981. Publications in Journal Chemical Physics, Journal Petroleum Tech, others; several patents. Chrmn, Gen Editorial Comm, SPE of AIME, 1973. *Society Aff:* AIME, AIChE, AIC, NSPE, NYAS.

Chesson, Eugene
Business: 130 DuPont Hall, Newark, DE 19716
Position: Prof *Employer:* Univ of DE *Education:* PhD/Civ (Struct)/Univ ov IL Urbana; MSCE/Civ (Struct)/Univ of IL Urbana; BSCE/Civ Engr/Duke Univ *Born:* 12/01/28 Born in Brazil, son of Eugene and Mary Foy Chesson. Married Marilyn Hershey Aug 1954. Served as officer in US Navy Civ Engr Corps 1950-1953. Refinery Engr, Standard Oil, Indiana, 1953. Asst and Assoc Prof, Civ Engrg, Univ of IL 1959- 1966. Prof and Chrmn, Civ Engrg, Univ of DE 1966-1975. Prof (U of D) 1975- present. Author of more than 50 papers and reports. Active at state and natl levels in several professional socs. A Epstein Award, Univ of IL, 1962. W E Wickenden Award, ASEE 1981. Outstanding Engr for DE, DSPE 1981. Reg PE IL and DE. Consit in structs and failures. *Society Aff:* ASCE, ASEE, NSPE, AISC.

Chester, Arthur N
Business: Bldg. E55/G200, PO Box 902, El Segundo, CA 90245
Position: Group VP & Mgr, Strat Systems Div *Employer:* Hughes Aircraft Co *Education:* PhD/Theoretical Physics/CA Inst of Tech; BS/Physics/Univ of TX-Austin *Born:* 8/5/40 Held research positions at Rockwell Int'l, Caltech, Hughes Research Labs (1963-1975), and Bell Telephone Labs (1965-1969). Deputy Dir Hughes Research Labs, 1975-1980. 1980-83, Asst Mngr Strategic Systems Div, Electro-Optical and Data Systems Group, Hughes Aircraft Co; also co-wide Program Mngr for Very High Speed Integrated Circuits, a major effort to develop design tools and VLSI circuits for next generation signal processors. 1983-85, Mgr Tactical Eng Div. 1985-present, Group VP & Mgr Space & Strategic Systems Division, responsible for space sensors & high energy Laser beam control product lines. IEEE Quantum Electronics and Apps Society: VP 1979, Pres 1980. Fellow of IEEE. Author of over 40 publications on quantum electronics and solid state physics. Co-Director, Int'l School of Quantum Electronics, Italy, 1980-present. *Society Aff:* IEEE, QEAS, APS, OSA, AAAS

Chew, Woodrow W
Business: Box 3066, Ruston, LA 71272
Position: Prof Emeritus *Employer:* LA Tech Univ. *Education:* MS/Chem Engg/OK State Univ; BS/Chem Engg/NM State Univ. *Born:* 1/29/13. Native of Alva, OK, Asst and Assoc Engg Public Works Administration 1938- 39. Taught Chemistry and Physics, Northeast Jr College, OK, 1939-40. Asst, Assoc, Prof and Head of Chem Engg Dept, LA Tech Univ 1940-75. Dev ChE Dept and graduate programs. Consultant, Petroleum refining and water quality. Active in AIChE. Mbr Research Adviser Committee, Natural Resources and Energy Dept, State of LA 1973-75. PE 1950-

Chew, Woodrow W (Continued)
Present. Retired LA Tech Univ, 1975. Presently, Pres, Woodrow W Chew and Assoc, Inc, Engg & Management Counsultants. Have agri and investment interests. Enjoy travel, music and reading. *Society Aff:* AIChE, ACS, ASEE, SIRE.

Chewning, Ray C
Home: 3414 Fairlomas Rd, San Diego, CA 92050
Position: Principal. *Born:* Aug 29, 1914; Univ of WA 1934-37. Worked for J Donald Kroeker Cons Engr 1938-40. Served in Army 1940-45; with Third Armored Div in Europe 1943-45. Decorations incl Bronze Star & Purple Heart. Rejoined Donald Kroeker 1945-74. Designed heat pump sys for first lrg commercial bldg, the Equitable Bldg in 1947 & co-authored papers on results. Equitable Bldg designated by ASME as "National Historical Mechanical Engineering Landmark" May 1980. Joined Dunn, Lee, Smith, Klein & Assoc 1974-1977. Now semi -retired. Served on following ASHRAE Cttes: Res 1955-58, Guide 1958-60, Honors and Awards 1970-73, Journal 1974-75. ASHRAE awards: Disting Serv Award 1965; Fellow 1970. Plays golf. *Society Aff:* ASHRAE

Cheyer, Thomas F
Business: One Ctr Plaza, Boston, MA 02108
Position: Sr VP., Director, Partner *Employer:* Camp Dresser & McKee Inc. *Education:* MSCE/Water Resources/MIT; BCE/Civ Engr/CCNY; BA/Math & Liberal Arts/Columbia College. *Born:* 5/15/35. Lectr in Civ Engg at the CCNY, 1962-63. Drainage Engr at King & Gavaris, NYC, 1962-63. Assoc with CDM since 1965. Joined the firm as a hydraulics & computer applications engr. Water supply engr and, ultimately, proj mgr on maj system redesign for Bangkok, Thailand. Became sr specialist in hydraulics in 1970; a VP of Camp Dresser & McKee, Inc in 1972. Appointed Sr VP, Partner and Dir in 1975. Mbr of firm's finance committee & technical review committee. Mbr of NEWWA & AWWA standpipe stds committee. Reg PE since 1967; reg in all New England & four other states. Diplomate, Amer Acad Environ Engrs. *Society Aff:* ASCE, AWWA, WPCF, AGU, XE, ΣΞ

Chhabra, Girdhari
Home: 90 Station Rd, Great Neck, NY 11023
Position: Pres. *Employer:* G Chhabra, Engr PC. *Education:* MS/Structural Engg/Univ of CA. *Born:* 8/10/36. Mr Chhabra has over 20 yrs experience in the design and construction of high rise bldgs, parking structures, bridges, water towers, and industrial bldgs. His previous assignments include Chief Design Engr and Resident Engr for multi-million dollar projs, Precast Plant Engr, Construction Mgr, and Chief Engr in a natl consultant firm. He has the unique advantage of having worked directly in construction as well as in design. He is regarded as one of the leading structural engrs. He specializes in design, precast prestressed, cast-in-place post-tensioned concrete, and high strength structural steel. *Society Aff:* ACI, PCI, NSPE.

Chi, Andrew R
Home: 3704 Chandler Dr, Fort Washington, MD 20744
Position: Staff Engr. *Employer:* NASA Goddard Space Flight Ctr. *Born:* Sept 1920; MA Columbia Univ 1945, BS Western MD Coll 1944. Taught phys at Cooper Union Sch of Engg 1946-53; Res Physicist at Army Signal Corps Engg Labs 1953-57 & Naval Res Lab 1957-63; Hd of Timing Systems Sect of Adv Dev Div 1965-71 & Staff Engr of Network Engg Div since 1971 at Goddard Space Flight Ctr. Chrmn IEEE's G-IM Tech Ctte on Frequence & Time, Pres of G-IM 1972, Mbr of Study Group 7 of CCIR since 1964, & Mbr IEC TC-49 since 1965. Guest Ed of Spec Issue on Frequency Stability of Proceedings of IEEE 1964, Cons Mbr of Apollo Navigation Working Group 1967-69, Tracking Scientist for OMEGA Pos Location Experiment-OPLE, Chrmn IEEE/NASA Symposium on Short Term Frequency Stability 1964 Editor, Trans. I & M since 1979. Disting Serv Award GSFC 1965, Scroll of Appreciation of Secy of St 1967, Skylab Achievement Award 1974 Moe I Schneebaum Memorial Award, 1978, Achievement Award Chinese Engineers and Scientists Ass'n of Southern California 1980. Fellow IEEE & Mbr Am Physical Soc, Philosophical Soc of WA & RESA.

Chi, Michael
Business: 956 N Monroe St, Arlington, VA 22201
Position: Pres. *Employer:* Chi Assocs, Inc. *Education:* ScD/Mechanics/George Wash Univ; MS/Hydraulics/LA State Univ; BSCE/CE/Univ of Tientsin. *Born:* 1/9/25. in China, naturalized US citizen in 1957. Taught in the Catholic Univ of America since 1958. Structural Res Engr with Natl Bureau of Stds. Deputy group Leader with Preload Engrs, Inc. Presently the Pres of Chi Assocs, Inc and reg PE in DC and VA. Certificate of Recognition. Natl Hgwy Traffic Safety Admin. Enjoy classical music and bridge games. *Society Aff:* ASME, WAS.

Chian, Edward S K
Business: School of Civil Engrg, Atlanta, GA 30332
Position: Prof *Employer:* GA Inst of Technology *Education:* ScD/Biochem Engrg/MIT; ChE/Chem Engrg/MIT; M.S./ChE/OK St; DIP/ChE/Taipei Inst of Tech *Born:* 5/24/35 in Shanghai, China. Research engr for Kimberly-Clark Corp, specializing in paper coatings. Senior research engr for Abcor, Inc., responsible for biochemical engrg design and membrane processes. Senior biochemical engr for Tenneco Chem, specializing in single cell protein research. Mgr of Technology for American Standard, responsible for the production of reverse osmosis equipment. Taught at the Univ of IL at Urbana-Champaign from 1971-78. Assume current responsibility as Research Leader and Prof in Environmental Engrg at GA Tech since 1978. *Society Aff:* ASCE, AIChE, ACS, WPCF, IAWPCR

Chiang, Donald C
Home: 9 Salem Pl, Terre Haute, IN 47803
Position: Prof *Employer:* Rose-Hulman Inst of Tech *Education:* PhD/Fluid Mech/Univ of MN; BS/ME/Taiwan Coll of Engrg *Born:* 1/29/31 Production engr for Taiwan Aluminum Co. Maintenance engr for Chinese Petro Corp. Teaching Mechanical Engrg since 1965. Naturalized US citizen. Registered PE in IN. Experienced in Technical translation (Japanese and Chinese). *Society Aff:* AIAA, ASME, ASEE

Chiang, Fu-Pen
Business: Dept of Mech Engrg, Stony Brook, NY 11794
Position: Leading Prof. *Employer:* State Univ of NY at Stony Brook *Education:* PhD/Engrg Science/Univ of FL; MS/Solid Mech/Univ of FL; BS/CE/Nat Taiwan Univ *Born:* 10/10/36 Joined SUNY Stony Brook as Asst Prof in 1967, promoted to Assoc Prof with tenure in 1970, Prof in 1974, leading Prof. 1987. Was visiting Prof (1973-1974) at Swiss Federal Inst of Tech (Lausanne) and Sr Visiting Fellow (1980-81) at Cavendish Lab, Univ of Cambridge. Specializes in optical methods of stress analysis and nondestructive testing. Contributed to the development of moire, holography, laser and white light speckle (random pattern) methods. Published extensively. Listed in Int'l Who's Who of Intellecturals, American Men & Women of Science, Who's Who in the East, Dictionary of Int'l Biography, Men of Achievement, Notable Americans of the Bicentennial Era, Who's Who in Tech Today, Who's Who of Sino-Americans, Who's Who in Frontier Sci & Tech, Dictionary of Distinguished Americans. Who's Who in the World 1985, 1986. Conslt to various governmental and industrial organizations. Currently also the Dir of the Lab for Experimental Mechanics Res. *Society Aff:* SEM, OSA, SPIE, AAM, AAAS, ASME, NYAS, ΣΞ

Chiang, Shiao-Hung
Business: Chemical/Petro Engrg Dept, Univ of Pittsburgh, Pittsburgh, PA 15261
Position: Prof *Employer:* Univ of Pittsburgh *Education:* PhD/ChE/Carnegie-Mellon Univ; MS/ChE/KS State Univ; BS/ChE/Nat Taiwan Univ *Born:* 10/10/29 He received his graduate education in the US. In addition to his teaching career, he has had extensive industrial experience in process design and analysis, involving a wide range of chemical industries and energy techs. He has served as a technical advisor/consultant to a number of industrial and governmental organizations. His research interests and publications are in the field of chemical engrg, fuel conversion process-

Chiang, Shiao-Hung (Continued)
es, industrial water use and coal processing. *Society Aff:* AIChE, ACS, ASEE, AIME, Filtration Soc.

Chiantella, Nathan A
Business: Dept 968/Bldg 630, P.O. Box 12195, Research Triangle Park, NC 27709
Position: Consultant *Employer:* IBM Corp *Education:* B/EE/NY Univ *Born:* Served in WWII in US Navy 1943-46, Radar Control Systems. From 1950 to present with IBM Corp. Positions include staff & management assignments in engrg development, production, field & marketing. At present is an IBM industry consultant in manufacturing systems. Recipient of an IBM outstanding contribution award for a model of a computer integrated manufacturing plant. SME offices held include: Pres of Computer & Automated Systems Assoc, Int'l Dir of SME (1981-82). Is a Certified Manufacturing Engr. Listed in Who's Who in Finance and Industry, Dictionary of Int'l Biography. *Society Aff:* IEEE, SME, CASA/SME

Chieri, Pericle A
Home: 142 Oak Crest Dr, Lafayette, LA 70503
Position: Prof & Hd ME Dept, USL (Retired) *Employer:* Dr. P. A. Chieri, PE Consulting Engr *Education:* Dr Aero E/Engrg (Aero.)/Univ of Rome-Italy; Dr Ing/Nav Arch & Mar Engrg/Univ of Genoa-Italy. *Born:* 1905. Chekiang prov. China; U.S. Citizen, nat. 1952; Married, 1938; Naval arch & mar eng, resch & exper div. Submarines, Ital Navy, Spezia, 1929-31; Naval arch Mar Engr Superintndt Libera Lines Ship Corp, Trieste, Genoa, It, 1931-35 Aero engr & Tech. Advisor, Chinese govt Commissn Aero Affairs, Nanchang, Loyang, 35- 37 Dir Mater Test Lab & Superintdt Tech Voc Instruction, Central Milit Aircraft Facty Nanchang, China, 1937-39; Aero Engr FIAT Aircft. Facty, Turin It, 1939; Aero Engr & Tech writer, Office Air Attache Italian Embassy, Washington, DC, 1939-41; Aero Eng Faculty, Tri-St Coll, Angola, IN, 1942; Aero Engr Helicopter Design, API, Aero. Products Inc, Detroit, MI, 1943-44; Sr. Aero Engr ERCO Engrg & Resch Corp Riverdale, MD, 1944-46; Assoc Prof Mech Engrg Univ Toledo, OH, 1946-47; Graduate Div Fac Assoc Prof Mech Engrg, Newark Coll Engrg, Newark, NJ, 1947-52; Prof & Head, Mechancial Engr Dept, Univ Southwestern LA, Lafayette, LA. , 1952-72; Consulting Engineer, Mem. CEC/L, Lafayette, LA., 1972-to date; Registered Professional Engrg: LA, NJ, SC, Italy, United Kingdom, (Ch. Eng.). *Society Aff:* AIAA, ASEE, ASME, SNAME, RINA

Childers, Donald G
Business: Dept of Elec Engg, Gainesville, FL 32611
Position: Prof. *Employer:* Univ of FL. *Education:* PhD/EE/Univ of S CA; MS/EE/Univ of S CA; BS/EE/Univ of S CA. *Born:* 2/11/35. in the Dalles, OR. BS 1958, MS 1959, PhD 1964 from Univ of S CA. Taught at Univ of S CA 1962-64, Univ of CA Davis 1964-65, Univ of FL 1965-present where now holds title of prof. Indus exper incl Aeronutronic 1958-60; 61-64 & Hughes Aircraft Corp 1960-61. Reg engr FL. ASEE southeastern sect award for teaching 1971-72 & res 1972-73. IEEE William J Morlock Award in biomedical electronics 1973. ASEE Geo Westinghouse Award 1975. IEEE Fellow 1976. Editor 1 bk. Co-author 2 bks. Dir Mind-Machine Interaction Res. Ctr. Enjoys golf. *Society Aff:* IEEE, ASA, AI, ACM, ASHA.

Childers, Mark A
Business: PO Box 61780, New Orleans, LA 70161
Position: Mgr of Domestic Operations *Employer:* ONECO, Inc. *Education:* BS/CE/VPI; MS/Hyd & Str/VPI. *Born:* 10/21/40. Formerly with Exxon Corp, U.S.A. and Exxon Production Research Co. Since joining ODECO in 1972, he has worked in many areas including new rig design; drilling and marine systems engineering, dynamic position, deepwater design, construction and operation; and currently is responsible for the operation of 34 MODU's in the Gulf of Mexico. He is the 1969 recipient of the SPE-AIME Cedrick K. Ferguson medal, a 1981-82 SPE-AIME "Distinguished Lecturer–, and has lectured in a number of Industry Schools. He holds a number of patents (both U.S. and foreign), and is a registered engineer in Texas and Louisiana. *Society Aff:* SPE-AIME.

Chilton, Ernest G
Business: Design Div - ME Dept, Stanford, CA 94305
Position: Prof. *Employer:* Stanford Univ. *Education:* PhD/ME/Stanford Univ;MS/Aero/Caltech; SB/Aero/MIT. *Born:* 5/3/19. Univ of Akron, Guggenheim Airship Inst 1942-45. Firestone Indust Products Div-Res 1945-1946 Stanford Univ 1946-1947. Shell Dev Co - 1947-1959. SRI Intl, Mgr Mechanics 1959-1969; Prof of ME, AZ State Univ 1969-1973; Prof of ME, Stanford Univ 1973-date. 3 books, 25 publications. Married, 3 sons. Present interest: teaching techniques, products liability consulting. *Society Aff:* ASME, ASEE, AAAS, ΣΞ.

Chin, Allan
Home: 854 Seale Ave, Palo Alto, CA 94303
Position: VP Engg & Mktg. *Employer:* Intl Nutronics Inc. *Born:* July 26, 1930. MS ChE MIT 1953. BS ChE MIT 1952. Varied engg pos with GE Co during 12 yrs. Mgr Prod Dev, Raychem Corp 3 yrs. VP past 7 yrs Intl Nutronics Inc respon for design of gamma facilities & marketing of radiation serv. Reg PE/Chem in CA. Mbr AIChE. Mbr ANS, HPS, Sigma Xi. Currently VP Gamma Fire Inc.

Chin, Gilbert Y
Home: 54 Roland Rd, Murray Hill, NJ 07974
Position: Dir *Employer:* AT&T Bell Labs *Education:* ScD/Met/MIT; SB/Met/MIT *Born:* 09/21/34 Native of China; came to US in 1950. Started work as Mbr of Tech Staff at AT&T Bell Labs in 1962 performing res on magnetic alloys. Appointed Dept Hd in 1973 and assumed current position of Dir of the Materials Res Lab in 1984. Mathewson Gold Medal, TMS-AIME 1970; Achievement Award, Chinese Inst of Engrs USA 1980; Achievement Awd, Chinese Engrs and Scientists Assoc of Southern CA 1983. Elected to Natl Acad of Engrg 1982. Active Episcopalian. *Society Aff:* ASM, TMS-AIME, ACerS, Mag Soc IEEE, AAAS, NAE, APS

Chinitz, Wallace
Business: 8 Haddon Hall, Melville, NY 11747
Position: Prof. *Employer:* Cooper Union. *Education:* PhD/Mech Engg/Polytechnic Inst of NY; MME/Mech Engg/Polytechnic Inst of NY; BME/Mech Engg/City College of NY. *Born:* 3/13/35. Native of NYC. Served in research engg capacities at Fairchild Engine Div and Republic Aviation Corp. Research Asst and Asst Prof of Mech Engg at the Polytechnic Inst of NY. Staff Scientist at Gen Applied Science Labs specializing in combustion problems in high-speed flows. Assoc Prof and, since 1975, Full Prof of Mech Engg at The Cooper Union, School of Engg. Consultant in energy, Air Pollution high-speed reacting flows, turbulent flows. Assoc Fellow, AIAA. Enjoy classical music and horticulture. *Society Aff:* AIAA, ASME, ASEE, CI.

Chinners, James E, Jr
Business: 363 North Belt-Suite 1600, Houston, TX 77060
Position: Corp Account Mgr. *Employer:* Occidental Chem Corp. *Education:* MBA/MGT/Cleveland State Univ; BS/ChE/Clemson Univ *Born:* 9/9/46 Native of Moncks Corner, SC. On the engrg staff of Firestone Synthetic Fibres Co in Hopewell, VA for 2 years before joining Diamond Shamrock in 1970. Technical Sales Rep selling PVC & PP resins for Diamond Shamrock from 1970 until 1974. Became PVC Dispersion Resin industry mgr in 1974, Business Mgr-Chlorowax in 1976. Business Mgr-Dry Caustic Soda and Potassium Chemicals in 1982, and assumed current responsibility in 1985. Diamond Shamrock Chemicals Company was bought by Occidental Chem Corp in Sept, 1986. SPE Pres (1981), Pres-Elect (1980), First VP (1979) and Secty (1978). Served on SPE's Plastic Promotion Ctte from its inception and was instrumental in seeing that the SPE film, "Plastics, the World of Imagination–, became a reality. Was responsible for SPE Liaison with SPI during 1982, 1983 and 1984. Mbr of SPE Council (1984-1987) Representing Mktg Div. Currently serving on SPE Mgmt Involvement Ctte. Enjoys sports of all types. *Society Aff:* SPE, NAMA.

Chinoy, Rustam B
Home: 226 Juniper Dr, Schenectady, NY 12306
Position: Consultant-Electro-Mech Measurements. *Employer:* GE Co. *Born:* Feb 1925 India. US citizen 1962. BS Mech 1945. Elec 1946 in Pakistan. Grad Aero engr from Northrop Aero Coll 1952. Worked as aircraft engr at Republic Aviation 1952-54; Mgr Kimball & Henzey cons engrs branch at Penn State 1954-55. Res Assoc Penn State Univ 1955-57; GE Co 1957-, starting as dev engr in Lrg Steam Turbine-Generator Dept 1957-65. Was appointed Mgr of Measurements Engg Unit of the divisional Mtls & Processes Lab 1965. Dev this org into a multimillion dollar sect of this div lab & managed it 1965-75. At present, hold the Staff Cons pos to the Dept Mgr of this lab of the Power Generation Div. Merit scholarship throughout India, & Rotary Intl Club. Enjoy arch, music & tennis. Kiwanian. Keenly interested in middle eastern commerce, esp fatherland Iran.

Chiou, Jiunn P
Business: Mechanical Engineering Dept, University of Detroit, 4001 W. McNichols Road, Detroit, MI 48221
Position: Prof *Employer:* Univ of Detroit *Education:* PhD/Heat Transfer/Univ of WI; MS/Heat and Power/OR State Univ; BS/Mech Engr/Natl Taiwan Univ *Born:* 07/29/33 Worked as Res Asst, Solar Energy Lab, Univ of WI, 1960-63. Served as Instructor, Mech Engrg Dept, Univ of WI, 1963-64. Worked as Engrg Specialist, AiRes Mfg Div, Garrett Corp, Los Angeles, 1964-69. Teach at Univ of Detroit since 1969; also serve as Dir, Heat Transfer and Solar Energy Lab of the Mech Engrg Dept. Serve as Chrmn, Climate Control Ctte, Passenger Car Activity, SAE and Secretary Solar Energy Div, ASME. Conslt for industry and government, specializing in Heat Transfer, Vehicular Heating/Cooling, Solar Energy, Computer Modeling and Simulation. Has 80 tech publs, and four books (editors). *Society Aff:* ASME, SAE, ISES

Chipouras, Peter A
Business: 1000 Western Ave, Lynn, MA 01910
Position: Lynn Product Engrg Dept *Employer:* GE Co. *Education:* MS/ME/Columbia; BS/ME/MIT *Born:* Mar 1925. BS MIT. MS Columbia. Native of Lynn, MA. Reg Engr MA. Served with USN 1943-46. With GE Co since 1955. Assumed respon as Mgr Engg 1968. Currently Gen Mgr Lynn Product Engrg Dept Respon for the Design, Dev, & Qualificiation of all Lynn Military & Commercial Engine Programs. Enjoy golf.

Chirlian, Paul M
Home: 15 Stephens Dr, East Brunswick, NJ 08816
Position: Prof. *Employer:* Stevens Inst of Tech. *Education:* EngrScD/Elec Engg/NY Univ; MEE/Elec Engg/NY Univ; BEE/Elec Engg/NY Univ. *Born:* 4/29/30. Native of NYC. Has been an Engg Educator since 1951. Instructor and Asst Prof at NY Univ 1951-1960. From 1960 to present has been at the Stevens Inst of Tech. Present position Prof of Elec Engg. Is the author of 25 innovative textbooks with wide adoptions which have had substantial influence on Elec Engg Education. Has authored over 60 res papers which have been published in learned journals. He is a Fellow of the IEEE and is Reg PE in NY and NJ. *Society Aff:* ASEE, IEEE.

Chisman, James A
Business: Industrial Engineering Dept, Clemson Univ, Clemson, SC 29634
Position: Prof Indus Engineering *Employer:* Clemson Univ. *Education:* PhD/Mgt Engg/Univ of IA; MS/Ind Engg/Univ of IA; BS/EE/Akron Univ. *Born:* 3/4/35. Native of Stow, OH. Prof of Indus Engg at Clemson Univ. Has been teaching at Clemson since 1963, within that time having instituted & coordinated the Systems Engg Prog 1967-74, having directed the Engg Tech Prog 1974-80 & having served as Acting Hd of Indus Engrg Dept 1983-84. Adjunct Professor Boston University Overseas Graduate Management program, 1980-81. Fulbright Scholar to University College, Cork, Ireland, 1987. Author of several articles published in operations res & ind engg journals. On editorial staff of various profl journals. Consultant with several natl ind firms. Owner of a real estate dev firm & an antique business. Hobbies are investing, collecting & restoring antique furniture and cars, art collecting, gadgeteering, tennis & snow skiing. *Society Aff:* TIMS, IIE, ASEE.

Chiswik, Haim H
Home: 1525 Thornwood Dr, Downers Grove, IL 60516
Position: Sr Scientist. *Employer:* Argonne Natl Lab. *Education:* ScD/Phys Met/Harvard Grad Sch of Engg; BS/Chemistry/Harvard College. *Born:* 11/29/15. Upon completion of grad studies 6 months before Pearl Harbor, joined the war res effort, first at Welding Res Council, then met at Chrleston Navy Yard and 1942-45 at Battelle evaluating enemy mtl. Res Engr 1945-49 at Air Reduction Co developing oxygen steel-making processes. Joined Argonne Natl Lab in 1949. Assoc Dir of Met Div 1955-1968 managing physical res prog in fuel elements, reactor mtls, neutron diffraction, radiation damage, and actinides. 1969 to present, Sr Scientist. Fellow ASM. Life member ASM. Editor of Journal of Nuclear Mtls 1969-1972. In "inactive" status since 1974 due to heart ailment. Retired Dec. 1981. Continuing as "guest scientist" at Argonne. *Society Aff:* ASM, ΣΞ.

Chittim, Lewis M
Home: 403 S Roberts, Helena, MT 59601
Position: VP, Marketing. *Employer:* Morrison-Maierle, Inc. *Education:* PE/Engr/MT Sch Mines. *Born:* 2/17/14. From 1938 to 1972, worked up from an instrumentman on survey crew for the MT Dept of Hgwys to Chief Engr and Chief Admin Officer. Worked for the Dept for 35 yrs. 1972 to Oct 1975 - Roy L Jorgensen Assoc as Proj Mgr on a large civ engg mgt proj in Vietnam. 1975 to present - VP of Mktg for Morrison-Maierle, Inc consltg engg firm. The firm has projs throughout the Middle East and in Zaire. Reg Engr and Land Surveyor, State of MT. Served on number of natl cttes of AASHTO while employed by MT Dept of Hgwys. *Society Aff:* NSPE, MSPE.

Chiu, Arthur N L
Home: 1654 Paula Drive, Honolulu, HI 96816
Position: Prof of Civil Engrg *Employer:* Univ of Hawaii at Manoa *Education:* PhD/Structural Eng/Univ of FL; SM/CE/MIT; BS/CE/OR State Univ; BA/Gen Sci/OR State Univ *Born:* 03/09/29 Univ of HI, Instructor-Assoc Prof 1953-62; Res Specialist Structural Mech, North American Aviation Inc 1962-63; Assoc Prof 1963-64; Civil Engrg Dept Chrmn 1963-66; Assoc Dean Res, Training & Fellowships, Grad Div 1972-76; Prof Civil Engrg Univ of HI at Manoa, 1964-present. Conslt to various engrs and government agencies; (on leave) Prof of Structural Engrg Asian Inst of Tech, Bangkok, Thailand 1966-68; Visiting Res Scientist, Naval Civil Engrg Lab, CA 1976-77. Res in dynamic response of structures to wind effects and earthquake forces, structural dynamics, computer applications for structural analysis. Numerous publications and reports. Natnl Pres, Chi Epsilon, Natl Civil Engrg Honor Society; Chrmn, Cttee on Natural Disasters, Natnl Res Coun, Natnl Reg Academy of Sciences., (HI), civil and structural branches. *Society Aff:* ASCE, ACI, NSPE, ASEE

Chiu, Chao-Lin
Business: Dept of Civil Engrg, Pittsburgh, PA 15261
Position: Prof *Employer:* Univ of Pittsburgh *Education:* PhD/Hydraulics/Cornell Univ; MSA/Engrg Sc/Univ of Toronto; BS/Hydraulic Engrg/Nat Taiwan Univ *Born:* 11/9/34 My professional activities include teaching, research, publication, and various activities for professional and scientific organizations. I teach courses in hydraulics and water resources. I have conducted scientific research under contracts and grants from federal agencies for the last 23 years. I have more than 50 publications, including 2 books which I have edited. I have organized 2 int'l conferences. I served as Chrmn of ASCE Committee on Stochastic Hydraulics, Chrmn of AGU Committee on Erosion and Sedimentation, and Chrmn of the Executive Committee of the Div on Experimental and Mathematical Fluid Mechanics of the Int'l Assoc for Hydraulic Research. I was awarded a Sr Fulbright Scholarship under the 1980-

Chiu, Chao-Lin (Continued)
81 program between the US and Germany. Certified Prof Hydrologist. *Society Aff:* ASCE, AGU, IAHR, AIH

Cho, Chun H
Business: 205 S Center St, Marshalltown, IA 50158
Position: Sr. Technical Consult *Employer:* Fisher Controls Int. Inc *Education:* PhD/ME-Cont/Univ of IA; MS/ME/Wichita State Univ; BS/IE/Wichita State Univ. *Born:* Jan 1934, Seoul Korea. PhD, ME Univ of Iowa BS, MS Witchita State Univ. Energy Cons Specialist on analog & digital sys applications, engg studies for sys justifications, sys anal & teaching adv control theory. His previous exper with res & dev of transducers, electro-hydraulic actuators & controllers for process control. Fellow of ISA & served as V.P. of Technology Dept & also a mbr of ASME. Reg PE CA. Has one patent on hydraulic amplifier. Credited with over 40 pub tech articles. *Society Aff:* AFIPS, ISA, ASME.

Cho, Soung M
Business: 8 Peach Tree Hill Road, Livingston, NJ 07039
Position: Director of Engineering *Employer:* Foster Wheeler Energy Applications, Inc. *Education:* Ph.D./Mech. Engrg./University of California, Berkeley; MS/Mech. Engrg./University of California, Berkeley; BS/Mech. Engrg./Seoul National Univ., Korea *Born:* 10/01/37 Born and raised in Korea. Had graduate studies in Univ CA, Berkeley. Served as Engrg Specialist at Garrett Corp and Staff Consultant at Rockwell International. Joined Foster Wheeler Corp 1973; currently Dir of Engrg at Foster Wheeler Energy Applications Group, responsible for engrg activities for nuclear and advanced technology products. Also Adjunct Prof of Mechl Engrg, Stevens Inst of Technology. Active in ASME; Ch of Nuclear Engrg Div, Assoc Ed of Applied Mechanics Reviews and Journal of Engineering for Gas Turbines and Power, Fellow of ASME. Recipient of the Republic of Korea Presidential Award (1962) and Fulbright Travel Scholar (1962). *Society Aff:* ASME, ANS

Chockie, Lawrence J
Home: 221 Condon Lane, Port Ludlow, WA 98365
Position: President *Employer:* Chockie Consulting Services *Education:* MS/Metallurgy/Colorado School of Mines; Engr. Met./Metallurgy/Colorado School of Mines; BS/Elec. Engr./Texas A&M College; Certificate/Court Reporter/Business College, Denver, Colo. *Born:* 02/02/23 President of his own consulting company, specializing in material application, nondestructive testing requirements and quality assurance for Codes & Standards in the development of inspection programs for operating nuclear power plants. On numerous cttees developing nuclear codes & standards for domestic & international use. VP ASME, chn of the bd on nuclear codes and standards. Mbr, bd of dir, Welding Research Council. Assigned by the UN to represent the US on International Atomic Energy Commission. Awarded the ASME Langer Medal, the ASME centennial medal, and a Fellow in the ASME. Has authored over two hundred publications from the creep of Zircaloy as influenced by neutron irradiation to historical reviews of the basic philosophy of codes and standards. A registered prof engr. *Society Aff:* ASME, ASTM, ANS, ASM, ASNT

Cholette, Albert
Business: 1324 Ave Lemoyne, Sillery Quebec, Canada G1S 1A3
Position: Prof. *Employer:* Univ Laval. *Born:* Oct 1918. ScD & MS MIT. B Eng with Hon from McGill Univ. Teaching Chem Engg since 1945. Chrmn of Chem Engg Dept until 1965. Pres Chem Engg Div of Chem Inst of Canada 1955-56 & Councillor "A" 1957-60. Fellow CIC 1958. Ed Can J Chem Engg 1959-62. Chrmn, Bd of Examiners, PE of Quebec 1968-80. Ed Chem Engg Sci since 1968. Fellow AIChE 1973. Awarded Archambault Medal of ACFAS 1973. R S Jane Meml Lect Award of Can Soc for CE 1974. Laureate, Sci Prize of Quebec 1975. Mbr Sigma Xi.

Chomitz, Morris A
Home: The Fairmont 603, Bala Cynwyd, PA 19004
Position: Sr VP. *Employer:* Day & Zimmermann, Inc. *Education:* MS/CE/Drexel Univ; BS/CE/MA Inst of Tech. *Born:* 12/5/25. Native of Phila. Served to Lt (jg) in USNR - 1943-46. Res & Dev Engr w/Allied Chem Co - Bridesburg, PA from 1946-49; Proj Engr w/Kuljian Corp, Phila, PA 1949-59; Plant Mgr of the Baldwin-Ehret-Hill, Inc Plant in Valley Forge, PA 1959-65. Came to Day & Zimmermann, Inc, Phila, PA in 1965 and since that time has served as Mgr - Proj Engg - 1972; VP Engg 1977-79 and is presently Sr VP - Engg. Reg PE in PA, NJ, NY, NC. Active mbr of Proj Mgt Inst.

Chong, Luis A
Business: 440 Franklin Turnpike, Manwah, NJ 07430
Position: VP-Engg. *Employer:* Otis Elevator Co. *Born:* May 1930. BSME Univ of IL, also attended Univ of MI. Design engr for Cummins Engine Co; var engg & mfg mgt pos with Curtiss-Wright Corp. With Otis Elevator Co since 1972, assumed current respon as VP 1973. Mbr AMA & NEMA. Enjoy tennis & skiing.

Chope, Henry R
Business: 1625 Bethel Rd, Columbus, OH 43220
Position: Pres, Dir-AccuRay Corp *Employer:* Chope Co *Education:* BEE/Elec Engrg/Ohio State Univ; SM/Engrg Sci/Harvard Univ; MS/Meteorology/CA Inst of Tech. *Born:* 07/19/21. MS Harvard 1950; MS Cal Tech 1948; BEE OH State Univ 1948. Electronic scientist rocket dev USAF 1949-50; co-founder, exec VP, dir, ind Nucleonics Corp, Columbus 1950-1981. Disting Alumnus Award 1961 Alumni Centennial Award 1970, OH State Univ; Tau Beta Pi-Pres, Chrmn Exec Ctte 1967-70; Fellow AAAS, Instrument Soc Am-Albert F Sperry Award 1972; Sr Mbr IEEE-Morris E Leeds Award 1967; Chambers of Commerce, US-Dir 1967-72, Chrmn Sci & Tech Ctte 1968-70; OH- Dir Riverside Meth Hosp, Columbus-Trustee, VP, Mbr Exec Ctte 1970-1984, OH Wesleyan Univ-Trustee 1970-1982. Patentee nucl energy, instrumentation, process control. *Society Aff:* ISA, AAAA, IEEE, TBII.

Chopra, Anil K
Business: Dept of Civ Engg, 709 Davis Hall, University of California, Berkeley, CA 94720
Position: Prof of Civ Engg. *Employer:* Univ of CA. *Education:* BS/Civ Engrg/Banaras Hindu Univ, India; MS/Civ Engrg/Univ of CA, Berkeley; PhD/Civ Engrg/Univ of CA, Berkeley. *Born:* 2/18/41. Native of India. Came to US 1961. Naturalized US citizen 1977. Asst Prof, Civ Engr, Univ of MN, 1966-67. Mbr, Faculty dept of Civ Engg, Univ of CA, Berkeley, since 1967. Prof of Civ engg since 1976. Chmn, ASCE-EMD Exec Ctte, 1986. ASCE-SD Ctte on Seismic Effects, Chmn 1983-86. Ctte on Natural Disasters, Natl Res Coun, 1980-85; Chmn (1982, 83). Dir, Applied Tech Coun 1972-74. EERI Steering Ctte, 8th World Conference on Earthquake Engrg, 1984. Mbr, Advisory Bd, MIT Press Series in Struct Mechs. Bd of Dirs, Seismological Soc of America, 1982-83. Awds and Honors: ASCE Huber Res Prize 1975, ASCE Norman Medal 1979. Certificate of Merit, Indian Soc of Earthquake Tech (1974). Banaras Hindu Univ Gold Medal (1960), Distinguished Alumnus Awd, 1980. Elected to Natl Acad of Engr, 1984. Honor Award, Association of Indians in America, 1985. Res interests: Structural Engg and Struct Mechs, Earthquake Engg. Over 130 pubs in struct dynamics and earthquake engrg. Conslt to private cos and govt agencies in earthquake engrg problems. *Society Aff:* ASCE, EERI, USCOLD, SSA.

Chopra, Randhir C
Home: 10100 El Pinar Dr, Knoxville, TN 37922
Position: VP. *Employer:* Chem Separations Corp. *Education:* BS/Chem Engg/Univ of TN. *Born:* 12/12/34. in West Punjab (now Pakistan) and migrated to India after partition in 1947. Attended BM College, Simla. In India, worked in DCM Chem & Bhilai Steel Plant. After coming to US grad from Univ of TN in Chem Engg and became citizen, 1974. Joined Chem Separations, 1963 in R&D for applications of Continuous Counter Current Ion Exchange for water treatment and pollution control. Promoted to Process Engg Mgr; Dir, Process & Commercial Operations; elected VP 1975. Most prominent papers written on Ion Exchange are "Chem-Seps Continuous Ion-Exchange Contactor and It's Applications to De-Mineralisation Process-

Chopra, Randhir C (Continued)
es–, 1969, Soc of Chem Industry, London; "A Closed Cycle Water System for Ammonium Nitrate Producers–, Internl Water Conference, 1971. Hold patents for Selective Removal of Chromates & Molybdate from Water. Contributor to "Ion Exchange for Pollution Control–, CRC Press, 1979. *Society Aff:* AIChE, ASTM, NWSIA, ASTM.

Chou, James C S
Business: University of Hawaii, Honolulu, HI 96822
Position: Prof *Employer:* Univ of HI *Education:* PhD/ME/OK State Univ; MS/ME/GA Inst of Tech; BS/Ordnance Engrg/Nat Inst of Tech - China *Born:* 1/13/20 Past Chrmn of ASME HI section. Past Dir of ASHRAE HI Chapter. Registered PE in mechanical and structural engrg branches in HI. Asst engr of the 21st Chinese Arsenal 1941-43. Design engr of H.C. & S. Co and Honolulu Iron Works 1951-59. U. H. faculty member since 1960. *Society Aff:* ASME, ASHRAE

Chou, Robert V
Business: Appliance Park AP2-234, KY 40225
Position: Mgr Microwave Range *Employer:* General Electric Co. *Education:* MBA/Marketing/Old Dominion Univ; MSEE/Elect/Univ of Notre Dame; BSEE/Elect/Univ of Notre Dame. *Born:* 6/6/39. Upon grad with MSEE in 1963, joined Gereral Electric Television Dept as Design Engg. In 1969, was app Mgr-Color Elect Design, and was responsible for the design of GEs first all-solid-state color TV chassis. In 1973, received MBA and was appointed Mgr-Engg of the GE TV plant in Singapore. In 1975, was app Mgr-Engg and Quality Control. In 1976, returned to US as Mgr Electronic Tuner Design. In 1977, was appointed mgr-evaluation Engg of the GE Range Dept (MABG). In Jan 1979, was appointed Mgr-Design Engg, responsible for total design of all cooking appliances with emphasis on applying solid state electronics. In Jan 1980, was appointed Mgr-Microwave Range Engineering responsible or all microwave product design, development, production, all electronic controls, and food technology. *Society Aff:* Tau Beta Pi, Eta Kappa Nu

Chou, Tsu-Wei
Business: Center for Composite Materials, Mechanical Engr Dept, Newark, DE 19716
Position: Prof *Employer:* Univ of DE *Education:* PhD/Matls Sci/Stanford Univ; MS/Matls Sci/Northwestern Univ; BS/Civil Engrg/Natl Taiwan Univ *Born:* 06/02/40 Prof, Univ of DE, 1978-present; Assoc Prof, 1973-78; Asst Prof, 1969-73; Liaison Scientist, US Off of Naval Res, London, 1983; Visiting Prof, DFVLR- Germany (1983), Argentina (1981), South Africa (1977), UK (1976), Argonne National Lab (1975). Sr Visiting Res Fellowship, British Sci Res Coun (1976). Frederick Gardner Cottrell Fellowship, 1970-71, Ford Foundation Fellowship, 1966-67, Stanford Univ Walter P Murphy Fellowship, 1964-65, Northwestern Univ. Author: "Composite Materials and Their Use in Structures–, Elsevier-Applied Sciences (London), 1975. North American Editor: Composites Science and Technology, Elsevier-Applied Science (London), 1985-present. *Society Aff:* ASM, ASTM

Chou, Wushow
Business: P.O. Box 8207, Raleigh, NC 27695-8207
Position: Professor *Employer:* North Carolina State Univ. *Education:* Ph.D./EE&CS/U. C. Berkeley; MS/EE/U. of New Mexico; BS/EE/ChengKung Univ., Taiwan *Born:* 02/12/39 Wushow Chou is recognized as one of those who created the technology of network optimization and evaluation. He has planned, analyzed, designed, and monitored communication networks for distributed computing, electronic banking, air traffic control, and a number of other areas. He is currently Dir of the Computer Studies Prog, and a Prof of Computer Science/Electrical and Computer Engrg at NC State Univ. He is also Pres of ACK Computer Applications, Inc., a Series Editor for Computer Science Press, & an Advisory Editor for Prentice-Hall. Previously, he was VP of Network Analysis Corp. and editor-in-chief of the Journal of Telecommunication Networks. Dr. Chou has published over 70 articles, edited 2 books, consulted for more than 30 large organizations and been a frequent invited speaker. He is an IEEE fellow. Dr. Chou received his Ph.D. in Electrical Engineering/Computer Science from U.C. Berkeley. *Society Aff:* IEEE, ACM

Chow, Ven Te
Home: 20 Sherwin Cir, Urbana, IL 61801
Position: Prof of Civ & Hydrosystems Engg. *Employer:* Univ of IL. *Education:* PhD/Hydraulic Engg/Univ of IL; MS/Struct Engg/PA State Univ; BS/CE/Chiao Tong Univ. *Born:* 8/14/19. Hon D Sc Andhra; Hon D Eng Yeungnam; D Hon C Strasbourg; D Eng Hon C Waterloo. Mbr Natl Acad Engg; Fellow Amer Acad Arts & Sci; Academician: Academia Sinica, China Acad; Hon TX Citizen; Hon Mbr Mexican Hyd Assoc; Hon Mbr KSCE; Hon Pres IWRA, Past Pres AGU Hydrology Sec; Past VP Intern Comm Surface water; Cons: UN & govt orgs; 23 countries. Author over 200 pubs; Ed: Advances in Hydrosci, Journal of Hydrology, Dev in Water Sci, Water Resources & Environmental engg, Water Intl Awards; Epstein; ASCE Res Prize; CIE Achievement; UNESCO Hon Cons; Freeman; ASEF Western Elec; Fulbright-Hays; CESASC Prof Achievement; ICID Commemorative Medal; IHD Endowment; Louis Pasteur Medal; Vincent Bendix Award. *Society Aff:* AGU, ASCE, ASEE, AAAS, AAM, AAEE, IAHR, ICID, IWRA, NAE, AAAS.

Chow, Wen L
Business: Dept. of Mech. & Ind. Eng, 1206 W. Green St, Urbana, IL 61801
Position: Professor *Employer:* University of Illinois *Education:* PhD/Mech. Eng./University of Illinois; MS/Mech. Eng. /University of Illinois; BS/Mech. Eng./National Central University of China *Born:* 03/06/24 Came to the U.S. for graduate study in 1949. After a year in Industry beyond PhD degree, came back to the Univ of Ill to do teaching and research in Gas Dynamics and Fluid Dynamics. Specialized in the area of Jet and Rocket Propulsion, Separated flows in incompressible, transonic and supersonic flow regimes. Recent interest also includes open channel flows. Naturalized in 1963. *Society Aff:* AIAA, ASME

Chrencik, Frank
Business: PO Box 7497, Birmingham, AL 35253
Position: Ret Vice Chrm of Bd & Dir Emeritus *Employer:* Vulcan Materials Co *Education:* BSCh Eng/Univ of IA; Adv Mgr Program/Harvard *Born:* 1/1/14. 1937-40 - Allied Chem Corp: Prod Supv various plants. 1940-1946 - US Army chem corps: Arsenal Operations last rank Lt Col. 1946-1972 - Diamond Shamrock Corp: Chem Plant Mgr then VP Mfg then Corp VP and Pres of a subsidiary. 1972- 1979 - Vulcan Mtls Co: exec VP then VChrmn of Bd and Dir. 1979 - Ret'd 5/1/79: became Dir Emeritus & Consultant. Intl Adv Com for: Encyclop of Chem Processing and Design. Distinguished Alumni Achievement Award - Univ of IA 1977. Alabama Outstanding Chem Eng Award 1983, Amer Inst of Chem Engrs. *Society Aff:* AIChE

Christ, John E
Home: 80 Orton Rd, West Caldwell, NJ 07006
Position: Traffic Engg. *Employer:* Essex Co Engg Dept. *Education:* Cert/H-Wat Traff Engg/Yale Univ Bur of Highway Traffic; BSCE/Civil Engg/Rutgers Univ. *Born:* 1/1/30. Reg Prof Engg in NJ. As Essex County Traffic Engg determines the need for, supervises and participated in the design for all types of traffic control devices. Reviews geometrics and plans by consultants. Gives court testimony. Privately provides consulting traffic engg services for various clients, including municipalities, insurance companies, citizen groups, and developers, at locations outside Essex County. Has taught Traffic Engg topics in the Rutgers Univ Continuing Engg Studies Program. Currently Pres of the NY and NJ Metropolitan Section of the Inst of Transporation Engrs. Recretion includes golf, chess, music, and travel with family. Army veteran. *Society Aff:* ITE.

Christensen, Bent A
Home: 10826 NW 15th Pl, Gainesville, FL 32601
Position: Prof. *Employer:* Univ of FL. *Education:* PhD/Hydraulic Eng/Univ of MN; MSCE/Civ Eng/Tech Univ of Denmark; Filosofikum/Philosphy/Univ of Copenhagen *Born:* 3/22/28. Born in Copenhagen, Denmark. Educated in Europe & the US.

Christensen, Bent A (Continued)
Profl background as researcher & educator specializing in hydraulics & water resources. Consulting activities in the US, Europe, Canada, Middle East & S Am. Since 1965 full prof in charge of hydraulics & water resources at the Univ of FL. Author of about 100 papers & articles & of numerous reports in the areas of sedimentation, groundwater flow & gen hydraulics. *Society Aff:* ASCE, ASEE, AGU, AWRA, IAHR, AIH.

Christensen, Palle S
Business: 502 Office Center Dr, Ft. Washington, PA 19034
Position: Pres *Employer:* Aydin Corp *Education:* MSEE/Telecomm/Copenhagen Inst of Electrical Engrg *Born:* 2/8/33 Born and educated in Denmark. Conducted research on microwave ferite devices, Danish Academy of Sciences 1979, Belt Northern Research Lab, Ottawa, designing Tropo and related communications equipment. 1961 Raytheon Co., Engrg Mgr guiding development of Mircowave Radio Relay equipment and other communications equipment. 1972 Philco Ford Mgr of Engrg commercial communications products. Since 1976, Aydin Corp, 1978 Pres of Aydin Monitor Systems and VP Aydin Corp. The division is a leading supplier of advanced Telemetry and Telecommunications products and systems. Technical Publications - 6 in various journals. *Society Aff:* IEEE

Christensen, Paul B
Home: 47 Roundabout Ln, Portland, ME 04102 *Employer:* Retired. *Education:* BS/ME/Stevens Inst of Tech. *Born:* 3/3/10. Native of NYC. Grad from Stevens 1931. VP & Chief Engr - Merchants Refrig Co, NYC 1940-1952. VP & Gen Mgr Northeast Cold Storage Corp 1952-1975. Pres ASRE (Am Soc of Refrig Engrs) 1951. Pres- Maine State Chamber of Commerce 1958, Pres- Greater Portland Chamber of Commerce 1972, Pres- Maine Chapter - Institute of Food Technologists 1973, Silver Beaver Award - Pine Tree Council Boy Scouts of of America 1974. *Society Aff:* ASHRAE.

Christensen, Ronald I
Business: W 115 Century Rd, PO Box 665, Paramus, NJ 07652
Position: VP Product Standards & Licensing Support *Employer:* El Paso Polyolefins Co *Education:* BSChE/CE/Univ of IA. *Born:* 10/5/35. Native of Davenport, IA. Served with US Army Corps of Engrs 1958-60. Dev engr for Monsanto Textiles Div in areas of spinning, polymerization & solvent recovery. With El Paso Polyolefins Co since 1967.. Assumed current responsibility in 1979. Co products are low density polyethylene & polypropylene. *Society Aff:* AIChE.

Christian, David B
Business: 241 W Maple Ave, Langhorne, PA 19047
Position: President *Employer:* D.B. Christian & Assoc. Inc. *Education:* BS/CE/Drexel Univ. *Born:* 4/16/38. Reg PE in PA, NJ and DE; Reg surveyor in NJ and PA. Adjunct Prof of Civ Engg, Drexel Univ, Phila, PA. Sole propietor of David B Christian & Assoc, 1973 to 1979. Pres. and chief engr of D.B. Chrisitan & Assoc., Inc. 1980 Formerly Assoc Mbr Dir, NJ Sec, ASCE. Past Pres of Bucks County Chapter, NSPE. Mbr of Rotary Intl. Mbr of Chi Epsilon (Natl Civ Engg Honor Fraternity). Employed as Proj Mgr for two firms in Southeastern PA prior to establishing private practice. Employed as resident engr on $1.9 million sewer proj in NJ. *Society Aff:* ASCE, ACSM, NSPE, PSLS.

Christian, Jack L
Home: 4726 Mt Frissel Dr, San Diego, CA 92117
Position: Staff Scientist. *Employer:* General Dynamics Convair. *Born:* May 1933. MS IA State Univ; BS N E MO State. Two yrs AF, 5 yrs teaching exper with Genl Dynamics Convair since 1959 working on R&D in cryogenic & adv composite mtls & proj support-missiles, aircraft & spacecraft. Author over 50 tech papers & articles. Reg PE & Mbr ASM & ASTM. Active natl ASM, ASTM, ASME, MPC & NMAB cttes. Invited lect at 3 intl confs. ASM hon award.

Christiansen, Donald D
Business: 345 East 47th Street, New York, NY 10017
Position: Staff Dir *Employer:* IEEE *Education:* B.E.E./Comm/Cornell Univ *Born:* 6/23/27 Birthplace: Plainfield, NJ. U.S.N., World War II. BEE, Cornell Univ. Experience: Engrg: Philco Corp and CBS Electronics Division of CBS. Publishing: Solid-state editor for Electronic Design; Editor-in-Chief, Electronics; Editor, IEEE Spectrum (since 1971) and Publisher (since 1975). Fellow, IEEE, 1980. Pres, Society of National Association Publications, 1981-1982. Chrmn, HKN Outstanding Young Engr Recognition Award, 1975-1979. Inter Dir, HKN, 1982-1984, Eminent Member, HKN, 1985. Recipient, Citation and medal for the advancement of culture, Flanders Academy of Arts, Science, and Letters, 1980, IEEE Centennial Medal, 1984. Commendation for Outstanding Accomplishment, CESSE, 1977. Member, Union Internationale de la Presse Radiotechnique et Electronique. Mbr, Franklin Inst (Phila), Royal Institution (London). Editor, several electronics books. Editor, Electronics Engrs Handbook, 3rd ed., McGraw- Hill. *Society Aff:* IEEE, NYAS, CESSE, SHOT, HKN, ASAE, AWA, RCA.

Christiansen, Ernest B
Home: 3025 S. 1935 E, Salt Lake City, UT 84106
Position: Prof. *Employer:* Univ of UT. *Education:* PhD/Chem Engg/Univ of MI; MS/Chem Engg/Univ of MI; BS/Chem Engg/Univ of UT. *Born:* 7/31/10. Native, Richfield, UT. Research, dev, and design engg and super for cellophane, rayon, nylon, and plutonium processes, flash and fluidized-bed coal pyrolysis, E I duPont, 1941-46; Prof of Chem Engg, Univ of Idaho, 1946-47, Univ of UT, 1947-date (Dept Chmn, 1947-75); AIChE: Organizer and Chmn, Great Salt Lake Section, Natl Chmn Humanities and Education Area, mbr Continuing Education Committee and others, Natl Dir, 1965-68, Fellow, 1971, Founders Award, 1978; Fellow, UT Academy of Sci, 1972; UT Engg Council Award, 1966; Univ of UT Distinguished Research Award, 1977; Research and publications: Newtonian and non-Newtonian tube flow and heat transfer, viscoelastic flow functions, protein from sewage, coal and oil shale pyrolysis. *Society Aff:* AIChE, ACS, ASEE, Am Soc Rheol.

Christiansen, Richard W
Business: 441 Clyde Bldg, Provo, UT 84602
Position: Prof *Employer:* Brigham Young Univ *Education:* PhD/EE/Univ of UT; MS/Physics/Univ of NM; BS/EE/Rutgers Univ *Born:* 05/18/39 Born Nephi, UT. Design Engr Minuteman Missile, Boeing 3 yrs. Res Asst, High Energy Neutron Experiments, Univ of New Mexico, 3 yrs Mgr High Voltage Lab, Electromagnetic sensor developments and Engrg Development Dept, EG&G Albuquerque Div, 4 yrs. Dir, Software Scis Corp, Salt Lake City, UT, 3 yrs. Res Asst, speech processing Univ of UT, 2 yrs. Dir of Applied Scis Div and Chief Scientist, Eyring Res Inst, Provo, UT, 2 yrs (120 employees). Prof, Electrical Engrg, Brigham Young Univ, 9 yrs. Mbr IEEE, Tau Beta Pi, Sigma Xi, Phi Kappa Phi. Enjoy running, hiking, skiing, classical music learning and teaching. *Society Aff:* IEEE

Christie, William B
Home: 660 Windermere Ave, Ottawa Ont Canada K2A 2W8
Position: Director Business Development *Employer:* Vickers Canada *Education:* BEng/EE/Nova Scotia Tech; Dip Engg/Gen Engg/Dalhousie Univ/Imperial Defense College, London, Eng.. *Born:* 9/20/19. Calois, ME USA. Educated in Nova Scotia, Canada. Served British Merchant Marine 1938-41. Royal Canadian Navy 1941-74 as Naval engr specializing in electrics & ship construction/repair. Ret as R/Adm from position as Assoc Deputy Minister (Mtl), responsible for all engg in Canadian Armed Forces. With Can Govt dept of Supply since 1974, firstly as Dir of Shipbldg Contracting, and now as Dir Gen Marine & Ind, responsible for all Canadian Govt contracting for Marine, special vehicles & Heavy Ind equip. Ret'd from govt service 1980. Currently with Vickers Canada as Director Business Development in Ohawa, Ont. *Society Aff:* SNAME, EIC, CSEE, APENS, MMI.

Christoffers, William H
Business: 4334 Secor Rd, Toledo, OH 43623
Position: Pres *Employer:* AVCA Corp *Education:* B/Engrg/Univ of Toledo *Born:* 10/3/44 Native of Toledo, OH. Formed AVCA Corp as Exec VP in 1973. Became Pres in 1980, responsible for administrative control of firms' activities, coordination of management team and overall leadership. Currently a licensed pilot with interests including Scuba Diving and Boating. Experienced in process plant and refinery mechanical systems from engrg development through construction. *Society Aff:* NSPE, ASCET

Christopher, Phoebus M
Business: Belknap Campus, Louisville, KY 40208
Position: Prof Chem Engg. *Employer:* Univ Louisville (Speed Scientific Sch). *Education:* MS/CE/Newark College of Engg; AB/Chemistry/Rutgers Univ. *Born:* 10/20/21. Although background is basically Chemistry, most of my teaching experience has been in Chem Engg. Have had approx 6 yrs of full-time ind experience (as a Chemist), & about 22 yrs of full-time teaching experience at the College/Univ level (with academic ranks in both Chemistry & Chem Engg). Much of my res interests have been involved with physiocochem properties (and struct), boron chemistry, electrochemistry, ultrasonics, etc. These interests have resulted in the publication of 19 technical articles (all in refereed journals). Have directed theses at the master's level, and have served on PhD thesis committees. Have taught all branches of Chemistry, mtls sci, stoichiometry, Chem Engg thermodynamics, environmental engg-oriented courses, etc, etc. *Society Aff:* ΣΞ.

Christy, Donald
Business: Box 278, 5th & Main, City, Scott City, KS 67871
Position: Chrmn of Bd *Employer:* First Natl Bank *Education:* MS Ag Eng/Water Resources/TX A&M Univ; BS Ag Eng/General/KS State Univ *Born:* 11/23/09 Son of Estes and Effie Christy. Married Helen G. Shedd, May 28, 1933. Children: Donald, Arthur, Alice A. Lay. Honors: Conservation Dist Supervisors; Individual of the Year ASAE (Midwest); KS State Grange, Service to Agriculture Award. KS Univ Distinguished Service Engrg - Agriculture Award. Profl Record: Southwest KS Senator 1968-76. Committees Agriculture; Energy and Natural Resources; U.S. Dept Agricultural Advisory Committee on Equip and Structures 6 years. KS State Bd Agriculture 12 yrs; Water Resources 13 yrs. Watershed 4 yrs; State Univ Engrg Advisory Council 23 yrs. Assoc Prof Irrigation drainage and soil conservation at TX A&M Univ 9 yrs. *Society Aff:* AAAS, WMA, AG ENG

Chryssafopoulos, Hanka Wanda S
Home: P.O. Box 6125, Boca Raton, FL 33427
Position: Pres *Employer:* HSce, Inc, Consulting Engineers *Education:* PhD/Soil Mech/Univ IL; MS/Soil Mech/Univ IL; CE/Civil Engg/Univ Rio Grande del Sul, Brazil. *Born:* Porto Alegre, Brazil of Stefan & Estacia Wilkoszynska. Came to US 1954; married Nicholas Cryssafopoulos 1956. Worked as civil engr Brazil; on faculty Univ CA Long Beach; Resch Asst Univ IL and IL State Geological Survey; worked as sr engr w Woodward, Clyde, Sherard & Assoc 1964-5; Dames & Moore 1968-74; currently Pres & owner, HSce, Inc., Consulting Engrs. Fulbright scholar & Fellow ASCE & GSA, sr mbr SWE. Listed in Who's Who in American Women, The World Who's Who of Women, Who's Who in the East, et al. Reg PE in Brazil. *Society Aff:* ASCE, ACEC, SWE, GSA, FICA, ΣΞ.

Chryssafopoulos, Nicholas
Business: P.O. Box 6125, Boca Raton, FL 33427
Position: Consultant *Employer:* HSce, Inc., Consulting Engrs *Education:* PhD, MS/Soil Mech/Univ of IL; BS/Civ Engg/Robert College, Istanbul *Born:* Istanbul, Turkey, of John and Despina (Hondropoulos) came to U.S. in 1951, was naturalized in 1959; married Hanka Wanda Sobczak in 1956. Worked as civil engineer on several construction projects in Turkey; was on the faculty of the Univ. of Illinois; worked as consulting engineer with Woodward, Clyde, Sherard and Associates from 1959 to 1968; and with Dames & Moore from 1968 to 1985. Self-emp & consult to HSce, Inc 1985- . Fulbright scholar and Fellow, ASCE and ACEC. US Dir on Bd UPADI. Listed in Who's Who in America. Reg PE CA, FL, IL, KS, MO, NJ, NY, PA, VA. *Society Aff:* ASCE, ACEC, ASTM, TRB, ΣΞ, NSPE, FES, FICE.

Chryssostomidis, Chryssostomos
Business: 77 Massachusetts Ave, Rm 5-325, Cambridge, MA 02139
Position: Assoc Prof. *Employer:* MA Inst of Tech. *Education:* PhD/Systems/MIT; Naval Engr/Naval Arch/MIT; SM/Naval Arch/MIT; BSc/Naval Arch/Durham Univ. *Born:* 10/27/42. Faculty mbr at MIT teaching design of ships and offshore structures. Interested also in the areas of seakeeping and Comp Aided Design. Past chrmn of SNAME (New England). Consultant to govt and ind. SNAME Capt Joseph H Linnard Prize, 1975. *Society Aff:* SNAME, RINA.

Chu, Jeffrey C
Home: 10 Baldwin Cir, Weston, MA 02193
Position: Chairman, President & CEO *Employer:* Sanders Technology, Inc. *Education:* BSEE Univ of MN/MS Univ of PA *Born:* July 14, 1919. Sr VP Wang Labs Inc 1973-75. VP Honeywell Info Sys Inc 1962-73. Dir of Engg Univac Div Sperry Rand Corp 1956-62. Sr Sci Argonne Natl Lab Univ of Chgo 1949-59. Sr Engr Reeves Inst Co 1947-49. Res Assoc Univ of PA 1943-47. Engr Philco Corp 1942-43. Fellow IEEE; Mbr Cie, Sigma Xi, Eta Kappa Nu, Res Soc of Am, Visiting Ctte Harvard Univ. Engg Achievement Award: Honeywell Res Ctr 1970, Chinese Inst of Engrs 1973. Trustee, Moore School, Univ of PA; Honorary Mbr, Chinese Acad of Social Sciences; PRC: Advisor, Natl Computer and Technology Commission, PRC; Mbr of the Bd, Shanghai Tiao Tung University.

Chu, Ju Chin
Home: 21 Yorktown, Irvine, CA 92714
Position: Pres. *Employer:* Tech Resources, Inc. *Education:* Doctor of Sci/Chem Engg/MA Inst of Tech; BS/Chem Engg/Natl Tsing Hua Univ. *Born:* 12/14/19. Sr Chem Engr, Shell Chem (1946); Asst Prof Wa Univ (1946-49); Assoc Prof (1949-54), Prof of Chem Engg (1954-67); Poly. Inst. of Brooklyn; Tech Dir, Chem Construction Corp (1955-57); Tech Advisor, Strategic Missiles Systems Div, Rockwell Intl (1966-70); Prof. of Chem. Engrg., VPI and State Univ (1969-72); Pres and Chrmn of Tech Resources, Inc (1970-to date). Served as cons for more than 30 maj chem, aerospace, nuclear firms and govt agencies. Published over 85 res papers, books, book chapters and US patents in chem engg and related fields. Prof Engrs of the State of MO since 1948. Represented US Twice at the NATO Propulsion Panel Conf. Achievement Award, CIE (1961); Medal of Honor, Univ of Liege', Belgium (1963). Fellow, American Institute of Chemical Engineers (since 1975); Fellow, American Association for the Advancement of Sciences (since 1955).; Achievement Award, the Ministry of Education, Republic of China (1963). *Society Aff:* AIChE, ASEE, ACS.

Chu, Richard C
Business: B02-701 Boardman Road, IBM Corporation, Poughkeepsie, NY 12602
Position: IBM Fellow *Employer:* IBM Corporation *Education:* MS/Heat Transfer/Purdue University; BS/Mechanical Eng./National Cheng-Kung University *Born:* 05/28/33 Was born in Beijing, China. Came to the US in 1958 after receiving BSME from National Cheng-Kung Univ in Taiwan. Studied heat transfer at Purdue Univ and received MSME in 1960. Joined IBM the same year. Had a variety of engrg and managerial assignments in the Poughkeepsie Lab - was Heat Transfer Technology Mgr in 1969, Product Technology Mgr in 1975 and Engrg Lab Mgr in 1980. Received 16 IBM Awards for Invention Achievements & related activities. Appointed an IBM Fellow, the company's highest technical honor in 1983 for sustained contributions & leadership in cooling technology for large scale computers. Elected ASME Fellow in 1982. Received ASME Heat Transfer Memorial Award in 1986, and Distinguished Alumnus Awards from both Alma Maters in 1984 and 1986 respectively. Elected a member of the National Acad of Engrg in 1987. *Society Aff:* NAE, ASME, AAAS, NYAS

Chu, Ta-Shing
Home: 112 Jumping Brook Rd, Lincroft, NJ 07738
Position: Technical Staff, Distinguished Member *Employer:* AT&T Bell Labs *Education:* PhD/EE/OH State Univ; MS/EE/OH State Univ; BS/EE/Natl Taiwan Univ *Born:* 07/18/34 Born in Shanghai, China. Res Assoc at Courant Inst of Math Scis, NY Univ, 1961-63. With Crawford Hill Lab of AT&T Bell Labs since 1963. Numerous Tech Pubs in the areas of Microwave Antennas and Wave Propagation. His work includes surface wave diffraction, precision gain standard, propagation through precipitation, dual polarization radio transmission, and the Crawford-Hill 7-meter offset Cassegrainian millimeter wave antenna. Bell Labs Distinguished Technical Staff award; Fellow, IEEE; Mbr, Intl Scientific Radio Union, Sigma-Xi, Pi-Mu-Epsilon. *Society Aff:* IEEE

Chu, Ting L
Business: Southern Methodist Univ, Dallas, TX 75275
Position: Prof. *Employer:* Southern Methodist Univ. *Education:* PhD/Chemistry/WA Univ; BS/Chemistry/Catholic Univ of Peking. *Born:* 12/26/24. Naturalized in 1961. With Westinghouse Res Labs 1956-67, specializing in semiconductor mtls and devices. Prof at Southern Methodist Univ since 1967, teaching and res in solid state electronics. Res activities during past fifteen yrs on solar cells for terrestrial applications and compound semiconductor technology. Served as consultant to Westinghouse, TX Instruments, Union Carbide, NCR, Monsanto, Poly Solar, and other corps. *Society Aff:* IEEE, ASEE, ISES, ECS.

Chu, Wesley W
Business: Computer Sci Dept, 3731K Boelter Hall, Los Angeles, CA 90024
Position: Prof. *Employer:* UCLA. *Education:* BSE/EE/Univ of MI; MSE/EE/Univ of MI; PhD/EE/Stanford Univ. *Born:* 5/5/36. Consultant to government agencies and private industries. He worked on seitching circuit design at Gereral Electric, 1961-62, on the design of large scale computers at IBM, 1964-66 and on res on computer comm at Bell Lab, NJ, 1966-69. He was Chmn of the ACM SIGCOMM (1973-77); Chmn of several Computer Conf, and he is an IEEE Fellow. He has authored or co-authored more than 70 articles on information processing systems, and has edited three widely used textbooks, *Advances in Computer Communications and* Networking, Artech House. *Centralized and Distibuted Date Base Systems* (Co-edited with P. Chen), IEEE Computer Society. *Distributed Systems*, Vol I *Distributed Processing Systems*; Vol II *Distributed Processing Systems*. Artech House 1986. He is mbr of the Editorial Bd on several computer journals, and an assoc editor of the Data & Knowledge Engineering. He has received a meritorious service award from IEEE in 1983. *Society Aff:* IEEE, ACM.

Chua, Leon O
Home: 955 Galvin Dr, El Cerrito, CA 94530
Position: Prof. *Employer:* Univ of CA. *Education:* SM/EE/MIT; PhD/EE/Univ of IL. *Born:* 6/28/36. Dr Chua's res interests are in the areas of gen nonlinear network & system theory. He is the author of the book, "Introduction to Nonlinear Network Theory–, published by the McGraw-Hill Book Co in 1969, & a co-author of the book "Computer-Aided Analysis of Electronic Circuits: Algorithms and Computational Techniques–, published by Prentice Hall in 1975. He has also published many res papers in the area of nonlinear networks & systems. Dr Chua is a Fellow of the IEEE. He has been awarded four patents & is a recipient of the 1967 IEEE Browder J Thompson Meml Prize Award, 1973 IEEE WRG Baker Prize Award, the 1973 Best Paper Award of the *IEEE Society on Circuits and Systems*, the Outstanding Paper Award at the 1974 *Asilomar Conference on Circuits, Systems, and* Computers, the 1974 Frederick Emmons Terman Award, & the 1976 Miller Research Professorship Award at the Univ of CA, Berkeley. Sr Visiting Fellowship at Cambridge Univ, England 1982. Alexander Humboldt Sr U.S. Scientist Award at Tech Univ of Munich 1983. U.S. Sr Scientist Award From Japan Soc for Promotion of Science, Waseda Univ, Tokyo, 1983. Hon Prof, Chengdu Inst of Radio Engr, Sichuau, Peoples Rep of China, 1983. Honorary Doctorate (Doctor honoris causa), Ecole Polytechnique, Lausanne, Switzerland, 1983. IEEE Centennial Medal 1984. Honorary Doctorate, Univ of Tokushima, Japan, 1984. *Society Aff:* IEEE.

Chuang, Hsing
Business: Mech Engg Dept, Louisville, KY 40292
Position: Prof. *Employer:* Univ of Lousiville. *Education:* PhD/Fluid Mechs/CO State Univ; MS/Hydromechanics/Univ of MN; BS/Hydraulic Engg/Natl Taiwan Univ. *Born:* 2/1/28. Born in Taiwan, naturalized US citizen; Design Engr Taiwan Power 1954-57, came to US in 1957, Postdoctoral Fellow at Harvard Univ 1962-64, Asst Prof and Engr CO State Univ 1964-66, Assoc Prof 1966-70 & Prof 1970-present. Faculty Res Participant & Visiting Scientist, Argonne Natl Lab 1978-79. Published papers on turbulent boundry layers, electrokinetics, plasma physics, MHD boundary layer control, heat and seed recovery in MHD power and computational fluid flow and heat transfer. Enjoy classical music & gardening. *Society Aff:* ASME, ΣΞ.

Chuck, Robert T
Business: PO Box 373, Honolulu, HI 96809
Position: Mgr-Chief Engr. *Employer:* State of HI. *Education:* M/CE/Cornell Univ; BS/CE/MI State Univ. *Born:* 10/1/23. Native of Honolulu, HI. Worked for several agencies in Territorial Govt as a profl civ engr. In 1960 assumed present position as Mgr-Chief Engr, Div of Water and Land Dev, Dept of Land and Natural Resources, State of HI. In charge of water resources for the State of HI. Selected as HI Engr of the Yr 1966; Pres of Natl Water Resources Assn 1975; Pres of Assn of Western State Engrs 1972; Natl Dir of American Water Works Assn 1975-78; Received Natl Water Res Assn President's Award 1980; Received Am Water Wks Assn Fuller Award 1981. Lectured in civ engg at Univ of HI, College of Continuing Education. Mbr of Tau Beta Pi, Phi Kappa Phi. *Society Aff:* ASCE, NSPE.

Chugh, Yoginder P
Business: Dept of Mining Engrg, Carbondale, IL 62901
Position: Prof and Chairperson *Employer:* Southern IL Univ *Education:* Ph.D./Mining Engrg/PA State Univ; MS/Mining Engrg/PA State Univ; BS/Mining Engrg/Banaras Hindu Univ-India *Born:* 10/6/40 in India. Worked for four years in coal industry and taught for a year juniors and seniors in India prior to coming to the USA in 1965. Post-Doctoral fellow at Columbia Univ, 1971; Research engr at IIT Research Inst, 1972-1974; Planning Engr, AMAX Coal Co, 1975-76; joined Dept of Mining Engrg at SIU-C, 1977. Very active in the national society of AIME, organized several professional conferences, at the natl and interntl level, author of over 50 technical publications and reports. Consultant to coal mining industry. Enjoy travelling and tennis. *Society Aff:* SME, ISRM, IBSM.

Chun, Michael J
Business: 1960 East-West Rd, Honolulu, HI 96822
Position: Assoc Prof of Environ Health. *Employer:* Sch of Pub Health-Univ of HI. *Born:* June 20, 1943. PhD & BS KS Univ; MS Univ of HI. Native of HI. Prof career includes environ engg consultation for Indonesia. Trust Territories of the Pacific Islands, State of HI, as well as industry. With Univ of HI since 1970, teaching both pub health & engg courses involving water supply, wastewater treatment, & solid waste mgt & control. Actively involved in statewide sewage treatment plant operator training. Current res int incl solid waste disposal and biological wastewater treatment. Enjoy surfing & competitive outrigger canoe paddling. Secy-Treas of HI Water Pollution Control Assn.

Chung, Benjamin T
Home: 766 Diandrea Dr, Akron, OH 44313
Position: Prof & Head *Employer:* Univ of Akron. *Education:* PhD/ME/KS State Univ; MS/ME/KS State Univ; MS/Math/Univ of WI; BS/ME/Cheng Kung Univ *Born:* 3/16/34. Benjamin T F Chung was born in China and came to the US in 1960. With Allis-Chalmers Mfg Co since 1962, as a Res Engr. Was awarded a fellowship by Allis-Chalmers for doctorate work. Joined the Univ of Akron as a full time faculty in 1970. A summer employee and consultant in Babcock & Wilcox Co.

Chung, Benjamin T (Continued)
Areas of interest include nonlinear heat conduction, turbulent convective heat transfer, heat transfer with phase change and radiactive heat transfer. Has authored and co-authored 100 technical papers in these areas. *Society Aff:* ASME

Chung, D S
Business: Dept. of Agri Engrg, Manhattan, KS 66506
Position: Professor *Employer:* KS State University *Education:* Ph.D/Grain Sci/KS St Univ; MS/ChE/KS St Univ; BS/ChE/Purdue *Born:* 3/20/35 Married in 1961. Wife: Okkyung, Daughters: Clara and Josephine. Teaching courses reated to agri process engrg and food engrg conducting res on food and feed grain storage, drying and handling areas, resulting in 80 tech publications. Received many grants and contracts. Outstanding educator award from the President, the Republic of Korea. Received outstanding paper award from America Society of Agri Engrs (ASAE). Chrmn of Mid-Central Region, ASAE (1973-1974); Chrmn of KS Section, ASAE (1972-1973); Member of Tau Beta Pi, Sigma Xi, Phi Lamda Upsilon, and Gamma Sigma Delta. Consulting on grain storage and drying for 15 countries. *Society Aff:* ASAE, AIChE, ASEE, IFT, AACC, AAUP

Chung, Deborah D L
Business: State University of New York at Buffalo, Department of Mechanical and Aerospace Engineering, Buffalo, NY 14260
Position: Professor *Employer:* State Univ NY *Education:* PhD/Mat Sci/MIT; SM/Mat Sci/MIT; MS/Engrg Sci/CA Inst of Tech; BS/Engrg & Applied Sci/CA Inst of Tech *Born:* 9/12/52 Native of Hong Kong. Asst Prof of Metallurgical Engrg and Materials Science and Electrical Engrg at Carnegie-Mellon Univ in 1977-82. Assoc Prof of Metallurgical Engrg and Materials Science at Carnegie-Mellon Univ in 1982-86. Professor of Mechanical and Aerospace Engineering, State University of New York at Buffalo since 1986. Current research on carbon fibers, carbon films and carbon composites. SAE Teetor Educational Award 1987. AIME Hardy Gold Medal 1980, Carnegie-Mellon Univ Ladd Award 1979, first woman graduate of CA Inst of Tech 1973, Josephine de Karman Fellowship 1972, Licentiate of the Royal Schools of Music in piano performance 1970. *Society Aff:* MRS, AIME, American Carbon Society, ASM

Chung, Jin S
Business: 1500 Illinois St, Golden, CO 80401
Position: Professor *Employer:* Colorado School of Mines *Education:* PhD/Engrg. Mechanics/University of Michigan, Ann Arbor; MS/Naval Arch./University of California at Berkeley; BS/Naval Arch./Seoul National University, Korea *Born:* 01/27/37 Born in Korea. Korea Electric Power Co. 1961-62. David Taylor Model Basin, Bethesda, MD 1964-66. Exxon Production Research Co. - Offshore Drilling Systems, Houston 1969-73. Lockheed Missiles and Space Co. - Deep-Ocean Mining, Sunnyvale, CA 1973-80. Prof. at Colorado School of Mines on offshore and arctic engineering since 1980. ASME Fellow, Eugene W. Jacobsen Award 1978, Founding Chairman of Offshore Mechanics and Arctic Engrg Div. 1985, Sr. Editor of J. Energy Resources Technology 1980-85. Amoco Foundation Outstanding Teaching Award 1983. Pres, Int'l OMAE Council 1987-. Initiator/Organizer/Chairman, Int'l Offshore Mechanics and Arctic Engineering Conference and Editor since 1982. *Society Aff:* ASME, ΣΞ, SPE, SNAJ, KCORE

Churchill, Alexander M
Business: 344 South Route 73 - Suite A, Berlin, NJ 08009
Position: Pres *Employer:* Alexander M. Churchill Assoc *Education:* MS/CE/Drexel Univ; BS/CE/Union Coll *Born:* 1/21/34 Native of Waterbury, CT, employed as a fire protection engr with Factory Insurance Association of Hartford CT in the Philadelphia office (1957 through 1962). Design engr with promotion to VP for John G Reutter Assocs, Civil Consulting Engrs, Camden, NJ, specializing in civil projects as engr and project mgr (1962 through 1976). In 1976, as Pres, formed the consulting engrg firm of Alexander M Churchill Assocs specializing in civil and municipal projects. Pres of Consulting Engrs Council of NJ, a functional section of NSPE, (1978 to 1979). Served as a Diplomat in the American Academy of Environmental Engrs. *Society Aff:* ASCE, NSPE, WPCF

Churchill, George H
Home: 65 Alfred St, Brantford, Ontario Canada N3S 5E8
Position: Pres. *Employer:* Churchill & Assoc. *Born:* Nov 4, 1912 Brantford Ont Canada. B Sc GMI & Univ of M. Cert Mfg Eng P Eng. A consultant for UNIDO & CESO to the developing countries in Africa, Arabia, and South East Asia. Dir SME 1966-75. OEA Award CA. S P Eng's 1973. Fishing and photography are enjoyment. *Society Aff:* Life Mem SME

Churchill, Stuart W
Business: 220 S 33rd St, Dept Chem Engrg, Phila, PA 19104
Position: Carl V S Patterson Prof. *Employer:* Univ of PA. *Education:* PhD/ChE/Univ of MI; MSE/ChE/Univ of MI; BSE/ChE/Univ of MI; BSE/Engg Math/Univ of MI *Born:* 6/13/20. Native of Imlay City, MI. Earned all his academic degrees at Univ of MI; Worked for Shell Oil, 1942-46, & for Frontier Chem, 1946-47. Began teaching at Univ of MI in 1950 & served as Chrmn, Dept of Chem & Met Engg, 1962-67. In 1967, became first Carl V S Patterson Prof at Univ of PA; receiving honorary MA in 1971 & S Reid Warren, Jr Award for teaching in 1976. Served as Pres of AIChE in 1966 & received their Profl Progress, William H Walker, Warren K Lewis & Founders Awards in 1964, 1969, 1978 & 1980. Max Jakob Award of AIChE & ASME in 1979. Active in res & consultation in combustion and heat transfer, & author of several books. *Society Aff:* AIChE, ACS, ΣΞ, COMB. INST,.

Chwang, Allen T
Business: Inst of Hydraulic Res, Univ of Iowa, Iowa City, IA 52242
Position: Prof *Employer:* Univ of IA *Education:* PhD/Mech Engrg/CA Inst of Tech; MS/Mech Engrg/Univ of Saskatchewan; BS/Mech Engrg/Chu Hai College *Born:* 11/7/44 Born in Shanghai. Came to the US in 1967. On faculty at Caltech 1971-78. With Inst of Hydraulic Res, Univ of IA since 1978. Earle C. Anthony Fellow 1969-70. John Simon Guggenheim Fellow 1974-75 at Univ of Cambridge, England. Main res interests in low-Reynolds-number hydromechanics, nonlinear water waves, earthquake effects on dams, general fluid mechanics and applied mathematics. *Society Aff:* ASCE, ASME, APS, AAM, IAHR, EERI, ΣΞ

Chynoweth, Alan G
Business: Morris Res & Engrg Ctr, 445 South St, Morristown, NJ 07960
Position: VP *Employer:* Bell Comm Res, Inc *Education:* PhD/Physics/U London Kings Coll; BS/Physics/U London Kings Coll *Born:* 11/18/27 Native of England. Doctoral thesis on solid state radiation detectors. Post-doctoral fellow, NRC, Ottawa, Canada, 1950-1952 (critical point phenomena; organic photoconductors). Mbr Tech Strff, Bell Labs 1953-(ferroelectrics, semiconductors). Head, Crystal Electronics Res Dept, 1960-65. Asst Dir, later Dir, Materials Res Lab, 1965-76. Exec Dir, Electronic Device, Process and Materials Div, 1976-83, VP Appl Res, Bell Comm Res Inc 1984-. Achievements include: first definitive demonstration of electron tunnelling in semiconductors; elucidation of origin of the Gunn effect (W.R.G. Baker Prize, IEEE, 1967), and demonstration of infrared detection using the pyroelectric effect. Active on Nat Acad and NATO Committees in Materials field. *Society Aff:* IEEE, APS, AIME

Cimino, Saverio Michael
Home: 1776 S Forge Mtn Dr, PO Box 452, Valley Forge, PA 19481
Position: Sales Mgr Natl Accounts. *Employer:* FMC Corp. *Education:* BS/Chem/Univ of CA. *Born:* 9/1/18. Native of San Francisco, CA. Reg Chem Engr, CA. Joined FMC, 1939, held responsible chem engg positions in R&D, Process Dev & Production. Mgr of large magnesia & phosphat facility, 1958; Resident Mgr of world's largest soda ash plant & underground trona mine, 1967; Mgr of Alkali Chem Div, 1972, responsible for all facets of the business; and, became Sales Mgr, Natl Accounts Dept, 1979. Mbr AIChE & AIME. Past mbr, Bd of Dirs, WY Ind

Cimino, Saverio Michael (Continued)
Dev Cooperation, & pst mbr, WY Ind Dept Planning & Economic Dev, (appointed by Governor). *Society Aff:* AIChE, AIME.

Ciricillo, Samuel F
Home: 3277 Somerford Rd, Columbus, OH 43221
Position: Dir Engg. & Administration *Employer:* Ranco Inc. *Education:* BSME/ME/Newark College of Engg; Indust. Design/Design/Newark School of Fine & Indust. Arts. *Born:* 11/14/20. Newark NJ - Various assignments GE Co, 1942-1952 on Test & Creative Design Progs. Worked Air Conditioning Div as Sec Engr on Controls. Joined Kellex Corp during WWII as Control Engr, Jersey City, Pilot Plant for Oak Ridge Facility. Awarded Presidential Citation for work. Chief Engr 1952-1958 Emerson Radio & Phono, Quiet Kool Div Air Conditioning. 1958-present - Joined Ranco Inc in Controls Div. Held various positions, including VP Engg & Res. Reg PE States of OH & NJ. Mbr ASME 1938: Fellow ASHRAE 1952; Awarded Distinguished Service Award 1979. Active in Scouting, Hobbies golf, painting & cars. *Society Aff:* ASME, ASHRAE.

Cisler, Walker L
Home: 1071 Devonshire Rd, Grosse Pte Park, MI 48230
Position: Chairman *Employer:* Overseas Advisory Assoc Inc. *Education:* ME/-/Cornell Univ. *Born:* 10/8/97. War Production Bd, 1941-43; Chief Public Utility Section, SHAEF, 1944; Exec Secy, AEC Ind Advisory Group, 1947-48; Founding mbr, Natl Acad of Engg; Past Pres and Fellow, ASME; Fellow, IEEE, AIM, ANS, SAME, IN Inst of Tech; Trustee emeritus, Cornell Univ, Marietta College; Trustee, MI Colleges Fdn; Honorary Dir, Suomi College; Honorary Chrmn Intl Exec Council of World Energy Conf; Dir, Economic Club of Detroit; Chrmn, Thomas Alva Edison Fdn; Dir, two corps; Chrmn of Bd, the Detroit Edison Co, 1964-1975. Recipient of 17 honorary degrees and of many decorations from US and foreign govts; also of many awards, citations, etc, from profl, bus and community organizations. *Society Aff:* NAE, ASME, IEEE, ESD, SAME.

Clabaugh, Jack H
Business: PO Box 15728, Baton Rouge, LA 70895
Position: VP. *Employer:* Bovay Engrs Inc. *Born:* June 3, 1925. BS OK State; AA from NE OK A&M. Native of OK. Served in Navy, Submarine Service 1943-45. With Dow Chem Co in TX, LA, FL & MI from 1951- 66. With Cabot Corp in Pampa, TX from 1966-68. With Bovay Engrs from 1968- present. Presently VP & Mgr of Bovay's Baton Rouge office with a total staff of 80 incl 24 PE providing engg serv to indus, util, municipalities, fed & state govt.

Clancy, Joseph W
Business: 1500 Fifth St, Sacramento, CA 95814
Position: Chief, Consultant Services. *Employer:* Office of the State Architect. *Born:* Dec 1920 Durand MI. BSME MI State Univ. CA Reg M 2911. Design engr spec in double duct & high velocity sys for Office of the State Architect 1947-60. Supr of Mech Design Sect 1960-75. Ch of Cons Services 1976-. Respon for selection & assignments to architects/engrs of State of CA bldg prog. P Pres ASHRAE, Sacramento Valley Chap, Fellow ASHRAE 1975. Hobbies: oil painting, art & golf.

Clapp, James L
Business: 101 Barrows Hall, Orono, ME 04469
Position: Dean. *Employer:* Univ of ME. *Education:* BS/CE/Univ of WI; BS/NS/Univ of WI; MS/CE/Univ of WI; PhD/CE/Univ of WI. *Born:* 3/14/33. Native of Madison, WI. Held position as Prof of Civil Engg at the Univ of WI-Madison from 1964 to 1978. Primary responsibilities in area of surveying, mapping and photogrammetry. Served as Dir of the Environmental Monitoring and Data Acquisition Group at the Univ of WI from 1970 to 1978. In 1978 accepted position at Univ of ME at Orono. Reg PE and Land Surveyor. *Society Aff:* ASEE, ASCE, NSPE, ACSM, ASP.

Clark, Alfred, Jr
Business: Dept Mechanical Eng, Rochester, NY 14627
Position: Prof. *Employer:* Univ of Rochester. *Education:* PhD/Applied Math/MIT; BS/Engg Sci/Purdue Univ. *Born:* 5/5/36. Joined Dept Mech Engineering 1965. Registered P.E., State of NY. Asst Prof 1964-67; Assoc Prof 1968-74; Prof 1974-; Chrmn of Dept 1972-77. Visiting Fellow Joint Inst for Lab Astrophys, Boulder, CO 1970-71. Maj res interest is theoretical fluid dynamics. Past res includes work on water waves, superfluid helium, solar magnetic fields, solar atmospheric waves, stellar & solar rotation, sunspot structure & circulations in large lakes. Present work on oxygen transport to tissue. *Society Aff:* ASME, AAAS, APS, AIBS, ISOTT, Microcircuitory Society

Clark, Arthur J, Jr
Business: Box 5800, Albuquerque, NM 87185
Position: Dept Mgr. *Employer:* Sandia Labs. *Education:* MS/Elec Engg/Univ of NM; MS/Mech Engg/Polytechnic Inst of Brooklyn; BS/Mech Engg/Cornell Univ. *Born:* 6/10/21. With Sandia Labs, Albuquerque, NM, since 1951. As Dept Mgr, engg assignments have included design of nuclear weapons, electromechanical component design, nuclear reactor design, aerospace nuclear safety, space isotope power, where he was in charge of the radioisotopic thermoelectric generator used to power the scientific experiments placed on the moon by the Apollo astronauts, and the design of new test facilites, the latest being the conceptual design of a 5 Mw(t) solar power tower test facility. Section Chrmn, NM Section of ASME in 1960 followed by many natl committee assignments and VP ASME 1975-79. *Society Aff:* ASME, IEEE, IES, CSE.

Clark, Brian M
Business: 1025 Cleveland Ave, Columbus, OH 43201
Position: Plant Met. *Employer:* Timken Co. *Education:* BMetE/Met Engg/OH State Univ; MBA/Bus/Xavier Univ. *Born:* 5/24/46. in London, England. Served as applications met at Timken's home plant in Canton, OH. Assumed position as Columbus operations plant met in 1976. Have served on various ASM committees, past mbr of ASM natl tech committee on career guidance (Secy of committee). Enjoy sports, music & sailing. *Society Aff:* ASM, ASNT.

Clark, Charles E
Home: 8 Norris Circle, Cottondale, AL 35453
Position: Advisor Assoc. *Employer:* C E Clark & Assoc *Education:* BS/IE/Univ of AL. *Born:* 12/30/23. BSIE Univ of AL 1950-Indus Engr; B F Goodrich 1947-53; Kirkhill Rbr 1953- 55; Gulf States Paper 1957-59; Assoc R R Wilson Cons 1955-57; Ernst & Whinney. Ernst and Ernst-Mgr 1959-83; C E Clark & Assoc - Co-owner 1984-86 Adviser Assoc 1987 to present; Adj Fac, UA Tuscaloosa etal Honors; 1986-Kentucky Colonel 1970 Engr of Yr Birmingham & Jeff Cty; 1972- Fellow Royal Soc of Health; Tau Beta Pi, Order of the Engr 1974-Fellow IIE; Bk Reviewer & Lectr IIE. Certification: PE AL & CA; Cert Mfg Engr (CMfgE), Cert Mgr (CM), Cert Mgmt Conslt (CMC); CPM Assns: IIE/AIIE, NSPE, AHA/HIMSS, Amer Arbitration Assoc, Capstone Engrg Soc, F&AM, Scottish & York Rites, Zamora Shrine Temple, NAA, BPOE, Civitans. Offices: IIE-Reg VP 3 terms; Exec VP - chapter operations. Current Area 3 V.P. Dir Mbrship Expansion, 2 terms; Honors & Awards, 2 terms, ABET Inspector-current. Exper: Shoe mfg to shopping ctrs, aerospace to health care; incentives to location feasibility, cost control to bldg feasibility. Personal: Lay Reader-Chalice Bearer *Society Aff:* IIE, ASPE/NSPE, HIMSS/AHA, NAA, AAA.

Clark, Clayton
Home: 798 N 1500 E, Logan, UT 84321
Position: Prof. *Employer:* UT State Univ. *Education:* PhD/EE/Stanford Univ; EE/Elec Engr/Stanford Univ; BS/Phys/UT State Univ. *Born:* 3/9/12. Joined EE faculty of UT State Univ, 1937 - Emeritus since 1977. US Army 1939-1946 (Col Sig C). Grad Sch, Stanford Univ, 1945-1946, 1965-1967. Prog Dir, Aeronomy, Natl Sci Fdn, 1966-67. Dir, Engg Experiment Station, USU, 1964-1975. Founder and Dir, Ctr for Atmospheric and Space Sciences, 1969-1977. Chrmn, UT Sec IEEE, 1967-70, Area Chrmn, 1971 and 1972. Public Service Award 1969, Fellow Award, 1970,

Clark, Clayton (Continued)
Life mbr 1979. Married Helen Brown 1933, two daughters. Publications and consulting in the field of the Ionosphere, Radio Wave Propagation, and microwave relay systems. *Society Aff:* IEEE, ASEE, ΣΞ, ΤΒΠ.

Clark, Donald E
Business: 101 Kendrick Ave, PO Box 2232, Gillette, WY 82716
Position: Exec VP. *Employer:* Cooper, Clark & Assoc. *Education:* BS/Civ Engg/Univ of CA; AA/Civ Engg/Fullerton Jr College. *Born:* 5/12/27. Native of Fullerton, CA. Worked in constn surveying with the Santa Fe RR, 1943-44. Served with the US Army 1945-48, 1950-51. Constn inspector with the City of Brea, CA, 1948-50. Soil engr with San Francisco Dist, Corps of Engrs, 1954-56. Field & office engr, San Francisco office, Dames & Moore 1956-58. Site engr, Honolulu office, FHA, 1962-65. 1958-62 & 1965- Cooper Clark & Assoc, originally Partner, now Exec VP & Corporate Consultant for Quality Control and Training. *Society Aff:* ASCE, ASTM, APWA, TRB, SME.

Clark, Ezekail L
Home: 4615 N Park Ave, Chevy Chase, MD 20015
Position: Consultant *Employer:* Self *Education:* BS/ChE/Northeastern Univ. *Born:* 6/29/12. Gomel, Russia, emigrated to US as child, US citizen. Educated Boston, MA, Public Latin Sch, Northeastern Univ, 1937. Career includes design & construction of petrochem plants; petrol refining; water chemistry. Specialist in coal conversion since 1945 at US Bureau of Mines; coal gasification & liquefaction, oil shale processing and high pressure systems. Last position at Dept of Energy in charge of col gasification development. Career included teaching at Israel Technion, Catholic Univ, PA State & Univ of Pittsburgh. Fellow of AIChE. *Society Aff:* AIChE, ACS.

Clark, Gail A
Home: 538 E. Lea Terrace, Mustang, OK 73064
Position: Consultant *Employer:* Gulfstream Aerospace Corp & others *Education:* BA/Chem/OH State U. *Born:* May 21, 1919. BA Chem OH State Univ 1940. Fabricated first transmitting radome fired on a rocket 1948. Repaired, prior to installation, the first all plastic wing flown on a military aircraft (AT6-1950). Fabricated first boron/polyimide struct component which was tested at 650F (1971). Pres SAMPE 1965-66. Rec highest award for ser to soc 1970. Ch SPI Prepreg Ctte 1961-66. Mbr SPI Policy Ctte 1961-66. George C Marshall SPC Certif for Contributions to the Tech Util Prog, May 1972. *Society Aff:* SAMPE.

Clark, George W
Home: 18 Washington St, Topsfield, MA 01983
Position: Consult in Lighting. *Employer:* Self *Education:* BS/EE/MIT. *Born:* 12/6/17. Native of Mills, MA. Signal Corps 1941-46 including Pacific Theatre of Operations, Rank of Lt Colonel. With Lighting Prod Grp of GTE Sylvania since 1948 in various positions of engg & marketing. Pres IES of N Am 1973-74, elected IES Fellow 1959; Pres US Natl Ctte of CIE 1975-79, Gold Medalist IES of N Am 1981. *Society Aff:* IES.

Clark, Gordon M
Home: 400 Longfellow Ave, Worthington, OH 43085
Position: Prof *Employer:* OH State Univ *Education:* PhD/Oper Res/OH State Univ; MSc/IE/USC; BS/IE/OH State Univ *Born:* 11/21/34 Served with Marine Corps 1957-60. Manufacturing engr with Gen Elec Co 1961. Sr Reliability Engr for Rocketdyne Div of NAA planning development programs for advanced rocket engines 1961-65. With the OH State Univ since 1965. Dir of research projects to develop combat simulations, improve effectiveness of urban goods movement and mine drainage pollution abatement, identify feasible commuter aviation routes. Teaching courses in simulation, stochastic processes, benefit- cost analysis, reliability. Member of MORS Bd of Dirs 1972-1976. Assoc editor of Operations Research 1975-1978. Am a marathon runner. *Society Aff:* IIE, TIMS, ORSA, SCS

Clark, Houston N
Business: Box 8789, Denver, CO 80201
Position: VP. *Employer:* Ideal Basic Ind Inc. *Education:* BS/Chem Engg/Rice Univ. *Born:* 11/28/18. US Navy 1944-46. Employed continuously with Potash Co of Amer or Potash Co of Amer Div of Ideal Basic Indus since 1941. Pos held incl res engg oper supervision & mgt. *Society Aff:* AIChE, AIME.

Clark, Hugh K
Home: 225 Lakeside Drive, Aiken, SC 29801
Position: Res Assoc *Employer:* E I duPont deNemours & Co *Education:* PhD/Physical Chem/Cornell Univ, 1943; AB/Chem/Oberlin Coll, 1939 *Born:* 01/22/18 Born St Louis, MO. Married Marie Folsom 1942. Two children Lawrence A Clark and Barbara Clark Ucko born 1945. Joined Radio Res Lab, Harvard U 1943. Joined duPont Co 1945 in Acetate Res Div. Transferred to Atomic Energy Div 1951. Spent two years at Argonne Natl Lab being trained. Transferred to Savannah River Lab in 1953. Mbr Phi Beta Kappa and Sigma Xi. Chrmn Nuclear Criticality Safety Div of ANS 1970-1971. Fellow of ANS. Recipient of Achievement Award of NCSD 1976. Author of several papers in field of criticality safety. *Society Aff:* ANS, ACS, AAAS

Clark, J Donald
Home: 5423 Queensloch, Houston, TX 77096
Position: Consultant, Reservoir Engr (Self-employed) *Education:* MS/Petrol Engg/Univ of OK; BS/Petrol Engg/NM Inst of Mining & Tech. *Born:* 1/1/18. Naval Aviator, WWII, Lt Cmdr. Ret Reserves, Air Medal. Inst Petrol Engr, Univ of OK 1947-48. Reservoir Engr, Stanolind 1948-52. Retired from Union Oil Co 7/1/80. Chief Reservoir Engr 1952-65, Reg Res Engr, 1965-80. Soc of Petrol Evaluation Engrs, Pres 1973, Dir 1967-68, 81, 82; Secy-Treas 1968 & 1972. Soc of Petrol Engrs of AIME, Received Soc of Petrol Engr, "Distinguished Service Award 1980" Dir 1973-76; Chrmn Editorial Com 1959, Sec Ch 1962; Ch Education and Accreditation, 1968. Soc of Profl Well Log Analysts, VP 1969-70, Pres Local Sec 1968. Am Men of Sci, Who's Who in Engrg in Southwest, Reg PE, Pi Epsilon Tau, Life mbr of Naval Reserve Assoc & Ret Officers Assn. Presently self employed, Reservoir Engrg Consultant. *Society Aff:* SPE of AIME, SPEE, SPWLA.

Clark, James A
Home: 181 Mendon Ctr Rd, Honeoye Falls, NY 14472
Position: VP Process & Development. *Employer:* Bausch & Lomb. *Education:* MS/ME/Univ of MD; BS/ME/Univ of MD. *Born:* 6/28/21. Native of Wash, DC. Completed education and military service 1942-48. Earned a BS & MS in mech engg from the Univ of MD; served four yrs in the US Army. Engg instr at Univ of MD 1946-48. With Bausch & Lomb since 1958 in various engg mgt positions. Presently, VP Process & Development. Hold 45 patents associated with various products such as microscopes, eye implant lenses, etc. Pi Tau Sigma, 1964. Pres, Rochester Engg Soc 1971-72; Pres, PE Soc 1976-77. In 1977, selected "Rochester Engineer of the Year." In 1978, selected for the "Distinguished Engineering Alumnus" award by the Univ of MD Engg Alumni Assn. Elected Fellow ASME 1980. Chosen for College of Engrg Innovation Hall of Fame, Univ. of Maryland 1987. *Society Aff:* ASME, NSPE, OSA.

Clark, James W
Business: 2500 Citywest Blvd, Suite 300, Houston, TX 77042
Position: Owner *Employer:* Cypraea Consulting *Education:* MBA/Corp Finance/NYU; MSChE/-/Univ of TX; BSChE/-/Univ of TX *Born:* 9/21/36. Native of Austin, TX. Process Engr at TX Butadiene & Chem Corp. Sr Process Engr & Process Mgr at Scientific Design Co. Mgr, Proj Dev, Commonwealth Oil. Mgr of Proposals, CE-Lummus. VP DeWitt & Co. Started Cypraea Consulting. Respon for worldwide studies on energy & refining, petrochemicals and petrochemical feedstocks. *Society Aff:* AIChE, ACS, CMRA.

Clark, John B
Business: Dept of Met Engrg, Rolla, MO 65401
Position: Assistant Dean & Prof Met Engg. *Employer:* Univ of MO-Rolla. *Education:* PhD/Met/Carnegie-Mellon; MS/Engg/Carnegie-Mellon; BASc/Engg/Univ of Toronto. *Born:* July 13, 1924. PhD & MS Carnegie Mellon, BA Sc Univ of Toronto. Met Engr-Dow Chem Co 1952-60; Principal Scientist-Scientific Lab Ford Motor Co 1960- 66; Prof Met Engg-Univ of MO-Rolla 1966-; Assoc Dean of Grad Sch at same inst 1973-1979. Assistant Dean-Mines & Metallurgy 1979-82. Mbr AIME, (mbr of Exec Ctte & Chrmn of Publications Ctte of IMD), ASM (mbr of Publication Ctte), ASM/NBS Data Program Category Editor for Binary Magnesium Alloys 1982-, & AAAS. Mbr Sigma Xi & Henry Marion Howe Medalist 1960. *Society Aff:* ASM, AIME, AAAS, NBS, ΣΞ.

Clark, John F
Business: Box 432, Princeton, NJ 08540
Position: Dir. Space Applic. & Technology *Employer:* RCA Corp. *Education:* PhD/Physics/Univ of MD; EE/EE/Lehigh Univ; MS/Math/Geo Wash Univ; BS/EE/Lehigh Univ. *Born:* 12/12/20. in Reading, PA. Commercial radio operator and announcer while student. Lehigh Asst Prof (EE) 1947-48. Naval Res Lab radar beacon dev 1942-47; airborn atmospheric electricity and ionospheric sounding rocket res 1948-58. NASA space sci prog dev and payload selection 1958-65. Goddard Space Flight Ctr Dir 1965- 76. GSFC has outstanding space sci, Earth observations and space communications progs. Since 1976 Dir, Space Applications and Tech, RCA Corp. Received NAA 1974 Collier Trophy for LANDSAT prog leadership and NASA medals for Distinguished Service, Outstanding Leadership and Exceptional Service. McGraw-Hill Consulting Editor, Space Tech. Author. Patentee. Radio amateur. Explorers Club. Fellow AAS, AIAA, IEEE. Chairman, FCC Industry Advisory committee on Direct Broadcast Satellite.. *Society Aff:* AAS, AGU, AIAA, IEEE, ΦBK, ΣΞ, TBΠ.

Clark, John P, Jr
Business: Box 368, Glenside, PA 19038
Position: Pres. *Employer:* John P Clark Co, Inc. *Education:* BS/St Jos Univ *Born:* 4/25/11. Native of Allentown, PA. Sales Engr in Met Equip and Products 10 yrs before starting present Mfrs Agency in 1946 for furnaces, ovens, heating equip, heating-elec elements, pollution control equip for air, water, solid wastes; solar energy equip. Trustee for American Soc for Metals 1965-67; life mbr, ASM, 1976; Fellow ASM, 1970; Wm Hunt Eisenman Award, Phila Chapter, 1978. *Society Aff:* ASM

Clark, John W
Business: Alcoa Labs, Alcoa Ctr, PA 15069
Position: Mgr Engg Properties & Design Div. *Employer:* Aluminum Co of Am. *Born:* July 1922. BS 1946 & MS 1947 Purdue Univ. PhD Univ of Pittsburgh 1954. Army Infantry 1943-45. Alcoa Labs as Res Engr 1947 & became div mgr 1970. Present pos is Mgr of Engg Properties & Design Div of Alcoa Labs. Has pub 35 tech papers relating to design & performance of aluminum structures. ASCE Rowland Prize 1957, ASCE Res Prize 1958, ASCE Croes Medal 1966. VChrmn Struct Stability Res Council. Mbr ASCE, SESA, Tau Beta Pi, Chi Epsilon & Sigma Xi.

Clark, Kenneth M
Business: 4800 East 63rd St, Kansas City, MO 64130
Position: Mechanical Dept, Mgr. *Employer:* Burns & McDonnell *Education:* B.S./ME/KS Univ *Born:* 2/12/43 Employed by Burns & McDonnell Engrg Co for total over sixteen years. Currently at Principal level, serving as Special Projects Division Mechanical Dept Mgr. Worked for Truog Nichols Inc., Mechanical Contractor, as chief engr from 1969 to 1975. Life long resident of the Kansas City area. Professional society involvement includes: P Pres of the K.C. Chapter & Regional Vice Chmn of ASHRAE and Past Pres, K.C. Chapter AEE. Other civic activities include: Mbr of the Code Bd of Appeals, City of Overland Park, KS; Member, Bd of directors for the Greater K.C. Epilepsy League and member of the Kansas City Chamber of Commerce Environment & Energy Committee. Enjoy snow skiing and jogging. *Society Aff:* NSPE, ASME, ASHRAE, AEE, NEBB

Clark, Lewis Gene
Home: 744 Fairway Ln, Columbia, SC 29210
Position: Cons Elec Engr-L Gene Clark PE (Self-emp) *Born:* July 23, 1934. Reg PE SC, NC, GA. Grad Univ SC SBEE. native Union SC. USAF 1953-57. Distribution Engr SC Elec & Gas Co until 1965, spec underground high voltage distrib sys. Cons engg field since 1965. Active in IES since 1966. Currently VChrmn SC State Chap IES. N Am IES Lighting Design Award of Merit- 1976. Amateur Radio Operator.

Clark, Melvin E
Home: 3200 Kiltie Ln, Birmingham, AL 35243
Position: Exec VP Chem Div. (Retired) *Employer:* Vulcan Mtls Co. *Education:* AMP/Bus/Harvard; EP/Bus/Columbia Univ; BS/Chem Engrg/Univ of CO *Born:* 10/02/16 Chem mfg exec; b Ord NE, Oct 2, 1916; s Ansel B and Ruth Joy (Bullock) C; BS in Chem Engrg, cum laude, Univ CO, 1937; grad exec prog Columbia Univ, 1952; grad Advanced Mgmt Prog Harvard Univ, 1961; m Virginia May Hiller, Sept 16, 1938; children - John Robert, Walter Clayton, Dale Eugene, Merry Sue. Asst editor Chem Engrg, McGraw-Hill, NYC, 1937-41; mktg staff Wyandotte Chem Corp (MI), 1941-53; chief program br War Prodn Bd, Washington, 1942-44; VP mktg Frontier Chem Co, Wichita, 1953-69; Exec VP Chems Div, Vulcan Materials Co, Birmingham, AL, 1969-81, VP planning chems and metals group, 1981-82 - Mgmt Conslt; Retired Vulcan, May 1982; Pres Chlorine Inst, 1977-80. Recipient Univ CO Alumni Recognition award, 1962; named Chem Market Res Assn Man of Yr, 1963; named Univ. CO Distinguished Engrg Alumnus, 1985. Mbr Am Inst Chem Engrs, Chem Mktg Res Assn, Tau Beta Pi, Pi Mu Epsilon. Republican. Mbr Christian Disciples. Clubs: Inverness Country, Relay House (Birmingham); Shoal Creek Country, Shoal Creek, AL. Contbr numerous articles to profl jours. *Society Aff:* AIChE, ACS, CMRA, NMA

Clark, Ralph L
Home: 4307 N 39th St, Arlington, VA 22207
Position: Retired *Education:* BS/EE/Physics/MI State College. *Born:* 6/2/08. Native of MI Radio Inspector (engineer) US Dept Commerce, Fed Radio Commission and FCC Drtroit, MI 1930-35; Engr FCC, Wash DC 1935-41; Consulting Engr, Ring & Clark Washington 1941-42, Lt to Cdr USNR 1942-46; Dir Pgms Div R&D Bd, Sec Defense 1946-49; Deputy Asst Dir, Sci Intelligence, CIA 1949-57; Dir Wash Office Stanford Res Inst 1957-59; Asst Dir Communication & c3 DDR&E, Ofc Soc Defense 1959-62; Spec Asst to Dir, Telecommunications Mgmt, Exec Office of Pres 1962-70; Sr US Rep NATO Joint Communciations-Electronics Committee 1968-70; Dir Wash Ofce IEEE 1972-75 Consultant 1975-Life Fellow IEEE, Fellow AAAS, Mbr Council on Foreign Relations and Amer Acad of Political and Social Science. 1984 Awarded IEEE Centennial Medal for Extraordinary Achievement. Retired 1976. *Society Aff:* IEEE.

Clark, Robert M
Home: 9627 Lansford Dr, Cincinnati, OH 45242
Position: Chief, Economic Analysis *Employer:* USEPA, Drinking Water Research Div *Education:* PhD/Civil and Environ Engrg/Univ of Cinn; MS/Civil and Environ Engrg/Cornell Univ; MS/Math/Xavier Univ; BS/Math/Portland State Univ; BS/CE/OR State Univ *Born:* 3/1/39 in Tuscaloosa AL, grew up in Portland, OR. Commissioned a Public Health Service Officer in June, 1961, currently holding rank of Sanitary Engrg Dir. Detailed to the US Environmental Protection Agency since it's inception. Currently responsible for directing research into the application of economic, systems management and operations research techniques in problems in drinking water research. Author of over 100 professional papers and publications, including two books in the field of environmental management. Recipient of ASCE's Walter L. Huber Research Prize, ASCE's Environmental Engrg Div Special Service Award,

Clark, Robert M (Continued)
and the US Public Health Service's Commendation Award. *Society Aff:* ASCE, AWWA, AAAS, TIMS

Clark, Samuel K
Business: Dept of Mech Engg & Appl Mech, Ann Arbor, MI 48109
Position: Prof. *Employer:* Univ of MI. *Education:* BSE/Aeronautical Engg/Univ of MI; MSE/Engrg Mech/Univ of MI; PhD/Engrg Mech/Univ of MI. *Born:* 11/3/24. A native of MI. After service in US Navy in WWII was employed in industry by Douglas Aircraft, Borg-Warner and Ford Motor Co. Taught engg at Case Inst of Tech, Cleveland, OH, 1952-1955. Join the Univ of MI in 1955 as Asst Prof and have remained there since. Have been active in Mechanics applied to the rubber industry with particular emphasis on tire mechanics. Active in consulting and research. *Society Aff:* ASME, SEM, SAE.

Clark, Stanley J
Business: Agricultural Engr. Dept, Kansas State Univ, Manhattan, KS 66506
Position: Prof *Employer:* KS State Univ *Education:* PhD/Agri Engrg/Purdue Univ; MS/Agri Engrg/KS State Univ; BS/Agri Engrg/KS State Univ *Born:* 09/22/31 Native of McPherson, KS. Served with USAF 1954-57 as pilot and installations engr. Res asst at KS State Univ, 1957-59-conservation tillage res; res asst Purdue Univ, 1959-61-development of lab soil bin and wheel traction res. Instructor Purdue Univ 1961-64-taught courses relating to farm equip design. Asst Agricultural Engr, CO State Univ 1964-66, fruit and vegetable crop mechanization. Assoc Prof Agricultural Engrg Dept, KS State Univ 1966-77- Teaching and res, tillage and alternative energy res. Acting Branch Chief of Agricultural and Food Processing Branch, Dept of Energy, Washington, DC, 1977 (on loan to DOE from Univ). Prof Agricultural Engrg Dept 1976 to present- Alternative energy and energy conservation res and teaching; President Tri- Valley Chapt at KS Engrg Soc 1979-80. 1970-Present - Forensic engrg relating to patents and product liability. *Society Aff:* ASAE, ASEE.

Clark, T Henry
Business: One Commerce Place, Suite 1925, Nashville, TN 37239
Position: Pres *Employer:* Ross Bryan Assoc, Inc *Education:* BE/CE/Vanderbilt Univ *Born:* 8/12/44 Native of Nashville, TN. Joined Ross H. Bryan, Inc, Consulting Engrs in 1967 as a design engr. Designed structures of all types and followed with construction administration. Performed inspections of prestressed concrete plants in PCI Certification Program. Appointed Architectural Projects Mgr in 1972. Developed projects from conception for architectural clients. In responsible charge of design projects. Developed client relationship with architect clients. Appointed VP (Principal) in 1974. Responsible for Architectural work. Directs Investigation of building failures. Appointed Exec VP in 1977. Elected Pres (Ross Bryan Assoc Inc) in 1983. Corp mgmt responsibilities include structural design projects, investigation of distressed structures and directs PCI certification inspection Prog. Elected to Bd of Dirs PCI-1980. Married with three children. *Society Aff:* ASCE, NSPE, ASTM, ACEC, SEAONC, IABSE

Clark, William G
Business: PO Box 1398, Pittsburg, CA 94565
Position: Mgr, Engineering Purchasing *Employer:* Dow Chem USA, Western Div. *Born:* 9/29/20 AA Diablo Valley College-Mech Engg & Tool Design Cert from Intl Correspondence Sch. Univ of CA Certificates: Struct Design, Cost Engg (Constr and Piping), Engg Admin. US Army 1943-46: Medical Corp & Air Rescue Serv. Pres Am Assn of Cost Engrs 1964 & Award of Merit Winner 1975. Pres Contra Costa Cty 4H Council 1958. With Dow since 1946. Served as Mech Designer, Design Supr, Maintenance Engr, Ch Estimator, Tax Cons, Admin Serv Mgr, Mgr of Cost Engg, Member Bldg Contractors & Western Council of Constr Consumers. Prepared & taught courses on Cost Engg & pub cost engg art in Chem Engg Journal in 1956, 1959, 1960 & 1965. Fellow AACE.

Clark, William R
Home: 129 Holly Rd, Seaford, DE 19973
Position: Retired *Education:* BS/CE/Lehigh Univ *Born:* 3/24/19 Native of Mifflintown, PA. Student engr with DuPont prior to WWII. Served with Army Corps of Engrs 1941-1945. Engr in maintenance, power and design with DuPont 1945-present; Chapter Pres of Kent Sussex Chapter of DE Society of PE's, served on council for DE Assoc of PE's; currently Pres of DE Society of PE's. Retired. *Society Aff:* NSPE

Clarke, Beresford N
Home: 3723 W Hamilton Rd-RR 6, Ft Wayne, IN 46804
Position: Pres. *Employer:* MPI Furnace Co. *Education:* BSME/Mech Engg/Purdue Univ. *Born:* Dec 1923. BSME Purdue Univ 1948. Rifleman 26th Infantry Div 3rd US Army 1942-45. Res & Dev Engr, Field Erection Engr, & Sales Engr Surface Combustion Corp 1948-59. Onc of founders & Pres 1959-85 Met Processing Inc, Ft Wayne Ind, commercial heat treating co. Reg PE IN 1957. Chrmn Ft Wayne Chap ASM 1963- 64. Pres Metal Treating Inst 1974-75. Active in promoting conservation & effective util of mtls & energy. Holder four U.S. patents, Pres MPI Furnace Co, Ft. Wayne, IN. *Society Aff:* ASM, MTI.

Clarke, Charles D
Home: Post Office Box 2471, Rancho Santa Fe, CA 92067
Position: Consultant (Self-employed) *Education:* BSME/Mechanics/Univ of S CA. *Born:* 2/18/11. Native of Spokane, WA. Exec VP & a Dir, 1967-1979, of Atlas Consolidated Mining & Dev Corp, Cebu, Philippines, the largest copper producer in Southeast Asia. Has been associated in senior operations and administrative positions, with large ferrous and non-ferrous mining and beneficiating operations in Bolivia, Peru, Chile & the Philippines for the past forty-three yrs. Received the William Lawrence Saunders Gold Medal from the AIME in 1975 and the Philippine Soc of Mining, Met & Geological Engrs named him the Most Outstanding Mining Exec of 1977. Mining Conslt since retirement in 1980. *Society Aff:* AIME-SME.

Clarke, Clarence C
Business: Hwy 41 N, Evansville, IN 47727
Position: Engr. *Employer:* Whirlpool Corp. *Education:* BSME/ME/Purdue. *Born:* 2/1/22. BSME Purdue 1944 (grad with distribution, 1st in Mech Engrg Sch). Cornell Midshipman's Sch (grad 1st in Engrg School) 1944. US Navy Electronics Off USS Estes (AGC-12) 1945-46; 1951-52; Radiological Safety Officer, 1st H-bomb at Eniwetok. Internatl Harvester Designer of 1st Residential Air Cond itioner, 1946-55. Whirpool Corp Chief Engrg Test & Evaulation Ordinance Div, 1957-68. Chrmn Artillery Div Self-Propelled Weapons & Recoil Systems-American Ordinance Assoc, 1968-69. Whirpool Corp PE, Refrigeration Div 1969-84, 4 Patents. Pres of SWISPE 1969-70; Engrg of Yr 1975. Pres Ind Soc of PE, 1976-77. Pres's Service Award 1973. IN Gov for Industry to NSPE, 1977-81. NSPE Dir, 1979-84. Engr Conslt. *Society Aff:* NSPE.

Clarke, Frank E
Home: 165 Williams Dr, Annapolis, MD 21401
Position: Consultant Corrosion (Self-employed) *Education:* AB/Chemistry/W MD College. *Born:* 12/26/13. Native of Annapolis MD. Taught secondary chemistry 1935-41. Successive positins with US Naval Engg Experiment Station 1941-61, including Hd of Process Branch & Chem Engg Div. With US Geological Survey 1961-71 & 1973-76, successive positions including asst chief hydrologist (res), assoc chief hydrologist, asst dir (engg) and sr scientist. Deputy under secy (engg), Dept of Interior 1971-72. Reg Chem Engr, MD. Corrosion consultant to United Nations, the USAID and eight countries. Pres ASTM 1974-75. Distinguished service awards ASTM, ACS (Best paper), and Depts of Interior, Navy & Commerce. *Society Aff:* ACS, AIChE, ASTM.

Clarke, Frederick J
Home: 4801 Upton Street N.W, Washington, D.C 20016
Position: Lieut. General, U.S. Army *Employer:* Retired *Education:*

Clarke, Frederick J (Continued)
MSCE/Structures/Cornell University; B.S./Gen. Engrg./U.S. Military Academy Born: 03/01/15 U.S. Army from 1937-73. Retired as Chief of Engineers in grade of Lieutenant General. After retirement acted as Exec Dir, National Comm on Water Quality, 1973-76 reporting to Congress on costs and benefits of cleaning up waters of U.S. Served as senior consultant to TAMS on water resources and energy 1973-83. Retired 1983. *Society Aff:* NAE, SAME, ASCE, AWRA, APWA, AIA

Clarke, John R
Home: 1733 Cricket Hollow, Austin, TX 78758
Position: Head, Dam Safety Unit *Employer:* TX Dept of Water Resources *Education:* BS/CE/TX A & M Univ *Born:* 4/10/20 at New York, NY. Childhood at Waco, TX. Service with Army, 2nd Engr Special Brigade 1942-45. Registered engr in TX. Employed by Army Corps Engrs (civilian) and consulting engrs 1948-62, on design and construction of earthwork and foundation projects. Design engr, City of Austin, TX 1963-67. With TDWR since 1968. Appointed Head, Dam Safety Unit in 1971. Am responsible for supervision of inspection and structural evaluation of all non-Federal dams in TX. *Society Aff:* ASCE, USCOLD

Clarke, Joseph H
Business: Box D, Providence, RI 02912
Position: Prof of Engg. *Employer:* Brown Univ. *Education:* PhD/Appl Mech/Poly Inst of NY; MAeE/Aeronautical/Poly Inst of NY; BAeE/Aeronautical/Poly Inst of NY. *Born:* 7/28/27. Professorships held at Poly Inst of NY, Politechnico di Torino, MIT, and Brown Univ. Awarded NSF Sr Postdoctoral, Guggenheim, and Fulbright Fellowships. Invited lectures presented all over USA and Europe. Consultant to numerous cos. Author of a book and many publications. Best known published research areas include wave drag reduction in supersonic lifting flight, uniformly valid second-order solutions in supersonic flow, reverse-flow integral methods for first- and second-order supersonic flow theory, photoionization upstream of a strong shock wave, gas dynamics with nonequilibrium radiative and collisional ionization, radiation gas dynamics with nonequilibrium/chem, physical gas dynamics, and gas dynamics with nonequilibrum condensation. *Society Aff:* APS, ASME.

Clarkson, Arthur W
Home: 916 8th Ave, Helena, MT 59601
Position: Sanitary Engr Consultant (Self-employed) *Education:* MS/Civ Engrg/U of MO at Columbia; BS/Civ Engrg/U of MO at Rolla. *Born:* 5/17/16. Native of St Louis, MO worked for MO Dept of Health from 1940 to 1950 with military leave from 1943-1946. During WWII served in the Army Sanitary Corps, after the war continued in the AF Reserve as a Bioenvironmental Engr until 1975 when retired from the reserve with the rank of Col. Moved to MT in 1950 to accept a position with the MT Dept of Health and Environ Scis from where I retired in 1980 to enter private practice as a conslt. Pres State Soc of Engrs 1964, Dir AWWA 1969-1972, Fuller Award from AWWA 1964, Life Mbr WPCF 1976, AWWA 1980, APHA 1981, NSPE 1981. Reg PE in MT & MO. *Society Aff:* AAEE, AWWA, WPCF, APHA, NSPE, ROA.

Clarkson, Clarence W
Business: 1-B Belmar Rd, Clearbrook, Cranbury, NJ 08512
Position: Consultant. *Employer:* C W Clarkson & Assoc. *Education:* BS/Elec Engrg/IL Inst of Tech *Born:* 10/2/12. Native of Chicago, IL. Engaged in Design & Sales of Theatrical Lighting Equip 1936-1940; Assoc Elec Engr War Dept, Office of Chief of Engrs 1941-1944; Instr course in Illumination, (eves) George Wash Univ 1943-1944; Sr Assoc Corning Glass Works 1944-1972; Dir Lighting Products Tech Services, Asg Inds, Inc, 1972-1977; consultant C W Clarkson & Assoc (Tech Glasses & Solar), 1977- 1984. Fellow Illuminating Eng Soc (Distinguished Service Award); Life (Sr) Mbr IEEE; Mbr Intl Solar Energy Soc, Inc, Christian Scientist, Patents, Publs in Field. Enjoy golf, travel, social rec. *Society Aff:* IES, IEEE, ISES

Clarricoats, Peter J B
Business: Queen Mary College, London E 1 England
Position: Prof Elec & Electronic Engg. *Employer:* Univ of London. *Born:* London, England 1932, ed Imperial Coll, Univ of London, awarded B Sc Engg 1953, PhD 1958, D Sc 1968. Prof of Electronic Engg Univ of Leeds 1963-68, Electrical Engg Queen Mary Coll 1968. Fellow of the Inst of Phys 1964, Fellow of the Inst of Elec Engrs 1967, and of the Inst of Elec & Electronic Engrs 1967. Mbr Council of Inst of Elec Engrs 1964-67, Chrmn of IEE Electronics Divisional Bd 1979, IEE Electronics Divisional Premiums 1961, 1962, Coopers Hill Mem prize 1964, Marconi premium 1974, founder editor of Electronics Letters.

Clary, Bobby L
Home: 1502 N Skyline Dr, Stillwater, OK 74074
Position: Assoc Prof. *Employer:* OK St Univ-Agri Engg Dept. *Born:* Aug 1938. PhD OK State Univ 1969. BS Univ of GA 1960. Native Odum, GA. Taught at State Univ of NY, Agri & Tech Inst, Alfred, NY & Polk Jr Coll, Winter Haven, FL from 1960-66. Joined Agri Engg Fac at OK State Univ in 1969. Current respon are teaching & res in food process engg. Actively engaged in cons on safe design of agri machinery. One yr on govt exchange prog with US Energy Res & Dev Admin developing an energy conservation R&D prog for Agri & the Food Indus. Active ASAE, IFT & NSPE. ASAE A W Farrall Young Educator Award 1975. ASAE Dinting Young Agri Engr of Southwest 1976. OSPE Young Engr of Yr 1974.

Clary, Eugene E
Home: 847 Redwood Ave, Santa Maria, CA 93454
Position: Acting Technical Dir. *Employer:* Space & Missile Test Organization. *Education:* BSEE/Electronics/Healds Engg College. *Born:* 3/3/20. Native of Hartford, CT. Served in Army AF 1943-45. Military Intelligence Analyst in Far East 1950-57. Deployed Matador & Mace tactical missiles to Far East 1957-60. Operations Analyst with HQ SAC 1960-62. Chief Scientist at Vandenberg AFB 1962-665. Currently Acting Technical Dir & Chief of Requirements and Evaluation Div. Chrmn, AIAA Vandenberg Section 1970. Chrmn PCFS&E 1971. Chrmn, Exec Committee, Range Commanders Council 1979. AF Exceptional Civilian Services Award 1954, 1974, 1979. Active in amateur radio & enjoys golf. *Society Aff:* AIAA, PCFS&E, AFA.

Clary, Leon H
Home: 50 Mendon Victor Rd, Mendon, NY 14506
Position: VP *Employer:* Sear-Brown Assoc, PC *Education:* MBA/Management/Univ of Rochester; B/CE/Clarkson Coll of Tech *Born:* 12/16/40 Native of Norwood, NY. Employed by State of CA's Div of Highways and Lewis Dickerson Assocs, Watertown, NY, before joining Sear-Brown Assocs. In 1979 assumed duties of VP of Sear-Brown Assocs, PC, with overall responsibility for all private clients. Served on several committees and held offices in the regional assocs of NSPE, ASCE, ASP & ACSM being a past VP of NYS Assoc of Professional Land Surveyors, member of the Curriculum Advisory Committee for the Construction Technologies Dept of the Nat'l Technical Inst for the Deaf, member Nat'l Assoc of Home Builders, listed in Who's Who in the East, registered PE and Land Surveyor in NY State. *Society Aff:* ASCE, NSPE, ASP, ACSM

Clauser, Francis H
Home: 4072 Chevy Chase, Flintridge, CA 91011
Position: Prof. *Employer:* CA Inst of Tech. *Education:* BS/Phys/Caltech; MS/ME/Caltech; PhD/Aero/Caltech. *Born:* 5/25/13. Kansas City, MO. In charge: Aerodynamic Res, Douglas Aircraft Co 1937-1946. Prof of Aero. The Johns Hopkins Univ. Chrmn dept of Aero 1946-1964. VChancellor, Univ of CA Santa Cruz 1965-1969. Chrmn Div of Engg and Appl Sci. Prof of Engg, CA Inst of Tech 1969-1980. Prof Emeritus 1980- . *Society Aff:* AIAA.

Clausing, Arthur M
Business: 1206 W Green, Dept of Mech Engrg, Urbana, IL 61801
Position: Prof of ME *Employer:* Univ of IL *Education:* PhD/ME/Univ of IL; MS/ME/Univ of IL; BS/ME/Valparaiso Univ *Born:* 8/17/36 Native of Palatine, IL.

Clausing, Arthur M (Continued)
Taught in Dept of Mech Engrg at Univ of IL since 1963. Consultant and researcher in convective heat transfer and various thermal aspects of solar energy. Developed & built variable temperature cryogenic heat transfer tunnel for the study of natural, forced & combined convection. Dir of solar energy program at Univ of IL. Enjoy photography, running, biking and hiking. *Society Aff:* ASME, ISES, ASHRAE, ASES.

Clay, Harris Aubrey
Home: 1723 Church Ct, Bartlesville, OK 74006
Position: Consulting Chem Engr. *Employer:* Self-employed. *Education:* ChE/Chem Engg/Columbia Univ; BS/Chemistry/Univ of Tulsa. *Born:* 12/28/11. Retired Phillips Petroleum Co after 34 yrs in R & D Dept, which included pilot plant supervision, process design, process engg, & plant start-up of petro and petrochem processes. Phillips tech rep to Fractionation Res Inc 7 yrs, 5 as chrmn of Tech committees. Co-author of techmagazine artilces on fractional distillation. Hold patents in field. Ref PE OK. Fellow AIChE, mbr Electrochem Soc, OSPE, NSPE, ACS. Cons spec in energy saving & separations processes. Enjoy photography & hunting. *Society Aff:* AIChE, ACS, ECS, NSPE, OSPE.

Clayton, Carl C
Home: 4701 Bonner Dr, Corpus Christi, TX 78411
Position: Sr Proj Engr. *Employer:* PPG Industries. *Education:* BS/CE/Univ of TX. *Born:* 5/23/19. Native of Texarkana, TX Tech Dept Southern Alkali Corp (PPG) 1942-44. US Navy 1944-46, Engg Officer (Ensign) on minesweeper. Dev Dept Columbia Southern Chem Corp (PPG) 1946-51. US Navy (Lt) 1951-53, Exec Officer on LST. PPG Industries 1953 to present. Responsible for preliminary design and economic evaluations on new chem processes, process improvements on existing plants, computer applications in related fields. Patentee in field. Founding Mbr of the AACE (1956), Natl Sec 1963-1966, Dir 1967, 1976-77. Charter mbr S TX Sec AACE, Pres 1971, 1976. Main hobby genealogy. Wife-Maxine Aaron, daughter-Carol. *Society Aff:* AACE.

Clayton, Joe T
Business: Dept of Food Engg, University of Massachusetts, Amherst, MA 01003
Position: Prof & Dept Hd. *Employer:* Univ of MA. *Education:* PhD/Agri Engg/Cornell Univ; MS/Agri Engg/Univ of IL; BSAE/Agri Engg/Univ of TN. *Born:* 10/2/24. Native of Etowah, TN. US Army 1943-46. Faculty positions at Univs of IL, CT, and MA. Prof of Agri Engg, Univ of MA, 1961. Natl Sci Fdn Sci Faculty Fellow, Cornell Univ, 1960-62. Appointed Hd, Dept of Food and agri Engg, Univ of MA, Jan 1966. Visiting Prof of Bioengg and NATO-NSF Sr Fellow in Sci, Univ of Reading (England), 1970-71. JSPS (Japan Society for the Promotion of Science) Fellow, University of Tokyo, 1981. Teaching and res interests include engg studies of plants and animals and their response to physical environment factors; engg properties of food stuffs, food storage, processing, and protection from mech damage. *Society Aff:* ASAE, IFT, ASEE.

Clayton, Robert J
Business: Hirst Res Ctr, East Ln, Wembly Middlesex, England HAP 7PP
Position: Technical Director. *Employer:* GE Co Ltd. *Education:* MA/-/Cambridge Univ *Born:* MA Cambridge Univ. GEC Res Labs 1937-50, res on communications, TV & radar, esp microwaves. 1950-61, GEC Applied Electronics Labs, finally as Mgr. 1961-66, Managing Dir GEC Electronics. 1966-68 Managing Dir GEC Res. 1968-, Tech Dir, the GE Co Ltd. 1971-83. Officer of Order of British Empire-OBE 1961; Commander of Order of British Empire-CBE 1970; Knighthood 1980; Prizes for papers read before Inst of Elec Engrs-IEE; Chrmn Electronics Div IEE 1968-69. Pres IEE 1975-76. Fellow IEEE 1976. Fellowship of Engg. VP 1980-83. Fellowship of Engg. Pres Inst Physics 1982/84. *Society Aff:* FENG, Inst. P, IEE, IEEE

Cleasby, John L
Business: 492 Town Engr Bldg, Iowa State Univ, Ames, IA 50011
Position: Prof & Anson Marston Dist Prof in Engrg *Employer:* IA State Univ. *Education:* PhD/Sant Engg/IA State Univ; MS/Sant Engg/Univ of WI; BS/Civil Engg/Univ of WI. *Born:* 3/1/28. Native of Madison, WI. Worked with Standard Oil Co, Indiana, and Consoer Townsend and Assoc, Chicago before joining IA State Univ in 1954. Research emphasis in filtration, coagulation and softening leading to over 70 pub, 4 book chapters and one textbook. Short term consultant to various industries and government agencies including short course teaching for WHO, PAHO, and the Economic Dev Inst of the World Bank. ASCE, Pres IA Section 1966; ASCE Rudolph Hering Medal in 1968, 1970 and 1983; AWWA, Chairman Iowa Section 1982; AWWA Publications Award 1962 and 1980; AWWA Water Quality Div Award 1970; AWWA GW Fuller Award 1978, ASCE Norman Medal 1980, Elected to Natl Acad of Engrg, 1983. *Society Aff:* ASCE, NSPE, WPCF, AEEP, ASEE, AWWA.

Cleaver, Oscar P
Home: 603 Mourning Dove Dr, Sarasota, FL 33577
Position: Retired. *Education:* Masters/EE/Yale Univ; BS/EE/GA Inst of Tech; Certificate/Gen/Oxford Univ; Certificate/Exec MOWW, Ret Officer, Dev/Cornell Univ; Certificate/Advances in Electrical Engineering and Physics for Research Executives/UCLA, Los Angeles Cal. *Born:* 1/8/05. in KY; Dist Engr for Mid-Western div, Westinghouse Lamp Co. Chief, Illuminating Eng Dept, Westinghouse Lamp Co prior to WWII. Served US Army Corps of Engrs 1942-1968. Specialized in res and dev in electrical generators, night vision, mine detection, personnel and military detection equipment. Became Technical Dir of Engr Res and Dev Labs, Ft Belvoir, VA. Retired as Colonel in 1968. Received Army two highest awards for contributions to dev in the Elec and Night Vision Military fields. Prior to Army career, was Chrmn of the Chicago and NY Chapters of IES, also Gen Secy of the Soc in 1941, and VP for the Northeastern Region prior to entering the Army in 1942. As grad of Yale Sch of Drama as well as of the Elec Engg Sch, my interests in retirement are dramatics and stage lighting with active participation in Sarasota Asolo Theater of Univ of Southern Florida. Numerous scientific articles in Electrical, Electronic and Theater fields. Published in 1980, Autobiography of activities during career. *Society Aff:* IEEE, IES, SAME, YUES, SAR, MOWW, ТВП, ΣΞ, ΦΚΦ.

Clegg, John C
Home: 1785 No 1500 East, Provo, UT 84604
Position: Prof *Employer:* Brigham Young Univ *Education:* PhD/EE/Univ of UT; MS/EE/Univ of UT; BS/EE/Univ of UT *Born:* 9/19/27 Gen Electric 1949-53 Sonar Systems, TRW Systems 1957-61 Intertial guidance, BYU 1961-present teaching electronics, control systems, industrial and commercial lighting. *Society Aff:* IEEE, IES

Cleland, Laurence Lynn
Business: PO Bos 808, L-30, Livermore, CA 94550
Position: Dep Director for Information Systems *Employer:* Univ of CA. *Education:* PhD/EE/Purdue Univ; MS/EE/Purdue Univ; BS/EE/Purdue Univ. *Born:* 10/22/39. Native of OH. Design Engr with Magnavox Co 1961-62. Proj Engr with Lawrence Radiation Lab specializing in controls & accelerators 1963-64. Consultant with Midwest Applied Sci Corp & CTS Microelectronics 1966-69. Asst Prof of Elec Engg at Purdue Univ 1968-69. Systems Res Engr 1969-70, Electromagnetics & Systems Res Group Leader 1970-71, Engg Res Div Leader 1971-78, Electronics Engg Deputy dept Hd 1978-79, NRC Prog Leader 1979-1985, Technical Services Prog. Leader 1985-86, Dep. Director for Information Systems 1986-present with Lawrence Livermore National Laboratory. Recent technical activities involve information and office automation systems. *Society Aff:* AAAS, IEEE.

Clema, Joe K
Business: Box 2691, MS D-42, W Palm Beach, FL 33402
Position: Exp Engr. *Employer:* Pratt & Whitney. *Education:* PhD/ME/CO State Univ; MS/Ind eng & Systems Analy/Univ Miami; BS/Chem/Univ NB. *Born:* 9/23/38. Native of Humboldt, NB. Taught at several univs & consulted for several

Clema, Joe K (Continued)
cos: 1969-present. Sr Software Dev Mgr of Simulation Tech, Inc 1977-1979. Won the sixth annual simulation symposium grant: 1972. Pres, Bd of Dirs of the 12th annual simulation symposium: 1979. Chrmn, IEEE comp soc technical committee on simulation stds. ACM special interest group chrmn: 1979-1981. Name appears in Who's Who in the West, Personalities of the West & Midwest, and Volume XVI of the Dictionary of Intl Biography. ACM National Lecturer 1980-1982. *Society Aff:* IEEE, AAAS, IIE, AFA, ACM, SCS.

Clement, Gordon M
Home: 117 Cold Hill, Granby, MA 01033
Position: Pres. *Employer:* H W Case Sales Co Inc. *Education:* BChemE/-/Cornell Univ; BS/CE/Cornell Univ. *Born:* 5/31/22. Native of and secondary schooling in Middletown, NY. Naval officer Pacific, WWII-Proj engr for Monsanto, Springfield, MA 47-54. Joined Case Sales Co, Springfield, MA as sales engr covering New England territory for engineered corrosion control systems. Pres '70. Enjoy travel, scuba and skiing. *Society Aff:* AIChE, TAPPI.

Clement, Richard W
Home: 1821 Emerson Dr, Deming, NM 88030 *Employer:* Retired *Education:* MS/ME/Harvard Univ.; BS/ME/Univ. of New Hampshire *Born:* 04/13/16 Lt Col, Asst G-2 Ninth Army, 1941-46. Consolidated Edison Co of New York, Inc., 1940-1976 (includes war leave). Design and Construction of power Generating stations, both conventional and nuclear. Worked through various engrg titles & responsibilities, ending my career as Asst. VP of Engrg. Life Fellow in ASME. Current PE license in NM. *Society Aff:* ASME

Clemente, Frank M, Jr
Business: 185 Park Row, New York, NY 10038
Position: Consultant *Employer:* Self Employed *Education:* PhD/Civ Engrg/Tulane Univ; MSCE/Civ Engrg/NY Univ; B.E. /CE/Cooper Union *Born:* 11/3/41 Fifteen years civil engrg experience with Parsons Brinckerhoff. Ten years in responsible charge for geotechnical engrg work in Honolulu Office. Additional assignments in New York, San Francisco and American Samoa. Licensed PE in NY, LA and HI. Principal Engr for design and implementation of the Keehi Interchange Field Testing Program which received the Grand Conceptor Award from CECH and a national Honor Award for Engrg Excellence from ACEC, 1979. Technical paper to ISSMFE, Stockholm, 1981. Lecturer, Univ of HI. Pres, HI Section, ASCE, 1979-80. Invited representative, ASCE People-to-People Delegation to Australia, New Zealand and Fiji, 1981. *Society Aff:* ASCE, NSPE, ASTM, EERI, DFI, ISSMFE.

Clements, Kevin A
Home: 46 Adams St, Westboro, MA 01581
Position: Prof. *Employer:* Worcester Poly Inst. *Education:* PhD/Systems Sci/Poly Inst of Brooklyn; MS/Systems Sci/Poly Inst of Brooklyn; BS/Elec Engg/Manhattan College. *Born:* 6/13/41. in Queens, NY. Worked for GE Co from 1963 to 1968 as a guidance and control engr. Proj engr specializing in aircraft and missile navigation systems for Singer-Gen Precision Co from 1968 to 1970. Joined Worcester Poly Inst in 1970 and is currently Prof of Elec engg. teaches in the areas of systems theory and power system operation and control. Conducts res on power system reliability and on-line control. Assoc Consultant with Power Technologies, Inc. *Society Aff:* IEEE.

Clements, Linda L
Business: Mtls Sci & Applications Office, Moffett Field, CA 94035
Position: Associate Professor/Res Scientist *Employer:* San Jose State University/NASA-Ames Res Ctr *Education:* PhD/Metallurgy & Engg/Stanford Univ; MS/Mtls Sci/Univ of PA; BS/Mtls Sci/Stanford Univ. *Born:* 10/6/45. My specialty is mech properties of mtls, with recent emphasis on high performance organic-matrix composite mtls. In addition to teaching materials engr at San Jose Univ. For the past 3 yrs I have been under contract to NASA/Ames Res Ctr. Before that I was with the Fiber Composites & Mechanics Proj at Lawrence Livermore Lab. I am currently active in ASM (Chrmn, Santa Clara Valley Chaphter, 1978-79), & ASTM (D30 cttee). I am principal author of 17 technical papers. My hobbies include backpacking, rock climbing, cross-country skiing, gardening, & remodeling. *Society Aff:* ASM, ACS, TMS-AIME, ASTM, SAMPE.

Clements, Wayne I
Business: N.J.I.T, Newark, NJ 07102
Position: Assoc Prof. *Employer:* NJ Inst of Tech. *Education:* MS/Biomed Engr/Univ of PA; MS/EE/Newark Coll of Engg; BS/EE/Newark Coll of Engg. *Born:* 2/11/31. in Hillsdale, NJ. Served as USN electronic technician. Employed at Bell Tel Labs and GE for radar and missile guidance systems dev. Electronic countermeasures experience with ITT Labs. Instrument design at Gulton Ind. Customer service engr for GE industrial products. Employed at Newark College of Engg since 1959. Developed biomedical engg prog. Recipient of Robert W Van Houten Teaching Excellence Award. Reg PE, engg consultant since 1967. Mbr of Tau Beta Pi, Eta Kappa Nu, Omicron Delta Kappa, Order of the Engr. *Society Aff:* IEEE, ASEE.

Clesceri, Nicholas L
Business: Dept of Chem & Environ Engrg, Route 102, Troy, NY 12181
Position: Prof *Employer:* RPI *Education:* PhD/Sanitary/Univ of WI; MS/CE/Univ of WI; B/CE/Marquette Univ *Born:* 9/13/36 in Chicago, IL. Design engr with City of Milwaukee, WI. Awarded NIH Postdoctoral Fellowship to study at ETH, Zurich Switzerland. At RPI since 1965. Chrmn, Joint Task Group on Phosphorus Analysis, Standard Methods. Member, Environmental Advisory Board, Chief of Engrs, US Army Corps of Engrs. Pres, Hudson River Environmental Society. Hobbies: skiing, boating. *Society Aff:* AIChE, WPCF, ASCE, AWRA, ASEE

Clevenger, William A
Home: 156 Meadow Lark Lane, Sequim, WA 98382
Position: Retired. *Education:* BS/Civ Engrg/Univ of WY *Born:* 9/12/19. Served with Army Corps of Engrs during WWII. With US Bureau of Reclamation for 10 yrs. Joined Woodward-Clyde in 1956 and assumed position of Chmn of the Bd in 1973 served until 1980. Have been consultant on over 300 dam projects, and geotechnical consultant on buildings, power stations, pipelines, canals, tunnels, marine structures, bridges, transmission lines. PPres, Amer Consulting Engrs Council, P. Pres. ASCE, Colo. section, Past Director WSCOLD. Current retired. *Society Aff:* ASCE, ACEC, USCOLD.

Cleverley, Morris L
725 Erie Blvd W, Syracuse, NY 13204
Position: Pres. *Employer:* Morris L Cleverley Engr, PC. *Education:* BS/Civ Engr/Rose Poly Inst. *Born:* 3/12/37. Engg Co Exec; B Oneonta, NY; BS Rose Poly Inst, 1962; Outstanding CE Award as student 1962 Ind ASCE; Postgrad Northwestern Univ 1963-64; Married W/2 children. Div Engg Mgr NYS Div Mobil Oil 1968-70, Dir Engr, VP Carrols Const Corp Syracuse, NY 1970-74; Pres Morris J Cleverley Engr, PE. Camillus NY 1972- , Pres Cleverley CM Assoc, Inc. 1978- ; Who's Who in East 1979-80; Mbr: ASCE, NSPE, PEPP, CSI, ACI, Sigma Nu, Terre Haute, Lodge 19 F&AM; Democrat, Lutheran, Bellevue CC-Director. Director 1981-88 American Lung Association Central NY. Who's Who in Technology Today 1981. *Society Aff:* ASCE, NSPE, PEPP, CSI, ACI.

Clewell, Dayton H
2954 E. Del Mar Blvd, Pasadena, CA 91107
Position: Retired *Education:* PhD/Physics/MA Inst of Tech; BS/Physics/MA Inst of Tech. *Born:* 12/15/12. From 1935 to 1938 was employed by C K Williams Co of Easton, PA as a Res Physicist working on optical properties of pigments. From 1938 to 1977 was employed by Mobil Oil Corp at Dallas, TX and NYC as follows: Dev of Gravity and Seismic Instrument for Oil Exploration, Dir of Field Res Lab for Res on Oil Exploration and Oil Production, Gen Mgr of Res for Processes and Products as well as exploration and production, Sr VP for res & engg. Has served

Clewell, Dayton H (Continued)
on several committees of Natl Res Council. Mbr of Bd of Cordis Co Miami, FL, Mbr Natl Acad of Engrg. *Society Aff:* NAE, SEG, IEEE, AAPG, AAAS.

Cliett, Charles B
Business: Drawer A, Mississippi State, MS 39762
Position: Hd, Aerospace Engr. *Employer:* MS State Univ. *Education:* MS/Aero Engr/GA Inst of Tech; BS/Aero Engr/GA Inst of Tech. *Born:* 7/10/24. Native of Clay County, MS. Served in US Navy 1943-46. Faculty, Dept of Aerospace Engg, MS State Univ, 1947-; appointed Hd, Dept of Aerospace Engg, 1960. Served as Natl Chrmn, Aerospace Dept Chrmn's Assn, 1979. Elder, First Christian Church, West Point, MS. Enjoy hunting and fishing. *Society Aff:* AIAA, NSPE, ASEE, ADCA.

Clifford, Eugene J
Business: 201 N Charles St-15900, Baltimore, MD 21203
Position: Agency Proj Engr. *Employer:* Daniel, Mann, Johnson & Mendenhall. *Born:* 1923 Boulder CO. Bach Civ Engg Catholic Univ of Am 1950. US Army 1943- 46. Automotive Safety Fdn Fellow Yale 1954-55; from draftsman to Cty Traffic Engr, Mont Cty MD 1950-58; City Traffic Engr Richmond VA 1958-59; Ch Transportation Services Div US Army Transportation Engg Agency 1959-63; Dir of Traffic Engg Baltimore Cty MD 1963-75; currently employed Daniel, Mann, Johnson & Mendenhall; formerly Instructor Essex Community Coll; Cons & guest lectr in traffic engg; VP MD Safety Council; Secy Theodore M Matson Meml Fund 1971-76; PE MD. Fellow Inst of Transportation Engrs, Mbr Bd of Dir; Mbr Natl Soc of PE.

Clikeman, Franlyn M
Business: School of Nucl Engineering, Purdue University, W. Lafayette, IN 47907
Position: Prof. of Nucl. Engrg. *Employer:* Purdue Univ. *Education:* PhD/Physics/Iowa State Univ; BS/Eng Physics/ MT State Univ. *Born:* 3/6/33. Native of Havre, Montana. Taught at MIT and Purdue Univ following a post Doctoral appointment in the Ames Lab of the Atomic Energy Comm. Res interests include experimental reactor physics, nuclear instrumentation and nuclear engg education. *Society Aff:* ANS, APS, ASEE, ΣΞ.

Cline, William E
Business: 65 Barrett La, Wyckoff, NJ 07481
Position: VP. *Employer:* C-E Lummus Operating Associates, Inc. *Education:* BS/Chem Engrg/NC State. *Born:* 8/11/19. Employed by DuPont at Belle, WV 1941-56 in various engg and supt positions. Was Capt Army Ordance Depat (1942-46). Joined Ethyl Corp, Baton Rouge, LA as Supr, chlorinated hydrocarbon operations. Moved to the Lummus Co (subsidiary of Combustion Engg), Bloomfield, NJ in 1962 as Mgr of Initial Operations. Was respon for all commissioning activities of the Lummus Group of Companies. Promoted within CE organization to VP C-E Lummus Operating Assoc in 1978. Has been an API Sub-Committee Chmn and a lecturer for Amer Mgmt Assoc. Retired in 1983. *Society Aff:* AIChE.

Clinebell, Paul W
Business: 211 N. Race St, Urbana, IL 61801
Position: VP *Employer:* CRS Group Engrs, Inc, Clark Dietz Div *Education:* MS/Environ Engrg/Univ of IL; BS/CE/Univ of IL *Born:* 10/11/27 Native of Springfield, IL. Taught at the Univ of IL, Civil Engrg Dept. Served in the Commissioned Corps of the US Public Health Service; current Inactive Reserves; Rank, Engr Office, Dir Grade, (equivalent to Col rank). Since 1957, member of Clark Dietz Engrs, progressively as design engr, chief environmental engr, VP with responsibilities in environmental engrg and new business development. *Society Aff:* WPCF, AAEE, AWWA, SAME, APWA

Clinedinst, Wendel W
Home: 951-A Argyll Cir, Lakewood, NJ 08701
Position: Retired. *Education:* ME/Power Plants-Marine Propulsion/Stevens Inst of Tech. *Born:* 7/23/95. NYC. ME Stevens Inst of Tech 1921. US Army Motor Transport Corps 1917-19. With Payne Dean Ltd, Field Engr 1921-24; Rust Engg co 1924. H Wheeler Mfg Co 1926-62; Bd of Tr Stevens Inst of Tech 1956-75. Pres Allied Artists of Am 1968- 70. Pres W Engg Corp 1936-61. PE CT 1956-63, NY 1925-. Mbr Exec Ctte Metro Sec ASME 1931-34; Medallist Stevens Alumni 1961 & 1969; Catholic Fine Arts Soc 1969; Lectr Bus Engg 1948-62. Patent West control for valves & bulk head door 1938. On Bds of many soc's. Fellow ASME. *Society Aff:* ASME, SNA & MarEngg.

Cloke, Thomas H
Business: Loebl Schlossman & Hackl, 845 N. Michigan Ave, Chicago, IL 60611
Position: Coord of Engg. *Employer:* Loebl Schlossman & Hackl. *Education:* BSME/Mech Engrg/Univ of IL. *Born:* 10/17/21. Served US Army 1943-46, Capt, Bronze Star. Experience: various levels of design, dev, administration and sales with architects Shaw, Naess & Murphy (1946-62), Jensen and Halstead (1962-64) and engrs, Neiler Rich & Bladen, Inc (1964-68). Formed Gritschke and Cloke, Inc, Consulting Engrs in 1968-79. Loebl Schlossman & Hackl - ARCH, (1979-Present). Mbr: ASHRAE, APCA, NFPA, AGA, SAME, Reg PE, IL. Served ASHRAE Chmn Research Promotion Committee and Research and Technical Committee, awarded ASHRAE Fellow. Designed mechanical systems Chicago- O'Hare Internatl Airport. Personal and civic interests include respon with Village of Glen Ellyn and US Power Squadron. *Society Aff:* ASHRAE, APCA, NFPA, AGA, SAME.

Cloud, Gary L
Business: MMM Dept, Michigan State University, E. Lansing, MI 48824
Position: Prof. *Employer:* MI State Univ. *Education:* Adv Study/Optics/Imperial College; PhD/Mechanics/MI State Univ; MS/Mechanics/MI Tech Univ; BS/ME/MI Tech Univ; Dipl/Pre-engr/Muskegon Community College. *Born:* 9/8/37. Native of MI. Active in Scouting. Taught 1969 at Univ of Zambia, Zambia, Africa. Sr Res Fellow at AF Mtls Lab, WPAFB 1975-76. Natl Sci Fdn Faculty Fellow for study of Phys at Imperial College, London 1976-77. reg PE-MI. Maintains consulting practice related to mech design, failure analysis, optics, etc. Holds several res grants in areas of strain & motion measurement, optics, fracture. Several summers glacier mechanics res in AK. Trained singer, performs recital and oratorio. Enjoys fishing, bldg, gunsmithing and shooting. *Society Aff:* SPIE, OSA, BSSM, SEM.

Clough, Ray W
Business: Rm 775 Davis Hall, Berkeley, CA 94720
Position: Prof of Civ Engg. *Employer:* Univ of CA. *Education:* ScD/Struct Engg/MIT; MS/Struct Engg/MIT; MS/Meteorology/CA Inst of Tech; BS/Civ Engg/Univ of WA. *Born:* 7/23/20. Native of Seattle, WA Served with USAF Aviation Engrs 1942-46. Mbr of faculty Univ of CA, Berkeley 1949-date. Chrmn SESM Div 1967-70, Dir Earthquake Engg Res Ctr (EERC) 1973-76, Asst Dir EERC 1976-1980. Teaching & res in structural dynamics & earthquake engg, analytical & experimental. ASCE Res Prize (1960), Howard Award (1970), Newmark Medal (1979). Moissieff Award (1980). Mbr Natl Acad of Engg (1968), Natl Acad of Sci (1979). Honorary Dr of Tech, Chalmers Univ, Sweden (1979) Univ of Trondheim (NTH) Norway (1982); Fellow ASCE. Mbr Earthquake Engg Res Inst, Seismological Soc of Am, Structural Engrs Assn of CA. *Society Aff:* ASCE, SSA, EERI, SEAOC.

Clough, Ronald J
Business: Computer Park, Albany, NY 12205
Position: Partner. *Employer:* Clough, Harbour & Assocs, Engrs & Planners *Education:* BE/CE/Univ of Queensland, Australia. *Born:* Aug 1928. BE Civ Engg Univ of Queensland, Australia. Native of Brisbane Australia. US citizen. Practiced Civ & Structural Engg in Australia & Canada prior to 1955 when became mbr of Clarkeson Engg Co, Cons Engrs of Boston & Albany; former Ch Engr, became partner of Clarkeson & Clough Assocs 1966 & successor firm Clough Assocs 1971, with offices in Albany NY, Coral Gables FL & Puerto Rico. Mbr ASCE, NSPE, ASTM, TRB, APWA, Inst of Engrs (Australia), Natl Defense Exec Reserve, Tau Beta Pi - Currently Sr Partner Clough, Harbour & Assoc. *Society Aff:* ASCE, NSPE, ASTM, TRB, TBΠ.

Who's Who in Engineering

Cloutier, Leonce
Home: 880 Lienard, Quebec Que G1V 2W5 Canada
Position: Prof. *Employer:* Universite Laval. *Education:* DSc/ChE/Universite Laval; BScA/ChE/Universite Laval. *Born:* 2/5/28. in Quebec City, Que, Can. Production engr at Canada Packers, Montreal, design engr at C D Howe Co, Montreal and full time prof at Laval Univ since 1955, in the Chem Engg Dept. Hd of the dept 1969-75, Visiting prof at Univ of CA in Berkeley, 1975-76, and appointed secy of the Faculty of Sci and Engg Jan 1980-Jan 86. Res and publications in mixing, simulation and process control, chem reactors, hydrocyclones, mineral processing. Special programme of teaching and consulting for Alcan in Arvida in 1965 and Demerara Bauxite in Guyana, SA in 1968. *Society Aff:* CSChE, CIC, OIQ.

Clum, James A
Business: T.J. Watson School of Engineering, SUNY-Binghamton, Binghamton, NY 13901
Position: Prof. of Mechanical Engrg *Employer:* State Univ. of New York at Binghamton *Education:* PhD/Met-Mtls Sci/Carnegie-Mellon Univ; MS/Met-Mtls Sci/Carnegie-Mellon Univ; BMetE/Met Engr/OH State Univ. *Born:* 7/7/37. Prof. Mech. Engr. Postdoctoral res at OH State Univ & at Cambridge Univ. During 1970-72 Assoc Dir of the Univ-Industry Res Prog at the Univ of WI. Ind experience in several labs of US Steel, also IBM & the Metal Sci Group of Battelle-Columbus. In 1973-74 on leave as an ASEE Resident Fellow in the Steel Div of Ford Motor Co; during 1977-78 Res Assoc Prof of Met at Vanderbilt Univ. 1979-82 Director Applied Research Outreach Program, Univ. of Wisconsin-Extension Engrg; 1982-84 Assoc. Prof Chemical & Metallurgical Engrg, Univ. of Alabama; 1985- Watson School of Engrg, State Univ. of New York-Binghamton. Mbr: ASM, ASTM, AmCerSoc, TMS-AIME and ISS-AIME, AAAS, AVS, ASEE. *Society Aff:* AAAS, AIME, ASM, ACS, AVS, ASTM, ASEE.

Coad, John D, Jr
Home: 15917 Heather Glen, Chesterfield, MO 63017
Position: Principal *Employer:* Coad & Rascovar, Consulting Engrs *Education:* BS/ME/WA Univ *Born:* 10/18/30 Native of St. Louis, MO. Attended private elementary and high schools in St. Louis. Served in Army in Japan 1951-54. Attended engrg schools of Purdue Univ, WA Univ and Catholic Univ, and special courses at Univ of MO, Sophia Univ, Tokyo and MIT. Was project engr for Nat'l Security Agency, Sr Engr and Mgr for space equipment and aircraft, training facilities design at McDonnell Douglas. Since 1972, principal and 50% owner of consulting engrg firm, mechanical and electrical, with building costs exceeding $100,000,000 annually. Hobbies: sailboat racing, skiing, guitar. Reg PE in 17 states, with project experience in Europe, Asia, Mid-East, Caribbean and Mediterranean. *Society Aff:* ASME, NSPE, ASPE, ISES

Coad, William J
Business: 7616 Big Bend Blvd, St. Louis, MO 63119
Position: Pres *Employer:* McClure Engg Assoc *Education:* BSME/Mech Engr/Washington Univ *Born:* 12/23/31 Electronics technician US Army. Mech Contractor 1957-62. Cons Engr (Pres McClure Engg Assoc), 1962 Pfes. Lectr and Affiliate Prof Mech Engg Wa Univ 1957-. Author num articles & papers on air cond, energy conversion & energy econ. Author book entitled "Energy Engineering and Management for Building Systems" published by Van Nostrand Reinhold and author monthly column Fundamentals to Frontiers' in Heating/Piping/Air Cond. Mbr CEC & ASHRAE. ASHRAE activities have incl: St Louis Chap Pres, Chrmn Energy Advisors Council, VChrmn Educ Ctte & Nomin Ctte, Co-Chrmn Presidential Ctte on energy resource eval, mbr & vchrmn var tech cttes, Mbr Bd of Dir 1976-79. Chrmn Bldgs Tech Adv Ctte of MO Energy Agency, & Mbr MO State Bldg Code Steering Ctte. Hon Mbr Pi Tau Sigma, 1985 Recipient of ASHRAE Crosby Field Award. *Society Aff:* ASHRAE, ACEC

Coates, Clarence L
Business: Sch of Elec Engg, W Lafayette, IN 47907
Position: Hd, Sch of EE. *Employer:* Purdue Univ. *Education:* PhD/EE/Univ of IL; MSEE/EE/Univ of KS; BSEE/EE/Univ of KS. *Born:* 11/5/23. Hastings, NB. Served US Navy 1944-46. On EE faculty, Univ of IL 1948-56. Res Engr for GE Res Lab, Schenectady, NY & mbr of adjunct faculty of RPI 1956- 63. Mbr of EE faculty of Univ of TX at Austin 1963-70; served as Chrmn of EE 1964-66 & Dir of Electronics Res Ctr 1966-70. Dir of Coordinated Sci Lab, Univ of IL-Urbana 1971-72. Became Hd, Sch of EE of Purdue Univ in 1972. Fellow IEEE, Fellow AAAS, served as VP for Publications & mbr of BOD of IEEE 1971 and 72. Mbr NSF Sci Info Council 1972-75 & Chrmn of Cosine Committee NAE 1970-73. *Society Aff:* IEEE, ASEE, AAAS.

Coats, Keith H
Home: 11335 Somerland, Houston, TX 77024
Position: Chrmn of Bd. *Employer:* Intercomp Resource Dev & Engg, Inc. *Education:* PhD/ChE/Univ of MI; MS/ChE/Univ of MI; MS/Math/Univ of MI; BS/ChE/Univ of MI. *Born:* 11/14/34. Asst Prof, Chem Engg, Univ of MI, 1959-61; Res Assoc, Esso Production Res, Houston, 1961-65; Assoc Prof Petrol Engg, Univ of TX, 1966-70; Bd Chrmn of Intercomp Resource Dev & Engg, Inc, 1968-. Active in dev and application of numerical, computer models for simulating oil & gas reservoirs and pipelines networks. Distinguished Lecturer for SPE, 1969-70. *Society Aff:* AIChE, SPE of AIME.

Cobb, Allen L
Home: 210 Thomas Ave, Rochester, NY 14617
Position: Corp Safety Dir (Retired) *Employer:* Self *Education:* Sci BA/Civil Engr/Mass. Inst. of Tech *Born:* 3/4/05 Engineer (Special Hazards) Engineering Div. Factory Mutual Fire Ins Co's Engineer and later. Director of Industrial Safety, Kodak Park Div. Eastman Kodak Co including Fire Protection, Safety and Security also consultant on these subjects to Tennessee Eastman, Texas Eastman Div and to Holston Defense Corp (Explosives) and Oak Ridge (atom bomb) WWII 1969-1971 Corporate Safety Director, Eastman Kodak Company Founder member and Honorary Member, SFPE Past President Natl Fire Protection Assoc. Have been a director, National Safety Council. Have served as chairman or member of a number of NFPA standards committees also member and chairman of review board, NY State Building Codes Council also chairman. Fire Protection & Safety Committee, manufacturing chemists assoc. *Society Aff:* NSPE, SFPE, ASSE

Cobb, James T, Jr
Business: Dept of Chem & Petrol Engg, Pittsburgh, PA 15261
Position: Assoc Prof. *Employer:* Univ of Pittsburgh. *Education:* PhD/ChE/Purdue Univ; MS/ChE/Purdue Univ; SB/ChE/MA Inst of Tech. *Born:* 3/9/38. Native of Cincinnati, OH. Served with US Army Chem Corps 1965-67. Engr for Esso Res Labs, Baton Rouge, LA with Univ Pgh since 1970. Tenured as Assoc Prof in 1975. Currently Assoc Dir, Mining & Energy Resources Div for Sch of Engg. Published papers in catalysis & reactor design. Past chrmn Pgh Sec, AIChE. Past chrmn Cont Educ Comm, AIChE. Mbr, Speakers Bureau, AIChE. Past chrmn, Commission on Certification, EJC. Past chrmn, Pgh Catalysis Soc. *Society Aff:* AIChE, ASEE, AEE, NSPE

Cobbs, James H
Business: 5350 E. 46, Tulsa, OK 74135
Position: Pres. *Employer:* Cobbs Engr Inc. *Education:* BS/Petrol Engr/Univ of OK. *Born:* 8/25/28. Grad study Univ of OK & Univ of Tulsa. With Tidewater Oil Co trainee to Div Reservoir Engr 1951-59. Private practice 1959-63. Fenix & Scisson Inc 1963-69. Private Practice 1970-, providing design & proj mgt for petro drilling & prod, drilled shafts, coal degasification, tunnel boring, deep well waste mgt, mined & solution storage of crude & oil high vapor pressure products, & special purpose down hole tools. 16 US pats. Elder, Bd Chrmn Harvard Ave Christian Church; 25 yr veteran Boy Scouts of Am, former Scoutmaster & com chairman. Fellow AAAS; Mbr NSPE, Soc of Petrol Engrs, AIME, Inst of Shaft Drilling Tech,

Cobbs, James H (Continued)
Nomads National Academy of Forensic Engineers. Registration: OK, CO, IL, IN, MO, MN, TX and VA. *Society Aff:* AAAS, NSPE, SPE, ISDT, NOMADS, NAFE.

Cobean, Warren R
Business: 800 Kinderkamack Rd, Oradell, NJ 07649
Position: VP. *Employer:* Burns and Roe, Inc. *Education:* AMP/Bus Admin/Harvard; BS/Naval Sci/US Naval Academy, Reactor Engineering/Bettis & U. Pittsburgh. *Born:* 6/26/23. Served in US Navy 1946-1971. Was mbr of original crew of USS Nautilus, World's first nuclear powered submarine. Active in nuclear power throughout naval career. Commanded four submarines. Deputy Dir, Proj Office responsible for dev, production and operational results of Polaris, Poseidon, Trident strategic weapons systems. Mgr - Nuclear Generation, Consolidated Edison 1971-1974. Proj Dir, then Div Dir and VP, Burns and Roe, Inc. Responsible for Nuclear Power Plant Projects - Three Mile Island, Forked River and Oyster Creek. Dir, Plant Modification TMI Recovery Organization. Responsible for operating plants (Nuclear) Service Division. Responsible for Coal Fired Power Plant Projects: Reid Unit 2&3. DB Wilson Unit 1&2, Winyah Unit 3&4, Cross Unit 1&2. VP and Dir of Burns and Roe, Inc responsibile for all Power Engrg Services, both domestic and intl. *Society Aff:* ANS, ASME.

Coberly, Camden A
Business: 1415 Johnson Dr, Madison, WI 53706
Position: Prof. Chemical Engr. *Employer:* Univ WI. *Education:* PhD/Chem Eng/Univ of Wis-Madision; MS/Chem Eng/Carnegie Inst of Tech; BS/Chem Eng/WV Univ *Born:* 12/21/22. Following PhD in 1949 employed Mallinckrodt Chem in St Louis, MO. Chem Engr 1949-55, Chief Engr, 1955-64. At Univ of WI-Madison, Prof of Chem Engg and Assoc Dir, Engg Experiment Station, 1964-68. Chrmn, Chem Engg Dept, 1968-71. Assoc Dean College of Engg, 1971-86, Prof of Chem Engr 1986-date. AIChE activities include Past-Chrmn, St Louis Sec; Awards Committee; Continuing Education Committee; Publications Committee; Technical Prog Chrmn, Natl Meeting, Kansas City, 1976; other committee mbrship. ASEE activities include committee chairmanships and mbrs bd, Chem Engg Div, and Dir of Engg Res Council. Mbr Sigma Xi, Tau Beta Pi, Phi Lambda Upsilon, Sigma Gamma Epsilon. Reg Engr - WI. *Society Aff:* ACS, AAAS, AAUP, NACE, ASEE, AIChE.

Coble, Hugh K
Business: 3333 Michelson Dr, Irvine, CA 92730
Position: Group Vice President-Marketing *Employer:* Fluor Engineers and Constructors, Inc. *Education:* BS/CE/Carnegie Mellon Univ; -/Bus Mgt/UCLA; -/Law/Univ of Houston. *Born:* 9/26/34. Started my industrial career with Gulf Res in Hamarville, PA while attending univ. Upon receiving BS degree, worked 6 yrs with Std Oil of CA in Refining, Engr & Operations. After 4 yrs in the chem ind I joined Fluor in 1966. Career has involved a 9 yr assignment in Europe, including position as Sr VP, Sales, responsible for bus dev activity of Fluor for Western Europe, Eastern Europe, Middle East, & Africa in 1978. Assumed present position in 1980 as Group VP, Mktg, responsible for worldwide business dev of Fluor Engrs and Constructors, Inc. *Society Aff:* AIChE, NSPE.

Cochran, Billy J
Business: Agri Engg Dept, Baton Rouge, LA 70803
Position: Prof. *Employer:* LA State Univ. *Education:* PhD/Agri Engr/OK State Univ; MS/Agri Engr/TX A&M Univ; BS/Agri Engr/MS State Univ. *Born:* 12/10/33. in Darbun MS. Grad from Columbia HS, Columbia, MS. Grad from Pearl River Jr College in 1953. Served in US Army 1953-1955. I was an instr in the agri engg dept of TX A&M Univ from 1958-1964. Conducted res in cotton production & machine design. In 1964 moved to LA State Univ as asst prof in agri engg. My work consisted of teaching farmpower & machine design & proj leader in sugarcane mechanization res. In 1970 I obtained sabatical leave, went to OK State Univ and received a PhD degree in 1972. In 1973 I was promoted to assoc prof & in 1976 promoted to full prof. *Society Aff:* ASAE, ASEE, ASSCT.

Cochran, Douglas E
Home: 11837 E Wagon Trail Rd, Tucson, AZ 85749
Position: VP Special Projs. *Employer:* Duval Corp. *Education:* BS/ChE/Rice Univ. *Born:* 08/06/25. m Frances Askins May 9, 1945; 1 son Stephen Douglas. Exec Dev Conf Univ of AZ 1971. Mbr NSPE, AAPG, AIChE, AIME. Reg PE TX. With Duval since Dec 7, 1951. Frasch sulphur mine pos of Chemist, Ch Chemist, Mine Engr, Asst Mgr, Res Mgr, Mgr Sulphur Div, VP Explor Non-Metallics; VP Spec Projs since 1970. Successfully applied Frasch Process to deepset sulphur deposit, & pioneered its application to W TX deposits. Dir group in discovery & dev of maj sulphur deposit in Culberson Cty, TX. *Society Aff:* NSPE, AAPG, AIChE, AIME.

Cochran, Robert G
Business: Dept of Nuclear Engr, Texas A & M Univ, College Station, TX 77843
Position: Prof of Nucl Engg. *Employer:* TX A&M Univ. *Education:* PhD/Phys/PA State; MS/Phys/IN Univ; BS/Phys/IN Univ. *Born:* July 12, 1919. BS Phys IN Univ 1948; MS Nucl Phys IN Univ 1950; PhD Nucl Phys PA State Univ 1957. Prof & Hd NE 1959-87, Assoc Prof NE PA State Univ 1954- 59; Consulting: USAEC, Univ of MI, Univ of MO, Teledyne Brown Engg Huntsville AL, US Army & USAF. Prof of Nuclear Engineering and Technical consultant to the Nuclear Industry. Mbr ANS, Am Phys Soc, Am Soc Engg Educ, Natl Soc of PE, Phi Kappa Phi, Sigma Xi. Reg PE TX. Past member: P Chrmn Subctte on Res Reactors, Natl Acad of Sci & Natl Res Council; P Mbr Bd of Dir Educ Div ANS & P Chrmn Educ Div. *Society Aff:* ANS, APS, ASEE, PE, ΦΚΦ, ΣΞ.

Cochrane, Gordon S
Home: 1630 Meadow Ln, Glen Mills, PA 19342
Position: Mgmt Consultant *Employer:* Self *Education:* BS/ChE/Drexel Univ; MS/Mgmt/MIT (Sloan School). *Born:* June 1929. MS Mgt MIT as Sloan Fellow; BS Chem Engg Drexel Univ. Joined Sun Oil Co 1952 as Tech Serv Engr at Marcus Hook Refinery. Army Chem Corps 1953- 55. Process Engr Sun Mfg-Engg Dept 1957. Mgr design & constr for an ammonia plant proj in Venezuela 1968. Products Mgr & VP Puerto Rico Sun Oil Co 1969. VP product sales & supply Sun Oil Trading Co 1974. Pres Sun Oil Trading Ltd Hamilton Bermuda, 1976; Pres, Sun Oil Trading Co, 1978; VP, Lubes, Sun Petro Products Co, 1980 to present. *Society Aff:* AIChE

Cockburn, Robert T
Home: 2111-15th Ave, San Francisco, CA 94116
Position: Exec Director Clean Water Program *Employer:* City & Cty of San Francisco. *Education:* BS/Mech Eng/Univ of Cal. Berk; MS/Civil Eng/Univ. of Cal. Berk. *Born:* 1942. BS, MS Univ of CA Berkeley. With City & Cty of San Francisco since 1968. Currently respon for all waste treatment and pumping operations. Past accomplisments incl co-author of 1971 Master Plan for Wastewater Mgt, Proj Engr for marine waste disposal studies and treatment studies since 1968. Dipl Am Acad of Environ Engrs; Mbr AWWA; Mbr WPCF; State of CA Reg C E, Grade V Waste Water Treatment Plant Operator and Grade IV Water Treatment Plant Operator. 1976 ASCE Wesley W Horner Award. *Society Aff:* AAEE, WPCE, AWWA.

Cockrell, William D
Home: 1512 Tuckahoe Rd, Waynesboro, VA 22980
Position: Consult (Self-employed) *Education:* EE/Electronics/Univ of FL; BSEE/EE/Univ of FL. *Born:* 9/2/06. Tallahassee, FL. Advanced engg GE Co until 1967. Over 40 patents in electronic field. Dir R&D at Rothermel Assoc until 1970. Author "Industrial Electronic Control–, Editor-in-Chief "Industrial Electronic Handbook" McGraw Hill. Also many papers and articles in field. 1977-78 recipient of Elfun Shenandoah Chapter and 17th Territory Award for outstanding performance in field of community service. Coffin Award. Fellow and Life mbr of IEEE. Mbr, Valley Community Serv Board (Mental Health and Retardation) 1979-84. Waynes-

Cockrell, William D (Continued)
boro Public Library Bd 1985-87, Waynesboro Chapter Virginia State Museum 1987- . *Society Aff:* IEEE.

Coda, Frank M
Business: ASHRAE, 1791 Tullie Cir, N.E, Atlanta, GA 30329
Position: Executive Director/Secretary *Employer:* ASHRAE *Education:* BS/Civ Engg/RPI *Born:* 4/8/40. Civil Engineer from 1961-67 with building and road and bridge contractors. Served with New Jersey National Guard (Corps of Engineers) from 1961-67, retiring as First Lieutenant. From 1966-73 worked for Perlite Institute, Inc., assuming responsibility as Managing Director. Served as Executive Vice President for Illuminating Engineering Society (IES) from 1973-80, a society consisting of approximately 10,000 members specializing in the advancement of illuminating engineering. Currently serving as Executive Director/Secy of ASHRAE; having responsibility for the Society's five staff divisions consisting of approximately 80 employees. ASHRAE currently has 55,000 members who share a common interest in the many aspects of human comfort and environmental control. *Society Aff:* IES, ASHRAE, ASAE, ASCE, CESSE.

Codlin, James B
Business: 3000 S 6th St, Springfield, IL 62710
Position: Dir Prod Safety. *Employer:* Fiat-Allis. *Education:* Bachelor/ME/Iowa State College. *Born:* 11/13/16. Started Student Engg, Allis-Chalmers Construction Machinery Div 1938, then various design capacities. Chief Engg, Tractomotive 1945, later their VP Engg. Gen Mgr - A-C Deerfield Wheel Loader Plant 1960; Mgr, Dev Engg 1963; Mgr Testing 1969. With formation of A-C/Fiat Joint Venture - Fiat-Allis, was made Product Safety and Regulatory Compliance Dir. Active Boy Scouts in early yrs, receiving their Order of the Arrow 1957. Lake Forest IL School Board; Chmn, Glenwood High School; American Field Service 1972. Served various SAE Committees: Chmn ORMTC, Con-Ag Council, Mbr Technical Board Chmn, US TAG ISO TC127 Earthmoving Machinery - 8 patents - enjoy camping fishing, travel. *Society Aff:* SAE, AOR.

Codola, Frank C
Home: 7 Nimitz Pl, Yonkers, NY 10710
Position: Assoc Prof. *Employer:* CCNY. *Education:* Master of Engg/-/Yale Univ; BME/-/RPI. *Born:* 4/22/25. in Bristol, RI where I also attended public schools through high sch. Served in the USNR during WWII and was discharged in 1946. Grad RPI in 1950 and Yale in 1951. Went to work with NASA in Langley Field, VA for one yr and then with Pratt & Whitney Aircraft in Hartford, CT, for three yrs. Taught Mechanics at New Haven Jr College in the evening session for about three yrs. Decided teaching was my profession, joined CCNY 1954 where I have been since. *Society Aff:* ASEE, AAUP.

Coe, Benjamin P
Home: 314 Paddock St, Watertown, NY 13601
Position: Exec Dir. *Employer:* Tug Hill Comm, N.Y. State *Education:* BS/Chem Eng. /MIT; AB/Phys/Bowdoin College. *Born:* 8/24/30. With Silicone Products Dept of GE, 1953-65 as Process Engr, Plant Start-up Supervisor; Process Economics Engr. Joined VITA (Volunteers in Technical Assistance) 1965-73, serving as Exec Dir, and VP. Currently Exec Dir of the Temporary State Commission on Tug Hill, NY State, concerned with future of 2000 sq mile area of critical environmental importance. Developed pioneering prog in rural land use planning, in which towns cooperate in making decisions but do not lose authority. Licensed PE, NY State. Former chrmn NENY Sec AIChE; Episcopal Church; married Margaret Butler; four children. *Society Aff:* ASPA.

Cofer, Daniel B
Business: Fertilla St, Carrollton, GA 30117
Position: Sr VP R&D. *Employer:* Southwire Co. *Born:* Nov 3, 1927 Wash, GA. Grad from Millen, GA High Sch 1944; BME from GA Inst of Tech 1953. US Marine Corps 4 yrs. Employed by Southwire Co 1953 as engr; elected to Bd of Dir 1959. Appointed VP of R&D in 1963. Named Sr VP 1974. Also Hd of Copper Div of Southwire Co. John W Mordica Award by Wire Assn 1971 for making the greatest contrib to the dev of the wire indus. Wire Assn medal for best tech paper publ 1964. 41 US pats & 300 foreign-issued patents.

Coffeen, William W
Home: 227 Kings Way, Clemson, SC 29631
Position: Retired *Education:* PhD/Ceramic Engrg/Rutgers; Prof Deg/Ceramic Engrg/Univ of IL; MS/Ceramic Engrg/Univ of IL; BS/Ceramic Engrg/Univ of IL *Born:* 8/13/14 Major activity 1935-1969, in industrial research in the following fields: Vitreous enamels for steel and cast iron, ceramic glazes, electronic ceramics, optical glass, zircon refractories, mineral recovery, tin smelting, antimony oxide production, thermite welding and metal production, arc welding rods and equipment. Major employer (1945-1969) Metal & Thermit Corp (now M & T Chemicals)--Dir of Research. 1969- Prof of Ceramic Engrg, Clemson Univ, teaching and research. Pres, Nat'l Inst of Ceramic Engrs, 1963. VP, American Ceramic Society, 1969. Greaves-Walker Award, 1980. Keramos, Sigma Xi. 23 Publications, 22 Patents. *Society Aff:* ACS, NICE

Coffin, Louis F, Jr
Business: 235 Met, KL Corp R&D, PO Box 8, Schenectady, NY 12301
Position: Mech Engr. Also Distinguished Research Professor Dept of Mechanical Engrg. *Employer:* GE Co. (Retired) now Consultant & Rensselaer Polytechnic Institute. *Education:* BS/ME/Swarthmore College; ScD/ME/MIT. *Born:* 8/30/17. Fellow of ASME, ASTM & ASM. Chrmn of Committee E-9 of ASTM, 1974-1978. Alfred E Hunt Award from ASLE in 1958. James Clayton Lectr (IMechE) in 1974. Carborundum's Award of Excellence in Iron & Steel in 1974. Elected mbr of Natl Acad of Engg in 1975. Dudley Medal & Award of Merit from ASTM. Coolidge Fellow of GE's Res & Dev Ctr. Distinguished Career Award from Hudson-Mohawk Chapter of AIME. Albert Sauveur Achievement Award, ASM; Nadai Award, ASME. Ten patents and 150 publications in plasticity, flow & fracture, fatigue, friction & wear. Also Distinguished Research Professor, Rensselaer Polytechnic Institute, Francis J. Clamer Medal, Frankoin Institute, jointly with S. S. Manson. *Society Aff:* ASME, ASTM, ASM, AIME.

Coggeshall, Ivan S
Home: 3 Murphy Cir, Middletown, RI 02840
Position: Asst VP. *Employer:* Western Union Telegraph Co. *Education:* Dr Engrg/Worcester Poly Inst. *Born:* 9/30/96. IEEE 1978 Haraden Pratt Award for "Outstanding service to IEEE–, IEEE Centennial Medal 1984, IEEE Communications Sys Group's 1963 Achievement Award "For leadership in the integration of wire & radio–. Hdquarters Staff Western Union Telegraph Co 1920-69; Dir Submarine Cables 1952; Asst VP 1969. Commander US Naval Res, retired; cons on overseas cables & radio Wash 1942-48. Author "Wire Telegraphy–, Radio Engg Handbook; "Telegraphy–, Ency Britannica 1974; "Compatible Tech of Wire & Radio–, Proc IRE; "Variations on a Theme by Oppenheimer," IEEE Spectrum 1984. Pres IRE 1951; Life Fellow IEEE. Dr of Engg Worcester Poly Inst *Society Aff:* IEEE.

Cohen, Alan
Business: Suite 300-2 Bala Plaza, Bala Cynwyd, PA 19004
Position: Consultant *Employer:* Self *Education:* SM in CE/Civil/MA Inst of Tech; BCE/Civil/Cornell Univ. *Born:* 4/15/32. Served two yrs in USAF; Proj Mgr for Moran, Proctor, Mueser & Rutledge, VP, J S Ward, Inc. Pres & founder of Site Engrs, Inc. VP of Day & Zimmermann, Inc. Pres Sonitrol Service, Inc. Consultant on Loss Prevention Construction Claims, expert witness services. *Society Aff:* ASCE, NSPE, AAA, AMA

Cohen, Arnold A
Home: 3517 W 39th St, Minneapolis, MN 55410
Position: Senior Fellow, the Charles Babbage Inst *Employer:* Univ of MN and the Charles Babbage Inst *Education:* PhD/Physics/U of MN; MS/Physics/U of MN; BEE/Elec Engg/U of MN *Born:* 8/1/14. Born Duluth, MN. With RCA as tube dev engr 1942-46. With Unisys predecessor cos 1946-71; managed systems dev, engg, planning operations; responsibilities in early yrs of computer industry included dev of magnetic drum storage techniques, system design of 1101 and 1103 computers. With Univ of MN Inst of Tech 1971-80 as Asst Dean for Industry and Profl Relations. With the Charles Babbage Institute for the History of Information Processing since 1980 as Senior Fellow. Valuable invention citation, Am and MN Patent Law Assns 1961. Fellow, IEEE 1964. Centennial Medal, IEEE 1984. Natl chrmn, IRE/PGEC (predecessor to IEEE Computer Soc) 1960-62. Foundingg exec committee, AFIPS (Am Fed Info Proc Socs) 1961-64. *Society Aff:* IEEE

Cohen, E Richard
Business: 1049 Camino Dos Rios, Thousand Oaks, CA 91360
Position: Distinguished Fellow. *Employer:* Rockwell Intl Sci Ctr. *Education:* PhD/Phys/CA Inst of Tech; MS/Phys/CA Inst of Tech; AB/Phys/Univ of PA. *Born:* 12/14/22. With Rockwell Intl (and its predecessor cos) since 1949. Served as Group Leader, Reactor Theory Group, Atomics Intl Div 1953-56; Res Advisor (1956-61); Assoc Dir, Res Dept (1961-62). Assoc Dir, Rockwell Intl Sci Ctr (1962-69); Mbr Technical Staff (1969-); Appointed Distinguished Fellow 1974. Contributed to the theory of neutron behavior in nuclear reactors & dev of advanced nuclear reactor concepts including only nuclear reactor to operate in space. Sr Lectr in Nuclear Engg, CA Inst of Tech 1962-65; Sr Res Assoc 1965-72. Recipient of E O Lawrence Award (Atomic Energy Commission) 1968. Mbr Intl Union of Pure and Applied Phys Commission on Atomic Masses & Fundamental Constants 1963-69; secy 1969-72; chrmn of Commission 1972-78; Commission on Symbols, Units and Nomenclature, 1978- ; IUPAP Delegate to CODATA (Committee on Data for Science and Technology) 1975- ; Chairman, CODATA Task Group on Fundamental Constants 1969- 85; mbr, 1985- ; Mbr CODATA Exec Ctte 1982-84. *Society Aff:* AAAS, ANS, APS, ACM.

Cohen, Edward
Business: Suite 1700, Two World Trade Ctr, NY, NY 10048
Position: Managing Partner and CEO *Employer:* Ammann & Whitney, Consulting Engrs. *Education:* MS/Structural/Columbia Univ; BS/Civ/Columbia Univ. *Born:* 1/6/21. Born Jan, 21. Mbr Natl Acad of Engg; Dept of Army Patriotic Civilian Service Award; Honorary Life Mbr NYAS; Wason Medal & Bloem Award ACI; Egleston & Illig Medals Columbia Univ; Laskowitz Res Award NYAS; Howard Medal, Ridgeway, State-of-the-Art & Reese Awards ASCE; Hon Mbr and Past pres ACI; Past Pres Met Sec ASCE; Honor Mbr Chi Epsilon, Rutgers Univ; Mbr Tau Beta Pi & Six Xi; Fellow ASCE and ACEC; Stanton Walker Lectr Univ of MD; Chrmn ASCE Reinforced Concrete Res Council; Chrmn ANSI A58 Committee on Loads. In charge of all projs pertaining to Restoration, bldgs, bridges, airports & spec struct incl designs for military, commercial, aerospace, communications, transportation & indus facils. Since 1950 has directed projs in res, dev & app of hardened design. Author over 100 Books, Manuals & tech papers on var of Engrg subjs. *Society Aff:* ASCE, NAE, ACI, NSPE, NYAS, SAME, ACEC.

Cohen, Gary
Home: 8927 Sylvia Ln, Phila, PA 19115
Position: Director of Engineering *Employer:* Ballinger Co. *Education:* BS/Mech & Elec Engg/Drexel Univ. *Born:* 8/25/34. 1958 to 1968: United Engrs & Construction, Inc designing heating, ventilating and air conditioning systems for industrial bldgs, chem plants, steel plants, fossil and nuclear power plants. 1968: Joined Arthur Engrs, Inc as Partner and Exec VP in charge of the mech engg. 1975: Joined Hassinger Schwam Assoc, Inc as VP; organized the firm of Energy Environmental Systems, which designs energy efficient bldg systems and solar systems. 1980: Founded Gary Cohen Assoc, a consulting engineering firm specializing in hospitals and high rise condominiums. 1981: Gary Cohen Assocs became a subsidiary of Robert D. Lynn Assoc, P.C., a firm specializing in the architectural and engrg design of hospitals. 1986: Joined Ballinger Co. As chief Mech. Engineer. The firm is an A/E Co of 100 people. Presently reg in six states: PA, NJ, DE, VA, MD, N.Y., N.C., & Florida. *Society Aff:* NSPE, ASHRAE, NFPA, ASPE.

Cohen, Ira M
Business: Dept of Mech Engg & Appl Mechanics, Univ of Pennsylvania, 111A Towne Bldg, Phila, PA 19104-6315
Position: Prof. *Employer:* Univ of PA. *Education:* PhD/Aero Engrg/Princeton Univ; MA/-/Princeton Univ; BAE/Aero Engrg/Polytechnic Univ. *Born:* 7/18/37. in Chicago. Attended NYC public schools. Married 1960 (Linda B Einstein); two daughters: Susan (1961), Nancy (1965). Asst Prof of Engg, Brown Univ, 1963- 66. Guest "ordinary" prof, Technische Hochschule Aachen, 1966. Faculty mbr, Univ of PA, 1966-. Tenured, 1967, Prof, 1976. Mbr Technical Staff, Sandia Labs Albuquerque (summers) 1971, 1974, 1977. Elected Assoc Fellow, AIAA, 1977. Secy, Greater Phila Section, AIAA, 1977-80, 1985-88. Consultant in fluid mechanics to local industry and attorneys. Author of about 50 papers on ionized gases and fluid mechanics. Squash player and bicyclist. *Society Aff:* AIAA, APS, ΣΞ, AAUP.

Cohen, Irving D
Business: 19 Copeland Rd, Denville, NJ 07834
Position: Pres. *Employer:* Enviro-Sciences Inc. *Education:* ME/ChE/NYU; BE/ChE/CCNY. *Born:* 5/12/45. Mr Cohen has been involved most recently with the mgt of environmental impact report progs for onshore/offshore natural gas pipelines in the northeastern & southeastern US, natural gas storage fields & proposed refineries. In addition, Mr Cohen has been responsible for environmental permit acquisition for the aforementioned as well as other projs & the design of air quality monitoring progs (PSD related) for many proposed projs. He has been responsible for the conceptual design of groundwater decontamination processes for chromium removal as well as the treatment of gaseous effluents from processing plants. He has defended the work of ESI as an expert witness at FERC, EPA & state hearings. Certified Environmental Professional. *Society Aff:* AIChE, AIHA, APCA, NAEP, IAPC.

Cohen, Jerome B
Business: The Technological Inst, Northwestern Univ, Evanston, IL 60201
Position: Prof. & Dean of The Technological Institute. *Employer:* Northwestern Univ. *Education:* ScD/Met/MA Inst of Tech; BS/Met/MA Inst of Tech. *Born:* 7/16/32. After receiving his doctorate at MIT he spent one yr as a Fulbright Scholar with A Guinier, Univ of Paris, one yr as Sr Scientist, AVCO Corp, Wilmington, MA, 6 months as 1st Lt US Army Ordnance Corps. Joined Northwestern Univ as Asst Prof in 1959; Full Prof since 1965. Liason Officer, US Navy (ONR), London, 1966- 67. Mbr, Tau Beta Pi, Sigma Xi and Phi Lambda Upsilon Honorary Societies, and the Royal Inst of Great Britain. Awarded the Hardy Medal, AIME, 1960 (Outstanding Metallurgist under 30). Appointed Faculty Fellow, Northwestern Univ Ctr for the Teaching Professions, 1971; the Northwestern Univ's Technological Inst's Teaching Award, 1972; Geo C Westinghouse Award, ASEE, 1976 (excellence in teaching and res); Fran C Engelhart Chair in Mtls Sci and Engg, Northwestern Univ, 1974. Elected fellow ASM, 1980; AIME 1981. Howe Medal AIME 1981 (best paper in 1980). Dept Chrmn, 1973-79. First Tech Inst Prof at Northwestern, 1983. Dean, Technological Institute, 1986- . Specialties: residual str measurement, phase transitions, catalysis, ordering, clustering (oxides and metals); fatigue, and defects. *Society Aff:* AIME, ASM, ACA, ASEE.

Cohen, Karl P
Home: 928 N California Ave, Palo Alto, CA 94303
Position: Consultant. *Employer:* Self. *Education:* PhD/Phys Chemistry/Columbia Univ; MA/Phys Chemistry/Columbia Univ; BA/Chemistry/Columbia College. *Born:* 2/5/13. Native of NYC. Lab asst to H C Urey 1937-40. Dir, Theoretical div, Manhattan Proj at Columbia 1940-44. Physicist, Std Oil Dev Co 1944-48. Technical Dir, H K Ferguson Co 1948-52. VP, Walter Kidde Nuclear Labs, 1952-55. Joined

Cohen, Karl P (Continued)
GE Co (nuclear energy div) 1955. Successively Mgr, Advance Engg (boiling water reactor dev & chrmn, GE Technological Hazards Council), Mgr, Advanced Projs Opration (fast breeders and ultracentrifuges), Gen Mgr, Breeder Reactor Dept, Mgr Operational Planning, and Chief Scientist, Nuclear Energy Group. Retired 1978. Elected to NAE, 1967. Pres ANS 1968, Alfried Krupp Energy Award 1977, AIChE Chem Pioneer 1979. *Society Aff:* ANS, APS, AAAS, NAE, IEEE.

Cohen, Louis A
Business: MOGADISHU (1D), c/o Dept of State, Washington, DC 20520
Position: Dir *Employer:* USAID Mission to Somalia *Education:* BS/CE/Purdue Univ; graduate work/Intl Rel/American Univ *Born:* 4/1/23 San Antonio, TX. Grew up in IN. Served AUS, 1943-46. Project Engr with IN State Hgwy Dept 1948-52. Consulting Engr Assoc, Indianapolis; Poughkeepsie, NY; Englewood, NJ, 1953-59. U.S. Intl Cooperation Agency/A.I.D. 1959; Hgwy Engr- Saigon, 1959-61; Chief Engr-Rangoon, 1961-64; Chief Engr-Bangkok, 1964-65; Capital Projects Officer for Southeast Asia-WA, 1965-70. Chief Engr for United Nations Mekong coodinating Committee Secretariat, 1970-74 (30 professional engrs & scientists). Deputy dir, AID Regional Economic Development Office-Bangkok 1974-76. Dir, USAID REDSO for East Africa (23 professional engrs and scientists)-Nairobi, 1976-79. Dir, USAID/Botswana-Gaborone, 1979-1982 ($55 million development program of projects). Dir, USAID/Somalia - Mogadishu, 1983- ($85 million/year program). *Society Aff:* NSPE, ASCE, USCOLD, AWRA, SAME, AFSA, ASA, IWRA.

Cohen, Morris
Home: 491 Puritan Rd, Swampscott, MA 01907
Position: Inst Prof, Emeritus. *Employer:* MA Inst of Tech. *Education:* ScD/Met/MA Inst of Tech; SB/Met/MA Inst of Tech. *Born:* 11/27/11. Maj research activities: phase transformations, strengthening mechanisms, mechanical behavior of metals, matls policy. Kamani Gold Medal, Mathewson Gold Medal, JIM Gold Medal, La Medaille Pierre Chevenard, Procter Prize, Albert Sauveur Achievement Award, Natl Medal of Science. Chrmn of NAS Survey on Matls Science & Engg. ASM Gold Medal, Clamer Medal, Howe Medal, Acta Metallurgica Gold Medal, New England Award of Engrg Societies of New England, Leadership Award of the Metallurgical Soc. *Society Aff:* AAAS, AIME, ASEE, ASM, NAE, NAS, INSA, BMS, JISI, JIM, IIM, KIM

Cohen, Norman A
Business: 1753 Cloverfield Blvd, Santa Monica, CA 90404
Position: Pres. *Employer:* Cohen & Kanwar Inc. *Education:* BE/Elect Engr/Univ of So CA *Born:* 12/4/25. US Navy 1943-46. Assoc Engr Los Angeles Dept of Water and Power, & Elec Designer with So Cal Edison Co until entering private cons engr practice in 1954 as self-proprietor. Formed Frumhoff & Cohen, Cons Engrs, a CA Corp in 1956. In 1973 formed Cohen and Kanwar, Inc - Pres since inception. Edwin F Guth Mem Award in Natl IES Lighting Competition 1968. *Society Aff:* IEEE, IES, NSPE, ACEC

Cohen, Paul
Home: 3024 Beechwood Blvd, Pittsburgh, PA 15217
Position: Consultant (Self-employed) *Education:* MS/Chem/Carnegie Mellon; BS/Chem/CCNY. *Born:* 8/11/12. Fuel engr for US Bureau of Mines 1936-49 spec in flow properties of coal- ash slag heat absorption in central-station boiler furnaces. Engr, Mgr, Cons for Westinghouse Elec Corp, nucl activities 1949-1977, Water Coolant Tech of Power Sys, Consult. "Water Coolant Tech of Power Reactors–. Fellow ASME; Mbr ASME Res Ctte on Water in Thermal Power Sys; Fellow ANS; Mbr ANS Bd of Dir; ANS Spec Award in Chem 1967; Mbr NACE; VChrmn NACE Tech Practice Ctte 11, Corrosion in Power Sys; Mbr Exec Ctte Intl Water Conf, Engrs' Soc of Western PA; Award of Merit 1976, Intl Water Conf; PE CA, Corrosion Engrg. *Society Aff:* ANS, ASME, NACE.

Cohen, Raymond
Business: Ray W Herrick Labs, School of Mechanical Engrg Purdue Univ, W Lafayette, IN 47907
Position: Prof & Dir. *Employer:* Purdue Univ. *Education:* PhD/Mech Engrg/Purdue Univ; MSME/Mech Engrg/Purdue Univ; BSME/Mech Engrg/Purdue Univ *Born:* Nov 1923. Native of St Louis MO. BSME, MSME, PhD Purdue Univ. PE IN. Served with US Army 1943-46. Teaching fac of Sch of Mech Engg Purdue Univ 1947-. Currently Prof of Mech Engg. Since 1971 Dir of the Ray W Herrick Labs, one of the maj res labs at Purdue Univ. Fellow ASME & Chrmn Shock & Vibration Ctte 1973-74; Mbr ASEE & Chrmn Engg Acoustics Comm 1972-74; Fellow ASHRAE & Chrmn Stds Comm 1971-73, chairman Research & Technical Committee 1986-87; Representative of ASHRAE on EAC of ABET 1978-82; NATO Sr Sci Fellow Inst of Sound & Vibration Res Univ of Southampton 1971. ASHRAE representative to the Bd of Dirs of ABET 1982-present. *Society Aff:* ASME, ASHRAE, ASEE, ASA, INCE, NSPE.

Cohen, Robert M
Home: 3822 Prince William Dr, Fairfax, VA 22031
Position: Pres *Employer:* Solar Energy Design, Inc *Education:* MSE/Naval Arch & ME/Univ of MI; BSE/Naval Arch & ME/Univ of MI *Born:* 7/5/44 Born and raised in Brooklyn, NY and Allentown, PA. Following coll, served 15 years active duty Navy and civilian employment with Div of Naval Reactors, USDOE. In 1977, formed own mechanical and electrical engrg co specializing in energy conservation and alternative energies including Solar. Fields of activity have included nuclear power plant engrg, naval architecture and marine engrg, heating ventilating and air-conditioning systems design, plumbing systems design, electrical systems design and energy consulting. Hobbies include micro-computers, photography, sailing and workshop. *Society Aff:* ASHRAE, NSPE

Cohen, Robert S
Business: 137 E Iron Ave, Dover, OH 44622
Position: Pres. *Employer:* Enterprise Chem Corp, Ltd. *Education:* MS/Physics/Univ of MI; BS/Chem Engg/Univ of MI; Hon/Assoc/Brandeis. *Born:* 9/23/23. Rochester, NY 1923; Married Roberta Joan Klar, children – Matthew and Mardah. Founder and Pres Dover Chem Corp, 1949-76; Merged into Ansul Co (NYSE) 1974; VP Res, ICC Industries, 1975 when Dover was acquired by ICC 1975-77; Mbr National Advisory Council, Consumer Products Safety Commission 1975-77; Fellow, Mellon Institute, 1947-49; Westinghouse, 1949-51; Pres, Anderson Refining, Palestine, TEX 1960-64; Currently Pres, Enterprise Chemical Corp, Dover; Chrmn Bd, Cripple Creek Trout Farm, Inc, Rural Retreat, VA; Pres Junior Achievement, Dover, 1966-67. US Army 1946-47. Meritorious Serv Award; Patents in chlorinated hydrocarbon field. Elected Fellow AIChE 1979. Mbr - Union C. C., Dover; Chemist's Club (NYC). American Chemical Society, American Institute of Chemical Engineers. *Society Aff:* AIChE, ACS, NSPE.

Cohen, Wallace J
Business: 2000 14th St. N.W, Wash, D.C 20009
Position: Dept Asst Dir *Employer:* DC Dept of Public Works *Education:* BSCE/-/Geo Wash Univ. *Born:* 9/9/36. Native of Wash, DC. Resident engr on maj bridge & tunnel construction projs. Currently responsible for public space and transportation policy. Pres, Natl Capital Section, ASCE. Active in church, civic and community affairs. Winner Jaycees Outstanding Young Man award 1967. Married to Joan Levy, 2 children Jay & Robin. *Society Aff:* ASCE.

Cohen, William C
Business: Tech Inst - Rm 2804, Evanston, IL 60201
Position: Prof of Chem Engg & Assoc Dean, Engg. *Employer:* Northwestern Univ. *Education:* PhD/ChE/Princeton Univ; MSE/ChE/Princeton Univ; BChE/ChE/Pratt Inst. *Born:* 5/30/33. Native of Brooklyn, NY. Taught chem engg at Univ of PA (1960-72) where was also Dir of Continuing Engg Education (1969-72). Teaching, res & consulting entered upon process dynamics, simulation & automatic control including the application of comp systems to control Assumed current responsibility

Cohen, William C (Continued)
as assoc dean for ind/academic affairs at Northwestern's Tech Inst in 1972. In position, hds progs in cooperative engg education, industrial associates & continuing engg education. Enjoys photography, tennis & students. *Society Aff:* AIChE, ΣΞ, ISA, ISPE, ASEE.

Coheur, Pierre M E O G
Business: Abbaye Du Val-Benoit, Liege 4000, Belgium
Position: Managing Dir & Univ Prof. *Employer:* Centre de Recherches Metallurgiques. *Born:* July 1913 Herstal Belgium. Grad civ engg & met 1937 & Agregee Ens Sup (PhD) 1942 both from Liege's Univ. Managing Dir Centre de Recherches Metallurgiques for the Benelux countries (CRM); also Prof of Met Liege Univ. Gold Medal A I Lg, ISI Bessemer Gold Medal London, Dr E h of RWTH Aachen. Hon VP ISI London, Dist Life Mbr ASM & AIME; Hon Mbr ILAFA, ISI of Japan, VDEh Germany, SFM Paris. Order of Merit of Italy, Austria, Luxemburg.

Cohn, Nathan
Home: 1457 Noble Rd, Jenkintown, PA 19046
Position: Consulting Engr. *Employer:* Self. *Education:* SB/Electromech Engg/MIT. *Born:* 1/2/07. Joined Leeds & Northrup on grad MIT 1927. Various field assignments, Phila, San Francisco, Chicago. VP, Technical Affairs, respon for R&D Engg, Pats 1958. Retired Exec VP 1972, from Bd of Dirs 1975. Bd mbr other cos. Author papers, book, pats power sys controls. PPres ISA, Scientific Apparatus makers Assn. Past Chrmn, Franklin Inst. Active Intl Federation Automatic Control, IFAC Life-Time Adv Award and Seal 1984, Fellow IEEE, ISA, AAAS. Mbr NSPE, Eta Kappa Nu, Sigma Xi, Tau Beta Pi. Mbr NAE 1969, IEEE Edison medal 1982, Lamme medal 1968, Franklin Inst Wetherill Medal 1968. ISA Sperry Medal 1968, Hon DEng RPI 1976, Hon ISA 1976. SAMA Award 1978. Reg. P.E. AZ, CA, IL, PA. *Society Aff:* IEEE, ISA, AAAS, NSPE.

Cohn, Seymour B
Home: 300 S Glenroy Ave, Los Angeles, CA 90049
Position: Pres *Employer:* S.B. Cohn Assoc, Inc *Education:* PhD/Engrg Sci & Applied Physics/Harvard Univ; MS/Engrg Sci & Applied Physics/Harvard Univ; BE/EE/Yale Univ *Born:* 10/21/20 in Stamford, CT. Positions: Radio Research Lab, Harvard Univ, 1942-45, Research Assoc; Sperry Gyroscope Co, Great Neck, NY, 1948-53, Research Engr; Stanford Research Inst, Menlo Park, CA, 1953-60, Lab Mgr; Rantec Corp, Calabasas, CA, 1960-67, VP & Tech Dir; S.B. Cohn Assoc, Inc, Los Angeles, CA, 1967-present, Pres & Consultant. Awards: 1954-Annual Award from Yale Engrg Assoc, 1964-Microwave Prize by Microwave Theory and Techniques Society of the IEEE, 1974-Lamme Medal of the IEEE, 1979-Microwave Career Award by Microwave Theory and Techniques Society of the IEEE. Publications: Over 50 papers in various technical journals. Patents: Approximately 40 patents. 1984 Centennial medal of IEEE. *Society Aff:* IEEE, ΣΞ

Cohn, Theodore E
Business: Minor Hall, University of California, Berkeley, CA 94720
Position: Prof. *Employer:* Univ of CA. *Education:* PhD/Bioengg/Univ of MI; MS/Bioengg/Univ of MI; MA/Math/Univ of MI; SB/EE/MIT. *Born:* 9/5/41. in Highland Park, IL. Taught Physiological Optics in the Sch of Optometry, Univ of CA, Berkeley since 1970. Studied role of visual depth illusion triggered by escalator appearance in falls on escalators. Res also includes signal detection methods for analyzing nerve messages & application of engg models to the study of both the normal human visual system & visual sensory deficit. *Society Aff:* AAAS, ΣΞ, IEEE, OSA, ARVO, NS

Cohon, Jared L
Business: Office of the Provost, The Johns Hopkins University, Baltimore, MD 21218
Position: Vice Provost for Research/Prof *Employer:* Johns Hopkins Univ *Education:* PhD/CE/MIT; MS/CE/MIT; BS/CE/Univ of PA *Born:* 10/7/47 Taught and performed research in environmental systems analysis at Johns Hopkins since leaving MIT in 1973. Became Asst Dean of Engrg in 1981, Assoc Dean of Engrg in 1983, and Vice Provost for Research in 1986. On leave 1977-78 to be Legislative Asst for Environment and Energy to Senator Daniel Patrick Moynihan. Co-editor of Water Resources Research 1979-1983. Author of text, Multiobjective Programming and Planning (Academic Press, 1978) and technical papers on apps of systems analysis to water planning, power plant siting, fire station location, nuclear waste management and offshore oil and gas development. *Society Aff:* ASCE, AGU, ORSA, TIMS.

Coit, Roland L
Business: 3412 Hillview Ave, PO Box 10412, Palo Alto, CA 94303
Position: Proj Mgr. *Employer:* Elec Pwr Res Inst. *Education:* MS/ME/Univ of Pittsburgh; BS/ME/Univ of OK. *Born:* 6/29/23. Native of OK. Army Ordnance WWII. With Westinghouse Elec 1947-date. Contributed to dev of flash evaporators, gas turbines, nuclear steam generators. Currently on loan to Elec Pwr Res Inst. Working on reliability improvement of secondary plant systems. *Society Aff:* ASME.

Coker, Billy B
Home: 2280 Macland Rd, Marietta, GA 30064
Position: Dept Engr. *Employer:* Lockheed-GA Co. *Born:* Currently serving a second term as Intl Pres of the Soc of Allied Wt Engrs and served as Exec VP 1974-75 & was active in Govt-Indus affairs for sev prior yrs. Hon pos of Fellow. Currently Dept Mgr Stds & Gen Wt & Producibility Engg at Lockheed-GA Co; career spans 25 yrs with exper in all areas of Mass Properties. In recent yrs, has held a var of mgt pos respon for design-to-cost, structures & design activities as well as mass properties.

Colabella, Alfred V Jr
Business: 138 Farnsworth Ave, P.O. Box 187, Bordentown, NJ 08505
Position: Pres *Employer:* AV Colabella, Engrs *Education:* ME/Mech Engrg/Stevens Inst of Tech *Born:* in Newark, NJ. Commissioned Ensign, US Naval Reserve, 1942; achieved rank of Lieutenant; served as Naval Aviator; resigned 1946. Employed in various engrg capacities from 1946 to 1955 by industrial firms in NJ and PA. Licensed PE in several states. Consultant in civil, electrical and mechanical engrg since 1955. Independent candidate for Governer of NJ in 1973. Formerly served as Commissioner of Urban Redevel, Mun Engr, and Plan Bd Mbr for City of Bordentown, NJ. Recipient of NJ Society of PE's Engr Award and Edwin F. Guth Memorial Design Award in 1977. Selected "Engineer of the Year" by PE Soc of Mercer Cty and awarded 1984 Cert of Merit by Engrs Club of Trenton. Fellow of American Consulting Engrs Council. Past Pres P E Soc of Mercer Cty and Engrs Club of Trenton. Chaired numerous tech soc cttes. Biography in Who's Who in America. Advocator of development of ultra-deep drilling tech to extract geothermal energy from Earth's mantle. *Society Aff:* ACEC, ASME, NSPE, TBΠ.

Colandrea, Thomas R
Home: 12988 Angosto Way, San Diego, CA 92128
Position: QA Dir *Employer:* GA Technologies *Education:* MBA/Mgmt/Western New England Coll; MS/Engrg Sci/RPI; BS/Met Engrg/U of MO at Rolla *Born:* 05/07/38 Extensive background in the cost-effective application of Qualtiy Assurance principles. Applied these techniques as QA Dir and, on a consulting basis, through litigation support, expert witness, mgmt reviews, trouble-shooting quality problems, etc. Recognized for expertise in the field of Quality Assurance: ASQC Certified Quality Engr, ASQC Certified Reliability Engr, Reg PE in CA and CT. In 1981, voted the ASQC Energy Div "QA Person of the Year–. In 1984, elected as an ASQC Fellow. Past Chrmn of the ASQC Energy Div, and currently a mbr of the ASME Nuclear QA Main Ctte and the Subctte on Nuclear Waste Mgmt. *Society Aff:* ASQC

Colangelo, James M
Home: 35 Bramble Dr, Brigantine, NJ 08203
Position: Pres *Employer:* Consulting Engr Service *Education:* B/CE/George Washing-

Colangelo, James M (Continued)
ton Univ *Born:* 9/1/30 in Summit, NJ; attended GWU due to their reputation in Structures. Received Scholarship from ASCE. Pres of Student Chapter ASCE, Member of Honorary Engrg Fraternity OT. Worked as Bridge Engr for Michael Baker, Jr, Inc, and other Consulting Firms specializing in Structures, until founded Consulting Engr Services in 1964. Firm employs 30 professionals, semi-professional draftsmen, and surveyors. Specialize in Wastewater Treatment, Highways and Bridges and Land Planning. *Society Aff:* ASCE, NSPE, WPCF, ACI, PCI

Colbert, Ted P
Business: 936 Enterprise Dr, Sacramento, CA 95852
Position: Prin & Pres. *Employer:* Terra Engg. *Education:* BSCE/Civ Engg/CA State Univ. *Born:* 5/6/40. CalTrans Engr - Materials & Res Dept, 1961-1966. Mgr of computer services for 200-person civ consulting firm, 1966-1971. Founded Terra Engg, 1971. Assumed Presidency of Terra Engg, 1979. Past Pres of ASCE-AMF & Past Sec Dir. *Society Aff:* ASCE, SAME.

Colby, Starr J
Home: 45 Linden Ave, Atherton, CA 94025
Position: Dir, Business Dev *Employer:* R&D Div, Lockheed Missile & Space Co. *Education:* MS/Aero Eng/Univ of MI; BS/Aero Eng/Univ of MI. *Born:* Apr 1, 1923. BS Aero E & MS Aero E from Univ of MI. Aerodynamics & Advance Design with Douglas Aircraft 1951-63; Asst Dir Space Tech with DOD-ODDRE 1963-65; Mgr & Dir of var Advanced Progs with LMSC 1965 to date - Mgr Advanced Remotely Piloted Vehicle Sys. VP Tech of AIAA 1973-74; AIAA Tech Dir & Ctte Chrmn 1965-72; AIAA Fellow. Mbr AAS, AFA, LWV, AOC AUSA, ADPA, AUYS; Mbr PSAC Ground Warfare Panel 1969-72; Mbr DSB RPV Task-force 1971-72; Mbr ARPA/DNA Long Range R&D Planning Prog 1973-75; Pres Assn for Unmanned Veh. Syst. 1980-81. *Society Aff:* AIAA, AUSA, AOC, ADPA, AUVS, LWS.

Colclaser, R Gerald Jr
Home: 407 Manor Rd, Delmont, PA 15626
Position: Prof *Employer:* Univ of Pittsburgh. *Education:* DSc/EE/Univ of Pittsburgh; MS/EE/Univ of Pittsburgh; BS/EE/Uiv of Cincinnati. *Born:* 9/21/33. BSEE Univ of Cincinnati; MSEE & DSc EE Univ of Pittsburgh. Native of Pittsburgh. Began engg career with Westinghouse Electric 1956. Spec in Design, Dev & Test of High Power SF(6) Breakers (21 patents). Dev Westinghouse Design Sch 1966. Joined Dept of EE Univ of Pittsburgh 1970, Prof & Chrmn 1974-84. Power related res in Arc Interruption, Transient Recovery Voltage. IPA Appt Air Force Weapons Lab, Albq, NM, in Pulsed Power 1985-86. Cons Westinghouse R&D in Pulsed Power Circuits. Fellow IEEE. PE PA, OH. Mbr IEEE High Voltage Circuit Breaker Subctte 1971-82; Served on Transient Recovery Voltage, Capacitor Switching, Out-of-Phase Switching, and Synthethic Test Circuit Working Groups. *Society Aff:* IEEE.

Cole, Eugene E
Business: 1515 River Park Dr, Sacramento, CA 95815
Position: Pres *Employer:* Cole, Yee, Schubert & Assoc *Education:* BS/Arch Engrg/CA State Polytechnic Coll-San Luis Obispo *Born:* 3/15/31 in Baker, OR. Married, with 5 children. Member of Preservation Bd, city of Sacramento. Member State Seismology Committee, SEAOC. *Society Aff:* SEAOC, CEA, ACI, ICBO

Cole, Jack H
Business: Mech Engg Dept, Fayetteville, AR 72701
Position: Prof, ME. *Employer:* Univ of AR. *Education:* PhD/ME/OK State Univ; MS/ME/OK State Univ; BS/ME/OK State Univ. *Born:* 6/7/34. Native of Tulsa, OK. Held engg positions with Am Airlines, Gen Dynamics, and North Am Rockwell before assuming current teaching position at the Univ of AR. Courses taught include machine design, system dynamics, control theory and fluid logic. Res area includes instrument and controls dev. Served seven yrs as faculty adviser for ASME student chapter. During 1978-1979 held position as Mgr, Instrument Systems Engg Branch of the Instrumentation Div of EG&G ID, Inc. Enjoy family and outdoor activities. *Society Aff:* ASME, SAE, ASEE, ΣΞ, FPS.

Cole, Jon A
Business: School of Engineering, College Place, WA 99324
Position: Prof of Engg. *Employer:* Walla Walla College. *Education:* BSCE IIT; PhD Univ of WI *Born:* 4/14/39 Native of Wheaton IL. Reg PE, Land Surveyor IL, WA, OR. Employed as civ engr, surveyor Harold F Steinbrecher 1958-63; Walla Walla Coll 1964-; civil/environmental engr and land surveyor 1970- ; NASA/ASEE Fac Fellow 1972; water qual consultant Walla Walla Dist Corps of Engrs 1975-80, hydrologic engineer 1981; Fulbright Lectr Ege Univ, Izmir Turkey 1976-77. Mbr ASCE (Pres Columbia Sect 1974-75), WPCF, ASEE, AWWA, Sigma Xi (Pres Whitman Coll-Walla Walla Coll Club 1972-73, secty 1979-), State of WA Adv Bd for Surveys & Maps 1975-, Seventh-day Adventist Church. Enjoy classical music, skiing, swimming, backpacking. *Society Aff:* ASCE, WPCF, ASEE, AWWA, ΣΞ, AAF, FAA

Cole, Julian D
Business: RPI, Troy, NY 12180-3590
Position: Prof. *Employer:* Dept of Math Science *Education:* PhD/Aeronautics/Cal Tech; BME/Mech Engrg/Cornell. *Born:* 4/2/25. PhD Caltech; BME Cornell. Teaching & res Caltech 1950-67, UCLA 1969-, Rensselaer Poly Inst ONR 1956-57, Boeing Co 1968. Cons RAND Corp, RDA, Boeing, North Amer Rockwell. Res in Fluid Mech esp Transonic Aerodynamics, on Applied Math esp Perturbation and Similarity Methods, on Math Physiology esp Cochlear Mech. Guggenheim Fellow 1963; Distinguished Alumni Award Caltech 1971; Lady Davis Fellow Technion, Haifa 1975; Fellow Amer Acad of Arts 6 Sci 1975; Natl Acad of Engg, Natl Acad of Sci 1976; V Karman Prize SIAM, 1984. *Society Aff:* APS, AIAA, SIAM.

Cole, Nancy Clift
Home: 4565 N Ravenwood Dr, Chattanooga, TN 37415
Position: Project Dev Manager *Employer:* Combustion Engg Inc. *Education:* BS/Met Eng/Univ of TN. *Born:* 8/4/40. BS Met Engg Univ of TN 1963. 1st female engg co-op; 1st female met engg grad; Tau Beta Pi. 1963-72 Oak Ridge Natl Lab, spec in corrosion, welding & brazing. Respon for GTA & EB welding of refractory metals; for dev improved filler metals for austenitic steels 1972-82, Combustion Engg, dev new and improved Fluxes, filler materials for welding products. 1982-84 Non-destructive testing, 1984- Responsible for Metallurgical Res and Dev Contracts. 3 pats on brazing. Author 22 pubs. Served Natl Cttes of AWS, ASM, WRC. Chairperson AWS Brazing & Soldering Committee 1983-86. Selected Outstanding Engg Alumnus Univ of TN 1973; Outstanding TN Woman 1974. Listed in "Outstanding Young Women of Amer" 1976. Husband: Leon; sons Douglas & Andrew. *Society Aff:* ASM, AWS, ASNT, WRC.

Cole, Phillip L
Business: CH2M Hill, 200 S. W. Market St, 12th floor, Portland, OR 97201
Position: Senior Engineer *Employer:* CH2M-Hill *Education:* BS/Civ Engg/OR State Univ. *Born:* 5/30/22. Over 30 yrs' experience in the design and construction of maj water resource projs in the northwest part of the US. Currently Hydropower Department manager with the Consulting firm of CH2M Hill with firm wide responsibilities for marketing, planning, design & Construction of Dams & Hydropowr facilities, Previously Chief, Engg Div of the N Pacific Div of the Corps of Engrs exercising responsibility for technical planning and design of all civil & military work accomplished in all or parts of five northwest states and AK. Coordinator for the Columbia River Treaty with Canada. Previously Chief, Construction Engr on the Libby Dam proj in northwest MT, the Green Peter, Foster, and Hills Creek Dams in OR. Served with the Army Corps of Engrs 1942-1946. *Society Aff:* ASCE, SAME, USCOLD.

Cole, Ralph I
3705 S. George Mason Dr, Apt. 1515 South, Falls Church, VA 22041
Position: Independent Tech/Education/Consultant. *Employer:* Self employed.

Cole, Ralph I (Continued)
Education: MS/Phys/Rutgers Univ; BS/EE/WA Univ. St. Louis, MO. *Born:* 8/17/05. After receiving BS Degree in EE (WA Univ) '27 entered in to Govt Service ultimately becoming Technical Dir, Rome Air Dev Ctr and resigning in '52 to join Industry (Melpar, Inc & WABCO). During war yrs on active duty as Col USAF and received "Legion of Merit–. Served industry as Exec Engr until '63 leaving to join Am Univ, Wash DC ultimately as Assoc Prof and Dir of Grad R&D Mgt Prog. Retired, 1973. Since that date Tech/Education & Mgt Consultant serving Academia and Govt. Long active career in profl socs and in publishing. PE, VA. *Society Aff:* AAAS, AIAA, IEEE.

Cole, Robert
Business: Potsdam, NY 13676
Position: Prof. *Employer:* Clarkson University *Education:* PhD/Chem Engr/Clarkson College; MChE/Chem Engr/Clarkson College; BChE/Chem Engr/Clarkson College. *Born:* 12/31/28. Native of Milford, CT. Served four yrs in US Marine Corps. Fundamental res in heat transfer at the NASA Lewis Research Ctr prior to joining the Clarkson faculty. Numerous refereed publications concerning bubble dynamics and boiling nucleation . Co-author of the recent (1979) book "Boiling Phenomena–. Guest Prof in 1971/72 and 1978/79 in the Appl Physics Dept of the Technical Univ of Eindhoven, the Netherlands. Co-Investigator for the current NASA space shuttle prog in materials sci (Physical Phenomena in Containerless Glass Processing). *Society Aff:* AIChE, ASME, OSA, ACERS.

Coleman, Edward P
Business: 10556 Strathmore Dr, Los Angeles, CA 90024-2541
Position: Pres & CEO *Employer:* Univ CA, Los Angeles; Sys Corp of Amer *Education:* PhD/Math Statistics, Indust Engrg/Columbia Univ; MS/Math/Univ of IA, IA City IA; BS/Mech Engrg & Math/MS State Univ *Born:* 07/09/10 Reg PE CA, NY, MI. PhD Columbia. Fellow and past Natl Dir, ASQC. Mbr, ASME, IIE, Inst Math Statistics, and over 15 other natl orgs. Natl VP Tau Beta Pi Engrg Honor Assoc, 1962-66. Prof Engrg, UCLA since 1953. Previously with Hughes Aircraft, Aberdeen (MD) Proving Ground, West Point, Aerojet Genl Corp, Lockheed, Bechtel. Outstanding Engr of So CA, 1957, ASQC Shewart Medal, 1967. Has made substantial contributions to the tech and engrg lit in fields of Quality, Reliability, Risk Evaluation and Professional Engrg Leadership. *Society Aff:* ASME, ASQC, IIE, IMS, ASEE, SAME

Coleman, Leonard F
Business: 343 State St, Rochester, NY 14650
Position: Gen Mgr, Mktg and V P Motion Picture and Audiovisual Products Div. *Employer:* Eastman Kodak Co. *Education:* BS/Physics & Math/Univ of Rochester. *Born:* Nov 27, 1930. BS Univ of Rochester. Native of Rochester NY. Began working for Eastman Kodak Co 1948. Var pos in Mfg dealing with the testing & usage of prof motion picture film. Dir of Intl Services, Eastman Kodak for 3 yrs, visiting 42 diff countries, giving lectures to mbrs of motion picture & tv industry. Ch Engr for Motion Picture & Audio-Visual Mkts, Midwest Region. Var mgt pos within Eastman Kodak, currently Gen Mgr, Marketing, and V Pres. Motion Picture & Audiovisual Products Div. SMPTE: VChrmn Intl Papers Ctte; Governor of Soc; Fellow; Secy & Mgr Dallas, Chgo & Rochester Sects. Hobbies: collecting antique cameras & US postage stamps. Additional memberships: British Kinematographic, Sound & TV Society, Fellow; Academy of Motion Picture Arts and Sciences; American Society of Cinematographers; The American Film Institute. *Society Aff:* SMPTE, BKSTS.

Coleman, Paul D
Home: 710 Park Ln, Champaign, IL 61820
Position: Prof EE. *Employer:* Univ of IL. *Education:* PhD/Phys/MIT; MS/Phys/PA State Univ; BS/Phys/Susquehanna Univ. *Born:* 6/4/18. Dr Coleman was first employed at WADC in Dayton, OH where he worked on countermeasures, aircraft antennas, microwaves and broadband impedance matching. Since 1946, his main interest has been on the generation and detection of electromagnetic energy in new spectral regions. From 1946-60, he did res on megavolt electronics (now called free electron lasers), and from 1960-79 he has been engaged with quantum electronics sources, molecular phys, nonlinear optics, and detectors. He has published over 100 papers, on topics ranging from free electron Doppler & parametric sources, coherent Cernkov generators, metal-oxide- metal detectors to molecular and chem lasers. *Society Aff:* IEEE, APS, OSA.

Coleman, William S
Home: 6771 Cottonwood Knoll, W Bloomfield, MI 48033
Position: Gen Mgr. *Employer:* Eaton Corp. *Education:* MS/ME/MIT; BS/ME/MSU. *Born:* 8/26/24. Originally from Grand Rapids, MI. Joined G M Res in 1946 in materials res. Joined Minneapolis-Moline in 1961 as Chief Admin Engr. Increasing responsibility positions with co Presidency in 1970. Dir Product Group Admin 1971 to 1975. Gen Mgr-Engg & Res Ctr 1976 to present. Married, 3 children. Named ESD Young Engr of Yr 1954. Bd of Dir - SAE 1972-75 Southfield Chamber of Commerce Bd of Dirs. Intnl Pres of SAE 1987-88. Honorary Fellow, Engrg Society of Detroit. *Society Aff:* SAE, ESD.

Coles, Donald E
Business: 1201 E California Boulevard, Pasadena, CA 91125
Position: Prof. *Employer:* CA Inst of Tech. *Education:* PhD/-/Ca Inst of Tech; MS/-/CA Inst of Tech; B Aero E/-/Univ of MN; E Mech/-/Boeing School of Aeronautics. *Born:* 2/8/24. Educ Boeing School of Aeronautics (E mechanic's license, 1947), Univ of MN (B Aero E, 1947), CA Inst of Tech (MS, 1948, PhD 1953). Mbr Caltech faculty (Prof Aeronautics, 1964 -). Mbr National Committee for Fluid Mech Films (produced 23 education films, 1960-65). Mbr organizing committee, Stanford Conference on Computation of Turbulent Boundary Layers (1968). Major papers on friction in supersonic flow (1953; AIAA Sperry award), similarity laws for turbulent flow (1956, 1962), shock-tube design (1962), transition in circular Couette flow (1965), boundary-layer transition (1975, 1979), stalling of airfoils (1978), vortex shedding from cylinders (1983), coherent structures in turbulence (current). *Society Aff:* AIAA, APS, NAE

Coli, Guido John
Business: Enka, NC 28728
Position: Pres. *Employer:* American Enka Co. *Education:* PhD/Chem Engg/VA Polytech Inst & State Univ; MS/Chem Engg/VA Polytech Inst and State Univ; BS/Chem Engg/VA Polytech Inst & State Univ. *Born:* 9/12/21. Native of Richmond, VA. Served to Lt USN 1943-46. Mobile Oil Co 1949-50. With Allied Chem Corp 1950-72; group VP Corp 1968-72; dir 1970-72. VP American Enka Co 1972-74; Exec VP 1974-79; Pres Jan 1979-Present; elected to Board of Dirs March 1979. Reg PE, NY, VA. Mbr Sigma Xi, Phi Lambda Upsilon, Tau Beta Pi, Phi Kappa Phi, Alpha Kappa Psi, Virginian of the Yr, Virginia Press Assoc 1968. Dir St Joseph's Hospital, Asheville, NC. *Society Aff:* AIC, AIChE, ACS.

Colleran, Robert J
Home: 27 Rimwood Dr, Lincroft, NJ 07738
Position: Pres. *Employer:* Dravo Van Houten *Born:* Feb 1933. BSCE Univ MD. PE NY & NJ. Presently Pres Dravo Van Houten Inc, respon for mgt, planning, design & constr of maj harbors, facils for the oil & gas industry, military facils, commerical & indus facils for intl & domestic projs having a constr value exceeding a half-billion dollars. Before joining Dravo Van Houten was assoc with a marine contractor and also served in USAF where respon for a num of design & constr projs carried out under the military constr prog.

Collier, Courtland A
Home: 830 NW 22 Terrace, Gainesville, FL 32605
Position: Assoc Prof. Emeritus *Employer:* Univ of FL. *Education:* ME/CE/Univ of FL; BE/CE/Yale Univ. *Born:* 7/29/25. Since 1961 taught Civ Engg at Univ of FL. Responsible for initiating the Construction option in Civ Engg, as well as the Sur-

Collier, Courtland A (Continued)
veying Tech prog. Authored 2 texts *Engineering Cost Analysis, Construction* Funding and over 20 published papers. Prior to 1961, worked in CE design and construction in 5 states and 4 foreign countries. Served 12 years as City Commissioner and 1 year as Mayor of City of Gainesville, Fla. *Society Aff:* ASCE, NSPE, FES, AAA.

Collier, Donald W
Business: Suite 1008, 307 N Michigan Ave, Chicago, IL 60601
Position: Lecturer & Consultant *Employer:* Self-employed *Education:* PhD/Phy Chem/Princeton Univ; ChE/Chem Engr/Princeton Univ; AM/Phy Chem/Princeton Univ; BACh/Appl Chem/Catholic Univ of Am. *Born:* 6/5/20. Wash, DC. Res Chemist & Chem Engr Sharples Corp 44-51. Dir Res to VP, Thomas A Edison Inc 51-57. VP to Pres, T A Edison Res Lab Div, McGraw-Edison Co 57-60. VP Res, Borg Warner Corp 60-75; VP Tech 75-78; Sr VP Corp Strategy 78-83. Advisory Council Plastics Prog, Princeton Univ 68-76, Chrmn 73-76. US Commerce Dept Technical Advisory Bd 74-81, VChrmn 75-76 VPres, AAAS 1968. Pres, Industrial Res Inst 66-67. Pres, Assoc of Res Dirs 58-59. Pres, Res Dirs Assoc Chicago 64-65. Chrmn, Dirs of Industrial Res 77-78. Annual Alumni Achievement Award, Engg Res, Catholic Univ of Am '79. Assoc, The Strategic Planning Inst 1983-85. Chmn of the Bd Guardian Interlock Systems Inc 1986- . *Society Aff:* AIChE, ACS, AAAS, DIR.

Collier, Samuel L
Home: 11002 Huntwyck, Houston, TX 77024
Position: Former Mgr R&D (Cons. Eng) *Employer:* TRW *Education:* B.S./M.E./Carnegie-Mellon Univ. *Born:* 04/10/22 Collier, author of 42 technical papers and the Industry Bible, a book - "Mud Pump Handbook–, was elected a Fellow of ASME in 1980 and received the Worthington Medal in 1985 for "Eminent Achievement in Pumping Machinery–. He holds 35 patents and is an expert in pumps - centrifugal and reciprocating - has done significant research in cavitation, fluid knock and valve bang, developed computer math modeling for above systems. Is an expert in air (gas) and liquid percussion drilling machines including computer simulation programs. Invented reciprocating pump computer diagnostic system. Conceived and built one of 2 in world- check valve transient flow lab. Collier is active in ASME - former member of Petroleum Div. exec cttee; he is 87-88 rep to Inventors Hall of Fame. As a consultant Collier is pursuing upgrades in pump and check valve testing and performance standards. Collier received the Hansen Award for Editorial Excellence in 1980 for best article in Plant Engineering. *Society Aff:* ASME.

Collin, Robert E
Business: 10900 Euclid Ave, Cleveland, OH 44106
Position: Prof *Employer:* Case Western Reserve Univ. *Education:* PhD/Elec Engg/Imperial College Univ of London; BSc/Engg Physics/Univ of Saskatchewan. *Born:* 10/24/28. in Western Canada. After grad work in England returned to Canada in 1954 and served as a Scientific Officer at the Canadian Armament Res and Dev Establishment. Immigrated to the US and became a naturalized citizen in 1964. Joined faculty at Case Inst of Tech in 1958. Served as chrmn of the Elec Engg and Appl Physics Dept from 1978-1981. Author of Field Theory of Guided Waves, Fdns for Microwave Engg, Antennas and Radiowave Propagation. Co-author with Prof Plonsey of Principles and Applications of Electromagnetic Fields, co-editor with F Zucker and contributor to Antenna Theory (2 volumes). Married with three children. *Society Aff:* IEEE, URSI.

Collins, Charles E
Business: 1704 Adolphus Tower, Dallas, TX 75202
Position: VP. *Employer:* Barnes & Click, Inc. *Education:* BS/CE/TX A&M. *Born:* 11/7/32. Native of Bonham, TX, served in US Army Corps of Engr 1955-57. Employed by Mobil Oil Corp 1957-60 as gas engr, Graff Engg Corp 1960-65 as plant design engr, by Northern Natural Gas Co as chief process engr 1965-67. Entered engg consulting 1967 specializing in technical & economic feasibility studies of oil/gas processing facilities for clients in the US, S Am, Europe, & N Africa. Assumed present position in 1977. Past chrmn of Dallas Sec of Am Inst of Chem Engrs & recipient of Outstanding Engr Award presented by Dallas Sec of AIChE in 1976. *Society Aff:* AIChE, SPE of AIME, AACE, NGS of NT.

Collins, Donald R
Home: 294 E Shore Dr, Massapequa, NY 11758
Position: Pres. *Employer:* Tele-Cine Inc. *Born:* Oct 1, 1930. Native of Ithaca (S Lansing) NY. BSEE Cornell Univ 1952. Started broadcasting career in military as pioneer (OIC) of Army Signal Corps TV. Joined ABC-TV Network in NYC, genl Engg Staff. Engg Mgr 5 yrs, first NYC ETV Station. Ch Engg 12 yrs Videotape Productions, NYC's first & largest Producton Co. Formed Tele-Cine Inc - North Am agent for Schneider tv lenses, sole owner & Pres. Chrmn NYC Sect SMPTE, invented "Editec" editing sys, respon for tech oper of first "Telestar" telecast. hobbies: boating, sailing, photography.

Collins, Henry E
Business: 23555 Euclid Ave, Cleveland, OH 44117
Position: Mgr-Matls' Engg. *Employer:* Argo-Tech Corp. *Education:* PhD/Met Engg/Carnegie-Mellon Univ; MS/Met Engg/Carnegie-Mellon Univ; MBA/Exec Prog/Baldwin-Wallace Coll; BS/Met Engg/Drexel. *Born:* June 1937. BS Met Engrg Drexel Univ; MS & PhD Met Engrg Carnegie-Mellon Univ; MBA Baldwin-Wallace College. Joined TRW Inc 1965 after completing PhD work. Assumed current pos as Mgr Matls Engrg 1976 with respon for providing metallurgical and chemical support to manufacturing divisions (Became Argo-Tech Corp, Oct 1986). Mbr ASM, SAMPE, Tau Beta Pi, Sigma Xi, SAE-AMS Ctte. Mbr ASM Handbook Ctte 1973-76, AIME High Temp Alloys Ctte 1973-77, IMD Publications Ctte 1972-78. *Society Aff:* ASM, SAE, TBΠ, ΣΞ, SAMPE

Collins, Jack A
Business: 206 W 18th Ave, Columbus, OH 43210
Position: Prof *Employer:* OH State Univ *Education:* PhD/ME/OH State Univ; MSc/ME/OH State Univ; BME/ME/OH State Univ *Born:* 11/23/29 Native of OH. Research Assoc, The OH State Univ Research Foundation 1952- 63. Assoc Prof of Engrg, AZ State Univ, 1963-72. Prof of Mechanical Engrg and Chrmn of Mechanical Design Section at The OH State Univ since 1972. Consultant to Babcock & Wilcox Research Ctr, Gen Elec Co, Air Research Mfg Co, Owens- Corning Fiberglas, others. Author of Failure of Materials in Mechanical Design: Analysis Prediction, Prevention, Wiley, 1981. Vice Chrmn of ASME Design Engrg Div 1981, Secy, ASME Policy Board Research Committee on Reliability, Maintainability and Failure Analysis. Bd of Dirs Seal of OH Girl Scout Council. Commissioner Central OH Boy Scout Council. *Society Aff:* ASME, ASTM, SESA, ASEE

Collins, James J
Home: 27 Deerhaven Dr, Nashua, NH 03060
Position: Dir *Employer:* Kollsman Instrument Co *Education:* MBA/Marketing/Adelphi Univ; PE/Elect/New Hampshire; B/EE/Manhattan Coll. *Born:* 6/16/36 Native of NY City. Design Engr Supervisor Airborne Instruments Labs, Div of Cutler Hammer 1957-62. Responsible for design of Electronic Intelligence Surveillance Systems. With Kollsman since 1962. Engrg consultant on Star Trackers for Space Program and ATE for US Navy and Air Force. As Engrg Mgr, directed R & D and new product development. Obtained one patent and authored several papers and articles. Program Mgr for multi-million dollar contracts; Mfg Engrg; Department head since 1979. Dir of Mfg Engrg 1980-85 (Multi-Plant responsibilities). Co presently second largest defense contractor in NH; Acquired by Sun Chemical 1986-87 Dir Avionics Engrg 1987-present Dir Programs Avionics Div. *Society Aff:* SME, AMA, IEEE, AUSA, Association of Old Crows

Collins, James P
Home: 161 Branford Rd, Rochester, NY 14618
Position: President *Employer:* James P. Collins & Assoc *Education:* SM/Civil Engg/MIT; MS/Chem Engg/Columbia Univ; BE/Chem Engg/Yale Univ. *Born:* 12/22/29. Employed, Olin-Matheson Chem Co 1953-55. Officer, Destroyer Force Atlantic Fleet 1955-58. Research Staff, MIT 1958-59. Principal Soils Engg, Hayden Harding & Buchanan, Consulting Engrs, Boston, MA 1959-60. VP, LeMessurier Assoc, Inc, Consulting Engrs, Boston, MA 1960-63. Pres, James P Collins & Assoc, Consulting Geotechnical Engrs, Rochester, NY 1963-1981. President, Wagner Associates, P.C. Consulting Engineers 1981-1982. Pres, James P. Collins & Assoc, Conslt Geotechnical Engrs, Rochester, NY, 1982-date. *Society Aff:* ACEC, ASCE, NSPE, ASTM.

Collins, Jeffrey H
Business: Advanced Robotics Research Institute, PO Box 19045, University of Texas At Arlington, Arlington, TX 76019
Position: Dir, Automation & Robotics Research Inst and Prof Elec Engrg *Employer:* The Univ Texas (UTA) *Education:* MSc/Math/Univ of London; BSc/Physics/Univ of London *Born:* 4/22/30 Luton, England. Engr at GEC and Ferranti, UK, 1951-57, working on microwave antennas and tubes. Sr lecturer, Univ of Glasgow, 1957-66, teaching electromagnetics and researching microwave magnetics. Research engr, Stanford Univ 1966-68. Dir of Physical Sciences, Rockwell Int'l, 1968-70. Prof and Head of Dept of Electrical Engrg, Univ of Edinburgh 1977-84; Dir of Automation and Robotics Inst and Prof of Elect Engrg, The Univ of Texas at Arlington, 1987-present; Dir of Advent Tech Plc and SETg Ltd (1984-1986). Co-recipient of Hewlett-Packard Europhysics Prize 1979 and elected F/IEEE 1980, F Engr 1981. *Society Aff:* IEEE, IEE, IOP, FRSE, FEng, IERE

Collins, John J
Business: 2 Penn Plaza-Rm 1005, New York, NY 10001
Position: Gen Mgr - Buildings Management & Constr *Employer:* NYNEX-NY Telephone Co *Education:* BS/Engg/US Nav Acad; -/EE/Syracuse Univ; -/Indus Engg/Columbia. *Born:* 12/9/21. BS US Naval Acad; studied elec engrg at Syracuse Univ & indus engrg at Columbia, Native of Syracuse NY. UNS (incl USNA) 1942-47. With NY Tele Co 1947- 83, except AT&T 1960-62, NYNEX-NY Tele Co 1984-present.. Assumed current respon for planning, design, constr, engrg, leasing & property mgt, operation & mtnce & protection of all Co real estate in NY State for NYNEX-NY Tele Co on 1/1/84. Life member NY Bldg Congress 1979. Chrmn NY Bldg Congress Council of Business & Labor for NY 1984-present. Founder 1971; Sectry NY Const Users Council, 1984- present. Bd of Gov Bldg Indus Employers of NY State 1972-present. *Society Aff:* IEEE.

Collins, John S
Business: 1211 Connecticut Ave, NW, Washington, DC 20036
Position: VP Marketing & Sales. *Employer:* De Leuw, Cather & Co. *Education:* BS/Civ Engg/Bucknell Univ. *Born:* 1/2/29. Held various engg positions in track maintenance on PA Railroad 1950-1964. 1964-1965 was Supervisor, Track & Tunnel Maintenance, Port Authority Trans Hudson Corp, NYC, & prepared 15-yr facilities improvement prog. 1965-1968 was Engr Maintenance of Way, Lehigh Valley Railroad. 1968-present, with De Leuw, Cather & Co, initially as Chief Railroad Engr (1968-71) in Washington, DC; Dir of Bus Dev (1971-73), VP Bus Dev (1973-78); then VP, Marketing & Sales, Corporate & Intl (1978-present). *Society Aff:* AREA, ASCE, NSPE.

Collins, John S
Business: 630 E Ninth St, Tucson, AZ 85705
Position: Pres. *Employer:* John S Collins & Assocs Inc. *Education:* BS/CE/Univ of AZ. *Born:* July 1936. BS CE Univ of AZ 1958. Field Engr for Southern Pacific Co 1958-60. Design Engr for Blanton & Co Arch & Engrs 1960-63. With Collins & Assocs since 1963. Presently Pres & ch oper officer for cons firm involved in civil & sanitary engg. Pres Southern Chap AZ Soc of PE 1967-68. Engr of Yr Southern Chap AZ Soc of PE 1970. Pres AZ Water & Pollution Control Assn 1972-73. Chrmn AZ Sec Water Pollution Control Fed 1972-73. Chrmn AZ Sect Am Water Works Assn 1972-73. Presently Pres AZ Cons Engrs Assn. Enjoy traveling, community & church activities.

Collins, Michael A
Home: 9427 Arborhill Ln, Dallas, TX 75243
Position: Prof of Civ Engr. *Employer:* Southern Methodist Univ. *Education:* PhD/Hydrodynamics/MIT; MSCE/Civ Engg/GA Tech; BCE/Civ Engg/GA Tech. *Born:* 7/6/42. Oak Park, IL. Grad GA Tech with honor under cooperative plan. Worked summers with Camp, Dresser, and McKee, Consulting Engrs, while completing PhD in hydrodynamics. Dissertation work cited in *Advances in Hydroscience* as state of art groundwater modeling study. Joined faculty of engg school, SMU, 1970, obtaining full professorship in 1979. Teach and conduct sponsored res in hydraulics, hydrology, and water resources engg. Have published in engg and scientific journals. Consultant to govt & private industry. Mbr water policy and technical advisory ctts, North Central TX Council of Govts. Chairman, Publications Committee, Sessions Program and Executive Committee, Pipeline Division, and Chairman, Sessions Program and Water Resources Planning Ctte, Water Resources Planning and Management Div, Amer. Soc. of Civil Engineers. Member Amer Soc. Civil Engineers, Amer Geophysical Union Amer Water Resources Assoc., Amer Water Works Assoc., International Association for Hydraulic Research, Water Pollution Control Federation, Natl Water Well Assoc. Sigma Xi. Reg PE, Cert Prof Hydrologist by Amer Inst of Hydrology, Pres of Eaumac, Inc, hydraulics & hydrology conslt firm. *Society Aff:* ASCE, AGU, AWWA, WPCF, NWWA, AIH, IAHR, AWRA, FRESHWATER SOCIETY, ΣΞ

Collins, Ralph E
Home: 1123 Joliette Rd, Bon Air, VA 23235
Position: VP *Employer:* Johnson & Higgins *Education:* BSEngg/Fire Protection/Univ of MD. *Born:* 12/27/39. Native of Wash, DC area. Grad Univ of MD. Field Engr, Factory Ins Assn specializing in Fire Protection. Asst Mgr/Sr Engr Fire Protection Services, Kennedy Space Ctr, FL. Assumed current responsibility as VP, Property Loss Control, Johnson & Higgins. President-Elect, Soc of Fire Protection Engrs; First Past Pres and Founding Mbr, SFPE, VA Chap; SFPE Committees - Symposium Planning (Past Chrmn), Long Range Planning (Past Chrmn), Honors, Measuring Fire Phenomena (Past Secy); Membership (Chrmn), Membership Criteria (Chrmn), Annual Program (Chrmn), Past Pres, ASSE Colonial VA Chap. Adjunct faculty various Community Colleges - Fire Sci. Mbr NFPA, SSS, BOCA, NFPA and NFPA Detection Devices Committee. Enjoy music, sports, square dancing. *Society Aff:* SFPE, ASSE, SSS.

Collins, Robert C
Home: 1725 Mora Ct, Los Altos, CA 94022
Position: VP-Engg. *Employer:* United Airlines. *Education:* M Ind Mgt/-/MIT; B of Arch E/-/WA State Univ. *Born:* 1/2/29. Seattle Wash. B of Arch & WA State 1952. M of Ind Mgt MIT 1965. Airborne Infantry 1946-47. Started with UA after grad in 1952. Served as Test Pilot for approx 10 yrs prior to entering engg mgt. Assumed current pos in Oct 1970. Remains qualified Air Pilot in all UA aircraft types. Mbr SAE, Chrmn Aerospace Council; Assoc Fellow AIAA; Reg PE CA. *Society Aff:* SAE.

Collins, Robert H, III
Home: 2387 Kimridge Road, Beverly Hills, CA 90210
Position: Pres. & CEO *Employer:* GSF Energy Inc. *Education:* MBA/Bus Adm/Stanford Univ. Advanced Mgmt Coll; BSME/Mec Engg/Stanford Univ *Born:* 5/5/35. Native of CA, and mbr of the Soc of CA Pioneers, and Executive Consultant to GSF Energy Inc. the first co in the world to purify methane gas from sanitary landfills. Prior to 1974 (when predicessor to GSF was founded) I was Pres if Inter-Tech Resources, Inc; Asst to the Sr VP of Cyprus Mines Corp, and Exec VP of Anvil Mining Corp, a subsidiary of Cyprus; Asst to the VP of Chief Engg of Best Industries and CA Ammonia Co. While in Canada with Inter- Tech, discovered one of the largest gas fields in the history of the Province of Ontario. When with Cyprus and Anvil I was instrumental in the dev of one of the world's largest lead, zinc and

Collins, Robert H, III (Continued)
silver mines. At leisure I enjoy boating and skiing, hiking, travel and photography. My wife, Emily Banks, is a Prof Model appearing in many commercials, and acting in movies for television and screen as well as on Stage. Emily is a Cert Clinical Psychologist in Family, & Marriage counseling. *Society Aff:* ASME, APE, AICE, SPE, ASME, APE, AICE.

Collins, S Kirk
Home: 1476 Summit Rd, Berkeley, CA 94708
Position: Safety Engr. *Employer:* Lawrence Berkeley Labs of Univ of CA *Education:* MA/Safety Admin/NY Univ; BA/Safety/Univ of CA. *Born:* 1/17/16. Native of New Orleans, LA, superintendent of enrg in eleven western states for Fireman's Fund Ins Group 1952-58; S Kirk Collins and Assoc, principal occupational safety consultant 1958-72; Lawrence Berkeley Lab/UC, construction safety engr 1972 to present. Retired LBL July, 86. Private industry & government occupational safety consult. Past pres, San Francisco Chapter ASSE, mbr of Soc bd of dirs and Region I VP. Re-Elected to bd of dirs of Certif Safety Profls 1980- 82. Certified Safety Profl 645 and reg profl safety engr in CA SF 01875. Expert witness in occupational safety and traffic safety cases. *Society Aff:* CSP, ASSE, SSS

Collins, William J, Jr
Home: 4230 N.W. 48th Street, Oklahoma City, OK 73112
Position: Chmn/Bd *Employer:* Collins-Soter Engineering, Inc. *Born:* May 1915, Iron River MI. Mech. Engrg MI Tech Univ 1935. Reg PE OK 1941. US Corps of Engrs 1942-45, Lt. Col. CERes. Chmn/Bd Collins-Soter Engrg, Inc, originally org 1946. Fellow & Pres. Mbr ASHRAE, Fellow Amer. Cons Engrs Council, Mbr Cons Engrs Council of OK, Natl Soc of PE, OK Soc of PE, Soc of Am Mil Engrs, Honorary Member MI Tech Univ Fund Bd of Tr. Pres's Club MI Tech Univ. ASHRAE Distinguished Serv Award 1964. Engineering Hall of Fame, Oklahoma State University 1981.

Collipp, Bruce G
Business: PO Box 2099, Houston, TX 77001
Position: Engrg Advisor *Employer:* Shell Oil Co. *Education:* SM/Nav Arch & Marine Engg/MIT; SB/Marine Trans/MIT. *Born:* 11/7/29. Inventor and developer of first semisubmersible floating drilling rig. Ocean engr design, construction and operation of offshore drilling rigs, fixed platforms, diving, subsea production, pipelines and oceanography. Awards Sigma Xi, Pi Sigma Phi (Propeller Club); Silver Cup (MIT); Holley Award ASME, OTC Indus Award 1971, 1984, Who's Who in Southwest; Guest Lecturer, Univ of TX; Visiting Committee MIT; Author of book "Buoyancy Stability and Trim–; Reg Engr TX; author of papers World Petroleum Congress, ASME, MTS, SNAME, SPE, Petrol Engr, Oil & Gas Journal, OTC, Offshore; Standardization Committee API on Marine Practices. *Society Aff:* API, MTS.

Colomb, Alain L
Home: 1, rue Jean Binet, Trelex, Switzerland, CH-1261
Position: Pres *Employer:* Energie Ouest Suisse, EOS *Education:* MS/Nuclear Eng/MIT; Diploma/Phys/Swiss Federal Inst of Tech. *Born:* 05/19/30 Native of Geneva (Switz). Res activities at the Swiss nuclear lab and at Oak Ridge Natl Lab. Since 1968 with EOS: power planning, nuclear power plant project. Former chrmn of the Swiss Nuclear Soc and of the European Nuclear Soc. Chrmn of the Swiss Atomic Indust Forum. Bd mbr of the Swiss Acad of Tech Sci and of Association Suisse des Electriciens. *Society Aff:* SNS, ANS, ASE, EPS

Coloney, Wayne H
Business: P.O. Box 668, Tallahassee, FL 32302
Position: Chrmn & CEO. *Employer:* Coloney Company Consulting Engineers, Inc. *Education:* BCE/CE/GA Inst of Tech. *Born:* 3/15/25. Bradenton, FL; m Anne Elizabeth Benedict, 1950; 1 dau, Mary Adore; Proj Engr, FL Rd Dept, Tallahassee, 1950-55; Hwy Engr, Gibbs & Hill, Inc, Guatemala, 1955-57, Proj Engr, Tampa, FL, 1957-59; Proj Engr, J E Greiner Co, Tampa, 1959-62, Assoc, 1962-63; Partner, Barrett, Daffin & Coloney, Tallahassee, 1963-70; Pres, Wayne H Coloney Co, Tallahassee, 1970-77; Chrmn & CEO, 1977-85; Pres, Sec, Tesseract Corp, 1975-85; Chrmn and CEO, Coloney Co Cons Engrs, Inc. 1978-; Dep. Chrmn Howden Airdynamics, Tallahassee, 1985-; V.P. Howden Coloney Inc., Tallahassee 1985-; Chrmn and CEO, Coloney Co Cons Engrs, Inc, 1978-; Mbr, Pres's Advisory Comm on Ind Innovation, 1978-79; White House Conf on Small Bus, 1978-79; P Mbr Tallahassee Chbr Commerce, FL State Chbr Commerce, Tallahassee 100 Club, Tallahassee Kiwanis Club, Tallahassee Popayan Friendship Comm, Advisory Com for Historic and Cultural preservation, Reg PE and Land Surveyor, AL, FL, GA, NC. *Society Aff:* ASCE, NSPE, FICE, FES, NCEE, NAFE, AAA

Colosimo, Joseph L
Business: 45 S Broadway, Yonkers, NY 10701
Position: Consulting Engr (Self-employed) *Education:* B.S./Engr/NY Univ *Born:* 11/24/27 Consulting Engr in private practice past 20 yrs-PE licensed in NY, NJ, CT, PA & NC-Engaged in various phases of construction industry & design of residents, commerical & consulting in structural & architectural engrg.- Foundations, piles, site planning etc. - Formerly chrmn Municipal Housing Authority city of Yonkers. Former member NYS Education Dept Advisory Council on Ethnic Studies. *Society Aff:* NYSSPE, NYSEPP

Colteryahn, Henry C
Home: 4926 Dorsie Dr, St Louis, MO 63128 *Education:* BS/EE/Univ of Pittsburgh. *Born:* 8/22/18. Native of Burgettstown PA. Spent 1 yr with Westinghouse Corp in E Pittsburgh. With Union Elec since 1940. Thirty six yrs in Transmission and Distribution, including operating, engg, construction and admin. Five yrs in Engg and Construction; VP of this responsibility April, 1978to Jan 30, 1981. Included elec engg, mech engg, civil engg, drafting, and construction, excluding nuclear. Enjoys wild turkey hunting, hiking, bird watching. *Society Aff:* NSPE.

Coltman, John W
Home: 3319 Scathelocke Rd, Pittsburgh, PA 15235
Position: Consultant *Employer:* Self employed *Education:* PhD/Phys/Univ of IL; MD/Phys/Univ of IL; BS/Phys/Case Inst of Tech. *Born:* 7/19/15. Native of Cleveland, OH. Joined Westinghouse Res & Dev Ctr in 1941. Res in microwave tubes & image amplification. Invented X-ray image intensifier, for which received Longstreth Medal of Franklin Soc in 1960 & the Roentgen Medal of Remscheid, Germany in 1970, and the Gold Medal of the Radiological Society of North America in 1982. Managed res in electronics, acoustics, mechanics, etc. Responsible for planning of program of R&D Ctr. Hobbies include playing, collecting & doing acoustics res on the flute. Retired Aug. 1980. *Society Aff:* IEEE, APS, NAE, ΣΞ.

Colville, James
Business: Dept of Civ Engg, College Park, MD 20742
Position: Prof. *Employer:* Univ of MD. *Education:* PhD/Civ Engg/univ of TX; MS/Civ Engg/Purdue Univ; BS/Civ Engg/Purdue Univ. *Born:* 6/13/36. in Coatbridge, Scotland. Naturalized US Citizen in 1963. Structural Engr with NAHB Res Inst lab, MD, 1960-1963. Structural engr with Stone & Webster, Boston, MA, 1963-1967, with responsibilities in power plant design and construction. With Hill Engg, Inc, Dalton, MA, 1967-1968 as Chief Structural Engr. Joined faculty of Univ of MD, Dept of Civ Engg, Aug 1970. Visiting Res Fellow & Lectr at Univ of Edinburgh, Scotland, 1976-1977. Attained current rank of Prof in 1979. Reg PE in NY, MA, & MD. Pres, Natl Capital Chapter of ACI, 1977. *Society Aff:* ASCE, ACI, TMS.

Colwell, Gene T
Business: School of Mechanical Engrg, Atlanta, GA 30332
Position: Prof *Employer:* GA Tech *Education:* PhD/Engrg Sci/Univ of TN-Knoxville; MS/ME/Univ of TN-Knoxville; BS/ME/Univ of TN-Knoxville *Born:* 8/3/37 Native of Chattanooga, TN. Research engr, Oak Ridge Nat'l Lab 1959-1962. Instructor, Univ of TN-Knoxville 1962-65. Design specialist, Oak Ridge Nat'l Lab

Colwell, Gene T (Continued)
1955-66; Prof, School of Mechanical Engrg, GA Inst of Tech, 1966-present; Assoc Dir, Sch of Mech Engrg 1983-85. Consultant to many companies and government organizations. Registered engr. Enjoys tennis & reading classics. *Society Aff:* ASME, ΣΞ, ΠΤΣ, ASEE

Comann, David H
Home: 4909 Woodland Apt D, West Des Moines, IA 50265
Position: Executive Director *Employer:* Iowa Ready Mixed Concrete Association *Education:* BS/Civ Engg/IA State Univ. *Born:* 8/17/25. US Naval Res WWII. Survey officer US Mil Govt Guam 1946. Design Engr & Hd of Civ Dept Stanley Cons, spec in municipal engg proj's. Field Engr 8 yrs & Dist Engr 8 yrs Portland Cement Assn IA. VP, Bus Dev, Shive-Hattery & Assoc. 11 yrs. With Iowa Ready Mixed Concrete Assn since 1980 as Exec Dir. Fellow Amer Soc of Civ Engrs (P Past Pres Tri-City & IA Sects Dist 17 Council, Civil Engr. of the Year 1985, IA Sect.), Natl Soc of PE, IA Engg Soc, Amer Pub Wks Assn (PPres IA Chap), Fellow Amer Concrete Inst. Reg PE IA & TX. *Society Aff:* NSPE, ASCE, APWA, ACI, IES.

Combs, Robert G
Business: Dept of Elec Engg, Columbia, MO 65211
Position: Prof, EE., Dir of Graduate Studies *Employer:* Univ of MO. *Education:* PhD/Elec Engg/Univ of FL; MSEE/Elec Engg/Univ of MO; BSEE/Elec Engg/Univ of MO. *Born:* 11/17/30. raised and mostly educated in MO. Served as Mineman in US Navy from 1948- 52. Married to Gertrude Minnick, 1958. Have two children, William and Melanie. Have taught Elec Engg at NC State, Univ of NB, Univ of FL and at Univ of MO- Columbia since 1965. Developed MYOCOM as part of work in communication-aids for non-verbal handicapped individuals. Also researching effects of competency based instruction. Served on many committees in MSPE, including chrmn of PEE Pres of Central Chapter MSPE, 1970. Pres of MO Academy of Sci, 1979. Likes golfing, bowling, fishing and singing barbershop harmony. *Society Aff:* ASEE, IEEE, NSPE

Comeaux, R J
Business: PO Box 3766, Houston, TX 77001
Position: VP, Petrochemicals *Employer:* Gulf Oil Chems Co. *Born:* Mar 1938. Attended Univ of TX in Austin & Lamar Univ in Beaumont TX, receiving BS in Chem Engg 1960. Indus exper incl 2 yrs with Texaco & 5 yrs with Goodrich-Gulf. In 1967 joined Gulf Oil Chem Co US. In 1972 appointed VP Petrochems, Korea Oil Corp (Gulf/Korean Govt joint venture). Late 1974 appointed VP Mfg & Raw Mtls for Gulf Oil Chem Co. Appointed VP Petrochemicals Gulf Oil Chemicals in July 1977. Mbr AIChE & received natl ward from AIChE 1960 in chem engg student design contest. Reg PE TX; Mbr Soc of Chem Indus.

Comella, William O
Home: 40 Water Oak Dr, Hilton Hd Island, SC 29928
Position: Retired. *Education:* BE/CE/Johns Hopkins Univ. *Born:* 5/8/13. in Pittsburgh, PA, raised in MD. Employed US Bureau of Public Rds, dev of State-wide Hgwy Planning and Transportation Surveys (1936-43), Admin Mgr, State- wide Hgwy Planning Surveys, DE and MD. WWII Service, US Naval Reserve, Civ Engr Corp, 71 St CB. Post-war employment with US Fed Hgwy Admin, Div Engr WY, Deputy Regional engr, Region 3 (Construction), Regional Engr, Region 15, directed construction of senic pkwys (Blue Ridge Pkwy), developed Res and Dev Demonstration Prog of natl scope. Regional Fed Hgwy Administrator Region 3. Retired 1973, continue active in Church and Civic affairs. *Society Aff:* ASCE.

Comer, David J
Home: 1375 Coldren Rd, Paradise, CA 95969
Position: Prof Head EE Dept. *Employer:* CA State Univ. *Education:* PhD/EE/WA State Univ; MS/EE/Univ of CA/BS/EE/San Jose State Univ. *Born:* 1/10/39. Patent holder with IBM Corp 1959-64. Asst Prof EE, Univ of ID 1964-66. Assoc Prof EE, Univ of Calgary 1966-69. Prof and Chrmn, Div of Engg CA State Univ, Chico 1969-74. Prof of EE, CSU, Chico 1974-present. hd, EE Dept CSU, Chico 1979-1981. Prof of EE, BYU 1981-present. Author of five textbooks including *Modern Electronic Circuit Design* and *Electronic Design with Integrated Circuits*. Author of 20 technical articles consultant, digital systems, Video Computer Systems and Microdesign Associates. *Society Aff:* IEEE.

Comingore, Edward G
Home: 142 Brookfield Dr, Moraga, CA 94556-1747
Position: Sr Industrial Engr *Employer:* US Postal Service *Education:* MS/Engrg Adm/Univ of UT; BIE/OH State Univ *Born:* 1/30/26 Raised in Washington, DC. Served in the US Navy as a Naval Aviator from 1943-1953. Worked in various engrg and production management assignments with Kimberly-Clark Corp, Thiokol Chemical Corp, Matson Navigation Co, and Leslie Salt Co from 1957-1970. Joined the US Postal Service as a Sr Industrial Engr in fall of 1970. Assigned to the Western Region Headquarters engrg staff in 1972. Responsible for regional program development and management for major US Postal Service programs. Regional VP IIE (IIE) 1980-1982. Member Nat'l Ski Patrol System. *Society Aff:* IIE

Comings, Edward W
Home: 509 Windsor Dr, Newark, DE 19711
Position: Prof Emeritus. *Employer:* Univ of DE. *Education:* ScD/ChE/MA Inst Tech; BS/ChE/Univ of IL. *Born:* 2/24/08. Chem engr (solvent extraction, dewaxing lubricating oils) the TX Co 1933- 35. Taught at NC State College lyr, Univ of IL 15 yrs concurrently assoc dir Munitions Dev Lab for Office of Scientific Res & Dev during WWII on smoke generators. Naval Ordnance Dev Award. Hd Sch Chem & Met Engg Purdue Univ 19561- 59. Dean College of Engg Univ of DE 1959-73. After retirement, Prof & Chrmn chem engg Univ of Petrol & Minerals Saudi Arabia 1974-78. Walker Award 1956 from AIChE. Author "High Pressure Technology" 1956 & some 60 articles in technical journals. Reg PE DE. Care giver for Alzheimer's Disease - spouse. *Society Aff:* AIChE, ASEE, ACS.

Comparin, Robert A
Business: Virginia Polytechnic Inst and State Univ, Blacksburg, VA 24061
Position: Dept Hd *Employer:* Cleveland State Univ *Education:* PhD/ME/Purdue Univ; MS/ME/Purdue Univ; BS/ME/Purdue Univ *Born:* 7/25/28 Instructor Purdue Univ 1954-1960. Asst Prof Univ of Maine 1962-1964, Prof VA Polytechnic Inst 1964-1974, Prof and Chrmn of Mechanical Engrg NJ Inst of Tech 1974-1977. Dean of Engrg CSU 1977-83; Test Engr Gen Elec summer 1953 and 1954, staff engr IBM 1960-1962, USAF 1946-49 and 1950-51, Registered PE NJ, Prof and He of Mech Engrg Dept. *Society Aff:* ASME, ASEE

Compton, Ralph T, Jr
2015 Neil Avenue, Columbus, OH 43210
Position: Prof *Employer:* OH State Univ *Education:* PhD/EE/OH State Univ; MSc/EE/OH State Univ; SB/EE/MIT *Born:* in St. Louis, MO. After receiving PhD in 1964, was Asst Prof of Electrical Engrg, OH State Univ 1964-65 and Asst Prof of Engrg, Case Inst of Tech, 1965-67. During 1967-68, held a Nat'l Science Foundation Postdoctoral Fellowship at the Technical Univ, Munich, Germany. Since 1968 has been with OH State Univ, first as Assoc Prof and then since 1978 as Prof of Electrical Engrg. 1983-84, at Naval Research Laboratory, Washington DC. Major interests include antennas, electromagnetics and communications. *Society Aff:* IEEE

Concordia, Charles
Home: 629 Alhamhra Rd, Venice, FL 33595
Position: Consulting Engr. *Employer:* Self. *Education:* ScD/-/Union College. *Born:* 6/20/08. With GE, 1926-1973. Consulting Engr on: elec power system dynamic performance, control (voltage, speed, power, frequency), protection, designed and built GE's first Transient Network Analyzer. AIEE Lamme medal, 6 patents, 115 & technical papers (elec machines, self-excitation, centrifugal compressors, wind-tunnels, power system stability, speed governing, voltage transients, computing machines, reliability). 1973-present, private consulting engr; 1960- 69, Chrmn of CIGRE Intl

Concordia, Charles (Continued)
Study Comm-Power System Planning & Operation. On Advisory Panels for studies of nearly all maj power blackouts. *Society Aff:* ASME, IEEE, NSPE, CIGRE, NAE, AAAS, ΣΞ, TBΠ.

Conderman, Charles W
Home: 40 Sunset Trail, Fairport, NY 14450
Position: Vice Pres *Employer:* Wagner Assocs *Education:* BSCE/Civil Engg/State Univ of NY at Buffalo; AAS/Engg Sci/State Univ on NY Tech Inst at Alfred. *Born:* 10/13/42. Reg PE in NY State with over 17 yrs of experience in the consltg engrg prof. Currently VP for business dev in NY State for the Wagner Group, a multi-discipline engrg-Architecture firm with principal offices in Reading, PA, Metuchen, NJ, and Rochester, NY. As a mbr of the Adjunct Faculty, Rochester Inst of Tech, Coll of Continuing Educ, taught courses in surveying and structural analysis and design. Pres (1977-78), Rochester Sect, Amer Soc of Civil Enggs. *Society Aff:* ASCE, CEC, NSPE

Condit, Carl W
9300 Linder Ave, Morton Grove, Illinois 60053
Position: Prof Emeritus Hist, Art Hist, Urban Affairs. *Employer:* Northwestern Univ. *Education:* BS/Mech Eng/Purdue Univ; MA/English/Univ of Cincinnati; PhD/English/Univ of Cincinnati; Post-doct Fellow/History of Science/Univ of Wisconsin. *Born:* Sept 29, 1914 Cincinnati OH. BS Purdue Univ; MA, PhD Univ of Cincinnati; Postdoc Fellow Univ of WI. Chief courses taught; hist of bldg tech, hist of sci, hist of city. Author of following bks: *The Rise of the Skyscraper* 1952; *Amer Bldg Art 19th & 20th Cent* 1960, 1961; *The Chicago Sch or Arch* 1964; *Amer Bldg: Materials & Techs* 1968; *Chicago 1910-70: Bldg, Planning & Urban Tech* 1973-74; *The Railroad and the City: A Technological and Urbanistic History of Cincinnati* 1977; *The Port of New York: A History of the Rail and Terminal System* 1980-81. Hon Dr of Letters Univ of Cincinnati 1967; Hist & Heritage Award, Am Soc of Civ Engrs 1971; Leonardo da Vinci Medal, Soc for the Hist of Tech 1973; Distinguished Service Award, Am Institute of Architects, 1980; Hon Doctor of Letters, Knox College, 1981. Hon Doctor of Letters, De Paul Univ, 1983. Mbr: History of Sci Soc; Soc of Architectural Historians; Society for the History of Technology. *Society Aff:* HSS, SAH, SHOT.

Condrate, Robert A
Business: Alfred Univ, Alfred, NY 14802
Position: Prof *Employer:* NYS Coll of Ceramics *Education:* BS/Chem/WPI; PhD/Chem/IIT *Born:* 1/19/38 in Worcester, MA. Worked up the professional ranks 1967-1981 to full prof in the Engrg & Science program of the NYS Coll of Ceramics at alfred Univ. Hd academic leaves at Los Alamos Scientific Lab and the research and development group at GTE Sylvania (Towanda, PA). Published over eighty publications in referred journals and various books. My group was the first to make a solid polycrystalline form of calicium phosphates (apatite) for potential application as a bioceramic replacement for bones or teeth. We are currently investigating the structure, spectral and related properties of glass compositions that have optical fiber applications & superconducting materials. *Society Aff:* ACHEMSOC, ACerS, APS, SAS, RSC, NICE, MRS, ΦΚΦ.

Conkey, David R
Business: 1500 Foshay Tower, Minneapolis, MN 55402
Position: Pres. *Employer:* Conkey Assoc Inc. *Education:* BS/Mech Engr/Univ of MN *Born:* 8/22/22. Army Corps of Engrs 1943-45. Engrg, Fegles Construction Co Ltd 1947-51. Mech & Structural Engrg, General Mills, Inc 1951-56 for rehabilitation and modernization of flour mills & other plant facilities. Founder and Pres, Conkey and Assoc, Inc, Consulting Engrs, since 1956; specializing in industrial plant design with emphasis on commodities handling & food processing facilities. Annual construction volume in excess of 50 million annually. Reg in 21 states. Projects completed in 16 foreign countries. *Society Aff:* ACEC, ASME

Conklin, Clement L
Business: PO Box 5837, Orlando, FL 32855
Position: Eng Dir - Battlefield Interdiction Systems *Employer:* Martin Marietta Corp. *Education:* BS/Mech Engr/IIT *Born:* 5/8/28. Native of Chicago, IL. Early employment as mech engg in the canning machinery and steel industries. Joined Martin Marietta as mech design engg in 1956. Res for design of structures and flight control systems on sev aircraft and missile programs. Later Engg Mgr resp for all mech engg activities on several major missile systems, then Dir of the Mech Systems Lab respon for all Orlando Div mech engg activities, then Engrg Dir of Pershing 1A and Pershing II programs. Currently Dir of Engrg for the Battlefield Interdiction Systems Programs. Reg PE FL. Enjoys fishing, boating.

Conly, John F
Home: 6478 Bonnie View Dr, San Diego, CA 92119
Position: Chrmn *Employer:* San Diego State Univ *Education:* PhD/EM/Columbia; MS/ME/Univ of PA; BS/ME/Univ of PA *Born:* 9/11/33 Native of Swarthmore, PA. Was instructor at Univ of PA, research asst at Columbia Univ. Won Daniel & Florence Guggenheim Fellowship. Prof at San Diego State Univ since 1962, Chrmn of Dept of Aerospace Engrg & Engrg Mechanics from 1977-84. Chrmn of San Diego section of AIAA, 1970, won award as outstanding section in US. *Society Aff:* AIAA

Conn, Arthur L
Business: 1469 East Park Pl, Chicago, IL 60637
Position: Pres. *Employer:* Arthur L. Conn & Associates Ltd. *Education:* SM/Chem Eng Practice/MIT; SB/Chem Eng/MIT; Cert/Inst for Mgmt/Northwestern Univ. *Born:* 4/5/13. Founder, Arthur L Conn & Assoc, Ltd, an independent professional consulting firm in 1978. With Std Oil Co (Indiana), and Amoco Oil Co (subsidiary), Res and Dev 1939-78. Directed Boron Isotope Separation proj for Manhattan District; supervised process dev, including fluid catalytic cracking, Ultraforming and other petroleum processes in various division director positions leading to Dir- Process Dev, Res Coordinator, Sr Consulting Engr, and Dir of Govt Contracts. Consultant, Atomic Energy Commission, Office of Coal Res, and Energy Res and Dev Administration. Committees of Natl Acad of Engg, Coal Liquefaction, and Refining Coal and Shale Liquids (Chrmn); AIChE, Bd of Dir, 1966-71, Pres 1970, Fellow founders Award; AAAS Fellow, Bd 1970-73. *Society Aff:* ACS, AIChE, AAAS.

Conn, Harry
Business: 1109 Arden Ave, Rockford, IL 61107
Position: Man-Tech Consultant. *Employer:* W A Whitney Corp. *Education:* PhD/Appl Sci/CO Polytech. *Born:* Aug 1914. Studied mech engg at Lewis Inst, Armour Inst, IL Inst of Tech. Lectured on engg & econ for SME, ASME, AISC, AWS, IIE & over 50 univ's in over 250 cities in US, Canada, Mexico, Europe & Orient. Has also written over 200 tech papers & written articles for over 75 tech & sci journals. Was employed by John Deere, Intl Harvester, Studebaker, LaSalle Engg, Scully-Jones Corp (as Ch Engr for 13 yrs). Was awarded 1975 SME Intl Gold Medal & ASME 1975 Edwin F Church Medal. Chrmn of Bd of W A Whitney Corp, Rockford, IL. Is retained as a mgmt-tech conslt by W.A. Whitney Corp, Anderson Consolidated Ind and Correct Craft Inc, and lectures widely on "Ethics and Morality for Mgmt-. On May 19, 1986, awarded a Charter Fellow Award and Medal by SME. *Society Aff:* SME.

Conn, Hugh G
Business: Kingston, Kingston Ont K7L 3N6
Position: Prof Emeritus. *Employer:* Queen's Univ. *Education:* MS/Thermodynamics/Univ of MI; BSc/ME/Queen's Univ. *Born:* 10/6/08. Education in Ontario & Queen's Univ (ME '31). 1931-32 - Demonstrator Phys Dept, Queen's. 1932-37 - Plant Engr, Procter & Gamble Co, Hamilton. 1937-39 - Mech Engg Dept, Queen's. 1939-45 - Armed Forces - Lt Col, RCEME. 1946 - MS (Mech) Univ of MI. 1946-63 - Hd of the Dept of Mech Engg, Queen's. 1955-64 - Dean, Appl Sci, Queen's. 1965-70 - VPrincipal (Admin) - Queen's. 1970-74 - Prof, Mech Engg, Queen's. 1974-to date - Prof Emeritus. *Society Aff:* ASME, EIC, CSME, FIMechE.

Connally, Harold T
Business: 5735 Pineland Dr, PO Box 31049, Dallas, TX 75231
Position: VP, Refining. *Employer:* Dorchester Gas Corp. *Education:* MS/ChE/Univ of MI; BS/ChE/Univ of OK. *Born:* 10/19/27. Norman, OK native for 21 yrs. Previously with Conoco, Inc at Ponca City, OK involved in refinery construction, engg & mgt for eleven yrs. VP, refining for Crystal Oil Co Shreveport, LA from 1974 to 1978. Currently VP, Refining for Dorchester Gas Corp in Dallas, TX. Avid hunter, fisherman & football fan. *Society Aff:* AIChE, NPRA.

Connell, James Paul
Business: 1023 Corporation Way, Palo Alto, CA 94303
Position: Mgr - Mining Engg. *Employer:* W A Wahler & Assocs. *Born:* Jan 1928 St Louis MO. BS Mining Engg & Geol Engg MT Sch of Mines. Grad studies PA State Univ. Qual Control Supr Purex Corp 1952-57. Mining Engr, Indus Engr, Res (Mining Methods) Engr, Sys Engr, Ch Planning Engr MT Oper the Anaconda Co 1959-71. Advanced to Coordinator of Corp Mtls & Inventory Sys in NYC 1971. Joined Wahler & Assocs 1972, advanced to Principal in Firm in 1975. Served as bus mgr, presently directs all mining engg planning, bus dev, proj completion, & reporting for cons geotech org. Served as Secy-Treas MT Sect AIME 1964-71. Chrmn Oper Res Ctte SME 1973-74. Also served on Mbrship & Publ cttes of SME.

Connell, Richard M
Business: 630 Sansome St, San Francisco, CA 94111
Position: Div Engr. *Employer:* US Army Corps of Engrs, S Pac Div. *Born:* Sept 13, 1925 Erie PA. Commissioned in Corps of Engrs 1949 upon grad from US Mil Acad; MSCE MIT; grad of Command & Genl Staff Coll, Army War 1974. Previously was Deputy Ch of Staff Logistics & Engr US Army Training & Doctrine Command. Between 1970 & 1973 was Dist Engr Walla Walla WA, during a part of the constr of John Day, Little Goose, Lower Granite, Ririe, & Dworshak Dams.

Connell, Terence J
805 SW 84th CT, Portland, OR 97225
Position: Dir, Engg and Drafting *Employer:* Blaesing Granite Co *Education:* MS/Phys/Purdue U; BS/Military Engr/USMA, West Point *Born:* 11/4/33 Native of Muskegon, MI. Military assignments have included staff and battalion command, and Dist Engr of Portland, OR Dist during construction of Bonneville Second Powerhouse, MT St Helen's emergency flood control and navigation, and construction of Applegate Dam. Asst Prof of Phys, USMA. Overseas military engrg duties in Korea, Vietnam and Philippines. Currently Dir of Resource Mgmt, Office Chief of Engrs, US Army. Pres, SAME Mobile, AL (1973) and Portland, OR (1981). *Society Aff:* SAME, ASCE

Connelly, John R
Home: 223 Harvard Ave, Palmerton, PA 18071
Position: Retired. *Education:* ME/ME/Univ of IL; MA/Education/Lehigh Univ; MS/ME/Univ of IL; BS/ME/Univ of IL. *Born:* 10/15/05. Faculty of Lehigh Univ 1929-1942. Capt Ordnance, Springfield Armory 1942- 1946. Asst Ch Engr, Ch Engr Kelly Springfield Tire Co 1946-1951. Asst Ch Engr, Ch Engr, Mgr of Engr NJ Zinc Co 1951-1970. Papers on Wear of Bearing Metals - ASME. Textbook on Mfg Processes-McGraw-Hill 1943. Patents concerning mfg, testing, & shipping small arms. News paper articles and slide story at Bicentennial on early history of east central PA, correcting many earlier articles of misinformation (since retirement). *Society Aff:* ASME.

Conner, David A
Home: 3406 Tequesta Lane, Birmingham, AL 35226
Position: Prof and Chrmn *Employer:* Univ of AL in Birmingham *Education:* PhD/EE/GA Inst of Tech; MS/EE/Auburn Univ; B/EE/Auburn Univ *Born:* 8/4/39 Native of Pensacola, FL. Taught at Auburn Univ, 1961-63. Engr for IBM Corp, 1963-65. Taught at GA Tech, Univ of TN at Chattanooga, and Univ of Louisville, 1966-78. Served as Asst Dean for Research of Speed Scientific School at Univ of Louisville. Prof and Chrmn of Electrical Engrg at Univ of AL in Birmingham since 1978. Consultant to over 40 corporations and governmental agencies. Registered PE in AL, GA, KY, and TN. Past Chrmn, AL State Bd of Registration for PEs. Active in IEEE at Section and National Levels. Authored two books and numerous articles and reports. *Society Aff:* IEEE, ASEE.

Conner, James L
Business: 3315 N Oak Trafficway, Kansas City, MO 64116
Position: Sr VP. *Employer:* Farmland Ind, Inc. *Education:* BS/Engg/Millikin Univ. *Born:* 4/5/28. Methods & Planning Engg - OGMC, Redstone Arsenal - 1950-1952. Began in fertilizer industry in 1955 as Process Engr. Plant Superintendent & Asst Mgr, V- C Chem Co 1958-1965. Gen Mgr, Farmland Inds' FL Operations - 1965-1974. VP, Phosphate Mfg, Farmland Inds - 1974-1975. VP, Fertilizer Mfg, Farmland Inds - 1975-1979. Current Sr VP, Fertilizer & Agri Chemicals Div, Farmland Inds. *Society Aff:* AIChE, ACS.

Connor, John R
6472 Via Benita, Boca Raton, FL 33433
Position: President *Employer:* Self *Education:* BS/CE/Univ of PA *Born:* 11/27/16 Engr. Panama Canal 1939-1940. Officer USN-Civ Engr Corps 1940-1945. Proj Engr and Proj Mgr Gahagan Constr Corp 1946-1949. Proj Mgr., Regional Mgr (South America), Intl Mgr Fred'k Snare Corp 1949-1961. Kaiser Indus and Kaiser Engrs. VP International Pres Consorcio Guri (Kaiser Joint Vent-Hydro Elec Project Venezuela) 1961-1975. Pres CEO Fredk Snare O'seas Corp. & VP Dir Constr Precomprimido-Snare (Venezuela) 1975-present. Pres. Connor Assoc (Consulting & Constr Mgmt). *Society Aff:* ASCE, NSPE, USCOLD, FES, ABA

Connor, Myles J, Jr
Business: P.O. Box 101, Florham Park, NJ 07932
Position: Retired *Employer:* Exxon Research & Engrg Co *Education:* MS/ME/Columbia Univ; BS/ME/Rutgers Univ *Born:* 5/5/23 Served in the US Air Force (2-1/2 years in England). Began Exxon career in 1948 with the Standard Oil Development Co, holding various positions, including Dir of Project Engrg and Asst Gen Mgr of Esso Engrg. Was elected VP of Exxon Int'l in November, 1968, VP of Esso Middle East in December 1971, and VP of Esso Europe in August, 1973. Mr. Connor assumed current responsibility as Exec VP of Exxon Research & Engrg in June, 1981. Enjoys tennis and skiing. *Society Aff:* AIChE

Conover, John S
Business: 2600 Denali St, Suite 305, Anchorage, AK 99503
Position: VP *Employer:* Century Engrg, Inc *Education:* B.S./CE/Univ of AK *Born:* 11/23/36 Registered PE (Civil) in AK, CA and HI. Past Secretary, Treasurer and VP of Fairbanks Branch A.S.C.E. Past State VP N.S.P.E. Assumed present duties July 1980 as VP and Mgr of AK Operations for Century Engrg, Inc. of Baltimore, MD. Graduate study in structures at Cornell Univ and the Univ AK. *Society Aff:* ASCE, NSPE, SEASC

Conrad, Albert G
Home: 4591 Camino Del Mirasol, Santa Barbara, CA 93110
Position: Dean Emeritus. *Employer:* Univ of CA. *Education:* EE/EE/Yale; MS/EE/OH State; BS/EE/OH State. *Born:* 5/19/02. Faculty Yale Univ 1928-62; Chmn EE Yale Univ 1943-62. Dean of Engg, Univ of CA Santa Barbara 1962-68. Dean Emeritus UCSB 1968-date. *Society Aff:* AIEE, ASEE.

Conrad, E T
Business: 11260 Roger Bacon Drive, Reston, VA 22090
Position: Exec VP. *Employer:* SCS Engineers *Education:* BS/Engg/Univ of CA. *Born:* 2/19/36. and raised in San Francisco. As Field Engr for the FAA, traveled throughout the West establishing radar facilities for 4 yrs. Then supervised field engrs for 2 yrs. Proj Engr (environmental), Bd mbr and officer for Ralph Stone & Co, Consulting Engrs, Los Angeles for 6 yrs. Founded SCS Engrs in 1970 and opened Reston, VA office in 1971. Provides comprehensive engg services with emphasis on all as-

Conrad, E T (Continued)
pects of solid waste mgt, landfills and landfill gas, hazardous waste control, economic and policy analysis. Enjoys athletics -- jogging, racketball, swimming, biking, skiing. *Society Aff:* ASCE, NSPE, WPCF, GRCDA, NSWMA.

Conrad, Hans
Business: North Carolina State Univ, Materials Engg. Dept, Raleigh, NC 27650
Position: Prof & Head. *Employer:* North Carolina State Univ. *Education:* DEngg/Met/Yale Univ; MEngg/Met/Yale Univ; BS/Met Engg/Carnegie Inst Tech. *Born:* 4/19/22. Konradstahl, Germany; came to US 1926, naturalized, 1944; student Wash and Jefferson Coll 1940-42; B.S. Carnegie Inst. Tech. 1943, M.Engr and D.Engr Yale Univ. 1951, 1956. Res Metallurgist Chase Copper & Brass Co, Waterbury, CT, 1953-55; supervisory engr Westinghouse Res Labs, Churchill Boro, PA, 1955-59; sr res specialist Atomics Intl, Canoga Park, CA, 1959-61; hd dept phys Aero Corp, El Segundo, CA, 1961-64; tech dir Franklin Inst Res Labs, Phila 1964-67; Prof & Chrm Metallurgical Engg. & Mat. Sci. Dept. Univ. KY 1967-81; Head Materials Engg. Dept, N.C. State Univ. 1981-86, Prof. 1986- . Recipient Univ KY Res award, 1971; US Sr Scientist award Alexander von Humboldt-Stiftung, 1974; Japan Sec Promotion Sci Visiting Prof, 1976, Visiting Prof. Am. Univ. Cairo 1984, Soviet Acad. Sci 1984, Chinese Ministry of Metallurgical Ind. 1986, Fellow ASM 1977. Mbr Sigma Xi, Tau Beta Pi, Contbr articles to profl jours and books. *Society Aff:* AIME, ASM, ASTM.

Conry, Thomas F
Business: Dept of Gen Engrg, 104 S Mathews St, University of Illinois, Urbana, IL 61801
Position: Prof *Employer:* Univ of IL *Education:* PhD/ME/Univ of WI-Madison; MS/ME/Univ of WI-Madison; BS/Engrg Mech/PA State Univ *Born:* in West Hempstead, NY. Married, three daughters. Product Design Engr for AC Electronics Div, Gen Motors Corp, 1963-66. Research Engr for Detroit Diesel Allison Div, Gen Motors Corp, 1969-71. With Univ of IL since 1971. Currently Prof of Gen Engrg and Mechanical Engrg. Research, publications, and consulting in areas of tribology, dynamics of rotating machinery, and system design optimization. Tech Ed of ASME Journal of Vibration, Acoustics, Stress & Reliability in Design, Oct 1984-Oct 1989. Chrmn, ASME Design Engrg Div 1979-80. Enjoy photography and bicycle riding. *Society Aff:* ASME, ΣΞ, ASEE

Constant, Clinton
Home: PO Box 1217, Hesperia, CA 92345
Position: Air Quality Engineer (cons) *Employer:* APCD, San Bernardino County *Education:* PhD/ChE/Western Reserve Univ-Cleveland; BSC/Chem/Univ of Alberta, Edmonton, Canada. *Born:* 3/20/12. PE Chem Engg, PE Mfg Engg; Natl Cert Chem Engr/Chem Achievements; engg design for equipment and process, R&D, mfg operations in the hydrofluoric industry and inorganic products, Napalm, phosphatic chemicals/fertilizers, envirn. Pats. 1936: R&D engr/mfg superintendent; anhydrous HF, fluorides, specialty chemicals-Harshaw, Cleveland 1950: mgr of engg; designed most automated Napalm plant in USA- Ferro, Cleveland. 1952: chief engr/proj mgr; phosphatic chemicals production, R&D, plant design- Armour, FL and Atlanta 1970: environ engr; EPA waste treatment plant projs, 201 plans, O&M manuals-Robert and Co, Atlanta, 1979: chemical engr; environ projs-Almon Associates, Atlanta. 1980: Chem engr, environ Projs - Engr Serv Assoc, Atlanta; 1981: VP of Engrg - ACI, Inc. 1984: Sr VP of Engrg & Chf Engr, Micronic Technologies, Inc. Hesperia CA. 1986: Air Quality Engineer (consultant)-APCD, San Bernardino County, Victorville, CA. Fellow AIChE, AIC, AAAS, NYAC, Assoc fellow AIAA (50-yr pioneer) P chrmn AIChE Atlanta 1978. Hobbies: pianist, astronomer. *Society Aff:* AAS, ACS, ASP, RASC, AIC, AIChE, AIAA, AAAS, NYAS, NSPE, AWWA, WPCF, APCA.

Constantinides, Alkis
Business: Dept of Chem & Biochemical Engg, PO Box 909, Piscataway, NJ 08854
Position: Prof & Grad Dir. *Employer:* Rutgers Univ. *Education:* DESc/Chem Engg/Columbia Univ; MS/Chem Engg/OH State Univ; BChE/Chem Engg/OH State Univ. *Born:* 1/29/41. Dr Constantinides is Prof and Grad Dir of the Dept of Chem and Biochemical Engg at Rutgers Univ, New Brunswick, NJ. He is responsible for teaching graduate and undergraduate courses and for conducting research in the areas of fermentation, enzyme tech, microbial engg, process optimization and computer application. Dr Constantinides is the author of several papers published in leading biochemical engg journals. For his work, he has received extensive support from the Natl Science Fdn in the form of research grants. Dr Constantinides has industrial experience with Exxon Research and Engg, Procter and Gamble, and Borg-Warner Corp. He also has extensive consulting experience with several other companies. *Society Aff:* AIChE.

Conta, Lewis D
Home: 738 Calle del Ensalmo, Green Valley, AZ 85614
Position: Prof Emeritus, Univ of RI; Adj Prof, Univ of AZ *Employer:* Self Employed Consultant *Education:* PhD/ME/Cornell Univ; MS/ME/Univ of Rochester; BS/ME/Univ of Rochester *Born:* 9/16/12 Taught at Cornell Univ, Univ of Rochester (Chrmn, Engrg Div & Assoc Dean for Graduate Studies) and Univ of RI (Dean, Coll of Engrg, and Dir, Div of Engrg, Research & Development). Non Academic employment included the Air Reduction Co Research Labs, the Nat'l Science Foundation, and the Engrg Societies Commission on Energy. Consultant to many companies and governmental agencies, author of many papers & research reports and patentee in field. Major offices in ASME included Chrmn, Diesel & Gas Engine Power Div, VP, Region III, and VP, Research. *Society Aff:* ASME, ASEE

Conti, James J
Business: Polytechnic University, L. I. Campus, Route 110, Farmingdale, NY 11735
Position: VP for Educational Dev. & Director, L.I. Campus *Employer:* Polytechnic Univ *Education:* DChE/Chem Engg/Poly Inst of Brooklyn; MChE/Chem Engg/Poly Inst of Brooklyn; BChE/Chem Engg/Poly Inst of Brooklyn. *Born:* 11/2/30. in Coraopolis, PA. Industrial experience with Westinghouse Atomic Power Div, Mobil Oil Corp and a number of other short term and consulting arrangements. With Poly since 1959, starting as Asst Prof of Chem Engg. Was dept head for 6 yrs and Provost for 8 yrs. In current position am responsible for admin and further dev of branch campus having undergrad, grad and res progs in engg. *Society Aff:* AAAS, ASEE, AIChE.

Contini, Renato
Home: 357 S Curson Ave, Los Angeles, CA 90036
Position: Sr Dev Engr. *Employer:* Univ of CA (Retired) *Education:* BSME/Aeronautics/NYU. *Born:* 6/10/04. Executive in Aeronautical and Railcar mfg organizations from the early thirties till after WWII. In charge of the design of the Super Chief, El Capt and other railcars from 1936-39. NYU, Sch of Engg and Sci in 1946 as Sr Res Scientist and Res Coordinator, Special Projects. Initiated first large scale res prog in Human Factors Engg and was elected first natl pres, Human Factors Soc, 1958. Pioneered res in Bio-engg. Retired from NYU in 1970. On staff of Div of Orthopaedics, UCLA, organizing res progs in Bioengg 1971-1981. Retired.. *Society Aff:* ASME, HFS, TBΠ, ΣΞ.

Conway, Lawrence
Business: 6001 S Westshore Blvd, Tampa, FL 33616
Position: Engg Mgr. *Employer:* Westinghouse Elec Corp. *Education:* PhD/ME/Univ of Durham; BS/ME Univ of Durham. *Born:* 5/28/36. Native of Newcastle upon Tyne England. Machinist with C A Parsons England 1952-56. Emigrated to USA in 1964. Mgr of Heat Transfer Analysis then Instrumentation Systems Dept at Brown Engg Co, Huntsville, AL 1964-67. Hired as LMFBR Systems Analyst by Westinghouse in 1967, then transferred to Tampa Nuclear Steam Generator fabrication facility as a stress analyst. Successively promoted to Mgr of Engg at Tampa Plant. Hobbies are golf and running. *Society Aff:* ASME, IMechE.

Conway, Richard A
Business: PO Box 8361, S Charleston, WV 25303
Position: Corp Dev Fellow *Employer:* Union Carbide Corp. *Education:* SM/San Engrg/MA Inst of Tech; BS/Public Health/Univ of MA. *Born:* 11/10/31. Served as Asst Preventive Medicine Officer at Fr Meade, MD 1954-5. Graduated MIT SM in Environmental Engineering. Joined Central Engg Dept of the Eng Manufacturing & Tech. Srv. Div of Union Carbide Corp. Appointed Corp Dev Fellow in 1981. Chrmn of Exec Committee of ASCE's Environmental Engg Div 1975. Reg PE. Co-authored books on ind waste treatment. Edited books on environmental risk analysis and solid waste testing. Co-recipient of WPCF's 1967 George Gascoigne Medal and 1974 and 1983 Willem Rudolfs Medal and ASCE's 1974 Rudolph Hering Medal and 1975 State-of-the-Art of Civ Engg Award and ASTM's 1984 Dudley Medal and Chemical Engg Magazine's 1986 Personal Achievement Award. Adjunct Assoc Prof at the WV College of Grad Studies. Mbr Governing Bd of IAWPRC. Mbr EPA Sc. Adv. Bd. Mbr NRC Board on Environmental Studies and Toxicology and Water Science and Technology Board. Elected National Acad of Engg 1986. Born in Weymouth, MA on Nov 10, 1931. *Society Aff:* AEEP, IAWPRC, ASCE, ASTM, WPCF, AAEE, SETAC, NAE, NWWA, AGWS&E

Conway, William B
Business: 1055 St Charles, New Orleans, LA 70130
Position: Managing Partner *Employer:* Modjeski & Masters *Education:* MS/CE/Thayer School-Dartmouth; BA/CE/Dartmouth College *Born:* 1/15/31 Mr. Conway joined the firm of Modjeski and Masters in 1957, following employment with other firms, an instructorship at Dartmouth College, and service with the US Navy Construction Battalion. Assigned to the Harrisburg office during his first 4 years with the firm, participated in planning and design of large bridge projects including first Newburgh-Beacon Bridge, Brent Spence Bridge, the second IA-IL Memorial Bridge, Acme Bend Bridges and Little River Bridge. 1961, appointed Engr in charge of firm's New Orleans design office; 1969 became the Resident Partner in New Orleans. Has led the New Orleans office in the dev of steel box girder design and in urban multiple use and design team studies. Principal in charge of design of the MS River Bridge at Luling and design of the Greater New Orleans Bridge No. 2. Since 1986, Managing Partner of firm. *Society Aff:* ASCE, ACI, AREA, IBTTA, IABSE, AISC.

Conwell, Esther M
Business: Xerox Webster Research Ctr - W114, Webster, NY 14580
Position: Research Fellow *Employer:* Xerox *Education:* PhD/Physics/Univ of Chicago; MS/Physics/Univ of Rochester; BA/Physics/Brooklyn Coll, CUNY *Born:* 5/23/22 During and after PhD, spent 5 years on the faculty at Brooklyn Coll. After a year on leave at Bell Labs, joined the technical staff at GTE Labs. Remained there, working on solid state physics (semiconductors) and then integrated optics, until 1972, except for a year (1962-63) as a Visiting Prof at the Ecole Normale Superieure in Paris. At GTE, was mgr of the Physics Dept from 1963 to 1972. During 1972, was the Abby Rockefeller Mauze Prof at MIT, after which she joined the staff at Xerox Webster Research Ctr as a principal scientist. Has worked there on integrated optics and on the solid state physics of quasi-one- dimensional organic materials. Has published over 100 papers and a book, *High Field Transport in Semiconductors.* Recreation: ballet and jogging. *Society Aff:* IEEE, APS

Conybear, James G
Business: 900 N Westwood, Toledo, OH 43696
Position: Mgr Met. *Employer:* Midland-Ross Corp. *Education:* MS/Engg Sci/Univ of Toledo; BS/Met Engr/IL Inst of Tech. *Born:* 2/2/43. in Chicago, IL, Mr Conybear joined Midland-Ross Corp in 1966. In his present position, he is responsible for equip dev related to heat treatment of metals. He serves as a consultant to Midland-Ross and their customers in matters related to heat treatment, and is responsible for identification of long range needs in heat treatment. Currently is a mbr of the ASM Advisory Technical Awareness Council, Chrmn of the Vacuum Heat Reat Subcommittee of ASM, and a mbr, Aerospace Metals Engg Committee, SAE. Active in Toledo Area Council, Boy Scouts of Am. *Society Aff:* ASM, AIME.

Coogan, Charles H, Jr
Home: RR 1, 735 Storrs Road, Storrs, CT 06268
Position: Emeritus Prof of Mech Engg. *Employer:* Univ of CT. *Education:* BS/Mech Engg/Tufts Univ; SM/Mech Engg/Harvard Univ; ME/Prof Degree/Univ of Penna. *Born:* 4/2/08. Mbr Tau Beta Pi, Pi Tau Sigma, Phi Kappa Phi. Reg PE CT. Instructor - Mech Engg Univ of Penna 1931-42. Asst Prof Mech Engg - Univ of CT 1942-46. Assoc Prof - Mech Engg Univ of CT1947-48. Prof - Mech Engg Univ of CT 1948-78. Emeritus Prof - Mech Engg Univ of CT 1978-Date. Dept Head - Mech Engg Univ of CT 1947-68. CT State Board of Reg for prof engg & land surveyors. Secretary & member 1957- 72, chmn & mbr 1972-73. Research commission state of CT - mbr - 1965-72. Consultant to industry 1955-1968. VP ASME 1959-63. *Society Aff:* ASME.

Cook, Barton B, Jr
Home: 1623 Old Trenton Rd, RR4, Trenton, NJ 08691
Position: Exec VP & Mbr of Bd of Dir. *Employer:* Transamerica Delaval Inc. *Born:* Aug 1920. BS 1943 Univ of MI. Joined Gibbs & Cox 1943. Respon incl Hd of Machinery, Scientific. 1955 joined De Laval Turbine as Marine Engr; 1961 Mgr Marine Dept; 1967 Mgr Engg; 1968 VP Mkting; 1972 VP & Genl Mgr Turbine Div; 1973 Mbr Bd of Dir; 1974 also VP & Genl Mgr Compressor Div; 1976 Group VP. 1979 Exec VP. Mbr SNAME, ASNE, ASME, AGMA, CAGI, NEMA, ABS, Newcomen Soc, Inst of Marine Engrs, Propeller Club, Hydraulic Inst, Shipbldrs Council. Author tech papers. Formerly VP & Mbr of Council SNAME & Bd of Dir Shipbldrs Council.

Cook, Billy C
Business: 5000 N Willamette Blvd, Portland, OR 97203
Position: Asst Prof *Employer:* Univ of Portland *Education:* PhD/CE/TX Tech Univ; MS/CE/Univ of So CA; BS/CE/TX Tech Coll; BS/Ed/TX Tech Coll *Born:* 3/7/30 Native of Odessa, TX. Undergraduate education at TX Tech in Lubbock, TX. Field engr for Civil Aeronautics Adm 1958. Design engr for Kenneth E. Esmond and Assocs 1959-62. Assoc Engr CA Dept of Water Resources 1962-1978. Asst Prof Civil Engrg 1978-present. Chrmn Civil Engrg program 1980 to present. *Society Aff:* ASCE, ASEE

Cook, Charles E
Business: Burlington Rd, MS N210, Bedford, MA 01730
Position: Sr Staff. *Employer:* MITRE Corp. *Education:* MEE/EE/Poly Inst of NY; SB/Phys/Harvard College. *Born:* 10/27/26. Res Engr for the Sperry Corp from 1951 to 1971. Performed basic res on the dev of complex signal processing techniques for radar systems. Holds several fundamental patents in this field. Coauthor of *Radar Signals: An Introduction to Theory and Application* (Academic Press, 1967; Soviet Radio Publishing, 1971). Coeditor of Spread Spectrum Communications, (IEEE Press, 1983), contributing author to three books on radar sys analysis. Elected Fellow of IEEE in 1972 "for contributions to signal processsing theory and radar design." Employed by the MITRE Corp since 1971. Current res interests involve the vulnerability and survivability of military communications systems. Author of technical journal papers in the fields of radar, communications, air traffic control. *Society Aff:* IEEE, ΣΞ.

Cook, Charles W
Home: 1180 Daleview Dr, McLean, VA 22102
Position: Deputy Asst Secty. *Employer:* US Air Force. *Education:* PhD/Physics/Cal Inst Tech; MS/Physics/Cal Inst Tech; BA/Physics/Univ of S. D. *Born:* 9/27/27. Hd, Nuclear Physics, Convair-San Diego, 1957-60; Chief, Ballistic Missile Def Br, Advanced Res Proj Agency, 1960-61; Corporate Dir-Electronics Res and Dev, North American Aviation, El Segundo CA 1961-67; CIA, WA, DC 1967-71; Asst Dir, Def Res and Eng (def sys), Wash DC 1971-74; Deputy under Secty (Space Sys). US Air

Cook, Charles W (Continued)
Force 1974-1979; Deputy Asst Secty (Space Plans & Policy) US Air Force 1979-present Consultant Mcgraw-Hill, Inc, NY, NY 1962-1976. Grad: Calif Inst Tech, Phd, M.S.; Oak Ridge School of Reactor Tech. Phi Beta Kappa, Sigma Pi Sigma, Sigma Xi; Secty of Def Meritorious Civilian Service Award 1974; Received Secty of Def Distinguished Service Award in 1977. Outstanding Service Awards, 1975-77. Air Force Exceptional Civilian Service award 1981, 1982; 1987; Distinguished Alumni Award Univ S. D., 1982, Superior Performance (AF), 1983. Board of Gov, National Space Club, 1983-Present; AIAA Fellow 1987. *Society Aff:* ΦBK, ΣΠΣ, ΣΞ, BΘΠ

Cook, George E
Home: Rt 6, 1203 Hood Dr, Brentwood, TN 37027
Position: Prof *Employer:* Vanderbilt Univ *Education:* PhD/EE/Vanderbilt Univ; MS/EE/Univ of TN; BE/EE/Vanderbilt Univ *Born:* 4/4/38 Subject and research specialties are welding automation, robotics, automatic controls and signal processing. Prof of Electrical Engrg at Vanderbilt Univ. Has been granted over 60 US and foreign patents on automatic controls and automatic welding inventions. Has presented numerous papers at conferences, symposiums, etc, in US and Europe. Has published over 120 technical articles. Member of Phi Eta Sigma, Tau Beta Pi, Eta Kappa Nu, Phi Kappa Phi, and Sigma Xi honorary societies. *Society Aff:* IEEE, NSPE, AWS, SME, ASM, ΣΞ, ASME

Cook, George J
Business: 1230 NE 3rd St, Bend, OR 97701
Position: Bd (Chrmn). *Employer:* High Desert Engg, Inc. *Education:* BS/Forest Engg/OR State Univ. *Born:* 11/20/31. Native of Coopersville, MI. Korean War Veteran. Grad of Forest Engg, OR State Univ. Worked as Logging Engr in timber industry until 1964. Owner/Mgr of George Cook Engg, Inc. Principal firm in the land dev of Central OR. Formed High Desert Engg, Inc in 1979. State Chrmn, Profl Land Surveyors, 1974. Appointed OR State Bd of Engg Examiners, 1975; NCEE, 1975, NCEE Land Survey Committee 1975- 1979. *Society Aff:* NCEE, ASCM, PEPP.

Cook, Harry M, Jr
Business: 2623 East Pershing Rd, Decatur, IL 62526
Position: Exec VP *Employer:* Blank, Wesselink, Cook & Assocs, Inc *Education:* BS/Gen Engrg/Univ of IL *Born:* 3/4/43 1966 graduate of the Univ of IL. After graduation, short term with Allied Chem, Solvay Process Div in Syracuse, NY as a Mechanical Engr, 3 years with the US Army in the Corps of Engrs. From 1969 to date, engrg consultant assignments for Blank, Wesselink, Cook & Assocs, Inc. These assignments were predominatly in Process and Industrial Operations ranging from small projects to those up to 20 million dollars construction cost. *Society Aff:* ASME, ISPE, AIPE, NSPE

Cook, John P
Home: R R 1 Mine Rd, Pennington, NJ 08534
Position: Sr Proj Engr. *Employer:* Ingersoll Rand Res Inc. *Born:* 1943. Native Phila area. BS Met Engg Drexel Univ 1966. 3 yrs E I duPont de Nemours Inc, Explosive Bonding R&D; 5 yrs Hoeganaes Corp, Powder Metal Forming R&D - Supervised forging lab, presented & published many tech papers on powder forming; currently Ingersoll Rand Res Sr Proj Engr. Corresponding Secy Phila APMI Chap; Mbr ASM Metalworking Ctte Powder Met.

Cook, Joseph O
Home: 1023 Dead Run Dr, McLean, VA 22101
Position: Admin, Coal Mine Safety & Health. *Employer:* Govt - Dept of Labor. *Education:* BS/Mining Engg/VPI & SU. *Born:* 9/2/30. Mr Cook worked for about 15 yrs as a laborer, mining, engr, chief mining engr, and mine mgr. He went to work for the Govt in 1968 and has held the position of mining engr, supervisory mining engr, and Asst Admin in Coal Mine Safety and Health. He has a degree in mining engg and is a Reg Profl Safety Engr, Reg Profl Mining Engr, and Certified Mine Foreman. *Society Aff:* AIME.

Cook, Leland B
Business: PO Box 1526, Tupelo, MS 38802
Position: VP. *Employer:* Cook Coggin Engrs Inc. *Education:* BSCE/-/MS State Univ *Born:* June 1920 Lee Cty MS; Bach of Sci CE MS State Univ 1942. Ingalls Shipbldg Corp 1942-44. USN 1944-46 in ship repair. Assoc with William E Johnson, cons engr of Jackson, MS 1946-47; private practice of engg at Tupelo, MS 1948; Principal of Cook Coggin Engrs Inc, cons of Tupelo since 1950. Principal of Fdn Services Inc and Planning Consultants Inc. P Mbr of State Bd of Reg for PE. P Pres Cons Engrs Council of MS. P Pres of state PEPP branch of NSPE. P Pres MS State Univ Engg Alumni Assn. *Society Aff:* NSPE, ASCE, AWWA, ACEC

Cook, Leonard C
Business: P.O. Box 97, Rt 250, Frank, WV 24937
Position: Dir of Pollution Control *Employer:* Howes Leather Co, Inc *Education:* MS/Engrg/Univ of Pittsburgh *Born:* 8/28/49 Native of Irwin, PA. Taught for PA Dept of continuing education preparing sewage treatment plant operations for State Certification. Studied Civil and Sanitary Engrg at Univ of Pittsburgh. Consulting sanitary engr for private consulting firm - Duncan, Lagnese and Assoc for 6 years (1974-1980). Registered PE in PA, MA, and WV. Currently responsible for directing all pollution control activites within The Howes Leather Organization at The Leather Tanneries in PA, MA and WV; and the development of Lotus 1-2-3 and Symphony Templates for engrg, operation and maintenance, industrial time study, and environ monitoring and reporting applications. Enjoy sailing, downhill skiing and computer programming. *Society Aff:* NSPE, WPCF

Cook, Michael C
Business: PO Box 529100, Miami, FL 33152
Position: Group VP *Employer:* FL Power & Light Co. *Education:* MBA/Mgt/CUNY; BChE/CE/CUNY. *Born:* 9/28/39. Native of NYC. Proj Engr with US Atomic Energy Commission, 1960-1965. Contract administrator in Engg Dept of Mobil Oil Corp, 1965-1967. Investment analyst, investment banker & consultant, 1967-1972. Joined FL Power & Light Co in 1972 & served as Treas until 1977. Elected VP in 1977 Group VP in 1984, responsible for Fuel, Corporate Dev, contracts, power supply & sys planning & Nuclear Engg. Dir & officer of Jr Achievement of Greater Miami. *Society Aff:* AIChE, FEI.

Cook, Nathan H
Business: Mtl Processing Lab, Cambridge, MA 02139
Position: Prof of Mech Engg. *Employer:* MIT. *Born:* 1925 Ridgewood NJ. 3 war yrs USN 1943-46 serving on a Destroyer in the Pacific Ocean. SB 1950, SM 1951, ME 1954, ScD 1955 MIT. Appointed Asst Prof of Mech Engg MIT 1953, Assoc Prof 1959, Full Prof 1965. Teaching, res & cons have been primarily in the genl areas of mtls, mtls processing, metal removal methods, machine tools, controls, vibration, applied mech, friction, wear, and instrumentation. Author or co-author of books & some 45 publ tech papers. Mbr Transactions Ctte SME; Adv to the Birla Inst of Tech & Sci, Pilani, Rajasthan India & Univ of Benin Nigeria. Mbr ASME, ASEE, SME, Sigma Xi, Pi Tau Sigma, Tau Beta Pi. In 1974, won the ASME Blackall Award. Reg PE MA & a Cert Mfg Engr. Since 1970, he & family have lived in a MIT dorm where he serves as "Housemaster–.

Cook, Robert L
Business: 704 Porter Waggoner Blvd, West Plains, MO 65775
Position: Pres. *Employer:* NA - The Sunni Co. *Education:* Masters/Construction-Mgt/Univ of MO; BS/Civ Engg/Univ of MO. *Born:* 8/6/48. As Pres of Sunni Co, design and construct steel and conventional bldg projs with a current staff of thirty employees. Specialty: Commercial and ind bldgs, dairy facilities, and bridges. Need for Engg/Construction services in a high growth, rural area provides challenging projs. Design Engr for MSHD - Bridge Div and consulting engr

Cook, Robert L (Continued)
for construction cos in metro areas prior to forming the Sunni Co. Personally enjoy fishing and tennis in rural area. *Society Aff:* NSPE, ASCE.

Cooke, Francis W
Business: 301 Rhodes Bldg, Clemson, SC 29631
Position: Dept Head *Employer:* Clemson Univ *Education:* PhD/Mat Sci/RPI; BS/Metallurgy/Notre Dame Univ *Born:* 11/3/34 After receiving BS degree from Univ of Notre Dame, in 1957, received an appointment in the Metallurgy Div of Oak Ridge Nat'l Lab 1957 to 1961: materials selection, design and failure analysis. Research Asst RPI 1961 to 1965: thermodynamics and electrical properties for ferroelectric ceramics. Mgr of Metallurgy Lab, Franklyn Inst Research Labs, Philadelphia, PA 1965 to 1970: thermomechanical processing of nonferrous metals, bio-materials. Head, Dept of Interdisciplinary Studies, Clemson Univ, 1970-82; Prof of Bioengrg & Materials Engrg 1982-present: biomaterials orthopedic implants, prosthesis attachment and bioethics. *Society Aff:* SFB, ASM, ORS, BES.

Cooke, James L
Business: 4400 Pt Arthur Rd, Beaumont, TX 77710
Position: Prof. *Employer:* Lamar Univ. *Education:* PhD/EE/Northwestern; MS/EE/TX; BS/EE/TX Tech. *Born:* 9/20/29. Profl Experience: Elec Engg Dept, Lamar Univ, Beaumont, TX. (Asst Prof 1956, Assoc, 1960, Prof 1962-). Elec power, machinery and automatic control; Consulting Engr, Gulf States Utilities Co, Beaumont, TX 1956-. Computer Applicattions to: (a) power system, (b) economic systems operation; Consultant, Texaco, Pt Arthur, TX 1967-71. Digital control applications to refinery problems; Instr US Army Signal corps, Camp Gordon, GA, 1954, FM Radio Theory. Engr, Southwestern Public Service Co, Amarillo, TX, 1952. Relay coordination, load-flow studies. Awards: Regents' Prof (1972), Sigma Xi, Tau Beta Pi, Eta Kappa Nu; listed Am Men of Sci. *Society Aff:* IEEE.

Cooke, Norman E
Business: PO Box 10, Complexe Desjardins, Montreal, Quebec H5B 1C8 Canada
Position: Manager of Tech Dev Dept and Chief Process Engr *Employer:* SNC Inc *Education:* ScD/ChE/MIT; MASc/ChE/UBC; BASc/ChE/UBC. *Born:* Aug 30, 1922. ScD MIT; BA Sc, MA Sc UBC. Assoc Scientist Fisheries Res Bd Canada 1946-50. Lt Engrs Korea 1950-52. Principal Chem Engr 1956-72, Tech Dev Mgr 1972-74 for CIL. Ch Process Engr SNC Inc 1975. Mgr Technical Dev 1979-. Fellow CIC, AIChE. EIC, Mbr CSChE, ORSA, MEAC. Sigma Xi. Reg Engr Quebec. Offices CSChE: Secy-Treas 1964-67, VP 1967-68, Pres 1968-69. Chrmn Ed Bd CJChE 1970-74. Served on 17 (Chrmn of 4) Govt & assn cttes dealing with environ. Auxilary Prof, Chem Engr McGill 1963-. Indus Waste Medal WPCA 1960. 37 papers, 2 pats.

Cooke, Robert S
Home: 2519 Humble, Midland, TX 79705
Position: Petroleum Conslt *Employer:* Self employed *Education:* BS/Chem Engg/Rice Univ. *Born:* 8/3/18. A native of Kansas City, MO. Was with the Carter Oil Co in various petroleum engrg assignments in several mid-continent locations from 1939-42 and 1946-52. Was a meteorologist in the Army Air Corps from 1943-45. Joined Union Oil Co of CA in 1952, serving as Regional Engrg Mgr from 1966-1981 and as Engrg Mgr of Union Oil Co of Great Britain in London 1982-1983. Now an independent conslt following retirement from Union. A past Dir of AIME and Soc of Petroleum Engrs of AIME. *Society Aff:* AIME.

Cooke, William T
Home: 5901 Dovetail Dr, Agoura Hills, CA 91301
Position: Retired. *Education:* MA/Phys/Univ of MI; BA/Phys/Univ of MI. *Born:* 11/3/03. 1921-24 Oberlin College. 24-26 Univ of MI. 26-28 Columbi Univ Lectr, Res. 28-34 Columbia Res, Hunter College Instr. 34 Published Thesis, Joined Sperry Gyroscope as Special Proj Engr. 38-41 Stanford Univ & Sperry Lab, Dev of Klystrons, microwave radar, aircraft landing. 41-45 Dir Radar Engg. Produced design for first gunsight microwave radar for Navy aircraft. 45-58 Dir for Radio & Radar Engg. 58-65 Asst to VP Res & Dev in charge Progs & Budget. Retired 1965. *Society Aff:* APS, IEEE, ΣΨ, EX.

Cookson, Albert E
Business: 320 Park Ave, New York, NY 10022
Position: Sr VP, Genl Tech Dir. *Employer:* IT&T Corp. *Born:* Oct 1921. BSEE Northeastern; MSEE MIT; ScD (Hon) Gordon Coll. With ITT since 1951. Currently Sr VP & Genl Tech Dir. 1942-46 Lt jg USN Radar Officer 1947-51, Res Lab of Electronics MIT. Tau Beta Pi, Sigma Xi; Natl Space Club; AFCEA; Northeastern Univ Natl Council; Indus Panel Sci & Tech, Natl Sci Fdn; Fellow IEEE.

Cooley, Wils L
Business: Electrical Engrg Dept, P.O. Box 6101, Morgantown, WV 26506-6101
Position: Prof *Employer:* WV Univ *Education:* PhD/EE/Carnegie-Mellon Univ; MS/EE/Carnegie-Mellon Univ; BS/EE/Carnegie- Mellon Univ *Born:* 8/23/42 Taught in Electrical Engrg and Biotech programs at Carnegie-Mellon Univ from 1968 through 1973. Joined faculty of WV Univ in 1973, was promoted to Prof in 1980, and recently assumed position as Assoc Chrmn of Electrical and Computer Engrg. Performed extensive research in mine power system safety. Author, over 40 papers on engrg topics, including two engrg design guides for mine power system construction. Currently studying mine electrical accidents and developing ways to reduce electrical hazards. Teaching energy designed and project mgmt. *Society Aff:* IEEE, ASEE, ΣΞ

Coon, Arnold W
Home: 3679 Oakview Dr, Salt Lake City, UT 84117
Position: Chrmn of the Bd. *Employer:* Coon, King & Knowlton. *Education:* BSCE/Civ Engg/Univ of UT. *Born:* 9/24/25. Native of Salt Lake City. USN Air Corps 1943-46. Asst Proj Engr for UT Constr Co on tunnel proj 1949-50. Struct Engr for H C Hughes 1950-54. Estab own struct cons firm in 1954. Formed ptnrship with C V King 1956 as cons civ engrs. Formed corp of Coon, King & Knowlton, Cons Engrs in 1962 in order to expand & provide greater service. Pres CEC UT 2 terms. Dr CECUS 2 terms. Recipient of Engg excell award for UT 1975. Formed company known as "Forensic Engrg" in March 1982. Maj work at present time involved investigation of accidents, structural failures & product liability - acting as expert witness in related litigation. *Society Aff:* ACEC.

Coonrod, Carl M
Business: 650 Westdale Dr, PO Box 9166, Wichita, KS 67277
Position: Chrmn of the Bd, Chief Exec Officer *Employer:* Coonrod and Walz Const Co, Inc *Education:* BS/Arch Engr/KS State Univ *Born:* 06/03/25 Native of KS. Served US Army 1943-45. Mayor of Maize, KS 1957. Present AGC of KS 1966, Chrmn of lst Natl Bank of Derby, KS 1978 to present. Outside interest include ranching hunting fishing and golf. *Society Aff:* AGC

Cooper, Arthur W
Business: 2590 Windy Hill Place, Auburn, AL 36830
Position: Agricultural Engr (Self-employed) *Education:* PhD/Ag Engg/MI State U; MS/Ag Engg/Auburn; BS/Ag Engg/Auburn. *Born:* Mar 1918. PhD MI State Univ; BS & MS Auburn Univ. Native of AL. USN 1945-45. Res & teaching Auburn Univ 6 yrs; Purdue Univ 3 1/2 yrs. Affiliated with soil conservation serv & Agri Res Serv USDA since 1949. Currently Agricultural Engr - Consultnt in Agri Res Adm & Agri Engr. John Deere Medal 1969. Hon VP ASAE representing ASAE to 12 countries 1964. Fellow ASAE; Fellow AAAS; Mbr Sigma Xi, Sigma Pi Sigma, Gamma Sigma Delta, Tau Beta Pi. *Society Aff:* ASAE.

Cooper, Bernard E
Business: 5017 Wythe Ave, Richmond, VA 23226
Position: Engr (Self-employed) *Education:* BS/Mech/VPI; BS/Elec/VPI. *Born:* 1/18/22. Petersburg, VA; 1922 US Army Coast Artillery 1943-44; Corps of Engrs 1944- 46 (Capt) QMC-Res 1948-55. BS in EE and BS in ME VPI, 1947. Employed

Cooper, Bernard E (Continued)
by Experiment Inc 1948-50 as Design Engr Carneal & Johnston, Architects & Engrs as design engr (1950-62); Chief mech design (1954-62). Began practice as Bernard E Cooper; Cons Engr 1962. Engaged in mech and elec engg design for bldgs and structures. Mbr: ASME, ASHRAE, ASPE, ACEC, NSPE. Reg Engr: VA, WV, MD, PA, OH; NCEE Cert. State Air Pollution Tech Advisory Bd. State Bldg Code Tech Review Bd. Forensic Engr Expert Witness. *Society Aff:* ASME, ASHRAE, ASPE, ACEC, NSPE.

Cooper, Earl D
Home: 3004 Emerald Chase Drive, Herndon, VA 22071
Position: Management Prog Chrmn, National Capital Region and Research Prof Medical Research Inst *Employer:* Florida Inst of Tech *Education:* DPA/Mgt/Nova Univ; BEE/EE/Geo Washington Univ. *Born:* 4/16/26. BEE from GWU; Grad work in EE & Mgt Univ of MD & GWU; DPA from Nova Univ, Native of Arlington VA. 1948-55 Journeyman Engineer in Navy Bureau of Ships. 1956-65 Started as Proj Engr in Navy Bureau of Ordnance for a guided missile dev proj & rose to Tech Dir Air Weapon Sys, Bureau of Naval Weapons. 1966 to 1979; Tech Dir Advanced Sys, Naval Air Sys Command. 1979 to 1983; Technical Director for Research and Technology, Naval Air Sys Command; 1983 to 1984 Ass't Commander for Res and Tech, Naval Air Systems Command. Member of the Senior Executive Service. Exper Summary: Over 36 yrs as a civilian employee of the Navy Dept with 23 yrs plus exper in a supergrade job featuring the engg mgt of a broad spectrum of lrg res & dev proj's, respon for lrg expenditures of money & leadership of lrg, complex terms of Navy Hdqtrs & lab, & private indus people. 1985-Date Management Program Chairman and Research Prof, Medical Research Institute, Florida Inst of Tech. *Society Aff:* ASPA.

Cooper, Franklin S
Home: 5 Parsell Lane, Westport, CT 06880
Position: Assoc Research Dir *Employer:* Haskins Labs, Inc *Education:* PhD/Physics/MIT; BS/Eng Physics/Univ of IL; DSc/Honorary/Yale Univ. *Born:* 4/29/08 From doctoral work at MIT, Cooper went to Gen Elec Research Labs (high voltage engrg), then joined Caryl P. Haskins in founding the Haskins Labs (1939). During WWII, Cooper was in, and later headed, the Liaison Office, Office of Scientific Research and Development. At war's end, he launched a program of research on speech perception and production that is still underway at the Haskins Lab. Public service included work with United Nations Atomic Energy Commission Group; advisory services to Nat'l Research Council, Nat'l Inst of Health, and various agencies in the Dept of Defense, also, adjunct affiliation with Columbia, Yale and the Univ of CT. *Society Aff:* ASA, IEEE, NAE, AAAS, CASE

Cooper, Geoffrey F
Home: 220 St Germain Ave, Toronto, Ontario M5M 1W1 Canada
Position: Supervisor, Harvesting Sys Lab. *Employer:* Massey-Ferguson Ltd. *Education:* MA/Eng Sc/Oxford Univ, Eng; BA/Eng Sc/Oxford Univ, Eng. *Born:* 1923. Oxford Univ England; BA Engg Sci 1944 & MA 1948. Sea-going Engr Officer Royal Navy 1944-47. Inspecting Engr 1947-52 London, on railway, harbor and automotive equip. Emigrated to Canada 1952, joined Massey-Harris Co Winnipeg as serviceman; to Engg Dept Toronto 1953. Var pos concerned with res, design, dev, & testing of harvesting equip, particularly crop-processing mech of combines. Reg PE Ontario. Mbr ASAE, Chrmn Power & Machinery Div 1974-75. Mbr Inst of Agri Engrs. Other ints: amateur music-making, boat-bldg, language study, sports. *Society Aff:* ASAE, Inst AgE, PEng Ont.

Cooper, George R
Business: P.O. Box 4620, Norton AFB, San Bernardino, CA 92409
Position: Prof Emeritus of Elec Engg. *Employer:* Purdue Univ. (Retired) *Education:* PhD/EE/Purdue Univ; MSEE/EE/Purdue Univ; BSEE/EE/Purdue Univ *Born:* Nov 1921. BSEE, MSEE & PhD Purdue Univ. Retired fac mbr Purdue Univ since 1949 and formerly Prof of EE & Coordinator of the EE Grad program. Teaching & res are in area of communication theory. Has had extensive cons exper in indus & govt. Author or co-author of 7 bks & num tech papers. Mbr Tau Beta Pi, Eta Kappa Nu, Sigma Xi, Sigma Pi Sigma, ASEE, & Fellow IEEE. Currently a consultant to industry. *Society Aff:* ASEE, IEEE, TBΠ, ΣΞ, ΣΠΣ, HKN.

Cooper, H Warren, III
Home: 7211 Windsor Ln, Hyattsville, MD 20782
Position: VP Operations *Employer:* Historical Electronics Museum *Education:* MS/EE/Stanford Univ; BS/EE/NM State Univ. *Born:* 7/19/20. WWII Hd OSS radio Comm station in Kandy (Ceylon). Microwave & antenna engg at Airborne Instruments Lab 1948-53; hd antenna sect, then Dir R&D for MD Electronic Mfg Corp (became Litton AMECOM) 1954-58; with Westinghouse-Baltimore (1958-Ret'd 1986) as Mgr R&D Prog, Mgr Strat opers, Mgr Space Systems, & Mgr Electromagnetic Tech Lab devoted to R&D in microwave ckts & semiconductor techniques, acoustic waves at microwave freqs, aircraft instrument landing systems. Elected Fellow IEEE 1970. President, 1975, IEEE Microwave Theory & Techniques Society. VP Operations & mmbr Bd Dir Historical Electronics Museum, Baltimore, Maryland from 1986 to present. President 1986-1987 IEEE Aerospace & Electronic Systems Soc. 15 patents in microwaves, surface acoustic waves, aircraft navigational systems, as well as papers. Interested in sailing & woodworking. *Society Aff:* IEEE.

Cooper, Hal B H
Home: 4234 Chevy Chase Dr, Flintridge, CA 91011
Position: Pres. *Employer:* Cerco Inc. *Education:* PhD/CE/IA State Univ; MS/CE/MIT; BS/CE/IA State Univ. *Born:* 11/4/14 Early life & educ in Panama Canal Zone. Taught in CE Dept at IA State Univ prior to WWII. Profl exp has been with following organizations: Am Cyanamid Co 1942-52 Tech Dir & Asst Ch Chem Engr.; Colgate-Palmolive Co 1952-55 Asst Dir R&D Dev.; Am Potash & Chem Co 1955-63 Dir Dev & Engrg.; Chemanox Inc 1963-73 Pres and Founder; Cerco Inc. 1973-date Pres & Technical Dir. Currently engaged in the dev & promotin of new tech in power plant emissions control, electrochem, combustion & minerals processing fields. *Society Aff:* AIChE, ACS, APCA, AAAS

Cooper, Harrison R
Home: 1450 S 1425 E, Bountiful, UT 84010
Position: Pres. *Employer:* Harrison R Cooper Systems Inc. *Education:* PhD/CE/Univ S CA; MS/CE/Univ of Houston; BS/CE/Lamer State College; BS/Chemistry/Univ of KY. *Born:* 11/12/29. Native of Milwaukee, WI. Employed successively by B F Goodrich Chem Co, Fluor Corp, IBM Corp, & Kennecott Copper Corp prior to forming mfg & service organization in 1970. Harrision R Cooper Systems specializes in equip for process control & automatic process measurements, mainly applicable to the mineral processing industries. The principle products are automatic sampling equip & on-stream mineral slurry analysis. *Society Aff:* SocMinEngg, AIME, AIChE.

Cooper, Martin
Home: 100 Beach Rd, Glencoe, IL 60022
Position: VP, Dir R&D. *Employer:* Motorola, Inc. *Education:* MSEE/EE/IL Inst of Tech; BS/EE/IL Inst of Tech. *Born:* 12/26/28. in Chicago, IL; Officer in US Naval Reserve, active duty 1950-53. Res Engg, Teletype Corp, 1953-54. Motorola, Inc, 1954-present. VP, 1960--; Gen Mgr, Communications Systems Div, 1977-78; Dir of R&D, 1978--. Fellow -IEEE (for contributions to telephony). Pres, Vehicular Tech Soc. Mbr, Bd on Computer/Telecommunications Applications of Nat Res Council. Patented in field (6 patents). Contributions in spectrum management, quartz crystal tech, mobil and portable FM 2-way radio. *Society Aff:* IEEE, HKN, NRC.

Cooper, Norman L
Home: 1403 Pennington Lane, Bowie, MD 20716
Position: Policy Staff *Employer:* US Dept of Transportation. *Education:* Master of Arts/Government/IN Univ; BS/Civil Engg/Univ of TX. *Born:* 10/20/36. From 1971, Office of the Secretary, US Dept of Transportation: Coordinate development and implementation of national transportation legislation, regulation, and other policy on land use and air quality. Chairman, Transportation Research Board Air Quality Committee. Past President, National Capital Section, American Society of Civil Engineers. Elected to Bowie, Maryland City Council, from 1978. Except for 1967 service in Israel as United Nations International Consultant, earlier career was in field and headquarters of Federal Highway Administration: Co-author of urban policy book on highway planning, location, and design, published by American Assoc of State Highway and Transportation Officials and adopted as Federal policy. *Society Aff:* ASCE.

Cooper, Peter B
Business: Dept of Civ Engg, Seaton Hall, Manhattan, KS 66506
Position: Prof. *Employer:* KS State Univ. *Education:* BS/Civ Engrg/Lehigh Univ; MS/Civ Engrg/Lehigh Univ; PhD/Civ Engrg/Lehigh Univ. *Born:* 3/30/36. Native of Colebrook, CT. Engr in Naval Architecture Dept, Elec Boat Div, Gen Dynamics 1957-58. Res Asst, Lehigh Univ 1958-60. Res Instr, Lehigh Univ, 1960-65. Res Asst Prof, Lehigh Univ 1965-66. Joined faculty at KS State Univ 1966. Teach courses primarily in structural analysis & design. Many published papers on strength & behavior of plate girders & beams with web openings. Chrmn, Structural Div Exec Ctte, ASCE, 1980. Chrmn Mgmt Group B, ASCE, 1986. Reg P E in KS. Favorite pastime - barbershop singing. *Society Aff:* ASCE, ASEE, SSRC

Cooper, Robert H
Home: 301 N Joslin, Charles City, IA 50616
Position: VP Production. *Employer:* Salsbury Labs. *Education:* BS/CE/IA State Univ. *Born:* 12/31/17. Native of Boone, IA. Married with 3 children, 2 sons, 1 daughter. Joined Salsbury Labs upon grad from IA State as Chem Engr. After variety of production assignments, named VP Production in 1961. Responsible for Mfg of Chem and BIological Products, Warehousing, Shipping, Purchasing, and Plant Engg. Was a founder of the IA Chapter of AIChE and its first pres. Named Fellow of AIChE in 1977. Have been active in community affairs including Chamber of Commerce and school bd. Enjoy most sports and playing bridge. *Society Aff:* AIChE.

Cooper, Thomas D
Home: 542 Rader Dr, Vandalia, OH 45377
Position: Chief, Mtls Integrity Branch. *Employer:* Mtls Lab, AF Wright Aeronautical Labs *Education:* MS/Met Engg/OH State Univ; MetE/Met Engg/Univ of Cincinnati. *Born:* 4/7/32. Engr, Westinghouse Elec Corp, 1955-1956. Officer, USAF, 1956-1958, AF Mtls Lab, remained as civilian, 1958-present. Served as Proj Engr, Sec Chief, Technical Mgr. Currently Chief, Mtls Integrity Branch, Systems Support Div. Served as Chrmn of the following committees: AIME High Temperature Alloys, AIME Structural Mtls, Dayton ASM Chapter, ASM Mtls Testing and QC Div, ASM Handbook Ctte, AIAA Mtls Technical Committee, ASNT Aerospace Ctte, SME Engg Processes Subdiv, SAE Aeronautical Materials Div. Fellow, ASM, ASNT and Assoc Fellow, AIAA. Received Distinguished Alumnus Award, College of Engg, Univ of Cincinnati (1972). Reg PE, State of OH. Technical specialties; Aero Mtls Applications, Nondestructive Evaluation, Failure analysis. *Society Aff:* ASM, AIME, AIAA, ASNT, ΣΞ.

Cooper, William E
Business: Teledyne Engrg Services, 130 Second Ave, Waltham, MA 02254
Position: Consulting Engr. *Employer:* Teledyne Engg Services. *Education:* PhD/Engrg Mechanics/Purdue Univ; MS/ME/OR State; BS/ME/OR State. *Born:* 1/11/24. GE (KAPL) 1952-63; responsible for structural integrity of naval reactor power plants. Lessells and Assoc and Teledyne since 1963, consulting in the design and analysis of mech systems and structures, was VP 1968-76. Purdue Distinguished Engg Alumnus, 1973; Fellow-ASME, 1972; ASME Pressure Vessel and Piping Medal 1983; ASME B F Langer Nuclear Codes & Stds Award, 1978; SESA William M Murray Lectureship, 1977; Pressure Vessel Res committee Certificate of Appreciation. ASME activities include Council on Codes and Stds 1981- , Sr VP and Chrmn (1981-84). Policy Bd - Codes and Stds (1972-81) Chrmn 1980-81), Nuclear Codes and Stds Committee (1974-80) (Chrmn 1975-77); Committee on Budget (1978-80) (Chrmn 1979-80); Ctte on Legal Affairs (1983-)(Chairman 1986- .) Boiler and Pressure Vessel Committee (1955-80) Honorary Member, 1981. ASME Codes & Standards Medal, 1986; US National Academy of Engineering, 1985. *Society Aff:* ASME, SEM, NAE.

Cooperrider, Neil K
Business: Mechanical Engineering Dept, Tempe, AZ 85287
Position: Prof *Employer:* Arizona State Univ *Education:* PhD/ME/Stanford Univ; MS/ME/Stanford Univ; BS/ME/Stanford Univ *Born:* 12/18/41 Native of CA. GE Res and Development Ctr 1968-70. Rensselaer Polytechnic Inst as Asst Prof, Machines and Structures Div 1970-73. Joined AZ State Univ in 1973 as Assoc Prof, promoted to Prof in 1977. Res interests in rail vehicle dynamics, vehicle mechanics, nonlinear sys dynamics, and control sys. Teaching activities in design, dynamics and control theory. Conslt to indus and government through Acorn Assocs, a firm he co-founded in 1974. Humboldt Fellow with DFVLR in Oberpfaffenhofen, W. Germany, 1978-79. Bd mbr, IAVSD. Author or co-author of more than 40 technical papers and reports. *Society Aff:* ASME, IAVSD

Copeland, Norman A
Business: 1007 Market St, Wilmington, DE 19898
Position: Dir. *Employer:* Du Pont Co. *Education:* PhD/CE/Univof DE; MS/CE/Univ of DE; BS/ME/MIT. *Born:* 8/16/15. 1973-1978 Sr. V.P., Exec Cttee, Bd of Dirs; 1978-present Bd of Dirs, Du Pont Co; 1970-73 - Chief Engr, Du Pont; 1965-70 - Asst Chief Engr; 1963-1965 - Asst Gen Mgr, Film dept; 1951-1963 - Tech Supt, Plant Mgr, Asst Dir Mfg, Dir of Res, Film Dept; 1937-1951 - Dev Engg, Design Proj Mgr, Engg Dept. Born Aug 16, 1915, Mercer County, OH. Married Gladys Tucker, 1949. Two sons - Eric and Terry. 1978 to present Bd of Dirs. *Society Aff:* AAAS, AIChE, NAE.

Copeland, William D
Business: Grad Sch, Golden, CO 80401
Position: Dean of Grad Sch. *Employer:* CO Sch of Mines. *Born:* Mar 1934. BA Math Carleton Coll 1956; PhD Met Engg Univ of MN 1966. US Naval Officer 1956-60. Joined CO Sch of Mines as Asst Prof of Met Engg 1966. Presently Dean of the Grad Sch & Prof of Met Engg. Am Council of Educ Admin Fellow 1970-71. Chrmn CO Ctte on Student Loan Prog 1972. Presently Chrmn CO Tech Ctte on Formula Budgeting.

Coppelman, Daniel P
Business: Deer Park Office Plaza, Katonah, NY 10536
Position: VP/Principal *Employer:* William A. Keane Assoc, P.C. *Education:* BS/CE/Northeastern Univ *Born:* 12/29/48 Dir of Operations of a consulting engrg firm dealing chiefly in Civil, Sanitary, and Environmental Engrg Projects with an emphasis on design relating to Land Development (Residential, Commerical, Institutional, Industrial). Work history includes project engrg for major condominium developments (in excess of 1,000 units), retail shopping centers, corporate headquarters, tract subdivisions, advanced waste water treatment facilities, water resource and treatment, storm water management programs, feasibility studies, and expert testimony at meetings and hearings. *Society Aff:* NSPE, ASCE, NCEE, WPCF, APA

Coppinger, John T
Business: Engrg Design Graphics Dept, College Station, TX 77843
Position: Prof *Employer:* TX A&M Univ *Education:* DED/Arch/TX A&M Univ; MS/ME/TX A&M Univ; BS/ME/TX A&M Univ *Born:* 1/28/43 Dr. Coppinger has worked as a project engr in the oil tool industry where his primary responsibilities were machine design and testing. In 1968, he joined the staff at TX A&M Univ where he has developed and taught such diverse courses as Computer Graphics,

Coppinger, John T (Continued)
Machine Design & Robotic Applications. He has published numerous articles and graphics workbooks. He has held nat'l offices in both the American Society for Engrg Education and the American Inst for Design and Drafting. In the State of TX, he is a registered PE, a Master Electrician and a Peace Officer. His hobbies include photography, pistol shooting and soccer. *Society Aff:* ASEE, ASME, SME.

Copulsky, William
Business: 1114 Ave of the Americas, New York, NY 10036
Position: VP. *Employer:* W R Grace & Co. *Education:* BS/Indus Chemistry/NYU; PhD/Bus Admin/NYU. *Born:* 4/4/22. US Army 1942-46. Consultant 1946-51. VP Office for Operations W R Grace & Co 1951-. Author Practical Sales Forecasting 1971; Co-author Entrepreneurship & the Corp 1974; Dir Chem Indus Assn; Ed Bd Indus Mkting Mgt; Adj Assoc Prof Baruch College 1966-76. *Society Aff:* AIChE, ACS, ASA, AMA.

Corbalis, James J, Jr
Business: 8560 Arlington Blvd, P.O. Box 1500, Merrifield, VA 22116
Position: Engr-Dir *Employer:* Fairfax County Water Authority (VA) *Education:* B/CE/Sanitary Engrg/Manhattan Coll *Born:* 7/13/19 In current position since 1958, directing provision of public water supply service in northern VA segment of metropolitan Washington, DC area. 1953-58, Engr-Dir, City of Alexandria, VA Sanitary Authority directing design, construction and operation of wastewater collection and treatment facilities for City and adjacent areas of Fairfax County. 1941-53, Sanitary Engr, VA State Dept of Health and Fairfax County. During WWII, served in Civil Engr Corps, US Navy. Currently, VA Commissioner on Interstate Commission on the Potomac River Basin. Registered PE in VA. *Society Aff:* NSPE, ASCE, AAEE, AWWA, WPCF, APWA, INSA.

Corbato, Fernando J
Business: 545 Tech Sq, Cambridge, MA 02139
Position: Assoc Head for Computer Science and Engrg *Employer:* MIT-Elec Engg & Comp Sci Dept. *Education:* PhD/Physics/MIT; BS/Physics/Cal Inst Tech *Born:* July 1, 1926. BS Caltech 1950; PhD MIT 1956 Phys. Dept of EE & Comp Sci MIT: Assoc Prof 1962-65; Prof 1965-; Assoc Hd for Comp Sci & Engg 1974-78, 1983- ; Cecil H. Green Prof of Comp Sci & Engg 1978-80; Director of Computing and Telecommunication Resources 1980-83; Founding Mbr MIT Lab for Comp Sci 1963- MIT Computation Ctr 1956-66. Developer of Compatible Time-Sharing Sys (CTSS) & Multics Sys. Natl Acad of Engg 1976. Am Acad of Arts & Sci 1975. Inst of Elec & Electronics Engrs, Fellow 1975. Assn for Computing Machinery. Am Phys Soc. Sigma Xi. AAAS. Tau Beta Pi. W W McDowell Award IEEE 1966; Harry Goode Memorial Award (AFIPS) 1980. Computer Pioneer Award (IEEE Computer Society) 1982. *Society Aff:* AAAS, AAAS, ACM, IEEE, NAE, ΣΞ

Corbitt, Robert A
Business: 2000 Clearview Ave, NE Ste 200, Atlanta, GA 30340
Position: Division Mgr *Employer:* Jordan, Jones & Goulding, Inc. *Education:* MS/Sanitary Engg/GA Inst of Tech; B/Civ Engg/GA Inst of Tech. *Born:* 9/15/45. From 1967 to 1971, GA Water Quality Control Bd working with all maj GA ind groups. Principle Sanitary Engr, Teledyne Browning Engg, for ind & environmental investigations. From 1972 to 1974, Prog Mgr, Municipal Engg Service, GA Dept of Natural Resources, responsible for all GA municipal wastewater abatement progs. Since 1974, Mgr, JJ & G including municipal & ind progs. Currently manages General Practice Division Reg engr in eight Southeastern States. Instrumental in two statewide technical periodicals and author of numerous articles. Certificates of Appreciation, 1970, GA/ASCE. Pres's Outstanding Service Award, 1975, CEC/GA. EPA Region IV Advisory Committee. Serv Appreciation Awards, ASCE Environ Engrg Div, 1980 and 1984. Certificate of Appreciation, Atlanta Reg Commission, 1983. *Society Aff:* ASCE, AWWA, SAME, AAEE, SAVE, WPCF.

Corcoran, William H
Home: 8353 Longden Ave, San Gabriel, CA 91775
Position: Inst Prof of Chem Engg. *Employer:* Caltech. *Education:* PhD, MS, BS/Chem Engg/Caltech. *Born:* Mar 11, 1920 Los Angeles CA. Cutter Labs: Chem Engr 1941 & 42, Dir Tech Dev 1948-52. Chem Engr & Res Supr, Rocket Ordnance, OSRD 1942-46. Caltech: Assoc Prof 1952-57, Prof 1957-79, Institute prof 1979-, Exec of Chem Engrg 1967-69, VP Inst Rel 1969-79, Am Hosp Supply Corp: Cons 1952-, VP & Sci Dir of Don Baxter Inc (Subsidiary) 1957-59. ABET, VP 1981. Mbr & Chrmn Adv Comm, AF Inst Tech 1974-78. Fellow AIChE, Founders Award, and Pres 1978. Lamme Award 1979. Western Elec Fund Award. National Acad of Engrg. Corresponding Mbr Natl Acad of Engrg of Mexico. Mbr Assembly of Engrg, NRC. Fellow AAAS. *Society Aff:* AAAS, AIAA, AIChE, ASME, NSPE, ACS, AIC.

Cordill, John J
Business: 16 W 260 83rd St, Hinsdale, IL 60521
Position: Mgr, Corp Qual Assur & Met. *Employer:* Intl Harvester Co. *Born:* June 1920. Attended St Ambrose Coll Davenport IA & Univ of MN. Employed by Intl Harvester Co since 1939 in var supervisory pos in mfg incl Foreman, Gen Foremen, Asst Ch Inspector, Ch Inspector, Div Genl Supr, Product Quality. Presently Mgr Corp Qual Assur & Met with respon for staff coord of qual control, reliability & met activities. Sr Mbr ASQC since 1952, P Chrmn Automotive Div & Dir at Lrg ASQC. Currently VP ASQC.

Cordovi, Marcel A
Home: 125-10 Queens Blvd, Kew Gardens, NY 11415
Position: Vice President *Employer:* Inco United States, Inc. *Education:* ME/ME/Poly Inst Brooklyn; MME/Met Engg/Poly Inst Brooklyn; BME/ME/Poly Inst-Univ MI. *Born:* 11/17/15. Profl career began in 1943 with a res assignment at Welding Res Council. In 1944 joined Babcock & Wilcox Co as Res Metallurgist and in 1951 became Chief Matallurgist of its Atomic Energy Div. Appointed Adjunct Prof of Metallurgical Engg at Poly Inst of Brooklyn in 1950 and taught grad courses until 1970. Joined Intl Nickel Co in 1958; became Mgr of Appl Engg in 1967, Asst to the Pres in 1974, and Dir Special Projs, Inco Ltd, in 1976. Assumed current position as VP, Inco US, Inc in 1980 with corporate responsibility for Inco's worldwide activities in nuclear power and other advance energy conversion systems. *Society Aff:* ANS, ASM, ASME, ASTM, NACE.

Corell, Edwin J
Business: 309 Dixie Terminal Bldg, Cincinnati, OH 45202
Position: Pres. *Employer:* J & E Inc. *Education:* BChE/Chem Engg/OH State Univ. *Born:* 3/6/13. Native of Cincinnati OH. Worked 23 yrs for PPG Industries, the last 8 of which was as Supt of Operations of their largest plant, 2000 people, 3000 tons of chemicals per day. Then did consulting work for some 13 yrs, in the USA and abroad in France, Germany, Italy, Switzerland, England, Mexico, South Africa, Pakistan, New Zealand, Brazil and Argentina. Also do consulting work for the United Nations out of Vienna. Last 10 yrs have been Pres of J&E Inc doing consulting work and as full time and now part time mgr of a personnel placement firm. Named Distinguished Alumnus of the College of Engg at OH State in 1962. Given special commendation and award by Brazilian Govt for work done there in 1972. Mbr Tau Beta Pi. *Society Aff:* AIChE.

Corelli, John C
Business: Jonsson Engg Ctr Rm 5050, Troy, NY 12181
Position: Prof *Employer:* Rensselaer Poly Inst. *Education:* BS/Phys/Providence College; MS/Phys/Brown Univ; PhD/Phys/Purdue Univ. *Born:* 8/6/30. Prof of nuclear Engg at RPI since 1962. Physicist at GE Knolls Atomic Power Lab, 1958-1961. Res interests include the effects of high fluence neutron irradiation on ceramic mtls for nuclear fusion reactor applications, the effects of ion implantation on the elec & optical properties of silicon, use of ionizing radiation to alter the chem and physical properties of polymer mtls. Consultant to GE Co & Westinghouse Elec Corp. Over one hundred res papers published in res fields. *Society Aff:* ANS, APS.

Corey, Richard C
Business: 1820 Dolley Madison Blvd, McLean, VA 22102
Position: Sr Energy Systems Engr. *Employer:* Mitre/Metrek Corp. *Education:* BS/CE/Brooklyn Poly Inst. *Born:* 12/5/08. For some 35 yrs, have been involved in conducting and managing res and dev in coal combustion, gasification, carbonization, and liquefaction; flue gas desulfurization; waste disposal via incineration; and fireside corrosion in large boilers. Res Dir of Bureau of Mines Pittsburgh Energy Res Ctr for 22 yrs followed by sr staff engg work with energy res & dev admin and dept of energy. Author or coauthor of 65 technical papers, author of chapters in four engg handbooks, and of a book on incineration. Recipient of ASME Percy Nicholls Award. *Society Aff:* ASME, APCA.

Corlew, Philip M
Business: 707 N Main St, Edwardsville, IL 62025
Position: Pres. *Employer:* M B Corlew & Assoc, Inc. *Education:* MS/Civ Engr/OK State; BS/Civ Engr/OK State *Born:* 6/28/38. Native of Edwardsville, IL. Pres/CEO of consulting firm, M B Corlew & Assoc, Inc (staff 20-25). Active in community affairs; legislative liaison for state engg socs, mbr of various regional planning agency advisory committees. Hobbies include music, hunting, fishing, golf, flying. Instrument-rated private pilot. Past chapter, pres of ISPE, IRLSA. Also mbr of Rotary, CECI, AWWA, ACSM, AOPA. Publications: "Development of a Rural Water District–, Public Works Mag. May 1981; "The Surveyrs Financial Statement–, Feb 1981 paper to IRLSA State Convention, "Improving Surveyor-Title Company Communications–, June 1980, paper to IL Land Title Assoc. State Convention. *Society Aff:* NSPE, ACEC, ACSM, AWWA, APWA

Corley, William G
Business: 5420 Old Orchard Rd, Skokie, IL 60077
Position: VP *Employer:* Construction Technology Labs, Inc. *Education:* PhD/Struct Engg/Univ of IL; MS/Struct Engg/Univ of IL; BS/Civ Engg/Univ of IL. *Born:* 12/19/35. VP, Construction Technology Labs, Inc. PhD, MS, BS from Univ of IL, Urbana, Native of Shelbyville IL, Jr Engr with Shelby Cty, IL Hgwy Dept 1958, Res Asst and Teaching Asst, Univ of IL 1958-61. US Army Corps of Engrs Res & Dev Coordinator 1961-64, Portland Cement Assoc 1964-86 with title of Divisional Dir since 1979, became VP, CTL 1987, Chrmn ACI, committee 443 Concrete Bridge Design 1973, ACI Wason Medal for Res 1970, ACI Bloem Award 1978, PCI Martin Korn Award 1978, ASCE TY Lin Award 1979, RCRC Arthur J. Boase Award 1986, ACI Reese Structural Res Award 1986. Reg Struct Engr, and Reg PE. *Society Aff:* ASCE, ACI, PCI, NSPE, RILEM, EERI, IABSE, BSSC.

Corliss, William K, Jr
Home: 600 York St, Mechanicsburg, PA 17055
Position: Proj Engr, Hydraulic Div. *Employer:* Gannett Fleming Corddry & Carpenter. *Born:* May 1939. Raised in Gloucester MA. Grad from Norwich Univ, VT with BS in CE 1962 & BA in Phys 1963. Officer US Army Corps of Engrs for 2 yrs. Joined Water Works Sect of Hydraulic Div, Gannett Fleming Corddry & Carpenter Inc 1965. Promoted in 1969 to present pos of Proj Engr for lrg water treatment plant and water supply projs. Became partner in firm in 1974. Lic PE PA; Mbr Tau Beta Pi, Am Water Wks Assn & Water Wks Operators Assn of PA.

Cormack, William J
Business: Caterpillar, Inc, 600 W Washington St, Bldg W, East Peoria, IL 61630
Position: Process Engrg Mgr *Employer:* Caterpillar Tractor Co. *Education:* BS/Met Engrg/London Univ; ARTC/Met/Royal Tech Coll., Glasgow. *Born:* 05/10/28. BS Met London Univ; Associateship in Met Royal Tech Coll Glasgow. Joined Colvilles Ltd, Scotland as met engr, principal respon being Res Sect Leader. Joined Caterpillar Tractor Co 1957 at their Glasgow, Scotland plant as Plant Met & transferred to their Peoria, IL hdqtrs where present pos is Process Engrg Mgr with respon for dev of robotics, mfg process, metallurgy, assembly, quality assurance and intergrated mfg in corporate office. Mbr Am Soc for Metals, Mbr SME. *Society Aff:* ASM, SME.

Corneliussen, Roger D
Business: 32nd & Chestnut, Philadelphia, PA 19104
Position: Prof of Materials Engrg *Employer:* Drexel Univ *Education:* PhD/Phys Chem/Univ of Chicago; MS/Phys Chem/Univ of Chicago; BA/Chem/Concordia Coll *Born:* 08/01/84 Dr. Corneliussen has been Asst Prof of Chem at Luther Coll, Decorah, IA, a Chemist with the Research Triangle Inst, The North Star Research Inst and Pres of Research Services, Inc. He is a past Pres of the Philadelphia section of SPE and a past chrmn of the Nat'l Education Committee. He has also been visiting Prof of the Ford Motor Co in Dearborn, MI and the Bell Telephone Lab in Columbus, OH. His research deals with the morphology and failure processes in polymeric solids. In 1983, he received the Man of the Yr awd for the Philadelphia Sect of Soc of Plastics Engrs. *Society Aff:* SPE, ACS, SAMPE, AAAS

Cornell, C Allin
Business: Dept. of Civil Engineering, Stanford, CA 94305
Position: Prof Research of Civ Engg. *Employer:* Stanford University *Education:* PhD/Civil Engrg/Stanford Univ; MS/Civil Engg/Stanford Univ; AB/Architecture/Stanford Univ. *Born:* Sept 1938. AB, MS, PhD Stanford Univ. Taught at MIT 1964-1982; holder of MIT Winslow Chair 1971-74. Visiting pos at Univ of Mexico, Univ of CA at Berkeley & Natl Civ Engg Lab in Lisbon, the last while holding Guggenheim & Fulbright Fellowships; teaching, res & cons activities in the application of probability to struct engg, esp earthquake engg. President, SSA, 1986-7. ASCE Huber Res Prize 1971. ASCE Moessieff Award 1977; ASCE Norman Medal 1983; Natl Acad of Engr, 1981. *Society Aff:* ASCE, SSA, NAE, EERI.

Cornell, Donald H
Home: 819 Huntingdon Dr, Schenectady, NY 12309
Position: Mgr, Power Plant Equip. *Employer:* GE Co. *Education:* B of ME, University of Akron, 1939 *Born:* Dec 1915. Native of Akron OH. Joined B F Goodrich Co, conducted res on rubber bearings, rubber torsion springs & Mgr Engg Services, Res Ctr Brecksville OH. Joined GE, Knolls Atomic Power Lab 1955. Assoc with var aspects of in-pile irradiation testing of fuels, poisons & struct mtls. Following this, made Mgr of Power Plant Equip, Reactor Facilities. Retired June 1976. Held num pos in ASME, local & natl, incl VP Mbrship 1970-74.

Cornell, Holly A
Business: PO Box 428, 2300 NW Walnut Blvd, Corvallis, OR 97339
Position: Senior Consultant *Employer:* CH2M Hill, Inc. *Education:* ME/Struct/Yale Univ; BS/Civ Engg/OR State Univ. *Born:* 4/5/14 Apr 1914, Boise, ID. BS from OR State Univ; ME Yale Univ. Design engr with Std Oil of CA, struct dept. During WWII was exec officer of an engg combat group, European Theater. Participated in maintenance and repair of the Remagen Bridge over the Rhine in 1944 & 1945. Received the Bronze Star. Founding partner Cornell, Howland, Hayes & Merryfield (now CH2M Hill, Inc), now a maj consulting firm employing 2,500. Dir of Profl Services directing 7 technical disciplines through which CH2M Hill insures technical excellence. Pres from 1974 to 1978. Chrmn of the Bd 1978 to 1980. *Society Aff:* AWWA, ASCE, EERI, AAEE.

Cornett, Jack B
Business: 5300 S. Yale, Tulsa, OK 74135
Position: President *Employer:* W R Holway & Assoc. *Education:* MS/CE/Univ of OK; BS/Civ Engg/Univ of OK. *Born:* 3/15/20. USAF 1941-45; Fighter Pilot ETO, Capt; Air Medl & Distinguished Flying Cross. BSCE 1948, MCE 1964 Univ of OK. 1948-53 with var engg & constr firms; 1954-59 in private practice; 1959-, with W R Holway & Assocs, cons in all phases water resources & hydro power. Presently President Holway & Assocs (Div of The Benham Group Inc). Sr VP, The Benham Group Inc. Principal-in-Charge of engrg, sev large-scale civ works projs, incl water pump stas, aqueducts and plants to $85 mil. Principal-in-Charge & Ed num reports on water supply, treatment & distrib; ASCE Fellow; ACEC (OK Pres 1969-70, Alternate Natl Dir 1969-70); WPCF (Natl Dir 1968-71); AWWA; AAEE; NSPE; Soc of Sigma Xi; Serv Award OK Water & Pollution Control Assn; listed Who's Who in

Cornett, Jack B (Continued)
South & Southwest; Reg Engr OK, KS, AR, TX. *Society Aff:* ASCE, NSPE, WPCF, AWWA, AAEE, ACEC.

Cornforth, Derek H
Business: 7440 SW Hunziker Rd, Tigard, OR 97223
Position: Pres *Employer:* Cornforth Consultants, Inc *Education:* PhD/Soil Mech/London Univ (Imperial Coll); MS/Soil Mech/Northwestern; BS/CE/Durham Univ - England *Born:* 7/14/34 in Whitby, Yorkshire, England. Derek Cornforth founded his own consIt geotech firm in 1983 after 15 years (1961-67; 1974-83) with Shannon & Wilson, Inc; the past 7 years as VP. Cornforth Consultants practice emphasizes landslides, dams, legal work. Hobbies include numismatics, photography, golf, tennis, travel. US Citizen. *Society Aff:* ASCE, CECO, ISSMFE.

Cornforth, Robert C
Business: P.O. Box 54828, Oklahoma City, OK 73154
Position: Owner. *Employer:* Cornforth Assocs. *Education:* MS/Arch Engr/OK State Univ; BS/Arch Engr/OK State Univ *Born:* 2/18/37 Bach & Master of Arch Engg from OK State Univ. Native of Coyle, OK. Served 1961-62 US Army as CO of Co E 317th ASA Bn. Design Engr with Sorey Hill Binnicker 1963-67 & with Benham Blair & Affil 1968-70. Founded Cornforth Assocs 1970. Outstanding Alumni Award 1967 from Phi Delta Theta at OSU. Elected VChrmn Okla City Bldg Code Comm 1976. Mbr AIA, ASCE, PCI, CEC. Structural Engrg Council of Ok, Pres 1981 & 82. Reg PE OK, TX, AR, AZ, ALA, CA, IND, MN, MO, FL, CO, NV, LA, KS, NC, NM, UT. TEA. Adult 4M (Married) Nichols Hills Baptist Church, Oklahoma City, OK. American Concrete Institute - Pres OK Chap Chap 1979-80; VP. 1986-87, Mbr Professional Advisory Cttee - School of Architecture - Oklahoma State Univ., Secretary 1986-87 - Engg Alumni Association-OSU. *Society Aff:* AIA, PCI, SECO, ACI

Cornish, George H
Business: Commiss Office-Bx 2100, Calgary, Alberta T2P 2M5 Canada
Position: Chief Commissioner *Employer:* City of Calgary. *Education:* BSc/Elec Engg/Univ of British Columbia *Born:* May 25, 1931. BA Sc Elec Engg Univ of British Columbia 1956. Native Calgarian. Field of engg expertise - Elec Utility engg, incl outdoor lighting speciality. Was Asst Genl Mgr City of Calgary Elec Sys 1962-67, and Commissioner of Planning & Transpor for City of Calgary (pop 560,000) 1973-1981. Currently Chief Commissioner utils (Chief Adminstrative Office for City) for City of Calgary, reporting on behalf of City Adminstration to City Council. Pres. Illuminating Engg. Soc. of N. America. 1974-75. *Society Aff:* IESNA

Cornog, Robert A
Business: 6914 Canby Ave, 109, Reseda, CA 91335
Position: Consultant *Employer:* Pacific Infrared *Education:* PhD/Physics/Univ of CA -Berkeley; MA/Physics/Univ of CA -Berkeley; BS/ME/Univ of IA *Born:* 7/7/12 Physicist and engr. Worked in aerospace development and atomic energy for the past 35 years. Atomic bomb development during WWII. Co-discoverer (with L.W. Alvarez) of tritium at Berkeley, and head of mechanical engrg group at Los Alamos. Asst Prof of Mechanical Engrg, Univ of CA, Berkeley, 1947-1950. Research on combustion stabilization in moving gas streams. Worked on performance studies of ICBM systems at Space Tech Labs (now TRW Systems), and of geostationary communication satellites at Aerospace Corp. Currently a consultant on product design and performance to several mfg companies. *Society Aff:* IEEE, ASES.

Corns, Charles F
Home: 5713 Guy Place, Springfield, VA 22151
Position: Consltg Engr *Employer:* Charles F. Corns, Inc *Education:* BCE/Structural Engrg/Akron Univ *Born:* 10/9/17 Consulting Engr. Formerly Chief Structural Engr in Civil Works Directorate, Office, Chief of Engrs, US Army, with responsible charge for all structural features of the Corps of Engrs water resources development program, including the promulgation of design criteria and direction of the Corps' structural research program. Directed Corps' activities authorized by the Nat'l Dam Inspection Act, PL 92-367, including development of the "Recommended Guidelines for the Safety Inspection of Dams–. Since retirement from the Corps in 1977, consulting in the fields of Dam Safety and structural engrg for water projects. Chrmn of USCOLD, 1975-76; Emeritus Member RCRC (Reinforced Concrete Research Council). *Society Aff:* USCOLD, ASCE, ACI, EERI

Cornwall, Harry J
Business: Cornwall Engineering, 381 E Anaheim St, Long Beach, CA 90804
Position: Pres *Employer:* Cornwall Engineering, Inc. *Born:* 9/7/21 Project engr. Douglas Aircraft Corp, Los Angeles, 1941-46; project supt C. A.C., Chgo., 1947-57; chief engr. TRE Corp, Los Angeles, 1958-62; pres ConServ Inc, Long Beach, CA, 1962-. Cornwall Turbo-Transmission Corp, Long Beach, 1967-, partner several small bus.'s; advisor 1968-; CA State Univ Long Beach, 1971-; active PTA, YMCA. Registered PE, CA. Mem. Nat. Tech. Services Assn. (nat VP 1974- , Bd of Dir 1976-), Soc Mfg Engrs (chrmn 1958-59), CSPE, NSPE, IIE, AIPE, ISES, Long Beach C of C, Marine Tech Soc, Intl Oceanographic Found., Nat Audubon Soc, CA. Clubs: Century, 49'ers CA State Univ Long Beach, Mason (Shriner). Patentee in field. Certified Mfg. Engr., Plant Engr. *Society Aff:* SME, NSPE, MTS, IIE, IOF, AIPE, CSPE, ISES

Corotis, Ross B
Business: The Johns Hopkins University, Dept of Civil Engrg, Baltimore, MD 21218
Position: Prof *Employer:* Johns Hopkins Univ *Education:* PhD/CE/MIT; SM/CE/MIT; SB/CE/MIT *Born:* 1/15/45 Nat'l Science Foundation Fellow at MIT. Member of Northwestern Univ faculty 1971-81. Head of Civil Engrg program and chaired professor at the Johns Hopkins Univ since 1981. Research interests in structural reliability, including wind and load modeling and nonlinear structural behavior, and in probabilistic analysis of civil engrg problems. Member of ASCE Engrg Mechanics Committee on Probabilistic Methods and Structural Committee on Safety of Buildings (chair), and ACI Committee on Structural Safety (chair). Chrmn of ASCE Loads Standards Subcommittee. Author of 85 publications. Registered PE and Structural Engr. VP of ASCE Maryland Section. *Society Aff:* ACI, ASCE, ΣΞ, ΤΒΠ, ΧΕ

Corporales, George W
Business: 13215 E Penn St, Whittier, CA 90602
Position: Gen Mgr. *Employer:* Ferro Corp Productol Div. *Education:* MS/CE/Univ S CA; BS/Applied Chem/CA Inst of Tech. *Born:* 6/16/31. Although born in Holyoke MA, now qualify as a basic Californian, having worked for the Productol Div since grad in '54. Early interest in Res & Dev matured into product marketing & sales & eventually to gen mgt of div. Emphasis as centered about dev & commercialization of specialty chems for worldwide chem industry. Along the way merged with Suzon Corporales as wife & invented three children Carym, Brad & Carol, while also discovering golf, skiing & tennis. *Society Aff:* AIChE, ACS, PEA.

Corradi, Peter
Business: 2905 Gulf Shore Blvd N, Naples, FL 33940
Position: Consultant (Self-employed) *Education:* ScD(Hon)/CE/NY Univ; BSCE/CE/NY Univ. *Born:* 11/24/10. Brooklyn, NY. Employed various engg capacities on Triboro Bridge and NY Board of Water Supply. Commissioned Lt(jg) US Navy Civil Engg Corps. Rose through grades to Rear Admiral. Appointed Deputy Chief Civil Engrs 1957, Chief Civil Engrs 1962. Retired USN 1965. Became Pres Gibbs & Hill Inc, Consulting Engrs, 1966. Became Exec VP Raymond Intl Inc 1969; Chmn of the Board 1971. Retired from Raymond in 1977 and became consultant to that company and continued as dir. Retired from Raymond Bd in 1981. Continued as dir DMJM Inc & conslt to US Agencies. Retired from DMJM 1985. *Society Aff:* SAME, ASCE, NSPE.

Correa, Jose J
Home: 845 Hickorywood, Houston, TX 77024
Position: Pres *Employer:* Jose Correa, Inc. *Education:* BSCE/Traffic-Transportation/TX Univ. *Born:* 11/7/21. Native of Nicaragua, Central Am, grad from the Univ of TX at Austin with a BS in Civ Engg. Did post-grad work in Traffic & Hgwy Engg. Fellow of ASCE, Reg PE in TX & naturalzed citizen USA. VP, Latin Am Div of the Houston-based firm of Bernard Johnson Inc. Proj Mgr TX Tpk Authority and Pres of Jose Correa Inc Traffic and Transportation Conslts. Have supervised feasibility studies & final design of hgwys & bridges in Panama, Nicaragua & Spain. Author of several papers on Civ Engg. Author of "Normas de Diseno Geometrico de Carreteras–. Immediate past Pres of TX Section ASCE - Immediate past Chrmn Exec Committee Hgwy Div ASCE. Mbr of the Bd of Dir of ASCE. Dir Dist 15 (TX, Mexico, NM, OK). Recipient of the TX Sec Award of Honor, Houston Branch Award of Merit, and the Hwy Div Award (ASCE). Enjoy cooking, fishing & hunting. *Society Aff:* ASCE, SAME, SHPE, ITE.

Corrigan, Brian
Home: 33 Sunny Hill Dr, Madison, CT 06443
Position: Retired. *Education:* ME/Mech Engg/Stevens Inst of Tech. *Born:* 4/1/15. Worked in Res and Tech Dept of Texaco Inc from 1940-70. Titles included Technologist, Sr Technologist, Asst to Mgr, Technical Mgr - Indianapolis, Indiana, Denver, Colorado and lastly at Detroit, MI. Elected Fellow Mbr of ASME in 1967 and Life Fellow in 1975. *Society Aff:* ASME.

Corry, Andrew F
Business: 800 Boylston St, Boston, MA 02199
Position: Sr VP. *Employer:* Boston Edison Co. *Education:* Adv Mgt Prog/Gen Mgt/Harvard; BS/EE/MIT. *Born:* 10/28/22. Native of Lynn, MA, Army Signal Corps 1942-46, with Boston Edison Co since 1947, specialist in res & dev of elec transmission tech, assumed current respons in 1979 as Sr VP in charge of operations & engrg including planning, res, nuclear activities, & environmental affairs. Elected to the Natl Acad of Engrg, 1978; Fellow IEEE since 1974; US Tech Ctte CIGRE since 1971. *Society Aff:* NAE, IEEE, CIGRE.

Corsiglia, Robert J
Business: 2655 Garfield St, Highland, IN 46322
Position: C.E.O. *Employer:* Hyre Electric Co of IN. *Education:* BSEE/EE/IIT. *Born:* 1/22/35. Native of Chicago, IL. Became C.E.O. of Hyre Electric Co. of IN and Exec VP of Hyre in Chicago in 1987. Mbr, The Chicago Presidents Organization. *Society Aff:* IEEE, NSPE, NCEE.

Corten, Herbert T
Business: 216 Talbot Lab, 104 S. Wright St, Urbana, IL 61801
Position: Prof. *Employer:* Univ of IL. *Education:* MS/Mechanics/Univ of IL; BS/ME/IL Inst of Tech. *Born:* 11/20/25. in Oak Park, IL. Navy V-12 prog, 1944-46. After teaching mechanics at Wayne State Univ, Detroit, returned to Univ of IL, in 1948. Became Prof of Theoretical and Appl Mechanics in 1957. Spent a yr leave at Oak Ridge Natl Labs, Oak Ridge, TN in 1967-68 planning res to insure pressure vessel safety. Author of res articles on ductile, brittle and fatigue fracture of metals and composites. Mbr Subcommittee on Properties of Metals. Chrmn of Subgroup on Toughness of ASME Boiler and Pressure Vessel Code Committee. Consultant to US Govt agencies and ind cos. *Society Aff:* ASME.

Cortese, Anthony D
Home: 41 Pershing Rd, Jamaica Plain, MA 02130
Position: Commissioner. *Employer:* MA Dept of Environmental Quality Engg. *Education:* ScD/Env Health/Harvard; MS, BS/Env Eng/CE/Tufts. *Born:* 1/26/47. Native of Boston, MA. Served as an air pollution control engg with the US Public Health Service 1969-72. Environmental mgr and operations research analyst for the US Environmental Protection Agency 1973-76. Dir of MA Air Quality Program 1976-78. Assumed present position as MA Environmental Commissioner since early 1979. Responsible for air and water pollution control, protection of water supplies, water ways and wetlands and regulation of solid and hazardous waste. Enjoy reading, classical music and physical activities. *Society Aff:* AAAS, WPCF, APCA, ΤΒΠ, ΣΞ.

Cortese, Sara H
Home: 41 Pershing Rd, Jamaica Plain, MA 02130
Position: Mgmt, Sci, Manager *Employer:* Digital Equipment Corp. *Education:* MBA/Genl Mgmt/Harvard Grad Sch Bus Adm; SBCE/Civil Engg/MA Inst of Tech. *Born:* 6/4/49 Worked 5 yrs as Sanitary Engr, US Environmental Protection Agency. Was Management Intern, 1974-75. Left to return to grad school full time in 1977. Was Secretary of Boston SWE in 1977. Section Pres, 1979-80. Currently working in Field Service area of DEC. Won National titles in women's lightweight sculling in 1973, 1974 and 1975. Other hobbies: squash, cross-country skiing and jogging. *Society Aff:* SWE.

Cortright, Edgar M
Business: Burbank, CA 91520
Position: Pres *Employer:* Lockheed CA Co *Education:* Dr of Sci//George Washington Univ; Dr of Engrg//RPI; M/Aero Engrg/RPI; B/Aero Engrg/RPI *Born:* 7/29/23 Served aboard USS Saratoga in WWII. Joined NACA in 1948; conducted research at Lewis Flight Propulsion Lab in propulsion aerodynamics until 1958. Member of original team to set up NASA. Organized and directed unmanned space programs in NASA Headquarters 1958-67; manned space programs 1967-68. Dir NASA Langley Research Ctr 1969-1975. Technical Dir Owens-IL 1975-78. VP Science and Engrg, Lockheed Corp 1978. Pres Lockheed CA Co 1979 to date. Past Pres AIAA. Member NAE. Numerous awards including NASA Distinguished Service Medal (2), AAS Space Flight Award, Fleming Award, ASME Spirit of St. Louis Medal. *Society Aff:* AIAA

Corum, James F
Business: WV Univ./Electrical Engg Dept, Morgantown, WV 26506
Position: Assoc Prof *Employer:* Dept of Elec Engr *Education:* BS/EE/Univ of Lowell; BS/EE/Tufts Univ; MSc/EE/OH State Univ; PhD/EE/OH State Univ *Born:* 8/15/43 From 1970-84 held various EE faculty positions at the OH Inst of Tech and WV Univ. Received 7 outstanding teaching awards. Prior to this, worked as Electronics Engr for the Natl Security Agency. Profl res and primary publications are in relativistic electromagnetics, general relativity, antennas and radiating structures. Serve as a consultant for a variety of private industries and governmental agencies involved with Commercial broadcasting (AM- FM-TV), Satellite communications (TVRO earth stations), and RF Engrg. Invented the Toroidal Helix Antenna (2 US and 6 foreign patents). Past Chrmn of Upper Monongahela Subsection of IEEE. *Society Aff:* IEEE, AAPT, ASA, ΣΞ, AAAS, ASEE, SMPTE.

Cory, Lester W
Home: 45 Summit Ave, Tiverton, RI 02878
Position: Prof *Employer:* Southeastern MA Univ *Education:* MS/EE/Northeastern Univ; M of Ed/Inst Media/Bridgewater State Coll; BS/EE/Southeastern MA Univ *Born:* 7/25/39 Prof Cory is the President and a co-founder of the Share Foundation, Inc. A non profit corporation which provides equipment and services to enable physically immobilized voice impaired people to communicate. With Prof B.B. Hardy (SMU), he is the co-author of *Electrical Measurements for Engrs*, a publication dealing with the fundamentals of electrical measurements. Col Cory is a career officer in the RI Air Nat'l Guard. From 1966 to 1978 he was the commander of the 143rd Communications Flight. Since 1979, he has served as the commander of the 281st Combat Communications Group. He is married to the former Patricia L. Barrett of East Liverpool, OH. They have 5 children. *Society Aff:* IEEE, ΣΞ

Cory, William E
Business: 6220 Culebra Rd, PO Box 28510, San Antonio, TX 78284
Position: VP. *Employer:* Southwest Res Inst. *Education:* MS/Engg/UCLA; BS/EE/TX

Cory, William E (Continued)
A&M Univ. *Born:* 4/5/27. Native of Dallas, TX, married Doris Garlington, two children. Worked with USAF in secure communications 1950-57, Lockheed Aircraft Corp in Communication, Navigation & Identification fields 1957-9. Joined Southwest Res Inst in 1959, appointed VP in 1972. Maj technical fields are electromagnetic effects and compatibility, communications, physical security and automation. Received IEEE Fellow Award, 1971; elected Dir, IEEE Region 5, 1972-3; IEEE Bd of Dirs 1972-3; Pres IEEE Electromagnetic Compatibility Society 1974-5; Gen Chrmn, 1975 IEEE Intl Symposium on Electromagnetic Compatibility; Midcon Bd of Dirs 1977-present; Reg PE, TX. *Society Aff:* IEEE, BEMS, AOC, NSPE.

Corzilius, Max W
Business: 520 Cypress Ave, Venice, FL 33595
Position: President. *Employer:* Emcee Electronics, Inc. *Education:* MChE/Chem Engg/Univ of DE; BS/Chem Engg/Grove City Collete. *Born:* 3/19/24. Served USAF 1943-45 CBI Theatre, Employed E I DuPont 1951-61, Emcee Electronics, Inc (Pres) 1961- , Crown Chem Corp (Pres) 1965-76, Emcee Industries, Inc (Chrmn) 1976-80, Airline Transport Pilot (Sels, Mels, Glider). *Society Aff:* AIChE, SAE, ISA, ASTM.

Cosens, Kenneth W
Home: 2620 Chester Rd, Columbus, OH 43221
Position: Retired *Education:* MS/Sanitary Engrg/MI State Univ; BS/CE/MI State Univ; CE//MI State Univ *Born:* 6/22/15 Born and educated in MI. Prof of Civil Engrg at MI State Univ, Univ of TX, and OH State Univ over 30 year career. Retired as partner, Chief of Public Works Div of Alden E. Stilson & Assoc, Consulting Engrs, over 10 year period. Supply Admin City of Columbus, Div of Water. Honorary fraternities include: Tau Beta Pi, Chi Epsilon, Phi Kappa Phi, Sigma Xi. Received Fuller Memorial Award 1980, AWWA OH Section. Author, or co-author of 40 publications in technical journals. Married to Arlene M. Four grown children. Hobbies: golf, photography and woodworking. *Society Aff:* ASCE, NSPE, AWWA, WPCF, AAEE

Cosgriff, Robert L
Business: Anderson Hall-Rm 556, Lexington, KY 40506
Position: Prof. *Employer:* Univ of KY. *Born:* Feb 1923 Big Timber MT. BEE, MS, PhD OH State Univ. US Army 1943-46. Sr Engr Curtiss Wright 1947-50. Staff EE Dept OSU 1950-67. Univ of KY 1967-. Dir CCSL Lab 1960-67 (OH State). Chrmn EE Dept Univ of KY 1967-71. Dev microwave instrumentations sys now commercially distributed. Dir collection and compiled radar reflection data in Terrain Return Handbook. Pioneered in design & eval of an automatic automobile & remotely controlled vehicles. Author: "Nonlinear Control Sys–. Former Chrmn Columbus Sect IRE & Ad Com of Automatic Control Group IEEE. Mbr sev honoraries. Reg PE OH. Disting Alumnae recipient OH State Univ.

Cosgrove, Benjamin A
Home: 18602 Brittany Dr SW, Seattle, WA 98166
Position: VP Engg *Employer:* Boeing Co. *Education:* BS/Aero/ Univ of Notre Dame *Born:* BSAE Notre Dame. Native of Detroit MI. Served with Navy 1944-46. Joined the Boeing Co 1949 as Struct Design Engr. Appointed to mgt 1960. Was chief proj engr 1977 of 767 & later respon for admin & production design & dev of the Model 747 & 767 airplanes & their derivatives. Assumed current Position of V.P. of Engg in 1985. Enjoy skiing, swimming & tennis. *Society Aff:* SAE, AIAA

Cosgrove, Donald G
Business: PO Box 12404, Ft Worth, TX 76116
Position: Pres. *Employer:* NDE-AIDS Inc. *Born:* 1932; native of E St Louis IL. Served in US Army during 1952 & 1953. BS in Engg Admin from Millikin Univ; Cert in Met, Metals Engg Inst 1962. Metallogapher Caterpillar Tractor Co 1956-63, duties in Met & NDT. With Convair Aerospace 1963-73 as a Sr NDT Res Engr. Currently Pres of NDE-AIDS Inc. Presented tech papers at natl mtgs of ASNT, & sev at educ seminars around the US. P Mbr of Bd of Dir ASNT, P Pres North TX Sect of ASNT, Fellow of the Soc. Reg PE.

Cospolich, James D
Home: 14 Theresa Ave, Kenner, LA 70062
Position: VP *Employer:* Waldemar S Nelson & Co, Inc *Education:* MS/EE/LA State Univ; BS/EE/LA State Univ *Born:* 12/19/44 Native of New Orleans, LA. Served in United States Coast Guard Reserve 1964-1972. With Waldemar S. Nelson & Co, Inc since 1967. Assumed responsibility as mgr of electrical engrg in 1974 and became VP in 1979. Specializing in construction, maintenance, design, and project management of industrial power generation and distribution systems, instrumentation, control and automation systems for gas processing plants, offshore and onshore oil and gas production and processing facilities, drilling rigs and petrochemical plants. Electrical design in hazardous areas in compliance with OSHA, NEC, USCG, USGS, ABS and API. Registered PE in CA, LA, and TX. *Society Aff:* IEEE, NSPE, ISA, IES, GPA, LES

Cost, James R
Business: Sch of Mtls Engg, W Lafayette, IN 47907
Position: Prof. *Employer:* Purdue Univ. *Born:* Sept 5, 1928; married, four children. Undergrad degree Psychology Univ of WI 1951. USAF 1951-55. MS & PhD Phys Met Univ of IL 1959-62. Res Scientist Ford Scientific Lab, behavior of point defects in metals 1961 to 1965. Adj Staff Univ of MI 1962 & 1963. Sch of Mtls Engg Purdue Univ 1965, Assoc Prof, Full Prof 1970. Alcoa Fdn Chair 1970-74. Summer pos at Aerospace Corp, Argonne Natl Lab & Lawrence Livermore Lab. Res ints: gases & interstitial solutes in metals, radiation damage, point defect behavior, elastic properties of composites.

Cost, Thomas L
Home: 1113 Northwood Lake, Northport, AL 35476
Position: Prof. *Employer:* Univ of AL. *Education:* PhD/AE/Univ of AL; MS/AE/ Univ of IL; BS/AE/Univ of AL. *Born:* 12/24/37. Dr Cost is currently a Prof of Aero Engg and serves concomittantly as the Pres of Athena Engg Co, a consulting engg organization. He is a mbr of several Govt advisory bds including the DOD Structures and Mech Behavior Committee and the AIAA Structures Steering Committee. He conducts res in the area of solid mechanics and structures and is the author of over fifty technical papers. *Society Aff:* ASEE, ASME, AIAA, AAM.

Costello, Charles V
Home: 30 Allendale Rd, Binghamton, NY 13903
Position: Dir. *Employer:* Costello's Lab Inc. *Education:* BS/Chemistry & Phys/Univ of Scranton. *Born:* 1/11/15. Native of Binghamton NY, City of Binghamton Bur of Water 1940-1978 Supt of Water 1963 Retired Apr 1978, Water Supply Div Govt of Liberia W Africa Consultant 1960-1962, Published Articles Je 1965: Monitoring & Control of Flouridation of Municipal Water Supplies, Mgt Training Short Course AWWA journal July 1970. Lectr Technicon Natl Symposium Chem Div 1964. Water Consultant, Town of Kirkwood NY 1979-1980. Editor of NYS Sec AWWA Dir of Costello's Lab Inc. Consulting & Analytical Service Water & Sewage Analysis, George Warren Fuller Award 1978 Life Mgr AWWA, George Warren Fuller Award 1978 Life Mgr AWWA. *Society Aff:* AWWA.

Costello, Frederick J
Home: RFD 1, 7 Sunset Ave Box 3040, Montpelier, VT 05602
Position: Director, Agency Facilities. *Employer:* State of VT. *Education:* BS/CE/New England College; AS/CE/Wentworth Institute. *Born:* 5/13/42. Native of Lynn, MA. Employed as Civ Engr in Dept of Hgwys 1963-69. In 1969 was appointed Chief Engr for the Dept of Forests & Parks, then in 1973 was made Chief Facilities Engr for the Agency of Environmental Conservation due to state govt reorganization. Appointed Director of Agency Facilities in 1986. Was VT's Acting Natl Dir in 1973 & Natl dir 1975-1977 in NSPE. Past-Pres of the VT Soc of PE (1973) & the VT Sec of the Am Soc of Civ Engrs (1976). Received "VT Young Engr of the Yr" award 1973. Registered Professional Engineer-Maine, NH, & VT. *Society Aff:* NSPE, ASCE.

Costello, George A
Business: 104 S Wright, Urbana, IL 61801
Position: Prof *Employer:* Univ of IL *Education:* PhD/Theoretical and Applied Mech/ Univ of IL; MS/Theoretical and Applied Mech/Univ of IL; B/CE/Manhattan Coll *Born:* 10/30/33 I have been teaching and doing research in the area of solid mechanics at the Univ of IL since 1955. I was appointed an Asst Prof in 1959, and Assoc Prof in 1964 and a Full Prof in 1968 in the Dept of Theoretical and Applied Mechanics. In the last ten years, I have been doing research in the area of wire rope. This research is concerned with a determination of the stresses in the individual wires in the rope when the rope is subject to both axial and bending loads. *Society Aff:* ASME, ASCE

Costello, Lawrence S
Home: 1459 Greenmont Ct, Reston, VA 22090
Position: VP. *Employer:* CH2M Hill Inc. *Born:* Aug 1939. MBA Finance Geo Wash Univ 1973. BSCE 1962, MSCE 1963 Univ of CA at Berkeley. Commissioned Officer Army Corps of Engrs 1963-65. Post-grad studies at MIT 1965-66. Data Processing Engr with Jacobs Assocs 1967. Water resources analyst with Water Resources Engrs Inc Walnut Creek, CA & Springfield VA. Dir water resource prog for Metro WA COG 1971-73. With CH2M Hill since 1973. Outstanding serv award No VA Council of PE 1974.

Costes, Nicholas C
Business: Structures Dynamics Lab, ED42, Marshall Space Flight Ctr, AL 35812
Position: Aerospace Tech; Senior Research Scientist *Employer:* NASA Geo C Marshall Space Flight Center. *Education:* PhD/MSCE/NC State Univ; ME/-/Harvard; AM/-/Harvard; AB/MSCE/Dartmouth *Born:* 9/20/26 Native of Athens, Greece. In US since 1948. Married, 3 children. AB Dartmouth 1950. Masters degrees from Dartmouth, NC State, Harvard. PhD NC State 1965. Mat Eng NC State Hwy Comm 1953-56. Res Civil Engr US Army SIPRE/CRREL 1956-62. Instructor, Ford Foundation Fellow NC State 1962-64. Since 1965 Sr Res Sci NASA MSFC. PE NC, IL. ASCE TCAS/Aerospace Div Chrmn Program Cttee 1973-75; Exec Comm 1977-1982. Chrmn Exec Comm 1980-81. Dartmouth Soc Engg Prize 1951; num NASA awards; ASCE Norman Medal 1972, Assn of Civil Engrs Greece; Hon Mbr 1973; AIAA, AL Sect Outstanding Aerospace Engg for 1976. 1979 Martin Schilling Award. Phi Kappa Phi, Sigma Xi, Chi Epsilon. Author, Co-author 60 pubs. *Society Aff:* ASCE, AIAA, AGU, AAAS, NSPE.

Costrell, Louis
Home: 10614 Cavalier Dr, Silver Spring, MD 20901
Position: Physicist *Employer:* Natl Bureau of Stds. *Education:* BS/EE/Univ of Maine; MS/EE/Univ of Maryland. *Born:* June 26, 1915 Bangor, ME. BSU Univ of ME 1939; postgrad Univ of Pittsburg 1940. MS Univ of MD 1949. With Elliot Co Ridgeway PA 1940; Westinghouse 1940-41; Navy Dept 1941-46; Bureau of Stds 1946-. Ch Radiation Instrumentation Sect Bureau of Stds. Tech Advisor to US Natl Ctte for intl Electro Tech Comm 1962-; Chrmn ERDA Natl Instrumentation Methods Comm. Recipient Meritorious Serv Award Commerce Dept 1955; Disting Serv Award 1968; Spec Serv Award 1963; IEEE Harry Diamond Mem Award 1975; IEEE Nucl & Plasma Sci Soc Merit Award 1975. Fellow IEEE and Washington Acad Acad of Sci. Mbr Am Phys Soc, Chrmn ANSI Ctte on Radiation Instruments 1960-. IEEE Stds Bd. Mbr Tau Beta Pi, Phi Kappa Phi. *Society Aff:* IEEE, APS.

Cota, Harold M
Business: Civil & Environmental Engg Dept, San Luis Obispo, CA 93407
Position: Prof. *Employer:* CA Polytechnic State Univ. *Education:* PhD/Chem Engg/ Univ of OK; MS/Chem Engg/Northwestern Univ; BS/Chem Engg/Univ of CA. *Born:* 4/16/36. Native of San Diego CA. Worked for San Diego Gas & Elec five summers while attending school. With Lockheed's electochemistry res group from 1960-62. Joined the faculty of CA Poly in 1966. Dir of the air pollution control training program since 1969. Teaching responsibilities in noise and water pollution control. Appointed to the CA Central Coast Regional Water Quality Control Bd (1970-84). Chrmn 1976-77. Air Resources Board Research Screening Comm. (1985-). APCA Dir 1979. Editor of WCS-APCA newsletter 1979-86. Chair WCS-APCA (86-88). Diplomat Amer Acad of Environ Engrs. Received APCA's Lyman A. Ripperton Award in 1984. Active in church, Boy Scouts of America, and enjoy swimming. *Society Aff:* APCA, AIChE, AIHA, INCE, CWPCA.

Cotellessa, Robert F
Home: 13 Bradley Dr, Potsdam, NY 13676
Position: V P for Acad & Student Affairs Prof of Elec & Computer Engrg *Employer:* Clarkson Univ *Education:* PhD/Physics/Columbia Univ; MS/Math & Phys Sci/ Stevens Inst of Tech; ME/Elec Engg/Stevens Inst of Tech. *Born:* 6/7/23. Native of Ridgewood, NJ. Served with US Naval Reserve 1943-46. Instructor (Mathematics), Stevens Inst of Tech 1946-48. Elec engr, A B Dumont Labs 1949-51. Tech Coordinator for Electrical Engrg 1951-55; Prof, 1955-68; Dir, Lab for Electroscience Res 1962-68; NYU. Prof & Chrmn of Elec and Comp Engrg, Clarkson College of Tech, 1968-80. Provost and Prof of Electrical Engrg and Computer Science, Stevens Inst of Tech 1980-84. Prof of Electrical and Computer Engrg Clarkson Univ 1984-present VP Acad & Student Affairs 1986-present. VP, IEEE, Publications (1973), Tech Activities (1974-75). Exec VP, IEEE, 1976. Fellow IEEE; IEEE Centennial Medal, 1984; IEEE US Activities Bd Citation of Honor, 1979. Haraden Pratt Award 1987. Fellow AAAS. Enj boating. *Society Aff:* IEEE, ASEE, APS, AAAS, HKN, ΣΞ, TBΠ, NSPE

Cotten, William C, Jr
Home: 1302 Belmont Pkwy, Austin, TX 78703
Position: Pres-Principal *Employer:* W. C. Cotten, Jr., Inc. *Born:* 10/22/10 Native TX, began Engrg career with TX Hwy Dept, 1930; reg PE, 1938; spent 3 yrs, 1940-1943 as Construction Engr & Gen Superintendent on Military Base Construction; US Navy Civil Engr Corps (CB) South Pacific Theatre, 1943-1946; Chief Engr and Exec VP for heavy construction company on bridge, expressway, dam, and hgwy construction, 1946-1954; Gen Civil and Mech practice to present, including many large lite devs, large bridges, expressways, fdns, stream hydraulic analysis, and forensic practice to the Legal profession. Travis Chapter TSPE Engr of the Yr, 1977. *Society Aff:* ACEC, CECT, NSPE, TSPE, AREA, AWS

Cotterill, Carl H
Home: 6030 Corland Ct, McLean, VA 22101
Position: Staff Assistant/MDA *Employer:* US Bureau of Mines. *Education:* Met Engr/Profnl/Univ MO-Rolla; MS/Bus Adm/Washington U (St Louis); BS/Chem Engr/Univ MO-Rolla. *Born:* 11/27/18. Lamar, MO. Employed by Am Zinc Co in positions of plant engr, res engr, corporate mgt, feasibility studies & exec Asst to the Chrmn & CEO over a period of 31 yrs. Served in US Armed Forces Europe & N Africa in WWII & in Army Reserves as Lt Col. Employed with US Bureau of Mines since 1971 in Mineral Policy, mineral supply/demand, strategic and Critical Minerals, & Mineral Investigation of Public Lands. Served as expert on US Delegation to U.N. International Preparatory meeting on Copper and US Delegation to International Lead-Zinc Study Group, 1975-6. Author book "Industrial Plant Location-Its Application to Zinc Smelting " 1950, and Co-Editor Proceedings "AIME World Symposium on Mining & Metallurgy of Lead & Zinc" 1970. Currently Staff Asst/ MDA, USBM. *Society Aff:* AIME, AIChE, IMM, MMSA, CIMM, ΣΞ, WAS.

Cottingham, William B
Business: 1700 W Third Ave, Flint, MI 48502
Position: Pres. *Employer:* GM Inst. *Education:* BSME/Engg/Purdue Univ; MSME/ Engg/Purdue Univ; PhD/Engg/Purdue Univ. *Born:* 12/1/33. Native of Chicago IL. 1956-75 Purdue fac except for 3 yrs with Bell Tel Co 1960-63, where assoc with res group studying microwave interaction with plasmas. Became Assoc Prof ME 1963; Full Prof 1966; Hd Sch of ME 1970. P Pres Symphony Orchestra Lafayette IN. Currently mbr Flint Inst of Music & mbr Exec Ctte. Guest Researcher Medisch Fysisch Instituct-TMO, Utrecht the Netherlands Feb 1970-Aug 1970. 1975-76 Dean of

Cottingham, William B (Continued)
Academic Affairs GM Inst; 1976-, Pres GM Inst. Co-author book "Phys Design of Electronic System–. Shares pat entitled "Cryostat–. Hobbies: music, flying, gardening, fishing. *Society Aff:* SAE, APS, ASEE, ASME.

Cottom, Melvin C
Business: Dept of EECE, Durland Hall, Manhattan, KS 66506
Position: Asst Prof. *Employer:* KS State Univ. *Education:* MS/Elec Engg/Univ of KS; BS/Elec Engg/Univ of KS. *Born:* 10/11/24. Native of Coffeyville, KS. Taught elec engg at KS Univ, 1945-50. Elec design engr at Black and Veatch, Consulting Engrs, 1950-55, specializing in power plant design. Asst Prof in Elec Engg Dept, KS State Univ 1955 to date. Sr Mbr, IEEE. Chrmn, KS City Sec of IEEE, 1971-72. Reg PE, States of KS & MO. Hobbies are photography and music. *Society Aff:* IEEE.

Cotton, Frank E, Jr
Home: P.O. Box 5235, Bloomington, IN 47402
Position: Consultant. *Employer:* Self. *Education:* PhD/Economics/Univ of Pittsburgh; MLitt/Management/Univ of Pittsburgh; BS/EE/MS State Univ; BS/Sci (Math & Phys)/MS State Univ. *Born:* 8/14/23. Reg PE. Electronics Engg Officer, USN, WWII. Ind Engr with Westinghouse Elec Corp, E Pittsburgh, 1947-51. Petro Economist with Gulf Oil Corp, Pittsburgh, 1951-58. Dir, Engg Extension Service, MS State Univ, 1958-68. Prof and Hd, Dept of Ind Engg, MS State Univ, 1962-86. Fellow, American Inst of Ind Engrs; Pres 1970-71. Mbr, Cost Study Ctte, Independent Petro Assoc of Amer. Profl activity in over 35 countries. Conslt in petro economics and in profitability evaluation of petro investments. Enjoy reading, travel, hiking *Society Aff:* IIE, ASEE, SPE, IPAA, AAPG, AПM, TBП.

Cotton, Jack L
Business: 5615 Corporate Blvd, Baton Rouge, LA 70808
Position: Assoc. *Employer:* Howard, Needles, Tammen & Bergendoff. *Born:* May 1930. BSCE Univ of Notre Dame 1951. USN 1951-54; HNTB since 1954, doing Hgwy Design & planning initially & Transpor Planning & Admin in ensuing yrs. Named Assoc in Jan 1974. Currently assigned in Baton Rouge LA. In charge of firm's extensive LA assignments.

Cotton, John C
Home: 4105 Stoconga Dr, Beltsville, MD 20705
Position: Consultant. *Employer:* Self Employed. *Education:* CE/Civil Engr/RPI. *Born:* 10/4/09. 1933-39 Agricultural Engg, USDA on Construction at Beltsville, MD and Res at WA office. 1939-43 Soil Conservation Service, USDA-Special hydraulic and drainage studies in MS, TX, IN and OH. 1943-54 SCS- Proj Engr for Drainage & Flood Control structures in MD and MS. 1954-72 Design Engr for Drainage and Flood Control structures in MD and DE. Retired in 1972. Now Hydraulic Consultant specializing in design and construction of Flood Control and Storm Water Mgt Structures. *Society Aff:* ASCE, NSPE, SCSA.

Cotton, Kenneth C
Home: 216 Sugar Hill Rd, Rexford, NY 12148
Position: Consltg Engr *Employer:* Self employed *Education:* BS/ME/PA State; Equivalent to Masters/Mech Engrg, Fluid Mech/GE Advanced Engrg Program *Born:* 2/8/22. US Navy 1943-46. Joined GE Test Prog 1946 and was assigned responsibility for turbine air testing dev lab 1950. Responsible as Mgr of Turbine Performance Engg from 1956 to 1984; includes testing and evaluation of new design steam turbine components in dev labs & complete turbine generator units in power plants. Retired from GE in 1984. Currently an active Consulting Engr. Past Mbr and Chrmn ASME PTC Policy Bd, VChrmn PTC 6, USA Technical Advisory and US Ctte Chrmn for IEC/TC 5 Steam Turbines. PE, ASME Fellow 1971, ASME Centennial Award, 1981 Prime Movers Award, 1981 GE Steinmetz Award, 1983 James Harry Potter Gold Medal Award, 1984 Performance Test Code Gold Medal Award and 1984 Power Movers Award. *Society Aff:* ASME.

Cottrell, Alan Howard
Business: Jesus College, Cambridge, CB5 8BL, England
Position: Research Associate. *Employer:* Cambridge Univ. *Education:* ScD/-/Univ of Cambridge; PhD/-/Univ of Birmingham; BSc/-/Univ of Birmingham. *Born:* 7/17/19. Prof of Physical Met, Birmingham Univ 1949-55. Deputy Hd of met Div, UKAEA, Harwell, 1955-58. Goldsmiths' Prof of Met, Cambridge Univ, 1958-65. Deputy Chief Scientific Adviser & Chief Scientific Adviser, HMG, 1968-74. Fellow of the Royal Soc. Recreation: music. Several awards including: Albert Sauveur Achievement Award, American Society for Metals, 1969; James Douglas Gold Medal, American Institute of Mining, Metallurgy & Petroleum Engineers, 1974; The Rumford Medal of the Royal Society 1974; Harvey Prize (Technion, Israel), 1974; Acta Metallurgica Gold medal, 1976; The Guthrie Medal & Prize, Inst. of Physics, 1977; The Gold Medal of the American Society for Metals, 1980; The Brinell Medal of the Royal Swedish Academy of Engineering Sciences, 1980. *Society Aff:* AAAS, NAS, ASM, NAE.

Couch, Frank B, Jr
Business: P.O. Box 1070, Nashville, TN 37202
Position: Chief, Geotechnical Branch *Employer:* US Army Corps of Engrs *Education:* MS/Soil Mech/Harvard Univ; BE/Civil Structural/Vanderbilt Univ *Born:* 7/17/37 Native of Nashville, TN. Served as Commissioned Officer with US Corps of Engrs 1960-62. Materials Engr at Barkley Dam 1962-63. Nashville District Corps of Engrs Soils Engr, Chief, Soils Section and Chief, Geotechnical Branch, 1963-present. Direct responsibility for design of Cordell Hull, Laurel and Bay Springs Dams; remedial work at Wolf Creek Dam; and TN Tombigbee Waterway Divide Cut. Consultant on numerous building foundations, 1968 to present. Nominated ENR Man of Year 1977. ASCE A.P. Greensfelder Prize 1980. *Society Aff:* ASCE, GSA, USCOLD.

Couch, George R
Home: P.O. Box 163, c108 Goldfinch Meadow, Allamuchy, NJ 07840
Position: Private Investor *Employer:* Retired *Education:* MS/Chem Engrg/Newark Coll of Engrg; BS/Chem Engrg/Univ of MO, MO Sch of Mines & Metallurgy; Prof Deg/Metallurgy/Univ of MO, Rolla; AMP/Advanced Management Program/81st AMP Harvard Business School *Born:* 10/9/19. Retired 1985. VP AMAX, Inc; Pres of Subsidiary AMAX Specialty Metals Co. Pres TMS-AIME, 1979; VP AIME, 1978-80. Prior engg and management experience with Kennecott Copper Corp, Western Mining Div in Salt Lake City, UT; VP, Molybdenum Corp of America in New York City. Prior engg respon with Titanium Div and Magnesium Reduction Co & National Lead Co, US Steel Corp, Southworks, Chicago, IL. Served with US Army, Tech Det in WWII. Grad MO School of Mines & Metallurgy, Univ of MO 1941. P.E. Chemical Engineering. Harvard Business School, AMP-81. Born in St Louis MO. *Society Aff:* AIME-TMS, ASM, M&MS.

Couch, James G
Business: 1 Diamond Hill Rd, Murray Hill, NJ 07974
Position: VP. *Employer:* C F Braun & Co. *Education:* BS/Mech Engr/WV Univ. *Born:* 2/20/20. BSME WV Univ. Native of Charleston WV Army Corps of Engrs 1942-46. Engr with duPont 1946-53 in appl dev work & tech supr assoc with urea, ammonia, methanol, polyethylene & nylon processes. Joined Natl Distillers & Chem Corp 1953. Assumed pos of VP Engg in 1969 with respon for process design, engg dev & proj engg of new facilities. In 1973 joined C F Braun & Co, in current pos respon for mgt of Eastern Div which provides engg & const services to the process indus. Enjoys music, tennis & sailing. *Society Aff:* ASME, EJC, TBП.

Couchman, Peter R
Business: Coll of Engrg, P.O. Box 909, Piscataway, NJ 08854
Position: Professor *Employer:* Rutgers Univ *Education:* PhD/Materials Sci/Univ of VA; MSc/Materials Sci/Univ of VA; BSc/Physics/Univ of Surrey (U.K.) *Born:* 1/5/47 in London, England. Research and teaching experience in Metals Physics, Physics of Polymers, Materials Science, Math, Physical Chem of Surfaces. Postdoctoral fellowships at Univ of Bristol (U.K.), Univ of MA; Asst Prof of Macromolecular Science at Case Western Reserve Univ; at present Prof of Mechanics and Materials Science, Rutgers Univ. Principal areas of research: polymer physics, esp glass transition behavior of multicomponent systems, equations of state, miscibility; also surface tension of simple and chain liquids. Fellow, American Physical Society. *Society Aff:* ACS, NATAS, SPE, NYAS, APS

Coughlin, Robert W
Home: 49 Storrs Heights Rd, Storrs, CT 06268
Position: Prof & Dept Head. *Employer:* Univ of CT. *Education:* PhD/Chem Engg/Cornell Univ; BS/Chemistry/Fordham Univ; Postdoctoral/Chemistry/Univ of Heidelberg (Germany). *Born:* 6/18/34. After early and undergrad (Fordham) education in NYC, Coughlin did grad work at Cornell and spent a postdoctoral yr doing res as a Fulbright Fellow at Heidelberg Univ Germany. At Cornell he held Univ and AEC fellowships. His industrial career includes service at Exxon Res & Engg Co, Isotopes-Teledyne and a wide variety of consulting assignments for a number of different cos in the chem processing and petroleum refining industries. He rose through the academic ranks to prof of chem engg at Lehigh Univ (1965-1977) and became prof and head of the Chem Engg Dept at the Univ of CT in 1977. His maj interests include kinetics, catalysis, surface chemistry and the processing and refining of fossil fuels. *Society Aff:* AIChE, ACS, NACS, AAAS.

Coulman, George A
Business: 1983 E 24th St, Cleveland, OH 44115
Position: Prof *Employer:* Cleveland State Univ. *Education:* PhD/Chem Engg/CIT; MS/Chem Engg/Univ of MI; BS/Chem Engg/CIT. *Born:* 6/29/30. Broad industrial and academic experience. Dev engr with Dow Corning Corp. Mgr of Dev American Metal Products Engg, Sci Div. Asst Prof for 3 yrs Univ of Waterloo, Ontario. Promoted to Prof during 12 yrs at MI State Univ. Was Co-prin and/or prin investigator on many res grants and contracts from govt and industry. Became Prof and Chrmn Dept of Chem Engg at Cleveland State Univ in 1976. Assumed Acting Chair in Metallurgical Engg in 1977. Responsible for extensive development of res & expansion of grad program for working engrs in cooperation with regional industry. *Society Aff:* AIChE, ASEE, CES.

Coulter, James B
Business: Dept of Natural Resources, 580 Taylor Ave, Tawes St Office Bldg, Annapolis, MD 21401
Position: Secy of Dept of Natural Resources. *Employer:* State of MD. *Education:* MS/Sanitary Engg/Harvard Univ; BS/Civ Engg/KS Univ. *Born:* 8/2/20. in Vinita, OK, 1920, Secy Coulter served in the US Army Combat Engrs, 1940- 45. He served with the US Public Health Service from 1950 through 1966 as a res engr and administrator. From 1966-69, he was Asst Commissioner for Environmental Health, MD Dept of Health. From 1969 through 1971, he was Deputy Secy of Natural Resources and Secy from 1971 to present. *Society Aff:* WPCF, AAEE.

Coulter, Kenneth E
Home: 511 W Meadowbrook, Midland, MI 48640
Position: Adjunct Consultant Prof. *Employer:* Univ of MI. *Education:* BS/ChE/Univ of MI; MS/ChE/Univ of MI. *Born:* 2/4/17. Employed by the Dow Chem Co 1937-1978. Served as Plant Supt, Plant Mgr, Dir of Res for MI Div, Mgr of Engg Construction & Maintenance, Process Consultant. Pres of AIChE in 1975. Mbr of John Fritz Medal Committee. Fellow of AIChE & received Founders Award in 1977. Retired from Dow Chem 1978. Adjunct Prof of Chem Engg 1978 ,1979,1980.Part time as Process Consultant. *Society Aff:* ACS, AIChE, ASFMRA, MSFMRA

Cound, Dana M
Home: 1730 Hastings Mill Rd, Pittsburgh, PA 15241
Position: Dir - Oper Audit. *Employer:* Rockwell Intl. *Born:* June 1931. BS CA State Univ, Long Beach CA. Reg PE CA. Fellow Am Soc for Qual Control. VP Educ & Dev of ASQC 1973-76. Chrmn Qual Assurance Ctte of Aerospace Indus Assn 1975. In 1953 joined N Amer Aviation which was subsequently merged with Rockwell to eventually form Rockwell Intl. Served as Aerospace Group Dir & Corp Dir of Qual Assurance. In Mar 1976 became Corp Dir of Oper Audit respoon for audit & improvement of production oper.

Counts, Cecil P
Business: 600 108th Ave NE Suite 405, Bellevue, WA 98004
Position: Assoc. *Employer:* Howard Needles Tammen & Bergendoff. *Education:* BS/Civil Engrg/Univ of NM. *Born:* 9/28/25. An Assoc of HNTB since 1979, Mr Counts became Engrg-in-Charge of the Seattle office in 1970, after being project mgr for transportation since 1961. During 31 yrs with HNTB in Kansas City and Seattle, he has experience in civil, transportation, structural and environmental engg. This includes design, plan supervision, specification writing, and construction inspection for major highway and bridge projects. A reg PE in 11 states, he grad from the Univ of New Mexico in 1950. He is a mbr of the American Public Works Assoc, a fellow of the American Soc of Civl Engrs and a member of the International Bridge, Tunnel and Turnpike Association. *Society Aff:* ASCE, APWA, IBTTA

Couper, James R
Business: Dept of Chem Engg., 3202 BEC, Univ of Arkansas, Fayetteville, AR 72701
Position: Prof *Employer:* Univ of AR. *Education:* DSc/Chem Eng/Washington Univ; MS/Chem Eng/Washington Univ; BS/Chem Eng/Washington Univ. *Born:* Dec 10, 1925 St Louis MO. BSChE 1949, MSChE 1950, ScD 1957 WA Univ St Louis. Merchant Marine 1944-46. Res Chemist Presstite Engg Co St Louis 1950; Res Engr MO Portland Cement Co St Louis 1950; Monsanto Co St Louis 1952-57 Process Design Engr & 1958-59 Production Supervisor. Taught Chem Engg Univ of AR 1959-. Dept Hd 1968 to 1979. Chrmn 1974, Chrmn-Elect 1973, Secy 1970-72 I & EC Div ACS; Chrmn 1971-72, Chrmn-Elect 1970-71, Secy 1969-70, AIChE Dept Chrmn Group; Tau Beta Pi, Sigma Xi, Omicron Delta Kappa. elected Fellow, AIChE (1986) Enjoy golf, choral singing and gardening. Chrmn-Elect 1980, Chrmn 1981, Ch.E. Division, ASEE; Reg PE-MO, AR. Vice-President 1978 to 1981, Pres 1981 to 1986, Omega Chi Epsilon. *Society Aff:* AIChE, ACS, ASEE, AACE, Soc of Rheology

Courage, Ken G
Business: 346 Weil, Gainesville, FL 32611
Position: Assoc Prof Civ Engg. *Employer:* Univ of FL. *Born:* Dec 1939. MSCE TX A&M 1968; BSEE Univ of Manitoba 1962. Native of Winnipeg, Canada. Traffic Signals Engr Winnipeg 1961-66. Asst Res Engr TX Transp Inst 1967-68. Sr Transp Engr Kelly Sci Corp 1969-71. Assumed present pos 1971. Principal areas of int incl advanced traffic control sys, automated traffic data collection & analysis, freeway surveillance & control & transp sys mgt. Transp Res Bd Award 1972 & Inst of Transp Engrs P Pres Award 1973.

Couret, Rafael M
Home: 1727 NW 51 Terrace, Gainesville, FL 32605
Position: Electrical Mgr *Employer:* Ch2M Hill *Education:* B/EE/Univ of FL; B/IE/Univ of FL *Born:* 6/17/41 Native of Pinar Del Rio, Cuba. Electrical Design Engr for Phillips Petro Co from 1964-69 specializing in plastic plants. Chief electrical engr for DSA Engrs from 1969-1974 specializing in commercial and institutional buildings. Joined Black, Crow and Eidsness (BCE) in 1974 as Chief Electrical Engr; assumed additional duties in 1975 as Chief Instrumentation Engr; when BCE merged into ChzM Hill. Became Chief Electrical Engr for the Gainesville Region and Electrical Coordinator for the Eastern District. NSPE 1970. PE in 10 states. Enjoys music and fishing. Active in Catholic Church Ministries. *Society Aff:* NSPE

Coursey, W
Home: 2222 Central Ave, Dubuque, IA 52001
Position: Dir *Employer:* Pam Imports *Education:* BS/Chem Engg/MT State Univ *Born:* 4/1/34. Grad from Montana State Univ in 1957 with a Bachelor Degree in

Coursey, W (Continued)
Chem Engg. Employed in the Material Engg Dept of the John Deere Dubuque Works 1957-86. Sr Chem Engg with respon for directing analytical, oil and environmental analysis groups. Additionally respon for the selection and dev of OEM and Manufacturing Process Lubricants; Retired May 1986. ASLE Chmn Metalworking Fluid Committee 1972-74. ASLE Midwestern Regional VP 1977-81. ASLE Director 1981-83, ASLE Pres 1984-85. Currently owns and directs "Pam Imports" an automotive, motorcycle dealership. Also currently consults as a Lubrication Engr to industry and manufactures. *Society Aff:* ATLE

Courtney, John C
Home: 10326 Hackberry Ct, Baton Rouge, LA 70809
Position: Prof of Nuclear Engrg *Employer:* LA State Univ *Education:* DEngr/Nuclear Engr/Catholic Univ of Amer; MNE/Nuclear Engr/Catholic Univ of Amer; BCE/Civ Engr/Catholic Univ of Amer *Born:* 06/11/38 John C Courtney was born and raised in Washington, DC. He served a three year tour of active duty in the US Air Force where he was involved with nuclear res for classified applications. After leaving active duty, he joined Aerojet Nuclear Sys Co in Sacramento. He was a Phys Specialist and Supervisor of Radiation Shielding Analysis on the nuclear rocket engine proj, NERVA. He joined LA State Univ in Baton Rouge in 1971. Since 1975, he has been a Visiting Scientist at Argonne Natl Lab in Idaho. He is a Reg Professional Nuclear Engr and Certified Health Physicist. He is married to the former Peggy Roberts of Rockwood, TN. *Society Aff:* ANS, HPS

Courtney, Thomas H
Business: Department of Materials Science, Thornton Hall, University of Virginia, Charlottesville, VA 22901
Position: Prof, Dept Chairman *Employer:* Univ. of Virginia *Education:* ScD/Phys Met/MA Inst Tech; MS/Engg Phys/Cornell; BS/Met/MA Inst Tech. *Born:* 9/26/38. Res Met at Babcock & Wilcox Co, Alliance, OH, 1964-66; Res Assoc, MIT, 1966-68; Asst Prof, Assoc Prof and Prof, Dept of Mech Engg, Univ of TX at Austin, 1968-1975; Prof, Dept of Met Engrg, 1975-86, Dean Grad Sch 1983-86, MI Tech Univ., Houghton MI; Prof & Chairman, Dept. Matls Science, Univ. of VA 1986- . Have worked in res on superconducting mtls, solidification, microstructural stability, mech behavior of composites and other two-phase alloys, and liquid phase sintering. Enjoy sports and reading. *Society Aff:* TMS-AIME, ASM.

Courtney-Pratt, Jeofry S
Business: President Physics Inc, 101 Wigwam Road, Locust, NJ 07760
Position: retired *Employer:* AT&T Bell Labs *Education:* ScD/Applied Physics/Univ of Cambridge-England; PhD/Applied Physics/Univ of Cambridge-England; BE/Engineering/Univ of Tasmania-Australia *Born:* 1/31/20 Worked for CSIR, Australia and later seconded to Admiralty, London during WWII. Univ of Cambridge as doctoral student 1945-49, and later until 1958 as Asst Dir of Research. Fellow of Gonville & Caius Coll 1949-1957. Consultant on instrumentation, ballistics, friction and wear, optics, high speed photography. Came to USA in 1958 to work for Bell Telephone Labs in mechanics, optics, acoustics, instrumentation, and more recently in all forms of communications-- audio, video and graphics, particularly for teleconferencing. Honorary Member of SMPTE. DuPont Gold and Progress Medals from SMPTE, and many other awards. Over 110 technical publications and more than 30 patents. Since 1985, Pres of Physics Inc. consulting in Physics and Engineering. *Society Aff:* SMPTE, SPIE, OSA, Inst of Physics (UK), Inst of Mech Engrs (UK)

Courtsal, Donald P
Business: 4800 Grand Ave, Neville Island, Pittsburgh, PA 15225
Position: Corp VP & Gen Mgr-Engr Works Div. *Employer:* Dravo Corp. *Education:* MS/Naval Architect & Marine Engg/MIT; BS/ME/US Coast Guard Acad. *Born:* 12/30/29. Native of New Haven, CT. Married with two children. Served three yrs with US Coast Guard. Spent eight yrs with Shipbldg Div, Bethlehem Steel Corp in the Central Technical Dept. Joined Engg Works Div of Dravo Corp in 1965. Named Engg Mgr in 1974. In 1975 became Gen Mgr of the Engg Works Div & in 1976 was elected a VP of Dravo Corp. Now serving as a VP, Mbr of Council, & as a Chrmn of the Advisory Public Service Committee of SNAME. Enjoy sailing, music, reading & hiking. *Society Aff:* SNAME.

Cousineau, Robert D
Business: 420 S Pine St, San Gabriel, CA 91776
Position: Pres. *Employer:* Soils Intl. *Education:* AA/CE/Los Angeles City College; Univ. of Conneticut Special Courses/Soil Mech/USC; Special Courses/Fnd Engr/UCLA. *Born:* 9/8/16. Reg Civ Engr, CA, OR, WA. Experience in Soils and Fdn Engr beginning in 1940. Corps of Engrs - Airfield investigations in six western states; flood control, dams, testing and control. Donald R Warren Co - Investigations for chem plants, shipyards, ind plants, steel mills, CA, AZ & TX. Frederick J Converse - Consulting Fdn Engrs Gen Mgr, 6 yrs. Converse Fdn Engr Co - Co-Founder and Partner, 10 yrs. Maurseth Howe Assocs, Principal, Pres, 11 yrs. Soils Intl, 13 yrs. *Society Aff:* ASCE, SAME, ASFE, EERI

Coustry, Eric E
Business: P.O. Box 24, Frostburg, MD 21532
Position: Pres *Employer:* CECA, Inc *Education:* MS/CE/IL Inst of Tech; BS/CE/IL Inst of Tech; BA/Lib Arts/Univ of Montreal *Born:* 1/28/48 in Belgium, raised in Chevy Chase, MD. Teaching Assistant for 2 years at IL Inst of Tech. Design Engr for Harza Engrg Co for 4 years specializing in soil Mechanics and Foundations. Field Engr for GAI Consultants for 2 years during construction of Pleasants Power Station. Chief Engr for Delta Intl for 4 years. In 1979, assumed additional position of Officer Mgr. In July 1980, started my own consulting E/A firm, CECA, specializing in Civil engrg, Energy Studies, Coal Investigations and Permits, and Architecture. Enjoy woodworking and playing piano. *Society Aff:* NSPE

Coutinho, John de S
Home: 602 Westgate Rd, Aberdeen, MD 21001
Position: Gen Engr. *Employer:* US Army Mtl Systems Analysis Activity *Education:* DrIng/Aerospace/Berlin Tech Univ; MAE/Aero Engg/NYU. *Born:* 8/9/13. Dr Coutinho is a Gen Engr, US Army Mtl Systems Analysis Activity, Aberdeen Proving Ground where he developed the Army's first Battlefield Damage Assessment and Repair Technical Manuals for combat vehicles, missiles, chemical defensive material, and an attack helicopter. Prior to 1972 he worked for 33 yrs at Grumman Aerospace as a stress analyst, structural designer, control systems designer; in 1954 he became the co's first reliability analyst & developed & prepared the first failure-effect & mode analyses in the US. In 1958 he organized the Reliability Control Dept & designed & directed the reliability & maintainability progs for many aircraft & space systems; in 1962 he was appointed Relaibilty Dir, Lunar Module; in 1966 he became Special Asst to the VP - Lunar Module. Author, Advanced Systems Development Management, Wiley, 1977, and many o publications, recipient of numerous awards. *Society Aff:* ASME, ASQC, NYAS 60Dr.

Covalt, Robert B
Business: 2 N Riverside Plaza, Chicago, IL 60606
Position: Pres. *Employer:* Morton Chem, Div of Morton Thiokol Inc *Education:* MBA/-/Univ of Chicago; BSChE/Chem Engrg/Purdue Univ. *Born:* 11/8/31. Originally employed with B F Goodrich Chem Co as a Dev Engg. Served as an instructor in electronics in the USAF (1/Lt-1954-56). Served as Dev Engg, Project Engg, Process Engg. Became Asst to Pres 1967, Dir of Engg 1968, VP Engg 1972, VP Manufacturing & Engg 1973, Group VP 1978, Pres 1979. Received BSChE from Purdue Univ, 1953 and MBA from the Univ of Chicago, 1967. *Society Aff:* AIChE, ACS, SCI.

Covarrubias, Jesse S
Business: 3838 NW Loop 410 Suite 200, San Antonio, TX 78229
Position: Owner. *Employer:* Structural Engg Assoc. *Education:* Grad Work/Civ Engg/Univ of CA; BSCE/Civ Engg/Univ of TX. *Born:* 6/26/40. Was born and raised in San Antonio, Texas. Has been employed by US Bureau of Reclamation. CA Div of Hgwys, Bridge Dept; TX Hgwy Dept, Bridge Div and various structural engg consulting firms. Holds structural engg licenses in several states. Established own private practice in 1976. Is also a principle (pres) of an architectural and engg consulting firm. Has designed and constructed a multitude of maj bridges including Laredo Intl Bridge & Miguel Aleman Intl Bridge. While in private practice designed several high rise and low rise bldgs. *Society Aff:* NSPE, TSPE, PCI, NFIB, PEPP.

Covault, Donald D
Business: Atlanta, GA 30332
Position: Prof *Employer:* GA Inst of Tech *Education:* PhD/CE/Purdue Univ; MS/CE/Purdue Univ; BS/CE/Purdue Univ *Born:* 4/19/26 in Fort Wayne, IN. Prior to coming to GA Tech in 1958, he worked at the Univ of CO, WI Highway Commission and Purdue Univ. He has acted as a consultant to numerous agencies and has authored numerous articles dealing with Highway Engrg, Traffic Engrg and Transportation Planning Problems. Active in several committees in nat'l societies. Plays classical guitar and enjoys photography. *Society Aff:* ASCE, TRB, ITE, ARTBA

Cover, Thomas M
Business: Dept of Elec Engg, Stanford, CA 94305
Position: Prof - EE & Statistics. *Employer:* Stanford Univ. *Education:* PhD/EE/Stanford; MS/EE/Stanford; BS/Phys/MIT. *Born:* Aug 7, 1938 San Bernardino CA. BS Phys MIT 1960; MS, PhD EE Stanford niv 1961 & 1964. Prof Depts of EE & Statistics at Stanford Univ. Res areas: communication & info theory, pattern recog, complexity & statistics. on fac at Stanford since 1964. Cons to Stanford Res Inst, Bell Labs & Sylvania. Served in the Visiting Lectr Prog in Statistics 1971-74. Visiting Assoc Prof EE MIT and a Vinton Hayes Res Fellow Harvard 1971-72. Pres IEEE Info Theory Group 1972. Book Review Editor & Assoc Ed for the IEEE Transactions on Info Theory. Assoc Ed for Pattern Recog and Am. Math. Stat., I T Paper Award 1972. Mbr IMS, IEEE, AMS, ACM, Sigma Xi. Fellow IEEE and IMS. *Society Aff:* IMS, IEEE, AMS, ACM, ΣΞ, DSI, ASA.

Covert, Eugene E
Business: Room 37-401, 77 Massachusetts Ave, Cambridge, MA 02139
Position: Prof of Aeronautics *Employer:* MA Inst of Tech *Education:* ScD//MIT; SM/Aerodynamics/Univ of MN; B/Aero Engrg/Univ of MN *Born:* 2/6/26 in Rapid City, SD. US Navy 1943-47. Preliminary Design Aerodynamics, Naval Air Development Ctr, Johnsville, PA 1948. Prof of Aeronautics and Astronautics, MIT, 1968. Visiting Prof, Technical Univ of Berlin, 1968. Chief Scientist, US Air Force 1972-73. Technical Dir, EOARD, 1979-1980. Fellow American Inst for Aeronautics and Astronautics, Royal Aeronautical Society. American Association for the Advancement of Science. Member Nat'l Academy for Engrg, Chrmn USAF Scientific Advisory Bd, AGARD Power and Energetics Panel. Bd of Dirs, Megatech Corp,, Rohr Industries Inc, Engrs Council for Professional Development 1970-73. Exceptional Civilian Sci Award - 1973; 1986. The American Society for Aerospace Education Univ Educator of the Year-1980. NASA Public Service Award - 1981. MIT Graduate Student Council Outstanding Teacher Award-1985. Consultant to several Aerospace Companies. Served on numerous Air Force, Navy, NASA, OSTP, Nat'l Research Council committees, and on President's Commission to investigate the Challenger accident. *Society Aff:* AIAA, AAAS, RAS.

Covert, Paul D
Business: 170 N High St, Columbus, OH 43215
Position: Past Chmn. *Employer:* Stilson & Assoc., Inc. *Education:* BSCE/Struct/OH Northern Univ. *Born:* 10/28/29. Born in Seneca County, OH. Served in US Army in Japan 1946-48. Started as survey crew chief for Alden E Stilson & Assoc following grad in 1952. Proj engg on a variety of civil and sanitary projects. Mgr of Branch Office. Taken in as Partner 1968. Elected Chmn 1978. PPres of WV Soc of PE, WV Section of ASCE, OH Council, & District 9 Council ASCE. Mbr of Kiwanis and Masonic bodies. *Society Aff:* ASCE, NSPE, WPCF, AWWA, APWA.

Covlin, Robert J
Home: 2230 South Vrain, Denver, CO 80219
Position: Sr. Pet. Engr Assoc *Employer:* Amoco Production Co *Education:* B.S./Petro Engr/MT Tech. Univ *Born:* 11/3/29 Native of North Dakota. Served 2 years with Army at Redstone Arsenal as engr working on initial satellite. With Amoco since 1955 specializing in drilling, completion and production of oil and gas in the Rockies. Instrumental in the development and application of massive hydraulic fracturing, (MHF) of low permeability gas sands. Active member of SPE serving various positions locally and nationally - Co-General Chrmn of the SPE/DOE 1981 Symposium on Low Permeability Gas Reservoirs; Chrmn of Meetings Policy Committee; Technical Editor of JPT. Authored several technical papers. Active politically and a Denver businessman in Travel and Art. Hobbies - art, music, hunting, fishing, gardening. *Society Aff:* SPE of AIME, CEC

Cowan, Everett H
Business: 2147 Belcourt Ave, Nashville, TN 37212
Position: Pres *Employer:* Miller/Wihry/Lee, Inc *Education:* BCE/Univ of Louisville *Born:* 4/4/46 Born in McKenzie, TN. Attended high school in Louisville, KY. Graduated from the Univ of Louisville with a Bachelors of Civil Engrg in 1969. Employed by Miller/Wihry/Lee, Inc. (MWL) in Louisville, KY, In 1966 as Draftsman. Became Engr Proj Mgr in 1969. Transferred to MWL's Nashville office in 1970. VP and Managing Dir of the firm's Nashville branch with full responsibility for operation of the office. Joined MCI Consltg Engrs, Inc, in 1982 as VP and Mgr of the firm's Nashville office. Currently serving as Pres of the firm, Mr. Cowan directs a multidisciplinary staff encompassing engrs, planners, landscape architects, hydrogeologists, land surveyors, and support staff involved in a broad range of engrg and land planning projs. Sec/Treas Nashville section ASCE 1980-82, State Dir for TSPE, 1981-82, VP Nashville Section ASCE 1982-83, Pres Nashville section ASCE 1983-84, Treas CET 1984-85. Reg PE in TN, KY, AL, VA, GA, SC, LA, OH, NV, and IN. *Society Aff:* NSPE, ASCE, TSPE.

Cowan, John D, Jr
Business: 2015 Neil Ave, Columbus, OH 43210
Position: Prof Emeritus (ret July 1, 1983) *Employer:* The OH State Univ *Education:* MS/EE/OH State Univ; BS/ME/OH State Univ *Born:* 9/7/18 Prof of Elect Engrg, The OH State Univ. Residence 233 Montrose Way, Columbus, OH 43214. Married Natalie Brewka 1940, children John Dale III, 1943, Natasha Eve, 1947. Tool room machinist 1940-43, USAAF, B-29 Engr 1943-45. Develop & Supervise OSU Experimental Instrument Shop 1946-83, teach OSU 1950-83, Emeritus July 1983, supervise various circuit & control courses, develop electronic instruments. Book *Intro to Circuit Analysis* 1961. Summer 1977, USAID control program in Damascus, Syria. Registered PE, consulting 1955 to present, systems, Forensic Engrg, private and contract with OH Arson Crime Lab. Numerous published and un-published articles, Seminars on Electrical Fires. *Society Aff:* IEEE, HKN, ΣΞ, IAAI.

Cowart, Kenneth K
Home: 4515 Davenport St, NW, Washington, DC 20016
Position: retired *Education:* MS/ME/Univ of CA-Berkeley; BS/Marine Engrg/US Coast Guard Academy *Born:* 1/16/05 Vice Admiral, USCG, Retired 1959. Native of Twin City, GA. Served on various Coast Guard ships and stations in engrg, line, staff and command positions for 20 years after graduation in 1926 from USCG Academy. Asst Engr in Chief 1946-1950; Engr in Chief 1950-58; Member, Nat'l Council and Steering Committee SNAME, 1955-58; Registered PE, District of Columbia, 3261; Editor Encyclopedia Britannica, Shipping Dept since 1956; Pres, Propeller Club Port of Washington, DC, 1954-55; Pres, Amer. Society of Naval Engineers, 1956; Grand Paramount, Military Order of the Carabao, 1967. Member, The Army and Navy Club, Columbia Country Club and American Newcomen Society. *Society Aff:* ASNE, SNAME, SAME

Cowherd, David C
Business: PO Box 51, Dayton, OH 45401
Position: Pres & Chief Geotech Engr. *Employer:* Bowser-Morner Testing Assoc Inc. *Education:* MS/Civ Engg/Univ of KY; BC/Civ Engg/Univ of KY. *Born:* Jan 25, 1940. Native of Greensburg KY. BSCE & MSCE Univ of KY. Res Engg Assoc KY Hgwy Res Lab 1963-65. Joined Bowser-Morner 1965. Advanced to Ch Geotech Engr 1972. Pres in 1985. Respon for expanding areas of service & qual control of all geotech & related work in corp & branch offices. Reg engr 8 states. Nationwide geotech cons with present emphasis on innovative approaches to design of earthfill embankments. Speaker at V & VII OH River Valley Soils Seminars & Seminar on Lateral Pressures Generated by Pipes, Piles, Tunnels, Caissons. *Society Aff:* ASCE, ACEC, NSPE, NACE.

Cowin, Roy B
Home: 3061 Mt. Holyoke Rd, Columbus, OH 43221
Position: JETS-Executive Dir Emeritus *Employer:* Retired *Education:* J.D./Jurisprudence/Univ of Toledo; A.B./Geol/Lehigh Univ *Born:* 8/13/21 Served in Europe as a tank unit commander during WWII. Industrial experience from 1946-72 includes management positions in glass technology, quality control, and market research with either Corning Glass Works and/or Owens-IL in product lines varying from powdered glass to TV bulbs. Was a member of several engrg and technical councils of EIA. Teaching experience includes Instructor of Industrial Statistics at Capital Univ, Production Management Program and numerous Quality Control courses for Columbus Foremen's Club, Columbus and Toledo ASQC Chapters, and the AMA. Served as pres of Columbus and Toledo ASQC sections. From 1973 until 1986 served as Executive Dir of JETS and also until 1983 as ECPD's (now ABET) Guidance Dir. *Society Aff:* ASQC, ASEE

Cowin, Stephen C
Business: Dept Biomed Engrg, Tulane Univ, 6823 St. Charles Ave, New Orleans, LA 70118
Position: Alden J. Laborde Professor of Engineering *Employer:* Tulane Univ *Education:* PhD/Engrg Mech/PA State Univ; MS/CE/Johns Hopkins Univ; BSE/CE/Johns Hopkins Univ *Born:* 10/26/34 Prof S.C. Cowin has been active in teaching and research in mechanics and biomechanics since receiving his Ph.D. in 1962. In 1963 he went to Tulane Univ where he has made significant contributions to the understanding of the mechanical behavior of such complex materials as granular media, wood, and bone. In 1985, he assumed his current position as Alden J. Laborde Prof of Engrg in the Dept of Biomedical Engrg. He also holds an appointment as Adjunct Professor in the Department of Orthotropic Surgery in the School of Medicine and Professor of Applied Statistics in the Graduate School. *Society Aff:* ASME, AAAS, ΣΞ, AAM, BMES, ASR

Cowles, Walter C
Home: 55 FieldStone Dr, Morristown, NJ 07960
Position: Consultant *Employer:* self *Education:* BSE/Naval Arch/Univ of MI *Born:* 8/25/19 Employed by The American Ship Building Co, Cleveland, OH, as design draftsman. Appointed Chief Hall Draftsman in 1951 and Naval Architect in 1957. Joined Esso Intl Co in 1963. (Now Exxon Intl Co). Experience with Exxon includes design of Ocean going tankers of all sizes, liquified natural gas carriers, floating drilling rigs and other miscellaneous types. Retired; consultant. *Society Aff:* SNAME, ASNE, NECI, USNI

Cox, Allen L
A-10 Dorado Del Mar, Dorado, P.R 00646
Position: Cons Engr (Prinicpal). *Employer:* Aqua-Terra Engrs Inc. *Born:* Nov 1934. BS & MS LA State Univ. Asst Prof of Agri Engg at LSU 1963-66. Asst Res Engr hydraulics LA Dept of Hgwys 1966-69. Asst Bridge Design Engr hydrology & hydraulics LA Dept of Hgwys 1969-73. Principal owner & cons engr Aqua-Terra Engrs Inc Baton Rouge 1973-. Spec in drainage, erosion control, hydrology, hydraulics, & irrigation. Mbr Transpor Res Bd Ctte on hydraulics, hydrology, & water qual. Am Soc of Civ Engrs, Am Soc of Agri Engrs, Am Council of Cons Engrs, & LA Engg Soc. Prof registration in LA & TX.

Cox, Alvin E
Business: 2 World Trade Ctr-9528, New York, NY 10048
Position: VP. *Employer:* J J Henry Co Inc. *Education:* BS/NA&ME/Webb Inst of Naval Arch. *Born:* 5/25/18. Grad of Oak Ridge Sch of Reactor Tech; completed mgt course Carnegie-Mellon Inst. USN, Lt 1944-46. Joined Newport News Shipbldg as a naval architect. Left Atomic Power Div as Sr Design Supr to hd CVAN-65 design group loaned to BuShips. Officially commended. Appointed Asst Ch Naval Architect 1958. Also served as Tech Dir FDL Prog. Appointed LHA Proj Mgr 1966. Upon completion of assignment, appointed DLGN 35 & 36 Prog Mtr. In 1969 assigned respon as Sr Prog Mgr to coord all prog mgrs. In 1973 assigned respon as Mgr of all Merchant Ship Constr & Repair. In 1975 joined J J Henry Co Inc as Asst to the Pres. Mbr Naval Arch Ctte of Amer Bureau of Shipping. Served 2 terms on Ship Struct Ctte Natl Acad of Sci/Natl Res Council. Council Mbr 1970-76 Soc of Naval Architects & Marine Eng. *Society Aff:* ASNE, SNAME.

Cox, Carl M
Business: PO Box 1970, Richland, WA 99352
Position: Mgr, Space Reactor Test Site *Employer:* Westinghouse Hanford Co *Education:* DSc/Nuclear Engrg/Univ of VA; BS/Mech Engrg/Univ of VA *Born:* 12/10/38 Dr Cox has managed nuclear tech organizations at Westinghouse Hanford Co since 1971. He is currently mgr of SP-100 Ground Engrg. System test site, a nuclear reactor test facility being constructed to support devel of 50 to 1000 kwe space nuclear power supplies. He previously directed design and irradiation testing of nuclear core components, hot cell examinations, dev of nuclear ceramic prods, and assurance of proper performance of reactor core components. He managed the extensive developmental fuels irradiation and testing prog for the Fast Flux Test Facility and the startup testing and operation of FFTF core components. Earlier experience included engrg and mgmt positions at Oak Ridge Natl Lab and the Savannah River Plant. *Society Aff:* ANS, ΣΞ

Cox, Don Emery
Business: PO Box 4162, Corpus Christi, TX 78469
Position: VP & Gen Mgr. *Employer:* Refinery Terminal Fire Co. *Education:* BS/ChE/OK State Univ. *Born:* 6/12/21. With PPG Industries, Chem Div 35 yrs in R & D, Planning, Marketing, Engg. Projects in USA, Mexico, Southern Africa, Canada. Chem raw materials, processes, products, market dev, plant locations studies, economic analysis. Reg PE Tx, patentee, author. Has also applied engg principles to solving community problems: developing regional water supply, promoting intergovernmental cooperation, improving environmental quality. Enhanced relationships between engrs of Southern TX and Northern Mexico. Chmn of Coastal Bend Section AIChE; National Service to Society Award, AIChE 1978. Fellow AIChE. Pres of Lower Nueces Rvr Water Supply Dist 1972 to date. Mbr Exec Ctte of Coastal Bend Council of Govt's 1966 to date. Presently managing co providing fire fighting services to 27 industries around Port of Corpus Christi. Also consultant on engg, economic, environmental matters. *Society Aff:* AIChE, ACS, APCA.

Cox, Donald C
Home: 24 Alden Ln, Tinton Falls, NJ 07724
Position: Div Mgr *Employer:* Bell Comm Res *Education:* PhD/EE/Stanford Univ; MS/EE/Univ of NB; BS/EE/Univ of NB. *Born:* 11/22/37. Res & Dev Officer in USAF 1960-63, Lt; microwave communication system design. Stanford Univ, Res Asst 1963-67, Res Assoc 1968; res in microwave amplifiers, microwave measurement techniques, & microwave radio propagation. Bell Labs 1968-83; res in mobile radio propagation, mobile & earth-satellite communication systems, microwave amplifiers, microwave & millimeter wave measurement techniques, microwave & millimeter wave earth-space radio propagation; Proj Supervisor for Comstar Satellite-Beacon propagation experiment. Bell Comm Res, Div Mgr 1984-; res in radio and

Cox, Donald C (Continued)
Satellite Comm Sys and propagation. Fellow IEEE; Recipient Guglielmo Marconi Prize in Electromagnetic waves Propagation, Inst. Internat. Communications (Italy) 1983; IEEE Morris E. Leeds Award 1985; Hon Dr of Sci from Univ of NB. Reg PE two states; mbr: US Comm B,C & F of URSI; Sigma Xi; Sigma Tau, Eta Kappa Nu, Fellow AAAS. *Society Aff:* IEEE, URSI, AAAS, ΣΞ.

Cox, Elmer J
Business: 49th St & AVRR, Pittsburgh, PA 15201
Position: Pres. *Employer:* Pittsburgh Comm Heat Treating Co. *Education:* BS/Metturgical Engg/Carnegie Tech. *Born:* 6/23/22. *Society Aff:* ASTM, ASM, AFS.

Cox, Ernest A, III
Business: Suite 243, 133 West Oxmoor Rd, Birmingham, AL 35209
Position: VP *Employer:* ATEC Assocs Inc *Education:* MSE/Soil Mech/Univ of FL; BS/CE/Univ of FL *Born:* 4/30/47 Native of Chipley, FL. After graduating from Univ of FL in 1971, was employed by Law Engrg Testing Co in Atlanta, GA as a staff geotechnical engr. Promoted to senior engr on rapid transit project in 1975. Joined ATEC Assocs, Inc as Atlanta district office chief engr in 1977. Became VP and district mgr of ATEC's Birmingham, AL district office in 1978. Treasurer, Cobb County, GA branch of GSPE, 1977. State membership chrmn, AL ACEC, 1978. Enjoys all outdoor sports and daughter and son. *Society Aff:* ACEC, ASCE, NSPE

Cox, Geraldine V
Business: 2501 M St, NW, Washington, DC 20037
Position: VP & Tech Dir *Employer:* Chemical Manufacturers Assn. *Education:* PhD/Environmental Science/Drexel University; MS/Environmental Science/Drexel University; BS/Biological Sciences/Drexel Institute of Technology *Born:* 01/10/44 Assumed current position of VP-Technical Dir, Chemical Manufacturers Assoc, in 1979. Direct technical activities for chemical industry, i.e., toxic substances, occupational health and safety, environmental programs (air, water, solid waste), transportation and distribution. Also direct management of National Chemical Response & Information Center, including, CHEMTREC, CHEMNET, and CRC; and Community Awareness & Emergency Response (CAER) program. Member of numerous professional associations and advisory panels. Honors received include: One of Ten Outstanding Young Women in America, 1975; White House Fellow, 1976-77; SWE Engineering Achievement Award, 1984; Engineering and Science Award, Drexel University, 1987. Hobbies are photography and mineralogy. *Society Aff:* ASTM, SWE, WPCF

Cox, J Carroll
Business: Box 491, Spartanburg, SC 29304
Position: VP & Corporate Legal Officer. *Employer:* Lockwood Greene Eng Inc. *Education:* BS/Architectural Engg/Clemson Univ. *Born:* 6/2/33. in Greenville, SC, joined Lockwood Greene in 1955 as structural designer, promoted to mgr of structural engg in 1966, VP & Chief Engr in 1973. Corporate Director in 1980. Corporate Legal Officer, 1984. Served in US Army Ordnance Corp 1957, served on Civ Engg Advisory Bd for Clemson Univ 1974 & 1975. Served as Pres of Carolinas Chapter ACI in 1975. Mbr ACI, PCI, ASCE, CESC. Reg PE in 11 states. *Society Aff:* ACI, PCI, ASCE, CESC.

Cox, J Robert G
Business: Satellite & Aerospace Systems Div, Quebec, H9X 3R2, Canada
Position: Mgr Sys Engrg Satellite & Aerospace Sys Div *Employer:* Spar Aerospace Limited, Ste Anne de Bellevue. *Education:* B Eng/Eng Physics/McGill Univ. *Born:* 4/14/27 in Montreal Canada. Educ Univ of British Columbia & McGill Univ, receiving B Eng (Engrg Phys) 1949. Member and formerly Dir & Mbr of Exec Ctte, Ordre des ingenieurs du Quebec. Mbr Exec Ctte Canadian Natl Org of CCIR. Mgr Sys Engr, Satellite & Aerospace Systems Div., Spar Aerospace Ltd. Ch Engr G&CS Div RCA Ltd Montreal 1970-77; Mgr Aerospace Engg 1965-70. Mgr Mil Sys, Sperry Gyroscope Co of Canada; previously Mgr Dev Engg 1955-65. Proj Engr IT&T Canada 1951-55. Engr Canadian Marconi Co 1950-51. Res Scientist Defense Res Bd 1949-50. *Society Aff:* OIQ, AIAA.

Cox, Jim E
Business: 1825 K Street, NW, Suite #215, Washington, D.C 20006
Position: Director, Government Affairs *Employer:* ASHRAE *Education:* Ph.D./M.E./Oklahoma State University; MSME/M.E./Southern Methodist University; BSME/M.E./Southern Methodist University *Born:* 08/29/35 Native of Texas. Reg PE TX. Fellow ASME. 1975 Congressional Fellow with the Science & Technology Cttee, U.S. House of Representatives. Prof of Mechl Engrg, Univ of Houston 1963-81 specializing in heat transfer and thermodynamics; recipient of both the College of Engrg and the Univ Teaching Excellence Awards. Dir of Govt Affairs for ASHRAE 1981 to present. Responsible for interacting with the U.S. Congress and the Federal Executive Branch on technical issues and for serving as liaison for ASHRAE with the Washington offices of industry, trade associations and other tech & professional societies. *Society Aff:* ASME, ASHRAE

Cox, Joe B
Business: 244 Perimeter Ctr Pkwy, NE, Atlanta, GA 30346
Position: Dir of Corp Engrg and Supply *Employer:* Gold Kist Inc. *Education:* MBA/Business/Emory Univ; MS/Mech Engr/GA Inst of Tech; BS/Mech Engr/Clemson Univ. *Born:* 3/20/37. Native of Greenville, SC. Engr with Douglas Aircraft 1959-60. US Army Transportation Res and Engg Command 1960. Sr Res Engr, aerospace projs, Brown Engg Co, 1962-1966. Dir of Corp Engrg, and Supply, Gold Kist Inc 1967-present. Mbr Tau Beta Pi, Beta Gamma Sigma. Participates in NSPE-PEI. Reg PE GA, FL, AL, AR, TX. Mbr Dunwoody Baptist Church. Wife Carolyn Boothe Cox, son Brad age 15, daughter Susan age 9. Enjoy golf, flying, swimming, travel. *Society Aff:* ASME, SAE, NSPE, AIAA.

Cox, Philip L
Business: 108 E Green St, Ithaca, NY 14850
Position: City Engr. *Employer:* City of Ithaca. *Education:* MSCE/Sanitary Engg/Wayne State Univ; BSCE/Civ Engg/MI Tech Univ; AAS/Civ Tech/Broome Tech Comm Col. *Born:* 10/15/46. Native of Binghamton, NY. After OCS, served in Army Corps of Engrs in Detroit as officer & civilian. Hydraulic Engr there on Great Lakes projs from 1969 to 1975. Since 1975 have been City Engr in Ithaca, NY, in charge of engg div whose work includes all facets of public works. Responsibilities include supervision of design & construction administration & supervision of consulting profls when used. Pres, Ithaca Sec ASCE, 1979-80, & past-secy, Detroit Post, SAME. *Society Aff:* ASCE, APWA.

Cox, Ronald B
Home: Chattanooga, TN 37401
Position: Dean. *Employer:* Univ of TN at Chattanooga. *Born:* Sept 1943. PhD Rice Univ; BS & MS Univ of TN. Native of Chattanooga TN, MBA, Vanderbilt Univ. E I duPont de Nemours Co 1965-66. Later served as Dir of Engg with Indus Boiler Co involved with design, res & dev of asphalt plant equip. With Univ of TN since 1970. have served as Dir of Coop Engg Prog & Prof of Engg. Director, of Res Prog spon by US Dept of Transp - 1 of 5 progs in nation, Dir of Engg Res. , Var cons assignments. Engg and Management. Eminent Engr Tau Beta Pi; Sigma Xi; Order of the Engr; Enjoy sports of all kinds. Married, 3 children. Listings: Who's Who in America, Who's Who in South, Outstanding Young men of America, Notable Americans. Member: NSPE, TSPE, ASME, ASEE, Chattengrs Club, Chatt Chamber of Commerce. Reg. Prog Engr. (PE).

Cox, William A, Jr
Business: Consulting Forensic Engineer, 2309 Broad Bay Road, Virginia Beach, VA 23451
Position: Consulting Engr/Principal *Employer:* Self *Education:* BS/ME/VA Tech *Born:* 06/17/13 Reg PE VA & FL. Consltg Forensic engrg in VA Beach. 1946-

Cox, William A, Jr (Continued)
Present. Pres Cox- Powell Corp, Mech & Genl Contractors Norfolk 1946-1984. Founder Mbr (Diplomate) & Pres 1985 Natl Acad of Forensic Engrs (NAFE). Founder Mbr, Charter Bd Mbr, Past Pres (1953-54), Past Natl Dir VA Soc of Prof Engrs (VSPE). Engr of the Year Tidewater Chapter, VSPE 1972. Engr of Yr VA 1973, 1981. Life Mbr, Past-Pres 80/81 Natl Soc of Prof Engrs; Founder Mbr PEC Pract Div. Mbr (1972-82) Hampton Rds Sanitation Dist Comm & Past Chrmn (75-77). Past Pres: Kiwanis Club of Norfolk (1959), Blds & Contractors Exchange (1956), Tidewater Chapter VA Tech Alumni Assoc. Mbr Bd of Dir (1962-present) & Past Chrmn Bldg Comm VA Beach Genl Hosp. Life mbr ASHRAE (Fellow). Mbr: ASME, Engrs Club of Hampton Rds, Order of the Engr, VA Assn of the Prof (VAP), Reserve Officers Assn (Life), The Retired Officers Assn, US Power Sqaudrons, Langley Yacht Club, Boat/US, Bahamas Air Sea Rescue Ass'n (BASRA). Past Commodore Cavalier Yacht Club, P V-Commodore Bay Harbor Yacht Club. Mbr: Ctte of 100 Coll of Engrg, Hokie Club & Century Club of VA Tech. "The Most Outstanding Mbr of the Assoc Professions, 1980" awd of VAP. "The Hd Pecker Awd" 1980-81 of the PE in Construction (PEC), Practice Div of NSPE. Honored as "Distinguished Alumnus" College of Engineering, VA Tech, Annual Commencement June 8, 1985. Twice commissioned a "KY Colonel–. US Army WWII 1941-46, ILT - Col, Artillery (AA), included ETO Service on Staff of General George S Patton, Jr's Third US Army, rec'd Bronze Star w/Oak Leaf Cluster, Fourraguerre (Belgium), Croix de Guerre (France). Colonel AUS-Ret. P Comdr VA Beach Post VFW. Wife Sue Hume Cox; 3 sons. Specialization: forensic consltg (mechanical). *Society Aff:* NSPE, NAFE, ASHRAE, ASME

Cox, William E
Home: Route 2, Box 272, Denver, NC 28037
Position: Exec VP *Employer:* Talbert, Cox & Assoc, Inc *Education:* BS/CE/Univ of SC *Born:* 07/07/28 Experience includes five yrs with the DuPont Co early in career and six yrs with the Federal Aviation Agency. Past twenty seven yrs have been spent in private consltg work specializing in airport planning and design. Served as Principal in Charge or Proj Mgr for over 400 airport planning and design projs ranging from small gen aviation airports to complete new intl air carrier airports, both domestic and foreign. Expertise includes new airports, runways, taxiways, lighting sys, navigational aids, hangars, cargo bldgs, terminal areas, master planning and environmental impact assessments. Testified before Congressional Committees on transportation matters. Past Pres of Airport Conslts Council. *Society Aff:* ACEC, ACC, NSPE, SAMA

Coyle, Harry M
Business: Dept of Civil Engrg, College Station, TX 77843
Position: Prof *Employer:* TX A&M Univ *Education:* PhD/CE/Univ of TX-Austin; MS/CE/MIT; BS/CE/US Military Academy (West Point) *Born:* 1/7/27 Served in the Army Corps of Engrs 1950-1962. Taught at Univ of TX-Austin 1962-65. Teaching and research at TX A&M Univ 1965-present. Teaching and research in Geotechnical Engrg specializing in soil-structure interaction, deep foundations, and retaining structures. Currently, Head of Geoengrg Group and Dir of Ctr for Marine Geotechnical Engrg. Registered PE (TX) and Pres, TX Section, ASCE 1980-1981. *Society Aff:* ASCE, ASTM

Coyne, James E
Business: 105 Madison St, Worcester, MA 01601
Position: VP - Tech Dir. *Employer:* Wyman-Gordon Co. *Born:* Sept 25, 1925 Springfield, MA. BS Met from Univ of Notre Dame 1953; MS RPI 1957. Worked at Pratt & Whitney Aircraft, United Tech Corp, E Hartford CT for 7 yrs in Mtl Dev Lab before coming to work at Wyman-Gordon. Mbr ASM, AIME, FIERF. Reg PE MA. Principal ints in titanium & nickel-base alloys in the area of microstructural control & the effects of plastic deformation on mech properties. Has authored many papers & is a frequent lectr at tech mgts. Most recent ints are centered around the powder met assoc with turbine hardware.

Cozzarelli, Frank, Jr
Home: 3 Van Reyper Pl, Belleville, NJ 07109
Position: Patent Manager *Employer:* Union Carbide Corp. *Education:* Juris Dr/Law/Seton Hall Univ; MS/ChE/NJ Inst of Tech; BS/ChE/NJ Inst of Tech. *Born:* 12/13/25. Native of Belleville, NJ. Employed by Union Carbide Corp since 1952 with experience as polyolefins Patent Manager, mgr plant engg, product distribution mgr and bus team mbr for ethylene oxide derivatives. Speciality includes plant design for chem and plastics. Elected a fellow in AIChE and served as past chrmn North Jersey Sec AIChE. Mbr US Navy 1944-1946. A mbr of the State of NJ Bar and a patent attorney assoc with the law firm of Cozzarelli, Mautone & Nardachone, Belleville, NJ; also patent manager for Unipol Systems/Polyolefins Div of Union Carbide Corp. *Society Aff:* AIChE, ABA, NJBA, NJPLA.

Crabb, William A
Business: P. O. Box 8405, Kansas City, MO 64114
Position: Partner *Employer:* Black & Veatch, Consulting Engrs *Education:* BSE/Arch/KS State Univ; *Born:* 3/11/25 Native of KS. Served in the Army Air Corps 1943-46 and U.S. Air Force 1951- 52. Joined Black & Veatch Consulting Engrs in 1948. Initial years with the firm included structural design engrg assignments. Since 1953, responsibilities have been in the Management Services Division of the firm related to engagements including property valuations, utility rate studies, financial feasibility analyses and other economic and financial matters. Many of the studies are presented in testimony as an expert witness before regulatory bodies and courts. Partner in the firm since 1964. Registered PE in 12 states. *Society Aff:* NSPE, ASCE, MSPE, AWWA

Crabtree, Samuel E
Business: P.O. Box 987, Antioch, CA 94509
Position: Owner *Employer:* Crabtree Engrg *Education:* BS/CE/Univ of CA-Sacramento; AA/Engrg/LA City Coll *Born:* 3/1/35 Native of CA. Worked for the City of Woodland, CA in 1960; CA Div of Highways (now CALTRANS) 1960-63; City of Walnut Creek, CA, 1964. In private practice for W.J. Hargreaves in 1964. Owner and chief engr of Crabtree Engrg since 1964. Also a partner in S & J Engrg Contractors, underground contractor, 1967-70. Has held various positions in the societies of which he has been a member. *Society Aff:* NSPE, ASCE, ACSM, EGCA

Cragon, Harvey G
Business: P.O. Box 226015, M/S 238, Dallas, TX 75266
Position: TI Sr Fellow. *Employer:* TX Instruments Inc. *Education:* BS/EE/LA Poly Inst. *Born:* 4/21/29. At TX Instruments for 20 yrs and have contributed to the design & dev of several digital computers. These include: Minicomputers, TI 870, 960, 980, & 990; the Advanced Scientific Computer in use at GFDL Princeton Univ, Naval Res Lab, & TX Instruments. Formulated the TX Instruments Microprocessor Strategy of a microprocessor/minicomputer family & directed the execution of the strategy for two yrs. *Society Aff:* NAE, IEEE.

Cragwall, Joseph S, Jr
Home: 4901 English Dr, Annandale, VA 22003
Position: Professional Engr/Hydvologist (Retired) *Education:* BCE/General CE/Univ of VA. *Born:* 8/3/19. Native of Richmond, VA. Plant maintenance engg with Solvay Process Co, 1940-41. With US Geological Survey of Dept of the Interior since 1941 as hydraulic engg and hydrologist, culminating in positions of Chief Hydrologist, 1974-79, and Assoc Dir, 1979-80 (Retired), Mbr Virginia Water Control Bd 1982-86, Commissioner (VA) ORSANCO 1983-86, Recipient of the Dept of the Interior's Distinguished Service Award, 1976. Former officer of Chapters, Branches, and Sections in ASCE and NSPE since 1962. Engg-of-Yr Award, Fairfax Chapter, VA Society of Prof Engg, 1975, 1979 & 86. Eng of Yr Award, Chapter level 1975, 1979; Eng of Yr Award, State level, 1986. Reg PE, VA. *Society Aff:* ASCE, NSPE, AWRA, AGU.

Craig, Donald A
Home: 1606 Eudora St, Denver, CO 80220
Position: VP. *Employer:* Metal Treating & Res Co. *Born:* Mar 21, 1924 Denver CO. Met Engg CO Sch of Mines 1948, also grad work. Army Corps of Engrs 1944-46. ASTP Prog Rutgers Univ (CE). Dist Met Panhandle Dist Phillips Petrol Co 1948-52. Fabrication Supt the Dow Chem Co 1952-62. Was US delegate to the first Intl Conf on Beryllium London Eng Oct 1961. Joined Metal Treating & Res Co 1962. Active in church work, CO Sch of Mines Alumni Assn - Bd of Dir. Hobbies incl photography - CO ghost towns. Reg PE CO.

Craig, Edward J
Home: 25 Pepper Hollow, Clifton Park, NY 12065
Position: Dean of Engrg *Employer:* Union Coll *Education:* ScD/EE/MIT; BS/EE/Union Coll *Born:* 7/17/24 in Springfield, MA. Served as a navigator (B-29) in the Army Air Corps in WWII. Teaching asst and instructor in EE at MIT 1949-53; Asst/Assoc Prof EE at Northeastern Univ, 1953-56; Union Coll since 1956, Worked part time as a design engr for Gen Elec Co in microwave tubes, 1956-70. Research in communications systems at Northeastern Univ 1953-56. Dean of Engrg since 1984. PE in NY. *Society Aff:* IEEE, ΣΞ, HKN, TBΠ, ASEE.

Craig, Harold O
Home: 22162 Longeway Rd, Sonora, CA 95370
Position: Owner *Employer:* Osborne Engrg Assocs *Education:* AS/CE/Coll of Marin *Born:* 4/24/23 in San Francisco, CA; Ship Design and Construction 1939-42; US Naval Aviation 1942-45; Asst Resident Engr, Bon Tempe and Cherry Valley Dams 1947-55; Resident Construction Engr, Canyon-Cherry Power Development 1955-60 and San Francisco Intl Airport 1960-64; Project Mgr for contractors on varied highway and power projects, 1964-68; Project Engr, BART Balboa Park Station 1968-70; Quality Control Engr, New Melones Project 1970-73; In 1973, established own business "Osborne Engrg Assocs" in Sonora, CA. Registered Civil Engr-CA. *Society Aff:* ASCE, ASTM, USCOLD, CCSCELS

Craig, John H
Business: 227 Church St, New Haven, CT 06510
Position: VP-Network. *Employer:* S NE Tel Co. *Education:* SMEE/EE/MIT; SBEE/EE/MIT. *Born:* 10/12/16. Started June 1939 at Bell Tel Labs. Stayed until May 1958 having worked on equip design of radar, toll and local telephone equip, and systems engg. Transfered to AT&T Co as private line & TWX Engr. 1961 I joined OH Bell as dir of marketing. Yr later I joined Southern New England Tel Co as VP-Engr, later as VP-Network operations and currently as VP-network. In 1961-62 I was natl pres of Eta Kappa Nu. Senior Member of IEEE. *Society Aff:* IEEE.

Craig, William A
Business: 708 Leeward Ave, Beachwood, NJ 08722
Position: Pres *Employer:* William A Craig, Inc *Born:* 05/06/18 CA PE license, ASQC Certified Quality Engr, Fellow of IQA (UK), Sr Mbr of ASQC. Set up courses, wrote curricula for Metrology, NDE and Quality Assurance Engrg and taught them for two years at NWTI, Green Bay, WI. Specialization was Quality & Reliability Control. Served as mbr of ASQC Pubs Mgmt Bd. Have been active with the Inspection Div of ASQC since 1975. Was chrmn of the Inspector's Handbook Ctte, Treas for two terms, V-Chrmn of Tech Affairs for two terms, and served as V-Chrmn of Regional councilors. In 1978 chapter Secy and in 1979 was chapt V-Chrmn of ASNT. *Society Aff:* IQA, ASQC

Crain, Cullen M
Business: 1700 Main St, Santa Monica, CA 90406
Position: Hd Engrg & Appl Sci *Employer:* Rand Corp. *Education:* PhD/EE/Univ of TX; MS/EE/Univ of TX; BS/EE/Univ of TX. *Born:* 9/10/20. Native of TX. Worked on airborne radar dev at Philco after receiving BS in EE from Univ of TX in 1942. Instr at Univ of TX 1943-44. Active duty in Naval Res with Bu Ord and ORI 1944-1946. Asst and later Assoc Prof of EE Univ of TX 1946-57 and Mbr of staff of Elec Engg Res Lab. Group leader and presently Dept Hd of Engg and Appl Sci, Rand Corp Santa Monica, CA 1957-. Inventor of airborne microwave refractometer. Life Fellow IEEE. Mbr JGAC 1975-1979 Chrmn 1978- 1979. Mbr of num gov adv bds and councils. Member of National Academy of Engineering, Who's Who In Amer. *Society Aff:* IEEE.

Crain, Richard W, Jr
Business: Dept of Mech Engg, Pullman, WA 99164
Position: Chrmn, Mech Engr. *Employer:* WA State Univ. *Education:* PhD/Mech Engr/Univ of MI; MS/Mech Engr/Univ of WA; BS/Mech Engr/Univ of WA. *Born:* 7/2/31. Grew up in Seattle, WA. Acting Instr, Gen Engg, Univ of WA, 1954-55. US Navy, 1955-61; ship repair and new construction, 1957-60. Ford Fdn Fellowship, Univ of MI, 1961. Instr, Univ of MI, Dearborn Campus, 1964-65. With Wa State Uiiv since 1965. Chrmn, Mech Engg since 1976. Registered PE, Acting Dean, College of Engrg, Jan 1983 to Aug 1984. Summer employment, Battelle NW Labs, 1966, 1967 and Lawrence Livermore Labs, 1968. Active on faculty committees including Univ Senate Chrmn, VChrmn and Exec Secretary. ABET accreditation visitor for engg progs representing ASME 1977-82. Boy scouting activities include scoutmaster since 1966, Silver Beaver and Eagle Scout awards. *Society Aff:* ASME, ASEE, AAAS, ΣΞ.

Cranch, Edmund T
Business: 242 Carpenter Hall, Ithaca, NY 14853
Position: Dean, College of Engg. *Employer:* Cornell Univ. *Born:* Nov 15, 1922; m. 1946; c. 3. BME Cornell 1945; PhD 1952. Mbr tech staff Appl Mech Bell Tele Labs 1947-48; Asst Prof to Assoc Prof Mech & Mil Engg Cornell univ 1951-56, Prof & Hd of Dept 1956-62, Prof Mech Engg 1962-, Assoc Dean Engg 1967-72, Dean Engg 1972-. NSF Fac Fellow Stanford 1958-59, Sr Fellow Swiss Fed Inst Tech 1964-65, Cons Calspan 1955-61, Aerojet-Genl Corp CA 1957, Bausch & Lomb NY 1961, IBM 1963-66, Battelle Mem Inst 1963-65; USN 1943-46. Fellow ASME; Mbr Soc Exper Stress Analysis; Am Soc Engg Educ Bd of Dir, Tool Steel Gear & Pinion Co Cinn, LASPAU Cambridge MA, Tompkins Cty Trust Co, Albany Med Ctr.

Crandall, Clifford J
Home: 715 Glenwood Dr, Picayune, MS 39466
Position: Dir, Hydrographic Dev Div. *Employer:* US Naval Oceanographic Office. *Born:* Dec 1927. Bach Syracuse Univ. Native of Ft Ann NY. Enlisted Navy 1945. Began career with CofE 1951. 1957 joined Autometric Corp as proj engr on radar recon sys. From 1959-64 with Army Engr R&D Lab. 1964 joined staff, Ch of Engrs and proved feasibility of topographic radar mapping through cloud cover. Became Dir Hydrographic Dev Div, Naval Oceanographic Office 1968. Elected Potomac Region Pres ASP 1968. Elected 2nd VP Am Soc of Photogrammetry 1976. Mbr Am Congress on Surveying & Mapping; Mbr Working Group on Oceanic Cartography, Intl Cartographic Assn; Mbr SAME. Tennis, guitar enthusiast. Elder Presby Church.

Crandall, John L
Business: Protection Mutual Ins. Co, 300 So. Northwest Hwy, Park Ridge, IL 60068
Position: VP, Dir of Und *Employer:* Protection Mutual Ins Co *Education:* BS/FPE/IL Inst of Tech *Born:* 04/17/27 Native of Chicago, IL. Served US Navy Air Corp 1945-46. Grad of IL Inst of Tech, Jun, 1951 with BS in Fire Protection Engrg. Employed by FIA (now IRI) 1951 to 1965 as inspector/engr and underwriter of HPR (Highly Protected Risks) properties, inspected, recommended fire protection, approved plans, and final installation of fire protection. Worked for Kemper Ins 1966-1971 inspecting, recommending fire protection, underwriting & sales of insurance for HPR properties. Employed by Protection Mutual Ins Co 1971 as asst ot Dir of Underwriting, 1973 VP Underwriting, 1978 to present VP-Dir of Underwriting. Received IIA (Ins Inst of Amer). Certificate in Gen Ins, 1969; CPCU (Chartered Property & Casualty Underwriter) Designation, 1972; Assoc in Mgmt Cert IIA, 1976. Charter Mbr of the Soc of Fire Protection Engrg and the Chicago Chapt. Charter Mbr of Suburban Chicago Chapt CPCU. Served as Treas 1978/79, Secy

Crandall, John L (Continued)
1978/80, Pres 1980/81, Dir 1981/82. National Dir, Soc CPCV 1987. *Society Aff:* SFPE, CPCU

Crandall, L LeRoy
Business: 711 N Alvarado St, Los Angeles, CA 90026
Position: Pres. *Employer:* LeRoy Crandall & Assoc. *Education:* BS/Civ Engg/Univ Calif Berkeley *Born:* 2/4/17. Native of San Diego, CA. With Dames & Moore 1941-54, Partner 1947-54. Pres LeRoy Crandall & Assoc, Cons Geotech Engrs 1954-present. Reg CE CA, AL, AZ, CO, FL, GA, HI, IL, NV, NM, OK, TN, TX, UT, WA. Commissioner CA Seismic Safety Comm; Chrmn CA Strong Motion Instrumentation Ctte; Bd of Dir LA YMCA & Hollywood Presby Hosp; Mbr Tau Beta Pi, Chi Epsilon, Rotary Club, & Town Hall. 1983- LA C of C "Achievement Award–; 1982- ASCE "Martin S. Kapp Fd Engrg Award–; 1976- CCCELS "Dist Service Award–; 1974- IAE "Engr of the Year Award–. *Society Aff:* ΤΒΠ, ΧΕ, ASCE, ACEC, EERI, AAAS, APWA, ASTM, SSA, SEA.

Crandall, Stephen H
Business: MIT, 3-360, Cambridge, MA 02139
Position: Prof. *Employer:* MIT. *Education:* PhD/Math/MIT; ME/Mech Engrg/ Stevens Inst of Tech *Born:* 12/2/20. Entire career since '46 in Mech Engg dept MIT, appointed prof '58, Ford Prof of Engg '75. Specialist in appl mechanics, random vibration, acoustics, and rotor dynamics. Active in teaching, res, and consulting. Published 8 books, more than 100 technical papers. Received Centennial Medal '70 from Stevens, Worcester Reed Warner Medal '71 and Centennial Medal '80 from ASME, Trent-Crede Medal '78 from ASA, Theodore von Karman Medal '84 from ASCE, Recipient of Festschrift volume "Random Vibration - Status and Recent Developments–, commemorating 65th birthday. Visiting appointments at Imperial Coll, London '49, Faculte des Sciences, Marseille '60, Univ of CA, Berkeley, '64-65, Univ of Mexico '67, Harvard Univ '71-72, Ecole Natl Superieure de Mecanique, Nantes '78. Lady Davis Visiting Professor, Technion, Israel, 1987. Chrmn US Natl Committee Theoretical and Appl Mechanics '72-74. VP ASME '78-80. Foreign Mbr. Polish Soc Theoretical and Appl Mechanics Society Aff; NAE, ASME, NSPE, ASEE, SIAM, ASA, AMS, AAM, AAA&S, AAAS. *Society Aff:* NAE, ASME, NSPE, ASEE, SIAM, ASA, AMS, AAM, AAA&S, AAAS

Crane, Jack W
Business: New Holland, PA 17557
Position: Res. & Dev Coordinator *Employer:* Ford New Holland. *Education:* MS/Agri Engg/MI State Univ; BS/Agri Engg/MI State Univ. *Born:* 9/21/32. Profl experience with Ford New Holland Sub. of Ford Motor Co. Advanced to present position of Res and Dev Coordinator for North America. Active in the America Soc of Agri Engrs where he served on various committees and as state chrmn for the PA sec. Also, numerous natl level committees including chrmn of the Power and Machinery Div. and Dir of ASAE. Reg Profl Mech and agri Engr in the State of PA and MI. A mbr of the Natl Soc of PE and their local chapter. Chaired several committees for the chapter and served as vp. Active on Farm and Ind Equip Inst committees and a past mbr of the bd of dirs of Council of Agri Sci and Tech. *Society Aff:* ASAE, FIEI, NSPE. ‡TTL‡Mr

Crane, L Stanley
Business: Consolidated Rail Corporation, 6 Penn Center Plaza, Philadelphia, PA 19103-2959
Position: Chairman & CEO *Employer:* Consolidated Rail Corporation (Conrail) *Education:* BS/Civ Engg/Geo Wash Univ. *Born:* 9/7/15. in Cincinnati, OH. Starting with Southern in 1937, as lab asst, he advanced steadily through posts of increasing responsibility to the position of asst chief mech officer before joining the PA Railroad in May, 1963. He returned to Southern as vp, Engg & Res in Jan, 1965; was elected exec vp, Operatons, in 1970; & pres & chief administrative officer on Mar 1, 1976. He was named pres & chief exec officer on Feb 11, 1977, and was elected chairman in 1979. 10-1-80 Retired from Southern Railway. 1-1- 81 named Chairman & CEO of Consolidated Rail Corporation. In 1978, he was elected a mbr of the Natl Acad of Engg of the USA, a Trustee of Geo Wash Univ, his Alma Mater. *Society Aff:* NAE, ASME, ASTM, SAE, AREA, ASCE, ASTT.

Crane, Robert K
Business: Thayer School of Engrg, Hanover, NH 03755
Position: Research Prof of Engrg *Employer:* Dartmouth Coll *Education:* PhD/EE/Worcester Polytechnic Inst; MS/EE/Worcester Polytechnic Inst; BS/EE/ Worcester Polytechnic Inst. *Born:* 12/9/35 Research Prof of Engrg, Dartmouth Coll working on Radar Meteorology and Propagation Pnenomena affecting satellite communications. Div Sr Scientist and Deputy Div Mgr, Environmental Research & Tech, worked on Radar Meteorology, Hazards to Aviation, and Satellite Communications 1976-1981. MIT Lincoln Lab, Tropospheric and Ionospheric Propagation Research, 1964-1976. Research on Refraction effects, the MITRE Corp 1959-64. Fellow IEEE, Vice Chrmn Commission F, URSI, Chrmn USNC-URSI. *Society Aff:* URSI, USNC, IEEE, AMS, AGU

Crane, Roger A
Business: College of Engg, Tampa, FL 33620
Position: Assoc Prof. *Employer:* Univ of S FL. *Education:* PhD/ME/Auburn Univ; MS/ME/Univ of MO; BS/ME/Univ of MO. *Born:* 11/27/42. Served as mech test engr at Newport News Shipbldg, 1964-65. Thermal analyst for Babcok & Wilcox 1966-74. Joined faculty of Univ of S FL 1974. Active in broad range of energy problems with FEA, ERDA, FL Energy Ctr, State Energy Office, Procter & Gamble, Tampa Electric & FL Power Corp. Offered continuing education seminars through FL Soc of PE & McGraw Hill. Promoted to Assoc Prof 1977. Tenured 1978. Res in heat transfer includes conduction in granular systems, contact conductance, & enhanced evaporation. *Society Aff:* ASME, ASEE, ΣΞ, AAAS.

Craven, John P
Business: 2540 Dole St, Honolulu, HI 96822
Position: Dean of Marine Prog. *Employer:* Univ of HI. *Born:* Oct 1924. PhD Univ of IA; JD Geo Wash Univ; MS Caltech; BS Cornell. Ocean tech & authority on Marine Affairs. State's Marine Affairs Coord. Prior to this appt spent a yr as Visiting Prof of Ocean Engg & Political Sci at MIT. On leave of absence during that yr from his pos as Ch Sci of the US Navy's Strategic Sys Proj & its Deep Submergence Sys Proj. The Strategic Sys Proj has respon for the dev of the Polaris-Poseidon Fleet Ballistic Missile Sys, and the Deep Submergence Sys Proj has the reson for the Man-in—the—Sea Prog and spec deep submersibles. Among num awards: Arthur S Fleming Award; the Parsons Award; the Disting Civilian Serv Award of the USN & the Disting Civilian Serv Award of the Dept of Defense. Has been named a mbr of the Natl Acad of Engg. Pres MTS 1971-72.

Cravens, Dennis C
Business: 524 Lagonda Ave, Lexington, KY 40505
Position: Pres & Gen Mgr. *Employer:* Cravens & Cravens, Inc. *Education:* -/-/Marine Blueprint School, Evansville IN; -/-/Aviation Pre Flight School, San Antonio, TX; -/-/MO State Teachers College, Springfield, MO; -/-/ University of Kentucky *Born:* 6/4/18 Reg PE in KY & TN; Mbr NSPE and KSPE; P Dir Blue Grass Chapter of KSPE; First state chmn of PE in KY; PVP of NSPE Prof Engrs in Construction; First Recipient of KSPE Award of Engg Achievement in construction; Dir of Blue Grass Chapter of Associate General Contractors; Mbr Salvation Army Advisory Board; Mbr of Adm Bd of Epworth United Methodist Church; Mbr of Masonic Lodge; Active in Scouting program for 18 yrs. Director Lexington Chamber Commerce; Trustee Kentucky, Wesleyan College. *Society Aff:* NSPE, KSPE, AGC.

Crawford, Frederick W
Business: Inst for Plasma Res, via Crespi, Stanford, CA 94305
Position: Prof. *Employer:* Stanford Univ. *Education:* PhD/EE/Liverpool Univ; MSc/ Math/London Univ; DipEd/Educ/Liverpool Univ; BSc/EE/London Univ. *Born:* 7/28/31. J Lucas Ltd 1948-52, NCB Mining Res Establishment 1956-57, CAT Birmingham 1958-59. At Stanford Univ since 1959. Dir, Ctr for Interdisciplinary Res and Assoc Dean of Grad Studies 1973-1977, Chrmn, Inst for Plasma Res 1974 to 1980. Leave at French AEC 1961-62 and Oxford Univ 1977-78. Awarded DEng, Liverpool Univ and DSc, London Univ, for publications on plasma physics. *Society Aff:* IEEE, APS, IEE, InstP, IMA.

Crawford, George L
Business: 100 Progress Pkwy, Suite 129, Maryland Heights, MO 63043
Position: Pres. *Employer:* George L Crawford & Assoc Inc. *Education:* BS/Civil Engg/Univ of IL. *Born:* 3/6/28. Davenport, IA. Served with Army 1945-1947. District Traffic Engr and Sr Field Traffic Engr, IL Div of Hgwys. In transportation consulting field since 1964. Pres, George L Crawford & Assoc since 1973. GLC is engaged in traffic and transportation consulting services to the public and private sector, primarily in Midwest part of USA. Intl Dir, ITE 1977 to 1980. Enjoy travel and golf. *Society Aff:* ITE, NSPE.

Crawford, Horace R
Business: PO Box 2820, Dallas, TX 75221
Position: Dir Process & New Prod Dev. *Employer:* NIPAK-sub of ENSERCH Corp. *Born:* Mar 1928 Haskell TX. PhD, MS Univ of TX; BS TX Tech. Engr & Plant Chemist for Sun Oil Co. Taught at Univ of TX & Army; Commendation Medal. Humble Oil Fellow. Mgr of Contract R&D for the Western Co. With ENSERCH since 1969. Created Solid Waste Mgt Div. Now respon for ammonia-related capacity increases. Mbr Tau Beta Pi, Sigma Xi. Chrmn, Dir Dallas AIChE. "Outstanding Achieve in Chem Engg" Award 1973. Pres SW Chap Am Soc of Gas Engrs; Natl Dir. Chrmn Northrail Dist of Boy Scouts; 1975 Award of Merit.

Crawford, Leonard K
1212 S. Grand Ave W, Springfield, IL 62704 *Employer:* Retired *Education:* BSCE/Civ Eng/Univ of ND. *Born:* 5/7/13. 1937-41 - Wood, Walraven & Tilly, consulting Engrs, Springfield, IL. 1941- 45 - US Army. Included service in European Theater. 1946-63 - Partner, Crawford, Murphy & Tilly, consulting Engrs, Springfield, IL. 1964-1979 - Pres, Crawford, Murphy & Tilly, Inc, consulting Engrs, Springfield, IL. 1980-present, Chairman, Crawford, Murphy & Tilly, Inc. Reg PE, IL, MO, IA, WI. Reg Struct Engr, IL. Diplomate, Am Acad of Environmental Engrs. Served as Secy, VP, Pres Elect & Pres of Am Consulting Engrs Council & as 1st Pres of IL affiliate of ACEC. Active in consulting Eng Practice particularly in Waterworks & waste treatment fields. *Society Aff:* ACEC, ASCE, NSPE, AWWA, WPCF.

Crawford, Lewis C
Business: 631 East Crawford, Salina, KS 67401
Position: Partner *Employer:* Wilson & Co *Education:* BE/CE/Yale Univ *Born:* 12/7/25 Native of Salina, KS. Joined Wilson & Co, Engrs & Architects in 1947 following one year with the Cemenstone Corp of Pittsburg, PA. Became an Assoc of the Co in charge of structural design in 1956 and a partner in 1967. Authored papers on computer management in the design office; capacity testing of vaulting poles; and beam deflections by electronic computer. Registered Civil Engr with NCEE and eight State Bds. *Society Aff:* ASCE, NSPE, ACEC, ACI

Creagan, Robert J
Home: 2305 Haymaker Rd, Monroeville, PA 15146
Position: Consulting Scientist. *Employer:* Westinghouse Elec Corp. *Education:* PhD/Phys/Yale Univ; MS/Phys/Yale; BS/Engg/IL Inst of Tech. *Born:* 8/24/19. Harvard-Electronics MIT-Radar; USNR 1943-6; Argonne Natl Lab 1946; Westinghouse 1948-58 Engg Mgr Atomic Power-Commercial; Bendix Corp-Dir Nuclear Prog 1958-1961; Westinghouse 1961-present Asst Dir of Nuclear Power Engg; Proj Mgr LMFBR Prototype; Dir Tech Assessment-Power Systems, Chrmn Power Div ANS 1969; Fellow ANS, Charter Mbr ANS, Mbr Tau Beta Pi, Sigma Xi, National Academy of Engrg. *Society Aff:* ANS.

Creed, Frank C
Business: PO Box 190, Hubbards, Canada BOJ ITO, Nova Scotia
Position: H. V. Consultant *Education:* PhD/EE/Univ of London; BSc/EE/Queen's Univ Canada. *Born:* 4/3/21 b Canada. ScB Queen's Univ Canada. PhD Univ of London Eng. Maj res work devoted to the generation & measurement of High Voltage Impulses. Have participated extensively in dev of Natl & Intl Stds in this area. Retired from Natl Res Coun of Canada Dec 1979. Canadian Rep on var IEC & CIGRE Cttes & Working Groups. In IEEE, have been Sect Chrmn, Chrmn of Power Sys Instrumentation & Measurements Ctte & Regional Dir for Canada. IEEE Awds - Morris E Leeds Centennial Medal. Power Soc Prize Paper. *Society Aff:* IEEE, APEO.

Creed, Michael W
Business: 201 N Front St, Wilmington, NC 28401
Position: Principal *Employer:* McKim & Creed Engrs *Education:* BS/CE/NC State Univ *Born:* 6/5/48 Native of Winston, NC. Prior to coll I worked for three years as a structural draftsman at a local steel fabricator graduated from NCSU with high honors in 1973. Did post graduate work through 1977 while employed with J.E. Sirrine Engrs. Worked with Sutton-Kennerly Engrs in Greensboro as Project Engr for one year prior to forming McKim & Creed Engrs in May of 1978. Married the former Linda Livengood of Winston-Salem & have two children, Thomas Derek age 13 & Catherine Elizabeth age 8. Active in the formation of Covenant Moravian Church in Wilmington since 1978. *Society Aff:* ASCE, NSPE

Creighton, Donald L
Home: 1307 Woodhill Rd, Columbia, MO 65203
Position: Prof. *Employer:* Univ of MO. *Education:* PhD/ME/Univ of AZ; MS/Mech Engg/Univ of KS; BS/Mech Engg/Univ of KS. *Born:* 1/3/32. Married 1953 to Monica Ann Price. Son Christopher Price born 1961. Served with US Navy, 1954-1957. Teaching experience: Univ of KS, Univ of MO. Industrial experience: Central KS Power Co, Pittsburgh Plate Glass Co, Minneapolis- Honeywell, North American Aviation. Reg PE: MO. Fields of Res include mtls, heat transfer, thermodynamics and design. *Society Aff:* ASTM, ASM, AWS, ASB, ASEE, ASME, ISA.

Cremens, Walter S
Home: 5220 Green Oak Ct, NW, Atlanta, GA 30327
Position: Sr Staff Engr, Mfg. Engr. Tech *Employer:* Lockheed-GA Co. *Education:* ScD/Met/MIT; MS & BS/Met/MIT; MBA/Finance/GA State Univ. *Born:* 8/14/26. BS, MS, & ScD in Met from MIT. MBA from GA State, '74. Met engr with GE at Schenectady Works Lab & Res Lab & At Thompson Lab (A/C Gas Turbines) in Lynn, MA. Civilian Sci attache to USAF Hq in Europe '57-'60. Hd, Powder Met res at NASA-Lewis '60-'67. Sr Staff Scientist & Mgr Graphite Composites Task Force, Lockheed-GA Res Lab '67-'71. Mtls Scientist and engr in Advanced Structures Dept '71-'84, Lockheed-GA Co. Chrmn, AIAA Mtls Technical Committee '82. Awarded 1980 SAE Wright Bros. medal for best aeronautical engineering paper. Fluent in German & French. *Society Aff:* ASM, SAMPE.

Cremers, Clifford J
Business: Mech Engg Dept, 242 Anderson Hall (00461), Lexington, KY 40506
Position: Prof *Employer:* Univ of KY. *Education:* PhD/Mech Engg/Univ of MN; MS/Mech Engg/Univ of MN; BS/Mech Engg/Univ of MN. *Born:* 3/27/33. Native of Minneapolis, MN. Worked as Res Fellow and Instructor at the Univ of MN. Joined ME Dept at GA Tech as Asst Prof in 1964. Became Assoc Prof of Mech Engg at the Univ of KY in 1966 and Prof in 1971. Held the position of Chrmn 1975-84. Have served as mbr of the Technology and Opportunities Planning Comm. and several comm in the Heat Transfer Div of ASME, including the Executive Committee. Res interests include heat transfer in frost and elec arcs and also thermophysical property measurement. Enjoy outdoor sports, gardening and classical music. *Society Aff:* ASME, AIAA, ASEE, AAAS, ΣΞ.

Cremisio, Richard S
Home: 50 Creekwood, Cincinnati, OH 45246
Position: Pres. *Employer:* RESCORP/Metech. *Education:* DSc/Engg/Univ of Pitt; MS/MetEng/RPI; BS/Physics/Siena College. *Born:* 12/22/27. Has held major corp engg and mgmnt post in the specialty steel ind. For past 11 yrs, Pres of RESCORP serving major US and European clients in metal and alloy selection, use and manufacture, Also active in patent mgmnt, failure analysis and expert witness test. Adjunct prof at NYU, Suracuse Univ. and Pitt Teaching Grad and Mgmnt Cournes. Has expertise in melting, casting, hot-working, welding, heat-treatment and application of superalloys, tool-steel, stainless and high strength steels, high performance non—ferrous alloys and castings of all types. A licensed pilot with instrument rating served in Army Chem Corps and US Navy. *Society Aff:* AIME ASM, SME.

Crenshaw, Paul L
Business: PO Box 21, Tulsa, OK 74102
Position: Mgr, Treatment Engr & Dev. *Employer:* Dowell Div, Dow Chem USA. *Education:* BS/Petr Geology/TX Tech Univ. *Born:* 5/19/33. P.O.L. & Nuclear Weapons Analyst US Army Corps of Engrs., Army Map Serv Wash DC, 2 yrs. Employed by Dowell since 1956: authored their Cementing Manual, held pos as Dist Mgr & Area Engr & Reg. Engr Mgr. SPE of AIME as local Sec Chrmn & Mbr of Natl SPE Bd of Dir. Reg PE TX & OK. *Society Aff:* SPE of AIME.

Cresci, Robert J
Business: Route 110, Farmingdale, NY 11735
Position: Prof. *Employer:* Polytechnic Inst of NY. *Education:* PhD/Applied Mech/Polytechnic Inst of Brooklyn; MAeE/Aero Engg/Polytechnic Inst of Brooklyn; BAeE/Aero Engg/Polytechnic Inst of Brooklyn. *Born:* 10/9/32. Native of NYC. Worked as aerodynamicist for North American Aviation and as consultant for various companies including GE, Marquardt, Republic Aviation, Grumman Aircraft, Allied Chemical Corp, Portman Assoc, Aerotech World Trade Corp. Primary tech interest include: fluid dynamics, wind tunnel facility design, industrial aerodynamics, heating and ventilation systems, fire safety. *Society Aff:* AIAA.

Cressman, Russell N
Business: Homer Res Labs, Bethlehem, PA 18016
Position: Mgr. *Employer:* Bethlehem Steel Corp. *Education:* BS/EE/Lehigh Univ. *Born:* 3/1/31. Native of Easton, PA. Reg PE (PA). Assumed position of Mgr of Instruments and Measurements in 1971 (30 prof, 35 non prof). Responsible for instrument dev primarily of the type used in automation of steel mill and raw mtls processes- including nondestructive testing, physical properties, temperature and analytical measurements. Several patent and technical papers. Hobbies — music, fine arts, woodworking. *Society Aff:* IEEE, ISA, ASNT, TBΠ, HKN.

Crews, Paul B
Home: 2300 Telequana Drive, Anchorage, AK 99017
Position: Retired, Former Chrmn of Bd of Crews, Mac Innes & Hoffman Consulting Engrs. *Education:* B.S./ME/WA State Univ *Born:* 8/8/17 Alaskan resident for 30 years. Bomb Disposal Officer in South Pacific in World War II. Division engr with General Petroleum Corp 1946 to 1950. Construction Engr with AK Plumbing and Heating Co 1951 to 1957. Founded firm of Crews, Mac Innes & Hoffman, Consulting Engrs in 1957, present name CMH Consultants. Pioneered inovative Heating, Petroleum and utilities applications in the Arctic environment. Received the "Engr of the Year" award from Anchorage Engrs Week Committee, in 1976. Major hobbies are skiing and sailing. *Society Aff:* ASME, ASHRAE, NSPE

Crews, R Nelson
Business: PO Box 22718, Houston, TX 77227
Position: Pres, COO, & Dir. *Employer:* Raymond Intl Inc. *Education:* BE/CE/Tulane *Born:* 2/28/24. Texline, TX. USNR WWII. BECE Tulane Univ 1948. Exec VP & Dir J Ray McDermott 1948-74; Exec VP & Dir Raymond Intl 1975-78. Prof civ engg licenses LA & TX. Previously served: Bd of Adv Tulane Univ Sch of Engr; Bd of Trustees Metairie Park Country Day Sch; Panel of Adv to the Commandant of USCG; Dir the Offshore Co 1961-67. Formerly Chrmn Offshore Operators Ctte. Interests: Cattle ranching, classical music & fine art. Mbr Natl Ocean Indus Assoc, Bd of Dir. The Moles. Dir - First City Bank-Highland Village, Dir - CFA Group, Inc.

Cribbins, Paul D
Home: 3416 Noel Ct, Raleigh, NC 27607
Position: Prof-Civ Engr. *Employer:* NC State Univ. *Education:* PhD/CE/Purdue Univ; MS/CE/Purdue Univ; BS/CE/Univ of AL; BS/Marine Transp/US Merchant Marine Acad. *Born:* 6/24/27. Native of Jacksonville, FL. Served in US Merchant Marine 1945-49 and Navy 1952-55. With NC State Univ since 1959. Assumed current responsibilities as Prof of Civ Engg in 1966. Pres, NC Sec ASCE, 1971; Pres, NC Div ITE, 1968; ASEE Western Elec Fund Award, 1971; NCSU Alumni Distinguished Prof, 1975-78. *Society Aff:* ASCE, ITE.

Crilly, Eugene R
Home: 18646 Ludlow St, Northridge, CA 91326
Position: Engrg Specialist *Employer:* N Amer Aircraft Oper, Rockwell Intl *Education:* MS/Physical Chemistry/Univ of PA; MS/Chemical Phys/Stevens Inst of Tech; ME/Mech Engg/Stevens Inst of Tech. *Born:* 10/30/23. Involved in aero mtls and process engg since 1954 specializing in the areas of composites and structural adhesive bonding. Associated with F-86, F-100, F- 107, Snark, Apollo, SST, C-5A, L-1011, B-1 and other aircraft, missile and space progs. Pioneered in the dev of Kevlar 49 for secondary aircraft structures and served as principal non-metallics engr on the Highly Maneuverable Aircraft Tech (HiMAT) prog. Chrmn, Los Angeles Chapter, Soc for the Advancement of Mtl and Process Engg (SAMPE), 1978-79, Natl Dir, 1979-86, Natl Treas, 1982-85 and Gen Chrmn, 1981 Natl Symposium and Exhibition. Served in Naval Reserve, 1942-1975, retiring with rank of Commander. *Society Aff:* SAMPE, NRA, USNI, ASC, SME-COG, MOWW, NIP, AFIO, VFW.

Crippen, Reid P
Home: 951 79th Ave N, St Petersburg, FL 33702
Position: Ret. (Consulting Elec Engrg) *Education:* MS/Elec Engg/Univ of CA; BS/Elec Engg/Univ of CA. *Born:* 6/10/97. Native of Tarpon Springs, FL. Served in Navy in WWI. Relay and system operating work for Great Western Power Co and Carolina Powr and Light Co. Asst to VP & GM of TN Public Service Co. Asst Elec Engr with Ebasco Services. Consulting Elec Engr with Ebasco Intl Corp. Served on Power Generation committee of AIEE; Elec Equip Committee of EEI; Triple Joint Committee of AIEE, EEI and ASME on hydrogen cooled generators; Engrs Joint Council; and Atomic Ind Forum. *Society Aff:* IEEE.

Crisp, John N
Business: 3013 Learned Hall, Lawrence, KS 66045
Position: Prof & Chrmn Mech Engrg *Employer:* Univ of KS *Education:* PhD/ME/Carnegie-Mellon; MSE/ME/Akron Univ; BME/ME/GA Tech *Born:* 9/9/33 Native of Hiawiasse, GA. Held several engrg positions with Timkens Roller Brg Co from 1958-1969. Chrmn of Mechanical Engr Dept at Tri-State Coll 1969- 1973. Assoc Prof of Mechanical Engrg at the Univ of Dayton 1973-1979. Assumed current responsibilities as Prof and Chrmn of Mechanical Engrg at the Univ of KS in 1979. Dir of the Univ of KS Energy Analysis and Diagnostic Center since 1980. Outstanding Professional Achievement Award, Engrg and Science Foundation, 1979 VP, Pi Tau Sigma, 1980. *Society Aff:* ASME, AIAA, ASEE, ASLE, ΠΤΣ

Crisp, Robert L, Jr
Home: 422 Atwood Dr, Marietta, GA 30064
Position: Consultant *Education:* Post Grad Studies/Soil Mech/GA Inst of Tech; Post Grad Studies/Soil Mech/MA Inst of Tech; BS/CE/GA Inst of Tech *Born:* 12/18/23 1948-1949 - Civil Engr, VA Dept of Hgwys. 1949-1979 - Chief Geotechnical Engr and Special Assistant to Division Engr, South Atlantic Division, Corps of Engrs. Responsible for Geotechnical investigations, design and construction consultant on such projects as NASA Space Center, Cape Kennedy, FL, lateroceanic Sea Level

Crisp, Robert L, Jr (Continued)
Canal Studies, Aiken, S.C. Atlantic Energy Plant, more than 30 large dams such as Carters Rockfill Dam and Richard B. Russell Dam, and as special Assistant for all phases of the TN-Tombigbee Waterway. Retired 1979-1980. VP for Geotechnical Division - SSI 1980-present-Geotechnical Consultant-Dams, Slurry Trench and Foundations for Power Companies, large industrial organizations and various governments. *Society Aff:* SAME, ASCE, ISSMFE, USCOLD

Crisp, Robert M
Business: Engg Bldg 309, Fayetteville, AR 72701
Position: Prof. Computer Sci. Engrg *Employer:* Univ of AR. *Education:* PhD/ME/Univ of TX; MSIE/IE/Univ of AR; BSIE/IE/Univ of AR. *Born:* 8/20/40. Performed extensive res and publication: Approximately 40 papers and PI on 25 sponsored projects totaling several million dollars. Received awards for teaching and research excellence and as reviewer/editorial bd for several journals. Native of AR. Current research is on simulation models of buildings to minimize energy consumption. Hobbies & other interests include racquetball, official for high sch football. *Society Aff:* IIE, AEE, ASEE, ORSA, TIMS, IEEE

Crist, Robert A
Northbrook, IL 60062
Position: VP *Employer:* Wiss, Janney, Elstner Assoc, Inc. *Education:* PhD/CE/Univ of NM; MS/CE/Univ of NM; BS/CE/Univ of NM. *Born:* Sept 1935 El Paso TX. PhD, MS & BS Civ Engg Univ of NM. Worked as cons struct engr 1957-61. Employed by Univ of NM at AF-owned lab 1961-71, where involved in res in struct dynamics. Joined NBS 1971. Respon for dir of progs in Struct Sect. Managing Dir, Publications and Technical Affairs ASCE, 1977-82. Joined Wiss, Janney, Elstner Assoc., Inc. 1982 in his current position as VP. Provides expertise in failure analysis, stuctural dynamics and experimental mechanics. Author over 50 pubs in field of struct behavior. Mbr Sigma Xi, ASCE, ACI. Among recipients of 1974 ASCE State-of-the- Art of Civ Engg Award & 1976 ASCE Raymond C Reese Res Prize. *Society Aff:* ASCE, ACI, ΣΞ.

Cristal, Edward G
Home: 10755 Mora Dr, Los Altos, CA 94022
Position: Proj Mgr *Employer:* Hewlett-Packard Co *Education:* PhD/EE/Univ of WI; MS/EE/WA Univ; BS/EE/WA Univ; AB/math/WA Univ *Born:* 1935 in St Louis, MO. With the Stanford Research Inst, Menlo Park, CA 1961-1971 specializing in microwave applied research and development. Taught electrical engrg at McMaster Univ, Hamilton, Ontario 1971-1973. Joined the Hewlett-Packard Co in 1973. Current position: Proj Mgr, Stanford Park Division, Hewlett-Packard Co. Recipient of the IEEE Microwave Applications Award in 1973. IEEE Fellow in 1980. *Society Aff:* IEEE

Cristy, Nicholas G
Home: 64 Chiltern Rd, Rochester, NY 14623
Position: Consultant (Self-employed) *Education:* BSAeE/Aero/NY Univ. *Born:* 12/27/21. Native of Pleasantville, NY. Served with Army Corps of Engrs 1944-46. VP and Chief Engr, The Macton Corp, for 22 yrs. Holder inpart or wholey of 10 patents. Responsible for design of most revolving restaurants & some 1500 turntables, including special turntables for IBM in NYC & Poughkeepsie, McGraw Hill Bldg, NYC, Time-Life Bldg, NYC & Susquehana Univ, Selins Grove, PA. Designed mech support & drives for swing bridges. Designed moving stage equipment for MGM Grand Hotel, Ren, NV, Metro Opera House, NYC, special lifts for many stages including Pikes Peak Ctr for the Arts, CO Springs, CO, SUNY at Bingham, & many other stages. *Society Aff:* NSPE, PEPP, NYSSPE, USITT.

Criswell, Marvin E
Business: Dept of Civ Engg, Colorado State University, Ft Collins, CO 80523
Position: Prof *Employer:* CO State Univ. *Education:* PhD/CE/Univ of IL; MS/CE/Univ of IL; BS/CE/Univ of NB. *Born:* 10/31/42. Native of Chappell, NB. With CO State Univ since 1970. Res Structural Engr with Corps of Engrs Waterways Experiment Station, 1967-70. Teaches structural design and behavior, mechs, and matls. Res areas include wood engrg, wind engrg, structural reliability, reinforced concrete slabs, and fiber reinforced concrete. Mbr ASCE-ACI Cttes Shear and Torsion, Connections in Monolithic Concrete; ASCE Cttes Timber, Practical Reliability Concepts, Curricula and Accreditation, ACI Ctte on Fiber Concrete, Chrmn, Rocky Mt Sec of ASEE, Past Chrmn of Civil Engrg Div of ASEE, ABET Accreditation Visitor (Civil). Tau Beta Pi, Sigma Xi, Chi Epsilon, Phi Kappa Phi, ASTM. Reg PE, CO. *Society Aff:* ASCE, ACI, ASEE, ASTM, FPRS

Criswell, Milton P
Business: 7th & D St. S.W, Washington, DC 20590
Position: Director, office of Development *Employer:* Dept of Transp, Fed Hgwy Admin. *Born:* Sept 1929 Brooklyn NY. MA Engr Admin Geo Wash Univ; BS Civ Engr Denver Univ. US Marine Corps 1946-48. Since 1953, with Fed Hgwy Admin, formerly Bureau of Pub Rds. Since 1980 - Director, Office of Development, FHWA with respon for managing national program for the Development and implementation of useable research results, 1973-80 Ch, Implementation Div, FHWA, Prior pos bridge. 1973- 1980 Ch, Implementation Div FHWA. Prior pos incl Planning and Res respon in FHWA Region 1 & Wash Office 1962-73; ara engr respon in NH, NJ & MA 1957-62; PE Wash DC. Hobbies: sports, bridge.

Crites, Joseph D
1400 North Lake Dr, Ishpeming, MI 49849
Position: Mgr. *Employer:* Tilden Mine, Cleveland-Cliffs Iron Co. *Education:* BS/Mining/Missouri Sch of Mines & Met; PE/Mining/Missouri Sch of Mines & Met. *Born:* 1/6/26. Native of Farmington, MO. During WWII served in the US Navy as a line officer in the Pacific. Employed in South America for three yrs with Bethlehem Chile Iron Co. Employed by Cleveland-Cliffs Iron Co since 1953. Currently employed as Mgr of the Tilden Mine Complex & have been mgr of the Republic and Empire Complexes. In 1973 was awarded PE of Mines Degree at the MO Sch of Mines with citation reading "outstanding contributions to the engineering profession." Past mbr of the Bd of Dirs and Past VP, Central Region, SME-AIME and mbr of the Engrs and Scientists Joint Ctte on Pensions. Past VP, Central Region, SME-AIME and mbr o Currenlly hold position of VP, Central Region, SME-AIME and am a mbr of the Bd of Dirs. *Society Aff:* SME-AIME.

Crocker, Burton B
Business: 800 N Lindbergh Blvd, St Louis, MO 63167
Position: Dist Engg Fellow *Employer:* Monsanto Co. *Education:* SMChE/CE/MA Inst of Tech; BSChE/CE/GA Sch of Tech. *Born:* 01/01/20 VP APCA (1974-5) & (1972-5); Fellow AIChE, Dir Environmental Div (1976-8), AIChE Environ Award (1982); Dir Particulate Solid Res Inc and Chrmn Tech and Tech Adv Committees (1978-85); Mbr Drying and Gas Cleaning Panels, separation processes service, UK (1981-); Mbr Tech Adv Cttee, Intl Fine Particle Res Inst (1982-84); Dist Engg Fellow, Monsanto Co, Corp Engg Dept, Consultant in gas- solids and solids processing & air pollution control; PE (MA); Lectr, air pollution control, source sampling & monitoring; past chrmn, APCA Chem Ind Committee, AIChE Air Committee; Mbr AIChE Air Committee; Mbr AIChE Dynamic objectives committee (1975) and Ad Hoc Committees; 43 publications, 3 books, 5 US patents. *Society Aff:* AIChE, APCA.

Crocker, John W
Home: 3602 Branigan Lane, Austin, TX 78759
Position: Former Assist. To President *Employer:* Retired Stone & Webster Eng Corp. *Education:* BSc./M.E./Heriott Watt University Edinburgh Scotland *Born:* 02/28/13 Ret from Stone & Webster Engrg Corp. 1978 after 25 years service as assist to Pres, dir of cost control, Mgr of Marketing (Houston, TX), Consultant - pulp & paper mill design chief engineer - Sandy Hill Corp 8 yrs service-complete charge of 80 man engrg dept designing paper machines and pulp & paper equipment. Project engr - research and development dept. Elliott Co, gas turbine devel.

Crocker, John W (Continued)
for marine application, positive displacement high speed compressors and the Manhattan Project atomic bomb. Eng. design-drafting and shop apprenticeship. Active in technical societies. P.E. Texas & Mass. *Society Aff:* ASME, NSPE, TSPE, TAPPI, ACES

Crocker, Malcolm J
Business: Dept of Mech Engrg, Auburn, AL 36849
Position: Prof (Full). *Employer:* Auburn Univ *Education:* PhD/Acoustics/Liverpool Univ; MSE/Noise Vibration/Southampton Univ; BS Aero/Aero Engrg/Southampton Univ *Born:* 9/10/38. Came to USA in 1963, worked first on Acoustics of space vehicles (1963- 1967). Joined Purdue in 1969. Became full prof in 1973 and was named as asst dir of Herrick Labs (for acoustics & noice control) in 1977. In 1983 moved to Auburn Univ as Hd of the Dept of Mech Engrg. Consultant to industry. Gen Chrmn of sevl acoustics conferences including Inter-noise 72, Wash DC, & Noise con 79, Natl Conf on Noise Control Engrg 1979. Published over 200 articles & several books. Editor-in-chief of US refereed journal "NOISE CONTROL ENGINEERING JOURNAL" since 1973. *Society Aff:* ASA, INCE, IOA, ASEE, ASME.

Crom, Theodore R
Business: 250 S W 36th Terrace, Gainesville, FL 32607
Position: Chmn of the Board. *Employer:* The Crom Corp. *Education:* BS/Civil Engg/Univ of MD; one Yr Grad Study/Struct Engg/Univ of FL. *Born:* 7/4/20. Asst Supt Construction, Univ of FL 1948-50; Chief Building Inspector, Gainesville, FL, 1950-51. Pres The Crom Corp, 1953-73; Chmn of the Bd, 1973- Present, specializing in design and construction of prestressed concrete liquid storage vessels; Fellow of ACI, FES, NAWCC; BHI; Tau Beta Pi; ODK, active in many areas with Univ of FL; PE accredited with NCEE; horology gymnastics, and tool collecting hobby; author of several tech book and number of papers; patent holder, tech speaker. *Society Aff:* ASCE, NSPE, FES, FCES, ACI, AWWA.

Cromack, Duane E
Business: Univ of MA, College of Engg, Amherst, MA 01003
Position: Prof. & Assoc. Dean *Employer:* Univ of MA. *Education:* Dr of Engr/Aero/RPI; ME/Mech/Yale Univ; BSME/Mech/Univ of MA. *Born:* 10/28/30. Prof Cromack joined the Mech Engg Dept as an Asst Prof in Sept 1964 & was promoted to Prof in Sept 1978 & appointed Assoc Dean of the college of Engr. in 1984. He has taught primarily in the gen thermal/fluids field & has done extensive res & consulting in the areas of applied aerodynamics. Most recently, he has served as dir of the Wind Power Prog at the Univ of MA & conducting res on wind turbine blade design & performance. He has also served as a consultant on wind power to govt & ind and was elected to the Bd of Dirs of the Wind Energy Div of the Am Sec of the Intl Solar Energy Soc. *Society Aff:* ASME, ASEE, AWEA.

Cromer, Sylvan
Home: 123 Windham Rd, Oak Ridge, TN 37830
Position: Energy Consultant. *Employer:* Self-employed. *Education:* MSE/Petrol Engg/Univ of OK; BS/ME/Univ of OK. *Born:* 10/19/06. Native Marshall OK. Taught Mech Engr Univ of OK and Petrol Engr LA State Univ. War Res on Uranium Enrichment at Columbia Univ. Assignments while employed by Union Carbide in the Nuclear field include Hd of Engg Dev during Oak Ridge Gaseous Diffusion Plant start up. Supt Plutonium Fabrication Plant, Los Alamos, NM. Chief Engr responsible for Process Design of post war Uranium Enrichment Plants at Oak Ridge, TN, Paducah, KY and Portsmouth, OH. Dir Aircraft Reactor Engg Div (ORNL). VP Nuclear Div and Dir of Engg, Mining and Metals Div, responsible for design and construction of mining and milling facilities for uranium and other ores. Dir of Engg Process Plants, Oak Ridge, TN. Pi Tau Sigma- ASME Richards Meml Award 1955. Fellow ASME. *Society Aff:* ASME, AIME, NSPE, ΣΞ, ΤΒΠ, ΠΤΣ.

Cromwell, Leslie
Home: 27661 Via Granados, Mission Viejo, CA 92692
Position: Dean Emeritus Sch of Engg *Employer:* CA State Univ Los Angeles *Education:* PhD/Engg/UCLA; MS/Engg/UCLA; MScTech/EE/Univ of Manchester; BScTech/EE/Univ of Manchester. *Born:* 4/2/24. Various positions in engg in England 1943-1948. Lecturer in Engg UCLA 1948- 1953. Since 1953 Asst, Assoc, and Prof of Engg CA State Univ, Los Angeles. Also Dept Chair 1957-1964, 1968-1973, and Dean of Engg 1973-1980. Co-author: "Biomedical Instrumentation and Measurements" Prentice-Hall, 1973, and 1979. "Medical Instrumentation for Health Care-, Prentice-Hall, 1976. Pres Inst for the Advancement of Engg 1977-1978. CSULA Outstanding Prof Award 1968; Distinguished Service Award, US Jr Chamber of Commerce 1961; Eminent Engr Tau Beta Pi; Fellow of the IAE Mbr, Engg Liaison Committee; State of CA, 1959-1963, and 1974-1978. Mbr Natl Ad Com IEEE, Engg in Medicine and Biology Group 1970-76. Reg Profl Elec Engr 1954. *Society Aff:* ASEE, IEEE, IAE.

Cromwell, Thomas M
Home: 1514 Park Somerset Dr, Lancaster, CA 93534
Position: Retired *Employer:* United States Borax Chem Corp. *Education:* AB/Chem/Occidental College. *Born:* 1/2/23. Previously VP . Sr. VP, Operations, US Borax & Chem Corp, previous positions include VP and Gen Mgr. Boron Operations, Res Mgr of Allan Polash Mines, Saskatechewan Canada, Dir of Engg operations US Borax, Asst Res Mgr Boron ops, Mgr of Chem Engg Res, US Borax. Graduate studies in physical chem, UCLA and Oregon State Univ. WWII-Naval Aviator. Married, one daughter, one grandson. Born New Orleans, LA. *Society Aff:* AIChE, ACS.

Cronan, Calvin S
Home: 25 Midbrook Ln, Old Greenwich, CT 06870
Position: Editor-in-Chief, Retired *Employer:* Chem Engg Magazine-McGraw-Hill. *Education:* BS/ChE/Northeastern Univ *Born:* July 12, 1917 Hartford CT. BS Chem E Northeastern Univ 1940. Wife Jean Ferguson; children: Susan, Christopher, Andrew. Indus exper: res engr Dennison Mfg Co, Dir Applications Engg Ctr Bird Machine Co. Chrmn Process Indus Div ASME; Chrmn NY Sect AIChE; Chrmn Ed Bd McGraw-Hill Publ Co; Chrmn Ed Exec Ctte Am Bus Press (trade org); AIChE Fellow, Natl Dir, Chrmn Public Relations Ctte; Mbr Taskforce drafting Prof Employment Guidelines & Mbr of subsequent Implementation Ctte; Secy, VChrmn Prof Dev Ctte; Mbr Energy Ctte. Member John Fritz Medal Bd of Award; Coordinator, Employment Guidelines Committee, Engineering Affairs Council, AAES; Winner of 1981 Crain Award of the American Business Press for a Distinguished Editorial Career; Winner of 1981 Tyler Award for outstanding contributions to the profession of chemical engineering, New York Section, AIChE. *Society Aff:* AIChE

Cronk, Alfred E
Home: 727 N Rosemary, Bryan, TX 77801
Position: Prof Emeritus. *Employer:* Retired from TX A&M Univ. *Education:* MS/Aero Engr/MN; BS/Math-Physics/College of St Thomas *Born:* 7/1/15. Taught Aero Engrg at MN as Instructor, Asst Prof. Assoc Prof 1943-56. At TX A&M Univ since 1956 as Prof & Dept Hd of Aero Engrg. Also serves as Dir of the Balloon Engrg Lab, part of the TX Engrg Exper Station. *Society Aff:* AIAA

Cronvich, James A
Business: Dept of Elec Engg, New Orleans, LA 70118
Position: Hd, Dept of Elec Engg. *Employer:* Tulane Univ. *Education:* SM/EE/MA Inst of Tech; MS/EE/Tulane Univ; BE/Mech-Elec Engg/Tulane Univ. *Born:* 10/26/14. James A Cronvich, the son of James Cronvich and Louise Lester, was born in New Orleans, LA on 10/26/14. After undergrad and grad studies at Tulane Univ, he attended the MA Inst of Tech on a Genradco Fellowship. He served consecutively as instr, asst prof, assoc prof, prof, and hd, Dept of Elec engg at Tulane Univ since 1938, except for 1941-42 when he was asst then assoc elec engr on the Third Locks Proj, Panama Canal. He has been involved in biomedical engg (instrumentation for cardiovascular studies with the Tulane Dept of Medicine) since 1946. Was asst editor of the *American Heart Journal. Society Aff:* IEEE, ASEE, ΣΞ, HKN.

Crook, Lawrence
Home: 8 Westmeadow Close, Willingham Cambridgeshire, England
Position: Sr Weight Engr. *Employer:* Hawker Siddeley Dynamics Ltd. *Born:* UK Aerospace Ind 1954/68 & 1973/74. Weight Engr, working upon aircraft, helicopters, missiles & spacecraft. 1968/73 tracked Hovercraft Ltd Proj Weight Engr. High speed ground trans. In local govt since 1974 as a highway engr. Joined the Soc of Allied Weight Engrs 1962. Dir of European Chap 1967/71. Sec UK Chap since 1971. Fellow Award 1940. Leisure pastimes: gliding, walking & active SAWE mbr.

Crook, Leonard T
Home: 3355 Yellowstone Dr, Ann Arbor, MI 48105
Position: Owner. *Employer:* Leonard T Crook & Assoc. *Education:* BSCE/Hydraulics/Univ NM. *Born:* 11/12/13. Soil Conservation Service; Bureau Reclamation; Office Engr, Corps Engrs, 2 yr; Sec & Branch Chief, Third Locks & Sea Level Canal I, Studies, Panama Canal, 8 yr; Special Asst, Acting Technical Dir, Developer, Conductor nation-wide Planning Associates Training Prog, Bd Engrs Rivers & Harbors, 19 yr; mbr fac grad sch, developer master & doctoral water resources prog, Catholic Univ, 3 yr; Exec & Planning Dir, Great Lakes Basin Comm, 11 yr; presently, water resources consultant. Pres, DC -- Dir, Natl Soc PE; Chrmn, Council Engr & Arch Socs -- Joint Bd Sci Educ. *Society Aff:* ASCE, NSPE, PIANC, DCSPE.

Croom, Richard D
Business: 340 Victoria Rd, Asheville, NC 28801
Position: Division Dir *Employer:* Asheville-Buncombe Tech Coll *Education:* BS/CE/NC State Univ *Born:* 11/5/37 PE and RLS in NC. Amateur radio operator. Cartoonist. Active in amateur and professional theatre (acting) in Asheville and Mars Hill. Editor for Bulletin, the NC Section ASCE Newsletter. Native of Fayetteville, NC. Past Pres of Western Branch and Past Pres of NC Section, ASCE. Sanitary Engr 1961-63, design engr at American Enka Corp 1963-66. With A-B Technical Coll since 1966. *Society Aff:* ASCE

Crosbie, Alfred L
Business: Mechanical Engr., University of Missouri-Rolla, Rolla, Missouri 65401
Position: Professor *Employer:* Univ. of Missouri-Rolla *Education:* PhD/Mech. Engrg. /Purdue University; MS/Mech. Engrg./Purdue University; BS/Mech. Engrg./ University of Oklahoma *Born:* 08/01/42 Born in Muskogee, OK. Joined faculty of Univ of Missouri-Rolla in 1968 where promoted to Prof of Mech Engrg in 1975. Active researcher in radiative heat transfer since 1964. Edited two books & authored over sixty papers in archival journals. Member 1976-78 and chairman 1984-86 of AIAA Thermophysics Technical Cttee. Tech Prog Chairman of AIAA 15th Thermophysics Conference 1980. Associate Editor of AIAA Journal 1981-83 and Journal of Quantitative Spectroscopy and Radiative Transfer 1979-90. First Editor-In-Chief of Journal of Thermophysics and Heat Transfer 1987. Assoc. Fellow of AIAA. Fellow of ASME. AIAA Thermophysics Award 1987. *Society Aff:* ASME, AIAA, OSA

Crosby, Alfred R
Business: PO Box 101, Florham Park, NJ 07932
Position: Division Manager & Sr Engg Advisor *Employer:* Exxon Res & Engg Co. *Education:* BS/ME/NJ Tech Inst. *Born:* Nov 1920. Naval Arch - Postgrad Sch, US Naval Acad. USN constr officer 1942-46, ship supt (hull) aircraft carriers USS Antietam & Valley Forge, commendation medal. With Exxon Res & Engg Co since 1947. Mtls procurement, engg stds, cost engg & proj mgt. Sr. Engg Advisor. Since 1942 Mgr of Contracts Engg Div providing services to Exxon affiliates, worldwide, that lead to award of construction contracts for new capital projs. AIChE; Charter Mbr Engg & Constr Contracting Ctte, Chrmn Exec Ctte 1974-75, P Chrmn 1975-76. Sailing, skiing, swimming. *Society Aff:* AIChE.

Crosby, Ralph
Business: 5500 Florida Blvd, Baton Rouge, LA 70806
Position: President *Employer:* Mayers & Crosby, Inc *Education:* BS/Mech Eng/LA State Univ *Born:* 2/21/21. Native of Bogue Chitto, MS. Multi-Engine Pilot during WWII. Began private practive May 1959. Have served as VP and Pres of my Firm. Mbr of Kiwanis Intl. Past Mbr of Bd of Governors of Consulting Engrs of LA. Mbr Organization of American Consulting Engrs Council having served as Secy, VP, Pres, Alternate Natl Dir and Natl Dir to ACEC Representing LA. Served two terms on Bd of Governors for Baton Rouge Chapter of ASHRAE. *Society Aff:* ASHRAE, ACEC

Cross, Edward F
Home: 626 SE Fourth St, College Place, WA 99324
Position: Emer Dean of Engg. *Employer:* Walla Walla College. *Education:* DE/-/Walla Walla College; MA/Math/Columbia Univ; ME/ME/Stevens Inst of Tech. *Born:* 11/16/08. Dr of Engg (Honoris Causa), Walla Walla College, 1974. 1929-1932 Mech Engr, Columbia Gas & Elec System. 1932-35 Mech Engr, Multi-Needle Engg Corp. 1935-47 Teaching (math, sci, drafting), NYC system and YMCA Evening Technical Sch. 1947- 79 Founded Engg dept, Walla Walla College and later retired as Emeritus Dean, Sch of Engg. Consultant to City of College Place, 1952-65. Designed maj campus bldgs at Walla Walla College. ASEE-Western Elec Fund Award, 1965; Distinguished teacher Award, 1967. PE WA, 1948; OR, 1952. Who's Who in Education; Who's Who in the West; Tau Beta Pi. In retirement: engaged in consulting services (energy conservation, structural/mechanical areas), occasional teaching and activities related to personal hobbies and interests. *Society Aff:* ASEE, NSPE, WSPE.

Cross, Leslie Eric
Business: Univ Park, PA 16802
Position: Prof Elec Engr & Assoc Dir Mtls Res Lab. *Employer:* PA State Univ. *Education:* PhD/Phys/Leeds Univ; BSc/Phys/Leeds Univ. *Born:* 8/14/23. Leeds, England. M 56; c 5. Exp off electronics, Brit Admiralty, 43-46; asst lectr phys, Leeds, 49-51; Imp Chem Indust fel 51-54; sr res assoc, Phys, Brit Elec Res Assoc 54-61; PA State Univ, 61-63, assoc prof 63-68, Prof, 68 to present, Assoc Dir Mtls Res Lab, 69 to present. Mtl sci; ferroelectric and antiferroelectric properties of titanates and niobates; thermodynamics of ferroelectricity; high permittivity mtls; dielectric measuring techniques; dielectric properties of glass systems. Address: Mtls Res Lab, Rm 251A,, the PA State Univ, Univ Park, PA 16802. *Society Aff:* APS, ACS, IEEE, NICE, JPS, OSA.

Cross, Ralph E
Home: 50 N Deeplands Rd, Grosse Pointe Shores, MI 48236
Position: Bd Chrmn. *Employer:* Cross & Trecker. *Education:* Dr/-/Lawrence Inst of Tech. *Born:* 6/3/10. Born in Detroit. Student MIT 1933. Pres Cross Co, Fraser MI; Chrmn of Bd, Pres Cross Intl A G Switzerland 1965-68; Pres Cross Export Corp; Dir Roberts Corp, Enshu-Cross K K 1970-75; Asst Admin Bus, Def Services Admin US Dept of Commerce 1954; spec cons to Asst Secy AF for Mtl 1955-59. Mbr Corp Dev Comm MIT 1970-. Recipient engg citation Am Soc Tool Engrs 1956. Mbr Natl Acad Engg, Soc Automotive Engrs, Soc Mfg Engrs, Engg Soc Detroit, Pres Natl Machine Tool Bldrs Assn 1975. *Society Aff:* NAE, SAE, SME, ESD.

Cross, Robert C
Home: 4020 Howard Ave, Western Springs, IL 60558
Position: Retired. *Education:* BME/Northeastern Univ. *Born:* 10/18/02. H B Smith Co 1925-1929 1930-1935. US Bureau Mines Pittsburgh Sta - 1930. Battelle Meml Inst, Col5,O - 1935-1940. Sears, Roebuck & Co, Chicago, IL 1941- 1954. Am Soc Refg Engrs 1954-1959. Am Soc Htg Refg & AC Engr 1959-1967. *Society Aff:* ASME, ASHRAE.

Cross, Willis J
Home: 61 Paxon Hollow Rd, Media, PA 19063
Position: President *Employer:* TEK Associates Inc *Education:* BS/ChE/VA Tech SU. *Born:* 12/15/23. I started work professionally for the Atlantic Refining Co in 1945

Cross, Willis J (Continued)
after obtaining a BS in CE from VPI. After one yr of fluid cracking pilot unit work, I joined Houdry Process Corp as a technical service engr starting up new catalytic plants and troubleshooting refinery problems. During the following ten yrs, I worked in process design, dev & proj engg. Became Dir of Technical Service in 1961, Mgr of Sales & Service in 1967, & VP of Sales & Service in 1969. Thru acquisition of Houdry by Air Products & Chems, Inc, became the Gen Mgr of Processes & Catalysts in 1971. Formed TEK Associates Inc March 1981- . Consulting firm covering catalytic process technology in the refining, petrochemical and automotive emissions control industries. *Society Aff:* AIChE, ACS, API, CS, AIC

Crossett, Frederick J
Home: 1680 Viewpoint Dr, Fayetteville, AR 72701
Position: Prof Emeritus. *Employer:* Univ of AR. *Education:* MS/Economics/Purdue Univ; BS/EE/Drexel Univ; BS/Indus Mgmt/Univ of AR. *Born:* 1/19/10. Res Engr, Elec Service Mfg Co, Phila, 1933-40: Design & dev of circuit protective devices. Officer, Army Ordnance Corps, 1940-63: Instr, Ordnance Sch, Command & Gen Staff Sch, & Purdue Univ; Engg Officer, Pentagon, supervising product engg & test at Frankford Arsenal & Aberdeen Proving Ground; Planning Officer, Pentagon, over all Ordnance Depots; Advisor to the Chief of Ordnance of the Peruvian and the Korean Armies; Staff & Command of battalions & groups. Prof of Elec Engg, Univ of AR 1963-75 (Emeritus). PE, PA. Sr Mbr, Sec Chrmn 1970-71, Faculty Counselor 1965-75 of IEEE. *Society Aff:* HKN, ΣIE

Crossley, F R Erskine
Home: 282 Pine Orchard Rd, Branford, CT 06405
Position: Prof of Civil Engrg, Emeritus *Education:* D Engrg/Mech Engg/Yale Univ; MA/Mech Sci/Cambridge Univ; BA/Mech Sci/Cambridge Univ. *Born:* 7/21/15. Born 21 July 1915 and educated in England, Harrow and Cambridge UNiv.; College oarsman, winning Ladies' Plate, Henley Royal Regatta, 1937 emigrated to USA 1937. Married Mary Eleanor DeLacy Coyne most fortunately 1941. Two children. Taught engg at Univ of Detroit, Yale Univ, GA Tech and Univ MA (Amherst); became Emeritus 1980. Started the *Journal of Mechanisms*, Editor 1966-72. Chmn, ASME Mechanisms conf, 1968-69. Mbr of original organizing committee, International Federation for Theory of Machines and Mechanisms, its first VP, 1969-75 and made Hon Mbr 1983. Fulbright Lectureship at T H Munich, Germany, 1962-63; Visiting Fellow, Manchester Inst of Sci and Tech, England, 1965; visiting res prof, Warsaw Polytechnic Univ, Poland, 1975; von Humboldt Sr Sci Award, Aachen Germany, 1975-76. Fulbright Lectureship in Bucharest, Rumania, 1976. Life Fellow, ASME, Received Centennial Medal ASME, 1980. Awarded honorary Membership as Corresponding Member, Verein Deutscher Ingenieure, 1983. Undertook research into pre-commercial forest harvesting methods for the Dept of Energy, visiting Finland and Sweden, 1976-81. ASME Legislative Fellow, to the Connecticut State Legislature, 1981-1983. Chrmn of town cttee for Resource Recovery (Garbage Disposal) 1986. *Society Aff:* ASME, VDI, ASEE.

Crossley, Frank A
Home: 7575 Woodborough Drive, Roseville, CA 95661 *Education:* PhD/Met Engg/IIT; MS/Met Engg/IIT; BS/ChE/IIT. *Born:* 2/19/25. Born in Chicago, IL. Served to Ensign USNR during WWII. Taught at IL Inst of Tech 1947-49. Prof & Dir of Foundry Engg at TN A&I State Univ 1950-52. Served to Sr Met at IIT Res Inst, Chicago 1952-66. Lockheed Missiles & Space Co, Sunnyvale, CA, since 1966 in the following capacities: Sr Mbr, Res Lab 1966- 74; Mgr, Producibility & Stds Dept, Missiles Systems Div 1974-78; Mgr, Missile Body Mech Engg, MSD 1978-79; & Consulting Engr 1979-86. Research Director-Propulsion Materials at Aerojet Propulsion Research Institute, Sacramento, CA 1986-87. Published more than fifty papers with more than forty of these dealing with titanium. Inventor of the "Transage" titanium alloys. Fellow of the Am Soc for Metals. *Society Aff:* TMS-AIME, ASM, SAMPE, AIAA, ΣΞ

Crotts, Marcus B
Home: 10 Gomar Ln, Winston-Salem, NC 27106
Position: Partner. *Employer:* Crotts & Saunders Engrg, Inc *Education:* BS/Mech Engrg/NC State Univ; MS/Mech Engrg/Univ of IL *Born:* 08/06/31 Native Winston-Salem, NC, served in Korean War as Lt in Aircraft Maintenance, Dir of SME, served as VP ASME 1967, served as Pres of the NC Soc of Engrs 1979-80, received 1979 Archimedes Engg Achievement Award from CA Soc of PE, rec'd 1976 Meritorious Service Award from NC State Univ, rec'd SME Award of Merit. Mbr Tau Beta Pi - 1980 Dist Governor of Rotary Intl - Rec'D University of Illinois Distingushed Engineers Award for 1980 - Received North Carolina State University " Distinguished Engineering Graduate for 1981 - Served as Pres of the American Machine Tool Distributors Association - 1981-82 Dir Citizens Natl Bank, Dir Objective Industries. 1987- Joseph A. Sieger Award from SME. *Society Aff:* NSPE, SME, SAE, ASME

Crouse, Philip C
Business: Suite 3120, 2001 Bryan Tower, Dallas, TX 75201
Position: Pres *Employer:* Southwest PetroCorp, Inc *Education:* MBA/Finance Management/Univ of TX-Austin; BS/Petro Engr/Univ of TX-Austin *Born:* 7/14/51 Founder of Southwest PetroCorp, a petroleum development co; Petroleum Management Consultant; Registered PE in the State of TX; Author-SPE8194-Crude Oil Pricing Regulation-Considerations for the Evaluation of US Petroleum Co Acquisition Candidates- 1979: SPE JPT published- A Method to Organize and Manage the Evaluation of Petroleum Co Acquisition Candidates-Feb 1980: World Oil Feb 1987- The Industry in 1987-OPEC Reorganization Could Spell Relief; miscl publications in World Oil. Past Pres, Exec VP-the Petroleum Engrs Club of Dallas, Held the offices of Sec Chrmn, Dir, Scholarship Chrmn, Second VP (Education) for the Dallas Section of SPE of AIME. Career: Atlantic Richfield, Champlin, Sun. 1980-worked as senior advisor on the US Senate Committee on the Budget Staff (Conference Bd Congressional Assistant Program). Recipient of SPE Young Mbr Outstanding Service Award - 1984, SPE Distinguished Lecturer 1984-1985, SPE Distinguished Member-1984, SPE Regional Service Award-1987. *Society Aff:* SPE, Inc.

Crow, John H
Business: 811 Westheimer, Houston, TX 77006
Position: VP. *Employer:* Trentham Corp. *Education:* BS/ChE/TX Tech Univ. *Born:* 11/14/22. Native of TX. Served with Army Ordnance Dept 1942-1945. Grad from TX Tech 1949. Joined Graff Engg in 1951 specializing in gas treating, sulfur recovery and gas processing. Participated in design, procurement, construction and commissioning of plants as well as mgt of the contracting engr organization. Very active as official in Dallas Sec AIChE 1960-1975. Married with two sons. *Society Aff:* AIChE.

Crowe, John J
Home: 715 Westfield Ave, Westfield, NJ 07090
Position: Consulting Engr (Self-employed) *Education:* BS/-/Geo Wash Univ. *Born:* 1/3/86. PE NY & NJ. Ten yrs - Natl Bureau of Stds Apprentice to Asst Physicist. Ten yrs - USN - Physical Met - Boston & Phila Navy yards. Fifty yrs - Airco Inc Engr in Charge of Apparatus Res & Dev - Asst VP - Consultant. Expert witness in a number of law cases involving compressed gases and failure of pressure vessels. Safe use of compressed gases - welding & cutting. *Society Aff:* AIME, ASME, ASM, AWS, CGA, NFPA.

Crowell, A Leavitt
Home: 10353 Hampshire Green Ave, Fairfax, VA 22032
Position: VP *Employer:* C.R.S. Group Engrs *Education:* BS/CE/WA Univ (St Louis) *Born:* 11/11/22 Native of Sandwich, MA. field engr, supt, proj mgr on major tunnels and mine development projects throughout the US. Currently VP C.R.S. Group Engrs. Responsible for construction engrg and construction management for tunneling projects in MD, GA, OH and ND. *Society Aff:* ASCE, AIME

Crowley, Francis X
Business: Teal Rd, Wakefield, MA 01880
Position: Pres. *Employer:* Natgun Corp. *Education:* BS/Bldg Engg&Con/MIT, Cambridge. *Born:* 12/15/25. Twenty-five yrs experience in design and construction of prestressed concrete tanks. Mbr of ACI Committee 344 "Circular Prestressed Concrete Structures" and AWWA Committee for Prestressed Concrete Water Storage Tanks. Designed and built shotcrete-steel diaphragm tanks, precast concrete tanks, underground tanks, partially buried tanks with balanced and unbalanced backfill, standpipes, tanks with flatslab roofs, cast-in-place concrete domes and precast concrete domes. Originated stack precasting and was the first to utilize this technique for tank walls and for precast domes of both the isotropic and orthotropic types. *Society Aff:* AWWA, ACI, BSCE.

Crowley, Joseph M
Home: 16525 Jackson Oaks Dr, Morgan Hill, CA 95037
Position: Prof. *Employer:* Univ of IL. *Education:* PhD/EE/MIT; MS/EE/MIT; BS/EE/MIT. *Born:* 9/9/40. Native of Phila. Recipient of GM Scholarship, AEC Fellowship and NATO Fellowship, the last for postdoctoral work at the Max Planck Inst for Fluid Mechanics. Joined the faculty at IL in 1966. Currently Prof of Elec Engg and of Mech Engg and Dir of the Appl Electrostatics Res Lab. Profl interests include electrostatics, electromechanics, energy conversion and feedback control, with applications in computer peripherals, copying machinery, bioelectric effects and controlled fusion. Presently on leave of absence from Univ of Illinois to start up company in California, Electrostatic Applications. Also active in local church and sch organizations. *Society Aff:* IEEE, APS, ESA.

Crowley, Thomas H
Business: 600 Mountain Avenue, Murray Hill, NJ 07901
Position: Exec Dir. *Employer:* Bell Lab. *Education:* PhD/Math/OH State Univ; MA/Math/OH State Univ; BA/EE/OH State Univ. *Born:* 6/7/24. Exec Dir, Computing Tech and Design Engg Div. Since joining Bell Labs in 1954, I have engaged in res on magnetic logic devices, switching theory, sampled -data systems, and computer-aided logic design. I was Appointed Dir of the Computing Sci Res Ctr in 1865, Exec Dir of the SAFEGUARD Design Div in 1968, and assumed my present position on June 1, 1979. I have been granted 4 patents on magnetic devices and digital circuits, and have written a number of books and articles on computers and magnetic devices for several tech journals. I am a Fellow of the IEEE. *Society Aff:* IEEE, AAAS.

Croxton, Frank C
Home: 1921 Collingswood Rd, Columbus, OH 43221
Position: Consultant. *Education:* PhD/Phys Chem/OH State Univ; MA/Phys Chem/OH State Univ; BA/Chem/OH State Univ.;DSc/Hon./Denison Univ. 1965. *Born:* 6/26/07. in Wash, DC, a native of Columbus, OH. Became a res chemist, Std Oil Co (IN) in 1930. To Battelle Meml Inst, Columbus, OH in 1939 as res chemist. In charge of res in chemistry, chem engg, and biosciences 1947-1952. In charge Battelle operations in Europe 1952-1957. Asst Dir and mbr Sr Technical Council 1958 until retirement in 1972. Active in AAAS (P 1957 and 1958), Sigma Xi, the Scientific Res Soc, (Pres 1971 & 1972), AIChE (Dir 1966-1968, Chrmn Central OH Sec 1948), and Am Chem Soc (Chrmn Columbus Sec 1963, Councillor 1964-1969, Gen Chrmn 11th Central Regional Meeting 1979). Hobbies: Gardening, reading. *Society Aff:* AIChE, ICHEME, CE, ΣΞ

Crozier, Ronald D
Home: 4 Daisy Ln, Ridgefield, CT 06877
Position: Consulting Engineer *Employer:* Self *Education:* PhD/CE/Univ of MI; MSE/CE/Univ of MI; BSE/CE/Univ of MI; BSE/Met Engg/Univ of MI. *Born:* in Chile, studied chemistry in Glasgow Univ, taught Chem and Met Engg at Univ of MI 1954-56. Process design engr CA Res Corp 1956-58. Joined res dept, textile fibers, Dow Chem Co 1958-64 as group leader process dev; business mgr Dow Chem Europe, 1966-69. VP Operations, Soquimich, Chile, 1969-72 in charge of four Sodium Nitrate mines and solar evaporation complex; Pres Minerec Corp, Baltimore 1972-76, producer of reagents for mineral processing; Managing Dir Tecnmonin Ltda, Santiago Chile 1977-82, res, dev and production of mineral processing reagents. International Consulting Eng 1982-. Patentee in acrylic fiber processes, catalytic organic synthesis, flotation reagents and phosgene derivatives. Publications in heat transfer, distillation, rheology, and mineral processes. Elected a Fellow of the Inst of Mining and Metallurgy (London) in 1980. Chartered Engineer (UK). Elected to IMM Council in 1987. *Society Aff:* AIChE, ACS, AIME, IMM.

Cruice, William J
Business: 200 E Main St, Rockaway, NJ 07866
Position: VP *Employer:* Hazards Res Corp. *Education:* MS/Chemistry/St. John's Univ; BA/English/St. John's Univ; BS/Chemistry/St. John's Univ. *Born:* 8/22/37. NYC. Res Chemist Reaction Motors Div Thiokol Corp 1963-69, rocket propellants, incendiaries, catalysis, corrosion, fires & explosions. Asst VP Engg US Banknote Corp 1969-70, proj mgt, corp diversification. Co-founder, VP, Tres Hazards Res Corp 1970-, eval of fire & explosion hazards for chem processors, users, gov agencies. Process review, design engg, lab & pilot scale studies of dust, mist, vapor explosions, thermal explosions, detonations, suppression sys. Dir HRC 1969-, Dir Deltronic Crystal Indus Inc 1971-. *Society Aff:* AIChE, ASTM, ACS, NFPA, AIC, CI.

Crum, Floyd M
Business: PO Box 10029, Lamar Univ Station, Beaumont, TX 77710
Position: Regents' Prof of EE *Employer:* Lamar Univ *Education:* MS/EE/LA St Univ; BS/EE/LA St Univ *Born:* 4/28/22 Born and reared in Baton Rouge, LA. Married Betty (Cook); have five children. Verteran of WWII, Honorable Discharge, three Combat Stars, Good Conduct Medal, Meritorious Service Award, ETO Service Ribbon. Graduate degree from LA State Univ; taught there in EE Dept eight years. Teach Lamar Univ EE Dept 1955 - date. Registered PE states of LA, TX. Member of Tau Beta Pi, Eta Kappa Nu, ARRI, and extra class license WSEL. Consultant to NRL 20 years; received awards of merit from Navy for services in conjunction with Scorpion and Threasher searches. Hobbies- amateur radio, music. *Society Aff:* IEEE, ASEE, ISA, ΣΞ

Crusinberry, Thomas F
Home: 4682 E. Windsor Ln, Columbus, IN 47201
Position: Govt Regulations Engr. *Employer:* Ford Motor Co - Tractor Oper. *Education:* BS/Mech Eng/SD State Univ *Born:* Sept 1921 Sioux Falls SD. BS Mech Engg SD State Coll 1947. Infantry Officer 1942-46. Mbr Ford Tractor Oper since 1947. Supr Soil Engaging & Combine Sect 1962-73 in tillage, planting & grain harvesting equip design. Currently as Govt Regulations Engr respon for monitoring & disseminating info on worldwide tractor regulations. Active in std dev for Am Soc of Agri Engrs, Am Natl Stds Inst & Intl Stds Org. Mbr Adv Ctte on Agri Oper for MI State Dept of Labor. Chrmn MI Sect Am Soc of Agri Engrs 1974-75. Retired from Ford Tractor 6/30/80. Served as VP of Engineering/Manufacturing from 7/1/80 to 4/30/84 for Claas of America, Inc., Columbus, IN. 5/1/84 until present. Serve as Consultant for Claas of America, Inc. and Cummins Engine Co. for Agricultural Applications. *Society Aff:* ASAE, SAE

Crutch, Kenneth Dean
Business: 1600 E Robinson St, Orlando, FL 32803
Position: FL Mgr. *Employer:* Franklin Consultants Inc. *Born:* June 1937. BCE OH State Univ. Native of Bryan OH. Hgwy Engr with Burgess and Niple Cons Engrs Columbus OH 1961-65. Field Engr Portland Cement Assn Columbus OH 1971-73. With Franklin Cons Inc since 1973. Mbr Am Soc of Civ Engrs and ASCE Natl Ctte on Student Chapters 1972-75. ASCE Adv Ctte to the Alfred Noble Joint Prize Ctte 1976. ASCE FL Sect, Chrmn Student Activities Ctte 1975- 76. ASCE E Central Branch Conact Mbr to FTU Student Chap 1975-78, Secy-Treas 1976-77. FL Engg Soc - Sr Mbr. Reg PE OH & FL. Pres OSU Alumni Club of Central FL 1975-76. Mbr Univ Club of Orlando.

Crutcher, Harold L
Home: 35 Westall Ave, Asheville, NC 28804
Position: Owner *Employer:* Self *Education:* PhD/Meteorology/NY Univ; MS/Meteorology/NY Univ; BS/Chem-Math/Soueastern State Univ OK; BA/Latin-Educ/Okla *Born:* 11/18/13 Born, Cheraw, CO. Attended primary and secondary schools, Colorado, South Dakota, Oklahoma. Attended Southeastern State Univ OK 1930-1934, OK Univ (Summers) 1935-1938, New York Univ 1950, 1955-1959. Taught secondary schools OK 1935-1938. US Weather Bureau - Natl Weather Service, 1939-1977. Assignments to US State Dept, Mexico 1942-1946, Peru 1947-1950. Observer, Forecaster, Inspector-Trainer, climatologist, statistician with Weather Service. Final position Sci Advisor to Dir Natl Climatic Control Ctr, Asheville, NC. Retired 1977. Consltg Meteorology, Reliability and Quality Control, Hydrology. Fellow, Amer Inst of Chemists; Fellow, Amer Soc for Quality Control. *Society Aff:* AMS, ASQC, APCA, ASTM, AGU, ACS, AIC, ASA.

Cruz, Jose B
Home: 14 Canyon Ridge, Irvine, CA 92715
Position: Prof and Chair *Employer:* Univ of California, Irvine Dept of Elec Engrg *Education:* PhD/EE/Univ of IL; SM/EE/MA Inst of Tech; BS/EE/Univ of the Philippines. *Born:* 9/17/32. in the Philippines. Instr, Univ of the Philippines, 1953-1954. Res Asst, MIT, 1954-1956. With Univ of IL 1956-87. Professor and Chair, Electrical Engrg, Univ of California, Irvine, 1986-present. Visiting Assoc Prof, Univ of CA, Berkeley, 1964-1965. Visiting Prof, MIT and Harvard, 1973. Pres, Dynamic Systems, Author or Ed of 6 books, 170 papers. Editor, IEEE Transactions on Automatic Control, 1970-1972. Pres IEEE Control Systems Soc, 1979. Mbr IEEE Board of Directors, 1980-1985; IEEE VP for Tech Activities, 1982- 1983; IEEE VP for Publ Activities 1984-85. Elected Fellow, IEEE, for significant contributions in circuit theory and sensitivity analysis of control systems. 1972 ASEE Curtis McGraw Res Award. Elected to National Academy of Engineering, 1980. Founding Member, Phil-American Academy of Science and Engineering, 1980. Registered Professional Engineer in Illinois. *Society Aff:* IEEE, NSPE, ASEE, AAAS.

Crynes, Billy L
Home: College of Engineering, Oklahoma University, Norman, OK 73019
Position: Dean, Engg *Employer:* Oklahoma Univ. *Education:* PhD/CE/Purdue Univ; MS/CE/Purdue Univ; BS/CE/Rose Poly Inst. *Born:* 3/16/38. Billy L Crynes was born on Mar 16, 1938, in Worthington, IN & attended elementary schools & high sch in Terre Haute, IN. For two yrs he was then in the US Marine Corps before entering Rose Poly Inst. Following his grad with a BS Degree in Chem Engg in June, 1963, he entered full time employment with E I Dupont Co in the Elastermers Dept at Louisville, KY. In Sept, 1964, he entered Purdue Univ for advanced schooling in Chem Engg. He was grad with his MS & PhD degree in Sept, 1967, and was employed at OK State Univ, Stillwater, OK, until May, 1987. In June, 1987 he became Dean of the College of Engg at Oklahoma Univ. He has extensive ind experience through full time & summer employment and consultation practices. He has over fifty technical publications & numerous presentations, and has authored or co-edited three books. He & his wife, Mary, have three children: Lawrence, David & Stephen. *Society Aff:* ACS, AIChE, OSPE, ASEE, ΣΞ, ΩXE, NSPE.

Csermely, Thomas J
Business: Electr & Comp Engg Dept, 111 Link Hall, Syracuse, NY 13244-1240
Position: Associate Prof. *Employer:* Syracuse Univ. *Education:* PhD/Phys-Sci Teach/Syracuse Univ; Dipl Ing/ME/Poly Univ of Budapest. *Born:* 6/25/31. Native of Hungary. US citizen, naturalized in 1964. Served as Res Engg with Carrier Corp, Syracuse, NY. Faculty appointments include Inst of Theoretical Physics, Polytechnical Univ of Budapest; Physiology Dept, SUNY Upstate Medical Ctr; Phys Dept, Le-Moyne College; Electr & Comp Engg Dept, Syracuse Univ. Won Wolverine - ASHRAE Diamond Key Award for Outstanding Publication in 1965. Author of 13 technical reports & over 20 papers. Main interest: analysis of physiological system dynamics by math modeling & comp simulation. Listed in: Who's Who in the East; Dictionary of Int Biography; Am Men & Women of Sci; Who's Who in the World; Who's Who in Technology. *Society Aff:* IEEE, AAAS, APS, AAPT, Biophys Soc., NY Acad. Sci.

Cuckler, Raymond E
Business: Room 203, Drumheller Bldg, Walla Walla, WA 99362
Position: Owner *Employer:* R. J. Consultants *Education:* BS/CE/Univ of NB *Born:* 11/13/23 Civil, structural engr for US Army Corps of Engrs for 30 years. Was design engr of Dworshak Dam and of the first to utilize finite element analysis for high head gravity dams. Chief of Design Walla Walla District for 10 years and responsible for all aspects of large multipurpose projects, with construction budgets of $140,000,000 per year. Received Meritorious Civilian Service medal. Established consulting firm in 1979 and was consultant to B.C. Hydro on Revelstoke project. Also consulted as construction specialist for WPPSS on Nuclear projects 3 and 5, with major responsibilities on civil activities. *Society Aff:* SAME

Cudmore, Russell D
Business: 1965 North Park Place, Atlanta, GA 30339
Position: VP *Employer:* Aldrich Co, Inc *Education:* B.S./CE/Northeastern Univ *Born:* 1/27/45 Native of Natick, MA. Started with Aldrich Co (member of the Carlson Group) in 1969 as project engr. Assumed position of mgr of contract engrg Sept. 1973. Responsible for managing the engrg dept in development of engrg design, working drawings and specifications. Elected VP of firm January 1979 responsible for company's engineering projects in Southeastern United States. Promoted to present position of VP in Charge of Aldrich Company's Atlanta office Sept. 1980. Responsible for engrg documents for construction, manpower utilization, cost control methods, code and regulatory compliance, sound engrg practices and value engrg methods. Presently registered in 17 states. *Society Aff:* ASCE, NSPE, SAME, NFPA, CSI

Culberson, Oran L
Business: Chem Tech Div, Oak Ridge, TN 37830
Position: Chem Engr *Employer:* Oak Ridge Nat Lab *Education:* PhD/ChE/Univ of TX; MS/ChE/Univ of TX; BS/ChE/TX A&M. *Born:* 5/2/21. Houston, TX. Capt, Infantry, WWII. Worked in process design & economics at Gulf Res & Dev Co, 1950-53, & at Celanese Chem Co, 1953-56, Mgr of Computing Ctr, Celanese Chem, 1956-62. Mgr of Operations Res, Celanese Corp, 1962-65. Assoc & Full Prof of Chem Engg, Univ of TN, 1965-1981. Sr Devel Staff Mbr, Oak Ridge Natl Lab, 1981-present. Chrmn: Coastal Ben Sec, Knoxville-Oak Ridge Sec, Natl Machine Computations Committee, Chm Engg Educaton Projs Committee, Personnel Supply and Demand Committee AIChE. Fellow AIChE. Univ of TN Alumni Assn Outstanding Teacher, 1978. *Society Aff:* AIChE, ΣΞ.

Culberson, Reid T
Business: 9250 W Flagler St, Miami, FL 33174
Position: Chief Engg. *Employer:* Florida Power & Light Co. *Education:* Exec Pro/Bus Adm/Stanford Univ; BS/Elec Engg/Univ of Miami. *Born:* 2/3/28. Native of Washington, DC. Joined FL Power & Light Co in 1953, and after four yrs of holding various commercial and operations positions, assumed the responsiblity for substation engg. Became Mbr of Engg Services and then Asst Chief Engg in 1970. Current position as Chief Engg since 1972, with responsibility for transmission, substations, protection and arch engg. Active in Electric Power Research Inst for new sources of energy. VChmn of the FL Governor's Council for Metric Conversion. Hiking, canoeing, and square dancing are outside interests. *Society Aff:* IEEE, NSPE, FES.

Culick, Fred E C
Home: 1375 E Hull Ln, Altadena, CA 91001
Position: Prof. *Employer:* CA Inst of Tech. *Education:* PhD/Aero & Astro/MIT; SM/Aero & Astro/MIT; SB/Aero & Astro/MIT. *Born:* 10/25/33. I have been on the faculty of Caltech since 1961; I am presently prof of appl physics and jet propulsion.

Culick, Fred E C (Continued)
My principal res interests are combustion, fluid mechanics and acoustics in propulsion systems; the history of early aeronautics; applied aerodynamics; the stability and control of aircraft; and robotics. *Society Aff:* AIAA, AIP, AAAS.

Cullen, Alexander L
Business: Dept of E & E Engg, Torrington Pl, London WC1E 7JE England
Position: Hon Res Fellow Univ College London *Employer:* Consultant *Education:* DSc/Elec Eng/Univ of London; PhD/Elec Eng/Univ of London; BSc(Eng)/ Elec Eng/Univ of London. *Born:* Apr 1920. ScD Engg, PhD, ScB Engg London Univ. Worked on microwave radar at Royal Aircraft Estab, Farnborough 1940-46. Lectr (later Reader) at Univ Coll London 1946-55. Prof & Hd of Dept of Elec Engg Sheffield Univ 1955-67. Pender Prof of EE & Hd of Dept of Electronic & Elec Engg Univ Coll London 1967-1980. 1980-83 Sci Res Council Sr Fellow (at UCL), 1983- Hon Res Fellow UCL and Consultant. Fellow IEEE, IEE, Inst of Phys, FRS, F Eng, City & Guilds Inst. Appointed OBE 1960. Author "Microwave Measurements" (jointly with Prof H M Barlow) (Constable 1950); about 70 papers, mostly on microwaves, in Proc IEE & elsewhere. *Society Aff:* (F)IEE, (F)IEEE, FRS, F Eng.

Cullen, Thomas M
Business: 911 W Main St, Chattanooga, TN 37402
Position: Mgr, Mtls Evaluation. *Employer:* Combustion Engg Inc. *Born:* Dec 16, 1935 Birmingham AL. MS & PhD Univ of MI, BS Univ of AL. Raised in Wash DC. Res Engr with Univ of MI 1963-66. With C-E since 1966. Engaged in characterization of creep, fracture toughness & corrosion behavior of alloys used in steam generation service. Dev first fully computerized creep lab in this country. Chrmn Applied Res Subctte, ASTM-ASME-MPC Joint Ctte on Effect of Temp on Properties of Metals. Fellow ASM. Avid golfer.

Culler, Floyd L, Jr
Business: 3412 Hillview Ave, Palo Alto, CA 94303
Position: Pres. *Employer:* Electric Power Res Inst. *Education:* BS/Chem Engg/Johns Hopkins Univ. *Born:* 1/5/23. Beginning with R&D in nuclear energy, a WWII proj, my profl career has led to mgt positions at the Oak Ridge Natl Lab where I had responsibility for R&D in broad areas of engg, chem tech, chemistry, phys, biology, environmental sciences, fusion, health phys and radiation, instrumentation and controls, computer sciences, metallurgy, and related fields, including all types of energy. In 1978, I became pres of Elec Power Res Inst, where I am responsible for mgt of R&D progs for the elec utilities. I serve in advisory capacity to various univ engg groups, the White House Offices of Tech Assessment and Sci and Tech Policy, and other government, business groups. I am the USA mbr of the Scientific Advisory Committee to the Intl Atomic Energy Agency.. *Society Aff:* AAAS, AIChE, ANS, ACS, NAE

Cullinan, Harry T, Jr
Business: P.O. Box 1039, Appleton, WI 54912
Position: VP Acad Affairs & Dean *Employer:* The Institute of Paper Chemistry *Education:* PhD 1965/Chem Eng/Carnegie Inst of Tech; MS 1963; Chem Eng/Carnegie Inst of Tech; BS 1961/Chem Eng/Univ of Detroit *Born:* 5/27/38 Raised on Long Island, NY. Worked in process development with Avon Cosmetics. Process Engineer with Westinghouse Research. Joined Chemical Engineering Department at SUNY/Buffalo in 1964. Became Department Chairman in 1969. Visiting Professor at University of Manchester, UK in 1972-73. Appointed Dean of the Institute of Paper Chemistry in 1976. Vice President-Academic Affairs in 1977. Enjoy tennis, squash, basketball and coaching youth soccer. *Society Aff:* AIChE, TAPPI

Cullinane, Murdock J, Jr
Home: 11500 Hollow Oak Ct, Austin, TX 78759
Position: Research Civil Engr *Employer:* US Army Corps of Engrs *Education:* JD/Law/MS Coll School of Law; MS/Sanitary Engrg/MS State Univ; BS/CE/MS State Univ *Born:* 11/22/46 Native of Gulfport, MS. Served on active duty with the US Army Corps of Engrs 1970-1972. Engr through project mgr with the firm of Clark, Dietz specializing in planning and design of municipal and industrial wastewater treatment facilities. 1972-1978. Research civil engr US Army Waterways Experiment Station specializing in wastewater collection and treatment research 1978-Present. Pres, MS Water Pollution Control Assoc 1980. Elected Dir to Water Pollution Control Federation 1981-1984 term. Recipient US Environmental Protection Agency Bronze Medal, 1981. Major, US Army Reserve. *Society Aff:* WPCF, AWWA, SAME

Cullinane, Thomas P
Business: Aerospace and Mech Engg, Notre Dame, IN 46556
Position: Assoc Prof. *Employer:* Univ of Notre Dame. *Education:* PhD/Ind Engg/VPI; SM/Environmental Health/Harvard Univ; MSIE/Ind Engg/Northeastern Univ; BS/Ind Engg/Boston Univ. *Born:* 12/24/42. and attended schools in Brockton, MA. Taught at Northeastern Univ, VPI and the Univ of MA. Have served as a consultant for many nationally known firms specializing in the fields of mtl handling and plant layout. Div Dir for IIE facilities planning and Design Div (1977-1979); ASEE Dow Outstanding Young Faculty Award 1976; NIOSH and NSF fellowships to support educational endeavors. Published several technical articles in journals. *Society Aff:* ASEE, IIE, AIDS, TIMS.

Cullison, William L
Business: P.O. Box 105113, Atlanta, GA 30348
Position: Executive Director *Employer:* TAPPI. *Education:* MBA/Bus/Florida Atlantic Univ; LLB/Law/LaSalle Univ; BS/Marine Engg/US Merchant Marine Academy. *Born:* 8/26/31. Native of Baltimore, MD. Sailed as Marine Engr with US Navy and Petroleum companies aboard ocean going tankers. Joined American Petroleum Institute Staff as Research Coordinator 1957. Served as res & engg coordinator, Asst Dir-Sci and Tech and Asst Dir-Air & Water Conservation. Resp for fundamental res, engg, petroleum measurement, medical res and world petroleum congress. Since 1968 with TAPPI as Technical Secretary then Dir of Tech Operations. Assumed present position 1982. Elected Treasurer 1975, then Dir of Operations. Responsible for all operations of worldwide tech society. PPres of GA Soc of Assoc Execs. Certified Assoc Exec of American Soc of Assoc Executives; Secretary/Treasurer ASAE. *Society Aff:* ASME, SRA.

Cullivan, Donald E
Business: One Ctr Plaza, Boston, MA 02108
Position: Pres-Intl Div. *Employer:* Camp Dresser & McKee Inc. *Education:* MS/Sanitary Engr/Harvard Univ; BS/Civ Engr/Northeastern Univ. *Born:* 10/14/29. Reg PE in IL, MA, MI, & NY. Mbr of the Inst of Engrs, Singapore. Served as water supply advisor in Northeast Brazil for US Agency for Intl Dev. Advisor to Bangladesh govt for natl water supply and sewerage prog. Proj mgr for water supply, sewerage and drainage master plans, design and construction in Bangkok, Thailand. Pres of Intl Div of Camp Dresser & McKee Inc with responsibility or maj environemental engg projs in 15 countries. Chrmn of Intl Engg Committee of American Consulting Engrs Council and active in affairs of the Intl Fed of Consulting Engrs. *Society Aff:* ASCE, AWWA, WPCF, ACEC, AAEE.

Culp, Neil J
Business: 101 W Bern St, Reading, PA 19601
Position: VP - Tech *Employer:* Carpenter Tech Corp. *Education:* BS/Met Engrg/Lehigh Univ. *Born:* 2/12/30. in Reading, PA; attended public schools in PA and NJ. Grad from Lehigh Univ; began prof career with Carpenter Tech Corp, Reading, PA, in 1952 as a research metallurgist. Became Supervisor of Tool & Alloy Steels at Carpenter in 1955 and was promoted to Supervisory Met-Alloy Dev in 1957. Went on to become Asst Mgr of Research in 1959 and then Mgr of the Research Lab in 1960; became Dir of Research in 1969. Was appointed Asst VP of R&D in 1971 and then Asst VP of Corp Dev in 1975; Dir-Planning & Economic Analysis in 1980; Dir-Tech Dept in 1983 (just briefly) and then VP - Tech 1983. *Society Aff:* APMI, IRI, AISI, ASM, AISE, MPCI.

Culpepper, Fred C
3506 Loop Rd, Monroe, LA 71201
Position: Consultant. *Employer:* Ford, Bacon & Davis Inc. *Education:* BS/Civ Engg/VA Military Inst. *Born:* 11/16/18. Native of Monroe, LA. 101st Airborne Div, 1942-45. Joined Ford, Bacon & Davis in 1941. Rose from proj engr to Pres and Chief Operating Officer of Ford, Bacon & Davis Inc. Now consultant to that firm. Past chrmn Natl Constructors Assn. First VP Pipe Line Contractors Assn. Recipient of LA Engg Soc A B Patterson medal for accomplishments in the field of mgt; Leo M Odom award for the distinguished service to the engg profession. Mbr, LA State Bd of Reg for PE & Land Surveyors. Past president Louisiana Engineering Society. Board of directors National Council of Engineering Examiners, Vice President of Southern Zone, NCEE. Chrmn & Chief Exec Officer Breck Construction Co.; Pres Faberect of Louisiana; Pres Culpepper Enterprises LTD; Chrmn, R.C. Boswell. Inc.; Pres Amer Council for Construction Education; Chrmn Advisory Council, Northwest Louisiana Univ School of Construction. *Society Aff:* ASCE, NSPE, NCEE

Culver, Everett E
Business: P.O. Box 38, Henning, TN 38041
Position: Pres. *Employer:* Revico, Inc. *Education:* BSChE/-/Univ of TX, Austin. *Born:* 9/15/35. Tech asst - Chlorine & Caustic - PPG Chem Div - Lake Charles, LA. 1958-62. VP - Industrial Management Club of Lake Charles - 1961. Process Engg & Group Leader - Jefferson Chem Co - TX - 1962-65. Engg Mgr - Chapman Chem Co - Memphis - 1965-67. Contract Div Mgr - Chapman Chem Co - 1967-69. Tech Service Engg - Drew Chem Co - Memphis area, 1970-72. VP - Revico, Inc engg & water treatment - 1972-74. Pres - Revico, Inc - Memphis - 1974-present. Pres - A-T Contractors, Inc - industrial cleaning & coating - Mphs 1975-present. Mbr of Pilots International Assn, United Meth Church, Rolling Hills Country Club. Mfrs Rep for Il Water Treatment Co. Son, Hunt C Culver - senior at Univ of FL; full scholarship - 3 engg fields, now 3.43/4.00 GPA; elected Cadet Commander-AFROTC, 1981-82. *Society Aff:* AIChE.

Culvern, Frederick E
Home: 4115 E 44th St, Tulsa, OK 74135
Position: Sr Consultant. *Employer:* Willbros Butler Engrs, Inc *Education:* MSCE/Civ Engrg/U of CO; BSCE/Civ Engrg/U of NC *Born:* 4/28/10. Native of Carolina but have lived in mid-west since grad. Engr with US Bureau of Reclaimation for 7 yrs. Engr & Asst Chief Engr for Panhandle Eastern Pipeline Co. 14 yrs & 22 yrs as Engg Consultant to the petrol & natural gas industry. Currently Sr Consultant for pipeline design, pipeline hydraulics and pipeline safety. Received an award from the NC Sec ASCE for best technical paper from a sr - 1934, & in 1979 was awarded the Stephen D Bechtel Pipeline Engg award. Life mbr ASCE, honorary mbr Pipeliners Club of Tulsa. Mbr Tau Beta Pi, honorary engg fraternity. *Society Aff:* ASCE

Cummings, George H
Home: 5108 Doe Valley Lane, Austin, TX 78759
Position: Tech Mgr-ret. *Employer:* Rohm & Haas. *Education:* PhD/Chem Engg/Penn State Univ; SM/Chem Engg/MA Inst Tech; BS/Chem Engg/Penn State Univ. *Born:* 12/28/13. With Rohm & Haas 1941 to 1976. From 1941 to 1951 in various chem engg capacities at Bristol and Phila, PA locations. From 1951 to 1976 Technical Mgr at Houston Plant. From 1976 to 1979 Instr (part time) San Jacinto College, Pasadena, TX. Author of several technical articles, editor of two books and co-author of patents on solvent extraction. Reg PE in TX. Dir AIChE 1967-69. *Society Aff:* AIChE, ACS.

Cummings, Samuel D, Jr
Business: 700 Wallace Bldg, Little Rock, AR 72201
Position: Principal. *Employer:* Pettit & Pettit Cons Engrs, Inc. *Education:* BS/Mech Engg/Univ of AR. *Born:* 10/31/47. Native of North Little Rock, AR. Assoc with Pettit & Pettit Cons Engrs, Inc., mech-elec cons, for 16 yrs. Responsible for HVAC design and proj coordination. Reg engr in AR, OK, KS & CA. Pres AR Chap ASHRAE, 1975-76. Pres AR Chap Prof Engrs in Priv Prac, 1979-80. Pres. AR Engr. Foundation, 1980. Pres AR Amer Cons Engrs Council, 1985-86, Chmn ASHRAE Educ & Chap Programs Ctte, 1987, Mbr AR Bd of Registration for Professional Engrs, 1987. Married, three children. Enjoy hunting, baseball and football. *Society Aff:* ASHRAE, NSPE, PEPP, ASME, ACEC, ΦΗΣ.

Cunniff, Patrick F
Home: 7006 Partridge Place, University Park, MD 20782
Position: Professor *Employer:* Univ of MD. *Education:* PhD/Engr Mechanics/VPI & SU; MS/Struct Engr/VPI & SU; BCE/CE/Manhattan College. *Born:* 10/25/33. Native of NY, NY. Served with the US Public Health Service 1956-58. Res mech engr for the Naval Res Lab 1960-63. Joined Mech Engg Dept, Univ of MD, in 1963. Served as dept chrmn from 1975-82. Author of three textbooks, several journal articles and book reviews. Teach a broad range of courses in the curriculum and conducts research on shock and vibration problems. *Society Aff:* ASME, ASEE.

Cunningham, Donald J
Business: 420 NW 13th St, Suite 200, Oklahoma City, OK 73103
Position: Pres. *Employer:* Cunningham Consultants, Inc. *Education:* BS/Civ Engg/TX A&M Univ. *Born:* 2/10/34. Native of Houston, TX. Spent 4 yrs with Continental Oil Co in Ponca City, OK, where he was involved in engg design for grading, drainage, pavement, sewage treatment & fire water systems as well as boundary, topographic, & construction surveys for Conoco's refinery properties. Entered private practice in 1966 & has been responsible for engg, design & related surveying & mapping services. Completed the flood plain hydraulics & hydrology short courses for HEC 1 & HEC 2 under Leo Beard at Austin, TX. Formed Cunningham-Judd & Assoc in May, 1973, of which he is Pres. July, 1979, the name of the firm was changed to Cunningham Consultants, Inc. Enjoys camping & working in the church. *Society Aff:* PE, LS, OSLS, ASCE, ACEC.

Cunningham, Floyd M
Business: Rm 208, Engr Mech Dept, Rolla, MO 65401
Position: Assoc Prof. *Employer:* Univ of MO. *Education:* PhD/Theoretical & Applied Mech/IA State; MS/Agri Engr/IA State; BS/Agri Engr/IA State. *Born:* 11/13/31. Served in US Army Corp of Engrs 1954-56 as platoon leader. Analyzed and designed spreaders for granular mtls in res projs t VPI from 1960 to 1966 as Assoc Prof of Agri Engg. Taught courses in Farm Power & Machinery and advised 6 MS students in this area. As Assoc Prof of Engg Mechanics at Univ of MO-Rolla since 1966, res has been in the area of plate & shell vibrations. *Society Aff:* ΣΞ, ASME.

Cunningham, John E
Business: PO Box X, Oak Ridge, TN 37831
Position: Assoc Dir of Metals & Ceramics Div. *Employer:* Oak Ridge National Lab. *Education:* MS/Metallurgical Engg/Univ of TN; BS/Metallurgical Engg/Univ of IL. *Born:* 3/18/20. Chicago. MS Univ Tenn, BS Univ IL, Metallurgical Engrg. Army Corps of Engrs 1944-46. Oak Ridge Natl Laboratory 1964-present; Mgr prototype facility for manufacture of fuel, control and related components of first generation res reactors, Asst Dir and currently Assoc Dir, Metals and Ceramics Div. Two patents. US delegate 1955 Geneva Atoms for Peace Conference and 1957 Paris Fuel Element Conference. Chrmn 1958 and 1962 Intl Fuel Element Conferences, Gatlinburg, TN. ANS Fellow, Mbr Board of Directors 1976-79, Chrmn Honors and Awards Ctte 1982-84. ASM Fellow. Chrmn Editorial Adv Ctte, ANS-ASM Nuclear Engrg Materials Handbook. *Society Aff:* ANS, ASM.

Cunningham, Richard G
Home: 900 Outer Dr, State College, PA 16801
Position: VP for Res & Grad Studies Prof Mech Engrg. *Employer:* PA State Univ. *Education:* PhD/-/Northwestern Univ; MS/-/Northwestern Univ; BS/-/Northwestern Univ. *Born:* 9/23/21. Native of Flora IL. Engg officer US Navy WWII. Ind exper incl Pure Oil Co as res engr, & Shell Oil Co as res engr, res group leader, & sr res engr. Joined Penn State Univ 1961 as Prof of Mech Engg. Named Hd of Dept 1962; to current pos of VP for Res & Grad Studies 1971. Pubs & pats in fluids engg &

Cunningham, Richard G (Continued)
educ. Chrmn Fac Senate 1967-68. ASME 1974 Moody Award for fluids engg paper; Outstanding Mech engr 1974 Central PA Sect ASME; ASME Policy Bd, Educ 1967-76 & VP Educ 1974-76; Bd of Dir ECPD 1969-75. VP ECPD 1976-78, Pres 1978-80. Chairman Accreditation Coordination Committee 1980-82. Bd of Tr Centre Community Hosp 1969-75. Mbr Phi Eta Sigma, Pi Tau Sigma, Tau Beta Pi, Sigma Xi, Fellow-AAAS. *Society Aff:* ASME.

Cunningham, Walter Jack
Business: Box 2157, Dept of Electrical Engineering, New Haven, CT 06520
Position: Prof. *Employer:* Yale Univ. *Education:* PhD/Engg Sci & Appl Phys/Harvard; AM/Phys/Univ of TX; AB/Phys/Univ of TX. *Born:* 8/21/17. Born in Comanche, TX. Did res and teaching at Harvard during WWII. At Yale Univ since 1946. Prof, Elec Engg, 1956-63; Engg and Appl Sci, 1963-81; Elec Engg 1981-.Assoc Chrmn Dept 1969-72. Author technical papers on engg problems, Nonlinear Analysis (1958). Mbr, Standing Comm on Accreditation, CT Comm for Higher Educ 1968-74; Chrmn 1973-74 Bd of Editors, American Scientist, 1955-81; Assoc Ed J Franklin Inst, 1962-75. Current interests: systems theory, energy problems, acoustics. *Society Aff:* IEEE, ASEE, ASA, ΣΞ.

Cunningham, William A
Home: 8303 Summerwood Dr, Austin, TX 78759
Position: Prof Emeritus. *Employer:* Univ of TX. *Education:* PhD/ChE/Univ of TX; MS/ChE/Univ of TX; BS/ChE/Univ of TX. *Born:* 5/22/04. Born & raised in TX. Following MS in CE in 1929 worked in petrol refining, sulfur production, geological scouting & mineral processing res. Joined chem engg at the Univ of TX 1941. Departmental chrmn 1942-45, 1947-49 & 1953-55. Retired as Prof Emeritus in 1971. Currently Assoc Editor, Encyclopedia of Chem Processing & Design, Marcel Dekker, Inc Named Distinguished Graduate, Col. of Eng. U. Tx, 1974. *Society Aff:* AIChE, ACS.

Cunny, Robert W
Business: PO Box 1367, Vicksburg, MS 39180
Position: Senior Associate *Employer:* The Geotechnical Associates *Education:* MS Ind E/Ind Engg/Purdue Univ; BS/Civ Engg/Purdue Univ. *Born:* 7/22/24. Native of Chicago, IL. Mbr of the Marine V-12 program in WWII. Served in Korea with the Army 434th Eng const Bn 1950-52. Res engr with Portland Cement Assoc 1952-54. First joined Waterways Experiment Station (WES) as soil mechanics engr in 1948; rejoined WES in 1954 and has served as Chief, Soil Dynamics Div; Chief, Earthquake Engg Div; and Dir, Soil Mechs Info Analysis Center. Currently, senior associate with the Geotechnical Associates. Pres of the MS section ASCE 1976-77. Reg PE in MS. *Society Aff:* ASCE, NSPE, SAME.

Cuomo, Dominick M
Business: 4513 Western Ave, Lisle, IL 60532
Position: Dir of Engg. *Employer:* Western Electric Co Inc. *Born:* Oct 1925 NYC. Grad Am Radio Coll 1943. US Army Field Artillery 1944-46. BSEE Manhattan Coll 1951. Began Western Elec career 1951 as mfg & dev engr. Currently Dir of Engg with respon for intro of new electronic switching sys into production & field installation. Former Mbr NYSPE, IEEE, Burlington Chamber of Commerce, Human Relations Council of Alamance Cty NC, Bd of Dir of Jr Achieve & Chrmn - Conservation Ctte of Baltimore MD Chamber of Commerce. Current mbr Telephone Pioneers of Am, Western Soc of Engrs & DuPage Area Council of BSA.

Cupit, William L
Business: 1360 Oil & Gas Bldg, New Orleans, LA 70112
Position: Area Sales Mgr *Employer:* Magcobar Div Dresser *Education:* BS/Petro Engrg/Univ of TX *Born:* 3/10/30 Born and raised in the oilfields of TX. Navy vetran 1951-54. With Dresser Industries since 1964. Involved with drilling and completion of oilwells in gulf coast area last 15 years. Served on Editorial Review Committee of "Journal of Petroleum Tech" (SPE-AIME) and "Society of Petroleum Engrs Journal" 1982-83. Enjoy picking guitar, singing country and western music, skiing, and tennis. *Society Aff:* SPE of AIME

Curran, Elton C
Business: PO Box 805, Lake City, FL 32055
Position: Genl Mgr *Employer:* St Johns Chem Corp. *Education:* BS/CE/Univ of MO. *Born:* 6/28/31. Native of Aliquippa PA. Served as naval officer with US Navy 1954-57. Process Engr with Columbia-Southern Div PPG. Asst Mgr of Met Refining, Kennametal, Inc. Plant Engr, AL Binder Co Plant Mgr, Stauffer Chem Co. Genl Mgr St Johns Chem Corp. Enjoy boating. *Society Aff:* AIChE.

Curran, Robert M
Home: 330 Columbia St, Cohoes, NY 12047
Position: Mtls Engg. Consultant *Employer:* Self *Education:* MS/Met Engg/RPI; BS/Met Engg/RPI. *Born:* 5/23/21. US Navy Electronics Officer, WWII. Welding Res at RPI '46 to '48. Welding Res & Dev, GE 1948 to 1955. Mgr, Mtls Engg Large Steam Turbine-Generator Dept 1955-84. Materials Engineering Consultant 1984-to date. Former Chrmn Metals Properties Council Technical Advisory Committee, Former Chrmn of Subcommittee 3 on Fatigue, Mbr of Joint ASTM-ASME-MPC Committee on Effects of Temperature on the Properties of Metals, Mbr of Appl Res Panel. Author of more than forty technical papers on Turbine Mtls Behavior. *Society Aff:* ASME, ASM, ASTM.

Current, K Wayne
Business: Electrical Engrg Dept, Univ Calif, Davis, CA 95616
Position: Prof *Employer:* Univ of CA *Education:* PhD/EE/Univ of FL; MS/EE/Univ of FL; BS/EE/Univ of FL *Born:* 2/7/49 In 1975, K. Wayne Current joined the Microelectonics Center of TRW Systems Group as a Member of the Technical Staff doing LSI ECL design and device modeling. In 1976, Dr Current joined the faculty of the Electrical Engrg Dept of the Univ Of CA at Davis where he is now Prof. He is presently directing some of the pioneering work in multiple valued logic integrated circuits. His other interests include high performance integrated circuit design, device modeling, and computer aids to integrated circuit design. Prof Current has published over 100 technical papers and reports on many aspects of electronic circuits. *Society Aff:* IEEE, ΣΞ, ΤΒΠ, HKN, ΦΗΣ

Curreri, John R
Home: 10 San Carlos Court, Toms River, NJ 08757
Position: Prof of ME. *Employer:* Poly Inst of NY. *Education:* MS/ME/Poly Inst of Brooklyn; BS/ME/Poly Inst of Brooklyn. *Born:* 7/20/22. Currently, Prof of ME, Poly Inst of NY. Design Engr, Manhattan Proj, 1944- 1947. Asst to Assoc Prof 1948-1955, Poly Inst of Brooklyn. Hd, Dynamics, Environmental Res & Dev, Arma Corp 1955-1957. Prof of ME 1957 to present. Hd of ME Dept, Poly Inst of NY 1966 to 1974, Consultant to Govt & Industry, currently Consultant, Brookhaven Natl Lab, 1974 to present. Co-author text, Vibration Control. More than 50 papers in field of linear, and non-linear vibrations, Chrmn various technical meetings. Mbr of ASME, ASEE, Sigma Xi, Tau Beta Pi, Pi Tau Sigma. Licensed PE, NY. *Society Aff:* ASME, ASEE, ΣΞ, ΤΒΠ, ΠΤΣ

Currie, Malcolm R
Business: 3E1006-Pentagon, Washington, DC 20301
Position: Dir of Defense Res & Engg. *Employer:* Pentagon. *Born:* Mar 1927. PhD, MS, AB Univ of CA Berkeley. Taught Elec Engg at Univ of CA Berkeley & UCLA. Joined Hughes Aircraft Co in Culver City CA 1954 as mbr o the tech staff & later became Dir Hughes Res Labs at Malibu, Corp VP & Mgr Res & Dev Div 1966. Joined Beckman Inst Inc 1969 as VP R&D. In 1973 received the Pres appt of Dir Defense Res & Engg for the US Dept of Defense. In 1971 was elected to mbrship in Natl Acad of Engg. Also Mbr IEEE & Assoc Fellow AIAA. Pub some 20 tech papers & holds 9 pats.

Curry, Elmer L
Business: PO Box 35832, Tulsa, OK 74135
Position: Pres. *Employer:* Corrosion Mitigation Systems, Inc. *Education:* Certificate/Elec Engg/Tech OK State Univ; Basic & Career/Army Engg/Ft Belvoir. *Born:* 9/9/25. Native of OK. Served with Navy communications 1943-46. Served with Army Engrs 1950-52. Sr corrosion engr, Cities Service Gas Co, 1952-68. Chief engr, Western div, General Corrosion Services, Inc, 1968-1977. Pres, Corrosion Mitigation Systems, Inc, 1977-present. Consultant to oil and gas cos and utility cos, world wide. Retired, Army Engr Reserves (Maj). *Society Aff:* NACE, NSPE, OSPE.

Curry, Norval H
Home: 227 Campus Ave, Ames, IA 50010
Position: Retired *Born:* Oct 1914. Native of St Francis KS. MS & BS IA State Univ. Field Engr Struct Clay Products Inst Ames IA 1940-44. Res Assoc to Prof, Agri Engg Dept IA State Univ 1944-59, teaching & mgr of arch services for IA Agri Exper Station. Private cons practice 1959-1977, primarily design of agri exper stations and facils for lrg, specialized agri operations. Founder and President, Curry-Wille & Associates, Consulting Engineers P.C., Ames, Iowa, 1978-1980. Chrmn Farm Struct Div ASAE 1953-54. Natl Bd ov Dir ASAE 1963-65. Pres ASAE 1969-70. Fellow ASAE 1974. ASAE Cyrus Hall McCormick Medalist, 1980. Enjoy all sports & active in Lions Club.

Curry, Roger W
Business: 501 3rd Ave, Moline, IL 61265
Position: Mgr, Mfg Engg. *Employer:* John Deere Plow & Planter Works. *Education:* MS/Agri Engg/Univ of IL; BS/Agri Engg/Univ of IL. *Born:* 4/27/42. Raised on a farm in western IL. Served two yrs in the US Army stationed at the Yuma Proving Ground as a test engr for Usatecom. Employed by John Deere for 20 yrs (incl. school & army time) and have held the following positions: Engr, Sr Engr, Admin engr, Proj Engr & Div Engr all in Product Engg. Since Dec 1978 I have held the position of Mgr, Mfg Engg. Honored by the Quad City Sec ASAE as Young Engr of the Yr 1975 and received the FMC Young Designer Award in 1976. Currently serving as an elder in the Rapids City Christian Church. And President of Citizens Advisory Council to the Riverdale School District.. *Society Aff:* ASAE.

Curry, Thomas F
Home: 2403 Beekay Ct, Vienna, VA 22180
Position: Chief Scientist *Employer:* E-Systems, Inc. Ctr for Advanced Planning & Analysis *Education:* PhD/EE/Carnegie Mellon Univ; MS/EE/Penn State; BEE/EE/GA Tech. *Born:* 11/22/26. Signal Corps, ILT WWII & Korea. BTL Fellowship in Elec Communications, Carnegie Tech; Mbr Technical Staff, BTL, later Dir, Electronics Res Lab, Syracuse Univ Res Corp. Chief Engr, Applied Electronics, Melpar; Technical Advisor to the Pres, E-Systems, & VP, Microwave Systems, Inc beford joining Dept of Defense as Asst Dir, Systems Evaluation (SIGINT) in 1976. Elected Fellow, IEEE, for "the Dev of Remotely-controlled Electronic Reconnaissance Systems–. (Assoc. Deputy Asst. Secy of Navy 1980-1983). Past Mbr of Natl Bd of Dirs, Assn of Old Crows (Electronic Warfare), Past Chrmn, No VA Sec, IEEE. Reg PE, VA & PA. Centennial Award, IEEE, 1984. *Society Aff:* IEEE, AOC, ΣΞ, HKN, ΤΒΠ, AAAS

Curtis, Arthur S
Business: 14 Rielly Rd, Frankfort, KY 40601
Position: Environmental Engr. *Employer:* Natural Resources & Environ Protection Cabinet, Div of Water *Education:* BSCE/-/U of KY *Born:* 8/5/33. Mr Curtis began his engg career with J Stephen Watkins, Consulting Engr in the early Interstate Hgwy Design. In 1961 he shifted to the sanitary field with employment by the Natl Clay Pipe Inst. Mr Curtis took a yr's leave of absence in 1970 to serve as Exec Dir of the KY Soc of PE. During 1972 he joined the firm of Parrott Ely & Hurt, Consulting Engrs. Currently employed by the Commonwealth of KY, he works in Water, a Div within the Dept for Environmental Protection. *Society Aff:* NSPE, ASCE, WPCF, APWA, AAEE

Curtis, Kenneth S
Home: 2204 Happy Hollow Rd, W Lafayette, IN 47906
Position: Prof Surveying & Mapping. *Employer:* Sch of Civ Engg - Purdue. *Education:* MSCE/CE/Purdue Univ; BSCE/CE/Purdue University. #AFI‡SAME, ASCE, ASPRS, CISM, ASEE, NSPS, ACSM, AAGS, ACA, HS. *Born:* Oct 10, 1925. Native of W Lafayette IN. BSCE & MSCE Purdue Univ. Advanced study OH State Univ. Taught at Purdue since 1946, spec in surveying & mapping. Sev full-time work exper with private concerns & govt agencies esp US Coast and Geodetic Survey. Reg land surveyor. 20 yrs as Exec Secy of IN Soc of Prof Land Surveyors & Ed of "Hoosier Surveyor–. As surveying & an outstanding grad prog in geodesy & photogrammetry. Served Surveying & Mapping Div ASCE as newsletter editor, journal editor, & 5 yrs as Secy of Exec Ctte and received the Soc's Surveying & Mapping award 1976. Served ACSM as Natl Dir on 3 separate occasions & received the Soc's Fennell Award 1975. Enjoys historical cartography & tech lit retrieval.

Curtis, L B
Business: PO Box 2197, Houston, TX 77252
Position: VP, Prod Engrg Services Exploration and Production, Intl *Employer:* Conoco, Inc. *Education:* BS/Petrol Engg/CO Sch of Mines. *Born:* 6/18/24 BS Co Sch of Mines 1949. 4 yrs in US Armed Services during WWII. Employed by CONOCO 1949 as Petrol Engr & assigned var engg & supervisory pos in US & Intl oper. Assigned to current pos in 1971, present title is Vice President. Production Engg Services is part of CONOCO's Worldwide Production oper. Mbr Soc of Petrol Engrs of AIME and has served in many capacities professionally, incl Pres in 1971 & most recently as Distingudished Lectr for 1975-76 term. Chrmn of Natl Petroleum Council Subctte on Enhanced Oil Recovery 1982-84. Named to status of SPE Dist Mbr in 1984. Colorado Sch of Mines Dist Ach Award - 1980. *Society Aff:* SPE.

Curtis, Lamont W
Business: URS Corp, 5606 Virginia Beach Blvd, Virginia Beach, VA 23462
Position: VP *Employer:* URS Corp. *Education:* BSCE/Civ Engrg/Univ of ME. *Born:* 9/14/37. Native of Brewer, ME; Capt USAR ret; taught Univ of Dayton & Cleveland State Univ; Engr with Miami Conservancy Dist, Dayton OH 1960-1965; Assoc with Havens and Emerson - Cleveland, OH 1965-1977; Dir Environmental Engg - Dalton Dalton Newport 1977, elected Principal 1978, currently Vice Pres URS Corp, Assoc Prof Old Dominion Univ; mbr Bay Village City Council; Pres Cleveland Sec ASCE 1977; Pres OH Council ASCE; Chmn Dist 7 ASCE; mbr & Chrmn ASCE Environmental Engg Div, Sessions Prog committee 1970-1974 & Exec Committee 1974-1978, TAC Mgt Group D; active in WPCF; presented & published several non-technical & technical articles. *Society Aff:* ASCE, AWRA, WPCF, AWWA, SME, APWA.

Curtis, Richard H
Home: 141 Skyland Dr, Lakeland, FL 33803
Position: Senior VP of Engg. *Employer:* Davy McKee Corporation *Education:* BApplSc/CE/Univ of Adelaide. *Born:* 4/23/31. Balaklava, S Australia. PE, Chartered Engr UK. Fellow Inst of Gas Engrs UK. Mbr ASME, Mbr AIChE, Mbr NSPE. Joined Powergas Corp, Melbourne, Australia 1953; transferred to Stockton-on-Tees England office & served 17 yrs including Chief Engr 1964-68, and Divisional Technical Dir 1968-71. Transferred to Wellman- Powergas (now Davy McKee Corporation) 1971. Assumed current responsibility as Senior VPof Engg in 1980. Enjoys travel, golf, & scuba diving. *Society Aff:* ASME, AIChE, NSPE, IGasE.

Curtright, Donald E
Home: 10415 Belinder Rd, Leawood, KS 66206
Position: Chrmn of Bd & Chief Exec Officer *Employer:* Greb X-Ray Co. *Education:* BS/EE/KS State Univ *Born:* 11/15/26 Native of Lyons, KS. Served in the US Navy in WWII. Attended KS State Univ after the war and was graduated with BSEE in 1950. Employed by Greb X-Ray Co. upon graduation to sell and service medical X-Ray equip and design Radiology depts. Have served as a consultant on design of medical X-Ray equip for Picker X-Ray Corp. since 1963. Became VP of Greb X-Ray in 1963 and then Pres in 1978. Currently serve on the Coll of Engrg Adv Council at KS State Univ Received the Distinguished Service Award in Engrg from Kansas State Univ 1987. *Society Aff:* HKN, ΣΤ, ΦΚΦ

Cusack, John J
Home: 106 Monatiquot Ave, Braintree, MA 02184
Position: Dir of Business Dev. *Employer:* Purcell Assocs. *Born:* Apr 11, 1922. BSCE & BBA Northeastern Univ. Native of Boston MA. USN 1942-46. Presently hold rank of Capt in the Naval Reserve Civ Engg Corps. With Purcell Assocs, Engrs, Archs & Planners since Jan 1974. Placed in charge of all Bus Dev for the firm in 1975 & respon for all design work in the Boston office. Worked for Edwards & Kelcey Cons Engrs 1955-74 & was the Oper Mgr of the Boston office. Presently the Am Soc of Civ Engrs Dist 2 Dir representing the approx 4000 mbrs who live in New England.

Cusack, Joseph G
Business: One Echelon Plaza, Voorhees, NJ 08043
Position: Pres *Employer:* SITE Engrs, Inc *Education:* B/CE/Manhattan Coll *Born:* 11/20/43 Native of NY City. Registered PE and Certified Professional Geologist. Served as Officer in NYARNG for 14 years. Field Engr on Tunnel Projects for Walsh Construction, Project Engr for Woodward-Clyde and Mgr-Environmental Projects for Envirosphere Co (EBASCO Services). Pres of SITE engrs, (a division of Day & Zimmermann, Inc) since 1980. Specializing in Geotechnical Engrg and Environmental Engrg, Geology, with emphasis on Utility and Industrial fields. Responsible for Technical, financial, marketing, and administrative areas of co. Enjoy all sports and Model Railroading. *Society Aff:* SAME, AEG, BTS

Cushen, Walter E
Home: 6910 Maple Ave, Chevy Chase, MD 20815
Position: Supv. Operations Research Analyst. *Employer:* Office of Sec Defense. *Education:* PhD/Logic/Univ of Edinburgh (Scotland); BA/Math/Western MD College. *Born:* 3/21/25. Hagerstown, MD. Married Helen Lingenfelter 1948. Daughter, Donna Gail Michlin, B 1954; son, Mark Edward, B 1969. US Army AK Communications System WWII. Elec Engg training TX A & M. Programmer Bell Relay Computer and Univac I. Policy studies in logistics, combat, intelligence, and R&D mgt for army and DOD. Assoc prof operations res, Case Inst. Created technical analysis div at Natl Bureau of Stds. Asst Dir, Fed Pwr Commission OFC of Energy Conservation; VP Mathematica (1975-1981); Studies at LMI (Logistics Management Institute) of overseas depot maintenance and military base operations (1982-1984). Office of Sec Defense 1984-present. Pres, Operations Res Socy Amer, 1970-71. Honorary SCD, Western MD College. Fellow, AAAS since 1965. *Society Aff:* ORSA, AAAS, TIMS, AAPSS.

Cussler, Edward L
Business: Dept of Chem Engrg and Materials Sci, Minneapolis, MN 55455
Position: Prof *Employer:* Univ of MN *Education:* PhD/ChE/Univ of WI; MS/ChE/Univ of WI; BE/ChE/Yale Univ *Born:* 3/23/40 After receiving his PhD in Chem Engrg, Dr Cussler began his work with 3plus post doctoral years in chem at the Univ of Adelaide, South Australia, and at Yale Univ. He then taught for thirteen years at Carnegie-Mellon Univ in Pittsburgh, PA. He also briefly worked in research for Unilever (London) and Gen Elec. In 1980, he moved to his present position. Dr Cussler has published two books and over 100 papers. His work centers on diffusion, especially diffusion coupled with chem reaction. He has developed selective membrane separations for metal ions and has made extensive studies of detergent kinetics, particulary as connected with the dissolution of gallstones. He has studied diffusion and reaction in porous solids, with particular applications to the decay of teeth. He has also done research on psychophysics as applied to the texture of food. Dr Cussler is the recipient of the Alan P. Colburn Award from the American Inst of Chem Engrs, a Research CareerDevelopment Award from the National Inst of Health and the Ryan Award from Carnegie-Mellon Univ. *Society Aff:* AAAS, ACS, AIChE, ΣΞ

Custer, Warren L
Home: 804 Rte 271 N, Ligonier, PA 15656
Position: Pres *Employer:* Warren L. Custer Co., Inc. *Education:* BS/ME/VPI *Born:* Oct 22, 1920. BS VPI, WWII Burma & China as guerrilla with US Army. Elected Pres H F Lenz Co 1968, in respon charge of design of Mech & Elec Sys for all types of bldgs & dev including military, educ, indus, commercial, housing, hospital, mental health & inst proj's. Active in ctte work with Natl and Penn Soc of PE. Active in design & promotion of Energy Conservation. Johnstown Chap PSPE: Pres, VP, Exec Ctte, Education, Mbrship & Public Relations Cttes. Retired from HF Lenz Co. in 1983 and started second career in Private Conslt Engrg. *Society Aff:* NSPE/PSPE/ASHRAE.

Cutchins, Malcolm A
Business: Dept of Aerospace Engrg, Auburn University, Auburn, AL 36849
Position: Prof of Aerospace Engrg *Employer:* Auburn Univ *Education:* PhD/Engr Mech/VA Polytechnic Inst; MS/Engr Mech/VA Polytechnic Inst; BS/CE/VA Polytechnic Inst *Born:* 3/27/35 Native of Franklin, VA. Engr with Lockheed-GA Co, 1956-66. On leave to serve with USAF 1957-59 and to attend graduate school 1962-66. Left Lockheed as Sr Mechanical Engr to become Assoc Prof of Aerospace Engrg at Auburn Univ, 1966. IR-100 Nat'l Award 1973, Patent 1976. Research for USAF, NASA and others. Received Outstanding Faculty Award, Auburn Univ School of Engrg 1967, 1976, 1981. Natl VP of Sigma Gamma Tau 1979-82, Natl Pres 1982-85. Engr. of Year, Auburn Chapter of ASPE, 1985. Faculty Sec of Auburn Circle of Omicron Delta Kappa. 1984-present. Natl AIAA Structural Dynamics Tech Ctte 1983-87, Chairman 1985-87. Currently Prof of Aerospace Engrg. Areas of interest include aeroelasticity, vibrations, structures, computers. Enjoys basketball, jogging, and fishing. *Society Aff:* AIAA, NSPE, ASEE, ΣΓΤ, ΩΔΚ, ΣΞ.

Cutler, C Chapin
Business: Ginzton Lab, Stanford, CA 94305
Position: Prof Emeritus *Employer:* Stanford Univ *Education:* BS/Gen Sci/Worcester Polytechnic Inst *Born:* 12/16/14 Bell Lab 1937-1978 Dir of Electronic and Computer Systems Research 1960- 1978. Professor, Applied Physics, Stanford Univ 1979-present. Honors & Awards: Honorary Doctor of Engineering, Worcester Polytech; Edison Medal, IEEE 1981. Centennial Medal, IEEE 1984, Robert H Goddard Alumni Award Natl Acad of Scis; Natl Acad of Engrg. *Society Aff:* IEEE, APS, ASA, OSA, ARRL

Cutler, Leonard S
Home: 26944 Almaden Ct, Los Altos Hills, CA 94022
Position: Dir Superconductivity Lab *Employer:* Hewlett-Packard Co. *Education:* PhD/Phys/Stanford; MS/Phys/Stanford; BS/Phys/Stanford. *Born:* Jan 1928. PhD, MS & BS Phys, Stanford Univ. USN 1945-46. VP Engg for Gertsch Prod Inc until 1957 specializing in frequency meters & precision ratio transformers; Hewlett-Packard 1957-, working in areas of precision oscillators, atomic frequency stds, magnetic bubble memories & superconductivity. Assumed respon as Dir of Physical Res Lab Hewlett-Packard 1969. Fellow IEEE 1974-; Electorate Nominating Ctte AAAS. Served on Adv panels to Natl Bureau of Stds 1968-78. Presently Dir Superconductivity Lab, Hewlett-Packard Co. Morris Leeds Award from IEEE in 1984. National Academy of Engg, 1987. *Society Aff:* AAAS, IEEE, ΣΞ.

Cutler, Verne C
Home: 8630 N Spruce Rd, Milwaukee, WI 53217
Position: Prof. *Employer:* Univ of WI-Milwaukee. *Education:* PhD/Engg Mech/Univ of WI; MS/Structural Engr/KS State Univ; BS/Civ Engr/KS State Univ; -/Civ Engr/Wichita State Univ. *Born:* 1/2/26. Native of Brookings, SD. Draftsman Boeing Airplane Co-1943, served in Air Force 1944-1946, Taught in Civ Engg Dept, KS State

Cutler, Verne C (Continued)
Univ 1950-51, Design Engr Boeing Airplane Co 1951. Taught in Mechanics Dept Univ of WI 1951-1963. Asst Prof Dept of Mechanics, Univ of WI 1960-63. Chrmn, Dept of Mechanics Univ of WI - Milwaukee 10 yrs to 1974. Proj Dir $173,000 NSF COSIP grant 1969-1973. Chrmn- Mechanics Div of Amer, Soc. for Engrg Education 1982-83. Presently Prof of Engg Mechanics, Univ of WI-Milwaukee, materials, fatigue, stress analysis. Enjoy fishing, hunting, woodwork, tennis. *Society Aff:* ASTM, ASEE.

Cutler, W Gale
Business: Whirlpool R and E Center, Benton Harbor, MI 49022
Position: Staff VP, Univ Relations *Employer:* Whirlpool Corp *Education:* PhD/Physics Egrg/PA State Univ; MS/Physics/PA State Univ; BA/Math/Monmouth Coll; Sr Exec Prog/MIT (Sloan School) *Born:* 1/31/22 Native of IL. Taught engrg physics and mathematics at PA State Univ, Monmouth Coll and Mankato State Coll; Physics Dept Head at Mankato State Coll. Joined Whirlpool Corp in 1957; served as Research Physicist, Mgr of Mechanical Engrg Research, Research Dir, Dir of Corp Research and became Staff VP, Univ Relations in 1984. Active in technical societies, member of Bd of Dirs of Industrial Research Inst, 1981-84. Member of Governing Bd of Monmouth Coll, 1980-84. Awarded Whirlpool Chapter, Sigma Xi, Elisha Gray Award in 1981 for outstanding management contributions to Whirlpool. *Society Aff:* ΣΞ, ACS, APS, AOCS, TIMS

Cutlip, Michael B
Business: Box U-139, Dept of Chem Engr, Storrs, CT 06268
Position: Professor and Head of Chemical Engineering *Employer:* Univ of CT. *Education:* PhD/Chem Engr/Univ of CO; MS/Chem Engr/OH State; BChE/Chem Engr/OH State. *Born:* 9/21/41. in OH. Postdoctorate at CO after PhD degree 1967-68. Became Asst Prof at CT in 1968. Consultant with United Technologies and Control Data Corp. Sabbatical Leave at Cambridge, England 1974-75 and 1983. Res interests include Catalysis/Chem Reaction Engg, Electrochemical Reaction Engg and Computer Based Education. *Society Aff:* AIChE, ACS, CACHE, ASEE

Cutting, Charles L
Business: 417 Montgomery St, San Francisco, CA 94104
Position: VP & Corp Princ, Public Works *Employer:* Sverdrup Corp *Education:* MS/Engrg/Stanford Univ; BS/Civ Engrg/Stanford Univ. *Born:* 05/17/26. MS & BS Stanford Univ 1948. Has been with Sverdrup & Parcel San Francisco since completing educ in civ engrg. Involved in engrg of high-voltage transmission lines, hydroelec projs, airports, mass transit facilities, hgwys, bridges, piers, & commercial & indus bldgs. As Corp Principal, Public Works has been or is respon for all S & P hydrpower and port projects. Previous responsibilities include a maj expansion airlines facility, IRS regional computer ctr, $330 million expansion S F airport, & sev multi-million-dollar jobs in SE Asia and elsewhere overseas. Reg 6 states. Mbr ASCE. *Society Aff:* ASCE, ASTM, ACEC.

Cutts, Charles E
Home: 4599 Ottawa Dr, Okemos, MI 48864
Position: Prof. Emeritus *Employer:* MI State Univ. *Education:* PhD/CE/Univ of MN; MSCE/CE/Univ of MN; BSCE/CE/Univ of MN. *Born:* 5/15/14. Taught and conducted res at the Univ of MN, Univ of FL and MI State Univ. Res Admin at Natl Sci Fdn, Wash DC. Served as Prof and Chrmn, Dept of Civ Engg, MI State Univ 1956-1969; Prof 1969-84; Prof Emer. 1984- ; Consultant, Univ of Minnesota, Morocco Project 1986. Natl VP of ASEE 1970-72. EE & A Committee, ECPD. Co-author of textbook "Structural Design in Reinforced Concrete–. Pres, MI Section, ASCE. *Society Aff:* ASCE, ASEE, ACI, NSPE.

Cywin, Allen
Home: 1126 Arcturus Ln, Alexandria, VA 22308
Position: Consulting Engineer & Retired Sr Sci Advisor EPA *Employer:* Self *Education:* BCE/Civ Engg/RPI. *Born:* 8/31/25. Private Consulting Engineer and Retired Sr Sci Advisor for Water & Waste Mgt of EPA after creating & directing the Effluent Guidelines Div responsible for establishing all natl ind waste water stds. Directed entire Water Pollution R&D Prog during the first yr of EPA after creating & directing the R&D progs for ind. & non-point source waste water control. Asst Commissioner of the Community Facilities Admin (now HUD), Asst Dir for Demonstration Plants in the Office of Saline Water. Award for Achievement in Water Quality Control 1969 ASME. Distinguished Career Gold Medal, Superior service medals 1975 and 1976 EPA. Hatfield Award 1978 WPCF. Over 60 publications & 4 patents. Cmdr USNR Ret. Reg PE (Mech). *Society Aff:* ASCE, ASME, WPCF.

Czyzewski, Harry
Business: 2245 SW Canyon Rd, Portland, OR 97201-2499
Position: Pres. *Employer:* OR Tech Serv Ctr, Inc *Education:* MS/Met Engrg/U of IL; BS/Met Engrg/U of IL. *Born:* 2/13/18. Now an independent sr consltg engr, Czyzewski was co-founder and president of MEI-Charlton, Inc, an engg and consulting firm, 1946-1983. Much of his work relates to use of mtls in large structures. He has authored more than 50 technical papers, including a presentation on metal corrosion to an Intl Congress, Brazil, 1978, Germany 1981. Recent awards include Consulting Engrs Council of OR Pres's Citation for Extraordinary Service, 1979; Univ of IL, College of Engg Alumni Honor Award (Gallo Medal) for Distinguished Service in Engg, 1979; Am Consulting Engrs Council Engg Excellence Award, 1977; Natl Pres of Am Council of Independent Labs, 1978-80. *Society Aff:* AIME, ASM, ASME, ACEC, NSPE, ΦΚΦ, ΣΞ, ΤΒΠ, ΑΣΜ, ACIL, AFS.

Dabkowski, Donald S
Business: 600 Grant St, Rm 2026, Pittsburgh, PA 15230
Position: Manager-Metallurgy & Quality Assurance, Tubular Products *Employer:* USS A Division of USX Corporation. *Education:* BS/Metallurgical Engg/Carnegie Inst of Tech. *Born:* Jun 1932. After serving in Marine Corp joined the Research Lab of the US Steel Corp working in high-and-low-alloy steel development for heavy product applications, induction heat-treatment of heavy products and control-rolling of high-strength low-alloy steels. Is presently Manager-Metallurgy & Quality Assurance, Tubular Products. Registered Prof Engr Pennsylvania. ASM Henry Marion Howe Medal 1965. Charles Hatchett Award & Medal of the Metals Soc 1979. *Society Aff:* ASM, TMS-AIME, API.

Dacey, George C
Business: Sandia National Laboratories, Albuquerque, NM 87185
Position: Pres. *Employer:* Sandia National Laboratories *Education:* PhD/Physics/CA Tech; BS/Elec Engg/Univ of IL. *Born:* Chicago, Ill. During WWII worked on radar countermeasures at Westinghouse Research Laboratories. Joined Bell Laboratories in 1952 working on transistor device feasibility research. From supervisor of transistor development work roce to Assistant Director of Solid State Electronics Research Laboratory in 1958 & Director two yrs later. In 1961 took a leave of absence to become VP of Research of the Sandia Corporation returning to Bell Laboratories two yrs later as Executive Director Telephones and Power Division. Elected VP Customer Equipment Development in 1968, VP Transmission Sys in 1970, and VP Operations Sys Network Planning in 1979. He assumed his present responsibilities in August, 1981. Received E2A2 Distinguished Alumnus Award in 1970 from Univ of Illinois EE Alumni Association. Elected to Natl Acad of Engg. *Society Aff:* IEEE, APS.

Dackis, William C
Business: 300 Park Ave, New York, NY 10022
Position: VP. *Employer:* Crane Co. *Education:* BSME/Mech Engg/Duke Univ. *Born:* 8/28/24. Born in Durham, NC. Served with US Navy 1944-1946. With Natl Advisory Comm for Aeronautics as Aeronautical Res Scientist before joining The Trane Co - last six yrs as VP, Heat Transfer Sales. With Crane Co since 1965, first as VP Res & Dev; currently, VP Adm. Duties include corporate planning, coordination of R & D, energy conservation and stds activities. Trustee, the Crane Fund, and CF&I

Dackis, William C (Continued)
Fund. Reg PE, WI. Phi Beta Kappa, Tau Beta Pi. *Society Aff:* ASME, NACE, ADPA.

Dadras, Parviz
Business: School of Engrg, Dayton, OH 45435
Position: Assoc Prof. *Employer:* Wright State Univ. *Education:* PhD/Mech Engg/Univ of DE; MS/Mech Engg/Univ of DE; BS/Mech Engg/Abadan Inst of Tech. *Born:* 9/26/40. Worked in Oil Industry 1964-66. Taught in Metallurgical Engg Dept, Arya Mehr Univ of Tech, Tehran, Iran, 1972-77. Was Adjunct Asst Prof in the Dept of Mtls Sci, and Engg, Univ of FL, 1977-78. Assumed current position in Sept 1978. Have taught and performed res in Solid Mechanics, Mech Properties, and Deformation Processings. *Society Aff:* ASM, TMS-AIME, ASME.

Daehn, Ralph C
Business: Midwest Materials & Engineering Consultants, Inc, 864 W. Stearns Rd, Bartlett, IL 60103
Position: President *Employer:* Midwest Materials & Engineering Consultants, Inc *Education:* MS/Mtls Sci & Engr/Northwestern; BS/Met Engr/IIT; BS/ME/IIT. *Born:* 8/21/33. A native of Chicago's Western suburbs. Service with the US Army during the cold war. Mgr of met engg for Danly Machine Corp, responsible for mtls & processes used. Mtls lab mgr for Cummins Engine. Managed key dev progs related to welding and adhesive bonding of cans for Continental Can. Transferred Continental's aluminum can making tech to the UK & Australia. Staff Conslt with Packer Engrg Assoc. Founder & Pres, Midwest Materials & Engineering Consultants, Inc., Wayne, Ill. Chrmn, Chicago-Western Chapter, ASM, 1974-75. mbr NSPE, SME, ASTM and ASME. PE State of IL. Many outdoor sports activities. *Society Aff:* ASM, ASME, NSPE, ASTM, SME

Dague, Delmer C
Business: 1 General St, Akron, OH 44329
Position: VP QA *Employer:* Gen Corp *Education:* BS/Mgmt Eng/Carnegie Inst; BA/Phys/Wash and Jefferson Coll *Born:* Native of Wheeling, W VA. Served in Army Air Force (1942-45). Quality Engrg Supvr for the Hoover Co 1951-62. Employed by Gen Corp, Inc (formerly the General Tire and Rubber Co) in 1962. Assumed current respon as VP, Quality Assurance in 1979. P Pres (1966) of Akron Coun of Engrg and Scientific Socs, Past Chrmn, Akron-Canton Section, Amer Soc for Quality Control, Past Chrmn, Automotive Div, ASQC. Recipient of Automotive Div, Roth Award, 1982. Fellow Mbr, ASQC. *Society Aff:* ΤΒΠ, ASQC

Dague, Richard R
Business: Professor and Chairman, Department of Civil Engineering, Iowa State University, Ames, IA 50011
Position: Prof, Civ and Environ Engrg *Employer:* University of Iowa. *Education:* PhD/Environ Engg/Univ of KS; MS/Sanitary Engg/IA State Univ; BS/Civil Engg/IA State Univ. *Born:* Feb 1931. Served with US Navy Seabees 1950-53. Taught in Department of Civil Engrg at Iowa State 1960-62, & at Kansas State 1966-67. At the University of Iowa since 1967 except for 3 yrs as Senior Environmental Cons Henningson, Durham & Richardson, Omaha Nebr. Mbr of Phi Kappa Phi & Chi Epsilon Honor societies. Served as Pres Iowa Water Pollution Control Assn & Natl Dir to WPCF from Iowa. Special cons to US Army Corps of Engrs & to US Energy R&D Administration. Consult to US Agency for Intl Dev. Married, four children. *Society Aff:* ASCE, AWWA, WPCF, AEEP.

Dahl, Robert E
Business: Manhattan, KS 66506
Position: Engrg Dept Head *Employer:* KS State Univ *Education:* MS/Struct Engrg/KS State Univ; BS/Arch Engrg/KS State Univ *Born:* 03/24/27 in Garnett, KS. Served in WWII. Taught in Civil Engrg Dept, KSU, 1952-54. Twenty-one yrs indus experience; structural engrg, architectural engrg design and construction admin. Returned to teaching in 1976, assumed current position in 1979. *Society Aff:* NSPE, ASHRAE, ASEE, AIC

Dahl Roy E
Home: 5050 S. Olympia, Kennewick, WA 99337
Position: Mgr Fuel Cycle Plant *Employer:* Westinghouse Hanford Co *Education:* PhD/Nuclear Engrg/NC State Univ; MS/Nuclear Engrg/Univ of WA; BS/Chemistry/Gonzaga Univ *Born:* 11/07/30 Native of Spokane, WA. Served in U.S. Army - Infantry Officer. Worked at Hanford since 1955 as an engineer and engineering manager in reprocessing, reactor design, materials development and fuel fabrication. Current assignment is Manager, Chemical Processing Programs with Westinghouse Hanford Company. Member of the American Institute of Chemical Engineers, American Nuclear Society, and the National Management Association. A licensed Professional Engineer, publications include 75 journal articles and reports, holds three patents. Active in community and church affairs. Enjoys golf, tennis, travel, and working in apple orchard. *Society Aff:* ANS, AIChE

Dahlberg, E Philip
Home: 4058 Dumbarton Rd, Houston, TX 77025
Position: Sr Metallurgical Consultant. *Employer:* Metallurgical Consultants, Inc *Education:* PhD/MET Engr/U Florida; MS/Matls Sci/NorthWestern U; BS/Sci-Engr/NorthWestern U; BA/Liberal Arts/Shimer College *Born:* 8/8/37. in Chicago, IL. Married 1959, 4 children. Res Met Naval Res Lab Wash DC 1960-63 & 1966-69; Assoc Res Coord UOP Inc Des Plaines IL 1969-76; Mgr Mat Tech Exxon Nuclear Co Inc Richland WA 1976-78; Sr Met Consultant Failure Analysis Assoc Houston, TX 1978-82; Sr Met Consit Metallurgical Conslts, Inc Houston, TX 1982- . Reg PE IL, TX & WA, Mbr Handbook Comm ASM Spec in mech met & failure analysis, SEM fract mech & stress-corrosion, product liability tech consltg, lecturer Univ Houston ME Engr Dept. *Society Aff:* ASM, ASTM, AIME-TMS, ANS, NACE, ADA, AAAS, TSEM, HAMAS.

Dahlen, Dean M
Business: P O Box 101, Florham Park, NJ 07932
Position: Asst Gen Mgr. *Employer:* Exxon Res & Engg Co. *Education:* BS/Mech Engg/Univ of DE. *Born:* 6/6/33. BS-ME Univ of DE 1955. Joined Esso Engg the same yr. Initial assignment in Cost Engg, later, fractionation res. Managed Engg Res Lab, Heat Transfer Engg, Cost Engg, Offsites and Util design, and Chem Proj Mgr. Asst Mgr- Esso Engg- Europe, Sen Refining Advisor-Exxon Corp, Mgr-Centrifuge/Enrichment Tech and Centrifuge Manufacturing-Exxon Nuclear Co, Mgr Proj Management-Esso Exploration and Production UK, Ltd./Esso Europe Producing Dept. Asst Ger Mgr-Project management Dept-Exxon Res and Engg Co, Graduate Cornell Exec Dev Prog. Licensed PE in NJ. Currently Sr Project Dir - Exxon USA Baytown Refinery Upgrade Proj.

Dahlgren, Shelley D
Business: Battelle Northwest Blvd, Richland, WA 99352
Position: Mgr, Metallurgy Res Section *Employer:* Battelle Memorial Inst. *Born:* 07/-/37 PhD Univ of CA Berkeley; BS & MS Univ of Wash. Native of Seattle WA. Metallurgist for Boeing Co, Commerical Airplane Div, 1961, served the US Army at the Tank Automotive Command 1961-63, Res Sci at Battelle-Northwest 1966-1977, specializing in basic res on synthesis of materials by high-rate sputter- deposition. Mgr, Surface Sci Sect 1977-79, Mgr, Metallurgy Res Sect, at Battelle Northwest, 1979- . Enjoys sailing, guitar, tennis, hunting, fishing & sports events.

Dahlke, Walter E
Business: Bethlehem, PA 18015
Position: Prof Elec Engrg. *Employer:* Lehigh University. *Education:* PhD/Physics/Berlin Univ; Dr Habil/Physics/Jena Univ. *Born:* Aug 1910. Dr Phil, Univ Berlin, Dr Habil, Univ Jena, Germany. Res Scientist, Hd Microwave Lab, German Aviation Res Estab, Berlin 1940-45. Senior Scientist, Head Applied Research Tube & Semiconductor Div & Solid State Res Group, Telefunken, Ulm, W Germany 1949-64. Lecturer at Univ Heidelberg, Stuttgart, Karlsruhe 1955-64. Hon-

Dahlke, Walter E (Continued)
orary Prof Univ Karlsruhe 1961. Natl Science Foundation Sr Foreign Scientist Fellow 1964 & Prof Elec Engrg 1965, Lehigh University. Fellow IEEE, Mbr EKN, German Phys Soc & Inst of Elec Engrs. Four dozen tech publications. *Society Aff:* IEEE.

Dahlstrom, Donald A
Business: Univeristy of Utah, Chemical Engrg Dept, Salt Lake City, UT 84112
Position: Res Prof *Employer:* Univ of Utah *Education:* PhD/Chem Engg/Northwestern Univ; BSChE/Chem Engg/Univ of MN. *Born:* 1/16/20. AIChE: Dir 1960-62, V P 1963, Pres 1964. Founder's Award 1972; Environ Award, 1977; Fellow. American Institute of Mining, Metallurgical & Petroleum Engrs: Bd/Dir 1973-75; V P 1975. Rossiter W. Raymond Award 1954; Richards Award 1976. Krumb Lectr, 1980. Taggart Award, 1983. Natl Academy of Engrs, Honorary Member, 1986. The Filtration Soc London. Named one of thirty as Emminent Engr at the 75th Anniversary Meeting of AIChe *Society Aff:* AIChE, AIME, ACS, NAE, MMSA, ASEE.

Dahm, J Richard
Home: 34 Mary Ann Drive, Pittsburgh, PA 15227
Position: Research Supervisor. *Employer:* LTV Steel Co. *Education:* BSEE/Elec Engg/Carnegie Mellon. *Born:* 2/16/32. Subsequent graduate work in automatic control. In current position is respon for process computers & control systems development, particularly the application of computer control technology to steel plant processes; prior experience in steam power plant instrumentation & control with Duquesne Light Co. Has published several papers on measurement & control of heating & rolling processes, & holds two patents covering instruments to measure cold rolled strip shape. Recipient of ISA R A Francy Memorial Award in 1970, & the ISA Instrumentation Technology Award in 1971. Sr Mbr ISA & Mbr IEEE. *Society Aff:* IEEE, ISA.

Dailey, Ralph R, Jr
Business: PO Box 1159, Harvey, LA 70059
Position: VP, Oper. *Employer:* Loop Inc. *Education:* BS/ChE/TX A&M Univ. *Born:* 6/23/23. Native of Pt Arthur, TX. Served with Army Corps of Engrs 1943-1946 with service in Europe in 1944-1945. With Texaco Inc since 1947. Served in various refining, engg, & managerial positions in several locations in US, Central & S Am & Europe. Currently on loan to the LA Offshore Oil Port (LOOP) of which Texaco is a shareholder as VP, Operations. Main hobby activities are centered around Amateur Radio. *Society Aff:* AIChE.

Daily, Eugene J
1114 Lincolnshire, Champaign, IL 61820
Position: Consulting Engineer. *Employer:* Self. *Education:* Prof Civ Engr/-/Univ of MO, Rolla; MS/CE/Univ of IL; BS/CE/MO Sch of Mines. *Born:* 8/18/13. An engr for more than 40 yrs, experience includes service with US Army Corps of Engrs, prof of engg at Univ of IL, and principal of consit engg firms. Specialties are hydrological and structural engg; projs including use of detentinn basin concept for urban stormwater control (Shelby Co, TN) and design memo for complex railway/hgwy re-routing in connection with navigation proj. (Kaskaskia R, IL). Active in profl engg organizations; past pres, Consltg Engrs Council of IL (CECI); past pres, Central IL ASCE Pres Illini Post S.A.M.E., Mem Acad civil Engr. U.M.R.. *Society Aff:* ACEC, ASCE, AREA, NSPE, APWA.

Dajani, Jarir S
Home: PO Box 814, Abu Dhabi United Arab Emirates
Position: Tech Adv *Employer:* ABU DHABI Fund for Economic Development *Education:* BEngg/Civ/American Univ of Beirut; MSc/Engg Econ Planning/Stanford Univ; Phd/Urban Systems/Northwestern Univ. *Born:* 4/5/40. Resident Engr, Associated Consulting Engrs; Lebanon, Jordan and Saudi Arabia, 1961-65. Transportation Planner, DeLeuw, Cather & Co, Chicago 1968-71. Assoc Prof of Civ Engg and the Policy Sciences, Duke Univ, Durham, NC 1971-76. Assoc Prof of Civil Engrg, Stanford Univ, Stanford, CA 1976-82. Research and Evaluation officer, USAID, Amman, Jordan 1981-82 (on leave from Stanford Univ) Consultant to Intl Bank for Reconstruction and Dev, Agency for Intl Dev. Sr. Technical Advisor, Abu Dhabi Fund for Economic Development, P.O. Box 814, Abu Dhabi, UAE 1982- . *Society Aff:* ASCE, TRB, API.

Dajani, Walid Z
Business: 306 S 19th St, Philadelphia, PA 19103
Position: Chief Bridge Engr. *Employer:* Urban Engineers Inc. *Education:* BSc/CE/Univ of MO-Rolla; MSc/CE/Univ of KS. *Born:* Dec 1929 Jaffa, Palestine. US Citizen & res Cherry Hill, N J. BSc Univ of Missouri-Rolla; MSc 1962 U of Kansas, thesis: Stress Distrib in Orthogonally Stiffened Plates; Prof Engr in N Y, PA, NJ & DE; prior experience included work for major Mid-West & East Coast Consultants; Chief Bridge Engr for Urban Engineers, Inc since 1972 having total responsibility for planning, design and in-depth inspection of all bridge projects; also responsible for hydrologic & hydraulic analysis projects & flood insurance studies; pres of Ithaca NY ASCE Sect in 1971-72; Mbr ASCE Committee on steel bridges and a fellow of ASCE; Enjoys bridge, golf & internatl travels. *Society Aff:* ASCE.

Dalal, Jayesh G
Home: 46 Sussex Court, Naperville, IL 60540
Position: Engineering Specialist *Employer:* GTE Communication Systems Corp. *Education:* PhD/ME/U of WI (Mad); MS/ME/U of WI (Mad); BE/ME/Gujarat U (India); BE/EE/Gujarat U (India) *Born:* 07/25/37 Resident of Naperville, IL. Worked at Allis-Chalmers from 1963-66, specializing in computer modeling and computer-aided design. Worked at Inland Steel Co from 1970-86, specializing in experimental design, statistical data analysis, process modeling, and optimization. Participated in the planning and dev of Inland's quality improvement process, and in the training for the same. Chrmn, Metals Tech Ctte, ASQC, 1980-82; Chrmn, Metals Group, CPID/ASQC, 1983-85; V-Chrmn, 1982 AQC Program, ASQC; Councilor- Region 12, CPID-ASQC, 1985-; AIME McKune Award Winner, 1976. Worked as the Mgr of Statistical Process Control, Amphenol Products, 1986. Since 1987 employed at GTE Communication System Corp. as Engrg Specialist in the Quality Assurance Dept. *Society Aff:* ASQC, ASA

Dale, James D
Business: Dept of Mech Engg, Edmonton Alberta T6G, CANADA
Position: Prof. *Employer:* Univ of Alberta. *Education:* PhD/ME/Univ of WA; MSc/ME/Univ of Alberta; BSc/ME/Univ of Alberta. *Born:* 12/5/39. Native of Edmonton, Alberta. Worked with Res Council of Alberta 1963-64. With Univ of Alberta since 1969. Res work has been in fundamentals of convection heat transfer, cycle improvements in gas turbines with steam injection, emissions from spark ignited engines, pulsed plasma jet & laser ignition systems for spark ignited engines, low temperature starting of Diesel engines, energy conservation in houses. Faculty advisor for ASME & CSME; mbr of Natl Students Committee, CSME 1971-75; 1979 SAE Ralph R Teetor Award. Enjoy running, x-country skiing, raquetball, antique automobiles. *Society Aff:* ASME, Comb Inst, CSME, SAE.

Dale, John C
Business: 4660 Wilkens Ave, Suite 201, Baltimore, MD 21229
Position: Pres *Employer:* Jack Dale Assoc Inc *Education:* BS/EE-ME/U.S. Naval Acad; *Born:* 5/25/24 Pres, Jack Dale Assocs, Inc, a consulting engrg firm specializing in energy management and energy conservation. Mr. Dale has authored four books on energy conservation including one now used as a text by several MD Colleges. He is a retired Captain, U.S. Naval Reserve. Past Pres of the Annapolis, MD chapter of the MD Society of PE, Past Pres of the MD Soc of PE and past Dir of the Natl Soc of PE. At the present time he serves as energy consultant to numerous commercial and Industrial firms. He is listed in Who's Who in the East, Who's Who in Ecology, International Who's Who in Engrg, Personalities of the South, Dir of Dist Americans and Intrntl Roll of Honor. *Society Aff:* NSPE, ASHRAE

Daley, Joseph F
Home: Corey Lane, Mendham, NJ 07945
Position: Retired 1984. *Employer:* Warner Lambert Co. *Education:* MS/Chem Eng/NY Univ; BS/Chem Eng/NJ Inst of Tech. *Born:* Oct 1918 N J. Initially employed by The Flintkote Co in product dev followed by plant management. With Warner-Lambert Co since 1957, currently respon for all internatl production operations manufacturing primarily pharmaceutical & confectionary products. Mem AICHE & one of the founding members of the New Jersey Section. Mbr Tau Beta Pi. Enjoy golf, tennis & gardening. Retired 1984. *Society Aff:* AIChE.

Dalia, Frank J
Home: 43 Imperial Woods Dr, New Orleans, LA 70123
Position: Prof. *Employer:* Tulane Univ. *Education:* PhD/Engg Economics/Tulane Univ; MS/Structural Engg/Tulane Univ; BS/Civil Engg/Tulane Univ. *Born:* 11/10/28. Officer in Civil Engg Corps, USNR, 1952-55; currently retired with rank of Commander. Joined faculty of Tulane Univ as asst prof in Civil Engg Dept specializing in structures, 1955. Granted fellowship in 1961 to obtain PhD in Economics. Returned to Tulane faculty in 1964, becoming full prof in 1968. Engaged in private consulting engg practice concurrently with teaching. Designer of various types of structures, including commercial, industrial and public buildings, municipal plants, and specialized fdns. Served as forensic expert in engg, providing investigations and testimony. Married, four children. Native of New Orleans, LA. *Society Aff:* ΤΒΠ, ASCE, NSPE, LES, STE.

Dallas, James P
Home: 8511 Vicksburg Ave, Los Angeles, CA 90045
Position: Principal. *Employer:* Self employed. *Education:* -/Chem Eng/Univ of WA. *Born:* Dec 1904. Fellow IEEE. Chmn Subctte on Aerospace Elec Insulation; Pubs: 'Recommended Practice for Aircraft', 'Missile & Space Equip Electrical Insulation Tests', IEEE Standard 135; 'Bibliography on High Temperature Electrical Insulation for Flight Vehicles', IEEE publication 68c 37AES. 'Administrative Ctte IEEE Professional Technical Group on Aerospace' 1963-66. Member of 'Advisory Staff on Aircraft & Missile Electrical Systems' sponsored by USAF Research & Development Command jointly with the US Navy Bureau of Aeronautics 1951-59. Mbr of Tech Staff of Hughes Aircraft Co 1941-70. Author or coauthor of 17 technical papers or articles. California Engrg Licenses Elec 2567 and Mech 6228. *Society Aff:* IEEE, AES, AIP.

Dalle Donne, Mario
Home: Kiefernweg 2, Stutensee-Blankenloch, Germany 7513
Position: Division Head. *Employer:* Euratom. *Education:* PhD/Engg/Bologna Univ. *Born:* Oct 1932 Bologna, Italy. 1957-59 heat transfer work at Agip Nucleare. 1959-63 seconded by Euratom to OECD High Temp Gas Cooled Dragon Reactor Proj, England. Since 1963 delegated from Euratom to Karlsruhe Nuclear Center, Germany. Head of Gas Heat Transfer Division & project leader of Gas Cooled Fast Reactor. 1973-77 European Tech Coordinator of OECD-NEA Coordinating Group on Gas Cooled Fast Reactor Development. Mbr German Nuclear Society. Fellow Amer Nuclear Soc. 1976-77 Chmn of ANS Local Sect in Central Europe. Enjoy philately, classical music, reading history, gardening & sauna. *Society Aff:* ANS, KTG.

Dalley, Joseph W
Home: 1911 Woods Dr, Arlington, TX 76010
Position: Assoc Dean Engrg. *Employer:* University of Texas. *Education:* PhD/Engr Mech/Univ of TX, Austin; MS/Engr Mech/Univ of TX, Austin; BS/Aero Engr/Univ of TX, Austin. *Born:* Aug 1918 Aberdeen Idaho. USAF 1940-46, Ret Res Lt Col Aircraft Dir of Maintenance. Instr & Asst Prof UT Austin 1948-59, Prof & Hd Aero Engrg Dept Univ of Wichita 1959-60, Prof & Hd Dept Engrg Mech 1960-69, Prof Aero Engrg & Engrg Mech 1970; Assoc Dean Coll of Engrg, UT Arlington since 1970. Assoc Fellow AIAA, Mbr Sigma Xi. V P Soc for Experimental Stress Analysis 1973-75, Pres SESA 1975- 76. Fellow SESA Enjoy fishing, hunting, and horticulture. *Society Aff:* AIAA, SESA, TSPE, NSPE, ASEE.

Dallin, Gary W
Home: 1254 Havendale Blvd, Burlington, Ont L7P 4A7 Canada
Position: Genl Supr Metallurgist. *Employer:* Stelco Inc. *Education:* BSc/Met Eng/Univ of Alberta. *Born:* 11/27/43. Raised in Edmonton. Grad from Univ of Alberta 1965; Have been employed by Stelco Inc. since 1965. Have held positions of Metallurgist, Galvanize; Supervising Metallurgist, Tinplate; and Supervising Metallurgist, ROD & BAR. Current position is Genl Supervising Metallurgist, Shapes & Bars, respon for quality control in Stelco's Primary Mills, Rod & Bar Div. ASM: Mbr since 1964, P Chmn of Young Mbrs Ctte, P Sec of the Chap Mbrship Council, Chmn of Ontario Chap. 1980-81. AIME/ISS: Mbr since 1976, Mbr Bar & Semi-Finished Products Ctte, Mech Working & Steel Processing Div. *Society Aff:* ASM, AIME/ISS.

Dallos, Peter
Business: Northwestern University, 2299 Sheridan Rd, Evanston, IL 60201
Position: Prof *Employer:* Northwestern Univ *Education:* PhD/Biomedical Engr/Northwestern Univ; MS/EE/Northwestern Univ; BS/EE/IL Inst of Tech; -/EE/Tech Univ Budapest *Born:* 11/26/34 Native of Budapest, Hungary where he studied electrical engrg between 1953 and 1956. Came to the US in 1956 and became a citizen in 1962. After receiving the PhD from Northwestern he remained on the faculty and specialized in the biophysics and physiology of hearing. Founding Chrmn of the Univ dept of Neurobiology and Physiology. Currently John Evans Prof of Neuroscience. Author of book: The Auditory Periphery, and over 80 research papers. Member of numerous national committees and editorial bds. Recipient of Beltone Award (1977), a Guggenheim Fellowship (1977/78) to do research at the Karolinska Inst in Stockholm, the Jacob Javits Neuroscience Investigator Award (1984- 1990), and the Amplifon International Prize (1984). Collector of contemporary American art. *Society Aff:* IEEE, ASA, AAAS, NS, ARO, Collegium OLAS.

Dally, James W
Business: Coll of Engrg, College Park, MD 20742
Position: Prof, Mech Engrg *Employer:* Univ of MD *Education:* BS/Mech Eng/Carnegie Mellon; MS/Mech Eng/Carnegie Mellon; PhD/Mech/IL Inst Tech *Born:* 8/2/29. Began engg career in 1951 with Mesta Machine Co in Homestead, PA. Served as a sr res engr at Armour Res Fdn, Chicago, IL from 1953-58. Entered the field of engg educ as an Asst Prof at Cornell Univ 1958-61. Returned to Armour Res Fdn as an Asst Dir of Res 1961-64. Appointed Prof and Dir of the Experimental Stress Analysis Labs, IL Inst of Tech 1964-71. Served as Chmn of the Mech Engg Dept, Univ of MD 1971-75 and Prof from 1975-79. Dean of the Coll of Engg, Univ of RI 1979-82, Mgr of Mech Dev, IBM 1982-84, Prof of Mech Eng 1984-present. Pres of SESA 1970-71, Frocht Award 1976, Murray Lecturship 1979. Hon Mbr SESA 1982, Fellow ASME 1978, Fellow Amer Acad of Mech 1983, Fellow SESA 1978. Chmn US Natl Ctte of Theoretical and Applied Mechs 1983 and 84. Elected to Natl Acad of Engrg 1984. PE in MD. *Society Aff:* ASME, SEM, AAM, NAE

Dalman, G Conrad
Business: Phillips Hall, Cornell University, Ithaca, NY 14853
Position: Prof *Employer:* Cornell university *Education:* DEE/EE/Poly Inst of Brooklyn; MEE/EE/Poly Inst of Brooklyn; BEE/EE/City College of NY. *Born:* 4/7/17. Electron device engr for RCA, Bell Telephone Labs, & Sperry Gyroscope Co 1940-56. Joined Cornell's School of Electrical Engrg in 1956 as Prof; served as Director. Research interest: microwave and mm-wave solid state devices. Active as a Consultant to industry. 1980-81 Sabbatical at TRW, Redondo Beach, CA, 1969- 70 at Cayuga Assoc. (NARDA), 1962-63 at Chiao Tung University, Hsinchu Taiwan. Fellow IEEE & AAAS. Mbr, Eta Kappa Nu, Tau Beta Pi, & Sigma Xi. Certificate of Distinction from Polytechnic Institute of Brooklyn (1957). Cornell EE Sch Excellence in Teaching Award (1977). 5 US Patents Awarded; Co-author textbook on electromagnetic theory. *Society Aff:* IEEE, AAAS, HKN, ΣΞ, ΤΒΠ.

Dalphin, John F
Business: 2101 Coliseum Blvd East, Fort Wayne, IN 46805
Position: Dean *Employer:* Purdue Univ *Education:* PhD/Math-CS/Clarkson Coll; MS/Math/Univ of NH; B/ME/Clarkson Coll *Born:* 5/7/40 Raised in Brooklyn, NY. Developed interest in computers in undergraduate school, employed by IBM after graduate school. Served as Captain, Instructor in Data Processing USAAGS, Fort Harrison 1965-66. Chrmn Computer Tech Dept IUPUI 1967-71. NSF Science Faculty Fellow Clarkson Coll 1971-72. Chrmn and Prof Computer Science Dept and Dir Computing Services SUNY Coll at Potsdam 1973-77. Appointed Dean School of Engrg, Tech & Nursing at IU-Purdue at Fort Wayne 1977. Member PiTauSig, TauBetPi, TauAlpPi. Subsidiary interests include philately (APS, ATA), photography, silver culture (AFA,NNGR) and woodworking. *Society Aff:* AAAS, AAHE, ACM, ASEE, MAA

Dalrymple, Gordon B
Business: 2600 Century Ctr Pkwy NE, Atlanta, GA 30345
Position: Chrmn of Bd. *Employer:* Law Engg Testing Co. *Education:* MS/Civil Engr/GA Inst of Tech; BS/Civil Engr/Univ of IL. *Born:* 12/2/24. Engg co exec born in Williamsburg, KY. Obtained degrees in civ engg from Univ of IL and GA Inst of Tech and attended advanced mgt prog at Harvard Business School 1968-69. Joined Law Engg Testing Co in 1952 as jr engr and successively vp, exec vp, chief exec officer and pres, chrmn. Served in USN 1943-46 and active in Christian Church at state and loca levels. Pres of the GA Soc of PE and Engr of Yr in GA 1977. Founding Pres and Dir of GA Engg Fdn. Listed in Who's Who in America. *Society Aff:* ASCE, NSPE, ASTM.

Dalton, Charles
Business: Col. of Engineering, 4800 Calhoun, U. of Houston, Houston, TX 77004
Position: Prof of Mech Engr & Assoc Dean of Engineering *Employer:* Univ of Houston. *Education:* BS/ME/Univ of Houston; MS/ME/Univ of Houston; PhD/Mechanics/Univ of TX (Austin) *Born:* 6/17/35. Native of Houston, TX, has served as a faculty mbr at the Univ of Houston since 1965. Consultant for maj oil and construction firms on a long-term basis. Author of over 40 papers in recognized profl journals. Currently is Assoc Editor of the Journal of Fluids and Structures. Pres of the consulting firm Dalton Hydrodynamics Inc. Is licensed commercial pilot and licensed Scuba diver. Mbr of the Exec Ctte of ASME's Fluids Engr Div. Registered Prof. Engineer in Texas. Fellow of ASME. Assoc Fellow in AIAA. *Society Aff:* ASME, AIAA, APS, ASEE

Dalton, Frank E
Home: 100 E Erie St, Chicago, IL 60611
Position: General Superintendent *Employer:* Metro Sanitary Dist of Gr Chicago *Education:* MS/IL Inst of Tech; BS/-/Purdue Univ. *Born:* 11/14/28. in Chicago. BS Purdue, MS IL Inst of Tech. Army Corps of Engrs 1951-53. Proj Engr with industry & several consulting firms 1953-63. With MSDGC since 1963. Current respon as Gen. Supt. which incl the Tunnel & Reservoir Plan. Former VP No Shore Chap ISPE Mem US Natl Com on Tunneling & Technology of the Natl Acad of Sciences/Natl Res Council. Radabaugh Award from Central States WPCA. Linn H Enslow Award from NYWPCA. George Bradley Gascoigne Medal from WPCF. 1976 Superior Service Award for Outstanding Supervisory Employee. Charles Walter Nichols Award from APWA - 1985. The Golden Beaver Engineering Award - 1986. *Society Aff:* ASCE, WPCF, NSPE, NAUFMA.

Dalton, John H
Home: 1900 Ashwood Dr, Akron, OH 44313
Position: Mgr Product Group Marketing *Employer:* Goodyear Aerospace Corp. *Education:* BA/Physics/Williams College; BS/Aero Engg/MIT; BS/Bus & Engg Admin/MIT. *Born:* May 1926 Philadelphia, PA. MK48 Torpedo Project-Programm Dir 1964-66 respon for all tech decisions & mgmt. Subroc Weapon Syst-Proj Engr 1966-68 all engrg respon for 70 prof Engrs & Analysts. Engrg Mgr-Advanced Dev 1966-68 all Defense Syst Div R&D. Mgr Engrg 1972-74 all Divisional Engrg-80 units; Acting Past Ch E 1972. 1974- , Mgr Product Group Marketing. Past Chmn Ctte, Anti- Submarine Warfare Ctte, the Natl Security Industrial Assn (NSIA). Leisure: Skiing, Tennis. *Society Aff:* AIAA.

Dalton, Robert L
Home: 6850 E 14th St, Wichita, KS 67206
Position: Weight Engr. Specialist-Consultant *Employer:* INCONEN *Education:* MS/Math/Wichita State Univ; BA/Math/Wichita State Univ. *Born:* Sep 1930. Served US Navy & Commissioned Off in Air Force. Taught math & sci in Wichita Pub Schools & Boys Detention Home. Weight Control Engr at Cessna 3 yrs, Northrop 1 yr, Boeing 13 yrs, Swearingen 1 yr, Beech 7 yrs, Gates Learjet 4 yrs & presently a conslt at INCONEN 5 yrs. Supervisor in Weights at Boeing & Chief of Weights at Swearingen. Currently Weight Engr Specialist in R&D. Elected Dir of SAWE Wichita 4 yrs. Weight Control & Preliminary Design experience in Light Business Aircraft, Large Bombers, Commercial Aircraft and Missiles. Hobbies: bass fishing & stamps. *Society Aff:* SAWE.

Dalton, Robert W
Business: 121 Manly St, Greenville, SC 29601
Position: President *Employer:* Dalton & Neves Co, Inc, Engineers *Education:* BS/Civil/Clemson Univ. *Born:* Mar 1924. Attend Wingate C, Furman U, Clemson U. Corps of Engrs WWII. With Dalton & Neves Co: Field Engr 1948-54, Design Engr 1954-58, VP Co 1958-64, 1964- President replacing father and founder of Co since 1897. All types Civil, Sanitary & Consulting Engrg. Mem South Carolina Society of Engineers, South Carolina Society of Professioanl Engrs, Natl Soc of Prof Engrs, Prof Engrs in Private Practice, Consulting Engrs of South Carolina, American Congress Surveying & Mapping. South Carolina Soc of Registered Prof Land Surveyors, Natl Soc of Cons Engrs. Robert William Dalton, PE. *Society Aff:* PEPP, NSPE, SCSE, NSCE.

Daly, Charles F
Business: 1365 Westgate Center Drive, Suite F-1, P.O. Box 24026, Winston-Salem, North Carolina 27114-4026
Position: President *Employer:* DSA Group Inc *Education:* BSE/Mech Engr/Univ of NC at Charlotte. *Born:* 8/18/38. Charles F Daly, PE, has practiced as a consulting engr in NC for twenty years and is currently president of a 52 person consulting engineering firm. A mbr of the NSPE, he has served as Pres of the Central Piedmont Chap and has been Past Pres of the Engg Advisory Council for UNC-Charlotte. He is past chairman of the national ASHRAE Technical Committee (TC 9.4) on Water to Air Heat Pumps. He attended Univ of FL and UNC-Charlotte where he received his BS in engg. An author of several articles & handbooks, he is a reg PE in the States of NC, SC, WV, KY, TN, GA, VA, AL, FL, NJ & IN; certified by National Council of Engineering Examiners (NCEE). Received 1st Place Regional Award from ASHRE in 1983 and 1986 for design of projects recognized for energy conservation. Holds mechanical contractors construction license in NC including commercial refrigeration certificate. *Society Aff:* NSPE, ASHRAE.

Daly, Edward A
Business: 11855 SW Ridgecrest Dr, Room 201, Beaverton, OR 97005
Position: Prin *Employer:* Daly Engrg Co *Education:* MS/Mech Engg/MI State Univ; BS/Mech Engg/NM State Univ. *Born:* 6/7/25. Raised near Columbia South Dakota. Received BSME from NM State Univ, 1950; MS MI State Univ, 1951. On the Mechanical Engrg faculty of South Dakota State University 1953-57 and Oregon State University 1957-69. Private consultant from 1969 until present. Registered as Mech Engr, Nuclear Engr, & Acoustical Engrg. Practice in the field of acoustical engrg *Society Aff:* ASA, ASHRAE, NSPE, CEC/O, NCAC, ASTM.

Dalzell, Robert C
Home: 2548 N Vermont St, Arlington, VA 22207
Position: Retired. *Education:* ScD/Metallurgy/Harvard; BE/Elec Engg/Johns Hopkins; MSc/Non-ferrous Metallurgy/Harvard. *Born:* Oct 1906 Beaumont TX. Tech

Dalzell, Robert C (Continued)
Advisor, Plant Metallurgist, General Mgr, Chief Tech Advisor-Revere Copper & Brass Inc. Chief Metallurgical R&D; Chief Engrg Development; Chief Foreign Activities; Asst Dir Reactor Devel Div USAEC. Representative of USAEC on ASTM Nuclear Policy Ctte, ASA Nuclear Standards Board, ICPUAE Geneva, ISO Conf on Nuclear Standards. Chmn Nuclear Engrg Div; Chmn Program Ctte; Founder Hist and Heritage Cotte ASME (Secy 1970/82 & Chmn 1982-87). Life Mbr ASM, Life Fellow ASME. Mbr Cosmos Club Wash DC. *Society Aff:* ASME, ASM Int.

Dalziel, Charles F
Business: Dept Elec Engrg, Berkeley, CA 94720
Position: Professor Emeritus. *Employer:* Retired. *Education:* EE/EE/UCB; ME/EE/UCB; BS/EE/UCB *Born:* 1904 San Francisco. Teaching Dept EE at UCB 1932-Prof Emerit 1967. Supr Engr Sci & Mgmt War Training 1941-44; Fulbright Prof IENGF Turin Italy 1951-52. Visit Prof U of Manchester England Fall 1966. 'Power Life Award' IEEE-PES 1975. 'Achievement Award' IEEE-IGA 1970. Hon Mbr ASSE 1969. Fellow AIEE 1957. Cert Commendation CA Disaster Office 1964. 1st Prize IEEE Paper 'Re-evaluation of Lethal Electric Currents' 1969, 1st Prize AIEE Paper 'Underfrequency Protection of Power Systs' 1960, 1st Prize AIEE Paper 'Effects of Electric Shock' 1959. 2nd Prize AIEE Paper 'Effect of Frequency on Perception Currents 1950. Ch Tech Aide NDRC-OSRD Div 13 P44-P45. Pubs: 154 tech papers worldwide. P E CA. US Pat: Mini Diff Circuit Breaker-The GFCI 1965. 1st Prize paper Nat Safety Coun 1980. *Society Aff:* ΣΞ, TBΠ, HKN

Daman, Ernest L
Home: 435 Wychwood Rd, Westfield, N.J 07090
Position: Sr. Vice President *Employer:* Foster Wheeler Corp *Education:* BME/Polytechnic Inst. of Brooklyn *Born:* 03/14/23 Graduated from the Polytechnic Inst. of Brooklyn with a BME in 1943; entered U.S. Army and served in the Pacific Theater with the 77th Division; joined Foster Wheeler Corporation in 1947 as a Development Engineer; involved in the development of advanced Naval Steam Power Propulsion Plants; developed a combined cycle plant for Navy antisubmarine ships; led Foster Wheeler's development in Breeder Reactor Components, fluidized bed combustion, and combined cycle power generation. I am currently Senior V.P. for Research, Foster Wheeler Corporation; Chairman, Foster Wheeler Development Corp; Pres-elect, ASME; Member, AAAS; Fellow, Institute of Energy - U.K.; enjoy sailing and tennis. *Society Aff:* ASME, AAAS

D'Amato, Salvatore F
Business: 70 Broad St, NY, NY 10004
Position: Pres *Employer:* Amer Bank Note Co *Education:* MS/ME/Columbia Univ; BS/ME/B'klyn Polytechnic Inst *Born:* 7/28/28 *Society Aff:* ASME.

Dames, Trent R
Business: 445 S Figueroa St, Los Angeles, CA 90071
Position: Founding Partner *Employer:* Dames & Moore *Education:* MS/Civ Engrg/CA Inst of Tech; BS/Civ Engrg/CA Inst of Tech. *Born:* 10/06/11 Born Brooklyn, NY. Held positions with Labarre & Converse, 1934-35; US Bureau of Reclamation, Denver, CO, 1935-36; R V Labarre, Los Angeles, 1936-38. Co-founded Dames & Moore with William W Moore, Los Angeles, 1938. Served as Exec Partner 1950-75; Founding Partner and Exec Ctte Mbr 1975 to present. Patented Soil Samplers and Methods of Determining Driven Friction Pile Capacities. Soils & Fdn Consltg Engrs who expanded into Environmental Scis Consltg, was instrumental in the domestic and intl growth of the firm into a multi-natl partnership. ASCE Honorary Mbr, VP-Zone IV, 1970-72, Dir 1960-62; Construction Man of the Yr, L A Chamber of Commerce, 1975; Tau Beta Pi, Sigma Xi. *Society Aff:* ASCE, SSA, CEASC, SEASC.

Damiani, A S
Business: 12050 Baltimore Ave, Beltsville, MD 20705
Position: Pres *Employer:* Com. Site Int'l. *Education:* MEA/Eng Adm/The Geo. Wash. Univ.; BS/Comm. & Eng./Drexel U *Born:* 02/28/38 A.S. "Migs" Damiani, CPE is Pres & Chief Operating Officer of Com-Site International. Mr. Damiani was previously Dir of Corporate Facilities with Planning Research Corp. in McLean, VA. His prior experience has included Dir of Facilities and Services with Montgomery Cty, Md, and Mgr of Facilities with Fairchild Industries. A certified plant engineer and author of more than forty magazine articles, Migs holds a B.S. degree in Commerce & Engrg from Drexel Univ and a Master's in Engrg Administration from The George Washington Univ. He has received over thirty national awards in his distinguished career, including being named Plant Engineer of the Year in 1983 and a Fellow in 1984 by the AIPE and an honorary assoc mbr of the Amer Inst of Architects in 1974. He was also honored by Drexel Univ with their Silver Salute Award in 1986. Mr. Damiani is very active in community and civic organizations. He currently serves on the Bd of Trustees of Holy Cross Hospital in Silver Spring, Md, and is a past bd member of the Lung Assoc of Mid-Maryland and past pres and bd member of the Montgomery Cty, Md, Chamber of Commerce. *Society Aff:* AIPE, AIAA

Damon, Richard W
Business: 1601 Trapelo Rd, Waltham, MA 02154
Position: Dir of Tech *Employer:* Sperry Corp *Education:* PhD/Appl Physics/Harvard Univ; MA/Appl Physics/Harvard Univ; BS/Physics/Harvard Univ. *Born:* 5/14/23. With Sperry Corp since 1962, currently Dir of Tech responsible for operation of Corp Tech Ctr and strategic tech planning. Served as Dir. Appl Physics Lab, Sperry Res Ctr, from 1970-82, responsible for progs in materials, elec devices, optics and electromagnetics. 1981 IEEE Pres; 1977-78 IEEE Dir/Delegate of Div IV. Fellow of IEEE and APS. Conslt mbr of Advisory Group on Electron Devices, Off of the Under Secy, Defense Res and Engrg (1973-80). Also served on NAS/NRC Advisory Panel to Natl Bureau of Stds (1969-73); Natl Bureau of Stds Elec Study Group (1972-42); and NASA Advisory Comms on Basic Res, Electrophysics and Electronic Materials (1967-71). Visiting Lecturer in Appl Physics, Harvard Univ, 1961-62; 1969 IEEE Natl Microwave Lecturer; Mbr Advisory Ctte to Lab for Surface Sci and Tech, Univ. of ME, 1981-84; Bd of Dir, Matec Corp 1981-. Prior employment included: Mgr, Control Devices Dept, Microwave Assoc (1960-62); Res Assoc, Gen Elec Res Lab (1951-60); Engr, Ratheon Co (1948- 49); Teaching Fellow and Res Asst, Harvard Univ (1946-48 and 1949-51). USNR, to Lt (jg) (1943-46). *Society Aff:* IEEE, APS, AAAS, ΣΞ

Damrell, Charles B
Business: 800 Boylston St, Boston, MA 02199
Position: VP. *Employer:* Boston Edison Co. *Education:* BS/EE/Northestern Univ; Mgt Dev Certificate/Bus/Harvard Bus Sch. *Born:* 12/8/32. Native of the Boston, MA metropolitan area. Worked for Boston Edison Co for 29 yrs in positions of succeedingly more responsible engg positions. Chief Engr 1973 to 1979. Elected VP of Engg & Distribution in May, 1979. Responsible for engg, planning & construction of generation (excluding nuclear), transmission, substation & distribution facilities; operation & maintenance of transmission & distribution facilities & the Co's environmental affairs. Chrmn of the Boston Section of IEEE 1968-1969. Recipient of the MSPE Young Engr of the Yr Award in 1967, the IEEE Centennial Medal in 1984, and the Boston Section, IEEE "Laurence F Cleveland Award" in 1984. Outside interests include music, tennis & woodworking. *Society Aff:* IEEE.

Danchak, Michael M
Business: The Hartford Graduate Center, 275 Windsor St, Hartford, CT 06120
Position: Dean, Eng & Sci *Employer:* The Hartford Grad Ctr *Education:* PhD/Nuclear Engrg/RPI; MS/Nuclear Sci/RPI; BS/Mech Engrg/Princeton Univ *Born:* 03/28/44 After receiving the BS degree, served in the USMC from 1965-1969. Upon completion of the doctoral program, worked for Combustion Engrg with primary design respon for the computer generated color display sys used in advanced power plant control rooms and the attendant human factors aspects. Since 1978, associated with The Hartford Grad Ctr; first as Chrmn of Computer Sci and then as

Danchak, Michael M (Continued)
Dean of the School of Engrg and Sci. Also active in human-computer interaction res. *Society Aff:* IEEE, ACM, HFS, AAEE

Dancy, Terence E
Business: Place du Parc, 300 Leo Pariseau, PO Box 2000, Succ. La Cite, Montreal, Quebec H2W 2S7
Position: VP Technology *Employer:* Sidbec *Education:* PhD/Fuel Tech/Imperial College London; BSc/Chemistry/Imperial College London. *Born:* 3/5/25. Educated Highgate School & Imperial College, London. British Iron & Steel Research Assoc, BISRA, 1947-53 - blast furnace research. British Embassy, Washington, DC 1953-54. BISRA 1955-56. Jones & Laughlin Steel Corp, Pittsburgh 1956-70 - research & engrg on new technology in steel industry. Numerous technical publications and co-editor of several books. Mbr Metals Society UK, AIME, ASM, AISE, AISI, & CIM. AIME R.W. Hunt Medal 1960 - oxygen in blast furnace. ASM Pittsburgh Night Lecturer 1967. ASM Fellow 1974. AISE Kelley First- Award 1964 - contin casting. AISI Institute Medal - hot strip & HSLA steels 1965 and - large-scale integrated steel-making using direct reduced iron 1980. CIM Airey Award 1983 - direct reduction 1971 to present - VP Engineering & Development Sidbec-Dosco to VP Technology Sidbec respon for development & met for integrated steel prod using dir reduc & elec furnaces. Mbr/Bd Sidbec- Normines Inc *Society Aff:* AIME, ASM, CIM, Met Soc.

D'Angelo, Vincent J
Business: 23875 Commerce Park Rd, Cleveland, OH 44122
Position: Pres *Employer:* Woodruff, Inc Consulting Engrs *Education:* BS/CE/Purdue Univ *Born:* 11/12/40 Native of Jamestown NY. Engr with Howard Needles Tammen and Bergendoff from 1963 to 1966, Wheeler & Melena 1966 to 1967, Williams and Hach 1967 to 1973. Joined Woodruff, Inc as Chief Structural Engr 1973. Became VP and Principal 1976 and Pres 1980. Chrmn of ASCE Cleveland Section Structural Committee 1978, 79 & 80. Presently secretary of Cleveland Section ASCE. *Society Aff:* ASCE

Daniel, David E
Business: Univ of Texas, Dept. of Civil Engrg, Austin, TX 78712
Position: Associate Prof of Civ Engrg *Employer:* The Univ of TX at Austin *Education:* BS/Civil Engr/Univ of TX; MS/Civil Engr/Univ of TX; PhD/Civil Engr/Univ of TX. *Born:* 12/20/49 PHD, MS, BS from the University of Texas at Austin in 1980, 1974, & 1972, respectively. Has directed a variety of research proj related to disposal of hazardous and radioactive wastes and has studied the movement of water and contaminated liquids in the ground around waste disposal facilities. Conslt to more than a dozen companies and corporations on waste related problems. Norman Medal and Croes Medal, ASCE, 1975 and 1984, respectively. *Society Aff:* ASCE.

Daniel, Isaac M
Home: 6708 N Francisco Ave, Chicago, IL 60645
Position: Professor. *Employer:* Northwestern University. *Education:* PhD/Civil Eng/IIT; MS/Civil Eng/IIT; BS/Civil Eng/IIT. *Born:* Oct 1933 in Salonica, Greece. Attended National Technical University in Athens & IIT in Chicago. Graduated with distinction. Received ASCE award, two Hetenyi awards (SEM best papers), Society of Plastics industry best paper award, SEM (Soc for Exp Mechanics) B. Lazan Award (distinguished contributions). Listed in American Men & Women of Science. Mbr of Chi Epsilon, Tau Beta Pi, Sigma Xi, ASTM & SESA (Past Chmn of Papers Review Ctte, Fellow). Active in experimental stress analysis & material engrg, fracture mechs, composite materials and nondestructive testing. Lectured at home & abroad. Author of over 130 publications including chapters in 7 books and a Monograph. Served on committee on characterization of National Materials Advisory Board. Chaired ASTM Conference on composites. Knowledge of Greek, Spanish, French, Italian, German & Hebrew. With IIT Research Institute from 1959 to 1982. From 1982 to 1986 Prof and Dir of Exper Stress Analysis, Mech and Aero Engrg Dept, IIT. Visiting Prof at Univ of Poitiers, France, 1984; Professor of Theoretical and Applied Mechanics at Northwestern University, 1986- . *Society Aff:* SEM, ASTM, AAM.

Daniel, Kenneth R
Home: 3212 Brookwood Rd, Birmingham, AL 35223
Position: President (Retired) *Employer:* American Cast Iron Pipe Co *Education:* DSc Hon/Mech Engr/Univ of AL; ME/Mech Engr/Univ of AL; BSME/Mech Engr/Univ of AL. *Born:* 1913 Milford Co. BSME 1936, Prof ME 1957 Univ of AL. US Army 1941-46 to Lt Col; Acting Ordnance Officer 1945, Third Army; Exec to Theater Chief of Ord 1945-46. European Theater ribbon, 5 campaign stars; French Croix de Guerre with gold star; Bronze Star medal & Legion of Merit. Joined Amer Cast Iron Pipe Co 1936; Ch Engr 1948; VP Engrg 1955; present position 1963-78. Named 1967 Engr of the Yr by B ham Engrg Council; Mbr AL State Bd of Registration for Prof Engrs & Land Surveyors 1967-85; 1974 recipient of Amer Foundrymens Soc Thos W Pangborr gold medal; Hon Mbr ASME. Trustee & Natl Pres FEF 1964-65; Pres B'ham Area Cham of Commerce 1969. Past Dir: Seaboard Coast Line Indus; First AL Bancshares; Past Dir Amer Iron & Steel Inst. Mbr Natl Advisory Council, Salvation Army 1976-80. Trustee & Mbr Executive Ctte Southern Res Inst 1978 recipient of Natl Mgmt Assn Exec of Yr Award, 1976 recipient of Henry Laurence Gantt Mem Medal Awarded by Amer Mgmt Assns & ASME, 1980 Hon. Dr of Sci Univ of Ala 1981 Sesqui Centennial Hon Prof Univ of Ala. 1986 Joan Hodges Queneau Palladium Medal Joint AAES & Natnl Audubon Soc. *Society Aff:* ASME, AISE, AFS.

Daniel, Richard A
Business: 1 Main St, Chatham, NJ 07928
Position: V P-Tech *Employer:* Celanese Specialties Oper *Education:* PhD/Chem Engg/Columbia Univ; MS/Chem Engg/Columbia Univ; BS/Chem Engg/Columbia Univ; BA/Chem/Columbia College. *Born:* 4/26/34. V.P. Manufacturing & Technical, Celanese International Division (1980-1984) (1976-1980) VP & Gen Mgr of Engineering Resins-Plastics Division of Celanese V. P. Manufacturing Chem Ca. Division of Celanese (1974-75), VP and Dir of R& D for Chemicals Division of Celanese (1971-74), Laboratory Director of Celanese Plastice Division (1969-71) and Tech Dir of the Resins Div (1967-69). Current position is V.P. Tech Celanese Specialities Oper (1984-). *Society Aff:* AIChE, SPE, ACS, NAS.

Daniels, John H
Home: 3139 Patterson Drive, Bethlehem, PA 18017
Position: Professor of Civil Engrg. *Employer:* Lehigh University. *Education:* PhD/Civil Engg/Lehigh Univ; MS/Civil Engg/Univ of IL; BSc/Civil Engg/Univ of Alberta. *Born:* July 1931. Native of Calgary, Alberta. Resident & Design Engr with Bridge Branch, Alberta Dept of Hwys 1955-58. Head of Structural & Soil Mechanics Division of Associated Engrg Services Ltd, Consulting Engrs, Edmonton Alberta 1959-64. Gra duate & undergraduate teaching plus research on fatigue of bridges & behavior of bridges & building at Lehigh University since 1967. Mbr ASCE, IABSE & TRB. Past Pres of Lehigh Valley Section ASCE. Mbr Sigma Xi. Enjoy classical music, piano & gardening. *Society Aff:* ASCE, IABSE, ΣΞ TRB

Daniels, Marjorie A
Home: 3417 Emelye Dr, Mobile, AL 36609
Position: Senior Manufacturing Engr. *Employer:* Ingalls Shipbuilding Corp. *Born:* Jul 22 1925 Fairhope AL. Married P E Daniels Jr; parents Joe & Madeline Schneider Greensboro NC. BS Texas Tech 1945; Avr U. Prof exper incl: Asst Claim Investigator ATSF Rwy Los Angeles CA 1945-48; USAF Indus Engr Mobile AL 1949-67; Sr Indus Engr Ingalls Shipbldg 1968- . IIE: Mbrship Chmn Natl Div Indus & Labor Relations 1973-77, Bd/Dirs Local Chap 1973-76, Pres 1971-72, num other offices 1966-67, Del to Natl Conf 1963-76, num speeches to var groups. Mbr Garden Club; Le Krewe de Bienville; Dauphin Way Meth. Hon: Mobile Woman of Day 1972, Ala Most Dist, Personalities of the South, Who's Who in Al, Outstanding Perf Award;

Daniels, Marjorie A (Continued)
Listed as source of Info on Indus Engrg 'Your Future In Indus Engrg' by Ross Hammonds.

Daniels, Orval W
Business: PO Box 2391, 535 N Washington, Wichita, KS 67201
Position: Partner. *Employer:* Engg Testing Co. *Education:* BS/CE/KS State Univ. *Born:* 2/14/21. near Bronson, KS. Structural Engr at Consolidated Vultee Aircraft, Ft Worth, TX 1943-1945. Structural Engr at Boeing Airplane Co, Wichita, KS, 1946. Structural Engr at Beech Aircraft Corp, Wichita KS, 1947. Mgr and Civil Engr of Wichita Branch, Tulsa Testing Lab 1947 to 1958. Mng partner and geotechnical civil engr - Engg Testing Co, Wichita, KS, 1958 to present. Reg PE in KS and OK, Reg LS in KS. *Society Aff:* ASCE, ASTM, AAPT, NSPE.

Daniels, Raymond D
Business: 100 East Boyd, Norman, OK 73019
Position: Prof. *Employer:* Univ of OK. *Education:* PhD/Metallurgy/Case Inst of Tech; MS/Physics/Case Inst of Tech; BS/Physics/Case Inst of Tech. *Born:* 2/14/28. Native of Cleveland, OH. Employed as Physicist with Natl Bureau of Stds, 1950-51. Engr, Linde Div, Union Carbide Corp, 1954-55. With the Univ of OK since 1958. Prof Chem Engg and Mtls Sci since 1964. Dir, Sch of Chem Engg and Mtls Sci 1963-65, 1969-70. Associate Dean of Engg, 1965-68. Halliburton Dist Lecturer, Coll of Engrg, 1983-88. Natl Sci Fdn Fellow, Univ of Neuchatel, 1968-69. Dir, the Univ of OK Res Inst 1971-83. Chrmn, S Central Region NACE, 1976. Natl Dir, NACE, 1980-83. Principal res interests: hydrogen in metals, corrosion. Co-author of book on res admin, 1977. *Society Aff:* TMS-AIME, ASEE, NACE, ASM, ASTM, AIChE.

Danly, Donald E
Home: 4470 Old Spanish Trail, Pensacola, FL 32504
Position: Dir of Tech. *Employer:* Monsanto. *Education:* PhD/Chem Engg/Univ of FL.; BSChE/Chem Engg/Cornell Univ. *Born:* 6/14/29. Currently Dir of Fiber Intermediates Tech for Monsanto Chem Intermediates Co at Pensacola, FL. Involved for past 20 yrs in dev and commercialization of processes for production of the monomers and auxiliary chem used in synthetic fibers. Project leader of team of engineers assigned the task of scaling up Monsanto's electrochemical adiponitrile process in the 1960's. Author of numerous articles on scaleup of electroorganic processes. *Society Aff:* ACS, AIChE, ΣΞ, ECS.

Danne, Herbert J
Home: 5823 E 57 St, Tulsa, OK 74135 *Education:* BS/ChE/OK St U *Born:* 3/11/26. Kingfisher, OK. 1956-1951, Sales Engr with Std Magnesium Corp in their formative days. 1951-1958, Sales Engr for Western Supply Co - Heat Transfer Equip. 1958-1963 - Ind Fabricating Co, VP Sales & Engg-Heat Transfer Equip. 1963-1978, Exec VP for Ind Fabricating Co with direct responsibility for admin of co. 1978-1982, Pres of Am Machine and Supply Co, Mfg of Process & Production Components. 1982-83 VP Energy Exchanger Co. Mbr of AIChE. Past Pres of Tubular Exchanger Mfgrs Assoc. Past Dir of heat transfer res. Listed in Marquis' Who's Who in the Southwest; Who's Who in Sci. Enjoy fishing, hunting and boating. *Society Aff:* AIChE

Dannenberg, Warren B
Business: Prudential Ctr SE Tower, Boston, MA 02199
Position: VP (Reitred 1/1/78). *Employer:* Chas T Main Inc. *Education:* BS/EE/MIT. *Born:* Jun 1936. BS from MIT in Elec Engrg. Native of Mass. 1936-42 Consulting Engrg with Arthur L Nelson Engrs Boston Mass. 1942-45 Refinery Engrg with E B Badger Co Boston Mass. Joined MAIN in 1945. Progressed thru Power Plant Design Engrg, Project Engrg, Project Mgmt, Ch Projects Dir & Asst Mgr of Thermal Div. Mbr ASME, IEEE, NSPE. Mbr Prof Practices Ctte ASME 1970-75. Prof Reg in MA, RI, KS, MI, DE, MO, DC, MD. Enjoy sailing & golf. *Society Aff:* ASME, IEEE, NSPE.

Danner, Ronald P
Business: 163 Fenske Lab, Univ Park, PA 16802
Position: Prof - Chem Engr. *Employer:* PA State Univ. *Education:* PhD/ChE/Lehigh Univ; MS/ChE/Lehigh Univ; BS/ChE/Lehigh Univ. *Born:* 8/29/39. in New Holland, PA. Sr Res Scientist at Eastman Kodak Co, Rochester, NY 1965-67. On faculty at Penn State since 1967. Faculty Fellow at US Gen Accounting Office 1974-75. Visiting Prof, Tech Univ of Denmark, 1983. Primary res publications are on the adsorption of gases and on the physical and thermodynamic properties of fluids. Co-author of Tech Data Book - Petrol Refining published by the American Petrol Inst and Data Prediction Manual and Data Compilation: Tables of Properties of Pure Compounds published by the Amer Inst of Chem Engrs. A class B squash player and a ragtime piano enthusiast. *Society Aff:* AIChE, ASEE, ΣΞ.

Danofsky, Richard A
Business: 261 Sweeney Hall, Ames, IA 50011
Position: Prof *Employer:* IA State Univ *Education:* PhD/Nuclear Engrg/IA State Univ; MS/Nuclear Engrg/IA State Univ; BS/ME/IA State Univ *Born:* 4/11/31 Native of IA. Taught at IA State Univ since 1957 in depts of Engrg Mechanics and Nuclear Engrg. Became Prof of Nuclear Engrg in 1970. Major research and teaching interests are in the fields of nuclear reactor dynamics safety, and application of artificial intelligence techniques to nuclear reactor operation. Registered PE in the State of IA. *Society Aff:* ANS, ASEE, NSPE, IES

Danpour, Henry
Business: PO Box 395, Walker Dr, Upton, MA 01568
Position: VP. *Employer:* Sure Chem Corp. *Education:* MS/ChE/MIT; BS/-/Northwestern Univ. *Born:* 9/19/53. Grad with a BS degree from Northeastern Univ June 1975 & MS from MA Inst of Tech in 1977. Thesis was on water pollution control in DE River Basin. Presently am vp of chem mfg co which specializes in detergents & sanitizers for the dairy industry & distributes animal health products. *Society Aff:* AIChE.

Danzberger, Alexander H
Business: 1250 Broadway, New York, NY 10001
Position: Vice President. *Employer:* Hydrotechnic Corporation. *Education:* BS/ChE/MIT. *Born:* March 1932. BS MIT. 1953-60 Staff Mbr Arthur D Little Inc, Tech Economics Group, except 1956-58 with LOA Project Off Army Chem Corps; 1960-70 Engrg Mgr Union Carbide Cryogenics. Var respon for new process & new prod dev. Subsequent cons exper as Ch Engr Foster D Snell Inc, sub of Booz-Allen-Hamilton & Principal Marcom Inc for process, energy res & pollution control. 1975 Hydrotechnic Corp respon for dev advanced tech for pollution control. Pubs. Mbr: AIChE (Fellow), ASME, API, WPCF, AWWA, DIPL AAEE PE MA. *Society Aff:* AIChE, ASME, WPCF, AWWA, Dip AAEE.

Danzinger, Edward
Business: 1000 Pacific Trade Center, Honolulu, HI 96813
Position: Loss Control Mgr *Employer:* Firemans Fund Ins Co *Education:* BS/CE/Northeastern Univ *Born:* 2/18/23 Chrmn-Fire Advisory Bd-city & county of Honolulu. Member-High Rise Task Force-city & county of Honolulu. Pres-HI Chapter-Society of Fire Protection Engrs. Past Pres-HI Chapter-American Soc of Safety Engrs. PE Registration-MA & CA. Dir-HI State Little League. *Society Aff:* ASCE, ASSE, SFPE

Dappert, George F
Home: 57 Allwood Rd, Darien, CT 06820
Position: Dir of Operations, Lignin Chems Div. *Employer:* American Can Co. *Education:* BE/ChE/Yale Univ; MChE/ChE/Poly Inst of NY. *Born:* Mar 1922. BE Yale 1943; MChE Brooklyn Polyy 1950. Lic Prof Engr NY. Mbr AIChE, NY Acad of Sci, SChI, Fellow AIC, Assoc Mbr Sigma Xi. Employed by Chas Pfizer & Co in mfg & process engrg 1946-60; GAF Corp 1961-79. Mgr of Process Engrg 1961-66; Plant Mgr, Dir of Mfg, Genl Mgr of Indus Photo Div 1966-70; Dir of Process Engrg

Dappert, George F (Continued)
(Corp) 1970-71; Dir Mfg Chem Group 1972-75; Dir Mfg Internatl Group 1975- ; V P GAF 1974-79. . *Society Aff:* AIChE, AIC, SCI, NYAS, ΣΞ.

D'Appolonia, Elio
Business: 10 Duff Road, Pittsburgh, PA 15235
Position: President. *Employer:* D'Appolonia Cons Engrs Inc. *Education:* PhD/Civil Engg/Univ of IL. *Born:* 1918 Coleman, Canada. Consultant, US Army Corps of Engrs, Alaska/northern Canada 1942-45; Res Assoc Univ of Illinois 1946-48; Asst Prof Civil Engrg Carnegie-Mellon Univ 1948-56. Began own consulting firm 1956. Consultant, US/NRC Advisory Ctte on Reactor Safeguards. Active in many natl/internatl professional socs: ASCE, AUA, ISSMFE, DFI, ASTM, ICOLD, NSPE, AICE, AWRA, ISRM, IABSE, AEG etc. O'Keefer Medal Engrg Inst of Canada 1948; Middlebrooks Award, ASCE 1969; Pittsburgh Civil Engr of the Year 1972; Metcalfe Award, Engrs' Soc of Western PA, 1981; Distinguished Alumnus Award, Univ of IL, 1981. Num publications. Enjoys skiing, tennis, scuba diving. *Society Aff:* AUA, DFI.

Darby, Joseph B, Jr
Business: 9700 Cass Ave, Argonne, IL 60439
Position: Assoc Dir. *Employer:* Argonne National Laboratory. *Education:* PhD/Metallurgy/Univ of IL; MS/Metallurgy/Univ of IL; BS/Metallurgy/VA Polytechnic Inst & State Univ; BS/Mathematics/College of William & Mary. *Born:* Dec 12 1925. USMC Air Corps 1944-46. Chemist Allied Dye & Chem Corp 1948-49; Metallurgist Union Carbide Corp 1951-53; with Argonne Natl Lab 1958- : 1966-72 Group Leader of 25 scis & engrs concerned with alloy props, 1973 Planner Lab Dir Off, 1973-74 Asst Dir Mat Sci Div, assumed pres pos as Assoc Dir Controlled Thermonuclear Fusion Res in 1974. Co-edit Journal of Nuclear Mats 1972- ; Edit Adv Bd Journnal of Less Common Metals 1970- ; Associate Editor, Materials Letters 1981- Fellow ASM 1975; Sr Fellow U of Birmingham-England 1970- 71; Nuclear Met Ctte AIME 1974- ; Alpha Sigma Mu, Tau Beta Pi, Sigma Xi. Interests incl antique collecting, woodworking & gardening. *Society Aff:* ASM, AIME, ASTM, AAAS.

Darby, Ronald
Business: Chem Engrg Dept, Coll Station, TX 77843
Position: Prof *Employer:* TX A&M Univ *Education:* PhD/ChE/Rice Univ; BS/ChE/Rice Univ; BA/ChE/Rice Univ *Born:* 9/12/32 Three years active duty, US Navy (1955-58); NSF Postdoctoral Fellow, Cambridge Univ (1962); Senior Scientist, LTV Research Center (1962-65); Asst Prof, Assoc Prof, Full Prof, Chem Engrg, TX A&M Univ (1965-present); Visiting Prof Ruhr Univ, Bochum, W Germany (1983). 40 Technical Publications in Refered Journals, Book "Viscoelastic Fluids–, Marcel-Dekker Pub; former Students Association Award for Teaching, 1971; S TX AIChE Best Publication Awards, 1967, 1971, 1976, 1984; Minnie Stevens Piper Prof of 1981; Registered PE; Research in Viscoelastic Fluids, Coal Slurries and Suspensions, Applied Electrochemistry. *Society Aff:* AIChE, SR, NACE, ASEE

Darden, Arthur D
Business: Internatl Trade Mart, New Orleans, LA 70130
Position: President *Employer:* Arthur D Darden Inc *Education:* BSE/Naval Arch & Marine Engg/Univ of MI. *Born:* 9/22/20. Lt USN Cruiser, USS Quincy 1943-45; Engr Welding Shipyards, Todd Shipyards, Am Shipbldg Co 1945-52. Started cons Naval Arch 1953- . V Chmn SNAME Gulf Sec 1963-64, Chmn SNAME Gulf Sec 1971-72, Pres, Port of New Orleans, Propeller Club of the US 1978-79. Pres-Ctte for Sch of Naval Arch and Marine Engrg at Univ of New Orleans. Mbr Awards Ctte SNAME, 1980-82. Registered PE LA No 18767, TX No 12262 *Society Aff:* SNAME, ASNE.

Darden, Catherine H
Home: 2914 Creekstone Dr, Acworth, GA 30101
Position: Sect. Mgr, Estimating & Planning. *Employer:* Loral Systems Group Engineered Fabrics Division. *Education:* Bach/IE/GA Tech. *Born:* Sep 1945. BIE Georgia Tech. President Atl Chapter IIE 1973-74. Chairperson 1978 Systems Conference IIE. With Carter Co from 1967-68. Promoted from Engr to Mgr Textile Quality Control in 1972. Respon for overall Quality Program for Fabric Mfg group. Established Flame Retardant Testing Program for fabrics. With Goodyear Aerospace Engineered Fabrics Div., Rockmart Ga. as Quality Assurance Engineer from Oct 1980. Promoted to Sect. Mgr. Estimating & Planning in 1985. With Loral Systems Group, Engineered Fabrics Division, Rockmart, Ga as Sect. Mgr. Estimating & Planning Mar 1987 to present. Enjoy waterskiing, motorcycles, backpacking & PC's. With Foreman-Geneva, Div of Motor Wheel Corp as Personnel Mgr, Cartersville Plant from Aug 1978 to Nov 1979. With Geneva Wheel and Stamping Corp from Nov 1979 to Sept 1980 as Quality Control and Personnel Mgr, being respon for overall quality program and personnel. *Society Aff:* IIE.

Dare, Harold A
Business: PO Box 1251, Aptos, CA 95001
Position: Principal *Employer:* Dare Engrg *Education:* B.A.Sc./Civil/Univ of Br. Columbia *Born:* 11/4/26 Served in various construction quality assurance roles from 1951 to 1959 in Jamaica, Pakistan and Canada. In building construction industry 1959 to 1974 in Alberta & CA. From 1974 served in senior construction quality assurance positions in Bangladesh, Iran, Argentina and Alaska. Recently est own firm, Dare Engrg. Projects include hydroelectric, irrigation & flood control, industrial plants, industrial & commercial buildings. Recent position was contracts, current positions in 1980 Mgr on Susithna Hydroelectric Proj. Currently principal of own firm. Served as Editor of C.S.P.E. 1968-1969. Enjoys golf, bridge & chess. *Society Aff:* ASCE, NSPE, CSPE, NFC.

Darlington, Sidney
Home: 8 Fogg Drive, Durham, NH 03824
Position: Adjunct Prof., University of N.H. *Employer:* Bell Tele Labs Retired *Education:* PhD/Physics/Columbia Univ; BSEE/Elec Communication/MIT; BS/Physics/Harvard College. *Born:* 1906. Mbr Tech Staff Bell Telephone Labs 1929-71; Dept Hd 1960-71; retired 1971; Adj Prof EE U of NH 1971- . Tech observer and expert cons to US Army 7 months 1944-45. Mbr NAE; Mbr NAS; Fellow IEEE; Assoc Fellow AIAA. Mbr US Comm VI of URSI 1959-75; Del to Genl Assemblies: 1960, 63, 66, 69. Recip US Army (Cvl) Medal of Freedom; IEEE Edison Medal; IEEE Medal of Honor. Univ of NH Hon DSc. IEEE CAS Soc. Award. Enjoy hiking, canoeing, gardening. *Society Aff:* IEEE, AIAA, NAE, NAS.

Daron, Walter S
Business: 1710 H Street, N.W, Washington, DC 20006
Position: VP Central Staff. *Employer:* The C & P Telephone Co. *Education:* MEA/Engr/George Washington Univ; BS/Geog/Penn State. *Born:* Oct 1932. Reg Engr MD, DC. Army Signal Corps 1954-56; With C&P Telephone Co 1956- , except 2 yrs with Bell Tele Lab; Ch Engr for C & P Telephone Co-MD 1969-75; present VP Central Staff in DC respon for Corp planning, etc. P Mbr Bd/Tr MD Acad/Sci, Exec Ctte & Dir of Balto Area Council-BSA; Sec of Balto Mayor's Engrg Cons Eval Bd; Engrg Soc of Balto.

Darragh, Robert D
Business: 500 Sansome Str, San Francisco, CA 94111
Position: Partner *Employer:* Dames & Moore. *Education:* MSCE/CE/Caltech; BCE/CE/RPI; BS/-/US Naval Acad. *Born:* April 1926. Received Bachelor of Science Degree from US Naval Academy 1947, Bachelors degree in Civil Engrg frm Rensselaer 1949, & Masters from Cal Tech 1954. Served with US Navy Civil Engr Corps 1947-53; discharged with rank of Lt. Joined Dames & Moore Consulting Engrs 1954, admitted to partnership 1964. Primarily engaged in practice of foundation & earthquake engrg for highrise structures, petrochemical refineries, offshore drilling platforms & oil and gas pipelines and terminals. Chrmn Exec Ctte Geotechnical Engrg Div, ASCE. Thomas A. Middlebrooks Award ASCE. *Society Aff:* ASCE, ASSMFE.

Dart, Jack C
10101 Gary Rd, Potomac, MD 20854-4109
Position: Owner. *Employer:* J C Dart & Assoc. *Education:* MSE/CE/Univ of MI; BSE/CE/Univ of MI; AB/Liberal Arts/Albion College. *Born:* 8/14/12. 1912, M 1940, Rachel Cecilia Henderson. S, James Laurence. Daus, Dianne Cecilia, Linda Lenore & Janis Jennette. Career: CE, Am Oil Co 1937-43, Magnolia Petrol Co 1943-44, Esso Res & Engg Co 1944-47; Houdry Process Corp, Dir of Dev 1947-52, Mgr, Res & Dev 1952-55, VP Chem Div 1955-58, Sales & Service Div 1958- 62; Partner, Weinrich & Dart 1962-63; Owner, J C Dart & Assoc 1963-; Dir Houdry Process Corp Katalysatorenwerke Houdry-Huels & Pontecassino, SPA 1955-62. Publications: numerous technical papers & patents in the fields of alkylation, hydrocarbon synthesis, catalytic cracking & catalytic reforming. Distinguished Service Award, Natl Capital Sec, AIChE, 1974. AIChE Founders Award of 1981. Distinguished Alumnus Award, Albion College 1982. *Society Aff:* ACS, AIChE.

Darveniza, Mat
Business: Dept of EE, Univ of Queensland, St Lucia, Australia 4067
Position: Prof & Head Dept of Elect Engrg *Employer:* Univ of Queensland. *Education:* BE/EE/Univ of Queensland; PhD/EE/Univ of London; DEng/EE/Univ of Queensland. *Born:* 11/3/32. Born Innisfail, Australia. Elec engr with Southern Elec Authority of Queensland 1953-1955; Academic in the Dept of Elec Engg from 1959, appointed Prof (personal chair) 1979 and Head 1983; visiting appointments with Westinghouse R&D Ctr (1966), Univ of La Plata, Argentina (1971), Univ of Southampton & the Technical Univ of Munich (1973), Univ of FL (1973 & 1978/9). Res interests in elec power engg, particularly in elec insulation, high voltage tech, lightning protection, and solar power plants. Has published widley in IEEE & IE Aust Journals, & has written a book "Electrical Properties of Wood and Line Design" (Univ of Queensland Press, 1979). Elected Fellow IE Aust (1974), Fellow IEEE (1979). Elected Fellow AATA (1983). Recreational interests, music, tennis, squash & jogging. Married, three children. *Society Aff:* IEEE, IE Aust, AATS.

Darwin, David
Business: Dept Cvl Engrg, University of Kansas, Lawrence, KS 66045
Position: Prof Civil Engg *Employer:* University of Kansas. *Education:* PhD/Civil Engg/Univ of IL at Urbana-Champaign; MS/Structural Engg/Cornell Univ; BS/Civil Engg/Cornell Univ. *Born:* Apr 17 1946. Native of Syosset NY. Served in Army Corps Engrs 1967-72, one yr in Vietnam. NSF Graduate Fellow 1967-68, 1972-74. Phi Kappa Phi Fellow 1967-68. Joined Univ of Kansas in 1974. Presently Prof Civil Engg engaged in teaching & res in fields of plain & structural concrete and composite construction. VP & Dir Kansas Chapter ACI 1975, 1977-78, 1981-84; Pres Kansas Chapter ACI 1976. Fellow of the Amer Concrete Inst, 1981; Mbr, Uniform Building Code Bd of Appeals, Lawrence, KS 1978-84. Author of over 50 articles on struct engrg and engrg materials. ASCE Huber Research Prize, 1985, ACI Bloem Award, 1986. PhD Civil Engg UIUC, MS Struct Engg & BS Civil Engg Cornell Univ. *Society Aff:* ACI, ASCE, PTI, PCI, AISC, ΣΞ, AAAS

Das, Khirod C
Business: 4010 W Broad St, Richmond, VA 23230
Position: Dir (PRO-DWCM) *Employer:* State Water Control Bd *Education:* Ph.D./Hydraulics & Hydrology/Purdue Univ; M.S./Soil & Water Conserv./MO Univ; Post-doctoral work/Sanitary & Environ/Rutgers Univ *Born:* 6/3/33 Khirod C. Das, now a US citizen, was born in India in 1933. He received his bachelor's degree in 1960 from India. Shortly after graduation, he spent two yrs with OUAT, Orissa, India. In 1962, he joined the Univ of Missouri, under a grant from the US AID, and received his MS degree. After his stay in Missouri, he returned to OUAT as an Assoc Prof and Dept Hd, but he soon got itchy feet again. At the end of 1971, he joined the VA State Water Control Bd and has made substantial contributions in the water resources field. He has also worked actively and productively in the area of water quality modeling. By means of clever but simple calculations, he has contributed significantly in the maintenance of improved water quality. *Society Aff:* WPCF, AGU, IWRA, USCOLD

Das, Mihir K
Business: Mech. Eng. Dept, Bellflower Blvd, Long Beach, CA 90840
Position: Prof *Employer:* Cal. State Univ *Education:* Ph.D/ME/Birmingham Univ-Eng; MS/ME/Ranchi Univ-India *Born:* 11/2/39 Well known for his research work in advanced design and manufacturing and computer applications to teaching and research in engrg. Author of over 65 publications: research papers, monographs and intl patents. Originator of the theory of dynamic cuttin which is currently taught in MS courses on machine tool tech. Lectured in many intl conferences all over the world. Mbr of several professional societies: ASME, ASEE, NCGA and I.Mech.E (England). Present Editor- in-Chief, ASME Publications (Orange County). Current interests: CAD-CAM, CIM and Teaching Innovation involving high-tech. *Society Aff:* ASME, ASEE, NCGA, I.Mech.E.

Das, Mukunda B
Business: Elec Engg Dept, University Park, PA 16802
Position: Prof. *Employer:* PA State Univ. *Education:* PhD/Electronics/Imperial College; DIC/Elec Engg/Imperial College; MS/Physics/Dacca Univ; BS/Physics/Dacca Univ. *Born:* 9/1/31. in Khulna, Bangladesh, on Sept 1, 1931. He was educated in Dacca Univ with BS and MS degrees in applied physics and in elec engrg and electronics in London Univ, Imperial College. He served as a Lecturer in Elec Engr, Imperial. College, from 1960-62; as a sr res officer at the Pakistan CSIR Lab in Decca, from 1962- 65; as a principal scientific officer and groupleader in MOS IC Tech at the Hirst Res Ctr, Wembley, England, from 1965-68; and he was apptd as assoc prof in elec engrg in 1968 and became a prof in 1979. He holds sevl Britich and US pats and one is in ionimplanation Mosfet's. He is involved in silicon and GaAs device and IC dev and characterization and also in CdTe solar cell characterization. *Society Aff:* IEEE, IEEE, EDS, ΣΞ.

Das, Sankar C
Home: 4621 Lakewood Dr, Metairie, LA 70002
Position: Assoc Prof *Employer:* Tulane Univ *Education:* Ph.D./Structural Engr/Univ of MO-Columbia; M.S./Structural Engr/Univ of MO- Columbia; M.S./Solid Mech/Brown Univ; B.E./CE/Calcutta Univ *Born:* 12/26/40 16 years experiences of teaching, research and consulting in the structural mech area involving structural dynamics, riser dynamics, ice-mechanics, numerous publications in Journals, National and Intl conferences. Consltg affiliations with some of the major US indus. Currently tenured assoc prof at Tulane Univ since 1977 and part time engrg conslt at offshore indus. Enjoys music, fishing, traveling etc. *Society Aff:* ASCE, AAM, AAUP.

Dashiell, Thomas R
Home: 504 Thomas Ave, Frederick, MD 21701
Position: Dir Environ and Life Sciences *Employer:* Dept of Defense. *Education:* BS/Bio/Western MD College; BSChE/Chem Engg/John Hopkins Univ. *Born:* 9/9/27. Native of MD. Served in Army Air Forces in WWII. Began government career at Fort Detrick, Frederick, MD in 1951. Performed in various cap from Bacteriologist, Lab Sup, Sup Chem Engg, completing work there in 1970 as Asst Sci Dir for Dev and Engg. Joined as staff spec for Chem Tech and currently serve as Dir Environ and Life Sciences in the Office of the Under Secretary of Defense (Res and Engg), Off of the Sec Def. Rec the Meritorious Civilian Svc and Spec Svc Award for directing the Army Biological Demilitarization Program, Rec Pres Meritorous Exec Award in 1982. Author of technical publications and listed in Whos Who in the South and MD. *Society Aff:* AIChE, ACS, NYAS, ASM, ΣΞ.

Date, Raghunath V
Home: Box 194, Trumbull, CT 06611
Position: Prof *Employer:* Sacred Heart Univ. *Education:* PhD/ChE/Yale Univ; MBA/Mgmt/Sacred Heart Univ; MPhil, MS/ChE/Yale Univ; MChEngg/ChE/Univ of OK; BChEngg/ChE/Univ of Bombay. *Born:* 10/22/37. Pioneer in intradisciplinary applications of tech, business principles, computer techniques & communication

Date, Raghunath V (Continued)
fields Inventor of novel extended surface heat exchanger - Olin Corp, computer simulation of multi stage desalination equip - AMF, basic & applied plastics res - celanese, pioneering work in analysis & computer simulation of plastics processing - TPT Corp (VP). Consultant to plastics & computer industry. Dir Plastics Dev, Natl Can. Rgtd Patent Agent, VP Omkar Inc, Prof of Mgmt Dept, Sacred Heart Univ. *Society Aff:* AIChE, SPE, APICS.

Datta, Subhendu K
Business: Dept of Mech Engrg, Campus Box 427, Boulder, CO 80309-0427
Position: Professor *Employer:* Univ Colo *Education:* PhD/Applied Math/Jadavpur University, Calcutta; M.Sc./Applied Math/Calcutta University; B.Sc./Mathematics (Honors)/Calcutta University *Born:* 01/15/36 Born in Howrah in the State of West Bengal, India. Became a naturalized citizen of the USA on May, 1, 1985. Was Fulbright grantee in 1962 and 1986. Elected Fellow of ASME in 1985 and received the College of Engrg & Applied Science Research Award in 1985. Was Chmn of the Wave Propagation Cttee of the Applied Mechanics Div 1981-86. Acting Chmn of the Dept of Mech Engrg, Univ Colo, in 1979 and 1984-85. Have published extensively on earthquake engrg, mechanics of composites, and ultrasonic nondestructive evaluation. *Society Aff:* ASME, AAM, SSA, EERI, SIAM

Datta, Tapan K
Business: 667 Merrick, Detroit, MI 48202
Position: Prof & Chrmn Civ Engg. *Employer:* Wayne State Univ. *Education:* PhD/Civ Engg/MI State Univ; MS/Civ Engg/Wayne State Univ; DTRP/Town & Regional Plng/Bengal Engg College; BE/Civ Engg/Bengal Engg College. *Born:* 9/3/39. in India but settled in the US since 1967. Worked five yrs as a Civ Engr for the Kujian Corp in their Calcutta office. Worked as Chief Transportation Engr for Goodell-Grivas, Inc, Southfield, MI from 1968 to 1973. Since 1973, working fulltime at Wayne State Univ. Also consultant and a stock holder of Goodell-Grivas, Inc. Have published over 30 papers in various journals. Prin investigator of several researches in the area of hgwy safety. *Society Aff:* ASCE, ITE, APWA.

Daubert, Henry
Home: 1648 S Boston Ave, Tulsa, OK 74119
Position: Exec VP *Employer:* Mansur-Daubert Strella Inc *Education:* BS/CE/CO State Univ *Born:* 4/19/24 Native of Ft Collins, CO. Served with US Army during World War II. Field Engr with USBR 1950-1952. Soils and Materials Engr, Phillips Petroleum Co 1952- 1955. In Consulting Engrg since 1955. With M-D-S Inc and predecessor firms. Became VP and Chief Engr in 1963; Executive VP in 1972. Member ASCE, ACEC, NSPE, APWA. Pres CEC/O 1979-80; Pres OK, Section ASCE 1980-81; Dir ACEC from CEC/O, 1981-1984; Director, DIstrict 15, ASCE, 1987-90. *Society Aff:* ASCE, NSPE, APWA, ACEC, ULI, NAHB.

Daubert, Thomas E
Business: 165 Fenske Laboratory, University Park, PA 16802
Position: Professor of Chemical Engrg. *Employer:* The Pennsylvania State University. *Education:* PhD/ChE/PA State Univ; MS/ChE/PA State Univ;l BS/ChE/PA State Univ. *Born:* 10/14/37. Mbr of Penn State Faculty since 1961; now Prof of Chem Engrg. Res in applied thermodynamics; prediction of thermodynamic, transport & physical properties of hydrocarbons, organics, and inorganics; vapor-liquid critical and equilibrium properties, characterization of petroleum. Co-author of *API Technical Data Book-Petroleum Refining, AIChE Data Compilation and Data Prediction* Manual Author of *Chemical Engineering Thermodynamics* textbook. *Society Aff:* AIChE, ACS, ASEE.

Daugherty, Tony F
Home: 8022 Sunnyvale, Houston, TX 77088
Position: Sr. Energy Analyst *Employer:* Tenneco, Inc. *Education:* PhD/Engg Statistics/Univ of Houston; MSIE/Engg/Univ of AR; BSIE/Engg/Univ of AR. *Born:* 1/24/49. Native of Little Rock, AR. Attended Univ of AR from 1967-73. Employed in Energy & Mtls Dept of E I du Pont de Nemours & Co, Inc. until 1975. Attended Univ of Houston where res on energy modeling was conducted. Held position of Asst. Professor at GM Insitutute, until Dec 1979, where major areas of work included statistical analyses, quality control, and reliability. Assumed current position of Senior Energy Analyst at Tenneco in Jan 1980. Presently work in areas of energy supply/demand modeling, forecasting, and simulation. Reg PE. Enjoy reading, running, tennis, and music. *Society Aff:* APM, TBΠ, IIE, ASA

Daum, Donald R
Business: 204 Agri Engg Bldg, Univer Park, PA 16802
Position: Prof. *Employer:* PA State Univ. *Education:* MS/Agri Engg/PA State Univ; BS/Agri Engg/PA State Univ. *Born:* 12/25/33. Native of Clarion County, Pa. Reg PE (IL). Taught and conducted res in Agri Engg at Univ of IL 1959-1966. Extension Agri Engr at the PA State Univ since 1966, responsible for Extension educational prog in farm power and machinery, including design, evaluation, energy conservation, safety and economics. Coordinator, Agri Engrg Ext Prog. Received 13 ASAE Blue Ribbon Awards for Educational Aids. Secy, pres of PA Sec; secy-treas of North Atlantic Region; mbrship on 13 natl committees (nine offices); secy, chrmn of Education and Res Dept, ASAE. Hon Soc: Sigma Xi, Gamma Sigma Delta, Epsilon Sigma Phi, and Alpha Epsilon. Sr mbr, ASAE. Community activities include Lutheran Church, 4-H and Lions. Hobbies are gardening, photography, hunting, and snowmobiling. *Society Aff:* ASAE, CAST.

Dautel, John D
Business: Sperry Syst Mgmt, Great Neck, NY 11020
Position: Asst Mgr Strategic Syst Group. *Employer:* Sperry Rand Corp. *Education:* BSME/ME/Stevens Inst. *Born:* Feb 22 1933. BSME 1955 Stevens Institute of Technology; Graduate Studies Hofstra University. Professional: Employed by Sperry since graduation in 1955. Specialized in Navigation Equipment for Submarines. Worked as a Design Engr, Product Engr & currently Systems Engr. 1965 promoted to Supervision & currently Asst Mgr for Strategic Sys Group of Sys Mgmt Div. Respon for dev & production of all equipment used in the Navigation Systems of the Polaris, Poseidon & Trident Submarines & for the Integrated Logistics Support of these equipments. Direct a Staff of over 200 prof employees. Pat C W Surveillance Radar Syst 3 815 131 6- 4-74.

Davee, James E
Business: Old Maysville Rd, Commerce, GA 30529
Position: VP of Engr *Employer:* Roper Pump Co *Education:* MS/ME/Stevens Inst of Tech; BS/ME/Univ of ME *Born:* 3/17/28 Raised in Tenafly, NJ. Tau Beta Pi at Univ of ME. 1st Lt US Army Signal Corps 1951-1953. Design, Product and Chief Engr, Rotary Pumps, at Worthington Corp. Design Engr, Precision Mechanical Products, Hamilton Standard. Joined Roper Pump Co, Manufactures of Positive Displacement Rotary Pumps in 1972. Assumed VP of Engrg 1977. Past Chrmn of Rotary Pump Section, Hydraulic Inst. Chrmn, Membership Committee, Hydraulic Inst. Member ANSI B215 Rotary Pump Committee. Member Manufacturer's Task Group to Develope API Standard 676, "Positive Displacement Rotary Pumps–. Chrmnm ASME PTC 7.1 Committee. *Society Aff:* ASME, HJ.

Davenport, Granger
Home: 6 Gordon Pl, Montclair, NJ 07042
Position: Retired. *Education:* BS/ME/NJIT; ME/-/NJIT. *Born:* Jul 11 1903 Brooklyn NY. Thesis 'Testing the Strength of Gear Mats'. With Gould & Eberhard Inc Irvington NJ, mfgrs of gear-cutting Mach & Shapers 1923-59. Ch Engr 1950-59 & at Norton Co G&E section to 1961. 17 US pats. PE in NJ, NY, MA. Mbr ASME since 1925, currently Life-Fellow. Formerly active on 8 ASME stds cttes incl 5 yrs on Standarization Ctte, 3yrs on Bd/Codes & Stds. Formerly on ANSI com B6, Gears, Mbr ASM 25 yrs, Co Rep Amer Gear Mfgrs Assn 1935-61, recd Edward P. Connell Medal for yrs of work on gearing stds & made Honorary Mbr. Wrote sections on gear milling & hobbing for Amer Mchst Hdbk in 1940, 53, et seq; ditto for T1 Engrs Hdbk 1959 et seq. 50 yrs Mbr F&AM. In scouting 11 yrs. Hobbies: Adirondack Mtn camp, canoeing, hiking, bird watching, classical music, stock market. *Society Aff:* ASME, ASM, AGMA.

Davenport, Wilbur B, Jr
Business: 35-214 MIT, Cambridge, MA 02139
Position: Prof Dept Elec Engg & Computer Sci. *Employer:* MIT. *Education:* BEE/Elec Engg/AL Poly Inst; SM/Elec Engg/MIT; ScD/Elec Engg/MIT. *Born:* Jul 1920. Joined MIT faculty in 1949. Worked on telecommunications at the MIT Lincoln Lab from 1951-60, & from 1963-65; held position there of Group Leader, Div Head & Asst Dir. Was Dir of MIT Center for Advanced Engrg Study from 1972-74. Was Head, Dept of Elec Engrg & Computer Science at MIT from 1974-78. Fellow of IEEE, AAAS & Amer Acad Arts & Sci. Mbr Natl Adacemy of Engrg. *Society Aff:* IEEE, AAAS, AAAS, NAE.

Daverman, Edward H
Home: 2101 Onekama Str SE, Grand Rapids, MI 49506
Position: Retired *Employer:* Daverman Associates, Inc *Education:* BS/ChE/Univ of MI. *Born:* May 1915. US Navy 1941-45 Ordnance Ship Superintendent with rank of Lt Commander. 1945-49 Survey Ch, Resident Engr on power line projects. 1949-1980, Partner/Director - Daverman Associates in charge of Civil, Mechanical Depts on architectural & engrg projects, 200 man office. General administration respon on office production, personnel. Mbr NSPE, Reg Prof Engr 10 states, Mbr ASCE. On Bd of Systems Planning Corp & Mid Continent Tele Co of MI. Retired from Daverman Associates 12/13/79. *Society Aff:* NSPE, ASCE.

David, Edward E, Jr
Business: 10 Gould Center, Rolling Meadows, IL 60008
Position: Executive Vice President. *Employer:* Gould Inc. *Born:* Jan 25 1925 Wilmington NC. BS from Georgia Tech 1945, ScD from MIT 1950. Bell Telephone Labs 1950-70, Science Adviser to the Pres, The White House 1970- 72. Gould Inc since Jan 1 1973. 6 hon doctorates. Originator of 'the Man-Made World', new high school course concerning principles behind tech. Chmn/Bd The Aerospace Cor. Mbr of the Corp MIT. Cons Natl Sec Council. White House Group on Contrib of Tech to Economic Strength. ERDA Task Force.Mbr NAS & NAE. Fellow: AAAS, ASA, AAS, IEEE, Audio Engrg Soc. Chmn US-France Cooperative Science Program, State Dept. Co-author of 2 books.

David, Pierre B
Home: 77 Rue des Martyrs, Paris, France 75018
Position: Ingenieur Gen Hon des Telecomm. *Employer:* Retired. *Education:* PhD/-/Ecole Polytechnique, Paris. *Born:* 10/18/97. 1916-18 WWI Lt. 1921 French Military Radiotelegraphy; 1938 Natl Radio Lab; 1939-56 French Navy Hd of Radio-Electron Lab; 1956-62 Sci Adv SHAPE. Retired. Fellow IEEE-USA; Chmn French Sect IEEE 1965-68; Mbr 'Academie de Marine'-France.

David, Thomas C
Business: 1022 Tenth Street, Alexandria, LA 71301
Position: Board Chairman. *Employer:* Pan American Engineers, Inc. *Education:* BS/Civil Eng/Univ of Notre Dame. *Born:* Nov 1912, Alexandria, Louisiana. Early Employment with contractors, Louisiana Highway Dept & Louisiana Dept of Public Works. In 1942 organized consulting firm of Pan American Engrs Inc, presently serv as Bd Chmn & operating one of two offices engaged in consulting work in Louisiana & Mississippi. Fields of experience cover design of civil & public commercial & industrial fields. ASCE Fellow, Life Mbr Louisiana Engrg Soc, Mbr WPCF, CEC. *Society Aff:* ASCE, WPCF, CEC.

Davida, George I
Business: Dept of EE & CS, Milwaukee, WI 53201
Position: Prof *Employer:* Univ of Wisconsin *Education:* PhD/EE/Univ of IA; MS/EE/Univ of IA; BS/EE/Univ of IA *Born:* 8/2/44 Dr. Davida joined the Dept of Electrical Engrg and Computer Sci at the Univ of Wisconsin in 1970. He was program Dir of Theoretical Computer Sci at NSF in 1978-79. In 1980-81 he was with the School of Information & Computer Science, Georgia Inst of Tech. Dr. Davida served on the governing bd of the IEEE Computer society. His research interests include cryptography and privacy protection. *Society Aff:* IEEE, ACM.

Davidson, Arthur C
Home: 106 Bideford Ave, Downsview Ont M3H 1K4 Canada
Position: Assoc Professor, Retired 30 June 1981 *Employer:* Dept Cvl Engrg U of Toronto. *Education:* MASc/Structures/Univ of Toronto; BScCE/Civil/Univ of Manitoba; BScEE/Elec/Univ of Manitoba. *Born:* 1914, Calgary, Alberta. Served in Royal Canadian Engrs 1941-46, Staff Capt procuring earth moving equipment, R&D A/Veh A/Pers mine detection; prior to war Field Engr for Dominion Bridge & for Canadian Inspection, design, construction, & inspection; after war, designer, checker, specification writer, inspector, coordinator for Hydro Electric Power Comm Ontario, Canadian Brazilian Services, Defense Construction, Central Mortgage & Housing Corp, Toronto Transit Comm. Co-author text in Statics, Res on vibrating crystals as pressure intensity measurers, Standard for compression members; Fellow Engrg Inst of Canada, Councilor; Examiner Assn of Prof PEs of Ontario; Hobbies: Canadiana, Photography. *Society Aff:* LMEIC, APEO, FEIC, LM CSCE

Davidson, Bruce R
Home: 1 Summerall Rd, Somerset, NJ 08873
Position: President *Employer:* Information Inc. *Education:* MBA/Exec Mgmt/Pace Univ; BS/Chem Engg/Lehigh Univ. *Born:* 9/1/37. *Society Aff:* AIChE, ACS, NACE.

Davidson, Edwin A
Home: 2510 Grand Ave, Kansas City, MO 64108
Position: VP, Chief Engr. *Employer:* M J Harden Assoc, Inc. *Education:* BS/Engr/MO School of Mines & Metallurgy. *Born:* 6/19/24. Mapping Co exec; born St Louis, June 19, 1924; S William Edwin & Willie Anne (McCombs) Di, BS in Ceramic Engg, MO Sch Mines & Metallurgy, 1949; m Mary Jeanette Forquer, Oct 30, 1953; children - William Joseph, James Edwin. Photogrammetric Engr US Geologicl Survey, Rolla, MO, 1949-53; field engr Pressite Engr Co, St Louis 1953-55; Sales Engr Hubbard Co, San Francisco Co, CA 1955; Materials Engr & Field Engr, MO Hwy Dept, Jefferson City, 1955-57; Chief field Surveys, Sales Engr, Chief Engr, M J Harden Assoc - VP & dir 1957-80, Kansas City, MO; US Air force 1943-45. Decorated Air Medal with 4 Oak Leaf Clusters; D F Cross; Reg PE MO, KS, AR, & IA; Am Soc Civil Engrs; Am Soc Photogrammetry; MO & Natl Soc Prof Engrs; Prof Engr in private practice. Home: 804 San Francisco Tower, 2510 Grand Ave, Kansas City, MO 64108. Reg P.E. also in New Mexico. *Society Aff:* ASCE, ASP, MSPE, NSPE, PEPP.

Davidson, Frederic M
Business: The Johns Hopkins Univ, Dept. of Electrical & Computer Engr, Baltimore, MD 21218
Position: Prof of Electrical Engrg *Employer:* Johns Hopkins Univ *Education:* PhD/Physics/Univ of Rochester; BE/Engr Physics/Cornell Univ *Born:* 2/11/41 in Glens Falls, NY. Received a Bachelor of Engrg degree from Cornell Univ, (1964) and a PhD degree from The Univ of Rochester (1969). Served as an Assistant Prof of Electrical Engrg at The Univ of Houston, Houston, TX (1968- 70), and as Assistant (1970-75), Assoc (1975-80) and Prof of Electrical Engrg (1980-) at The Johns Hopkins Univ in Baltimore, MD. Professional interests are optical communications and quantum electronics. Has received research grants from The National Science Foundation and The National Aeronautics and Space Administration. *Society Aff:* OSA, IEEE

Davidson, Harold O
Home: 445 Rookwood Dr, Charlottesville, VA 22901
Position: Principal. *Employer:* Davidson Associates *Education:* PhD/Operations Res-IE/OH State Univ; MS/Ind Engg/GA Inst of Tech; BSME/Heat-Power/GA Inst of Tech. *Born:* 12/10/18. Principal, Davidson Assoc, Charlottesville, VA; Bd. of Direc-

Davidson, Harold O (Continued)
tors, Pedras USA Inc Partner (ret 1979) Arthur Young & Co; Pres - Davidson, Talbird and McLynn, Inc 1964-68; VP - Operations Res Inc 1959-64; Prof GA Inst of Tech 1958-59; Sr Scientist, Hq US Army Europe 1956-58. Guest Lectr - Harvard Bus Sch, Cornell Univ, US Civ Service Commission, Army War College. Author "Functions & Bases of Time Standards–, IIE; Johns Hopkins monograph on gamma radiation effects; & articles in Journal of Ind Engg, Harvard Bus Review, & McGraw-Hill "Handbook of Business Problem Solving–. Former Editorial Bd & Monograph Editor, Jo of Ind Engr. Married Georgia Lee Stone, children Lee & Scott.

Davidson, J Blaine
Business: Dept of Ocean Engg, Boca Raton, FL 33431-0991
Position: Prof of Ocean Engg. *Employer:* FL Atlantic Univ. *Education:* MS/Appl Physics/UCLA; BS/Engg Electronics/US Naval Postgrad Sch; BS/Engg/US Naval Acad. *Born:* 11/10/23. Native of Oklahoma City, OK. Commissioned Ensign in USN in 1946. Served on destroyers in ASW and Electronics positions. Designated an Engg Duty (Electronics) Officer (1954) with following tours in Japan, in submarine construction at Mare Island Naval Shipyard, and as Dir Undersea Progs in the Office of Naval Res as most significant assignments. Retired as Commander in 1967 and joined the faculty of FL Atlantic Univ in the Dept of Ocean Engg. Chrmn 1976-77. Consultant in underwater acoustics, beach restoration and deep ocean manganese nodule mining. Visiting Professor Norwegian Technical Institute, Trondheim, Norway 1981-82. Leader, People to People, coastal zone management group visit to People's Republic of China, 1985. *Society Aff:* ASEE, ASA, MTS.

Davidson, Robert C
1329 Beachmont St, Ventura, CA 93001
Position: VP. *Employer:* Pharm-Eco Labs, Inc. *Education:* BS/Appl Chem/CA Inst of Tech. *Born:* Jun 19 1916. Discovered rehydroxylation of natural cracking catalyst & adapted it to catalytic cracking of petrol. Dev sev other indus catalysts. Mgr Nalco Catalyst Div & Pres of Katalco, a joint venture of Nalco & Imperial Industries. Involved in spec chem oper as post-retirement act. Enjoy golf, sailing, woodworking. *Society Aff:* AIChE.

Davidson, William W
Business: Coll of Tech, Berrien Springs, MI 49104
Position: Prof. of Engineering, Dept. of Architecture. *Employer:* Andrews University. *Education:* PhD/Civil/PA State Univ; Ms/Civil/OH Univ; BSCE/Civil/OH Univ. *Born:* Jul 1930 Chauncey Ohio. Served US Army during Korean conflict 1951-53. 10 yrs bldg const supr. Taught engrg at Ohio U, Penn Sta, Kanawha Grad Ctr W Va U, Andrews U. Appointed Dean of newly estab C of Tech 1974 with respon for dev of programs in engrg & indus technology. MI. Registered PE Colorado, Michigan, Pennsylvania. Member ASCE, ASEE, Sigma XI & Chi Epsilon. Presently Prof. of Engineering, Dept. of Arch, Andrews Univ and Field Engineer for Maranatha Flights International. *Society Aff:* ASCE, ASEE, ΣΞ, XE.

Davies, Clarence W
Home: 37 Middlesex Road, Fredonia, NY 14063
Position: Consulting Eng. *Employer:* Self *Education:* BS/Structures/Union College. *Born:* 4/20/12. Native of Schenectady, NY. Designer of Wiring Devices, Weber Electric Co Schenectady NY 1934-36. Struc Designer, Genl Electric Co 1936-41. Struc Designer, Field Inspector to Proj Mgr, Amer Locomotive Co, Schenectady NY, Latrobe PA, Auburn NY, Chicago Hghts IL 1941-52; Beaumont TX, Sharonville OH, Dunkirk NY Design Engr, Stress Analyst, Proj Mgr 1952-62. NY Sta Thruway Authority Buffalo NY Asst Div Engr 1962-78. Pres C W Davies Cons Engr Dunkirk NY 1967-69. Mbr ASCE, Sect Sec 1952, Chmn Pipeline Plan Ctte 1968, Dir Dist III 1974-76, Chmn Ctte on Pubs 1976. Chrmn Committee on Technology Assessment 1976- 81 Co Chrmn Specialty Conference on Social & Economic Impact of Earthquakes on Utility Lifelines. Mbr NSPE, Chap Pres NYS PE, Lic Prof Eng NY. Consulting engr 1978-present. *Society Aff:* ASCE, NSPE, NYSPE.

Davies, James A
Home: 7506 Prestwick, Houston, TX 77025 *Education:* MS/CE/CA Inst Tech; BS/Appl Chem/CA Inst Tech. *Born:* 9/2/14. Joined Engg Dept of Texaco in NY upon acquiring MS degree in Chem Engg in 1936. Was appointed Gen Mgr of Dept in 1973. Dept contracts maj process units constructed around the world for Texaco. Published many papers on bubble trays. Holds Fellow grade of AIChE. Retired in 1979. *Society Aff:* AAAS, ACS, AIChE.

Davies, Richard G
Home: 156 S Waverly, Dearborn, MI 48124
Position: Senior Scientist. *Employer:* Ford Motor Company. *Education:* PhD/Metallurgy/Univ of Birmingham, England; BSc/Metallurgy/Univ of Birmingham, England; MBA/Finance/Univ of MI. *Born:* Nov 1934 in Congleton, Cheshire, England; naturalized US citizen, married with 3 children. PhD, BS from University of Birmingham, England; MBA from University of Michigan. Visiting Metallurgy Professor at Institute of Physics, SC de Barcloche, Argentina 1959-61. With Ford Motor Company since 1962. Major areas of study have included ordered alloys, high temperature Ni-base superalloys, martensitic & dual-phase steels & the influence of high rate deformation upon the strength of materials; published over 55 papers; 4 patents. Mbr ASM & AIME; awarded ASM Howe Gold Medal in 1972 for the work on martensitic alloys. *Society Aff:* AIME, ASM.

Davis, Cabell S, Jr
Home: 1673 Trap Rd, Vienna, VA 22180
Position: Consultant (Self-employed) *Education:* MS/ME/US Naval Postgrad School; BS/ME/US Naval Postgrad School; BS/Engrg/US Naval Academy *Born:* 7/20/26 Lakeland, FL. Attended Elementary, Secondary Schools Charleston, WV. Grad US Naval Acad 1947. Promoted to Rear Admiral 1975. Served as Engrg Duty Officer. Various shipbd duties '47-'55. Ship Supt Norfolk Naval Shipyard '55-'58. Atlantic Fleet Maintenance Officer Staff '58-61. Tender Repair Officer '61-63. Acquisition & Maintenance of Ships BUSHIPS '63-68. Amphibious Force Staff Engr '68-71. LHA Proj Mgr '71-'74. Supvr Shipbldg. Pascagoola, MS '74-'75. Commander Charleston Naval Shipyard '75-'78. Depty CDR Naval Sea Systems Command for Industrial Facilities Management '78-'79. Marine Industry Consultant since retirement from Navy in '79. Council Member ASNE National Office '71- '74 and '81-'85. Mbr Smithsonian, Natl Geographic. Hobbies; golf, fishing, philately. *Society Aff:* ASNE

Davis, Charles S
Business: 220 Bagley, Suite 700, Detroit, MI 48226
Position: Pres *Employer:* Charles S Davis & Assoc, Inc *Education:* Ph.D./CE/Univ of MO-Rolla; M.S.C.E./Structures/Univ of WA; B.S./C.E. /Prairie View A & M Univ *Born:* 6/6/39 Charles Simon Davis is Pres of Charles S. Davis and Assocs, Inc, a Consulting Engrg Co offering professional and technical services in structural engrg, civil engrg, automotive design and computer science. Prior to 1978, he worked as a Structural engr in the Aircraft (North American, Boeing and Lockheed) and Automotive (Ford Motor Co) industry. He was an officer in the United States Army. Publications have covered a wide spectrum of engrg topics and papers presented at the Column Research Council Annual Meeting, Specialty Conference on Cold-Formed Steel Structures, Structural Engrg Research Session of ASCE and the Society of Automotive Engrg Annual Meeting. Local Professional affiliations includes, MI Society of PE, Engrg Society of Detroit, Conslt Engrg Council/MI and Society of Engrs and Applied Scientist. Listed in Who's Who in the Midwest, 19th Edition. He is a member of Alpha Phi Fraternity, Inc. He is an avid tennis player. *Society Aff:* ACEC, NSPE, WPCF.

Davis, David C
Home: 2485 Midvale Forest Dr, Tucker, GA 30084
Position: Genl Manager-Switching & Design. *Employer:* Southern Bell Telephone & Telegraph. *Education:* MSEE/Elec Engg/GA Tech; BSEE/Elec Engg/GA Tech; BBA/Mgmt/GA State Univ. *Born:* Mar 1929. MSEE & BSEE Georgia Tech Atlanta, BBA

Davis, David C (Continued)
Georgia Sta Univ. Served in Signal Corp 1951-53. Employed by Southern Bell since 1949 except for 3 yrs when employed by American Telephone Co in New York. Assumed current respon in 1972. Respon for policy & capital improvements in switching communications facilities for GA involving the direction of over 400 engrs.

Davis, Donald E
Business: 2605 Routh St, Dallas, TX 75201
Position: Consulting Engineer. *Employer:* Donald E Davis Cons Engrs. *Education:* BS/CE/Citadel. *Born:* 12/8/29. Native of Dallas, Texas. Taught in Southern Methodist University School of Engineering. Have been in private practice as a cons engr since 1957. Have served as Chapter First VP & as Pres of American Soc of Plumbing Engrs as well as mbrship on the National Technical Ctte. Have served as Chap Sec & on the Bd/Dirs of the Amer CEC. Previous Natl First VP & on Bd/Dirs ASPE Res Found Inc. Holder of a Natl Council of Engrg Examiners Cert & Licensed P E in 12 states. *Society Aff:* CEC, ACI, ASCE, ASPE.

Davis, E James
Business: P.O. Box 1039, Appleton, WI 54912
Position: Dir. Engrg Div. *Employer:* Inst for Paper Chem *Education:* PhD/ChE/Univ of WA; BS/ChE/Gonzaga Univ. *Born:* 7/22/34. Native of St Paul, MN. Taught at Gonzaga Univ & was Dir of the Comp Ctr there until 1968. At Clarkson College served as Chem Engg Dept Chrmn & Assoc Dir of the Inst of Colloid & Surface Sci. From 1978 until 1980 Chrmn of Chem & Nuclear Engg at the Univ of NM. Winner of the ASEE/Leeds & Northrup Predoctoral Fellowship (1964-65) at Univ of WA & London Univ, respectively. Admin of the AIChE Design Inst for Multiphase Processing & author of over 60 publicatins on two-phase flow, aerosol phys & chem engg sci. Since 1980, Director of Engineering Division at Institute for Paper Chemistry. *Society Aff:* AIChE, ACS, ASEE.

Davis, Edward J
Home: 1103 Glourie Dr, Houston, TX 77055
Position: Exec VP, Chief Oper Off *Employer:* Bernard Johnson Inc *Education:* MS/CE/Rice Univ; BS/CE/Rice Inst; BA/CE/Rice Inst *Born:* 3/6/37 Native of Brussels, Belgium moving to US in 1940 and raised in San Antonio, TX. Moved to Houston, TX as permanent resident upon graduation from Rice Inst in 1958. Served as reserve officer in US Army Corps of Engrs holding ranks of 2nd Lt through Captain. Entire professional career with Bernard Johnson Inc. Married former Myrna Bieberdorf in 1958 and have two sons, Scott Winfield and Matthew Lloyd. Hobbies include travel, photography, white water rafting. *Society Aff:* ASCE, ACEC, SAME

Davis, Edward M
Business: PO Box 1498, Reading, PA 19603
Position: Group VP. *Employer:* Gilbert Assocs, Inc. *Education:* BS/Elec Engg/Bucknell Univ. *Born:* 12/20/17. Native of Naticoke, PA. Employment: Gen Elec Co--Applications Engr (1939- 1946); Firestone Tire and Rubber Co--Staff Engr (1946-1947); Gilbert Assoc, Inc- -1947 to present in the following capacities: Proj Elec Engr (1947-1959), Chief Elec Engr (1959-1969), Engg Mgr (1969-1970), VP and Gen Mgr, Central Operations Div (1970-1973), Admin VP (1973-1975), Group VP (1975 to present). Present responsibilities are: As Chief Operating Officer, am responsible for operations of the Environmental, Industrial, Facilities, and Environmental Systems Div Other current affiliations: Commonwealth Assoc-VP; Gilbert/Commonwealth Intl, Inc-Bd of Dir; Gaico, Inc-Bd of Dir; Laurel Run Civil Assoc-Bd of Dir. *Society Aff:* IEEE.

Davis, Evan A
Home: RD 4 Box 52, Slippery Rock, PA 16057
Position: Consulting Mechanical Engineer. *Employer:* Westinghouse Electric Corp - Ret. *Education:* MS/Mech Engg/Univ of Pittsburgh; BS/Mech Engg/Univ of Pittsburgh. *Born:* Dec 10 1904 Johnstown PA. At Westinghouse Res Labs 1933-70. Worked on Problems of Creep, Relaxation, Elastic-Plastic Stress Distrib & Theories of Strength. ASTM Dudley Medal (with M J Manjoine) 1953. ASME NADAI Award 1976. Fellow ASME. Former mbr ASTM & Soc of Rheology (AIP). *Society Aff:* ASME.

Davis, Frank W
Home: 6328 Curzon Avenue, Fort Worth, TX 76116
Position: Retired. *Employer:* General Dynamics Corporation. *Education:* BS/ME/CA Inst Tech. *Born:* Dec 1914 Charleston West Virginia. Married 1941 Frances Pfeifer. US Marine Corps Aviator. Joined Vultee Aircraft 1941 Engrg Test Pilot. 1948 Ch Design Engr Consolidated Vultee. 1954 Chief Engr Convair Fort Worth Div. 1961 Pres Fort Worth Div General Dynamics. 1970 Pres Convair Aerospace Div of General Dynamics including Fort Worth & San Diego operations. Dir Canadair Ltd. Mbr National Academy of Engrg. Honorary Fellow Soc of Experimental Test Pilots. Fellow American Institute of Aeronautics & Astronautics. Honorary DSc University of West Virginia. Cal Tech Disting Alumni Award. Registered Engr Texas, California. Sports Illustrated Silver Anniversary All America. Dir Langley Corp, Kanawha Manufacturing Co. Cohu Inc Bd of Overseers, University of California at San Diego. *Society Aff:* AIAA, SAE, SETP, NAE.

Davis, Franklin T
Home: 7750 West 64th Avenue, Arvada, CO 80004 *Education:* MS/Metal/Univ of WA; BS/ME/NM School of Mines. *Born:* 10/27/18. Engrg Officer US Merchant Marine 1943-45. Metallurgist Magma & San Manuel Copper Corporation 1945-53; Dir Metallurgy Div, Colorado School of Mines Research Institute, 1953-70; Mgr Environmental Research, Hazen Research 1970-72; Dir Environmental Division, Colorado School of Mines Research Institute 1972-80. 1980 to present Conslt Engr. Registered Engr Colorado; Past Chmn Colorado Section AIME; Dir & Western VP SME; Dir and Vice-Pres, AIME; Mining & Metallurgical Soc of America; Distinguished Member Award, SME. Distinguished Service Award, New Mexico Tech. Editor 'Unit Process in Hydrometallurgy' 1964. Numerous Technical Publications. *Society Aff:* AIME.

Davis, H Ted
Business: Dept of Chem Engrg, Minneapolis, MN 55455
Position: Prof and Head *Employer:* Univ of MN *Education:* PhD/Chem Physics/Univ of Chicago; BS/Chem/Furman Univ *Born:* 8/2/37 Hendersonville, NC, married 1960; NSF fellow, Univ Brussels, 1962-63; from asst prof to assoc prof 63-69, Prof and Head (1980-), Chem Eng & Materials Science, Univ of MN, Minneapolis 1969-. Concurrent Position: Sloan Found fellow 67-69, Guggenheim fellow 69-70; consultant to food and petroleum industry. Research: Theoretical and experimental studies of transport processes; statistical mechanical studies of fluids; electronic structure and radiation chem of hydrocarbon liquids and gases; molecular theoretical studies of low tension interfaces and micellar forms; theoretical and experimental studies of multiphase fluid displacement in porous media & oil recovery technology. *Society Aff:* AIChE, SPE, ACS, APS, AAAS.

Davis, Harmer E
Business: 109 McLaughlin Hall, Berkeley, CA 94720
Position: Emeritus Professor & Director ITT. *Employer:* University of California. *Education:* MS/CE/Univ of CA-Berkeley; BS/CE/Univ of CA- Berkeley. *Born:* Jul 1905 Rochester NY. Began career of teaching & res at U of Ca at Berkeley in 1930. Assoc with work of University's Engrg Mat Lab on dev work relating to concrete, foundations, bridges & pavement until 1948. As Dir of Inst/Transport & Traffic Engrg 1948-73. Hd dev of educ & res progs relating to hgwy, air & urban transport. Adv to gvmt agencies & legislative groups on planning, financing & dev problems in transport field. Chmn U of Ca Dept Cvl Engrg 1954-59; Chmn Hgwy Res Bd of Natl Res Council 1959. President's Task Force on Hgwy Safety 1969. Books on engg matls; num reports & papers on transport subjects. James Laurie Prize 1967 ASCE, G S Bartlett Award 1970 HRB, Berkeley Citation 1973. Univ of Calif Berke-

Davis, Harmer E (Continued)
ley, T.M. Matson Award, 1974, Inst. of Transp Engrs. Prof Emeritus of CE & Dir Emeritus of ITTE, 1973- . *Society Aff:* ASCE, ASTM, APWA, NAE, ACI.

Davis, Hunt
Home: 21 Briar Hollow, Houston, TX 77027
Position: Senior Staff Engr, Chief Engr Dept *Employer:* M.W.Kellogg Co. *Education:* MS/Mech Engr/Univ of Pittsburgh; BS/Haverford College. *Born:* 8/12/20. Res Engr Westinghouse Res Labs 1941-44; dev gas centrifuge. Aerodynamics Div Engr Elliott Co 1944-52, dev & des centrifugal compressors, turbochargers, gas turbines. Ch Engr, Mgr of R & D, Sr Engrg Consultant Worthington Corp 1952- 75, dev & des centrifugal & recip compressors, pumps, engines. Sr Staff Engr M. W.Kellogg 1975- , consultant on rotating equipment in process industries. ASME: Past Chmn Westmoreland Section, Past Exec Comm Pittsburgh, No Jersey, Buffalo Sections, Past Chmn Performance Test Code Subcttes. Jr Award 1948. Fellow 1972. Var pubs & pats. Ltd private cons turbo machinery design & operations. Dir of Continuing Education Course in Compressor and Turbine Technology. *Society Aff:* ASME, NSPE, ΦBK, ΣΞ

Davis, J Franklin
Home: 2684 Lakeside Drive, Erie, PA 16511
Position: Manager-Engrg. *Employer:* General Electric Co. *Education:* BSEE/-/Univ of IA. *Born:* Apr 14 1922 Unionville Missouri. Served as officer US Naval Reserve 1944-46. Joined G E Co 1943 with var engrg assignments relating to d-c motors & generators in Erie Pa and Lynn Ma to present. Became Manager of Engrg Sect, D-C Motor & Generator Dept in 1968. Senior mbr IEEE, past chmn Erie Sect, past chmn D-C Subctte of Rotating Machinery Ctte. Active in motor standards dev in NEMA, ANSI, & IEC. Author of several articles on d-c machines. *Society Aff:* IEEE.

Davis, Jeff W
Business: 6 South Second St, Ste 720, Yakima, WA 98901
Position: Partner *Employer:* Davis/Scheible Engr *Education:* MS/Environ Engr/Univ CA-Berkeley; BS/CE/OR State Univ *Born:* 7/24/46 Project mgr for Gray & Osborne, Inc PS, Consulting Engrs from 1971 through 1979, responsible for industrial and municipal wastewater projects. Established Davis/Scheible Engrg in 1980 to provide environmental management consulting services to the food processing industry. *Society Aff:* AWWA, WPCF

Davis, John M
Home: 812 Linkhorn Dr, Virginia Beach, VA 23451
Position: Consult Eng *Employer:* Self *Education:* MSA/Engr Mgmt/GWU; BS/Mine Engr/VPISU *Born:* 6/9/27 Danville, VA early years SC. Bob Jones Coll 1943-45. Army Engr 1945-47. VPISU 1947-51. Asst Mine Supt, Standard Lime and Stone Co, Martinsburg, WV 1951-52. Joined Atlantic Div, NAVFAC in 1952, as Construction Engr and Sr Proj Mgr. Promoted to Design Dir 1967 & subsequent position as Sp. Engr. Asst 1983-86. Design & Construction of Navy and DOD facilities in Middle Atlantic states and numerous overseas locations in Europe, Africa, Caribbean, Central and South America for 35 years. Navy and NSPE Federal Engr of the Year 1979. Private consult in Oct. 86. Enjoy rockhounding, railroading and amateur photography. *Society Aff:* NSPE.

Davis, Luther, Jr
Home: 2 Winthrop Terrace, Wayland, MA 01778 *Employer:* Retired 7/31/87. *Education:* PhD/Physics/MIT; BS/Physics/MIT. *Born:* Jul 12 1922 Mineola NY. MIT Radiation Laboratory 1942-45. MIT Research Laboratory of Electronics 1945-49. Raytheon Co, Res Div since 1949 & General Manager, Res Div 1969-1987. Served as cons to USAF Scientific Advisory Board, to the NSF & as member of cttes of the NAS-NRC Materials Advisory Board. Fellow of the IEEE. *Society Aff:* APS, IEEE.

Davis, Milton W, Jr
Home: PO Box 242, Columbia, SC 29202
Position: Prof. *Employer:* Univ of SC. *Education:* PhD/Chem Engr/Univ of CA; MS/Chem Engr/Univ of CA; BE/Chem Engr/Johns Hopkins Univ. *Born:* 4/5/23. After graduation from Johns Hopkins in 1943 I joined the US Naval Reserve as an Ensign. I was honorably discharged in 1946 with the rank of Lt (sg) having served as Chief Engr on two destroyers in the Pacific, (USS the Sullivans and USS James E Kyes). After grad school I joined the DuPont Co Atomic Energy Div in 1951 and participated in the design of the Savannah River Plant. I left DuPont in 1962 to join the Univ of SC as a Prof. *Society Aff:* AIChE, ACS.

Davis, Philip K
Business: Dept of Civil Engineering and Mechanics, Southern Illinois University at Carbondale, Carbondale, IL 62901
Position: Professor Dept of Civil Engrg and Mechanics. *Employer:* Southern Illinois University at Carb. *Education:* BS/Mech Engg/Univ of TX at Austin; MS/Mech Engg/Univ of TX at Austin; MSE/Engg Mechs/Univ of MI; PhD/Engg Mechs/Univ of MI. *Born:* 8/29/31. Reg PE Sta of IL. Native of central IL. Boeing Airplane Co 1959-60; NASA Langley Res Center Summer 1967. Asst Prof to Prof at SIU-C 1964-71. Chair, Department of Engr Mechs & Materials 1971-78, 1979-84. Chair, Dept of Civil Engr and Mechanics 1984-87. Acting Dean, Coll of Engrg & Tech, 1978-79. Chrmn of SIU-C Grad Council 1973-75; Pres SIU-C Chap Sigma Xi 1977. Pres Elect/Vice President SIU-C Chap Phi Kappa Phi, 1987. Publications in the areas of fluid mechs, vibrations, rheology & biomedical. *Society Aff:* ASEE, ASCE, ASME, AAM, ACDM, ΣΞ, IMI, SME, FPS, ΦΚΦ, ΤΒΠ

Davis, Robert L
Business: 101 Engrg Res Lab, Rolla, MO 65401
Position: Dean *Employer:* Univ of MO *Education:* PhD/Mech Engrg/Univ of MD; MS/Mech Engrg/Univ of MD; BS/Mech Engrg/Univ of Evansville *Born:* 10/23/36 BS Mech Engrg, Univ of Evansville, 1958; Staff Engr, Naval Ordinance Lab, 1958-1962; MS Mech Engrg, Univ of MD, 1962; Instructor, Univ of MD, 1962-1965; PhD Mech Engrg, Univ of MD, 1965. Asst, Assoc, Full Prof Engrg Mech, Univ of MO- Rolla, 1965-present; Asst Dean of Engrg, Univ of MO-Rolla, 1978-1979; Dean of Engrg, Univ of MO-Rolla, 1979-present. Consltg work for numerous industrial organizations and law firms, Mbr, Amer Soc of Engrg Education, Amer Soc of Mech Engrs, Natl Soc of PE, MO Acad of Sci, Phi Kappa Phi, Tau Beta Pi, Pi Tau Sigma, and Sigma Xi. *Society Aff:* ASME, ASEE, NSPE, ΣΞ

Davis, Robert W
Business: P.O. Box 683, Houston, TX 77001
Position: VP. *Employer:* Columbia Gulf Transmission Co. *Education:* BME/Fluid Flow - Hydraulics/OH State. *Born:* 5/9/22. Barre, VT. Served in AF 1942-1946 with Air Inspectors office specializing in Auditing Aircraft maintenance (B-17). Employed by Columbia Gas (1950) as Cadet engr. Specialized in high pressure gas measurement. Became US expert. Served on Intl Units sub-committee, Chrmn AGA Gas Measurement Committee; 1967 became Chief Engr in charge of large Construction prog; 1976 became VP of Engg, Planning and Purchasing. Responsible for planning, design and construction of approximately $50 million Annual budget. Also responsible for ext natural gas exchange and transportation agreements as well as operating and maintenance agreements of joint facilities. *Society Aff:* ASME, AGA, SGA, API.

Davis, Ruth M
Business: 1625 I St, NW, Suite 1017, Washington, DC 20006
Position: President *Employer:* Pymatuning Group, Inc *Education:* PhD/Math Univ of MD; MA/Math/Univ of MD; BS/Math/American Univ *Born:* 10/-/28 Dr Davis is Pres and founder of the Pymatuning Group Inc. Dr Davis has most recently been Asst Secy of Energy for Resource Applications 1979-81, and Deputy Under Secy of Defense for Res and Advanced Tech 1977-79. Prior to that she served as Dir of the Inst for Computer Scis and Tech at the Natl Bureau of Standards; as the first Dir of the Natl Center for Biomedical Communications in the Dept of Health, Educ and

Davis, Ruth M (Continued)
Welfare; and as Staff Asst for Intelligence and Reconnaissance in the Office of the Secy of Defense. Her first two jobs were working for Admiral Rickover in developing the first computer programs for nuclear reactor design, and establishing the Navy's first Command and Control Tech Organization. Dr Davis has received numerous awards. *Society Aff:* NAE, NAPA, AIAA, SID, AAAS

Davis, Stephen R
Business: 1700 W Third Ave, Flint, MI 48502
Position: Provost and Dean of Faculty *Employer:* GMI Engrg & Mgmt Inst *Education:* PhD/Mech Engg/Univ of IL; MSME/Mech Engg/Univ of DE; BSME/Mech Engg/Drexel Univ. *Born:* 3/19/23. Prior to joining GMI Engrg & Mgmt Inst in Jan 1984, Dr Davis was Dean of Engrg at Lawrence Inst of Tech, Assoc Dean of Grad Studies and Res, Coll of Engrg, Wayne St Univ. He served as Dir of Tech Planning, Cummins Engine Co and held engrg positions at Ford Motor Co and Westinghouse. Has been on the faculties of the Univ of DE and the Univ of IL in the Mech Engrg Dept. Mbr of several sci and prof soc. Actively engaged in conslt work in combustion and energy mgmt with Dept of Defense, Cameron Iron, DuPont and other major co. "Gold Award" 1979 (Outstanding Sci and Engr in MI). *Society Aff:* SAE, ASEE, SME, ASME, ASM.

Davis, Thomas W
Business: P.O. Box 644, Milwaukee, WI 53201-0644
Position: Dean of Academics and Research *Employer:* Milwaukee Sch of Engg. *Education:* BS/EE/Milwaukee Sch of Engg; MS/EE/Univ of WI. *Born:* 3/14/46. Began in 1966 as a Student Instructor in the Physics Dept at MSOE and was promoted to a full time faculty mbr in 1968. Since that time, have been to Full Prof and Hd of the Computer Engg Tech Prog and Chairman of EECS until assuming responsibilities of the Dean of Academics and Research in 1984. Have done consulting work for 150 local cos and heads the Applied Technology Center at MSOE. Have authored several textbooks including "Introduction to Interactive Programming–, coauthor of "Computer Aided Analysis of Electrical Networks" and Experimentation in Microprocesor Applications (Reston). Have authored over 30 papers in the area of microprocessors, robotics and their importance in education. Is a senior member of IEEE and SME/RI. *Society Aff:* IEEE, ASEE, SME/RI, ACM, NSPE

Davis, W Kenneth
Home: 209 Fairhills Dr, San Rafael, CA 94901
Position: Consultant *Employer:* self *Education:* MS/Chem Engrg/MIT; BS/Chem Engrg/MIT *Born:* 07/26/18 Am an independent conslt on mgmt and engrg. Served as Deputy Secy of Energy from 1981 to 1983. Prior to that served as a corp VP of Bechtel Power Corp and its predecessor companies for nealy 23 years. Before that was Dir of Reactor Dev of the Atomic Energy Comm. At the present time am immediate past Chrmn of the US Ctte for the World Energy Conference and Chairman, Programme Cttee of the intl organization. Served as VP of the Natl Acad of Engrg, Pres/Chrmn of The Atomic Indust Forum, and Pres of the Amer Inst of Chem Engrs. Am a Dir of the Mgmt Analysis Co and the Atlantic Coun and Adj Prof of Engrg at UCLA. *Society Aff:* AAAS, AIChE, ANS, ASME, ACS

Davis, Wilbur M
Business: Box 421, Ottumwa, IA 52501
Position: Division Engr *Employer:* Self *Education:* BS/Agri/Kansas St Univ *Born:* 6/26/22 Native of Belleville, KS, Graduate of Kansas State University with BS in Agr. Engr. Served in European Theatre in WW II, decorations include Silver Star, Bronze Star w/Cluster, Purple Heart and Combat Infantrymen's Badge. Professional engineering experience with Deere & Co. include agricultural tractors particularly with hydraulic, implement hitch and transmission. Also Division Engineer on Forage Equipment directing exploratory designs on future concepts, characteristics of forage materials and introduction of new models of equipment into production. Extensive experience in industry standards served on Farm and Industrial Equipment Institute Engineering Committee and numerous Power and Machinery Technical Division Committees of the American Society of Agricultural Engineers. Served three years as Technical Vice President and member of the Executive Committee and is currently President Elect of the Society (ASAE). Currently self employed as consulting engineer in New Product Development and as an Expert Witness in Product Litigation. Also a member of Society of Automotive Engineers (SAE) and the American Alpine Club. Hobbies, mountain climbing, skiing, tennis. *Society Aff:* ASAE, SAE, Steel Ring, American Alpine Club.

Davis, Wm C, Jr
Home: 13 Cone St, Wellsboro, PA 16901
Position: Pres/Owner *Employer:* Tioga Engrg Co, Inc *Education:* B.S./Physics Math/St. Bonaventure Univ *Born:* 5/18/21 Native of Millport, PA. Ordnance officer, U.S. Army Ballistic Research Labs, 1944-1946. Faculty, St. Bonaventure Univ, 1947-1951. Chief, Small-Arms Branch, Chief Engr, Infantry & Aircraft Weapons Division, D&PS, Army Aberdeen Proving Ground, 1951-1959. Superintendent, NATO North American Regional Test Center for Ammunition, 1959-1964. Engrg Mgr for Military Products, Colt's Firearms Division, 1964-1965. Chief, Technical Division, U.S. Army Project Mgr for Rifles, 1965-1969. Chief, Small-Caliber Ammunition Engrg Lab, U.S. Army Frankford Arsenal, 1969-1972. PE in private practice (PA) 1972-present. Consultant, Battelle Labs, U.S. Army Armamant R&D Command, National Rifle Association. Author of numerous articles and engrg reports, "Ammunition" section of the Encyclopaedia Britannica (1974), book "Handloading" published National Rifle Association 1981. *Society Aff:* NSPE, ASME, SAME

Davison, Beaumont
Home: 201 W Park St, Angola, IN 46703
Position: Pres *Employer:* Tri State Univ *Education:* PhD/EE/Syracuse Univ; MEE/EE/Syracuse Univ; BE/EE/Vanderbilt Univ. *Born:* May 1929, Atlanta, Georgia. Instructor & Res Associate, Syracuse University 1951-56. Asst Prof Case Institute of Technology 1956-59. Executive VP Industrial Electronic Rubber Co 1959-67. Chmn Dept of Electrical Engrg 1967-69; Dean of Engrg & Technology 1969-71; VP for Regional Higher Education 1971-74, Ohio University; Dean of Engrg California State Polytechnic Univ Pomona 1974-83 Pres Tri State Univ since 1983. Sr Mbr SME; ASEE, Sigma Xi, Tau Beta Pi, Eta Kappa Nu, Phi Kappa Phi. *Society Aff:* ASEE, SME.

Davison, Edward J
Business: Dept of Elec Eng, Univ of Toronto, Toronto, Ont, Canada M5S1A4
Position: Prof. *Employer:* Univ of Toronto. *Education:* ScD/Control Sys/Cambridge Univ; PhD/Control Sys/Cambridge Univ; MA/Applied Math/Univ of Toronto; BASc/Engg Physics/Univ of Toronto; ARCT/Piano/Royal Conservatory of Music. *Born:* 9/12/38. Presently prof of Elec Engg at Univ of Toronto. He was an associate editor of the *IEEE Trans on Auto Control*, a Mbr of the IDC during 1974-76, a Consulting Editor of IEEE Trans on Auto Control in 1985, and is presently an assoc editor of the IFAC Journal *Automatica & Large Scale Systems: Theory & Applications* and mbr of the Editorial Bd for *Optimal Control Appl & Methods*. He was an elected Mbr of the Administrative Committee of the IEEE Control Sys Soc for the Period 1977-83, was VP (Technical affairs) of the IEEE control sys soc 1980-81 and was Pres-Elect and Pres of the IEEE Control Sys Soc in 1982 and 1983 respectively. He was Vice Chairman, IFAC Theory Committee 1978-87 and is Chairman; IFAC Theory Committee 1987-90. He was an Athlone Fellow 1961-63, received The Natl Research Council of Canada's E.W.R. Steacie Meml Fellowship during 1974-77 & received a Killam Res Fellowship 1979-80, 1981-83 was awarded the IEEE Centennial Medal in 1984, was elected Distinguished Member of the IEEE Control Systems Society in 1984, and was elected Honorary Professor of Beijing Institute of Aeronautics and Astronautics in 1986. His main res interests are in linear sys theory, large scale sys theory and computational methods. He is a fellow of IEEE since 1978, a fellow of the Royal Society of Canada since 1977, and a designated

Davison, Edward J (Continued)
consultant of the Association of Professional Engineers in the Province of Ontario since 1979. He is a Director of Electrical Engineering Con sociates LTD. (Toronto) since 1977. *Society Aff:* IEEE.

Davison, Joseph E
Business: 300 College Park, Dayton, OH 45469
Position: Res Met. *Employer:* Univ of Dayton. *Education:* PhD/Met/IA State Univ; MS/Phys/IA State Univ; BS/Phys/St Louis Univ. *Born:* 11/23/32. Conducted research on metal matrix composite materials, phase equilibria in multi component alloy systems, the application of refractory metals to gas turbine engines, phase change thermal energy storage materials. Presently conducting research on graded index thin film optical filters. *Society Aff:* ASM.

Davison, Joseph W
Home: 8505 East Temple Dr., #461, Denver, CO 80237
Position: Senior Vice President, Planning & Develop. *Employer:* Phillips Petroleum Company. *Education:* BS/Engg/KS Univ. *Born:* Nov 1921. Joined Phillips as Process Design Engr Refining Dept 1943. Served USNR 1944-46. Re-joined Phillips in R&D 1946. Process Evaluation Mgr 1956-64. Dir R&D Div 1964-71. Vice Chmn Company Operating Ctte 1971-73; Chmn 1973-75. Mgr R&D 1975-76; VP 1976-80. Senior VP-Plns. & Develop. 1980- Coordinating Res Council Bd; Natl Assn of Conservation District Business Advisory Ctte; Industrial Res Inst. Fellow AIChE Civic Bd Mbrships: Presbyterian Church (P Pres); County Chap of KU Alumni Assn; KU Sch of Engg Advisory Bd & Exec Ctte; Cherokee Area Council of Boy Scouts. *Society Aff:* SAE, AIChE, ASTM.

Davison, Wellen G
Business: 1215 Wilbraham Rd, Springfield, MA 01119
Position: Prof Mech Engg. *Employer:* Western New England College. *Education:* BS/Mech Engg/WPI; MS/Mech Engg/RPI; PhD/Mech Engg/Univ of CT. *Born:* Apr 1925. Served in Army Air Corps 1943-45 as Navigator. Test & design engr in refrigeration with Westinghouse 1949-58. With Western New England College as part time inst since 1951. Full time teaching since 1958. Dept Chrmn 1965-78. Assumed current position in 1965. Mbr ASME, served in Sectional & ASME Region I Vice President Regional Offices. Assoc Mbr Sigma Xi. *Society Aff:* ASEE, ASME, ASHRAE, ΠΤΣ, ΣΞ, AEE.

Davisson, Lee D
Business: Elec Engg Dept, College Park, MD 20742
Position: Prof. *Employer:* Univ of MD. *Education:* PhD/Engg/UCLA; MSE/Engg/UCLA; BSE/Elec Engg/Princeton. *Born:* 6/16/36. First Lt, US Army, 1958-1960. Res Engr, Aeronautics, 1960-62. Mbr of technical staff, Hughes Aircraft Co, 1962-64. Assoc Prof, Princeton Univ 1964- 70. Prof, Univ of Southern CA, 1969-1976. Prof, Univ of MD, 1976-present. Chairman, Electrical Engineering, Univ. of Md 1980-1985. Eta Kappa Nu Outstanding Young Elec Engr in US (honorable mention), 1968. Treas IEEE Info Theory Group 1973-78. Second VP IEEE Info Theory Group 1978. First VP IEEE Info Theory Group 1979. Pres. IEEE Info. Group 1980Co-chrmn, 1979 IEEE Intl Symposium on Info Theory. Fellow, IEEE, 1976. Mbr of Editorial Bd SIAM 1972- 1980. 1976 Prize Paper Award of IEEE Group on Info Theory (with R M Gray). *Society Aff:* IEEE.

Davisson, Melvin T
Home: 14 Lake Park Rd, Champaign, IL 61821
Position: Consulting Engineer *Employer:* Self employed *Education:* PhD/Civil Engg/Univ of IL; MS/Civil Engg/Univ of IL; BCE/Civil Engg/Univ of Akron. *Born:* Dec 1931. Exper in construction & structural engrg prior to engagement in res & teaching of foundation engrg. Cons to industry in foundation engrg, deep foundations, excavation & bracing, waterfront, marine projs, earth & rock structures. Over 40 publications. Raymond Award 1958, Collingwood Prize ASCE 1964. Mbr: ASCE, ACI, AREA, ASTM, ISSMFE, NSPE, DFI. Reg Prof Engr Ohio & Illinois; Structural Engr Illinois. Chmn or Mbr of 6 tech cttes. *Society Aff:* ASCE, ACI, ASTM, AREA, NSPE, ISSMFE, DFI.

Davy, M C W
Business: Garratt Blvd., Downsview, Ontario, Canada M3K 1Y5
Position: VPres., Engrg *Employer:* deHavilland Aircraft of Canada Ltd. *Education:* -/Aeronautical Eng/London Univ *Born:* 4/10/24 R.A.F. Pilot during WW II. Still an active Pilot. Joined A. V. Roe 1949 as Junior Engr on the 102 Jetliner & CF 100 Fighter Programs. Joined de Havilland Aircraft of Canada 1951-56 as Designer Otter, Caribou Buffalo & Twin Otter Airplanes. Joined Sierracin Corp, Sylmar, Calif. 1965, position: design & development. Joined Canadair Ltd 1967-73 as Mgr of Design. Rejoined de Havilland Aricraft of Canada 1973 - Dash 7 Program. 1979 VP Engrg. *Society Aff:* CASI

Davy, Philip Sheridan
Business: 115 South 6th St, La Crosse, WI 54601
Position: President. *Employer:* Davy Engineering Co. *Education:* MS/Civ Engrg/Univ of WI Madison; BS/Civ Engrg/Univ of WI Madison *Born:* July 1915. BSCE, Univ WI, 1938 with high honors; Awarded Res Fellowship 1938; BS in Civ & Sanitary Engrg, 1938; Tau Beta Pi, Chi Epsilon, Phi Kappa Pi. US Army 1941-46, 1st Lt to Lt Col Corps of Engrs. Army Commendation Award w/Cluster 1946. VP & Pres Davy Engineering Co. Consulting Engrs La Crosse, WI; primary fields: water works, sewerage, treatment plants, & land development. Trustee & Chmn Wisconsin AWWA 1957-61. Dir Natl Soc of Prof Engrs 1966-69. Director WSPE 1963-1965, 1973-1976. VP, Pres Wisconsin Soc of Prof Engrs 1974-75. Honors: George Warren Fuller Award 1985. Member, Governor's cttee on water resources 1965-6. Diplomate AAEE. Engr of Year Wisconsin Soc of Prof Engrs 1970. Distinguished Service Award Wis. Sect ASCE, 1979. Fellow ASCE, AWWA. Diplomate Environmental Engrs Intersociety Bd. Mbr AWWA, WPCF, APWA, WSPE, NSPE, AAAS, Res Off Assn, AAAS, NSPE; Reg PE WI, MN, IA, IN, MI. *Society Aff:* NSPE, AWWA, WPCF, WSPE, APWA, ASCE, AAAS, ROA, TROA, AAEE

Dawkins, Mather E
Business: 520 North Semoran Blvd, Orlando, FL 32857
Position: Pres/Chief Engr *Employer:* Dawkins & Assocs, Inc Consulting Engrs *Education:* B/CE/Univ of FL; MSE/Sanitary Engrg/Univ of FL *Born:* 12/11/21 Sanitary Engr, Hq SAC, USAF 1951-52; Sanitary Engr, FL State Bd of Health 1953-55, Sanitary Engr, Reynolds, Smith & Hills (AE) 1955-62; Partner, Flood & Dawkins, Consulting Engrs, 1972-1964. Pres and Chief Engr, Dawkins & Assocs, Inc 1964 to present. Fellow Member (1974) Chapter Pres 1973-74, State Dir 1975-79, FL Engrg Society (NSPE); FES Engr of Year 1979; Award for Distinguished Service to FES 1980; WPCF honors: FL Assn Pres 1960-61, Federation Dir 1965-69, Bedell Award, 1960; Environmental Protection Award, Univ of Central FL 1976; Honorary Life Member FL Water and Pollution Control Operators Association. *Society Aff:* ASCE, NSPE, WPCF, AWWA, APWA

Dawson, David C
Business: Bet 8109, Philadelphia, PA 19101
Position: Pres & CEO. *Employer:* ESB Ray-O-Van. *Education:* Bach/Met Eng/Cornell Univ; Masters/Bus Admin/Duguesne Univ. *Born:* Feb 1933. Bachelor Metallurgical Engrg from Cornell U; MBA from Duquesne; New Jersey native. Naval Officer 1955-58. With Inco Limited since 1958. Assumed respon as VP in 1975. This covers Diversification; Corporate Dev. AFS, ASM (Pittsburgh Chapter Offices), AIME. Apptd Pres in April 1978 & CEO in July. Beta Gamma Sigma (Bus Honorary). *Society Aff:* AFS, ASM, AIME.

Dawson, J Gordon
Home: Mildmay House, Apethorpe, Peterborough, England PE8 5DP
Position: Consultant *Employer:* Self *Education:* BSc/ME/Aberdeen Univ; BSc/EE/Aberdeen Univ *Born:* 2/3/16 1938-46 Rolls Royce Ltd, finally Development Test Engr. 1946-55 Shell Research Ltd, finally Chief Engr. 1955-66 Perkins Engines Ltd,

Dawson, J Gordon (Continued)
Technical Dir. 1966-69 Dowty Group Ltd, Dir, Industrial Div. 1969-81 Zenith Carburetter Ltd, Managing Dir and Chrmn. 1979-80 Pres, Institution of Mechanical Engrs. *Society Aff:* FSAE, FEng, Hon FIMechE, MRAeS.

Dawson, Thomas F
Business: 555 Technology Square, Cambridge, MA 02139
Position: Staff Engineer *Employer:* C.S.Draper Lab., Inc. *Education:* BSEE/Elec. Engr./Virginia Polytechnic Institute; Engr. Flight Test/Electronics/U.S. Naval Aviation Flight Test School *Born:* 12/11/30 Served in U.S. Navy during the Korean war, 1949-1953. Hazeltine Corp - Aircraft Flight Test Engr for high performance military aircraft. M.I.T. Lab for Nuclear Science and M.I.T. Center for Space Research. Served as cognizant Project Engr/Dept. Mgr for a series of scientific plasma and x-ray astronomy spacecraft experiments flown on the Mariner Mars, Mariner Venus, Pioneer 6 and 7, O.S.O. 7, HEAO, Apollo, S.A.S. and Voyager Spacecraft. Involved in transfer of interial guidance technology for Trident and MX Peacekeeper Missile Progs at C.S. Draper Lab., Inc. Author of several papers in scientific & technical journals. Received U.S. & foreign patents in the field of automation/mfg. Past Pres AISES; received Sequoyah Fellows Award, and is Chairman Emeritus. Bd of Dir Winds of Change magazine. Cited for contributions to space science in Journal of Geophysical Research; Science; & Nature publications. Biography included in text-book Concepts and Challenges in Physical Science. Bd of Governors AAES; served as a US Delegate to UPADI *Society Aff:* IEEE, AISES, AAES

Day, C LeRoy
Home: 504 Crestland Ave, Columbia, MO 65201
Position: Chmn Dept Agri Engrg. *Employer:* University of Missouri-Columbia. *Education:* PhD/Engg/IA State Univ; MS/Agri Engg/Univ of MO, Columbia; BS/Agri Engg/Univ of MO, Columbia. *Born:* Oct 1922. Mbr of faculty at University of MO at Columbia since 1945. Prof of Agricultural Engrg since 1962. Dept Chmn since 1969. Visiting Prof at University of Thessaloniki (Greece) 1972. Pres of PEN-REICO INC 1968-79. Fellow; American Society of Agricultural Engrs. Elder in the Church of Christ. Mbr ASEE & MSPE/NSPE. *Society Aff:* ASAE, ASEE, MSPE, NSPE.

Day, Emmett E
Business: Mech Engrg Dept FU-10, Seattle, WA 98195
Position: Prof of Mech Engrg. *Employer:* Univ of Wa *Education:* MS/Mech Engg/MIT; BS/Mech Engg/MIT; BA/Chemistry/East TX State Univ. *Born:* Jul 1915. Native of Texas. War Production Bd prior to WWII. Joined faculty of ME Dept of University of Washington 1947, full prof 1954. Received Pi Tau Sigma Gold Medal from ASME 1954. Served as VP for Region IX & mbr of Council ASME 1962-66. Cons for many firms & labs, author of many technical papers in the field of stress analysis. VP of Soc for Experimental Stress Analysis (SESA) 1972-74, Natl Pres SESA 1974-75. Fellow ASME, Fellow SESA. Registered Prof Engr in Washington, mbr of Sigma Xi. Enjoys snow skiing, cruising Puget Sound & travel. *Society Aff:* ASME, SEM.

Day, Ivor J
Home: PO Box 279, Halfway House, Transvaal, R of South Africa 1685
Position: Consultant (Self-employed) *Education:* PhD/-/Cambridge Univ(England); MSc/Mech Engg/Natal Univ (Durban)RSA; BScMech Engg/Natal Univ (Durban) RSA. *Born:* 11/13/48. Obtained Bachelor and Masters degrees in Mech Engg at the Univ of Natal, Durban, South Africa. Lectured in Fluid Mechanics for a semester. Awarded Charles Rylott Foster Scholarship and proceeded to England to complete a PhD at Cambridge (1976). A new phase lock sampling technique for studying rotating stall in axial compressors was pioneered. The results led to the pub of three elated papers and to the proposal of a new model for compressors operating in stall. Along with E M Greitzer and N A Cumpsty was awarded the 1977 ASME Gas Turbine Award. In 1978 was awarded the George Stephenson Research prize by the British Inst of Mech Engrs. Returned to South Africa and is now working for Brown Boveri, SA. *Society Aff:* SACPI, ASME.

Day, Robert W
Business: Mech Engr Dept, Amherst, MA 01003
Position: Prof. *Employer:* Univ of MA. *Education:* MME/ME/Rensselaer Poly Inst; BSME/ME/Univ of MA. *Born:* 2/7/24. in Worcester, MA. Served with Army Corps of Engrs during WWII. Did grad work and taught Mech Engg courses at Rensselaer Poly Inst from 1948 to 1954. Joined Mech Engg Dept at Univ of MA in 1954. From 1952 to 1968 consulted with GE Co and other cos in the field of heat transfer and thermodynamics. Was active in the dev of the fuel cell batteries used in the Gemini Space Prog. Also contributed to the design and dev of other aircraft accessory power units. Currently a Prof and Assoc Hd of Mech Engg at the Univ of MA and a recipient of the Western Elec Award for excellence in instruction of engg students. *Society Aff:* ASME, ASEE, ΣΞ, ΤΒΠ.

Day, Roger W
Home: 12318 Cobblestone, Houston, TX 77024
Position: VP. *Employer:* Geosource. *Education:* BChE/CE/Cornell Univ. *Born:* 3/18/26. Equip Mfg Exec; Bird Machine Co Sales Engr '55-'60, Regional Sales Mgr '61- '68, Planning Mgr '68-'69; Pioneer Centrifuging Co Gen Partner '69-'73, Managing Partner '73-'75; Picenco Intl Inc Pres/Dir '75-'78; Geosource VP Drilling Equip Div '78-; Author numerous technical papers. Co-author Hydrocyclone Separation Chapter in "Handbook of Separation Techniques for Chemical Engineers." Inventor/Patent In-line Mixer Device; Served in USNR '44-'46. BChE Cornell Univ '49. Sigma Chi. Republican. Presbyterian. Houston Club. *Society Aff:* AIChE, ASME, SPE, Filtration Soc, IADC.

Day, William C
Home: 559 Brevoort Rd, Columbus, OH 43214
Position: Ret. (Consultant) *Education:* BME/Mech Engg/OH State Univ. *Born:* Sep 1909. Native Columbus Ohio. Army Air Corps 1941-46. Columbia Gas System 40 years in design, construction, operation & maintenance of natural gas storage, transmission & distribution facilities. As Mgr of Gas Engrg, supervised Syst operational procedures incorporating new standards, materials & advanced methods; initiated compressor station operational efficiencies. Natl Defense Executive Reserve as Dir - Natural Gas Pipeline Operations, Emergency Petroleum & Gas Administration. Served on numerous industry & code cttes. Chmn DGP Div ASME 1968-69; AGA Award of Merit 1967; SGA Disting Service Award 1972; ASME Fellow 1973. Ohio Registration. Lt Col USAF Retired. *Society Aff:* ASME.

Daykin, Robert P
Home: 3174 Menomonee R Pkwy, Wauwatosa, WI 53222
Position: VP - Quality & Tech Retired *Employer:* Ladish Co, Cudahy Forgings Div. *Education:* AB/Chem/Oberlin College; MSc/Met Engg/Univ of WI. *Born:* 10/11/20. Contributed to met dev in deformation processing of heat resistant alloys and reactive & refractory metals. Have contributed to the alloy dev & physical met of high strength steels. Reg PE. ASM Fellow. *Society Aff:* ASM.

Dealy, John M
Home: 305 Grosvenor Ave, Montreal Quebec, Canada H3Z 2M1
Position: Professor of Chemical Engineering. *Employer:* McGill University. *Education:* PhD/Chem Engg/Univ of MI; MSE/Chem Engg/Univ of MI; BSChE/Chem Engg/Univ of KS. *Born:* 1937 Waterloo IA. Prof of Chem Engrg since 1964 at McGill University. Dir of res program on polymer rheology & plastics processing, consultant to plastics industry. Res grants received from Sci & Eng. Res Counc Can & ACS; res contracts rec from Nat Res Counc Can & Alberta Oil Sands Technology & Res Auth. Author of technical papers, reviews and a book on polymer melt rheology. Holder of several patents on plastics testing devices. Mbr of AIChE, ACS, Soc of Rheology (Pres. 1987-89), Soc Plastics Engrs, Can Soc Chem Engrg; Prof Engr in Quebec. Pres of STOP, Montreal citizen's env group 1975-78. Pres McGill Assoc U of Profs 1968-69, Pres McGill Chap Sigma Xi 1973-74 & Pres McGill Fac Club

Dealy, John M (Continued)
1973-74; Natl VP Theta Tau Engrg Frat 1956-58. Mbr Tau Beta Pi, Phi Lambda Upsilon, Sigma Xi, Omicron Delta Kappa. *Society Aff:* AIChE, ACS, SPE, CIC, SRheol, CRG, BSR.

Dean, Albert G
Home: 219 N Wynnewood Ave, Narberth, PA 19072
Position: Consulting Engr (Self-employed) *Education:* BS/Aeronautical/MIT. *Born:* May 1909. Worked for The Budd Company from 1931 until retirement, September 1969. During first yr & WWII worked with aircraft in structural design, performance & flight test. Otherwise involved with stainless steel railway passenger cars. Respon for structure & performance, then running gear, brakes, equipment, interiors, laboratory & road testing, crash worthiness; then Ch Engr design. 120 patents, mostly in railway field. Fellow ASME. Continuing similar work as cons engr including Budd International Licensee consultation. 27 technical papers & talks, domestic & foreign. Hobbies; hiking, mountain climbing, skiing, canoeing, gunsmithing. *Society Aff:* ASME.

Dean, Charles E
Home: 15301 Pine Orchard Dr, Apt 3-A, Silver Spring, MD 20906
Position: Retired. *Education:* PhD/Physics/Johns Hopkins; MA/Physics/Columbia (NYC); AB/Engg Scis/Harvard. *Born:* May 1898. With Hazeltine Corp 1929-63, in various capacities especially Dir of Publications. Licensed PE in NY State 1934. Officer or editor, Radio Club of America 1939-47. Awarded US Navy Certificate of Commendation for outstanding civilian WWII work at Hazeltine, 1947. One of 6 panel chmn of Television Allocations Study Organization in industry-wide examination of UHF TV channels 1957-59. VP Communications, American Inst of Elec Engrs 1961-62. Author or editor of books: Principles of Television 1944; Color Television Receiver Practices 1955; Principles of Color Television 1956. Technical Editor, Scripta Technica, Inc 1963-77 . *Society Aff:* IEEE.

Dean, Donald L
Business: Pyramid Electronics, Sarasota, FL 33581
Position: Owner & Consultant Pyramid Electronics *Employer:* Self employed *Education:* PhD/CE/Univ of MI; MS/CE/Univ of Mo, Rolla; BS/CE/Univ of MO, Rolla. *Born:* 11/25/26. PhD at University of Michigan; BS & MS at University at Rolla. Native of Litchfield, Illinois. Taught at U of Missouri from 1949-55; taught at U of Kansas 1955-60; Chmn Cvl Engrg U of Del 1960-65; Hd Cvl Engrg NC Sta U 1965-78. Dean of Engg, IL Inst of Techn 1978-80. Owner Pyramid Electronics, Personal Computers & Consulting, 1982-. Journal Editor ASCE Engrg Mech Div 1961-64, as Div Chmn 1965-66, as Mbr of Structural Div Comm on Method Analysis 1966-69. Received ASCE Walter L. Huber Res Prize 1967. Registered prof engr in Missouri, Kansas & Virginia. Special interest in design & mathematical analysis of large & unusual structural systems. Enjoy skin diving & piloting private plane *Society Aff:* ASCE, IASS.

Dean, Eugene R
Business: 105 Drood Lane, Pittsburgh, PA 15237
Position: Consultant *Born:* Aug 1916. Specialty is blast furnance iron & ferroalloy res & production & raw materials evaluation. Introduced first successful injection of fuel through tuyeres of commerical blast furnace to the ironmaking industry 1959. Directed dev of equipment & method to wet clean ferromanganese blast furnace gas with total water recycle in 1956. Chmn, Allegheny County, Pennsylvania Smoke Control Research Ctte for The Steel Industry 1964-65. Chmn American Iron & Steel Institute Technical Ctte on Blast Furnaces 1965-67. Mbr Assn Iron & Steel Engrs. Dir AIME 1971-74. WWII Veteran, Cpt, Field Artillery, Battery Commander. *Society Aff:* AIME-ISS.

Dean, Robert C, Jr
Business: SYNOSIS Inc, P.O. Box 318, Norwich, VT 05055
Position: President *Employer:* SYNOSIS Inc. *Education:* SCD/ME/MIT; MS/ME/MIT; BS/ME/MIT *Born:* 4/13/28 Married, E Nancy Hayes, Sept 22, 1951; Proj engr Ultrasonic Corp, 1949-51; hd advance engrg dept Ingersoll-Rand Co, 1956-60, dir res Thermal Dynamics Corp, 1960-61; dir Ecol Sci Corp, 1968-70; founder and pres Creare Inc 1961-75; pres Ecol Res Corp 1968-70; cofounder, chrmn & principal engr, Creare Innovations 1976-79; founder, pres, Verax Corp, 1979-83, Chrmn and Dir of Sci & Tech 1983-1987, Founder & President, SYNOSIS Inc., 1987- . Asst prof mech engrg MA Inst Tech, 1951-56; prof engrg Thayer Sch Enrg, Dartmouth, 1960- (now adjunct). Master Designer award Product Engrg mag, 1967. Dir Turbomachinery Inst 1968-80. Thurston Lecturer 1977, ASME Fellow 1978, Fluids Engrg Award 1979, Natl Acad of Engrg 1977. *Society Aff:* ASME, NAE, ACS, AIChE, ASM

Dean, Russell K
Business: Mechanical & Aerospace Engrg Dept, Morgantown, WV 26506
Position: Assoc Prof & Graduate Program Director. *Employer:* WV Univ *Education:* PhD/ME/WV Univ; MS/ME/WV Univ; BS/EE/WV Univ *Born:* 8/26/52 Has been teaching undergraduate and graduate courses in engrg mechanics for five years. Conducting research in engrg education for several years and currently active in "problem-solving" area of educational research. Interest in robotics and energy management. Served as chairman of Educational Research Methods Division and the North Central Section of ASEE Received award as "Outstanding Teacher" at WV Univ during 1980-81 academic year & received the 1986 WVU College of Engineering Outstanding Teaching Award. *Society Aff:* ASME, ASEE, IEEE, TBП, EKN

Dean, Stephen O
Business: 2 Professional Dr, Suite 248, Gaithersburg, MD 20879
Position: Pres *Employer:* Fusion Power Assocs *Education:* PhD/Phys/Univ of MD; SM/Nuclear Engrg/MIT; BS/Phys/Boston Coll *Born:* 05/12/36 Initiated Laser Fusion program at US Naval Res Lab (1968); Managed (as Div Dir) AEC/ERDA/DOE Magnetic Fusion Confinement Sys Div 1972-1979. Founded Fusion Power Assocs 1979 and serves as CEO (1979-present). Edited book "Prospects for Fusion Power" (1981). Assoc. Ed. Journal of Fusion Energy (1985-present). *Society Aff:* ANS, APS, AVS, ASAE, FPA

Dean, Walter A
Home: 3623 Muirfield Drive, Sarasota, FL 33583
Position: Retired. *Employer:* Aluminum Co of America *Education:* PhD/Chem Engg/Cooper Union Inst Tech; MS/ Metallurgy/RPI; BS/Metallurgy/ RPI. *Born:* Feb 1905, Bridgeport, Conn. Offices: Chmn Advisory Ctte to Metallurgy Div Natl Bureau of Standards 1955-58; Chmn Forgings & Castings Panel Materials Advisory Bd 1958-61; Chmn Non Ferrous Metallurgy Section, American Ordnance Assoc 1960-63; Mbr Governors Materials Advisory Panel of Pennsylvania 1965-68. Societies: Fellow of ASM, mbr of AIME, mbr various cttes ASM & AIME, Chmn IMD of AIME 1952, VP & Dir AIME 1955-58, Sigma Xi & Scientific Res Soc of America. Publications: Papers on machining aluminum alloys & powder metallurgy. Contributor to books & Soc Proceedings on metallurgical subjects. 85 patents on aluminum alloys & heat treating practices. Honors: American Ordnance Assoc Bronze Medal & Citation 1964, ASM Zay Jeffries Lecturer 1966, Gano Dunn Medal from Cooper Union, professional achievement, 1967. *Society Aff:* ASM, AIME, ΣΞ, RESA.

Deane, Richard H
Home: 1171 Nogales St, Lafayette, CA 94549
Position: Clerk of Court. *Employer:* US Court of Appeals. *Education:* PhD/Ind Engr/Purdue; MSIE/Ind Engr/MS State Univ; BSIE/Ind Engr/MS State Univ. *Born:* 10/8/44. Native of MS. Taught at GA Tech College of Engg for 4 yrs. Earned Law Degree at night. Worked 4 yrs as Asst Dir, Admin Office of the US Courts. Was involved in comp planning. Now Clerk of Court for the US Court of Appeals for the Ninth Circuit. Reg PE and Mbr GA Bar. *Society Aff:* IIE, TBП.

DeAnnuntis, William J
Business: 1626 Mount Pleasant Rd, Villanova, PA 19085
Position: President. *Employer:* DeAnnuntis & Associates. *Education:* Dip/-/Drexel Inst. *Born:* Apr 1926. Native of Phila, Pa. Served in US Navy 1943-46 in Pacific. Mech engr for 8 yrs des heating, ventilating, air conditioning, plumbing & drainage systems with various firms. Founded DeAnnuntis & Associates, Consulting Engrs in 1956 to date, engaged in designing mech systs for all types of buildings & NASA R&D projects. Treasurer CEC of Phila 1975-76 & Secy 1977. Enjoy art, water color & sketching. Pres Phila Sketch Club 1968-70 & on the bd 1971- 75. *Society Aff:* ASHRAE, CSI, ACEC, CEC/PA, NSPE, PEPP.

Deardorff, Mark E
Business: 9696 Businesspark Ave, San Diego, CA 92131
Position: Pres *Employer:* Deardorff & Deardorff *Education:* BS/CE/Univ of So CA *Born:* 6/10/53 Married. No children. Pres, CA Society of PE San Diego Chapter; Chrmn, San Diego National Engrs Week, 1981; Member, CA Republican Central Committee, 1974- 1978; Member, Chi Epsilon; Univ of Southern CA Outstanding Senior Award, 1975; Who's Who in Finance and Industry; Who's Who amoung American Coll and Univ Students; Outstanding Young Men of America. *Society Aff:* ASCE, NSPE, ACI, TMI, ICBO

Deatherage, James H
Business: 3801 East Indian Sch Rd, Phoenix, AZ 85018
Position: Owner. *Employer:* Self-employed Hastain & Deatherage,. *Education:* BCE/Civil Engg/Univ of MN. *Born:* Jul 1927. Bachelor of Civil Engrg degree from University of Minnesota 1950; employed by The Bureau of Reclamation on the Wellton-Mohawk irrigation project near Yuma, Arizona, Lescher & Mahoney, Architects & Engrs on hospital construction in Phoenix, & Allison Steel Manufacturing Co. Formed Hastain & Deatherage, Cons Engrs in 1959; Past Pres Arizona Section ASCE, Phoenix Branch Ariz Section ASCE, & Ariz PEPP Section NSPE. Mbr ASCE, NSPE, ACI, ASTM, CSI, Arizona Structural Engrs Assn & Cons Engrs Assn of Arizona. Techical specialities include structural collapse investigations & expert witness testimony. *Society Aff:* NSPE, ASCE, ACI, ASTM, CSI.

Deatrick, Warren J
Business: 3852 L. B. McLeod Rd, Orlando, FL 32805
Position: VP & Chief Engr *Employer:* Universal Engr Testing Co *Education:* MS/Management/Rollins Coll; BS/CE/Univ of KY *Born:* 9/27/35 in Louisville, KY. Registered PE in FL & GA. Chief Engr for Universal Engrg Testing Co., a major FL Geotechnical & Consulting Engrg firm. Six offices provide subsurface soils investigations, construction testing & failure analysis. Also Pres of Orlando Concrete Contractors, Inc., a concrete subcontracting firm specializing in concrete & masonry construction for commercial & industrial structures. Projects ranged from FL to GA to TX. Former VP & General Mgr of Conco, Inc., the supplier of all concrete to Walt Disney World theme park. Currently VP of the Central FL Chapter of the American Concrete Inst. *Society Aff:* ASCE, NSPE, ACI, ASCC

Deb, Arun K
Business: Roy F. Weston Inc, Weston Way, West Chester, PA 19380
Position: VP Environmental Systems *Employer:* Roy F Weston, Inc. *Education:* PhD/Civil Engg/Univ of Calcutta; MS. Envir Engg/Civil Engrg/Univ. of Wisconsin; BS/Civil Engg/Univ of Calcutta. *Born:* 5/1/36. Native of India. Was responsible for dev and teaching of Sanitary engg courses at the Univ of Calcutta. USAID Fellow at the Univ of Wisconsin. As Sr Research Fellow at the Univ College London, dev methodology of dual water supply analysis for urban water systems. Visiting prof at the Univ of Notre Dame. Since 1974 responsible for Weston's project management and high technology projects in the area of urban water and wastewater management. Pioneered concept of dual water supply and water reuse. Mbr of various prof committees. Pub/presented more than 70 papers. Inst of Engrs PHE medal 1973. Diplomate, AAEE. Past Editor ASCE Journal Environmental Engrg. Chrmn ASCE-EED sessions program Cttee. *Society Aff:* AAEE, ASCE, AWWA, WPCF.

DeBell, Gary W
Home: 619 Teresi Lane, Los Altos, CA 94022
Position: Manager Optics Division. *Employer:* Spectra-Physics. *Born:* Aug 1943. AB Physics University of California, Berkeley 1966. MS 1970 & PhD 1972 from Institute of Optics, University of Rochester. Technical specialization in optical thin films. Sr optical engr for Spectra-Physics since 1972. Assumed respon for business & technical mgmt in 1974. Mbr of OSA & SPIE. Mbr of SPIE technical council.

DeBra, Daniel B
Business: Dept Aero & Astro, Stanford, CA 94305
Position: Prof of Aeronautics & Astronautics. *Employer:* Stanford University. *Education:* PhD/EM/Stanford; SM/ME/MIT; BS/ME/Yale *Born:* 6/1/30. Married Esther Crosby DeBra, 6 children. Fellow of AIAA & VP Education 1972-75; Mbr: NAE; IFAC Chrm Space Comm; Pilot. Registered Elec Engr in CA. Education: BE Mech Engrg Yale 1952; MS in Mech Engrg MIT 1953; PhD in Engrg Mechanics, Stanford Univ 1962. 75 pubs. *Society Aff:* AIAA, AAS, SME, ASME, IEEE, SAE, AUVS, AGU, ION.

DeCamp, Robert A
Business: 1500 Meadow Lake Pkwy, Kansas City, MO 64114
Position: Partner. *Employer:* Black & Veatch, Consulting Engineers. *Education:* BSME/Power/Purdue Univ; MSME/Power/Purdue Univ. *Born:* Feb 1924. Native Kansas City. Joined Black & Veatch 1950 & served as Project Engr & Project Mgr in Power Div. Exper includes major thermal electric generating programs for private utilities, municipal electric systems & a large overseas client, The Electricity Generating Authority of Thailand. Has been in responsible charge of over 6 million kilowatts of generating capacity covering coal, gas & oil units. Tau Beta Pi, Natl Soc Prof Engrs, Registered Prof Engr Past Chmn Kansas City Section ASME. *Society Aff:* ASME, NSPE.

DeCario, Victor N
Home: 9520 SW 25th Dr, Miami, FL 33165
Position: Pres *Employer:* Victor N. DeCario & Associates, Inc. *Education:* BS/CE/Northeastern Univ *Born:* 1/2/42 Civil Engr with Metropolitan Dade County FL, 1965-69. Partner in Irani Assocs, Inc, Consulting Engrs, 1969-73. Principal in Post, Buckley, Schuh & Jernigan, Inc (PBSJ), 1973-1987. Pres Victor N. DeCairo & Associates, Inc. since 1987. Nationally recognized expert in the application of computers and data processing techniques to engrg applications. Author, lecturer, and consultant in the effective use of computers in the consulting engrg environment and the administration of engrg offices. Pres CEPA 1977-78. Pres FES 1981-82. Engr of the year 1986. *Society Aff:* ASCE, FES, NSPE, CEPA

Decker, Erwin L
Business: PO 2040, 2040 Ave C, Bethlehem, PA 18001
Position: VP, Operations *Employer:* Fuller Company. *Education:* MSME/Power & Machine Design/Lehigh Univ; BSME/Power & Machine Design/Lehigh Univ. *Born:* 7/12/25. Native of Allentown, PA. Served with Army Air Force 1943-45. Reg PE in State of PA-Mech Engrg. Employed by Fuller Co since 1949. Service in Engrg Dept 1949-80. In 1958 appointed Asst Ch Engr; 1966 Mgr of Engrg Services; 1966 Mgr of Engrg; & 1973- VP Engrg. Respon for engrg of Fuller covering a broad line of indus process equip, process & application engrg, design & proj engrg & tech support of Fuller Co foreign subsidiaries & licensees. Awarded sev pats. In 1980 appointed VP of Gen Products Div. Responsible for Sales, Engrg, Mfg and Services of Broad Line of heavy duty industrial equipment. In 1983 appointed Sr VP and Gen Mgr of Intl. In 1985 appointed V.P. Sales & Marketing including R&D. In 1986 appointed V.P. of Operations including Enginrg & 3 Mfg facilities and responsible for Indian joint venture company in India.

Decker, Gerald L
Home: 213 Wimbledon Rd, Walnut Creek, CA 94598
Position: VP. *Employer:* Kaiser Alum & Chem Corp. *Education:* MS/Nuclear Phys/Univ of MI; BS/Phys & Chemistry/Alma College. *Born:* 10/11/21. Native Midland, MI. Worked for Dow Chem Co in Power Engg & Mgt. Assumed present responsibility for Kaiser Aluminum & Chem Corp in 1978 as Corp VP & Dir of Energy. Received Secy of Commerce Medal in 1974 for directing Commerce Technical Advisory Bd Panel on Proj Independence blueprint. Received local chapter AIChE & Natl Energy Resources Organization "Service to Society" awards in 1978 & in 1979 received the Natl Service to Society" Award from AIChE. Served, Chrmn, AIChE Energy Coordinating Committee in 1977 & 1978. Served, Chrmn, House Sci & Tech Committee Panel, "Synthetic Fuels–, 1979. *Society Aff:* ASME, AIChE, NSPE.

Decker, Howard E
Home: 19902 Encino Brook, San Antonio, TX 78259
Position: VP. *Employer:* Jud Plumbing & Heating Co. *Education:* Grad Study/Solar Studies/Trinity Univ; BS/ME/TX A&M Univ. *Born:* 2/13/22. Native of El Paso, TX. Served as an Engg Officer with the Army AF in WWII. With Jud Plumbing & Heating Co since 1960. Assumed present position of VP in 1976 with mgt & engg responsibilities. Also Principal in Applied Solar Engg Co with responsibility for engg design. Dir & Regional Chrmn of ASHRAE 1978, 1979 & 1980. ASHRAE Distinguished Service Award 1982. Made Fellow of ASHRAE 1983. Dir At Large ASHRAE 1983, 1984 & 1985. Pres TX Chpt of Natl Environmental Balancing Bureau 1984 & 1985. *Society Aff:* ASHRAE.

Decker, Raymond F
Home: 3065 Provincial, Ann Arbor, MI 48104
Position: Pres & CEO *Employer:* University Science Partners, Inc. *Education:* PhD/Met Engg/Univ of MI; MS/Met Engg/Univ of MI; BS/Met Engg/Univ of MI. *Born:* 7/20/30. In Afton, NY served to 1 Lt. Army Ordnance 1952-54, INCO Ltd. V.P. 1978-82. Michigan Technological University V.P. 1982-86. ASM International President 1986-87. University Science Partners, President and CEO 1986- . Board of Directors Lindberg Corporation 1987- . Co-inventor of maraging steels. Research publications on these and nickel superalloys and alloy design. Coined Ausaging Maraging and Microduplex. Adj. Prof. Polytechnic Institute of Brooklyn and NYU in 1960's. I.R. 100, University of Michigan Sesquicentennial, R.F. Mehl AIME Gold Medal Award 1973. IMD Lecturer of AIME. Member of National Academy of Engineers, National Materials Advisory Board 1982-87. Chairman of Board of Michigan Energy Resources and Research Association, 1985-87. ASM International (Fellow, 1981 Gold Medal, 1985 Campbell Lecture). *Society Aff:* ASM, IRI, WRC, ASTM, AIME, AIChE, AAAS, ΣΞ.

Decker, Wallace D
Business: PO Box 808, Livermore, CA 94550
Position: Ass't To The Assoc. Director For Engineering *Employer:* Lawrence Livermore Lab. *Education:* BS/Mech Engr/Univ of CA Berkeley *Born:* 6/30/23. Mech Design Engr in nuclear energy R&D at the Lawrence Labs at Berkeley and Livermore, CA beginning in 1943. Hd of 1100 person Mech Engg Dept at Lawrence Livermore Lab (LLL) from 1962 to 1971. Staff Specialist in LLL Director's Office since 1971 on education, organization, and mgt matters. Active since 1968 in developing continuing educ techniques and in TV services and in-house progs. Published papers and articles about profl personnel practices, and engrg careers and educ in the work place. Serving on IEEE careers ctte since 1977. Bd membr and chmn of Assoc for Continuing Ed (ACE) since 1981- an industry consortium providing televised educ. *Society Aff:* ASME, ASEE, IEEE

DeCorte, Robert V
Home: 7441 Emerson, Canton, MI 48187
Position: Traffic Engr *Employer:* Auto Club of MI *Education:* BS/CE/Detroit Inst of Tech; Surveying and Drafting/Pre-Engrg/Ferris State Coll; Gen Bus/Bus/Univ of Detroit *Born:* 12/10/41 I attended the Detroit Inst of Technology where I was active in the Society of Automotive Engrs and the American Society of Civil Engrs. Upon graduation with a Civil Engrg degree in 1970, I became assistant city engr in Mount Clemens, MI. I then became supervisor of the Traffic Engrg Section of the Safety and Traffic Engrg Dept of the Automobile Club of MI. After joining the Inst of Traffic Engrs, I worked and chaired many national and local committees up through Pres of the MI Section. I have been responsible for several educational films, books, seminars and papers. *Society Aff:* ITE, NSPE

Decossas, Kenneth M
Home: 6529 General Diaz St, New Orleans, LA 70124
Position: Chem Engr *Employer:* US Agri Res Serv *Education:* BE/Chem Engrg/Tulane Univ of LA *Born:* 08/14/25 Native of New Orleans, LA. With US Agri Res Service's Southern Regional Res Center since 1944, specializing in chem engrg cost analysis and design of new and improved processes and plants for producing new and improved products from cotton, cottonseed, peanuts, rice, sugarcane, pine gum, tung and other southern- grown commodities. Supvr of Industrial Analysis Unit 1954-61 and Hd of Cost and Design Investigations 1961-74. Since 1961, have been respon for cost engrg res program at the Center and have been awarded three Agri Res Serv Certificates of Merit for outstanding performance. Authored and published 59 research papers. Reg PE in LA and MS. *Society Aff:* AIChE, NSPE, AACE, TBΠ, ACS, AOCS, ΣΞ, AXΣ FRATERNITY

De Cwikiel, Casimir A
Home: 782 Robinhood Rd, Rosemont, PA 19010-1240
Position: Cost Engr *Employer:* Currently Allstates Eng. Co. w/ Dupont as Contract Eng. *Education:* BS/Structural Eng/Drexel Univ *Born:* 06/12/23 Native of Phila, PA. Employed at Sun Oil Co in structural design of refinery units from 1942 to 1960. Served in U.S. Army, 1943-46. Sr Cost Engr at Sun (Oil) Co specializing in engrg and construction cost control of capital refinery projects, 1960-83. Retired from Sun Co in 1983. President of the American Association of Cost Engrs, 1984-85. AACE Awd of Recognition, 1982. Registered prof engr in State of Pennsylvania. Mbr of Toastmasters, International. Certified Cost Engineer - AACE; AACE Fellow, July, 1 1987. *Society Aff:* AACE, NSPE

Dedrick, Allen R
Business: 4331 E Broadway, Phoenix, AZ 85040
Position: Agri Engr. *Employer:* USDA-ARS *Education:* PhD/Agri Engg/UT State Univ; MS/Agri Engg/Univ of NB; BS/Agri Engg/Univ of NB. *Born:* 8/16/39. Native Red Cloud, NE. Employed by US Dept of Agri as agri engr since 1958. Researched water harvesting systems with emphasis on flexible membranes for seepage and evaporation control. Has developed floating covers to control evaporation from tanks. Is currently working to improve surface irrigation water mgt procedures, emphasizing the use of level-basin irrigation with automatic controls. Enjoy hunting and fishing. *Society Aff:* ASAE.

Dedyo, John
Business: 355 Main St, Armonk, NY 10504
Position: Prin. *Employer:* Charles R Velzy Assoc, Inc. *Education:* MBA/Mgt/Lone Island Univ; BCE/Transportation/Rensselaer Poly Inst. *Born:* 5/11/31. Resides in Irvington, NY. Employed in Venezuela with Chicago Bridge & Iron Co, from 1953 to 1955 as field engr and with El Dorado Engg Corp, from 1957 to 1967, managing oil industry related construction projs. Served in US Army Corps of Engrs at Ft Belvoir from 1955 to 1957. Joined Velzy Assoc in 1967 and has completed projs in areas of water distribution, municipal/industrial waste water treatment, sludge disposal and solid waste/energy from refuse. Past Pres Westchester County Chap, NY State Soc of PE; Chi Epilson; Avid golfer; enjoys bowling. *Society Aff:* AAEE, WPCF, NSPE, ASCE.

Deegan, William Carter, Sr
Home: 7099 East Isleway Ct, Fairfield Plantation, Villa Rica, GA 30180
Position: Manager. *Employer:* Southwire Co. *Education:* MBA/Mgt/W GA College; BS/Geology/GA State Univ. *Born:* 12/17/38. Currently Manager of R&D Cost/Proposals Group for Southwire's worldwide system of continuous casting of nonferrous elec conductor rod, refineries and wire & cable mfg plants. Previously Sr Engr at Lockheed GA specializing in weight, value & inertia engg on L-1011, C-5, C-141 & C-130 aircraft progs. Born in Savannah, GA, married Gloria Loncon, five children. Attended GA Tech - Civ Engg, grad GA State Univ - BS Geology & W GA College - MBA Mgt. Pres, Atlanta Sec AACE. Certified Cost Engr by AACE. Previously active in SAWE, recipient of Personal Excellence Award while at Lockheed. Interests in golf, football & soccer. *Society Aff:* AACE.

Deen, Robert C
Home: 708 Old Dobbin Road, Lexington, KY 40502
Position: Director, Transportation Research Program *Employer:* Univ. of Ky, College of Engineering *Education:* BSCE/Civil Engg/Univ of KY; MSCE/Civil Engg/Univ of KY; PhD/Civil Engg/Purdue Univ; JD/Law/Univ of KY *Born:* May 1929, Henderson, Ky. Graduate study in meterology at University of Chicago; JD, Univ KY; PhD, CE, Purdue; MSCE, Univ KY; BSCE Univ KY. Fellow of ASCE, Pres of Ky Sec ASCE 1965. Mbr Triangle Fraternity; Natl Pres 1967-69. Chmn 1971-73 & Treasurer 1973-81 UK Campus Ministry. V Chmn 1973-75 & Chmn 1976, Ky Comm on United Ministries in Higher Education. Treasurer 1976 & 1977 & Chairman 1978-1980 Voluntary Action Center. Mbr Rotary Internatl. Served as a meteorologist (cpt) in USAF in Japan 1951-55. Res engr & head of Soils Section for Ky Dept of Hyws 1955-63; Asst Dir for Research for KY Dept of Transportation 1963-81; Assumed present position in 1981. Currently respon for technical aspects, budget, & adm of a comprehensive transportation res program. Assoc Prof of Civil Engrg, Univ of KY. Instructor in Geotechnical Engrg, Transportation Engrg and Engrg Law. Registered Prof Civil Engr & Land Surveyor in KY. *Society Aff:* ASCE, KSPE, NSPE, ASTM, RI, TRB

Deevy, William J
Business: 1130 Petroleum Bldg, Beaumont, TX 77701
Position: Exec VP. *Employer:* Deevy & Shannon, Inc. *Education:* BS/ME/Polytech Inst of Brooklyn *Born:* 11/16/21 in NY City. Registered PE in TX, NY, and NJ. Served on Bd of Directors of NSPE, chrmn NSPE Bd of Ethical Review, Vice-chrmn NSPE-PEPP, Chrmn consulting engrs section, TSPE, and on numerous committees at local, state, and national levels, engr of the year award by Sabine chapter, TSPE in 1973. Was with the Lummus Co., M.W. Kellogg Co. and James P. O'Donnell, Engrs. Started partnership of Deevy and Shannon, consulting engrs in 1958. In 1975 was named pres of Deevy & Shannon, Inc. Hobbies include tennis, writing, and theology. Elected to Tau Beta Pi, 1982 (eminent engr). Elected to Tau Beta Pi, 1982 (eminent engr). *Society Aff:* NSPE, ASME, CSI, TSPE.

DeFalco, Frank D
Business: Worcester Polytech Inst, Civil Engg Dept, Worcester, MA 01609
Position: Civil Engg Prof. *Employer:* Worcester Polytech Inst. *Education:* PhD/Civil Engg/Univ of CT; MS/Civil Engg/Worcester Polytechnic Inst; BS/Civil Engg/Worcester Polytechnic Inst. *Born:* 4/25/34. Prof of Civil Engineering, Worcester Polytechnic Institute. Reg PE and Land Surveyor. Pres of DeFalco Engg, Inc, a civil engg consulting firm (Est 1974). Geodetic Engr for Western Elec Co, 1960. Fulbright Prof in the Middle East, 1965-1967. Recipient of the WPI Trustees' Award for Outstanding Teaching 1977 and Lincoln Arc Welding Fdn, Structural Div Faculty Awards(16) Special Commendation 1985. Professional Dev Lecturer for NSPE and US Defense Civil Preparedness Agency. Natl Honor Societies - Chi Epsilon and Sigma Xi. *Society Aff:* ASCE, ASEE, NSPE, ACI.

Deffeyes, Robert J
Business: Subsidiary of Carlisle Corporation, 6625 Industrial Park Blvd, North Richland Hills, TX 76118
Position: President. *Employer:* Graham Magnetics Incorporated. *Education:* BS/Appied Chemistry/Caltech. *Born:* Aug 1935; Dow Chemical Co 1957-63, Process Dev Engr, Technical Service Market Dev; 1963-69 Memorex Corp, Mgr of Manufacturing Tech Servs, Magnetic Tape; 1969- , VP Tech Dev, Graham Magnetics Inc, man of precision magnetic recording tapes, 1972 Pres Cobaloy Div-magnetic pigments & conductive pigments. 1978 Pres & Chief Exec Officer Graham Magnetics, Div of Carlisle Corp-Computer Tape Manufacture, & Related Prods. Fellow AICh, accredited. Mbr AIChE; Nationally certified AAU Swimming Official. *Society Aff:* AIChE, AICh.

DeFigh-Price, Cherri
Business: P.O. Box 800, Richland, WA 99352
Position: Mgr Engrg Analysis Unit *Employer:* Rockwell Hanford Operations. *Education:* BS/CE/WA State Univ *Born:* 1/19/53. Lic PE (Civ) in WA State. Pres (1977-78) & Natl Sec Rep (1978-79) of the Eastern WA Sec of the Soc of Women Engr. Pres-Elect for the Columbia Sec of the Am Soc of Civ Engrs. Atlantic Richfield Hanford Co's (ARHCO's) Young Career Woman in 1976; published article on "Nuclide Migration" in AIChE Journal in 1978. Prepared a metrication prog for ARHCO. Was lead rock mechanics engr for Rockwell Hanford Operations Basalt Waste Isolation Proj, which is studying the feasibility of final disposal of nucl waste in a deep basalt flow & is presently manager of the Engrg Mech group. Soc of Women Engrs "Distinguished New Engineer" in 1983. (national award), Treasurer of the Tri-Cities Sectin, NSPE. *Society Aff:* SWE, ASCE, NSPE

deFigueiredo, Rui J P
Business: Dept of ECE Rice Univ, Houston, TX 77251-1892
Position: Prof of EE & Mathematical Sciences. *Employer:* Rice University. *Education:* PhD/Appl Math/Harvard; SM/Elec Eng/MIT; SB/Elec Eng/MIT *Born:* Born Panjim Goa India. SB, SM from MIT; PhD from Harvard. Served first as advisor & then as principal investigator with the Junta de Energia Nuclear, Portugal 1956-62. Assoc Prof at Purdue Univ 1962-64 & at the Univ of IL at Champaign-Urbana 1964-65. Since 1965 in the faculty of Rice Univ as Assoc Prof 1965-67 & Prof 1967 to date, with joint appt in Depts of EE, CE & Math Scis. In 1972-73 was Visiting Prof at Math Res Ctr of Univ of WI, Madison & at Univ of CA, Berkeley. Visiting Prof, Inst of Telecomm, Swiss Fed Inst of Tech ETH, Zurich, Switzerland in the summer of 1981. Mbr SIAM. Fellow IEEE. Mbr of various IEEE cttes. Former Assoc Editor of IEEE Trans on Circuits & Sys. Author of about 150 pubs including two bks. Gen Chmn of the 1980 IEEE Intl Symposium or Circuits and Sys. Currently, Technical Editor for space applications of the IEEE Journal of Robotics and Automation. Also, member of the Editorial Board of *Electrosoft* and of *Circuits, Systems, and Signal Processing*. Hobby: Music in which received the diploma of Licenciate from the Trinity Coll of Music, London. *Society Aff:* IEEE, SIAM, NYAS.

De Forest, Sherwood S
Home: 106 Fitzrandolph Rd, Coraopolis, PA 15108
Position: Agri Engineering Consultant. *Employer:* Self employed. *Education:* BS/Agri Engr/IA State Univ; MS/Agri Engr/IA State Univ. *Born:* 9/20/21. Army Air Force engrg off & flight engr 1943-46. IA State Univ Instructor Agri Engrg 1946 & Extension Agri Engr from 1947-52. Became Engrg Editor, Successful Farming magazine 1952. In 1959-77 was agri engr Market Dev US Steel to dev mkts for steel products in the agricultural industry. From 1964-70 was Mgr Agri Equipment Mktg. Recipient of 3 patents; contributor to 1960 USDA Yearbook of Agri & Steel In Agri, High Authority of European Coal & Steel Community 1966. Dir Am Soc of Agri Engrs 1961-62, 1969-70, 1974-77; Pres 1975- 76. Received ASAE-MBMA Award 1964. Fellow ASAE, Mbr Soc Prof Journalists, Reg prof engr, VP Montgomery Assoc 1977-81; 1977-78 4-month assignment, Visiting Associate Professor, Agricultural Engineering, Univ. of Illinois, Champaign, Il 1980-81-One year assignment, Visiting Assoc Prof, Agri Engrg, Purdue Univ, to work as technology transfer program leader, Northern Agricultural/Energy Center, USDA-SEA, Peoria IL, Publisher Travelhost of Pittsburgh 1981-83, De Forest Agri-Services, Coraopolis, PA 1977-. Con-

De Forest, Sherwood S (Continued)
sultancies, Pakistan 1984, Portugal 1985 and 1986, Member Industrial and Professional Advisory Commitee, College of Engineers, Pens State 1966-71. Member N.E. Reg Agri Research Plan Cttee 1970-72. Author or Co-author of over 130 articles. *Society Aff:* ASAE, SPJ.

DeFraites, Arthur A, Jr
Business: 1700 Grand Caillou Rd, Houma, LA 70360
Position: Pres *Employer:* Gulf South Engrs, Inc. *Education:* MS/CE/Tulane Univ; BS/CE/Tulane Univ *Born:* 5/28/32 Native of New Orleans, LA. NROTC graduate, regular Naval Officer Commissioned, August, 1953. Served on Bd USS Maury (AGS 16), Hydrographic Survey Ship, 1953-56. Completed graduated studies June, 1958. Employed DeFraites Associates Consulting Engrs. Made Firm partners December, 1959. Changed co organizational structure, 1973, became Pres of Gulf South Engrs, Inc. a wholly owned subsidiary of DeFraites Assocs, Inc., a position presently held. NSPE/PEPP Governor 1976-77. PEPP Southwest Regional VP 1979-80. PEPP Chrmn-elect 1980-81. PEPP Chrmn 1981-82. VP NSPE 1981-82. *Society Aff:* NSPE, LES, ASCE, ACI.

Degenhardt, Robert A
300 National Press Building, Washington, DC 20045
Position: Dir-Mech/Electrical Dept *Employer:* Ellerbe Assocs, Inc *Education:* MS/ME/Univ of NB; BS/ME/Univ of NB *Born:* 5/29/43 After graduation and service in the US Army Ordnance Corps as a commissioned officer, worked for Sunstrand Aviation from 1970 to 1973 as a Project Engr responsible for design and development of turbo-hydraulic systems for aircraft and missiles. From 1973 to 1977 was Dir of Engrg at Davis/Fenton/Stange/Darling A/E firm responsible for the Mechanical and Electrical Engrg Depts. Then spent three years at Durrant Engrs as VP of Mechanical Engrg managing a 15-man dept. At present, am dir of the Mechanical/Electrical Dept of Ellerbe Assocs, Inc, responsible for the overall leadership, direction, management, planning and coordinating of a 130plus man dept in a 600plus man A/E firm. *Society Aff:* ASHRAE, NSPE, MSPE

Degenkolb, Henry J
Business: 350 Sansome St, San Fancisco, CA 94104
Position: Structural Engr *Employer:* H J Degenkolb Assoc Engrs. *Education:* BS/Civil Engg/Univ of CA. *Born:* 7/13/13. H J Degenkolb, Chmn of the Board, H J Degenkolb Assoc, has been a lecturer in the Univ of CA's Engg Extension program and in the Univ's College of Engg. He is also the author or co-author of a number of tech papers dealing with structural design problems, including reports on damage caused by earthquakes in Kern County CA (1952), Alaska (1964), Venezuela (1967), and Nicaragua (1972), Guatemala (1976). He is an Honorary Mbr of ASCE, a mbr of the SSA, SAME, ACI, ASTM, Tau Beta Pi and Chi Epsilon. He is a mbr and PPres of the Structural Engrs Assoc of CA, Consulting Engrs Assoc of CA, San Francisco Section of ASCE and has served on the Exec Ctte, Structural Div, ASCE. He was Secretary of the Technical Council on Codes and Standards, ASCE. He is also a mbr and PPres of the Earthquake Engg Research Inst and a mbr of the National Academy of Engg. He is a former mbr of the State of CA, Bldg Std Commission and a former mbr of the Seismic Safety Commission currently serving on an advisory ctte to that commission. *Society Aff:* ASCE, ASTM, SAME, ACEC, EERI, ACI, SSA

DeGuise, Yvon
Home: 205 St Catherine Rd, Apt 702, Outremont, Quebec, Canada H2V-2A9
Position: Retired *Education:* BApSC/Ecole Polytechnique-Mtl; CE/Ecole Polytechnique-Mtl *Born:* 3/10/14. Graduated in engrg Ecole Polytechnique of Montreal in 1937. After 20 yrs in various respon with the Quebec Hydroelectric Comm, was appointed Commissioner in 1965 & became VP in 1974. 1970-76 executive dir of Atomic Energy of Canada Limited. Chmn of Canadian Ctte of the World Energy Conf 1976-79 & Chmn of the Canadian Ctte of Internatl Comm on Large Dams 1974-77. Fellow mbr of Engrg Inst of Canada, mbr of IEEE, Prof Engrs of Quebec. Retired from Quebec Hydro in 1976 to become Principal Cons Energy The Lavalin Group, Cons Engrs, Montreal. Retired from the latter in 1983. *Society Aff:* IEEE. FEIC, PEQ

DeGuzman, Jose P
Home: 22645 Oak Canyon Rd, Salinas, CA 93908
Position: Pres. *Employer:* De Guzman Export Management Co. *Education:* MBA/Marketing Mgt/Univ of Chicago; MSChE/Chem Engg/Purdue Univ; BSChE/Chem Engg/Univ of Santo Tomas. *Born:* 2/10/32. Born and raised in Manila. Came to the US on a 1955 Fulbright Scholarship. Plant Engg in Westinghouse's Mica Products Dept. Res Engg in DuPont's Textile Fibers New Products Div involved in R & D of Dacron nonwoven product (Reemay). Project management of new business ventures in Sohio's Chems and Plastics Div (Cleveland). Market res and market dev work in petrochemicals and fertilizers. Manager of Res & Dev resp for dev of new utility and energy-related products. Pres of export management co'specializing in dev export markets for high tech firms in electronics, chems, computers, instruments, advanced processes and equip, pharmaceuticals, software, plastics, textiles, specialty surfactants, and pulp and paper. *Society Aff:* AIChE, SCC, TAPPI.

De Haven, Eugene S
Business: Dow Chemical U.S.A, PO Box 1398, Pittsburg, CA 94565
Position: Insur & Loss Mgr. *Employer:* Dow Chem USA. *Education:* BS/ChE/Univ of CA., Berkeley. *Born:* 9/18/26. Past Chrmn of Safety & Health Div of AIChE. Former mbr of Loss Prevention Prog Committee of AIChE. Former Pres San Francisco Bay Area Engineering Council. Field of concentration is chem hazard identification & special designs for chem reaction hazards. Former Engg Mgr, Process Design Supervisor, & Sr Process Engr for Western Div, Dow Chem USA, concerned with process design, proj engg, & construction of process plants, usually first of a kind. Recent papers on hazard identification; past papers on non-Newtonian flow analysis & equip design (three patents). Patent also on fiber processing devices & on a process for uranium recovery. *Society Aff:* AIChE, AAAS, NFPA, ΣΞ.

DeHaven, William K
Business: 341 White Pond Dr, Akron, OH 44320
Position: Senior Officer Glaus, Pyle, S. *Employer:* chomer, Burns & DeHaven Inc. *Born:* Apr 1924. BS with Honors Wisconsin University. Resident of Akron, Ohio. Served US Army Air Corps 1942-46, Radar Officer. With Babcock & Wilcox as Stress Analyst 1948-51. Joined Beiswenger & Hoch 1951-61. VP Highway & Bridge Design. Went into private practice 1961- . Sr Officer respon for Civil & Structural Engrg of firm. P Pres Akron Section ASCE, P Chmn Ohio Council ASCE, Pres Elect Akron Section NSPE, Chmn City of Akron Urban Design & Fine Arts Commission. Mbr NSPE, ASCE, ASME, ITE. Hobbies: golf & gardening.

Dehne, Manfred F
Business: 14350 Chrisman Rd, Houston, TX 77039
Position: Pres. *Employer:* Polutrol Industries *Education:* DiplIng/CE/TU Magdeburg-Germany. *Born:* 3/31/35. Abitur at A H Francke Gymnasium/Halle/Germany 1953. Diplom-Engr at TU Magdeburg 1959. R&D Engr with CFK Cologne (BASF Group) involved both in organic and inorganic process devs (1959-1961). Proj Engr with GEA/Germany, 1961-1967, specializing in design of large scale heat transfer equip. Worldwide travel. 1967-1971 Mgr of Advanced Engg at Perfex Co, Milwaukee, WI. Founder & Pres of Polutrol Industries, Houston - specializing in dev & fabrication of ultrahigh performance particulate collection devices & systems. Holder of numerous patents. Author of several technical papers (heat transfer & fluid mechanics). *Society Aff:* AIChE.

DeHoff, Robert T
Business: Dept Mat Sci & Engrg, Gainesville, FL 32611
Position: Professor. *Employer:* University of Florida. *Education:* PhD/Metallurgy/Carnegie-Mellon; MS/Metallurgy/Carnegie-Mellon; BE/Metallurgy/

DeHoff, Robert T (Continued)
Youngstown State. *Born:* 1/15/34. Teach phys metallurgy & related topics U of Florida 1959- . Mbr AIME, ASM, IMS, Sigma Xi, Tau Beta Pi, ISS 1976 ACerS. VP for the Americas, ISS. Primary interest: quantitative microscopy (Textbook). Res Areas: phase transformations, sintering, multicomponent diffusion, math of evolving microstructs. Cons Bausch & Lomb, Inst for Gas Tech, Argonne, Oak Ridge & Hanford Labs. 80 pubs. *Society Aff:* AIME, ASM, ISS, ΣΞ, IMS, ACerS, AMS, ISSS.

Deissler, Robert G
Business: 21,000 Brookpark Road, Cleveland, OH 44135
Position: Staff Scientist *Employer:* NASA Lewis Research Center *Education:* MS/ME/Case Institute of Technology; BS/ME/Carnegie Institute of Technology *Born:* 08/01/21 Fluid dynamicist. Married with four children. Aeronautical research scientist NASA Lewis Research Center, Cleveland, 1947-52; Chief, Fundamental Heat Transfer Branch, NASA Lewis Research Center, 1952-70, staff scientist and scientific consultant (fluid physics) 1970-present. Contributed articles to professional journals. Max Jacob Memorial Award ASME/AIChE, 1975; NACA/NASA Exceptional Service Award, 1957, and Outstanding Publication Award, 1978; ASME Heat Transfer Memorial Award, 1964. Fellow AIAA and ASME; member APS and Sigma Xi. Areas of research: fluid turbulence, turbulent heat transfer, turbulent solutions of equations of fluid motion, vortex flows, and radiative heat transfer in gases. *Society Aff:* AIAA, ASME, APS, ΣΞ

DeJarnette, Fred R
Business: Mech & Aero Engg, N.C. State University, Raleigh, NC 27695-7910
Position: Prof *Employer:* NC State Univ. *Education:* PhD/Aero Engg/VPI & SU; MS/Aero Engg/GA Tech; BS/Aero Engg/GA Tech. *Born:* 10/21/33. Native of Rustburg, VA. Aerodynamic Engg with Douglas Aircraft Co in Longbeach, CA & Charlotte, NC 1958-61. Asst Prof at VA Tech 1961-63, Aerospace Engg. Aerospace Engg at NASA, Langley 1963-65, research in computational fluid flow fields. Assoc Prof at VA Tech 1965-70 (acting head of aerospace engg 1967- 69). Prof, NC State Univ 1970-present (Graduate Administrator 1973-83, Associate Head 1980-83). *Society Aff:* AIAA, ASEE.

Dekema, Jacob
Business: PO Box 81406, San Diego, CA 92138
Position: District Director of Transportation. *Employer:* State of California Dept of Transpor. *Education:* BS/Civil Engg/USC. *Born:* 1915, Indonesia. Educated Holland; Canada; USA. Started work for State as Student Engrg Aide 1936; rose to District Engr 1955. Honors: Selected as one of Top Ten Public Works Men of Year by APWA 1972. State Senate Resolution- Congrats 1972. Certificates: D M Wilson Assoc & Dept of CE USC 1973; Kiwanis Club 1973; United Community Services 1972; San Diego Chamber of Commerce 1964; City Councils of Chula Vista, Calexico, El Centro 1964. State Assembly Resolution/Meritorious Service 1964. Award: San Diego Hgwy Dev Assn/Outstanding Service 1964; Associated General Contractors 1958 & 1979; San Diego County Bd/Supervisors. Keys to Cities San Diego - Natl City 1964. Articles published: California Highway Public Works Magazine & Highway User. Pres ASCE San Diego Sect 1974. Retired December, 1980. Interstate 805 named the "Jacob Dekema Freeway" by California State Legislature 1981. *Society Aff:* ASCE, APWA.

DeLancey, George B
Business: Stevens Institute of Technology, Castle Point Station, Dept of Chem Engg, Hoboken, NJ 07030
Position: Prof. *Employer:* Stevens Inst of Tech. *Education:* PhD/CE/Univ of Pittsburgh; MS/CE/Univ of Pittsburgh; BS/CE/Univ of Pittsburgh. *Born:* 10/19/40. in Cresson, PA. Res engr for Jones & Laughlin Steel Corp 1962-65. Post doctoral study in applied math at Natl Bureau of Stds 1967-68. Res in math models of soaking pits, computer models of decision making, diffusion through membranes, math of nonisothermal diffusion with chem reaction, analysis of diffusion data, multicomponent gas absorption with chem reaction, optimal preparation of catalytic pellets, thermal effects in gas absorption, reactor optimization for exothermic reactions, catalyst effectiveness & hysteresis in SO2 oxidation, storage of thermal energy in chems, kinetics of C6H12 dehydrogenation over Pt, heat transfer with homogeneous reactions. On leave as dept chrmn for Polymer Engg, Algerian Petrol Inst 1976-78. Academic Adv in Intl Programs Office at Stevens Inst of Tech. 1979-87 & Prof of Chemical Engrg. *Society Aff:* AIChE.

Delaney, Francis H
70 Kerry Lane, Kingston, NY 12401
Position: Pres *Employer:* FHD Ltd. *Education:* B.S./EE/Univ of CT *Born:* 9/25/24 Served in the United States Army 1943-1946. Employed by IBM Corp from 1948-1979 with positions in Field Engrg, Engrg and Manufacturing. Assignments encompassed Quality Assurance, Manufacturing Engrg, Manufacturing Research and Industrial Engrg with emphasis on process and equipment design. General Mgr of two manufacturing locations in the U. S.A. and manufacturing operations in the United Kingdom. Retired from IBM in 1979. Registered as a PE, Manufacturing Engrg, CA. Formed a Manufacturing and Management Consultant firm, F.H. Delaney Ltd., in 1980. Assumed responsibility as VP/Gen Mgr, Commercial Div, EG & G Rotron in 1984. Returned to consultant business as of 2-87. *Society Aff:* NSPE, NYSPE.

De Lange, Owen E
Home: 14 Auldwood Lane, Rumson, NJ 07760
Position: Member Technical Staff. *Employer:* Bell Telephone Labs. *Education:* MA/Physics/Columbia Univ; BS/EE/Univ of UT. *Born:* Aug 1906 in Utah. Mbr of Technical Staff of Bell Telephone Labs. Activities were largely res on electronics as applied to communications. F M , P C M, microwaves, M M waves, propagation etc. Did radar R&D during WWII. Res on Optical Communications Sys. Fellow IEEE. Granted 25 US Patents. Technical papers published in all of above fields. *Society Aff:* IEEE.

DeLapp, Kenneth D
Home: 1828 San Felipe Circle, Sante Fe, NM 87501
Position: President. *Employer:* DeLapp Engg Corp (Owner). *Education:* MSE/Civil Engg/Univ of MI; BSE/Civil Engg/Univ of MI. *Born:* Nov 8 1925 Frederick SD. Cvl Engr & Struc Engr 1946-52 with Giffels & Valet Detroit and Ayres, Lewis, Norris & May Ann Arbor MI. Partner Wood & DeLapp C Es Santa Fe NM 1953-65. Owner DeLapp Engrg Corp Cons Struc Engrs Santa Fe NM 1965- . Reg Land Surveyor NM. Reg Prof Engr NM, MI, CO, TX, AZ, HI, NV. NCEE Cert. Mbr: Tau Beta Pi, ACI, NM Sect of ASCE: Dir 1976-77, Pres-elect 1977-78, Pres 1978- 79, CEC/NM, NM SPE (Sta Dir 1973-75, VP 1975-76, Pres-Elect 1976-77, Pres 1977- 78, NMSPE Natl Dir 1979-80). *Society Aff:* NSPE, ASCE, ACEC, ACI, TBΠ, EERI.

Delbridge, N G, Jr
Business: DAEN-ZB, Washington, DC 20314-1000
Position: Maj Gen USA Deputy Chief of Engrs *Employer:* US Army Corps of Engineers. *Education:* MS/CE/IA State Univ; BS/Mil Sciences/US Military Acad, West Point, NY; - /CE/MI Technological Univ. *Born:* 1/4/28. Registered Prof Engr, Iowa; 1953 Grad US Military Academy; Co Commander, 10th Engrs, 3d Infantry Div, Korea; Asst Prof of Military Science, Michigan State University; Asst Area Engr & Ch of Construction, Mediterranean Div. Cdr of 4th Engr Battalion, 4th Infantry Div, Vietnam; Ch of Engr Assignments, DA Office of Personnel Operations; Ch, STANO Div, Office of Ch of R&D, Dept of Army, Wash DC; District Engr, Pittsburgh, PA; Cdr, Div Support Command, 3d Armored Div, Germany; Div Engr, European Engr Div, Frankfurt, Germany; Div Engr, US Army Engr Div, South Pacific, San Francisco, CA; Asst Chief of Engrs, Washington, DC; Deputy Chief of Engrs, Wash, DC. *Society Aff:* SAME.

De Leeuw, Samuel L
Business: Dept of Civil Engrg, University, MS 38677
Position: Prof of Civil Engrg *Employer:* Univ of MS *Education:* PhD/Applied Mechs/MI State Univ; MS/Applied Mechs/MI State Univ; BS/Civil Engg/MI State

De Leeuw, Samuel L (Continued)
Univ. *Born:* Aug 1934. Asst Prof at Yale University 1960-65. Present position at Univ of Miss since 1965. Co-authored textbook, Digital Computations & Numerical Methods McGraw Hill. Published several papers in fields of solid mechanics & transportation. *Society Aff:* ASCE, ASME, ASEE, NSPE.

Delgass, W Nicholas
Business: School of Chem Engg, Purdue University, West Lafayette, IN 47907
Position: Prof. *Employer:* Purdue Univ. *Education:* PhD/Chem Engg/Stanford Univ; MS/Chem Engg/Stanford Univ; BSE/Chem Engg, Math/Univ of Michigan. *Born:* 10/14/42. After a postdoctoral yr at Berkeley, began academic career as an Asst Prof in the Dept of Engg and Applied Sci at Yale, 1969. Moved to Purdue in 1974. Now Prof of Chem Engg. Other experience includes summer jobs with Dow, American Cyanamid, Humble Oil, and North American Aviation and consulting. Res interests are in Heterogeneous Catalysis. Spectroscopic tools such as XPS, SIMS, and the Mossbauer effect are used to probe the surface chemical origins of catalytic behavior. *Society Aff:* AIChE, ACS, ΣΞ.

Delgrosso, Eugene J
Home: 1468 Tuttle Ave, Wallingford, CT 06492
Position: Chief, Adv Mtls. *Employer:* United Tech Corp. *Education:* MS/Met/Univ of PA; BS/Chemistry/Yale Univ. *Born:* 12/14/29. Met, 1952-53, at Frankford Arsenal, Phila. Served in AF, 1954-56. Res Met for Kennecott Copper Corp, 1956-58. With United Tech Corp since 1958: Canel Div of Pratt & Whitney Aircraft until 1964; then Hamilton Standard. Particpated in dev of new columbium alloys, "Cryoshock" strengthening of alloys, boron/aluminum tape and composites, new powder metal alloys; awarded twelve patents in these areas. Presently responsible for advanced mtls devs & applications. Published seven papers in Am & British journals. Mbr of ASM: advisory technical awareness committee, 1979-1981. Interested in opera, classical music & wildlife conservation. *Society Aff:* AAAS, AIME, ASM, ASTM, YSEA.

Dell, Leroy R
Business: 245 East Lakewood Boulevard, Holland, MI 49424-2066
Position: Pres *Employer:* Dell Engrg, Inc. *Education:* BS/CE/MI State Univ; Assoc of Applied Sci/Hgwy Tech/Ferris State Coll *Born:* 4/30/43 Native of Stanton, MI. Responsible for design and construction of numerous major water and wastewater facilities. Registered Engr in MI, WI, KY, IN, OH, IL & MO. Registered Land Surveyor in MI. Certified water and wastewater plant operator. Founder of Western MI Environmental Services, Inc., a full-service environmental-analytical lab. Founded Dell Engrg, Inc. an environmental firm specializing in industrial air, water, solid and hazardous waste consulting services, waste treatability studies, treatment plant design, sanitary landfills, hydrogeological studies and industrial process design. Licensed pilot. Enjoy golf, fishing and canoeing. *Society Aff:* ASCE, ASTM, AWWA, WPCF, NSPE, AIPE, AAEE.

Della Giovanna, Ciro M
Home: 15 V le delle Rimembranze di Lambrate, Milano Italy 20134
Position: Technical Consultant. *Education:* Doctor in Industrial Engineering *Born:* Dec 1916. Doctor in industrial engrg from Genoa University in 1938. Native of Potenza, Italy. Served in different technical positions in Siemens & Safar from 1938-50. With FACE Standard from 1950 to 1977 as Lab Mgr, Asst to Managing Dir & Tech Dir. In the past Dir/Bd of IATC Antwerp Belgium, VP of Eurocae Steering Ctte, expert in CCITT & CCIR, teacher in specialization courses for graduate engrs in Milan Politechnic & ISPT. Presently Fellow of IEEE, Mbr of Fitce & AEI. Holder of patents in telecommunication & writer of several papers. *Society Aff:* AEI, IEEE, FITCE

Della Torre, Edward
Business: Dept of Elect & Computer Engrg, Detroit, MI 48202
Position: Chairman *Employer:* Wayne State University *Education:* D.E. Sc./EE/Columbia Univ.; M.Sc./Physics/Rutgers Univ; M.Sc./EE/Princeton Univ; B/EE/Brooklyn Polytech Inst *Born:* 3/31/34 Taught at Rutgers 1956-1967. Mbr of the tech staff of Bell Telephone Lab, Murray Hill, NJ 1967-1968. Taught at Mc Master University 1968-1979, was chrmn of dept 1972-1978, joined Wayne State Univ 1979 as chrmn of the Electrical and Computer Engrg Dept. Has carried out res in the theory of operation of magnetics devices including bubble memories. Is the author of two books: *The electromagnetic Field* with C. V. Longo and *Magnetic Bubbles* with A. H. Bobect. Also he has written over 100 technical papers and holds over 15 potents. *Society Aff:* IEEE, ADS, ΣΞ, ΤΒΠ, HKN.

Dellon, Alfred L
Home: 804 Saddle River Rd, Saddle Brook, NJ 07662
Position: Pres *Employer:* Dellon Associates, Inc. *Education:* BA/Social Sciences/Fordham Univ; Certif/Engg Mgt/PA State Univ; Certif/Advance Refinery Dev/Manhattan College; Diploma/Mech Inst/Architectural Design & Estimating; Continuing Ed. Units/Cost Engg/WV Univ; Continuing Ed. Units/Cost Engg/Univ of Miami, FL. *Born:* 2/11/15. Over 35 yrs of diversified experience in engg & construction mgt & a mgt background encompassing more than 20 yrs as a proj mgr, construction resident mgr, consultant to private industry, state & fed agencies, utilities & intl projs. Served as tech advisor to the USAF & Navy on the construction & admin of high energy fuel plants. Broad spectrum of experience includes projs in the US & abroad of heavy construction projs, power field-combined cycle, fossil & nucl plants, pharmaceuticals, chemicals, petro-chemicals, electro-mech & met facilities, hospitals, airports & wastewater treatment plants. A recognized authority on project management. Expert consulting services in support of litigation of claims. Univ Lecturer, course director, author, faculty mbr of the Ctr for Profl Advancement, NJ. Fellow AACE and Certified Cost Engineer. Pres, VP and mbr of community Bd of Educ. Recipient of "Silver Beaver" award for "Distinquished Serv" to the Boy Scouts of America. *Society Aff:* ABEEC, AACE, AAES, ASEE, APA, SAM, PMI, SMIEC, FEPIEC, UPADI, BSA, CMAA, ICEC.

Dellon, Robert E
Business: 4550 Montgomery Ave, Bethesda, MD 20014
Position: Utility Mgmt Consultant. *Employer:* Booz, Allen & Hamilton. *Education:* PhD/Mgmt/CA Western Univ; MBA/Mgmt/Fordham Univ; BS/Industrial Admin/Univ of CT. *Born:* 10/19/46. Native of Saddle Brook, NJ. 14 years utility industry experience. Field Engr with Ebasco. Youngest Project QA Mgr with Burns & Roe. Joined WPPSS in 1972. Youngest Corporate QA Mgr in Nuclear Utility Industry. Youngest utility executive with WPPSS, Asst Dir (VP) Mgmt Services. Other Mgmt positions held with WPPSS: Mgr Personal Programs, Project Engg Mgr WNP 2, Mgr Mgmt Sys, Mgr Organization Performance Operation. Assumed current respon as Utility Mgmt Consultant in Nov 1979. Conducts mgt studies re: project/construction mgt, organizational development, operations, corporate services; strategic planning & QA. Mbr ASME, ANS, & ASQC; WPPSS Representative on EEI QA Task Force. Has worked on 12 Nuclear Projects, 5 Fossil & has visited 90 Nuclear Plants. Was selected for IAEA assignment. Lectures nationally and internationally on project/construction management techniques, quality assurance principles and human resource programs. Enjoys snow skiing & scuba diving. *Society Aff:* ASME, ANS, ASQC.

De Loach, Bernard C, Jr
Business: 600 Mountain Ave, Murray Hill, NJ 07974
Position: Department Head. *Employer:* Bell Laboratories. *Education:* PhD/Physics/OH State Univ; MS/Physics/Auburn Univ; BS/Physics/Auburn Univ. *Born:* Feb 19 1930 Birmingham AL. Joined Bell Labs in 1956 & is Head of the Solid State Materials & Devices Dept at Bell Labs, Murray Hill, N J. Author of a number of published technical articles & has been granted 11 pats for investigations. Mbr IEEE, Pi Mu Epsilon Sigma Xi, Sigma Pi Sigma. Fellow Ohio Sta 1955-56 & IEEE Jan 1972. Recipient of IEEE David Sarnoff Award 1975 & Stuart Ballentine Medal of the Franklin Inst. *Society Aff:* IEEE, ΠΜΕ, ΣΞ, ΣΠΣ.

DeLoach, Robert E, II
Home: 6800 Z3 Peachtree Industrial Blvd, Atlanta, GA 30360
Position: VPres *Employer:* Flood Assoc Inc *Education:* MS/SE/GA Inst of Tech; B/CE/Univ of FL *Born:* 11/18/37 Career consulting engr to industry and government. Chief of Operations, Georgia Water Quality Control Bd 1965-69. With Flood Assoc Inc. since 1969. Presidential Exchange Executive assigned to Corps of Engrs Hdquarters 1977-1979. VPres in charge of Atlanta, Georgia regional office operations and marketing since 1979. Mbr Who's Who in American Universities and Colleges 1964, University of Florida Hall of Fame 1964, Florida Blue Key 1964, Personalities of the South 1971, Notable Americans of the Bicentennial Era, Who's Who in the South and Southwest 1971-1972. Awarded University of Florida Pres Award, Engrg Leadership Award. Married. Three Children, Baptist. *Society Aff:* WPCF

DeLong, Carl E
Home: PO Box 51591, Durham, NC 27717-1591
Position: Civil Engr *Employer:* Self *Education:* MS/CE/Univ of MO-Rolla; BS/CE/Univ of MO-Rolla *Born:* 7/9/38 Active in professional societies. Past secretary/treasurer of IL Society of PE. Past Pres of the Central IL Section ASCE. Past Pres of local ISPE and Toastmaster chapters. With Corps of Engrs since 1969. Currently Principal Investigator with CERL investigating computer development and applications in construction management. *Society Aff:* ASCE, SAME, NSPE

DeLong, William T
Home: 730 Dogwood Cir, York, PA 17403
Position: VP, Corp Dev. *Employer:* Teledyne McKay. *Education:* BS/Met Engrg/Lehigh Univ *Born:* 12/9/21. About 25 yrs of res in the field of welding consumables, plus 10 yrs of gen mgt. Over 15 patents & 15 publications, the latter primarily in Cr-Mn austenitic stainless steels, ferrite measurement & control in stainless steel weld metal, and welding safety & health. Pres 81-82 of the Am Welding Soc. About 10 yrs of activities in the Intl Inst of Welding. A Fellow of the Am Soc for Metals. *Society Aff:* AWS, ASM

Del Re, Robert
Home: 10130 White Trout Ln, Tampa, FL 33618
Position: Mgr Const Servs, South Region *Employer:* Camp Dresser & McKee, Inc. *Education:* BS in Petroleum/-/Marietta College/BS/Pet Engrg/Marietta College *Born:* 6/3/30. Virginia (Mother) Higginbotham Del Re; m Joyce Maley, Aug 20, 1971. Engineer, Camp Dresser & McKee Inc, South Region Manager of Construction Services, Tampa FL 1983-present. Engr, Greeley and Hansen, Chgo, 1957-71, Asso, 1971, Resident Engr, Knoxville, TN, 1963-64, Resident Engr, City of Panama, Republic of Panama, 1964-69, Resident Rep, Tampa, FL, 1971-83. Patron Tampa Museum of Art, Tampa Historical Society. Served with CEC, USNR, 1952-55. Reg PE, CA. Diplomat American Academy of Environmental Engineers, FL Engg Soc, Natl Soc of Profl Engrs, Soc Am Mil Engrs, FL Pollution Control Assn, AM Arbitration Assn, Earthquake Engg Res Inst, Inter-Am Assn Sanitary Engg, Am Water Works Assn, Am Soc Qual Control, Pi Epsilon Tau, Alpha Tau Omega. Clubs: Univ (Tampa), Univ (Chgo). Fellow ASCE V Chrmn, Ctte on Contract Adm; Author, "The Resident Engineer: Intermediary Between owner and contractor–, ASCE vol 108, 1982, Journal Const Div Article. 1986 Engr of Yr, West Coast Branch, Florida Section, ASCE. *Society Aff:* ASCE, ASQC, NSPE, SAME, WPCF, AWWA, AAEE, AIDIS, EERI, AAA

Delve, Frederick D
Home: 102 Lansdowne Dr, Coraopolis, PA 15108
Position: Manager of Quality Control. *Employer:* Jones & Laughlin Steel Corp. *Born:* Oct 13 1923. BS Met Engrg Carnegie Inst of Technology. Mbr Phi Kappa Phi Scholastic Honorary. Mbr of ASM, AIME, AISE, ASTM. Chmn Pittsburgh Chapter ASM 1965-66, Chmn Pittsburgh Section AIME 1968. Registered Prof Engr Pa. Joined Vanadium Corp of America as Res Engr, transferred to Jones & Laughlin Steel Corp as Res Engr 1953. Series of promotions in Metallurgical Dept culminating in present position 1975. Worked mainly process metallurgy area & written papers on Ti solubility in blast furnace iron, blast furnace thermochemical models, kinetics of oxygen removal in the Dortmund-Hoerder vacuum degassing process, hardenability of boron treated steels. Enjoy golfing & classical music.

Delyannis, Leonidas T
3030 Clarendon Blvd, Arlington, VA 22201
Position: Pres. *Employer:* L T Delyannis & Assoc. *Education:* MSCE/Structures/Univ of IL; BSCE/Civ Engg/Greek Technical Mil College; BS/-/Greek Military Acad. *Born:* 11/8/26. Native of Athens, Greece. Served as officer of the Corps of Engrs, Greek Army & taught at Engg Ctr. Chief structural engr Ben Dyer Assoc 1958-1960, Chief Bridge engr David Volkert & Assoc 1960-1970, Pres L T Delyannis & Assoc 1971-. Fellow of ACI. Founding mbr of ACI Committee on Concrete Bridge Design. Prog Chrmn of two Intl Symposia on concrete Bridge Design 1967 & 1969. Author of technical papers. Reg PE in several states. Mbr Arlington County Planning Commission: (1976 - 1980). "Mbr. Virginia Board of Geology" to Mbr Virginia Board of Geology: (1981-1982). Speaks & reads French & Greek. Resides in Arlington, VA with wife Georgia & sons Theodore & Harry. *Society Aff:* ASCE, NSPE, SAME, ACI, IABSE, PCI, CRSI.

Demand, Lyman D
Business: Texas Commerce Bldg, 2201 Cinis Cir, Amarillo, TX 79109
Position: Owner *Employer:* Demand & Assoc *Education:* BSME/Mech Engrg/Univ of OK. *Born:* 5/9/25. Native of OK City, OK. Served three yrs in US Marine Corps Reserve. Attended Univ of OK before & after WWII, received BSME 1948. Worked for Phillips Petrol Co, Foster Wheeler Corp, John Deere Chems, Lone Star Gas (Nipak), Cominco Am Inc. Was appointed to present position of Pres & Chief Exec Officer of Cominco Am Inc at annual bd meeting May 1979. Baptist (Southern). Golf (duffer). Enjoy music, fishing, hunting, & flying. Retired from Comino American Dec 83. Independent Oil Operator & Consltg Engr. *Society Aff:* ASME, AIChE.

Demaree, David M
Home: 3030 E Puget, Phoenix, AZ 85028
Position: Pres. *Employer:* Demaree & Assoc. *Education:* BS/ME/Rose-Hulman Inst. *Born:* 11/29/20. Born in Bloomington, IN. Student Engr with GE Plant Engg with Goodyear. Chief EE at Hacienda Cartavio, Peru for W R Grace. Founded Demaree & Assoc, Elec Consulting Engrs in Phoenix, AZ 1951. Designer of elec systems & lighting systems for many large projs mainly in southwestern USA. Reg PE in AZ, CA, NV & NM. Bd Mbr of AZ Consulting Engrs Assn for seven yrs. Sr Active Mbr of Rotary. Life Mbr of Theta Xi Fraternity. *Society Aff:* NSPE, IEEE, IES.

Demarest, Donald M
Business: 6901 Elmwood Ave, Philadelphia, PA 19142
Position: Manager HVDC Project Engineering. *Employer:* General Electric Company. *Education:* MBA/Bus Admin/Amer Inst College; BSE/Elec Engg/Cornell Univ. *Born:* Nov 28 1921. 35 yr career has been with the General Electric Co in technical & managerial positions concerned with High Voltage Systems. The last 17 years has been Managing Engr respon for the dev design & installation of highly reliable thyristor (SCR) utility type High Volt DC Conversion Systems. Was a major contributor to the first commercial All Thyristor HVDC Conversion System in the world. The sys established new worldwide industry standards for performance. Holds 10 US patents on HVDC, serves on several IEEE Working Groups & has received several co awards for outstanding achievements. Is a Steinmate Award Nominee.

DeMaria, Anthony J
Business: Silver Ln, E Hartford, CT 06108
Position: Asst Dir of Res for Elec & Electro-Optics Tech *Employer:* United Tech Res Ctr. *Education:* PhD/EE/Univ of CT; MS/Phys/RPI; BS/EE/Univ of CT. *Born:* 10/30/31. Specialist in res & dev in acousto-optics, electro-optics, lasers, nonlinear optics & in the interaction of coherent radiation with matter. Reported the first use of acousto-optic devices to control the output of lasers (1961), was the first to report

DeMaria, Anthony J (Continued)
& recognize the ability of passive optical saturable absorbers to Q-switch & mode-lock lasers for the generation of picosecond pulses (1966). Have pub in excess of 35 papers in referred journals, have been issued 28 patents, received the 1980 IEEE Morris N Liebmann Meml Awd & medal, the 1978 Univ of CT Distinguished Alumni Awd; the RPI Davies Medal & Awd in 1982, the Univ of CT Engrg Alumni Awd 1983; the IEEE Outstanding Service Medal in 1980; Fairchild Distinguished Scholar of Calif. Institute of Technology, 1983-84; & the AF Distinguished Service Medal in 1986. Assoc Editor of the IEEE *Journal of Quantum Electronics* from 1969-77, Editor 1977-1983; VP of the Optical Soc of AM (1979), Pres-Elec (1980) & Pres (1981). Conslt to the Dept of Defense & the Natl Bureau of Stds, Elected to the Natl Acad of Engrg (1977), elected to the CT Acad of Sci & Engrg (1979). *Society Aff:* OSA, IEEE, APS, SPIE, NAE, CASE.

Demas, Nicholas G
Business: Tenn Tech Univ, Cookeville, TN 38501
Position: Assoc Prof. *Employer:* TN Tech Univ. *Education:* PhD/Nuclear Engrg/U of WI; MS/Physics/U of Pittsburgh; BS/Physics/West Liberty St Coll. *Born:* 10/24/34. Presently teaching Nuclear Engg courses in the Elec Engg dept at TN Tech Univ and is also engaged in res in Nuclear design methods. Has accumulated eleven yrs of Nuclear Design experience since he worked in 1957 on the first commercial Nuclear power plant at Shippingport, PA. Spent six yrs in San Juan, PR during which time he did Environmental studies and Nucelar Reactor studies for the PR Water Resourses Authority. He is an avid (but not too successful) hunter and fisherman. *Society Aff:* ANS, IEEE.

Dembowski, Peter V
Business: Materials Research Corporation, Route 303, Orangeburg, NY 10962
Position: VP & Gen. Mgr. *Employer:* Materials Research Corporation *Education:* MS/Met Engg/Poly Inst of Brooklyn; BS/Met Engg/Poly Inst of Brooklyn. *Born:* 9/24/46. Res Met & Proj Leader, Watervliet Arsenal, 1969-1976; Functional & Gen Mgr. GE Co, 1976-1987; MRC, 1987- . Profl activities have centered in Business dev, semiconductor/specialty materials, dev and implementation of metalforming and metalcutting processes, composite structure fabrication, effects of ultra high pressure on the structures and properties of mtls, and mtls applications & stds for nuclear systems. Chrmn, Eastern NY Chapter, ASM, 1976-1977. VChrmn, AIME Shaping & Forming Committee, 1976-1978. Mbr, ASM Process Modeling Committee, 1977-, and ASME Production Engg Div, 1977-. *Society Aff:* ASM, SME.

De Mello, F Paul
Business: POBox 1058, Schenectady, NY 12027
Position: Vice President. *Employer:* Power Technologies Inc. *Education:* MSEE/Power/MA Inst Tech; BSEE/Power/MIT. *Born:* Jul 1927 Goa Port India. MS, BS in EE from MIT 1948. Engr Sys Planning & Special Studies Rio Light & San Paulo Light & Power Co's 1948-55. Analytical Engrg Genl Electric Apparatus Sales Div 1955-63. Sr Engr General Electric Utility Engrg 1963-69 respon for power sys dynamic analysis & controls. Principal Engr, Dir & Sec/Tres Power Technologies Inc 1969-73. VP PTI since 1973. Cordiner Award 1963 (G E). Fellow IEEE 1974. Mbr CIGRE. Sr Mbr ISA. *Society Aff:* IEEE, ISA, CIGRE.

De Merit, Merrill W
Home: 11302 Kley Rd, Vandalia, OH 45377
Position: Director of Engineering, Aeronautical Equipment *Employer:* Aeronautical Systems Division USAF. *Education:* BSEE/Electrical Engrg/Univ of WI. *Born:* Mar 1924. BSEE Wisconsin 1949. Graduate 3 yr General Electric Advanced Engrg Program. Single Engine Fighter Pilot in WWII. Mgr Airborne Communications, General Electric, Missile Dept 1957. Dir, Advanced Product Center, General Precision, Link Div 1963. Deputy Eng Dir, McDonnell Douglas Electronics Co 1973. Technical Advisor for Avionics ST-855, USAF in Jan 1974, Director of Eng, Aeronautical Equip, 1980 Respon for support of airborne weapon systems in avionics life support, auto. test equip, producibility. Mbr IEEE, RESA, Tau Beta Pi, Eta Kappa Nu, Phi Kappa Phi, Phi Eta Sigma, Sigma XI, and Assoc. of Ex-Prisoners of War. Author Servo Sect of Automation Handbook. Past Boy Scout Council, Cub Scout Master, Boys Club Dir, Pres PTA. *Society Aff:* IEEE, RESA, AXPOW, ΣΞ.

Demmy, Richard H
Business: Weston Way, West Chester, PA 19380
Position: Executive VP *Employer:* Roy F Weston, Inc *Education:* BS/EE/Cornell Univ *Born:* 6/2/22 Thirty-two years of diversified experience in electric utility engrg, construction and operation, in coal operations, and in energy utilization and conservation. Functional areas of responsibility included design, construction supervision, operation supervision and management, maintenance supervision, long-range facilities planning, new-ventures planning, and general top- management policy making. Numerous assignments including: investigation of the technical and marketing feasibility of a $500,000,000 5,000-ton/day plant for production of methanol from coal; member of a 4-man group that developed the 1974 national energy plan. *Society Aff:* IEEE, NSPE, WPCF, API, ACEC, NAM

DeMoney, Fred W
Home: 325 Cardinal Way, Freeland, WA 98249
Position: President. *Employer:* Self-Consultant. *Education:* PhD/Metallurgy/Univ of MN; MS/Metallurgy/Univ of MN; BS/FPE/IL Inst Tech. *Born:* Nov 1919 Oak Park IL. Res Assoc U of MN 1951-54; Res Met Dow Chem Co Midland MI1954-55; Res Engr, Branch Hd, Tech Supr Kaiser Aluminum Corp Spokane WA 1955-69; Tech Supr, Prog Mgr Kaiser Pleasanton CA 1969-72. Pres Montana Tech 1972-1985. Named Engr of Yr Inland Empire 1962. Mbr AIMME, ASME (Cert of Merit 1969), ASM (Chap Chmn, Natl Long Range Planning Ctte 196162). Mbr Salamander, Tau Beta Pi, Sigma Xi, NY Acad of Scis; Pres Montana Energy Research Inst. Adv bd Inst for Intl Education Adv bd Canadian Inst of Mgmt. Named Man of the Year, 1985, Button Silver Bow. *Society Aff:* ASME, AIMME, ASM, TBΠ, ΣΞ, NYAS.

Demopulos, Chris
Business: 600 Petroleum Tower, Shreveport, LA 71101
Position: Corporate Officer. *Employer:* Demopulos & Ferguson Inc. *Education:* BS/Engineering/Texas A&M *Born:* 10/30/24. US Army Air Force 1943-45. Lic PE Tx, LA, AK, OK, PA. NCEE Cert. of Qualification. Pres.1972-73, Dir 1974-75 CEC LA., Ntl Past P Shreveport Chap ASCE, SAME. P Dir Shreveport Cham of Commerce. Past Bd/Dir Metro YMCA Shreveport, Past mbr of LA Bd of Registration for Professional Engs and Land Surveyors. Recipient: 1979 and 1985 Engr of Yr Award from Engrg & Scientific Council of Shreveport; 1980 AE Wilder Jr Award from Consltg Engrs Council of LA. *Society Aff:* ACEC, ASCE, NSPE, SAME

DeMott, Alfred E
Business: Blvd 22CE, 1285 Boston Ave, Bridgeport, CT 06611
Position: Manager Design Engineering. *Employer:* General Electric Co. *Education:* MSME/Mech Engg/Lehigh Univ; BSME/Mech Engg/Lehigh Univ. *Born:* Feb 1925. Native of Allentown, Pennsylvania. Served in US Navy 1943-46. Design Engr Heavy Equipment for Bethlehem Steel Co 1949-52. Joined General Electric Housewares Div 1952. Have held dev, design, administrative & managerial positions in Engrg of small appliances. Mgr-Engrg since 1965. 7 patents, Bd/Governors Lehigh Valley Engrs Club 3 yrs, Corporate Technical Recruiting 6 yrs. Active in School Authority, Chamber of Commerce, Church Trustee. Enjoy golf, tennis, bowling, woodworking.

Dempsey, Barry J
Business: 111 Talbot Laboratory, 104 S. Wright St, Urbana, IL 61801
Position: Prof. *Employer:* U of Illinois-Dept C E. *Education:* PhD/Tranportation Matls/Univ of IL; MS/Tranportation Matls/Univ of IL; BS/Civil engg/Univ of IL. *Born:* March 17 1938. Resident Engr Illinois Dept of Transportation 1960-61 & 1963-64; Commissioned Officer in US Army Corps of Engrs 1961-63; Res Asst 1964- 69 & Faculty & Staff 1969- , Dept of Civil Engrg University of Illinois. Cons to various governmental agencies, consultants & industry. Field of interest is highway materials with major emphasis on the investigation of the influence of climatic factors on pavement systems. Published approx 80 tech journal papers. Mbr ASCE, ASTM, SSSA & TRB (Chmn Ctte A2L06 & Mbr 2 other cttes, A W Johnson Mem Award, received the K B Woods Award twice). Holder of 4 patents for Geocomposite drainage materials and installation for Highway and Airport pavements. Reg Prof Engr IL. *Society Aff:* ASTM, TRB, SSSA, ASCE.

Dempsey, J Ned
Business: 1444 NW College Way, P.O. Box 1174, Bend, OR 97701
Position: President *Employer:* Century W Engrg Corp *Education:* MS/Envir Engrg/San Jose State Univ; BS/Civ Engrg/San Jose State Univ. *Born:* 12/06/42 Major stock holder, Century West Engrg Corp, Bend, OR. Reg engr in OR, WA, CA and NE. ID Seventeen yrs experience beginning with US Geological Survey as proj engr in CA Water Resources Div. Private sector engrg in water and wastewater mgt 1970-1973. Became owner in Century West Engrg Corp in 1973. Firm provides consltg services in water & wastewater mgt, aviation community and industrial engrg dev, ground and surface water dev, geological engrg, environmental sci and hazardous waste. Century Testing Laboratories is a state-of-the-art, completely certified EPA contract lab. Office also in Spokane & Kennwick WA, Portland OR and Woodland, CA. Native of Witchita, KS. Lived in Midwest until attended San Jose State Univ earning BS in Civ Engrg 1969 and MS in Sanitary Engrg 1970. *Society Aff:* WPCF, AWWA, PNWA, NSPE, ASCE

Denbrock, Frank A
Business: 209 East Washington Ave, Jackson, MI 49201
Position: VP *Employer:* Gilbert/Commonwealth *Education:* BS/EE/IN Inst of Tech *Born:* 7/6/25 Resident of Jackson, MI. PE with Gilbert/Commonwealth since 1949, VP and general mgr for energy transport facilities since 1969 (400 employees). Registered PE in 19 states. IEEE Fellow; MSPE Engr of the Year (1979) and State VP (1980-82); IEEE T&D Standards Chrmn (1975-82); IEEE Substations Chrmn (1982-84); ANSI C2, National Electrical Safety Code, Chrmn, Strengths and Loadings, and Interpretations Committee; US Representative CIGRE (Substations); PES Technical Delegation to People's Republic of China; US State Dept HVdc Technical Exchange to Russia (1976); US DOE/Italy, Joint UHV Program (1980-85). Enjoys fishing, golf. *Society Aff:* IEEE, Ansi C2, UNSC of Cigre, MSPE NSPE, ASCE

de Nevers, Noel H
Business: Rm 3062 MEB, Salt Lake City, UT 84112
Position: Prof Chem Engrg *Employer:* Univ of UT *Education:* PhD/ChE/Univ of MI; MS/ChE/Univ of MI; BS/ChE/Stanford *Born:* 5/21/32 Worked for research arms of Standard Oil Co of CA, 1958-63; Univ of UT since then, with summers and sabaticals elsewhere. Univ of UT Distinguished Award, 1969. EPA Citizen Participation Award, 1980. *Society Aff:* AIChE, APCA

Denham, Frederick R
Business: 2300 Yonge St, Toronto Ontario, Canada M4P 1G2
Position: VP *Employer:* Thorne Stevenson & Kellogg *Education:* PhD/Mech Eng/University of Durham; MBA/Bus Admin/Univ of Buffalo; BSc/Appl Sci/Univ of Durham *Born:* 10/21/29. Defense Res Bd 1953-54; Ford Motor Co of Canada 1954-55; Union Carbide Canada 1956-61. Joined Thorne Stevenson & Kellogg 1961, appointed Principal in 1964, elected Dir 1966, VP 1968. Currently in charge of Mgmt Scis. Mbr Assn of Prof Engrs of Ontario. Fellow Engrg Inst of Canada. Fellow, Inst of Directors. Fellow Inst of Mgmt Conslts of Ontario. *Society Aff:* APEO, IMCO, EIC

Denham, Roy S
Business: 4 Research Place, Rockville, MD 20850
Position: Vice President, International Activities *Employer:* NUS Corp. *Education:* ChE/Chem Engg/ Univ of Cincinnati; BS/Chem Engg/Univ of Cincinnati. *Born:* 12/29/32. in Baldwin, NY; married 1960; 4 sons. Joined NUS Corp 1963. Currently Vice President, responsible for business development and management of company activities in Mid-East and Latin America.Previously employed by E I DuPont and Rockwell International. Mbr, Bd of Dir, Air Pollution Control Assoc; Past.Chmn, APCA Finance Committee and mbr of APCA Government Affairs, International, and Energy/Environment Interactions Committees. Special interest in environmental impact of energy prod, effect of legislation and reg on industry and economy, and energy conservation. Author or Co-Author ofmultiple papers on these subjects, Specific prof interests include industrial site selection and evaluation, with particular emphasis on nuclear and coal-fired power stations. *Society Aff:* APCA.

Denison, John S
Home: 1203 Merry Oaks, Coll Station, TX 77840
Position: Prof of Elec Engrg *Employer:* TX A&M Univ *Education:* MS/EE/TX A&M Univ; BS/EE/NM State Univ *Born:* 6/18/19 in Waco, TX, son of Frank & Emma Denison. Graduated as Class Salutatorian, Waco High School, 1935. Surveyor of rural electric lines, 1936-1942. USAAF, 1942-45. Pilot, Southwest Pacific Area. Instructor-Prof of Electrical Engrg, TX A&M, 1949-date. Acting Head of Dept 1966-67. Organizer and Dir, Electric Power Inst, A Power System Research Group in EE Dept, 1964-1976 and 1981-date. Owner and Pres, Electric Power Engrs, Inc, a consulting engrg firm serving primarily municipal and rural electric systems. *Society Aff:* IEEE

Denkhaus, Walter F
Home: 132 E Golf View Rd, Ardmore, PA 19003
Position: Retired. *Education:* EE/Elec Engg/Swarthmore College; BS/Elec Engg/Swarthmore College. *Born:* Jun 1905. Native of Colwyn, Pennsylvania. Served with Army Signal Corps 1942-46 as Dir of Production Div Philadelphia, Signal Corps Procurement District 1945 & as Ch of Industrial Mobilization Branch, Office of Ch Signal Officer 1946 Lt Col 1946. With Bell Telephone Co of Pennsylvania starting in 1928. Assumed Ch Engr respon in 1960, then General Engrg Mgr Western Area 1962. Pres, Liberty Bell Chapter, Telephone Pioneers of America 1954. Natl Dir of Publications Dept AIEE 1962. Founder & Trustee of Penna Engrg Foundation 1970-84. Fellow IEEE. *Society Aff:* IEEE, NSPE.

Denman, Eugene D
Home: 13402 Taylorcrest, Houston, TX 77079
Position: Prof *Employer:* Univ of Houston *Education:* DSc/EE/Univ of VA; MS/Physics/Vanderbilt Univ; BS/EE/WA Univ (St Louis) *Born:* 3/15/28 Farmington, MO. US Signal Corp 1946-1948. Jr Engr Magnavox Co, 1951-1952. Engr with Sperry Gyroscope Co, 1954-1956, Midwest Research Inst, 1956-1960. Res Scientist with Univ of VA, 1960-1963. Assoc Prof Elec Engr and Asst Prof Medicine, Vanderbilt Univ, 1963-1969. Prof at Univ of Houston 1969-Present. Consultant to Boeing Co, 1965-1967, Oak Ridge National Labs, 1967, Lockheed Electronics, 1974-1986. Author/Co-author of two books, Coeditor of two books and Author/Co-author of more than 20 papers. Academic fields are networks, systems, controls and computer software. *Society Aff:* IEEE

Denn, Morton M
Business: Dept Chem Engrg, Univ. of California, Berkeley, CA 94720
Position: Prof *Employer:* Univ CA, Berkeley *Education:* PhD/ChE/Univ of MN; BSE/ChE/Princeton Univ *Born:* 7/7/39 Native of NJ. At Univ of DE from 1964 to 1981, as the Allan P Colburn Prof from 1977 to 1981. Prof at Univ CA, Berkeley since 1981 and Program Leader for Polymers, Center for Adv Materials, Lawrence Berkeley Lab since 1983. Editor of AIChE Journal since 1985. Was a Guggenheim Fellow in 1971-72, a Fulbright Lecturer at the Technion-Israel Inst of Tech in 1979, Visiting Chevron Energy Prof at CA Inst of Tech in 1980, Visiting Prof at the Univ. of Melbourne in 1985. Received AIChE Professional Progress Award in 1977, CA Tech Lacey Lectureship in 1979, Notre Dame Reilly Lectureship in 1980, AIChE William H. Walker Award in 1984, LSU Bicentennial Commemoration Lectureship in 1984, Society of Rheology Bingham Medal in 1986, Purdue Kelly Lectureship in 1987. Elected to Natl Acad of Engg in 1986. Books "Optimization by Variational

Denn, Morton M (Continued)
Methods" (1969), "Introduction to Chem Engrg Analysis" (with T.W.F. Russell, 1972), "Stability of Reaction and Transport Processes" (1975), "Process Fluid Mechanics" (1980), "Process Modeling" (1986). *Society Aff:* AIChE, SOR, BSR, PPS.

Dennard, Robert H
Business: P.O. Box 218, Yorktown Heights, NY 10598
Position: Staff Member *Employer:* IBM Research Lab *Education:* PhD/EE/Carnegie Inst of Tech; MS/EE/So Methodist Univ; BS/EE/So Methodist Univ *Born:* 9/5/32 in TX. Joined IBM Research in 1958 to work on digital circuit techniques. Active in MOSFET device and circuit research since 1963. Authored basic patent on dynamic one-device memory cell. With co-workers, introduced scaling concept for miniaturizing FET's. Appointed IBM Fellow in 1979. IEEE Fellow since 1980. Active in tennis and Scottish Country Dancing. *Society Aff:* IEEE

Denney, Roger P, Jr
Business: 6531 River Clyde Dr, Winnetka, Highland, MD 20777
Position: Exec VP *Employer:* Block, McGibony, Bellmore and Assoc, Inc *Education:* MSE/IE/Johns Hopkins Univ; BES/IE/Johns Hopkins Univ *Born:* 4/26/37 Native of Frederick, MD After coll, joined Westinghouse Electric, first as engrg recruiter, then manufacturing planner, from 1961-65. Served as Editor-in- Chief and Dir of Publications for IIE from 1965-70.Served successively as Senior Consultant, Dir of Community Health Ser, VP for Govt Ser, Sr VP for Eastern Region, Executive VP for Operations, MEDICUS Sys Corp, 1970-84. Joined current employer in 1984. Mbr of Editorial Advisory Bd for Wiley Hanebood of Indus Engrg, 1978-82. VP-Publications, 1981-83; Pres-Elect and Chief Financial officer, 1983-84; President, 1984-85, IIE. Registered PE. *Society Aff:* IIE

Denning, Anthony J
Business: P.O. Box 909, Piscataway, NJ 08854
Position: Exec Officer, Indust Engrg Dept *Employer:* Coll of Engrg Rutgers Univ (Retired) *Education:* IE/IE/Columbia univ; MS/IE/Columbia Univ; BS/ME/Newark Coll of Engrg *Born:* 11/17/20 Native of Newark, NJ. Aircraft design, Curtiss-Wright Corp St. Louis, MO. Served U.S. Navy WWII (U.S. Retired Reserve). Manufacturing Engr, Western Electric Co. Methods & Standards Engr, E.I. DuPont Co. Major Contributor to New Plant Design, Otis Elevator Co 1956-67 Rutgers Univ 1967 to date, chrmn of New Industrial Engrg Dept 1981-83. Exec officer 1983-87. Retired Prof Emeritus 1987. Raritan Valley Chapter (IIE), Pres 1971, present member of Bd of Trustees Vice Chrmn, Industrial Engrg Section ASEE Mid-Atlantic Section 1973 to date. Registered PE in NJ and CA. Maintains active part time consulting practice in manufacturing, chmn zoning Bd of Adjustment, Franklin Township, NJ Married, four children. Interests photography and woodworking. *Society Aff:* ASME, IIE, ASEE, AAUP, SME.

Dennis, John F
Business: 406 South Prospect Ave, Clearwater, FL 33516
Position: VP & Dir *Employer:* Briley, Wild & Assocs, Inc *Education:* BS/CE/MIT *Born:* 04/01/29 Native of New England; PE-FL, NY, PA, MA, ME, and NH. VP and Dir, Briley, Wild & Assocs. Supervises engrs and other personnel on investigation design, construction and report projs in Western FL. Projs include water supply, sewage interceptor and wastewater treatment facilities. Mbr FL Engrg Soc. *Society Aff:* NSPE, WPCF

Dennis, William E
Business: PO Box 490, Aliquippa, PA 15001
Position: Dir-Quality Control-Eastern Div. *Employer:* Jones & Laughlin Steel. *Education:* PhD/Phys Chemistry/London Univ; DIC/of Steelmaking/; BS/Met/Durham Univ. *Born:* 10/4/26. W E Dennis first worked for the British Atomic Energy Authority and later served as Chief Metallurgist for GEC Atomic Energy Div. In the steel industry, he served as Deputy Gen Mgr at Redbourn Works and later as Deputy Chief Technical Officer of Richard, Thomas and Baldwin. Coming to the USA and Jones & Laughlin Steel in 1966, he has served as Dir of Res, Genl Mgr of Quality Control and is presently Dir of Quality Control for the Eastern Div. A Fellow of ASM, he is the author of many articles and the holder of 7 patents. *Society Aff:* AIME, ASM, AISI.

Dennison, Robert A, Jr
Home: 2602 Village Dr, Brewster, NY 10509
Position: Sr VP *Employer:* Cashin Assocs, P.C. *Education:* Juris Doctor/Law/Geo Washington Law School; B/CE/RPI *Born:* 6/21/26 Mohawk, NY - Served US Navy 1944-45 - 11 years for NYSDPW as Supervising Constr Engr - 3 yrs US Bureau of Public Roads as Chf Prel Engr Branch, Fed Highway Prov Div - Comm of Public Works, Westchester County, NY 1971-1974 - Prof of Civil Tech, Westchester Community Coll 1974-1981- With Cashin Assoc, P.C., as Sr VP, Secretary, and Dir since 1977 - VP, Area 3, NYS SPE 1977-1981 - Founding Trustee and first Pres of Practising Inst of Engrg 1980-1981.- Licensed PE in NY and FL - Licensed Land Surveyor in NY - Member of NYS Bar and admitted to practice US Court of Appeals, 2nd Circuit. *Society Aff:* NSPE

Denno, Khalil I
Home: 68 Ridgeway Ave, W Orange, NJ 07052
Position: Prof. *Employer:* NJ Inst of Tech. *Education:* PhD/EE/IA State Univ; MEE/EE/RPI; BSc/EE/Univ of Baghdad. *Born:* 12/25/33 US citizen, prof of Elec Engg at NJ Inst of Tech in Newark, NJ, & his maj fields are elect power, electromech energy conversion & direct energy conversion. The author of large number of technical papers in the above- mentioned fields, published in US & Europe, Licensed PE in the state of NJ & chartered engr in Great Britain & all British Commonwealth Countries. Sr Fellow in IEEE, Fellow in the IEE (England) & mbr in the Am Assn of Engg Education. Received the Excellence for Res Award in 1982, mbr in the Amer Nuclear Soc, honored with the Award of the Cultural Doctorate in Phil of Engrg by the World Univ Roundtable in 1983, and honored with numerous listings in the natl and intl biographies and honor socs. Author, "Power System Design and Applications of Alternative Energy Sources," Prentice-Hall 1987. *Society Aff:* IEEE, IEE, ASEE, ANS.

Denny, James O'H III
Business: 149 Engrg Sciences Bldg, Coll of Engrg, Morgantown, WV 26506
Position: Asst to the Dean *Employer:* WV Univ *Education:* ML/Bus Adm/Univ of Pitt; BS/Mgt Engr/Carnegie-Mellon Univ *Born:* 8/2/19 1941, taught women Design Engrg at H. H. Robertson Co after working three years as Structural Design Engr. Joined US Army Corps 1942, held positions of pilot, airplane commander, and flight leader. Graduated from Carnegie-Mellon Univ 1948. Taught undergraduate classes, Coll of Engrg. Rejoined H.H. Robertson Co as estimator, field design engr, and sales engr 1948. Accepted position with J.S. McCormick Co 1951. Held positions in quality control, inspection, production, sales, and service. Was first Dir of Res. Later became VP of Development. Retired from the business world 1971, joined the Industrial Engrg Dept Coll of Engrg (COE), WV Univ. Sec of the Pittsburgh Chap of the American Foundrymen's Society and Treasurer of the Monongahela Chapter of the IIE. *Society Aff:* ASEE, IIE, AFS

Denton, Richard T
Business: Packard Laboratory 19, Lehigh University, Bethlehem, PA 18015
Position: Department Head & Prof *Employer:* Lehigh Univ. *Education:* BS/EE/Penn State; MS/EE/Penn State; PhD/EE/Univ of Mich. *Born:* Jul 13 1932. Served as Mbr of the Steering Ctte for INTERMAG in 1962, Mbr of Program Ctte for NEREM 1968 & Mbr of Education Ctte of the North Jersey Section of IEEE. Employed by Bell Labs since 1954 with respon to 1968 for dev of magnetic, acoustic, & optical solid state devices. From 1968-85 has supervised dept with respon for dev of new signal processing equipment for military systems. NSF fellowship 1954, NEC fellowship 1957. In 1985, he joined the faculty of Lehigh Univ and was appointed Prof of Comp Sci and Elect Engg. *Society Aff:* IEEE, APS.

Deo, Narsingh
Business: Computer Sci Dept, University of Central Florida, Orlando, FL 32816
Position: Millican Chair Professor *Employer:* Univ of Central Florida *Education:* PhD/Elec Engg/Northwestern Univ; MS/Elec Engg/CalTech; Diploma IISc/Elec Tech/Indian Inst of Sci; BSc (Hons)/physics/Patna Univ (India). *Born:* 1/2/36. Native of India, first came to USA for grad studies; worked as electronics engg for Burroughs for three yrs and for Jet Propulsion Lab for five yrs as mbr of tech staff. Taught at UCLA, Univ of IL (Urbana), and Univ of Nebraska. Was Prof of Electrical Engg at Indian Inst of Tech, Kanpur from 1971-77. Was Prof of Computer Science 1977-86, and Chairman from 1980. Assumed current position as an endowed chair professor in computer science at Univ of Central Florida in 1986. Author of four textbooks (Prentice-Hall) and 60 research papers. Holds three US patents on computer hardware. Recipient of Drake Scholarship at Cal Tech and Apollo Achievement Award from NASA. *Society Aff:* ACM, IEEE.

DePhillips, Fred C
Business: 601 Bergen Mall, Paramus, NJ 07652
Position: Assoc. *Employer:* N H Bettigole Co. *Education:* BCE/Civ Engg/Manhattan College. *Born:* 2/5/32. Mr DePhillips has had an engg career covering both the private sector as well as the public arena. Beginning as a Jr Engr in 1955 for a NJ consulting engg firm, he went on to ten yrs of private practice before becoming the full-time municipal engr for the town of Leonia, NJ, in 1964. This was followed by an appointment as Bergen County Engr & Dir of Public Works in 1968, & then Asst Commissioner of Transportation for the State of NJ in 1971. After nine yrs in govt service, he returned to the private practice of consulting engg in 1973. *Society Aff:* ASCE, NSPE, SAME, WPCF.

Depp, Oren Larry, Jr
Business: 2625 Frederica St, Owensboro, KY 42301
Position: President. *Employer:* Johnson Depp & Quisenberry CEs. *Education:* BSCE/Civil Engr/Univ of KY. *Born:* Aug 3 1931. Native of Owensboro, Kentucky. Served in USAF upon graduation as Air Installation Engrg Officer in Narsarsuak, Greenland. Joined Johnson, Depp & Quisenberry as a Design Engr in June 1956. Principal in the firm in 1957. Pres 1970. Firm is actively involved in all phases of Civil Engrg with 76 employees currently. Registered Prof Engr in states of Kentucky, Indiana, Illinois, Tennessee, West Virginia & Florida & holds NCEE registration. Has served as Pres & Natl Dir of Kentucky Soc of Prof Engrs. Pres & Natl Dir of Consulting Engrs Council of Kentucky. Received Distinguished Service Award from KSPE in 1975. *Society Aff:* KSPE, NSPE, CECKY, ACEC, ASCE.

Derenyi, Eugene E
Business: Dept. of Surveying Eng, P.O. Box 4400, Fredericton, N.B, Canada, E3B5A3
Position: Professor of Surveying Engineering. *Employer:* The University of New Brunswick. *Education:* PhD/Photogrammetry/Univ of New Brunswick; MSc/Geodesy/Univ of New Brunswick; Dipl Ing/Surveying/Tech Univ of Sopron. *Born:* 1932 Sopron Hungary. Registered Prof Engr in the Province of Ontario. Was Lecturer & Res Assoc, University of Sopron until 1956; Asst Surveys Coordinator Ontario Hydro Commission Toronto until 1961. Appointed to the faculty, University of New Brunswick in 1963 & presently holds the rank of Professor. Photogrammetry and remote sensing specialist. Author of 60 scientific publications. Recipient of the Talbert Abrams Grand Trophy Award 1972. American Soc of Photogrammetry. *Society Aff:* ASPRS, CISM, CASI.

Deresiewicz, Herbert
Business: Dept. of Mechanical Engineering, Columbia University, New York, NY 10027
Position: Prof *Employer:* Columbia Univ *Education:* PhD/Applied Mech/Columbia Univ; MS/CE/Columbia Univ; BS/ME/Coll of City of NY *Born:* 11/5/25 Served with US Army, 1946-47. Senior staff engr, Applied Physics Lab, The Johns Hopkins Univ, 1950-51. Member of faculty of Columbia Univ since 1951, prof of mechanical engrg since 1962, chrmn of dept since 1981. Fulbright senior research scholar, National Inst of Applied Mathematics, Rome, Italy, 1960-61. Fulbright lecturer, Weizman Inst of Science, Rehovoth, Israel, 1966-67. Visiting prof of mechanics, Israel Inst of Tech, Haifa, and visiting prof of mechanical engrg, Ben-Gurion Univ, Beersheba, Israel, 1973-74. Fields of research are stress analysis, vibrations, elastic contact, wave propagation, mechanics of granular and of porous media. *Society Aff:* AAAS, SSA

DeRoze, Donald G
Home: 13103 Tamarack Rd, Silver Spring, MD 20904
Position: Spec Asst for Dod Reporting. *Employer:* Defense Intell Agency. *Education:* MS/ME/RPI; ME/ME/Univ of Cincinnati. *Born:* 11/21/34. Taught Mech Engg courses at sr undergrad level (Univ of Dayton) & performed post grad work at the Univ of MD. Worked in aerospace res at United Aircraft Corp for three yrs, obtaining a patent in space cabin environmental control. Worked in military intelligence on foreign weapons & tech for 21 yrs. Latest position for eight yrs requires authorship & verbal briefings on scientific & technical intelligence to key govt technical personnel including the under secy of defense for res & engg. Enjoy tennis, running, fishing, hunting. Active in church work. *Society Aff:* ΤΒΠ, ΠΤΣ.

Derr, Jack
Home: 16 Woodland Park Dr, Parkersburg, WV 26101
Position: VP of Engg. *Employer:* Borg Warner. *Education:* BS/CE/Penn State. *Born:* 11/29/27. *Society Aff:* AIChE, NSPE.

Dertinger, Ellsworth F
Business: 204 Edison Way, Reno, NV 89520
Position: Chrmn & Ch Exec Off. *Employer:* Lynch Communication Systs Inc. *Education:* BSEE/Elec Eng/Univ of AL. *Born:* Sep 1920. BSEE from University of Alabama. Native of Baltimore, Md. Served in US Army Signal Corps 1943-46. 30 yr background in electronics field with respon for design engrg, reliability engrg, quality control, manufacturing & administration. With Arma, Bendix Missile & Radio Divisions, Raytheon & ITT. Taught reliability engrg for 7 yrs at Northeastern University. Conducted employee motivation & quality control seminars throughout the US & Canada. Elected Pres of Lynch Communication Systems in 1972. Fellow Mbr in both Institute of Electrical & Electronic Engrs 1964 & American Soc for Quality Control 1967. Enjoy golf, swimming & reading. *Society Aff:* IEEE, ASQC.

Dertouzos, Michael L
Home: 15 Bernard Lane, Waban, MA 02168
Position: Dir & Prof MIT Lab for Computer Sci *Employer:* MIT. *Education:* PhD/EE/MIT *Born:* 11/5/36. in Athens, Greece. PhD EE MIT 1964. MIT Faculty 1964- . Fellow IEEE; Best Educator (Terman) Award ASEE 1975; Best Paper (Thompson) Award IEEE 1968; Auth 5 bks, sev papers, pats in computer field; Founder, Bd/Chmn Computer Inc 1968-75. Tr Athens C; Bd Chmn Boston Camerata, Early music. Mbr Athens Academy. *Society Aff:* IEEE

DeRusso, Paul M
Business: RPI Engrg, Troy, NY 12181
Position: Assoc Dean of Engrg. *Employer:* Rensselaer Polytech Inst. *Education:* ScD/EE/MIT; MEE/EE/RPI; BEE/EE/RPI. *Born:* Sep 9 1931. Principal employment, Rensselaer Polytechnic Inst: Asst Prof 1959-61; Assoc Prof 1961-64; Prof 1964- ; Chmn, Systems Engrg Div 1967-74; Assoc Dean of Engrg 1974- ; E I du Pont de Nemours, Engrg Dept, Year-in-Industry Prof 1966-67. Cons for NYSDE, Robotics, Albany Med Ctr, USARO, G E, others. Reg P E in NY. Sr Mbr IEEE. Co-Auth 'State Variables for Engrs' & num tech papers & reports. Enjoy golf, bowling, horserace handicapping, hi-fi & card games. *Society Aff:* IEEE, ASEE, NY AcSci.

Desai, Drupad B
Business: 201 N. Charles St, Suite 1900, Baltimore, MD 21201
Position: Assoc VP *Employer:* DMJM *Education:* MS/Structural Engr/Univ of WI-

Desai, Drupad B (Continued)
Madison; BS/CE/S. V. Univ-India; Intermediate Sci/Math/Bombay Univ *Born:* 01/08/40 Presently Assoc VP for DMJM 1968 to present. Proj Mgr for Vancouver ALRT Chief Engr for Balto Subway Sys sections A & B, Proj Mgr for Buffalo LRRT's S. Campus Station, Construction Mgr for Buffalo LRRT's Surface Section Chief Engr for Cornell Collider Proj, Principal Design Engr for WMATA's tunnels and subway structures. Licensed as PE in MD, PA, VA, IL, FL, DC, NY, and NJ. Over 20 yrs experience in this field. Published papers for MA-DOT on Charles Center Slurry Walls, 1980; ASCE, PA Section on Concrete Tunnel Liners; etc. *Society Aff:* ASCE, NSPE, USNCTT, UTRC, ISRM

DeSalvo, Joseph M
Home: 264 Lynn Dr, Franklin Lakes, NJ 07417
Position: Senior Vice President *Employer:* Converse Ward Davis Dixon *Education:* MS/Civil Engg/Columbia Univ; BS/Civil Engg/Rutgers Univ. *Born:* Jul 4, 1931. Instructor & Asst Prof of CE at The Cooper Union, New York City, 1952-57. With Joseph S Ward & Associates from 1952 to 1978, becoming partner in 1960, sharing respon for technical & administrative functions. Sr VP of Converse Ward Davis Dixon following a 1978 merger of Joseph S Ward & Assoc & Converse Davis Dixon Assoc. Active in ASCE cttes incl P Chmn Executive Ctte of Geo-Technical Engrg Div; Past Chmn Information Advisory Ctte of ISSMFE; active in Cons Engrs Council & Assn of Soils & Foundation Engrs; P Trustee of Engrg Index; Deep Foundation Institute. Hobbies incl music & golf. *Society Aff:* ASCE, ASFE, ACEC, DFI, ISSMFE.

De Serio, James N
Business: 86 W Chippewa St, Buffalo, NY 14202
Position: Pres *Employer:* De Serio-Abate Engineers, P.C. *Education:* BS/CE/Northeastern Univ *Born:* 12/19/10 Vastogirardi, Italy. Employed Truscon Div of Republic Steel Co 1935-1941 as District Engr, Buffalo, NY; Chief Engr Bero Engrg & Construction Corp, 1942- 1943; Construction Engr for Line Air Prods Co, Tonawanda NY, 1943-1944 on Manhattan Project; 1944 to present, self employed as Consulting Engr on Civil and Structural Engrg projects. Taught Professional Engrg refresher courses for 21 years; Authored book *Notes on Structural Planning & Design*; had published 16 Technical papers on Engrg subjects in National Technical Publications. Licensed in states of NY, PA, MN, AL. Registered with National Council of Engrg Examiners. Member of NY State Bldg Code Council, 1967-1979; Chrmn of State Bldg Code Bd of Review, 1974-1979. VP of CSI 1966; Member National Council of Northeastern Univ, 1960 to present. Chairman Civil Engg Technology at Erie County Community College, 1985 to present. *Society Aff:* ACI, ACEC, ASCE, CSI, NSPE

De Shazer, James A
Business: Dept of Agriculture, Univ. of Nebr. - Lincoln, Lincoln, NB 68583-0726
Position: Prof of Ag Engr *Employer:* Univ of NB *Education:* PhD/Bio & Ag Engr/ NC State Univ; MS/Ag Engrg/Rutgers Univ; BS/ME/Univ of MD; BS/Gen Agr/ Univ of MD *Born:* 7/18/38 Directed the First Intl Livestock Environment Symposium - 1974 Lincoln, NE. Awards include National Metal Bldg Manufacturer's Association Award - 1979, Outstanding Young Agricultural Engr for Mid Central Region (MO, KS, IA, NE) - 1975, and Outstanding Young PE (NE) - 1975. He has been Chrmn of the Mid Central Region of ASAE - 1979 to 1980 and National ASAE Committee Chrmn of the Environmental Group - 1980 to 1981; Environmental Physiology - 1973 to 1974, Struct and Environ Div - 1984-85. He has developed design criteria for livestock housing. He has published extensively in areas of scientific research, in engrg education, in extension service publications and in the farm press. *Society Aff:* ASAE, NSPE, ASEE, CAST.

DeShazo, John J, Jr
Business: 330 Union Station, Dallas, TX 75202
Position: President *Employer:* DeShazo, Starek & Tank, Inc. *Education:* BS/CE/TX A&M Univ; Cert/Transportation/Yale Univ; Grad/Urban Exec Course/MIT. *Born:* 8/23/27. Married: Bonner Burke Dec. 22, 1962; Son Charles. Served US Army Corps of Engrs; Design Engr, TX Highway Dept 1950-56; Dir of City Planning & Traffic Engrg, City of Amarillo 1957-66; Dir of Traffic Control, City of Dallas 1966-75; VP Young Hadawi DeShazo Inc, Cons Engrs 1975-80; Pres DeShazo, Starek & Tang, Inc., Engrs/Planners 1980-. P Pres TX Sect Inst of Trans Engrs, Dallas Chapter TSPE, Panhandle Chapter TSPE, High Plains Branch ASCE. P Mbr ITE Tech Council, Texas Transportation Engineer of the Year - 1984, Col USAR. *Society Aff:* NSPE, ITE, ASCE

De Silva, Ananda SC
Business: 470 Park Ave South, New York, NY 10016
Position: Principal *Employer:* Ewell W Finley, P.C. *Education:* MS/Ind Management/Polytechnic Inst of Brooklyn; BS/Engrg/Univ of Ceylon *Born:* 12/4/39 Native of Sri Lanka. Graduated with honors, Univ of Ceylon, 1965. Recipient Fulbright-Hays Scholarship, 1969. Joined Finley & Madison Consulting Engrs 1969. Principal-in-charge structural engrg, Ewell W Finley, PC, Ewell W Finley & Partners, Inc 1973 to present. Directs structural engrg designs for firm, fiscal administration. Recipient co Award for Engrg Excellence, Honor Award, Structural Engrg Design Services for Atlanta Life Ins Corp Headquarters, Atlanta, GA, from NY Association of Consulting Engrs, May 16, 1981. Enjoys classical music, stamp collecting, tennis, swimming. PE-NY, NJ, GA, PA and MA. *Society Aff:* ASCE, ACI, AISC, CRI, ICE, NYSPE, NYACE, SAME

De Simone, Daniel V
Business: 415 Second St NE, Washington, DC 20002
Position: Exec Dir *Employer:* AAES *Education:* JD/Law/NYU Sch of Law; BS/Elec Engrg/Univ of IL *Born:* 05/04/32 Bell Telephone Labs Staff (1956-62). Conslt to US Dept of Commerce (1962- 64). Dir, Office of Invention and Innovation, Natl Bureau of Standards (1964- 69). Dir, US Metric Study (1969-71). White House Office of Sci and Tech (1971- 73). Exec Dir, Fed Council for Sci and Tech (1972-73). Deputy Dir, Congressional Office of Tech Assessment (1973-80). President, Innovation Grp, Inc (1980-84). Exec Dir, AAES (1984-), First VP, World Federation of Engineering Organizations (1987-). Served in the Force during the Korean War. Mbr Eta Kappa Nu, Tau Beta Pi, NYU Law Review Alumni Assoc, IEEE, AAAS, Cosmos Club of Washington, DC. Published works include *Technological Innovation: Its Environment and Management* (1967); *Education for Innovation*, with J. H. Hollomon, et al. (1968); *A Metric America* (1971). *Society Aff:* IEEE, AAAS

Desoer, Charles A
Business: Dept EECS, Berkeley, CA 94720
Position: Professor. *Employer:* University of California. *Education:* ScD/EE/MIT; Dsc (Hon.)/-/Univ. of Liege. *Born:* 1/11/26. Charles A. Desoer worked at Bell Telephone Labs, Murray Hill, 1953-58. From 1958 – present, Dept of Elec Engrg and Computer Scis, Univ of CA, Berkeley. 1967-68, Miller Inst. 1970-71, Guggenheim Fellowship. Best Paper Prize (with J Wing), JACC 1962. Medal of the Univ of Liege, 1970. Distinguished Teaching Awd, Univ of CA, Berkeley, 1971. 1975 IEEE Education Medal. Prix Montefiore 1975. Amer Automatic Control Ed Awd 1983. 1986 Control Systems Science and Engineering Award. Fellow of AAAS. Member of Natl Acad of Engrg, 1977. One of his papers received Honorable Mention from IEEE Control Soc 1979. His interests lie in sys theory with emphasis on control sys and circuits. Author, with L A Zadeh, *Linear System Theory* (1963); with E S Kuh, of *Basic Circuit Theory* (1969); of *Notes for a Second Course on Linear Systems* (1970); with M. Vidyasagar, of *Feedback Systems: Input-Output Properties* (1975); with F M Callier, of *Multi-variable Feedback Systems* (1982) with L.O. Chua and E.S. Kuh, of Linear and Nonlinear Circuits (1987). Dr Sc (hon.) Liege 1976. *Society Aff:* IEEE, AMS, SIAM, MAA.

deSoto, Simon
Business: 1250 S Bellflower Blvd, Long Beach, CA 90840
Position: Prof of ME. *Employer:* CA State Univ. *Education:* PhD/Engg/UCLA; MME/ME/Syracuse Univ; BME/ME/CCNY. *Born:* 1/8/25. in NYC. Taught in Engg Schools at Syracuse Univ (48-50) UCLA (53-69), CA State Univ, Long Beach (69-present). Received Outstanding Faculty Award at UCLA (1962) & at CA State Univ, Long Beach (1971, 73, 76). VChrmn of Mech Engg Dept at CA State Univ, Long Beach. Res Engr at Stratos Fairchild (50-53) developing air cycle machines & heat transfer equip for jet aircraft. Turbomachinery consultant, & res specialist on high speed heat transfer for Lockheed Missile Systems Div 1953-56, sr tech specialist at Rocketdyne Div of Rockwell Intl 1956- 69. Dev of rocket engines. *Society Aff:* AAAS, WFS, ΤΒΠ, ΠΤΣ.

DeSpirito, Victor P
Business: 232 Alice Ave, Solvay, NY 13209
Position: Plant Engineer. *Employer:* Syroco-Div Dart Indus Inc. *Born:* 5/6/22. Native of Syracuse, New York. Industrial Engr for Easy Washing Machine Co Syracuse NY 1943-63. Industrial Engr of McMillan Book Co Syracuse NY 1963-65. Plant Engr for Syroco, Div of Dart Industries Inc Syracuse NY 1965-76. Also respon for complete Safety Program (OSHA) at Syroco. Have specialized in methods, facilities planning, plant layout, plant engrg & safety. Have been Pres of the local chapter of the following socs: IMMS 1974-76, IIE, Easy Mgmt Club cert by IMMS in both Mat Handling & Mat Mgmt. Enjoy golf, bowling, fishing, music & gardening. *Society Aff:* IIE, IMMS, AIPE, ASSE.

Despres, David R
Home: 1924 Meadowfield Dr, Grand Rapids, MI 49505
Position: Dir Dept Public Works. *Employer:* County of Kent-Michigan. *Education:* MSCE/Civil Engg/Univ of MI; BSCE/Highway Eng/Univ of MI. *Born:* Dec 21 1926 Grand Rapids MI. BSCE 1955, MSCE 1956 U of Michigan. Reg Engr MI. Pres MI Sect ASCE 1974-75. Field Service Engr, Marketing Div of Standard-Vacuum Oil Co White Plains NY 1956-57. Design & Project Engr, Brewer Engrg Owosso MI 1957-60. Field contact engrg in wastewater collection & trtmt, Michigan Dept Public Health Lansing MI 1960-63. Dir, Kent County Dept of Public Works Grand Rapids MI 1963- . *Society Aff:* AWWA, ASCE, WPCA, NSWWA.

Dessauer, John H
Business: P.O. Box 323, 37 Parker Dr, Pittsford, NY 14534
Position: Retired. *Employer:* Xerox Corp. *Education:* Doctor/Chem Engg/Inst of Technology, Aachen Germany; Master/Chem Engg/Inst of Technology, Aachen Germany/ BS/Chem Engg/Inst. of Technology Munich Germany. *Born:* May 13, 1905 Germany. Employed: Agfa-Ansco Binghampton NY 1929-35, Haloid-later Xerox Corp 1935-70. Org Xerox R&D effort, discovered Carlson & his xerographic invention. Respon for dev of xerographic process to commercially useful machines. Elect VP Res & Engrg, Exec VP & 1968 V Chmn/Bd. Dir until 1973 Mbr NAE; Tr Emeritan Fordham U. Sev Pats & Pubs incl *My Years with Xerox* 1971 & *Xerography & Related Processes* jointly with H E Clark 1965. *Society Aff:* NAE.

DeTemple, Thomas A
Business: 1406 West Green St, Urbana, IL 61801
Position: Prof of EE. *Employer:* Univ of IL. *Education:* PhD/EE/Univ of CA; MS/Phys/CA State Univ; BS/Phys/CA State Univ; AA/Sci/Skagit Valley College. *Born:* 9/2/41. Native of NY, past employment includes the Navy Electronics Lab, San Diego and the Univ of AZ, Tuscon. Presently at the Univ of IL presuing res in infrared & far infrared gas lasers, strong light-matter interactions & gaseous electronics & plasma kinetics. *Society Aff:* IEEE, APS.

Deutsch, George C
Home: 8303 Whitman Dr, Bethesda, MD 20034
Position: Dir Materials & Structures Div. *Employer:* NASA. *Education:* BS/Metallurgy/Case-Western Reserve Univ. *Born:* Apr 19 1920 Budapest Hungary; came to US 1924; married to Ruth A Amster Oct 4 1942; 3 children: Fred, Harvey & Marilyn. With Copperweld Steel Corp 1942- 44; materials res NASA Cleveland Ohio 1946-60, Washington 1960, Dir Mat & Struc Div 1970-81 Dir, Res and Technology Div. 1981-present-Conslt Engr. USNR 1944-46 (to Lt-jg). Recipient Except Serv Medal NASA 1964. Fellow ASM; Mbr AIME. *Society Aff:* ASM, AIME.

Deutsch, Sid
Business: Electrical Engr. Dept, University of South Florida, Tampa, FL 33620
Position: Visiting Professor. *Employer:* University of South Florida. *Education:* DEE/Electrical Engrg/Poly. Inst. of Brooklyn; MEE/Electrical Engrg/Poly. Inst. of Brooklyn; BEE/Electrical Engrg/Cooper Union. *Born:* Sep 19 1918 NYC. BEE 1941 Cooper Union; MEE & DEE 1947 & 1955 from Polytech Inst of Brooklyn. Started 1935 as elec motor tech, designer of elec- mech equip, in US Navy, electron engr at Poly R&D Co, Microwave Res Inst, Cons for Budd Electrons, Affil of Rockefeller Inst. Since 1943, taught physics at Hunter C, Radio & TV at CCNY, Polytech Inst of Brooklyn. Prof Bioengrg Rutgers Med School & Adj Prof EE Rutgers U. Fellow IEEE & Soc for Information Display. Was Director, Biomedical Engineering Program, Tel Aviv University. Now Visiting Prof of EE. at Univ. of South Florida. *Society Aff:* IEEE, SID.

DeVelis, John B
Business: Turnpike Rd, N Andover, MA 01845
Position: Prof of Physics. *Employer:* Merrimack College. *Education:* PhD/Physics/Boston Univ; MA/Physics/Boston Univ; BA/Physics/Boston Univ. *Born:* 1/24/35. Fellow of the Optical Soc of America. Fellow of the Soc of Photo-Optical Instrumentation Engrs. Contributor, author and/or editor of numerours papers, books, and articles in the fields of holography, physical optics, classical and quantum coherence, and optical instrument design. Former editor of "Optical Engineering–, former Dean of Sci and Eng at Merrimack College and past governor of the Soc of Photo-Optical Instrumentation Engrs. *Society Aff:* SPIE, OSA.

Devereux, Owen F
Business: Box U-136, Storrs, CT 06268
Position: Prof. *Employer:* Univ of CT. *Education:* PhD/Metallurgy/MIT; MS/Metallurgy/MIT; BS/Metallurgy/MIT *Born:* 8/23/37. Native of Lexington, MA. Subsequent to formal education employed as Res Chemist at Chevron Res Co (La Habra, CA) 1962-4, Corning Glass Works (Corning, NY) 1964-6, Chevron Oil Field Res Co (La Habra, CA) 1966-8. Joined Univ of CT in 1968. Presently Prof and Head, Dept of Metallurgy and Mbr Inst of Materials Sci. *Society Aff:* ECS, NACE, AIME, AAUP

Devey, Gilbert B
Business: 1800 G Str NW, Washington, DC 20550
Position: Head, Middle East Section. *Employer:* National Science Foundation. *Education:* BS/Elec/MA Inst Technology. *Born:* Jan 1921. BS Mass Institute of Technology 1946. Native of East McKeesport, PA. Attended Carnegie Inst Tech 3 yrs prior WWII. Served as radar officer, USS SARATOGA; attended US Naval Acad PG School 1941-45. Antenna systems engr, Navy Bu Ships & USNUSL, New London, CT 1946-49. Electronic scientist, Off Naval Res 1949-52. Field engr, product mgr & marketing mgr, special products, Sprague Elec 1953-65. NSF Program Dir in Electrical Sciences, Instrumentation Technology & Internatl Programs, 1965-67; 1969- . Natl Acad Engrg 1967-69. AAMI Foundation Award 1975; Fellow, IEEE 1976. Amateur radio & photography hobbies. *Society Aff:* ISEE, AIHM, SAMS, BMES.

Devitt, Timothy W
Business: Chester Square Towers, Cincinnati, OH 45246
Position: Executive VP *Employer:* PEI Associates, Inc. *Education:* BS/Chem Engg/ Univ of CA, Berkeley. *Born:* 8/2/44. Chem Engr, Univ of Calif, Berkeley; Environ engg studies, Univ of Cincinnati. Worked as a Chem Process Engr with Stauffer Chem Co. In 1968, joined the US Environmental Protection Agency. Served as a

Devitt, Timothy W (Continued)
Proj officer directing EPA sponsored res on advanced methods for controlling atmospheric emissions. Joined PEI Associates 1970. Currently serving as Executive VP with respon for proj execution & corp dev. Served as Cons to the US EpA on natl & internatl control technology dev programs. Diplomat-, American Academy of Environmental Engineers, Mbr of the Technical Ctte of the Air Pollution Control Assn; received the High Score Award in Professional Engrs Examination in chemical Engg in OH. *Society Aff:* AIChE, APCA, OSPE.

Devonshire, Grant S
Business: 605-5th Ave. SW, PO Box 3333 Station M, Calgary Alta, Canada T2P2P8
Position: Manager Supply (West). *Employer:* Texaco Canada Inc. *Education:* BSc/Chemistry/Univ of Alberta. *Born:* Nov 1929. Native of Banff, Alberta. Joined Texaco Canada 1951 as Dev Chemist-Lubricants. Later Co experience has been in areas of sales, product application, technical service, and supply & distribution. Currently respon for crude and refined product supply in Western Canada. Reg Prof Engr Provinces of Alberta & Ontario. Past Regional V P, Dir, Coord Indus Councils, ASLE. Past Chairman Alberta Section ASLE. *Society Aff:* APEO, APEGGA, CIM.

DeVries, Alfred L
Home: 22 Goodwin Ave, White Plains, NY 10607
Position: Chief Engr. *Employer:* Stauffer Chem Co. *Education:* BChE/CE/GA Inst of Tech. *Born:* 3/2/23. USN in WWII, 10 yrs production mgt, twenty plus yrs in Process & Proj Engg of Chem plants in various supervising & mgt levels. Mbr of Chem Cargoes Committee of Am Bureau of Shipping & previously involved with Chlorine Inst of Safe Trans Comm. Has been Chief Engr, Licensing since 1969 responsible for technical activities of licensing chem & tech. *Society Aff:* AIChE, ABS.

deVries, Douwe
Home: 802 Piney Point, Houston, TX 77024
Position: Pres *Employer:* Oilfield Systems, Inc *Education:* MS/Mech Engg/Tech Univ of Delft; BS/Mech Engg/Tech Univ of Delft. *Born:* 10/25/22. Born and educated in Holland. Naturalized US citizen living US since 1952. Worked for Shell Oil Co 16 yrs and was mbr of special Offshore Dev Group pioneering offshore oil and gas exploration tech. Has 13 patents in this field. Had own Consulting Firm "Project Engineering, Inc" which was later sold to Stewart & Stevenson and subsequently to NL Industries. Was in charge of design, mfg and sales of NL Subsea Production Systems. Now pres of his own co, Oilfield Sys, Inc; founded in 1981, mfg high pressure hydraulic equipment and offering oilfield equip cert, design and conslt serv specialized in the floating drilling and subsea prod fields. Mbr Deep Sea Drilling & OTEC Committees Natl Res Council & NSF. ASME Holley Medal 1979. Elder in Presbyterian Church. Enjoy golf, tennis, boating and music, arts. *Society Aff:* ASME, NSPE, AIME, API, MTS.

DeVries, Kenneth L
Business: Mechanical and Industrial Engrg, Salt Lake City, UT 84112
Position: Prof Assoc. Dean of Engineering *Employer:* Univ of UT *Education:* PhD/Physics/Univ of UT; BS/ME/Univ of UT; AS/CE/Weber Coll *Born:* Engr - Convair 1957, Assistant Prof, Assoc Prof, then Prof of Mechanical Engrg 1962 - present. Dept Chrmn, Mechanical and Industrial Engrg 1969-1981. Assoc. Dean of Engrg 1984-present. Head Polymer Program at National Science Foundation 1975-1976. Member NSF Advisory Council Materials Research Labs 1978- 1979, member Advisory Committee. NSF Division for Materials Research 1967-1979, NBS Materials Advisory Bd 1976-1979. Organizing Chrmn various professional societies national meetings. Research Interest - Mechanical properties of materials, polymers, medical and dental materials, rock mechanics. Published one book and numerous papers on research. *Society Aff:* ASME, ASTM, APS, ACS, SES, SEM.

De Vries, Marvin F
Home: 901 Tompkins Dr, Madison, WI 53716
Position: Prof of Mech Engg. *Employer:* The University of Wisconsin-Madison. *Education:* PhD/Mech Engg/Univ of WI-Madison; MSME/Mech Engg/Univ of MI; BSME/Mech Engg/Univ of MI; BS/Pre Engg/Calvin College. *Born:* 10/31/37. Native of Grand Rapids, MI. Prof at Wisconsin' since 1966 teaching & res in the manufacturing processes area. More than 60 publications. Registered PE in Wisconsin & Certified Manufacturing Engr. 1985-86 President of the Soc of Manufacturing Engg; ASTME Scholar 1960; SAE Teetor Award 1967; Full Member of the Internatl Institution for Production Engrg Res, CIRP. Fulbright Sr Lecturer, Visiting Prof, Cranfield Inst of Technology, England, 1979-80. Dir of the UW-Madison Manufacturing Systems Engrg program since 1983. 1984 recipient of the NASA and Rocketdyne Rockwell Intl Space Shuttle Tech Award. Fellow of ASME and SME. *Society Aff:* SME, ASME, CIRP.

De Witt, David
Home: 109 Via Teresa, Los Gatos, CA 95030
Position: IBM Fellow. *Employer:* IBM. *Education:* MS/EE/Columbia Univ; AB/Econ/Columbia Univ. *Born:* Jun 1915. Engr at Radio Receptor Co NY 1939-56. VP Semicons 1951-56. Award from War Dept for radar in WW2. With IBM 1956- . Semicon design, manufacturing, application. Co-author of 3 texts on semiconductor devices, mbr of SEEC, NIC, ISSCC, editor of IEEE Spectrum, Fellow IEEE, IBM Fellow, Visiting Prof UCB. *Society Aff:* IEEE, APS.

DeWitt, David P
Home: 3033 Georgton Rd, W Lafayette, IN 47906
Position: Prof. *Employer:* Purdue Univ. *Education:* PhD/Thermophysics/Purdue Univ; MS/ME/MIT; BS/ME/Duke Univ. *Born:* 3/2/34. Bethlehem, PA. Received ASME 75th Anniversary Outstanding Student Award (1955) Whitney Fellow 1955-56 (MIT) & Overseas Summer Fellow 1956 (College of Aero, England). Instr, Duke Univ, 1957-59. Following PhD, Res Physicist, Natl Bureau of Stds, 1963-65. During 1965-72 served as Deputy Dir, Thermophysical Properties Res Ctr (Purdue Univ). Visiting Physicist at the Physikalish- Technische Bundesanstalt, W Germany, 1970-71. Assoc Prof (1972) and Prof (1978) in Mech Engg, Purdue Univ. Res & consulting interests in heat transfer, thermo- optical properties/instrumentation and radiation thermometry. Holder of several patents. Author of numerous scientific articles & several books. *Society Aff:* ASME, AIAA, ASTM.

DeWitt, Frederick W
Home: 7419 Armstrong Rd, Orlando, FL 32810
Position: Vice President. *Employer:* Amick Const Construction Co Inc. *Education:* BS/CE/Swarthmore College. *Born:* April 6 1925. Partner Building Construction in Conn 1946-53. Project Engr Florida State Rd Dept 1954-57. County Engr Orange County Florida 1958-69, dual role as Mgr of County Sanitary Sewer Districts. Engr with Amick Construction Co Inc since 1969; perform estimating, billing & engrg coordinating duties at Amick. P Chmn & Mbr, Orange County Road Construction Advisory Bd. P Chmn & Mbr Orange County Underground Utilities Contractors Licensing Bd. Sen Mbr NSPE. Fellow Florida Engrg Soc. Fellow ASCE. Pres Florida Section ASCE. Charter Mbr, Florida Council of Engrg Societies. Pres, FL Council of Engg Society; Chairman, Engrs in Construction, FL Eng Society. Hobbies: golf & bowling. District 10 Director, ASCE. *Society Aff:* ASCE, FES, NSPE.

DeWitt, John D
Business: 601 Jefferson Str, Houston, TX 77002
Position: President. *Employer:* DeWitt & Co Inc. *Education:* BS/ChE/Vanderbilt Univ. *Born:* Jun 12 1938 Richmond Kentucky. Lived most of young life in Nashville Tennessee. Attended Vanderbilt University & graduated with a bachelor of engrg in chemical engrg. Employment 1963-67 Union Carbide, New York & Houston, in petrochemicals marketing. 1967-73 Coastal States Gas Producing Co, Houston, initially Mgr Commercial Dev, later VP Petrochemical Marketing. 1973- , Founder & Pres of DeWitt & Co Inc, Internatl Petrochemical Marketing Cons. Mbr AIChE (Marketing Program Chmn 1970-77, Dir Fuels & Petrochemicals Div 1974-77, Mbr Executive Ctte 1974-77, Mbr Steering Ctte Houston Petrogroup, Pollution

DeWitt, John D (Continued)
Solution Ctte, ACS, CMRA, EVAF, Southwest Chemical Assn. *Society Aff:* AIChE, NPRA, ACS, EVAF, CDA.

DeWitt, John H, Jr
Business: 3602 Hoods Hill Rd, Nashville, TN 37215
Position: Consulting Engr (Self-employed) *Born:* Feb 20 1906 in Nashville Tennessee. Attended Vanderbilt University Engineering School 1928. Received Distinguished Alumnus award 1974. Received IEEE Fellowship award 1951 for 'Achievements in field of broadcast radio engrg & for demonstration of radar reflections from the moon'. In 1964 received The Natl Assn of Broadcasters Engrg Achievement Award. Lt Col US Army Signal Corps WWII, Dir Evans Signal Laboratory 1944-45. Awarded Legion of Merit for invention of radar mortar locator. 1947-68 Pres WSM Inc, radio & TV stations, until retirement 1968. Registered Prof Engr Tennessee. Hobbies, Astronomy & Sailing. *Society Aff:* IEEE, ΣΨ.

Dexheimer, Wallace D
Home: 1216 W. Jackson Blvd, Spearfish, SD 57783
Position: Owner/Pres (Self-employed) *Education:* BS/CE/SD School of Mines *Born:* 9/9/29 Native of SD. Spent some thirteen years in CA in a consulting office doing design & field observation work. After spending a couple years in the CO area, moved back to SD where established own firm in 1973. Hobbies include music & outdoor sports as fishing & backpacking. *Society Aff:* ASCE, NSPE, PCI, ACI, SEAOC, NCEE

Dhanak, Amritlal M
Business: Dept of Mech Engg, E Lansing, MI 48824
Position: Prof, Mech Engg. *Employer:* MI State Univ. *Education:* PhD/ME/Univ of CA; MEngg/ME/Univ of CA; BS/EE/Univ of CA; BSc/Phys- Math/Royal Inst of Sci. *Born:* 7/13/25. in Bhavnagar, India and naturalized as US citizen in 1958. From 1950-56 I taught and did res in the field of heat transfer at the Univ of CA in Berkeley as sssoc in mech engg and grad res engr. From 1956-58 I was employed as proj engr for thermal design at the GE Co in Schenectady, NY. From 1958-61 I held the position of assoc prof of mech engr at the Rensselaer Poly Inst in Troy, NY. Since 1961, I have held the position of prof of mech engg at MI State Univ, E Lansing, MI. Classical music & backpacking are my favorite recreational activities. *Society Aff:* ASME, ISES, ΣΞ.

Dhir, Vijay K
Business: 2445 22rd St, Santa Monica, CA 90405
Position: Prof. *Employer:* Univ of CA. *Education:* PhD/ME/Univ of KY; MTech/ME/IIT; BSc/ME/Punjab Engg College. *Born:* 4/14/43. Vijay K Dhir obtained his PhD degree in Mech Engg in 1972. He joined the faculty of school of engg & applied sci at UCLA in 1974. At UCLA, his res activities include heat transfer related to phase change, thermal hydraulics of nucl reactors and reactor safety. He has authored or co-authored about 100 papers in refereed technical journals & proceedings of national & international conferences. He is married to Komal Khanna and has two children Vinita and Vashita. *Society Aff:* ASME.

DiAiso, Robert J
Business: 2594 Riva Rd, Annapolis, MD 21401
Position: Principal *Employer:* Dewberry & Davis *Education:* BS/Gen Engr/U.S. Naval Acad; M/CE/NYU; M.U.R.P./Urban Plan/Univ of Pittsburgh; PhD(abd)/Public Admin/Univ of Pittsburgh *Born:* 1/3/40 Registered PE (civil) in several states, certified planner and professional planner (NJ), current in charge of Annapolis Division of Dewberry & Davis with a staff of 70 persons offering services in architecture, civil and environmental engrg, planning, landscape architecture, surveying, and construction inspection, mapping & cartography. Acted as project planner, project engr & project mgr on a variety of projects ranging from new town in town, to large residential communities as well as regional shopping centers & institutional facilities. Particular expertise in real estate development, economics & finance. Married - one child - active in community - Trustee, Anne Arundel Community Coll (1974- 1980) Reappointed (1980-1986); Chrmn (1977-79); Pres Crofton Civic Association (1973), Business Leader of the Year (1982-83) Anne Arundel Trade Council; Dir, Scotts Seaboard Corp (1978-present); Pres, Dir, Property Improvement Collaborative; Dir, Bay Natl Bank. *Society Aff:* NSPE, ASCE, NAR, APA

Diamond, David J
Business: BNL-475B, Upton, NY 11973
Position: Nuclear Engr *Employer:* Brookhaven Natl Lab *Education:* PhD/Nuclear Engrg/Massachusetts Inst of Tech; MS/Nuclear Engrg/Univ of AZ; BS/Engrg Phys/Cornell Univ *Born:* 12/31/40 At Brookhaven Natl Lab since 1968 (Grp Ldr since 1974) analyzing nuclear reactor plant transients and reactor core performance under contract for the US Nuclear Regulatory Commission. Also involved with the devel of neutronic-thermal-hydraulic computer programs for this analysis. Consultant to The Singer Co, Link Simulation Sys Div (since 1982) and The Elec Power Res Inst (since 1984). Lecturer in Korea (1984) and Italy (1986) for the International Atomic Energy Agency and at the Massachusetts Inst of Tech (1982, 1983). Adjunct Prof at Poly Inst of NY (1977-78). Has given lectures to the public and participated in debates on nuclear energy. Author or co-author of more than 35 reports and 50 publications or presentations. *Society Aff:* ANS, ΣΞ, CSE

Diamond, Fred I
Home: 1715 Lincoln Lane, Rome, NY 13440
Position: Chief Scientist *Employer:* Rome Air Dev Ctr-Griffiss AFB. *Education:* PhD/EE.Syracuse Univ; MEE/EE/Syracuse Univ; SB/EE/MIT. *Born:* 12/13/25. Native of New York State. Served with Army Corps of Engrs 1944-46. With RADC since 1950; Chief Scientist. Directs R&D programs for Air Force command, control, communications and intelligence. Chrmn AGARD (Nato) Avionics Panel. Executive Chairman, Communications Subgroup, the Technical Cooperation Program (Australia, Canada, New Zealand, UK, US) Fellow IEEE, AAAS. Air Force Decoration for Exceptional Civilian Service. Air Force Meritorious Civilian Service Medal. Mbr MIT Educational Council, ABET Ad Hoc Visitor. *Society Aff:* IEEE, AAAS, NYAS

Diamond, John
Business: Boeing Vertol Company, P.O. Box 16858, Philadelphia, PA 19142
Position: Dir Product R&D. *Employer:* Boeing Vertol Company. *Born:* Sep 4 1928. BS Aeronautical Engrg from MIT 1948. Joined Boeing Vertol Co in 1948 & served in various engrg capacities of preliminary design, final design & flight test. Current assignment is to develop advanced subsystems which will provide competitive advantages for new programs & for improvement of existing products.

Diamond, Sidney
Business: School of Civil Engineering, Purdue University, West Lafayette, IN 47907
Position: Prof of Engg Mtls. *Employer:* Purdue Univ. *Education:* PhD/Engg Mtls/Purdue Univ; MS/Forest Soils/Duke Univ; BS/Forestry/Syracuse Univ. *Born:* 11/10/29. Res engr, FHWA-(US DOT) 1953-1965, soil and highway chem res. Assoc Prof, Purdue 1965-1969, Prof 1969-present. Active res in cement chem and concrete tech, including hydration, scanning electron microscopy, alkali-aggregate reactions, admixture responses, flyash utilization, silica fume effects, fiber reinforcement problems, and other areas. Past mbr, Bd of Trustees, Amer Ceramic Soc Chrmn, Theme 6, 7th Intl Congr on Chem of Cement, Paris, 1980; Scientific Bd, 8th Intl Congr on Chem of Cement, Rio de Janeiro 1986; Chrmn, Flyash Symposium, Materials Res Soc 1981. Copeland Award, Amer Ceramic Soc, 1983. Everard Dist Lectureship, Natl Bldg Res Inst, S Africa, 1985. Pres, Sidney Diamond and Assoc, Inc, conslt, engrg materials. *Society Aff:* MRS, ACS, ASTM, ACI.

Dibelius, Norman R
44 Hyde Blvd, Ballston Spa, NY 12020
Position: Consultant *Employer:* Self *Education:* MSME/Mech. Eng./Rensselaer Polytechnic Inst.; BME/Mech. Eng./Polytechnic Inst. of Brooklyn *Born:* 12/24/22 Consultant: gas turbine combustion, emissions, and environmental impact (1984--

Dibelius, Norman R (Continued)
present). Held a variety of engrg and managerial positions in G.E.'s Corporate R & D Center and the Gas Turbine Div, mainly involving devel and design of combustion systems for gas turbines. Past member ASME International Gas Turbine Inst Exec Cttee, 1979-84, Chairman 1983-84. US delegate to CIMAC 1975-84. ASME Fellow, ten issued patents, over two dozen technical papers, received Industrial Research 100 Award, 1965; received G.E.'s prestigious div and sector engrg awards for professional activities 1977 and 1983. Veteran WW II, European theater, National Ski Patroller, teach handicapped people to ski. *Society Aff:* ASME, CIMAC

Dibner, Bern
Business: Electra Sq, Norwalk, CT 06856
Position: Director. *Employer:* Burndy Corporation. *Education:* EE/Polytech Univ.; DEng/Polytech Inst of NY; DSc/Brandeis Univ; DST/Technion; DHL/Univ. of New Haven; D. Sc./Wesleyan Univ. *Born:* 8/18/97. Grad Polytech Inst of NY, EE cum laude; post-graduate studies Columbia & Univ of Zurich. Elec Engr at Adirondack P & L, Amer & Foreign Power. Founded Burndy Corp 1924, then Pres, now Chmn Emeritus. Served SATC Plattsburgh in WWI & as Lt Col Air Forces in ETO in WWII. Trustee Univ of Bridgeport, Fellow Brandeis Univ, Fellow Pierpont Morgan Libr, Fellow IEEE, Fellow Amer Acad Arts & Scis. Awarded D Eng in 1959 & Hon Prof in 1976 by Polytech Univ. D Sci in Tech by Technion of Israel, DSc by Brandeis Univ in 1977, DSc by University of Bridgeport in 1981 & DHL Univ of New Haven in 1984. Given Eli Whitney Award in 1973, Leonardo da Vinci Medal in 1974 & Smithsonian Gold Medal, George Sarton Medal in 1976, Sir Thomas More Medal by Univ of San Francisco in 1983. Founded Burndy Library 1936, Dibner Lib at Smithsonian Inst 1975. Pres Soc History of Tech 1971-72, Chmn Elec Historical Foundation, VP Amer Technion Soc *Society Aff:* IEEE, AAAS, HSS, SHOT, PML.

di Cenzo, Colin D
Home: 28 Millen Ave, Hamilton Ontario, Canada L9A 2T4
Position: Chrmn, Colpat Entprs *Education:* MSC/Electronics/U New Brunswick; DIC/Control/Imperial Coll; BSc/Electrical/U New Brunswick *Born:* 7/26/23. Royal Canadian Navy Reserve 1966-79; Professor & Dean, McMaster Univ 1965-80; Prof and Dean, Memorial Univ of Newfoundland 1980-82. Chmn, Colpat Entprs 1983-. Decorated Order of Canada & Canadian Decoration thrice; Recipient Centennial Medal; Queens Silver Jubliee Medal; Fellow, Engrg Inst of Canada (Julian C Smith medal, 1977; Pres 1979-80; Fellow , Inst of Elec & Electronics Engrs; Mbr CSEE (Pres 1976-78); Fellow, Canadian Academy of Engineering. Mbr, Assoc Profl Engrs of Ontario (Eng Medal 1977). Canadian Mbr Commonwealth Engrs Coun 1979-80); Canadian Dir, (UPADI) 1980-83; Mbr Ctte Engrg Info WFEO 1980-82 Canadian Delegate, WFEO 1979, 1981. Prof Emeritus, McMaster Univ 1980. *Society Aff:* FIEEE, FEIC, MCSEE, FCAE

Dick, Richard I
Business: Sch of Civ & Environmental Engg, 118 Hollister Hall, Ithaca, NY 14853
Position: Joseph P Ripley Prof of Engg. *Employer:* Cornell Univ. *Education:* PhD/Environmental Engg/Univ of IL; MS/Sanitary Engg/Univ of IA; BS/Civ Engg/IA State Univ. *Born:* 7/18/35. Res & teaching water quality control with emphasis on wastewater treatment and sludge treatment utilization & disposal. Past pres of Assn of Environmental Engg Profs & mbr of Governing Bd & Exec Committee of the Intl Assn on Water Pollution Res. Recipient of WPCF Harrison Prescott Eddy Medal, AEEP Distiquished Lecturer, ASCE Rudolf Hering Medal, AEEP Outstanding Publication Award. Formerly employed by US Public Health Service (1958-1960), Clark, Daily & Dietz Consulting Engrs (1960-1962), Univ of IL (1962-1972), & Univ of DE (1972-1977). *Society Aff:* WPCF, AEEP, IAWPRC, ASCE, IWPC, AWWA, AAAS, APHA.

Dickens, James W
Business: NC State Univ, Raleigh, NC 27650
Position: Agricultural Engr *Employer:* USDA-ARS *Education:* MS/Agric Engr/NC State Univ; BS/Agric Engr/NC State Univ *Born:* 9/28/25 Graduate of NC State Univ with a MS degree in agricultural engrg. Worked with NC State Univ in tobacco and peanut curing research from 1951 to 1957. Since 1957 worked with the USDA-ARS in research on harvesting, curing, marketing, Market Quality and Handling Research, located at NC State Univ where he is Prof in the Dept of Biological and Agricultural Engrg. He is Technical Adviser on Toxic Constituents of Food and Feed, USDA-ARS. Elected Fellow in the American Society of Agricultural Engrs in 1981. *Society Aff:* ASAE, APRES

Dickens, Lawrence E
Home: 5501 Leith Rd, Baltimore, MD 21239
Position: Advisory Engg. *Employer:* Westinghouse Electric. *Education:* DrEngg/EE/Johns Hopkins Univ; MSE/EE/Johns Hopkins Univ; BSEE/EE/Johns Hopkins Univ. *Born:* 12/8/32. Born North Kingstown, RI, Dec 8, 1932. After military service (1950-53) joined Bendix Radio, Field Engg, then design R&D in radar and communications. From 1960-65, at the Carlyle Barton Lab of the Johns Hopkins Univ he pursued res on microwave and millimeter wave receiving systems. From 1965-69, on the staff of Advanced Tech Corp, and since at Westinghouse Electric Corp (Baltimore MD) as an Advisory Engg, he has been engaged in R&D of semiconductor components, circuits and subsystems. Mbr and officer of Baltimore Section, IEEE-MTT/AP. Committee person, IEEE. Elected Fellow, IEEE in 1976. *Society Aff:* IEEE, ΣΞ.

Dicker, Daniel
Home: 228 Forest Dr, Jericho, NY 11753
Position: Prof of Engg & Applied Mathematics. *Employer:* SUNY at Stony Brook *Education:* Doc of Engg Science/Engg Mech/Columbia Univ; MCE/CE/NYC; BCE/CE/CCNY. *Born:* 12/30/29. Associated with cons engrs 1951-58., Columbia Univ 1958-62: Quincy Ward Boese Fellow, Res Asst, Instructor Instructor & Asst, Prof of Engg. Since 1962 at Stony Brook. Prof of Engg & cons 1968- ; Asst Dean of the Graduate Sch 1965- 69; Exec Officer, College of Engg 1970-71. Nato-NSF Fellow 1969-70 at Imperial College London. Honorary Fellow, Harvard University, 1978-1979; Visiting Professor, University of the West Indies, Barbados, 1986. Major works have been in the theory of aeroelastic stability of suspension bridges, and transient free surface flow in porous media, heat transfer and method for solutions of partial differential equations of engineering and applied science. ASCE: Fellow, Norman Medal 1967, A M Wellington Prize 1972; Elect Fellow NY Acad of Sci 1974; P E in NY, NJ & NC. *Society Aff:* ASCE, NYAS, NSPE, SIAM, ΣΞ

Dickerson, Ronald F
Home: 377 Springs Dr, Columbus, OH 43214 *Employer:* Retired *Education:* MS/Met Engg/VPI & SU; BS/Met/VPI & SU. *Born:* 1/27/22. Born in Thompson Ridge, NY, grad from Middletown High School, Middletown NY 1939, served in armed forces WWII, highest rank attained Capt, Inf, MS from VPI & SU in 1948, BS from same inst 1944. Employed at Battelle Meml Inst since 1948 as div chief, asst dept Mgr at Battelle Columbus, Staff mgr at the Pacific Northwest Labs of Battelle, Marketing Dir of the Battelle Intl Operations and VP Battelle Dev Corp. Retired Jan 1987. Married with two sons. Active Goodwill, Kiwanis, Boy Scouts, and Church. *Society Aff:* ASM, ΣΞ.

Dickey, John M Jr
Business: Box 217, 108 Main St, Brookville, PA 15825
Position: Consulting Engg. *Employer:* Self. *Education:* MS/-/OH State Univ; BC/Cer Engg/OH State Univ. *Born:* 10/19/24. Born and raised in Columbus, OH. Resident of Brookville, PA past 31 yrs. Worked for National Clay Pipe Research Assoc, Engg Experiment Station, OSU, 1946-48. Ceramic Engg for Hanley Co, Inc, 1948-55. Plant Superintendent for Hanley Co-Face Brick, Glazed Brick, and Glazed Tile Producer, 1955-59. Consulting Engg, Private Practice, 1959-80. Respon for all types of civil engg and surveying projects, and ceramic consulting projects. Borough Engg, Borough of Brookville, for past 20 yrs. Maintain office and staff in Brookville, PA.

Dickey, John M Jr (Continued)
Ceramic projects included national and overseas projects. Enjoy golf, tennis, and family. *Society Aff:* NSPE, PSPE, ACS, NICE, PEPP.

Dickey, Paul S
Home: 3828 Crayton Rd, Naples, FL 33940
Position: Retired. *Employer:* Bailey Meter Co. *Education:* BS/ME/Purdue; Hon deg/Engg/Purdue. *Born:* 9/21/03. With Bailey Meter Co 1925-70, VP 1947-55, Chmn, Dir & Pres 1955-70; also Dir Bailey Meter Co Ltd Montreal Canada. Past Trustee & Chmn Fenn Educational Foundation, P Chmn Euclid General Hospital Assn. Fellow Purdue Res Foundation, P Pres Scientific Apparatus Makers Assn. Fellow ASME; Instrument Soc of America, Pi Tau Sigma, Sigma Xi. Patented some 65 patents covering instruments & control systems. *Society Aff:* ASME, ISA, ΠΣ, ΣΞ.

Dickey, Robert, III
Business: One Oliver Plaza, Pittsburgh, PA 15222
Position: Retired. *Education:* BS/ME/Princeton Univ. *Born:* 1/28/18. Mr Dickey joined Dravo in 1948 as a salesman for towboats & barges, later becoming sales mgr. He transferred to a subsidiary, Union Barge Line Corp (now Dravo Mechling Corp) where he held successive posts as engg mgr; vp, sales; and pres. He rejoined the parent co in 1962 as exec vp; was elected pres in 1964, & was named chief exec officer in 1965. He was elected chrmn in July 1974, retaining pres & chief exec officer. Mr Dickey is a director of Gulf Oil Corp, Joy Mfg Co, Pittsburgh Natl Bank, the Regional Ind Dev Corp of Southwestern PA & is on Executive Committee of the Allegheny Conf on Community Dev. He also is chrmn of the Public Auditorium Authority, a dir of the Pittsburgh Symphony Soc, a trustee of Carnegie-Mellon Univ, Carnegie Inst, & Shadyside Hospital in Pittsburgh. He was recently selected by Pres Carter to be a Mbr of the Pres's Export Council, & named "Exec of the Yr" for 1979 by the Natl Mgmt Assoc. Retired 1983. *Society Aff:* AISI.

Dickie, H Ford
Home: 16 Crescent Rd, Riverside, CT 06878
Position: Retired *Education:* BS/Ind Engg/Univ of KS; Certificate/Grad Bus Ad/Stanford Univ. *Born:* 8/2/14. With the Gen Elec Co since 1935. Mgr of Aeronautic & Ordnance Systems Div Mfg. Corporate Mgr of Production & Inventory Control. Mgr Engg & Mfg, Intl GE Co. Mgr Bull-GE Corp Proj, Paris, France. Corporate Mgr, Mfg Mgt, Education & Personnel Staff Exec, Corporate Production Resources. Retired 1977. Consultant- Industrial Management. Lectr - Columbia Univ, Grad Sch of Engg. Adjunct Prof - NYU Grad Sch of Engg. Fellow - IIE 1977; Fellow APICS 1977. Honorary mbr Alpha Pi Mu; Mbr Tau Beta Pi; Theta Tau; Phi Delta Theta. IIE rep to ECPD 1970-73 & 1978-81. Mbr, Riverside Yacht Club & Nutmeg Curling Club. Mbr of Bds of Dirs: U. S. Sm Bus Adm, S.B.D.C. Stamford, CT, Small Bus Resource Ctr. Greenwich, CT. Y.M.C.A. *Society Aff:* IIE, APICS.

Dickinson, Bradley W
Business: Department of Electrical Engineering, Princeton University, Princeton, NJ 08544
Position: Professor of Electrical Engrg *Employer:* Princeton University *Education:* Ph.D./EE/Stanford University; M.S./EE/Stanford University; B.S./Engineering/Case Western Reserve Univ. *Born:* 4/28/48 Bradley W. Dickinson is Prof of Electrical Engrg at Princeton Univ where he has been a faculty member since 1974, engaged in teaching & research in systems & signals. Elected a Fellow of IEEE in 1987 "for contributions to system theory and its application in signal processing–. Active in the IEEE Control Systems Society, Information Theory Group, and Acoustics, Speech and Signal Processing Society. A founding editor of the journal Mathematics of Control, Signals, and Systems, and a founding editor of the Electronic Newsletter of Systems, Control, and Signal Processing. *Society Aff:* IEEE, ΣΞ, SIAM, TBΠ

Dickinson, James M
Business: MS 770, PO Box 1663, Los Alamos, NM 87545
Position: Consultant. *Employer:* Los Alamos National Lab. *Education:* PhD/Chem-Met/IA State Univ; BS/Chemistry/IA State Univ. *Born:* 7/31/23. Native of Waterloo, IA - NM resident since 1953. Served in the South Pacific and China with the US Marine Corp 1942-1946. Employed at the Los Alamos Scientific Lab since 1953 with responsibilities in mtls engg, fabrication & joining, particularly of unusual mtls such as uranium, refractory metals, pure oxide ceramics, glasses, graphite & graphite nuclear fuel elements. Assumed mgr position in the Mtls Tech Group at LANL in 1974. Retired LANL 1986. Presently consulting in materials and management areas. Past Mbr AIME Refractory Metals Committee, past chapter chrmn ASM. Married, three daughters. Non-profl activities include hunting, fishing, hiking, skiing & square dance calling. *Society Aff:* ASM, AIME, ADPA, SAMPE.

Dickson, George H
Home: 3911 NE 100th St, Seattle, WA 98125
Position: Pres *Employer:* G. H. Dickson & Assoc, Inc *Education:* BS/Econ & Engrg (electrical option)/MIT *Born:* 8/20/28 Cons. elec. engr; b. Tacoma, WA; m. Lavonne Irene Schuler, June 7, 1951; children--James Morton, Rodney Glenn, Geoffrey David; Sigma Chi Fraternity; Kiwanis Intl; Presbyterian Elder. Distribution engr Puget Sound Power and Light Co, Bellevue, WA 1950-58, design engr. Bouillon, Griffith et al, Seattle, WA 1958-61; partner Miskimen and Dickson, Seattle 1961-67; dir engrg, Naramore Bain Brady & Johnson, Seattle 1967-71; assoc/exec. engr R. W. Beck & Assoc, Seattle 1971-78; vp-elec. engring Wood/Harbinger, Inc, Kirkland, WA 1979-80; pres G. H. Dickson & Assoc, Inc, Seattle 1981--. U.S. Army 1951-53. Registered PE WA, AK, OR, ID, MT, WY, CA, and VA. *Society Aff:* IEEE, NSPE, IES

Dickson, James F
Business: Room 701-H, 200 Independence Ave, SW, Wash, DC 20201
Position: Asst Surgeon Gen. *Employer:* Dept of Health and Human Serv *Education:* AB/Chemistry/Dartmouth College; MS/Instrumentation/Drexel Unv; MD/Medicine/Harvard Medical Sch; Special Fellow/Control Engg/MIT. *Born:* 5/4/24. Educated as a thoracic & cardiovascular surgeon & a control systems engr. From 1965 through 1984: Sr Consultant for Health with the Pres's Commission on Tech, Automation & Economic Progress. Dir of Health for the Pres's Advisory Council on Mgt Improvement. Deputy Asst Secy for Health. Acting Asst Secy for Health. Asst Surgeon Gen, Dept of Health and Human Serv. Mbr, Inst of Medicine, Natl Academy of Sciences. *Society Aff:* IEEE, ABS, IOM/NAS, ACS.

Dickson, James G
Business: 120 S Parking Pl, Lake Jackson, TX 77566
Position: Pres. *Employer:* Dunbar & Dickson, Inc. *Education:* BSCE/Civ Engg/Univ of TX. *Born:* 5/2/22. Native of Freeport, TX. Resident Engr with TX Hgwy Dept, 1946-48. Pres of Dunbar & Dickson, 1948-50, 1953-64 and 1975 to present. Pres of Limbaugh Engrs, Inc of NM, 1970-74. Dir of First Freeport Natl Bank, 1955-78. Chrmn of Bd, First Freeport Corp, 1978 to present. Served with Corps of Engrs, AUS, 1944-46 and 1950-53. Chrmn of Technical Council on Aero, ASCE, 1973-74. Reg PE in TX, LA, AZ, KS, OK. *Society Aff:* ASCE, NSPE, MTS, AWWA, ASPO.

DiDomenico, Mauro, Jr
Business: Bell Communications Research, 435 South St, Morristown, NJ 07960
Position: Div Mgr *Employer:* Bell Communications Res *Education:* PhD/EE/Stanford Univ; MS/EE/Stanford Univ; BS/EE/Stanford Univ. *Born:* 1/12/37. Joined Bell Labs in 1962. Res and dev on lasers and optoelectronic devices for optical communications systems, as well as mtls unique to these devices. Hd of Optical Device Dept (1970-1980) responsible for dev of devices and laser transmitters and photodiode receivers for lightwave sys 1980-1982 Hd of integrated Circuit Customer Service Dept coordinating IC developments. 1982-1984 Div Mgr AT&T Strategic Planning. Joined Bell Communications Res Jan 1984. Presently Div Mgr Tech Liaison Office. Fellow, IEEE; Fellow, Am Physical Soc; Mbr, Tau Beta Pi & Sigma Xi *Society Aff:* IEEE, APS, AAAS, NYAS

Diehl, Douglas S
Business: 8989 Westheimer, Ste 111, Houston, TX 77063
Position: Pres *Employer:* ERM-Southwest, Inc *Education:* MS/Sanitary Engr/Purdue Univ; BS/CE/Purdue Univ *Born:* 10/5/43 Spent three years in US Navy Civil Engr Corps, 1966-69 (Public Works Dept, NAS Pt Mugu & Instructor, US Naval Academy); Ten years with Roy F Weston, Inc, Environmental Designers-Consultants, as project engr, project mgr, VP-Southwest Region & Dir. Co-founded ERM-Southwest, Inc, Environmental Consulting Engrs, Houston in 1979. Founded SW Closures, Inc (Special Clousure Hazardous Waste Sites) 1984. Pres & CEO of ERM-SW as well as SW Closures. Reg PE in six states. *Society Aff:* AAEE, ASCE, WPCF

Diener, Robert G
Business: Agri Engrg Bldg, Morgantown, WV 26506
Position: Prof- Agri Engr. *Employer:* West Virginia University. *Born:* Apr 12 1938. BS Agri Engrg Penn State 1960, Test Engr Internatl Harvester Co 1960-62; MS Agri Engrg Penn St U 1963; PhD Agri Engrg Michigan St U 1966; Agri Engr ARS, USDA-Michigan St 1966-68, fruit & vegetable harvesting res; 1968- , Agri Engr-Prof WV Univ. Mbr ASAE, Sigma Xi, Alpha Epsilon; Student Branch Advisor WVU; Former Chmn PM-48, FE-03, A216, A117 Cttes ASAE; former Chmn Student Activities Ctte, North Atlantic Region ASAE; winner of ASAE technical paper awards 1969, 70 & 71; presently Past Dir, North Atlantic Region, ASAE.

Dienes, Andrew J
Home: 831 Ensenada St, Berkeley, CA 94707
Position: Professor *Employer:* U. of California *Education:* PhD/Physics/Calif. Inst of Tech; MSc/Electr Eng/Calif. Inst of Tech; B. Eng./Electr/McGill Univ, Montreal Canada *Born:* 10/22/39 in Hungary. Went to Canada in 1957, to USA in 1962. After receiving PhD worked at Bell Lab, Holmdel, NJ where he specialized in dye lasers and ultrashort pulses. Was one of the researchers to develop the first CW mode- locked dye lasers. Pursues research in the same field at both Davis and Berkeley campuses. Hobbies: classical music, sailing, hiking.

Dienhart, Arthur V
Home: 5740 Morgan Ave S, Minneapolis, MN 55419
Position: VP. - Retired *Employer:* Northern States Power Co *Education:* BCE/Civil Engg/Univ of MN. *Born:* 7/23/20. Born 1920 in Sioux Falls, SD. Professional experience began with Dravo Corp, Pittsburgh, in construction mgt for shipyard facilities, 1942-45. Commissioned officer in Civil Engr Corps, US Naval Reserve, engaged in overseas shore facility construction, 1945-46. Joined Northern States Power Co, Minneapolis, in 1946; engaged continuously in mgt of plant facilities design and construction for electric utility serving a four-state area. Retired July 1985. Reg PE in MN, WI, ND. *Society Aff:* ASCE, NSPE, ANS, TBΠ, ACI, XE

Diercks, Frederick O
Home: 9313 Christopher St, Fairfax, VA 22031
Position: Consultant. *Employer:* National Ocean Survey-NOAA. *Education:* MSCE/Civil Engg/MA Inst of Technology; MS/Photogrammentry/Syracuse Univ; BS/Civil & Military Engg/US Military Acad. *Born:* Sep 1912. Commissioned 2nd Lt Army Corps of Engrs 1937. Commanded 656th Engr Topographic Battalion, European Theater of Operations, WWII; Ch Topographic Engr Far East Command Tokyo 1948-49 & of European Command Heidelberg 1953-56. Commanding Off US Army Map Serv Washington 1957-61. 1962-63 Act Asst Dir for Mapping, Charting & Geodesy Defense Intelligence Agency. Deputy Engr 8th US Army Korea 1963-64. Dir US Army Coastal Engrg Res Ctr Wash 1964-67. Retired from Army 1967-Col. 1967-74 Assoc Dir for Aeronautical Charting & Cartography US Coast & Geodetic Survey, Dept of Commerce. Luis Struck Award 1969 ASP, Legion of Merit 1967 USA, President 1970-71 ASP, VP 1969-70 ASP, Dir Cartography Div 1968-70 ACSM; Colbert Medal, SAME, 1972; Bronze Medal, Dept of Commerce, 1974; Fellow ASCE, SAME; Sigma Xi; Reg Profl Engr DC. *Society Aff:* ASCE, SAME, ASP, ACSM, ΣΞ.

Dieter, George E
Home: 1 Locksley Ct, Silver Spring, MD 20904
Position: Dean of Engr. *Employer:* Univ of MD. *Education:* DSc/Metallurgy/Carnegie-Mellon Univ; BS/Met Engg/Drexel Univ. *Born:* 12/5/28. Native of Phila, PA. Served at Ballistics Res Lab, APG, MD 1953-55 and then employed by DuPont Co at Engg Res Lab. Res on fatigue, powder metallurgy and shock hardening of metals. Went to Drexel in 1962 as Dept Hd, Metallurgical Engg. Dean of Engg at Drexel, 1969-73. Prof of Engg and Dir, Processing Res Inst, at Carnegie-Mellon Univ 1973-77. At Univ of MD as Dean of Engrg and Prof of Mech Engrg since 1977. Active in AIME, Fellow ASM. Trustee of ASM, and Federation of Materials Societies. Mbr Natl Mtls Advisory Bd. *Society Aff:* AIME, ASM, SME, ASEE, NSPE, AAAS.

Dietrich, Edwin J
Business: 1500 City W Blvd, Ste 1003, Houston, TX 77042
Position: Pres *Employer:* Dietrich Engineers Inc *Education:* JD/LAW/SOTX Law; BSCE/Civil Engrg/Univ of TX. *Born:* BS Civil Engrg University of Texas; Doctor of Jurisprudence South Texas Law. Native of Shreveport, Louisiana & a resident of Houston, Texas since 1949. Wife is Adele Marie Odom of Shreveport; daughters are Shelle & Melanie. Pres of Dietrich Engineers Incorporated, Conslt Engrg of Houston, Texas. Served as Director of the Houston Engrg & Scientific Soc, the Texas Cons Engrs Council & is a Fellow in the American Soc of Civil Engrs and Amer Conslt Engrg Council; Houston Bar Assoc. *Society Aff:* ASCE, ACEC, NSPE, SAME.

Dietsche, William O
Home: 11 Marshall St, East Longmeadow, MA 01028
Position: Exec VP *Employer:* O R Cote Co *Education:* BS/Chem Engg/Univ of NH. *Born:* 12/4/35. Hooker Chem 1957-59, Geigy Proj Engg 1959-63; D L Thurmott Co Equip sales 1963-67; Hayes Pump & Mach, sale of pumps Dist Mgr, 1967-73; O R Cote Co Equip Sales, Indust Sales Mgr 1973-77, VP 1977-79, VP & Gen Mgr, 1979, Exec VP '80 to present. Reg PE RI, & MA,Town of East Longmeadow Planning Bd 1970-80, AIChE RI Sect PChmn, Western MA sect, PChmn. *Society Aff:* AIChE.

Dietz, Albert G H
Home: 19 Cambridge St, Winchester, MA 01890
Position: Prof Emeritus *Employer:* MIT *Education:* ScD/Mat/MIT; BS/Bldg Const/ MIT; AB/Humanities/Miami Univ (Ohio) *Born:* 3/7/08 At MIT: Prof of Bldg Engrg, founder Plastics Research Lab, member committees on advanced degrees in materials and space heating by solar energy. World War II: office of Field Service, OSRD; consulting senior engr, Forest Products Lab. chrmn, Bldg Research Advisory Bd and chrmn, Committee on Micromechanics of Materials, National Academy of Sciences. Dir, ASTM, SPE. Senior Research Fellow, East-West Center. Visiting Prof, Australia, Germany, England. Intl Gold Medal, SPE; Derham Intl Award, Plastics Inst of Australia; Marburg Lecturer, Templin Award, Voss Award, ASTM. Honorary Member ASCE, ASTM. Enjoy photography, swimming, reading, classical music. *Society Aff:* ASCE, ASTM, ASME, SPE, AAAS, AAAS, SPI, FPRS

Dietz, Daniel N
Business: Laan Copes van Cattenburch 93, 2585 EW The Hague, The Netherlands
Position: Professor, Retired *Employer:* Independent Conslt *Education:* -/Engr Phys/ Delft U of Tech *Born:* 11/7/13. Worked on groundwater problems 1941-46 for Dutch Govt & on reservoir engrg in oil fields office & lab 1946-72 for Royal Dutch Shell. Distin Lecturer SPE 1972-73. Currently engaged in Teaching, dev of tertiary oil recovery & of unconventional mining techniques, study of groundwater pollution by oil, underground disposal of radio-active waste 1972-84 for Delft U. of Tech. Publications on surface water, groundwater, unstable oil displacement, thermal drive, pressure build up analysis, in situ coal gasification. John Franklin Carll Award 1970 SPE, Chmn 1965-71 Hydrologisch Colloquim, Chairman 1973-75 Netherlands Section SPE. *Society Aff:* KIVI, SPE

Dietz, Jess C
Home: 469 Village Place, Longwood, FL 32779
Position: Special Consultant *Employer:* Conklin, Porter & Holmes-Engineers, Inc. *Education:* PhD/Civ Engg/Univ of WI; MS/Sanitary Engr/Univ of WI; BS/Civ Engg/ Univ of WI. *Born:* 10/10/14. Native of WI. Res Asst-Instr, Univ of WI, 1940-42 & 1945-47. Served US Army (2nd Lt to Col) 1942-45. Awarded Bronze Star & Croix de Guerre. Prof of Sanitary Engg, Univ of IL 1947-57. Prof Emeritus 1974-present. Joined present firm as Principal in 1957. Diplomat, Am Acad of Environmental Engg. Mbr, Sigma Xi, Tau Beta Pi, Chi Epsilon, Gamma Alpha, Pi Mu Epsilon. Awarded Distinguished Service Citation (1976) Univ of WI. Author papers on indus waste and sanitary engg. *Society Aff:* WPCF, AWWA, ASCE, NSPE, APWA.

Dietz, John R
Business: P.O. Box 1963, PA 17105
Position: Chrmn *Employer:* Gannett Fleming Corddry & Carpenter *Education:* BS/CE/Drexel Univ. *Born:* 1912 Carbondale PA. Designer PA Highway Dept; Designer/Resident Engr PA Turnpike Commission; Designer Army Air Base Trinidad, Caribbean Architect Engrs; Ch Designer Andrews Air Field Washington D C; Flood Control Study, Potomac River Basin, US Engrs. With GFCC Pres 1964-76, Chmn since 1970. Registered PE 26 states & DC, NCEE. Life ASCE; Sr Mbr ASHE; Mbr: NSPE, PEPP, PSPE, ARTBA, ACEC, Transportation Res Bd. P Pres, Central PA Chapter ASCE & PA Hwy Info Assn. Engr of Year, Harrisburg Chapter PSPE 1965. A J Drexel Paul Award, most outstanding alumni 1973. *Society Aff:* ASHE, NSPE, PEPP, ACEC, ASCE, ARTBA.

Dietz, Robert J
Business: Gannett Fleming Corddry et al, PO Box 196, Harrisburg, PA 17105
Position: Vice President, Chief Mass Transit. *Employer:* Gannett Fleming Corddry et al *Education:* BS/Civil Engg/Lehigh Univ; BA/Applied Science/Lehigh Univ. *Born:* Oct 1943. Involved with planning, design & testing of the Philadelphia- Lindenwold Rapid Transit Line PATCO. With Gannett Fleming since 1965. Principal & Mgr of the Detroit Office 1973-76. Assumed respon of Ch, Mass Transit in Jan 1976 and Vice President in March 1981. Respon for transportation planning & urban transit programs. Project Director for transportation planning & engrg programs including preliminary engineering of a Light Rail System for Southeastern Michigan, final design management for the Detroit Downtown People Mover and preliminary design of the Miami Metiomover. Chmn Publication Ctte of American Public Transit Assn for the publication 'Guidelines & Principles for Design of Rapid Transit Facilities'. Chmn ITE committee on Activity Center Circulation. Member of the American Public Transit Association commitees on Light Rail Transit and Advanced Technology. *Society Aff:* ASCE, NSPE, ITE, APTA, ARTBA.

Differt, Douglas H
Business: Rm 408, Transportation Bldg, St Paul, MN 55155
Position: Deputy Commissioner. *Employer:* MN Department of Transportation *Born:* Aug 27 1933. Served with US Army in Europe 1953-55. Graduate of the Bureau of Highway Traffic, Yale University 1965. BSCE University of North Dakota 1961. Registered Civil Engr in Minnesota. Employed by Minnesota Dept of Highways since 1961. Have held the positions of District Traffic Engr, Metropolitan Planning Engr & Engrg Standards Engr. Asst Commissioner of Transportation Planning & Programming, District Engineer, Asst. Commissioner of Program Mgmt. Presently Deputy Commissioner. Pres of NCITE 1975. Secy District IV ITE 1976. Chmn of NCHRP Ctte 8-18 1976-77. Board of Directors for P.M.I. (Project Mgmt Institute).

DiFranco, Julius V
Home: 32 Candlewood Path, Dix Hills, NY 11746
Position: Res Dept Head *Employer:* UNISYS Corporation *Education:* MS/EE/Columbia Univ; BS/EE/Columbia Univ. *Born:* 1925. Served with the US Navy in 1943-45. Servo systems engr with Liquidometer Corp from 1950-58. With UNISYS/SPERRY Corp since 1958. Assumed current respon for sys group in Systems Design in 1973. Prepare proposals, conduct research studies & dev of new technology systems. Received Fellow Award in Jan 1975. Taught graduate courses at BPI for a period of time. Enjoy literature, music and art. *Society Aff:* IEEE, AES, TBΠ.

Digges, Kennerly H
Home: 2510 Rocky Branch Rd, Vienna, VA 22180
Position: Assoc Administrator for R&D *Employer:* US Dept of Transportation. *Education:* PhD/Mech Eng/OH State Univ; MS/Mech Eng/OH State Univ; BS-hons/ Mech Engg/VA Polytechnic Inst; Res Fellow/Bio Eng/Oxford Univ. *Born:* 10/23/33. Native of Northern VA. Served in USAF 1956-57. Nineteen yrs experience of aircraft res and dev at Wright Patterson AFB. Dir dev of graphite composite components, high performance windshields and air cushion landing gear for aircraft. With National Highway Traffic Safety Admin, US Dept of Transportation since 1975. Directs res which serves as the tech basis for federal standards for auto safety and fuel economy. Major programs include trauma res to define human injury tolerance, and the dev of experimental auto which are clean, fuel efficient and safe. Awarded DOT Silver Medal for Meritorious Svc, 1979, GEICO Public Service Award, 1980. *Society Aff:* ASME, ΣΞ, SAE

Di Giacomo, William A
Business: 1133 Ave of Americas, New York, NY 10036
Position: Pres & Chrmn. *Employer:* W A Di Giacomo Assocs, PC. *Education:* BSME/Mech Engg/NY Univ College of Engg. *Born:* Jan 1928. Native of New York City. Capt with Army Corps Engrs 1949-50. Sr Mech Project Engr with three firms 1950-55. Founder cons mech-elec engrg practice W A Di Giacomo Assocs since 1956 with offices in NYC & San Francisco Project engineer design respon for over 100 major commercial & institutional bldg projects. Engrg licenses 15 states & NCEE certification. Fellow member ASHRAE & Honorary Mbr NAPE. Listed Whos Who in the East. Golfer. *Society Aff:* ACEC, NSPE, ASHRAE, NAPE.

DiGioia, Anthony M, Jr
Business: 570 Beatty Rd, Monroeville, PA 15146
Position: President. *Employer:* G A I Consultants Inc. *Education:* PhD/Civil Engg/ Carnegie-Mellon Univ; MS/Civil Engg/Carnegie-Mellon Univ; BSCE/Civil Engg/ Carnegie-Mellon Univ. *Born:* Aug 24 1934 Pittsburgh PA. BSCE, MSCE, PhD CE Carnegie-Mellon. Taught Cvl Engrg Dept Carnegie-Mellon U. Served US Army Corps of Engrs Ft Belvoir VA 1961-62. V P, then Pres General Analytics Inc from its inception to 1974. Formed G A I Cons Inc 1974 & serves as Pres. Exper in geotech engrg. Pres Pittsburgh Sect ASCE 1973; Dir District 4 ASCE 1974-77; Regional V P SAME 1974-76; Young Engr of Yr Award Pittsburgh Sect ASCE 1970. Eng of the Year Pittsburgh Sect. ASCE 1981; Carnegie-Mellon Univ Alumni Service Award-1978; Enjoy golf, tennis & bowling. *Society Aff:* ASCE, SAME, CIGRE, IEEE, ASTM, NSPE.

DiJulio, Roger M, Jr
Home: 18743 Dylan St, Northridge, CA 91324
Position: Assoc Prof. *Employer:* CA State Univ. *Education:* PhD/Struct Mechanics/ UCLA; MS/Mgt/UCLA; MS/Struct Mechanics/UCLA; BS/Gen Engg/UCLA. *Born:* 1/22/47. Mbr of faculty at CSUN since 1974. Teaches in the areas of appl mechanics, civ engg and engg economics. Chrmn dept of mechanics and mtls (1976-1977), Assoc dean, school of Engg and Comp Sci (1977-1978), Asst Dean (1979). Reg Civ Engr, State of CA. Consultant in the areas of structural analysis and vibration test and analysis. Res in the area of structural dynamics. Chairman, Dept of Mechanics, Civil and Industrial Engineering (1980-1981). *Society Aff:* SEAOC; EERI.

Dilger, Walter H
Business: Dept Cvl Engrg, U of Calgary Alberta, Canada T2N 1N4
Position: Professor of Civil Engineering. *Employer:* Univ of Calgary Dept of Civil Engg. *Education:* Dr Ing/Structures/Tech Univ Stuttgart; Dipl Ing/Civil Engg/Tech Univ Stuttgart. *Born:* Sep 19 1934. 1954-59 Tech U Stuttgart W-Germany Dipl-Ing

Dilger, Walter H (Continued)
Cvl Engrg. 1959-60 Design Engr G Drueckler Cons Engr Stuttgart W-Germany. 1960-64 Res Engr Otto-Graf-Institut at the Tech U Stuttgart W-Germany. 1964-65 Sci Asst to Chair of Prof Leonhardt & 1965 Dr-Ing Degree from Tech U Stuttgart W-Germany. 1966-67 Res Assoc Dept Cvl Engrg (Dr Neville) U of Calgary Alberta Canada & since 1967 Prof Dept Cvl Engrg at U of Calgary. 1974 Martin P Korn Award PCI; 1976 T Y Lin Award ASCE. Author & coauthor of 60 papers & 1 book. Involved in analysis & check of design of the Ekofisk Oil Tank & the Frigg Structs in the North Sea. Mbr ACI, ASCE, PCI, IABSE, APEGGA. *Society Aff:* ACI, ASCE, IABSE, PCI, APEGGA.

Dill, Ellis H
Business: P.O. Box 909, Piscataway, NJ 08854
Position: Dean of Engrg *Employer:* Rutgers, The State Univ *Education:* PhD/CE/Univ of CA-Berkeley; MS/CE/Univ of CA-Berkeley; BS/CE/Univ of CA-Berkeley; AA/Engrg/Grant Tech Jr Coll *Born:* 12/31/32 in McAlester, OK. Assistant Prof to Prof and Chrmn of the Dept of Aeronautical Engrg at the Univ of WA, 1956-1977. Author of numerous research papers and reports on structural mechanics, structural analysis, and continuum mechanics. A. A., Grant Tech J.C.; BS, MS, PhD, in Civil Engrg, Univ of CA Dean of Engrg, Rutgers Univ, 1977-present. *Society Aff:* SNP, ASEE, ASCE.

Dill, Frederick H
Home: RR1, Box 144, S Salem, NY 10590
Position: Research Staff. *Employer:* IBM Corp. *Education:* PhD/EE/Carnegie Inst Tech; MS/EE/Carnegie Inst Tech; BS/Phys/Carnegie Inst Tech. *Born:* 3/1/32. In the early 1960's Dr Dill worked primarily on exploratory semiconductor devices and was co-inventor of the semiconductor injection laser. He has managed groups exploring very high speed integrated circuits, has done important research in photolithography, has worked on computer controlled measurement tools and info display techs. He is presently mgr of the Info Arch Group in advanced semiconductor Mfg at IBM East Fishkill, NY. He taught at Berkeley and MIT in 1958 and 1964 respectively. He was Pres of the IEEE Electron Devices Soc 1982-83. *Society Aff:* IEEE, EDS, AAAS.

Dill, Hans G
Home: Rue Deleynes 21A, St. Blaise, Switzerland 2072
Position: Tech Dir Microelectronic Components. *Employer:* Ebauches Electronics. *Education:* PhD/EE/Swiss Federal Inst of Technology, Zurich; MS/EE/Swiss Federal Inst of Technology, Zurich. *Born:* May 1927 in Switzerland. 1957-59 IBM, digital circuit des; 1959-74 Hughes Aircraft, microelectronics, from design engr to dept mgr. Since 1975 mgr in charge of manufacturing integrated circuit & display components for electronic watches, Ebauches SA, Marin, Switzerland. 27 US patents & num papers in the microelectronic field. 1970 Hughes-Hyland patent award. 1973 Fellow IEEE. 1964 naturalized citizen. Wife May, children Corinne, Ellen, Martin. Enjoy classical music, photography, hiking. *Society Aff:* IEEE.

Dillard, Joseph K
18 Anchor Point, Anderson, SC 29621
Position: Gen. Mgr. Adv. Tech Div (Ret) *Employer:* Westinghouse Electric Corp. (Retired) *Education:* BS/EE/GA Inst of Tech; MS/EE/MA Inst of Technology. *Born:* 5/10/17. AMP, Harvard Grad. School of Bus Admin. Native of Westminster S C. Taught at MIT until 1950. Joined Westinghouse Electric Corp as electric utility engr. Presently Retired. Published 70 technical papers. Registered Prof Engr. Mbr NSPE. Fellow IEEE. Pres IEEE 1976. Mbr Natl Acad of Engrg, Mbr Engrg Soc of Western Penna. *Society Aff:* IEEE, NSPE.

Dillaway, Robert B
Home: 1306 Ballantrae Ct, McLean, VA 22101
Position: Consultant *Employer:* Dillaway & Assocs Ltd *Education:* PhD/Mechs/Univ of IL; MS/Physics/Univ of IL; BS/Math/Univ of MI; BS/Mech Eng/Univ of MI. *Born:* 11/10/24. Carrier Corp R&D 1945-46. Engrg Res Assocs. R&D in acoustics & rocket propulsion 1946-48. Taught mech engrg U of IL 1948-53. North American Aviation Engr & Mgr 1953-68. Rocket Engine & Space Propulsion Direction to 1964, Corp R&D direction 1964-68. Spec Asst US Navy Sec 1968-69. Deputy to Commanding General, US Army Material Command, for R&D & Lab Mgmt 1969-75. Sr VP Cons Diesel Elect Co 1975-76. Dir Engrg & Marketing. On leave Jan 1976-77 serving as sci cons to US Congress, House Sci & Tech Ctte, organizing new oversight staff for special investigations. Global Defense Products Inc, Defense & Electronics Mktg 1977-79. Dillaway & Assocs Constls in Tech & Mgmt 1979- . Dadico Systems Inc 84- pres. Intiatives Inc 1985-Enec. VP. Pronimity Inc 1986-VP. *Society Aff:* ASME, AIAA

Dillon, John G
Business: PO Box 3965, 50 Beale St, San Francisco, CA 94119
Position: VP & Dir. *Employer:* Bechtel Civil & Minerals Inc.. *Education:* BS/Civ Engg/WA State Univ; Dipl/Logistics/Naval War College. *Born:* 12/11/19. Native of Tacoma, WA. Served 31 yrs in Navy Civ Engr Corps. Retired as Rear Admiral. Joined Bechtel 1973. Currently Dir, VP, & Gen. Mgr - Hydro & Community Facilities Div which is responsible for heavy civ engg & construction in Bechtel group of cos on worldwide basis. Felow, former Dir & Mbr of Exec Committee, SAME. Former Dir, Treas & Mbr of Exec Committee, NSPE. Mbr of Bd of Dirs, EJC. Authored numerous technical publications. *Society Aff:* NSPE, SAME.

Dilpare, Armand L
Business: Jacksonville Univ, Jacksonville, FL 32211
Position: Dir Engrg Prog *Employer:* FL Inst of Tech *Education:* PhD/ME/Columbia Univ; MS/ME/Columbia Univ; BS/ME/City Coll of NY *Born:* 8/12/32 and raised in NY City; relocated to FL 1975. Twenty years of industrial experience (including Bell Telephone Labs, Republic Aviation, Lundy Electronics and RCA) focused on conception, research, and development of advanced electronic and aerospace systems. Recent interests include robotics, which combines long term backgrounds in mechanisms and computers. Ten years of univ teaching (Columbia Univ, FIT); currently Dir of Engrg Programs responsible for initiating and administering electrical and/or mechanical engrg. *Society Aff:* ASME, ASEE

Dilworth, Paul B
Business: 6415-2 Viscount Rd, Mississauga, Ontario, Canada L4V 1K8
Position: Consultant (Self-employed) *Education:* BA/MEngg/Univ of Toronto. *Born:* 1/31/15. in Toronto, Canada; Dir Dilworth-Secord-Meagher & Assocs Ltd Cons Engrs Toronto. Fellow of Engg Inst of Canada & of Canadian Aeronautics & Space Inst. Assn of PEs of Ontario. Mbr Natl Club & Lambton Golf & Country Club, Toronto. *Society Aff:* FEIC, FCASI.

DiMaggio, Frank L
Business: 616 Mudd Bldg, New York, NY 10027
Position: Prof. *Employer:* Columbia Univ. *Education:* PhD/Civil Engg/Columbia Univ; MS/Civil Engg/Columbia Univ; BS/Civil Engg/Columbia Univ. *Born:* 9/2/29. Served in Army 1954-56 in Armed Forces Special Weapons Proj. Prof at Columbia since 1956. Chrmn, Dept of Civil Engg and Engg Mech, 1975-1978. Consultant to Weidlinger Assoc since 1956. Technical Consultant to Implements and Ball Comm, United States Golf Assoc, 1958-1966. Amer Soc of Civil Engrs, Engr Mechanics Div: Exec Cttee, Mbr 1981-1984, Chrmn 1983; Advisory Cttee, 1984-87, Chrmn 1986. Specialty: Structural Dynamics and Fluid-Structure Interaction. Natl Sci Fdn Sr Postdoctoral Fellow, 1962, at Turin Polytechnic Inst, Turin, Italy. Guest Scholar, Kyoto Univ. 1986. *Society Aff:* ASCE.

DiMascio, Angelo J
Home: 4466 Dale Blvd, Woodbridge, VA 22193
Position: Deputy Commander *Employer:* Naval Air Systems Command. *Education:* DBA/R&D/Science & Tech/George Washington Univ; MSEM/OR/SA/Drexor Univ; BSME/Aero Engr/Drexor Inst of Tech *Born:* 6/14/38. Native of Phila, PA. Served

DiMascio, Angelo J (Continued)
with Army Corps of Engrs 1964-1966. Design engr/proj engr/supv engr at Phila Naval Shipyard (specializing in missile sys & hydraulics), 1966-1968. Prog mgr (Aviation support equip) at Naval Air Engr Ctr, 1968-1971. With Naval Air Sys Command (Mtl Acquisition Group, Res & Tech Group, & Test & Evaluation group) since 1971, Tech Dir for Test and Evaluation, 1977- 1979. Exec Dir of Acquisition mgmt, 1979-1982; Corp Mgmt Conslt to Commander, 1982-1983; Assumed current position of Deputy Commander in 1983. *Society Aff:* PiTau Sigma, Beta Gamma Sigma

DiMatteo, Frank A
Home: 3300 Durbin Place, Falls Church, VA 22041
Position: Dir. *Employer:* AID-State Dept. *Education:* BSCE/CE/Univ of OK. *Born:* 5/25/23. Grad from Univ of OK in 1951 as a Civ Engg and served in the US Navy during WWII as a flyer and in Korean War as a Civ Engg. Served as Proj Mgr and Chief Engg on overseas heavy construction projs for the US construction industry, before joining AID/State Dept in 1962. Served in SE Asia as a Foreign Service Officer as AID Dir of Engrg, joined the US Army Corps of Engrs in 1980 as spec asst to the Chief of Engrg on Intl Affairs, and have had supervisory engg repson for over thirty yrs. Am a PE, fellow ASCE and Mbr of SAME, PIANC and APWA. Married to former Eugenia Caruso, three children Louise, Guy and Gia. *Society Aff:* NSPE, ASCE, SAME, PIANC, APWA.

Dinlenc, Sadi
Home: 157 All Angels Hill Rd, Wappingers Falls, NY 12590
Position: VP *Employer:* Fishkill Engrg Assoc *Education:* M.Sc.M.E./Heat Transfer/Syracuse Univ, NY; B.Sc.M.E./Heat Transfer/Istanbul Tech Univ *Born:* 6/22/39 Native of Turkey, worked there at USAF installations as mech. engr after graduation. Wrote articles on HVAC topics, taught heating/refrigeration courses in trade schools. Resigned from IBM in 1973 to accept overseas assignments. Established own co 1978 for private practice. Registered PE in NY, NJ. *Society Aff:* PEP, NSPE

Dinos, Nicholas
Business: Dept of Chem Engg, Ohio University, Athens, OH 45701
Position: Chrmn. *Employer:* OH Univ. *Education:* PhD/ChE/Lehigh Univ; MS/ChE/Lehigh Univ; BS/ChE/PA State Univ. *Born:* 1/15/34. in Tamaqua, PA of Greek immigrant parents. Married (1955) to Lillian Margaret Gravell. Three children: Gwen (1958); Christopher (1961) Janet (1966). With E I Dupont (1955-1964) in atomic energy prog. OH Univ (1967-), tenured 1970; Prof 1972; Chrmn (1977-present), NASA faculty fellow (1972, 1974). Visiting Prof, Japan (1976); Danforth Fdn Assoc (1975); elder, Presbyterian church. *Society Aff:* AIChE, ASEE, ACS, AAAS, SHOT

DiPippo, Ronald
Business: Southeastern Mass. Univ, Mech. Engin. Dept, N. Dartmouth, MA 02747
Position: Prof and Chairman of M.E. Dept. *Employer:* Southeastern MA Univ *Education:* PhD/Engrg/Brown Univ; ScM/Engrg/Brown Univ; ScB/ME/Brown Univ *Born:* 6/2/40 grew up, educated in Providence, RI. First to earn Bachelors, Masters and Doctors degrees in engrg consecutively at Brown Univ. Research interests: transport properties of fluids; design of energy conversion systems; applications of geothermal energy; energy-related topics. Author of Geothermal Energy as a Source of Electricity: A Worldwide Survey of the Design and Operation of Geothermal Power Plants. Editor and contributor to Sourcebook on the Production of Electricity from Geothermal Energy. Author of numerous reports and articles on geothermal energy. Previous employment: ITT-Grinnell; US Naval Underwater Systems Center. Presently also Consultant to industry and governments. Three children. Enjoy photography. *Society Aff:* ASME, ASEE, GRC, ΣΞ

Director, Stephen W
Business: Dept of Elec and Computer Engr, Carnegie Mellon University, Pittsburgh, PA 15213
Position: Whitaker Prof and Head, Dept of Elec and Computer Engr and Prof of Computer Sci *Employer:* Carnegie Mellon Univ. *Education:* PhD/Elec Engg/Univ of CA, Berkeley; MS/Elec Engg/Univ of CA, Berkeley; BS/Elec Engg/SUNY-Stony Brook. *Born:* 6/28/43. Joined Univ of FL in 1968, promoted to Prof in 1974 of Elec Engrg. Joined Carnegie Mellon Univ as a Prof in 1977 and appointed U A and Helen Whitaker Prof of Electronics and Electrical Engrg in 1980 and Prof of Comp Sci in 1981. Appointed Head, Dept of Elec & Comp Engr Oct, 1982. Was Visiting Scientist at IBM Watson Res Ctr Sept 1974-Aug 1975. Fellow IEEE, Mbr Sigma Xi & Eta Kappa Nu. Has been Pres-Sec-Treas and Mbr ADCOM and Chrmn of Computer-Aided Network Design Ctte of IEEE CAS Soc & guest editor & assoc editor of IEEE Trans CAS and Proc of IEEE. Written 3 texts, received the 1970 IEEE CAS Soc Best Paper Award the 1976 ASEE Frederick Emmons Termar Award, the 1978 IEEE W R G Baker & Prize Paper Award, 1984 Outstanding Alumni Award, SUNY Stony Brook, and IEEE Centennial Medal. *Society Aff:* IEEE, HKN, ΣΞ.

Dirscherl, Rudolf
Business: PO Box 451, St Louis, MO 63166
Position: Chief Metallurgist. *Employer:* Nooter Corp. *Education:* MS/Metallurgical Engg/Univ of MO; BS/Civ Engg/KS State Univ. *Born:* 5/2/33. Native of Mannheim, Germany. Immigrated to US in 1953. With Nooter Corp since 1960. Production engr 1960-68; Metallurgical Engr 1968-1977; Chief Metallurgist since 1977. Supervise testing lab and responsible for solving mtl related problems. Mbr of two subgroups within the ASME Boiler & Pressure Vessel Committee. Past Chrmn of the St Louis Sec of ASM. Enjoy travel and gardening. *Society Aff:* ASM, AWS.

Disantis, John A
Home: 11 Greenview Dr, Jeannette, PA 15644
Position: Material Utilization Engineer. *Employer:* Elliot Co Div of United Techs Corp. *Education:* BS/Met E/MI State Univ. *Born:* May 1920. Native of Cleveland, Ohio. Married, 3 children. US Navy 1942- 45. Mgmt trainee, General Motors, Cleveland, Ohio. Technical Specialist, Corrosion-Metallurgy, SOHIO, Cleveland, Ohio. Sen Metallurgical Engr, Res Div, Carrier Corp. Material Utilization Engr Elliott Co. ASM: Handbook Cttes Vol 1 1958-59, Vol 2 1962-63; 1st in No America to receive MEI Ferrous Met Ext Diploma 1963; Res Applications Prog Ctte 1967-71, Sec for 3 yrs; Syracuse Chap Chmn 1970. ASTM Subctte X of A-10 & Sec 1962-63. 1st VP The Tech Club of Syracuse NY 1974-75; NMA Director, 1971-75. *Society Aff:* ASM, AWS, NMA.

Dischinger, Hugh C
Home: Box 472, Gloucester, VA 23061
Position: Owner, Consulting Engrg Firm. *Employer:* Hugh C. Dischinger, Engineers & Surveyors *Education:* BS/CE/VA Military Inst. *Born:* Mar 1924. Served as Fighter Pilot, US Army Air Corps 1943 to 1947 with overseas duty in the Pacific Theater. With Stone & Webster Engrg Corp two yrs before entering private practice in 1952. Married, 4 children. Served as Pres of local chapters ASCE & VSPE Hampton-Newport News Area; Pres Virginia Section ASCE. Appointed by Governor of Virginia 1966 to Mbr of Virginia Marine Resources Study Comm. Pres of Local Council, BSA Winner of Silver Beaver Award. *Society Aff:* ASCE, NSPE, WPCF, ACSM.

Disher, Jerrold W
Business: 1400 Rymal Rd E, Hamilton Ontario Canada
Position: Vice President *Employer:* C C Parker Consultants. *Education:* BSc/Civ/Queen's. *Born:* 2/12/28. Grad Engr Queens 51, mbr the Assn of PE of Ontario, Fellow of the Inst of Traffic Engrs. Past Pres of the Canadian Soc for Civ Engg. Now Vice President and Treas of the Consulting Engg firm of C C Parker Consultants Limited of Hamilton, Canada, & mgr of the Transportation & Municipal Depts of that firm. Am a private pilot & also enjoy tennis & fishing. *Society Aff:* CSCE, EIC, TRB, ITE.

Disher, John H
Home: 8407 Whitman Dr, Bethesda, MD 20817
Position: Pres *Employer:* Avanti Sys *Education:* Advanced Mgt Prog/Mgt/Harvard

Disher, John H (Continued)
Bus Sch; BS/ME/Univ of ND. *Born:* 12/23/21. Pres, Avanti Sys, an Aerospace Consltg firm 1981-Date. Sr Engr, Ital Space Agency, Rome, Italy, 1981-82. Dir of NASA Advanced Progs 1974-1980, in charge of advanced studies & devs for prospective new space projs in areas of space transportation & manned space flight. Deputy Dir Skylab Prog 1965-74. Dir Apollo Prog Test Div 1963-65. Asst Dir for Apollo Spacecraft Dev 1961-1963. Proj Engr Mercury Prog 1958-1960. Hd Free Flight Res Sec, Lewis Res Ctr 1948-1958. Res Engr 1943-48. Author numerous Technical Papers in fields of Hypersonic Heat Transfer, flight testing, ramjets, solid rockets. Received NASA's highest honor - the Distinguished Service Medal in 1974 for work on skylab prog and Univ of ND "Sioux Award" for outstanding achievement the same year. Mbr Indianapolis "500" Technical Committee. *Society Aff:* AIAA, IAA, BIS

Disney, Ralph L
Business: Dept of IEOR, Blacksburg, VA 24061
Position: Charles O Gordon Prof. *Employer:* VA Polytechnic Inst & State Univ. *Education:* DrEng/Oper Res/Johns Hopkins Univ; MSE/Oper Res/Johns Hopkins Univ; BE/Ind Engr/Johns Hopkins Univ. *Born:* Feb 1928. Native of Baltimore Maryland. Taught at Lamar Univ 1956-59, Univ of Buffalo 1959-62, Univ of MI 1962-77, Charles O Gordon Prof of IEOR, VPI & SU 1977- . Visiting Prof I T A Sao Jose dos Campos, Brazil 1970-71; Distin Visiting Prof, Ohio State University 1974-75. Tau Beta Pi Outstanding Teaching Award 1967, Alpha Pi Mu Outstanding Teaching Award, 1977, Univ of MI. Siam Visiting Lecturer 1975- . Sr Ed AIEE 'Transactions', ORSA Council 1978-81, Author of many papers & co-author of 'Probability & Random Processes for Scientists & Engrs'. David F Baker Dist Res 1972 IIE, Operations Res Div, IIE "Special Citation" for Distinguished Res and Teaching,1979. Fellow IIE, 1981.. *Society Aff:* IIE, ORSA, SIAM, AMS, AAAS, TIMS, MAA.

Disque, Robert O
Business: AISC, 400 N. Michigan Ave, Chicago, IL 60611
Position: Asst. Director of Engineering *Employer:* A.I.S.C. *Education:* MS/Civil Engg/Drexel Univ; BS/Civil Engg/Northwestern Univ. *Born:* 8/28/26. Served in US Navy in WWII. Structural design engg for Day & Zimmermann from 1950-54 and for Moran, Proctor, Mueser & Rutledge from 1954-59. Joined American Inst of Steel Construction in 1959. Served as Chief Engg from 1963-79. Appointed Assoc Prof of Civil Engg, Univ of ME at Orono in 1979. Rejoined AISC in 1980 as Assistant Director of Engineering. Author of Textbook *Applied Plastic Design in Steel*, *Structural Steel Design and Construction* chapter of McGraw-Hill Standard Handbook for Civil Engrs and various papers on structural engg. Lecturers to prof audiences and consults to industry on structural steel. *Society Aff:* ASCE, AWS.

Ditmore, Dana C
Home: 13020 Oak Valley Dr, Morgan Hill, CA 95037
Position: Vice President, Engineering *Employer:* Applied Materials, Inc. *Education:* MS/ME/Univ of CA; BS/ME/Univ of CA. *Born:* 12/31/40. Dana C. Ditmore attended the University of CA at Berkeley where he obtained by a BS degree (1962) and an MS degree (1966) in ME. He is a grad of GE Advanced Engg Training Prog-ABC Courses. He worked for the GE Co in commercial nucl power in San Jose from July 1962 until December 1979. While with GE he worked in the boiling water reactor (BWR) fuel & core design area for approximately ten yrs. From 1974 through 1979 he was responsible for engineering management in the area of overall BWR system design & performance. His last position at GE was as Mgr of Proj Engg from Dec 1977 to Dec 1979. In December of 1979 Dana left GE to join Applied Materials, Inc., as Vice President of Engineering. In this role Dana is responsible for the management of engineering activities of the development of semiconductor wafer processing equipment. Dana is a reg PE in the State of CA in Mech & Nucl Engg. *Society Aff:* ASME.

DiTrapani, Anthony R
Home: 11500 Drop Forge Ln, Reston, VA 22091
Position: Dir for Civilian Personnel Policy/EEO *Employer:* Asst Secy of the Navy for Manpower & Reserve Affairs *Education:* BS/ME/Univ of WI. *Born:* 4/1/37. A mech & electronics engr, Mr DiTrapani is currently the Navy's Dir for Civilian Personnel Policy/EEO, for the Asst Secy of the Navy. Previously, he was the Deputy Proj Mgr for the Navy's largest shipbldg prog, the FFG-7 guided missile frigate ship acquisition prog, since its inception in 1970 to May 1980. Other responsibilities included: dir of ship silencing & ASW systems for the DD963 (Spruance class) destroyers for the first four yrs of that prog; asst proj eng for the SQS-26 sonar system; Deputy Asst Deputy Chief of Naval Material for Acquisition Control; 1981-86 was the Navy's Dir for Shipbldg, in the office of the Asst Secy of the Navy (Shipbldg & Logistics). *Society Aff:* ASNE.

Dittfach, John H
Business: Mech Engg Dept, Amherst, MA 01003
Position: Prof ME. *Employer:* Univ of MA. *Education:* MSME/ME/Univ of MN; BSME/ME/Univ of MN. *Born:* 4/17/18. Currently Assoc Hd responsible for undergrad prog at Univ of MA. Prior teacher and responsible for supervising ME Instrumentation Labs. *Society Aff:* SAE, ASEE.

Ditzler, Harold E
Business: 4020 N 59th St, Phoenix, AZ 85018
Position: Priesident Principal Engineer. *Employer:* Benham-Blair-Ditzler & Elling. *Education:* BS/Civil/OK Univ. *Born:* Apr 1926. BS Civil Oklahoma University. Served Marine Corps WWII. Designed all types heavy Cvl Engrg Projects, principally water & waste water facilities. Employed Benham-Blair et al 1954- . VP Rocky Mountain Region SAME 1976, Pres AZ CEs Assn 1978, Dir Papago Chap ASCE, Chmn AZ JC of Design Professions 1973. Reg Prof Engr 5 states. Enjoy hunting & fishing. *Society Aff:* ASCE, NSPE, SAME, AIME.

Dix, Rollin C
Business: 10 W 31st St, Chicago, IL 60616
Position: Assoc Dean *Employer:* Ill. Inst. of Tech. *Education:* PhD/ME/Purdue U.; MS/ME/Purdue U.; BSME/-/Purdue U. *Born:* 02/08/36 After schooling at Purdue and service with US Army Ordnance Corps, appointed Sr Engr with Bendix Mishawaka Div, 1962-64. Joined MAE Dept, IIT, in 1964 and appointed Professor in 1980. Since 1980, has served as Assoc Dean for Computing, responsible for all academic computing. Holds two patents, is a director of the Bimet Corporation, and consultant to industry. Appointed Fellow, ASME, in 1986. *Society Aff:* ASME, SME, ASEE

Dixon, John R
Home: Field Hill, Conway, MA 01341
Position: Professor. *Employer:* University of Massachusetts. *Education:* PhD/ME/Carnegie Mellon; MS/ME/Carnegie Mellon; BS/Civ Engg/MIT. *Born:* Oct 1930 Youngstown OH. US Army 1953-55. Plant Engr Jarl Extrusions Rochester NY 1955-57. Proj Engr Jos Kaye Co Cambridge MA 1957-58. PhD Carnegie Tech 1960. Asst Prof Carnegie Tech 1960-61. Assoc Prof Purdue 1961-64. Assoc Prof Swarthmore Coll 1964-66. Prof & Hd ME Dept 1966-71 U of Mass-Amherst. Prof U of Mass 1971- . Author: 'Intro to Probability' Wiley 1964, 'Design Engrg' Mc Graw-Hill 1966, 'Thermodynamics' Prentice-Hall 1975. 1976: Elect Fellow ASME & Ralph Coates Award ASEE. *Society Aff:* ASME, ASEE, ASHRAE.

Dixon, John R
Home: 9919 High Dr, Leawood, KS 66206
Position: Chmn *Employer:* HSD, Inc. *Education:* BSEE/EE/Univ of KY *Born:* 9/5/32. Sr mgmt and ownership positions held in the aviation, aerospace & construction indus. Pres & Dir of Kansas City based cons engrg mgmt firm operating internatly. Mbr MENSA. Dir Cons Engrs Coun of MO. Reg PE in 28 states & with the Natl Coun of Engrg Examiners. Airline Transport Pilot and Certified Flight Instructor active in civil aviation. HSD, Inc. is an A/E FRM specializing in Agri Bus and in industrially-oriented design-build projects. The Co operates internatly and is

Dixon, John R (Continued)
currently included in projs in Central America, the PRC, and in the Middle East. *Society Aff:* IEEE, NSPE, NCEE

Dixon, Marvin W
Business: Mech Engg Dept, Clemson, SC 29631
Position: Prof. *Employer:* Clemson Univ. *Education:* PhD/Mech Engr/Northwestern Univ; MS/Mech Engr/LA State Univ; BS/Mech Engr/LA State Univ. *Born:* 10/5/41. Native of Baton Rouge, LA. Taught Mech Engg at LA State for three yrs prior to Clemson appointment. During 15 yrs at Clemson, served as Assoc Dept Head, as chairman of the Mech Systems Group, as lab coordinator, and as advisor for the ASME student sec. Dev Mech Systems Lab prog, 8 new grad & undergrad courses, high school visition prog, and over $500,000 funded res during the last 5 yrs. Directed the res of PhD & MS students. Mech Engg Graduate Student Coordinator since 1983 during which time graduate enrollment increased by 280%. Active in ASME at the regional & sec level. Listed in Who's Who in the Southeast and in Who's Who in Education. *Society Aff:* ASME, ASEE, SCAS, IFTOMM, NSPE.

Dixon, Richard W
Business: 600 Mtn Ave, Murray Hill, NJ 07974
Position: Dir *Employer:* Bell Labs. *Education:* PhD/Applied Phys/Harvard Univ; MA/Applied Phys/Harvard Univ; AB/Applied Phys/Harvard Univ. *Born:* 9/25/36. in Hubbard, OR grad summa cum laude from Harvard College. Following post-doctorate work with Prof N Bloembergern at Harvard, joined Bell Labs in 1965. Became concerned with the interaction of light & elastic waves in solids and liquids. Subsequently worked on the dev of semiconductor light emitting diode devices. From 1973 to 1979 was supervisor of semiconductor lightwave lasers with responsibility for the device aspects of gallium arsenide communications laser dev. Hd of Optoelectronic Devices Dept 1979-83. Dir, Lightware Devices Lab, 1983-87. Dir, Adv Tech Dev Lab, 1987-present. Author of more than 50 technical papers. Fellow, IEEE. *Society Aff:* IEEE, APS, AAAS.

Dixon, William T
Home: 524 Gainesboro Rd, Drexel Hill, PA 19026
Position: Retired *Employer:* Formerly Atlantic Richfield Co *Education:* SM/ChE/MIT; BS/Chem & Science/Princeton Univ *Born:* 1/20/03 Hanover, NH. Taught in Practice School of Chem Engr 1926-27 (Buffalo Station). Joined Atlantic Refining 1927 and progressed thru various jobs to Dir of Engrg. Joined Engrg Dept in 1952 as Mgr of Design continued in this position until retirement in 1967. Dir of AIChE 1952-1956. Enjoy golf. *Society Aff:* AIChE

Dizer, John T
Home: 10332 Ridgecrest Rd, Utica, NY 13502
Position: Prof & Dean Emeritus; retired *Education:* Ph.D./Ind.Eng./Purdue Univ.; MS/Ind.Eng./Purdue Univ.; BS/Mech.Ind.Eng./ Northeastern Univ. *Born:* 11/07/21 Native of Bellows Falls, Vermont. Naval officer WWII and Korean War. Engr and supervisor for E. I. du Pont de Nemours and Cummins Engine Co. Prof, Dept Head and Dean, Engrg Technology and Business, Mohawk Valley Community College, 1959-85. NSF consultant for science education in India, 1969. Currently consultant in engrg technical education. Author of Tom Swift & Company and numerous papers and articles. Life Fellow, ASME. ASME Centennial medal, BSA Silver Beaver award and numerous engrg and civic awards. Cited in various American and foreign biographical publications. *Society Aff:* ASME, IIE, SME, ASEE, SIA

Dlouhy, John R
Business: 2620 Mishawaka Ave, South Bend, IN 46615
Position: Pres *Employer:* Lumm Engrs, Inc *Education:* MS/BA/IN Univ; BS/ME/Christian Brothers Coll *Born:* 6/9/45 After graduation from coll, spent two years with US Navy as Sonar Specialist. Worked for Bendix Corp for ten years as facilities engr responsible for HVAC, plumbing, pollution control and facilities upgrade of mechanical systems. Came with Lumm Engrs in 1979 as chief mechanical design engr for HVAC and plumbing group. Assumed present position in 1981 as Pres and Treasurer, responsible for overall job coordination, contract supervision, client relations, and work scheduling for co. *Society Aff:* NSPE, ASHRAE, APCA

Doane, Douglas V
Home: 2302 Vinewood, Ann Arbor, MI 48104
Position: Conslt *Employer:* Self *Education:* MSE/Metallurgical Engg/Univ of MI; BS/Chem Engg/Wayne State Univ. *Born:* 11/11/18. Elected to Tau Beta Pi & Sigma Xi. Res Assoc Univ of Michigan 1943-45. Metallurgical Engr M W Kellogg Co 1946-51 specializing in applications for corrosion-resistant & highstrength steels. Joined Climax Molybdenum Co of Michigan Res Lab in 1951; Res Supervisor in 1952, Mgr of the Res Lab 1960-69, Mgr Ferrous Metallurgy Res 1970-1979, Assoc Dir of Res 1979-1982. Retired 1982. Res in molybdenum metal powders, oxidation-resistant coatings, but major area of interest in physical metallurgy of steel & cast iron. Consulting metallurgist. MBR ASM, AIME, AFS, SAE. Publications in journals of these organizations, as well as active national ctte mbrships *Society Aff:* TMS/AIME, ASM, SAE, AFS.

Dobbins, Richard A
Home: 11 President Ave, Providence, RI 02906
Position: Prof of Engrg *Employer:* Brown Univ *Education:* PhD/Aeronautical Engrg/Princton Univ; MS/Phys/Northeastern Univ; BS/Engrg Sci/Harvard Univ *Born:* 07/15/25 Indus Exper: Eastern Inspection Bureau 1948-50; Arthur D Little Inc 1950- 53; Sylvania Electric Products 1953-56. Presently Chrmn of the Exec Div of Engrg 1983 present; Prof Engrg Brown Univ 1960- . Visiting Teaching & Res Appointments helds at CA Inst of Tech 1967-68; Univ of Essex England 1971; Abadan Inst of Tech Iran 1975. Cons to gvmt & industry. Mbrships: ASME, The Combustion Inst, Amer Physical Soc. Honorary Mbrships: Sigma Xi, Tau Beta Pi, Who's Who in America, Who's Who in Aviation. Author of numerous tech & scientific papers. Textbook *Atmospheric Motion & Air Pollution: A Textbook for Students of Engg & Science*, J Wiley & Sons 1979. *Society Aff:* ASME, APS, AAAR

Dobie, John B
Home: 3320 N El Macero Dr, PO Box 3064, El Macero, CA 95618
Position: Agricultural Engr, Emeritus, P.E. *Employer:* Univ of California. *Education:* BS/Agri Engg/WA State Univ; MS/Agri Engg/WA State Univ. *Born:* Native of Yakima, Washington. Served as Rural Service Engr with Pacific Power & Light Co for 3 yrs & Rural Electric Investigator at WSU for 3 one-half yrs. With Agricultural Engrg Dept Univ of Calif, Davis since 1946. Retired in 1983. Res interests incl application of electricity to agriculture, livestock feed processing & mechanization, densification & packaging of forage & collection, handling & utilization of agricultural residues. ASAE Journal Paper Award 1954, ASAE Fellow awarded 1974. Enjoy golf & travel. *Society Aff:* ASAE, ΦΚΦ, ΤΒΠ, ΣΤ, ΣΞ.

Dobras, Quentin D
Business: Nela Park-Noble Rd, Cleveland, OH 44112
Position: Manager Product Planning. *Employer:* General Electric Company. *Education:* BSEE/Elec Engg/Purdue Univ. *Born:* Oct 30 1928 in Lorain Ohio. Came to General Electrics Lamp Div headquarters at Nela Park after graduation as an Application Engr in the Large Lamp Dept. From 1960-64 worked in Advanced Application Dev & served as color cons for the Dept. 1967 appointed Sr Prod Planner for fluorescent lamps & in 1971 was appointed Mgr, Product Planning for the Fluorescent Lamp Dept. Hold three patents on lamp & luminaire designs. Fellow of the Illuminating Engrg Soc. Reg Prof Engr in St of Ohio. *Society Aff:* IES.

Dobrovolny, Jerry S
Business: 117 Transportation Bldg, Urbana, IL 61801
Position: Prof & Hd Genl Engrg Dept. *Employer:* U of Illinois - Urbana. *Education:* MS/Mech Engg/Univ of IL; BS/Genl Engg/Univ of IL. *Born:* Nov 1922. Grad work in mech engrg and engrg geology. Joined staff of College of Engrg at Univ of Illinois in 1945. Became Prof & Head of Dept of Genl Engrg 1959. Under leadership, the

Dobrovolny, Jerry S (Continued)
curriculum in genl engrg has changed significantly & today is recogni zed as one of the leading interdisiplinary engrg design progs in the country. Since 1957 has been a leader in identifying needed changes in the preparatory programs for teachers of technical education & has written num papers pertaining to engrg technology education. Registered Prof Engr & P Pres of the Illinois Soc of Prof Engrs. His activitites in various professional & civic organizations are many & varied. *Society Aff:* NSPE, ASCE, ASEE, ATEA.

Dobson, David B
Business: 6411 Chillum Pl NW, Washington, DC 20012
Position: Managing Director *Employer:* McGregor & Werner Inc. *Education:* BEE/Elec Eng/Rensselaer Poly. *Born:* 9/22/28. Worked at US Army Signal Corps Labs Fort Monmouth NJ. Served as Audio/Radio Mbr of US Army Psychological Warfare Bd Ft Bragg NC. Participated in dev of semiautomatic & computer-controlled test equipment & participated in dev of APOLLO/LEM electronics with RCA. With McGregor & Werner Inc since 1971; respon for Mktg; co supplies professional, tech & scientific socs. Has authored many articles & lectured in the fields of automated testing & Communicating, publishing. Mbr IEEE- AES Bd/Governors 1969-81, Also IEEE Publications Bd, Standards Bd & Finance Committee. Assoc Fellow Soc for Tech Communication; mbr, Soc for Scholarly Publ; Reg PE. *Society Aff:* STC, SAME, WSE, SSP, IEEE, NSPE, DCSPE, AWA.

Dobyns, Samuel W
Home: 10 Sellers Ave, Lexington, VA 24450
Position: Retired *Employer:* VMI. *Education:* MS/CE/Lehigh Univ; BS/CE/VMI. *Born:* 3/7/20. 1941 - Carnegie IL Steel Corp. Practice Apprentice. 1941-46 - US Army, Cavalry, Air Corps, Pilot. 1946- Sr. Instrumentman, EI DuPont DeNemours 1946-1985 Faculty, Civl Engg, VMI, 1954-64 mbr, VA State Bd of Architects Prof Engrs and Land Surveyors. 1960-NSF Faculty Fellow 1963-NSF Faculty Fellow. 1967-NASA Faculty Fellow. 1969- OCD Faculty Fellow. PE VA. CLS VA. 1969-LTC USAF (Retired). 1946-85 - Conslt Engr 1985 - Retired. *Society Aff:* ASCE, ASEE, ACSM, ASP.

Dockendorff, Jay D
Business: 1717 N. Naper Blvd, Napperville, IL 60540
Position: Pres *Employer:* H P D Inc. *Education:* BS/ChE/IA State Univ. *Born:* Nov 1939 Burlington IA. Worked for Chicago Bridge & Iron Co for 9 yrs. In the Horton Process Division, gained exper in genl areas of evaporation, crystallization, prilling & fluidized bed processing, spec in the design of caustic evaporator systs. Asst Sales Mgr & Sales Mgr of the Div's prod lines. With H P D Inc since 1970 as Pres. Ch Engr & now involved in the dev of designs for Process Systs & in the execution of all process equip contracts. *Society Aff:* AIChE.

Dockendorff, Ralph L
Home: 1704 E Texas Ave, Baytown, TX 77520
Position: Consultant. *Education:* ScD/CE/MIT; BS/CE/Tufts. *Born:* 9/12/12. 41 yrs with Exxon, now retired, and consulting in area of safety & fire protection. 17 yrs in process design/process control at Baytown Refinery; 1953- 1962, engg mgt; 1963-1964, asst mgr mfg, Esso Std Italiana, at Genoa; 1965-1970, engg mgr, Exxon Chem, Baytown; 1970-1979, Sec Hd, Safety/Fire Protection, Exxon Engg, Florham Park, NJ; 1973-1977, S/FP consultant to Baytown Ref'y, in particular, the refinery expansion proj. Consulting since retirement, Aramco Services Co, Houston, 14 months. *Society Aff:* AIChE, ACS, ΤΒΠ, ΣΞ, NFPA.

Dodd, Caius V
Business: P.O. Box X, Oak Ridge, TN 37830
Position: Research Engineer. *Employer:* Oak Ridge National Lab. *Born:* Jan 1936. Native of Memphis, Tennessee. BS in Engrg Physics from University of Tennessee in 1959. Joined Oak Ridge National Lab in 1960, doing dev work in nondestructive testing, with emphasis on electromagnetic testing. Concurrently as part time grad student received MS in 1965 & PhD in physics from UT in 1967. A Fellow in & Sec of Electrical & Magnetic Methods Ctte of American Soc for Nondestructive Testing, on Editorial Advisory Ctte of Internatl Journal of Nondestructive Testing, Mbr Sigma Xi & authored over 50 pubs.

Dodds, Walter B
Home: 8262 Fremont Ct, Greendale, WI 53129
Position: VP. *Employer:* Milwaukee Forge. *Education:* BS/Metallurgy/Grove City College. *Born:* 12/13/41. Native of McKeesport Penna. Graduated Grove City College 1963, BS Metallurgical Engg. Has Held various production, QC, and metallurgical positions with Inland Steel Co, Kaiser Aluminum and presently Milwaukee Forge. Elected a VP of Milwaukee Forge in 1974. PChmn 1978-79 Milwaukee Chapter, ASM. Mbr of Forging Industry Assoc Technical Committee. *Society Aff:* ASM, SAE, ASTM.

Dodge, Donald D
Home: 22827 Buckingham Ave, Dearborn, MI 48128
Position: Sr Res Engr. *Employer:* Ford Motor Company. *Education:* BS/EE/Univ of MI. *Born:* 3/3/26. Served with navy Air Corps 1943-45. Machine des & process dev engr for Owens IL Glass Co in manufacture of television picture tubes 1948-55. With Ford Motor Co since 1955 in var capacities. Presently respon for dev of non-destructive tests for manufacturing operations. Author of NDT sect of Mark s Mechanical Engrs Handbook. Author of NDT section of McGraw Hill Encyclopedia of Science and Technology. Holder of 5 patents. Natl Dir of ASNT 1968-71 & 1973-76. Fellow of ASNT 1973. ASNT Tutorial Citation 1979. ASNT Lester Honor Lecturer 1980. Chmn SAE Handbook Ctte Div 25. SAE Vice-Chmn. of ISTC. Taught Community Colleges 1971-74. Enjoy yacht racing, skiing, & the arts. *Society Aff:* ASNE, ASTM, SAE, ASM.

Dodge, Franklin T
Business: 6220 Culebra Rd, San Antonio, TX 78284
Position: Institute Engineer *Employer:* Southwest Research Institute *Education:* PhD/Mech. Engrg/Carnegie Mellon Univ.; MS/Mech. Engrg./Carnegie Mellon Univ.; BS/Mech. Engrg./Univ. of Tennessee, Knoxville *Born:* 11/11/36 Born in Uniontown, PA & attended public schools in WV, KY & TN. Employed by Southwest Research Inst since 1963 except for a two-year period, 1971-73, at the Univ of TN. Assumed present position of Institute Engineer in 1982. Nationally known specialist in low-gravity fluid dynamics and spacecraft propellant dynamics. Also part of the advisory group directing the long-term research program at SWRI. Fellow, ASME, and Associate Fellow, AIAA. Chairman of ASME National Fluid Transients Cttee, 1984-86. Interests include jogging, golf, and music. *Society Aff:* ASME, AIAA

Dodgen, James E
Home: 130 Briarcrest Pl, Colorado Springs, Colorado 80906
Position: Pres. *Employer:* Dodgen Engg Co. *Education:* BS/Chem Engg/GA Inst of Tech. *Born:* 9/15/21. Native of Anniston, AL. Pres and founder of Dodgen Engg Co, in 1974 which spec in process engg of propellant, explosives and related chem plants in the US and Europe. Engr Duty Officer-Commander-US Navy, spec in res, test and evaluation, and production of weapon systems (1951-68). Sen Chem Eng, Penn Salt Co, specializing in design, construction and operation, heavy and fine chem plants (1946-51). Prod Dir and Process Engg Mgr Olin Corp, res for operations and plant improvement (1969-72). Program Mgr Aerojet Chem Co responsible for dev high performance rocket motors and commercial specialty chem (1972-74). Registered PE CO & CA. *Society Aff:* AIChE, ACS, NACE, AUSA.

Dodington, Sven H M
Home: 1 Briarcliff Rd, Mountain Lakes, NJ 07046
Position: Avionics Consultant. *Employer:* Self *Education:* AB/Engrg/Stanford Univ. *Born:* 5/22/12. 1935-41 television res London, England. 1941- , ITT on radar countermeasures & 1945-85, radio aids to navigation. Inventor of tacan navigation system for which received New Jersey outstanding patent award 1967. Holds 45

Dodington, Sven H M (Continued)
patents. Volare award of Airline Avionics Inst 1967. Author of radio navigation chapters in 'Aviation Navigation Systems' Wiley 1969 & 'Electronics Engineers Handbook' McGraw-Hill 1975. Second Ed 1982. Technical advisor to Radio Technical Commission for Aeronautics 1969- . Fellow IEEE. IEEE Pioneer Award 1980. Inventor of the Year award, NY Patent Law Assoc 1983; RTCA Achievement Award 1984; Conslt Editor, McGraw-Hill Encyclopedia of Sci and Tech 1984 *Society Aff:* ION, IEEE, AIAA.

Dodson, Charles R
Business: 539 San Ysidro Rd, Santa Barbara, CA 93108
Position: Pres. *Employer:* Dodson & Assoc, Inc. *Education:* MS/Mech Engr/MA Inst of Tech; BS/Mech Engr/Univ of MD; 2 yrs Grad Courses/Mech Engr/Univ of CA. *Born:* 9/1/07. 11 yrs Std Oil Co of Ca, Dir Prod Tech Labs, Div Petrol Engr; 5 yrs Prof, Hd Petrol Engg Dept USC; Consultant 6 yr VP Oil Dept Citibank, NY; 11 yrs Sr VP Natural Resources Dept, First Interstate Bank of Calif, Los Angeles; 5 yrs Pres IDS-Oil Progs, VP Investors Diversified Services; 5 yrs Exec. VP Ogle Resources Inc; 3 yrs Pres, Dodson & Assoc. Inc; Treas & Dir AIME, 1977 Recipient De Golyer Medal; SPE Distinguished Lectr; Chrmn Trust Committee, Tau Beta Pi; Advisory BD SW Legal Fdn; Reg. Petrol Engr. CA *Society Aff:* SPE, SPEE, API.

Dodson, Roy E, Jr
Home: 1226 Alexandria Dr, San Diego, CA 92107
Position: Sanitary Engr. *Employer:* State of California. *Education:* BS/Engg/Univ of CA. *Born:* Aug 1913 San Diego CA. Public Health Engr in Contr Costa County CA & with Oregon State Bd/ Health 1939-44. USPHS Reserve WWII. Joined City of San Diego Water Utilities in 1944 as Sanitary Engr; retired as Dir in 1972 to accept appointment to State Water Resources Control Bd, a full-time 5-Mbr Bd which regulates water rights & water pollution in CA, 1972-77. Consultant to CA Dept Health Services in Wastewater Reclamation, 1977-1982. Appointed to CA Water Commission 1979; reappointed 1982-86. (CA Water Comm is public sounding-board on State Water Project; advisory to Dir of Water Resources, to Governor and Legislature.) Special interests have been in training & certification of treatment plant operators & water reclamation & demineralization. Retired 5/1979 from full-time State Service. Fellow ASCE. Life Mbr AWWA; National Dir 1969-72; Fuller Awardee 1965; George A Elliott Awardee 1950. Diplomate, Amer Academy of Environ Engrs. Reg PE - Civil and Chem (CA). *Society Aff:* ASCE, AAEE, AWWA.

Dodson, William H
Business: PO Box Y, Oak Ridge, TN 37830
Position: Div Dir. *Employer:* Martin Marietta Energy Sys *Education:* BS/CE/Univ of TN. *Born:* 11/14/25. Native TN. After military service in WWII, obtained Chem Engrg Degree at Univ of TN in 1950. Total ind career has been in mtls dev, processing, characterization, and evaluation of mtls used in fabricating nuclear weapons components at the Martin Marietta operated Y-12 Plant at Oak Ridge, TN. Maj effort has been in uranium and uranium alloys. Served as dev engr, group leader, dept superintendent, and currently as div dir with responsibilities for the full range of mtls: metals, ceramics, plastics, chem compounds, etc. Charter mbr of local ASNT and active in ASM for 25 yrs. *Society Aff:* ASM.

Doe, Peter W
169 Valley Forge Drive, Mystic Shores, Tuckerton, NJ 08087
Position: Consultant *Employer:* Havens and Emerson, Inc *Education:* B.Sc./CE/Manchester Univ, Eng *Born:* 6/12/26 St. Annes, England. Served in British Army 1947-1949, Commissioned Royal Engrs. Engaged in Roads, Sewers, Structural and Solid Waste for Blackpool Borough, Lancashire 1949-1956. Later headed team designing major aqueducts and filtration plants for Fylde Water Bd. Sludge disposal research lead to unique freezing process for dewatering. Emigrated 1968 to join Havens and Emerson, Mgr of Special Projects. Specialized Water Resources work in NJ, CT, GA, VA including Sludge Disposal. Project Mgr, NJ State-wide Water Supply Master Plan 1976-1980. Authored over 25 papers; active professional circles. Chrmn several AWWA Committees. Enjoys golf, tennis, painting, traveling. *Society Aff:* ICE, AWWA, WPCF, IWSA, AAEE

Doeppner, Thomas W
Home: 8323 Orange Ct, Alexandria, VA 22309
Position: Prof, Engrg Mgnt *Employer:* Defense Systems Management College *Education:* MS/EE/Univ of CA; BS/EE/KS State Univ.; -/Mathematics/McPherson College *Born:* 5/22/20. Native of Berlin, Germany. Served 29 yrs in US Army Signal Corps, primarily in communications res. Proj Mgr, Advanced Res Projs Agency, 1965-69, directing radio propagation res in tropics; established "lateral wave" component as maj carrier of VHF-UHF energy in tropical forests. Dir, Electronics Directorate, US Army Gen Staff, 1969-73; concentrated on spectrum mgt & electro-magnetic compatibility prog. Dir, Communications Res, Gen Res Corp, specializing in ILS planning for command, control, & telecommunications systems, 1973-77; Dir, Logistics Engrg Oper, 1977-84, Prof, Engg Mgmt, Defense Systems Management Coll, directing and teaching courses in systems engrg, logistics engrg, 1985 to date. Legion of Merit with Oak Leaf Cluster, other military decorations. Fellow, IEEE, Centennial Medal Award, IEEE. Fellow, Wash Acad of Sci; McPherson Coll Citation of Merit; IEEE Patron Award. *Society Aff:* IEEE, WAS

Doering, Herbert K
Business: Melatener Strasse 25, 5100 Aachen, West Germany
Position: Prof High Frequency Technique. *Employer:* Rheinisch-Westfalische Techn Hochschule. *Education:* Habil/-/Techn Univ, Stuttgart; Dr Techn/-/Techn Univ of Vienna; Dipl Ing/- /Techn Univ of Vienna. *Born:* Feb 10 1911 Vienna. Dipl-Ing Elec Engrg 1934; Dr-Tech Technical University Vienna 1936. 1936-52 Indus activities as res engr incl: radio receivers, transmitters, transit time effects in electron tubes, klystrons, travelling wave tubes, passive devices & microwave measurement technique. Lecturer & habilitation on microwave techniques & ultrashort-wave tubes at Stuttgart Technical University 1949. Since 1952 Full Prof & Dir of the Institute of High Frequency Techniques at Aachen Institute of Technology. Curator of the Physikalisch-Technische Bundesanstalt 1963. Head of the 'Nachrichtentechnische Gesellschaft in VDE' 1967 & 69. VP of the 'Deutsche Forschungsgemeinschaft' 1969-74. Mbr 'Rheinisch-Westfalische Akademie der Wissenschaften' & 'Osterreichische Akademie der Wissenschaften'. VDE hon ring 1972; Fellow IEEE 1974; Wilhelm Exner Medal Vienna 1975. Enjoys classical m. *Society Aff:* ITG, DPG, IEEE.

Doering, Robert D
Home: 2520 Lauder Dr, Maitland, FL 32751
Position: Prof of Engrg *Employer:* Florida Tech University. *Education:* PhD/Engr/Univ of Southern CA; MS/Civil Engr/Univ of Southern CA; MS/Industrial Engr/Univ of Southern CA; BE/Mechanical Engr/Univ of Southern CA. *Born:* Feb 23 1925. Native of Orlando, Florida. Served with US Navy WWII Electronics Technician. Sales Engr with Allis Chalmers, General Machinery Div. Union Oil of Calif: Field Engr, Refinery Construction; Plant Engr, Res Center; Operations Supervisor, Distribution Terminals. Honeywell Inc Ordnance Div, Ch Engr Space Simulation. Hughes Aircraft Sr Staff Head. Joined Univ of Central FL College of Engrg 1969, Dept of Indus Engrg & Mgmt Systs. Mbr Tau Beta Pi, Pi Tau Sigma, Phi Kappa Phi, Alpha Pi Mu, NSPE, FES, IIE, AMA, SAFSR. Pres local chap FLorida Engrg Soc 1977-78 Regional VP Alpha Pi Mu. 1976-83 Elected National Chrmn, 1984, Dir Central FL Crime Watch Inc, Pres Omega Systems, Inc. Registered Prof Engr Florida & California. Named Engr of Yr by FES 1976, Elected Fellow Mbr 1983. *Society Aff:* ΤΒΠ, ΤΣ, ΦΚΠ, ΑΠΜ, NSPE, FES, IIE, AMA, SAFSR

Doherty, William H
Home: 18 Belted Kingfisher Rd, Hilton Head Island, SC 29928
Position: Retired. *Education:* ScD/Science/Catholic Univ of Amer; SM/Engg/Harvard Univ; SB/Elec Communication Engg/Harvard Univ. *Born:* Aug 21 1907

Doherty, William H (Continued)
Cambridge From 1929 spent 40 yrs with Bell System. Assignments included Dir of Electronic & Television Research Bell Labs; Asst VP AT&T; Patent License Mgr Western Electric; Asst to Pres Bell Labs. Received Morris Liebman Memorial Prize IRE 1937 for invention of a high-efficiency amplifier for high-power broadcast transmitters. During WWII supervised design of Navy fire-control radars at Bell Labs. Author of numerous papers on electric communication. Awarded 108 U.S. & foreign patents. Was made fellow of IEEE 1944 for contrib to the design of radio transmitters. Fellow AAAS 1978. Dir IEEE 1951-53 & 1958-60. Bd Chrmn Engg Soc Lib (NY City) 1967-69. After retirement in 1970, co-authored A History of Engineering & Science in the Bell System published 1976. *Society Aff:* IEEE, AAAS, TBΠ, ΣΞ.

Dohse, Fritz E
Business: Muggardt 6, 7840 Muellheim 16, West Germany
Position: Retired *Employer:* Univ of New Orleans. *Education:* PhD/Theoretical & Applied Mechs/Univ of IL, Urbana; MS/Engr Mechs/LSU, Baton Rouge. *Born:* 2/6/25. Native of Goettingen Germany. Undergrad education: Tech University of Berlin. Taught engrg mechanics at LSU & Univ of New Orleans. Chrmn of Engrg Sciences Dept, LSUNO 1967-73. Founding Dean of the College of Engineering 1973- 1984. Registered prof. engineer (ME). Retired August 31, 1984 and now lives in West Germany. Hobbies: classical music, electric model trains, home repair, travel. *Society Aff:* ASEE

Doigan, Paul
Home: 2242 Pine Ridge Rd, Schenectady, NY 12309
Position: Conslt-Engrg Manpower *Employer:* Self *Education:* PhD/Physical Chem/NYU; MS/Electrochem/Univ of MA; BS/Meteorology/NYU; BS/Chem/Univ of CT *Born:* 06/08/19 Born in Greenfield, MA. Married Edna Levy; three children, Rhoda Sherman, Lloyd, Amy. Served with USAF 1942-46 as Meteorologist-rank of Major, on faculty of Univ of Conn 1950-51. With Gen Elec 1951-1984. Initial work in Engrg High Temperature Insulation Materials. (Obtained patent on insulation structure) and then successively more responsible technical and mgmt positions in Electronics and Aerospace Businesses. Assumed responsibilities in 1967 for company-wide doctoral and intl recruiting and assisting components in their overall technical recruiting activities. Elected to Sigma Xi, Phi Lamda Epsilon and Sigma Pi Sigma Honorary Societies. Served as Chrmn Coll-Indus Council and Relations with Indus Div of ASEE; ASEE Bd of Dirs; Chrmn Engrg Manpower Commission of Pres 1977-79. Currently consltg on engrg. Education, engrg manpower and immigration problems. *Society Aff:* IEEE, AAES, AIC, ASEE

Dolan, John E
Business: 180 East Broad Street, Columbus, OH 43215
Position: V-Chrmn Engg & Const. *Employer:* American Electric Power Serv Corp. *Education:* BSME/Mech Engg/Columbia. *Born:* May 1923. ILLIG Medal. Tau Beta Pi. Advanced Progs in Steam Design, Nuclear Technology, Mgmt, Power Systems Engrg. WWII Pilot 1942-46. With American Electric Power Service Corp since 1950 as Staff Engr, Ch Design Engr, Ch Mech Engr, Ch Engr. Presently V-Chrmn, Engrg & Construction for 7-State System, including 1300 MW Supercritical-Pressure Generating Units & 765-kv Transmission. Registered Prof Engr in 7 states. Fellow ASME. Dir of 22 AEP companies. *Society Aff:* ASME, TBΠ.

Dolan, Roger J
Home: 2369 Lariat Lane, Walnut Creek, CA 94596
Position: Gen Mgr-Chief Engr *Employer:* Central Contra Costa Sanitary District *Education:* MS/Environ Engrg/Harvard Univ; BS/CE/Northeastern Univ *Born:* 3/22/41 Manage Sanitary District responsible for wastewater treatment for a population of approximately 350,000; wastewater collection for approximately 240,000 people. Annual O & M budget of $22 million, 240 employees. From 1969 to 1977, Mgr of Technical Services, East Bay Municipal Utility District. Pres, CA Water Pollution Control Association (CWPCA) 1977; Chrmn of Bay Section, CWPCA, 1975; Dir, Water Pollution Control Federation 1980-82; Pres, CA Association of Sanitation Agencies (CASA) 1983. Married; 2 children. *Society Aff:* WPCF, ASCE, AAEE, AWWA.

Dolan, Thomas J
Home: 510 S Highland Ave, Champaign, IL 61821
Position: Prof Emeritus T&AM Dept. *Employer:* Self-Employed. *Education:* MS/CE/Univ of IL. *Born:* Dec 1906. Head T & A M Dept 1952-70. Capt directing mfg Army Ordnance 1942-46. Over 130 technical publications on res on fracture & properties of metals, pressure vessels, pipe lines, rocket engines, failure analysis etc. Mbr num cttes: SAE, ASTM, ASM, PVRC, ISO. Bd/Dir Packer Engrg Assoc. Awards: R L Templin & Dudley Medal ASTM, Murray Lecture SESA, Outstanding Educators of America 1972, Univ of Illinois Disting Service in Engrg 1974. Cons to industries in failure analysis & in product liability litigation & in design problems to fed agencies (such as AEC). Spec interest in technical photography & golf. VP ASME 1958-60; Bd/Dir ASTM 1962-65; Pres SESA 1952. *Society Aff:* ASME, ASTM, SESA.

Dolch, William L
Business: CE School, W Lafayette, IN 47907
Position: Professor of Engineering Materials. *Employer:* Purdue University. *Education:* BSChE/ChE/Purdue Univ; MS/Chem/Purdue Univ; PhD/Chem/Purdue Univ. *Born:* Jul 1925. Prof of Engrg Materials 1964- , Hd Materials Div School of Civil Engrg Purdue. Mbr ASTM cttes on cement & concrete, ACI. Sigma Xi. ASTM Dudley Medal 1966. ACI Wason Medal 1968. Res & cons in cement, concrete, aggregates. ASTM Award of Merit (Fellow), 1977. ACI Anderson Award, 1984. ACI Fellow, 1985. *Society Aff:* ASTM, ACI, ΣΞ.

Dolecki, Stanley
Home: 9113 Hatton Dr, Crestwood, MO 63126
Position: Civil-Environmental Engineer *Employer:* Black & Veatch *Education:* BSCE/Civil Engr/Univ of MO, Rolla. *Born:* Jul 15 1928. Native St Louis, Missouri. US Air Force 1951-53. Cons Engr Russell & Axon 1953-55. Planning Engr St Louis County Plan Comm 1955-57; Dir of Public Works Ferguson MO. 1957-60; Cons Engr Vonder Bruegge & Gresowski 1960-62; Field-office Engr Portland Cement Assn 1962-65; Ch Engr St Louis Office Harland Bartholomew & Assocs 1965-81; Assoc Partner 1968. Civil-Environmental Engineer, Black & Veatch, 1981-. Mbr NSPE, MSPE (State Dir & P Chapter Pres), ASCE(Past Sect Pres), APWA, AWWA, ITE, WPCF, SAME, Engrs Club of St Louis (P Dir) Missouri Assoc Reg Land Surveyors, Chi Epsilon (Chapter Honor Mbr). Registered Engr 22 states; Registered Land Surveyor 2 states. *Society Aff:* NSPE, ASCE, APWA, AWWA, ITE, WPCF, SAME, XE.

Doll, Paul N
Home: 933 Moreau Dr, Jefferson City, MO 65101
Position: Profl Engrg Advisor (Self-employed) *Education:* LLB/Honor/Univ of MO; MS/Agric Engrg/Univ of MO; BS/Agric Engrg/Univ of MO; AA/Sci/Kidder Jr. Coll *Born:* 4/4/11 m. Mary Ruth Choplin, 1939; ch.: Mary Beth, Anne Colby, Robert Paul, 1937- 44, Univ MO, agric engr, co ext agent, agric. ext. serv.; 1944-47, MO div. Resources & Dev., chg. agric. indus.; 1947-54, MO Limestone Producers Assn, mgr. ; 1954 to 1976, Ed., MO Engr.; 1954 to 1976 exec. dir, MO Soc. Profl. Engrs. Danforth fellow, Univ MO, Mystical Seven, Univ MO; Knight of St. Patrick, Summa Cum Laude, Univ MO Recipient, Univ of MO awards: Dist. Fac.-Alum.- 1973. Dist. Service to Engrs-1969. Dist. Service to Univ - 1979. Dist. Service to Agric - 1981. Dist. Service to NSPE-1986. Inventor, prefabricated farm bldgs, lecturer. Author numerous pubns., Ed. MO Engr. (monthly profl. engrg jl.) Dir, 1949-52, D-V Equip. Co., Doll-Redman Min. Co.; Soil Nutrients Inc., farm owner. Univ MO; dir Agric. Engrs of MO; secy. MO Chem Council; dir MO Assoc Reg. Land Surveyors. 1949 to date, dir, Boy Scouts of Am., Lake of Ozarks Council; 1956 to

Doll, Paul N (Continued)
date, treas., MO Presby. Ch. Synod.; Elder, First Presby. *Society Aff:* NSPE, ASAE, ASCET, AE, TBΠ.

Doll, William O
Business: Alexander Dr, Wallingford, CT 06492
Position: Sr V P & Ch Engr. *Employer:* Cahn Engrs Inc. *Born:* Jun 1933. Native of Trenton New Jersey. BSCE University of Utah. 2 yrs service with USMC. With Cahn Engrs Inc since 1961. Named Principal of firm in 1964. Currently holds position of Sr VP-Ch Engr. Respon for technical excellence of engrg design for projects with construction values exceeding 75 million dollars annually. Active Mbr of many tech & prof socs; holds certification by the Natl Council of Engrg Examiners; Registered Prof Engr in 11 states. Enjoys tennis & motorcycling.

Dollar, David L
Business: Box 676, Greeneville, TN 37743
Position: Process Metallurgist. *Employer:* Ball Metal & Chemical-Div Ball Corp. *Born:* May 1944. MS from University of Wisconsin 1971; BSME from General Motors Institute 1967. Worked as project engr for AC Electronics Div GMC until 1969. Joined Ball Corp, Technical Div as Staff Metallurgist in 1969 & later promoted to Mgr of Metallurgy R&D. Transferred 1974 to the Metal & Chemical Div of Ball Corp. Presently respon fpr process dev involving Hazelett continuous casting for zinc based alloys & the characterization of zinc alloy formability. Mbr of SME & National Career Guidance Council of ASM. Author of 4 US Patents.

Dombeck, Harold A
Business: H2M Group, 575 Broad Hollow Road, Melville, NY 11747
Position: Exec VP *Employer:* H2M Labs Inc/Holzmacher, McLendon & Murrell, PC *Education:* M/CE/NYU; B/CE/NYU *Born:* 3/23/41 Exec. VP of H2M Labs Inc./ Holzmacher, McLendon and Murrell, P.C. since 1970. 1963-65 1st Lt USAF; '79-81 VP NYSSPE; VP, LI Chapter ACEC; Pres Suffolk County Chapter NSPE; '79-82 Dir NYWPCF; Pres '77-79 LI Chapter NYWPCF. Received ACEC and CEC/NYS Design Awards 1973 and 1975. NYSSPE Young Engr of the Year 1974; Diplomate AAEE. With H2M since 1965 serve as Dir of Environmental Engrg of H2M until 1981. Presently Dir of Finance and Marketing. Trustee NSPE Political Action Ctte 1984- 90; Pres NYSSPE 1983-84; Dir NSPE 82-85; Pres LI Chapter ACEC 82-84; Dir NYSACEC 82-87; Prof Engr NY, NJ, CT; Fellow ASCE; Trustee, Founder, NYSSPE Political Action Ctte 1981-87. *Society Aff:* AAEE, ASCE, NSPE, WPCF, ACEC, AWWA, XE

Dome, Robert B
Home: 645 Terry Rd, Syracuse, NY 13219
Position: Consulting Engineer. (Ret) *Education:* MSEE/Elec Engg/Union Univ; BS/EE/Elec Engg/Purdue Univ. *Born:* Oct 1905, Tell City, Indiana. Entire prof life with General Electric from student engr 1926 to retirement 1968 when Cons Engr Television Receivers. Assignments included Head of Advance Dev Radio & Television Transmitters, later same position in Receivers. Hold 115 US patents. Served on national cttes on transmitting standards for Monochromatic Television, Color Television & FM- Stereo. Authored num technical papers & book Television Principles . Sigma Xi honorary fraternity, Morris Liebmann memorial prize IRE, Fellow IRE, two GE Coffin awards. Enjoy travel & operating W2WAM. *Society Aff:* IEEE, ARRL, ΣΞ, QCWA

Domian, Henry A
Business: Beeson St, Alliance, OH 44601
Position: Research Specialist. *Employer:* Babcock & Wilcox Co. *Education:* MS/Met Engr/Univ of MI; BS/Met Engr/Syracuse Univ. *Born:* Jul 1926. Raised in Worcester, Mass. Served in the Army Infantry for two yrs & 6 yrs in the Air Force as Engrg & R&D Officer. Employed as Scientist with Ford Scientific Lab Dearborn MI for 8 yrs, doing fundamental res in diffusion & phase transformations in metals. Since 1965 employed at the Alliance Res Center of the Babcock & Wilcox Co as Supervisor & Res Specialist in metallurgical res. Enjoy fishing. *Society Aff:* AIME, ASM.

Domingos, Henry
Business: Clarkson Univ, Potsdam, NY 13676
Position: Prof *Employer:* Clarkson Univ *Education:* PhD/EE/Univ of WA; MS/EE/Univ of S CA; BS/EE/Clarkson Univ. *Born:* 9/17/34. Industrial experience has been with Union Carbide Corp, Hughes Aircraft Co, Atomics Intl, Rome Air Dev Ctr, and Gen Elec. Taught at the Univ of NV, the Univ of Wa, and Clarkson Univ. Res work has been concerned with VLSI design, new integrated circuit technologies, elec overstress failure, and reliability. *Society Aff:* IEEE, ASEE.

Domingue, Emery
Business: PO Box 52115, Lafayette, LA 70505
Position: President. *Employer:* Domingue Szabo & Assocs Inc. *Education:* MS/Civil Engg/Univ of IL; BS/Civil Engg/Univ of Southwestern LA. *Born:* Jan 9 1926 Louisiana. Served in European Theatre in WWII. Worked for construction company as field engr & then taught Civil Engrg at USL for 10 yrs. He is Pres of Domingue Szabo & Assocs Inc, organized in 1957. Served as Pres of Lafayette section of Louisiana Engrg Soc; Pres of Baton Rouge Section of ASCE; served as Bd/Dir of Louisiana Engrg Soc & is Past Pres of the Louisiana Intercoastal Seaway Assn. Recipient A.E. Wilder Award from Consulting Engineers Council/Louisiana. *Society Aff:* ACEC, ASCE, NSPE, ACI, ACSM, APWA, AME.

Dominguez, Renan G
Business: 306 Railroad Ave, S San Francisco, CA 94080
Position: Structural Engr. *Employer:* Dominguez Assoc. *Education:* BS/Civ Engr/Univ of IL. *Born:* 6/2/26. in Merida, Yucatan, Mexico. Educated in Merida until 1943. Grad BSCE at Univ of IL, Urbana in 1947 with Honors and was awarded Rotary Fdn Fellowship for advanced studies 1947-48. Practiced in Merida until 1955. In Decatur, IL 1955- 57. In San Francisco, CA from 1957 to date. Started his own firm in 1966, Dominguez Assoc. Reg to practice in CA, AZ, CO & WA. Has designed bridges for hgwys and innumerable bldgs, including 60 ft high masonry shell walls & 200 ft trunkated dome. Consultant for Taliesin Architects of the Frank Lloyd Wright Fdn. Involved in subdiv work. *Society Aff:* ASCE, NSPE.

Dominicak, Robert H
Home: 222 Stanley Court, Rapid City, SD 57701
Position: Owner *Employer:* Crow Dominicak Associates *Education:* BSCE/Civil Engg/SD School of Mines & Tech. *Born:* 2/26/41. Engr and VP of Thomas & Lockwood Consulting Engrs, Rapid City, 1968-76. VP & Prin Thomas, Erickson, Dominicak & Crow Consulting Engrs, Inc 1976-80. Partner, Crow Dominicak Assocs Consulting Engrs, 1980. Owner, Crow Dominicak Assoc Consulting Engrs, July 1981-present. Reg PE in SD, MT, CO, WY; Reg Land Surveyor in SD. Past Pres of Dakota Chapter of Amer Concrete Inst 1980, PPres of Black Hills Branch and SD Section ASCE, 1978; Secy-treasurer SD Section ASCE, 1975-76. Mbr Methodist Church. Mbr Exec Board and P VP of Admin Black Hills Council Boy Scouts of Amer, Council Commissioner Black Hills Council BSA. *Society Aff:* ASCE, ACI, NSPE, ACSM.

Donachie, Matthew J, Jr
Home: 296 Porter St, Manchester, CT 06040
Position: Staff Materials Project Engr. *Employer:* Pratt & Whitney. *Education:* ScD/Metallurgy/MA Inst Tech; Met Engr/Metallurgy/MA Inst Tech; SM/Metallurgy/MA Inst Tech; BMetE/Metallurgy/Renss Poly Inst. *Born:* Oct 1932. Native Holyoke, Mass. Employed at UTC Res Center, Electric Boat & Chase Brass & Copper prior to joining P&W. Adjunct Prof of Metallurgy, employed at Hartford Graduate Center since 1958. Extensive technical publications & presentations. Primary fields of interest: superalloys & high temperature metallurgy; titanium alloys; materials data retrieval; X-Ray & electron metallographic analysis. Active on ASTM & ASM cttes & in local AIME positions. Am Vac Soc Best Paper Award 1971. Hon Mention

Donachie, Matthew J, Jr (Continued)
1954 ASM Metallographic Competition. Active in Boy Scouts. Enjoy hiking, camping & photography. *Society Aff:* ASM, ASTM.

Donahue, D Joseph
600 N Sherman Dr, Indianapolis, IN 46201
Position: V Pres, Sr Scientist. *Employer:* GE/RCA Consumer Electronics *Education:* MS/Phys Chem/Univ of Michigan; MS/Phys Chem/Univ of Michigan; PhD/Phys Chem/Univ of Michigan. *Born:* May 1926. Native of Lemont, IL. Served with US Navy 1944-46. Joined RCA in 1951 as engr working on dev of mask etching & screening for color picture tubes. 1958: transferred to Solid State Div & held various engrg mgmt & general mgmt positions. 1970 became Div VP of Europe, Middle East & Africa Solid State located in London. 1973 returned to US as Div VP of Integrated Circuits & 1974 returned to London as Div VP of Internatl Solid State. 1975 Became Div VP Engg RCA Color Picture Tubes. 1977 apptd Div VP operations, RCA Consumer Electronic, Indianapolic, IN. 1981 appointed VP & Gen Mgr RCA Consumer Elec, Indianapolis, IN. In 1987 became VP Sr Scientist, GE/RCA Consumer Electronics, Indianapolis, IN. *Society Aff:* IEEE.

Donaldson, Bruce K
Business: Dept of Aerospace Engg, College Park, MD 20742
Position: Prof. *Employer:* Univ of MD. *Education:* PhD/Astro & Aero Engg/Univ of IL; MS/Aero Engg/Wichita State Univ; MS/Math/Wichita State Univ; BS/Civ Engg/Columbia Univ. *Born:* 8/5/32. USN pilot 1955-58. Structural dynamics engr with Boeing-Wichita 1/59-1/62 and then same with Beech Aircraft, Wichita 6/62-8/63. Grad study at Univ of IL, Urbana, 9/63-1/68. Asst Prof, then Assoc Prof, then Prof at the Univ of MD, College Park, 1/68 to present, teaching in the areas of solid mechanics and structural dynamics. Res in the area of numerical methods.

Donaldson, Coleman
Home: PO Box 279, Gloucester, VA 23061
Position: Consultant *Education:* PhD/Aero Engrg/Princeton Univ; MA/Aero Engrg/Princeton Univ; B/Aero Engrg/RPI *Born:* 9/22/22 Philadelphia, PA. Served with Army Air Corps, 1945-46. Head, Aerophysics Section, National Advisory Committee for Aeronautics, 1946-52. Formed Aeronautical Research Assocs of Princeton, Inc (NJ) in 1954; served as Pres until 1979. Received AIAA Dryden Research Lecture Award, 1971; Certificate of Commendation, US Marine Corps, 1978. General Editor, 12-volume Princeton Series on High-Speed Aerodynamics and Jet Propulsion, 1955-64. Member, Pres's Air Quality Advisory Bd, 1973-74. Chrmn, Lab Advisory Bd for Air Warfare of Naval Research Advisory Committee, 1972-77; member, Marine Corps Panel of NRAC, 1972- 77. Chrmn, Advisory Council, Dept of Aerospace and Mechanical Sciences, Princeton Univ, 1973-78; Member, Naval Research Advisory Ctte 1986-. *Society Aff:* NAE, AIAA, APS

Donaldson, Merle R
Home: 1833 Almeria Way S, St Petersburg, FL 33712
Position: Prof Elec Engrg Dept. *Employer:* University of South Florida. *Education:* PhD/EE/GA Tech; MS/EE/GA Tech; BEE/EE/GA Tech. *Born:* Apr 1920. Native of Grainola, Oklahoma. Served in Navy 1940-46 in radio & radar. Taught Elec Engrg at Georgia Tech 1946-50. Design cyclotrons at Oak Ridge Natl Lab 1950-57. Staff Engr to Lab Dir in communications, antennas & propagation R&D at Electronic Communications Inc 1957-63. Assumed current respon for creating a new Electrical Engrg Dept at University of South Florida in 1964. Fellow IEEE 1970. Editor IRE Trans on Comm Systems 1959-60. Registered Prof Engr Florida. Sigma Xi, Eta Kappa Nu, Tau Beta Pi, Phi Kappa Phi. *Society Aff:* IEEE, ΣΞ, ΦΚΦ, ΤΒΠ, HKN.

Donaldson, Robert M
Home: 60 Columbia Ave, Hampton, VA 23669
Position: Retired. *Education:* BS/Nav Arch & Mar Eng/Webb Inst. *Born:* Aug 1915. With Newport News Shipbuilding & Drydock Co from 1936-72 as Designer, Hydraulic Engr, Asst Mgr of Atomic Power Div & Mgr of Industrial Products Div. Products were hydraulic turbines, pumps, pump-turbines, paper drying & other heavy machinery & the power plant for the aircraft carrier Enterprise . VP of Planning & Control, Offshore Power Systems Inc 1972-75. Planning for production of floating nuclear power plants. Cofounder & P Pres of Hampton Roads Academy. Registered Prof Engr Virginia. Fellow ASME. Married to Harriet Houston. Twin daughters, Margaret & Anne. *Society Aff:* ASME, NSPE.

Dondanville, Laurence A
Home: 717 Westgate Rd, Deerfield, IL 60015
Position: Vice President *Employer:* De Leuw Cather & Company. *Education:* BS/Civil Engg/Purdue Univ 1950. *Born:* Mar 1928, Moline, Illinois. Served with Army Corps Engrs 1951-52. With De Leuw, Cather & Co since 1953 serving first as project engr & later as Sr Project Mgr on transportation projects incl studies of feasibility of high level bridge and rail tunnel crossing, Bosphorus Straits, Istanbul; Innermetropolitan freeway system, Sydney, Australia, Rapid Transit Integration, Taipei, Taiwan & major highway & public transportation projects in US. VP & Dir of De Leuw, Cather Internatl 1966-1980. Dir of De Leuw, Cather of Australia Pty Ltd since 1968-1980. Sr VP & Dir De Leuw, Cather & Co since 1972-1980. VP and Director of Quality Assurance and Transportation projects, DeLeuw Cather & Company since 1980. Registered Professional Engineer in Illinois and six other states. Internatl Pres Inst of Transport Engrs 1974. Disting Engrg Alumnus Purdue U 1973; Chairman Deerfield Illinois Sesquicentennial 1985; Purdue Alumni Association Director 1985-1988. Married to Rose Marie Lacey, Ebensburg PA. *Society Aff:* NSPE, ASCE, ITE, IEA, SAME, ARTBA.

Dondanville, Leo J, Jr
Business: 1525 S Sixth St, Springfield, IL 62703
Position: Board Chairman & President *Employer:* Hanson Engrs, Inc. *Education:* MS/Civ Engg/Univ of IL; BS/Civ Engg/Univ of Notre Dame. *Born:* 6/25/30. Served with USAF 1954-56. Joined Hanson Engrs in 1956 as a specialist in soil mechanics and fdn engg; has supervised projs for co clients in all 50 states and in several foreign countries; reg in 26 states and holds Natl Engg Certification; recipient of the Univ of IL Civ Engg Alumni Distinguished Alumnus Award, 1978; recipient of IL Soc of PE Distinguished Service Award, 1978, Recipient of Consulting Engineer Council/Illinois Distinquished Service Award, 1980; CEC/I (Pres, 1979-80); ACEC (Natl Dir 1980-81); 1986 Honor Award, College of Engineering, University of Notre Dame. *Society Aff:* ASCE, NSPE, ACEC, USCOLD.

Dong, Stanley B
Business: 4514 Boelter Hall, Civil Engineering Dept, Los Angeles, CA 90024
Position: Prof. *Employer:* Univ of CA. *Education:* PhD/Civ Engr/Univ of CA; MS/Civ Engr/Univ of CA; BS/Civ Engr/Univ of CA. *Born:* 4/2/36. Sr Res Engr (1962-1965) Aerojet-Gen Corp, Sacramento, CA. Asst Prof (1965- 1969), Assoc Prof (1969-1973), Prof (1973-present) Univ of CA Civil Engrg Dept Sch of Engg & Appl Sci Los Angeles, CA. ONR-AIAA Res Scholar in Structural Mechanics (1970-71). Achievement Award - Chinese Engrs & Scientists Assn of Southern CA 1978. *Society Aff:* ASCE, ASME, AAAS, ΤΒΠ, XE, ΦΤΦ.

Donnalley, James R
Business: 2730 Burr St, Fairfield, CT 06430
Position: VP, Corporate Environmental Issues Project *Employer:* Genl Elec Co., Fairfield, Conn. *Education:* BS/ChE/PA State Univ; PhD/ChE/Cornell Univ. *Born:* Jun 1918, Camden, New Jersey. Attended public schools in Prospect Park, PA. Joined General Electric Co Res Lab in 1943. Transferred to Chemical & Metallurgical Div in 1946. Group Leader for process dev & plant des work on silicones. Named Plant Mgr of Waterford, New York plant in 1948; Mgr Manufacturing Silicone Products Dept in 1952. Was appointed Genl Mgr of Insulating Materials Dept in 1960, Genl Mgr of the Semiconductor Products Dept in 1966 & Mgr Technical Resources for the Consumer Products Group in 1971. Mgr Lighting Res & Technical Services Operation in 1974. Assumed current responsibilites as Vice President,

Donnalley, James R (Continued)
Corporate Environmental Issues Project in 1980. Hobbies are golf, bowling & curling. *Society Aff:* AIChE, ACS, AAAS, CES, AXE.

Donnell, Lloyd H
Home: 656 Lytton Ave, Apt D307, Palo Alto, CA 94301
Position: Adj Prof. *Employer:* U of Houston. *Education:* BS/ME/Univ of MI; PhD/Mechanics/Univ of MI; L1D (Hon)/-/IL Inst of Tech *Born:* 05/20/95 Engr Franklin Auto Co, Dodge Bros 1915-22. Quaker Unit-Red Cross 1918-19. Small mfg bus 1922-25. Taught Univ of MI 1923-30. Fellow CA Inst of Tech 1930- 33. Chg of structures Goodyear Zeppelin Corp 1933-39. Chg. structures dev, Chance Vought Aircraft 1942-44. Prof, Acting Chrmn Mech Dept, Prof Emerit IL Inst of Tech 1939 to date. Visiting Prof Univ of MI 1960, Univ of Houston 1966. Fellow ASME, Amer Acad of Mechs. Founding editor Applied Mech Reviews 1948-50. Chrmn Appl Mech Div ASME 1951. Gen Chrmn First US Natl Cong Appl Mech 1951. Worcester Reed Warner Medal ASME 1960. Sesquicentennial Awd, Univ of MI 1967. Von Karman Medal ASCE 1968. ASME Medal 1969. Honorary L1D, 1968, IIT Hall of Fame 1978, IL Inst of Tech. Author of many res papers on dymanics, wave transmission, elasticity, theory of shells, buckling problems etc; book "Beams, Plates and Shells–, Eng Soc Monograph, MaGraw Hill Book Co, 1976; Russian translation, NAUKA Publisher, 1982. Commemorative volume, "Theory of Shells–, to celebrate 70th birthday, Univ of Houston, 1967. *Society Aff:* ASME, AAM, AAUP, ΣΞ, ΤΒΠ

Donnelly, John I, Jr
Business: 212 West Hubbard St, Chicago, IL 60610
Position: Pres *Employer:* Anderson Elevator Co *Education:* BS/EE/Inst of IL *Born:* 7/12/27 Consulting Engr, Elevators - J.I. Donnelly, J.K. Dawson and Assocs, and Metropolitan Elevator Co. Chief Engr - C.J. Anderson Co. Dir of Education Soc. IUEC Local 2 Chicago 1962-1972. Trustee for National Elevator Industry Educational Program 1967-1973. Certified forensic engr on elevators. Legal expert on elevators. PE State of IL. National dir for IL Society of Professional Engrg. Served all offices of Chicago Chapter. Served and chaired numerous committees at local and state level. Chrmn engrs group Union League Club of Chicago. *Society Aff:* NSPE/ISPE, IEEE, SAME, WSE, PE

Donnelly, Lloyd W, Jr
Business: Daniel Bldg, Greenville, SC 29607
Position: Group VP Engrg. *Employer:* Daniel Internatl Corp. *Education:* BS/Mech Eng/TX A&M Univ. *Born:* Mar 1927. Served in Navy in WWII. Associated with Monsanto Co in Texas, Massachusetts, Ohio & Missouri doing maintenance, construction, engrg & mfg assignments. Served as Dir of Engrg for Geigy Chemical prior to entering private practice. Joined Daniel International Corp in 1971 & served as VP of the Chemical Div prior to assuming current duties of Group VP-Engrg with respon for mgmt of US based engrg offices as well as those located in San Juan Puerto Rico, Glasgow Scotland, Brussels Belgium. Reg Engr TX, MA, MI, LA. Mbr ASME, AIChE & NSPE. *Society Aff:* SCSPE, NSPE, AIChE, ASME.

Donnelly, Ralph G
Business: Martin Marietta Energy Systems, P.O. Box 628 NS 1223, Piketon, OH 45661
Position: Section Hd. *Employer:* Union Carbide Corp, Nuclear Div. *Education:* MS/Met Engg/Univ of TN; BS/Met Engg/Case Inst of Tech. *Born:* 1/13/38. Native of Cleveland, OH. Joined Nucl Div of Union Carbide Corp at Oak Ridge Natl Lab in 1959 specializing in mtls joining & refractory metals R&D. Pioneered & managed R&D progs on mtls for radioisotope-powered thermoelectric generators for space satellites, and more recently, mtls for ind & residential energy conservation. Presently hd of Metals & Ceramics Div's nuclear fuels and mtls processing R&D sec at ORNL. Chrmn of newly established ORNL committee responsible for internal energy conservation policy. Past Chrmn of East TN Sec, AWS, & 1978 ASM Fellow. Avid sailor & bicyclist. *Society Aff:* ASM, ECS.

Donoghue, John L
Home: 1427 Blackthorne Dr, Glenview, IL 60025
Position: Chmn *Employer:* Ralph Burke Assoc. *Education:* MSE/Aero Engg/MA Inst of Tech; BSCE/Civ Engg/IL Inst of Tech. *Born:* 2/23/19. Hd of consulting firm specializing in planning and engg of airports, parking facilities, park and recreation facilities. Chrmn, Founding Committee, Airport Consultants Council; Founding Mbr and later Chrmn, Parking Consultants Council. Past Pres, North Shore Chapter, Natl Soc of PE. Former VP and Dir, Consulting Engrs Council - IL. Active mbr of leading profl organizations in the airport and parking fields. Recipient, Eminent Conceptor Award, CEC-I, for planning the redevelopment of Norfolk (VA) Intl Airport. Navy service: active duty, WWII, engg officer South Pacific, later Bureau of Aeronautics; Capt, USNR, Ret. Former Trustee, Village of Glenview, IL. *Society Aff:* ASCE, NSPE, ACEC.

Donohue, David A T
Business: 137 Newbury Street, Boston, MA 02116
Position: President. *Employer:* Internatl Human Resources Dev Corp. *Education:* BS/Petroleum Engrg/OK Univ; PhD/Petroleum Engrg/PA State Univ; JD/Law/Boston Coll Law Sch. *Born:* April 11 1937 Montreal Canada. Prof Engr in Penna. Employment with Exxon & Gulf Oil Corp (1957-59, 1963-64, 1966-67) Assistant and Associate Professor, Pennsylvania State Univ (1964-68); Pres IHRDC (1969-present); Pres Honeoye Storage Corp (1973-present); Partner Donohue Anstey & Morrill (1976-present). Pres Arlington Exploration Co (1978-present). Cedric K Ferguson Award of Soc of Petroleum Engrs of AIME (1968), Distinguished Mbr SPE (1984). *Society Aff:* SPE, AAPG, ABA.

Donovan, James
Business: 73 Pond St, Waltham, MA 02154
Position: President *Employer:* Artisan Industries Inc. *Education:* BS/Chem Engg/MIT. *Born:* Oct 26 1906. Mbr, Fellow AIChE; IFT; ACS; AWS. Reg Prof Engr Commonwealth of Mass. Mb of Commonwealth of Mass Bd of Registration of Prof Engrs & of Land Surveyors. Mbr Natl Council of Engrg Examiners. Has been associated with the dev of processes, design & manufae of equipment, installation & operation of small & medium sized plants for the process industries.

Donovan, Robert L
Home: 2135 Herbert Dr, Bethlehem, PA 18018
Position: Dist Bridge Engr *Employer:* PA Dept of Transp. *Education:* MS/CE/Lehigh Univ; BS/CE/Lehigh Univ. *Born:* 8/7/29. Native of Scranton, PA. Served with the Naval Civ Engr Corps 1953-56. Bridge Engr for DC Dept of Hgwys, 1956 & for West & Wells Engrs, 1956-1963. Tower Designer for Bethlehem Steel Corp 1963-65. Joined PA Dept of Hgwys in 1965 as Bridge Engr in Allentown. Assumed Asst Dist Engr position in 1967 in operations, then planning & predesign & in 1979 in design. Assigned as Dist Bridge Engr in 1982. Am responsible for developing and receiving all bridge design plans and managing the bridge inspection program for all bridges in the District. Pres, Lehigh Valley Sec, ASCE, 1976-1977. Pres, Lehigh Valley Post, SAME, 1972-1973. Pres East Penn Section, ASHE 1982-83. Enjoy walking. *Society Aff:* ASCE, ASHE, ACECO.

Doolittle, Jesse S
Business: Mechanical & Aero Engr Dept NC State Univ, Raleigh, NC 27650
Position: Prof Emeritus and Conslt *Employer:* North Carolina State University. *Education:* MS/Mech Engr/Penn State Univ; BS/Mech Engr/Tufts Univ. *Born:* Sep 1903. Engr General Electric Co 1925-27. Instructor Case Inst of Tech 1927-31. Instructor, Asst Prof, Assoc Prof Penna State University 1931-47. Prof of Mech Engrg North Carolina State Univ since 1947. Semi-retired 1973. Author of six textbooks. Registered Engr North Carolina. Life Fellow American Soc of Mech Engrg 1968. Western Electric-ASEE, Southeastern Section Award for excellence in engrg education 1965. Burks-ASEE Award excellence in mechanical engrg education in US & Canada 1970. Life Fellow AAAS, ASME, ASEE, Fellow, Pi Tau Sigma, Tau Beta Pi,

Doolittle, Jesse S (Continued)
Sigma Xi, Phi Kappa Phi. *Society Aff:* ASME, ASEE, AAAS, AAUP, ΠΤΣ, ΤΒΠ, ΣΞ, ΦΚΦ.

Dopazo, Jorge F
Home: 30 Clinton Rd, Garden City, NY 11530
Position: Gen. Mngr. Eastern Office. *Employer:* Systems Control Inc. *Education:* MS/EE/Univ of Havana; -/Power Systems. *Born:* 4/2/21. Native of Havana, Cuba. Joined the Cuban Elec Co in 1945, grad in 1950 & became Hd, System Operation in 1955. In 1961, joined Am Elec Power where he directed res, dev & implementation of maj comp applications in the areas of Engg & Mgt sci. In 1984, joined Systems Control as Gen. Mngr., Eastern Office. Has developed analytical methods for the solution of power systems problems. Pioneered the application of State Estimation to power system monitoring & directed the implementation of the first operational system in the industry. Has published over forty technical papers & received IEEE, PES, Prize Paper Award 1981 for "Real-Time External System Equivalents for on-line Contingency Analysis–. Lectured at univs in the USA & abroad. Fellow of IEEE. *Society Aff:* IEEE, PES, ΣΞ.

Dorato, Peter
Business: Eece Dept, University of New Mexico, Albuquerque, NM 87131
Position: Prof *Employer:* Univ of NM. *Education:* BEE/EE/CCNY; MSEE/EE/Columbia Univ; DEE/EE/Polytechnic Inst of Brooklyn *Born:* 12/17/32. Native of NY, NY. Taught at the Poly Inst of Brooklyn for twelve yrs & the Univ of CO for five yrs. He was Chrmn of the Dept of Elec & Comp. Engr at the Univ of NM from 1976 to 1984. His areas of specialization include optimal & Robust control theory & solar-energy control system design. He is a past assoc editor of the "IEEE Transaction on Automatic Control" & assoc editor of "Automatica–. He is a mbr of the Bd of Governors for the Control Sys Soc of the IEEE and is a Fellow & "Distinguished Member" of the IEEE. *Society Aff:* IEEE, ASEE.

Dore, Stephen E, Jr
Home: 33 Birchcroft Rd, Canton, MA 02021 *Employer:* Coffin & Richardson Inc (Retired) *Education:* ScB/CE/Brown Univ *Born:* 4/1/18 in Providence, RI - After graduation worked as a surveyor in Hartford CT and a draftsman - designer for the US Navy at Quonset Point, RI - 1942-1946 - US Army Corps of Engrs - Highest rank attained - Captain. In 1946-1947 worked as highway draftsman, hydraulic engr, and structural designer for various organizations. Joined Coffin & Richardson Inc in 1947 as a civil engr for water supply and waste water facilities. Became VP in 1962, Executive VP in 1972 and Pres in 1979; Retired in 1983. Authored and presented several technical papers which were printed in engrg journals. *Society Aff:* ASCE, ASTM, SAME, ACEC, AWWA

Doremus, Robert H
Business: Rensselaer Poly. Inst, Troy, NY 12180
Position: Prof *Employer:* Rensselaer Poly Inst *Education:* PhD/Phys Chem/Cambridge Univ; PhD/Phys Chem/Univ of IL; MS/Phys Chem/Univ of IL; BS/ChE/Univ of CO *Born:* 9/16/28 Worked sixteen years (1955-71) at the Gen Elec Research Lab in Schenectady, NY, as a physical chemist in precipitation in low-carbon steels, crystal growth from solution, formation and optical properties of thin metallic films, and diffusion, crystal growth, strength, and synthesis of glasses NY State Prof of Glass and Ceramics at Rensselaer since 1971. Research interests in glass science, sintering of ceramics, crystal growth, and biomaterials. Author of books "Glass Science" and "Rates of Phase Transformations" and over 150 research publications. Fellow, winner of Purdy award, and Glass Division Chrmn of America Ceramic Society. Consulting with many different companies and government labs. *Society Aff:* AAAS, Am. Cer. Soc.

Doret, Michel R
Home: 20 E 9th St, New York, NY 10003
Position: Chief Engr. *Employer:* Doret Product Design & Dev. *Education:* PhD/French/George Wash Univ; MA/French/NYU; BS/LS/Univ of the State of NY; BA/French/Pace College of NY. *Born:* 1/5/38. Foreign born. Grad Civ Engr (ICS). Diploma in Architecture (Universidad Nacional de Colombia). Served in US Army Ordnance Corps (1963-1965). Former draftsman, designer and engr in various firms in NYC. Created Doret Product Design in 1969 for the intl dev of several new products. Writer of nine books add various literary and scientific artilces. Teacher (secondary and college level). Inventor - Technical translator. Mbr of Pi Delta Phi and Phi Sigma lota. Fluent in several languages. *Society Aff:* NSPE, ASCE, MTS.

Dorf, Richard C
Business: Electrical Engr Dept, Davis, CA 95616
Position: Prof of Elec and Computer Engrg. *Employer:* University of California. *Education:* PhD/Elec Engg/US Naval Postgrad Sch; MS/Elec Engg/Univ of CO; BS/Elec Engg/Clarkson College of Technology. *Born:* Dec 1933, New York City. BSEE Clarkson, MS University of Colorado & PhD US Naval Postgraduate School 1961. Prof & Chmn University of Santa Clara 1963- 69. VP Ohio University 1969-72. Dean of Extended Learning, University of California 1972-1981 . Res: Control Systems, Energy Resources & Policy, Engrg Mgr Technology Assessment. Author of 13 textbooks Modern Control Systems' 4th Ed 1985. Natl Officeholder: American Soc for Engrg Education. Fellow IEEE. Married, 2 daughters. Wife is a Presbyterian clergywoman. *Society Aff:* IEEE, ASEE, AAAS.

Dorfman, Myron H
Business: Dept of Pet Eng, Univ of Texas, Austin, TX 78712
Position: Chrmn, Dept of Petroleum Engrg *Employer:* Univ of TX *Education:* PhD/Pet Engrg/Univ of TX; MS/Pet Engrg/Univ of TX; BS/Pet Engrg/Univ of TX *Born:* 7/3/27 Engr & Geologist, Sklar Oil Co, Shreveport, LA, 1950-56; VP, Sklar Oil Co, 1956-59; Consulting Engr and Geologist, Dorfman Oil Properties, 1959-1974; Faculty of Dept of Petroleum Engrg, Univ of TX, 1974- ; Chrmn of Dept of Petroleum Engrg, Univ of TX, 1978 - ; Director, Tex Petroleum Research Committee 1982-86; Dir of Geothermal Studies, Center for Energy Studies, Univ of TX, 1974- ; Soc of Pet Engrs. Distinguished Lecturer 1978-79; SPE Distinguished Author, 1982. Chrmn, Geothermal Resources Committee, Interstate Oil Compact Commission, 1976-80; Chrmn, various Advisory Committees for Geothermal Energy, Dept of Energy, Lawrence Berkeley Lab, and others; over 50 technical articles on geothermal energy and petroleum engrg. W.A. Moncrief Jr. Centennial Chair in Petroleum Engrg, 1984-; Distinguished Author, Soc of Pet Engrs, 1982. *Society Aff:* SPE, GSA, SPWLA, AAPG

Dorko, Ernest A
Business: Attn: AFIT/ENP, Wright-Patterson AF Base, OH 45433
Position: Prof *Employer:* Air Force Inst of Tech *Education:* PhD/Chem/Univ of Chicago; MS/Chem/Univ of Chicago; B/ChE/Univ of Detroit *Born:* 9/16/36 Native of Detroit, MI, served with Army Ordnance Corp 1964-1966, highest rank Captain, received Army Commendation Medal, Vietnam Service Ribbon, worked at Redstone Arsenal, AL for US Army Missile Command 1966-1967 as spectroscopist. Assumed present position as prof in the School of Engrg, Air Force Inst of Tech. Am currently Prof of Chem. Serve as consultant for Air Force Avionics Lab and Air Force Weapons Lab. Recieved Air Force Scientific Achievement Award. Enjoy camping and reading. Am married to former Betty Jane Kurtz, have a son, Thomas. *Society Aff:* ACS, AAAS, CI

Dorman, Albert A
Business: 3250 Wilshire Blvd, Los Angeles, CA 90010
Position: Chrmn & Ch Exec. Off *Employer:* Daniel Mann Johnson & Mendenhall. *Education:* MS/Civil/Univ of So CA; BS/Mech/NJ Inst of Tech. *Born:* 4/30/26. in Philadelphia, PA. Reg Prof Engr 9 states; Reg Architect, CA & OR. Joined DMJM, Los Angeles in 1965. Assumed current respon as Ch Exec Officer of DMJM, one of largest cons firms in US, in 1977 (also named chrmn in 1983) & as Pres in 1974. Past dir of CHB Foods and Past dir of Financial Insts. Tau Beta Pi; Chi Epsilon; distin Civil Engr Alumnus Award (USC) 1976; P Chmn ASCE Natl Land-Use

Dorman, Albert A (Continued)
Policy Planning Ctte; ASCE Harland Bartholomew Award 1976; Outstanding Engineer Merit Award (Inst. for the Advancement of Engrg) 1980; Dir and Past Pres CEAC and Dir of ACEC; Past Pres ASCE Los Angeles Section; Edward F. Weston Medal for Distinguished Professional Achievement by an Alumnus (New Jersey Institute of Technology) 1986. Author of 13 papers; active in education & civic affairs. *Society Aff:* ASCE, AIA, APWA, AWWA, SAME, ΤΒΠ, ΧΕ, CEAC, ACEC

Dorman, William H
Home: 7 Orchard Dr, Corning, NY 14830
Position: Proj Engr & Sr Assoc.-Retired *Employer:* Corning Glass Works. *Education:* Engr Physics/Physics/Lehigh Univ. *Born:* Oct 18 1918 Corning New York. US Army 1941-46. Joined Corning Glass Works 1949 as a Prod Engr. Specialized in Lighting System Design & Photometry. Issued 16 patents for Lighting Systems. Fellow IES, active on IES Roadway & Testing Proceedures Cttes. Mbr of Optical Soc of America & the Amer Inst of Physics. Married Helen Devenport in 1942, 3 children, 4 grandchildren. Retired from Corning Glass Works 1984. *Society Aff:* IES, OSA, AIP.

Dorning, John J
Business: Reactor Facility, Charlottesville, VA 22901
Position: Whitney Stone Prof of Nuclear Engrg & Prof. of Engrg Phys *Employer:* Univ of VA *Education:* PhD/Nucl Sci & Engr/Columbia Univ, NYC; MS/Nucl Sci & Engr/Columbia Univ, NYC; BS/Marine Engr/US Merchant Marine Acd, Kings Point NY *Born:* 04/17/38 Ensign, US Navy, 59-60; Marine Engr, Merchant Marine, 60-62. AEC Fellow, Nucl Sci & Engr 63-66. Asst Physicist, reactor theory 67-69; Assoc Physicist & Grp Ldr, Brookhaven Nat Lab 69-70. Assoc Prof 70-75, Prof 75-84, Nucl Engr Prog, Univ of IL. Whitney Stone Prof of Nucl Engr & Prof of Engr Phys, Univ of VA 84- Pres. Concurrent positions: Consultant, Brookhaven Nat Lab, Argonne Nat Lab, Los Alamos Nat Lab, Sci Applications, Inc, Schlumberger-Doll Res, ARCO Oil & Gas Corp; Nat Res Council Visiting Prof Math Phys. Italian Nat Res Council 76, 81, 85; Physicist, Plasma Phys, Lawrence Livermore Lab, 77-78; Internat Prof of Nucl Engr, Italian Ministry of Public Educ 83, 84, 86; Fellow, Am Assoc Adv Sci 86; Fellow, Am Phys Soc 82; Fellow, Am Nucl Soc 78; Mark Mills Awd, Am Nucl Soc 67; Mbr, Soc Ind & Appl Math; Res: Nonlinear Dynamical Systems and Chaos, computational methods for particle transport, computational methods for continuum flow, mathematical physics, Fission reactor theory, neutron transport theory, fission reactor kinetics, plasma phys, kinetic theory of gases. *Society Aff:* ANS, APS, SIAM, AAAS

Dorny, C Nelson
Business: 371 Moore School, Philadelphia, PA 19104
Position: Prof & Chrmn, Sys Engrg *Employer:* Univ of Penn *Education:* PhD/EE/Stanford Univ; MSEE/EE/Stanford Univ; BSE/EE/Brigham Young Univ. 2r23729925acountic waves!acoustic waves*sys theory!control of multi-robot sys
Born: 01/20/37 Dr Dorny was appointed a White House Fellow by Pres Nixon for 1969-70. In this capacity he served as Special Asst to the Secy of Agriculture and participated in activities involving all depts of the federal government. His present research is in high resolution imaging with microwaves and ultrasonic acoustic waves and in multi-sensor, multi-robot control strategies. He also has experience in computer-integrated mfg, in computer-aided design of high-voltage equipment, and in the design of high-resolution radar sys. He has published numerous scientific articles and one book, entitled *A Vector Space Approach to Models and Optimization* *Society Aff:* IEEE, ASEE.

Dorothy, Walter W
Business: 100 S 19th St, Omaha, NB 68102
Position: VP. *Employer:* Northwestern Bell Telephone Co. *Education:* BS/ME/Univ of NB. *Born:* Aug 1924. Native of Fullerton, Nebraska. Joined Northwestern Bell in 1947. Held positions in var depts in Grand Island, Fremont & Omaha. With AT&T in NY 1959 assisted in early studies on satellite communications. Returned to Northwestern Bell-Omaha 1961 in marketing capacity. Appointed Asst VP Engrg in 1966. Assumed present respon Jan 1976. Served as Dir of Nebraska Alumni Assn for University of Nebraska & P Pres of Omaha Chapter. Mbr & P Chmn of Advisory Council of College of Engrg & Tech at University. Served as Dir of NSPE; Mbr & former Pres of Prof Engrs of Nebraska. Mbr of Inst of Electrical & Electronic Engrs & SAME. Mbr and P Chrmn of NB State Bd of Examiners for Professional Engrs & Architects. Mbr of City of Omaha Committee for Selection of Consultants. *Society Aff:* IEEE, NSPE, NCEE.

Dorrance, William H
Business: 1925 Pavline Plaza, Ann Arbor, MI 48103
Position: Pres. *Employer:* Organization Control Services, Inc. *Education:* Dr of Sci/Phys/Occidental Univ; MSE/Aero Engg/Univ of MI; BSE/Aero Engg/Univ of MI. *Born:* 12/3/21. Native of Ann Arbor, MI. Hd-Aerodynamics Dept Univ of MI Aero Res Ctr 1948- 51, Chief of Aero Atlas Missile Dev, Convair, San Diego, CA 1951-55, Sr Staff Scientist, Convair Gen Offices 1955-61, Group Dir, Aerospace Corp, 1961-1964, Corporate VP Conductron Corp, Ann Arbor, 1964-69 Chrmn, Chief Operating Officer Interface Systems Corp, 1969-70, Chrmn & Pres, OCS Inc 1971-present. Past Dir, Ann Arbor Chamber of Commerce, Past Pres-Barton Hills Country Club Ann Arbor 1978-79. Avid golfer - 3 books, 4 patents, 22 papers. *Society Aff:* AIChE, ACS, AAAS.

Dorros, Irwin
Home: 34 Hillside Pl, Fair Haven, NJ 07701
Position: Asst VP. *Employer:* AT&T. *Education:* EngScD/EE/Columbia Univ; SM/EE/MIT; SB/EE/MIT. *Born:* 10/3/29. Served in Army Signal Corps 1950-52. Joined Bell Labs in 1956 after receiving SB&SM from MIT. Worked on solid state circuit design for first electronic switching. Later on data communications & digital transmission of communications. Since 1966 engaged in network systems engg of Bell Network. Shifted from exec dir - network planning at Bell Labs to asst VP - network planning at AT&T 1978. Responsible for planning to utilize modern tech at AT&T 1978. Responsible for planning to utilize modern tech & growth. Fellow of IEEE. Enjoy golf, skiing, photography, stamps. *Society Aff:* IEEE.

Dorsett, Ronald
Business: PO Box 26473, Tempe, AZ 85282
Position: Pres. *Employer:* Dorsett Industries, Inc. *Education:* BSc/Marine Mech Engr/Royal Canadian Naval Acad. *Born:* 12/1/24. Native of Vancouver, BC, Canada. WWII Service in Royal Canadian Navy on N Atlantic Submarine Patrol, on Assignment to RCAF Flying as Command Pilot of PBY 5A Patrol Bomber on Convoy Submarine Patrol, Royal Canadian Naval Acad. Post Grad work in Forestry, Forest Engr, Wood Sci & Tech at Univ of BC, Canada. Pres & Managing Dir of cos established by myself for past 26 yrs engaged in Consulting Engg, Res & Dev, & Mfg for Process & Mfg industries. Reg PE in cA (MF 1182) & Natlly in Canada (7906290) Certified Manufacturing Engineer, Certified Plant Engineer. Full voting mbr in 12 Profl Engg Socs. Many processes, products & machines conceived, engineered and/or mfg. Current Pilot & mbr of Confederate AF restoring & Flying WWII Aircraft. *Society Aff:* NSPE, ASME, SME, IIE, AIPE, ASPE, CSPE, AIChE, SAE, SPIE,.

Dorsky, Lawrence R
Business: 184-10 Jamaica Ave, Hollis, NY 11423
Position: Corp Dir-Qual Assurance. *Employer:* Ideal Toy Corp. *Born:* Nov 29 1923. BSEE from Newark College of Engrg. Home in Springfield NJ. Served with Army Signal Corps 1943-46. Prior to Ideal Toy, served as a Dir Quality Assurance with SCM Corporation. Have spoken frequently for the American Mgmt Assn on Scope, objectives, organization & mgmt of quality control. Former Executive Sec, present VP American Soc for Quality Control & Fellow of the Soc. Panelist at Annual Meeting of NAE, Comm of Engrg Manpower Comm & many offices in ASQC.

Doshi, Mahendra R
Business: PO Box 1039, Appleton, WI 54912
Position: Pres *Employer:* Doshi & Associates, Inc. *Education:* PhD/ChE/Clarkson University; MS/ChE/Clarkson University; BS/ChE/Bombay Univ *Born:* 05/06/41 Res interests are in recycled paper, solid-liquid separation, mmembrane processes-reverse osmosis and ultrafiltration, fluid mechs, heat and mass transfer, mathematical modelling and waste water treatment. Born in India. Moved to USA in 1965 and became a citizen in 1977. Taught at the State Univ of NY at Buffalo for seven yrs and at the Institute of Paper Chemistry for nine years starting his own consultation/education/research service in 1986. Organizes conferences, seminars, continuing ed courses and on-site short courses. Hobbies include: puzzles, brain teasers and bridge. *Society Aff:* AIChE, TAPPI, ΣΞ

Dossett, Jon L
Home: 1702 Boulder Dr, Darien, IL 60559
Position: Supr Met Process Control. *Employer:* Internatl Harvester Co. *Born:* Jan 1940. BS Met E Purdue. Prof Engr Indiana & Illinois. Assumed position of Met Mgr for Warner Gear Div of Borg-Warner in 1970. Respon incl materials engrg, metallurgical process control & metallurgical standards. Joined Internatl Harvester in 1974 as Supervisor Metallurgical Process Control with respon for policy & standards dev for the control of foundry & heat treat processes. Served on ASM Handbook Cttes on metallurgical definitions & ductile iron. Currently Chmn Heat Treating Div of ASM. Enjoy tennis & bowling.

Doten, Herbert R
Home: 71 Purinton Ave, Augusta, ME 04330
Position: Project Manager *Employer:* Kleinschmidt Associates *Education:* MS/CE/Univ of ME; BS/CE/Univ of ME. *Born:* Dec 1932. Civl Engr of Maine Dept of Transportation for 13 yrs. Lt Army Corps of Engrs. COL in USAR. Instr of CE at Univ of Maine. Transportation Engr for E C Jordan Co Portland ME. Started own firm in 1972 with practice in CE & Land Surveying. Joined Hunter Ballew Assocs in 1982. Prof Socs: PPres ME Sect ASCE; P State Pres ME SPE/NSPE; Past V Chrmn, NE Region PEPP; Bd/Dirs from ME NSPE Soc of Engrs; PPres Cons Engrs of ME (ACEC). Married Patricia Fortier; children: Debora, Dorothy, Kathleen, Carol, Daivd. On the Engineers Joint Contract Documents Committee. *Society Aff:* ASCE, NSPE, ACEC.

Dotson, Billy J
Home: 611 Turner Blvd, Grand Prairie, TX 75050
Position: Sr. Res. Physicist Retired *Employer:* Mobil Res & Dev Corp *Education:* MA/Physics/Amherst Coll; BA/Phys, Chem, Math/Southwestern Coll *Born:* 9/11/20 Physicist. Born Liberal, KS September 11, 1920. B.A. Physics, Chemistry, Mathematics, Southwestern Coll 1942. M. A. Physics, Amherst Coll 1943. Married Rowena V. Knisely December 4, 1943. Children: Brian Mark, Marsha Gale, Darlene Joy. Instructor physics and math Amherst Coll 1942-4. Research physicist Mobil Research and Development Corp, Dallas, TX, 1944-84. Retired. Now Consultant. Patentee and author in petroleum engrg. Member Soc. Petr. Engrs, of AIME; Research Society of America. SPE committee on symbols and metrication; technical symbols reviewer for Soc. Petr. Engrs Jour. Publications. Office: 3600 Duncanville Rd, Dallas, TX 75236. *Society Aff:* AIME, RESA, SPE, RESA

Dotson, Clinton
Business: 8000 Arlington Expressway, Jacksonville, FL 32211
Position: Chief Engr Struct Engg. *Employer:* Offshore Power Systems. *Education:* BS/Naval Arch & Marine Engg/Univ of MI; 5-yr Certif/Struct Designer/Newport News Shipbldg Apprentice Sch. *Born:* 5/24/34. Entered Newport News Shipbldg Apprentice Sch in 1953 and attended college on a co sponsored scholarship. With Newport News Shipbldg 1961-72 progressing through assignments of engr, supervisor, sr supervisor & engg sec mgr involving engg responsibilities for both Navy & commercial vessels. With Offshore Power Systesm since 1972 involved in the design of Floating Nuclear Plants for electricity generation. Mgr, Naval Architecture from 1972-78. Presently Chief Engr, Struct Engg, responsible for directing the design of all structures assoc with the FNP. Author of several technical papers. Mbr of technical committees in SNAME & ANSI; Chrmn of SNAME, Southeast Sec 1976-77. *Society Aff:* SNAME.

Dotterweich, Frank H
Business: POBox 2084, Kingsville, TX 78363
Position: Prof of Chemical & Natural Gas Engrg. *Employer:* Texas A&I University. *Education:* BE/Gas Engg/Johns Hopkins; PhD/Physical Chem & Engg/Johns Hopkins. *Born:* Jun 1937. Native of Baltimore, MD. Asst Instr Baltimore Polytechnic Inst & Instr & Coach Johns Hopkins U prior to WWII. Served as Tech Cons Natural Gas, Natural Gasoline, Gas, Petrol Admin 1942-45. With Texas A&I U since 1937. Assoc Prof 1937, Prof 1941, Hd/Dept 1945, Dean Sch of Engrg 1965-71. Presently Prof Chem & Natural Gas Engrg. 60 pubs, 4 textbooks (Natural Gas). Amer Olympic Lacrosse Team 1928; Fellow AIChE, Sigma Tau & Tau Beta Pi, Cert of Merit Petrol Admin for War, Engr of Yr Nueces Chap, Texas Chap SPE. Enjoys hunting. *Society Aff:* NSPE, AIChE, SPE, AIME, ΣΞ, ΤΒΠ, APCA.

Dotts, Homer W
Home: 112 Hiwon Dr, Conroe, TX 77304
Position: Assistant Program Manager (Retired). *Education:* BS/Aero/Univ of MI. *Born:* Sep 1910. Native of Chicago, Illinois. 1938-49 Engr with Curtiss Wright Airplane Div. Designer, Group Engr, Proj Engr, Ch Design Engr. 1950-61 Columbia R&D Corp VP & Ch Engr. 1961-- NASA: Deputy Spacecraft Mgr Gemini Program, Mgr Systems Engrg Skylab Program, Asst Mgr Space Shuttle Program. Received num NASA awards. Authored sev technical papers. One of small group respon for structuring Shuttle Mgmt Organization. *Society Aff:* AIAA.

Doty, Clarence W
Home: 2527 Twin Creeks Dr, San Ramon, CA 94583
Position: Principal Engr *Employer:* Rockwell Intl *Born:* 1/18/27 Native of Pecatonica, IL. US Army Coast Artillery Anti Aircraft 1945-1946. Chief Bn Administrative NCO 1946. 1947-1949 Spartan School of Aeronautics aviation mechanics/airport administration graduate. W. F. & John Barnes Co, machine tool builder and nuclear reactor components 1950-1962. Machinist/assembler apprenticeship, nine years production control-process engrg. US Dept of Energy Rocky Flats Plant with Dow Chem Co 1962-1975 and Rockwell Intl 1975 to present in Product Engrg major nuclear project and program engr assignments. Temporary exchange engr assignment from plant to Lawrence Livermore National Lab Aug 1979 to April 1982. SME Dir since May 1979. *Society Aff:* SME

Doty, Coy W
Business: Coastal Plains Soil & Wtr Con Res. Ctr, PO Box 3039, Florence, SC 29502
Position: Agricultural Engg. *Employer:* USDA-ARS *Education:* MS/Agri Engg/SD State Univ; BS/Agri Engg/Auburn Univ. *Born:* 9/18/31. Native of Blount County, AL. Served in US Army 1952-54. Grad of Auburn Univ, 1958. South Dakota State Univ 1968. Joined USDA in 1958 at USDA Sedimentation Lab, Oxford, MS, working in erosion control res until 1964. Moved to Brookings, SD in 1964 and worked in erosion control and water conservation res until 1968. Joined USDA's Coastal Plains Soil & Water Conservation Res Ctr, Florence, SC in 1968. Res has involved tillage, irrigation and drainage. Received ASAE Natl Paper Award in 1979. Prominent in the dev of a Controlled and Reversible Drainage (CaRD) System, a single system which drains excess water from the soil during wet periods and supplies crop water needs through subirrigation during drought. Dev stream water level control in water resource proj which provides water for irrigation and increases crop yields from storage of shallow ground water which is recharged by rainfall. *Society Aff:* ASAE, SCSA, ICID.

Doty, Keith L
Home: 4813 NW 19th Pl, Gainesville, FL 32601
Position: Prof. *Employer:* Univ of FL. *Education:* PhD/EE/Univ of CA; MS/EE/MIT; BS/EE/MIT. *Born:* 8/28/42. Native of Enid, OK. Academic & ind experience in minicomputer & microprocessor hardware & software systems design. Served on the faculty of the Univ of CA at Berkeley & Santa Barbara, Univ of FL. Active consultant. Author of numerous technical papers & one text book, "Fundamental Principles of Microcomputer Architecture–, Matrix Publishers (July 1979). Served as a technical dir of an IBM res proj aimed at producing design concepts for high level language machines. Currently Assoc Dir of the Ctr for Intelligent Machines & Robotics (CIMAR). Actively engaged in microcomputer & software education, presenting lectures & tutorials intlly. *Society Aff:* IEEE.

Doty, William D
Business: Box 98243, Pittsburgh, PA 15227
Position: President & Principal Consultant *Employer:* Doty & Associates, Inc. *Education:* PhD/Metallurgy/Rensselaer Polytechnic Inst.; M. Met. E./Metallurgical Engr./Rensselaer Polytechnic Inst.; B. Met. E/Metallurgical Engr./Rensselaer Polytechnic Inst. *Born:* 03/11/20 Native of Rochester, NY. Graduate research received national award from AWS. Joined U.S. Steel in 1947 and held various research and supervisory positions at their Technical Center. Chief of Bar, Plate and Forged Products Div, 1958-66. Research Consultant, Steel Products Dev, 1966-73. Sr Research Consultant, Product Engrg, 1973-83. Sr Metallurgical and Product Consultant, 1983-85. Widely known for research & publications in steel product dev & welding. Co-author of book on "Weldability of Steels–. Member of Main Cttee of the ASME Boiler and Pressure Vessel Cttee. Ch, 1967-73, of Pressure Vessel Research Cttee of the Welding Research Council. Received AWS Spraragen Award 1966, and ASME J. Hall Taylor Award 1984. Elected to Sigma Xi in 1944. Reg PE in PA. Upon retirement from U.S. Steel in 1985; formed consulting firm of Doty & Assoc. *Society Aff:* AIME, ASME, ASM, AWS, TWI

Doubleday, Laurence W
Home: 8833 Karlen Rd, Rome, NY 13440
Position: Chief. *Employer:* Rome Air Dev Ctr-USAF. *Education:* MSSM/Mgmt/Univ of South CA; BEE/Electronics/Rensselaer Polytechnic Inst. *Born:* May 25 1936. BEE from RPI in 1958, MS in Systems Mgmt from USC in 1973. Held office in Rome-Utica chapters of AFCEA & IEEE. He served as a full time cons to NATO military budget & infrastructure cttes & as the US Mbr of 'NATO Working Group of Natl Communications experts'. He currently is a Branch Ch at the Rome Air Dev Center managing USAF dev in areas of VLF, HF, Satellite, Tropo, Microwave, & Optical Communications. *Society Aff:* AFCE, IEEE.

Dougal, Arwin A
Home: 6115 Rickey Dr, Austin, TX 78731
Position: Prof of Elec & Computer Engrg *Employer:* University of Texas-Austin. *Education:* PhD/EE/Univ of IL; MS/EE/Univ of IL; BS/EE/IA State Univ. *Born:* Nov 1926 Iowa. Air Force 1946-49. Radio Engr Collins Radio Co 1952. Entered engrg education as Asst & Assoc Prof EE at University Illinois 1957-61. Since 1961 Prof EE Univ Texas at Austin; Dir electronics res. Fed executive service 1967-69 as Asst Dir of Defense Res & Engrg (Res), Defense Dept. Since 1969 continuing as Prof EE at Univ Texas Dir of Electronics Res Ctr 1971-77. Reg PE TX. Fellow IEEE & APS. Elected to Bd of Dir, IEEE 1980-81. Centennial Medal, IEEE, 1984. Hobby is flying. *Society Aff:* IEEE, APS, ASEE, OSA.

Dougherty, Grafly C R
Home: 150 Dayton Ave, Somerset, NJ 08873
Position: Prof. Emeritus *Employer:* Rutger Univ. *Education:* B.S./Ch.E./Univ. of Penna.; -/Graduate Study/Iowa State College *Born:* 12/29/15 Taught in the engrg drawing dept. at Iowa State and in Mech Engrg There for eight years. Left as an Asst. Prof. in 1946. Taught thermodynamics, air conditioning, and fuels and combustion for thirty years at Rugters. Retired in 1976. Worked numerous years in the summer at Pratt & Whitney Aircraft. *Society Aff:* ASME, ASEE

Dougherty, John J
Home: 4766 Calle De Lucia, San Jose, CA 95124
Position: Consultant *Employer:* Retired *Education:* BSEE/EE/Villanova Univ. *Born:* Aug 1924. Native of Philadelphia, PA. Served in Submarine Service during WWII. After brief period with US Signal Corps Procurement Agency, joined Philadelphia Elec Co 1951. Specialized in insulated cables, synthetic insulations & dc transmission. Managed an Edison Electric Inst Res Proj 1964-69, on loan from Philadelphia Elec Co. Joined the Elec Power Res Inst in 1975 as Dir of the Electrical Sys Div. Fellow IEEE, Mbr of CIGRE. *Society Aff:* IEEE, CIGRE, IEC.

Dougherty, Robert A
Business: POBox 8149, Prairie Village, KS 66208
Position: President. *Employer:* Dougherty & Associates. *Education:* BSME/Ind Engr/Univ of Notre Dame. *Born:* 5/3/28. Sales Engr & District Mgr in St Louis & KS City for Robert R Stephens Machinery Co from 1952-72. Formed Dougherty & Assocs, machine tool & heat treat equip distribution 1972. Mbr ASM. Cert Mfg Engr; Reg Prof Engr. SME: Intl Sec 1974, Intl Tres 1975, Intl VP 1976-80, Intl Pres 1980-81, Bd/Dirs 1971-81. IProdE Fellow. Enjoys golf, bowling & coaching youth basketball. *Society Aff:* ASM, SME, IProdE.

Dougherty, Thomas A
Business: 3939 Fabian Way, Palo Alto, CA 95118
Position: Spacecraft Engineering Manager *Employer:* Ford Aerospace and Communications Coporation *Education:* PhD/Physical Chemistry/IA State; MS/Chemistry/Wichita State; BS/Chemistry/Wichita State. *Born:* 12/31/37. Matls & Processes Engr with The Boeing Co (7 years) and with Ford Aerospace (6 years). Specializing in fiber reinforced plastic structures. Currently manager of structures, power, antenna and prepulsion subsystems on commercial satellite programs. Num pats & pubs. *Society Aff:* SAMPE, ACS.

Doughty, Eric R
Business: Edison Dr, Augusta, ME 04336
Position: Coordinator Commercial & Agri Services. *Employer:* Central ME Power Co. *Education:* BS/Agri Engg/Univ of ME. *Born:* 3/8/26. Native of Topsham, ME. Served US Navy Submarine Service 1943-46. Taught Univ of CT Agri Engg Dept 1952-54. Supv at Agri Sales and Service CMP Co 1954 to date. Founder and Co-Chrmn ME Farm Days 1968-1979. Chrmn Acadia Sec ASAE 1979- 80. Friend of Agri Award by ME Farm Days. 1975 by New England Extension Agents Assoc 1978. Currently responsible for Agri Services, Solar Res CMP Co. Enjoy Family Life, grandchildren canoeing including building canoes. Reproducing early tin ware. Past pres Square Dance Club. Active church and affiliates progs. *Society Aff:* ASAE.

Doughty, Julian O
Business: PO Drawer ME, Tuscaloosa, AL 35487
Position: Prof Mechanical Engrg *Employer:* The Univ of AL *Education:* Ph.D./Engrg Sci/Univ of TN; M.S./Aero Engrg/MS State Univ; B.S./Aero Engrg/MS State Univ *Born:* 6/11/33 Engrg design with McDonnell Aircraft in 1956-57. Academic positions with MS State Univ (1957-60), the Univ of TN (1960-66) and The Univ of AL (1966-87). Academic positions include aerospace engrg, mechnaical engrg and engrg mechanics. Research and publications in viscoelastic fluids, fluid mechanics and aerodynamics. Positions on the AL faculty include three years as Dir of the departmental Thermal-Fluids Division and three years as Chrmn of the Aerospace Engrg Program. Service to engrg societies include membership on the AIAA Student Activities Committee and Faculty Representative for the AIAA Southeastern Regional Section. PE in AL. Currently serving as mech engrg grad prog super and fac adv for SAE student branch. *Society Aff:* AIAA, ASME, ASEE, SAE, ΣΞ, ΣΓΤ

Douglas, Bruce M
Home: PO Box 8121 Univ Station, Reno, NV 89507
Position: Chrmn CE Dept. *Employer:* Univ of NV. *Education:* PhD/Engr Mechanics/Univ of AZ; MSCE/Engr Mechanics/Univ of AZ; BSCE/Civ Engg/Univ of Santa Clara. *Born:* 10/13/36. Eureka, CA - Married - four children. With the Univ of Nevada since 1964. Prof of CE 1973, Chrmn CE Dept. 1976-1984, Dir of Cte for Civil Engrg Earthquake Res (CCEER) - 1983 to present. Cosultantships in area of seismic analysis. Strong res interests in area of earthquake engrg with emphasis on dynamic testing of full scale bridges. Author of a number of papers in this field. Participant in several earthquake engineering workshops in foreign countries. Named "Engineer of the Year" in 1983 for the State of Nevada. Co-author of one US patent. *Society Aff:* ASCE, SSA, ASEE, EERI, AAA, IABSE, NZSEE.

Douglas, Charles A
Home: 7315 Delfield St, Chevy Chase, MD 20015
Position: Physicist. *Employer:* Retired. *Education:* MS/Physics/Univ of ID; BS/Physics/Univ of ID. *Born:* Sept 12 1911. Instructor Math & Physics Coeur d'Arlene Jr College 1934- 35. With National Bureau of Standards at Seattle 1935-40; at Washington D C since 1941 specializing in photometry & aviation lighting & marking. Developed atmospheric transmissometer used to determine runway visual range at airports. Mbr Visual Aids Panel, Internatl Civil Aviation Organization 1960-1976. Mbr Natl Res Council 1958-64. Dept of Commerce Meritorious Service Award 1954; Fed Aviation Admin Extraordinary Service Award 1974. Retired 1976, Illuminating Engrg Soc of N Amer Gold Medalist, 1983. *Society Aff:* IESNA, OSA, AIP

Douglas, Fred R
Home: P.O. Box 193, Glenham, NY 12527
Position: Technologist *Employer:* Texaco Inc *Education:* MS/Chem Eng/Poly Inst of NY; BS/Chem Eng/NJ Inst of Tech *Born:* 04/25/24 Native of Newark, NJ. MS in Chem Engrg, Poly Inst of NY 1949, BS in Chem Engrg, NJ Inst of Tech 1945. Served with US Army Corps of Engrs 1945-47. Ass't Prod'n Mgr Bristol-Myers Corp Hillside NJ. Engr with Jefferson Chem NYC. With Texaco, Inc since 1953, specializing in res and corp planning. Author and co- author of several tech papers. Lecturer in Profitability Analysis. VPres Amer Assoc of Cost Engrs 1985-present. Served as Dir in 1983-84, Secy in 1968-69. Past officer of Res Soc of Amer. *Society Aff:* AACE

Douglas, James
Business: Civil Engrg Dept, Stanford, CA 94305
Position: Emeritus Professor Cvl Engrg. *Employer:* Stanford University. *Education:* PhD/Civ Engrg/Stanford Univ; MCE/Civ Engrg/RPI; BCE/Civ Engrg/RPI; BSc/Engrg/USNA. *Born:* Oct 1914. Native of San Antonio, Texas. Served as officer in Navy Seabees 1941-61. Regt Comdr in charge of construction at Cubi Point, Philippines 1951-53. Dir Seabee Div Washington D C 1954-58. Antarctic Operations 1957-58. Retired as Capt (CEC) USN 1961. Faculty Stanford Univ 1963- . Author Construction Equipment Policy' (McGraw-Hill) & numerous technical papers & reports. Chmn ASCE Construction Equipment Comm 1965-70. Chmn Comm on Construction Mgmt Trans Res Bd 1970-76. Awards: ASCE Thomas Fitch Rowland Prize 1969, Construction Mgmt Award 1975. Enjoy backpacking & fishing. Retired as Emeritus Prof. of Civil Engr., Stanford Univ. 1980. Consulting Engineer, 1980-. *Society Aff:* ASCE, ΤΒΠ, ΣΞ, ΧΕ.

Douglas, Jim, Jr
Business: Dept. of Mathematics, Purdue University, Mathematical Sciences Bldg, West Lafayette, IN 47907
Position: Loveless Disting Prof of Computational Math. *Employer:* Purdue University. *Education:* PhD/Math/Rice Univ; MA/Math/Rice Univ; MSCE/Civ Engr/Univ of TX; BSCE/Civ Engr/Univ of TX. *Born:* 8/8/27. *Society Aff:* AMS, SIAM.

Douglas, Robert A
Business: 208 Mann Hall, Raleigh, NC 27650
Position: Prof of Civ Engg. *Employer:* NC State Univ. *Education:* PhD/Mechanics/Purdue Univ; MS/Mechanics/Purdue; BS/Mechanics/Purdue. *Born:* 12/4/25. Has published reports, papers, and one book in the gen area of Solid Mechanics. Holds US patent for machine to form optical diffraction gratings directly into the surfaces of metals where used as strain gages for extreme environments. Principal res area concerned with wave propagation and gross deformation in solids as result of damaging impact. Secondary res effort explores the mechanical condition of human bone and uses ultrasonic waves to determine in vivo condition in a non-intrusive way. Pres of Alden-Roberts Inc specializing in dynamic transients. Deputy Chief of Party, US Engg Team, Afghanistan, 1964-1966. *Society Aff:* AAAS, ASCE, SESA.

Douglass, David L
Business: 6531 Boelter Hall, Los Angeles, CA 90024
Position: Prof *Employer:* UCLA. *Education:* PhD/Met Engg/OH State Univ; MS/Metallurgy/Penn State Univ; BS/Metallurgy/Penn State Univ. *Born:* 9/28/31. Gen Elec Co Schenectady 1958-60 & Vallecitos CA 1960-66. Exchange Scientist AEC/Euratom Mol Belgium 1963-65. Stanford Res Inst Head Corrosion Sciences Sect 1966-68. UCLA Prof 1968-. Ch Editor *Oxidation of Metals* 1969-. Chrmn Gordon Conf Corrosion 1975. P Chrmn S Cal Met Soc AIME. Cons to numerous companies. Author *Metallurgy of Zirconium*. 72 publications. *Society Aff:* NACE.

Douglass, Jack L
Business: POBox 1934, Oklahoma City, OK 73101
Position: Environ Control Engr. *Employer:* Mobil Oil Corporation. *Born:* Jun 1928. BSEM from West Virginia University 1954. Native of Clarksburg West VA. Aviation Electronic Technician with US Navy 1946-49. With Mobil Oil Corp (& predecessor company) since 1954 in various assignments incl field operations & reservoir engrg. Assumed respon for environmental engrg for all Mobil producing operations in a ten-state area in 1971. SPE of AIME Bd/Dir 1975- 78; Mbrship Chmn 1969-70. Active in many industry environmental cttes.

Douglass, Nelson L
Home: 190 Bristol Blvd, Jackson, MS 39204
Position: VP *Employer:* Cooke-Douglass-Farr Ltd. *Education:* BS/ME/MS State Univ. *Born:* 4/8/34. Grad study at Univ of TN. Design Engr for Union Carbide Nuclear Co. For the past 26 yrs Partner in Cooke-Douglass-Farr Ltd, Architects-Engrs. Past Pres of Mississippi Engrg Soc and Conslt Engrs Council/Miss. Natl Dir of NSPE and CEC/US. *Society Aff:* NSPE, CEC, ASHRAE, ASEE

Douglass, Thomas E
Business: PO Box P K-1580 MS 596, Oak Ridge, TN 37830
Position: Mgr Plant Engrg *Employer:* Union Carbide Corp-Nuclear Div. *Education:* BS/ME/MS State Univ; BS/Bus Admin/MS State Univ. *Born:* 2/15/36. Native of Grenada. Pres of Engrg Sch Student Body. Employed since graduation by Union Carbide Nuclear Div. Has served in areas of production inspection of nuclear warheads, producibility dev related to the gas centrifuge process for isotopic environment & engg design for a broad array of equip, processes, experiments & facilities. Active in Tech & Prof Socs & formerly Pres of the TN Soc of Prof Engrs. Currently Chrmn of Oak Ridge Section of ASME. Enjoys golf and formerly Pres of Oak Ridge Country Club. Mbr of Bd of Dirs, Oak Ridge Chamber of Commerce. *Society Aff:* NSPE, ASME, SME, ASEE.

Douma, Jacob H
Home: 1001 Manning Street, Great Falls, VA 22066
Position: Consulting Hydraulic Engr. *Employer:* US Army Corps Engrs-Off/Ch/Engrs - Retired. *Education:* BS/Civil Engg/Univ of CA, Berkeley. *Born:* May 30 1912 Hanford California. Employed with US Army Corps of Engrs 1935-79, except with US Bureau of Reclamation 1936-39. Presently Consulting Hydraulic Engg Washington D C. Hydraulic cons to 30 private engrg firms & foreign govts. Fellow ASCE; Mbr

Douma, Jacob H (Continued)
Executive Ctte Hydraulics Div 1868-74, Chmn 1973-74. Mbr internatl: Comm on Large Dams; Irrigation, Drainage & Flood Control; Assn of Hydraulic Res & Permanent Assn of Navigation Congresses. Mbr National Acad of Engrg. Author of 25 publications & technical papers presented at Natl & Internatl meetings. *Society Aff:* ASCE, ICOLD, USCID & FC, IAHR, PIANC.

Doumas, Basil C
Business: Hibiya Chunichi Bldg, 1-4, Uchisaiwaicho 2-Chome, Chtyoda-Ku, Tokyo 100 Japan
Position: Mgr, Mfg & Engg. *Employer:* Dow Chem Japan Limited. *Education:* PhD/ChE/VPI; MSc/ChE/VPI; BSc/ChE/VPI *Born:* 12/31/33. Native of Fredericksburg, VA. Entered VA Tech in 1950 & earned the MSc degree in 1955. Employed by the Natl Lead Co of OH, working on processes to recover uranium from scrap mtls. After earning PhD degree, joined Dow's TX Div in Freeport Worked in various res, dev, production & engg assignments. Significant assignment was the Styrene Monomers Tech Ctr in which responsibility was for all technical phases of global ethylbenzene/styrene tech. Joined Dow Chem Japan Limited in 1979 as Mgr of Mfg & Engg. Serving as Natl AIChE Dir, 1979-81. Reg PE in TX since 1967. *Society Aff:* AIChE, ACS, ΣΞ, ΦΛΥ.

Douty, Richard T
Home: 1412 Ridgemont Ct, Columbia, MO 65201
Position: Prof of Civ Engg. *Employer:* Univ of MO. *Education:* PhD/Structures/Cornell Univ; MS/Structures/GA Inst of Tech; BSCE/Civ Engg/Lehigh Univ. *Born:* 6/12/30. in Williamsport, PA. Served in US Navy 1950-54. Mgt trainee for the Fabricated Steel Div of the Bethlehem Steel Corp 1957-58. Joined the faculty of the Univ of MO at Columbia in 1962. Specializing in the use of computers for structural design as reflected in teaching and res interests, and as consultant to industry. Over twenty papers presented and published in profl journals. Pres, Engg Design programmers, Inc. *Society Aff:* ASCE.

D'Ovidio, Gene J
Business: 285 Riverside Ave, Westport, CT 06880
Position: Principal *Employer:* Marketing Corp of America *Education:* MBA/Marketing-Mgt Sci/Boston Univ; BS/Ind Engg/Northeastern Univ. *Born:* 2/3/49. Native of New England. The Stop & Shop Cos, Inc corporate staff from 1973- 1977 specializing in distribution & store operations. Assumed position with Booz, Allen in early 1977, specializing in logistics & strategic planning. Joined the Dev Agency, a consltg div of MCA which specializes in new business/new product dev, in 1984. Div Dir, IIE 1979-81, VP, NCPDM 1981-83. Enjoy yacht racing and competitive squash *Society Aff:* IIE, NCPDM, IIE.

Dow, John D
Business: Dept of Physics, Notre Dame, IN 46556
Position: Frank M. Freimann, Prof of Phys *Employer:* Univ of Notre Dame *Education:* PhD/Physics/Univ Rochester; BS/Physics/Univ Notre Dame *Born:* 11/6/41 Native of Paterson, NJ, NSF postdoctoral fellow (1967), instructor (1968), Assistant Prof at Princeton Univ (1969). Assoc (1972) and full Prof (1975) Univ of IL. Assoc of IL Center for Advanced Study (1976-77). Fellow of the American Physical Society (1976). Frank M. Freimann Prof of Physics at Univ Notre Dame (1983). *Society Aff:* APS

Dow, Thomas A
Business: Mech Engr Dept, P O Box 7910, Raleigh, NC 27695-7910
Position: Prof *Employer:* NC State Univ *Education:* PhD/ME/Northwestern Univ; MS/ME/Case Inst of Tech; BS/ME/VPI. *Born:* 7/3/45. After graduating from Northwestern, Dr Dow joined the Tribology Group at Battelle's Columbus Labs. At Battelle his res covered broad areas of friction, lubrication, and wear. He conducted experimental & analytical progs concerned with cold rolling, wire drawing, journal & rolling-element bearings, gears, seals, brakes, & elec brushes. Dr Dow joined the faculty at NC State Univ. in 1982. He is the Director of the University's Precision Engrg Center. He was past chrmn of the Metalworking Lubrication Ctte of ASME & is currently on the planning ctte of the Joint ASME/ASLE Lubrication Conf and the Chairman of the Tribology Div of ASME. His outside interests include tennis & photography. *Society Aff:* ASME, ΣΞ, SPIE.

Dow, William G
Home: 915 Heatherway, Ann Arbor, MI 48104
Position: Prof Emerit. *Employer:* The University of Michigan (1926-1971). *Education:* MSE/Elect Eng/Univ of MI; EE/Elect Eng/Univ of MN; BSE/Elect Eng/Univ of MI; Honorary Doctor of Science/Univ. of Colorado *Born:* Sep 30 1895 Faribault MN. Industrial positions 1919-26; electrical engrg faculty Univ of Michigan 1926-65, Dept Chmn 1958-64, Space Res 1966-71; Div Head Radio Res Labs Harvard Univ 1943-45. Mbr: Vacuum Tube Dev Ctte NDRC 1943-45; US Rocket Res Panel 1946-60; Sci Adv Comm Diamond Ordnance Labs 1953-64; Bd/Trustees Environmental Res Inst of Michigan 1972- . Reg Prof Engr Michigan. Fellow IEEE, ESD & AAAS. Mbr: European & Far Eastern Surveys of Physical Electron Res Space Research & Engrg Educ, Cosmos Club Washington, ASEE, AAAS, AAS, AIAA, AAUP, AWS, AGU, NSPE, APS, NYAS, Tau Beta Pi, Sigma Xi, Eta Kappa Nu, Phi Kappa Phi. IEEE Medal in Engrg Education 1963. Distinguished Alumni Achievement Award, Univ. of Minnesota 1961. Articles in Journals. Author 'Fundamentals of Engrg Electronics' Wiley & Sons 1937, rev 1952. Patent holder in electrical welding and in controlled nuclear fusion (Jan 1981). *Society Aff:* IEEE, NSPE, ASEE, AIAA, AWS, AGU, APS, NYAS, AAAS, AAUP, AAS, MSPE

Dowdell, Roger B
Home: 16 Gull Rd, Naragansett, RI 02882
Position: Prof. *Employer:* Univ of RI. *Education:* PhD/Civ Engr/CO State Univ; ScM/Mech Engr/Brown Univ; BE/Mech Engr/Yale Univ. *Born:* 3/18/25. W-Engr Officer US Navy. 1946-1948 Aircraft Gas Turbine Div Gen Elec Co. 1948-52 Brown Univ ScM. 1952-1960 Fluid Mechanics Engr B-I-F Industries, Prov RI. 1960-1964 Assoc Prof Univ of Bridgeport. 1966 PhD CO State Univ. 1966-pres Prof Mech Engg Univ of RI. 1966-pres Consultant in Hydrodynamics US Navy Underwater Systems Ctr. 1973-pres Chrmn ASME Res Comm on Fluid Meters. Reg PE State of RI. *Society Aff:* ASME, ASEE, ΣΞ.

Dowdey, Wayne L
Home: 1315 Villa Nueva Dr, P.O. Box 1047, Litchfield Park, AZ 85340
Position: Sr Consultant *Employer:* Eimco Process Equipment Co. *Education:* None/-/Berry College; None/-/Auburn Univ; None/-/Samford Univ. *Born:* 3/17/18. Native of AL, joined Republic Steel in 1937 and became superintendent of their Spaulding, AL iron ore concentrator in 1944. Joined Eimco Corp, now a div of Envirotech, n 1945, advanced to VP of Sales in 1961. Became Pres of Eimco Process Machinery Div in 1969; Pres of the Envirotech Process Equip Group in 1971; Exec VP of Envirotech in 1973; Sr VP of Envirotech in 1978. Being an active mbr of AIME since 1945, have served as Chrmn of the Mineral Processing Div, SME-AIME; AIME Dir; Pres of SME-AIME, and Pres of AIME (1978). Retired in 1983 and is now Sr Consultant Eimco Process Equipment Co, a Div of Baker Mine Equipment Co. *Society Aff:* AIME, MMS of Am., CIM, Aus. IMM.

Dowell, Dr Douglas
Business: Dept of Mechanical Engineering, Cal Poly University, Pomona, Pomona, CA 91768
Position: Prof. *Employer:* CA State Poly Univ. *Education:* PhD/Eng Mech/IA State Univ; MS/Nucl Engr/AF Inst of Tech; BS/Civ Engr/Univ of IA. *Born:* 5/31/24. Presently Prof of Mech Engrg at CA St Poly Univ, Pomona. Presently Pres of Dowell Engrg designing facilities and equip for the testing and development of inter-ocular lenses. Former Dir of CA State Univ Statewide Energy Consortium. Former Energy Conservation Mgr for CA State Univ system of 19 campuses. Former Assoc Dean and Acting Dean of the CA Poly Sch of Engg. Former Dir of Inst & Educational

Dowell, Dr Douglas (Continued)
Services for CA Poly. Former Dir of the KATE proj in Greece involving the dev of five schools of tech in Greece. Former prof at the AF Acad. *Society Aff:* AEE.

Dowell, Earl H
Business: School of Engineering, Science Dr, Durham, NC 27706
Position: Dean, School of Engrg *Employer:* Duke University *Education:* ScD/Aero & Astro/MIT; SM/Aero & Astro/MIT; BS/Aero & Astro/ Univ of IL. *Born:* 11/16/37 Earl Dowell, a leading authority in the field of aeroelasticity, is the 1980 recipient of the major res awd in his field, the Amer Inst of Aeronautics and Astronautics (AIAA) Structures, Structural Dynamics and Materials Awd. He has served as vpres, publications and mbr of the exec ctte of the Bd of Dirs of AIAA. He is a fellow of the Amer Acad of Mech and also the Amer Inst of Aeronautics and Astronautics. He is a mebr of the US Air Force Scientific Advisory Bd. In addition to being author of over 80 res articles, Dean Dowell is the author of two books, *Aeroelasticity of Plates and Shells* and *A Modern Course in Aeroelasticity.* He is widely sought after as a conslt to local, state and fed govt and ind. His teaching and interests span the component disciplines of acoustics, aerodynamics and structural dynamics. *Society Aff:* AIAA, ASME

Dowell, James C
Home: 109 Overhill Rd, Salina, KS 67401
Position: Assoc *Employer:* Wilson & Co *Education:* BS/Gen Engrg/Stanford *Born:* 3/31/50 After graduation in 1972, Mr Dowell joined Wilson & Co as a member of the Environmental Division. His experience began in the field of sanitary engrg, assisting in the design of sewage collection and treatment systems for several municipalities. Mr Dowell has been Project Engr on several major wastewater collection and treatment projects. In April 1980, he was appointed Head of the Wastewater Section of the Co's Environmental Division. Mr Dowell is currently Pres-Elect of the Smoky Valley Chapter of the KS Engrg Society and Chrmn of the KS Water Pollution Control Association's Seminars and Workshops Committee. *Society Aff:* WPCF, NSPE

Downey, Joseph E
Business: State Office Bldg, Div of Pub Water Supplies, Montgomery, AL 36130
Position: Dir. *Employer:* AL Dept of Public Health. *Education:* MS/Sci & San Engg/ Univ of MI; BS/Civil Engg/Auburn Univ. *Born:* 3/28/34. Employed with the Al Highway Dept until Jan 1955, then served with the US Army for 2 yrs. Received BS Civil Engg, Auburn Univ, 1960. Served as Public Health Engg with the AL Dept of Public Health for 4 yrs, beginning Jan 1961. Received MS, Sci & San Engg, Univ of MI. 1965. Appointed Dir of Div of Public Water Supplies, in 1970 and is presently serving in that capacity. Married to the former Modean Lett, father or 4 children. Mbr of Morningview Baptist Church. Active participation with Dixie Youth Baseball League and "Y" Football. *Society Aff:* ASCE, WPCF.

Downing, Mason L
Home: 17 Olmstead Rd, Morristown, NJ 07960
Position: Asst Gen Manager-Proj Management. *Employer:* Exxon Res & Eng Co. *Education:* MS/Chem Engg/MA Inst of Tech; BS/Chem/MIT. *Born:* 6/3/20. Native of North Andover, MA. Served in Chem Warfare Service 1941-46. Instructor MIT School Chem Engg Practice 1947. Employed by Exxon Res and Engg Co initially in enng res and process design. In 1954 began career in Project Management. From 1957-59 was res in France with respon for building a grass roots refinery and simultaneous major expansion of existing refinery. In 1965-67 managed major expansion of an Italian refinery. Currently Asst General Manager of Exxon Engrs Proj Management Dept resp for supr of project teams on many of Exxons major engg and construction projects world-wide. *Society Aff:* API.

Dowson, Duncan
Business: Dept of Mech Engg, Leeds England LS2 9JT
Position: Prof of Engg Fluid Mechanics & Tribology. *Employer:* Univ of Leeds. *Education:* DTech/-/Chalmers Univ of Tech; DSc/-/Univ of Leeds; PhD/-/Univ of Leeds; BSc/ME/Univ of Leeds. *Born:* 8/31/28. Kirkbymoorside, York, England. Educated Univ of Leeds (Mech Engg). Res Engr, Aircraft industry 1952-55. Univ of Leeds Lectr (1955), Sr Lectr (1963), Reader (1965), Prof of Engg Fluid Mechanics & Tech (1966). Dir Inst of Tech (1967). IMechE James Clayton Fund Prize (1963). Thomas Hawksley Gold Medal (1966), James Clayton Prize (1978), Tribology Gold Medal (1979). ASME's Lubrication Div's Best Paper Awards (1975), (1976). Mayo D Hersey Award (1979). ASLE's Natl Award (1973). British Soc of Rheology Gold Medal (1969). Joint author of books on elastohydrodynamic Lubrication (1966, 1977) and Ball Bearing Lubrication (1981). Author of book *History of Tribology* (1979). Interested in Archaeology and history of tech. *Society Aff:* CEngFIMechE, Fellow ASME, Hon Mbr ASLE

Doyle, Frederick J
Business: USGS Natl Ctr 516, Reston, VA 22092
Position: Research Cartographer *Employer:* US Geological Survey-Natl Mapping *Education:* Dr. Eng (hc)/Photogrammetry/Tech Univ Hannover FRG; MSc/ Photogrammetry/Int Tng Ctr Delft Neth; B Civ Engrg/Civil Engrg/Syracuse Univ. *Born:* 4/3/20. Assoc Prof Photogrammetry OH State Univ 1952-60. Ch Scientist Raytheon Autometric performing res on NASA & DoD reconnaissance sys 1960-69. Res Scientist on mapping sys for USGS 1969- . Exec Ctte Div of Earth Sciences & Chmn Cartography Panel of Advisory Ctte on Space Programs of Natl Acad of Scis. NASA medal for Exceptional Scientific Achievement for Apollo Photography and mapping. Pres 1969, Fairchild Award 1968, Hon Mbr 1974 Amer Soc of Photogrammetry. Fellow AAAS 1975, Meritorious Service Award 1977, Distinguished Service Award 1981, Department of Interior Secretary Genl 1976-80 President 1980-84, Brock Award 1984 Intl Society for Photogrammetry and Remote Sensing. Dr. Engrg (hc) Techn Univ Hannover FRG 1976. Dr. Science (hc) OH State Univ 1986. Dr. Science (hc) Univ Bordeaux-France 1987. *Society Aff:* ASPRS, ACSM, AGU, AAAS.

Doyle, J Edward
Business: 3901 Industrial Blvd, Indianapolis, IN 46254
Position: Secretary-Treasurer *Employer:* Reid, Quebe, Allison, Wilcox & Assoc, Inc *Education:* BS/CE/Purdue Univ *Born:* 7/18/36 in southern IN at Cannelburg, graduated from Purdue Univ in January, 1959. Since graduating from Purdue I have been working with a consulting engrg firm. Primary responsibilities have been in the transportation area with design work on interstate highways and toll roads. Also worked on other design projects including airports, water and sewer projects. Pres of Metropolitan Indianapolis Branch of ASCE in 1973, Pres of IN Section, ASCE in 1979-80, Dir of Consulting Engrs of IN, 1979 thru 1981. Enjoy coll football and vacations with family. *Society Aff:* ASCE, ISPE, NSPE, ACEC, ITE, IBTTA

Doyle, James A
Business: 100 N Main Bldg, Memphis, TN 38103
Position: Exec V P. *Employer:* W R Grace & Co-Agri Chem Group. *Born:* Nov 6 1921. BS Chem Engrg University of Arkansas; Advanced Mgmt Program Harvard. Employment with W R Grace & Co 1953- ; held positions of Superintendent, Plant Mgr, Branch Production Mgr, Branch Mgr, Operations Mgr. Presently Executive V P in Agricultural Chemicals Group of W R Grace & Co in charge of Mining & Production Operations. Respon incl everthing assoc with operations reporting to me, incl design, engrg & procurement for old & new projects, some logistical respon & all respon for marine transportation of three cryogenic anhydrous ammonia tankers. Responsible for 9 different operations, incl two in foreign countries.

Dracup, John A
Business: Civil Engr Dept, UCLA, 405 Hilgard Ave, Los Angeles, CA 90024
Position: Prof. *Employer:* Univ of CA. *Education:* PhD/CE/Univ of CA, Berkeley; MS/CE/MIT; BS/CE/Univ of WA *Born:* 7/14/34. Dr Dracup is currently a prof in the Sch of Engg & Applied Sci at the Univ of CA, Los Angeles. He has been a mbr of the Univ of CA, Los Angeles engr faculty since 1965 & formerly was chrmn of

Dracup, John A (Continued)
the Engg Systems Dept. His profl interests & expertise are in the fields of engg hydrology & water resource systems engg. His recent work has been connected with the engg economics of water resources, the statistical analysis of hydrologic drought, the analysis of the conjunctive use of ground water & surface water systems, the modeling of groundwater systems, the impact of climatic change on hydrology and water supplies.. *Society Aff:* ASCE, AGU, AAAS, AWRA, IWRA

Drake, James H
Business: 2244 Walnut Grove Ave, Rosemead, CA 91770
Position: VP. *Employer:* S CA Edison Co. *Education:* MS/EE/Caltech; BS/EE/ Caltech. *Born:* 12/28/24. Phoenix, AZ. Received MS, EE, Caltech 1947. Instr in electrical engg Univ of CA at Los Angeles 1947-48. Joined S CA Edison Co 1949. Elected VP 1970. Currently VP, engg and construction. Served to Lt JG, USN, 1943-46. Reg PE State of CA. *Society Aff:* ΣΞ, IEEE, AIF, CIGRE, AEIC.

Drake, L Kenneth
Business: 1600 W 135th St, Gardena, CA 90249
Position: Mgr Engrg Services *Employer:* HITCO/Defense Prod Div. *Born:* 12/9/33. Mech Engrg major, Business Admin minor, working on Master's degree (MBA) through 'American Graduate University'. Native of Southern California. Specializing in estimating, costing, pricing & proposal preparation with HITCO/Armco Corp since 1970. Areas incl new business proposals, contract change proposals, termination claims & Mbr of negotiating team. Worked in composites, adhesives, polymer advanced reinforced plastic & sandwich structures since 1955. Manage the Engrg Services, Tool Design & Estimating Dept in costing/pricing of materials, engrg, manufacturing, subcontracts, facilities etc. Mbr SAMPE Executive Ctte, Chmn Natl Education & Career Dev Ctte 1975-77, Chmn LA Chapter 1976-77. Enjoy tennis, bridge, golf, fishing & music. *Society Aff:* SAMPE.

Drake, Richard J
Home: 5439 Hallford Dr, Dunwoody, GA 30338
Position: Unit Project Manager. *Employer:* Metro Atlanta Rapid Transit Auth. *Education:* BS/Civil Engg/UT State Univ. *Born:* Jan 1930. Native of Illinois. BSCE in 1952 from Utah State University; construction survey crew for Alaska Rd Comm; Installations Engr U S Air Force 1952-54; Highway & Drainage Design Engr Howard-Needles-Tammen & Bergendoff 1954- 59; Sr Highway Planning & Design Engr Meissner Engrs Inc 1959-61; DeLeuw-Cather & Co Project Engineer, Project Mgr & Chief Engineer of Atlanta office, specializing on planning & design of traffic & transportation projects 1961-76; 1976 to Date, Unit Project Mgr for MARTA rail transit system planning, design & construction. Mbr ASCE, SAME; Registered PE in MA & GA. Land Surveyor-In- Training in GA. *Society Aff:* SAME, ASCE.

Drake, Robert L
Business: Tulane Univ. Dept. Elec. Eng'g, New Orleans, LA 70118
Position: Prof Elec Engrg *Employer:* Tulane Univ *Education:* Ph.D./EE/MS State Univ; MS/EE/Tulane Univ; BS/EE/Tulane Univ *Born:* 6/24/26 B.S., M.S. in electrical engrg Tulane Univ. Ph.D. MS State Univ. Member IEEE. Now prof of electrical engrg at Tulane Univ. Formerly in plant engrg at Procter and Gamble and member of technical staff at Space Technology Labs. Current research in digital computer applications in control systems and signal processing. Currently consulting with industry and electric utility companies and with attorneys. *Society Aff:* IEEE

Drake, Robert M, Jr
Home: 648 Tally Rd, Lexington, KY 40502
Position: Consulting Engineer. *Employer:* Self Employed. *Education:* PhD/Heat Transfer-Fluid Mech/Univ of CA, Berkeley; ME/Heat Transfer-Fluid Mech/Univ of CA, Berkeley; BS/ME/Univ of KY, Lexington. *Born:* 12/13/20. Army Air Corps 1942-47. Faculty Univ. of CA-Berkeley 1947-55; Faculty Princeton Univ 1956-63, Chmn of Mech Engrg 1957-63; Faculty Univ of KY 1964- 1977, Dean of Engrg 1966-72, Special Asst to Pres 1975-77; VP Technology Studebaker-Worthington Inc 1977-78; Pres Univ Investment Co Inc 1978-79; Private Consulting Practice 1979. Co-founder 1981 and Chrmn of the Bd Projectron Inc 1983-85 also Director 1981-1985. Employed Genl Electric Co 1954-56; Arthur D Little Inc 1963-64; Corporate VP R&D Combustion Engrg Inc 1971-75. Co-founder & Dir Intertech Corp 1962-72; Dir Magnetic Corp of Amer 1974-81. Sometime cons Genl Electric, Rand Corp, McGraw-Hill Book Co, Arthur D Little Inc, Combustion Engrg Inc, Natl Sci Foundation, General Telephone and Electronics. Mbr NAE; Fellow & honorary member ASME; Registered KY. *Society Aff:* NAE, ASME.

Drange, Robert O
Home: 5010 W 67th St, Shawnee Mission, KS 66208
Position: Secretary-Treasurer *Employer:* Advanced Brake & Clutch Co. Inc *Education:* BSCE/Civil Eng/IA State Univ *Born:* 3/14/24. Native of Watertown, SD. US Navy (CEC) 1943-46 and 1952-54. Field Engr Rust Engg Co 1946-48. Howard Needles Tammen & Bergendoff 1948-78. Staff engr in Kansas City 1948-65. Partner 1966-78. Phila office 1966-74. Exec admin in Kansas City 1974-78. Partner ESE 1982-86. Consultant Corporate Finance Assoc 1986-present, Advanced Brake & Clutch-Secy-Treas, 1987-present. Pres of ARTBA Planning & Design Div 1972. Pres of Phila Sec ASCE 1973-74. Chrmn Comm on Fed Procurement of Architect-Engr Services 1977. ARTBA Guy Kelcey Award 1979. Seminar lecturer Univ of MO 1978-79. Listed in Who's Who in America. *Society Aff:* ASCE, NSPE, CEC.

Drangeid, Karsten E
Business: 4 Saumerstrasse, CH 8803 Rueschlikon, Switzerland
Position: Dir Res Lab. *Employer:* IBM Corp-Zurich. *Education:* Dipl Ing/Elec Engg/ Swiss Federal Inst of Technology, Zurich. *Born:* Mar 24 1925. MSEE at Swiss Federal Inst of Technology Zurich 1951. 4 yrs Elec Engr Brown Boveri Baden Switzerland. Joined IBM Res Lab in 1956. Was first involved in thin magnetic films & superconductivity res, later Mgr IBM Zurich semiconductor tech effort, which realized microwave Schootkybarrier field effect transistors limiting frequencies above 30 GHz. Dir IBM Zurich Res Lab 1971- . Fellow IEEE, Mbr ACM, Swiss Electrotech Soc, Swiss Phys Soc, Zurich Phys Soc.

Draper, Alan B
Business: 207 Hammond Bldg, Univ Park, PA 16802
Position: Prof. *Employer:* PA State Univ. *Education:* PhD/Mech/Univ of IL; MS/Ind Engr/Syracuse; SB/Bus & Engr Adm/MIT. *Born:* 1/3/22. Newton, MA. Apprentice Machinist, Tool & Die Maker during WWII at Brown and Sharpe Mfg Co. Mfg & Paint & Metal Finishing Engr at Carrier Corp 1947-1951. Instr of Ind Engg at Syracuse Univ 1951-1957. Taught Production Control, Engg Economy, Plant Layout & Mfg Processes; set-up Process Engg Labs. Asst Prof, Assoc Prof & Prof Industrial Engg at PA State Univ 1957-present. Taught most undergrad courses in IE but specialize now in Mfg Engg. Teach grad courses in Metal Casting, Sand Tech, Die Casting, Metal Melting, Cupola Melting. *Society Aff:* AFS.

Draper, E Linn
Business: 350 Pine St, Beaumont, TX 77701
Position: Vice Chmn, Pres & CEO *Employer:* Gulf States Utilities Co *Education:* PhD/Nuclear Engrg/Cornell Univ; BS/Chemical Engrg/Rice Univ; BA/Chem Engrg/ Rice Univ *Born:* 02/06/42 Vice Chmn, Pres and CEO for Gulf States Utilities. Since joining GSU in Feb 1979, has also served as Vice Chmn, Pres and COO, Exec VPres-External Affairs and Prod, Sr VP- External Affairs, Sr VP-Engrg & Tech Servs, VP-Nuc Tech, and Tech Asst to the Chrmn of the Bd. Prior to Feb 1979, was the Dir of the Nuc Engrg Prog at the Univ of Texas at Austin for about ten years. Past Pres of Amer Nuc Soc and has served as conslt to over thirty utility companies, several federal and state agencies, and a number of indust concerns. Has presented testimony on energy issues in ten states as well as before cttees of the US Congress. Author or co-author of more than seventy tech pub and ed of two books. Has served on the Natl Acad of Sci Comm on Radioactive Waste Mgmt. Is past Chrmn of the Utility Nuclear Waste Mgmt Grp and past Chrmn of AIF Nuclear

Draper, E Linn (Continued)
Waste Policy Act Oversight Comm. Served on the Secretary of Energy's Adv Panel on Alternative Means of Financing and Managing Radioactive Waste Facilities. Is currently serving on the Texas Radiation Advisory Bd. Received BA in 1964 and BS in Chem Engrg in 1965 from Rice Univ and PhD in Nuclear Engrg from Cornell Univ in 1970. Fields of specialization are nuclear waste mgmt and public understanding of energy. *Society Aff:* ANS, APS

Draper, John C
Home: 605 Orchardhill Drive, Pittsburgh, PA 15238
Position: Environmental Specialist *Employer:* Duquesne Light Co *Education:* BSEM/Mining Engrg/WV Univ; AB/English/WV Univ *Born:* 1/28/22 Worked for Island Creek Coal Co, Penowa Coal Co and Harmar Coal Co. Thirty years in charge of mining engrg for Duquesne Light Co. Have presented 19 technical papers, published 19 items. Responsible for hazardous and residual waste disposal. Executive Committee of Coal Division of SME. Distinguished Member of SME, 1980. Active in Keystone Bituminous Operators Assoc and in National Coal Association, Environmental Committee. Past Chrmn of AMC, Roof Control Committee and of AIME, Coal Division Utilization Committee. Past Pres of WVA Univ Mining Alumnae. Ordained Deacon and Order of St George from Episcopal Church. Silver Beaver, Woodbadge and Order of the Arrow from Boy Scouts. *Society Aff:* AIME, AMC, CMIA, NMRA, SME

Drapes, Alex G
Home: 312 1st Ave. N, Great Falls, MT 59401
Position: Chairman of Board *Employer:* Drapes Engrg. *Education:* BS/Engg/CA Inst of Tech. *Born:* Jan 31 1924 Montana. Grad of Great Falls High School 1942. Served in European Theatre WWII. Registered Prof Engr Montana, Wyoming, Washington. Founded Drapes Engrg 1954, providing Mech & Elec Design Services for bldgs & systems. P Pres Montana Tech Council, Cons Engrs Council of Montana, Great Falls Lions Club. City/County Planning Bd, Mbr of Univ of Montana Advisory Council, Pres Bd/Dirs of Deaconess Medical Ctr & Diocese of Montana Exec Council. Mbr of: ACEC. *Society Aff:* ACEC.

Drenick, Rudolf F
Home: 35 Melody Ln, Huntington, NY 11743
Position: Pres *Employer:* Poly Syst Anal Corp *Education:* PhD/Theo Phys/Univ of Vienna. *Born:* 8/20/14. With the GE Co as engr (1946-49), with the Radio Corp of Am as group mgr (1949-57), and with the Bell Labs as res mathematician (1957-1961). He was prof at the Poly Inst of NY from 1961 to 1984, with a one-yr leave to the Technical Univ in Munich, German, in 1964 another to the Natl Sci Fdn, Wash, DC, in 1977. He specialized initially in the math treatment of problems in flight mechanics and later in those of electronics. More recently, his involvement has broadened to other areas of engg as well. He is a fellow of the IEEE, was chrmn of its Info Theory Group and was active on the Ad Coms of the IEEE/SMC & Control Soc.'s. *Society Aff:* IEEE, SIAM, IMS, TIMS.

Dresden, Anton
Home: 2430 SW Summit Ct, Lake Oswego, OR 97034
Position: Principal Engr & Owner. *Employer:* Anton Dresden - Consulting Engineer. *Education:* BS/Mech Engg/MTS Amsterdam. *Born:* Sep 1915 Holland Europe. J & L Machine Co Vermont 1938-46 tool engrg; Mining Equipment Corp NYC 1946-53 purchasing engrg; Wiliamette Iron & Steel Co Oregon 1953-62 ending as Ch R&D Engr valves; Cons Engr 1962- . Pres Cons Engrs Council Oregon 1974-75 Life Member, Engrs & Architects Council Oregon 1972, Portland Chapters IIE 1961-62, ASME 1968-69. Pres OR Bd/Engrg Examiners 1979-81 Mbr, PEO, NSPE; Fellow ASME; Awards: CEC-USA design excellence 1968, CECO 1972, AIEE Engr of Yr 1970. Patentee, contributor prof publications. Ardent sailor, magician, play piano. *Society Aff:* NSPE, ASME, IIE, CECO, ACEC, AGC.

Dresher, William H
Home: 13 Singing Woods Rd, Norwalk, CT 06850
Position: Pres, Intl Copper Research Assoc *Employer:* International Cooper Research Association, Inc. *Education:* PhD/Metallurgy/Univ of UT; BS/Chem Engg/Drexel Univ. *Born:* 3/15/30. in Philadelphia, PA. 1956-71 Union Carbide Corp 1971- Dean College of Mines, Dir State of AZ, Bureau of Geology & Mineral Technology. 1981 Pres, Intl Copper Research Assoc, Inc. Mbr AIME, The Metallurgical Soc Bd/Dirs. Mbr: ACS, Amer Soc for Advancement of Science, Mining and Metallurgical Soc of Amer. *Society Aff:* AIME, TMS, SME, ACS, MMSA, AMA, AAAS.

Dresselhaus, Mildred S
Business: 13-2090, Cambridge, MA 02139
Position: Abby Rockefeller Mauze Prof of EE. *Employer:* MIT. *Education:* PhD/Physics/Univ of Chicago; MA/Physics/Radcliffe Coll; AB/Liberal Arts/Hunter Coll. *Born:* 11/11/30. Fulbright Fellow Cambridge England 1951-52. PhD (Physics) Univ of Chicago 1958; Hon D Engrg WPI 1976. Hon D Sci Smith Coll 1980. Hon Dr. Sci Hunter Coll 1982, Hon Dr. Sci Newark Coll Engr, 1984. NSF postdoctoral fellow Cornell Univ 1958-60, at MIT since 1960, Staff Lincoln Lab 1960-67, Abby Rockefeller Mauze Visiting Prof 1967-68, Prof Elect Engrg 1968- , Mauze Chair 1973- , Assoc Dept Hd Elec Sci & Engrg 1972-74. Dir, Ctr of Materials Sci and Engrg 1977-83. Res in solid state physics & materials engrg. Chrmn, Steering Ctte of Evaluation Panels of Natl Bureau of Standards 1978-84. Hunter Coll Hall of Fame 1972, Radcliffe Alumnae Medal 1973, Natl Acad of Engrg 1974, Mbr NAE Council 1981-87. Am Acad of Arts & Scis 1974, Corresponding Mbr Brazilian Acad of Sci 1976. Achievement Award, Soc of Women Engrs, 1977. Married to Gene Dresselhaus, 4 children. Amateur musician. Fellow IEEE Pres, Amer Physical Soc 1984. *Society Aff:* APS, IEEE, MRS, ACS.

Drew, John
Home: 2577 S Ponte Vedra Blvd, Ponte Vedra Beach, FL 32082
Position: Pres. *Employer:* Drew Forest Chem Lab, Inc. *Education:* MS in Engg/Indust Eng/Univ of FL; BSChE/Chem Engg/GA Inst of Tech. *Born:* 1/24/15. Tampa, FL. R&D with Hercules, Inc; Mgr R&D and Engg with Crosby Chem Co; Div Engg, Plant Manager, Mgr of Manufacturing, and Dir of Dept with SCM Corp, Organic Chem Div. Started existing consulting and testing lab July 1978. Governor s appointment to Air Pollution Authority 1967, Served local Environment Protection Board 11 yrs. Served on panel of CORRIM - National Academy of Science, 1975-76. Authored or Co-authored 15 patents in field of naval stores processing. Consulted with pulp mills of 13 foreign countries. Pub articles in six national trade journals. Engg of Yr, FL Engg Soc 1974. FL Sect ACS Civil Award 1978. Boss of Yr, National Sect 1970. *Society Aff:* AIChE, ACS, PCA.

Drew, Thomas B
Home: 34 Glen Dr, Peterborough, NH 03458
Position: Prof Chem Engrg-Emeritus. *Employer:* MIT *Education:* MS/Chem Engrg/MIT; BS/Chem Engrg/MIT. *Born:* Feb 9 1902 Medford Mass. After teaching briefly at Drexel Inst & at MIT joined DuPont Co as res & then proj engr 1934-40 & later recalled from Columbia Univ as design engr for Hanford plutonium plant 1943-44 & still later as cons on Savanah River plant. Assoc Prof then Prof of Chem Engrg Columbia University 1940-65; Head of Dept 1948-57. Tech Dir AEC Heat Transfer Facility at Columbia 1951-62. Program Dir for MIT Ford Foundation proj for establishment of Birla Inst of Tech in India 1962-66. Prof of Chem Engrg MIT 1965-67; Prof Emeritus 1967- . Editor Advances in Chemical Engineering 1956-81. Charter Mbr of ASME Heat Transfer Div & Chmn thereof 1938-42. William H Walker Award AIChE 1937; Hon Mbr ASME 1959; Max Jacob Mem Award ASME & AIChE 1967. Author, 1961, "Handbook of Vector and Polyadic Analysis–. Honors: Elected to Natl Academy of Engrg, 1983; Designated "Eminent Chem Engr" By AIChE at their 75th Anniversary Meeting 1983. *Society Aff:* AIChE, ASME, ACS, ΣΞ, AAAS

Drewry, William A
Business: Department of Civil Engineering, Old Dominion University, Norfolk, VA 23508
Position: Prof of Civil Engrg *Employer:* Old Dominion Univ. *Education:* PhD/Civ Engr/Stanford Univ; MSCE/Civ Engr/Univ of AR; BSCE/Civ Engr/Univ of AR; AS/Civ Engr/AR Tech Univ. *Born:* 10/23/36. Asst Prof, Univ of AR 1965-67; Assoc Prof and Prof, Univ of TN 1968-75. Prof Dept of Civ Engr, Old Dominion Univ, 1976-present, Chrmn 1976-1984. Reg PE, AR, TN, and VA. Mbr Chi Epsilon, and Tau Beta Pi. Pres for VA Sec of Amer Soc of Civ Engrs 1983-84. Pres of the VA Soc of PE 1984-85. Vice-Chrmn, SE, NSPE practice div for PE's in Ed, 1982-present. Contract res and conslt to numerous indus, governmental agencies, and conslt firms in water and wastewater problems. Public service as Bd Mbr to several local and st agencies. *Society Aff:* ASCE, ASEE, NSPE, WPCF, AWWA.

Drexel, Roger E
Business: 13000 Du Pont Bldg, Wilmington, DE 19898
Position: VP. *Employer:* E I duPont de Nemours. *Education:* ScD/ChE/MIT; BS/CE/Univ of Rochester *Born:* 2/10/20. Native of Rochester, NY. Joined DuPont Engg Dept in 1944. Specialized in agrichemicals & process res, 1949-1961. Worked in mfg functions, finally as Dir of Ind & Biochemicals Dept's Mfg Div until appointed Asst Gen Mgr of Dept in 1967. Headed Ind & Biochemicals Dept unitl becoming VP - Polymer Products in 1979. *Society Aff:* ACS, AIChE, ΦBK, ΣΞ.

Drickamer, Harry G
Business: SCS-1209 W. California St, Urbana, IL 61801
Position: Prof Chem Engrg, Chemistry and Physics *Employer:* University of Illinois. *Education:* BSE/ChE/Univ of MI; MS/ChE/Univ of MI; PhD/ChE/Univ of MI. *Born:* Nov 1918 Cleveland Ohio. ChE Pan Amer Ref Corp 1942-46. With University of Illinois 1946- . Res area is the effect of very high pressure on the electronic behavior of matter. Buckley Prize APS 1967; Colburn Award 1947; Alpha Chi Sigma Award 1967; Walker Award 1972 AIChE; W.K. Lewis Award-AIChE-1986; A. Von Humboldt Award-Fed Rep. Germany-1986; Welch Prize in chemistry-Welch Foundation-1987; Ipatieff Prize 1956; Langmuir Award 1974 ACS; Peter Debye Award-ACS 1987; Bendix Prize 1968 ASEE; Mbr NAS; Fellow AAAS. Hobbies incl hiking & reading ancient & medieval history. P W Bridgman Award - Int Assoc for High Pressure Sci & Tech (AIRPT)-1977 Michelson-Morley Award-Case-Western Res Univ 1978. Mbr - Natl Acad of Engg, Natl Acad of Sci, Amer Phil Soc, John Scott Award - City of Philadelphia 1984. *Society Aff:* NAS, NAE, APHS, AAA, APS, ACS, AIChE, AGU, AIC.

Dries, David A
Business: 1527 Starks Bldg, Louisville, KY 40202
Position: Project Engr *Employer:* Chevron USA Inc *Education:* MBA/Mgmt Finance/Univ of Louisville; ME/CE/Univ of Louisville; BS/CE/Univ of Louisville *Born:* 1/31/52 Current (1981) Pres of the KY Section of the American Society of Civil Engrs. Employed as project engr with Chevron USA Inc, East Central Division, Louisville, KY. Formerly technical VP with Schimpeler-Corradino Assocs, Consulting Engrs. Past recipient of the ASCE Daniel V Terrell Award, The Speed Scientific School Alumni Award, and The Louisville Engrg and Scientific Societies Council Award. *Society Aff:* ASCE, NSPE

Driggers, L Bynum
Business: Box 7625, Raleigh, NC 27695-7625
Position: Extension Prof *Employer:* NC State Univ-Agri Extension Serv. *Education:* MS/AG Eng/VPI, BS/AG Eng/Clemson *Born:* in Manning, South Carolina. M.S. from VPI and B.S. from Clemson. Assistant Professor at VPI 1957-66. Asst. Professor, Assoc. Professor and Professor at N. C. State University 1966-present. Over 100 research and extension publications on the design, construction, controlled environment and waste management of livestock and poultry housing systems. Chairman, N.C.Section ASAE 1975-76. ASAE Engineering Achievement Young Extension Man Award, 1975. ASAE blue ribbon educational awards 1973, 1975, 1978, 1980, 1983, 1984. Outstanding Extension Service Award, N.C.S.U., 1974. Personalities of the South, 1976. Superior Leadership Award, N.C.S.U., 1978. Metal Building Manufacturers Association Award, ASAE, 1981. Senior Member, ASAE. Registered P.E. *Society Aff:* ASAE, NSPE, ΕΣΦ

Driggers, William J, Sr
Business: P.O. Box C-50, Little Rock, AR 72203
Position: President *Employer:* Garver & Garver, P.A. *Education:* BS/CE/Univ of AR *Born:* 1/22/30 Native of AR. Served in US Army 1948-1954. Began working for Garver & Garver upon graduation Univ of AR in 1959. Elected to Bd of Dirs; named VP in 1968. Professionally responsible for foreign and domestic projects including highways, bridges, drainage facilities, bldg structures, ports and railroads. Past Pres AR Society of PE; Past National Dir NSPE. Honor Awards: Tau Beta Phi, Life Member; Chi Epsilon, Member Pi Mu Epsilon, Member; Project Designer, White River Bridge in Des Arc, AR; AR Society of PE "Engr of Year, 1979–; AR Academy of Civil Engrg, Charter Mbr, past Pres. *Society Aff:* NSPE, SAME, ASCE, ACI, ASPE, ACEC.

Dripps, William N
Home: 2602 Ryegate Lane, Alexandria, VA 22308
Position: Codes Engineer. *Employer:* Natl Conf of States on Bldg Codes & Stds. *Education:* MSCE/Structures/MIT; BE/Civil/Univ of Toledo. *Born:* 8/20/22. in Toledo, OH. MSCE from MIT. BE from Univ of Toledo. Served in USA Corps of Engrs 1943-46. Structural Designer for Office, Municipal Architect, District of Columbia. Transferred in 1954 to DC Dept of Licenses & Inspections at Supt Instpection Div. Became Ch Bureau of Bldg Housing & Zoning for DC Government in 1968. Retired from govt service in 1976. Now with Natl Conf of States on Bldg Codes & Stds as Codes & Stds Coordinator. *Society Aff:* ASCE, NSPE.

Driskell, Leslie R
Business: 455 Greenhurst Dr, Pittsburgh, PA 15243
Position: Consultant-Control Systems. *Employer:* Independent. *Born:* Aug 1916. After graduate work in industrial instrumentation at the University of Louisville, a career devoted to control systems engrg began at Seagram-Calvert in 1939. Instructor at Purdue Univ 1941-42. Riggs Distler & Co Constructors 1942-44. E I DuPont Co 1944-47. Dravo Corp Chemical Plants Div (formerly Blaw-Knox Co) from 1947-79. Section Ch, Instruments-Electrical. Fellow of Instrument Soc of America, Pres Pittsburgh Section 1954-55, Natl Chmn Final Control Elements Ctte 1958-66. Chmn ISA Stds Ctte SP75, Control Valves. Author 33 publications. Instructor 89 Short Courses on control valves. Instrumentation Tech Award 1968. *Society Aff:* ISA.

Drnevich, Ronald J
Business: POBox 1963, Harrisburg, PA 17105
Position: Chief Transit Facilities Design. *Employer:* Gannett Fleming Corddry & Carpenter. *Education:* BS/Civil Engg/Univ of Notre Dame. *Born:* 1/26/42. Grad studies Carnegie Mellon Univ 1964. With Gannett Fleming Corddry & Carpenter Inc since 1964 & a partner in the firm since 1973. Asst VP and Chief of Facilities Design. Respon for Final Designs of mass transit & transp related projs. Dir of computer dev for Transp Div. Reg in PA. Recipient of PSPE Harrisburg Chapt Young Engr of Yr Award in 1973. Pres of ASCE Central PA Section 1976. ASCE Natl Dir 1977-80. Enjoy golf & fishing. *Society Aff:* ASCE, NSPE.

Drnevich, Vincent P
Business: 212 Anderson Hall, Lexington, KY 40506
Position: Prof of Civil Engg *Employer:* Univ of KY. *Education:* PhD/Civ Engg/Univ of MI; MS/Civ Engg/Univ of Notre Dame; BS/Civ Engg/Univ of Notre Dame. *Born:* 8/6/40. Native of Pittsburgh. Joined faculty at Univ of KY in 1967. Prof of CIv Engg since 1978. Dept. Chairman 1980-84. Summer & sabbatical experiences include: soils engr for Ove Arup & Partners, London, England; proj engr for E D'Appolonia Consulting Engrs, Pittsburgh; & visiting engr with Bureau of Reclama-

Drnevich, Vincent P (Continued)
tion, Denver. Res focused on field & lab measurement of soil properties. Published numerous papers & been involved with soil testing equip dev, holding a patent on the Long-Tor Resonant olumn apparatus. Consultant to industrial, governmental, & engg organizations. Received the 1973 Norman Medal of ASCE (with Bobby O Hardin), 1976 Engr of the Yr in Education from KY Soc of PE, and the 1979 Hogentogler Award of ASTM (with Bobby O Hardin & David J Shippy), and the 1980 Walter L. Huber Research Prize of ASCE, and the 1985 Harold T. Larson Award of Chi Epsilon. *Society Aff:* ASEE, ASCE, ASTM, NSPE, ISSMFE, EERI, TRB.

Drobile, James A
Home: 401 Audubon Ave, Wayne, PA 19087
Position: Partner. *Employer:* Schnader, Harrison, Segal & Lewis. *Education:* LLB/Law/Temple Univ; SM/ChE/MIT; BS/ChE/Villanova Univ. *Born:* 9/29/27. Engg positions, 1950-56; patent agent, then attorney, Sun Oil Co, 1956-61; assoc, then partner (Managing Partner, 1968-71; Exec Cttee, 1983-86), Schnader, Harrison, Segal and Lewis, 1961-date; reg engr, PA; Natl Soc of PE Budget Cttee, 1983-date; PA Soc of PE Legislative & Govt Affairs Committee, 1971-76 (Chrmn, 1972-73); Task Force on Engg Contract Awards, 1973-74; Task Force on Engrs' Reg Law, 1975-date; Engrs Club of Phila (Pres, 1970-71); Phila Engg Fdn (Treas, 1967-1983; Vice Chrmn of Bd of Mgrs, 1984-date) Morehouse Award, Villanova Univ College of Engg, 1974; Trustee, Villanova Univ, 1976-86; Secy, Radnor Township Civ Service Commission, 1969-85; Tau Beta Pi; Who's Who in Am (43rd ed); Who's Who in World (6th ed). *Society Aff:* NSPE, AIChE, PSPE.

Dromgoole, James C
Business: POBox 1729, Houston, TX 77001
Position: Vice President. *Employer:* Maintenance Engrg Corp. *Education:* BS/Chem Eng/Univ of TX. *Born:* Jul 1938. University of Texas 1961. Native of Runge Texas. Two yrs res at Ft Dietrick Maryland. Marine Engr for Maintenance Engrg from 1963-65. Promoted to Field Engr in land plants until 1973. Promoted to Chemical Dept Head in 1973. In 1974 made Vice President of Res & Marketing. Enjoy rodeoing, hunting & fishing with family. *Society Aff:* AIChE, NACE, AWWA, CTI, IWC.

Drost, Edward J
Business: EC23, Huntsville, AL 35812
Position: Supr Electron Engr. *Employer:* G C Marshall Space Flight Ctr. *Education:* BSEE/Elec Engg/Univ of MI. *Born:* May 6 1923. BSEE U of Michigan. With G C Marshall Flight Ctr Huntsville AL since 1960. Presently a lead engr respon for the dev & application of measuring sensors for ground & flight tests of Space Shuttle & other sys; Previously Ch of a section providing instrumentation support for structural & environmental tests of guided missile, space vehicle & payload systems & components. Fellow ISA, won Phillip T Sprague Applications Award 1974. *Society Aff:* ISA.

Drosten, Fred W
Home: 945 Old Bonhomme Rd, St Louis, MO 63132
Position: Mtl Engg. *Employer:* Retired, Army Aviation Res & Dev. *Education:* BS/CE/WA Univ; -/Reactor Design/Oak Ridge Natl Lab; -/Physical Met/Univ of TN. *Born:* 11/24/09. Native of St Louis MO. Plant & Product Chemist Natl Lead, St Louis 1934- 1941; Chief Chemist Emerson Elec Gun Turret Plant 1942-1945; Dev Engg for Nuclear Fuel, Oak Ridge Natl Lab 1946-1951; Dir Res Nickel Processing Co, Nicaro, Cuba 1951-1954; Dir Metal Res Crane-Republic Steel's Titanium Plant, Chattanooga, TN 1954-1958; Reactor Engg, AEC, Oak Ridge 1958. Dir Met Res Vitro Chem, Chattanooga, 1958-1964; Mtl Engg Boeing Co Huntsville AL (Saturn 5 Prog), 1964-1968; Mtl Engg, US Army Aviation Res & Dev Command, St Louis MO 1969-1979. Improved Nickel Recovery from Laterite Ores of Cuba, devised changes in Titanium Sponge Production increasing yields & quality. Since 1976 has served on NACE Tech Practices committee for Corrosion Reduction. *Society Aff:* AAAS, ASM, NACE.

Drouillard, Thomas F
Home: 11791 Spruce Canyon Cir, Golden, CO 80403
Position: Sr NDT Engr. *Employer:* Rockwell Intl-Rocky Flats Plant. *Education:* BA/Chemistry/Kent State Univ; BS/Pre-Med/Kent State Univ. *Born:* 4/16/29. Native of Lakewood OH. Supervisor of Ultrasonic Testing for Babcock & Wilcox Co Boiler Div 1952-64. Assoc Engr at IIT Res Inst Metals Div 1965-67 specializing in nondestructive testing. Came to Rocky Flats Plant in 1967 under Dow Chem USA 1967-75 & under Rockwell Intl since 1975. Working in R&D in acoustic emission and modal testing to characterize matls & structures for incipient failure analysis & nondestructive evaluation. Charter Mbr AEWG (Acoustic Emission Working Group): Sec-Treas 1973-74, V Chmn 1974-75, Chmn 1976-78, Charter Fellow AEWG 1982. Mbr ASNT; P Chmn CO Sect ASNT, Fellow ASNT 1974. Reg PE CA. Author of book, *Acoustic Emission: A Bibliography with Abstracts*. Received AEWG Inaugural Publication Award 1983. Assoc Editor Journal of Acoustic Emission. *Society Aff:* AEWG, ASNT.

Drucker, Daniel C
Business: Dept Engineering Sciences, 231 Aerospace Engr Bldg, Gainesville, FL 32611
Position: Graduate Res Prof of Engrg Sci *Employer:* U of FL *Education:* PhD/Appl Mech/Columbia Univ; CE/Civil Engrg/Columbia Univ; BS/Civil Engrg/Columbia Univ. *Born:* 1918 New York City. Supvr mechanics of solids Armour Res Fdn. Taught at Cornell, ITT & Brown Univ. Dean of Engrg at U of IL at Urbana-Champaign, 1968- 84. Graduate Res Prof of Engrg Sci U of FL, 1984-. Presently Vice-Pres, Intl Union of Theoretical & Appl Mechanics (IUTAM). P Pres of IUTAM, ASEE, ASME, SESA & Amer Acad of Mechanics. Author of one text & over 150 papers. Received Lamme Medal ASEE; von Karman Medal ASCE; M M Frocht Award SESA; Guggenheim & Fulbright Fellowships. Mbr of NAE, & Amer Acad of Arts and Sci. Reg Prof Engr in RI & IL. Hon Doc of Engrg, Lehigh Univ, 1976; Doc of Sci in Tech, Tech Israel Inst of Tech, 1983, Hon Doc of Sci, Brown Univ, 1984; Northwestern Univ 1985; Thomas Egleston Award, Columbia Univ Sch of Engrg & Appl Sci, 1978; Gustave Trasenster Award, Univ of Liege, Belgium, 1979; Chmn of Sec M (Engrg), Fellow, Amer Assoc for the Adv of Sci; Foreign Mbr, Polish Acad of Sci; Wm Prager Medal, Soc of Engrg Sci, 1983; Timoshenko Medal ASME, 1983; John Fritz Medal 1985. *Society Aff:* ASEE, ASME, AAM, SEM, AAAS, ASCE, NSPE, SOR, AIAA.

Drucker, Jules H
Home: 4 Bonnie Ct, Hicksville, NY 11801
Position: Conslt Engr *Employer:* Self-employed *Education:* MS/Mech Engg/Stevens Inst of Tech; MS/Mgt Engg/Long Island Univ; ME/Mech Engg/Stevens Inst of Tech. *Born:* 8/9/21. Lic PE (NY, NJ). Lic merchant marine Chief Engr (USCG), steam vessels unlimitd. Conducted cooperative prog of industrial res as chief admin officer of Pressure Vessel Res Comm, Engg Fdn, (1954-7). Marine Engr in charge of operations and field inspection for fleet of ocean-going dredges and other floating plant, Marine Div, US Army Corps of Engrs, (1957-60). Chief, Safety Branch, North Atlantic Div, C of E, (1960-2). Prof of Engg, US Merchant Marine Academy, Kings Point, NY (1962-82) Conslt Engr, Mech and marine engg, product safety, mech failure analysis. President, LI Maritime Services, Inc. *Society Aff:* ΤΒΠ.

Drumwright, Thomas F, Jr
Business: Alcoa Technical Center, Alcoa Center, PA 15069
Position: Senica Technical Specialist *Employer:* Aluminum Co of America. *Education:* BS/Elec Engrg/Virgina Military Inst; MS/Welding Engineering/Ohio State Univ. *Born:* Sep 1928. MS in Welding Engrg specializing in Nondestructive Testing from Ohio State University 1957. BSEE from Virginia Military Inst 1951. Native of Newport News Virginia. First Lieutenant Army field artillery 1951-53. Staff Trainee & Welding Engr for Newport News shipyard 1953-55. With Aluminum Co of America-Alcoa Labs 1957- . Respon for dev of NDT techniques, procedures & instrumentation for Alcoa nondestructive testing programs. Elected Fellow ASNT 1976. P Chmn of Pittsbugh ASNT, SEC ASTM Ctte E07 on nondestructive testing, Sec ASTM Subctte E07.06 on ultrasonic testing, Sec ASNT Ctte on ultrasonic testing. Awarded the ASTM Charles W. Briggs award, 1987. Sigma Xi, Active in church & community affairs. Married, two children. *Society Aff:* ASTM, ASNT

Druyvestein, Terry L
Business: 1018 Burlington, Missoula, MT 59801
Position: Pres *Employer:* Stensatter Druyvestein & Assoc *Education:* MS/CE/SD State Univ; BS/CE/SD State Univ *Born:* 3/29/41 St James MO; Married, Loretta T in 1962; 2 children, Paul 15, Keneth 13; Graduated with BSCE, SD State Univ 1963, MSCE 1970, worked for US Forest Service from 1963-1973, Assistant Forest Engr, Lolo Natl, Forest, Principal in Civil Engrg Consulting Firm 1973-present. Has held several offices and is active in ASCE. Currently is Pres of MT Section, American Society of Civil Engrs. Past Western Branch Pres & Member of Pacific Northwest Council. *Society Aff:* ASCE, AWWA, WPCF, CEC

Dryden, Robert D
Business: Dept of Ind Engrg, Blacksburg, VA 24061
Position: Dept Hd. *Employer:* VPI. *Education:* PhD/Ind Engg/TX Tech Univ; MS/Ind Engg/OK State Univ; BS/Ind Engg/OK State Univ; Assoc in Sci/-/Northern OK College. *Born:* 2/12/42. Native of Ponca City, OK. Taught at the Univ of TX at Arlington 1968- 1972, 1973-1977; TX Tech Univ 1972-1973. Chrmn Ind Engg Dept at Wichita State Univ 1977-1979 and Hd of the Ind Engg Dept at VPI and SU since 1979. Consultant occupational injuries, Res and Productivity in construction, Aerospace Mfg and pipeline construction. Sr mbr IIE, mbr HFS, ASEE & APM; Regional Dir and Mbr of the Exec Council of Alpha Pi Mu 1974-1976, Natl Exec Dir 1976-1979; Natl Pres 1982-1984; Currently Exec Dir; Reg PE in TX, OK, KS & VA. *Society Aff:* APM, HFS, ASEE, IIE, RESNA

Dryer, Frederick L
Business: D329 Engrg Quadrangle, Dept of Mech & Aerospace Engrg, Princeton Univ, Princeton, NJ 08544
Position: Assoc Dean *Employer:* Princeton Univ *Education:* PhD/Mech & Aerospace/Princeton Univ; BAE/Aeronautical Engrg/RPI *Born:* 11/3/44 Dr Dryer joined the professional Research Staff of Princeton in 1972 as a member of the Fuels Research Group and was appointed as a tenured Assoc Prof in the Dept in 1981. His principal research interests are in the fundamental combustion sciences with particular emphasis in high temperature combustion chem of hydrocarbons, formation/ignition/secondary atomization/liquid phase chem of fuel droplets, and fire safety-related phenomena. Dr Dryer has published over 75 articles in both archival and trade journals, and he has lectured and consulted extensively on these and related subjects. On two separate occasions, he has contributed invited presentations to the intl symposia of The Combustion Inst. Dr Dryer is Assoc Editor of the intl journal, Combusting Science and Tech, and a member of Tau Beta Pi, Sigma Gamma Tau, Sigma Xi, and the Combustion Institute. Presently, Assoc Dean, School of Engg and Applied Science in charge of Undergrad and Grad affairs.

Drysdale, David D
Business: Fire Safety Engineering Unit, Kings Bldgs, Mayfield Rd, Edinburgh, EH9 3JL, Scotland, UK
Position: Lecturer *Employer:* Univ of Edinburgh *Education:* PhD/Combustion Chem/Cambridge Univ (England); BSc (Hons)/Phys Chem/Edinburgh Univ (Scotland) *Born:* 09/30/39 Native of Dunfermline, Scotland. Res lecturer at the Univ of Leeds (1967- 1974), specializing in the measurement and evaluation of rate data for gas phase reactions in combustion systems, air pollution, etc. Since 1974, lecturer, in Fire Safety Engrg at Edinburgh Univ. Visiting Prof at the Centre for Fire Safety Studies, Worcester Poly Inst, MA, 1982. Elected SFPE 'Man of the Year' in 1983. Author of "Introduction to Fire Dynamics" (John Wiley and Sons, Chichester, 1985). Enjoys hillwalking, squash, running, and classical music. *Society Aff:* RSC, CI, SFPE, FSS, IFE, SFSE.

Duane, James T
Business: 1285 Boston Ave, Bridgeport, CT 06602
Position: Mgr Computer Tech *Employer:* General Electric Co. *Education:* AMP/Business/Harvard Bus. School; SB/EE/MIT; SM/EE/MIT. *Born:* 8/24/28. Elem, Sec Education in Phoenix, SBEE, SWEE MIT 1954; AZ; PE PA. USN 1946-48; Employed GE since 1954. Engr, Mgr Engrg Aerospace Motors-Generators 1954-65. Mgr Adv Lab 1965-66. Genl Mgr Speed Variator Products 1966-73. Mgr Special Purpose Computer Center 1974-77, Mgr Production Resources Planning 1978. Present position 1983. Mgr Computer Operation 1978. Six AIEE/IEEE Papers. 3 Patents. Mbr Tau Beta Pi, Eta Kappa Nu, Sigma Xi, ACM. Fellow IEEE 1971. EKN Award Outstanding Young Elec Engr 1962. Activities incld JC s, United Way (Campain Chmn 1974), Boy Scouts, Mfg Assn, Golf. *Society Aff:* IEEE, ACM, ΣΞ.

Dubal, Gajen P
Business: 8074 Military Ave, Detroit, MI 48204
Position: Special Projects Engr. *Employer:* Park Chemical Co. *Born:* 1945. Native of Sanand India. MS from Montana State University. BS from University of Bombay. Registered Prof Engr in State of Michigan. With Park Chemical Co since 1969. Park develops & supplies speciality chemicals to heat treating, automotive & metal finishing industries. My specialty is application of molten salt in metallurgical & chemical operations. Areas of respon are technical admin, process dev, procurement of lab & plant equipment, providing customer assistance on safety & environmental aspects of products & processes. Mbr AIChE, ASM. Enjoy reading & photography.

Dube, Rene L
Home: 13 Fairview Rd, Wilbraham, MA 01095
Position: Professor & Chmn Elec Engrg. *Employer:* Western New England College. *Education:* PhD/EE/Univ of CT; MS/EE/Univ of PA; BS/EE/Univ of MA. *Born:* Jul 1931. PhD from University of Connecticut; MSEE from University of Pennsylvania; BSEE from University of Massachusetts. Native of Chicopee Massachusetts. Served with USAF 1952-56. Microwave Engr for RCA Systems Planning Engr for USN. With Western New England College since 1964. Assumed respon for dev Elec Engrg Curriculum & Labs. Chmn IEEE (Springfield Section) 1974-75. PE Mass. Enjoy squash, tennis, bridge etc. *Society Aff:* IEEE, ASEE, ΤΒΠ, ΣΞ.

Duberg, John E
Home: 4 Museum Dr, Newport News, VA 23601
Position: Prof *Employer:* George Washington Univ. *Education:* PhD/Civil Engg/Univ of IL; MS/Civ Engg/VPI; BS/Civ Engg/Manhattan College. *Born:* 11/30/17. in NYC & educated there. Began a career in structural res at VPI & then Univ of IL. Began govt service in 1943 in structural res at Langley Meml Aero Labs, NACA; ultimately rose to Assoc Dir of Lab in 1969 with intermediate excursion to industry (SO of IN 1946-48; Ford Aero 1955-56) & teaching (Univ of IL 1957-59). Was responsible for basic res, univ relations, profl staff education. Retired. Co-dir of Joint Inst for Advancement of Flight Sciences with Geo Was Univ. Active in community affairs; Assoc Fellow, AIAA; DeFlorez Award, AIAA; Govt Council Dir, ASEE. Retired from NASA 1980, since 1981 Prof (Engrg Adm) George Washington Univ at Joint Inst of Flight Sciences. *Society Aff:* AIAA, NSPE, AAAS, ASEE, ΤΒΠ, ΦΚΦ, ΣΞ.

Dubin, Eugene A
Business: 10 S Wabash Ave, Chicago, IL 60603
Position: Cons Engr & Architect. *Employer:* Self-employed. *Education:* BSAE/Engg/Univ of IL. *Born:* Oct 24 1908 Chicago Illinois. Fellow ASCE. Mbr ACI (President Chicago Chapter 1976); Tau Beta Pi; Married Julia Lipow 1932. Office Supervising Architect Washington D C 1930-36, Jr & Asst Struct Engr; PBB-FWA 1936-41 Assoc Struct Engr; 1941-46 Lt Cmdr US Coast Guard Ch Struct Engrg; Lecturer G W University 1943-45; Cons Struct Engr Chicago Illinois 1946- . Principal Eugene A Dubin Struct Engrs. Contrib sev articles to Tech Journals. Cons AISC

Dubin, Eugene A (Continued)
Special Ctte for Computer Design Aids 1973-76. Reg in 5 states & Natl Council Engrg Examiners. *Society Aff:* ASCE, ACI.

Dubin, Fred S
Home: 1 Seaside Pl, East Norwalk, CT 06855
Position: President. *Employer:* Dubin-Bloom Assocs PC. *Education:* MA/Arch/Pratt Inst; BS/ME/Carnegie Inst of Techn. *Born:* 1/31/14. Prof Engr 24 states. Retired Lt Cmdr on active duty US Navy WWII. Prof Sch of Architectures. Lectured more than 30 architect & engrg univs in US & 11 countries abroad. Engr of Yr Award 1975 by 'Engrg News Record'. Selected twice as 'Amer Specialist' by US Dept of State to lecture abroad Consultants to num Fed & State Agencies. Fellow ASHRAE & ACEC. Chmn Natl Energy Ctte ACEC. Native of Hartford Ct. Major Designs-Salk Inst, La Jolla CA, Solar Energy Res Inst, Golden, CO. Dir AMER Section, Solar Energy Society. *Society Aff:* ACEC, ASHRAE, NSPE, ASES.

DuBois, J Harry
Business: PO Box 346, Morris Plains, NJ 07950
Position: Proprietor. *Employer:* J. Harry DuBois Co. *Education:* BS/Elec Eng/Univ MN. *Born:* 9/18/03. Fifty years in Plastics. Employment at General Electric Co, Gorham Co, Shaw Insulator Co, Plax Corp, Mycalex Corp of America, Mykroy Corp, Tech Art Plastics Co. Engr Editor Plastics World 12 years. Now Technical Editor Industry Media magazines. Recipient of Int'l Award in Plastics Science and Engg in 1966, Charter member of Plastics Hall of Fame. Active industry plastics consultant on product design, marketing, processing and material selection. Author of numerous testbooks on plastics. *Society Aff:* SPE, SPI.

Dubois, William D
Business: 5150 E 65th St, Indianapolis, IN 46220
Position: Exec VP *Employer:* ATEC Assocs, Inc *Education:* MS/CE/MI State Univ; BS/CE/Tri-State Coll *Born:* 10/21/44 ATEC Assoc, Inc, Executive VP-Regional Mgr, Senior Project Engr, 1971 to present. ATEC Assocs, Inc, Material Engrg Mgr, 1969-1971. ATEC Assoc, Inc, Geotechnical Engr, 1968-1969. PE-IN, IL, FL, GA, AL, VA, SC *Society Aff:* ASCE, NSPE, ASTM, ACI, CSI

DuBose, Lawrence A
Business: 457 E Gundersen Dr, Wheaton, IL 60187
Position: Pres. *Employer:* Testing Service Corp. *Education:* PhD/Civ Engg/TX A & M Univ; MS/Civ Engg/TX A & M Univ; BS/Civ Engg/TX A & M Univ. *Born:* 8/19/20. Native of TX. Served as US Army officer almost five yrs. Taught engg two yrs at the Univ of Al. For ten yrs taught and did res at TX A & M Univ and TX Engg Experiment Sta. Pres, Testing Service Corp since 1958. Presented twelve series of lectures on "Soil Mechanics and Foundation Engineering" to Engrs, Architects, and Contractors. Several engg publications. Intl consultant. Hobbies: forestry, bee-keeping, fishing. Honor Societies: Sigma Xi, Tau Beta Pi. *Society Aff:* ASCE, ASTM, NSPE, SAME, ACI.

Dubowsky, Steven
Business: 4731 Boelter Hall, Los Angeles, CA 90024
Position: Prof. *Employer:* Univ of CA. *Education:* ScD/ME/Columbia Univ; MS/ME/Columbia Univ; BME/ME/RPI. *Born:* 1/14/42. in NYC. Educated at RPI & Columbia Univ. Employed 1964 to 1971 by the Optical Group of the Perkin Elmer Corp, Wilton CT as a Sr Engr. Responsible for design & analysis of complex electromech systems. Jointed faculty of Univ of CA, Los Angeles in 1971. Published over 30 articles on the dynamics & controls of electromech systems, robotics & design problems. Principal Investigator of NSF grants, Assoc of Danforth Fdn, Chrmn of ASME Mechanisms Committee (1978-80), Assoc Editor of Mechanisms & Machine Theory, & consultant to various ind firms. Reg engr, State of CA. *Society Aff:* ASME, IEEE, ΣΞ.

DuBroff, William
Business: 9313 Walnut Dr, Munster, IN 46321
Position: Staff Scientist. *Employer:* EG&G Idaho, Inc. *Education:* PhD/Metallurgy/Columbia Univ; MS/Metallurgy/Columbia Univ; AB/Chemistry/Columbia Univ. *Born:* 10/1/37. Joined Inland Steel as a Res Engr in 1967. Progressed to Supervising Res Engr responsible for the areas of magnetic and mech properties, phase transformations, and recrystallization in steels, as well as microstructural analysis. Assumed position of Assistant Director of Research in 1976, Associate Dir of Res in 1980 & Dir of Research in 1984. In this capacity, had responsibility for res in the areas of raw mtls (coal, coke, iron-oxide pellets, and iron- bearing mtls), blast furnaces, refractories, and environmental control. In 1985, joined EG&G Idaho as a Staff Scientist with responsibility for the DOE Steel industry related research, including near net-shape casting programs. Has been chrmn, AIME long range planning committee, editoral bd of Ironmaking and Steelmaking. Also served on several cttees of AIME. *Society Aff:* AIME, ASM, ΣΞ.

Ducatman, Fred P
Home: 886 Village Green, Westfield, NJ 07090
Position: VP Finance & Adm. *Employer:* El Paso Poly Co. *Education:* MS/ChemEng/Columbia Univ; MastBA/Finance/Rutgers; BS/ChemEngg/Columbia Univ. *Born:* 10/2/25. 12 Yrs with Merck & Co as Pilot Plant Engg, Asst Factory Hd, Sen Fin Analyst & Budget Dir. With present co 16 yrs (Dart until 8-31-79 when div sold to El Paso Co) as Mgr. Marketing Services Controller, Pres of div, VP of Engg and Construction and presently as VP Finance & Adm of El Paso Polyoletins Co. Served in US Navy 1944-46. Enjoy tennis and swimming. *Society Aff:* AIChE; CDA

Du Chateau, Joseph E
Business: 2900 N 117th St, Milwaukee, WI 53222
Position: Vice President. *Employer:* Ring & Du Chateau Inc-Cons Engrs. *Education:* BCE/CE/Marquette Univ. *Born:* Aug 1929 Luxemburg Wisconsin. Grad from Marquette University College of Engineering with Bachelors Degree. Worked for US Navy Bureau of Ships upon grad. Subsequently employed by private cons in Milawukee until organizing Ring & Du Chateau Inc Cons Engrs 1961. Served as V P of firm respon for Heating, Ventilating & Air Conditioning design. Reg as a Prof Engr in California, Indiana, Ohio & Wisconsin. Residence in Elm Grove Wisconsin with wife Anita & 5 children. Enjoys golf. *Society Aff:* ACEC, ASHRAE.

Duchscherer, David C
Business: 2320 Elmwood Ave, Buffalo, NY 14217
Position: President *Employer:* Duchscherer Oberst Design, PC. *Education:* MSCE/Structural/State Univ of NY; BSCE/Civ Engr/Union College. *Born:* 2/12/45. Native of Buffalo, NY. Initially worked in Pilot Sewage Treatment Plant design with Morrell Vrooman Engrs & subsequently with the Bridge Design Group for the City of Seattle. Presently, Pres of Duchscherer Oberst Design, PC, charged with management, special projs, reports, transportation facilities, & marketing. Visiting Lectr of Structural Engg at SUNYAB. Buffalo Sec, ASCE 1980 Pres, 1981 Chrmn-Natl ASCE Committee on "Younger Members–. Hobbies include sailing & snow skiing. *Society Aff:* ASCE, NSPE.

Duckworth, Winston H
Business: Battelle Institute, 505 King Ave, Columbus, OH 43201
Position: Res Leader & Mbr, Res Council Office of the Dir *Employer:* Battelle Memorial Inst. *Education:* MSE/Ceramic Engg/OH State Univ; BChE/Chem Engg/OH State Univ. *Born:* Oct 1918. Native of Greenfield Ohio. Served with Army Corps Engr & AF 1941-46. Col USAFR (retired). Registered Prof Engr in Ohio since 1947. With Battelle Memorial Inst since 1946. Dev & dir Battelle's ceramic res operations 1951-66. Fellow & Res Leader 1966- . Over 90 technical publications, mostly in ceramic field. NICE Pres 1965, Permanent Secy 1977- ; ACers: VP 1976, Tr 1968-74, Cramer Award 1975, Fellow 1956. Amer Ceramic Soc - Distinguished Life Member 1985; NICE - Greaves-Walker Award 1987. *Society Aff:* ACS, NICE, AAAS, ROA.

Duda, John L
Business: Department of Chem Engg, 122 Fenske Lab, Univ Park, PA 16802
Position: Prof. *Employer:* PA State Univ. *Education:* PhD/Chem Engg/Univ of DE; MChE/Chem Engg/Univ of DE; BSc/Chem Engg/Case Inst of Tech. *Born:* 5/11/36. Native of Donora, PA. Res Engg with the Dow Chem Co from 1963-71; worked in areas of polymer processing. With the PA State Univ since 1971. Active research in the areas of molecular diffusion in polymeric systems, num analysis of transport processes, and lubrication/tribology. *Society Aff:* AIChE, ACS, SPE.

Dudderar, Thomas D
Business: Mountain Ave, Murray Hill, NJ 07974
Position: Mbr of Tech Staff. *Employer:* Bell Laboratories. *Education:* PhD/Engg/Brown Univ; ScM/Engg/NY Univ; BSME/Engg/Lehigh Univ. *Born:* Jan 1936. PhD Brown University, MS New York University, BS in Mech Engrg Lehigh University. Native of New Jersey. Employed by Bell Labs since 1958. Worked originally in Electro-Mechanical design incl Telstar Project. Last 10 yrs have been doing res, mostly studies of materials properties, especially mechanical properties. Involved in early dev of holographic testing techniques holographic interferometry, dynamic scattered light speckle photography and remote metrology using fiber optics. Mbr SESA & SES; Active with ASTM. Holds several patents on testing devices & have numerous publications in fields related to the above listed areas of interest. Co-recipient of the Hetenyi Award in 1971 SESA. Fellow of SESA (now SEM) in 1983. Co-recipient of IEEE-CHMTS Best 1983 Conference Paper Award *Society Aff:* AAAS, SEM, ΣΞ.

Dudek, Richard A
Business: Dept. of Ind Engr, Texas Tech Univ, P.O. Box 4130, Lubbock, TX 79409
Position: P.W. Horn Prof of Indust Engrg *Employer:* TX Tech Univ *Education:* PhD/IE/Univ of IA; MS/IE/Univ of IA; BS/IE/Univ of NB *Born:* 9/3/26 Dr Dudek's experience includes several years in industry, in his last position as Division Industrial Engr he supervised an IE program for twenty-one plants. Teaching and research positions were held at the State Univ of IA, Univ of NB and Univ of Pittsburgh. He has been professionally active through IIE, Region IX VP; ASEE, IE Division Dir; TIMS and Sigma Xi. Interests include management, sequencing and scheduling, tech assessment, and work design. He has directed considerable research in these areas, authored many articles and given presentations at several national and intl meetings and at Univ in Europe and Asia. *Society Aff:* IIE, ASEE, NSPE, ASME, TSPE, WFS, HFS, TIMS, ΣΞ, ΦΚΦ, APM, ΤΒΠ, ΠΤΣ

Duderstadt, James J
Business: Coll of Engrg, Ann Arbor, MI 48109
Position: Dean *Employer:* Univ of MI *Education:* PhD/Engrg Sci & Physics/Caltech; MS/Engrg Sci/Caltech; B Engrg/EE/Yale Univ *Born:* 12/4/42 Present Position: Dean of Engrg, Univ of MI; Past Positions: Asst Prof (1969-72), Assoc Prof (1972-75), Prof of Nuclear Engrg (1975-81), Univ of MI. Research Interests: Nuclear Reactor Theory, Thermonuclear Fusion, Kinetic Theory and Statistical Mechanics, Computer Simulation. Author of 6 textbooks and 50 papers in areas of nuclear fission and fusion power, mathematical physics, and engrg. *Society Aff:* ANS, APS, AAAS, ΤΒΠ, ΣΞ

Dudukovic, Milorad P
Business: Dept. Chem Eng, W.U. Box 1198, Washington University, St Louis, MO 63130
Position: Prof and Dir, Reaction Engineering Laboratory *Employer:* Washington Univ. *Education:* PhD/Chem Engg/IIT; MS/Chem Engg/IIT; BS/Chem Engg/Univ of Belgrade. *Born:* 3/25/44. in Belgrade, Yugoslavia. Received advanced engg degrees at IIT, Chicago and taught there as instructor 1970-72. Taught at OH Univ, Athens, as asst prof 1972-1976. Joined WA Univ in 1974 as assoc prof and assumed additional responsibility as dir of the Chem Reaction Engg Lab. Primary res interests and publications in the area of gas-solid reactions, multiphase reactors, tracer kinetics, chemical vapor deposition and crystal growth. Consultant to a number of chem companies. *Society Aff:* AIChE, ACS, ASEE, ΣΞ, AACG, AAAS.

Dudzinski, Sharon D
Home: 21981 Heatherbrae Way S, Novi, MI 48050
Position: Test & Dev Engr. *Employer:* Chrysler Corp. *Education:* MS/Radiation Phys/Univ of CO Med Ctr; BS/Phys/Wayne State Univ. *Born:* 11/22/51. Native of MI. Employed by Chrysler Corp since 1973. Present position is in the engine & emission systems dev area. Natl Sci Fdn Summer Res Fellow-1972. Attended the Engg Fdn's Summer Seminar on non-invasive diagnostic techniques (in medicine) - 1976. SWE (Detroit Chapt), Secy - 1979-80. *Society Aff:* SWE, AWIS, AAPM.

Duer, Beverley C
Business: 1172 Park Ave, New York, NY 10128
Position: Proprietor (Self-employed) *Education:* DENG/Engrg/Univ of CA-Berkeley; APC/Finance/NYU Grad School of Bus; Geol Engr of Mines/Mining geol/CO School of Mines *Born:* 4/7/28 s, Beverley and Julia Mary (deForest) D; Geol Engr, CO Sch Mines, 1953; D. Eng, U CA, 1962; advanced prof certificate in Fin, NYU, 1974; m. Helen Crandell Feb. 10, 1962; children John, Alexandra, Staff mem. Arthur D. Little, Inc, 1962- 64; supr. Lybrand, Ross Bros & Montgomery, NYC, 1965-67; mgr ops. research Corporate Mgmt. Sci, CPC Intl Inc, Englewood Cliffs, NJ 1967-75; owner, mgr B.C. Duer, indsl. Consult. NYC, 1975-; dir, Lexington Goldfund, Inc. 1978- ; dir, Lexington Global Fund Inc. 1987- ; Strategic Planning Inst, Cambridge, MA, 1974-75, research fellow/asst. , 1972-75; dir 1172 Corp 1970-87, pres, 1973-75. Served with C.E. U.S. Army 1953-55. Registered PE, NY. Episcopalian. Club Union. *Society Aff:* ASME, ASHRAE, IIE, ORSA, ΣΞ, ACM, NSPE

Duffee, Floyd W
Home: 3426 Blackhawk Dr, Madison, WI 53705
Position: Retired (Emerit Prof Agri Engrg-Univ of WI.) *Education:* S/Mech Engg/Univ of LO; BSC Agriculture (Major Agri Engrg) Ohio State Univ; BS/Mech Engg/Univ of LO. *Born:* 1893. Connecticut Agri Coll 1915-18; University of Wisconsin Dept of Agri Engrg 1918-63; Dept Chmn 1937-62; Internatl Harvester Co 1963-65. Res: ensilage cutters, brush plows, seed corn drier, forage harvester, dairy farm electrification, total dairy farm mechanization, grass silage, seeding grass seed, crushing hay, stored feed vs pasture. Established first full time farm safety program in America. Ten awards incl Cyrus Hall McCormick by ASAE & hon Doctors Degree by Univ of Hohenheim, Stuttgart Germany. *Society Aff:* ASAE, ΑΓΡ, AZ.

Duffey, Dick
Business: Nuclear Engr Dept, College Park, MD 20742
Position: Prof of Nuclear Engr. *Employer:* Univ of MD. *Education:* PhD/Chem Engr/Univ of MD; MS/Chem Engr/Univ of IA; BS/Chem Engr/Purdue Univ. *Born:* 8/26/17. Native of IN. Worked as: Chem engr Union Carbide 1940-42; US Army 1942-47; Nuclear engr Atomic Energy Commission at St Louis, MO, Hanford, WA, and Wash, DC 1947-54; Nuclear engr MIT nuclear engr proj, summer 1954. Initiated the nuclear engr prog at the Univ of MD in 1954 and the nuclear reactor proj in 1957 serving as Nuclear Reactor Dir until 1967; now prof of nuclear engrg. Technical interests are nuclear reactor design, construction, and operation and neutron uses. *Society Aff:* ANS, APS, AIChE, ACS, AGU, HPS.

Duffey, Loren A
Home: 1397 Arrowhead Dr, Placentia, CA 92670
Position: Member Technical Staff *Employer:* Rockwell Internatl-Autonetics Strategic Systems Division *Education:* BS/Civil Engg/Univ of IA; BA/Engg/Cornell College. *Born:* Apr 1919. Native of Manchester Iowa. Served with US Air Force 1945-46. Weights Engr for Curtiss Wright, specializing in Flight Test weights & Actual weights. With Boeing as Preliminary Design Weight Engr & Weights Specialist in Prod Res Office. At Rockwell Internatl with Preliminary Design Weights respon & supervisor weights area on various contractual efforts. SAWE: VP 1969 & 1970, Fellow 1971, Hon Fellow 1973. Enjoys photography & gardening. *Society Aff:* SAWE, IAE.

Duffie, John A
Business: Engg Res Bldg, 1500 Johnson Drive, Madison, WI 53706
Position: Prof of Chem Engrg. *Employer:* University of Wisconsin-Madison. *Education:* PhD/ChE/Univ of WI; MChE/ChE/RPI; BChE/ChE/RPI. *Born:* 3/31/25. Instructor in Chem Engrg RPI; Res Asst at UW; Res Engr Electro-Chems Dept of DuPont; Scientific Liaison Officer at Office of Naval Res. AT UW since 1954 in various positions; as Proj Assoc, Asst, Assoc & Prof, Dr Solar Energy Lab, Asst Dir Engrg Experiment Sta, Dir Univ-Indus Res Prog & Assoc Dean of the Grad Sch. 1964 was a Fulbright Res Scholar & Guggenheim Fellow at Univ of Queensland Australia & at Mech Engg Div of CSIRO Australia. Co-author with W A Beckman of the books *Solar Energy Thermal Processes* 1974 and *Solar Engrg of Thermal* Processes 1980 and with W A Beckman & S A Klein of the book *Solar Heating* Design 1977. P Pres of International Solar Energy Soc & recipient of ISES Amer Sect Abbott Award for 1976. 1976-77 was a Fulbright Res Scholar at Mech Engrg Div of CSIRO Australia. 1984 Honorary Senior Research Fellow, at Univ of Birmingham, UK. *Society Aff:* AIChE, ASES, AAAS, ISES.

Duffy, Jacques
Business: Div of Engg, Providence, RI 02912
Position: Prof. *Employer:* Brown Univ. *Education:* PhD/Appl Mechanics/Columbia Univ; MS/Mech Engg/Columbia Univ; BS/Mech Engg/Columbia Univ; AB/Mathematics/Columbia Univ. *Born:* 7/1/22. Served in US Army, in European Theater of Operations, 1943-1946. Finished studies after the war and worked in industry. Since 1954 on the faculty at Brown Univ. Field of interest is applied mechanics and has published about 70 papers in following areas: mech behavior of granular media, photoelasticity, bioengg in orthopedics and neuromuscular behavior, dynamic plasticity of metals and dynamic fracture. Recipient of Guggenheim Fellowship, 1964, and Engg Fdn Fellowship 1978; Fellow ASME 1979; Honorary DSc Univ of Nantes, France, 1980. TAC Award for Excellence in Teaching, 1987. *Society Aff:* ASME, ΣΞ, ASTM, ASM, SEM.

Duffy, Robert A
Business: 115 Indian Pipe Lane, Concord, MA 01742
Position: Director *Employer:* C.S. Draper Lab, Inc *Education:* BS/Aero Engr/GA Inst of Tech *Born:* 9/9/21 Br.G.Gen: LSAF (ret.) served as an engrg officer in the US Air Force (and predesser Air Corps, US Army) for 32 yrs. Awarded Distinguished Service Medal Legion of Merit. Member Nat'l Academy of Engr, Fellow, AIAA. Past Pres Inst of Navigation. Awarded Thurlow Award, ION; Thos D White, USAF Space Trophy, Nat'l Geographic Society. Member International Academy of Astronautics. *Society Aff:* AIAA, ION, NAE, IAA

Duffy, Robert E
Business: Dept. of Mech. & Aero. Engr, Rensselaer Polytechnic Inst, Troy, N.Y 12181
Position: Prof. *Employer:* Rensselaer Poly Inst. *Education:* PhD/Fluid Mech/RPI; M Engr/Aero Engr/RPI; BS/Aero Engr/RPI. *Born:* 5/27/30. Married Ann Silver 11/26/53; three children, Brian, Patricia, Suzanne; Aero Engg, US Govt 1951-1953. Consultant to numerous firms including GE, Grumman, Xerox, US Forestry Service. Principle investigater on contracts dealing with wind energy systems, non steady aerodynamics and high-temperature heat transfer. Treas and Bd of Trustees of N Greenbush Library; Chrmn of the Planning Bd of Town of N Greenbush NY since 1973. Technical dir, Panaflight Corp. Chrmn of Aero Engg Dept at RPI, 1968-1974. *Society Aff:* AIAA, ASME, ISA.

Dugan, John M
2335 Shawnee Blvd, Springfield, OH 45504
Position: President *Employer:* Rolltech-Pittsburgh *Education:* BMetEng/Metallurgy/OH State Univ; BA/Math/Wittenberg Univ. *Born:* 1909 Springfield Ohio. Ch Metallurgist Ohio Steel Foundry Co 1946-59, V P Operations 1959-66. V P for Res Blaw-Knox Co Pittsburgh PA 1967-83. Pres, Rolltech Pittsburgh 1983-. Reg Prof Engr. Author numerous technical papers. Patentor var casting & heat treating inventions. Dev of Differential Heat Treating Practice for Mill Rolls. Received Distin Alumnus Award from Coll of Engrg Ohio St U in May 1970. Mbr Bd/Dir Marchal Ketin SA Belgium & Aceros Tepeyac Co Mexico. *Society Aff:* AIME, AISE, AWS, NSPE, ASTM, ESWP.

Dugat, Reginald E
Business: 1333 W Loop S, Houston, TX 77027
Position: Gen Mgr. *Employer:* Exxon Chem Co USA. *Education:* BS/CE/Rice Univ. *Born:* 12/27/21. Native of Freeport, TX. Served in USNR 1942-46, principally in the Pacific Area. With Exxon since 1946, in refining, marketing, corporate planning, and chemicals. Assumed current responsibility as Gen Mgr of the Specialty Intermediates Div of Exxon Chem Co USA in 1972. Also serve on the Worldwide Mgr Committee for the Specialty Intermediates' businesses. Enjoy golf & tennis. *Society Aff:* AIChE.

Duggan, Herbert G
Home: 400 Virginia Rd, Oak Ridge, TN 37830
Position: Proj Manager. *Employer:* Union Carbide Corp-Nuclear Div. *Education:* BS/Mech Engg/Univ of TN. *Born:* Jan 1919. Graduate studies through MS level 1950-53. Received Prof Engr license Tennessee 1950. 1943 Design Engr with Buick Motor Div. Served US Navy 1944-46 with rank of Lt (jg). Since 1946 Union Carbide Corp Oak Ridge Natl Lab as: 1946-1955 Design Engr & Group Leader, 1955-74 Head Experimental Engrg Dept, 1974-1979 Y-12 Plant, Hd Mech Design Engg, 1979- Oak Ridge Gaseous Diffusion Plant, Proj Mgr on Centrifuge program, and Manager, Engineering Subcontracts. Respon for engrg studies, tech reports, analysis & preliminary & final design for a wide area of special nuclear equip items, remote handling systs, reactor experiments, fuel reprocessing facilities, & centrifuge machines for the nuclear enrichment program. Technical advisor for Information Ctr on Nuclear Standards. ANS Standards Ctte. Fellow Am Nuclear Soc, Chmn Remote Systems Tech Div 1964-65; Fellow ASME, VP 1969-71; Mbr NSPE. *Society Aff:* ANS, ASME, NSPE.

Duggin, Michael J
Business: 308 Bray Hall, Coll of Environmental Science & Forestry, Forest Engrg Dept, Syracuse, NY 13210
Position: Prof *Employer:* State Univ of NY *Education:* Ph.D./Physics/Monash Univ; BSc./Physics/Math/Melbourne Univ *Born:* 7/30/37 M.J. Duggin received his BSc degree in Science at Melbourne Univ in 1959 and his Ph.D. in physics at Monash Univ in 1965. Since then he has been continuously engaged in research activity and for the past 20 years has been conducting and directing research in the physics of remote sensing, sensor design and image analysis. He is at present a prof at the State Univ of NY, Syracuse, where he teaches the physics and engrg of remote sensing and directs research in the area. He is the author of a large number of publications, and is a fellow of several learned societies. *Society Aff:* AIAA, ASP, RAS, OSA, SPIE, Institute of Physics (London), RAS, IEEE, RSS.

Duke, Robert E
Home: 111 S Wilshire Ln, Arlington Heights, IL 60004
Position: Vice President. *Employer:* Fire Control, Inc. *Education:* BS/Fire Prot Engr/IL Inst of Tech. *Born:* 4/15/26. Employed Atomic Energy Comm as fire protection engr to 1955. Engr with Viking Corp & Sprinklers Contractors Inc for 3 yrs. Hydraulic specialist & product dev with Fire Protection Co for 5 yrs. Currently VP & Dir of Engrg. Serve on Engrg & Standards Ctte of Natl Automatic Sprinkler & Fire Control Ctte. Sub-ctte Chmn on technical cttes of Natl Fire Protection Assn. P Pres of Chicago Chapter Soc of Fire Protection Engrs. Currently Mbr Bd/Dirs of SFPE (Natl) & on Qualifications Bd. Prof Engr by Exam Illinois. *Society Aff:* SFPE.

Duker, George H
Business: 350 N Sherman St, York, PA 17403
Position: General Manager. *Employer:* Molycorp Inc. *Education:* BS/Chem Engg/IA State Univ. *Born:* Nov 1923. BS from Iowa State University 1950. Native of Fort

Duker, George H (Continued)
Madison Iowa. Served in US Army Air Force 1943-45. Production Supt with former Lindsay Chemical Co 1951-60; Sr Process Engr 1960-64; Process Engr with Molycorp Inc 1964-66, Genl Mgr York Plant since 1966. Involved in processing of rare earth chemicals for 25 yrs. Mbr AIChE, NACE. Enjoy golf, growing roses, travel. *Society Aff:* AIChE, NACE.

Duket, Steven D
Business: 500 Sagamore Pkwy W, West Lafayette, IN 47906
Position: VP *Employer:* Pritsker & Assocs, Inc *Education:* MS/IE/Purdue Univ; BS/IE/Purdue Univ *Born:* 9/17/51 Steven D Duket has been VP of Pritsker & Assocs since 1977. He is codeveloper of SAINT, a combined continuous discrete network simulation language for the system analysis of integrated networks of tasks. He was the prime architect and developer of BETHSIM, a simulation support system for modeling steel operations. Mr Duket has been involved in over 25 industrial applicatons of simulation and has demonstrated through his writings and seminars the procedures by which productivity is improved through analysis and planning using simulation. The development and use of BETHSIM at Bethlehem Steel Corp is a major industrial engrg contribution. *Society Aff:* Tau Beta Pi, IIE

Dukler, Abraham E
Business: Cullen College of Engg, Chemical Engin. Dept, Houston, TX 77004
Position: Prof. Chemical Engin. *Employer:* Univ of Houston. *Education:* PhD/ChE/Univ DE; MS/ChE/Univ DE; BS/ChE/Yale. *Born:* 1/5/25. Dev Engr, Rohm & Haas Co, 1945-8; Res Engr, Shell Oil Co, 1950-52; Asst Prof to Prof, Univ of Houston, Chem Engg Dept, 1932-; Chrmn, 1967-73; Exec Dir, TX State Energy Council, 1973-5; Dean of Engg, 1976-82. Res & publication in fluid mechanics & heat transfer in gas-liquid flow. AIChE Alpha Chi Sigma Res Award; ASEE Chem Engg Res Lectureship Award; Fellow, AIChE; Mbr, Natl Acad of Engg. Consultant, Schlumberger, Shell, DOE, Brookhaven Natl Lab, others. *Society Aff:* AIChE, ASME, ACS.

Dulis, Edward J
Business: POBox 88, Pittsburgh, PA 15230
Position: Pres Crucible Res Ctr *Employer:* Colt Industries - Crucible Inc. *Education:* BS/Met Eng/Univ of AL; MS/Sci/Stevens Inst of Tech. *Born:* 10/30/19. Graduate work at New York Univ. Mgmt programs AMA & Columbia Univ Grad School of Mgmt. Res metallurgist Naval Air Station Philadelphia & US Steel Fundamental Res Lab; Supervisor Stainless, Elevated Temperature Steels, Enameling Steels & Hot Dip Coatings Res at US Steel. With Crucible Inc since 1955. Assumed current position as Pres Crucible Res Ctr in 1971. Author of over 50 technical publications on physical metallurgy & dev of steels, superalloys, titanium alloys and high-performance powder met. products and processes. Inventor of 12 patents. Hobbies: tennis, golf, bridge & classical music. ASM Fellow; P Chmn AIME IMD Pittsburgh Chapter; P Mbr AISI Gen Res Ctte, ASM; P Mbr Bd of Trustees, ASM, Exec. Com. Natl Materials Advisory Bd *Society Aff:* ASM, AIME, APMI, NMAB, AMA

Dumack, Ralph C
Home: 102 Blue Spruce Ln, Levittown, PA 19054
Position: Pres. *Employer:* Ralph C Dumack, PE & Assoc. *Education:* BS/Civ Engg/Drexel Univ. *Born:* 3/26/27. in Phila, PA and attended public schools. Served in US Navy electronics prog 1945-46 and as a commissioned officer in the US Navy Reserve CEC 1949 to 1960. Employed by consulting firms of Andrews & Clark and Day & Zimmermann. Entered private practice as structural consultant in 1964. Became a Fellow of ASCE in 1971. Pres of Bucks County Chap of PSPE 1971-72. Established consulting firm of Ralph C Dumack, PE & Assoc as a professional corp in 1973. Awarded engr of the yr 1976 of Bucks County Chap Licensed in 33 states, & WDC. Assistant District Commissioner, Boy Scouts of America. *Society Aff:* NSPE, ASCE, ACI, PCI.

Dumin, David J
Business: Riggs Hall - Elec & Comp Engg Dept, Clemson, SC 29631
Position: Prof. *Employer:* Clemson Univ. *Education:* PhD/EE/Stanford Univ; MSEE/EE/Purdue Univ; BSEE/EE/Johns Hopkins Univ. *Born:* 10/6/35. Over 20 yrs experience in high speed electronic circuits and in electronic matls. Work performed in cryogenics, III-V compounds and devices, silicon material and devices, thin film materials. One of the pioneers in silicon-on- sapphire materials and devices. Present interests include physics of very small devices and improvements in electronics education. Author of over 50 technical publications. *Society Aff:* IEEE, APS.

Dummer, Geoffrey W A
Home: 27 King Edwards Rd, Malvern Wells Worcester, England WR14 4AJ
Position: Electronics Consultant & Author. *Employer:* Self. *Born:* 1909. Educated Manchester England. After 15 yrs in Radio Industry joined Royal Signals & Radar Estab 1939, closely assoc with design of 1st PPI to be used in radar. Designed radar synthetic trainers, for which awarded MBE. Also awarded Amer Medal of Freedom n 1946. 1944 began R&D on components & wrote 6 books on this subject. Pioneered reliability work in the UK. Initiated thin film circuit res & in 1952 put forward first ideas on semiconductor integrated circuits. In early 1960s initiated the majority of all British Govt res in microelectronics & awarded the Wakefield Gold Medal by Royal Aeronautical Soc in 1964. Fellow of IEEE, IEE & IERE. Now full time Author & Cons Awarded IEEE Cledo Brunetti Award 1979. *Society Aff:* IEEE, IEE, IERE.

Dunbar, Robert A
Business: Dunbar Geotechnical Engrs, 1286 W. Lane Ave, Columbus, OH 43221
Position: Owner & Principal Engr. *Employer:* Dunbar Geotechnical Engrs *Education:* MCE/Trans-Soils-Geol/Cornell Univ; BCE/Civil/Cornell Univ. *Born:* Sep 1928. Native of Eveleth Minn. After grad Airphoto-Analyst of engrg soils including location of Brasilia. Proj Engr 1955-58, Photographic Survey Corp Toronto, performing engrg reconnaissance studies; period incls yr in Ceylon with Canadian Colombo Plan. From 1958-60 Ch Geological Engr Photronix Inc Columbus soil engrg studies. Taught airphoto interpretation at Ohio State Univ 1960-64. Since 1960 owner & principal engr Dunbar Geotech Engrs, spec in appl of soil mech & geol to civil engrg practice, Company services include material testing, laboratory and field inspection of construction. Scottish Rite Mason & Priv Pilot. Tr 1968-69 CECO. Dir Great Lakes Region ASP 1962-64. Dir, Central Ohio Chapter ACI, 1984-87. *Society Aff:* ASCE, NSPE, ASTM, ACI, GSA, AAP.

Duncan, Charles C
Home: 255 Elderfields Rd, Manhasset, NY 11030
Position: V Chmn;Internatl Communications Cons. *Employer:* Group 800 N V;Self. *Education:* BS/EE/Washington Univ. *Born:* Feb 1907, Missouri. AMP Harvard Graduate Business School. ATT Long Lines 1927-72 incl V P, Ch Engr, Pres Eastern Tel & Tel Co, Transoceanic Communications, Transpacific Communications, Transoceanic Cable Ship Co & Cuban Amer Co. Exper incl dir supervision of 18 submarine cable projects totaling over 43,000 miles incl 5 Transatlantic, Transpacific, Alaska, Bermuda, Puerto Rico, St Thomas, Venezuela, Jamaica & Panama cables. 1972- , Internatl Communications Cons for Spanish, French, Italian & Israel communication administrations, ITT & Page Iran. V Chmn Group 800 N V. Fellow IEEE; Disting Service CrossEpiscopal. Alumni Citation Washington University & Gold Medal Finnish Telecommunications. *Society Aff:* IEEE, OES.

Duncan, Charles W
Business: 5728 LBJ Freeway, Dallas, TX 75240
Position: Partner *Employer:* Black & Veatch *Education:* MS/Envir Health Engrg/Univ of KS; BS/CE/IA State Univ *Born:* 05/22/33 Resident partner in the Dallas Regional Office of Black & Veatch. He directs a staff of approximately 70 providing engrg services to clients in TX and the southwest. He joined Black & Veatch in 1956 as an engr-in-training. Following a two yr period of military service, he rejoined Black & Veatch in 1958 as a design engr. Projs included field inspection, survey, and preparations of plans and specification for water and wastewater projs.

Duncan, Charles W (Continued)
He was assigned as a proj engr in 1971 and proj mgr in 1974. In 1977 he was transfered to the Dallas Office, being name mgr in 1980. Duncan was named a partner in Black & Veatch in Jan, 1981. *Society Aff:* ASCE, NSPE, AAEE, AWWA, WPCF

Duncan, Donald M
Business: Consltg Geotech Engr PC, 12321 Baltimore, Kansas City, MO 64145
Position: Principal & Pres *Employer:* DMD CGE PC *Education:* AB/Pre-Engrg/Ripon Coll; BSCE/Civ Engrg/MIT; MSCE/Civ Engrg/MIT. *Born:* 1/6/29. BS & MS from MIT. Native of Sheboygan, WI. Registered Prof Engr in NY, MO, KS. With Woodward Clyde Cons 25 years experience in engg for projects throughout the US and "foreign nations–. Dev of underground space & earth dams are special interests. Author of engg papers on shale properties and mine space use. 1974 awarded on ACEC Engg Excellence Award for rock mechanics proj at Crown Center Hotel, KC City. P Pres ASCE, KC. Section, and CEC-MO. 1982 established consltg practice in geotech engrg. Problem evaluations & solutions provided for underground construction sequences. *Society Aff:* ASCE, NSPE, ISSRM.

Duncan, George A
Business: 201 Agr Engg Bldg, Rose St, Lexington, KY 40546-0075
Position: Ext Prof *Employer:* Univ of KY. *Education:* PhD/Agri Engg/Univ of KY; MS/Agri Engg/Univ of KY; BS/ Agri Engg/Univ of KY. *Born:* 12/27/39. Native of Auburn, Ky. Reared on family farm. Res exper with Univ of KY and USDA on Crop Harvesting and Curing Systems, 1961-64. Served as officer, Signal Corps, US Army, 1964-66. Extension Specialist in Agri Engg, 1966-present. Emphasis on design, blueprints, publications, and educational programs on agri structures, environment, crop harvesting mechanization, storage, and curing systems, energy conservation, computers.ASAE Young Extension Man of the Year Award, 1978. Man of the Year in Agriculture Award for Kentucky - 1983, by Progressive Farmer. *Society Aff:* ASAE.

Duncan, James M
Business: 437 Davis Hall, Berkeley, CA 94720
Position: Professor of Civil Engineering. *Employer:* University of California. *Education:* PhD/Civil Eng/Univ of CA, Berkeley; MS/Civil Eng/GA Tech; BS/Civil Eng/GA Tech. *Born:* Jan 1937. PhD from University of California-Berkeley; BS & MS from Georgia Tech. Worked for Corps of Engrs & private cons through 1962. With Univ of Calif-Berkeley since 1965. Cons to govt & private engrg organizations since 1965. Main res & cons activities in area of soil stability & deformation problems. Recipient of Walter Huber Res Prize from ASCE 1973 & Collingwood Prize from ASCE 1972. *Society Aff:* ASCE, ASTM, ACI.

Duncan, John D
Business: 1810 N Main, Miami, OK 74354
Position: Pres & Owner. *Employer:* Consulting Engrs of Miami, Inc. *Education:* MEd/Gen Engg/Univ of IL; BS/Ind Tech/KS State Univ. *Born:* 3/15/42. Native of Miami, OK. Taught at Cameron Univ for six yrs. Past Pres of Wichita Mtn Chapter of OK Soc of PE and past State Dir. Past VP of M G Fuller and Assoc, consulting engrs, and Poe and Assoc, consulting engrs. Assumed current responsibility as Pres and Owner of Consulting Engrs of Miami, Inc in Aug 1977. Designed the largest Rural Water District in OK, over 600 miles. *Society Aff:* NSPE, OSPE, PEPP.

Duncan, Richard H
Business: US Army White Sands, White Sands Mis Rge, NM 88002
Position: Technical Director & Chief Scientist. *Employer:* Department of Army. *Education:* PhD/Physics/Univ of MO; MS/Physics/Univ of MO, Rolla; BS/Elec Engg/Univ of MO, Rolla. *Born:* Aug 1922. Served with Army Air Corps 1942-46. Student & grad student 1947-54. Prof of Physics & EE New Mexico State Univ 1954-65. VP for Res New Mexico State Univ 1965-69. Technical Dir & Ch Scientist White Sands Missile Range 1969- . Publications in electromagnetic theory. Member, Technical Advisory Committee for the New Mexico Energy Research and Development Institute. *Society Aff:* IEEE, ΣΞ, ADPA, AUSA.

Duncan, Robert C
Home: 5109 Yuma St. NW, Washington, DC 20016
Position: Asst. Secretary of Defense (Research and Technology) and Director, DARPA. *Employer:* Dept. of Defense *Education:* ScD/Instrumentation/MIT; SM/Aeronautical Engg/MIT; BS/Aeronautical Engg/US Navy PG School; BS/Elec Engg/US Naval Academy. *Born:* 11/21/23. Commissioned Ensign, USN, 1945, advanced through grades to Commander 1960. Asst to Dir Defense Res and Engg 1961-1964. Chief guidance and control div NASA Manned Spacecraft Ctr, Houston 1964-1967. Asst Dir NASA Electronics Res Ctr 1967-1968. Polaroid Corp VP 1968-1985. Pres, Polaroid Fdn 1978-82. Appointed Director, Defense Advanced Research Projects Agency 1985, Appointed Assistant Secretary of Defense (Research and Technology) 1986. Dir of Charles Stark Draper Lab 1972-1985, Trustee Emeritus of Forsyth Dental ctr, and Corporator, Boston Museum of Sci. Elected to National Academy of Engineering. Awarded Legion of Merit, NASA Exceptional Service Medal, Inst of Navigation Hayes Award, Silver Beaver. Dist Eagle Scout Award. Author of "Dynamics of Atmospheric Entry–, McGraw-Hill, 1962. *Society Aff:* NAE.

Duncan, Samuel W
Home: 716 Ramona Place, Godfrey, IL 62035
Position: President. *Employer:* Duncan Foundry & Machine Works Inc. *Education:* BSChE/Washington Univ, St Louis, MO. *Born:* Oct 13, 1909. Employed by Duncan Foundry & Machine Works Inc Aug 1931- . Involved in the Co production of steel castings, metallurgical machinery, coal burning chain grate stokers & municipal incinerators. *Society Aff:* ASChE, AICHE.

Dundurs, John
Business: Civil Engg, Northwestern Univ, Evanston, IL 60201
Position: Professor of Civil Engineering. *Employer:* Northwestern University. *Education:* PhD/TAM/Northwestern; MS/ME/Northwestern; BSME/ME/ Northwestern. *Born:* Sep 1922. Native of Riga Latvia. Educated in Latvia, Germany & USA. Faculty mbr of Northwestern Univ since 1958, Prof since 1966. Res in theory of elasticity, structural mechanics, dislocations. Numerous publications & mbr of sev editorial bds. Mbr ASCE, P Chmn Engrg Mechanics Div; Fellow ASME & Amer Acad of Mechanics. *Society Aff:* ASCE, ASME.

Dundzila, Antanas V
Home: 7621 Tremayne Pl, No. 202, McLean, VA 22102
Position: Systems Manager *Employer:* American Red Cross *Education:* MS/Mech Engrg/CA Inst of Tech; BS/Mech Engrg/Univ of IL. *Born:* 7/13/32. Dir, Data Systems & Services Dept, Purdue Univ, Calumet Campus 1972-74; Sr Engrg Analyst-Programmer Sargent & Lundy Engrs 1974-82; Sr Scientist, DHR, Inc. 1984-85. Syst. Manager, American Red Cross 1985-. Visiting Lecturer, Data Processing & Mathematics, various Chicago and Washington, DC area Colleges & Universities 1965- . Sr US Fulbright Lecturer Abroad, Fall Semester 1981. Mbr ACM, AIAA. ACM Chicago Chapter Chmn; ACM Ombudsman Ctte Chmn; ACM Mbrship Ctte Chmn; Genl Chmn 1973 ACM North Central Regional Conference. *Society Aff:* ACM, AIAA

Dunham, Roy H
Business: 889 Ridge Lake Blvd, Memphis, TN 38119
Position: Mgr of Engrg *Employer:* Bechtel Energy Corporation *Education:* BS/Mech Engg/Univ of MO, Rolla *Born:* 8/29/23. US Army Air Corps 1943-45. Allis-Chalmers Co 1947-49. Instructor of Mech Engg Univ of AL 1949-50. TVA Div of Engg Design 1950-80. Involved in various aspects of power generating plant engg design. Chief Mech Engg 1967-71. Dir and Mgr of the div 1973-80 with overall resp for all arch and engg design work for TVA electric power generating capacity additions and other TVA facilities. Since March 1980, Bechtel Energy Corp, as Mgr of Eng. Resp for all arch and engr design work and operating plant support services for the corporation's clients. *Society Aff:* ASME, NSPE, TBΠ.

Dunham, Thomas E
Business: Appliance Park AP3-236, Louisville, KY 40225
Position: Engg Dept Gen Mgr. *Employer:* General Electric Co. *Education:* PhD/Mettallurgy/OH State Univ; BMetEng/Mettallurgical Engg/OH State Univ; BA/Chem/OH Wesleyan. *Born:* 8/24/41. Tom was born in North Canton, OH in 1941. After completing his undergraduate degree at OH Wesleyan, he received his Bachelor and PhD Degrees in Metallurgy from OH State Univ. He joined the General Electric Co as a Res Metallurgist for the Refractory Metals Product Dept of the Lamp Bus Div. He later assumed the position of Manager of Refractory Metals Res and Dev within the same organization. In 1974 he became Manager of the Chem Products Plant of the Quartz and Chem Products Sect in the Lamp Business Div. He was appointed Manager-Manufacturing, Quartz & Chem Products Dept in 1976. In June 1979, he became Gen Mgr of Dishwasher & Disposal Engr Dept. He was apptd to his current position in June 1980.

Dunman, Leonard J, Jr
Business: Genl Elec Co AP 1-135, Louisville, KY 40225
Position: Mgr Occupational Safety etc. *Employer:* GE Co-Home Laundry Mfg Dept. *Education:* MS/Mech Engg/Univ of Louisville; BS/Mech Engg/Univ of Louisville. *Born:* Nov 1926 Hunstville Alabama. BS & MS Mech Engrg University of Louisville. Certified Safety Professional 1970. Lt Col US Army Corps of Engrs Retired Reserve. Resident safety engr, Michigan Mutual Liability Co 1951-64. First Dir of Occupational Safety, Commonwealth of Kentucky 1964-66. With General Electric since 1966. Mgr of Occupational Safety, Healty & Environmental Pollution Control, Home Laundry Dept since 1972. Prof Mbr ASSE. Regional VP & mbr of Bd/Dir ASSE 1977-81. Bd/Dir Southern Safety Conf 1970-74. Favorite recreation is golf, fishing & camping. *Society Aff:* ASSE.

Dunn, Andrew F
Home: 734 Eastbourne Ave, Ottawa Ontario, Canada K1K 0H7
Position: Chief Electricity Section. *Employer:* National Research Council of Canada. *Education:* PhD/Physics/Univ of Toronto; MSc/Physics/Dalhousie Univ; BSc/-/ Dalhousie Univ. *Born:* Jan 17, 1922 Sydney Nova Scotia. Experimental Physics. Natl Res Council of Canada Asst Res Off 1950-54, Assoc Res Off 1954-63, Sr Res Off 1963- . Ch Elect Section 1971- . Cdn Army 1942-45, Capt. Fellow IEEE; Cdn Assoc of Physicists; Reg Prof Engr. Chmn 1970 IEEE Inst & Meas Group; VP 1954-64, Pres 1964-71 Dunn Const Co Ltd, Baddeck, NS. Precision electrical measurements; natl primary standards; absolute determination of electrical quantities. *Society Aff:* IEEE, CAP.

Dunn, Cullen L
Business: United Amer Plaza, Suite 1801, Knoxville, TN 37929
Position: Exec VP *Employer:* Russell & Axon Inc. *Education:* MS/Civil Engg/ Vanderbilt Univ; BS/Civil Engg/TN Polytechnical Inst. *Born:* 10/12/41. Joined the TN Dept of Public Health, Div of Sanitary Engg in 1964. In 1967 opened a regional office for the Div. Respon in the region office were to operate a public water & sewerage sys program for one-third of the State. Work incl design review, const surveillance & sys operation review as well as other regulatory functions. Joined Russell & Axon Engrs, Planners, Architects, Inc in 1973. Worked as Regional Office Mgr for the TN Region for 7 yrs. Presently is Exec VP and mbr of the Exec Ctte. *Society Aff:* AWWA, WPCF, NSPE, ASCE.

Dunn, Donald A
Business: Dept of Engg-Economic Systems, Stanford, CA 94305
Position: Prof. *Employer:* Stanford Univ. *Education:* PhD/EE/Stanford; LLB/Law/ Stanford; BS/EE/Cal Tech *Born:* 12/31/25. Engaged in the study of public policy in relation to telecommunications and info systems. Res has included studies of policy issues in the areas of comp communications, communication satellities, and cable TV. Current issues include economics of info services in natl and intl markets with associated policy issues. Author of Models of Particles and Moving Media (Academic Press 1971). *Society Aff:* IEEE, APPAM

Dunn, Floyd
Business: Dept of Elec & Computer Engg, 1406 West Green St, Urbana, IL 61801
Position: Prof. *Employer:* Univ of IL. *Education:* PhD/EE/Univ of IL; MS/EE/Univ of IL; BS/EE/Univ of IL. *Born:* 4/14/24. From Asst to Full Prof, Univ of IL, 1957-1965. Currently, Prof of Elec Engr and Prof of Bioengg, College of Engr, and Prof of Biophys, Dept of Physiol and Biophys, College of Liberal Arts & Sciences; Chrmn, Bioengr. Faculty, 1977-1982; Dir, Bioacoustics Res Lab, 1977-present. Visiting Prof, Dept Microbiol, Univ College Cardiff, 1968-9; Visit; Sr Scient, Phys Dept, Inst of Cancer Res, Sutton, 1975-6 & 1982-3. Mbr FDA Tesprssc, 1973-7; Mbr NIH Rad Study Sec, 1977-1982. Assoc Ed, J Acoust Soc Am; Mbr Ed Bds. Ultrasound Med & Biol, Ultrasonics, Rad. Environ. Biophys. Author of over 150 articles on interaction of ultrasound and living systems. 1980/81, VP, 1985-6, Pres Acoustical Society of America. Fellow: Acoustical Society of America, IEEE, AAAS, Amer. Inst. of Ultrasound in Medicine, Institute of Acoustics (UK). Mbr Natl Acad Engrg; Visiting Prof Tokoku Univ, Sandai, Japan 1982; Visiting Prof, Univ Nanking, People's Repub of China 1983. *Society Aff:* AAAS, IEEE, ASA, AIUM, BS, Institute of Acoustics (UK).

Dunn, Frank W
Business: 1 Bidwell Rd & Rt 5, South Windsor, CT 06074
Position: President. *Employer:* Metals Testing Co Inc. *Born:* Apr 13, 1923. Bentley College; Norwich St Tech College; Mitchell College; Seminars & short courses. 24 yrs experience in NDT. Prior positions held: NDT Group Leader at Pratt & Whitney Aircraft & Supervisor of Tech Training at Electric Boat Co. Socs: ATS; ASQC; ASTD; ASM; HPS; ASTM & ASNT. Offices Held: Dir of ASNT; V Chmn of Education Council of ASNT; Fellow ASNT & Prof Engr, Quality, State of California.

Dunn, Parker S
Home: 3332 Quail Creek Rd, Oklahoma City, OK 73120
Position: Consultant. *Employer:* Kerr McGee Chemical Corp. *Education:* MSChE/Chem Engg/MIT; BChE/Chem Engg/OH State Univ. *Born:* Aug 1910 Portsmouth Ohio. Employed Mead Corp 1931-33, Pittsburgh Plate Glass Industries 1933-41, Potash Co of America 1941-51 the last five yrs as Resident Mgr, Kerr-McGee Chemical Corp subsidiaries 1951-71 as VP, Pres & Chmn/Bd of Kerr-McGee Chemical Corp. VP Kerr McGee Nuclear Corp 1969-75 (retired). At present Consultant Kerr-McGee Chemical Corp. Benjamin Lamme Engrg Gold Medalist OSU 1966. Enjoy walking, fishing & photography. *Society Aff:* AIChE, AIME

Dunn, Robert M
Home: 25 Orchard Rd, Woodbridge, CT 06525
Position: V.P., Dev & Eng. *Employer:* Summagraphics Corporation *Education:* MSEE/CIS/Univ of PA; BA/Math/Rutgers. *Born:* May 1937 Chicago Ill. Worked in physical chemistry, numerical methods, computer systems software design, computer systems architecture design, design of integrated computer-telecommunication networks, software engrg mgmt, computer graphics & interactive systems.dev, computer-based tools for human use. Assumed current position for development of interactive CAD Sys in 1979; past Natl Chmn ACM-SIGGRAPH. Enjoy contact sports, classical music, jazz. *Society Aff:* ACM, IEEE.

Dunnavant, Guy P
Home: 201 Girard Ave, Dothan, AL 36301
Position: Ch Utilities Div etc. *Employer:* US Army Aviation Ctr. *Born:* Apr 1924. BSME from Auburn University. Native of Alabama. Served in US Navy, WWII & Korean War. Field Engr for Babcock & Wilcox 1955-57, Plant Engrg Monsanto Co 1957-65. Currently serves as Ch, Utilities Div, Directorate of Facilities Engrg, US Army Aviation Ctr. Respon to evaluate, identify & eliminate air, land & water pollution problems. Respon to plan & execute energy conservation plan for the Ctr. Serves as technical advisor to Dir of Facilities Engrg on all matters concerning planning, construction, operation & maintenance of utilities systems. ASME 75th Anni-

Dunnavant, Guy P (Continued)
versary Oustanding Student Award. Sustained Superior Performance Award 1972. Toastmasters.

Dunne, Edward J, Jr
Home: 4939 Pepperwood Dr, Dayton, OH 45424
Position: Chair and Prof Dept of Decision Sciences *Employer:* Univ of Dayton *Education:* PhD/IE-Ops Rsch/Univ of IL; MS/Mat Engrg/Air Force Inst of Tech; BS/Aeronautical Engrg/St Louis Univ *Born:* 4/21/41 Native of St Louis, MO. Worked as a test program engr at the Air Force Special Weapons Center and an R&D mgr at Air Force Aeronautical Systems Division. Served one year as an operations research analyst in Southeast Asia. 1973-82, on the faculty of the School of Engrg, Air Force Inst of Tech and since 1982 on the faculty of the Univ of Dayton. Have been active in research, consulting, and publishing (some 30 articles and papers) in the areas of military operations research, techniques of engrg and project management, and management information systems. *Society Aff:* DSI.

Dunsmore, Robert L
Home: 4 Centre Str, Kingston Ontario, Canada K7L 4E6 *Education:* LLD/BCS/Kingston Onario; BSc/Civ Engrg/Queens Univ. *Born:* 1893, Seaforth, Ontario. Doctor Commercial Science Laval University Quebec (honorary). LLD Queens Univ Kingston, Canada. WWI: Major Royal Canadian Engrs 1914-19, Military Cross 1916. WWII: Cdr Royal Canadian Navy Dir of Fuel. 40 yrs Petroleum Industry: Constr Engr, Refinery Mgr, Executive & Pres Champlain Oil Products Ltd. Life Mbr & Fellow Engr Inst of Canada; Life Mbr Corp Engrs Quebec; Chmn Montreal Bd/Trade 1955; Chmn Can Broadcasting Corp 1958-63; Mbr Bd/Trustees 1953-74. Montreal Medal Distin Service Award, Queens Univ Kingston Canada. Diversions painting, tennis. *Society Aff:* EIC

Dupies, Donald A
Home: 4733 N Cumberland Blvd, Whitefish Bay, WI 53211
Position: Partner *Employer:* Howard-Needles-Tammen & Bergendoff. *Education:* BCE/Structures/Marquette Univ. *Born:* 4/17/34. Tau Beta Pi & Chi Epsilon. Served with US Army Corps Engrs 1957-59 in Iceland & US Reserves from 1959-64. With HNTB since 1959, respon for the operations & mgmt of a major design office involved in all phases of engrg, architecture & planning for highways, bridges, airports, building; environmental, municipal, electrical & mechanical engrg projects; urban & regional planning. Prof Engr in Minnesota Michigan Ohio Wisconsin & Illinois. ASCE (Natl Dir 1982-85, Pres Wisc Sec 1976); Active in University, Civic & charitable organizations *Society Aff:* ASCE, APWA, WPCF, ITE, TRB, AMA, CECW, ESM, IBTTA, AWP.

Dupuis, Russell D
Business: Room 7C204, 600 Mountain Ave, Murray Hill, NJ 07974-2070
Position: Distinguished MTS *Employer:* AT&T Bell Laboratories *Education:* PhD/Elect. Eng./University of Illinois at Urbana-Champaign; MS/Elect. Eng./ University of Illinois at Urbana-Champaign; BS/Elect. Eng./University of Illinois at Urbana-Champaign *Born:* 07/09/47 Russell D. Dupuis was born on July 9, 1947 in Kankakee, IL. He received his BS degree in Electrical Engineering with Highest Honors from the Univ of IL in 1970. He received his MSEE degree in 1971 and his PhDEE in 1973 also from the Univ of IL. After graduation, he joined the Semiconductor Research and Dev Lab of Texas Instruments in Dallas. In 1975 he joined the Electronics Research Center of Rockwell International in Anaheim, CA where he was the first to demonstrate high-quality epitaxial films grown by metalorganic chemical vapor deposition (MOCVD). In 1979, he joined the Materials Science Research Dept of AT&T Bell Labs in Murray Hill, NJ where he is presently continuing his research on MOCVD. *Society Aff:* IEEE, APS, ECS

Duquette, David J
Business: Mtls Engg Dept, Troy, NY 12181
Position: Prof of Met Engg. *Employer:* Rensselaer Poly Inst. *Education:* PhD/Metallurgy/MIT; BS/Engg/US Coast Guard Acad. *Born:* 11/4/39. Native of Springfield, MA, four yrs in USCG LT,. Res Asst MIT, 65-68. Res Assoc Adv Matl Res & Dev Lab P&W Aircraft 68-70. Asst Prof 70-73. Assoc Prof 73-76. Prof 76-. Perform & supervise res on corrosion, environment/mech properties. High temp metallurgy. Listed in Who's Who in the East, Who's Who in Education. AM Men of Science. Alcoa Fdn Awards for Excellence in Res 1978, 1979, Case-Western Reserve Centennial Scholar, 1980 Sr. Scientist Prize, A. von Humbolt Foundation, 1983, Fellow ASM, Int., 1986. *Society Aff:* AIME, ASM, NACE, ECS.

Durant, John H
Home: 42 Summer St, Weston, MA 02193
Position: Product Mgr *Employer:* Vacuum Industries, Inc *Education:* AB/Phys Sciences/Harvard Univ; Courses/Met/MIT; Courses/Met/NYU. *Born:* 6/7/23. Native of Cambridge, MA. Acquired early experience in high vacuum processes and equipment dev at Natl Res Corp, where he conducted res in vacuum reduction of alkali metals, dev in vacuum melting & coating equipment. Moved into field engg large plants, preparing technical manuals, sales promotion and public relations. Cofounder of Am Vacuum Soc, 1954. 1957-1961 with Engelhard Industries fabrication of nuclear fuel elements, proj engr Experimental Breeder Reactor Core I. With Vacuum Industries since 1961 as mgr, std products, Gen Sales Mgr, VP since 1969. Author of technical papers & articles in vacuum equip & processing. Trustee, Museum of Transportation since 1968; Trustee Brimmer & May School since 1971. Enjoy photography, dramatics, wood working. Active in Amer Soc for Metals, variously Chrmn Adv Tech Awareness Cttee, Chrmn, Govt & Public Affairs Cttee. 1984, 85, 86. *Society Aff:* ASM, AVS, AVEM.

Durante, Raymond W
Business: 1 Farragut Square, NW, Washington, DC 20006
Position: VP *Employer:* Schneider Group *Education:* MSIE/Indus Eng/Stevens Inst of Tech; ME/Mech Engr/Stevens Inst of Tech; Armed Forces Ind Coll/(Certificate)/ US Coll of Armed Forces *Born:* 06/08/28 35 yrs prof & govern experience, currently VP Mktg Schneider Grp. 1970-1984 VP Adv Energy Progs Westinghouse Elec Co. Respon for Energy Related Tech Progs with Federal Govt & Utilities. 1966-1970 Presidential Exchange with US Dept Interior Proj. Mgr for Nuclear Power and Desalination Projects and other large sea water conversion proj. 1960-1966 Dir Planning Aerojet General Corp. 1950- 1960 Proj Engr for Nuclear Submarine Prog, Savannah River H-Bomb Proj, Nuclear Reactors for res, test, power prod. *Society Aff:* ANS, APCA, SAME, NERO, APS

Durbetaki, Pandeli
Business: School of ME, Atlanta, GA 30332
Position: Prof *Employer:* GA Inst of Tech *Education:* PhD/ME/MI State Univ; MS/ ME/Univ of Rochester; BS/ME/Robert Coll Engrg School *Born:* 5/31/28 Native of Istanbul, Turkey became naturalized citizen 1959. Draftsman and designer in Rochester, NY and then Instructor and Assistant Prof at Univ of Rochester, Instructor at MI State Univ and NSF Science Faculty Fellow. At GA Tech since 1964 and promoted present position 1977. During tenure at three academic institutions taught classical, statistical and irreversible thermodynamics, heat transfer and combustion. Research interest and activity in combustion, pyrolysis, flammability, air pollution and charge stratification of internal combustion engines. Published more than seventy journal articles, papers and research reports. Enjoy classical music and bowling. *Society Aff:* ASME, CI, AAUP, ΣΞ

Durelli, August J
Business: Mech. Eng.-Univ of MD, College Park, MD 20742
Position: Professor *Employer:* Univ. of Maryland *Education:* Doctor in Engineering/ University of Paris (Sorbonne); Doctor in Social Sciences/Catholic University of Paris; Civil Engineer/University of Buenos Aires *Born:* 04/30/10 Prof. at Illinois Inst. of Tech & Catholic Univ., Oakland Univ (chair), Univ. of MD, Univ of Mexico (chair). Author of 300 scientific papers, 200 papers on educational and

Durelli, August J (Continued)
social problems, 10 books. Honorary mbr of SEM; Gold Medal ASEE. Courses and lectures in Europe, So. Am and Asia. *Society Aff:* ASME, SEM, N.Y. Ac. of Sc

Durelli, August J
Business: Mech Engrg, College Park, MD 20742
Position: Prof *Employer:* Univ of Maryland *Education:* CE/-/Univ of Buenos Aires; Dr Eng/-/Univ of Paris; Dr Soc Sc/-/Catholic Univ of Paris. *Born:* 4/30/10. Guggenheim Fellow 1941. Visiting Prof Ecole Polytech Montreal. Head Buenos Aires City Testing Lab. Prof CE at IL Inst Tech 1956-61. Prof at Catholic Univ 1961-75. John F Dodge Prof in Engrg at Oakland Univ Rochester MI 1975-80. Nabor Carrillo prof at Univ of Mexico 1980-81. 10 books in Theory of Elast Exp Stress Analysis, social & political philosophy & educ. 250 papers in English & French, about 100 in Spanish. Cons to US Govt agencies & industrial co s. Hon Mbr SESA. Fellow ASME. *Society Aff:* ASME, SESA, ΣΞ, NYAS.

Durham, Charles W
Business: 8404 Indian Hills Dr, Omaha, NB 68114
Position: Chrmn of the Bd *Employer:* Henningson, Durham & Richardson, Inc *Education:* Prof Deg/CE/IA St Univ; BS/CE/IA St Univ; BS/Gen Engrg/IA St Univ *Born:* 9/28/17 in Chicago. Elementary and coll education, Ames, IA. Married, one son and three daughters. Post Engr and U.S. Public Health Service-WW II. Civil Engr with Henningson Engrg '46-'50. Pres Henningson, Durham & Richardson, Inc. 1950-1976. Chrmn of Bd 1976 to present. Responsible for administration and policies of intl consulting engrg firm with 21 offices and 1200 employees providing services throughout the United States and twenty foreign countries. Active in civic, state and national affairs; Bd of Junior Boy Scouts. Past Pres, Chief Executives Forum. Member Masons, Shrine and Jesters. Award of Merit, IA State Univ Coll of Engrg, 1978, Engr of the Year, National Society of PE, Newcomen Society honor, 1978. Received Marston Medal-IA State College, 1984. Enjoys tennis, golf, flying planes. *Society Aff:* NSPE, CEC, ASCE, APWA, SAME, AAEE

Durrani, Sajjad H
Home: 17513 Lafayette Dr, Olney, MD 20832
Position: Research & Planning Manager, NASA Communications Division *Employer:* NASA-Goddard Space Flight Ctr *Education:* ScD/Elec Engrg/U of NM; MSc Tech/Elec Engrg/Coll of Tech, Manchester UK; BSc Eng (Hons)/Elec Engrg/ Eng Coll, Lahore, Pakistan; BA/Math/Govt Coll Lahore Pakistan. *Born:* Aug 1928. Taught in Pakistan (Chmn EE Dept, Engrg Univ, Lahore 1964-65) & Kansas State Univ; Visiting Prof Univ of MD. Sen Engr GE, RCA Space Ctr; Mbr, Tech Staff & Branch Mgr, Comsat Labs; Sr Scientist, ORI. With NASA since 1974 including Chief Comm Scientist NASA HQ (1979-81). Cons. Office of Telecomm Policy, Exec Office of the Pres & Natl Telecomm & Info Admin. Author of numerous papers on space comm. Fellow, IEEE; Assoc Fellow, AIAA; Fellow, Washington Acad of Scis. Editorial Bd, IEEE Spectrum & AES Transactions, & Telecomm magazine. Member, Executive Advisory Board, Encyclopedia of Physical Science & Technology, Academic Press; EASCON Bd/Dir 1975 & 1978-81; IEEE Educational Activities Bd 1974-75. AES Bd of Governors 1976- . Pres, IEEE Aerospace & Elec Sys Soc, 1982, 1983. Dir IEEE Div IX, 1984, 1985. Citation of Hon, IEEE US Activities Bd, 1980. Outstanding Mbr Awd IEEE Region 2, 1982. NASA Special Achievement Awds, 1977 and 1978. *Society Aff:* IEEE, AIAA.

Durrenberger, J E
Home: P.O. Box 1655, Las Cruces, NM 88004
Position: Opto-Mechanical Engr *Employer:* Self, consultant *Education:* BS/ME/NM State Univ *Born:* 9/8/16 Native of Greenville, NJ. Engaged in Industrial/ Commercial photography prior to WW II. Chief of US Army Signal Corps photo lab ETO. Project Engr, US Army White Sands Missile Range, 1951-81, specializing in photo-optical instrumentation of extra-terrestrial phenomena. Consultant to Smithsonian Astrophysical Observatory. (IGY) Opto-Mechanical Engr (Special Projects), Physical Science Laboratory, NM State Univ. 1983 to date. Fellow, Life Mbr, Historian SPIE, Gordon, Governors' and Pezzuto Awards, Assoc Editor Optical Engrg 1972-80, SPIE secretary, VP, Governor 1964-81. Enjoy optics astronomy, photography, pecan culture, educational pursuits. *Society Aff:* SPIE, ASME, ADPA

Durrett, Joseph B, Jr
Business: 333 West Loop North, Houston, TX 77024
Position: Pres. *Employer:* Sergeant Oil & Gas Co, Inc. *Education:* BS/CE/Univ of MS; Post Baccalaureate/Bus Admin/Univ of Houston. *Born:* 7/27/35. Attended public schools in Aliceville, AL. Grad Univ of MS in 1957 - Pres of Sch of Engg, Pres of Phi Delta Theta, Mbr of Omicron Delta Kappa, Editor of Univ Annual "Ole Miss" in 1956. Profl career began at Union Carbide with experience in Technical Service, Mfg, Purchasing, & Marketing/Sales. VP, Sales/Marketing of Hybrid Systems, Inc - 1969. Established & Pres of United Home Delivery of Houston, Inc - 1971-84. Pres of Sergeant Oil & Gas Co, Inc - since 1976. Teacher of Sunday School & Bible Study Classes since 1957. Hobbies include hunting, fishing, all sports. Local Sch Ctte Mbr - 1979-81. Trustee Univ of Mary Hardin-Baylor -1980-86, Chairman, Second Baptist School Foundation, Inc. *Society Aff:* AIChE.

Duscha, Lloyd A
Home: 11802 Grey Birch Pl, Reston, VA 22091
Position: Deputy Director of Engineering & Construction *Employer:* Headquarters, U.S. Army Corps of Engineers *Education:* BCE/Civil Engg/Univ of MN. *Born:* 3/18/25. in Foley, MN. Federal Exec Inst. Cdr, Civil Eng Corps, US Naval Reserve. With US Army Corps of Engrs since 1946, serving through assignments as structural engr on multi-purpose dams, as Ch Engg Br on construction of various ICBM projs, as Ch Engg Div Philadelphia District, as Ch Engg Div MO River Div, as Ch, Engg Div, Dir of Civil Works, Office of the Chief of Engrs. Assumed current responsibility as Deputy Director of Engineering & Construction, Office Chief of ENgineers in 1983. Involved in directing execution of design and construction responsibilities of a worldwide program of civil works and military construction. National Academy of Engineers NSPE, ASCE, & USCOLD, SAME Wheeler Medal 1975, President's Meritorious Executive Rank 1980 & 1985, Department of Army Decoration for Meritorious Service Chairman, U.S. Section, Columbia River Treaty Permanent Engineering Board Registered Professional Engineer (Minnesota) Tau Beta Pi. and Chi Epsilon (Honorary Engineering Fraternities). *Society Aff:* ASCE, NSPE, USCOLD.

Dussourd, Jules L
Home: 14 Cleveland Rd. W., RD. #2, Princeton, NJ 08540
Position: President *Employer:* Jules L. Dussourd & Assoc. *Education:* BSME/Mech. Eng./City College of New York; MS in ME/Mech. Eng./ Columbia University; Sc. D./Mech. Eng./Massachusetts Institute of Technology *Born:* 12/21/24 Teaching Experience: City College of NY and Arizona State U. (3 yrs.) Aerospace and Industrial Experience: General Dynamics, Fort Worth, TX (Senior Propulsion Engineer, 3 yrs.) AiResearch, Garrett Corp., Phoenix, Ariz. (Senior Eng. Specialist, 9 yrs.) Ingersoll-Rand Research Center, Princeton, NJ (Assistant Director and Department Head, 23 yrs.) Aerospace/Industrial Consultant and R & D Contractor: Jules L. Dussourd & Associates, President (2 yrs.) ASME Fellow, Past Chairman of ASME Fluids Eng. Division. Recipient of Institution of Mechanical Engineers' C. Stephenson Award. 22 Journal publications, 8 patents. Born in New York City, married, two children. Hobbies: tennis, camping, lumberjack, do-it-yourself. *Society Aff:* ASME, SAE

DuTemple, Octave J
Business: 555 N Kensington Ave, LaGrange Park, IL 60525
Position: Exec Dir. *Employer:* American Nuclear Soc. *Education:* MBA/Business/Northwestern; MS/Chem Eng/MI Tech Univ; BS/Chem Eng/MI Tech Univ *Born:* 12/-/20. MS MI Tech Univ. MBA Northwestern Univ. 1958 to present: Exec Dir. Am Nucl Soc. 1949-58: Argonne Natl Lab - Chem Engr. R & D in Chem Processing. 1941-48: Flying instr. Served one yr active duty with US Army Air Corp. 1961- Edited and published book on "Prehistoric Copper Mining in the

DuTemple, Octave J (Continued)
Lake Superior Region–. Other interests: farming, archeology, geology, economics. Distinguished Service Award 1978 ANS, ASAE Management Award-1972, Octave J. Du Temple Award-1983 ANS; Michigan Tech Univ. Board of Control Silver Medal for Outstanding Career Achievements-1987. Bd Mbr 1972-77 CESSE, Pres 1975-1976 CESSE, Exec Dir, American Nuclear Society 1958-Present. *Society Aff:* ACS, WSE, ISPE, AIChE, AAAS, ASAE, CESSE.

Dutta, Subhash
Business: Div of Chem Engg, 323 High St, Newark, NJ 07102
Position: Asst Prof. *Employer:* NJ Inst of Tech. *Education:* PhD/Appl Chem/Indian Inst of Tech; Post-Grad Dipl/CE/Tokyo Inst of Tech; MS/CE/Calcutta Univ. *Born:* 1/8/44. Native of Calcutta, India. Recipient of Silver Medal from Calcutta Univ in 1965, UNESCO Scholarship from the Govt of Japan in 1970 and Natl Scholarship from the Govt of India in 1973. Taught at the Indian Inst of Tech, New Delhi, before joining the WV Univ, Morgantown, in 1973, as a post-doctorate. Assumed the current position of Asst Prof of the Div of Chem Engg at NJIT in 1977. Author of more than a dozen res publications and reports in the fields of Coal Pyrolysis, Combustion and Gasification reactions, Non-catalytic Heterogeneous Reaction Kinetics and Fluidization, and a book-chapter in a grad text on Coal Conversion Tech. *Society Aff:* AIChE.

Dutta, Sunil
Home: 3996 Tennyson Lane, North Olmsted, OH 44070
Position: Materials Engr *Employer:* NASA Lewis Research Center. *Education:* PhD/Ceramics/Sheffield Univ; MS/Ceramics/Sheffield Univ; MS/Applied Chem/Calcutta Univ, IN; MBA/Finance/Babson Coll *Born:* 11/2/37. Post-Doctoral Fellow Lehigh Univ. Sr Res Ceramic Engr US Army Materials Res Ctr 1968-76. Wr matls Engrg with NASA-Lewis Res Ctr since 1976. Respon for R&D programs & projs on ceramic materials dev for gas turbine, diesel engine & sturling engine and other structural applications. Fellow ACS & Inst of Ceramics. Full Mbr: NICE, CEC, & Sigma Xi. Serves as session Chrmn, program chmn for various basic sci, DOD, & ACS meetings. Res areas: processing, characterization, high temperature properties of high performance ceramics for structural applications. *Society Aff:* AMER CER SOC, NICE, INST OF CER, ΣΞ, CEC.

Duttenhoeffer, Richard
Home: 8 Willow La, Kings Park, NY 11754
Position: Sr VP and Partner *Employer:* Parsons Brinckerhoff Quade and Douglas, Inc *Education:* B/CE/NYU *Born:* 1/26/28 Native of Kings Park, Long Island, NY. Served in the US Navy between 1946 and 1948. Attended NYU where he received his B of CE in 1955. He is a registered PEin NY, NJ, CT, VA, OH, TX, FL, DE, and MA. With Babcock and Wilcox Co from 1948 to 1949. Joined Parsons Brinckerhoff Quade & Douglas, Inc in 1949 and became the technical dir of hgwy and civil engrg and also serves as mgr for the firm's Atlantic Region. He is also a Dir on the Bd of several of Parsons Brinckerhoff's companies and a Senior VP of Parsons Brinckerhoff Quade and Douglas, Inc as well as a Partner in Parsons Brinckerhoff Quade and Douglas (a partnership). He enjoys golf, fishing, gardening and music. *Society Aff:* ARTBA, ASCE, SAME, IBTTA, IRF

Dutton, Granville
Home: 9810 Ridgehaven Dr, Dallas, TX 75238
Position: Senior Vice President *Education:* BS/Engg/US Naval Acad; LLB/Law/Southern Methodist Univ. *Born:* 10/23/24. Served in US Navy 1941-48. Currently-Sr.VP-Sabio Oil & Gas, Inc. Initially employed by Sun as a Petroleum Engr in 1949 & served as Conservation Engr & Attorney 1957-70 & Mgr of Joint Operations, and VP of Canyon Reef Carriers, Inc. 1970-78 and Dir Govt Relations Sun Co, Inc. 1978-81. VP & Mbr of Bd/Amer Inst of Mining, Metallurgical, & Petroleum Engrs for whom he also serves as Chmn of the Governmental, Energy & Mineral Affairs Ctte. Formerly on the Bd/ Soc of Petroleum Engrs for whom he served as a Distin Lecturer & Chmn of the Technical Info Ctte. Registered Professional Engineer in Texas. A trustee of the Reformed Theological Seminary & a Ruling Elder in the Presbyterian Church, US. Married to the former Carol Sullivan of Navasota Texas & the father of 3 sons *Society Aff:* SPE-AIME, AAPL.

Dutton, Roger W
Business: 1500 Meadowlake Pkwy, Kansas City, MO 64114
Position: Partner *Employer:* Black & Veatch Consulting Engr *Education:* BS/ME/KS State Univ *Born:* 4/19/33 A native of Parsons, KS. Joined Black & Veatch Consulting Engrs in 1956. Served in the US Air Force 1956-1959 as Aircraft Maintenance Officer for the 524th Fighter Squadron. Joined the partnership of Black & Veatch in 1977. Serves as project mgr on major electric generating stations. *Society Aff:* NSPE

Duttweiler, David W
Business: College Station Rd, Athens, GA 30613
Position: Lab Dir. *Employer:* US EPA Environmental Res Lab. *Education:* PhD/Sanitary Engg/Johns Hopkins Univ; MSE/Sanitary Engg/Johns Hopkins Univ; BSE/Civ Engg/Univ of MI. *Born:* 9/15/27. Structural Engr, T H McKaig Consulting Engrs, NY, 1948-49; Sanitary Engr Officer (2nd Lt to Lt Col), US Army Medical Service Corps, 1949-69; Dir, Environmental Res Lab, US Environmental Protection Agency, 1969-. Reg PE, TX, 1953; Diplomate, Am Acad Environmental Engrs, 1966-; Res ssoc, Univ of GA, 1969- ; Adjunct Prof, Clemson Univ, 1971-. Mbr, Commission on Environmental Health, Armed Forces Epidemiology Bd, 1970-74; Univ of FL Dept of Environmental Engg Sciences, Visiting Committee; Univ of NC Water Resources Res Inst Advisory Committee; Athens Rotary Club. Profl interests: math models in sanitary engg; res mgt; water quality control; heat exchange in natural waters; environmental health. *Society Aff:* ASCE, WPCF, AAAS, IAWPR.

Duttweiler, Russell E
Home: 3220 South Cove Court, Haineville, OH 45039
Position: Mgr Technology Integration *Employer:* GE-Aircraft Engines *Education:* BS/Met Engg/Univ of IL *Born:* 5/12/38. in Canton, IL. s Lester Edwin and Gladys Juanita Duttweiler; m Sherry Lee Rector, Dec 24, 1957; children-Mark, Jeff. BS Met Engrg Univ of IL 1960. With Gen Elec Co, 1960- , mgr matls engg, aircraft engine grp, Cin, 1973-75, Mgr, Mfg, Tech Labs, Cin, 1976-77, purchased matls quality, aircraft engine grp, Cin, 1977-80. Mem Am Soc Metals (Elected Fellow 1979), Alpha Sigma Mu. Republican. Roman Catholic. Mgr Matls Dev 1981-84, Mgr Technology Integration 1985- . *Society Aff:* ASM-International

Duwel, Paul F
Business: 4747 Spring Grove Ave, Cincinnati, OH 45232
Position: Superintendent of Commercial Div. *Employer:* Cincinnati Water Works. *Born:* Feb 5, 1930 Harrison Ohio. Educated at Elder High Sch (Cincinnati) & the University of Cincinnati. Served with the Army Medical Corps 1943-46, Tech 5 Sgt for the 325th Station Hospital. Started with the Cincinnati Water Works as a Junior Account Clerk in 1946. Promoted through 7 different titles to present title on May 1, 1976. Mbr of Amer Water Works Assn since 1962. Served as Sec/Treas Ohio Section AWWA from 1964-68; served as Trustee, Ohio Section, AWWA from 1968-72; served as Chmn of the Ohio Section, AWWA 1972-73. Presently serving as Sec/Treas, Southwest District Ohio Section AWWA. Married to Lee for 30 yrs, with 3 children. *Society Aff:* AWWA.

Duwez, Pol E
Business: Pasadena, CA 91125
Position: Prof of Applied Physics etc. *Employer:* California Institute of Technology. *Education:* DSc/Physics/Univ of Brussels Belgium; -/Metallurgical Engg/Sch of Mines, Mons Belgium. *Born:* Dec 1907, Mons, Belgium. Metall Engr Mons, Belgium 1932, DSc University Brussels, Belgium 1933. Res Fellow Calif Inst Tech 1933-35. Res Engr Calif Inst Tech 1941-45, Jet Propulsion Lab 1945-54, Assoc Prof Matls Sci 1947-52, Prof Appl Physics & Materials Science 1952- . Teaching & Res in materials science, structure & properties of alloys, glassy alloys, magnetic & superconducting alloys. Mbr Nat Acad of Sci, Natl Acad of Engg, Am Acad of Arts & Sciences, Am Physical Soc, Fellow Metall Soc AIME, AAAS, Am Soc Metals, Am Ceramic Soc, French Civil Engrg Awards; Charles B Dudley ASTM 1951, Mathewson Gold Medal AIME 1964, F J Clamer Medal Franklin Inst 1968, Albert Sauveur Achievement Award ASM 1973, Prix Conez Belgium 1973, P Lebeau Medal, France 1974. American Physical Soc Intl Prize for New Materials, 1980. Emil Heyn Medal of the Deutsche Gesellschaft fur Metallkunde, 1981. W. Hume-Rothery Award Metallurgical Soc of AIME, 1982. *Society Aff:* NAS, NAE, AAAS, APS, AIME, ACS, ASM, ISF.

Dwinnell, James H
Home: 12424 Holmes Pt Dr NE, Kirkland, WA 98034 *Education:* MS/Aeronautics/Caltech; BS/AE/U of WA. *Born:* 6/20/15. Assoc Prof of Aeronautical Engrg Univ of WA 1941-50. Author of "Principles of Aerodynamics" McGraw-Hill 1949. Respon for much of Boeing tech dev of B52-G, Minuteman, Gas Turbines to 1966. Respon for Boeing Commercial Airplane Co technology res, labs & computing during following 12 yrs. Retired Jan 1981. AIAA Assoc Fellow & Past P NW Sect V Chmn, Sec & Council Mbr as well as mbr of natl comms.

Dwivedi, Surendra Nath
Business: Mech Engg Tech, Norfolk, VA 23508
Position: Asst Prof. *Employer:* University of Wisconsin, Milwaukee *Education:* PhD/Manufacturing Engg/Birla Inst of Tech, India; MASc/Mech Engg/Univ of British Columbia, Canada; Master of Engg/Machine Design/Univ of Roorkee, India; BSc/Engg/Banaras Hindu Univ, India. *Born:* 7/15/45. Native of India. Taught in many reputed insts of India and Canada from 1971-78. Worked as res engg in IIT Res Inst Chicago, 1978-79. 1979-81 Asst Prof in Old Dominion Univ. Presently Asst. Prof., University of Wisconsin, Milwaukee. Published 14 res papers in the field of manufacturing design and production. Invented a new method of dynamic visio-plasticity for study of metal working processes. Discovered the latest concept of frictional conditions during high speed deformation of metals. Worked for developing a new method of internal shear forging. Guided many engg proj. Enjoy country music and religious study. Received outstanding faculty award in 1980 and 1981 at Old Dominion University. Chaired many international conferences. Committee member of Robotics International. *Society Aff:* ASME, SME, ASM, ASEE.

Dwon, Larry
Home: 103 Hudson Hills Rd, Rt 4, Box 103, Pittsboro, NC 27312
Position: Consultant for Energy Industries *Employer:* Amer Elec Power Service Corp - Retired. *Education:* EE/Elec Engg/Cornell Univ; MBA/Mgmt/NYU Grad Sch of Bus. *Born:* 5/2/13. PE NY State 1941. Taught evenings at Prat Inst of Tech & Polytechnic Inst of Brooklyn; designed motors Diehl Mfg Co; lighting design Holophane Lighting Inc; dev & evaluated communication countermeasurers with Office of Sci & res Dev; 40 yrs engg & mgmt at AEP Service Corp. Lond Assn with engg education, accreditation, guidance & manpower matters. Numerous published papers & articles. Chmnsip ctte work with IEEE, ECPD, EEI, IES, ASEE & Eta Kappa Nu, (Natl Pres 1958-59). Publ History of Eta Kappa Nu in 1976, Official Historian, Dis Service Awards: Eta Kappa Nu 1976. Power Engg Educ Comm (IEEE) 1977, Edison Elec Inst 1978. IEEE Fellow 1968, Life Mbr 1978. Fred Plummer Lectr 1976 AWS. Special interest: Consulting career dev, proper utilization tech human resources; energy audits for industry, teaching seminars for North Carolina State University, good music. *Society Aff:* IEEE, CSE.

Dworsky, Leonard B
Home: 8 Winthrop Pl, Ithaca, NY 14850
Position: Prof. *Employer:* Cornell Univ. *Education:* MA/Public Admin/Am Univ; BS/Civ Engg/Univ of MI. *Born:* 1/5/15. Native of Chicago, IL. IL Div Sanitary Engg 1936-1940; Lt Col. AUS 1941-46; Engr Dir US Public Health Service, Water Pollution Control Prog 1946-1964; Mbr, Chrmn Columbia Basin Interagency Committee 1958-62; Prof Civ Engg Cornell Univ 1964-to date; Water Resources Staff Asst to Pres Sci Adviser, Exec office of the Pres 1967-1968; Environmental Consultant Rockefeller Fdn 1973-1976; Dir Cornell Univ WRMSC 1964-1974. Mbr ASCE Natl Water Policy Committee 1968-1972; and Natl Energy Policy Comm 1973-1977; Mbr Sci Advisory Bd. Intl Joint Commission 1972- 1977; Diplomate EEJC; Reg PE in IL. Chrmn, Federal Ctte on Water Resources Res, Federal Coun on Sci and Tech, 1967-68; Mbr Pres Sci Advisory Ctte, Panel on Environ 1969-72; Natl Acad of Sci EW Studies B6 Panels 1970-74; Author Vol 1 Conservation in the US Water and Air Pollution, 1970; Unified Riv Basin Mgmt 1979-81. *Society Aff:* ASCE, XE.

Dybczak, Zbigniew W (Paul)
Home: 129 Arrowhead Dr, Montgomery, AL 36117
Position: Prof *Employer:* Tuskegee Inst *Education:* PhD/ME/Univ of Toronto, Canada; BSc/ME/Univ of London, England *Born:* 6/27/24 Native of Poland, served with Polish Air Force in England during WW II. Thirty years in engrg education in England, Canada and US; twenty one years as Dean of Engrg at Tuskegee Inst. Initiated graduate education, dual-degree programs, cooperative education and new facilities. Responsible for numerous initiatives to recruit and retain more minorities in engrg. Member or officer of many committees, task forces and consortia. ASME Centenial Medal, ASEE Vincent Bendix Minorities in Engrg Award. Hobbies: Golf, Tennis, Music and Travel. *Society Aff:* ASME, ASEE.

Dyer, Garvin H
Home: 11715 Markham Rd, Independence, MO 64052
Position: Retired *Education:* Prof/CE/Univ of MO at Rolla. *Born:* 1905 Ash Grove MO. Professional Civil Engrg Degree University of Missouri at Rolla. 1923-47 Engr, Asst Superintendent & Sec Springfield City Water Co. 1947-71 VP, Mgr, Ch Engr, Dir Missouri Water Co (Retired 1971). Was in charge of planning, rebuilding, operation & maintenance of water utility now serving approx 225,000 people. Hon Mbr Amer Water Works Assn (Fuller Award 1958, Mgmt Div Award 1964 & 65); Mbr: Missouri Soc of Prof Engrs (Pres 1953-54), NSPE (Pres 1957-58), ASCE (Fellow); Advisory Dir of First Natl Bank of Independence MO. *Society Aff:* ASCE, NSPE, AWWA.

Dyer, Harry B
Business: 2015 Spring Rd, Suite 550, Commerce Plaza, Oak Brook, IL 60521
Position: Regional Mgr Tech Service. *Employer:* Republic Steel Corp. *Education:* BS/Met Engg/Univ of KY. *Born:* 4/7/21. Native of Dekoven, KY. Received BS in Met Engg at Univ of KY in 1943. After grad, worked on Low Temperature Impact Res NRC for one yr. Since 1944, employed by Republic Steel in various locations and capacities as follows: S Chicago Testman, Met, Technical Service Rep, Dir of Process Control, Asst Chief Met, Chief Met. Union Drawn Div Republic Steel, Massillon, OH - Asst Chief Met. Oakbrook-Regional Mgr Technical Service. Enjoy all sports and classical music. *Society Aff:* ASM, AIME.

Dyer, Ira
Business: Dept of Ocean Engrg, Cambridge, MA 02139
Position: Prof & Hd of Dept Ocean Engrg. *Employer:* Mass Inst of Tech. *Education:* PhD/Physics/MIT; SM/Physics/MIT; SB/Physics/MIT. *Born:* Jun 1925. SB, SM, PhD from MIT. Native of NY City. Served with Army Air Force 1944-45. With Bolt, Beranek & Newman Inc 1951-70: served as VP, Dir of Physical Sciences Div, Dir of its Program for Advanced Studies & as Pres of its subsidiary General Oceanology Inc. At MIT since 1970 having served as Dir of MIT Sea Grant Program and Head of Dept Ocean Engrg, as well as in the above listed category. Res specialization in underwater acoustics & sonar engrg with genl res interests in marine resources & offshore engrg. Elected to Natl Academy of Engrg 1976. Chmn Res Advisory Ctte U S Coast Guard 1975. VP Acoustical Soc of Amer 1973-74; Fellow & Biennial Award Winner ASA, Fellow IEEE and AAAS. Meritorious Public Service Award, US Coast Guard 1979. Distinguished Technical Contribution Award 1982 IEEE. Visiting Fellow Emanuel College & Visitor Engg Dept, Cambridge Univ 1979-80. Addicted to sailing. *Society Aff:* ASA, IEEE, AGU, AAAS.

Dyer, Thomas K
Home: 13 Demar Rd, Lexington, MA 02173 *Education:* BS/Civil Engg/MA Inst of Technology. *Born:* Jan 1922 Medford Massachusetts. Served as Navy Pilot 1943-45. Engrg Dept of Boston & Maine Railroad 1946-63, Ch Engr 1960-63. Organized Thomas K Dyer Inc Cons Engrs 1963, specializing in studies & design of rail transportation systems, Pres & Tres 1963-86. Pres & Tres Kencomp Inc, computer software subsidy incorporated in 1970. Presently is a rail transportation consultant. Registered Prof Engr Massachusetts, New Hampshire, Vermont, Connecticut. Mbr: ASCE, AREA, ACI, AWS, BSCE, NSPE. *Society Aff:* ASCE, AREA, ACI, AWS, BSCE, NSPE.

Dykes, Glenn M, Jr
Business: 2600 Blair Stone Rd, Tallahassee, FL 32312
Position: Professional Engr. Administrator Div. of Operations *Employer:* FL Dept of Environmental Regulation. *Education:* MSE/Sanitary Engg/Univ of FL; BCE/Civ (Sanitary) Engg/Univ of FL. *Born:* 3/25/31. Native of Tampa, FL. Served in Army Corps of Engrs 1954-56, presently a COL. in Retired Reserve. Employed by State of FL, as Sanitary Engr in drinking water program in 1957. Elevated to Administrator of FL's Drinking Water Prog in 1972 & served in that position with Dept of Environmental Regulation until 1986. Presently Engr Administrator overviewing restoration of state's contaminated drinking water supplies. Served as Pres Jacksonville Chapt of Engg Soc, Chrmn Engrs in Govt, VP Fla Engrg Soc, Chrmn & Natl Dir of FL Sec AWWA. Recipient of AWWA's Geo Warren Fuller & Ambassador Awards. *Society Aff:* NSPE, AWWA, FES

Dy Liacco, Tomas E
Home: 651 Radfod Dr, Highland Heights, OH 44143
Position: Pres *Employer:* Dy Liacco Corp *Born:* Nov 12 1920 Naga Philippines. BSEE, BSME University of the Philippines. MSEE Illinois Inst of Tech. PhD Case Western Reserve Univ. Intl consultant in realtime computer control of electric power systems. A Fellow of IEEE & Mbr of sev IEEE & indus Cttes. Pub many papers for IEEE & internatl conference proceeedings. Received IEEE Power Group Prize 1967 for 'The Adaptive Reliability Control Syst'. Married Nijole Abraitis of Marijampole Lithuania. Have 3 sons, 1 daughter.

Dym, Clive L
Business: Dept of Civ Engg, Univ. of Massachusetts, Amherst, MA 01003
Position: Prof of Civil Engg. *Employer:* Univ of MA. *Education:* BCE/Civ Engg/Cooper Union; MS/Applied Mechanics/Poly Inst of Brooklyn; PhD/Aeronautics and Astronautics/Stanford Univ. *Born:* 7/15/42. in Leeds, England. Naturalized citizen US. Asst Prof, State Univ of NY at Buffalo, 1966-1969. Res Staff, Inst for Defense Analyses, Arlington, VA, 1969- 1970. Assoc Prof of Civ Engg, Carnegie-Mellon Univ, 1970-1974. Sr Scientist, Bolt Beranek and Newman, Cambridge, MA, 1974-1977. Assumed current position Aug 1977. Visiting positions at Technion-Israel Inst of Tech (1971), Inst for Sound and Vibration Res at Univ of Southampton (1973), and Xerox PARC and Stanford Univ (1983-84). Founding Editor of Artificial Intelligence for Engg Design, Analysis and Mftg. Awarded NATO Sr Fellowship in Sci (1973), Fulbright-Hays Grant (1977, declined), ASCE Walter L Huber Research Prize (1980), ASEE Western Electric Fund Award (1983). Reg in four states, author/co-author of seven books, more than 60 res articles and 30 tech reports. *Society Aff:* ASA, ASCE, ASME, AAAS, AAAI.

Dyment, John T
Home: 4005 Bayview Ave, Apt 715, Willowdale Ontario M2M 3Z9
Position: Self-employed *Education:* LLD/-/Univ of Toronto; BASc/ME/Univ of Toronto. *Born:* 11/23/04. As a stress analyst & performance predictor with Dept Natl Defense for eight yrs, as Chief Engr of Air Canada for its first 30 yrs, as an aviation consultant for eight yrs, have been involved in all aspects of Civ Aviation such as aircraft selection, modification, maintenance, operations & economics of airtransports, airwothiness requirements; airport site selecting & requirements; troubleshooter of troubled airlines; R&D requirements for industry; fuel conservation analyist; on numerous intl aviation committees and recipient of many honors. *Society Aff:* SAE, CASI, EIC, RAeS.

Dysart, Benjamin C, III
Business: 401 Rhodes Research Center, Clemson, SC 29634-0919
Position: Prof of Envir Engrg *Employer:* Clemson Univ *Education:* PhD/CE/GA Tech; MS/Sanitary Engrg/Vanderbilt Univ; BEngrg/CE/Vanderbilt Univ *Born:* 02/12/40 Native of Columbia, TN. Staff engrg for 3 yrs with Union Carbide Corp. Joined Clemson faculty in 1968, teaching grad course work and conducting water quality, nonpoint-source pollution, and energy-related research since then. Dir, Water Resources Engrg grad prog, 1970-75. Consultant to numerous federal agencies and corporations. Served as Scientific Advisor for Civil Works, Office of Secy of Army 1975-76. Pres, AEEP, 1981. Mbr, OCS Advisory Bd, US Dept of Interior, 1979-81. Mbr, Science Advis Bd, US Environmental Protection Agency, 1983-date. Pres & Chrmn of Bd VP, Natl Wildlife Federation, 1983-85. Trustee, Rene Dubos Center, 1985-date. Sr. Fellow Conservation Fund, 1985-date. *Society Aff:* AGU, ASCE, AEEP, WPCF, XE, ΣΞ, ΦΚΦ, ΩΡ

Dziurman, Theodore D
Business: 34400 Glendale Ave, Livonia, MI 48150
Position: Prin & VP *Employer:* Soil and Matls Engrs *Education:* MSIE/Indust Engrg/Wayne State Univ; BCE/Civil Engrg/Univ of Detroit *Born:* 02/10/39 A registered professional engr in MI and OH. Joined Soil and Matls Engrs, Inc, in 1981 where assumed current respons of Prin and VPres. Am respons for SME's mktg prog and proj mgmt of pavement and quality control mgmt systems. Prior to SME, served as Dir of Admin Services and Sr Assoc for Neyer, Tiseo & Hindo, LTD, from 1977 to 1980 and Dir of Facilities Engrg for County of Oakland/MI from 1963 to 1977. Currently hold the positions of National Director of MI Soc of Prof Engrs and Natl Treasurer of Natl Academy of Forensic Engrs. *Society Aff:* ASCE, NSPE, CSI, NAFE

Eachus, Joseph J
Home: 4935 Stevens Dr, Sarasota, FL 33580
Position: EDP Consultant *Employer:* Self-employed. *Education:* PhD/Math/Univ of IL; MA/Math/Syracuse Univ; AB/Math, Physics/Miami Univ (OH). *Born:* Nov 1911 Anderson, Ind. Fellow AAAS & IEEE. USNR WWII, civilian DOD 1948-55. With Honeywell Info Sys & processores employee R&D capacitites 1955-76. Approx 20 pats computer related electronics. Retired from Honeywell end 1976 to become employee of Raytheon Co. Became an independent consultant in 1982 *Society Aff:* AMS, AAAS, IEEE, ACM.

Eagan, Constantine J
Business: PO Box 160187, San Antonio, TX 78280-2487
Position: VP *Employer:* ERT Lighting & Sales, Inc. *Education:* BSEE Univ. of Illinois (1949) *Born:* Mar 3, 1926 Joliet, IL. With IL Power Co 1950-72. Directed co-wide tech serv for sales dept. Taught courses in lighting and HVAC. Was involved in pioneering designs of infrared comfort heating. Also involved in designing sys for recovery & utilization of heat generated by lighting sys. Joined Wide-Lite in 1973. Managed for 9 years, Application Engg Dept Regional Manager 1 1/2 years, Export Director for 3 years. As Export Director, Directed Mfr. by Lic, Export of Co. products and conducted technical seminars in N. & S. america, Europe, Asia and Australia. Since January, 1987 V. Pres of ERT Lighting, Mfrs, Agents in San Antonio, Tx. Mbr IES since 1948. Midwest Regional VP, Soc Dir. Has served on sev natl ctts incl: Office Lighting, Sports Lighting, Regional Conferences, Programs, & Disting Serv Award. Also served on Sports Lighting Committee of C.I.E. (Int'l Commission on Illumination). *Society Aff:* IES.

Eagan, James R
Home: 30485 Woodgate, Southfield, MI 48076
Position: Sales Director *Employer:* Panax Trading Co Ltd *Education:* MBA/Bus Adm/Wayne State Univ; BS/Met Engg/MI State Univ; BA/English Lit/Wayne State

Eagan, James R (Continued)
Univ. *Born:* 7/14/24. Live in Southfield, MI. Served in AF Weather Service in WWII. Employed by Ryerson Steel, Eaton Mfg Co & Teer Wickwire prior to working in Chevrolet Engg Mtls Group. Based on work at this time was selected to be a Fellow in ASM for maj contributions to automotive mtls tech. Moved to Inco in 1963, becoming Dist Mgr & in 1974, the Regional Mgr in Charge of Automotive Market Dev. In 1976 put in charge of marketing new Inco-developed plateable plastic through MPD Tech. Retired in 1981. Worked four years as self-employed manufacturers' Rep selling to the automotive indus. Currently Sales Director for Panax Trading Co, Ltd. Active in SAE. Served as Chrmn Publ Ctte, and 3 yrs on SAE Bd of Dir. Reg PE, State of MI. In 1975, selected for Detroit Chapter ASM John Shoemaker Award of Distinction as Outstanding Exec Metallurgist. Enjoy golf & fishing. *Society Aff:* ASM, SAE.

Eager, George S, Jr
14 Bellegrove Dr, Upper Montclair, NJ 07043
Position: President *Employer:* GRJ Consulting Services Inc *Education:* BEng/Electrical/Johns Hopkins Univ; DrEng/Electical/Johns Hopkins Univ. *Born:* 9/5/15. Native of Baltimore, MD. Served with Army Signal Corps 1941-46. Res Engr, Asst Dir Res, Dir of Res & Assoc Dir of Res for Genl Cable Corp Res Ctr. Pres, GRJ Corp. President Cable Technology Laboratories, Inc. P Secy, V Chmn, Chmn CEIDP, Natl Res Council; P Secy, V Chmn, Chmn ICC, IEEE. Fellow IEEE, Play golf. *Society Aff:* IEEE, CEIDP.

Eager, Walter J
Business: 3550 NW Glenridge, Corvallis, OR 97330
Position: Principal Engr. *Employer:* Eager & Assoc. *Born:* Apr 1931. MS from Naval Post Grad Sch; BVS from Caltech. Designed amphibious vehicles for Ford Motor Co. Designed process equip. Served USN 1954- 74. Developed transportation & maintenance progs. Start-up engr on nucl-elec power plant. Performed and taught Nucl Defense Engg. Designed, developed ocean facilities & construction systems. Originated, developed, directed Navy's Ocean Facilities Engg & Construction Prog. Awarded Meritorious Service Commendation & Legion of Merit medals. ASNE Gold Medal 1971. Private Cns Engr. Principal consultant to maj deep ocean mining group on mineral collection equip, collector deployment systems & deployment operations planning. Reg Engr in OR & CA.

Earl, Christopher B
Business: 6075 S Florida Ave, (P O Box 5260), Lakeland, FL 33803
Position: VP *Employer:* Bearden-Potter Corp *Education:* HNC/ChE/Borough Polytech, London *Born:* 12/10/37 London, England. Lead Process Engineer 1962-65, Lummus Co., Ltd., U.K. Process Mgr with: Welllman-Lord (now Davy-McKee Corp.) 1965-72, with specialization in fertilizer plant design. Patended and commercialized Davy- Mckee's flue gas desulfurization tech. Dir Business Development Davy-McKee 1972- 79. Promoted to VP 1980. Joined Bearden-Potter Corporation as VP in January 1981. PE-Chemical Engr-State of FL. AIChE "Service to Soceity Award" receipient 1980. *Society Aff:* AIChE

Earle, Emily A
Home: 1846 Stonehenge Dr, Birmingham, AL 35215
Position: Proj Engr. *Employer:* Square D Co. *Education:* BS/Mech Engg/Auburn Univ. *Born:* 11/02/55. Cooperative Education Student 1974-76. NASA-MSFC in Data Acquistion, reduction, and solar res. Summer, 1977, employee GTE-AE Huntsville, AL. Reduced waste in telephone receiver assembly by 50%. Summer, 1978, Employee IBM-OPD Lexington, KY. Modified print module in electronic typewriter eliminating customer complaints concerning printing voids. Employed by profs, 1975-76, 1978- 79 in nuclear and solar res. Grad June, 1979. Employed by Square D Co, Leeds, AL as Proj Engr in elec transmission equip design. Secy ASME 1976-1978, Pres ASME 1978-79, 1st Natl C T Main Leadership Award 1978. Hobbies: reading, horseback riding, dancing. *Society Aff:* ASME.

Earley, Duane E
Home: 7700 Essington Cir, Dayton, OH 45459
Position: Pres. *Employer:* Science Unlimited, Inc. *Education:* BS/Chem & Chem Eng/Wright State Univ. *Born:* Sept 1930. Baccalaureate & grad studies Univ of Dayton & Wright St Univ. Over 45 publications in thermodynamics, x-ray diffraction & energy. Consultant on waste disposal processes. Designed & built incinerator for disposal of toxic and hazardous mtls. Currently involved in Environmental & solar energy areas. Profl Artist. Hobbies-woodworking & organic gardening. Cofounder & co-chrmn of a series of Natl Confs on Energy & the Environment. Founding Pres & Chrmn of Bd, Engrg Sci Hall of Fame. *Society Aff:* AIChE.

Earlougher, R Charles
Business: 6600 S Yale, Bldg 2, Tulsa, OK 74136
Position: Chrmn, Petroleum Consult Division *Employer:* Williams Brothers Eng Co. *Education:* Petrol Engr/-/CO Sch of Mines. *Born:* 5/6/14. In charge of Secondary Oil Recovery Prod Lab, Sloan & Zook, Bradford, PA, 1936-38. Owned & operated core & water analysis lab, Tulsa, OK, & consultig eng service with emphasis on water injection for increased oil recovery, 1938-70. Responsible for evaluation of many oil reservoirs for adaptability to water injection, including design & supervision of installation & operation of numerous projs, domestic & foreign. Sold Earlougher Eng, Inc 1970-75 to D E Godsey. Joined Williams Brothers Engrg Co 1976. Recipient of Distinguished Service Award by CO Sch of Mines, by SPE of AIME, & by Mid-Continent Dist API. Dir 1967-68 & VP 1968-69 of AIME. Distinguished Lecturer 1967-68, SPE. Recipient of AIME's Anthony F Lucas Gold Medal, Feb 1980; Honorary Mbr AIME & SPE 1984. *Society Aff:* AIME, SPE, OSPE, SPEE, NSPE.

Earlougher, Robert C, Jr
Business: P.O. Box 120, Casper, WY 82602
Position: Division Reservoir Engr. *Employer:* Marathon Oil Company *Education:* PhD/Petro Engrg/Stanford Univ; MS/Petro Engrg/Stanford Univ; BS/Petro Engrg/Stanford Univ *Born:* 6/26/41 1966/72-Research Engr, Marathon Oil Company, Littleton, CO. Research and development of thermal recovery methods, reservoir engineering, well simulation, and transient well testing. 1972- Senior Consultant, Scientific Software Corportion. Application of reservior simulation to petroleum reservoirs. 1973/77-Senior Research Engr, Marathon Oil. Reesearch and development in fields of oil recovery, transient well testing, and simulation applications. 1977/81-Engineering Department Mgr, Marathon Oil Company Research Center. 1981/Casper Division Reservoir Engr, Marathon Oil Company. Author of SPE Monograph, Advances in Well Test Analysis. *Society Aff:* SPE/AIME

Early, David D
Business: PO Box 3707, Mail Stop 5E-10, Seattle, WA 98124
Position: Mgr. *Employer:* Boeing Co. *Education:* MS/Mtls Sci/Vanderbilt Univ; BE/Mech/Vanderbilt Univ. *Born:* 6/18/43. Kingsport, TN native. Undergrad Natl Sci Fdn fellow; cum laude grad. Fabrication R&D with Dow Chem, developing production criteria for nuclear mtls. Ten yrs with the Boeing co including aircraft mtls & process design consultation; met lab support for mtls characterization; resident met at remote facility. Assumed current Boeing resonsibility as mgr of subcontractor process surveillance in 1978; serves as corporate focus for this activity. Reg WA State PE. Technical & admin contributor to the Am Soc for Metals, local & natl. Participates in local community affairs; enjoys mtn sports. *Society Aff:* ASM.

Early, James M
Home: 740 Center Dr, Palo Alto, CA 94301
Position: Consultant. *Employer:* Self. *Education:* PhD/EE/OH State Univ; MS/EE/OH State Univ; BS/Pulp & Paper Mfg/New York State College of Forestry. *Born:* 7/25/22. Married to Mary Agnes Valentine, Dec 28, 1948; children: Mary Elizabeth, Kathleen R, Joan T, Rhoda A, Maureen M, Rosemary, James M, Margo. Res assoc, Instr, OH State Univ, Columbus, OH, 1946-51; mbr technical staff, Bell Tel Labs, Murray Hill, NJ, & Allentown, PA, 1951-69; supr, 1953-57, dept hd, 1957-62, lab dir, 1962-69; vp res & dev div, Fairchild Camera & Instrument Corp 1969-1971.

Early, James M (Continued)
Dir of VLSI Res 1972-1983, Scientific Advr 1983-1986. Recipient Texnikoi Outstanding Alumnus award, OH State Univ, 1967. Fellow IEE: J.J. Ebers Award of IEEE Electron Device Society, 1979. Distinguished Alumnus Award, OH State Univ, 1985. *Society Aff:* IEEE, ECS, APS, IPA.

Earnest, Samuel A
Business: 1501 Alcoa Building, Pittsburgh, PA 15219
Position: Sales Mgr. *Employer:* Aluminum Co of America. *Education:* BS/Met Eng/Case Inst; MBA/Bus/Western Reserve Univ. *Born:* Oct 1928. Native of Cleveland area but have also resided in N J. Currently Sales Manager-Technology Marketing for Alcon Located at Company headquarters in Pittsburgh. *Society Aff:* ASM, AIME

Eason, Glenn A
Business: Box 3337, Durham, NC 27702
Position: Structural Engr (Self-employed) *Education:* MSCE/Structures/NCSU; BS/CE/NCSU *Born:* 11/27/30 Smithfield, NC. Graduated NCSU 1953. Member Tau Beta Pi and Chi Epsilon. Served with 13th Engr Combat Batallion in Korea 1953-55 as platoon leader and as engrg intelligence officer. Awarded commendation medal. Taught engrg mechanics NCSU 1955-58. Graduate work included study in plastic design of steel. Employed as structural engr with various consulting engrg firms until 1971. Formed Glenn A Eason, Consulting Engrs 1971. Work has included design of bldgs and various structures throughout NC and in other states. Past pres NC Section, ASCE. Officer CEC/NC. Registered engr NC, SC, VA, FL, GA. *Society Aff:* ASCE, ACEC, ACI, PCI

Easterday, James R
Business: 12890 Westwood Ave, Detroit, MI 48223
Position: Mbr - Application Engr. *Employer:* Kolene Corp. *Born:* June 23, 1926. BS & MS Wayne St Univ, Met Engg. Engg Supr, Mtls Evaluation Group, Chrysler Missile Div. Joined Bower Roller Bearing Div in 1959 as Supr, Met R&D. In 1967 Pres & Gen Mgr, Bosworth Steel Treating Corp. Assumed current responsibility in 1976. P Chrmn Detroit Chap ASM. P Chrmn: Iron & Steel Tech Ctte SAE & Hardenability Activity, Heat Treat Div, ASM. Reg PE, MI.

Eastman, Jay M
Home: 25 Arlington Dr, Pittsford, NY 14534
Position: Exec VP. *Employer:* Photographic Sciences Corp. *Education:* PhD/Optcs/Univ of Rochester; BS/Optcs/Univ of Rochester *Born* 6/09/48 Native of Seneca Falls, New York. Mgr of Engrg, Optics Dept, Spectra Physics, Inc. 1974-1975. Optical Engr Laboratory for Laser Energetics (LLE) Univ of Rochester (U of R) 1975-1976. Project Mgr for the construction of the $15, 000,000 OMEGA Laser System at LLE 1976-1980. Dir, Division of Engrg, LLE 1980- 1981. Dir, LLE 1981 to 1982; Pres Optel Systems Inc 1983-1986. Exec. VP Photographic Sciences Corp., 1986-present. I am a consultant to several industrial firms and have held offices in all societies of which I am a member. *Society Aff:* OSA, SPIE, Roch Optcl Soc

Eastman, Lester F
Business: 425 Phillips Hall, Ithaca, NY 14853
Position: John L. Given Found Prof of Engg. *Employer:* Cornell Univ. *Education:* PhD/EE/Cornell Univ; MS/EE/Cornell Univ; BS/EE/Cornell Univ. *Born:* 5/21/28. Faculty Mbr, Sch of EE, Cornell Univ. One-yr leaves at Chalmers Tech Univ, Gothenburg, Sweden; RCA Res Lab, Princeton, NJ; Cayuga Assoc, Ithaca, NY; Lincoln Lab of MIT, Lexington, MA, IBM Watson Lab, Yorktown Heights N.Y. Res on microwave electron devices and compound semiconductor matls. Founding Mbr of Natl Res and Resourse Facility on submicron structures and Founding Dir of Joint Series Electronics Program at Cornell. Founder and Organizer of microwave semiconductor work shop; and IEEE Biennial Conference at Cornell, Mbr US Gov Advisory Grp on electron devices and consultant to Lincoln Labs, Honeywell, General Electric (US), General Electric (England), Siemens, Sanders, United Technology, Martin Marietta and Alpha, and member of National Academy of Engineers. Fellow of IEEE. *Society Aff:* IEEE

Eastman, Robert M
Home: 600 S Glenwood Ave, Columbia, MO 65203
Position: Prof of Ind Engg. *Employer:* Univ of MO. *Education:* PhD/Ind Engg/PA State Univ; MS/Ind Engg/OH State Univ; AB/Accounting/Antioch Univ. *Born:* 4/18/18. Prof Ind Engg, Univ of MO-Columbia, 1958-present, Chmn 1958-68 Prof Mech Engg 1955-1958. Previously on faculties of GA Tech, Penn State and OH State. Visiting Prof Middle East Technical Univ, Ankara, Turkey, 1969-1971. Fulbright Lecturer, Bucaramanga, Colombia, 1962-1963. Consultant to US Depts of Commerce and HEW, AMF, Gen Motors, Sandia and Southwestern Bell Tel and several other govt agencies and private industries. Labor arbitrator. Over 70 publications, reports and presentations. NSF Summer Fellowship in resource recovery NASA/ASEE Design Fellowships at Langley Res Ctr. NASA/ASEE Research Fellowships, Marshall and IPL, Navy/ASEE Research Fellowships, Naval Shipbuilding R&D Center. Res in resource recovery, industrial dev, health care systems. Married, three children. *Society Aff:* ASEE, AAAS, IMMS, IIE.

Eaton, Thomas E
Business: PO Box 1100, Nicholasville, KY 40356
Position: Consulting Engr. *Employer:* Eaton Engg Co. *Education:* Doctor of Sci/Nuclear Engr/MA Inst of Tech; Nuclear Engr/Mech E/MA Inst of Tech; MS Nuclear Engr/Mech E/MA Inst of Tech; MS Mechanical Engr/Mech E/Univ. of Mo-Rolla; BS Mechanical Engr/Mech E/Univ of Mo-Rolla. *Born:* Consulting Engr, Specialist in Accident/Loss Investigations; Owner - Eaton Engg Co, Nicholasville, KY. Consulting experience involving machinery, heavy equip, motor trucks, rrs, mining and drilling equip, heating and AC systems, power plants and boilers, industrial electrical systems, fire and explosion investigation, Diesel engines, equip repairs. Engg res on nuclear reactor thermal-hydraulic analysis, reactor containment, cooling tower performance, power plant design and performance, nuclear fuel cycle analysis, spent nuclear fuel storage, fire causation. US AEC Fellow 1970-73. Asst Prof, Mech Engr, Univ of KY, 1975-78. Reg PE - KY, VA, OH, IN, CO, WV, TN. *Society Aff:* ASME, ANS, AIChE, AREA, SME, ASHRAE, SAE, ASM.

Eaves, Elsie
Home: 18 Third Ave, Port Wash, NY 11050
Position: Civ Engr (Self-employed) *Education:* BS/CE/Univ of CO. *Born:* 5/5/98. Office engr, Herbert S Crocker, 4 yrs; dir market res, Engg News-Record, 6 yrs; mgr business news dept, 31 yrs; advisor on housing costs Natl Commission Urban Problems, 1968-69; consultant, construction cost indexes to Plan and Budget Organization, Tehran, Iran, 1974; mbr intl editorial bd, "Engineering Costs and Production Economics–, Amsterdam, 1977-81;Honorary Mbr, ASCE; Honorary Life Mbr, AACE; Award of Merit, AACE; Certificate of Recognition for Pioneering, SWE; Service-to-Country Award, Intl Exec Service Corps; Distinguished Engg Alumnus Award, Univ of CO; George Norlin Silver Medal, Univ of CO; Tau Beta Pi; Chapter Honor Mbr, Chi Epsilon. Honorary Mbr, Altrusa, Intl; Trustee, North Shore Sci Museum, Plandome Manor, NY. *Society Aff:* ASCE, AACE, NYSSPE, SWE, CSE.

Ebaugh, Newton C
Home: 2737 SW 4th Place, Gainsville, FL 32607
Position: President, 1950-1978 *Born:* Oct 1907 New Orleans. Tulane University BE 1928, ME 1930; Ga. Tech MS 1932 Inst.-Asso. Prof. ME Ga. Tech 1929-35. Ch. M.E. Dept. Univ of Florida 1935- 50. Pres. Ehangh & Goethe, Inc. Consulting Engineers 1950-78. Director Marineland, Inc. Dir. Century Banks, Inc. Dir. The Charter Co. Dir. Financial Computer Center; Dir. Century Banks of Gainesville & Palatka, Fl. Author - Handbook of Air Conditioning, W.R.C. Smith Pub. Co 1936, 2nd 2ed 1946 - Engineering Thermodymanics, D. Van Nostrand Co. 1937, 2nd ed. 1952 - 50 Technical Papers in ME subjects. Memberships - Life Fellow ASME; Fellow FES; Life member AWWA & ASHRAE & NSPE - Phi Kappa Phi; Pi Tau Sigma; Tau Beta Pi; Sigma Xi - Newcomer Society; Rotary Club. Professional Engineer Florida, formerly Ga. Texas & Cal.

Ebeling, Dick W
Business: 10150 SW Nimbus Ave, Suite E2, Portland, OR 97223-4306
Position: Pres Chief Engr. *Employer:* Dick W Ebeling Inc. *Education:* MS/CE/OR State Univ; BS/CE/OR State Univ. *Born:* 7/2/19. Native of Portland, OR. Grad in 1941. Taught at OR State, worked for Curtiss-Wright Corp, Corps of Engrs, Timber Structures. Became assoc with Cooper and Rose' and Assoc in 1952 becoming a partner in 1956. Left to start my own practice of consulting struct engg in 1962. Ex-mbr of ASCE Struct Div Committee on Wood, writing part of their recent books. On AITC's Technical Advisory Committee group that revised the Timbe Construction manual. Mbr of Natl Forest Products Assn's special technical advisory group on the updating of NDS. Licensed in 18 states mostly western. *Society Aff:* ASCE, IABSE, EERI, ACI, AISC, AITC, ICBO, ACEC (CEO).

Ebert, Ian O
Business: EE Dept, E Lansing, MI 48824
Position: Assoc Prof of EE. *Employer:* MI State Univ. *Education:* MS/EE/Univ of IL; BS/EE/IA State Univ *Born:* 3/13/20. Parents: Albert J & Grace M (Adkins); Wife: Doris E (Fuller); Children: James Ian & Barbara Elizabeth. Res Scientist, Naval Res Lab, Wash DC 1943-48. USNR Ens to CDR 1943-1971. Advisor, Prof Poona Engg College, India 1962-63. Engg Consultant, Reg PE (MI). *Society Aff:* AAAS, ASEE.

Ebert, Lynn J
Business: Dept Met & Mat'l Sci, Cleveland, OH 44106
Position: Prof Met & Mat'l Sci. *Employer:* Case Western Reserve University. *Education:* BS/Met Engg/Case Sch; MS/Met Engg/Case Sch; PhD/Met Engg/Case Inst. *Born:* 4/17/20. Work Experience: 1941-53 Various res positions, Case Sch and Case Inst. 1953-present various profl position; presently Prof of Met and Matls Science. Areas of Profl Activities: Teaching - engg mechanical behavior of matls, both undergraduate and graudate. Special courses in Failure Analysis. Matls in Sports, High Temperature Behaviors. Research - Mechanics behavior of matls, flow and fracture, forming, matls processing matls treatment, fiber composites, hi gh temperature alloys, golf club design. Consultation - Failure analysis, metal forming, designing with matls, fabrication of structures, mechanical performance of mtls, joining of metals. Publications: Approx 70 papers, 2 ref vols, pts of bks. *Society Aff:* ASM, AIME, NYAS.

Eby, Martin K, Jr
Business: 610 North Main, PO Box 1679, Wichita, KS 67201
Position: Pres & Chrmn of Bd *Employer:* Martin K. Eby Const Co, Inc *Education:* BS/CE/KS State Univ *Born:* 04/19/34 Joined Eby Construction Co in 1956 as field engr. Subsequently served as superintendent, proj mgr and VP. Elected pres in 1967, and chrmn in 1979. Mbr and past chrmn of the KS State Univ Engrg Dean's Advisory Council. Past Chrmn of Construction Indus Political Action Ctte of KS. Mbr of Bd of Trustees, Wesley Medical Endowment Foundation; Bd of Dirs First Natl Bank of Wichita; and Chrmn of Wichita Area Economic Education Foundation. *Society Aff:* ASCE, NSPE, KES, WPES

Eccles, Richard M
Business: 207 Russell Rd, Princeton, NJ 08540
Position: Pres *Employer:* Princeton Process Engrs *Education:* BChemE/Chem Engg/Cornell Univ. *Born:* 10/18/31. Twenty-four yrs. exp. with Exxon at two refineries, two hdquarter locations and the engg center, spec in coal liquefaction dev, heavy oil upgrading, refinery planning and tech mgmt. Former hd of process engg dept at Exxon, Baton Rouge Refinery. Five years at Hydrocarbon Res, as VP, for liquid-phase hydrogenation using ebullated bed reactor tech. Activities included marketing support and tech coord and supervision of engg R&D and commercial plant planning. Currently heads own conslt firm in use of advanced hydrocarbon process tech with emphasis on heavy oil upgrading. *Society Aff:* AIChE.

Echelberger, Wayne F, Jr
Home: 1709 E Cedar St, South Bend, IN 46617
Position: Prof Public & Environmental Affairs. *Employer:* IN Univ. *Born:* Oct 23, 1934. BSCE SD Sch of Mines & Tech 1956. MSE, MPH, PhD Univ of MI 1959, 1960, 1964. Instructor & Res Assoc Univ of MI 1964-65. Asst Prof 1965-67, Assoc Prof 1967-73, Univ of Notre Dame. Prof IN Univ 1973-. Bd/Dirs, Assoc Environmental Engg Profs 1971-73; Secy-Treas 1972-73. Water Pollution Control Federation Harrison Prescott Eddy Medal 1973. St Joseph Valley Chap, IN Soc of PE, Engr of the Yr 1976. Listed in Am Men & Women of Sci & Who's Who in the Midwest. PE IN, IL, MI, OH. Secy-Treas Ten Ech Environemtnal Consultants Inc 1969-.

Eck, Bernard J
Business: 200 W Monroe, Chicago, IL 60606
Position: Dir Prod Engg. *Employer:* Griffin Wheel Co. *Education:* BS/Ceramic Eng/MO School of Mines and Met *Born:* 5/2/28. Native of Springfield, IL. Served with Army Counter Intelligence Corps 1951-53. Previously technologist with US Steel Corp, specializing in steel plant refractories and defects in steel products. With Griffin since 1957 concerned with steel plant processes, mtls and product defects. Assumed present position 1972, Responsible for railroad wheel design testing, tooling fabrication, wheel specifications. Technical Asst to customers, technical rep to ASTM & AAR. Past Chrmn of Rail Transportation of ASME. Chrmn Subcommittee A01.07 of ASTM. *Society Aff:* ASME, ASTM.

Eck, Ronald W
Business: Dept of Civil Engrg, PO Box 6101, WVU, Morgantown, WV 26506-6101
Position: Prof *Employer:* WV Univ *Education:* Ph.D./CE/Clemson Univ; BS/CE/Clemson Univ *Born:* 5/11/49 in Allentown, PA. Joined WV Univ in 1975; assumed current position in 1984. Teach graduate and undergraduate courses and conduct research in highway and traffic engrg. Have authored over 60 papers and research reports. Involved in program development and committee activity at Coll and Univ level. Chair, Faculty Senate 1986-87; WVU Board of Advisors, 1986-87; Pres, WV section, ASCE 1980; Chrmn, Civil Engrg Division of ASEE, 1983- 84; Chrmn, Prof. Interest Council I of ASEE, 1986-88; Chrmn, ASCE Engrg Applic. of Remote Sensing Ctte, 1985-1987; SAE R. R. Teetor Award 1977; Dow Outstanding Young Faculty Award 1980; National Defense Executive Reserve. Enjoy tennis and backpacking. *Society Aff:* ASCE, ITE, ASEE, ASPRS, TRB

Ecker, H Allen
Business: P.O. Box 105600, Atlanta, GA 30348
Position: VP-Corporate Dev *Employer:* Scientific-Atlanta, Inc. *Education:* PhD/EE/OH State Univ; MSEE/EE/GA Inst of Tech; BEE/EE/GA Inst of Tech. *Born:* 10/22/35. in Athens, GA. From 1959 to 1962, he was an officer in the USAF working in radar & electronic countermeasures res for the Navigation & Guidance Lab, Wright-Patterson AFB, OH, & from 1962 to 1966 as a civilian, he directed in-house sys analysis projs for the AF Sys Command. From 1966 to 1976 he was with the Eng Experiment Station, GA Inst of Tech, where he supervised a wide range of res and dev activities in radar & associated tech areas as Chief of the Radar Div & Dir of the Applied Engrg Lab. He has been with Scientific-Atlanta since 1976, where he served as Dir or Res V Pres for Res & Dev, V Pres-Telecommunications, and currently is V Pres-Corporate Dev. His responsibilites at Scientific-Atlanta include strategic planning, tech dev, new product dev and dev of new business activities. *Society Aff:* IEEE, HKN, TBΦ, ΦΚΦ,

Eckert, Charles A
Business: Dept of Chem Engg, 297 RAL, Urbana, IL 61801
Position: Prof of Chem Engrg *Employer:* Univ of IL. *Education:* PhD/CE/Univ of CA; SM/CE/MIT; SB/CE/MIT. *Born:* 12/13/38. Teaching in Chem Engg at Univ of IL since 1965. Res in molecular thermodynamics, chem kinetics, phase equilibria, high pressure, high temperature; also in comp teaching of engg. NATO Post-doctoral Fellowship, 1964; Guggenheim Fdn Fellowship, 1971; Alan P Colburn Award of AIChE, 1973; Ipatieff Award of ACS, 1977. Mbr, NAE. Active in techni-

Eckert, Charles A (Continued)
cal consulting & presentation of short courses in ind. *Society Aff:* AIChE, AIME, ASEE, ACS, CS, AIRAPT, NAE.

Eckert, Ernst R G
Home: 60 W Wentworth Ave, W St Paul, MN 55118
Position: Regents Prof. *Employer:* Univ of MN. *Education:* DrEnggHabil/Heat Transfer/Tech Univ Danzig; DrIng/ME/German Tech Univ Prague; Ing/ME/German Tech Univ Prague. *Born:* 9/13/04. Doctors degrees from German Inst of Tech, Prague and Tech Univ, Danzig. After 1945 associated with USAF, Wright Field, OH; Natl Advisory Committee for Aero. With Univ of MN since 1951. There he founded Heat Transfer Lab and became its director. 1966 named Regents' Prof of Mech Engg. Scientific contributions and books in heat transfer. Mbr Natl Academy of Engg, Honorary mbr or fellow of profl societies. Recipient of Max Jakob Award in USA; gold medal in France; honorary doctors degrees from Technical Univ, Munich, Germay; Univ of Manchester, England; Poly Inst, Iassi, Romania; Purdue and Notre Dame Univs, USA. *Society Aff:* AIAA, ASME, NYAS, DWGLR, NAE.

Eckert, Jerry W
Business: 7400 Trail Blvd North, Naples, FL 33940
Position: Regional Mgr *Employer:* Post, Buckley, Schuh & Jernigan, Inc *Education:* BS/CE/Purdue Univ *Born:* 10/21/46 Native of IN. Have held various positions in consulting and construction engrg. VP of Bruce Green & Assocs, Inc 1978-80. Joined Post, Buckley, Schuh & Jernigan, Inc as Project Mgr 1980. Opened and manage Naples PBS & J office since December 1980. Will assume Regional Mgr responsibilities Oct. 1981. Chrmn FL Engrg Society Chapter Activities Committee 1975-76. Pres FL Engrg Society Calusa Chapter 1979-80. Listed in Who's Who in South and Southwest. Enjoy golf and racquetball. *Society Aff:* NSPE, ASCE

Eckert, Richard M
Business: 80 Park Plaza - T4B, P.O. Box 570, Newark, NJ 07101
Position: Sr. VP - Energy Supply & Engg. *Employer:* Public Service Elec & Gas Co. *Born:* Nov 1929. MS in Engg from Union Coll, Schenectady; BME from Univ of Louisville, KY. Served with USN 1952-55. Started with PSE&G in 1955. Attended Oak Ridge Sch of Reactor Tech in Idaho Falls, ID 1957. Following this, obtained practical experience at Natl Reactor Testing Station. After an active career in the design & construction of many generating stations, assumed current responas VP in charge of engg & construction of combustion turbines and nucl power plants in 1974. This includes land based nucl plants and offshore nucl plants. Mbr of Am Nucl Soc, m Soc of Mech Engrs, & Lic PE in the State of NJ. Enjoy sailing & skiing. In July 1977, he became Sr. Vice Pres., Energy Supply & Engineering.

Eckhardt, Carl J
Home: 4100 Jackson Ave, Apt. 360, Austin, TX 78731
Position: Professor Emeritus, ME. *Employer:* University of Texas at Austin. *Education:* MS/ME/Univ of TX; BS/ME/Univ of TX. *Born:* Oct 1902. Prof Emeritus of Mechanical Engrg, The University of Texas. Served this institution also as Supt of Power Plants, Supt of Utilities, & Director of Physical Plant. Served ASME as Chairman, South Texas Section; Speaker, Regional Delegates Conference; and VP. Served Engrs Council for Profl Dev as Chrmn Ctte on Student Selection and Guidance; Chrmn, Natl Nominating Ctte. Mbr of Pi Tau Sigma (honorary), Tau Beta Pi, Sigma Xi, and Phi Kappa Phi. Life Mbr, ASEE; Life Fellow, ASME. Recipient 75th Anniversary ASME Medal. Author of books, technical papers and bulletins. Licensed Profl Engr. *Society Aff:* ASEE, ASME.

Eckhoff, David W
Home: 4720 S Ichabod St, Salt Lake City, UT 84117
Position: Chrmn, CE Dept. *Employer:* Univ of UT. *Born:* Dec 1937. PhD 1969 & MS 1963, Univ of CA, Berkeley. BS CA State at Sacramento. CA Div of Hgwys 1959-62. Res on biodegradable detergents at UC Berkeley. PhD thesis on Biological Waste Treatment Systems. Employed by Engg- Sci, Inc 1966-71. Combined sewer overflow investigations, regional water quality mgt, and waste treatment. Mgr of NY Regional Office 1968-71. Directed Jamaica Bay Study for the City of NY, solid waste mgt, & sanitary landfill investigations. Chrmn 1973. Regional water quality mgt & county-wide solid wastes mgt. Currently, Dir of Salt Lake County 208 Water Quality Proj.

Eckhoff, N Dean
Business: Ward Hall, Dept. of Nuclear Engrr, Manhattan, KS 66506
Position: Prof & Head *Employer:* Kansas State Univ *Education:* PhD/NE/KS State Univ; MS/NE/KS State Univ; BS/NE/KS State Univ; AA/PE/Pratt Jr. Coll *Born:* 4/10/38 Have worked for Boeing and the Litwin Corp., Wichita; AEC in Oak Ridge, TN. Consulting work has included several industries and government agencies. Registered PE in CA; Prof and Head, Dept. of Nuclear Engrg at Kansas State Univ. *Society Aff:* ANS, ASEE

Eckmann, Donald E
Business: 20 N Wacker Dr, Chicago, IL 60606
Position: Partner. *Employer:* Alvord Burdick & Howson. *Education:* BSCE/Civ Engg/Univ of IL. *Born:* 11/11/34. Employed by Alvord, Burdick and Howson in 1956; partner 1969. Engr on water works for Danville, IL; Gary, IN: Detroit, MI; Green Bay, WI; Chicago, IL; Du Page Water Commission, IL; on sewerage for ST Louis, MO; Milwaukee, WI; Fulton, NY; Galesburg, IL; Roanoke, VA; Waukesha, WI. Expert witness on Lake MI Allocation Hearings, several technical papers; Diplomate in Am Acad of Environ Engrs; Reg PE in 11 states. Mbr ASCE (P Pres IL Sec), AWWA, NSPE, WPCF, WSE & ASTM. *Society Aff:* ASCE, ASTM, NSPE, WPCF, AWWA, AAEE, ASEE.

Eckrose, Roy A
Home: 3214 Windsor Lane, Janesville, WI 53545
Position: VP *Employer:* Donohue & Associates, Inc. *Education:* BA/Liberal Studies/Western IL Univ *Born:* 2/16/41 Currently VP of Donohue and Associates, Inc. Previous- 1975-1979 Dir of Public Works, Janesville, WI (pop. 51,000); 1970-75 Ass't Dir of Public works, Janesville, WI; 1961-1970 Asst City Engr, Janesville, WI. Registered PE-WI, IA, MO, TX, and AL. Recipient: "Young Engr of the Year" - Wisconsin Society of PE. *Society Aff:* WSPE, NSPE, APWA, AEE

Eckstein, Marvin
Home: 8210 SW 203 St, Miami, FL 33189
Position: Project Mgr *Employer:* Post, Buckley, Schuh & Jernigan, Inc. *Education:* MS/CE/The City Coll of NY; BS/CE/The City Coll of NY *Born:* 12/11/35 and educated in New York City. Employed by The New York City Transit Authority for 15 years. Served as Administrative Engr in charge of Construction Mamagement of a portion of NYCTA new route expansion. For 16 years taught Construction Cost Estimating evenings at The Institute of Design and Construction. Relocated in 1978 to Miami to become The Kaiser Transit Group Construction Mgr for The South Corridor of The New Metropolitan Dade County Rapid Transit System. An amateur photographer who enjoys darkroom work. *Society Aff:* NSPE

Economides, Leander
Business: 110 E 30th St, New York, NY 10016
Position: Partner. *Employer:* Economides & Goldberg. *Education:* BSME/Mech Engg/CCNY. *Born:* Jan 31, 1917. Chief Mechanical Engineer, Voorhees Walker Smith Smith & Haines 1955-65. Notable Projects; National Bureau of Standards, Caterpillar Research Center, NASA Goddard Flight Center, Bethlehem Steel Research Center, Argonne Natl Lab. *Society Aff:* NSPE, ASME, ASHRAE, AIA.

Eddy, Lial N
Business: PO Box 401, Oklahoma City, OK 73101
Position: Gen Mgr Transpor & Shops. *Employer:* OK Natural Gas Co. *Born:* Oct 1922, Davis OK. BSEE from the Univ of OK 1954. Grad study in law at the Okla City Univ. Served in USN 1943-46. Previous work has been in hgwy bridge con-

Eddy, Lial N (Continued)
struction, oil & gas well drilling & production & pipe line construction. Employed by OK Nat Gas Co in 1954 as engr in training. Served as Dist Engr, Asst Mgr Transportation & promoted to Gen Mgr of Transportation and Shops in 1962. P Pres of OK Soc of PE & Dir of the Natl Soc of PE. P Chrmn of Am Gas Assoc Automotive & Mobile Equip Ctte. Award of Merit AGA 1969. Mbr Am Welding Soc, Soc of Automotive Engrs, Fleet Supervisors, Consistory & Chamber of Commerce.

Eddy, Robert A
Business: Worcester St, North Grafton, MA 01581
Position: Mgr Technical Services. *Employer:* Wyman Gordon Co. *Born:* July 1921. BS & MS from MI Technological Univ. Native of Lake Linden, MI. Served with US Army Air Corps 1942-46. Specialized as a P-51 Pilot, Fighter Gunnery Instr. Joined Wyman Gordon as a Met & have held many supervisory & Managerial pos including Plant Mgr, Dir of Product Quality, & Mgr Technical Services-Eastern Div. Currently responsible for all outside Quality & Tech contacts. Have served on cttes for AMS & ASTM & am now on a Ctte on NDT of Aerospace Systems, Natl Mtls Advisory Bd. Enjoy skiing & swimming.

Eddy, W Paul
Business: 41 Neptune Dr, Groton, CT 06340
Position: Owner. *Employer:* Ind Mgt & Engg Consultant. *Education:* BS/CE/Syracuse Univ. *Born:* 12/3/99. Career: Supv, Crucible Steel Co; Chief Met, Geometric Tool Co; Chief, Mtls and Service Engg, GMC Truck & Coach; Chief, Mtls Engg, the Chief, Engg Operations, Pratt & Whitney Aircraft, East Hartford, CT. Author of 40plus technical papers on various eng and met subjs. Past Pres and Honorary Mbr, Soc of Automotive Engrs Chrmn New Haven and Detroit Chapters, and Honorary Life Mbr, Am Soc for Metlas. Pres, Hartford Engrs Club. Pres, SECT Chamber of Commerce. Pres, SECT Economic Dev Corp. Chrmn, SECT Water Authority. Pres, ECT Council, Navy League of US. Pres, CT Council, Automotive Organization Team. *Society Aff:* SAE, ASM, IOF, AOT, NL.

Edelson, Burton I
Home: 116 Hesketh St, Chevy Chase, MD 20015
Position: Assoc Administrator *Employer:* NASA *Education:* PhD/Met/Yale; MS/Met/Yale; BS/Engg/US Naval Acad. *Born:* 7/31/26. Native of E Lansing MI. Served 20 yrs in US Navy as Engg Duty Officer. Service included Bureau of Ships, Natl Aero & Space Council, & Office of Naval Res. Retired 1967. Joined COMSAT as Asst Dir of COMSAT Labs, 1968; Dir, COMSAT Labs, 1973; VP, 1979. Senior VP, 1980. Appointed Assoc Admin of NASA for Space Service and Applications, 1982. Fellow of AAAS, AIAA, IEEE & BIS. Many soc committees, many publications. Received Howe Res Medal, ASM, 1963; Legion of Merit, US Navy, 1965; Wilbur Cross Medal, Yale University, 1984; Government Industry Science Award, IEEE, 1985; Doctor of Science (Honorary), Capitol Institute of Technology, 1986. *Society Aff:* AAAS, AIAA, IEEE, ΣΞ, BIS.

Eden, Edwin W, Jr
Home: 5375 Sanders Road, Jacksonville, FL 32211
Position: Consulting Engr (Self-employed) *Education:* CE/Civil Engg/Rutgers Univ; MS/Natl Res Dev/Univ of MI; MS/Hydraulics/Univ of IA; BS/Hyds/Rutgers Univ. *Born:* 6/4/11. 1950-65 Chief Jacksonville District-Corps of Engrs. Respon for Planning, Direction, Coordination & Execution-Engg and Economic Investigations covering Flood Control. Navigation, Beach Erosion and other water related programs. 1965- 70 Respon Charge Interoceanic Sea-level Canal including Construction Methods- Nuclear or conventional. 1970-75 Consultant Atlantic-Pacific Interoceanic Canal Study Commission & Jacksonville District, Corps of Engrs. 1975-76 Reisdent Proj Mgr, Bicol River Basin Water Resource Dev Study-Philippines. 1976 to date consulting Engr. Numerous papers. ASCE Wellington Prize; Dept of Defense, Disting Civilian Servie Award. *Society Aff:* NSPE, ASCE, PIANC.

Edgar, C E, III
Business: 424 Trapelo Rd, Waltham, MA 02254
Position: Div Engr *Employer:* US Army Corps of Engrs *Education:* MS/CE/IA State Univ; BS/CE/VA Military Inst *Born:* 01/15/36 Career Officer, Corps of Engrs. Served in various engr capacities overseas and in US including: Area Engr, USA Engr District, Louisville, KY; Asst Dir Civil Works, Lower MS Valley, Office, Chief of Engrs; District Engr, USA Engr District, Little Rock, AR. Currently, Div Engr, USA Engr Div, New England, Waltham, MA, supervising staff of 600 with more than 85 PE. Reg PE, KY and AL; Recipient of numerous US and foreign military decorations; Mbr, American Soc of Civil Engrs and the Soc of American Military Engrs; Mbr, US Army Coastal Engrg Res Bd; Mbr, Intl St Croix River Bd of Control, US-Canadian Intl Joint Commission. *Society Aff:* ASCE, SAME

Edgar, Thomas F
Business: Dept of Chemical Engineering, EPS 211B, Austin, TX 78712
Position: Prof *Employer:* Univ of TX *Education:* PhD/ChE/Princeton Univ; BS/ChE/Univ of KS *Born:* 04/17/45 Prof and Dept chrmn of Chem Engrg, Univ of TX, Austin, TX. Profl experience includes employment with Continental Oil Co and consltg for Fisher Controls. Profl activities include Pres of CACHE Corp 1981-84; Chrmn, CAST Div of AIChE, 1986; Dir, American Automatic Control Council 1978-80; Reg PE - TX; and Editor of *In Situ*. Awards include Outstanding Student Chapt Counselor (AIChE-1974); Outstanding Young Man, AIChE South TX Section 1976; Colburn Award for res (AIChE-1980); and Katz Lecturer, Univ of MI (1981). Author of three books inc. *Optimization of Chemical Processes* (McGraw-Hill) and of numerous articles, book chapts, papers, and reports. Res interests in computer process control and mathematical modeling. *Society Aff:* AIChE, ISA, ASEE

Edgerley, Edward, Jr
Home: 582 Brookhaven Ct, Kirkwood, MO 63122
Position: Pres. *Employer:* SITEX Corp. *Education:* PhD/Sanitary Engg/Univ CA; SM/Sanitary Engg/MIT; BS/Sanitary Engg/PA State Univ. *Born:* 3/8/31. Native of Lancaster, PA. Served as officer in AF, 1954-57. Faculty mbr of Environmental Engg Grad. Prog & Civ Engg Dept, WA Univ, St Louis 1957-73, Asst Dean 1968-69. One of founders of consulting engg firm Ryckman, Edgerley, Tomlinson & Assoc & Pres 1975-78. Upon merger with Envirodyne Engrs became Sr VP 1978-79. Formed new environmental engg firm of SITEX Corporation in 1984. Diplomat, Am Acad of Environmental Engr. Reg PE in five states. *Society Aff:* WPCF, ASCE, AWWA, APCA, AIHA, ACEC.

Edgerton, Harold E
Home: 100 Memorial Dr, Cambridge, MA 02142
Position: Prof. *Employer:* MIT. *Education:* DSc/EE/MIT; MS/EE/MIT; BS/EE/Univ of NB. *Born:* 4/6/03. One of the founding partners of EG&G Inc. Best known for his dev of the modern stroboscope, which makes possible high speed photography of moving objects. He has designed flash lamps & cameras to go to the bottom of the deepest oceans & presently is dev sonar devices for sub-bottom exploration of geology & archacoloy. Morris E. Leeds Award 1965, IEEE; Richardson Medal, OSA 1968; John Oliver LaGorce Gold Medal, NGS 1968; Albert A Michelson Medal Franklin Inst 1969. *Society Aff:* IEEE, SMPTE.

Edgerton, Robbie H
Home: 7704 Melody Lane, Dickinson, TX 77539
Position: Owner *Employer:* Edgerton Engineering & Res. *Education:* BS/ME Univ of TX/University of Texas, Austin. *Born:* 2/27/20 Native Texan, began engineering career 1941, graduated from the Univ of Texas 1947, licensed 1948, progressed with several concerns as Project Engr, Group Leader, Assistant Plant Engr, Primary Engr, Plant Mgr, Division Mgr, VP. Served on committees for inventory control, design standards, super pressure design, profitability. Did original work in fields of engine foundation resonance, tall stack vibration, predicting buckling of thin walled vessels above supports, and drip irrigation. Presently operating a consulting firm in the fields of management organization, investment analysis and feasibility, project

Edgerton, Robbie H (Continued)
management, plant design, and equipment design for petroleum, petrochemical, marine, industries. *Society Aff:* ASME, HESS

Edlund, Milton C
Home: 302 Neil St, Blacksburg, VA 24060
Position: Prof *Employer:* VPI & SU *Education:* PhD/NE/Univ of MI; MS/Physics/Univ of MI; BS/Physics/Univ of MI *Born:* 12/13/24 in Jamestown, N.Y. Research physicist with Union Carbide 1948-55. Published first book with S. Glasstone on nuclear reactor theory in 1952. With Babcock & Wilcox Co. from 1955-66-served as mgr of physics and mathematics, development dept. and assistant division mgr of Atomic Energy Division. Received E. O. Lawrence Award from US Atomic Energy Commission (AEC) 1965. Consultant to utilities, nuclear industry and AEC 1966-1970. Joined VPI & SU in 1970 as Prof of Nuclear Engrg. Research directed to improvement of nuclear fuel cycles. *Society Aff:* ANS, ASME, NAE, AIF.

Edmister, Wayne C
Home: 75 Summit Ave, San Rafael, CA 94901
Position: Retired. *Education:* Profl/CE/OK State Univ; MS/ME/Cornell Univ; BS/ME/OK State Univ. *Born:* 3/22/09. Native of Cleveland, OK; married, 34; res engr, Std Oil Co (IN) 34-43; process engr, Foster-Wheeler Corp 44-47; prof, Carnegie Inst Tech 48-51; res engr, CA Res Co 52-59; prof OK State Univ 58-71; visiting prof, UC-Davis, 72; visiting prof Federal Univ Rio & Petrobras (Brazil) 73-74; consultant, UNESCO (Venezula) 74- 75; visiting prof, Univ PR (Mayaguez) 77; visiting prof, Univ of Manchester (UK) 77, volunteer prof, Korea Advanced Inst of Sci 78; visiting prof, Tech Univ- Berlin 78; consultant, UN Ind Dev Organization proj in Argentina 79 and 80. Visting Prof UC-Santa Barbara 79; Lecturer East China Petroleum Institute in Peoples Republic of China 80. Author "Applied Hydrocarbon Thermodynamics–, Gulf Publishing Co. Houston, 2nd Edition: Vol 1, 84; Vol 2, 88. Fellow & past dir of AIChE; Richards Meml Award of ASME 57; Hanlon Award of Gas Processors Assn 66. *Society Aff:* AIChE, ACS.

Edwards, Arthur A
Business: 330 West 42nd St, New York, NY 10036
Position: President. *Employer:* Edwards & Zuck, PC. *Education:* BSME/Mech Engr/Duke Univ. *Born:* 9/18/26. Served in US Naval Res 1944-47 attaining rank of Lt. Joined Krey & Hunt, Consulting Engrs 1947, Partner 1957. Formed Arthur A Edwards, Consulting Engr 1967 as Sole Proprietor. Professional Corp formed 1976. Reorganized as Edwards & Zuck, PC 1978. *Society Aff:* ASME, SAME, ASHRAE, NSPE.

Edwards, Carl V
Business: 238 E South St, Grangeville, ID 83530
Position: Pres. *Employer:* Edwards*Howard*Martens, Inc. *Education:* BSCE/Civ Engr/Univ of ID. *Born:* 5/1/38. Worked 8 yrs for the Sawtooth Natl Forest in ID & the Klamath Natl Forest in CA as Engr designing & supervising the surveying & construction of forest roads. Employed 8 yrs by Morrison-Knudsen as Area Engr for ID & UT providing extensive experience in Hgwy construction including estimating, planning, quality control, and construction supervision of maj construction projs. Presently located in Grangeville, ID supervising a branch office of EdwardsHowardMartens. He also holds the position of County Engr for ID County, & City Engr of the City of Grangeville. Reg (Reg as a PE in both ID and NV). Reg as a Land Surveyor in ID (PE-LS 2098). Reg as a PE in NV (PE 4018). *Society Aff:* NSPE, CEC, ISPE, CEI.

Edwards, Charles E
Business: Bandag Center, Muscatine, IA 52761
Position: Pres and CEO *Employer:* Bandag, Incorporated *Education:* BS/ME/NC State Univ *Born:* 10/31/28 Born Norlina, NC, s. Clyde and Elizabeth (Hayes) Edwards; m. Linda Williams, Jun 14, 1950; children, Charles Elvin Jr., Brenda Kay. With Dayco, Southern, 1950-64; Dayco, Intl, 1964-67; Three Rivers Rubber Corp, 1967-69; joined Bandag, Inc, 1969; names Pres and Chief Exec Offr, 1980; also Chrmn of Bd: Bandag, Canada Ltd, Bandag N.V., Heavy Duty Parts, Inc, Muscatine Natural Resources, Vakuum Vulk, US; and Dir: Bandag, Inc, Bandag Comercio E. Industria Ltda, Bandag Equip, Bandag Europe N.V., Vitafrio Bandag S.A. de C.V., Mexico and Master Processing. Civic Assns: Muscatine Chamber of Commerce, and Rotary Intl. *Society Aff:* ASME, AMA, ACS, AIPE, IAS

Edwards, Donald K
Home: 13 Holly La, Irvine, CA 92715
Position: Prof, Assoc Dean *Employer:* Univ. Calif *Education:* PH.D./Mech. Engin./Univ. of Calif., Berkeley; M.S./Mech. Engin./Univ. of Calif., Berkeley; B.S. (Highest Honors)/Mech. Engin./Univ. of Calif., Berkeley *Born:* 10/11/32 Second son of Richmond CA manufacturer. Thermodynamics Engr., Lockheed Missile Sys. Div., 1958-59. Asst. Prof. Engrg., UCLA, 1959-63. Chmn., Gier Dunkle Instruments, 1963-66. Assoc. Prof., UCLA, 1963-68. Prof., 1968-81. Chmn., Dept. of Chemical, Nuclear, and Thermal Engrg., 1975-78. Prof., UCI, 1981- . Chmn., Dept. of Mech. Engrg., 1983-86. Assoc. Dean, Engrg. Grad. Affairs, 1986- . ASME Heat Transfer Memorial Award, 1973; Fellow 1981. AIAA First Thermophysics Award, 1976; Assoc, Fellow 1986. Assoc. Editor, ASME J. Heat Transfer, 1975-81; Solar Energy, 1982-85. Editorial Board, Int. J. Heat Mass Transfer, 1970- ; AIAA J. Thermophysics, 1986- ; Author, Radiation Heat Transfer Notes, 1981. Coauthor, Transfer Processes, 1973, 1979. Research in Thermal Radiation, Natural Convection, Evaporation, Heat Pipe Technology. Mem. Pi Tau Sigma, Tau Beta Pi, Phi Beta Kappa, Sigma Xi and above. *Society Aff:* ASME, AIAA, OSA, ISES

Edwards, Donald M
Home: 4557 Arrowhead Road, Okemos, MI 48864
Position: Chrmn and Prof *Employer:* MI State Univ *Education:* BS/Agri Engrg/S Dak St Univ; MS/Agri Engrg/S Dak State Univ; PhD/Agri Civ Engrg/Purdue Univ. *Born:* 4/16/38. Native of Tracy, MN Engr for Soil Conservation Service. Teaching & Res in Agri Engr, Univ of NB 1966-70. Assoc Dean, Col of Engrg and Tech - respon for engrg res & dev, cooperative educ program and student programs, Coll of Engg, & Tech Univ of NB, 1970-1980. Dir, Energy Res and Dev Ctr, Univ of Mich, 1976-80. Conslt and collaborator to several state & fed agencies, industries, and fdns. Currently, chmn, Dept of Agri Engrg, Michigan State Univ, Respon for teaching, res, extension and internatl programs in the Dept. Publ in the areas of irrigation, remote sensing, energy, water pollution, porous media, and engg education. 1970 received Univ of NB Disting Teaching Award. 1973 awarded Outstanding Young Educators Award by the Amer Soc of Agri Engg and Fellow in Amer Soc of Agri Engrg in 1983. BS and MS - S Dak State Univ - 1960 and 1961. PhD - Purdue Univ - 1966. *Society Aff:* ASAE, EAC, NSC, NSPE/MSPE, AAAS, CAST.

Edwards, Eugene H
Home: 69 Sleepy Hollow Ln, Orinda, CA 94563
Position: Chief Mtls Engr. *Employer:* Std Oil Co of CA. *Born:* July 1922. PhD & MS from Univ of CA, Berkeley; BS from Birmingham Southern Coll. Native of Birmingham, AL. Served in Navy 1943-46 as destroyer engg officer. Res asst, Inst of Engg Res, Univ of CA, Berkeley 1950-52. Joined Std Oil Co of CA in 1953. Assumed current responsibility of Chief Mtls Engr, 1969. ASM Fellow. Recipient 1956 of AIMME. Mathewson Gold Medal for Res. Mbr Sigma Xi. Reg Met Engr, State of CA. Enjoy hunting, fishing & photography.

Edwards, Harry W
Business: Department of Mechanical Engineering, Colo State Univ, Fort Collins, CO 80523
Position: Prof *Employer:* Colorado State Univ *Education:* Ph.D./Physical Chem/Univ of AZ; B.S./Chem/Univ of NV-Reno *Born:* 10/6/39 Harry W. Edwards was born in Syracuse New York. He attended public schools and universities in the western US. During summers of his school years, he worked for Stauffer Chemical Co. and American Potash in Nevada. He was a General Motors Corp. Scholar at Nevada and a General Electric Foundation Fellow at Arizona. In 1966, he became an assistant prof of mechanical engrg at Colorado State Univ; promotion to prof was in 1976. Courses taught: Air Pollution Control, Thermodynamics. Principal Investigator for research projects: Environmental Lead (NSF), Nitric Oxide (EPA), Aerosol Behavior (NASA). Chairman, CSU Environmental Engineering Committee, 1974-80. Member, APCA Education and Training Committee. Consultant on analysis of environmental data. Regional Editor for Environmental Technology Letters. Senior Visiting Fellow, Univ of Lancaster, England 1977-78 (Sabatical Leave). Summer 1984: DOE/ASEE Fellowship, Natl Bureau of Standards, Washington, DC. *Society Aff:* ACS, APCA, NYAS, AAAS

Edwards, James T
Home: 44040 Halcom, Lancaster, CA 93534
Position: Deputy Chief, Tech Div. *Employer:* AF Rocket Propulsion Lab. *Born:* Aug 1, 1933. MS in Mech Engg from Univ of S CA; BS from OK Univ. Native of Okla City. Served in AF 1956-58. Eighteen yrs in Rocket Propulsion R&D field. Current position includes responsibility for rocket combustion, exhaust plume, and chem R&D tech. Past supervisory positions have been in areas of solid rocket component dev & air augmented rocket dev. Received the AF Assoc Award as the Outstanding Civilian employee of the AF Rocket Propulsion Lab in 1973.

Edwards, Ralph P
Home: Belle Isle Villy 263, Georgetown, SC 29440
Position: Mgr-Technical Services *Employer:* Georgetown Steel Corp *Education:* BMetE/Metallurgy/RPI. *Born:* 11/14/35. in Scranton, PA. Worked for Youngstown Sheet and Tube after grad. Primarily worked in steel-making. Presently working for Georgetown Steel Corp. Experience has included stints as rod & wire metallurgist, customer contact metallurgist for hot roll & cold roll sheets and back to steelmaking and iron making. Active in AISI and ASTM Committee work, deeply involved with community volunteer work in mental health, social services, and ecumenical groups. *Society Aff:* ASM, AISI, AIME, ASTM, Wire Assoc..

Edwin, Edward M
Business: 85-80 188 St, Hollis, NY 11423
Position: Management Consultant/Engr *Employer:* Edward M. Edwin & Co. *Education:* PhD/Bus Mgmt/Heed Univ; MBA/Management Health Serv/St. John's Univ; Prof. I.E./Management Engineering/Columbia Univ; MS/Industrial Engineering/Columbia Univ; BME/Mechanical Engineering/New York University; BBA/Business Administration/City University of New York *Born:* 8/3/17 Certifications Professional Engineer (PE)-California, Certified Management Consultant (CMC)-Institute of Management Consultants, Certified Appraiser (ASA) American Society of Appraisers Machinery & Equipment, Business Enterprise Valuation, Certified Public Accountant (CPA) New York State. Prof York College (CUNY), and Adjunct Prof Hofstra Univ. Arbitrator at American Arbitration Assn in commercial & construction disputes. *Society Aff:* NSPE, ASME, IIE, AICPA, ΠΤΣ, ΒΓΣ

Edzwald, James K
Business: Dept of Civil Engr, Amherst, MA 01003
Position: Prof *Employer:* Univ of MA *Education:* PhD/Envr Engr/Univ of NC - Chapel Hill; MS/Envr Engr/Univ of MD; BS/Civil Engr/Univ of MD. *Born:* 6/27/42. Civil engr with US Army Corps of Engrs, WA Aqueduct Div, 1964-1966. Envr engr with the Fed Water Poll Control Admin, Washington, DC, 1968-1969. Asst Prof of Civil Engrg at the Univ of MO-Columbia, 1972-1974. Asst, Assoc, Prof of Civil Engrg at Clarkson Coll of Tech 1974-1984. Visiting Prof, Johns Hopkins Univ, 1982-1983. Teaching res position in environ engrg at the Univ of MA since 1984. Major teaching-res areas are water supply and drinking water treatment, aquatic chem. Reg PE, NY. *Society Aff:* ASCE, AWWA, WPCF, AAEE, IAWPRC, AEEP

EerNisse, Errol P
Business: 1020 Atherton Dr. Bldg C, Salt Lake City, UT 84123
Position: Pres *Employer:* Quartex, Inc *Education:* Ph.D./EE/Purdue Univ; MS/EE/Purdue Univ; MIA/Bus/Univ of NM; BS/EE/SD State Univ *Born:* 2/15/40 Rapid City, SD. Employed at Sandia Labs 1965-1979: Staff Member researching piezoelectric devices (1965-1968); Supervisor of Device Physics Division in area of radiation damage in semiconductor devices (1968-1971); Supervisor of Ion Implantation Physics Division involving ion implantation into metals, semiconductors, and insulators (1971-1975); Supervisor of Solid State Device Physics Division studying microwave, solar cell, and quartz resonator devices (1975-1979). Helped start Quartex, Inc in 1979 as an R&D co working on quartz resonator force transducers and related instrumentation. Fellow of the IEEE and the APS. Enjoy golf, hunting, fishing, and skiing. 1983 W. G. Cady Award from Frequency Control Symposium. 1985 Distinguished Engineer Award, South Dakota State Univ. *Society Aff:* IEEE, NSPE, APS

Efimba, Robert E
Business: Howard Univ/Civil Engg Dept, Washington, DC 20059
Position: Assoc. Prof. *Employer:* Howard Univ. *Education:* ScD/Structural Mech/MA Inst of Tech; CE/Structures/MA Inst of Tech; SM/Structures/MA Inst of Tech; SB/Civil Engg/MA Inst of Tech. *Born:* 11/1/39. Reg PE. Assoc Prof, Civil Engg, Howard Univ. Mbr Howard Univ Senate Council. Former Chmn Univ-wide Library Systems Committee. Howard Univ Faculty Resource person for Engrg Computer Appl. Born in Cameroon. Attended King's College, Lagos and Government Secondary School, Afikpo, Nigeria. Four CE degrees from MIT including ScD. Mbr Sigma Xi, Chi Epsilon, Tau Beta Pi Advisory Board, ASCE, NSPE, WSE. Several yrs consulting experience. Previously Lecturer Univ of Nigeria, Nsukka. Designer, Limbe bridge in Cameroon. Mbr PADC Architect Engr Evaluation Board. Pres, Queen of Peace Arlington Federal Credit Union. VP, D.C. Council of Engrg & Archit Socs. Dir, Washington Soc of Engrs. Former Dir and Chrmn, VSPE Registration/Qualifications Cttee. PPres NVA Chapt, NSPE. PPres, National Capital Sect, ASCE. PChrmn ASCE District 5 Reg Council. Mbr, ASCE Public Information Cttee & ASCE/NCS Policy Issues Cttee. 1985 Arlington Soccer Assoc award. 1986 Howard Univ Faculty/Staff doubles tennis second place trophy. Received six annual Outstanding Instructor Awards, 1976-1987. Mathcounts competition 1985 Northern Virginia Regional Coordinator. Former Secretary, Board Mbr, Dir, ASCE/NCS, 1977, 1980, 1981. Chmn Admin Div and Vice President, ASCE/NCS. HU Outstanding Instructor Awards. 1980 ASCE/NCS Meritorious Service Award. 1974 ASCE Moisseiff Award. Research: simplified finite element modelling and stability analysis of building systems including shear-walls and elastomeric roofing materials. *Society Aff:* ASCE, NSPE, TBPI.

Egami, Takeshi
Business: Dept. of Materials Science and Eng., University of Pennsylvania, 3231 Walnut St, Philadelphia, PA 19104-6272
Position: Prof Mat Sci & Engrg *Employer:* Univ of Penn *Education:* PhD/Met & Met Sci/U of PA; B Eng/App Phys/U of Tokyo. *Born:* 07/15/45 Fullbright Fellowship for Study in US PhD, PA, 1971; Thesis on Magnetic Domain Walls in Rare Earth Metals. Post-Doctoral Fellow, Univ of Sussex, and Max Planck Inst, Stuttgart. Asst Prof Univ of Penn 1973; Assoc Prof 1976; Full Prof 1980. Visiting Prof, Max Planck Inst, Stuttgart, 1979-80. Res interests: Magnetic materials; structure and properties of Amorphous Alloys; quasicrystals. Robert Lansing Haroy Medal, AIME, 1974. *Society Aff:* APS, TMS-AIME, ASM, IEEE, MRS

Eggen, Donald T
Home: 254 Sequoia Ct #21, Thousand Oaks, CA 91360
Position: Prof. Retired *Employer:* Northwestern Univ. *Education:* PhD/Phys/OH State Univ; BA/Phys/Whittier College. *Born:* 2/11/22. Native of Hemet, CA. Worked on Manhattan Proj in Oak Ridge, TN, during WWII. Prog Mgr for liquid metal components & tech, fast reactor design, phys, & engg at Atomics Intl (a div of Rockwell Intl) between 1949 & 1966. Joined the Liquid Metal Fast Reactor Prog office at Argonne Natl Lab as Prog Sect Mgr for Safety and later Core Design. Dev

Eggen, Donald T (Continued)
natl prog plan for LMFBR for the AEC. Joined Northwestern Univ in 1968 as prof of Nuclear Engg. Retired in 1987. Cons to Argonne Natl Lab on LMFBR core design, safety & fuel dev; US Gen Accounting Office on LMFBR; Gen Atomics on Gas Cooled Fast Reator Safety, Tech Advisor to IL Commission on Atomic Energy, Adv Comm to IL Energy Resources Comm. Argonne Universities Association, Nuclear Engg Educ Comm. and EBR-II Review Comm. Res on LMFBR & GCFR safety. Fellow Am Nuclear Soc; Board of Directors 77-81, mbr Planning., Chrm, CHI:SECT. 1976 Reg PE IL. and CA. Enjoy camping, stamps and fishing. *Society Aff:* ANS, ΣΞ, AAAS.

Eggener, Charles L
Business: 401 E Fireweed Lane, Anchorage, AK 99503
Position: Principal *Employer:* Quadra Engrg Inc *Education:* MS/Civ & Envir Engr/Univ of WI; BS/Civ Engr/Univ of WI. *Born:* 10/3/46. Began as an engrg technician while attending high school. Later worked as Proj Engr for consulting firm. Subsequently served 3 1/2 yrs as a commissioned officer with USPHS achieving the Dist Engr assignment for Southeast AK. Later hd of sanitary engrg for constr of the Trans-Alaska Oil Pipeline. Subsequently Chief Engr for the Anchorage Water & Sewer Utilities. Then Chief Engr for Tam Constr, Inc. Presently a principal of Quadra Engrg Inc. Currently mbr of the State of AK Water/Wastewater Works Advisory Bd. Published works in water distribution analysis & arctic water and wastewater treatment. *Society Aff:* WPCF, AWWA, AAEE, ASCE, APWA.

Eggers, Alfred J, Jr
Business: 260 Sheridan Ave, Suite 414, Palo Alto, CA 94306
Position: Pres *Employer:* Rann, Inc *Education:* Ph.D./Engrg/Stanford Univ; MS/Engrg Sci/Stanford Univ; BA/Math Physics/Univ of NB-Omaha *Born:* 6/24/22 Native of Omaha, NB. Served with Navy in WW II. Aerospace Scientist with NASA, Ames Research Center 1944-64. Served in NASA Hq. through position of Ass't. Adm. for Policy, 1964-71. Hunsaker Prof at MIT, 1969-71 & Ass't. Dir for Research Applications, NSF, 1971-1977. Dir Lockheed PARL, in Aerospace and Energy R&D including resource inputs and environmental impacts. Have served on numerous Engrg Society and National Advisory Bds. Awards include Reed Award of AIAA, 1961, membership in NAE, 1973, and the Pres's Award for Distinguished Federal Civilian Service, 1977. *Society Aff:* NAE, AIAA, AAAS

Egle, Davis M
Business: 815 Asp St, Room 212, Norman, OK 73019
Position: Dir, AMNE *Employer:* Univ of OK *Education:* Ph.D./ME/Tulane Univ; M.S./ME/Tulane Univ; B.S./ME/LA State Univ *Born:* 1/31/39 Native of New Orleans, Louisiana. Joined faculty at OKlahoma as Assistant Prof, 1965, Associate Prof, 1969, Prof 1973, Dir, 1981. Summer and sabbatical experience with NASA Langley Research Center (1966, 1967) and Lawrence Livermore Laboratory (1973, 1979). Received Oklahoma Regents Award for Superior Teaching (1968), ASNT Achievement Award (1980). Author of more than fifty technical publications on structural dynamics, nondestructive evaluation, wave propagation. *Society Aff:* ASME, ASA, SEM, AAM, ASNT, AEWG

Ehasz, Joseph L
Business: Two World Trade Center, New York, NY 10048
Position: Chief Consulting Civil Engr *Employer:* Ebasco Services Inc *Education:* MSCE/Foundation Engrg/Rutgers Univ Grad School; BS/CE/Rutgers Univ *Born:* 5/3/41 Native of Phillipsburg, NJ. Attended Rutgers Univ and obtained BS and MS degrees in Civil Engr. Worked in field engrg and construction during early career with Ebasco Services Inc., a major utility design - constructor, since 1965. Past positions inclulded responsibility for all geotechnical design aspects of all hydroelectric, fossil fueled and nuclear power stations designed and constructed by EBASCO. Assumed present position as Chief Consulting Civil Engr in 1980. Responsible for all civil and earth sciences activities necessary for feasibility and siting of power facilities. *Society Aff:* ASCE, ACI, USCOLD, ISSMFE

Ehrenberg, John M
Home: 912 Michigan Avenue, Evanston, IL 60202
Position: Pres *Employer:* BHP, Inc. *Education:* BS/Mech Engrg/Northwestern U; -/Law/Chicago Kent Coll of Law; -/Bus Admin/Northwestern U. *Born:* Sept 28, 1937. Graduate study at Northwestern and Chicago-Kent Coll of Law. Now Pres of BHP, Inc with respon for the design, manufacture, and world- wide mkting of lab equip for the professional motion picture industry. With Bell & Howell 1956-1982 where served as motion picture projector Design Engr, and Mgr of Engg, Mgr of Product Planning, Mgr of Mkting and Pres for the B&H Professional Equip Div. Fellow of Soc of Motion Picture & Television Engrs; Mbr British Kinematograph, Sound and TV Soc. Assoc mbr Assoc of Cinema and Video Labs. Enjoy Photography, bicycling, sailing, and golf. *Society Aff:* SMPTE, BKSTS, ACVL.

Ehrich, Fredric F
Home: 36 Marion Rd, Marblehead, MA 01945
Position: Staff Engr *Employer:* GE Co, Aircraft Engine Group. *Education:* ScD/ME/MA Inst of Tech; ME/ME/MA Inst of Tech; BS/ME/MA Inst of Tech. *Born:* 12/17/28. 1947-1948 Res Fellow at Delft Inst of Tech, Netherlands. 1951-1957 Engg and engg mgt positions at Westinghouse Aircraft Gas Turbine Div including 1955-1956 as rep at Rolls, Royce, Ltd. 1957-date Engg & engg mgt positions at GE Co Aircraft Engine Grp of aircraft engine analysis, res, tech, design and dev. Des Mgr T64 Engine. Part-time instr at Drexel, Univ of KS & GE engg ext education. Author of over 30 papers in fluid mechanics, appl mechanics (vibrations) and engine design and dev. Holder of 8 patents. Former Chrmn of ASME Design Engg Div and editor of Journal of Vibrations, Stress Analysis etc. Former ECPD visiting mbr. Reg PE. Fellow ASME; Assoc Fellow AIAA; Mbr AHS. *Society Aff:* ASME, AIAA, AHS.

Ehrlich, I Robert
859 Columbus Drive, Teaneck, NJ 07666
Position: VP-Academic Affairs *Employer:* Stevens Inst of Tech *Education:* PhD/Mech Engrg/Univ of MI 1960; MS/Engrg Sci/Purdue Univ 1956; BS/Military Engrg/US Military Academy 1950 *Born:* 09/01/26 PhD ME 1960; MS Engrg Sci, Purdue 1956; US Army 1950-60. Conslt, Grumman, 1961-69. Dept ME, Stevens Inst Tech, 1968-present; Dean for Res, 1974-83; Acting Head Dept ME, 79-83; VP for Sponsored Res, 1983-86. VP, Academic Affairs 1986-. Mbr Intl Soc for Terrain Vehicle Sys (Secy for US, 1964-67, Gen Secy 1967-78; VP 1978-81, Pres 81-84) ASTM, Natl Acad Sci, SAE, ASME, Assoc of US Army; NSPE. Author of 50 tech Papers, recipient of many honors and awards. *Society Aff:* SAE, ISTVS, ASME

Ehrsam, Otto, Jr
Business: 701 E Third St, Bethlehem, PA 18016
Position: Market Dev Engr *Employer:* Bethlehem Steel Corp. *Education:* BS/Met Engg/Lehigh Univ. *Born:* 4/20/30. Native Brooklyn, NY. Engg Officer USMCR 1951-53. Bethlehem Steel Met Div 1953-61 advancing Gen Foreman Labs, responsible physical testing, experimental heat treatment. Conducted first high temperature strain gauge application shrink fitting large billet extrusion containers. Since 1961, Commercial Res, Market Res, Sales Engg, & Mktg. Market development sheet products for appliances, office furniture & motors. Former Chrmn AISI/ILZRO joint task group Durability Coated Sheets, ASM Metal Progress Editorial Committee, Lehigh Valley Chapter Secy, Exec Committee, & Mgr Engrs Club Lehigh Valley. 1975 Governor's Review State Govt Mgt, Exec Committee, Team Leader. Bethlehem City Councilman, Chrmn Human Resources & Environment, Public Works, Public Safety & Finance Ctte. 1981 Lehigh Univ Alumni Award. *Society Aff:* ASM

Eibling, James A
Home: 1380 Camelot Dr, Columbus, OH 43220
Position: Energy Consultant (Self-employed) *Education:* BME/Energy/OH State Univ. *Born:* 11/22/17. B F Goodrich 1940-42, quality control, tires; Army Corps Engrs 1942-46 Battalion Commander and Div Engr, Europe, 29 & 69 Inf Divs,

Eibling, James A (Continued)
Army Commendation Medal, Bronze Star, three battle stars; Member the Retired Officer Assoc. Battelle Meml Int, Res Engr 1946-60, Ch Thermal Sys Div 1960-73, Mgr Solar Energy Res 1973-79; Intl Solar Energy Soc, VP 1975-77, Bd Dir 1966-74, 1978-81, Pres Am Sec 1973-74; Elected Solar Hall of Fame 1986; ASME (Fellow) Chrmn Solar Div 1960-64; ASHRAE, (Life Member) Heat Transfer Ctte 1962-68, Best Paper Award 1974; IIR US Natl Ctte, Natl Acad Sci 1970-78; Sigma Xi; PE State OH; over 60 tech pubs. Visited numerous countries on solar energy applications e.g. Australia, Japan, Soviet Union, India, Africa several European countries. Tech activities: solar energy, energy conversion, comfort control, desalination. Married, 2 sons. *Society Aff:* ASME, ASHRAE, ISES, TROA

Eichholz, Geoffrey G
Business: Nuclear Engineering & Health, Physics Program, Georgia Institute of Technology, Atlanta, GA 30332
Position: Regents' Prof *Employer:* GA Inst of Tech *Education:* D.Sc/-/Univ. of Leeds, Leeds, England; PhD/Phys/Univ of Leeds, Leeds, England; BSc/Phys/Univ of Leeds, Leeds, England *Born:* 06/29/20 After grad, worked for British Admiralty on radar devel 1942-1946. PhD res on microwave properties of magnetic matls. Asst Prof of Phys, Univ of British Columbia, Vancouver, Canada, doing research in nuclear phys. In 1951 joined Canadian Bureau of Mines to hd subdivision doing work on radiometric analysis of uranium ores, tracer work in metallurgical plants and devel of on-line instrumentation for mines and mills. At Georgia Tech since 1963, helped devel acad prog in nuclear engrg and health phys, as well as in architectural acoustics. Res activities have concentrated on radiation applications, radiation detector devel, environmental impact of nuclear power, radon, and subsurface migration processes of radioactive waste. Published books on Environmental Aspects of Nuclear Power, Radioisotope Engrg, and Nuclear Radiation Detection. *Society Aff:* ANS, APS, HPS, CAP, ΣΞ, IOP

Eichhorn, Roger
Business: Cullen Coll of Engrg, Houston, TX 77004
Position: Dean of Engrg *Employer:* Univ of Houston *Education:* PhD/ME/Univ of MN; MS/ME/ Univ of MN; BS/EE/Univ of MN *Born:* 4/1/31 Native of MN. Public Schools in Lake Wilson and St. Paul. Research Associate and Instructor, Mechanical Engineering, Univ of MN, 1955-59. Faculty, Princeton Univ, 1959-1967; Univ of Kentucky, 1967-1982. Chairman, Mechanical Engineering , Univ of Kentucky, 1967-1975; Acting Dean, College of Engineering, 1975-1976; Dean, 1979-1982. Associate Dean for Research, Graduate School, 1976- 1978; Univ of Houston, Prof and Dean of Engrg, 1982-present. Visiting Professor, Imperial College, London 1963-1964 and Spring, 1974. Fulbright Lecturer, Ecuador, Summer 1976. Visiting Professor, Assiut Univ Egypt, December 1978-1979. Editorial Boards-Journal of Heat Transfer (ASME), International Journal of Heat and Mass Transfer, Journal of Mathematical and Physical Sciences (madras) India, International Journal of Heat and Fluid Flow. *Society Aff:* ASME, AIAA, NSPE, AAAS, ASEE, AAUP, ΣΞ.

Eiden, Carl H
Business: 230 Park Ave, New York, NY 10017
Position: Partner. *Employer:* Meyer, Strong & Jones. *Born:* Aug 25, 1923 Cochem, Germany. Indus Mech Engg Cert Pratt Inst 1943; ME Stevens Inst of Tech 1948. USNR 1943-46. MS&J since 1950, being admitted as genl partner in 1964. Pres NY Assn Cons Engrs 1970-72. VP NY Bldg Congress 1975-. Fellow of Am Cons Engrs Council. Mbr ASHRAE & Am Arbitration Assn.

Eidt, Clarence M, Jr
Business: PO Box 101, Florham Park, NJ 07932
Position: VP *Employer:* Exxon Research & Engineering Company *Education:* MS/ChE/LA State U; BS/ChE/LA State U. *Born:* 2/4/35. Natchez, MS. BS & MS Chem Eng LSU. Employed June 1956 at Exxon Res. & Dev. Labs Baton Rouge, LA. 1963-68, Section Head, Proc. Dev.; 1972-73, Exxon Eng. Petroleum Dept., Florham Park, NJ as Asst Genl Mgr of Petroleum Eng. Returned to Baton Rouge Labs in 1973 as Mgr. 1976-78, Mgr. Regional Planning Logistics Dept, Exxon Corp., NYC. 1978-80, Mgr. Corp. Planning Coordinaton, Corp. Planning Dept. 1980-82, Genl Mgr, Exxon Engrg Petroleum Dept, Florham Park, NJ. Promoted to VP, Petroleum & Synthetic Fuels Res, 1982. State of LA PE. *Society Aff:* SAE, API, AIChE, CRC.

Eilering, John G
Business: P O Box 767, Chicago, IL 60690
Position: VP *Employer:* Commonwealth Edison Co. *Education:* BS/ME/Purdue Univ *Born:* 4/20/18 Joined Commonwealth Edison in 1940. Served in various positions and for 2 years; was in charge of the building of Dresden Nuclear Power Station. Elected VP May 17, 1966. He is a registered professional engineer. Is a member of the Economic Club of Chicago, and the Conservation and Energy Services Executives' Conference, a trustee of Western Society of Engineers, and Underwriters Laboratories, Inc., a dir and former Pres of the Electric Assoc and a dir of Junior Achievement of Chicago. He served as a major in the Field Artillery during World War II. *Society Aff:* ASME

Eisen, Edwin O
Business: Dept. of Chemical and Electrical Engineering, P.O. Box 976, Lake Charles, LA 70609
Position: Hd, Dept of Electrical and Chemical Engineering *Employer:* McNeese State Univ. *Education:* DEngrSc/Chem Engg/NJ Inst of Tech; MS/Chem Engg/NJ Inst of Tech; BS/Chem Engg/NJ Inst of Tech. *Born:* 3/12/40. in Newark, NJ. Asst Prof, Assoc Prof, Lamar Univ, Beaumont, TX, 1964-74; Sr Process Engr, Gulf Sci & Tech Co, Houston, TX, 1975-79; Prof and Hd, Dept of Chemical & Electrical Engrg, McNeese Univ, Lake Charles, LA, 1979-. Mbr Omega Chi Epsilon, Phi Kappa Phi, Tau Beta Pi. Natl Secretary, Exec Secretary, Omega Chi Epsilon, 1970-. Consultant to E I DuPont Co, Gulf Oil Corp, Educational Testing Service, Cities Service co, Conoco, Inc. *Society Aff:* AIChE, ASEE.

Eisenbach, Robert L
Business: G&E Engineering, Inc, P.O. Box 45212, Dept. 186, Baton Rouge, LA 70895
Position: Vice President. *Employer:* G & E Engineering, Inc. *Education:* BChE/Chem Engg/GA Inst of Tech; BS/Chemistry/Univ of AL. *Born:* 8/21/29. Native of NYC. Regular Army (Korea) 1951-54. Process Engr with Ethyl corp (1956-60); Proj engr with TN Products & Chem Corp (1960-62); Proj Engr subsequently has been involved with environmental. (1962-69) and Staff Environmental Engineer (1969-85) with Kaiser Aluminum & Chemical Corp.; Program Manager & Vice President with G & E Engineering (1985-present). Elected to Bd of Dirs of the Air Pollution Control Assoc (APCA) in 1977. Elected First VP in 1979; served as Pres of APCA in 1980-81. Founding mbr of the LA Sec of APCA; served as Chrmn 1973-74. Reg PE in LA. Enjoys philately and fishing. *Society Aff:* APCA.

Eisenberg, Lawrence
Business: School of Engineering & Applied Science, Dept of Electrical Engineering, Philadelphia, PA 19104
Position: Prof *Employer:* Univ of PA *Education:* D Eng Sci/EE/New Jersey Inst of Tech; MS/Math/NYU; BS/EE/Fairleigh Dickinson Univ *Born:* 06/09/33 He was previously a faculty mbr in the Dept of Elec Engrg at the New Jersey Inst of Tech. He has had industrial experience with Sys Development Corp, Intl Telephone and Telegraph, Gen Precision Aerospace, Electronics Assocs, Inc, and the Philadelphia Elec Co, where he is presently a conslt. Dr Eisenberg was a Natl Sci Fdn Faculty Fellow from June 1965 to June 1966. His major res interests are in the areas of energy sys, networks, and the theory and application of automatic control sys. He has authored several technical papers in these areas, and is a mbr of the IEEE, ISA, Sigma Xi, Eta Kappa Nu, etc. *Society Aff:* IEEE, ISA, ASEE.

Eisenberg, Martin A
Business: Dept of Engg Sciences, Gainesville, FL 32611
Position: Prof and Chairman. *Employer:* Univ of FL. *Education:* DEng/Mech Solids/Yale Univ; MEng/Mech Solids/Yale Univ; MS/Aero & Astro/NYU; BAeroE/Aero/NYU. *Born:* 3/8/40. Brooklyn, NY. On Res Staff of NYU Engg Res Div, 1958-61; Stuctures Engr at Sikorski Aircraft, 1962-64. DEng in Mechanics of Solids, Yale Univ, 1967. On Univ of FL Faculty since 1966; Prof of Engg Sciences since 1975, Associate Chairman since 1980, Chairman since 1986. During this period has held visiting appointments at Univ of IL, Yale, and GM Res Labs and has served as consultant to private ind & govt agencies Res interests include theory of plasticity & viscoplasticity, inelastic wave propagation, and composite mtls. author of *Introduction to the Mechanics of Solids*, Addison-Wesley, 1980. *Society Aff:* ASME, AAM, SES, ΣΞ, TBΠ, AIAA.

Eisenberg, Phillip
Business: 7210 Pindell Sch Rd, Laurel, MD 20810
Position: Pres. *Employer:* Hydronautics, Inc. *Education:* BSCE/Structural Engg/Wayne State Univ; -/Hydraulics/Univ of IA; CE/Fluid and Structural Mechanics/CA Inst of Tech. *Born:* 11/6/19. Native of Detroit, MI. Hd of hydrodynamics res group, US Navy's David Taylor Model Basin 1942-44, 1946-53. Active duty, US Navy Tech Mission in Europe (1944-45). Hd, Mechanics Branch, Office of Naval Res (1953-1959). Founder and Pres, Hydronautics, Inc (1959-present), res, engg, and advanced model testing in naval & industrial hydrodynamics, marine & ocean engg. Pres, SNAME (1973-74); Pres, MTS (1976-77); Honorary Mbr & Fellow, SNAME; Fellow, ASME, RINA; Elected Natl Acad of Engg (1974). David W Taylor Medalist, SNAME (1971), Gibbs Brothers Medalist, Natl Acad of Sci (1974). Mbr, Natl Res Council Marine Bd (1974-81) & Maritime Transportation Res Bd (1975-1979); Dept of Commerce Technical Advisory Bd (1978-80); Editor, Journal of Ship Res (1961-71). *Society Aff:* SNAME, ASME, ASNE, MTS, RINA, NSPE, AIAA, ASEE.

Eisenbud, Merril
711 Bayberry Dr, Chapel Hill, NC 27514
Position: Prof Emeritus. *Employer:* Inst of Env Med, NY Univ Medical Ctr. *Education:* ScD/Hon/Fairleigh Dickinson Univ; DHC/Hon/Catholic Pontifical Univ; BS/Elec Engg/NYU. *Born:* 3/18/15. Indus hygienist, Liberty Mutual Ins Co, 1936-47; Dir, Health & Safety Lab, US Atomic Energy Com, 1947-57; Mgr of Oper NY Oper Office, USAEC, 1957-59; Prof & Dir, Lab for Environ Studies, NYU Medical Ctr, 1959-85. On leave 1968-70 as first Admin, Environ Proj Admin, NYC. Adjunct Professor, Univ of N. Carolina, Dept of Env Eng of Science 1985- . Author of more than 180 technical articles, "Environmental Radioactivity," –The Environment, Technology and Health" (NYU Press 1978), from Industrial, Military, and Natural Sources, Academic Press, 3rd Edition 1987. *Society Aff:* NAE, HPS, NCRP, ANS, NYAS.

Eisenstein, Bruce A
Business: ECE Dept, Drexel Univ, 32nd and Chestnut Sts, Philadelphia, PA 19104
Position: Dept. Head *Employer:* Drexel University *Education:* Ph.D./E.E./Univ of PA; M.S./E.E./Drexel; B.S./E.E./M.I. T. *Born:* 9/10/41 in Phila, PA. Taught at Drexel and Princeton Univ. Became head of the Electrical and Computer Engineering Dept of Drexel in June, 1980. *Society Aff:* IEEE, ASEE, TBΠ, HKN, ΣΞ

Ekberg, Carl E, Jr
111 Lynn, Ames, IA 50010
Position: Prof. *Employer:* IA State Univ. *Education:* PhD/Civil Engr/Univ of MN; MS/Structural Engr/Univ of MN; B/Civ Engr/Uni of MN. *Born:* 10/28/20. Originally from Minneapolis, MN. Served as aircraft maintenance officer with USN 1943-46 on West Coast & Asiatic-Pacific Theater. Commenced academic career as instr of Mathematics & Mechanics at Univ of MN in 1946. Served as Asst Prof of Civil Engr at ND State 1951-53, Asst Prof & Assoc Prof at Lehigh Univ 1953-59. Taught structural engr conducted prestressed concrete res with particular emphasis on fatigue of bridges. Joined IA State as Prof & Hd, Dept of Civ Engrg in 1959. Maintained interest in structural concrete research while administering academic unit with 45 faculty positions, and over 700 students. Prof. at IA State 1985- . Elected Fellow of ACI and ASCE, Chrmn of Civ Engrg Div of ASEE 1971-72. Exec Council of PE in Educ of NSPE 1976-78. Pres Iowa Sec of ASCE in 1975-76. Pres of the Iowa- Minn Chap of ACI 1982. Vice-Chrmn of AREA Ctte 24-Ed in 1984-87, Chairman 1987-90. *Society Aff:* ASCE, ACI, ASEE, AREA, NSPE, PCI, PTI

El-Abiad, Ahmed H
Business: P.O. Box 144, Univ of Petroleum & Minerals, Box 1937, Dhahran, Saudi Arabia
Position: Prof of EE. *Employer:* Univ of Petroleum & Minerals *Education:* PhD/EE/Purdue Univ; MS/EE/Purdue Univ; BSc/EE/Cairo Univ. *Born:* 5/24/26. He has taught at Cairo & Ein Shams Univs, Purdue Univ & has been a visiting prof at the MIT, the Fed Univ of Rio de Janeiro, Brazil, Lund Inst of Tech, Sweden & Kuwait Univ, Kuwait. His consulting activities include work with Am Elec Power Service Corp, Ebasco Services, Consumers Power Co, the Edison Elec Inst, the Gen Egyptian Electricity Corp, System Control, Inc, Data Indexing Systems Corp, Education Dev Ctr & the Ctr for Profl Advancement. He also had been involved in consulting work in the areas of tech transfer, technical higher & continuing education & proj organization & mgt with the Ministry of Water & Energy, Iran, and the Ministry of Industry & Energy, Algeria. He is the author or co-author of over eighty technical papers & reports, contributor to the World Book Encyclopedia & McGraw-Hill Yearbook of Sci & Tech, & the co-author (with G W Stagg) of the book, *Computer Methods in Power System Analysis*, McGraw-Hill, 1968 & the editor of the book "Power Systems Analysis and Planning" Hemisphere Publishing Corp., 1983. *Society Aff:* IEEE, CIGRE

Elam, Edward E, III
Business: 1211 Old Albany Rd, Thomasville, GA 31792
Position: Plant Mgr TRW Thomasville *Employer:* TRW, Inc. *Education:* BS/Indus Mgt/TN Tech Univ *Born:* 7/7/29. Native of Nashville, TN. Upon completion of my education, I was a mbr of US Army Signal Corps stationed in Korea. Joined TRW as a QC Technician and thru successive job promotions became mgr, QC for the Compressor Components Div with the responsibility of formulating & implementing QC policies, system maintenance, & assurance. I am responsible for developing what is considered by many as the most unique quality award system in industry. I am now plant manager of TRW Thomasville. *Society Aff:* ASM, SME

El-Bayoumy, Lotfi E
Home: 6041 Peridot Ave, Alta Loma, CA 91701
Position: Mbr of Tech Staff *Employer:* Western Gear Corp *Education:* PhD/Solid Mech/NYU; MSc/Aero/Cairo Univ; BSc/Aero/Cairo Univ *Born:* 01/18/42 in Egypt; served as an instructor in Aero Engrg at Cairo Univ during 1964- 66 period; came to the USA for grad studies; became Asst Prof of Aero & Astro at NYU after finishing PhD; worked as Advanced Vibration Analyst at PWA, United Tech 1972-74. as a Sr Proj Engr then Group Engr at Sundstrand Corp 1975-80 serving as Engrg Conslt to NASA & as an Adjunct Assoc Prof at CSU, Long Beach beside working as mbr of tech staff at Western Gear Corp; has published numerous papers on Mechs. *Society Aff:* AIAA, AAM, NMA

Elder, John H, Jr
Home: 148 Arborvitae Dr, Pine Knoll Shores, Morehead City, NC 28557
Position: Retired. *Education:* BS/Civ Engg/VPI. *Born:* 6/14/20. in Richmond, VA. Served in US Army Corps of Engrs 1941-76 in US, Europe, Korea, Vietnam. Commanded 18th and 20th Engr Bdes in Vietnam. Dep Cmd, US Army Construction Agency, Vietnam. Retired 1976 as Dir Plans and Policy, Joint Chiefs of Staff in grad of Lt Gen. Enjoy golf, fishing, painting, civic activities of a planning nature. *Society Aff:* SAME.

Elder, Rex A
Business: 50 Beale St, San Francisco, CA 94105
Position: Engr Mgr *Employer:* Bechtel Civil & Minerals, Inc *Education:* MS/Hydraulics/OR State Coll; BS/Civil Eng/Carnegie Inst of Tech *Born:* 10/04/17 Native of PA. With the Tennessee Valley Authority from 1942 to 1973. Dir of TVA Hydraulic Lab for 13 yrs and of TVA Engrg Lab for 12 yrs. Have been Engrg Mgr for Hydraulic/Hydrology Group of Bechtel Civil & Minerals since 1973. Specialist in Hydraulic Model Testing and Hydraulic problems associated with dams, reservoirs and large hydraulic structures. *Society Aff:* ASCE, ASME, USCOLD

Eldred, Kenneth McK
Home: 722 Annursnac Hill Rd, Concord, MA 01742
Position: Director *Employer:* Kem Eldred Engineering *Education:* BS/Gen Engg/MA Inst of Tech. *Born:* 11/25/29. Native of Springfield, MA. Post grad studies at MIT, 1951-53 and at UCLA, 1960-63. Worked as Engg in charge vibration and sound lab Boston Naval Shipyard, 1951-54. Supervisory Physicist, Chief Phys Acoustics sect USAF, Wright Field, O, 1956-57. VP, Cons Acoustic Western Electro-Acoustics Labs, Los Angeles, 1957-63. With Wyle Labs, El Segundo, CA, 1963-73, as VP, Tech dir Sci Services and Systems Group. Joined Bolt Beranek & Newman Inc, Cambridge MA, 1973 as VP & dir div environ and noise control tech; Elected Principal Consultant 1977-81. Established Ken Eldred Engineering (KEE) Mar 1981. m. Helane Barbara Koerting Fischer May 31, 1957; 1 dau, Heidi Jean McKechnie. Enjoys sailing. *Society Aff:* ASA, NAE, INCE, SAE, SNAME

Eldridge, Bernard G
Business: 19006 La Puente Rd, Suite 121, West Covina, CA 91792
Position: Dir, Energy Systems *Employer:* The Bendix Corp *Education:* MS/ME/Univ of Durham, Eng; BS/ME/Royal Naval Tech Coll, Eng; BS/Naval Arch/Royal Naval Tech Coll, Eng; BS/EE/Univ of Birmingham, Eng *Born:* 11/2/34 Experience & Achievements include: Design, manufacturing, engrg & construction of engrg systems & process for a broad spectrum of industry. Development & implementation of energy conservation & renewable energy sources. Conceptual design analyses, engrg & construction of demonstration projects for DOE, veterans administration for solar HVAC, industrial process steam wind energy & photovoltaics. Presentation & publication of numerous technical & economic papers at national & intl conferences. Representing DOE/DOC & United Nations on trade/technology conferences in sixteen countries. Part-time lecturer at Carnegie Mellon Univ, Pittsburgh. *Society Aff:* ASHRAE, NSPE, ASME, ISES, AEE

Eldridge, John W
Business: Chem Engr Dept, Goessmann Lab, Amherst, MA 01003
Position: Prof Chem Engr. *Employer:* Univ MA. *Education:* PhD/ChE/Univ MN; MS/ChE/Syracuse Univ; BS/ChE/Univ of ME. *Born:* 8/22/21. Chem Engr, educator; b Nashua, NH, s Clarenc C & Grace (Hamor) E; BS, Univ ME, 1942; MS, Syracuse Univ 1946; Phd, Univ MN, 1949; m Lucille Eleanor Patten, May 26, 1942; Children - John Wm, Stephen Chapman, David Patten. ChE Carnegie-IL Steel Co, Clairton, PA, 1942; chem engr Semet-Solvay Co, Syracuse, NY, 1942-46; chem engr Barrett div Allied Chem & Dye Corp, Phila 1949-50; cons Albemarle Paper Mfg CO, Richmond, VA, 1951-64; prof Univ VA, 1950-62; prof Univ MA, Amherst, 1962-. Mbr Am inst Chem Engrs, Am Chem Soc, Am Soc Engg Edn, Tau Beta Pi, Sigma Xi, Alpha Chi Sigma, Phi Lambda Upsilon. Contbr articles profl jour. Patentee in field. *Society Aff:* AIChE, ACS, ASEE.

Elfant, Robert F
Business: OMEX, 2323 Owen Street, Santa Clara, CA 95051
Position: Vice President Systems *Employer:* OMEX *Education:* BS/EE/SMU; MEE/EE/NYU; PhD/EE/Purdue. *Born:* March 12, 1936. BTL 1957-59 Mbr of Technical Staff; IBM Res 1961-67; IBM Product Div in var mgmt rolls; OMEX Vice President System 1967 Eta Kapp Nu Outstanding Young Elec Engr; 1968 Distinguished Alumni Purdue Univ; 1971 Fellow IEEE. *Society Aff:* IEEE.

Elgerd, Olle I
Business: Dept of EE, Gainesville, FL 32611
Position: Prof. *Employer:* Univ of FL. *Born:* Mar 31, 1925. BSEE Orebro Tech Coll, Sweden; Diploma degree: Royal Inst of Tech, Stockholm, Sweden; DSc: WA Univ, St Louis, MO; Served in Swedish Army 1945-46; Design Engr, Sverdrup & Parcel; has held teaching positions with WA Univ, Royal Inst of Tech (Stockholm), Univ of CO. Since 1956, Prof of EE, Univ of FL, coordinator for EE Energy Prog. Author of three textbooks in controls and energy-environmental interaction. Fellow, IEEE (1971). Hobbies: Mountaineering, canoeing.

Elgin, Joseph C
Home: c/o Bishop, 166 Wilson Rd, Princeton, NJ 08540
Position: Retired (Dean of Engrg) *Education:* PhD/Chem Engg/Princeton Univ; PhD/Phy Chem/Princeton Univ; MS/Phy Chem/ Univ of VA; ChE/Chem Engg/Univ of VA. *Born:* 2/11/04. Nashville, TN. Mbr Faculty Princeton University, 1929--; Prof Chem Engg, 1939-72; Dean of Engg, 1954-71, Emeritus, 1972. During WWII served with NDRC, OSRD; chief polymer dev, Rubber Div WPB, 1940-44; div chief SAM Labs, Manhattan Proj, Columbia, 1944-45. Recipient William H Walker Award, AIChE, 1958; Founders Award, 1972; Warren K Lewis Award 1975; Lamme Award, ASEE, 1964. Fellow, AICh/w (Dir); mbr Bd Trustees, Assoc Univ, Inc, 1950-62, 1968-71, chmn Bd, 1957-58. *Society Aff:* AIChE, ACS, ASEE.

Elias, Peter
Business: 77 Mass. Ave., Rm NE43-317, Cambridge, MA 02139
Position: Prof *Employer:* MIT *Education:* Ph.D./Applied Physics/Harvard; M. Eng. Sci/App. Phys. /Harvard; M.A./App. Phys/Harvard; S.B./Indust. Mgmt/MIT *Born:* 11/26/23 I was brought up in Manhattan, Graduated from the Walden School in 1940, Attended Swarthmore Coll 1940-1942 and MIT 1942-44, Served in the Navy 1944-46, Did graduate work at Harvard 1946-50, Married Marjorie Forbes in 1950, was Harvard Junior Fellow 1950-53. Joined MIT as Asst Prof 1953, Head of Electrical Engineering Dept. 1960-66, Cecil H. Green Prof of E.E. 1970-72, Edwin S. Webster Prof of E.E. 1974-Present. My research has been in information Theory in both Communications and Computation. *Society Aff:* IEEE, AAAS, IMS, ACM, NAS, NAE, AAAS

Elias, Samy E G
Business: 600 Fifth Street, NW, Washington, DC 20001
Position: Dir, Office of Transit Engrg & Safety *Employer:* Washington Metropolitan Area Transit Auth *Education:* PhD/Ind Engg/OK State Univ; MS/Ind Engg/TX A&M Univ; BS/Aero Engg/Cairo Univ. *Born:* 6/28/30. Emigrated from Cairo, Egypt in 1956; Served as Exec Asst to the Chrmn of the Bd of the Organization of Military Factories in Egypt; Taught at TX A&M Univ, KS State Univ and WV Univ, where he served as Chrmn of the Ind Engg Dept, as Benedum Professor of Transportation, as Director of the Harley O. Staggers National Transportation Center, and as Special Asst to the Univ Pres for Personal Rapid Transit (PRT); Is credited for the initiation of the PRT, the first system of its kind in the world, which was selected by the PE Soc as one of the top 10 outstanding engg achievements in 1972; Served as Dir of the Transportation Div of IIE & Chrmn of the Ind Engg Sec of ASEE. *Society Aff:* IIE, SAVE, SCS, NSPE.

Eligator, Morton H
Business: 79 Madison Ave, New York, NY 10016
Position: Partner. *Employer:* Weiskopf & Pickworth Cons Engrs. *Education:* BS/Civ Engg/Columbia Univ *Born:* 6/28/24 BS in Civ Engg from Columbia Univ. Lives in Briarcliff Manor, NY. Joined Weiskopf & Pickworth in 1948; admitted to partnership in 1960. Respon for design of many maj projs, including James Forrestal Bldg in Wash, DC; VA Hospitals in Gainesville, Nashville, & DC; Chase Manhattan Bank Hdqtrs in NYC; Corp Hdqtrs for Procter & Gamble in Cincinnati; Mt. Sinai Hospital in NYC. Mbr of numerous technical and profl organizations. On Natl Panel of Arbitrators. Co-author of section "Multi-Story Buildings" in Structural Engg Handbook. *Society Aff:* ACEC, NSPE, ASCE, AWS

Elijah, Leo M
Home: 235 Main St. (4c-1), East Hartford, CT 06118
Position: Consulting Engrg (Self-employed) *Education:* B.S./Chemistry/Univ of Bombay; M.S./Met. Eng. /Univ of WI; Post- Grad Dipl/Management & Tech/Nat Foundry Coll *Born:* 3/30/28 As registered prof eng of both USA and Canada, have delivered/published twelve papers on heat-treating, alloying, casting, management, maximizing productivity, including contributor to the Metals Handbook (ASM). Received intensive training in management/marketing and mechanization/automation/computerization techniques for profit maximation at Ford Motor, I.B.M, etc, that boosted productivity over ten million dollars annually overall. Numerous inventions. Represented colleges for track, field and hockey. Areas of specialization include: Materials, Heat-Treating, Casting, Alloying, Machining, Grinding Design. (Ferrous & Non-Ferrous) Marketing, Productivity, Automation. *Society Aff:* NSPE

El-Kareh, Auguste B
4415 Brady Road, Colorado Springs, CO 80915-1307
Position: President *Employer:* ABEK Inc. *Education:* Dr of Sci/EE/Univ of Delft; Diplng/EE/Univ of Delft. *Born:* 7/9/32. Married, 2 children. Undergrad education, Univ of London (England). Grad education, Univ of Delft (Netherlands). Received Dip Ing and DSc Degrees from Univ of Delft. Mbr Technical Staff RCA Labs Princeton 1960-1963, Faculty Mbr Univ of PA, Penn State Univ, Syracuse Univ, between 1963 and 1971. 1971-1981 with Univ of Houston, Prof of Elec Engg and Assoc Dean for Grad Progs and Res. June 1981, President ABEK R & D Company. Consultant to GE, Honeywell, & Burroughs. Sr mbr of IEEE. Author of three books on electron beam tech & engg physics. *Society Aff:* IEEE, ASEE.

Elkins, Lincoln F
Home: 3412 N Preston, Oklahoma City, OK 73122
Position: Petroleum Consultant *Employer:* Self *Education:* Prof Deg/Petrol Eng/CO School of Mines *Born:* 2/12/18 Native of Denver, Colorado. Research Engineer, Amoco Production Co., 1941- 45. Production Eng, Continental Oil Co., 1945-47. Special Projects Engr., Chief Engr., Tech Advisor to V.P., Senior Consulting Engr., Sohio Petroleum Co., 1947- 83, Petroleum Consultant 1983-present. Frequent contributor of reservoir related papers to SPE literature. SPE Distinguised Lecturer 1962-63. SPE De Golyer Distinguished Service Medal 1971. AIME Honorary Membership 1977. National Academy of Engineering 1980. Mbr of Bd of Adjustment of Oklahoma City 1973-82. SPE/DOE EOR (Enhanced Oil Recovery) Pioneer 1986. *Society Aff:* SPE of AIME, NAE

Elkins, Lloyd E
Home: 2806 East 27th St, Tulsa, OK 74114
Position: Petroleum Consultant. *Employer:* Private Practice. *Education:* PE/Petroleum Production/CO School of Mines. *Born:* 4/1/12. Wheatridge, CO. Prof Pet Engg Degree, CO Sch of Mines, 1934. Employed June 1934-1977 by Amoco Production Co. Mandatory retirement from Amoco Production Co April 1, 1977. Opened Petroleum Consulting office 4/1/77. Mbr of API, SPE of AIMME, AAPG & NSPE (reg in OK). Offices in AIME; Chmn, AIME Petroleum Branch (1949-50); National Dir (1945-50), National Pres of AIME 1962-63. Honors: CO Sch of Mines Disting Achievement Medal 1961; OK State Univ Engrs Hall of Fame 1961; AIME Anthony F Lucas Gold Medal 1966 & Hon Mbrship 1969; Univ of Tulsa Hall of Fame of the College of Engg & Physical Sci 1975; Natl Acad of Engg 1976. *Society Aff:* SPE-AIME, AAPG, API, NSPE.

Ellefson, George E
Business: 1200 Summit Ave, Little Rock, AR 72202
Position: Owner (Self-employed) *Education:* BSEE/EE/Univ of AR. *Born:* 6/19/29. M Dorothy Claire Stannus. Appts incl Elec Assocs Erhart, Eichenbaum, Rauch, Blass-Archts 1956-60; Engr Leo L Landauer and Assocs Inc 1960-61; Prin G E Ellefson & Assocs 1961-70; Gen Ptnr Ecol Dynamics Assocs, Little Rock, Dallas, 1970-73; Present assn 1973; Mbrships incl IEEE (Sr); IES (M); Am Mngm't. Assoc.; NY Acad of Sci; Honors: Engg Excellence Honorable Mention AMA, 1968; Engr of Distinction, EJC, 1974; Hobbies: swimming, traveling. Co-founder Brock, Ellefson, & Assocs, Inc, a minority bus enterprise, mgt consultants, engrs. PO Box 653 Tahlequah, OK 74464 (918) 456-4782. *Society Aff:* IEEE, NFPA, AEE, IES, NYAS, AAAS, AMA.

Ellenberger, William J
Home: 6419 Barnaby Street, N.W, Washington, DC 20015
Position: Eng Consultant (Self-employed) *Education:* BS/EE/George Washington Univ; BS/ME/George Washington Univ; Cert/Eng & Bus/Gen Elec Co *Born:* 1/14/08 59 yrs. prof eng in public utility, industrial, teaching (EE), Government (military and civilian) and consulting work. Public Utility: engineering service to large customers. Col (Ret) US Army Signal Corps including service on Army General Staff as advisor on aeronautical and R&D facilities. 5 yrs. Plant Superintendent, National Bureau of Standards: design, construction, maintenance of physical plant and facilities. 7 yrs. staff supervision of Army Chemical Corps installations. 7 yrs. responsible for preparation and defense Army military Construction Program. Now consultant largely engaged in design power supplies for communications and computer systems. WSE Award 1972. *Society Aff:* IEEE, ASME, NSPE, WSE

Eller, Thomas J
Business: USAF Academy, CO 80840
Position: Prof & Head *Employer:* Dept of Astro & Computer Sci *Education:* PhD/Aerospace Engr/Univ of TX-Austin; MS/Aero/Astro/Purdue Univ; BS/Engr Sci/USAF Acad *Born:* 10/19/37 Born in Pelham, Georgia. Attended public schs in Pickens, Sc and Furman Univ. Continuous active duty in USAF since 1961 advancing to rank of Colonel. Served as pilot/aircraft commander in Military Airlift Command C-130s and Pacific Air Forces C-7a (Vietnam) until 1967. Faculty mbr in various capacities at USAF Acad from 1969 to the present. Conslt on Global Positioning Sys 1974 to present. Became head of dept of Astronautics and Computer Sci in 1979. Pres and Chrmn of the Bd of Dirs of the Association of Grads of the USAF Acad. *Society Aff:* AIAA, ACM

Ellingwood, Bruce R
Home: 605 Wilton Road, Baltimore, MD 21204-7615
Position: Professor *Employer:* Johns Hopkins University *Education:* PhD/CE/Univ of IL at Urbana-Champaign; MS/CE/Univ of Il at Urbana- Champaign; BS/CE/Univ of IL at Urbana-Champaign *Born:* 10/11/44 Native of Evanston, IL. Structural eng for Naval Ship Research and Development Center, 1972-1975, specializing in structural fatique and fracture reliability. Joined NBS in 1975, as research eng, where research activities have included the application of methods of probability and statistics to structual engineering problems, analysis of structural loads, and development of structural design criteria. Since 1986, Professor of Civil Engineering at Johns Hopkins University, Baltimore, MD. Author of numerous publications and reports. Member or officer of numerous national tech or standard committees. Recipient of ASCE Walter L. Huber Engineering Research Prize, DC Joint Council of Engineering and Architectural Societies Engineering Achievement Award, Department of Commerce Silver Medal, ASCE State of the Art in Civil Engrg Award, and ASCE Norman Medal. NSPE Engineer of the Year of the Department of Commerce, 1986. *Society Aff:* ASCE, ASTM, ΣΞ, TBΠ

Elliott, Daniel P
Home: 515 Tanacrest Dr, Atlanta, GA 30328
Position: Mgr. *Employer:* Simons-Eastern Co. *Education:* MBA/Mgt/GA State Univ; BS/Mech Engr/LA State Univ. *Born:* 8/13/25. Native of Shreveport, LA. Worked as Proj & Div Engr for W R Grace & Co. Presently Mgr of the Engg Support Dept of Simons-Eastern Co. In this position assists Corporate, proj and dept mgt in the dev of progs and systems, also responsible for the application of these progs and systems to aid in the mgt and control of a wide range of projs. PE, Certified Cost Engr, author of papers presented to AACE, TAPPI, and PMI. Mbr of AACE Cost Index, Continuing Education and Cost Control Committees. Founder and past pres of the

Elliott, Daniel P (Continued)
Atlanta Area Chapter of AACE. Natl dir of the AACE 1979-81, Technical VP 1981. Am interested in sailing and woodworking. *Society Aff:* NSPE, AACE, ASME.

Elliott, David L
Business: Dept of Systems Sci & Math, St Louis, MO 63130
Position: Prof *Employer:* WA Univ. *Education:* PhD/System Sci/UCLA; MA/Math/USC; BA/Math/Pomona College. *Born:* 5/29/32. in Cleveland, Educated in CA. Worked as patent trainee 1955, Navy mathematician 1955-1969. With WA Univ since 1971, specializing in mathematicl systems theory, control lab. Visiting appointments at Brown Univ. 1979, UCLA 1987-88; Assoc editor for *Mathematical System Theory*, Fellow, IEEE. Enjoy listening to chamber music and have promoted free summer concerts. Science fiction enthusiast. *Society Aff:* IEEE, AAAS, AMS, SIAM.

Elliott, John F
Business: 77 Mass Ave.-Room 4-138, Cambridge, MA 02139
Position: Prof of Metallurgy. *Employer:* MA Inst of Tech. *Education:* BMetE/Met/Univ of MN; ScD/Met/MIT *Born:* 7/31/20 Education: Univ of MN 1939-42; MIT, 1946-49. Employment: US Steel Corp 1949-50; Inland Steel Co, 1950-55; MIT, 155-present. Served with US Navy 1942- 46. Fields of interest - Process and Extractive Metallurgy, hot corrosion, high temperature physical chemistry. Active in profl societies and affairs related to natl resources policies. Publish extensively in profl journals. Honorary Mbr: JIM; JISI; AIME; Fellow: The Met Soc, AIME; AIChE; ASM; AAAS; Distinguished Prof 1982-5, Am Iron & Steel Inst; Mbr Natl Acad of Engrg. *Society Aff:* AIME, AIChE, ASM, CIM, ECS, AAAS, IMM, JISI, JIM, Met. Soc.

Elliott, Kendall C
Business: 124 Agri Sci Annex, West Virginia Univ, Morgantown, WV 26506
Position: Assoc Prof. *Employer:* WV Univ. *Education:* MS/Agri Engg/WV Univ; BS/Agri Engg/Univ of VT. *Born:* 2/6/32. Have been responsible for mtls handling work in baled hay and tobacco. Designed two picking aids for strawberry production. Currently engr with mech harvesting proj for peaches and apples. One half responsibility for harvester component design. One patent application for hydraulic soil sampler (1979). Conceptual design for frost temperature indicator strobe lights and tree spacing indicator for mechanical tree planter. *Society Aff:* ASAE.

Elliott, Kenneth M
Business: PO Box 1026, Princeton, NJ 08540
Position: VP-Engg. *Employer:* Mobil Res & Dev Corp. *Education:* BS/ChE/Univ of TN; -/64th AMP/Harvard Bus Sch. *Born:* 11/8/21. Native of Athens, TN. Joined Mobil Oil Corp in 1942 in Res Dept. 1964 named Gen Mgr of Engg. 1967 named VP-Engg, Mobil Res & Dev Corp. Active in Am Petrol Inst progs. Received 1976 API Certificate of Appreciation. Mbr of Bd of Dirs and Exec Committee of NJ State Safety Council. Mbr of Chem Engg Consultor Committee of Manhattan College, Bronx, NY. *Society Aff:* AIChE, AASA.

Elliott, Martin A
Home: 13623 Alcheste Ln, Houston, TX 77079
Position: Energy Consultant (Self-employed) *Employer:* Self *Education:* PhD/Gas Engg/Johns Hopkins Univ; BE/Gas Engg/Johns Hopkins Univ. *Born:* 2/21/09. Worked in mfg gas plant, Baltimore Gas & Elec Co, 1934-38. Held various positions at Bruceton Station of US Bureau of Mines eventually becoming Chief, Synthetic Fuels and Dev Branch. At IL Inst of Tech, served as Res Prof of Mech Engg; Dir, Inst of Ga Tech; Academic vp; 1952-1967. Was Corporate Scientific Advisor with TX Eastern Transmission Corp, 1967-74. On retirement became private energy consultant. Authored more than 100 technical articles. *Society Aff:* AIChE, ASME, SAE, ACS, AAAS, INST, NAE.

Elliott, Paul C
Home: 10810 Oak Creek, Houston, TX 77024
Position: Pres & CEO *Employer:* Concord International Corp *Education:* PMD/Bus/Harvard Grad Sch of Bus; BS/Chem Eng/Rose-Hulman Inst of Tech *Born:* in Terre Haute, IN. Grad 1955 with BS Chem Engg, Rose Poly Inst. Attended Bus Schools IN Univ & Wharton Sch. Grad Harvard Grad. School of Business 1971 PMD 21. Career in industry began 1955 with Socony Mobil Oil, Res Div, Paulsboro, NJ 1955-60, Commercial Solvents Corp 1960-62, Marathon Oil Co, Findlay, OH 1962-74 in various exec positions including: proj engr & refinery Construction Interntatioanl Refining & Marketing, & Chemical products marketing. Served as VChrmn AIChE, Wabash Valley. Mbr of Tau Beta Pi. In July, 1974, named Pres and CEO of Tampimex Petroleum Corp and later Tampimex, Inc in Houston, TX. Left Tampimex in 1978 to form Concord Petroleum Corp and serve as Pres and CEO of Concord. Also Pres of Concord Int'l Corp. Formed in 1979. Serves as Dir of Corps: Kors, Marlar & Assocs, Concord Petroleum, Concord Int'l, & Chesapeake Corp. *Society Aff:* AIChE, API, NPRA, TBΠ.

Elliott, Robert S
Business: 7732 Boelter Hall, Los Angeles, CA 90024
Position: Professor. *Employer:* Univ of Calif. *Education:* PhD/EE/IL; MS/EE/IL; MA/Economics/UC Santa Barbara; BS/EE/Columbia; AB/English Liter/Columbia. *Born:* 3/9/21. in Brooklyn, NY. AB English Lit, 1942 and BS Elec Engg, 1943 from Columbia. MS, 1947 and PhD, 1952, both in Elec Engg, from IL. MA in Economics, 1971 from Univ of CA, Santa Barbara. Jr Engr, Appl Physics lab, 1943-46. Asst Prof, IL, 1946-52. Lt Jg USN, 1952-53. Hd, Antenna Res, Hughes, 1953-56. Founder and Tech Dir, Rantec Corp, 1956-57. Prof, UCLA 1957-present. Mbr: Tau Beta Pi, Sigma Xi, NY Acad of Sciences. Fellow: IEEE. *Society Aff:* IEEE, URSI.

Elliott, Roger W
Business: Engineering & Industrial Experiment Station, College of Engg, Gainesville, FL 32611
Position: Associate Director, EIES. *Employer:* Univ of FL. *Education:* BA/Math/TX Western College; PhD/Math Comp Sci/Univ TX. *Born:* 10/5/35. Native of Northwood, IA. Worked as Comp Engr, Autonetics, Downey CA two yrs. Served on faculty of Ind Engg Dept, TX A&M Univ for 10 yrs. Served as Chairman of CIS Dept. at University of Florida for 9 years. *Society Aff:* ACM, IEEE, ASEE, DPMA.

Ellis, Charles W
Business: PO Box 16858, Philadelphia, PA 19142
Position: VP - V-22 Joint Program Office *Employer:* Boeing Vertol Co. *Education:* MS/Aero Engg/MIT; BS/Aero Engg/MIT. *Born:* 10/8/27. Worked for Kaman Aircraft Corp from 1952-65 in various engg mgmt postions. Involved with development & qualification of the World's first droned helicopter (DASH), USAF HH-43B and USN UH-2 helicopters plus testing of K-16 tilt wing deflected slipstream VTOL aircraft. Joined Boeing Vertol Co in 1965 as Dir of Preliminary Design respon for new bus efforts in the helicopter field. Was Dir of Tech respon for aerodynamics, structures, propulsion, electrical/mech and weapon sys analysis. From 1968-72 was Dir of Engg respon for the supervision of all engg efforts as well as direction of flight test, technology and product assurance. Current assignment since 1982 - Program VP for overall direction of Bell-Boeing V-22 (Tiltrotor aircraft for joint USMC/USN/USAF/USARMY usage). Holds numerous pats in rotary wing filed and has pub several art and papers. Mbr of the Army Scientific Adv Panel from March 1970 to June 1976. Hon Fellow AHS, Fellow AIAA. *Society Aff:* AHS, AIAA, ΣΞ

Ellis, Harry M
Business: 4012 Myrtle St, Burnaby BC, Canada V5C 4G2
Position: Dir of Res & Dev. *Employer:* B C Hydro & Power Authority. *Born:* Jan 1923 in Vancouver B C, Canada. PhD (EE & Phys) & MSc from CA Inst of Tech, BA (EE) from Univ of British Columbia. With B C Hydro & its predecessor cos since 1956. In 1974 was Chrmn of Task Force on Future Generation & Transmission Requirements for B C Hyrdo. Appointed Dir, Res & Dev, Oct 1975. Fellow IEEE, Canadian Rep of CIGRE (Intl Conf on Large High Voltage Elec Systems)

Ellis, Harry M (Continued)
Study Ctte 41 on Future of Elec Power Transmission & Sys; Chrmn R&D Ctte of Canadian Elec Assn; mbr IEEE Transmission & distrib Ctte. Author a var of tech papers.

Ellis, Lynn W
Business: 320 Park Ave, New York, NY 10022
Position: Dir-Res. *Employer:* ITT Corp. *Born:* Feb 1928. BEE Cornell 1948, MS Stevens Inst of Tech 1954. ITT Corp: Fed Tele Labs 1948-55; ITTE Espana 1955-58 & 1961-62; Std Tele & Cables, London 1958-61; Std Tele & Cables Pty, Sydney 1962-66; Headquarters 1966 to date. Fellow Inst of Elec & Electronics Engrs Inc 1974; Fellow AAFAS 1976. Certificate of Commendation, US Dept of Commerce 1975; Chrmn, Telecommunications Equip & Tech Adv Ctte to Dept of Commerce 1973-75; Mbr Ctte on Telecommunications, Natl Acad of Engg (subsequently Natl Res Council) 1973-76.

Ellis, Richard G
Home: 2700 Pencoyd Ln, Charlotte, NC 28210
Position: Agent *Employer:* The Teague Group *Education:* BS/CE/Univ of IL. *Born:* Sept 1930. BS Univ of IL 1952. Native of Mt Zion, IL. Corps of Engrs 1952-54; the Texas Co 1954-56; GE Co 1956; Harry Balke Engrs 1956-64; Am Inst of Steel Const 1964-71; Mgr, Bldg Div, Carolinas Branch Associated Gen Contractors of Am 1971-81. 1981-86 Exec. VP Carolinas Branch Associated General Contractors of Am. Reg PE OH, NC & SC. P Pres NC Sec ASCE; P Chrmn NC Chap AWS; P. Pres NC Soc of Engrs; Prof Const Estimators Assn of Am, SCSPE, PENC, NSPE, Dir ASCE 1976-78.

Ellis, Robert M
Business: P O Box 8405, Kansas City, MO 64114
Position: Partner *Employer:* Black & Veatch *Education:* BS/ME/Univ of KS; Engrg/Kansas City Jr Coll *Born:* 10/25/23 and raised in Kansas City. Enlisted in Navy and served as pilot 1942-45. Graduated from Kansas Univ in 1948 and joined Black & Veatch that year. Serving as partner and project manager on design and construction management of coal- fueled electric generating units with all support facilities. *Society Aff:* NSPE, ASME

Ellis, Robert W
Home: 35945 Fredericksburg Rd, Farmington Hills, MI 48018
Position: Dean of Engrg, Prof *Employer:* Lawrence Inst of Tech *Education:* PhD/Engg Mechanics/VPI; MS/Engg Mech/VA Tech; BS/Met Engr/VA Tech. *Born:* 10/16/39. in Richmond, VA. Served as prof & asst dean of engg, Univ of S FL, 1965-69- -developed mtls prog & related res. Res Fellow NASA Langley R&D Ctr 1969. Asst VP, 1970-72--responsible for grad study, sponsored res, & academic admin. Founded Sch of Tech, FL Intl Univ, 1972-78. Chief academic & operating officer, Detroit Inst of Tech, 1978-1981. Sr Engr, US Army Tank Automotive R&D Ctr, 1981- 1984. Chrmn Relations with Industry Div, ASEE, 1980; Pres. Michigan Professional Engineers in Education 1987-88; Pres, Detroit Chapter Michigan Society of Professional Engineers 1987-88; Engineer of the Year in Michigan 1987; Fellow of the Engineering Society of Detroit 1987; General Chairman, ASEE/CIEC conference 1987. ABET visitor committee, Am Soc for Metals; consultant in engg & academic fields. *Society Aff:* ASM, NSPE, ASEE, SAE, SME, SAMPE.

Ellis, William W
Business: P.O. Box 541, Princeton, NJ 08542
Position: Exec Dir *Employer:* Univ Assoc *Education:* Bachelor of Sci/Chem Engrg/Univ of MO-Columbia *Born:* 07/29/25 Exec Dir Univ Assoc, P.O. Box 541, Princeton, NJ 08542. b. July 1925. BS. Univ of MO. Served as commissioned officer US Navy in Pacific during WWII. Proj & process engr exp with DuPont & Owens Ill. Assoc for 11 yrs with M & R Dietetic Labs. Resp for various spray dryer & other food processing res & eng design activities in Holland & US. Dir of non-degree progs of prof educ for engrs and scientists first at Carnegie-Mellon Univ and then Princeton Univ 1964-78. Conslt to indus & govt on dev progs for professionals. *Society Aff:* AIChE, ASQC, ASEE

Ellsworth, Donald M
Business: 140 N Corner Ave, PO Box 1822, Idaho Falls, ID 83403
Position: Pres - Gen Mgr. *Employer:* Ellsworth Engg, Inc. *Education:* BS/CE/Univ of UT *Born:* 10/4/29. Native of ID. Design engr, City of Salt Lake 1956-57. Asst City Engr & City Engr City of ID Falls, ID 1957-64. Established consulting engg and land surveying firm in 1964. Mbr of ACEC, ASCE, AWWA, NSPE & Consulting Engrs of ID. Past chrmn Intermountain Section AWWA. Pres Consulting Engrs of ID 1979, Natl Dir to ACEC 1968-71 & 1984-86. ACEC Contract Negotiation Ctte. Licensed in ID, CA, MT, NV, WY & AZ. Aircraft Pilot. Enjoy flying and hunting. Mbr of the ID State Bd of Engrg Examiners and NCEE. *Society Aff:* ACEC, ASCE, AWWA, NSPE, NCEE.

Ellsworth, William M
Business: 6110 Executive Blvd. Suite 315, Rockville, MD 20852
Position: VP *Employer:* Engineering & Science Assocs. *Education:* MS/Fluid Mechanics/State Univ. of Iowa; BS/Civil Engr./State Univ. of Iowa *Born:* 11/05/21 48-58 David Taylor Model Basin; Hydromechanics. Lab. Project Engr, Physical Science Coordinator, and Head, Towing Problems Branch; R&D on ships. Model tests, analysis, and full-scale at-sea trials. Included destroyer condenser scoops, submarine snorkels, submarine superstructure flooding holes, periscope vibration, boundary layers, mine countermeasures, undersea surveillance systems, mooring and towing systems, etc; 58-64 PneumoDynamics, Inc., Systems Eng Div, Washington; Hydrodynamics Specialist, Manager Marine Systems, Div General Mgr, and Corporate VP. R&D contracts with government and industry in marine and aerospace systems; 64-83 Naval Ship R&D Center; Technical Mgr, Navy Hydrofoil dev, Assoc. T.D. for Systems; Head, Systems Dev Dep, a Sr Exec Service position. Developed hydrofoil ships, hovercraft, assault craft, planing craft, small waterplane area twin-hull (SWATH) ships, advanced submarine concepts, and shipboard material handling and transfer systems. Retired from federal service in January 1983; 83-Date Engrg and Science Assoc, Inc., Rockville MD. VP & Ch, Bd of Dir. Consultant in Naval Engrg and R&D mgt to govt & industry. Professional Recognition: 67- David W. Taylor Medal; 72- Navy Superior Civilian Service award; 73- ASNE Engineer of the Year award with gold medal; 80- Navy Distinguished Civilian Service award; 80- Senior Executive Service Meritorious Presidential Rank award; Honorary Life Member ASNE; Life Fellow ASME; Dir. Intl. Hydrofoil Society; Prof Engr; Author and co-author of more than 50 technical reports and papers; Author of Hydrofoil section, McGraw-Hill Yearbook of Science and Technology; Chairman, Edit. Committee, Special Issue (Feb 1986), ASNE Journal; Author, "History of the U.S. Navy Hydrofoil HIGHPOINT-Twenty Foilborne Years." *Society Aff:* ASME, ASNE, IHS, USHS

Elmblad, Thomas R
Business: GWF Power Systems, Signal Companies, 17900 Skypark Circle, Irvine, CA 92714
Position: Pres. *Employer:* Whiting Corp. *Education:* BS/ChE/Univ of MI. *Born:* 8/2/29. in Hancock, MI, & attended high sch in L'Anse, MI. Grad of the Univ of MI. Has been with the Whiting Corp since 1951 in various sales positions and was Pres from 1975 thru 1982. Since 1982 been Pres of GWF Power Sys. Irvine, CA, a unit of the Sign Companies. Active in the field of Crane Mfg Assn of Am, Inc, Machinery and Allied Products Inst, Mtl Handling Inst, Inc. Has been elected to the bd of dirs of Suburban Fed Savings & Loan Assn, Ingalls Meml Hospital, Greater Chicago Safety Council, and Midwest Ind Mgt Assn. Wife, Trudy, & two daughters, Linda & Christine, reside in Crete, Ill. Enjoys golfing, hunting, fishing, & skiing. *Society Aff:* AIChE.

Elmendorf, Charles H, III
Home: 34 Cross Gates Rd, Madison, NJ 07940
Position: Retired, Asst V P Engg *Employer:* AT&T *Education:* BS/Phys/CA Inst of

Elmendorf, Charles H, III (Continued)
Tech; MS/EE/CA Inst of Tech. *Born:* 7/1/13. 1936-1966 Bell Labs - Transmission. 1966-1978 AT&T - Asst VP, Engg. 1978-1985 Tech Mgmt Consultant. *Society Aff:* IEEE, NAE.

Elmer, William B
Business: 2 Chestnut St, Andover, MA 01810
Position: Consulting Reflector Designer *Employer:* Self employed *Education:* BSC/Electr. Eng'g/Mass. Inst. of Technology *Born:* 01/22/01 1922 Boston: Supervisor High Voltage Laboratory, the Div Head, Transmission & Distribution Dept; 1942 Westinghouse Electric, Sharon & Cleveland: Designed Mark 31 torpedo-mine. Redesigned complete street lighting line including enclosed oval luminaire; 1946 Hart Manufacturing (Canada) Ltd, Aurora, Ontario: Managing Dir, heater switch plant; 1949 Wheeler Reflector Co., Boston: Mgr, Street Lighting Dept. Designed open surban reflector used extensively; 1960 Reflector Design Consultant, Boston; Many designs, including reflector in walking-stick camera on moon since 1969; 1974 Authored "The Optical Design of Reflectors-, world technical classic. Second edition published by John Wiley, N.Y., 1980. Basis for Titanic undersea lighting. *Society Aff:* IEEE, IES, OSA

Elms, James C
Home: 112 Kings Pl, Newport Beach, CA 92663
Position: Consultant *Employer:* SDIO *Education:* MA/Physics/Univ of CA; BS/Physics/CA Inst of Tech. *Born:* 5/16/16. Began as stress analyst Consolidated Aircraft San Diego 1940. Air Force officer 1942-1946. Res Assoc Geophysics UCLA 1948-50. Served in mgt positions at N Amer Aviation, Martin, AVCO and Ford Aeronutronic 1950-1962. Joined NASA 1962 as Deputy Dir Manned Spacecraft Ctr Houston. Later returned to industry as Corp VP and Gen Mgr Space Div, Raytheon. Rejoined NASA as Deputy Assoc Administrator for Manned Spaceflight 1965. Then Dir Electronics Res Ctr Cambridge 1966-1970. Later Dir DOT's Transportation Systems Ctr Cambridge 1970- 75. Retired 1975. Consultant to Administrators NASA and ERDA and mgts of aerospace and energy industries. 1975-85. Consultant to director SD10 1985-Present. Fellow IEEE, Assoc Fellow AIAA, Amer Physical Soc. Mbr of Natl Academy of Engrg. Mbr Explorers Club, Life Assoc of CA Inst Tech registered PE, CA. Enjoy soaring, skiing and sailing. *Society Aff:* IEEE, AIAA, APS.

Elovitz, David M
Business: 26 Robinhood Rd, Natick, MA 01760
Position: Consultant. *Employer:* Energy Economics Inc. *Education:* BS/ME/Worcester Poly Inst. *Born:* 11/19/31. Equip Engr, Aircraft Nuclear Propulsion prog, Pratt & Whitney Aircraft. Mgr of Facilities, High Temperature Mtls, Inc: develop, install and operate thermal processing equip. Chief Engr, BTU Engg Corp: design and manufacture of industrial furnaces. Dir, Facilities and Services, Computer Control Div, Honeywell, Inc: facilities planning and construcion, maintenance, admin services. Chief Engr, then VP BALCO Inc: energy systems engrs and contractors. Exec VP, Medical Area Service Corp: shared services including central power plant, energy conservation, maintenance engg, urban planning for maj Boston multi-institutional complex. Independent consultant (Energy Economics): energy conservation, energy systems troubleshooting, planning and economic evaluation, expert testimony. *Society Aff:* ASHRAE, NAFE, NSPE.

El-Ramly, Nabil A
Business: 2404 Maile Way, Honolulu, HI 96822
Position: Prof, Dept of Decision Scis *Employer:* Univ of HI. *Education:* PhD/Engg/UCLA; M.S.M.E./Mech Engg/IIT; B.Sc./Mech Engg/Cairo Univ. *Born:* 1936. Attended Cairo Univ MSME 1958, IIT MSME 1962, & Univ of CA, Los Angeles PhD Engg 1970. Held engg pos with Kaiser Engrs, Bechtel, & the Alexandria Portland Cement Co. Conslt on water & power devs to the US Off of Water Res & Tech, US Bur of Reclamation, the UN, & to the Govts of Kuwait and Saudi Arabia. Pres, Tech Economic Services of Honolulu. Held profl pos in Engg as well as in Bus Adm at UCLA, Univ of OR, & WV Univ, & the Japan-Amer Inst of Mgmt Sci. He has published sev articles & monographs on the economics of desalination, geothermal & nucl power, & other new technologies. Editor-in- Chief, the NWSIA Journal. Recipient of S D Bechtel Merit Award for Communication of Technical knowledge, and the Eugene L. Grant Award of ASEE for best article in the Engrg Economist in 1976. *Society Aff:* NWSIA

Elsayed, Elsayed A
Business: College of Engg, Piscataway, NJ 08855-0909
Position: Prof and Chrmn *Employer:* Rutgers Univ. *Education:* PhD/Ind Engg/Univ of Windsor; MSc/Mech Engg/Cairo Univ; BSc/Mech Engg/Cairo Univ. *Born:* 12/29/47. Prof Experience: Lecturer of Mech Engg, Cairo Univ '69-'73; Teaching Asst of Industrial Engg, Univ of Windsor '73-'76; Teaching and res assoc, Univ of UT, '76-'77; Asst Prof of Ind Engg, Rutgers Univ, '77-'81. Assoc Prof, 1981-85. Prof 1985-present. Chairman of Ind Engg. 1983-present. Concurrent positions: Principal investigator, Rutgers Res Council, '78 and Sea- Land Service Inc res grant, '78-Mbr: NSF, FAA, EPA, IIE, ASEE; Res area: Production Planning and Control, stochastic processes with specialization in queueing theory & Reliability Engrg. Co-author "Analysis and Control of Production Systems-, Prentice-Hall, 1985. *Society Aff:* IIE, ASEE, ASME, SME.

Elshout, Raymond V J
Business: PO Box 2650, Pasadena, CA 91105
Position: Owner *Employer:* Energy Systems Engineering *Education:* MS/Chem Engg/Univ of MI. *Born:* 3/12/41. Reg Chem Engr-CA & mbr of AIChE. Founder of Energy Systems Engrg, a consulting firm specializing in computer programs for Heat Exchanger Network Simulation & optimization. Formerly associated with Jacobs Engg, Ralph M Parson Co & the Fluor Corp & Union Oil Res Co. Dev of computer programs to simulate Heat Exchange networks, sour water strippers, sulfur plants, process engg consulting in petroleum refining specializing in energy conservation, heat exchange and distillation. *Society Aff:* AIChE.

Elspas, Bernard
3464 Janice Way, Palo Alto, CA 94303
Position: Staff Scientist. Retired. *Education:* PhD/EE/Stanford Univ; MEE/EE/NYU; BEE/EE/CCNY. *Born:* 7/26/25. Native of NYC; Taught Elec Engg at CCNY (1946-49); Visiting Lecturer at Stanford Univ and Univ of HI; Mbr Technical Secretariat, Panel on Electron Devices (1949-51); SRI Intl 1955-86 (Staff Scientist since 1968-1986), specializing in Info Theory and Computer Science; over 20 papers in profl journals; contributor to three books; Chrmn, IEEE Prof Group on Info Theory (1967-68); Hobby: musical composition. IEEE Fellow (since 1976). *Society Aff:* IEEE, ACM, ΣΞ.

El-Sum, Hussein M A
Business: 74 Middlefield Road, Atherton, CA 94025
Position: Pres & Chief Executive *Employer:* El-Sum Consultants *Education:* PhD/Physics/Stanford Univ; MSc/Physics/CA Inst of Tech; BSc/EE/Cairo Univ (Egypt) *Born:* 10/24/18 Naturalized American since 1957. Established El-Sum Consultants in 1963 for International Technical & Management Consulting Services mainly in fields of optics, electro-optics, holography, nondestructive testing, acoustics, ultrasonics, electronics and related fields. Professional teaching, lecturing and research positions held at universities in the US and abroad (Stanford; Univ of MI; Redlands; Cavendish Lab of Cambridge, England, Oslo, Norway; consultant to governments & Industries (IBM, Xerox, Lockheed, MacDonell-Douglas, Spectra Physics, NASA, Electro-Equipment, etc.) Listed in Who's Who in the West, American Men of Science, etc. *Society Aff:* AAAS, AIP, OSA, OSNA

Eltimsahy, Adel H
Business: 2801 W Bancroft, Toledo, OH 43606
Position: Prof. *Employer:* Univ of Toledo. *Education:* PhD/Elec Engg/Univ of MI; MSc/Elec Engg/Univ of MI; BS/Elec Engg/Univ of Cairo. *Born:* 6/10/19. Adel H El-

Eltimsahy, Adel H (Continued)
timsahy received the BSEE degree from Cairo Univ, Cairo, Egypt. He received the MSEE and PhD degrees in elec engg from the Univ of MI, Ann Arbor, in 1961 and 1967, respectively. In 1958, 1959 and 1962 he was with the Natl Res Ctr. From 1958 to 1959 he also taught at Cairo Univ. During his PhD work at the Univ of MI he worked as a Res Asst and a Teaching Asst. In 1967 he joined the Elec Engg Dept of the Univ of TN, Knoxville. Since 1968 he has been with the Elec Engg Dept of the Univ of Toledo, Toledo, OH, where he is presently a Prof. He taught and developed new courses in robotics and control systems. He is actively engaged in the optimal control of robotic systems res. Dr Eltimsahy is a mbr of the Inst of Elec and Electronics Engrs, Sigma Xi, Eta Kappa Nu and Phi Kappa Phi. He is also listed in American Men and Women of Sci and in Who's Who in Tech Today. *Society Aff:* IEEE.

Eltinge, Lamont
Home: Box 251, Northville, MI 48167
Position: Dir of Res. *Employer:* Eaton Corp. *Education:* PhD/Mech Engg/IL Inst of Tech; MS/Mech Engg/IL Inst of Tech; BS/Mech Engg/Purdue Univ. *Born:* 5/9/26. Native of Chicago, IL. USNR during WWII. Trainee and Foreman for Electro-Motive Div, GMC; Automotive Engr and Sec Leader - American Oil (Std Oil Co-IN) working on aviation, motor, and diesel fuels; Dir Automotive Res for Ethyl Corp with emphasis on engine modifications to achieve clean exhaust; VP of Res & Tech for Cummins Engine Co with responsibility for dev and application of advanced specialized tech; Dir of Res for Eaton Corp since 1973 where attention is focused upon identification and assimilation of new tech that will help make Eaton products better. SAE - 1973 Horning Memorial Award; SAE-Bd of Dirs 1977- 79, Chairman-Detroit Section 1980. IRI - Bd of Dirs. *Society Aff:* ASME, SAE, ΣΞ, IRI.

Ely, Berten E
Home: 30 Wood Rodge Rd, Thornton, PA 19373
Position: Consultant (Self-employed) *Education:* MA/Chem Engrg/Cornell U; BA/Chem Engrg/Cornell U *Born:* 1/3/23. Florham Park NJ. S/Sgt 89th Inf in WWII Purple Heart Oak Leaf Cluster. Dev Engr for DuPont Plastics Dept. Patents & articles on Teflon. Sec ASTM D-9 for 3 yrs. Started Penntube Plastics Co 1955. Ex VP Dixon Ind 1969. Started Penn Dixon 1965. Pres both cos until 1978. Activities include sch bd sch authority, boy scouts, Methodist Church, tennis, sub committee chrmn of ASTM and SPI. Started 7plus Corp 78 Retired Jan 1980 from Penntube & Penndixon & Dixon industries. Became Consultant on Fabrication & New Products made from Fluoropolymers Started White Birch Racquet Club New London N.H. April 1980 Sec of 7plus & White Birch Member ASTM D-20 & F-17 & Subcommitte Chairman. *Society Aff:* SPI, ASTM, AIChE, ACS

Ely, George M, Jr
Business: P O Box 22738, 620 Euclid Ave, Lexington, KY 40522
Position: Pres *Employer:* Parrott, Ely & Hurt, Cons. Engs. *Education:* B.S./CE/Univ of KY *Born:* 12/28/27 Lynch, KY, attended North Georgia Coll, Dahlonega in 1945 as ASTRP student; was graduated from Univ of KY 1955, BSCE; was with H. K. Bell Engineers, Lexington, KY 1953-1967, was resident eng on municipal water and sewerage project in Kentucky and Tennessee; in 1966 served as project manager on Master Plan for Lexington-Fayette County Sewerage System. A founding principal, Parrott, Ely and Hurt, in 1968. In 1969, honored by KSPE for engineering achievement in private practice. Awarded Arthur Sidney Bedell award in 1981 from Water Pollution Control Fed. Past President, Kentucky Society PE; Past Pres, Kentucky-Tennessee Water Pollution Control Assoc; Past President, Kentucky Chapter American Public Works Association; Registered Eng in five states; Diplomate, American Academy Environmental Engineers; Past Natl Dir KY-TN Water Pollution Control Assoc; Natl Dir KY Soc of P E's. Awarded Distinguished Service Award by Kentucky Society PE in 1985. *Society Aff:* NSPE, ASCE, WPCF, AAEE, ACEC.

Ely, John F
Business: Box 7908, North Carolina State Univ, Raleigh, NC 27695
Position: Prof CE *Employer:* NC State Univ *Education:* PhD/Theoretical & Applied Mech/Northwestern Univ; MS/Theoretical & Applied Mech/Northwestern Univ; BS/CE/Purdue Univ *Born:* 03/20/30 *Society Aff:* ASCE, NCSE

Elzinga, Donald Jack
Business: Industrial & Systems 303 Weil Engg, 303 Weil, University of Florida, Gainesville, FL 32611
Position: Prof & Chrmn. *Employer:* Univ of FL. *Education:* PhD/Chem Engg/Northwestern Univ; MS/Chem Engg/Northwestern Univ; BE/Chem Engg/Univ of WA. *Born:* 1/16/39. Worked at Hanford Atomic Works during summer 1959. Chem Engr at Shell Dev, Emeryville, CA, 1960-61. Peace Corps Volunteer, Colombia, SA, 1961-63. BSChE, Univ WA (1960) MS, Am Northwestern Univ (1965) PhD, Northwestern Univ (1968). Taught at the Johns Hopkins Univ 1967-79 in the Depts of Chem Engg, Operations Res and Industrial Engg, and Mathematical Sciences. Chrmn of Dept of Industrial and Systems Engg, Univ of Fl, 1979-present. Dir of Manufacturing Systems Engrg Program, Univ of Fl, 1984-present. Consultant to NSF, OMB, HEW and private firms. *Society Aff:* ORSA, IIE, MPS, ΣΞ.

Emanuel, Alexander E
Business: Worcester Poly Inst, Worcester, MA 01609
Position: Prof. *Employer:* Worcester Poly Inst. *Education:* Dr Engg/EE/Technion Israel Inst; MSc/EE/Technion Israel Inst; BSc/EE/Technion Israel Inst. *Born:* 3/8/37. in Bucuresti, Romania. Obtained all Engg Degrees from Technion IIT Haifa where he specialized in Electromech Energy Conversion, Powr Electronics and High Voltage Tech. 1963-1969, he taught and engaged in res at the Technion IIT. In 1969, he joined High Voltage Engg, Burlington, MA, where he lead R&D of EHV Equip. Since 1974, he is associated with WPI, Worcester MA, teaching and supervising res teams. He was Investigator in Res Contracts for NEES, NSF, Quebec Hydro & EPRI. Dr Emanuel did pioneering work in the field of Power Factor Improvement in circuits with nonsinusoidal currents. He authored more than 60 papers and holds several patents. He is listed in Am Men & Women of Sci and Who's Who in Tech Today. He is a reg PE. *Society Aff:* IEEE, ΣΞ, SRBE.

Emanuel, George
Business: 865 Asp Ave, Rm 212, Norman, OK 73019
Position: Prof *Employer:* Univ of OK *Education:* PhD/Aero/Stanford Univ; MS/ME/USC; BA/Math/UCLA *Born:* 4/3/31 Joined Laboratory Division of the Aerospace Corporation in 1963, where I performed the first modeling of the CW and pulsed chemical laser and where I directed the chemistry and kinetics program. Joined TRW Systems Group in 1972 as senior scientist for the High Energy Laser program, where I managed several research programs. Joined LASL in 1976 as staff to the Applied Photochemistry division. Moved to the School of Aerospace, Mechanical and Nuclear Engineering of the Univ of OK in 1980. Over 60 archival publications in chemical lasers, fluid mechanics, radiative transfer theory, chemical kinetics and isotope separation theory. Author of two books, "Geodynamics: Theory and Applications', 1986 and "Advanced Classical Thermodynamics', 1987; both published by the AIAA. *Society Aff:* AAAS, AIAA, APS

Emanuel, Jack H
Business: 307 Butler-Carlton Civil Engrg Hall, Rolla, MO 65401
Position: Prof of Civil Eng *Employer:* Univ of Missouri-Rolla *Education:* Ph.D./Structural Eng/IA St Univ; M.S./Structural Eng/IA St Univ; B.S./ Arch Eng/IA St Univ *Born:* 9/26/21 Centerville, Iowa, parents Wilbur Harold and Ufah Relma (McGhee) Emanuel; married Marie Emilie Mammen, 1946; Children, Stephen James and Linda Marie; Army of the US, 1944-47; Registered Professional Eng, Missouri; Comm. A2C01, Transportation Research Bd, Natl Research Council; Bd of Building Appeals and Fire Bd of Appeals, Rolla; Research and more than 35 technical papers and reports on bridge supporting and expansion devices, dynamic behavior and thermal stresses in bridges and concrete; Coauthor two-volume structural engrg text; Dir numerous conferences ad short courses; Tau Beta Pi (Eminent

Emanuel, Jack H (Continued)
Eng Initiate), Chi Epsilon (Faculty Mbr Initiate), Soc of Sigma Xi; Faculty Adviser, Student Chapter ASCE; Past Master, Masonic Lodge; Consistory; former Unit Leader, Explorer Post. *Society Aff:* ACI, ASCE, PCI, MSPE, NSPE

Emberson, Richard M
Home: 3588 Spring Blvd, Eugene, OR 97405
Position: Consultant. *Education:* PhD/Phys/Univ of MO; MA/Phys/Univ of MO; BA/Phys/Univ of MO. *Born:* 4/2/14. 1936-9 - post-grad Fellow, Harvard College Observatory, steller radiometry; 1940 - instr, Medical Sch, Pittsburgh, cyclotron proj; 1941-6 - Radiation Lab, MIT, dev antennas, land/seaborne systems, chief engr Res Construction Corp; hd Fisher's Island experimental station. 1946 - Naval Res Lab, planning Combat Info Ctrs. 1947-51 - Res Dev Bd, DOD, staff for mixed civilian/military committees on principal areas (aero, biological/chem, electronics, guided missiles, human factors, etc). 1951-62 - Associated Univs, Inc - asst to Pres, asst corp secy, civ defense Proj E River, establishment Natl Radio Astronomy Oservatory. ITU- CCIR Frequency Allocations for Res - 1962-78 - IEEE staff director technical activities; 1978 gen mgr/exec dir; 1979-retired. IEEE Dir Emeritus. *Society Aff:* AAAS, IEEE, AAS, APS, AAPT, NYAS, ΣΞ.

Embleton, Tony F W
Business: National Research Council, Montreal Rd, Ottawa Ont, Canada K1A 0R6
Position: Prin Res Officer. *Employer:* Natl Res Council. *Education:* DSc./Acoustics/University of London (England); PhD/Acoustics/University of London (England); B.Sc./Physics/University of London (England). *Born:* Oct 1929. Hornchurch Essex, Eng. D Sc, PhD, B Sc Univ of London. Postdoc Fellow Natl Res Council of Can 1952. Sci Staff of NRC 1954-, Principal Res Officer Phys Div 1974-. Respon for Canadian acoust stds, also involved with noise reduction of machinery by tech means & control of community noise by tech & regulatory means. Fellow Acoust Soc of Am; Biennial Award ASA 1964, Chrmn ASA Tech Ctte on Noise 1964-67, Assoc Ed Journal ASA 1970-74, VP ASA 1977-78, President 1980-81. Arch T Coldwell Award Soc Auto Eng 1974. John Wiley Jones Lectures Rochester Inst Tech 1976. Fellow Royal Society of Canada 1978-. Foreign Associate, National Academy of Engg 1987-. *Society Aff:* ASA, INCE, NAE

Emerick, Harold B
Home: 479 Salem Dr, Pittsburgh, PA 15243
Position: VP-Retired *Employer:* Jones & Laughlin Steel Corp (now LTV SteelCorp) *Education:* Non-degree/Metallurgy/Carnegie-Mellon Univ *Born:* 07/06/13 Studied metallurgy at Carnegie Inst of Tech night Engrg Sch 1932-38. Employed by Jones & Laughlin Steel Corp 1935-1972. Last position held was VPres- Res and Quality Control. Served as Pres, the Metallurgical Soc of AIME, 1965. VPres AIME 1965-66. Fellow-grade mbr ASM and TMS/AIME. Author of more than 20 tech publications on Tech of Steel Prod and Steel Proc. Following retirement served as volunteer conslt in steel plants located in the Philippines, South Korea, Mexico and Peru under auspices of International Exec Serv Corps (IESC). *Society Aff:* AIME, ASM, AISI

Emerson, C Robert
Business: Dept of Ind Engg/Comp Sci, Roberts Hall, Bozeman, MT 59717
Position: Prof. *Employer:* MT State Univ. *Education:* PhD/Ind Engg/Purdue Univ; MS/Ind Admin/Purdue Univ; BS/Ind Engg/Purdue Univ. *Born:* 12/3/42. in MA, raised in MI, educated in IN. Taught at Purdue Univ and MT State Univ until 1974. Joined the Collins Radio Group of Rockwell Intl as the Mgr of Mtl and later as the Mgr of Systems Dev. Returned to MT State Univ and promoted to full Prof in 1979. Dir of the Engg Economy Div of IIE (1973-1979), and Chrmn Elect of the Ind Engg Div of ASEE (1979). Enjoy teaching and developing inovative methods for teaching. Outside of work enjoy the family, coaching volleyball, and the out-
e doors. John 3:16. *Society Aff:* ASEE, IIE.

Emerson, Warren M
Home: 241 Mainsail Court, Foster City, CA 94404
Position: VP *Employer:* International Engineering Co. (IECO) *Education:* BS/CE/Univ of Ut *Born:* 1/10/29 Served with US Navy Seabees 1948-52. Joined IECO 1955; engaged in design and engrg mgmt of major hydro power and water resources development projects, both domestic and foreign. Was Chief Civil Eng of IECO Brazilian affiliate 1969- 72; Ass't. Hydro Division Mgr, Brazil 1972-77; IECO Water Resources Division Chief Eng 1977-81; elected to vice-presidency 1981. Member: Comite Brasileiro de Grandes Barragens. *Society Aff:* USCOLD

Emery, Donald B
Business: 3750 Wood St, Lansing, MI 48906
Position: Pres. *Employer:* Emery & Porter, Inc. *Born:* Feb, 1931. BS & MS from MI State Univ 1952 & 1956. Tau Beta Pi & Chi Epsilon Engg Honor Fraternities. Army Ordnance Corps Officer 1952-54. Bridge Design Engr with Hazelet & Erdal 1956-59. Bridge Squad Leader with MI Hgwy Dept 1959-65. Hd Bridge Engr with MI Br of Brighton Engg Co 1966-73. Pres & Principal in Charge of Engg with Emery & Porter, Inc from 1973 to present. Reg PE in MI since 1958. Mbr of ASCE & NSPE, Pres of MI Sec of ASCE in 1973.

Emery, Michial M
Business: 4125 Carlisle Blvd N. E, Albuquerque, NM 87107
Position: VP *Employer:* Bohamnan Huston, Inc *Education:* MS/Environ Health Eng/Univ of TX at Austin; BS/CE/Univ of TX at Austin *Born:* 9/12/44 Bio Environmental Eng with the US Air Force, served from 1968-1971. Joined the firm Bohamnan Huston, Inc. in 1971. Employed as design eng, project mgr, division mgr and is presently VP in charge of Community Development and Planning Divison, and Business Development Coordinator. President Albuquerque Chapter NMSPE. Avid hunter. *Society Aff:* NSPE

Emery, Willis L
Business: Dept of Elec Comp Engg, Urbana, IL 61801
Position: Prof of EE. *Employer:* Univ of IL. *Education:* PhD/EE/IA State Univ; MS/EE/IA State Univ; BS/EE/Univ of UT. *Born:* 11/23/15. Native of Salt Lake City, UT. Taught at UT and IA State Univ before coming to the Univ of IL. Electronics Engr at the Naval Res Lab during WWII. Visiting Prof at the Indian Inst of Tech, Kharagpur, India 1960-1962. Prof of Elec Engg at the Univ of IL Urbana-Champaign and served as a departmental grad study coordinator until retirement Aug 21, 1982. Hobbies include art, photography and crafts. *Society Aff:* APS, IEEE.

Emkin, Leroy Z
Business: School of Civil Engineering, Atlanta, GA 30332
Position: Prof *Employer:* Georgia Inst of Techngy *Education:* PhD/CE/MIT; MS/CE/GA Tech; BS/CE/GA Tech *Born:* 8/15/43 Native of Brooklyn, NY. On the Civil Engineering Faculty of Georgia Tech since 1969 and currently Prof of Civil Eng and Dir of the GTICES Systems Laboratory. Numerous publications and presentations in the field of advanced and effective uses of computers in engrg. Current research included the development of large scale, integrated engrg software systems. Pres, IUG 1973, 1974; Dir of IUG International Board of Directors; ASCE Moisseiff Award 1973. Enjoys tennis, swimming and fishing. *Society Aff:* ASCE, ASEE, ACI.

Emling, Dale H
4258 S Alton St, Greenwood Village, CO 80111
Position: Consultant *Employer:* Self *Education:* MBA/Accounting/Univ of Chicago; BS/Mining Engg/MO Sch of Mines & Met. *Born:* 9/30/32. at Du Quoin, IL. Mining Engr Sahara Coal Co, 1954. Army Corps of Engrs 1954-56, 1st Lt. Mining Engr to Chief Mining Engr the United Electric Coal Cos, 1957-62 and 1965-1968. Tool Engr, Gen Motors Corp 1962-1965. Held Various engg and mgt positions in coal organizations of Union Pacific Corp 1969-1974. Began self-employment as mining engg and geologic consultant to coal industry in 1974. Inc D H Emling Co 1975-86 owner and pres of this coal consulting firm. Presently self employed coal consultant.

Emling, Dale H (Continued)
Married and 4 children. Recreational activities-camping, fishing, hunting. *Society Aff:* AIME.

Emling, John W
Home: 1217A Shetland Dr, Lakewood, NJ 08701
Position: Retired. *Education:* BS/EE/Univ of PA. *Born:* 6/3/04. Retired from Bell Tel Labs in 1966 as Exec Dir. of Transmission Systems Engg Div. A principal author of "History of Engineering and Science in the Bell System–, Vols I & II (1975, 1978). *Society Aff:* IEEE.

Emrick, Harry W
Business: Golden, CO 80401
Position: Dept Head *Employer:* CO School of Mines *Education:* PhD/Geod Sci/OH State Univ; MS/Geod Sci/OH State Univ; BS/Military Engrg/USMA-West Point *Born:* 8/15/32 Native of Freeport, IL. Graduated, USMA, at West Point, NY in 1954. Served with USAF 1954-1975 in ranks of 2nd Lt thru Colonel. Rated Command Pilot and Parachutist with over 6000 flying hours. Taught at USAF Academy in ranks of Instructor thru Prof and Dept Chrmn (1963-1974), Visiting Prof at National War Coll 1975, Faculty at CO School of Mines since 1976, Dir of Engrg Continuing Education 1978-1981, Basic Engrg Dept Head thru present time. NSPE scholarship selection committee, active in surveying and mapping at the national level. Enjoys computer technology, golf, jogging and classical music. *Society Aff:* NSPE, SAME, ACSM, AGU, ASP

Emrick, Jonathan E
Home: 6205 Hope Dr, Camp Springs, MD 20748
Position: Director, Quality Assurance *Employer:* VSE Corp. *Education:* BS/Mech Engrg/OH Northern Univ. *Born:* 12/28/26. Native of Wapakoneta, OH. Mbr of the US 30th TOPO Bn during WWII. Joined Westinghouse Elec after coll in 1952. With the US Govt 1958-86. Worked for AF, Defense Logistics Agency & with the Navy until 1986. Mbr of original cadre that established DCAS & specifically team mbr that dev the OA procedures for DOD. Dir of Contracting Mgmt Reviews for the NFCS until retirement in 1986, working to improve the efficiency and effectiveness of the Navy's contracting operations. Reg PE. Currently Dir. Quality Assurance - VSE Corp - a rapidly growing engg, development, testing, management, data processing & graphic communications firm. *Society Aff:* NSPE, ASQC.

Endahl, Lowell J
Business: 1800 Massachusetts Ave, NW, Wash, DC 20036
Position: Mgr R&D. *Employer:* Natl Rural Elec Co-op Assoc. *Education:* BS/Agri Engg/SD State Univ. *Born:* 7/3/22. Wessington Springs, SD. BS Agri Engg, SD State Univ, 1948. marine Corps pilot during WWII. Did engg & educational work for Tri-County Elec Co-op (1948- 51) & Sioux Valley Empire Elec Assn (1951-54) in SD. Joined NRECA in 1954. In 1969 was US rep on res team "Working Party on Rural Electrification" spon by US Econ Comm for Europe. NRECA-AID Rural Electrification Specialist to newly formed elec co-ops in Colombia & Equador in 1968 & Vietnam in 1970. Chrmn Md-D C Chap ASAE 1968-69. Chrmn Elec Power and Processing Div ASAE 1973-74. Fellow ASAE; ASAE George W. Kable Electrification Award, 1983; Dist Serv Award, Natl Food & Energy Council, 1984. *Society Aff:* ASAE.

Enell, John Warren
Home: 165 Lake Dr W, Wayne, NJ 07470
Position: VP for Res. *Employer:* Amer Mgt Assns. *Education:* EnggScD/Mgt Engg/NYU; MIE/Indus Engg/NYU; ME/Mech Engg/Univ of PA; BSME/Mech Engg/Univ of PA. *Born:* 6/24/19. Test engr & engrg supervisor, aircraft engine div of Curtiss-Wright Corp, 1940-46. Instructor/prof of industrial & mgt engrg, NYU, 1947-56. With Amer Mgt Assns since 1954, VP for res since 1967. On leave for USAID projs in Italy, 1952-53, and Vietnam, 1972; and IESC missions in Greece, 1973, and Colombia, 1978. Pres, Council on CEV, 1980-82; Secy/Treas, AAES, 1980; Treas EJC, 1979-; Treas ECPD, 1971-77; Pres IIE, 1968-69; Exec VP, 19967-68; VP-finance, 1965-67. Fellow: IIE. Life Mbr: Sigma Xi, Alpha Pi Mu. Author of book, Setting Standards for Executives, chapters of Quality-Control Handbook and Production Handbook, many res reports. PE, PA. *Society Aff:* IIE, ASME, ASQC, ASA.

Engdahl, Richard B
Home: 1750 W. First Ave, Columbus, OH 43212
Position: Consultant. *Employer:* Battelle's Columbus Labs(semi-retired). *Education:* BS ME/Mech Eng/Bucknell; MS ME/Mech Eng/U of Illinois. *Born:* MS, ME Univ Ill (UC) 1938; BS, ME Bucknell 1936; 1941-1976, Battelle, research fuels, combustion, air poll; 1979-80, WHO, envir. cons. Malaysia; 1980, Int. Exec. Serv. Corps., Turkey; 1981, WHO, envir. cons. P.R. China; waste-to- energy systems; 1983, air qual. deleg. P.R. China. *Society Aff:* ASME, ASHRAE, AAAS, AMS.

Engel, Joel S
Business: Crawfords Corner Rd, Holmdel, NJ 07733
Position: Dept Head *Employer:* Bell Telephone Laboratories *Education:* PhD/EE/Polytechnic Ins of Brooklyn; MS/EE/MIT; B/EE/City Coll of NY *Born:* 2/4/36 Native of New York City. Member of Staff, MIT Instrumentation Laboratory, 1957 to 1959. Joined Bell Telephone Laboratories in 1959; have worked in areas of data transmission, mobile radio communications, and video teleconferencing. Performed assignments at Bellcomm, on the Apollo space program, and at the American Telephone and Telegraph Company, in the Corporate Planning Dept. Currently responsible for systems planning of advanced residential services. Fellow of the IEEE; Paper of the Year award, 1969, 1973, IEEE Vehicular Technology Group. Pres, Congregation Bnai Israel, Rumson, N.J. *Society Aff:* IEEE, ΣΞ

Engelberger, John E
Business: 32 S Lafayette Ave, Morrisville, PA 19067
Position: Consulting Engr (Self-employed) *Employer:* Enerjee Intl *Education:* BS/ME/Drexel Univ. *Born:* 5/1/30. A 1953 grad of Drexel Univ with a BS in Mech engg. He is Reg in PA, NJ, NY, and DC. Presently a Consulting Engr in private practice, he specializes in Plant engg and Energy Conservation. Previously, he held Facilities and Plant Engg positions with such firms as McGraw-Hill, Inc, Pittsburgh Plate Glass, RCA, and CF&I Steel. He is active in many Engg Societies including: NSPE (Pres of local chapter), ASME, AIPE (Plant Engr of Yr 72-73 & Treas of local chapter), ASHRAE, AEE and Engrs Club of Trenton (President). He holds grade of Lt Commander in US Coast Guard Reserve. Selected as 1981 Engineer-of-the-yrr by Professional Engineers Society of Mercer County, NJSPE. Chrmn, NJSPE energy Ctte. 1981-4. *Society Aff:* NJSPE, NSPE, ASHRAE, ASME, AIPE, AEE, EC of T, ASPE, USNI.

Engelbrecht, Eugene W
Home: 3107 Woodhaven Dr, Cinnaminson, NJ 08077
Position: Vice Pres *Employer:* Richard A. Alaimo Assocs *Education:* BA/Art & Sci/Univ of PA *Born:* 11/03/27 PE specializing in the areas of Sanitary and Transp Engrg and Solid Waste Mgmt. From 1956-1974 held various positions from Engrg Dir to VP of Engrg with consltg firms: Albert C. Jones Assocs and Korman Corp. With Richard A. Alaimo Assocs, Consltg Engrs, since 1974. Serving as VP since 1978, responsible for Profl and Business Development, Engrg Coordination and Agency Liaison. Received NJSPE PE Recognition Award in 1976 and 1980. An avocational geologist, archaeologist and inventor, and a mbr of the NJ Water Supply Advisory Council. *Society Aff:* AAEE, ASCE, NSPE, ACEC, ITE, NACE, NJSPE, NJSME.

Engelbrecht, Richard S
Business: Dept Civ Engrg, Univ. of Illinois, 208 N Romine St, Urbana, IL 61801
Position: Prof. *Employer:* Univ of IL. *Education:* ScD/Sanitary Sci/MA Inst of Tech; MS/Sanitary Sci/MA Inst of Tech; AB/Zoology/IN Univ. *Born:* 3/11/26. Joined Dept of Civ Engg faculty of Univ of Il at U-C in 1954; Prof of Environmental Engg since 1959; Dir, Advanced Environmental Control Technology Research Center

Engelbrecht, Richard S (Continued)
since 1979. Proffly interested in environmental pollution - particularly microbiological (eg, bacteria, viruses) problems associated with water quality mgmt including water and wastewater treatment. Author of more than 120 articles in technical literature. Past Pres WPCF; President, and Mbr, Governing Bd and Exec Committee, IAWPRC; Commissioner from IL & Past Chrmn, OH River Valley Water Sanitation Commission; Mbr Natl Acad of Engrg; Mbr Water Sci and Tech Bd, NAS/NAE; Harrison Prescott Eddy Medal, WPCF; Arthur Sidney Bedell Award, WPCF; Honorary Mbrshp, WCPF; Geo Warren Fuller Award, AWWA; Publication Award, AWWA; Ernest Victor Balsom Commemoration Lecture, London; Abwassertechnische Vereiningung, Honorary Mbrship, W Germany; Eric H Vick Award, Inst of Public Health Engrs, London, UK; Benjamin Garver Lamme Award ASEE. *Society Aff:* AAAS, ASM, AWWA, IAWPRC, WPCF.

Engesser, Donald G
Home: 707 Fairmount Ave, Chatham, NJ 07928
Position: Pres *Employer:* PMEX, Inc *Education:* BS/Admin Engr/Lafayette College; MS/Mech Engr/Newark College Engr. *Born:* 5/5/27. Native of Chatham, NJ. Retired from Exxon Res & Engr Co in 1982 and founded conslt engrg firm specializing in proj mgmt. Currently working for numerous major firms world-wide. Direct experience all aspects proj work including proj execution planning and mgt of maj petrol and chemicals projs. Exxon position included Gen Mgr of engg Tech for Exxon Res and Engr Co. responsible for technical specialists and engg R&D. Asst Gen Mgr of Tanker Dept, Exxon Intl Co, responsible for ship design, construction and marine R&D. Sr Mgmt Advisor in connection with North Sea crude oil and gas production activities and Aramco projs in Saudi Arabia. Gen Mgr Project Mgmt Exxon Res & Engr Co. Enjoy skiing, golf, fishing, tennis *Society Aff:* AIChE, PMI.

Engin, Ali E
Business: 155 W Woodruff Ave, Columbus, OH 43210
Position: Prof. *Employer:* OH State Univ. *Education:* PhD/Mech/Univ of MI, 1968; MS/Mech/Univ of MI, 1966; BS/ME/MI State Univ, 1965 *Born:* 2/23/43. Native of Samsun, Turkey. Naturalized US citizen since 1974. Worked for two yrs as a res engr at the Hgwy Safety Res Inst of the Univ of MI after receiving PhD & one yr as a visiting Asst Prof at the Middle East Technical Univ, Ankara, Turkey. Accepted Asst Prof position at the Engg Mechanics Dept of the OH State Univ in 1971; promoted to Assoc in 1974 & to Full Prof in 1977. Served as United Nations Consultant in Summers 1977, 1980, 1982 & UNESCO Consultant in Summer 1979. Received the ASEE Best Paper Award for 1976-77 & the OH House of Rep Certificate of Commendation, 1978. Rcvd the Ervin G Bailey Award for meritorious achievement in 1974 and Sr Res award in 1982 of the OH State Univ. Coll of Engrg. Co-dir of NATO Advanced Study Inst, Portugal, 1983. Author of numerous scientific publications in mechanics & biomechanics. *Society Aff:* AAM, ASB, ISB, ASME.

Engl, Walter L
Home: 5 Zum Heiderbusch, Herzogenrath, W. Germany 5120
Position: Director *Employer:* State of Nordrhein Westfalen *Education:* Dr. Rer.Nat/Physics/Tech Univ of Munich; Diplom-Physiker/Physics/Tech Univ of Munich *Born:* 4/8/26 W.L. Engl was born in Regensburg, Germany. He worked in industry from 1950 to 1963. In 1963 he became Prof of Electrical Engineering at the Technical Univ of Aachen, where he is director of the Institute for Theoretical Electronics. He was a visiting Prof at the Univ of Arizona (1968), Stanford Univ (1970), Univ of Tokyo (1972, 1980). Dr. Engl is a Fellow IEEE, Member of The Academy of Science of North Rhine-Westfalia, Member URSI. *Society Aff:* IEEE, VDE, URSI

Englekirk, Robert E
Business: 3242 W 8th St Suite 200, Los Angeles, CA 90005
Position: Pres. *Employer:* Robert Englekirk, Consltg Smuctsane Engrs, Inc *Education:* PhD/Structural Engg/UCLA; MA/-/UCLA; BS/-/Tulane Univ. *Born:* 8/21/36. Pres of Robert Englekirk Consulting Structural Engrs, Inc Los Angeles CA, with offices in Newport Beach, CA, Oakland, CA, Honolulu, HI, Seattle, WA and Rome Italy. The design of multimillion dollar regional shopping ctrs & high-rise bldgs is the firms specialty. Subsidiary firm of Englekirk & Hart specializes in Special Engrg problems such as Wind & Earthquake engrg in Structural. Adjunct Prof at the Univ of CA, Los Angeles. Favorite past times are travel & tennis. *Society Aff:* SEAOC, ACI, PCI.

Engleman, Christian L
Home: Skamania Mines Rd, Washougal, WA 98671
Position: Prof Engr (Self-employed) *Born:* BS US Naval Acad, MS Harvard Univ, Advanced Mgmt Program Harvard Business School, Atomics Sandia, US Navy Post Graduate School Graduate. Numerous tech & exec positions in Armed Forces & in business & financial world. Numerous positions in IEEE prof groups & section activities.

Engquist, Richard D
Home: 5808 Flambeau Rd, Rancho Palos Verdes, CA 90274
Position: Head-Met Materials Engrg Sect. *Employer:* TRW Defense & Space Systems. *Education:* BMetE/Metallurgical Engg/Univ of MN. *Born:* 4/4/29. in MN. BMetE Univ of MN 1952. Dev of Ultra-high strength cast steels, res in shell & graphite permanent mold casting of steel, Amer Steel Foundries (1952- 59). Made worlds first laser weld, R&D on welded electronic packaging, ultra- lightweight high reliability welded antennae & pressure vessels, Hughes Aircraft Co (1959-67). Pioneered scanning electron microscope & anger spectroscopy in failure analysis of sophisticated electronic sys, TRW Sys (1967-), currently respon for metallurgical engg related to aerospace electronic sys. Reg prof engr CA. Mbr ASM, AIME. Married, 1 son, 2 daughters. Enjoys choral music & gardening. *Society Aff:* AIME, ASM.

Engwall, Richard L
Home: 560 Choptank Cove Ct, Annapolis, MD 21401
Position: Mgr. Systems Planning, Analysis & Assurance *Employer:* Westinghouse Electric Corp. *Education:* Master of Bus Admin/Production Mgmt/Northwestern University; Bachelors of Science/Industrial Eng./Northwestern University *Born:* 06/09/33 Native of Chicago, Ill. Mbr of IIE over 25 years, past Chapter Pres, Dir Aerospace Div, Chmn Special Productivity Projects Cttee, IIE to the AAES Coordinating Cttee on Productivity and Innovation. Have authored numerous papers, text book chapters, quarterly columns, serve on numerous task forces/cttees and am frequent speaker in areas of expertise. Have over five years of general mgmt, fifteen years of line & staff multi-plant operations, seven years of line & staff product marketing/project mgmt with mayor blue chip companies such as Honeywell, Bell & Howell, Burndy, Singer and Steelcase companies. Presently at Westinghouse Electric Corp, Defense & Electronics Center as Mgr Systems Planning, Analysis & Assurance & GetPRICE Prog Mgr responsible for mfg systems integration, advanced CIM systems/technology, cost benefits analysis, cost mgmt systems, and D&EC Productivity Improvement Prog. Hold national cttee chmn offices with Aerospace Industries Assoc in Work Measurement & Productivity, am Westinghouse representative and sponsors VChmn of Computer Aided Manufacturing Intl Cost Mgmt System Prog. *Society Aff:* IIE, SME/CASA, APICS

Enoch, Jay M
Business: School of Optometry, Berkeley, CA 94720
Position: Dean, School of Optometry *Employer:* Univ of California, Berkeley *Education:* PhD/Physiological Optics/OH Stat U; BS/Optics and Optometry/Columbia U; Post Doc/Natl Phys Lab/Teddington, Mddx, England; Post Doc/Applied Optics/Wasedau., Tokyo *Born:* 4/20/29. Prof Enoch serves as Dean and Prof, Sch of Optometry, Univ of CA, Berkeley. He conducts res in three areas. 1. Retinal Receptor Optics. He first demonstrated that photo receptors are optical waveguides & has recently presented evidence indicating that they may be phototropic as well. 2. Experiemental Perimetry. He has developed a test battery which allows point-by- point, layer-by-layer analysis of visual function. This allows finer diagnostic analysis than has been previously possible. 3. Correction of Sensory De-

Enoch, Jay M (Continued)
privation. He is attempting early visual correction in infants born with cataracts, without irises, etc. Early correction is essential for visual dev. He has rec'd the GA Fry & C Prentice Medals from the AAO & the F Proctor Medal from ARVO. *Society Aff:* OSA, ARVO, AAO, AOA

Ensign, Chester O, Jr
Business: Amax Center, Greenwich, CT 06830
Position: Exec. VP *Employer:* Amax Inc. *Education:* BSc/Geol/Univ of NC; BSc/Gen Commerce/Univ of NC *Born:* 10/23/24 1951-55 Davison Chemical Co., Bartow, Fla (now W.R. Grace) Supervisor Mines Planning & Exploration. 1955-61 American Metal Co. Ltd. (Now Amax Inc.) Sr. Geologist and Mgr-Mid Continent Exploration-Discovered high grade lead-zinc deposit in Southeast Missouri-now jointly owned by Amax Inc. & Homestake. 1961- 77 Copper Range Co.-Pres & CEO 1970-77 1977- Amax Inc. present Executive VP- responsible for all corporate activities in Strategic Planning, Business Dev, Exploration, R&D, Engrg & Mgmt, Environ Services and Mgmt Dev. *Society Aff:* AIME, SEG, GSA

Ensor, Walter Douglas
Business: 12368 Warwick Blvd, Newport News, VA 23606
Position: VP. *Employer:* Malcolm Pirnie Engrs, Inc. *Education:* BSCE/Civ Engg/ Newark College of Engg. *Born:* 4/17/35. Mr Ensor is VP of Malcolm Pirnie Engrs, Inc, Newport News, VA and Malcolm Pirnie, Inc, White Plains, NY. During his engg career he has worked in various fields of civ engg including environmental/ sanitary, transportation/hgwy, geotechnical/soils, municipal engg and land surveying. Mr Ensor heads the firms 50-man regional office in Newport News, VA. He is a licensed PE in VA, NC, NY, NJ and a licensed land survey in NY and NJ. Mr Ensor is married to the former Joan Elberfeld. They have one daughter and reside at 29 Indigo Dam Rd in Newport News. *Society Aff:* AWWA, ACEC/CEC/V, WPCF, ASCE, AAEE.

Entenman, Alfred M
Business: 1000 Marquette Bldg, Detroit, MI 48226
Position: Pres. *Employer:* Giffels Assocs, Inc. *Born:* Feb 6, 1921. BSCE Cornell Univ 1942, Grad Studies, Columbia Univ. Native of NY State. Capt US Army 1942-45 (European Campaign). Commenced employment with Giffels Assocs, Inc in 1951 as a proj exec with daily responsibility for the quality of design, cost control, scheduling & proj mgt. Apptd VP in 1963, exec VP in 1969, & Pres in 1972. As proj exec initial assignment was complete direction of the Ford Motor Co assembly plant at Toronto, one of the largest plants under one roof in Canada. Subsequent projs included glass works, steel mills, assembly & component mfg plants, warehousing, foundries & res facilities, in addition to maj projs for comml & inst clients. Delivered talks before the Natl Plant Engg & Maintenance Conf, Engg Soc of Detroit, Natl Construction Ind Ctte of the Am Arbitration Assn, & the first intl conf sponsored by the Am Inst of Plant Engr. Mbr Tau Beta Pi, Cornell Univ. Council, Dir of several local cos & associated with numerous engr & prof societies. Reg in 39 states & Canada.

Entzminger, John N, Jr
Home: RD 1 Valley View Rd, Clinton, NY 13323
Position: Branch Chief. *Employer:* USAF Rome Air Dev Ctr. *Born:* Dec 17, 1936. MSEE Syracuse Univ, BSEE Univ of SC. Native of Clinton, NY. Served as officer in USAF at RADC during 1960-63. Remained in civilian status in various capacities in command, control & communications R&D including VLF/LF/HF communications, ECCM communications, & TOS emitter location. Assumed current responsibility as Chief of Location & Control Branch in 1973 where responsible for directing R&D activities of 30-40 professionals in areas of target location, identification & strike & ECCM communications. Sr mbr IEEE, Phi Beta Kappa, Tau Beta Pi, AF Decoration for Exceptional Civilian Service 1973. Enjoy flying, lay religious ministry & camping.

Enyedy, Gustav
Business: Rt 1 Box 64, Gates Mills, OH 44040
Position: Pres. *Employer:* PDQS, Inc. *Education:* MS ChE/Chem Eng/Case Inst of Tech; BS ChE/Chem Eng/Case Inst of Tech *Born:* 8/23/24. President, PDQS, Inc. E I duPont, Rayon Technical Div, 1950-51, Grasselli Chems Div, 1951-54; Diamond Alkali/Diamond Shamrock, 1954-73; Consultant, 1973-85; Founder of PDQS, Inc (computerized Price & Delivery Quoting Service for chem process equip costs), 1976-. Author of PROVES (PROj Valuation & Estimation System) computer prog. Conslt, Aspen Project, at MIT for DOE, 1978-80. Taught design & cost engg courses at Fenn College & Cleveland State Univ. AACE: Fellow Mbr, Certified Cost Engr, Pres, 1969-70, Technical VP, 1966-68, author of COME (Cost Of Maj Equip) prog. AIChE: Fellow Mbr, Speakers Bureau, 1971-72. Over 50 publications. Reg in OH. *Society Aff:* AACE, AIChE

Ephremides, Anthony
Home: 3333 Univ Blvd, W-801, Kensington, MD 20895
Position: Prof. *Employer:* Univ of MD. *Education:* PhD/EE/Princeton Univ; MA/ EE/Princeton Univ; BS/EE/Natl Tech Univ of Athens. *Born:* 9/19/43. Native of Greece. Has taught at Univ of MD and conducted res there since 1971 on areas of communications and data networks. Has participated in adult engg education projs. Has been consulting with private indus, the Naval Res Lab and other govt agencies. Has served as officer of the info theory and control systems societies. Has participated in diverse intl scientific activities. Is the Pres of Pontos, Inc, a Washington DC based conslt firm. *Society Aff:* IEEE.

Eppler, Richard A
Home: 2113 Folkstone Rd, Timonium, MD 21093
Position: Scientist. *Employer:* Pemco Ceramics Group, Mobey Chemical Corp. *Education:* PhD/CE/Univ of IL; MS/CE/Univ of IL; BS/CE/Carnegie Mellon Uni. *Born:* 4/30/34. in Lynn, MA. Res Chemist at Corning Glass Works, 1959-1965. Assumed present position 1965. Responsible for all Res & Consulting activity at Pemco. Have 43 published papers and 16 US patents. Reg PE NY. Areas of expertise include porcelain enamels, ceramic glaze, pigments, glass, glass-ceramics, solar energy, high pressure, corrosion prevention. *Society Aff:* ACerS, NSPE, ASTM, NICE, ACS, ES.

Epremian, Edward
Business: 2101 Constitution Ave, NW, Washington, DC 20418
Position: Exec Dir. *Employer:* NRC/Natl Acad of Sci. *Education:* DSc/Metallurgy/Carnegie Inst of Tech; MS/Metallurgy/Rensselaer Polytech Inst; BSc/Metallurgy/MIT. *Born:* 9/3/21. in Schenectady, NY. DSc Carnegie Inst of Tech (1951), MS Rensselaer Polytechnic Inst (1947), BSc MIT (1943) Metallurgy. Res Assoc, Res Lab, Genl Elec Co (1943-47); Sci Liaison Officer (1951-52), Dep Sci Dir, US Office of Naval Res, London, Eng (1953-54); Chief, Metallurgy & Natsl Branch, Div of Res, US Atomic Energy Commission (1954-57); Sen Metallurgist, Res Lab (1957-59), Tech Coord, Technology Dept (1960-61), Mgr, New Product Mktg, Union Carbide Metals Co (1961-63); Asst Dir Res (1964-65), Genl Mgr, High Temperature Matls Co, Inc (a subsidiary) (1965-67), Genl Mgr, Advanced Matls Dept, Union Carbide Carbon Products Div (1965-70); Mgr, Tantalum Products (1970-71), Mgr Specialty Products (1971-71), Dir of New Ventures, Union Carbide Metals Div (1973-76); Exec Dir, Commisson on Sociotechnical Sys, Natl Res Council, Natl Academy of Sciences (1976-). Fel ASM (1970); Fel, AAAS (1978); Mbr Advisory Cmte, Schl of Met and Natl Sci, Univ of PA (1969-73); Mbr, Bd of Dir, Acta met (1971-73); Mbr, Bd of Trustees, Webb Inst of Naval Arch & Marine Eng (1976); Mbr, Adv Council, Col of Eng, Univ of MD (1978); Mbr of Chmn, various Matl cmtes of ASM, AIME, & AIAA (1954-). 2 pats 15 tech pubs on matls, book on "Metallurgy of Columbian and Tantaluim." *Society Aff:* AAAS, ASM, AIME.

Epstein, Henry David
Business: 600 Third Ave, New York, NY 10016
Position: Sr Group VP. *Employer:* Loral Corp. *Education:* MS/Elec, Bus Admin/ Harvard; SCB/Engrg/Brown *Born:* 11/5/27. From 1977 - Sr Group VP. Loral Corp (NYSE). Mfg & designs electronic warfare, telemetry & telecommunication systems. 1969-1977 Div mgr & asst vp control products div TX Instr Inc. 1959-1969 Mgr, Precision Control Dept TX Instr. 1950-1959 Mgr dev engg, Metals & Controls Corp. *Society Aff:* IEEE

Epstein, Howard I
Business: Box U-37, Dept. of Civil Engrg, Storrs, CT 06268
Position: Assoc. Prof *Employer:* Univ of CT. *Education:* Ph.D./Applied Mech/ Northwestern Univ; M.S./Applied Mech/Northwestern Univ///;BS/CE/The Cooper Union *Born:* 12/16/41 From 1967-9 he was an Assoc Research Eng at IIT Research Inst in Chicago. From 1969-76 he was on the faculty of the Univ of MN. Since 1976 he has been at the Univ of CT where he is an Assoc Prof. He has authored over fifty technical publications in areas such as soil-structure interaction, wave propagation, vibrations and various specialized structures such as floating roofs, scalloped tanks and space frames. He was on the task force which investigated the Hartford Civic Center collapse. He is a Fellow in ASCE. *Society Aff:* ASCE, AAUP, ASEE

Epstein, Norman
Business: Dept. Chem Engg, UBC, Vancouver BC, Canada V6T 1W5
Position: Prof. *Employer:* Univ of British Columbia. *Education:* EngScD/Chem Engg/NYU; MASc/Chem Engg/McGill Univ; BASc/Chem Engg/McGill Univ. *Born:* 12/6/23. 1923. Taught chem engg at UBC since 1951, with working leaves at Cambridge, Maitland, Oriente, Berkeley, Haifa, London, Harwell. About 100 res publications in multiphase, particulate and transport phenomena, with recent emphasis on three-phase fluidization, spouting and fouling of heat exchangers. Co-author with K B Mathur of Spouted Beds, Academic Press, 1974. Chem Inst of Canad Fellow since 1964. AIChE Fellow since 1973. Prog Co-chrmn, AIChE-CSChE Joint Conf, Vancouver, 1973. Killam Sr Fellow, 1975-76. Chrmn, NRC Assoc Committee on Heat Transfer, 1977-80. Pres, Canadian Soc for Chem Engg, 1979-80. Chrmn, Second Int'l Symposium on Spouted Beds, Vancouver, 1982. Editor, Canadian Journal of Chem Engrg, 1985- . *Society Aff:* CSChE, CIC, AIChE

Erdman, Carl A
Business: 301 Engineering Res. Center, College Station, TX 77843-3126
Position: Associate Dean *Employer:* Texas A& M Univ *Education:* PhD/Nuclear Engr/Univ of IL; MS/Nuclear Engr/Northwestern Univ; BS/Sci Engr/Northwestern Univ *Born:* 11/15/42 in Decatur, IL. Undergraduate COOP at Oak Ridge National Lab (1962-64). Assistant research engineer for IL Inst of tech (1965). AEC Fellow (1965-1969). Prof in Nuclear Engrg Dept at Univ of VA from 1970-1981 excepting year as Associate Nuclear Engr and Group Leader at Brookhaven Natl Lab. Asst Dean of Engrg, Univ VA 1979-81. Consultancies: US Nuclear Regulatory Commission, BNL, LANL, and NASA JSC. Joined Texas A&M Univ as Head of Nuclear Engrg in July, 1981. Associate Dean of engrg., Oct 1985 to present. Main interest: reactor safety, advanced reactors, systems engineering. *Society Aff:* ANS, ASEE, ASME

Erdoss, Bela K
Home: 2 Stoneleigh Alger Ct, Bronxville, NY 10708
Position: Chrmn Emeritus. *Employer:* Korfund Dynamics Corp *Born:* 7/3/03. PhD in ME from Lehigh Univ, ME in Heat-Power, Hydraulic & Elec Engg from the Royal Joseph Univ of Engg in Hungary; MS from Lehigh Univ, Engg M h c from Stevens Inst of Tech, Cert Master-Builder in Hungary; PE (ME) in PA. Byllesby Fellow to Assoc Prof (ME), Dir classified res porjs for the US Navy & AF, at Lehigh Univ 1942-47; Prof & Hd Dept of Fluid Dynamics, Dir Fluid Dyn Lab, Stevens Inst of Tech 1947-56. Consultant to Davidson Lab, E I Dupont and others 1943-56. Pi Tau Sigma, Tau Beta Pi, Sigma Xi. Mbr 3 Power Test Code Cttes, ASME, Consultant, Mbr, Sigma Xi. Mbr: ASME, AIAA, ASEE, SNA&ME, APS, SESA, ISA, Assoc Maritime & Aeronautique, NY Ac of Sciences, with Korfund Dynamics Corp since 1956, majority Stockholder 1968-75, responsible for gen mgt, res & product dev, production, sales, etc. Likes: photography, classical music, books, swimming & tennis. Retired in 1978 as Chm of the Bd emeritus of Korfund Dynamics Corp. *Society Aff:* ΣΞ, ASME, AIAA, SNAYME, APS, ISA.

Erganian, George K
Home: 7410 Central, Indianapolis, IN 46240
Position: Retired *Employer:* Self *Education:* MS/Sanitary Engg/Purdue Univ; BS/CE/ Purdue Univ. *Born:* 9/11/17. Grad work in sanitary engg and ind waste treatment. Res asst for the Natl Council for Stream Improvement of the Pulp and Paper Industry. Chief of the Sewage and Refuse Disposal Sect of the IN State Bd of Health. Principal and subsequently Pres of Henry B Steeg & Assoc Consltg Engrs from 1951-1973. A Partner of Howard Needles Tammen & Bergendoff from 1973-81. Retired Jan 1982. Appointed Indiana Commissioner to Midwest Low-Level Radioactive Waste Commission, 1983. Mbr Bd of Public Wks City of Indianapolis. Mbr Sigma Zi and Chi Epsilon honoraries. *Society Aff:* NSPE, ASCE, WPCF, AWWA.

Erickson, Arthur W
Business: PO Box 61, Belmont, CA 94002
Position: Owner *Employer:* Erickson Wood Products Co. *Born:* Mar 1926, San Francisco. Involved in all aspects of industrial packaging, wood, plastics, & corrugated since 1949. With local chapter of Soc of Packaging and Handling Engrs held various offices through Pres 1962-64. On Natl level, Gen Chrmn 1966 Natl Conf & Packaging Competition, San Mateo & judged in Natl Competition in other yrs throughout US. Represented Chap at SPHE Natl Bd of Dirs & in 1972 served as Western Regional VP. Elected Fellow Mbr of SPHE & currently serving as Nat'l SPHE VP, Publications. *Society Aff:* SPHE, ASTM

Erickson, Claud R
Business: 360 Hollister Bldg, Lansing, MI 48933
Position: Self-employed *Education:* BS/Eng/MI State; ME/Mech/MI State; EE/Elec/ MI State; CE/Civil/MI State. *Born:* 1/8/00. Listed in Men of Achievement 1973 London, Community Leaders of Amer 1970- 71, Int'l Who's Who in Community Service, Man of the Yr by Amer. Public Works Assn 1968, written many technical papers, publication award Amer Water Works Assn 1960, AWWA Fuller Award, admitted to US Supreme Ct Bar, 1963, MI Supreme Ct Cert No 1387. US Treasury Dept Savings Bond Div. County Chrmn since May 1, 1941, Delegate MI Constitutional Convention 1961-62, Mbr Lansing Charter Commission 1976-78; SATC WWI, Lt Officer Res Engr Corps 1922-27, Chrmn Gov Romney's Task Force on Water Rights, use and pollution control. Hobbies are stamps, music, travel and electric vehicles. *Society Aff:* ΤΒΠ, ΠΤΣ, HKN, XE, MES.

Erickson, Larry E
Business: Durland Hall, Dept Chem Engg, Manhattan, KS 66506
Position: Prof of Chem Engg. *Employer:* KS State Univ. *Education:* BS/Chem Engrg/ Kansas State Univ; PhD/Chem Engrg/Kansas State Univ. *Born:* 10/8/38. and raised near Wahoo, NB. Attended Luther Jr College and KS State Univ. Conducted post doctoral res in biochemical engg at Univ of PA and at the Inst for Biochemistry and Physiology of Microorganisms, USSR Academy of Sciences. Chem Engg faculty mbr at KS State Univ from 1964 to present. *Society Aff:* AIChE, ACS, ASM, ASEE, IFT, SIM, AWWA, WPCF.

Erickson, Larry L
Home: 169 Rainbow Ln, Cary, IL 60013
Position: Regional Sales Mgr *Employer:* The Austin Company *Education:* MS/CE/MT State College; BS/CE/Tri-State College. *Born:* 6/14/36. Formerly VP of Corporate Engg for Velsicol Chem Co. Formerly was Dir of Engg for MI Chem Co; VP of Mfg for E Shore Chem Co; Engg Mgr for Ott Chem Co; Process Design Supervisor & Proj Mgr for Monsanto; & Asst Prof of Chem Engg at Tri-State College. Native of Tustin, MI & an avid tennis player. Enjoy skiing. Currently Regional

Erickson, Larry L (Continued)
Sales Mgr for the Austin Co. Now Reg in the state of OH, WI, MN, MI, IN. *Society Aff:* AIChE.

Ericsson, Eric O
Home: 222 South Garden St, Bellingham, WA 98225
Position: Consultant. *Education:* BS/Chem Engr/Univ of Washington. *Born:* 12/10/12. Lic Engr. Native of Bellingham, WA. VP of Operations, Bellingham Div; Genl Mgr Chem Div; Portland, Or; Genl Mgr Pulp Div; Samoa, CA; Genl Mgr Paper Div, Toledo, OR; Genl Mgr Pulp & Chem Div, Bellingham, WA; all GA-Pacific Corporation. Mbr ACS, AIChE, TAPPI Fellow. *Society Aff:* ACS, TAPP, AIChE.

Eriqat, Albert K
Business: 4443 30th St, San Diego, CA 92116
Position: Pres. *Employer:* AKE Profl Engrs. *Education:* MS/EE/Univ of S CA; BS/EE/Univ of CA. *Born:* 8/16/27. Over 30 yrs elec & mech engg. Reg PE in CA, EE & ME, CO, AZ, NV, NM, TX, OR, OH, HI. Dept of Defense: Fallout Shelter Analyst, Certified Nucl Defense engg, Environmental engg, Certified Natl Council of Engg Examiners, Panelist-Am Arbitration Assn. Pres of AKE Profl Engrs.. *Society Aff:* IEEE.

Ernst, Edward W
Business: College of Engineering, 1308 W. Green St, Urbana, IL 61801
Position: Prof & Assoc Dean *Employer:* Univ of IL. *Education:* PhD/Elec Engg/Univ of IL; MS/Elec Engg/Univ of IL; BS/Elec Engg/Univ of IL. *Born:* 8/28/24. Born in Great Falls, MT; attended elementary and high schools in Ardmore, OK. Res engr, GE Co, Syracuse, NY, 1955; res engr, Stewart Warner Electronics, 1955-58. With the Dept of Elec Engg, Univ of IL, Urbana-Champaign since 1958; Assoc Prof, 1958-68, Prof, 1968-. Assoc Hd, 1970-85. Assoc Dean, 1985- . Fellow, IEEE, Chrmn, IEEE Accreditation Committee, 1977-1980, IEEE Vice-president for education, 1981-82, IEEE Board of Directors, 1981-82, IEEE Executive Committee, 1981-82; Bd of Dirs, ABET, 1983-, ABET Engineering Accreditation Commission, 1979-87, Vice Chrmn, 1981-85, Chairman, 1985-86. *Society Aff:* IEEE, ASEE.

Ernst, Robert H
Business: 2139 Main Ave, P O Box 177, Durango, CO 81301
Position: Engineer *Employer:* LWP Services Inc. *Education:* BS/CE/Univ of NM *Born:* 11/17/29 Certified Consulting Engr and Registered Professional Engr and Land Surveyor in Colorado and New Mexico. Bachelor of Science degree in Civil Engrg from the Univ of New Mexico after attending New Mexico Military Institute . From 1955 to 1963, mgr of the Durango office of San Juan Engrg Co. In 1963 formed the Ernst Engrg Co and has been its President since its incorporation in 1972. Services provided included consulting civil engrg, land surveying, material and concrete design, testing and quality control. Joined LWP Services Inc, a subsidiary of Pettit-Morrey Co in Risk Management for insurance underwriters. *Society Aff:* NSPE

Errera, Samuel J
Business: Steel General Office, Bethlehem, PA 18016
Position: Tech Cons *Employer:* Bethlehem Steel Corp. *Education:* PhD/Structural Engrg/Cornell Univ; MS/Civ Engrg/Univ of IL; BS/Civ Engrg/Rutgers Univ. *Born:* 1/7/26. Native of Hammonton, NJ. Served with US Army Infantry & Signal Corps 1944- 46. Married, 2 Children, 3 grandchildren. Engr of tests, Fritz Lab, Lehigh Univ, 1951-62. Mgr of structural res, Cornell Univ, 1962-70. Bethlehem Steel Corp 1970-present; Chrmn SSRC. Chrmn of Advisory Group on AISI Specification for Design and Cold-Formed Steel Structural Mbrs. Several published papers on cold-formed steel design. Enjoy golf & travel. *Society Aff:* ASCE, SAE, SSRC.

Ershaghi, Iraj
Business: Dept Petro Engrg-USC, Los Angeles, CA 90007
Position: Assoc Prof./Acting Chmn *Employer:* University of Southern California. *Education:* PhD/Petroleum/Univ of SCA; MS/Petroleum/Univ of SCA; BS/Petroleum/Univ of Tehran. *Born:* 8/1/42. Prior Employment Signal oil and Gas, State Lands Commission. Married. Wife's name Mitra. MS in Bus Adm. 3 children Marsha, Minta, Millad. Res Engineering, enhanced oil recovery, well Logging. Geothermal Engr. Distinguished Facility Award Soc of Petroleum Engrs (1984). *Society Aff:* SPE, SPWLA, AAES.

Ersoy, Ugur
Business: CE Dept, Ankara Turkey
Position: Prof/METU *Employer:* Middle East Technical Univ. *Education:* PhD/Structures/Univ of TX; MS/Structures/Univ of TX; BS/Civ Engrg/Robert Coll. *Born:* 1932, Mersin, Turkey. Received Civ Engrg degree from Robert Coll, Istanbul 1955, Masters from Univ of TX 1956 & PhD 1965. Worked as a design engr with R C Reese 1956-57. Since 1959 with the Middle East Technical Univ. Served twice as the dept chrmn & Asst Pres. Presently Prof of Civ Engg. Assumed responsibility as Dean for founding a new Campus in Mersin. Author of several books & res papers on reinforced concrete. Mbr of ACI & European Concrete Ctte. Received ACI Wason Medal for res 1967 & was listed in Engrs of Distinction 1970. *Society Aff:* ACI, CEB.

Ervin, Fred F
Business: 621 South Westmoreland, Los Angeles, CA 90005
Position: Pres *Employer:* Ervin Engineering *Education:* BS/CE/Univ of So CA *Born:* 10/27/30 Served as City Engr and Dir of Public Works for the City of Palos Verdes Estates from 1957 to 1960. He entered private practice as the Division Engr in charge of Design for a Los Angeles based civil engrg firm, and served in this capacity until 1963 when he formed the firm of Ervin Engrg. He is actively involved in day to day engrg activities with the firm, and is currently serving as District Engr for the Carpenteria Sanitary District. *Society Aff:* ASCE, CEAC, CCCELS

Erzurumlu, H Chik M
Business: School of Engineering (EAS), Portland State Univ., Box 751, Portland, OR 97207
Position: Dean, Sch of Engrg and Applied Sci *Employer:* Portland State Univ. *Education:* PhD/Civ Engg/Univ of TX at Austin; MS/Civ Engg/Univ of TX at Austin; Profl Degree/Civ Engg/Istanbul Technical Univ. *Born:* 3/7/34. After receiving profl degree in civ engg (1957), worked as structural engr until 1960 when began grad work. In 1962, joined Portland State Univ as Instr in Engg with promotions in rank to Asst Prof (1965), Assoc Prof (1969), Prof (1972). Served as Hd of Civ-Structural Engg at Portland State Univ (1975-79), was selected to hd the Division of Engg and Applied Science in 1979 and was named the first Dean of the Sch of Engrg and Appl Sci in 1982. Reg professional engr and technical consultant to industrial/consulting firms, and govermental agencies in the field of structural engg. Res interests include tubular structures and steel bridges. Author of numerous publications in these fields, currently serving on ASCE Administrative Committee on Bridges. *Society Aff:* ASCE, ASEE, NSPE.

Eschenbrenner, Gunther Paul
6 South Cheska, Houston, TX 77024
Position: Vice President - Engineering *Employer:* The M.W. Kellogg Comp *Education:* MSME/Mech Engr/Columbia Univ; Diplom Engr/Mech Engr/Tech Univ Darmstadt, Germany *Born:* 3/3/25. Native of Hamburg, Germany. Educated in Germany, England and USA. With M.W. Kellogg since 1952, Vice President and Director of Engineering Operation since June 1981. Author of technical, published papers in field of Pressure Vessel Tech, several US patents. Mbr ASME since 1952, Fellow 1973, VP 1978. Govenor 1982. Present and past activities include Intl, Natl and Local ASME Committees. Mbr Sigma Xi - Kellogg Branch, AIChE, NSPE, Houston Engr & Scientific Soc. Active in several outdoor-oriented organizations and societies. Biographic Ref: Who's Who in TX; Who's Who in the Southwest. *Society Aff:* ASME, AIChE, NSPE, ΣΞ

Eschman, Rick D
Home: 8903 Echo Valley, Houston, TX 77055
Position: Owner. *Employer:* Eschman Engg Co. *Education:* MS/ME/Univ of TX; BS/ME/Univ of TX. *Born:* 6/24/44. Native of W TX. Staff engr & VP of John & Glasgow Consulting Engrs, Corpus Cristi, TX from 1968-70. Owner of Eschman & Assoc, Consulting Engrs, San Angelo, TX from 1970-77. VP of CGS Engrg, Inc, Houston from 1977-79. Presently owner of Eschman Engrg CO, Houston. Have performed mech/elec design of over 1100 bldg systems; including hospitals, nursing homes, governmental, institutional & industrial bldgs. Received R C Baker Fdn Fellowship, 1967. Performed ASHRAE sponsored res in High Velocity Duct Design. Contributed to ASHRAE handbook of fundamentals, 1972. Pres, W TX Chapter of NSPE, 1976. Enjoy hunting, fishing, skiing. *Society Aff:* NSPE, ASHRAE, TSPE.

Escoffier, Francis F
Home: 2759 Brookwood Dr, Mobile, AL 36606
Position: Retired. *Born:* 3/19/08. Attended Arts and Science Sch, Loyola Univ, New Orleans, 1928-30; did not graduate. Employed by US Army Corps of Engrs in 1929; attained prof level in 1937. During 1940-43 served on staff of Beach Erosion Bd. During 1943-65, served as hydraulic engr in Mobile District of Corps, being chief of Hydraulic Section 1953-65. with Marden B Boyd, received Karl Emil Hilgrad Prize of ASCE, 1965. Held rating of GS 13 at retirement in 1965. Consultant in coastal Engg 1966-75. Now Retired. *Society Aff:* ASCE, SAME.

Esgar, Jack B
Home: 5097 Whitethorn Ave, N Olmsted, OH 44070
Position: Ret *Education:* BS/ME/Univ of CO. *Born:* 12/1/20. Spent 4 yrs as Experimental Test Engr at Pratt & Whitney Aircraft in E Hartford, CT after grad from Univ of CO in 1943. At NACA/NASA in Cleveland, OH since 1947 conducting res in gas turbine cooling, spacecraft structures, & full scale gas turbine engines at successive levels of supervision up to Chief of the Airbreathing Engines Div. In 1977 assumed position as Dir of Engrg Services resp for all facilities & engrg design at the NASA Lewis Res Ctr at Cleveland, OH. Ret in 1980. Presently involved in many volunteer activities including score (Service Corps of Retire Execs) with the small business admn. Enjoy gardening, tinkering, jogging, dancing, & water skiing. *Society Aff:* AIAA.

Eskinazi, Salamon
Home: 1316 Broad St, Syracuse, NY 13224
Position: Prof. *Employer:* Syracuse Univ. *Education:* DrEngg/ME/Johns Hopkins Univ; MS/ME/Univ of WY; BS/ME/Robert College. *Born:* 11/25/22. Born in Izmir, Turkey. Instr at the Johns Hopkins Univ 1948-55. Assoc Prof at Syracuse Univ 1955-60. Prof to date. Teaching Fluid Mechanics, Thermodynamics, Turbulence, Environmental Fluid Mechanics, Wind energy. Twice Fulbright Scholar 1963 & 1972. Prof Assoc at the Univ of Poitiers, France 1963, 1972, 1980. Chrmn of dept of ME 1965-71. Extensive ind consulting 1948-present. Principal in a consulting firm: Associated Engg Consultants. Decorated by the French Ministry of Education for excellence in teaching & res with title of Knight of the Order of Academic Palmes in 1975. Listed in Who's Who in the East, Who's Who of Authors Intl Who's Who in Community Service, Dictionary of Intl Biographies, Am Men of Sci, ASME Honors Chrmn 1960-62. Lectr at thirty univs across the world. Author of four books & many technical papers. *Society Aff:* AAAS, AWEA, AAM, ASME.

Esmay, Merle L
Home: 1272 Scott Drive, East Lansing, MI 48823
Position: Prof Agricultural Engrg. *Employer:* Michigan State University. *Education:* PhD/Agri Engr/IA State Univ; MS/Agri Engr/IA State Univ; BS/Agri Engr/SD State Univ. *Born:* 12/27/20. in IA. Wife Katherine and 2 children. Army Corps of Engr, Prof of Agri Engr MI State Univ since 1957 with specialization in the Structures & Environment field. Authored 200 technical articles and 4 books. Has served as an advisor in many Asian and latin Amer countries since 1962. Major advisor for 45 MS & PhD graduates. Received the ASAE Metal Bldgs, Manufacturers Assoc. Award in 1966 and was named "Engineer of the Yr" in 1966 by the MI Section of ASAE. Named a Fellow of ASAE in 1975. Listed in the 40th edition of Who's Who in America in 1978. Lead an ASAE technical Delegation to China in 1979. First Dir of ASAE Internatl Dept 1978-80. Distinguished Alumnus Award South Dakota State Univ. 1980, Served on International Consultancy Assignment in 15 different countries and participated in 26 international symposia. Received the ASAE Kishida Intl Award in 1982. Named Distinguished Engr, South Dakota State University, 1983. Received the Distinguished Faculty Award from Michigan State Univ., 1985. *Society Aff:* ASAE, ASEE.

Esogbue, Augustine O
Business: School of Industrial and Systems Engrg, Atlanta, GA 30332
Position: Prof *Employer:* GA Inst of Technology *Education:* Ph.D./Engrg-Op Research/Univ of So CA; MS/Ind. Engrg & Op Research/Columbia Univ; BS/EE/UCLA *Born:* 12/25/40 Tenured prof, School of Industrial and Systems Engrg, GA Inst of Technology. Interests include dynamic programming-theory and applications, fuzzy sets, operations research in health care, water resource, urban systems, and development problems of developing countries. Co-author or author of 2 books and over 40 technical papers in referred journals. Serve on editorial bd of several journals including the Journal of Mathematical Analysis and Applications, Intl Journal of Fuzzy Sets and Systems, Intl Journal of Computer Science and Its Applications. Pres and Principal AESO Systems, Inc. Consultant to the National Academy of Sciences (National Research Council), member, Executive Committee, Water Resources Advisory Council, Atlanta Regional Commission. Listed in numerous Who's Who. *Society Aff:* AWRA, ORSA, IMS, AAAS

Espelage, John J
Business: 11541 S Champlain Ave, Chicago, IL 60628
Position: Gen Mgr - Chicago Site. *Employer:* Sherwin Williams Co. *Born:* Aug 1938. ChE 1961 Univ of Cincinnati, MS ChE 1965 Wayne State Univ, MBA 1974 Univ of Cincinnati. Native of Cincinnati. Process Dev Engr at Parke Davis & Co, Detroit 1961-65; Process Engr at Monsanto Port Plastics, OH 1965-67; With Sherwin-Williams since 1967. Assumed current duties 1974. Respon for gen mgt of the Chicago Site & direct mgt of the support groups serving several mfg & technical groups located on the Site. Mbr AIChE. Active in several Business & Community Organizations.

Esposito, Raffaele
Business: Via Tiburtina KM 12,400, Rome Italy 00100
Position: General Manager *Employer:* Selenia SPA. *Education:* Libera Docenza/Communications/Univ of Rome; Dr Elec Engg/Electronics/Univ of Rome. *Born:* 8/11/32. Native of Rome, Italy. Had my education at Univ of Rome-Fulbright scholar at Univ of AR, Fayetteville 1959-1960. Res Scientist at Raytheon, Waltham, MA. From 1960 to 1973 served as a consultant on systems problems for all divs of Raytheon. Did res on communication theory & served as editor for IEEE transactions on communications. Joined Selenia in 1973. Technical Dir from 1975. Vice Direttore Generale from 1981. Responsible for all technical activities. IEEE Fellow. Direttore Generale from 1984. *Society Aff:* IEEE, ASEE, AEEI.

Esselman, Walter H
Home: 1141 Buckingham Dr, Los Altos, CA 94022
Position: Consultant (Retired) *Employer:* Elec Power Res Inst. *Education:* BS/EE/Newark Coll of Engg; MS/EE/Stevens Inst of Tech; DEE/EE/Poly Inst of Brooklyn. *Born:* 3/19/17. Westinghouse Elec Corp, 1939-1949 - Dev of various control systems including elevator & anti-aircraft dir systems. 1950-1955 - Bettis Atomic Power Lab. Progressive positions on the dev of the nuclear plant for the USS Nautilus, including reactor & plant control, plant systems, operations & testing. 1955- 1959 - Mgr of Advanced Dev. 1960-1969 - Dev of Nuclear Rocket. Position including Mgr of Nerva Proj. 1970-1972 - Dev of Breeder Reactor. Pres of Westinghouse Hanford, Dir Hanford Engg Dev Lab. 1973-1975 - Dir Strategie Plan-

Esselman, Walter H (Continued)
ning Nuclear Systems. EPRI 1975-81 - Dir of Strategic Planning. 1982-86 Dir of Engrg Assessment and Anal; 1986-Retired, Consultant. *Society Aff:* IEEE, ANS, AAAS, AIAA.

Estcourt, Vivian F
Business: P O Box 3965, San Francisco, CA 94119
Position: Consulting Engr. *Employer:* Bechtel Power Corp. *Education:* AB/ME & EE/Stanford Univ. *Born:* 5/31/97. Native of London England. Retired Mgr of Steam Generation Dept, Pacific Gas and Electric Co. With Bechtel Power Corp as independent power consultant since 1962. Career in fossil and nuclear power generation has included engg design, operational and technical mgmt, res, and engg education. Present emphasis is on R&D in air pollution control tech, plant availability and safey protective controls and engg safeguards. Fellow and Hon Mbr ASME; Newcomen Gold Medal 1965, & Hon Life Mbr Newcomen Soc; Mbr. Ntl Academy of Engrg. Enjoys world travel. *Society Aff:* ASME, IEEE/PES, ASTM, ISA, ISA, NFPA.

Esterbrooks, Robert C
Business: 3325 W Durango St, Phoenix, AZ 85009
Position: Asst County Mgr, Dir of Public Works & County Engr *Employer:* Maricopa County. *Education:* Masters/Engr/UCLA; BSCE/CE/Univ of TX *Born:* Glendale CA. MS Engg from UCLA, BS in Civ Engg from Univ of TX. Served in Navy in WWII & Korean War. Reserve Officer currently retired with rank of Rear Admiral, Civil Engr Corps, USNR. Currently responsible for all public works depts within County admin. Formerly Pres AZ & San Diego Sections of ASCE; San Diego & Phoenix posts SAME; AZ Chapt APWA. Past Pres of APWA & Past Natl VP at large of SAME. Selected as one of APWA's Top Ten Men of Yr in 1973. Nat'l Bd of Dirs, ASCE (1983-86). Nat'l Dir "Keep America Beautiful–. NACE, Natl. Urban Engineer of Year; 1982. *Society Aff:* ASCE, APWA, SAME, IPWP, AAEE, NACE

Estes, Edward R Jr
Business: College of Engineering & Technology, Old Dominion University, Norfolk, VA 23529-0236
Position: Assoc Dean of Engrg *Employer:* Old Dominion Univ. *Education:* MS/Appl Mechanics/VPI; BE/Civ Engg/Tulane Univ. *Born:* 3/2/25. Native of Richmond, VA; Reg PE; Ensign, Civ Engr Corp, US Naval Reserve, WWII; Asst Prof, Engg, Univ of VA, 1948-55; Res Engr, Am Inst of Steel Construction, 1955-60; Dir of Engg, FL Steel Corp, 1960-66; Dir of Engg, M-B Co, 1966-68; Chief Res Engr, Am Iron & Steel Inst, 1968-69; Engg Mgr, Republic Steel Corp, 1969-72; Estes & Assoc, Consulting Engrs, 1972-78; Chrmn, Dept of CET, ODU, 1978-84; Assoc Dean of Engrg, ODU, 1984. Chrmn, Res Council on Riveted and Bolted Structural Joints of the Engg Fdn, 1974-79; Am Welding Soc's AF Davis Silver Medal, 1964; Past pres, Skyline Chap NSPE, 1954-55. Past Pres, Norfolk Branch, ASCE, 1983-84; Chairman, Tidewater Section, American Welding Society, 1986-87; Accreditation Bd for Engrg and Tech, Ad Hoc Visitor, Civil Engrg Tech. Sigma Xi, Chi Epsilon, Methodist. *Society Aff:* ASCE, AWS, ASTM, AREA, ASEE, NAC

Estrin, Gerald
Business: 3732 Boelter Hall, Comp Sci Dept, Los Angeles, CA 90024
Position: Prof & Chair *Employer:* Univ of CA. *Education:* PhD/EE/Univ of WI; MS/EE/Univ of WI; BS/EE/Univ of WI *Born:* 9/9/21. Native of NY, NY. Mbr of von Neumann Comp Engrg group which developed first machine at Princeton Inst for Advanced Studies 1950-1956. Dir of Elec Comp Group which developed WEIZAC at the Weizmann Inst of Sci, Rehovot, Israel 1954-55. Mbr of engrg faculty at UCLA since 1956 & Chrmn, Comp Sci Dept 1979-82, 85- . Guggenheim Fellow 1963, 1967. IEEE Fellow 1968. Governor, Weizmann Inst of Sci since 1971. Dir Sys Engrg Labs, Ft Lauderdale, FL 1977-80. Distinguished Service Citation, Univ of WI, Sch of Engrg, 1975. Dir Electronic Sys Business, Gould Inc, Rolling Meadows IL, 1981-86. *Society Aff:* IEEE, ACM, ASEE, ΣΞ, ΤΒΠ, AAAS.

Estrin, Thelma
Business: Sch of Engg & Appl Sci, Boelter Hall - 6722, Los Angeles, CA 90024
Position: Prof, Engrg *Employer:* Univ of CA at Los Angeles. *Education:* PhD/EE/Univ of WI; MS/EE/Univ of WI; BS/EE/Univ of WI. *Born:* 2/21/24. Biomedical Engr, Neurological Inst, NYC (1951-1953); Electronic Computer Proj, Weizmann Inst of Sci, Israel (1954-1955); Brain Res Inst, Univ of CA at Los Angeles, combining res in neuroscience, electronic instrumentation & computers (1960-80). Dir of the Data Processing Lab of the Brain Res Inst (1970-1979). Prof of Computer Science, UCLA 1980-present. 1982-1984 - Natl Sci Foundation, Dir of the Elec, Computer and Systems Engrg Div; 1984 - Asst Dean, UCLA Sch of Engrg and Applied Science and Dir, Dept of Engrg, Science and Mathematics, Univ Extension. Mbr Bd of Trustees of Aerospace Corp (1978-82), Army Science Bd 1981-83, Honors include: Fellow IEEE & AAAS; Distinguished Service Citation from the Univ of WI, College of Engg, Madison, 1976; and Outstanding Engr of the Yr Merit Award, 1978. Achievement award Soc Women Engrs, 1981. Active in profl societies: Past Pres, IEEE Engg in Medicine and Biology Soc (1977); IEEE, BOD 1979-1980), *Society Aff:* IEEE, SWE, AAAS, ASEE, AAAAI.

Esvelt, Larry A
Home: E 7905 Heroy Ave, Spokane, WA 99212
Position: Consultant. *Employer:* Esvelt Environmental Engg. *Education:* PhD/Engg (Sanitary)/Univ of CA; MS/Civ (Sanitary) Engg/Univ of CA; BS/Civ Engg/WA State Univ. *Born:* 10/19/38. MS & PhD, Univ of Calif Berkley, BSCE Washington State Univ. Owner & principal of Esvelt Environmental Engg, a consulting firm specializing in water quality, water & wastewater treatment & ind wastewater mgt since 1976. Previously sr specialist with Bovary Engrs, Inc (1973-1976). Sanitary Engr with STR-Esvelt, 1972-73; Sanitary Engr with Gray & Osborne Consulting Engrs, 1964- 1969. Notable projs have included: Food Processing Wastewater Treatment, Reclamation and Reuse; Toxicity assessment of & removal from wastewaters; design of conventional and advanced wastewater treatment plants; ground water quality studies & aquifer protection, sanitary engineering in cold regions. Past Pres of Spokane Section, ASCE; Past Chrmn of WA State Univ Civ Engg Advisory Council; Reg in 8 states; Spokane Section ASCE Engineer of Merit, Diplomate AAEE, Mbr ASCE, WPCF, AWWA, IAWPRC. *Society Aff:* ASCE, WPCF, IAWPRC, AWWA.

Etherington, Harold
Home: 84 Lighthouse Dr, Jupiter, FL 33469
Position: Consulting Engr (Self-employed) *Education:* ARSM/Metal Eng/Royal School of Mines; BSc/Eng-Metallurgy/Univ of London (England) *Born:* 1/7/00 London, England. Mechanical Engr, Allis-Chalmers Mfg Co, 1937-46. Dir, Power Pile Div, Oak Ridge Natl Lab, 1947. Dir, Naval Reactor and Reactor Engrg Divs, Argonne Natl Lab, 1948-1953. VP, Nuclear Products-Erco Div; ACF Industries, 1954-59. Gen Mgr, Atomic Energy Div, Allis-Chalmers Mfg Co, 1959- 1963. Consulting Engr 1963 to present. Member Emeritas, Nuclear Regulatory Commission Advisory Committee on reactor safeguards. Registered PE in the state of WI, member ASME, member Natl Academy of Engrg. Atomic Energy Commisson Citation and Gold Medal for 1974. Co-author *Modern Furnace Tech.* Editor *Nuclear Engrg Handbook. Society Aff:* ASME, NAE

Ethington, Robert L
Business: 809 N E 6th Ave, Portland, OR 97232
Position: Dir *Employer:* Pacific NW Forest & Range Experiment Station *Education:* PhD/Theoretical & Applied Mechanics/IA State Univ; MS/Theoretical & Applied Mechanics/IA State Univ; BS/Forestry-Wood Tech/IA State Univ. *Born:* 2/13/32. Native of IA. Proj Leader for Res on Physical & Mech Properties of wood, US Forest Products Lab, Madison, WI 1963-1974. Asst Dir of same lab, 1974-1976. Natl authority on stress grading of wood products for structural use. ASTM Fellow & L J Markwardt Award winner. 1976-1979 Dir of all res progs in forest products & engg for USDA Forest Service. Since 1979, Dir of Forestry Res for USDA Forest Service in Oregon, Washington, Alaska. *Society Aff:* ASTM, ΣΞ, FPRS, SWST, SAF.

Ettles, Christopher M McC
Business: Mechanical Engrg Dept, Troy, NY 12181
Position: Assoc. Prof *Employer:* R.P.I. *Education:* DSc/Lubrication/London Univ England; PhD/ME/Imperial Coll London England; BSc/ME/Imperial Coll London England *Born:* 5/12/37 in London, ICI Research Fellow at Imperial Coll, London, then Lecturer 1967-80. Speciality Hydrodynamic Lubrication, with emphasis on industrial applications. A Chartered Mechanical Engr. Emigrated to US in 1980. Enjoys skiing, sailing, and tennis. *Society Aff:* ASLE, IME

Eubanks, Robert A
Business: 1806 S. Vine St, Urbana, IL 61801
Position: Owner *Employer:* Eubanks and Moody, Consultants *Education:* PhD/Mechanics/IIT; MS/Mechanics/IIT; BS/Math/IIT. *Born:* 6/3/26. Born in Chicago, IL. Chief Warrant Officer, AUS, 1942-46. Asst Prof of Mech, IIT, 1950-54. Sr. Res Engr, Bulova R&D, NY, 1954-55. Res Engr, AMF, Chicago, 1955-56. Scientist, Borg-Warner Corp, Des Plaines, IL, 1956-60. Sci Advisor, IIT Res Inst, Chicago, 1960-64. Prof of CE & of Theoretical & Applied Mechanics, Univ of IL, 1965 to 1986. Owner, Eubanks and Moody Consultants, 1985 to date. Visiting Distinguished Prof of CE, Mech and Aerospace Engg, & Math, Univ of DE, 1973-74. Dir, Afro-Am Studies Commission, Univ of IL, 1968-71. Natl Councilmbr, AAUP, 1974-76. Exec Committee, Natl Consortium for Grad Degrees for Minorities in Engg, 1977-82. VP 1981-82. Editorial Bd, Journal of Elasticity, 1971 to date. *Society Aff:* ASCE, ASME, AMS, ASEE, AAUP, AAM

Eubanks, William H
Business: PO Box E G, MS State, MS 39762
Position: Prof & Hd Engg Graphics. *Employer:* MS State Univ. *Education:* MS/Indust Ed/MS State Univ; BS/Aero-space Engg/MS State Univ. *Born:* 12/13/21. born in Columbus, MS. Married to Juanita Patrick. 3 children, Dr W Hunter, Jr, Kenneth B and Diane Schmidt. Attended MS State and Louisiana State Univ. Majoring in Mech and Aero Engg, ATO Social Fraternity. Mbr of ASEE since 1950. Reg PE no1609-MS. Industrial Experience with 4-County Elect Power Assoc. Mobil Dist Corp of Engg, TCI Div of US Steel Corp. Military service WWII overseas European Theatre of operation - Army Engg Corp. Employed MS State faculty 1947 - Mbr ASEE, Dir Zones Design Graphics Div - mbr MS and National Society PE, Rotary International. President Starkville - Miss State Rotary Club 1981-82. *Society Aff:* EDG-ASEE, MS/SPE, NSPE.

Euffa, Chauncey
Business: Rm 6C-101, Whippany, NJ 07981
Position: Tel Off Planning & Engrg Sys. *Employer:* Bell Telephone Labs. *Born:* Brooklyn Polytechnic Inst 1942. Bell Telephone Labs 1942- . Bell Tel Labs, F B Jewett Tel Pioneers Council Award 1976. Bell Tel Labs Speakers Bureau 1974- . Standards Engrs Soc Positions & Awards; Asst Secy Treas 1953-68, Nat'l Treas. 1975-1980 Outstanding Section mbr service award 1973, Dir 1973-75, Historian 1971- , chmn NY Sect 1976-77. toastmasters Internatl Secr-Dist 1970 to 1975, and 1979-,Facil Chmn-Dist 1970- ,Hall of Fame Award 1970-71, Abel Toastmaster Award 1973, Toastmaster of Yr Award 1978 & 1976. Adv Bd Mbr 1973- (United Way) and Vice Pres. 1979- Ret Sr Volunteer Program. (R.S.V.P.) *Society Aff:* SES.

Eunpu, Floyd F
Home: 5837 River Dr, Lorton, VA 22079
Position: Deputy Engr-Dir *Employer:* Fairfax County Water Authority. *Education:* BCE/Sanitary Engg/RPI. *Born:* 8/17/29. Experience: 1953-1960 - Engr, Greeley & Hansen, Consulting Engrs Chicago, IL. 1960-85 - Dir, Engg & Construction Div Fairfax County Water Authority. Principal Duty: Responsible for the design & construction of all facilities relating to the Water Authority's capital improvement progs. 1985-present - Deputy Engr-Dir Fairfax County Water Authority. Personal: Married. Wife, Sally, is elementary schoolteacher. Children. One daughter, three sons. *Society Aff:* AWWA, ASCE, APWa.

Eustis, Robert H
Business: Mech Engg Dept, Stanford, CA 94305
Position: Prof. *Employer:* Stanford Univ. *Education:* ScD/Mech Engg/MA Inst of Tech; MS/Mech Engg/Univ of MN; BME/Mech Engg/Univ of MN. *Born:* 4/18/20. Native of Minneapolis, MN. Grad from Univ of MN and taught there to earn MS degree. Worked on jet engine res at NASA during WWII. Taught at MIT while earning ScD and then moved to industry in 1951. Chief Engg of Thermal Res and Engg Corp until 53 and Head of Heat and Mechanics at SRI Intl until 55. Joined Stanford Univ and teach courses in fluid mech, heat transfer, thermodynamics, and energy conversion. Dir of High Temperature Gasdynamics Lab involving res on magnetohydrodynamics, combustion and particulate clean-up until 1980. Assoc. Dean of Engg, 1984-present. Received Soviet Science Academy Medal, 1973, Woodard Professorship, 1978, Fellow AIAA, 1981, Fellow ASME, 1983, and Fellow AAAS, 1986. *Society Aff:* ASME, ASEE, AIAA, CI, AAAS.

Evans, Anderson P
Business: 555 W 57 St, New York, NY 10019
Position: Dir, TV Facilities Planning. *Employer:* CBS Inc. *Education:* BEE/Electronics/Univ of FL. *Born:* 7/8/24. St Augustine FL. North TX State Univ (ASTP). Signal Corp, WWII, served to Capt. Joined CBS Radio Network, 1949. Transferred to CBS TV Network, 1954. Planned, directed, or participated in design of most current CBS TV prog production and broadcasting facilities. Made innovative contributions to TV audio and communications sys. Active in indus stds. Papers in JAES, JSMPTE, IEEE broadcast transactions. Fellow, AES. Sr Mbr, IEEE. Mbr SMPTE. Sigma Tau. Avocations writing, painting, music, recreational math. *Society Aff:* AES, IEEE, SMPTE.

Evans, Bernard M
Home: 105 Berrywood Dr, Severna Park, MD 21146
Position: Partner *Employer:* Blunt & Evans, Consltg Engrs *Education:* MS/Structural/Rensselaer Polytech Inst; BS/CE/Clarkson Coll of Tech *Born:* 10/29/24 Former Regional Dir, NY State Dept of Transp for 23 yrs. Former State Highway Administrator, MD State Highway Admin. Presently, Partner - Blunt & Evans Consltg Engrs. *Society Aff:* ASCE, NSPE, ASBPA, ARTBA, SAME

Evans, Bob O
Business: 1000 Westchester Ave, White Plains, NY 10604
Position: IBM VP & Pres Sys Comm Div. *Employer:* Intl Bus Machines. *Born:* Aug 1927. IBM VP & Pres System Communications Div, heads a team responsible for designing, developing & mfg terminal & computer communications related products. Took part in development of IBM's first large scale computers. Was promoted in 1962 to VP, Dev, DSD responsible for dev of IBM Sys/360. In Jan 1965 was named pres of FSD, which dev adv products for the natl interest. In Oct 1969, was promoted to pres of SDD, elected an IBM VP in June 1972, & named to present post in May 1975. Received BSEE from IA State Univ in 1949. In June 1971 was awarded IA State's Professional Achievement Citation in Engg. Fellow IEEE, served on the Defense Sci Bd, mbr of NAE, RPI's Bd/Trustees, MIT's EE Visiting Ctte & the Charles Stark Draper Lab.

Evans, David C
Business: 580 Arapeen Dr, Salt Lake City, UT 84108
Position: Pres *Employer:* Evans & Sutherland Computer Corp *Education:* PhD/Physics/Univ of UT; BS/Physics/Univ of UT *Born:* 2/24/24 Adjunct Prof of Electrical Engrg and Computer Science, Univ of UT. Fellow, IEEE. Pres and cofounder of Evans & Sutherland Computer Corp, 1968 to present; Prof/Chrmn of Computer Science, Univ of UT, 1965-1977; Prof Electrical Engrg and Computer Science, Univ of CA-Berkeley, 1962-66; Dir of Engrg, Computer Div, Bendix Corp, 1953-1962. In April, 1978, Was elected a member of the Natl Academy of Engrg. Was named "Engr of the Year" by the UT Engr Council in Feb, 1979. Contributions to the computer field are in the fields of interactive computing, computer

Evans, David C (Continued)
graphics and computer aided design and manufacture. *Society Aff:* IEEE, NAE, ACM

Evans, David F
Business: 200 S.W. Market St, Suite 110, Portland, OR 97201
Position: Pres *Employer:* David Evans & Assocs, Inc *Education:* BS/CE/Heald Engrg Coll *Born:* 4/7/38 Pres and Founder of David Evans & Assocs, Inc with offices in Portland, OR; Bellevue, WA; Kennewick, WA. 15 years of prior consulting experience in Municipal & Site Dev. Engrg. Formerly VP and Office Mgr for Wilsey & Ham. Registered in OR, CA, WA, and HI. Recognized expert in land development for housing. Spoke at National Conventions of NAHB, 1976 and 1981, and previous work shops. Responsible for design of bldg sites for 12,000 residents of Metro Portland Area. Mbr. ACEC, NAHB, CEO, NSPE, and PEO. *Society Aff:* ACEC, NAHB, CEO, NSPE, PEO

Evans, David J I
Home: Box 4 - Site 9 - R R 6, Edmonton Alberta T5B 4K3 Canada
Position: Asst VP, Tech. *Employer:* Sherritt Gordon Mines Ltd, Alberta. *Born:* Sept 1928 Margate, England. Caterham Sch. Imperial Coll, London Univ, BSc, Met 1949, PhD, Mineral Engg 1953. Am Cyanamid Field Engr 1951-53 spec in flotation & gravity concentration. Immigrated to Canada in 1953 & worked for Sherritt Gordon, Ft Saskatchewan in var pos from Sr Res Met to Asst VP Tech, since 1953. Work has comprised a broad coverage of all aspects of res dev, innovation, operations & bus in fields of extractive, powder & physical met. Personal preference is for metallurgical process design. Jules Garnier Prize 1966. 37 papers; 26 patents issued. Prof Soc Cttes: AIME, Hydromet 1966-73, Chrmn 1969-71; AIME, Papers & Pubs 1971-73, Co-Chrmn 1973 Intl Hydromet Symposium, Co-Chrmn 1979 Intl Symposium on Laterites; ASM, Transactions 1966-72. Others incl PE of Alberta; Can Res Mgt Assn; Alberta Res Council.

Evans, Ersel A
Business: Westinghouse Hanford Company, P.O. Box 1970, Jadwin Bldg, Richland, WA 99352
Position: VP & Laboratory Technical Dir. *Employer:* Westinghouse-Hanford. *Education:* PhD/Chemistry/OR State Univ; BA/Chemistry/Reed College. *Born:* 7/17/22. Native of NB. Naval Aviation Cadet 1943-1945. Managed res & dev in ceramics and met processing assignments for nucl reactors & isotope heat sources with GE Co 1951-67. Directed wide range of both nucl & non-nucl mtls progs with Battelle Memorial Inst (PN2) 1967-70. Since 1970 have had increasing responsibilities as mgr then technical dir of the Hanford Engineering Development Laboratory & VP of Westinghouse-Hanford Co. This included directing dept of energy tech mgt for US fuels & mtls progs. Present assignment: Sams. Member & former Chrmn Univ of WA College of Engg visiting Committee. Fellow ANS, ACS, ASM, & AIChe; mbr NAE. *Society Aff:* ANS, ASM, ACS, AIChe.

Evans, Garrett H
Business: 111 W Kingsley, Ann Arbor, MI 48103
Position: Prin. *Employer:* Stoll, Evans, Woods & Assoc. *Education:* MS/Civ Engg/Univ of MI; BS/Civ Engg/Univ of MI. *Born:* 10/17/38. Proj Engr at Ayres, Lewis, Norris & May, 1960-66. Responsible for coordination and financing of civ engg construction. Joined U W Stoll in 1966 and co-founded Stoll, Evans, Woods & Assoc in 1972. Consults in field of geotechnical engg including feasibility studies, design and optimization. Mbr of Education Committees for ASCE from 1972 and chaired Exec Committee 1977-79. Mbr of Bd of the Accreditation Board for Engineering and Technology since 1977. Served on Bd of Review for City of Ann Arbor from 1976. Appointed to teach grad course in fdn design at the Univ of MI, 1978. *Society Aff:* ASCE, ASEE, ACI, ISSMFE.

Evans, George F
Business: 5755 Granger Rd, Cleveland, OH 44131
Position: Chrmn. *Employer:* Evans and Assoc, Inc. *Education:* BS/Mech & Aero Engg/Univ of Pittsburgh. *Born:* 9/17/22. Hometown: Erie, PA. Engr with GMC during WWII. Prof at Penn College of Engg for 10 yrs, Specialty-thermodynamics. Pres of Evans & Assoc Inc, consulting engrs since 1947. PPres of the OH Soc of Prof Engrs, P Dir of NSPE, immediate P Chmn of the OH State Bd of Engg Regis, NCEE 1976 meritorious service award recipient; Mbr ASHRAE, Tau Omega, Sigma Tau and Pi Tau Sigma. Most recent work in applied solar energy sys. Special bass tenor sax, flute, skiing and sailing. *Society Aff:* NSPE, ASHRAE, ECPD, ACEC, NCEE.

Evans, Gerald W
Business: Dept of Indus Engrg, Louisville, KY 40292
Position: Assoc Prof *Employer:* U of Louisville *Education:* PhD/Indus Engr/Purdue U; MS/Indus Engr/Purdue U; BS/Mathematics/Purdue U *Born:* 09/29/50 Native of New Albany, IN. Sr Res Engr for Gen Motors Res Laboratories 1978- 81. Asst Prof, Univ of Louisville 1981-86. Assumed current position in 1987. Teaches courses in Operations Research, Simulation Modeling, and Engrg Economics. Known for his res in multiobjective decision making. Has served as conslt to several organizations and has published articles in mathematical programming, simulation modeling, project management, energy planning, quality control, and facilities planning in several journals including *Management Science, IIE Transactions, International Journal of Production Research*, and *IEEE Transactions on Engineering Management*. *Society Aff:* IIE, ORSA, TIMS, DSI, ASEE

Evans, J Harvey
Home: 8 Doran Farm Lane, Lexington, MA 02173
Position: Emeritus Prof. *Employer:* MA Inst of Tech. *Education:* BEng/Naval Arch/Univ of Liverpool. *Born:* 5/1/14. Native of Rochester, NY. Ten yrs' experience in ship design with Bethlehem Steel Co concluding as a supervisor. Joined MIT faculty in 1947 as Asst Prof. Retired in 1978 as Prof of Naval Arch. Honorary Member, Fellow & Honorary VP of SNAME & recipient of its Davidson Medal for "outstanding scientific accomplishment in ship research–. Fellow of RINA. Chrmn 6th ISSC meetings in Boston in 1976. Official commendation from Seoul Natl Univ, Korea, for technical services rendered during 1957. Author of papers on gen and ship structural design before natl and intl socs. Author and editor of "Ship Structural Design Concepts" & "Ocean Engineering Structures–. Was honored as the dedicatee at a Ship Structure Symposium, sponsored by the Interagency Ship Structure Ctte and the Soc of Naval Architects and Marine Engrs, Wash, DC Oct, 1984. 1985-88, Chairman of the Committee on Marine Structures, National Research Council. *Society Aff:* SNAME, RINA.

Evans, James L
Business: 116 E. Elm, West Union, IA 52175
Position: Civ/Proj Engr/Area Engr. *Employer:* USDA - Soil Conservation Service. *Education:* BS/Agri Engg/Univ of MO. *Born:* 7/15/49. Native of MO. Educated at the Univ of MO. One yr of grad work in Soil and Water Engg. Active in ASAE - past chrmn of both the student and regular State Chapters in MO. Mbr natl Dam Safety Committee, ASAE. Employed by SCS 1972 to present. Worked in design and have been responsible for the construction of works of improvement under PL-566, Watershed Protection and Flood Prevention Act, in NE MO. Currently serving as Area Engineer for the SCS. Responsible for quality of engineering designs and installations, all programs, in N.E. Iowa. Licensed Professional Engineer in MO. *Society Aff:* ASAE, SCSA, AE.

Evans, John V
Business: Wood St, Lexington, MA 02173
Position: Asst Dir. *Employer:* Lincoln Lab MIT. *Education:* PhD/Phys/Univ of Manchester; BSc/Phys/Univ of Manchester. *Born:* 7/5/33. Dr J V Evans came to Lincoln Lab in May 1960, where he has conducted res into the radar reflection properties of the moon, planets, meteors, the upper atmosphere and ionosphere. He is the coeditor (with T Hagfors) of the book *Radar Astronomy* and has published over seventy scientific papers on these topics. In Sept 1975, Dr Evans was appointed Asso Hd of the Aero Div, and in May 1977, to the post of Asst Dir of the Lab. In this position he is responsible for the progs at the Lab in Advanced Electronics Res. In Jan 1980 he took on the additional task of directing the North East Radio Observatory Corporation (NEROC) Haystack Observatory, located in Tyngsboro, Mass which is operated by MIT on behalf of 13 educational institutions. Dr. Evans was appointed a Professor in the MIT Meteorology Dept at the same time, and continues to oversee the prog of high-power radar studies of the upper-atmosphere and ionosphere being conducted at the Millstone Hill Field Station. *Society Aff:* IEEE, AGU, URSI, IAGA, AAAS.

Evans, Lorn R
Business: 2132 Pankratz St, Madison, WI 53704
Position: VP. *Employer:* Alster & Assoc, Inc. *Born:* June 1926. BS in Geology, minors in CE & History. Grad from Univ of KS, June 1950. From 1950-52, schooling in photogrammetry, surveying, mapping & instructor at the Engr Sch, Ft Belvoir, VA. 1952-, employed by & VP, part owner of Alster & Assoc, Inc, Photogrammetric Engrs. Reg Engr in WI. Fellow of Am Soc of Civ Engrs, Am Congress of Surveying & Mapping. Mbr of Am Soc of Photogrammetry, Natl Soc of PE, Am Consulting Engrs Council, WI Soc of PE. Chrmn, Prof Activities Ctte of Am Soc of Photogrammetry, P Chrmn WSPE-PE in Private Practice Sec. P Pres, SW Chap of WSPE. P Bd Mbr & Treas, Consulting Egrs Council of WI.

Evans, Ralph A
Home: 804 Vickers Ave, Durham, NC 27701
Position: Consultant (Self-employed) *Employer:* Self *Education:* PhD/Physics/U CA, Berkeley; BS/Engrg Physics/Lehigh U *Born:* 2/2/24. Consulting, teaching, writing, & litigation-assistance in reliability, quality assurance, & engg statistics. Editor of *IEEE Transactions on Reliability* 1969-1985. Editor of *ASQC Reliability Review* 1981-1986. Fellow of ASQC & IEEE. ASQC-certified quality & reliability engr. Mbr of mgt committee of Annual Reliability & Maintainability Symposium since 1965. Formerly a sr physicist at Res Triangle Inst in NC & Dir of Link-Belt Res Lab. Reg PE in CA (Elec & Quality) & NC. *Society Aff:* ASQC, IEEE, ASM

Evans, Richard A
Home: 403 Ash St, Richland, WA 99352
Position: Chief Engr *Employer:* Vitro Engrg Corp *Education:* BS/Engrg/CA Polytechnic State Univ; BA/Math/Pomona Coll *Born:* 9/3/34 Current Pres Mid-Columbia Chapter of ASHRAE. Registered PE. Co-holder of 4 patents. Background in Contracting, Mfg and Consulting. Member ANS Speakers Bureau. Published papers on Solar Energy and Energy Conservation. Was Chief Systems Engr for Astro Science Corp, Chief Engr for Divs of Lear Siegler and Frick, and mgr of New Product Development for Div of Rucker. Specialized fields include heat transfer, control systems, refrigeration and air conditioning. Opened a Vitro office in MD, devoted to Solar Engrg and Energy Conservation. *Society Aff:* ASHRAE, ANS, NSPE, WSPE

Evans, Robert C
Business: PO Box 297, Bluffton, IN 46714
Position: Pres. *Employer:* PCP of IN. *Education:* BAgriEngr/Agri Engr/OH State Univ. *Born:* Findlay, OH, Oct 14, 1924, parents Roland R and Alice Crone; B Agri engr, OH State Univ, 1950, married Jane B Evans, June 12, 1946. With Intl Harvester Co 1951-55, New Idea Farm Equip, 1955-68, Chore Time Equip, Inc 1968-70, Fairfield Engg and Mfg 1970-72, Bacon Bin Inc 1972-75, Central Soya Co 1975-78, Pres of PCP of IN 1978-81. CEO of Agricultural Technologies of America, 1981-1985. Presently Pres. Swiss Properties, Inc and Swiss Am, Inc 1986- . Mbr Am Soc Agri Engrs, Gamma Sigma Delta, Presbyn (elder) Mason, Kiwanian. Holds several patents. *Society Aff:* ASAE.

Evans, Robert J
Home: 8615 Stockton Pkwy, Alexandria, VA 22308
Position: Conslt *Employer:* Self-employed *Education:* BS/ME/IL Inst of Tech *Born:* 3/13/17. 1947-56 Proj Engr in various mfg organizations. 1956-61 Chief Engr - Air-Conditioning - A O Smith Corp. 1961-1982 Air-Conditioning & Refrig Inst, as Sr Engr, Asst Dir of Engg, Dir of Engg & Dir of Special Projs. 1982- present Conslt Engr. ASHRAE Bd of Dirs, 1970-73. ASHRAE Distinguished Service Award 1975. ASHRAE Fellow 1979. Honorary Societies - Tau Beta Pi, Pi Tau Sigma. Lt Col, USAF (Ret'd). *Society Aff:* ASHRAE, NSPE

Evans, Robert R
Business: 1105 Williams, Great Bend, KS 67530
Position: Partner. *Employer:* Evans-Bierly-Hutchison & Assoc. *Born:* June 1921. BE in CE from OK Univ, 1943. CEC-USNR WWII, Staff of CEC Midshipman Sch, Camp Peary, VA & Davisville, RI - Pontoon Assbly Detachment South Pacific. City of Dallas Public Works Dept; construction engr Colwell Constr Co, Dallas. Formed civ engg consulting firm, Great Bend, KS 1951. Offices in Great Bend, Lawrence & Garden City, KS. Pres KS PEPP, 1963-64; Pres KS Engg Soc 1966-67; VChrmn NSPE/PEPP 1973-75. Other ints include ser on volunteer and elective civic bds & cttes, travel & sports.

Evans, Thomas C, Jr
Business: CE Dept, Charleston, SC 29409
Position: Prof of CE. *Employer:* Citadel, The. *Education:* PhD/Engg Mech/Univ of AL; MS/CE/Carnegie Inst of Tech; BS/CE/Citadel. *Born:* 11/17/36. Attended High School in Pittsburgh, PA. After grad from Carnegie Tech, worked for Rust Engrg Co; spent short hitch with US Army Corps of Engrs as 2nd Lt; Attended the Univ of AL; taught four yrs at LA Tech Univ; joined the Citadel faculty in 1971. Been active in SE Sec of ASEE and local Civ Engrs Soc. V Pres of SE Architects: Engs: Planners, Inc. Enjoy fishing, flying, soaring, classical music, and sailing. *Society Aff:* ASCE, ASEE, ACI.

Evces, Charles R
Box 2908, Tuscaloosa, AL 35487
Position: Prof. *Employer:* Univ of AL. *Education:* PhD/Engg Mechanics/WV Univ; MS/Mech Engg/Univ of Notre Dame; BS/Mech Engg/Univ of Notre Dame. *Born:* 12/31/38 Native of E Liverpool, OH. Industrial experience with Great Lakes Steel Corp American Cast Iron Pipe Co and General Motors Corp. Previous academic appointments at the Univ of Notre Dame and WV Univ. Mbr of mech engg faculty at The Univ of AL since 1969, specializing in vibrations, noise control and dynamic phenomena. Co-author of two textbooks in engg mech. AL Reg PE. *Society Aff:* ASME, ASEE, SAE.

Everard, Noel J
Home: 1505 Freeman Ct, Arlington, TX 76013
Position: Prof *Employer:* Civil Engg Dept. *Education:* BS/Civil Engg/LA State Univ; MS/Civil Engg/LA State Univ; PhD/Civil Engg/TX A&M Univ. *Born:* 12/24/23. Prof Civil Engg Dept, Univ of TX at Arlington. Consulting Engg, computer methods for structural analysis and design. Fellow-ASCE, ACI. Mbr-ACI-318, National Building code. ASCE-ACI-343, Bridge Design. Former Chmn ACI—340, mbr ACI-TAC. Former Assoc David W Godat, Consulting Engrs, New Orleans. Computer methods consultant to Industry. Author, Schaums Outline, "Theory and Problems in Reinforced Concrete–. Captain, Corps of Engrs, US Army, Retired Reserve. Lecturer on Reinforced Concrete throughout USA, Peru, Argentina, Brazil, Taiwan, Hong Kong, Malaysia, and Thailand. Reg PE. Joe W Kelly Award of the Amer Concrete Inst. Distinguished Service Award, Amer Concrete Inst. Editor Ctte 21-A Tall Buildings and Urban Habitat-UNESCO.. *Society Aff:* ASCE, ACI.

Everett, James L, III
Business: 2301 Market St, Philadelphia, PA 19101
Position: Pres. *Employer:* Phila Electric Co. *Born:* July 1926 Charlotte NC. BSME 1949, MSME 1949, PA St Univ; MS Ind Mgt MIT 1959. Joined Phila Elect Co 1950; Exec VP 1968-71; Pres 1971-; Dir, Mbr Exec Ctte, parent co; Dir Subsidia-

Everett, James L, III (Continued)
ries. Chrmn Bd Radiation Mgt Corp; Dir, Bellevue Stratford Co, Fidelity Mutual Life Ins Co, Martin Marietta Corp, Phila Natl Bank, Tasty Baking Co; Trustee, Drexel Univ. Reg PE PA; Fellow, ASME; Mbr ANS, IEEE, SAME, NSPE, Natl Acad of Engg. Outstanding Young Man of Yr, Phila Jr Cham Comm 1961; Annual Engg Tech Award Temple Univ 1963; Engr of Yr DE Valley 1972; Geo Wash Medalist, 1974; Outstanding Layman of Yr, Phila. YMCA 1975.

Everett, Robert R
Business: Burlington Rd, Bedford, MA 01730
Position: President and Chief Executive Officer *Employer:* The MITRE Corp. *Education:* BS/Elec Engg/Duke Univ; MS/Elec Engg/MIT. *Born:* 6/26/21. With Servomechanisms Lab of Mass Inst of Tech, 1943-51. Assoc Dir Digital Computer Lab 1951. Assoc Div Head Lincoln Lab 1951-56; Div Head, 1956-58. Tech Dir The MITRE Corp, Bedford, MA 1958-59; VPres Tech Operations 1959-69, Exec VPres 1969; Pres 1969- . Fellow IEEE; Mbr Assn of Computing Machinery, AAAS, Phi Beta Kappa, Sigma Xi, Tau Beta Pi. Senior Scientist, USAF Scientific Advisory Bd; Mbr Def Comm Agency Scientific Advisory Group; Mbr Natl Academy of Engrg, FEMA Advisory Bd, US Information Agency, Voice of Amer, Engrg Advisory Cttee; Duke Univ Distinguished Engrg Alumnus Award, D.D Medal For Exceptionally Distinguished Public Service. Contributed articles to a number to tech journals and awarded patents in the fields of magnetic drum memories & display devices. Pres residing in Concord, MA; Married 6 children. *Society Aff:* NAE, IEEE, AAAS, ACM.

Everett, Wilhelm S
Business: 3098 Solimar Dr, PO Box 1535, Ventura, CA 93001
Position: Owner *Employer:* Everett Assoc *Education:* MS/Mech Engg/Univ of CA; BS/Phys/CA Inst of Tech. *Born:* 10/2/14. Native of Ventura, CA. Served in WWII, Lt CDR, USNR in charge of heavy machinery inspection at the office of the Naval Mtls, San Francisco, CA. Pres and founder of Pulsation Controls Corp, which sold high technical noise, vibration and pulsation control products to industry and defense. In addition, developed surge control equip for Public Works market. Holds over 30 patents in these fields. Life Fellow ASME, Reg Profl Mech Engr in CA. Pres Venture County Engrs Club, Dir Ventura Rotary Club and active in many community affairs. *Society Aff:* PEA, ASME.

Everett, William B
Business: PO Box 18514, 3400 Democrat Rd, Memphis, TN 38118
Position: Partner. *Employer:* Bowman - Everett & Assoc. *Education:* BSME/ME/Princeton Univ. *Born:* 5/17/25. Native of Memphis, TN. Principal of A T Kearney & Co Mgt Consultants 1951- 1963. Developed physical distribution concepts with resulting wide practice. 1963-1966 Group VP, IMC Industries with Profit Responsibility for production of lawnmowers, builders hardware, 1967-1969. VP Operations Ryder Truck Line. Partner Bowman-Everett since 1969. Ind Engg Consultants specializing in labor cost control, product dev, mfg engg, value analysis, safety, warehouse and plant location and layout. *Society Aff:* NSPE, IIE, ACEC.

Everett, Woodrow W, Jr
Home: 1101 Massachusetts Avenue, St. Cloud, FL 32769
Position: Chrmn of the Bd. *Employer:* Southeastern Ctr for Elec Engg Education (SCEEE). *Education:* PhD/Ind & Labor Rel/Cornell Univ; MS/Ind & Labor Rel/ Cornell Univ; BEE/Communications-Elec/Geo Washington Univ. *Born:* 10/11/37. Proj Engr, Scott Paper Co (1959). Active duty, USAF (1959-1962). Proj Engr, Ithaca Res Labs, Atlantic Res Corp (1962-1964). Rome (NY) Air Dev Ctr Post-doctoral Prog dir (1964-1975). Chrmn, SCEEE (1975-present). Chrmn, Northeast Consortium for Engg Education (1976-present). AF Systems Command Award for Scientific Achievement (with prize) - 1966. SCEEE Distinguished Service Award - 1977. Author/inventor. Who's Who in the East. Who's Who in the South and Southwest. Who's Who in Finance and Industry. Notable Americans. Community Leaders of Am. Dictionary of Intl Biography. Who's Who in the World. Personalities of the South. The American Scientific Registry. Men of Achievement, Who's Who in America, Personalities of America, International Who's Who in Engineering. *Society Aff:* IEEE, ASEE.

Everhart, James G
Home: 902 Eastmont Dr, Centralia, MO 65240
Position: Retired *Education:* BS in EE/Power/PA State Univ. *Born:* 8/29/15. Native of Pittsburgh, PA. Retired Pres, Pitman Div Emerson Elec Co. and Exec VP, A B Chance Co, having previously been VP and Gen Mgr of its Utility Systems Div. Before joining Chance in 1970, had many yrs experience in various engg and exec capacities with McGraw-Edison's Power Sytesm Div, and its predecessor Line Mtl Industries. Fellow, AIEE; Former mbr American Mgt Assns (trustee 1973-75, Distinguished Service Award 1975, President's Council). Former Dir, Power Equip Div, Natl Elec Mfg Assn. Author articles and book chapters; nine US patents. Eta Kappa Nu, Tau Beta Pi, Sigma Tau. *Society Aff:* IEEE.

Everhart, Thomas E
Business: Office of the Chancellor, 601 E John St, Champaign, IL 61820
Position: Chancellor *Employer:* Univ of IL at Urbana-Champaign *Education:* PhD/Engr/Cambridge Univ; MSc/Applied Phys/UCLA; AB/Physics/Harvard *Born:* 2/15/32. Born in Kansas City, MO. Received following honors: Wm Scott Gerrish Scholarship, Harvard College, 1949-53; Phi Beta Kappa, Harvard College, 1953; Marshall Scholar, Cambridge Univ, 1955-8; Sigma Xi, Univ of CA, Berkeley, 1958; Natl Sci Fdn Sr Postdoctoral Fellow, 1966-7; Fellow, IEEE, 1969; Miller Res Prof Univ of CA, Berkeley, 1969-70; John Simon Guggenheim Mem Fellowship 1974-5; Natl Acad of Engg, 1978. Lawrence Berkeley Lab Scientific and Ed Advisory Ctte, 1978-85, chmn 1980-85; General Motors Sci Advisory Ctte, 1980- , Chairman, 1984- . Served as conslt for several firms including: Westinghouse Elec Corp, Ampex, Hughes Res Labs, IBM Res Lab & Bell Labs. Asst, Assoc, & Prof Dept of Elec Engg & Comp Sci, Univ of CA at Berkeley, 1958-1978; Dept Chrmn 1972-77. Joseph Silbert Dean of Engg & Prof, Elec Engg, Cornell Univ, 1979-1984. Chancellor, Univ of IL at Urbana-Champaign, 1984- . *Society Aff:* NAE, IEEE, AAAS.

Everitt, William L
Business: 106 Engg Hall, 1308 W. Green St, Urbana, IL 61801
Position: Dean Emeritus. *Employer:* Univ of IL. *Education:* PhD/Physics/OH State Univ; MS/Elec Engg/Univ of MI; EE/Elec Engg/Cornell Univ. *Born:* 4/14/00. Former pres of the IRE, the ASEE and the ECPD. A founding mbr of the Natl Acad of Engg, he received the Medal of Honor from the IRE, the Lamme Medal from the ASEE, the Education Medal from the AIEE and the Kelly Telecommunication Medal from the IEEE. The author of the text Communication Engineering and editor of over 100 volumes in the Prentice hall Elec Engg Series, he was the recipient of ten honorary doctorates. *Society Aff:* IEEE, ASEE, NSPE, AAAS.

Evert, Carl F
Home: 3163 Bellewood Ave, Cincinnati, OH 45213
Position: Prof Emeritus Elec & Comp Engg *Employer:* Univ of Cincinnati. *Born:* Dec 1923. PhD from Univ of WI; BSEE & MSEE from Purdue Univ. Ser in USN WWII as Chief Radio Tech. Taught at Univ of Cincinnati, Dept of EE since 1950. Cons to industry in power systems, biomedical engg, computers & instrumentation. Bd Dir Ohmart Corp, Cin, OH. P Dir, Univ Computing Ctr, Univ of Cincinnati. Sr mbr IEEE, mbr SCS, IMACS, Sigma Xi, Tau Beta Pi, Eta Kappa Nu. Reg Engr OH. Visiting Prof Dartmouth College, NSF Cons for Educ in India, UNESCO Cons for Computer Sci in Romania. NSF Lect, Cancer Inst, Melbourne, Aust.

Ewen, Harold I
Business: 60 Beaver Rd, Weston, MA 02193
Position: Pres. *Employer:* Ewen Knight Corp. *Education:* PhD/Physics/Harvard; MA/Physics/Harvard; BA/Math/Amherst College. *Born:* 3/5/22. Pres, Ewen Knight Corp since 1952 and Ewen Dae Corp since 1958, specializing in microwave and millimeterwave radiometric applications. Taught mathematics and astronomy at Amherst College 1943 and Radio Astronomy at Harvard 1952-56. Served in England as a lieutenant in WWII with Navy Patrol Bombing Squadron VPB112. Scientific Advisor to Cincinnati Electronics Corp since 1975 for the Aif Force Global Solar Radio Telescop Network. Contributing author to "Electromagnetic Sensing of the Earth from Satellites" Polytechnic Press (1965), "Advances in Microwaves–, Volume V(1979) Academic Press, and "Geoscience Instrumentation" (1974)-John Wiley & Sons. Co-discoverer of 21 centimeter interstellar hydrogen line (1951). IEEE Morris E Leeds Award (1965). *Society Aff:* AAAS, IEEE, ΣΞ, ΦBK.

Ewing, Benjamin B
Business: 408 S Goodwin, Urbana, IL 61801
Position: Dir and Prof, Emeritus *Employer:* Univ of IL. *Education:* PhD/Sanitary Engg/Univ of CA; MS/Civ Engg/Univ of TX; BS/Civ Engg/Univ of TX. *Born:* 4/4/24. Donna, TX. Married Elizabeth Malone 1947. Three children. Instr, Asst Prof, Univ of TX, Austin 1947-1955. Assoc in Civ Engg, Asst res engr, Univ of CA, Berkeley, 1955-58. Assoc prof, Prof, Univ of IL, Urbana-Champaign, 1958-85; Prof. Emeritus, 1985-present. Dir, Water Resources Ctr, 1966-1973. Dir, Inst for Environmental Studies, 1972-85. Consulting engr, 1959-present. trustee, Urbana & Champaign Sanitary Dist (1974-80). Public mbr of IL Water Resources Commission 1975-84. Served to Lt (jg), Civ Engr Corp, US Naval Reserve 1943-1946. Recipient of Harrison Prescott Eddy Award for Noteworthy Res, WPCF & Epstein Award, Dept of Civ Engg UIUC 1961. *Society Aff:* ASCE, WPCF, AGU, AAEA, AWWA, AEEP

Ewing, Ronald L
Business: Dibble & Associates, 3625 N 16th St, Ste 128, Phoenix, AZ 85016
Position: Partner *Employer:* Dibble & Associates *Education:* BS/Civ Engr/CA State Univ. *Born:* 1/31/46. Native of Merced, CA. Joined Boyle Engg Corp - Ventura, CA in 1969 serving in numerous capacities including resident construction engr on two maj wastewater treatment facilities. Appointed Asst Office Mgr - Phoenix, AZ in 1974 specializing in sanitary engg applications particularly land treatment. Developed sanitary engg applications and has prepared numerous profl papers and publications including presentation at UNESCO Conference in Amsterdam - 1977. Received AZ NSPE Young Engr of the Yr 1978 Award. Private Pilot, enjoys skiing and outdoor recreation. Established RLE Consulting Engineers in 1978. Became partner in Dibble & Associates in 1980. *Society Aff:* ASCE, AWWA, ACEC, NSPE, WPCF.

Ewing, William C
Business: PO Box 1647, Midland, MI 48640
Position: Supt-Water. *Employer:* City of Midland. *Education:* BSCE/Sanitary Engr/ Purdue Univ. *Born:* 10/27/31. Past-Chrmn of MI Section of AWWA, Past-Pres of Saginaw Valley Chapter of MI Soc of PE. Presently Chrmn of AWWA natl committee on "Water Service Line Fittigs–. *Society Aff:* ASCE, APWA, AWWA, NSPE.

Eyerman, Thomas J
Business: 30 W Monroe St, Chicago, IL 60603
Position: Gen Partner. *Employer:* Skidmore, Owings & Merrill. *Education:* MBA/Business/Harvard Grad Sch of Bus Admin; BA/Archt/OH State Univ. *Born:* 6/11/39. Trustee: Ctte for Economic Dev (CED) Washington. Mbr of Res and Policy Ctte CED. Fellow American Inst of Architects. Trustee: Chicago Symphony Orchestral Assn. Dir: Harvard Business Sch Club of Chicago govering mbr: the Art Inst of Chicago. Mbr: Pres's Club - Univ of Chicago. Mbr: Dean's Fund, Harvard Bus Sch. Trustee Field Museum of Natural History - Chicago, Pres Chicago Chpt Amer Inst of Architects, Dir Chicago Commerce and Industry Assoc. *Society Aff:* AIA.

Eykhoff, Pieter
Business: Eindhoven Univ of Technology, PO Box 513, NL-5600 MB Eindhoven Netherlands
Position: Prof, Former Dean *Employer:* Netherlands govt *Education:* PhD/Control/Univ of CA Berkeley; MSc/Elect Eng/Delft Univ of Tech *Born:* 4/9/29. Born in the Hague, Netherlands 1929. Appointments at Delft Univ of Tech, 1949-1964, visiting Res Fellow US Natl Acad of Sci 1958-1960; with Eindhoven Univ of Tech since 1964. Teaching control engg; res in system identification/parameter estimation with engr and bio-medical applications. Books: System Identification, Wiley 1974, also translated into Russian, Rumanian, Polish & Chinese, Edit: Trends and Progress in System Identification, Pergamon 1981, and other publications. Dean EE Dept, 1977-1980. Officer in the Intl Federation of Automatic Control (IFAC), eg Honorary Editor 1971-1975; Mbr Exec Council 1975-1981, Mbr IFAC Publications Managing Board since 1976. Chrmn IFAC Publ Ctte since 1984. Stays & lectures in many European countries, including USSR, as well as USA, Canada, Japan, PR China, Chile. Among interests: classical music, archeology, Eastern cultures. Fellow IEEE. Honorary Professor Xian Jiaotong University, PR China. *Society Aff:* IEEE, KIVI, ΣΞ

Ezzat, Hazem A
Business: Res Labs, GM Tech Ctr, Warren, MI 48090
Position: Head, Power Systems Research *Employer:* General Motors Corporation. *Education:* PhD/Mech Engg/Univ of WI; MS/Mech Engg/Univ of WI; BSc/Mech Engg/Univ of Cairo. *Born:* 7/12/42. Field engr with the Suez Canal Authority 1963-65. Instructor of Theory of Machines and Vibrations, Univ of Cairo 1965-66. Joined General Motors Res Labs 1970: 1970-1981 respons in the area of "Friction and Lubrication Research" 1981- 1984 Responsibility in areas of acoustics and vehicle noise and vibration; 1984- present, responsibility in vehicular chassis and powertrain system dynamics and control, and advanced powertrain synthesis. Mbr: ASME; SAE; ESD; Sigma XI. ASME Henry Hess Award 1973. Publications in lubrication Theory, Dynamics and Optimal System Design. *Society Aff:* ASME, SAE, ESD, AMER ACAD MECH, ΣΞ

Fabrycky, Wolter J
Business: Department of Ieor, Vpi and State University, Blacksburg, VA 24061
Position: Prof, Indust Engrg and Operations Research *Employer:* VPI and State University. *Education:* PhD/Engg/OK State Univ; MS/Indust Engr/Univ of AK; BS/ Indust Engr/Wichita State Univ. *Born:* 12/6/32. Taught at AK, OK State, and VPI & State Univ; Chrmn Sys Engr, Assn Dean of Engrg, 1970-76, Dean, Research Div 1976-81 at VPI & State Univ. Co- author of 5 Prentice-Hall textbooks (Engr Economy, Economic Decision Analysis, Proc & Inventory Systems, Applied OR/MS, & Systems Engrg and Analysis). Co-editor of the Prentice-Hall Intl Series in Indust & Sys Engrg. Exec VP IIE, and Mbr of the BD of Trustees, 1982- 84. Chrmn, Engrg Res Council, ASEE, VP & Mbr of the Bd of Dirs, 1976-78. Registered Professional Engr in AK & VA. Fellow and mbr IIE and Fellow and mbr AAAS. *Society Aff:* IIE, ASEE, ORSA, AAAS.

Fadum, Ralph E
Business: 408 Mann Hall, Box 7908, Raleigh, NC 27650-7908
Position: Dean Emeritus of Engg & Prof Civil Engg. *Employer:* North Carolina State Univ. *Education:* DEng, Hon./-/Purdue Univ; SD/Soil Mechs/Harvard Univ; MS/ Soil Mechs/Harvard Univ; BSCE/Civil Engg/Univ of IL. *Born:* 7/19/12. in Pittsburgh, PA. Taught at Harvard and Purdue. In 1949 apptd CE Dept Hd & CE Prof, NCSU. From 1962-78, NCSU Dean of Engg & CE Prof. Consultant to civic, defense, and scientific agenices; 19 yr mbr of Army Scientific Advisory Panel. 3 yr mbr of Army Sci Bd. Mbr of Bd of Dirs, Natl Driving Center & Highway Safety Res Center of the Univ of NC. Chrmn, Bd of Dirs of UNC Water Resources Research Inst. Mbr of Natl Acad of Engg. Hon Mbr ASCE. Hon Mbr ASEE, mbr ECPD, NSPE, PENC, NC Soc of Engrs & Natl Council on soil Mechs & Fnd Engg. Dept of Army Outstanding Civilian Service Award 1973 & 77; Disting CE Alumnus Univ of IL, 1969; Outstanding Engg Achievement Award NC Soc of Engrs, 1971; & ASCE Outstanding Civil Engr of NC 1971; Chi Epsilon Natl Hon Mbr, 1978;

Fadum, Ralph E (Continued)
Award of Merit, Alumni Assoc, NC State Univ 1979. *Society Aff:* ASCE, NSPE, ASEE.

Faeth, Gerard M
Business: 218 Aerospace Engineering Bldg, University of Michigan, Ann Arbor, MI 48109-2140
Position: AB Modine Professor of Aerospace Engingering *Employer:* University of Michigan *Education:* PhD/Mech Engr/PA State Univ; MS/Mech Engr/PA State Univ; BME/Mech Engr/Union College. *Born:* Jul 1936. Native of New York City. With The Pennsylvania State University 1958-1985 as teacher and in research, retiring with the rank of Professor Emeritus of Mechaniacl Engineering. With the University of Michigan since 1985 as teacher and in research, and also serving as Head of the Gas Dynamics Laboratories. Research is in the areas of turbulent combustion and multiphase flow. Author or co-author of over 140 articles, holds one patient. *Society Aff:* ASME, AIAA, CI, AAAS.

Fagerlund, N David
Business: Municipal Bldg, Oakland, NJ 07436
Position: Dir-Public Works. *Employer:* Borough of Oakland. *Born:* Feb 23, 1931 New York City. Attended Newark College of Engrg, Rutgers Univ, Cornell Univ. USMC 1946-47; US Navy Submarine Service 1948-56; Dir-Public Works Oakland, New Jersey 1956-present. Professional Societies: N J Water Pollution Control Assn; North Jersey Water Conference - Past President & Trustee; AWWA - Past Chairman 1970-72 New Jersey Section - International Director N J 1974- 1977. Dir NJ, 1974-75. Articles & Awards - AWWA Water Utility Advancement for Public Relations 1965; American City Magazine 1965, Study of Cathodic Protection Water Tanks; American City Magazine 1969, Use of Meter Maids in Water System & Results; AWWA Journal 1969, Better Relations with Consumers; Enjoy Boating & Fishing.

Faget, Maxime A
Business: Lyndon Johnson Sp Ctr, Houston, TX 77058
Position: Dir Engrg & Development. *Employer:* Lyndon B Johnson Space Ctr, NASA. *Education:* BS/Mech Engg/LA State Univ. *Born:* Aug 1921. Born in British Honduras of American parents. Served on US Navy Submarines in WW II. Research Scientist at NACA Langley Research Ctr. in transonic aerodynamics, rocket & ramjet propulsion, & hypersonic heating & boundary layer phenomena. Originated Mercury design concepts. With NASA since initiation of space program in manned spacecraft engrg. Hon Doctorate of Engrg degrees from Univ of Pittsburgh and Louisiana State Univ. Spacecraft Design Award 1970 AIAA, Award for Outstanding Accomplishment 1971 IEEE, William Randolph Lovelace II Award 1971 AAS, 1973 Daniel & Florence Guggenheim Intl Astronautics Award, 1975 Amer Soc of Mech Engineers' Gold Medal, 1976 Inst of Elec & Electronics Engineers' Harry Diamond Award, 1976 Amer Astronautical Soc's Space Flight Award, 1976 Instrument Soc of Amer's Albert F Sperry Medal, 1979 Ame Inst of Aeronautics & Astronautics' Goddard Astronautics Award. *Society Aff:* NAS, IAF, AIAA, AAS.

Faherty, Keith F
Business: Sch of Engg, Fresno, CA 94740
Position: Prof of Civil Engineering *Employer:* CA State Univ. *Education:* PhD/Structures/Univ of IA; MS/Structures/Univ of IL; BS/Civil Engg/Univ of WI-Platteville. *Born:* 12/7/31. Native of Platteville, WI. Design engr & construction foreman for US Gypsum Co from 1954-57. Joined the faculty of Univ WI-Platteville in 1957. Held ranks of Instructor, Asst, Assoc prof from 1957-66. Apptd Dept of Civil Engg in 1966. Promoted to Prof of Civil Engg in 1973. 1979 Apptd Prof of Civil engg at CA State Univ, Fresno. Pres Madison Branch, Brahcn rep, dir and VP WI section Bd of ASCE. Chmn of sub-cmtte on Tech Sessions of Cmtee on Wood, ASCE. *Society Aff:* ASCE, ASEE, ACI.

Fair, James R
Home: 2804 Northwood Rd, Austin, TX 78703
Position: Ernest & Virginia Cockrell Chair *Employer:* Univ of TX at Austin. *Education:* PhD/Chem Eng/Univ of TX; MS/Chem Eng/Univ of MI; BS/Chem Engg/GA Tech; DSC//WA Univ. *Born:* 10/14/20. Charleston, MO. Secondary schools, Little Rock, Arkansas, m Merle Innis, Port Lavaca, TX. DSc (Hon), WA Univ, Shell Dev Co, 1954-56; Monsanto Co, 1942-52 and 1956-79. Presently Professor of Chem Engg, Univ of TX at Austin & Consultant. Registered prof engr; Fellow AIChE (and former dir); Natl Acad of Engrg. Awards: Walker (AIChE), Chem Engrg Practice (AIChE), Founders (AIChE), Inst Lecturer (AIChE), Personal Achievement (Chem Engrg Magazine), Distinguished Grad of Univ of TX. Over 100 articles published. *Society Aff:* AIChE, ACS, NSPE, ASEE.

Fairbanks, Gustave E
Business: Agri Engr Dept, Scaton Hall, Manhattan, KS 66506
Position: Prof Emeritus *Employer:* Kansas State Univ *Education:* MS AgE/AG Engr/KS State Univ; BS/AG Engr/KS State Univ *Born:* 6/10/15 Reared in KS where I have spent my entire life except for a period in the Army during WWII. Served overseas in Pacific area. Discharged as captain. Maintained mbrshp in US Army Reserves. Retired in grade of Colonel with Legion of Merit Medal. Employed by Agri. Engrg. Dept at KS St Univ from Dec 1946 to 1983. Duties involved teaching and res in the Power and Machinery area. Students have won five Natl Agri Machinery Design Championships. I retired July 1, 1983. *Society Aff:* NSPE, ASAE, NCEE

Fairbanks, John W
Home: 4717 Jasmine Dr, Rockville, MD 20853
Position: Mechanical Engineer. *Employer:* Dept of Energy. *Education:* BSNS/Marine Eng/Maine Maritime Acad; BSME/Mech Engg/Stanford Univ; MSME/Mech Engg/Univ of Santa Clara. *Born:* 11/21/31. Taught at TX A&M and Univ of MD. Served with US Navy 1954-57 in Pacific. Res Engr with Hiller Aircraft on annular ejector and designed high speed bearing and shaft test stand for XC-142A. At Philco Ford worked on advanced space power sys. With NASA-Goddard in 1967 as power sys engr sevl space craft including Orbiting Astronomical Observatory. With Naval Ship Engg Ctr as Program Engr for FT9 Marine Gas Turbine Dev & Ceramic Demonstrator Gas Turbine from 1971-77. Also, co-ordinated the Navy's gas turbine matl dev. Organized first 2 Gas Turbine Matls in Marine Environment Conferences, US participation in US Navy/Royal Navy Conference '76, 1st NATO Workshop Coatings for Heat Engines: Italy '84 and two DOE/EPRI Conference on Advanced Matls for Alternative Fuel Capable Directly Fired Heat Engines. Organized Interagency Workshop Coatings for Advanced Heat Engines '87. Since 1977 with Dept of Energy. Responsibilities included management of advanced gas turbine compressor, turbine airfoil cooling, disel engine alternative fuel capability, applied materials heat engines (26 contracts). Currently Program Mgr Advanced Diesel Engine Development-Transportation. Authored or co-authored 68 tech papers. Outstanding Alumni Award - Marine Acad - 1976, winner John C. Niedermair Award - Best Tech Paper ASE 1973 and 1975. Chairman ASME Washington Sect, ASME National Nominating Cttee 86-87, Chairman Energy Materials Co-ordinating Ctte. P. Pres. Maine Maritime Academy W.A. Sect. Captain US Naval Reserve. *Society Aff:* ASME, AVS, NRA, ASE, SAE.

Fairchild, Jack E
Home: 136 Varsity Circle, Arlington, TX 76013
Position: Prof, Aerospace Engrg. *Employer:* Univ of Texas, Arlington. *Education:* PhD/Eng Sci/Univ of Ok; MS/Aeronautical Engg/Univ of So CA; BS/Aeronautical Engg/Univ of TX, Austin. *Born:* Oct 25, 1928. Native of Houston, Texas. Served as Naval Aviator in carrier fighter squadron in Korean War 1947-51. Aerodynamics Engineer specializing in stability and control and performance with Bell Helicopter, Chance Vought Aircraft, 1953-56. Lecturer in Aerospace Engineering, Aviation Safety Division, U of S Calif 1956-60. Instructor and Research Engineer, Univ of Oklahoma, aerodynamics and gas turbine vibration analysis 1960-64. Assoc Prof to

Fairchild, Jack E (Continued)
Prof, U of Texas at Arlington, 1964-present. UMTA sponsored research on transit bus performance and fuel efficiency. 1974-76. Important consulting work: The Western Corp, ordnance trajectory analysis; LTV Corp, stability and control; Bell Helicopter, experimental research on new-concept tail rotor systems; The VLM Corp, aerodynamic design of new concept VTOL aircraft; TCA Co, aerodynamic design of energyefficient truck cooling fan. Numerous forensic engr. investigations of aircraft accidents: General Dynamics/Ft. Worth, stability/control adv. aircraft; Chairman, N Texas section AIAA 1973-74. Dir., Region IV AIAA, 1980- . Consulting N Texas section AIAA 1973-74. Dir., Region IV AIAA, 1980- . Consulting. *Society Aff:* AIAA, NSPE

Fairclough, Dennis A
Business: 468 CB, Provo, UT 84602
Position: Assistant Prof *Employer:* Brigham Young Univ *Education:* PhD/Computer Engrg/Brigham Young Univ; MS/EE/Electronics/Univ of Santa Clara; BS/EE/Univ of UT *Born:* 7/24/35 Assistant Prof of Electrical Engrg at Brigham Young Univ. Teaches digital synchronous and asynchronous state machines, microprocessor fundamentals and architecture, and advanced computer architecture. Senior Staff Consultant for Novell Data Systems. Work includes high speed microprocessor controlled disk subsystems and high performance local networks. *Society Aff:* IEEE, ACM, ΦΚΦ, ΤΒΠ, HKN

Fairclough, Hugh
Home: 1624 42nd Ave. E, Seattle, WA 98112
Position: Cons Chem, Mech & Nuclear Engr. *Employer:* Self Employed. *Education:* ME/ME/Case Sch of Applied Science; BS/ME/Case Sch of Applied Science. *Born:* Sept 1904. Registered Professional Engr, Washington. Fellow AIChE, Fellow SAME, Fellow AAAS. Life Member, NSPE, Assoc. Mbr. U.S. Naval Institute; Past Pres, Puget Sound Engrg Council (PSEC). Specialist in Petrochemical, Mechanical & Nuclear Engineering. 30 years in Chemical Engineering, 18 with Dow Chemical Co as R&D Engr, Engrg Asst to V P; 12 with E I duPont de Nemours & Co as Senior Process Engineer, Resident Design Engineer. 11 years with The Boeing Co. as Design Specialist. Several Design Firsts including Cracker to produce Ethylene from Residual Fuel Oil, Mass Polymerizer to produce Polystyrene, Large Scale U S Heavy Water Plant & Tritium Plant. Congregationalist, Mason, Shriner, Rotarian. Lt Col Wash Wing Civil Air Patrol. Grad of USAF Air War College. (Correspondence) Knight of the York Cross of Honour. *Society Aff:* AIChE, SAME, AAAS, NSPE, US Naval Institute.

Fairhurst, Charles
Business: Dept of Civil and Mineral Engrg, University of Minnesota, 500 Pillsbury Drive S.E, Minneapolis, MN 55455-0220
Position: Prof *Employer:* Univ of MN *Education:* PhD/Mining Engrg/Univ of Sheffield, UK; B/Mining Engrg/Univ of Sheffield, UK *Born:* in Widnes, England. He worked for over 3 years in coal mining production operations with the National Coal Bd, England. He has conducted research in rock drilling and rock mechanics at the Univ of MN since 1956 and was appointed to his present position in 1972. He was Program Chrmn, 3rd Intl Congress on Rock Mechanics, Denver 1974 and has held various committee offices in both AIME and ASCE. He is a mmbr of the Bd on Radioactive Waste Management of the US Natnl Acad of Sci. He was awarded the 1971 Best Research Paper award (with B Haimson) by the US National Committee on Rock Mechanics, AIME's Outstanding Achievement Award for Rock Mechanics and in 1979 was elected a Fellow of the Royal Swedish Academy of Engrg Sciences. He is senior editor of "Tunnelling and Underground Space Technology'; and past-pres American Underground-Space Association and Editorial Advisor to the Journal of Rocks Mechanics and Mining Sciences. He is married, with 7 children. 1983 he was awarded special award of U.S. Natl Ctte on Rock Mechanics for 25 years of distinguished service, and he was appointed E.P. Pfleides Prof of Rock Mechanics and Mining Engrg, Univ of Minnesota. *Society Aff:* ASCE, AIME, SAIMM, ΣΞ

Fairstein, Edward
Home: 228 Outer Drive, Oak Ridge, TN 37830
Position: Conslt *Employer:* Self *Education:* B of EE/Elec Engrg/City Coll of NY of the City Univ of NY. *Born:* Dec 14, 1922 Brooklyn N Y. Two sons. Sr. design engr Oak Ridge Nat'l Lab., Oak Ridge, TN 1946-59. Founder, Tennelec, Inc. 1960. Seven patents (assigned to U S Govt), approx 24 papers, chapter in 'Nucl. Instr. and their Uses' (Wiley, 1962). Fellow IEEE. Mbr edit board, Rev. Sci. Instr. 1958-61. Member several standards groups 1958- . Member Prog. Comm. IEEE Nucl. Sci. Symposium. Directorships: Tennecomp Systems, Inc 1967-1976, Hamilton 1st Natl Bank 1968-74, Mbr Nucl Inst and Det Comm IEEE, City Ind. Dev. board 1970-76 (Oak Ridge, TN). Enjoy literature, music, motorcycle, woodworking. *Society Aff:* IEEE, AAAS, RESA

Falco, James W
Business: Office of Environ. Processes and Effects Research (RD-682), Washington, DC 20460
Position: Office Director *Employer:* U.S. EPA *Education:* Ph.D/Chem E/University of Florida; MS/Chem E/University of Florida; BS/Chem E/University of Tennessee *Born:* 05/14/42 Born in Chicago. Experimental Test Engr with Pratt & Whitney 1964-67. Since 1971, served in various positions of increasing responsibility with the U.S. Environmental Protection Agency (EPA) with the exception of one year with the Waterways Experiment Station, U.S. Army Corps of Engineers, Vicksburg. Research at the EPA Environmental Research Lab, Athens, GA (1971-81) and at Vicksburg (1973-74) concentrated on dev of mathematical and physical models for ecosystem simulation, behavior of chemicals in the environment, and exposure assessment. From 1981-85, Exec Dir, Exposure Assessment Group, EPA, Washington, DC. Responsible for developing EPA policy on and conduct of exposure assessments. Since August 1985, Dir of the Office of Environmental Processes and Effects Research, EPA, Washington, DC, responsible for directing the research of approximately 430 staff assigned to 6 Research Labs and managing an extramural research program with an annual expenditure of approximately $45 million. *Society Aff:* ACS, AIChE, AAAS

Falcocchio, John C
Home: 102-12 72 Ave, Forest Hills, NY 11375
Position: Prof *Employer:* Polytechnic Univ *Education:* PhD/Trans/Polytch Inst of Bklyn; MS/Trans/Polytch Inst of Bklyn; Cert in Hgwy Traffic/Traffic Eng/Yale Univ Bureau of Hgwy Traffic; B/CE/Polytch Inst of Brooklyn *Born:* 8/29/37 Native of Tagliacozzo, Italy. Came to the US in 1955. Construction engr with the Arthur A Johnson Corp (1960-63). Transportation engr with TAMS (1964- 66). Senior transportation engr with the DE Valley Regional Planning Commission (1966-68). With the Polytechnic Inst of NY since 1968. Prof of Transportation Engrg, specializing in transportation system planning, transportation planning methods, travel demand analysis and forecasting, travel needs and travel behavior of disadvantaged groups. Sr partner and principal with Urbitran Assoc and Urbitran Assoc. Inc. *Society Aff:* ASCE, ITE, NYAS, ΣΞ

Falcone, Philip A
Business: 67B Mt Blvd Ext, P.O. Box 4039, Warren, NJ 07060
Position: Principal *Employer:* Paulus, Sokolowski and Sartor, Inc. *Education:* MS/CE/MIT; BS/CE/Univ of Detroit *Born:* 9/14/41 Native of East Orange, NJ. Current resident of Westfield, N.J. Attended St Peter's Coll, Jersey City, NJ. Completed Engrg curriculum at Univ of Detroit, Detroit, MI. Held positions as Teaching/Research Assistant at MIT, 1965-66. Published "Small Scale Steel Frameworks," MIT, 1966. Served with US Army Corps of Engrs, 1967-69. Batallion Engr in Vietnam, specializing in road/bridge construction. Joined Paulus, Sokolowski and Sartor in 1970. Assumed position as Principal in 1976. Serves as VP in Charge of Civil Engrg Services. Major responsibilities are marketing, contract negotiation and client

Falcone, Philip A (Continued)
contact. Licensed as PE in NJ, MA, ME, PA, NY, CT, MD, GA. Active in civil, church and alumni organizations. Past Pres Parish Council and past Pres MIT Club of NJ. Enjoy sailing and tennis. *Society Aff:* NSPE, ASCE, ACEC, SAME

Falivene, Pasquale J
Business: 105 Hudson St, Jersey City, NJ 07302
Position: Mgr - Admin Svcs. *Employer:* Colgate-Palmolive Co. *Education:* BChE/Chem Engg/CCNY. *Born:* 9/7/26. BChE from CCNY in June 1948. With Foster Wheeler Corp 6 yrs, design and construction of prototype atomic energy plant in Arco, ID. Then 8 yrs at Picatinny Arsenal, on US Army Ordinance Corps Res & Dev projs. Advanced to Sec Hd of Small Arms Unit which designed and tested Antipersonnel mines, rifle ammunition, etc. Past 22 yrs with Colgate-Palmolive Corporate Res & Dev Dept. Responsible for product and process dev for consumer products such as Dynamo, Palmolive Dishwashing Liquid, Handi-Wipes, etc. Since Aug '78 Mgr of Admin Svcs, responsible for Employee Relations, Pilot Plant Adm, and Clerical Staff. Enjoy opera, travel, and reading.

Falk, Martin Carl
Home: 508 Montclair Street, Pittsburgh, PA 15217
Position: Owner *Employer:* Martin C. Falk, Prof Engineer *Education:* Dr OEC/Engrg Econ/German Univ; MS/Metallurgy/British College; BS/EE & ME/German Univ; Cert. Tool Engrg *Born:* 6/2/12. German & British Univs, BSEE, BSME 1932; MS Met Engg 1933; PhD Engg Economics 1934. Cert Tool Engg Case 1942. Reg P.E. PA, IL. Life Fellow ASME, AM. Arbitration Association (Arbiter), Score, AISE. ME Republic Steel to 1941; Dev Engineer. Jones & laughlin Steel to 1945. Group Leader Res & Dev PPG to 1951: Asst Chief Engineer Copperweld Steel to 1954; Chief Engr R&D Yoder to 1957; Manager Tube Mills, Mannesmann-Meer to 1960; Exec Proj. Engr Schloeman to 1963; Corp VP, PA Engrg Corp to 1970; Sr Staff Engr, Koppers Co, E&C Div. Retired at Age 65 in 6/77. 7/77 to 3/1/83: (Exec) Asst to the Pres Pittsburgh Testing Lab 3/1/83 to now: Owner of the firm Martin C. Falk, Professional Engr. Community College of Allegheny County: Lecturer and Seminar Leader (Dept. of Labor Studies.) *Society Aff:* ASME, AISE, AAA, SCORE

Falkie, Thomas V
347 Echo Valley Lane, Newtown Square, PA 19073
Position: Pres. *Employer:* Berwind Natural Res Co. *Education:* PhD/Mining Engg/Penn State Univ; MS/Mining Engg/Penn State Univ; BS/Mining Engg/Penn State Univ. *Born:* Sep 1934. PhD, MS, BS Penn State University. Native of Mt. Carmel, Pa. Held various scholarships and fellowships. With International Minerals & Chemical Corporation (1961-69) in various management and engineering positions. Head of Mineral Engineering Department, The Pennsylvania State University (1969- 74)-Jan 1977. Respon for academic programs, research and continuing education in mining engineering, petroleum and natural gas engineering, and mineral engineering management. Consultant for various industrial and government organizations. Served on various governmental commissions. Director, US Bureau of Mines, February 1974-Jan 1977. Responsible for research and development, mineral factfinding, and policy programs funded at approximately $212 million annually. Pres, Berwind Natl Res Co Feb 1977-present. Pres-Elect AIME, Distinguished Mbr and 1985 Pres Soc of Mining Engrs, Mining and Metallurgical Society of America. Dir Natl Coal Assoc 1980-present. Mbr Bd Mineral and Energy Resources, NRC, 1982- present. Mbr Boards: Fook Mineral Co., National Coal Association. *Society Aff:* AIME, SME, MMSA.

Faller, Harold E
Business: 508 Coolidge St, New Orleans, LA 70121
Position: President. *Employer:* Faller & Assocs, Cons Engrs. *Born:* October 8, 1924. BEME from Tulane University. Native of New Orleans, La. Served with US Infantry 1943-45. Sales Engineer for Air Conditioning Contractor 1948. On staff of A R Salzer Jr, Consulting Engineer, 1948-58. Served as Vice President, Salzer & Associates, 1954-58. Assumed current responsibility as President, Harold E. Faller & Associates, Consulting Engineers in 1959. Served as member of Louisiana Engineers Selection Board and of State Firm Marshall's Committee High Rise Buildings and Proposed Code Changes. Past President New Orleans Chapter ASHRAE, Past Interim Pres and Pres of ASPE, New Orleans Chapter Past Pres of New Orleans Chap Consulting Engrs Council of Louisiana. Past President Consulting Engineers Council La. Past President of Rotary Club of Carrollton (N.O.).

Fan, Liang-Shih
Business: 140 West 19th Ave, Columbus, OH 43210
Position: Assoc Prof *Employer:* Dept of Chem Engrg The OH State Univ *Education:* PhD/ChE/WV Univ; MS/ChE/WV Univ; MS/Statistics/KS State Univ; BS/ChE/National Taiwan Univ *Born:* 12/15/47 in Taiwan, came to the US in 1971. Postdoctoral Research Asoc, KS State Univ (KSU), 1975-76. Visiting Assistant Prof, KSU, 1976-78, and Univ of Tokoyo, summer, 1978. Served as a research engr for Amoco Oil Co, summer, 1979 and a Research Assoc at Argonne National Labs (ANL), summer, 1980. Presently serve as a consultant for ANL and Battelle Memorial Inst (BMI). Now employed as an Assoc Prof (tenured) of Chem Engrg, OH State Univ. Recipient Research Award, BMI, 1980; OH State Univ, 1983; Sigma Xi OH State Univ Chapter, 1983. NSF Grantee, 1979, 1980-84. Has over 80 publications on fluidization and process modeling and simulation. *Society Aff:* AIChE, SCEJ, ΣΞ

Fan, Liang-tseng
Business: Dept. Of Chemical Engg, Durland Hall, Kansas State University, Manhattan, KS 66506
Position: University Distinguished Professor and Head *Employer:* Kansas State University. *Education:* PhD/Chem Engg/W VA Univ; MS/Chem Engg/KS State Univ; MS/Mathematics/W VA Univ; BS/Chem Engg/Natl Taiwan Univ. *Born:* August 7, 1929. Production Engineer, Koahsiung Agricultural Chemical Works, Formosa, 1951-52. Research Specialist, West Virginia Univ (1/2 time) 1954-58. Engineer, U S Bureau of Mines (1/2 time) 1956-58. Instructor, Kansas State Univ, 1958-59. Assistant Professor, Kansas State Univ, 1959-61. Assoc. Prof, Kansas State Univ, 1961-63. Professor, Kansas State Univ, 1963-1984. University Distinguished Professor, Kansas State Univ, 1984-present. Visiting Research Professor, Cambridge University, England, Feb. 1965 to Aug 1965. Visiting Professor, The University of Sydney, Australia, 1979, Kansas Power & Light Co., Professorship, 1967-73. Head, Chemical Engrg Dept, Kansas State U, 1968 to present, Fellow of AIChE, 1983-present, Fellow of AAAS, 1985-present. *Society Aff:* ACS, AIChE, AWRA, ASEE, AAAS, AAUP, NYAS, BEI, CHMT, PEF, SES, IEEE.

Fancher, George H
Home: 600 East 32nd Street, Austin, TX 78705
Position: Petroleum Consultant (Self-employed) *Education:* DSc/Petroleum Engg/CO Sch of Mines; MSc/Chem Engg/Univ of MD; BS/Ind Chem/Univ of Southern CA. *Born:* September 1901. 2 yrs graduate work Chemical Engrg Univ of Mich Native of San Francisco. Taught U of Mich, Colorado School of Mines, Penna St U, U of Texas 1926-60. Pioneer research in cracking, core analysis, waterflooding. Prof Petroleum Engrg, U of Texas, 1935-60, Chairman 1955-60; Director Texas Petroleum Res Ctte 1950-60; V P & Board of Directors, Sinclair Oil and Gas Co, 1960-67. John Franklin Carl Award of SPE of AIME 1968, Distinguished Achievement Silver Medal, Colorado School of Mines 1963. Listed Who's Who in America, Rotarian, Tau Beta Pi, SAR, Society Mayflower Descendants, American Arbitration Assn, American Institute of Mgmt. Author 3 books and numerous technical articles. George Homer Fancher Presidential Scholarship in Petroleum Engg in perpetuity, Univ of TX 1979. Distinguished Mbr, Soc of Petrol Engrs, 1983. *Society Aff:* AIME, SPE, API, ACS.

Fang, Pen J
Business: Bliss Hall, Univ RI, Kingston, RI 02881
Position: Assoc Prof Civil/Environmental Engrg. *Employer:* University of Rhode Island. *Education:* PhD/Structures/Cornell Univ; MS/Hydraulics/OK State Univ; BS/Hydraulics/Natl Taiwan Univ. *Born:* July 1931. Structural engr for I C Hillman & Assoc (Chicago) for over 2 years and Senior Research Engr at US Steel for 3 yrs. Taught at U of Pittsburgh (p t) & Sir George Williams Univ. Teaching CE at URI since 1970; extensive domestic and international consulting. Visiting Prof of Post-Graduate Engrg Prog, Universidade Federal do Rio de Janeiro & Univ do Sao Paulo; concurrently as staff consultant for Promon Engenharia SA, Brazil 1976-78; consultant for Electroconsult to Brasil since 1976 and for Natron (Rio de Janeiro) since 1980. P E RI, Struct Engr Ill. Collingwood Prize ASCE 1967. Enjoy classical music & photography. *Society Aff:* ASCE, ASEE, ACI, AAM.

Fanger, Carleton G
Business: PO Box 751, Portland, OR 97207
Position: Prof Emeritus *Employer:* Portland State Univ. *Education:* MS/Aero Engg/OR State Univ; BS/ME/OR State Univ. *Born:* 3/22/24. Native of Huron, SD. Completed secondary education in Medford, OR. Class Salutatorian. Served with USNR during WWII as ETM. BS 1947, MS 1948 from OR State Univ. Consolidated Vultee Fellowship Award. With Portland State Univ since 1948. Chrmn of Engg 1954-60. Six summers as ME Design Engr, US Army Corp of Engrs. Reg Mech Engr, OR, since 1954. Pres PSU Chapter AAUP 1958, 1962. Pres State Fed of AAUP 1960-61. Published 2-vol Engg Mechanics, CE, Merrill, 1970. Chrmn ASME, OR Sec 1975-76. Acting Hd, Mech Engg, PSU, 1979-80. Member PNW Section ASEE Governing Bd, Council of Sections, 1979-1983. Chrmn PNW Section ASEE, 1981-82. Pres Engrg Coord. Council of Oregon, 1982-83. Married, three children, five grandchildren. *Society Aff:* ASME, ASEE, NSPE

Fangmeier, Delmar D
Business: Agricultural Engr. Dept, Univ of Arizona, Tucson, AZ 85721
Position: Prof. *Employer:* Univ of AZ. *Education:* PhD/Engg/Univ of CA; MS/Agr Engg/Univ of NE; BS/Agr Engg/Univ of NE; BS/Agr/Univ of NE. *Born:* 10/27/32. NE native. Served in US Army 1954-1956. Agr Engr US Dept of Agr, 1960-61. Asst Prof at Univ of WY, 1966-68. Assoc Prof, Prof, and Agr Engr working in irrigation engg at the Univ of AZ since 1968. ASAE activities include Chrmn - Irrigation Group, 1976-78, VChrmn and Chrmn-Soil and Water Div 1978-80, Co- Chrmn, Natl Irrigation Symposium, 1978-1980. Co-Chrmn, Fourth International Conference on Guayule Research and Development, 1985. *Society Aff:* ASAE, ASCE, ASEE, AAAS, ICID, CAST, GRS.

Fanning, Daniel P
Home: 175 Hamilton Allenton, No. Kingstown, RI 02852
Position: President. *Employer:* CFI Construction Co. *Education:* MS/Civil Eng/Univ of MI; BCE/Civil Engr/RPI; BA/Physics/Williams College. *Born:* Aug 28, 1937. MS Civil Engrg Univ of Missouri; BCE RPI; BA Williams College. Superintendent, Raymond Concrete Pile, New York, N Y and Fanning & Doorley Construction, Prov R I. Founder and President CFI Construction and related corporations 1966. Member ASCE; P Pres R I Sect RISPE, NSPE; Pres R I Section Williams College Alumni; Dirctor Providence Country Day School. Married Susan Collard Fanning, Buffalo, N Y; 4 sons. Enjoy golf, tennis, squash, gardening and fishing. *Society Aff:* ASCE, NSPE.

Fano, Robert M
Business: NE43-511, MIT, Cambridge, MA 02139
Position: Ford Prof of Engrg., Emeritus. *Employer:* Dept/Elec Engrg & Computer Sci, MIT. *Education:* ScD/EE/MIT; SB/EE/MIT. *Born:* Nov 1917. Native of Torino, Italy. Member MIT faculty since 1947. Member Radiation Lab, R L E, Group Leader Lincoln Lab. Ford Prof 1962, Emeritus 1984. Founding Director of Project MAC 1963-68, Assoc Head EE Dept 1971-74. Author 'Transmission of Information', co-author 'Electromagnetic Fields, Energy and Forces' and 'Electromagnetic Energy Transmission and Radiation'. Worked in network theory, microwave circuits, electromagnetism, communications, computer science. Mbr Natl Acad of Sciences. Mbr Natl Acad of Engineering, Fellow American Academy of Arts & Sciences, Fellow IEEE, Member ACM, Sigma Xi. *Society Aff:* NAS, NAE, AAAS, IEEE, ACM, ΣΞ

Fante, Ronald L
Home: 26 Sherwood Rd, Reading, MA 01867
Position: Asst. Vice President *Employer:* AVCO Systems Division *Education:* PhD/EE/Princeton Univ; MS/EE/MIT; BS/EE/Univ of Penna. *Born:* 10/27/36. Native of Phila Pa. Consulting Engg, Avco Corp (1964-71). Engaged in res on reentry physics. Res physicist, AFCRL, Bedford, MA, specializing in res on propagation in random media. Sr scientist at Rome Air Dev Ctr (1971-80) studying problems on optics, antennas and radar. Asst. V.P./Engrg, Textron Defense Systems 1980-present. Received Atwater Kent prize (1958), Marcus ODay Prize (1975), and USAF achievement Award (1974). IEEE Antennas & Propagation Society Best Paper Award (1980) Fellow of IEEE, mbr Adcom of Antennas & Propagation Soc, Chmn of Central NE Council (APS) of IEEE. Adjuct Prof, Univ of MA, Director of Electromagnetics Society, Fellow of Optical Soc of Amer; Ed-in- Chief of IEEE Transactions on Antennas and Propagation; Mbrshp Chrmn of Intl Union of Radio Sci. *Society Aff:* IEEE, URSI, OSA.

Fanucci, Jerome B
Home: 1313 Anderson Ave, Morgantown, WV 26505
Position: Dept of Mech and Aerospace Engrg *Employer:* West Virginia Univ. *Education:* PhD/Aerospace Eng/Penn State Univ; MSAE/Aerospace Eng/Penn State Univ; BSAE/Aerospace Eng/Penn State Univ. *Born:* Oct 7 1924 Glen Lyon PA. 1964-81 Prof & Chrmn Dept of Aerospace Engrg. 1959-64 RCA Sr Res Sci; 1957-59 GE Missile & Space Vehicle Div, Research Engr; 1952-57, Penn State U, Asst Prof of Aerospace Engrg. Mbr: Pi Tau Sigma, Sigma Xi, Sigma Gamma Tau, AIAA Assoc Fellow, ASEE, ARRL. Co-designer of first circulation controlled flying aircraft (STOL); author of over 60 technical papers in Aerospace Engrg. Married Janice Bovitz; children: Dr Jerome Paul, aerospace engg grad MIT and Karen Marie, student in medicine. Hobbies: amateur radio (K8JF) and hunting. *Society Aff:* AIAA, ASEE.

Farber, Erich A
Home: 1218 NE 5th St, Gainesville, FL 32601
Position: Prof & Research Prof, Director. *Employer:* Univ of Florida, Solar Energy. *Education:* PhD U of Iowa; MS & BS U of Missouri *Born:* Sep 1921. Ed in Europe; Taught at U of Missouri, Iowa, Wisc, Fla. Dev 'Boiling Curve' (heat transfer), 'Critical Mass' concept for liquid rocket propellants; started & built the Solar Energy & Energy Conversion Lab into one with internatl reputation. Battlefield Commission US Army WW II, Silver Star, Purple Heart wCs, etc. Over 400 pubs, co-author and contributor to 6 bks. Solar Energy Citation from Air Force; ASME Worcester Reed Warner Gold Medal 1966, Missouri Citation & Gold Medal 1968, Fellow ASME 1971; NASA/NSF Panel 'Solar Energy as a Nat'l Energy Source', US/USSR Solar Energy Work Group, Blue Key Distinguished Fac Award 1973, Coll Eng. Outstanding Ser Award 1974, Dir Int'l Solar Energy Soc, former Chmn ASME Solar Energy Div, appointed by Gov to Fla Energy Ctte, also Energy Task Force. Since 1976 Dr Farber has acted, by request of the State Dept, as advisor on energy problems to such countries as the Phillipines, Peru, Morocco, etc and has helped to establish energy centers in these countries.

Farber, Jack D
Home: 100 W 96th Terr, Kansas City, MO 64141
Position: VP *Employer:* Burns & McDonnell Engrg Co, Inc *Education:* BS/EE/KS Univ *Born:* 8/25/25 Lifetime resident of Kansis City, MO except for five years of service with the US Navy during World War II and the Korean Conflict. All 35 years of employment have been with Burns & McDonnell Engrg Co. Performed as a design and resident engr on large electric generating station and transmission line projects, and as project mgr on transmission line and substation projects. Served in the capacities of Electrical Dept Mgr, Senior Mgr of Design, and currently VP and

Farber, Jack D (Continued)
Deptuy Mgr of the firm's Power Division. Responsible for administrative management as well as division's design functions. Member of firm's Bd of Dirs. A golfing enthusiast. *Society Aff:* IEEE, NSPE, MSPE

Fargo, James T
Business: 390 Washington Ave, Nutley, NJ 07110
Position: Engr Tech Conslt *Employer:* ITT Avionics Division *Education:* MS/EE/-/NJIT; BS/Phys/Math/-/Upsala Coll *Born:* 07/01/33 Born in Irvington, NJ. Son of Joseph J. and Alice V. (Lott) Fargo, Engr Tech Conslt with ITT Avionics Division, Nutley, NJ. (1972-), involved in the dev of advanced electronic warfare systems for the military; Supvr Engr, Honeywell, Inc. (1971-72) responsible for dev of helicopter avionics systems. Principal Engr Lockheed Electronics Co (1959-71), involved in the dev of advanced ground based and airborne radar systems. Served with US Army, Redstone Arsenal, AL (1957-58), Weapon Systems Evaluation Group. *Society Aff:* IEEE

Farhat, Nabil H
Business: 200 South 33rd, Philadelphia, PA 19104
Position: Prof *Employer:* Univ of Pennsylvania *Education:* PhD/EE/Univ of PA; MSc/EE/Univ of TN; BSc/EE/Technion *Born:* 11/03/33 Joined the faculty of the Moore Sch of Electrical Engrg, Univ of PA in 1964 where he is now Prof in Electrical Engrg and heads the Electro-Optics and Microwave-Optics Lab. His current res interests are in the areas of imaging radars, microwave holography and electro-optics in which he has numerous publications. He is teaching courses in EM Theory, Electro-Optics and Holography on both grad and undergrad levels. While Assoc Prof, Dr Farhat held the Ennis Chair in Electrical Engrg. He is a recipient of the Lindback Foundation award for destinguished teaching. He is a Fellow of the Inst of Electrical and Electronics Engrs, and is a mbr of Sigma Xi, Eta Kappa Nu, the NY Acad of Sci, the American Inst of Physics. He has served on the Natl Bd of Dirs of Eta Kappa Nu and has been in RCA conslt since 1969. He has also served as Editor of *Advances in Holography* and is Assoc Editor of *Acoustical Imaging and* Holography. *Society Aff:* IEEE

Farison, James B
Business: 2801 W Bancroft St, Toledo, OH 43606
Position: Professor of Elec Engrg *Employer:* University of Toledo. *Education:* BSEE/Elec Engg/Univ of Toledo; MS/Elec Engg/Stanford Univ; PhD/Elec Engg/Stanford Univ. *Born:* 5/26/38. Phi Kappa Phi & NSF fellowships. Engineering experience with Bell Telephone Laboratories, Ford Motor Company, Stanford Electronics Laboratories and Medical College of Ohio. Joined UT as assistant professor in 1964. Advanced to associate professor 1967, professor 1974. Served as assistant dean 1969-70, acting dean 1970-71 and dean 1971-80. Selected US Outstanding Young EE of 1970 by Eta Kappa Nu, Toledo's Outstanding Young Man of 1971 by local Jaycees, 1973 Ohio Young Engineer and 1984 Outstanding Engrg Educator of Ohio by OSPE, and Toledo's 1984 Engr of the Year. Married, two sons. P E Ohio *Society Aff:* NSPE, IEEE, ASEE, ISA, ΣΞ, TBΠ, HKN, SME/MVA, ΦΚΦ

Farkas, Emery
Home: 48 Oakwood Road, Newtonville, MA 02160
Position: VP, Construction Div. *Employer:* W R Grace & Co. *Education:* MS/Chem Engg/Polytech Univ of Budapest. *Born:* Oct 28, 1925 Budapest, Hungary. Chem Co executive. s. Imre and Margit (Egerhazy) F; m. Elizabeth Tersztyanszky Sept 17, 1949; Children: Lillian Theresa & Leonora Marguerite. Came to U S in 1956, naturalized in 1962. MS in Chem Engrg Polytech Univ, Budapest 1948. Research engr at State Hygienic Institute, Rockefeller Foundation, Budapest 1948-56. Asst of Chem, Harvard Univ, Cambridge, Mass 1957. Research chemist W R Grace and Co, Dewey & Almy Chem Div, Cambridge 1957-60, group leader 1960-65. Director Tech Serv 1965-70. Sr Vice Pres Materials and Constr Mgmt Consultants 1970-72. General Manager Construction Products Div, W R Grace & Co, Cambridge, Mass 1972-77, VP 1977-present. Pres, Hungarian Club of Boston Inc 1966-67. Member American Society for Testing and Materials, American Concrete Inst. Member Board of Directors. Reg PE in the Commonwealth of MA. President of American Concrete Institute 1985-86. Board of Direction of American Society for Testing and Materials 1987-89. *Society Aff:* ACI, ASTM.

Farley, Charles S
Business: Ste 404B, Pembroke One Office Park, Virginia Beach, VA 23462
Position: Pres *Employer:* Farley Engrg, P.C. *Education:* BS/ME/Univ of VA *Born:* 4/28/39 Extensive experience in governmental, industrial, commercial and institutional mechanical systems design, estimating, construction supervision, construction inspection, trouble shooting, system balancing and engrg analysis. Experienced in preparation of pollution control reports, life cycle costing analysis, energy conservation analysis and audits, sound and vibration analysis, automatic temperature control and energy conservation controls and energy recovery systems. Mechanical design accomplished for GSA, HEW, NASA, US Navy and US Army Corps of Engrs, Federal Agencies and for city and state agencies for stateside and intl projects. *Society Aff:* NSPE, ASHRAE

Farmer, Herbert E
Home: 6387 W 80th St, Los Angeles, CA 90045
Position: Professor of Cinema. *Employer:* Univ of Southern Calif. *Education:* MA/Cinema/Univ of So CA; BA/Cinema/Univ of So CA. *Born:* 3/31/20. in Buffalo, NY. MA & BA from U S CA. Faculty U S CA 1942-43 & 1946-present. Photographic officer US navy during WWII. Now Professor of Cinema specializing in motion picture engg & business. Fellow, Society of Motion Picture Engrs & Audio Engg Society; SMPTE Editorial VP 1963-66; SMPTE Educational VP 1970-73; Award of the Audio Engg Society 1957; Pres, Univ Film Producers Assn 1952-54; Trustee, Univ Film Fnd, 1961-present. Eastman Kodak Gold Medal awarded by SMPTE 1976, Chrmn, Intl Electrotechnical Commission, IEC/Tech Com 84. *Society Aff:* SMPTE, AES, UFA.

Farmer, Larry E
Business: P O Box 3, Houston, TX 77001
Position: Vice President. *Employer:* Brown & Root, USA Inc. *Education:* PhD/CE/Univ of TX-Austin; MS/CE/Univ of TX-Austin; BS/CE/Univ of MO-Rolla. *Born:* October 4 1939. MS & PhD from University of Texas-Austin. BS from Missouri School of Mines & Metallurgy. Native of Ash Grove, Missouri. Was Asst. Prof of CE Univ of Missouri-Rolla, 1965-67. President L E Farmer Inc 1967-73. Director of Consulting Services, Compu-Serv Network Inc 1973-76. Joined Brown & Root in 1976 as specialist in offshore concrete structures. Co- recipient of the ACI Wason Medal for Res 1967. Proj Engg Mgr for N Sea offshore platform projs 1978-82. Chief Engineer Brown & Root (UK) responsible for engg. activities in Europe and Africa 1982-85. President Brown & Root Marine Ltd responsible for engg. and construction activities in Canada, 1985 to date. Also Vice President Brown & Root USA responsible for marine engg. *Society Aff:* ASCE, ACI, SPE.

Farmer, Thomas S
Business: 200 So Michigan Avenue, Chicago, IL 60604
Position: Vice President. *Employer:* Borg-Warner Corporation. *Born:* September 1931. BS Chemical Engrg from Tulane. MS Chemical Engrg from Princeton. Native of New Orleans, Louisiana. Joined Borg-Warner in 1971 as President of the Chemicals and Plastics Group. Had been Vice President of Essochem Europe Inc, Brussels, Belgium since 1968 at which time was responsible for directing the marketing, manufacturing and distribution of various product lines in Europe and Africa. Currently responsible for Borg-Warner's worldwide chemical operations and Borg-Warner Educational Syustems. Enjoy reading and sailing.

Farnell, Gerald W
Home: 1509 Sherbrooke St West, Montreal, Quebec, Canada H3G 1M1
Position: Professor of Engineering. *Employer:* McGill University. *Education:* PhD/Elec Engg/MCGill UNiv; SM/Elec Engg/MIT; BASc/Elec Engg/Univ of Toronto. *Born:* 1925 Toronto Canada. McGill University 1950 to present: Currently with academic rank of Professor of Engineering Physics; Nuffield Fellow at Clarendon Laboratory, Oxford 1960; Chairman of the Department of Electrical Engineering for six years from 1967; appointed Dean of the Faculty of Engineering in 1974. Research interests have included successively multichannel pulse telemetry systems, diffraction in microwave lenses, paramagnetic relaxation, electroacoustic interactions in semiconductors and most recently surface acoustic wave phenomena and devices. Fellow of the Institute of Electrical and Electronics Engineers (1970). *Society Aff:* IEEE, CSEE, OEQ, ΣΞ.

Farnham, Charles P
Home: 8629 NE 6th, Bellevue, WA 98004
Position: Pres. *Employer:* Lighting Tech Inc. *Born:* Dec 1932. BA Ohio Wesleyan University. Raised in Cleveland Ohio. Joined G E 1956. Lamp Division Application Engineer. Field Sales Engr Fresno Calif. Promoted to District Engineer, Seattle Wash 1963. Responsible for all applications of light sources including such diverse fields as aerospace, equipment design and application, medical equipment, TV and stage lighting and infra-red u v applications. Presently responsible for application and design of roadway, sports and industrial lighting for G E in the Pacific Northwest. Publications: 'Economics of Plant Growth Lighting' (co-author) 1971; 'Lighting Economics Applying Energy Rate Increases and Present Value Costing Concepts' 1975. Offices held: President Puget Sound Section IES 1969-70; Regional Vice President IES 1972-74. Instructor in Illumination Univ of Washington, for IEEE; Have organized and taught. Since 1977 Mr Farnham has been Pres of Lighting Technology, Inc. Lighting Technology is engaged in lighting design, maintenance and energy mgmt. LTI is the lighting arm of Energard, which offers total energy mgmt, for schools, office bldgs, commercial and industrial accounts. LTI is also engaged in res, to dev more efficient luminaires. It holds a pat on the "retroflector–. A device which allows users to remove 2 lamps from 4 lamp luminaries and obtain 70% of the initial illumination. Mr Farnham is also a mbr of the IES. *Society Aff:* IES, AEE, NSPE.

Farnsworth, David E
Business: P O Box 185, Houston, TX 77001
Position: President. *Employer:* Eddy Refining Company. *Education:* BS/ChE/Rice Univ. *Born:* 9/8/20. in Ranger, TX, Sept 8, 1920; son of Madison (deceased) & Martha (Fokert) Farnsworth; Process design engg, Union Carbide, 1942-43. Genl Mgr, Secy-Treas, Eddy Refining Co, 1946-58, Pres 1958- ; Pres, Key Oil Co, 1950-79. Served as Lieutenant, USNR (1943-46), Pacific Theater of Operations. Serving as a Bd Mbr of the TX Commerce Medical Bank & as a Mbr of Amer Soc of Chem Engrs, Chamber of Commerce, TX Assoc of Bus, and other civic & bus organizations. Mbr of River Oaks Country Club, Petroleum Club, & other social clubs. *Society Aff:* A.I.C.H.E.

Farnsworth, George L
Business: 2709 McGraw Drive, Bloomington, IL 61701
Position: Chairman of Board *Employer:* Farnsworth & Wylie, Cons Engrs. *Education:* MS/CE/Univ of IL; BS/CE/Univ of IL. *Born:* Oct 1917. 1941, Sanitary Engr Ill Dept of Public Health; 1942-46 US Navy; 1946-52 Engr, J J Woltmann; 1952-54 Partner, Farnsworth & Conley Cons Engrs; 1954- , Partner Farnsworth & Wylie Cons Engrs. Pres Ill Engrg Council 1964; Secy Ill Soc of PE; Natl Dir NSPE. IL Award IL Soc of Prof Eng 1979. *Society Aff:* ASCE, NSPE, AWWA, ACSM, WPCF, ISPE, CEC.

Farr, W Morris
Business: Engg 112C, Tucson, AZ 85721
Position: Assoc Prof. *Employer:* Univ of AZ. *Education:* PhD/Nuc Engg/Univ of MI; MS/Nuc Engg/Univ of MI; BA/Engg/Rice Univ. *Born:* 10/20/38. Currently employed as Assoc Prof of Nuclear Engrg at the Univ of AZ. Res mainly in the field of controlled thermal nuclear fusion, with some work in the area of disposal of radioactive waste mtl. Two children, Jeffry & Katy. Moved to Tucson in 1969 from Oakridge, TN. *Society Aff:* AAAS, ANS, APS.

Farrall, Arthur W
Home: 1858 Cahill Dr. East, East Lansing, MI 48823
Position: Prof & Chmn Emeritus-Agri Engrg. *Employer:* Michigan State Univ-ret. *Education:* Dr of Egg/Agri Engg/Univ of NB; MS/Agri Engg/Univ of NB; BS/Agri Eng/Univ of NB. *Born:* 2/23/89. M. Luella Buck June 1923; c Margaret (Longnecker), Robert Arthur. Instructor and Asst Prof Univ of CA 1922-29; Dir Douthitt Res Lab Chicago 1929- 32; Res Engr and Dir Research Cremery Package, Chicago 1932-45; Chrmn Agri Engg Dept, MI State Univ 1945-64. Massey-Ferguson Gold Medal Award 1971; Food Engg Gold Medal Award ASAE and Dairy & Food Industries Supply Assn 1972. Fellow & Pres ASAE 1962-63; Mbr Tau Beta Pi, ASEE, Sigma Xi, Phi Kappa Phi. Author 7 books, many tech paper & bulletins; consultant to India, Brazil, WA DC; 7 patents. Pres. E. Lansing Sr. Citizens 1980-81 Chrm. Mich. State Univ. Retiree Service Corp. 1980- . Oct 5, 1984 Bd of Trustees renamed MI State Univ Agr Engrg Bldg the Arthur W Farrall Agr Engrg Hall; April 1984, Natl Bd of Trustees of Phi Kappa Phi awarded distinguished mbr status to Arthur W Farrell elected Intl Who's Who of Intellectuals, Cambridge England, Aug 1984. *Society Aff:* ASME, AAAS, TBΠ, ΦΚΦ, ΣΞ, IFT.

Farrand, William A
Home: 632 Lakeside Dr, Fullerton, CA 92635
Position: Fulltime Faculty (Retired) *Employer:* California State Univ Fullerton Coll of Engrg & Computer Sci *Education:* PhD/Engr/UCLA; MS/Engr/UCLA; BA;Chem/UCLA. #TTL‡Dr *Born:* May 6, 1922. BA, MS & PhD from UCLA. Member engrg staff UCLA 1945-48. North American Aviation engr for computers since 1950. Developed the early computers and controls for inertially guided missiles, including the first transistor computer and the first disc memories. Retired from Rockwell Int. & Cal State Fullerton Coll of Engr. & Computer Sci. PGEC - a founder and board member LA Chapter 1950's. WJCC - committee chairman 1957-58. DDA - council chairman 1960-64 board member 1958-64. SCI - board member 1961-65. SIMULATION - associate editor 1963-68. SCI Life member (elected). 17 patents and over 25 publications.

Farrell, Keith P
Home: 1150 Albany Dr, Ottawa Ontario K2C 2L3
Position: Private Conslt *Employer:* German & Milne Inc. *Education:* Master/Naval Arch/RN College; BSc/Engg/London Univ; Dipl/Bus Adm/McGill Univ. *Born:* 8/28/19. Born in Portsmouth, England. Trained with Royal Navy. Served in Royal Corps of Naval Constructors 1942-1949. Served in constructor branch of Royal Canadian Navy from 1949-1974. Responsible for design of maj warships. Dir, Ottawa Div of British Columbia Consulting Architect firm 1974-1977. VP Bus Dev with German & Milne Inc, responsible for design and dev projs in the field of ship design and ocean engrg. 1984 - Private Consultant. *Society Aff:* RINA, SNAME, PEO, C.I.Mar.E.

Farrell, Robert J
Business: 3650 Mayfield Rd, Cleveland, OH 44121
Position: Staff Cons Physical Distribution. *Employer:* The Austin Company. *Education:* BSEE/Elec Engg/Cleveland State Univ; BEE/EE/Cleveland State Univ. *Born:* October 1, 1922. BSEE 1944 and BEE 1948 Fenn College. Resides in Hudson, Ohio. Electrical Division Officer Light Cruiser WW2. Taught Electrical Engineering Fenn College 1947-50. Application Engineer Reliance Electric 1950- 62. Assistant Div Manager Control Systems Division Beloit Corporation specializing in Process Computer Control of Papermaking 1962-68. Reliance Electric Manager of Systems at Toledo Scale Division Integrating Weighing and Computer Technology 1968-72. The Austin Company since 1972. Specializing in Financial and Technical Justification of Physical Distribution Centers including Associated Material Handling & Control Systems. Reg P E Ohio. Fellow Tech Assn Pulp & Paper Indus, P Chmn

Farrell, Robert J (Continued)
Engrg Div. Sr Mbr IEEE, P Chmn Pulp & Paper Indus Ctte. Mbr Natl Council of Physical Distribution Management. *Society Aff:* IEEE, TAPPI, NCPDM, IIE.

Farrell, William C
Home: 5940 Pinebrook Dr, Boca Raton, FL 33433
Position: Retired *Education:* BE/EE/Johns Hopkins Univ. *Born:* 1/16/25. After three yrs in the Army, Mr Farrell returned to Johns Hopkins Univ. He received a BE degree in 1948. He started working in the Engg Dept of the C&P Telephone Co and progressed through various levels of engg resp in the Bell System to Div Manager-Transmission Engrg from which job he retired on 7/16/84. Mr Farrell s engg soc activities began as a student mbr of AIEE. He has been a Sr mbr of IEEE since it was formed. His yrs of service to the IEEE has culminated in the election as Dir-Reg II. He has also served on the Engineer s Week Council, the Engg Council of MD, and the Board of the Engg Soc of Baltimore (currently a past Pres). *Society Aff:* IEEE.

Farrelly, Richard J
Home: 339 Conestoga Rd, Wayne, PA 19087
Position: General Manager. *Employer:* General Electric Company. *Born:* Oct 9, 1931 New York City. BSEE from Manhattan College; MSEE from Univ of Penna. Registered Professional Engineer in State of Pa. Member of AIAA, MENSA. Senior Member of IEEE. Mbr of Technical Staff at Bell Telephone Laboratory 1953-56. With General Electric Co since 1956. 15 years aerospace engrg management. 23 years aerospace electronics experience. Section Manager, Electronics Engrg with product and system responsibility, directing 350 professionals. Manager, Systems Development and Technology Marketing including technical direction of Independent Research & Development Program. Directed systems engineering and product development on major aerospace programs. Program management responsibility for NASA-Pioneer Venus Re-entry Probe and Viking Systems Support Engineering. Department General Manager for 5 years. Presently General Manager, Advanc.

Farrin, James M
Home: 4750 S Ocean Blvd-303, Highland Beach, FL 33431
Position: Rear Admiral US Navy (retired). *Employer:* US Navy and Aerojet-General Corp. *Education:* MS/Naval Arch & Marine Engg/MA Inst of Tech; BS/-/US Naval Acad. *Born:* July 16, 1908. Advanced through ranks to Rear Admiral in 1957. Professor of Naval Architecture, MIT; Director of Ship Design, Navy Department; Commander of Naval Shipyards in Philadelphia, Pearl Harbor and Norfolk; Asst Chief Bureau of Ships, Navy Dept. After retirement from US Navy in 1965 was Shipyard Consultant for Aerojet-General Corp and Asst General Manager, Surface Effect Ships Division of Aerojet until 1973. Was Council Member, Vice President, and Honorary Vice President for Life of the Society of Naval Architects and Marine Engineers. *Society Aff:* SNAME, ASNE.

Farrington, David E
Home: PO Box 88, Subiaco, AR 72865
Position: Chrmn of Bd *Employer:* Farrington & Assoc, Inc. *Education:* MPA/Public Admin/Texas Christian Univ; BS/Civil Engg/TX A&M. *Born:* June 20, 1937. Asst City Engineer San Angelo, Texas 1960-64. City Engineer Cleburne, Texas 1964-65; City Manager, Cleburne Texas 1965-66; City of Fort Worth, Texas 1966-72 (Asst Public Works Director, Asst City Manager, Director of City Planning); Director of Engineering Flower Mound New Town 1972- 73; President of Farrington & Associates Inc 1973-1983. Chrmn of Bd, Everage Smith Farrington & Assoc Inc 1983-86. Currently Pres of Farrington & Assoc. F ASCE (Aesthics and Land Use Committees). TSPE Secy-Treas San Angelo, Texas 1962-64, Secy-Treas Fort Worth, Texas 1968. M CEC, AM AIP. *Society Aff:* NSPE, APA, ASCE, ICMA.

Farst, James R
Business: One PPG Place, Pittsburgh, PA 15272
Position: Works Mgr. *Employer:* PPG Ind, Inc. *Education:* MS/CE/OH State Univ; BS/CE/OH State Univ. *Born:* 10/18/33. A native of Barberton, OH, J R Farst joined PPG Industries at the co's Barberton plant in 1956, following grad from OH State Univ where he received BS & MS degrees in chem engg. His initial assignment at Barberton was Operations Engr. He served in several supervisory capacities there before transferring to the Bartlett, CA, plant as Plant Mgr in 1966. Farst transferred to Houston Chem Co (a subsidiary of PPG Industries) in Beaumont, TX as Plant Superintendent in 1967 & was promoted to Mgr of Mfg for that plant in 1973. He transferred to the Lake Charles plant on Nov 1, 1974, as Asst Works Mgr and served in that capacity until his promotion to Works Mgr on Jan 1, 1978. Have participated in community activities in Beaumont prior to moving to Lake Charles & have served as a Dir of the Greater Lake Charles Chamber of Commerce. Have been active in United Appeal work in both Beaumont and Lake Charles and currently am on the Bd of Dir of the LA Chemical Assoc. *Society Aff:* AIChE.

Fassnacht, George G
Home: 302 Poplar Road, Indianapolis, IN 46219
Position: Chief, Water Supply Section(ret). *Employer:* Indiana State Board of Health (ret). *Education:* MCE/Sanitary Engr/NY Univ; BSCE/Civil Engr/Purdue Univ. *Born:* July 1909 South Bend, Ind. Road survey, design and drafting with US Indian Service. Div of Sanitary Engineering, Indiana State Board of Health 1940. Responsibilities in school construction, plumbing, and general sanitation programs prior to becoming Chief, Water Supply Section in 1947. In latter position was responsible for public water utility surveillance & swimming pool program. Assisted in operator training and development of codes and standards. Director for AWWA 1971-73; Sect Secy 1950-54; Chairman 1956; Fuller Award 1961. Director AAEE 1972-74. Now travelling. *Society Aff:* AWWA, WPCF, NSPE, APHA, AAEE.

Fast, C Robert
Home: 4504 Gran Tara, Afton, OK 74331
Position: Pres. *Employer:* Fast Engg Co. (Fasco) *Education:* BS/Petroleum Engg/Tulsa Univ. *Born:* Feb 1921 Tulsa., Okla. Research Engineer and Supervisor for Amoco 1943- 77. Married, 4 children. Currently engaged in consulting and teaching drilling, well completion, stimulation and production mechanics. Chmn Mid-Cont Dist API 1965-66. SPE of AIME Uren Award 1967. API Citation for Service 1968. Author SPE Monograph 'Hydraulic Fracturing 1970'; SPE Distinguished Lecturer 1970-71; SPE Distinguished Member 1983; Chmn API Committee on Perforating 1971; J C Slonneger Petro Engrg Award 1972. Author 22 technical papers and 35 U S Letters Patents. Hobbies: sailing, golfing, hunting & fishing *Society Aff:* SPE, API.

Fateley, William G
Business: P.O. Box 688, Manhattan, KS 66502-0688
Position: Pres *Employer:* D.O.M. Assoc, Intl; Prof. Dept Ch., KS State *Education:* PhD/Chem/KS State Univ; D.Sc/Honorary/KS State Univ; A. B./Chem- Math/Franklin Coll *Born:* 5/17/29 Editor, J. Applied Spectroscopy, 1974-present. Editor, Raman Newsletter, 1975-1980. Coblentz Award, 1965. SSP Award, 1976. H.H. King Award, 1979. Phi Beta Kappa (Honorary), 1976. Who's Who in America, 1972. Co-Director, NATO Advanced Study Inst, 1980. Tour Speaker for American Chem Society, 1974-87. Honorary D.Sc, 1965. Senior fellow at Mellon Inst, 1965-1972. Prof at Carnegie- Mellon Univ, 1966-1972. Head, Dept of Chemistry, 1972-1979. Dist Grad Faculty Award, 1984 from KSU Univ. Hon Mbrshp, Coblenly Soc 1987, Hon Mbrshp, Soc. for Applied Spectroscopy 1987, Distinguished Service Award, Soc. for Appl. Spectro 1987, Gold Medal, Eastern Analytical Symposium, NY Section 1987. *Society Aff:* SAS, ACS, OSA

Fatic, Vuk M
Business: Division of Electrical and Computer Engineering, 5500 Wabash Ave, Terre Haute, IN 46703
Position: Assoc Prof *Employer:* Rose-Hulman Inst of Tech *Education:* PhD/Elec Engg/VPI & SU; MS/Elec Engg/VPI & SU; Dipl Ing/Elec Engg/Belgrade Univ. *Born:* 3/22/32. Born in Yugoslavia. Has lived in USA since 1970. From 1961 to 1970 worked as a res engr in "Elektroinstitut" and INTDI (Belgrade), and an asst in control engg at Novi Sad Univ. Taught elec engg at VPI & SU, Union College (Schenectady, NY), Tri-State Univ and Rose-Hulman Inst of Tech. Wrote 11 technical reports and published 28 papers. The main profl interest is in the area of variational formulation for dissipative networks and systems. Related interests in electromagnetic fields, electric power, control theory, mathematics and theoretical physics. Non-professionally addicted to jazz music, science fiction and gastronomy. *Society Aff:* IEEE, HKN, ΦΚΦ, ΤΒΠ.

Faucett, Robert E
Business: 3386 Longhorn Road, Boulder, CO 80302
Position: President. *Employer:* Independent Testing Lab Inc. *Education:* MS/Elect Engrg/Case Western Reserve; BS/Elect Engrg/Cleveland, OH. *Born:* Nov 21, 1926. BS and MS in EE, Case Western Reserve University 1947-51. Full member Sigma Xi, National Honorary Research Society. Elected to 'Fellow' status by the Illuminating Engrg Society (IES), 1962. His research on the evaluation of discomfort glare at low adaptation levels has become an important reference in the field. Among many of his architectural lighting designs, the most prominent is the programmed color sequence spectacular at Grand Coulee Dam, Washington. His participation in National Technical Committees of the IES includes Photometric Testing Procedures, Aviation, Searchlight, Sign, Outdoor Productive Areas, Progress, Industrial and he has served as Chairman of the Sports Lighting and Aviation Industry Lighting Committees. His participation in the International Commission on Illumination has included Sound & Light Spectacles, Sports,. *Society Aff:* IES, ΣΞ.

Faucett, Thomas R
Business: 107 Mech Engr Bldg, Rolla, MO 65401
Position: Prof of ME. (Emeritus) *Employer:* Univ of MO. *Education:* PhD/ME/Purdue Univ; MS/ME/Purdue Univ; BS/ME/Univ of MO. *Born:* 8/22/20. Hatton, MO. Educated-Public schools, Fulton, MO. Grad-Class of '42-Mech Engg-Univ of MO. Served as Design Engr & Physicist-Cleveland Diesel Div, GMC until 1946. Received MS & PhD Degrees-Purdue Univ while serving as Instr of Mech Engg to 1952. Assoc Prof-Mech Engg, Univ of Rochester (1952-1960), Prof-Mech Engg, MO Sch of Mines & Met (1960-1962), Prof-Mech Engg & Dept Hd, State Univ of IA (1962-1965), Prof & Chrmn-Mech Engg, Univ of MO-Rolla (1965-1978), Prof-Mech Engg, Univ of MO-Rolla (1979-). Ind Consultant & Dir. Author of papers on vibration control & stress analysis. *Society Aff:* ASME, ASEE.

Faucher, Richard L
Home: 54 Giroux Ave, Quebec, H7N 3G9 Canada
Position: Director, Program Management *Employer:* Canadair Limited, Montreal. *Born:* Montreal 1928. Educated in Montreal Publc Schools and Montreal Technical Institute. Specialized in Plant Design and Construction for 15 years prior to assuming position as Director of Plant Engrg in 1967. Was Appointed to Present Position June 1980. Chrmn air Industries Assoc of Canada Committee on Energy 1977-79. Represent Air Industries Association of Canada on Federal Energy Conservation Task Force for Transportation Manufacturing Sector. Article on budgeting published in Plant Management and Engineering Magazine 1972. Article on Financial Control in Plant Engrg published in Plant Engineering Magazine 1975. Vice President Canada Region & Vice President Finance and Publications 1971-74, American Institute of Plant Engineers. Canadian Plant Engineer of the Year, 1976. Cert Plant Engr, AIPE, 1978. Elected Fellow AIPE, 1980. *Society Aff:* AIPE.

Faulkner, C Shults
Home: 5618 Longmont, Houston, TX 77056
Position: Pres & CEO *Employer:* C.S. Faulkner, Inc. *Education:* Postgrad work/ChE/Univ of TX; BS/ChE/Univ of TX. *Born:* 10/20/19. Process Engr, operating cos (oil refining); Process Design Engr, eng & construction cos (oil refining & nat gas processing) 1943-1949. VP/Chief Engr, Grebe & Doremus Process Co (E & C) 1949-1952. VP, Refining, McBride Oil & Gas Co 1952-1954. Pres & CEO PONA Engrs, Inc & subsidiaries, 1954-1981 with specialization in design/construction of oil refining, petrochem and natural gas processing plants as well as consulting & environmental eng. Pres & CEO, CS Faulkner, Inc, Conslt Engrs, 1981-date. Reg PE, TX & WY. Hobbies, golfing/hunting. *Society Aff:* AIChE, ACS, TSPE, NSPE.

Faust, Delbert G
Home: 7902 E Bonnie Rose Ave, Scottsdale, AZ 85253
Position: VP (Ret) *Education:* BSEE/EE/Lehigh Univ. *Born:* June 1912. BS EE, Lehigh Univ, Have held job titles of Chief Engr, VP Engrg & Research, VP Marketing, VP International Operations, VP & General Manager. Employed by Atwater Kent Mfg Co, Walter Kidde Co, C A Norgren Co. Hold 12 U S patents and corresponding foreign patents. Author of many tech articles and lecturer. Past member of AIEE, ASLE, ASM, ASTM, AOS, NFPA, NSPE. Currently Life Fellow ASME. PE License NY & Colorado. Retired September 1974. *Society Aff:* ASME.

Faust, J W, Jr
Home: 2455 Robincrest Dr, W Columbia, SC 29169
Position: Prof of Elec Engg *Employer:* Univ of SC. *Education:* PhD/Phys Chem/Univ of MO; MS/Phys Chem/Univ of MO; BS/Chem Engr/Purdue Univ. *Born:* 7/25/22. Engg Officer, 1942-46 with US Navy, Westinghouse Res Labs as engr in semiconductor group 1951-59, built up Mtls Characterization lab as mgr, 1959-1965, Mgr crystal growth 1965-67. Prof Solid State Sci, Penn State 1967-69. Prof of Engg USC 1969-present. Research Physicist, Naval Research Labs, 1980-81. Held various Local and Natl offices in ECS, ASTM, AIME and ISHM. Mbr of Sigma Xi, Tau Beta Pi, and Eta Kappa Nu Hon Soc. Co-editor of two books, many papers and patents in fields of semiconductors and metals. Co-chrmn Intl Committee on Silicon Carbide 1969-1974. Ran two intl confs, several natl confs and symposia. Consultant to industry and govt in crystal growth, surface phenomena, mtl characterization, and fabrication processes in microelectronics. *Society Aff:* AIME, ASM, ECS, FAIC, ACS, ISHM, ICCG, AACG

Faust, Josef
Home: 2717 Pembroke Terrace, Oklahoma City, OK 73116
Position: Owner (Self-employed) *Education:* Diplom-Ingensiur/Mining/Technische Hochschule Berlin, Germany; Doctor/Geolog/Technische Hochschule Berlin, Germany *Born:* 2/6/02 in Meggen-Lemmestadt, Germany. Immigrated into U.S.A. - 1929. Has been independently active since 1931. Was counsultant for a number of companies. Has been very active in the Mid Continent Region of America's oil-fields, doing research on oil mining. Pres of Mar-Mar Minerals Corp, Josef Faust, Inc. *Society Aff:* AAG, AAPG, AAPL

Faustman, D Jackson
Business: 2415 L Street, Sacramento, CA 95816
Position: Consulting Traffic Engr. *Employer:* Self Employed. *Education:* DEng/Transp Engr/Univ of CA, Berkeley; MEng/Transp Engr/Univ of CA, Berkeley; BS/Civil Engr/Univ of CA, Berkeley; Cert Transp/Transp Engg/Harvard Univ. *Born:* 8/7/16. BS, M Engrg, & D Engrg from Univ of California, Berkeley; Certificate in transportation, Harvard. Asst planning, traffic engr, Sacramento 1938-40. Genl Mgr Sherrill Mfg Corp, Peru Ind. 1946-47. Transp engr. Calif Pub Util Comm 1947- 48. City Traffic Engr Sacramento 1948-56. Cons. Traffic Engr 1956-present. Adjunct prof transp engr, Cal State U Sacramento 1973-75, lectr traffic engr, Northwestern U, Yale U, UC Berkeley & Davis. Mbr Sacramento Bd of Education 1964-, Pres 1966, 1971-87, 1980; Fellow, Institute Transportation Engrs (President North Calif Sect). Fellow Amer Soc CE. Sacramento Camellia Soc (Pres '66). Harvard Soc Scientists & Engineers, Phi Beta Kappa, Sigma Xi, Chi Episilon, Phi Delta Theta, Sacramento Valley Engineer of the Year 1971. Rotary and Sutter Clubs, Hon Citizen of Matsuyama, Japan, VP. Sacramento-Matsuyama Sister City Corp. Trustee, Crocker Art Museum, 1983-. Service World War II, Corps Engrs, US Army to Major, Retired Colonel, AUS. Registered PE Calif-civil, mechanical and traffic. *Society Aff:* ASCE, ITE.

Favreau, Romeo R
Business: 11722 Sorento Val Rd, San Diego, CA 92121
Position: Consultant. *Employer:* Sorrento Valley Associates. *Education:* MS/Aero Eng/UCSD; BS/Physics/MIT. *Born:* Sept 1925. BS Physics MIT 1945 as Aero Engrg UCSD 1974. Mbr Tech Staff Hughes Aircraft Co 1947-54. Dir of Computating Electronic Assocs Inc 1954-58. Vice Pres Res & Computation 195867. General Manager Computer Div 1968. Mbr of the Board of Directors EAI 1958-70. Tech Dir Oceanographic Engrg Co Dillingham Corp 1968-70. Pres Dillingham Environmental Co 1970-72. Private cons Sorrento Valley Associates 1974 - present. Mbr of Sigma Xi, IEEE, Dir Publication Society of Computer Simulation. *Society Aff:* ΣΞ, IEEE, SCS, ISA.

Favret, Andrew G
Business: School of Engineering & Arch, CUA, Washington, DC 20064
Position: Dean *Employer:* Catholic Univ *Education:* DEng/El Engrg/Catholic Univ; MSEE/El Engrg/Univ of Penn; BS/Engrg/US Military Acad *Born:* 05/09/25 Regular Army Officer, 1945-54; MIT, Lincoln Lab, 1954-55; Alexandria Div, AMF, 1955-59; Sr Scientific Adv, Army Intelligence, Pentagon, 1959-63. Catholic Univ faculty since 1963 incl Dir, Computer Ctr, 1969-73. Intelligence Officer, CIA, 1973-77. Mbr Cosmos Club, Army Sci Bd. *Society Aff:* IEEE, ACM, ASEE

Favret, Louis M
Business: 1010 Common Street, New Orleans, LA 70112
Position: Exec VP-Business Integration Group *Employer:* Babcock & Wilcox Company. *Education:* BS/Mech Engrg/OH State Univ; MS/Mech Engrg/OH State Univ. *Born:* 6/5/28. Native of Columbus, OH. With Babcock & Wilcox since 1951. Held pos incl serv engr, serv pts engr, boiler sales engr & mgr of mkting res. 1965 - Proposal mgr for nuclear power plants. 1966 - Proposition, planning & commercial mgr. 1970 - Genl mgr of the nuclear equip div. 1971 - VP of nuclear equip div. 1973 - Corp VP of nuclear div's, with respon for all nuclear operations for the power generation group. 1979 - Exec VP Power Genration Group. 1980 - Exec. V. Pres - Business Integrated Group. Mbr of AIF, ASME. Distinguished Alumnus of OH State Univ-1974. *Society Aff:* AIF, ASME, ANSI.

Faw, Richard E
Business: Nuclear Engg Dept, Kansas State University, Manhattan, KS 66506
Position: Prof. *Employer:* KS State Univ. *Education:* PhD/Chem Engg/Univ of MN; BS/Chem Engg/Univ of Cincinnati. *Born:* 6/22/36. Born 1936, Rural OH. Parents: Robert H Faw and Mary E (Baird) Faw. Educated Univ of Cincinnati (BS, 1959) and Univ of MN (PhD, 1962). Married (1961) to Beverly Giltner of Marion, IN. Two children: Jennifer (1962) and Andrew (1968). Served in US Army, 1963-1965. Mbr of Nuclear Engg Faculty, KS State Univ, since 1962. Dir, KSU Nuclear Engg Shielding Facility, 1968-80. Dept Hd, 1972-1977. Dir, KSU Nuclear Reactor Facility, since 1977. Held various offices in American Nuclear Soc and KS Engg Society. Licensed PE, KS & OH. *Society Aff:* ANS, ASEE, AAEE, HPS.

Fawcett, Sherwood L
Home: 2820 Margate Rd, Columbus, OH 43221 *Education:* PhD/Physics/Case Inst of Technology; MS/Physics/Case Inst of Technology; BS/Engg Physics/OH State Univ. *Born:* Dec 1919. Navy Officer, active duty 1941-46. Joined Battelle 1950. Early work on development of nuclear reactors for power and naval propulsion. In 1964, selected by Battelle to establish its Pacific Northwest Laboratories, Richland Washington. Chief Exec of Battelle 1968-84. As such, headed organization with 7300 scientists, engineers and supporting specialists and 1983 research expenditures of $433 million. A director of the Atlantic Council of the US; former Vice Pres of AAAS. Trustee of several universities; six honorary doctorates. *Society Aff:* ANS, APS, AIME, ASM, ASME, NSPE.

Fay, Charles B
Home: 341 E Knowlton Rd, Media, PA 19063
Position: Director-UTTAS Technologyk. *Employer:* Boeing Vertol Company. *Education:* MS/Aero Eng/MIT; BA/Bus Econ/Middlebury College. *Born:* 8/6/32. AB Bus Econ from Middlebury College; MS Aero from MIT. 1956 - Flight Res Engr with Sikorsky Aircraft. Joined Vertol Div Boeing Co in 1957 as Res Engr in Design Analysis. Progressed through areas in Aerodynamic Technology, Preliminary Design, Tilt Wing Technology & Product Aircraft (CH-47) Technology. Deputy Dir Technology Dept 1968-72. Dir-UTTAS Technology 1972-76. Technology Manager-New Products 1976-present. Mbr Amer Helicopter Soc since 1957, Chrmn of Phila Sect 1969-70. Publisheld papers in various AHS Natl Forums, NY Acad of Scis annals, Aircraft & Missiles Magazine & the Intl Helicopter Forum. Registered Professional Engr-State of PA. *Society Aff:* AHS.

Fay, Richard J
Home: 8 Meadowview Ln, Littleton, CO 80121
Position: Pres. *Employer:* Fay Engineering Corp. *Education:* MS/ME/Univ of Denver; BS/ME/Univ of Denver. *Born:* in St Joseph, MO. Lived in Denver, CO since 1943. Design Engr for Denver fire Clay Co 1957-60. Design & Proj engr for Silver Engg Works 1960-63. Res Engr & Lectr of ME, Univ of Denver 1963-75. Adjunct Prof CO Sch of Mines 1975-78. Adjunct Prof Univ of Denver since 1977-80. Part time consultant 1971-75. Formed Richard J Fay & Assoc in 1975. Name changed to Fay Engineering Corp in 1981. Firm has 18 employees. Chrmn ASME CO WY Sec, 1979-80. Past Chrmn SAE 1977-78. ASME Region VIII Coll Relations Chrmn 1981-87. VP ASME Region XII. *Society Aff:* ASME, SAE, NSPE, AIME, AAAM

Fazio, Anthony F
Home: 205 Garfield St, Haworth, NJ 07641
Position: VP. *Employer:* Chem Air Pollution Con Corp. *Education:* AChe/Chem Engg/Manhattan College. *Born:* 11/27/40. Joined Chemico in 1965 from Hercules, Inc prior to assuming current resp as VP. Proj was involved in marketing, proj mgmnt, proj engg, contract admin, cost engg, proposals and estimating for the co. I enjoy tennis, softball, basketball, jogging, Chmn local scout committee. *Society Aff:* AIChE.

Feagan, Wilbur S
Home: 2128 S Cedarbrook, Springfield, MO 65804
Position: Pres. *Employer:* F & H Food Equip Co. *Education:* Profl Degree/Engg Mgt/Univ of MO Rolla; MS/Engg Mgt/Univ of MO Rolla; BS/Civ Engg/Univ of IL. *Born:* 9/9/13. Collinsville, IL - Sanitary Engr, St Louis MO Dairy Commission. Dairy Plant Engr, St Louis MO Health dept, responsible for review and redesign of milk and food handling equip to assure sanitation and product quality. Milk Specialist MI State Health Dept. Public Health Engr, Kansas City, MO. Sanitary Engr, Klenzade Products Co. Founder and Pres of F & H Food Equip Co, specializing in application and sale of food handling equip. Limited consulting practice sanitary engg mgt and investment. *Society Aff:* NSPE.

Fear, J V D
Business: 1801 Market St, Phila, PA 19103
Position: VP, Fuels *Employer:* Sun Refining & Mktg Co *Education:* MS/ChE/Univ of Louisville; BS/ChE/Univ of Louisville. *Born:* 11/7/25. Native of Morgantown, WV; lived in Phila area for 39 yrs. Served in US Naval Reserve 1943-46 final rank Lt (jg). With Sun since 1948. Started as: res engr; Sec Chief (1960), Mgr, Process Dev (1966) in R & D; Mgr, Special Projs (1970); Pres, Suntech, Inc (1975) a subsidiary of Sun Co; and VP, Marketing, Sun Petrol Products Co (1977) a subsidiary of Sun Co. VP, Fuels Sun Petrol Products Co (1980), VP, Fuels Sun Refining and Mktg Co (1981) a subsidiary of Sun Co. Married, three children. Work with Boy Scouts of Am; Bd of Dir Citizens Crime Comm; enjoy hunting, hiking, boating. *Society Aff:* AIChE, API.

Fearnsides, John J
Home: 1502 Chain Bridge Court, McLean, VA 22101
Position: VP, Civil Systems Div *Employer:* The Mitre Corporation *Education:*

Fearnsides, John J (Continued)
PhD/EE/MD; MS/EE/Drexel; BS/EE/Drexel. *Born:* May 1934. Cryptographer in US Army 1955-57. Electrical Engineering Department faculty member Drexel Univ 1962-64 & Univ of Maryland 1964-68. Member of Tech Staff Bellcomm Inc. and Bell Labs 1968-72 specializing respectively in spacecraft automatic control. U S Dept of Transportation 1972-79. Initiated & managed advanced R&D program 1972-75, Spec Asst to Asst Secretary 1973-74, Chief of R&D Policy Div 1974-75, Exec Asst to Deputy Secretary 1975-78 Acting Adm Res & Special Programs Adm, 1978-79, Deputy Under Sec & Chf Scientist, 1979-80. 1980 Dir., Transportation Analysis & Dir, Planning & Policy Studies; 1982-1986 Tech Dir Air Trans Sys Eng Div. 1986-VP Civil Systems Div The MITRE Corp, 1982-present, Adjunct Prof of Engrg & Public Policy, Carnegie Mellon Univ. 1980-present. Senior Member, IEEE. Member Sigma Xi, AIAA. Assoc Editor IEEE Trans on Auto Control 1974-1976. Mbr, IEEE R&D Ctte. Authored several technical papers. *Society Aff:* IEEE, ΣΞ, AIAA

Fedler, Richard E
Business: 4738 N 40th St, Sheboygan, WI 53083
Position: VP *Employer:* Donohue & Assocs, Inc *Education:* MS/CE/Johns Hopkins Univ; B/CE/Marquette Univ *Born:* 2/3/43 Native of Milwaukee, WI. Engr with Harza Engrg Co, 1966-1969, specializing in water quality investigations and regional wastewater planning. With Donohue & Assocs since 1969 as project engr (1969-73), project mgr (1973-76), assoc and dept head (1976-80). VP since 1980. Responsible for supervising the work of 60 professionals in the fields of solid waste and and hazardous waste disposal, hazardous waste site cleanups, groundwater contamination evaluations, environmental assessments, and laboratory analyses. Pres, ASCE, Fox River Valley Branch, 1979-80. *Society Aff:* ASCE, NSPE, WPCF, AWWA, NWWA, AMA, HMCRI, HWAC

Fedorochko, John A
Business: 3198 Chestnut St-5837A, Philadelphia, PA 19101
Position: Manager, Product Support Engrg. *Employer:* Genl Elec Co. *Education:* BIE/-/NYU. *Born:* October 14, 1929. BIE New York University, Bronx N Y. Native of Yonkers, N Y. Served in US Army 1948-49. Began in 1951 as a Components Engr for Radio Receptor Co Inc 1951-53. Associate Section Head Standards Engrg for GPL Div, General Precision Inc 1953-61. With G E Co Re-Entry & Environ Systs Div 1961- . Assumed current position as Manager, Product Support Engineering responsible for Reliability, Configuration Mgmt, Design/Drafting, Standards, Specifications, and Interactive Graphics. International President-Standards Engineers Society; Vice- President (Technical)-Standards Engineers Society; Corporate Representative- National Aeropace Standards Committee; Elected Fellow-Standards Engineers Society (Sept. '73). Nationally recognized in the Standards Field. *Society Aff:* SES, NASC.

Fee, John R
Home: 2152 Canyon Road, Arcadia, CA 91006
Position: Chmn of the Bd Emeritus *Employer:* J M Montgomery, Cons Engrs Inc. *Education:* BS/Engrg (Civ)/CA Inst of Tech; BS/Bus (Accounting)/Univ of KS. *Born:* Dec 1922. Native of Kansas, moved to CA in 1947. Served as Major in Army Coast Artillery Corps during WW II 1942-46. With James M Montgomery, Consltg Engrs Inc since 1951. Assumed current respon of Chmn of the Bd in 1983. Is also respon for the planning & design of major water facilities, including water treatment plants reservoirs, pumping stations, and mains. Is a registered PE in CA, FL, HI, ID, NV, and VA. Treas Emeritis, Caltech Alumni Assoc since 1959, Pres 1978. Pres IAE Coll of Fellows 1977. Pres IAE 1978. Commissioner Planning Commission. Arcadia, CA 1979-87. Dir Consltg Engrs Assoc of CA 1979-85, Secy - Treas 1981-82, VP 1982-83, Pres 1983-84. Enjoy photography. Rotary International 1975-present; Pres, Arcadia Rotary Club, 1984-85; Gov Rotary Dist 530, 1986-87. Methodist Hospital of Southern CA, Foundation Bd of Dir 1986-89, Pres. 1987. BSA, San Gabriel Valley Council, Exec Bd 1984-present. *Society Aff:* ASCE, AWWA, CWPCA, AAAS, AAEE, IAE.

Fee, Walter F
Home: 43 Tee Ln, Wethersfield, CT 06109
Position: Exec Vp Engg & Operations. *Employer:* Northeast Utilities. *Education:* BS/EE/Purdue Univ. *Born:* 11/6/21. Native of Providence, RI. Served in USN from 1942 to 1945. Held various positions in the utility engg field between 1946 & the present. With Northeast Utilities & its predecessor cos since 1954. Assumed current responsibilities as Exec VP-Engg & Operations on May 1, 1978. Responsible for all Engg & Operations, including nuclear, for Northeast Utilities. Held various positions with Power Engg Soc of IEEE, becoming Pres in 1969. Former Dir of IEEE. Became a Fellow of IEEE in Jan 1977. *Society Aff:* IEEE, ANS.

Feeley, Frank G, Jr
Home: Barnstown Rd, Camden, ME 04843
Position: Consulting Mech Engg. *Employer:* Retired. *Born:* 5/12/12. Education: Deerfield Academy and Massachusetts Institute of Technology. Marine Engineer 1932-36 in American Flag Merchant Vessels. Service Engineer, Babcock and Wilcox Co 1936-42. Served in US Navy during WWII as Assistant Machinery Superintendent, New York Navy Yard. 1946-55 Utilities Engr for Union Carbide Corp. 1956 to 1977 principal mech engr Olin Corp. Since 1977 in private consulting practice specializing in industrial energy systems design. Life Fellow American Soc of Mechanical Engrs, Member Industry Cttee of American Power Conference *Society Aff:* ASME, CES, APC.

Feely, Frank J, Jr
Business: P O Box 101, Florham Park, NJ 07932
Position: Execuitve VP (July 1, 1979). *Employer:* Exxon Research & Engrg Co. *Education:* BS/Mech Eng/Univ of MI. *Born:* 8/26/18. Resident of Mountainside, NJ. Employed by Exxon Research & Engrg Co since graduation in a variety of positions. Early in career helped develop fluid catalytic cracking process. Later directed research on brittle fracture in steel which led to new specifications for storage tanks & pressure vessels. Assumed responsibility as VP for Engg in 1966, but spent three yrs, 1971-74, as manager, logistics Operations for Exxon Corp during intervening period. Named Executive VP July 1, 1979. Active in industrial standardization for much of career, as Chrmn of API Central committee on Engrg & Amer Natl Standards Inst Bd, currently Pres of ANSI. Mbr, Natl Acad of Engineer; Mbr ASME - Fellow. Enjoys mountain climbing, boating & tennis for relaxation. *Society Aff:* ASME, ANSI, NAE.

Feerst, Irwin
Home: 368 Euclid Ave, Massapequa Park, NY 11762
Position: Consultant *Employer:* Feerst Electronics Corp *Education:* MS/EE/Polytechnic Inst of NY; M/EE/NY Univ; B/EE/City Coll of NY *Born:* 11/18/27 Electronics engr (circuit design) - 1949-1962; Asst Prof, Physics and Electronics, Adelphi Univ 1962-69; Independent conslt (circuit design, ECM, radar systems, signal processing, electronic typesetting, medical electronics) 1969-present. I consider my most valuable contribution to the profession is made in my role as gadfly. I have consistently proclaimed that IEEE (and other engrg societies) primarily serve academics and corporate executives, and to a far lesser degree the American working engr. I have run, unsuccessfully, for IEEE pres and I publish a monthly newsletter espousing these views. *Society Aff:* IEEE, CCEE

Feeser, Larry J
Business: Vice Provost for Computing & Inform. Technology, Pittsburgh Building, Rensselaer Polytechnic Institute, Troy, NY 12180
Position: Vice Provost for Computing & Information Tech *Employer:* Rensselaer Polytechnic Institute. *Education:* PhD/Civil Engg/Carnegie-Mellon; MS/Civil Engg/Univ of CO-Boulder; BS/Civil Engg/Lehigh Univ. *Born:* February 23, 1937, in Hanover Pennsylvania. Instructor, Asst Prof, Assoc Prof, Prof Department of Civil Engineering University of Colorado 1958-74. Prof and Chairman of Civil Engineering

Feeser, Larry J (Continued)
at Rensselaer Polytechnic Institute, 1974- 1982. Assoc Dean of Engrg, 1982-85. Vice Provost for Computing & Information Technology, 1985-present. PE in Colorado and New York. *Society Aff:* ASCE, ACI, ASEE, NSPE.

Fegley, Kenneth A
Home: 115 Park Avenue, Paoli, PA 19301
Position: Prof of Systems Engrg; Chair, Department of Systems *Employer:* Univ of PA. *Education:* PhD/Elec Engg/Univ of PA; MSEE/Elec Engg/Univ of PA; BSEE/Elec Engg/Univ of PA. *Born:* 2/14/23. Son of Henry Stanley and Bertha (Malone) Fegley; married Virginia Ruth Weaver Sept 1, 1951; children: Alan Donald, John David, Paul Andrew. Consultant to industry and gov; IEEE Publications Board 1983-1984. Chairman, Phila. Sect. IEEE, 1981. Served with USNR 1946-48. Fellow IEEE; Fellow AAAS, Amer Assn of Univ Professors, ASEE. Presbyterian. Registered PE. *Society Aff:* IEEE, AAAS, AAUP, ASEE, HKN, TBΠ, ΣT, ΣΞ

Fehribach, William J
Business: 5160 E 65th St, Indianapolis, IN 46220
Position: VP. *Employer:* A & F Engg Co, Inc. *Education:* BS/Civ Engr/Purdue Univ. *Born:* 8/4/34. Native of Jasper, IN. Served with the USAF 1953-57. Following grad from Purdue Univ served as Asst Traffic Engr for the City of Indianapolis 1961-66. Formed Consulting Engg Partnership as A & F Engg Co 1966, then became VP when firm was changed to a corp 1970. Am responsible for all Traffic & Transportation Engg functions of the Firm. Past Pres, IN Sec ITE 1969. Past Intl Dir, Dist 3, ITE. Reg P E in the state of IN. Enjoy sports & fishing. *Society Aff:* ITE.

Fei, Fames R
Business: 1 Poston Rd., Suite 300, P.O. Box 300001, Charleston, SC 29417
Position: CEO & Chmn *Employer:* Life Cycle Engrg, Inc *Education:* MS/Ocean Engrg/Univ of HI; BS/ME/Univ of So CA *Born:* 5/24/47 James R. Fei was born in Tucson, AZ. He received a B.S. degree in Mech Engrg from USC. He earned his M.S. in Ocean Engrg from the Univ of HI. From 1969-1973 was Proj Mgr for the Pearl Harbor Shipyard Pre-Overhaul Test and Inspection Program. From 1973-1977 was Proj Engr/Mech Engr for Naval Sea Systems Command. From 1977-Present Chmn & CEO, Life Cycle Engrg, Inc., Charleston, S.C ., with offices also in Arlington, VA, San Diego, CA, Norfolk, VA, and Portsmouth, N.H. Responsible for technical overview and direction of projects to ensure program objectives and contractual requirements are met. Provides technical evaluation and input in areas of mech engrg and fluid power and control systems. Currently a mbr of the ASME and NSPE. *Society Aff:* ASME, NSPE

Feinstein, Joseph
Home: 2398 Branner Dr, Menlo Park, CA 94025
Position: Conslt in Elec *Employer:* Self *Education:* PhD/Physics/NY Univ; MA/Physics/Columbia Univ; BEE/Elec Engrg/Cooper Union Inst of Tech. *Born:* July 1925. National Bureau of Standards (Radio Propagation Lab) 1949-54, Bell Telephone Labs (Electron Tube Dev, Murray Hill) 1954-59, Varian Associates (S F D Labs, Union N J 1959-64; Palo Alto Calif - Vice President, Corporate Research, 1964-79). Dept. of Defense (Director of Electronics, 1980-83). Consltg Prof, Dept of Elec Engrg Stanford Univ, 1983-). Consultant Member Advisory Group on Electron Devices, DOD. Panel Chairman, Committee on Energy and the Environment, National Research Council. Fellow IEEE; Member, National Academy of Engineering. *Society Aff:* IEEE, NAE.

Feintuch, Howard M
Home: 46 Nicholas Ave, West Orange, NJ 07052
Position: Mgr, Proc Design Ser *Employer:* Foster Wheeler USA Corp *Education:* PhD/Chem Engrg/NYU 1973; ME/Chem Engrg/NYU 1967; BS/Chem Engrg/Poly Inst of Brooklyn 1964 *Born:* 04/06/43 Joined Foster Wheeler in June 1964 as a Process Engr. Assumed current position as Mgr, Proc Design Ser in 1981. He is respon for the activities of energy conservation, environmental control, coordination of lab operations and coordination of computer applications and tech data. Previously he held the positions of energy conservation mgr and asst petroleum proc mgr. In addition he is adjunct assoc prof of chem engrg at Manhattan Coll and has also taught at Stevens Inst of Tech. He is an author of various tech articles including two chapters in McGraw-Hill's Handbook of Petroleum Refining Processes. *Society Aff:* AIChE

Feisel, Lyle D
Business: Dept of Elec Engrg, Rapid City, SD 57701
Position: Prof & Dept Hd, Elec Engrg. *Employer:* S D School of Mines & Technology. *Education:* PhD/EE/IA State; MS/EE/IA State Univ; BS/EE/IA State Univ. *Born:* 10/16/35. Native of Tama IA. US Navy 1954-58. With SD of Mines & Technology since 1964, Dept Head 1976. Natl Visit Prof, Cheng-Kung Univ, Tainan Taiwan 1969-70. Chrmn of the Faculty 1973-74. Outstandin Teacher Award 1972; Danforth Assoc 1972; ASEE Zone IV Campus Coordinator of the Yr 1973; Western Elec Fund Award 1974; ASEE Ctte on New Engg Aduc Affairs, Chmn 1974-77; ASEE Campus Activity Coordinator Affairs Ctte 1974-77, V Chmn 1977, Chmn 1978- . IEEE Sr Mbr; IEEE Educ Soc AdCom, Pres 1978-79. IEEE History Ctte; IEEE Mining Industry Ctte. S O Renewable Energy Assoc, Chrmn 1979- . Elec Engg dept Hds Assn, Sec-Trans 1979- 80. *Society Aff:* IEEE, AVS, ISHM, AUSA.

Fejer, Andrew A
Home: 122 LeMoyne Parkway, Oak Park, IL 60302
Position: Sr Advisor, Gas Turbine Res. *Employer:* Inst of Gas Technology. *Education:* PhD/Aeronautics & Physics/Caltech; MS/Aeronautics/Caltech; Dip Engr/Mech Engg/Tech Univ of Prague. *Born:* June 4, 1913. Instructor Caltech 1940-45, Research Engr Packard Motor Car Co Toledo Ohio 1945-48, in charge of Advanced Aerothermodynamic Design of Gas Turbine Engines. Professor & Chairman Aeronautics Dept University of Toledo 1948-58, Prof of Mechanical & Aerospace Engineering Illinois Institute of Technology 1958-78 & Chairman of Dept 1958-71. Research activities involve aerodynamics of jets, flows in turbo-machines, aerodynamics of wind energy converters, microclimate of urban areas. Energy Conservation sys in Industry. Fellow ASME (1973), Associate Fellow A1AA, Associate Editor, Journal of Applied Mechanics (1972-76), mbr Closed Cycle and Process Industries Committee, ASME Gas Turbine Div. Awarded order of Academic Palm leaved by government of France June 4, 1981. Visiting Professor Ecole Nationale Superieure de Cachan 1987. *Society Aff:* ASMe, AIAA, ASEE, AAAS.

Felbeck, David K
Business: 2250 G.G. Brown Laboratory, Ann Arbor, MI 48109-2125
Position: Professor of Mechanical Engineering. *Employer:* University of Michigan. *Education:* ScD/Mech Eng/MIT; MS/Mech Eng/MIT; BME/Mech Eng/Cornell Univ. *Born:* April 2, 1926. Registered professional engineer in Michigan and District of Columbia. US Fulbright Lecturer, Technical Univ of Delft, Holland, 1952-53. Assistant Professor MIT 1953-55. National Academy of Sciences 1955-61. University of Michigan 1961-present. Adams Memorial Membership, American Welding Society, 1963; Wilson Award, American Society for Metals, 1973. National Chairman, Materials Division of ASME, 1974-75. Specialist in analysis of failures involving deformation and fracture of metals. Consultant for industry and as expert in product liability litigation. *Society Aff:* ASME, ASM, AIME, SAE, ASTM, NSPE.

Feldman, David
Business: 555 Union Blvd, Allentown, PA 18103
Position: Dir, Film & Hybrid Technology Lab. *Employer:* AT&T Bell Laboratories *Education:* MS/EE/NJ Inst of Tech; BS/EE/NJ Inst of Tech. *Born:* October 16, 1927 New York City. BSEE 1947, MSEE 1949 Newark College of Engrg; DEE studies 1951-56 Polytechnic Institute of Brooklyn N Y. Asst Prof Electrical Engrg 1949-54 Cooper Union School of Engrg. 1954-56 Senior Research Engr Polytechnic Research and Development Co. With Bell Laboratories since 1956. Presently Technical Director, Hybrid Integrated Circuits, Components and Thin Film Technology Laboratory. IEEE activities include Magnetics Committee 1953-62; General Chairman

Feldman, David (Continued)
1959 Magnetics Conference. Electronics Components Conference 1964-74. General Chairman 1972. President IEEE - Parts, Hybrids & Packaging Group 1972-74. Consultant Aerospace Industries Assn, Electronic Industries Assn; Prof Engr New York State; IEEE-ECC Best Technical Paper Award 1968; elected IEEE Fellow 1972. IEEE Contributions Award for PHP Soc Activities 1977; IEEE Centennial Medal 1984. *Society Aff:* IEEE.

Feldman, Edwin B
Home: 1023 Burton Dr, Atlanta, GA 30329
Position: Pres. *Employer:* Service Engg Assocs. *Education:* BS/IE/GA Inst of Tech *Born:* Founder and Pres, Service Engrg Assocs, Inc, a firm of PE specializing in physical facilities consltg and training: Custodial, Maintenance Mgt. Author of "Industrial Housekeeping" (MacMillan), "How To Use Your Time To Get Things Done" (Fell), "Housekeeping Handbook" (Fell), "Building Design For Maintainability" (McGraw-Hill), Supervisors Guide to Custodial and Bldg Maintenance Operations (Harris Communications) and "Energy Saving Handbook" (Fell). Co-author (with George B Wright) The Supervisors Handbook (Fell); Editor Programmed Cleaning Guide (Soap & Detergent Assn). Contributor "Handbook of Coll and University Administration" (McGraw-Hill) and "Facilities Management and Plant Engineering Handbook" (McGraw-Hill), author of over 200 articles in periodicals. *Society Aff:* AIPE, IIE, NSPE.

Feldman, Harry R
Business: 2586 Forest Hill Blvd, West Palm Beach, FL 33480
Position: President. *Employer:* Harry R. Feldman Inc & VP *Education:* BEE/Elec Engrg/Northeastern Univ. *Born:* 7/27/10. Civil Engrg & Land Surveying. Started prof career with the Engrg Dept in Framingham; served a short term with the US Army Engrs on Cape Cod Canal; worked as const engr with the Met Dist Water Supply Comm of Mass; joined a large cons firm in Boston 1940-43; served as Lt - Civil Engrg Corp in US Navy during WW II; entered private practice in 1946 spec in engrg & geodetic surveying, mapping & photogrammetry. Pres Mass Soc Prof Engrs 1962-63; Dir Natl Soc 1966-74. VP Natl Soc Prof Engrs 1974-76; For (FIG) Intl Federation of Surveyors, he served as Deputy Auditor - Auditor and at the last Congress meeting was elected Chrmn of Commission 6. Hobbies: photography & electronics. Elected to Hon Mbrship ACSM March 1984. Retired-. *Society Aff:* NSPE, FES, NSPE, ASCE, ACSM, ASP.

Feldman, Melvin J
Business: P O Box X, Oak Ridge, TN 37830
Position: Manager, Engineering Systems. *Employer:* Oak Ridge National Laboratory. *Education:* MS/Metallurgical Engg/Univ of TN; BS/Metallurgial Engg/Purdue. *Born:* 1/6/26. Native of South Bend IN. Employed Oak Ridge Natl Lab 1950-56 specializing in radiation damage to metals & fuels. 1956-60 radiation damage studies Westinghouse Elec Corp. Joined Argonne Natl Lab 1960, involved in design, construction & operation of fuel cycle facility through 1964. Manager EBR-II Fuel Cycle Facility & Hot Fuels Examinaltion Facility (ID Falls) 1965-72. ASsoc Dir ANL-WEst 1973-75. Present employment (Manager, Engg Sys, Consolidated Fuel Reprocessing Program, ORNL). Active in Amer Nuclear Soc: Office held include VP & Pres (1975-76), Chrmn, natl Program Committee; Chrmn Nominating Committee & Mbr Bd of Dirs & Exec Committee. Fellow ANS; Chrmn; Public Policy Committee of ANS; Mbr, Sigma Xi - Research Soc of America. Present employment-Senior Advisor to the Director, Consolidated Fuel Reprocessing Program, ORNL. *Society Aff:* ANS, RESA, ΣΞ

Feldman, Rubin
Home: 46 Rio Vista Dr, St Louis, MO 63124
Position: Pres. *Employer:* Thermal Sci Inc *Education:* MS/Engg/Johns Hopkins Univ; BS/Chem Engg/WA Univ. *Born:* 7/25/25. Pres of TSI Inc-a St Louis Co specializing in the dev and manufacturning of subliming fire protective coatings and ablative coatings for protection from aerodyhamic heating. With the Co as the Chief Exec and Operating Officer since 1967. Holds basic patents on fire protective and high temperature resistive coatings in the US, Canada, United Kingdom, Japan and Sweden. Also patents on rocket engine concepts, nucleonic sensors and packaging equipment in the US. From 1956-67 Manager of the Thermal Systems Dept-Emerson Electric. Served with the US Army 1953-54. Mbr of the Advisory Council-College of Engg- WA Univ. *Society Aff:* NSPE, MSPE, AIChE, AIAA, NFPA.

Feledy, Charles F
Home: C/O John Feledy, 9227 Solon Dr, Cincinnati, OH 45242
Position: VP, Offshore Sourcing. *Employer:* Harris Corp. *Education:* MBA/Bus Admin/Univ of MI; BS/ME/Cornell Univ; BBA/Bus Admin/Univ of NB- Omaha *Born:* 7/12/35 Won national awards from Assoc of Energy Engrs and also National Energy Resource Organization for having best coporate energy management program in U.S. Selected as "Energy Man of the Year" (1981) by Modern Industrial Energy magazine. Previously Dir, Corporate Energy Programs for United Technologies Corp and also its Dir, Operations Analysis. Formerly was VP and General Mgr, Tuner Div., General Instrument Corp and previously its Corproate Dir of Material Control. Elected to Phi Kappa Phi and Betta Gamma Sigma, national scholastic honorary societies. Currently, VP, Offshore Sourcing for Harris Corp. Established Corporate Asian Purc Off in Taipei, Taiwan. Respon for buying electronic prod and components in all of Asia and developing new vendors. *Society Aff:* SME, AEE

Fellers, Rufus G
Business: College of Engrg, Columbia, SC 29208
Position: Prof of Elec Engrg *Employer:* Univ of South Carolina. *Education:* PhD/EE/Yale Univ; BSEE/EE/Univ of SC. *Born:* Sept 1920. US Naval Research Laboratory: Electronic Scientist, Section Head 1944-55. US Navy 1945. Professor of Electrical Engrg University of S C 1955- ; Dean of Engineering 1960-69. Bell Telephone Laboratories 1970-71. Fellow of IEEE, IEEE Director 1966-67, IEEE Region III Outstanding Engineer 1972. Delegate to General Assemblies of International Union of Radio Science 1954 (Hague), 1957 (Boulder), 1960 (London), 1963 (Tokyo), 1966 (Munich), 1969 (Ottawa), 1972 (Warsaw), 1975 (Lima) 1981 (Washington), 1984 (Florence).Chairman S C Sect IEEE 1963. Registered Engr S C. *Society Aff:* IEEE, ASEE.

Fellinger, Robert C
Business: Dept of Mech Engg, IA State Univ, Ames, IA 50011
Position: Prof & Div Leader, Thermodynamics & Energy Utilization *Employer:* Iowa State Univ of Sci and Tech. *Education:* MS/ME/IA State. *Born:* August 10, 1922. Graduate ASTP program in Mechanical Engrg Univ of Minnesota 1944. ISU 47-48. US Army Corps of Engrs 1943-46. Manhattan Project 1944-46 at U of Chicago & MIT. Registered PE, Iowa 2521. Member Tau Beta Pi, Pi Tau Sigma, Sigma Xi, Phi Kappa Phi. Life Fellow ASME. Faculty Iowa State Univ since 1947. ALCOA Professor of Engineering Iowa State Univ 1966-68. Iowa State Alumni Assn Faculty Citation 1968. AMOCO Outstanding Teacher Award (College Engrg) 1974. Elected (By Students) M.E. Prof of the Year, 1976. Consultant fuels, fires, explosions. Prof & Acting Chrmn, Dept Mech Engg 1979-80. *Society Aff:* ASME

Fellows, John A
Home: 3201 San Gabriel Ave, Glendale, CA 91208
Position: Metallurgical Consultant (Self-employed) *Education:* ScD/Phys Metallurgy/MIT; MS/Physics/MIT; AB/Physics/Williams College. *Born:* July 27, 1906 Greenfield Mass. m. 1931; c. 3; m. 1958. AB Williams Coll 1928; Indus Res Fellow MIT 1930-37; graduated Cum Laude from Williams College - mbr Phi Beta Kappa. Indus exper in foundry dev work with Amer Brake Shoe Co (now ABEX) 1937-53 except for time with Manhattan Project 1943-46 & Union Carbide 1946-47. Dev work with uranium metal Mallinckrodt Chem Wks 1953-66. ASM Staff 1966-70 in tech programming. Cons ed Metals Handbook 1971-75. Genl Met Consulting 1976- . President American Soc for Metals 1964-65. Joint recipient Henry Marion Howe Medal ASM 1944. Fellow Inst of Metallurgists ASM. Hon Mbr Iron & Steel Inst of

Fellows, John A (Continued)
Japan. Past Chairman Inst of Metals, Dir of the Met Soc of AIME. *Society Aff:* AIME, ASM, ASTM, ANS.

Felsen, Leopold B
Business: 333 Jay St, Brooklyn, NY 11201
Position: Inst Prof. *Employer:* Polytechnic Institute of New York. *Education:* DEE/Electophysics/Polytech Inst of Brooklyn; MEE/Electrophysics/Polytech Inst of Brooklyn; BEE/Electrophysics/Polytech Inst of Brooklyn. *Born:* May 1924. Professor of Electrophysics since 1961; Dean of Engineering since 1974-78. Inst Prof since 1978. Professional area & teaching: electromagnetics, optics. Author or coauthor of more than 130 technical journal articles, and of two books. Professional societies: Institute of Electrical and Electronics Engineers (IEEE) (Fellow); International Union of Radio Science (former Chairman of US Commission 6, current Chairman of International Commission B); Opt Soc of America (Fellow) Acoustical Soc of America; Amer Soc for Engrg Education. Honors and awards: 3 best paper prizes from IEEE; Invited Guest of Soviet Academy of Sciences 1967 and 1971; Citation for Distinguished Research (Society of Sigma Xi) 1973; John Simon Guggenheim Memorial Fellow 1973- 74; Van der Pol Gold Medal URSI 1975. Election to Natl Acad of Engg 1977. Humboldt Senior Scientist Award 1979 Honoray doctorate from Technical University of Demark, Lyngby, Denmark, 1979. Recreation: music, photography.

Felske, Armin
Business: Volkswagen AG, FE-Forschung Messverfahren, D-3180 Wolfsburg, W Germany
Position: Physical Scientist. *Employer:* Volkswagenwerk AG. *Education:* Dr rer nat/ PhD/Univ of Munster; Diplom Phys/Experimental Phys/Univ of Kiel. *Born:* 4/9/36. Native of Konigsberg. Taught in public schools in Kiel after WWII. Studied in Germany Univs of Kiel & Munster. Since 1963 physical scientist in German Inst of Spectrochemistry in Dortmund. Specializing in laser techniques. With VW Corp since 1970, specializing in optics, holography, vibration, nondestructive testing, physics. Arch T Colwell Merit Award of SAE 1979. Enjoy music, traveling. *Society Aff:* DGAO.

Feltner, Charles E
Business: 24500 Glendale, Detroit, MI 48239
Position: Mgr; Mftg Sys & Oper Engg Dept. *Employer:* Ford Motor Company. *Education:* PhD/Theoretical & Applied Mechs/Univ of IL; MS/Theoretical & Applied Mechs/Univ of IL; BS/Mech Engg/Univ of NC. *Born:* June 1936. PhD, MS Univ IL; BS Univ of NC. Native of Raleigh, NC. With Ford Motor Co since 1963. 1963-70 carried out research programs, plastic deformation & fatigue behavior, materials in Met Dept; 1970-73, Mgr respon, materials research & development. 1973-75, Mgr, Prod Qual Off, respon for investigation & coord, Ford's position in prod liability investigations. 1975-81 Mgr in the Mfg Devel Off respon for mfg feasibility of new casting processes & high energy laser processing. 1981-83, Mgrf, Mfg Sys & Machining Dept, Manu Proc Lab. Since 1983, Mgr, Mftg Systems and Oper Engg Dept, Mftg Dev Center, focusing on such automotive mftg problems as schedule stability, option complexity JIT, vehicle sequencing, assembly line balancing facility design and plant floor information systems using computer modeling, analysis and simulation techniques. Current assignment is to help implement advanced Computer Integrated Manufacturing (CIM) concepts in Ford's powertrain manufacturing and vehicle assembly plants. Enjoy Computers, scuba diving, boating, skiing, tennis, music, reading. *Society Aff:* IIE, CASA/SME.

Felton, Dean R
Business: 320 West Port Plaza, St Louis, MO 63141
Position: Pres *Employer:* D.R. Felton & Associates, Inc. *Education:* B/CE/Univ of IL *Born:* 2/28/26 Native of Kewanee, IL. Served with US Army, 1944-1946. Graduate of the Univ of IL, 1951. Assistant Hgwy Engr, CA Division of Hgwys, 1951-1955. Went from Project Mgr in 1955 to Executive VP in 1976 for the firm of Warren and Van Praag, Inc. Past-pres Consulting Engrs Council of MO; National Dir of ACEC. Opened own firm in 1976. Registered civil engr in seven states: CA, IA, NB, IL, MO, FL, MI. Holder of Certificate of Qualification from National Council of Engrg Examiners. Registered Land Surveyor in State of MO. *Society Aff:* ACEC, ASCE, AWWA, NSPE

Felton, Kenneth E
Business: Dept of Agri Engg, College Park, MD 20742
Position: Prof *Employer:* Univ of MD. *Education:* MS/Agri Engr/PA State Univ; BS/Civ Engr/Univ of MD; BS/Agri/Univ of MD. *Born:* 8/18/20. Native of Parsons WV, Deputy Tax Assesor of Tucker County 1938-42. Armed services 1942-45. Fire Protection Engr for Factory Mtl Engg Div 1951-52. Sanitary Engr with Interstate Commission on the Potomac River Basin 1952-54. Irrigation Design Engr with S States Cooperative 1954. Faculty Mbr Dept of Agri Engg, Univ of MD 1954 to present. Responsibilities in res have included: irrigation, pneumatic handling of forages, lateral stability of farm structures, evaporative cooling of broiler houses, broiler response to varying temperatures, utilization of solar energy in broiler production. Have served as a consultant to the Republic of Korea, the Govt of Somalia & the Fed Ext Service (USDA). Ext Agricultural Engr in the area of farmstead engg Univ of MD. *Society Aff:* ASAE, ISES, NFBA.

Felton, Lewis P
Business: 4531 BH, School of Engg, Los Angeles, CA 90024
Position: Assoc Prof. *Employer:* Univ of CA, LA. *Education:* PhD/Civil Engg/ Carnegie Inst of Tech; MS/Civil Engg/Carnegie Inst of Tech; BCE/Civil Engg/ Cooper Union. *Born:* 12/14/38. *Society Aff:* AIAA, ASCE, ΣΞ.

Femenia, Jose
Home: Fort Schuyler, Bronx, NY 10465
Position: Chairman, Engineering Dept. *Employer:* SUNY Maritime College. *Education:* MS/Mech Engr/CCNY; BE/Marine Engr/SUNY Maritime. *Born:* May 24, 1942. Federal license: Third Assistant Engr, Steam and Motor Unlimited Horsepower. A native of New York City. Commenced teaching at Maritime in 1964. Advanced from Assistant Instructor to Prof and Chairman of the Engrg Department. Responsible for the administration of the Electrical Engrg, Marine Engrg & Naval Architecture, at the college. Visiting Prof, World Maritime Univ, Malmo Sweden, 1983-present. Adjunct Assistant Professor and Research Associate at Webb Institute of Naval Architecture 1970-74. Member of SNAME, ASNE, ASEE and the Society of Marine Port Engineers N Y, N Y. VP of SNAME and CH of SNAME Education Cttee. Enjoys sailing. *Society Aff:* SNAME, ASNE, ASEE, SMPE.

Fenech, Henri J
Home: 500 Via Hierba, Santa Barbara, CA 93110
Position: Prof. *Employer:* Univ of CA. *Education:* ScD/Nucl Engg/MIT; MS/Nucl Engg/MIT; AEM/Engg Sci/Arts-et-Mefiers. *Born:* 3/14/25. Faculty mbr, Dept of nucl engg at MIT (1959-69). VChrmn for nucl engg at Univ of CA Santa Barbara (1969-74). Presently prof of nucl engg at UCSB. Consultant to utilities & nucl industry on nucl power systems safety. Thermal and hydraulics, nucl safety, nucl fuel m. Author of 70 publications & co-author of 3 books on above subjs. Fullbright Fellow (1955), NATO Senior Fellowsip, (1968-1970). Fellow ANS (1982). Exec bd mbr of Power Div, ANS. Member Education & Research Council, Associate Western Universities (1970-present). *Society Aff:* ASME, ANS.

Feng, Tse-yun
Business: Dept. of Electrical Engineering, The Pennsylvania State University, University Park, PA 16802
Position: Binder Prof of Computer Engrg *Employer:* The PA State Univ *Education:* PhD/Computer Engg/Univ of MI; MS/Electrical Engg/OK State Univ; BS/Electrical Engg/National Taiwan Univ *Born:* 2/6/28. in Hangchow, China. Now, a US citizen. Served with Taiwan Power Co, China; & Ebasco Services, Inc, NY. Taught at Univ of MI, MI; Syracuse Univ, NY; Wayne State Univ, MI; Wright State Univ, OH & The Ohio State Univ, OH. Presently, Binder Prof of Computer Engrg and Director of Computer Engineering Program, The PA State Univ, PA. Served as a consultant to Transidyne Gen, Syracuse Univ, Pattern Analysis & Recognition, Inc, New York State Dept. of Education. Current profl activities include Past Pres, IEEE Computer Soc Chrmn, Distinguished Visitors Program; Program Evaluator, Computer Science Accreditation Commission; Dir, Northeast Consortium for Engg Ed; Gen chrmn, Intl Conf on Parallel Processing. Numerous technical publications. Patentee in field. *Society Aff:* IEEE, ACM.

Fenn, Raymond W, Jr
Home: 13428 Carillo Lane, Los Altos Hills, CA 94022
Position: Manager, Materials & Process Control Ctr. *Employer:* Lockheed Missiles & Space Co Inc. *Education:* D Eng/Metallurical Engg/Yale Univ; M Eng/MEtallurgical Engg/Yale Univ; B MetE/Metallurgical Engg/Rensselaer Polytechnic Inst. *Born:* Feb 1922. Native of Torrington Conn. Employment: Met Engr General Electric Co West Lynn Mass 1943-44; USNR 1944-46; Supr Test & Instrumentation Lab, Met Lab Dow Chem Co Midland Mich 1949-61; Consulting Scientist Mat Sci Lab 1961-69; Mgr Mat & Prod Systems Engrg - Space Sys Div 1969-74, Mgr Mfg Research 1974-79, Mgr, Mat'l & Process Labs 1981-87, Mgr Mat'l & Proc Control Ctr 1987- . Lockheed Missiles & Space Co Sunnyvale Calif. Reg P E Calif. Technical Soc activities: Amer Soc for Testing & Materials (Natl Dir 1966-69), R E Templin Award 1961; Amer Soc for Metals (Natl Tr 1969-71), elected to Fellow 1976; Soc for Advancement of Material & Process Engrg (Natl Dir 1974-76), Pres 1978-79, 1st VP 1977-78 2nd VP 1976-77; Amer Nat'l Mat'ls Advisory Bd, Materials Tech Ctte 1974-76; Natl Mgmt Assn; Sigma Xi; Contrib articles to professional journals. *Society Aff:* ASM, SAMPE, NMA.

Fenn, Rutherford H
Business: Walter H. Wheeler Jr. Dr, Stamford, CT 06926
Position: Dir, Corp Standards *Employer:* Pitney Bowes Inc *Education:* BME/Polytechnic Inst of NY *Born:* 7/15/21 Native of Brooklyn, NY. Aviation Cadet - Yale Univ, commissioned 2nd Lt Aircraft Engrg Officer, USAAF. Active duty 1943-46. Presently Lt Colonel USAF (Ret). Joined Pitney Bowes, Stamford, CT 1947 as engr in Production Engrg. Current appointment as Dir, Corporate Standards with responsibility of co's worldwide standardization programs. Has given numerous talks and written papers on industrial standardization and co metrication. Presently Eastern Regional Dir and Fellow, SES. Enjoys travel and antique collecting. *Society Aff:* SES, ASME, SME.

Fenner, William G
Home: 30 Keyes St, Ashburton, Victoria, Australia 3147
Position: Manager, Manpower Studies. *Employer:* Victorian Railways. *Education:* BE/Mech/Univ of Adelaide; FSAIT/Mech/SA Inst of Tech. *Born:* 3/11/22. Born Adelaide, South Australia. Worked: BHP 2 yrs; Dept of Mines 3 yrs (geophysics); ICI Australia 12 yrs (proj engg); Comalco 15 yrs (proj management). Currently Manager, Manpower Studies, with Victorian Railways-This work is aimed at raising productivity throughout the system. Considerable international travel & negotiations. Prof int: Cost Estimating; Project Management: Productivity. Currently Immed. Past Pres of Victorian Sect of AACE. Hobbies: Travel, Macro-economics, human geography, natural history. *Society Aff:* AACE, MIEAust, FSAIT

Fenster, Saul K
Home: 524 Bernita Dr, River Vale, NJ 07675
Position: Pres. *Employer:* NJ Inst of Tech. *Education:* PhD/ME/Univ of MI; MS/ ME/Columbia Univ; BME/ME/CCNY. *Born:* 3/22/33. Native of NYC. Taught mech engg, City College of NY, 1953-1956. Teaching and Shell Oil Fellow, Univ of MI, 1957, 1958. With Sperry-Rand Corp, 1959-1962. With Fairleigh Dickinson Univ 1962-1978 as Chrmn of Physics, Chrmn of Mech Engg, Asst to Dean, Assoc Dean, Exec Asst to Pres, Provost of Rutherford Campus. Served as Mbr, 1972-1978 and Chrmn of the Bd, 1978, Assn of Independent Colleges and Univs in NJ. Chrmn, 1978-1979, Statewide Task Force on Programmatic Mission Differentiation. Appointed to present position, 1978. Co-author of two texts. Author, numerous articles. 1981, 1985, Chairman Citizens Task Force on Water Management Emergency (Appointed by Governor); 1979-1980. Member, Hudson River Waterfront Study, Planning and Development Commission (Appointed by Governor); 1980 - Member Board of Directors, New Jersey Association of Colleges and Universities; 1981 - New Jersey Committee and Bd Dir, Regional Plan Association; 1981- , Mbr, NJ Water Supply Auth (App by Gov); 1982-1984, Gov Comm on Sci and Tech (Appointed by Gov); (1985-) New Jersey Commission on Science and Technology (Appointed by Gov). *Society Aff:* AAAS, ASME, ASEE, ΣΞ, SME

Fenton, Robert E
Business: 2015 Neil Ave, Columbus, OH 43210
Position: Prof (elec eng) *Employer:* Dept Elec Engr, OH State Univ *Education:* PhD/EE/OH State Univ; MSc/EE/OH State Univ; B/EE/OH State Univ *Born:* 9/30/33 Native of Brooklyn, NY. Served in USAF 1957-60. Joined OH State Univ (OSU) in 1960 and am currently Prof of Electrical Engrg. Involved in research on various communication and control aspects of automated ground transport. Active in IEEE-VTS as Assoc Editor for Transportation Systems of IEEE Transactions on Vch Tech (1970-75), Treasurer 1981-83, Vice President 1983-85, President (1985-87), and in Transportation Research Bd (comm on hgwy communications) (1968-83). Gamma chapter of Eta Kappa Nu Teaching award 1963, OH Society of Prof Engrs Neil Armstrong award 1971, and IEEE-VTS Technical paper award 1980, Fellow IEEE 1986, and Fellow, Radio Club of America 1987. *Society Aff:* IEEE, RADIO CLUB OF AMERICA, ΣΞ

Fenves, Steven J
Business: Dept of Civil Engrg, Carnegie-Mellon Univ, Pittsburgh, PA 15213
Position: Sun Company Univ Prof of Civil Engrg *Employer:* Carnegie-Mellon University. *Education:* PhD/Civ Engrg/U of IL; MS/Civ Engrg/U of IL; BS/Civ Engrg/U of IL *Born:* June 1931. Native of Yugoslavia; immigrated 1950; naturalized 1953. Draftsman & designer for various firms 1950-57. On faculty of Dept of Civil Engrg from 1957-71, with visiting appointments at MIT, National University of Mexico, & Cornell. At Carnegie-Mellon University since 1972 (Dept Hd 1972-75; Univ Prof 1978-86; Sun Company Univ. Prof since 1986). Teaching, research and consulting in computer aided engineering, incl expert systems for structural design, engineering databases, standards processing, and design theory. Mbr NAE, ASCE (var offices held), and other professional and honorary organizations. ASCE Research Prize, 1966. *Society Aff:* ASCE, NAE.

Feorene, Orlando J
Home: 12 Old Farm Circle, Pittsford, NY 14534
Position: Dir, Mgmt Services Div. *Employer:* Kodak Park Div, Eastman Kodak Co. *Education:* BIE/OH State Univ. *Born:* 1/29/18. Exec Committee: Finger Lakes Health Sys Agency. Advisory Committees: OH State Univ, OK State & TX Tech Univ. Hospital Association of NY State. Chrmn Civic Dev Council, Rochester Chamber of Commerce. P Pres Rochester Chapter IIE. P Natl VP IIE. Mbr: Council on Industrial Engrg, Inst of Mgmt Science, Rochester Engrg Society, Industrial Engrg Society of Rochester. Honors: Tau Beta Pi, Alpha Pi Mu; Outstanding Alumnus OH State Univ. Captain, Corps of Engrs WWII. Visiting Lecturer Air Force Inst of Technology, Univ of Rochester, Dept of Defense Mgmt Training Ctr. Bd of Dirs, OH State Univ Research Fnd., Bd of Trustees, American Institute of Industrial Engrs. *Society Aff:* IIE, TIMS, RES, IES.

Ference, George
Home: 120 Poplar Drive, Wilmette, IL 60091
Position: Director Corp Plans & Business Dev. *Employer:* Internatl Minerals & Chemicals Corp. *Education:* BS/Chem Engr/Purdue. *Born:* January 1920. BSChE from Purdue Univ. Native of Wilmette, Ill. Held supervisory and managerial positions in manufacturing prior to current employment. With IMC since 1957. Re-

Ference, George (Continued)
sponsibilities have included corporate engrg, VP of operations for corporate subsidiary and business development. Assumed present responsibility in 1975. Member AICHE and ACS author of technical & economic articles for journals & reference books. *Society Aff:* ACS, AIChE.

Ferguson, Arnold D, Jr
Home: 159 Linda Court, Gretna, LA 70053
Position: Chief Engr Marine Division. *Employer:* Prager Inc. *Education:* BA/Bus/Rice Univ; BS/Naval Sci/Rice Univ. *Born:* 1923. BS in Naval Science and BA Rice University. Native of Houston, Texas. Served in US Navy 1941-46. Foreman, Superintendent, & General Manager Majors Tool Co Inc. Joined Prager Inc 1974. Presently Chief Engineer-Marine Division. Received Section Performance Award as Chairman New Orleans Section, American Welding Society, 1973. Currently District Director AWS. Enjoy golf & fishing. *Society Aff:* AWS, SPE, AIME.

Ferguson, Colin R
Business: Sch of Mech Engg, W Lafayette, IN 47907
Position: Assoc Prof. *Employer:* Purdue Univ. *Education:* PhD/ME/MIT; MS/ME/MIT; BS/ME/Univ of CA; AA/Engr/Sierra College. *Born:* 1/3/50. Colin is a native of N CA. His education began in that state's Jr College system and terminated at MIT where he received a PhD degree. He assumed his current position at Purdue in 1978 and has developed labs for studying combustion processes occuring in diesel engines. He has taught thermodynamics, heat transfer, combustion and internal combustion engines. In 1979 he was a recipient of the SAE's Ralph K Teetor Award. In 1986 his textbook entitled *Internal Combustion Engines: Applied* Thermosciences was published by John Wiley and Sons, New York. *Society Aff:* SAE, CI.

Ferguson, Don E
Business: P.O. Box X, Oak Ridge, TN 37830
Position: Sr Tech Adv *Employer:* Martin Marietta, Oak Ridge Nat L *Education:* MA/Math/Univ of TN; BS/Chem Engrg/TN Tech Univ. *Born:* October 1923. Native of Roswell, NM. Served as engrg officer in US Navy 1944-6, Joined ORNL in 1946. Engaged in Res and Dev, primarily in the field of radiochemical separations. Assumed present position in 1983. Responsible for the tech aspects of waste mgmt at ORNL. Also respon for the Consolidated Edison uranium solidification program. Fellow of the Amer Nuclear Soc and served on Bd of Dirs. Enjoy outdoor sports. *Society Aff:* ANS.

Ferguson, Donald M
Home: PO Box 1076, Marblehead, MA 01945
Position: Chief of Engrg (retired). *Employer:* Air Force Cambridge Res. Labs. *Education:* MS/Engg Mgmt/Northeastern Univ; BS/Mech Engg/Univ of MI. *Born:* New York City, Industrial engrg assignments at Yale & Towne, 1939-40, and as Lieut, Ordnance Dept, US Army (Springfield Armory, Aberdeen Proving Ground, & OFDAP) 1940-6, and at E.D. Jones & Sons 1946-7. Reseach engrg at United Shoe Mchy., 1947-51. Chief of Engineering , Air Force Cambridge Research Labs, 1951k-73. Retired June 1973. 'Fellow' American Society of Mechanical Engrs 1969- . Member American Geophysical Union. Patentee, research equipments. Registered Professional Engr Mass. 5749. Enjoying art (sculpture & drawing) and travelling. *Society Aff:* ASME, AGU.

Ferguson, George
Business: 2300 6th st.N.W, Washington, DC 20059
Position: Professor. *Employer:* Howard Univ. School of Engrg. *Education:* BS/Physics/Howard Univ; MS/Physics/Howard Univ; PhD/Physics/Catholic Univ. *Born:* 5/25/23. Awards: AEC Fellow (Univ PA), Thomas Edison Fellow (Naval Res Lab). Mbr: Bd of Dirs Amer Nuclear Soc (1973-76), Nuclear Div, Amer Soc for Engg Education. *Society Aff:* ANS, APJ, AAAS, AIPT, AAPT

Ferguson, George E
Home: 1737 Brookside Lane, Vienna, VA 22180
Position: Historical documentation (Volunteer Status) *Employer:* US Geological Survey, Water Resource. Retired. *Education:* BS/CE/Univ. of Minn. *Born:* April 1906. Native of Stillwater, Minn. Hydraulic Engr with Water Resources Div., US Geological Survey from 1928-retirement, 1972, with service in Ohio 1928, Texas 1929-31, Hawaii 1931-7, Maryland-Delaware 1938-40, Florida 1940-7, Nat'l Hqs. Staff 1947-55 and as Regional Hydrologist for Atlantic Coastal States 1955-72. Life member AGU and ASCE. AWWA Director from Florida 1946-8; Honorary Member 1974. US Dept of Interior Distinguished Service Award 1970. *Society Aff:* AWWA, AGU, ASCE.

Fernandes, John H
Home: 294 Riverside Dr, Tiverton, RI 02878
Position: VP-Energy Div *Employer:* Maguire Group, Inc. *Education:* Doctorate/Sci/Calvin Coolidge Coll; MS/ME/Lehigh Univ; BS/ME/Univ of RI. *Born:* 8/21/24. Registered Prof Engr in NY, PA, NJ, CT MA, RI, & KS. Joined Combustion Engg 1949 after grad from Univ of Rhode Island and then returned to college to teach & do grad work in Fall of 1950. Taught at Lafayette College until 1960, when apptd hd of the Dept of Mech Engg at Manhattan College. In 1963 returned to Combustion Engg, serving as the New Products Div's Chief Proj Engr, Genl manager of Raymond Div, the Industrial Group's Dir of Technical Activities, Corp Coordinator of Environ Control sys, Dir of Corp Technology Transfer. In May of 1984, joined Maguire Group Inc., as VP and Dir of the Company's Energy Div managing this consulting firm's activities in the energy field and directing the company's Elec and Mech engrs. Mbr of Air Pollution Control Assoc, Natl Society of Professional Engineers, Amer Society of Mech Engrs (Fellow Grade), Amer Soc of Engg Education, Diplomat of American Academy of Environmental Engrs and Amer Assoc for the Advancement of Sci. Awarded ASME's performance test code medal 1986. *Society Aff:* ASME, APCA, NSPE, ASEE, AAEE, AAAS

Fernandez, Rodolfo B
Home: 4314 East-West Hgwy, University Park, MD 20782
Position: Associate *Employer:* RJN Environmental Associates, Inc. *Education:* BSE/CE/Princeton Univ *Born:* in Havana, Cuba; immgrated to NY City after the revolution. Technical mgr with NY Telephone 1970-72; Consulting Engr with Brokaw Engrg Assoc 1972-79 Elson T. Killan Assocs 1979-82, KJN Environmental since 1982, named Associate 1987. Professional experience primarily in infiltration/inflow; involvement began while in coll by compiling a state-of- the-art file of abstracts as part of EPA's initial study of nationwide I/I. Currently in charge of the firm's Maryland and Virginia offices. *Society Aff:* NSPE, ASCE, WPCF, APWA

Ferrante, John A
Business: Dept 982 Bldg 330-74A, E Fishkill, NY 12533
Position: Advisory Engr *Employer:* IBM. *Education:* MS/Phys/St Marys College; BS/Met Eng/Lafayette College. *Born:* 11/23/32. Native, Maple Heights, OH. Lab Metallurgist, Heat Treatment Engr for Natl Screw and Mfg in Cleveland and Los Angeles. West Coast Metallurgist, specializing in stainless steel strip and tubing for Wallingford Steel. With IBM since 1961. Assignments in Rochester, MN (Heat Treatment Dev, Mgr Quality Mtls/Electrical labs), Vimercate, Italy (Quality Liaison), Toronto (Quality Engr) and E Fishkill, NY (Substrate Process Dev and Product Engg). Chapter Chrmn, MN ASM - 1968, Mid-Hudson SME - 1977-80. Chrmn, SME Quality Council 1972- 76, SME Engineer Council 1976-78. Product Engr PE - state of MN. Enjoy skiing and duplicate bridge. *Society Aff:* ASM, SME.

Ferrara, Norman
Home: 24 Spring Tree Lane, Yardley, PA 19067
Position: Consultant (Retired) *Employer:* Colgate Palmolive Co *Education:* Mechanical Engineer/M.E/Rutgers University; MS/M.E/Steven Institute of Technology; B.S./C.E/Rutgers University *Born:* 02/24/13 Retired engrg consultant. Former VP of Engrg Center, Inc., Edison, N.J. Former Assoc Dir of Engrg for the Colgate-

Ferrara, Norman (Continued)
Palmolive Co, N.Y.C. in charge of international plants and equipment. Former Dir of Engrg for the Revlon Co, Edison, N.J. Former Project Engineer for Hercules Co. Parlin, NJ. Former Development Engineer for Johnson & Johnson, New Brunswick N.J. Has 55 years experience as an engineer-in-industry and has extensively traveled through the USA, S America and Europe supervising plant construction, equipment installation, and expansion. *Society Aff:* ASME, NSPE, USNL-UPADI

Ferrara, Thomas C
Business: Div of Engrg, Chico, CA 95926-0930
Position: Prof and Chair of CE *Employer:* CSU, Chico *Education:* PhD/Civ Engr/Univ of CA; Davis Campus; MS/Civ Engr/Univ of CA; BS/Civ Engr/Univ of CA; Davis Campus *Born:* 3/27/47. Born and raised in Sacramento, CA. Served on Civ Engg Faculty at CSU, Chico since 1971. Concurrently Consulting Traffic Engr on Traffic Safety, Planning, and Parking studies. Selected as Hd of Civ Engg Prog at CSU, Chico in 1978. Served as faculty advisor to award winning ASCE Student Chapter. Chrmn of ITE Committee - Planning for bicycle transportation and mbr of other natl committee relating to bicycle transportation. Married with 2 children. *Society Aff:* ASEE, ASCE, ITE, TRB

Ferrari, Domenico
Business: Computer Science Division, EECS Dept/University of California Berkeley, CA 94720
Position: Professor *Employer:* Univ. of California, Berkeley *Education:* Dr. Ing./Electronics Eng./Politecnico di Milano, Italy *Born:* 08/31/40 Native of Gragnano Trebbiense (Piacenza), Italy. Asst Prof at the Polytechnic Institute of Milan in 1967, was awarded the Libera Docenza in computer science in 1969. In 1970, joined the Dept of Elect Engrg and Computer Sci, Univ of CA, Berkeley, where he is now Prof and leader of the Berkeley UNIX and DASH Projects. For his contributions to the evaluation and improvement of computer systems performance, was elevated to the grade of IEEE Fellow in 1987. Author of Computer Systems Performance Evaluation (Prentice Hall, 1978) and coauthor of Measurement and Tuning of Computer Systems (Prentice Hall, 1983). *Society Aff:* IEEE, IEEE-CS, ACM, CMG

Ferrari, Harry M
Home: 144 W Swissvale Ave, Pittsburgh, PA 15218
Position: Consulting Engineer. *Employer:* Westinghouse Electric Corp. *Education:* PhD/Met Eng/Univ of MI; MS/Met Eng/Univ of MI; BS/Met Eng/Wayne State. *Born:* May 1932. Native of Detroit Michigan. Technical Advisor to Euratom in 1960-61. With Westinghouse Electric since 1958. Assumed current position as Consulting Engr in Nuclear Fuel Division in 1973 after a series of managerial positions. Executive Ctte of ANS, Fellow of the ANS in 1972, Mbr of ASME Nuclear Fuels Subcommittee. Over 50 publications & 20 patents in the nuclear fuel area. Enjoys tennis and squash. *Society Aff:* ANS, ASM, AIME, ASME

Ferrero, Joseph L
Business: Ferrero & Associates Inc, 600 So. Washington Road, McMurray, PA 15317
Position: Partner. *Employer:* Fayette Engrg Company. *Education:* BSCE/Civil Engg/Univ of Pittsburgh; MSE/Sanitary Engg/Univ of MI. *Born:* July 7, 1937 - 42 years old. BSCE Univ of Pittsburgh 1959. MS in Sanitary Engrg Univ of Michigan 1962. Worked Pa Health Dept, Gannett Fleming Corddry & Carpenter. In 1968 assumed current position as Partner Fayette Engrg Company. President WPCA of Pa (1975-76), Sludge Shovelers Award WPCAP (1971) AM, ASCE, & Member Assn Waste Engrs of Pittsburgh. Arthur Sidney Bedell Award (1978)-FWPCA. Considers golf, accordion, water sports, & family enjoyable pastimes. *Society Aff:* ASCE, WPCAP, FWPCA.

Ferrigni, George P
Business: 62 Glenmere Dr, Chatham, NJ 07928
Position: Conslt *Education:* MChE/Chem Engg/Polytechnic Inst of Brooklyn; ChE/Chem Engg/Cooper Union Inst of Tech; BChE/Chem Engg/Cooper Union Inst of Tech. *Born:* 7/2/19. Officer in US Navy 1944-46. With American Cyanamid since 1941 serving in a number of mfg & engg positions. Dir of Engg & Construction Div 1969-79. 1979-83 Served as VP mfg & engg for Cyanamid Europe, Mid-East & Africa with responsibilities for Cyanamid's plants in these areas producing agricultural and industrial chemicals, pharmaceuticals, sutures and Formica brand decorative laminates. In 1983 became VP Mfg & Engg for Cyanamid International with added responsibilities for plants in Asia, Australia, New Zealand as well. Retired from Cyanamid in Aug 1984. Conslt in domestic and foreign manufacturing, engrg and project mgmt. *Society Aff:* AIChE, NSPE.

Ferris, Robert R
Home: 150 Jonathan Drive, McMurray, PA 15317
Position: Pres. *Employer:* Glass Gorham Co. *Education:* PhD/Strat Plann/CA Western Univ; MBA/Fin/OH Univ; BS/Chem Engg/Penna State Univ. *Born:* 1/31/38. Pres - Glass Gorham group of manufacturing companies serving the lighting industry with facilities in USA, Canada, England, Germany, Venequela, Mexico. Previously (1971-78) Pres - Armco Composites, manufacturer of fiber reinforced plastics. Also, Manager-Bus Dev and mbr of strategic planning group for Corp Armco (Steel) Inc (1970-71). Earlier employed (1968-70) as GM-Sargent Art Co, a div of Mead Paper Co, manufacturing artist paints and supplies. From 1959-68, worked for MOBAY Chemical Co (Bayer) in engg and production of urethane plymers, isocyanates, and polucarbonate plastics. Active in numerous prof, industry and commumity groups. *Society Aff:* AIChE, SPE, NSPE.

Ferrise, Louis J
Home: 531 E Manoa Rd, Havertown, PA 19083
Position: Consultant *Employer:* Wyeth Lab., Inc. *Education:* MS/Ch.E/NYU; BS/Ch.E/Lehigh Univ. *Born:* 12/12/21. After grad from Lehigh Univ in 1950, joined Merck & Co as process dev engr, and spent nine yrs in various assignments in pilot plant, new plant start-up, and process engg design. In 1958, joined Cardinal Chem Co as plant mgr & plant engr. After 1.5 yrs, left Cardinal to join Wyeth Intl Ltd in 1960. Initially assigned as proj engr, for design and construction of fine chem/steroid plant in India. Moved into Intl Production Mgt in 1965, and appointed VP for mfg and technical activities in 1974. Promoted to Group V.P. Tech. Affairs in 1980. Transferred to Wyeth Labs. Inc as VP of Operations in 1985. Retired in 1987. Consultant for Wyeth. *Society Aff:* AIChE.

Ferry, David K
Business: Ctr for Solid State Electronics Res, Tempe, AZ 85287-6206
Position: Prof of EE & Dir, CSSER *Employer:* AZ State Univ *Education:* PhD/EE/TX; MS/EE/TX Tech; BS/EE/TX Tech *Born:* 10/25/40. Teaching and res asst at TX Tech (1962-63); Res engr and Instr at Univ of TX (1963-66); NSF Postdoctoral Fellow at Univ of Vienna (1966-67); Asst and Assoc Prof, TX Tech (1967-73); Scientific Officer at Office of Naval Res (1973- 77); Prof and Hd of Elec Engg at CO State Univ (1977-83); Prof of Elec Engrg and Dir of Ctr for Solid State Electronics Res, AZ State Univ. Fellow of American Physical oc; Fellow of Inst. Electrical and Electronic Engrs. Res activities in the areas of semiconductor transport, surface physics, and submicron semiconductor devices. Mbr of NAS/NRC Solid State Sci Panel; Consultant to various DOD agencies. *Society Aff:* IEEE, APS

Fertig, Marcel M
Home: 230 N Craig St, Pittsburgh, PA 15213
Position: Consulting Engr. *Employer:* Fertig Engg Co. *Education:* MS/Civ Engg/Carnegie-Mellon Univ; BS&CE/Civ Engg/Royal Technical Univ of Budapest. *Born:* 7/11/95. From 1921 to 1944 held responsible positions with industrial and consulting firms, preparing engg plans and specifications for construction projs. In 1945 started my engg firm offering consulting on industrial plants, bridges, and hgwys. In 1948, Fertig Engg Co was founded, because of expanded engg activities. Reg PE in PA, reg structural engr in IL and reg civ engr in WV. Pres, Pittsburgh chapter PSPE, vp of state society PSPE, dir SAME. In 1954, Governor of PA appointed as

Fertig, Marcel M (Continued)
engr mbr of the State Arbitration Bd of Claims. Mbr: ESWP, Pittsburgh Athletic Assn, Edgewood Country Club, Kiwanis, 32 degree Mason, Shrine. *Society Aff:* ASCE, NSPE, SAME.

Fertl, Walter H
Business: Dresser Atlas, P.O. Box 1407 (DC-21), Houston, TX 77251
Position: VP Marketing, Dir of Intptn & Field Dev *Employer:* Dresser Atlas, Inc *Education:* PhD/Petrol Engr/Univ of TX; MS/Petrol Engr/Univ of TX; Dr. Mont/Mining/Mining Univ of Leoben; Dipl Ing/Petr Eng/Austria *Born:* 3/16/40 Sixteen years worldwide experience in formation evaluations. Author of 220 technical papers, 19 US Patents, book on Geopressures, technical contributor to 5 other books. Recipient of two Best Technical Awards by SPWLA, Technical Creativity Award by Dresser Industries, Gold Medal by SPWLA (81), SPE of AIME Distinguished Lecturer (80/81), awards by governments of Iran (76) and Thailand (81). Speaker at over one hundred technical schools and seminars on all but one of the continents. *Society Aff:* SPE of AIME, SPWLA, SEG, AAPG

Ferziger, Joel H
Business: Dept of Mech Engg, Stanford, CA 94305
Position: Prof - ME. *Employer:* Stanford Univ. *Education:* BChE/Chem Engrg/Cooper Union; MSE/Nuclear Engrg/Univ of MI; PhD/Nuclear Engrg/Univ of MI. *Born:* 3/24/37. Brooklyn NY. Married, three daughters. At Stanford since 1961; Prof since 1972. Worked previsouly in nuclear engrg and kinetic theory of gases 1959-72. Current res is numerical simulation of turbulent flows. Conslt. Author 3 books, approx 100 papers. *Society Aff:* ASME, AIAA, SIAM, APS.

Fetters, Karl L
Business: 7099 Oak Dr, Poland, OH 44514
Position: Consulting Engr (Ret) *Education:* BS/Metallurgical Engg/Carnegie Inst of Tech; DSc/Metallurgical Engg/MA Inst of Tech. *Born:* Nov. 1909. 36 years engineering and executive in steel industry. 3 years Asst Prof Metallurgy, Carnegie Tech. Vice Pres Reasearch Youngstown Sheet & Tube Co 13 yrs. President AIME 1964. Fellow ASM. Fellow Met Society of AIME. Mbr. National Academy of Engrg. *Society Aff:* ASM, AIME, NAE, AISE, AISI.

Fettweis, Alfred L M
Home: Im Koenigsbusch 18, D-4630 Bochum 1, W Germany
Position: Prof of Communication Engrg. *Employer:* Ruhr-Universitaet Bochum. *Education:* Dr Appl Sci/El Engrg/Cath Univ of Louvain; Ing Civ Elect/El Engrg/Louvain, Belgium. *Born:* Nov 27, 1926 Eupen Belgium. Ingenieur civil electricien 1951 & Docteur en sciences appliquees 1963, both University of Louvain, Belgium; m. Lois Jane Piaskowski, BA; children: Luise, Maria, Gerhard, Gerlinde, Joerg;. Development engineer, telephone transmission and electronic switching (circuit design, systems planning), Bell Telephone Mfg Co, Antwerp, Belgium (1951-54 & 1956-63) and Federal Telecommunication Labs, Nutley N J (1954-56), both ITT subsidiaries; Professor theoretical electricity, Eindhoven University of Technology, Netherlands (1963-67); Professor communication engineering, Ruhr-Universitaet, Bochum W Germany; recipient honorary doctorate (Dr. h.c.) from University of Linkoping (Sweden) 1986, Prix Acta Technica Belgica 1963, Prix 1980 of the Fondation George Montefiore (Belgium), 1980 Darlington Prize Paper Award of the IEEE circuits and sys soc, IEEE Centennial Medal 1984, and VDE Ehrenring of the Verband Deutscher Elektrotechniker 1984 Fellow IEEE, Vice President IEEE circuits and Systems Soc. 1987, Member Rheinisch Westfaelische Akademie der Wissenschaften, Informationstechnische Gesellschaft, Sigma Xi, Eta Kappa Nu, European Association for Signal Processing, Societe Royale Belge des. Ingenieurs des Telecommunications et d'Electronique. *Society Aff:* ITG, SITEL, EURASIP.

Few, William E
Home: 4520 Murray Hills Drive, Chattanooga, TN 37416
Position: Mgr Lab Serv, Met & Matls Lab-Retired *Employer:* Combustion Engrg Inc. *Education:* BS/Met Engg/OH State Univ. *Born:* June 1923. Native of Niagara Falls, New York. Served with Army Corps of Engineers, Manhattan Project 1944-46. Principal Metallurgist Battelle Memorial Inst, 1947-55. Technical Staff AssistantCramet Inc. Republic Steel and Crane Company 1955-58. Metallurgical Consultant-includes staff of NMAB, NRC, National Academy Sciences, 1959-63. With Combustion Engineering, Inc since 1964. Presently Manager of Laboratory Services, Metallurgical and Materials Laboratory (Power Systems). Member AIME, ASM. Served on National Committees of ASM including Past Chapter Chairman and Past Chairman Southern Metals Conference. *Society Aff:* ASM, AIME.

Fey, Robert T
Business: 2702 Monroe St, Madison, WI 53711
Position: Pres *Employer:* Foth & Van Dyke of Madison, Inc. *Education:* BS/CE/Univ of WI *Born:* 12/9/28 Native of Madison, WI. Employed by Shell Oil Co in 1951. Entered Consulting Engrg in 1956, specializing in wastewater, water supply, hgwy and municipal design. Serves as Pres of Carl C. Crane, Inc, Consulting Engrs, responsible for administration of business and review of all major projects including flood studies, wastewater, water supply and municipal engrg. Currently concentrating on development of alternative systems for medium and small unsewered communities. *Society Aff:* NSPE, ASCE, ACEC.

Fey, Willard R
Business: Sch of Indus & Sys Engrg, Georgia Tech, Atlanta, GA 30332
Position: Assoc Prof *Employer:* Georgia Tech *Education:* MS/EE/MIT; BS/EE/MIT; BS/MGT/MIT *Born:* 6/29/35 Native of Cincinnati. Studied EE, Management and Economics at MIT, then Assistant Prof of Management 1964-67. 1966 E. M. Baker Award (Outstanding Teacher of MIT Undergraduates). Original member of Industrial Dynamics Group 1957-67 that developed System Dynamics field. Researcher at MITRE Corp 1967-68 before joining GA Tech ISyE School 1969-81 to establish System Dynamics graduate program. Research has been supported by the Law Enforcement Assistance Administration, US Air Force and US Forest Service. Consulted for major (Kodak, SCM, Whirlpool) and smaller corps and published Ecosystem Succession, co-author L Gutierrez, (MIT Press 1980) and many articles and papers. Selected for Leadership Atlanta 1977. *Society Aff:* IEEE, TIMS, AAAS, SDS, NYAS, ΣΞ

Fiala, Richard J
Business: 1604 S Wash St, Grand Forks, ND 58201
Position: Senior Vice-President. *Employer:* KBM, Inc. *Education:* BSCE/Civil Engg/Univ of ND. *Born:* September 1928; Native of N D; USAF officer 1952-54; EIT K B MacKichan & Assocs Inc 1954-58; Chief Civil Engr 1958-69; Vice-Pres CE 1969-75; Senior VP 1976. Headed design team for first rural water sys in N D 1970, $3.5 million construction cost; responsible for coordination of all Depts of Firm; total firm strength is 80; registered in 8 states; land surveyor N D; Fellow ASCE; Past Pres NDCEC; Past Pres N D ASCE; Past Pres Chap One NDSPE; married, 4 children; enjoys golf & public service activities. *Society Aff:* ASCE, NSPE, ACEC, WPCF, AWWA.

Fiander, Allan D
Business: Suite 200, 335 Queen St, Box 207, Fredericton, New Brunswick, Canada E3B 1B1
Position: Pres *Employer:* Fiander-Good Assoc Ltd. *Education:* MSc/Trans Plng/Univ of Birmingham (England); BSc/Civil Engg/Univ of New Brunswick. *Born:* 4/5/45. Since 1980 he has been pres of Fiander-Good Assoc Ltd., a firm of engrs, planners and economists providing consulting services to govt and industry. Previously he served as a dir and sr transportation engr for a regional engg firm. In this position his responsibilities included promotion and technical mgmt of the firm's planning and engineering consulting activities dealing with all modes of transportation. During the early 1970's he served on the faculty of the Transport Group of the Dept of Civil Engg, Univ of New Brunswick. In the mid 1970's he also served as

Fiander, Allan D (Continued)
special advisor on transportation policy to the New Brunswick govt. *Society Aff:* ITE, CTRF, RTAC, APENB.

Fiaschetti, Rocco L
Home: 13628 E Camilla, Whittier, CA 90601
Position: Quality Consultant *Employer:* Self *Education:* BA/Chemistry/Hobart College. *Born:* June 21, 1918. Held various technical and management positions at Ansco Division, GAF, including Manager of Quality Control until 1962. Spent two years as member Corporate Reliability & Quality Staff at Aerojet General. Since 1964 with Space Div Rockwell International in management positions in Quality & Reliability Assurance. (Lecturer Quality Control and Management). Member ACS since 1940. Fellow of American Society for Quality Control; Vice President 1962-64, President 1964-66. Mbr Board of Directors ASQC 1955-68, 1971-75. Certified Quality Engr (ASQC). Registered Professional Engr (Quality) CA Edwards Medal, ASQC, 1977. Certified Mgr, Academiciam, International Academy for Quality - 1977. Currently Quality Consultant. *Society Aff:* ASQC, IAQ.

Ficalora, Peter J
Business: 419 Link Hall, Syracuse, NY 13210
Position: Prof. *Employer:* Syracuse Univ. *Education:* PhD/Mtls Sci/PA State Univ; BS/Chemistry/Manhattan College. *Born:* 4/9/38. in NYC. Received PhD from the PA State Univ in 1965. Post-doctoral fellow at Rice Univ. Joined faculty at Syracuse Univ 1967. Served two terms as faculty chrmn of engg college at SU. Present res interests include gas-solid interactions & their relation to the mech properties of the solid. *Society Aff:* TMS-AIME.

Fiddes, Basil A
Home: P.O. Box 878, Asuncion, Paraguay
Position: General Coordinator *Employer:* Government of Paraguay *Education:* CE//LA State Univ *Born:* 5/19/14 1942-1945, Corps of Engrs U.S. Army. All types of Engrg Work. 1945-1948, Project Engr for J.A. Jones Contractors in Ecuador. 1948-1951, Creole Co in Venezuela. Project Engr in charge of the installation "Oil pipeline" Catia de la Mar-Caracas. 1951-1954, Bolivia: Macco-Pan Pacific Project Engr in charge of construction of roads and bridges. 1954-1976, Paraguay: Ministry of Public Works. 1954-1960, Pres and Technical Dir of the Official commission of 1954-1967 Cons Gen, Charge d'Affairs, Counselor Minister of the Republic of Panama in Paraguay. Executive VP and Chief Operating Officer of Chesapeake Intl Corp. August 1953 Official Paraguayan Delegate to the Panamerican Hgwy Congress in the Republic of Panama. August 1957 Official Paraguayan Delegate to the Panamerican Hgwy Congress in Bogota, Columbia. Two years of Post-grad work in CE at Loyola Univ. *Society Aff:* ASCE, SAME, USCOLD, ASTM, FESAE, PCI, ACI

Fiedler, George J
Home: 100 Pierce St-Apt 805, Clearwater, FL 33516
Position: Principal Engineer. *Employer:* Sverdrup & Parcel & Assocs Inc. *Education:* MS/EE/KS Univ; EE/Professional/KS State Univ; BS/EE/Kansas State Univ. *Born:* March 1904 at Bushton, Kansas. Graduate work towards PhD at Univ of Michigan. Electronics Engr for G E, RCA & US Navy Electronics Lab; Prof of Electrical Engrg at Univ of California Berkeley & Montana State Univ. Technical papers on automatic control systems published in procedings of the AIEE, IRE, ISA, ASME and International Federation of Automatic Control Moscow Russia 1960 and Basle Switzerland 1963. Fellow awards from IRE, AIEE, ISA, IEEE and American Assn for Advancement of Scien ce for contributions to electronic process control systems. *Society Aff:* IEEE.

Fiedler, Ross A
Business: Mechanical Engrg, UW-Platteville, Platteville, WI 53818
Position: Prof *Employer:* Univ of WI-Platteville *Education:* PhD/ME/Univ of WA; MS/ME/Univ of CA-Berkeley; BS/ME/Univ of MN *Born:* 7/1/35 Native of Osseo, WI. Served with US Navy, CEC, 1958-60. Previously employed by United Tech Ctr, Sunnyvale, CA, The Boeing Co, Seattle, WA, The Univ of WI- Stout. Currently employed by The Univ of WI-Platteville, Coll of Engrg, Dept of Mechanical Engrg, Dept Chrmn 1980-83. Chrmn of Rock River Valley Section of ASME, 1981-82. Mbr of ASME, ASEE, NSPE, Sigma Xi. Dept Chrmn of Gen Engg, 1985-87. *Society Aff:* ASME, ASEE, NSPE, ΣΞ.

Field, A J
Business: 900 Wilshire Blvd, Los Angeles, CA 90017
Position: Consulting Engr (Self-employed) *Education:* MS/CE/Stanford Univ; BS/Engr/CA Inst of Tech *Born:* 1/6/24 CA. Officer US Navy Seabees WWII - Drilling and Petroleum Engr. Union Oil Co, 1947-1958. During this period participated in the invention and development of drilling systems for deep water oil drilling from floating vessels, 1959- 1977. Founder-Pres - Global Marine Inc. During this period pioneered deep water exploration drilling from ships in all major oil producing regions of the world. Also developed and operated Glomar Challenger - deep water scientific drilling ship. 1977-1981 - Consultant - associated with Santa Fe and in design & development of large deep water scientific drilling system for the National Science Foundation & the oil industry. *Society Aff:* NAE, API, AIME-SPE, AAPG

Field, Joseph H
Business: 1319 Pinewood Dr, Pittsburgh, PA 15243 *Employer:* Self *Education:* MS/CE/Univ of Pittsburgh; BS/CE/Carnegie Tech. *Born:* 5/29/20. 1941-1944 & 1946-1968 US Bureau of Mines during the last ten yrs of this time I was proj coordinator of several projs in fischer tropsch synthesis of liquid fuels & high BTU gas, coal gasification & removal of sulfur gases. Co- discoverer of the Benfield gas purification process. Since 1968 I have been VP of Benfield Corp supervising design & dev of gas purification plants for the NH3, petrol, gas & chem industries. Natl Chrmn of fuels div of ACS in 1968 and winner of Storch Award for coal res. Semi retired as of 1981. Consultant in gas purification. *Society Aff:* AIChE, ACS.

Field, Michael
Business: 3980 Rosslyn Drive, Cincinnati, OH 45209
Position: Chief Executive Officer. (Retired) *Employer:* Metcut Research Assocs Inc. *Education:* PhD/Physics/Univ of Cincinnati; MS/Mech engg/Columbia Univ; BS/Mech Engg/City College of NY. *Born:* February 1914. Chief Exec Officer Metcut Research Assocs Inc Cincinnati. Member numerous technical societies and active as officer, committee chairman and committee member at local and national levels. Over 120 publications in field of machining and metal cutting. Registered PE in Ohio. Member National Academy of Engineering. Fellow ASM. Member Sigma Xi. '1966 Engineer of the Year'. Cincinnati; 1968 Gold Medal SME; Joseph Whitworth Prize 1968, Institution of Mechanical Engrs, England; 1969 Distinguished Alumnus Award Cincinnati 1978; AM Award, Amer Machinist; 1980 HOnorary Mbr, Amer Soc of Mechanical Engineers. *Society Aff:* ASM, SME, SAE, CIRP, ASME.

Field, Richard
Business: Woodbridge Ave, Edison, NJ 00837
Position: Chief, Storm & Combined Sewer Sec, Office of R&D. *Employer:* US Environmental Protection Agency (EPA). *Education:* MCE/Sanitary Engg/NYU; BCE/CE/CCNY. *Born:* 11/22/39. Sanitary des, Velzy, Assoc, 1963-64. Chief, Interceptor Maint and Plant Services Sec, NYC, 1964-66. Special Assis for Envir Engr, Eastern and Northeastern Div, Naval Fac Engr Command respon for all envir engr probs for Naval estab in NE, US, 1966-70. With EPA's Natl Storm and Combined Sewer Poll Control R&D Prog since 1970 and Prog Chief since 1971. Natl and intl leader and maj contributor urban stormwater poll control. Over 300 R&D reports. Over 100 confer lectures. WPCF Res Comm respons for annual JWPCF liter review on urban runoff; Urban Water Resources Coun; VWRC, ASCE, 1972-present; ASCE UWRC Task Ctte on Manual of Practice, Design and Construction of Storm Drainage Sys, 1981- present. ASCE/WPCF Task Force on Sewer Rehabilitation Manual of Practice; Tech Advisory Ctte for NYC Dept of Envir Protection Admin. Combined Sewer overflow regulator improvement prog, 1983-present. Numerous awards incl: EPA's Bronze Medal, 1973; ASCE State-of-the-Art in CE Award. 1976; NY Water

Field, Richard (Continued)
Poll Cont Assoc Kenneth Allen Award for outstanding contrib to the advmt of wastewater treat, 1977; and First Level EPA Scientific and Tech Achiev Award in Control Systems, 1979. *Society Aff:* WPCF, ASTM, XE.

Field, Sheldon B
Home: 195 Kensington Rd, Garden City, NY 10530
Position: Pres, Flume Stabilization Sys. *Employer:* John J McMullen Assocs Inc. *Born:* September 1931. BS US Merchant Marine Academy, Graduate Study at Columbia University and Stevens Institute of Technology. Native of New York City. Worked as project naval architect with M Rosenblatt and Son & H Newton Whittelsey Inc. Joined John J McMullen Associates Inc in 1959. Appointed V P of Ship Motions Div in 1966. Joined US Lines as Vice President, Market Research and Development in 1969. Asst to the Pres, American Export Industries 1970. Rejoined John J McMullen Assocs Inc as V P Ship Motions Div 1973. Appointed Pres of Flume Stabilization Systems, a div of John J McMullen Assocs Inc 1976. Received USMMA Alumni Award for Distinguished Professional Achievement, 1968.

Fieldhammer, Eugene L
Business: 1139 Olive St, St. Louis, MO 63101
Position: Consultant, Retired. *Employer:* Booker Assocs, Inc *Education:* BS/CE/KS State Univ *Born:* 2/11/25 February 1925. Native of NY City, NY. Served in U.S. Navy, 1943-46. B.S. in Civil Engrg in 1950 from KS State Univ, Manhattan, KS. Entered consulting engrg practice in 1950 with Edwards & Kelcey Engrs, Newark, NJ. Joined Booker Assocs, Inc. of St. Louis, Missouri in 1963 as Mgr, Bridge Design Dept, promoted to VP in 1965, Senior VP in 1968, Executive VP in 1973 and Pres in 1980. Retired 1985. Serves on Bd of Directors and as Consultant. Registered PE in MO, IA, NY; also reg PE in TX, UT and CA. Reg struc engr in IL; registered land surveyor in ME. *Society Aff:* SAME, NSPE, ASCE.

Fielding, Raymond E
Business: Houston, TX 77004
Position: Prof. & Dir of the Sch of Communication *Employer:* Univ of Houston. *Education:* PhD/Communication/USC; MA/Cinema-TV/UCLA; BA/Theater Arts/UCLA. *Born:* 1/3/31. Prof of communication, specializing in motion picture and tv tech. Asst and Assoc Prof, UCLA, 1957-65 where also hd of special cinematographic effects lab; Assoc Prof, Univ of IA, 1965-69; Prof, Temple Univ, 1969-78; Prof, Univ of Houston, 1978-. Author, *The Technique of Special Effects Cinematography* (1965); *A Technological History of Motion Pictures and Television* (1967), and other books and articles on communication tech. Elected Fellow of the SMPTE, 1976. Elected Active Mbr of the Academy of Motion Picture Arts & Sciences, 1981. VP for Educational Affairs, SMPTE, 1977-79. Trustee, American Film Inst, 1973-79, and Univ Film Fdn, 1967- , (President, 1985-). Pres, Univ Film Assoc, 1967-68; Pres, Industry Film Producers Assoc, 1961-62. Consulting Director of Research & Development, Zoetrope Studios (Hollywood) 1980-81.. *Society Aff:* SMPTE, UFVA.

Fifield, Charles D
Business: McNamee, Porter & Seeley, 3131 So State St, Ann Arbor, MI 48104
Position: Indus Proj Dept Hd *Employer:* McNamee, Porter & Seeley *Education:* BS/CE/Wayne State Univ *Born:* 11/22/45 Process design engr with Greeley and Hansen, Chicago, specializing in wastewater treatment plant design. Joined McNamee, Porter and Seeley in 1975. Currently, Proj Mgr and Indus Proj Dept Hd for Industiral Services Div. Responsible for management of firms initial planning phase on all indus water and wastewater projects. Pres MI Society PE's, Ann Arbor Chapter 1983, State Dir 1984. Mbr of Tau Beta Pi & Chi Epsilon Engrg Honorary Frat. Active in church and social organizations and enjoys outdoor sporting activities. *Society Aff:* NSPE, WPCF, AWWA, ASCE.

Figge, Kenneth L
Business: P.O. Box 70, Carlyle, IL 62231
Position: VP *Employer:* Henry, Meisenijeimer & Gende, Inc *Education:* BS/CE/Univ of IL *Born:* 11/19/38 Registered PE in IL, MO, and CO; Registered Structural Engr in IL. Member, Structural Engrs Assoc of IL. Primary specialties; bridge design, structural analysis, investigations and reports. *Society Aff:* NSPE, ACI.

Figueroa-Laugier, Juan R
Home: 601 Austral Street, Rio Piedras, PR 00920-4225
Position: Consulting Engr. *Employer:* Self-Employed. *Education:* BS/Civil Engg/Univ of PR. *Born:* 12/14/31. BSCE with High Honors in 1953 from Univ of PR. Native of Caguas PR. Served with Army Corps of Engr 1953-54, with Eighth US Army in Korea 1954-55. At present a Retired Reserve Colonel and a Consulting Engr after retiring from PRWRA (Puerto Rico Water Resources Authority) in 1982 after a 30 yr government practice as Structural Design Engr, Civil Engr VI, General Office Engr, Prin Engr & hd of the Specifications & Procurement Dept. Mbr PR Bd of Examiners of Engrs, Architects & Surveyors 1967-72; VP 1963-66; 1973-74, Pres PRSPE (NSPE) 1974-76; Natl Dir 1978-date; Fellow Mbr, Pres PR Sect ASCE 1962; Charter Mbr 1968, Dir 1969-70, VP of the PR Chapter American Public Works Assn; Mbr, Engrs and Surveyors Assoc of PR; Treasurer 1984-85, Director 1985-date; Pres, Institute of Civil Engrs of P.R., 1984-86, Delegate 1986-date. *Society Aff:* NSPE, ASCE.

Filbert, Howard C, Jr
Home: 515 E Seminary Ave, Baltimore, MD 21204
Position: Pres & Chrmn. *Employer:* MRC Corp. *Education:* ME/ME/Univ of MD; MS/ME/Univ of MD; BS/ME/Univ of MD. *Born:* 2/7/17. Native of Baltimore, MD. Grad from the Univ of MD, College of Engg with highest honors. Employed initially by the Naval Ordnance Lab, Silver Spring, MD, starting as a res engr & finally as Chief of the Engg Div with MRC Corp, in 1952, and as Pres & Chrmn until retired, March 31, 1987, at age 70. Now employed by MRC as Consultant. MRC is a producer of engineered systems for ind & govt automation & mtls handling. Received Navy Meritorious Service Medal, & ADPA Bronze Medal for contributions to natl defense. Mbr Tau Beta Pi, Pi Tau Sigma, & Omicron Delta Kappa, Phi Kappa Phi, Phi Kappa Phi. *Society Aff:* ASME, AIAA, NSPE, ADPA.

Filer, William A
Business: Box 7055, Ancaster, Ontario Canada L9G 3L3
Position: President. *Employer:* Filer Consultants Ltd. *Education:* BASc/Civ Engrg/Univ of Toronto. *Born:* 1926 Toronto Ontario. Consulting Engr since 1956. Principal, Filer Consultants Limited, specializing in structural engrg design for industrial and commercial clients, and historical building renovation. Active in engineering organizations since graduation. Fellow Engrg Inst of Canada 1972. Councillor, Assn of Prof Engrs of Ontario 1976-79. Natl Chrmn Professional Development Program EIC 1960-61. Councillor EIC 1963-66. Director Ontario Chamber of Commerce 1971-75. Continuous activity in local and provincial organizations for culture, politics and local improvement. Pres of Royal Hamilton College of Music 1973-79. Engrg 1982-83. Fellow Conslt Soc for Civ Engrg, 1984. Officer Order of Sons of Martha, 1982. *Society Aff:* FEIC, FESCE, OSM

Files, John T
Business: PO Box 61529, Houston, TX 77208
Position: Chmn & CEQ. *Employer:* Merichem Co. *Education:* MS/Chem Engg/Univ of TX at Austin; BS/Chem Engg/Univ of TX at Austin. *Born:* 8/13/18. in Austin, TX. Served on engg faculty at Univ of TX from 1940-42. While doing res at the Univ, dev a process for recovery of magnesium from sea water by ion exchange and in 1942 joined Dow Chem Co to develop this into a commercial process. Became Asst Chief Engg at Dow and was respon for the design and installation of serval of their chlorinated hydrocarbon processes. Left Dow in 1945 to start the Merichem Co in Houston. Merichem is a chem manufacturing co producing cresols, phenols, cresylic acids and soda chem from petroleum. In 1968 he was named a Dist Grad of the College of Engg at the Univ of TX. Now serves on the Engg Foundation at the Univ, also on Bd of Dir of several assoc. He and his wife Barbara enjoy snow

Files, John T (Continued)
skiing, scuba diving and tennis and are active in civic and church work. *Society Aff:* AIChE, ACS, API, SCI, TSPE, SOCMA.

Files, Wilber D
Business: 900 S Glenn Cir, State College, PA 16803
Position: Consultant (Self-employed) *Education:* BS/EE/PA State Univ *Born:* 4/20/22 I am a native of Centre County, PA and a Registered PE since 1954. I have been serving as State Dir from Central PA for The PA Society of PEs from 1975 to the present. I am self employed as a consultant in Communications and Broadcast Engrg. Prior to 1972, I was employed by HRB-Singer, Inc for 25 years as a Staff Engr in communication and electronic surveilance system design. During WWII, I served in the US Army Signal Corp, Second Signal Service Bn. *Society Aff:* NSPE

Filley, Richard D
Home: 2445 W. Pecos Ave, Mesa, AZ 85202
Position: Director, Technology Transfer Programs *Employer:* Department of Industrial & Management Systems Engineering, Arizona State University *Education:* BS/Indus Engrg/U of WA *Born:* 07/28/55 Began engrg career in 1978 in aerospace indus. At Boeing, was co-author of the book *Industrial Engineering in the Boeing* Company, and at Garrett AiResearch and Sperry Flight Systems was involved in facilities planning and manufacturing systems design. Joined staff of *Industrial* Engineering magazine in 1981. As people and tech ed, achievements included authoring over 50 articles for *Industrial Engineering* (including 13 cover features) and other engineering publications. On the staff of the Dept of Ind & Management Systems Engrg at Arizona State Univ since 1985. Unique technology transfer activities include co-founding College of Engrg's Industrial Fellows Prog, and work with the aerospace and maquiladora industries. Enjoy skiing, mountain climbing, travel and reading. *Society Aff:* IIE, AAC, SME-CASA, SPJ-SDX

Finch, Rogers B
Home: 12 Sherwood Rd, Little Silver, NJ 07739
Position: Exec VP (Retired) *Employer:* Illuminating Engrg Soc of N Amer *Education:* ScD/HighPolymer Mechs/MA Inst of Tech.; MS/Textile Technology/MA Inst of Tech; BS/ME/MA Inst of Tech. *Born:* 4/16/20. Broadalbin NY. Prof MIT 1946-53; director, US Foreign Aid Mission Rangoon, Burma, 1953-54; Dir of Research Rensselaer Polytechnic Institute 1954-61, Assoc Dean, Hartford grad Ctr 1963-66, V P for Planning, 1966-72 Dir of Univ Relations, Peace Corps, Washington, DC, 1961-63; Exec Dir and Secretary, Amer Soc of Mech Engrs 1972-80. Consultant 1981-82; Exec VP, Illuminating Engrg Soc of N Amer 1982-1987. Past Pres, Council of Engrg and Scientific Soc Exec; Served from 2nd Lt to Major AUS 1941-46, to Brig Gen. US Army Commendation Medal, Legion of Merit. License P E Mass and N Y. Fellow ASME, I Mech E (UK), Inst Engrs (Australia), AAAS; Mbr IESNA, SAE, AWS, AIAA, Amer Soc Assn Execs, Sigma Xi, ASEE,EIC, Canadian Soc for Mech Engrg; Tau Beta Pi. *Society Aff:* ASME, AAAS, SAE, AWS, AIAA, ASAE, ASEE, IESNA.

Finch, Stephen J
Home: Box 2891, Setauket, NY 11733-0788
Position: Assoc Prof *Employer:* SUNY at Stony Brook *Education:* Ph.D./Stats/Princeton Univ; MA/Stats/Princeton Univ; BS/Math/St. Louis Univ *Born:* 3/20/45 Born and raised in St. Louis, MO. Joined faculty of Stony Brook in September, 1974 and holds a joint appointment with the Dept of Community and Preventive Medicine. Was visiting Assoc Prof of Statistics, Dept of Mathematical Statistics, Columbia Univ, Spring, 1981. Consultant to Brookhaven National Lab, Biomedical and Environmental Assessment Division, June 1975 to 81, to Brookhaven National Lab, Dept of Nuclear Energy, Containment Systems Group, Thermal Reactor Safety Division, June 1979 to 81, to Columbia Univ Sch of Social Work, 82-present, and to various corporations and NY State agencies. *Society Aff:* ASA, AAAS, APHA, NYAS.

Fincher, James R
P.O. Box 785, Fairburn, GA 30213
Position: Pres. *Employer:* quo modo, Inc. *Education:* Master's/Structures/GA Tech; Bachelor/Civ Engg/GA Tech. *Born:* 8/4/28. Native of LaGrange, GA. Married former Jane Hart. Four children. Served with Army Corps of Engr 1946-52. Faculty and Grad Faculty, GA Tech 1957-67. Founder and Pres quo modo, Inc since 1968. Firm specializes in analysis, design and investigation of complex structures. Conceived and designed unique, 350' span space frame roof with integrated 100' high wall truss for Atlanta arena (Omni). Holds patent on structural elements. Reg engr in several states. Dir, GA Sec ASCE 1977; VP, 1978; Pres, 1979. Several Natl Committees. Twelve yrs Boy Scout leader at Troop, Dist, and Council levels. Licensed Ham Radio Operator. *Society Aff:* ASCE, NSPE.

Finchum, Willis A
Home: 4900 Nogales, Atascadero, CA 93422
Position: Prof of Engg Tech *Employer:* CA Poly St Univ. *Education:* MS/Elec Engg/UT State Univ; BS/Elec Engg/UT State Univ. *Born:* 5/27/21. Native of Indianapolis, IN. Served as Elec Tech in US Navy during WWII Engr exp Sandia Corp, Bell Aircraft Corp, Raytheon Man Co, Radio Plane Co, Coleman Engr Co, Ramo-Wooldridge, Boeing, Lawrence Rad Lab, Assoc Prof EE at UT State Univ 1960-68, Assoc Prof Purdue Univ 1968-69, Prof and Head EE Dept Univ of the Pacific 1969-74, Prof & Head Elect Tech Dept at Purdue Univ 1974-76. Prof and Head Engg Tech CA Poly St Univ 1976 to 1981. Sr mbr IEEE. PE in IN and UT. *Society Aff:* IEEE, ASEE.

Findley, William N
Business: Div of Engrg - Bx D, Providence, RI 02912
Position: Prof of Engrg, Emeritus *Employer:* Brown University. *Education:* DS/Hon/IL Coll; MS/Mech/Cornell; BSE/Mech Engrg & Math/U of MI; AB/Physics & Chem/IL Coll. *Born:* February 1914. Instructor George Washington Univ, Instructor to Assoc Prof Univ of Illinois, Professor Brown Univ Emeritus 1984 to date. Life Fellow ASME. Mbr Scientific Advisory Council Picatinny Arsenal from inception for 10 years. Consultant to Lawrence Livermore Lab 1962-78. Director Central Facility for Mechanical Testing 1965-68. Member Organizing Ctte Joint International Conf on Creep. Recipient Charles B Dudley Medal Amer Soc Testing & Matls; twice recipient Richard L Templin Award Amer Soc Testing & Matls; prize for paper Soc Plastics Engrs (twice). Author book on Nonlinear Creep 1976; chapters in 6 books; 155 papers; baritone soloist; yachtsman. Office of Naval Res-Amer Inst of Aeronautics & Astronautics (ONR-AIAA) Res Scholar in Structural Mechs for 1978. *Society Aff:* ASTM, SESA, ASEE, ASME, AAM.

Fine, Morris E
Business: Tech Inst, Northwestern Univ, Evanston, IL 60201
Position: Technological Institute Professor. *Employer:* Northwestern Univ, Dept of Matls Sci & Eng. *Education:* PhD/Physical metallurgy/Univ of MN; MS/Metallurgical Engg/Univ of MN; BMetE/Metallurgical Engg/Univ of MN. *Born:* 4/12/18. Jamestown, North Dakota. Member of Technical Staff, Bell Telephone Labs, Murray Hill New Jersey 1946-54. Presently W P Murphy Prof, Matls Science and Engineering Department, First Chairman of Department, Northwestern University, Technological Inst, Evanston Illinois. Fellow and former Director TMS-AIME and former, Chairman IMD. Fellow American Society for Metals, Fellow American Physical Society, Fellow Am Assoc. Adv. Sci and Fellow American Ceramic Society, Honorary Member Japan Institute of Metals, Former Chairman Membership Committee, National Academy of Engrg; former member ASEB; and Materials Advisory Board of the NRC. Sigma Xi, Tau Beta Pi, Alpha Sigma Mu 1979 ASM Campbell Memorial Lecturer; 1981 TMS-AIME Mathewson Gold Medal; 1982 AIME James Douglas Gold Medal; 1986 ASM Gold Medal; Chicagoan of the Year in Sci, 1960. Over 150 technical and scientific articles. Presently visiting Prof Univ Texas at Austin. *Society Aff:* ASM, AIME, NAE, MRS

Fine, Morton S
Business: 211 1/2 N First St, PO Box 5000, Seneca, SC 29678
Position: Exec Dir. *Employer:* NCEE. *Education:* BS with high distinction/Civil Engg/Worcester Poly Inst. *Born:* 6/3/16. Forty-four yrs of increasing responsiblity in Civil Engg, Landscape Architecture, Land Surveying, Business Mgt and civil and professional activity. Prin and owner of established professional firm, engaged in Civil Engg, Planning, Landscape Architecture and Land Surveying from 1950 to 1975. 1976 to present Exec Dir of Natl Council of Engg Examiners. "Pres 1974-75; V-Pres 1971- 73; Member 1962-74" 1978 elected "Eminent Engr" mbr Tau Beta Pi. Enjoy music, stamp and coin collecting, golf and tennis. NSPE, "Natl Chrm NSPE/PEPP 1970-71-. ASCE, "Fellow." *Society Aff:* NSPE, ASCE, ACSM, ASEE, ACEC.

Fine, Terrence L
Business: Sch of Elec Engg, Cornell Univ, Ithaca, NY 14853
Position: Prof of EE. *Employer:* Cornell Univ. *Education:* PhD/Applied Phys/Harvard Univ; SM/Applied Phys/Harvard Univ; BEE/EE/CCNY. *Born:* 3/9/39. Lectr & res assoc, DEAP, Harvard Univ 1963-64. Miller Inst Fellow, UC Berkeley, 1964-66. Visiting Prof, Info Systems Lab, Stanford Univ, 1972-73, 1979-80. With the Sch of Elec Engg, Cornell Univ since 1966, where I am a mbr of the grad fields of applied math, elec engg, statistics history & Philosophy of Science and technology. Past Chrmn, IEEE, Ithaca Sec, VP Board of Governors IEEE Info. Theory Group, fellow of IEEE Consultant to several systems- oriented cos generally with regard to statistical & methodological issues. Res interests focus on the fdns of, and new approaches to, uncertainty & decision making. *Society Aff:* IEEE, IMS.

Finerman, Aaron
Business: 3210 Computing Center, Univ of Michigan, 1075 Beal Avenue, Ann Arbor, MI 48109-2112
Position: Prof, Elec Engg and Comp Sci; Special Assoc in Information Tech *Employer:* Univ of MI Computing Ctr. *Education:* ScD/Structural Engg/MA Inst of Tech; SM/Structural Engg/MA Inst of Tech; BCE/Civil Engg/Coll of City of NY. *Born:* 4/1/25. Born: 4/1/25 in New York City. He has been a Prof (Electrical Engin & Computer Sci) at The Univ of MI since 1978 and was the Dir of the Computing Ctr from 1978-86; he is now also Special Associate in Information Technology. He was a Prof at SUNY at Stony Brook from 1961-78, Dir of the Ctr from 1961-69 and Chrmn of the Dept of Computer Sci from 1975-77. He was Mgr of Office of Computing & Info Sys at the Jet Propulsion Lab from 1971-73 while on leave from Stony Brook. He was Natl Treas & Council Mbr of the Assoc from Computing Machinery, the professional computing society; he served on the Exec Committee and Bd of Dirs and was Chrmn of the Publications/Committee of the Amer Federation of Info Processing Societies. Currently, he is Editor-in-Chief of *Computing Reviews* published by ACM; he is a mbr of the Space Applications Advisory Committee for the National Aeronautics and Space Administration. *Society Aff:* ACM, AAAS, AAUP, TBΠ, ΣΞ.

Finfrock, M Frank
Business: PO Box 105062, Atlanta, GA 30348
Position: VP-Industrial Chemicals Div. *Employer:* Cities Service Company. *Education:* BSME/-/Purdue Univ. *Born:* 11/8/29. in Fort Wayne, IN. Has held various positions with Cities Service Co since 1951. Assumed current responsibility as VP, Industrial Chems Div in 1979. Served with the US Marine Corps 1952-53. *Society Aff:* ASME, AIChE.

Fink, Daniel J
Business: 3135 Easton Turnpike, Fairfield, CT 06431
Position: Sr VP Corporate Planning & Development *Employer:* General Electric Company. *Born:* Dec 13, 1926. BS, MS Aeronautical Engineering 1948 MIT. GE Sr VP Corporate Planning and Development, previously VP and Group Executive, Aerospace Group and VP and General Manager, Space Division. Formerly served as Deputy Director of Defense Research and Engineering, DOD. Pres AIAA 1974; Honorary Fellow AIAA, Vellow AAAS, Sr. Consultant - Defense Science Board, Department of Defense; Scientific Advisory committee - Army; Member - National Academy of Engineering. Honors: Recipient Distinguished Public Service Award - DOD 1967; Collier Trophy 1974. Clubs: Cosmos

Fink, Donald G
Home: 103B Heritage Hills, Somers, NY 10589
Position: Director Emeritus, IEEE. *Employer:* Retired, formerly Exec Dir, IEEE. *Education:* MSc/Elec Engg/Columbia; BSc/Elec Engg/MIT. *Born:* November 1911. Native of Englewood, NJ. Editorial staff 'Electronics' 1943-52, Editor-in-Chief 1946-52. Director Research, Vice President-Research Philco Corporation 1952-62. Editor Proceedings IRE 1956-57; Pres IRE 1958. General Manager and Executive Director IEEE 1963-74; Executive Consultant IEEE 1975-76. Executive Ctte WFEO 1972-76. Operations Director ACE 1975-76. Author & editor of 15 books in fields of electronics, television, radar and computers including 'Standard Handbook for Electrical Engineer' and 'Electronics Engineers Handbook'. Eminent Member Eta Kappa Nu 1965. Member National Academy of Engineering 1969. Hobbies: Amateur radio and astrophotography. Recipient of IEEE Founder's Medal, 1978 and SMPTE Progress Medal, 1979. Honored by establishment of "IEEE Donald G Fink Prize Paper Award" 1980. *Society Aff:* IEEE, SMPTE, RCA.

Fink, George B
Business: 1023 Corporation Way, P.O. Box 10023, Palo Alto, CA 94303
Position: Chrmn *Employer:* Wahler Assocs *Education:* MS/CE/CA Inst of Tech; MS/Intl Affairs/George Washington Univ; BS/Engrg/US Military Academy *Born:* 1/1/25 Chrmn of consulting geotechnical engrg co practicing nationally and internationally. Has had 38 years of continually increasing responsibility, both technical and managerial, in both the Corps of Engrs and private engrg practice. In last Corps assignment, as a brigadier general, was a member of the Natl Bd of Engrs for Rivers and Harbors, the Coastal Engrg Research Bd, and headed the Corps South Pacific Div. In private practice for past 10 years, was Regional VP of Metcalf and Eddy before assuming present duties. *Society Aff:* ASCE, SAME

Fink, Harry A, Jr
Home: 602 32nd Avenue, E Moline, IL 61244
Position: Senior Engineer. *Employer:* John Deere Harvestor Works. *Born:* Jan 1933. BS Agri Engrg Purdue Univ 1951. Native of Lafayette Ind. Sr Res Engr, Deere & Co Moline Illinois 1955-66; Sr Engr John Deere Harvester Works, E Moline Ill 1966- . Quad City Sect ASAE Chmn 1971-72. Ill-Wisc Region ASAE Secy-Treas 1974-75. ASAE Natl Continuing Educ Comm Chmn 1975-76. Co-owner of two patents. Taught computer programming Black Hawk Jr College, Moline Ill 1973-74 & 1976. Developed computerized sys for ASAE Yearbook Product Directory 1972. Currently employed as sci computer coordinator in Prod Engrg Dept. Listed in 1970-72 'Who's Who in the Midwest'.

Fink, James E
Business: 3777 Gaines Street, San Diego, CA 92110
Position: President. *Employer:* San Diego Aircraft Engrg Inc. *Education:* BSME/Aero/Univ of SC. *Born:* July 21, 1930, Glen Cove NY. Extensive background in technical management, particularly related to Aerospace Sys & Products. 1951-56 North American Aviation, Ryan Aeronautical Co. 1956-66 General Dynamics/Convair: Program Manager and Development Project Engr. 1966-present, San Diego Aircraft Engineering Inc (Sandaire): President and Director. Society of Automotive Engrs: Director 1976; chairman various committees and national meetings. Member and officer of several technical societies and defense-oriented organizations. Aviation Week Award 1964; 'Engineer of the Year', American Institute of Aeronautics & Astronautics 1964. Enjoy sailing, staff Commodore, San Diego Yacht Club. *Society Aff:* SAE, AIAA, NL, AFA, AAA.

Fink, Lester H
Home: 11304 Full Cry Court, Oakton, VA 22124
Position: Chrmn of Board *Employer:* Carlsen and Fink Assoc. Inc. *Education:*

Fink, Lester H (Continued)
MS/EE/Univ of PA; BS/EE/Univ of PA. *Born:* 5/3/25. Native of Philadelphia PA. US Army Infantry, European Theatre 1943-46. Employed by Philadelphia Elec Co 1950-74 in Res Div, spec in modeling, simulation, and control of generating plants & of individual & interconnected power sys. Employed by US Govmt 1974-79, respon for conceiving, dev, and managing a coordinated natl res program applying Sys theory to the structuring and operation of electric power sys. Left Dept of Energy in 1979 to found consulting firm Sys Engg for Power, Inc (SEPI). Founded Carlsen and Fink Assoc Inc in 1983 3 US patents. Fellow of IEEE & ISA. Reads widely & enjoys classical music. *Society Aff:* IEEE, ISA, CIGRE, NYAS.

Fink, William B
Home: 17 Vincek Lane, Saratoga Springs, NY 12866
Position: Energy Advisor. *Employer:* N.Y State Energy office *Education:* BChE/Ch Engr/Brooklyn Poly Inst; BSc/Economics/Rensselaer Polytechic Inst; -/College Prep/Polytechnic Prep CD Sch. *Born:* 8/4/12. Brooklyn NY. s. William L. & Veola (Schiff) F.; m. Edna Madeline Buck June 6, 1942; children - Wm Bruce, Jr, Edward Martin. Chemist Allied Chem & Dye Co Phila 1940-44; Supr Charles Pfier & Co Brooklyn 1944-47; Proj Engr Caltex Oil Co NYC 1947-50; Sr Engr Air Reduction Co NYC 1950-62; Facils Engr C E Co Schenctady NY 1962-63; Proj Mgr 1963-66; Ptnr Teeling Spindler & Fink Engrs Inc. Albany NY 1966-72, VP 1970-72; Managing Ptnr W B Fink & Assocs 1972-81. Pres. Fink & Fink P.C. Advisor, N.Y. State Energy Office 1981- , Active Boy Scouts of Amer. St George's Lodge 6 F&AM P Cons Engrs Council NY 1969-70. Scouters Key 1968; Phi Lambda Upsilon Key 1942. Reg PE NY, MA, VT, NH. Mbr ACS, NY Soc PE, Cons Engrs Council. Methodist (Steward 1958- 76). Dir Industries for the Blind 1981-86, Founding Member & Crew Chief- Wilton Emergency Squad. *Society Aff:* ASHRAE, ACS, AIChE, CEC NYS.

Finkel, Edward B
Business: 333 N Broad St, Elizabeth, NJ 07208
Position: Consulting Engr (Self-employed) *Education:* MEngg/Struct Engg/Yale Univ; BSCE/Struct Engg/Univ of IL. *Born:* 9/6/28. Born in Elizabeth, NJ. Formerly with consulting firms: Parsons, Brinckerhoff, Hall & MacDonald; Saul Shaw & CO; Weinberger, Frieman, Liechtman, & Quinn. In private practice as consulting structural engr since 1962. Mbr of Natl Soc of PE & Consulting Engrs Council. Proflly lic in NJ, NY, MA, IL & FL. Married: three children, one grandchild, Profl musician. Hobbies: jazz music & swimming. *Society Aff:* ASCE, ACI, NSPE, ACEC

Finley, J Browning
Business: Box 2139, Kingsville, TX 78363
Position: Prof. *Employer:* TX A&I Univ. *Education:* PhD/CE/OK State Univ; MS/CE/Univ of S LA; BSChE/CE/Univ of S LA; BS/Chemistry/Univ of S LA. *Born:* 7/18/19. From 1942 to 1958, practiced chem engg as researcher in petrol industry and in private consulting business. Returned to college and received adv degrees in chem engg. Has taught since 1963. Outside consulting works has been in heat transfer, mass transfer and corrosion. *Society Aff:* AIChE, ASEE, ΣΞ, TBΠ.

Finley, James H, Jr
Home: 283 Sherbrooke Ave, Williamsville, NY 14221
Position: Senior VP *Employer:* Cannon Design, Inc. *Education:* BME/Mech Engg/Rensselaer Poly Inst. *Born:* 1/13/28. Native of Buffalo, NY. Served with Army Corps of Engrs 1951-53. Three yrs process engg, three yrs plant engg, 26 yrs architectural engg experience. With Cannon Design, Inc since 1963. Became principal in 1967, assuming present responsibility as Dir of Engg. Specialization in medical facility environmental design and energy conservation. Reg in NY, PA, and MA, Maine, New Jersey Natl Engg Certificate. *Society Aff:* AEE, ASHRAE, NSPE.

Finlon, Francis P
Home: 1134 William Street, State College, PA 16801
Position: Professor of Engrg Research. *Employer:* Applied Res Lab, State College Pa. *Education:* MS/EE/Penn State Univ; BS/EE/Penn State Univ. *Born:* September 2, 1924 Carbondale Pa. Member, Institute of Electrical Engrg, American Academy of Science. Honors: Tau Beta Phi, Eta Kappa Nu, Pi Mu Epsilon, and Sigma Xi. Professor of Engineering Research and Head of Acoustics Department of the Applied Research Laboratory, the Pennsylvania State University. Applied research and development of sonar, underwater communications, and diagnostic medical ultrasonic systems. Several papers and patents. Consultant to US Navy Dept and the Commonwealth of Pennsylvania. Hobbies: golf and square dancing. *Society Aff:* TBΦ, HKN, ΠME.

Finn, James Francis
Home: 48 Hampshire Rd, Wash Township, Westwood, NJ 07675
Position: Partner. *Employer:* Howard Needles Tammen & Bergendoff. *Education:* BS/Civil Engrg/Newark Coll of Engrg *Born:* 7/11/24. Jersey City, NJ. Asst Engr and Sr Engr with NJ Dept of Conservation, Newark, 1949-1956. Hgwy Design Engr with HNTB, NY, 1956-1964. Engg Mgr and Gen Partner, HNTB, Fairfield, NJ, 1965. VP HNTB Intl, Inc VP Kencomp, Inc, Clerk Thomas K. Dyer Inc. Served as Officer CE, US Army, 1943-1946. Mbr American Rd Builders, Assn, Intl Bridge Tunnel & Turnpike Assn, The Moles and Transportation Res Bd. Listed in *Who's Who in America* and *Who's Who in the East, Who's Who in Engineering, Who's Who in Finance & Industry*. HNTB S de R L - Partner & Admin, HNTB Inc (MO)--VP, HNTB, PC (DC)--VP, HNTB, Inc, PS (WA)--VP, HNTB, PC (OR)--VP, HNTB A & E, Inc (FL)--VP, HNTB Inc, (AK)--VP, HNTB, AE&P, PC (NJ)--VP, HNTB Int'l., Inc.--VP. *Society Aff:* NJSPP, ASCE, NSPE, ACEC

Finnegan, Thomas J
Home: 66 Summer Street, Buffalo, NY 14209
Position: Consultant (Self-employed) *Education:* BSChE/Polytechnic Inst of Brooklyn. *Born:* June 11, 1901. 1938-41 - Research Assoc Consolidated Edison Co of N Y, studying corrosion and water treatment problems. 1942-70 Chemical Engr Niagara Mohawk Power Corp. Established Chemical Engrg function at NMP, including laboratory design and supervision. Work took on many mechanical engrg features. Was a pioneer in the replacement of evaporators by deionizers for the preparation of make-up water in power plants. Was instrumental in making design changes to combat air and water pollution. Work embraced both fossil fuel and nuclear power stations. Author of many papers in the field of corrosion and water treatment. Registered PE state of New York. Fellow ASME & AIC. Emeritus Mbr American Chem Society. Participant American Nuclear Society. Mbr & former Chairman Chemistry Subcommittee Prime Movers Ctte, Edison Electric Institute. *Society Aff:* ACS, ASME, NYSSPE, AIC.

Finneran, James A
Business: 3 Greenway Plaza E, Houston, TX 77046
Position: Vice President. *Employer:* Pullman Kellogg, Div of Pullman Inc. *Education:* BS/Chem Engg/Univ of Notre Dame. *Born:* 1923 in the Bronx, New York. BSChE University of Notre Dame, 1943. Instructional Asst MIT 1943. Operating Engr, the Kellex Corp, 1944-45. Process Engr, Carbide & Carbon Chemical Corp, Oak Ridge, Tennessee 194546. Process Engr Hydrocarbon Res Inc, 1947-54. With Pullman Kellogg since 1954, as Process Engr, Project Manager, and Director of Process Engineering. Present position with Pullman Kellogg is Vice President & Dir of Process Operations. *Society Aff:* AIChE, ΣΞ.

Finneran, James A
Business: Three Greenway Plaza E, Houston, TX 77046
Position: VP. *Employer:* The M. W. Kellogg Company *Education:* BS/ChE/Univ of Notre Dame *Born:* 9/7/23. Employed as Chem Engr by the Kellex Corp 1944-45. Worked for Union Carbide and Chems Corp Oak Ridge, TN, 1945-47. Employed as Process Engr by Hydrocarbon Res Inc, 1947-54. Joined M W Kellogg Co in 1954 as Process Engr. Present position is VP & Dir of Process Operations. Holder of 10 US patents & is the author of numerous technical publications. *Society Aff:* AIChE, ΣΞ.

Finney, Essex E, Jr
Business: Agri Res Ctr, USDA, Beltsville, MD 20705
Position: Assistant Director *Employer:* US Dept of Agriculture, ARS. *Education:* BS/Agri Engg/VA Polytechnic Inst; MS/Agri Engg/Penn State Univ; PhD/Agri Engg/MI State Univ. *Born:* May 1937. Native of Powhatan, Virginia. Served in the US Army Transportation Corps, Denver Colorado, 1963-65. Research engr with the Agricultural Research Service, US Dept of Agriculture, Beltsville Md, 1965-77. Responsible for developing instruments and methods for measuring the quality of agricultural products. Chairman, Agricultural Marketing Research Inst, Beltsville Md, 1972-75. Princeton Fellow in Public Affairs, Princeton Univ N J, 1973-74. Board of Directors and Chairman Food Engrg Div, American Society of Agricultural Engrs (ASAE) 1970-72. ASAE Fellow, 1984. ASAE Journal Paper Award, 1969. Town Councilman, Glenarden, Md 1975. Sr. Policy Analyst, Office of Sci and Tech Policy Exec Office of the Pres, Washington DC 1980-81. Outstanding Alumni Award College of Engg Penna State Univ, 1985. Asst Dir Beltsville Agricultural Research Center, U S Dept of Agriculture, since 1977. *Society Aff:* ASAE, IFT, TBΠ.

Finney, William G
Home: 1212 Northwood Lane, Muscatine, IA 52761
Position: private practice *Employer:* self *Education:* BS/EE/Univ of NE *Born:* 2/23/24 Native of Chadron, NE. Served in US Naval Air Corps 1942-46 as pilot and navigator. Field engr for Buell & Winter for construction of HV transmission lines and substations. Project Engr for Stone & Webster Engrg for HV and EHV transmission lines and substations. VP of Engrg and part owner of 60 man firm operating the New England area. Started steel pole firm, mfg poles for HV and EHV transmission lines. Currently in private practice, consulting on transmission and distribution systems. Also Manufacturers Representative for several products used by electric utilities. Member of Natl Transmission & Distribution Committee. Enjoys golf and participates in local politics. *Society Aff:* IEEE, PES

Finnie, Iain
Business: Dept of Mech Engrg, Berkeley, CA 94720
Position: Prof of Mechanical Engrg. *Employer:* University of California. *Education:* BSc/Mech Engrg/U of Glasgow; DSc/Mech Engrg/U of Glasgow; MS/Mech Engrg/MIT; ME/Mech Engrg/MIT; ScD/Mech Engrg/MIT. *Born:* July 1928. BSc and DSc from Glasgow Univ; MS, ME and ScD from Mass Inst of Technology. Shell Development Co 1954-61; Assoc Prof 196163 and Professor 1963 to present at University of Calif Berkeley. 1965 Visiting Prof Catholic University of Chile; 1967-68 Guggenheim Fellow; 1976 Invited Prof at Ecole Polytechnique Federale, Lausanne Switzerland; 1975-78 Technical Editor of ASME Journal of Engrg Matls and Tech. Hon Mbr ASME and Fellow I Mech E (London). Mbr, Natl Acad of Engg. Specialty - Mechanical Behaviour of Materials (especially Fracture, Creep, Erosive Wear and Failure Analysis). *Society Aff:* NAE, ASME, SEM, IMechE.

Fino, Alexander F
Home: 51 Hillside Dr, Warren, PA 16365
Position: VP *Employer:* Struthers Wells Corp *Education:* BS/ME/Univ of Pittsburgh *Born:* 5/25/22 Native of Warren, PA. Mgr Engrg for former Hammond Iron Works, now Pittsburgh Des Moines Steel, 1947-1961. Pres Dorcon, Inc 1961-69. VP of Struthers Wells Corp since 1976. Authority on petro conservation devices and high pressure layered vessel tech. Developed in 1955 first commercially accepted soft seal for large floating roofs for oil storage tanks. Has received 23 patents for conservation designs used worldwide to prevent petro evaporation losses. Authored articles for Welding Journal, Welding Research Council, API Bulletin 2516, Erdol and Kohle, Cost Engrg and Pressure Vessels and Piping. Active in community affairs. *Society Aff:* NSPE, ASME, AWS, AWWA, API

Finsterwalder, Ulrich
Home: Pagodenburgstr. 8, 8000 Munchen 60, Fed Repub of West Germany 8000
Position: Beratender Ingenieur. *Employer:* (Independent) *Born:* Dec 1897 Munich Germany. Dr Ing Tech Univ Munich. Son of the Privy Councillor Dr h c Sebastian Finsterwalder (plus1951), Prof at the Tech Univ in Munich. 1923 grad as a Diplom-Ingenieur at the Munich Tech Univ. Dec 1923 entered the firm Dyckerhoff & Widmann; summer 1930, Doctor's degree at the Munich Tech Univ; Dec 1930 appointed Ch Engr, Sept 1934 Proxy, Dec 1938 Dir, Jan 1941 mbr of the Tech Mgmt of Dyckerhoff & Widman K G, 1949 personally liable partner; July 1950 appointed Dr Inc E h of the Tech Univ of Darmstadt; 1968 Dr Ing E h of the Tech Univ of Munich. Hon Mbr 1964 ACI. Charles S Whitney Award 1967 ACI; Freyssinet Medal 1970 FIP.

Fintel, Mark
Home: 1833A Wildberry Dr, Glenview, IL 60025
Position: Director/Engrg Services. *Employer:* Portland Cement Assn. *Born:* 1922. MSc Munich Inst Technology 1950. Since 1961 with PCA, named Director, Engrg Services Dept 1967. ACE WASON Medal in 1971 (with Fazlur Khan). Chairman several ACI and ASCE committees dealing with high rise design. Contributed in several developments of special problems in high rise buildings. Recent concentration of effort in earthquake engrg. Heads PCA Earthquake Investigation Team; visited and reported on most major quakes in last fifteen years. Published many papers; edited Handbook of Concrete Engrg (Van Nostrand, 1974). Named one of Engineering News-Record's Men Who Made a Mark in 1975.

Fiore, Nicholas F
Business: Box E-Met E & Matl Sci, Notre Dame, IN 46556
Position: Professor and Dept Chairman. *Employer:* University of Notre Dame. *Education:* BS/Met Engg/Carnegie Inst of Tech; MS/Met Engg/Carnegie Inst of Tech; PhD/Met Engg/Carnegie Inst of Tech. *Born:* Sept 1939, Pittsburgh Pa. Served with Army Signal Corps, 1964-66. At Notre Dame since 1966, becoming Chairman in 1969. Research interests include internal friction studies of reaction in solids, non-destructive testing, wear, hydrogen embrittlement. Have consulted for Union Carbide, Miles Laboratories, Corning Glass Works, Ford Motor Company, Cabot Corporation, Argonne National Laboratory. Adams Award (AWS), 1971. President, Alpha Sigma Mu, 1976-77. *Society Aff:* ASM, AIME, AAUP.

Fiorentino, Robert J
Home: 836 Singing Hills Ln, Worthington, OH 43085
Position: Program Manager. *Employer:* Battelle's Columbus Labs. *Education:* MS/Met Engg/OSU; BS/Met Engg/RPI. *Born:* July 1930. MS from Ohio State Univ; BS from RPI in metallurgical engrg. Native of Utica, NY. WIth Battelle's Columbus Labs since 1952. Involved in dev of advanced metal-working techniques, including improved form of hydrostatic extrusion (Hydrafilm), hydrostatic wire drawing (Hydraw), & hydrostatic sheet forming. Have several patents in this area. Also involved in new devs in isothermal rolling, cold extrusion, die designs for extrusion of brittle materials, & extrusion cladding. Became Mgr of Battelle's Metalworking Sect in 1970, & Prog Mgr of Metalworking Systems since 1974. Chmn Emerging Processes Div SME; Mbr Extrusion & Drawing Ctte ASM; Mbr Bd of Dirs, Amer Tube Assoc. Enjoy golf, art, music, & fishing. *Society Aff:* ASM, SME, ATA.

Fiorillo, Michael
Business: NAVAIR HQS (PMA-259A), Washington, DC 20361
Position: Deputy Program Manager. *Employer:* Department of Navy NAVAIR. *Born:* August 1923. MBA from Loyola Univ, New Orleans; BSAE from Univ of Oklahoma. Formerly an Executive with Chrysler Corp, Space Div. Apollo Saturn Program. Presently, Deputy Program Manager on the Sidewinder/ Chaparral Missile Systems with NAVAIR. In addition, an Executive Dir of NAVAIR for the Assoc of Scientists and Engrs of the Naval Air & Sea Systems Commands and am a Registered PE (Louisiana). Mbr of Pi Tau Sigma, Tau Omega engrg honor societies and am licensed with NASD. Served as a pilot and navigator with the USAF during WWII and Korea. As an ardent Skier, am VP of the Air Force Ski Club, Pentagon, and am an enthusiastic lover of Grand Opera, classical music, and sailing.

Fire, Philip
Business: Office of Naval Research, Branch Office-London, Box 39, FPO New York, NY 09510
Position: Liaison Scientist *Employer:* US Navy *Education:* PhD/EE/Stanford Univ; MS/EE/MIT; BS/EE/MIT *Born:* 12/18/25 in Paterson, NJ. Served with US Navy 1944-46. Member of Technical Staff of MIT Lincoln Lab, 1952-54. Member of Technical staff of GTE Sylvania, Inc Systems Group Labs in Mountain View, CA, 1955-1980. Presently, on leave of absence from position as Sr Scientist at GTE Sylvania, with the London Branch Office of US Navy's Office of Naval Research. Research and development interests include error-correcting codes used in communication systems and analysis and synthesis methods in electronic signal processing systems. *Society Aff:* IEEE

Firmage, D Allan
Business: 1079 Ash Ave, Provo, UT 84602
Position: Principal Engineer *Employer:* Firmage & Assoc *Education:* MS/Sfr Eng/MIT; BS/Civil Eng/Univ of UT *Born:* 02/15/18 Engrg educator, researcher, design engr, author and consultant for over 44 years. Pres, UT Section, ASCE, 1967-68, Intl contact Dir, ASCE, 1977-80. UT Engr of the Year 1980. Western Electric Fund Award for excellence in teaching 1971- 72, Rocky Mt Section, ASEE. Chrmn UT Engrg Council 1973-74. Chrmn ASCE Natl Technical committee on bridges 1977-82. Author of 4 engrg books and numerous papers. Lectures on bridge design in 12 countries. Consultant on bridge design to several consulting firms. *Society Aff:* ASCE, IABSE

Fischer, Frederick K
Home: 155 Valley Road, Wawa, PA 19063
Position: Manager, Dev Engrg Dept-Steam Divs-Retired *Employer:* Westinghouse Electric Corp. *Education:* MS/Engrg/Univ of Pittsburgh; EE/-/Rens Poly Inst. *Born:* Sept 1906. Native of Fosterdale, NY. Entire engrg career with Steam Divisions of Westinghouse Electric Corporation as: a heat transfer, steam turbine and gas turbine design and developmment engr; consultant to Electric Utilities on selection and application of steam turbines and heat transfer apparatus, responsible for design of a closed cycle gas turbine propulsion plant for the Navy; Manager of Applied Research and Development Dept for five major Divisions of Westinghouse. Fellow and Life mbr of ASME. Mbr of Sigma Xi and Tau Beta Pi. Have eight US patents. Enjoy golf, gardening, and great outdoors. *Society Aff:* ASME

Fischer, Irene K
Home: 301 Philadelphia Ave, Takoma Park, MD 20912
Position: Retired *Education:* MA/Math/Univ of Vienna, Austria; MA/Descriptive Geom/Vienna Inst of Tech, Austria; Dr. Eng/Geodesy/Tech Univ *Born:* 7/27/07 Taught Math, Descriptive Geometry, Engrg Drawing at Austrian and US secondary schools and Bard Coll, NY; Mathematician and later Supervisory Research Geodesist at Army Map Service, now Defense Mapping Agency (1952-77). Research in "Figure of the Earth," history of Ancient Geodesy and related topics. Over 100 internationally well known technical publications. Honorary Doctorate of Engrg from Technical Univ Karlsruhe, Germany. Distinguished Civilian Service Award from Defense Dept, Decoration for Exceptional Civilian Service, and twice Meritorious Civ Serv Award from Dept/Army. AGU Fellow. Natl Civil Service League Award, Federal Retiree of the Year Award, Mbr of Natl Academy of Engrg, a.o. Past officer at Intl Assoc of Geodesy. *Society Aff:* AGU, WSE, NAE

Fischer, James R
Business: Bldg T-12, Agri Engr Dept, Columbia, MO 65211
Position: Res Engr. *Employer:* US Dept of Agri. *Education:* PhD/Agri Engg/Univ of MO; MS/Agri Engg/Univ of MO; BS/Agri Engg/Univ of MO. *Born:* 2/23/45. Conducting res on pollution control for USDA - 1969-72. From 1973 to present eveloped alternative energy system for agri. Responsible for the dev and implementation of the first automated anaerobic digester integrated in an agri production system. Pres, MO Sec ASAE, 1977; Vchrmn, Mid-Central Region ASAE, 1979. *Society Aff:* ASAE, ΣΞ, AE, ΓΣΔ.

Fischer, Jerome M
Business: One Rockefeller Plaza, New York, NY 10020
Position: Exec VP - Dir Pres *Employer:* Planning Research Corp PRC Engrg, Inc *Education:* B/CE/NC State *Born:* 3/15/24 Native New Yorker. Career began in 1948. In 1952, managed Central American operations of TAMS. Joined Frederic R. Harris in 1960 as VP; managed their European and Middle Eastern activities. In 1967, named Exec VP and Dir and 1972, Chrmn. 1974-present, Dir and Exec VP of Planning Research Corp. In 1981, Pres of PRC Engrg. Registered engr in US, Great Britain, Netherlands. Decorated by Belgian Government with Order of the Crown, holds Spanish Medal of Honor, and Interpetrol Award from Italian Government, all for outstanding engrg accomplishments. *Society Aff:* ASCE, NSPE, SAME, ACEC, IRF

Fischer, Peter A
Home: 1812 N Furness St, Maplewood, MN 55109
Position: consulting civil engineer *Employer:* self *Education:* MS/Structural Engrg/Univ of MN; B/CE/Univ of MN; BS/Univ of MN *Born:* 11/2/31 Life-long resident of St. Paul, MN area. Member of Chi Epsilon and Tau Beta Pi. Worked for the US Bureau of Public Roads (1953) and MN Highway Dept (1954). St. Paul Dist. Corps of Engrgs 1955-1987. Formerly Chief Engrg Div responsible for the management of water resources planning and engrg in Upper MS, Souris and Red River Basins. Self employed consultant in Hydrologic and Hydraulic design of reservoirs, dams, channels, hydraulic structures and dam safety. Member ASCE Hydraulic Structures Committee 1976-1981, Chrmn 1980. Chrmn, ASCE Task Committee, Design and Performance Reversible Flow Trash Racks 1980-1983. Mbr NSPE, SAME. Pres Mpls-St. Paul Post SAME 1983. Reg PE MN, ND, WI. *Society Aff:* ASCE, SAME, NSPE, USCOLD, USCIDFC.

Fischer, Stewart C
Home: 9615 Lantana, San Antonio, TX 78217
Position: Director of Traffic & Trans. *Employer:* City of San Antonio. *Education:* Cert/Highway Transportation/Yale Univ; BS/Civil Eng/Univ of TX-Austin. *Born:* January 1924. Certificate in Hwy Traffic, Yale University. Native of New Braunfels, TX. Served in US Army in Europe in WWII. Employed by Texas Highway Dept as structural designer. Employed by City of San Antonio since 1952; Dir of Traffic and Transportation since 1962. Fellow, Institute of Transportation Engineers; International Director, 1972. Texas Traffic Engr of the Year for 1975. Past President San Antonio Lions Club. Married to Myra Katt; 2 children: Nancy (Mrs. Donald Shindler) and Kenneth. Listed in Who's Who in South and Southwest. *Society Aff:* ITE.

Fisher, C Page
Home: P.O. Box 8761, Durham, NC 27707
Position: Principal *Employer:* Fisher Assocs *Education:* PhD/CE/NC State Univ; SM/Soil Mech/Harvard Univ; BS/CE/Univ of VA *Born:* 9/24/21 Early education in Richmond, VA; Employed by VA Highway Commission 1940; Served in US Navy 1941-1945; Employed as engr in various capacities by W.W. LaPrade & Bros, Richmond, VA; Metcalf & Eddy, Thule Greenland; Robertson & Assocs, Baltimore, MD, H.S. Porter Assocs, Richmond, VA 1946-1955; Instructor to Assoc Prof NC State Univ, 1955-1969; Principal and Pres, Geotechnical Engrg Co, Raleigh and Research Triangle Park, NC 1961-1978; Principal, Gardner-Kline Assocs, 1968-1971; Principal C. Page Fisher Consulting Engr, Durham, NC 1978 to date. Fellow ACEC: Past Pres and Natl Dir CEC/NC, Fellow ASCE: Chrmn ASTM Sub Committee D18.01, Registered PE in VA, NC, SC, MD, TN; NCEE Certificate. Secretary, Troxlier Electronics Corp. Member TRB Comm. AZH0I *Society Aff:* ASCE, ASTM, ACEC, NSPE, AAA

Fisher, Cary A
Business: Dept of Engr Mech, Colorado Springs, CO 80840
Position: Dept Head *Employer:* US Air Force Academy *Education:* PhD/Mech/Univ of OK; MA/Gov/Univ of NM; MS/Aeronautics/CA Inst of Tech; BS/Engrg/US Military Academy *Born:* 12/29/41 Serving as Prof and Head of the Dept of Engrg Mechanics and Chairman of the Engineering Division at the US Air Force Academy. He is a graduate of the US Military Academy and his graduate degrees include a Master of Science in Aeronautical Engrg from the CA Inst of Tech; a Master of Arts in Government from the Univ of NM, and a PhD in Engrg Mechanics from the Univ of OK. He has served in a variety of Air Force engrg and combat support roles in addition to his career as a military educator and Asst Dean. His Sabbatical Assignment include Chief Sci for the European office of Aerospace Res and Dev and Sci off in the Amer Embassy, London. He is a member of the American Academy of Mechanics and has served on the Bd of Dir of ASEE. *Society Aff:* ASEE, AIAA, ASME, AAM, SES

Fisher, David L
Business: 11237 Pellicano, El Paso, TX 79935
Position: VP. *Employer:* Saltech Corp. *Education:* BS/ChE/NM State Univ. *Born:* 5/20/48. Helped form Saltech Corp in 1976 for the purpose of mfg water and waste treatment equip. Most equip is of the membrane separation and ion exchange type. Dir of all engrg activities. Previously work for El paso Environmental Systems and Continental Water Conditioning Corp. A native of the southwest. Actively enjoy nearly all types of recreational activity including jogging and soccer. *Society Aff:* AIChE.

Fisher, Dennis K
Business: P.O. Box 808, Livermore, CA 94550
Position: Associate Director for Engineering *Employer:* Lawrence Livermore National Laboratory *Education:* PhD/M.E./University of Michigan; M.S./M.E./Cornell University; B.S./M.E./Cornell University *Born:* 09/23/43 Employed Lawrence Livermore National Lab, 1970-. Assumed current position as Assoc Dir in June 1986; responsible for Electronic and Mech Engrg support to the Lab's scientific programs, Engrg Research activities, and the Technology Transfer and Precision Engrg programs. Previously headed Materials Fabrication Div, 1982-86 and Magnetic Fusion Engineering Div, 1979-84. Technical specialties include design, analysis, and control of dynamic systems; digital data acquisition, control, and signal processing systems. Mbr, Bd of Visitors, College of Engrg, Univ of California-Davis. Reg PE Society affiliations: Pi Tau Sigma, Tau Beta Pi. *Society Aff:* ASME, IEEE, SME, ΤΒΠ, ΠΤΣ

Fisher, Edward R
Business: College of Engg, Detroit, MI 48221
Position: Assoc Dean. *Employer:* Wayne State Univ. *Education:* PhD/Chem Engr/Johns Hopkins; BSc/Chem Engr/UC Berkeley *Born:* 3/24/38. in Detroit; grew up in Southern CA. While attending Berkeley, worked at Stauffer Chem Co. Before beginning grad study and during summers, worked at Lawrence Radiation Lab, Livermore. Spent post-doctoral yr at Univ of Copenhagen working on fundamental collision problems. Worked at Space Sciences Lab, GE Co for two yrs before coming to Wayne. I am Assoc Dean for Res and Grad Progs, Dir of Res Inst for Engg Sciences, Dir of Ctr for Industrial Safety and Security, and Prof of Chem Engg. *Society Aff:* APS, AIChE, ASEE

Fisher, Edwin E
Home: 3319 Maxwell, Midland, TX 79703
Position: Pres. *Employer:* Vapor Compression, Inc. *Education:* BS/ChE/Univ of TX, Austin. *Born:* 5/22/34. Process Engg with El Paso Gas Products Co 1958-60. Process Sup with Union Carbide Chem Co 1960-63. Chief Engg National Sulfur Co 1963-65. VP; Engg and Dev, Elcor Chem Co 1966-70. Chmn of Permian Basin Chapter of AIChE, VP, Western Oil Shale Corp, 1971-77. Pres, Vapor Compression, Inc. 1977-present. Active in consulting and design in the area of gas processing, especially vapor recovery and petroleum conservation activities. *Society Aff:* AIChE, NSPE, ACS.

Fisher, Franklin G
Home: 1942 Palm Street, Reading, PA 19604
Position: VP-Rail Trans & Mass Transit. *Employer:* STV, Inc. *Born:* Dec 1916. PE 1953. Wyomissing Polytechnic Inst & Penn State University. Native of Reading, Pa. 35 years with Reading Company, Class I Railroad, 1938-73; held position last 20 years: Chief Engineer, Chief Mechanical Officer, General Manager of Operations, and Vice President Operations and Maintenance. Assumed current responsibility, Vice President, Rail Transportation & Mass Transit Division, STV Inc, an A/E consulting firm, May 1, 1973. AM responsible for all marketing, design and operations of the Transportation Division regarding rail and bus transportation contracts: Federal and State Departments of Transportation, Railroads, Transit and Bus Authorities. ASME Fellow. Enjoy all sports. *Society Aff:* ASME, NSPE, AREA.

Fisher, Garland F
Home: 603 Maiden Lane, Muscatine, IA 52761
Position: VP *Employer:* Stanley Consultants, Inc *Education:* BS/EE/IA State Univ *Born:* 5/4/24 Joined Stanley Consultants in 1946. 3 years designing rural electric lines, power plants and electric service for other facilities. 1 year inspecting construction of steam electric generating station. 2 years of electric design of power plants, substations and lines. 4 years in new business development and administration of training program for new engrs. Project Management experience on electric system studies, diesel and steam power plants, substations, transmission and distribution lines, steam heating plants, chilled water plants and wastewater treatment facilities. BS, 1946; Chrmn, IEEE, Bd of CEC (IA); Committees of NSPE, VP, Stanley Consultants, 1968; Registered Engr in nine states. *Society Aff:* NSPE, IES, IEEE, PEPP

Fisher, Gary D
Business: 2700 South Post Oak, Houston, TX 77056
Position: Vice President. *Employer:* ChemShare Corporation. *Education:* PhD/ChE/Johns Hopkins; BS/ChE/Univ of TX. *Born:* August, 1934. PhD from Johns Hopkins, 1965; BSChE from Univ of Texas. Project Engr for AEC, Naval Reactors Div, 1957-61. Asst Prof, Rice Univ, 1964- 69, with research interests in thermodynamic properties of fluids. VP with ChemShare Corp, 1969-76. Major responsibilities involve development of distillation and thermodynamic property methods for computer aided process design. Mbr AIChE. *Society Aff:* AIChE, ASEE.

Fisher, Gordon H
23431 No 82nd St, Scottsdale, AZ 85255 *Employer:* Retired from Kenecott Copper Corp. *Education:* MS/Petroleum Engrg/Univ of TX; BS/Petroleum Engrg/Univ of TX. *Born:* March 12, 1918 in Gainesville, Texas. Served as Petroleum Engr for Phillips Petroleum Company in reservoir engrg work (1939-1941). Employed by Gulf Oil Corp as Geologist, Geological Technician, Chief Petroleum Engr, Manager of Supply & Services, Executive Asst to VP and Manager of Production in Fort Worth, Texas Production Division (1941-55). In 1955, joined Plymouth Oil Company as VP and General Manager and Director. Was responsible for all exploration and production activities on a world-wide basis. In 1963, joined Kennecott Copper Corp as Asst to the President. Elected VP-Planning in 1968. In 1978 elected VP-Asst to Chrmn Also served as a Director of Peabody Coal Company (Chmn - 1971-72), Quebec Iron and Titanium Corp and Quebec Columbium Ltd. Director of Engineers Joint Council 1971-72. Interested in lapidary, gardening and sports.

Fisher, John W
Business: Fritz Engrg Lab 13, Lehigh University, Bethlehem, PA 18015
Position: Prof CE; Assoc Dir Fritz Engrg Lab. *Employer:* Lehigh University. *Education:* PhD/Structures/Lehigh Univ; MSCE/Structures/Lehigh Univ; BSCE/Civil Engg/WA Univ, St Louis. *Born:* 2/15/31. Native of Scott City, MO. Served with Army Corps Engrs 1951-54. Asst Bridge engr with NAS-HRB at AASHO Road Test 1958-61. At Lehigh Univ since 1961. Apptd Professor of Civil Engg in 1969,

Fisher, John W (Continued)
Assoc Dir of Fritz Engg Lab in 1971-1985, respon for operations & planning. Co-chairman, Dept. of Civil Eng. 1984-85, Director of NSF Eng. Res. Center, Advanced Tech. for Large Structural Systems in 1986. Specialist in behavior, fatigue, & fracture of structural details, & composite steelconcrete mbrs. Specialist in Connection Behavior. Consultant to industry & government. Author of 150 articles in scientific journals & co-authors of 5 books. ASCE Walter L Huber Research Prize, 1969; AWS Adams Mbrship Award 1975; AISC T R Higgins Lectureship Award 1977; ASCE Ernest E Howard Award 1979; Lehigh Valley PSPE Engr of Year 1980; ASCE Raymond C. Reese Res Prize 1981; Civil Coll Eminent Overseas Speaker, Inst of Engrs, Australia 1983; ABCD Outstanding Member Award, 1983; Welding Inst of Canada Education Award, 1983. Cited by Engineering News Record for improving Bridge design and helping streamline inspection in 1984; Named Senior Visiting Scholar to the People's Republic of China in 1985; Co-recipient of the Eleanor and Joseph F. Libsch Research Award at Lehigh University for outstanding research in 1985; Elected a member of the National Academy of Engineering in 1986, Named "Cnstruction's Man of the Year" by Engineering New Record in 1987; Presented with the 1987 Engineering Alumni Achievement Award from Washington University, MO, in 1987. *Society Aff:* ASCE, NSPE, AWS, IABSE, ASEE, PSPE, AREA, AISC.

Fisher, Joseph F
Home: 1517 Powder Mill Lane, Wynnewood, PA 19096
Position: Retired. *Employer:* Ford Aerospace & Communications Corp. *Education:* Grad/Elec Eng/Drexel Univ/-/Diploma-Elect Eng (1936)/-/ *Born:* 2/28/11. Elec Engrg, Drexel Univ (Day School) 1929-34; Drexel Univ (Evening School) 1934-36, Graduate. Engrg Sect Mgr & Res Engr with Philco, Philco-Ford, & Ford Aerospace & Communications Corp, 1934-76. Author of over 20 articles on television published in tech journals. Awarded five patents in Television field, Respon for design & implementation of High Resolution TV Command & Control System installed in NASA Mission Control, TX. Life Mbr of SMPTE 1976., elected to grade of Fellow IEEE in 1961 for contribs to color television; Former Mbr of the Natl IEEE Ctte G2.1 and G2.1.1 on Audio/Video Techniques; Chmn of Phila Sect IEEE Awards Ctte 1974-75. Awarded Phila IEEE Sect Award Feb 1976: "For his important pioneering contribs in applying television principles to space, medical & instrumentation applications & his leadership in the IEEE'. Life Fellow of IEEE 1980. Mbr of Natl SMPTE Committee on Television Video Tech (1984- 87). Co-Author of two chapters of second ed of "Television Engrg Handbook" published by McGraw Hill 1986. Mbr Lower Merion Township, PA, "Cable TV Evaluation and Advisory Ctte" (1981-1984). *Society Aff:* SMPTE, IEEE.

Fisher, Lawrence E
Home: 149 Sedgwick Road, West Hartford, CT 06107
Position: Owner. *Employer:* Fisher Engineering. *Education:* BSEE/Univ of MI. *Born:* March 1905. Fellow IEEE. Engrg employment includes Square D Co, Hudson Motor Car Co, Bull Dog Electric Co in Detroit Mich, and General Electric Co in Plainville Ct. Presently, consulting engr registered in CT and Mich. He has been a leader in using relays to protect equipment and people from destructive and dangerous arcing on low-voltage systems. He is listed in Who's Who in Engineering, International Blue Book, and American Men of Science. He is the author of many IEEE technical papers, and the inventor of 52 patents. *Society Aff:* IEEE.

Fisher, Lawrence R, Jr
Home: 7808 S Florence, Tulsa, OK 74136
Position: Pres. *Employer:* Williams Brothers Engg Co. *Education:* MSCE/Water Resources/Univ of AR; BSCE/Water-Wastewater/Univ of AR. *Born:* 6/26/41. Native of Benton AR - Proj Engr for Exxon Co; USA 1965-1970; Proj Engr for Pyburn & Odom, 1970-1972; Sec Mgr Ford, Bacon & Davis 1972-1976; Mgr, Water Tech, Williams Brothers Engg Co 1977 to 1980. Dir Env Engr 1980 to 1982; VP Process Div 1982-83; VP Engg. Div 1983-84; Pres. 1984 to present. *Society Aff:* ASCE, NSPE, SAME, WPCF.

Fisher, Will S
Business: Nela Park, Cleveland, OH 44112
Position: Technical Relations *Employer:* GE. *Education:* BS/EE/Vanderbilt Univ. *Born:* 6/27/22. Mr Fisher was Pres of the Illuminating Engg Soc of N Am in 1978-79, VP for Technical Activities in 1975-77, Dir 1972-75. He has performed original work and written many papers about the integration of lighting, heating and cooling in bldg design. He presented papers at the Intl Commission on Illumination (CIE) in Amsterdam 1983, Kyoto 1979, London 1975, Barcelona 1971, Vienna 1963; is US expert to CIE Division 3 (Interior Environment and Lighting Design). Was a mbr of Panel 9 (Lighting) of ASHRAE 90-75, and the Proj Committee on ASHRAE 100.4 (Industrial Bldgs) '79. Has been Chapter Chrmn for Lighting on all editions of the IEEE Gray Book since its inception. During World War II, served as 1st Lieutenant in the US Army Engrs on the Manhattan Project (Atomic Bomb). *Society Aff:* IES, ASHRAE, IEEE.

Fisk, Edward R
Business: 1224 E. Katella, Suite 105, Orange, CA 92667
Position: VP and Vice P-Western Region *Employer:* Lawrance, Fisk & MacFarland, Inc. and Construction Consulting Group, Inc. *Education:* cert/Bus Adm/Education/Univ of CA *Born:* 7/19/24 Over 38 yrs in engrg and construction. VP. Western Region, Construction Consulting Group, Inc., and VP of Lawrance, Fisk & McFarland, Inc., Orange, CA. Reg PE in 12 states; land surveyor in two states; and licensed contractor. Construction Mgmt conslt. Adjunct Prof of Construction Mgmt. Also teaches seminars for Univ of CA, Berkeley, and for ASCE nationwide. Past pres and state dir of CSPE/NSPE. Fellow of ASCE and Past Chrmn of Construction Div Exec Ctte. Past Chrmn, ASCE Natl Ctte on Inspection. Diplomate, Natl Acad of Forensic Engrs. Member Tau Beta Pi, National Engrg Honor Soc. Author of several books, one adopted by over 60 univ as a text. Author of 25 technical papers. Lifetime teaching credential. Studied CE at Marquette Univ. He is a candidate for a BS in mgmt and a MBA in bus adm at CA-Western Univ. *Society Aff:* ASCE, USCOLD, AAA, ΤΒΠ, NSPE.

Fitch, Bryant
Business: Auburn-Univ, Auburn, AL 36830
Position: Adjunct Professor. *Employer:* Department of Chemical Engineering. *Education:* BS/Chem Eng/Cal-Tech; MS/Chem Eng/Univ of CT. *Born:* 4/30/10. In production, then R&D for Amer Potash & Chem Copr, 1933-44. With Dorr- Oliver Inc, 1944-75, holding various positions incl Research Dir & Chief Scientist. Visiting Indus Professor, Carnegie-Mellon Univ 1973-77. Publications largely in field of sedimentation. currently doing research on thickening under grant from NSF. Fellow of AIChE. *Society Aff:* AIChE.

Fitch, Lewis T
Business: Electrical and Computer Engrg Dept, Clemson, SC 29634
Position: Alumni Professor (EE) *Employer:* Clemson Univ *Education:* PhD/EE/OH State Univ; MS/EE/NC State Univ; BS/EE/Duke Univ *Born:* 12/29/32 Born in Columbus, OH. Worked for RCA, Vitro Labs as design engr. Army Signal Corps Instructor, NC State Coll. Became Electronics Engr and Research Assoc at the OH State Univ Radio Observatory. With Dr. John Kraus discovered the astronomical object with the highest red shift ever observed. Currently Prof of Electrical Engrg at Clemson Univ, and engaged in research in the behavior of solid state electronics at high temperatures. Enjoys playing musical instruments, electronic and traditional, classical and folk, and is active in amateur radio. *Society Aff:* IEEE, ASEE

Fitch, W Chester
Business: Depreciation Programs, Inc, 5320 Holiday Terrace, Oshtemo, MI 49077
Position: Dean, College of Applied Sciences. *Employer:* Western Michigan University. *Born:* Nov 12, 1916 Billings, Montana. BSIE, Montana State Coll 1938; MS in Engrg, Valuation Iowa State Univ, 1939; PhD, 1950. Teacher of engrg at Iowa State

Fitch, W Chester (Continued)
Univ, 1939-45; Asst Prof Indus Engrg, Montana State Coll, 1945-46; also Prof at Iowa State Univ 1947-52. Asst Dir Valuation Div, Gannett Fleming Corddry & Carpenter, cons engrs, Harrisburg Pa 1952-58, 1959-64. Prof, Head Mech Engrg Dept Utah State Univ 1958-59; Mich Tech Univ 1964-68; Prof, Chmn Engrg & Tech Dept at Western Mich Univ 1968-72, also Assoc Dean Coll of Applied Sciences, 1972-73 & Dean 1973- . Cons engrg 1952- . Pres Depreciation Programs Inc 1978- . Reg PE Ia, Pa. Mbr ASME, ASEE, NSPE, IIE, Tau Beta Pi, Phi Kappa Phi, Pi Mu Epsilon, Pi Tau Sigma, Lambda Chi Alpha. Dir Depreciation Prog 1967-77. m. Manzella L Groth 1946; 1 son, David P Presbyn. Rotarian.

Fitchen, Franklin C
Home: College of Engg/Univ of Bridgeport, Bridgeport, CT 06602
Position: Professor of Electrical Engineering *Employer:* University of Bridgeport. *Education:* DEng/EE/Yale Univ; MSEE/EE/Northeastern Univ; BSEE/EE/Univ of RI. *Born:* June 15, 1928 New Rochelle, NY. GE Co, Lynn, Mass. 1950-56; Assoc Prof Univ of RI 1956-65; Chmn, Elec Engrg, South Dakota State Univ, 1965-72; Dean of Engineering, University of Bridgeport 1972-80. GE Co, Bridgeport CT, 1980-81. Professor of Electrical Engineering, Univ of Bridgeport 1981- . Director, Brookings Municipal Utility Board, 1971-72. Served with US Army Ordnance Corps. Author: Transistor Circuit Analysis & Design, 1960, 2nd ed 1966; Electronic Integrated Circuits & Systems, 1970, 2nd ed 1980; Low-Noise Electronic Design, 1973; numerous foreign translations; numerous journal articles; editor: World Directory of Engineering Schools, 1980. *Society Aff:* ASEE, NSPE, IEEE.

Fithian, Theodore A
Business: 227 S Div St, Zelienople, PA 16063
Position: Pres. *Employer:* Sybron/Leopold. *Education:* BCE/Sanitary/Rensselaer Poly Inst. *Born:* 3/30/28. Bridgeton, NJ. Served with Army Airborn Engrs 1946-1949. Joined the Chester Engrs, Sanitary Engr Consultants in 1953-Resident Engr 1953-1955, Proj Engr 1955-1964, Partner 1964-1969, Proj Mgr 1969-1972-, Dir SE Office 1972-1977. Became Pres Leopold Div, Sybron Corp, in 1977-Mfg of Water and Waste Treatment equip. Borough Councilman 1965-1972. Pres RPI Alumni Assn 1977-1978. Reg Engr in 9 States. *Society Aff:* NSPE, ASCE, AAEE, AWWA, WPCF.

Fittz, Ronald D
Business: 1144 Petro Bldg, Beaumont, TX 77701
Position: Principal *Employer:* Fittz, Crozier & Shipman *Education:* M of Engrg/CE/TX A&M Univ; BS/CE/McNeese Univ *Born:* 11/16/48 Center, TX. Engrg Technician, Vernon F. Meyer & Assocs, Consulting Engrs, Lake Charles, LA, 1966-70. Research Asst, TX Transportation Inst, College Station, TX, 1970-71. From 1971-74, with Leonard Shoemaker & Assocs, Consulting Engrs, Houston, as project engr for over 40 projects. From 1975-79, VP and Beaumont mgr for Houston-based Walter Moore & Assocs, Consulting Engrs. Managed office of 10-15 engrs and support staff, responsible for average of 30 projects and $200,000 gross fees per year. Founded present firm in 1980, currently with 15 employees. Presently, on Bd of Trustees, Beaumont Independent School District. Deacon and Sunday School Dir, First Baptist Church, Beaumont. Enjoy camping and jogging. *Society Aff:* ACEC, NSPE, ASCE

Fitzgerald, Dennis J
Home: 11 Princess Lane, Loudonville, NY 12211
Position: Exec Dir./Gen. Mgr. *Employer:* Capital District Transportation Auth. *Education:* BCE/-/RPI; MCE/-/RPI. *Born:* April 1939. USN Line Officer and Civil Engr Corp; Society of American Military Engineers Award, CECOS. Worked for New York Telephone after graduate school, then Bertram D Tallamy Associates, consulting engineers. With CDTA since 1971, assumed present position in 1980. President, Mohawk-Hudson Section, ASCE 1974. Serve on RPI Civil Engineering Curriculum Advisory Council. Registered Prof. Engr, NYS. Regional Director of American public Transportation Association.

Fitzgerald, Edwin R
Business: 20 Latrobe Hall, Baltimore, MD 21218
Position: Prof. *Employer:* Johns Hopkins Univ. *Education:* PhD/Phys/Univ of WI; MS/Phys/Univ of WI; BS/EE/Univ of WI. *Born:* 7/14/23. Physical Res Dept of B F Goodrich Co, Akron, OH 1944-46; Teaching Asst, Univ of WI, Phys 1946-48; Math 1947-48; Res Asst Chemistry 1948-51; Res Assoc, Chemistry 1951-53 Asst Prof of Phys, PA State Univ 1953-56; Assoc Prof of Phys 1956-59; Prof of Phys, 1959-61. Prof (Mechanics Dept) the Johns Hopkins Univ 1961-. Fellow Am Physical Soc. Former Chrmn and mbr of exec committee of High Polymer Div of Am Physical Soc. Mbr Acoustical Soc of Am, Mtls Res Soc, Eta Kappa Nu, Tau Beta Pi, Sigma Xi, Phi Beta Kappa. Reg PE (ME, EE) MD No 11,669. Author "Particle Waves and Deformation in Crystalline Solids" (Wiley, 1966), numerous technical articles on elec and mech properties of mtls, including polymers, metals, alloys, crystals, etc. *Society Aff:* APS, ASA, MRS.

Fitzgerald, J Edmund
Business: Sch of Civil Engg, Atlanta, GA 30332
Position: Dir & Prof, Civil Engg. *Employer:* Georgia Inst of Technology. *Education:* DSc/Math & Physics/Natl Univ of Ireland; MSc/Math & Physics/Natl Univ of Ireland; MS/Civil Engg/Harvard Univ. *Born:* September 1923. Director, School of Civil Engineering at Tech since 1975. Previously Assoc Dean and Chairman CE, Univ of Utah since 1966. Director Research and Engrg, Lockheed Propulsion Co, prior to 1966. Over 100 articles and monographs in applied mechanics, polymer and propellant behavior. Active consultant in aerospace field. Many government and professional cttes. Alexander von Humboldt Senior Fellow, 1973-74; Fellow ASCE and Inst Physics. *Society Aff:* ASCE, AIAA, AAM.

FitzGerald, James E
Home: 231 Woodlawn, Glencoe, IL 60022
Position: Pres. *Employer:* Clayton Mark *Education:* BS/CHE/Univ of MI *Born:* 10/28/21. Native Muskegon MI; Plant Mgr-Div Mgr 1950-1960; VP Intl Operation 1960- 1972 Brunswich Corp. Pres Intercraft Ind 1972-1974; Pres Div Sargent Welch 1974- 1979. Pres Clayton Mark 1980- . Scout master 1939-1984. Mbr 1st Russian Cultural Exchange Group. Active in church & community. Dir Biosynergy Corp, Steria Corp; Pres James FitzGerald & Assoc. *Society Aff:* AMA, ACS, AICHE, WSC, NWWA

Fitzgerald, John E
Home: 255 Seaside Avenue, Westbrook, CT 06498
Position: Consulting Engr (Self-employed) *Education:* BS/Civil Eng/Univ of Ct. *Born:* September 1928. Native of New Haven, Conn. Served with Army Airborne 1946-48 specializing in Air Transport. Sr Safety Engr Liberty Mutual Ins 1952- 62. Town Engr Southington, Conn 1962-64. Private practice since 1964. Served on US Consumer Product Safety Commission for Arch Glass Standards. Vice Pres, Secy, Pres of Conn Soc PE; Natl Dir NSPE. Hobbies are golf and fishing. Conn Engineer of the Year 1982. *Society Aff:* NSPE, ASCE

Fitzgerald, Joseph P
Business: PO Box 5000, Cleveland, OH 44101
Position: Principle Sys Engr *Employer:* Cleveland Electric Illuminating Co. *Education:* BS/Elec Engg/PA State Univ. *Born:* September 1923. Native of Sheffield, Pa. Served with Infantry in Germany 1943-46. Grad Electrical Engr Penn State Univ 1950. Employed since then by the Cleveland Electric Illuminating Company. Since 1982, Principal Sys Engr, System Operations & Planning Dept. Formerly Mbr IEEE Standards Board and Joint Committee on Nuclear Power Standards; Past Chairman, IEEE Switchgear and Standards Coordinating Committees; VChrmn EE I ES&E Ctte; Head, EL&P Delegation to ANSI C-50; Chrmn 1980-81 AEIC Ctte on Electric Power Apparatus; IEEE/PES Ad Com. Received 1972 IEEE Standards Medal; Fellow, IEEE 1975. *Society Aff:* IEEE, AEIC, EEI.

Fitzpatrick, Edward B, Jr
Home: Wishing Well Farm, Old Brookville, NY 11545
Position: Pres & Chief Exec Officer. *Employer:* Clark Fitzpatrick Constr. Corp *Education:* BS/Civil Engg/Univ of Notre Dame. *Born:* January 1933. PE in New York, New Jersey, Georgia & Michigan. Fellow of ASCE and mbr of The Moles. Mbr of Advisory Council, College of Engrg, Univ of Notre Dame. President & CEO of the Fitzpatrick organization for the past 12 years during which the firm constructed over $300,000,000 of new heavy construction projects such as Battery Park City, NYC Subway, Atlanta Subway, Dade County Transit System, Jamestown, R.I. Bridge, Coney Island Viaduct, Hoosick St. Bridge, Northeast Marine Terminal, Nassau County Outfall, Meadowbrook Parkway and many more. Lives with his wife Elizabeth and their nine children in Old Brookville. His oldest Daughter, graduated with a degree in Civil Engg from Notre Dame. *Society Aff:* ASCE, MOLES, AGC

Fitzpatrick, Thomas C
Business: 555 S Front St, Columbus, OH 43215
Position: Pres. *Employer:* Elford, Inc. *Education:* BCE/-/OH State Univ. *Born:* 4/6/21. Worked for Elford, Inc during high sch & to pay way thru OSU, & after three yrs in US Navy, returned to Elford, Inc as superintendant, chief engr in 1952, exec VP in 1956, pres in 1965, & owner in 1976. This co builds all types of commercial, ind, & inst construction since 1910, including all of Battelle Meml Inst in Columbus & the BMI Nuclear Ctr at W Jefferson, OH. Mbr of several professional organizations, Active in Civic Affairs, Winner of numerous awards.

Fitzroy, Nancy D
Home: 2125 Rosendale Rd, Niskayuna, NY 12309
Position: Mgr, Energy & Environ Programs *Employer:* Gen Electric Co, Turbine Marketing & Projects Operation *Education:* BChE/Chem Engg/Rensselaer Poly Inst; D.Sc. (hon)/New Jersey Inst. Technology. *Born:* Oct 1927. Native of Pittsfield, MA. Engr, Gen Elec Corp Res, 1950-76, specializing in heat transfer. Editor and author GE Heat Transfer and Fluid Flow Data Books 1955-74. ASME President, 1986-87; ASME Bd of Governors, 1983-85; ASME, Senior VP, Council on Public Affairs 1981-83; ASME Section Chrmn 1963-64; Secy Region III 1966-68; Policy Bd Education 1974-77; Natl Nominating Cttes 1977-78; VP, Policy Bd Prof and Public Affairs 1978-81. Papers published in founder societies journals. Natl Sci Found Advisory Comm for Res 1972-75. Natl Res Council, Assembly of Engrg, Mbr Bd on Engrg Manpower and Educational policy 1974-76. Am Inst Chem Engrs: Fellow, The Institution of Mechanical Engineers, Great Britain. Charter Mbr Rensselaer Council, 1972 - present. Two patents. Reg PE, NY. Soc Women Engrs Achievement Award, 1972 (hon). RPI Demers Medal, 1975. Fellow ASME 1978, Power Systems Sector, Gen Elec Co, Engrg Award, 1979. Federation of Prof Women Achievement Award 1984. Conference Chrmn Career Guidance for Women Entering Engrg, 1973. Pilots helicopter and personal twin-engine Cessna 310. Who's Who in Amer; Who's Who Amer Women; Who's Who East; World's Who's Who, etc. Married to Roland V. Fitzroy, Jr. *Society Aff:* ASME, AIChE.

FitzSimons, Neal
Home: 10149 Cedar Lane, Kensington, MD 20895
Position: Principal. *Employer:* Engineering Counsel. *Education:* BCE/Civ Engrg/Cornell Univ. *Born:* 9/28/28. Army service in WWII and Korean War. Researcher & teaching asst at Cornell and Lehigh Univs. Overseas work in design & construction of roads, airfields, bldgs, & pub utilities in Europe, Africa and Asia. R&D and mgt of programs in nuclear testing & protective structures. Res in strategic defense sys including economic, social and political aspects. Civil Engrg studies of performance of structures & sys. Extensive activities in engrg history and historic preservation. Presently hd of own firm, Engg Counsel. Author and editor of numerous articles and several books on engrg & history. P Pres. ASCE (NCS) Chmn Ctte on Damaged & Failed Structures (ASCE); Chrmn History Ctte (ASCE); Dir. ACI (NCC), Chmn Adv Bd of AEPIC (Univ/MD); Chmn Ctte on Engrg Performance (ASCE). National Dir (ASCE). *Society Aff:* ASCE, NSPE, ACI, SAME, ASTM, IABSE, CSI, CSCE.

Fladung, Jerome F
Business: PO Box 8405, Kansas City, MO 64114
Position: Proj Mgr *Employer:* Black & Veatch Consulting Engrs. *Education:* Grad Sch/Environ Engrg/Univ of KS; BS/Civ Engrg/KS State Univ. *Born:* Aug 1925 Emporia, Kansas. PE, LS. ASCE Fellow; NSPE; MSPE, AWWA; Trustee WPCF; Bd of Control 1981-84, Arthur Sidney Bedell Award; Diplomate, AAEE; MWPCA, Pres 1976-77; MWCPA Award of Merit. APWA Overland Park Bd of Zoning Appeal, Former Bd of Governors Jo Co Mental Retardation Center. Active with Heart of America Council BS of A. United Way. K C Chamber of Commerce, Former Mbr Pres Club; Mbr of Tau Beta Pi and Sigma Tau. Former Chrmn Johnson County, KS, Republican Central Ctte. *Society Aff:* ASCE, NSPE, MSPE, AWWA, WPCF, MWPCA, APWA.

Flaherty, Hugh C
Business: One Columbus Plaza, New Haven, CT 06510
Position: Chrmn of the Bd *Employer:* Flaherty Giavara Assocs, P.C. *Education:* BS/Civ Eng/Tri-State. *Born:* 7/3/29 Born Waterbury, Ct., son of Michael J. and Josephine M. Bagley; BSCE, 1951 Tri St Coll, Angola, Indiana; married Dorothy M. Peters, Oct 1, 1955; children - Paula, Martha and Ann. Proj Mgr Cahn Engrs, Inc., New Haven, Ct; 1966-1969; owner Hugh C. Flaherty Assoc., cons. engrs. and land surveyors, North Branford, CT, 1969-1970; partner Flaherty Giavara Assocs, environmental design consult, West Haven, CT, 1970-1976; 1976 to Present, Chrmn of the Bd, Flaherty Giavara Assocs, P.C., engrs, planners, and environmental scis; 1970 to Present, Secretary and Treas, Environmental Labs, Inc., W Haven, CT; 1973 mbr Connecticut Bd of Reg of Profl Engrs and Land Surveyors. Served with USAF, 1951-1955. *Society Aff:* NSPE, ASCE, APWA, AWWA.

Flaherty, Joseph A
Business: 555 W 57th St, New York, NY 10019
Position: VP, Engrg & Development. *Employer:* CBS, Inc. *Education:* BS/Physics/Univ of Rockhurst, MO *Born:* Degree in physics, Univ of Rockhurst, Kansas City MO in 1952. Began his TV career at WDAF-TV in that city. Served with US Army Signal Corps 1953-54. Joined NBC-TV in New York City in 1955. In 1957 became assoc with CBS TV Network. In 1959, became network's Dir of Tech Facilities Planning. In 1967 promoted to Genl Mgr and subsequently appointed VP Engrg & Dev. Responsible for all engrg & dev activities for CBS TV. Fellow SMPTE; held position of VP for TV Affairs, Treas Financial VP, Exec VP. Recipient of Emmy Awd Citation for the CBS Minicam Color Camera in 1969, and was recipient of the David Sarnoff Gold Medal for Progress in TV engrg 1974; received the 1975 CBS Tech Emmy Awd for Electronic News Gathering; 1979 Montreux Achievement Gold Medal; 1983 Natl Assoc of Broadcasters Engrg Award. Frequent lecturer and author on TV tech. Mbr Royal TV Soc UK; Fernseh-Kingtechnischen G, FDR; Soc Des Electriciens (SEE) Fr. *Society Aff:* SMPTE, IEE, RTS, SEE, FKTG

Flammer, Gordon H
Business: CEE-Utah State Univ, Logan, UT 84322
Position: Prof of CE. *Employer:* Utah State University. *Education:* PhD/Fluid Mech/Univ of MN; MS/Irrig and Drng/UT State Univ; BS/Civil Engg/UT State Univ. *Born:* June 1926. Have taught at Utah State 26 years, Univ of Minnesota 3 years, Asian's Inst of Tech 2 years, Stanford 1 year; served on Board of Directors of the American Society for Engineering Education 1972-74, four terms on Board of the Education Research and Methods Division of ASEE and one term as chairman of that division. Teaching Excellence Award, Utah State Univ for 1972; Utah Engineers Council Outstanding Educator Award 1974; Western Electric Teaching Award 1974. V. Pres for Mbr Affairs & Mbr of the Bd of Dirs of the ASEE 1980-82. Utah State University Teacher of the Year 1987-88. *Society Aff:* ASCE, ASEE.

Flanagan, H Russell
Home: 1203 Montvale Rd, Maryville, TN 37801
Position: Pres. *Employer:* Ruskat, Inc. *Education:* MSc/Organic/OH State Univ; AB/

Flanagan, H Russell (Continued)
Chem-Math/Univ of Vermont. *Born:* 12/21/27. in Worcester, MA. A grad of Mt Hermon Prep School and veteran of US Navy serving in Pacific Theater, 1945-47. A graduate (1953) of The University of Vermont with a Bachelor of Arts degree (Chemistry & Mathematics), did graduate work under Dr. Melvin Newman at Ohio State. Receiving an M.Sc. in 1956 (Organic Synthes) went with Naugatuck Chemical to work on heterogeneous polymerization development. Transfered later to MAP in Malaysia to study Chemical-biochemical interactions in latex processing. Returned to join Marbon Chemical in 1961, later Borg-Warner Chemical. Started Ruskat, Inc in 1976. Enthusiastic fly fisherman and computer bug *Society Aff:* NSPE, ACS, AIIM

Flanagan, James L
Business: AT&T Bell Laboratories, Murray Hill, NJ 07974
Position: Director, Information Principles Research Laboratory *Employer:* AT&T Bell Laboratories. *Education:* ScD/EE/MIT; SM/EE/MIT; BS/EE/MS State Univ. *Born:* Aug 1925. Faculty of Engrg, Mississippi State Univ, 1950-52. US Army, 1944-46. Bell Laboratories, since 1957. Specialization in digital communications, speech transmission and acoustical signal processing. Responsible for research programs in digital communications, robotics, artificial intelligence, man-computer communications and electro-acoustics. Author, 120 technical publications, 40 patents, 2 books. Has served as President, Acoustical Society of America. Pres, IEEE Acoustics, speech and Sig Proc Soc; and on Bd of Governors, American Inst of Physics. Fellow, IEEE. Fellow, Acoustical Society of America. Mbr, Natl Acad of Engg. Mbr, Natl Acad Scis, Achievement Award, IEEE Acoustics Speech and Sig Proc Soc, 1970. Society Award, IEEE Acoustics, Speech and Sig Proc Soc, 1976. IEEE Centennial Medal, 1984. L.M. Ericsson International Prize in Telecommunications, 1985. IEEE Edison Medal 1986. Acourtical Society of America Gold Medal 1986. Sigma Xi, Tau Beta Pi. Swims year-round, likes salt- water fishing. Is an instrument-rated pilot, amateur trumpet player; reads biographies. *Society Aff:* IEEE, ASA, NAE, NAS.

Flaschen, Steward S
Business: 320 Park Ave, New York, NY 10022
Position: Sr VP *Employer:* ITT. *Education:* PhD/Geochemistry/Penn State; MS/Chemistry/Miami; BS/Chemistry/Univ IL. *Born:* 5/28/26. Native of Chicago, IL. Served in Naval Air Corps 1944-46. Res Scientist with Bell Tel Labs 1952-59 specializing in mtls res. Dir of R&D with Motorola semiconductor Div 1959-64. With ITT since 1964. Assumed current responsibility of corporate Sr vp tech in 1982. Mbr bd of dirs Natl Retinitis Pigmentosa Fdn and Natl Voice Fdn. *Society Aff:* AAAS, IRI, NYAS, AIC

Flatt, Douglas W
Business: 6111 SW 29th St, Topeka, KS 66604
Position: Pres. *Employer:* Cook, Flatt & Strobel Engrs, PA. *Education:* BSCE/Civ Engg/ND State Univ. *Born:* 6/11/29. Born and raised on a ND farm. The 10 yrs after coll grad were spent with the OR & CA Hgwy Commissions, officer in the AF and with a consulting firm. In 1961, organized engg firm which currently has a staff of 50 persons providing a wide range of services in civ and environ engg. Served as Chrmn, KS Bd of Technical Professions. Golf and hunting are my prime outside activities. *Society Aff:* ASCE, NSPE, ACEC, WPCF, KCE.

Flawn, Peter T
Business: Univ of TX, Austin, TX 78712
Position: President. *Employer:* University of Texas. *Education:* PhD/Geology/Yale Univ; MS/Geology/Yale Univ; BA/Geology/Oberlin College. *Born:* Feb 1926. Director, Bureau of Economic Geology, Univ of Texas at Austin 1960-70; Academic Vice Pres, Univ of Texas at Austin 1970-72; Exec VP, Univ of Texas at Austin 1972-73. President, Univ of Texas at San Antonio 1973-77. Prof Geological Sciences and Public Affairs, Univ of Texas at Austin 1970- ; Pres, The Univ of TX at Austin, 1979- . Elected National Academy of Engrg 1974. Field: application of geology to engineering, environmental, natural resources, and public affairs problems. *Society Aff:* AASG, AIME, GSA, NAE, SEG.

Flay, George F, Jr
Home: RDS Milltown Rd, Brewster, NY 10509
Position: VP-Engrg. *Employer:* Grove Shepherd Wilson & Kruge, Inc. *Education:* Civil Engg/Structures/Polytechnic Inst of Brooklyn. *Born:* July 25, 1912, New York City. Post Graduate Structure & Hydraulics, PIB; Post Graduate Naval Architecture Tankers, Univ of Alabama. PE New York, Penna, Conn, Ill, Mich, Ohio & Wisc. Author Technical Construction Articles: underpinning, cofferdams, bridge foundations, moving heavy structures. Consultant on construction methods, tunnels, foundations, marine structures, cofferdams, moving and raising structures. Fellow American Society of Civil Engineers, Director 1975-77; Mbr National Society of Profl Engrs; Mbr ACI, SAME. Mbr of The Moles. Mbr Chi Epsilon. Joined present business affiliation 1963. Engaged in engineering construction planning and evaluations of heavy construction and building. *Society Aff:* ASCE, NSPE, ACI.

Fleckenstein, William O
Business: Bell Tele Labs, Holmdel, NJ 07733
Position: V Pres, Switching Systems. *Employer:* Bell Tele Labs Inc. *Education:* BS/EE/Lehigh *Born:* Apr 1925 Scranton RA. USN 1943-44. Dupont Memorial Prize, Lehigh, 1949. Bell Tel Labs 1949, Communications Dev'l Training Prog Valedictorian. Specialized in switching, data communications, advanced to Executive Dir. 1968 Western Elect Co in charge of Engrg Center. 1970 Bell Labs, Exec Dir Switching Sys Engrg Div. Assumed present post V Pres, Switching Sys April 1973. Phi Theta Kappa, Eta Kappa Nu (1959 Hon Mention Outstanding Young Amer Engr). Fellow IEEE. 1970 Chairman Tech Prog Ctte IEEE Convention; 1971 IEEE Bd of Dir, Exec Ctte. Dir editing of 'Physical Design of Communications Equipment. Member Lehigh University Board of Trustees. Chmn, Lehigh Computer Center Visiting Committee.m Jean H. Swarts, 2 daughters. Enjoy golf, fishing, swimming.

Fleischer, Gerald A
Home: 3281 Donnie Ann Rd, Los Alamitos, CA 90720
Position: Prof. *Employer:* Univ of S CA. *Education:* PhD/Ind Engg/Stanford Univ; MS/Ind Engg/Univ of CA; BS/Ind Engg/St Louis Univ. *Born:* 1/7/33. Native of St Louis, MO. Served with the US Navy 1954-57. Operations Analyst, Consolidated Freightways 1958-60. With AID-sponsored proj in Brazil, 1963-64. At Univ of Southern CA since 1964. Served as UNESCO Expertin Engg Economics, Venezuela, 1969. Fulbright Scholar, Ecuador, 1974. Honors include election to grade of Fellow, IIE, 1978, USC Univ Marshal, 1982-87, Region XII VP, IIE, 1984-86. Pres, USC Faculty Senate, 1986-87. *Society Aff:* ASEE, TIMS, IIE.

Fleischer, Henry
Business: 1450 N. Milford Rd, Highland, MI 48031
Position: Dir. of Engr. *Employer:* Numatics, Inc. *Education:* MSc/ME/Columbia Univ; BME/ME/Coll of the City of NY *Born:* 11/20/23 Residence: Grand Blanc, Michigan. Assumed current responsibility as Dir of Engrg. for Numatics in 1968. Prior employment-V.P. Engrg., Flair Mfg Corp., Chief Engr. Finkel Outdoor Products, Ass't Chief Engr. Swingline & Project Engr. Eversharp. Served with Army Corp of Engrs. 1942-46., Written numerous articles for Nat'l publications and a manual on optimum pneumatic component sizing & optimum energy utilization. Patentee in field. Am a Prof Eng & a certified Mfg. Engr. Presented many papers at National and Regional Conventions. Received a BME in 1948 from CCNY & MS in Mech Eng from Columbia Univ in 1952. Awarded the 1985 "Outstanding Engr in Industry" for the State of Michigan. Awarded the 1986 "Clovis Award" by the Michigan Soc. of Prof Engrs. Mbr of NSPE (State Dir.) & other Societies. Married to Rhoda Brotman and have two sons, Niles and Bruce. *Society Aff:* NSPE,FPS,ESD,SME

Fleischman, Marvin
Business: Dept. of Chem Eng, University of Louisville, Louisville, KY 40292
Position: Prof *Employer:* Univ of Louisville. *Education:* PhD/ChE/Univ of Cincinnati; MS/ChE/Univ of Cincinnati; BChE/ChE/City College of NY. *Born:* 5/19/37. Currently, I am tenured Full Prof of Chem Engrg, having started as an Asst Prof in 1970. My res interests have included biomedical engrg and water/wastewater/waste treatment and health effects. I have over 25 publications and have received more than 10 grants, 3 of them from the Natl Sci Fdn. I also have 8 yrs of profl experience in industry, govt & consulting. This included res, dev and design in various areas, with a strong emphasis on environmental problems. I have taught a wide variety of chem and environmental engrg courses, and have taken numerous continuing education courses, and was Dir of Engrg Extension Services. *Society Aff:* AIChE, ASEE, TBΠ, HMCRI

Fleisher, Donald R
Home: 435 Berryhill Rd, Harrisburg, PA 17109
Position: Bridge Eng. *Employer:* PA Pub Utility Commission. *Education:* BS/CE/PA State Univ. *Born:* 3/14/26. BS from Penn State Univ. Native of Central Penn. Joined Brookhart & Tyo, Consulting Engrs in 1948. Served as structural design engr, bridge engr and Project Engr for highway transportation projects. Became Partner of firm in 1962. Assumed Presidency of Tyo and Fleisher, Inc upon formation in 1973. Employed by Bureau of Rail Transportation, PA.PUC in 1977. *Society Aff:* ASCE, ASHE, NSPE, PSPE.

Fleisher, Harold
Business: D/C14, B/704, Box 390, Poughkeepsie, NY 12602
Position: IBM Fellow. *Employer:* IBM DSD Lab. *Education:* PhD/Physics/Case Western; MS/Physics/Univ of Rochester; BA/Physics/Univ of Rochester. *Born:* 1921. Staff Mbr MIT Radiation Laboratory 1943-45, Senior Engr Rauland Corporation 1945-46, Instructor Case Western 1946-50. With IBM since 1950. Manager of Advanced Technology 1964-74, appointed IBM Fellow, 1974. Visiting Professor, Vassar College, Vassar Fellow, 1976. Mbr of AAAS, Amer Physical Soc, and Sigma Xi, and Fellow of IEEE. Received certificate of merit from OSRD, and outstanding invention award from IBM. Past mbr of Admissions and Advancement Committee IEEE, Vice Chairman (Computers) of IEEE Intercon 75, Eastern Region Chairman IEEE Computer Society 1974-75. Holds 37 patents. *Society Aff:* IEEE, APS, ΣΞ, AAAS.

Fleming, John F
Business: Dept of Civil Engrg, University of Pittsburgh, Pittsburgh, PA 15261
Position: Prof of Civil Engrg *Employer:* University of Pittsburgh *Education:* PhD/CE/Carnegie Inst of Tech; MS/CE/Carnegie Inst of Tech; BS/CE/Carnegie Inst of Tech *Born:* 12/15/34 Assoc Prof of Civil Engrg, Northwestern Univ 1960-1965. Project Eng, General Analytics Inc, Pittsburgh, Pa 1965-1969. Assoc Prof of Civil Engrg, Univ of Pittsburgh, 1969 to 1983, Prof, 1983 to present. Area of speciality-computer analysis of structural systems. Engaged in research and consulting on static and dynamic analysis of structural systems. Author of two books on structural analysis on computers. *Society Aff:* ASCE

Fleming, Paul D III
Business: Gen Corp, Research Division, 2990 Gilchrist Rd, Akron, OH 44305
Position: Group Leader *Employer:* Gen Corp *Education:* PhD/Chemical Physics/Harvard Univ; AM/Physics/Harvard Univ; BSc/Physics/Ohio State Univ *Born:* 6/1/43 Raised in Columbus, OH. Research Associate in Chemistry, Columbia Univ, 1970-71 and Brown University 1971-74. With Production Branch, Phillips Petroleum since 1974. Reserach Physicist 1974-79, Research Associate 1979-1985. Senior Research Specialist 1985-86. With Gen Corp Research, Applied Mathematics and Computation Department, Group Leader 1986-present. Technical Editor, Society of Petroleum Engineers of AIME. Author of over 35 articles published in technical journals and 1 US patent. Speciality in mechanisms of OilRecovert and their simulation utilizing Thermodynamics, Statistical Mechanics, and Fluid Mechanics. Also Numerical Methods, Computer Graphics and Viscoelasticity. *Society Aff:* SPE of AIME, APS

Fleming, Richard
Home: Box 117 Windfields, Zionsville, PA 18092
Position: Pres *Employer:* Richard Fleming Associates, Inc. *Education:* BChE/Chem Engg/Pratt Inst. *Born:* 6/15/24. in NY City. BS Chem Engg Pratt Inst 1944. Attended Princeton Univ, MIT, and Univ of DE for advanced courses in electronics & bus admin. MS in Chem Engg NY Univ 1949. Served 2 yrs as Ensign, US Navy. 1946-62 wide variety of assignments in res, engg & new product dev. 1962-69 Pres and CEO Avisun Corp, 1969-77 Group VP, 1977-1980 Exec VP. Air Products & Chems, Inc., Chrmn Bd of Dirs Chemical Industry Institute of Toxicology 1976-1980, Director & Member of Executive Committee American Industrial Health Council 1977-1980, Executive Committee Society of Chemical Industry 1980 to 1983, Pres & Dir GAF Corp 1980-1981 President of Board and CEO Lehigh Valley Hospital Center 1978-1984 Pres Richard Fleming Associates Inc 1981 to present. President & CEO of Catalytica Inc, 1985 to Present. Member Board and Executive Cmte of Health East, Inc. 1984 to Present. Vice Chairman of Chemical Industry Institute of Toxicology 1986 to Present. *Society Aff:* ACS, AIChE, SCI.

Fleming, William H, Jr
Business: Route 73 and Ramblewood Parkway, Mount Laurel, NJ 08054
Position: VP *Employer:* Gerald E. Speitel Associates *Education:* BS/Geo/Univ of Notre Dame; BS/CE/Univ of Notre Dame *Born:* 9/6/49 Began engineering practice with CE Maquire, Wethersfield, CT in 1971. Served with John G. Reutter Associates, Camden, NJ in 1972. Instrumental in building Gerald E. Speitel Associates, Consulting Environmental Engineers since 1973. Presently responsible for development and completion of various wastewater, water and solid waste projects. VP New Jersey Section, ASCE, 1979- 1981 and President South Jersey Branch, ASCE, 1981-1982. Executive Committee NJCEC, 1981-1982. *Society Aff:* ASCE, AWWA, ACEC, WPCF

Fleming, William W, Jr
Business: Pembroke Six Bldg., Suite 208, Virginia Beach, VA 23462
Position: VP *Employer:* Wiley & Wilson *Education:* B/CE/Univ of VA *Born:* 12/25/30 Native, Middleburg, North Carolina; Graduate, Folk Union Military Academy, 1949, Univ of VA, 1956. engr, Southeastern Underwriters Association 1957-59 specializing in municipal water and fire protection systems. Subdivision-Drainage Eng, Henrico County, Virginia 1960-70 responsible for development standards and compliance, design, and construction of urban drainage improvements. Director of Public works, City of VA Beach, VA, 1970-76 responsible for five major divisions;Highways, Engineering, Sanitation, Building Codes, and Utilities with over 500 employees. Vice President, Talbot & Associates, Architects/Engineers, 1976-80 responsible for municipal clients and business development. VP, Wiley & Wilson, 1981- Mgr Tidewater VA Office. Registered Professional Eng, VA, NC, SC, NH & TN. *Society Aff:* NSPE, ASCE

Flemings, Merton C
Business: 77 Mass Ave/Rm 8-309, Cambridge, MA 02139
Position: Head, Dept of Mats Sci & Engrg; Toyota Prof *Employer:* Massachusetts Inst of Technology. *Education:* ScD/Metallurgy/MIT; SM/Metallurgy/MIT; SD/Metallurgy/MIT. *Born:* Sept 1929. Metallurgist Abex Corporation 1954-56. Joined faculty of MIT in 1956. Abex Professor of Metallurgy 1970-1975, Ford Professor of Engrg 1975-1981. Dir, Matls Processing Center, MIT, 1979-82. Co-author of 206 papers and two books in the fields of solidification science and engrg, materials processing and foundry technology. Received Simpson Gold Medal AFS, Mathewson Gold Medal AIME, Henry Marion Howe Medal ASM. Overseas Fellow Churchill College Cambridge. Mbr, National Academy of Engrg, AM. Academy of Arts & Sciences. Toyota Prof of Mtls Processing 1981-present. Head,. Dept of Mtls Sci and Engrg, MIT, 1982-present *Society Aff:* ASM, AIME, AFS.

Flemming, Frank L
Business: 855 N. Cahuenga, Hollywood, CA 90038
Position: Director of Technical Operations *Employer:* Vidtronics Co. *Education:* BS/EE/Univ of Buffalo. *Born:* April 1928 Los Angeles, California. Production engr and television equipment design engr Sylvania Electric Products 1950-54. Joined CBS Television Network in 1954 as equipment and systems design engr. In 1964 became Director of Plant Systems Engrg for CBS responsible for broadcast system engineering projects. In 1967 with Visual Electronics as Chief Engr responsible for the engineering and manufacture of a broad line of broadcast equipments. In 1969, joined NBC Television Network as Vice President Engineering responsible for the design and installation of all major technical systems as well as r.f. spectrum management. Joined Vidtronics company in 1981 as director Technical Operations. Responsible for daily technical operations and long range capital planning for this video tape post-production editing firm. Mbr of the Audio Engineering Society & IEEE and a Fellow of the Society of Motion Picture and Television Engrs. Past VP Television Affairs, Society of Motion Picture & Television Engrs 1977-78. *Society Aff:* SMPTE, IEEE, AES.

Flesher, Cranston W
Business: 954 E First Natl Ctr, 1340 Mid-America Tower, OK 73102
Position: Pres. *Employer:* Flesher & Assoc, Inc. *Education:* BS/Petrol Engr/Univ of OK; BS/Geol Engr/Univ of OK. *Born:* 2/2/24. and reared in Edmond, OK. Educated in TX and OK. Served in U.S. Navy 1942- 1946. Worked in Engg of oil and gas operations in OK, KS, NB, WY, TX, OH & NY. Employed by OH Oil Co in Northwestern US for 3 yrs. Did Jr and Sr engg work for consulting firm in Oklahoma City covering the Mid-Continent Area. Established firm of Flesher & Assoc, Inc in 1962 to supervise drilling-completion operations and evaluation of oil & gas properties. Participate in developing solutions to drilling & completion problems and techniques of application of processes. Leasure time spent on lake and family. *Society Aff:* AIME-SPE, AAPG, SIPE.

Fletchall, Lyle R
Business: 1728 Central Ave, Fort Dodge, IA 50501
Position: President & Principal Engineer. *Employer:* Associated Engineers, Inc. *Born:* June 1929. BSCE from Iowa State Univ 1954. Native of Webster City Iowa. Post Engineer US Army Corps of Engrs 1954-55. EIT, Assistant Resident Engr and Resident Engr Iowa Highway Commission 1956-65 with Associated Engineers Inc since 1966 (1966-68 Vice Pres and Treasurer; 1969- , President). Responsible for direction of Professional Services Firm. On Board of Directors Fort Dodge Chamber of Commerce, United Methodist Church Board, Past President Fort Dodge Chapter of IES, President Iowa Engineering Society.

Fletcher, Alan G
Business: PO Box 8155 Univ Station, Grand Forks, ND 58202
Position: Dean, Sch of Engr. *Employer:* Univ of ND. *Education:* PhD/Civ Engg/Northwestern Univ; MS/Civ Engg/CA Inst of Tech; BASc/Civ Engg/Uni of British Columbia. *Born:* 1/2/25. Employed with the British Columbia Elec Co, 1948-56: Engr-in-Training 1948- 52. Hydraulic Designer in the Hydraulic Dept, 1952-56, holding a variety of responsibilities in hydroelectric power proj planning, design, and construction. Supervisor of Hydroelectric Planning with the British Columbia Engg Co, 1956-59. The maj proj was the dev of the Columbia River in Canada. Employed by the Univ of ID as Asst and Assoc Prof of Civ Engg, 1959-62. Walter P Murphy Fellow in Civ Engg at Northwestern Univ, 1962-64. Assoc Prof of Civ Engg at the Univ of Utah, 1964-69. Consultant on fluid mechanics and dam stability problems. Dean of Engg at the Univ of ND, 1969 to present. *Society Aff:* ASCE, NSPE, ASEE, NDSPE.

Fletcher, Clarence T
Business: PO Box 509, Lockport, NY 14094
Position: Technical Mgr-Tool and Alloy Steels *Employer:* Guterl Special Steel, Inc *Education:* BS/Metallurgical Engg/Carnegie Inst of Tech. *Born:* 3/25/20. Native of New Kensington, PA. Research Metallurgist of Braeburn Alloy Steel 1948-59, then Chief Metallurgist until 1971. Chief Corp Metallurgist, UTD Div, Litton Industries 1971-73. Since 1973 with Guterl Steel Corp, as Technical Mgr Tool and Alloy Steels, I am responsible for processing, quality-and sales development of tool and alloy steel products. Chmn, Pittsburgh Charter of ASM 1966-67. Chmn, Tool Steel Technical Com, AISI, 1970. Play at golf and enjoy other sports and music. *Society Aff:* ASM, AIME, ASTM.

Fletcher, Edward A
Home: 3909 Beard Ave S, Minneapolis, MN 55410
Position: Prof. *Employer:* Univ of MN. *Education:* PhD/Chemistry/Purdue; BS/Chemistry/Wayne. *Born:* 7/30/24. Naval aviator, WWII. Hd, flame mechanics and propulsion chemistry sections, NASA Lewis Res Ctr, 1952-9. Visiting scientist, USSR Acad, Sci, 1964. Visiting Prof, Univ Poitiers, 1968, Natl Acad Sci Committee on Fire Resistant Hydraulic Fluids, 1977-, Chrmn, Central States Sec, Combustion Inst. First to establish combustion in supersonic airstream. NASA Special Award for High Altitude Rocket Ignition System. Publications in profl jours and books. *Society Aff:* AAAS, CI, ACS.

Fletcher, Leonard J
Home: P.O. Box 6087, Carmel, CA 93921
Position: VP (Ret) *Education:* LLD/-/Bradley Univ; BS/-/IA State Univ. *Born:* May 1891, on Nebraska homestead. Learned sheet metal trade. BS in Agricultural Engrg Iowa State Univ 1915; LLD Bradley Univ, honorary 1956. Head, Dept of Agri Engrg Univ of California 1920-27; Caterpillar Tractor Co 1927, employed in foreign and domestic technical sales; in charge education and training and community relations. Elected Vice President 1951; retired 1956. Honorary Mbr and Past President ASAE; awarded Society's Cyrus Hall McCormick Medal, 'For Exceptional and Meritorious Engineering Achievement in Agriculture'. In 1954, recipient Univ of Missouri award, ' For Distinguished Service in Engineering'. Since retirement, lectured before technical, business and college groups; active in community affairs. *Society Aff:* ASAE.

Fletcher, Leroy S
Business: College of Engineering, Texas A&M University, College Station, TX 77843
Position: Associate Dean, College of Engineering *Employer:* Texas A&M University *Education:* PhD/Mech Engrg/AZ State Univ; Engr/Mech Engrg/Stanford Univ; MS/Mech Engrg/Stanford Univ; BS/Mech Engrg/TX A&M Univ. *Born:* 10/10/36. Res Scientist at NASA-Ames Res Ctr 1958-63. Instructor at AZ State Univ 1964-68. Involved in teaching & res at Rutgers Univ 1968-75; Assoc Dean, Rutgers Univ Coll of Engrg 1974-75. Chmn of the Mech & Aero Engrg Dept, Univ of VA 1975-80. Assoc Dean, Coll of Engrg, TX A&M Univ, 1980-.Assoc Dir. Texas Engineering Experiment Station, 1986-. Involved in teaching & res in programs involving experimental heat transfer & fluid mechs as well as energy sys analysis. PE in AZ, NJ, VA & TX. ASEE Bd of Dirs 1974-77, 1978-80; ECPD/ABET Bd of Dirs 1979-80, 82, 83-89; SCORE Bd of Dirs 1976-80. ASME VP, Educ, 1980-81, Sr VP and Chmn, Council on Educ 1981-83, Bd of Governors 1983-85, Pres-1985-86; AIAA Bd of Dirs 1981-84. AAAS Section M Chairman-Elect, 1987-88. Fellow, AIAA, AAAS, I. Mech E, AAS, ASEE; Recipient SAE Ralph Teetor Awd, 1970; ASEE Dow Awd, 1972; ASEE George Westinghouse Awd 1981; ASEE/AIAA Aerospace Educ Awd, 1982; ASME Charles Russ Richards Mem Awd, 1982; ASEE Ralph Coats Roe Awd, 1983; AIAA Energy Sys Awd, 1984, Arizona State University, Distinguished Alumni Award, 1985; ASME Engineering Leadership Award, 1985; ASEE Donald E. Marlowe Award, 1986. *Society Aff:* ASME, ASEE, AIAA, ABET, AAS, AAAS, I. Mech. E. (London)

Fletcher, Robert C
Home: 25 Dorchester Road, Summit, NJ 07901
Position: Exec Dir/Intergrated Circuit Design Division *Employer:* Bell Laboratories.

Fletcher, Robert C (Continued)
Education: PhD/Phyiscs/MIT; BS/Physics/MIT. *Born:* 5/27/21. Began Bell sys career in 1949. Early work in res on traveling wave tubes, crossed field devices & semiconductors. 1964 elected VP, Res, Sandia Corp - Western Elec Co subsidiary. Returned to Bell Labs 1967 as Exec Dir Military Sys Res Div & later Exec Dir Ocean Sys Div in March 1971. Responsible for Common Subsystems Lab, Integrated Circuit Customer Service Lab, Computer Aided Design & Test Lab, Electronics Technology Planning Center, and LSI Design Lab. Since June 1982 responsible for the Computer Tech and Design Engrg Div. Fellow of Amer Phy Soc & IEEE. Mem Amer Assoc for Adv of Science & Sigma Xi. Author numerous publications & holds 6 patents. *Society Aff:* IEEE, APS, AAAS.

Fletcher, Stewart G
Home: 4407 Tournay Rd, Bethesda, MD 20816
Position: Retired *Education:* ScD/Metallurgy/MIT; BS/Met Engg/Carnegie Inst of Technology. *Born:* 1918. Native of New Kensington Pa. Research metallurgist at Alcoa, research associate at MIT during WWII. Joined Latrobe Steel Co in 1945, becoming Vice President and Technical Director in 1957, responsible for all technical activities. Came to AISI in 1973 to head Division of Research and Manufacturing as Senior Vice President. Retired 1980. Active in American Society for Metals National President in 1965-66, Fellow in 1972. Received ASM Howe Medal in 1945 and 1949. Professional engr. Member Cosmos Club, and many technical societies. *Society Aff:* ASM, AIME-ISS, TMS.

Fletcher, William J
Business: 444 N Michigan Ave, Chicago, IL 60611
Position: Agricultural Safety Engr. *Employer:* National Safety Council. *Education:* BS/AGri/Univ of IL. *Born:* October, 1925. Native of rural Belvidere, IL. Served in WWII as bombardier instructor in Army Air Force. Sales experience with electric utility. Management trainee in retail farm equipment. Joined Doane Agricultural Service, Inc in 1952 as farm structures engr and consultant. Engineering Editor Doane Publications 1959-61. Engrg and Agri Industries Editor of Successful Farming 1961-66. First editor of Farm Building News until 1968, then joined National Safety Council as Farm Safety Engineer. Have held numerous committee and leadership positions at the chap, regional and national level in the American Society of Agricultural Engrs. Former Chairman IL-WI Region ASAE. Currently Chmn, Safety Goals and Plans Committee of ASAE. Honorary Mbr Alpha Epsilon. Honored Mbr for 1975, Chicago Section ASAE. Mbr, Northern IL Conf United Methodist Church. *Society Aff:* ASAE.

Fleuelling, Lewis E
Business: One International Drive, Monroe, MI 48161
Position: V P North Amer Operations. *Employer:* Monroe Auto Equipment Company. *Education:* BS/Engg/Univ of MI. *Born:* 8/24/20. in St Thomas Ontario Canada. Grad Univ of MI, BS in Engg. Employed by Monroe Auto Equipment Company-holding positions in the past in engg, sales & mgmt. P Chmn Detroit Section, Soc of Automotive Engrs. Pres SAE 1979-80. Mbr of Natl Finance Ctte, SAE. *Society Aff:* SAE, ESD.

Flikke, Arnold M
Business: Agri Engrg Dept, St Paul, MN 55108
Position: Prof & Head, Agricultural Engrg Dept. *Employer:* Univ of Minnesota. *Education:* PhD/Agri Engg/Auburn Univ; MS/Agri Engg/MN; BS/Agri Engg/WI. *Born:* July 1919. University of Minnesota. 1946-present. US Navy 1944-46. Specialized in Electric Power and Processing. Publications on electric motors, grain drying, instrumentation and equipment. Department Head 1973-present. President North Central Region ASAE, 1973. Regional Director ASAE 1975-77. Mbr ASAE, ASEE, IEEE. Reg Agricultural Engr, Minnesota. Fellow, ASAE. *Society Aff:* ASAE, IEEE, ASEE.

Fling, Russell S
Home: 477 E Dominion Blvd, Columbus, OH 43214
Position: President. *Employer:* Consltg Engrg *Education:* BS/Arch Engrg/OH State Univ. *Born:* Columbus Ohio, Nov 11 1926. m. Dona A Thompson Oct 2, 1954. Children: Russell T, Karen and Paul. Pres R S Fling & Partners Inc, Consulting Engrs 1953- 1982. Reg P E 9 states. Reg Architect. Mbr ASCE. Pres Cons Engrs of Ohio 1963-65. Pres Franklin Cty (Ohio) Chapter NSPE 1971-72. Pres ACI 1976. ACI Hon Mbr 1982. Meritorious Service Award, Ohio Soc of Professional Engrs, 1968. Author of 12 technical papers and one book. *Society Aff:* NSPE, ACI, ASCE.

Flinn, Richard A
Home: 140 Underdown, Ann Arbor, MI 48105
Position: Prof of Mat & Met Engrg. *Employer:* Univ of Michigan. *Education:* ScD/Metallurgy/MIT; MS/Metallurgy/MIT; BSc/Chem Eng/CUNY. *Born:* May 1916. Research metallurgist INCO 1937-39. Asst Ch Met ABEX 1941-52. Developed new cast railroad car wheel, austenitic ductile iron for aircraft, abrasion resistant irons, basic cupola melting. Prof Materials and Met Engrg Univ of Michigan 1952- . In charge of Cast Metals Lab. Publications 80 in fields of cast metals, basic oxygen steelmaking, austenite transformation, alloy development. Consultant: Chrysler, McLouth Steel, INCO, Hastings, USN, etc. Visiting Prof U Catolica de Chile 1968-69. Books: Fundamentals of Metal Castings, A-W 1964; Copper, Brass and Bronze Castings NFFS 1962; Engineering Materials & Their Applications, Houghton-Mifflin 1975, 3rd Ed. 1986. Awards: Howe Medal ASM 1944 and 1963; Simpson Gold Medal AFS 1947. Fellow & Life Mbr ASM, Hoyt Lecturer AFS 1981. *Society Aff:* AFS, ASM.

Flipse, John E
Business: PO Box 486, Gloucester Point, VA 23062
Position: Chmn, Pres & Ch Exec Officer. *Employer:* Deepsea Ventures, Inc. *Born:* Feb 1921 Montville N J. BS MIT, MME NYU. Lic P E Vir and N Y. Served in US Merchant Marine, WWII. Faculty of NY Maritime College 1946-55. Proj Engr for Ship Stabilization 1956-57 wth Sperry. Dir of Research and Administrative positions with Newport News Shipbuilding & Dry Dock Co 1957-68. Founded Deepsea Ventures, Inc to develop deep ocean hard mineral mining and metal winning 1968 to date. Mbr, Interagency Law of the Sea Advisory Committee, Dept of Commerce Ocean Petroleum and Minerals Advisory Ctte. Fellow, Marine Tech Soc and Mbr and Past Chmn of the Tech and Research Ctte of SNAME. Enjoy foxhunting, sailing, and farming.

Flock, Henry H
Home: 2659 Sparta Ct, Olympia Fields, IL 60461
Position: President. *Employer:* Tuthill Pump Co. *Education:* MBA/-/Univ of Rochester; BSME/-/Univ of IL. *Born:* Jan 2, 1935. US Army 1955-57; Ch Engrg Parker Hamifin Corp 1962-68; Asst to Pres, Garlock Inc 1968-73; Operations Manager, Federal Signal Corp 1973-74; Pres Tuthill Pump Co 1974- ; Technical Articles, 'Cushioning in Pneumatic Cylinders', 'Exclusion Devices in Earth Moving Equip', 'Proper Gasket Installation and Designs' and 'A Change to Direct Costing in a Multi product Co'. Pres Hydraulics Inst, Beta Gamma Sigma Honor Soc, Past Chrmn of US Delegation to ISO on Sealing devices, Mbr of US Delegation on Fluid Power Products; Director of ALSIP industrial district ASSN, Christ Hospital. *Society Aff:* ΒΓΣ

Flood, H William
Business: 1 Univ Ave, Lowell, MA 01854
Position: Associate Prof Chem Engrg *Employer:* Univ of Lowell *Education:* BS/Chem Engg/Univ of MO-Rolla; Ch.E./Chem Engg/University of Mo-Rolla. *Born:* 10/7/22. Production and maintenance supervision, Freeport Sulphur, Nicaro Nickel, Mutual Chemical, Dewey and Almy Chemical 1943.56. Consulting minerals & metals processing projects, Arthur D Little 1956-70. Kennecott 1970-80. Functioned as Manager, Process Engrg and Economic Evaluation until 1975. Then assignments to identify new technical and business opportunities in traditional and non- traditional areas. Resource Engrg Inc 1980-1983. Univ of Lowell Chem Engrg faculty

Flood, H William (Continued)
since 1983. Acting Head and Associate Professor 1987. AIChE, Chmn National Prog Ctte 1976; Kngfish (Chmn) Boston Sect 1972-73. Dir 1978-80. AIME. Prog Ctte International Symposium on Hydrometallurgy 1973. Area Director MSM-UMR Alumni Assn 1970-1980 Chmn Board of Appeals Town of Acton NMAB Ad Hoc Panel on Manganese Recovery Tech 1974-75. Enjoy choral singing & woodworking *Society Aff:* AIChE, AIME, AEE, ASEE, ACS.

Flood, Paul E
Business: 4421 Harrison Street, Hillside, IL 60162
Position: President. *Employer:* A & H/Flood Eng Div, Prof Service Indus, Inc *Education:* BS/ChE/IL Inst of Tech. *Born:* May 7, 1926. Reg PE Ill, Ind, Mich, Wisc. Reg Struct Engr Illinois. US Naval Reserve 1944-54. Employed by Precision Scientific Co, Chicago 1946-56; Walter H Flood Co (Foundation Engrs and Materials Lab) 1956-1982. President since 1972. Pres, A & H/Flood Engrg Div, Prof Service Indus, Inc, 1982- . Pres Ill Soc of PE's 1975-76. Mbr ASTM, ACI, ASNT, APWA, ACS, CSI, WSE, NSPE. Pres IIT Alumni Assn 1978-79; Chicago District Council ASTM. VP CECI 1979-81, Trustee, Ill Inst of Tech 1980-Director, Drexel National Bank 1979-, Director NSPE 1979-82, Dir. Prof Service Indus, Inc. 1982-. *Society Aff:* NSPE, ACI, AAPT, ASTM.

Floreen, Stephen
Business: INCO Res & Dev Ctr Sterling Forest, Suffern, NY 10901
Position: Research Fellow. *Employer:* Intl Nickel Co Inc. *Education:* PhD/-/Univ of MI; MS/-/Univ of MI; BS/-/MIT. *Born:* July 19, 1932. USAF stationed at NASA Lewis Lab 1955-57. With International Nickel Co since 1960. Supervisor, Managing Steel Sect 1962-64, Research Assoc 1964-73, Research Fellow since 1973. Current responsibilities are research and development work in stress corrosion cracking, creep failure, brittle fracture & fracture mechanics techniques, including material characterization and creating materials with improved resistance to crack growth. Responsible for development of new grades of cast air-meltable high strenght steel & nickel bas alloys for nuclear & sour gas well appls. *Society Aff:* ASM, AIME, ΣΞ.

Flores, Ivan
441 Redmond Road, South Orange, NJ 07079
Position: Pres. *Employer:* Flores Assoc. *Education:* PhD/Ed Psych/NYU; MA/Math/Columbia Univ; BA/Math/Bklyn Coll *Born:* 01/03/23 From 1950-60 worked as a Proj Engr in the comp indust for firms such as UNIVAC. Established Flores Assoc in 1960 & have been consltg for maj computer firms ever since. Have written 20 books on hardware, programming, operating systems, data structures, peripherial devices etc for Prentice Hall and Allyn & Bacon. All have been translated into German, Spanish, Russian, Hungarian & Japanese. Data Base Architecture 1981; Word Processing Handbook 1982; Microcomputer Systems 1983; Handbook of Computers and Computing 1984; Professional Microcomputer Handbook 1985, publ by Van Nostrand Reinhold. Author of 80 papers in prof journals. Organized 2 Conf on Computers & the Humanities. Chaired maj prof meetings. Former editor of JACM, Modern Data System Editor Journal of Computer Languages. Conslt, US Army Scientific Adv Panel, UN Dev Prog. Contributed to Encyclopedia Americana & Encyclopedia of Comp Sci. Assoc Prof EE, Stevens Inst 1965-67. Prof Comp & Info Sci, Baruch Coll, City Univ of NY 1967 to present. *Society Aff:* IEEE, ACM

Florio, Pasquale J, Jr
Business: 323 High St, Newark, NJ 07102
Position: Assoc Prof ME *Employer:* NJ Inst of Tech. *Education:* PhD/ME/NYU; MME/ME/NYU; BSME/ME/Newark College of Engg. *Born:* 5/24/36. Engr NASA, Geo Marshal Space Flight Ctr, Huntsville, AL, 1961-62. Assoc Prof Mch Engg, NJ Inst of Tech, Newark, 1966-. Teaching and res in the area of Thermodynamics, Fluid Mech, Heat Transfer and the application of boundary element methods to heat transfer and fluid flow design problems. Received Ph.D. in M.E. from New York Univ-1967, MME degree from New York Univ-1961 and BSME from Newark College of Engg-1959. Was a ASEE-NASA Summer Faculty at Lewis Res Center-Summers 78 & 79. *Society Aff:* ASME.

Florman, Samuel Charles
Business: 97 Montgomery St, Scarsdale, NY 10583
Position: VP, General Manager. *Employer:* Kreisler Borg Florman Const Co. *Education:* MA/English/Columbia; CE/Civil Engg/Dartmouth; Bs/Civil Engg/Dartmouth; Hon. Dr. of Sci./Manhattan College; Hon. Dr. of Sci/Clarkson University *Born:* 1/19/25. BS (summa cum laude, Phi Beta Kappa) & CE from Dartmouth College. MA from Columbia Univ. Hon Dr of Sci Manhattan Coll 1983 and Clarkson Univ. 1986. Ensign, Seabees 1945-46 in the Philippines and Truk. 1948 Field Engr Hegeman Harris Co in Venezuela. 1949-54 Constr Engr Thompson Starrett Co and Joseph P Blitz Inc NY. 1955 to date, principal and Vice President, Kreisler Borg Florman Construction Company. Mbr Board of Overseers, Thayer School of Engrg, Dartmouth College 1971-77. Past President Dartmouth Society of Engrs. Dir Engrg Info Inc., VP Hospital for Joint Diseases Orthopaedic Inst. Author 'Engineering and the Liberal Arts' 1968, 'The Existential Pleasures of Engineering' 1976, 'Blaming Technology' 1981, 'The Civilized Engineer', 1987 and more than 100 articles in Harper's and other periodicals. *Society Aff:* ASCE, NSPE, ASEE, ASME, IEEE, SAME

Flowers, A Dale
Business: Dept of Operations Research, Case Western Reserve Univ, Cleveland, OH 44106
Position: Assoc Prof *Employer:* Case Western Reserve Univ *Education:* DBA/Prod Management/IN Univ; MBA/Prod Management/IN Univ; BS/Op Research/IN Univ *Born:* 3/21/46 Dr. Flowers' professional service has focused on the successful application of quantitative modeling via computers to solve Operations and Industrial Engrg problems. He is a principal in the highly successful Storm Software Project for IBM-PC and compatible microcomputers. His areas of specialization focus on integrated manufacturing control systems. His corporate clients have included General Motors, IBM, General Electric, TX Instruments, Sherwin-Williams, Keithley Instruments, TRW, Inc, & others. He received the 1981 IIE Transactions Application and Development Award. He has published widely and served on the Material Advisory Board for the General Motors Manhattan Engine Project, & the Bd of Directors for Storm Software Inc. & MIMS Mutual Funds, Inc. *Society Aff:* IIE, TIMS, ORSA, ASQC, IIF, APICS, SME.

Floyd, Dennis R
Business: PO Box 464, Golden, CO 80401
Position: Dept Mgr, Product R&D. *Employer:* Rockwell Intl. *Education:* PhD/Met Engg/CO Sch of Mines; MS/Met Engg/CO Sch of Mines; PE/Met Engg/Co Sch of Mines. *Born:* 10/3/41. Individual technical contributor for 12 yrs at DOE Rocky Flats Plant - specialized in beryllium tech. Last four yrs managing R&D efforts. Currently lead 225 person R&D dept doing high tech work with fabrication, assembly, coating, welding, inspection, NDE processes as applied to plutonium, uranium, beryllium, stainless steel, & many non-ferrous metals. Co-editor Berylliu, Sci & Tech, Vol II, 1979. Held all offices local ASM chapter, Past-Pres.Wheatridge Chamber of Commerce. Active in several community, church & athletic organizations. Listed in Who's Who in the Am West. *Society Aff:* ASM, AAAS, ΣΞ, NMA.

Flugum, Robert W
Business: 20 Mass Ave, NW, Washington, DC 20545
Position: Mgr, Res and Adv Tech. *Employer:* Chas T Main, Inc, Boston, MA. *Education:* BS/Elec Engg-Pwr/Univ of WI. *Born:* June 1924, Madison Wisconsin. In USAAF during WWII. BSEE Univ of Wisconsin, 1948. Joined Westinghouse and served as electric utility engr for eight years in Midwest. Moved to Bloomington, Indiana in 1957 as Advisory Engr on many types of power apparatus. Chief US Delegate IEC for surge arresters 1965-67, & active on other U S & IEC Standards Cttes. Was with Ohio Brass 1967- 75 as Dir, Applied Research. Directed US Government research in electric power transmission 1974-79. Has published over 30 tech papers

Flugum, Robert W (Continued)
incl 3 IEEE PES prize papers. Fellow of IEEE, Chmn IEEE Lightning & Insulator Sub-Ctte. V Chmn IEEE Trans and Dist Committee. Former Chmn NBS Electricity Div Advisory Panel, and Mbr CIGRE, Rec'd Dist Service Citation. Univ of WI 1976. *Society Aff:* IEEE/PES.

Fluhr, Wallace E
Business: DFCEM, USAF Acad, Colorado Springs, CO 80840
Position: Prof & Head, Dept CE. *Employer:* US Air Force Academy. *Education:* BSCE/Civil Engg/Univ of KY; MS/Structural Engg/Univ of IL; PhD/Structural Dynamics/Univ of IL. *Born:* Jan 1932. PhD Univ of Ill, MS Univ of Ill, BS Univ of Kentucky. Base Engr, Gunter AFB Ala 1954-57. Dir of Facilities Design, Air Force Ballistic Missile Prog Los Angeles 1960-63. Faculty of Air Force Acad 1963- , with academic rank of Permanent Prof. Mbr ASCE, ASEE. Former Chmn of ASCE ctte on Experimental Analysis, Engrg Mechanics Div; former Chairman of ASCE Research Committee, Structural Division; mbr ASCE Tech Council on Res. Mbr of ASEE, Civil Engrg Div, Ctte on Educational Policies. *Society Aff:* ASCE, ASEE.

Flurscheim, Charles H
Home: 20 Moore Str, Chelsea London SW3 2QN, England
Position: Consulting Engr. *Education:* ScD. BA/-/Cambridge Univ, England. *Born:* Feb 4, 1906. Metropolitan-Vickers professional apprenticeship. Switchgear Engr G E Phila 1932-33. 1949 Chief Engr Switchgear, Metropolitan- Vickers. 1957-Dir of Engrg, MV and later AEI Ltd. Part Chrmn Associated Nuclear Services (1970-77). Now Consulting Engr. Past Chmn Power Div IEE. Past Mbr of Council, Royal College of Art, London, FEng. Fellow IEE, IEEE and I Mech E. Editor of Power Circuit Breaker Theory & Design (Peregrinus - IEE 1975). Author many tech papers High Voltage Circuit Breakers, Aircraft Electro-Mechanical Components and seventy patents on devices in these fields. Hopkinson award of IEE and Power Division Premium for papers. Author, *Engrg Design Interfaces*, (Design Council 1977). Editor, Industrial Design in Engineering (Design Council 1983). *Society Aff:* FEng, FIEE, FIMechE, IEEE.

Flynn, John E
Business: 104 Engg A Bldg, University Park, PA 16802
Position: Prof of Architectural Engg. *Employer:* PA State Univ. *Education:* Bach-Architecture/Architecture/Univ of MI. *Born:* 1/10/30. Licensed architect in OH and presently Prof of Architectural Engg at the PA State Univ. Corporate mbr of AIA and a Fellow in the Illuminating Engg Soc. Served as a mbr of the IES Bd of Dir since 1974; served as Sr VP in 1978-79; and served as Natl Pres of IES in 1979-80. Mr Flynn is also a mbr of the US Natl Comm of the Intl Commission on Illumination (CIE), and presently serves as intl chrmn of the Comm on Environmental Lighting. Author of two books on lighting and architecture, and has authored numerous technical papers and articles. *Society Aff:* IES, CIE, AIA, ISCC, IALD.

Flynn, John J
Business: Empire Bldg, Bedord, NY 10506
Position: Owner. *Employer:* John J Flynn, Consulting Engrs. *Education:* BCE/Engg/Villanova Univ. *Born:* Native of Brooklyn, NY. Construction Engr with US Army Corps of Engrs in NY from 1958 to 1960. Technical Service Engr for Shell Chemical in Union NJ 1960- 61, Sr Bldg Construction Engr, NYS div of Housing, Bldg Code Bureau 1961-62; Field Engr, George A Fuller Co, NY 1962-64, Supervising Engr, Robert W Hunt Co, NY 1964-66. In private practice since Jan 1977. In addition, Instr, Inst of Design & Construction, Brooklyn, NY 1968-1971. *Society Aff:* NYSPES, NYCPES, ASA, ASHRACE, NFPA.

Flynn, Peter C
Business: Suite 805, 605-5th Ave SW, Calgary, Alberta, Canada T2P 3H5
Position: VP *Employer:* Nova, An Alberta Corp *Education:* PhD/ChE/Univ of Alberta; MSc/ChE/Univ of CA; BSs/ChE/Univ of DE *Born:* 08/01/46 Spent six yrs with Syncrude Canada Ltd, an oil sands mining co, in a variety of res engrg and operations positions. In 1980 joined Nova, An Alberta Corp to head up their participation in Canstar Oil Sands Ltd, a NOVA/Petro- Canada subsidiary, which plans to develop a fourth commercial Athabasca mining oil sands proj. Active: as a dir of The Chemical Inst of Canada; as Chrmn of the Mbrship and Local Sections Committee, Canadian Soc for Chem Engrg; and as mbr of the Publications Committee, CSChE. *Society Aff:* CSChE, CIC, APEGGA

Focht, John A
Home: 2823 Salado, Austin, TX 78705
Position: Prof CE/Asst Dean Col of Engrg-ret. *Employer:* Univ of TX, Austin (Retired). *Education:* MS/Civil Engr/Univ of TX; CE/Civil Engr/Univ of TX. *Born:* Feb 1894 Sweetwater Texas. Asst City Engr and City Engr Sweetwater, Texas. Lieutenant Co B 315th Engrs two Years WWI. Asst Co Engr and Co Engr Rockwall Co, Texas 5 years. Co Engr Nolan Co Texas, 3 years. Prof of Civil Engrg, Asst Dean and Counselor, College of Engineering, Univ of Texas, 38 years. DSC for action in 1918, Silver Beaver Boy Scouts 1958; National Honor Member Chi Epsilon Fraternity 1972; Honorary Mbr ASCE 1975. Registered PE Texas. *Society Aff:* ASCE, CHI EPS, TBΠ, ASEE

Focht, John A, Jr
Business: PO Box 740010, Houston, TX 77274
Position: Exec Vice President. *Employer:* McClelland Engrs, Inc. *Education:* MS/Civil (Soil Mech)/Harvard Univ; BS/Civil/Univ of TX. *Born:* August 31, 1923. Soils Engr at Waterways Experiment Station, Vicksburg, Miss, 1947-53 except for 2 years in Korea as Captain with the US Army Corps of Engrs. With McClelland Engrs, Inc since 1953, Vice President for Engrg 1955-71, Exec Vice President since 1972. President Texas Sect ASCE 1971. National Director ASCE Dist 15 1980-1983. Natl VP ASCE Zone III, 1983-1985. Recipient of ASCE Thomas A Middlebrooks Award, 1957 & 1976; the James Laurie Prize, 1959; the State-of-the-Art of Civil Engrg Award, 1971 & 1979; Named Distinguished Engineering Graduate by the Univ of Texas, 1964. Elected National Academy of Engineering, 1986. Named Engineer of Year, 1987, Zone IV Tex. Soc. of Prof. Engr's. *Society Aff:* ACEC, NSPE, ASCE

Focke, Arthur E
Home: 7799 E Galbraith Rd, Cincinnati, OH 45243
Position: President. *Employer:* A E Focke Corporation. *Education:* PhD/Metallurgy/OH State Univ; MS/Metallurgy/OH State Univ; BS/Met Engrg/OH State Univ. *Born:* June 17, 1904. Spec Met, Cleveland Wire Div GE 1927-29; Ch Engr P R Mallory 1929-31; Res Met & Ch Met Diamond Chain Co 1931-51; Mgr Matls Dev & Met Aircraft Nuclear Prop Dept GE 1951-61; Assoc Prof & Prof Met U of Cin 1962-74; Prof Emer Met U of Cin 1974- ; Pres A E Focke Corp 1962- ; Natl Pres ASM 1950; ASM Fellow 1970; Mbr AIME, ASTM, Charter Mbr ANS, Metals Soc UK. Lic P E Ind & Oh. Ints: Wear & Wear Testing; High Temp, Nuclear & Superhard Matls; Career Awareness. *Society Aff:* ASM, AIME, ANS, ASTM.

Foderberg, Dennis L
Business: Stanley Consultants, Inc, 8501 W. Higgins Rd., Suite 400, Chicago, IL 60631
Position: Vice President & Regional Manager *Employer:* Stanley Consultants, Inc *Education:* BS/Civil Engr/Univ of IA; MS/Structural/Univ of IA. *Born:* 9/27/41. Native of Iowa. Taught at the Univ of Iowa & worked in research under a NASA Traineeship & a Ford Fdn Fellowship. Previously employed by three other consulting engrg firms in Iowa developing computer programs, performing structural design, managing projects and developing new business. Assumed duties with the Stanley Consultants, Inc in 1980 as Project Mgr with State and Local Market responsibilities. Currently a Vice President and Regional Manager responsible for the Chicago Office. Mbr of ASCE, ASTM, NSPE, ISPE, APWA, SMPS, Chi Epsilon, Tau Beta Pi, and SEAEI. ASCE Iowa Sect Pres 1974, IES Dunlop- Woodward Award for outstanding achievement before the age of 35 in 1975. *Society Aff:* ASCE, NSPE, ISPE, ASTM, APWA, SMPS, WPCF, NAFE.

Foerster, George S
Business: NL Permanent Mold Castings, PO Box 0788, Greeneville, TN 37744
Position: Sr. R & D Engineer *Employer:* Doehler-Jarvis, Div. of Farley Industries *Education:* BS/Chem Engg/Tulane Univ. *Born:* Sept 29, 1929 (New Orleans). BS Chem Engrg Tulane Univ 1980. Joined Dow Chem as student trainee & advanced through several res levels to Assoc Scientist 1961. Improved the performance of magnesium and other non-ferrous metals through alloy and process dev. Designed creep resistant magnesium diecasting alloys needed for VW's Superbeetle. Joined N L Indust in 1972 as Prin Res Scientist and have continued innovative R&D on new prods and processes. Created and helped commercialize DeSulf/DeFlake, improved additives for desulfuring steel and making ductile iron. Increased recovery of Mg, Al & Pb from scrap & dross. Devised highly efficient process for producing pure antimony trioxide fume & pure Pb from Pb-Sb scrap. Designed new Pb alloys for battery gidmetal, body solder, and shot. Transferred to Doehler-Jarvis Division of Farley Industries as sr process engr in 1979. Improved core properties and quality of Al castings. Named ASM Fellow 1972. Holder of over 80 patents and author of several tech papers. Enjoy sports, bridge and security markets. *Society Aff:* ASM, TBΠ

Foerster, Richard E
Business: 233 Broadway Suite 1069, New York, NY 10279
Position: Partner *Employer:* Greeley and Hansen *Education:* BS/CE/IL Inst of Tech *Born:* 7/24/23 With Greeley and Hansen since 1948. Involved in major sanitary engrg projects in Chicago, Tampa and Dade Co., Florida, Panama, Brazil, New York City and State, and New Jersey. Resident representative for Greely and Hansen in Tampa, FL 1954 to 1961 and in New York City since 1962. Elected to the Partnership in 1968. Past Director and Chairman of Met. Section WPCF *Society Aff:* AAEE, NSPE, ASCE, WPCF, AWWA

Fogel, Irving M
Business: 373 Park Ave So, New York City, NY 10016
Position: President. *Employer:* Fogel & Assoc Inc. *Education:* BSCE/Civil/IN Inst of Tech. *Born:* Apr 1929 Gloucester Mass. Reg PE in several states and Israel. Fellow, ASCE; Mbr, NSPE, CSI, and other professional societies. Articles have appeared in several professional publications. Lecturer at Harvard Univ, Inter-University Center, Chicago; Newark College of Engrg; MIT; Rutgers Univ, New Jersey; Ohio State Univ; & AGC, AAA, ABC, STA, etc. Founded Fogel & Assocs Inc 1969. Previously engaged in design, construction supervision, and management of major construction projects in the United States and abroad. Listed in 'Who's Who in the East' 'Who's Who in the World' and 'Who's Who in Bus and Finance'. *Society Aff:* ASCE, NSPE, AACE, SAME, AAA

Fok, Thomas D
Home: 325 S Canfield-Niles Rd, Youngstown, OH 44515
Position: Chrmn. *Employer:* Thomas Fok & Assocs, Ltd. *Education:* PhD/Civ Engg/Carnegie-Mellon Univ; MBA/Finance/NYU; MS/Civ Engg/Univ of IL; BEng/Civ Engg/Natl Tung-chi Univ, China. *Born:* 7/1/21. Structural Designer, Lummus Co, NYC, 1951-53; Design Engr, Richardson, Gordon & Assoc, Consulting Engrs, Pittsburgh, PA, 1956-58; Assoc Prof of Engg, Youngstown State Univ, 1958-67; Dir Computing Ctr, 1963-67; VChrmn, Mosure-Fok & Assoc, Youngstown, OH 1956-76; Pres Computing Systems & Tech, Youngstown, OH, 1967-72; Chrmn, Thomas Fok & Assoc, Ltd 1977. Trustee, Youngstown State Univ, 1975-84, chrmn, 1981-1983, Pres, Youngstown Branch, ASCE, 1970; Pres, Mahoning Valley Tech Soc Council, 1972; Pres, Mahoning Valley Soc of PE, 1979. Consulting Engr for State and Municipalities in Mahoning Valley, 1958-. Enjoy classical music and water color painting. *Society Aff:* ASCE, NSPE, OSPE, ACI.

Fok, Yu-Si
Business: Dept. of C.E, 2540 Dole Street, Honolulu, HI 96822
Position: Prof *Employer:* Univ of Hawaii *Education:* PhD/CE/UT; MS/Irrig § Drain. Eng/UT; BS/Argic Eng/Nat Taiwan Univ *Born:* 1/15/32 Worked as Asst Prof and Asst Research Eng in Dept of Agricultural plus Irrigation Engineering and Utah Water Research Laboratory, Utah State Univ, Dec 1963. toJune, 1966. Hydrologist and Research Associate for Illinois State Water Survey and Univ of ILL at Urbana, June 1966 to July, 1968. Head of the Basin Hydrology Section, Texas water Rights Commission, July 1968 to August, 1970. Assoc Prof, Dept of Civil Engineering and Water Resources Research Center, Univ of Hawaii at Manoa, Sept. 1970 to June, 1977. Promoted to full professor July 1977. Enjoy reading, fishing, and gardening. Published more than 100 papers and reports. *Society Aff:* ASCE, ASAE, AGU, AWRA, ΣΞ, IWRA

Foland, Donald L
Home: 429 French Ave, Fort Wayne, IN 46807
Position: Chief Water Engr & Admin Engr. *Employer:* Fort Wayne Water Utility. *Education:* BS/ME/IN Inst of Technology. *Born:* Apr 1927 Fort Wayne Ind. Married, 2 daughters. Navy Reserve 1945-46. Employed Aug 1949, Fort Wayne Water Utility, Jr Engr; Chief Engr 1972. City Administrative Engr 1976. Registered P E. Certified Water Works Operator. American Water Works Assn: State Chairman, District President, national ctte to review distribution system manuals - presented papers at national and state meetings; National - Indiana Society of Professional Engineers, Chapter President, State Director, administrator of professional engineering review course. Fort Wayne Engineer's Club. Five state, county, local review boards - committees since 1969. Three papers, one article published in AWWA, municipal engineering periodicals. Became Fuller Awardee of AWWA in 1978. *Society Aff:* AWWA, NSpE, APWA.

Folds, Donald L
Home: 4427 Fletcher Street, Panama City, FL 32401
Position: Deputy Head, Research & Tech Dept. *Employer:* Naval Coastal Systems Center. *Education:* BS/Physics/Florida State U. *Born:* July 1935. Air Force 1954-58. Field Engr Philco Corp spec in radar & electronic countermeasures equip. With Naval Coastal Sys Center since 1962, conducting res in underwater acoustics & signal processing related to high resolution sonar & ultrasonic imaging. Assumed current respon as Deputy Head Research & Tech Dept 1986. ASNE Solberg Award 1972. Fellow, Acoustical Soc of America. Life Mbr Amer Soc of Naval Engrs, Assoc Ed, "Journal of Underwater Acoustics–. *Society Aff:* ASA, IEEE, ASNE.

Folsom, Jack H
Home: 14712 NE 164 St, Woodinville, WA 98072
Position: Asst Dir Engrg-707/727/737 Div. *Employer:* Boeing Company. *Education:* BS/Aero Engg/IA State Univ. *Born:* Dec 1922. Native Des Moines Iowa. Served as pilot US Army Air Corps 1942-45. BSAE Iowa State Univ. Joined Boeing Airplane Company 1949 as structural design engr; specialized in fuselage design, Models XB52, 367-80, KC-135, 727, and 737. Appointed to management in 1958. Chief Project Engr on 727 from 1968 to present (now acting). Assumed current position Assistant Director of Engrg for 707, 727, and 737 production models in 1975. Assigned chief proj engr structures and payloads new airplane program model 767, 1977-present. Active in Christian lay work and enjoy outdoor activities. *Society Aff:* AIAA.

Folsom, Oliver H
Home: 1518 Allegheny Drive, Sun City Ctr, FL 33570
Position: Const Engr, Water Resources. *Employer:* ret - Part-time Cons - Self employed. *Education:* BE/Gen Engg/Univ of So CA; ME/Civil Engg/Univ of So CA. *Born:* June 1909; BS Gen Engrg U of So Calif 1934; Office Engrg Met Water Dist of So Calif 1934-37; Inspect, Jr Engr, Asst Engr Parker Dam & Central Valley USBR; 2Lt to Capt AUS Corps of Engrs Oper Off Const Camp Cook Calif. Engr Planning London & Oran, S-4 Const Bn Italy Engr Intell Off of Ch Engr 1941-45. USBR Off of Comm Wash DC, Field & Resident Engr Delta Mendota Canal, Asst Reg Plan Officer 1945-51. Contract const Proj Mgr Travis AF Base 1952; Dir Proj Serv Portsmouth Atomic Plant for Peter Kiewit 1952-55; US St Dept ICA River Dev

Folsom, Oliver H (Continued)
Advisor India 1955-60, Ch Engr ICA Mission to Lebanon 1960; loaned to Jordan Gov to create & operate Central Water Auth as Dir Gen 1960-64. AID Ch Mission Engr at Dacca E Pakistan 1964-66; AID Ch Engr for Near East, S Asia in Wash DC 196670. Retired July 1970. Fellow ASCE & Inst of Engrs. (India) Mbr of ICID & Tau Beta Pi Short term consultation with trips to and reports submitted- U.S. Army Engrs (1970) Vietnam - A.I.D. 1971-1975 Pakistan (2) Jordan, Bolivia, Syria. Retired with rank of Colonel U.S. Army Engineers 1969. Chartered Engineer in India and U.K. *Society Aff:* ASCE, ICID, TBΠ, ROA

Folsom, Richard G
Home: 585 Oakville Crossroad, Napa, CA 94558
Position: Pres Emeritus. *Employer:* RPI - retired. *Education:* PhD/ME/CA Inst of Technology; MS/Mech Engr/CA Inst of Technology; BS/Mech Engr/CA Inst of Technology. *Born:* Feb 3, 1907 Los Angeles Calif; 1932-33 Engr, Water Dept, City of Pasadean Calif; 1933-53 Asst to Prof, Univ of Calif Berkeley; 1953-58 Dir Engrg Research Inst, & Professor University of Michigan Ann Arbor; 1958-71 President Rensselaer Polytechnic Inst; 1972-73 Pres, 1965 Wright Lectr, Honorary Mbr, Amer Soc of Mech Engrs; 1971 Lamme Award, Honorary Mbr, Amer. Inst. of Chem Engrs., Honorary Mbr Amer Soc of Engrg Educ; 1975- 78 Mbr of Council, Natl Acad of Engrg; former, corporation director for Airco Inc, American Electric Power Co Inc, Bendix Corp Arthur D. Little, Inc. Others. 1966 Alumni Distinguished Service Award, Calif Inst of Tech; Honorary degrees from Northwestern Univ, Union College, Rose Polytechnic Inst, Lehigh Univ, Rensselaer Polytechnic Inst, Albany Medical College; PE in Calif. *Society Aff:* NAE, ASME, ASEE, AIChE, AIAA.

Fonash, Stephen J
Business: 127 Hammond Bldg, PA State Univ, Univ Park, PA 16802
Position: Prof. *Employer:* PA State Univ. *Education:* PhD/Engrg/Univ of PA; BS/Engrg/PA State Univ. *Born:* 10/28/41. Distinguished Alumni Prof of Engrg Sciences at the PA State Univ. Activities include teaching, res, & consulting. Res interest in mtls & device phys. Specific areas include microelectronics, solar cells, and solid state sensor res & dev; mtls dev (amorphous, polycrystalline); solid state device res. Fellow of IEEE. Married; two children. Enjoy traveling, wood-working, gardening. *Society Aff:* IEEE, APS, AVS, ECS.

Fonda, LeGrand B
Home: 704 W Gore Blvd, Erie, PA 16509
Position: Corp Secy *Employer:* Bartell Metallurgical Group, Inc. *Education:* BS/Ind Chemistry/Univ of WI. *Born:* 10/20/12. Troy, NY; Allegheny Ludlum Steel 1936-42. Mtls & Proccesses Engr, Lab and QC Mgr GE 1942 to retirement Oct 1977, at Lynn & Everett, MA; Tokyo; Munich, Detroit; Cincinnati; & Erie, PA. Consultant to Sifco Industries, Cleveland, Union Carbide Corporate Technical Center. Tarrytown, NY, Chief Met US Welding Corp,Bartell Metallurgical Group Inc., Meadville, PA. and insurance companies since retirement. Helped design & mfg first jet engine built in US. Developed early superalloys; established sound melting & forging techniques for producing heavy superalloy jet engine components; responsible for locomotive diesel engine mtls engg; served 20 ys on two intl mtls specification committees. Elected Fellow ASM 1978: NWPA Chapter ASM "Man of the Year" 1979. *Society Aff:* ASM, ADPA, THE Ξ

Fondahl, John W
Home: 12810 Viscaino Rd, Los Altos Hills, CA 94022
Position: Prof of C.E. *Employer:* Stanford Univ. *Education:* MS/CE/Thayer Sch, Dartmouth College; BS/CE/Dartmouth College. *Born:* 11/4/24. in Washington, DC. US Marine Corps, Pacific Theatre, WWII. American Bridge Co, Structural Detailer, 1948. Univ of HI, Instr and Asst Prof, 1948-51. Winston Bros Co, Construction Engr, 1951-52. Proj Engr-Nimbus Dam and Powerplant proj, 1952-55. Stanford Univ since 1955, full prof since 1966, in charge of grad prog in Construction Engg and Mgt. Golden Beaver Award (Heavy Construction)-1975. ASCE Construction Mgt Award-1976. Phi Beta Kappa. Reg Engr CA Pres and Chrmn, PMI, 1974-77. Dir, Caterpillar Tractor Co since 1976 and Scott Co of CA (Mech Contractors) since 1963. *Society Aff:* ASCE, PMI.

Fondy, Philip L
Home: 444 Schenck Ave, Dayton, OH 45409
Position: Sr VP. *Employer:* Chemineer, Inc. *Education:* MSc/ChE/OH State Univ; BS/ChE/OH State Univ. *Born:* 3/3/35. Native of Columbus, OH. Began engg career in 1957 & have held numerous engg and mgt positions, all with Chemineer, Inc, Dayton, OH, in design & mfg of fluid agitation & mixing equip. Named VP, Engg, in 1966. Currently Sr VP in charge of all marketing, engg, & mfg operations of Chemineer. Mbr AIChE since 1957. Also mbr of several community service organizations. *Society Aff:* AIChE.

Fong, Arthur
Business: 1501 Page Mill Rd, Palo Alto, CA 94304
Position: Corp Engrg Mgr, Engrg Design. *Employer:* Hewlett-Packard Co. *Education:* BSEE/EE/Univ of CA, Berkeley; MS/EE/Stanford Univ. *Born:* Feb 1920. MSEE Stanford Univ 1968; BSEE Honors, Univ of Calif Berkeley. MIT Radiation Labs 1943-45; R&D microwave generation, spectrum analysis, power measurements. Browning Labs 1946: designed 1st commercial 88-108 MHz AM/FM receiver. Hewlett-Packard Co 1946- : work incls electronic instruments, generators & impedance devices to 11GHz; wide sweep spectrum analyzers to 40GHz; low cost satellite TV receivers; computer-aided art work & thermal analysis- design. 6 pats; 17 pubs. IEEE Fellow - 1971, for contribs in microwave. Reg P E Calif. Mbr IEEE, Sigma Xi, Tau Beta Pi, Eta Kappa Nu, Amer Soc Engrg Educ, Internatl Electrotech Comm. Enjoys skiing, fishing, hiking. *Society Aff:* TBΠ, HKN, ΣΞ, IEEE.

Fontana, Robert E
Home: 6679 Statesboro Rd, Dayton, OH 45459
Position: Professor Emeritus, Dept of Elec Engrg. *Employer:* Air Force Inst of Tech. *Education:* PhD/Elec Engg/Univ of IL; BS/Elec Engg/NY Univ; MS/Elec Engg/Univ of IL; *Born:* Nov 1915. Native of Brooklyn New York. Served with the Air Corps during WWII as radar officer. Research Leader with Sandia Corp specializing in microwave systems 1950. Special Asst for Nuclear Development, HQ USAF, 1954. Director of Plans, Office of Aerospace Research 1958. Dir, Aerospace Res Laboratories 1961, Head, Dept of Elec Engrg AFIT, Wright-Patt AFB 1966-1984, Currently Professor Emeritus. Fellow IEEE; Chmn, Bd of Dir, Natl Aerospace Electronics Conf 1971-74. Distinguished Alumnus Award, Univ of Illinois 1973. Legion of Merit, USAF 1966, 1969; IEEE Meritorius Service Award 1983, Dept. of Air Force Exceptional Civilian Service Award, 1985. *Society Aff:* IEEE, ASEE.

Foote, Bobbie L
Home 1314 Barkley Ave, Norman, OK 73071
Position: Prof *Employer:* Univ of OK. *Education:* PhD/Indust Engr/Univ of OK; MS/Math/Univ of OK; BS/Math/Univ of OK. *Born:* 1/24/40. Born and educated in OK. Taught math and coached debate at NEO A&M Jr Coll. Dir. of modeling studies for predictions of cost of elec power and future of nuclear powrr in TX through 2000 AD. Director School of I.E. 1977-81. MRP Analyst for ASO, US Navy, Phil, Penn. Merrick Foundation teaching award 1985. Currently Prof, School of I.E. *Society Aff:* IIE, ORSA, TIMS, NSPE.

Footlik, Irving M
Home: 1548 Tower Rd - Sylvia Ln, Winnetka, IL 60093
Position: Chairman *Employer:* Footlik & Assoc. *Education:* BS/Mech Engr/IL Inst of Tech. *Born:* 2/7/18. Irving M Footlik, licensed PE, is an author, college instr, matl handling and plant layout consultant, and a practical engr all rolled into one. He grad as a mech engr from the IL Inst of Tech. His background covers over 40 yrs' practical diversified experience in all types of industries. He is especially known for his outstanding contributions to the matl handling industry as well as his involvement in civic and community activities. He is a founder of the APMHC and IMMS and also a mbr of ASME, NSPE and ISPE. He is listed in Who's Who in the World,

Footlik, Irving M (Continued)
Who's Who in Finance and Industry, and Who's Who in the Midwest. *Society Aff:* ASME, NSPE, ISPE, APMHC, IMMS, SME.

Footlik, Robert B
Business: 2521 Gross Point Road, Evanston, IL 60201
Position: Pres *Employer:* Footlik and Assoc *Education:* BS/IE/IL Inst of Tech *Born:* 10/29/46 Named as Illinois' Outstanding Young Eng in 1972, specializies in plant layout, materials handling systems design, and facilities planning, internationally. In addition to above affiliations, he has served as VP and Secretary of I.M.M.S., Secretary of A.P.M.H.C., Chairman of the Ethics & Practice Committee of I.S.P.E. Currently participates in the Standards writing process on both the A. N.S.I. MH 10 (Carton Standardization) and MH 14 (Dockboard) Committees (Vice-Chairman). As Material Handling Educational Director of the Midwest Industrial Mgmt Assoc., developed and taught supervisory and technical courses in materials handling. Over 150 articles have appeared in various trade and professional journals. Is also active in civic functions, serving as school board member, VP of North Shore Ass'n. for the Retarded, and as an advisory Dir of the First Oak Brook Bank of Addison, IL. *Society Aff:* IIE, NSPE, IMMS, APMHC, SME, ASME

Foott, Roger
Business: 34 School St, Action, MA 01720
Position: Consulting Soil Engr. *Employer:* Self Employed. *Education:* ScD/Soil Mech/MIT; BSc/Civil Engg/Univ of Birmingham. *Born:* 9/30/46. in England. BSc from Univ of Brimingham, ScD from MIT. 1967-69 with Binnie & Ptnrs London. 1972-73 Instrucltor at MIT. 1973-75 with Louis Berger Inc as Dir of Nigeria Fnd Services & Soils Res Co & in Cyprus & Senegal. 1975- , consulting soil engr working individually in USA & intlly. Awards: Norman Medal (ASCE) 1976; Co-reporter, IX Intl Conf Soil Mech & Found Eng, 1977; Tucker Voss (MIT) 1972; Sloan Res Traineeship (MIT) 1971; Kennedy Meml Scholarship 1969 & 1970, Coultas Prize (U of B) 1967; Taylor Woodrow Prize (U of B) 1966. *Society Aff:* ASCE, ICE, ΣΞ.

Forberg, Richard A
Business: 5303 DuPont Cir, Milford, OH 45150
Position: VP *Employer:* Intl Tech Group Inc *Education:* BChE, Univ of MN; BBA/Ind Mgt/Univ of MN. *Born:* 6/27/15. Directed Corp Indus Engrg Div 1954-68. Mgmt Systems Div 1969-80. Responsibilities included indust engrg, data processing & communications. Employed (6-30-80 to 5-30-83) by Structural Dynamics Research Corp in Engrg Consltg and Software Dev-Business partically owned by Gen Elec Co. Co-founder, Intern'l TecheGroup, Inc. 1983; Engg & Engg Software Co. Active in Cincinnati, psychiatric care organizations and the public schs. VP 1970-72 Engrs Council for Profl Dev. Pres 1972-73 ECPD. Bd Mbr at Large 1969-73 Engrs Joint Council. Mbr of IIE, Tau Beta Pi. *Society Aff:* IIE, ΤΒΠ.

Forbes, Leonard
Business: Dept of Elec Engg, Davis, CA 95616
Position: Assoc Prof. *Employer:* Univ of CA. *Education:* PhD/EE/Univ of IL; MS/EE/Univ of IL; BSc/Engg Phys/Univ of Alberta. *Born:* 2/21/40. A native of the province of Alberta, Canada, he received his BSc degree in engg physics in 1962 from the Univ of Alberta at Edmonton & his MS & Phd degrees in elec engg in 1963 & 1970 from the Univ of IL at Urbana. He has held faculty positions at Howard Univ & the Univ of AR & was with IBM Corp for two yrs. A mbr of IEEE & the Am Assn of Univ Profs, he is author or co-author of about 50 papers, articles, & reports. His res at UC is in GaAs mtls, devices, and circuits. *Society Aff:* IEEE, AAUP.

Force, Richard W
Business: 3131 South State Street, Ann Arbor, MI 48108
Position: Partner *Employer:* McNamee Porter and Seeley *Education:* BS/CE/MI State Univ *Born:* 5/28/39 Native of Williamston, Michigan. Worked with U.S. Geological Survey during Coll. Held position with State of CA Dept. of Water Resources in 1962. Served with U.S. Army Corps of Engrs. 1962-64 joined McNamee Porter and Seeley in 1964 aa Project Eng. Promoted to Associate in 1974. Became Partner in 1977. Duties include overall client responsbilities, Director of Professional Development and Assistant Director of Business Operations. Active in NSPE and ASCE as President of Local Chapters in 1978-79 and 1981-82 respectively. Co-authored Epa Paper on "Combined Sewer Overflow Monitoring and Assessment" for Manistee, Mich. Active in Water Pollution Control Federation. *Society Aff:* NSPE, ASCE, WPCF, AWWA

Ford, C Quentin
Home: 1985 Crescent Dr, Las Cruces, NM 88001
Position: Assoc Dean of Engrg. *Employer:* New Mexico State Univ. *Education:* PhD/Mech Engr/Mich State Univ; MS ME/Mech Eng/Univ of MO; BSME/Mech Engr/NM State Univ; BS/Marine Engr/USMMA. *Born:* 8/6/23. Instr Univ of Mo 1949-50; Instr WA State Univ 1950-53; Asst Prof 1953-56; Instr Mich State Univ 1956-59; Prof NM State Univ 1959- ; Hd Dept Mech Engrg 1960-70; Assoc Dean of Engrg 1974- . Prin Ford & Assocs. Served to Lt USNR 1942- 46. Mbr Amer Soc ME, Fellow AAAS, Amer Soc Engrg Educ, NM Prof Engrs (Outstanding Engr 1964), Sigma Xi, Phi Kappa Phi, Pi Tau Sigma, Tau Beta Pi, Pi Mu Epsilon, Presbyn, Mason, Kiwanian. Editor: Space Tech and Earth problems, Vol 23 Science and Tech Series 1969. Member NM State Bd of Reg for Prof Engrs and Land Surveyors 1978-. V.P.-western zone NCEE-1986-88. *Society Aff:* ASME, ASEE, NSPE, AAAS, NCEE.

Ford, Curry E
Home: 143 Parish Rd South, New Canaan, CT 06840
Position: Consultant. *Education:* MSE/Harvard Univ; BSCE/Purdue Univ. *Born:* 8/23/11. Joined Union Carbide in 1937. Served in a number of positions with Carbon Products Div in areas of sales, new products, mktg and dev. Appointed VP, Technology, Carbon Products 1965; Exec VP, Matls Sys Div 1967; VP & Genl Manager, Carbon Products 1968. Director, Fiber Materials, Inc; AIChE (Fellow). Pres, Federation of Matls Societies 1976; Pres, United Engrg Trustees, Inc 1984-86; Distinguished Engrg Alumnus, Purdue Univ 1968; AICHE Materials Engineering and Sciences Division Award-1979; Visiting Lecturer, Sch of Industrial Mgmt, Purdue Univ; Chrmn, Environ Comm, New Canaan, CT 1982-; Retired from Union Carbide 9/1/76. *Society Aff:* AIChE.

Ford, Edwin R
Business: 398 East San Antonio St, New Braunfels, TX 78130
Position: Pres *Employer:* Ford Engrg, Inc *Education:* BS/CE/TX A & M Univ *Born:* 5/26/43 Native of Brownfield, TX. U.S. Army Transportation Corps, O.I.C., TMA-MAC V, V.N. 1967-1969. Deisgn engr for Urban Engrg specializing in municipal and environmental engrg 1970-1977. Past Pres of Region Three American Society of Civil Engrs, Corpus Christi, TX. Young Engr of the Year 1976 Nueches Chapter, TX Society of PE. January 1978 until the present, responsible for the municipal and environmental design consulting service with Ford Engrg, Inc. Pres, Ford Engrg, Inc. Registered PE, State of TX. Registered Public Surveyor, State of TX. Enjoys soccer and sailing. *Society Aff:* ASCE, NSPE, TSPE, PEPP, TSA

Ford, Frank F
Business: P O Box 19652, Sta N, Atlanta, GA 30325
Position: President. *Employer:* Frank F Ford & Assocs. *Education:* BE/Metallurgy/Yale Univ. *Born:* Apr 26, 1914 Augusta Georgia. Permanent location, Atlanta Ga. Metallurgical Engr, Sales Engr, specialty alloys steels, Great Lake Steel Division, National Steel Corporation Detroit Mich 1948. Assoc Prof of Chem Engrg (Met) Georgia Inst of Tech. 1949-76 President Frank F Ford & Assocs Inc, Consultants and Sales Engineers, specializing in automated production facilities. 1971-72 President of Society of Manufacturing Engrs. P E and Cmfg E 1966, SME Joseph A Siegel Memorial Award. *Society Aff:* SME, AIME, ASM, AWS, NSPE, ASEE.

Ford, Hugh
Business: Exhibition Rd, London SW7 England
Position: Senior Research Fellow - Professor Emeritus *Employer:* Imperial College. *Education:* DSc/Mech Engg/Univ of London; PhD/Mech Engg/Univ of London; BSc/Mech Engg/Univ of London. *Born:* 7/16/13. in Northamptonshire England. Apprentice to locomotive engg and after 5 yrs at Imperial College, London Univ, res engr in Imperial Chem Ind on dev of the first high pressure polyethylene plant. Steel ind 1942-47, in charge of engg res into metal working processes. Researches led to the dev of the first successful automatic gauge control system for cold strip mills. After further ind posts, became prof of appl mech at Imperial College, the prof and hd of dept of mech engg until 1978, pro-rector of the college 1978-80. Now Senior Research Fellow. Pres, IMeche 1976-7. Chrmn of Ford & Dain Partners Ltd (consulting and contract res co) 1971-present, knighted for services to engg education and res, 1975 Vice President, Fellowship of Engineering 1981 -. *Society Aff:* ASME, FIMechE, FICE, FEngg, FRS.

Ford, Maurice E, Jr
Home: 4 Sunriver, Irvine, CA 92714
Position: Regional VP. *Employer:* Boyle Engrg Corp. *Education:* BS/MEch Engg/CA Inst of Technology; AA/Engg/Fullerton Jr College. *Born:* Feb 1924. Native of Fullerton Calif. BS from Calif Inst of Tech. Commissioned Officer-US Navy in WWII. Registered Mechanical EngrCalif. Registered Civil Engr-Calif. 1947-51 Industrial Engr, Columbia Steel Corporation. 1951-56 Manager, Yorba Linda Water Co. 1956-63 Manager, Escondido Mutual Water Co. 1963- , Consulting Engr Boyle Engineering Corporation. Currently Regional VP respon for Quality Control and San Diego and Water Resources Branch Offices. Previously served as manager of Latin American operation for firm. Speak and read Spanish. Publications include: 'Air Injection for Control of Reservoir Limnology', Journal AWWA, March 1963; 'Elimination of Thermal Stratification of Reservoirs and Resulting Benefits', US Geological Survey Water Supply Paper 1809-M. Energy Optimization in Grandwater Development & Production; US Dept of Energy ANL/EES-TM-96. *Society Aff:* ASCE, SAME, NWWA, AWWA.

Fordyce, Samuel W
Home: 6716 Selkirk Court, Bethesda, MD 20817
Position: Chief Engineer *Employer:* Riparian Research Corp. *Education:* MS/EE/WA Univ; SB/Engrg Scis/Harvard Coll. *Born:* Feb 28, 1927 Jackson, Miss. Served with the US Navy during 1944-46. From 1949-58, worked for the Emerson Elec Manufacturing Co in St Louis, specializing in airborne radar sys. Joined NASA in 1962 as the Chief of Comm and Tracking in the Office of Sys on the Apollo Program. Retired from NASA in Nov '82. Currently Chief Engineer of the Riparian Research Corp. *Society Aff:* IEEE, AIAA, Prof Engri.

Foree, Edward G
Home: Commonwealth Technology, Inc, 2520 Regency Road, Lexington, KY 40503
Position: President *Employer:* Commonwealth Tech, Inc *Education:* PhD/Civ Engr/Stanford Univ; MS/Civ Engr/Stanford Univ; BS/Civ Engr/Univ of KY. *Born:* 2/24/41. Native of Sulphur, KY. Began academic career Univ of KY Civ Engg Dept in 1968. Taught grad and undergrad courses and conducted res in water quality control engg, served as dir of water quality labs and dir of grad studies for the Civ Engg Dept. Was responsible for computer modeling and simulation associated with river basin planning studies for KY. Now full-time as pres of Commonwealth Tech, Inc, a firm specializing in water quality control and the environmental effects of coal mining. Extra-profl activities include church, fishing, hunting, gardening, farming, and collegiate spectator sports. *Society Aff:* ASCE, NSPE, WPCF, AAEE, AWWA.

Foresti, Roy J
Home: 301 Willington Drive, Silver Spring, MD 20904
Position: Staff Engineer, Test Dept *Employer:* Vitro Laboratories *Education:* PhD/Fuel Technology/Penn State; MS/Chem Engg/Carnegie-Mellon; BE/Chem Engg/Johns Hopkins. *Born:* Mar 1925. Combustion Engr wth US Bureau of Mines 1951-53; Group Leader Engrg Res Div Monsanto, directing studies in mixing, catalysis & electrostatics 1953-59; Sr Engr, Univ of Dayton Res Inst, consulting for Gov contracts 1959-61; Assoc Prof Chem Engrg Univ of Conn 1961-63; Chmn Chem Engrg Dept Catholic Univ 1963 to 1980. Reg Prof (DC). Cons for misc gov agencies & private corp, officer on numerous sect & natl AIChE committees, Mbr BOD for Jr Engr Tech Soc. Staff Engr, Vitro Corp. consult. on subjects of Corrosion, thermo., heat trans., air pollution 1980-present. *Society Aff:* ACS, AIChE, NACE, ASHRAE

Forger, Robert D
Business: Society of Plastics Engineers, 14 Fairfield Drive, Brookfield Center, CT 06805
Position: Exec Dir. *Employer:* Soc of Plastics Engrs. *Education:* BS/Chemistry/Norwich Univ. *Born:* 5/24/28. Native of Westport, CT. Joined Dorr-Oliver Inc in 1949 upon grad from college serving as res and dev engr and mgr of technical publicity. With SPE since 1959 as dir of mbr activities. conf. mgr, publisher and since 1971 as exec dir. Retired Lt Col, US Army Reserve. Grad of US Army Command and Gen Staff College. Married, two grown sons. Former Chrmn, Westport (CT) Public Housing Authority. Former Pres, Norwich Univ Alumni Assoc. Trustee, Norwich Univ. 1987-1988 Pres, Council of Engrg and Scientific Soc Executives. *Society Aff:* AIChE, SPE.

Forlenza, Gerard A
Home: 75 Llewellyn Rd, Montclair, NJ 07042
Position: President-COO *Employer:* Beker Industries Corp *Born:* Mar 25, 1923 NYC. BS & ChE degrees in Chemical Engineering from Columbia Univ. Joined American Cyanamid in 1947 as research engr. Named Asst Dir, Engr & Const Div in 1960; Asst General Manager, Organic Chemicals Div 1963. In 1967, Dir of Engrg & Const Div; Jan 1969, Dir of Commercial Development Div; July 1969 Genl Manager of Plastics Div; 1973 President, Industrial Chemicals and Plastics Div; 1976 President, Industrial Chemicals Div. Mbr AIChE; 1979 to present- President-Beker Industries Corp . Licensed P E in N Y, N J and Texas. Resides in Montclair New Jersey with wife and five children.

Forman, George W
Business: 4007 Vintage Court, Lawrence, KS 66046
Position: Professor Emeritus *Employer:* Univ of Kansas. *Education:* MS/Mech Engg/Univ of KS; BS/Mech Engg/Univ of IL. *Born:* Dec 1919. Design Engr & Supervisor of Engrg Training, Hamilton Standard 1941-46; Mgr of Mech Engrg, The Marley Co, 1946-53; Mgr of Research, Butler Mfg Co 1953-55; Mbr of the faculty Univ of Kansas since 1955. Fellow ASME 1970; mbr Tau Beta Pi, Pi Tau Sigma, Sigma Xi; Consultant to a wide variety of industry and govt agencies. *Society Aff:* ASME.

Forman, J Charles
Business: FACTS, 77 Stanton Rd, Darien, CT 06820
Position: Exec Dir & Secretary. *Employer:* American Inst of Chem Engrs. *Education:* PhD/Chem Engrg/Northwestern; MS/Chem Engrg/Northwestern; SB/Chem Engrg/MIT *Born:* 12/22/31. J Charles Forman is Exec Dir and Secretary of the AIChE, a professional soc of 60,000 mbrs based in NY. He was appointed to this position in June of 1978 and was elected Secretary six months later. A native of Chicago, IL, he received his SB degree in Chem Engg from MIT and his MS and PhD from Northwestern Univ. Dr Forman joined Abbott Labs of N Chicago in 1965 doing process dev work in chem and fermentation mfg. He later advanced to Sec Mgr and Proj Mgr charged with constructing new antibiotic plants both here and abroad. After serving as Operations Mgr in charge of Corporate Procurement, he was elevated to Dir of Mfg Operations in the Agricultural and Veterinary Prod Div, a position he held until joining the AIChE as Assoc Exec Dir in 1977. An AIChE Fellow, Dr Forman has served on both the Career Guidance & the Education & Accreditation

Forman, J Charles (Continued)
Committees of the Inst. He is a Reg PE in IL. *Society Aff:* AIChE, ACS, AAAS, NSPE, CESSE, ABET, SCI

Fornander, Sven G E
Home: Asgaton 8, Grythyttan, Sweden 71060
Position: Retired. *Education:* Fil Dr/-/Univ of Gothenberg; -/-/Bergsingenior Royal Inst of Technology, Stockholm. *Born:* Dec 19, 1913 Hagfors Sweden. Bergsingenjor, Royal Inst of Tech, Stockholm 1939; Research Asst Inst of Metals Research, Stockholm 1939-41; production engr, later res engr, Surahammars Bruks AB 1941-52; dir of res, Jernkontoret (Swedish Ironmasters Assn), Stockholm 1952-57; Tech Dir Surahammars Bruks AB 1957-68, V Pres same company 1968, Retired 1977. Fellow Swedish Academy of Engrg, Life Mbr ASM, AIMME. Fil Dr h c Univ of Gothenburg. Publ papers on iron & steelmaking in Swedish, American & English journals. *Society Aff:* AIME, ASM, VdEh.

Forney, Bill E
Business: 6011 E 57th Str, Tulsa, OK 74135
Position: Owner. *Employer:* Self (Consulting). *Education:* BSME/Mech/OK Univ; BSCE/Structural/OK Univ. *Born:* June 1921. BSME and BSCE Oklahoma Univ. Continuous study since. PE registration 1948. Air Corps Service WWII. Twenty-two years Div Manager CE- Natco. Manager of Engrg Maloney-Crawford Tank Corp. President Southern Supply and Valve Corp. Owner BEF Engrg Company. Author. Patentee. Expert Witness. Mbr numerous Engrg Societies. *Society Aff:* ASME, ASCE, AISC, EST, NSPE, NAFE, AAFS.

Forney, G David, Jr
Business: 20 Cabot Blvd, Mansfield, MA 02048
Position: VP *Employer:* Codex Corp. *Education:* SCD/EE/MIT; MS/EE/MIT; BS/EE/Princeton. *Born:* Mar 6, 1940 New York City. MS ScD MIT 1965. Author 'Concatenated Codes', MIT Press, 1966. Joined Codex Corporation 1965; became Vice President and Director 1970. Visiting Scientist at Stanford Univ 1971-72. Mbr IEEE Information Theory Group Board of Governors 1970-76 1987- ; Editor, IEEE 'Transactions on Information Theory' 1971-74. IEEE Fellow 1973; IEEE Browder J Thompson Award 1972; IEEE Information Theory Group Prize Paper Award 1970. Member, National Academy of Engineering, 1984. *Society Aff:* IEEE, AAAS

Forney, Larry J
Home: 4029 Menlo Way, Doraville, GA 30340
Position: Assoc Prof *Employer:* GA Inst of Technology *Education:* PhD/Engrg Sci/Harvard Univ; ME/ME/MIT; MS/ME/MIT; BS/Engrg Sci/Case Inst of Tech *Born:* 11/1/44 Attended public schools, Seville, OH. Research Engr for General Electric, 1966. Research Assistant at M.I.T. Fluid Mechanics Lab, 1966-68. Research Engr for Norton Research Corp., 1968. EPA Fellow, M.I.T., 1969. EPA Traineeship, Harvard Univ, 1969-74. Assistant Prof, Univ of IL, 1974-79. Awarded NSF Initiation Grant, 1975-77, SCEEE Fellow, 1982. Assumed current position of Assoc Prof of Chem Engrg, GA Inst of Technology, 1979. Consultant: Walden Research Div. of Abcor Inc, 1972-74; Commercial Union Insurance Corp, 1976; Lockheed GA Co, 1982-83; Sverdrup Tech Inc, 1982-1984. Over 30 publications in the fields of aerosol mechanics, instrumentation, pipe mixing, dynamics and chemistry of buoyant jets and power plant plumes. *Society Aff:* AIChE, AAAR

Forney, Robert C
Home: Centerville, DE 19807
Position: Exec VP *Employer:* E I duPont De Nemours & Co Inc. *Education:* PhD/ChE/Purdue Univ; MS/Ind Engg/Purdue Univ; BS/ChE/Purdue Univ. *Born:* 3/13/27. PhD, MS & BS from Purdue Univ. Native of Chicago, IL. Joined Du Pont, Textile Fibers Dept, 1950. After a series of assignments in res, manufacturing, technical & mktg, became VP & Genl Mgr in 1975. VP of the Plastics Products and Resins Dept in 1977. Assumed respon as Sr VP in 1979, Exec VP 1981. Mbr Amer Chem Soc, AIChE, & Sigma Xi honor scientific res society; Mbr Soc of Chem Industry *Society Aff:* ACS, AIChE, ΣΞ.

Forrester, A Theodore
Business: Buelter Hall-Rm 7731, Univ. of Calif, Los Angeles, Los Angeles, CA 90024
Position: Prof of Engrg & Physics. *Employer:* Univ of California. *Education:* PhD/Physics/Cornell Univ; AM/Physics/Cornell Univ; AB/Physics/Cornell Univ. *Born:* Apr 1918 N Y. U Calif Berkeley (Manhattan Dist) 1942-45; RCA Labs Princeton 1945-46; Asst-Assoc Prof Phys Univ So Calif Los Angeles 1946-55; Westinghouse Res Labs Pgh 1955-58; Atomics Intl & Rocketdyne Divs of North Amer Aviation 1958-59; Electro-Optical Systems, Pasadena Calif 1959-65; Prof Phys U Cal Irvine 1965-67; Prof Engrg & Physics U Cal Los Angeles 1967- ; Visiting Prof Astronomy U Utrecht (Netherlands) 1971; Assoc, Culham Lab England 1974; Visiting Prof Physics Technion (Israel) 1977. Fellow IEEE & Amer Phys Soc Research Award, Amer Rocket Soc (now Amer Inst Aeronautics & Astronautics) 1962. Research in optical heterodyning, ion propulsion, isotope separation, superconductivity and now, primarily, ion beams for fusion plasma application. *Society Aff:* APS, IEEE, AAAS, AAUP.

Forrester, Jay W
Business: Bldg E40-294, Cambridge, MA 02139
Position: Germeshausen Prof of Mgmt. *Employer:* Sloan Sch of Mgmt - MIT. *Education:* DEng (hon)/-/Univ of NE, 1954; DSci (hon)/-/Boston Univ, 1965; DEng(hon)/- /Newark Coll of Engrg, 1971; DSci (hon)/-/Union Coll, 1973; DEng (hon)/-/Univ of Notre Dame, 1974; DPolSci (hon)/-/Univ of Mannheim, 1979; SM/EE/MIT, 1945; BSc/EE/Univ of NE, 1939 *Born:* July 1918. Inventor random access magnetic core memory for computers. Prof of Mgmt 1956. Developed field of Sys Dynamics for policy design in social sys. Received Valdemar Poulsen Gold Medal, Danish Acad Tech Scis, 69, Medal of Honor, 72, IEEE & Howard N Potts Awd, 74 Franklin Inst. Natl Inventors Hall of Fame, 1979. Author *Industrial Dynamics & Urban Dynamics* MIT Press, Cambridge MA 1961 & 1969; & *Principles of Systems, World Dynamics, & Collected Papers of Jay W Forrester* MIT Press, Cambridge Mass 1968, 1971, & 2nd ed 1973 & 1975. Inventor of Yr, 68; Sys, Man & Cybernetics Awd for Outstand Accomp, IEEE, 72; New England Awd, Engg Socs N England, 72; Henry Goode Mem Awd, 77; Am Fed Info Proc Socs, 77; Common Wealth Awd, Disting Serv, 79; Computer Pioneer Awd, IEEE Comp Soc 82. *Society Aff:* NAE, IEEE, AAAS, AM, AIP, SME, IMS, AEA

Forstall, Walton
Home: 124 Woodland Dr, Pittsburgh, PA 15236
Position: Prof of Mech Engrg *Employer:* Carnegie-Mellon University. *Education:* ScD/Mech Engg/MIT; ME/Mech Engg/Lehigh; MS/Mech Engg/Lehigh; BS/Mech Engg/Lehigh. *Born:* June 26, 1909. Asst Test Engr, Philadelphia Electric Company; Design Engr, The Franklin Inst; Asst Supervisor of Industrial Gas Servicing, Philadelphia Gas Works Company; Project Engineer, Manhattan Project, Oak Ridge; Asst Dean of Engrg and Science, Professor & Associate Dept Hd of Mechanical Engrg, CarnegieMellon University. Mbr Phi Beta Kappa, Tau Beta Pi, and Sigma Xi. Service to ASME includes: Chmn of Constitution & By-Laws Ctte & four years as Vice President, Region V. *Society Aff:* ASME, ASEE, APS.

Forster, William H
Business: 320 Park Ave, New York, NY 10022
Position: VP & Prod Group Mgr - Telecommunications & Elect *Employer:* IT&T Co *Education:* AB/Physics/Harvard College; -/Adv Mgt Program/Harvard Bus Sch. *Born:* July 1922 Phila Pa. 1943-65 Philco Corp Phila. Engineering & research included Director of Solid State Electronics Research, Technical Director of the Communications & Electronics Div primarily defense communications. Since 1966 with ITT successively as Staff Asst to the President, Vice President and Technical Director of ITT Europe. In current position since April 1976. Fellow of the IEEE, Physical Society UK. Previously served as Vice Chairman of original RMA (now EIA)

Forster, William H (Continued)
Committee on Microwave Relays & as Chmn of JTC-14 (Semiconductor Devices). *Society Aff:* IEEE, ΣΞ, ΦBK.

Forsyth, Raymond A
Home: 5017 Pasadena Ave, Sacramento, CA 95841
Position: Prin Transp Engr *Employer:* Calif D.O.T. *Education:* M/CE/Auburn Univ; BS/Engr/San Jose Stae Univ *Born:* Native of Nevada. Served as Civil Engrg officer, USAF 1954-56. Instructor in civil Engrg, Aunburn Univ 1957-58. Currently Chief of the CA D.O.T. Lab. Directed the development of the CA D.O.T. flexible pavement overlay design procedure and earthwork reinforcement sys (MSE). Chrmn, TRB Committee on Embankments and Earth Slopes (1976-1982). Chrmn, CA State Council, ASCE, 1980- 81, Chrmn TRB Section K (Soil Mechanics) 1982- . Chmn AASHTO Joint Task Force on Pavements 1987- . *Society Aff:* ASCE, TRB, ASTM, AASHTO

Forsyth, T Henry
Business: 9921 Brecksville Rd, Bracksville, OH 44141
Position: Research Associate *Employer:* BF Goodrich Chemical *Education:* PhD/Chem Engg/VA Poly Inst and St Univ; MS/Chem Engg/VA Poly Inst and St Univ; BS/Chem Engg/Univ of KY. *Born:* 11/8/42. in KY. Employed by Dow Chem from 1967 to 1970 for res on polymers. Employed by The Univ of Akron from 1970 to 1980 in the Chem Engg Dept. Maj teaching responsibilities have been Materials Sci, Polymer Engg and Water Pollution. Grad res has produced 15 publications. Employed by BF Goodrich since 1980 for Polymer Reactor Research. Active Nationally in the Materials Div of AIChE (Dir, Secretary-Treasurer), regionally for ASEE (Chrmn for N Central conference). Organized a natl Conf on Polymerization Reactors. *Society Aff:* AIChE, SPE, ACS.

Forsythe, Peter
Home: 1209 Foxhill Drive, Clinton, MS 39056
Position: State Eng *Employer:* Soil Cons. Service *Education:* BS/CE/MT State Univ *Born:* 8/23/32 Native Montanan staring with Soil Conservation Service in 1955. Worked at Three locations in Montana before promotion in 1966 to Assistant State Eng in IN. Promoted to State Conservation Eng in MS with responsiblity for total SCS Engrg Program. Presented five papers to American Society of Agricultural Engineers and one to ASTM. Experienced in all types of Irrigation, Flood Control, Construction, Dispersive Soil Treatment and Special Stream Studies. Recognized in 1977 as Eng of the Year and Chairman Elect in 1980 by MS Section ASAE. Chairman, Planning Committee, Water Resource Structural Group ASAE. *Society Aff:* ASAE, NSPE

Fort, Tomlinson
Business: California Polytechnic State University, San Luis Obispo, CA 93407
Position: Prof of Chem and Matls Sci *Employer:* California Polytechnic State Univ *Education:* Postdoctoral/Surface Chem/Univ of Sydney Australia; PhD/Surface Chem/Univ of TN; MS/Chem/Univ of TN; BS/Chem/Univ of GA. *Born:* April 16, 1932. Res Chem Sr Res Chem, and Proj Ldr at E.I. duPont de Nemours & Co 1958-65, Asst Prof, Assoc Prof and Prof of Chem Eng at Case Western Reserve Univ 1965-73. Prof of Chem Eng and Chemistry and Head, Dept of Chem Eng at Carnegie-Mellon Univ 1973-79. Prof of Chemistry and Chem Eng and Provost, Univ of Missouri-Rolla, 1980-82. Prof of Chem and Materials Sci and Provost, CA Polytechnic State Univ 1982-86. Summer appointment as Visiting Prof, National Univ of Mexico, 1973. Summer appointment as Visiting Prof, Univ of Copenhagen, 1977 and 1980. Pres, Frances Fort Brown Realty Co, 1970- . Chrmn, Eastern N C Sec of the ACS, 1962-63. Chrmn, 43rd Natl Colloid Symposium 1969. Chrmn, ACS Div of Colloid and Surface chemistry 1977. Chrmn, Gordon Conf on chemistry at Interfaces, 1977. National Councilor, ACS 1978- . Univ administration; teaching, research and consulting on interfacial phenomena. *Society Aff:* AIChE, ACS, ΣΞ, ASEE, IACIS.

Forte, Vincent A
Business: 9107 Interline Ave, Baton Rouge, LA 70809
Position: Bd Chrmn *Employer:* Forte & Tablada, Inc. *Education:* BS/EE/LA State Univ. *Born:* 5/5/28. Native of Plaquemine, LA. Worked as an elec engrg with various consulting engrg firms for ten years prior to forming Forte and Tablada, Inc in 1961, now serving as Bd Chrmn. Also Pres of F&T Services Corp, F&T Energy Corp and Aero Construction, Inc. Is responsible for gnl mgmt activities of these companies. Past pres of LA Engrg Foundation, Past Pres. LSU Engr-Alumni, Baton Rouge Chpt-LA Engrg Soc, and Past Pres of state soc. Mbr C.E.C. Enjoys fishing and gardening. Mbr, Bd of Dir, Plaquemine Bank & Trust Co. *Society Aff:* IEEE, NSPE, LES, LEF, IES, CEC.

Fortenbaugh, Charles E
518 Hiawatha Ave, Inverness, FL 32652
Position: Retired. *Employer:* Self. *Born:* Apr 1909 Cleveland. s. Charles Henry and Helena Theodore (Boll) F; student Case Inst Tech 1930-31, Western Reserve Univ 1938, 1942, Cleveland Coll 1944, 1946; m. Edna Marlan Jackson April 20, 1934; children: Marlan (Mrs R Barrett), Ann (Mrs William Kingzett), Charles P, Kathleen, Thomas Jr. Engr City Cleveland 1937-41; Republic Steel Corp 1941-44, Cleveland, 1944-50; cost engr, asst supt Diamond Shamrock, 1950-65; sr proj estimator Lummus Co 1967- ; Exec Dir AACE 1970- . *Society Aff:* AACE

Fosdick, Ellery R
Business: PO Box 561, Port Townsend, WA 98368
Position: Cons Engr. *Employer:* Self. *Education:* EE/Elec Engg/WA State Univ; BS/Hydrolec Engg/WA State Univ. *Born:* Feb 1900. Asst Engr for Wash Water Power Co in Spokane, WA 7 years. Chief of electric utility investigations for Wash State PUC 5 years & for Federal Power Comm in Wash D C 7 years. Asst Dir of Power for Bur of Rec in Wash D C 5 years. Dir of Utilities for First Naval Dist in Boston, Mass 1 year. Administrator of very large construction projects for Atomic Energy Comm in Wash DC 5 years. Cons Engr self employed 30 years working in many parts of the US and in Central and S America, the Carribean, Europe, Asia and the Near East for electric utilities, foreign governments, engrg consultants, the U N, U S Gov, & conservation organizations. Registered engr. Fellow of IEEE and ACEC. *Society Aff:* ACEC, IEEE.

Fosdick, Lloyd D
Home: 276 Acorn Ln, Boulder, CO 80302
Position: Prof. *Employer:* Univ of Colorado. *Education:* PhD/Physics/Purdue Univ; MS/Physics/Purdue; BS/Liberal Arts/Univ of Chicago; PhD/Liberal Arts/Univ of Chicago. *Born:* Jan 18, 1928. BS 1948 Univ of Chicago; MS 1950, PhD 1953, Purdue Univ. Res Assoc Control Systems Lab, U of Illinois, 1953-54; US Army 1954-56; Assoc Dir Computer Center, MURA, 1956-57; On faculty, Dept of Computer Science, U of Illinois 1957-70; Prof, Dept of Computer Science U of Colorado, 1970-present. Purdue Fellowship 1951-53; Guggenheim Fellowship 1964-65. Editor, Algorithms, Communications of the ACM, and Transactions on Mathematical Software, 1969-76. Board Mbr, Special Interest Group on Numerical Mathematics of the ACM 1979; V Chmn, International Federation for Information Processing Working Group on Numerical Software, 1974- ; Sec-Treas, Computer Science Board 1976-78. Mbr Sigma Xi, SIAM, ACM.

Fosholt, Sanford K
Home: 1208 Northwood Lane, Muscatine, IA 52761
Position: Retired (Exec VP) *Education:* BS/EE/IA St Univ *Born:* 5/11/15 B. Rudd, IA, Past President CEC/USA. Life Fellow ACEC; Life Fellow ASME; Past President and Life Member Consulting Engineers Council of IA; Life Member IEEE, Life Mbr NSPE, Life Mbr. IA Engineering Society, served as USA delegate to International Federation of Consulting Engineers and as CEC representative to National Council of Engrg Examiners; Member Advisory Board Coll of Engrg, Univ of IA; Distinguished Service Award of IA Engineering Society, Registered Eng, Consultant, Who's Who in America, and in the World, Retired 1981 from Executive VP, Stan-

Fosholt, Sanford K (Continued)
ley Consultants, Inc., Muscatine, IA, Mbr Bd of Dirs Muscatine Community Sch Fnd, and Vesterheim Norwegian-Amer Museum, Decorah, IA. *Society Aff:* ACEC, IES, IEEE, ASME, CEC/Iowa.

Foss, Edward W
Business: 499 E. Palmetto Park Rd, Boca Raton, FL 33432
Position: Dir of Agr. Programs *Employer:* Access/International *Education:* MSA/Agr Engg/Cornell Univ; BSA/Gen Agr/Univ of NH. *Born:* 12/04/14 Native of Laconia, NH. Taught Voc Agr & Ind Arts in high schools in Averill Park, NY & Walpole, NH (1936-42). Instr, Applied Farming & Agr Engr Univ of NH (1945-49), Ext Agr Engg, Univ of ME, 1945-49, Prof, Agr Engr, Cornell Univ 1945- present. Visiting Prof of Agr Engg Univ of Ibadan, Nigeria 1968-1970, Visiting Prof, Univ of HI (Hilo) 1977-78. Consultant, USDOL-OSHA. Agri Machine Gurding; Consultant: Agri Safety Code Protection. 1974-77. Trustee, Village of Lansing, 1972-74 Agri Safety Code Committee, NYSDOL. 1975-77 Campground Code Committee. NYSDOH. 1963-68 Ithaca Sch Bd. 1966-68 Bd of Cooperative Educational Services. 1951-63 Secy Tompkins County Fire Advisory Bd. 1945-79 450plus Engg, Safety, Housing, etc res, ext & private media publications, movies, slide sets, including one book. *Society Aff:* ASAE, NFPA, ΣΞ, ΕΣΦ.

Foss, James B
Home: 304 Linwood Ave, Bel Air, MD 21014
Position: Retired *Employer:* Frederick Ward Assoc, Inc *Education:* BS/CE/Univ of IL *Born:* 7/1/29 Native of Bridgeport, IL. Graduated from Univ Of Ill in 1951 with BS in Civil Engrg. Served as lieutenant in U.S. Army during Korean conflict, 1951 - 1953. Design and Project Engr with Warren and Van Praag, Inc, till 1963. Project Mgr with Frederick Ward Assocs, Inc, until firm was inc by Fred Ward and Jim Foss in 1968. Co-owner, vp and chief engr, and responsible for all firm engrg activities. unti retirement June 30, 1984, NEC Certificate No. 4416. Registered PE in MD, VA, PA, DC, DE, WV, SC, IL and MO. *Society Aff:* ASCE, NSPE, WPCF, AWWA, SAVE.

Foss, John F
Business: 200 Engr. Bldg, East Lansing, MI 48824
Position: Prof *Employer:* Mech Engr., Mich. State U *Education:* PhD/ME/Purdue Univ; MS/ME/Purdue Univ; BS/ME/Purdue Univ *Born:* 3/24/38 Research interests: basic mechanics of turbulent flows, primarily free shear flows, allied interests in advanced techniques for hot-wire anemometry including "direct" measurements of the transverse vorticity. Pedagogical interests: fluid mechanics with a strong interest in lecture/laboratory coordinated instruction. Co-author (M.C. Potter) of a basic text: Fluid Mechanics (Ronald Press Co. 1975, presently published by Great Lakes Press, Okemos, MI 48864). Research leaves supported by Sloan Fellowship-The Johns Hopkins Univ. (1970-71) and the Alexander von Humboldt Foundation-The Univ. of Karlsruhe, West Germany (1978-79) (and the Univ. of Erlangen (1985-86)). Research support from Department of Army, ONR, NSF, NASA, Lockheed-Georgia, Co., (United Technologies Research Center, Ford Motor Co. and the Whirlpool Corp.) and the Whirlpool Corporation. *Society Aff:* APS, ASME, AIAA, ASEE, ΣΞ

Foss, Mark B
Business: 1701 Westview Dr, PO Box 2823, Fargo, ND 58108
Position: Chmn Bd *Employer:* Foss Associates Archs & Engrs & Interiors *Education:* Mas of Sci/Civil Engg/Univ of ND; Bach of Sci/Civil Eng/Univ of ND. *Born:* 11/18/27. Native of Fergus Falls, MN. Served US Army Special Services, 1945-47 & US Army Corps of Engrs, Japan & Korea 1951-53; Struct Designer with S T DeRemer, Grand Forks ND 1953-55; Struct Designer, Struct Engr, Proj Engr, HD of Struct Engg Dept, Hd of Admin Dept & Pres & Chmn of Bd Foss Associates Fargo, ND & Moorhead, MN. PPres NDACEC; PPres ND Branch ASCE; Mbr NSPE & ASCE; Reg Civil-Struct Engr & Land Sur; 1966 DOD & AIAA Natl Fallout Shelter Design Award; AISC special Citation Award, 1974;. Member, U.N.D. Sch. of Engr. Advisory Council. *Society Aff:* ASCE, NSPE, SMPS

Fossum, Robert R
Business: 1400 Wilson Blvd, Arlington, VA 22209
Position: Dir, Defense Advanced Research Projects Agency *Employer:* Dept of Defense *Education:* PhD/OR State Univ; MS/Univ of OR; BS/Univ of Idaho *Born:* Nov 1928, El Paso Texas. US Navy Officer, 1951-54. Engr & Engrg Mgr, Sylvania Electronic Defense Labs, Mountain View, Calif 1957-69. Vice Pres Systems Tech, ESL Inc, 1969-74. Dean of Res US Naval Postgrad School, 1974-76. Dean of Sci & Engrg, 1976-1977. Phi Beta Kappa, Sigma Xi. Enjoys aviation, sailing & classical music. 1977-present, Director, Defense Advanced Research Projects Agency (DARPA). *Society Aff:* IEEE.

Foster, Albert L
Home: 5308 Tidwell Hollow Rd, Nashville, TN 37218-4027
Position: Plant Engineer *Employer:* Vanderbilt University *Education:* BET/Mechanical/University of North Carolina at Charlotte; AAS-MET/ Mechanical Design/Central Piedmont Community College *Born:* 10/30/50 Native of Charlotte, N.C. After service in U.S. Navy worked myself through engrg school by supervising steam plant operations at UNC at Charlotte. Elected first non-faculty president of Univ Senate at UNCC. Asst Dir of Physical Plant at Western Carolina, 1981-85. Headed Plant Engrg Section at Vanderbilt Univ since 1985, where I direct programs of energy conservation, utility plant operation, deferred maintenance and capital budgeting. Am published in areas of utility metering and facilities mgt. Selected as AIPE's South East Regional Plant Engr of 1981. Am presently VP, Education and Training, AIPE. *Society Aff:* AIPE, APPA

Foster, Asa B
Home: Rt 7-Irvin Bridge Rd, Conyers, GA 30207
Position: Dir Air & Hazardous Matls Div. *Employer:* Environ Protection Agency. *Education:* BCE/Civil Engr/GA Inst Tech. *Born:* 1/29/30. Bach of Civil Engrg, Ga Tech 1952. Native of Gibson County, TN. Lt Corps of Engrs 1952-54 (Kores). Reg Eng - GA, KY, TN, AL, NC, SC, MS. Fellow ASCE - served Bd of Dirs & Pres - GA Sect ASCE. Employed by Robert & Co Assocs & Davis & Floyd Inc before joining Fed service in 1961. Have been in various capacities in the Environ Protection Agency's Atlanta Office. Currently a special assistant to the Regional Admin and projects that reduce the Rlais attention. *Society Aff:* ASCE.

Foster, Charles K
Home: 5706 Highland Hills Cir, Austin, TX 78731
Position: Dir Water Hygiene Div. *Employer:* TX Dept of Health. *Education:* BS/CE/TX Univ *Born:* 11/24/20. Sipe Springs, TX. Employed Imperial Irrigation Dist, Imperial, CA prior to WW II. Served with Navy Seabees; Ret CEC Res. Employed TX Health Dept, Dallas County Health Dept & TX Dept of Health since 1949. Dir Environ Engrg TDH. Respon for statewide solid waste mgmt & drinking water programs since from 1967-76. Dir Water Hygiene Div, 1976-81. Chief, Bureau of Environ Health, 1981. Past Pres, Travis Chap, TSPE; Diplomate, Amer Acad of Environ Engrgs. Past Chmn, TX Sect, AWWA; Past Pres, TX Public Health Assn; Secy TX Water Utilities Assn. Served on numerous cttes. Active in church & masonic work. *Society Aff:* NSPE, AAEE, AWWA, WPCF.

Foster, Derrell V
Business: 1701 21st Ave South, Nashville, TN 37212
Position: Pres *Employer:* Advanced Digital Products *Education:* PhD/CS & EE/Univ of TX-Austin; MS/CS/Univ of Houston; BS/Math & EE/Univ of TX-Arlington *Born:* 9/12/46 After receiving PhD, taught Computer Science at Duke Univ until 1977. Then taught Computer Science at Vanderbilt Univ School of Engrg specializing in operating systems, computer engrg, and in digital communications. Assumed current position in 1981 to design/develop/market a desktop VAX running the UNIX operating system with a totally transparent local computer network. *Society Aff:* ACM, IEEE

Foster, Edward T, Jr
Business: 4360 Nicholas St, Omaha, NE 68131
Position: Pres/Chrmn of Bd *Employer:* Nicholas Indus/Foster-Western/Nicholas Const *Education:* PhD/Engg/Univ of CA; SM/Nucl Engr/MIT; SB/Civ Engr/MIT. *Born:* 3/11/41. Native of Omaha, NE. Spent 1968-1970 as Capt, US Army Gen Staff, Operations Res Analyst. Served as VP, HDR Systems, computer subsidiary of Henningson, Durham, and Richardson during 1970-1975 except for five months service in 1971 as Fulbright-Hays Visiting Prof, Inst of Earthquake Engg and Engg Seismology, Skopje, Yugoslavia. From 1975 to present in gen mgt of design and construction including marketing, admin, and proj mgt. Reg PE in nine states, plus NCEE Certificate and ICCP Certificate in Data Processing. Mbr of NE State Bd of Examiners for PEs and Architects. Enjoys classical music, sailing, classic automobiles, and model railroading. *Society Aff:* NSPE, ASCE, ACM, OEC, AICCP.

Foster, Edwin P
Business: Engg, 222 Grote Hall, Chattanooga, TN 37402
Position: Prof and Dir of Civil Engrg *Employer:* Univ of TN. *Education:* PhD/Structural Engr/Vanderbilt Univ; MS/Structural Engr/Vanderbilt Univ; BE/Civ Engr/Vanderbilt Univ. *Born:* 5/7/42. Prof and Dir, Univ of TN at Chattanooga 1979-, Pres Nashville Sec of ASCE 1979, Univ of TN at Nashville 1968-1979, NASA- ASEE summer faculty fellowship 1977, 78, Avco Aerostructures - stress analysis engr 1968, Brown Engg Co 1965, 66. Am Bridge Div of US Steel 1964. PE 10820 TN, Alumni Outstanding Teaching Award, 1977, teacher excellence Award of TSPE student chapter 1973, 1984, 1987, Vanderbilt Univ adjunct Assoc Prof 1975, NSF Traineeship 1965, res published in ASCE structural journal, computers & structures, & AIAA Journal of Aircraft, presentations of res at intl confs in US, Canada, England, Czechoslovakia, Germany. *Society Aff:* ASCE, ASEE, NSPE, TSPE, AAM.

Foster, Elmer P
Home: 2613 Drayton Drive, Wilmington, DE 19808
Position: Material Coordinator. Retired *Employer:* E I DuPont De Nemours & Co. *Education:* MSE/ChE/Univ of MI; BSE/ChE/Univ of MI; BSE/Math/Univ of MI. *Born:* Dec 1919. Native of Flemingsburg, Ky. Served with US Army Sig Corps 1941-46, US Army Reserve (Sig Corps & Mil Intell) 1947-74. Grad US Army War College. Col US Army Res (Retd). With DuPont Co from 1947 with tech assignments & production supervision. Rtd 1982. Founding officer NJ Sect & Savannah River Sect, AIChE. AIChE Admissions Comm Chmn 1976-77, Member 1956-77. *Society Aff:* AIChE.

Foster, Eugene L
Home: 3316 Wessynton Way, Alexandria, VA 22309
Position: Pres *Employer:* UTD, Inc. *Education:* ScD/Mech Eng/MIT; Mech E/Mech Eng/MIT; MS/Mech Eng/Univ of NH; BS/Mech Eng/Univ of NH. *Born:* Oct 9, 1922 at Clinton, Mass. Taught Mech Engrg subjects at UNH (2 yrs) & MIT (6 yrs). Founder & Chmn, Foster-Miller Assocs Inc, an Engrg R&D co. Served as Pres of Foster-Miller 1956-72. Cons to Office of the Secr, U S Dept of Transportation, 1972-74 where he planned & implemented $100 million R&D program in Tunneling Tech. Founded UTD Corp in 1974 (incorporated in 1976) to develop new tech for underground applications. Mbr: U S Natl Ctte on Tunneling Tech, Natl Acad of Engrg. Reg Prof Engr New Hampshire, Mass, Va, Md. Mbr: ASME, ASCE, ASEE, NYAS, NSPE. *Society Aff:* ASME, ASCE, NSPE, ASEE.

Foster, John Stanton
Business: 275 Slater St-20th Floor, Ottawa Ontario, K1A OS4
Position: Pres & Ch Exec Officer. *Employer:* Atomic Energy of Canada Ltd. *Born:* June 1921. BEng (Mech) & BEng (Elec) from N S Technical College; Governor-General's Medal. Served in Royal Canadian Navy 1943-45. Joined Montreal Engrg 1946 & worked on engrg of thermal power plants in Canada & Central & S America. 1953 became involved in nuclear reactor design. 1954 seconded to group which did feasibility study for Canada's first nuclear power plant (NPD), then became head of CGE Design Engrg Group for NPD. 1958 Deputy Mgr, Nuclear Power Plant Div, Atomic Energy of Canada Ltd. 1959 Mgr Douglas Point NPP Project. 1966 resigned Meco & appointed Vice Pres, Power Projects AECL. Pres of AECL since Dec 31, 1974. Hon degrees of D Eng from N S Tech College & Carleton Univ. Fellow Royal Soc of Canada, Engrg Inst of Canada.

Foster, Leroy E
Business: P O Box 2500, Daytona Beach, FL 32015
Position: Manager-Tech & Production Operations. *Employer:* General Electric Company. *Born:* Troy NY 1931. Joined G E's Engrg Lab Schenectady 1949. Completed co's Adv Engrg Prog A, B & C courses 1956. Moved to Phila & was sys engr in Re-Entry Sys Dept. Apptd Mgr Communications Sys Analysis 1961. Mgmt positions in Cocoa Beach, Daytona Beach, Huntsville, and Syracuse. Wrote two engrg textbooks published by John Wiley. In 1970, moved to Daytona Beach as Manager of Program Management and subsequently to the Manager of Engrg - Daytona Beach Progs. Appointed to present position in 1975.

Foster, Lowell W
Home: 3120 E 45th St, Minneapolis, MN 55406
Position: Pres ' Dir Lowell W. Foster Associates, Inc. *Employer:* Manage Own Co. *Education:* BS/USCG Acad *Born:* Oct 1919, Minneapolis, MN. US Coast Guard 1941-46. Grad USCG Academy, attended U of Minn. WW II Atlantic, Pacific, Far East. Employed Honeywell Inc 1946; Designer, Supervisor, Tool Design, Process, Production, Standards; Sr Principal Engr; Corporate Exec responsible worldwide standardization. Author 35 texts, lecturer, teacher on product definition, standards, metrication; 100 papers, presentations; conducted 1200 training seminars. Active natl and internatl standardization; leader U S and internatl cttes to USSR, Japan, Europe, U K. SES: Fellow 1970, Leo B Moore Award 1973, Dir 1971-73. Registered PE; Certified Manufacturing Engrs SME; Certified Teacher Minnesota; Cert Stdization Engr; Who's Who America, Centennial Award ASME 1980. *Society Aff:* SME, SES, AIDD, NSPE, MSPE, ECM.

Foster, Norman
Home: 4300 Stony River Dr, Birmingham, MI 48010
Position: Pres, Chief Exec Officer. *Employer:* Oxy Metal Industries Corp. *Born:* Apr 19, 1938. MEA & BS Chem E from Washington U, St Louis, Mo. Native of St Louis. Chemical Corps Officer US Army, Ft McClellan, Ala. Functional engrg and plant management responsibilities with Monsanto Co in St Louis (J F Queeny Plt) & Nitro, W Va. Business Controller & Financial Analysis Manager in corporate staff role with Monsanto in St Louis Mo. Tenure with Monsanto from June 1960-May 1972. President & Chief Operating Officer of Pan American Chemical Co in Toledo Ohio from June 1972-August 1973; a basic producer of pentaerythritol and sodium formate. With Oxy Metal Industries from Oct 1973, initially as Dir of Planning & Dev. Promoted to V P of Finance & Planning, then to Exec V P. Became Pres & CEO Aug 1975 of this $260 million div of Occidental Petroleum.

Foster, Walter E
Business: Black & Veatch, PO Box 8405, Kansas City, MO 64114
Position: Partner *Employer:* Black & Veatch *Education:* BSCE/Sanitary Engrg/Univ of KS. *Born:* 5/28/30 With Black & Veatch Constgl Engrs since 1955. Manage environ engrg work for private and public sectors primarily involving water, wastewater, solid and hazardous waste. Also other types civil works including transportation and marine projects. Example project is the Green Bay, Wisconsin, wastewater treatment plant, winner of first annual engrg design achievement award given by WATER & WASTES ENGRG magazine. Dir of Black and Veatch office in Detroit, MI 1978 to 1982. Coauthor of "Sewage Pumping" section in McGraw-Hill PUMP HANDBOOK. Mbr ASCE, APWA, AWWA, HMCRI, NCEE, NSPE, SAME and WPCF. Also diplomate, Amer Academy of Environ Engrs. *Society Aff:* ASCE, NSPE, SAME, AAEE, AWWA, APWA, NCEE, WPCF, HMCRI

Fouladpour, K Danny
Business: PO Box 2863, 707 North 27th St, Boise, ID 83701
Position: VP. *Employer:* H & V Engineering, Inc. *Education:* MS/Sanitary Engg/WA State Univ; BS/Civil Engg/Univ of ID. *Born:* 2/25/38. in Iran, entered US in 1957, to attend college. Worked for ID Dept of Highways several yrs. Returned to Iran in 1967 serving as Sanitary Engg consultant to Iranian government. Returned to States 1968, accepting position with WA State Univ on a solid waste management demonstration project. Joined ID Dept of Water Resources in 1970, active in water resource planning. Joined Hamilton & Voeller Consulting Engrs in 1972, establishing the Boise Office. Became part owner and VP in 1974, forming H & V Engg, Inc, with three offices in ID and two in NV. *Society Aff:* NSPE, ASCE, WPCF, ACEC.

Foulke, Donald G
Home: 455 Johnston Dr, Watchung, NJ 07060
Position: Director. *Employer:* Laboratorium. *Education:* PhD/Chem/Rutgers Univ; MS/Chem/Rutgers Univ; BS/Chem/Juniata College. *Born:* Aug 1912, Burnham Pa. Taught Rutgers, Beaver, Brooklyn Poly. Cons A K Graham & Assocs, Foster D Snell and Laboratorium. Industry: Republic Steel, Houdaille Hershey (Manhattan Project), Hanson-Van Winkle-Munning & Sel-Rex Corp (U S & Swiss). Res in electroplating applications and processes (25 U S patents and over 50 papers). Editor Electroplaters' Process Control Handbook. Exec Secy AES & Bd of Dir of AES and ECS. AES Gold Medal (3), Precious Metal Plating Award (2), Award of Merit and Proctor Award. Elected Fellow of ASM in 1974 & AES Honorary Mbr in 1979. Enjoy travel, gardening and building. *Society Aff:* ACS, ASM, AES, ECS.

Foundos, Albert P
Home: One Dorchester Dr, Muttontown, NY 11545
Position: Pres. *Employer:* Fluid Data, Inc. *Education:* MBA/Bus Mgt/City College of NY, Baruch Sch; BCHE/Chem Engg/City College of NY, Engg. *Born:* 5/23/35. in Albania. Emigrared to the US in 1948 after a four yr stay in Greece. Served in the US Army Corps of Engrs and thereafter in the Res. Worked for MW Kellogg Co four yrs as instrument engg. Joined manufacturers representative co in 1961 as sales application engg for process analyzers specializing in Process Chromatography and became co-ower in 1969. Dev a unique sample conditioner for sampling outlet gas of cracking furnaces for ethylene production. Founded Fluid Data, Inc in 1973 to manufacture the Sample Conditioners and Process Analyer systems with emphasis on control instrumentation for energy conservation. Patent No. 4,259,867 Gas Sample Conditioning Apparatus. *Society Aff:* AIChE, ISA, API-SOI.

Fourney, Michael E
Business: 4532-J Boelter Hall, Civil Engineering Dept, UCLA, Los Angeles, CA 90024-1600
Position: Prof *Employer:* Univ of CA at Los Angeles (UCLA) *Education:* Ph.D./Aeronautics/CA Inst. of Tech; M.S./Aeronautics/CA Inst. of Tech; B.S. /Aeronautical Eng/WV Univ *Born:* 1/30/36 Native of WV. Registered professional eng in states of WA And ST of CA. Past President SESA. Past professional experience: Eng, General Electric Co; Douglas Aircraft Co; Boeing Aircraft Co. 1963-64 Eng, Boelkow Entwicklungen KG, Muchen, Germany. Research Asst Prof, Univ of WA. Dir of Engrg, Mathematical Science. Research Associate Prof, Univ of WA. Univ of WA Faculty Award/1972-73. B.J. Lazan Award (SESA award), 1975. Treasurer Eight U.S. National Congress of Applied Mechanics, UCLA 1978. Enjoys sailing. *Society Aff:* ASME, OSA, SESA, AIP, RILEM.

Foushi, John A
Business: 3210 Watling St, East Chicago, IN 46312
Position: Supervising Eng *Employer:* Inland Steel Company *Education:* BS/ME/IL Inst of Tech *Born:* 6/13/28 Past Pres of American Assoc of Cost Eng, authored numerous papers in the field of cost engineering, certified cost engr and is a Supervising Eng, Project Control Group, Engrg Dept, Inland Steel Company, with over 30 years of experience in all phases of cost engineering. Responsible for the development of capital budgeting, feasibility estimates and various phases of planning for the Steel Divison of Inland Steel. Developed the Inland Steel Company construction cost indices for new construction and blast furnace relines. Also School Board Member and VP of the Board of Education of Illinois School 170 (1969-85); Member of Inland Managment Association and Inland Athletic Association; Past AAES Bd of Govs Mbr (1980-83); past Vice-Chairman and Chairman of the Chicago Heights Community Caucus; 4th Degree-K of C Lansing 3540; Chairman and member of various church organizations. *Society Aff:* AACE

Fouss, James L
Business: 401 Olive St-Bx 1047, Findlay, OH 45840
Position: V P, Res and New Product Dev. *Employer:* Hancor, Inc. *Education:* PhD/Agri Engg/OH State Univ; MSc/Agri Engg/OH State Univ; Bach Agr Eng/Agri Engg/OH State Uhhiv. *Born:* Feb 1936. BAE, MS, PhD Ohio State Univ 1959, 1962, 1971. Employed as Agri Engr with U S Dept of Agri - Agri Res Serv, & stationed at OSU 1960-72. In 1972 transferred to Coastal Plains Soil & Water Res Ctr, Florence S C; became Dir in 1974. Assumed current pos as V P for Res & New Prod Dev with Hancor Inc in 1976. Conducted original U S R&D on corrugated plastic drainage tubing, a plow for its installation, & the prototype laser beam grade control sys. USDA Awards: Spec Serv 1972, Outstanding Performance 1968 & 1972, Disting Serv 1972. ASAE Awards: FMC Young Designer 1972, & Outstanding ASAE Paper 1973. OSU Alumni Award: Disting Alumnus, Coll of Engrg 1975. *Society Aff:* ASAE.

Fowler, Alan B
Business: IBM Res Div, Yorktown Hts, NY 10598
Position: Res Staff Mbr. *Employer:* IBM. *Education:* PhD/Applied Phys/Harvard; MS/Phys/Rensselaer; BS/Phys/Rensselaer. *Born:* 10/15/28. After US Army Service (1946-48, 1952-53) & employment at Raytheon Res, he joined IBM Res Div. His career has been primarily directed to the study of semiconductor electronics including work in GaAs lasers & photodiodes. His main interest has been in the physics of semiconductor surfaces & especially in the transport properties of electrons in MOS structures. With his associates he demonstrated quantum effects in an inversion layer, studied mobility & hot electron effects & showed impurity band conduction in the surface. He has managed various applied physics groups studying device properties. He is the author of numerous papers & patents. *Society Aff:* APS, IEEE, AVS, AAAS.

Fowler, Byron H
Home: 2276 Berryessa Ln, Santa Maria, CA 93455
Position: Engineer. *Employer:* Martin Marietta Corp. *Education:* MBA/Bus Admin/ Pepperdine Univ; BSAE/Aeronautical Eng/Northrop Univ; Cert AE/Aeronautical Eng/Northrop Inst of Tech. *Born:* Feb 1931. Certificate in Aeronautical Engrg from Northrup Inst of Tech, 1951; BS Aeronautical Engrg Northrup U 1970; Master of Bus Admin Pepperdine U 1978. Mass Prop Engr Northrup Aircraft 1951-56. Martin Marietta Corp 1956- ; Group Engr. Elected Fellow Soc of Allied Wt Engrs 1969, Hon Fellow 1970. *Society Aff:* AIAA, SAWE.

Fowler, Charles A
Home: 15 Woodberry Rd, Sudbury, MA 01776 *Employer:* C. A. Fowler Associates *Education:* BS/Eng Physics/Univ of IL. *Born:* 12/17/20. Graduate work in EE at Polytechnic Inst of Brooklyn and in Applied Scis at Adelphi Univ. C.A. Fowler Associates 1986-present. Previously Senior VP MITRE Corp 1976-1985. VP of Raytheon Co's Equipment Dev Labs Sudbury, MA 1970-76; Deputy Dir of DDR&E Dept of Defense 1966-70; Dept Hd AIL Div Cutler Hammer 1946-66; staff mbr MIT Radiation Lab 1942-46. Fellow IEEE, Fellow AIAA, Fellow AAAS, Mbr NAE. Mbr Defense Sci Bd 1972- , Chrmn 1984- , AF Sci Advisory Bd (1971-77), and Mbr Defense Intelligence Agency Advisory Ctte 1971- . Chrmn 1976-82. *Society Aff:* AIAA, IEEE, AAAS, NAE, AOC.

Fowler, Delbert M
Home: 5708 Willow Ln, Dallas, TX 75230
Position: Coordinator, Architect-Eng Services. *Employer:* Dallas Independent School District. *Education:* MS/Intl Affairs/Geo Wash Univ; MS/Civ Engg/TX Agri & Mech Univ; BS/Military Engg/US Military Acad; Grad/-/Ind College of the Armed Forces. *Born:* 9/14/24. Ladonia, TX. Coordinator AE Services, Dallas Independent School District since 1985; Central Exprwy, Dallas, TX 75204, since 1979; PE; Govt Exec; Regular Army Officer. Educ: BSc, US Mil Acad 1945; MSc, TX A&M Univ 1953; MSc Intl Affairs, Geo Wash Univ 1965; Grad, Indl Coll of Armed Forces, 1965. M Betty Alouise Reichey, 1948, 1s, 2d Military service included duty in Austria, Korea, Fed Rep of Germany, & Vietnam; ret'd as Col, 1972. Proj Mgr, design and constr, Westinghouse 1973; Regional Administrator, Fed Energy Admn, for states of TX, LA, OK, NM & AR 1973-77. Energy Conservation Consultant, Planergy Inc, Austin, TX 78-79, Pres. Blum Energy consultants, Blum Consulting Engg, 1979-85, Cert Energy Mgr; Natl Defense Exec Reservist. Decorations incl: Legion of Merit; Bronze Star; Air Medal. Fed Exec Fellowship, the Brookings Inst, 1972. Hobbies: genealogy; tennis. *Society Aff:* NSPE, AEE, ASHRAE, IES, SAME, IPA.

Fowler, Earl B
Business: Navy Dept, Elec Sys Cmd, Washington, DC 20360
Position: Commander, Naval Elec Sys Command. *Employer:* US Navy. *Born:* 1925 Jacksonville Fla. Enlisted Navy V12 prog 1943. Commissioned Ensign 1946. BSME Georgia Tech 1946; BSEE MIT 1949. Harvard Advanced Management Program 1971. Twenty years shipboard and shore duty preceded promotion to Rear Admiral in August 1974. Named Vice Commander Naval Electronic Systems Command July 1975. Became Commander in September 1976. Manages diverse electronic programs to assure Navy readiness. Scope includes res, development, procurement, installation and maintenance of command and control communications, surveillance, intelligence and electronic warfare systems. Mbr Delta Tau Sigma, IEEE and American Soc of Naval Engrs.

Fowler, Hardy B
Home: 5935 Coliseum St, New Orleans, LA 70115
Position: Engrg Consultant. *Employer:* Self. *Education:* 2 yr/Engg/Rice Univ; BS/ EnggUS Naval Acad; 2 yr/Marine/NY Univ. *Born:* Oct 1920. Two years engrg Rice Univ, BS U S Naval Acad; two years post grad work New York Univ. Chief Engr Navy Destroyer 1943-46 WWII. Alcoa S/S Co & Aluminum Co of America 1946-50 in Engrg & Construction Div. W Horace Williams Co, New Orleans 1950-56 Proj Engr and Dir. Pres and Founder H B Fowler & Co, Engrs & Contractors 1957-72. Company engaged in design and construction of piers, docks, bulk material and liquid handling facilities in No & So America, Caribbean area and Africa. Company sold to The Offshore Co. Presently acting as independent engrg consultant. *Society Aff:* CEC, LES.

Fowler, Harlan D
Home: P.O. Box 484, Solvang, CA 93463
Position: Aero Consultant. *Education:* Civil Engineering *Born:* 6/18/95. Native of Sacramento, CA. Invented & developed the Fowler Flap. In use on over 100 different forms of aircraft: transports, bombers, fighters, cargo planes. Patented May 22, 1928. Developed first all-cargo aircraft, Patented July 19, 1949. Also an aluminum container to go with it, Patented June 1, 1948. Conceived laminar flow glove to reduce turbulence drag over a wing. Patented May 6, 1958. Recently AF extended contracts to two aircraft cos to test the same form of device. Conceived, built & had tested wind-tunnel model at Ames Res Ctr, Published as NASA TN D-6323, 1971. Concept now used on two STOL aircraft contracted by AF. Patents 3,093,347 & 3, 312,426. *Society Aff:* SAE-Franklin Institute.

Fowler, Jackson E
Home: 1137 Millington Rd, Schenectady, NY 12309
Position: Mgr, Fluid Mech Engrg (Retired) *Education:* BSc/Mech Eng/Univ of Utah. *Born:* Dec 1915. Native of Salt Lake City. Joined G E 1943 in educ prog. Completed G E Co ABC courses 1947. Now Retired from Lrg Steam Turbine-Generator Dept, where was responsible for internal aerodynamics of lrg steam turbines 1951- . Have made significant contribs to turbine internal efficiency & reliability. Jr Mbr ASME 1943, Mbr 1949, elected Fellow ASME 1967. Chmn Fluids Engrg Div ASME 1972-73. Formerly Mbr-at-Large, Policy Board, Basic Engrg Dept, and was Mbr Policy Bd Communications, Past Secy Natl Nominating Comm, ASME. Enjoy sailing and sailboat racing, travel and reading. *Society Aff:* ASME.

Fox, Arthur J, Jr
Business: 1221 Ave of Americas, New York, NY 10020
Position: Editor, ENR. *Employer:* McGraw-Hill Inc. *Education:* BCE/-/Manhattan Coll; DSc (Hon.)/-/Manhattan Coll. *Born:* Sep 1923. Served Combat Engrs 1943-45, European Theatre. Early experience structural design Sanderson & Porter; instructor engrg drawing at Manhattan College. Joined Engineering News-Record editorial staff 1948; editor- in-chief since 1964. Served on boards: Engrg Societies Library, Engrs' Council for Professional Development, Engrs Joint Council, Bd for Engrg Cooperation. President of ASCE 1975-76. VP of EJC 1978-80. Tau Beta Pi, Chi Epsilon, American Acad of Environmental Engrs, Bldg Res Advisory Bd, The Moles. Recipient: 1975 Award of Merit from Fellows of American Consulting Engrs Council, 1971 Distinguished Alumni Award, 1981 honorary Doctor of Science. Manhattan College. 1986 Hon Mbr, Amer Inst of Arch, 1987 Hon Mbr Amer Subcontractors Assoc, 1987 We Dig Amer Award from Natl Utility Contractors Assoc. *Society Aff:* ASCE, AAEE, USCOLD.

Fox, Bruce L
Business: 2850 Gravois, St Louis, MO 63118
Position: CB & Pres. *Employer:* Meridian Eng & Assoc Inc. *Education:* PhD/MMet/Meridian Environmental & Tech Mgt Inst; BS/Mech/Univ of Miami. *Born:* 4/20/23. Virginian; WWII ASTP-Engr; Designer; Const Supt; Proj Engr; Prod Engr; Engrg Supt; Asst Chief Dsn; Proj Mgr; Proj Dsn Mgr; Asst to VP & Gen Mgr; Mgr Projs; VP Chem & Design Engrg, for: Gov't, Olin, Air Products & Exxon; Init & Pres of ME & AI Mgmt-Engrg Firm; Pres-METMI Mgmt-Envir & Techno Inst. Leadership Processes; Engrn'd Systems, Operations & Performance Control Materials & Tools Proj, Plan, Estim, Sched, Assign & Report System. Published book on Mgmt, Life, Organization & Career Dev "Naturality-Tomorrow is Where You Live." Initiating organization guide and control books for Companies, Physical Installations, Programs and Projects. *Society Aff:* AIChE, ASME, AIPC, ASBMC, ISA

Fox, Frederick M, Jr
Business: 4765 Independence St, Wheat Ridge, CO 80033
Position: Pres. *Employer:* F M Fox & Assoc, Inc. *Education:* Profl Degree/Geological Engr/CO Sch of Mines. *Born:* 8/13/30. Native of Springfield, MA. Served as an officer in the US Army 1954-56. Geologist for the Shell Oil Co 1957-59. A private consultant since 1959. Founded F M Fox & Assoc, Inc, Consulting Engrs & Geologists in 1963 & is now Pres & Chrmn of the Bd of F M Fox & Assoc, Inc (Denver); F M Fox & Assoc of the Southwest, Inc (Albuquerque); & Fox & Assoc of AZ, Inc (Phoenix). Served as Dir of Jefferson Chapter, CO Soc, NSPE 1954-58. Enjoys flying & spectator sports. *Society Aff:* AEG, ASCE, NSPE.

Fox, Gerard F
Home: 3 Whitehall Blvd, Garden City, NY 11530
Position: Partner. *Employer:* Howard Needles Tammen & Bergendoff. *Education:* BCE/Structures/Cornell Univ. *Born:* Jan 1923. Served with US Air Force 1942-46. Joined Howard Needles Tammen & Bergendoff, New York, NY 1948. Elected to present position of partner in 1967. Also Adjunct Prof at Columbia Univ. Specialty is applied mechanics and structural engrg, including foundations, for all types of structures, esp long span bridges. Skilled in use of computers & programming. Chmn U S Group and V.P. of IABSE and Vice Chmn of the Structural Stability Res Council. 1980 EE Howard Award, ASCE; 1986 Roebling Award, Met. Sec. ASCE;

Fox, Gerard F (Continued)
1987 JA Roebling Medal, Eng. Soc of Western Penn. Fellow of ASCE and mbr ACI, AREA, SAME, IABSE, Moles and Natl Acad of Engrg. Avid surf fisherman. *Society Aff:* NAE, ASCE, ACI, AREA, SSRC, IABSE, PCI, AISC, SAME.

Fox, John A
Business: 201 Carrier Hall, University, MS 38677
Position: Chrmn & Prof M.E. *Employer:* Univ of Mississippi *Education:* PhD/Aero/PA State Univ; MS/Aero/PA State Univ; BS/Aero/Unif of MI; BS/Math/Univ of MI *Born:* 02/08/24 Toronto, Canada. Emigrated to US in 1927. Married to Catharine Sauer; 4 children. Teaching: Penn State 1949-61; Univ of Rochester 1961-67; Univ of MS 1967-present. Consltg: Piper Aircraft 1952-61, HRB Singer 1956-61. RASA, 1963- 67; USDA Sedimentation Lab, Oxford, MS, 1968. Private business since 1968. Res: Plates & Shells (NACA), Hydrodynamic Stability ONR, ARL Boundary Layer, Stability and Control: NASA, MHD: NSF High Speed Hydrodynamic Stabilty; US Navy- DTNSRDC Fluid Mechs. Teaching, res 1949 to date for NACA, NASA, NSF, US Navy. Interests: Primarily Fluid Mechs, including boundary layer, potential flow in rotating systems. *Society Aff:* ASME, AIAA, ASEE, SNAME, FPS

Fox, John H
Home: Box 249, Bowmanville, Ontario, Canada
Position: *. *Employer:* *. *Education:* BASc/-/Univ of Toronto. *Born:* May 24, 1903 in Toronto. Elementary and Secondary Schooling Toronto. Graduated 1927 Univ of Toronto, Bach Appl Science. Served 1939-46 rank of Lt Col in England and Northwest European Theatre in Canadian Army (Royal Canadian Elec & Mech Engrs). Decorated-Order of British Empire. 1930-67 V P Honeywell Controls Ltd. Mbr Assn of Prof Engrs of Ontario, Councillor & Past President 1957; Fellow & Past President ASHRAE; Fellow Inst Mech Engrs London Eng & Charter Engr, Great Britain: Fellow & Life Mbr Engrg Inst of Canada. Toronto. Proprietor-John Holloway Fox Engg Ltd - Consultin Engrs - Toronto. *Society Aff:* FEIC, FASHRAE, FIMechE, CEB

Fox, Joseph M, III
Home: 3396 Angelo Str, Lafayette, CA 94549
Position: Manager Process Technology. *Employer:* Bechtel National Inc. *Education:* BS/ChE/Princeton; MS/ChE/Princeton. *Born:* Nov 20, 1922. Phila Penna. Was tech serv engr with Pan Amer Refining from 1944-46. Sect. Hd for M W Kellogg Co in pilot plant development, process evaluation, research planning from 1947-66, Proc Mgr. With Bechtel since 1966 in process design and development. Dir of American Inst of Chem Engrs from 1975-78. Delegate Calif Legislative Council for Prof Engrs and Bay Area Engrg Council. Specialties include petroleum refining processes and natural gas processing to produce LNG and liquid transportation fuels. Wife Elizabeth Larkin, 6 children. *Society Aff:* AIChE, ACS.

Fox, Michael R
Home: 1606 Amon Dr, Richland, WA 99352
Position: Staff Engineer *Employer:* Rockwell Hanford Ops *Education:* PhD/Phys Chem/Univ of WA; BS/Math & Chem/St Martins Coll (Lacey WA) *Born:* 12/31/36 Held staff positions in Engrg at the Idaho Natl Engrg Lab. Activities in nuclear fuels reprocessing as well as in solidification and storage of nuclear wastes. Hold a patent in the area of water desalination. Staff engr at Hanford (State of Washington) specializing in waste mgmt and plutonium proc devel. Have held several mgmt positions in nuclear waste mgmt, and geologic repository engrg. An authority on regional and natl energy supplies and demands, and adv state and Congressional leaders in this area. Am a mbr of the Washington State Adv Bd on Low Level Wastes. Have delivered more than a hundred speeches on energy issues. Enjoy handball, tennis, travel, jogging, reading/studying. *Society Aff:* ANS, AIChE

Fox, Portland P
Home: 500 Hiwassee St. N.E, Cleveland, TN 37311
Position: Consultant (Self-employed) *Education:* B.S./Geo/Univ of NC *Born:* 8/10/08 Principal work has been engrg Geology from 1934 to 1947 for the Tennessee Valley Authority, U.S. Bureau of Reclamation, and from 1947-1953 as chief geologist for Sao Paulo Power & Light Co. in San Paulo Brazil. From 1953 to Present (May 87) as an independent consulting geogolist on numerous major projects. *Society Aff:* ASCE, GSA, AEG, TAS, AIPG, ASCE, GSA, AEG

Fox, Robert W
Business: Sch Mech Engg, W Lafayette, IN 47907
Position: Prof & Assoc Hd. *Employer:* Purdue Univ. *Education:* PhD/Fluid Mechanics/Stanford Univ; MSME/ME/Uni of CO; BSME/ME/RPI. *Born:* 7/1/34. Joined faculty Mech Engg Purdue 1960, Asst Prof. Promoted to Assoc Prof 1963 and Prof 1966. Primarily involved in teaching and res until 1971. Since 1971 in addition to teaching has devoted a large portion of time to the dev and admin of undergrad progs, working both in Mech Engg and in Office of Dean of Engg (Asst Dean for Instruction 1972-75). Served as Acting Hd of Mech Engg 1975- 1976; since 1976 Assoc Hd of ME and Chrmn of Curriculum Committee. Recipient of Std Oil and Harry L Solberg outstanding teacher awards. Chrmn of Univ Senate 1971-72. *Society Aff:* ASME, ASEE, ΠΤΣ, ΤΒΠ.

Fox, William R
Business: P O Box 5465, Mississippi State, MS 39762
Position: Agri & Bio Engrg Dept Head. *Employer:* Mississippi State Univ. *Education:* PhD/Agri Engg/IA State Univ; MS/Agri Engg/Univ of TN; BS/Agri Engg/Univ of TN. *Born:* Jun 15, 1936. Native of Sevierville, Tennessee. Instructor of Agri Engrg 1960-62 ISU. Asst and Assoc Prof Agri Engrg, Miss State Univ 1962-67. Head of Agri and Biological Engrg Dept 1967-present. Mbr of ASEE, AAAS, and ASAE. Chmn of Southeast Region 1972. Education & Res Ctte Chmn 1973. Curriculum & Course Content Ctte Chmn 1978 (accreditation group). Mbr of Sigma Xi, Phi Kappa Phi, Tau Beta Phi, Gamma Sigma Delta, CAST, Bd of Dirs, Security State Bank, Starkille, MS 1977-present. Enjoy outdoor activities. *Society Aff:* ASAE, ASEE, CAST.

Foye, Robert, Jr
Home: 10531, Idlebrook Dr, Houston, TX 77070
Position: Manager of Waste Management Div *Employer:* Woodward-Clyde Consultants *Education:* PhD/CE/TX A&M Univ; M Eng/CE/TX A&M Univ; BS/Military Engrg/US Military Acad *Born:* 12/31/37 Born in NC. Raised in NC, FL and GA prior to entrance into West Point in 1956. Served with Army Corps of Engrs after grad from 1960-81. Served in various Command and Staff positions in US and Pacific in Combat, airborne and construction engrg responsibilities. Field engr on Bangkok By-Pass Highway (Thailand), Res Engr (TX A&M Univ), Proj Mgr for Lines of Communication (LOC) Program (Vietnam). Last duty with Crops of Engrs as Asst Prof of Civil Engrg, USMA. With Woodward-Clyde Conslts since Aug 1981 as Associate and Vice-President, Waste Management Div., Houston, TX. Military awards include LOM, BSM, ACM for military engrg achievement and service. *Society Aff:* NSPE, ASCE

Fraas, Arthur P
Home: 1040 Scenic Dr, Knoxville, TN 37919
Position: Consultant. *Employer:* Self Employed. *Education:* BS/ME/Case Inst of Tech; MS/Aero Engg/NY Univ. *Born:* Aug 1915. Native of Lakewood, O. Aircraft power plant test engr at Wright Aeronautical Corp and Packard Motor Car Co (engine and supercharger performance, cooling, vibration, etc) and taught at NYU, Case Inst of Tech, and Instituto Technologico de Aeronautica in Brazil 1938-50. Joined Oak Ridge Natl Lab in 1950 as Principal Design Engr on the Aircraft Nuclear Propulsion Proj 1950-57; Assoc Dir of the Reactor Div respon for developing advanced concepts for fission and fusion power plants for central stations, spacecraft, and marine applications 1957-75; Mgr of High Temperature Systems 1975-76. Retired from ORNL in 1976 to engage in private consulting practice. *Society Aff:* ASME, AIAA, RESA, ANS.

Fradkin, Edward J
Business: 2 Park Ave, New York, NY 10016
Position: Vice President. *Employer:* Scientific Design Co . *Education:* SM/Chem Eng/MIT; BChE/Chem Engg/CCNY. *Born:* Sep 1924, New York City. Mbr AIChE (Fellow), ACS, Sigma Xi, Tau Beta Pi, NSPE, PE NY & TX. Vice Pres Scientific Design Co, involved in marketing engrg services for chemical and petrochemical plants. Formerly Ch Engr & Dir of SD London office; also rep in Japan. Previously Ch Engrg Halcon Internatl Inc; also with Chem Construction Corp and Hydrocarbon Res Inc. Served in US Navy. Resides in NYC and Cold Spring, N.Y. with wife Eileen. *Society Aff:* AIChE, ACS, NSPE.

Frame, John W
831 W. Paseo Potrerro, Green Valley, AZ 85614
Position: Retired. *Education:* PhD/Met Engr/Lehigh Univ; MS/Met Engr/Lehigh Univ; BS/Met Engr/Univ of MO. *Born:* 9/28/16. Native of Rolla, MO. Started work for Bethlehem Steel in 1938 at Lackawanna Plant. During war was responsible for quality control of approx 40,000 tons/month of carbon shell steel. Transferred in 1946 to Res Dept as an engr. Eventual position - mgr product res with about 100 engrs & 100 technicians reporting. Responsible for res progs & budget in alloy dev, heat treating, corrosion, paints, metallic coatings, welding, fatigue & fracture, & forming. Chrmn Mech Working & Steel Processing Committee of AIME in 1967. Fellow of ASM 1978. Retired Nov 1978. *Society Aff:* ASM, SAE.

France, Jimmie J
Home: 110 Howard St, Roanoke, IL 61561
Position: ch Engr Mech Mining Trucks. *Employer:* Wabco/Cmeg. *Born:* 11/1/26. Attended Marshall Univ Huntington, W Va. Reg PE in OH & IL. Sr Design Engr at Marion Power Shovel Co from 1951-59. Ch Engr at Ulrich Mfg 1959-65. Design respon for earth moving attachments & positive displacement pumps. Have been empl at Wabco/Cmeg now Dresser Industries, Haulpak Div. from 1986- . Respon for Mechanical Mining Trucks, Wabco Technical Respresentative in Litigation Suits on Mechanical Trucks & was Wabco Technical Respresentative on Mechanical Trucks in China. Now occupies the position of Chief Engineer Quality Assurance, Dresser Industries, Haulpak Division. Enjoys good music, plays & good wine. *Society Aff:* SAE, FPS.

France, W DeWayne, Jr
Business: G M Tech Center, Warren, MI 48090
Position: Asst Dept Head. *Employer:* General Motors Res Labs. *Education:* PhD/Matls/Rensselaer Poly Inst; BE/Metallurgy/Yale Univ. *Born:* Nov 1940. Joined G M after receiving PhD from RPI & BE from Yale U. Reg P E Mich, accredited Corrosion Specialist (NACE). Publ 25 papers, co-edit book Recognition for Outstanding Contributions 1974 NACE, Award of Appreciation for outstanding service as ASTM Chmn 1976, Mbr Sigma Xi hon soc. Mbr ASM, Mbr ASTM, Mbr ACS Analytical Div. Mgrm of dept respon for analysis, identification & characterization of inorganic & organic engrg materials - Analytical Chem Dept. Mbr, ASTM, BD of Dir, Mbr, FMS Bd of Trustees (Federation of Material Sec), ASTM Award of Merit & Fellow, 1982. *Society Aff:* ASTM, ASM, FMS, NACE.

Franceschi, Bruno J
Business: 6165 N Green Bay Ave, Milwaukee, WI 53209
Position: Pres. *Employer:* Franceschi Assoc, Inc. *Education:* Bachelor/Civil Engg/Marquette Univ/Continueing Education, U.W. Milwaukee, U.W. Madison M.S.O. E., Marquette University. *Born:* 4/29/31. Entered private practice in 1962 - Inc in 1967. Corp name change in 1979 to Franceschi Assoc, Inc Architects, Engrs, Planners. Registered Architect. Registered Engineer. *Society Aff:* AIA, WSA, ASCE, ESM, PEPP, NSPE, WSPE

Francis, Gerald P
Business: Office of the Provost, 349 Waterman Bldg, Univ of Vermont, Burlington, VT 05405
Position: Vice-Provost *Employer:* Univ of Vermont *Education:* PhD/ME/Cornell Univ; MME/ME/Cornell Univ; BME/ME/Univ of Dayton. *Born:* 2/15/36. Seattle, Washington. Married: Anne Virginia Stewart. Children: Timothy P., Michael S., and Peter L. Faculty: Cornell University, 1961-64; Georgia Institute of Technology, 1964-66; State University of New York at Buffalo, 1966-77; United States Merchant Marine Academy, 1977-80; University of Vermont, 1980-present. Administration: Chairman of Mechanical Engineering, SUNY/B; Head of Engineering, USMMA; Dean, Division of Engineering, Mathematics & Business Adminstration, interim VP for Academic Affairs, Vice-Provost, Univ. Vermont. Awards: ASEE/Western Electric Award, 1970; US Dept. of Commerce Silver Medal, 1979; Fellow of the Institute of Marine Engineers. Registration: P. E. - Ohio; Charatered Engineer, United Kingdom. *Society Aff:* ASME, ASEE, IME, NSPE, VSE, ΠΤΣ, ΣΞ, ΤΒΠ, ΦΚΑΡΡΦ.

Francis, John E
Business: 865 Asp, Norman, OK 73019
Position: Prof & Assoc Dean of Engrg *Employer:* Univ of OK. *Education:* PhD/Engr Sci/Univ of OK; MS/ME/Univ of OK; BS/ME/Univ of OK. *Born:* 3/14/37. Native of Kingfisher, OK. Employed as an engr for Allis-Chalmers Mfg & taught at the Univ of MO at Rolla prior to present employment. Joined the Univ of OK in 1966, served as Asst Dean of the Grad College 1968-71, attained rank of Prof in 1974. Assoc Dean of Engrg 1981. Responsible for undergrad progs for the college. Mbr of AIAA Thermophysics committee 1975-78, 1984-87, chrmn 1978-80. Assoc Editor AIAA Journal 1984-86, Associate Editor AIAA Journal of Thermophysics and Heat Transfer 1986-present. *Society Aff:* ASME, AIAA, ASEE, NYAS

Francis, Lyman L
Business: 201 Mining Bldg, Rolla, MO 65401
Position: Prof. *Employer:* Univ of MO. *Education:* MS/ME/Univ of MO; BS/ME/Univ of MO. *Born:* 5/17/20. Native Missourian. BS 1944. Infantry WWII-Europe. Two yrs industry then MS 1950. Taught Univ of MO Columbia, WA State Univ. Now at Univ of MO Rolla as Prof & Dir of Engr Tech Div of ME Dept. 15 yrs service to ECPO-includes 2 yrs Chrmn Engr Tech Committee, 6 yrs Bd of Dir Representing SME, 2 yrs Chrmn ECPO Council. Chrmn SME Accrediation Committee and Mbr SME Bd of Dir. Co-authored a book - authored several publications. Summers in industry, consultant. Prefers to relax outdoors. *Society Aff:* SME, ASME, ASEE, NSPE.

Francis, Philip H
Business: 1301 E. Algonquin Rd, Schaumburg, IL 60196
Position: Dir, Advanced Manufacturing Technology *Employer:* Motorola Inc. *Education:* Ph.D/Mechanics/Univ. of Iowa; MBA/Management/St. Mary's Univ. of Texas; MS/Mech. Engr./Univ. of Iowa; BS/Mech. Engr./Calif. Polytechnic State Univ. *Born:* 04/13/38 Stress Analyst with Douglas Aircraft Co. (Santa Monica), 1960-62. At Southwest Research Inst. (San Antonio, TX), 1965-79, engaged in applied research in fracture and materials. Pro and Ch, Dept. of Mech and Aerospace Engrg, Il Inst. of Tech., 1979-84. Dir of Flexible Inspection & Assembly Lab. at Indust Technology Inst. (Ann Arbor, MI), 1984-86. At Motorola Inc. (Schaumburg, IL) as Director of Adv. Manufacturing Tech., General Systems Group, 1986-present. Reg PE (TX), Gustas Larson Award of the ASME/Pi Tau Sigma (1978). Mbr of Army Science Bd, and Editor-in-Chief, Manufacturing Review, published by ASME. *Society Aff:* ASME

Francis, Richard L
Business: Dept. of Industrial & Systems Engineering, 303 Weil Hall, Univ. of Florida, Gainesville, FL 32611
Position: Professor. *Employer:* Univ of Florida. *Education:* PhD/Indus Engg & Mgmt Sci/Northwestern Univ; MSc/Indus Engg/GA Inst of Techn; BSc/Indust Engg/VA Polytechnic Inst. *Born:* 6.9/38. Sys Engr for Western Elec Co 1962-63. Asst Prof 1966-68. Assoc Prof 1968- 71, Dept of Indus Engg OH State Univ. Prof Dept of Indus & Sys Engg, Univ of FL 1971- . Teach courses & carry out res in the areas of facilities planning, facilities location, applied optimization & modeling of emergen-

Francis, Richard L (Continued)
cy evacuation problems. Prog Chmn of Joint ORSA/TIMS Puerto Rico meeting 1974. Amer Program Chrmn of TIMS XXVI Intl Copenhagen Meeting, 1984. Served on editorial bds of IIE Transactions & Operations Res. Recipient of IIE-O.F. Baker Distinguishted Res Award, 1977. Co-recipient of IIE-H B Maynard Book of the Year Award 1975. *Society Aff:* IIE, ORSA, TIMS, RSA, ΤΒΠ.

Francis, Warren T
Home: 9 Coleridge Rd, Short Hills, NJ 07078
Position: President. *Employer:* Warren T Francis Assocs. *Education:* BS/Che/Bucknell Univ. *Born:* 6/22/24. Patent Engr BuShips 1946. Colgate-Palmolive 1947. Tech Editor Reinhold 1948-56. Sales & Commercial Dev Air Producst & Chemicals 1956-60. Program Mgr United Technologies Corp 1961-68. VP Airco/BOC Cryogenic Plants Corp 1968-71; VP Airco Indus Gases Div 1972-76, respon for turnkey LNG plants & for introducing cryogenic sys for gaseous radwaste removal at nuclear power & fuel reprocessing plants & for oxygen wastewater treatment sys. 1976- cons in indus gas & related tech & bus areas. AIChE Sect Chmn 1958. Designer and Builder of Liquid Carbon Dioxide Facilities - Production, Storage, Handling and Vaporization. Manager of Manufacturing. *Society Aff:* AIChE, ACS.

Franck, Kurt G
Business: c/o Johns Manville, 214 Oakwood Ave, Newark, OH 43055
Position: Engg Consultant. *Employer:* Holophane, Div of Johns-Manville. *Education:* Bach of Sci/Elec Engg/Tech Univ, Berlin, Germany. *Born:* Jul 1911. Diploma in EE Tech Univ Berlin, Germany. Engrg Exec with Holophane Div of Johns-Manville. Licensed P E state of Ohio. Holder of over 60 patents. Mbr Amer Optical Soc, Pres Illuminating Engrg Soc of N America. Mbr Tech Ctte TC 1.5 Internatl Comm on Illumination (CIE). Contributor of various papers to Scientific & Trade journals in the field of elec engrg & illumination. Lutheran. Enjoys traveling & music. *Society Aff:* IES, CIE.

Frandina, Philip F
Home: 75 Chatham Pkwy, Buffalo, NY 14216
Position: Commissioner of Public Works *Employer:* Erie County Div of Public Works *Education:* BS/Civil Engg/State Univ of NY, Buffalo. *Born:* 6/4/28. Formerly with City of Buffalo Div of Engrg. Joined Erie County Div of Highways in 1959. Responsible for design, construction and maintenance of all county buildings, the 1200 mile county highway system, Asst Prof in Engrg at Erie Community Coll. Officer of the Erie-Niagara Regional Planning Board, Founder and past president of the Western New York Association for Bridge Construction and Design, First Presiding officer of the ASCE Natl Council of Presidents, Director of NYSSPE; named "Engineer of the Year" for 1987 by the Erie-Niagara Chapter, Member of several civic and academic advisory boards, Enjoys music, chess, & gardening. *Society Aff:* ASCE, NSPE, ABCD, NYSPLS, ABET, APWA, XE, NACE, TNS.

Frandsen, John P
Home: 3318 Bragg Dr, Wilmington, NC 28403
Position: Mfg. Eng/Quality Control *Employer:* Industrial Consultant. *Education:* Bat Met Eng/-/Ransselaer Poly Inst. *Born:* 2/4/20. Native of Troy NY. Met Engr Amer Locommotive Co (Secretary Welding Ctte). 1947 jointed Genl Electric (Knolls Atomic Power Lab); left in 1957 as Mgr Mfg Engrg, Plant Mgr G E Aircraft Engines Blade Mfg Plant Rutland Vt, then Mgr of Nuclear Quality Control Knolls Atomic Power Lab. Sr Engr and Consultant Nuclear Fuel Mfg Dept, Wilmington NC. Fellow ASNT, P E CA, Certified Quality Engr ASQC, Certified Mfg Engr SME, Certification Appeals Comm. SME. Life Member AWS, ASM. many civic & service organizations Senior Instructor Cape Fran Technical Institute. *Society Aff:* ASM, ASNT, ASQC, SME, AWS.

Frank, Howard
Business: 130 Steamboat Rd, Great Neck, NY 11024
Position: Pres. *Employer:* Contel Information Systems *Education:* PhD/EE/Northwestern Univ; MS/EE/Northwestern Univ; BSEE/EE/University of Miami. *Born:* 6/4/41. Since 1982 has been Pres of Control Information Systems. Prior '82, was Pres of Network Analysis Corp. Formerly Assoc Prof of EE & Comp Sciences, Univ of CA, Berkeley. Was full time consultant in Exec Office of the Pres, 1968-69. IEEE Communications Soc Award 1969 for best paper of the yr. Won Honorable Mention, the Lanchester Prize 1972 for book "Communications, Transmission and Transportation Networks." Has authored over 100 papers & articles. White House Advisory Committee 1979-80. Fellow of IEEE, 1978. Became Pres of Network Analysis Corp in 1970. Has lectured widely throughout US, Canada, & Europe. *Society Aff:* IEEE, ORSA, AAAS, NYAS.

Frank, Otto
Home: 6 Barberry Rd, Convent, NJ 07961
Position: Consultant *Education:* MSChE/Chem. Eng'g/Princeton; BChE/Chem. Eng'g/Clarkson *Born:* 09/13/25 For 26 years worked as a process supervisor for Allied Corp. Accepted early retirement end of 1986. Since then have consulted for such companies as Drew Engineering, BASF and Air Products. Active member of North Jersey AIChE. Have held every administrative office, including chairman of the section. Also active on a number of national cttees. Elected Fellow in 1985. Reg P.E. in NY State. Published a number of articles in technical magazines. Reviewing engineering texts for McGraw-Hill. *Society Aff:* AICHE

Frank, Robert L
Home: 30795 River Crossing, Birmingham, MI 48010
Position: Elec/Electronics Cons. *Employer:* Self. *Education:* MS/EE/MA Inst of Tech; BS/EE/Univ of MI. *Born:* Jun 1917 Detroit, Mich. BullDog Elec Products Co (since merged with Gould Inc) 1939-42 as res engr in industrial power distribution. US Navy as Lt USNR 1942-46; radar engrg, then radio navigation (Loran) at Bureau of Ships Wash DC. Sperry Gyroscope Div Sperry Rand Corp 1946-74 as Sr Res Sect Supervisor in radio navigation & allied communications, electronic instrumentation & computers. m. Mary Brown 1985. Navy commendation 1946, Fellow IEEE 1964, Pioneer Award IEEE Aerospace & Electronic Sys 1971. Medal of Merit Wild Goose Assn (a navigation soc) 1977, Fellow Eng. Soc Detroit, 1981; Trustee, Detroit Sci. Center. P E New York and MI, 28 U S Patents, many tech papers. Interested in study & teaching of creativity in engrg. *Society Aff:* IEEE, WGA, ION, RIN

Franke, Ernest A
Home: 10687 E Black Forest Dr, Parker, CO 80134
Position: VP - R&D. *Employer:* Alpha Electronics. *Education:* PhD/EE/Case Inst; MS/EE/TX A&I Univ; BS/EE/TX A&I Univ. *Born:* 10/22/39. TX A&I Univ: Assoc Prof 1967-1969; Prof 1969-1979; Chrmn, EE, 1969-1971; Dean, College of Engg 1971-1979; Consultant: Minicomputers, microcomputers, control systems; VP-R&D, Alpha Electronics, 1979-. *Society Aff:* IEEE, ASEE, TSPE.

Franke, Milton E
Business: AFIT/ENY, Wright-Patterson AFB, OH 45433
Position: Prof. *Employer:* AF Inst of Tech. *Education:* PhD/ME/OH State Univ; MSME/ME/Univ of MN; BME/ME/Univ of FL. *Born:* 4/7/31. Engr with Westinghouse 1952. Active duty with USAF in res and dev in fuel systems and propulsion at Wright-Patterson AFB 1954-57. With Dupont's Petrol Lab in fuels res 1957-59. Since 1959 engrg faculty (aeronautical and mech engg) at the AF Inst of Tech. Full prof since 1970. Res projs with AF labs. Also, colonel, USAF Reserve, Retired. Sr Reservist for AF Labs, 1978-82. Chrmn, Dayton Sec ASME, 1968-69. Chrmn natl ASME Fluid Control Sys Panel 1975-79. Assoc editor of ASME Journal Dynamic Sys, Meas, and Control, 1979-84. Exec Ctte, ASME Dynamic Sys and Control Div, 1983- . ASME Publications Ctte, 1984- . AIAA Liquid Propulsion Tech Ctte, 1982-85. ASME Board on Communications, 1986- . AIAA Air Breathing Propulsion Tech. Committee, 1986- . *Society Aff:* AIAA, ASME, ASEE, ADPA, AFA, ROA.

Frankel, George J
Home: 26 Fountain Ln, Jericho, NY 11753
Position: Principal Engr, Engrg Test Operations *Employer:* Grumman Corp. Aircraft Systems Div *Education:* BME/Mech Engg/CCNY. *Born:* 1/3/23. BME CCNY 1944. Post grad PIB 1968. Product test engr, Arma Corp 1944-45. Corp Secretary & Ch Engr Metaplast Process Inc 1945-60. Production Mgr Brillium Metals Corp 1960-62. Ch, Space Environ, Republic Aviation Corp 1962-65. With Grumman Corp since 1965. Currently Principal Engr, Engrg Test Operations & Corp Metrication Officer. Certified Advanced Metrication Specialist, U.S. Metric Assn. (USMA). Vacuum metallizing cons. General Chmn AIAA/IES/ASTM/NASA Space Simulation Conf 1973; Tech Program Chmn 1975. Assoc Fellow AIAA (Chmn Long Island Sect 1976-77); Sr Mbr IES; mbr Natl Fire Protection Assn (Chmn Tech Ctte on Fire Hazards in Oxygen-Enriched Atmospheres 1966-); Amer Vacuum Soc; Pi Tau Sigma. Elected AIAA Director, Region I 1981-84, Vice President-Member Services 1985-1989. Chairman of Aerospace Sector Ctte of American Natl Metric Council, and Member of ANMC Board of Directors (1986-1989). Recipient of AIAA Dist. Service Award 1981. Contributed to numerous professional journals. Married, Miriam Josephson Frankel. Children: Paul J. Frankel, MSEE, AT&T Information Sys, Alice (Mrs. Donald F. Pratt); Lee J. Frankel DPM. 4 grand- children. Listed in Who's Who *Society Aff:* AIAA, NFPA, IES, USMA, ANMC, ΠΤΣ.

Frankel, Sidney
Business: 1165 Saxon Way, Menlo Park, CA 94025
Position: Tech Consultant (Retired) *Education:* PhD/Math/RPI; MS/Math/RPI; EE/Elec Engg/RPI. *Born:* Oct 6, 1910 New York City. Instructor in math RPI 1931-34. Engr & Proj Mgr at ITT 1938-50. Assoc Hd of Microwave Lab, Hughes Aircraft Co 1951-54. Dir of Engrg Sierra Electronic Corp 1958-60. Self employed 1960-68. Mbr of Sigma Xi, Eta Kappa Nu; Fellow of IEEE. Author of 17 tech papers. Asst Prof, Elec Engg Dept, San Jose State Univ, CA. Hobbies are recording classical music & gardening. *Society Aff:* IEEE.

Frankel, Victor H
Business: Dept/Orthopaedics, Seattle, WA 98195
Position: Prof & Chmn, Orthopaedics. *Employer:* Univ of Washington. *Born:* May 14, 1925. BS Swarthmore College; MD Univ of Pennsylvania; PhD Univ of Uppsala Sweden. Positions held-Founder & Dir of Biomechanics Lab, Hospital for Joint Diseases & Case Western Reserve Univ 1960-75. Prof & Chmn Dept of Orthopaedics Univ of Wash 1976-81. Dir of Orthopaedic Surgery Hospital for Joint Diseases Orthopaedic Inst and Prof of Orthopaedic Surgery Mt Sinai Sch of Medicine NYC 1981- . Chmn ASTM Ctte F-4 on Medical & Surgical Materials & Devices 1974- . Mbr ASTM Bd/Dir 1975- . Chmn, Ctte on Biomedical Engrg, American Acad of Orthopaedic Surgeons 1972- . Coauthor of textbook: Orthopaedic Biomechanics; co-author of numerous papers in field of biomechanics as applied to orthopaedic surgery.

Franken, Peter A
Business: Univ. Ariz, Tucson, AZ 85721
Position: Dir, Optical Sciences Center *Employer:* Univ of Arizona *Education:* Ph.D./Physics/Columbia Univ; M.S./Physics/Columbia Univ; B.S. /Physics/ Columbia Univ *Born:* 11/10/28 On the Physics faculty at Stanford University 1952-56, Univ of MI 1956-73, and Univ of AR 1973-present. Visiting Professorships at Oxford (1959) and Yale (1963). Director of the Optical Sciences Center at the Univ of AZ (1973-83). Prof. of Physics and Optical Sciences, Univ. of Ariz. (1973-present). Deputy Dir and Acting Dr of Defense Advanced Research Projects Agency (1967). Inventor of metastable helium magnetometer, co-discover of modern nonlinear optics and level-crossing spectroscopy. Sloan Foundation Fellow 1958- 62. Winner of the American Physical Society Prize (1967) and the Wood Prize of the Optical Society of America (1979). *Society Aff:* AAAS, APS, OSA, RSNA, SPIE

Frankfurt, Daniel
Business: 800 Second Ave, New York, NY 10017
Position: Pres *Employer:* Daniel Frankfurt, P.C. *Education:* MS/Structures/Columbia Univ; B/CE/City Coll of NY *Born:* 5/28/25 His consulting firm, Daniel Frankfurt P.C. Consulting Engrs, was formed in 1968 as Carroll and Frankfurt to perform civil and structural engrg for public works projects. Mr. Frankfurt, Chief Engr of the firm, is responsible for all technical aspects of the work. The firm specializes in hgwys, bridges, sewers, streets, water distribution systems, transportation facilities, structures, bldg rehabilitations, traffic studies, park rehabilitations, industrial bldgs, and other urban projects. Mr. Frankfurt was Dir of Technical Operations and Chief engr of Brill Engrg Corp; Principal Civil Engr of Burns and Roe, Inc; and Chief Civil Engr of McConothy, Hoffman and Assocs. *Society Aff:* NSPE

Franklin, Gene F
Business: Dept of EE, Stanford, CA 94305
Position: Prof of EE. *Employer:* Stanford Univ. *Education:* DEngSci/EE/Columbia; SM/EE/MIT; BSEE/EE/Geo Tech. *Born:* 7/25/27. Prof Gene F Franklin has been at Stanford Univ since 1957 & is currently Prof of EE. His res interests include computer-aided design progs, system identification techniques, & algorithms for dual identification & control. He is co-author of *Sampled Data Control Systems* (with Ragazzini - 1958), and of *Digital Control of Dynamic Systems* (with Powell-1980) and of *Feedback Control of Dynamic Systems* (with Powell and Ememi, 1986). He is also a Fellow of the IEEE. *Society Aff:* IEEE, SIAM.

Franklin, J Stuart, Jr
Business: 612 First Fed Bldg, Roanoke, VA 24011
Position: Partner. *Employer:* Sherertz, Franklin, Crawford, Shaffner *Education:* BS/Civ Engr/VPI & SU. *Born:* 5/2/20. Richmond VA. Field Engr Mason and Hanger - 1941, 1942. 30th Engr TOPO BN US Army - 1942-1946. Structural design for schools, churches, hospitals, public and industrial bldg - Eubank and Caldwell - 1948-1955. Structural designer American Bridge Div, US Steel Corp - 1955-1958. VP and Gen Mgr Cates Bldg spec - 1958- 1960. Partner present firm since 1961. Reg PE in VA, Mbr VSPE, PEPP. Fellow American Society civ Engrs, Pres VA Sec 1968, Served on Natl Committee for student chapters Pres Roanoke Kiwanis 1967. Pres Roanoke Valley Chamber of Commerce 1977. Enjoy intl travel and photography. *Society Aff:* ASCE, NSPE.

Franklin, William B
Home: 5005 Glen Haven, Baytown, TX 77521
Position: Retired Engrg Exec. *Employer:* Humble Oil & Refining Co. *Education:* PhD/Chem/Univ of TX; MS/Chem Eng/Univ of TX; BS/Chem Eng/Univ of TX. *Born:* 10/7/08. Natchez, Miss. Instructed in Chem Engrgistry & Chem at Univ of TX. Spent 39 yrs with Humble Oil Co (now Exxon Co) until retirement in 1972 in various engrg & mgmt positions in refining & chem mfg, including Mgr of Tech & Res & Engrg Mgr of the Baytown Refinery. Very active in AIChE, serving as Dir 1959-61, V P 1964, Pres 1965, P Pres 1966. Honors include Distinguished Service Award of South Texas Sect & Fellow AIChE & Distinguished Engrg Grad Award from Univ of Texas. *Society Aff:* AIChE, ACS.

Franks, L E
Business: Dept. of ECE, University of Massachusetts, Amherst, MA 01003
Position: Prof. *Employer:* Univ of MA. *Education:* PhD/EE/Stanford Univ; MS/EE/Stanford Univ; BS/EE/OR State Univ. *Born:* 11/8/31. Bell Tel Labs 1958-69. MTS specializing in circuit design until 1962. Supervisor, Data Systems Analysis Group until 1969. Joined faculty of Dept of Elec & Comp Engg at the Univ of MA in Amherst in 1969. Served as Dept Chrmn 1975-78. Active in res & teaching in Communication Systems & Signal Processing. Author of textbook on Signal Theory & several papers on related topics. Mbr of admin committee of IEEE Circuit Theory Group 1965-70. Assoc Editor for Communications for the IEEE Trans on Info Theory 1977-80. IEEE Fellow Award in 1977. *Society Aff:* IEEE.

Franques, J Thomas
Home: PO Box 16000-NH, Temple Terrace, FL 33687
Position: Asst Prof. *Employer:* Univ of S FL. *Education:* BSCE/Civ Engg/LA State Univ; MSCE/Civ Engg/LA State Univ; PhDCE/Water Resources/LA State Univ. *Born:* 3/16/42. Native of Baton Rouge, LA. Hydrologist with Water Resources Div, US Geological Survey, Baton Rouge, LA, 1966-71 Engg faculty at Univ of S FL, Tampa, FL, since 1971. Faculty Advisor to ASCE Student Chapter since its inception in 1973. Pres of W Coast Branch, FL Sec, Am Soc of Civ Engrs, 1978-79. Author of "A Two-Dimensional Analysis of Backwater at Bridges," *Journal of the Hydraulics Division, ASCE,* 1975; paper was first reported application of finited element analysis to surface water hydraulics. Consulting practice in hydrology & environmental impact assesment in Tampa Bay Area since 1973. *Society Aff:* ASCE, AWRA.

Frantti, Gordon E
Business: Dept of Geol & Geol. Eng, Houghton, MI 49931
Position: Dept. Head *Employer:* MI Tech Univ *Education:* MS/Geophysics/MI Tech Univ; BS/Engrg Physics/MI Tech Univ *Born:* 7/28/28 Native of Upper MI Served in U.S. Navy from 1946-48. Educated in Eng. Physics/Geophysics at Michigan Tech from 1949-54. Employed as Mining Engr for Copper Range Co. and as Geologist/Geophysicist for CCI Co. from 1954-56 and as Research Geophysicist for U.S. Bureau of Mines from 1956-59. After one year as Assistant Prof of Physics at Michigan Tech, joined Univ of Mich as Assoc Research Geophysicist where advanced graduate study was carried out in seismology from 1960-65. Returned to Michigan Tech as Assoc Prof of Geophysics. Accepted current position, Head, Dept of Geol & Geol Eng April, 1987, after serving as Acting Head since May 1981 and promotion to Professor of Geophysics. *Society Aff:* SEE, SSA, SEG, AGU.

Frantz, Joseph F
Business: 2500 Tanglewilde, Suite 110, Houston, TX 77063
Position: President. *Employer:* The Frantz Co. *Born:* Feb 21, 1933 McComb, Miss. Arkadelphia, Arkansas High School 1951; BS Chem Engrg LSU 1955; MS Chem Engrg LSU 1956; PhD Chem Engrg LSU 1958. Joined Monsanto Co in July 1958. Worked in Res in El Dorado, Arkansas & St Louis, Missouri. Transferred to Engrg Dept in Texas City, Texas 1962. Joined South Hampton Co in Houston Texas Feb 1965 as Mgr of Dev & Marketing. Started the Frantz Co (a cons & engrg design firm) in April 1958.

Franzini, Joseph B
Business: Dept of Civil Engg, Stanford, CA 94305
Position: Prof of Civil Engg. *Employer:* Stanford Univ. *Education:* PhD/Civil Engg/Stanford Univ; MS/Civil Engg/CA Inst of Tech; BS/Civil Engg/CA Inst of Tech. *Born:* 11/10/20. Native of Las Vegas, NM. US Navy 1944-1946. Teaching Asst, Stanford Univ 1946-49; Mbr Stanford faculty 1949-; Prof 1962-; Assoc Chrmn of Dept of Civil Engg 1963-1980; Special Consultant to Nolte and Assoc Consulting Civil Engrs, San Jose, CA 1958-; Co-author (with Ray K Linsley) of Water Resources Engineering; Co-author (with Robert L Daugherty and E. John Finnemore) of Fluid Mechanics with Engineering Applications; consultant to governmental agencies and private enterprise; Reg Engr State of CA. Mbr Caltech Alumni Assoc, First Congregational Church of Palo Alto. *Society Aff:* ASCE, AGU, ASEE, ΣΞ, AIH.

Franzoy, C Eugene
Business: P O Box 27381, Tempe, AZ 85282
Position: V P, Southwest District Mgr. *Employer:* Tech Consultants Inc. *Born:* Nov 1940. BS in Agri Engrg New Mexico State Univ; MS in Agri Engrg Univ of Arizona. Engr & Sr Engr Salt River Proj 1965-71. With Technical Consultants Inc since 1971. Currently respon for dev & mgmt of irrigation cons service in Southwestern U S. P E in Arizona, Agri Engr & C E. Chmn ASAE Arizona Sect.

Frashier, Gary E
Business: Waters Associates, 34 Maple St, Milford, MA 01757
Position: Pres Waters Associates *Employer:* Millipore Corp *Education:* MS/Mgt Sci/MIT; BS/CE/TX Tech Univ. *Born:* 7/2/36. President & CEO of leading scientific Co. specialty Liquid chromatography, analytical instruments & spec. chemicals; previously Pres & CEO of leading specialty adhesives & sealants. Co, Loctite Corp of Newington, CT. Previously VP & Gen Mgr Latin Am, & VP Mfg; Dir of Bus planning, Ind Products, Rockwell Intl; Mgr of Production, Western Hemisphere, Cabot Corp; Plant Mgr, Tuscola IL, Cabot Corp. Registered Professional engineer Texas and LA; member Young Presidents Organzation, City Club of Hartford, Avon Golf Club, Algonquin Club of Boston, and Annisgram Yacht Club (Mass). *Society Aff:* AIChE, ISA, ASM.

Frasier, Charles L
Business: 2701 St Julian Ave, Norfolk, VA 23504
Position: Sales Engr. *Employer:* Seven Santini Brothers. *Born:* Jun 1948. BS from Michigan State Univ 1970. Native of Port Huron Mich. After graduation spent 2 yrs as field engr, Detroit Metro Water Supply. Next 4 yrs with General Electric Internatl in New York, specializing in protection of proj equipment for internatl shipment. Current respon include sales engg & design of packaging for commercial & industrial equipment for export. Secretary & Exec V P of Society of Packaging & Handling Engrs NY-NJ Chap. Also Natl Competition Chmn 1975. Cert a Professional in Packaging 1975, SPHE..

Fratessa, Paul F
Business: 11881 Skyline, Oakland, CA 94619
Position: Pres. *Employer:* Paul F. Fratessa Assoc, Inc. *Born:* San Francisco. Raised in Carmel. MS San Jose State Univ. 9 yrs with H J Degenkolb & Assoc, San Francisco. 5 1/2 yrs dir of Ruthroff & Englekirk Oakland & Seattle offices. Founded Paul F Fratessa Assoc 1978. Dir of SEAONC. Mbr Rotary Club. Numerous SEAOC papers on seismic analysis of shear walls. ASCE State of the Art Award 1974. ASCE Raymond Reese Award 1976. P Pres of the Oakland Exchange Club. Enjoys family & golf.

Frazier, John Earl
Business: 436 E Beau St Box 493, Washington, PA 15301
Position: Chairman & Treasurer. *Employer:* Frazier-Simplex Inc. *Education:* DSc/Cer Sci/Univ of Brazil; SM/Chem Engg/MIT; BS/Chemistry & Math majors/Washington & Jefferson College. *Born:* Jul 1902. Reg P E & Prof Chemist-accredited. BS W & J College; MS MIT; Dr Sci Univ of Brazil. 1924-26 Engr Owens Ill Glass Co; 1926-28 Fuel Engr Frazier-Simplex Inc; 1928-30 Asst Secy Asst Treas, ibid - 1930-38 Secy & Treas ibid - 1938-45 V P & Treas - 1945-66 Pres & Secy - 1966- , Pres & Treas. Mbr Phi Beta Kappa, Phi Chi Mu, Keramos, Druids, Sigma Xi. FAAS, FACERS: FSGT, FAIC, FRSA. Hon NICE & Life Fellow AIC 1968; V P 196768, Treas 1968-69, Pres Elect 1969-70, Pres 1970-71. A Cer S, Greaves Walker Roll of Honor by Keramos Fraternity 1967. Albert Victor Bleininger Memorial Award 1969, John Jeppson Award 1976 of A Cer S, Hon Life Mbr 1978 of A Cer S; Natl Acad of Engg 1978. *Society Aff:* ACS, ACERS, AIC

Frazier, John W
Business: 1646 Knollinood Drive, Topeka, KS 66611
Position: Mgr *Employer:* Finney & Turnipseed, Cons Engrs. *Education:* BSCE/Structures/KS State Univ. *Born:* May 20, 1913 Columbus Ohio. Kansas Hwy Comm 1935-46. Finney & Turnipseed 1946- . Mbr Kansas Engrg Soc, SecyTreas 1949-50. Mbr Amer Welding Soc. ASCE: Fellow; Secy-Treas, V P, Pres, Kansas Sect; Dir Dist 16 1967-70; V P Zone III 1971-73. Hon Mbr Chi Epsilon. 1974 Civil Gov Award ASCE, 1975 Edmund Friedman Prof Recog Award, ASCE. Tr Kansas State Univ Foundation; P Pres Kansas State Univ Alumni Assn. Mbr Kansas State Bd of Educ 1969-79; Chmn 1971- 77. Commissioner Educ Comm of the States 1972-79, Citizenship Award of Kansas Engrg Soc, 1981, Distinguished Service Award, Coll of Engrg, Kansas State Univ 1983. *Society Aff:* ASCE, AWS.

Frazier, Richard H
Business: 77 Mass Ave, Cambridge, MA 02139
Position: Consulting Engr (Self-employed) *Education:* SM/EE/MIT; SB/EE/MIT. *Born:* May 1900 Bellevue Pa. Taught subjects in elec circuitry & electromechanical control components there 1925-65. Prof of Electromechanics Emeritus 1965- . Cons engr in area of electromechanical control. Reg P E in Mass. Cons on rod controls for submarine Nautilus, Westinghouse Atomic Power Div & for various reactors at Reactivity Measuring Facility, Natl Reactor Testing Station. Cons on design of controls for Apollo vehicles, Polaris missiles & submarines & var other missiles, Charles Stark Draper Lab Inc. Author or co- author num publ articles, books & reports on educational, tech or scientific subjects. *Society Aff:* IEEE, ASEE.

Fread, Danny L
Home: 9417 Bulls Run Parkway, Bethesda, MD 20034
Position: Research Hydrologist. *Employer:* Off of Hydrology, Natl Weather Serv. *Education:* PhD/Civil Engg/Univ of MO-Rolla; MS/Civil Engg/Univ of MO-Rolla; BS/Civil Engg/MO Sch of Mines & Metallurgy. *Born:* Jul 1938. Native of Lovington, Ill. PhD & MS Civil Engrg from Univ of Missouri Rolla; BS Civil Engrg with the Highest Honor from Missouri School of Mines & Met. Sr Engr with Texaco, Inc Lawrenceville, Ill 1961-67 specializing in design of pressurized piping & sewer systems. Res Hydrologist with Hydrologic Res Lab, Natl Weather Service NOAA since 1971. Respon for river mechanics res to dev efficient operational math models of unsteady flow, temp & sediment transport for real-time flow forecasting of U S rivers. ASCE Huber Prize 1976; ASCE J C Stevens Award 1976. Dept of Commerce Gold Medal Award 1979.. *Society Aff:* ASCE, AGU, AWRA, ΣΞ.

Freberg, Carl Roger
Home: 846 S Hudson Ave, Los Angeles, CA 90005
Position: Prof of Mech Engrg. *Employer:* Univ of Southern Calif. *Education:* BS/Mech Engg/Univ of MN; MS/Mech Engg/Univ of MN; PhD/Mech Engg/Purdue Univ. *Born:* Mar 17, 1916. Taught at Univ of Minn & Purdue Univ prior to 1945. Industrial experience at Carrier Corp, Southern Res Inst, US Naval Civil Engrg Lab & Borg Warner Central Res Lab through 1957. From 1957 has been Prof of Mech Engrg Univ of Southern Calif, 1957-66 as Dept Chmn. Specialties are mech vibrations & design. Has held var positions in ASME including Chmn Los Angeles Sect. Mbr ASEE, Sigma Xi, Pi Tau Sigma & Tau Beta Pi. Author of two vibration texts. *Society Aff:* ASME, ASEE, ΣΞ, ΠΤΣ, ΤΒΠ.

Freche, John C
Business: 21000 Brookpark Rd, Cleveland, OH 44135
Position: Retired (Chief, Matls Div) *Education:* BS/Petrol Engg/Univ of Pittsburgh; MS/Mtls Sci/Case Western Reserve Univ. *Born:* 4/29/23. Responsible for technical & admin mgt of mtls & structures res for aerospace propulsion at NASA Lewis Res Ctr. Author of over 100 technical papers on nickel and cobalt base superalloys, powder met, ceramics, metal fatigue, & turbine cooling. Holder of numerous ptents for advanced superalloys & turbine blade cooling concepts. Awarded NASA Medal for Exceptional Scientific Achievement for outstanding contributions in advancement of high temperature mtls tech and sci of metal fatigue. Received three IR-100 awards for: ferromagnetic alloy, high temperature nickel base alloy & superplastic forming process. PE, state of OH. *Society Aff:* AIME, AIAA, SESA, ASTM, ASM, AIC.

Frechette, Van D
Home: 22 S. Main St, Alfred, NY 14802
Position: Prof *Employer:* NY State Coll of Ceramics *Education:* BS/Ceramic Engrg/Alfred Univ; MS/Ceramic Engrg/Univ of IL; PhD/Ceramic Engrg/Univ of IL *Born:* 01/05/16 Born in Ottawa, Canada, was married 1940, has 5 children, is a naturalized US citizen; Res Physicist for Corning Glass Wks, 1942-1944; Prof of Ceramic Sci, NY State Coll of Ceramics, since 1944, also Dir of Res since 1970, PE, NY. Sr Fulbright Fellow (Res) 1955; Gordon Res Conferences Award 1955; George Westinghouse Award ASEE, 1968. Distinguished Educator Awd, AM Ceram Soc 1983. Guest prof Gottingen 1955-56; Max Planck Institut f. Silikatforschung 1965; Univ Erlangen-Nurnberg 1973. Eighty papers, one book; editor on nine books. Present res and consltg interests are in fractography, fracture mechs. *Society Aff:* ACS, NICE, CEC, DKG

Fredendall, Gordon L
Home: 969 Gravel Hill Rd, Southampton, PA 18966
Position: Retired. *Employer:* *. *Education:* PhD/Elec Engg/Univ of WI. *Born:* Dec 20, 1909. BS, MS, PhD Univ of Wisconsin. Mbr Tau Beta Pi, Sigma Xi. Fellow AIEEE. Retired from RCA Labs, Princeton NJ 1936-75. Sr Mbr Research staff. Received 5 achievement awards from RCA Corp for contributions to the field of television communication. Mbr of num IEEE cttes from 1947-75 concerned with Television Standards. The AIEEE Fellow citation read-'For his application of network analysis & synthesis to television system problems. Interested in municipal government; mbr Bd of Township Supervisors Bucks County Pa. Philadelphia Sect AIEEE Award 1981 "for his pioneering contributions to transient analysis in filtering in color television equipment. *Society Aff:* AIEEE.

Frederick, Carl L, Sr
Home: 3560 Chiswick Ct, Silver Spring, MD 20906
Position: Scientist-Consultant *Employer:* Self *Education:* DSc/-/NB Weslyn UNiv; MA/Physics/George Washington Univ; AB/Physics/NB Wesleyan. *Born:* Jul 1903. Native of Memphis, Nebraska. Taught Physics at G W Univ; Mbr Tech Staff Bell Telephone Labs; Asst Dir Res Dictaphone Corp; Head Physics Dept Cornell Aeronautical Lab; Supervisor Naval Ordnance Lab; President Frederick Res Corp; Fellow Scientist, HRB Singer Inc; cons to Vitro Labs, Melpar Inc; Scientific Warfare Advisor to Secretary of Defense; Inst Scientist Southwest Res Inst; over 10 patents & disclosures. Published over 25 tech papers & four volume set of books on Electromagnetic Compatibility & Radio Interference. Assoc Fellow AIAA, Life Fellow IEEE, Life mbr Electromagnetic Compatibility Soc. *Society Aff:* IEEE, AIAA, EMCS

Frederking, Traugott, H K
Home: 11 314 Homedale St, Los Angeles, CA 90049
Position: Prof *Employer:* UCLA *Education:* Ph.D./Cryogenics/Ch.E./ETH (Swiss Fed Inst Tech) Zurich; MSc/Mach Eng./Inst of Tech/Hannover (Tech. Univ.) W Germany *Born:* 6/21/26 Industrial R & D work at BASF from 1954 to 1957, after completion of education at Hannover. Assumed involvement in low temperature work in 1957 at ETH Zurich, Helium Lab. Started educational career and continued University research in cryogenics in 1961. Cryog. Eng. Conf. Board; R.B. Scott Memorial Award for Cryogenic Engineering Research 1971. Fellow AIChE 1985. *Society Aff:* AIChE, APS, IIR, CEC, VDI.

Fredrich, Augustine J
Home: 1108 Glen Moor Ct, Evansville, IN 47715
Position: Prof. *Employer:* Univ of Southern Indiana *Education:* MS/Civil Engrg/CA St Univ Sacramento; BS/Civ Engrg/Univ of AK *Born:* 9/12/39. Native of Little Rock, AR. Employed by Corps of Engineers, Little Rock Dist, on hydroelectric planning and design from 1962-66. With Corps Hydrologic Engg Center 1966-72, serving as Chief of Res Branch from 1969-72. Consultant to UNESCO in Porto Alegre, Brazil, in 1970-71. Congressional Fellow on staff of US Senator John L McClellan 1972-73. Employed in office, Chief of Engineers, and as Dir of Corps Inst for Water Resources 1973-79. Prof of Civil Eng Tech 1979-1981. Chmn, Engrg Tech Div 1981-present. Chmn, ASCE Tech Council on Computer Practices 1976-77. Chmn, ASCE Water Resources Planning and Management Div 1979-80. Exec Secretary, US Committee for IWRA. Reg PE. *Society Aff:* ASCE

Free, John Martin
Business: 8600 Indian Hills Dr, Omaha, NE 68114
Position: Sr V P, Corp. Dir, Operations. *Employer:* Leo A Daly Co. *Education:* BS/Arch-Engr/IA State Univ. *Born:* Oct 1923. Native of Minn. Served with Army

Free, John Martin (Continued)
1943-45; ETO. Decorated Bronze Star. Mgmt Dir of Tech Projs US & overseas Leo A Daly Co Omaha: 1953-60 Assoc, VP 1960-66, Sr. VP/Exec Dir Omaha 1966- , Managing Dir Pacific 1970-. Sr VP & Exec Dir/Hdqtrs Div 1982-86. Sr. V.P. Corp. Dir. Operations 1987- . Mbr ACEC Intl Engrg Comm; Mbr Advisory Council U NB Coll Engrg & Arch; Mbr Advisory Council IA State U; Reg P E IA; Reg Arch 14 states; Mbr AIA; Natl Soc P E's; NCARB; Archs Regis Council UK; SAME; Nebr Cons Engrs Assn; Consltg Engrs Council US; Royal Inst British Arch (RIBA); Inst Engrs Singapore; Fellow The Royal Australian Inst of Arch: Nebr Assn Commerce & Indus; Received 'Professional Achievement Citation' IA State U 1974, Mbr Republic of Singapore Bd of Archs.

Freed, Charles
Business: 244 Wood, Lexington, MA 02173
Position: Sr Staff. *Employer:* MIT Lincoln Lab. *Education:* EE/EE/MIT; SM/EE/MIT; BEE/EE/NYU. *Born:* 3/21/26. Charles Freed joined MIT's Lincoln Lab as a staff mbr in 1962, & was appointed to the position of Sr Staff in 1978. Since 1969 he has also been lectr in the EE Dept of MIT. He has written technical papers on noise, electron beams, varactor diode parametric amplifiers, photoelectron counting statistics, frequency stability, sealed-off stable co2 & co lasers, stabilization of gas lasers, precision heterodyne calibration in the infrared region, varactor photodiode detection, determination of co2 isotope laser transitions, & fundamental, quantum phase noise limited linewidths in tunable, lead-salt diode lasers. In 1979 he has been elected a Fellow of the IEEE. He enjoys classical music, literature, skiing, skating, swimming, & sailing. *Society Aff:* IEEE, ΣΞ, ΤΒΠ, HKN.

Freedman, Jerome
82 Shady Hill Rd, Weston, MA 02193
Position: Asst Dir. Emeritus *Employer:* Retired. Private Consultant *Education:* MS/EE/Poly Inst of Brooklyn; BS/EE/CCNY. *Born:* 8/16/16. Engr for the Army Signal Corps prior to WWII. Served as Radar Officer with US Army 1942-1946. Engr for Watson Lab (AF) specializing in radar res & dev. With Lincoln Lab, MIT 1952-1987. Assumed position of Asst Dir in 1968. Was responsible for res & dev in strategic offense & defense systems & high energy lasers. Assisted in guidance of Lab mgt & policies. Retired; now private consultant. Author of many technical papers. served on many govt advisory committees including the Army Sci Bd. and Defense Sci Bd. Received citations from SAC and NATO. Fellow IEEE, Assoc Fellow AIAA. Editor's Award, IEEE 1952. *Society Aff:* IEEE, AAAS, AIAA.

Freedman, Steven I
Business: 8600 West Bryn Mawr Ave, Chicago, IL 60631
Position: Executive Scientist *Employer:* Gas Research Institute *Education:* Ph.D./Mechanical Eng./Mass. Inst. of Tech.; Mechanical Eng/Mechanical Eng./Mass. Inst. of Tech.; S.M./Mechanical Eng/Mass. Inst of Tech; S.B./Mechanical Eng./Mass. Inst. of Tech *Born:* 06/05/35 Dr. Freedman is Executive Scientist of the Gas Research Institute where he is responsible for the engineering, scientific and technical aspects of the advancement of gas utilization equipment. Previously he was Asst Dir for Combustion and Advanced Power in DOE and ERDA where he was most noted for leading the Government's R&D efforts which resulted in the commercialization of the fluidized bed boiler. He has also had responsible R&D positions in the Budd Company, the General Electric Company, and I. U. International. He has taught on the faculty of MIT. *Society Aff:* ASME

Freeland, Manning C
Business: 216 South Pleasantburg Dr, Greenville, SC 29615
Position: VP & Chief Engr *Employer:* CRS Sirrine, Inc. *Education:* B/CE/Clemson Univ *Born:* 12/12/25 Native of SC. Served in US Navy 1943-1946. Specialized in Pulp and Paper Mill Engrg Design with J. E. Sirrine Co since 1951. Presently, Mgr of Pulp and Paper Staff Engrg, J. E. Sirrine Co, VP, Chief Engr and Dir of J. E. Sirrine Co. Enjoy tennis. VP & Chief Engr, CRS Sirrine, Inc. *Society Aff:* NSPE, TAPPI.

Freeman, Arthur H
Home: 14 Arrowhead Ln, Huntington, CT 06484
Position: Director Engrg. *Employer:* Black & Decker *Education:* MS/ME/Lehigh Univ; BS/ME/CCNY. *Born:* 1927 Brooklyn, New York. US Army service 1944-46. BSME-CCNY 1951 & MSME Lehigh Univ 1956. Joined G E Test Engrg Program 1951. Dev & Design Engr 1952-59; Unit Engr Motorized Dev 1959 -1965; Mgr Design Engrg 1965-69 & Mgr Engrg 1969- at Brockport NY Housewares plant, Mgr Evaluation Engg & Technical Support and Mgr Engg - Bridgeport, CT. Received P E licenses NY & PA 1958 & Cordiner Award 1962. Holds 12 patents. Authored articles on Electric Knife Dev with presentations on that subject to ASEE, SPE & GE Professors Conferences. Taught courses for GE Apprentice Program. Presently Director Engrg for Black & Decker Household Products.

Freeman, Herbert
Home: 7 Woodview Dr, Cranbury, NJ 08512
Position: Prof. *Employer:* Rutgers Univ *Education:* BSEE/EE/Union College; MS/EE/Columbia Univ; DrEngSc/EE/Columbia Univ. *Born:* 12/13/25. Affiliated with Sperry Gyroscope Co 1948-60; visiting prof at MIT, 1958-59. Prof at NYU, 1960-75. Chrmn, IEEE Tech Comm on Machine Intelligence and Pattern Analysis. Pres, Intl Assn for Pattern Recognition, 1978-80. US Delegate to IFIP, Am Assn of Info Processing Socs (AFIPS), 1978. NSF Postdoctoral fellow, 1966; Guggenheim Fellow, 1972-73; IEEE Fellow, 1966. State of New Jersey Prof of Comp Engg and Dir, CAIP Center, Rutgers University. *Society Aff:* IEEE, ACM, IAPR, PRS.

Freeman, Mark P
Business: 77 Havemeyer Ln, Stamford, CT 06904
Position: Sr Scientist. *Employer:* Dorr-Oliver Inc. *Education:* PhD/Phys Chemistry/Univ of WA; BS/Phys/Univ of WA. *Born:* 6/9/28. Military Service 1945-1949, 1950-1951. Instr in Chemistry, Univ of CA (Berkeley) 1956-58, Res Assoc with Central Res Labs Am Cyanamid Co 1958-1972. Visiting Prof Chem engr at MA Inst of Tech 1967-68. Since 1973, Sr Scientist with Dorr-Oliver. Has served on Res Committee of AIChE since 1975 and is past chrmn. Also, past chrmn and counselor of Western CT ACS. On Engineering Society Library Bd since 1978. Prefers to think of himself as Chem Engg Scientist. Enjoys fencing (Epee) and bicycling. *Society Aff:* AIChE, ACS, APS.

Freeman, William C
Business: One Center Plaza, Boston, MA 02108
Position: Sr Vice President. *Employer:* Camp Dresser & McKee Inc. *Education:* MS/Civil Engg/Northeastern Univ; BS/Civil Engg/Northeastern Univ. *Born:* Oct 1931. MS in Sanitary Engrg Northeastern Univ 1954; BS in C Engrg Northeastern Univ 1962. Assoc exclusively with Camp Dresser & McKee since 1957. Named partner in 1970 & Sr V P of Camp Dresser & McKee in 1971. In these latter capacities, respon for planning, design & supervision of constructiion of a wide variety of sanitary engrg proj's, including some of CDM's largest proj's involving collection, treatment & disposal of wastewater. Diplomate American Academy of Environ Engrs. *Society Aff:* AWWA, FWPCA, AAEE.

Freeman, William C
Home: 40 Little Harbor Road, Tiverton, RI 02878
Position: Engrg Consultant. *Employer:* W C Freeman & Associates. *Education:* BS/Naval Architecture & Marine Engg/MIT. *Born:* 10/29/20 Mgmt courses Amer Mgmt Assn. Tech experience-design, construction, testing & operation of marine power plants, naval & merchant ships. Mgmt experience-genl mgmt, intl sales, domestic sales, budgets, financial/production planning, personnel, organization. Genl Mgr Marine Div Combustion Engg, 16 yrs. Engr Officer, Merchant marine WWII. Reg P E, Mbr Engg Ctte, Amer Bureau of Shipping. Served on Bd/Dir Shipbuilders' Council of Amer; Mbr SNAME, ASNE, Mbr of Council SNAME. *Society Aff:* SNAME, ASNE.

Freeman, William R, Jr
Business: 500 Terrace Plaza, 475 Steamboat Rd, Greenwich, CT 06820
Position: VP Corp Quality Assoc & Tech Advisor. *Employer:* Howmet Turbine Components Corp. *Education:* BS/Metallurgy/MA Inst of Technology. *Born:* Sep 16, 1927. 1945-47 US Merchant Marine Acad; 1947-51 MIT; 1967 - Cornell Executive Dev Prog. Work History: 1951-53 Lt USAF, Proj Engr Wright Patterson AFB - monitored titanium programs & conducted res in phase diagrams; 1953-72 AVCO Lycoming Div, dev engr to Dir Matls & Processes Tech Labs - respon for all aspects of matls/ processes dev, selection, qual control & failure analysis - principal product gas turbines, rocket motor cases & reentry vehicles; 1972-77, V P & Tech Dir Howmet Turbine Components Corp - respon for R&D, tech services for investment casting concern - activities incl ceramics, intermetallic coating (high temp), casting & P/M dev, production support. Sev pats & publs in field. 1977-VP CQA & TD-Define & Coordinate Quality Related Matters Corp, Prod Reliability, Advise on Tech Programs, special studies. *Society Aff:* ASM, AIME.

Freer, Robert K
Home: 1031 Anna Road, Huntingdon Valley, PA 19006
Position: Project Eng *Employer:* Gigliotti Corporation *Education:* MBA/Accounting/Drexel Inst; B/CE/Cornell Univ *Born:* 9/15/29 Ithaca, NY ; Photogrammetric eng and chief Estimator-Aero Service Corp., Phila,PA 1952-64; Mgr coordinate scaling and plotting operations-Graftek, Inc., Princeton NJ 1964-65; Chief Eng-Air Survey corp., Reston, VA 1965-66; Project Eng, Data Processing Division Mgr-A.W. Martin Assoc.King of Prussia, PA 1966-76; President Diamond Engineering Co., Warrington, PA 1976-77; Data Processing Manager-Libson Contractors Inc., Danboro, PA 1977-81; joined Gigliotti Corp., Langhorne, PA in 1981 as civil engineer engaged in design and construction management on land development projects in PA and NJ. Registered PE in NY, PA, NJ, IL, RI, CA, FL; Registered Professional land surveyor in PA; Fellow, ASCE; Member, NSPE, and PSPE; Valley Forge Chapter PSPE Treasurer, Past President; Advanced class amateur radio operator (W3YLT) *Society Aff:* ASCE, NSPE

Freers, Howard P
Home: 536 Whitehall Rd, Bloomfield Hills, MI 48013
Position: Chief Engineer Engrg Tech Services *Employer:* Ford Motor Co. *Education:* MAE/Automotive Engg/Chrysler Inst of Engg; BSME/ME/Rose Poly Inst W/Honors. *Born:* 8/13/26. Native of Indianapolis, IN. USN 1945-46 ETM 2/C. Chrysler Engg 1948-1955 Supervisor, Ordnance Engg. Ford Motor Co 1955 to present: Exec Engr 1962, Chief Engr Car Powertrain Systems 1967, Chief Engr Light Car 1968, Chief Car Engr, Chief Powertrain & Chassis Engr 1975, Chief Body & EE 1978-1983. Chief Engr ETS 1983- . SAE - Chrmn Detroit Sec 1973-74, Bd of Dirs 1977-1979. SAE Fellow 1982. ESD - Fellow 1974 - Bd of Dirs 1975-81. Rose Hulman Inst of Tech Bd of Mgrs 1975 - continuing Tau Beta Pi - Eminent Engr 1978. Mason - Scottish Rite - Shriner - Enjoy bowling; alpine skiing. *Society Aff:* SAE, ESD, ASBE.

Freese, Howard L
Home: 140 Cabell Way, Charlotte, NC 28211
Position: VP Bus Dev. *Employer:* Luwa Corp. *Education:* MBA/Marketing/Syracuse Univ; BS/Chem-CE/Columbia Univ. *Born:* 12/9/41. Native of Charleston, WV. BS, Chem/ChE, Columbia Univ, MBA, Mktg, Syracuse Univ. Theta Tau, Soc of Dumbbells, Order of Owls, Beta Theta Pi, Football, Varsity "C–, Glee club. Beta Gamma Sigma Bus Honorary. Union Carbide Corp 1962- 64, process engr, Corning Glass Works 1964-72, process engr/applications engr. Luwa Corp 1972-date, Marketing Mgr, mgr of engg, mgr of Polymer gear-pump prog 1975. Chrmn Central Carolinas Section AIChE 1974, Charlotte Chamber of Commerce 1975, lectr Ctr for Profl Dev "Evaporation Technology" short course 1977-81, "Fermentation Technology" 1980-81, Co-author seven papers on new venture teams, evaporation, and radioactive waste mgt methods. PE NC 1979. *Society Aff:* AIChE, ACS, CMRA.

Freeston, Robert C
Home: 128 S Mansfield Blvd, Cherry Hill, NJ 08034
Position: Engr. *Employer:* Am Water Works Service Co. *Born:* 8/20/16. Joined Honeywell in 1940 as a Sales Engr for instrumentation, then diaphragm control valves. In 1956, became the Sr Application Engr for the Water & Waste Ind, writing many of Honeywell's Bulletins for that ind. In 1963 transferred to Fischer & Porter, with the same industry responsibilities. In 1966, joined Am Water Works Service Co, Inc as their first & only instrumentation specialist. Presented many papers at ISA & AWWA Confs, with three of them published in natl technicl journals. Received AWWA's Best Paper Published award in 1971. A mbr of AWWA's Automation & Instrumentation Committee since 1972. Attended Drexel Univ & PA State. *Society Aff:* AWWA, ISA.

Freeston, W Denney, Jr
Business: Georgia Institute of Technology, College of Engineering, Atlanta, GA 30332
Position: Assoc Dean Coll of Engrg *Employer:* Georgia Inst of Tech. *Born:* 5/8/36. BS Mech Engrg Princeton Univ 1957; PhD from same 1961 as Fellow of the Textie Res Inst. Joined the FRL Div of Albany Intl 1960 as Sr Res Assoc; Assoc Dir in 1969. Joined GA Tech in 1971 & became Dir of the Sch of Textile Engrg in 1972 and Assoc Dean, Coll of Engrg in 1980. Received Fiber Soc Award for Distinguished Achievement in Basic & Applied Fiber Sci 1969. Currently treas and Dir GA Tech Res Corp.

Frei, Ephraim H
Home: 3 Klausner St, Jerusalem-Talpiot, Israel
Position: Professor Emeritus Dept of Elec & Visiting Prof Hebrew Univs *Employer:* Weizmann Inst of Science. *Education:* PhD/Physics/Univ Vienna, Austria. *Born:* 1912 Vienna Austria; married, two children. Dr Phil Univ of Vienna. 1936-38 B Scharff & Co, battery mfg Vienna. 1942-46 British Army & British Embassy Athens; Broadcasting Engr in Palestine, Egypt & Greece. 1946-48 Hebrew Univ Jerusalem; Asst. 1950-52 Inst for Advanced Study Princeton NJ. 1953-. Dept of Electronics, Weizmann Inst of Sci; Prof & Dept Hd 1960-1977. Presently Prof Emerit. Also visiting prof Hebrew Univ Jerusalem. 1975-80 Yeda, Res & Dev Co; Chmn of Bd & Scientific Dir; 1968-1973, Jerusalem Coll of Tech; Mbr Advisory Bd. 1970-1980 IEEE Tech Ctte: Magnetics in Life Scis. Chmn XII Internatl Conf, Medical & Biolog Engrg. Delegate to URSi. Awards-Weizmann Prize 1956. Fellow IEEE 1967. Computer Pioneer NCC1975. Life Mbr Magnetic Soc, 1980. Hon Fellow Israel soc Med & Bio Engr, 1981. "Golden Doctors-Diploma" Vienna Univ, 1986. *Society Aff:* IEEE, ISMBE.

Frein, Joseph P
Home: 4312 Edgemont St, Boise, ID 83706
Position: Const. Consult. *Employer:* Self Employed *Born:* 5/22/04 My background after technical study mentioned included the positions of Project Office Engr for the City of St. Louis, Project Engr for W.E. Callahan Construction Co on three large dams and several tunnels. Since joining Morrison Knudson Co in 1942 I advanced from Estimating Engr, to District Engr, to World Wide Assistant Chief Engr, to Chief Engr, to VP-Engrg, to VP-Charge of Foreign Operations. I was a member of the Bd of Directors of MK, organizer and chrmn of the Construction Committee for U.S.C.O.L.D, and long time member of the Beavers, and presently a Construction Consultant. Studied Civil Engrg for 5 1/2 years at night. WA Univ-St. Louis, MO Extension Section. *Society Aff:* USCOLD, Beavers

Freise, Earl J
Business: University of Nebraska-Lincoln, 303 Administration Bldg, Lincoln, NE 68588-0430
Position: Asst Vice Chancellor for Research & Prof of Mechanical Eng. *Employer:* Univ of Nebraska - Lincoln. *Education:* PhD/Met/Univ of Cambridge; MS/Mtls Sci/Northwestern Univ; BS/Met Engg/IIT. *Born:* 12/30/35. Native of Chicago, IL. Assoc Prof of Mtls Sci at Northwestern Univ from 1962 through 1977. Also Asst Dir of Res office at NU from 1969 through 1977. Director, Office of Research and Spon-

Freise, Earl J (Continued)
sored Programs & Adjunct Prof of Mech Eng at Univ of North Dakota 1977-81. Presently, Assistant Vice Chancellor for Research at the Univ of Nebraska - Lincoln responsible for Office of Research & Sponsored Projects Services. Also Full Prof in Department of Mechanical Engineering, & as such teaches courses in Metallurgy & Materials Science. Held a Ford Fdn Residency in Engg Practice Award at Western Elec Co during 1968. Recipient of Fulbright Award 1960-1962 & Western Elec - ASEE Award for Excellence in Engg Education in 1969. NCURA-Distinguished Service Award 1987. Previously chrmn of Chicago Chapter of ASM & NCURA, Region IV, & Secy-Treas of IL-IN. Section of ASEE. P Pres of NCURA. Presently on Editorial Board of Grants Magazine, SRA Journal & NCURA Research Management Review. *Society Aff:* ASM, AIME, ASEE, NCURA, SUPA, SRA.

Freitag, Dean R
Business: Tennessee Tech Univ, Cookeville, TN 38505
Position: Prof *Employer:* Tennessee Technological University. *Education:* PhD/Engrg/Auburn U; SM/Soil Mech/Harvard U; BS/Civ Engr/IA State U. *Born:* Oct 1926 Ft Dodge IA. Employed by U Army Corps of Engrs 1951-1981. Conducted R&D at Waterways Experiment Station (WES) Vicksburg, MS in fields of soil stabilization, trafficability of soils & vehicle mobility. NASA consultant on Lunar Rover. 1970-1972 Asst Tech Dir of WES. 1972-1981 Tech Dir, US Army Cold Regions Res & Engrg Lab. 1981-present Prof Tenn Tech Univ, Geotech & Trans Engrg Consltg Engr, Geotech, Soil-Vehicle Mechanics & Cold Regions Engrg. Mbr ASCE & ASAE; founder mbr & past pres Internatl Soc for Terrain-Vehicle Sys. Past Chrmn ASCE Tech council on Cold Regions Engg. PE, TN. *Society Aff:* ASCE, ASAE, ISTVS.

Fremon, Richard C
Home: 32 Barn Owl Dr, Hackettstown, NJ 07840
Position: Director of the Computer Center *Employer:* Centenary College *Education:* MS/EE/Columbia Univ; BS/EE/Columbia Univ; AB/Math, Physics/Columbia Univ *Born:* 5/28/18 Bell Telephone Labs 1941-1981: Dir, Personnel Planning 1954-1967; Dir, Salary Administration 1967-1973; Dir, Administrative Systems (data processing) 1973-1981. Developed statistical tools for salary administration and an early personnel date base system. Directed planning, development, and operation of a large business/management information system. Dir. Computer Center, Centenary College 1981- . Engrg Manpower Commission: member since 1965, Vice Chrmn 1977-1979, Chrmn 1979-1981 member Emeritus 1986. Ex-officio member of Engrg Affairs Council of American Association of Engrg Societies 1980-1981. *Society Aff:* IIE, DPMA.

Fremouw, Gerrit D
Home: 3000 Spout Run Pkwy, Arlington, VA 22201
Position: Deputy Asst Secretary. *Employer:* Dept of Health, Education & Welfare. *Education:* Civil Engr/Civil/Clarkson Coll of Tech; BSCE/Civil/Clarkson Coll of Tech. *Born:* Jan 1909. Native of New York & retired A F Colonel (1969). Served as Engr HQ.14 Air Force China, during WWII & 17 yrs at HQ SAC as Deputy Ch of Staff, Civil Engrg. Respon for facility engrg at SAC bases worldwide totaling $5.4 billion. Awarded distinguished service medal, Legion of Merit and Bronze Star Medal. Established & heads HEW facilities engrg & construction, respon for design & construction of $10 billion in program of Fed assisted & direct Fed special purpose facilities. Mbr of Building Res Advisory Bd, Natl Acad of Sciences. Mbr of Tau Beta Pi 1942. Life Fellow, ASCE 1975. Special Citation AIA 1972. Distinguished Service Award Clarkson Coll 1976. Special Citation from Sec DHEW 1977 Sponsored adv design & construction technology which has resulted in cost avoidance of over $700 million in 6 yrs. *Society Aff:* ASCE, ΤΒΠ.

French, Clarence L
Business: 28th St. and Harbor Drive, San Diego, CA 92138
Position: Chrmn & CEO *Employer:* National Steel and Shipbuilding Company *Education:* BS/ME/Tufts Univ; BS/NS/Tufts Univ *Born:* 10/13/25 in New Haven, CT, Married to Jean Ruth Sprague on June 29, 1946. Three sons: Craig Thomas; Brian Keith; Alan Scott. Military Experience: 1943-53, Lieutenant, USN. With NASSCO since 1967, assumed current position as president and chief operating officer in 1977. Prior occupational experience was with Bechtel Corp. (1967-77) as supervising engineer; Kaiser Steel Corporation (1956- 64) as staff engineer and assistant superintendent BOSP; Bethlehem Steel Corporation (1947-56) as foundry engineer. *Society Aff:* SNAME

French, Heyward A
Business: 492 River Rd, Nutley, NJ 07110
Position: Dir International Marketing *Employer:* ITT Defense Communications Div. *Education:* BEE/Elec Engg/CCNY. *Born:* 8/20/21. Resides in Ridgewood, New Jersey. Joined ITT Res & Dev Lab in 1944. Engaged in design & dev of all aspects of electronic communications systems & equipment for the military. Respon for initiating many innovative designs in UHF & SHF microwave & tropospheric scatter transmission systems & equipments. Awarded 6 patents & authored 9 articles. Fellow IEEE. *Society Aff:* IEEE.

French, Orval C
Home: 4680 S A1A, Melbourne Beach, FL 32951
Position: Prof of Agri Engrg, Emeritus. *Employer:* Ret from Cornell Univ 1973. *Education:* MS/Agri Engg/KS State Univ. *Born:* Jan 1908, Geneseo Kansas. Dev Engr Black-Sivalls & Bryson Kansas City Mo 1930. Mbr Faculty Agri Engrg Dept, Univ of Calif 1931-47. Res Engr Univ of Calif Radiation Lab 1942-45. Prof of Agri Engrg & Head of Dept Cornell Univ 1947-72. Visiting Prof Univ of the Philippines 1958-59. Life Fellow ASAE, Life Mbr ASEE, Fellow AAAS; Mbr Sigma Xi, Phi Kappa Phi, Tau Beta Pi. ASAE Gold Medal Award 1975. Dir ASAE 1965-68, Pres 1966-67. Dir EJC 1969-70 & 1972-75. *Society Aff:* ASAE, ASEE, ΣΞ, ΦΚΦ, ΤΒΠ.

French, Robert D
Business: Arsenal St, Watertown, MA 02172
Position: Chief Metals and Ceramics Lab *Employer:* Army Mtls & Mechanics Res Ctr. *Education:* PhD/Engg/Brown Univ; MS/ME/Northeastern Univ; BS/ME/ Northeastern Univ. *Born:* 10/26/39. Instr of Met, Brown Univ, 1967. Served with US Army Corps of Engrs, 1967- 69. Res Met with Army Mtls & Mechanics Res Ctr (AMMRC) 1970-73; Group Leader, Physical Met 1973-78, Branch Chief, Prototype Mtls, 1978-79. Division Chief, Prototype Development 1979-80. Assumed current position in 1980. Past Chrmn of Boston Sci-AIME, 1973-74. Member, Board of Directors, TMS-A1ME, 1979-81. Mbr - Sigma Xi Honor Soc. *Society Aff:* TMS-AIME, ASM, ACA.

French, William D
Home: 2216 N Toronto St, Falls Church, VA 22043
Position: Exec Director & Secretary-Treasurer. *Employer:* American Soc for Photogrammetry & Remote Sensing. *Education:* BSCE/Civil Engg/Northeastern Univ. *Born:* May 24, 1937. Native of Boston Mass. From 1963-76 held various positions on staff of ASCE; Exec Asst 1968-72, Editor Tech Publications 1963-68; Asst Secy 1972-76; Dir Support Services 1972-76. Served on RNRF Board and Excom 1985-1986. Mbr of ASCE, ASPRS, ASAE; GWSAE Northeastern Univ Alumni Club. Soc of Industrial Archeology, Soc for the History of Technology. Friends of Cast-Iron Architecture, Amer Topical Assoc Exec Dir ASPRS 1977-present. Executive Director International Geographic Information Foundation. *Society Aff:* ASCE, ASPRS, ASAE, SHOT, SIA, RNRF.

Freudenstein, Ferdinand
Home: 435 W 259th St, Riverdale, NY 10471
Position: Prof of Mech Engg. *Employer:* Columbia Univ. *Education:* PhD/Mech Engg/Columbia Univ; SM/Mech Engg/Harvard Univ. *Born:* 5/12/26. Prof Freudenstein received his MS degree from Harvard Univ in 1948. Following several yrs in industry and further studies, he received his PhD in mech engg from Columbia Univ in 1954. He has served on the faculty of Columbia Univ since 1954 and has

Freudenstein, Ferdinand (Continued)
been active as a consultant to industry, including Bell Telephone Labs, IBM, and Gen Motors. His res interests are in mechanisms, kinematics, dynamics, and engg design. Prof Freudenstein received the Great Teacher Award from Columbia Univ in 1966, the ASME Machine Design Award in 1972, the Mechanisms Comm Award in 1978 and the Charles Russ Richards Memorial Award in 1984. He is a mbr of the Natl Academy of Engg and a Fellow of ASME. *Society Aff:* ΣΞ, ASME.

Freund, Lambert Ben
Business: Division of Engineering, Brown University, Providence, RI 02912
Position: Prof of Engrg *Employer:* Brown Univ *Education:* BS/Engrg Mech/Univ of IL; MS/Appl Mech/Univ of IL; PhD/Appl Mech/Northwestern Univ *Born:* 11/23/42 Native of McHenry, IL. Engr-in-Training with Intl Harvester, F. G. Hough Division 1961-65. Member of Brown Univ faculty since 1968; Prof of Engrg since 1975. Chrmn of the Executive Committee, Division of Engrg 1979-83. Over 90 research papers and review articles published in areas of stress waves in solids, fracture mechanics, seismology. ASME Henry Hess Award 1974; Fellow of ASME 1980; Fellow of AAM 1983; ASTM George R. Irwin Medal 1987. Editor-in-chief of ASME Journal of Applied Mechanics 1983- , Mbr, US National Cttee/THM 1985- . Visiting Prof, Stanford Univ 1974-75, Visting Scholar, Harvard Univ 1983-84, Engrg conslt. *Society Aff:* ASME, AGU, AAM

Freund, Louis E
Business: 475 Potrero Ave, Sunnyvale, CA 94086
Position: Sr VP. *Employer:* Whittaker Medicus Systems *Education:* PhD/Ind Engr/ Univ of MI; MSIE/Ind Engr/WA Univ; BSIE/Ind Engr/WA Univ. *Born:* 12/30/40. Upon completion of military service, was appointed to faculty of Ind Engr Dept at Univ of MO-Columbia in 1970. Promoted to Assoc Prof in 1973. Directed Health Systems Trng Grant between 72-75. Co PI of "Optimal Nursing Assignments based on Difficulty 73-75. Joined Medicus Systems in 75 as Regional Mgr. Apptd Dir of Nursing Services Div & VP in 77. Apptd Sr VP for Dev of Medicus Microsystems in 78. Apptd SR VP Whittaker Medicus Systems in 81. Past Div Dir, Health services Div IIE. Also past council mbr, Health Application Section, OPNS Res Soc of Am. Past Bd Mbr, Hospital Mgt Systems Soc, Am Hosp Assn. *Society Aff:* IIE.

Freund, Richard A
Business: Quality Planning Services, 155 Yarkerdale Dr, Rochester, NY 14615
Position: Pres *Employer:* Quality Planning Services *Education:* MS/Indus Eng/ Columbia Univ-Sch of Engrg; AB/Pre Engrg/Columbia Univ- Columbia Coll *Born:* 11/14/24 In addition to managing a conslt service in quality mgmt and planning, works with the Ctr for Quality and Applied Statistics of the Rochester Inst of Tech as Conslt for curricula and contract training planning. Served with the Army Air Corps 1942-46. Sr staff consult for Quality, Kodak Park Div, Eastman Kodak Co 1949-83. Pres ASQC 1972-3, Fellow ASQC, ASTM, AAAS, ASA, IQA (Great Britain), Academican Int'l Acad for Quality, Chrmn Ctte E-11 ASTM, Chrmn USTAG to ISO/TC176 on Quality Assurance. Shewhart Medal (ASQC), Dodge Awd (ASTM), Brumbaugh Awd (ASQC). *Society Aff:* ASQC, ASTM, AAAS

Freund, Robert J
Business: 300 W Wash Str, Chicago, IL 60606
Position: Principal & Owner. *Employer:* Robert J Freund, Cons Engrs. *Education:* MS/Engg Admin/IL Inst of Technology; BS/Elec Engg/IL Inst of Technology. *Born:* 8/23/25. BS & MS Ill Inst of Tech. Served USNR Ltjg. Principal & Owner Robert J Freund, Consultg Engrs since 1957. P E IL, IN, IA, MI, NY & WI. OH, TX, PA, LA. Sr Mbr IEEE. Mbr ISPE, NSPE, ACEC, ARRL, IAEI. Mbr Panel of Arbitrators, Amer Abritration Assn. *Society Aff:* IEEE, NSPE, ACEC, ARRL, IAEI.

Frevert, Richard K
Home: 5052 N Swan Rd, Tucson, AR 85718
Position: Prof Agri Engrg. *Employer:* Univ of Arizona. *Education:* PhD/Agri Engg/ IA State Univ; BS/Agri Engg/IA State Univ; MS/Agri Engg/IA State Univ. *Born:* Feb 1914. Native of Odebolt Iowa. Instr through Prof Iowa State Univ 1937-58. Asst Dir Iowa Agri Experiment Station 1952-58. Dir Arizona Agri Exp Station 1958-74. Assoc Dean Col of Agri Univ of Arizona 1972-74. Currently Prof of Agri Engrg Univ of Arizona. US Army Air Corps as Weather Officer 1942-46, with discharge as Captain. Res & teaching specialities in Soil & Water Conservation Engrg. Coauthor of two textbooks & author of numerous scientific papers. Mbr of Council 1952-55, Fellow 1962 & Admin Dir 1976-78 of Amer Soc of Agri Engrs. *Society Aff:* ASAE, AGU, AMS, SCSA.

Frey, Carl
Business: 19 W Harding Ln, Westport, CT 06880
Position: President. *Employer:* Carl Frey Associates, Ltd. *Education:* MA/Education/Columbia; BME/Mech Engg/NY Univ. *Born:* 2/1/27. Carl Frey is the president of Carl Frey Associates, Ltd, a firm involved in technology related management and information projects. Previously he was executive director of the American Association of Engineering Societies, the Engineers Joint Council and was staff head of the Engineering Manpower Commission. Other activities include: vice president and founding partner of Consultation Networks, Inc, founding publisher of Who's Who in Engineering, organizer of the U.S./Egypt Alliance for Engineering Cooperation, and director of the Study of Engineering Utilization. Frey holds a bachelor of science degree in mechanical engineering from New York University and a masters degree in education from Columbia University. *Society Aff:* AIAA, ASEE, AAAS, ASAE, CESSE, SSA.

Frey, Donald N
Business: 7100 McCormick Rd, Chicago, IL 60645
Position: Chmn of Board, President, CEO *Employer:* Bell & Howell Co. *Education:* PhD/Met Engrg/U of MI; MS/Engrg/U of MI; BS/Engrg/U of MI. *Born:* Mar 13, 1923 St. Louis MO. Asst Prof U of Mich 1950-51. Engrg & exec positions For Motor Co 1951-67. Pres Gen Cable Corp 1968-71. Chmn & CEO Bell & Howell Co 1971- . Hon doctorates in engrg from U of Mich & U of MO at Rolla. Dir Cincinnati Milacron Inc. Clark Equip Co, Springs Mills Indus & Mbr, Natl Acad of Engrg (mbr council 1972), Amer Motors Corp, Engrg Soc of Detroit (ESD). *Society Aff:* ASME, NAE, ESD.

Frey, Howard A
Home: 1302 Woodshole Rd, Towson, MD 21204
Position: Retired. *Employer:* *. *Education:* BS/EE/Johns Hopkins Unvi. *Born:* Oct 1905. With General Elec Co from 1926-68. Held various engrg & mgmt positions, including Mgr-Engrg & Gen Mgr at Insulator Dept of G E in Baltimore Md. Principal engrg activity in field of high voltage & insulators. Author of number of tech papers & holder of several patents in this field. Served on number of cttes of IEEE, NEMA & ASA, sometimes as chmn. Former Chmn Md Sect AIEE. Fellow IEEE, Mbr NSPE, Engr Soc of Baltimore. Presently retired but doing occasional cons work. *Society Aff:* IEEE, NSPE.

Frey, Jeffrey
Business: Dept of Electrical Engrg, Univ. of Maryland, College Park, MD 20742
Position: Prof. *Employer:* Univ. of Maryland *Education:* PhD/EE/Univ of CA; MSc/ EE/Univ of CA; BEE/EE/Cornell Univ. *Born:* 8/27/39. Jeffrey Frey was born in NYC & educated at Cornell Univ & the Univ of CA, Berkeley. After working at the Watkins-Johnson Co in Palo Alto, CA on microwave semiconductor devices & integrated circuits, he was a NATO Postdoctoral Fellow at the Rutherford Lab in England, & worked on ion implantation at the UK Atomic Energy Res Establishment in Harwell. He taught at Cornell 1970-86, working on res in high-speed semiconductor devices & integrated circuits, & set up Cornell's teaching & experimental prog in the latter field. In 1976 he was a Fellow of the Japan Soc for the Promotion of Sci at the Univ of Tokyo, and subsequently was a participant in the US-India Exchange of Scientists Prog, and a guest researcher at the Technical Univ, Aachen, Germany. In 1984-85 he was a visiting professor at the Univ. of Tokyo, & is now a Prof. of Elec. Engrg at the Univ. of Maryland.

Frey, Muir L
Home: 5951 Orchard Bend, Birmingham, MI 48010
Position: Consultant. *Employer:* Self. *Education:* BS/Met Engg/Univ of MO at Rolla. Formerly MO Sch of Mines; MS/Met Engg/Univ of MO at Rolla. Formerly MO Sch of Mines. *Born:* May 1900 Ill. Mbr Tau Beta Pi, Phi Kappa Phi, Theta Tau. Asst Ch Met, Holt Mfg Co Peoria Ill 1925-26; Ch Met Gerlinger Steel Casting Co Milwaukee Wisc 1925-26; Ch Met John Deere Tractor Co Waterloo Ia 1926-36; Service Met Republic Steel Corp 1936-40; Aircraft Met Packard Motor Car Co 1940-45; Asst to Gen'l Works Mgr Tractor Plants Allis Chalmers Corp, Milwaukee Wisc 1945-65. Mbr AIME since 1923, Chmn Detroit Sect 1942-44; ASM since 1924, Fellow 1970; Wm Hunt Eisenman Award 1974, chmn sev natl cttes 1955-63, Milwaukee Chap 1950; SAE since 1946 Fellow 1981, Distinguished Service Award, Tech Board 1957, V P 1957, Bd/Dir 1957 & 1961-62, Mbr Tech Bd 1958-60, Chmn Genl Matl Council & its Publ Policy Comm 1958-63; Mbr Iron & Steel Comm since 1946 and sev of its panels & divs, chmn in 1950; Mbr Advisory Ctte on Naval Gun Mounts, 1946. Mbr Advisory Ctte on Welding of Armor, 1952-63.

Frey, Stuart M
Home: 3970 North Darlington, Birmingham, MI 48010
Position: Vice President, Car Engineering *Employer:* Ford Motor Company. *Education:* SM/Indus Mgmt/MA Inst Tech; BS/Mech Eng/Univ of MI. *Born:* Feb 13, 1925. Have been with Ford Motor Co since 1953. Have contributed to the design, engrg & dev of domestic passenger cars while moving from Exec Engr to Ch Body Engr. Had respon for product planning & res programs as Ch Car Planning & Res Engr. Also had respon for advance pre-programs & supplier res programs. As V.P. Car Engrg now respon for Component Design as well as vehicle design & dev engrg, testing operations, vehicle material dev & reliability programs. *Society Aff:* SAE.

Friar, John II
Home: 537 Montgomery St, Fall River, MA 02720
Position: Pres. *Employer:* Qualitex Inc. *Education:* BS/ChE/Univ of MA *Born:* 5/29/42. After 2 yrs as res engr in Titanium Dioxide Div of Cabot Corp, entered Rubber Latex Ind as process dev engr for Globe Mfg in Fall River, MA. Transfered to position of latex chemist, then chief chemist in 1977. Joined Heveatex corp in 1978 as gen mgr, & Qualitex, Inc in 1979 as Pres. Served as appointed chrmn of local water supply bd for 6 yrs. Currently chrmn of Wastewater Commission, overseeing construction of $55 million treatment facility. Also chrmn of local United Way budget panels & bd mbr of local cooperative bank. Named 1978 citizen of yr. 1982, moved to Corp Hdquars of Qualitex's Parent Co, Lugano, Switzerland. *Society Aff:* AIChE

Friberg, Emil E
Home: 3406 Woodford Dr, Arlington, TX 76013
Position: Pres. *Employer:* Friberg Associates Inc. *Education:* BSME/Mech Engg/Univ of TX at Austin. *Born:* 4/11/35. Native of Wichita Falls, TX. After ten yrs experience with TX Elec Service Co in Wichita Falls and Ft Worth joined present firm as Prin in 1969. Elected Pres of firm in 1973. Firm specializes in mech and elec design services related to bldgs. Friberg specializes in energy mgt for bldgs. Dir and Regional Chrmn VIII, ASHRAE 1975-78; Dir at Large, ASHRAE, 1979-82; Pres, Consulting Engrs Council of TX, 1979-80; Pres, Ft Worth Chap ASHRAE, 1974; Pres, Ft Worth Chap, TX Society of Professional Engrs, 1973; Rotary Intl; Ft Worth Club. *Society Aff:* ASHRAE, ASME, AIME, NSPE, ACEC.

Frick, John P
Home: 6924 Middle Cove Dr, Dallas, TX 75248
Position: Sr Res Engr. *Employer:* Mobil R & D Corp. *Education:* PhD/Met/PA State Univ; MS/Met/PA State Univ; BS/Met Engg/Univ of KS. *Born:* 1/20/44. Dr Frick received his PhD degree in 1972. Since that time he has been associated with the oil & gas industry. He has presented papers before the AFS & NACE. Because of his efforts in the area of oil country tubular goods, he was appointed Chrmn of the Tubular Products Session of the 20th Mech Working and Steel Processing Conf of the AIME. He has been selected by the ASM to edit the 1984 Edition of the handbook, *Engineering Alloys.* Dr Frick has been an active mbr of ASM & has held offices of Secy & VP of chapt of that organization. *Society Aff:* ASM, AIME.

Frick, Pieter A
Home: 12180 NW Reeves, Portland, OR 97229
Position: Prof & Head Elect. Eng *Employer:* Portland State University *Education:* PhD/Control Systems/Univ of London; DIC/Control Systems/Imperial Coll, London; MEng/EE/Univ of Stellenbosch, South Africa; BScBEng/EE/Univ of Stellenbosch, South Africa *Born:* 2/11/42 in Heidelberg, Transvaal South Africa. Received his education at the Univ of Stellenbosch (SA) (1960-66) and the Univ of London (Imperial College) (1968- 71). Was a technical officer in the South African Air Force 1964-67, research eng with S.A. Iron and Steel Co., and has taught at Oregon State University (1973-80) and Portland State University 1980- . Has held visiting faculty positions at UCLA 1972, University of Rome 1976 and the Oregon Graduate Center 1979-80. *Society Aff:* IEEE, SIAM, ΣΞ, HKN

Frick, Thomas C
Business: 731 Meadows Building, Dallas, TX 75206
Position: Consultant, Petroleum & Natural Gas. *Employer:* Self. *Education:* BS/Petroleum Engg/Univ of Tulsa. *Born:* Apr 1909. Native of Caney Kansas. Worked 4 yrs in the oil fields & natural gas industry before entering college. Worked for Phillips Petroleum Co for 3 yrs & taught Petroleum Prod Engrg & was head of dept at U of Tulsa 1936- 41. Held var positions with Atlantic Richfield Co from 1941 until retiring in 1972 as VP of the Central Region (Production) & Natural Gas Dept. Editor of two- volume Handbook on Petroleum Prod. Enjoy fishing & hunting on Texas Gulf Coast. Pres of AIME 1965. Hon Mbr AIME 1970. DeGolyer Disting Service Medal 1973 SPE/AIME. *Society Aff:* AIME, SIPES, AAAS.

Frick, Warren A
Business: 5 N Country Club Rd, Decatur, IL 62521
Position: Partner. *Employer:* Homer L Chastain & Assocs. *Education:* BCE/Civil Engg/George Washington Univ. *Born:* 4/4/21. Native of IL. Reg PE IL & GA. US Bureau of Public Roads: 1951-53, Engr Training Program & 1953-55 Hwy Engr assigned to IL District Office. IL Dept of Transp: 1955-61 Geometric Design Engr, 1961-66 Engr of Location & Roadway Planning & 1966-71 Engr of Traffic. Also 1966-71, Commissioner for IL and Mbr of Natl Exec Ctte of Vehicle Equip Safety Comm & 1968-71, Secy of Governor's Traffic Safety Ctte. Since 1971 Partner in cons engrg firm. Also Col US Army Reserves (Retired), Past VChrmn of Natl Joint Ctte on Uniform Traffic Control Devices, past Mbr and Chrmn of Development Ctte of Group Council of Transp Res Bd, past. Mbr of TRB Cttes, 25 yr Mbr of AASHTO, Fellow of Inst of Transp Engrs & Past VP of Cons Engrs Council of IL. *Society Aff:* ITE, NSPE, AREA, ACEC, ARTBA.

Fricke, Arthur L
Home: 11 Edgewood Dr, Orono, ME 04473
Position: Gottesman Prof *Employer:* Univ of ME. *Education:* PhD/CE/Univ of WI; MS/CE/Univ of WI; ChE/CE/Univ of Cincinnati. *Born:* 3/6/34. Native of W VA. Res & dev positions with Shell Cos 1961-1967. Mbr ChE dept at VPI&SU 1967-1976 with primary interest in polymers. Chrmn of ChE dept at ME 1976 to 1981. Gottesman Res Prof of Chem Engrg, 1981 to present. Ind experience related to plastic foams, fibers, plastic films, paper, pigments, & ind chem through full or part-time employment with six cos. Have held supervisory positions in ind & academia. Reg PE. Active as an ind consultant. Primary current res interest is polymer processing. Married with three children. *Society Aff:* AIChE, SPE, ACS, TAPPI, PIMA.

Fricke, William G, Jr
Business: Alcoa Tech Center, Alcoa Center, PA 15069
Position: Sr. Scientific Associate *Employer:* Aluminum Co of America. *Education:* PhD/Metallurgy Eng/Univ of Pittsburgh; BS/Metallurgy/Penn State; MS/Metallurgy/Penn State; Assoc/Electronics/Penn State. *Born:* May 1926. Native of Pittsburgh Penna. Served in Army (Infantry) 1944-46 & 1951-52. Employed by Alcoa (Res) since 1952. Sr. Scientific Assoc in Structures Group. Metallography, microbeam analysis, fatigue, crystallographic texture. ASTM Templin Award 1955. *Society Aff:* ASM, AIME, ASTM, MAS, ISS, SQM, ΣΞ

Fridley, R B
Business: Aquaculture & Fisheries, Univ. of Calif., Davis, CA 95616
Position: Dir., Aquaculture & Fisheries Program *Employer:* Univ. of California, Davis *Education:* PhD/Ag Engrg/MI State Univ; MS/Ag Engrg/Univ of CA - Davis; BS/Mech Engrg/Univ of CA - Berkeley; -/AA/Sierra Coll *Born:* 06/06/34 Mgr, R&D Dept. (1977-85), Weyerhauser Company, Tacoma WA. Director of Aquaculture & Fisheries at the Univ. of Calif., Davis 1985-present; Specialist, Univ CA, Davis, 1956-61; asst prof, assoc prof, 1961-69 prof, 1969- 77, chrmn agri engr dept, Univ CA, Davis, 1974-76; NATO visiting prof Univ Bologna, Italy, 1975; visiting prof, MI State Univ, 1970-71. Recipient Charles G Woodbury Award, Am Soc Hort Sci, 1966. Fellow Am Soc Agri Engrs (FIEI Young Researcher Award, 1972, concept of yr award, 1976, outstanding paper awards, 1966, 68, 69, 76). Patentee, contributor articles in Fields of engr in Forestry and Fruit harvesting and handling; b Burns, OR; m Jean Marie Griggs, June 12, 1955; 3 sons James Lee, Michael Wayne, Kenneth Jon. *Society Aff:* ASAE, ASEE, ΣΞ, ΦΚΦ, WAS, AFS.

Fridy, Thomas A, Jr
Business: P. O. Box 491, Spartanburg, SC 29304
Position: VP & Dir of Projects *Employer:* Lockwood Greene Engrs, Inc *Education:* BCE/Structures & Hgwys/Univ of FL; Graduate Studies/Structural Engrg/Univ of FL *Born:* 12/21/23 Native of Palatka, FL. Worked in early part of career for the FL State Hgwy Dept in Bridge Design Office and for Harwood Beebe for six years in Municipal Engrg. Served in Armed Forces actively for three years during World War II. Have been with Lockwood Greene since 1956 occupying various positions from Designer, Project Mgr, Dept Mgr, Dir and VP of the Co. Presently responsible for all projects in Lockwood Greene's Spartanburg Office and consultant for all marine work done by the co on a worldwide basis. Graduate study in Structural Engrg at the Univ of FL. *Society Aff:* ASCE, AWWA, NSPE, PIANC.

Fried, Benjamin S
Home: Bantam Lake Rd, Bantam, CT 06750
Position: Plant Superintendent. *Employer:* Cerro Wire & Cable Co. *Education:* BS/Math/Brooklyn College; Adv Courses/Ind Eng/NY Univ. *Born:* Aug 2, 1924. Grad study at NYU. Presently Plant Superintendent Cerro Wire & Cable Co. Previously held position as Ch Indus Engr with same company. Sr Mbr IIE. P Pres & Mbr of Bd of Governors IIE Chap 5. Edited Monograph 3 'Film Index of Work Measurement & Methods Engineering Subjects'. P Dir of W M & M E Div of IIE. *Society Aff:* IIE.

Fried, Erwin
Home: 53 Terrace Court, Ballston Lake, NY 12019
Position: Engineering Consultant *Employer:* Self *Education:* MS/Mech. Engrg./Union College; BS/Mech. Engrg./Columbia University *Born:* 08/24/22 Spent entire engrg career with General Electric Co. 1952, Development Engineer in General Engrg Lab in heat transfer and fluid mechanics. 1961, Thermal Control Consulting Engr in GE Space Division. Performed pioneer work in Contact Heat Transfer applied research and published numerous papers and book articles. Served on AIAA Thermophysics Cttee. 1971, Thermal and Hydraulics Consultant on Navy Nuclear power plants 1986, Retired from GE. Chairman, ASME Heat Transfer Div. 1977-8. Fellow ASME 1980; ASME Dedicated Service Award 1984; National Nominating Cttee 1985-7; ASME Board on Public Information, 1986-present. Editor of "Handbook of Hydraulic Resistance" translation. *Society Aff:* ASME

Fried, George
Business: 519 Eighth Ave, New York, NY 10018
Position: VP Engg & Mfg. *Employer:* Manostat Corp. *Education:* MS/Engg Mgt/NYU; BS/ME/Univ of WI. *Born:* 9/11/25. VP of Engg and Mfg of Manostat Corp since 1969, Mfg of Lab Instruments and Scientific Apparatus. Noted authority on peristaltic pumping, featured in Product Engineering Magazine; wrote on this subj in Machine Design Magazine, lectured at Columbia, CUNY, WI, Cornell and other colleges. Formerly mgr of engg for Bowling Products Div of AMF, Inc. Holds eleven US and foreign patents. Mbr of ASME for 26 yrs, VP since 1979, chrmn of its design engg div in 1973-74 and active in div for over 15 yrs. Mbr of Infusion Devices Std Committee for AAMI. *Society Aff:* ASME, AAMI.

Fried, Walter R
Home: 11321 Brunswick Way, Santa Ana, CA 92705
Position: Proj Tech Dir *Employer:* Hughes Aircraft *Education:* MS/EE/OH State Univ; BS/EE/Univ of Cincinnati *Born:* 4/10/23 Did early work on Doppler navigation systems at Wright Air Development Center from 1948 t0 1958. Pioneered in the development of a new type of FM-CW Doppler radar at General Precision Laboratories, which became the standard approach for Doppler radars in civiland military aircraft. U.S. Delegate to the International Civil Aviation Organization. Chairman of IEEE Chapters in New York and Los Angeles. Worked extensively on Relative Navigation systems at Hughes Aircraft C ompany. Editor and co-author of the book Avionics Navigation Systems and co-author of Airborn Radar. Elected a Fellow of the IEEE in 1980 *Society Aff:* IEEE, ION

Friedberg, Arthur L
Business: 65 Ceramic Dr, Columbia, OH 43214
Position: Exec Dir. *Employer:* American Ceramic Society. *Education:* PhD/Ceramic Engg/Univ of IL; MS/Ceramic Engg/MIT; BS/Ceramic Engg/Univ of IL. *Born:* 3/25/19. Prof and Hd of the Dept of Ceramic Engg, Univ of IL, Urbana, 1963-79; Prof of Ceramic Engg, 1955-63. Exec Dir, Amer Ceramic Soc 1979-present; Exec Dir, Natl Inst of Ceramic Engrs, 1979-present; Adjunct Prof, Dept of Ceramic Engrg OH St Univ 1979-present.

Frieden, B Roy
Business: Optical Sciences Ctr, Tucson, AZ 85721
Position: Prof *Employer:* Univ of AZ *Education:* PhD/Optics/Inst of Optics, Univ of Rochester, NY; MS/Physics/Univ of PA; BS/Physics/Brooklyn Coll *Born:* 9/10/36 B. in Brooklyn, NY. Received PhD in Optics in 1966 from the Inst of Optics, Univ of Rochester. Joined Univ of AZ Optical Sciences Ctr as Asst Prof in 1966, became Assoc Prof in 1972, Prof in 1976. Main research interest is signal processing, with emphasis on optical images. Invented maximum entropy image processing and other computer algorithms for restoring and enhancing images. Produced the first recognizable features on a picture of a moon other than our own (Ganymede, 1976). Strong interests in information theory, probability and statistics. Enjoy analyzing pictures of strange phenomena to verify authenticity. Hobbies are hiking and bird keeping. *Society Aff:* OSA, SPIE, ΣΞ

Friederich, Allan G
Business: 342 Mech Engrg Bldg, Urbana, IL 61801
Position: Prof of Mech Engrg. *Employer:* Univ of Ill. *Education:* MS/ME/Univ of IL; BS/ME/Univ of IL. *Born:* Jan 1, 1922 Wood River, Ill. Attended Bradley Univ. 3 yrs with CurtissWright Corp during WWII BS & MS from Univ of Ill, 1947 & 1953. Progressed through ranks to Prof of Mech & Indus Engrg at Univ of Ill 1972. V P ASME 1969- 71. Principal interest Machine Design. Mbr Pi Tau Sigma, natl scholastic honorary for Mech Engrs. Hobbies-woodworking, automotive repair, travel. Prof Eng, Reg in WI.. *Society Aff:* ASME, ΠΤΣ.

Friedkin, Joseph F
Home: 3821 Hillcrest Dr, El Paso, Texas 79902
Position: Consult *Employer:* Self *Education:* BS/ME/Univ of TX-El Paso; grad work/Hydraulics/Miss St Univ. *Born:* 10/18/09 Engineering Consultant. United States

Friedkin, Joseph F (Continued)
Commissioner on Intl Boundary and Water Commission, United States and Mexico 1962-85. Entrusted jointly with Mexican Commissioner by treaties with responsibility for resolving water and boundary problems along the 2,000-mile border, requiring cooperative action by the two countries. Served as Technical Adviser to the Dept of State in the negotiation of the Treaty of 1963 for Settlement of the Chamizal Boundary Dispute, the Boundary Treaty of 1970 with Mexico, in 1973 for a Permanent and Definitive Solution to the Intl Problem of the Salinity of the CO River. *Society Aff:* ASCE, NSPE, ICID.

Friedlaender, Fritz J
Business: Purdue University, Sch of Elec Engrg, W Lafayette, IN 47907
Position: Prof of Elec Engrg. *Employer:* Purdue Univ. *Education:* PhD/EE/Carnegie Inst of Technology; MS/EE/Carnegie Inst of Technology; BS/EE/Carnegie Inst of Technology. *Born:* May 7, 1925 in Freiburg, Germany. During 1954-55 was Asst Prof at Columbia Univ. Since 1955 has been at Purdue Univ where is now Prof of Elec Engrg. Spent sabbatical years at the Max-Planck-Institut fur Metallforschung in Stuttgart, W Germany 1964-65 & Ruhr Universit at Bochum, W Germany 1972-73 Universitat Regensburg l98l-82. Received the Sr Scientist Award from the Alexander Von Humboldt Foundation. Japanese Society for the Promotion of Science Senior Research Fellowship, Aug & Sept. 1980. Cons for var orgs & active in res on magnetic devices & materials with current interests in high gradient magnetic separation & amorphous & garnet magnetic films & micromagnetic memories (VBL) Pres of IEEE Magnetics Soc 1977-78; has held var positions in the Soc & on editorial boards of magnetics publications as well as the Proceedings of IEEE. Fellow of IEEE & mbr of numerous prof & hon societies. *Society Aff:* IEEE, APS, ASEE, AAUP, ΣΞ, ΤΒΠ, HKN, ΦΚΦ

Friedland, Alan L
Home: 40 Rockland Dr, Jericho, NY 11753
Position: Principal. *Employer:* Alan L Friedland, Cons Engr, P.C. *Education:* MSME/Mech Engrg/Columbia Univ.; BME/Marine Engrg/NYS Maritime Coll. *Born:* Oct 4, 1929. BME NYS Maritime Coll; MS ME Columbia Univ. Engaged in design of all types of mech & elec sys for office buildings, commercial, institutional, educational, industrial & health-medical facilities. Served 2 yrs as Ch Engr aboard a naval vessel. Licensed as P E in NY, NJ, Penna, Conn, Mass & Fla. P Pres of L I Chap Amer Cons Engrs Council. Mbr NSPE, NYSSPE, ASME, ASHRAE, IES, ACC. *Society Aff:* NSPE, ASME, ASHRAE, IES, ACC.

Friedland, Bernard
36 Dartmouth Road, West Orange, NJ 07052
Position: Research Manager *Employer:* The Singer Company *Education:* A.B./Pre-engr./Columbia University; B.S./EE/Columbia University; M.S./EE/Columbia University; Ph.D./EE/Columbia University *Born:* 05/25/30 Born and educated in New York City. Taught electrical engrg at Columbia Univ where he received his A.B., B.S., M.S. and Ph.D degrees, prior to joining The Singer Company (then General Precision, Inc.) in 1962 as Mgr of Systems Research. Also Adjunct Professor of Electrical Engrg of Polytechnic Univ. Author of three textbooks and numerous technical papers on applications of modern control theory to navigation, guidance and control in aerospace field. Holds ten patents for navigation instruments and systems. Fellow of IEEE and ASME; ASME Oldenburger Medal, 1982; IEEE Control Systems Society Distinguished Member Award, 1985. *Society Aff:* IEEE, ASME, AIAA

Friedland, Daniel
Home: 1046 Field Ave, Plainfield, NJ 07060
Position: Mgmt Consultant (Self-employed) *Education:* PhD/Chem Engg/Brooklyn Polytech; MChE/Chem Engg/Brooklyn Polytech; BChE/Chem Engg/CCNY. *Born:* 4/5/16. in Columbus, OH. Work Experience: M W Kellogg Co 1937 Res. Associate - ; Citites Services Oil Co Chem Engr, - Refinery Superintendent 1938-46. Chem div UOP Inc from 1946. (Sometime Assoc Prof Chem Engg. Author many tech papers. Mbr ACS, AIChE, CDA & American Inst of Chemists) to 1976 as VP operations & then as VP mkting. From 1976-77, Pres of Crompton & Knowles Org Chmn Div Since 1977. Mgmt Consultant. *Society Aff:* ACS, AIChE, CDA, AmInstChem.

Friedlander, Benjamin
Home: 703 Coastland Dr, Palo Alto, CA 94303
Position: Dir *Employer:* Saxpy *Education:* Ph.D./Elec. Eng./Technion, Israel; M.S./Elec. Eng./Stanford University; M.S./Statistics/Stanford University; B.S./Elec. Eng./Technion, Israel *Born:* 02/24/47 Director of Advanced Technology at Saxpy Computer Corp since 1985. Mgr of the Advanced Technology Div, Systems Control Technology, 1976-85. Fellow of the IEEE. Member of the ADCOM of the ASSP Society. Recipient of the 1983 Senior Award of the ASSP and the 1985 Best Paper Award from ERASIP. Specializes in the theory and applications of signal processing. Past associate editor of the IEEE Trans. Automatic Control. Mbr of the technical cttee on Spectrum Estimation of the ASSP. Vice Chairman of the Bay Area Chapter of the ASSP Society. *Society Aff:* IEEE, ΣΞ

Friedlander, Sheldon K
Business: Chem, Nucl, & Thermal Engg, Boelter Hall, Los Angeles, CA 90024
Position: Prof of Engg & App Sci VChrmn, CNTE. *Employer:* Univ of CA. *Education:* PhD/CE/Univ of IL; MS/CE/MIT; BS/CE/Columbia Univ. *Born:* 11/17/27. Dr S K Friedlander is Prof of Engg & App Sci & VChrmn, Chem Engg at UCLA. He has worked in the field of aerosol dynamics & its applications to explaining the origins & behavior of particulate pollution. Dr Friedlander has been a consultant to the Los Angeles Air Pollution Control Dist, chrmn of the NRC Panel on the Abatement of Particulate Emissions from Stationary Sources & chrmn of the NRC Sub-committee on Photochem Oxidants & Ozone. He received the Colburn Award from the AIChE in 1959 & was a Fulbright Scholar in 1960 & Gugenheim Fellow in 1969, both at the Univ of Paris. In 1974, Dr Friedlander was the recipient of the Alpha Chi Sigma Award from the AIChE, & in 1979, he received the AIChE Walker Award. He was elected to the Natl Acad of Engg in 1975 in recognition of his work on the origins & control of particulate pollution. He is the author of *Smoke, Dust and Haze: Fundamentals of Aerosol Behavior* and of over 90 scientific papers. He is currently chrmn of the EPA Clean Air Scientific Advisory Comm & a mbr of the EPA Sci Adv Bd Exer Comm. He is Director of the Center for Intermedia Transport Research established by EPA at UCLA (1980). *Society Aff:* AIChE, ACS, APCA.

Friedly, John C
Business: Dept of Chem Engg, Rochester, NY 14627
Position: Prof. *Employer:* Univ of Rochester. *Education:* PhD/CHE/Univ CA (Berlekey); BS/CHE/Carnegie Tech. *Born:* 2/28/38. in Glendale, WV. Res Engr for GE Co, 1969-67. Asst Prof of CE at the Johns Hopkins Univ, 1967-68. With Univ of Rochester since 1968. Served as Chrmn of Interdepartmental Prog in Engg & Applied Sci 1973-75. NATO Sr Fellow at Univ of Oxford 1975-76. Served as Assoc Dean for Grad Studies 1979-81. Became Chairman of Chemical Engineering Department in 1981. *Society Aff:* AIChE, ACS, ASEE.

Friedman, Edward A
Business: Stevens Inst. of Tech, Hoboken, NJ 07030
Position: Prof of Management *Employer:* Stevens Inst of Tech. *Education:* PhD/Physics/Columbia Univ; SB/Physics/MA Inst of Technology. *Born:* 9/29/35. Married Arline J Lederman 1963. Two sons. Info Tech Appl in Higher Education & Training, Mossbauer & neutron scattering studies of magnetic matls. Laser scattering studies of polymers (with R Andrews) - awarded Stevens Inst of Tech Res Award 1970. Faculty mbr at Stevens 1963-current rank of Prof. Visiting Prof Afghanistan 1965-67. Gov of Afghanistan education medal recognized accomplishments while Kabul Univ Eng Coll Dev Prog Ch of party 1970-73. Dean of the Coll at Stevens 1973-1983, VP Acad Affairs 1983-85, Professor of Management 1985-present. Chrmn, Council for the Understanding of Tech in Human Affairs 1979-

Friedman, Edward A (Continued)
1983. Mbr Bd of Dirs of the Assoc of Independent Coll and Universities of NJ 1978-1983. Activities include computers in education and training and infor for public regarding tech policy issues. Writing Text on Information Tech. Honorary Master of Engrg, Stevens Inst of Tech 1983. Co Ed Intl Journal *Machine-mediated* Learning, 1983- (with Howard L. Resnikoff). *Society Aff:* ASME, ASEE, AAAS, APS.

Friedmann, Peretz P
Business: 5732 Boelter Hall, University of California, Los Angeles, CA 90024
Position: Professor *Employer:* Univ of CA. *Education:* DSc/Aero & Astro/MIT; MSc/Aero/Technion-Israel Inst of Tech; BSc/Aero/Technion-Israel Inst of Tech. *Born:* 11/18/38. Native of Rumania, came to US in 1969, naturalized US citizen. BS Aeronautical Eng (1961) Technion, Israel Inst of Tech; MS Aeronautical Eng (1968) Technion, Israel Inst of Tech; DSc Aeronautics (1972), MIT, Cambridge, Mass. Dr Friedmann has been an engr officer in the Israeli AF (1961-65), sr engr at Israel Aircraft Industries (1965-69) & res asst at the Aeroelastic & Structures Res Lab at MIT (1969-72). He has been with the Univ of CA since 1972, when he joined the faculty as an asst prof. Dr Friedmann has been engaged in res on rotary & fixed-wing aeroelasticity as well as structural dynamics & has published extensively. He has served on the AIAA structural dynamics technical committee & the dynamics committee of the AHS. He is the recipient of the ASME Structures and Materials Award (1984) & the Editor-in-Chief of Vertica, the International Journal of Rotorcraft & Powered Lift Aircraft. *Society Aff:* AIAA, ASME, AHS.

Friels, David R
Business: 4161 Ridgemoor Ave, Memphis, TN 38118
Position: Principal/Geotechnical Manager *Employer:* Test, Inc. *Education:* MS/Civ Engg/OK State Univ; BS/Civ Engg/OK State Univ. *Born:* 10/23/42. Reared SW OK, began career as construction engr - Phillips Petrol Co, served 4 yrs as engg officer, USAF. Returned to grad sch working as soil mechanics res asst. Joined Test, Inc in 1971 as geotechnical proj engr, promoted to princial with duties of Dir of Engg & Testing in 1974. Responsible for geotechnical testing & engg for numerous projs including a $12,000,000 ICGRR Bridge over the TN-Tombigbee Waterway. Past Pres Mid-South Sec, ASCE. Mbr Fdn Inspection & Testing Committee, Deep Fdns Inst. Author of several technical papers & reg PE in four states. *Society Aff:* ASCE, NSPE, DFI.

Friend, William L
Home: PO Box 401, Kentfield, CA 94914
Position: President *Employer:* Bechtel National, Inc. *Education:* Masters/ChE/Univ of DE; Bachelors/ChE/Polytechnic University *Born:* 6/17/35. 1957-72 Lummus Co. 1969-72 Gen Mgr, founded Lummus GmbH Wiesbaden. Mgr, Process Engrg & Asst Mgr, Lummus Tech Center. Dev Lummus ethylene process. 1973-77 Pres, Pritchard Group of Companies. 1977-80 Mgr, Area Office Ops, Refinery & Chem Div, Bechtel, Inc. 1980-84 VP & Mgr, San Francisco Div, Bechtel Petroleum, Inc. 1984-85 VP & Mgr, Houston Div, Bechtel Petroleum, Inc. 1985-86 VP & Mgr, Adv Technology Div, Bechtel, Inc., San Francisco. 1986-Present Pres, Bechtel National, Inc., San Francisco. Director, Bechtel Group, Inc. Fellow, American Institute Chemical Engrs & Polytechnic University. *Society Aff:* AIChE, ACS, ΤΒΠ, ΣΞ, ΦΑΥ, ANS, API.

Friesen, Clarence W
Home: 620 G Str, SW, Washington, DC 20024
Position: Ch, Prog Mgmt Div. *Employer:* Federal Hwy Admin. *Born:* Feb 1928. BSCE from Univ of New Mexico. Native of Wyoming. Served with Marine Corp 1946-52. Awarded the Silver Star for Action in Korea. With FHWA since 1956. Worked as Planning & Res Engr in Colorado, Iowa & Maryland. Currently am Ch of the Prog Mgmt Div in the Headquarters Office. Respon for Statewide Planning & dev of information for Natl Policy Dev. Am a Reg P E in Colorado. Enjoy fishing & backpacking.

Frisby, James C
Business: 216 Agr Engg Bldg, UMC, Columbia, MO 65211
Position: Prof. *Employer:* Univ of MO. *Education:* PhD/Agri Engr/IA State Univ; MS/Agri Engr/IA State Univ; BS/Agri Engr/Univ of MO; BS/Education/Univ of MO. *Born:* 10/22/30. in Bethany, MO (Harrison Cty). Parents-Jackson C Gladys Frisby. Wife-Hazel M (Kallenbach) Frisby. Children-None. Armed Service-Army Artillery 1952-54 (Released as 1st Lt). IA St Univ-1961-66; Mgr Farm Ser Dept, Faculty Univ of MO-1966-present; Asst Prof, Assoc Prof with Tenure, Prof with Tenure. Reg prof Eng in MO E9795. *Society Aff:* ASAE, ASEE, GSD, NACTA, ΣΞ.

Frisch, Ivan T
Home: 390 Latham Ln, E Williston, NY 11596
Position: Prof *Employer:* Polytechnic Univ *Education:* PhD/EE/Columbia Univ; MS/EE/Columbia Univ; BS/EE/Columbia Univ; BS/Phys/Queens College. *Born:* 9/21/37. 1962-69, Asst & Assoc Prof, Univ of CA, Berkeley. Leave of absence, Bell Labs, 1965. 1969, Founding mbr, Network Analysis Corp. 1969-80, VP, Network Analysis, 1980-86 VP Contel Business Networks. Have been Visiting Prof, App Math & Statistics, SUNY & Adjunct Prof, Comp Sciences, Columbia. Published over 80 papers & book "Communication, Transmission and Transportation Networks–, Addison Wesley, 1971. Was Managing Editor, "Columbia Engineering Qaurterly–, Assoc Editor "IEEE Transactions on Circuit Theory–, Founding Editor-in-Chief, "Networks–, Wiley Consulting Ed., Journal of Telecommunication Networks. Honors: Fellow IEE, Guggenheim Fellow, Ford Fdn Resident, Honorable Mention ORSA Lanchester Prize for best publication in operations res, 1971. *Society Aff:* IEEE, AMA, NYAS, ΦΒΚ, ΤΒΠ, HKN.

Frische, James M
Business: 6400 E 30th St, Indianapolis, IN 46219
Position: Plant Mgr *Employer:* RCA Records *Education:* MBA/OR/Butler Univ; MSIE/Stat & Sys/Purdue Univ; BSIE/OR & Sys/Purdue Univ. *Born:* 6/15/41. Joined RCA Engg organization in 1966. Became Mgr of Quality Control in 1971 & Mgr of Div Mgr & Quality Controls in 1972. Assumed respon as hd of total Engg organization for Record Div in 1973. Apptd Plant Mgr of Record Div Manufacturing Operations in 1978. Appointed Dir of Manufacturing Operations in 1980. Mbr of Ame Inst of Indus Engrs, Audio Engg Soc, Tau Beta Pi & Alpha Pi Mu Engg Socs & Hons. Selected as Outstanding Young Man of America. *Society Aff:* IIE, AES.

Fritz, Kenneth E
Business: 1 River Rd, Schenectady, NY 12345
Position: Sr Matls Engineer. *Employer:* General Elec Co. *Education:* MS/Metallurgical Engg/Univ of WI; BS/Metallurgical Engg/OH State Univ. *Born:* Oct 1918. Reg P E in Wis. Served the Navy as civilian & as a Naval Officer 1942-45 at Naval Res Lab. Employed by Bucyrus-Erie Co in South Milwaukee, Wis for 10 yrs, then by General Elec Co in Gas Turbine Engrg & Mfg Dept - specializing in indus gas turbines, steam turbines & generators. Several tech papers published in this field. *Society Aff:* ASM.

Fritzlen, Glenn A
Home: 306 Rue De Maison, Kokomo, IN 46902-3636
Position: Asst Prof. *Employer:* Purdue Univ. *Education:* BS/Met Engr/Purdue Univ. *Born:* 1919. BS Met Engrg Purdue 1941. Served as Army Air Corps Officer WWII; Asst Ch Metals Branch, Materials Lab WPAFB 1942-45. With Stellite DivCabot Corp for 38 yrs. Held positions of Tech Dir & Mgr of Quality Assurance. Authored many tech articles & dev numerous process innovations & controls for nickel- & cobalt-base alloys. Served on NASA subctte & 14 other gov materials adv panels. Served 3 yrs each on Transactions, Finance & Metals Handook Cttes of ASM. Fellow & Life Member of ASM, Amer Men of Science, Who's Who in Midwest. Senior Member, AIME; Asst. Prof. School of Technology, Purdue Univ (1985 to Present). Reg P E Engr. *Society Aff:* ASM International, AIME, ASTM, ASEE.

Frohling, Edward S
4370 So Fremont Ave, Tucson, AZ 85714
Position: President. *Employer:* Mntn States Mineral Enterprises, Inc. *Education:* BS/Metallurgy/MA Inst of Technology; Grad/-/Lawrenceville Sch Combat Engrs US Army. *Born:* Mar 26, 1924 Princeton, New Jersey. 1942-45 US Army Combat Engrs, European Theatre 1948. BS Metallurgy MIT. 1948 res engr AEC; 1948-51 plant metallurgist Climax Molybdenum Co; 1951-54 field sales engr Western Machinery Co; 1954-57 Mill Supt St Lawrence Fluorspar Inc; 1957-62 Regional Sales Mgr Western Machinery Co; 1962-69 V P & Mgr of Bus Dev, Parsons-Jurden Corp; 1969- , Pres Mountain States Mineral Enterprises Inc & Mountain States Engrs Inc. Pres Dev Authority for Tucson's Economy; Dist Citizen Award by the Univ of Ariz 1975; Man of the Year Tucson 1975; Author. Mbr AIME, SME of AIME (Dir); Mining Club of New York, Mining Club of Southwest, Univ Club of NYC. Dist Mbr Award by the Society of Mining Engrs of AIME 1979; 1st VP of Assoc Builders & Contractors, Inc 1982; Recipient of the 1982 Charles F Rand Memorial Gold Medal Award-AIME. *Society Aff:* AIME, SME, AMC.

Frohmberg, Richard P
Business: P O Box 1159, Kansas City, MO 64141
Position: Mgr, Plastics & Materials Engrg *Employer:* Bendix KC Div, Allied Bendix Aerospace *Education:* PhD/Met Engr/Case Inst of Technology; MS/Met Engr/Case Inst of Technology; BS/Met Engr/Case Isst of Technology. *Born:* Mar 1920 Cleveland Ohio. USAF 1942-45 Engrg Officer. Case Inst of Tech, Grad student, Instructor, Res Assoc 1948-55. Bendix Kansas City Div 1970- present. Mgr, Plastics & Materials Engr. Rocketdyne Div Rockwell Internatl 1955- 70. Ch Production Dev Lab; Mgr Materials Res. Listed Amer Men & Women of Science, Mbr Soc of Sigma Xi, Amer Soc for Metals, Reg Prof Metallurgical Engr Ohio & Calif. Henry Marion Howe Award ASM 1955. Former Mbr NASA R&T Advisory Subctte, Natl Mgmt Assn, AAAS & Assoc Fellow AIAA Elected Fellow ASM 1978. Enjoy backpacking, bicycling, hunting & shooting. *Society Aff:* ASM, ΣΞ.

Frohrib, Darrell A
Home: 2144 Princeton Ave, St Paul, MN 55105
Position: Prof & Head of the Design & Mfg. Division. *Employer:* Mech Engg Dept, Univ of MN. *Education:* SB/ME/MIT; SM/ME/MIT; PhD/Mechanics/Univ of MN. *Born:* 6/25/30. Dr Frohrib is responsible for the Design and Control Div & is Director of Graduate Studies of the Mech Engrg Dept. He served as proj engg at Sperry Gyroscope Co, Lake Success, NY, 1955- 1959, and res engr at the Budd Co, Phila, 1954-1955. His engg interests are in innovative and shock & vibration. He is consultant to Dacomed Corp, Minneapolis in design and Biomedical Engrg. He is author of 42 papers in his field. *Society Aff:* ASME, ΣΞ.

Froid, Stanley H
Business: 1071 Via Del Pozo, Los Altos, CA 94022
Position: President *Employer:* Froid Consultants, Inc. *Education:* BS/Civil Engr/ Univ of CA, Berkeley. *Born:* Jan 18, 1923 Valley City N Dakota. Reg P E in AL, CA, HI, & OR & reg with Natl Council of State Boards of Engrg Examiners. Joined Tudor Engrg Co in 1951. Has directed feasibility studies, planning, design & construc supervision of transit systems, highways, bridges, ports, military installations & tramways. As Associated Consultant served as member of Boards of Control for various Tudor Engineering Company joint venture projects. Is Pres. of Froid Consultants Inc. Is Fellow in ASCE, mbr of SAME. *Society Aff:* ASCE, SAME

Fromherz, Frank C
Business: 4747 Earhart Blvd, New Orleans, LA 70125
Position: President. *Employer:* Fromherz Engrs, Inc. *Education:* MS/San Engrg/ Harvard; BE/Civil Engrg/Tulane. *Born:* Sep 1921. Native New Orleanian. Commissioned USPHS WWII. Civil & Sanitary Engr, Alvin M Fromherz, Cons Engr 1948-52 after one year as Public Health Engr Louisiana State Board of Health. Partner, Fromherz Engrs-Cons Engrs 1952-75. President Louisiana Sect ASCE 1960. Since 1952 respons have incl developing & dir the professional activities of the firm in the performance of assignments for federal, state, parish (cty) & local governments; industries; developers; & architects. *Society Aff:* ASCE, AWWA, NSPE, ACEC, ARTBA.

Fromherz, Thomas A
Business: 1539 Jackson Ave, New Orleans, LA 70130
Position: Exec V Pres. *Employer:* Fromherz Engrs, Inc. *Education:* BE/Civil Eng/ Tulane Univ. *Born:* 2/11/24. 70115/1524 Leontine St/New Orleans, LA. *Society Aff:* ASCE, ACEC, NSPE, ACI.

Fromm, Eli
Home: 2604 Selwyn Dr, Broomall, PA 19008
Position: Prof. *Employer:* Drexel Univ. *Education:* PhD/Bioengg/Physiology/Jefferson Medical College; MS/Engg/Drexel Univ; BSEE/ EE/Drexel Univ. *Born:* 5/7/39. Primary technical interests in Biomedical Engg, physiologic instrumentation, biotelemetry. Engr Missile & Space Div GE Co 1962; Engr Appl Phys Lab E I DuPont 1963; NIH Special Fellow 1963-67; Prof Drexel Univ 1967- present. Assoc Dir Biomed Engg & Sci 1973-4; Consultant; Lectr Univ of PA Sch of Medicine. Grantee NIH, NSF, private fdns. Recipient Drexel Univ Res Achievement award; Sr Mbr IEEE; Congressional Science and Engineering Fellow award; Pres IEEE Engg in Medicine & Biology Soc. Mbr Eta Kappa Nu, Tau Beta Pi, Sigma Xi. Invited speaker & panel participant scientific meetings, legislative committees. More than 40 publications. Married, 3 children, active in community affairs. *Society Aff:* IEEE

Frosch, Robert A
Business: General Motors Research Laboratories, Warren, MI 48090-9055
Position: VP *Employer:* Gen Motors Corp *Education:* PhD/Theoretical Physics/ Columbia; BA/-/Columbia; MS/-/Columbia. *Born:* 5/22/28. Hudson Labs of Columbia Univ: Scientist 1951, Asst Dir 1955, Dir 1956-63. Advanced Res Projs Agency, Office of the Secy of Defense, Dir for Nuclear Test Detection 1963, Deputy Dir 1965. Asst Secy of the Navy (Res & Dev) 1966-73. Asst Exec Dir, United Nations Environ Programme 1973-75. Assoc Dir for Applied Oceanography, Woods Hole Oceanographic Inst 1975-77. Mbr Natl Academy of Engrg 1971, Fellow IEEE, Fellow Acoustical Soc of Amer Mgmt of R&D, Sys, Policy Formulation. Admr, Natl Aeronautics and Space Adm 1977-81. Pres, Amer Assoc of Engrg Societies, Inc. 1981-82. VP in charge of Res Labs, Gen Motors Corp 1982- present. *Society Aff:* NAE, AAAS, ASA, IEEE, AIAA, AAS, APS, SSA, MTS, SNAME, SEG, AGU, SAE

Frost, Brian R T
Business: 9700 S Cass Ave, Argonne, IL 60439
Position: Dir, Technology Transfer Center. *Employer:* Argonne Natl Lab. *Education:* BS/Metallurgy/Univ of Birmingham (UK); PhDMetallurgy/Univ of Brimingham (UK); MBA/Business/Univ of Chicago. *Born:* 9/6/26. Joined Atomic Energy Res Estab. Harwell England 1949 & worked on fission & fusion reactor materials in staff & mgmt functions. Joined Argonne Natl Lab as Assoc Dir, Materials Sci Div 1969. Became Dir of Div in 1973. 1984: Sr Scientist with special responsibility for interactions with industry. 1985: Director Technology Transfer Center. Fellow of Amer Nuclear Soc & inst of Metals (UK). Mbr ASM, AIME & ACerS. Enjoys sailing, golf, classical music *Society Aff:* ANS, Inst Metals, ASM, AIME-TMS, ACerS.

Frost, George C
Home: 25 McHugh St, S Glens Falls, NY 12801
Position: Pres & CEO *Employer:* Rist-Frost, Assocs, PC *Education:* BS/ME/Northeastern Univ. *Born:* 1/25/28. Assoc 1960-80!Assoc 1960-80. 1980-84 Exec VP & Dir of Tech Services, Rist- Frost Assoc, PC since 1984 Pres & CEO, Rist-Frost Assoc, PCNSPE, AAAS;!NSPE; NY!NY, NH, ME *Society Aff:* ACEC, NSPE, ASHRAE, ASME, IES.

Frost, Paul D
Home: P.O. Box 297, Hartland, VT 05048
Position: Pres *Employer:* Frost Assoc *Education:* BMetE/Metallurgy/Rensselaer Polytechnic Inst. *Born:* Aug 9, 1916. Graduate courses Ohio State Univ. Republic Steel Corp 1940- 43. Curtiss-Wright Corp 1943-45. Taught Cornell Univ Extension Courses, Buffalo New York 1941-45. Battelle Memorial Inst 1946-1983. Head Light Metals Sect 1953- 61. Head Metal Product Planning Sect 1961-70. Sr Tech Advisor, Tech-Economic Res since 1970. Dir res on aluminum, magnesium & titanium resulting in more than 60 published papers & gov publications. Now engaged in techno-econ projs in the metals industry in USA & Mexico. *Society Aff:* ASM.

Froula, James D
Business: PO Box 8840, Univ Station, Knoxville, TN 37996-4800
Position: Secy-Treas & Editor *Employer:* Tau Beta Pi Assn *Education:* MS/Mech Engrg/Univ of TN-Knoxville; BS/Mech Engrg/Univ of TN-Knoxville *Born:* 05/17/45 Native of Oak Park, IL. Married in 1968 to Barbara Leftwich and has 2 children. 1968-70 Lieutenant in US Army Ordance Corps. Service in Vietnam, Bronze Star, 101 Airborne Div. Employed as assoc, sr assoc, proj and dev engr by IBM Corp, 1970-1982; first invention achievement award, 1976; outstanding innovation award, 1982. Dist Dir, Tau Beta Pi, 1977-82. Assumed current respon as Secry-Treas & Editor of the Tau Beta Pi Assn in 1982. Respon as exec officer for all progs and activities of Tau Beta Pi, natl engg honor soc. Reg PE in TN. Mmbr, exec cttee, Assoc of College Honor Societies, 1987-. Enjoy mtn climbing and photography. *Society Aff:* ASME, NSPE, AAAS, ASEE, ΟΔΚ, TBPI, ΦΚΡΗΙ, ΠΤΣ

Froyen, Hugo G
Business: Rue Belliant No 7, 1040 Brussels, Belgium
Position: Loss Control Consultant *Employer:* Am. Int. Group. *Education:* Bachelor/Chemistry (Eng)/Superior Institute of the State for Nuclear - Industries. *Born:* 05/06/49 Performs loss prevention surveys with regard to property damage & business interruption of American multi-national companies having outfits in Europe, Middle East and Africa. Working with AIG since 1972, with one year for Johnson & Higgins Insurance Brokers. Was reserve officer in Belgian Army. Currently Pres, SFPE European Chapter. *Society Aff:* SFPE, AII Br.

Frumerman, Robert
Business: 218 So Trenton Ave, Pittsburgh, PA 15221
Position: President. *Employer:* Frumerman Assocs Inc. *Education:* MS/Che/Carnegie-Mellon Univ; BS/ChE/Univ of Pittsburgh. *Born:* Aug 1924 Rochester, Penna. U S Army Signal Corps 1944-46. Process engr with Elliott Co, Blaw-Knox Chem Plants Div. Proj engr Koppers Co E & C Div. Mgr of Chem Processing NUMEC. Founded engrg firm in 1962 & operated it until incorporation as FAI in 1968. Pres FAI, an engrg & cons firm since inception. Cons on dev of processes & sys for chem, met, coal conversion & nuclear applications. U S patents & publications. Reg in Penna, N Y, N J & Fla. Teaching CMU. Fellow AIChE. Hobbies include photography, riding, & painting. Married: Marcia; children: Bruce & Julie. *Society Aff:* AIChE, APCA, AACE, AMC.

Fry, Donald W
Home: Coveway Lodge, 25 Bowleaze Coveway, Weymouth Dorset, England
Position: Director, Retired. *Employer:* U K Atomic Energy Authority. *Education:* MSc/-/London Univ. *Born:* Nov 1910. Native of Weymouth, England. Res Engr G E C Labs 1932-36; Airborne Communications, Gov Service 1936-40. Radar Dev 1940-46. Joined U K A E A 1946 leading a res team investigating methods for producing high energy particles. Appointed Deputy Dir A E R E Harwell 1958, Dir Atomic Energy Estab, Winfrith Dorset 1959. Ret 1973. Dudell Medal Inst of Physics & Physical Soc 1950. Fellow IEEE 1960. Decorated Commander of order of the British Empire 1970. Elected Fellow of IEE (UK) 1954. Hon Freeman of Borough of Weymouth & Melcombe Regis 1959. Enjoy classical music & travelling. *Society Aff:* IEE, IEEE, AIP.

Fry, Glenn A
Business: College of Optometry, The Ohio State University, Columbus, OH 43210
Position: Prof Emeritus of Optometry. *Employer:* The Ohio State Univ. *Education:* PhD/Psychology/Duke Univ; MA/Psychology/Duke Univ; AB/Psychology/Davidson Coll. *Born:* Sep 10, 1908 Wellford S C. Licensed Optometrist since 1937. Taught optics, optometry & illumination at Ohio State Univ since 1935. Present rank is Prof Emeritus. Dir School of Optometry (1937-66). Co-dir Inst for Res in Vision (1949-56). Mbr IERI Tech Advisory Ctte on Light & Vision. IES Ctte on Quantity & Quality of Illumination. IES Fellow & Gold Medalist. Mbr of the U S Natl Ctte of the Internatl Comm on Illumination *Society Aff:* OSA, IESNA, AAO, AOA.

Fry, James P
Business: Computer Information Systems, Grad School Bus Admin, Ann Arbor, MI 48109-1234
Position: Professor, Res Scientist. *Employer:* Univ of MI, Grad School Business *Education:* BS/Engr/Univ of MI; MS/Engr/Univ of MI; MS/Ind Engr/Univ of MI *Born:* 5/2/39. Listed in Who's Who in Computer Educ & Res & Who's Who in the Midwest & Technology Today has been active in the following orgs: Assn for Computing Machinery 1965- (Natl Chmn SIGMOD - Spec Int Grp/Mgmt of Data, Mbr of SIGBDP Spec Int Grp/Bus Data Processing, & SIGPLAN - Spec Int Grp/ Programming Languages; CODASYL - Conf on Data Sys Lang 1967- , mbr of the Sys Ctte, Chmn of the Stored-Data Definition & Translation Task Grp (SDDTTG) 1969-74; ASCE Authored over 30 prof articles & published one book "Design of Database Structures–. *Society Aff:* ACM, CODASYL

Fry, Thornton C
Home: Del Mesa Carmel 46, Carmel, CA 93921
Position: Retired. *Education:* AB/Arts/Findlay College; MA/Math/WI; PhD/Math, Physics & Astronomy/WI; DSc/Hon/Findlay College. *Born:* Jan 7, 1892. Bell Tel Labs 1916-56: organized & until 1944 directed Math Res Dept; Dir Switching Res & Engrg 1944-49; Asst to Pres 1949-56. By courtesy of BTL: Visiting Lectr MIT 1926, Princeton 1929; Dept Ch & Deputy Div Ch Office of Sci R&D 1939-44 for which awarded Pres Certificate of Merit. V Pres Sperry Rand Univac 1957-60. Cons on R&D Mgmt 1956-74 ITT, NCAR, Boeing Res Labs, GranvillePhillips Co & others. Mbr, Fellow and/or officer of prof soc's incl Fellow IEEE & APS. Author papers & books, especially *Probability & its Engrg Uses*, 1st publ 1928, in print for 50 yrs. Mathematical Assn of Amer Award for Distinguished Service to Mathematics, 1982. *Society Aff:* AAAS, AAS, ASEE, AMS, APS, LMS, ASEE, MAA.

Frye, Alva L
Business: 703 Murfreesboro Rd, Nashville, TN 37201
Position: Vice Pres Research & Development. *Employer:* Aladdin Industries Inc. *Education:* BS/Che/IA State College. *Born:* July 19, 1922 Gray Okla. Married 1943, 2 children. Control Chemist Shell Chem Co Texas 1943-44; Process Engr Natl Synthetic Rubber Corp Ky 1944-47; Sect Leader Minn Mining & Mfg Co 1947-57, Mgr Cent Res Pilot Plant 1957-62, Tech Dir Paper Prod Div 1962-68; V Pres Res & Commercial Dev Inmont Corp 1969-70; V Pres R&D Aladdin Indus 1970- . Mbr co representative Ind Research Inst Amer Chem Soc, AIChE, Amer Mgmt Assn. R&D Council i.e. still a member 1970- synthetic polymers, process dev, flouro-chem & pilot plant, specialty papers & new products. *Society Aff:* ACS, AIChE, AMA, IRI.

Frye, John C
Home: 4470 Chippewa Dr, Boulder, CO 80303
Position: Executive Director. *Employer:* Geological Society of America Inc. *Education:* PhD/Geology/St Univ IA; MS/Geology/ST Univ IA; AB/Geology/ Marietta College; ScD/Marietta College. *Born:* July 25, 1912 Marietta Ohio. Employed by US Geol Survey 1938-42; Univ of Kansas Asst Prof to Prof, Asst Dir to Exec Dir, & State Geologist, State Geol Survey 1942-54; Ch, Ill St Geol Survey & Prof Geol Univ of Ill 1954-74; Exec Dir Geol Soc of Amer 1974- . Mbr NAE; Fellow Geol Soc of Amer; Amer Geophys Union; AAAS; Hon Mbr SEPM; Assoc Amer State Geologists Mbr AAPG; SME of AIME; Soc of Econ Geol. Pres Amer

Frye, John C (Continued)
Geol Inst 1966. Mbr & Chmn various cttes of NAS-NAE-NRC 1955- . Recipient US Dept of Interior Public Service Award 1972. Mbr Cosmos Club Wash DC. *Society Aff:* NAE, GSA, SME, AIME, AAPG, SEPM, SEG, AGU.

Frye, John H Jr
Home: 1520 High Forest Dr N, Tuscaloosa, AL 35406
Position: Prof Met Engr *Employer:* Univ of AL *Education:* Doctor Phil/Phys Sci/ Oxford Univ, England; MS/Met/Lehigh Univ; AB/English/Howard Coll *Born:* 10/01/08 Asst and Assoc Prof of Met Engrg, Lehigh Univ; civilian employee, office Sci R & D; and awarded Lincoln Gold Medal of Amer Welding Soc in Period 1940-1944. Engr, Bethlehem Steel Co 1944-1948. Dir Metals & Ceramics Div, Oak Ridge Natl Lab (TN) 1948-1973. Between 1948 and 1973: served as dir, Bank of Oak Ridge and tech adviser at Second Intl Conference on Peaceful Uses of Atomic Energy, Geneva, 1958; elected as fellow of Amer Soc Advancement Sci, of Amer Soc Metals, and of Met Soc of AIME. Prof, Met Engrg, Univ of AL, 1973-. *Society Aff:* AIME, ASM, AAAS

Fthenakis, Emanuel
Business: Fairchild Industries, P.O. Box 10803, Chantilly, VA 22021
Position: Chrmn of the Bd & Chief Exec Officer *Employer:* Fairchild Industries *Education:* MS/EE/Columbia Univ; Diploma/ME/Tech Univ of Athens; Diploma/ EE/Tech Univ of Athens *Born:* 1/30/28 Native of Salonica, Greece; came to US in 1952, naturalized in 1956. Early career with Bell Telephone Labs as member of technical staff. Subsequently served as Engrg Mgr, Space and Missile Div of Gen Elec Co; then as VP and Gen Mgr of the Space Div of Philco-Ford Corp. At Fairchild Industries, have line responsibility for communications and electronics group consisting of several divisions and/or subsidiaries. Direct responsibility for Fairchild Space Co; Fairchild Communications & Electronics Co; American Satellite Co (partnership with Continental Telephone Corp); and SPACECOM (partnership with Continental Telephone). Adjunct Prof, Univ of MD. *Society Aff:* AFCEA, IEEE, AIAA.

Fu, King-sun
Business: Sch of Elec Engrg, West Lafayette, IN 47907
Position: Goss Distinguished Prof of Engrg. *Employer:* Purdue University. *Education:* PhD/EE/Univ of IL; MASc/EE/Univ of Toronto; BS/EE/Natl Taiwan Univ. *Born:* Oct 2, 1930. Guggenheim Fellow 1971-72. Fellow IEEE 1971; Mbr NAE 1976. Herbert N McCoy Award for Contributions to Sci Purdue Univ 1976. Teaching at Purdue 1960-. ASEE Senior Research Award, 1981, IEEE Education Medal, 1982. AFIPS Harry Goode Award, 1982, IEEE Centennial Medal, 1984. Sigma Xi Faculty Res Award-Purdue Univ, 1982. *Society Aff:* IEEE, ACM, ASEE.

Fu, Li-Sheng W
Business: 155 W. Woodruff Ave, Columbus, OH 43210
Position: Assoc Prof *Employer:* OH State Univ *Education:* PhD/Theo Appl Mech/ Northwestern Univ; MSc/Micromech/Northwestern Univ; BSc/CE/Natl Taiwan Univ. *Born:* 9/17/39 Native of Hu-pei Province, China. Have taught at OH State Univ since Oct, 1967. Worked during summers at Naval Res Lab (1969), GE Res & Dev Ctr (1974), and Air Force Mtls Lab (1977, 1978). Specialties: solid mechanics and structural mechanics. Patent: Combined load testing device (1971). Award: US Navy Dept (1970, 1972). Enjoys Basketball, table tennis and golf. Visiting Prof at Hua Zhong Univ, PROC (Summer, 1981), and at Natl Taiwan Univ, ROC (Autumn 1983 - Winter, 1984). *Society Aff:* ASME, ASEE, AAM, AIAA.

Fuchs, Henry O
Business: Mech Engrg Dept, Stanford, CA 94305
Position: Professor Emeritus. *Employer:* Stanford University. *Education:* Dr Ing/ ME/Tech Univ Karlsruhe; Dipl Ing/ME/Tech Univ Karlsruhe; Bach/Philosophy/ Univ of Strasbourg. *Born:* 1907. Baccalaureat Strasbourg France; Dipl Ing & Dr Ing Karlsruhe Germany. Design Engr GM 1933-45; Ch Res Engr Preco Inc 1945-54; Pres Metal Improvement Co 1954-61; Prof 1961-76. Author of numerous papers on metal fatigue & self stresses; on use of case histories in engrg educ. Carlson Award ASEE for innovation in educ 1974. AMSE Da Vinci Medal for innovative Design 1980. ASME Machine Design Award 1981. *Society Aff:* ASME, ASTM, SAE, SEM.

Fuchsluger, John H
Home: 906 Breezewick Cir, Towson, MD 21204
Position: Manager, R&D *Employer:* Kaydon Ring & Seal, Inc. *Education:* BS/Phys/Loyola College. *Born:* 8/1/27. Married, four children. High sch Baltimore Poly Inst, Loyola College BS Phys maj, grad studies in met at Univ of MD. Elected to Alpha Sigma Mu, honorary met society. Served on exec committee of Baltimore Chapter ASM for past 30 yrs, chrmn twice. Served on ASM Natl Nominating Committee. Entire career in Res & Dev. Co-author of a number of ASME & ASLE technical papers on wear resistance of metals & plastics. Interests include physical fitness, travel, & christian fellowship. *Society Aff:* ASM, ASTM, WSE.

Fucik, E Montford
Home: 57 South Deere Park Dr, Highland Park, IL 60035 *Employer:* Retired. *Education:* MS/Engrg/Harvard; BS/Civ Engrg/Princeton. *Born:* Jan 1914. MS Engrg Harvard Univ 1937; BSCE Princeton 1935. Phi Beta Kappa. With Harza Engrg Co Chicago 1938-40; Foundation Engr Panama Canal 1940- 42; Lt Cmdr USNR 1943-45; with Harza 1945-85, Pres 1963-74, Chmn of Bd 1969-79. Chrmn Emeritus 1979; Reg Prof Engr: Ill, Ind, 4 other states. Fellow ASCE, Natl Dir 1968-71, Chmn Ctte on Convention Policy & Practice, Mbr Subctte on Code of Ethics; Thomas Fitch Rowland Prize 1953; Western Soc of Engrs, Pres 1972-73; Soc of Amer Military Engrs, Pres Chicago Post 1975-76; Mbr AAAS, NSPE, ACEC, Natl Acad of Engrg. Hon. Mbr Western Soc. of Engrs - Rickey Medal ASCE 1979 Goethals medal, Soc American Milt. Engr 1980. *Society Aff:* ASCE, WSE, SAME, NSPE.

Fuentes-Sanchez, Gustavo E
Business: 216 Las Marias Ave, Hyde Park, Rio Piedras, Puerto Rico 00927
Position: Principal *Employer:* Gustavo E. Fuentes-Sanchez, PE Conslt *Education:* BS/CE/Univ of Puerto Rico-Mayaquez *Born:* 07/01/31 Fallout Shelter Analyst 1 TT-955-63 Grad Sch of Planning Univ of TN; Shelter Design and Analysis Office of Civil Defense-Staff Coll Battle Creek, MI. After a couple yrs in government service joined one of the most prestigious A/E firms in Puerto Rico. In short became Chief Engr with over 40 people (architects engrs, designers, technicians, etc). Under direct/indirect supervision, in charge of production and technical aspects of all A/E/P/ projs. In 1967 formed the firm Lebron, Sanfiorenzo & Fuentes which became one of the most professionally recognized A/E/P firm in Puerto Rico and abroad. This Firm grew to over 80 people. Was in charge of production and in later yrs handled coordination of projs and also did marketing and sales. In 1981 separated from this firm and established a consltg office handling A/E/P/ projs. SAME Civil/Military liasion, active PTA mbr, enjoy boating (cruising) and stamp collecting. *Society Aff:* NSPE, SAME

Fuerstenau, Douglas W
Business: Univ of California, Hearst Mining Bldg, Berkeley, CA 94720
Position: P. Malozemoff Professor of Mineral Engineering. *Employer:* University of California. *Education:* BS/Metallurgical Eng/SD Sch of Mines and Tech; MS/ Mineral Engr/Montana Sch of Mines; ScD/Mineral Engrg/MIT *Born:* Dec 6, 1928 Hazel S D. Asst Prof Dept of Metallurgy MIT 1953-56; Sect Leader Metals Res Labs Union Carbide Corp Niagara Falls N Y 1956-58; Mgr Mineral Engrg Lab Kaiser Aluminum & Chem Corp Permanente Calif 1958-59; Assoc Prof 1959- 62, Prof 1962- , Chmn Dept of Matls Sci & Engrg 1970-78, Univ of Calif Berkeley. Awards: Robert Lansing Hardy Gold Medal AIM 1957, Rossiter W Raymond Award AIME 1961, Robert H Richards Award AIME 1975, Antoine M Gaudin Award AIME 1978, Distinguished Mbr Soc of Mining Engrs 1975. Mbr NAE 1976; Chmn Minerals Beneficiation Div AIME 1966, Mbr Board of Directors Soc of Mining Engrs AIME 1967-70. Reg Met Engr: CA; Mbr, Bd of Dirs, Homestake Mining Co Co-ed. Min Indus Educ Awd, AIME, 1983 Alexander Von Humboldt

Fuerstenau, Douglas W (Continued)
Prize, W. Germany, 1983. Chrmn, Faculty of the Coll of Engrg, Univ. of CA. Berkeley, 1982-84. *Society Aff:* AIME, AIChE, ACS, IMM

Fuerstenau, Maurice C
Business: SD Sch of Mines & Tech, Rapid City, SD 57701
Position: Distinguished Professor Metallurgical Engrg *Employer:* S D Sch of Mines & Technology. *Education:* ScD/Metallurgy/MIT; MS/Metallurgy/MIT; BS/Geol Engr/SD Sch of Mines *Born:* June 6, 1933 Rapid City S D. ScD & MS MIT; BS S D Sch of Mines & Tech. Res Assoc MIT; Res Engr Beryllium Corporation; Res Engr N M Bureau of Mines; Prof Colo Sch of Mines; Prof Univ of Utah; Prof & Head Dept of Met Engrg S D Sch of Mines & Tech 1970- . Chmn Mineral Processing Div, Soc of Mining Engrs AIME 1975; Mbr Board of Dir Soc of Mining Engrs AIME 1974-76 and 1980-83; Mbr Board of Directors AM. Inst. of Mining, Metallurgical & Petrolleum Engrs 1980-83; President, Society of Mining engrs AIME 1982;. Editor 'Flotation'. A M Gaudin Memorial Volume. Numerous tech articles for publ. Mbr: Amer Inst of Mining & Met Engrs, Sigma Xi. Editor, "Gold, Silver, Uranium and Coal Mining, Geology, Extraction and the Environ–, Co-author, "Chem of Flotation–. *Society Aff:* AIME, ΣΞ

Fugate, Charles R
Home: 712 Ihler Rd, Jefferson City, MO 65101
Position: Field Mtls Engrg Supervisor. *Employer:* MO State Hgwy Dept. *Education:* BS/Civ Engg/Univ of KS. *Born:* 8/13/35. Native of Pomona, MO. High school & pre-engg at Rockhurst, Kansas City, MO. Employed with State Hgwy Dept since grad serving in various capacities in Construction & Mtls Divs including Dist Mtls Engr in Kansas City, MO. Responsible for asphalt & portland cement concrete mix designs, preparation or supervision of preparation of specifications for design & control of asphaltic concrete, new mtl & new equip, & functional responsibility over 10 Dist Mtls Engrs. MSPE North Chapter Young Engr of the Yr - 1970. Past Pres Jefferson City Chapter MSPE & Mid-MO Sec ASCE. Active in church, fraternal & civic organizations. MSPE Treasurer 85-86, Secretary 86-87, Vice President effective July 1, 1987. Fellow, ASCE. Reg PE MO & KS. *Society Aff:* ASCE, NSPE, AAPT.

Fuhriman, Dean K
1457 Cherry Lane, Provo, UT 84604
Position: Professor Emeritus *Employer:* Brigham Young Univ *Education:* PhD/Hydrology-Civil Eng'g/Univ of WI; MS/Irrigation-Civil Eng'g/UT State Univ; BS/CE/UT State Univ *Born:* 6/6/18 in Ridgedale, Onieda County, ID. Married Alta Christensen. Five children, Susan, Julie, Mark, Christine and Jana. Prof of Engrg 30 years at UT State, CO State, & Brigham Young Universities. Engr/Consultant on water resource development in US, Puerto Rico, and countries of the Middle East. Research on lake evaporation, groundwater evaluation, flood control, irrigation methods and water use - domestic, agricultural and industrial. For total of 15 years, Pres of 3 consulting engrg firms. Author of many technical publications. Registered Engr in UT and NV. *Society Aff:* ASCE, IWRA, ASEE, ICID

Fuhrman, Brian L
Business: W 705 First Ave, P.O. Box TAF-C4, Spokane, WA 99220
Position: Sr Engr-Appraiser *Employer:* Federal Land Bank of Spokane *Education:* BS/Agric Engr/MT State Univ *Born:* 7/12/53 Native of Northeastern MT. Started with the Federal Land Bank as a loan officer in Southeastern ID. Moved to Spokane as an Engr Spring of 1978. Cover the states of MT, ID, WA, OR and AK on matters dealing with water law, ground water hydrology and irrigation system design. Chrmn elect for the Inland Empire Section of ASAE. *Society Aff:* ASAE

Fuhrman, Robert A
Business: Pres and Chief Operating Officer, Lockheed Corporation, 4500 Park Granada Blvd, Calabasas, CA 91399
Position: Pres and Chief Operating Officer. *Employer:* Lockheed Corp *Education:* MSE/Fluid Mech & Dynamics/Univ of MD 1952; BS/Aero Engrg/Univ of MI 1945. *Born:* 1925 Detroit. MSE Univ of Md; BSAE Univ of Mich. Flight Test Engr USN Air Test Center 1946-53; Ryan Aeronautical Co 1953-58; Lockheed Corp 1958, Pres Lockheed-Ga Co 1970-71, Pres Lockheed-Calif Co 1971-73, Exec V Pres Lockheed Missiles & Space Co 1973-76, Pres LMSC 1976-83, Chrmn 1979-. Mbr NAE; Former V Chrmn Aero & Space Engg Bd, Hon Fellow AIAA. Dir Amer Defense Preparedness Assn. Awards: Secy Of Navy Certificate of Commendation for Polaris Task Group Service, John J Montgomery Award NSAP for Polaris role, Soc of Manufac Engrs' Indus Tech Mgmt Award, Natl Mgmt Assn Silver Knight of Mgmt Award. Von Karman lecture AIAA 1978, Eminent Engr Award, Tau Beta Pi, Univ. of CA at Berkeley 1983.. *Society Aff:* NAE, AIAA, AAS, SME

Fuhs, Allen E
Home: 25932 Carmel Knolls, Carmel, CA 93923
Position: Distinguished Prof. *Employer:* Naval Postgrad School. *Education:* PhD/Mech Engr/CA Inst of Tech; MS/Mech Engr/CA Inst of Tech; BS/Mech Engr/ Univ of NM. *Born:* 8/11/27. Naval Postgrad School since 1970 Chief Scientist, AFAPL, WPAFB, 1968-70. Professor, NPS, 1966-68. Staff Scientist, Aero Corp, 1960-1966. Mbr of Tech Staff, TRW Systems, 1959-60. Asst Prof, Northwestern Univ, 1958-59. Lecturer, CA Inst of Tech, 1957-58. Officer, USN, 1951-1954. Fellow, VP, and President, AIAA. Fellow, JILA, 1964-65. Consultant to govt and industry. Service on fed govt bds. USN Rep to AGARD, NATO, 1969-76. Author of two books; editor of four books. Editor-in-Chief, Journal of Aircraft, 1974-1979. 142 technical papers. Areas of interest: lasers, space systems, and gas dynamics. *Society Aff:* AIAA, ASME, ASNE, SNAME, OSA, SAE

Fujisawa, Toshio
Business: Faculty of Engg Sci, Toyonaka Japan 560
Position: Prof. *Employer:* Osaka Univ. *Education:* PhD/EE/Osaka Univ; BS/EE/ Osaka Univ. *Born:* 12/27/28. in Kobe, Japan. Teaching, res and/or visiting positions at Kinki Univ, Univ of Osaka Prefecture, Osaka Univ, Conductron Corp, the Univ of MI, & Univ of CA, Berkeley. Prof of Osaka Univ since 1965. Awarded 1974 Guillemin-Cauer Prize Award of IEEE Circuits & Systems Soc. IEEE Fellow. Having studied elec circuit theory, particularly the analysis of large-scale networks. *Society Aff:* IEEE.

Fukunaga, Keinosuke
Home: 542 Avon St, W. Lafayette, IN 47906
Position: Prof of EE. *Employer:* Purdue Univ. *Education:* PhD/EE/Kyoto Univ; MS/ EE/Univ of PA; BS/EE/Kyoto Univ. *Born:* 7/23/30. After 13 yrs in Central Res Lab & Comp Div of Mitsubishi Elec Co in Japan, joined Purdue Univ in 1966, & currently Prof of EE, working in the area of pattern recognition & pattern processing. Fellow of IEEE. *Society Aff:* IEEE.

Fulford, P James
Business: Sch of Nuclear Engg, W Lafayette, IN 47907
Position: Assoc Prof. *Employer:* Purdue Univ. *Education:* PhD/Nuclear Engg/Purdue Univ; MSc/Reactor Physics/Birmingham Univ; Grad Cert/Bus Admin/London Sch of Economics; BSc/ME/Univ of Manitoba. *Born:* 10/18/35. Winnipeg Manitoba 1935. Educated in Canada, UK, and USA. Worked as Engr for Atomic Energy of Canada, Ltd, and Dilworth, Secord, Meagher Assoc. Consulted in reactor economics and design for the power industry. Taught at Univ of Manitoba and Purdue Univ. Univ Gold Medallist at Manitoba, Athlune Fellow (57-59) and ASEE Engg Fellow. *Society Aff:* ANS, ΣΞ, AAAS.

Fuller, Dudley D
Mink Hollow Road, Lake Hill, NY 12448-0219
Position: Prof. *Employer:* Columbia Univ. *Education:* MS/Mech Engg/Columbia Univ; BME/Mech Engg/City College of NY. *Born:* 2/8/13. Concerned with teaching, res and consulting in mech engg with primary emphasis on bearing design and lubrication, friction, wear and tribology. About 60 publications in the field including

Fuller, Dudley D (Continued)
book, "Theory and Practice of Lubrication for Engrs–, now available in four languages. Have received a number of honors and awards including the gold medal from the Intl Tribology Council (London) 1978. *Society Aff:* SAE, ASME, ASLE, I MECH E.

Fuller, Robert H
Business: 2901 No High St, Columbus, OH 43202
Position: Pres *Employer:* Robert H. Fuller & Assoc, Inc *Education:* BS/ME/OH State Univ *Born:* 7/1/41 Quality Control Engr, Eastman Kodak 1965 to 1968. VP, W.E. Monks and Co, Consulting Engrs, 1968 to 1974. Technical Dir, OH State Univ Energy Conservation Div, 1974 to 1979. Pres, Robert H. Fuller & Assocs, Inc, 1979 to present. Testified on the Natl Energy Plan before the US Senate Committee on Interior and Insular Affairs (1976), and US House of Representatives Sub-Committee on Energy & Power (1977). Publications in ASHRAE Transactions & Journal; ASME Technical Digest, CEFP Journal, Heating/Piping/Air Conditioning, Energy and Bldgs, and Contributor to numerous training manuals on Energy & Mech Sys publ by TPC Training Systems. Panel leader for Amer Council for an Energy Efficient Economy 1934 summer study. Chrmn of ASHRAE'S Special Proj Ctte for revision of standard 100.5 energy conservation in existing bldgs-institutional. Distinguished Service Award, Council of Educational Facilities Planners, Intl, 1978 & 1986. *Society Aff:* ASME, ASHRAE, NSPE, CEFP/I.

Fullerton, John H
Home: 24 Cortland Ln, Lynnfield, MA 01940
Position: Retired *Education:* BE/CE/Yale Univ. *Born:* 12/5/22. Following grad from college in 1944 spent two yrs in US Navy Civ Engr Corps including one yr in Pacific Area. Presently Lt Commander retired. Have spent entire profl career in Boston with Jackson & Moreland and United Engrs & Constructors following their merger. Was pres of MA Sec ASCE in 1961. Was chrmn of planning bd, town of Lynnfield, MA, 1972. *Society Aff:* ASCE, NSPE, SAME, TBΠ, ΣΞ.

Fullman, Robert L
Business: G E Res & Dev Center, Schenectady, NY 12301
Position: Metallurgist. *Employer:* General Electric Co. *Education:* D Eng/Metallurgical Engg/Yale; B Eng/Metallurgical Engg/Yale. *Born:* Sept 1922 Sewickley Pa. USN 1942-45. Instr New Haven YMCA Jr Coll 1947- 48; GE E Co Res Assoc 1948-55, Mgr Materials & Processes Studies 1955-59, Mgr Metal Studies 1960-63, Mgr Fuel Cell Studies 1964-65, Mgr Properties Branch 1965-68; Mgr Planning & Resources Materials Sci & Engrg 1969-72, Metallurgist 1972- . RPI Visiting Lectr 1951-56, Adjunct Prof 1956-65. Secy-Treas Board of Governors Acta Met 1965- . Fellow ASM; Mbr AIME. 24 tech papers; 8 patents. Res fields: interfacial energies in solids, crystal growth, origin of microstructures, recrystallization & grain growth, relationships between microstructure & properties, quantitative metallography. *Society Aff:* AIME, ASM.

Fulton, Robert J
Business: 701 E Third St, Bethlehem, PA 18015
Position: Sr Mgr *Employer:* Bethlehem Steel Corp. *Education:* BS/Chemistry/Muhlenberg College; MBA/Bus/Lehigh Univ. *Born:* 2/15/43. in NJ. Began working after college in 1965 for Bethlehem Plant-Bethlehem Steel Corp. Have worked in a variety of capacities within the Met Div of the Bethlehem Plant. Current assignment as Sr Mgr involves Mgmt of Plants Manufacturing Operations. Past chrmn of Lehigh Valley chapter of ASM. Past Mbr natl long range planning ctte. Past mbr of AM Iron and Steel Inst, Steel Fellows Prog. *Society Aff:* ASM.

Fulton, Robert P
Business: 1235 Jefferson Davis Highway, Suite 700, Arlington, VA 22202
Position: VP, Hd of Washington Div. *Employer:* Gibbs & Cox, Inc. *Education:* BSME/Mech Engg/Polytech Inst of Brooklyn. *Born:* 4/18/28. in Queens County, NY. US Army 1946-47 on occupation duty in Korea. From 1951-71 with Gibbs & Cox, Inc naval architects & marine engrs. From 1971-74 Pres of Modular Sys Inc, a subsidiary of Warren Pumps Inc. Rejoined Gibbs & Cox Inc 1974 & currently serving as VP & Div Hd of Wash DC Div. Chmn 1975-76 NY Metropolitan Sec Soc of Naval Architects & Marine Engrs. Chmn annual Dinner Dance Ctte SNAME. Reg P E in NY, NJ & VA. Mbr Bd of Education, N Shore Sch District No 1, Glen Hd, NY 1964-69. *Society Aff:* ASME, SNAME, ASNE.

Fung, Yuan-Cheng B
Business: 5028 BSB M-005 Ames, University of California, La Jolla, CA 92093
Position: Prof of Bio-Engrg & Appl Mech. *Employer:* Univ of Calif San Diego. *Education:* PhD/Aero & Math/CA Inst of Tech *Born:* Sep 15, 1919. Res Asst to Prof Caltech 1946-66. Prof at UCSD 1966- . Author over 240 original papers & books 'Theory of Aeroelasticity' John Wiley Inc 1955, Revised Dover Publications 1969; 'Foundations of Solid Mechs' Prentice-Hall Inc 1965; 'A First Course in Continuum Mechanics' Prentice Hall 1976, 2nd ed 1979; Biomechans: 'Mechanical Properties of Living Tissues', Springer-Verlag New York, 1981, "Biodynamics, Circulation–, Springer-Verlag, NY, 1984, Ed of 'Biomechanics: Its Foundations & Objectives' 1972 & 'Thin Shell Structures: Theory, Experiment & Design' 1974. Guggenheim Fellow 1958. Achievement Award Chinese Inst of Engrs 1965, 1968. Landis Award, Microcirculatory Soc 1975. Von Karman Medal ASCE 1976, Lissner Award, AMSE, 1978. Member U.S. National Academy of Engineering. *Society Aff:* ASME, APS, BMES, IBS

Funk, William U
Home: Twin Oaks E-RD 3, Cambridge, OH 43725
Position: Pres. *Employer:* Packaging Materials, Inc. *Education:* BS/Chem Engg/Case Inst of Tech. *Born:* Jun 1926. Bd mbr ASBE 1959. Formed Packaging Mtl's Inc 1970 for mfg of flex pkg. *Society Aff:* AIChE, SPE.

Fuquay, Garth A
Business: 2501 S El Camino Real, Apt 310D, San Clemente, CA 92672
Position: Retired (Consulting self-emp) *Education:* SMCE/Soil Mech & Fdn/Harvard Univ; BSCE/Structures/TX Tech Univ. *Born:* 4/5/17. in TX & grad from TX Tech in Civ Engg. WWII, Army Corps of Engrs, 1943- 1946, worked as a civilian civ engr with Corps of Engrs 1941-1943, 1946-1956, 1963-1978. Was Chief, Fdns & Mtls Branch 1964-1970, Pittsburgh Dist, Chief Design Branch 1970-1972, Chief Engr Div Los Angeles Dist 1972-1976, Chief Engg Div, N Central Div 1976-1978. Worked for Tippets-Abbett-McCarthy-Stratton, Chief Soils Engr, in Thailand, Quebec & Colombia 1956-1963. Retired to consulting in 1978. Married, with three grown children. Enjoy travel, golf and fishing. *Society Aff:* ASCE, SAME, USCOLD, TBΠ.

Furfari, Frank A
Home: 117 Washington Rd, Pittsburgh, PA 15221
Position: Service Mgr, Ind Equp & Services Group. *Employer:* Westinghouse Elec Corp. *Education:* BSEE/WV Univ *Born:* 2/14/15. July 1938 joined Koppers Coal Co, elec inspector, WV mining operations. 1940 to retirement Nov 1981 Westinghouse Elec Corp. Early career at Westinghouse includes 2 yrs Rotational Test Course at E Pittsburgh Works & field service eng & applicaton engg assignments, Cleveland & Toledo, OH. 1951-71, mgt positions in product service operations, dist & hdquarters locations. In last position since 1972. IEEE Sec Chrmn, Toledo 1949-50, Pittsburgh 1968-69. 1976 Pres IEEE-Industry Applications Soc. 1979 recipient, IAS Outstanding Achievement Award. IEEE Bd of Dirs 1979-80. Profl mbr Eta Kappa Nu. Reg Engr, States of OH and PA. Married: three sons Anthony D, Mark V, Ross V. *Society Aff:* IEEE, IAS, SSIT.

Furgason, Robert R
Business: Administration 202, University of Nebraska-Lincoln, Lincoln, NE 68588-0420
Position: Vice Chancellor, Acad Affairs *Employer:* University of Nebraska *Education:* BS/ChE/Univ of ID; MS/ChE/Univ of ID; /-/ChE/Northwestern. *Born:* Aug 1935 Spokane Wash. P E Idaho. Teaching Experience: 1961-1984 Instr through Prof in Chem Engrg, Dept Chmn 1965-74 at Univ of Idaho; Dean College & Engr 1974-78 at Univ of ID, 1973-74 NSF prog at Escuela Politecnica Nacional, Quito, Ecuador to teach courses in computer applications, modeling & economic analysis, to help establish computer ctr. Indus Experience: 1969-70 B F Goodrich Chem Co - dev cons full time in computer simulation, polymer processing & economic analysis. Other Experience: Boeing, Phillips Petroleum & Martin Marietta. Outstanding Teacher Award U of Idaho; Outstanding Young Engr ISPE; offices in AIChE, ABET, ASEE & ISPE; private pilot; Vice Pres Academic Affairs and Res, Univ of Idaho 1978-1984; Vice Chancellor for Academic Affairs University of Nebraska 1984-; Prof of Chem Engrg Univ of Nebraska 1984-. *Society Aff:* AIChE, ACS, NSPE, ISPE.

Furland, Loren P
Business: Suite 200, 2600 Century Pkwy, NE, Atlanta, GA 30345
Position: Principal Environmental Engr *Employer:* Stanley Consultants, Inc *Education:* MS/Environ Engrg/Univ of IA; BS/CE/IA State Univ *Born:* 11/22/45 Registered PE, IA, IL and GA. Employed since 1968 by Stanley Consultants, Inc, in Muscatine, IA and Atlanta, GA. Professional experience includes design management and technical responsible charge for major wastewater treatment; wastewater collection, water treatment and water transmission facilities for municipal, industrial and intl clients. Co-author of technical papers presented at Purdue Univ Industrial Waste Conference and Water Pollution Control Federation natl conference. *Society Aff:* WPCF, NSPE, ASCE

Furlong, Richard W
Business: Dept of Civil Engg, Ernest Cockrell Hall 4.6, Austin, TX 78712
Position: Prof. *Employer:* Univ of TX. *Education:* BSCE/Civil Engr/Southern Methodist; MSCE/Civil Engr/Washington Univ; PhD/Structures/Univ of TX at Austin *Born:* 3/30/29. Public school education Norwalk, OH. Coop engr - Southern Methodist included experience as Inspector TX Hgwy Dept, laborer and draftsman Austin Steel Inc, Stress Analyst McDonnell Aircraft one yr, steel detailer and checker 2 yrs, structural engr, F Ray Martin, Inc 3 yrs. Univ of TX at Austin since 1959. Consultant part-time, industrial, sanitary, architectural structures. Res and teaching applications in steel and concrete, design aids, handbooks, composite construction. Dir American Concrete Inst 1977, 79-81. Exec Secretary TX Sec ASCE 1979-85. Donald J. Douglass Centennial Professorship 1983-1987, Univ. of Texas at Austin. Enjoy golf, pre-1900 music, travel. Married Helen Prince 1951, 2 children, 2 grandchildren. *Society Aff:* ASCE, NSPE, ACI, CRSI, SSRC, CSCE, ΣΞ, NEWCOMEN SOCIETY.

Furman, Thomas deS
Home: Route 3, Pickens, SC 29671
Position: Prof Emeritus CE Engrg & Environ Engrg Sci. *Employer:* Univ of Fla, Gainesville. *Education:* BS/Civil Engr/The Citadel; MSE/San Engrg/Univ of FL *Born:* Nov 1915. S Carolina Hwy Dept 1936-41. Harland Bartholomew & Assocs 1941-42. US Navy (CEC) 1943-46. Quattlebaum Engrg Co 1946-51. Fac of Univ of Fla 1951-80. Retired from Univ of FL Faculty Jun 30, 1980. Design cons Black, Crow, Eidsness, Cons Engrs 1954-70. Partner & Dir of Water & Air Res Inc 1970-80. Dir at large Water Pollution Control Federation 1964-67. Dir Water Pollution Control Federation 1974-77. Dir Natl Soc of Prof Engrs 1970-74. Pres Fla Pollution Control Assn 1963-64. Pres Fla Engrg Soc 1966-67. Hobbies-Genealogical Res Conslt. Camp Dresser & McKee, Consltg Engrs 1982- .. *Society Aff:* ASCE

Furr, Howard L
Business: Texas A&M Univ, College Station, TX 77843
Position: Prof of Civil Engrg. *Employer:* Texas A&M Univ. *Education:* PhD/Civil Engr/Univ of TX; MS/Civil Engr/TX A&M Univ; BS/Civil Engr/MI State Univ. *Born:* 1915 Pontotoc County Miss. US Army 1941-45; Struct design with Sverdrup & Parcel Inc 1949-52; summers with Humble Oil & Refining Co; ARO Inc, Applied Physics Lab. Prof Univ Miss 1952-59; Univ Missouri Rolla 1959-62; Texas A&M Univ 1962- . Res & publications in reinforced & prestressed concrete with Texas Transportation Inst 1962- . Southern Fellowships Award 1956; General Dynamics Award for Excellence in Engrg Teaching 1969; Coawardee T Y Lin Award ASCE 1974. *Society Aff:* ASCE, ACI, ASEE, TBΠ, AAA.

Furry, Ronald B
Business: Dept of Agri Engr, Riley-Robb Hall, Ithaca, NY 14853
Position: Prof. *Employer:* Cornell Univ. *Education:* PhD/Agr Engr/IA State Univ; MS/Agr Engr/Cornell Univ; BS/Agr Engr/Cornell Univ. *Born:* 10/22/31. Native of Niagara Falls, NY. Served as Ext Agri Engr responsible for dairy systems prog at Cornell Univ to 1962, then assumed full time res & teaching functions to present. Teach upper & lower div undergrad, & grad courses. Recipient of 1979 Outstanding Faculty Award by Cornell Student Branch ASAE. Past dir of Agri Engg grad prog at Cornell, Departmental Coordinator of Res, among other functions. Principal current res responsibility deals with long duration controlled atmosphere/temperature posharvest storage of agricultural mtls; gen res interest areas include structures & environemtns for plants & animals & application of similitude methodology. *Society Aff:* ASAE, ASEE, ΣΞ, ΓΣΔ, AZ.

Furter, William F
Business: Dean of Graduate Studies and Research, Royal Military College of Canada, Kingston, Ontario, Canada K7K 5L0
Position: Dean and Prof *Employer:* Royal Military Coll of Canada *Education:* PhD/ChE/Univ of Toronto; SM/ChE/MIT; BASc/ChE/Univ of Toronto; Dipl RMC/ChE/Royal Military Coll of Canada *Born:* 04/05/31 Born in North Bay, Ontario. Resident of Kingston, Ontario. Sr Technical Investigator, R&D Dept, Du Pont of Canada, 1958-60. Headed RMC's chem engrg degree program, 1960-80. Was Dean of the Canadian Forces Military Coll (RMC's continuing education div), 1980-84. Currently Dean of Graduate Studies and Research, and Prof of Chem Engrg. Many scientific res papers and book contributions published. Consltg engr to several major corps. Dir, CSChE. Fellow, CIC. Attended Natl Defence Coll of Canada, 1969-70. Various medals and awards, including Engrg Inst of Canada Prize. Listed in Who's Who in Amer. Reg PE, Ontario. Married; three daughters. *Society Aff:* CSChE, CIC, CNA, ANS, APEO, CAGS, OCGS

Fusfeld, Herbert I
Business: 329 Shimkin Hall, Washington Sq, New York, NY 10003
Position: Dir, Sci & Tech Policy Ctr. *Employer:* NY Univ. *Education:* PhD/-/Univ of PA; MA/-/Univ of PA; BS/-/Brooklyn College. *Born:* Feb 1921. Native of Brooklyn, NY. Joined Frankford Arsenal 1941, Head Physics & Math Div. Dir of Res, Amer Machine & Foundry Co 1953. Joined Kennecott Copper Corp as Dir of Res 1963- . Serves on Advisory Bd School of Materials Science Univ of Penn. Pres of IRI 1973. Is Chmn Nonferrous Subctte on Energy Conservation AMC. Mbr AIP Governing Bd 1968-71. On Advisory Bd Inst of Materials Res NBS. Mbr of Natl Materials Advisory Bd, US-USSR Joint Comm for Scientific & Tech Cooperation & Advisory Ctte on Transnational Enterprises, State Dept. Serves on Council of AAAS, 1979-80 Visiting Prof of Sci and Techn Policy, NY Univ. Dir of Ctr for Sci & Techn Policy.

Gabel, Richard
Business: POBox 16858, Boeing Center, PA 19142
Position: Manager, Rotor Tech. *Employer:* Boeing Vertol Co. *Born:* 1928. BS Mech Engrg & MS Aeronautical Engrg Drexel University, registered P E Pennsylvania. Tech Engr Boeing Co spec in struct & dynamics. Held pos of Ch Dynamics Engr, Ch of Structures. Currently Mgr of Rotor Tech respon for rotor tech support, helicopter dynamics & aeroelastic stability. Author of numerous helicopter dynamics papers. Chmn AHS Dynamics Ctte 1968, Chmn Philadelphia Chap 1975, Chmn AHS Rotor Sympos 1976.

Gabelman, Irving J
Business: 225 Dale Road, Rome, NY 13440
Position: President. *Employer:* Technical Associates. *Education:* PhD/EE/Syracuse Univ; MEE/EE/PIB; BEE/EE/CCNY; BA/Physics/Brooklyn College. *Born:* Nov 1918; Engr US Army Engr Office NYC 1941-45; electronics Engr Watson Labs USAF Eatontown NJ 1945-51; electronics scientist Rome Air Dev Ctr Griffiss AFB NY 1951-59, dir advanced studies 1959-69, ch plans 1969-71, ch scientist 197175; Pres Tech Assocs - Engrg Consultant 1975- . Chmn avionics panel AGARD NATO 1971-73, US Natl Coordinator 1971-76. Recip Air Force Exceptional Civilian Service award 1974. Fellow IEEE (chmn systems com nat group on computers 1966), Fellow AAAS. Editor Displays for Command and Control Centers 1969; Techniques for Data Handling in Tactical Systems 1969; Storage & Retrieval of Information " A User-Supplier Dialogue 1969; Data Handling Devices 1970. Patents: air traffic control device. Mbr Council SUNY Utica/Rome 1974-Advisor, Div Adviso. *Society Aff:* IEEE, AAAS.

Gada, Ram
Business: 1030 Soo Line Bldg, Minneapolis, MN 55402
Position: Pres *Employer:* Gada & Assoc, Inc *Education:* MS/ME/ND State Univ; BS/ME/Gujarat Univ, India *Born:* 11/16/40 Has designed HVAC systems for bldgs and managed construction projects since 1965. Has performed energy conservation studies since 1973 and has developed a comprehensive Energy Management Program for bldgs. This includes a preventive maintenance program, a cost/benefit study of conservation measures, and advising management of bldgs on conservation goal-setting and implementation. Has also developed a computerized monthly Energy Accounting System in use for over 150 bldgs throughout the US. Pres of ASHRAE-MN for 1981-82. *Society Aff:* ASHRAE

Gadd, Charles W
Home: 218 Pheasant Run, Hendersonville, NC 28739
Position: retired *Education:* B.S./Automotive engineering/Mass. Institute of Technology *Born:* 04/06/15 After graduation, MIT, in 1937, joined Research Labs., G.M. Corp., carrying on original R & D in strength of materials, vibration, and acoustics, as applied to automotive, aircraft engine, and military vehicle fields. Supervised in these fields. From 1955 to retirement (76) supervised auto safety and related biomechanics R & D, receiving in 1980 the US Dept. of Transporation Safety Award. Received ASME Certificate for 25 years distinguished service as reviewer for Applied Mechanics Reviews. 1985, made Life Fellow, ASME. Since retirement has specialized in Stringed Instrument construction and tone research. *Society Aff:* ASME

Gaddy, Frank L
Business: Suite 709, 421 Tenth St, Huntington, WV 25701
Position: Pres. *Employer:* Gaddy Engr Co. *Education:* MS/Mining Engr/VPI; BS/Mining Engr/VPI. *Born:* 6/3/22. Coal Mining Consultant. Mining Engr Warner Collieries Co, Mammoth, WV, 1948-53; health & safety Engr, US Bureau of Mines, Norton, VA 1953-59; Asst Engr coal properties Chesapeake & OH RR Co, Huntington, WV 1959-67, Pres Gaddy Engg Co 1968-. Mgr & chief engr various corps owning coal properties. Reg PE WV, KY, OH, VA. Mbr American Mining Congress (Chrmn Roof Control Committee 1959-69). Sigma Gamma Epsilon, Mason. *Society Aff:* NSPE, AIME.

Gaddy, James L
Home: 964 Arlington Ter, Fayetteville, AR 72701
Position: Prof & Head, Dept. of Chem. Eng. *Employer:* Univ of Arkansas *Education:* PhD/ChE/Univ of TN; MS/ChE/Univ of AR; BS/ChE/LA Tech Univ. *Born:* 8/16/32. Author of over 230 technical presentations & 110 publications in the areas of chem process optimization & biological production of fuels & chems from biomass & coal. Res awards of $2420M supported by USDA, DOE & private industry. Consultant to twelve cos, as well as United Nations on alternative energy supplies for remote areas. Invited as Visiting Prof, Swiss Federal Inst Tech 1977-78. Eleven yrs of ind experience with Ethyl Corp & AR La Gas. Reg PE AR. *Society Aff:* AIChE, ACS, AAAS, ASEE, ΤΒΠ, ΣΞ, ΩΧΕ, ΑΧΣ.

Gaddy, Oscar L
Home: 9 Carriage Way, Champaign, IL 61820
Position: Prof *Employer:* Dept of Elec Engrg U of Illinois. *Education:* PhD/EE/Univ of IL; MS/EE/Univ of KS; BS/EE/Univ of KS. *Born:* Jul 1932. Native of St Joseph Mo. Mbr of IEEE & APS. Elect Fellow IEEE 1974. Mbr of the faculty of the Dept of Electrical Engrg University of Illinois since 1962, engaged in teaching & res in electronics, electrooptics and quantum electronics. Presently part-time assoc head of dept. *Society Aff:* IEEE, HKN, ΤΒΠ, ΣΞ.

Gaden, Elmer L, Jr
Business: Dept of Chem Engg, Thornton Hall, Charlottesville, VA 22901
Position: Wills Johnson Prof. *Employer:* Univ of VA. *Education:* PhD/ChE/Columbia Univ; MS/ChE/Columbia Univ; BS/ChE/Columbia Univ. *Born:* 9/26/23. Dept of ChE & Appl Chemistry, Columbia Univ (1949-74); Chrmn 1960-69, 72- 74. Teaching responsibilities in chem engg, bioengg, & history. Dean, College of Engr, Math, & Bus Adm, Univ of VT (1975-79). Wills Johnson Prof of Chem Engg, Univ of VA, 1979-. Born, Brooklyn, NY (1923); educated at Columbia Univ. Mbr, Natl Acad of Engr; Food & Bioengg Award, AIChE; Chem Engr Lectureship Award, ASEE; Egleston Medal, Columbia Univ. *Society Aff:* AIChE, ACS, AAAS, ASEE

Gadomski, Richard T
Business: 2930 Airways Blvd, Memphis, TN 38116
Position: Pres. *Employer:* PSI-Process Systems Inc. *Education:* MSME/Dynamics & Controls/Univ of S CA; BS/Engg Chem/CBC. *Born:* 4/29/40. Chief exec officer of PSI-Process Systems Inc, a design engg & construction mgt firm specializing in the chem & food processing industry. Licensed PE in ten states. Founder of firm which began operation in May 1974 which now has a permanent staff of approx 80 personnel. Former Pro Mgr with BASF Wyandotte Corp- Parsippany, NJ - specializing in translating German tech in capitol projs for construction in the US. Handled proj responsibilities in dyestaffs, intermediate, polystyrene, & pigments. Had responsibilities with Homko Products div of Kroft Foods as Proj Engr specializing in edible oil plant operations. *Society Aff:* AIChE, ISA, AOCS, GEAPS, NCEE.

Gafarian, Antranig V
Home: 1115 23rd St, 1, Santa Monica, CA 90403
Position: Professor. *Employer:* U of So California, Sch of Engrg. *Education:* PhD/Prob & Math Stat/UCLA; BSE/Math/Univ of MI; BSE/Mech Eng/Univ of MI. *Born:* DEC 1924. Mgr, Transportation & Telecommunications Dept 1969-73, at Sys Dev Corp of Santa Monica CA 1959-73. Taught at Ca St U at Northridge. Chmn ORSA, Western Reg 1975-76. *Society Aff:* IIE, ORSA, IMS, ASA.

Gaffney, Francis J
Home: 205 S Pebble Beach Blvd, Sun City Center, FL 33570
Position: Consultant. *Employer:* Self. *Education:* BS/Elec Engg/Northeastern Univ. *Born:* Jun 27, 1912. Grad studies Tufts, MIT, Brooklyn Polytechnic. Head, Test Equip Group MIT Radiation Lab 1940-45. Genl Mgr PRD Co 1946-52. Dir Engrg Fairchild's Guided Missiles Div, VP Engrg Teleregister Corp. Since 1960 No Am Philips Co serving as Tech Dir, Dir of Product Planning & Pres North Am Philips Communications Corp. Cons since 1975. Fellow IEEE. Mbr Tau Beta Pi, Sigma Xi. Certificate of Appreciation, War & Navy Depts 1947. Enjoy Golf & Music. *Society Aff:* ΤΒΠ, ΣΞ.

Gaffney, Jerome J
Business: P.O. Box 14565, 1700 SW 23 Drive, Gainesville, FL 32604
Position: Res Agri Engineer. *Employer:* ARS, US Dept Agri. *Education:* MS/Agri Eng/PA State; BS/Agri Eng/IA State. *Born:* April 1942. Native of Winthrop Iowa. Res Engr working in areas of heat transfer & thermal & optical properties of foods with Agri Research Ser, US Dept of Agriculture 1965- . USDA Certificate of Merit

Gaffney, Jerome J (Continued)
1972. Reg Prof Engr Florida. Mbr ASAE; Chmn FL State Sect ASAE 1975-76; Chmn Natl ASAE Food Engrg Div 1977- 78. Mbr ASHRAE. Prof Mbr IFT. Mbr of 5 Natl Tech Cttes in ASAE & 2 in ASHRAE. Author of num tech art, papers & presentations before prof socs. Co recipient of ASHRAE Best Symposium Paper Award 1976. FL. Sect. ASAE Special Award for Outstanding Service, 1981. *Society Aff:* ASAE, ASHRAE, IFT.

Gaffron, John M C
Home: 9577 Doliver Dr, Housten, TX 77063
Position: Retired. *Education:* BS/Mining/Columbia Univ; EM/Mining/Columbia Univ. *Born:* May 1920. BS & EM Columbia U. Native of NYC; Joined Freeport Minerals Co as Jr Mining Engr 1942. US Naval Reserve 1943-46. Discharged with rank of Lt. Returned to Freeport 1947 as Mining Engr; in Oil & Gas Div 1949; VP Freeport Oil 1964-76; Recovered Sulphur Advisor 1976. Joined Conoco 1978; retired March 1987 as Sr Staff Engineer. Reg Prof Mining & Petrol Engr La. Bd/Dir AIME(SPE) 1962, 63, 68-70; V P 1964, Pres 1969; Dir AIME 1968-69, V P 1970. SPEAIME Distin Service Award 1974. *Society Aff:* SPE, AIME.

Gage, Elliot H
Business: 210 Chicago St, West Chicago, IL 60185
Position: Consultant. *Employer:* Self. *Education:* BS/Mechanical Engr/IL Inst of Tech; JD/Law/DePaul Univ. *Born:* Jan 1923. BSME from IIT 1944; JD from DePaul U 1951. Officer Submarine Service WW II. Engr Internatl Harvester, Engr S J Reynolds Co Inc, VP & Engr Hunter Clark Ventilating System Co, VP & Engr Alexander Gammie Plumbing & Heating Co. P Pres of Ill Chap ASHRAE, Chicago Chap of CSI, VP & Dir of Cons Engrs Council of Illinois. P Pres Chicago Engineers Club. North Central Sect Dir of CSI, Fellow of CSI. Established & Instruct Air Test & Balance Sch for Ventilating & Air Conditioning Contractor's Assn of Chicago. P Mbr Zoning Bd & Zoning Bd of Appeals West Chicago. Chrmn - Engrs Joint Contracts Document Committee. Mbr Illinois P.E. Registration Committee. *Society Aff:* ASHRAE, ACEC, NSPE, CSI, AAA, NACE, BOCA, ASPE, ASTM, APCA.

Gaggioli, Richard A
Home: 6425 Western Ave NW, Washington, DC 20015
Position: Prof of Mech Engrg *Employer:* Catholic Univ of America *Education:* PhD/ME/Univ of WI, Madison; MS/ME/Northwestern Univ; BS/Engrg Sci/Northwestern Univ *Born:* 12/3/34 Highwood, IL. Parents: Gustavo and Tina (Mordini) Gaggioli. Wife: Anita (Sage), Children: Catherine, Michael, Daniel, Edward, Mary. Positions: Engr, Abbott Labs, Postdoctoral Fellow in ChE, Univ of WI; research member, US Army Math Res Ctr; Asst & Assoc Prof of ME, Univ of WI; visiting Fellow, Battelle Memorial Inst; Chrmn & Prof of ME, Marquette Univ; Mgr, Professional Engrg Consultants, Washington; Prof of Engrg & Dean of Engrg & Arch, Cath. Univ of America. Author of 60 articles on Theoretical & Applied Thermodynamics, Applied Math. Co-author of textbooks, *Thermodynamics*, with E. F. Obert. Editor of reference book, *Thermodynamics: Second Law Analysis*, *Efficiency & Costing: Second Law Analysis of Processes*. Honors and Awards: Steiger Award for Distinguished Teaching, U of WI, 1965; NSF-Soc for Indus and Applied Math Natl Lecturer, 1969-71; Pere Marquette Faculty Excellence Award, Marquette U, 1976. *Society Aff:* ASME, ASHRAE, AIChE

Gaggstatter, Henry D, Jr
Business: 3400 SW 3rd Ave, Miami, FL 33145
Position: VP *Employer:* Bradley, Whitworth Assocs, Inc *Education:* BSE/ME/Univ of MI *Born:* 12/15/25 Columbus, GA. US Navy Air Corp in WWII. Chief Engr, Air Conditioning Div, Miami Roofing & Sheet Metal Co 1953-54. Secty, R.L. Duffer Assoc, Inc, consulting engrs, 1958-65. VP, Bradley, Whitworth Assocs, Inc, consulting engrs, 1965 to present. Specialized in hospital design since 1967. Pres, Miami Chapter, ASPE 1979-80. Hobbies are fishing and tennis. *Society Aff:* ACEC, NSPE, FES, ASHRAE, ASPE, SCA

Gagnebin, Albert P
143 Grange Ave, Fair Haven, NJ 07701
Position: Dir. *Employer:* Inco Ltd. *Education:* MS/Met/Yale Univ; BS/ME/Yale Univ. *Born:* 1/23/09. Native of Torrington, CT. Assn with Intl Nickel Co of Canada Ltd started in 1930 at Huntington WV alloy plant. From 1932 to 1949, undertook met res in Bayonne, NJ lab on steels and cest irons -- co-inventor of Ductil Iron. Field dev work on Ductil Iron 1949-55. Mgr of nickel sales and marketing 1956-64. VP 1960, Exec VP 1964, Pres 1967, Chrmn 1972 - retired 1974 but continue as Hon Dir. *Society Aff:* AIME, ASME, AFS, ASM.

Gaige, C David
Business: P.O. Box 173, Kansas City, MO 64141
Position: Project Mgr & Group Leader *Employer:* Burns & McDonnell *Education:* MS/ME/CO State Univ; BS/ME/TX Tech Univ *Born:* 7/12/47 Served with Army Corps of Engrs 1970-72. Engr for CO Air Pollution Control Div, involved with automobile control program and permit review. Obtained MS Degree and began work as a Consultant with Rust Engrg in AL. Began work with Burns and McDonnell in 1975. Assumed position as Group Leader of the Air Pollution/Meteorology Group in 1980. Promoted to Mgr for the Planning and Environ Analysis Div in 1982. *Society Aff:* ASME, NSPE, APCA

Gaither, Robert B
Business: 237 Mech Engrg Bldg, Gainesville, FL 32611
Position: Prof & Chmn Mech Engrg. *Employer:* U of Florida. *Education:* PhD/Mech Engg/Univ of IL; MSME/Melch Engg/Univ of IL; BME/Mech Engg/Auburn Univ. *Born:* Aug 1929. U S Navy Off 1951-54. Ford Foundation Fellow 1959-62. Extensive exper in mech engrg instruction, res & admin. Natl Chmn ME Dept Heads 1971-72; VP Educ of ASME 1976-80; President of ASME 1981-82; ECPD Bd/Dirs 1976-78; U S Army JSHS Bd/Dirs 1974-76; Chmn Bd/ Dirs FL Foundation for Future Scis 1973- . Enjoy classical music & golf. *Society Aff:* ASME, ASEE.

Gajewski, Jan
Business: 1011 Kennedy Blvd, Manville, NJ 08835
Position: President. *Employer:* Manville Rubber Products Inc. *Education:* BS/Chem Engg/NJ Inst of Technology. *Born:* Poland 1919. US Citizen. Holder of Diplome De Batchelier Du Lycee France. Served in Artillery, France & in the RAF during WWII. Escaped from Stalag V11-A, Germany. Studied French & Literature at Universite de Grenoble, Textiles, University of Nottingham, Pure Science, University of Bristol. BS degree in Chem Engrg from Newark C of Engrg, grad courses Columbia U. Pres of Sterling Laboratory & Development Co from 1964-70. Dev prototype internal rubber insulation for Polaris, Posidon, Moon Rangers, & Sprint missiles. Organizer & Pres of Manville Rubber Prod Inc since 1970. Sr Mbr ACS, AIChE. Hobby: Gardening, playing violin, tennis, alpine skiing. *Society Aff:* ACS, AIChE.

Gajewski, Walter M
Home: 3103 S. Everett Pl, Kennewick, WA 99337
Position: Asst. Mgr., Eng. Dept. *Employer:* Westinghouse Hanford Co. *Education:* MS/Eng/University of Connecticut; BS/EE/University of Connecticut *Born:* 04/04/23 Served with U.S. Army 1943-46. Graduated Univ 1950 and started with Westinghouse at Bettis Atomic Power Lab. Involved in and managed design of first of kind nuclear propulsion systems for submarines, cruisers and aircraft carriers. Managed engrg of Fast Breeder Reactor FFTF. Currently Asst Dept Mgr for central engrg encompassing 500 engineers. Chaired Advanced Reactor Cttee of Nuclear Engrg Div of ASME. Fellow (ASME) and Professional Engineer (Nuclear). *Society Aff:* ASME, ANS, IEEE

Gajjar, Jagdish T
Home: 221 Steinmetz Hall, Schenectady, NY 12308
Position: Prof EE/CS *Employer:* Union Coll *Education:* PhD/EE/Univ of Houston; M/EE/Univ of OK; BE/ME/Univ of Bombay; BE/EE/Univ of Bombay *Born:* 5/23/40 Native of Bombay, India. Taught at Univ of OK, Univ of Tulsa, and Univ

Gajjar, Jagdish T (Continued)
of Houston. On the faculty of Union Coll since 1970. Chrmn div of engrg and applied science (1979-82). Consultant to industry in electronics, communications, energy systems, optics and instrumentation. Chrmn IEEE, Student Activities Committee. Region 1 - 1975-77, Chrmn IEEE Schenectady Section (1981-82). Fulbright Exchange Scholar 1985-86. *Society Aff:* IEEE

Gakenheimer, David C
Business: P O Box 9695, Marina del Rey, CA 90291
Position: Program Manager. *Employer:* R & D Associates. *Education:* PhD/Mechs/Caltech; MS/Mechs/Caltech; BES/Mechs/Johns Hopkins Univ. *Born:* Aug 1943. Staff Scientist with The Rand Corp 1968-72. With R & D Assocs since 1972. Presently program mgr for laser system application studies. Past res activ incl seismic wave propagation from explosions & particle erosion of reentry vehicle materials. ASME Henry Hess Award 1972. Chmn Jr Awards Ctte of Applied Mechanics Div of ASME 1973-79. *Society Aff:* ASME.

Galambos, Theodore V
Business: Civil and Mineral Eng. Dept, Univ. of Minnesota, Minneapolis, MN 55455
Position: Prof Civil Engg. *Employer:* Univ of Minnesota *Education:* PhD/CE/Lehigh; MS/CE/UND; BS/CE/UND. *Born:* Apr 1929. Taught at Lehigh 1959-65. Prof CE Wash U 1965-1981, Chmn/Dept 1970-78. Appoint H D Jolley Prof of Cvl Engrg 1969. Primary res interest in struc steel behavior & design. Since 1965 cons Engr for Steel Joist Institute. Served 1970-74 as chmn of Column Research Council of the EJC. Received ASCE Walter Huber Research Prize 1964; ASCE Moisseieff Award 1968. Author of text 'Structural Members and Frames.' Elected to Natl Acad of Engg, 1979. Appointed J. L. Record Prof of Structural Engrg, Univ of Minnesota. Sept 1981. ASIC T.R. Higgins Lectureship Award 1981; ASCE Norman Medal, 1983; Honorary Doctorate, Tech. Univ, Budapest; 1982. *Society Aff:* ASCE, ASEE, SSRC, IABSE.

Galandak, John
Home: 8 Riverview Dr W, Upper Montclair, NJ 07043
Position: Consultant *Employer:* Camp Dresser & McKee *Education:* MBA/Mgt/NYU; MS/Sanitary Engg/IL Inst of Tech; BS/Civ Engg/IL Inst of Tech. *Born:* 12/4/15. Early employment in water and wastewater equip design and mfg cos. Seventeen yrs at Graver Water Conditioning Co. Positions included design, res and dev, field service, mgr of engg. Was asst prof, environmental and hydraulic subjects at NJ Inst of Tech for four yrs. Past sixteen yrs engaged in all phases environmental engg at Alexander Potter Assoc, operating unit of Camp Dresser McKee. Was proj mgr design of advanced wastewater treatment plants. Positions held: Assoc, VP, Pres, Chrmn Exec Bd. Held officer positions ASCE, Environmental Div. Published articles in advanced wastewater treatment, wastewater sludge mgt. Enjoy sailing. *Society Aff:* ASCE, WPCF, AAEE, XE.

Galas, David E
Home: 29504 Whitley Collins Dr, Rancho Palos Verdes, CA 90274
Position: Consultant *Employer:* Self *Education:* MS/Elec Power/Univ of IL; MA/Intl Affairs/Univ of Geo Wash; BS/Sci & Military/USMA; Certificate/Econ/Natl & Intl Power & Military/USAF War Coll *Born:* 10/19/19 in Evanston, IL. Completed enlistment in the US Army Air Corps. Admitted to the US Military Academy-graduated as pilot. Developed special weapons safety guide, responsible for Air Force aeronautical electrical equipment, supported-in plans and safety aspects-special weapons tests, assigned to the FAA for SST development tasks in maintainability, airport compatibility and ground support equipment. Was the Air Force Plant Representative at the Northrop Corp, Aircraft Div. Senior proj engr with Hughes Aircraft Corp. Held several offices in the System Safety Society including 2 years as technical editor of the Journal- Hazard Prevention. Now consulting in safety, liability cases and investigations, specialize in aviation. *Society Aff:* SSS

Galazzi, Joseph A
Home: 810 Red Mill Drive, PO Box 336, Tecumseh, MI 49286
Position: Vice President. *Employer:* Tecumseh Products Co. *Education:* BS/Mech Engr/Tufts Univ *Born:* March 1914. Native of Somerville Mass. Asst Dir Engrg Labs York Corp York PA, specializing in refrigeration & air conditioning. Ch Engr American Instrument Co. Successively, Ch Engr, Liquid Carbonic Co, Morrison IL; Internatl Harvester Co, Evansville IN; Tecumseh Products Co, Tecumseh MI. Assumed current respon as V P, Foreign Oper Div, Tecumseh Prod Co, then to Pres, Internatl Div in charge of exports & lic program. Holder of num pats in refrig. Fellow Amer Soc of Heating, Refrig & Air-Cond Engrs Inc. Enjoy gardening, world travel, good music.

Galerman, Raphael
Home: 302 Woodland E Dr, RR 7, Greenfield, IN 46140
Position: Supervisor, Process Metallurgy *Employer:* Bridgeport Brass Corp *Education:* BA/Chemistry/Butler Univ; BS/Chem Engg/Purdue Univ. *Born:* December 8, 1923. Native of Indianapolis, Ind. Served with US Navy 1943- 46 and with Naval Reserve until 1954. Design & Supervision of metallurgical processes & copper base alloy dev in the Bridgeport Brass Co, Ind Plant since 1948. P Chmn of Ind Chap ASM. Served 3 years as a mbr of the Soc's Chap Adv Ctte. Have traveled extensively in US, Europe & Middle East. *Society Aff:* ASM.

Galindo-Israel, Victor
Home: 3221 Emerald Isle Dr, Glendale, CA 91206
Position: Sr Scientist. *Employer:* Jet Propulsion Lab. *Education:* PhD/EE/Univ of CA; MS/EE/Univ of CA; BEE/EE/NYU. *Born:* 9/14/28. Dr Victor Galindo-Israel (PhD, UC Berkeley, 1964) has contributed to the field of applied electromagnetics for nearly 35 yrs. His work includes significant contributions to the synthesis & analysis of phased array antennas and periodic structures, reflector antennas, microwave devices, numerical analysis, & other related areas. This work is reflected in more than 70 scientific papers including a book on "The Theory and Analysis of Phased Arrays" (Wiley 1972). He was recently (1977) elected Fellow to the IEEE for his work in this area & others, & more recently received the first prize for best paper published in the 1977 IEEE Antennas & Propagation Soc transactions. He has served as full prof of EE at VPI & SU & at Ben Gurion Univ of the Negev (tenured), Israel. *Society Aff:* IEEE, URSI, ΣΞ.

Gall, William R
Home: 103 Antioch Drive, Oak Ridge, TN 37830
Position: Consultant *Employer:* W.R. Gall, Inc. *Education:* M Eng/Mech Eng/Yale Univ; BS/Mech Eng/Univ of Tenn. *Born:* May 20, 1913. Native of Rogersville TN. Asst Prof Mech Engrg U of Louisville 1941-44. Design Engr in Nuclear Programs at Oak Ridge 1944-64, Sect Ch in Nuclear Reactor Design 1949-60. Engrg Cons to Oak Ridge Natl Lab Engrg Div 1964-78. Chmn ANSI B31.7, 1961-78. Mbr Subctte on Nuclear Power of ASME Boiler & Pressure Vessel Ctte 1969-79 & Chmn Subgroup on Materials. Fellow ASME 1966. ASME J Hall Taylor Medal 1972, TSPE Award 1972 as outstanding Engr of East TN. Mbr Tau Beta Pi, Phi Kappa Phi. *Society Aff:* ASME, NSPE, ASTM.

Gallager, Robert G
Business: Room 35-206, Cambridge, MA 02139
Position: Prof *Employer:* MIT *Education:* ScD/EE/MIT; SM/EE/MIT; BS/EE/Univ of PA *Born:* 5/29/31 in Philadelphia, PA. Worked at Bell Telephone Labs and served in Army Signal Corps before doing graduate work at MIT. Has been on the electrical engrg and computer science faculty at MIT since 1960 and is currently also Assoc Dir of the Lab for Information and Decision Systems. Received the IEEE Baker Prize Paper in 1966, was elected a Fellow of the IEEE in 1968 and became a member of the Natl Academy of Engrg in 1979. Bd of Governors of IEEE group on Information Theory and Chrmn of group in 1971. *Society Aff:* NAE, IEEE

Gallagher, Richard H
Business: Worcester Polytechnic Institute, Worcester, MA 01609
Position: Provost and V.P. Academic Affairs. *Employer:* Worcester Polytechnic Inst *Education:* PhD/CE/State Univ of NY at Buffalo; M/CE/NYU; B/CE/NYU *Born:* 11/17/27 Spent the first 17 years of his career in industry, including a 12-year period with Bell Aerospace Corp (1955-67). Became Prof of Civil Engrg at Cornell Univ in 1967, and for 9 years served as Chrmn of the Dept of Structural Engrg. Dean of Engrg, Univ of AZ, 1978-84. Became VP at WPI in 1984. At Bell Aerospace undertook research in the finite element method and has been an active contributor to its development for 25 years. Has authored 3 books on this topic, published over 100 papers, and edits the Intl Journal for Numerical Methods in Engrg. *Society Aff:* ASME, ASCE, AIAA, ASEE, AAM, IACM.

Gallagher, Ronald P
Business: 130 Jessie Street, San Francisco, CA 94105
Position: Vice President. *Employer:* URS/John A Blume & Assocs, Engrs. *Education:* MS/Structural Engg/Univ of CA; BS/Civil Engg/Univ of CA. *Born:* Apr 30, 1940. MS Struc Engrg & BS Cvl Engrg Univ of Calif, Berkeley. Mgr of the Nuclear & Licensing Div. Extens experience in struc engrg design, seismic analysis, & earthquake engrg. Dir or participated in analysis of structures, piping & equipment for more than 30 nuclear facilities, incl power plants, waste trtmt & storage facilities, & test & experimental reactor facilities. Prev exper incl planning & design of high-rise, hospital & indus bldgs, air craft hangars, military facilities, & marine structures. Mbr ASCE, ASME, ANS, EERI. *Society Aff:* ASCE, AME, ANS, EERI.

Gallagher, William L
Business: 315 Powers Bldg, Rochester, NY 14614
Position: Prin. *Employer:* Phillip J Clark, PE Engrs and Consultants. *Education:* B of CE/Civ Engrs/Catholic Univ of America. *Born:* 2/20/42. Native of Pittsford, NY. Licensed Engr in NY & PA State. VP of Clark Engrs in charge of transportation and municipal engrg for local govts in the western NY area. Major accomplishments include design of surface structures for the lake and tunnel sys in Rochester, NY, Hwy and bridge proj for NYSDOT, and st and utility proj for the city of Rochester and other municipalities in western NY. *Society Aff:* ASCE, NSPE, WPCF, APWA.

Gallardo, Albert J
Home: 3814 Moddison Ave, Sacramento, CA 95819
Position: Exec Dir *Employer:* Alameda County Transportation Authority *Education:* BS/Civil Engg/UC Berkeley. *Born:* 4/8/27. BS from UC Berkeley 1950. Reg Civil Engg CA. Fellow ASCE. With Fed Highway Admin 1950-83. 1976 Pres Sacramento Sec ASCE. Am serving as Natl Dir for ASCE and on ctte for "Construction, Maintenance and Operation of Highways–. Have received Silver Medal for Superior Achievement from Secretary of Transportation in 1978. Have been accorded Hon Mbrship in "Professional Engrs in CA Government" (PECG) in 1978. Enjoy history & handiwork hobbies. Currently am enployed as Exec Dir of the Alameda County Transportation Authority charged with the disbursement of sales tax funds for transportation improvements. *Society Aff:* ASCE, APWA.

Gallen, Donald R
Business: 111 West Port Plaza, St Louis, MO 63141
Position: Sr Project Mgr. *Employer:* Arthur G McKee & Co. *Born:* Nov 1925. BS Chem Engrg Fenn Coll (now Cleveland St U). Grad work at Case Inst of Tech & U of Calif. Diversified exper as Genl Mgr, Proj Mgr, Super Engr, Cons & Sr Process Engr in engrg & const of chem, power & petroleum plants, pipelines & mfg facilities. This exper incl dept organization & const coordination. Managed projects involving computerized design techniques, business & control procedures & automation of plant & pipeline operations. Design Engr for Babcock & Wilcox 1950-51; Arthur G McKee & Co 1951-57; Bechtel Corp 1957-67; Mobil Res & Dev 1967-69; Treas G E Co 1966. Enjoy bridge & golf.

Galliett, Harold H, Jr
Home: 1616 Garden Street, Anchorage, AK 99504
Position: Cons Civil Engr. *Employer:* Galliett & Assocs, Engrs & Surveyors. *Education:* BS/Civil Engg/Univ of CA. *Born:* 8/22/23. BS Univ of CA 1950. Res Anchorage AK 1954-79. USAAF WWII 1942-45, Navigator Southwest Pacific. Cvl Engg officer USAF CA, Iceland, OH, AK 1951-55. Priv Pract Cvl Engg AK 1956-79. Mbr Alaska Bd Engrs & Arch Examiners 1960-63. Reg CE AK, CA, HI. Fellow ASCE. Mbr NSPE, AWWA, AREA, SAME, ASP. Papers "Cadastral Surveys by Flare Intersection" 1966; "The Radio Theodolite" 1973; "A Plan for Alaska" 1973; "The Great Coal Giveaway–, a newspaper series 1975-76. Candidate AK Legislature 1976, 1978. *Society Aff:* ASCE, NSPE, AWWA, AREA, SAME.

Galligan, James M
Home: 75 Sawmill Brook Lane, Willimantic, CT 06220
Position: Prof *Employer:* Univ of CT *Education:* BS/Metallurgy/Poly Inst of Brooklyn; MS/Metallurgy/Univ of IL; PhD/Metallurgy/Univ of CA-Berkeley *Born:* 5/13/31 in NY City, NY, educated in NY. Staff scientist, US Steel Fundamental Research Lab, Monroeville, PA. US Steel Fellow, Univ of CA-Berkeley. Assoc Prof, Columbia Univ, 1963-67. Group Leader - Brookhaven Natl Lab. Prof, Univ of CT 1972 to present. Consultant to Brookhaven Natl Lab, Stanley Tool Co. Expert witness on failure analysis in metals and materials. Visiting Prof at Max Planck Inst, Stuttgart, CA Inst of Tech, D. Ben Gurion Univ. *Society Aff:* APS, AIME, AAAS, NYAS, ΣΞ, ΦΚΦ

Gallin, Herbert
Home: 39 Chestnut Ridge, Dobbs Ferry, NY 10522
Position: Pres. *Employer:* Herbert Gallin Assoc, PC. *Education:* BME/ME/CCNY; BCE/Civ Engg/NYU, JD/PACE U. School of Law. *Born:* 11/3/23. NYC 1923: Army Corps of Engrs 1944-46: NYC Bd of Transportation 1947-50 Mech Installation Design: PE, NYS 1949: Pres Apex Contracting Co Inc 1951- Overall design & completion of bldg alteration & renovation: Organized Herbert Gallin Assoc PC in 1973 to engage in construction consultation & analysis & representation of clients before Bldg Depts & other Admin Agencies: Have been appointed umpire by NY State Supreme Ct: Have been appointed Surveyor of Unsafe Bldgs by NYC Bldg Dept: Admitted NY State Bar 1981. *Society Aff:* ASME, ASCE.

Gallo, Frank J
Business: 1983 E 24th St, Cleveland, OH 44115
Position: Chrmn, Tech. *Employer:* Cleveland State Univ. *Education:* MSCE/Civil Engg/Cleveland State Univ; BStrE/Struct Engg/Fenn College (now CSU). *Born:* 7/29/21. Native of Cleveland. Served in Army Air Force 1945-46. Worked in Fisher- Aircraft Div GMC, 1943-1945 as Stress Analyst. Taught in Fenn College (now Cleveland State Univ) from 1946 to present in Civil Engg Dept. Became chrmn of Tech Dept in 1977. Worked 27 years on part-time basis as consulting engr with Barber & Hoffman, Inc, Cleveland, OH Consulting Structural Engrs. Held ASCE, Cleveland Sec, offices: Treas 1976 & 77, VP 1978, Pres 1979-80. Secretary-Treasurer & Board Member of National Order of the Engineer, Inc, 1980-present. Hobby - Photography. Publication: Small residential structures, John Wiley & Sons 1984, co-author Regis I Campbell. *Society Aff:* ASCE, NSPE, ASEE, OE.

Gallotte, Willard A
1311 Maple St. Sp 5, Wenatchee, WA 98801
Position: Manager-Engg. *Employer:* Puget Sound Power & Light Co. *Education:* BS/EE/Worcester Polytechnic Inst *Born:* 1902 Manette WA. Distrib & Transmission engrg Niagara Falls P Co & Puget Power prior to WWII. In Army Ordnance Corps WWII 1942-46. With Puget Power to retirement - construction & operations to 1951. Genl engrg dept to 1967, last 14 yrs as dept hd. Respon for overhead, underground & submarine transmission & distrib & substation engrg & const budget, also liaison with architects & genl contractors on major projs, & sys planning. Ctte work IEEE, Seattle; Fellow AIEE 1962. P.E. 1952 Hosp tr 10 yrs; cvl service comm 6 yrs; Army res 21 yrs; LtC AUS Ret. Lay missionary & famine relief, Ethiopia 3 yrs re-

Gallotte, Willard A (Continued)
tirement; occasional consultant work. incl. Metro Transit, Seattle 1978-79.. *Society Aff:* IEEE.

Galloway, William J
Home: 19343 Olivos Dr, Tarzana, CA 91356 *Education:* PhD/Physics/UCLA; MS/Applied Physics/UCLA; BS/Applied Physics/UCLA *Born:* 9/15/24 US Army 1942-46, 1951-53. Bolt Beranek and Newman, Inc 1953-82. Consultant 1983- . Primary professional interests are in the developments of methods for measurement and prediction of noise produced by aircraft, motor vehicles, and other sources of community noise, and the development of predictive models for relating such noise to community response. Active participant in development of natl and intl acoustical standards. VP and Dir, INCE 1978-1980; Standards Dir, ASA 1979-83; SAE Arch T. Colwell Merit Award, 1973. Active airplane pilot. *Society Aff:* NAE, ASA, AIAA, INCE, ΣΞ

Gallup, Robert B
Business: R.W. Beck and Associates, 2121 Fourth Avenue, Seattle, WA 98121
Position: Chief Consulting Engr. *Employer:* R W Beck and Assocs. *Education:* BS/Elec Engg/Univ of WA. *Born:* 10/27/19. Native of Seattle, WA. Engr for Rural Electrification 1941-42, 1946-51, spec in power supply matters. Served with Signal corps & Army Intelligence 1942- 46. Since 1951 con engr for R W Beck and Assoc, Partner 1954-1985, Mgr Seattle Office 1968-1979. Chief Consulting Engr 1979-1985. Executive Engineer since 1985. Cons to elec utility sys; feasibility reports in connection with the issuance of revenue bonds to finance acquisition & const of hydroelec projects, coal-fired steam-elec power projs & nuclear power projs. Respon for preparation of const engr's reports in connection with over 10 million kW of generating capacity. *Society Aff:* IEEE, NSPE, CEC

Galson, Edgar L
Business: 6601 Kirkville Rd, East Syracuse, NY 13057
Position: Pres, Galson & Galson. Galson Technical Services - Sect, Treasurer, Galson Research Corp - Sect Trea *Employer:* Galson & Galson, PC, Cons Engrs *Education:* MME/Mech Engr/Purdue Univ; BME/Mech Engr/Cornell Univ *Born:* 11/10/25 PE NY, MA, IL, RI, NJ, TX, NCEE. GE Test Engr 1950. Heat Transfer Spec 1951. Joined Galson & Galson 1952. Partner 1954. Pres 1980. Firm specializes in all mech, elec facilities for major bldgs. With brother Allen Galson started environmental consulting firm, Galson Tech Servs, Inc 1971. Galson Technical Services has 4 major divisions; Industrial Hygiene, Laboratory, Air Quality and Shell Testing. These firms are located in Syracuse with branch offices in Rochester, NY, Plymouth Meadows, PA and Oakland, CA. Galson Control Co. founded in 1987. With Allen Galson and Bob Peterson formed Galson Res Corp 1983. Adjunct prof at Harvard Grad Sch of Design 1972-74. Lecturer at Syracuse & Cornell. Active in ASHRAE, CEC, APEC. Lecturer at ASHRAE, IES, and other tech societies. Community affairs activities include Past Pres Dunbar Center, a Family Service Agency; Scoutmaster; Past Pres Coun on Urban Renewal; skier, wilderness camping, canoeing, tennis, gardening, naturalist-activities family centered. Married -- 4 children. *Society Aff:* ACEC, IES, ASHRAE

Galstaun, Lionel S
Business: P.O. Box 2166, Houston, TX 77001
Position: Prin. Process Engr. *Employer:* Bechtel Group, Inc. *Education:* PhD/Physical Chem/MA Inst of Tech; SM/Physical Chem/MA Inst of Tech; BS/ChE/Univ of Dayton. *Born:* 12/17/13. in Java, high sch in Japan, and since 1929 a resident of the US. Res Chem Engr with Tidewater Oil Co at Avon, CA 1936-1948. Instr of Electronics, Univ of CA War Time Ext Courses, 1943-45. Leader of Chem Industry Team of US Strategic Bombing Survey in Japan, 1945. Supervisor of Res with Tidewater Oil, 1948-58. With Bechtel Corp 1958-present in various engg mgt positions including Mgr of Process Design Depts in NY & Houston offices, 1966-79. Mbr: Committee for Supplement to the Chemistry of Coal Utilization under sponsorship of Natl Res Council. *Society Aff:* AIChE, AIC, ACS, ASTM.

Galt, John K
Business: POBox 5800, Albuquerque, NM 87115
Position: VP. *Employer:* Sandia Labs. *Education:* PhD/Physics/MIT; AB/Physics/Reed College. *Born:* Sept 1, 1920 Portland OR. Teaching Fellow & Res Assoc MIT 1941-43 & 1945-47. Civilian with Office of Scientific R&D 1943-45. Natl Res Council Fellow Bristol, England 1947-48. Mbr Tech Staff Bell Telephone Labs 1948-74. Dir Solid State Electronics Research 1961-74. Dir Solid State Sciences Research, Sandia Labs 1974-78. VP 1978- . Fellow IEEE & Am Physical Soc. Mbr Air Force Studies Board (Natl Academy) 1971-76. Mbr Air Force Scientific Advisory Board 1975- . *Society Aff:* IEEE, APS, AAASΦBK, AAAS.

Galvin, Cyril
Business: Box 623, Springfield, VA 22150
Position: Principal Coastal Engr *Employer:* Self *Education:* PhD/Geology/MA Inst of Tech; SM/Geology/MA Inst of Tech; BS/Geol Engg/St Louis Univ. *Born:* 6/16/35. Consulting Coastal Engr specializing in the action of waves and currents on engineered structures and in sediment movement in coastal waters. Formerly Res Asst, MIT Hydrodynamics Lab, 1959-1963; Oceanographer, Beach Erosion Bd/Coastal Engg Res Ctr, 1963-1970; Chief, Coastal Processes Branch, CERC, 1970-1978. Reg PE, VA, MD, & NJ. Huber Res Prize (1969) and Norman Medal (1970) of ASCE. Native of Jersey City, NJ. Started out in geology, took up fluid mechanics as a means of putting geologic studies on a physical basis, and got interested in coastal res and then coastal engg. Other interests in history of sci, literature. *Society Aff:* ASCE, IAHR, AGU, SEPM, AAAS, WEDA, AAPG.

Gamble, William L
Business: 2209 Newark Civ Eng Laboratory, 208 N. Romine St, Urbana, IL 61801
Position: Prof. *Employer:* Univ of IL. *Education:* PhD/Civ Engg/Univ of IL; MS/Civ Engg/Univ of IL; BS/Civ Engg/KS State Univ; Assoc of Arts/Pre-Engg/Dodge City (KS) Jr College. *Born:* 11/25/36. Specializing in analysis and design of reinforced and prestressed concrete structures. Res work on most aspects of concrete design and behavior. Special emphasis on reinforced concrete floor slabs and on long-term behavior of prestressed concrete hgwy bridges. Teaching in all areas of reinforced and prestressed concrete, and fire resistance of structures. One book on floor slabs plus about 60 papers and reports. Consulting work on distressed reinforced concrete structures and special design problems *Society Aff:* ASCE, ACI, PCI, NFPA, ASTM.

Gambrell, Carroll B
Business: P O Box 25000, Orlando, FL 32816
Position: Professor of Engineering. *Employer:* Univ of Central FL. *Education:* PhD/Indus Engr/Purdue Univ; MSE/Indus Engr/Univ of FL; BS/Engr/Clemson Univ; BA/History/FL Southern Coll. *Born:* Dec 1924. BS Clemson Univ 1949; MS Univ of Florida 1952; PhD Purdue Univ 1958. BA FLA Southern Coll 1977. Active duty Army 1943-46. Lt Col USAR Ret. Major cons employment: C & D Battery Co, Western Elec Co, AvSER, Sperry Phoenix Co, US Army Test & Eval Command, US Army Sci Adv Panel. Res Asst, Univ of Fla 1951-52; Instructor Clemson Univ 1949-51; Asst Prof Lamar Univ 1952-55; Asst Prof Purdue 1955-59; Prof, Dept Chmn Ariz State Univ 1959-67; V Pres for Acad Affairs, Fla Tech Univ 1967-78 . ECPD Accreditation Visitor; Bd of Tr, Embry Riddle Aeronautical Univ, Winter Pk Mem Hosp. V Pres Publications IIE, 1965-67; Disting Alumnus Award Purdue Univ 1969; Fellow Award IIE 1969; Engr of Yr, Fla Engrg Soc 1969. *Society Aff:* ASEE, IIE.

Gambrell, Samuel C, Jr
Business: Box 2908, Tuscaloosa, AL 35487
Position: Prof. of Engrg Mech *Employer:* Univ of AL. *Education:* PhD/Theo & Appl Mechanics/WV Univ; MS/Agri Engr/Clemson Univ; BS/Agri Engr/Clemson Univ. *Born:* 9/15/35. Native of Owings, SC, Grad of Clemson Univ. Served in US Army Air Defense Artillery 1957-59. Grad Sch and Teaching at Clemson and WV Univs, 1959-65. Appointed Asst Prof of Engg Mechanics, the Univ of AL, in 1965; Prof,

Gambrell, Samuel C, Jr (Continued)
1973; Asst dean of Engg, 1976-83. Teaching interests in Solid Mechanics. Research interests in stress analysis of highway bridges, wire rope, and mine roof support systems. Pres, Southeastern Sec ASEE, 1982-83. VP, Res Southeastern Sec ASEE, 1978-79. Col, US Army Res., Retired. *Society Aff:* ASEE, ASME, SEM, ΣΞ, NSPE, ASPE

Gambro, A John
Home: 511 Spring Valley Rd, Paramus, NJ 07652
Position: Project Dev Specialist *Employer:* Self Employed. *Education:* BChE/Chem Engg/NY Univ. *Born:* 2/7/27. Chem E New York Univ. Native of Yonkers, NY. Employed by Lummus for 26 yrs. Have been involved in design & startup of petro & petrochem processing facilities. Pos held incl process engr, project engr, proj mgr, start-up mgr, mgr of process design, mgr of tech, genl mgr of commercial activities & VP, mgmt dev. VP, Bloomfield Div. Employed by Arusuisse of Amer for 5 years. Responsible for all business activities of engg group in the Western hemisphere. Have contributed the growth of indus in less developed countries by plant install & training of nationals in required tech. Currently involved in dev alternate energy and cogeneration projects for clients. Enjoy language, travel & writing. *Society Aff:* AIChE.

Gambs, Gerard C
Home: 1725 York Ave-33C, New York, NY 10128 *Education:* BEM/Min Engg/OH State Univ. *Born:* May 1918. Native of Columbus, Ohio. Mining engr with Pittsburgh Coal Co 1940-42. Major, Corps of Engrs Manhattan Dist, Atomic Bomb Proj 1942-46. Asst Prof Ohio State Univ 1946-47. Asst to V Pres Consolidation Coal Co 1947-69. Mgr of New Bus Dev, Gibbs & Hill 1969-70. V Pres Ford, Bacon & Davis Inc 1970-1983. Consulting Engineer 1983-. *Society Aff:* AIME, ASME, ANS

Games, Donald W
Home: 6946 Parmelee Dr, Mentor, OH 44060
Position: Chmn *Employer:* Production Machinery Corp. *Education:* BSME/Mech Engr/OH State Univ (1947) *Born:* 1/28/24. Native of Brown County, OH. Secondary sch Aberdeen, OH. Served in the Army Air Corp & 44th Infantry Div 1943-45. Tau Beta Pi 1947. Reg PE - OH, 1952. Helped found Production Machinery Corp Oct 1955. Served as Sales Engr, VP- Operations, & became Chief Executive in March, 1976, Chrmn, 1982- . Production Machinery Corp supplies finishing lines for coiled flat rolled metals to Producers & Fabricating Plants around the World. A wholly owned subsidiary re-manufactures pipe line valves. *Society Aff:* AISE.

Gammill, Robert C
Business: Comp. Sci Dept, Fargo, ND 58105
Position: Prof of Computer Sci & EEE *Employer:* ND State Univ. *Education:* PhD/Computer Sci/MIT; MS/Meteorology/MIT; BS/Physics/Univ of Rochester. *Born:* 2/12/38. Served as USAF Lt from 1959-62 at Global Weather Central, SAC Hdqurtrs, Omaha. PhD Comp Sci MIT 1969. 1969-73 at Univ of CO as Asst Prof of Computer Sci & consult at National Ctr for Atmos Res. Rand Corp from 1973-78 as res scientist in computer languages & interfaces. Also Assoc Prof Adj of EE & Computer Sci at USC 1976-78. Chmn ACM Mbrs and Chapters Bd 1976-78. Dir of Computer Sci Program at NDSU 1978-84. ACN Ntl. Lecturer 1980, 1981. Chmn of ACM SIGPC 1982-84. Dir of NDSU Computer Systems Inst, 1984-. *Society Aff:* ACM, IEEE, ΣΞ

Gandhi, Om P
Business: Dept of EE, Salt Lake City, UT 84112
Position: Prof. *Employer:* Univ of UT. *Education:* ScD/EE/Univ of MI; MSE/EE/Univ of MI; Post-grad Dipl/Elec Communication Engg/Indian Inst of Sci; BS/Phys/Univ of Delhi India. *Born:* 9/23/34. 32 yrs of profl & 24 yrs of managerial experience, prof since 1973, holds 6 US patents, author of Microwave Engrg and Applications (Pergamon Press, 1981) & over 200 journal articles, internationally recognized authority in microwave dosimetry, principal investigator on 26 federally funded res projs since 1970, consultant to several governmental agencies & private industries, guest Editor of *Special Issues on Electromagnetic Biological Effects & Medical Applications Proceedings IEEE* Jan 1980 & IEEE Engg in Medicine & Biology Magazine March 1987, has served on several profl panels including Bd of Dirs of Bioelectromagnetics Soc; honors include Fellow of IEEE, Distinguished Res Award of Univ of UT & Special Award for outstanding Technical Achievement from IEEE - UT Sec. Listed in Who's Who in Tech Today, Who's Who in America and Who's Who in the West. Amer Men and Women of Science, Interntl Who's Who in Engrg (UK) Directory of World Researchers (Japan). *Society Aff:* IEEE, BEMS.

Gangarao, Hota V S
Business: Rm 507, E S Bldg, Morgantown, WV 26506
Position: Prof. *Employer:* WV Univ. *Education:* PhD/Structural Engg/NC State Univ; MS/Structural Mechanics/NC State Univ; BTech/Civil Engg/IIT, Madras, India. *Born:* 1/12/45. Native of Andhra Pradesh, India. Grad from IIT, Madras with a bachelors degree in Civil Engg before joining NC State Univ at Raleigh NC in 1965. With WVU since 1969. Assumed current responsibility as Prof of Civil Engg in 1979. Worked on several res projs, published over 100 tech papers in various journals, and serving on several committees of natl recognition. Have a couple of patents in the areas of bldg systems. Enjoy reading and jogging. Fellow of the American Society of Civil Engineers. *Society Aff:* ASCE, FIE

Gannon, Philip E
Home: 86 Broadacres Rd, Atherton, CA 94025
Position: Chrmn Emeritus *Employer:* Keller & Gannon *Education:* BS/Elec Engrg/Univ of Wash *Born:* 4/1/16 With Corps of Engrs & Navy 3 yrs; co-founder of Keller & Gannon 1941. Engaged in military design during WW II; K&G in joint venture participated in reconstruction of Guam, Sangley, Philippines 1946-53. Engaged in design of commercial, institutional & indus projects since WW II. Performed power- energy/air cond studies on world-wide basis for DOD, NASA & others. Merged with Lester B. Knight & Asso in 1969. Pres of San Francisco Bay Area Engrg Council; Pres of SAME, San Francisco Post 1980; IEEE Mbr 1950; Chmn Reliability Sub-Cttee 1976-81; Illuminating Engr Soc Mbr 1950; Amer Defense Preparedness Assn Mbr 1976; Cons engr Assn of Calif & ACEC. Prof reg in Calif, Nevada, Arizona, Missouri & Guam. Enjoys golf & swimming. *Society Aff:* IEEE, IES, SAME, CEAC

Gans, Manfred
348 Highwood Ave, Leonia, NJ 07605
Position: President. *Employer:* Technology and Education Development Associates, Inc. *Education:* SM/Chem Eng/MIT; BSc Tech Hon/Chem Eng/Manchester (England) Univ *Born:* Apr 1922. MS, MIT; B Sc Univ of Manchester. British Army Parachute Commandos 1940-46. Joined Scientific Design (SD), a Halcon -SD Group Inc Division, as Process Engr 1951. Became operations engr 1957; Asst Dir of Oper 1959; Asst V P Operations 1963. In 1967, became V P, Process Engrg & Oper, and in 1970 V P, Process & Project Eng & Operations. 1978-1985 Sr VP for evaluation & dev of new technology. Played a vital role in guiding the design & start-up of most of the plants developed by SD. Publications incl: 'Planning the Successful Startup', 'Effective Process Engrg'. Fellow AIChE. Speaks German, French, & Dutch. Mbr UNIDO Tech Asst Ctte for Argentina, Chrmn UNIDO Tech Asst Ctte for Turkey. Presently also President of GTM, Inc, a company manufacturing advanced polymer composites. *Society Aff:* AIChE

Gant, Edward V
Home: 16 Eastwood Rd, Storrs, CT 06268
Position: Univ Prof of Civil Engrg *Employer:* Univ of CT *Education:* MS/CE/CA Inst of Tech; BS/CE/Vanderbilt Univ *Born:* 3/13/18 Native of Ethridge, TN. Graduate of Public Schools, Nashville, TN. Engr, American Bridge Co 1940-42. Member of Univ of CT Civil Engrg Dept since 1942. Univ Provost and Academic VP 1965-74; Acting Pres 1969, 1972-73, 1978-79. Univ Prof of Civil Engrg since 1974. Pres CT Section ASCE, 1960. Chrmn of New England Council ASCE, 1959. Honorary

Gant, Edward V (Continued)
Alumnus, Univ of CT, 1973. Numerous special awards from Univ of CT Bd of Trustees and Alumni Assoc. Benjamin Wright Award, CT Section ASCE, 1980. *Society Aff:* ASCE, ASEE

Gantz, Gordon F
Business: P.O. Box 310, (4400 W National Ave), Milwaukee, WI 53201
Position: Cost Engr *Employer:* Harnischfeger Corp *Education:* BS/ME/Marquette Univ *Born:* 11/4/25 Work experience includes 7 years in Application Engrg in overhead crane division of the national leader in this field. Designed the component cost system currently in use at this co for reporting costs on all industrial and electrical division manufacturing operations. Designed and operated the first in-house crane cost control system (an I.E. Standard Cost System); including computerizing this system for job and stock order variance determination, manufacturing performance and operations monitoring. This system has been applied to all shop operations. Also designed and operated all Quality control test fixtures for a stack type oil burner control line for White-Rodgers Electric Co, St. Louis, MO. *Society Aff:* ASME, AACE, IIE

Ganzhorn, Karl E
Home: Gluckstr. 1, Sindelfingen, West Germany D-7032
Position: Former member of Manging Board *Employer:* IBM Germany *Education:* PhD/Physics/Univ of Stuttgart; Diplom/Physics/Univ of Stuttgart *Born:* 04/25/21 1952-86 employed with IBM; Mbr Managing Board IBM Germany 1963-86. Dir IBM Laboratories in Germany, Austria, Sweden & France 1964-78. Dir Science & Technology IBM Europe 1973-75. Mbr, Science Council of Fed. Republic of Germany 1978-87. Pres of German Physical Society 1969-71. Hon. Prof at Karlsruhe Univ 1960-87, lecturing Info Processing Technologies. Dr. honoris causa Univ of Stuttgart, Honorary Senator of Techn. Univ of Munich. Member of the Bd of the German Inst of Standardization (DIN) 1980-86. Author of 3 books, 50 publications and 49 patents. *Society Aff:* DPG, IEEE, VDE, GI

Garbarini, Edgar J
Home: P.O. Box 3221, San Francisco, CA 94119
Position: Consultant (Self-emp) *Employer:* Self *Education:* BS/Civ Engrg/Univ CA at Berkeley. *Born:* 8/1/10 1933 to 1940 - Several employers in research design and construction engrg. 1940 to 1975 - Bechtel Group of Companies; progressed from Field Construction Engr to Vice Chrmn and Dir of the Bechtel Group of Companies and Pres of Bechtel Power Corp. 1975 to present - Practicing consultant. *Society Aff:* NAE, ASCE, AIME.

Garber, Charles A
Business: PO Box 342, W Chester, PA 19380
Position: Pres. *Employer:* Structure Probe, Inc. *Education:* PhD/Engrg/Case Inst of Tech; MS/Chem Eng; BS/Chem Eng/U of IL (Urbana) *Born:* 5/23/41. After working 2 1/2 yrs for DuPont, formed structure Probe, Inc, a private lab analytical service. Also an active mbr of the Am Council of Independent Labs & has testified several times before cttees of the US Congress on subjs involving small tech based films and, most recently, tech innovation in small firms. In January 1980, served as an elected delegate to the White House Conference on Small Business. Formerly served as VP, Association of Consulting Chemists and Chemical Engrs, New York. Mbr American Chemical Society, American Inst of Chemical Engrs, American Society for Metals, American Acad of Forensic Science, American Association of Small Res Companies, Soc of Cosmetic Chemists, Amer Soc for Artificial Internal Organs, and Int'l Soc for Hybrid Microelectronics. *Society Aff:* ACS, ASM, AIChE, ISHM

Garber, Dwayne C
Business: 707 North 3rd Ave, Marsalltown, IA 50158
Position: Pres *Employer:* Clapsaddle-Garber Assoc *Education:* BS/CE/IA State Univ *Born:* 5/18/37 Native of the Marshalltown, IA area, BSCE I.S.U. in 1961, employed by Clapsaddle Engr. From '61 to '62 U.S. Forest Service '63 to '67, Clapsaddle-Garber Assoc from 1967 to present. Past Pres of IA Consulting Engrg Council. Past Chrmn of the Bd of Advisors to the Center For Industrial Research & Service, IA State Extension Service, IA State Univ. Current member of the Beaman-Conrad-Liscomb Community School Bd. *Society Aff:* NSPE

Garcia, Raymond G
Home: 2 Schroeder Ct, Savannah, GA 31411
Position: Consultant. *Employer:* Self *Education:* BE/Mech Engg/GA Inst of Technology; BIE/Industrial Engg/GA Inst of Technology. *Born:* Nov 1927. Native of Florida. Educated in Florida public schools. Served with Spec Engrg Detachment, US Army 1945-47. Employed Florida phosphate with IMCC, 1952. Progressed through Engineering, Maintenance & Oper to Oper Mgr. In 1971 moved to Consolidation Coal Co as Dir of New Projects, later V Pres of MidWestern Div. Since Jan 1973 V Pres, Mining-Agrico Chem Co, respon for all phosphate mining operations. He was promoted in 1977 to Sr VP of Agrico Mining Co. Married, 4 children. Enjoys golf, fishing, sailing. Mbr Bd of Dir 1969-71 SME, Eastern Region V Pres 1969-71 SME. Retired October 1986. Currently consulting on a limited basis. *Society Aff:* SME, AIME.

Gardiner, Keith M
Business: Manufacturing Systems Engrg Program, Harold S. Mohler Building #200, Lehigh Univ, Bethlehem, PA 18015
Position: Prof *Employer:* Lehigh Univ. *Education:* PhD/Met/Univ of Manchester, UK; BSc/Met/Univ of Manchester, UK *Born:* 3/30/33. Currently a professor at Lehigh University. Most recently a staff member and program manager with IBM Corporate Technical Education. Adjunct Prof at Univ of VT 1972-80 Joined IBM 1966 involved in many aspects of mfg engg and with relationship of semiconductor reliability to mfg process. Previously with Rolls-Royce developing mfg methods for gas turbines; Sec Leader nuclear fuel mtls dev with English Elec Co. Published in all these areas plus bioengg. Former Chrmn of Electronics Mfg Council Soc of Mfg Engrs, former Chrmn Intl Prog Committee SME, mbr Editorial Bd Journ Mfg Systems, Education ctte SME, various ASME and IEEE cttes, certified Mfg Engr, reg PE (CA). Constl to Brookings Institute Nat Sci Fnd, Nat Acad of Sci. *Society Aff:* SME, ASME, SRS

Gardner, Chester S
Business: 1406 West Green St, Dept of Elec Engr, University of Illinois, Urbana, IL 61801
Position: Professor & Assoc Dean *Employer:* Univ of IL. *Education:* PhD/EE/Northwestern Univ; MS/EE/Northwestern Univ; BS/EE/MI State Univ. *Born:* 3/29/47. Jamaica, NY and raised in Kittanning, PA. Mbr of Technical Staff, Bell Telephone Labs 1969-71, specializing in software dev of No 4 ESS. EE Prof at Univ of IL since 1973. Involved in federally sponsored res on laser ranging, laser radar, atmospheric optical propagation. Consultant in electro-optics to Caterpillar Tractor Co, US Army, Northern Illinois Gas Co, Deere & Co, and McGraw-Hill Book Co. Chrmn, Central IL Sec of IEEE 1979-80. *Society Aff:* IEEE, OSA, AGU, AAAS, ΣΞ.

Gardner, Leonard B
Business: Automated Integrated Manufacturing, PO Box 1523, Spring Valley, CA 92077
Position: Senior Partner *Employer:* Automated Integrated Manufacturing *Education:* ScD/EE/Golden State Univ; MSc/Appl. Comp. Sci/Augustana Coll; BSc/Appl. Physics/Univ of CA; LLB/Law/Univ of MI *Born:* 2/16/27 Elected Fellow of IEEE 1981 for "Pioneering concepts and leadership in the development of computerized control systems used in process simulators and in automated production lines." Senior Partner, Automated Integrated Manufacturing (AIM) Consultants to industry with practice limited to industrial automation. Author of Principles Automated Integrated Manufacturing AIM engaged in assisting industry to implement factory automation. Dir of center for AIM at local colleges. Nationally recognized authority, lecturer, and teacher in AIM, (authored over 100 technical articles). Registered PE in CA

Gardner, Leonard B (Continued)
& Cert Mfg Engr. See Who's Who in America, 1982 edition for further details. *Society Aff:* IEEE, SME, ASTM, NSPE, ΣΞ

Gardner, Robin P
Business: Box 7909, Nuclear Engrg Dept, Raleigh, NC 27695-7909
Position: Prof *Employer:* NC State Univ *Education:* PhD/Fuel Tech/PA State Univ; MS/ChE/NC State Univ; B/ChE/NC State Univ *Born:* 8/17/34 Charlotte, NC, s. Robin Brem and Margaret (Pierce) G.; BChE, NC, State Univ, 1956, MS 1958; PhD, PA State Univ, 1961; m. Linda Jean Gardner, Oct 21, 1976. Scientist, Oak Ridge Inst Nuclear Studies, 1961-63; research engr, asst dir measurement and controls lab Research Triangle Inst, Research Triangle Park, NC, 1963-67; research prof nuclear engrg and chem engrg, dir Center Engrg. Applications of Radioisotopes, NC State Univ, 1967. Cons Oak Inst Nuclear Studies, Research Triangle Inst, Oak Ridge Nat Lab, Intl Atomic Energy Agy. NASA, AEC. Served to Lt AUS, 1956. Author: (with Ralph L. Ely, Jr.) *Radioisotope Measurement Applications in Engrg*, 1967. Contbr articles to sci jours. Recip of Radiation Industry Awd in 1984 from Isotopes and Radiation Div of the ANS. *Society Aff:* ANS, AIChE, FPS, SME of AIME

Garetson, Paul K
Business: 7454 E 46th St, Tulsa, OK 74145 *Employer:* Chem Equip Corp. *Education:* -/CE/OK A&M College; Third Officer Lic/Nav Sc/USMMA; -/CE/OK Inst of Tech; -/ME/Tulsa Univ. *Born:* 10/7/25. Paul was interested in chemistry at a very early age, having established a home lab where many analysis of local minerals were performed. During WWII he was a Midshipman in the Merchant Marine. While at the Acad, he met & married his wife Stannie Glowinski. They claim their greatest success in life has been their four children, all with Drs degrees or studying for one. He worked seventeen yrs for Std Oil Res Ctr before establishing his own co, Chem Equip Corp. He is a 1980 candidate for the natl office of Pres Elect of the Am Chem Soc. *Society Aff:* AAAS, NSPE, AIChE, AIC, ACS, ASME, ASTM.

Garfinkel, David
Business: Moore Sch of Elec Engg, Philadelphia, PA 19104
Position: Prof. *Employer:* Univ of PA. *Education:* PhD/Biochemistry/Harvard; MA/Honorary/Univ of PA; BA/Biochemistry/Univ of CA. *Born:* 5/18/30. Have been active in both comp & biological res since the mid-fifties. Principal interest is in the comp simulation of biological systems, in an interdisciplinary fashion, emphasizing metabolic systems, but also including physiological & pharmacological ones, and in development of expert systems related to these using artificial intelligence techniques. Secondary interests in comp sci developments related to principal interest: solving stiff differential equations, simulation languages, data base mgt, etc. *Society Aff:* IEEE-CS, IEEE-EMB, BMES, SCS, ACM.

Garg, Devendra P
Business: Sch of Engrg, Durham, NC 27706
Position: Prof of Mech Engrg. *Employer:* Duke University. *Education:* PhD/ME/NY Univ, NY; MS/ME/Univ of WI, Madison; BE/ME/Univ of Roorkee; BS/- /Agra Univ. *Born:* Mar 1936. Taught as Asst, & Assoc Prof at MIT, Chmn Engrg Projects Lab MIT, Prof Mech Engrg Duke Univ since 1972. Director of undergraduate studies. (1977-86) Recipient of T C M Award & N Y U's Founder's Day Award for outstanding scholastic achievement; Chairman, Vice-Chairman secretary and member executive committee-Dynamic Systems and Control Div ASME; Guest Ed of Two Spec Issues of ASME Transactions, one on Ground Transportation (June 1974) & the other of Ecological & Socio-Econ Sys (Mar 1976); Author of 2 bks & num tech papers. *Society Aff:* ASME.

Garing, Robert S, Jr
Business: 141 S Elm, Arroyo Grande, CA 93420
Position: Chrmn *Employer:* Garing, Taylor & Assoc, Inc. *Education:* BS/CE/Univ of AZ. *Born:* 6/23/20. Flagstaff AZ - Robert Sweet & Lucy May. Education: Elementary, High Schools, & Univ of AZ, Tucson, AZ. Married: Sarah Helen Kyser, 1943 Sons - Robert James and Russel Samuel Daughters - Mary Helen and Rose Ann. Non Profl Experience: Douglas Aircraft, subforman template loft US Navy South Pacific WW2 MOMM 2/C. Profl Experience: Asst Dist Design Engr, Asst Dist Matls Engr CA Div of Hgwy, Dist. V Chrmn Garing, Taylor & Assoc, Inc, Past Consulting City Engr, Arroyo Grande, Guadalupe, Lompoc, Paso Robles, Pismo Beach, Pres. District Engr, Lucia Mar Unified Sch Dist, CA, subdivisions, drainage, water, sewer, EIR handbook, Planning, Surveying. PE CA, AZ, NV - Past Chrmn CSPE - PEPP. Past Pres CSPE Chap Eleven Past Western Regional Vice Chrmn PEPP, NSPE Kiwanis. *Society Aff:* NSPE, PEPP.

Garland, Eric C
Business: Rm 104 Old Arts Bldg-UNB, Fredericton, New Brunswick, Canada E3B 5A3
Position: Assoc. VP (Administration) *Employer:* Univ of New Brunswick. *Education:* Engr/Struct Engg/Stanford; MS/Struct Engg/Stanford; BSc/Mech Engg/Univ of New Brunswich; BSc/Civil Eng/Univ of New Brunswick. *Born:* Aug 1931. Native of Moncton, N B, Canada. On staff of U N B since 1955 and holds pos of Dir of Planning, Prof of Civil Engrg & assumed current pos of Asst V P Admin in 1974. Elected fac mbr of U N B Bd of Governors 1968-73. Prof organizational pos: Pres, Canadian Council of Prof Engrs 1975-76; Pres, Canadian Soc for Civil Engrg 1974-75; V Pres, Engrg Inst of Canada 1972-74; Hon Pres, Engrg Undergraduate Soc 1975-77. Pres, Assoc of Prof Engrs of New Brunswick 1972-73. Elected Fellow of Engrg Inst of Canada 1975 Awarded Queen's Silver Jubilee Medal 1977 Appointed Commissioner-N. B. Electric Power Commission 1981-. Awarded C.C. Kirby Award 1986 & CCPE Meritorious Award for Professional Service to Canadian Engineering 1987. *Society Aff:* CSCE, EIC, SCUP, FPRS, AIR, AEE.

Garlow, Gary M
Business: P.O. Box 3333, 916 South 17th St, Wilmington, NC 28406
Position: Exec VP *Employer:* Talbert, Cox & Assocs, Inc *Education:* BS/CE/NC State Univ; BA/Math & Economics/NC Wesleyan Coll *Born:* 11/15/42 Mr Garlow's professional responsibilities with the firm, and prior to joining the firm, have provided a broad background of experience in many phases of engrg and planning, including water and sewer engrg, streets and drainage, airports, planning and studies, recreation facilities, city planning, and municipal operations. He has directed major water and sewer projects, ranging in size from the very small to in excess of 26 million. Before joining the firm in 1976, Mr. Garlow was employed by J.A. Jones Construction Co and the City of Rocky Mount, NC, where he worked in varied positions, gaining experience in a wide field of engrg and planning activities. Hobbies: Sports, wood working and fishing. *Society Aff:* ASCE, NSPE, APWA, AME, ACEC

Garmire, Elsa
Business: Center for Laser Studies, DRB 19, Los Angeles, CA 90089-1112
Position: Prof *Employer:* Univ of Southern CA *Education:* PhD/Physics/MIT; AB/Physics/Harvard-Radcliffe *Born:* 11/9/39 Dir, Center for Laser Studies, and Prof of Electrical Engrg and Physics, USC. After her PhD in nonlinear optics under C.H. Townes at MIT, she was researcher at CIT, studying electro-optics, integrated optics, GaAs devices, and crystal growth. Garmire came to USC in 1974 and is presently developing waveguide switches, bistable optical devices, GaAs lasers, infrared waveguides and fiber optics. Fellow of IEEE, Fellow of OSA, Sr mbr, SWE. Author of over 100 papers and 5 patents, she is Assoc Editor of *Fiber and Integrated Optics*, on the Intl Commission for Optics, and on Board of OSA. She was the first woman on the USC Engrg Faculty. *Society Aff:* OSA, IEEE, SWE, AWIS

Garner, Harvey L
Business: Moore Sch-Elec Engrg, Philadelphia, PA 19174
Position: Prof. *Employer:* Univ of Pennsylvania. *Education:* PhD/EE/Univ of MI; MS/Physics/Univ of Denver; BS/Physics/Univ of Denver. *Born:* Dec 1926. Mbr of group that designed MIDAC & MIDSAC at Univ of Michigan. Res specialty is Computer Arith. Faculty mbr Dept of Elec Engrg & Dept of Communication &

Garner, Harvey L (Continued)
Computer Sci at Univ of Mich 1958-70. Dir & Prof Moore School of Elec Engrg, Univ of Penn 1970-77. *Society Aff:* IEEE, AAAS, ΣΞ, HKN.

Garoff, Kenton
Home: 49 Winding Way, Little Silver, NJ 07739
Position: Project Director. *Employer:* Palisades Institute for Research. *Education:* BA/Chem & Physics/Brooklyn College. *Born:* June 1918. Engr, Army Electronics Command 1942-74 in electron device R&D & mgmt pos. Retired after 15 yrs as Dir Electron Tube Div & currently Project Dir at Palisades. Army mbr Adv Group on Electron Devices to Dept of Defense for 15 yrs & served on other AdHoc study groups. Authored sev tech papers in microwave tubes. Mbr IEEE AdCom Group Electron Devices, Standards Comm, Ch Tube Techniques Conf, Program Comm Electron Device Res Conf, & Internatl Electron Device Meeting. IEEE Fellow 1968. Fellow Ctte 1972-76. *Society Aff:* IEEE.

Garr, Carl R
Business: 750 Third Ave, New York, NY 10017
Position: Vice Pres-International Operations *Employer:* ACF Industries, Inc. *Education:* PhD/Metallurgical Engr/Case Inst of Tech; MS/Physics/Case Inst of Tech; BS/Physics/Kent State Univ. *Born:* Apr. 4, 1927, Olean, N.Y.; S. Frederick H. J. and Mary Magdalene (Zimmerman) G.; m. Arlene Crawford, Dec. 20, 1947; children--Christine Weber, Elizabeth Reese, Anne H.; Engring. supr. Bettis plant Westinghouse Co., 1956- 58; Supt. tech. services, nuclear fuel ops. Olin Mathieson Chem. Corp., 1958- 62; Dir. engring. and research Albuquerque div. ACF Industries, Inc., 1962- 68; V.P. Research and Devel. N.Y.C. 1968-70; Pres., Chief Exec. Officer Polymer Corp. subs. ACF Industries, Inc., Reading, Pa., 1970-1976; V.P. ACF Industries, Inc., N.Y.C., 1976; Trustee of Albright College, Reading, Pa. 1971-1976; Dir of Bank of PA Carpenter Technology Corp, Reading; Polypenco Ltd., United Kingdon; Polypenco B.V., Holland; Polypenco S.R.l., Italy; N.V. Polypenco S.A. Belgium; Nippon Polypenco Ltd., Japan; Polypenco (Pty) Ltd., South Africa; Polydrop, S.A ., Spain; Carter-Weber, Inc. U.S.A. Served with USN 1944-46. *Society Aff:* ASM, AIME, SAE, NSNA, ΣΞ.

Garrelts, Jewell M
Home: 15 Brook Rd, Tenafly, NJ 07670
Position: Asst to Dean engr. *Employer:* Columbia Univ. *Education:* Doctor of Sci (hon)//Doane College; MS/Civ Engr/Columbia Univ; BS/Civ Engr/Valparaiso. *Born:* 10/25/03. McPherson, KS. Taught Cornell Univ 1925-27, Columbia Univ since 1927. Became Prof and Chrmn, Dept of Civil Engg and engg Mechanics at Columbia in 1946. Named Assoc Dean for grad work in engg in 1957. Retired from full-time service and became Special Asst to the Dean for Combined Plan prog in 1972. Profl experience includes consultantships with Waddel & Hardesty on the design of several notable bridges such as Marine pkwy Lift Bridge, St George's Tied Arch and Rainbow Arch Bridge at Niagara. ASCE, Dir 1955-57; VP 1966-67. IABSE. I retired from Columbia June 30, 1983. *Society Aff:* ASCE, ASEE, IABSE.

Garretson, Owen L
Business: P O Box 108, Farmington, NM 87401
Position: President. *Employer:* Consultant *Education:* BS/ME/IA State Univ *Born:* 2/24/12. Salem Iowa. Bailey Meter Co, Phillips Petroleum Co, engrg & mgmt pos. Pres Genl Tank Corp, United Farm Chem; Exec VP, Arrow Gas, Roswell, NM. and retired as Chrmn of the Board, Plateau, Inc Natl Propane Corp NY. Pres since 2/60, Plateau, Inc (petro refining), Farmington, NM and was Chrmn of the BD, Plateau, Inc. Dir Suburban Propane Gas Corp, Whippany, NJ. VP Garretson Equip, Mt Pleasant, Iowa. Dir Western Bk. Pres Independent Refiners 1975-76. NLPGA since 1954. NM LP-Gas Commission 1955-1975. Mbr Iowa Governor's Trade Commission to Europe 1970. Mbr NSPE, ASME, SAE, AIChE, ASAgE, Tau Beta Phi. 42 U S patents issued. *Society Aff:* NSPA, ASME, SAE, AIChE, TBΠ, ASAGE

Garrett, Donald E
Business: 911 Bryant Pl, Ojai, CA 93023
Position: President. *Employer:* Saline Processors, Inc. *Education:* PhD/Chem Engg/OH State Univ; MS/Chem Engg/OH State Univ; BS/Chen/Univ of CA, Berkeley. *Born:* Jul 5, 1923. R&D Engr, Group Leader, Army Spec Engrg Detach, Los Alamos 1943-46; Dow Chem Co Pittsburg Calif 1950-52; Union Oil Co, Brea Calif 1952-55; Mgr of Res, Amer Potash & Chem Corp, Trona, Calif 1955-60; Pres Garrett Res & Dev Co, La Verne Calif 1960-75; Exec V Pres, Res & Dev, Occidental Petro Corp 1968-75; Pres Saline Processors, Garrett Energy Res & Engrg, Ojai, Calif 1975- . Mbr AIChE, ACS, Tau Beta Pi, Sigma Xi; Kirkpatrick Chem Engrg Achievement Award 1963; Merit Award 1971; Ohio State Outstanding Alumnus 1971; Lamme Medal, 1975, Engr of Month, AIChE 1964; Mbr Genl Tech Adv Comm, Fossil Fuel Div, DOE, ERDA; Engrg Adv Comm, U of Calif; Over 200 patents & tech pubs. *Society Aff:* ACS, AIChE.

Garrett, Luther W, Jr
Home: 537 Virginia Ave, San Mateo, CA 94402
Position: Pres. *Employer:* Garrett Assoc, Inc. *Education:* BS/ChE/Univ of TX; 30 credits MS work/ChE/Poly Inst of Brooklyn. *Born:* 4/26/25. Birthplace Corsicana, TX. USNR service 1944-45. Joined M W Kellogg Co 1947 as chem engr. Selected for team to engr Sasol Synthetic Fuels complex, 1951. First process design engr, subsequently proj mgr. From 1957-63 various engg mgt positions at Kellogg. Was Mgr of projs, then vp operations Swindell-Dressler 1963-69. With Bechtel Corp as engg Mgr 1969-74. Joined Fluor Utah as sr vp operations in 1974, serving as exec vp & pres to 1978. Is pres of Garrett Assoc, Inc, consulting engrs, 1978 to present. Is patentee in field and reg PE. *Society Aff:* AIChE, AIME.

Garrett, Richard Marvin
Business: 1116 Eighth St., B, Manhatten Bch, CA 90266
Position: President. *Employer:* Margar Co, Div of Margar Engrg Corp. *Education:* MBA/Sales Mgmt/Univ of CA at Berkeley; BS/Chem Engr/Ohio St Univ; Grad work/Sales Mgmt/Univ of Chicago-Ohio St *Born:* Jun 1921. Production engr for Rohm & Haas: 1943-44. Naval Officer 1944- 46. Instrument sales engr: 1948-52 with Taylor Instrument Co. Co-founder Silver Plastic Co, Chem-Nickel Co, Carbon Wool Corp 1952-54. Founder & Pres Margar Co & Heatcon Corp from 1953-83, involving sales & distrib of chem engrg process materials & equip. President, Margar Engineering Corp, 1983 to present. P Chmn The Southern Calif Sect of AIChE. P Pres of Los Angeles Council of Engrg Soc. Pres of Amer Cetacean Soc. *Society Aff:* AIChE, NACE, ACS

Garrett, Roger E
Business: Dept of Agri Engr, Davis, CA 95616
Position: Prof *Employer:* Univ of CA. *Education:* PhD/Agri Engr/Cornell Univ. *Born:* 9/23/31. Worked with the John Deere Tractor Works in Waterloo, IA, before joining the Univ of CA staff in 1962. He teaches design and metal working and does res leading to mechanized systems for aquaculture, plant tissue culture, and crop production, harvesting and handling. He is a former Dir of the ASAE and former Chrmn of the UCD Agri Engg Dept. *Society Aff:* ASAE.

Garry, Frederick W
Business: 3135 Easton Turnpike, Fairfield, CT 06431
Position: VP *Employer:* GE *Education:* BSME/ME/Rose-Hulman Institute of Technology *Born:* 07/12/21 Native of Stratford CT. Graduated from Rose-Hulman Inst of Tech & joined GE's Aircraft Engine Group (AEBG) in 1951. Elected VP AEBG Technical Div 1968. Doctor of Engrg Degree (honorary) Rose-Hulman 1968. Joined Rohr Industry as Pres 1974; became Chrmn and CEO 1976. Rejoined GE in 1981 as VP Corp Engrg and Manufacturing. Elected to National Acad of Engrg in 1981. Serve as Trustee, Clarkson Univ and mbr Bd of Mgrs, Rose-Hulman. *Society Aff:* SAE, SME.

Gartner, August J
Business: 225 W. Randolph St - HO 29D, Chicago, IL 60606
Position: Asst VP *Employer:* IL Bell Tel Co *Education:* MS/Industrial Mgmt/MIT; BS/EE/Univ of IL *Born:* 10/11/26 in Highland, IL. Attended grade and high school in Edwardsville, IL. Served 2 years in the US Navy. Began career with IL Bell in 1950. Has had assignments with Bell Telephone Labs and AT&T. Received a Sloan Fellowship in 1962 for graduate study at MIT. A Registered PE in IL. A member of the Executive Club of Chicago, Economic Club of Chicago and the IEEE. A member of the Bd of Dirs of the Western Society of Engrs. Married to Lola Etzkorn. Three children - Nancy Witt, Warren, Diane Boley. Resides in Park Forest, IL. *Society Aff:* WSE, IEEE

Gartner, Joseph R
Home: 147 Davis Rd, Storrs, CT 06268
Position: Prof. *Employer:* Univ of CT. *Education:* PhD/ME/Univ of WI; MS/ME/Univ of WI; BS/ME/Univ of IL. *Born:* 11/7/30. Native of Chicago, IL. Consultant to several New England cos in the field of structural vibrations. In addition to educational background, one yr (1966) post doctoral res at Univ of Birmingham (England), six months res activity at Mech Engg Faculty, Univ of Belgrade (Yugoslavia) (1979), six months res activity at Mech Engg Faculty, Edvard Kardelj Univ (Ljubljana, Yugoslavia) (1985), & eight months res activity at the Technical University Munich (West Germany). *Society Aff:* ASME, AGMA, ASEE, SEM, AAUP.

Garver, Robert V
Business: 2800 Powder Mill Rd, Adelphi, MD 20783
Position: Sup Physicist *Employer:* Harry Diamond Labs *Education:* M/EA/George Wash Univ; BS/Physics/Univ of MD *Born:* 6/2/32 Employed as Harry Diamond Labs 1956 - present. 15 years research on microwave diode switches, limiters, and phase shifters. 5 years project engr for the Minirad Mod A proximity fuze (now on the Mark XII A warhead). 5 years heading up assessment teams for the survivability of major defense communications systems to nuclear electromagnetic pulse environments. 5 years directing the U.S. Army High Power Microwave Hardening Technology development. Published 34 govt reports, 24 journal articles, 1 book, and issued 12 patents. IEEE fellow. IEEE assignments: MTT Adcom 1971-1973, Solid-State Circuits Council 1966-1969, assoc Ed Journal of Solid-State Circuits 1969-1973, Technical Program Committee Chrmn 1971 Intl Microwave Symposium. *Society Aff:* IEEE

Garvey, James R
Home: 812 North Ridge Dr, Pittsburgh, PA 15216 *Education:* Bach Eng/Mining/OH State. *Born:* Jun 8, 1919 Pittsburgh, Pa. Married Bettina Margaret Craig 1945; children--Craig Randall & Lynn Garvey Handford; mining engr Pittsburgh Coal Co 1941-42; Captain USAAF 1942-46; Capital Airlines, Washington 1946; project engr to pres Bituminous Coal Res Inc, Monroeville, Pa 1946-80; Retired 1980; exec v p, Natl Coal Assn Washington 1971-73. Percy Nicholls award ASMEAIME 1963; disting alumnus Ohio State Univ 1972. Mbr ASME, AIME, N Y Acad of Sci, Amer Acad of Sci, Amer Assn for Advancement of Sci, Internatl Comm Coal Res. Author of papers on coal res & associated subjects. *Society Aff:* AIME, ASME.

Garvin, Clifton C, Jr
Business: 1251 Ave of Amer, New York, NY 10020
Position: Chmn of Bd & Ch Exec Officer. *Employer:* Exxon Corp. *Education:* MS/Chem Engrg/VA Polytech Inst; BS/Chem Engrg/VA Polytech Inst *Born:* 12/22/21 Portsmouth, VA. Bach Chem Engrg 1943 & Masters Chem Engrg 1947 VA Polytech Inst; Engr-exec-oil & chem oper, Exxon Co USA, Baton Rouge, LA, & Houston, TX 1947-64; Exec Asst to Pres and then Chrmn Exxon Corp NYC 1964-65; Pres Exxon's Domestic and then Intl Chem Co NYC 1965-68; Dir, VP Exxon Corp NYC 1968, Exec VP 1968-72, Pres 1972-75, Chrmn of Bd & Chief Exec Officer 1975- . Dir: Citicorp & Citibank, N A; Pepsi Co, Inc; Johnson & Johnson, Amer Petro Inst; Chrmn The Bus Coun (until Feb 1985); VChrmn Bd of Mgrs Sloan-Kettering Inst for Cancer Res; VP Bd of Trust Vanderbilt Univ; Trustee Ctte for Econ Dev; Mbr Bd of Gov United Way of Amer; Sr Mbr The Conference Bd; Mbr: The Bus Roundtable, Coun on Foreign Relations, Inc; Natl Petro Coun; Labor-Mgmt Group; ASC; AIChE; Soc of Chem Indus, Amer Sect. Holds honorary degrees from NY Univ and Stevens Inst of Tech. *Society Aff:* ACS, AIChe, SCI

Gary, James H
Business: Colorado School of Mines, Dept of Chem Engg, Golden, CO 80401
Position: Prof. *Employer:* CO Sch of Mines. *Education:* PhD/ChE/Univ of FL; MS/ChE/VPI; BS/ChE/VPI. *Born:* 11/18/21. Native of Victoria, VA. Served Lt to Maj. Army Anti-aircraft Artillery, Pacific Theater, 1942-46; Technical service engr & hd of pilot plant dev lab, Std Oil Co (OH), 1946-52. Asst Prof ChE, Univ VA, 1952-56; Assoc Prof & Prof, Univ AL, 1956-60; Prof Chem & Petrol - Refining Engg, CO Sch of Mines, 1960-; Hd Dept chem & Pet Ref Engg, 1960-72; VP Academic Affairs & Dean of Faculty, 1972- 79; Trustee, CSM Res Inst 1970-72, 1980-83; Prin Investigator, Synthetic Fuels Res, 1960-; Co-author "Petroleum Refining Technology--, Dekker Pub Co, 1984; Chrmn, oil shale committee, OTA, 1978-80; Dir, CSM Oil Shale Symposia, 1964- Fellow, AIChE; Halliburton Award for Professional Achievement, 1981; Fellow, AAAS; George R. Brown Gold Medal for Outstanding Contributions to Engineering Education, 1987. *Society Aff:* AIChE, AIME, ACS, ASEE, AAAS.

Gasich, Welko E
Business: 1840 Century Park East, Los Angeles, CA 90067
Position: Exec VP-Programs *Employer:* Northrop Corp *Education:* Professional Degree/Aero Engr/CA Inst Tech; MS/Aero Engrg/Stanford Univ; AB/ME/Stanford Univ. *Born:* 3/28/22 Aerodynamacist - Douglas Aircraft Co 1943-44, Supervisor-aeroelastics 1947-51; Chief aero design - Rand Corp 1951-53; Chief prelim design-aircraft div - Northrop Corp 1953-56, Dir Adv Systems 1956-61, VP & Asst gen mgr-technical 1961-66, Corp VP & Gen Mgr - Northrop Ventura Div 1967-71, Corp VP & Gen Mgr - Northrop Aircraft Div 1971-76, Corp VP & Group Exec - Northrop Aircraft Group 1976-79, Sr VP Adv Projects 1979-85, Exec VP-Programs 1985. *Society Aff:* AIAA, SAE, NAE.

Gaskell, David R
Business: Materials Engineering, Purdue University, West Lafayette, IN 47907
Position: Prof *Employer:* Purdue Univ *Education:* PhD/Metallurgy/McMaster Univ-Canada; BSc/Metallurgy & Tech Chem/Univ of Glasgow-Scotland *Born:* in Glasgow, Scotland. Employed 1962-64 as a metallurgist with Laporte Chems Co Luton, England specializing in corrosion and chem plant failure. Joined the Univ of PA in 1967. Areas of specialization; Chem and Extraction Metallurgy with emphasis on the thermodynamics and kinetics of metallurgical refining reactions. Author of *Introduction to Metallurgical Thermodynamics*-Hemisphere Publishing Co. 1st edition 1973, 2nd edition 1981. Recipient of the 1977 McMaster Univ Distinguished Alumni Award. Joined Purdue Univ in 1982. *Society Aff:* TMS-AIME, ASM, CIM, ISIJ

Gaskill, Jack D
Business: Optical Sciences Center, Tucson, AZ 85721
Position: Prof. *Employer:* Univ of AZ. *Education:* PhD/EE/Stanford Univ; MS/EE/Stanford Univ; BS/EE/CO State Univ. *Born:* 12/9/35. Jack D Gaskill received his BS degree from CO State Univ in 1957, and his MS & PhD degrees from Stanford Univ in 1965 and 1968, respectively, all in electrical engg. During 1957 he was an electronics engr with Motorola, Inc, and from 1958 to 1963 he served with the USAF as an instructor pilot and academic instructor. In 1968 he joined the faculty of the Univ of AZ as an asst prof in the Sciences Ctr, where he is currently Asst Dir, Academic Affairs and Prof of Optical Sciences. Dr Gaskill is a Fellow and Governor of SPIE - the Intl Soc for Opt Engg, and serves as Ed of that society's journal, Optical Engineering. He is also a Fellow of the Optical Soc of Amer. *Society Aff:* SPIE, OSA, IEEE.

Gasner, Larry L
Business: Dept of Chem & Nucl Engg, College Park, MD 20742
Position: Assoc Prof. *Employer:* Univ of MD. *Education:* PhD/Biochem Eng/MIT; MS/CE/MIT; BS/CHEME/Univ of MN. *Born:* 5/17/43. Environmental Res & Consulting Engr, Betz Labs, 1970-1974 in wastewater treatment, chemistry, process & plant innovation & design. Private consultant 1974-present. Joined faculty, Dept of Chem Engr, Univ of MD, 1974-present. Currently Assoc Prof with tenure. Res areas: Biochemical Engg, Microbial Polysaccharides, Fluidized bed combustion of coal, gas-solid chem reaction kinetics & reactor optimization. Coal conversion catalysis including combustion, liquefaction & gasification. Consultant to industry, govt, & private institutes. Reg PE in PA. *Society Aff:* AIChE, ACS, AAAS.

Gasser, Eugene R
Business: 17904 Georgia Ave Ste 316, Olney, MD 20832
Position: Pres. *Employer:* Gasser Assoc, Inc. *Education:* BS/Chem Engg/Case Inst of Tech. *Born:* 8/5/25. Native Pennsylvanian. Has been associated with Armstrong Cork, Gen Elec, Allis-Chalmers, Southern Nuclear Engg and presently is Pres of Gasser Assoc, Inc. Actively engaged in nuclear energy since 1952, having specialized in systems design including reactor protection and control systems, for light water, liquid metal, and gas-cooled reactors. Attained position of VP with SNE prior to forming Gasser Assoc. Serving as educational advisor for Case Inst of Tech in promoting and encouraging engg profession among high school students. Mbr of ISA, Alpha Chi Sigma and Sigma Alpha Epsilon. Reg PE in CA, MD and WI. *Society Aff:* ISA, ACS, SAE, NSPE.

Gasser, Robert F
Home: 104-78 114 St, Richmond Hill, NY 11419
Position: Exec VP. *Employer:* Ecolotrol, Inc. *Education:* MCE/Civ Engg/NYU; BCE/Civ Engg/Manhattan College. *Born:* 10/9/44. Native of Queens, NY. Worked at Manhattan College's Environmental Engg Lab performing res studies on the biological treatment of various municipal & industrial wastewaters. In 1969 started up Ecolotrol, Inc, a co selling pollution control equip for municipal & ind wastewater treatment. Present position is exec vp & have responsibility for sales. Licensed Engr in NY. *Society Aff:* WPCF.

Gassett, Richard B
Business: 1042 US Highway 1, North, PO Box 607, Ormond Beach, FL 32074
Position: Dir of Engrg *Employer:* Briley, Wild & Assocs, Inc *Education:* MS/CE/Univ of ME; BS/CE/Univ of ME *Born:* 06/15/37 Native of Massachusetts, Chief of Wastewater Sys Engrg, Metcalf & Eddy, Boston, MA 1964-74; author *Foam Separation of Surface Active Agents*; PE Maine, PE FL; Dir of Engrg for Briley, Wild & Assocs, Inc, Ormond Beach, FL 1974-87; specializing in feasibility planning and engrg design in water and waste water treatment and related fields, municipal, industrial, commercial, etc. Directed award winning THM control ozone water treatment plant projects in Belle Glade and Ormond Beach, FL. Pres, Florida Inst of Consulting Engrs, National Dir PEPP. Enjoys golf and bowling. *Society Aff:* NSPE, ASCE, WPCF, PEPP, ASME, AAEE

Gassner, Robert H
Home: 2324 Camino Escondido, Fullerton, CA 92633
Position: Conslt Metall Engr *Employer:* CNS Inc. *Education:* BS/Chemistry/Queens College. *Born:* 9/2/23. Native of NY. Bomber pilot with 35 WWII Combat missions. 36 yrs in Mat and Process Engg at Douglas Aircraft; supv 29 yrs. Retd 12/31/84. Responsibilities have include design consultation, failure analysis, mtl and process selection and procurement, specifications, mfg problems, heat treating, non-destructive testing and corrosion engg. Honorary member (active) of SAE/AMD/AMS Commodity Committees and Aero Metals Engg Ctte. ASM Fellow and mbr for 31 yrs; 1964-65 chrmn of LA Chapter. LA Consltg Editor for *Metal Progress* (1962-1986). Play tennis and golf. Since 1972, have managed, for ASM, the largest golf tournament and summer party in the world. *Society Aff:* ASM, SAE.

Gates, Albert S, Jr
Home: PO Box 163, New Smyrna Beach, FL 32070
Position: Consultant (Self-employed) *Born:* Jan 4, 1909. Univ of Maine, MIT, George Washington Univ. Navy Bureau of Ships, respon for submarine air cond & environ control until 1961. Natl Inst of Health, Ch Engrg, Environ Services until 1972. Presently cons for Environ Air Control Inc; Amer S F Products and others. Disting Civilian Serv Award, US Navy. Fellow ASHRAE. Formerly chmn Physiology & Human Comfort Ctte, mbr sorption, shelters & heat recovery cttes. ASHRAE Disting Serv Award. Sailor. PE District of Columbia *Society Aff:* ASHRAE.

Gates, David W
Home: 117 Abby Lane, Portland, ME 04103
Position: Mgr Civil & Construction Services *Employer:* E. C. Jordan Co., Portland, ME *Education:* BS/Civ Engg/Univ of ME. *Born:* 6/9/32. Native of Melrose, MA. BSCE Univ of ME, 1954. Jr Hgwy Engr US Bureau Public Rds, Missoula, MT 1954; Platoon Leader US Army Signal Corps 1955-1957; Hgwy Engr SSV&K, NY 1957-1959; Design Engr Levitt & Sons, Inc, Willingboro, NJ 1959-1961; Proj Engr, Proj Mgr, Assoc, VP, Chief Engr, Proj Mgr, James P. Purcell Associates, Inc, Glastonbury, CT 1961-1979. Manager Civil & Construction Services, E. C. Jordan Co, Portland, ME, 1980-81. Past Pres CT Section ASCE (1977), Past Pres CT Soc Civ Engrs (1977). Married, 4 children. *Society Aff:* ASCE, WPCF, NSPE.

Gates, Leslie C
Business: Post Off Drawer AF, Beckley, WV 25802-1854
Position: Pres *Employer:* Leslie C. Gates, Limited *Education:* BS/CE/VPI. *Born:* Nov 1918. Received CE degree from VPI 1940. US Army 1941-46. Discharged rank of Major. Chmn and CEO Gates Engrg Co. Firm is active in all phases of Mining & Civil Engrg & Architecture. Became Pres 1962, owner 1958-62, partner 1955-58, assoc 1946-54 Retired 1982. Chmn, L. A. Gates Company, 1986- . Pres NSPE 1974-75, Mbr AIME, ASCE & AMC. Mbr W Vir Registration Bd for Prof Engrs 1965-80; mbr W Vir Chamber of Commerce. PPres, WVA Chamber of Commerce, Pres WVA C of C - 1978-79. Other mbrships incl AWWA, W Vir Coal Mining Inst, Amer Rd Bldrs Assoc, Colo Mining Assn, Ill Mining Assn, Amer Concrete Inst, Amer Arbitration Assn, ASEE. Dir Cardinal State Bank, Trustee Engrg Index. Mbr American Mining Congress 1981 Resolution Comm; Visiting Ctte - West Virginia Univ - College of Engrg; Ctte of 100 - Virginia Polytechnic and State Univ - Coll of Engrg. *Society Aff:* NSPE, WVSPE, AIME, ASCE, AWWA, AMC.

Gates, Lewis E
Business: 5870 Poe Ave, Dayton, OH 45414-1123
Position: VP Engg & Development *Employer:* Chemineer Inc. *Education:* BCE/CE/OH State Univ; MS/CE/OH State Univ. *Born:* 4/10/41. Native of Gallipolis, OH. Author or coauthor of six articles on fluid agitation, mixing, & heat transfer. Currently responsible for product engg, res & dev, market communications, technical sales, & customer service depts of Chemineer Inc. Ind experience with DuPont & Chemineer. Reg PE in OH. *Society Aff:* AIChE.

Gates, Marvin
Business: 1014 Wethersfield Avenue, Hartford, CT 06114
Position: Vice President. *Employer:* Gates-Scarpa & Assocs Inc. *Born:* 10/29/29. in New York City. Registered Professional Engineer, Planner, and Landscape Architect; Certified Cost Engineer and Cost Analyst. Recipient Conn. Society Civil Eng. Elwood Nettleton Award 1960; ASCE James Laurie Prize 1962; ASCE Walter Huber Civil Engineering Research Prize 1971; Tau Beta Pi Eminent Engineer 1975; Chi Epsilon honor Engineer 1976. Dir, So.W.E. Section AACE; Arbitrator for Am. Arb. Assoc.; Past-Chairman ASCE Construction Division Executive Committee. Past-Director Tau Beta Pi and Connecticut Society of Civil Engineers. Formerly distinguished visiting Professor and Head, Construction Engineering Programs, Univ. of Hartford. Guest Lecturer and Seminar Leader on Cost Engineering throughout United States. Author numerous papers construction economics. Since 1961 consulting civil engineer with assignments throughout the world. Attended

Gates, Marvin (Continued)
Cooper Union School of Engineering. *Society Aff:* NSPE, ASCE, ASCE, IEEE, ASME

Gatewood, Buford E
Home: 2150 Waltham Rd, Columbus, OH 43221
Position: Prof Emeritus. *Employer:* OH State Univ - Retired. *Born:* Aug 1913. MS & PhD Univ of Wisconsin, BS La Tech Univ, PE state of Ohio; 1939-42 asst prof mathematics La Tech; 1942-46 stress analyst to acting chief structural engr McDonnell Aircraft Corp; 1947-60 assoc Prof to Prof & head Mech Dept, to res coordinator, at Air Force Inst of Tech; 1960-1978, Prof Aerospace Engrg at Ohio State Univ. Cons for var aircraft & airline companies. Author 'Thermal Stresses', & about 45 tech papers. Mbr ASME, AIAA, ASEE, SESA, MAA, Sigma Xi. Retired, Prof Emeritus OSU.

Gathy, Bruce S
Business: Engg Dept, New London, CT 06320
Position: Prof. *Employer:* US Coast Guard. *Education:* PhD/Nuclear Engg/IA State Univ; MS/Heat-Power Engg/Stevens Inst of Tech; BS/Gen Engg/US Coast Guard Academy. *Born:* 8/2/34. Commissioned Officer US Coast Guard, 1956-59; Civilian teaching faculty, Coast Guard Academy, 1961-; Ford Fdn Grantee, 1966-67; Tenured prof, 1974-; Chief-Marine, Nuclear, Ocean sec, 1973-76; Developed marine (ECPD accredited) and nuclear progs; Natl Academy of Sci Risk Analysis Panel, 1969-75; Chrmn, Committee on Relations with Colleges and Univs, Region I, ASME, 1974-77; PE, State of IA; Visiting Prof, IA State Univ Engg Res Inst, 1977-79; Designed and installed microprocessor-controlled heat-pump system for ISU Energy Res House, utilizing outdoor air/solar/water as energy sources to heat and cool with 17 different modes. *Society Aff:* ANS, ASEE, ASME, ASNE, ΣΞ.

Gatti, Emilio
Home: Via Lambro 32, Lesmo Italy 20050
Position: Full Prof Hd Physics Dept. *Employer:* Politec of Milano Ministry of Pub Ed. *Education:* Master/Electronics/Politecnico Torino; Grad/EE/Univ of Padova. *Born:* Mar 18, 1922 Torino, Italy. Grad in elec engg Padova Univ 1946. Master in electronics at Politecnico of Torino 1947. Assoc prof 1951, full prof 1957, Vice Dean of Politecnico of Milano 1969-72. Fellow IEEE 1972. Ed "Alta Frequenza–, the scientific journal of AEI (Associazione Elettrotecnica ed Elettronica Italiana). Winner of the prizes Bianchi (AEI 1956), Righi (1972 AEI), Della Riccia (1968), Pubblicazioni (1982 AEI), Feltrinelli (1986 Accademia dei Lincei). Mbr of "Instituto Lombardo di Scienze e Lettere–. Mbr of "Accademia Nasionale delle Science detta dei XL–. Invited sr scientist at Brookhaven Natl Lab 1973, 74, 75, 76, 77, 78, 79, 80, 81, 82, 83, 84, 85, 86. Main int: electronic instrumentation for nuclear physics experiments, radiation detectors, info & noise as related to limit in measurement accuracy. Enjoy mountaineering, music & arts. Married with Laura Semenza. Sons: Gabriella, Aldo, Carlo, Annapaola. *Society Aff:* SIF, AEI.

Gatti, Richard M
Home: 618 Warren Blvd, Broomall, PA 19008
Position: Engineer In Charge, Communications & Controls *Employer:* Phila Elec Co. *Education:* MEES/EE/Penn State Univ; BEE/EE/Villanova Univ. *Born:* 1/20/47. Married, 2 children. Reg PE, PA. Phila Elec Co. 1969-present. Engr-In-Charge, Communications & Control. Responsibilities include management and administration of engineers and technicians charged with testing and maintaining all protective relays, communications and supervisory contol systems throughout the PE Co service territory. Adjunct faculty member at Villanova Univ; College of Engrg. Recipient of Villanova's "Gallen Award" as Outstanding Engrg Alumnus. Civic, social and sporting activities. Enjoy music and photography. *Society Aff:* HKN.

Gattis, Jim L
Business: College of Engineering, University of Arkansas, Fayetteville, AR 72701
Position: Assoc Dean *Employer:* Univ of AR *Education:* PhD/EE/Purdue Univ; MS/EE/Univ of AR; BS/EE/Univ of AR *Born:* 8/11/43 A native of Ozark, AR, and a graduate of the Univ of AR and Purdue Univ. Has worked for Phillips Petro Co and System Development Corp. Has worked in the area of image processing and pattern recognition since 1967. Joined the Univ of AR faculty in 1972. Currently serves as Associate Dean of Engineering at the Univ. of Ar. *Society Aff:* IEEE, ASEE, ΣΞ, HKN

Gaudy, Anthony F, Jr
Business: Dept of Civil Engrg, Newark, DE 19711
Position: H Rodney Sharp Prof of Civ Engrg. *Employer:* Univ of DE. *Education:* PhD/Sanitary Engrg/Univ of IL; MS/Sanitary Engrg/MIT; BS/Civil Engrg/Univ of MA. *Born:* Jun 16, 1925. Army Air Corps 1943-46. Genl civil engrg prior to 1955; indus wastes investigations, NCSI, pulp & paper indus prior to 1957. Educator since 1959. Asst Prof Univ of Ill 1959-61; Assoc Prof Oklahoma St Univ 1961-63; Professor Oklahoma St Univ 1963-1979. Edward R Stapley Prof of Civil Engrg, Oklahoma St Univ 1968-79. H Rodney Sharp Prof of Civ Engrg 1979- . Chrmn Dept Civ Engrg Univ of DE 1979-84. Over 200 pubs in pollution control area. Active mbr 13 prof orgs. Dir WPCF 1962-65; Dir AEEP 1973-75; Chmn Environ Engrg Div ASEE 1971; Fellow ASCE. Trustee Amer Acad Environ Engrs 1977-82. Harrison P Eddy Award 1967 (WPCF). VP, Thomas, Gaudy, McCaskill Inc, Memphis, Tenn. VP, Rozich & Gaudy Inc. Newark, Dela. *Society Aff:* ASCE, ASM, ACS, NSPE, WPCF, AWWA, ASEE, IAWPR, IWRRA, AEEP, AIChE, AAEE, ISSE.

Gaum, Carl H
Home: 9609 Carriage Rd, Kensington, MD 20895
Position: Chief Engineer *Employer:* Gaum & Associates *Education:* BS/CE/Rutgers Univ; Graduate Programs MS equivalent/University of Oklahoma; /Water Resource Planning & Management/US Army Corps of Engineers Board of Engineers for Rivers and Harbors. *Born:* 7/29/22 Served in Air Corps, India, Burma 1942-45. Groundwater hydrologist, USGS. Consultant; geologic engrg, site planning, economic evaluation. Responsible for planning and design of water resources, river and harbor projects, shore protection, pollution abatement, environment for U.S. Army Corps of Engrs. Retired from Office Chief of Engrs as Chief of Central Planning. Past Dir of engrg review and evaluation, earth sciences section Greenhorne & O'Mara Inc. First chrmn ASCE Water Resources Planning and Management Division, Pres, Trenton, NJ Branch, VP Cincinnati Section, Dir & VP & Pres National Capital Section. Dir NJSPE, Pres Mercer County Chapter. Chrmn of Publications Committee, Pianc, Leningrad, Edinburgh, Brussels and Oseka Navigation Congresses. Currently Chief Engr., Gaum & Associates. Chairman, Dist 5 Council ASCE Zone 1. *Society Aff:* ASCE, AAAS, NSPE, PIANC.

Gaumer, Lee S
Business: PO Box 538, Allentown, PA 18105
Position: Tech Dir Govt Sales *Employer:* Air Products & Chem. *Education:* BS/Chem Engr/Penn State Univ. *Born:* 5/27/26. Native of Lehighton, PA. Served in Army Corps of Engrs 1946-47, 1954-55. Manhattan proj Univ of Chicago. White Sands Proving Ground. With Argonne Natl Lab 1948-52. As Chem Engr responsible for fuel separation pilot plants. With Air Products & Chem since 1952. Qualified PE state of PA. Supervised up to 50 chem engrs in process design 1958-1976. Dept hd process systems group R&D, 24 professionals, 1976-1980. Gen Mgr Tech Planning & Research Services, 50 professionals 1980-1984; Tech Dir Corporate Govt Sales, 1984-present. 15 patents. Chem Engg, Engr of the Yr, 1976-Enjoy music, sports, reading. *Society Aff:* AIChE.

Gautreaux, Marcelian F, Jr
Home: 1662 Pollard Pkwy, Baton Rouge, LA 70808
Position: Sr VP Consultant (Ethyl Corp.) *Employer:* self *Education:* PhD/ChE/LA State Univ; MS/ChE/LA State Univ; BS/ChE/LA State Univ *Born:* 1/17/30 Nashville, TN. Married Mignon Thomas 1952. 4 children. With Ethyl Corp 1951-55. Taught at LA State Univ 1955-58. Returned to Ethyl in R & D Dept, 1958. Responsible for R & D Dept 1969-1980. VP 1969. Sr VP 1974-86. Member of Bd

Gautreaux, Marcelian F, Jr (Continued)
1973-present. Advisor to Executive Committee, 1981. Elected to NAE 1976, Natl Awards: 1968, Personal Achievement in Chemical Engrg, 1978, CMRA Memorial Award, 1983, CMRA Honorary Mbr, two best paper awards in 1952 and 1980. Also, recipient of local Charles E. Coates Memorial Award. Dir of SES, 1972-75. Fellow AIChe, 1981 and Charter Member of the LA State Univ Engrg Hall of Distinction. Sci Advisory Chair, LSU, named in his honor. Inst for Amorphous Studies. Hobbies, golf and piano. *Society Aff:* AIChE, CMRA, SES, SCI, NAE

Gauvenet, Andre J
Home: 31, rue Censier, Paris 75005 France
Position: Prof *Employer:* Ecole Nationale Superieure de Techniques Avancees (ENSTA) *Education:* Professorship degree/Phys/Ecole Normale Superieure De St-Cloud (PhD equiv) *Born:* 03/31/20 I have been trained as a Phys Prof. I shifted early to Res in Engrg first in Electronics (1945-1948). I came back afterwards to Univ work with the Ecole Normale Superieure de St-Cloud. From 1954 to 1956 I was Sci Attache with the French Embassy in the US. Returning to France in 1956 I was appointed as an Engr to the Commissariat a l'Energie Atomique (French AEC). In 1963 I became Asst to the High Commissioner. I soon specialized in Radioprotection and Safety, becoming Dir for Safety & Radiological Protection. From 1982 to 1985 I was "Inspecteur General" at Electricite de France (the French government owned utility) for Nuclear Safety & Security. I retired in April 1985 and am now Prof of Nuclear Safety at the Ecole Nationale des Techniques Avancees (linked to Ecole Polytechnique). *Society Aff:* ANS, APS

Gauvin, William H
Business: 240 Hymus Blvd, Pointe Claire, Quebec Canada H9R 1G5
Position: Dir of R&D. *Employer:* Noranda Res Ctr. *Education:* PhD/Physical Chemistry/McGill Univ; MEng/CE/McGill Univ; BEng/CE/McGill Univ; DEng/-/Waterloo Univ. *Born:* 3/30/13. Native of France. Immigrated to Canada in 1929. Assoc Prof, Dept of Chem Engg, McGill Univ, prior to joining Pulp & Paper Res Inst of Canada, & Noranda Mines Limited. Dir of Res &Dev for Noranda Res Ctr, & Sr Res Assoc & Dir of Plasma Tech Group, McGill Univ. Over 130 papers in the fields of electrochemistry, high-temperature heat & mass transfer, fluid mechanics & particle dynamics. Patents in high-temperature chem processing. Companion of Order of Canada (1975). *Society Aff:* AAAS, AIChE, AMA, TAPPI, ΣΞ.

Gavert, Raymond B
Business: Bldg 207, Ft Belvoir, VA 22060
Position: Prof. *Employer:* Defense Sys Mgt College. *Education:* PhD/Met Engr/Univ of WI; MS/Met Engr/Univ of WI; BS/Met Engr/Purdue Univ. *Born:* 9/27/32. Native of Jamestown, NY. Twenty yrs experience in mtls engg & high tech products. Supervised resolution of mtls problems for isotope generator used in Apollo lunar experiments. Promoted in 1969 to GE Co Corporate Engg & R&D offices. Mtls information consulting & special nuclear & ind studies. From 1976- 1978 given Wash assignment with US Congress-OTA. Developed critical mtls policy alternatives & assessed impacts of tech for the Congress. He is a Prof of Technical Mgt at the Defense Systems Mgt College resolving Dept of Defense res, dev, engg and nuclear systems problems. PE (PA), NASA Tech Transfer Award molybdenum dev. Enjoys civic affairs. *Society Aff:* ASM, AAAS, WFS.

Gavin, Joseph G, Jr
Business: Grumman Corporation, 1111 Stewart Ave, Bethpage, NY 11714
Position: Senior Management Consultant. *Employer:* Grumman Corp. *Education:* MS/Aero Eng/MIT; BS/Aero Eng/MIT *Born:* 9/18/20. Somerville, Mass. USNR 1942-46. Grumman Aircraft Engrg Corp 1946. Lunar Module Dir, Apollo Program Grumman. Dir: Grumman Corp. Mbr: Admin and Exec and Special Projects Ctte of Grumman Corp. President and Director Grumman Corporation Jan. 1975 - Feb. 1985. Dir: Pine Street Fund; Chrmn, Ctte on Intl Coop in Magentic fusion of the Natl Res Coun. Mbr of The Corp and Dir of the Charles Stark Draper Lab, Inc.; Member: MIT Corp. Natl Acad Engrg. Trustee Huntington Hosp. *Society Aff:* AIAA, AAS, NAE.

Gavlin, Gilbert
6500 Kenton Ave, Lincolnwood, Ill 60646
Position: President. *Employer:* Gavlin Assoc, Inc. *Education:* BS/Chem Engr/Univ of IL; PhD/Organic Chem/Cornell Univ *Born:* Jan 12, 1920. Married Feb 1947 Carolyn Epting, 3 children-Suzanne, Nancy, Patricia. Manhattan Dist, Columbia Univ, Res Chemist 1943-45; Tenn Eastman Corp, Oak Ridge, Tenn, Res Chem 1945-46; Armour Res Foun, Chgo, Res Chemist 1946-47; Cornell Univ, Ithaca, N Y, Res Asst 1947-48; Armour Res Foundation, Chgo, Proj Leader, Dir Natl Register Rare Chemicals 1948-54; Richardson Co, Melrose Pk, Ill, Mgr Chem Res 1954-64; Poly-Synthetix Inc, Chgo, Pres 1964-69; Custom Organics Inc, Chgo, Pres 1969- . Mbr Amer Chem Soc, Amer Inst of Chem Engrs, Tau Beta Pi, Sigma Xi, Phi Lambda Upsilon; B.S. Univ of Ill. 1941, PhD Cornell Univ. 1948. *Society Aff:* ACS, AIChE, ASTM

Gayer, George F
Home: 26486 N Oak Highland Dr, Newhall, CA 91321
Position: Plant Manager. *Employer:* Westinghouse Elec Corp-ret. *Education:* MS/ME/OSU; EDP/Bus/Stanford; BS/ME/OSU. *Born:* May 22, 1908. EDP Stanford Bus School 1952. Grad Student 1929 E Pittsburgh, Apprentice S Phila. Westinghouse, Exp. Lab Technician, Marine Engr, Sect. Engr. Marine Div. 1935, Chief Test Engr. Todd Ship, Tacoma 1940, Chief Engr. Joshua Hendy Iron Works-Marine Turbines, Gears & Gas Turbines 1945. Engr. Mgr. Westinghouse, Asst Plant Mgr. 1953, Plant Mgr. Manufacturing Div. SunnyvaleSteam, Electrical & Marine Apparatus, Lrg Machinery, Tullahoma Compressor, Nuclear Submarines & Ship Propulsion, Deep diving submarines, Astronomical & Radio Telescope mounts, Large valves & Hydraulic Gates. ASME activities: Vice Chmn Seattle Sect, Chmn Santa Clara Valley Sect, Fellow Award 1967, History & Heritage Chmn Reg IX 1970-72, presently a mbr of the San Fernando Valley Sect. Enjoy Lapidary, Jewelry work, Gunsmithing & Black Powder shooting. *Society Aff:* SNAME, ASME.

Gaylord, Edwin H
Home: 27 G H Baker Dr, Urbana, IL 61801
Position: Prof emeritus. *Employer:* Univ of IL. *Education:* MSCE/Civ Engr/Univ of MI; BSCE/Civ Engr/Case Western Reserve Univ; AB/Math/Wittenberg Univ. *Born:* 1/16/03. Youngstown, OH. Teaching & res in structural engg: OH Univ until 1956; Univ of IL 1956-71, prof emeritus 1971-. Consultant on steel structures & author various papers. Coauthor "Design of Steel Structures–, "Design of Steel Bins for Storage of Bulk Solids" & coeditor "Structural Engineering Handbook–, all with C N Gaylord, & coeditor with R J Mainstone of volume on Criteria & Loading of Tall Bldg Monograph. Fellow, Life member ASCE, mbr IABSE, SSRC (Chrmn 1962-66). Mbr Committee on Specifications AISC & Advisory Committee to Sr Technical Committee of Aluminum Assn. DSc (hon) Wittenberg Univ 1959. ASEE Western Elec Award 1971. *Society Aff:* ASCE, IABSE, SSRC.

Gaylord, Thomas K
Business: Sch of Elec Engg, Atlanta, GA 30332
Position: Prof *Employer:* GA Inst of Tech. *Education:* PhD/Elec Engr/Rice Univ; MS/Elec Engr/Univ of MO; BS/Physics/Univ of MO. *Born:* 9/22/43. Received BS & MS degrees from Univ of MO-Rolla in 1965 and 1967. Received PhD in Elec Engg from Rice Univ in 1970. Since 1972 with Sch of Elec Engg at GA Tech. Author of some 130 technical journal publications in areas of electro- optics, optical data processing, solid state, education, and instrumentation. Industrial consultant. Reg PE. Received six Sigma Xi res awards, 1987 "Engineer of the Year in Education Award" from GA Soc of PE, 1979 "Curtis W McGraw Research Award" from ASEE, 1984; "Outstanding Teacher Award" from GA Tech; 1985 honorary Professional Degree of Electrical Engineer from Univ of MO-Rolla. Fellow of IEEE and Optical Soc of Amer. Editor of *IEEE Transactions on Education*. *Society Aff:* AAAS, AIP, IEEE, OSA, ASEE.

Gaynes, Chester S
Business: 1642 W Fulton St, Chicago, IL 60612
Position: President. *Employer:* Gaynes Testing Labs Inc. *Education:* BS/Mech Engr/IL Inst of Technology. *Born:* Oct 1920. Native Chgo, Ill. Tool & Die Designer 1941-44; US Navy 1945- 46. Formed Gaynes Engrg with Father 1947, manufacturing all types of equip & conveyorized sys. Started producing Vibration, Drop & Incline Impact Testers for pkging field in 1953. Helped organize Independent Testing Lab in 1956, to conduct environ engrg, material & product evals & container appraisals. Has given 50 formal presentations on Pkging, Product & Seismic Analysis. Fellow Mbr & Former Chmn ASTM Ctte D-10; Fellow Mbr & former Natl Treasurer SPHE. Also mbr ANSI, PI, TAPPI, IES, NIPHLE, ACIL, ITEA. *Society Aff:* ASTM, IES, ANSI, PI, SPHE, TAPPI.

Gaynor, Joseph
Home: 1407 Oaklawn Place, Arcadia, CA 91006
Position: President. *Employer:* Innovative Technology Assoc. *Education:* PhD/Phys Chem/Case-Western Reserve Univ; MS/Chem Engg/Case-Western Reserve Univ; BChE/Chem Engg/Poly Inst Brooklyn. *Born:* Nov 1925. NYC native. US Army Chem Warfare Serv 1944-46. Genl Elec Co 1955-66: Res engr to Mgr-Imaging Materials Sys; Bell & Howell Co 1966-72: V P- Res, Bus Equip Group, Dir-Graphic Materials Res, Corp Labs; Horizons Res Inc 1972-73: Mgr, Comm Dev Grp, Mbr-Pres Office Innovative Tech Assoc-1973 to present: President. Fellow AAAS, AIC. Mbr Tau Beta Pi, Sigma Xi, Phi Lambda Upsilon. Ed Bd Photographic Sci & Engrg Journal, Ed Adv Panel-Chem Week. Chmn: Gordon Res Conf, Internatl Conf on Electrophotography, Bus Graphics Symposium & Engrg Sect SPSE. Invited Plenary Lec Internatl Congress of Photographic Sci (Moscow). 70 pats & pubs. Lic Prof Engr Ohio. Two Indus Res Awards (IR-100) 1963 & 1965. Sr Mbr - SPSE; Chmn - Org. & Prog. ctte.: 2nd Intl Bus. Graphics Symposium, 1st Intl Congress - Advances in Non-Impact Printing (NIP) Tech, Advisory Bd - 2nd Intl Cong. Adv. in NIP Tech, Ed - Advances in Non-Impact Printing Tech - 1983, Van Nostrand - Reinhold Inc. Ed.- *Advances in Non-Impact Printing Technologies*, Vol. II. *Society Aff:* AIChE, ACS, AAAS, SPSE, SID, SPIE.

Gaynor, Lester
Home: 152 Burgess Ave, Westwood, MA 02090
Position: Environ Engrg Conslt *Employer:* Self (Semi-Retired) *Education:* MS/Civil & Sanitary/Harvard Univ; BS/CE/Tufts Univ *Born:* 12/22/14 in Brockton, MA. Boston Latin School. Served with Air Corps Weather Service and Corps of Engrs 1939-1945. Officer Instructor Engr OCS 1942-43. Captain, 4 battle stars. Graduate Fellow Tufts 1949-1950. Instructor, Northeastern Univ 1950-54; Boston Univ 1954-56. Jr Engr, Whitman & Howard 1950-53. Sr Engr, Anderson, Nichols 1953-59. Partner and Sr VP SEA Consultants, Inc and predecessors 1959-82.Sewer Commission, Westwood, MA 1961-82, chrmn 1965-82. Environmentalist, Pacific history author. Recipient, Tufts Dept of Civil Engrg Distinguished Service Award 1983. *Society Aff:* WPCF, NEWWA, AMS, SAME, BSCE, ΤΒΠ, ΣΠΣ, AAEE.

Gazis, Denos C
Home: Lake Rd, R D 2, Katonah, NY 10536
Position: Asst Dir, Semiconductor Sci & Tech Dept. *Employer:* IBM. *Education:* PhD/Eng Mechs/Columbia Univ; MS/Civil Engg/Stanford Univ; BS/Civil Engg/Tech Univ of Athens, Greece. *Born:* 9/15/30. Sr Scientist Genl Motor Res 1957-61. Joined IBM Res 1961; served as Dir of Genl Sci 1971-74 & Cons to VP & Dir of Res 1974-75; Tech Adv to VP & Chief Scientist at Corp Hdquarters 1975-77; Mbr Res Review Bd 1977-79. Asst Dir, Computer Sciences Dept 1979-1983, Asst Dir Semiconductor Sci & Tech Dept. since 1983. Visiting Prof Yale 1969-70. Chmn Transportation Sci Sect, Oper Res Soc (ORSA) 1965-66; Member Comm on Eng and Tech Systems, National Research Council 1980-83; Member, Building Res. Board, Nat. Res. Council 1983-; Lanchester Price of ORSA 1959. Editor, Transportation Sci; Assoc Editor, Res Policy; Advisory Editor, Networks. *Society Aff:* ORSA, BRB/NRC, IPA

Gear, Charles W
Business: Dept of Comp Sci, 1304 W. Springfield, Urbana, IL 61801
Position: Prof. & Head *Employer:* Univ of IL. *Education:* PhD/Math-EE/Univ of IL; MS/Math-EE/Univ of IL; MA/Math/Cambridge Univ; BA/Math/Cambridge Univ. *Born:* 2/1/35. in England. Engr for IBM British Labs 1960-62. Joined faculty of Univ of IL in 1962. Author of seventeen books in programming & numericl problem solving. Awarded the ACM's Special Interest Group in Numerical Math's (SIGNUM) Forsythe Award, 1979. Chrmn of SIGNUM, 1973-75. President SIAM 1987-88. Fellow, IEEE, AAAS. Managing Editor, SIAM J. of Scientific and Statistical Computing 1985- . *Society Aff:* SIAM, ACM, IEEE, AAAS.

Gebhardt, Wilson A
Home: 650 Allegheny Dr, Sun City Ctr, FL 33570
Position: Exec Assistant, now Consultant. *Employer:* Bendix Corp, now Retired. *Education:* BSME/Auto Mach Design/Case Sch of Applied Science; ME/Auto & Mach Design/Yale Univ; MSME/Auto & Mach Design/Yale Univ; LLB/-/Blackstone College. *Born:* Aug 26, 1910, Cleveland, Ohio. Instructor in Mech Engrg at Case 1934-36. With The Bendix Corp from 1936 until retirement in 1975, serving successively as Res Engr, Patent Attorney, Western Group Patent Counsel, Mgr of Engrg, Mgr of Engine Equip Sect, & Exec Asst to VP, in charge of Patent & Legal Affairs with emphasis on Product Liability Litigation. Natl Pres of SAE in 1974. Now serving as Cons in Product Liability Litigation stemming from aircraft accidents. *Society Aff:* ΤΒΠ, ΣΞ.

Gebhart, Benjamin
Business: Meam Towne 1D3, Philadelphia, PA 19104
Position: Samuel Landis Gabel Prof of ME *Employer:* Univ of PA *Education:* PhD/ME/Cornell Univ; MSE/ME/Univ of MI; BSE/ME/Univ of MI; MA/Arts/Univ Penna *Born:* 07/02/23 Born in Cincinnati. Apprentice toolmaker in Detroit. Active duty in Marine Corps, Pacific, 1942-5. Consltg engr 1948-50. Taught at Univ of MI, Lehigh and Cornell. Also at State Univ of NY at Buffalo, Chrmn of Mech Engrg 1976-9. Since 1980 Samuel Landis Gabel Prof of Mech Engrg at Univ of PA. Res and publication in areas of heat transfer, buoyancy induced flows, arctic ice transport: Heat Transfer Div of ASME Heat Transfer Memorial Award 1972, Fellow ASME, ASME Freeman Scholar 1978, NAVSEA Res Chair Prof at the Naval Postgrad Sch 1980-81. Special interests include nature and travel. *Society Aff:* ASME, ASEE, AGU

Geddes, Leslie Alexander
Business: Potter Bldg, W Lafayette, IN 47907
Position: Prof Dir Biomedical Eng Ctr *Employer:* Purdue Univ. *Education:* B Eng/EE/McGill; M Eng/EE/McGill; PhD/Physiology/Baylor Med College *Born:* 5/24/21. Born in Scotland & educated in Canada, Dr Geddes has published over 500 scientific papers & 6 books. He is a consultant to NIH, NSF, NASA, & the USAF. Dr Geddes is currently a consulting editor of IEEE Transactions on Biomedical Engg, Medical Instrumentation, Journal of Electrocardiology & Mecial and Biological Engineering 8 Computing and others.. Who's Who, Am Men of Sci, Leaders in Am Sci & Who's Who in Engg all carry listings of his activities. *Society Aff:* IEEE, AAMI, NYAS, BMES

Geddes, Ray L
Home: 825 Monticello Ct, Cape Coral, FL 33904
Position: Retired Cons Engr. *Employer:* Stone & Webster Engg Corp. *Education:* PhD/Physical Chemistry/OH State Univ; MS/Organic Chemistry/OH State Univ; BS/Ind Chem/KS State Univ. *Born:* 1/27/05 Native of Wellington, KS. Fractional distillation and process res for Std Oil (IN) in Whiting refinery, 1930-1936. Petrol plant design with Alco Products, Inc, 1936-1939. With Stone & Webster Engg Corp for period of 1939-1970: Fractionation, chem-petrochem plant designs; process Consulting Engr, 1956-1970, Chem plant investigations & reports. *Society Aff:* AIChE, ACS.

Gedney, David S
Home: 9326 Boothe St, Alexandria, VA 22309
Position: Vice-President *Employer:* Deleuw Cather & Co. *Education:* MSCE/Civil Engg/Purdue Univ; BSCE/Civil Engg/Tufts Univ. *Born:* Mar 10, 1938. Native of Boston, Mass. Staff engr with Haley & Aldrich Inc, Cons, Boston. With FHWA since 1964. Served 2 yrs as Regional Engr of Eastern Region; 1 yr as Chief, Construction Div, Headquarters; 5 yrs as Dir of Engrg for Northeast Corridor Rail Passenger Improvement Program. Respon for engrg of Corridor up-grading program to be accomplished by 1981 with $2.5 billion Fed expenditure. Chmn ASCE Hwy Div Exec Ctte 1978; Chmn Construction Sect, TransRes Bd; mbr Natl Coop Hwy Res Program Bd. Enjoy all sports, good music, reading, theater, Reg Prof Engr-MD and now V.P., Deleuw Cather Co, responsible for Management of International work. *Society Aff:* ASCE.

Gee, Louis S
Business: 502 N Willis, Abilene, TX 79603
Position: Pres. *Employer:* Tippett & Gee, Inc Consulting Engrs. *Education:* BS/Mech Engr/TX A&M Univ. *Born:* 11/2/22. Native of Baird, TX. Air Corps Fighter Pilot WWII. Engr, W TX Utilities Co, 1948-1954. From 1954 to date: Chief Exec Officer, Tippett & Gee, Inc, Consulting Engrs, Specializing in design of maj elec power stations utilizing coal and low grade lignite as fuels. Author of technical papers pertaining to the design of fossil fuel power stations which have been delivered at maj engg conf and published in Natl and Intl Periodicals. Mbr, Advisory Committee for Lignite Res, Dev and Demonstration, TX Energy Advisory Council. Councilor, TX A&M Univ Res Fdn. Director First State Bank, Abilene Texas. Director, Independent Bankshares, Inc. *Society Aff:* IEEE, NSPE.

Geer, Ronald L
Business: P.O. Box 440335, Houston, TX 77244-0335
Position: Sr Mechanical Engrg Consultant *Employer:* retired *Education:* B/ME/GA Inst of Tech *Born:* 9/2/26 Native of West Palm Beach, FL. Served US Merchant Marine in WWII. Principal contributor to Shell Oil Co's overall pioneering offshore tech efforts in development of deep sea floating drilling vessels and associated hydrocarbon drilling and production equipment systems. Past activities include coordination of Shell Oil Co's Production Dept's overall mechanical engrg R&D programs including offshore engrg-related non-Artic and Artic engrg projects. Member of various professional organizations and natl level ocean engrg and tech advisory committees. Past Chrmn of Marine Bd of Natl Research Council. Enjoy sailing and antique cars. *Society Aff:* ASME, MTS, API, NAE

Geers, Thomas L
Business: 3251 Hanover St, Palo Alto, CA 94304
Position: Sr Staff Scientist. *Employer:* Lockheed Palo Alto Res Lab. *Education:* PhD/Appl Mechs/MIT; MS/Mech Engg/MIT; SB/Physics/MIT. *Born:* 11/16/39. in El Paso, TX. Raised in St Louis, MO. Served as Ensign, Ltjg USNR at Naval Ship R&D Ctr, Carderock, MD 1961-63, conducting exper res in underwater shock. Discharged from naval Res 1972 as Lt. Staff Engr & Sr Engr with Cambridge Acoustical Assocs, Cambridge, MA 1963-66, & took present pos in 1966; directs, conducts & publishes results of analytical res in struct dynamics & acoustics; reviews for var technical journals; coordinates tech symposia; mbr of Shock and Vibration Comm and Appl Mech Div Prog Comm of ASME. Active church mbr; plays piano, guitar. Received Henry Hess Award ASME 1970. *Society Aff:* ASME, ASA, $\Sigma\Xi$

Geffe, Philip R
Business: 3845 Pleasantdale Rd, (ATL-18), Atlanta, GA 30340
Position: Staff Engr *Employer:* Scientific Atlanta, Inc *Born:* 10/22/20 in CA. Studied network synthesis under Ernst Guillemin at UCLA in 1952, and has been working in network theory and filter design ever since. Was Chief Filter Engr at Triad Transformer Corp, Fellow Engr at Westinghouse Elec Corp, and is currently Staff Engr at Scientific-Atlanta, Inc. Has twice been a guest editor for *IEEE Transactions on Circuits and Systems*. Is the author of *Simplified Modern Filter Design* (Hayden, 1963), and of more than 50 papers. Was recently elected a Fellow of IEEE, "for contributions to computer-aided filter design." *Society Aff:* IEEE, AAAS

Geffen, Ted M
Business: 812 Philtower Bldg, Tulsa, OK 74103
Position: Petroleum Consultant *Employer:* Independent Consultant *Education:* BS/Petro Engg/OK Univ. *Born:* Feb 22, 1922 Calgary, Alberta, Canada. Employed as field engr with Albert Petro & Nat Gas Conservation Bd 1943-45 & Calif Standard Co 1945-46. Joined Stanolind Oil & Gas Co (now Amoco Production Co) in 1946 as Res Engr in Tulsa, Okla. Retired from Amoco 1981. Now a Petroleum Consultant. Served as Chmn of API Oil Recover Tech Domain Ctte & the API-Gov Res Liaison Ctte. Active in affairs of SPE of AIME, serving as Program Chmn for 1966 Fall Mtg, Region IV Dir on SPE Bd 1966-70, Genl Chmn of 1st Improved Oil Recovery Symposium 1969, Chmn 1975-76 of Implementation Ctte for Pub of 'Data on Field Tests of Improved Methods for Oil Recovery'. Received AIME Sect Serv Award in 1970, was 1974 recipient of John Franklin Carll Award for Disting Contributions to Petro Engg, was Chmn of the 1979 Gordon Res Conference on "Fluid Flow in Porous Media." Mbr of the Bd of Dir of AIME, (1979-81), Vice-President 1980. Member of BD of Dir and Ex Ctte of The Engrg Societies Commission on Energy (1981). *Society Aff:* SPE, AIME.

Geier, John J
Home: 3 Dogwood Glen, Rochester, NY 14625
Position: Engineer Associate-Indus Engr *Employer:* Eastman Kodak Co. *Education:* BSME/Mech Engrg/Duke Univ. *Born:* May 1, 1926 Rochester, N Y. US Navy 1943-47, 51-53. Employed at Kodak Park 1947-81 in var manufacturing areas generally concerned with facil planning & econ analysis. Elected V P Region V of The Amer Inst of Indus Engrs 1974-77 & served on sev cttes as a mbr of the Bd of Trustees. Pres Rochester Chap IIE- 1974. Elected Fellow of IIE 1981, Mbr NSPE, NYSSPE, MCPES.

Geiger, David H
Business: 500 5th Ave, New York City, NY 10036
Position: President. *Employer:* Geiger Berger Assoc P C. *Born:* May 29, 1935. PhD Columbia Univ; MS Univ of Wisconsin; BS Drexel Univ. Adj Prof at Columbia Univ; and NY Univ. Lic engr 18 states. Mbr ACI, ASCE, IASS, BRI/BRAB, CIB & others. Received ASCE NY Sect Award in 1971 & 1972; 1974 Dickson Prize for Outstanding Contribution to Sci, 1975 NSPE Outstanding Engrg Achievement among others. Internationally recog as a foremost auth of air struct. Respon for all the long span air supported structures built in the U S to date. Holds sev patents & has written num art on air structures.

Geiger, Gordon H
Home: 2118 Evans Rd, Flossmoor, IL 60422
Position: Executive *Employer:* Inland Steel Co. *Education:* BE/Metallurgy/Yale Univ; MS/Metallurgy/Northwestern Univ; PhD/Met & Mat Sci/Northwestern Univ. *Born:* Apr 1937. Indus exper with Allis-Chalmers Mfg Co, Jones & Laughlin Steel Corp, U S Steel Corp, Hanna Mining Co, Cabot Corp, Spec Metals, ITT-Harper & others. Teaching exper at Univ of Wics 1965-69; Univ of Ill at Chgo Circle 1969-73; & Univ of Arizona (Dept. Head, 1973-80), Co-author texts 'Transport Phenomena in Metallurgy', and 'Handbook of Material and Energy Balance Calulations in Metallurgical Processes', pub tech papers on Metallurgical thermodynamics, transport phenomena, & oper res applications to met process analysis. Received Bradley Stoughton Award from ASM & Campbell Award from NACE. Served on many TMS-AIME & ISS AIME cttes & on Engineering accreditation commission of ABET. *Society Aff:* ISS-AIME, TMS-AIME, ASM, AREA, NACE, ABET

Geiger, James Edward
321 Forest Park Blvd, Knoxville, TN 37919
Position: Cons Elec Engr. *Employer:* James E Geiger & Assoc. *Education:* BS/EE/Univ of TN. *Born:* Jan 2, 1931 Chattanooga, Tenn. Attended public schools

Geiger, James Edward (Continued)
in Chattanooga, Univ of Chattanooga, & received Bach of Sci in Elec Engrg. Has served as Guest Lecturer in School of Architecture at Univ of Tenn. Formed his own Elec Engrg Cons Firm in 1960 & is reg in 10 Southeastern states as well as in Ohio, Missouri, & Ill. Has served as Pres of the Constr Specifications Inst & of the local chap of the Inst of Elec & Electronic Engrs. Has done extensive work in legal area in relation to Elec Engrg. Past Pres of consulting engr of TN. Past-Southeastern Regional V Chrmn NSPE-PEPP. *Society Aff:* IEEE, NSPE, CSI, ACEC.

Geils, John W
Home: PO Box 432, Basking Ridge, NJ 07920
Position: Engg Dir. *Employer:* A T & T Co. *Education:* BEE/Communications/RPI. *Born:* 1/2/21. AT&T Engg Mgr - Responsible for administration of R&D work of BTL, 1964 to 1968. Dean, Bell System Ctr for Technical Education, 1968 to 1969. Engg Dir - Responsible for Engg Dept staffing, Budget and Expense Controls, Bell Labs R&D Prog & Budget, Technical Suggestions Studies, Safeguarding Proprietary Info, Mgt Dev, Manpower Selection, Utilization & Organization Planning for Engg Dept at AT&T & 23 Operating Tel Subsidiaries, 1969 to 1978. Dir - Network Dept Administration - Minorities and Women in Engg Progs for Bell System,, Budget & Expense, Profl Soc & Engg Licensing matters, progs for HEW of Bell System people in engg & technical jobs. *Society Aff:* IEEE, ASEE, NSPE, AAAS, AAES, TBP

Geis, A John
Business: 4615 Forbes Ave, Pittsburgh, PA 15213
Position: Dir, Technical Services. *Employer:* Graphic Arts Technical Foundation. *Education:* MBA/-/Ohio State Univ; BSME/IE/Purdue Univ. *Born:* 10/08/31. US Army Corps of Engrs 1953-55. Employed as Mech or Indus Engr in Petro & Graphic Arts Indus for 10 yrs, Indus Engrg Mgr in Graphic Arts for 10 yrs. 3 yrs VP mfg in printing, & in present pos for 10 yrs. P Pres of Columbus Chap IIE, Graphic Arts Div Dir IIE, & Fellow of IIE. Int incl golf, bowling & bridge. *Society Aff:* IIE

Geisser, Russell F
Business: 120 Pershing St, E Providence, RI 02914
Position: Pres. *Employer:* R F Geisser & Assoc. *Education:* MBA/-/Univ of RI; BSCE/-/Univ of RI. *Born:* 11/11/23. Native of E Providence, RI. US Infantry, WWII, Adjunct Prof, Engg, Roger Williams College, Northeastern Univ, Specializing in Engg & Testing in field of civ, structural, mech (problem solving), Licensed Pilot, enjoys sailing. *Society Aff:* NSPE, RISPE, ACI, ACIL, ASTM, AWS, ASCE.

Geist, Jacob M
Home: 2720 Highland St, Allentown, PA 18104
Position: Pres *Employer:* Geist Tec *Education:* PhD/Chem Eng/Univ of MI; MS/Chem Eng/Penna State Univ; BS/Chem Eng/Purdue Univ. *Born:* 2/2/21. Army service WW II. with Geist Tec since 1982 consulting engr cryogenics, industrial gases, speciality gases, distillation, heat transfer membranes, safety, semi-conductor industries. With Air Products & Chemicals Inc 1955-1982: Ch Engr Process Sys Grp 1970-1982, respon for review & approval of new engrg designs & concepts; had been Assoc Ch Engr, Assoc Dir R&D, Process Mgr. Also lecturer and adjunct prof at Lehigh Univ since 1960. Previously taught at Technion, Israel Inst Tech, & MIT. Reg Engr; V P Comm A-3 Int Inst of Refrig; V Chmn Cryogenic Engrg conf; Hon Fellow Indian Cryogenic Council. Fellow AIChE; Fellow AAAS, MBR ACS, NSPE, Tau Beta Pi, Phi Lambda Upsilon, Sigma Xi; AIChE Award Chm eng Practice 1976, Laureate Intl Inst of Refrigeration 1983, Member National Academy of Engineering 1980. Outstanding Engineering Alumnus - Pennsylvania State University, (1985), Honorary Sc. D. Technion, Israel Institute of Technology, 1987. *Society Aff:* AIChE, ACS, AAAS, NSPE.

Geisthoff, Hubert G
Home: 1M Wiesengrund 14, 5204 Lohmar, West Germany
Position: Dir *Employer:* Jean Walterscheid GMBH *Education:* Diploma/Agri-Engr/Univ of Agri-Machinery, Cologne *Born:* 10/30/30 German by birth. Grown up on a farm. Taught in schools, during and after WWII. Mechanic in 1951. I am a 1955 graduate of the German Univ of Agricultural Machinery, where I received an engrg degree. With Jean Walterscheid GMBH since 1955. Mgr research and development since 1960. 1980 Doerfer Engrg concept of the year award winner. Enjoy agricultural history and tennis. *Society Aff:* ASAE

Gelb, Arthur
Business: 6 Jacob Way, Reading, MA 01867
Position: Pres & Tech Director. *Employer:* TASC (The Analytic Sciences Corp). *Born:* Sep 1937 NYC. BEE City College of New York 1958, Magna Cum Laude; SM Applied Physics Harvard Univ 1959; ScD Instrumentation MIT 1961. The 2 latter as NSF Fellow. Founder TASC 1966 (The Analytic Sci Corp); TASC's Pres & Tech Dir since its inception. IEEE Fellow, Control Sys Soc AdCom mbr 1974-80. AIAA Fellow. Pat in med instrumentation. Has pub 25 tech papers (inertial navigation, nonlinear control, optimization). Is co-author of 2 tech bks, incl 'Applied Optimal Estimation', MIT Press 1974. Enjoys music, tennis, games & puzzles.

Gelders, Morris V
Home: 1300 Pinecrest Rd, Spartanburg, SC 29302
Position: VP *Employer:* Lockwood Greene Engrs, Inc *Education:* BS/EE/GA Tech *Born:* 12/22/17 Native of Fitzgerald, GA. Graduated from GA Tech in 1940. Employed by National Theatre Supply Co (Division of General Precision Equipment Corp) in Memphis, TN 1940-41. Employed by Special Engrg Division of Panama Canal, C. Z. 1941-42. Served in USAAF 1942-46. Employed and associated with Lockwood Greene Engrs, Inc in Spartanburg, SC 1946-present. Elected to Bd of Directors of Lockwood Greene 1964 and made a VP in 1974. Principle responsibilities have been in the design and management of design groups for industrial plants, especially for textiles. Active in church and civic organizations. *Society Aff:* NSPE, PEPP

Geller, Seymour
Business: Dept of Electrical & Computer Engrg, Campus Box 425, Boulder, CO 80309
Position: Prof *Employer:* Univ of CO *Education:* PhD/Physical Chem/Cornell Univ; BA/Math & Chem/Cornell Univ *Born:* 3/28/21 Native of NYC. Junior Chem, Picatinny Arsenal, US Army Ordnance Dept 1941- 43. Private to 1st Lt, US Armed Forces 1943-46. Research Chem, Benger Lab, duPont, Waynesboro, VA 1950-52. Member Technical Staff, Physical Research Dept, Bell Labs, Murray Hill, NJ 1952-1964. Group Leader, Rockwell Intl Science Center, Thousand Oaks, CA 1964-71. At Univ of CO since 1971. Research interests are relation of physical properties to arrangement and types of atoms in crystals; magnetic materials, solid electrolytes, superconductors, pressure- induced phases, phase transformations, magnetic structures, crystal structures of inorganic compounds. Did pioneering work on ferrimagnetic garnets (seven patents and numerous publications) and other areas of solid state science. Fellow IEEE, APS, MSA. *Society Aff:* IEEE, APS, MSA, $\Sigma\Xi$

Gelles, Stanley H
Business: 2836 Fisher Rd, Columbus, OH 43204
Position: Pres *Employer:* S H Gelles Assoc. *Education:* ScD/Met/MIT; SM/Met/MIT; SB/Met/MIT. *Born:* 9/12/30. Since 1976, Pres of Gelles Lab, Inc. Presently engaged in res on mtls processing in space, beryllium res & dev, failure analysis, scanning electrn microscopy asbestos analysis & met consulting. Associated with Battelle Columbus Labs as Sr Met & Assoc Sec Mgr from 1965-1976, with Ledgemont Lab of Kennecott Copper Corp as a Sr Met from 1961-1965 and with Nuclear Metals Inc as Prog Mgr from 1956-1961. Former mbr of the Mtls Advisory Bd Committees on beryllium and high pressure tech. Secy-Treas Boston Sec AIME, 1966-1968, Chrmn MAB Subcommittee on Impurity Analysis in Beryllium, 1964-1965, VChrmn ASTM Committee on Non-Ferrous Metals, 1973-1975, Pres Columbus Chpt ASM, 1985-86, Treas Ohio Asbestos Council 1987-88. *Society Aff:* ASTM, AIME, ASM, EMSA, AWS, NAC.

Gelopulos, Demos P
Business: EE Dept, Valparaiso, IN 46383
Position: Prof of Elect Engr *Employer:* Valparaiso Univ *Education:* PhD/EE/Univ of AZ; MS/EE/Notre Dame; BS/EE/Valparaiso *Born:* 4/24/38 Teaching Elect Engr since 1960. Various consulting activities in Power Systems & Aerospace. Principle Investigator for EPRI Research on computer simulation of large Power Grids while member of the faculty at AZ State Univ (1967-1980). *Society Aff:* IEEE, PES, SCS

Gempler, Edward B
Box 347, Mt Blanchard, Ohio 45867
Position: Staff Metallurgist. *Employer:* United Aircraft Products Inc-. *Education:* MS/Physical Metallurgy/Univ of IL; BS/Metallurgical Engg/Univ of IL. *Born:* Oct 1921. Native of N J. Alpha Sigma Mu (Natl Met Hon), Sigma Xi (Natl Res Hon). P Chap Chmn Syracuse, NY, ASM. P Mbr & Acting Chmn ASM Natl Mbrship Ctte. Bd of Governors, Tech Club, Syracuse N Y. US Navy 1942-45. Design Engr, Commerical Eng (Gov't Sales) GE Schenectady, Tube Dept. Sr Met Engr, Carrier R&D Co Syracuse, N Y (Air Cond). Joined United Aircraft Products, Dayton Ohio 1971 as Staff Metallurgist. Company-wide respon, materials, welding, brazing, processes, Failure analysis nuclear ORings, Met cons. Natl AWS C3 Brazing & Soldering Ctte. and Vice Chrmn, Chmn C3c Sub-Ctte on Education & Information. C3f Sub Ctte on Safety, C3d Sub-Ctte on Applications, C3x Executive Sub-Ctte. Findlay, Ohio Rotary Club. Elder 1st. Pres. Ch, Advisory Bd. Hancock Co. Home Health Care Agency. *Society Aff:* AWS, ASM.

Genereaux, Raymond P
Home: The Devon Unit 514, 2401 Pennsylvania Ave, Wilmington, DE 19806
Position: Manager. *Employer:* DuPont Company-retired. *Education:* AB/Chemistry/Stanford; Chem Eng/Chemical Engineering/Columbia *Born:* Sep 1902. With DuPont: chem engrg res, chem plant design incl original plutonium facilities at Hanford Engr Works; mgmt of res, design & engrg cons. Author of arts on fluid flow, engrg design, & engrg practice. Retired 1967; now cons engr. Fellow & former Dir AIChE & recipient of Founders Award 1963; Pres UET 1968-69; V P UET 1967; Chmn 1980-2 Dir EF 1967-82; Trustee EI 1968-75. *Society Aff:* AIChE.

Genheimer, J Edward
Business: 3290 W Big Beaver Rd, Troy, MI 48084
Position: Pres & Chrmn of Bd *Employer:* Ellis/Naeyaert/Genheimer Assoc, Inc *Education:* BS/Arch Engrg/OH Univ *Born:* 8/17/29 Native of OH. Joined Giffels & Assocs, Detroit, in 1951 as design engr specializing in industrial projects. Advanced and served as Assistant Dir of Industrial Engrg from 1958 to 1961; Assistant Dir of Development until 1962 and Dir of Development until 1967. Became stockholder and Dir in 1967, serving as VP/New Business Development from 1967 to 1969. Joined present firm in 1969 as VP, became stockholder and was elected to Bd of Dirs. Elected Pres in 1977. Elected Chrmn of Bd 1986. Dir of First of America-Wayne Oakland Bank, Royal Oak, MI. Holds a certificate of qualifications in NCEE, registered PE in seventeen states. *Society Aff:* NCEE, TBΠ

Genin, Joseph
Home: 2012 Crescent Drive, Las Cruces, NM 88005
Position: Exec VP *Employer:* Corner Stone Enterprises *Education:* PhD/Engg Mechanics/Univ of MN; MS/Struct Engg/Univ of AZ; BS/CE/CCNY. *Born:* 9/9/32. Bridge design for Ammann and Whitney, NY 1952/1955; Own struct design firm (bldgs, bridges), Tucson, AZ, 1955-1959. Instr, Engg Mechanics Dept, Univ of MN, 1959-1963; Sr struct Analyst (Aeroelasticity), Gen Dynamics, Ft Worth, TX 1963-1964; Prof, Sch of Aero, Astro & Engg Sci, Purdue Univ, 1964-1972; Prof of Mech Engg Purdue Univ 1972-1974; Prof & Chrmn of Engg Mechanics Purdue Univ 1974- 1981; Dean of the Coll of Engrg and Prof of Mech Engrg New Mexico State Univ 1981-85; EVP, Corner Stone Enterprises 1985. Specialize in dynamics, vibrations and dynamic stability. *Society Aff:* ASME, ASEE, NSPE.

Gennaro, Joseph J
Home: 3 Wells La, Warren, NJ 07060
Position: Prof of Civil Engrg *Employer:* Stevens Inst *Education:* MS/Structures/Columbia Univ; B/CE/City Coll of NY *Born:* 4/21/19 Civil Engr, Educator, Author, Licensed PE and Land Surveyor. Author of *Advanced Structural Analysis*, D. Van Nostrand, 1959, *Computer Methods in Solid Mechanics*, MacMillan 1965, *Modern Structural Analysis*, D. Van Nostrand, 1969, Consultant on Structural Analysis and Design, Computer Aided Design. *Society Aff:* NJSPE, NSPE, ASCE

Genovese, Philip W
Home: 39 Maple Ave, North Haven, CT 06473
Position: Pres *Employer:* Genovese & Assoc, Inc *Education:* MBA/Bus Ad/New Haven Univ; BE/Civil/Yale Univ *Born:* 1/22/17 Field Engr - Bridgeport Hydraulic - Dam, tunnel and hgwy relocation prior to WWII; Lt. Cmdr, CEC, USNR, Underground Fuel Storage, then Public Works Officer 2 Naval Air Stations; Designer, Project Engr to Partner Newman E Argraves and Assoc, Sewage Treatment Plants, groundwater development, Monmouth Park. Jockey Club House, Norwich State Hospital Power Plant; (CEO Genovese & Assoc, 1955 to Present), airplane hangar Bradley Field, major state hgwys and bridges, water system studies and supply development, many sewage systems and treatment plants, many bridges in Flood Recovery Progam, FEMA Flood Insurance Studies 15 Towns, Separation Storm from sanitary sewers Derby, New Haven, CT, 65D 650' steel box girder bridge, Trumbull airport runway upgrade, shore protection works. Immediate Past Pres Conn. Engrs Private Practice - 2 years; Past Pres Conn Soc Prof Engrs; Former National Dir - NSPE (10 years); Advisory Dir Conn Natl Bank; North Haven Conservation Comm. *Society Aff:* NSPE, ASCE, WPCF, ACEC.

Gensamer, Maxwell
Home: 1057 Forest Lakes Dr Apt. 309, Naples, FL 33942
Position: Howe Professor Emeritus. *Employer:* Columbia University *Education:* PhD/Met/Carnegie-Mellon Univ; MS/Met/Carnegie-Mellon Univ; BS/Met/Carnegie-Mellon Univ. *Born:* 6/3/02. Metallurgist, Page Steel & Wire Co, 1924-1929. Metals Res Lab, CMU 1929- 1945. Asst, Assoc & Prof, CMU 1935-1945. Hd, Div of Mineral tech, PA State Univ, 1945-1947. Asst Dir of Res, Carnegie-IL Steel Co, 1947-1950. Howe Prof of Met, Columbia Univ, 1950-1970 (Emeritus 1970-) Campbell & Howe Meml Lectr Consultant GE, US Steel, US Atomic Energy Commission, Westinghouse, Esso, etc.-NASA, Mtls Subcommittee-Chrmn IMD (AIME)-Fellow ASM & AIME (TMS). Numerous papers, lectureships, awards, etc. *Society Aff:* AIME, ASM.

Gentry, Donald W
Home: 6590 Ridgeview Dr, Morrison, CO 80465
Position: Dean of Engineering and Undergraduate Studies *Employer:* CO School of Mines. *Education:* PhD/Mining Engg/Univ of AZ; MS/Mining Engg/Univ of NV; BS/Mining Engg/Univ of IL. *Born:* 1/18/43. Industrial experience with Kennecott Copper Corp, Anaconda Co and NL Industries in new property evaluations and applied rock mechanics. Taught courses in Mining Engg at the Mackay School of Mines, Univ of AZ and CO School of Mines. Specialty areas include applied rock mechanics and the financial aspects of project and new mine property valuation. Active in res associated with longwall mining and mineral economics. Has presented numerous short courses in US and abroad. Industrial and governmental consulting on mining-related issues. Vice Chrmn-Operations of the EAC of ABET. SME/AIME representative to the ABET Bd of Dirs, Mbr, CO Section-AIME. Chairman of the Engineering Accreditation Commission of ABET, Past-Chairman of the Colorado Section-AIME, AIME Henry Krumb Lecturer for 1987. *Society Aff:* ASEE, ABET, SME, EAC.

Genzlinger, Bryce S
Home: 9118 N. Miami Ave, Miami Shores, FL 33150
Position: Dir, Computers & Communications *Employer:* City of Miami, FL *Education:* MS/Ind Engr/VPI & SU; BS/Ind Engr/OKLA State Univ. *Born:* 4/17/28.

Genzlinger, Bryce S (Continued)
Native of Bryn Athyn, PA. Asst Prof Ind Engr, Syracuse Univ 1956-60; Assoc Prof Ind Engr Drexel Univ 1960-63; Mgr, Operations Analysis, Burroughs Co, 1963- 67; VP Systems Design & Computer Operations. Banco Credito, Puerto Rico, 1967- 70; Dir, Mgt Advisory Services, Franklin Assoc 1970-72; Sr VP Operations and Admin Group, Farmers Bank, DE, 1972-76; V.P. Management Consultant, BEK Industries Inc., Brynathyn, PA 1976-78 Dir, Admin Services & Mgt Advisory Services, New Castle County, DE, 1978-80; Dir, Computers & Communications, City of Miami, FL 1980-Pres.. Natl Sci Fdn Fellow, Ford Fdn Fellow, Drexel Fellow; Dir & Treas Alpha Pi Mu, Inc; Dir & Treas 1st Am Sav & Loan Assn; Dir IIE COmp Syst Div; Trustee, Midwestern Acad; Trustee; Bryn Athyn Church. *Society Aff:* ALPHA Pi Mu, Sigma PHi Epsilon, IIE, ORSA, TIMS, ACM

Geoffrion, Louis P
Home: 79 Carpenter St, Foxborough, MA 02035
Position: Dir, Quality *Employer:* ITT Royal Elec Div *Education:* MBA/Mgmt/Boston Univ; AB/Phys Sci/Harvard Univ *Born:* 08/21/38 Strong experience in prod liability sys, expert testimony and failure analysis. Statistical process control and quality sys design, implementation and mgmt as well as direction of quality engrg activities major component of professional activities. Elected Fellow of ASQC in 1983, Sr Mbr of SME, Chrmn- Elect Boston Sect ASQC. *Society Aff:* ASQC, SME, ASTM, IAQC

George, Albert R
Business: Mech & Aerospace Engg, 105 Upson Hall, Ithaca, NY 14853
Position: Prof & Dir. of Mech. & Aerospace Engg. *Employer:* Cornell Univ. *Education:* PhD/Aero & Mech Sci/Princeton; MA/Aero & Mech Sci/Princeton; BSE/Aerospace Engg/Princeton. *Born:* 3/12/38. in NY, NY. Visiting Asst Prof, Univ of WA, Seattle 1964-5. Asst Prof, Assoc Prof & Prof at Cornell Univ 1965-present. Visiting Sr Fellow, Univ of Southampton, England 1972-1977. Since 1977 Prof & Dir of Sibley Sch of Mech and Aerospace Engg. Res, teaching & consulting in fluid dynamics, acoustics, aerodynamics, & automotive engg. Res publications in areas of aerodynamics, sonic boom, helicopter noise, acoustics & turbulence. Married to Carol Frerichs George; children: Albert F, David K, & Amy M George. Active in Congregational Church. *Society Aff:* AIAA, ASME, SAE, ASEE, AAUP, AHS

George, John A
Business: Aerospace Engg Dept, Cahokia, IL 62206
Position: Chrmn Aerospace Engg Dept. *Employer:* Parks College of St Louis Univ. *Education:* PhD/Phys/St Louis Univ; MS/Aero/Caltech; BS/Aero/Parks College of St Louis Univ. *Born:* 9/10/34. Joined the faculty of Parks College in 1959 and currently Prof of Aerospace Engg. Assumed the position of Chrmn of Aerospace Engg in 1977. Teaching in areas of fluid mechanics, aerodynamics and vehicle performance. Res interests are in the area of aircraft condition monitoring. Mbr of the Detection, Diagnosis and Prognosis Committee of the Mech Failures Prevention Group of the Natl Bureau of Stds. *Society Aff:* AIAA, ASEE.

George, Rogers E, Jr
Home: 188 E Emerson, W St Paul, MN 55118
Position: Dir of Construction Services. *Employer:* Hammel Green & Abrahamson Inc. *Born:* Jan 1917. Educated in public schools, Univ Minnesota 1935-39. Military Service: WW II 1942-46, Command & Genl Staff College 1947, Korean Conflict 1950- 53, Bronze Star, Lt Col. Subsequently practiced in both design & constr segments of the constr indus. Officer in arch engrg firm. Respon for contracts, specifications, constr admin, prof supporting services. Fellow Soc of Amer Military Engrs; Mbr Construction Specification Inst, Scientists & Engrs Technological Assessment Council, Environ Balance Assn of Minn, U S Power Squadrons. Lecturer-Panelist: Univ of Wisc, Univ of Minn, var conventions, institutes, symposiums, etc. Papers pub: 'Construction Specifer', 'Northwest Architect'. Dir SAME 1969 through 74. Dir C S I 1969, 70, 71.

Georgian, John C
Business: Dept of Mech Engg, Washington Univ, Box 1185, St Louis, MO 63130
Position: Prof. *Employer:* WA Univ. *Education:* MS/Mech Engr/Cornell Univ; BS/Mech Engr/Univ of MN. *Born:* 12/26/12. Native of Minneapolis, MN. Instr machine design Cornell Univ 1938-40. Engr in charge of steam turbine calculations, Allis-Chalmers Mfg Co, 1940-46. Vibration Engr, Nordberg Mfg Co 1946-49. Assoc Prof Mech Engg WA Univ 1949-55, full prof to 1981, Prof Emeritus to date.. Part time consultant to Sverdrup and Parcel Inc on turbomachinery design for atomic-powered aircraft 1952-54. Head, sponsored research for engrg controls, St. Louis MO. Designed small highspeed steam turbine utilizing steam from diesel engine exhaust; tested diesel engine waste heat exchangers; tests binary vapor power plant. 1958-1963. Designed for brace research inst a small high speed steam turbine for a solar power plant. 1965. at Bengal Engg College, Calcutta, India, 1954-55, summer 1956 and 1961-62. Mbr ASEE mission to India winter 1958-59. Guest prof at Instituto Tecnologico de Aeronautica, Sao Jose dos Campos, Brazil, 1965-67. Guest prof at Faculty of Engg, Kabul Univ, Kabul, Afghanistan 1971-73. Registered Prof engr MO. *Society Aff:* ASME, ASEE, AGMA, AAUP

Gerard, Roy D
Home: 1418 Castlerock, Houston, TX 77090
Position: Gen Mg. *Employer:* Shell Development Company *Education:* MS/ChE/LA State Univ; BS/ChE/LA State Univ. *Born:* 9/14/31. Native of New Orleans, LA. Reg engr in the state of LA. Served in US Army Corps of Engrs. Upon discharge from service returned to LA State Univ to obtain masters degree in chem engg. Employed by Shell Oil Co, assignments include engg. & mgt jobs in mfg, res & dev, and commercial sales, followed by the gen Managership of engineering, Presidency of Saudi Petrochemical Company and more recently the Gen managership of Shell Development Company - Westhollow Research Center. Outside activities include fishing and woodworking. *Society Aff:* AIChE, CCR, IRI, SCI

Gerber, Eduard A
Home: 11 Community Drive, West Long Branch, NJ 07764
Position: Consultant. *Employer:* US Army Electronics Command. *Education:* PhD/Physics/Tech Univ Munich; Dip Physicist/Physics/Tech Univ Munich. *Born:* Apr 1907. Native of Fuerth, Germany. Mbr of scientific staff, Carl Zeiss Works, Jena, Germany, res & dev in piezoelectric crystals 1935-46. Cons for frequency control, Signal Corps Engrg Labs, Ft Monmouth N J 1947-54. Dir Frequency Control Div 1954-63 & Dir Electronic Components Lab, US Army Electronics Command, 1963-70, all at Ft Monmouth N J. In this pos, respon for res & dev of components & devices for US Army. Since 1970, cons US Army Electronics Tech & Devices Lab. Fellow Inst of Elec & Electronics Engrs, author of 35 scientific & tech papers in the field of frequency control & related subjects. Editor and/or Coauthor of 5 books.. *Society Aff:* IEEE.

Gerber, George C
Business: MMP Intl Inc, 3222 N St N.W, Washington, DC 20007
Position: Pres *Employer:* MMP Intl Inc *Education:* BS/CE/Duke Univ *Born:* 9/10/32 Cum Laude graduate-Phi Beta Kappa-Leadership Honaries. Civil Engrg officer, U.S. Air Force, 1954-1956, AK. Chief Civil/Structural Dept and Partner, Engr- Architect firm; Commissions include major lab, office bldg, advanced waste water treatment facility, schools; 1956-1974. Chief Engr, Architect-Engr firm; Project Dir for $100 million medical center including owner's point of contact and direction of multi-discipline architect-engr team; 1974-1979. Established civil/structural and cost & project management firm in 1979; Pres and co-owner; Engr of record, Univ Inter-cultural Center featuring largest bldg mounted solar power cogenerator. Professional engrg practice-Washington DC area over 35 years. *Society Aff:* NSPE, ASCE, CIAA, TBΠ.

Gerber, H Bruce
Home: 309 E Elmwood Ave, Mechanicsburg, PA 17055
Position: President. *Employer:* Gannett Fleming Corddry & Carpenter, Inc.

Gerber, H Bruce (Continued)
Education: BS/Chem Engg/PA State Univ. *Born:* Oct 3, 1925; BS in Chem Engrg Pennsylvania State Univ 1948; Reg Prof Engr 33 states. With GFC&C since 1950. Assumed current respon as VP of GFC&C in 1972 & Dir of Pollution Control Div in 1975. Diplomate, Amer Acad of Environ Engrs; P Pres Water Pollution Control Assn of Penn; P Dir Water Pollution Control Fed; Trustee, Harrisburg Area Community College; Trustee Seidle Mem Hosp; P Pres Bd of Ed, Mechanicsburg Area School Dist; Mbr Governor's Adv Health Bd-Penn; Water Pollution Control Federation Bedell Award 1969; Water Pollution Control Assn of Penn Haseltine Award 1976. *Society Aff:* NSPE, WPCF.

Gerberich, William W
Home: 21035 Radisson Inn Road, Christmas Lake, Shorewood, MN 55331
Position: Full Prof. *Employer:* Univ of MN. *Education:* PhD/Matl Sci & Engg/Univ of CA, Berkely; MS/Indust Engg/Syracuse Univ; BS/Engg-Sci/Case Inst of Tech. *Born:* 12/30/35. Married: Susan Elizabeth Goodwin, August 15, 1959. Children: Bradley Kent, Brian Keith, Beth Clarice. 1959-61 Research Engg, Jet Propulsion Lab, Col Tech, Pasadena. 1961-64-Research Sci, Aeron, Newport Beach, CA. 1964-67-Engg Research Specialist, Aerojet General, Sacramento, CA. 1967-71-Lecturer & Research Metallurgist, LRL and Univ of CA, Berkeley. 1972-Prof & Dir of Materials Sci, 1980-Associate Head of Chemical Engrg and Materials Sci, Univ of MN. William Spraragen Award for best research paper, *Welding Journal*, 1968 contributed over 150 articles to tech journals and international symposia. 1984 - Bd of Publications, Acta Metallurgica. 1986-Vice Chairman, Maetallurgical Transactions Board of Review. 1986-Chmn of the Bd. of Publications, Acta Metallurgica. *Society Aff:* AIME, ASM, ASTM, ΣΞ, MRS.

Gerdes, Walter F
Home: P O Box 585, Lake Jackson, TX 77566
Position: Technical Specialist. *Employer:* Dow Chemical Co, Texas Div. *Born:* Nov 1916 Seely, Calif. BS EE Texas A&M Univ. Worked as engr for Gulf Oil Corp prior to WW II. Served with Signal Corps 1941-45. Worked as control engr for Genl Elec 1945-48. Worked in instrument field with Dow Chemical Co since 1948. Received ISA INSTRUMENTATION TECHNOLOGY Award in 1967 for paper on the STREAMING CURRENT DETECTOR. Received ISA ARNOLD O BECKMAN Award in 1968 in recognition of dev of the STREAMING CURRENT DETECTOR.

Gere, James M
Business: Dept of Civil Engrg, Stanford University, Stanford, CA 94305-4020
Position: Prof. *Employer:* Stanford Univ. *Education:* PhD/Appl Mechanics/Stanford Univ; MCE/Civil Engg/Rensselaer Poly Inst; BCE/Civil Engg/Rensselaer Poly Inst. *Born:* 6/14/25. Born and raised in Syracuse, NY; moved to CA in 1952. Married Janice M Platt in 1946; three children. Bombsight mechanic, US Army Air Force, WWII. NSF Grad Fellowship, 1952-54. Prof at Stanford since 1954 (teaching and res). Chrmn of the Dept of Civil Engg, 1967-1972. Assoc Dean, School of Engrg, 1960-70. Co-Dir of the John A Blume Earthquake Engg Ctr since 1975. Lecturer and consultant. Author seven textbooks and many articles and reports. Reg CE, CA and NY. Ardent hiker and backpacker in Sierra Nevada and Grand Canyon. Mbr of numerous conservation organizations. *Society Aff:* ASCE, ASEE, EERI, ΣΞ, ΤΒΠ

Gergely, Peter
Business: Civil and Environmental Eng, Hollister Hall, Cornell University, Ithaca, NY 14853-3501
Position: Prof & Dept. Chrmn & Director School of Civil & Environmental Eng. *Employer:* Cornell Univ. *Education:* PhD/Civ Engg/Univ of IL; MS/Civ Engg/Univ of IL; BEng/Appl Mechanics/McGill Univ. *Born:* 2/12/36. Born in Hungary & attended the Technical Univ of Budapest. MS, PhD, U of Illinois. Joined the faculty of the Dept of Structural Engg of Cornell Univ in 1963. Has been consultant to a number of ind & governmental offices. Worked for the Dominion Bridge Co of Montreal (1959 & 1960) and the Pittsburgh-Des Moines Steel Co (1964 & 1969). Bd of Directors, ACI. Distinguished service award, ACI (1981); co-receipient ASCE awards (1977, 1978). Fellow of ACI and ASCE. Executive Comm. National Center for Earthquake Engrg Research. Maj res & profl interests: reinforced concrete, earthquake engg, structural dynamics, & shells. Reg PE in NY State. *Society Aff:* ASCE, EERI, ACI, IABSE.

Gerhard, Earl R
Business: Speed Scientific School, Louisville, KY 40292
Position: Dean-Speed Scientific Sch *Employer:* Univ of Louisville. *Education:* PhD/Chem Engr/Univ of IL; MS/Chem Engr/Univ of LO; BS/Chem Engr/Univ of LO. *Born:* Aug 1922. PhD Univ of Illinois; BS Chem Engr. & MS Chem Engr. Univ of Louisville. Prof rank at Univ of Louisville since 1951. Chmn Dept of Chem Engrg 1969-73; Assoc Dean 1973-80. Dean 1980- Shell Oil Co 1943-44; US Army 1944-46. Cons & Process Design with C&I Girdler & Chemetron Catalyst. Prof Engr KY. Full mbr AIChE, ACS, ASEE. Honor soc Phi Kappa Phi, Omicron Delta Kappa, Phi Lambda Upsilon, Sigma Xi, Tau Beta Pi. 18 pubs in Kinetics, heat & mass transfer & environ areas. Major res areas are environmental assessment & energy applications. *Society Aff:* ASEE, AIChIE, ACS, NSPE

Gerhardt, Lester A
Business: Elec & Systems Engg Dept, New York, NY 12181
Position: Prof - Elec, Computer & Systems Engg (Computer Science) Director - Computer Integrated Manufacturing *Employer:* RPI. *Education:* PhD/EE-Comm/SUNY; Master/EE/SUNY; Bachelor/EE/CCNY. *Born:* 1/28/40. Born in the Bronx NY, & attended CCNY. Following grad, he joined Bell Aerospace Co in Buffalo NY, where he became hd of info & signal processing, and asst to the dir of advanced systems res. Concurrently, he obtained his grad degrees at SUNYAB. Joining RPI in 1969, he was selected chrmn of elec and systems engg in 1975, a position held until 1986. He is now Director of Computer Integrated Manufacturing at RPI. He was a visiting faculty at Univ of CA in 1979 on sabbatical leave. Areas of interest include digital signal processing, interactive graphics, pattern recognition & adaptive systems. He is a fellow of the IEEE among other honors & organizations. *Society Aff:* IEEE, NYAS, ΣΞ.

Gerhardt, Ralph A
Business: John Deere Rd, Moline, IL 61265
Position: Mgr, Engr Planning *Employer:* Deere & Co *Education:* MS/ME/IA Univ; BS/ME/IA State Univ *Born:* 10/8/34 Was design mgr at Motorola, Inc Phoenix AZ until 1960. Obtained MSME 1961 Univ of IA. With John Deere since 1961 specializing in product design, holding various positions including divisional responsibility in the US and as head of design dept in factory in France. In 1977 assumed responsibility in Deere Co product planning area for agricultural eqpt. Specializing in hay and forage. Current responsibility since 1980 is on corp staff as mgr of product engrg planning coordination worldwide. Enjoy sports and traveling. Active in professional societies. Currently Natnl Awards dir for ASAE. *Society Aff:* ASAE, ASME, SAE, NASCP

Geri, Don W
P.O. Box 1121, San Carlos, CA 94070
Position: Pres. *Employer:* Geri Engg, Inc. *Education:* BSME/ME/OR State Univ. *Born:* 9/20/33. Native of Portland, OR. Commissioned in the Navy in 1958 and served with the Polaris Dev team in Wash DC. On return to industry in 1962 various design and mgt positions were held in the tape recorder, microfilm and marking equip fields with responsibilities up to director of product dev. Formed Geri Engg, Inc in July 1974 to provide a specialized design and prototype service in the areas of sophisticated machine design and air control systems. *Society Aff:* SAE, SPE, NSPE.

Gerke, John F
Home: 7414 NE Par Lane, Vancouver, WA 98662
Position: Pres *Employer:* Columbia Sentinel Engrs Western Inc *Education:* BS/EE/Univ of N.S.W. *Born:* 3/4/28 Native of Sydney, Australia. Project mgr, assistant dir of engrg and dir of cost recovery for Lockheed Shipbuilding and Construction Co, Seattle, WA. Co- founder of consulting engrg firm of Columbia-Sentinel Engrs established in 1973. Assumed current responsibility of Pres when firm was inc in 1976. Has had broad experience in developing both critical path method and deterministic pert networks for program cost and schedule control. Also has prepared construction claims for contract changes. *Society Aff:* IEEE, SNAME, NSPE

Gerken, John M
Home: 185 Willow Lane, Chagrin Falls, OH 44022 *Education:* PhD/Metallurgy/RPI; M Met E/Metallurgy/RPI; B Met E/Metallurgy/RPI. *Born:* 9/3/20. Native of Kearny NJ. Served in Army Signal Corps 1942-46. Res Assoc at Knolls Atomic Power Lab 1955-60. Dev adv welding tech for nuclear fuels. Respon for welding R&D at TRW Materials & Manufacturing Tech Center for aircraft, automotive & nuclear components. Active on sev Welding Res Council Cttes, past chmn of Reactive & Refractory Metals Ctte and Heat Resistant Alloy Sub. Ctte of High Allows Ctte. Served 7 yrs as Dist 10, Dir & 2 yrs as Dir at Large AWS & mbr of natl bd of dir of AWS. Mbr AWS Tech Papers Ctte. Hobbies incl restoring antique automobiles. Currently serving as President, AWS. Retired from TRW, performed consulting work 10/1/86 to 6/15/87. Working for Lincoln Electric Co, starting 7/1/87 in development of new welding processes. *Society Aff:* AWS, ASMI, IIW, WRC, EWI-IAB, ΣΞ

Gerlich, James W
Business: P.O. Box 9009, 4250 Glass Rd N.E, Cedar Rapids, IA 52409
Position: VP *Employer:* Howard R. Green Co *Education:* BS/CE/Univ of IA *Born:* 11/5/24 Responsible for directing research and project development programs, pilot studies and design. Since joining the co in 1953, his responsibilities have included serving as resident engr on construction, design engr for civil projects, and project mgr on major water pollution control, water treatment and other sanitary engrg projects. He has published numerous technical and research publications in the field of environmental engrg. His previous experience includes serving as a sanitary engr with the IA State Dept of Health and as design engr for the Des Moines Public Works Dept. *Society Aff:* NSPE, AIPE, WPCF, AAEE, NAFE

Gerliczy, George
Business: 17 Mileed Way-Box 10, Avenel, NJ 07001
Position: Vice Pres-Operations. *Employer:* Wall Trends, Inc. *Born:* Feb 1929. PhD (Dr Sc Tech) from ETH Zurich 1954; Chem Engrg Degree ETH 1951. Native of Budapest, Hungary-naturalized US citizen 1960. 195657 res chem E I DuPont de Nemours Inc, Wilmington Del. Associated with Solvay & Cie subsidiaries in USA since 1957 (Hedwin Corp 1959-70, The Solvay Amer Corp 1970- 73) in var staff pos. Assumed present pos 1974 as V P Operations for Wall Trends, Inc, an importer & distributor of wallcoverings, and a subsidiary of Griffine Marechal S A Paris, France (a subsidiary of Solvay & Cie, S A Brussels, Belgium).

Germain, James E
Business: P.O. Box 1498, Reading, PA 19603
Position: VP & Gen Mgr *Employer:* Gilbert Assocs, Inc *Education:* MS/CE/MI State Univ; BS/CE/MI State Univ *Born:* 4/15/33 Born and educated in MI. His career has been in Civil Engrg and Water Pollution Control. After 4 years' experience as a consulting engr, he was employed by The Dow Chem Co as a pollution control engr. After moving to PA in 1965, his assignments during 12 years with Roy F. Weston, Inc, included process engrg, project management, and Divisional VP. During the last 7 years, he was a member of the Bd of Dirs. At Gilbert Assocs, Inc, he is a VP & General Mgr of the Environmental Division. His family include wife, Pat; and two daughters. *Society Aff:* AAEE, WPCF, ASCE, NSPE

German, John G
Home: 19 Holloway Dr, Lake St Louis, MO 63367-1357
Position: President *Employer:* John German Railway Engineering Consultants, Ltd *Education:* B.S./Mech. Engr./Case School of Applied Science *Born:* 09/22/21 Transportation consultant; b. Devils Lake, N.D. 1921; m. Mary Alice Chambers, 1973; 1 son, John R. B.S.M.E., Case Sch. Applied Sci., Cleve., 1943. With Gt. No. Ry. Co., 1943-61; with Mo. Pacific R.R. Co., 1961-83, chief mech. officer, 1961-66, asst. v.p. engring., 1966-75, v.p. engring., St. Louis, 1975-83; exec dir International Heavy Haul Association; 1983--; pres. John German Ry. Engring. Cons. Inc., 1983--. Fellow ASME; mem. Air Brake Assn., Am. Ry. Bridge and Bldg. Assn., Am. Ry Engring. Assn., Car Dept. Officers Assn., Locomotive Maintenance Officers Assn., Railway Fuel and Operating Officers Assn., Assn. Am. Railroads, Roadmasters and Maintenance of Way Assn. Republican. Lodges: Masons; Shriners; Scottish Rite. *Society Aff:* A.S.M.E.

German, Randall M
Business: Materials Engrg Dept, Troy, NY 12181
Position: Prof *Employer:* RPI *Education:* PhD/Mat Sci/Univ CA; MS/Metallurg Engr/OH State Univ; BS/Mat Sci/San Jose State Univ *Born:* 11/12/46 On the faculty at RPI since 1980. Metallurgist who emphasizes processing as a science with attention to powder metallurgy. Previous employment in contract and nuclear research (Bettelle and Sandia) and as dir of research for Mott Metallurgical Corp as well as J. M. Ney Co. Author of 110 technical papers and one textbook "Powder Metallurgy Science" and currently retained as a consultant to industry and government labs; responsible for new developments in such diverse fields as nuclear ordinance and dentistry. Journal editor, national and regional awards, patents. *Society Aff:* AIME, ASM, APMI, AADR, SAE

Gernert, Marvin L
Business: P O Box 2357, Batesville, AR 72501
Position: President. *Employer:* Ark Eastman Co-Div Eastman Kodak Co. *Education:* MChE/-/Univ of Louisville. *Born:* Jun 1924. Native of Louisville. With Tenn Eastman Co, Div Eastman Kodak Co since 1947. Had respon for patents, licensing, supply, & distrib prior to current pos. Currently respon for constr & oper of plant to manufacture organic chemicals. Enjoy gardening & golf. *Society Aff:* AIChE.

Gerstein, Melvin
Business: USC - OHE 430, Los Angeles, CA 90089-1453
Position: Prof of Mech Engr *Employer:* Univ of Southern Calif. *Education:* BS/Chemistry/Univ of Chicago; PhD/Chemistry/Univ of Chicago *Born:* 5/8/22. PhD & BS Univ of Chicago. Manhattan Proj 1944-45; Navy Fuels Res 1945-46; NASA (NACA) 1946-59, Section Chief Combustion Fund. Sect, Branch Chief Comb Chem Branch, Asst Div Chief Propulsion Chemistry Div; JPL 1959-60, Chief Phys Sci Res Div; Dynamic Sci Corp 1960-67, VP-Tech Dir, Pres; Univ So CA 1966-present; Chmn ME Dept 1966-69, Assoc Dean Engg 1970-1981; Interim Dean Engrg 1981-1983; Prof Mech Engrg 1983-present; NATO (AGARD) Advisory 1952-64. Outstanding Young Engr Award, Cleve Tech Soc Council 1957. USC Alumni Assoc Outstanding Fac Award 1974. Combustion Inst, Sigma Xi, Pi Tau Sigma, Tau Beta Pi. Listed in Who's Who in America, Amer Men of Sci, Outstanding Educators. *Society Aff:* CI, ΣΞ, ΠΤΣ, ΤΒΠ

Gerstle, Kurt H
Business: Dept of Civil Engrg, Boulder, CO 80302
Position: Prof of Civil Engrg. *Employer:* Univ of Colorado. *Education:* BS/CE/Univ of CA, Berkeley; MS/CE/Univ of CA, Berkeley; PhD/CE/Univ of CO, Boulder. *Born:* 11/11/23. US Army 1943-46. Engrg & Cons, San Francisco & Denven areas. Instructor to professor, Civil Engr, Univ of CO, Boulder 1952- . Study at Fed Inst of Tech, Zurich, Switz & Brown Univ, Providence, RI. Visiting Prof SEATO Grad Sch of Engrg, (1963-64), Tech Univ, Munich, (1970-71), Trondheim, Darmstadt (1979), Res in Struct Engrg; about 70 pub papers; Co-recipient, Wason Medal of ACI for best paper of 1964. Co-Recipient, Moisseif Award of A.S.C.E., 1984 Humboldt Senior Scientist Award, 1978. Author of 2 textbooks. Pres, Colo Section ASCE 1973-74. *Society Aff:* ASCE, ACI.

Geselowitz, David B
Business: 217 Elec Eng West & PG, University Park, PA 16802
Position: Prof of Bioengineering. *Employer:* Pennsylvania State Univ. *Education:* PhD/EE/Univ of PA *Born:* 5/18/30. Grad Central HS, Philadelphia. Mbr of fac of Univ of PA from 1951-71 with appointments in Moore School of Elec Engrg & Sch of Medicine. Visiting assoc prof MIT 1965-66. With PA State Univ since 1971 as Hd of Bioengrg Program. Major res int is in cardiac electrophysiology. and artificial hearts. ED IEEE Transactions on biomedical Engrg 1967-72. Fellow Amer College of Cardiology 1975. Chmn, Amer Heart Assn Ctte on Electrocardiography 1976-81. Fellow IEEE, 1978. Guggeheim Fellow and visiting Professor of Biomedical Engg, Duke Univ, 1978-79. IEEE Centennial Medal, 1984. *Society Aff:* IEEE, BMES, AAAS, AAPT

Gessner, Gene A
Business: 321 East Market St, Iowa City, IA 52240
Position: Pres *Employer:* Gene Gessner, Inc *Education:* BS/ME/Univ of NB *Born:* 1/22/30 Native of Lincoln, NB. Began engrg career in mechanical contracting management with Omaha, NB firm. Soon promoted to Assoc in Charge of Mechanical and Electrical with a NY based Consulting Firm. In 1967 organized Gene Gessner, Inc and has served a Principal/Pres for past fourteen years. Have undertaken various projects throughout the US in electrical/mechanical consulting engrg. Married, has 3 sons, active in outdoor sports, especially golf, swimming, jogging. Likes to travel, attends sports events and the theatre regularly. *Society Aff:* ACEC, ASHRAE

Gessow, Alfred
Home: 7308 Durbin Terr, Bethesda, MD 20817
Position: Professor & Chairman, Dep't of Aeros Engrg *Employer:* Univ. of Maryland *Education:* M Aero Eng/Aero Engg/NY Univ; B Cvl Eng/Civil Eng/CCNY. *Born:* Oct 1922. With NASA, Hampton Va, & Wash, 1944- , asst dir res, Washington 1966-70, chief aerodynamics & fluid mech 1970-80. Adj prof aeros & astronautics NYU 1968-69, Catholic Univ, Wash 1969-70; cons helicopters France 1956-60, Germany 1956-60. Honory Fellow Amer Helicopter Soc (p Tech Dir). Alexander Nikolsky Honorary Lectureship, Am. Hel. Soc. 1985. Author: Aerodynamics of the Helicopter 1951. Contribute profl journals, encys. Founding ed Jour Amer Helicopter Soc 1955. NASA Exceptional Service Medal 1974, Fellow Amer Inst Aeronautics & Astronavtics, Dir Center for Rotocraft Education & Research, Univ. of Md. Member U.S. Army Science Board 1986- , Commission on Professionals in Science & Technology; AIAA Career Enhancement Committee; Am. Hel. Soc. Education Committee. *Society Aff:* AHS, AIAA, ASEE

Gesund, Hans
Business: Dept of Civil Engg, Univ of Kentucky, Lexington, KY 40506-0046
Position: Prof of Structural Engg., and Chairman, Dept. of Civil Engg. *Employer:* Univ of KY. *Education:* DrEngg/Civil Engg/Yale Univ; MEngg/Civil Engg/Yale Univ; BEngg/Civil Engg/Yale Univ. *Born:* 9/18/28. Grew up in New Haven, CT. Served in US Army 1950-52, 1961-62, retired with rank of Maj, Corps of Engrs, US Army Reserve. Employed as Instructor, Civil Engg, Yale Univ 1955-58. Asst Prof, Univ of KY 1958-59, Assoc Prof 1959-65, Prof of Structural Engg 1965 to present. Director of Graduate Studies in Civil Engineering 1984 to present, Chairman, dept. of Civil Engineering, 1987 to present. Author or co-author of approximately 50 technical papers. Elected Fellow of ACI and ASCE, and mbr of Sigma Xi, Tau Beta Pi and Chi Epsilon. Reg PE in CT and KY with considerable consulting experience in industry and govt. *Society Aff:* ASCE, ACI, IABSE, ASTM, ASEE.

Getsinger, William J
Business: Clarksburg, MD 20734
Position: Sr. Scientist *Employer:* Comsat Labs *Education:* Engr/EE/Stanford; MS/EE/Stanford; BS/EE/Univ of CO *Born:* 1/24/24 in Waterbury, CT. From 1950 to 1957 worked with microwave tech at Technicraft Labs and Westinghouse Electric Co. From 1957 to 1962 in microwave research at Stanford Research Inst and MIT Lincoln Lab. In 1969 joined COMSAT Labs as a dept mgr in the microwave lab, and in 1981 became sr scientist. Elected a fellow of the IEEE in 1980. Guest editor of IEEE transactions on microwave theory and techniques special issue on computer-oriented microwave practices (August 1969). Project mgr of Centimeter Wave Beacons Orbited on four COMSTAR communication satellites. His published technical papers are well known to microwave engrs. *Society Aff:* IEEE

Getter, Gustav
Home: 21 Bayberry Lane, New Rochelle, NY 10804
Position: President. *Employer:* Gustav Getter Assoc P C. *Education:* MCE/Civil Structural/Polytech Inst of Brooklyn; BSE/Civil/Brown Univ. *Born:* Aug 1926. US Navy 1944-46 & 1951-53; highest rank LT CEC USNR. Field Engr & office engr with internatl Cons Engrg firms. 1946-63 respon charge for design of military facilities incl missile bases & airfields, water & waste treatment plants, highways, & bridges & petro chemical facilities. Estab own practice in 1963 as multi-discipline engrg firm operating on a natl basis serving Gov & indus. Fields of specialty incl: ground support facilities, aircraft, engines & missiles; indus bldgs; instrumentation & process control; total sys engrg. *Society Aff:* ASCE, ACEC, SAME, NSPE, AWS.

Getting, Ivan A
Business: 312 Chadbourne Ave, Los Angeles, CA 90049
Position: Consultant. *Employer:* Self-employed. *Education:* DSc/Hon/Northwestern Univ; D Phil/Artrophysics/Oxford Univ, ESg; SB/Physics/MIT. *Born:* Jan 1912 NYC. Jr Fellow Harvard 1935-40, MIT Radiation Lab 1940-45, Prof Elec Engrg MIT 1945-50. V Pres Engrg & Res Raytheon 1950-60, Pres & Bd mbr The Aerospace Corp, 1960-78. Served as Asst for Dev Planning USAF, mbr USAF Scientific Adv Bd, Naval Studies Bd NRC. President's Scientific Adv Ctte, Panels & cons to Natl Security Council. Fellow APS, IEEE, AIAA, Amer Acad Arts & Sci. Mbr Natl Acad of Engrg. Received President's Medal of Merit, Naval Ordnance Dev Award, AF Exceptional Serv Award. Mbr of Bd of Northrop, Environ Inst of MI (ERIM), Verac, Registered PE. *Society Aff:* IEEE, AIAA, AAAS, NAE, APS

Geyer, John C
Home: 710 Bosley Rd, Cockeysville, MD 21030
Position: Prof Emeritus. *Employer:* Johns Hopkins Univ-ret. *Education:* BSCE/Civil/Univ of MI; MSE/Sanitary/Harvard; DrEng/Sanitary/Johns Hopkins Univ. *Born:* 8/11/06. Employed U of N C 1934-37, Johns Hopkins Univ 1937-76 with time for USNR 1943-46 & for WHO in Chile 1954-55. Mbr num prof & hon societies incl the Natl Acad of Engrg. Many consultancies & service on many cttes, commissions & councils incl: AEC-ACRS, AEC Panel of Examiners & PESAC Panel on the Environ. Res: water filtration, nuclear wastes, storm runoff, ground water, water & waste sys & cooling water. Author of some 60 articles & sev text-books, the latter being jointly authored with G M Fair & Daniel Okun. *Society Aff:* ASGE, AWWA, FWPG, AAAS.

Ghaboussi, Jamshid
Business: 3106 Newmark Civ Engg Lab, Urbana, IL 61801
Position: Prof *Employer:* Univ of IL. *Education:* PhD/St Engrg/Univ of CA Berkeley; MEng/St Engrg/Nova Scotia Tech Coll; BSc/Civ Engrg/Univ of Tehran. *Born:* 7/24/41. in Iran. Worked on structural design of some important bldgs in Iran. Experience in North Am includes working for Dames and Moore, Consulting Engrs in Toronto on Geotechnical problems and after receiving PhD from UC, Berkeley worked with Agbabian Assocs in Los Angeles on computer methods in struct mechanics. Before joining the faculty at CE Dept at Univ of IL in 1973, spent one yr as Res Fellow in Trondheim, Norway. Res interest includes: earthquake engg, soil dynamics, computer methods in struct mechanics and geomechanics. *Society Aff:* ASCE, EERI, AAM

Ghandakly, Adel A
Business: Dept of Elect Engineering, Toledo, OH 43606
Position: Assoc Prof *Employer:* Univ of Toledo *Education:* PhD/EE/Univ of Calgary; MSc/EE/Univ of Calgary; BSc/EE/Univ of Alexandria *Born:* 03/15/45 Native of

Ghandakly, Adel A (Continued)
Egypt. Practiced Elec Engrg in both Canada and US since 1971- . Taught at Univ of New Orleans (LSU) and Univ of Toledo. *Society Aff:* IEEE, HKN, APEGA

Ghandhi, Sorab K
Business: Troy, NY 12180
Position: Prof *Employer:* Rensselaer Poly Inst *Education:* PhD/EE/Univ of Ill; MS/EE/Univ of Ill; BSc/EE&ME/Benoves Hindu Univ *Born:* 1/1/28 Mbr - Advanced Circuits Grp, General Electric Co., Syracuse, NY 1950-; Manager - Electron Devices Group, Philco Corp., Philadelphia, PA 1960-1963; Professor of Electrophysics, Rensselaer Polytechnic Institute, Troy, NY 1973- present; Author - *Theory and Practice of Microelectronics*, 1968; *Semiconductor Power Devices* 1977; 1983; 90 papers in the area of semiconductor materials, processes, and devices. Present interest - wine, women and song.

Ghanime, Jean
Business: 795 First Ave, Lachine Quebec, Canada H8S 2S6
Position: National Sales Mgr Electric Utility Ind *Employer:* General Electric Canada Inc. *Education:* BS/Elec Engg/Ecole Polytechnique of the Univ of Montreal. *Born:* Jun 1927 in Middle East. Prof Engr, mech-elec engrg, Ecole Polytechnique (Univ of Montreal). Completed Test course Canadian Genl Elec in 1952. Three yrs design with Canadian Genl Elec Meters & Instruments dept. Sales Engr from 1956- 74. Indus Heating Eqt of C G E (Process heating). Spec in residential, commercial & indus heating C G E Montreal 1974-75. 1975-80 Quebec Sales Mgr- Hydraulic turbines Dominion Engrg Works Ltd (Affiliate Co of C G E). 1986 Mgr Electric Utility Sales - General Electric, Canada Inc. 20 yrs as RCEME Major with Canadian Forces (Militia) & retired in 1974. Chmn ASM-Montreal Chap 1967-68, Chmn of Canadian Council of ASM 1973-75. Appointed P Chmn Representative in 1976. Mbr of Adv Ctte in metallurgical Techniques-Dept of Ed-Province of Quebec. VP of Montreal Electrical Club, Mbr Canadian Electrical Assn Mbr of Order of Engrs of Quebec. *Society Aff:* ASM, OIQ, CEA.

Ghausi, Mohammed S
Business: Coll of Engrg, Davis, CA 95616
Position: Prof & Dean. *Employer:* Univ of CA, Davis. *Education:* PhD/Elect Eng/Univ of CA (Berkeley); MS/Elect Eng/Univ of CA (Berkeley); BS/Elect Eng/Univ of CA (Berkeley). *Born:* 2/16/30. From 1960 to 1972, Mohammed S Ghausi was on the faculty of NY Univ, NY, as Asst Prof, Assoc Prof and Full Prof. From 1972 to 1974, he was the Sec Head of Elec Sciences and Analysis Sec of the Engg Div at Natl Science Fdn in Washington, DC. From 1974-1977 he was Prof and Chrmn of the Elec and Computer Engg Dept at Wayne State Univ, Detroit, MI. From Jan 1978-Oct 1983 he was Dean of Engr, Oakland Univ, Rochester, MI. Dr Ghausi also held the John F Dodge Prof of Engg. Since October 1983 he has served as Dean of Engrg at the Univ of California, Davis. He is the author of five text books, and numerous publications. Dr Ghausi is a member of Sigma Xi, Phi Beta Kappa, Eta Kappa Nu, Tau Beta Pi, Circuits and Systems, Systems, Man and Cybernetics Societies. He was the VP of Circuits and Systems Society (1972-1974) and Pres of the Society (1976). He also served as the Chrmn of Fellows Award Comm (1977-81) and the Profl Activities Coordinator of Div I of the IEEE (1978-80). He is mbr of Res & Dev Ctte of IEEE. Dr Ghausi is a Fellow of IEEE. Dr. Ghausi is the winner of IEEE Centennial Medal (1984) as well as the winner of the Alexander von Humboldt prize from the government of West Germany, (1985). *Society Aff:* IEEE, ESD.

Ghiotto, Robert A
Business: 1454 US 19 South, Clearwater, FL 33516
Position: VP *Employer:* CH2M-Hill *Education:* B/CE/Univ of FL *Born:* 2/14/33 Mulberry, FL-served US Army 1955-57, graduated Univ of FL 1959, Sanitary Engr for development co 1959-60; Ass't City Engr and Utilities Dir City of Pompano Beach, FL 1960-63; Ass't Utilities Dir Broward County, FL 1963-64, Regional Mgr Black, Crow & Erdsness 1964-72, VP 72-77, purchased by CH2M-Hill 1977. Currently Regional Mgr, VP member of Bd of Dirs CH2M-Hill. *Society Aff:* AAEE, NSPE, AWWA

Ghormley, Edward L
Business: 75 W Green St, Suite 2, Pasadena, CA 91105
Position: VP. *Employer:* I Sheinbaum Co, Inc. *Education:* PhD/CE/Univ of S CA; MS/CE/MIT; BS/CE/Univ of WA; Certificate/Engg Mgt/Univ of CA. *Born:* 10/21/23. Thirty yrs experience in the chem engg field. Managed a number of construction projs in the oil refining & petrolchem industries as well as in the chem & geothermal industries. Experience includes process dev, facility operations, engg design, proj mgt, scheduling & estimating. Has served many local contracting cos, the aerospace industry, govt & the Navy. Has had many publications & patents. Is presently Chrmn Elect of S CA Sec of AIChE. *Society Aff:* AIChE, ΣΞ, RSA.

Ghose, Rabindra Nath
Business: 696 Hampshire Rd, Westlake Village, CA 91359
Position: Chairman of Bd of Dir. *Employer:* American Nucleonics Corp. *Education:* PhD/Engg/Univ of IL; MA/Math/Univ of IL; EE/Engg/Univ of IL; MS/Engg/Univ of IL; DIIS/Engg/Ind Inst of Science. *Born:* Sep 1, 1925. Fellow IEEE, IEE (London), Amer Physical Soc, Inst of Physics (London), Amer Assoc for the Adv of Sci, Wash Acad of Sci. Mbr Sigma Xi, Eta Kappa Nu, Pi Mu Epsilon, State Bar of Calif., Amer Bar Assoc., Patent Attorney, US Patent and Trademark Office. Charter Engr, British Commonwealth, Prof Engr State of Calif. Winner of Asoka Medal, Engr of Yr 1972 SFV Engrg Council, Los Angeles, Exec & Prof Hall of Fame, Commendation, State Assembly, Calif. Res & Tech Adv Ctte (Guidance, Control & Info) Natl Aeronautics & Space Admin (1974-76), Div Adv Group, Scientific Adv Bd, US Air Force. *Society Aff:* IEEE, IEE, AAAS, APS, IP.

Ghosh, Amit K
Business: Rockwell International Science Ctr, 1049 Camino Dos Rios, Thousand Oaks, CA 91360
Position: Manager, Metals Processing *Employer:* Rockwell Intl Sci Ctr. *Education:* PhD/Metallurgy & Materials Sci/MIT; MS/Metallurgical Eng/Univ of IL; BE/MET E/Univ of Calcutta *Born:* 11/18/45. Came to the US from native India in 1966 with BE degree in Metall Engg & officers training in steel plant tech. Following grad work at M.I.T., joined the Res Labs of Gen Motors Corp in 1972. Directed and conducted res related to sheet metal forming, with emphasis on high strength/weight ratio mtls for energy conservation. Developed sheet metal quality control test (1975). Received ASM Grossman award for best paper in Metall Transactions (1976). Joined Rockwell Sci Ctr in 1976 and since working in the areas of high temperature deformation, metal matrix composites, fracture and superplastic forming problems. Currently manager of Metals Processing Department. Responsible for several innovative methods of grain refinements, superplastic forming and metal matrix composite fabrication. Author of over 62 technical papers including an article in Scientific American (1976), and 4 patents. Several Rockwell R & D Awards (1982, 1983), ASM Quad Chapter Award 1986. Served on the Bd of Dir for The Metallurgical Society. *Society Aff:* AIME, ASM

Ghosh, Arvind
Home: 5222 Moss Glenn, Houston, TX 77088
Position: Conslt Engr *Employer:* Self-employed *Education:* BS/CE/Calcutta Univ-India; BS/Physics/Calcutta Univ-India; Diploma/Prestressesd Concepts/French Courts; Diploma/Ocean Engrg/TX A&M Univ. *Born:* 10/6/24 Native of India, now naturalized US citizen. Educated in India, England France and the US. Senior engr with Societe Anonyme Hersent, Paris, France; chief engr of DeLong Corp, NY, NY. VP, Structural and Marine Division of Worley Engrg, Inc. Fellow of ASCE; Fellow of IStructE(London). Publication: "Fabrication and Installation of Production Platforms in Shallow Open Sea Areas–, presented at the Energy Technology Conference and Exhibition, Houston, TX, Nov 5-9, 1978. PE, NY and TX. *Society Aff:* ASCE, SNAME, SPE of AIME

Ghosh, Mriganka M
Home: 560 Lanceshire Lane, State College, PA 16803
Position: Prof. *Employer:* Pennsylvania State Univ *Education:* PhD/Civil Engg/Univ of IL; MS/Civil Engg/Univ of IL; B Tech (Honors)/Civil Engg/Indian Inst of Tech Kharagpur, India. *Born:* 11/5/35. Born and raised in India. Immigrated to the USA in 1968. Taught at the Univ of ME and Univ of MO for eight years each before coming to PA in 1984. Was a visiting prof at the Swiss Fed Inst of Tech in Zurich in 1974-75. Have authored or co-authored more than eighty papers on various aspects of environmental engrg in various professional journals and bulletins. Res expertise in physiocochemical processes for water quality control and aquatic chemistry of natural systems. Chaired the Environmental Engrg Res Council, ASCE. Past editor of the Journal of the Environmental Engrg Division, ASCE. Past Exec Secy of the Environ Engr Div, ASCE, Chairman, Exec Cttee, Environ Engrg Div, ASCE-Enjoy jogging, classical music, and traveling. Mbr, USA National Committee for International Association for Water Pollution Research and Control. *Society Aff:* ASCE, ACS, AWWA, IAWPRC, WPCF, AEEP.

Ghublikian, John R
Home: 192 Commonwealth Ave, Boston, MA 02116
Position: Adjunct Prof Chem Engrg *Employer:* Tufts Univ *Education:* BS/ChE/Tufts Univ.; Dip/Mgmt Dev/Northeastern Univ *Born:* 11/7/17. Wife Leona L. PhD, Case Western Reserve Univ. Children: John R Jr & Ann. Fellow AIChE, Mbr ACS, Tau Beta Pi. Reg Prof Engr MA 3038. Awards: ASME 1976 Outstanding Leadership, Mgmt Energy Resources. Mbr: Univ Club, Boston; Mbr Algonquin Club, Boston; New Seabury Tennis & Country clubs. Retired Dir: The Badger Co Inc (Sr VP Western Hemisphere); Badger Pan Amer Inc (Pres); Badger Amer Inc (Pres); Bader BV B Netherland (Managing Dir 1970-74). Chairman Board of Trustees, Wentworth Institute of Technology, Boston. *Society Aff:* TBΠ, AIChE, ACS.

Giacobbe, John B
Business: Superior Tube Company, P.O. Box 191, Norristown, PA 19404
Position: Vice Pres-Metallurgy. *Employer:* Superior Tube Co. *Education:* MS/Metallurgical Eng/Lehigh Univ; BS/Chem Eng/Leigh Univ. *Born:* Oct 11, 1917 Dudley, Pa. Joined Superior Tube Co 1941 as Asst Metallurgist; Plant Metallurgist 1945; Dir-Nuclear Products Div 1955; Ch Metallurgist 1964; promoted to V P-Metallurgy May 1967. Chmn: Phila Chap ASM 1960-61; Metals-Material Exposition ASM 1964; Ed & Dev Council 1969-71; Fellow Amer Soc for Metals; Bd of Trustees ASM 1970; Chmn ASM Study Ctte on Cert & Council for Prof Interest 1974-76; Delaware Valley Man of Yr, Phila Chap ASM 1967. Mbr Tech Adv Ctte, Metals Properties Council; VP & Trustee ASM 1979-80. Pres & Trustee ASM 1980-81; Immediate Past Pres & Trustee ASM 1981-82. *Society Aff:* ASM, ASTM, NSPE.

Giacoletto, Lawrence J
Home: 4465 Wausau Rd, Okemos, MI 48864
Position: Prof Emeritus of Elec Engrg. *Employer:* Michigan State Univ. *Education:* BS/EE/Rose-Hulman Inst of Techn; MS/Physics/State Univ of IA; PhD/EE/Univ of MI. *Born:* Nov 14, 1916. US Army Signal Corps 1941-46; Res Engr RCA Labs 1946-56; Mgr Electronics Dept Ford Motor Co Scientific Lab 1956-61; Prof of Elec Engrg, Michigan State Univ 1961- . Fellow & Past Mbr of Bd of Dir IEEE; Fellow AAAS; Mbr Soc of Sigma Xi; P Mbr of Bd of Trustees, Natl Electronics Conf; Mbr, Editorial Advisory Bd, Solid-State Electronics. Contributor, Transistors I, Methods of Experimental Physics; Author, Differential Amplifiers; Ed, 2nd Edition, Electronics Designers' Handbook. *Society Aff:* IEEE, AAAS, ΣΞ.

Giaever, Ivar
Business: Box 8, Bldg K-1, 3C32, Schenectady, NY 12301
Position: Biophysicist. *Employer:* G E Co Research & Dev Ctr. *Education:* PhD/Physics/RPI, Troy; Siv Ing/Mech Eng/Norwegian Inst of Tech. *Born:* Apr 5, 1929. Native of Norway, naturalized U S citizen in 1964. Immigrated to Canada in 1954 & to US in 1956. Staff mbr at G E R&D Ctr since 1958. Recipient Olive Buckley Award 1964, Nobel Prize in Physics 1973, Zworykin Award 1974. Hobbies are camping, hiking, skiing, tennis & windsurfing. *Society Aff:* NAS, NAE, APS, IEEE.

Giamei, Anthony F
Home: 54 Virginia Drive, Middletown, CT 06457
Position: Sr Consulting Scientist. *Employer:* United Technologies Research Center *Education:* PhD/Mat Sci/Northwestern Univ; BE/Metallurgy/Yale Univ. *Born:* Oct 1940. G M Scholar, Grad w/highest hon. NSF Fellow 1963-65. Native of Corning N Y. Metallurgist with G M summers 1960-62. With Pratt & Whitney Aircraft, Commercial Products Div, United Technologies Corp 1966-1981. Currently Sr Consulting Scientist in Materials Technology, United Technologies Research Center. Work covers physical metallurgy of high temp alloys & solidification R&D. Thirty-Fivetech papers & sixteen patents. New England Regional Conf Outstanding Paper Award in 1970. United Tech. Mead Medal for Engineering Achievement in 1981. Active mbr AIME & ASM Fellow 1977. Family man with 2 children. Hobbies incl bowling & fishing. *Society Aff:* ASM, AIME, ΣΞ.

Giampaoli, Donald A
Business: Chief, Energy Division, Inter-American Development Bank, 1300 New York Ave., N.W, Washington, D.C 20577
Position: Chief Energy Division *Employer:* Inter-American Development Bank *Education:* MCE/Structures/Catholic Univ of America; BCE/Structures/Univ of Santa Clara. *Born:* 1/4/32. Native of San Francisco. Served with US Army Corps of Engrs 1955-57; Hwy Engr for D C Hwy Dept 1957-59; Assoc Design Engr NCMA 1959-60; Dir of Heavy Constr, Dir of Legislation & Spec Programs, & Assoc dir of Labor Relations AGC 1960-75; from Jan 12, 1975 to June 18, 1978, Asst Commissioner for Resource Dev, Reclamation, Interior; from June 18, 1978 to March 24, 1980, Dir Office of Dam Safety Engg and Science; from Mar 24 1980 to July 15, 1984, Dep. Dir. Div. of Hydropower Licensing, July 15, 1984 to Dec 1986, Dir, Hydropower Analysis, FERC; Jan. 1, 1986 to present, Chief, Energy Div. Inter-American Development Bank. Policy respon incl hydropower, construction, design, & mining petroleum projects. Mbr ASCE; USCOLD. 1985 Special Achievement Award Fed. Emergency Management Agency; Award 1974 Annual Alumni Engrg Achieve Award, Catholic Univ; Prof Contribution Cert, Gov Procurement Commission; ASCE-NCS Disting Serv Award. Reg PE, Cert Contracts Mgr, Profl *Society Aff:* ASCE, USCOLD.

Giampaolo, Joseph A
Business: 3907 N Rosemead Blvd, Rosemead, CA 91770
Position: Pres. *Education:* MBE/Bus Admin/Claremont Univ; BS/EE/Gonzaga Univ. *Born:* 3/20/46. I am experienced in general construction since 1964. I have been in private practice as a consulting engr since 1968 & am Pres & Proj Engr for elec & mech design for all types of bldgs in the construction industry locally & nationally including hotels & casinos, convention centers to heavy ind facilities. My co is in a continual growth pattern necessitating expansion of office space in present facilities which is a bldg privately owned by myself. A maj function of the firm, Giampaolo & Assoc, Inc, recently is energy conservation studies & reports performed by competent personnel researching & keeping abreast of new info. *Society Aff:* IEEE, ASHRAE, NSPE, CSPE, ASPE.

Giancarlo, Samuel S
Business: 4801 Woodway, Suite 251 W, Houston, TX 77056
Position: VP & Gen Mgr. *Employer:* GKN Birwelco (US), Inc. *Education:* MBA/Bus/SUNY; BS/CE/SUNY. *Born:* 6/11/42. Engr trainee Rep Steel Corp, Buffalo, 1961-63; dev engr Allied Chem Corp, Buffalo, 1965-67; mgr engg, asst sales mgr Luwa Corp, Charlotte, NC, 1968-73; contract engr engg & constrn J F Pritchard & Co, Kansas City, MO, 1973; proj mgr M W Kellogg Co, Houston, 1973-74; engg & constrn sales exec A G McKee & Co, Houston, 1974-79; vp & gen mgr GKN Birwelco (US), Inc, Houston, TX 1979-. Pack exec com Boy Scouts Am; cub master Cub Scouts, Leader Webelos 1975-77; active YMCA Indian Guide Prog, 1973-77.

Giancarlo, Samuel S (Continued)
Licensed PE, TX; Mbr Am Inst Chem Engrs, Natl, TX soc PE, U Buffalo Alumni Assn, Beta Gamma Sigma, Tau Kappa Epsilon. Roman Catholic. Club: Univ Club of Houston. *Society Aff:* AIChE, NSPE, TSPE.

Gianelli, William R
Business: 973 Pioneer Rd, Pebble Beach, CA 93953
Position: Consltg Civil Engr *Employer:* Self Employed *Education:* BS/CE/Univ of CA-Berkeley *Born:* 2/19/19 Stockton, CA, Married, two daughters. Commissioned Officer, Corps of Engrs, 1941-45. Served in HI, Saipan, Okinawa and Korea. Joined CA State Government in various positions in State Engrs Office and Dept of Water Resources until 1960. Senior partner, Gianelli and Murray, Consulting Engrs, 1960-67. Director of CA Dept of Water Resources 1967-1973 (Member/Chrmn of Western States Water Council and Member of National Commission on Water Quality). Consulting Engr, 1973-81. Assistant Secretary of the Army (Civil Works) 1981-84; Consulting Civil Engr, Chrmn Panama Canal Commission, 1984- ; Awards: Construction Man of the Year, 1973; Public Works Man of the Year, 1973; ASCE Royce J Tipton Award, 1973; Distinguished Service Award, Soil Conservation Soc of Amer, 1973; US Dept of Interior, Bureau of Reclamation, Citizen Award, 1975; US Dept of Defense Distinguished Public Service Award, 1984; US Dept of the Army Decoration for Distinguished Civilian Service, 1984. *Society Aff:* ASCE, AWWA

Gianniny, Omer Allan, Jr
Home: 2411 Jefferson Park Ave, Charlottesville, VA 22903
Position: Prof *Employer:* Univ of VA. *Education:* EdD/Educ Psychology/Univ of VA; MEd/Communications/Univ of VA; BME/Mech Engrg/Univ of VA. *Born:* 12/5/25. Native of Charlottesville, VA. Worked as refinery engr, ESSO Std Oil Co, Linden, NJ, 1947-1951. Engr officer in USNR, 1945-6, 1951-53. Res engr in fluid flow, especially mixed phase, 1953-1956. Began teaching technicl communications in 1955 at Univ of VA; mgt communications in 1961-5, role of engr in soc, tech and western culture, in 1963; taught history of tech & values in engg. Consultant to Newport News Shipbldg & Drydock & to Norfolk Naval Shipyard in technical communications. Chrmn, humanities faculty in engg sch, Jan, 1979-Sep 1980. Bd. Directors ASEE, 1979-1981. PE (VA) 1955. *Society Aff:* ASEE, SHOT, AAAS, TBΠ.

Gianola, Umberto F
Business: AT&T-Bell Labs - Room 14A-383, 1 Whippany Road, Whippany, NJ 07981
Position: Govt Prog Planning Ctr. *Employer:* AT&T Bell Labs *Education:* PhD/Electron Physics/Univ. of Brimingham, UK; BSc/Physics/Univ. of Brimingham, UK. *Born:* 10/29/27. Post-doct Fellow 1951-53 Univ of Brit Columbia. Fellow IEEE. Early work in Electron Optics & Nuclear Radiation Detectors. Joined Bell Labs at Murray Hill, N.J. in 1953 as mbr of Communications Res Dept. Appointed supervisor 1960 of dev of memories for Electronic Switching Sys. Appointed hd Fundamental Memory Components Dept in 1963, initiated the dev of Magnetic Bubble dices. Appointed head Ocean Res Dept at Whippany, N.J. in 1969, respon for oceanographic studies. Appointed Dir, Ocean Systems Studies Ctr in 1971, respon for sys engrg, res & dev of adv sonar signal & data processing sys. Assigned to tech liaison at Sandia Natl Labs, Albuquerque, N.M. in 1984. Appointed Dir, Defense Systems Ctr, Summit, N.J., in 1984, respon for systems engrg for defense against strategic and tactical missiles. Appointed Dir, Government Program Planning Ctr, Whippany, N.J., in 1985, respon for management information systems for planning and coord of all AT&T - BL contract R&D for Government depts. & agencies. *Society Aff:* IEEE, ΣΞ, AAAS, ADPA

Gianoli, Napoleon H
Business: Jr. Callao 268-24/42, Lima 1, Peru
Position: Pres *Employer:* Ofic. Tech. de Ingenieria *Education:* MS/EE-ME/Univ NAC. De Ingenieria. *Born:* 6/16/24 Native of Lima (Peru). Taught in La Salle and Escuela Nacional de Ingenieros (1932-1946). Chief Engr Chimbote Branch CPS (1947-1949). Electrical Engrg Prof Escuela Militar (1950-1952) Electrical Instalations Prof Architecture Faculty (1961-1967). UNI Lighting Engrg Prof (1978- .) Consulting Engr Association VP (1967). IEEE Peru Section Pres (1968) Mechancial & Electrical Chapter Colegio de Ingenieros Del Peru President (1970-1972). Instituto de Urbanismo General Secretary (1981- .) IESA Mgr (1951-1956). NEISSER Chief Engr (1957-1958). CPS Pres (1972). INIE Dir (1973). CIER Pres (1974-1975) Electricity Rate Tariff Commission Pres (1969-1972). Lima Metro Politan Technical Commission Mbr (1980-1981). Independent Contractor (1959-1968). Independent Consultant (1974- .) Ten children and 15 grandchildren. Electricity Rate Tariff Comm Mbr (1983-1985) CIE Natl Peruvian Lighting Ctte (Pres). *Society Aff:* CIP, APLE, AFE, IES.

Giavara, Sutiri
Business: 1 Columbus Plaza, New Haven, CT 06510
Position: Pres *Employer:* Flaherty Giavara Assoc Inc *Education:* BS/CE/IN Inst of Tech; Cert/Military Engr/Fort Belvoir *Born:* Directs FGA, a multioffice and multidisciplined engrg, architectural, planning, environmental Labs Inc CT, Partner: Santiago Vazquez Flaherty Giavara Assoc Puerto Rico. Wide range of natl and intl experience on many major projects in the Environmental and Civil engrg fields. Registered PE; in all New England States, NY, NJ, WV, IN. Diplomate, Amer Academy of enviormental Engrs, Fellow, Amer Soc of Civil Engrs; Sr VP CT Engrs in Private Practice, mbr; of over twenty Tech Soc, bd mbr of several civic institutional organizations. *Society Aff:* ASCE.

Gibala, Ronald
Business: Dept. of Materials Science & Engineering, The University of Michigan, Ann Arbor, MI
Position: Prof of Metallurgy & Mat Sci. *Employer:* Case Western Reserve Univ. *Education:* PhD/Met Eng/Univ of IL; MS/Met Eng/Univ of IL; BS/Met Eng/Carnegie Inst of Tech. *Born:* Oct 3, 1938 New Castle, Pa. Asst Prof Metallurgy & Materials Science, Case Western Reserve Univ 1964-69. Assoc Prof CWRU 1969-76. Prof CWRU since 1976. Alfred Noble Prize ASCE 1969. Outstanding Young Mbr Cleveland Chap ASM 1971. Tech Achieve Award, Cleve Tech Soc Council 1972. Visiting Scientist, Centre d'Etudes Nucleaires d'Grenoble, France 1973-74. Chmn Cleveland Chap ASM 1975-76. Director, Materials Research Laboratory, CWRU 1981- Metallurgical cons for more than 50 indus, publishing, & legal firms since 1964. Board of Directors, TMS-AIME. Served on sev ASM & AIME natl cttes. *Society Aff:* ASM, AIME.

Gibble, Kenneth
Business: P.O. Box 362, Old Saybrook, CT 06475
Position: Pres *Employer:* Besier Gibble & Quirin *Education:* B/Arch Engrg/PA State Univ *Born:* 1/16/39 Chrmn of ASCE Exec Committee of Engrg Management Div. Editorial Bd of Structural Engrg Practice Journal of Analysis, Design, Management. Past dir and treasurer of CT Engrs in Private Practice. Visiting faculty member of Yale Univ School of Architecture in fire safety design. Author of *Organizational Development for small consulting firms,* and *Improving the effectiveness of engrg organizations: Six case histories,* both published in Issues in Engrg Journal of Professional Activities, ASCE. Who's Who in Industry and Finance. PE in CT, NY, VT. *Society Aff:* ASCE, ASEM, NSPE

Gibboney, James A
Business: AGMC/MAW-Newark AF Sta, Newark, OH 43055
Position: Resources Mgmt Div Chief. *Employer:* Aerospace Guidance & Metrology Ctr. *Born:* Nov 24, 1931. Native of Fletcher, Ohio. Assoc of Sci Degree from Univ of Dayton (Ohio). Reg Prof Indus Engr Ohio. US Navy 1950-54. U S Civil Service since 1954. Held pos as Ch of Plant Engrg, Ch of Indus Engrg, & since 1972 as Ch of Resources Mgmt Div. Current respon incl Dev of Depot Maintenance Work Load Capability, Negotiation of Depot Maintenance Work Load, Manpower Re-

Gibboney, James A (Continued)
quirements & Distribution, Financial Mgmt, Depot Test Equip & Industrial Facilities. Recipient of the AF Meritorious Civilian Service Medal in 1971.

Gibbons, Eugene F
Home: 585 Chauncey Ln, Cedarhurst, NY 11516
Position: Pres *Employer:* Gibbons and Hyland, P.C. *Education:* BS/CE/Manhattan Coll *Born:* 11/2/14 My early experience was as engr for construction companies, and involved both field and design work. During 1955-1956, I was Chief Engr for the NY State Dept of Public Works. I then spent six years with Raymond Intl as VP and Chief engr of its African subsidiary. In 1962, I became Commissioner of Public Works and County Engr for Nassau County, NY. I entered private practice in 1966, and as VP of McFarland-Johnson-Gibbons, directed the company's metropolitan NY activities. In 1972, I founded my present practice, and am Pres of Gibbons and Hyland, P.C. *Society Aff:* ASCE, NSPE, SAME, WPCF, ACEC

Gibbons, Harry De R
Business: Eng-Science, Cleveland, OH 44119
Position: Group VP *Employer:* Eng-Science *Education:* MSCE/Fdns-Structures/MIT; BSCE/Structures/Univ of TX. *Born:* 1/24/24. Reg PE in OH and MI. Worked for industry for twenty yrs and rose through the sales and engg ranks to become mgr of a product div. For the last ten yrs has served as the Midwestern Regional Mgr for Engg Science a natl and intl consulting engg firm currently ranked 43rd in the ENR listing. As a Group VP is responsible for two regions and four subsidiary companies. *Society Aff:* ASCE, NSPE, SAME, WPCF.

Gibbs, Charles V
777 108th Ave., NE, PO Box 15000, Bellevue, WA 98004-2050
Position: Sr VP, Northwest District Manager *Employer:* CH2M HILL Inc. *Education:* MS/Civil Engr/Univ of WA; BS/Civil Engr/Univ of WA *Born:* Jul 1931. MS, CE Univ WA; BS, CE Univ of WA. Dist Engr, WA State Pollution Control Comm 1956-58. Seattle Metro, Chief, Water Qual Div of Metro 1959-64, respon water qual monitoring & analysis prog. 1964-66 Dir of Tech Services for Metro's wastewater operations. Served as Exec Dir Seattle Metro 1967-74. Respon admin & implementing policies & progs for water pollution control and transit King Cty. Joined CH2M HILL in mid-1974; & in present pos as Northwest District Mgr, Seattle office. Awds: Charles Walter Nichols 1969 APWA; Top Ten Public Works Men-of-Yr 1971 APWA; Recipient, commendation from Pres for exceptional srvc in environ prot effort. Mbr AAEE, WPCF; Engr of Yr, Wash SPE, 1973; Outstanding Pub Official, Seattle-King Cty Municipal League, 1974; ASCE Wesley W. Horner Awd, 1974. *Society Aff:* AAEE, APWA, ASCE, PMI, WPCF

Gibbs, William R
Business: 1500 Meadow Lake Parkway, Kansas City, MO 64114
Position: Partner. *Employer:* Black & Veatch, Cons Engrs. *Education:* BS/Civil & Environ Engg/Sch of Engg Univ of KS. *Born:* 7/21/18. Prof CE, Univ of MO, Rolla (Hon). In 1940, joined Black & Veatch, Cons Engrs, Partner; VP, Black & Veatch Intl. Major, Corps of Engrs, US Army WWII. Practice in civil & environ engg. Dir, VP, & Pres 1977 of ASCE. Pres of local & state groups, chmn of Bd of Ethical Review & Natl Dir of NSPE. Dir, Engrs Joint Council. Natl Dir of Planning-Design Div of ARTBA. Mbr, mgmt Advisory Group to EPA. Delegate of Union of Panamerican Assocs of Engrs. Mbr AWWA, WPCF, AAEE, APWA, SAME, MARLS, Tau Beta Pi, Sigma Tau, Chi Epsilon. Active in civic orgs. Wife Dee; daughter Linda; sons Richard, Donald, & William Jr. *Society Aff:* ASCE, NSPE, AWWA, AAEE, WPCF, APWA, SAME, ARTBA, COFPAES, ACEC.

Giberson, Harry F
Business: 2601 Teepee Dr, Stockton, CA 95205
Position: Manager, Maint. & Engr. *Employer:* California Cooler *Education:* MS/Eng. Mgt./Northeastern University; BS/Ch.E/Northeastern University *Born:* 02/13/38 Graduated from Northeastern Univ in 1961 BS Ch.E. Spent next two years in areospace, then joined General Foods in 1963. In 1968, joined Norton Co. as Sr. Engineer; progressed through numerous engrg and maintenance management positions. Left Norton in 1985 as Plant Engineer. Assumed current responsibility as Dir of Maintenance and Engrg for California Cooler in 1986. President Ga. Soc. of Prof. Engrs 1984-85. Plant Engr of the Year Atlanta Chapter & Southeast Region 1983. Engr of the Year in Industry - Atlanta Metro Engrs and Ga. Soc. of Prof. Engrs 1984. Present Chairman - Prof. Eng. in Industry for Ca. Soc. of Prof. Engrs. *Society Aff:* AIChE, AIPE, NSPE, PMI, IFMA, SME

Gibian, Thomas G
Home: PO Box 219, Sandy Spring, MD 20860
Position: Pres *Employer:* Henkel Corporation *Education:* DSc/Chem/Carnegie Mellon Univ; BS/Chem/Univ of NC. *Born:* 3/20/22. Born Prague, Czechoslavakia. Came to USA 1940 US citizen. 1948-51 Atlantic Refining Co Chemist. 1951-74 W R Grace & Co (hyc) Plant Mgr, VP Organic Chemicals Div, Pres Res Div, Corp. VP & Tech Group Executive 1974-76. Pres Chem Construction Corp 1976-1980. Chem and Pres TGI Corp Engg Design & Construction 1980-Present, Pres & CEO Henkel Corporation. Married, four children. Mbr Union League Club (NYC), Cosmos Club (Washington DC), 1942-45 Royal Air Force Pilot. *Society Aff:* AIChE, ACS, IRI.

Gibney, James J, Jr
Home: 1712 Providence Ave, Schenectady, NY 12309
Position: Manager. *Employer:* General Electric Co. *Born:* Nov 1920. BME The City College of New York. With the G E Co since 1942. In 1954 the first Manager of the Steam Turbine & Generator Product Dev Lab. Since 1960, mgmt pos with respon for thermal & mech design & performance, as well as procurement of pressurized water reactor plant components used in Navy nuclear ships. Since 1974, Mgr of Oper Planning & Auditing for dept. Enjoy community activities, golf & tennis.

Gibson, Charles A
Business: P.O. Box 6169, Tuscaloosa, AL 35487
Position: Prof of Elect Engrg *Employer:* The Univ of AL *Education:* PhD/EE/Univ of FL; MS/EE/MS State Univ; BS/EE/MS State Univ *Born:* 2/11/31 in Greenville, MS. Served in the US Navy 1951-54. Prof of Electrical Engrg at the Univ of AL where he has taught since 1959. Research and consulting in electric power system planning and operations. Teaching primarily in power systems and automatic control systems. Industrial experience as a transformer design engr, relay engr, planning engr, and operating system development engr. Registered PE in the State of AL. Senior member in the Inst of Electrical and Electronics Engrs. He is also a member of the American Society for Engrg Education, Tau Beta Pi, Eta Kappa Nu, and Sigma Xi. Hobbies are golf, hunting, and fishing. *Society Aff:* ASEE, IEEE, TBΠ, HKN, ΣΞ

Gibson, David W
Business: Room 346 Weil Hall, University of Florida, Gainesville, FL 32611
Position: Assoc Prof *Employer:* Univ of FL *Education:* MS/CE/Univ of Miami; BS/CE/Univ of Cinn *Born:* 10/05/44 He has been an Assoc Prof with the Univ of FL since 1974, developing, teaching, and directing programs in Surveying & Mapping. Awards and distinctive service includes co-authorship of a best selling text *Route Surveying and Design*, exam consultant to the state of FL, ACSM Professional Journalism Award, FSPLS Pres's and Land Surveyor of the Year awards, member ABET Bd of Dirs, exam consultant to NCEE for land Surveying content. *Society Aff:* ACSM, ASPRS.

Gibson, George T
Business: 800 N Harvey, One Bell Central, Oklahoma City, OK 73102
Position: Genl Manager-Network. *Employer:* Southwestern Bell Telephone Co. *Education:* BA/Math & Physics/Univ of Tulsa; AA/Math/Muskogee Jr College. *Born:* 6/23/33. Native of Tulsa, OK. Worked at Bell Labs 1958-60; developed data training sch at Cooperstown, NY 1961-62. Served in wideband data mktg at AT&T in 1962. With the Bell sys since 1955. Assumed current respon as hd of the Network Dept for the State of OK in 1969. Is a Sr Mbr of IEEE & a Reg P E. Receive the "Outstanding Engr in Mgmt" award from the OK Soc of Prof Enggs in 1973. Under his guidance, the Engr Dept was awarded the Natl Indus Prof Dev award for 1976 from the NSPE. Past Pres of the Okla Engr Foundation. Instructor at Okla Univ. Enjoys Ranching, tennis, canoeing and fishing. Mbr Amer Soc of Engg Ed, IEEE, NSPE, OSPE, OETGCO, OEF. Member Board of Advisors at Oklahoma University, Tulsa University and Oklahoma Christian College and is a Director of Academy Computing Corporation. *Society Aff:* IEEE, NSPE, OSPE, ASEE, ASEM.

Gibson, Harold J
Home: 28505 Inkster Rd, Farmington Hills, MI 48018
Position: Eng Consultant *Education:* MS/Mech Eng/Univ of MI; BS/Mech Eng/Univ of MI. *Born:* Jun 1909. Native of Wixom, Mich. Joined Ethyl Corp in 1930. Specialized in dev of fuels, engines, & lubricants for power, econ, durability & low emissions. Was Manager, Detroit Res Labs of Ethyl Corp until retirement in June 1976. Now Tech Cons. Was Chmn Motor Fuels Div of Coord Res Council, Sci-Engrg & Transportation Sys Activities of Soc of Automotive Engrs. Received Horning Mem Award from SAE 1976. Also Fellow of SAE 1979. Mbr Sigma Xi, Pi Tau Sigma & Engrg Soc of Detroit of which he is a Fellow & Former Dir. *Society Aff:* SAE, ESD, ΣΞ, ΠΤΣ.

Gibson, Harry G
Home: 2367 Wilshire Ave, W. Lafayette, IN 47906
Position: Assoc Prof Forest Engg. *Employer:* Purdue Univ. *Education:* MS/Forest Engg/WV Univ; BS/ME/WV Univ. *Born:* 9/10/38. Native of Morgantown, WV. From 1962 to 1965, worked as Res Proj Engr with US Bureau of Mines researching flow measurement of 2-phase (gas-solids) heat transfer media. Obtained patent for 2-phase meter. Worked as Res Engr with US Forest Service from 1965 to 1978. Proj Leader of Timber harvesting Res Unit from 1969 to 1978. Patent obtained on stability indicator for off-road articulated vehicles. Was plantation harvesting dept mgr for Jari Florestale e Agricupari Ltda in the Amazon region of Brazil for 2 yrs & presently Assoc Prof & Coordinator of Forest Engrg at Purdue Univ. *Society Aff:* ASAE, SAE, ISTF, SAF.

Gibson, John E
Business: Thornton Hall, Charlottesville, VA 22901
Position: Commonwealth Professor of Engrg. *Employer:* Univ of Virginia. *Education:* Phd/EE/Yale; MS/EE/Yale; BS/EE/Univ of Rhode Island. *Born:* 6/11/26. As an engrg educator, was appointed Instructor of Elec Engrg, Yale 1952. Asst Prof Yale 1956; Assoc Prof of Elec Engrg Purdue Univ 1957; Prof Purdue 1960. Was founding Dean of Engrg, Oakland Univ 1965; John F Dodge Prof, Oakland 1972; Most recently has been Dean of Engrg & Applied Sci & Commonwealth Prof at the Univ of VA 1973-83. Has written numerous articles & reports in the field of automatic control & is co-author of books on conrol system components, nonlinear automatic control, & intro to engrg design. Is author of DESIGNING THE NEW CITY: A SYSTEMIC APPROACH John Wiley & Sons; *Managing Research and* Development, John Wiley & Sons, 1981. Is Fellow of Institute of Electrical and Electronic Engineers; Professional Engineer, Commonwealth of Virginia. *Society Aff:* IEEE, ASEE, AAAS.

Gibson, John Sevier
Home: 813 Parkway Drive S.E, Smyrna, GA 30080
Position: Advanced Program Manager *Employer:* Lockheed-Georgia Co. *Education:* BS/Physics & Math/Stetson Univ. *Born:* Dec 1934. 1st Lieutenant, Project Officer, US Army, Redstone Arsenal Army Rocket & Guided Misisle Agency 1956-58. With Lockheed-Georgia Co since 1959. Became Group Engr in 1965 in charge of structural vibration & sonic fatigue, interior soundproofing, & engine noise programs on the C-5A airplane project. 1971 thru 1978 primarily involved in R&D contract technical & mgmt work and U.S. and international noise regulations. 1979-Present, advanced program management with current emphasis on Airborne Surveillance Systems. Mbr Acoustical Soc of Amer, Georgia Chap Pres 1969-70, & on the Ctte on Noise 1976- 79. Elected Fellow Amer Inst of Aeronautics & Astronautics in 1986. Mbr of the Aero- Acoustics Tech Ctte 1975-77 & Mbr of Bd of Dirs 1978-83. *Society Aff:* ASA, AIAA.

Gibson, Kenneth A
Business: Rm 200, City Hall, 920 Broad St, Newark, NJ 07102
Position: Mayor. *Employer:* City of Newark. *Born:* 5/15/32. in the small town of Enterprise, AL. He moved with his parents and younger brother to Newark in 1940. He attended Newark public schools and was grad from Central High School. He served in the Army & later earned a bachelor's degree in civ engr from NJ Inst of Tech, then known as Newark College of Engg. He also has honorary degrees from several colleges. A licensed PE, Gibson worked in the 1950s and 1960s for the NJ State Dept of Transportation, Newark Redevelopment & Housing Autority, & Newark Bureau of Bldgs. Mayor Gibson a widower lives in Newark's West Ward. They have 5 adult daughters and three grandchildren. Gibson was a saxophonist in his younger days, and his principal hobby in recent yrs has been jogging. He has competed in long distance runs in Newark, NY & other cities, and was the first mayor ever to finish the 26-Mile Boston Marathon.

Gibson, Richard C
Business: Box 105, Surry, ME 04684
Position: Retired. *Education:* ScD/Instrmentation/MIT; SM/EE/MIT; SB/EE/MIT. *Born:* 12/31/19. Officer USAF 1942-67 with var R&D assignments. Prof of Astronautics USAFA 1960-65. V Commander Natl Range Div 1965-67. Chmn Dept of EE U of ME 1967-77. Professor of EE 1967-80. Mbr USAF Scientific Adv Bd 1970-81. Active in sailing & skiing. *Society Aff:* ASEE, ΣΞ, ΦΚΠ.

Gibson, Richard T
Business: Amtac Aquatech - Hawaii, P.O. Box 596, Kekaha, HI 96752
Position: Mgr *Employer:* Amfac Aquatech - Hawaii *Education:* MS/Agri Engg/Univ of HI; BS/Agri Engg/CA Poly State Univ. *Born:* 8/4/47. Returned to college after serving in USAF, 1965-70. Specialized in Aquacultural Engr. Directed the drafting of the Nation's first comprehensive, state-wide Aquaculture Dev Plan. Directed the Nation's first State Aquaculture Dev Prog. Participated in drafting Natl Aquaculture Plan. Responsible for identification & recommendation of investment in HI-based aquaculture bus for large, NYSE listed co. Currently industrializing Hawaiian Prawn Farming. *Society Aff:* ASAE, WMS.

Gibson, Robert C
Business: 740 Duke St, Suite 200, Norfolk, VA 23510
Position: Partner/Sr VP. *Employer:* Clark, Nexsen, Owen, Barbieri, Gibson. *Education:* BS/ME/VPI & SU. *Born:* 11/6/37. Native of Norfolk, VA. Past-Pres of the VA Soc of PE, & the Tidewater Chapt VSPE. Was Engr-of-the-Yr in VA - 1979. Served four terms as NSPE treasurer. Presently is Chrmn, NSPE Budget Committee, and NSPE Southeastern Regional VP. Is reg PE in VA, NC, MD & WV. He is a mbr of Soc of Am Military Engrs & VA Assoc of Professions. Enjoys family & traveling. *Society Aff:* SAME, NSPE.

Gidaspow, Dimitri
Business: IIT Center, Chicago, IL 60616
Position: Prof *Employer:* IL Inst of Tech *Education:* BChE/ChE/CCNY; MChE/ChE/Polytechnic Inst of Brooklyn; PhD/Gas Eng /IIT. *Born:* 6/4/34 Started teaching chem engrg at IIT in 1962. From 1963 to 1977 was teaching gas engrg and doing research at the Inst of Gas Tech in Chicago. The research is described in 100 journal publications and in 9 US patents. Directed 25 doctoral theses. Assumed present position in 1977. Served as consultant to several national labs and to industry. Served as papers chrmn for AIChE Heat Transfer and Energy Conversion Division (1973-4), member of its executive committee (1976-9), chrmn of two phase flow subcom-

Gidaspow, Dimitri (Continued)
mittee (1981-5). Was program chrmn for 1979 Intersociety Energy Conversion Engrg Conference. *Society Aff:* ACS, AIChE, FPC.

Giddens, Don P
Business: School of Mechanical Engineering, Georgia Institute of Technology, Atlanta, GA 30332
Position: Regent's Prof *Employer:* GA Inst of Tech *Education:* PhD/Fluid Dynamics/GA Inst of Tech; MSAE/Aerodynamics/GA Inst of Tech; BAE/Aerospace Engrg/GA Inst of Tech *Born:* 10/24/40 Regent's Prof of Mechanical Engrg, GA Inst of Tech. Co-Dir, Emory/Georgia Tech Biomed Tech Rsch Center. Research in cardiovascular fluid dynamics, ultrasonics in medicine, turbulence and biomedical engrg. *Society Aff:* ASME, ΣΞ

Giedt, Warren H
Home: 2280 Hastings Drive, Belmont, CA 94002
Position: Prof Emeritus *Employer:* Univ of California-Davis 95616. *Education:* PhD/Mech Engrg/Univ of CA, Berkeley; MS/Mech Engrg/Univ of CA, Berkeley; BAS/Mech Engrg/Univ of CA, Berkeley. *Born:* Nov 1, 1920. Instructor Air Force Inst of Tech 1946-47. Univ of Calif, Berkeley 1947-65. Univ of Calif, Davis 1965-83. Chmn, Dept of Mech Engrg 1965- 69. Assoc Dean-Grad Study of College of Engrg 1972-80. Chmn 1964 Heat Transfer & Fluid Mech Inst. Tech Ed ASME 'Journal of Heat Transfer' 1967-72. Author 'Principles of Engrg Heat Transfer' 1957 & 'Thermphysics' 1971 and tech journal papers. Fulbright Prof Univ of Tokyo 1963. Awards for outstanding teaching and prof contributions. Prof Emeritus 1983-. *Society Aff:* ASME, ASEE, AWS.

Giessner, William R
Home: 2201 Coronet Blvd, Belmont, CA 94002
Position: Sr Sanitary Engr. *Employer:* City & Cty of San Francisco. *Education:* BS/Civil Engg/Univ of CA, Berkeley. *Born:* Sep 1924. Native of Cassel, Calif. Attended school in Long Beach & maritime academy at Vallejo. Marine Engr during W W II. Resident Engr, Designer, Project Engr, Branch & Div Hd with San Francisco Department of Public Works. Spec in sewer, treatment plant & outfall design. Conceived, managed computer rain gauge-sewer stage data acquisition network. Dev analysis corrolating rainfall & facility sizes to volume & frequency of overflows. Co-authored 'San Francisco Master Plan, Sept 1971', authored Contingency Operation Manual for North Point Treatment Plant. Mbr ASCE, Diplomate AAEE, recipient of 1976 Wesley W Horner Award. Enjoy gardening & abalone fishing. *Society Aff:* ASCE, AAEE.

Giffin, Walter C
Home: 4277 Kenmont Pl, Columbus, OH 43220
Position: Prof *Employer:* OH State Univ *Education:* PhD/IE/OH State Univ; MS/IE/OH State Univ; B/IE/OH State Univ *Born:* 4/22/36 Native of Walhonding, OH. Worked as a research engr for General Motors Research Labs, Warren MI 1960-1961. With OH State Univ since 1961. Prof of industrial engrg specializing in inventory control, forecasting and stochastic modeling. Research interest in aircraft pilot behavior and air transportation systems. Author of three engrg textbooks. Registered PE in OH. Active in Experimental Aircraft Assoc as a builder of homebuilt aircraft. Licensed commerical pilot, flight instructor and aircraft mechanic. *Society Aff:* IIE, ASEE, ORSA, ΣΞ, TIMS

Giglio, Richard J
Business: Dept of Industrial Engrg &, Operations Research, 114 Marston H, Amherst, MA 01003
Position: Prof & Head *Employer:* Univ of MA *Education:* PhD/IE/Stanford Univ; MS/IE/Stanford Univ; BS/Aero Eng/MIT *Born:* 08/27/37 Native of Hartford, CT. Worked on stability and control problems for United Aircraft after receiving BS. After receiving PhD worked four yrs for Exxon Res doing engrg-economic planning in the US and Europe. At the Univ of MA, Amherst, he has conducted res and received funding from corps and the government in two major areas: the planning capacity of large integrated sys (e.g. elec power sys) and the mgmt of health care sys. Head of Dept of Industrial Engrg and Operations Res since 1978, Acting Dean of Engrg in 1982, Assoc Dean for Computing Services, 1983 to present. *Society Aff:* IIE, ORSA, TIMS.

Gigliotti, Michael F X
Home: 498 Washington St, Gloucester, MA 01930
Position: Pres. *Employer:* MGA Engg Inc *Education:* ME/Gen Engg/Stevens Inst of Tech. *Born:* 1/31/21. Also Pres of Polytechnique Ltd, producer and supplier of Unique Polymer Technologies. Also Pres of Michael Gigliotti and Assoc, Inc, an intl technical mgt consulting group. With Monsanto Co 35 yrs, variety of technical mgt assignments, Dir of R & D Cycle Safe-Lopac Packaging (1968-1976) Dir of Plastic Process Tech and Engg (1962-1968). Pres (1966-1976) Dir (1961-present) JETS Inc the Jr Engg Technical Soc. Currently Dir and Chrmn, Res Committee, Plastics Inst of America. Has been aDir of the Plastics Education Fdn, the Bldg Res Inst and the Plastics Advisory Bd for Underwriters Labs. Has chaired committees in MCA, and SPI. Active in guidance activities with ECPD, ASEE, NSPE, AIChE, SME and SPE. Mbr of Assn of Consulting Chemists and Chem Engrs. Reg PE-TX and MA. *Society Aff:* NSPE, SPE, ASEE, AIChE

Gilat, H Isaac
Home: 26 Snir St, Ramat Hasaron 5, Israel 47226
Position: Executive Director. *Employer:* Transportation Sys Cons Ltd. *Education:* ME/Transp Engg/Univ of CA; BSc/Civil Engg/Technion, Haifa, Israel. *Born:* 1934 Tel-Aviv, Israel. Hwy Engr 1958-60; Transportation Planner with Parsons, Brinckerhoff, Quade & Douglas participating in such projects as BART, & other rapid transit projects in Chgo, Atlanta & Baltimore USA 1961-64; Ch Transportation Planner & Traffic Engr with Colin & Zahavi, Tel-Aviv 1964-67. Study Dir of the Jerusalem Area Transportation Master Plan 1967-71. Estab cons firm 1971. Mbr of the Israel Assn of Engrs & Architects; Israel Inst for Environ Planning & Inst of Transportation Engrs (ITE). *Society Aff:* ITE, AEAI.

Gilbert, Joseph
Business: 400 Commonwealth Dr, Warrendale, PA 15096
Position: Executive Vice President *Employer:* Soc of Automotive Engrs. *Born:* Dec 1920. Joined SAE staff 1946 as Tech Ed of the SAE Jrnl. Later became Mng Ed, then was named Mgr of the Soc's Tech Div. 1957 appointed Asst Gen Mgr, and to present pos 1960. During WWII Indus Engg Officer with the Army Air Forces, involved in mfg of aircraft engines, accessories & components. Pres, CESS 1964.

Gilbertson, Conrad B
Business: 305 Plant Industry Bldg, Univ of Nebraska, Lincoln, NE 68583-0816
Position: Agricultural Engr. *Employer:* Agri Rese, USDA *Education:* BS/Agri Engg/NDSU; MS/Agri Engg/SDSU. *Born:* 4/11/38. Born in Aneta, N Dak. Wife, Carolyn; Children, Taunja, Tausha, Milissa, Susan, Cheryl, Conrad Brian II, Carrie Renae, Catherine Britt. ASAE, NSPE, Sig Xi, Gamma Sig Delta, Alpha Epsilon. Served as Ext Agri Engr NDSU 1963-66; Dev Engr USMARC, Clay Ctr Nebr 1967-68 & currently Res Engr AR-USDA, Lincoln Nebr. Registered PE Served as Chmn, Nebr Sect ASAE; V Pres Midcentral ASAE, served as Chmn, and Natl Tech Dir Structures and Environment Division, National ASAE and Chmn of sev Natl ASAE Cttes; Pub Chmn, Secy, & V Pres of PEN. V. Pres, Pres. Elect, Pres, and Dir. of SE chptr NSPE. Recipient of Reserve Champion award, NDSU 1960; Outstanding J C of Month Award 1967; Outstanding Young Engr of Yr Award SE Chap PEN 1972; Outstanding Young Engr of Yr Award PEN 1973, & Young Engr of Yr Award, Mid-Central Reg ASAE 1973. Hobbies incl woodworking & most sports *Society Aff:* ASAE, NSPE, ΣΞ, ΓΣΔ, AE

Gilbreath, Sidney G, III
Home: Rt 9 Box 14, Cookeville, TN 38501
Position: Pres Hytec Systems, Inc (Self-employed) *Education:* PhD/Indus Engg/GA Inst of Tech; MS/Indus Engg/Univ of TN; BS/Indus Engg/Univ of TN. *Born:* 8/11/31. Formerly Gnl Mgr, Stonega Div, Westmoreland Coal Co. Prof & Chrmn, Dept of Indus Engg, TN Tech Univ. Asst Prof, Indus Engg, VA Tech & Instructor, GA Tech. Field engr, indus engr & cons, Presently Chrmn Dept of Indus Engrg, TN Technological Univ. Mbr, SPE, IIE: VP, Region VII 1975-77, Exec VP, Pres 1978-79. Dir, Chap Dev 1981 Outstanding IE Fac. Award-VA Tech & State Univ 1968; H P Emerson Outstanding I E Alum Award 1969 Univ of TN. *Society Aff:* IIE, NSPE, IMMS, ASEE.

Gilchrist, Bruce
Business: 612 W 115th St, New York, NY 10025
Position: Senior Advisor for Information Strategy. *Employer:* Columbia Univ. *Education:* Ph.D./Meteorology/Univ London; B.Sc./Mathematics/University London *Born:* 1930. Ed Univ of London; BSc Maths 1950; PhD Meteorology 1952. Staff mbr Electronic Computer Project, Inst for Advanced Study 1952-56; Dir of Computing Ctr, Syracuse Univ 1956-59; IBM Corp 1959-68; Exec Dir, Amer Fed of Info Processing Soc (AFIPS) 1968-73; Dir of Computing Activities, Columbia Univ 1973-85, Sr. Advisor for Information Strategy, Columbia University, 1986-. AFIPS, Pres 1966-68; ACM, V Pres 1962-64; ACM, Secy 1960-62. Current interests incl the social implication of computer usage, gov involvement in the computer indus, & manpower planning for computer related jobs. *Society Aff:* IEEE, ACM, AMS, AAAS, ΣΞ

Gilden, Robert O
Home: 1277 Similk Bay Rd, Anacortes, WA 98221 *Employer:* Consultant. *Education:* MS/Agri Engr/WA State Univ; BS/Agri Engr/WA State Univ. *Born:* 12/4/22. and raised on an island in the Puget Sound. Navigator in WWII - shot down, spent 1 yr in German prison camp. Investigator in WA State Farm Struct Res Fdn, Instr in Agr Engg at WSU, Asst Prof on Agri Engg. Ext at Univ of Wyoming, Ext Farm Machinery specialist for Ext Service USDA, Wash, DC. Farm Bldg & Housing Specialist for USDA, Coordinator Ext Agri Engr. Progs, USDA, Safety Prog Leader USDA. Consultant World Bank, AID, UNDP, Univ of NB, Asia, Africa & So Am. *Society Aff:* ASAE.

Giles, George W
Business: 2600 Wade Ave, Raleigh, NC 27607
Position: Prof Emeritus-NCSU. *Employer:* N Carolina State Univ. *Education:* MS/Agri Engr/Univ of MO; BS/Agri Eng/Univ of NB. *Born:* Mar 1910 Nebr. Fac N C State Univ 1936; Hd of Dept of Agri Engrg 1948- 60. Cons in India 5 yrs, Pakistan, Iran, Ghana, & Egypt. Dev in India one each, strong programs in undergrad, grad & Dev Ctr. First to document yield increases from mechanization in India. Mbr ASAE, Indian Soc of Agri Engrs (Fellow), Sigma Xi, ASEE (formerly), Phi Kappa Phi, Gamma Sigma Delta. Mbr Subpanel U S President's 'World Food Study' 1967-68. U S delegate to 'Programs for UNDP, United Nations 1969. Listed in 'Amer Men of Sci' 1960. Recipient of The Univ of Missouri Award for Distinguished Serv in Engrg 1969; John Deere Gold Medal Award ASAE 1971, Kishida International Award ASAE 1980. *Society Aff:* ASAE, ISAE, ΦΚΦ.

Giles, William S
Business: 1455 East 185th Street, Cleveland, OH 44110
Position: Manager Product Engrg. *Employer:* TRW Inc Valve Div. *Born:* Oct 19 1934. BS ME 1957 Cornell U. Engrg Mgr Program 1975 Carnegie Mellon U. Native of Cleveland OH. Joined TRW Valve Div in July 1957 as Jr Test Engr in Engine Test Lab. Subsequently, worked as Test Engr & Dev Engr. Joined Product Engrg Dept in 1963 & became Mgr in 1971, with respon for customer engrg liaison and inhouse product design. Mbr SAE & ASME. Active on governing bd of Cleveland SAE Sect. Received Russell S Springer Award 1972 for SAE paper on Valve Problems with Lead Free Gasoline. Interests incl sailing, skiing and fishing.

Gill, Arthur
Business: EECS - Comp Sci Div, Berkeley, CA 94720
Position: Prof. *Employer:* Univ of CA. *Education:* BS/EE/MIT; MS/EE/MIT; PhD/EE/Univ of CA. *Born:* 4/18/30. in Israel, 1930. Raytheon Co (Waltham, MA), 1956-57. Univ of CA, Berkeley, 1959-date. *Society Aff:* ACM.

Gill, Charles B
Home: 405 Monroe St, Easton, PA 18042
Position: Prof & Head, Metallurgical Engrg Dept *Employer:* Lafayette College. *Education:* PhD/Metallurgical Engg/Univ of MO; MSc/Metallurgical Engg/Univ of MO; BASc/Metallurgical Engg/Univ of Toronto. *Born:* 4/8/21. in Sudbury, Ontario, Canada. Res Metallurgist Univ of MO School of Mines 1952-55, Technical Superintendent Deloro Smelting and Refining 1955-57, Prof at Lafayette College 1957-present, in crge of extractive metallurgy courses. Department Head 1986-present. *Society Aff:* AIME, ΣΞ, ASEE, ΑΣΜ

Gill, James M
Business: 451 Florida St, Baton Rouge, LA 70801
Position: Sr VP. *Employer:* Ethyl Corp. *Education:* MS/CE/LA State Univ; BS/CE/LA Tech Univ. *Born:* 9/2/22. Native of Ruston, LA. Served in USN WWII. Exec Officer LSM. With Ethyl Corp since 1948. Positions held in mfg, R&D, Maintenance & Construction, Marketing. VP 1966; Sr VP & mbr Bd of Dirs 1969. Bd of Dir - Chem Mfg Assn 1972 to 1975 *Society Aff:* API, CMA.

Gill, Lowell F
Business: 180 E First S. Salt Lake, UT 84139
Position: Mgr, Rates & Industrial Engg. *Employer:* Mtn Fuel Supply. *Education:* Bachelor/Industrial Engg/Univ of UT. *Born:* 8/24/42. Native of Nebraska, Reg PE in UT and WY. Presently serving as Vice President of Regulatory & Consumer Affairs for the Distribution Division of Mountain Fuel Supply Company. This position encompasses the regulatory and consumer affairs activities of the Company as well as direction of the Industrial Engineering and planning processes for the Distribution Division. *Society Aff:* IIE, ASHRAE.

Gill, William H
17 Hopland Court, Sacramento, CA 95831
Position: Consultant/Owner *Employer:* William H. Gill, Civil Engrs *Education:* MS/Civ Engg/CA State Univ Sacramento; BS/Civ Engg/Worcester Poly Inst. *Born:* 4/6/40. Born in Boston, lived in Shrewsbury, MA, resides in Sacramento, Ca. Graduate of Worcester Poly Inst in 1961 and CA State Univ, Sacramento. Design engr on CA Aqueduct Proj with CA Dept of Water Resources. Dir of Environmental Services & Dir of Special Projs, 1967-78 with the Dev and Resources Corp with maj interests in water resources planning & dev, drainage, & flood control. Co-founder of Gill & Pulver Engrs Inc of Sacramento, 1979-86 and Consultant in water resources and hydrology, 1987- , with specialized analysis and research in natural stream systems and urban flood control. Pres, Sacramento Sec ASCE, 1977; Chrmn, CA State Council of ASCE, 1977. *Society Aff:* ASCE, APWA, AWRA.

Gillen, Kenneth F
Home: 581 Welshire Dr, Bay Village, OH 44140
Position: Salesman *Employer:* Bethlehem Steel Corp *Education:* BS/CE/Carnegie-Mellon Univ *Born:* 8/5/29 Career since coll has been with Bethlehem Steel Corp in various capacities. This included engrg, fabrication and erection of major fabricated steel buildings and bridges. Past 12 years have been in sales and administration of contracts for structural steel, reinforcing bars and piling. Pres of Cleveland Section, ASCE in 1975. Chrmn of District 9 Council, ASCE in 1981. *Society Aff:* ASCE

Giller, Edward B
Home: 825 Mackall Avenue, McLean, VA 22101
Position: Program Manager Arms Control. *Employer:* Pacific-Sierra Research Corp. *Education:* BS/Chem Engg/Univ of IL; MS/Chem Engg/Univ of IL; PhD/Chem Engg/Univ of IL. *Born:* July 1918. Native of White Hall Ill. Comdr 2nd lt USAF

Giller, Edward B (Continued)
1942 to Maj Gen USAF ret 1972. served as pilot 1942-50; Chief Radiation Branch armed forces Spec Weapons Proj 1950-54; Dir of Res Directorate Air Force Weapons Ctr, Albuquerque NM 1954-59; Off of Aerospace Res 1959-64; Dir USAF Off of Sci & Tech 1964-67; Asst Gen Mgr for Military Application AEC 1967-72; Asst Gen Mgr for Natl Security 1972-75; Dept Asst Admin for Natl Soc 1975-77; JCS Rep to CTB negotiations 1977-84; Program Manager. Arms Control Pacific-Sierra Res. C. Mil decor & awards incl Silver Star, the DSM, Legion of Merit with oak leaf cluster, Distinguished Flying Cross with oak leaf cluster, Air Medal with 17 oak leaf clusters, Purple Heart, and the Croix de Guerre (France). Enjoy hunting & fishing. *Society Aff:* AIChE, AAAS, AIC.

Gillespie, Daniel C
Business: P.O. Box 9312, 77 Havenmeyer Ln, Stamford, CT 06904
Position: VP, Bus Development *Employer:* Sohio Chemicals & Industrial Products Co *Education:* BS/ChE/PA State Univ; MS/ChE/Univ of MI. *Born:* 09/22/22. Native of Shamokin, PA. Served with US Army on Manhattan Proj, Los Alamos Lab, NM 1944-46. Joined Dorr-Oliver Inc in 1948 and worked in various technical and sales positions. Became VP-Marketing, Exec VP, Pres & CEO 1976-82. Appointed VP-Bus. Dev't. of Sohio Chem & Int'l Prod Co in 1982. Fellow of AM. Inst. of Chem. Engrs. *Society Aff:* AIChE, AIME, PEMA.

Gillespie, James W
Business: POBox 300, Tulsa, OK 74102
Position: Mgr Marine Dept & Sr VP. *Employer:* Grand Bassa Tankers, Inc & Cities Service Co. *Education:* PhD/Engg/OK State Univ; MS/Civil Engg/OK State Univ; BS/Genl Engg/OK State Univ. *Born:* 1/30/35. BS in Gen Engg, MS in Cvl Engg, PhD in Engg from OK State Univ. Assoc with Cvl Engg Fac at OK State Univ 1959-64. Mbr Mkt Tech Serv US Steel Corp 1964-68. Ch Tech Advisor Cities Service Co 1968-78; Mgr Marien Dept 1978- . Pres OK Sec ASCE 1970. Pres Tulsa Chap OSPE 1976. Young Engr of Year Tulsa Chap OSPE 1974. Mbr num tech cttes, ASCE. Outstanding Engr Tulsa Chap OSPE 1979. Outstandin Engr OSPE 1979. Adm VP OSPE 1978, Exec VP OSPE 1979, Pres Elect OSPE 1980. Dir ASCE 1976-79. Dir NSPE 1977, 1980p-81. *Society Aff:* ASCE, NSPE, ASEE.

Gillespie, LaRoux K
Home: 1300 E 109 St, Kansas City, MO 64131
Position: Sr Proj Engr. *Employer:* Allied Bendix Aerospace *Education:* MS/Mfg Engg/UT State Univ; MS/ME/KS Univ; BS/ME/KS Univ. *Born:* 11/11/42. Responsible for supervisory manufacturing engrg of precision large turned case sections of thin wall aluminum & titanium, improving quality, and supporting expert system development for computer automated process planning. Previously responsible for Robotic and Flexible Automation applications & development for aerospace products, engineering coordination of purchased case, forged, powder metal, ceramic & machined metal parts for defense applications. With Allied Bendix since 1966. Prolific author on precision machining and deburring subjects. Dir, SME 1977-1981. ASME Arthur L. Williston medal & award 1965. Bendix Outstanding Technical award, 1977. SME Albert M. Sargent Progress Award 1984. SME fellow 1986. *Society Aff:* ASME, SME, IPRODE.

Gillett, John B
Business: 1111 N Charles St, Baltimore, MD 21201
Position: Partner. *Employer:* Whitman, Requardt and Assoc. *Education:* BS/Civil Engg/NC State Univ. *Born:* 10/14/27. 1948-1951 - Office Engr - Duke Power Co, Charlotte, NC. Construction of steam generating stations. 1951-1981 - Whitman, Requardt and Assoc. In charge of civil, sanitary and public works projects in MD, DE and VA. In charge of utility valuation, cost-of-service & rate design for clients in 15 states. Baltimore Jr Assoc of Commerce, 1975-63; Pres 1961-62. Baltimore Cty Bd of Health 1962-69; V Chrmn 1967-69. Pres, Consulting Engrs Council of MD 1978-79. Officer & committee chrmn of various profl societies. Author of articles in *Civil Engineering, Public Utilities Fortnightly, Management Accounting, and Baltimore Engr. Society Aff:* ASCE, NSPE, WPCF, AWWA, ACEC, ESB.

Gillette, Jerry M
Business: 205 W Randolph St, Chicago, IL 60606
Position: Pres *Employer:* Consulting Consortium, Inc *Education:* BS/EE/IL Inst of Tech *Born:* 8/17/39 Former Assoc of C.F. Murphy Assocs (Murphy/Jahn), architects & engrs: now pres of Consulting Consortium, Inc, consulting engrs with offices in Chicago, IL and Gary, IN, licensed PE in states of IL, WI, IA and MN; received "Outstanding Professional Achievement" award for 1979 from (NTA) National Technical Association; member of "Electrical Industry Evaluation Panel," Electrical Construction and Maintenance Magazine (McGraw-Hill); published article "Power Distribution in the All-Electric Bldg" 1967, in actual specifying magazine (Cahners). *Society Aff:* IEEE, ISA, NTA, WSE, NSPE/ISPE

Gillette, Leroy O
Business: Tech Ctr, Lancaster, PA 17604
Position: Dir Indus Engrg *Employer:* Armstrong World Industries, Inc. *Education:* BSIE/Indus Engg/VA Poly Inst & State Univ. *Born:* 11/27/22. in Arlington, VA. Married Jane A Kirk 1948. Children L Kirk, Thomas L, William J, Barbara J. Major USAR ret. Employed by Armstrong World Industries 1947- . Fellow & Past Pres, IIE. Sr Mbr ASQC. Mbr Council on Indus Engg. *Society Aff:* IIE, ASQC.

Gillette, Robert W
Home: 110 Captain Small Rd, So Yarmouth, MA 02664
Position: Div Engr. *Employer:* Consolidated Edison Co of NY Inc. *Education:* BS/EE/Worcester Polytech Inst. *Born:* Oct 1905. Employed by General Electric Co 1927-28; then by Consolidated Edison Co of NY 1928-70. Became Div Engr 1962. In cable engrg field since 1940. Respon for engrg of cable syst for overhead & underground distrib & transmission up to 345 kV. Fellowship in IEEE 1962 for 'contributions in the field of power cable engrg'. Chmn IEEE Power Group 1965-67. Fellow AAAS. Dir IEEE 1967-68. Dir UET 1969-70. *Society Aff:* IEEE, AAAS.

Gillette, Roy W
Home: 6111 Glennox Lane, Dallas, TX 75214
Position: Consultant (Ret) *Born:* 03/21/19 BS Texas A&M University. Served in US Army Corps of Engrs WWII-Ret Res Col USAF. Spec in design, construction & maintenance of airfield pavements while with Portland Cement Assn. Respon for production, quality control, & construction of precast & prestressed concrete products for past 10 yrs with HB Zachry Co & Texas Industries Inc. Have been cons for proj in Central America & Middle East. Reg Prof Engr. Enjoy fishing & hunting. Conslt on Plant Mgmt and Construction.

Gilley, James R
Business: Agri Engr Dept, Lincoln, NB 68583
Position: Professor *Employer:* Univ of NB. *Education:* PhD/Agri Engr/Univ of MN; MS/Agri Engr/CO State Univ; BS/Agri Engr/CO State Univ. *Born:* 3/19/44. Native of CO. Began profl career at the Univ of MN in 1968, teaching courses in irrigation and drainage and conducting res on uses of waste products in agri. Joined the Agri Engg Dept at the Univ of NB in 1975. Primary responsibilities include res on energy conservation in irrigated agri, teaching grad courses in irrigation engg and advising engg grad students. Author of 30 technical articles in the irrigation area. Have served as technical consultant on several irrigation projs, federal agencies and irrigation companies. Chrmn, NB Sec of ASAE 1978; Natl ASAE paper award 1978; served on several ASAE natl committees. *Society Aff:* ASAE, ASA, SSSA, IA.

Gilliam, Charles W
Business: 1750 New York Ave, N.W, Washington, DC 20006
Position: Vice Chairman *Employer:* MMM Design Group *Education:* BS/CE/IN Inst of Tech *Born:* 2/24/28 Native of Searcy, AR. Surveyor and Construction Engr with L&NRR, 1951-1953. Engr with C&EI RR, 1953-1956. With Stanley Consultants 1956-1967 progressing from Design Engr, Specification Engr, Resident Branch Mgr

Gilliam, Charles W (Continued)
in Ghana and Nigeria to Area Mgr for Africa in the Intl Division. Joined McGaughy, Marshall & McMillan in Rome, Italy in 1967 progressing from Assistant to Pres to VP in the Intl Div. in Athens, Greece, Senior VP of Washington, DC office for MMM Design Group and President of MMM Intl in Dublin, Ireland. Currently Chairman of MMM Design Group Intl and Vice Chairman of MMM Design Group Inc. Enjoys travel, tennis, and classical music. *Society Aff:* ASCE, SAME, NSPE.

Gilliland, Bobby E
Business: College of Engg, Clemson, SC 29634-0915
Position: Special Asst to the Pres. & Prof *Employer:* Clemson Univ. *Education:* PhD/Instr Sci/Univ of AR; MS/Instr Sci/Univ of AR; BS/CE/LA Tech Univ. *Born:* 8/6/36. ISA and GA Pacific Cor Grad Fellow, 1964-66. Asst Prof to Prof of Elec & Comp Engg, 1967-76; Asst to Dean of Engg 1973-79. Assoc Dean of Engg. for Research and Planning 1980-86. Assumed current responsibility as Special Asst to the Pres in charge of special projects. Active in comp-based instrumentation res, especially as applied to textile mfg quality control. Author of over thirty technical papers, presentations and reports. Chrmn, ASEE, Instrumentation Div, 1977; Newsletter Ed, 1980-83 and Mbr, Bd of Dir, 1984-present, ASEE Relations with Indus Div; Pres, Western Carolinas Sec of ISA, Bd of Dirs, ISA Textile Div, 1975-79; Outstanding Educators of Amer, 1974- 75; Peer Reviewer for several Fed Agencies and tech journals. Reg PE. Private Pilot. Enjoy woodworking, boating and travel. *Society Aff:* IEEE, ISA, ASEE, ΣΞ

Gillin, James
Home: 8 Breeze Knoll Dr, Westfield, NJ 07090
Position: Pres MSD AGVET; Retired July 1, 1987. *Employer:* Merck & Co, Inc. *Education:* PhD/Chem Eng/Cornell Univ; BChE/Chem Eng/Cornell Univ. *Born:* 9/16/25. Joined Merck & Co Inc in 1949 and held num pos in the Res Labs since that time, most recently to pos as VP of Dev in 1971. Current position-since July 1979, Pres, MSD AGVET div of Merck & Co Inc. Mbr Tau Beta Pi, AAAS, NYAS, ACS, AIChE. Sev pubs & patents. Enjoy swimming & golf. *Society Aff:* ACS, AIChE, AAAS, NYAS.

Gillis, Manly E
Business: (Exxon Chemical Americas), P.O. Box 3272, Houston, TX 77253-3272
Position: Pres *Employer:* Exxon Chemical Americas *Education:* BS/ChE/Univ of TX. *Born:* Jun 15 1930 Fentress TX. 16 yrs in tech oper & plant mgmt at Baytown Chem Plant Exxon Chem Co USA. Since 1969 with var affil of Exxon Chem Co in product line & genl mgmt assignments in NY, Brussels & Houston. Long time mbr of AIChE & So Texas Sec AIChE. *Society Aff:* AIChE.

Gillis, Peter P
Business: Metallurgical Engg, 763 Anderson Hall, Lexington, KY 40506
Position: Prof of Materials Sci. *Employer:* Univ of KY. *Education:* PhD/Engg/Brown Univ; ScM/Engg/Brown Univ; ScB/Engg/Brown Univ. *Born:* 12/23/30. Newport, RI. Three yrs, US Navy; worked for Fram Corp, Leesona Corp, and taught at RI Sch of Design; Grad Sch at Brown Univ. Came to the Univ of KY and is now Prof of Materials Sci. Res interests ctr on theories of the mech behavior of materials, especially dislocation theories, and how these relate to time- dependent effects. Has been a consultant to Spindletop Res, the Lawrence Livermore Lab, the AEC Directorate of Regulatory Operations and the Los Alamos Scientific Lab. Recipient of Fulbright Res Fellowship in 1970. Fellow of ASM (1981). Author of approximately 100 technical articles. *Society Aff:* AIME, ASM, ASME, FAS, VITA.

Gillmor, Robert N
Home: 1124 Greengate Rd, Fredericksburg, VA 22401
Position: Retired (GE Co.) *Education:* BS/Met/SD Sch of Mines & Tech. *Born:* 12/25/06. For 40 yrs a consultant & hd of a Mtls & Process Lab involved in the broad fields of Physical Met, Metal Joining, Instrumentation, & Inorganic & Organic Chem. In the fields of high temperature met; authored & present technical papers. Elected a Fellow of the Am Soc for Metals. PE, NY state. *Society Aff:* ASM.

Gillott, Donald H
Business: 6000 J Street, Sacramento, CA 95819
Position: Dean School of Engrg. *Employer:* CA State University. *Education:* PhD/Elec Engg/Univ of Pittsburgh; MS/Elec Engg/Univ of Pittsburgh; BS/Elec Engg/Univ of Pittsburgh. *Born:* Aug 25 1931. Native of PA. BS, MS, PhD from U of Pittsburgh. Taught at U of Pittsburgh 1957-68. Chmn Dept EE CA State U 1968-76. Cons IBM, Westinghouse, NAS. Pub num articles on hysteresis effects on flux penetration. Pat on Electron Controlled Med Ventilator. Chmn num cttes incl Sacramento Sec IEEE. Sr Mbr IEEE. Mbr ASEE. Reg Prof Elec Engr CA. Outstanding fac award at CA State University by Phi Kappa Phi. *Society Aff:* IEEE, NSPE, ASEE.

Gillum, Jack D
Business: 4251 Kipling, Wheat Ridge, CO 80033
Position: Exec V.P. Chief Operating Officer *Employer:* KKBNA Inc. *Education:* BS/Arch Engr/KS Univ. *Born:* 11/21/28. Grad studies at CO Univ & Denver Univ. Reg Struc Engr in CA & 23 other states. Army Corps of Engrs 1950-52. Founded Jack D. Gillum & Assoc in 1955. Pioneered struc comp prog for design/cost var. Authored pubs in prestressed concrete, struc steel and membrane struc. Received in 1980 an Award of Merit from the Struc Engrs Assn of IL for 1980 Olympic Fieldhouse; 1979 Eng Excellence Award from Consltg Engrs Council for Learning Resources Ctr, St Louis Univ Sch of Medicine; the 1979 and the 1980 Excellence in Engrg Award from the IWEA of Washington, DC for the projs Marriott Corp Intl Hdquarters and Mobil Corp, Fairfax, VA, respectively; Honorable Mention from the Consltg Engrs Council Engrg Award in 1979; Second Place in Engrg Excellence for the Mercantile Bank Bldg, Kansas City, MO from the Consltg Engrg Council of MO in 1975; Spec Citation Award 1975 from American Inst of Steel Construction & the 1974 Award from Prestressed Concrete Inst for Crown Ctr Hotel Proj, Kansas City. *Society Aff:* SEANC, SEAI, ACI, NSPE, PCI.

Gilman, John J
Business: PO Box 400, Naperville, IL 60566
Position: Mgr, Corporate Research *Employer:* Standard Oil Co (Indiana) *Education:* PhD/Metallurgy/Columbia Univ; MS/Metallurgy/Illinois Inst Tech; BS/Mech Eng/Illinois Inst Tech. *Born:* 12/22/25. Presently Mgr of Corp Research Standard Oil Co (Indiana) in Naperville, IL. Previous assoc with the Allied Corp 1968-80, Gen Elec Res Lab 1952-60, the Crucible Steel Co of Amer 1948-52 & faculty mbr at Brown U 1960-62, and the U of IL 1963-68. His area of res interest incl res mgmt, metallic glasses, optical materials, plasticity, fracture and dislocations. Pub 150 tech papers & 5 books. Awards incl Fellow ASM & APS; the Raymond Award (American Inst of Met Engrs); the Geisler Award ASM; the Mathewson Gold Medal AIME; ASM Campbell Memorial Lecture 1966; Distinguished Service Award (IL Inst of Tech Alumni Assn). Mbr of NAE. *Society Aff:* ASM, APS, AIME, AACG.

Gilman, Stanley F
Home: 505 Cricklewood Dr, State College, PA 16801
Position: Prof Architectural Engg. *Employer:* Pennsylvania State U. *Education:* BSME/Mech Eng/Univ of ME; MS/Mech Eng/Univ of IL; PhD/Mech Engg/Univ of IL. *Born:* March 1921. Reg Prof Engr in PA & NY. Pres ASHRAE 1971-72, Fellow 1974. Mbr NSPE, ISES, ASEE, ASHRAE, ASME & IIR. Joined PA State U after 20 yrs engrg exper in indus, primarily with Carrier Corp, manufae of heating, vent & air conditioning products & systems. Prior Asst Prof of Mech Engrg U of IL. Author of over 60 pub in field. Mbr Tau Beta Pi, Sigma Xi & Pi Tau Sigma. *Society Aff:* ASHRAE, IIR, ASEE, ASME, ISES, NSPE

Gilmont, Roger
Business: 401 Great Neck Rd, Great Neck, NY 11021
Position: President. *Employer:* Gilmont Instruments Inc. *Education:* DrChemE/Chem Eng/Brooklyn Polytechnic; MChemE/Chem Eng/Brooklyn Poly-

Gilmont, Roger (Continued)
technica; BChemE/Chem Eng/Cooper Union. *Born:* Dec 1916 NYC. Brooklyn Tech HS Chem Engr for Quartermaster Corps 1941; Phys Chem for NBS 1942-43; Instruc Chem Eng CCNY 1943-44; in charge Chem Instrum General Foods 1944-47; Instrument Design for Emil Greiner as VP & Manostat Corp as Pres 1947-61; Pres Roger Gilmont Instrument 1961- ; Adj Prof Chem Engrg Polytech Inst of NY 1947- . Book on Thermodynamics (P-H 1959), many papers & patents for lab instruments. Hobbies: Baroque Music & Baseball. *Society Aff:* AIChE, ACS, AIC, ISA, AAAS.

Gilmore, Robert B
Business: 400 1 Energy Square, Dallas, TX 75206
Position: V Chmn of the Exec Comm. *Employer:* DeGolyer & MacNaughton. *Education:* BS/Petr Prod Engg/Univ of Tulsa, OK. *Born:* Jul 1913 Tulsa OK. Employed by Shell Oil 1934-41. With DeGolyer & MacNaughton since 1941. Officer & Dir since 1949. Elect Sr Chmn Bd in 1972. Reg in Petrol Engrg & Geology in TX. and CA. Pres SPE of AIME 1955. Awarded DeGolyer Medal SPE 1975. Distinguished Alumni U of Tulsa 1972. Dir Whitehall Corp, Dallas; Former Dir. Valero Energy Corp, San Antonio. Cooper Industries, Houston; LTV, Dallas. *Society Aff:* SPE, AAPG, NSPE, AIPG.

Gilmour, Charles H
Home: 4013 Kirby Street, So Charleston, WV 25309
Position: Consultant (Self-employed) *Education:* MS/Chem Engg/Syracuse Univ; BS/Chem Eng/Syracuse Univ; MS/Chem Engg/Carnegie Mellon Univ. *Born:* 1902 Yonkers NY. Instruct Syracuse U, Carnegie Mellon, MIT, WVU. Chem Engr Union Carbide 1933-67. Self-employed Consultant. Pub in Heat Transfer 1952- 65. Fellow AIChE; AIC; Mbr ACS. Chmn B78 ANSI Standards Ctte for Heat Exchangers. ASME Heat Transfer Div Memorial Award 1967. Charleston Sec AIChE Distinguished Service Award 1971. 1977 Donald Q Kern Award, Heat Transfer & Energy Conversion Div Amer Inst of Chem Engrs.. *Society Aff:* AIChE, AIC, ACS.

Gilruth, Robert R
Home: 5128 Park Ave, Dickinson, TX 77539
Position: Consultant (Self-employed) *Education:* MS/Aero Engg/Univ of MN; BS/Aero Engg/Univ of MN. *Born:* Oct 1913. Honorary: DS U of MN, IN Inst of Tech, & George Washington U. D Engrg MI Tech U. Dr LL NM State U. Dir of US's first manned spaceflight program, Proj Mercury. As the space program broadened in 1961, with Pres Kennedy's decision to fly Americans to the Moon, was placed in chg of creating new space ctr in Houston TX, a facility designed spec for manned spaceflight. Here the Gemini Prog & the Apollo spacecraft were conceived & managed. Other activities: Astronaut selec & tng, spacecraft design & testg, mission contr, flight ops & lunar sci. In Jan 1972, after svg for over 10 yrs as Dir of the Ctr, assumed new pos as Dir of Key Personnel Devmt for NASA, reptg to Deputy Admin in Wash DC. Has resp for identifying near & longer range potential cands for key jobs in the Agency, & for creating plans & proceds which will aid in devmt of thes. Retired from NASA in 1975. Now independent consultant to NASA and various aerospace corporations. *Society Aff:* AIAA, IAA, AAS, NAE, NAS.

Gilson, Richard D
Business: 2160 West Case Rd, Columbus, OH 43220
Position: Full Prof & Former Dir of Aviation *Employer:* The OH State Univ *Education:* PhD/Human Factors Psych/Princeton Univ; MA/Human Factors Psych/Princeton; Univ; BS/ME/Univ of CT. *Born:* 11/8/43 After three years of service as an officer in the US Navy, Dr. Gilson rose through the academic ranks to achieve full prof in two depts at the age of thirty-five. In his capacity (1978-1982), Dr. Gilson has as Dept Chrmn spearheaded the first degree granting program in Aviation (BS Degree), since aviation's inception at OH State Univ in 1917. As Airport Dir, and as it's chief administrative officer he in three years turned a substantial yearly loss into a consistently profitable position. During the same period he was instrumental in acquiring 2.3 million for the airport and dept through grants, gifts, and an endowed prof. Dr. Gilson is a FAA Designated Pilot Examiner and holds an airline transport pilot certificate with a citation jet type rating, a multi-engine rating, a flight instructor certificate with airplane, instrument, and multi- engine ratings and a flight instructor certificate in helicopters. *Society Aff:* ASEE, ISASI, HFS, APA, Aero Med. Assn.

Giner, Jose D
Business: 14 Spring St, Waltham, MN 02154
Position: Pres. *Employer:* Giner Inc. *Education:* Dr Sci/Electro Chem/Univ of Madrid; Diplom/Chem/Univ of Valencia. *Born:* 8/4/28. Dr Giner received a Dipl in Chemistry from the Univ of Valencia (Spain) in 1951 & a doctorate from the Univ of Madrid in 1954. After a brief period in the Consejo Superior the Investigaciones Cientificas in Madrid, he did electrochem res at the Univs of Erlangen & of Bonn (W Germany), from 1956 to 1961. From 1961 to 1965, he headed a fuel cell res group at Pratt & Whitney Aircraft. From 1965 to 1973, Dr Giner worked for Tyco Labs where he became Asst Dir of Res. In 1973 he founded Giner, Inc to perform res & dev in Electrochemistry. *Society Aff:* AIChE, ES, ISE, ACS.

Gingrich, James E
Business: 7035 Commerce Circle, Pleasanton, CA 94566
Position: VP. *Employer:* Nuclepore Corp *Education:* BS/CE/OR State Univ. *Born:* 10/21/30. Lafayette, IN. Production engr, Alcoa; chem process engr, Mobil Oil Co; & proj engr on numerous nucl energy projs with Rockwell Intl. Mgr energy conversion & res projs at GE Co. Co-founder & Dir Terradex Corp. Author numerous technical papers on energy conversion, radiation detection & minerals exploration. (Outstanding ANS technical paper 1973). *Society Aff:* AIChE, ANS.

Ginsburg, Charles P
Business: 401 Broadway, Redwood City, CA 94063
Position: VP Advanced Dev. *Employer:* Ampex Corp. *Born:* Jul 1920. Attended U of CA at Berkeley & Davis, & San Jose State C. BA Math from San Jose 1948. Studio & transmitter engr in San Francisco Bay Area radio stations from 1942-52. Employed by Ampex Corp in 1952 to investigate recording of t.v. signals on magnetic tape, & became leader of team which succeeded in this dev. Awards incl David Sarnoff Gold Medal of the SMPTE in 1957, Vladimir K Zworykin Television Prize of the IRE in 1958, Valdemar Poulsen Gold Medal of the Danish Acad of Tech Scis in 1960, Howard N Potts Medal of Franklin Inst in 1969, John Scott Medal of the City of Philadelphia in 1970. Fellow SMPTE, IEEE; Mbr Franklin Inst, Mbr NAE.

Ginsburg, Seymour
Business: Computer Sci Dept, Los Angeles, CA 90007
Position: Professor. *Employer:* U of Southern CA. *Born:* Dec 12 1927. PhD math from U of MI. BS from CCNY. Taught math 1951-55 at U of Miami. In industry (computers) 1955-71. At USC since 1966 as professor of computer sci. Author over 90 papers & 3 books in math, automata theory, formal languages, grammar theory and databases. Fellow IEEE. Guggenheim fellow 1974-75. *Society Aff:* ACM, IEEE, AMS, SIAM.

Ginwala, Kymus
Business: 39 Olympia Ave, Woburn, MA 01801
Position: Pres. *Employer:* Northern Res & Engg Corp. *Education:* BS/ChE/MIT; SM/ChE/MIT *Born:* 3/27/31. Born in the Republic of S Africa. Secondary sch education in S Africa & India 1954. Engr with S E Ginwala Filhos in Mocambique 1955-59. Chem Engr work Arthur D Little, Inc, Cambridge, MA. Joined Northern Res & Engg Corp in 1959 as Sr Engr. Became Pres in 1974. NREC has been a subsidiary of Ingersoll-Rand Co since Feb 1978. The co provides res & dev services in mech engg, turbomachinery, gas & steam turbines to over 150 clients in US & overseas & to US govt agencies. The company also develops and markets software for mechanical engineering design for machining and for data acquisition and control of software. Included in the company's products are air dynamometers for aircraft engine testing. *Society Aff:* AIAA, AIChE, ACS, ADPA, NCGA.

Giordano, Anthony B
Business: 333 Jay St, Brooklyn, NY 11201
Position: Prof & Dean Emeritus *Employer:* Polytechnic University *Education:* BEE/EE/Polytech Inst of Brooklyn, 1937; MEE/EE/Polytech Inst of Brooklyn, 1939; DEE/EE/Polytech Inst of Brooklyn, 1946 *Born:* 02/01/15. Native of NYC. Joined Polytechnic 1937, became Prof of EE 1953, Dean of Grad Studies 1960, Assoc Provost 1983, Prof and Dean Emeritus, 1986. Fellow IEEE; Fellow AAAS; Charter Fellow ASEE; Active Mbr Intl Sci Radio Union; former Assoc Ed URSI Radio Sci; Mbr Sigma Xi, Eta Kappa Nu, Tau Beta Pi, Pi Tau Sigma. Participant in 1962 Space Sci Study of Natl Acad of Sci. Selected for educ teams doing evals in France 1964; Pacific Ocean areas 1965; Venezuela 1966. Cairo Egypt 1977. During WWII, in microwave res, dev precision waveguide attenuators which became intl stds. IRE Bd 1960-62; Chrmn NYC Bd of Educ's Engrg Adv Ctte 1962-63; Chrmn Tech Program IEEE Intl Convention 1966-67. Chrmn IEEE External Awds Ctte 1975-77; Chrmn IEEE Edison Medal Ctte 1978; Mbr IEEE Nominations and Appts Ctte 1974-76; Secy IEEE Communications Soc 1964-75; VP at Large ASEE. Chrmn IEEE Education Medal Ctte 1981; Chrmn ASEE Sr Res Ctte 1979-81; VChrmn Hoover Medal Ctte 1981; Dir ASEE Zonel 1974-75; Chrmn ASEE Campus Liaison Bd 1975; Chrmn ASEE Middle Atlantic Sect 1971-72; Chrmn Proj Ctte of Engrg Fdn 1981; Chrmn of Bd of Engrg Fdn 1980; Mbr IEEE Awards Bd 1981. Recipient, ASEE W. Leighton Collins Distinguished and Unusual Service Award 1987. IEEE Centennial Medal 1984; Achievement Award, IEEE Education Soc 1983; ASEE-Western Elec Award for Excellence in Engrg Teaching 1981; Meritorious Award, IEEE Communications Soc 1976; Distinguished Alumnus Award of Polytechnic 1967. *Society Aff:* ASEE, IEEE, URSI, AAAS

Giordmaine, Joseph A
Business: 600 Mtn Ave, Murray Hill, NJ 07974
Position: Member of Technical Staff *Employer:* AT&T Bell Labs. *Education:* PhD/Phys/Columbia Univ; AM/Phys/Columbia Univ; BA/Phys & Chemistry/Univ of Toronto. *Born:* 4/10/33. in Toronto, Canada. Instr in Phys, Columbia Univ 1959-1961; Mbr of Technical Staff, Bell Labs, 1961 to present. Visiting Prof of Phys, Technical Univ, Munich, 1966. Lab. Dir. 1971-87. Res on lasers, quantum electronics, nonlinear optics. Coinventor optical harmonic & tunable coherent light sources, high speed optical pulse devices. Married Mary A Mills, 1958. Children, Paul, Anne, & Claire. Optical Society of America R.W. Wood Prize, 1986. IEEE Microwave Theory & Techniques Society, National Lecturer, 1983. *Society Aff:* APS, IEEE, OSA, AAAS, AAS.

Gipe, Albert B
Business: 416 Goldsborough St, PO Box 1147, Easton, MD 21601
Position: President. *Employer:* Gipe Associates, Inc. *Education:* BEE/Elec Engg/Marquette Univ; BNS/Naval Science/Marquette Univ. *Born:* June 21, 1925. US Navy 1943-46. Allis Chalmers Mfg Co, Switchgear Design & Dev 1946-48. US Army Med Res 1948-51. Whitman, Requardt Assocs 1951-56. Partner Miller, Scherholz, Gipe 1956-63. Pres Albert B Gipe & Assocs 1964-1978. Sr V P Kidde Cons 1975-78. Pres Gipe Assocs 1978- . Taught elec courses 1951-57. Reg P E Md, DC, DE, W Va. Mbr ACEC, NSPE, IEEE, IES, AWWA, WPCF, ASHRAE . Has written papers, magazine articles. Pres Bldg Congress & Exchange 1976-77, Baltimore Cty Adv Bd.,GSA Public Adv Panel, Pres. Cons. Engr Council of Maryland 1962. Chairman ACEC Procurement Committee 1986-1988. *Society Aff:* ACEC, NSPE, IEEE, IES, AWWA, WPCF, ASHRAE.

Girifalco, Louis A
Business: Univ. of Pennsylvania, 3231 Walnut St, Phila, Pa 19104
Position: University Professor of Materials Science & Engineering. *Employer:* Univ of Penna. *Education:* Ph.D./Chemical Physics/University of Cincinnati; M.S./Chemistry/ University of Cincinnati; B.S./Chemistry/Rutgers University. *Born:* July 3 1928 Brooklyn NY. Married 1950, 8 children. Materials Science & Engineering. BS Rutgers 1950; MS Cincinnati 1952; PhD (phys chem) 1954. Res Chem E I DuPont de Nemours & Co 1954-55; solid state physicist NASA 1955-59; Hd solid state physics sect Lewis Res Ctr Ohio 1959-61; Assoc Prof Met Engrg U of PA 1961-65; Prof Met & Mat Sci 1965-81. Dir Lab Res Struct of Matter 1967-69. Pres Cara Corp 1969-70. Assoc Dean C of Engrg & App Sci U of Pa 1974-79. Vice-Provost for Res, Univ of PA 1979-81, Acting Provost, Univ. of PA, 1981, Univ Prof Mat Sci & Engrg 1981-, AAAS, APS, AIME, Met & Petrol Engrg. Electronic theory of metallic solution thermodynamics; The metallic bond & interatomic potentials in metals; The dynamics of technological change. 79 articles published in scholarly journals & author of 3 books. *Society Aff:* AAAS, APS, ASM, TMS/AIME, ΣΞ

Givens, Paul E
Home: Staplcoth, Box 547, Greenwood, MS 38930
Position: Mgt Specialist Assoc Prof. *Employer:* MS State Univ. *Education:* PhD/Ind Engg/Univ of TX; MBA/Bus Admin/Creighton Univ; BS/Ind Engg/Univ of AR. *Born:* 8/12/34. Native of Pawhuska, OK (reg Osage Indian). Designed and build pipeline systems as a Pipeline Engr for Service Pipeline Co (Std Oil of IN) 1957-1963. Economics Engr and Chief Ind Engr, Pipeline Operations, and Dir of industrial Relations for Northern Natural Gas Co, 1963-1969. dir of Manpower, the Western Co of NA (Oilwell Service - Offshore Drilling) 1969-1972. Pres, Paul E Givens and Assoc, States Army Ordinance Sch, APG MD, 1958-1960. MS State Univ, 1974- present. Pres, Omaha Chapter, IIE, 1969 and current Dir of Mgt Div IIE, 1977- present. *Society Aff:* IIE, ASEE, MAS.

Gizienski, Stanley F
Business: 3467 Kurtz St, San Diego, CA 92110
Position: Sr Consulting Principal & Chmn of the Bd *Employer:* Woodward-Clyde Consultants *Education:* M.Sc/CE/Harvard Univ; BS/Gen Engr/Univ of MA *Born:* 7/3/21 Combat and construction officer, Army Corps of Engrs 1943-47 and as civil engr, 1948-55. With Woodward-Clyde Consultants since 1955. Special consultant earth dam design and geotechnical problems involving groundwater seepage and construction on expansive soils. Co-author of technical book, "Earth and Earth Rock Dams." Author several articles in geotechnical engrg. Broad experience in geotechnical engrg on variety of projects. In charge of design of number of earth dams including Lopez, Virginia Ranch, La Costa, Patagonia, and Matanzas. Registered PE CA, NB, NV, and AZ. Dist engineer alumnus, 1987, Univ of Mass. Current Chmn, California OSHPD Bldg Safety Bd. Served as Pres & Chmn of San Diego County Planning Commission, California Legislative Council of Prof Engrs, and San Diego Chapters of ASCE, CCCELS and Downtown Lions Club. Dir of SAME and Consulting Engrs Assoc of California. *Society Aff:* ASCE, SAME, ACEC, NSPE, AWWA, USCOLD, ACI, ASTM.

Gjostein, Norman A
Home: 544 S Claremont, Dearborn, MI 48124
Position: Director Materials & Design Analysis Laboratory. *Employer:* Ford Motor Co. *Education:* PhD/Met Eng/Carnegie-Mellon Univ; MS/Met Eng/IL Inst Tech; BS/Met Eng/IL Inst Tech. *Born:* 5/26/31. in Chicago, IL. Past-Chmn of the Phys Met & chem & Phy of Metals Ctte of AIMe. Member Advisory Board, MAE Dept., Princeton Univ. Chairman Education Committee, Detroit Chapter, ASM presently Dir, Materials and Design Analysis is respon for co-ordination, planning & execution of LRR programs in a variety of fields of automotive technology: computer-aided engineering analysis, expert systems, advanced materials, sensors and IC devices and energy storage/conversion. *Society Aff:* ΤΒΠ, ΣΞ, ΦΛΥ, ASM, AIME, IEEE, ESD.

Glandt, Eduardo D
Business: Dept. of Chem. Eng, U. of Pennsylvania, Philadelphia, PA 19104-6393
Position: Prof *Employer:* Univ of PA *Education:* PhD/ChE/Univ of PA; MSE/ChE/Univ of PA; BS/ChE/Univ of Buenos Aires *Born:* 03/04/45 Native of Buenos Aires, Argentina. After BS, worked for 5 yrs 1968-73 as conslt in mineral and inorganic

Glandt, Eduardo D (Continued)
separations for INTI, a branch of the government of Argentina, while also serving as an adjunct prof at the Univ of Buenos Aires. Moved to the US in 1973. Obtained PhD in 1977. Faculty mbr at Univ of PA since then. Presently, Prof. Res work in Thermodynamics, Statistical Mech, Phase Equilibrium, Gas Adsorption, Random Media, Recipient of the 1979 Victor K. LaMer Award, American Chem Soc. Also, 1977 S. R. Warren Award, and 1980 Lindback Award, both for distinguished teaching. *Society Aff:* AIChE, ACS, APS, AAAS

Glaser, Peter E
Business: Acorn Park, Cambridge, MA 02140
Position: Vice Pres. *Employer:* Arthur D Little Inc. *Education:* PhD, MS/ME/ Columbia; BS/ME/Charles Univ, Prague *Born:* 9/5/23. Joined Arthur D. Little 1955, VP 1973; Dir of projects on cryogenic thermal protection systems, Apollo lunar science experiments on lunar heat flow, lunar gravity & earth-moon distance, solar & arc imaging furnaces, solar heating & cooling, solar photovoltaic conversion, solar power satellites & commercial activities in space; Pres Sunsat Energy Council, March 1978-pres, Dir AAS 1978-1982, Director L5 Soc. 1979-Pres. Pres Intl Solar Energy Soc, 1968, Dir Am Sect, 1975; VP IIR, 1974; Mbr cttes of ASME, ASHRAE, AIAA, ISES, IIR, Recipient of Kayan Medal, Columbia, 1974 Farrington Daniels Award Intl Solar Energy Soc., 1983; Editor-in-Chief Journal of Solar Energy, 1973-1985 Mbr Cosmos Club, 1978-Pres. *Society Aff:* ASME, ASHRAE, AIAA, ISES, L5.

Glasford, Glenn M
Business: 231 Link Hall, Syracuse, NY 13210
Position: Prof of Elec & Computer Engrg. *Employer:* Syracuse Univ. *Education:* MS/EE/Iowa State Univ; BS/EE/Univ of Texas. *Born:* Nov 1918, Arcola,Texas. Staff Mbr, Radiation Lab, MIT, 1942-45; Head Adv Dev Group, Allan B DuMont Labs, 1945-47; Faculty Syracuse Univ since 1947, Prof from 1955; Occasional cons to government & ind, SignalCorps, Brookhaven Natl Lab, IBM, Genl Elect. Author, 'Fundamentals of Television Engrg, Linear Analysis of Electronic Circuits', 'Analog Electronic Circuits', 1986, 'Digital Electronic Circuits', 1988; contributor, 'Radiation Laboratory Technical Series', 'Encyclopedia of Science and Technology', 'Encyclopedia Americana'. Numerous technical & educational journal papers. Natl Elec Conf & IEEE Broadcasting Group Natl Awards. Fellow, IEEE & AAAS; Mbr, ASEE, AAUP, SMPTE. *Society Aff:* IEEE, ASEE, SMPTE, AAAS, AAUP.

Glaspell, W Leon
Business: 2750 ABW/DE, Dayton, OH 45433
Position: Dep Base Civ Engr. *Employer:* WPAFB. *Education:* MBA/-/Wright State Univ; BSAE/-/WV Univ. *Born:* 7/25/43. Manage the largest Base Civ Engg function in the AF. Responsible for over 8,000 acres, 900 admin bldgs, 2,335 housing units, 110 miles of rds, & fire protection; all of which support over 26,000 Govt employees. These functions are accomplished by a work force of 700 craftsmen, 100 engrs, & 400 support personnel. We average over $30,000,000 in projs under construction at all times and have an annual operating budget for intl maintenance & operations of $26, 000,000. *Society Aff:* SAME.

Glasscock, Green B
12 Point Circle, Mill Pond Acres, Lewes, DE 19958
Position: VP Chem. Oper. & Eng. *Employer:* Barcroft Co. *Education:* MS/ChE/Univ of TX; BS/ChE/Univ of TX *Born:* 1/4/22. Served as a Naval Aviator in WWII. Employed by Phillips Chem Co as a Process Engr; by the Lord Corp in Erie, PA as a Res Engr, Process Engr & Plant Mgr. Served briefly with W R Grace Res Div in Clarksville, MD. Joined Bacroft Co at the start of plant construction as a Proj Engr 6 now operate as VP Res & Engrg. *Society Aff:* AIChE, NSPE.

Glasser, Julian
Home: 3400 Glendon Dr, Chattanooga, TN 37411
Position: Pres. *Employer:* Chem & Met Res, Inc. *Education:* PhD/Physical Chemistry/PA State Univ; MS/Chemistry/Univ of IL; BS/Chemistry/Univ of IL. *Born:* 5/23/12. and raised thru high sch in Chicago, IL; Res Engr in Electromet at Battelle Meml Inst, 1938-42; Chief Chemist at Olin-Aluminum Div, Tacoma, WA, 1942-45. Dir of Res at Gen Abrasive Co, 1945-47; Supv in Extractive Met, IIT Res Inst 1947- 53; Consultant, Office of Naval Res, 1952-53. Consultant, Natl Res Incuncil, 1951-53 & 1958-60 Technical Dir, Cramet, Inc, Chattanooga, TN 1953-58. Pres, Chem & Met Res Inc. 1960-to date. *Society Aff:* AIME, ASM.

Glassey, C Roger
Business: Dept of Indus Engrg & Operations Res, Berkeley, CA 94720
Position: Prof *Employer:* Univ of CA *Education:* PhD/Op Research/Cornell Univ; MS/Applied Math/Univ of Rochester; Post Grad Cert/Ind Admin/Univ of Manchester (England); BME/IE/Cornell Univ *Born:* 7/12/30 Born in Glen Bridge, NJ, raised in upstate NY. After graduation from Cornell, studied in England as a Fullbright Scholar. Officer in US Navy from '54-'57. Industrial Engr and applied Mathematician for Eastman Kodak in Rochester, NY for 6 years. Studied and taught in Univ of Rochester applied math evening program. After PhD from Cornell, joined faculty of Univ of CA Berkeley. '78-'80 on leave as assistant administrator for applied analysis in the Energy Information Administration, US DOE. Chairman of Department of IE for '80-'86. Married, 3 children. *Society Aff:* AAAS, TIMS, ORSA, IIE, IEEE

Glassman, Irvin
Business: Dept. of Mech. & Agri. Eng, Princeton University, Princeton, NJ 08544
Position: Prof *Employer:* Princeton Univ *Education:* Dr Engr/ChE/Johns Hopkins Univ; BE/ChE/Johns Hopkins Univ *Born:* 9/19/23 Joined Princeton in 1950. Now Prof of Mechanical and Aerospace Engrg. American Cyanamid Prof of Environmental Science, 1972-1975. Dir, Center for Environmental Studies 1971-1979. Executive Office, Fac Comm of Guggenheim Aerospace Propulsion Labs 1963-1970. Member and Former Chrmn, Propulsion and Energetics Panel, AGARD/NATO. Member NAS Committee on Motor Vehicle Emissions and Climatic Impact Committee. Visiting Prof - Univ of Naples (1966, 1978), Technion (1967) and Stanford (1975). Author of two books - *Combustion* and *Performance of Chemical* Propellants. Edgerton Medalist. Combustion Inst.; Roe Award, ASEE. Consultant UTC and Chrysler. *Society Aff:* ACS, AAUP, C.I. FAS

Gleason, James G
Home: 1139 Sunset Drive, Fayetteville, AR 72701
Position: Consultant, Mech. & Aero Engrg. *Employer:* Self *Education:* MS/ME/Univ of AR; BS/Aero Engg/AL Poly Inst. *Born:* 3/24/15. Native of Hammondsport, NY. Employed in Mech Engrg Dept at AL Poly Inst 1938-1940 and at Univ of AR 1940-85. Duties include teaching and res. Consulting Mech. & Aero. Engrg, 1985-present. Summers spent in industry include 8 at Boeing Airplane Co, Power Plant Staff Unit, 7 at P&WA Advanced Combustion Group, 2 at Beech Airplane Co Engine Unit, 1 at EG&G Nuclear Res Unit in ID Falls doing res. During past 19 yrs, served on SAE Mid-Continent Sec Governing Bd (V Chrmn 1976) (Chrmn 1977) including 4 yrs on SAE Natl Committee. SAE Natl. Fuel & Lubricant Meeting General Chairman, 1981. Faculty Advisor for SAE Student Branch 20 yrs. Reg Private Pilot. Reg PE AR & KS. *Society Aff:* ASME, SAE, ASEE.

Glen, Thaddeus M
Home: 3324 Pelham Rd, Toledo, OH 43606
Position: Prof of Ind Engr. *Employer:* Univ of Toledo. *Education:* PhD/Ind Engr/ Univ of MI; MME/ME/Univ of DE; BSIE/Ind Engr/Wayne State Univ. *Born:* 5/18/27. Has taught at the Univ of Toledo since 1960. Has also had extensive engg and mgt experience in several cos including Ford, DuPont, Seabrook Farms, and the Frick Co. Served as a consultant for the Asian Productivity Organization in Japan, Philippines, and India. Spent two summers in the NASA-ASEE co-sponsored prog at the Manned Spacecraft Ctr in Houston. Has presented numerous papers on productivity, human factors engg, applications of operations res, habitability, and work design for the physically and mentally handicapped. *Society Aff:* IIE, ASEE, HFS, ORSA, AIIM, ΣΞ, TBΠ.

Glenn, Gerald M
Business: Daniel Bldg, Greenville, SC 29602
Position: VP. *Employer:* Daniel Construction Co. *Education:* BS/Civ Engg/Clemson Univ. *Born:* 8/20/42. Native of Greenville, SC. Began career with Daniel on summer jobs while an undergrad. Have progressed through the mgt ranks of engg & construction to present position of Evp, responsible for Worldwide Marketing & Sales. Played an active role in expandig Daniel's engg capabiities from a humble beginning to a current status in excess of 1000 people & serving a wide spectrum of industries. *Society Aff:* AIChE, TAPPI

Glenn, Joe Davis, Jr
Business: PO Box 12154, Norfolk, VA 23502
Position: Pres. *Employer:* Joe D Glenn & Assoc Inc (Formerly Glenn-Rollins Assoc) *Education:* BS/Civil Engg/Clemson Univ; MS/Civil Engg/Univ of TN. *Born:* 8/12/21. Native of Fair Play SC. Capt, Corps of Engrs 1942-46. Asst Prof Clemson Univ 1946-55. Structural Engr, Tidewater Const Corp 1955-60. Cons Lt. Engr 1960-75. Pres, Glenn-Rollins & Assoc Inc 1975-1983. Pres. Joe D. Glenn & Assoc 1983-. Pres, Tidewater Chap Va Soc of Prof Engrs, 1967; Pres, Engrs Club of Hampton Rds, 1974; Pres, Va Soc of Prof Engrs 1975-76, Dir, Natl Soc of Prof Engrs 1976-80, Engr of the Yr, Va Soc of Prof Engrs 1976; Pres, Kiwanis Club of Norfolk- Princess Anne, 1963. Mbr NSPE, VSPE, ASCE, ACEC & Soc of Am Military Engrs. Former Elder Presbyterian Church *Society Aff:* NSPE, VSPE, ACEC, SAME, ASCE

Glenn, Roland D
Business: 50 E 41st St, New York, NY 10017
Position: Pres. *Employer:* Combustion Processes Inc. *Education:* MS/Chem Eng/ MIT; BS/Chem Eng/MIT. *Born:* 3/22/12. Sloan Program in BA in 1939. With Union Carbide Corp Chem & Plastics Div at So Charleston, W Va 21 years & NY 13 years; positions included Plant Mgr, VP Plastics Dev, VP Utilities Operations & VP Production. VP Pope, Evans & Robbins in charge of fluidized bed boiler dev, Alexandria, Va 1969-71. Since 1971, Pres Combustioin Processes, Inc, New York, specializing in fluidized-bed combustion, coal-water slurries & specialized coal cleaning operations. AIChE, ACS, Assoc of Cons Chem & Chem Engr, Chemists Club (NY) & Sandy Bay Yacht Club (Rockport, Mass). Reg Prof Engr in NY, Conn & Va *Society Aff:* AIChE, ACS.

Glennan, T Keith
Home: 11400 Washington Plaza W, Apt 903, Reston, VA 22090
Position: Retired. *Education:* BS/EE/Yale. *Born:* Sept 1905. 16 honorary doctorates, 1 hon MS. Indus exper incl exec positions with Elec Res Prods in early days of talking motion pictures incl 2 yrs serv in Eur; 5 yrs Paramount Pictures, 2 yrs Samuel Goldwyn Studios as oper & studio mgr; 2 yrs with Ansco, Bingham NY. Apptd 4th Pres, Case Inst of Tech, Cleveland Ohio 1947-66. On lv 1950-52 as US Atomic Energy Cmsnr & 1958-61 as 1st Admin of NASA. Svd during the war 1942-45 as Dir, US Navy Underwater Sound Lab, New London CT. Since retiring from Case, svd over 3 yrs as Pres Associated Univs Inc & as Asst to Chmn of the Urban Coalition. Former Dir or Trustee of 8 natl corps & of Rand Corp, Cleveland Clinic Found & as Chmn of the Bd of The Aerospace Corp. Fellow, of NAE, 1967. US Medal for Merit 1946. NASA Disting Serv Medal 1967. Disting Hon Award US Dept of State 1973. Retired Fellow AAAS. Retired Benj Franklin Fellow, Royal Society of Arts in London, Eng. 1970-73 served as US Representative to IAEFI personal rank of Ambassador *Society Aff:* TBΠ, ΣΞ, AAAS.

Glennon, James M
Home: 1335 Bluebird Dr, Mt Pleasant, SC 29464
Position: VP *Employer:* Enwright Assoc Inc *Education:* MS/EE/VA Polytechnic Inst; BS/CE/VA Polytechnic Inst *Born:* 12/15/43 Native of VA Beach, VA. Served with US Army, Medical Service Corps 1967-70. Dir of continuing educations, env engr, Clemson Univ 1970-71; Chief Engr, SC Pollution Control Authority 1971-74. Joined Enwright Assocs as project mgr and design engr for environmental projects in 1974. Assumed current responsibility as VP for Coastal Area Offices in 1979. Nominated to "Outstanding Young Men of America" in 1972. Appointed to the SC Bd of Certification for Environmental Systems Operators in 1971. Has served on and chaired several past committees for AWWA and SCWPCA and has co-authored two technical training publications. *Society Aff:* WPCF, NSPE, CESC, CCEC

Glicksman, Leon R
Home: 8 Maiden Lane, Lynnfield, MA 01940
Position: Prof *Employer:* MIT. *Education:* BS/Mech Engg/MIT; MS/Mech Engg/ Stanford; PhD/Mech Engg/MIT. *Born:* 5/12/38. Chicago; raised in San Antonio, TX. Capt US Army, 1964-66. Taught at MIT since 1966 as Lecturer, Assoc Prof, Sr Res Scientist Prof. Currently also mbr of MIT Energy Lab. Have conducted res in glass forming, end-use of energy, cooling towers fluidized beds. Winner of ASME Melville Medal, 1969. Married to Judith (Kidder), children: Shayna, Eric & David. *Society Aff:* ASME.

Glicksman, Martin E
Business: Matls Engrg Dept, Troy, NY 12181
Position: Prof. *Employer:* Rensselaer Polytechnic Inst. *Education:* PhD/Metallurgical Engrg/Rensselaer Poly Tech; BMetE/Metallurgy/Rensselaer Poly Tech. *Born:* April 4, 1937. Chosen as a Natl Acad of Sci - Natl Res Council Post- doctoral Assoc 1961-63 & studied transport properties of liquid metals in the Metal Phys Branch of the Naval Res Lab. In 1963 joined the staff of the Met Div at the Naval Res Lab & was appointed Acting Head, Metal Phys Branch 1966-67; Hd, Liquid Metals & Solidification Sect 1967-69; Head, Transformations & Kinetics Branch 1969-75; Assoc Supt, Mat Sci Div 1975. In Sept 1975, was appointed Prof & Chmn of the Mat Engrg Dept at RPI, and served as such until 1986, when he was appointed the John Todd Horton Distinguished Professor of Materials Engineering. RESA Award in Pure Science 1968, Arthur S Flemming Award 1968, Outstanding Young Man of America 1969, NRL Res Paper Awards, 1969, 1971, 1972, M A Grossmann Award of the Amer Soc for Metals 1971 & the Achievement Award in Physical Sci of the Washington Acad of Sciences 1973, Intl Metallographic Society Award 1979; NASA Award for Tech Excellence 1980, 1982, 1983; NSF Res Creativity Award, Div of Mtls Res 1981; Fellow Amer Soc for Metals 1982; Intl Metallographic Soc Award 1982; Stanley P. Rockwell Medal 1982; VanHorn Distinguished Lectureship, Case Western Reserve Univ 1984. *Society Aff:* AIME, ASM, AAAS, AIAA, RESA, AACG.

Glicksman, Maurice
Business: Box 1862, Brown University, Providence, RI 02912
Position: Provost *Employer:* Brown U. *Education:* PhD/Physics/Univ of Chicago; ScM/Physics/Univ of Chicago. *Born:* Oct 1928. undergrad work at Queens U. Native of Toronto Ontario; U S Citizen. Res in nuclear physics, Atomic Energy Proj Chalk River Ontario 1949-50; taught phys at Roosevelt U 1953-54. Res at RCA Labs Princeton NJ 1954-69 (Dir of RCA's Tokyo res lab 1963-67). University Prof & Prof of Engrg at Brown from 1969, Dean of Grad School 1974-76, Dean of Fac & Acad Affairs 1976-78; Provost & Dean of Fac 1978-86; present pos since 1986. Visiting Scientist, Physics Dept, MIT, 1983-84 Chmn Ctte on Mats for Radiation (electromagnetic) Detection Devices, NAS 1971-74, Vice Chmn, Bd, Univ Res Assoc, 1987- . *Society Aff:* IEEE, APS, ΣΞ, ΦBK.

Gloge, Detlef C
Business: Optical Sys Res Dept, Holmdel, NJ 07733
Position: Dept Head. *Employer:* Bell Tele Labs. *Education:* Dipl Ing/-/Univ of Braunschweig; Dr Ing/-/Univ of Braunschweig. *Born:* Feb 2, 1936. Dipl Ing 1961, Dr Ing 1964, Tech Univ of Braunschweig, Germany; Bell Labs since 1965. Dept Head, Long Haul Lightwave System Development Dept. Worked on the design & testing of various optical transmission media. Presently engaged in systems studies related to optical fiber telecommunications; Mbr Optical Soc of America; Fellow, Inst of Elec & Electronics Engrs. *Society Aff:* IEEE, OSA, APS.

Glover, Charles J
Business: Dept of Chem Engrg, Texas A&M University, Coll Station, TX 77843
Position: Assoc Prof *Employer:* TX A&M Univ *Education:* PhD/ChE/Rice Univ; B/ChE/Univ of VA *Born:* 4/6/46 Native of Arlington, VA. Served with the US Army, 1968-1970. Research engr in chem flooding techniques with Exxon production Research Co in Houston, 1974- 1977. Joined TX A&M in 1977, working in multicomponent polymer solution vapor-liquid equilibrium (VLE) and chromatographic measurement of VLE and diffusion. Recent activities also in asphalt chemical characterization and performance; AIChE student chapter advisor, 1978-1981. Recipient of the SPE of AIME Cedric K. Ferguson Medal for 1980. Distinguished Mbr, SPE of AIME. Registered PE, TX. *Society Aff:* AIChE, SPE of AIME, SOR

Glover, Douglas
Business: 3980 Res Park Dr, Ann Arbor, MI 48104
Position: Mgr-Metallurgical Engineering. *Employer:* Federal-Mogul Corporation. *Education:* BS/Met Engg/MI Tech Univ. *Born:* July 1930. Served as communications officer in USAF 1954-56. Res metallurgist, Owens Illinois specializing in glass mold materials, 1956-57. Joined Continental Aviation Toledo Ohio 1957. Was Chief Metallurgist responsible for materials engrg and special manufacturing processes for gas turbine engines. Started with Federal-MOGUL 1968 as Manager of Metallurgical Engineering. Present position is Manager of Research for the Ball and Roller Bearing group of Federal-MOGUL Corporation. Responsible for development of materials, manufacturing methods and new design of ball and roller bearings. Major contributions include development of powder metallurgy bearings and ceramic roller bearings. Chmn Toledo ASM 1965-66. Active SAE & ASTM. Hobbies: sailing, fishing and vegetable gardening. *Society Aff:* ASM, SAE, ASTM.

Glover, Frederic L
Home: 1044 Armada Dr, Pasadena, CA 91103
Position: Chems Mgr. *Employer:* Getty Oil Co. *Education:* MBA/Bus/Harvard Grad Sch Bus Adm; BSE/CE/Univ of MI. *Born:* 6/25/33. Presently, (1977-) Chems Mgr, Getty Oil Co & serves as Chrmn of the Bd, Hawkeye Chem Co, Clinton, IA; Dir of Chembond Corp, Eugene, OR; & Dir of Nucl Fuel Services, Inc, Rockville, MD. Also serves on Policy Committee, Chemplex Co, Rolling Meadows, IL. Formerly, (1974-76) Flurochems Div Pres, Pennwalt Corp. BSE in Chem Engg, univ of MI, 1956; MBA Harvard univ, 1964. Born, Detroit, June 25, 1933. Married Jane Hobday, Jan 3, 1959; children-Justine (19), David (18). Mbr of Am Inst of Chem Engrs, Natl Petrol Refiners Assn, Sierra Club. Republican, Episcopalian. Home: 1044 Armada Dr, Pasadena, CA 91103. Office: 3810 Wilshire Rd, Los Angeles, CA 90010. *Society Aff:* AIChE.

Glover, John W
Home: 3008 Churchill Rd, Raleigh, NC 27607
Position: Ext Prof *Employer:* NC State Univ. *Education:* BS/Ag Engg/NC State Univ. *Born:* 4/30/28. Native of NC; Field Rep for Intl Harvester 1950 with time out for military service in 1951 & 1952. Design & sales engr for DARF Corp 1963 to 1955. Assumed present position in Nov, 1955. Maj responsibilities include ext work in the field of agri crop processing & storage, machinery selection, & teaching in the field of electricity & farm electrification. Hobbies include mechanics and carpentry. *Society Aff:* ASAE.

Glower, Donald D
Business: The Ohio State University, College of Engineering, 2070 Neil Avenue, Columbus, OH 43210
Position: Dean-College of Engg. *Employer:* The Ohio State University. *Education:* BS/Marine Engg/US Merchant Marine Acad; BS/-/Antioch College; MS/Mech Engg/IA State Univ; PhD/Nuclear Engg/IA State Univ. *Born:* July 1926 Shelby Ohio. Asst Engrg Officer Grace Lines Inc 1947-49. Res engr Battelle Mem Inst 1953-54. Asst Prof Coll Engrg Iowa St Univ 1954-58. Mbr res staff Sandia Corp 1961-63. Hd, Radiation Effects Dept Genl Motors Corp 1963- 64. OSU Prof Mech Engrg 1964-68, Dept Chmn 1968-76, Dean, College of Engineering, The Ohio State University 1976- . Cons num indus & univs. Mbr Bd/Dir Ohio TRC, Orton Foun, ITI, Bd/Gov CICT. Indus Tech. Enterprise Bd. Ohio. Author & mbr num tech soc's. Chmn Nuclear Engrg Div 1971-72 ASEE; Pres 1973-74 CTC. Mbr Bd/Dir - Natl Res Inst; OSU Dev. Fund; OSU Engg Exp Station. Outstanding Bus Achievement Alumni Award, US Merchant Marine Acad Outstanding Prof Achievement Award, IA State Univ. Ohio Society of Prof. Engrs Citation 1985. *Society Aff:* ASME, ASEE, ANS, NSPE

Gloyna, Earnest F
Business: Cockrell Hall 10.310, College of Engrg The Univ of Texas, Austin, TX 78712
Position: Dean of Engg. *Employer:* Univ of TX. *Education:* Dr of Engg/Sanitary Engr & Water Resources/Johns Hopkins Univ; Masters/CE/Univ of TX; BS/CE/TX Tech College. *Born:* 6/30/21. Dean & Smith Chair Univ of TX at Austin; mbr Natl Acad of Engrg, US, corresponding mbr Natl Acad Scis Venezuela, Natl Acad of Engrg Mexico; Eddy, Fair Medals and Hon Mbr Award WPCF, Water Resources Award AWAA, Natl Conserv Medal, Order of Henry Pittier, Venezuela, Hon. Member Award, Award of Honor, Simon W. Freese Environmental Engrg Award, ASCE; EPA Educator Award; Conservation Achievement Award in Science, Natl Wildlife Fed; Natl Environ Devel Award, JJ King Award, Univ TX, Pres, Water Poll Ctrl Fed, Past Pres & Diplomate, Amer Acad of Environ Engrs, Pres. & past Natl Dir, TX Soc of PE; Bd of Dirs, various organizations. Author 7 books & 300 publications & professional reports. *Society Aff:* NAE, ASCE, NSPE, AIChE, AAEE, ASEE, AWWA, WPCF.

Gnesin, Albert J
Business: Box M-1, Short Hills, NJ 07078
Position: V P - Maintenance Coatings. *Employer:* Mobil Chem Co. *Education:* BSChE/Chem Engg/Syracus Univ. *Born:* July 1924. Native of N Y. Process Engr Standard Oil Co Ind 1949-52. Sales Engr Davison Chemical Co, Div of W R Grace & Co 1953-58. Petrochemical Sales Manager Delhi-Taylor Oil Corp 1958-63. Marketing Planning Manager, Petrochemical Div, Mobil Chemical Co 1963-64. Marketing Manager, Mobil Chemical International 1964-68. President, Mobil China Allied Chemical Co, Taipei Taiwan 1968-71. V Pres International, Chemical Coatings Division, Mobil Chemical Co 1972-76. V Pres Maintenance Coatings, North American Affiliates & Foreign Licensing, Mobil Chemical Co 1976 to date. *Society Aff:* AIChE, ACS.

Go, Mateo L P
Home: 2415 Ferdinand Ave, Honolulu, HI 96822
Position: Chrmn *Employer:* Univ of HI-Manoa *Education:* PhD/CE/Cornell Univ; MS/CE/MIT; BS/CE/Cornell Univ *Born:* 9/17/18 in China; naturalized US citizen. In the Philippines 1947-56; Pres of Mateo L. P. Go Construction Co and Hamilton Furniture Co; Tech consultant for Go Occo & Co 1957-59, assist prof, Syracuse Univ. Since 1959, assoc prof and later prof at Univ of HI, Manoa. 1969-72, and currently, Chrmn, Dept of Civil Engrg. Civil Defense consultant in fallout shelters, protective construction and disaster mitigation design. *Society Aff:* ASCE, ACI.

Goad, Cecil G
Business: Office of Chief of Engineers, 20 Mass Ave, NW, Washington, DC 20314
Position: Chief, Construction Operations Div. *Employer:* US Army Engr Div, S Atlantic Div, Corps of Engrs. *Education:* BS/Civil Engr/VA Tech. *Born:* 11/5/28. in Floyd County, VA, began engg career with VA Dept of Hgwys in 1951. Served four yrs, including two yrs active duty with Army in the Office of Chief of Engrs in Washington, DC. Joined Corps of Engrs as a Civil Engr, civilian employee, in 1955. Served in field construction organization around Washington as proj, office, resident, and area engr for six yrs and in Office of Chief of Engrs as construction mgt engr for 13 yrs. Became Chief of Construction Div for Baltimore Dist 1974 and Chief of Construction-Operations Div, South Atlantic Div, Atlanta, GA, 1977. Promoted to Chief, Construction-Operations Div, Directorate of Civil Works, Office

Goad, Cecil G (Continued)
Chief of Engrs, Wash, DC, Aug 81, with staff mgmt responsibility for the Cons of Engrs nation-wide civil works construction and operations program. *Society Aff:* ASCE, SAME.

Gochenour, Donal L, Jr
Business: Room 739 Industrial Engrg Dept, Morgantown, WV 26506
Position: Prof & Assoc Chrmn *Employer:* WV Univ *Education:* Ph.D./Engrg WV Univ; MS/IE/WV Univ; BS/IE/WV Univ *Born:* 4/15/43 Prior to joining the WV Univ Industrial Engrg Dept in 1968, served as a research Industrial Engr for the U.S. Forest Service. Is currently prof and assoc chrmn of the (WVU) Industrial Engrg Dept, directing graduate and research programs. Area of specialization is simulation modeling and management information systems. Research has resulted in the design and implementation (for WV) of statewide (1) energy management and (2) health care delivery information systems. Division Dir, IIE, 1979-81; Executive Dir, ATTM, 1981-present; WVU Teaching award, 1970-71; SAE Teaching Award, 1979. Registered PE in WV. *Society Aff:* IIE, NSPE, ASEE, ORSA, TIMS

Goddard, George W
Home: 310 S Ocean Blvd 405, Boca Raton, FL 33432
Position: Ret. (Brigadier General USAF) *Born:* June 1889 England; naturalized 1917. WW I Army Pilot, active 35 yrs. Honorary BS Boston University; BL Keuka College. 22 yrs Director, R&D Aerial Photo Lab, Dayton. Mbr Fed Board Survey Maps, 1921. First Night Aerial Photo, Rochester, Nov. 22, 1925. Extensive Aerial Surveys California, Mexico 1923; Philippines 1927-29; Alaska 1934. Dir Tech Photo School, 1930-36. Developed First Shutterless Aerial Stereo Strip Camera 1942. Used by Navy, Okinowa, 1945. Air Corps Normandy BeachesInchon Korea. Photo Disarmament Officer, Germany 1945. Dir Photo Recce NATO 1952. Retired 1953. Distinguished Service Medal, 5 patents, Hon Mbr ASP, Master Photographer PAA, National Geographic. Aviation Hall of Fame, July 24, 1976 - International Aerospace Hall of Fame, July 30th, 1976.

Goddard, Joe D
Business: University of Southern California, Los Angeles, CA 90089-1211
Position: R.J. Fluor Prof Chem Engrg. *Employer:* Univ of Southern Calif. *Education:* PhD/Chem Engg/Univ of CA, Berkeley; BS/Chem Engg/Univ of IL, Urbana. *Born:* 7/13/36. Johnson County Illinois; married Shirley M Keltner 1957, 5 children. Univ & Dept Scholarships Awards, Kennecott Copper Scholar, and Merck A'wd in Ch.E. U of Ill 1956-67; Malinckrodt Chem Co St Louis Mo summer 1956. Columbia Southern Chemical Co 1957. Dow and NSF Graduate Fellow U of Calif 1957-61. NATO Fellow, Laboratoire d'Aerothermique du CNRS & L'Institut Francais du Petrole, Paris 1961-62. Univ of Michigan, Ann Arbor, Asst to Full Professor Chemical Engrg 1963-76. NSF Senior Postdoctoral Fellow Dept Applied Maths & Theo Physics Cambridge Univ England 1971. R.J. Fluor Professor, 1976- , and Chmn, 1976-86, of Chemical Engrg, University of Southern California Los Angeles Fulbright Res and Visiting Scholar, Univ Leuven, Belgium, 1984. Mbr Amer Inst Chem Engrs, Amer Inst Phys Soc. /Rheology, Amer Chem Soc; Amer Phys Soc Div Fluid Dynam, Amer Math Soc., Amer Acad Mechanics, Alpha Chi Sigma, Tau Beta Pi, Pi Lambda Upsilon, Sigma Xi. *Society Aff:* AIChE, ACS, AIP, APS, AMS.

Goddard, Thomas A
Home: 6 Great Hill Rd, Darien, CT 06820
Position: Coordinator *Employer:* Exxon International *Education:* MS/Shipping & Shipbuilding Mgmt/MIT; BS/Naval Architecture & Marine Engg/MIT. *Born:* 2/11/42. Served in US Navy 1964-68. With Exxon Intl Inc 1968-77; 1968-72 Project Eng, R & D Div, Tanker Dept; 1972-75, Planning Analyst Planning Div, Tanker Dept; 1975-77, Mktg Analyst, Marine Sales Dept 1977-1981 Exxon Corp, Sr Analyst, Petroleum Product Dept. 1981- Exxon International, Coordinator, Industry Analysis Div, Mbr Sigma Xi, SNAME. Winner SNAME Linnard Prize for best paper presented at 1972 Annual Meeting. *Society Aff:* SNAME, ΣΞ.

Goddard, William A
Home: 2325 Northridge Drive, Modesto, CA 95350
Position: Plant Industrial Engr. *Employer:* Hershey Chocolate Co. *Education:* BSME/Mech. Engr./Fresno State Univ.; MBA/Business/Fullbrton State Univ *Born:* 05/23/24 I am a California Reg. PE in Industrial and Manufacturing Engrg. I am a Certified Plant Engr and a Fellow in AIPE. I am also a Certified Manufacturing Engr. I organized and was the first Pres of the AIPE chapter 155 in Modesto, CA. I have also been Pres of two IIE Chapters, *38 in Orange County & 169 in Modesto, CA. I am currently the AIPE Northwest Region IV VP/Dir for the past 2 3/4 years.* *Society Aff:* AIPE, SME, IIE

Godfrey, Albert L, Sr
Business: Godfrey Engineering Associates, P.O. Box 91, Winthrop, ME 04364
Position: Pres. *Employer:* Godfrey Engg Associates. *Education:* Certif/Traffic Engg/Northwestern Univ; BSCE/Civ Engg/Univ of ME. *Born:* 4/8/35. Native of Winthrop, ME. Recipient of New England Rd Builders Award for technical paper on Photogrammetry - 1957. Area Traffic Engr - ME State Hgwy Commission - 1958-64. Asst State Traffic Engr 1964-68. State Traffic Engr 1968- 76. Dir, Bureau of Safety, ME Dept of Transportation 1977-80. Governor's Hgwy Safety Representative, ME Dept of Public Safety 1980-87. Pres, Godfrey Engg Associates 1987-present. Chrmn, Natl Assn of Governor's Hgwy Safety Representatives 1983-85. Past Pres New England Sec ITE. Past Deputy Dist Governor - Lions Intl. Reg Civ Engr, Land Surveyor & Site Evaluator - ME. Married - 5 children. Enjoy cooking, hunting and sports. District Deputy, Maine State Council Knights of Columbus 1986-87 *Society Aff:* ITE.

Godfrey, Kenneth E
Home: 2361 Leeward Shore Dr, Virginia Beach, VA 23451
Position: Dir of Engrg & Design *Employer:* Atlantic Division, Naval Facilities Engrg Command *Education:* MS/Engrg Admin/George Washington Univ; BS/CE/VA Military Inst *Born:* 9/26/36 Native of Norfolk, VA. Infantry officer. Held positions of construction, civil, and structural engr with Army Corps of Engrs. Served as project mgr and Head of Project Management, Atlantic Division, Naval Facilities Engrg Command (LANTDIV NAVFAC). Selected as Dir of Construction, LANTDIV NAVFAC 1975 & current position Dir of Engrg & Design 1983. Received numerous NAVFAC outstanding performance awards including Navy Meritorious Civilian Service and Outstanding EEO Supvr Awards. Tidewater Chapter VA Society PE (VSPE) Engr of the Year 1977. Presented two VSPE Outstanding Service awards. Held numerous VSPE positions including Tidewater Chapter Pres and State VP. Member, Methodist Church. *Society Aff:* NSPE

Godleski, Edward S
Business: Cleveland State Univ, Dept. of Chem Engr, Cleveland, OH 44115
Position: Deputy Chair/Chem Engr *Employer:* Cleveland State Univ *Education:* PhD/ChE/OK State Univ; MS/ChE/Cornell Univ; BS/ChE/Lehigh Univ *Born:* 11/16/36 Deputy Chairman and Professor of the Department of Chemical Engineering, Fenn College of Engineering at Cleveland State University, Associate Dean of Engineering for Undergraduate Affairs, 1977-1985. Author of several research papers. Areas of interest: effective teaching, student learning styles, Myers-Briggs Type Indicator. *Society Aff:* AIChE, ASEE, WPCF, APT.

Godsey, Dwayne E
Business: 6600 So. Yale, Tulsa, OK 74177
Position: Sr. VP *Employer:* Williams Bros Engr Co *Education:* BS/IE/OK Univ *Born:* 10/27/31 Native of Tulsa, OK. Petroleum Engr for Carter Oil Co (now Exxon) 1954- 1957, Petroleum and District Engr for Tenneco Oil Co 1957-1965. Responsible for all engrg activities in Mid-Continent area. Consultant for and eventual owner of Stiles-Godsey Engrg Co, Inc 1965-1970. Purchased Earlougher Engrg, merged into Godsey-Earlougher, Inc 1970-76. Sold co to Williams Brothers Engrg Co 1976. Sr. VP in charge of all oil and gas consulting and operations 1976 to pres-

Godsey, Dwayne E (Continued)
ent. Bd of Dirs of both Society of Petroleum Engrs of AIME and Society of Petroleum Evaluation Engrs. *Society Aff:* SPE of AIME, SPEE, NSPE

Goehler, Donald D
18811 SE 42nd St, Issaquah, WA 98027
Position: Mgr - 7J7 Materials Tech *Employer:* Boeing Commerical Airplane Co. *Education:* BS/Met Engg/MT College of Mineral Sci & Tech. *Born:* 4/11/30. Starting at age twelve at Father's foundry in Portland, OR; through mines, mills & smelters while at sch in MT; including current responsibilities as Mgr-7J7 Materials Tech to Boeing Airplanes; have maintined "hands on" approach to applicatin of Mtls Sci to engg activitie. Successfully completed mgt of $2.5 million R&D contract with USAF titled "Cast Aluminum Structures Technology–. Authored eleven (11) articles published in range of technical journals & have lectured in eight states & Canada. Currently serving on Nominating Ctte-ASM Intl. Hobbies: wine making, gardening & sailing. *Society Aff:* ASM.

Goel, Amrit L
Business: 327 Link Hall, ECE Dept. Syracuse Univ, Syracuse, NY 13244-1240
Position: Prof. *Employer:* Syracuse Univ. *Education:* PhD/Mech Engg/Univ of WI, Madison; MS/Mech Engg/Univ of WI, Madison; BEngr/Mech Engg/Univ of Roorkee, India; BS/Math/Agra Univ India. *Born:* 3/4/38. A native of India, Has been residing in the USA since 1962. Was a Mech Engg with the Atomic Energy Commission of India before starting grad work at the Univ of Wi, Madison in 1962. Currently pursuing research and doing teaching in modelling of computer systems including software reliability, exact methods for reliability assessment using sequential tests, and improved methods for quality control. Has published extensively in the tech publications, and is the recepient of the 1979 and 1980 P K McElroy Best Technical Paper Awards in reliability and Maintainability (sponsored be IEEE ASQC ASME, etc) Organizer and Chairman of Minnowbrook Workshop on Software performance assessment 1978, 1979, 1980, 1981. Active in professional societies-nationally and internationally. *Society Aff:* ASQC, IEEE, IIE, ACM, RSS, ORSA.

Goepfert, Detlef C
Business: 4520 SW Water Ave, Portland, OR 97201
Position: VP *Employer:* Robert W Zanders & Assocs, Inc *Education:* Diploma/Tech/Borsig Berufsschule (Trade school) Berlin, and Wiesbaden Volkshochschule (Community Coll)/Technology *Born:* 06/23/27 Born 1927 in Berlin, Germany, educated at Berlin's Borsig Berufsschule and Wiesbaden Volkshochschule. Since 1957 building environ sys design and engrg for educational, res, commercial, indust, institutional and govt facilities, energy conservation studies, proj mgmt; VP, RW Zanders & Assocs, Portland, OR. Past pres Engrs Coordinating Coun of OR; Engrg Tech Advisory Ctte, Portland Community Coll. Dir ASHRAE and Regional Chrmn of chapters in OR, WA, AK, ID, and four Western Provinces of Canada. Mbr ASPE, NSPE, AAAS, USMA, and Deutscher Kaelte und Klimatechnischer Verein (Germany). ASHRAE Distinguished Service Award and Regional Awards of Merit; Portland Convention Award. *Society Aff:* ASHRAE, NSPE, AAAS, DKV, USMA, ASPE

Goering, Carroll E
Business: 360 Agri Engrg Sci Bldg, 1304 W Pennslyvania Ave, Urbana, IL 61801
Position: Prof. *Employer:* Univ of IL. *Education:* PhD/Agri Engr-Engr Mech/IA State Univ; MS/Agri Engr/IA State Univ; BS/Agri Engr/Univ of NB. *Born:* 6/8/34. Born and educated in NB. Following grad from the Univ of NB, he did advance implement design for Intl Harvester Co. After earning a PhD degree from IA State Univ, he joined the Univ of MO where he taught courses in power and machinery design and conducted res on pesticide application. In 1970, he received a faculty-alumni award for distinguished service. In 1977, he joined the Univ of IL where he teaches courses on farm tractors and does res on alternative fuels. He has published 35 articles in scientific journals and authored one textbook, *Engine and Tractor* Power. *Society Aff:* ASAE, SAE, ASTM, ASEE.

Goering, Gordon D
Business: 17 Phillips Bldg, Bartlesville, OK 74004
Position: Sr VPres Petroleum Prod Grp *Employer:* Phillips Petroleum Company *Education:* BS/Chem Engr/KS St Univ *Born:* 5/3/22 Native of Pretty Prairie, KS. Graduated from KS St Univ with a degree in Chemical Engineering. Joined Phillips Petroleum Company in 1946 as an engineer at the company's KS City refinery. Held various technical and supervisory positions in Phillips domestic refineries and starting in 1966 held a series of overseas managerial positions. Was elected vice president in charge of refining operations in 1978, and in 1980 was elected senior vice president of the Petroleum Products Group, consisting of petroleum refining, crude supply, NGL, transportation and marketing. *Society Aff:* NSPE, API, NPRA, OSPE

Goering, William A
Business: Central Lab, 15000 Century Dr, Dearborn, MI 48121
Position: Manager-Central Lab. *Employer:* Ford Motor Company. *Education:* ME/Metallurgy/Yale Univ; BE/Metallurgy/Yale Univ. *Born:* 7/16/32. Ford Motor Co 1956- . Sci Res Staff 1956-71; Product Testing Lab 1971-72; Supr Matls Engrg 1973-76. Vehicle Matls Engrg Mgr 1977-78, Central Lab Mgr 1979- present. Taught co courses in Analysis of Failures. Present position Mgr of Central Tes ting Lab. Mbr ASM (Education Ctte), SAE (Member-Ferrous Cttes). Executive VP Amateur Fencers League of Amer. *Society Aff:* SAE, ASM, ACS.

Goerner, Joseph K
Home: 1532 S Gessner, Houston, TX 77063
Position: VP Marketing *Employer:* Texaco Chemical Co *Education:* BS/ChE/Rice Univ. *Born:* March 1925. Native of Houston, Texas. Served in Navy 1945-46. Began with Jefferson Chemical in 1946 as research chemical engineer. Subsequently worked in process design, economic evaluations, plant start-up and technical service, marketing and sales. Named Genl Mgr R&D in 1967 and V Pres Res & Planning in 1974. When Texaco acquired Jefferson in Dec 1974, became Mgr, Res and Dev Div, Petrochemical Dept for Texaco. When Texaco Chemical Co. was formed in April 1980, became Vice President, Res and Dev. In July 1982 became VP Marketing. Major hobby is sailing. *Society Aff:* ACS, AIChE.

Goethert, Bernhard H
Business: Univ TN Space Inst, Tullahoma, TN 37388
Position: Dean Emer; Prof Aerospace Engrg. *Employer:* Univ of Tenn Space Inst. *Education:* PhD/Aerospace Eng/Tech Univ of Berlin; MSAE/Aerospace Eng/Tech Univ of Danzig; BSME/Mech Eng/Tech Univ of Hannover *Born:* Oct 1907 Germany; naturalized 1954. PhD Berlin 1938; Aeronautical Res Inst Berlin, high speed aerodynamics 1934-45; US Air Force Centers Wright Field & Tullahoma Tenn: Chief, aerodynamic & propulsion; Dir Engrg, Vice Pres Res, Chief Scientist 1945-64; Chief Scientist US Air Force Systems Command 1964-66. Space Inst Univ of Tenn Tullahoma: Prof Aerospace Engrg, Dean, Dean Emeritus 1964- ; Prof Tech Univ Aachen Germany 1961- . US Air Force Scroll of Appreciation 1959; AF Award for Meritorious Service 1966; Plaque of Honor Tech Univ Aachen 1972. Fellow AIAA 1958; Hon Mbr German Assn for Aeronautics & Astronautics 1975; 1st recipient AIAA Award 'Simulation & Ground Testing' 1976; Humboldt Award by Humboldt-Fnd on behalf of German Government, 1977. June 1978: Plaque of Tribute from Aerospace Indus for leadership in founding the Univ of TN Space Inst in 1964 and its guidance as its first Dir and Dean. 1982: Akademischer Ehrenburger (Academic Hon Citizen) der Tech Univ, Aachen, West Germany. 1982: "B. H. Goethert Professorship" at the Univ of TN Space Inst, Tullahoma, TN. 1982: "The B. H. Goethert Graduate Study Schship Awd" at the Univ of TN Space Inst, Tullahoma, TN. Award by AIAA for outstanding contribution to airbreathing propulsion education, 9 April 1987. (American Institute of Aeronautics and Astronautics)

Goethert, Bernhard H (Continued)
Ludwig Prandtl Ring award Bonn Godesberg, Federal Republic of Germany 1984. *Society Aff:* AIAA.

Goetschel, Daniel B
1514 Steinhart Ave, Redondo Beach, CA 90278 *Education:* PhD/Structures/UCLA; MS/CE/CA State-Northridge; BS/Ocean/UCLA *Born:* 07/28/48 Mech Engr for Electrodynamics Div of Bexdix Corp. Working anti-submarine warfare in 1972 and 1973. AISC Fellow in 1975-76 while at CA State Northridge. Worked on computer analysis of atomic bomb cratering for CA Res and Tech Inc as a res engr in 1980. Joined the Mech Engrg Dept of R.P.I. in 1981 as an asst prof. Working in the composite materials program there. Specialize in structural finite element analysis. Enjoy swimming and micro- computers. *Society Aff:* ΣΞ

Goetz, John L
Home: 482 Woodward Blvd, Pasadena, CA 91107
Position: Sr Vice Pres. *Employer:* Southwest Portland Cement Co. *Born:* August 1923 Wapakoneta Ohio. BS Chemistry St Joseph's (Indiana). Grad School of Engrg Harvard Univ 1943 (US Army). Served Dec 1942Oct 1945. Employed by Southwestern Portland Cement Co 1948. Chemist, Chem Engr, Dir of Labs, Chemical Dir, Dir of Manufacturing, V P & Mgr of California Div, Vice Pres of Manufacturing Operations. Now Sr Vice Pres responsible for R & D, quality control, environmental matters, fuel, energy. Mbr of Amer Concrete Inst (Vice Pres), ASTM Ctte C-1, Portland Cement Assn, Genl Technical Ctte. Enjoy classical music, walking, water sports.

Goetzel, Claus G
Home: 250 Cervantes Rd, Portola Valley, CA 94025
Position: Consultant (Self-employed) *Education:* Doctorate/Met/Columbia Univ; ME/Mat. Testg./Technical Univ. Berlin, Germany *Born:* 7/14/13. Consulting Scientist. Has been Sr mbr of Res Lab with Lockheed 1960-1978. Associated with Mtls Sci Dept Stanford Univ since 1961. Lecturer: High temperature mtls and Powder Met. Recipient 1978-79 of Alexander von Humboldt Award by Fed Republic of Germany. Guest Prof: Technical Univ of Karlsruhe, Germany 1978-1979. Holder of over 30 patents. Author of text on Powder Met, 5 volumes. Numerous publications, domestic and foreign. *Society Aff:* AIAA, ASM, AIME.

Goff, John A
Home: 623 Righters Mill Rd, Narberth, PA 19072
Position: *. *Employer:* Retired. *Education:* PhD/ME-Math/Univ of IL; MS/ME/Univ of IL; BS/ME/Univ of IL. *Born:* Oct 1899. Asst in Mech Engrg to Prof of Thermodynamics; Univ of Pennsylvania: Dean of Engrg & Dir of ME, Prof of Thermodynamics, Prof Emer; Natl Defense Res Ctte Official Investigator Central Engrg Lab; Dir Univ of Pennsylvania Thermodynamic Res Lab; Chairman Faculty Senate 1956, Founding Pres Faculty Club. Societies: ASME Chairman Applied Mechanics Div. Chairman Standing Cttes on Prof Divs, Chmn Heat Transfer Div Ctte on Thermo-Physical Properties, Chairman Special Res Ctte on Properties of Gases & Gas Mixtures, Chairman Bd on Honors. AMS (Amer Mathematical Society) Mbr Bd of Collaborators Applied Mathematics Quarterly. ASHVE-ASHRAE Mbr Res Exec Ctte, of Technical Advisory Ctte on Instruments & Sorbents, Publications Ctte, & produced definitive values thermodynamic properties of moist air adopted as intl std by the Intl Meteorological org (IMO) 1947. Mbr Subcommission on Physical Constants & Functions, Aerological Commission IMO. *Society Aff:* ASME.

Goff, Kenneth W
Home: 2143 Horace Ave, Abington, PA 19001
Position: VP Advanced Technology. *Employer:* Performance Controls, Inc. *Education:* ScD/EE/MIT; MS/EE/MIT; BS/EE/WV Univ. *Born:* 6/14/28. Native of Salem, WV. Consulting Engr, Bolt, Beranek & Newman, Inc, Cambridge, MA 1954-56; Proj Mgr, Gruen Precision Labs, Cincinnati, OH 1956-57. Mgr, Systems Analysis, Leeds & Northup Co, North Wales, PA 1957-69, responsible for the application of hybrid simulation, data analysis techniques & digital control algorithm design concepts to the application of process control computers. Mgr, Systems Dev, Leeds & Northrup 1969-85. Responsible for design & dev of hardware & software for digital comp based systems for process control applications. VP Advanced Technology, Performance Controls, Inc. 1985-present. Development of motion control equipment, particularly brushless motor drives and microprocessor-based controllers. Elected Fellow ISA 1973 & Fellow IEEE in 1979. Enjoy tennis, boating & fishing. *Society Aff:* IEEE, ISA.

Gogan, Harry L
Home: 2913 Charleston St NE, Albuquerque, NM 87110
Position: Technical Dir, Special Systems *Employer:* BDM Corp *Education:* MPA/Sys Analysis/Harvard Univ; BS/Aero Engg/Univ of AL. *Born:* April 1922. Graduate study at Univ of Virginia, Univ of New Mexico & Federal Exec Inst. Native of Buffalo, NY. Fighter pilot in USAAF in World War II. Aeronautical Res Scientist with NACA 1948-51. With Air Force Spec Weapons Ctr 1951-76, where as Technical Dir responsible for flight testing, air-launched missile testing & other programs. 1976-80 Technical Dir, Directorate of Aerospace Studies, Air Force Systems Command, conducting System Analyses. Former State Pres Air Force Assn, Assoc Fellow in Amer Inst of Aeronautics & Astronautics, Tau Beta Pi. Theta Tau, Phi Eta Sigma. Registered Prof Engg in NM. New Techical Dir, Special Systems, The BOM Corp; Conducting Systems analysis operational test and evaluation; test range engrg. *Society Aff:* AIAA, AFA, TBΠ, ΘT.

Gogarty, W Barney
Business: 7400 South Broadway, Littleton, CO 80122
Position: Assoc Research Dir *Employer:* Marathon Oil Co *Education:* PhD/ChE/Univ of UT; BS/ChE/Univ of UT *Born:* 4/23/30 Provo, UT. Married with five children. Started with Marathon Oil Co at their Denver Research Center in 1959. Presently Assoc Research Dir for Production Research. Work primarily in areas of research and management related to field of new oil recovery techniques. Is one of inventors of Marathon's Maraflood TM oil recovery process. Holds 57 US patents and many foreign patents and has published 33 technical papers. *Society Aff:* AIChE, SPE OF AIME, ΣΞ, TBΦ

Gogate, Anand B
6112 Sedgwick Rd, Worthington, OH 43085
Position: Pres. & Assoc. Prof. *Employer:* Anand Gogate Engineers & Ohio State Univ. *Education:* PhD/Structures/OH State Univ; MS/Structures/Univ of IA; BE/Civil/Poona Univ. *Born:* Jan 1935 Rangoon, Burma. Native of Indore, India. US Citizen 1970. Structural Engr with Peterson & Appell, DesMoines, Iowa 1963-65. Design Engr PDM DesMoines, Iowa 1966-67. Ch Structural Engr AE Stilson & Assoc 1967-72. Ch Structural Engr Elgar Brown Cons Engrs 1972-85. President Anand Gogate Engineers 1985 to present. Assoc Prof Dept of Architecture, Ohio State Univ. 1981-84. Awards: OSPE Award for obtaining highest grade in the PE Exam 1968 in the State of Ohio. ASCE State of the Art Award 1975. ASCE Raymond C. Reese Award 1976. Student Chapter of A.I.A. Outstanding Faculty Award 1983-84. Mbr of Joint ACI, Inst. of Engrs. Joint ACI-ASCE Ctte 426 Shear & Diagonal Tension Fellow, Institution of Engineers, India. ACI Ctte 350 Sanitary Engrg Structures. Joint ACI-ASCE ctte 445 Shear and Torsion ACI ctte 435 Deflections, Maharishi Award Winter 1976, Co- author of the book titled *Elements of Reinforced Concrete.* Interested in Vedic studies and higher states of conciousness. *Society Aff:* ASCE, PCI, ACI, Inst of Engrs (India)

Goglia, Mario J
Home: 3066 Arden Rd NW, Atlanta, GA 30305
Position: Engrg Conslt *Employer:* Self employed *Education:* PhD/Mech Eng/Purdue Univ; MS/Mech Eng/Stevens Inst of Tech; ME/Mech Eng/Stevens Inst of Tech. *Born:* March 1916. Taught at Univ of Ill, Purdue Univ, Notre Dame (Dean of Engrg), Georgia Tech (Regents Prof of ME & Dean Grad Div) - with Bd of Regents Univ Syst Ga as Vice Chancellor for Res. Retired December 1981 - presently Engi-

Goglia, Mario J (Continued)
neering Consultant. Cons to industry, institutions of higher education, government. Served on governing bds of natl prof & educ societies. Reg Prof Engr. WWII Naval Ordnance award. Author textbooks, contributor to prof & educational journals. Fellow AAAS. *Society Aff:* ASME, ASEE, AAAS.

Goguen, Joseph A
Business: SRI International, 333 Ravenswood Ave, Computer Science Lab, Menlo Park, CA 94025
Position: Sr Staff Scientist. *Employer:* SRI. *Education:* PhD/Math/UC, Berkeley; MA/Math/UC, Berkeley; BA/Math/Harvard *Born:* 6/24/41. Semantics of programming; software engg; Massively parallel computer architectures; declarative and multi-paradigm languages; specification and abstract data types; social systems and linguistic methods; fuzzy sets. Prof Comp Sci, UCLA, 1974-81; Sr Staff Scientist, Stanford Re Inst; Managing Dir, Structural Semantics. Principal mbr of the Ctr for the Study of Language & Information, Stanford. Author of over 90 papers, on software engineering, computer architecture, program semantics, artificial intelligence, linguistics. *Society Aff:* ACM, AMS, MAM, IEEE

Going, E Jackson, Jr
Business: 820 Morse St, San Jose, CA 95126
Position: Pres *Employer:* E. Jackson Going Jr., Inc. *Education:* BCE/Civil Engg/Univ of Santa Clara. *Born:* 2/22/28. Native Houston, TX. Worked as Engg Aide & Jr Engr, Airport Soc, Public Works Dept San Jose CA, to Sept 1949. Employed as office mgn for consulting engg firm James & Wates & Assocs. In 1952 became partner of same firm under the new name of Waters, Ruth & Going. In 1956 acquired firm, currently under the name of Ruth and Going, Inc. Architects, Engrs, Planners 1975. Until July 1983 served as Chrmn of the Bd-Treas. At that time sold interest in Ruth and Going, Inc. and incorporated. Presently retained as conslt for business development at Ruth and Going, Inc. as well as conslt to Univ of Santa Clara. Served as Pres CEAC 1975. *Society Aff:* NSPE, ACEC, ASCE, APWA

Goland, Martin
Business: P O Drawer 28510, San Antonio, TX 78284
Position: President. *Employer:* Southwest Res Inst. *Education:* LLD/Honoris Causa/St Mary's Univ; Prof of Research/Honoris Causa/St Mary's Univ; ME/summa cum laude/Cornell Univ. *Born:* 7/12/19. in NY. Instructor Cornell 1940-42; Hd Applied Mechs Curtiss-Wright, 1942- 46; Midwest Res Inst 1946, Dir Engrg Sci 1950-55; Southwest Res Inst 1955, pres 1959- ; Pres Southwest Fdn for Res & Education 1972-1982. Awards: ASME Spirit St Louis Jr Award 1944; ASME Jr Award 1946; ASCE Alfred Nobel Prize 1946. Fellow, P Pres & Dir AIAA; Fellow AAAS; Hon Mbr & Past VP ASME; Mbr natl Acad of Engrg; Mbr NASA Aeronautics Adv Ctte & Chmn Subctte on Matls, Structures & Structural Dynamics Tech; Mbr NRC Ctte on Transportation Panel on Innovation in Transportation; Trustee, Univs Research Assoc, Inc; VChmn & Dir, SMU Fdn for Sci and Engg. *Society Aff:* ASME, AIAA, AAAS, NAE, ΣΞ.

Golay, Marcel J E
Home: Creux de Corsy 94, CH-1093 La Conversion, Switzerland
Position: Sr Scientist. *Employer:* Perkin-Elmer Corp-Norwalk Ct. *Education:* BSc/Gymnase de Newchatel; Lic Elec Engr/EPFZ; PhD/Physics/Univ of Chicago *Born:* May 1902 Neuchatel Switz. Bacc Sc Gymnase de Neuchatel 1920. Lic Elec Engr EPFZ 1924. PhD in physics Univ of Chicago 1931. BTL 1924-28. Signal Corps Labs 1931-55. Cons to Perkin-Elmer Corp & Philco Corp 1955-61. Inauguration of Chair of Sci of Analogies Technological Univ of Eindhoven 1961-63. Sr Scientist of Perkin-Elmer Corp 1963- . Over 100 publications in the fields of: Physics, Chemistry, Communication Engrg, Naval Engrg, Math & Philosophy. Over 50 U S & foreign pats. Fellow OSA & IEEE. Harry Diamond Award of IEEE (then IRE) 1951. Sargent Award of ACS 1961. Disting Achieve Award ISA 1961. Jimmie Hamilton Award ASNE 1971. Tswett Award 1976, Doct. h.c. EPFL, 1977. Tswett Mem. Med. All-Union Chromat. Coun of the Acad Sc. USSR. Supelco Award of ACS 1981. S.Dal Nogare Awd of the Chrom. Forum Del Val. 1982. *Society Aff:* OSA, IEEE

Gold, Albert
Business: 333 Jay St, Brooklyn, NY 11201
Position: Provost. *Employer:* Polytechnic Inst of NY. *Education:* PhD/Physics/Univ of Rochester; BS/Engg Physics/Lehigh Univ. *Born:* 7/2/35. Native of Phila, PA. Mbr, State Legislative Comm, Commission on Independent Colleges and Univs; Mbr, School Governing Bd, NY Province, Religious of the Sacred Heart, 1971 - 1978; Trustee, Cheshire Academy, 1972-1974; Cons in tech and admin areas; Dir, Scientific Calculations, Inc (Rochester, NY) 1969-74. Director, University Patents, Inc. 1979- present; Held various admin posts at Rockefeller Univ including VP 1973-78, Special Asst to the Pres and Dir of Postdoctoral Affairs 1969-73. Served as Assoc Dean of Engg and Applied Sci, Univ of Rochester 1966-68. Served on faculty of Univ of Rochester, Inst of Optics 1962-69. Assumed present position 1978. *Society Aff:* ASEE, ΣΞ, ΤΒΠ, ΦΒΚ, CICU.

Gold, Bernard
Business: Wood St - P O Box 73, Lexington, MA 02173
Position: Group Leader-Digital Processors. *Employer:* Lincoln Lab MIT. *Born:* April 1923. PhD from BPI 1948; BEE from CCNY 1944. Native of New York City. Staff Mbr Hughes Aircraft Co 1950-53, specializing in radar signal processing. Staff Mbr MIT Lincoln Lab 1953-72, specializing in pattern recognition, noise theory, speech analysis & synthesis & digital signal processing. Group Leader, Lincoln Lab 1972 to present, specializing in speech communications. Fulbright Fellow Italy 1955-56. Visiting Prof of Elec Engrg MIT 1965-66. Fellow Acoustical Society of America. Fellow IEEE.

Gold, S H
Business: P O Box 800, Rosemead, CA 91770
Position: Ch of Apparatus Engr. *Employer:* So Calif Edison Co. *Born:* Sept 1923. Native Tucson Arizona. BSEE Univ of New Mexico, Grad Courses USC, UCLA, UCR, U of Idaho, Stanford & Westinghouse Sch in Elec Util Engrg. Calif Elec Power Co 1951-63 Asst Ch Engr. Southern Calif Edison Co 1963- , Ch Apparatus Engr. Professional Activities: CIGRE-USNC Tech Ctte & Study Ctte 13; EEI Past Chairman Elec System & Equip Ctte; Chairman Task Force on D-C Breaker R&D; IEEE Fellow Grade PES, Admin Ctte & Council Chairman, Awards & Recog Dept Past Chairman Royal Sorenson Fellows; EIC; LA Elec Club (Scholarship Ctte). Honors: Sigma Tau, Kappa Mu Epsilon, IEEE Council Recognition Award, Fellow IAE. Reg Elec Engr. Enjoy fishing, horseback riding & classical music.

Goldberg, Alfred
Business: P O Box 808. L342, Livermore, CA 94550
Position: Sr Scientist. *Employer:* UC Lawrence Livermore National Lab. *Education:* PhD/Metallurgy/Univ of Berkeley, CA; MS/Metallurgy/Carnegie-Mellon Univ; BEng/Metallurgy/McGill Univ. *Born:* Oct 1923. Res Engr at UCB prior to PhD. 11 years on faculty at US Naval Postgrad Sch. Fulbright Prof Engrg Univ, Lima Peru. Internatl Atomic Energy Agency Prof at Inst of Physics, S C deBariloche Argentina. U C Lawrence Livermore Lab since 1964. Past res activities include mechanical properties, phase transformations, calorimetry & corrosion. Presently program mgr on steel res. Chairman ASM Santa Clara Valley Chapter 1969-70. Interested in interactions with Latin Americans: Chairman Livermore Sister City Org 1972-73. Chmn ASM/LARC 1973-74. Mbr Exec Ctte Inter-American Conferences on Materials Technology. Enjoy music, dancing, traveling. *Society Aff:* ASM Intl.

Goldberg, David C
Business: PO Box 10864, Pittsburgh, PA 15236
Position: Mgr, Adv Prog. *Employer:* Westinghouse Elec Corp. *Education:* BS/CE/Antioch College. *Born:* 6/27/21. Pioneered application of titanium and cast high temperature alloys for aviation gasturbines in early 50's. Led the dev of columbium, tantaium, and other refractory alloys for space systems. As engg mgr admin

Goldberg, David C (Continued)
the transition from a space-nuclear oriented activity to a div responsible for the dev of alternate energy systems-solar thermal, photovoltaics, wind turbine generators, hydrogen, MHD, nuclear waste mgt. Currently responsible for div strategic planning, product dvpt, and marketing. Served as chrmn and mbr of numerous natl mtls advisory bd. ASM, and other soc committees. *Society Aff:* ASM, AIAA, ASTM, ΣΞ.

Goldberg, Harold
Home: 311 S Hollybrook Dr 303, Pembroke Pines, FL 33025
Position: Retired. *Education:* PhD/EE/Univ of WI; MS/EE/Univ of WI; BS/EE/Univ of WI; PhD/Physiology/Univ of WI. *Born:* 1/31/14. VP Engineering, Rockwood Systmes Corp., 1979-84. Chief WPNS Engr US Naval Sea Sys Command 1974-79. Exec VP, Dir Riker-Maxson Corp 1969-73. VP, Dir, Genl Mgr E Sys Inc 1964-69. VP Engg & Res Raytheon Co 1963-64. Exec VP Emertron Inc, VP Emerson Radio 1954-63. Ch, Ordnance Electronics Div Natl Bureau of Stds 1948-54. Prin Res Engr Bendix Radio Div 1945-48. Sr Res Engr Stromberg-Carlson Co 1941-45. Univ of WI faculty positions 1935-41. Dir First Lexington Corp., 1969-86. Exceptional Service Award US Dept of Commerce. Distinguished Service Citation, Univ of WI. Author - "Principles of Guided Missile Design–, D. Van Nostrand, 1956. Author - "Extending the Limits of Reliability Theory–, John Wiley, 1981. Inventor - over 35 patents. Fellow, IEEE. Member, American Physical Society. Exceptional Service Award US Dept of Commerce. Distinguished Service Citation, Univ of WI. V.P. Engineering, Rockwood Systems Corp., 1979-84. *Society Aff:* IEEE, APS.

Goldberg, Harold S
Business: 66 Cherry Hill Dr, Beverly, MA 01915
Position: VP *Employer:* Aerosystems Corp *Education:* M/EE/Polytechnic Inst of Brooklyn; B/EE/Cooper Union *Born:* 1/22/25 An Electronics Engr who, for the past 30 years has been involved in instrumentation, first as Design Engr, then Chief Engr, mgr and recently founder and leader of what has become the third largest DMM manufacturer in the world, DATA PRECISION CORP. Merged in 1978 with ANALOGIC. He led DATA PRECISION as a Division, into other more exotic instrumentation fields. These include digital computing oscilloscopes and automatic test equipment. He presently, in addition to remaining as VP of Analogic, has joined Aerosystems as its Pres. Aerosystems designs and manufactures data acquisition and control systems for use with personal computers. He is also an avid writer of technical articles and for years has been an active contributor and officer of IEEE. *Society Aff:* IEEE, NSPG, ISA

Goldberg, John E
Home: 1805 Western Dr, W Lafayette, IN 47906
Position: Prof Dr. Emeritus. *Employer:* Purdue Univ. *Education:* BSCE/Civil Engrg/NW Univ; MSCE/Civil Engrg/NW Univ; PhD/Engrg Mech/Ill Inst of Tech *Born:* Sept 1909, Seattle, Wash. Corps of Engrs & City of Chicago 1932-42. Structures Engr WACO Aircraft & Convair 1942-47. Res Asst Prof of Mechanics & Dir of Mech Res IIT, 1947-50. Asst Editor 'Applied Mechanics Review' 1947-50. Assoc Prof Purdue Univ 1950-55, Prof & Academic Head of Structural Engrg 1955- 75. Visiting Prof Univ of Ill Chgo Circle 1975-76, Georgia Tech 1976-77. Ruhr Univ 1977-78, Southern Methodist Univ, 1978-79, TX A&M Univ 1979-80. Prog. Mgr Natl Sci Fdn 1980-83. Mbrship Research Ctte of Natl Acad/Natl Res Council 1963-77, & Chmn 1973-77. Mbr Exec Ctte Engrg Mech Div ASCE 1964-68 & Chmn 1967-68. Mbr Sigma Xi, Tau Beta Pi, Chi Epsilon, IABSE. Fellow ASCE. First recipient of the Newmark Medal of ASCE. US Sr-Scientist Award von Humboldt Fnd. Tech ints: structural mechanics, continuum mechanics, numerical methods, nuclear structures, tall bldgs, shells, ships, submersibles, vibrations, stability, forensic engineering. *Society Aff:* ASCE, ΣΞ, XE, ΤΒΠ, IABSE, SSRC

Goldberger, Marvin L
Business: 1201 E California Blvd, Pasadena, CA 91125
Position: Pres. *Employer:* CA Inst of Tech. *Education:* ScD/Hon Degree/Carnegie-Mellon Univ; ScD/Hon Degree/Univ of Notre Dame; BS/-/Carnegie Inst of Tech; PhD/Physics/Univ of Chicago. *Born:* 10/22/22. in Chicago, IL in 1922. Married, 2 children. Acted as advisor to various govt agencies on natl sec affairs. He was one of the founders (in 1959) of the Jason Group, originally assoc with the Inst for Defense Analyses and now with SRI Interntl. This group of about 35, composed mainly of physicists, works for the Dept of Defense and other agencies on both classified and unclassified problems involving advanced technological concepts. Mbr of President's Science Advisory Com from 1965-69. Chmn of Strategic Military Com Chmn of Federation of Amer Scientists, 1972-73. Was a Mbr of Mitre Corp study *Nuclear Power - Issues and Choices*; 1963-69, Chmn of High Energy Physics Commission of Internat'l Union of Pure & Applied Physics (IUPAP) and US Rep to IUPAP. May 1972, headed scientific delegation to the People's Republic of China which arranged for the first visit to US of a group of Chinese scientists. *Society Aff:* NAS, AAAS.

Golden, Robert G
Home: 330 Rahway Rd, Edison, NJ 08820
Position: Assoc Prof *Employer:* NJ Inst of Tech. *Education:* MDiv/Theology/New Brunswick Theological Seminary; AM/Guidance & Personnel/Seton Hall Univ; BSME/ME/Newark College of Engg, N.J.I.T.; AB/Philosophy/Little Rock College College. *Born:* in Big Heart, OK. Served an apprenticeship as a toolroom machinist at the Singer Mfg Co in Elizabethport, NJ. Attended night sch in engr and worked as an engr for Babcock & Wilcox, the Turner Construction Co, NJ Dept of Transportation and Wright-Patterson AFB. Earned a Master of Divinity Degree at the New Brunswick Theological Seminary. Faculty Fellow in both Design & Research at Geo Marshall Space Flight Ctr, Huntsville, ALA. and a visiting Fellow in Sch of Engr, Princeton Univ. Taught since 1959 at NJ Inst of Tech, Department of Mechanical Engineering. Reg PE in NJ. *Society Aff:* ASME, ΠΤΣ, CBA, SAR.

Goldey, James M
Business: 2525 N. 12th Street, Reading, PA 19612
Position: Director. *Employer:* AT&T Bell Labs. *Education:* PhD/Physics/MIT; BS/Physics/Univ of DE *Born:* Married 1951; 2 children. Home address 3930 Azalea Rd, Allentown Pa. BS Delaware 1950; indus fellow & PhD physics MIT 1955. IEEE Fellow Jan 1969; Dir, Integrated Circuit Customer Service Lab, 1981. Currently director, linear and high voltage integrated circuit laboratory AT&T Bell Laboratories. *Society Aff:* OME DEL KAP, ΦΚΦ, ΠΜΕ, IEEE

Goldman, Alan J
Business: 34 & Charles Sts, Baltimore, MD 21218
Position: Prof of Math Sci *Employer:* The Johns Hopkins Univ *Education:* PhD/Math/Princeton Univ; MA/Math/Princeton Univ; BA/Math Physics/Brooklyn Coll *Born:* 3/2/32 Following 1956 PhD (topology), joined National Bureau of Standards, becoming Deputy Chief of Applied Math Division and Chief of Operations Research Division. Directed applied & basic research projects involving distribution, transportation, information, military, and public-sector service systems. WA Academy of Sciences Award for Achievement in Math (1971), Commerce Dept Gold Medal (1976). Assumed current Johns Hopkins position in 1979, research interests emphasize optimization (esp of facility location), game theory, graph theory, ecological models. Assoc editor, Transportation Science, Naval Res Logistics Quarterly, past council member, Operations Research Society. *Society Aff:* ORSA

Goldman, Allan E
Business: PO Box 6016, Cleveland, OH 44101
Position: Applications Manager. *Employer:* Union Carbide - Carbon Prod Div. *Education:* BS/MET/UTEP. *Born:* 12/17/26. Native of Olean, NY. Served with US Army 1944-46. Metallurgis with US Steel 1951-53. Cornell Aeronautical Lab 1954-55. Proj Metallurgist, Oak Ridge National Lab 1955-68 with foreign assignment to OECD Dragon Proj 1964-66. With Carbon Prod Div since 1968; responsible for marketing specialty graphites on Domestic and International Basis. Mbr US/UK

Goldman, Allan E (Continued)
Info Exchange Panel on Gas Cooled Reactors. Mbr ASM, ASTM, ANS. *Society Aff:* ASTM, ASLE, ANS, ASM.

Goldman, Jay
Business: School of Engineering, University of Alabama at Birmingham, University Station, Birmingham, AL 35294
Position: Dean *Employer:* Sch of Engrg, Univ of AL at Birmingham *Education:* DSc/Indust Engrg/WA Univ; MS/Mech Engrg/MI State Univ; BS/Mech Engrg/Duke Univ *Born:* 04/15/30 Dr Jay Goldman is currently Dean and Prof, Sch of Engrg, Univ of AL at Birmingham. He was formerly Prof & Chrmn of Indust Engrg at the Univ of MO, Columbia and prior to this was at NC State Univ and WA Univ. He has authored or co- authored over 60 tech pubs, made numerous presentations to business and professional organizations, conslt with local and natl organizations as well as the Fed Govt and has served on a number of Fed, State and Univ natl advisory panels and study sections. He served as Exec VP for the Inst of Indust Engrs, and currently serves as mmbr, Accred Bd for Engrg & Tech. *Society Aff:* IIE, ASEE, NSPE, ASPE

Goldman, Kenneth M
Home: 2223 Shady Ave, Pittsburgh, PA 15217
Position: Advisory Scientist, Core Materials *Employer:* Bettis Lab, Westinghouse. *Education:* DSc/Met Eng/Carnegie Mellon Univ; BS/Met Eng/Carnegie Mellon Univ. *Born:* 1922. Native of Pittsburgh Pa. National Tube Co, student engr. US Army, Manhattan Proj Los Alamos, New Mexico, Jr Scientist. Metals Res Lab Carnegie-Mellon Univ, Res Asst. Positions held at Bettis Atomic Power Lab: Sr Engr Materials & Metallurgy, Supervisory Engr Corrosion & Fuel Element Dev, Mgr Irradiation Test & Fabrication Dev. Fields of interest: physical chemistry of steelmaking, radioactive tracers applied to metallurgy, metallurgy of zirconium & uranium. Co-inventor of the Zircaloys. Hunt Award AIME 1952, Engrg Materials Achievement Award ASM 1972, Fellow ASM 1976. Presently Advisory Scientist, Core Materials, Bettis Lab, Westinghouse Electric Corporation. *Society Aff:* ASM, AIME.

Goldman, Manuel
Home: 1342 Winhurst Dr, Akron, OH 44313
Position: Retired. *Employer:* Goodyear Tire & Rubber Co. *Education:* MS/Phys Chem/Tufts Univ; BS/Phys Chem/Boston Univ; ASTP Dipl/EE/OH State Univ. *Born:* 9/22/23. Native of Revere, MA. Served in US Army Signal corps 1943-46. Res Asst Harvard Univ 1949-50. Physical Met Branch Watertown Arsenal Lab 1950-52. Ferrous Met Dept Battelle Meml Inst 1952-53. Past Chrmn Akron Dist Chapter ASM, Former special instr the Univ of Akron. Taught courses on "Heat Treatment of Metals–, "Inspection of Metals–, and "Selection of Metals in Design–. Presently retired from the Goodyear Tire & Rubber Co. Was responsible for metals selection & application in radar, aerospace, anti-submarine warfare, missile components, and off-hgwy wheels. *Society Aff:* ASM.

Goldmann, Theodore B
Business: PO Box 11581, Fort Worth, TX 76109
Position: VP. *Employer:* Southwest Silicone Co. *Education:* BS/Chem Engr/TX A&M Univ. *Born:* 10/5/23. Born in Dallas, TX. TX A&M Univ 1940-42 and 1946-48. Served in US Army 1943-45, including ASTP at Purdue Univ and 3 European campaigns as paratrooper in 101st Airborne. With Dow Corning Corp 1948-1973 in a variety of sales and management positions. Originated many silicone applications in refinery processing and petro-chemicals, especially antifoam and non-adhesion uses. Started companies in Brazil and France; was market manager at European headquarters start-up. Took early retirement in 1973 and joined Maremont Corp. Started own firm in 1976. Major contributions to delayed coking. Published 15 papers. *Society Aff:* AIChE, AIME(SPE) SRG.

Goldreich, Joseph D
Business: 257 Park Ave South, New York, NY 10010
Position: Partner *Employer:* Goldreich, Page & Thropp *Education:* MCE/Soils/NYU; BSCE/Structures/Union Coll *Born:* 10/8/25 Native New York City. Served to Lt. (j.g.) Civil Engr Corps, U.S. Navy, World War II. construction surveyor 1946-47. Structural designer, major NYC consulting engr firms 1947-60. Partner Goldreich, Page & Thropp, Cons. Engrs since 1961. Published in Consulting Engr Magazine & A/E Concepts in Wood Design. Volunteer leader Boy Scouts of America since 1965, district commissioner 1979- 81. Marshal Memorial Day Parade, Town of New Castle, NY since 1975. Past dir, secretary, VP of NY Assn. of Cons. Engrs. Recipient of BSA Shofar Award 1973, Silver Beaver 1977; Award of Merit Concrete Industry Bd of NY 1979. *Society Aff:* ACI, ACEC, ASCE, SAME

Goldschmidt, Victor W
Business: Sch of Mech Engg, W Lafayette, IN 47907
Position: Prof. *Employer:* Purdue Univ. *Education:* PhD/ME/Syracuse Univ; MSME/ME/Univ of PA; BS/ME/Syracuse Univ. *Born:* 4/20/36. Native of Uruguay. Worked with Honeywell's Control Valve Div 1957-1960. Involved with res on Basic Aspects of Turbulence; Energy Conservation in Bldgs, and Performance of Heating & Air Conditioning Systems. Past dir of Engg Fellowship prog in Latin Am, past chrmn Fluid Mechanics Committee of ASME & presently V-Chmn Intl Activities Ctte of ASHRAE. Freeman Fellowship to England 1971; Fulbright Sr Scholar to Australia 1979. Chrmn various conferences related to Energy Conservation in Appliances & Bldgs. Recipient ASHRAE Award of Merit (1975) & Engg News-Record Recognition for Contributions to the Construction Industry (1979). ASHRAE Fellow (1986) & E.K Campbell Award (1987). *Society Aff:* ASME, ASHRAE, ASEE, APS.

Goldsmith, Arthur
Home: 4303 Wynnwood Dr, Annandale, VA 22003
Position: Consultant (Self-employed) *Education:* PhD/Elec Engg/IL Inst of Tech; MBA/Bus Admin/Stanford Univ; MS/EE/IL Inst of Tech; BS/EE/IL Inst of Tech. *Born:* May 1915. Native of Chicago Ill. Asst in Elec Engrg, Ill Tech, 1937-41. Served with US Navy 1941-45 and 1947-58. Captain, USNR Ret. Between 1960-68 was Director of Engrg, Wilcox Electric Co; Ch Engr, Military Communications Lab, Motorola Inc; & Dep Director, Communications Applications Div, Communications and Systems, Inc. Chief, Technical Div, Office of the Secy of Transportation 1968-79. President IEEE Engrg Mgmt Soc 1968-69. Chmn, Inter-Soc Ctte on Transportation 1974-75. Tau Beta Pi, Sigma Xi, Eta Kappa Nu. Reg Prof Engr. Treas, IEEE Engrg Mgmt Soc. 1981- . Treas, IEEE Vehicular Tech Soc. 1983-. *Society Aff:* IEEE, ASEE, USNI, ION, RCA, WGA.

Goldsmith, Werner
Home: 6129 Etcheverry Hall, Berkeley, CA 94720
Position: Professor & Cons Engineer. *Employer:* University of California. *Education:* PhD/ME/Univ of CA, Berkeley; MS/ME/Univ of TX, Austin; BS/ME/Univ of TX, Austin. *Born:* May 23 1924 Germany. US immigration 1938. Married Penelope Alexander 1973. BSME, MSME U of Texas 1944, 1945. PhD ME U of CA Berkeley 1949. Engr Westinghouse Corp 1945-47. Berkeley faculty mbr since 1947. Visiting Prof in Europe since 1968. Cons engr, expert witness to govmt, industry & bar. Chmn West Coast Applied Mech Ctte ASME 1960. Chmn Head Injury Model Const Ctte NIH 1966- 70. Managment Chmn, 3rd Intern. Fulbright Alumni Assoc. Convention, 1980 Current cons Naval Weapons Ctr. Reg Mech Engr CA, Reg Safety Engr CA. Fellow ASME. Guggenheim Fellow 1953-54. Fulbright Res Scholar Greece 1974-75, 1981-82 Lady Davis Fellow, Technion, Haifa, Israel 1986 Pub monograph 'Impact', 160 tech articles. Specs: collisions, wave propagation, experimental mech, material behavior, biomechanics, rock mechanics. *Society Aff:* ASME.

Goldspiel, Solomon
Home: 732 Gerald Court, Brooklyn, NY 11235
Position: Dept of Envir Prot-Bureau of Water Resource Devel *Employer:* Dept of Envir Pro-Bureau of Water Res Develop *Education:* Chem Eng/Chem Eng/City College of NY; BS/Engrg/City College of NY; MS/Metallurgy & Physics/Polytechnic

Goldspiel, Solomon (Continued)
Inst of NY *Born:* June 1913. Licensed Professional Engineer Metallurgist, Physicist, Nondestructive Test (NDT) Specialist with Navy 33 yrs, NY City Bureau of Water Resource Development 11 yrs. Retired in 1982 and became part time conslt engr and nondestructive test specialist. Specs: foundry tech, metal fabrication, radiography, spectroscopy, X-ray diffraction, reference radiographs. Cons part time; coll evening teaching 25 years. Mbr ASTM, ASNT, ASM, SAS, MI, ANSI. ASNT: Fellow, Mehl Honor Lecturer, Kahn Mem Award; Sigma Xi & RESA; mbr NRC MTRB Adv Bd on Metals, Fabrication & Inspection; Pres CCNY Engrg & Architect Alumni; rep ASTM at VIIth Internatl NDT Conference. Present num tech talks & papers. Family Data: Former wife Mrs Fannie (Stern) Goldspiel deceased 12/15/77; daughter Mrs Gloria (Goldspiel) Muskat; Alfred A Goldspiel, PE son; current wife Mrs Iboja (Cybuch) Goldspiel. Chrmn of E07.02 ASTM on Reference Radiographs; Mbr of ASTM A01, G07 and F16 on Steel, Corrosion and Fasteners, respectively. Holder of ASNT NDT Level III Cert in RT, MPT, UT and PT since Aug 29th 1977. Recipient of 1978 ASTM "Award of Merit–, and "ASTM Fellow" and Honorary Membership for Life in 1980;. consultant to Jerusalem College of Technology and Israel Institute of Science and Halachad. *Society Aff:* NSPE, ASTM, ASNT, ANSI

Goldstein, Alexander
Home: Hertensteinstr 50, 5400 Ennetbaden, Switzerland
Position: Dir *Employer:* Brow Boveri Ltd *Education:* Dr Sc Techn/EE/Swiss Fed Inst of Tech, ETH Zurich; Dip Engrg//Swiss Fed Inst of Tech, ETH Zurich *Born:* 10/6/16 Native and taught in Zurich, Switzerland. With Brown Boveri 1941 - 1981. 1941 Development High Current Mechanical Rectifier. 1945 Powerline Carrier and Telemetering Dept. 1951 Head Dielectric Heating Dept. 1954 Head Special R&D Dept Transient Phenomena, Lightning Arresters, Electronic Relays, Computer Applications. 1959 Mgr Power Transformer Design Dept. 1962 Asst Technical Directorate High Voltage, Transformers. 1966 Pres MICAFIL Ltd, Zurich. 1970 General Mgr Transformer Division Brown Boveri. 1974-1981 Dir Corp Staff. 1982 retired, Consulting Engr, Advisor to American firms in Europe. 1951-1958 Assistant Prof ETH Zurich, El. Utilities Telecomm. 1967-1979 VP SIA. Enjoy rose gardening and birdwatching, art and theater, travelling. *Society Aff:* SEV, SIA, IEEE

Goldstein, Edward
Business: 295 N Maple Ave, Basking Ridge, NJ 07960
Position: Asst Financial Officer. *Employer:* American Telephone & Telegraph Co. *Education:* BS/Elec Engg/Univ of MN. *Born:* Jan 1923. Army Signal Corps 1944-48, 1950-52. Joined Bell Tele Labs 1949. Var assigns in switching systems dev, computer dev, data communications systems engrg, military communications systems engrg. Last BTL assgin: Dir Military Communications Systems Engrg 1964-66. Joined AT&T 1966 Engrg Dir Equip & Bldg. N Y Telephone 1970 V P - Mkting. Returned AT&T 1972. Pres assign: Assistant Financial Officer. Mbr Tau Beta Pi, Eta Kappa Nu. Fellow IEEE. Chmn Sci Adv Group Defense Communications Agency. Likes bicycling, sailing.

Goldstein, Irving R
Business: 21 Janet Ln, Springfield, NJ 07081
Position: Prof Emeritus *Employer:* NJ Inst of Tech. *Education:* MS/ME/Stevens Inst of Tech; BS/ME/Newark College of Engg. *Born:* 4/28/16. Ind engr, Wm Bal Corp, 1939-40; Maidenform Co, 1940-41; Inspector of Ordnance, US Army, 1941-3; Army anti-aircraft Artillery, 1943-46; Joined Dept of Ind & Mgt Engg, Newark College of Engg in 1947, Appointed Prof 1970. Assoc Chrmn of Dept, 1960-63, Faculty advisor, Student Chapter, IIE 1962-1977; Retired 7/81. Pres (1977-8), Chrmn, Bd of Governors (1966-68), (1981-83), Sec'y (1983-84) Distinguished Service Awards 1970, 1976, & 1985 all of Metropolitan NJ Chapter, IIE; Dir (1970-3), Work Measurement and Methods Engg Div, IIE, Phil Carroll Achievement Award (1975); Fellow, IIE (1978); Alpha Pi Mu; Pi Tau Sigma; Order of the Engr (1978); PE-NJ & CA; Vice Director, N.J. Engineers Committees for Student Guidance (1981-83), Treas (1983-86); Consulting Engineer (1981-). *Society Aff:* IIE, ASME, NSPE, TIMS, ORSA, NYAS.

Goldstein, Kenneth M
Business: PO Box 3395, Univ Sta, Laramie, WY 82071
Position: Staff Systems Analyst. *Employer:* US Dept of Energy. *Education:* PhD/Systems Engg/AF Inst of Tech; MSE/Systems Engg/AF Inst of Tech; BSE/ChE/AZ State Univ. *Born:* 9/15/46. Chem Engr, Shell Chem 1968-69. Sr Engr, Motorola SPD 1969-71. AF Command Pilot & Communications Engr 1971-74. Energy Consultant & Systems Engr 1974-75. Nuclear Engr, ITT 1975-77. Pres, KLG Energy Systems, Inc 1977-79. Assumed current position as Systems Analyst, Staff to Dir, Laramie Energy Tech Ctr in July, 1979. Responsible for long-range strategic planning for natl dev & utilization of oil shale, tar-sand, & underground coal gasification. Dir, Arizonans for Jobs & Energy since 1978; Mbr, Synthetic Fuels Presidential Task Force. Teach judo & fencing; enjoy metal sculpture, stained glass, & profl photography. Married with three children. *Society Aff:* AIChE, SPE, ACS, AIME, AJE, ANS.

Goldstein, Lawrence
Business: Best Lock Panama, Box 3119, Panama 3, Panama
Position: Conslt, Door Hardware Washroom Accessories Remote Sensing Techniques Latin Amer *Employer:* Self *Education:* MS/CERS/Purdue; BS/CEC/NC State Univ. *Born:* 8/19/36. Postgrad: Remote Sensing Purdue Univ. Native of Mooresville NC. US Army Corps of Engrs 1954-57 in SE Asia. With DMAIAGS since 1962; Inter-American Geodetic Survey 1962-80. Techniques & Services as remote sensing specialist; field exper in Peru, Ecuador, Bolivia, Guatemala, Nicaragua, Honduras, Venezuela, Panama Thailand & Japan. Mbr ASCE, AAAS, SAME & ASP. Secy, VP & Pres Panama Region of ASPR. Prior natl Dir for Panama-Central America Region of ASPR. Coordinator for 1st Pan Amer Symposium on Remote Sensing 1973. Gave prof papers at COSPAR Germany 1973 & AAAS Mexico 1974, Paigh, Guatemala 1977, IGAC Columbia 1977 & SPIA Seminar Panama 1982. Enjoy water skiing, photography, fresh/salt water fishing/travel. Serve as constl to Govt & private agencies in remote sensing. Consultant in Avionics, Giftware and Security Hardware and Arch Accessories. *Society Aff:* ASPR, ACSM, SAME, ALOA, DHI

Goldstein, Moise H, Jr
Business: Barton Hall, 34th & Charles Sts, Baltimore, MD 21218
Position: Prof. *Employer:* Johns Hopkins Univ. *Education:* ScD/Elec Engg/MA Inst of Tech; SM/EE/MA Inst of Tech; BS/EE/Tulane Univ. *Born:* 12/26/26. Native of New Orleans, LA. Undergrad studies were at Tulane Univ. received ScD in 1957 at MIT where I remained doing teaching and res until 1963. Have been at Johns Hopkins Univ since then, now as Edw. J. Schaefer Prof of Elec Engg with a joint appointment in Biomedical Engg. Director of the Speech Processing Laboratory in the ECE Department at Hopkins. Work on Aids for deaf children and human and machine processing of speech. Enjoy sailing and squash. *Society Aff:* AAAS, ASA, IEEE, SN, BME.

Goldstein, Paul
Business: 1910 Cochran Rd, P.O. Box 330, Pittsburgh, PA 15230
Position: Vice President & Genl Mgr. *Employer:* NUS Corp- Cyrus Wm Rice Div. *Education:* BS/Marine Eng/US Merchant Marine Acad. *Born:* Dec 1934. Test Engr Large Steam Turbine Generator Div, General Electric Co 195657. Ltjg Engrg Officer US Navy 1957-59. Marine Operating Engr United Fruit Co 1959. Service Engr, Sr Res Engr Foster Wheeler Corp 1959-65. Sr Res Engr Combustion Engrg Inc 1965-68. Current respon for cons, conceptual design & lab services in the field of indus water & wastewater mgmt. Received ASME Prime Movers Award 1969. Mbr ASME; Policy Bd Res; Natl Nom Ctte; Chmn Res Ctte on Water in Thermal Power Systems; Mbr ASTM; Mbr WPCF; Mbr Engrs Soc of W PA, Genl Chmn Internatl Water Conference. Mbr TAPPI. Hold patents on boilers & nuclear steam generating equipment & present num papers on water & wastewater tech. Operate horse farm

Goldstein, Paul (Continued)
as primary non-engrg activity. Received Engineer's Society of Western Pennsylvania Award of Merit, 1980. *Society Aff:* ASME, ASTM, WPCF, TAPPI.

Goldstein, Richard J
Business: 111 Church St SE, Minneapolis, MN 55455
Position: Prof & Hd. Mech. Engr. *Employer:* Univ of MN. *Education:* PhD/Mech Engg/Univ of MN; MS/Physics/Univ of MN; MS/Mech Engg/Univ of MN; BS/ Mech Engg/Cornell Univ. *Born:* 3/27/28. Instr. U. Minn., Mpls., 1948-51, instr., research fellow, 1956-58, mem. faculty, 1961--, prof. mech. engring., 1965--, head dept., 1977--; research engr. Oak Ridge Nat. Lab., 1951-54, Lockheed Aircraft, 1956; asst. prof. Brown U., 1959-61; vis. prof. Imperial Coll., Eng., 1984; cons. in field 1956--; chmn. Midwest U. Energy Consortium; chmn. Council Energy Engrg. Research; NSF sr. postdoctoral fellow, vis. prof. Cambridge (Eng.) U., 1971-72. Served to 1st lt. AUS, 1954-55. NATO fellow Paris, 1960-61; Lady Davis fellow Technion, Israel, 1976; recipient NASA award for tech. innovation, 1977. Fellow ASME (Heat Transfer Meml. award 1978, Centennial medallion 1980, BEG v.p. 1984--, pres. assembly for internat. heat transfer confs. 1986--); mem. AAAS, Am. Phys. Soc., Am. Soc. Engring. Edn., Minn. Acad. Sci., Nat. Acad. Engring., Sigma Xi, Tau Beta Pi, Pi Tau Sigma. Research, publs. in thermodynamics, fluid mechanics, heat transfer, optical measuring techniques. *Society Aff:* AAAS, APS, ASEE, ASME.

Goller, Carl H, Jr
Home: Rt 1 Box 85B, Defiance, MO 63341
Position: Chrmn of Board-Retired *Employer:* Bendy Engrg Co-Earth City Mo. *Education:* BS/Chem Engg/Univ of MO-Rolla. *Born:* May 1917. WWII 1941-45 Army Air Corps Pilot. Missouri Portland Cement 1949-55 as lab analyst, finally ch chemist. Bendy Engrg Co, cons engr since 1955. Performed as process engr, project engr, mgr of cement & pyro-processing projects until VP respon for project mgmt 1969. Spec in project economic analyses and process operating trouble-shooting & personnel training. Exec VP & Treasurer 1974. President 1976, Chrmn of Bd, 1980. Mbr AIChE, ASTM, APCA, Tau Beta Pi. Enjoy gardening, boating, dog obedience & showing. *Society Aff:* AIChE, ASTM, APCA, ΤΒΠ.

Golomski, William A J
Home: 1228 Middlebury Lane, Wilmette, IL 60091
Position: Principal. *Employer:* W A Golomski & Assoc. *Education:* MSEM/Engg/Milwaukee Sch of Engg; MS/Math/Marquette Univ; MBA/Accounting, Fin/Univ of Chicago; BS/Math/Univ of WI, Stevens Point; BA/TV/Columbia College. *Born:* 10/14/24. Taught at var univs prior to 1957 Corporate Exec '57-71; 1971-pres Tech & Mgmt Cons. Mbr Indus Adv Comm U of MI-Dearborn/Honorary Member, Center for Quality & Productivity U of WI-Madison. P Natl Pres ASQC 1966-67. Specs: Productivity & Invest Studies, Process and Quality Control. Fellow NYAS, Edwards medal, ASQC 1976, W.A. Golomski Award of ASRC Started 1986, Fellow AAAS, Fellow ASA, 1984; Award for Excellence, Quality Control & Reliability Engrg Div. IIE 1981 Div. Chm 1987-88; PE CA. *Society Aff:* AAAS, IIE, ASQC, NYAS, ASA, ORSA.

Golze, Alfred R
Home: 1508 La Sierra Dr, Sacramento, CA 95825
Position: Consulting Engineer. *Employer:* Self-employed. *Education:* BS/Civil Engg/ Univ of PA. *Born:* Jul 6 1905 in DC. Reg Cvl Engr DC & CA. With Federal Govnt 1930-61 (ICC, Bureau of Budget, Bureau of Reclamation). Engrg assignments Bureau of Reclamation from Asst Engr to Asst Commissioner. 1961 Ch Engr, 1967 Dep Dir CA St Dept Water Resources, charged with desing, const, oper St Water Proj. Ch Water Resources Engr for Burns & Roe Inc 1971-74. Water Resources Engr Cons 1975- . Fellow ASCE. Pres Natl Cap & Sacramento Sects ASCE; Chmn Comm Engr Educ 1959-60; Chmn Natl Water Policy Comm ASCE 1972-73. Mbr US Comm on Large Dams; Internatl Ctte on Irrigation & Drainage; Earthquake Engrg Res Inst, SAME. Honors: Disting Service Award Dept of Interior 1962; Engrg - Award, Engrg Council Sacramento Valley 1966; Toulmin Medal SAME 1964; Hon Mbr ASCE 1976, D. Robert Yarnall Award U of PA 1979, Author *Reclamation in US*, Editor, Handbook of Dam Engrg & num prof papers. Clubs, Cosmos, Wash. D.C., Engrs. Sacto. CA. *Society Aff:* ASCE, SAME, USCOLD, ICID.

Gomberg, Henry J
Business: Ann Arbor Nuclear Inc, 500 City Center Bldg, Ann Arbor, MI 48104
Position: President & Chief Exec Officer. *Employer:* Ann Arbor Nuclear Inc. *Education:* PhD/EE/Univ of MI; MS/EE/Univ of MI; BS/EE/Univ of MI. *Born:* Apr 16 1918. Born in NYC. Started in elec engrg but became involved in nuclear energy during WWII. Eventually Prof & 1st Chmn of Dept of Nuclear Energy U of MI, also Dir of MI Mem Phoenix Proj for Res on peacetime use of nuclear energy. Moved to PR in 1961 to become Dep Dir & then Dir of the Nuclear Ctr, a USAEC Lab. In 1971 returned to Ann Arbor as Pres of KMS Fusion Inc, created for dev on a private basis of lasardriven fusion power systems. Fellow ANS. Resigned December 1980. Founded Ann Arbor Nuclear Inc for development of synthetic fuel processes using nuclear energy (fission or fusion) as the driving energy source. Exec Bd 1970 & Bd/Dir 1970-72 ANS. Fellow AAAAS. Mbr APS, IEEE, ASEE. Fellow APHA. *Society Aff:* IEEE, ANS, ASEE, APS, AHPA AAAS.

Gomez, Rod J
877 South Alvernon Way, Tucson, AZ 85711
Position: Pres *Employer:* RGA Consulting Engrs *Education:* BS/CE/Univ of AZ *Born:* 3/14/24 Native of Jerome, AZ, Commander, CEC, US Naval Reserve (Retired). Design engr for Phelps-Dodge Corp 1950-1954, Convair, Division of General Dynamics 1954-1957; structural engr for Tucson architect 1957-1960; chief engr for Boduroff, Meheen & Gomez 1960-1963. Formed Rod Gomez & Assocs 1963 and have since expanded the structural consulting services to include civil, environmental, land surveying, mechanical, and electrical. As Pres, am responsible for overall direction and accountability for client relations, staff utilization, quality control, and profitability for RGA, which is currently ranked as the tenth largest minority-owned consulting engrg firm in the US. Engr of the Year, Southern Chapter AZ Society of PE 1979. Chrmn, AZ State Bd of Technical Registration 1979-1980. Chrmn, City of Tucson Bd of Appeals. Chrmn, City of Tuscon Bd of Adjustment. Mbr, State Coliseum Bd Dir, UA Engr Alumni Assoc Recd Distinguished Alumni Citizen Award 1983-UA Coll of Engr. *Society Aff:* NSPE, SAME, ACEC, APWA.

Gomezplata, Albert
Home: 513 Powell Drive, Annapolis, MD 21401
Position: Professor/PT. *Employer:* University of Maryland. *Education:* PhD/Chem Engr/Rensselaer Polytech Inst; MChE/Chem Engr/Rensselaer Polytech Inst; BChE/ Chem Engr/Brooklyn Polytech Inst. *Born:* Jul 2 1930 Colombia SA. Became a US naturalized citizen in 1948. Indus exper incl a yr as a process engr for General Chemical Delaware, a yr in process eval at Picatinny Arsenal with US Armed Forces, a yr with the Engrg Dept duPont in their 'Year in Industry' Program. Joined MD faculty in 1958, Prof in 1968, Act Chmn 1973, Chmn 1975-78. Assoc. Dean 1982-1985. Reg Prof Engr Md. Mbr Sigma Xi, Phi Lambda Upsilon, Tau Beta Pi & Omega Chi Epsilon. Maj res area is heterogeneous flow systems. Author. *Society Aff:* AIChE, ΣΞ.

Gomory, Ralph E
Business: POBox 218, Yorktown Heights, NY 10598
Position: V P & Dir Res. *Employer:* I B M Corp. *Education:* PhD/Math/Princeton Univ; BA/Physics, Math/Williams College. *Born:* May 1929 Brooklyn NY. Grad Williams College; studied at Cambridge U; PhD math Princeton 1954. U S Navy. Higgins Lecturer math Princeton. Author 42 papers on linear programming & nonlinear differential equations. Joined I B M 1959: I B M Fellow 1964; Dir Math Sci Dept 1965; Dir Res 1970; V P of I B M 1973. Mbr NAS; NAE. Fellow AAAS; Econometric Soc. Was Andrew D White Prof-at-large Cornell U 1970-76. Mbr: Visit Ctte MIT Sloan School of Mgmt; Adv Council Dept Math Princeton. Hon DSc William College. *Society Aff:* NAS, NAE, AAAS.

Gompf, Arthur M
Business: Box 155, Jarrettsville, MD 21084
Position: President. *Employer:* Gompf Engrg Services Inc. *Education:* BE/Mach-Elec Engrg/Johns Hopkins Univ; MLA/Lib Arts/Johns Hopkins Univ *Born:* Jan 24, 1909 Pikesville Md. Balt Coll Commerce 1925-29; B.E., Mach-Elec. Engrg., Johns Hopkins Univ., M.L.A., Lib Arts, Johns Hopkins Univ. m.Margaret Jane Purdam Aug 24, 1940; c. Arthur Purdum, Henry Lewis. Esso - Standard Oil Co of NJ, Balt 1934-38; Secy-Treas Egli & Gompf Inc Balt 1938-50, Pres 1950-72, Chmn/Bd 1972-75; Dir Progress Fed Sav & Loan Assn 1952- , Chmn/Bd 1972-79; fac Johns Hopkins 1936-40; Pres Bldg Congress & Exchange of Balt 1955; Chmn Air Qual Control Adv Council Md 1966-67; Tr McDonogh Sch 1965-79, V P 1971-75; Engrg Prof Merit Serv Award 1968; Founders Award Engrg Soc of Balt 1974; Hon Life Mbr Bldg Cong & Exchange 1975; Tau Beta Pi, Reg P E Md, WV, Fla, DC, Va; Life Fellow ASME; Life Mbr IEEE, Life Mbr ASHRAE, Engrg Soc Balt (Pres 1961), NSPE (V P 1963-66, Treas 1966-70). Contrib articles to mags. Office: Gompf Engrg Serv Inc, Street Md 21154. *Society Aff:* ASME, IEEE, ASHRAE, NSPE.

Gonser, Bruce W
Home: 1301 Arlington Ave, Columbus, OH 43212
Position: retired *Employer:* Battelle Columbus Labs. *Education:* SD/Metallurgy/Harvard Univ; MS/Metallurgy/Univ of Utah; BS/Chem Eng/Purdue Univ; ChE/Professional Deg/Purdue Univ; Hon. Dr. Eng./-/Purdue Univ. *Born:* 9/9/99. Native of Hudson, IN. Taught Country School. Signal corps in WWI. Res metallurgist, ASARCO; Durango, CO; Amarillo, TX; Selby, CA; Omaha, NB; Salt Lake City, UT; Perth Amboy, NJ. Natl Radiator Corp, Johnstown, PA 1933-34 developing electrolyticiron process, Battelle Meml Inst, 1934-retirement, at Columbus, OH, as res metallurgist, chief, asst dir, and technical dir. Specialties - non ferrous metallurgy, vapor phase deposition, metallic coatings, tin, and unusual metals. Activities abroad in W Germany, Switzerland, Spain, etc. With United Nations in Argentina in 1935. Honorary Dr of Engg, Purdue, in 1967, James Douglas Medal from AIME, 1970. Married; three children. *Society Aff:* AIME, ASM, ASTM, ES

Gonseth, Alan T
Business: 1 World Trade Center, New York, NY 10048
Position: Assistant Manager, Transportation Planning Div *Employer:* Port Authority of NY & NJ. *Born:* Nov 23 1934. BSCE Clarkson C of Tech 1956. Cert of Highwy Engrg Yale U in 1959. NYU Grad School of Public Admin 1959-62. Reg Prof Engr NY; Mbr ASCE; Fellow Inst Transportion Engrs. Nom for VP ITE 1976 and 1978; Dist I Dir ITE 1970-71; Chmn Dist I 1975; VP & Pres Metro Sect ITE 1967-69; Dir Metro Sec ITE 1974-75; Port Authority's Exec Dir's Unit Citation Award 1971 and 1980. *Society Aff:* NSPE, ASCE, ITE

Gonzales, John G M
Business: 431 Prater Way, Sparks, NV 89431
Position: Sanitary Engr *Employer:* City of Sparks *Education:* MS/Sanitary Engrg/ Univ of NV; BS/CE/Univ of NV *Born:* 4/5/48 Native of Northern NV. Graduated with Honors, Univ of NV, Distinguished Military Service Award, Commissioned US Army Corps of Engrs, 1971, Water and Wastewater Systems Engr, South Lake Tahoe Tertiary Wastewater Treatment Plant, 1972-76. Consulting Engr 1976. Sanitary Engr, City of Sparks, NV since 1977, Mgr, Reno-Sparks Wastewater Treatment Facility since 1977. Registered PE, NV and CA. Grade V Water and Wastewater Plant Operator Certificates in CA and NV. Certified Community Coll Instructor - CA and NV. Pres, NV Water Pollution Control Assoc. Enjoy snow skiing, sailing, and gardening. *Society Aff:* ASCE, APWA, WPCF, NSPE

Gonzalez, Rafael C
Business: Elec Engg Dept, Knoxville, TN 37916
Position: IBM Prof *Employer:* Univ of TN. *Education:* PhD/EE/Univ of FL; MS/EE/ Univ of FL; BS/EE/Univ of Miami *Born:* 8/26/42. In addition to his present position, Dr Gonzalez is Dir of the Image & Pattern Analysis Lab at the Univ of TN and is a consultant to govt & industry on problems dealing with pattern recognition, image processing, & machine learning. He has received the 1978 UTK Chancellor's Res Scholar Award, the 1980 Magnavox Eng. Prof. Award and the 1980 M.E. Brooks Distinguished Prof. Award for his contributions to these fields, including co-authorship of three text books. Dr Gonzalez is a mbr of numerous profl & honorary socs & is an assoc editor for the *IEEE Transactions on Systems, Man and Cybernetics* and the *International Journal of Computer and Information Sciences.* He is a Fellow of the IEEE. *Society Aff:* IEEE, PRS, ΤΒΠ, ΣΞ

Gonzalez, Raul S
12258 Buckingham Ave, Baton Rouge, LA 70815
Position: President. *Employer:* Gonzalez Engineering Co. *Education:* BSCE/CE/LSU; MS/CE/LSU. *Born:* Oct 1934. BS & MS La State Univ. Native of Colombia SA; US citizen. US Army Corps of Engrs, 1953-56. Over 26 years of civil & struc engrg & proj mgmt with various Louisiana firms. In 1972 left Pyburn & Odom as Asst Ch Engr to form own co which grew around struct & cvl engrg, const, const mgmt & surety engrg with specialties in marine dev & marine structures, municipal engrg, mathematical modeling & drainage & san engrg. In 1976 broadened engrg to full range of design service. Pres of Gonzalez Engineering Company, Inc. Internatl speaker & author. *Society Aff:* ASCE, LES, PCI, ACI, ΞE.

Gonzalez-Karg, Sergio
Business: Ing Sergio Gonzalez-Karg, Monte Athos No 165, Lomas De Chapultepec, Mexico 10, DF
Position: Dir-Genl. *Employer:* Gonzalez-Karg y Asociados. *Born:* 7/3/25. Native of Mexico City Mexico. Profl studies at the Natl Univ of Mexico, post-grad Harvard. prof of Bus Admin in the UNAM's Engrs Faculty & in the Iberoamerican Univ. PPres ASCE Mexican Sect 1972-74 & of sev other assn, USA Alt Dir in Mexico for EJC, Corresponding Mbr IEP, Delegate to Mexico's Pres during the XIX Olympic Games in Acapulco, Gro Const work in genl. Owner Gonzalez-Karg y Asociados SC, a cons engrs & architects co. Enjoys sailing, water & snow skiing, biking & motorcross & all types of outdoor sports; classical music & languages. Member AAES USNC-UPADI, International Contact Director, ASCE 1980-83. *Society Aff:* ASCE.

Good, Melvin F
Business: 8977A Complex Dr, San Diego, CA 92123
Position: Pres Alega Corp. *Employer:* Alega Corp, Melvin Good Assocs. *Education:* BSME/Mech Engr/Univ of TX; MBA/Bus Adm/Natl Univ; MSSM/Bus. Mgt/USC. *Born:* 1/24/31. in Hanover, PA. Const Engr for Natkin & Co, Pneumatics & Hydraulics Sr Design Engr & Field Engineer for Genl Dynamics Corp. Since 1966, Pres & Principal Engr of Cons. Engr firm spec in Engineering & Architectural Services, Reg Control Syst Engr, Natl Council of Engg Examiners Cert, Panel Mbr Amer Arbitration Association. *Society Aff:* ASME, ASHRAE, SAME, NECC, AAS

Goodbar, Isaac
Home: 93-02 211th St, Queens Village, NY 11428
Position: Chief Engr. *Employer:* Edison Price Inc. *Born:* Dec 1918 Argentina. Prof Civil Engrg Univ of Buenos Aires. MS MIT Elec Engrg. Awarded fellowship to MIT. In Argentina: first employed by Catita; Later as Tech Dir & ptnr of sev co's, manufactured elec motors, generators, signalling devices, lighting equip (own pat); Theatrical lighting cons (Auditorium, San Martin theaters). In US: Illuminating Engrg (Holophane 1945, Century 1946); Ch Engr of Edison Price Inc since 1957. Lighting cons to Illuminating Engrg Res Inst, Elec Testing Labs, City of Buenos Aires. US Delegate to CIE. Fellow 1968 IES; Fellow 1969 IES Great Britain. Patents & publications. U.S. Delegate to International Congresses (CIE) Barcelona (1971), London (1975), Kyoto (1979).

Goodell, Paul H
Home: Apartment 108, 1,600 SE. St. Lucie Blvd, Stuart, FL 33494
Position: Retired *Born:* Native of Stuart, Florida. Reg Ohio EE. 1934-81 U of Cincinnati. Fellow IEEE 1959, Mbr. Manhattan Proj WW II; foreign engrg assignments: Italy 1964-65, Australia 1973-74. Granted US pats: 2,504,516 - 2, 688,685 - 3,332, 179. Successive US assignments as: Prod Line or Engrg Mgr to Vice Pres Perfection Products, Inc., Hale & Kulgren (Consulting Engineers), Fostoria Corp, General Electric Co, Catalytic Combustion Corp. Pioneering contrib in fields of Air Pollution Control, Safety, Industrial Process Heating & Prod Finishing with numerous published papers in Journals of AIEE, ASME, APCA, Soc of Manufacturing Engrs, Indus Gas, Indus Heating, Indus Finishing. Quarter century Mbr Natl Fire Protection Assn. Safety Standards Ctte on Indus Ovens & Furnaces. Hobbies: Boating, Photography, Order of the Blue Gavel (Past Yacht Club Commodores) and Shrine Mariners. *Society Aff:* IEEE, ASME, IES, APCA, AAAS, USPS

Goodemote, DeWitt F
Home: 112 Densmore St, Buffalo, NY 14220
Position: Sr Mfg Engr Group Leader. Ret. *Employer:* Moog Inc Controls Div. *Education:* AAS/Mech Engg/Rochester Inst of Technology. *Born:* Sept 1915. Native Western NY. Grad Rochester Inst of Tech; Ext - State Univ of NY at Buffalo, Erie Comm College & Cornell Univ. Tool Engr, Supervisor Houdaille Indus Inc, Hydraulic Div Buffalo. Joined Moog 1956 - originate, plan, evaluate cost & schedule mfg processes and tooling requirements for complete programs of mfg & supervise mfg engrs, technicians & tool designers. BS Mbr SME since 1957 with Certification. Presently Certified Mfg Engr, and Reg P.E. SME Dir 1974-1982. SME citation 1964 for Public Rel; Pres Award 1964. Member Woodside United Methodist Church. Mbr Free & Accepted Mason-33 degree, Past Master, Past Grand Lodge Officer. Enjoy golf, travel. *Society Aff:* SME.

Gooden, Charles D
Home: 6218 Heatherbloom, Houston, TX 77085
Position: Owner. *Employer:* Charles D Gooden Cons Engrs. *Born:* Aug 30, 1942. BS Arch Engrg Prairie View A&M Univ 1965. Native Madisonville Texas. Worked in Aircraft Indus for about 2 years with Boeing & Lockheed Aircraft Co 1965-67. 2 years US Army 1967-69. Sr Assoc & Asst Proj Mgr Walter P Moore & Assocs Inc 1969-75. Formed own firm Jan 1976, providing civil & struct engrg serv. Mbr of Gulf Coast Soc of Prairie View Engrs, NSPE, TSPE, ACEC, CEC-T, HESS.

Goodenough, John B
Business: Center for Materials Science & Engineering, ETC II 5.160, Univ. of Texas At Austin, Austin, TX 78712
Position: Professor. *Employer:* Oxford Univ. *Education:* PhD/Physics/Univ of Chicago; AB/Math/Yale Univ. *Born:* July 25, 1922 Jena, Germany of Amer parents. Grad Groton School, magna cum laude in classics 1940; AB Yale Univ, summa cum laude in math 1943; MS & PhD Univ of Chicago in physics 1951-52. Served as meteorologist in USAAF 1942-48 (Captain). Westinghouse res engr 1951-52; MIT Lincoln Lab 1952-76, Leader Electronic Matls Group; Prof of Inorganic Chem Oxford Univ England 1976-1986; Virginia H. Cockrell Centennial Chair of Engineering, Univ. of Texas at Austin, 1986-. IEEE Award London 1957; Docteur Hon Causa Univ of Bordeaux 1967; Centenary Lect British Chem Soc 1976; Natl Acad of Engrg 1976; Solid State Chem Award Roy Soc Chem 1982. Mbr Solid State Sci Panel & Natl Matls Advisory Bd 1975-76. Assoc Editor Materials Res Bulletin, Journal Solid St Chemistry, Solid St Ionics, & Structure & Bonding; Eec. Editorial Board J. Applied Electrochemistry Co-Editor International Scr of Monographs on Chemistry (Oxford Univ). *Society Aff:* APS, JPS, ACS, RSC, NAE, IAS.

Gooderham, Ronald M
Home: R R 1, Terra Cotta, Ontario LOP 1NO Canada
Position: Retired. *Education:* BASc/ME/Univ of Toronto. *Born:* April 1903 Toronto Can. Graduated EE Univ of Toronto 1926. Recipient of Mech Engrg Degree 1943. Established European activities of Lincoln Elec Co of Canada Ltd, London England. Welding Advisor British & Canadian Governments World War II 1939-45. Welding Cons Engr in private practice. Estab Canadian Welding Bureau 1947 & became first Genl Mgr. Implemented the welding standards of certification & approval & originated & dev educational programs at all levels to enable fabricators & personnel from the operator to prof engr to meet designated specifications. Conceived & founded Canadian Welding Dev Inst in 1972. Fellow of Engrg Inst of Canada & of the Welding Inst of Great Britain. *Society Aff:* EIC, WIC, WI.

Goodfriend, Lewis S
Business: 7 Saddle Rd, Cedar Knolls, NJ 07927
Position: Acoustical Engr. *Employer:* Lewis S Goodfriend & Assoc. *Education:* ME/-/Stevens Inst of Tech; MEE/-/Poly Inst of Brooklyn. *Born:* 5/21/23. Consulting engr in the areas of architectural acoustics; noise surveys & control measures for industry, communities & airports; electroacoustics & electronics; air conditioning noise control, & product dev. He is Fellow of the ASA & the AES. He was Pres of INCE, & sr mbr of IEEE as well as several other socs. Mr Goodfriend was editor of Sound & Vibration magazine & of Noise Control a publication of the Acoustical Soc of Am. Mr Goodfriend serves on the Bd of Trustees at Stevens Inst of Tech. Mr Goodfriend is a reg PE in NY, NJ, CA, & several other states. *Society Aff:* ASA, IEEE, INCE, ASHRAE, AIHA

Goodgame, Thomas H
Business: Env. & Chem. Consulting Engrs., Inc, P.O. Box 914, Alamogordo, NM 88310
Position: Pres & CEO *Employer:* Environmental & Chem Conslt Engrs., Inc. *Education:* ScD/ChE/MIT; MS/ChE/LA State Univ; BS/ChE/LA Tech Univ *Born:* 5/23/21. Camden Ark. Engr ALCOA 1942-43; Submarine Officer US Navy 1943-46; Asst Prof LTU 1947-49; R&E group leader Cabot Corp 1953-55; head R&E, TB&CC 1955-57; Assoc Prof Georgia Tech 1957-59; Sr Proj Engr Cabot 1959-62; Gen Mgr Cabot Titania 1962-64; Sr Engr & Res Mgr Whirlpool Corp 1964-70; Dir Environ Control Whirlpool Corp 1970-84; Pres. & CEO Env. & Chem. Conslt Engrs., Inc., 1985- . Reg P E Michigan, New Mexico. Diplomate (APCE&SE) AAEE. Chrmn Environmental Div AIChE 1976. Over 70 papers & pats in pollution control, mass/heat transfer, engrg economy, chem prod. Captain USNR-ret. Addl accomplishments since last report Natl Dir AIChE 1979-81; Recipient of AIChE Environmental Awd 1978; Natl Dir of Sigma Xi - The Scientific Res Soc 1974-80; 1983 President's Awd of Porcelain Enamel Inst.; Designated an Arkansas Traveler by Gov. Clinton; Various Advisory Cttes and Panels for U.S. Office of Tech Assessment, U.S. EPA and Ohio EPA. *Society Aff:* AIChE, ACS, AAAS, ΣΞ, ΑΧΣ, ΤΒΠ

Goodjohn, Albert J
Business: P O Box 81608, San Diego, CA 92138
Position: Dir, High Temp Gas-Cooled Reactor Prog Div. *Employer:* General Atomic Co. *Education:* PhD/Physics/Queen's Univ; MSc/Physics/Univ of Alberta; BSc/Physics/Univ of Alberta. *Born:* 2/18/28. in Calgary, Canada. Naturalized US citizen 1966. Reactor physicist Canadair Ltd until 1959. Joined Genl Atomic Co 1959. Participated in sev progs related to gas-cooled reactor technology. Assumed current respon for mgmt of HTGR Programs in 1978. Amer Nuclear Soc Fellow. Chrmn San Diego Sect ANS 1967. Enjoy outdoor activities, golf. *Society Aff:* ANS.

Goodkind, Michael N
Business: Alfred Benesch & Company, 233 North Michigan Ave, Chicago, IL 60601
Position: VP *Employer:* Alfred Benesch & Co *Education:* PhD/Structural Engrg/Northwestern Univ; M/BA/Univ of Chicago; MS/Structural Engrg/IA State Univ; BS/CE/Rutgers Univ *Born:* 8/4/43 Responsible for business development and project management. Previous positions included assignments as a structural engr, civil engr and rail transportation planner. Registered Professional and Structural Engr. Pres of IL Section, ASCE 1981-82. Grandfather, Morris, was National Dir ASCE and Pres NJSPE. Father, Herbert, and uncle, Donald, founded consulting firm Goodkind & O'Dea. Wife, Mary, member of ASCE; ANS, President of Chicago

Goodkind, Michael N (Continued)
Chapter 1983; and Heath Physics Society, Pres of Midwest Chapter 1977. Active in community affairs and local politics. *Society Aff:* ASCE, NSPE, PCI, APWA, ARTBA, SMPS, CECI, AREA, SAME.

Goodman, Alvin S
Business: 333 Jay St, Brooklyn, NY 11201
Position: Prof *Employer:* Polytechnic Univ *Education:* PhD/CE/NYU; MS/CE/Columbia Univ; B/CE/City Coll of NY *Born:* 3/14/25 Native of New York City. Served in U.S. Army (1944-47), First Lieutenant, Corps Engrs. Have taught at CCNY (1947-49), Northeastern Univ (1962-69), and NYU (1969-73). At PINY since 1973, where am head of Civil & Environmental Engineering Dept. and Coordinator of Water Resources Program. Principal Investigator and doctoral advisor for research projects in water resources systems, hydrology, and energy development. Affiliated between 1951 and 1985 with New York consulting firm of Tippetts-Abbett-McCarthy-Stratton, earlier as Project engr and, after 1962, as Staff Consultant Member and officer of several committees of ASCE-WRPM Division. *Society Aff:* ASCE, AWRA, WPCF, AGU, ASEE, IWRA, NYAS, AIH

Goodman, James R
Business: Engr. Data Management, 736 Whalers Way Bldg G, Ft Collins, CO 80525
Position: President of Engineering Data Management, Inc. *Employer:* Colorado State Univ *Education:* PhD/Struc. Eng & Mech/Univ of CA-Berkeley; MS/CE/CO State Univ; BS/CE/Univ of WY *Born:* 5/23/33 Native of WY. Worked as a Bridge Engr with the WY Highway Dept following graduation for two years prior to joining the Civil Engrg Faculty at CO State Univ in 1957. Active in teaching and research since that time. Pres of CO Section of ASCE 1971. L.J. Markwardt Award for contributions to wood engrg ASTM 1974. ASCE Technical Committee Chairman. Past Chi Epison Advisor. Registered PE and consultant to industry. Member of National Science Foundation research team, cooperative project with Hungary 1978-81. Author of numerous papers and editor of ASCE book on Wood Structures. PE and Pres of Engrg Data Mgmt, Inc. *Society Aff:* ASCE, FPRS, ΣΞ

Goodman, Joseph W
Business: Dept of Elec Engrg, Stanford, CA 94305
Position: Prof of Elec Engrg. *Employer:* Stanford University. *Education:* PhD/EE/Stanford; MS/EE/Stanford; AB/Appl Phys/Harvard *Born:* 2/8/36. in Boston, MA. Raised in Boston area. At Stanford since 1958. Engaged in res & teaching in Fourier & Statistical Optics. Post-doctoral Fellow Norwegian Defence Res Establishment 1962-63. Visiting Prof Univ of Paris XI, Inst d Optique 1973-74. Fellow OSA, IEEE, SPIE, F E Terman Award ASEE 1971, Max Born Awd, OSA 1983, Education Medal, IEEE 1987. Member, National Academy of Engineering, 1987. Author introduction to Fourier optics statistical optics & approximately 150 tech papers. Past editor, *Journal of the Optical Society of America*. Past Mbr, Bd of Dirs OSA, Bd of Governors SPIE. VP, International Comm For Optics (ICO). *Society Aff:* OSA, IEEE, SPIE, NAE.

Goodman, Lawrence E
Home: 1589 Vincent St, St Paul, MN 55108
Position: Prof. *Employer:* Univ of MN. *Education:* PhD/Appl Mechanics/Columbia Univ; MS/Civ Engg/Univ of IL; BS/Civ Engg/Columbia Univ; AB/Math/Columbia Univ. *Born:* 3/12/20. Employed as mech engr, Johns Hopkins Appl Physics Lab, on design of radio- proximity fuse. Served in US Navy Pacific Fleet 1942-45 and introduced blind firing gun dir. Subsequently Natl Res Council Fellow, Prof of Civ Engg at Univ of IL (Urbana-Champaign) and at Univ of MN. Hd, Dept of Civ and Mineral Engg, Univ of MN, 1965-1972. James L. Record Professor of Civil Engg. 1980- . Sr Fellow Natl Sci Fdn, Distinguished Teaching Award, Univ of MN. Author monographs on Statics, Dynamics and journal articles on stress analysis and structural analysis. Fellow ASME and Fellow ASCE. Honorary Mbr, Chi Epsilon Society. *Society Aff:* ASCE, ASME, Med. Acad. Am.

Goodman, Leon
Home: 187 I U Willets Rd, Searingtown, NY 11507
Position: Manager, Transportation Planning Div. *Employer:* Port Authority of NY & NJ. *Education:* Prog for Mgmt Dev/Harvard; Cert in Transp/Yale; Bachelor Civil Engg/CCNY. *Born:* 1935 NYC; Cert in Transp Yale, Prog for Mgmt Dev Harvard, BCE CCNY; With Port Auth since 1956 as Traffic engr, Prin Transport Planner & since 1976 Mgr- Transp Planning Div. In 1970 coordinated one-way tolls prog & received Exec Dir's Award for dir of Exclusive Bus Lane Proj. Teaching transport at Pratt Inst since 1963 ' at Stevens Inst from 1970-78, at City Univ. of NY from 1987; mbr ASCE, ITE, NSPE; Tech papers for ASCE, ITE, IBTTA, TRB, AHONAS, Austral Rd Res Bd, Pan Am Hwy Congress. Secy ASCE Hwy Div 1966-71; Pres ITE Met Section 1975, Chrmn ITE District One 1978, ITE Intl Bd 1981-1983; ITE Transp Engr of Year 1983; member, ITE U.S. Legis Comm; Reg PE Ny. Interests: sports, photography, bridges. *Society Aff:* NSPE, ASCE, ITE.

Goodman, Theodore R
Home: 21 Chapel Pl, Great Neck, NY 11021
Position: Retired *Education:* PhD/Aero Engg/Cornell Univ; MS/Aero Engg/Caltech; BS/Aero Engg/RPI. *Born:* 3/21/25. Native of Brooklyn, NY. Staff positions with Cornell Aeronautical Lab & Allied Res Assoc, Inc. VP & founder Oceanics, Inc 1962-76. Stevens 1976-82. Presently retired. Author of over forty published papers in aerodynamics, hydrodynamics & heat transfer. Contributory author in "Advances in heat Transfer I," academic press 1954. Reviewer for applied mechanics reviews 1950-82. Enjoy classical music & contract bridge. *Society Aff:* AIAA, ASME.

Goodno, Barry J
Business: Atlanta, GA 30332
Position: Prof *Employer:* Georgia Institute of Tech *Education:* PhD/CE/Stanford Univ; MS/Structural Engrg/Stanford Univ; BS/CE/Univ of WI- Madison *Born:* 7/19/47 Native of Madison,WI; Previous positions with John A Blume and Associates, San Francisco, Arnold and O'Sheridan, Consulting Eng, Madison; Wisconsin Division of Hwys; National Science Foundation. Teaching Fellow at Stanford Univ. Currently a Member of ASCE (EMD, SD), EERI, SSA, PCI and Sigma Xi. Faculty Advisor to Georgia Tech Student Chapter of ASCE 1975-85. Registered PE, GA. Evans Scholar graduate of Univ WI. *Society Aff:* ASCE, PCI, EERI, ΣΞ.

Goodrum, John W
Business: Driftmier Engrg Center, Athens, GA 30601
Position: Assoc Prof *Employer:* Univ of GA *Education:* PhD/ChE/GA Inst of Tech; MS/Chem/Univ of GA; BS/Chem/Univ of GA *Born:* 12/27/42 He did graduate research in high temperature chem. His PhD research dealt with the microstructure of materials. He conducted a four-year program of solid- state chem research at the Air Force Cambridge lab at Bedford, MA. For five years he was employed the GA Tech Engrg Station where he headed the Thermal Conversion Branch. He served as chrmn of the GA Section of the American Society for Metals, 1979-80. Currently Prof Goodrum is researching the application of alternative energy sources. *Society Aff:* ASAE, ASM, ACS, NATAS, ASCG

Goodson, Raymond Eugene
Business: A A Potter Engg Ctr, W Lafayette, IN 47906
Position: Assoc Dean of Engrg, Dir Engrg Exp Station *Employer:* Purdue Univ. *Education:* PhD/Engg/Purdue; MSME/Mech Engg/Purdue; BSME/Mech Engg/Duke; AB/Economics/Duke. *Born:* 4/22/35. As Assoc Dean and Dir of EES/IIES Dr. Goodson is responsible for coordinating $19 million of research programs; responsible for fourteen mission research laboratories and centers, and for initiating new research centers and laboratories in areas of national and regional importance (started centers on automotive transportation, energy policy, coal research, and CAD/CAM); fostering growth of existing centers; and equipping and managing the new A. A. Potter Engineering Research Center. He is also Chairman of the Midwest Universities Energy Consortium with responsibilities including forming the structure of the

Goodson, Raymond Eugene (Continued)
consortium and steering program through its initial two years. *Society Aff:* ASME, ΠΤΣ, ΣΞ, NSPE.

Goodson, Raymond L, Jr
Business: 3409 Oak Grove, Dallas, TX 75204
Position: President. *Employer:* Raymond L Goodson Jr Inc. *Education:* MCE/-/NY Univ; BS/CE/So Methodist Univ. *Born:* Dec 11 1918. Native Dallas Texas; m. Ann Meriwether; c. James, David & Sally. US Navy 1942-46 Lt Cdr; Design Engr Myer, Noyes & Assoc; Asst Prof CE, SMU; founded Raymond L Goodson Jr Inc, Cons Engrs in 1953; CHI EPSILON; Pres Dallas Branch ASCE 1967; Pres Cons Engrs Council of Texas 1963; named Civil Engr of Year Dallas 1971; NSPE; Pres Texas Sect ASCE 1975-76; Engrs Club of Dallas; St Matthews Cathedral (Episp), Lakewood Country Club, Eng of Yr, Dallas 1976; Rotary Club. *Society Aff:* ASCE, NSPE, ACEC.

Goodwin, Gene M
Business: P O Box X, Oak Ridge, TN 37830
Position: Group Leader *Employer:* Oak Ridge Nat'l Lab *Education:* PhD/Mat/RPI; B/METE/Metallurgy/RPI *Born:* 12/02/41 Native of York, Maine. With Oak Ridge National Laboratory since 1968. Assumed current position as Group Leader-Welding and Brazing in 1976. NASA Fellow, 1965-68. AWS McKay-Helm Award, 1979 and 1980, William Sparagen Award, 1974, Committee Service Award 1977. Approximately 60 publications. Director-at- Large, Welding Society, 1979-81. *Society Aff:* AWS, ASM, ΣΞ, WRC

Goodwin, James G
Business: PO Box 3208, Ogden, UT 84409
Position: Mgr, Product Assurance. *Employer:* Western Zirconium. *Education:* MS/Met Engg/Univ of Pittsburgh; BS/Met/Penn State. *Born:* 9/30/29. Native of Pittsburgh, PA. Spent 27 yrs at Westinghouse Electric's Bettis Atomic Power Lab. Specialized in nuclear mtls. Engg & mgt assignments included dev of alloys, fabrication methods, powder met, ceramic tech, heat treatments, welding, & advanced nuclear core components; corrosion res, mech property studies, & process quality control. Published 78 articles & contributed to several books on nuclear mtls. Assumed current duties as Quality Assurance Dept Mgr with Western Zirconium (a Westinghouse subsidiary) on Sept 1, 1978. Elected to Am Men of Sci, 1967; listed in Who's Who in the East. Enjoy tennis, skiing, swimming, bowling, golfing, classical music, opera & photography. *Society Aff:* ASM.

Gor, Vishnu J
Business: 14465 S Waverly Ave, Midlothian, IL 60445
Position: Pres. *Employer:* Poly Enviro Chem Co. *Education:* BS/CE/Univ of MO; BSc/Chemistry/St Xavier College. *Born:* 10/21/40. Native of Malpur, India. Res Chemist for Burgess Cellulose Co. Sr Polymer Chemist, Paint Res Asst Lab 1966-1968. Res Engr Continental Can Co Mgr Surface Phenomena - Group Continental Can Co 1968-1979. Specialty of polymers & lubricants formulation & process design. Able Toastmaster Award 1975. Patentee in the field of coatings, polymers & lubricants. Pres of Poly-Enviro Chem Co 1979-cont. VP of Dober Lubricants Inc 1979-cont. Enjoy public speakings & classical music. *Society Aff:* AIChE, ACS, ASLE, FSCT.

Gordon, Bernard B
Business: 40 Bower Place, Danville, CA 94526
Position: Consulting Eng (Self-employed) *Education:* MS/CE/Soil Mechanics/Harvard Univ; BS/CE/MIT *Born:* 3/19/15 in Boston MA. Career has spawned from 1936 to date. Major interest has been in Applied Soil Mechanics and Foundation Engrg, particularly in Embankment Dam Design and Construction. Recent years in Earthquake Resistant Design. Worked with private firms (Woodward-Clyde & Jackson& Moreland) federal agencies (Corps of Eng & US Navy) and state (Calif Dept. of Water Resources) and municipal (East Bay Municipal Utility District) Self-employed as a consulting engr since 1980. Have served as a lecturer or a guest lecturer over the years in graduate training at UC Berkely, UC Davis, Sacramento State Univ and Univ of MO at Rolla. *Society Aff:* ASCE, ASTM, SAME, ISSMFE, ICOLD

Gordon, Daniel I
Business: 4405 East-West Hwy, Suite 512, Bethesda, MD 20814
Position: Staff Scientist/Conslt, Electronics/Physics *Employer:* Vela Associates *Education:* MEngg/EE/Yale Univ; BEngg/EE/Yale Univ. *Born:* 8/12/20. Res Physicist, Naval Surface Weapons Ctr (NSWC), White Oak, MD, 1942-80. Visiting Scientist, Weizmann Inst of Sci & Technion - Israel Inst of Tech, 1968- 69. Res in magnetic mtls, devices & measurements. Developed low noise magnetometer sensors used by Apollo 16 scientists on surface of moon & by Voyager Jupiter Saturn experimentors. Contributed to dev of radiation tolerant magnetic devices. Program Manager, Very High Speed Integrated Circuits (VHSIC), Naval Sea Systems Command, 1980 to 1984. Vela Assoc, 1984 to date. Pres, Magnetics Soc of IEEE, 1979, 1980. Gen Chrmn, Intl Magnetics Conf, 1980, Editorial Bd, IEEE Transactions on Magnetics. *Society Aff:* IEEE, APS, SMAG.

Gordon, Donald T
Business: P.O. Box 269, Littleton, CO 80160
Position: Dept. Mgr *Employer:* Marathon Oil Co. *Education:* BS/CHE/Univ of TX *Born:* 8/15/31 Joined Marathon Oil CO in 1953 as Junior Petroleum Eng. Have held various enrg and eng supervisory positions. Currently Mgr of Reservoir Mgmt. Dept, Applied Technology Division, Denver Research Center, Marathon Oil. Co-Author of two SPE papers in reservoir description and reservoir engrg;. Former local section chairman of SPE. Served 2 yrs as Technical Editor on Technical Review Committee for Journal of Petroleum Technology of SPE. Contributing author to *Modern Reservoir Description for Improved Oil Recovery*, IOCC Vol. 3. *Society Aff:* SPE

Gordon, Eugene I
Business: 600 Mountain Ave, Murray Hill, NJ 07974
Position: Dir Optical Devices Lab. *Employer:* Bell Lab. *Education:* PhD/Physics/MIT; BS/Physics/CCNY. *Born:* Sep 14 1930 NYC. With Bell Labs since 1957. Member of the Advisory Group on Electron Devices for the Office of the Dir of Defense Res & Engrg. Mbr of APS, Fellow of the IEEE, mbr of Phi Beta Kappa & Sigma Xi. Author of over 50 published articles & holds 23 pats. 1975 IEEE Vladimir K Zworykin Award. *Society Aff:* NAE, IEEE, ΦΒΚ, ΣΞ.

Gordon, Gerald M
Home: 5358 Maretta Dr, Soquel, CA 95073
Position: Mgr Fuel and Plant Materials Technology. *Employer:* General Electric Co. *Education:* PhD/Metallurgical Engg/OH State Univ; BS/Wayne State Univ; Detroit, MI. *Born:* May 1931. Native of Detroit Mich. Sr Metallurgist at Stanford Res Inst from 1959-63, serving as proj leader on progs related to high temp refractory metal alloys. With General Electric Co since 1964. Assumed present position as Mgr Fuel and Plant Matl Tech in 1979 working on Corrosion & qualification of materials & processes for nuclear reactor sys. Served on Exec Ctte of ASM & AIME. Fellow ASM 1975. Reg P E Calif. Numerous publications & pats in the Metallurgical Engrg field. Chairman NACE Ctte on Corrosion in High Purity Power Plant Water 1975-78. Enjoy classical music, hiking. Remarried on 9/15/79 to Priscilla M Gordon. *Society Aff:* ASM, AIME, NACE, AAAS

Gordon, Paul
Business: 233 Perlstein Hall, Chicago, IL 60616
Position: Prof. *Employer:* IIT. *Education:* DSc/Met Engr/MIT; SM/Met Engr/MIT; SB/Met Engr/MIT *Born:* 1/1/18. Five yrs (1942-1947) as Metall Group Leader for Manhattan Proj (atomic bomb proj) and Atomic Energy Commission. Thirty yrs in univ teaching, res, and admin (MIT, Univ of Chicago, IIT); books and many published res papers on metall transformations, properties, and fracture. Since 1954, extensive consulting work in industry and product liability involving investigation of over 600 failures in metal structures, devices, and machinery, including court testimony in many cases. *Society Aff:* ASM, AIME, ASTM

Gordon, Raymond G
Home: PO Box 1078, Atascadero, CA 93423
Position: Dept Head. *Employer:* CA Poly State Univ. *Education:* PhD/Mech Engr/Univ of CA, Santa Barbara; MSME/Mech Engr/Univ of MI; BSME/Mech Engr/Western New England College. *Born:* 2/15/42. Industrial experience was obtained at such cos as Universal Design Inc, Duchess Design and Dev, Oriole engg Co, and Combustion Engg. Teaching experience was obtained at CA Poly State Univ at San Luis Obispo since 1967. I was Mbrship Chrmn 1973-75, VChrmn 1975-76 and Chrmn 1976-77 of the Central Coast Group of the LA Sec of ASME. Secretary 1979-80 of the Central Coast Sec of the CA Society of Professional Engrs. My publication and res area is Bioheat transfer. *Society Aff:* ASME, NSPE, ASEE, ΣΧ, ΤΒΠ.

Gordon, Richard H
Business: Beaver Falls, PA 15010
Position: Chrmn-Engg. *Employer:* Geneva College. *Education:* PhD/Higher Educ/Univ of Pittsburgh; MS/Ind Engg/Columbia Univ; BEE/Elec Engg/RPI. *Born:* 12/30/25. Native of Clifton, NJ. Served with US Navy 1944-1946. Supervisor with Bigelow Sanford Carpet Co, 1947-1950. Helped establish the Afghan Inst of Tech, Kabal, Afghanistan 1950-1952. Proj Engr with Congoleum-Nairn Inc, 1952-1958. Served as a missionary educator in Nigeria, W Africa 1959-1966. With Geneva College since 1966. Assumed current responsibility as Chrmn of the engg dept in 1975. *Society Aff:* IIE, ASEE.

Gordon, Ruth V
Business: 124-Beale St, Mezzanine Fl, San Francisco, CA 94105
Position: District Structural Engr. *Employer:* State of CA, Office of the State Architect. *Education:* MS/Structures/Stanford Univ; BS/Civ Engg/Stanford Univ. *Born:* 9/19/26. Native of Seattle, WA. With State of CA, OSA, since 1956. Currently responsible for supervision of construction of public sch & hospital bldgs for compliance with state earthquake safety regulations in six counties. First woman reg structural engr, state of CA; first woman mbr, SEAONC; Outstanding Service Award, Golden Gate Sect, SWE, 1979, San Francisco Working Woman of Archievement Finalist, Working Woman Mag, 1983, Profile in "The Women's Book of World Records & Achievements–, Anchor Books, 1979. 1975, 1982 Woman of Achievement, Union Sq Bus & Profl Women's Club; 1984-1986 Dir, SEAONC; Past Pres, Golden Gate Sec, SWE; VChair, 1979 Natl Convention, SWE; Past Pres, San Francisco Bay Area Engr Council; Engr Mbr, CA Bd of Architectural Examiners Examinations Revision Proj Adv Panel. Structural Engr & Civil Engr, State of CA. Active in career guidance to encourage young women to enter technical careers. *Society Aff:* ASCE, SEAONC, AWIS, EERI, BPW, AAUW.

Gordon, Walter S
Home: 12411 Lakeholme Rd SW, Tacoma, WA 98498
Position: Consulting Engineer *Education:* BS/EE/Univ of WA (Seattle). *Born:* Nov 1904. Native of Seattle Wash. Railroad electrification: Milwaukee, Seattle 1925-28; Lackawanna, Hoboken 1928-31; PRR: Philadelphia 1931-34; Baltimore 1934-35; Downingtown 1936-38; International Panel Iran 1975. Elec Supt (WPA) Philadelphia Navy Yard 1935-36. Sr Elec Engr Fort Lewis & McChord Field, Army Constructing Quartermaster 1938-41. Army Corps of Engrs Cpt - Lt Col, air base & highintensity runway lighting const: 11th AF Alaska, Aleutians; 20th AF Guam, Tinian, Saipan, Iwo Jima, 1942-46; Legion of Merit. Cons Engr office Tacoma 1945-63; Gordon & Cross Engrs 1964-80. Consulting Engineer 1980 - present. Washington State Power Comm 1953-57. IEEE Fellow, Life Mbr. Chapter & State Pres, Natl Dir Wash Soc of Prof Engrs. Reg P E Penna, Wash, Oregon, Calif. *Society Aff:* IEEE, AREA, SAME, IES, Sigma Xi, WSPE, NSPE, IAEI, ROA, TROA.

Gordon, William E
Business: Provost, Houston, TX 77251
Position: Provost and Vice President *Employer:* Rice University. *Education:* PhD/Elec Engg/Cornell Univ; MS/Meteorology/NY Univ; MA/-/Montclair State College; BA/-/Montclair State College. *Born:* 1918 Paterson New Jersey. Joined res staff at Cornell Univ 1948; became Assoc Prof of EE in 1953, Prof in 1959 & Disting Prof in 1965. Between 1960-65 conceived, planned & directed the design & const of the world's largest radar- radio telescope, Arecibo Ionospheric Observatory & directed the Observatory during first years of operation. In 1966 became Dean of Sci & Engrg & Prof of EE & Space Physics & Astronomy at Rice Univ; 1975 became Dean of the School of Natural Sci; 1980 became Provost and Vice President. Mbr National Acad of Sci & Natl Acad of Engrg; recipient of Balth. van der Pol Award for distinguished res in radio sci 1963-66, presented by Internatl Scientific Radio Union; Amer Meteorological Soc Medal 1969; Guggenheim Fellow 1972-73; Fellow IEEE; Fellow AG; Nat'l Academy of Sciences Arctowski Medal 1984; IEEE Centennial Medal 1984. *Society Aff:* NAS, NAE, URSI, IEEE, AAAS, AMS, AGU.

Gore, Willis C
Business: Dept of Elec Engg, Baltimore, MD 21218
Position: Prof. *Employer:* Johns Hopkins Univ. *Education:* DrEng/Elec Engr/Johns Hopkins Univ; BE/Elec Engr/Johns Hopkins Univ. *Born:* 5/20/26. Currently Prof in the Dept of Elec Engg at the Johns Hopkins Univ with res interests in error-correcting codes. Has been a consultant for AAI Corp, Aerojet Gen Corp, ARINC Res Corp, Beckman Instruments, Electronic Communications, Inc, Fairchild Stratos Corp, Goddard Space Flight Ctr, HF Communications, Leeds and Northrup Co, Litton Industries, Martin-Marietta Corp, and Westinghouse Elec Corp. *Society Aff:* IEEE, HKN, ΤΒΠ.

Gorham, Robert C
Home: 1061 Findley Dr, Pittsburgh, PA 15221
Position: Prof Emeritus EE. *Employer:* Penna State Bd Prof Engrs. *Education:* MS/EE/Univ of Pittsubrgh; BS/EE/Cornell Univ; BA/Physics/NB Weslayan Univ. *Born:* Apr 11, 1893 Smith Ctr Kansas. Taught in Nebr public schools prior to WW I. Served as Reg Electrician USN 1917-19. Pres Pgh Chapter PSPE 1953-54. Mbr of Penna State Bd for Prof Engrs Registration 1951-58; 6 years in charge of all written exams, covering 13 disciplines given Jan & June each year. Received Distinguished Serv Certificate from NCSBEE 8-1-58 for Loyal & Intelligent Service to NCSBEE & to the Engrg Profession. *Society Aff:* IEEE, NSPE, ASEE.

Gorman, James E
Business: 400 - 108th Ave NE, Bellevue, WA 98004
Position: Pres. *Employer:* Constructioneering NW, Inc. *Education:* BS/Engg Sci/Univ of Portland. *Born:* 10/4/40. Pres of Constructioneering Northwest, Inc, Bellevue, WA; engg, construction mgt, & construction firm that specializes in proj mgt & cost engg work. Past Chrmn of WA state, NSPE/PE in Construction. Presently is the Western Regional VChrmn for the Natl Soc of PE in Construction. Instr on "Proj & Construction Mgt" for Seattle Univ's Dept of Civ Engg, Adult Continuing Education Prog. Grad of the Univ of Portland (Magna Cum Laude); BS Degree in Engg Sci. Author of numerous articles. Also author of text: "Simplified Guide to Construction Management for Architects and Engineers–, Cahners Publishing Co, Boston, MA, 268pp, 1967. Lic PE in WA, OR, HI, DC. Married with two children. Enjoys fishing, hunting, jogging, & racket ball. *Society Aff:* NSPE, NSPE/PEC, NSPE/PEPP.

Gorman, Paul F
89 Briarcliff Rd, Brokton, MA 02401
Position: Chrmn *Employer:* Chas T Main, Inc. *Education:* MS/Mech Eng/Northeastern Univ; BS/Mech Eng/Worcester Polytechnic *Born:* Oct 1923. Native of Bridgewater MA. Jackson & Moreland/United Engrs & Constructors Inc 1946-75. Last pos as Dir of Jackson & Moreland Internatl, V P - Mgr of Boston Office & V P

Gorman, Paul F (Continued)
Boston Power Dept, respon for all study, design and/or const from Power Div Boston Office - large nuclear & fossil fueled elec generating stations. Chas T Main Inc, 1975 to date - currently Chrmn; Pres Metropolitan Chap of MSPE 1963. V Chmn ASME Boston 1965. Cert of Qualifications with Natl Council of Engrg Examiners. Enjoy tennis & golf. *Society Aff:* AIF, ASME, NSPE, ACEC

Gornowski, Edward J
Business: P O Box 101, Florham Park, NJ 07932
Position: Exec Vice Pres. *Employer:* Exxon Res & Engrg Co. (Retiring 7/1/81) *Education:* PhD/Chem Eng/Univ of PA; BChE/Chem Engr/Villanova Univ. *Born:* Feb 27, 1918 Wilmington Del; married, resides Scotch Plains NJ; 3 sons & 1 daughter. Dir & Exec VP Exxon Res & Engrg Co. Joined Exxon Baton Rouge LA Labs 1942; Exxon Res & Engrg Co Linden NJ 1945; transferred to Exxon Co USA's Bayway Refinery Linden NJ, Mgr Chem Prods 1964; Mgr Corp Planning Dept Exxon Corp 1965; became VP Logistics Esso Europe London 1966; present pos 1969. B Chem E Villanova Univ 1938; PhD Univ of Penna 1943. 18 U S pats. Mbr NAE, ACS, API, Sigma Xi; Fellow AIChE; Assoc Tr Univ of Penna; Fellow AAAS; Received J. Stanley Morehouse Award for outstanding engineering achievement from Villanova Univ. 1979. *Society Aff:* NAE, ACS, API, AAS, AIChE, ΣΞ, ΑΧΣ.

Gorrill, William R
Business: 10 Boardman Hall, University of Maine, Orono, ME 04469
Position: Prof. of Civil Engrg. *Employer:* Univ of ME at Orono. *Education:* MS/Civ Engg/Univ of ME; BS/Civ Engg/Northeastern Univ. *Born:* 10/1/21. Asst Soils Engr and Soils Engr, ME State Hgwy Comm 1951-57. Consulting Engr 1957-68 and pres of successor firm 1968-1972. Prof of Civ Engg at Univ of ME for 24 yrs. Dir of Sch of Engg Tech and Prof of Civ Engg, teaching part-time 1975-83. *Society Aff:* ASCE, NSPE, ASEE.

Gorton, Robert L
Business: Durland Hall, Dept of Mech Engrg, Manhattan, KS 66506
Position: Prof, Mech Engr *Employer:* Kansas State Univ *Education:* PhD/ME/KS State Univ; MS/ME/LA State Univ; BS/ME/LA Tech Univ *Born:* 10/19/31 Worked as Engr for AR-LA Gas Co, Shreveport, LA and as Field Engr for Schlumberger, Morgan City, LA. Taught at LSU, and is currently Prof of Mech Engrg of KS State Univ. Served as conslt to USDA, GM, Oak Ridge Natl Labs and Black & Veatch and Wilson Co Consltg Engrs. Accomplished res projs sponsored by DOE, DOD, US Navy, NSF, ASHRAE. Concentration in areas of Energy Systems, Production and Transport, and in Air-Conditioning Sys. Serves on ASHRAE Technical R&T and Program Committees and is ASME Fellow. Recipient of ASME Centennial Medallion. *Society Aff:* ASME, ASHRAE

Gorup, Ivan L
Home: 621 Crossridge Terrace, Orinda, CA 94563
Position: VP. & General Manager *Employer:* Liquid Air Engg Corp. *Education:* BE/Chem/McGill Univ. *Born:* 7/13/32. Over 30 yrs of experience in cryogenic engg & related fields. Was involved in projs for the separation of air, CO-H2, separation of hydrocarbons, helium and hydrogen, production of ethylene & ammonia synthesis gas, LNG & liquefaction of hydrogen, carbon dioxide & other gases, waste water treatment. Was process engr with L'Air Liquide, Montreal (1956), Mgr - Plant Operations (1962), Mgr - Field Opertions (1964), Mgr of Engg (1966) for Am Air Liquide, NY. At Lotepro, NY was Mgr of Engg (1971), VP of Engg & Construction 1975. Joined Liquid Air Engg in 1985. Since 1986 is VP & General Mgr for Liquid Air Engg (San Francisco & Montreal). *Society Aff:* AIChE, ANSI/ASME, ASTM

Goryl, William M
Home: Corey Lane, Mendham, NJ 07945
Position: Asst. Project Executive *Employer:* Exxon Res & Engg Co. *Education:* ME/Mech Engg/Stevens Inst of Tech. *Born:* 4/4/24. Served as naval officer (engg) WWII. Joined Std Oil Dev Co (now Exxon Res & Eng Co), 1946. Dev improved design of large diameter piping expansion joints for fluid catalytic cracking units. Asst Dir Products Res Div 1961 and Gen Engg Div 1962. Holder of two US patents. In 1963 assisted ER&E VP in establishing Marine Research Program. Project Manager of complete new refinery for Esso France, 1965. Manager of Exxon Engg European Office, 1967-69. Engg Res Coord, 1974-78. Currently Project Dir for major new res center for ER&E. SME & Tau Beta Pi. *Society Aff:* ASME.

Gosden, John A
Home: 111 Fourth Ave, Suite 10-N, NY, NY 10003
Position: Assoc Prof of Information Systems. *Employer:* NYU - GBA. *Education:* MA/Maths/Cambridge UK; BA/Maths/Cambridge, UK. *Born:* 3/-/30. Native of Folkestone England. Programmer Sys Software Developer, LEO Computers London 1953-60. Editor Std EDP Reports, Project Mgr Auerbach Corp 1961-66. Dept Hd Info Sys MITRE Corp 1966-69. With Equitable 1970-1986, as VP & Dept Hd Tech Services until 1971, Mgr CAPS Proj 1971-75. VP Corp Compter Sr 1975-78, VP Telecommunications 1978-80, Technology Officer 1981-86. Since 1986 Adjunct Assoc Prof of Information Systems, and Research Fellow at NYU Graduate School of Business Administration; also independant consultant. Fellow of British Computer Soc. Publication Bd Chrmn & Council Mbr ACM 1973-76 & involved in stds since 1956. Mbr Washington Affairs Committee of AFIPS 1977-1986, chrmn 1980-86. *Society Aff:* ACM, BCS.

Gosman, Albert L
Home: 244 Bonnie Brae, Wichita, KS 67207
Position: Professor *Employer:* Wichita State Univ *Education:* PhD/ME/Univ of IA; MS/ME/Univ of CO; BS/ME/Univ of MI *Born:* 5/27/23 Native, Detroit MI. Served Army Engineers. Assist Prof Mechanical Engrg, CO School of Mines 1950-56. Senior Research Engr, Advanced Development, Northrop Aircraft 1956-58. Researching Environmental Controls for Missile Guidance. Assistant Prof, CO School of Mines, 1958-62. Staff ME Dept, State Univ of IA, while completing PhD, 1962-65. Associate Prof, Wayne State Univ, Detroit 1965- 67. Prof and Head, ME Dept., Wichita State University, 1967-71. Associate Dean Engrg and Assoc. Dir Research, Wichita State Univ, 1971-80. Returned Teaching, Prof ME Dept., Wichita State Univ 1980-present. Engr, National Bureau of Standards, CO, Summers, 1961-69. Outstanding Teacher Award 1952, 1953, 1967, 1981. *Society Aff:* ASME, ASEE, ΣΞ, ΦΚΦ

Gosnell, Gary J
Business: 5801 Pecos St, Denver, CO 80221
Position: Regional Engrg Mgr. *Employer:* Prestressed Concrete of Col Inc. *Born:* Sept 1942. BS VMI; Masters & Doctorate Univ of Virginia. Native of Monroeville Pa. Army Corps of Engrs 1968-70. Joined Prestressed Concrete of Colorado 1971 as Design Engr. Have since held pos of Sales Engr, Proj Mgr, Ch Engr & currently Rocky Mountain Regional Engrg Mgr. Ctte Mbr ACI-ASCE 426 - Shear & Diagonal Tension; & PCI- Prestressed Concrete Columns. Enjoy farming, skiing, hunting.

Goss, John R
Business: Dept of Agri Engg, Davis, CA 95616
Position: Prof. *Employer:* Univ of CA. *Education:* MS/Agri Engr/Univ of CA, Davis; BS/Engg/Univ of CA, Los Angeles. *Born:* 5/30/23. Winona County, MN. US Marine Corps June 1943-Aug 1949. Resigned first lt, regular commission. Appointed Specialist, Agri Engg Dept, Univ of CA, Davis, Aug 1953. Prof from July 1967. Half-time asst to the Chancellor for academic planning 1964-66. Agri Engg Dept Chrmn July 1968 to Jan 1974. Fellow, ASAE 1979. Fellow, AAAS, 1986. Res on combine harvester performance, forage harvesting, rice straw burning, and legume seed processing. Since Jan 1977 res on gasification of agri & forest residues & utilization of low Btu gas in boilers, burners for heated air, & in spark & compression ignition engines. *Society Aff:* ASAE, AAAS, ASEE, SAF, NSPE, NYAS

Goss, William P
Business: Mech Engr Dept, Amherst, MA 01003
Position: Prof. *Employer:* Univ of MA. *Education:* PhD/Thermal Sciences/Univ of CT; MS/ME/Univ of CT; BSE/ME/Univ of CT. *Born:* 5/23/38. Native of Milford, CT. Worked in industry for the Torrington Co (1961); Pratt & Whitney Aircraft (1962-64); The Mitre Corp (1967-68); Alfa-Laval Inc (1978-1982); and Owens-Corning Fiberglas (1983-1984). Taught at the Univ of CT (1964-67); VPI (1967-70); & the Univ of MA (1970-to-present). Currently Prof of ME Teaching courses & doing res in the areas of Heat Transfer, Applied Math, & Bldg Thermal Performance. Enjoy travel, music, hiking, running & reading. *Society Aff:* ASME, ASTM, ASHRAE, ASEE.

Gossick, Lee V
Home: 909 Country Club Dr, Tullahoma, TN 37388
Position: Vice President, Deputy General Manager *Employer:* Sverdrup Technology, Incorporated *Education:* MSc/Aero Engr/OH State Univ; BSc/Aero Engr/OH State Univ. *Born:* Jan 1920. BS, MS Ohio State. Native of Meadville MO. Military Service 1941, retiring from US Air Force as Major General in 1973. Cdr Arnold Engrg Dev Ctr 1964-67. V Cdr & subsequently Cdr Aeronautical Systems Div AFSC 1968-70. Dep Ch of Staff & subsequent Ch of Staff Headquarters Air Force Systems Command 1970-73. Asst Dir of Regulation US AEC 1973. Exec Dir for operations, US Nuclear Regulatory Commission - 1975-1980. SVERDRUP Corp - 1980 to present. Fellow AIAA. *Society Aff:* AIAA, SAME.

Gostelow, Jonathan Paul
Business: PO Box 123, Broadway, NSW 2007 Australia
Position: Dean of Engg. *Employer:* NSW Inst of Tech. *Education:* PhD/Turbomachinery/Liverpool Univ; MA/Modern History/War Studies/London Univ; MA/-/Cambridge Univ; BEng/Fluid Mechanics/Liverpool Univ. *Born:* 11/6/40. Leicester, England. Worked as postgrad student for Prof J H Horlock in aerodynamics of blade cascades. Sr Compressor Engr at GE Co, Evendale, OH working on advanced fan design & cascade aerodynamics, 1965-68. Asst Dir of Res, Cambridge Univ, UK, 1968-75 & Deputy Dir of the new Whittle Lab. Hd of Sch of ME, the New South Wales Inst of Tech, Sydney, Australia 1975-present. Dean of Engg, NSWIT, 1984-present. ASME Gas Turbine Award, 1976. Over 60 published papers and 2 books. Actively interested in turbomachinery, solar energy & engg education; more passively in soccer & classical music. *Society Aff:* ASME.

Goth, John W
Home: 15140 Foothill Rd, Golden, CO 80401
Position: Sr VP Retired *Employer:* AMAX Inc. *Education:* ME/Metallurgy/McGill Univ; BS/Metallurgy/SD Sch of Mines; DR Bus Adm//SD Sch of Mines (Honorary). *Born:* 4/22/27. Sr Exec VP of AMAX Inc responsible for all metals operations of the Corp 1982. Joined AMAX in 1954 and after a variety of assignments in metallurgy and sales, including Sales Mgr - Europe and VP - Sales, was appointed Pres of the AMAX Molybdenum Div and VP of AMAX in 1975. Designated Exec VP in 1981. Designated Group Exec in May 1978. Honorary Doctor of Bus Admin Degree from S Dakota Sch of Mines, 1981. Born in Ree Heights, SD. Obtained BSc MEt SD Sch of Mines, and MEng Met from McGill Univ. Designated Exec VP in Fed 1981. Resides in Greenwich, CT with wife. Retired from AMAX 1985. Now Dir of Dome Mines, Toronto, Canada; and Magma Copper Corp. San Manuel, Ariz. Consultant to the mining & metals industries. *Society Aff:* AIME, ASM, AISI.

Goto, Satoshi
Home: D5-403, Green Heights, 571 Unomori, Sagamihara, Japan 228
Position: Manager *Employer:* NEC Corporation *Education:* Ph.D/Elec. Engrg./Waseda University; MS/Elec. Engrg./Waseda University; BS/Elec. Engrg./Waseda University *Born:* 01/03/45 Born in Hiroshima, Japan in 1945. Got Ph.D, MS and BS all from Waseda Univ., in Japan. Joined Central Research Labs. of NEC Corp in 1970 and worked for VLSI design, AI applications and C & C Systems. Currently, a manager of C & C Systems Research Labs of NEC. Got best paper award from IEEE and is an IEEE Fellow. Enjoy tennis, golf and travelling. *Society Aff:* IEEE, IFIP

Gotschall, Donald J
Business: PO Box 50, Juneau, AK 99802
Position: Civ Engr. *Employer:* AK Power Administration. *Education:* BS/Civ Engg/CO State Univ. *Born:* 10/10/34. Early career was with Army Corps of Engrs and Bureau of Reclamation. Worked with Structural design, soil mechanics, and irrigation proj operation & maintenance from 1960-1966 in Denver. Reg PE in CO. Started with AK Power Administration in 1966 & developed a specialty in hydropower. Currently position is with oper and maint (O&M) of a hydropower plant. Formed and operated a corp to care for homeless teenagers. Pres of Juneau Branch, ASCE, Secy/Treas of ASCE, Pacific Northwest Council, & was Pres of AK Sec ASCE 1980. Enjoy hunting big game, fishing, taxidermy, boating, & house bldg. *Society Aff:* ASCE.

Gott, Jerome E
Home: 1604 Sweetbriar Dr, San Jose, CA 95125
Position: Sr Engineer. *Employer:* General Electric. *Born:* Sep 1935. MS from San Jose State U; BS U of Detroit. Native of Erie, Pa. With United Tech Ctr as materials engr, project engr & Mgr of Materials Engrg from 1962-69. Respon for dev of processes for large diameter one piece rocket motor cases. Joined General Electric (NED) in 1970; estab product of Penetration assemblies, proprietary plating processes for uranium. Present respon as sr cons materials & processes C & I Dept. Reg Prof Met Engr CA. Elect Natl Dir SAMPE 1972-75. Married Sandra Langley; 6 children. Enjoy backpacking, music, & travel.

Gottscho, Alfred M
Home: 348 Landis Ave, Millersville, PA 17551
Position: VP. *Employer:* Helme Tobacco Co. *Education:* MS/Chemistry/Franklin & Marshall College; BChE/ChE/CCNY. *Born:* 4/29/19 Received BChE 1940 & MS 1954. Military service 1944-46, infantry in Italy. Most of experience is in tobacco chemistry and process engg and dev of tobacco sheets. With General Cigar Co. since 1946, starting in lab on chemistry of nitrogn compounds in tobacco, and dev of tobacco sheets. Became dir of res in 1968, VP in 1977. In 1986 Helme Tobacco Co separated from General, Assumed same title and responsibility at that time with Helme. Received Distinguished Scientist Award from Cigar Industry Assn in 1973. Author and coauthor of numerous papers on tobacco chemistry and holder of patents on tobacco sheets. *Society Aff:* AIChE, AAAS, ACS.

Gould, Arthur F
Home: 835 Pinetop Drive, Bethlehem, PA 18017
Position: Prof *Employer:* Leigh Univ *Education:* MS/IE/Leigh Univ; SB/Eng Admin (ME)/MIT *Born:* 3/24/16 Worked as production engr prior to WWII with Congoleum-Nairn, Inc. Served with Ornance Dept, US Army, engr with Textron, Inc., 1945-1947. With Leigh Univ since 1947 as Prof, Dept Chairman, and Assoc Dean of Engrg. Elected Fellow in IIE in 1980 and a Fellow in APICS in 1977. IE Division Chairman of ASEE, 1956- 57. National Asian Productivity Organization, 1972 and 1976 with assignments in Hong Kong and Pakistan. *Society Aff:* IIE, ASME, APICS, ORSA, SME, ASEE

Gould, David S
Business: Caterpillar Tractor Co, 100 N.E. Adams St, Peoria, IL 61629
Position: Exec VP *Employer:* Caterpillar Tractor Co. *Education:* BS/Met/MO Sch of Mines; MS/Met/MO Sch of Mines; PhD/Univ of MO; MS/Ind Mgt/MIT; DEng (Honorius Causa)/-/Univ of MO-Rolla. *Born:* 8/28/26. in Decatur, IL. Indus career with Caterpillar Tractor Co incl pos in met & mfg; Ch Metallurgist 1964-69; Plant Mgr 1971-76. Mbr AFS, ASM & SAE. Reg Prof Engr. *Society Aff:* SAE, ASM.

Gould, Gerald G
Home: 2329 Magnolia Dr, Panama City, FL 32401
Position: Consultant *Employer:* Gerald Gould Associates, Inc. *Education:* MS/EE/Purdue Univ; BS/EE/NY City College. *Born:* Nov 1913. Started in design of high voltage equip. WWII Mine Warfare Officer in US Navy. R & D for torpedos, fire control, & launching systems for submarines. Instrumental in the dev of the Atlantic Underseas Test & Eval Ctr (AUTEC) in the Bahamas. Tech Dir of two of the Navy's major R&D Labs since 1955, dir the total tech efforts in ASW, mine, & coastal tech R&D. Assumed respon as Tech Dir, NCSL, 1972 to 1981. 1981-present Pres, Gerald Gould Assoc, Inc. Consultants in Engrg and Mgmt. Reg Prof Engr in NY, RI, & FL; Fellow IEEE. Received Navy League RADM Parsons Award in 1975 & IEEE Reg III Outstanding Engr Award in 1973. Navy Superior Service Award 1981. *Society Aff:* IEEE.

Gould, Merle L
Home: 1134 Lee Dr, Baton Rouge, LA 70808
Position: Chrmn & CEO *Employer:* Benedict & Myrick, Inc *Education:* BS/CE/NB Univ. *Born:* 3/10/19. Spent 32 yrs in petrochem industry. Over 16 yrs in R&D; mostly with Shell Dev Corp & Ethyl Corp. Responsible for maj concepts, process designs & mech designs in the fields of ethylene oxide, fluidized bed ethane chlorination, orthoalkylated phenols, aluminum alkyls, etc. A number of patents in reactor design, process design, separations, high velocity (high temperature) gas mixing, chem reactions, etc. Active in formation & early operation of P&P Div of AIChE. Involved in energy projs. They include coal gasefication to synthesis gas for combined cycle power generation & chemicals, studies on possible nonbiological sources of methane, solubility of methane in Geopressured Zones and Potential Meth in Natural Gas Recovery From Geopressured Zones. Temporarily CEO of an Electronic Sec Devices Co in a Turnaround situation. *Society Aff:* AIChE, ΣΞ, AAAS.

Gould, Phillip L
Business: Washington University, Box 1130, One Brookings Drive, St Louis, MO 63130
Position: Harold D. Jolley Prof & Chrmn. of Dept. of Civil Engrg. *Employer:* WA Univ. *Education:* BS/Civ Engg/Univ of IL; MS/Civ Engg/Univ of IL; PhD/Civ Engg/Northwestern Univ. *Born:* 5/24/37. Active in the field of hyperbolic cooling tower shells where he has conducted a number of res projs and has served as a consultant to industry and to governmental agencies. Published over 100 technical papers and authored four books. Founding editor of the journal Engineering Structures and has received a US Scientist Award from the Alexander V Humboldt Fdn in W Germany. Has lectured at many universities in Europe, Asia and Australia. Served as Visiting Professor at the Univ. of Sydney, Australia. *Society Aff:* ASCE, ASEE, IASS, AAM, ΣΞ, EERI.

Gould, Roy W
Business: M S 104-44, Pasadena, CA 91125
Position: Prof & Div Chrmn. *Employer:* California Inst of Tech. *Education:* BS/EE/CA Inst of TEch; MS/EE/Stanford Univ; PhD/Physics/CA Inst of Tech. *Born:* 4/25/27. Prof of Elec Eng & Physics CA Inst of Tech; 1955- , Exec Off for Applied Physics 1973-79, present Chrmn Div of Engg & Applied Science ; 1979-, Dir Div of Controlled Thermonuclear Res, US Atomic Energy Commission 1970-72. NSF Sr Post Doctoral Fellow 1963-64. Fields of interest plasma physics, controlled thermonuclear fusion, microwaves. Elect Fellow IEEE 1965, elect mbr NAE 1971. Elec Mbr NAS 1974. Chem Div of Plasma Physics 1973-74 APS. *Society Aff:* NAE, NAS, APS, IEEE.

Gould, William R
Business: 2244 Walnut Grove Ave, Rosemead, CA 91770
Position: Chrmn of the Board Emeritus *Employer:* So California Edison Co. *Education:* BS/Mech Engrg/Univ of UT; Post Grad/-/MIT; Post Grad/-/UCLA; Post Grad/- /Univ of ID *Born:* 10/31/19. in Provo, UT. US Navy 1942-47. With So CA Edison Co since 1948. Filled var pos in Engg, Const & Oper. Exec VP 1973-78/Pres 1978-80; Chairman/BD & CEO since 1980. Chrmn of the Bd Emeritus since Nov. 1984. Indust Forum 1973-74. Pres US Natl Ctte CIGRE since 1973. Chmn Western Sys Coord Council 1971-72. Chmn Inst for Advancement of Engg since 1974. Fellow ASME. Elect Engr of Yr IAE 1970; elect NAE 1973. Trustee CA Inst of Tech, Mbr Natl Advisory Ctte Univ of UT. Dir Kaiser Steel Corp, Union Bank Eyring Res Inst. Chrmn, Electric Research Power Institute, 1981. *Society Aff:* ASME, CIGRE.

Goulding, Randolph
Business: 2000 Clearview Ave NE, Atlanta, GA 30340
Position: Exec Vice Pres *Employer:* Jordan, Jones & Goulding Inc. *Education:* BCE/Civil Engg/GA Inst of Technology; BBA/Finance & Econ/GA State Univ. *Born:* 6/16/24. Reg PE FL & GA. Cert of Qualification Natl Council St Bd of Exams. Served in Air Force WWII 1942-46 & Korean War 1951-53. Proj Engr with cons engrg firm 1949-51, 1953-63. Principal J J & G since 1963. Chmn SE Sect AWWA 1970, Intl Dir AWWA 1973-76, Exec Ctte 1974-76. Pres CEC GA 1981-82. Dir ACEC 1982-83. Mbr AWWA Engrg and Construction Div, Mbr AWWA T&P Council, Mbr ACEC 708 Environmental Ctte. Enjoy golf & tennis. *Society Aff:* AWWA, WPCF, ACEC.

Gouse, S William
Business: 7525 Colshire Drive, Mclean, VA 22102
Position: Sr VP & GM *Employer:* The Mitre Corp *Education:* ScD/ME/MIT; SM/ME/MIT; SB/ME/MIT *Born:* 12/15/31 Received his Doctor of Science degree from MIT in 1958. Since then he has served on the faculty of the Mechanical Engrg Dept at MIT and at the Carnegie Inst of Tech. In addition, he has been employed with Atomics International, has served as Technical Assistant for Civilian Tech to the President's Science Advisor, Exec Office of the Pres, and as Associate Dean of the College of Engrg and the School of Public and Urban affairs at Carnegie-Mellon Univ. He was also the founding Dir of an Environmental Studies Institute at Carnegie-Mellon University. In the US Dept of the Interior he was the Dir of the Office of Research and Development, Science Advisor and Assistant to the Secretary of Interior as well as Acting Dir of the Office of Coal Research. In the Energy Research and Development Administration he served as Deputy Assistant Administrator for Fossil Energy. *Society Aff:* AAAS, AIAA, ASME, SAE, ASEE, IAEE.

Govier, George W
Home: 1507 Cavanaugh Pl NW, Calgary, Alberta, Canada, T2L 0M8
Position: Pres (Self-employed) *Education:* ScD/CE/Univ of MI; MSc/Chemistry/Univ of Alberta; BASc/CE/Univ of British Columbia. *Born:* 6/15/17. L1D, PE, was born in Nanton, Alberta in 1917. From 1940 to 1963 Dr Govier was associated with the Engg Faculties of the Univ of Alberta & the Univ of Calgary. He has been a mbr of Alberta's Energy Resources Conservation Bd since 1948 & was Chrmn of the Bd from 1962 to 1978 with a two yr leave of absence serving as Chief Deputy Minister of Energy & Natural Resources for the Govt of Alberta beginnning Oct 1, 1975 to Oct 1, 1977. He is Pres of his own firm, Govier Consulting Services Ltd, Chrmn of the Bd of Dirs of International Permeation Inc & is on the Bd of Dirs of Canadian Foremost Limited; Texaco Canada, Inc. Canadian-Mt Gas Co, Ltd; Canadian-Mt Pipe Line Co; Room Resources Supply Ltd; Stone & Webster Canada Ltd; Western Gas Marketing Ltd.; Cooperative Energy Development Corp. He is author & co- author of over 60 tech papers. He has received a number of awards and honors. *Society Aff:* AIChE, EIC, CIC, APEGGA, CIM.

Gowdy, John N
Business: Riggs Hall, Clemson, SC 29631
Position: Prof (Elec & Comp. Engr) *Employer:* Clemson Univ *Education:* PhD/EE/U of MO; MS/EE/U of MO; BS/EE/MIT *Born:* 2/7/45 Teaching and res in the areas of microcomputer applications and digital signal processing, NSF sponsored projects for automatic speech recognition and dev of a computer design lab. Industry-sponsored projects on process monitoring by computer. Extensive continuing education involvement. Grad Prog Coordinator for E & CE dept. at Clemson Univ. *Society Aff:* IEEE

Gowen, Richard J
Business: 500 St Joseph St, Rapid City, SD 57701
Position: President *Employer:* South Dakota School of Mining and Technology *Education:* Ph.D./Elect Engr/Iowa State University; M.S.EE./Elect Engr/Iowa State University; BS EE/Elect Engr/Rutgers University *Born:* 07/06/35 Engineer, educator, researcher, manager. 1957 Research Engr, RCA Labs; 1957-77 Air Force Officer; 1962-72, Air Force Academy, Faculty and Director, Air Force/NASA Space Medical Instrumentation Lab with five major astronaut medical experiments; Consultant, DOD, NASA, NIH, OMB and industries in management, health care, manpower, computers; 1976-86 IEEE Bd Officer & 1984 Centennial Pres; 1981- AAES Bd Officer & 1988 Bd Chairman; 1977-84, South Dakota Tech, VP & Dean of Engg; 1984-87, Dakota State College Pres, changed mission to center to enhance learning through computers in liberal arts, business & education; 1987- South Dakota Tech Pres, lead renewal & redirection; 1983- ETA systems, Founding Bd, new supercomputer company. *Society Aff:* IEEE, ASEE, AAAS, NSPE, RMBS

Graber, G F
Business: 3800 Stone School Rd, Ann Arbor, MI 48104
Position: President. *Employer:* Applied Dynamics Internatl. *Education:* BS/EE/Rutgers Univ. *Born:* June 1931. Grad study at Princeton U. 1st Lt Army Signal Corps 1951-54; Sales Engrg & Sales Mgmt, Electronic Assocs 1959-61; V P Marketing, Applied Dynamics 1962-69; past Dir of Soc for Computer Simulation; Pres of ADI since early 1975. *Society Aff:* SCS, IEEE.

Graber, S David
Home: 118 Larson Road, Stoughton, MA 02072
Position: Consulting Engineer. *Employer:* Self. *Education:* -/Civil Engg/MIT; SM/Mech Engg/MIT; BS/Mech Engg/Univ of Miami. *Born:* 1/12/42. San Engr Army Med Serve Corp 1967-69. Stationed Panama Canal Zone, Rank of Capt. Advisor to US Agency for Intl Dev on rural water supply. With Camp Dresser & McKee Inc from 1966-67, 1969-74 most recently as Dir Hydraulic Services. With Metcalf & Eddy Inc from 1974-77, most recently as Wastewater Div Tech Dir. Established Consulting Engg firm 1977. Principally work on planning & design of environmental engrg. facs. J C Stevens Award 1972 ASCE, Samuel A Greeley Award 1969 ASCE. *Society Aff:* ASCE, ASME, WPCF, TBΠ.

Grace, J Nelson
Business: Regional Administration, U.S. Nuclear Regulatory Comm, 101 Marietta St, Atlanta, GA 30323
Position: Regional Administrator (R-II). *Employer:* US NRC. *Education:* PhD/EE/Carnegie Inst of Technology; MS/EE/Carnegie Inst of Technology; BS/EE/Carnegie Inst of Technology. *Born:* 5/27/24. 1951-72 Westinghouse Bettis Atomic Power Lab serving Naval Reactors Program in var capacities in I&C, Reactor Plant Kinetics & Safety, Reactor Physics & Engg. 1972-74 Westinghouse Hanford Co serving as Mgr Reactor & Safety Engg Dept. 1974-79 Asst Dir for Technical Projs, Div of Controlled Thermonuclear Res AEC/ERDA/DOE. 1979-1983 Dir, Princeton Fusion Program Office DOE. 1983-84: Director, CRBR Program Office USNRC; 1984-85: Director, Division QA, Safeguards & Inspection Programs USNRC; 1985-present: Regional Administrator, R-II NRC; responsible for safety oversight & regulation of 33 operating power reactors, 5 under construction, 6 fuel facilities & other nuclear licensees in SE USA. Pub sev papers in reactor tech. Chmn Pittsburgh Sect IEEE & ANS. Elected Fellow ANS 1967. *Society Aff:* ANS, IEEE.

Grace, Richard E
Home: 2175 Tecumseh Park Lane, West Lafayette, IN 47906
Position: VP for Student Services *Employer:* Purdue University. *Education:* BSMetE/Met Engg/Purdue Univ; PhD/Met Engg/Carnegie Inst of Technology. *Born:* 6/26/30. Engrg educator; B.S. MET.E., Purdue Univ 1951; Ph.D., Carnegie Institute of Technology 1954; married Consuela Cummings Fotos Jan 29, 1955; children, Virginia Louise, Richard Cummings (deceased). Asst Prof Purdue U 1954-58, Assoc Prof 1958-62, Prof 1962- ; Head School of Materials Sci & Met Engrg 1965-72; Head Div of Interdisciplinary Engrg Studies 1970-82; Head Dept of Freshman Engrg 1981-87; Asst Dean of Engrg 1981-87; VP for Student Services 1987- . Also Registered PE NJ. P Chrmn Engg Educ & Accred Ctte of Engrs Council for Prof Dev, Mbr ASM (Teachers Award 1962, Fellow 1975), AIME, ASEE, Amer Assn of Univ Profs, Tau Beta Pi, Sigma Xi, Omicron Delta Kappa, Phi Gamma Delta, Rotarian, Elk. *Society Aff:* AIME, ASM, ASEE, AAUP

Grace, Thomas M
Business: 1043 E. South River St, Appleton, WI 54915
Position: Prof of Chem Engr *Employer:* Inst of Paper Chemistry *Education:* PhD/ChE/Univ of MN; BS/ChE/Univ of WI *Born:* 10/03/38 Native of Beaver Dam, WI. Started career with NASA Lewis Res Ctr working on dynamics of two-phase systems. Joined Inst of Paper Chem in 1965. Specialty is Chem Recovery Tech. Monitor recovery boiler explosions for the API Recovery Boiler Committee. Secy of the Emergency Shutdown Procedures Committee of the Black Liquor Recovery Boiler Committee. Actively researching black liquor combustion. Hobbies include cross-country skiing, hiking and running. *Society Aff:* TAPPI, AIChE, APCA

Grader, Jerome E
Home: 245 Jamestown Terr, Rochester, NY 14615
Position: Manager. *Employer:* Taylor Instru Process Control Div. *Education:* BS/Chem Eng/Purdue Univ. *Born:* 3/26/36. Rochester, NY. BS Chem Engrg Purdue Univ 1957. Grad study in control theory Cast Inst & Brooklyn Polytech. With Taylor Instrument Co (Rochester NY) since 1957, initially as application engr & later as sys engr. Author num tech articles on control theory & application. Mgr Sys Engrg 1970; Mgr Electron R&D 1972; Mgr prod Mkt 1976; Area Exec-Latin Amer Operations 1979. Mbr-AIChE, Sr Mbr ISA. Dir ISA Chem & Petrol Indus Div 1974; VP Industs & Scis Dept 1979. Interested in most sports, especially sailing. Active in community organizations dedicated to intl understanding; Pres Rochester Intl Friendship Council 1974. *Society Aff:* ISA, AIChE.

Grady, C P Leslie, Jr
Business: Environ. Sys. Engg, Rhodes Res. Center, Clemson University, Clemson, SC 29634-0919
Position: R.A. Bowen Prof of Environ Systems Engg *Employer:* Clemson University. *Education:* PhD/Environ Engg/OK State Univ; MS/Environ Engg/Rice Univ; BSCE/Civil Engg/Rice Univ; BA/Engr/Rice Univ. *Born:* June 25, 1938. Prof Exper: Charles R Haile Assocs Houston TX 1963; US Army Med Serv Corps 1963-65; Purdue U 1968- 1981; Visit Scholar U Texas - Austin 1975-76; Clemson U. 1981- Res Area: biological wastewater trtmt. Teaching Areas: phys & biochem unit opers; wastewater treatment process design. Author 80 tech pubs; 1 Textbook. Bd/Dir Assn of Environ Engrg Profs 1976-79. Merck Foundation Career Dev Award 1971. NSF Faculty Fellowship 1975-76. McQueen Quattlebaum Faculty Achievement Award, Clemson Univ., 1984, Service Award, Water Pollution Control Federation, 1983. *Society Aff:* WPCF, ACS, ASM, AEEP, ASEE, IAWPRC, AICHE

Gradziel, Albert Z
Home: 104 Arbor Place, Bryn Mawr, PA 19010
Position: Vice President *Employer:* ARCO Chemical Co. *Education:* BS/ChE/Univ of PA. *Born:* April 3, 1923. Native of Hartford Ct. US Army Infantry 1944-46. Atlantic Richfield Co 1943-67. Genl Mgr Oxirane Chem Co 1967-69. Oxirane Corp Oper V P 1969-76. Pres Oxirane Corp. 1976-1980 Mbr AIChE, Currently manufacturing V.P., ARCO Chemical Co., Alpha Chi Sigma. *Society Aff:* AIChE, ΣΤ, ΑΧΣ.

Graef, Luther W
Business: 6415 W Capitol Drive, Milwaukee, WI 53216
Position: Board Chairman & Sec *Employer:* Graef, Anhalt, Schloemer & Assocs.

Graef, Luther W (Continued)
Education: MSCE/Civil Engg/Univ of Wi; BCE/Civil Engg/Marquette Univ. *Born:* 8/14/31. Native Milwaukee WI, US Army Artillery 1953-56. Discharged Rank 1st Lt. Design Engr C Yoder & Assoc 1956-61. Started own cons firm 1961. Spec in Cvl & Struct Engrg. Respon for Proj Design, & Office Mgmt. P Pres ASCE WI 1969, CEC WI 1974-76. Chmn Indus Advisory Cttee UWM 1976-78. Pres Engrg Scientists of Milwaukee 1977. Active in local civic, church, educ & scout organizations. Past Rept, Distinguished Service Award Wis ASCE: Disting Alumnus Award Marquette 1982, Engr of Year WSPE 1983. *Society Aff:* NSPE, ASCE, APWA, ACEC, NSPE, SAME

Graeser, Henry J
Business: 5728 LBJ Freeway, Suite 300, Dallas, TX 75240
Position: Retired *Employer:* Self-employed *Education:* B.S/Sanitary Engr./Tex. A & M Univ. *Born:* Dec 1915. BS Cvl & San Engrg Texas A&M College 1938. San Corps 194245; Ch of Engrs US Army 1945-50; City of Dallas Water Utilities Dept 1950, Dir 1955- 75 (ret). Black & Veatch Cons Engrs 1976 (ret. 1986). Spec interest mgmt of water & wastewater utilities & water resources plan & dev. AWWA Pres 1967-68, Fuller Award, Diven Award, Hon Mbr, Distinguished Public Service Award. Dipl AAEE; APHA - Kiwanis Internatl, Public Works Man of the Year 1963; TSPE Dallas Chap Engr of the Year 1969; Outstanding Achievement Award, Civil Engineering, Dalls Chap ASCE 1976. Natl Clay Pipe Inst Dist Serv Award 1973; Environ Protection Agency Public Service Award 1976, EPA Environimental Quality Award Region VI 1977. *Society Aff:* AAEE, AWWA, ASCE, NAWC

Graf, Edward D
Home: 715 San Miguel Lane, Foster City, CA 94404
Position: V-P, Technical Services *Employer:* Pressure Grout Co *Education:* BS/Bus. Admin./UCLA *Born:* 12/31/24 Native of Los Angeles. Commanded a Sub-Chaser during WWII. Worked in heavy construction since apprentice carpenter in 1940. Pres of Grout Company for 31 years specializing in geotechnical grouting throughout the world. Issued 6 U. S. patents in the field of soil stabilization. Guest lecturer at C.E. graduate schools including Stanford, UCLA, U.C. Berkeley, Purdue, GA Tech, U of ILL, Northwestern, etc. in addition to geotechnical seminars. Five technical papers published (ASCE). Teaches course in Geotechnical Grouting at Stanford Univ. Graduate School of Engineering. Does consulting work regarding Geotechnical Grouting. Former pres of San Francisco Branch, ASCE. Active in technical committee work of ASCE, ACI, ASTM, TRB. *Society Aff:* ASCE, AIME, SAME, ACI, ASTM

Graff, Karl F
Business: Dept of Welding Engg, 190 W 19th Ave, Columbus, OH 43210
Position: Prof & Chmn. *Employer:* Ohio State University. *Education:* PhD/Eng Mech/Cornell; MS/Eng Sci/Purdue ; BS/Eng Sci/Purdue. *Born:* June 22, 1936. Entered Ohio State U as Asst Prof, becoming Prof of Engrg Mech in 1970 & Chmn of the Engrg Mech Dept from 1972-77. Chrmn of Welding Engg, 1979. Author of the book 'Wave Motion in Elastic Solids'. Res & cons interests in the areas of vibrations, impact, acoustics & ultrasonics. *Society Aff:* AWS, ASNT, IEEE.

Graham, A Richard
Business: Ctr for Prod. Enhancement, The Wichita State University, Wichita, KS 67208-1595
Position: Prof of Mech Engr & Dir, Center for Prod Enhancement *Employer:* Wichita State Univ. *Education:* PhD/Mech Engr/Univ of IA; MS/Mech Engr/KS State Univ; BS/Mech Engr/KS State Univ. *Born:* 8/5/34. Joined the Mech Engg faculty at Wichita State Univ in 1965 Dept Chrmn 1978- 84. Mbr of the ME Faculties at KS State Univ, Univ of MO-Rolla and Univ of IA at various times. Primary teaching interest in area of instrumentation and experimentation. Author of book "An Introduction to Engg Measurements–. Served as Program Chrmn and Chrmn of ASEE Div for Experimentation and Lab Oriented Studies & Instrumentation Div Chrmn, Midwest section (84-5); Chrmn, ASEE Zone III, 1987-89. Current activities include res in application of advanced manufacturing systems. *Society Aff:* ASME, ASEE, SME.

Graham, Beardsley
Home: P O Box 567, Yucca Valley, CA 92284
Position: Consultant. *Employer:* Self. *Education:* BS/Chem/UC Berkeley. *Born:* April 1914 Berkeley Calif. Grad study EE & Phys UC & Columbia 1939-42. Asst Dir Stanford Res Inst 1951-57; Pres Spindletop Res 1961-67; Cons 1967- . Found Sec & Dir Internatl Solar Energy Soc 1954-70; Appointed by Pres Kennedy as Incorporator, Communications Satellite Corp 1962-63 & Bd Mbr 1963-64; Group Chmn & Mbr Central Review Ctte, NAS's Space Applications Study 1967-69; Advisor Broadcasting & Cable TV, Ctte for Economic Dev 1973-75. Life Fellow IEEE; Assoc Fellow AIAA. *Society Aff:* IEEE.

Graham, Charles D Jr
Business: Dept of Mtls Sci & Engg, Univ. of Pennsylvania, 3231 Walnut St, Philadelphia, PA 19104-6272
Position: Prof *Employer:* Univ of PA. *Education:* PhD/Met/Univ of Birmingham; BMetE/Met Engg/Cornell Univ. *Born:* 10/15/29. Fifteen yrs at GE Res & Dev Ctr, Schenectady, NY, 1960-61, Guggenheim Fellowship at Inst for Solid State Phys, Univ of Tokyo 1978, Sci Res Council Fellow at Univ College, Cardiff, UK. 1985, Co-operative researcher, Research Development Corp of Japan, Inst. for Electric & Magnetic Alloys, Sendai, Japan. Dept. Chairman, Dept. Materials Science & Engr., U of Pa, 1979-84. Research Interests - magnetic materials & measurements. Five yrs as co-editor of transactions of Annual Conf of Magnetism & Magnetic Mtls. *Society Aff:* AIME, IEEE, APS, Phys Soc Japan.

Graham, Harry T
Business: 225 E. 6th St, Cincinnati, OH 45202
Position: Principal. *Employer:* Graham, Obermeyer & Partners Ltd *Education:* CE/Univ of Cincinnati. *Born:* Oct 1919. Field Engr on dams, tunnels & powerhouses with Utah Const Co & Morrison Knudsen Co. Bridge design with NY Central RR. Joined present firm 1949, became principal 1962. Main field is structure for industry. Pioneered in steel hyperbolic paraboloids (1959). Local pres ASCE 1972-73. Mbr, ACEC, AREA, Optimist Internatl. Reg OH, PA, NJ, WV, KY. *Society Aff:* ASCE, AREA, ACEC, OACE.

Graham, Jack H
Business: 2000 Classen Blvd, Oklahoma City, OK 73106
Position: Pres/Engr *Employer:* Graham & Assocs Professional Consltg Engrs, Inc *Education:* MBA/Finance/OK Univ; BS/EE/OK State Univ *Born:* 3/16/37 Native of OK City, OK. married JoAnn Lacy; 3 children; Pres of Graham & Assocs Professional Consltg Engrs, Inc an Elec/Mech consulting firm. Registered PE OK & AL. Elder Christian Church. Past Pres Lions. Treas OK S. Prof Engrs. Worked in Missile and Space Industry 11 years. Who's Who in South & SW. Graduated OSU 1959. Received MBA 1974. Received citations for business and civic contributions. Enjoys golf, whittling, reading, and grafting trees. *Society Aff:* NSPE, IEEE

Graham, Jack M
Home: 4372 Westdale Ct, Ft Worth, TX 76109
Position: Exec VP. *Employer:* Farrington & Assoc, Inc. *Education:* BS/Civil Engr/ Southern Methodist Univ. *Born:* 1/18/23. Native, Dallas, TX. Served with Army Air Force as Meterologist and with Corps of Engrs in Pacific Theater, 1942-1946. Air Force Reserve for 26 yrs, retired Lt Col, 1969. Design engr, Asst to City Mgr and Asst Public Works Dir, City of Dallas, 1947-1959. Public Works Dir, City of Corpus Christi, 1959-1966. Public Works Dir, City of Ft Worth, 1966-1978. Assumed present position as part owner of consulting firm, 1978. Pres, TPWA, 1962. Pres, Ft Worth Branch, ASCE, 1969. Bd of Trustees, Public Works Historical Soc, 1975-1978. Masonic orders. Elder, Christian Church. *Society Aff:* APWA, ACE, NSPE, TPWA.

Graham, James S
Home: 446 Park Ave, Rye, NY 10580
Position: Staff Engr *Employer:* Natl Timber Piling Council *Education:* MS/Civil/Univ of Pittsburgh; MA/Economics/Univ of Pittsburgh; BS/Civil/Univ of Pittsburgh *Born:* 12/22/32 Live in Rye, NY. Born and raised in Pittsburgh, PA. Served with Army Corps Engrs, 1954-56 in Kobe, Japan. Wrote draft and supervised final edition of US Steel Sheet Piling Design Manual. Contract negotiator for claims, changes and wrap-up on Washington, DC Metro Subway proj. Presently Staff Engr for Natl Timber Piling Council. This work involves pile foundations, bulkheads, bridge fenders, highways, and residential use of pressure treated wood. *Society Aff:* ASCE, IEEE, ASTM.

Graham, John
Business: Box 800, Richland, WA 99352
Position: Mgr Licensing *Employer:* Rockwell Hanford Operations *Education:* BSc (Hons First)/Math/Univ of Wales; Fulbright Fellow/Math/Univ of IL; Grad Study (doctoral) Theoretical Phys/Univ of London *Born:* 01/25/33 Educated in math, leading to quantum mechs and reactor kinetics, he applied this expertise in the evaluation of the safety of a number of different reactor types. These ranged from water to sodium cooled, fast and thermal variants. Most of this evaluation work was performed at the UK Atomic Energy Authority from 1958-1968. From 1968-1984 he led the Westinghouse effort in liquid metal fast breeder reactor safety first on the Fast Flux Test Reactor and then on the Clinch River Breeder Reactor. In addition he set divisional safety guidelines and initiated national safety policies. He was a member of the US delegation on successive visits to Japan, the Soviet Union and France in LMFBR safety work. Since 1984 he has managed the licensing effort and preclosure safety assessment for the Basalt Waste Isolation Proj at Rockwell Hanford Operations. *Society Aff:* ANS, ACSM

Graham, Lois
P.O. Box 221, Edwards, NY 13635
Position: Prof Emeritus *Employer:* Illinois Inst Tech; Retired *Education:* PhD/ME/IIT; MS/ME/IIT; B/ME/RPI *Born:* 4/4/25 Test engr Carriercorp 4/46-9/48 with IL Inst of Tech since 1948. Assist Chrmn, ME Dept 1950-1970. Dir Minorities in Engr Programs 9/78-9/81. Dir Women's Engrg Program 9/75-8/85. Secretary American Power Conference Fellow, Trustee Past President SWE, Fellow AAAS, Recipient 1980 Ralph R. Teetor Awards, 1980 IIT Alumni Professional Achievement Award, 1980 RPI Alumni Key Award, 1982 Chicago Associated Tech Societies Merit Award, Elected mbr of the Academy of CATS 1984. 1981 Metropolitan Chicago YWCA Outstanding Achievement Award. *Society Aff:* ASHRAE, ASME, SWE, ASEE.

Graham, Malise J
Business: P O Box 8405, Kansas City, MO 64114
Position: Proj Mgr *Employer:* Black & Veatch *Education:* MS/Environ Eng/KS Univ; MS/Structural Eng/MO Univ; BSc/CE/Cape Town Univ, South Africa *Born:* 11/14/30 After graduation worked for 3 years in South Africa and Rhodesia for consultants on water and sewerage projects. Following a year on construction site in London and a year in office structural engr in Glasgow, came to US as research assistence. Joined B & V on graduation and worked on environmental design projects for 2 years. Returned to Glasgow for further 2 years. After which returned to B & V. With B & V has specialized in design and feasibility reports for water supply and sewerage. Spent 3 years in Manila, RP and 2 years in Detroit. Reg PE in MO, KS, MI. *Society Aff:* ASCE, WPCF, AAEE, ICE, WPC, NSPE, AWWA, ΣΞ

Graham, Robert W
Home: 22895 Haber Dr, Fairview Park, OH 44126
Position: Chief Office of Tech Assessment *Employer:* NASA-Lewis *Education:* PhD/Mech. Eng./Purdue Univ.; MSME/Mech. Eng./Purdue Univ.; BSME/Mech. Eng./Case Inst. *Born:* 10/10/22 First joined the staff of the Lewis Research Center in 1948. Have specialized in turbomachinery, heat transfer and energy systems research. Co-authored a text book "transport processes in boiling & two phase systems–. I am the author or coauthor of approximately 100 technical papers. My technical society activities have been mostly in ASME. I have chaired the heat trans. div; chairman of Issues Management Bd, served as senior VP of the Council on Public Affairs. Currently I am a member of the Bd of Governors of ASME. My job assignment is Chief of the Office of Tech. Assessment for the Lewis Research Center. *Society Aff:* ASME, AIAA

Graham, Thomas A
Home: Gypsy Hill Rd, Gwynedd Valley, PA 19437
Position: Pres. *Employer:* Metric Mgt & Engg. *Education:* Bachelor of Engr/Civil/ Villanova Univ. *Born:* 6/8/32. As pres of Metric Mgt & Engg, Mr Graham has designed Power Plants in Sanna, Yemen, CPM Scheduling of Desalinization plants in Saudia Arabia and numerous energy conservation studies in the US. He has been the principal in charge of greater than 100 projs ranging from modest industrial and hospital conversions to vast complexes such as Phila's Airport, costing $120 Million and Bellevue Hospital in NYC costing $80 Million. He has been a Consultant in construction claims and is an Originator of computer applications to CPM Scheduling and to Energy Conservation Progs. *Society Aff:* NSPE.

Graham, W William, Jr
Business: 100 N Rodney Parham Rd, Little Rock, AR 72205
Position: President. *Employer:* W William Graham Jr Inc. *Education:* MSE/Sanitary Engr/Univ of AR; BSCE/Gen Civ/Univ of AR. *Born:* March 13, 1925 DeQueen Ark. Educ in Little Rock Public School Sys; attended U of Ark; grad in 1949 BS Cvl Engrg; MS Engrg Grad Inst of Tech at U of Ark at Little Rock 1974; P Pres of Ark CEC; represented Ark as a Natl Dir to ACEC. V P ACEC elected 1975 - 77. Mbr AWWA; Fed of Sewage & Indus Waste; ASCE. Served many natl cttes ACEC. Mbr of the Presbyterian Church having served as an Elder. Appointed as a Fellow in ACEC and a mbr of the Arkansas Academy of Civil Engrs. The engrg firm has designed and observed the construction of numerous rural water sys with a combined aggregate of over 2,000 miles of pipe line. Married and has 3 children and one grandson. Hobbies: building furniture and hunting. *Society Aff:* ACEC, AWWA, ASCE.

Grandle, Edward D
Home: 530 Bolton Pl, Houston, TX 77024
Position: President *Employer:* Grandle Associates, Inc. *Education:* BSCE/Civil Engrg/Univ of KS. *Born:* 1930. Postgrad studies at Delft Tech Inst Holland; Served in USAF 1951- 55. With Raymond Internatl Inc 1959-1983 as Sr. Vice President. Prior employment with MerrittChapman-Scott & with Texstar Gp of companies. Established Grandle Associates Inc. in 1983 offering construction consulting services to contractors, consulting engineers, and project owners. *Society Aff:* ASCE, The Moles

Grandy, Charles C
Business: 7525 Colshire Drive, McLean, VA 22102
Position: VP *Employer:* Mitre Corporation *Education:* MS/Physics and Math/ Northeastern Univ; MS/Mgmt Sci/Amer Univ; BS/Math/CO State Univ *Born:* 12/06/28 Native of Longmont, CO. US Army Security Agency 1946-48. Joined Digital Computer Lab at MIT in 1952; MIT Lincoln Lab 1953; Mitre Corporation in 1958. Assumed current position in 1979. Was VP WA Operations 1972-79. Background in Real-Time Computer Control Systems, Military Command, Control and Communications Systems and Air Traffic Control Systems. Currently has general mgmt responsiblity for (non-defense) work in Air Traffic Control, Advanced Info Systems, Criminal Justice Systems, Energy and Environment, and Resource Recovery. Active in community affairs and church. *Society Aff:* AAAS, AMA

Granger, John V
Business: Box 40, FPO, NY 09510
Position: US Embassy. *Employer:* Dept of State. *Born:* Sept 1918 Cedar Rapids IA.

Granger, John V (Continued)
MS & PhD Harvard U; BA Cornell College (Iowa) Natl Defense Res Council 1942-45. Asst Dir Engrg Res Stanford Res Inst 1949-56. Pres Granger Assocs 1956-70. Dep Dir Bureau of Internatl Sci & Tech Affairs, Dept of State 1971-75. Exec Secy Fed Council for Sci & Tech 1975. Acting Asst Dir for Sci, Tech & Internatl Affairs National Science Foundation 1976-77. Counselor for Sci & Tech Affairs, US Embassy, London, 1977-81. Pres IEEE 1970. NAE.

Granstrom, Marvin L
Home: 931 Oakwood Place, Plainfield, NJ 07060
Position: Prof II Emeritus *Employer:* Rutgers, The State Univ *Education:* PhD/Sanitary Eng/Harvard Univ; MS/Sanitary Eng/Harvard Univ; BS/CE/IA State *Born:* 9/25/20 My teaching at the graduate level for 30 years (Dept Chrmn for 12 years) was combined with res in water and wastewater treatment, chemistry of disinfectants, hydrology-hydraulics, sediment transport, regional water resource management, and hazardous waste treatment, and included supervision of 13 PhD theses. Consulting work on nuclear energy plant projects: floods and flood waves, droughts, emergency core cooling systems, hurricane surges and ocean waves, sea wall design, and environmental impacts - worked on some ten different plant locations in US and Brazil. Other consulting work includes: treament plant design and operation, flood, analysis and design of control structures, and engrg economic analysis. Retired July 1, 1983 from Rutgers Univ. However, continuing engr consulting. *Society Aff:* ASCE

Grant, Albert A
Business: 1875 Eye St. N.W, Suite 200, Washington, DC 20006
Position: Sr Transportation Advisor *Employer:* Metro. Wash. Council of Gov'ts *Education:* B/CE/Catholic Univ of America *Born:* 10/9/26 Employed by D.C. Dept of Hwys and Traffic 1949-1966. Served as Bridge Design Engr and Chief of Planning and Programming. Dir of Transportation Planning, Metropolitan Washington Council of Governments, 1966-87. Responsible for areawide highway, transit and aviation system planning. Registered PE in DC. National Pres, ASCE, 1967-68. 1981 recipient of ASCE Harland Bartholomew Award. Has been active in Transportation Research Board Committees and ABET accreditation activities. Has served as consultant to various Federal and State agencies. Academic experience includes teaching graduate course in Urban Transportation Planning at Howard and American Universities. *Society Aff:* ASCE, ASEE, ITE.

Grant, Arthur F, Jr
Home: 1432 Via Catalina, Palos Verdes Est, CA 90274
Position: Retired *Employer:* TRW Inc. *Education:* BS/Chem Engg/Univ of PA. *Born:* Aug 1921. BS U of Pennsylvania. Native of PA. Naval Off WWII Pacific Theatre. Chem Engr for Union Carbide Corp. Ch of the Power Plant Res Sect at the CA Inst of Tech JPL. With TRW since 1958. Currently VP & Genl Mgr Applied Tech Div. Respon for TRW's work in chem, chem engrg, fluid mechs, materials & advanced phys res; & for the laser, propulsion, space instrument, & energy process product lines. Mbr Alpha Chi Sigma, AIChE, ACS, Sci Res Soc of Amer, & Sigma Xi. Enjoy piano, hi-fidelity, & woodworking. *Society Aff:* AIChE, ACS, RESA, ΣΞ, ΑΧΣ.

Grant, Charles H
Home: Star Route, Box 3290, Tower, MN 55790
Position: Pres *Employer:* Energy Sciences & Conslt, Inc *Education:* Dipl/Mining Eng/WI Inst of Tech *Born:* 07/08/27 Chief Mining Engr J & L Steel MN Ore Div 1954-60. Mgr New Explosives Business in MN for Dow Chem Co 1960-65. Dir Res, Mgr Mfg, Sales Mgr to Gen Mgr Dow Chem Explosives Bus 1965-76. 1976 to date Pres, Dir, Treas Energy Scis & Conslts. Won AIME Peele Award 1964. 13 issued patents in field of explosives and use, 18 articles conventions and trade journals. Bd of Governers, IME. *Society Aff:* AIME, IME

Grant, Donald A
Business: 246 Boardman Hall, Orono, ME 04469
Position: Prof. *Employer:* Univ of ME. *Education:* PhD/Mech Engg/Univ of RI; MS/Mech Engg/Univ of ME; BS/Mech Engg/Univ of ME. *Born:* 1/3/36. Taught at the Univ of ME at Orono since 1956. Currently Prof of Mech Engg specializing in mech vibrations and solid mechanics. Consultant for several industries in stress analysis, mech properties of materials, weld analysis and mech vibrations. Employed as a flight test engr at the Naval Air Test Ctr, Patuxent River, MD upon grad. PE reg in the State of ME. Distinguished ME Prof Award in 1976. *Society Aff:* ASME, ΤΒΠ, ΣΞ, ΠΤΣ, ΦΚΦ.

Grant, Ian S
Business: 1482 Erie Bvd, Schenectady, NY 12305
Position: Manager/Software Products *Employer:* Power Technologies Inc. *Education:* ME/Power/University of NSW, Australia; BE/EE/University of New Zealand. *Born:* 04/01/40 Mr. Grant graduated BE 1962, ME in 1967. He joined the Electricity Commission of NSW in 1962. In 1969, he joined the GE HV Lab in Pittsfield, MA. Mr. Grant joined PTI in 1972, and was promoted to Dept Mgr in 1986. He directed the early power transmission research studies at PTI's Saratoga R&D Center. He is presently responsible for PTI's commercial software products. He is a Fellow of IEEE, Chairman of the IEEE Lightning & Insulator Subcttee, U.S. Member of CIGRE W.G. 33.01, and Convener of CIGRE W.G. 22.09. *Society Aff:* IEEE, CIGRE.

Grant, Jeffrey C
Business: 2318 Watterson Trail, Louisville, KY 40299
Position: President. *Employer:* Best Photo Industries Inc. *Education:* BS/Chem Engr/Univ of KY. *Born:* Sep 14, 1947. BS Chem Engrg from U of Kentucky. Joined Best Photo in 1972 as the Plant Mgr. In 1973 became VP. In 1975 became Pres of the Louisville KY plant that formulates & packages photographic & related chemicals. Mbr AIChE, Soc Photographic Scientists & Engrs. Hobbies: photography, backpacking & fishing. *Society Aff:* AIChE, SPSE.

Grant, John A, Jr
Business: 3333 No Federal Hwy, Boca Raton, FL 33432
Position: President & Ch Engr. *Employer:* John A Grant Jr Inc. *Education:* BSCE/-/TX A&M. *Born:* Jul 13, 1923. Mbr Fellow ASCE, Sr Mbr & PPres FES, P Pres FICE, Mbr NSPE & PEPP. Ch Engr & Pres John A Grant Jr Inc. Native of Crockett TX. Grad from North TX Agri Coll; US Army Ordnance 1942-46 with rank of 1st Lt. Grad from Tx A&M U 1947; Hwy Des Engr DE State Hwy Dept; Hwy Engr U S Bureau of Public Roads; Proj Engr Michael Baker Jr Inc; Ch Engr Arvida Corp, Ch Engr Private Cons Engrg Firm respon for extensive Turnpike & Hwy design & supr also land dev in FL. Enjoy golf & commercial banking. Reg. PE, Fla, GA, SC, NC, VA, KY, MD, DE & TX. *Society Aff:* ASCE, NSPE, ACSM, AWWA

Grant, Leland F
Business: 817 Broad St, Chattanooga, TN 37402
Position: Chief Geologist *Employer:* Hensley-Schmidt, Inc. *Education:* BS/Geol/Univ of TN *Born:* 10/30/13 in Etowah, TN, the son of Leland and Nettie Grant. Attended elementary school in Maryville and Alcoa. My wife, the former Nellie Pauline Schild, and I have two children, Leland Schild Grant and Annette G. Schwall. From 1936 through 1957 I worked at TVA assessing and evaluating foundation geologic conditions of major dams. Since 1957 I have been a principal and chief geologist for Schmidt Engrg Company and Hensley-Schmidt, Inc., providing geological consulting services to company clients. *Society Aff:* ASCE, APGS, CEC, GSA, SME, USCOLD

Grant, Nicholas J
Business: 8-305, Cambridge, MA 02139
Position: Prof Dept Matls Sci & engg. *Employer:* MIT. *Education:* ScD/Metallurgy/MIT; BS/Met Engg/Carnegie Tech. *Born:* Oct 21, 1915. Asst, Assoc & Prof of Met MIT from 1945; Dir Ctr for Mat Sci & Engrg 1968-77; ABEX Prof

Grant, Nicholas J (Continued)
of Advanced Materials 1975. Pres NE Materials Lab 1954-67. Distinguished Service Award Investment Castings Inst 1956; Annual AIME Powder Met Lect 1957; Merit Award Teach & Res Carnegie Tech 1967; ASM Fellow 1971; AIME Fellow 1980. Amer Acad Arts & Sci, 1972; J Wallenberg Award Swedish Acad Engrg Scis, 1978; Chrmn, US Side, US-USSR Electrometallurgy and Malts Prof, under US-USSR Joint Agreement in Sci & Tech 1977-; Natl Acad of Engr 1979; Author (with A W Mullendore) of *Deformation & Fracture at Elevated Temperatures*, MIT Press 1965; Editor (with B Giessen) Proceedings of the Second Internatl Conference on Rapidly Quenched Metals, Sect I MIT Press 1976; Sect II Mat Sci & Engrg 1976. *Society Aff:* ASME, AIME, ASTM.

Grant, Wallace R
Home: 121 George Ave, Edison, NJ 08817
Position: Dir of Maintenance *Employer:* NJ Turnpike Authority *Education:* BSCE/Civil Engr/Rutgers Univ *Born:* June 1926. BS CE from Rutgers U. Served US Navy 1944-46, 195152. With Franklin Contract Co Little Falls N J from 1953-75, eventually becoming VP. Natl Dir NSPE from 1969-73. Pres NJSPE 1975-76. Enjoys sailing, golf & swimming. *Society Aff:* NJSPE, NSPE

Grant, Walter A
Home: 10 Tenth St Apt 23, Atlantic Beach, FL 32233
Position: President. *Employer:* John B Pierce Foundation. *Born:* Aug 27, 1904 Brooklyn NY. AB, BS & ME Columbia U; Tau Beta Pi, Sigma Xi. Employed Carrier Corp 1928-67: VP Res & Dev 1954-57, VP & Dir Engrg 1958-62, VP Engrg Elliott Co Div 1962-67. John B Pierce Foundation: VP 1972-73, Pres 1973- 78. Fellow & Pres Mbr ASHRAE; recipient F Paul Anderson Medal 1967. Cons Engr 1968- ; Lic Prof Engr NY, NJ, PA. Co-author: 'Modern Air Conditioning, Heating & Ventilating'; author num tech papers, articles, pats.

Grant, Whitney I
Business: 2217 S First St, Milwaukee, WI 53207
Position: Exec VP. *Employer:* Vilter Mfg Corp. *Education:* BS/ME/Marquette Univ. *Born:* 10/1/26. After grad in 1949 from Marquette Univ with a BS in ME, joined Vilter Mfg Corp (mfg of ind refrigeration, air conditioning & heat exchange equip) in 1950. Held various positions including VP Mfg & Engg & Chief Engr; currently (since '77) Exec VP. Patentee in field. Reg PE in State of WI. Served with USAF, 1945. Was 1977 recipient of Marquette Univ Profl Achievement Award. Received ASHRAE's highest individual award - Fellow. Listed in several "Who's Who" biographical directories. Hobbies include foreign travel; camping; photography. *Society Aff:* W & NSPE, SAE, ASHRAE.

Grasso, Salvatore P
Home: 32 Elm Str, P.O. box 28, Milford, NH 03055
Position: Professor of Engrg. *Employer:* New England College. *Education:* MCE/CE/Univ of NH; BS/CE/Univ of NH. *Born:* Jul 1914. Native of Milford NH. Taught CE at UNH, U of Santa Clara & CCNY prior to WWII. With San Corps AUS 1944-46. Subcontract Mgr Hitchinger Mfg Co Inc 1954-56. Project & Ch Engr Anderson-Nichols & Co Inc Cons Engrs Concord NH 1956-65. Prof at NEC since 1965. VP Rollins King & McKone Inc Cons Engrs 1967-69. Short term cons for PAHO/WHO since 1972. Rep NH Genl Court since 1974. Pres NH Sect ASCE 1972, NHSPE 1973, & Chmn NE Sect AWWA 1970. Reg Prof Engr. *Society Aff:* ASCE, ASEE, SAME, AWWA, NSPE.

Gratch, Serge
Home: 32475 Bingham Rd, Birmingham, MI 48010
Position: Professor, Mech. Engrg. *Employer:* GMI Engrg. & Mgt. Inst. *Education:* PhD/ME/Univ of PA; MS/ME/Univ of PA; BS/ChE/Univ of PA *Born:* May 2, 1921 Monte San Pietro Italy. Naturalized; married to Rosemary A Delay 1951; 10 children. Asst Instr Mech Engrg Pa 1943-44, Instr 1944-47, Assoc 1947-49, Asst Prof 1949-51; Sr Sci & res Rohm & Haas Co 1951-59; Assoc Prof Mech Engrg Northwestern 1959-61; Supr Appl Sci Ford Motor Co 1961-62, Mgr Chem Process 1962-69, Asst Dir Engrg Sci 1969-72, Dir Chem Sci Lab 1972-83. Dir. Mat & Chem Sci Lab 1984-86; Prof Mech Engrg GMI EMI 1986- . Mbr AAAS, ACS, SAE, AIAA, ASEE, NYAS; Hon Mbr ASME, Fellow Engrg Soc Detroit; VP - Res ASME 1973-77; Member, ASME Bd of Governor, 1981-84; President ASME 1982-83; President Engrg Soc Detroit, 1986-87; Mbr, Natl Acad of Engr 1983- ; Reg Editor Int Jour Fracture, Mbr, Natl Alcohol Fuels Commission. Res interests: thermodynamic properties of gas mixtures; chem kinetics; polymerization kinetics; viscoelasticity; control of vehicle emissions; energy conversion. *Society Aff:* ASME, SAE, ACS, AAAS, ASEE

Gratiot, J Peter
Business: 39 Central St, PO Box 453, Woodstock, VT 05091
Position: Proprieter. *Employer:* Gratiot Engrg Co. *Education:* BS/Aeronautical/MIT; MS/Economics & Engrg/MIT *Born:* Feb 1921. Prof Engr VT, NH, ME & NY. Land Surveyor in VT. Employed deFlorenz Co NYC 1947-54 in design of automatic mach for graphic arts & aircraft flight simulators. 1954 present cons in VT in building, mech, electric & cvl work. Since 1970 primarily heating, air conditioning & fire protect work in institutional & commercial buildings with much work in Energy Conservation and Recovery in existing buildings and forensic engr Related to Accidents, and Fires in Buildings. Active in public school finance. Mbr of ASHRAE, Assoc of Energy Engrg, NE Solar Energy Assoc NEWWA, NFPA, BOCA & ACEC of VT. Honor Award 1971 ACEC Engrg Excellence competition. *Society Aff:* ACEC, ASHRAE, NFPA, BOCA

Graves, David J
Business: 311A Towne Bldg-D3, Philadelphia, PA 19104-6393
Position: Assoc Prof *Employer:* Univ of Penn *Education:* ScD/ChE/MIT; SM/ChE/MIT; BS/ChE/Carnegie Mellon Univ *Born:* 02/25/41 After receiving his doctoral degree under the direction of E. W. Merrill, Prof Graves spent two yrs as a captain in the Army Medical Service Corps at the Inst of Surgical Res, Ft Sam Houston. He then joined the Penn faculty, which has been his profl home since then. In 1976-77 he spent a yr in Germany and Sweden as an Alexander von Humboldt Fellow with the additional sponsorship afforded by a Fulbright-Hays grant. In addition to an active research prog on biochem and biomedical topics, he is an active conslt in the field, the author of numerous technical papers and three patents. *Society Aff:* AAAS, AIChE, ACS, ΣΞ, ΤΒΠ

Graves, Ernest
Home: 2328 South Nash Street, Arlington, VA 22202
Position: Sr Fellow *Employer:* Ctr for Strategic and Internatl Studies *Education:* PhD/Physics/MA Inst of Tech; BS/Engrg/US Military Academy *Born:* 7/6/24 Engr platoon leader in WWII, engr battalion commander in Korea, and engr group commander in Vietnam; Deputy Dist engr, Los Angeles Dist, US Army Corps of Engrs; Dir US Army Engr Nuclear Cratering Group, Livermore, CA; Div Engr, North Central Div, US Army Corps of Engr, Chicago, IL. Deputy Dir of Military Construction. Dir of Civil Works and Deputy Chief of Engr OCE, Washington, DC; Dir, Defense Security Assistance Agency, Washington, DC. *Society Aff:* SAME, USCOLD, PIANC

Graves, Gilman L, Jr
Home: 12621 Hanover Dr, Ocean Springs, MS 39564
Position: Project Coordinator. *Employer:* US Navy Ship Engrg Ctr. *Born:* May 23, 1925 Leominster Mass. BS from NYU 1949. US Army WW II-wounded in action Battle of Leyte Gulf. Project Engr 1949-62 & Head Engr 1962-68 for Gas Turbines Bureau of Ships. Respon for R & D Tech Programs of Ships Engrg Ctr in propulsion auxiliary, electrical, & deep ocean machinery 1968-80. Currently Dir, Advanced Progs & IRAD Ingalls Shipbldg. Active in ASNE (Chmn Flagship Sect 1965), Bureau of Ships Assn of Sr Engrs (Pres 1963), and ASME (Chmn Wash Sect 1959, VP for Reg III 1974-76). Elected Fellow Mbr ASME 1970. Active in community activities & Methodist Church.

Graves, Harvey W, Jr
Home: 7723 Curtis Str, Chevy Chase, MD 20015
Position: Pres, Energy Analysis Software Service, Inc *Education:* PhD/Nuclear Engg/Univ of MI; MS/Elec Engg/Dartmouth; AB/Engg Sci/Dartmouth. *Born:* Jun 18, 1927. Started career with Westinghouse Electric Corp 1951. Respon for Nuclear Engrg & Reactor Physics of first generation of commercial pressurized water nuclear reactors. Subsequently Mgr of Reactor Engrg in Westinghouse Advanced Reactors Div (LMFBRs). Mbr USAEC Advisory Ctte on Reactor Physics 1967-69. Taught Nuclear Engrg at U of MI 1968-73. Reg Prof Engr. Fellow ANS. Mbr Exec Ctte of Reactor Physics Div 1976-79, Standards Ctte, & Chmn Standards Working Group on Power Reactor Physics Measurements of ANS. Also wrote textbook titled *Nuclear Fuel Management* Published by John Wiley & Sons in 1979. *Society Aff:* ANS, IEEE-Computer Society.

Gray, Allen G
Home: 2741 Belvoir Blvd, Shaker Heights, OH 44122
Position: Tech Director. *Employer:* ASM Metals Park. *Education:* PhD/Chem & Metal/Univ of WI; MS/Metallurgy/Vanderbilt Univ; BS/Chemistry/Vanderbilt Univ. *Born:* Jul 1915. Teaching U of WI; 12 yrs with DuPont Co, incl work on Hanford-Manhattan Atomic Project during WWII. With ASM since 1958; Ed of Metal Progress magazine 1958; Dir Periodical Pubs 1961; assumed current respon as Tech Dir ASM 1973. Fellow AIC; Fellow ASM (1976); Eisenman Medal ASM Philadelphia Chap 1967, Who's Who in America, Who's Who in the world. Advisory Ctte on Tech Info AEC 1954- ; Chmn Natl Mat Adv Bd Ctte on 'Tech Aspects of Critical & Strategic Materials' 1969-72; Tech Adv Bd Metal Properties Council; Adv Bd Federation of Materials Socs; author Modern Electroplating 1953; author sect Alloy Steels Ency Brit; author sect on steels Ency Americana. Lect, writer, advisor to indus & government on materials availability & substitution & conservation of materials & energy in mfg practices. Issued pats. Home: 2741 Belvoir Blvd, Shaker Hts. *Society Aff:* ASM, ASTM, AES, EC, MSACS, ANS.

Gray, Glenn C
Business: 9233 Ward Pkwy - Suite 300, Kansas City, MO 64114
Position: VP, Treasurer *Employer:* Larkin Assocs. Consulting Engr's, Inc. *Education:* BS/Civ Engg/Univ of KS. *Born:* 3/2/24. Native of Eureka, KS. Served US Naval Airforce 1943-1946. Distinguished Flying Cross, Air Medal with two gold stars, Presidential Unit Citation. Student Instr Univ of KS School of Engg 1948-1949. Office Engr Bureau of Reclamations - 1950. Larkin & Assoc Consulting engrs 1950-85. Partner 1956. Managing Partner 1971-85. Larkin Assoc. Consulting Engr's, Inc, 1985 to present, VP & Treasurer. Enjoy reading, photography, sports. Honorary Organization - Tau Beta Pi, Sigma Tau, Professional organizations - American Society of Civil Engineers (ASCE)- National Society of Professional Engineers (NSPE)-American Water Works Association (AWWA) - Water Pollution Control Federation (WPCF) - Professional Engineers in private practice (PEPP). *Society Aff:* ASCE, NSPE, NSPE-PEPP, WPCF, AWWA.

Gray, Harry J
Home: 412 Colonial Park Dr, Springfield, PA 19064
Position: Prof. *Employer:* Univ of PA. *Education:* PhD/EE/Univ of PA; MS/EE/Univ of PA; BS/EE/Univ of PA. *Born:* 6/24/24. Prof of Elec Engg & Sci, & Comp & Info & Sci at the Moore Sch of EE of the Univ of PA. Ind experience: Univac, consulting: Philco, ITT, Burroughs, Curtiss- Wright Electronics, Xerox Corps; Patents: seven; author of two books and 33 papers in periodicals & conf records; proj experience: test & installation of ENIAC; design of EDVAC, UDOFTT (the first digital operational flight trainer); advanced dev of Univac's LARC circuits (first appearance of today's dominant clocked digital solid-state circuit system); inventor of multilist technique used for storage & retrieval of info. *Society Aff:* IEEE, ASEE, FRANKLIN INST, ΣΞ, ΤΒΠ.

Gray, Jon F
Business: 747 Alpha Dr, Highland Hgts, OH 44143
Position: Senior Design Engr. *Employer:* Allen-Bradley. *Education:* BS/EE/Case Western Reserve Univ. *Born:* 7/10/57. A Jan, 1979 grad of Case Inst of Tech of Case Western Reserve Univ, I am in my second yr of employment at the Allen-Bradley Co, PC Systems Div. I am a former mbr of the Natl Exec Council for Theta Tau, natl PE Fraternity, and a current mbr of the IEEE & its Comp, Software Engg, & Pattern Analysis & Machine Intelligence Societies. My current involvements are in the Software design of Ind Control Systems. *Society Aff:* ΘΤ, IEEE.

Gray, Paul E
Business: 77 Mass Ave, Cambridge, MA 02139
Position: President *Employer:* MIT. *Education:* ScD/Elec Engrg/MIT; SM/Elec Engrg/MIT; SB/Elec Engrg/MIT *Born:* SB, SM, ScD MIT; faculty MIT 1960-71; Chancellor 1971-80, Pres 1980- . Dir Shawmut Bank of Boston, N A, Cabot Corp, The New England, A D Little Inc. Trustee Wheaton College Norton MA; Trustee, mem corp Mus of Sci, Boston, Woods Hole Oceanographic Institute; Trustee, Whitaker Health Sci Fund, Cambridge, MA; Trustee, Kennedy Memorial Trust, London, England; trustee (ex officio) WGBH Ed Foundation, Boston; director, Natl Action Coun for Minorities in Engineering. Mbr NAE, AAAS, Sigma Xi, Eta Kappa Nu, Tau Beta Pi, Phi Sigma Kappa; Fellow Amer Acad of Arts & Sciences, IEEE. *Society Aff:* NAE, AAAS, IEEE

Gray, Robert M
Business: Durand 133, Electrical Engrg Dept, Stanford, CA 94305
Position: Prof *Employer:* Stanford Univ *Education:* PhD/EE/Univ of Southern CA; MS/EE/MIT; BS/EE/MIT *Born:* 11/1/43 Native of San Diego, CA. Electrical Engr at US Naval Ordnance Lab (MIT Cooperative Program) 1962-1965. Electrical Engrs, Jet Propulsion Lab, summers of 1968 and 1969. At Stanford since 1969. Currently Prof of Elect Engrg & Dir, Info Systems Lab. Associate Editor (1977-1980) and Editor (1980-) IEEE Transactions on Information Theory. Board of Governors, IEEE Professional Group on Information Theory (1974-1981). 1985-87 Co-Recipient of IEEE Information Theory Group 1976 Prize Paper Award and the IEEE ASSP Senior Award (1983), recipient of IEEE Centennial Medal (1984). Awarded fellowships from the Japan Society for the Promotion of Science (fall 1981) and the John Simon Guggenheim Memorial Foundation (Jan-Sept 1982). Amateur Radio License KB6XQ. *Society Aff:* IEEE, SIAM, IMS, AAAS

Gray, Robin B
Home: 1077 Spring Mill Lane, NE, Atlanta, GA 30319
Position: Regents' Prof *Employer:* Georgia Inst of Tech *Education:* PhD/Aeronautical Eng/Princeton Univ; MS/Aeronautical Eng/GA Inst of Tech; B/AE/Rensselaer Polytechnic Inst *Born:* 12/04/25 in Statesville, NC. Served in U.S. Navy 1943-46. Res Asst then Res Assoc, Princeton Univ, 1949-56. At Georgia Inst of Tech since 1956. Assoc Dir of Aerospace Engrg since 1967 and Regents' Prof since 1973. Teach undergraduate and graduate courses in fluid mechanics and helicopter performance and stability and control. Continuous research activity in rotary-wing aerodynamics and related topics since 1957. Served as Councilman, North Atlantic, Ga., 1963-64 and as Councilman and Vice-Mayor, 1964-65. *Society Aff:* AIAA, AHS, ΣΞ

Gray, Roland H, Jr
Home: 2128 Fernglen Way, Baltimore, MD 21228
Position: VP Operations *Employer:* Remal Information Corp *Education:* BS/ChE/Northwestern Univ. *Born:* 11/9/28. in Toledo, OH. 1951-59 Dow Chem R&D latex & fibers. 1959-76 W R Grace R&D mgmt organic & inorganic chemicals, pesticides, ceramics, sugar. Production mgmt latex & auto catalysts. 1976-77 ERDA Program mgr coal conversion, 1977- VP Operations-Micrographics Production. *Society Aff:* AIChE.

Gray, Truman S
Home: 22 Hayes Ave, Lexington, MA 02173
Position: Prof of Engg Electronics, Emeritus. *Employer:* Mass Inst of Technology.

Gray, Truman S (Continued)
Education: ScD/Elec Engg/MA Inst of Technology; SM/Elec Engg/MA Inst of Technology; BA/Physics/Univ of TX at Austin; BS/Elec Engg/Univ of TX at Austin. *Born:* May 3 1906 Spencer Ind. Educator Univ of Texas & MIT 1924- . Engr with Leeds & Northrup Co summer 1929, Genl Elec Co summer 1935 & Naval Ordnance Lab summer 1941. Cons to indus firms, gov labs & pat attorneys. Author 'Applied Electronics' 2nd edition 1954. Life Fellow IEEE (chairman instruments & measurements ctte 1945-47 & Boston section 1946-47). Mbr ASEE, Phi Beta Kappa, Sigma Xi, Tau Beta Pi, Eta Kappa Nu, Phi Mu Alpha & Pi Kappa Alpha. Enjoy music, silversmithing, sailing & fishing. *Society Aff:* IEEE, ASEE.

Gray, Victor O
Business: 2100 5th Ave Bldg, Seattle, WA 98121
Position: President. *Employer:* Victor O Gray & Co Inc. *Born:* June 1926. Married; 3 children. BS Gonzaga Univ Spokane WA 1950; MS Univ of Wash Seattle WA 1951. Principal Victor O Gray & Co Inc org in 1958; lic civil & struc engr & planner. Firm involved in struc design of buildings & indus facilities, distrib centers, schools, hospitals, parking garages, traffic & transportation planning, restor & rehab of struc, seismic analysis & feasabil studies. Pres Downtown Seattle Dev Assn 1973; Natl Vice Pres Amer Cons Engrs Council 1972-74; Pres Cons Engrs Council of Wash 1969; Engr of Yr CEC/Wash 1970. Mbr ASCE; Struc Engrs of Wash.

Gray, Wilburn E
Home: Route 10 - Box 498, Tyler, TX 75701
Position: President. *Employer:* Almeg Inc. *Born:* June 1937. BS Univ of Texas 1962. Chem Engr Texas Eastman R&D pilot plants prod, organ chemicals & plastics 1962-66. Process Engr ammonia plant design & nitrogen prod W R Grace Memphis Tenn 1966-67. Sr Engr Howe Baker Engrs 1967-71. Genl Mgr & Dir Howe Bakers European Div 1971-75. Responsible for Engrg & Sales Europe, Africa & Middle East: oil field prod equip & chem process plants. Man Dir Delgra Chem Co Zurich; Pres Almeg Inc Texas. Responsible for internatl sales of Oil Field Treating Chemicals 1975-present. Enjoy golf, bridge, hunting & fishing.

Graybeal, Paul E
Home: 524 Kerfoot Farm Rd, Wilmington, DE 19803
Position: Program Director-Major New Facilities *Employer:* Hercules Inc. *Education:* BChE/Chem Eng/Yale Univ. *Born:* May 31, 1919. Employed by Hercules Inc during war years in the manufacture of military propellants & explosives. Led pilot plant programs & plant designs on 2 very successful cellulose derivatives. Assigned to successive positions as production supt, plant mgr, dir of operations & asst genl mgr. Mbr ACE, AIChE. Also serve as Pres Haveg Industries, a subsidiary of Hercules. *Society Aff:* ACS, AICHE, AMA, ADPA.

Graybill, Howard W
Home: 5158 Don Matta Dr, Carlsbad, CA 92008
Position: Consultant (Self-employed) *Education:* BS/Elec Engg/Drexel Univ. *Born:* Aug 1915 Bareville Pa. Westinghouse Elec Corp 1937-45; on graduate student prog, then as design engr on disconnect switches. Joined I-T-E 1950. Held var pos in design & application engrg. In 1968-69 designed 1st 345 kV SF6 gas-insulated substation & 1st gas-insulated transmission systems in the world. As Mgr-gas-ins Prod Engg, 1968-78, Had overall respon for complete system design & installation supervision of approximately 85% of all gas-insulated substations & transmission systems in US, from 115 through 500 kV, as well as installations in Mexico & Canada. Fellow IEEE. Reg Engr in Pa. Mbr Tau Beta Pi, Eta Kappa Nu. Hold 30 U S Pats. Retired Aug 1980: Currently consultant on gas-insulated substations and transmission systems.. *Society Aff:* IEEE, PES.

Grayson, Lawrence P
Home: 9714 Carriage Road, Kensington, MD 20895
Position: IEEE Congressional Fellow. *Employer:* Congressman Jack Kemp U.S. House of Representatives. *Education:* PhD/Elec Engrg/Polytechnic Inst of Brooklyn; MEE/Elec Engrg/Polytechnic Inst of Brooklyn; BEE/Elec Engrg/Polytechnic Inst of Brooklyn *Born:* 5/16/37. Taught at John Hopkins Univ 1962-67, at Manhattan College 1968-69 & part-time at the Catholic Univ of America 1980 - 1985. Worked at IBM's Thomas J Watson Res Center 1967-68. Joined US Office of Education 1969 & assumed position as Inst. Advisor for Math, Sci & Tech in the Natl Inst of Educ 1982. Served as IEEE Congressional Fellow in Office of Congressman Jack Kemp, 1986-87. Served as President-elect of ASEE & as V Chrmn of the Educational Activities Bd of the Inst of Elec & Electron Engrs. Received the HEW Distinguished Service Awd 1976, US Army Achievement Awd 1962, Natl Assoc Pub Continuing & Adult Ed Special Awd on Soc Justice 1975, Second Century Awd from Polytechnic Inst of NY, 1980, Distinguished Alumnus Award Polytechnic University, 1987, ASEE's Distinguished Service Awd 1977, William Elgin Wickenden Award 1979, & Distinguished Service Citation 1981, ASEE's Continuing Prof Div Distinguished Service Citation 1982, & Certificate of Merit 1984. IEEE's, Centennial Medal 1984, & Ed Soc Achievement Awd 1979 & Testimonial Citation for achievement in educational Satellites from the Learning Channel, 1984. Hoover Medal for public service by an engineer jointly awarded by ASCE, ASME, AIChE, AIME & IEEE 1986. Reg PE. *Society Aff:* IEEE, ASEE, AAAS

Grayson, Leonard C
Home: 105 Shady Lane, Randolph, NJ 07869
Position: Sr Assoc/Chief Elec Engr *Employer:* Haines Lundberg Waehler *Education:* BS/EE/Cornell Univ; M/CS/Computer Arch./American Institute; PhD/EE/DePaul Univ. *Born:* 7/15/31 Thirty year of experience in engrg, with two patents for Control Sys. Specializes in Electrical (Power) Engrg, as well as Computer and Computerized Control Sys. Responsible for HLWs Electrical Dept and programs for latest state of the art in sys computerization. Enjoys classical music, fishing, and tennis *Society Aff:* NSPE, IEEE, ISA, ASPE, NACCC

Grebene, Alan B
Business: 750 Palomar, Sunnyvale, CA 94086
Position: VP. *Employer:* Exar. *Education:* PhD/Electronics/RPI; MSC/EE/Univ CA; BSc/Electronics/Robert College. *Born:* 3/13/39. Born in Istanbul, Turkey, in Mar, 1939. Emigrated to the US in 1961. Worked as Res Engr, Sr Engr & Engg Mgr at Fairchild Semiconductor Corp, Sprague Elec Co, & Signetics Corp, during 1963 through 1971. In June 1971, founded Exar Integrated Sytesm, Inc, as a joint-venture with Toyo Electronics of Japan, and is currently the VP of Engg & Marketing at Exar. Author of "Analog Integrated Circuit Design–, Van Nostrand Reinholt, 1972; & "Analog Integrated Circuits–, IEEE Press, 1978. Elected as a Fellow of the IEEE in 1979. Published over 40 technical papers & holds 7 US & foreign patents. *Society Aff:* IEEE.

Grecco, William L
Home: 7935 Corteland Dr, Knoxville, TN 37909
Position: Assoc Dean Coll of Engrg *Employer:* Univ of TN. *Education:* PhD/Transportation/MI State Univ; MSCE/Sanitary/Univ of Pittsburgh; BSCE/Civ Engr/Univ of Pittsburgh. *Born:* 8/28/24. Native of Butler, PA. Res and teaching at Univ of Pittsburgh (13 yrs), Purdue Univ (10 yrs) and Univ of TN since 1972. Chaired the Exec Committees for Urban Planning & Dev Div 1973; for Water Resources Planning & Mgt Div, 1977. Also the Committee on Curricula and Accreditation, 1978; Pres, TN Valley Sec, 1980; and adhoc Accreditation Visitor; all for ASCE. Previously consultant t Transportation Res Bd, Brown & Root Intl-Honduras, Fdn Assoc and Donald M McNeil Consultant engrs. Mbr - Bd of Dirs - Purdue - Calumet Dev Fdn, 1965-72. Reg PE (3 states). Alternate Member - Bd of Dir, ABET 1987; EAC/ABET Team Chair, 1981-86; ASCE, Bd of Dir, 1987-90. *Society Aff:* ASCE, ASEE ITE, APA, AICP. ΣΞ

Green, Carl E
Business: 5570 SW Menefee Dr, Portland, OR 97201
Position: Cons Engr. *Employer:* Self. *Education:* AB/Civil Engrg/Stanford Univ; AB/Engr, Civil & Sanitary Engrg/Stanford Univ *Born:* Apr 28, 1906. Mbr ASCE, Amer

Green, Carl E (Continued)
Cons Engrs Coun, Prof Engrs Oregon, Cons Engrs Coun Oregon, Water Pollution Control Federation, PNW Pollution Cont Assoc, Amer Pub Health Assn, Tau Beta Pi, Sigma Xi, Amer Water Works Assn. Awds & Offices: ASCE Ore Sect, V P, Pres; Pollution Control Fed Bedell Award; Pres PNW Pollution Control Assn; Chairman Portland Air Pollution Control Comm; Chmn Ore Air Pollution Auth, Sect & Ch Engr Ore Sanitary Auth. Experience: Ore State Sanitary Engr; Cunningham & Assocs, Cons Engrs - Partner; owner Carl E Green & Assocs, Cons Engrs; projs in cons work - 670. Reg OR & WA Dir at Large, Water Pollution Control For U.S. Fed. *Society Aff:* ASCE

Green, David G
Business: School of Engrg, Birmingham, AL 35294
Position: Lecturer *Employer:* Univ of AL in Birmingham *Education:* MSE/EE/Univ of AL in Huntsville;BSE/EE/Univ of AL in Huntsville *Born:* 9/6/54 Instructor of Electric Engrg at The Univ of AL in Huntsville (UAH) 1976- 1981. Chrmn of the Huntsville Section of IEEE 1979-1980. Counselor of UAH IEEE Student Branch 1976-1981. Outstanding IEEE Counselor/Advisor Award 1981. Chrmn of Alabama Section of IEEE's Industrial Application Society 1986-1987. Res Interests include computer applications, power electronics, and industrial control. Am an amateur radio operator and personal computerist. Counselor of UAB IEEE Student Branch 1982-present. Advisor to UAB HKN IOTA Alpha Chapter. *Society Aff:* IEEE, HKN, TBΠ, ΣΞ, ASEE

Green, David M
Business: 33 Kirkland St, Cambridge, MA 02138
Position: Prof of Psychophysics. *Employer:* Harvard Univ. *Education:* PhD/Psychology/Univ of MI; MA/Psychology/Univ of MI; BA/-/Univ of MI; BA/-/Univ of Chicago. *Born:* 6/7/32. Jackson Mich; m. Clara Lofstrom; wife died Jul 11, 1978. m. Marian Heinzmann Jun 7, 1980. c. Allan, Phillip, Katherine, George. Prof of Psychophysics Harvard Univ 1973-chrmn Dept of Psychology & Social Relations 1978-81. Prof of Psychology Univ of CA San Diego 1966-73. Assoc Prof of Psychology Univ of PA 1963-66; Asst Prof of Psychology Mass Inst of Tech 1958- 63; cons Bolt Beranek & Newman Inc 1958- . Biennial Award, AcSoc of Amer 1966; Guggenheim Fellowship, Overseas Fellow, St John's Coll Cambridge Eng 1973-74; elected to Natl Acad of Sciences 1978. Pres, Acoustical Soc of Amer, 1981. Distinguished Scientific Contribution Award, American Psychological Assn, 1981. Visiting fellow. All Souls Coll, Oxford England, 1981-82. *Society Aff:* NAS, ASA, APA, AAAS.

Green, Frank W
Point O'View, Wast, 504 Glorietta Blvd, Coronado, CA 92118
Position: Pres & Ch Engr Cons. *Employer:* Point O View Inc. *Education:* BS/Bus Admin/Boston Univ. *Born:* 4/14/16. Cons Engr in pkging, handling & phys distrib. Lecturer, educator, writer, Reg Prof Engr; Fellow Soc of Packaging & Handling Engrs; Cert Mgmt Cons; Edited 'Glossary of Packaging Terms' & author of sev books & more than 100 magazine articles & tech papers. Lecturer & Prog Chairman for credit course at Columbia, the Armed Forces Packaging School & instructor & lecturer at many colleges & orgs in USA, Canada, Germany & England. Clients Tokyo to Stuttgart incl natl indus corps; major air & surface carriers, marine underwriters. Pkging Sys & Cost Reduction Specialist. Expert witness in in state, federal and international Admiralty Courts for cases involving cargo, packages, containers, pilferage and handling. *Society Aff:* SPHE, PI, SPMC, CI.

Green, Lawrence J
Home: 10552 Mooring Rd, Longmont, CO 80501
Position: VP *Employer:* J.C. Zimmerman Engrg Corp. *Education:* MCRP/City Planning/Catholic Univ of America; B/CE/Marquette Univ *Born:* 4/5/30 Native of Oshkosh, WI. Served with U.S. Navy Civil Engr Corps from 1952 to 1971 attaining rank of Captain. Served as Executive VP and General Mgr of large intl consulting firm in southern CA. Acted as Construction Mgr for multi-million dollar heavy construction projects. Joined present firm as VP Construction Management in 1980. Presently serve on working commission, National Academy of Sciences called "Organization and Management of Construction." *Society Aff:* NSPE, ASCE, USNC/CIB, PMI

Green, Leonard A
Business: P O Box 6100 Station A, Montreal H3C 3H5, Quebec Canada
Position: Chief - Sound Div. *Employer:* National Film Bd of Canada. *Born:* 1925 England. 1941 British Broadcasting Corp. 1952 Canadian Broadcasting Corp. 1964 Natl Film Bd. Mbr of SMPTE, BKSTS, ISO Cttes.

Green, Paul E, Jr
Home: Roseholm Place, Mt Kisco, NY 10549
Position: *. *Employer:* Internatl Business Machines Corp. *Education:* ScD/EE/MIT; MS/EE/NC State Univ; AB/Physics/UNC *Born:* Jan 1924. Married 1948 Dorrit Gegan, 5 children. Nat Academy of Eng 1981. Native of Chapel Hill NC. Employed MIT Lincoln Lab 1951-69, Grp Leader R&D in communication theory applications to anti-jam, anti-multipath communications sys, radar astronomy & seismic discrimination. 1969-81 IBM Res Ctr Yorktown Hts NY, Senior Mgr Computer Sci Dept of teleprocessing sys performance modelling, speech & image processing. 1981-83 Corporate Technical Ctte. IBM Corporate HQ, Armonk, NY. Currently Research Staff, IBM Research, Yorktown Hts., NY. Hobbies: organ building, chamber music, tennis. *Society Aff:* NAE, IEEE, ACM, ΣΞ

Green, Richard S
Home: 9209 E Parkhill Dr, Bethesda, MD 20814
Position: Cons Engr (Self-employed) *Education:* SM/Engg/Harvard Grad Sch Engg; SB/Engg/Harvard College. *Born:* Mar 2 1914 Somerville MA. Career comm officer US Public Health Service over 32 yrs following earlier engrg work with Corps of Engrs, Mass Dept Pub Health, Panama Canal & U of Pa. Wide variety PHS assignments at home, abroad incl over 4 yrs as Ch Engr in Alaska Dept Pub Health & liaison officer with Army, Navy. Later became Dir, Indian Health Serv environmental progs, incl water supply, waste facilities const. Awarded PHS Commendation Medal & Distinguished Service Medal. Retired from PHS 1973 as Ch Sanitary Engrg Officer with rank of Asst Surgeon Genl. Mbr of sev engrg soc & served 1976 as Pres Amer Academy of Environmental Engrs. *Society Aff:* AAEE, ASCE, AWWA, WPCF, APHA, ROA, NAUS.

Green, Robert E, Jr
Business: Materials Science & Engineering, Maryland Hall/Room 102, The Johns Hopkins University, Baltimore, MD 21218
Position: Prof & Dir of Center for Nondestructive Evaluation *Employer:* Johns Hopkins Univ. *Education:* PhD/Physics/Brown Univ; MS/Physics/Brown Univ; BS/Physics/William & Mary College. *Born:* 1/17/32. Clifton Forge, VA. Fulbright Scholar, Aachen, Germany, 1959-60; Johns Hopkins Univ since 1960; Chrmn, Dept of Mechanics 1970-72; Chrmn, Dept of Mechanics and Materials Sci 1972-73; Chrmn, Dept of Materials Sci and Engr/Civil Engr 1979-82; Chrmn, Dept of Materials Sci and Engr 1982-85; Dir. Ctr. for Nondestructive Evaluation 1985-present. Ford Found. Resident Sr. Engr. RCA, Lancaster, Pa., 1966-67; Cons. U.S. Army Ballistic Research Labs., Aberdeen Proving Ground, Md, 1973-74; Physicist Center for Materials Sci., U.S. Nat. Bur. Standards, Gaithersburg, Md, 1974-81; Program Manager, Defense Advanced Res. Proj. Agency 18"981-82. Consultant to various industrial firms, govt facilities, and legal offices. Technical talks at numerous scientific meetings and lectured in the U.S., Australia, Austria, Canada, Egypt, England, France, Germany, Ireland, Israel, Italy, Japan, Portugal, Republic of China, Russia and Spain. Major res interests include materials sci and engr, nondestructive evaluation, mech properties of materials, ultrasonics, non-linear elastic waves, acoustic emission, light-sound interactions, x-ray diffraction, electro-optical systems, residual stress, synchrotron radiation, composites, electronic materials, advanced sensors,

Green, Robert E, Jr (Continued)
process control. Author of more than 140 scientific papers and 3 books. *Society Aff:* ASNT, ASM, ASA, IEEE, APS, MIM

Green, Robert H
Business: 550 High Point Ln NE, Atlanta, GA 30342
Position: Pres Consulting Engr. *Employer:* R H Green Engg Co. *Education:* BS/Struct Engg/Univ of FL; BS/Ind Engg/Univ of FL. *Born:* 7/3/35. Since 1970 Mr Green has been the principal of his engg firm specializing in structural designs of superstructures for commercial, institutional, and industrial bldgs. Also has done product designs with fiber reinforced plastics. His past employment was with engg design firms as Proj engr, as a Scientist Assoc doing experimental structural res, and as a Test Engr doing inspections and certifications. He has published technical papers and has obtained patents in his field of expertise. Presently GSPE state chrmn of engrs in private practice, Technical Advisor to ASCE's Structural Plastics Res Council, Reg PE in seven southeastern states. *Society Aff:* NSPE, ACI, Tau Beta Pi, CEC

Green, Robert P
Business: P O Box 2500, Chillicothe, OH 45601
Position: Senior Assoc *Employer:* Mead Central Res *Education:* MBA/Ind Mng/Miami Univ (OH); BS/Pulp & Paper/NYS College of Forestry *Born:* Mar 1925. Native of NYC. Served in US Army Infantry 1943-45. With Champion Paper 1949-61; Pulp Res Mgr, Asst Mgr of pulp prod. With Kimberly-Clark 1961-73: engrg, pulp sales, tech sales serv. Joined MCS 1973 with respon for mktg catalytic sys for white liquor oxidation. Mbr TAPPI, Pulp Mfg Div; Chmn, Mbr Bd of Dir 1970-73. Presently, senior assoc, Fiber Technology Dept., Mead Central Res. *Society Aff:* TAPPI

Green, Stanley J
Home: 3348 Middlefield Rd, Palo Alto, CA 94306
Position: Director, Steam Generator Project Office *Employer:* Electric Power Research Institute *Education:* PhD/Chem Eng/University of Pittsburgh; M.S.Ch.E./Chem. Eng/Drexel Institute of Technology; B.S.Ch.E/Chem Eng./College of the City of New York *Born:* 03/11/20 Currently, Dir, Steam Generator Project Office, Electric Power Research Institute. Previously was with Bettis Atomic Power Lab from 1954 to 1977 progressing to manager of reactor development and analysis. Earlier, was Chemical Engr for Acme Coppersmithing and Machine co, Fercleve Corp, and U.S. Bureau of Mines. Author of numerous publications on pressurized water reactor thermal and hydraulics and steam generator preformance. Chairman of ASME Heat Transfer Div. Received ASME Centennial Medallion, in 1980, ASME Fellow and recipient of Donald O.Kern Award in 1985. PE in PA. *Society Aff:* AIChE, ASME

Greenawalt, Jack O
Business: 3111 E Frank Phillips Blvd, Bartlesville, OK 74006
Position: Pres. *Employer:* Greenawalt-Armstrong, Inc. *Education:* BS/Elec Engr/KS State Univ; BS/Bus Adm/KS State Univ *Born:* 10/10/24. Native of Paola, KS. Served in Armored Force in US and Europe (1943-1946). KS State Univ attended 1946-1950. Field Engr for Aetna Insurance Co in IA and OH 1950-1955. Supervising Cathodic Protection or Corrosion Engr, Phillips Petroleum Co, 1955-1971. Started Greenawalt Engg Co in Bartlesville, OK in 1971. Formed Greenawalt-Armstrong in 1973, specializing in Land Planning, Dev, Surveying, and Civil Engg. Twenty seven yrs work in Little League Baseball, 16 yrs in DeMolay, and ten yrs in Scouting. *Society Aff:* NSPE, OSPE, ACSM, NACE, ASCE.

Greenbaum, Milton M
Business: 994 Longfield Ave, Louisville, KY 40215
Position: Secretary & Principal Engr. *Employer:* Milton M Greenbaum Assocs Inc. *Education:* BSCE/Civil Eng/City College of NY. *Born:* 2/10/30. Mr Greenbaum, principal engr of Milton M Greenbaum Assocs Inc since 1963, is expert in geophys engrg with regional recognition. He is a grad of City College of NY with post-grad studies at Newark Coll of Engrg & Catholic U of America. With Vollmer Assocs 1960-62 as Proj Engr & Mgr of Louisville office; Michael Baker Jr Inc 1956-60 Proj Engr; Louis Berger Assocs 1953-56 Ch Hydrol Engr; DeLeuw, Cather & Brill 1952-53 Asst Resident Engr. P Pres of: Ky C. E.C.; Louisville Chap KSPE; Engrs & Architects Club of Louisville; Engrs & Surveyors Assn; Ky Soil Mech & Found Group. Mbr ASCE, NSPE, ACEC, ASFE Soc. Active in design, investigation & consultation for site & found design & const qual control in a 6 state area. *Society Aff:* ACEC, NSPE, ASCE, ASTM, ASFE.

Greenberg, David B
Home: PO Box 21068, Cincinnati, OH 45221
Position: Prf & Grad Studies Advisor Chem & Nuc Engrg *Employer:* U of Cincinnati. *Education:* PhD/ChE/LA State Univ; MS/ChE/Johns Hopkins Univ; BS/ChE/Carnegie Inst of Tech *Born:* Nov 2, 1928 Norfolk Va. USNR 1947-49. Indus Res & Dev with RCA, USI, FMC 1952-56. Prof US Naval Acad 1958-61. NSF Fellow 1961. Prof LSU 1961-74. Esso Res Fellow 1964-65. Editor CEP Symp Series 1968. Program Dir Engrg Div NSF 1972-73. Assoc Edit Journal of Simulation 1970-74. Prof & Head Dept of Chem & Nuclear Engrg U of Cincinnati 1974-81. Fellow Am Soc for Laser Medicine and Surgery. Mbr AIChE, ASEE, ACS, Sigma Xi, Phi Lambda Upsilon, Tau Beta Pi. Res interests incl reactor kinetics, transport phenomena, laser chem, biochem, & biology. Reg Prof Engr. *Society Aff:* AIChE, ACS, ASEE

Greenberg, Herman D
Business: P.O. Box 355, Pittsburgh, PA 15230
Position: Dir Special Projects, Nuclear Fuel Div *Employer:* Westinghouse Electric Corp. *Education:* BS/Chemistry/Pittsburgh Univ; MS/Met Engg/Pittsburgh Univ. *Born:* 12/11/23. Over 40 years of service with Westinghouse Electric Corp: 19 years as metallographer and materials engineer at E. Pittsburgh Div., 8 years as Manager of Metallourgical Applications in the Research Labs, 6 years as Director of Advanced Mfg. Technology at the R&D Center and 8 years as Director of Special Projects in the Nuclear Fuel Division. Retired in 1982. Instructor at Carnegie- Mellon University in Materials Dept. Specialties on applications of materials, NDE, large forgings and weldments, automation, productivity improvement and isotope separation member ASM. Enjoy tennis, chess and bridge. *Society Aff:* ASM, ASME.

Greenberg, Joseph H
Home: 8801 W. Golf Rd, Apt. 5I, Des Plaines, IL 60016
Position: Vice President-Retired. Senior Consultant *Employer:* A T Kearney Inc. *Education:* BS/ME/MIT; MS/MetE/IIT. *Born:* June 1918. Native of Chgo Ill. Evening instruct at IIT in Met Engrg 1941-85. Ch Met Perfection Gear Co 1940-45. Ch Proj Engr Boynton Engrs 1945-61. VP Engrg Boynton Engrs 1961-64. Principal A T Kearney Inc 1964-72; VP 1972-82. Consultant for Metal Processing Industries, 1982- . Currently in charge of all met plant design, environ control, OSHA control & energy conservation engrg. Chmn Chicago Chap ASM-1960. Elected Fellow ASM-1970. Book to be published by ASM entitled "Industrial Heating Equipment Handbook–. *Society Aff:* ASM, AFS, AIME, AISE.

Greenberg, Leo
Home: 6711 No California, Chicago, IL 60645
Position: Assoc Prof. *Employer:* IL Inst of Technology. *Education:* PhD/Ind Sat/NY Univ; MME/Mech Eng/CUNY; BME/Mech Eng/CCNY. *Born:* Nov 1935 Poland. PhD (1969) NYU; MME (1965) CUNY; BME (1957) CCNY. Prof Engr NY, Cert Safety Prof. Mbr ASSE, AIHA, SSS, ACGIH. Worked as safety spec and/or indus hygienist for NY State, U of CA, etc. Fulltime university teaching OS&H at undergrad & grad levels for 5 yrs. Freelance cons since 1974, presently affil with Internatl Labour Office as Expert, assisting the Singapore Ministry of Labour in training & activity programming. NYU Founders' Day Award for Outstanding Scholarship (1970); Batsheba DeRothschild Award for Promising Young Scientists (1970); Third

Greenberg, Leo (Continued)
Prize (1970) & Second Prize (1973), Paper Competition ASSE. *Society Aff:* ARGIH, AIHA, ASSE, SSS.

Greenberg, Ronald David
Business: Visiting Prof of Taxation, Harvard Business School, 232 Baker, Boston, MA 02163
Position: Prof of Business Law and Taxation (Columbia Univ) *Education:* JD/Law/Harvard Univ; MBA/Bus/Harvard Univ, Grad. Sch. of Business; BS/Physics & Engrg/Univ of TX *Born:* 9/9/39 Preparatory education: public schools (San Antonio, TX). Honors: Chrmn, Taxation Committee, American Bar Association (General Practice Section); Visiting Prof, Stanford Business School (1978); Outstanding Prof Award (Columbia Business School); Silver Spurs (Univ of TX); Goodfellows, (Univ of TX); Counsel, Delson & Gordon (law firm, N.Y.C.). Bar Memberships: U.S. Supreme Court; U.S. Court of Appeals (2d Cir.); Federal District Courts (S.D. N. Y. and E.D. N.Y.) NY State Bar Assn.; American Bar Assn.; Assn. of the Bar of the City of NY. Military: Navy Lt. Corporate & engrg experiences: Exxon and Allied Chem. Publications: various articles (7), books (3), chapters (2), reviews (2), reports (8). *Society Aff:* AAAS, ASME, NSPE, NYSPE, TSPE, ΤΒΠ, ΠΤΣ, RA, NYAS

Greenberger, Martin
Business: 34th and Charles, Baltimore, MD 21218
Position: Prof of Math Sci *Employer:* Johns Hopkins Univ *Education:* PhD/Applied Math/Harvard; SM/Applied Math/Harvard; AB/Applied Math/Harvard *Born:* 11/30/31 Staff mbr at the Harvard Computation Lab working with the Mark I computer in early 1950's. Formed and managed IBM applied sci Cambridge, 1955-58. Taught at MIT 1958-67, helping to establish proj mac and leading its mgmt res group. Guggenheim Fellow at Berkeley 1965-66. Chrmn of NSF/OST Review Group of Cosati, 1972. Chrmn of Council and Mbr of Bd of Trustees. Educom Chrmn of AAS Section T on Info Computing, and Communications. Fellow AAAS. Mgr of Systems Program, Epri, 1976-78. Isaac Taylor Prof of Energy, Tech, 1978. At Hopkins since 1967 as Dir of Info Processing and Chrmn, Computer Sci, then current position. *Society Aff:* AAAS

Greene, Arnold H
Business: 6 Huron Dr, Natick, MA 01760
Position: Pres. *Employer:* Micro/Radiographs, Inc *Born:* May 27, 1922 Boston Mass. Educ in the Boston School Sys; attended Lowell Inst MIT, Harvard Business School, Franklin Inst. Reg Prof Engr Mass; Mbr ASME, ASTM (P Chmn NE Sect, Award of Merit, & Fellow); ASNT - Fellow; Amer Council of Independent Labs, Chmn Eastern Div; MA Area Council of Independent Test Labs - Pres & Founder; founded Arnold Greene Testing Labs Inc 1947; Arnold Greene Test Labs of Puerto Rico Inc 1968. *Society Aff:* ASME, ASTM, ASNT.

Greene, George R
Business: AF Rocket Prop Lab, Edwards AFB, CA 93523
Position: Dep Chief Test & Support Div. *Employer:* AF Rocket Prop Lab-Edwards AFB. *Education:* BS/Mech Engg/NC State Univ. *Born:* Jan 31, 1925. BS Mech Engrg North Carolina State U. Native of San Fran Calif. 1947-59 Proj Engr (12 yrs) & Sect Supr on jet aircraft propulsion subsystems Wright-Patterson AFB. 1959-69 Branch Supr & Asst to the Chief Liquid Rocket Div on rocket propulsion tech. Assumed current pos in 1969. Enjoy working with Boy Scouts, backpacking, camping & fishing.

Greene, Howard G
Business: 3637 Green Rd, Cleveland, OH 44122
Position: Partner *Employer:* Greene-Frantz & Associates *Education:* BS/Const Eng/Chicago Tech Coll *Born:* 6/22/22 and raised in Cleveland, OH. Served with the US Coast Guard in World War II-1942-1946. During the past ten years-involved in the dev, design, construction, operation and ownership of Medical Office Buildings in the Eastern portion of the United States. *Society Aff:* NSPE, OSPE, CES

Greene, Howard L
Business: Univ of Akron, Akron, OH 44325
Position: Prof, Dept Chem Engr. *Employer:* Univ of Akron. *Education:* PhD/Chem Engr/Cornell Univ; MChE/Chem Engr/Cornell Univ; BChE/Chem Engr/Cornell Univ. *Born:* 4/3/35. Raised in up-state NY. Summer employment with Allied Chem, DuPont, NASA, PPG, IBM, Sandia and Mobil. Taught Chem Tech at Broome Technical College in Binghamton while attending grad school. Joined Univ of Akron in Sept 1965 as Asst Prof of Chem Engg. Became Prof and Hd of the Dept in 1977. Teaching and res interests in Chem Reaction Engg, catalysis and mixing. Enjoy woodworking, boating, fishing. *Society Aff:* AIChE, ΣΞ, ΤΒΠ.

Greene, Irwin R
Home: 7318 N Kedvale Ave, Lincolnwood, IL 60646
Position: Mgr Package Engrg. *Employer:* Sears Roebuck & Co. *Born:* Involved with all types of Indus Packaging since 1949. Pioneered use of Expanded polystyrene & developed techs for the use of 'honeycomb' & built-up as packaging materials. Was Mgr of Package Engrg at Alden s Inc until 1968 when joined Sears & started the Pkg Engrg function as a part of the Sears Lab. Currently Chmn of ANSI MH-6 Ctte on Marking, on the ASTM D-10 Ctte on Packaging & the Electronics Indus Assn Ctte on Packaging. Certified as Prof in Packaging & Materials Handling by the Soc of Packaging & Handling Engrs. Vice Pres of Illinois Chapter of SPHE. Past Pres of the Illinois Region of the Muscular Dystrophy Assn of America. *Society Aff:* SPHE, PI, ASTM.

Greene, James H
Home: Oak Point, 555 N 400 W, W Lafayette, IN 47906
Position: Prof Emeritus of Ind Engr. *Employer:* Purdue Univ. *Education:* PhD/Ind Engr/Univ of IA; MS/Ind Engr/Univ of IA; BSME/Ind Engr/Univ of IA. *Born:* 3/12/15. Born in Elmwood, NB, 1915. Grad Univ of IA, BSME 1947; MSIE 1948; PhD 1957. He is the Author of Production and Inventory Control - Systems and decisions, 1974, Operations Planning & Control, 1967, pub Richard D Irwin; Production and Inventory Control Handbook, 2nd ed. 1987, McGraw-Hill Book Co. Oper Mgmt for Profit & Prod, 1984, Prentice Hall Publ Co. Awards: Fulbright Lectureship, Finland Inst of Tech. Presidential Award Am Production and Inventory Control Soc. Married: Barbara Holt Greene. Children: Robin Tower, Timothy J Greene. *Society Aff:* ASEE, APICS.

Greene, John H, Jr
Business: PO Box 1608, Port Arthur, TX 77640
Position: Senior Tech. *Employer:* Texaco Inc. *Education:* SM/ChemEngg/MIT; BS/ChemEngg/Univ of ALL; BS/MetEngg/Univ of AL; AB/Chem/Univ of A. *Born:* 6/22/19. in Tuscaloosa, Al, Son of John Herman & Mary Lee (Hollingsworth). Married Lovie Cathryn Milam on Nov 30. 1945; children-Charles, John, Cathryn. Engg at Texaco, Inc; Port Arthur, TX 1941-49; Staff Div Res NYC, 1950-55. Sr Proj Mgr, Port Arthur 1956-64. Exec div staff, NYC, 1965-70. Sup of prof dev, Beacon, NY, 1970-77. Sr Tech of Res, Port Arthur, 1977--. Fellow, Amer Inst Chem Engrs; (Chmn 1972 An Mtg Prog Com, chmn Natl Mbr Com, Speakers Panel, Dynamic Objectives, chmn Sabine Area Sect). API subcom tech data, Phi Beta Kappa, Tau Beta Pi, Gamma Sigma Epsilon,. Republican, Episcopalian. Patented only comm continuous grease process. *Society Aff:* AIChE, ΣΞ.

Greene, Joseph E
Business: 1101 West Springfield Ave, 137 Coordinated Sci Lab, Urbana, IL 61801
Position: Prof of Met. *Employer:* Univ of IL. *Education:* PhD/Mtls Sci/USC; MS/Elec Engr/USC; BS/Mech Engr/USC *Born:* 11/25/44. Native of CA. Married. One of the founders and dirs of Mtls Dev Corp, 1969- 71. Prof of Met at Univ of IL, Urbana, since 1971. Visiting Prof, Univ of Campinas, Brazil, 1976-77. Visiting Scientist, IBM Res Lab, summer 1977. Served as Asst Dean of Engg, Univ of IL, 1977-78. Visiting Prof. in Materials Science and Electrical Engineering, Virginia Polytech-

Greene, Joseph E (Continued)
nic Institute and State Univ., spring 1980. Visiting Prof, Physics Dept Linkoping Univ, Sweden Winter, 1984; Visiting Scientist, CSIRO Appl Physics Lab, Sydney, Australia, Fall, 1984. Consultant for several large corps. Listed in Outstanding Young Men of America, 1979. More than 100 scholarly publications in the area of crystal growth, surface sci, and thin films. Mbr of the Bd of Trustees of AVS, 1978-1980, Dir of the Electron Spectroscopy Soc 1976-1981, Mbr Bd of Dirs of American Vacuum Soc 1983-date, Assoc Editor of Journal Vacuum Sci and Tech 1983-date. *Society Aff:* AVS, APS, ASM

Greene, Joseph L, Jr
Business: Dept of Chemistry, Auburn, AL 36830
Position: Prof of Chemistry. *Employer:* Auburn Univ. *Education:* PhD/Organic Chemistry/Emory Univ; MS/Organic Chemistry/Auburn Univ; BS/Chemistry/Auburn Univ. *Born:* 5/5/24. Montgomery, AL. Service in US Army in WWII. Asst, Assoc, Sr Chemist, TN Eastman Co 1950-1955. Instr & Res Assoc, Emory Univ 1955-1957. Sr Chemist, Shell Dev Co 1957-1958. Sr Chemist & Sec Hd, Thiokol Chem Corp 1958-1960. Sr Chemist, Southern Res Ins 1960-1968. Adjunct Assoc Prof, Univ of AL in Birmingham 1965- 1968. Assoc Prof, Auburn Univ, 1968-date. Natl Secy Phi Lambda Upsilon 1975- 1981, Natl Pres. Phi Lambda Upsilon 1981-date. *Society Aff:* ACS, ΦΛΥ, ΣΞ, AAAS, ΣΧ.

Greene, Leroy F
Home: PO Box 254646, Sacramento, CA 95825
Position: Owner. *Employer:* Leroy F Greene & Assoc. *Education:* BSCE/Civil Engg/Purdue Univ. *Born:* 1/31/18. Newark, NJ Bridge Dept Ind Hwys, TVA Design of Structures. Designer of bridges, hydro-electric power plants, dams, commercial industrial bldgs, schools. Leroy F Greene and Assoc formed 1951 Sacramento CA. Mbr CA State leg 1963-present. Chmn Assembly Education Committee. Lincole Arc Welding award 1964. Amer Inst of Steet Constr award 1964. ASCE Civil Govt award 1978. *Society Aff:* ASCE.

Greene, Ray P
Business: 508 - S Byrne Rd, Toledo, OH 43609
Position: President. *Employer:* Ray Greene Indus. *Education:* BS/Mech Engg/OH State; BS/Ind Engg/OH State; Dist. Alumnus/-/OH State. *Born:* Jan 1913. Pres Ray Greene Indus 1976, Pres Ray Greene & Co Inc to 1973. m. Dorothy Owen Aug 1975. c. Kathleen Jan, Kristina Louise, Raymond Karl, Lance. Owner Ray Greeene Indus, builder, designer 1931. Mbr Mil Asst Adv Group Viet Nam 1961-62. Mbr Recreational Panel of the Ocean Sci & Tech Advisory Ctte (OSTAC) for NSIA 1970-72. Chmn BIA Sailboat Engrg Comm 1967-75. BIA Engrg Coord Comm 1967-75. ABYC Sailboat Design Comm 1971-74. Recipient Distinguished Alumnus Award in engrg Ohio U 1963; Mbr Sigma Chi, Rotarian. Comm for Tomorrow OH State Univ.

Greene, Robert L
Business: 1130 One Energy Squ, Dallas, TX 75206
Position: Pres & Owner. *Employer:* Greene & Assocs Inc. *Education:* MS/Chem Engg Practice/MIT; BS/Chemistry/William & Mary College. *Born:* Mar 1923 VA. Army Air Corps AUS 1943-47. Exxon Eng 1947-53; V P Purvin & Gertz Inc Cons Eng Dallas 1953-74; Cons in Engrg in Belgium, Germany, Bahamas, PR, Canada, Mexico, Venezuela, Nigeria, Mid-East. Expert petrol refining, natural gas processing, petrochems, oil & gas PL & terminals, fertilizer/synthetic gas/chems mfg. Expert feasibility studies & econom evals, proj mgmt, market surveys/forecasts, investigations, expert testimony. Reg TX, NH; Cert Natl Panel Am Arb Assn NY; AIChE; Phi Beta Kappa; Alpha Chi Sigma; Who's Who South & SW. Pat oil refining. Enjoys tennis, skiing, hunting. *Society Aff:* AIChE, AACCCE, AAA

Greenfield, Eugene W
4747 Oak Crest Rd-45, Fallbrook, CA 92028
Position: Elec Power Transmission Cons. *Employer:* Self-employed. *Education:* Dr Engg/Elec Engg/Johns Hopkins Univ; M Engg/Elec Engg/Johns Hopkins Univ; B Engg/Elec Engg/Johns Hopkins Univ. *Born:* 11/27/07. Balt, Md. Di-electrics under J B Whitehead. Post-docs at Brooklyn Poly & Columbia. Head elec res lab Anaconda W & Cable Co, cable R&D 1935-50. Supr Elec Res Lab Kaiser A & C Co, wire & cable R&D 1950-58. Dir Div of Engrg Res, Asst Dean & Prof Wash St Univ 1958-73. Elec Power Transmission Cons 1973- . Over 250 major reports on res findings; 52 arts & papers publ nationally. Author 2 books on dielectrics & cable engrg. V Pres AIEE 1962-64, Chairman IEEE High Power Testing Symposium 1971, Chairman Engrg Foundation Conf 1972; Chairman Elec Power R&D Center 1962-72; Chairman 6th Region IEEE Advisory Ctte 1970. Wash State Energy Policy Council 1974. Distinguished Engr Award IEEE 6th Region 1965. Awarded IEEE Centennial Medal and Citation, 1984. Major hobby, chamber music playing (flute) with wife Louise (cello) "Eng'r of the Year" - Spokone WA 1973; Founder and Director of Cable Failure Clinics 1968-. Member American Board of Arbitrators, 1986. *Society Aff:* IEEE, ΣΞ.

Greenfield, Irwin G
Business: Mech Engrg, U of DE, Newark, DE 19716
Position: Prof *Employer:* Univ of DE *Education:* PhD/Met Engrg/Univ of PA; MS/Met Engrg/Univ of PA; AB/Met/Temple Univ *Born:* 11/30/29 Unidel Prof of Materials Sci and Mech Engrg, 1984-. Dean, College of Engrg, Univ of DE, 1973-1984. Visiting Prof, Dept of Mech Engrg, Eindhoven Tech Univ, Netherlands, 1978. Acting Chrmn, Dept of Mech and Aerospace Engrg, 1970-1971. Visiting Prof, Oxford Univ, UK, 1970. Visiting Prof, Stanford Univ, 1969. Prof, Met and Materials Engrg, Mech and Aerospace Engrg Univ of DE, 1968. Assoc Prof, 1965-1968. Asst Prof, 1963-1965. PhD - Univ of Penn 1963, Sr Res Met, Franklin Inst Labs for Res & Dev, 1953-63. Met, Naval Air Experimental Stations, Phila PA, 1951-1953. Major res fields: Processing of Metal Matrix Composites, surface effects on mech properties, erosion and wear, electron microscopy, Auger Spectroscopy, fatigue and other mech properties. *Society Aff:* AIME, NSPE, ΣΞ, ASM INTL.

Greenfield, Seymour S
Business: 250 W 34th St, New York, NY 10119
Position: Chrmn, Bd of Dirs. *Employer:* Parsons Brinckerhoff Quade Douglas. *Education:* BME/Mech Elec Engg/Polytechnic Inst of Brooklyn. *Born:* July 1922 Brooklyn N Y. Partner, Chmn, Bd of Dirs Parsons, Brinckerhoff, Quade & Douglas Inc, Married, 2 children. Lt USNR active duty 1944-46; joined PBQ&D 1947; led M/E engrg efforts: NORAD-COC; Communications Ctr, Ft Ritchie, Keflavik AF Base Iceland. Since 1964 Ptnr PBQ&D dir many lrg tnransportation projs & mil const projs. Lic PE in num states; Mbr ASME, ASHRAE, NYSPE; Natl Pres Soc of Amer Military Engrs; Tr MOLES; Consult for Cvl Engrg Dept Manattan Coll NYC. *Society Aff:* ASME, APTA, ASHRAE, NYSPE, SAME, MOLES.

Greenkorn, Robert A
Home: 151 Knox Drive, West Lafayette, IN 47906
Position: Professor. *Employer:* Purdue University. *Education:* PhD/Chem Eng/Univ of WI; MS/Chem Eng/Univ of WI; BS/Chem Eng/Univ of WI. *Born:* Oct 12, 1928. USN - naval aviator 1946-51. Postdoc Fellow Norwegian Tech Inst 1957-58. Res Assoc Jersey Prod Res Co Tulsa 1958-63. Lecturer Univ of Tulsa 195863; Assoc Prof Mechanics Marquette Univ Milwaukee 1963-65. Assoc Prof School of Chem Engrg Purdue Univ West Lafayette Indiana 1965-67; Head School of Chem Engrg 1967-73; Prof 1967- ; Assoc Dean of Engrg 1976-1980; Dir of Engrg Experiment Station 1976-1980 ; Dir Inst for Advanced Interdisciplinary Engrg Studies 1972- 75. Acting Head Aeronautical & Astronautical Engrg 1973; Dir Environmental Engrg Center 1976-78, Acting Dir, Coal Res Lab, 1979-1980. Vice-President and Associate Provost 1980-86; Vice-President for Programs, Purdue Research Foundation 1980-, Vice-President for Research, 1986-. *Society Aff:* AIChE, SPE-AIME, ACS, ASEE, AGU.

Greenleaf, John W, Jr
Home: 1014 Manati Ave, Coral Gables, FL 33146
Position: Retired. *Education:* BSCE/Civil Engg/Northeastern Univ. *Born:* Oct 1908. Prof Engr Mass & Fla. Designer/Resident Engr san engrg projs until WW II. Served in Inst of Inter-American Affairs as Ch Engr of San Mission Asuncion Paraguay. Attained rank of Major San Corps. Priv practice since 1949 as principal in civil/environmental engrg. Major water & sewerage projs in Paraguay, Peru, Colombia, Costa Rica, Nicaragua & US. Natl hwy projs Ecuador & Panama. Extensive work in solid waste mgmt. Inventor. John C Vaaler Award 1964 CPI. ACEC 1974 Grand Conceptor Award. Extensive work in marine work related to recreational boating. *Society Aff:* ASCE, NSPE, AWWA, WPCF, APWA, AAEE.

Greenslade,.Wiliam M
7500 N. Dreamy Draw Dr #145, Phoenix, Arizona 85020
Position: Partner *Employer:* Dames & Moore *Education:* MS/Hydrology/Univ of NV Reno; BS/Geol Eng/Mackay School of Mines *Born:* 1/2/39 Cleveland, OH. Worked as Res Assist for Desert Res Center, Univ of NV while in graduate school. Presently Managing Partner of Denver Office of Dames & Moore. Conducted hydrogeologic investigations throughout the world. Recent technical work specialized in mine dewatering. Served as State Pres of AZ Section AIPG. PE in AZ. After Jan, 1982, will assume managerial responsiblity for firm's operations in Pacific, Far East and Australia. *Society Aff:* SAME, AEG, AIPG, NWWA

Greenstein, Teddy
Home: 3000 Ocean Pkwy, Brooklyn, NY 11235
Position: Prof of Chem Engg *Employer:* NJ Inst of Tech. *Education:* PhD/Chem Engg/NY Univ; MChE/Chem Engg/NYU; BChE/Chem Engg/City College of NY. *Born:* 3/16/37. Rating engr, Davis Engg, 1960; teacher, high sch, 1963-64; res asst Chem Engg NY Univ, 1964-67. With NJIT since 1967. Contbr articles to profl journals. Res areas: Low Reynolds number hydrodynamics; biochemical engg. Recipient Grants, NSF, 1964, Inst Paper Chemistry, 1965-67, Found Advanced Grad Study Engg, 1967-69, Excellence Initiative, 1987. *Society Aff:* AIChE, ASM, ASEE, ΣΞ, TBΠ, ΩXE, AAUP.

Greenwood, Allan N
Business: Troy, NY 12181
Position: Philip Sporn Prof of Engg, Chrmn Elec Power Engg. Dept. *Employer:* Rensselaer Poly Inst. *Education:* BA/Mech Eng/Cambridge Univ; MA/Mech Eng/Cambridge Univ; PhD/Elec Eng/Univ of Leeds *Born:* British subj. Served in the Royal Navy during WWII. Worked in British industry and taught at the Univs of Leeds and Toronto before joining the Power Delivery Group of GE in 1955. Became Sr Consulting Engr. Interested in power switching tech, especially vacuum, which he helped to pioneer. Prolific patentee and author of numerous publications, including books, "Electrical Transients in Power Systems–, "Vacuum Arcs–, (co-author) and "Adv Power Cable Tech–, Vols I & II (co-author). Assumed current position in 1972. Active consultant to industry and govt. *Society Aff:* IEEE, HKN, ΣΞ, CIGRE

Gregg, Dennis E
Business: Box 488, 4001 Stavanger, Norway
Position: Pres & Managing Dir *Employer:* Conoco Norway Inc *Education:* Petrol Engr//CO School of Mines *Born:* 1/26/28 Raised in CO. Employed by Conoco since then - in the US Gulf Coast to 1966, NY 1966-69, London 1969-1980 and now Stavenger, Norway. Work largely offshore- related since 1954. In New York coordinated dev of Fatch Field in Dubai using unique underwater and floating storage. In London participated in Viking gas field development offshore Southeastern England then was general proj mgr for Murchison Field dev in 512 feet of water in northern North Sea. Active in SPE throughout career, now Director for European region. *Society Aff:* AIME

Gregg, Henry T, Jr
Home: 766 Crooked Cr Rd, Hendersonville, NC 28739
Position: Engr-M&P. *Employer:* GE Co. *Education:* MS/Met Engr/VPI & SU; BS/Met Engr/VPI & SU. *Born:* 6/21/27. Native of Richmond VA. Began work for GE Co in 1950 on Chem-Met Training Prog. Have worked for GE in wire & cable, ordnance, avionic controls & lighting systems depts. Work has involved lab, design engg, adv mfgr engg, and adv engg. Have been in present position since 1971. Taught met courses at VPI while getting MS degree. Also taught at Blue Ridge Technical College. Served as chrmn of Southern Tier & Old South Chapts of ASM. Served on ASM Natl Mbrship Comm for 3 yrs. *Society Aff:* ASM.

Gregorits, John W
Home: 12 Hugenot St, East Hanover, NJ 07936
Position: Dir *Employer:* Engg Enterprises *Education:* MSME/Mech Engg/Newark College of Engg; BSME/Mech Engg/Newark College of Engg. *Born:* Sept 1926. BSME, MSME Newark Coll of Engrg. Designer & Proj Engr with various industrial organizations 1943-58. With the Dept of the Army in engineering and managerial positions, 1958-1987, Reg Prof Engr active in Natl Soc of Prof Engrs (past Chrmn-PEG) & the ASME. Founder and current Director of Engrg Enterprises, since 1973, a correspondence school offering Prof Engrs Examination Review Courses. Author of many publications & articles, the latest being books on the preparation for the Prof Engr examination. *Society Aff:* NSPE, ASME, ADPA.

Gregory, Bob Lee
Business: Sandia National Laboratories Org. 2100, P.O. Box 5800, Albuquerque, NM 87185
Position: Dir *Employer:* Sandia National Labs *Education:* PhD/EE/Carnegie Mellon Univ; MS/EE/Carnegie Mellon Univ; BS/EE/Carnegie Mellon Univ. *Born:* 9/30/38. He joined Sandia Labs in 1963 as a staff mbr in the Radiation Phys Dept, where he has been active in the dev of radiation-tolerant, high-reliability semiconductor tech. He currently is Dir of Microelectronics, and is responsible for the dev & production of integrated circuits & hybrid microcircuits for Sandia applications. Dr Gregory is a Fellow in the IEEE & is a past Chrmn of the Albuquerque Section. Dr. Gregory has been an active participant in the Annual IEEE Nucl & Space Radiation Effects Conf, & served as Technical Prog Chrmn in 1972 & Gen Chrmn in 1978. He is a past Assoc Editor of the IEEE Journal of Solid State Circuits and is a past mbr of the IEEE Electron Device Soc ADCOM. *Society Aff:* IEEE

Greif, Ralph
Business: Dept of Mech Engrg, UC Berkeley, Berkeley, CA 94720
Position: Professor *Employer:* Univ of CA, Berkeley *Education:* Ph.D.,/MA/Engineering/Harvard; MS/Engineering/UCLA; BSME/Mechanical Engineering/New York University *Born:* 11/28/35 Staff member, Hughes R & D Labs, Los Angeles, 1956-58. Faculty Member, Univ CA, Berkeley, 1963 to the present. Post-Doctoral Research Fellow, Harvard Univ, Cambridge, 1963. Visiting Scholar, Imperial College of Sci & Tech, London, 1969-70. Visiting Professor, Technion, Israel Inst of Tech, Haifa, 1977. Guggenheim Fellow, John Simon Guggenheim Memorial Foundation, 1969-70. ASME Heat Transfer Memorial Award, ASME, 1985. ASME, Fellow, 1986. ASME, Technical Editor, Journal of Heat Transfer, 1983 to the present. ASME, Honors and Awards Cttee, Heat Transfer Div, 1979-82, (Chairman, 1981-82). ASME, Heat Transfer Div, Technical Cttee on Aircraft and Astronautical Heat Transfer, 1967 to the present (Chairman, 1970-73). *Society Aff:* ASME

Greif, Robert
Business: Dept of Mech Engg, Tufts University, Medford, MA 02155
Position: Prof and Chrmn, Dept of Mech Engrg *Employer:* Tufts Univ. *Education:* PhD/Appl Mechanics/Harvard Univ; SM/Appl Mechanics/Harvard Univ; BME/Mech Engrg/NYU. *Born:* 1/17/38. Ph.D. (Applied Mechanics) Harvard Univ, 1963. Staff Scientist, Avco, 1963- 1965. Sr Staff Scientist, Avco, 1965-1967; Tufts Univ: Lectr 1966-67, Asst Prof 1967-70, Assoc Prof 1970-78, Prof 1978-, Chrmn 1981-. Visiting Res Fellow, Univ of Sussex, Sussex England 1974. Visiting Scholar, Harvard University, Cambridge, Mass, 1981. Consultant in Applied Mechanics. Associate Fellow, AIAA. *Society Aff:* ASME, AIAA, ASEE, TBΠ.

Greig, James
Home: Inch of Kinnordy, Kirriemuir, Scotland DD8 4 AA Great Britain
Position: Prof Emeritus. *Employer:* Univ of London. *Education:* PhD/-/Birmingham; MSc/-/London; BSc/-/London. *Born:* Apr 24, 1903 Edinburgh. 1923 Diploma of Heriot-Watt College Edinburgh. Exp in Telephone Engrg with Bell Telephone Co Canada 1924-26. 1928-33 Res on Power Amplifiers for the output stages of radio transmitters Res Lab GEC Wembley. 1933 MSc Univ Coll London. 1938 PhD U of Birmingham. 1936-39 Lecturer in the Univ of Birmingham. 1939-45 Head of the Elec Engrg Dept Northampton Poly London; engaged in Univ teaching & in dev work for Services instruction in radar. 1945-70 Prof & Hd Dept of Elec Engrg King s College Univ of London. Res on Tooth-Ripple Phenomena in Alternators. *Society Aff:* IEE, IRE.

Greiner, Harold F
Business: E G & G Sealol Inc, 15 Pioneer Ave, Warwick, R.I 02888
Position: Vice Pres & Ch Engr. *Employer:* EG&S Sealol, Inc. *Education:* BME/Mech Engrg/Syracuse Univ; MBA/Gen Bus/Byrant Coll. *Born:* May 20, 1922. Joined Sealol Inc 1950 as Proj Engr; held the following pos: Mgr Marine & Indus Div, Sales Mgr, Mgr Seal & Valve Div; presently V Pres & Ch Engr. Military: Capt USNR Retired, qualified submarine officer, engrg officer on various submarines. Lic aircraft pilot, single engine rating. Reg PE RI. Mbr Providence Engrg Soc. *Society Aff:* STLE, ASME, SNAME, ASTM, ASNE.

Greitzer, Edward M
Business: 31-265 Dept of Aero & Astro, Cambridge, MA 02139
Position: Prof, Dir Gas Turbine Lab *Employer:* MA Inst of Tech. *Education:* PhD/Mech Engrg/Harvard; SM/Engrg/Harvard; AB/Phys/Harvard. *Born:* 5/8/41. Res Asst, Harvard Univ, Div of Engg and Appl Physics, 1965-1969; Res Engr, Compressor Component Group, Pratt & Whitney Div of United Technologies Corp, 1969-1976; Ind Fellow Commoner, Cambridge Univ, Cambridge, England, 1975; Sr Res Engr, United Technologies Res Ctr, 1976-1977; Asst Prof, Dept of Aero & Astro, MA Inst of Tech, 1977-1979; Assoc Prof, 1979-1984. Royal Soc Guest Fellow and SERC Visiting Fellow, Cambridge Univ, Cambridge, England, 1983-1984. Prof and Dir, Gas Turbine Lab 1984-present. ASME Gas Turbine Award for Outstanding Technical Paper of 1975, 1977; IMechE (England) T. Bernard Hall Prize, 1976; ASME Freeman Scholar Award in Fluids Engg, 1980. Fellow, ASME, 1986- . *Society Aff:* ASME, AIAA, ΣΞ.

Grekel, Howard
Home: 8844 Meml Dr, Houston, TX 77024
Position: Pres. *Employer:* H G Consulting Assoc Inc. *Education:* MS/CE/MIT; BA/Organic Chemistry/UCLA. *Born:* 10/8/18. Pres, H G Consulting Assoc Inc, Houston, developed EVOP for Profit prog for improving profits and guided programs in 170 natural gas-processing and pipeline, LPG terminal, refining, & oil production operations. 1947-1975, Amoco Production Co, directed res in chem synthesis & separation, underground combustion for oil recovery, & Claus sulfur recovery; engg supervisor responsible for process design & modification of 30 gas processing plants including cryogenic, sweetening, & sulfur recovery plants. WWII, 1941- 1945, US Army, Infantry & AF, Pilot, Capt, Air Medal. Fellow of AIChE. Reg PE in OK & TX. *Society Aff:* AIChE, GPA, AXΣ, ΣΞ.

Grenga, Helen E
Home: 6139 Hickory Dr, Forest Park, GA 30050
Position: Asst VP *Employer:* GA Inst of Tech *Education:* PhD/Chem/Univ of VA; BA/Chem/Shorter Coll, GA *Born:* 4/11/38 Native of Newnan, GA. PhD Thesis on catalytic reactions on single crystals. Positions at GA Institute of Tech, School of Chemical Engrg: Postdoctoral, 1967; Asst Prof. 1968-72; Assoc. Prof. 1972-77; Prof 1977-Present. Assoc. Dean, Graduate Div, 1978-82. Dir, Grad Co-op Prog, 1983-present; Asst. VP, 1987-present. Established school laboratories for field-ion and emission microscopy. Research and publications on catalysis, corrison and oxidation. Chrmn of 22nd Intl Field Emission Symposium, 1975. GA Tech Sigma Xi President, 1978-79; Society of Women Engineers, National Pres, 1981-82; Member Phi Kappa Phi; Ga. Tech Phi Kappa Phi President, 1983-84. *Society Aff:* AIME/TMS, ASM, SWE, ASEE, TBΠ

Grethlein, Hans E
Business: Dartmouth College, Hanover, NH 03755
Position: Prof of Engg. *Employer:* Thayer Sch of Engg. *Education:* PhD/Chem Engg/Princeton Univ; BS/Chem Engg/Drexel Univ. *Born:* 5/16/34. Born in Plainwell, MI, raised in Phila, PA. Chem engr since 1957. Mgr of Systems Analysis at American cyanamid Co, Organic Chem Div, Bound Brook, NJ 1964-68. Prof of Engg at Thayer Sch since 1968. Maj profl interest in process dev with emphasis in membrane dev for reverse osmosis and biomass conversion, such as cellulose hydrolysis. Author of numerous papers, consultant, and register PE in NH, and register PE in NH. *Society Aff:* AIChE, ACS

Greve, Norman R
Business: 2042 Broadway, Santa Monica, CA 90404
Position: Structural Engr. *Employer:* Greve & O Rourke Inc/Sys Comp Corp. *Education:* BS/Civil Eng/Cal Tech; BAS/Civil Eng/Univ of CA (Berkeley); Numerous cont ed courses *Born:* 05/17/25 Native of CA. Lic Civil & Struc Engr CA with Greve & O Rourke Inc Santa Monica, CA spec in design of schools, hospitals & pub wks. Mbr Seismology Ctte of Struc Engrs Assn of CA; VP of Sys Comp Corp, developers of the Sys Prof Library of Computer Programs used worldwide by struct engg offices; Mbr ASCE & as such; P Chmn of Tech Council of Computer Practices. TCCP; P Chmn Computer Practices Ctte, respon for publication of 'Pricing Policy & Method for computer usage; Chmn Ad Hoc Ctte on ASCE policy towards Natl Inst of Computers in Engg (NICE); P Chmn Soc for Computer Applications in Eng Planning & Arch (CEPA), a group of over 200 Cvl Engg firms. Lect & author of papers concerning computer usage by engrs, principal author of 30plus struct eng computer programs and 7 vol of documentation for more than 70 programs. *Society Aff:* ASCE, NSPE, CEAC, CEC, PEPP, SEAOC, CEPA

Grewal, Manohar S
Business: Gillette Park, South Boston, MA 02106
Position: Group Manager. *Employer:* Gillette Co. *Education:* ScD/Matl Science/MIT Camb; MS/Matl Engg/RPI; BSc/Metallurgy/BHU. *Born:* Dec 1, 1935 Tatanagar India. ScD MIT Cambridge Mass. M Mat Engrg RPI Troy NY. BSc Met Engrg BHU India. Metallurgist in steel melting at Indian Iron & Steel Co India; Dev Metallurgist Avco, Lycoming Ct, & R&D Sci at Gillette Co Mass. Assumed present position in 1973. Published sev tech papers. Gold Medal Internatl Microstruc Soc & Lucas Award ASM; $1000 cash award for excell in Metallography in 1973 at the Internatl Metallographic Exhibit. *Society Aff:* AIME, ASM, MS, EMSA, APMI, IMS.

Grey, Jerry
Business: 370 L'Enfant Promenade, SW, Washington, DC 20024
Position: Dir., Sci & Tech. Policy *Employer:* AIAA *Education:* PhD/Aeronautics/Calif. Inst. of Technology; MS/Engg Physics/Cornell Univ; BME/Mech. Engg/Cornell Univ *Born:* 10/25/26 Tenured professor at Princeton 1951-67. Taught Aerodynamics, propulsion, nuclear powerplants and directed nuclear propulsion research laboratory. Pres of Greyrad Corp 1959-71 and Calprobe Corp. 1973-82. Pres of International Astronautical Federation 1984-86. Chairman, Solar Advisory Panel, Office of Technology Assessment; Chairman, Coordinating Cttee on energy, AAES. Director, Applied Solar Energy Corp; Dir, Scientists Institute for Public Information. Publisher, Aerospace America (1982-87). Author of Enterprise (Morrow, 1979), Beachheads in Space (MacMillan 1983), Aeronautics in China (AIAA 1981). V.P. International Academy of Astronautics. Deputy Secretary-

Grey, Jerry (Continued)
General, UN Conference on Exploration and Peaceful Uses of Outer Space 1982. *Society Aff:* AIAA, IEEE, AAAS

Gribaldo, Albert C
Business: 1900 Embarcadero Rd, Palo Alto, CA 94303
Position: President. *Employer:* Earth Systems Cons. *Born:* Sept 1925. BSCE Univ of Calif. Became Reg Cvl Engr Calif 1952; active reg in 8 other states. 1957-62 Mgr of Oper of Testing & Controls Inc. 1962-69 Pres & Oper Mgr of Gribaldo, Jones & Assocs spec in soil engrg & geology. Since 1969 duties have been as Pres & Oper Mgr of Earth Sys Inc & Earth Sys Cons. Fellow ASCE; Mbr NSPE, SAME, SEAC, CEAC. Citizen of Yr 1971 by City of Mountain View.

Gribben, J Clark
Business: 20 Vine St, Reno, NV 89503
Position: Pres. *Employer:* J Clark Gribben Consulting Engrs, LTD. *Education:* BSCE/Civ Engg/Univ of NV-Reno; MSCE/Structural Engg/Univ of NV-Reno. *Born:* 4/10/36. 29 yrs of experience in the structural engg field, twenty-three as a firm principal. Presently pres of a consulting engg firm. Past pres of local chap of NSPE and state sec of ASCE. Past or current mbr of Strong Motion, Education, Legislative, and Ethics Committees for ASCE, NSPE, and AGC. Past mbr on the bd of dir for NSPE, ASCE, AGC. Mbr of the City of Reno Bd of Appeals, mbr of Washoe County Planning Commission, Governor's (NV) Panel on Seismic Mitigation, and past mbr of the GSA Regional Panel on Architecture and Engg. Also an Arbitrator for the American Arbitration Assn. Married with two daughters. *Society Aff:* ASCE, NSPE, PEPP, ACI, CSI, AGC.

Grice, Harvey H
Business: Chemical Engineering Department, University of Missouri, Rolla, Rolla, Mo 65401
Position: Professor. *Employer:* U of Missouri-Rolla. *Education:* PhD/ChE/Ohio State Univ; MS/ChE/Ohio State Univ; BS/ChE/Ohio State Univ. *Born:* Sep 25, 1912. Native of St Clair MI. BChE 1937, MSc 1938, PhD 1941, from Ohio State Univ. Employment with General Foods Corp: Project Leader Central Labs 1942; Tech Asst to Mgr Mfg & Engrg 1948-52; Supt Processing & Power Diamond Crystal Salt Div 1946-48; Plant Mgr Diamond Crystal Salt Div 1952; Mgr Mfg & Engrg Kankakee Operations 1952-58. Col Chem Corps US Army, service with 42nd Inf Div Europe, Retired AUS 1972. Pres Graceland College, Lamoni IA 1958-64. Prof Chem Engrg U of Missouri-Rolla 1964-1978. Retired (Prof Emeritus, Chem Engg) May 1978. Elect Fellow AIChE 1976. Mbr NSPE, Sigma Xi, Phi Kappa Phi, Tau Beta Pi. *Society Aff:* NSPE, AIChE, ΣΞ, ΦΚΦ, ΤΒΠ

Gridley, Robert J
Home: 1400 Seymour Rd, Vestal, NY 13850
Position: Advisory Engineer. *Employer:* IBM Corp. *Born:* Dec 1931. BS, MS Case Western Reserve U. Native of Cleveland OH. Metallurgist for Clevite Res Ctr (now Gould Corp). Spec in Titanium Powder Metallurgy & Dev of Nuclear Fuel Components. With IBM Corp since 1959. Managed Mat Engrg Depts (Met, Chem, Plastics Activs) for last 15 yrs. Sec (1975) & mbr Advis Tech Awareness Council (197175) ASM. Enjoy tennis & sailing.

Griebe, Roger W
Business: P O Box 736, Idaho Falls, ID 83401
Position: Vice President. *Employer:* Energy Inc. *Born:* June 1943. PhD 1968 with BS & MS in Mech Engrg from Purdue U. Mbr Sigma Xi, ASME, ANS. Taught at Purdue U prior to joining the Atomic Energy Div of Phillips Petroleum Co at the Natl Reactor Testing Station 1968. Proj Engr for BWR-FLECHT, GE-BDHT for ID Nuclear Co & Aerojet Nuclear Co & part in a wide var of water reactor safety programs incl SECHT, PWR-FLECHT, Semiscale, LOFT. Joined Energy Inc in Fall of 1972 to dir the Syst & Analysis efforts of the company, building the prof staff to over 40 degreed engrs & scientists & a multi-million dollar cons program in 4 yrs.

Grier, William F
Home: 712 Cromwell Way, Lexington, KY 40503
Position: President. *Employer:* W F Grier & Assoc *Education:* BCE/Civil Eng/GA Tech. *Born:* 8/15/32. Native of Charlotte NC. Army Corps of Engrs in Korea; work primarily in Water Resources. With Southern Co on Coosa River Dev Proj & with R W Beck & Assoc on Columbia River Proj. Worked in Bangladesh on maj Flood Control & Irrigation proj near influence of Ganges and Bramaputra Rivers 1970 & Pres of KY Div of Mayes, Sudderth and Etheridge Inc in 1971. Supr a wide var of engrg projs. Formed Grier, Asher, Fuqua, Inc in 1979 and W F Grier & Assoc in 1981, as pres & Chief Engr. Engaged in consltg Engr including solid waste energy recovery projects, water projects, land dev, sanitary sewers and dams. *Society Aff:* ASCE, NSPE.

Grierson, Joseph B
Business: 6250 W Park, Suite 270, Houston, TX 77057
Position: Pres. *Employer:* Systems Appl Engr, Inc. *Education:* BS/CE/Univ of Houston. *Born:* 10/11/39. Sixteen yrs experience in the design of complete real-time computer hardware and softwater systems for dat acquisition & process control. Chem process engr, Monsanto Co, 1961-65 specializing in process simulation & ophmization studies. Control systems engrs, IBM Corp, 1965-68. Responsible for design, implementation & installating sensor-based process control systems in the oil, chem & utility industries. Co-founder & pres of systems Application Engg, Inc; a Houston, TX based engg & consulting firm providing programming services to the utility, oil & gas, process industries & mfg industries. *Society Aff:* AIChE ΤΒΦ.

Grieves, Robert B
Business: College of Engineering, Lexington, KY 40506
Position: Assoc Dean & Inst Dir. *Employer:* U of Kentucky. *Education:* PhD/Chem Engg/Northwestern Univ; MS/Chem Engrg/Northwestern Univ; BA/Russian/ Northwestern Univ. *Born:* 10/15/35. Asst & Assoc Prof Civil Engg at Northwestern Univ & IL Inst of Tech from 1961-67. With the Univ of KY since 1967: 1967-present Prof Chem Engg Dept; 1973- present Dir KY Water Resources Res Inst (Pl 95-467); 1976-79 Assoc Dean for Grad Programs & Res College of Engrg. 1979-present Assoc Dean for Administration and for Grad Affairs & Res College of Engg. Mbr ESWQIAC Ctte (PL 92-500), advisory to Administrator US Environ Protection Agency. Mbr Editorial Board of journal, Separation Science and Technology. Mbr Phi Beta Kappa & Tau Beta Pi. Res in physicochem processes-foam separations & membrane separations. Pub 150 tech papers. *Society Aff:* AIChE, WPCF.

Griffin, Daniel M
Home: 4120 Hickory Hill Blvd, Titusville, FL 32780
Position: Director Engr, Tvl *Employer:* McDonnell Douglas. *Education:* BS/EE/NC State. *Born:* 8/19/31. Native of Lenoir, NC. Served with USAF 1951 thru 1954. Designer of elec- electronic equip for Army tactical missiles 1959 thru 1967 for Douglas Aircraft Co. Systems Engr on anti-tank missile 1968 thru 1974 with McDonnell Douglas Astro Co. Design Dept hd 1974 thru 1978. Chief Engr, Titusville Div Jan, 1979 thru November 1980. Director of Engg, December 1980-August 1986. Presently Chief Program Engr, Dragon. *Society Aff:* AIAA, AUSA, ADPA.

Griffin, James H
Home: Box 422, Brookshire, TX 77423
Position: Chief Exec Officer. *Employer:* Rapsilver Internatl Inc. *Education:* MS/ChE/TX A&M; BS/ChE/TX A&M. *Born:* Jul 1917. Reg Prof Engr TX 6469. Supr of Prod Control for TX Div Dow Chem Co 1941-46. Independent Cons Engr 1946- . Mbr Prof Engrs in Private Practice, NSPE, San Jacinto Chap Texas Soc PE's, AIChE, So Texas Sect AIChE. Married Dorothy Rapsilver 1942 - 5 Sons - 2 Daughters - 15 Granchildren. Founded J.H. Griffin & Associates 1946 - International Engineering Design and Management Consultants. Projects: North and South America, Europe, Africa, Persian Gulf, India, Philippines, Orient. Several international patents Cereal Grain Field. Specialists in the Third World Developing Countries 1946 to present. Assistance to private and Governmental Entities for Food Production, including Conservation, processing, and Internal Distribution; Natural Resource Development, Tourist Development, and establish Trade Barter using available exports for needed imports. Founded Rapsilver International, Inc. 1957 as engineers for Design and Construction Management, whose services have been utilized worldwide. *Society Aff:* NSPE, PEPP, TSPE, AIChE.

Griffin, John J
121 Shore Line Drive, Downingtown, PA 19335
Position: Consulting Engineer *Employer:* J.J. Griffin, P.C. *Education:* BS/Chem/Mt St Mary's, MD. *Born:* 1/23/19. Education: BS/CHEM/MT. St. Mary's College, Md.; US Navy 1941-45. Amphibious Forces Europe-Africa 1943-45. Reg. Prof. Engr. NY, Penna. With Alcorn Combustion Co. and Petro Chem. Dev. Co. 1955-81. Entered private practice 1981, specializing in energy recovery and process heater design. Numerous patents, combustion and fired heater fields. Author numerous technical papers, process design. Past Chairman Fired Heater Subcommittee API. Member ASME, AICHE, API. Wife: Adele Harrington. Hobby: Boating. *Society Aff:* ASME, AIChE, API.

Griffin, Norman E
Home: 9713 Singleton Dr, Bethesda, MD 20817-2466 *Employer:* Retired *Education:* BS/IE/IL Inst of Technology. *Born:* 9/9/15. Added 42 grad credit hours. Native of Waukegan Il. 1958 transferred from Ch Reg Engr (Cincinnati) to Postal Headquarters as Proj Mgr for Mechanization, Assigned sev pats, USPS Industry Coordinator, Intl Postal Tech Liaison & USPS Program Mgr for Packaging & Mail Damage Reduction. Chap Pres IIE & NIPHLE/LIFE MEMBER. Cut to provide space. Chmn of Packaging Mat Ctte for EJC/TAP supporting OTA; res arm of Congress. 1979, Transportation Industry Award. Enjoys photography & POPS Concerts. *Society Aff:* NSPE, DSPE, IIE, NiPHLE, NARFE.

Griffith, James W
Home: 3921 Caruth Blvd, Dallas, TX 75225
Position: Engrg Consultant. *Employer:* James W. Griffith Inc. *Education:* BS/EE/SMU; MS/IE/SMU. *Born:* April 1922. Native of TX. USAAF 1942-46. Prof & Chmn of the Depts of Systems & Industrial Engrg SMU 1949-69. Cons to: LOF Co, Holophsne Co, HEW, NBS, DOE & the States of ME, NC, MD, NY, PA, CT, LA, FL, MN, TX & AK on Life Cycle Cost Benefit Analysis 1969-79. Pres IES 1971-72. US Expert on Daylighting to CIE 1957-79. Fellow IES & AAAS. Sigma Tau. Eta Kappa Nu. Who's Who Amer Men of Sci Internatl Bio. Mbr ASHRAE Panel 4-89. Mbr Tag of AIA/RC Beps Res Program for DOE.. *Society Aff:* ASHRAE, IES, AAAS, NSPE.

Griffith, P LeRoy
Business: 59 Prospect Ave, Montclair, NJ 07042
Position: Engg Consultant. *Employer:* Self-employed. *Education:* Columbia University *Born:* 7/24/08. US citizen. Spec in util opers & mgmt up to 1950 with Ebasco Services & Gilbertt Assocs. Financial, engrg, & mgmt posts with Union Securities, Sanderson & Porter, Gilbert Assocs & Walter Kidde Contractors Inc until 1959. VP James King & Son Inc 1960-67. CEO, Pres & Dir 1967-77. Lic Prof Engr NY, NJ, PA. Mbr ASCE, Fellow ASME, Mbr SAME, Chmn Urban Renewal Agency Montclair NJ 1960-65. Mbr Plan Bd Montclair 1960-65 & Defense Council 1960- . Pres Columbia U Club 1968- 70. Mbr Montclair Golf Club, Union League Club. Home: 59 Prospect Ave, Montclair NJ 07042. *Society Aff:* ASME, ASCE, SAME.

Griffith, Wayland C
Business: 3217A Broughton Hall, Raleigh, NC 27695-7910
Position: R.J. Reynolds Professor of Mechanical & Aerospace Engrg. *Employer:* North Carolina State Univ. *Education:* PhD/Engg Sci/Harvard Univ; MS/Mech Eng/ Harvard Univ; AB/Engg/Harvard Univ. *Born:* Jun 1925. Native of Urbana IL. AEC PostDoc Princeton 1950, Instr & Asst Prof Physics 1951-57, shock waves. NSF Fellow London 1956. Lockheed Missiles & Space Co 1957, Dir of Res, Tech Dir Polaris Missile, VP for Res. MIT Program for Sr Execs 1960. Cttes for NASA, NAS, NSF, NBS and State of N.C. NC State U from 1973 as R J Reynolds Prof of Mech Engrg. Chmn APS - Div of Fluid Dynamics. Royal Aero Soc, AIAA, Phi Beta Kappa, Sigma Xi, Pi Tau Sigma. Assoc Editor Journal of Fluid Mechanics, NC Reg Engr. *Society Aff:* APS, AIAA, AAAS, ΣΞ, ΦΒΚ, REAS.

Griffiths, Lloyd J
Business: Mail Code 0272, Dept of EE-Sys, Los Angeles, CA 90066
Position: Professor. *Employer:* Univ of S CA *Education:* BS/EE/Univ of Alberta CD; MS/EE/Stanford; PhD/EE/Stanford *Born:* 09/30/41 Staff sci for Barry Res Corp 1969-70 working on digital ionospheric sounder processor & display. EE Dept, Univ of CO 1970-84. 1984-present, Signal & Image Processing Inst, Dept of EE-Sys, Univ of Southern CA, Los Angeles. Applications of interest incl radar, biomedical, sonar, geophysics & adaptive arrays. Awards: IEEE Browder J Thompson Memorial Prize for published paper in 1971; Eta Kappa Nu Outstanding Young Elec Engrg Educator 1974; Dow Outstanding Young Faculty Award for the Rocky Mountain Sect of ASEE 1976; Fellow IEEE 1985. *Society Aff:* IEEE, SEG, HKN, ΣΞ.

Griffiths, Vernon
Business: MCMS&T West Park St, Butte, MT 59701
Position: Prof Dept of Metallurgy and Mineral Processing Engrg *Employer:* Montana Coll of Mineral Sci & Tech. *Education:* BSc/Metallurgy/Univ of Wales, Swansea; MSc/Metallurgy/Univ of Wales, Swansea; ScD/Metallurgy/MA Inst of Technology. *Born:* May 1929. 1955-59 Vancouver Canada, Sherritt Ltd/UBC powder metallurgy/ teaching. 1959- , Montana College of Mineral Sci & Tech. Met Dept Head 1959-74, 1983-86. Dir of Res 1974-84 & Exec Dir Montana Tech Foundation 1974-79 . MHD Proj Mgr 1974-79. . Summer res activities 1969-71 NASAJSC Houston, 1972 Ames Lab Iowa State, 1973 Douglas Labs Richland WA. Cons activities in failure analysis. French Gov Scholarship 1969 Ecole des Mines Paris. Dir of Res MCMSOT 1974-1983. Exec Dir, MT Tech Fnd 1974-79; MT Tech MHD Proj Mgr 1974-79. Dir, MT Mining & Mineral Resources Res Inst, 1978-1983. Director Graduate School 1978-86. *Society Aff:* TMS-AIME, ASM, ΣΞ, PMI, INSTITUTE OF METALS, MATERIALS RESEARCH SOCIETY.

Griggs, Edwin I
Business: Box 5213 TTU, Cookeville, TN 38501
Position: Prof. *Employer:* TN Tech Univ. *Education:* PhD/ME/Purdue Univ; MS/ ME/MS State Univ; BS/ME/MS State Univ. *Born:* 5/2/38. Native of western KY. Participated as a Cooperative engrg student with Union Carbide Corp in Paducah, KY throughout BS work. Served two yrs as Instr with TN Tech prior to PhD work. Subsequently, joined Tech's faculty upon completion of adv degree. Served as consultant to several industries on energy- related problems. Pursuing interest in energy-related problems. Current emphasis includes studies of pressure-losses and flow patterns in HVAC ducts. *Society Aff:* ASME, ASHRAE, ASEE.

Grigoropoulos, Sotirios G
Business: Department of Civil Engrg, University of Patras, GR-26110, Patras, Greece
Position: Prof of Civil Engrg *Employer:* Univ of Patras *Education:* SCD/Environ & Sanit Engrg/WA Univ; MS/Chem Engrg/ WA Univ; CHE/Chem Engrg/Natl Tech Univ of Athens *Born:* 3/24/33 Born in Athens, Greece. Came to the US in 1955, naturalized US citizen. Married to Marina Malamou in 1972, one son-Gregory. Chemist, Sigma Chemical Company, 1956-57; res engr, Ryckman, Edgerley, Tomlinson & Associates, 1958-60; research engineer, Metcalf & Eddy, Inc., 1960. Instructor in civil engrg, 1958- 60, visiting associate prof of civil engrg, 1961, WA Univ. Assoc prof of civil engrg, 1960-63, prof of civil engrg, 1963-81, director of Environmental Research Center, 1966-81, University of Missouri-Rolla (on leave 1979-81). Professor, chair of environmental engineering, 1979-82, prof of civil engrg, 1982-, Univ of Patras. *Society Aff:* ASCE, AIChE, AAEE, WPCF, AWWA, ASEE, ΣΞ, TEE(GREECE), XE, ΤΒΠ, ΦΚΦ.

Grill, Raymond A
Business: 300 Ellinwood Way, Suite 201, Pleasant Hill, CA 94523
Position: Consulting Engineer *Employer:* Rolf Jensen & Assoc. *Education:* Bachelor of Science/Fire Prot. Eng./Illinois Institute of Technology *Born:* 10/18/57 Currently Pres of the Northern CA/NV Chapter of SFPE, I have held the positions of VP, Treasurer, and Editor of the newsletter. I am active in the National Fire Protection Assoc and currently serve on the following NFPA Technical Cttees: 1) Cttee on Protective Signaling Systems, 2) Cttee on General Storage, 3) Cttee on Rack Storage, & 4) Cttee on Detection Devices. I am a Reg Mech Engr in the State of CA. *Society Aff:* SFPE

Grimes, Arthur S
Business: 2 Broadway, New York, NY 10004
Position: Consulting Mechanical Engr. *Employer:* Amer Electric Power Serv Corp. *Education:* BS/ME/Univ of Cincinnati. *Born:* Aug 1922. Native of W Va. Army Corps of Engrs 1943-46. Asst Ch Mech Engr 1972-77; Consulting Mech Engr since 1978. Received the ISA Philip Sprague Medal 1969. Chmn ASME Internatl Standardization Ctte 1974-76. *Society Aff:* ASME.

Grimes, Dale M
Business: 118 Elec Engr E, Elec Engg Dept, Univ Park, PA 16802
Position: Prof & Hd. *Employer:* PA State Univ. *Education:* PhD/EE/Univ of MI; MS/Phys/IA State Univ; BS/Phys/IA State Univ. *Born:* 9/7/26. Native of Marshall County, IA. Electronics Technician US Navy, 1943-1946. res Asst, IA State Univ, 1949-1951. Univ of MI, 1951-1976 as Res Assoc, Asst Prof, Assoc Prof and Prof. Chief Scientist Conduction Corp, 1960-1963. Consultant: Natl Bureau of Stds, Dow Chem, Gen Motors, and TRW. Prof and Chrmn, Elec Engg, Univ of TX at El Paso, 1976-1979. The PA State Univ, 1979-. Specialties are electromagnetic theory, quantum effects, automotive radar. *Society Aff:* AAAS, IEEE, APS, ASEE

Grimes, William W
Home: 2505 Dogwood Drive, Lima, OH 45805
Position: Refinery Mgr. *Employer:* The Standared Oil Co. (Ohio) *Education:* BCE/ChE/OH State Univ *Born:* 2/7/27. in Gorgas, AL. Raised in Cleveland, OH. Served in US Army 1946-47. Various process engg & engg specialists assignments with Std Oil (OH) from 1951-1970. Proj Mgr of maj expansion & modernization of BP Oil Inc (wholly-owned Subsidiary of Std Oil Co) refinery at Marcus Hook, PA. After 4 yrs as Mgr, Refinery Operations, became Refinery Mgr in 1978. Transferred to the Lima Refinery of Std Oil (OH) as Refinery Mgr in 1981. Holds six patents resulting from earlier engg activities. Received Distinguished Alumnus Award from College of Engg, OH State Univ in 1975. Mbr AIChE Council serving as a Dir (1984-86), Past Chrmn of the Exec Bd of the Natl Program Ctte, AIChE and Mbr of the Govt Programs Steering Ctte, AIChE. *Society Aff:* AIChE.

Grina, Kenneth I
Business: P O Box 16858, Philadelphia, PA 19142
Position: VP, Engg. *Employer:* Boeing Vertol Co. *Education:* BS/Aero Engg/Univ of MN. *Born:* 1923 Fergus Falls, Minn. Joined Vertol 1947. Apptd VP, Engg at Boeing Vertol Co in June 1979 with responsibilities for engrg, prod dev & res for Boeing Vertol aircraft. Joined Vertol as a stress analyst & successively attained positions as Proj Stress Engr, Ch of Stress, Ch of Struc, Ch Engr for Dynamic Components, design of rotors, hubs, blades, drive & transmission sys, and Ch Proj Engr CH-47 Helicopter; Dir of Engrg Proj & Dir of Engrg. *Society Aff:* AHS, AAAA, AUSA, HAA, AIAA

Grinnan, James A
Business: 2121 Corporate Square Blvd, Jacksonville, FL 32216
Position: Pres *Employer:* Grinnan Engrg, Inc *Education:* BS/ME/Univ of FL *Born:* 8/25/28 Graduated Univ of FL in 1950 with Bacheler of Science in Mechanical Engrg, Registered Engr in FL and GA, Fellow in FL Engrg Society, Received ASHRAE "Merit" award for distinguished service to the Society. Active participant in the establishment of ASHRAE, Standard 90-75 and Standard 100. Served on various ASHRAE technical committees and as vice-chrmn of TC 4.1 "Load Calculations–. Served as Mechanical Dept Head for the Firm of Reynolds, Smith and Hills of Jacksonville, FL for four years and as Senior VP and Mechanical Dept Head for Van Wagenen & Searcy, Inc. for eight years. Instituted Grinnan engrg, Inc in Jacksonville in July 1980. Notable projects include Tampa Intl Airport, Orlando Intl Airport, Independent Square office tower in Jacksonville, and Atlantic National Bank Bldg with design emphasis toward central chilled water plants, energy management, and control systems. *Society Aff:* ASHRAE, NSPE, ACEC, NSPE

Grinnell, Robin R
Business: Agri Engr Dept, Ca Poly State Univ, San Luis Obispo, CA 93407
Position: Prof *Employer:* CA Poly State Univ. *Education:* PhD/Agr Engg/Purdue Univ; MSAE/Agr Engg/Univ of MN; BSAE/Agr Engg/Purdue Univ. *Born:* 7/21/32. Native of Palo Alto CA. Grew up in WI, IN & IL. With US Army 1956-58. Taught at Univ of Guelph, Ontario, 1961-66, and at CA Poly State Univ 1967 to present. Reg PE (EE) in Ontario 62-66, Treas IEEE Sec 1966, Chrmn ASAE So CA Sec 1977-78, Mbr ISA Committee 8D-RP5.1, Am Stds Inst Sectional Committee to the Intl Organization for Standardization, subcommittee TC1/SC3. I received a certificate of Recognition for the Dev of RP51 - now S5.1. I enjoy classical and folk music, books, backpacking & outdoor activities, photography and working with Boy Scouts and Church groups. *Society Aff:* ASAE, IEEE, ISA.

Grinter, Linton E
Business: Grinter Hall-U of Fla, Gainesville, FL 32611
Position: Dean Emeritus. *Employer:* Educ Cons - Univ of Florida. *Education:* PhD/Engg/Univ of IL; MS/Civil Engg/Univ of IL; BS/Civil Engg/Univ of KS. *Born:* Aug 28, 1902 Kansas City, Mo. Prof Struc Engrg Texas A&M Univ 1928-37; Prof, Dean Grad School & V P Ill Inst of Tech 1937-52. Dean Grad School & VP Univ of Florida 1952-70. Author of 6 books & numerous papers on struc engrg & on educ. Chmn Report on Eval of Engrg Educ of ASEE 1956, Chmn Report on Engrg Tech Educ Study of ASEE 1972, Lamme Medalist 1958; Hon Mbr ASEE 1958, ASCE 1971. LLD degree Arizona State Univ 1962; DSc Univ of Akron 1969. Awarded Order of the North Star by King of Sweden 1970. Elected to Hall of Fame of Engrg Educ 1968. First recipient of Linton E Grinter Disting Serv Award of ECPD. Pres of ASEE 1953-54, Pres of ECPD 1965-67. *Society Aff:* ASCE, ASME, ASEE, ACI.

Grisewood, Norman C
Home: R.D.1, Box 200, Califon, NJ 07830
Position: Dir of Production. *Employer:* F.X. Matt Brewing Co. *Education:* B/ChE/Yale Univ *Born:* 4/4/29 Attended Yale Univ under N.R.O.T.C. program. Spent three years, 1950-1953, in U.S. Navy in Electronics and Engrg Billets. Joined AM Star Corp in 1953. Held various positions in production and quality assurance. Joined Schaefer Brewing Co in 1962 as Research and Development Engr. Became VP-Technical and Engrg with complete responsibility for quality control, product research and engrg activities in 1977. Also became a dir and member of Operating Committee ofSchaefer in 1977. On completion of merger with Stroh Brewery Co, resigned from all pos. Dir of Production-F.X. Matt Brewing Co. *Society Aff:* MBAA.

Griskey, Richard G
Home: 4633 N Cramer St, Milwaukee, WI 53211
Position: Prof. *Employer:* Univ of WI. *Education:* PhD/Chem Engg/Carnegie-Mellon Univ; MS/Chem Engg/Carnegie-Mellon Univ; BS/Chem Engg/Carnegie-Mellon Univ. *Born:* 1/9/31. Pittsburgh PA; Engr Gulf R&D Co Harmarville PA 1951; US Army Corps of Engrs (1/Lt 18 mos Korea, Japan) 1952-1953; Sr Engr, DuPont 1958-1960; Asst Prof Univ of Cincinnati 1960-62; Assoc and Full Prof VPI 1962-66; Dept Hd and Prof Univ of Denver 1966-68; Dir of Res and Prof Newark Coll of Engg, 1968-71; Dean of Engg and Prof Univ of WI-Milwaukee, 1971-; Visiting Prof at Polish Academy of Sci, 1971, Estadual Sao Paulo (Brazil), 1973; Manash U. (Australia), 1974 Algerian Inst Du Petrole 1975, 1976; Consultant 3M, Celanese,

Griskey, Richard G (Continued)
Phillips Pet, Monsanto; Author 185 Articles, 3 Monographs. Listed Who's Who in World and Who's Who in America. *Society Aff:* ACS, AIChE, ASEE, SPE, ASME

Grizzard, John L
Home: 8810 Teresa Ann Ct, Alexandria, VA 22308
Position: Dir of Engrg. *Employer:* VSE Corp *Education:* BS/IE/VPI & SU. *Born:* Dec 1930. Plant Indus Engr Richmond Dairy Co. Called into AF 1954 as Facilities Engr. Plant IE Lynchburg Foundry Co 1955. Plant IE H K Porter Co 1956. Recalled to AF 1957 & served 18 yrs as Prod Control Engr, Maintenance Engr, Ch IE, Ch Engr, Ch Const Engr, Dir of Engrg AF Hqs Command 1972. Transferred to Southern Cmd as Dir of Engrg 1975. Responsible for design, const, const mgmt, plant & facilities engrg, supervising 36 graduate engrs & 664 technicians & craftsmen. Founded & 1st Pres of Pikes Peak Chap IIE. Prog Chmn & 1st VP Natl Cap Ch IIE 1975. Spec Events Chmn IIE Natl Conference 1975. 1st VP NSPE Northern Va Ch 1975. VP NSPE Canal Zone Ch 1976. 1st VP Natl Cap ch IIE 1978. Pres Natl Cap Ch IIE 1979-80. 1st VP NSPE No VA Ch, Genl Chrmn Natl Conf 1981 IIE. Reg PE Calif & Colorado. Mbr Wash DC Bd of Reg for PE. Enjoy golf & boating. *Society Aff:* IIE, NSPE, AIPE.

Grocki, John M
Business: 605 Third Ave, New York, NY 10158
Position: Regional Mgr Technical Mktg *Employer:* VDM Technologies Corp *Education:* BS/Mtls Sci & Engg/Cornell Univ. *Born:* 8/17/46. Native of Hartford, CT. Process Control Engr for Kaman Aircraft before joining Kelsey Hayes Co, where principal responsibility was met control of purchased raw mtls. In 1971 promoted within Kelsey Hayes to casting met responsible for dev of casting processes, including involvement with all production aspects. In 1972 joined Howmet Corp as Asst Chief Met responsible for lab performing mech testing, metallography & experimental casting. In 1975 joined Uddeholm Corp as Applications Engr responsible for marketing of specialty stainless alloys, particularly UHB 904L. In 1977 promoted to Mgr - Technical Services with increased marketing responsibilities, & all technical support activities nationwide. In 1981 joined VDM Tech Corp as Regional Mgr Technical Mktg responsible for natl mktg of speciality nickel alloys & stainless alloys to the Pollution Control & Pulp & Paper Industries as well as technical support to regional commerical activities. *Society Aff:* ASM, NACE, ASTM, AWS, TAPPI, MPC.

Grodins, Fred S
Business: Biomedical Engrg, Los Angeles, CA 90089-1451
Position: Prof & Chmn Biomedical Engrg. *Employer:* Univ of Southern Calif. *Education:* MD/Medicine/Northwestern Univ Medical School; PhD/Physiology/Northwestern Univ; MS /Physiology/Northwestern Univ; BS/Chemistry/Northwestern Univ. *Born:* Nov 18, 1915 Chgo Ill; m. 1942. Asst in Physiology Northwestern 1938-40, Instructor 1942-44. Asst Prof Ill Coll of Medicine 1946. Assoc Prof Northwestern Med School 1947-50, Prof 1950-67. Prof Elec Engrg & Physiology Univ of So Calif 1967- . Chmn Biomedical Engrg 1970-86. Prof Emeritus, 1986- . Natl Inst of Health Career Res Awardee 1962-67. Biomed Engrg Training Ctte 1969- . Cpt Medical Corp 1944-46. AAAS Amer Physiological Soc, Soc Exper Biology & Medicine, Biomed Engrg Soc (Pres 197071). Biological Control Sys, cardiovascular & respiratory physiology, mathematical models of physiological systems. *Society Aff:* APS, BMES, AAAS.

Grogan, William R
Business: Inst Rd, Worcester, MA 01609
Position: Dean. *Employer:* Worcester Poly Inst. *Education:* MS/EE/Worcester Poly Inst; BS/EE/Worcester Poly Inst. *Born:* 8/2/24. Native of Lee, MA, wife: Mae K., US Navy 1943-46, 1950-52. Faculty mbr Worcester Poly Inst since 1946. Also associated with GE Co, Bell Telephone Labs and US Navy Dept, Missile Systems Office 1952-1967. Prof of Elec Engg, Wpl, 1962-. Dean of Undergrad Studies, 1970-. ASEE Chester F Carlson Award 1979. ASEE William E. Wickenden Award 1980, 1983 Fellow IEEE, 1986 IEEE Major Educational Innovation Award. *Society Aff:* IEEE.

Grohse, Edward W
Home: 75 Clifton Pl, Port Jefferson Station, NY 11776
Position: Consltg Chem Engr *Employer:* E W Grohse Assoc *Education:* PhD/Chem Engrg/Univ of DE; ChE/Chem Engrg/Cooper Union; BChE/Chem Engrg/Cooper Union *Born:* 12/05/15 *Professional Exper* Res Chem Engr, FMC Corp, 1940-45; Res Fellow, Asst Res Prof, Univ of DE, 1945-1949; Asst Prof, Carnegie-Mellon, 1949-1951; Sr Technologist, Monsanto Co, 1951-1952; Res Assoc, GE Res Lab, 1952-1958; Nuclear Engr, Knolls Atomic Power Lab, 1958-1960; Prof, Univ of AL in Tuscaloosa and Huntsville, 1960-1980; Conslt, Marshall Space Flight Center (NASA), 1970-1980; Sr Res Engr, Brookhaven Natl Lab, 1980-1982; Conslt, E W Grohse Assocs, 1982- present. *Honors* Elected Fellow of Amer Inst of Chem Engrs; Prof Emeritus, Univ of AL in Huntsville; Elected Mbr of Soc of Sigma XI (Hon Res Soc); Elected Mbr of Tau Beta Pi (Hon Engrg Soc); Chrmn, Air Pollution Control Bd, Huntsville, AL, 1974-1980; Reg PE, NY and AL. *General* Married, two children. Hobbies: tennis, sailing, reading, chess. *Society Aff:* AIChE(FELLOW), ACS, ΣΞ, ΤΒΠ, NYAS

Groover, Mikell P
Home: 3505 Nicholson Rd, Bethlehem, PA 18017
Position: Prof. *Employer:* Lehigh Univ. *Education:* PhD/Ind Engg/Lehigh Univ; BA/Appl Sci/Lehigh Univ; BS/ME/Lehigh Univ; BS/Ind Engg/Lehigh Univ. *Born:* 9/8/39. Mfg Engr for Eastman Kodak Co 1962-1964; Res Engr for Bethlehem Steel Co, 1966. Asst Prof of Industrial Engg, 1969-73; Assoc Prof, 1973-78; Prof, 1978-present; Dir, Mfg Tech Lab, 1978-present. Consultant to numerous industrial firms, 1969-present. Approx 40 technical articles and papers. Four books on automation, robotics, and CAD/CAM. Fellow of Inst. of Indust Engs. *Society Aff:* IIE, SME, ΑΠΜ.

Grosch, John F, III
Business: P O Box 66317, Baton Rouge, LA 70896
Position: Vice Pres. *Employer:* Woodward-Clyde Cons. *Education:* MS/Civ Engrg/Tulane Univ; BS/Civ Engrg/Tulane Univ. *Born:* Apr 2 1928. BS, MS Tulane Univ New Orleans LA. Native of New Orleans. Have been in cons engrg practice for over 20 yrs. Was VP & Mgr of Louisiana oper for Etco Engrs 1965-72 when firm became an affiliate of Woodward-Clyde Cons. Presently, Principal and VP of Woodward-Clyde & manager of Louisiana operations including geotechnical, geological, hydrological, and waste mgmt investigations and recommendations. Pres Louisiana Sect ASCE 1974; Chmn PEPP Sect Louisiana Engrg Soc 1970; Chmn Baton Rouge Sect CEC of Louisiana 1969, mbr ACEC and NSPE. Reg Prof Engr in Louisiana, Texas, Mississippi, Arkansas & Alabama. Registered Land Surveyor, Louisiana. *Society Aff:* ACEC, ASCE, NSPE, ASTM.

Grose, Gordon G
Home: 10643 St Lawrence Lane, St Ann, MO 63074
Position: Sect Mgr Tech-Aerodynamics. *Employer:* McDonnell Douglas Corp. *Born:* Mar 1925. BS Aero E Indiana Tech Univ. 1945-48 Chance Vought Aircraft. Joined McDonnell Aircraft in 1948 progressing from aerodynamics engr to sect mgr. Responsible for aerodynamics res in flight mechanics & external aerodynamics. Mgr of res contracts on reentry communications, trajectory optimization, transonic & viscous flow. Mbr AIAA. Wright Brothers Award 1974 SAE.

Grosh, Richard J
Business: 701 W 5th Ave, Columbus, OH 43221
Position: Chmn of the Bd & CEO. *Employer:* Ranco Inc. *Born:* 10/15/27. BS, MS & Doctorial degrees Purdue Univ. Matt 1946-Aug 1947 US Army CIC Japan. Asst Prof ME 1958, Dir Indus Dev & Hd of Sch of Mech Engg, Assoc Dean of Schs of Engg Purdue 1965 & Dean in July 1967. July 1971 Pres of Rensselaer Polytechnic Inst. Early 1976 appted Chmn & Pres of Ranco Inc. Author 45 tech papers. Mbr

Grosh, Richard J (Continued)
ASME & ASEE. Prof Engr NY & IN. Jan, 1979 titled Chrmn of the Bd and Chief Executive Officer. Serves on the Bd of Dirs for Sterling Drug Inc & Transway Intl Corp.

Gross, Donald
Business: George Washington Univ, Dept. of Operations Research, Washington, DC 20052
Position: Prof & Dept Chrmn. *Employer:* George Wash Univ. *Education:* PhD/Operations Res/Cornell Univ; MS/Operations Res/Cornell Univ; BS/Mech Engg/Carnegie-Mellon Univ. *Born:* 10/20/34. Field of specialization: operations res. First Lt, US Army Signal Corps, 1962-63. Operations Res Analyst, Atlantic Refining Co, 1961-65. From 1965 to present, the George Wash Univ: Asst Prof of Engg and Appl Sci, 1965-67; Associate Prof, 1967-74; Prof of Operations Res, 1974-; Chrmn, Dept of Operations Res, 1977-. Consultant to Inst for Defense Analyses, Res Analysis Corp, and US Govt. Coauthor of *Fundamentals of Queueing Theory*, John Wiley. Also author of numerous articles in the profl journals. Mbr of ORSA, TIMS, IIE, Sigma Xi, Tau Beta Pi, and Omega Rho. *Society Aff:* ORSA, TIMS, IIE, TBΠ, ΣΞ, ASEE

Gross, Eric T B
Home: 2525 McGovern Dr, Schenectady, NY 12309
Position: Philip Sporn Prof Emer of Engrg. *Employer:* Rensselaer Polytech Inst-Troy NY. *Education:* DSci/Eng/Tech Univ Vienna; EE/Elec Engg/Tech Univ Vienna. *Born:* 5/24/01. US Citizen. Prof CUNY, Cornell, IL Tech. Since 1962 Disting Prof RPI, occupant of Philip Sporn Chair, Chmn Elec Power Engg. Reg Engr IL, NY, VT. Fellow NY Acad of Sci, IEEE, AAAS, IEE London. Elected Natl Acad of Engg 1978, Power Engg Society-Power Education Committee Service Award 1978-Power Generation Committee Service Award 1977, Citation Edison Elec Inst 1976, Northeastern Reg Award IEEE 1976, Eminent Mbr Award Eta Kappa Nu 1975, Silver Plaque IEEE Brazil Council 1974, ASEE Award 1972, Citation Amer Power Conf 1972, Disting Faculty Award RPI 1972, VP Panamer Engg Congress, nel Amer Arbitration Assn. Mbr tech ctte IEEE & other soc's in power & energy. Over 100 tech papers published. Member Eta Kappa Nu, Tau Beta Pi, Sigma Xi. Award Austrian Cross of Honor in Science and the Arts First Class 1980. *Society Aff:* IEE, IEEE, AAAS.

Gross, John H
Business: 125 Jamison Lane, Monroeville, PA 15146
Position: Dir-Technology Implementation *Employer:* US Steel Corp. *Education:* BS/Met Engr/Lehigh Univ; MS/Met Engr/Lehigh Univ; PhD/Met Engr/Lehigh Univ. *Born:* Jan 27, 1923. Native of Bethlehem Pa. Officer US Navy WW II Pacific Theater. Taught Met E Lehigh Univ 1946-59. Joined US Steel 1959 as Asst Div Ch. Promoted to Div Ch, Asst Dir, Mgr & 1972 to Dir Res. 60 pubs, primarily mech metallurgy & weldability. Cons to to indus & gov. Sev pats. Numerous tech cttes including Chrmn, Welding Res Council. ASM Fellow. AWS Adams Lecturer 1967. Comm Lay Minister. Hobbies: raise & show collies, golf. *Society Aff:* WRC, AWS, ASM, ASME, IIW, ΣΞ.

Gross, Jonathan L
Business: Dept of Computer Sci, New York, NY 10027
Position: Prof and VChrm *Employer:* Columbia Univ. *Education:* PhD/Math/Dartmouth; MA/Math/Dartmouth; BS/Math/MIT *Born:* 6/11/41 in Philadelphia, PA. On Columbia faculty since 1969. Served as Acting Chrmn of Mathematical Statistics in 1979. Previously in Mathematics at Princeton Univ. Awarded postdoctoral fellowships by IBM and by Alfred P. Sloan Fdn. Active professional interests included combinatorial and probabilistic models, topological graph theory, design of higher-level computer languages, legal apsects of computer science. Consultant to Inst for Defense Analysis, Russell Sage Foundation, IBM, Oak Ridge Natnl Labs, Bell Laboratories, others. Author of many res papers and 10 books on computer science, mathematics, and anthropology. Publications bd for Columbia Univ. Press, Journal of Graph Theory, Computers and Electronics. *Society Aff:* AMS, SIAM, ACM

Gross, Joseph F
Business: Dept of Chem Engg, Tucson, AZ 85721
Position: Prof Chem Engg. *Employer:* Univ of AZ. *Education:* PhD/ChE/Purdue Univ; BChE/ChE/Pratt Inst. *Born:* 8/22/32. Fulbright Scholar, Germany 1956-57. Studied with Ernst Schmidt and Hermann Schlichting. Joined Rand Corp, 1957. Areas: multi-component boundary layer stability, hypersonic flow theory and magnetohydrodynamic flows. Post 1968 involved in bioengineering studies: flow mech and mass transfer phenomena in the microcirculation, pharmacokinetics of cytotoxic drugs, quantification of flow and mass transfer in developing human tumor microcirculatory systems. Joined Univ of AZ (1972), Prof, Chem Engg. PPres, Microcirculatory Soc. Secretary-Treasurer, International Insitute For Microcirculation. Humboldt Award 1979. *Society Aff:* AIChE, APS, AACR, MS, ASME.

Gross, Robert A
Business: School of Engrg & Applied Science, Columbia Univ, New York, NY 10027
Position: Dean of the Faculty of Engr. & Appl. Sci. *Employer:* Columbia Univ. *Education:* PhD/Appl Physics/Harvard Univ; MS/Appl Physics/Harvard Univ; BS/Mech Engr/Univ of PA. *Born:* 10/31/27. Dean of the Faculty of Engineering & Applied Science, Columbia Univ. Res area: plasma physics, controlled fusion energy, and high temperature gas dynamics. Founder of Columbia Plasma Lab. Consultant to industry and govt agencies. mbr AIAA, APS, AAAS, IEEE. Former vp AIAA. Honors: Guggenheim fellow, Fulbright-Hays Fellow, AIAA Pendray award, Waverly gold medal for res, NSF senior postdoctoral fellow, USSR exchange scientist; visiting fellow of Australian Academy of Sci. Columbia Great Teacher award; Assoc editor, Physics Fluids journal; editor in chief, AIAA Selected Reprint Series; Visiting faculty mbr-Leiden Univ, Fermi School of Physics, Sydney Univ, Flinders Univ. US delegate to IAEA meetings, Hudson Professor of Applied Physics, Columbia University. *Society Aff:* AIAA, APS, AAAS, IEEE.

Gross, Stanley H
Business: Long Island Ctr, Route 110, Farmingdale, NY 11735
Position: Prof *Employer:* Polytechnic Inst of New York *Education:* PhD/EE/Polytechnic Inst of Brooklyn; MS/EE/Polytechnic Inst of Brooklyn; BS/EE/Coll of the City of NY *Born:* 4/4/23 Faculty member, Polytechnic Inst of NY, 1967-present, prof since 1974. National Academy of Science Senior Assoc, 1980-81, at NASA Goddard Space Flight Center-Guest Investigator and Co-Investigator on NASA Satellite Programs. Consulting Activities for Industry and NASA. Consultant and Section Head of Space Sciences, Airborne Instruments Laboratory 1959-67. Project Engineer to Division Chief, Fairchild Astrionics Division, 1951-59. Project Engineer to Engineer, W. L. Maxson Co., Control Instruments Co., Hazeltime, 1945-51; Visting lecturer for American Institute of Physics, 1967-72; over 50 publications on atmosphere and ionospheric science in journals. *Society Aff:* AAAS, IEEE, AGU, APS, AAS, URSI

Gross, William A
Business: Dept. of Mechanical Engineering, Univ. of NM, Albuquerque, NM 87131
Position: Prof *Employer:* Univ. of NM *Education:* PhD/Applied Mechanics/U. Calif., Berkeley; MS/Applied Mechanics/ U. Calif., Berkeley; BS/Marine Engr/US Coast Guard Academy *Born:* 11/17/24 Three years Coast Guard engrg officer. Professor, University faculties: Calif, Berkeley; New Mexico; Iowa State; Depts Mech Engrg, Applied Mechanics, Dean of Engrg at UNM. Member of Technical Staff, Bell Labs (worked on under sea cables, maximally flat delay line filters) and IBM (with five others, accomplished breakthrough that made computer disk memories possible, developed design procedures still being used). VP, Dir of Research, Ampex (facilitated fundamental developments for low cost video recording; royalties accrue for all VCR's produced). Started Technological Innovation Prog, teach Entrepre-

Gross, William A (Continued)
neurial Engrg, UNM. Active in renewable energy technology; develop programs, consult in developing countries. *Society Aff:* STLE, ASME, IEEE, NSPE, AAAS, ASEE, ASHRAE

Grossberg, Arnold L
Business: 576 Standard Ave, Richmond, CA 94802
Position: Vice President. *Employer:* Chevron Research Co. *Education:* MS/Chem Eng/Univ of MI; BS/Chem Eng/CA Inst of Technology. *Born:* 1921. With Chevron Res Co 1943- . Ch Design Engr 1965-72. VP Process Engrg Dept 1972- . Fellow AIChE. Dir Fuels & Petrochem Div AIChE 1972-74. Task force mbr Federal Power Comm Natl Gas Survey. Dir & Pres Berkeley Bd of Education 1965-71. *Society Aff:* AIChE.

Grosser, Christian E
Business: 214 Sleepy Hollow Rd, Richmond, VA 23229
Position: Mech Engrg Design Cons. *Employer:* Self - Private Practice. *Education:* MS/ME/MIT; BS/ME/MIT. *Born:* April 1909. Early prof training in mech power transmission machinery with Watson-Flagg Mach Co (now Div of GE) & Vickers (Div of Speery-Rand). MIT staff WW II in defense res teaching. VP Engrg for Standard Machinery Co Providence RI. Prof in Applied Mech & Machine Design Syracuse Univ. Sect Head Computer & Controls Dept Hughes Aircraft Co-Missile Dev. Corp Ch Engr Philip Morris Inc. Cons Engr serving indus & gov since 1961. Fellow ASME. Virginia Lic Bd for Engrs. Avocations: sailing, tennis, handicrafts. *Society Aff:* ASME, SAE, NSPE, ASME.

Grossman, Alexander J
Home: 2297 Jones Dr, Dunedin, FL 33528
Position: Retired. *Employer:* *. *Education:* EE/-/Rensselaer Poly Inst. *Born:* Sept 1, 1904 NY. Received EE degree from Rensselaer Poly Inst 1925. Mbr Tech Staff Bell Telephone Labs Inc 1925-64. At time of retirement, Hd of the Transmission Circuits Dept. Awarded Fellow IRE 1962 'For contributions to circuit design'. Life Mbr IEEE. *Society Aff:* IEEE.

Grossmann, Elihu D
Home: 207 Avon Rd, Narberth, PA 19104
Position: Prof. Chem Engr *Employer:* Drexel Univ *Education:* PhD/CHE/Univ of PA; MS/CHE/Drexel Univ; BS/CHE/Drexel Univ *Born:* 11/29/27 Native of Delaware Valley. Served in US Army 1945-47. Teaching since 1952. Research areas in: drying, thermodynamics, process analysis, biomass conversion; waste agricultural products recycling and energy conversion, safety. Industrial and consulting experience in: process analysis, interior ballistics, propellant (process design); and manufacture; safety, fire and explosion investigation, hazardous chemical detoxification and disposal. Have served on professional society committees both national and local, ocassionally, as Chrmn. Mbr of Speakers Bureau of Natl AIChE. An occassional expert witness. *Society Aff:* AIChE, ACS, ASEE, NFPA

Groth, William R
Home: 277 Beechwood Crescent, Webster, NY 14580
Position: Sr Dev Engr. *Employer:* Eastman Kodak Co. *Education:* AAS/Elec/Rochester Inst Tech. *Born:* Oct 1926 Rochester NY. Served in US Navy 1944-46. Grad Rochester Inst of Tech in elec engrg, AAS degree 1951. Employed in Mfg Tech Div, (R&D) Eastman Kodak. Primary activity - instrumentation, measurement & control as it pertains to the operation & maintenance of pilot plants. Active in the Instrument Soc of Amer, P Pres of the Rochester Sect ISA, served as Dist IV VP Nov 1976-Oct 1978. W A Kate's Award 1973, conferred annually for the 5 most definitive reports of ISA sect activities during the year. Enjoy photography, hiking, camping. *Society Aff:* ISA.

Grout, Harold P
Home: 11025 Linda Vista Drive, Lakewood, CO 80215
Position: Consulting Hydrologist *Education:* EM/Mining Eng/CO School of Mines *Born:* 4/2/10 Native of Denver, CO. 1938-1941 employed as Hydrologist, CO Water Conservation Bd. Served in Army Corps of Engineers 1941-46. Assistant to Corps Engineer, XVIII Airborne Corps. 1946-1970 Army Reserve, Commanding Officer of 927th Engineer Aviation Group, retiring with rank of Colonel. Bureau of Reclamation 1946-1972 as Chief, Flood Hydrology Branch. Consulting Hydrologic Engineer 1972-1987, with assignments with World Bank in India and Malaysia. Also Consulting Hydrologist with Comision Federal de Electricidad in Mexico and Bechtel Corporation. 1983-1984-Consulting Hydrologic Engr For Water Resources Assoc in Phoenix, AZ on Safety of Dams. *Society Aff:* USCOLD

Grove, H Mark
Home: 2135 Springmill Rd, Kettering, OH 45440
Position: Deputy Director Computer & Electronics Policy *Employer:* Dept of Defense. *Education:* MSc/EE/OH State Univ; EE/EE/Univ of Cincinnati. *Born:* 8/2/33. Native of Highland County, OH. Currently responsible for computers, software, & electronics acquisition policies for Dept of Defense. 1974-78 Asst Chief Systems Avionics Div, USAF Avionics Lab; 1970-74 Chief Microwave Res, Walter Reed Army Inst of Res, managed Natl prog in bioeffects res; 1960-70 res engr for AF in laser, ECM, unconventional Weapons, advanced systems planning. Chrmn Dayton Sec, IEEE 1976; VChrmn IEEE Region 2 1979-80; Mbr Eta Kappa Nu, Tau Beta Pi, Sigma Xi Amer Assoc Adv Science; Reg PE, OH. *Society Aff:* IEEE.

Grover, John H
Business: 123 N Pitt St, Alexandria, VA 22314
Position: Vice Pres & Treasurer. *Employer:* Research Indus Inc. *Education:* MS/Chem Engg/MA Inst of Tech; BS/Chem Engg/MA Inst of Tech. *Born:* Sept 1927. Group Head for Atlantic Research Corp working on combustion, aerothermodynamics & solid-rocket propulsion. Have been VP, Treas & Dir of Res Indus Inc since its formation 1968. Also on the Bd of Dir of sev other companies. *Society Aff:* AIChE, ACS, AAAS, AIAA.

Groves, Richard H
Home: Qtrs 11-A, Ft Myer, VA 22211
Position: Dep Advisor to the Secy of Defense on NATO Affairs *Employer:* US Army. *Education:* MS/Civil Engg/Harvard; MA/Intl Affairs/George Washington Univ; BS/Military Engg/USMA; -/Math/Princeton. *Born:* July 1923. Comm in Army Corps of Engrs 1945; served in various positions with the Corps & Army since that time, incl co cmdr, battalion cmdr, group cmdr, div engr, Engr US Army Europe, Asst Secy Genl Staff HQ DA, Military Asst to the Secy of the Army, Ch of Staff US Army Europe & Dep Advisor to the Secy of Defense on NATO Affairs. Fellow ASCE, Fellow SAME; Mbr NSPE; VP SAME 1971-76; SAME Toulmin Medal 1973. Reg DC, NY. *Society Aff:* ASCE, SAME, NSPE.

Groves, Stanford E
Business: 3418 CEBA Bldg, Baton Rouge, LA 70803
Position: Associate *Employer:* LA State Univ, Hazardous Waste Res Ctr *Education:* BS/ChE/Purdue Univ *Born:* 7/1/28. in Ft Wayne, IN. Served in Marine Corp 1946-48. Employed by Freeport Sulphur Co 1952-60 & by Allied Chem Co 1960-63. Employed by Copolymer Rubber & Chem Co 1963 to 1982. Initially worked on dev of a polymerization system to make ethylene propylene elastomers. Patentee in that field. Later held positin of dir of engg. Assumed position as VP, Mfg in 1976. Reg PE in LA & TX. Retired from Copolymer 1982. Self employed consultant 1982-1984. Assumed current position of Associate at La. State Univ. Hazardous Waste Research Center 1984. *Society Aff:* AIChE, ACS, LES, NYAS.

Grow, Richard W
Business: Univ of Utah MEB 3054B, Salt Lake City, UT 84112
Position: Prof, Chmn, Dean. *Employer:* University of Utah. *Born:* 1925 Utah. BS & MS Univ of Utah. PhD Stanford 1955. Employed: Naval Res Lab 1949-51; Stanford Electronics Lab 1951-58; Univ of Utah 1958- . Dir Microwave Device & Phys Electronics Lab 1960- . Res on microwave & millimeter- wave tubes, quantum elec

Grow, Richard W (Continued)
devices, solid-state devices & rocket exhaust plasmas. Chmn EE Dept 1965- . Assoc Dean Coll of Engrg 1976- . Cons for G E, Microwave Electronics, Litton Indus, Eitel McCullough, Northrup & Varian Assocs. Fellow IEEE 1973. IEEE Educ Activ Bd 1971-75, IEEE EDS Admin Ctte 1972- , ECPD Ad Hoc Visit List 1972-76.

Gruber, Jerome M
Home: 711 Cheyenne Dr, Franklin Lakes, N.J 07417
Position: Consultant *Employer:* Waukesha Bearings Corp. *Education:* B.S./Mech. Engineering/Univ. of Wisc - 1941 *Born:* 07/21/19 Four published Society papers on Turbine Bearings & oil Lubricated Stern Tube Bearings and Seals. Co-authored Chapter XX - "Bearings and Lubrication" of "Marine Engineering" published in 1971 by Soc of Naval Arch & Marine Engrs. Nine patents on Bearings a&d Lubrication issued in 4 countries - USA, Germany, U.K. & Japan. Life Membership Award in SNAME; Life Fellow Award in ASME; Chairman, ASME Lubrication Div - 1971; Distinguished Service Award, ASME, 1972; U.S. Army Corps of Engrs Bronze Star, Five Campaign Awards 1944-45; Reg PE in MA, WI & NJ. Born - Chilton, WI, Married 1943. Six children. *Society Aff:* ASME, SNAME, ASNE, STLE (Formerly ASLE)

Gruenwald, William R
Business: 344 New Albany Rd, Moorestown, NJ 08057
Position: Director Marketing *Employer:* Northern Telecom. *Education:* BSEE/Electronics/Univ of MO. *Born:* 2/13/36. Mr Gruenwald grad from Univ of MO-Rolla with a BSEE in 1958. His early career was with GE's Missile & Space Div in various engg positions on satellite vehicle progs. He later transferred to GE's Comp Group as Prog Mgr-Large Comp Systems. He joined Simmonds Precision in 1973 as VP Engg where was instrumental in developing a microprocessor based airborne fuel mgt system now used on many commercial aircraft. He joined Norther Telecom as VP Marketing in 1978 & is now leading their marketing/sales activities for telecommunications test worldwide.

Grum, Franc
Home: 42 Shorecliff Dr, Rochester, NY 14612
Position: Res Lab Head-Photometrology Lab *Employer:* Eastman Kodak Co. *Education:* MS/Optics/Univ of Rochester; BS/Physics/Univ of Rochester; BA/Lib Arts/Univ of Ljubljana *Born:* 5/21/22 Born and educated in Slovenia-Yugoslavia. In 1950 he joined the Physics Division of Eastman Kodak Company's Research Laboratories. His responsibilities include research and development in spectrophotometry, colorimetry, optical radiation and image structure. In 1983 became Chaired Prof in Color Sci and Tech at Rochester Inst of Tech and Dir of Munsell Color Sci Lab. F Grum is past Pres of the Inter-Society of Color Council, Fellow of the Optical Soc of Amer. He is chrmn of CIE Tech Committee on Materials, he was Pres of the US Natl Committee of the CIE (1979-1983). Is listed in Who's Who in British biographical book. He is author of some 50 scientific papers and of 7 books. He is presently editing a Treatise entitled *Optical Radiation Measurements* this is a 5 volume Treatise two of which have been published by the Academic Press. *Society Aff:* OSA, SPSE, ISCC, USNC

Grumbach, Robert S
Business: EE Dept, 302 E Buchtel Ave, Akron, OH 44325
Position: Assoc Prof. *Employer:* Univ of Akron. *Education:* MS/EE/WV Univ; BS/EE/Case-Western Reserve. *Born:* 12/3/25. Native of Morgantown, WV. Served US Navy 1944-1946. Monongahela Power Co 1947. Instr, WV Univ 1947-1953. Sr Engr, Westinghouse Elec Corp, 1953-1957. Mbr, KY Contract Team, Bandung, Indonesia, 1957-1961. At Univ of Akron since 1961. Chrmn Akron Chap PES-IEEE 1977-1978. Chrmn Akron Sec IEEE 1981-1982; Mbrship Committee - PES; Mbr Eta Kappa Nu, Tau Beta Pi. Continued involvement in Boy Scouts since age 12. Active mbr of United Methodist Church. Social Fraternity, Theta Chi. Outstanding IEEE Student Branch Counselor 1983; IEEE Centennial Medal 1984. *Society Aff:* IEEE

Grun, Charles
Business: 111 Cedar Ln, Englewood, NJ 07631
Position: VP. *Employer:* Electromedia Corp. *Education:* BCHE/Chem Engrg/Poly Tech Inst of Bklyn. *Born:* 7/23/32. From 1958 to 1974 have worked in & supervised design - dev & mfg of high energy density batteries. Since 1974 as VP of Electromedia have designed metal- air batteries & developed high current density air electrodes. Avocation - Racquetball. *Society Aff:* AIChE, ACS.

Grunfeld, Michael
Business: 2525 Hyperion Ave, Los Angeles, CA 90027
Position: Principal & Owner/Cons Engrg Firm. *Employer:* Self-employed. *Education:* BS/Math/Louis-Le-Grand; MSc/Mech Engg/Federal Polytechnic; MBA/Economics/UCLA. *Born:* 1919 Istanbul Turkey. Resident of USA since 1958. Naturalized citizen. Lived in Los Angeles since 1959. Bach degree Math College Louis-le-Grand Paris France. Dipl Ing Federal Tech Univ, ETH, Zurich Switzerland in Mech Engrg 1941. Prior to coming to USA: Military serv in Turkish Air Force as 1st Lieutenant. Ch Mech Engr for Ministry of Public Works Div I. In US: Masters Bus Admin at Univ of Calif Los Angeles. Reg State of Calif in branches of Struc, Mech, Civil Engrg. Accredited by US Dept of Labor & State of Calif for inspection & certification of Cargo Gear, Cranes & material handling equip. Taught grad classes in diverse struc engrg subjects at Calif State Univ Los Angeles. Currently self-employed as cons engr spec in indus strucs, cranes, antennas, towers & their fnds & Expert witness & accident reconstruction. *Society Aff:* SEAOC, CEAC.

Gryna, Frank M
Business: 88 Danbury Rd, Wilton, CT 06897
Position: VP *Employer:* Juran Inst *Education:* PhD/Ind'l Engrg/U of IA; MS/Ind'l Engrg/NYU; BS/Ind'l Engrg/NYU *Born:* 08/29/28 Frank M Gryna, Jr, VP of Juran Inst Inc, has participated as Dr Juran's assoc since 1971. Prior to 1982 Dr Gryna was based at Bradley Univ where he taught indust engrg and also served as Acting Dean of the Coll of Engrg and Tech. He is now Distinguished Prof of Indust Engrg Emeritus. He co-authored *Quality Planning and Analysis* with J M Juran and as Assoc Editor of the Second and Third Eds of *Quality Control Handbook.* His res report on *Quality Circles* received the Book of the Year Award sponsored by joint publishers and the Inst of Indust Engrs. He is a fellow of the Amer Soc for Quality Control, a Certified Quality Engr, a Certified Reliability Engr, and a PE (Quality Engr). He has been the recipient of various awards including the E L Grant Award of ASQC. *Society Aff:* ASQC, IIE, АПМ, ТВП

Gschneidner, Karl A, Jr
Business: Ames Lab, Ames, IA 50011
Position: Distinguished Prof. *Employer:* IA State Univ. *Education:* PhD/Physical Chemistry/IA State Univ; BS/Chemistry/Univ of Detroit. *Born:* 11/16/30. Dir Rare-Earth Info Ctr, mbr Dept of Mtls Sci & Engg, group leader Ames Lab-DOE. Published 169 journal articles, 84 chapters in books, 27 reports, & 11 books on rare earth mtls. Prior to 1963 associated with Los Alamos Scientific Lab & Univ of IL. Current res: alloy theory; & physical met of rare earths. Recipient of the 1978 Wm Hume-Rothery Award of the Met Soc of AIME for his contributions to the sci of alloys; named Distinguished Prof in Sci & Humanities, IA State Univ in 1979. *Society Aff:* ASM INTL, TMS-AIME, ACS, AAAS, ACA, ΣΞ, IAS.

Gschwind, Gerry J
Business: PO Box 10, MC:1EC2, Longueuil, Quebec Canada J4K 4X9
Position: Project Manager *Employer:* Pratt & Whitney Aircraft of Canada Ltd. *Education:* BSc/Chemistry/Swiss Inst of Tech. *Born:* 6/15/25. in Switzerland. Arrived in Canada in 1951. Worked for Sorel Industries in Met-Chem Quality Control. Started at Pratt & Whitney Aircraft as technical investigator in 1953. Working his way up to Project Manager. In over twenty yrs as an active ASM mbr has held numerous positions including: Chrmn Montreal Chapter, Chrmn Canadian Council of ASM. Served in many committees such as: AWS. Brazing & Soldering committee. Also serves as Chrmn of the Consulting committee on Met-Manpower & Labour Dept. *Society Aff:* ASM, CCASM, AWS.

Guard, Ray W
Home: 1018 Blanchard, El Paso, TX 79902
Position: Prof, Met Engr. *Employer:* Univ of TX. *Education:* PhD/Met Engr/Purdue; MS/Met Engr/Carnegie Inst of Tech; BS/Met Engr/Purdue. *Born:* 11/28/27. After receiving the PhD, I worked for 12 yrs fro GE Co in Res & in Diamond Process Engr. After 5 yrs in mgt with Gen Precision and N Am Aviation, I joined MI Tech as Dept Chrmn, Met Engr. In 1970, I joined UT-El Paso as Dean of the Coll of Engg and in 1976, I became Prof Met Engr. I have worked in & published on high temperature mtls, high pressure processes and mtl characterizations as well as work in Extractive Met since 1970. Current res is on gold, silver, copper & their extraction. *Society Aff:* ТВП, ΦΚΦ, AIME.

Guarrera, John J
Home: 17160 Gresham St, Northridge, CA 91325
Position: Dir of Research; Dir of Center for Res & Services *Employer:* Sch of Engr & Comp Sci-CA State Univ, Northridge, CA *Education:* BS/EE/MIT; PE/EE/Univ of the State of NY Ed Dpt; PE/EE/CA State Bd of Engr Registration. *Born:* 3/4/22 Dir of Res, Dir of the Center for Res and Services, Prof of Electrical Engg, School of Engg and Comp Sci, California State Univ Northridge, CA 91330. Bd of Dir of: Wincon, Electronic Conventions Inc., Electronic Conventions Mgmt, San Fernando Valley Engrs Council, San Fernando Valley Child Guidance Clinic, California State Univ Northridge Foundation, Northridge Hospital Foundation, Healthwest Foundation. Fellow of IEEE and past Pres. Life mbr and past Pres of NCGA. Fellow of IAE, Mbr of ASEE, NSPE, CSPE, Tau Beta Pi. Commissioner, Citizens Commission for Pension Policy. Past Pres and former Commissioner of Los Angeles Dept of Water and Power. Mbr of IEEE Task Forces on Pensions and Age Discrimination. PE, registered in NY and CA. Awards: Dist Intl Interprofessionalism Award, IAE; Engg Professionalism, IEEE; Engr of the Year, San Fernando Valley Engineers Council; Service Awards from NCGA, IEEE, Wincon, Wescon, California Dept of Consumer Affairs, City of Los Angeles and County of Los Angeles. *Society Aff:* IEEE, ASEE, NSPE, NCGA, IAE, ТВП

Guba, Howard J
Home: 8118 Northumberland Rd, Springfield, VA 22153
Position: Resident Construction Engineer *Employer:* Ralph M Parsons Co *Education:* MS/Engg Mgmt/Univ of AK; MS/Civil Engg/TX A&M Univ; BS/Univ of MD; BS/Civil Engg/Polytechnic Inst of Brooklyn. *Born:* 12/15/34. US Army Corps of Engrs Officer 1956-1980 . Most recently served as const battalion Cdr; staff officer in R/D; Asst Dir of Construction Operations, Office of the Secretary of Defense. Respon for Construction execution of the Military Construction Program. Present position is with the Northeast Corridor Improvement Program. Dir SAME 1976; Fellow ASCE, SAME; PE District of Columbia *Society Aff:* ASCE, SAME.

Gubala, Robert W
Business: 24 Wolcott Hill Rd, Wethersfield, CT 06109
Position: Chief Eng *Employer:* CT Dept of Trans *Education:* PhD/Transp/Univ of CT; MS/CE/Univ of CT; BS/CE/Univ of CT; Certificate in Traffic Engineering/Yale Univ *Born:* 3/17/38 *Society Aff:* ASCE

Gubbins, Keith E
Business: Sch of Chem Engg, Ithaca, NY 14853
Position: T R Briggs Prof & Dir, Sch of Chem Engrg *Employer:* Cornell Univ. *Education:* BSc/Chem/London Univ; PhD/Chem Engrg/London Univ *Born:* 1/27/37. After receiving the PhD in 1962, I moved to the Univ of FL, Chem Engg Dept, as Postdoctoral Fellow (1962-1964), Asst Prof (1964-1968), Assoc Prof (1968- 1972), Prof (1972-1976). In Aug 1976 moved to Cornell Univ (Ithaca, NY) as T R Briggs Prof of Engg (1976-). Sabbatical leaves at Imperial College, London (1971/1972), Oxford Univ (1979/1980, 1986/1980), Univ of CA, Berkeley 1982. Author of approx 170 papers in scientific & engg journals and 2 books, *Applied Statistical* Mechanics, with T M Reed, McGraw-Hill, 1973, and *Theory of Molecular Fluids* with C G Gray, Oxford Univ Press, 1984. Guggenheim Fellow 1986-87. Alpha Chi Sigma Award, AIChE, 1986. *Society Aff:* AIChe, ACS, CS

Guenther, Richard
Home: 42 Hunt Rd, Sudbury, MA 01776
Position: President. *Employer:* Communication & Information Systems. *Education:* PhD/EE/Inst of Tech; MS/EE/Danzig *Born:* Sept 1910. Siemens e Halske AG Berlin Ger 1937-46 working on dev of voice multiplex & radio relay equip. With US Army Signal Corps Engrg Lab Fort Monmouth NJ 1947-52 as cons in communications & test instrumentation. With Bell Tele labs NY working on advanced systems studies 1952-56 and with RCA diteting Communication Sys Labs 1956-70 and as Comput-er Sys Architect in Computer Div Until 1972. 1972 founded cons firm CIS Inc, Pres. Work is in area of telecommunications & integration of computers & communication. Siemens Ring Stiftung 1935, Fellow IEEE 1965. Enjoy music, hiking, sailing. *Society Aff:* IEEE

Guentzler, William D
Home: 454 Goulburn Ct, El Cajon, CA 92020
Position: Prof, Dir Power & Transp *Employer:* San Diego State University. *Born:* May 1942. PhD Ohio St Univ; BS, MA Kent St Univ. Native of Lakewood Ohio. Taught at Valley Forge & Lakewood High Schools in fields of Transportation & Traffic Safety. Assumed current position as Dir of Power & Transportation for Indus Studies Dept San Diego St Univ 1968. Responsible for establishing a Rationale & Struc for Fluid Power in Indus Arts Educ'. Past President PDK; Univ Rep SDEIA; Univ Clubs' Ctte AIAA; Res on electronic ignitions for San Diego Cty. Enjoy water sports & antique cars. Current res includes 121.3 mpg diesel converted MG midget to be seen on "That's Incredible–, Author: *Fuel Economy Buyer's Guide* with 261 fuel economy products and services. Fuel efficient vehicles, diesel conversions, and fuel efficiency product development a specialty. Consultant services available for fleet or individuals.

Guernsey, Curtis H, Jr
Business: 3555 NW 58th, Oklahoma City, OK 73112
Position: Chrmn of Bd, Pres and CEO *Employer:* C H Guernsey & Co Consulting Engrs & Architects. *Education:* MS/Mech Engr/OK State Univ; BS/Mech Engr/Univ of OK *Born:* 11/25/24. Chrmn, Pres and CEO of C H Guernsey & Co, Consulting Engrs and Architects (established 1928). From 1946-75 served as proj mgr on thirty-five elec power generation and dist heating/cooling projs involving steam turbines, simple cycle and combined cycle combustion turbines, and dual fuel/Diesel engines as well as numerous elec power source/plant addition studies. Active mbr and deacon of Nichols Hills Baptist Church, OK City, and Gideons Intl. Past Pres of Oklahoma City Kiwanis Club. Retired Naval Reserve Officer, Civil Engg Corps. Past Pres and Past Natl Dir, Consulting Engrs Council of OK. *Society Aff:* ASME, NSPE, OSPE, SAME

Guernsey, John B
Business: P.O. Box 1975, Baltimore, MD 21203
Position: Tech Dir *Employer:* Eastern Stainless Steel Co *Education:* BS/Met Engr/Univ of Cincinnati. *Born:* 10/2/29. Native of Middletown, OH. Successively res tech for Lunkenheimer Co, res met for Rem-Cru Titanium, & Chief Met - Precision Parts Div for Ex-Cell-O Corp. Eleven yrs specializing in process res for Crucible Steel - last title Assoc Dir-Res. Became Dir of Met for Jessop Steel in 1971, & VP 1972-1984. Became Tech Dir for Eastern 1984. Responsible for all quality-oriented activities - met, chem, & inspection. Active in Am Iron & Steel Inst (Past Chrmnone Technical Committee), ASM (Past Chrmn Pittsburgh Chapter, mbr Natl committees, Fellow), and ASTM (mbr six subcommittees, Past Chrmn of one). Named Distinguished Alumnus by U of Cincinnati. Enjoy reading, sports (especially auto racing) & classical music. *Society Aff:* ASM, ASTM.

Guernsey, Nellie E
Business: 300 Broad - Box 10745, Stamford, CT 06904-1745
Position: VP. *Employer:* Mineral Systems Inc. *Education:* MBA/Exec Mgt/Pace Univ; BS/Earth Sci/Cornell Univ. *Born:* 3/25/35. in Mt Vernon, NY. Received secondary education in Mt Vernon, VP of Mineral Systems Inc in charge of all financial planning and responsible for dev and implementation of all financial systems applications in corporate and regional offices. Previous experience includes Sr Planning Analyst Exxon Corp, NYC; systems consultant and proj mgr, Mgt Controls & Info Tech Group, Kennecott Copper Corp, NYC; financial analyst, systems analyst, and proj mgr - Financial Dept, Kennecott Copper Corp, NYC; systems specialist, Eastern Data Ctr, Kennecott Copper Corp. Mbr SME of AIME; Chrmn Mineral Resources Mgt Committee of SME of AIME; mbr international organization for application of computers in the mineral industry (APCOM); NY mining club; adjunct prof Pace Univ. *Society Aff:* AIME, SME-AIME, APCOM.

Guild, Philip W
Business: 596 Belvedere Ct, Punta Gorda, FL 33950
Position: Retired *Employer:* US Geological Survey. *Education:* PhD/Geology/Johns Hopkins Univ; AB/Geology/Johns Hopkins Univ *Born:* Oct 1915. US Geol Survey 1939- : Geology & resources of chromite Alaska, Cuba, Guatemala; of iron ore Brazil, Missouri. ICAF resident course 1954-55. Minerals mobilization, stockpiling 1955-60; Ch, Br Base & Ferrous Metals 1960- 65. Currently hd of world metallogenic map effort to relate ore deposits to their geologic/ tectonic settings. Mbr NRC Bd on Mineral Resources 1974-78. Chmn M&E Div SME/AIME 1971; Dir & VP SME-AIME 1973-76. Hobbies: sailing & photography. *Society Aff:* GSA, SEG, SME/AIME

Guillou, John C
Home: 124 Maple Grove Lane, Springfield, IL 62707
Position: Pres *Employer:* Guillou & Assoc, Inc Cons Engrs *Education:* BE/Civil Engrg/Univ of Southern CA; MS/Hydraulic/Univ of IL. *Born:* June 1921. MS Univ of Illinois, BS Univ of Southern Calif. Taught at Univ of Southern Calif before & after WW II serv in Combat Engrs. Joined faculty of Univ of Ill 1947 doing work in teaching & res, & dev Hydraulic Engrg Lab. Ch Waterway Engr of State of Ill 1963-73 with responsibility for water resource work throughout the State. Received Top 10 Award from Amer Public Works Assn 1973. VP Metcalf & Eddy consultg engrs from 1974 to 1981 with responsibility for water resource dev and principal in charge of Springfield office. From 1981 to date, Pres of Guillou & Assoc, Inc, conslt engrs specializing in drainage, flood control, dams and reservoirs, groundwater studies and well design, hydraulic model studies, and hydrologic analyses. *Society Aff:* ASCE, ASME, AWWA, APWA, AGU.

Guinnee, John W
Business: 2101 Constit Ave NW, Washington, DC 20418
Position: Engr Soils, Geology, Foundations. *Employer:* Transportation Research Board *Education:* MS/Civil/Princeton; BS/Civil/Univ of MO. *Born:* July 1920. Native of Joplin Missouri. Soils Res Engr Missouri State Hwy Dept 1947-62. With Transportation Res Bd since 1962. Respon for 23 ctte's on Soils Mech, Properties, Foundations etc. Member Amer Rd Bldrs Assn Ctte on Stabilization; Chmn Nominating Comm Amer Soc for Testing & Materials Ctte D-18 on Soil & Rock for Engrg Purposes; Liaison to US Natl Ctte on Tunneling & US Natl Ctte on Rock Mech. Mbr Internatl Soc for Soil Mech & Foundation Engrg. Mbr Amer Soc of Civil Engrs Geotech Engrg Div. *Society Aff:* ARTBA, ASTM, ASCE, USNC/TT, ISSMFE.

Guins, Sergei G
Home: 4496 Dobie Rd, Okemos, MI 48864
Position: Transportation Consultant. *Employer:* Self employed. *Education:* BS/Aero/Univ of MI; MS/Appl Mech/Univ of MI. *Born:* Mar 1915 Petrograd Russia. Instructor Thermodynamics N Y Univ 1940-41. Stress Analyst Acrotorque Co on High Pressure hydraulic pumps & motors 1942-44. Transmission Res 1945-46. Chesapeake & Ohio RR 1947-70. Ret as Asst Dir of Res. Dir Shock & Vibration Res & worked on dev of new equip. Instructor of Machine Design Fenn College night school 3 years. Assoc Prof School of Packaging Michigan State Univ 1970-72. At present Transportation Cons on Car Dynamics & Ladding Damage Prevention Problems. Fellow ASME. P Chmn Cleveland Sect & Rail Transportation Div. Fellow ASTM Ctte D-10 & SPHE Pres Kaser Assoc Inc.. *Society Aff:* ASME, SPHE, ASTM.

Gulbransen, Earl A
Home: 63 Hathaway Court, Pittsburgh, PA 15235
Position: Research Prof *Employer:* Univ of Pittsburgh *Education:* PhD/Physical Chem/Univ of Pittsburgh; BS/CHE/WA State Univ *Born:* 1/20/09 Seattle, WA. National Research Council Fellow at Univ of CA, Berkeley 1934-1936. Instructor Chem Engrg Tufts Univ 1936-1940. ResearchAdvisory and Consulting Eng Westinghouse Research Labs. 1940-1974 Research Professor Dept of Materials Science & Engineering 1974. Over 200 papers in field of high temp oxidation and corrison. Whitney Award Nat. Assoc of Corrosion Engs; Pittsburgh Award, American Chemical Society; Atcheson Award and Prize, Electrochemical Society. Hobbies reading, classical music and travel. *Society Aff:* AIME, Electrochem. Soc., American Chem. Soc.

Gulcher, Robert H
Home: 30034 Via Borica, Rancho Palos Veroes, CA 90274
Position: VP, R&E *Employer:* Rockwell Intl. *Education:* BS/ME/US Merchant Marine Acad; BEE/EE/OH State Univ. *Born:* 8/26/25. Native of Columbus, OH. Served with Merchant Marine during WWII. Joined Rockwell Intl (formerly North Am Aviation) in 1951. Served in numerous design and mgt positions in engg. Was appointed chief engr of the Columbus Aircraft Div in 1969. Presently VP of Res & Engg of the combined North Am Aircraft Operations, with plants in Los Angeles CA, Palmdale CA, Columbus, OH, & Tulsa OK, headquarted in Los Angeles. *Society Aff:* AIAA, IEEE, NMA.

Gully, Arnold J
Home: 2406 Slide Rd, Lubbock, TX 79407
Position: Assoc VP. *Employer:* TX Tech Univ. *Education:* Dr of Philosphy/ChE/LA State Univ; Master of Sci/ChE/LA State Univ; Bachelor of Sci/ChE/Auburn Univ. *Born:* 7/8/21. Native of Mississippi. BS, Auburn; MS & PhD, LA State in Chem Engg. Eight yrs industrial professional employment as quality control engr, proj engr and research supervisor. Specialty: catalytic hydrogenation processes. Chem Engg Prof MS State, 1951-59. At TX Tech Univ since 1963. Prof of Chem Engg and successively Chrmn, Chemical Engg, Assoc Dean of Engg and Assoc VP for Research. Active in accreditation evaluations in engg technology and in chemical engg, Engrs Council for Prof Dev. Fellow, Amer Inst of Chemical Engrs. Mbr AIChE, ACS, ASEE, NSPE. *Society Aff:* AIChE, ACS, ASEE, NSPE.

Gully, Sol W
Home: 6 Iroquois Ave, Andover, MA 01810
Position: Sr. VP. *Employer:* Alphatech, Inc. *Education:* PhD/Sys Eng/Polytechnic Inst of NY; MS/EE-SYS Sci/Polytechnic Inst of NY; BE/EE/Youngstown State Univ. *Born:* 5/31/40. in Greenville, PA. USN 1958-62. Sys Dev Engr Genl Elec Co 1967-71. Controls Sys Cons MIT Inst Lab 1971-73. Doctoral Res MIT 1971-73. Honeywell from 1973-77. Worked on Adv AF & Space Shuttle Prog. Control Sys Dept Mgr at TASC from 1977- 79. Mgr of a dozen Govt contracts in controls. PI on lasers, fighters, gun & pwr sys. Co-founder & Dir of Alphatech in 1979. VP/Engr. Dir of 30 Govt Engr contracts in Sys Engrg. Engaged in Surveillance, Battle Management and C3 Contracts in the Strategic Defense Institute and The Air Defense Institute. Visiting scientist at MIT during summer 1979. Mbr AIAA & IEEE. Chmn AIAA Tampa Bay Sect 1976, 1977; Chmn 1976, 1978, 1981-82 IEEE Conf on Decision & Control. Space Shuttle sess Chmn for same Conf. Editor IEEE control sys

Gully, Sol W (Continued)
mag. PE LiC CA. Hobby: sailing, running marathons. *Society Aff:* IEEE, AIAA, ADPA

Gumb, Raymond Daniel
Business: CSUN, School of Engrg & Computer Sci, Northridge, CA 91330
Position: Prof of Computer Science *Employer:* CA State Univ, Northridge *Education:* SB/Ind Mgmt/MIT; MA/Philosophy/Emory Univ; PhD/Philosophy/Lehigh Univ *Born:* 10/20/38 Professional Experience: Computer Programmer, 1960-1970, with TRW, Raytheon, MIT, Bunker Ramo, GA Tech, Lehigh Univ. Coll Prof, 1970-present, LaFayette Coll, Temple Univ, CSUN. Books: *Rule-Governed Linguistic Behavior* (1972), *Evolving Theories* (1979), *Essays in Semantics and Epistemology* (1981). Papers: various on formal logic, computing and philosophy. *Society Aff:* ACM, ASL, ACL, APA, AAAI

Gunaji, Narendra N
Business: P O Box 3664, Las Cruces, NM 88003
Position: Commissioner, International Boundary and Water Commission, U.S. - Mexico *Employer:* U.S. Government *Education:* PhD/Civil Eng/Univ of WI; MSCE/Civil Eng/Univ of WI; BE/Civil/Univ of Poona. *Born:* Jan 1931. Univ of Bombay 1948; Univ of Poona India, BE with hon 1953; Univ of Wisc MSCE 1955, PhD 1958; Advanced Study MIT 1963-68; USAF Acad 1970; Univ of Wash 1973. New Mexico State Univ Dir, Engr Expt Sta 1966-1982; Prof of Civil Engrg 1960-87. Commissioner, International Boundary and Water Commission, US - Mexico, 1987-. ASCE, ASEE, NSPE, AGU, Internatl Assn for Hydraulic Res, WPCF, AAAS, Amer Met Soc; NM Soc Prof Engrs Outstanding Serv Award 1967; ASEE Western Elec Award 1974; Chi Epsilon, Tau Beta Pi, Sigma Xi; Recipient of Natl Prof Awards & Grants for Sci Res; Author of num prof pubs. *Society Aff:* NWWA, ASPRS

Gunderson, Morton L
Home: 1005 Kankakee, Lowell, IN 46356
Position: Mgr, Ind/Comml Gas Marketing. *Employer:* Northern IN Public Service Co. *Education:* BS/ME/Valparaiso Univ. *Born:* 5/18/32. Native of Strum, WI. Served two yrs in military service. Worked for Northern IN Public Service Co as Dist Engr, Ind Gas Sales Engr, Div Gas Sales Supvr, Goshen & Hammond dist. Presently Mgr, Ind & Comml Gas Marketing. Mbr of Midwest Ind Gas Council, Past Chapter Chrmn Calumet Chapt ASM, MEI Natl Committee for ASM, Toastmasters Intl, & Central Profl Ski Instr Assn. Reg PE, Lic in IN. Enjoy skiing, golf, & restoring old homes. *Society Aff:* ASM, NSPE.

Gungor, Behic R
Business: Electrical Engrg Dept, Mobile, AL 36688
Position: Assoc Prof *Employer:* Univ of South AL *Education:* Ph.D./Engrg/Univ of AR; MS/EE/Univ of AR; BS/EE/Istaubul Engrg & Arch Acad *Born:* 3/28/27 Worked as an instructor at IEAA, at the Naval Acad in Istanbul, as a design engr with ETI-BANK, Ankara, and as proj engr with the Engrg Group of Turkish Air Force in Ankara. Started teaching, at Univ of AR in 1960, moved to the Univ of South AL in 1975, in the areas of Network Theory Energy Conversion, Power Sys and Electronic Instrumentation. Is a mbr of Tau Beta Pi, Eta Kappa Nu, IEEE and ASEE. *Society Aff:* IEEE, ASEE

Gunnerson, Charles G
Home: 205 Pawnee Drive, Boulder, CO 80303
Position: Principal *Employer:* Kalbermatten Assoc *Education:* BA/Geology/UCLA. *Born:* June 1920. Oregon State, CE. US Army 1941-46/1950-53. City of Los Angeles 1947-50/1953-60; Calif Dept of Water Resources 1947-63; US PHS 1963-66, 1968-70. Ch Sanitary Engr, DAMOC, WHO water & sewerage proj Istanbul 1967-68. Stanford Res Inst 1970-71. Woodward-Clyde Cons 1971- 73. Dir Great Lakes Regional Office Internatl Joint Comm Windsor Ontario 1973-74. Envt'l Eng'g Advisor, National Oceanic and Atmospheric Admin 1974-81, 1985-86. Project Mgr/Project Officer, World Bank 1976-78, 1981-85. ASCE Horner Award 1975, Hering Medal 1967 & 1960. Spec in marine and estuarine environmental assessments and protection, appropriate tech for water & wastewater systems, enviromental data mgmt, resource recycling and utilization, ancient municipal systems. *Society Aff:* ASCE, MEI, MESA, IAWPRC, AIDIS, AAEE.

Gunness, Robert C
Business: 1 1st National Plaza, Chicago, IL 60603
Position: Corporate Dir. *Employer:* Self employed. *Born:* Jul 28 1911. S. Christian I & Elizabeth (Rice) G; BS Mass Sta College Amherst 1932; MS Mass Inst Tech 1934; D Sc 1936; m Beverly Osterberger 1936; Children Robert Ch, Donald A, Beverly A. Asst Prof chem engrg Mass Inst Tech 1936-38; res dept Standard Oil Co Ind. 1938-47, mgr res 1947-51, asst genl mgr mfg 1952-54, genl mgr supply & transp 1954-56, exec VP 1956-65, Pres 1965, retired 1975. Vice chmn R/D Bd Dept Def 1951. Trustee U Chgo; Life mbr Mass Inst Tech Corp; P Pres John Crerar Library. Mbr AIChE, council 1951, AChS, API; Sigma Xi, Phi Kappa Phi, Kappa Sigma. Clubs: Commercial Chicago, Chgo,; Chemists, NYC, Glen View. Fellow AIChE; Mbr of Council Natl Acad of Engrs; Mbr Amer Acad of Arts & Sci. Dir: Champion Internatl Corp, Inland Steel, Consolidated Foods Co, Foote Cone & Belding Comm Inc.

Gunnin, Bill L
Business: 4403 No Central Expressway, Suite 300, Dallas, TX 75205
Position: Exec VP *Employer:* Gunnin-Campbell Consulting Engrs, Inc *Education:* Dr of Philosophy/CE/Univ of TX-Austin; MS/CE/Univ of TX-Austin; BS/CE/TX Tech Univ *Born:* 1/19/43 Native of Dallas, TX. Taught at the Univ of TX at Austin and Southern Methodist Univ at Dallas, TX. Sr Structural Engr at Ellisor Engrs, Inc, Houston, TX, from 1970 to 1971. VP of Ellisor and Tanner, Inc, Dallas, TX from 1972 to 1979. Principal at Guinnin-Campbell Consulting Engrs, Inc., Dallas, TX, since 1979. Officer of Northeast TX Chapter, American Concrete Inst, and Chrmn of Structural Div, TX Section, American Society of Civil Engrs. Author of numerous technical articles. *Society Aff:* ASCE, ACI, CEC

Gunwaldsen, Ralph W
Home: 42 Temple Rd, Wellesley, MA 02181
Position: Independent Consulting Engr *Employer:* Self-Employed *Education:* B/CE/Polytechnic Inst of Brooklyn *Born:* 2/12/14 Since 1973, Independent Consulting Engr to electric utilities, engrg firms, law firms and individuals providing hydroelectric and water resources advisory services. Member Bd of Consultants several hydroelectric projects. Dam safety inspection several hydroelectric projects. 1957-1973: Asst Engrg Mgr and Chief Hydraulic Engr Stone & Webster Engrg. 1948-57: Consulting Civil Engr Ebasco Services. 1947-48 Asst Prof-Civil, Brooklyn Polytechnic Inst. 1946-47 Asst Civil Engr Parsons Brinkerhof Hogan Mcdonald. 1942-46: 2nd Lt-Major US Army. 1939- 1942: Jr Hydraulic Engr US Geological Survey and Corps of Engrs. Tau Beta Pi and Chi Epsilon Honory Engrg Societies. *Society Aff:* ASCE, AWRA, IAHR, USCOLD, BSCE

Gupta, Ajay
Business: 230 Truman St, NE, Albuquerque, NM 87108
Position: Chief, Sanitary Engr *Employer:* William Matotan & Assocs *Education:* MS/Enviro/Univ of IL-Urbana; BS/CE/Indian Inst of Tech Kanpor *Born:* 3/4/46 Native of India. Came to US in 1970. Did research at Univ of IL at Urbana Champaign while doing graduate work from 1970-74. Project Co-ordinator for Greeley & Hansen Engr, Chicago, designed wastewater treatment plants from 1974- 1980. Chief Sanitary Engr for William Matotan & Assocs, Albuquerque, responsible for management of environmental engrg dept and ongoing design projects, since 1980. Registered PE in the states of IL and NM. Enjoy camping and cooking. *Society Aff:* ASCE, WPCF, AWWA

Gupta, Ashwani K
Business: Dept. of Mechanical Engineering, University of Maryland, College Park, MD 20742
Position: Assoc Prof *Employer:* Univ of MD *Education:* D.Sc./Chem Engrg & Fuel Tech/Sheffield Univ; PhD/Chem Engrg & Fuel Tech/Sheffield Univ; MSc/Gas Dynamics/Univ of Southampton; BSc/Sci/Panjab Univ *Born:* 10/23/48 Born in India in 1948, obtained PhD from Sheffield Univ, England in 1973, at present with Univ of MD since 1983. Prior experience includes 6 years as Res Staff at MIT, Chem Engrg and Energy Lab, 3 years as Independent Res Worker and Sr Res Assoc at Sheffield Univ, conslt to many indust in UK, USA, Japan and Europe. Dr. Gupta was awarded D.Sc. in 1986 from Sheffield Univ U.K. for his international recognition in Engrg & Combustion Sci. Publ over 80 papers in various journals and symposia proceedings. Co- authored two textbooks. He is founder and co-editor of Energy & Engrg Sci Series. Chartered engr, fuel technologist and corp mbr of many professional socs. *Society Aff:* AIAA, CI, InstE

Gupta, Gopal D
Business: Enviresponse Inc, 110 South Orange Ave, Livingston, NJ 07039
Position: R&D Section Chief. *Employer:* Enviresponse Inc. *Education:* PhD/Mech Engg/Lehigh Univ, Bethlehem, PA; MS/Mech Engrg/Lehigh Univ, Bethlehem, PA; BTech/Mech Engg/IN Inst of Technology Kanpur, IN. *Born:* Aug 1946. Native of New Delhi India. Postdoctoral fellow & asst Prof at Lehigh 1970-73. Involved in R/D in Applied Mech & design proj in bioengrg. With Foster Wheeler since 1973. Assumed current responsibility as R&D section chief at Enviresponse, Inc (division of FW). Manager of Engrg Sci & Tech Dept of Res Div in 1980-87. Involved in R/D in energy related proj. Elected Fellow of ASME 1985. 1976 ASME Henry Hess Award. 1st prize in James F Lincoln Foundation natl student design competition 1969. Enjoy tennis. *Society Aff:* ASME, ΣΞ, BIS.

Gupta, Hem C
Business: 35 E Wacker Drive, Chicago, IL 60601
Position: President. *Employer:* Enviromental Systems Design Inc. *Education:* MS/ME/Univ of IL; BS/EE/Delhi Poly IN. *Born:* Jun 17 1931. BSEE Delhi Polytech; MSME Univ of Illinois. Reg Prof Engr in 20 states & DC. Proj engr Skidmore Owings & Merrill 1954-59; A Epstein & Sons Inc 1959-61; Proj engr Perkins & Will Partnership 1961-62; assoc 1962-63, sr assoc 1963-66; Ch Engr VP P&W Engrs Inc 1966-67; Pres Enviromental Systems Design Inv, mech & elec cons engrs, 1967-. Mbr ISPE, NSPE, Natl Council of State Bd of Engrg Examiners; ASHRAE. Am responsible for all major decisions in fields of mech & elec engrg, troubleshooting & energy analysis. *Society Aff:* NSPE, ISPE, ASHRAE.

Gupta, Krishna Chandra
Business: Prof, Dept of Mech Engrg, Univ. of IL at Chicago, Chicago, IL 60680
Position: Prof *Employer:* Univ of IL (Chicago) *Education:* PhD/Mech Engrg/Stanford Univ; MS/Mech Engrg/Case Inst; BTech/Mech Engrg/IIT Kanpur (India) *Born:* 09/24/48 Permanent resident of US. BTech/ME/1969, IIT Kanpur (India); MS/ME/1971, Case; PhD/ME/1974, Stanford. At Univ IL at Chicago since 1974; Asst Prof 1974- 79, Assoc Prof 1979-84, Prof 1984- , Dir Grad Studies 1982-84. Res contributions in kinematics, mechanism synthesis, robotics, numerical methods and design optimization. Past and current res under the sponsorship of Univ IL at Chicago, Natl Sci Fdn, and US Army Res Office. Assoc Editor, ASME Jour of Mech Design, 1981-82. Papers Review Chrmn, 1982 ASME Mechanisms Conference (Washington DC). Mbr, ASME Applied Mech Reviews Editorial Advisory Board. Faculty Adv, ASME Student Sect, 1978-80. Merit Scholarship, 1964-69. Res/Teaching Assistantship, 1969-74. Best Paper Awd, 1978 ASME Mechanisms Conference (Minneapolis); ASME Henry Hess (Lit) Awd, 1979. *Society Aff:* ASME

Gupta, Virendra K
Home: Murray Hill Apt 6, Mt Morris, NY 14510
Position: Dir of Engineering/Planning *Employer:* Town of Henrietta *Education:* BE/Civil Engg/Univ of Roorkee, MBA/U.O. Rochester. *Born:* 6/30/40. Worked for state gvmt engrg dept & taught at Univ of Roorkee. Held position of City Engr for City of DehraDun 1968-69. Came to US 1969. Worked for County of Onondaga NY 1969-72. Reg Prof Engr State of NY form 1972. Worked as Deputy Dir of Engrg & water Sys Engr for Town of Greece NY 1972-75. Worked for Livingston County Health Dept as Dir of Environmental Health at Mt Morris NY. 1975-80. Currently working for the Town of Henrietta as Dir of Engrg/Planning. *Society Aff:* ASCE, AAA

Gupton, Guy W, Jr
Home: 2405 Woodward Way NW, Atlanta, GA 30305
Position: President *Employer:* Gupton Engrg Assocs, Inc *Education:* BBA/-/Georgia State Univ; BE/Engg/Univ FL (ASTP) *Born:* 11/-/26 Native of Atlanta GA. Served with USAAF 1944-45. Studied Basic Engrg in ASTP Univ of FL 1944, mech engrg Georgia Tech. Began work in HVACR field 1945, cons engrg 1948. Formed Gupton Engrg Assoc 1976. Responsible for design & supr of bldg systems. Active in ACEC, Author of 3 books on HVAC controls operation and maintenance. ASHRAE, CSI. Reg PE in GA, NC, & TN. Former Lecturer at Georgia Tech. Speaker in US, Australia, Canada, England, Belgium & Germany on energy conservation. Author of tech articles on design, installation, commissioning & operation of bldg systems. ASHRAE Regional Award of Merit 1974, Fellow 1981, Dist Svc Award 1981. Mbr ODK. *Society Aff:* ASHRAE, CSI, ACEC

Gupton, Paul S
Home: 1100 Deats Rd, Dickinson, TX 77539
Position: Consulting Metallurgist *Employer:* Bell And Assoc. Inc. *Education:* BS/Engr/Lamar Univ; MS/Met/TX A&M Univ. *Born:* Dec 1934. Native of Houston TX. Welding Engr for the Hughes Tool Co; Res Engr for Texas Engrg Experiment Station; Sr. Monsanto Fellow Tech Sect Monsanto Co Texas City. With Monsanto for over 25 yrs. Presently Consulting Metallurgist Bell & Associates, Houston, TX. Prof affil incl ASM & mbr ASTM, ASNT & AWS. Hon organizations Pi Tau Sigma (Mech Engrg). P Chmn Houston Sect ASM & currently Tr ASM. Fellow of ASM. *Society Aff:* ASM, AWS, ASNT, ASTM.

Gurfinkel, German R
Home: 2510 S Prospect Ave, Champaign, IL 61820
Position: Prof of Civ Engr. *Employer:* Univ of IL. *Education:* PhD/Civ Engg/Univ of IL; MS/Civ Engg/Univ of IL; Ingeniero Civ/Civ Engg/Universidad de la Habana. *Born:* 9/14/32. Active in Structural Engg practice, res and academics since 1955. Taught at Civ Engg Dept of Univ of Havana, Cuba, 1959-61 and at Univ of IL at Urbana since 1962. Designer of multistory bldgs,tanks bridges and stadiums. Author of "Wood Engineering," –Reinforced Concrete Bunkers and Silos," and dozens of technical articles, coauthor of "Prestressed Concrete–. Awards by James F Lincoln Arc Welding Fdn for Structural Design received in natl competitions in 1973, 1977, 1978 1979, 1980, 1981, 1982 and 1983. Award in 1981 and First Prize in 1984 by Structural Engineers Assoc. of Illinois in most-innovative structural-design competition. Consultant to Natl Sci Fdn in 1967, 1968 and 1969 Summer Inst in India. Intl consultant on structural adequacy of silos and grain-containment structures. *Society Aff:* ASCE, ASTM, ACI.

Gurland, Joseph
Business: Div of Engg, Brown University, Providence, RI 02912
Position: Prof. *Employer:* Brown Univ. *Education:* ScD/Phys Metallurgy/MIT; ME/Metallurgy/NYU; BE/Chem Engr/NYU. *Born:* 1/26/23. BE (1944), ME (1947), NYU ScD (1951) MIT. US Army 1944-46. Res Engr at Battelle Memorial Inst, Columbus, OH, 1947-48. Res Engr and Mgr at Firth- Sterling, Inc, Pittsburgh, 1951-1955. Since 1955 at Brown Univ. Res interests: mech properties of engg solids, powder metallurgy, quantitative stereology, steels and cemented carbides. *Society Aff:* AIME, ASM, ASTM, IMS.

Gurnham, C Fred
Business: 223 West Jackson Blvd, Chicago, IL 60606
Position: VP & Consultant. *Employer:* Peter F Loftus Corp (IL). *Education:* DEngSc/Chem Engg/NY Univ; MChE/Chem Engg/NY Univ; BS/Chem Engg/Yale Univ. *Born:* Oct 19 1911 Ludlow MA. Industry 15 yrs: paper, textiles, chems, wire, plastics. Academic 20 yrs: Hd Chem Engrg Dept Tufts U, Mich State U (both depts accredited under his leadership); Hd Environ Engrg Dept Ill Inst of Tech. Cons 13 yrs (also while teaching): Pres Gurnham & Assocs Inc pollution control cons; Dir Peter F Loftus Corp; VP & Consultant, Peter F Loftus Corp (IL). 2 books on indus wastewater control, num journal articles. Reg Prof Engr; active on engrg cttes. 'Mr Pollution Control' 1970 AIPE. Philatelist (British Commonwealth). *Society Aff:* AAEE, ACS, AES, AIChE, AIC, ASCE, WPCF.

Gurther, H Louis
Business: 3443 Chicago Dr SW, Grandville, MI 49418
Position: VP. *Employer:* Haven Busch. *Education:* BSCE/Civ Engg/Purdue Univ. *Born:* 10/16/38. Engr Borne trained officer in US Army Corp of Engrs (1961-63) Engr with Haven Busch Co, specializing in engg, estimating, & sales of structural steel framing (1963-73). VP of Sales (1973-77). VP of Mfg (1973 to pres). VP of ASCE MI Section. Native of Plymouth IN. *Society Aff:* NSPE, ASCE.

Gustafson, Robert J
Business: Agri Engrg Dept, 590 Woody Hayes Dr, Columbus, OH 43210
Position: Prof & Chrmn *Employer:* Ohio State Univ *Education:* PhD/Agri Engrg/MI State Univ; MS/Agri Engrg/Univ of IL; BS/Agri Engrg/Univ of IL *Born:* 9/25/48 Native of Coal Valley, IL. Engaged in teaching and research at Univ of MN 1975-87. Currently, Prof & Chrmn of Agricultural Engrg. Author of text "Fundamentals of Electricity for Agriculture." *Society Aff:* ASAE, IEEE, NACTA

Gustafsson, Ross H
Home: 1725 Harrison Dr, Clinton, IA 52732
Position: VP Engg. *Employer:* Clinton Corn Processing Co Inc. *Born:* Jul 24, 1920 Bagley Minn & reared in agri-rich area of Red River Valley with farming & financial family business background. BS Chem Engrg 1943 U of N D. Ltjg USNR (Engrg-Diesel) 1943-46. 1946-48 Northern Regional Res Lab Peoria Ill, Process Dev Engr Oilseed Processing; 1948-50 Pillsbury Mills Inc Clinton Iowa, Production Supr for Soy Processing; 1950- Clinton Corn Processing Co Inc (formerly Clinton Foods Inc): 1950-57 Chem Engrg, 1957-61 Ch Chem Engr, 1961-63 Asst Mgr of Engrg/Chem Engrg, 1963 Planning & Coordination Mgr, 1963-66 Engrg Asst to Sr VP, 1966-68 Mgr Prod & Engrg, 196873 VP/Mfg, 1973-74 VP/Plan & Prod Dev, 1974-76 VP Tech, 1976-present VP/New Ventures Present VP Engg 8/15/79. *Society Aff:* AIChE.

Guth, John J, Jr
Business: 208 Milam Str, Shreveport, LA 71101
Position: President. *Employer:* John J Guth Assocs Inc. *Education:* BSEE/Power/Univ of TN. *Born:* Sep 3 1928. Armament Off USAF 2 yrs. With Carl M Hadra Assocs & Malahy & Guth Cons Engrs until 1966. Pres of John J Guth Assocs Cons Engrs 1966- . State Pres CEC LA 197576. Former Pres of Shreveport Chap ASHRAE, SAME, CSI & LA Engrg Soc. Reg Mech & Elec Engr in LA, Reg Prof Engr in AR, CO, FL, KS, MS, MO, OK & TX & reg with Natl Council of Engrg Examiners. *Society Aff:* NSPE, ASHRAE, ACEC, SAME.

Guth, Sylvester K
Home: 637 Quilliams Rd, South Euclid, OH 44121
Position: Consultant. *Employer:* Self-employed. *Education:* Prof EE/EE/Univ Wis; BS/EE/Univ Wis. *Born:* Dec 31 1908. Native of Milwaukee WI. Started with General Electric Co at Nela Pk as vision res in 1930; Hd Lighting Res 1950; Mgr Radiant Energy Effects Lab 1956; Mgr Applied Res 1969. Respon for res in light, vision, color, performance, vis comfort, physiological effects of radiant energy. Ret 1974. Now cons. IES Gold Medal 1967. American Academy of Optometry Charles Prentice Medal 1980. Pres Internatl Comm on Illumination 1975-79; American del at all sessions since 1951. Author of 75 tech & sci papers. Enjoy travel, golf & photography. *Society Aff:* IESNA, OSA, ISCC, ASPE, CIE.

Guthrie, Hugh D
Home: Apt C-34, 3557 Collins Ferry Rd, Morgantown, WV 26505
Position: Dir, Extraction Projects Mgmnt Div *Employer:* U.S. Govt Dept Energy Morgantown Energy Tech Center *Education:* BS/Chem Engrg/State Univ of IA *Born:* 05/11/19 Shell Cos 1943-76 Che R&D 1943-52, mfg, mktg, economics, mgmt 1952-76. Turbogrid pats. ERDA/DOE 1976-79, Dir Oil, Gas & Shale, SRI 1979-80 Dir Energy Center, Occidental 1980-86; Consultant 1986-87; DOE, Dir. Ext; Div. METC 1987-. Fellow AIChE; Dir 1965-68; Pres 1969; Founders Award; Tau Beta Pi, Sigma Xi, Chemists' Club NYC, UET Bd 1966-80, Pres 1978-79; Engg Fdn Bd 1966-79; Engg Socs Lib Bd 1968-76; Adv Bd Pittsburgh Univ Engg 1969-74; Adv Bd Univ IA Engg 1974-79; Adv Ctte Univ TX ChE 1976-79; Adv Bd Univ CA ChE 1979; Adv. Bd Tulsa Univ. ChE 1984-1987. *Society Aff:* AIME, AIChE, ACS, SPE, AAAS, TBΠ, ΣΞ

Gutierrez, Ariel E
Business: PO Box 13171, Santurce, PR 00908
Position: Pres *Employer:* Systems Concept Intl, Inc. *Education:* BSCE/-/GA Inst of Tech; MSCE/-/GA Inst of Tech. *Born:* 12/1/43. Honors Mbr of a Special Committee appointed by Governor Reuben Askew to study the condominum industry in the State of FL. Montilla, Latimer, Gutierrez & Gutierrez, San Juan, P.R. - Partner. *Society Aff:* ASCE, NSPE, GSPE, CIAPR.

Gutshall, Thomas L
Business: Syntex Corporation, 3401 Hillview Ave, Palo Alto, CA 94303
Position: Exec. Vice Pres. *Employer:* Syntex Corp. *Education:* BS/Chem Engrg/Univ of DE *Born:* Feb 24 1938. Grad work at WV U & U of MO. Native of Huntingdon PA. US Army Ordnance Corps 1961-62. Process Engr Union Carbide Corp 1960-69. Joined Mallinckrodt 1969 as Prod Supr for St Louis Plant. Plant Mgr of Raleigh General Chem Plant 1972. Assumed respon as Genl Mgr for Drug & Cosmetic Chem Div 1975. Named VP and Gen Mgr in 1978. Joined Syntex Mar 1981 as Group VP for worldwide chemical operations. Appointed Pres of SYVA Co, Sept 1983. Named Pres, Syntex Diagnostics, April 1984. Appointed Exec. Vice Pres, Syntex Corp, Jan. 1987. P Pres Kanawha Valley Alumni Chap Tau Beta Pi. Enjoy tennis, hunting & fishing. *Society Aff:* AIChE, TBΠ, ΣΞ

Guttmann, Michel J
Business: Centre Des Materiaux de L'ENSMP, Evry France 91003
Position: Maitre de Recherche. *Employer:* Ecole natl Superieure Des Mines de Paris (ENSMP). *Education:* DrSc/Met/Univ of Paris; Ing Civ des Mines/Met Engg/Ecole des Mines de Paris. *Born:* 2/19/47. Born in Neuilly S/Seine near Paris, France. Taught at Lycee Louis-Le-Grand and Ecole Nationale Superieure des Mines, Paris (ENSMP). Res in Physical Met at the Ctr des Materiaux of ENSMP since 1968. Res Group Leader since 1974. Specialized in grain boundary segregation & related phe-

Guttmann, Michel J (Continued)
nomena, & Auger Electron Spectrosopy. Rist award of the SFM (1976) & Marcus A Grossmann award of the ASM (1977). *Society Aff:* SFM, ASM, Met Soc.

Gutzwiller, F William
Business: Fairfield, CT 06431
Position: Mgr Technology Review. *Employer:* General Electric Co. *Education:* BS/EE/Marquette Univ. *Born:* Dec 8 1926. Early engrg pos with Allis-Chalmers, Schindler (Switzerland), Cutler-Hammer, Harnischfeger. Pos with General Electric since 1955 incl Mgr Application Engrg for Semiconductors, Mgr Engrg for Transport Equip, Mgr Engrg for Integrated Circuits, & Mgr Mech Drives Control Prod Sect. Performed critical role in dev & applic of first comm silicon rectifiers, controlled rectifiers, & triacs. 15 US pats. Author num papers & books on power semiconductors. Currently respon for tech reviews & strategy analysis for Consumer products. Fellow IEEE. PE NY. Steinmetz Award (GE). *Society Aff:* IEEE.

Guy, Arthur W
Business: Bioelectromagnetics Res Lab, RJ-30, Center for Bioengineering, Seattle, WA 98195
Position: Dir, BEMRL.; Prof and Assoc Dir Bioengg *Employer:* Univ of WA. *Education:* PhD/EE/Univ of WA; MS/EE/Univ of WA; BS/EE/Univ of WA. *Born:* 12/10/28. Native of Havre, MT. Served with the USAF 1947-50, 1951-52. Res engr, antennas and propagation, at Boeing Co, 1957-64. Res engr, Univ of WA, 1964-66. Joined the faculty, Dept of Rehabilitation Medicine, Univ of WA, 1966. Became Prof & Dir of the Bioelectromagnetics Res Lab in the Dept in 1974, Prof of Bioengg 1983, Assoc Dir of Bioengg and Grad Prog Dir 1987, where he is directing & conducting res as well as teaching grad students on the biological effects & medical applications of electromagnetic fields. Mbr of a number of natl committees involved with dev of safety stds, and dev & analysis of res projs concerning human exposure to electromagnetic fields 1971 to 1982. Chrmn, ANSI Subcommittee C95.4 for developing protection guides for human exposure to RF fields. *Society Aff:* AAAS, IEEE, IMPI, BEMS.

Guy, Louis L, Jr
Business: Guy & Davis, 5208 Rolling Rd, Burke, VA 22015
Position: Partner *Employer:* Guy & Davis, Consulting Engineers *Education:* BSCE/Civil Eng/VA Poly Inst. *Born:* 4/26/38. Native of Norfolk VA. BSCE VPI Grad study GWU. Lic PE in 5 states. Dipl (San Engrg) Amer Acad Environ Engrs. Lt Army Transport Corps 1959-62. Design engr/proj mgr on cvl/san projs 1962-70 with Wiley & Wilson; George, Miles & Buhr; and Langley, McDonald & Overman. Asst Mgr Tech Servs for WPCF 1970-73. Partner/Sr. V.P., Patton, Harris, Rust & Guy, 1973-81. V.P. SCS Engineers 1981- 83. Partner Guy & Davis Consulting Engineers 1983-. Fellow ASCE; Treas NSPE 1975-80; Pres NSPE Educ Found 1980-82; Chmn. NSPE-PAC 1983; NSPE Young Engr of the Year 1973. VA SPE Engr of the Year 1978; Pres. Amer Acad Environ Engrs 1986-87; mbr VA Water Study Commission 1977-82; Chrmn., Upper Occoquan Sewage Authority 1985-; Chrmn. Order of the Engineer 1987- ; Presidential appt. to the Bd of Directors, National Inst. of Bldg. Sciences 1987- . *Society Aff:* NSPE, AAEE, ASCE, WPCF, AWWA.

Guy, Warren J
Business: Dept of Electrical Engg, Easton, PA 18042
Position: Prof & Head. *Employer:* Lafayette College. *Education:* EngrScD/Elec Engg/NJ Inst of Tech; MA/Physics/Temple Univ; BSEE/Elec Engg/Drexel Univ. *Born:* 5/3/36. Joined Lafayette College in 1964 and assumed dept headship in 1972. Prior employment was with Philco Corp, Rohm and Haas, and the US Army (Signal Corps). Res interest is in instrumentation and control. *Society Aff:* IEEE, ASEE.

Guzzi, Louis A
Home: Rd 5, Box 251, Johnstown, PA 15905
Position: VP Engr Products *Employer:* VRACO Inc *Education:* BS/Mech Engr/Univ of Pgh; MS/Mech Engr/PA State Univ *Born:* 10/27/37 Graduated from Univ of Pittsburgh in 1959, worked as design engr for Crahe Co. Attended Penn State Univ to perform research on Bldg Products while completing my Master Degree. Employed as tech conslt by PA Elec Co for 5 years. Previously a Project Engr and Designer of Heating, Ventilating, Air Conditioning, Plumbing and Elec Sys for Commercial Inst and Indus Bldgs. Responsible for more than 80 million dollars of bldg design for each of the last 16 years. Mbr of Tech Adv Ctte to the AIA Res Corp for the development of Natl Bldg Energy Performance Standards. Mbr of Intl Ctte for environ Design Ctte of the AIA. Presently resp for applying engrd products to commercial bldgs. *Society Aff:* NSPE, ASME, ASHRAE

Gwynn, David W
Home: 26 Metekunk Dr, Trenton, NJ 08638
Position: Ch Engr Transp Operation. *Employer:* New Jersey Dept Transportation. *Education:* BSCE/CE/VMI; Cert/Trans/Yale; MSCE/CE/W VA Univ. *Born:* Nov 1936. Traffic Engr VA Dept of Hwys 1959-64. Supr Engr Bureau of Traffic & Safety Res & Dir of Res NJDOT 1965-75. Ch Engr Transportation Operations 1975. Internatl Pres Inst of Transp Engrs 1976. V P 1975. Chmn Tech Council 1972-74. Mbr Transp Oper & Maintenance Council Transportation Res Bd. P Pres Ewing Lions Club & Ewing Jaycees. Outstanding Young Man Ewing Twp 1969. Bd/Cons ENO Found for Transp Safety. Mbr Ewing Two Bd of Education-1977, Pres 1979-80 "Engineer of the Yr in NJ" 1978-selected by 'Central Jersey Engineers Council (SJEC), Selected as one of "Top Ten Public Works Officials for 1977" by Amer Public Works Assoc (APWA Married Lydia Melville; children: David, Daniel, Dennis & Susan. "Transportation Engr of Yr" 1980 selected by NY/NJ ITE Section. *Society Aff:* ITE, AASHTO, TRB, ARTBA

Gyftopoulos, Elias P
Business: MIT Room 24-109, Cambridge, MA 02139
Position: Professor *Employer:* MIT *Education:* Sc.D./Electrical Eng./MIT; Diploma in Mech. S/Elect. Eng/Technical University of Athens Greece *Born:* 07/04/27 Engrg educator & consultant, born in Greece, came to U.S. in 1953, naturalized in 1963. On teaching staff of MIT since 1955, Ford Professor of Engrg since 1970. Dir Thermo Electron Corp. and Thermo Instrument Systems, both in Waltham, MA. Ch, National Energy Council of Greece, 1975-78. Author: Thermionic Energy Conversion, Vol. 1, 1973, Vol. 2, 1979; Fuel Effectiveness in Industry, 1974; Editor, Energy Conservation Manuals, 1982. Fellow ANS, ASME, Am. Acad. of Arts and Sciences, Academy of Athens, and National Academy of Engineering. *Society Aff:* ANS, AAAS, ASME, APS

Haag, Kenneth W
Home: 27W681 Washington, Winfield, IL 60190
Position: Mmbr of Technical Staff *Employer:* AT&T Bell Laboratories *Education:* PhD/EE/IL Inst of Tech; MS/EE/IL Inst of Tech; BS/EE/IL Inst of Tech *Born:* 09/29/37 Native of West Chicago, ILL. Army Reserve with 6 months active duty. Faculty mbr of IL Inst of Tech 1960-85. Consultant for Inst of Gas Tech on instrmentation and SCR inverters 1960-64. Consultant for Univ of Chicago on signal processing systems for meteorological radar 1966-76. Part time Biomedial Engr with Hines VA Hospital working on applications of ultrasound to blood flow measurements and gait analysis. 1978-81. Acting Chairman Dept of EE, IIT 1980-82. Member of Technical Staff, AT&T Bell Laboratories, 1985-present.

Haan, Charles T
Business: Agri Engrg Dept, Oklahoma State University, Stillwater, OK 74078
Position: Professor, Agri Engg Dept. *Employer:* OK State Univ. *Education:* PhD/Agri Engg/IA State Univ; MS/Agri Engg/Purdue; BS/Agri Engg/Purdue. *Born:* 7/10/41. Teaching & Research in Hydrology & Water Resources at Univ of KY 1967-78; Hd, Agri Engg Dept, OK State Univ 1978-1984; Teaching and Research in Hydrology, OK State Univ 1985-present; Reg P E in KY & OK; Mbr ASAE, AIH; ASAE Board of Directors 1986-1988; Chmn Hydrology Group ASAE 1974-76; Vice Chmn 1975 & Chmn 1977 KY Sect ASAE; Mbr several SE U S regional hydrology committees; Research Paper Award 1969 ASAE; Tau Beta Pi teaching award Univ

Haan, Charles T (Continued)
of Ky 1970-71; Young Researcher of Year Award 1975 ASAE; Consultant in area of hydrology. Author 100 plus tech papers reports & 4 bks in area of speciality. *Society Aff:* ASAE, AIH.

Haar, Herbert R, Jr
Business: PO Box 60046, New Orleans, LA 70160
Position: Associate Port Director. *Employer:* Bd of Commiss/Port of New Orleans. *Education:* MS/City Planning/Univ of IL; BS/Civil Engr/VA Polytechnic Inst; Grad/Internatl Affairs/US Army War College. *Born:* April 1923. Native of Alexandria, Virginia. Served Army Corps of Engineers 1943-71. OIC Nicaraguan Canal Survey 1949-52. Engr Advisor-Peruvian Army 1959-62. Asst Engr Commissioner D C Govt 1965-66. US Army Engr-Thailand 1967-68. Dist Engr New Orleans Dist, CE 1968-71. Deputy Exec Port Dir for Port of New Orleans 1971- . Bd Chmn Natl Waterways Conf 1974- 76. Awarded Order of Ayacucho, Peruvian Gov 1962. Meritorious Pub Serv Award D C Government 1966. Cited by President Nixon 1969. Certificate of Merit from Governor of La 1972. Fellow SAME. Registered P E DC & LA. Full Mbr AICP. US Commissioner, Permanent Internatl Assoc of Navigation Congresses 1977- . Chrmn Internatl Assoc of Ports and Harbors Ad Hoc Dredging Task Force 1981. Observer for IAPH to the London Dumping Convention 1980-. Vice Chairman Bd of Directors Gulf Intercoastal Canal Assoc 1985-. Cited by President Carter in 1979. Awarded Dept. of Army Outstanding Civilian Service Medal 1984 and American Association of Port Authorities Important Service Award 1985. *Society Aff:* APA, AAPA, SAME, IAPH

Haas, John C
Home: 330 Spring Mill Road, Villanova, PA 19085
Position: V Chrmn of the Bd. *Employer:* Rohm & Haas Co. *Education:* BA/Chemistry/Amherst; MS/Chem Eng/MIT. *Born:* 1918. Other than 4 yrs in Naval Reserve, entire career wth Rohm and Haas Co. Serve as mbr of MIT Corporation and as Chairman (currently) of the Visiting Committee of the Chemical Engrg Dept. Received honorary degree from Amherst College 1975. *Society Aff:* ACS, AIChE.

Haas, Paul A
Home: 8000 Bennington Dr, Knoxville, TN 37909
Position: Sr Engr *Employer:* Oak Ridge Natl Lab *Education:* PhD/Chem Engr/U of TN-Knoxville; -/Nucl Engr/Oak Ridge Sch of Reactor Tech; MS/Chem Engr/Montana State Univ; BS/Chem Engr/U of MO-Rolla *Born:* 08/11/29 Born in Rolla, MO. Engr, Group Leader, and Sr Staff Mbr with the Chem Tech Div of the Oak Ridge Natl Lab from 1952 to date. Ten patents and numerous pubs concerning the nuclear fuel cycle including solvent extraction, fuel conversion, SOL-GEL, and waste treatment processes. Fellow of the Amer Inst of Chem Engrs. Reg Engr in TN. Hobbies are travel, duplicate bridge, fishing, and tennis. *Society Aff:* AIChE, ANS, ACS, TBΠ

Habach, George F
Business: 20 Cambridge Road, Glen Ridge, NJ 07028
Position: Consultant (Self-employed) *Employer:* Self-employed. *Education:* ME/Stevens Inst Tech; MME/Poly Inst NY *Born:* 8/2/07. Started with Worthington Corp 1929 at Harrison NJ plant - Dev Engr & successively Application Engr, Proj Engr, Ch Engr, Mgr Engg, Corp VP Engg & VP Admin, Studebaker Worthington Inc 1968 VP Admin, Retired 1971. Adj Prof Brooklyn Poly 1939-52. Life Fellow ASME 87th Pres. Life Mbr NSPE; NJ PE Lic No. 7917. Life Fellow ASME, Life Fellow Stds Engr Soc. Bds of Poly of NY, Stevens Inst Tech UET, & Burns & Roe Indus Serv Corp. Stevens Honor Medal. P Pres Stevens & Poly Alumni Assn. Hon Dr Engg-Stevens Inst of Tech 1977. ASME codes and stds medal 1980. Enjoy music & photography. Pres United Engrg Trustees 1982-1983, Mbr of Vet Bd 1977-1985. *Society Aff:* ASME, NSPE, SES

Haber, Bernard
Business: 1501 Broadway, New York, NY 10036
Position: Partner. *Employer:* Hardesty & Hanover. *Education:* BCE/Civil/CCNY *Born:* 1/13/29. BCE City College of NY 1951, Reg Prof Engr NY, 7 other states & Ontario. Mbr of ASCE, AREA, ITE, IBTTA, ACEC, NSPE, TRB USAF Korean War 1953. Retired as Lt Colonel 1975. With Hardesty & Hanover since 1953, Assoc Engr 1962, partner 1972. Specializing in bridge design, rehabilitation, inspec & transportation engg. Mj proj include West Side Hwy, & Henry Hudson Pkw,NYC; Pulaski Skyway NJ and I-80 NJ, 300 bridge inspections and rehabilitations, James River Lift Bridge VA. Past V. Chmn. Met Sect. ASCE Urban Planning & Transportation Committees. Past Chrmn, ACEC Transportation Committee; Chrmn, Legislative Committee and Past V. Pres., NYACE; Chrmn of NY City Planning Bd 11Q. Wrote series of papers on heavier truck load effects on bridge structures and bridge construction and community involvement in planning for cities. Member of N.Y.C. Commission For The Year 2000. *Society Aff:* ACEC, AREA, ITE, IBTTA, TRB, ASCE, NSPE.

Haber, Fred
Business: Moore Sch of Elec Engg, Phila, PA 19104
Position: Prof. *Employer:* Univ of PA. *Education:* PhD/EE/Univ of Penn; MS/EE/Univ of Penn; BS/EE/Penn State Univ *Born:* 7/1/21. New York, NY; Served US Army 1942-1946, Signal Corps 1944-1946; Arma Corp Engr 1948; RCA Engr 1948-1951; Univ of PA 1951-present, Prof; Gen Precision Inc Sr Staff Scientist 1962-3; Navy Underwater System Ctr & Aerospace Corp summers 1972,3,6; Visiting Prof Pahlavi Univ, Shiraz Iran 1968, Visiting Prof Einohoven Univ of Tech, Netherlands, 1984. Res & Teaching in Communication Theory, Adaptive Array Processing, Spectral Estimation, & Electromagnetic Compatibility; Chrmn, Phila Sec IEEE 1974-5, Elected Fellow 1977. *Society Aff:* IEEE, ΣΞ, HKN, ΠΜΕ

Haberl, Herbert W
Home: Old Scott Rd, Lakefield, Quebec JOV IKO Canada
Position: Consultant (Self-employed) *Born:* Denver Colo, 1902. Duquesne University, Carnegie Tech. Corporation Professional Engineers of Quebec. 1925 Engr Duquesne Light and Public Service NJ, 1929 Montreal Light Heat & Power, 1944-67 Hydro Quebec. Retired 1967 having held pos as Ch Engr, Asst Genl Mgr (Engrg). Presently self-employed Cons Spec Proj's. Fellow & Life Mbr IEEE, Life Mbr Engrg Inst of Can. Spec: Power Sys Design, Oper, & High Voltage Equip.

Haberman, Charles M
Home: 1432 Calle Grande, Fullerton, CA 92635
Position: Prof. *Employer:* CA State Univ. *Education:* Engr/ME/USC; MS/Aero Engg/USC; MS/ME/USC; BS/Engg/UCLA. *Born:* 12/10/27. Eight yrs in the Thermodynamics & Aerodynamics Depts of Northrop Aircraft, with lead engr & group engr responsibilities. Res specialist title at N Am and Lockheed Aircraft. Consultant for Royal McBee & Northrop. Thermal Control Dept of Aerospace Corp. Asst Prof, Assoc Prof, & Prof (since 1967) in the Mech Engg Dept of CA State Univ at Los Angeles, from 1959 to the present. Sole author of textbooks on Engg Analysis, Computer Applications, Aerodynamics, and Vibrations. *Society Aff:* AIAA, ASEE, AAM, CTA, NEA, CSEA, TBΠ, AAUP

Haberman, Eugene G
Home: 43647 21st Street W, Lancaster, CA 93536
Position: Director, Solid Rocket Division *Employer:* US Air Force Astronautics Lab. *Education:* MSA/Mgmt/George Washington Univ; BChE/Chem Engg/City College of NY; - /Mgmt/Indus Coll of the Armed Forces. *Born:* 8/18/33. Industrial College of the Armed Forces. With Air Force Rocket Propulsion Lab since 1954. Project engr on advanced liquid propellant rocket engines for piloted aircraft including X-15. 1960 Chief, Advanced Techniques Section, conducting engrg evaluations of advanced propellants & related concepts. 1964 Plans Staff, Ballistic Missile Propulsion. 1965 As Asst for In-house Operations, provided staff guidance and direction to lab inhouse program. 1971 Deputy Chief of Test & Support Division. 1972 Chief of Operations Office. 1975 Dir of Plans. 1977 Dir of Plans & Operations 1979 Chief of Propulsion Analysis Div 1981 Chief of Liquid Rocket Div Assumed cur-

Haberman, Eugene G (Continued)
rent position as Dir of Solid Rocket Div in 1983. Enjoy Skiing, reading, snorkeling and traveling *Society Aff:* AIAA.

Haberman, William L
Business: P O B 1723, Rockville, MD 20850
Position: Engrg Consultant *Employer:* W. L. Haberman Assoc *Education:* PhD/Appl Math/Univ of MD; MS/E/Univ of MD; B/ME/Cooper Union *Born:* 5/4/22 Served Army Corps of Engrs 1942-45. Employed by US Navy and NASA in various engrg and physical science positions in ship propulsion, aerodynamics, technology application, manned space flight programs. Engrg consultant in enery mgmt and conservation. Taught broad range of college courses; was chairman of ME Dept at Newark, NJ. Authored over forty technical papers and three textbooks in *Fluid Mechanics, Engineering Thermodynamics, Heat Transfer* Chairman, Rockville Energy Commission. Registered PE Certified Fallout Shelter Analyst. Member Tau Beta Pi, and Pi Tau Sigma Honor Societies. *Society Aff:* AEE, ASEE, ASME, NSPE, SNAME, AAUP, AGA, AIAA, APS, RSES, SIAM

Haberstroh, Robert D
Business: Dept of Mech Engg, Fort Collins, CO 80523
Position: Prof. *Employer:* Colorado State Univ. *Education:* ScD/Mech Engg/MIT; Mech Eng/Mech Engg/MIT; SM/Mech Engg/MIT; BS/Mech Engg/Carnegie. *Born:* 2/11/28. in Altoona, PA. Sales mgr and engr for Structures, Inc, mfr of fabricated structural steel products in Johnstown, PA, to 1956. Industrial Liaison Officer at MIT, 1956-59. Faculty mbr at Colorado State Univ since 1959. Research and teaching interests in drying of solids, natural convection, particle-to-particle dry heat transfer, applications of Second Law of Thermodynamics in society. Personal interests in community dev, active sports, and the outdoors. *Society Aff:* ASME, AIChE.

Habib Agahi, Karim
Home: 5230 Deerfield Ave, Mechanicsburg, PA 17055
Position: Chief, Soils Engr *Employer:* Gannett Fleming *Education:* BS/CE/Univ of IL; MS/CE/Univ of IL; Ph.D./Soil Mech./Univ of IL *Born:* 9/22/43 1971-1978: Held faculty positions here and overseas including Univ of TX at Austin, State Univ of NY at Buffalo and Pahlavi Univ in Iran. Supervising research and graduate students. Consultant to Pahlavi Univ Construction Office and the Iranian Geotechnical Inst. 1978-1979: Managing Dir of Iran Geotech, a soils and foundation engrg and material testing firm in Tehran, Iran. Consultant to Iran Atomic Energy Organization in design and construction of two nuclear power plants. 1979-now: Chief Soils engr at Gannett Fleming Corddry and Carpenter in Harrisburg, PA, involved with foundation investigations and designs of earth and rockfill dams, concrete gravity and buttress dams, tunnels, bridges, terminal structures and pollution control facilities. Registered PE in PA and DE. *Society Aff:* ASCE, USCOLD, ISSMFE, PSPE

Hacker, Herbert, Jr
Business: Dept of Elec Engg, Durham, NC 27706
Position: Assoc Prof. *Employer:* Duke Univ. *Education:* PhD/Elec Engg/Univ of MI; MS/Elec Engr/Princeton Univ; BS/Elec Engg/OH Univ. *Born:* 6/4/30. Served on the faculties of OH Univ (1959/60), Univ of MI (1964/65), and Duke Univ (1965-present). Served as Chrmn of Elec Engg at Duke (1975/79). Appointed as Visiting Assoc Prof and Res Physicist at the Univ of CA, San Diego (1971/72). Teaching and res interests in electromagnetic theory, solid state solid sci including electron spin resonance in amorphous materials, and physics of liquid crystals. Consultant to the Res Triangle Inst of NC on magnetic measurements and materials. Enjoys tennis. *Society Aff:* APS.

Hackman, E Ellsworth, III
Business: Box 2857, Wilmington, DE 19805
Position: President. *Employer:* NST/Engineers, Inc. *Education:* PhD/Chem/Univ of DE; MS/ChE/Univ of PA; BS/Chem/Juniata College. *Born:* March 22, 1928. Chemist and engr with Prismo Safety Corp, Sinclair Refining Co, US Army Chem Corps, ARCO and American Viscose Corp. R&D Project Manager with Thiokol Chem Corp. Mbr Maryland Governor's Science Advisory Council. With NST/ Engrs Inc since 1973. Responsible for designs, marketing & consulting in air and water pollution control devices; sulfonation and sulfuric acid indus manufacturing plants, fuel conversion, cryogenics, new technology investigations, and preparation of process and equipment operating manuals. Prepared PCBs waste treatment technology evaluation for EPA preparatory to the setting of indus emissions standards. Enjoy swimming, hiking, genealogy studies. Author of: *Toxic Organic Chemicals, Destruction and waste Treatment* (Noyes Data Corp, 1978) Reg PE, PA & DE. Cert Chem & ChE with natl Cert Bd (Amer Inst of Chemists) McGraw-Hill Seminar Leader. Air Force Office of Scientific Research grantee. *Society Aff:* ACS, AIChE.

Hackmann, Robert E
Business: 1901 Gratiot St, St Louis, MO 63166
Position: Gen Mgr-Customer Relations. *Employer:* Union Elec Co. *Education:* Profl Degree/EE/Univ of MO; BS/EE/Univ of MO. *Born:* 8/27/24. Presently Gen Mgr Customer Relations at Union Elec. Prior to that held positions of Mgr, Commercial & Ind Sales & Dist Mgr. In these capacities active in various civic organizations, including Pres United Way, Chamber of Commerce, Lions & Jr Achievement. Also proflly as Pres Sales & Marketing Exec. Currently Pres St Louis Rotary & VP Meml Hospital Bd of Dirs.

Hackney, John W
Home: 176-C Rossmoor Dr, Jamesburg, NJ 08831
Position: Consultant *Employer:* Self *Education:* CE/Civil/Carnegie Mellon Univ; MS/CE/Carnegie Mellon Univ; BS/CE/Carnegie Mellon Univ *Born:* 3/24/12 Hydraulic Engr and Construction Superintendent ALCOA, 1935 to 1946. Mgr of Cost and Construction Engrg, Diamond Shamrock, 1946 to 1959. VP, Pan-American Assocs 1959-61. Operations Research Mgr, M. W. Kellogg, 1961-65. Mgr of Cost Engrg, MOBIL, 1965-77. Principal, Consultant for Capital Projects, 1977 to present. Cost Engrg consultant to Dupont, MOBIL, STATOIL, Aspen Tech Inc, ARAMCO, etc. Author of numerous papers and articles. Book, "Control and Management of Capital Projects," published by John Wiley & Sons. Past Pres, Fellow, Life Member, and recipient of "Award of Merit–, Am. Assoc of Cost Engr. Fellow & Life Member, American Soc. of Civil Engrg. *Society Aff:* AACE, ASCE, AIChE, PMI, ISPA.

Haddad, Abraham H
Business: Sch of Elec Engrg, Atlanta, GA 30332-0250
Position: Prof *Employer:* GA Inst of Tech *Education:* PhD/EE/Princeton Univ; MA/EE/Princeton Univ; MSc/EE/Technion-Israel Inst of Tech; BSc/EE/Technion. *Born:* 1/16/38. A H Haddad joined the Univ of IL at Urbana in 1966, where he was Prof of Elec Engg and Res Prof at the Coordinated Sci Lab. During 1969-1979 he was an advisor ot the US Army Missile Res and Dev Command at Redstone Arsenal, AL. In 1979 he was a sr staff consultant with the Dynamics Res Corp in Wilmington, MA. From 1979 to 1983 he served as Prog dir for Systems Theory and Operations Res at the Natl Sci Fdn, (on leave from the Univ of IL during 1980 and 1981). He is now Prof of Elec Engrg and Assoc. Dir. of Computer Integrated Manufacturing Systems Program at the GA Inst of Tech in Atlanta. He served as Assoc Editor of the IEEE Transactions on Automatic Control, was Secy-Treas of the IEEE Control Systems Soc during 1979-1981, and is now the Ed of the IEEE Transactions on Automatic Control. *Society Aff:* IEEE, ORSA, AAAS.

Haddad, Anwar
Home: 3323-102nd Ave. N.E, Bellevue, WA 98004
Position: Retired *Education:* BS/CE//Univ of WA-Seattle *Born:* 4/11/09 Native of Seattle, WA. Retired after 34 years of Federal service, last 30 with US Corps of Engrs. Chief of Design Branch, Far East District, Japan and Korea, 4 years. Balance of time with Seattle District. Was Chief of Civil & Structures Section at time of retirement. Worked on 7 mutiple-purpose dam projects in Pacific northwest and two on Ohio River. Remained active until 1979. Last 2 1/4 years were in Buenos Aires

Haddad, Anwar (Continued)
with Charles T. Main Co., on design of Salto Grande Dam & Locks project. *Society Aff:* USCOLD, SAME, ΤΒΠ, ΣΞ

Haddad, George I
Business: Dept EE/Computer Engrg, Ann Arbor, MI 48109
Position: Professor and Dir Center for High-Frequency Microelectronics *Employer:* University of Michigan. *Education:* PhD/EE/Univ of MI; MSE/EE/Univ of MI; BSE/EE/Univ of MI. *Born:* April 1935 Aindara, Lebanon. Served successively as Instructor, Assistant Professor, Associate Professor & Professor in EE Dept at UM from 1960- present. Served as Director of Electron Physics Laboratory from 1968-75. Served as chairman of EECS Department 1975-1987. Currently serving as Dir, Center for High Frequency Microelectronics and for Solid State Electronics Laboratory. Has engaged in research on masers, parametric amplifiers, detectors, electron-beam devices, & presently on microwave solid-state devices. Mbr Eta Kappa Nu, Sigma Xi, Phi Kappa Phi, American Physical Society, American Society for Engineering Education. Fellow IEEE. Curtis W McGraw Research Award of ASEE 1970. Excellence in Research Award, College of Engineering, 1985. Distinguished Faculty Achievement Award, Univ. of Mich. 1986. *Society Aff:* ASEE, IEEE, HKN, ΣΞ, ΦΚΦ.

Haddad, Jerrier A
Home: 162 Macy Rd, Briarcliff Manor, NY 10510
Position: Ret. VP/Tech Personnel Dev; Consultant (Self-emp) *Education:* BEE/Elec Engg/Cornell Univ; ScD/-/Union College; ScD/-/Clarkson College. *Born:* 7/17/22. in NYC. Teaching asst at Cornell U before joining IBM in 1945 as lab technician. Held various engrg and mgmt positions. Assisted in design of company's first large scale electronic calculator and resp for company's first large scale electronic computer. Elected IBM VP, Engg, Programming & Technology in 1967; 1970 IBM VP & Dir Poughkeepsie Lab; 1972 IBM VP & Sys Products Div VP Dev; 1974 IBM VP, EP&T-Corp HQ; April 1977 assumed position IBM VP, Technical Personnel Dev. Ret 8/30/81. Holds 18 pats for inventions in computer & electronic field. Member Emeritus Cornell Coll of Enggt Advisory Council; Mbr Clarkson College Bd; Engg Fnd Bd; Mbr NAE, serving on NAE Education Advisory Board; Mbr Bd on Army Science & Technology; Chrmn of Ctte on Educ and Utilization of the Engr; Fellow of the IEEE; Member IEEE Education Activities Board, Member IEEE Awards Board, Chairman IEEE Honorary Membership Committee; Mbr Sigma Xi, Tau Beta Pi & Eta Kappa Nu; Bd/Mbr ADT. *Society Aff:* AAAS, ACM, IEEE, NAE

Haddad, Richard A
Home: Fawn Hill Rd, Tuxedo, NY 10987
Position: Assoc Dean *Employer:* Polytechnic Inst of NY *Education:* PhD/EE//Polytechnic Inst of Brooklyn; M/EE//Polytechnic Inst of Brooklyn; B/EE// Polytechnic Inst of Brooklyn *Born:* 11/26/34 Brooklyn, NY 1934; NYS Regents Scholarship 1953-6; G.E. Fellowship 1956-7; Jr. Engr, Columbia Univ 1955; Instructor in EE 1957-61, Polytechnic Inst of Brooklyn, Member of Technical Staff, Bell Telephone Labs 1962; Asst Prof 1961, Assoc Prof 1965, Assoc Dean of Westchester Center 1981-82. Dir, Westchester Grad Center, Polytechnic Inst of NY, 1982-. Head, Engrg Division, Institute National d'Electricite et d'Electronique, Boumerdes, Algeria, 1978-80. Consultant & lecturer to industry and government, 1963-81 in: systems analysis, simulation, and control, digital signal processing, radar signal smoothing and prediction. Elected to membership in Eta Kappa Nu, Tau Beta Pi, Sigma Xi, NY Academy of Science. *Society Aff:* IEEE

Haddadin, Munther J
Home: P O Box 1961, Amman Jordan
Position: Vice President. *Employer:* Jordan Valley Commission. *Born:* 1940, Native of Amman, Jordan. B.Sc. (1963) Alexandria University, Egypt, M.Sc. (1966) and Ph.D. (1969) University of Washington, Seattle. Field Engineer, (1963-65) in Suadi Arabia for highways, With Portland Cement Associaion, Skokie, Ill (1969-71) as research design engr. With the Royal Scientific Society (1971-73) as Director of Computer Systems. Vice President of the Jordan Valley Commission since 1973 engagined with planning and implementation of projects for intergrated economic and social development. Co- recepient of ASCE 1972 Raymond C. Reese Award.

Haddick, John S
Business: Box 1145, Dayton, OH 45401
Position: Pres *Employer:* Durion Co. Inc. *Education:* BS/ChE/IA State, Ames; AMP/-/Harvard *Born:* March 1929. Operations Engr and Ass't Dept Foreman PPG Chem Div Barberton Ohio in Chlorine & Perchlor plants 1950-1953. Sales Engr for Durion 1954-59. Product Mgr Durion 1960-67. Director Internatl Operations 1968-72. VPres in charge of Operations for 4 of Duririons's Divisions 1973-79. Group VP for US Operations. Presently Pres & C.O.O. Fellow AIChE. *Society Aff:* AIChE

Haden, Clovis R
Business: College of Engg & Appl Sci, Tempe, AZ 85281
Position: Dean. *Employer:* AZ State Univ. *Education:* PhD/EE/Univ of TX at Austin; MS/EE/Caltech; BS/EE/Univ of TX at Arlington. *Born:* 1940. Hometown Blooming Grove, Texas. Married with 3 children. Summers with Texas Instruments and Mesa Instruments. Asst Prof Univ of Okla 1965-68. Assoc Prof and Prof Texas A&M 1968-72. Director Inst for Solid St Elec 1969-72. Director & Prof School of EE Univ of Okla 1972-78. Dean, Coll Engr & Appl Sci, AZ State Univ, 1978-present. President Amer Mbr Okla City IEEE Exec Ctte, Region V Conf Ctte, and national IEEE committees. OKC IEEE Outstanding Contribution Award, 1976. Exec Ed, *Elec Power Sys Res*, 1977. Fellow, 1985, IEEE. *Society Aff:* NSPE, ASEE, IEEE.

Hadley, Henry T
Home: 4418 137th Ave, SE, Bellevue, WA 98006
Position: Consult Engr. *Employer:* Self *Education:* BE/ME/Univ of S CA. *Born:* 7/21/22. Native of WA state. Served with the US Marines in WWII & Korean wars. Met Engr for US Steel Corp 1946. Assumed Mtls, Met & Welding Engg repsonsibilities for ind & military products of PACCAR in 1960's until 1985. Chrmn ASM PUGET Sound 1962 - Mbr ASM Natl Committee 1970 Contributing author of ASM Metals Handbooks. Lectr, Northwest Regional ASM chapters. Life Member ASM Wife-Bette, Children - Victoria & Mark - all college grad. Enjoy golf & opera. Reg PE. *Society Aff:* ASM, ASTM, AWS, AUSA, ADPA, AIME.

Haefeli, Robert J
Business: N.H. Bettigole, P.A, Consulting Engineers, 601 Bergen Mall, Paramus, NJ 07652
Position: Chief Hydraulic Engineer *Employer:* N.H. Bettigole, P.A. *Education:* MS/Environ. Engg/Rutgers; BE/CE/Univ of MI *Born:* 9/7/26. 1948-52 Bureau Reclamation, Denver and Beirut, Lebanon. 1957-59 Ebasco Services; VPres, Intl 1960-66, Hydrotechnic Corp; Associate, Hazen and Swayer, 1966-69; consulting engr 1969-79; Pres, Frank & Haefeli Associates, P.A. 1972- 77; Pres, Haefeli Engrg, P.A., 1977-79; Water Resources Engr/Planner & Project Manager, Havens and Emerson, Inc, 1979-85; Chief Hydraulic Engineer, N.H. Bettigole, P.A., 1985-present; B.E. Civil Engrg, Univ Michigan; ASTP- Civil, Virginia Polytechnic Institute; US Corps of Engrs, Engrg School. Grad Studies Univ of CO; M.S. Environmental Engrg, Rutgers. PEPP/CECNJ NSPE Delegate 1975-77; Pres, 1970-71, Raritan Valley (NJ) Society of PE and Land Surveyors; Pres, 1972-74 Central NJ Pollution Control Association; Pres, 1972- 73, New Jersey Section Amer Water Resources Association; Pres 1981-82 Member Bd of Ethical Rev; 1984-present NJ Soc of rofessional Engineers. Diplomate, Amer Acad of Environmental Engineers, ASCE (Fellow). Mbr: NSPE, AWWA, AIDIS, WPCF; AGU, AWRA, US Comm on Irrigation Drainage & Flood Control, Chmn, NJ Clear Air Council 1978-80, Mbr 1968-80; Member New Jersey Clean Water Council, 1987-; Inter-American Soc of Sanitary Engrgs; NJ Alliance for Action; Licensed PE, NJ, NY, MA, WI, PA; Free and Accepted Masons, Lebanon Lodge No. 10 Beirut, Lebanon. *Society Aff:* NSPE, AWRA, AGU, AIDIS, AAEE, WPCF, AWWA, ICILD

Haensel, Vladimir
Business: Ten UOP Plaza, Des Plaines, IL 60016
Position: Vice President-Science/Technology. *Employer:* UOP Inc. and Univ of Mass, Amherst MA *Born:* September 1, 1914 in Frieburg, Germany. Came to US in 1930 and was naturalized in 1936. BS Northwestern University, 1935; PhD 1941; DSC (Hon) 1957; MS Massachusetts Inst of Technology, 1937; DSc. (Hon) Univ of Wisconsin 1979. Joined UOP Inc Des Plaines, Illinois in 1937. After serving for eight years as UOP Vice President and Director of Research, assumed position as VP Science & Technology in 1972. Retired Sept 1, 1979 and is presently a Consultant to UOP and Prof of Chemical Engineering, Univ. of Massachusetts; President of The Catalysis Society. Many awards including Perkin Medal for outstanding work in applied chemistry in 1967 and National Medal of Science, 1973. *Society Aff:* NAS, NAE, ACS, CS

Haentjens, Walter D
Business: 225 N Cedar St, Hazleton, PA 18201
Position: Pres. *Employer:* Barrett, Haentjens & Co. *Education:* MS/ME/Case; BME// Cornell. *Born:* 7/25/21. Designer of specialized centrifugal pumps for the chem and mining industry. Holds 15 patents relating to pump design and construction. Private pilot. Capt, Ordnance Dept, AUS, WWII. *Society Aff:* ASME, AIME.

Haertling, Gene H
Home: 3624 Colorado Ct NE, Alburquerque, NM 87110
Position: Officer of the Technical Staff. *Employer:* Motorola Inc. *Education:* PhD/Ceramic Engg/Univ of IL; MS/Ceramic Engg/Univ of IL; BS/Ceramic Engg/ Univ of MO. *Born:* 3/15/32. Native of Ste. Genevieve, MO. Dev engr with Ipsen Ceramics from 1954-58, specializing in thermal shock resistant alumina ceramics. Joined the res staff of Sandia Labs in 1961 emphasizing processing & fabrication of ferroelectric ceramics via chem co-precipitation & hot pressing techniques. Developed the first transparent electrooptic ceramic (PLZT) in 1969. Formed Optoceram, Inc in 1973 & served as pres. Optoceram subsequently acquired by Motorola, Inc; presently mgr of ceramic R&D group. Products include piezoelectri loud speakers and electrooptic shutters. Fellow, Am Ceramic Soc 1970; Fellow, IEEE 1979. NICE Profl Achievement in Ceramic Engg (PACE) Award in 1972. *Society Aff:* ACS, IEEE, NICE, SID.

Haestad, Roald J
Business: 37 Brookside Rd, Waterbury, CT 06708
Position: Pres. *Employer:* Roald Haestad, Inc. *Education:* BCE/Civ Engg/College of the City of NY. *Born:* 9/12/22. Established Roald Haestad, Inc, Consulting Engrs, in 1971. Formerly a Prin Engr with Malcolm Pirnie Engrs; and began engg career with Clinton Bogert Assoc in 1950. Maj engg projs in the field of water and sewage works, particularly design and construction supervision of new water supply dams, and inspection and remedial work on old dams. CT State Pres of NSPE in 1977, and Naugatuck Valley Chap Pres of CSPE in 1973. 1981 Chrmn of Review Comm to select the CT, American Water Landmark. Married to Jean Munson in 1952, and have three children, Randi, Cynthia and John. *Society Aff:* ASCE, AAEE, AWWA, NSPE, NEWWA, WPCF, ACI, ASTM, SAME, AISC, APWA, ASP, CSCE, ASDSO, NEADS, ASFPM, ISSMFE

Hagemaier, Donald J
Home: 9582 Scotstoun Dr, Huntington Beach, CA 92646
Position: Section Manager *Employer:* Douglas Aircraft Co. *Education:* AA/Genl Ed/ Pierce College. *Born:* February 1928 in Flushing, NY. AA in Gen Ed from Pierce College, 1968. native of Los Angeles, snce 1949. Instructor of Nondestructive Testing courses at Long Beach City & State Colleges. Nondestrcutive Test Specialist with Douglas Aircraft Company 1955-61; Rocketdyne Div of R.I. 1961-68. Presently, Sect. Mngr. NDE Section of M&PE at Douglas Aircraft Company. Mbr of ASNT and SAE Committee K. Recipient of 1973 ASNT Acheivement Award. Mbr of 1976 National Materials Advisory Board of NDT of Aerospace Components. Registered Professional Quality Engr State of CA. Qualified ASNT Level III in ET, PT, EC, RT, UT, & NRT. Fellow of ASNT. Presented Honor Lecture to Italian Soc For NDT in 1983. Presented ASNT Lester Honor Lecture in 1984. *Society Aff:* ASNT, SAE

Hagen, Vernon K
Business: 20 Mass Ave, Washington, DC 20314
Position: Chief/Hydraulics & Hydrology *Employer:* Office, Chief of Engrs. *Education:* M/CE//Catholic Univ; BS/CE//MT State Univ *Born:* 9/3/26 Employed by US Bureau of Reclamation (1949-53) in soils investigations and hydrology. Joined Corps of Engrs in 1953. Served with Fort Peck District (1953- 54) in hydrology. Served with Garrison District (1954-58) in hydrology, planning and hydraulic design. Assigned to Office. Chief of Engrs 1954 in hydraulics and hydrology. Assumed charge of hydrology in 1972 and charge of hydraulics and hydrology in 1979. Served on national committees in ASCE and USCOLD. Participated as member and chairman of several interagency committees and work groups. Author of many technical papers and lectures at numerous training courses. Responsible for many aspects of dam safety. *Society Aff:* ASCE, AGU, USCOLD

Hagerty, D Joseph
Business: Civ Engg Dept, Louisville, KY 40292
Position: Prof. *Employer:* Univ Louisville. *Education:* PhD/CE/Univ Louisville; MS/ CE/Univ Louisville; MEngg/CE/Univ IL; BCE/CE/Univ Louisville. *Born:* 12/5/42. Native of Louisville, KY. Dir of Geotechnical Engg, Vollmer Assoc, 1969-70. Prof of Civ Engg, Univ Louisville, 1979-. Geotechnical consultant to fed and state agencies, and industry. Waste mgt consultant to city, county and regional govt. Co-author of seven books and more than 40 papers. 1971 KSPE Young Engr of the Yr; 1973 ASCE Friedman Award. Eucharistic Minister, Archdiocese of Louisville. *Society Aff:* ASCE, ASEE.

Hagn, George H
Home: 4208 Sleepy Hollow Rd, Annandale, VA 22003
Position: Asst Director, Info Sci and Tech Center *Employer:* SRI International. *Education:* MS/EE/Stanford Univ; BS/EE/Stanford Univ. *Born:* 9/15/35. Born in Houston, TX (1935); educated Stanford Univ; joined Stanford Res Inst (renamed SRI Internatl) 1959, currently Asst Director in SRI's Info Scis and Tech Center. Work has included communications systems, radio propagation and noise, electromagenetic compatibility (EMI) and spectrum mgmt. Author of over 50 technical and scientific papers and over 150 reports. Reg PE (CA); Assoc Editor, *Radio* Science (1975-78); Chrmn, Internal Union of Radio Science (URSI) Commission E on "Electromagnetic Noise and Interference" (1978-81), and V Chrmn, US Natl Committee of URSI (1978-81). Elected Fellow, IEEE in 1979 for contributions to spectrum mgmt and EMC. Enjoys family, fishing, hunting,, and reading. *Society Aff:* URSI, IEEE, AGU, AAAS, AFCEA

Hahn, G LeRoy
Business: PO Box 166, Clay Ctr, NE 68933
Position: Agri Engr. *Employer:* US Dept of Agri. *Education:* PhD/Atmos Sci/Univ of MO-Columbia; MS/Agri Engg/Univ of CA-Davis; BS/Agri Engg/Univ of MO-Columbia. *Born:* 11/12/34. in Muncie, KS; educated in KS, MO, & CA. Employed since 1957 as res agricultural engr by Agri Res Service, US Dept of Agri in MO, CA, & NE. Also served in additional capacities of proj leader, res leader & technical advisor for res progs directed toward assessment of environmental influences on livestock stress and performance & resulting housing needs, & have written many technical and popular articles for journals & books. ASAE-MBMA Award, 1976; AMS Bioclimatology Award, 1976. Chrmn, Mid-Central Region ASAE, 1970-71; Regional Dir, ASAE, 1978-80; Chrmn, Structures and Environment Division, ASAE, 1982-83; Chmn., Structures and Environment Div ASAE Ctte on Technical Issues & Awareness, 1983-84; Co-Chmn, Animal Div, Intl Soc Biometeorology, 1983-84. *Society Aff:* ASAE, ASAS, AMS.

Hahn, Jack
Home: 1 Benvenuto Place, Apt 726, Toronto, Ontario, Canada M4V 2L1
Position: President *Employer:* Jack Hahn Assocs *Born:* June 1920. B Engrg McGill Univ. 1947. Partner, Surveyor, Nenninger & Chenevert, Montreal, 1959-66. VP and Dir SNC Inc 1966-80. Pres, Gen Engrg Co, Toronto, 1970-75. Chrmn, SNC Conslt 1975-75. Chrmn SNC-GECO 1978-80. Fellow, Engrg Inst of Canada, 1973. Chrmn Montreal Branch, 1964. Chrmn Toronto Branch 1980, Natl Pres 1981. Founding Mbr, Canadian Soc of Mech Engrs, Mbr, Assn of Consltg Engrs, Canada, Canadian Inst of Mining and Metallurgy. Mbr, Montefiore Club, Montreal & Natl Club, Toronto. Past Chrmn Engrg and Architectural Divs of Federated Appeal, Montreal United Appeal Toronto 1980. Hobbies: Sailing and Gardening.

Hahn, James H
Business: 10500 Kahlmeyer Dr, St Louis, MO 63132
Position: VP-Res. *Employer:* Interface Tech Inc. *Education:* PhD/EE/Univ of MO; MS/EE/Univ of Pittsburgh; BS/EE/MO Sch of Mines. *Born:* 6/28/36. Born E Prairie, MO, 1936. Native of Chaffee, MO. Engr, Westinghouse Elec Co 1959-1966. Engrg specialist, Monsanto Co, St Louis, 1967-71. Co-founder of Interface Tech, Inc, 1972. Currently vp res. Publications: two technical articles. Patents: one US, one Canadian. Reg PE, MO. Lic radio amateur, 20 yrs. Mbr: Tau Beta Pi, Eta Kappa Nu. *Society Aff:* TBΠ, HKN.

Hahn, Ottfried J
Business: Mech. Engr. Dept, U of Ky, Lexington, KY 40506-0046
Position: Prof *Employer:* Univ of KY. *Education:* PhD/Mech Engr/Princeton Univ; MA/Mech Engr/Princeton Univ; MS/Nuclear Engr/WV Univ; BS/Engr Phys/Univ of Alberta. *Born:* 6/21/35. Experience includes assignments with Bechtel Corp in the pipeline div and Can Gen Elec as a nuclear reactor designer and physicist. I joined the Mech Engr Dept, Univ of KY in 1967 as asst prof responsible for the nuclear engr prog and was promoted in 73 to assoc prof with increased responsibility as assoc dir for coal utilization in the Inst for Mining and Minerals Res. After an assignment with Bergbau Forschung in Germany I returned to Univ of KY and was promoted to full prof in 1984. Res interests include waste disposal, coal gasification, coal/liquefaction, cost studies and product liability. I am a native of Germany and am married to the former Joyce C King of Morgantown West VA. *Society Aff:* ASME, ANS.

Hahn, Ralph C
Business: 1320 South State Street, Springfield, IL 62704
Position: President *Employer:* Ralph Hahn & Assoc, Conslt & Design Engrs *Education:* MSCE/Structural/Univ of IL; BSCE/Civil/Univ of IL; Assoc/-/Springfield College in Illinois. *Born:* 11/9/27. Bridge Design Engr II Div Hwys 1952-61. Founded Ralph Hahn & Associates, Inc 1961, opened West Palm Beach, FL office, 1974. Chrmn Sangamon County Percy for Gov Ctte 1964; IL Youth for Nixon Ctte 1968. Elected Tr U of IL 1966, 1972, 1978 and 1984. Mbr State Universities. Cvl Service Merit Bd 1969-71 and IL Advisory Council to Small Business Admin 1969-71. Chrmn IL Architect-Engrs Council 1966. Mbr NSPE (Chapter Pres 1963-64), ACEC, ASCE. Pub articles on engrg educ. Mbr Bd of Dir, Springfield Symphony Orchestra; Member, Frank Lloyd Wright Historic House Foundation Board; Member, U of I Foundation & President's Council; Life Member, U of I Alumni Association; Member St Paul's Lodge *500 AF & AM Consistory & Shrine; Member Sangamo Club of Springfield; Member Tavern Club of Chicago. Honors & Awards: Alumni of Distinction from Springfield College in Illinois, 1964. Loyalty Award, U of I Alumni Association, 1975. Society Aff:* NSPE, ASCE, ACEC, ISPE, CECI, IAHE.

Hahn, Richard D
Home: 140 Fairview Ave, Frederick, MD 21701
Position: Assoc Dir, Office of Military Applications *Employer:* Dept of Energy. *Education:* MS/Nucl Engg/UCLA; BS/Met/OR State Univ; BS/Math/College of ID. *Born:* 10/9/33. In present position since July 1984. Responsible for: managing the Nation's nuclear weapons research, development & underground testing program; manages the Department's SDI research; & the Nation's Inertial Confinement Fusion Program. Previously served as Deputy Director, Office of Nuclear Materials Production (1979-84); Production Operations Div Dir for the Office (1976-1979); as Branch Chief for Clinch River Breeder Reactor Proj for plant components (1973-1976); AEC reactor engr (1971-1973); Mtls & Processes Engg Mgr for Litton Industries, Woodland Hills, CA (1967-1971); R&D mgr for San Fernando Labs, Pacoima, CA (1966-1967).

Hahn, Robert S
Business: Hahn Engg, Inc, 160 Southbridge Street, Box 311, Auburn, MA 01501
Position: Consulting Engineer *Employer:* Self Employed *Education:* DSc/Physics/Univ of Cincinnati; MSc/Physics/Univ of Cincinnati; ME/Mech Eng/ Univ of Cincinnati. *Born:* Nov 1 1916 NYC. Adj Prof Worcester Polytech Inst, U of Cinn. Reg Prof Engr MA. Chmn Prod Engrg Div ASME, Abrasive Process Ctte SME. Mbr: ASME Policy Bd-Res, NMBTA Res & Dev Ctte, CIRP (Coll Internatl pour l'Etude Scientifique de Production Mecanique), AAAS, SME, Tau Beta Pi, Pi Tau Sigma, Sigma Xi. Recipient: Master Design Award, ASME Blackall Machine Tool Award, SME Res Medal, Worcester Engrg Soc Sci Achievement Award. 30 pats, 45 tech papers. *Society Aff:* ASME, SME, NAE.

Haile, James M
Business: Chemical Engrg Dept, Clemson University, Clemson, SC 29634
Position: Prof *Employer:* Clemson University *Education:* PhD/ChE/Univ of FL; ME/ ChE/Univ of FL; BS/ChE/Vanderbilt Univ *Born:* 12/7/46 Native Nashville, TN. Served U.S. Navy 1968-72. Visiting Res Assoc, Physics Dept, Univ of Guelph, Ontario, 6/75-10/75. Visiting Res Assoc, Sci Res Lab, US Military Academy, West Point, 1/76-7/76. 7/77 Res Conslt, School of Chem Engrg, Cornell Univ, 7/78-8/78. Visiting Res Scientist, Neutron and Solid State Physics Branch, Chalk River Natl Labs, Chalk River, Ontario, 6/82-9/82. Visiting Assoc Prof of Chem Engrg, Cornell Univ, 9/82-5/83. Visiting Res Scientist, Inst. Thermo. Fluiddynamik, Ruhr Univ, Bochum, West Germany, 6/1986. Asst Prof of Chem Engr, Clemson Univ, 8/76-7/80. Assoc Prof of Chem Engrg, Clemson Univ, 8/80-7/84. Prof of Chem Engrg, Clemson Univ since 8/84. Clemson Univ McQueen Quattlebaum Faculty Ach Award 1981. Natl Science Foundation Presidential Young Investigator Award 1984. Co-editor (with Dr. G. A. Mansoori) of *Molecular Based Study of Fluids*, vol. 204 in ACS Advances in Chemistry Series, ACS, Washington, DC, 1983. Prof publ in Journal of Chemical Physics, Molecular Physics, Chemical Physics, Chemical Physics Letters, etc.plus *Society Aff:* ACS, APS, AIChE, HSS

Haimes, Yacov Y
Business: Dept. of Systems Engg, University of Virginia, Thornton Hall, Charlottesville, VA 22901
Position: Lawrence R. Quarles prof of Sys Engg & Dir of Center for Risk Mgmt of Engg Sys. *Employer:* Univ of Virginia *Education:* PE/Reg Prof Engr/Commonwealth of VA; PhD/Large Scale Sys/UCLA; MS/Sys Engr/UCLA; BS/Math, Phys, Chem/ Hebrew Univ, Jerusalem; PE/Reg Prof Engr/State of OH *Born:* 6/8/36 Dr. Haimes is Lawrence R. Quarles Prof Engg and Appl Sci and Dir of the Center for Risk Mgmt of Engg Sys at the Univ of Virginia. He was on the faculty of Case Western Reserve Univ in Cleveland Ohio since 1970 serving as Chmn of the Sys Engrg. Dept, Dir of the Center for Large Scale sys and Policy Analysis and Assoc Dean for Interdisciplinary Activities. His res, teaching, and consulting activities are in the area of risk management and of modeling and optimization of large-scale systems with emphasis on the hierarchical-multiobjective approach and its applications to water resources sys. He is a Fellow of the ASCE, AAAS, IWRA, and the IEEE, chmn of several natl and intl cttes. He is also the recipient of other honors and awards. He is former Pres of the Univ Council on Water Resources (1984-85), and the author and editor of thirteen books. *Society Aff:* AGU, AWRA, IWRA, ORSA, SRA, ASEE.

Haines, Roger W
Home: 7300 N Mingo, Maderira, OH 45243
Position: Project Manager. *Employer:* Ziel-Blossom & Assocs. *Education:* BS/Mech Engg/IA State Univ. *Born:* August 1916. 28 yrs experience in Cons Engrg during which I have designed mechanical systems including HVAC, Plumbing and Fire Protection for over 800 projects. Author of a widely used techincal reference book 'Control Systems for HVAC' and numerous technical papers and articles. Fellow, ASHRAE, and also recipient of the Society's Distinguished Service Award, have served on several committees, Hobbies: music, tennis, chess. Reg Prof Engr in 7 states. *Society Aff:* ASHRAE.

Hair, Charles W, Jr
Home: 2964 Murphy Dr, Baton Rouge, LA 70809
Position: Civil Engr *Employer:* Self *Education:* BS/CE/LA State Univ *Born:* 11/6/19 Commanded Engr Ponton Battalion in World War II - River Crossings in Europe. Retired USAR Engr Colonel. Engrg work on bldgs, chem plant and oil field construction. Employed by Pyburn & Odom, Consulting Engrs, 1950-60 - engrg pipe lines, river crossings and related. Baton Rouge Dept of Public Works, 1960-80, Chief Engr, sanitary sewers, streets, drainage and all municipal works. Presently consultant on Environmental and municipal engrg. Pres LA Section, ASCE, LA Water Pollution Control Association, LA Conference on Water Supply and Sewerage and Baton Rouge Chapter, LA Engrg Society. WPCF Bedell Award. US Army Legion of Merit. Kiwanian, Boy Scouts, Mbr LA Naval War Mem Comm, Baton Rouge Little Theater. Fellow, ASCE. *Society Aff:* ASCE, NSPE, SAME, WPCF, ACSM

Haisler, Walter E
Business: Aerospace Engrg Dept, College Station, TX 77843
Position: Prof and Head *Employer:* Texas A&M Univ *Education:* PhD/Aerospace Eng/TX A&M Univ; MS/Aerospace Eng/TX A&M Univ; BS/Aerospace Eng/TX A&M *Born:* 6/3/44 Native of Temple, TX. Joined Texas A&M Univ in 1970. Has taught courses in aerospace structures, numerical methods, finite element techniques, plasticity, and nonlinear structural mechanics. Research and consulting in the areas of nonlinear mechanics, inelastic structural response, constitutive modeling, and finite element methods. Developed several finite element computer programs (including SNASOR, DYNAPLAS, AGGIE) for predicting nonlinear, finite deformation and inelastic structural response. Published over seventy-five papers and reports on research conducted during past fifteen years. Hobbies include sailing, camping, woodworking and photography. *Society Aff:* AIAA, ASCE, ASEE, ASME

Hakimi, S Louis
Business: Dept. of Electrical and Computer Engineering, University of California, Davis, CA 95616
Position: Professor, EE and computer Science Dept *Employer:* Univ of California. *Education:* PhD/EE/Univ of IL, Urbana; MS/EE/Univ of IL, Urbana; BS/EE/Univ of IL, Urbana. *Born:* Dec 16, 1932. Asst Prof of Elec Engrg, Univ of Ill Feb 1959-Aug 1961. Came to Northwestern Univ as an Assoc Prof in 1961; Served on the Chmn Elec Engrg Dept, Northwestern Univ 1972-77. Joined the University of California at Davis in September 1986 as Professor and Chairman of electrical and computer engineering department. Res interests lie in applications of graph theory & combinatorics to curcuits & network theory, & coding theory. Mbr Adm Comm IEEE Circuits & Systems Soc 1966-69. Chmn, Midwestern Elec Engrg Dept Heads 1973-75. Accos Editor, IEEE Trans on Circuits & Systems 1975-77; Presently Assoc Editor, Journal of Networks. Fellow IEEE 1972; Mbr Sigma Xi, Tau Beta Pi, Eta Kappa Nu, Phi Kappa Phi, & AAUP. *Society Aff:* IEEE, SIAM, AAUP, ΣΞ.

Halacsy, Andrew A
Home: 1135 The Strand, Reno, NV 89503
Position: Prof Emeritus. *Employer:* Univ of NV. *Education:* PhD/Elec Engg/Engg Univ; DiplEng/Mech Engg/Engg Univ. *Born:* 5/15/07. 1929-48 asst and assoc prof, Engg Univ, Hungary elec engr (heavy construction) German AEG and Siemens Co. 1948-52 chief engr South East European Industrial Dev Co, Switchgear and Cowans Co, (England) resident engr (elec power generation and distribution) Colonial Dev Corp British West Indies. 1952-6 Designing transformers ITE Circuit Breaker Co (Canada, USA) chief engr Jeffries Transformer Co (Los Angeles), transformer design Delta Star and Radiation at Stanford, mgr corporate R & D, Fed Pacific Elec Co, consultant and pres Heat Magnetic Engg Inc, (CA and NV), prof Univ of NV (organized conferences on magnetic fields), consultant to CEPEL, Brasil. *Society Aff:* IEEE, NSPE, MMEV, IEE.

Halbouty, Michel T
Business: 5100 Westheimer Road, Houston, TX 77056
Position: Chairman of the Board and CEO *Employer:* Michel T. Halbouty Energy Co. *Education:* PhD (HC)/Engg/MT College of Mineral Sci & Technology; MS/Geol & Pet Eng/TX A&M Univ; BS/Geol & Pet Eng/TX A&M Univ. *Born:* June 21, 1909, Beaumont, Texas. Geologist & Petroleum Engr-Yount-Lee Oil Company, 1931-35. Chief Geologist & Petroleum Engr, Vice President & General Manager for Glenn H. McCarthy, Houston, 1935-37. Consulting Geologist & Petroleum Engr since 1937. Served in US Army 1942-45, detached with rank of Lt. Colonel. Author of over 280 scientific articles on geology and petroleum engrg; book entitled 'Petrographic Characteristics of Gulf Coast Oil Sands' 1937; Coauthor of 'Spindletop' 1952. Author of 'Salt Domes, Gulf Region, United States & Mexico' 1967, 2nd ed 1980. Co-author of 'The Last Boom' 1972. Chmn of Board & CEO Michel T. Halbouty Energy Co., Houston, TX. Distinguished Lecturer of SPE of AIME 1964-65 & American Assn of Petroleum Geologists (AAPG) 1965-66. Pres of AAPG 1966-67; First Distinguished Lecturer Emeritus of SPE 1983; AAES Hoover Medal 1982; Tex. Acad. of Sciences Distinguished Texas Scientist of the year 1983. *Society Aff:* NAE, AAPG, GSA, SPE, SEG, SEPM

Haldar, Achintya
Business: School of Civil Engineering, Georgia Institute of Technology, Atlanta, GA 30332
Position: Associate Professor *Employer:* Georgia Inst. of Technology *Education:* Ph.D./Civil Engrg./University of Illinois, Urbana; M.S./Civil Engrg./University of Illinois, Urbana; B.S./Civil Engrg./Jadavpur University, Calcutta, India. *Born:* 09/13/45 Born in Calcutta, India. Came to U.S. 1972, naturalized U.S. citizen, 1980. Worked for Bechtel Power Corp., 2 years, taught for one year at Ill Inst of Tech and at Georgia Tech since 1979. Research in risk-based design in Civil Engrg. Authored about 100 technical articles. Technical cttee member of ASCE's EMD, GT, and ST divisions. Registered PE in GA, CA, and IL. Consultant on many projects. Received Presidential Young Investigator Award, National Science Foundation, 1984; ASCE Huber Research Prize, 1987; Sigma Xi Research Award, 1982; Univ Gold Medal, 1968; Outstanding Young Man of America, 1974; Georgia Tech Outstanding Teacher Award, 1987; Georgia Tech ASCE's Outstanding Professor Award, 1982, 1987: *Society Aff:* ASCE, AISC, XE, ΣΞ, ΦΚΦ.

Hale, Bobby L
Home: 5239 Cripple Creek Ct, Houston, TX 77017
Position: President. *Employer:* Ventech Engineers, Inc. *Born:* January 31, 1935. BSChE University of Houston, 1957; BS Mathematics University of Houston, 1957. Pursued graduate studies toward an MSChE at University of Houston. Native of the oil fields of Texas. Process Engr for Texas Gas Corporation 1957-61. Research & Development Engr for Crown Central Petroleum 1961-62, Process Design Engr-Houston Res Inst 1962-64. Mgr of Engrg and Construction-Merichem Company 1964-67. Formed Ventech Engrs, Inc with 2 other engrs in 1967- . Complete services of Engrg and Construction of Natural Gas Processing and Treating Plants, Sulfur Recovery Plants, and Crude Oil Fractionation Systems. Specializing in Remanufacturing and relocating existing plants. Mbr AIChE. Enjoy studies in Nutritional Therapy and orthomolecular psychiatry.

Hale, Clyde S
Home: 148 Firwood Drive, Webster Groves, MI 63119
Position: Partner. *Employer:* Hale & Harvie Consulting Engineers. *Born:* November 8, 1920 Baldwin, Illinois. BSCE Univ of Illinois, 1942. Served with Army Corps Engrs before entering service in WWII. Ltjg US Navy Air Corps 1942-45, Pilot, Navigator, Bombardier, multiengine craft, Pacific area. Army Corps of Engrs 1945-56. Structural Designer, W.J. Knight and Co. Consulting Engrs St. Louis, MO. 1946-48. Structural Designer, Bank Building & Equipment Corp 1948-50. From 1950 to present Partner in the firm of Hale and Harvie Consulting Engrs, practicing in the structural field Design, Detail & Construction Observation of reinforced concrete and structural steel buildings. Mbr ASCE, ACEC, CECMo, MSPE & NSPE. P President, CECMo. Elder, Presbyterian Church, Board of Managers, YMCA. Married to Rena Eberspacher Hale; 3 children, Freda Hlavacek, Tom & John Hale; 3 granddaughters, Anne, Jeanne, Katie Hlavacek.

Hale, Edward B
Home: 1215 Sanders Street, Blacksburg, VA 24060
Position: Professor Emeritus, Agri Engrg *Employer:* Virginia Polytechnic Inst/State Univ. (Ret.) *Education:* MS/Agri Engg/Univ of TN; BS/Agri Engg/VA Tech. *Born:* December 1922. Native of Narrows, Va. Served in U.S. Army 1943-46. District Agri Engr, Appalachian Power Co., 1946-49. Extension Agri Engr, Univ of Tennessee, 1950-60. Active with VPI & State U 1960-82. Retired 1982. Responsible for educational programs throughout Va. in the areas of irrigation, rural water supply, agricultural waste management, and soil and water conservation. Received four ASAE Blue Ribbon awards for educational progs conducted & materials developed on agri engrg subjects. Received fellow and commendation awards from SCSA. Mbr Sigma Xi, Gamma Sigma Delta, ASAE, SCSA. Enjoy golf, fishing & hunting. *Society Aff:* ASAE, SCSA, ΣΞ, ΓΣΔ

Hale, Francis J
Business: Dept of Mech & Aerospace Engg, Raleigh, NC 27695
Position: Prof. *Employer:* NC State Univ. *Education:* ScD/Aero & Astro/MIT; SM/Aero & Astro/MIT; BS/Military Engg/US Military Acad. *Born:* 10/24/22. Corps of Engrs, 1944-48; parachute engr unit; surveyor & map-maker, atomic weapons prog, Sandia Base. USAF, 1948-65; pilot & flight instr; Dir of Aeroballisics Armament Ctr; Dept Dir of Thor & Minuteman; Hd, Dept of Astro, AF Acad. With NC State Univ since 1965. Visiting Prof in Turkey, 1973-74, Vis Prof of Mechanics, W Point, 1977-78. Author of Control Systems book. Assoc fellow, AIAA. Army Outstanding Civilian Service Medal, 1978. Author of Aircraft Performance book. Co-author of Thermodynamics book. Tech Dir of Wrightsville Beach Desalination Test Facility, 1981-82. Mbr, Order of Deadaliens. *Society Aff:* ASME, ASEE, AIAA, SAE

Hale, John K
Home: 591 Valley View Dr, New Holland, PA 17557
Position: Design Dir-N Amer Div. *Employer:* Sperry New Holland. *Education:* BS/Agri Engg/VA Tech; MS/Agri Engg/VA Tech. *Born:* 5/16/29. Native of Narrows, VA. Served in Army Chem Corps 1951-53. Joined Sperry New Holland 1954-58-Professional Engg. Became Design Mgr for Hay Tools 1964. In 1966, became Internatl Engineering Liaison Mgr. Became Design Mgr for Garden Prods, Round Balers & Manure Spreaders in 1972. 1974-79 European Engg Dir with respon for design of products built in France, Belgium and England. 1979- Prod. Design Director - N. American Div. Mbr of of ASAE, NSPE. Former mbr of Town Council, New Holland PA. Enjoy civil projs, tennis, swimming and participatory sports. Former Mbr Bd of CICR. *Society Aff:* ASAE, NSPE.

Hale, Nathan C
Home: 158 Winchester St, Warrenton, VA 22186
Position: Pres *Employer:* Nathan C Hale Assoc Ltd. *Education:* BS/CE/VA Polytech Inst *Born:* 9/28/25 Native of Bluff City, VA. Surveyor for USC&GS prior to World War II. Lt. US Air Corp. 1943-47. Chf. Engr Lublin, Mc Gaughy, WA, Partner, Pickett-Hale Assoc 1953-54. Owner, NCHA Assoc, 1956-78. Pres Nathan C Hale Assoc, Ltd., 1978-Pres. Inventor, Adjustable Manhole Frame. Listed in WHO'S WHO IN ENGRG-1964, WHO'S WHO IN SOUTH-SOUTHWEST 1965-66, WHO'S WHO IN COMMERCE & INDUSTRY-14th Edition. Registered PE in VA, MD, SC, DE, & DC. Member Tau Beta Phi & Chi Epsilon. Letters of Commendation from NAVFAC, U. S. Navy in 1961 & 1965 for Excellence in design work. Married Miriam Dickerson, 2 Children, Hilary & Melanie. Member- Fauquier Country Club. Hobbies, Golf. Graduate study in structural engrg at VA Polytech Inst. *Society Aff:* NSPE, ACEC, ASCE

Haley, Ted D
Home: 451 G Boone Trail, Danville, KY 40422
Position: Consulting Engineer *Employer:* Self *Education:* BS/Mining/Univ of KY; BA/Education/Murray St Teachers College. *Born:* 4/23/21. Coal mining experience in KY, TN, OH, OK; surveyor, Engr, superintendent to 1959. Technical service in blasting in contiguous United States 1960-1966. Prof and consultant in coal mining, blasting 1966 to 1981. Outside interests are instrumental music, tennis. 1981 to 1985, Sr., V.P., CSX Minerals, Inc. Self-employed Consulting Engineer, 1985-present. *Society Aff:* AIME, NSPE, SEE, KSPE.

Halfmann, Edward S
Business: 432 Parkview Dr, Wynnewood, PA 19096
Position: Consultant. *Employer:* Self-employed. *Education:* MS/Elec/MIT; BS/Elec/MIT. *Born:* 6/11/14. Employed by the Philadelphia Electric Co 1937; held numerous engrg positions specializing in aerial & underground transmission, T&D const standards, protective relaying & R&D; promoted to Dir of Res in 1966. Served in US Navy 1943-46; rose to the rank of Lt. Been active in the res progs of the Edison Electric Inst, Electric Res Council, Electric Power Res Institute, the Federal Power Commission's Natl Power Survey, & the Middle Atlantic Power Res Ctte. Reg PE in Penn. Fellow IEEE. Awarded US pat, 'Apparatus for Measuring Alternating Current in the Conductors &/or in the Sheaths of Electric Cables.' Retired from Phila Elec Co, Mar 1, 1979 Granted Amateur Radio License (KA3GGV) 8/26/80. *Society Aff:* IEEE, NSPE.

Halibey, Roman
Business: 2 Penn Plaza, New York, NY 10121
Position: VP *Employer:* Carlson & Sweatt-Monenco, Inc *Education:* MS/EE/Tech Inst - Munich *Born:* 9/7/19 Specialized in electrical distribution and power generation. Former co- owner of Carlson & Sweatt, P.C. and head of its Electrical Dept. In 1980 assumed present position of VP of Engrg with leading engrg firm. Past Pres (1967/68) of Ukrainian Engrs Society of America. Civic activist. PE licensed in NY, NJ and ME. *Society Aff:* IEEE, UESA

Hall, Albert C
Home: Wye Island, Queenstown, MD 21658
Position: Pres, Albert C Hall, PA. *Employer:* Self. *Education:* ScD/Elec Engr/MIT; MS/Elec Engr/MIT; BS/Elec Engr/Texas A&M *Born:* June 1914. Staff & faculty of MIT, 1937-50. Pioneer in servomechanisms. Director, Bendix Res Labs 1950-58. Ch Engr & Vice Pres Engrg, Martin Marietta 1958-63 & 1965-71. Assistant Secretary of Defense, Intelligence 1971-76. Mbr, National Academy of Engrg. Fellow, IEEE and AIAA. Mbr, Cosmos Club. Distinguished Intelligence Award, DIA; Meritorious Civilian Service Award, OSD; Exceptional Service Awd, Sec of Air Force; Distinguished Public Service Award (twice) Secretary of Defense. *Society Aff:* IEEE, AIAA

Hall, Allen S, Jr
Business: Sch Mech Engrg, Lafayette, IN 47907
Position: Prof of Mech Engrg *Employer:* Purdue Univ *Education:* PhD/Mech Engr/Purdue Univ; MSME/Mech Engr/Columbia Univ; BSME/Mech Engr/Univ of VT *Born:* December 1917. Native of Greensboro, VT. Teaching at Purdue since 1939, specializing in machine and mechanism design. Author of book 'Kinematics and Linkage Design' (Prentice-Hall); co-author of 'Machine Design' (McGraw-Hill), and

Hall, Allen S, Jr (Continued)
approximately 40 papers. Former mbr of Advisory Ctte to U.S. Munitions Command. Former chmn of Mech Engrg Division of ASEE. Visiting Professor UC Berkeley 1964-65, Univ of FL 1972. ASME Machine Design Award 1974. Author of "Mechanism Analysis" (BALT). ASME Life Fellow, 1984. *Society Aff:* ASME, ASEE

Hall, Carl W
Business: 1800 G. Street, N.W, National Science Foundation, Washington, DC 20550
Position: Deputy Asst Dir *Employer:* National Sci Foundation *Education:* PhD/Agr Engr/MI State Univ; MME/Mech Engr/Univ of DE; BAE & BS/Agr Engr/OH State Univ. *Born:* 11/16/24. Served with US Army, SSgt 99th Inf, 1943-46, European Theatre, Bronze Star; at Univ of DE, 1948-51; MI State Univ 1951-70; Dean Emeritus, Engr, WA State Univ and Prof Mech Eng, 1970-82. Natl Science Foundation, 1982- . Various consulting assignments. Served as Pres of Sec IV, and V Pres, CIGR 1967-74. Sec of ECPD, 1973-74; Pres of ASAE, 1974-75; Chairman, Engineering Accreditation Commission, 1979-80; Dir of NSPE 1975-79; Dir of ECPD, 1967-73; Fellow, ASAE; Life Fellow, ASME; ASEE; Life Fellow, AAAS, Tau Beta Pi. Assignments: Puerto Rico, USSR, Colombia, India, Peru, Ecuador, Nigeria, Brazil, Libya, Indonesia, China (PRC). Author/Co-author of 22 books on Drying, Energy, Food Engg, Errors, etc. Editor: Drying - An International Journal. Awards: Distinguished Faculty, MSU; Centennial Achievement Award, OSU; Massey-Ferguson Gold Medal, ASAE; Award du Merite, French Govt; Max Eyth Medal, German Engineering Society; La Medaille d'Argent, Paris; Cyrus Hall McCormick Gold Medal, ASAE; Dist Alumnus, OSU. *Society Aff:* ASAE, ASME, AAAS, NSPE, ASEE.

Hall, Donald R
Business: 300 Wamphanoag Trail, E Providence, RI 02914
Position: Exec VP. *Employer:* HART Corp. *Education:* BS/Chem Eng/URI; MS/Bus Admin/URI. *Born:* 2/28/43. BSChE and MBA from University of Rhode Island. Officer and Construction Engr with Corps of Engrs 1964-66. Field engr and estimator for chemical, power and waste treatment plants for Hart Engrg Company, 1966-69. Promoted to Project Manager-Chemical Projects in 1969-responsible for major chemical projects from negotiation of contracts and through start up. Appointed Manager of Engrg of Herzog-Hart Corp in 1973 & Vice President in 1974 responsible for all design and construction activities. Promoted to Exec VP of HART Corp, Parent Co of Herzog- Hart to Hart Engg, in 1979. Now respon for overall mgmt of the corp. Areas of particular technical interest and experience are pharmaceutical intermediate and fine chemical production facilities. Married with five children, active in community affairs and enjoys sailing and anthropology. *Society Aff:* AIChE.

Hall, Donivan L
Business: 350 W Wilson Bridge Rd, Worthington, OH 43085
Position: Dir of Engrg *Employer:* Toledo Scale Corp *Education:* MS/EE/OH State Univ; BS/Engg Phys/OH State Univ. *Born:* 10/11/23. Native of OH. Spent two (2) yrs in the army of the US after BS Engg Phys degree. Discharged as 1st Lt Corp of Engrs. Returned to OH State & obtained MSEE degree. Worked in industry since 1948 performing & managing product oriented res & dev. Named inventor or joint inventor on 23 US Letters Patents. Responsible for the technical direction of the Toledo Scale Corp. Mbr of the OH State Univ Ind-Education Advisory Council.

Hall, Francis E
Home: 432 S Wash Ave, Greenville, MS 38701
Position: Consultant (Self-employed) *Education:* CE/Concrete Design/Univ of MS; BE/Water, Sewer, Structural/Univ of MS. *Born:* 4/3/04. Native of Greenville, MS. BE degree; Grad Degree; "Civ Engr" Univ of MS, former visiting lectr Sch of Engg. Prepared enabling legislation MS for City Planning & Zoning, committee mbr for enacting Reg Laws for Engrs & Arch. WWII Lt Col Army Corps of Engrs, mbr twelve mn review bd BOCA Bldg Code. MS River Pkwy Planning Commission: "Distinguished Service Award". Included in "Who's Who in Engineering-, "Who's Who in Industry & Commerce-, Who's Who in the South and Southwest-. Reg PE in several states. *Society Aff:* ASCE, NSPE, ACI, AAEC, LES, MES.

Hall, Harold C
Business: Post Office Box 6, 49 State Street, Hart, MI 49420
Position: Consulting Engineer (Self-employed) *Education:* MS/Civil Engg/ Northwestern Univ; BS/Civil Engg/IA State Univ. *Born:* 11/17/27. Registered PE in 5 states. Consulting Engr as Employee or Owner, 1953- . Specializes in design, engrg reports & expert testimony pertaining to foundations, earthwork, structures, soil-related structural problems, soil- & groundwater-related waste disposal & environmental problems. Who's Who in the Midwest. Mbr of American Society of Civil Engrs, American Consulting Engrs Council, Tau Beta Pi, American Society for Testing and Materials, National Society of Professional Engineers, American Arbitration Association, and National Association of Environmental Professionals. Enjoy classical music, sailboating, & bicycle touring. *Society Aff:* NAEP, NSPE, ACEC, ASTM, ASCE, AAA.

Hall, Joseph C
Business: Bldg K-1401, MS367, Oak Ridge, TN 37830
Position: Superintendent-Plant Services *Employer:* Union Carbide Corp Nuclear Div-ORGDP. *Education:* BS/Mech Eng/OK State Univ. *Born:* 12/4/31. Lietenant Commander, USN, ret. Worked as Service & Erection Engr for Babcock & Wilcox Co central steam generation facilites & as Test Engr on new nuclear reactor hardward & sys. Currently working for Union Carbide Corp, having served as Equipment Dev Engr on large scale numerically controlled machine tool applications and Superintendent of Fabrication Shops. Present assignment is Superintendent -Plant Services. Primary interest concerns the improvement of productivity through innovation & the proper utilization of personnel & equipment. Award of Merit 1969 ASTME; Dir 1972-81 SME; Executive Committee 1973- 76 SME; VP 1976-79 SME; Pres 1979-80 SME. *Society Aff:* SME.

Hall, Kenneth R
Business: 301 ERC, Texas A & M, College Station, TX 77843
Position: Prof, Chem Engrg; Asst. Dir, Texas Engrg Experiment Station; Assoc. Dean, Engrg *Employer:* Texas A&M Univ *Education:* PhD/ChE/Univ of OK; MS/ ChE/Univ of CA; BS/ChE/Tulsa Univ *Born:* 11/5/39 After receiving Ph.D., I taught at the Univ of VA as an Asst Prof from 1967-70. In 1970 I worked for Chem Share Corp. as Asst to Pres and for Amoco Production Research as Sr. Research Engr. In 1971, I spent 1 year at KUL in Leuven, Belgium. During 1972-1974, I was at the Univ of VA and moved in 1974 to Texas A&M as Assoc Prof. In 1978, I became prof at Texas A&M, and in 1979, I also became Director of the Thermodynamics Research Center. In 1985, I became Asst Dir of the Texas Engrg Experiment Station and, in 1987, Assoc Dean of Engrg. *Society Aff:* AIChE, ACS, ASTM, ASEE

Hall, Newman A
Home: Town Hill Road, New Hartford, CT 06507
Position: Consultant. (Retired.) *Employer:* Self-employed. *Born:* June 1913. PhD Calif Inst of Tedchnology; AB & DSC (hon) Marietta College. Research Engr, United Aircraft Corp 1940-47. Prof Mech Engrg Univ of Minnesota 1947-55; Prof & Chmn Mech Engrg Yale Univ 1956-64; Exec Director, Commission on Education 1962-71. Science Advisor to Minister of Science & Technology, Republic of Korea, 1972-75. Cons in Educational Mgmt & Planning 1974-76. Mbr & Chmn, Ed & Accred Committee, Engrs' Council for Prof Development 1955-64. V Pres, ASEE 1960-62; V Pres Engrg, AAAS 1965-70. Internatl Combustion Inst Dir 1954 & Secy Chmn of Bd on Ed, ASME 1958-62. Author, 2 books on thermodynamics. Cons in nuclear power, combustion, education, & manpower planning. *Society Aff:* ASME, ASEE, ICI.

Hall, Robert N
Business: Genl Elec Res Lab Box 8, Schenectady, NY 12301
Position: Physicist. *Employer:* General Electric. Retired *Education:* PhD/Physics/CA

Hall, Robert N (Continued)
Inst of Technolgy. *Born:* 12/25/19. in New Haven, CT. Genl Electric R&D Center, Schenectady NY 12301, 1948- . Semiconductors, solidstate, junction devices, alloy process for making junctions, PIN power rectifiers, recombination via deep-level centers, tunnel diodes, GaAs injection laser. Ultrapure Ge for gamma-ray spectrometers, photovoltaics, Fellow Am Phys Soc, Fellow IEEE. Mbr Electochem Soc, Editorial bd Solid-State Electronics. IEEE David Sarnoff Award in electronics 1963. Genl Electric Collidge Fellow 1970, IEEE Morton Award 1976, Electrochemical Society Award in Solid State Science and Technology 1977, Mbr Natl Acad of Engg 1977, Mbr Natl Acad of Sci 1978. *Society Aff:* APS, IEEE, ECS, NAE, NAS, BPS.

Hall, Wilfred M
Business: Prudential Tower 46th Fl, Boston, MA 02199
Position: Chrmn of Bd and Chief Exec Off. *Employer:* Chas T Main, Inc. *Education:* BS/Civ Engg/Univ of CO; Honorary Doctorate/Engg/Tufts Univ. *Born:* 6/12/94. With C T Main, Inc: 1916-17, 1920-22, 1941-; Dir 1943-; VP 1953-57; Pres and CEO 1957-72; Chrmn Bd 1972-; Also Chrmn of Bd since 1972 of C T Main Intl, C T Main of MI, C T Main of VA, C T Main of NY, Buerkel & Co, Inc, Technical Services Co; Partner of Uhl, Hall & Rich 1953-62; Managing Partner 1963-80; Distinguished Engg Alumnus Award, Univ of CO 1967; Ralph W Horne Award for Public Service, Boston Soc of Civ Engrs, 1970; George Westinghouse Gold Medal, ASME, 1971; Past Dir US Committee on Large Dams; Past Pres NE Chapter of American Inst of Consulting Engrs. *Society Aff:* ASCE, ASME, NSPE.

Hall, William J
Home: 3105 Valley Brook Drive, Champaign, IL 61821
Position: Head and Professor of Civil Engineering. *Employer:* University of Illinois, Urbana. *Education:* PhD/Civil Eng/Univ of IL; MS/Civil Eng/Univ of IL; BS/Civil Eng/Univ of KS. *Born:* Sohio Pipe Line Co, 1948-49. University of Illinois Faculty, 1949- ; engaged in teaching & research. Author of books & articles in the fields of structural engrg, mechanics, soil & structrual dynmaics, shock, earthquake engrg, plasticity, fatigue, brittle facture mechanics, civil defense, transportation & education. Mbr, numerous technical organizations. Consultant to industry and governmental organizations. Elected Mbr National Academy of Engrg, 1968; American Welding Society, Adams Memorial Membership Award, 1967; U of Ill Engrg Coll Halliburton Engr Education Leadership Award 1980; University Scholar, 1986; ASCE, W L Huber Res Award 1963, N.M. Newmark Medal 1984, E. E. Howard Award 1984; Chmn, Structural Division ASCE, 1973-74, Technical Council on Life Line Earthquake Engrg, 1982-83 *Society Aff:* ASCE, ASME, AWS, ACI, SSA, ASTM, SESA, IABSE, ISPE, NSPE, AAAS, EERI

Hall, William M
Home: 1357 Massachusetts Ave, Lexington, MA 02173
Position: Consulting Engineer (Self-employed) *Education:* ScD/Elect Engr/MA Inst Tech; SM/Elect Engr/MA Inst Tech; SB/Elect Comm/MA Inst Tech. *Born:* July 1906 Burlington, VT. Instructor, Asst Prof MIT, 1930-46. Indicator Division Head, Radiation Laboratory, 1940-41. With Raytheon Company, 1941-71. Project Engr, first microwave surface-search radar for US Navy; Director of Development Engrg, 1968 until retirement. Specializing in radar design & performance. Fellow Acoustical Society of America, IEEE. *Society Aff:* ASA, IEEE, AAAS, ΣΞ.

Hallenbeck, John J
Business: 1485 Park Ave, Emeryville, CA 94608
Position: Pres *Employer:* Hallenbeck-McKay & Associates *Education:* BSC/CE/Univ of CA *Born:* 2/4/30 Raised in the Bay Area, graduated from Piedmont High School, and Studied for 2 1/2 yrs at Stanford Univ, after which he served three years in the US Coast Guard. He continued his education, after being discharged from the Coast Guard, at the Univ of CA. Upon graduation, he worked with the geotechnical engrg firm of Woodward, Clyde & Associates for 14 years, during which he gained experience in a vast number of projects and assumed the responsibility and title of Sr. Project Engr. In 1970, he formed the firm of Hallenbeck-Mckay & Associates, Geotechncial Engrs, located in Emeryville and Santa Clara. He now serves as president of the firm, which has a staff of about 20 professional engrs, geologists, and technicians. *Society Aff:* ASCE, SEA ONC, ABSCSE

Haller, Gary L
Business: 9 Hillhouse Ave, New Haven, CT 06520
Position: Prof, Deputy Provost. *Employer:* Yale Univ, Chem Engrg Dept. & Chem Dept. *Education:* Phd/Phys Chem/Northwestern Univ; BS/Chemistry/Kearney State College. *Born:* 7/10/41. Heterogeneous catalysis & surface chemistry are the main res interests. Recent studies of surface structure & adsorbed species have employed X-ray photoelectron extended X-ray fine structure analysis & NMR. The investigation of non-equilibrium translational & vibrational energy exchange & effects on surface reatin; the quantitative measurement of surface diffusion & the observation of gen correlation between catalyst structure & activity are also of current res interest. Mr Haller came to Yale following a NATO fellowship at Oxford Univ. He has held visiting res apptments at the Univ of Louvain, Belgium and the Univ of Edinburgh, Scotland. In addition to catalysis, the gen field of air pollution is an area of consulting & teaching interest. *Society Aff:* AAAS, AIChE, Royal Soc. of Chem., The Catalysis Soc.

Halligan, James E
Business: Box 3Z, Las Cruces, NM 88003
Position: Pres *Employer:* New Mexico State Univ *Education:* PhD/ChE/IA State Univ; MS/ChE/IA State Univ; BS/ChE/IA State Univ. *Born:* 6/23/36. Chem Engr with ind experience with Exxon Corp and El Paso Products Co. Academic experience at TX Tech Univ as a faculty mbr and dept hd; at the Univ of MO, Rolla, as Dean of Engg, Dean Engr Univ arkansas and currently serving as Pres of New Mexico State Univ. Res interests include: Mass transfer and membrane systems, sorbent oil spill cleanup systems, and the utilization of biomass residue employing pyrolytic reactors. Distinguished res award at TX Tech Univ; Distinguished teaching award at TX Tech Univ and the Univ of MO, Rolla. *Society Aff:* AIChE, ASEE, ACS, ТВП, ФКФ, NSPE.

Halloran, Joseph M
Home: 63 Mowry St, North Haven, CT 06473
Position: Mfg Rep. *Employer:* Halloran Equip Co. *Education:* BS/ME/Worcester Poly Inst. *Born:* 3/18/18. Native of New Britain, CT. Reg PE - MA. Mgr, Furnance Div, A F Holden Co to 1949. New England Rep, Ajax Elec Co, salt bath heat treating furnaces, Park Chem Co, heat treating supplies, to present. Past Chrmn, So CT Chapter, ASM, Past Chrmn New Hampshire Chapter, ASM, Interests in fire buffing & photography. Past Pres, W Ridge Volunteer Fire Assn, Treas, Box 22 Source IFBA. *Society Aff:* ASM, IFBA.

Halperin, Herman
Home: 275 Santa Margarita Ave, Menlo Park, CA 94025
Position: Retired *Education:* BC/Elec Engr/Cornell Univ; EE/ME. *Born:* June 21, 1898. 40 years, Commonwealth Edison Company in all fields relating to underground-power cable, load limits on all electric facilities, special investigations. Also responsible for determination of reserve generating capacity. Received William H. Habirshaw Award in 1962 from IEEE & elected to three honorary societies in 1943-47. Consultant 1961-81. Enjoy public affairs, music & travel. Life Fellow IEEE.
e *Society Aff:* IEEE, ΣΞ, ТВП, HKN.

Halpin, DW
Business: Division of Construction Engrg & Mgmt, Purdue Univ, W. Lafayette, IN 47907
Position: Full Professor *Employer:* Purdue Univ *Education:* BS/Engg/US Military Acad; MSCE/Civ Engg/Univ of IL; PhD/Civ Engg/Univ of IL. *Born:* 9/29/38. Native of KY. Grad of US Military Acad, W Point, 1961. Presently Prof & Hd of Div of Construction Engrg & Mgmt, Purdue Univ. Has been associated with maj construction efforts in the US, Europe & the Far East both as a mgr & in staff ca-

Halpin, DW (Continued)
pacities. Previous positions as Prof, School of Civil Engr, Georgia Inst of Tech, A.J. Clark Chair Prof U of Maryland, College Park. Res Asst Prof, Univ of IL, & Res Analyst, Construction Engg Res Lab, Champaign, IL. Winner of 1979 ASCE Huber Civ Engg Res Prize. Visiting Prof U of Sydney, 1981 & Swiss Federal Inst of Technology, Zurich, Switzerland, 1985. Author of 5 books & numerous papers on Construction Mgt. *Society Aff:* ASCE, ASEE, ΣΞ.

Halsey, George H
Home: 63 Shady Drive, Indiana, PA 15701
Position: President. *Employer:* Halsey Industrial Systems, Inc. *Born:* June 18, 1922. BME from Cornell University 1949. Native of Akron, Ohio. Merchant marine & apprentice machinist prior to WWII. Served in US Navy 1942-45. Technical & management positions with Goodyear, GE & McCreary Tire Companies. Formed own company in September 1967 for the purpose of developing nondestructive testing methods for rubber products with particular emphasis on ultrasonic & microwave methods. Served as chmn of rubber products sub committees for ASNT and ASTM. Elected a fellow of ASNT in 1976. Enjoy duplicate bridge, handball & writing poetry.

Haltiwanger, John D
Business: 3129 Newmark Lab, 208 N. Romine St, Urbana, IL 61801
Position: Prof of Civ Engg. *Employer:* Univ of IL. *Education:* PhD/Civ Engg/Univ of IL; MS/Civ Engg/Univ of IL; BS/Civ Engg/Univ of SC. *Born:* 6/10/25. B., 1925 in Irmo, SC. Served in USN Civ Engg Corp 1942-46. Began career as Instr in Civ Engg at Auburn Univ in 1946. Joined Civ Engg faculty at Univ of IL (in Urbana-Champaign) in 1951 as Instr; advanced to rank of Prof in 1959 & was appointed Assoc Hd of Dept in 1967. Appointed Distinguished Visiting Prof, USAF Academy, 1977-78. Public Service Commendation, US Coast Guard 1974. Mbr USCG Acad Adv Comm 1980-83. Recd SAME Bliss Medal 1980. Rec'd US Coast Guard's Meritorious Public Service Award in 1984. Rec'd 1984 Haliburton Educ Fdn Award of Excellence. 1983-84 served on USAF Scientific Advisory Bd ctte on the Role of Engrg and Services in the Employment of Air Power. *Society Aff:* ASCE, ACI, ASEE.

Halverstadt, Robert D
Business: PO Box 1649, New Canaan, CT 06840
Position: General Partner. *Employer:* AIMe Inc *Education:* BSME/Mech Engrg/Case Inst of Technology. *Born:* January 25th, 1920. 1940-51 Formal four-year apprenticeship course- Republic Steel Corp. 1951-63 Gen Elec Co Design Engr, Supvr Metal Working Laboratory, Mgr Thompson Engrg Laboratory. 1963-64 Continental Can Co Gen Mgr Oper Engrg. 1964-73 Booz Allen and Hamilton, Inc. 70-73 Group VP and Chrmn of Operating Bd for Products and Processes 64-70 Pres Design and Dev Inc (subsidiary) 64-73 Chief Exec Foster D. Snell Inc (subsidiary) 1973-74 Singer Co VP Tech 1974-84 Alleghny Intl 74-82 Pres Special Metals Corp (subsidiary) 81-83 Pres Metals Tech Group, Co-chrmn Titanium Metals Corp of America. Reg Prof Engr - NY, OH. 1984-Present AIMe Associates-General Partner. *Society Aff:* ASM-FELLOW, ASME, AIChE, NYAS, ΣΞ.

Ham, Inyong
Business: Dept of Ind & Mgt Systems Engg, Univ Park, PA 16802
Position: Prof. *Employer:* Penn State Univ. *Education:* PhD/ME/Univ of WI; MSc/ME/Univ of NE; BEng/ME/Seoul Natl Univ *Born:* 12/22/25. With PA State Univ since 1958 and currently Prof of Ind Engg. During 1960- 62, successively Dir of Ind, Asst Minister of Industry, Republic of Korea. Part time engr at Cushman Motor Works, NB (1954-55) and at Kearney & Trecker Corp, WI (1956-58). Specific fields of interest and res are: analysis of mfg processes & systems, group tech, computer aided mfg, etc, and conducted many workshop seminars on the subj here and abroad. Consultant to domestic cos and also intl organizations; Asian Productivity Organization, UNIDO, etc. Chrmn, Prod Engg Div of ASME (1977-78); Dir, Mfg Sys Div of IIE (1973-74); Pres of North Am Mfg Res Inst (1985-86); Pres of Korean Sc & Engg Assoc in Am (1974); Fellow of ASME (1984); Fellow of IIE (1985); Fellow of SME (1987); Coun Mbr of Intl Inst of Prod Eng Res; CAM-I Award (1978) by Computer Aided Mfg-Intl, Mfg Sys Div Award (1980) by IIE, Univ of Wis Dist. Alumuni Service Award (1985), SME Intl Educ Award (1986). *Society Aff:* ASME, IIE, SME, NAMRI, CIRP, ASEE

Ham, Lee E
Business: 1035 E. Hillsdale Blvd, Foster City, CA 94404
Position: Chrmn of the Bd *Employer:* Wilsey & Ham. *Education:* BS/Civ Engrg/Univ of CA-Berkeley *Born:* December 1919. Served in Army 1941-45. Joined firm of Wilsey & Ham as surveyor; became president 1957, chairman 1986. Firm involved in large-scale community planning & design engineering, development. Participated in design of Foster City, CA; Northstar-at-Tahoe; Las Positas New Town; Irvine; Puerto Montt, Chile; Anchorage, Alaska. Led other civil engrg projects as design of 500 Km of highway in Chile. Fellow in ASCE & ACEC. Licensed in 8 states. Participates on Board of Trustees of local hospital, school, and Boy Scouts of America. *Society Aff:* ASCE, ACEC, APWA

Hamada, Harold S
Home: 2084 Alaeloa St, Honolulu, HI 96821
Position: Prof *Employer:* U of Hawaii-Manoa *Education:* PhD/CE/Univ of IL; MS/CE/Univ of IL; BS/CE/Univ of HI *Born:* 11/1/35 in Honlulu, HI. Married Lucy Igawa. Two children (Kyle and Lee Ann). Served 3 years in US Air Force. Employed at Univ of HI-Manoa as Prof of Civil Engrg. Consultant in Structural Engineering, Past Pres Hawaii Section ASCE. *Society Aff:* ASCE, ΣΞ, SEAOH.

Hamann, Carl L, Jr
Home: 4365 NW Queens Ave, Corvallis, OR 97330
Position: Chief Process Engineer. *Employer:* CH2M Hill, Inc. *Education:* MS/Environ/Univ of KS; BS/Health Engg & Civil Engg/Univ of KS. *Born:* November 1937. Employed by Black & Veatch Consulting Engrs, Kansas City, MO in 1958. Became Project Engr in 1965. Joined CH2M Hill Inc. in 1972 as Eastern Regional Director of Water & Wastewater Treatment. Assumed current responsibility as firmwide Asst Dir of Water and Wastewater Treatment in 1981. Projects managed have included planning & design of water supply, treatment, & distribution facilities, & wastewater collection, treatment, & reclamation facilities. Author of numerous technical articles. Mbr of numerous professional socitites and technical committees. AWWA Publications Award, 1966. *Society Aff:* ASCE, AIChE, ACS, AWWA, WPCF, IOI, APHA, AAAS.

Hamann, Donald D
Business: North Carolina State Univ, Dept/Food Sci, Raleigh, NC 27695-7624
Position: Professor. *Employer:* North Carolina State University. *Education:* PhD/Engr Mechs/VA Polytechnic Inst; MS/Agri Engr/SD State; BS/Agri Engr/SD State. *Born:* July 1933. PhD from Virginia Polytechnic Institute; BS and MS from South Dakota State University. Native of Luverne, MN. Instructor at South Dakota State University, Asst Prof at VPI and Assoc Prof at NC State Univ prior to present position. Presently engaged in research on rheology of foods and other biomaterials and design of food processing equipment. National Science Foundation Fellow 1962-63. Editor, Food Engrg Div of ASAE, 1973-77. Editorial Bd of J. Texture Studies, 1982-present On Committee on Food Stability of Advisory Board on Military Personnel Supplies, National Res Council (1976-78). On Technical Advisory Board, Inst for Creation Research, San Diego, CA. *Society Aff:* ASAE, IFT, SR

Hamed, Awatef A
Business: Department of Aerospace Engg., ML 70, University of Cincinnati Cincinnati, , OH 45221
Position: Professor *Employer:* University of Cincinnati *Education:* Ph.D/Aerospace Engg./Univ. of Cincinnati; M.S./Aerospace Engg./Univ. of Cincinnati; B.S./Aeronautical Engg./Cairo University *Born:* 06/17/44 Worked as Aerodynamicist, in the Egyptian General Aero Organization 1965-67 in external aerodynamics and

Hamed, Awatef A (Continued)
engine Airframe integration. After obtaining PhD in 1972, joined the Aerospace Engineering Department at the Univ of Cincinnati, and is presently Dir of Graduate Studies. Over 150 publications in flow numerical modeling and non-intrusive measurements in turbomachines. Research experience in two phase flow in gas turbines, blade erosion and performance deterioration. ASME Fellow, Chairman of ASME Gas Turbine Education Committee 1984-86. Enjoy tennis and travel. *Society Aff:* ASME, AIAA

Hamel, James V
Home: 1992 Butler Drive, Monroeville, PA 15146
Position: Consulting Engr *Employer:* Hamel Geotechnical Consultants *Education:* PhD/CE/Univ of Pittsburgh; SM/CE/MIT; BS/CE/Univ of Pittsburgh *Born:* 4/21/44 Native of Penfield, NY. Research Engr, Univ of Pittsburgh 1967-69. Assist Prof. Civil Engr, South Dakota School of Mines & Technology, 1969-72. Consultant on soil & rock slope stability, MO River Divison, U.S. Army Corps of Engrs, 1970-73. Project Engr, General Analytics, Inc., Monroeville, PA 1972-73. Partner, Hamel Geotechnical Consultants, with practice as individual consultant, since 1973. Specialization in soil and rock slope stability, embankment and concrete dams, and solid (including hazardous) waste disposal with projects in the US, Canada, and the Philippines. Author or co-author of numerous research reports and technical papers. Enjoy Tang Soo Do (Korean Karate), old Volvos, pistol shooting, international travel. *Society Aff:* ASCE, AEG, AIME, ISRM, ISSMFE, USCOLD, IAEG

Hamer, Walter J
Home: 3028 Dogwood St NW, Washington, DC 20015
Position: Chemical Cons/Science Writer. *Employer:* Consultant. *Education:* PhD/Physical Chemistry/Yale Univ; DSc/Hon/Juniata College; BS/Chemistry & Math/Juniata College. *Born:* Nov 5, 1907. From 1935-72 was successively Research Chemist, Chief of Electrochemistry Section, and Director of the Electrolyte Center at the National Bureau of Standards. Chemical Consultant since 1971. With the Manhattan Project and OSRD during WWII. Battery Consultant to the Department of Defense 1951-53. President Electrochemical Society 1963-64; Honorary Mbr 1980. IEEE First Prize Award 1954. Superior Accomplishment Awards, 54,62,65; Gold Medal 1966, U.S. Dept of Commerce. Editor: 'The Structure of Electrolytic Solutions' and 'Electrochemical Constants'. *Society Aff:* ECS, ACS, IEEE, AIC, APS, AAAS, WAS, NAS, ΣΞ

Hamid, Michael
Business: Univ of Manitoba, EE Dept, Manitoba, Canada R3T 2N2
Position: Prof. *Employer:* Univ of Manitoba. *Education:* PhD/Elec Engrg/U of Toronto; M Eng/Elec Engrg/McGill Univ; B Eng/Elec Engrg/McGill Univ *Born:* 6/7/34. Prof Hamid served as an engr & as an engg consultant in a variety of ind and defense organizations. He has also been a prof of elec engg at the Univ of Manitoba in Winnipeg, Canada since 1965 & is a Fellow of IEE & IEEE. He is on the editorial bds of many scientific & engg journals, has served on many technical prog committees for engg confs, is a mbr of the Canadian delegation to URSI & has published over 200 tech papers, 150 conf papers, 25 reports and 30 patents. *Society Aff:* IEE, IEEE, URSI, CSEE

Hamielec, Alvin E
Business: Dept Chem Engrg, Hamilton, Ontario L8S 4L7
Position: Chmn/Prof, Dept Chemical Engrg. *Employer:* McMaster University. *Education:* PhD/Chem Engg/Univ of Toronto; MASc/Chem Engg/Univ of Toronto; BASc/Chem Engg/Univ of Toronto. *Born:* January 1935. Native of Toronto, Ontario. Worked as res engr with Canadian Industries Ltd, 1960-63. Professor of Chemical Engrg, McMaster University 1963 to present. Res and cons in polymer production technology. Experience with manufacture of polystyrene, polypropylene, polyethylene, polyvinychloride, polyacrylamide and copolymers with analytical techniques for measurement of MWD and PSD by GPC and light scattering. Awarded CSChE Erco Award in 1974 and CIC Protective Coatings Award in 1978. Awarded CIC Dunlop Award in 1987 and elected Fellow of the Royal Society of Canada in 1987. *Society Aff:* CSChE, AIChE, ACS.

Hamilton, Bruce M
Business: Box 271, 681 King St West, Hamilton Ontario, Canada L8N 3E7
Position: President. *Employer:* Slater Steels Corp *Education:* BSc (Hons)/Metallurgy/Queen's Univ. *Born:* April, 1920. Native of Hamilton, Ontario. Chief Metallurgist - Atlas Steels Corp, Welland Ont., responsible for manufacturing processes, customer technical service & product development to 1965. Director of Metallurgy - Crucible Steel Corp, Pittsburgh Pa, responsible for corporate policies and application of quality control and customer services to 1967. President Crucible Specialty Steel Division, Syracuse, NY to 1971. President Slater Steel Industries Limited, Hamilton, Ont to present, concerned with manufacture of steel bar products and pole line hardware. Chmn ASM Ontario Chapter 1962. ASM Fellow 1972. Medal in Metallurgical Engrg Queen's University 1943. Enjoy skiing, fishing, golf. *Society Aff:* ASM AISI

Hamilton, DeWitt C, Jr
Business: Mech Engrg Dept, Tulane Univ, New Orleans, LA 70118
Position: Emeritus Prof *Employer:* Tulane Univ *Education:* Ph.D./ME/Purdue Univ; MS/ME/Univ of CA-Berkeley; BS/ME/Univ of OK; BS/Petro Engr/Univ of OK *Born:* 12/4/18 Worked in Eastern OK Oil fields through coll. Heavy bomber pilot during World War II. Member of Mechanical Engrg Faculty at Purdue Univ from '47 to '51. Employee of Union Carbide Nuclear Co from '51 to '65: Head Engr in Reactor Division from '51 to '57; Principal engr in charge of Reactor Engrg at Oak Ridge School of Reactor Technology from '57 to '65. Head of Mechanical Engrg Dept, Tulane Univ, from '65 to '74; Prof of Mechanical Engrg since '74. Research and consulting in thermal radiation, energy conservation and conversion, and, recently, in solar applications and meteorology. Held all offices in local section of ASME. Retired 1986. *Society Aff:* ASME, ASHRAE, ASES.

Hamilton, Douglas J
Business: Dept of EE, Tucson, AZ 85721
Position: Prof. *Employer:* Univ of AZ. *Education:* PhD/EE/Stanford; MS/EE/UCLA; BS/EE/Case Western. *Born:* 12/6/30. Fellow, IEEE. *Society Aff:* IEEE.

Hamilton, Howard B
Home: 1422 Oak Street, Oakmont, PA 15139
Position: Prof Emeritus *Employer:* Univ of Pittsburgh *Education:* PhD/EE/OK State Univ; MS/EE/Univ of MN; BS/EE/Univ of OK *Born:* 10/28/23 Employed General Electric Co, 1949-54; Wichita State Univ 1955-65. Boeing Co 1958-60; Chief of Party, Univ of Pittsburgh at Univ Santa Maria Volparaiso Chile 1965-66. Chairman of Elec. Engr., Univ of Pgh 1966-73, Prof Univ. of Pgh 1973-86. Professor Emeritus 1986- . Consulting Engineer 1986- . Served as Dir, Reg II IEEE 1978-79. Member at Large USAB, IEEE, 1980-81; member at large, Power Engr Soc, IEEE 1980-81. *Society Aff:* IEEE

Hamilton, James F
Business: Sch of Mech Engg, Lafayette, IN 47906
Position: Prof Mech Engr. *Employer:* Purdue Univ. *Education:* PhD/Vibrations/Purdue Univ; MME/Mech Design/Cornell Univ; BSME/ME/Purdue Univ. *Born:* 4/8/27. Prof in the Sch of Mech Engg, Purdue Univ since 1960. Have taught at Cornell Univ and WV Univ and a Sr Res Engr at Clevite Res Labs, Cleveland, OH. Conducts res at the Ray W Herrick Labs in system modeling and simulation, optimization and design. Consulting experience with Lawrence Livermore Labs, Livermore, CA; Labeco, Mooresville, IN; Gilbarco, Greensboro, NC; General Electric Corp., Louisville, KY and others. Enjoys jazz and classical music, married with three children and four grandchildren. *Society Aff:* ASME, ASHRAE, INCE, IAMM.

Hamilton, Robert E
Business: 2380 SW Chelmsford Ave, Portland, OR 97201-2265
Position: Pres. *Employer:* Robert Hamilton & Co. *Education:* BS/ME/IA State Univ. *Born:* 5/29/25. Reg P E numerous states, provinces & territories. In 1955, established first prof mech engr office in Territory of HI. In 1963, joined John Graham & Co as Managing-Dir John Graham Conslts, Ltd with offices in Toronto, Ontario, Canada. Assumed current responsibility as Pres of Robert Hamilton & Co in 1965. In 1967, introduced computer-aided design procedures to State of HI for mech sys & spec. In 1976, appointed to proj ctte for dev of ASHRAE Standard 100. 3P, "Energy Conservation in Existing Buildings - Commercial–. Elevated to grade of Fellow by ASHRAE in 1978. Elected to various offices in Automated Procedures for Engrg Conslts 1976-1979. In 1980, moved co headquarters to Portland, OR and limited practice to specialization in consultation for computer-aided design sys and analysis. in 1982, admitted to Natl Forensic Ctr listing of expert witnesses. *Society Aff:* ASHRAE.

Hamilton, Stephen B, Jr
Business: 1285 Boston Avenue, Bridgeport, CT 06602
Position: Mgr Engrg, Wire/Cable Business Dept. *Employer:* General Electric Company. *Education:* PhD/Org Chem/Northwestern Univ; BS/Chem/ College of William & Mary. *Born:* October 1931. PhD from Northwestern Univ; BS from the College of William and Mary; Fulbright Fellow at Tubingen Univ, Germany 1959-60. Native of Norfolk, Va, US Navy 1950-52. With GE Research and Development Center 1960-68; developed new monomer synthesis and process for Noryl. Product Development Mgr for GE Silicones-RTV 1968-72. Assumed present position with responsibility for insulation development, wire and cable design, specifications, testing, and processing in 1974. Enjoy classical music, antiques, golf and sailing. *Society Aff:* IEEE, ACS.

Hamilton, Wayne A
Business: 101 Barrows Hall, Orono, ME 04469
Position: Assoc Dean *Employer:* Univ of ME *Education:* PhD/Structures/OK State; MS/Eng Mech/Case Inst of Tech; BS/CE/OH Northern Univ *Born:* 1/19/32 Native of Randolph, OH. Served in USMC 1951-53. Has been a member of the Faculty at Univ of ME since 1960. Chairman of Civil Engrg Department 1969-77; Associate Dean of Engrg since 1977. Past President of Maure Sechian ASCE, Member ACI, Committee 340, Design Aids for Reinforced Concrete Structures. *Society Aff:* ASCE, ACI, ASEE

Hamilton, William H
Box 613, Ligonier, PA 15658
Position: Consulting Engineer *Employer:* Self. *Education:* MS/Math & Physics/Univ of Pittsburgh; BS/Math & Physics/Washington & Jefferson College. *Born:* April 2, 1918, Greenville, Pa; Westinghouse Research Labs: Research Engr Electronics 1945-50; Westinghouse Bettis Atomic Power Lab-Engrg Mgr, Nuclear Reactor Plants 1950-65; Mgr, Operating Plants 1965-70; General Mgr 1970 to1979; Consulting Engineer, Westinghouse, 1979-1980; Consulting Engineer (private) 1980 to present; Lt Cmdr, US Navy, 1942-45; Mbr, American Nuclear Society, Institute of Electric and Electronic Engrs, American Standards Association, American Society for Metals; Awards: Westinghouse Order of Merit, 1958; Fellow in the Institute of Electrical and Electronic Engrs, 1966. Development, design, installation, and operation of nuclear reactor propulsion and power plants. *Society Aff:* ASM, IEEE, ANS.

Hamlin, Jerry L
Home: 7408 S Juniper, Broken Arrow, OK 74012
Position: Mgr of Supply *Employer:* Cities Service Oil and Gas Corp *Education:* PhD/Mgt/Columbia Pacific Univ/MBA/Mgt/Univ of Tulsa; MS/Ind Engg/Univ of TN; BS/Ind Engg/OK State Univ. *Born:* 8/1/43. Native of Stigler, OK. Served with USAF 1966-69. Ind engr for TN Eastman Co 1970-1972. With Cities Service Co since 1973, serving as Sr Technical Advisor responsible for ind engr (1973-76), proj engr responsible for design and construction of Tech & Computing Ctr (1977) and loaned Assoc to the AM Productivity Ctr (1977-79). Was instrumental in dev of a firm-level productivity measurement system. Sr conslt, Productivity Info Sys, performing productivity constlg throughout the corp (1979-80). Assumed current responsibility as Mgr of Supply, Marine Div and VP of Grand Bassa Tankers, Inc., a wholly arvered subsidiary. Past Dir, Transp & Distribution Div, IIE. Past Pres of Northeast OK Sec, AACE. Received Award of Excellence in Publications, AACE, 1979, as contributing author for *Cost & Proj Engr's Handbook*. Currently Chrmn, IIE Productivity Ctte. A Reg PE and Cert Cost Engr. *Society Aff:* IIE, AACE, NSPE, OSPE.

Hamm, William J
Business: One Camino Santa Maria, San Antonio, TX 78284
Position: Professor. *Employer:* St Mary's Univ, Dept of Physics. *Education:* PhD/Physics/WA Univ, St Louis, MO; MS/Physics/Catholic Univ, Washington, DC; BS/-/Univ of Dayton, Dayton OH. *Born:* July 1910. Taught in high schools of Chicago and St. Louis. On physics faculty of St. Mary's University, San Antonio: 1935-37, 1940-42, and 1946 to present. General conference chmn, 5th Annual SWIEECO Conference, San Antonio, 1953. Minnie Stevens Piper Foundation award- 'Piper Professor' 1958. Fellow Award of the Inst of Radio Engr, 1960, for contributions to teaching and research. Laser research participation, Stanford U, summers 1966, 1967, 1968. NIH-MBS res grantee, 1973. Enjoy: Jogging, amateur radio (W5FMG). *Society Aff:* IEEE, AIP, ARRL.

Hammack, Charles J
Business: 3501 W 11th St, PO Box 7395, Houston, TX 77248
Position: VP Thermal Engr. *Employer:* Engrs & Fabricators, Co. *Education:* BS/ChE/Univ of TX *Born:* 6/2/23. in Alamogorda, NM. Served in the Navy during WWII. After grad from college I was employed by Sheffield Steel Corp & Sinclair Refining Co before accepting employment with Engrs & Fabricators, Co (EFCO) in 1957. I was appointed VP of Thermal Engg at EFCO in 1977. In this position I am responsible for the thermal & hydraulic design & tube vibration analysis of the shell & tube heat exchangers that we design & fabricate. I am a lay leader in the Evangelism Explosion III Intl Prog at St Mark Lutheran Church. *Society Aff:* AIChE.

Hammer, Carl
Home: 3263 O St NW, Washington, DC 20007
Position: Director, Computer Sciences. Retired, 1981. *Employer:* Sperry Univac. *Education:* PhD/math Statistics/Univ of Munich, Germany; BSc/Mathematics/ Luitpold- Oberrealscshule Munich, Germay. *Born:* May 1914, Chicago IL; Statistician with Texas Co to 1943; with Pillsbury Mills 1944-47. Walter Hervey Junior College to 1951. Sr Staff Engr, Franklin Inst to 1955. As Director of Univac European Computing Center (1955-57) installed first Univac I in Europe. Next two years with Sylvania's Computer Operations. 1959-63 with RCA in charge of initial design of Minuteman Communications Systems. Since 1963 with Sperry Univac. Retired 1981. Fellow AAAS; Fellow IEEE; AFIPS Director 1973-75; ACM Council 1968-1977. DPMA Computer Sciences Man- ofthe-Year 1973; ACM Distinguished Service Award 1979.. *Society Aff:* ACM, IEEE, DPMA, AFIPS.

Hammer, Guy S, II
Home: 4626 River Road, Chevy Chase, MD 20015
Position: Manager Technical Development. *Employer:* Assn for Advancement of Med Instru. *Born:* Oct, 1943; Graduate Studies in Clinical Engrg, Johns Hopkins University School of Medicin; BSEE U of Md; Engrg/Executive with AAMI-Biomedical Standards Program since 1975. Staff- U of Md EE Dept 1970-74; US Army Nuclear Weapons 1966-69; Secretary- Washington Society of Engrs 1976; Editor- 'WSE News' and 'Journal'; Tech Editor- 'Medical Instrumentation'; Staff Contributor- 'Standards Engineering'. Contributing author to book: 'Clinical Engrg'. Tau Beta Pi; Eta Kappa Nu; Phi Kappa Phi; Who's Who Univ & Colleges. Certificates: US Army, U of Md Biomedical Lab, and Standards Engrs Society; Founded the TEK ART Institute in 1976 for technological artists and engineers.

Hammer, Kathleen M
Home: 4729 Winewood Ln, Milford, MI 48042
Position: Product Design Engr. *Employer:* Ford Motor Co. *Education:* MSE/EE/Univ of MI; PDED/Prof Dev/Univ of MI; BA/Phys/Wayne State Univ; BS/Math/Wayne State Univ. *Born:* 9/30/51. Native of Detroit, MI. Honors prog curricula while attending WSU; recipient of four yr Bd of Governors & State of MI Scholarships. Currently employed by Ford Motor Co; recent technical accomplishments in field of automotive electronics include circuit dev & production launch of 1980 Digital Speedometer and 1979 Graphic Display. Res asst under Dr Mary Beth Stearns (Phys Dept) on high resolution electron spectrometry & nuclear magnetic resonance while on FCG prog. Previously employed by DeVlieg Machine Co, working on microprocessor based design for numerical controller. 1979-80 Detroit Section SWE Pres, 1978-79 Rep, 77-8VP. *Society Aff:* SWE, IEEE, ESD, SAE.

Hammer, William E, Jr
Home: 3033 Meadowcrest Lane, Kettering, OH 45440
Position: Dir of Info Systems *Employer:* The Duriron Company, Inc. *Education:* MS/IE/OH State Univ; B/IE/Univ of Dayton *Born:* 10/25/39 Native of Dayton, OH. Planning Engr for Western Electric Co. 1962-67. With Duriron Company since 1967; Sr. Industrial Eng 1967-69, Systems Mgr 1969-77. Promoted to Dir of Information Systems in 1977 with responsibility for business and engrg systems development and data processing. Fellow of IIE; VP and Member of Board of Trustees, Past Dir of Computer Information sys Div, and Past Pres of Dayton Chapter. Member of Bd of Govrnors and President of Engs Club of Dayton. Indus Engrg Accredition Visitor for ABET. Listed in Who's Who in the Midwest and Directory of Top Computer Executives. Receipient of 2nd Annual Award of the Computer and Information Systems Division of IIE in 1980 for "significant accomplishments in the theory and applications of management information systems–. *Society Aff:* IIE, AIM.

Hammerschmidt, Andrew L
Home: 181 Ramblewood Road, Moorestown, NJ 08057
Position: VP (Ret) *Education:* BE/Elec Engg/OH State Univ. *Born:* 1914 Medina Ohio. Technical Supervisor WOSU 1938-41. Development Engr National Broadcasting Co, NYC 1941-48. Television Operations Supervisor WNBK Cleveland, Ohio 1948-52. Associate Director Color TV Systems Development NBC, NYC 1952-54. Associate Director TV Technical Operations NYC 1954-55. Vice President and Chief Engr NBC 1955-61. Chief Engr, RCA Missile and Surface Radar Division 1961-65. Manager, Operation Plans, RCA Broadcast and Communications Division, 1965-68. Vice-President RCA Broadcast Systems Division 1968-71. Vice- President Inductotherm Corp 1972-79. Retired 1980. Distinguished Alumnus Award, Ohio State University 1964. Married, 3 children. Presbyterian, Rotarian.

Hammes, John K
Business: 399 Park Ave, Fl 4, New York, NY 10043
Position: VP. *Employer:* Citibank, NA. *Education:* PhD/Mineral Engg/Univ of MN; MS/Mining Engg/Univ of MN; BA/Math-Geol/Univ of MO. *Born:* 11/28/34. Kansas City, MO. Served with US Marine Corps 1956-1959. US Bureau of Mines Fellowship 1962. Planning engr for Kennecott Copper 1965. With Citibank since 1968. Responsible for analysis of mining and met projs proposed for financing. VP - Dir, AIME 1977-current. Mbr of Mining Club of NY. *Society Aff:* AIME, CIMM, MMSA

Hammesfahr, Frederic W
Home: 25 Oak Place, Bernardsville, NJ 07924
Position: Director, Board of Directors. *Employer:* Hammesfahr, Winter & Assoc, Inc. *Education:* MS/Chem Eng/MA Inst of Tech; BS/Chem Eng/MA Inst of Tech. *Born:* 10/12/19. 1955-60 Consolidation Coal Co, Asst to VP, Res & Dev; 1960-64 J T Baker Chem Co, Bd/Dir & Dir of Commercial Dev; 1965-70 Allied Chem Co, Secy Corp Operations Commitee and Dir Intl Licensing; 1971-75 PVO Intl, Sr Exec VP; 1975- 77 Hydrocarbon Res Inc, Pres & Dir; 1977-present, pres and director of Hammesfahr Winter & Assoc Inc. Past Chrmn, NJ Energy Res Inst.; Advisor to US Congress on technical policy and on synthetic fuel. NJ Council for Res & Dev, Dir & Treas; Amer Inst of Chem Engrs, "Fellow" Past Chrmn, Govt Programs Steering Committee, Treasurer, Mgmt Div, New Technology Ctte, Patents and Publications *Society Aff:* AIChE, ACS, ΣΞ, ΤΒΠ.

Hammett, Cecil E
Home: P O Box 1069, Auburn, AL 36830
Position: Professional Engineer (Self-employed) *Education:* MS/Elect Engg/Univ of NB; BS/Elect Engg/KS State Univ. *Born:* Apr 6, 1906 Marysville Kansas. Graduate Study University of Michigan 1939-40. US Army Reserve major retired. Professor Auburn University 1949-52. Civil Service Engr, US Army Missile Command, retired 1972. Fellow ASME. Chmn, Professional Engrs in Government 1961 to 1977 and President 1965 of Alabama Society of Professional Engrs. Vice-Chmn NSPE Professional Engrs in Gov 1974- . Alternate Dir NSPE; Dir 1976-78 NSPE Ed Ctte 1974-77 . Spec: machine design, stress analysis, vibrations, pressure vessels, high pressure equipment. Member of The Society of Rheology since about 1942. Was a member of the NSPE Public Relations Committee for two years circa 1965, member of NSPE Committee on Education, 1975-78. registered in Alabama and Delaware. Hobbies: pistol, fishing, boats, electronics and shopwork. *Society Aff:* ASME, NSPE, AUSA, ADPA, ROA, NRA, SR.

Hammett, Robert L
Business: 1400 Rollins Rd, Burlingame, CA 94010
Position: Pres *Employer:* Hammett & Edison, Inc *Education:* MA/EE/Stanford Univ; BA/Engr/Stanford Univ *Born:* 7/11/20 At Radio Research Lab, Harvard Univ, on radar countermeasures 1943-1946. Consulting engr since 1946 in radio and television broadcasting. Practice involves frequency allocations, system design, construction, field services, and forensic engrg. Specialty is design of directional antennas and transmitting facilities. Founder of Hammett & Edison, Consulting Engrs, 1952; Elected to Tau Beta Pi and Phi Beta Kappa. Three U.S. Patents. Registered PE in CA, TX, and DC. *Society Aff:* IEEE, NSPE, AFCCE

Hamming, Richard W
Home: 1140 Sylvan Road, Monterey, CA 93940
Position: Adjunct Professor. *Employer:* Naval Postgraduate School. *Education:* PhD/Math/Univ of IL; MA/Math/Univ of NB; BS/Math/Univ of Chicago. *Born:* February 11, 1915. Employed Los Alamos 1945-46; Bell Labs 1946- 1976; Naval Postgraduate 1976- . President Assoc Comp Mach; Vice President AAAS; Fellow IEEE, Piore Award (IEEE), Member of National Academy of Engineering, UNI. Penna., Moore school Henry Peuder Award. IEEE RW Hamming Medal ($10,000 each year plus gold medal) named for me. *Society Aff:* ACM, IEEE, MAA, ASA.

Hammitt, Frederick G
Home: 1306 Olivia Avenue, Ann Arbor, MI 48104
Position: Full Professor of Mechanical Engrg. *Employer:* University of Michigan. *Education:* BS/ME/Princeton Univ; MS/ME/Univ of PA; MS/Applied Mech/Stevens Inst Tech; PhD/Nuclear Engrg/Univ of MI. *Born:* September 1923 Trenton N J. Served as engrg officer, US Navy 1945-46. Worked as engr in industry, particularly turbomachinery, until coming to Univ of Michigan in 1955. Fellow of IME, ASME & ASTM. Mbr of Phi Beta Kappa, Tau Beta Pi, and Sigma Xi. Currently Professor-in-charge, Cavitation & Multiphase Flow Laboratory, Mech Engrg, Univ of Michigan; previously professor of Nuclear Engrg. Author or co-author of over 450 papers or articles 2 handbook chapters, and 2 books. 5 patents. Now Prof.-Emeritus, Univ. Mich. 6 wk Lecture trips, China & Japan, 1982 & 1986. Prev. Sabbaticals, etc to France & Poland. *Society Aff:* Fellow ASME, ANS, IAHR, IM, ASTM.

Hammond, David G
Business: 3250 Wilshire Blvd, Los Angeles, CA 90010
Position: VP *Employer:* Daniel, Mann, Johnson, & Mendenhall *Education:* MS/CE/Cornell Univ; BS/CE/PA State Univ *Born:* 9/8/13 Native to Paterson, NJ. Commissioned Lt. US Army, 1937; advanced to Colonel, 1953; retired 1964. Assis-

Hammond, David G (Continued)
tant general mgr - Engrg & Operations, BART, San Francisco, 1964-1973. VP, Baltimore Rapid Transit System and Eastern Regional Transit Projects - Daniel, Mann, Johnson & Mendenhall, 1973-Present. Registered engr, NB, CA, MD, WV. Past chrmn, US National Committee on Tunneling Technology. Member, National Academy of Engrg. Member, The Moles. Named one of Top Ten Public Works Men of the Year by APWA and Kiwanis Intl. Listed in "Who's Who in the East" and "Who's Who in Railroading and Rail Transit–. Currently Proj Dir, MRTC (Metro Rail Transit Consultants), design engrs for the $3 billion LA Metro Rail System. *Society Aff:* NAE, ASCE, SAME, ITA, UITP, APTA.

Hammond, Donald L
Business: 1501 Page Mill Rd, Palo Alto, CA 94304
Position: Assoc Director, Hewlett Packard Labs *Employer:* Hewlett-Packard Co. *Education:* DSC/Eng/U of Bristol; DsC/Phys/CO State Univ; MS/Phys/CO State Univ; BS/Phys/CO State Univ. *Born:* 8/7/27. Native of CO; worked as civilian physicist at US Army Electronics Command 1952-1956; Dir of Res, Scientific Electronic Products 1956-1959; Hewlett-Packard Co 1959-present. Development & production of precision components until 1964. Founding Dir of HP Labs 1966, Director, Ctr for Physical Research 1978-84, Assoc Dir & part-time Acting Director Hewlett-Packard Labs 1986-present. *Society Aff:* NAE, IEEE, APS, AAAS.

Hammond, Joseph L
Business: 225 North Ave, Atlanta, GA 30332
Position: Prof *Employer:* Georgia Institute of Technology *Education:* PhD/EE/GA Inst of Tech; SM/EE/MIT; SB/EE/MIT *Born:* 10/16/27 Native of Birmingham, AL. Served in US Navy 1945-46. Research Engr for Southern Research Institute 1952-55. With GA Institute of Technology since 1955, advancing to present position of Prof of Electrical Engrg in 1966. Significant consulting projects for United Aircraft, Scientific Atlanta and Lockheed Ga Corp. Presently engaged in teaching and research in areas of communications, computer networks, and computational techniques. Published several dozen technical reports and papers; holds on US patent. Member of Sigma Xi and Tau Beta Phi honor societies. *Society Aff:* IEEE

Hammond, Ogden H
Home: 1365 Massachusetts Ave, Arlington, MA 02174
Position: Pres. *Employer:* Count Digital, Ltd. *Education:* PhD/-/MIT; SM/Chem Engg Practice/MIT; SB/Chem Engg/MIT. *Born:* 8/20/47. Taught at MIT from 1973-77. Participated in early stages of MIT Innovation Center and MIT energy Laboratory. Mbr MIT Policy Studies Group. Developed and taught courses on Invention, Electrochemistry, Synthetic Fuels, Developed heat transfer method of materials testing. Founded and subsequently sold Hetra Corp. Founded Count Digital in 1977. *Society Aff:* AIChE, ASTM.

Hammond, Robert H
Business: Rm 119 Riddick, Raleigh, NC 27607
Position: Dir Engg Student Services. *Employer:* North Carolina State University. *Education:* BME/Mech Engg/Cleveland State Univ. *Born:* July 17, 1916. Graduate studies in Machine Design, Purdue University, 1948-51. Served with Army Corps of Engrs, 1941-46 and 1951-66. Instructor US Military Academy 1945-46; Instructor Purdue Univ 1946-51; Assoc Professor US Military Academy 1951-66; Assoc Professor N C State University 1966-date. Co- author 'Engrg Graphics,' 2nd Edition, 'Introduction to FORTRAN IV'. 2nd Edition. Vice-Chmn, Engrg Design Graphics Division, ASEE, 1963; Chmn, 1964. Awarded Legion of Merit, US Army, 1966. Awarded Distinguished Service Award, Engrg Design Graphics Division, ASEE, 1976. *Society Aff:* ASEE.

Hammond, Ross W
Business: PO Box 888104, Atlanta, GA 30338
Position: Pres (1981 to present) *Employer:* Applied Research & Dev Assocs (ARDA) *Education:* BS/Elec Engrg/Univ of TX; MS/Ind Engrg/Univ of TX. *Born:* 3/20/18. Born in NYC. Served with Army Air Corps, 1943-46. Prin res scientist for GA Inst of Tech, 1963 to 1981, including respon for domestic and intl development programs. Contributor to engg and mgr handbooks, pubs and journals. Past exec dir and natl pres, Inst of Indust Engrs (Fellow). Tau Bet Pi, Eta Kappa Nu. International Development Consultant. *Society Aff:* IIE.

Hammontree, R James
Business: 5233 Stoneham Rd, N Canton, OH 44720
Position: Chrmn & Chief Executive Officer *Employer:* Hammontree & Assoc, Ltd. *Education:* BSCE/Civ Engg/Univ of Akron. *Born:* 12/16/33. Native of Akron, OH. Served with US Army Transportation Corps 1957-59. Hgwy Design Engr for Howard, Needles, Tammen & Bergendoff of Cleveland, OH, 1959-64. City Engr of N Canton, OH, 1964-66. Went into private consulting practice in 1966 and developed co of Hammontree & Assoc, Ltd of Akron & Canton, OH into present staff of 60 people doing work in Civ Engg in E OH and adjoining areas & Florida. Served as chrmn of co. Dir of NSPE 1978-80. Wife's name is Irene and have four children, Robert III 26, Charles F 24, Hope M 22, and Barbara H 20. Also operates branch offices in Pittsburgh PA, Lake Wales and Orlando Florida. R. James Hammontree is also Chrmn of Morris Knowles Inc, Environmental and sanitary engrs of Pittsburgh PA. *Society Aff:* NSPE, ASCE, ACSM, ACEC.

Hampton, Delon
Home: 12804 Brushwood Terrace, Potomac, MD 20854
Position: Pres & CEO *Employer:* Delon Hampton & Assocs., Chartered *Education:* PhD/CE/Purdue Univ; MS/CE/Purdue Univ; BS/CE/Univ of IL *Born:* 8/23/33 Married to Fay L. Myers-Hampton March 17, 1979. Instructor, Prairie View A & M College, Prairie View, TX 1954-55; US Army 1955-1957; Asst. Prof Civil Engrg, Kansas State Univ 1961-64; head, Soil Mechanics Research (on leave), E.H. Wang Civil Engrg Research Facility 1962-63; Sr Research Engr, IIT Research Institute, 1964-68 U.S. Naval Reserve 1967-1972; Prof of Civil Engrg, Howard Univ, 1968-85; President Gnaedinger, Baker, Hampton & Associates, chartered, 1971-73; President, Delon Hampton & Associates, chartered, 1973- present. *Society Aff:* ASCE, TRB, ASTM, SAME, ACEC, APTA

Hampton, Philip M
Home: 2440 Ostrum, Pontiac, MI 48055
Position: Pres. *Employer:* Hampton Engg Assoc. Inc. *Education:* BA/Geology/Berea College. *Born:* September 5, 1932. Advanced Studies at Michigan State University, Oakland University and Wayne State University. Reg Engr in Michigan, Ohio, Minnesota, West Virginia, Tennessee, Kentucky & Mississippi. Project Mgr, Director-PR, Director Corporate Services, Director-Ventures & International Business Development for Johnson & Anderson, Inc (J & A), Pontiac Michigan; Vice President J & A - 1965-71, Exec Vice President J & A - 1972-76; President Environmental Research Associate 1971-73; Mbr - Waterford Board of Education 1967-74 (President 1969-71); National Water Well Association (Chmn Tech Div 1969-71); American Consulting Engrs Council (Int'l Engrg Comm 1971- , V Chmn Public Relations Comm 1970-72); Chmn-ABA Procurement Code Committee, 1977- , Cons Engr's Council/Mich (Pres 1976-77), Chmn Cofpaes-ABA Model Code Taskforce, 1978-80 National Director, 1986- ; State Democratic Party (Alt State Central Comm 1972-); Precinct Delegate 1972-78; 19th Congressional District Coordinator, Geo McGovern for Pres. Mbr, Amer Arbitration Assoc Comm Panel, 1974- ; Mbr Genl Service Admin, Regl Pub Advisory Panel for A-E Services 1977; Treasurer, MI Consulting E. Founder and President-HMA Consultants Inc 1977-. Founder & Pres, Geointernational, 1976; Founder & Pres, Hampton Engg Assoc. Inc. 1985. *Society Aff:* ACEC, NWWA, AIPG, AGWSE, ITE.

Hampton Robert K, Sr
Home: P.O. Box 64, N. Myrtle Beach, S.C 29597
Position: Conslt *Employer:* Self *Education:* Cert/Civil Structural/Drexel Univ Evening Sch. *Born:* 11/20/19. Diploma Civil Engrg Drexel University, Evening School. Proj Engr Beaumont Birch Company 1936-42 and 1946-48. Proj Engr Hewitt

Hampton Robert K, Sr (Continued)
Robins 1948-49. U.S. Army Officer Field Artillery & Engrg Corps 1942-46 United States, France and Germany. New York Regional Engr Sales Manager Beaumont Birch Co 1949-57; Mfg Engrg Sales Representative 1948-1981; Associate, Charles R Velzy Assoc. Inc 1982-84; Conslt 1984- . PE Registered Professional Engr Member at large ASME Board on Research 1981-85, Fellow/ASME; Mbr. at large-ASME 1978-81 Liason Representative to ASME Policy Bds, Research & Communications, ASME Board Environ & Trans 1981-85 Chmn ASME Solid Waste Processing Div & 1972 ASME 1974-75 Incinerator Conference. Mbr NSPE, APWA, APCA, GRDCA, AAEE. Instructor Hofstra Univ Environmental Course. Publs-section in Plant Engrs Handbook. Papers presented at ASME Winter Annual meetings and ASME Natl Incinerator Conferences and other publs. Patents in Mtls Handling Field *Society Aff:* NSPE, APWA, ASME, APCA, GRDCA, AAEE

Han, Chang D
Business: 333 Jay St, Brooklyn, NY 11201
Position: Prof & Head of Dept. *Employer:* Polytechnic Inst of NY. *Education:* ScD/Chem Engrg/MIT; MS/Chem Engrg/MIT; MS/Math/NYU; MS/Elec Engrg/NJ Inst of Tech; BS/Chem Engrg/Seoul Natl Univ. *Born:* 9/28/35. After graduating from Seoul Natl Univ, he continued his education in US. Worked for American Cyanamid Co and Esso Res & Engg Co before joining Polytechnic Institute in 1967, where he served as Head of its Chem Engg Dept between 1974 and 1982. Active in research, he has published over 170 scientific papers dealing with polymer rheology, polymer processing, process control, and related areas; he has published two research monographs, "Rheology in Polymer Processing–, and "Multiphase Flow in Polymer Processing–. He serves as Chrmn of the Polymer Engg Prog, MESD, of the AIChE Natl Prog Comm. *Society Aff:* AIChE, ACS, SPE, SRAS.

Hanawalt, Joseph D
Business: Matls & Met Engrg Dept, Ann Arbor, MI 48109
Position: Professor Metallurgical Engrg. *Employer:* University of Michigan. *Education:* PhD/Physics/Univ of WI; MS/Physics/Univ of WI; AB/Physics/Oberlin College. *Born:* July 1902. Natl Res Fellow Univ Mich & Groningen Univ 1931. Dow Chemical Co Midland, Mich 1931-64, Hd X-ray & Spectroscopy, Met Lab, Mgr Magnesium Dept, V P. 1964-Prof Met Engrg Univ of Mich. 1976 Chmn/Bd Intl Centre for Diffraction Data, Swarthmore Pa. Mathewson Gold Medal AIME 1943. Hon Mbr & ASM Gold Medal 1964. Fellow & Award of Merit ASTM 1965, Hon Mbr Internatl Magnesium Assn 1968, Res Paper Award AFS 1972. Prof ints: magnesium met, x-ray diffraction analysis. Family man. Enjoy music & outdoor sports, summer & winter. *Society Aff:* APS, ACS, ACA, ASM, AIME, ASTM, AFS.

Hanbury, William L
Home: 492 N Owen Street, Alexandria, VA 22304
Position: Director, Interagency Coordination. *Employer:* Federal Emergency Mgmt Agency - US Fire Admin. *Education:* BSME/Mech Engg/GA Tech. *Born:* 11/5/26. Norfolk, VA. Reg P E in MA & CA. In fire protection and safety engg fields since 1950. Served in Navy, Army, GSA, NASA, and Commerce federal agencies in fire protection. Presently respon for serving as focal point for liaison in fire safety between federal fire safety organizations and the US Fire Admin. *Society Aff:* SFPE, IFSTA.

Hancher, Donn E
Business: Civil Engr. Bldg, Purdue University, W Lafayette, IN 47907
Position: Prof. *Employer:* Purdue Univ. *Education:* PhD/Constr Engr/Purdue Univ; MSCE/Civ Engr/Purdue Univ; BSCE/Civ Engr/Purdue Univ. *Born:* 6/17/43. Native of Elwood, IN. Currently Prof of Construction Engrg & Mgt at Purdue Univ. Special areas of interest are Construction Planning & Scheduling, Cost Engrg & Mgt Systems. Assoc Hd, Div of Construction Engrg and Mgt at Purdue. Very active in Profl Socs, Technical Committees, Student Assns & Continuing Education Progs for Construction. Pres. IN Sec, ASCE, 1977. Chrmn, Dist 9 Council, ASCE, 1975. Friedman Award, ASCE, 1975. IN Jr Engr of Yr, ISPE, 1969. Bd of Dir, ASCE, 1984-87. Bd. of Dir., ARTBA, 1986-89. *Society Aff:* ASCE, ASEE, ASEM.

Hancock, John C
Business: PO Box 11315, Kansas City, MO 64112
Position: Exec VP Corporate Development and Technology. *Employer:* United Telecommunications, Inc. *Education:* PhD/Elec Engg/Purdue Univ; MSEE/Elec Engg/Purdue Univ; BSEE/Elec Engg/Purdue Univ. *Born:* Oct 1929 Martinsville Ind. Res Engr, US Naval Avionics Facility 1951-57 (responsible for design of radar fire control equipment); 1957-65, Asst Prof to Prof of Elec Engrg (area of specialization-communications); 1965-72, Hd of School of Elec Engrg; 1972-84, Dean Schools of Engrg, Purdue Univ; 1984 to present: Executive VP - Corporate Development and Technology, United Telecommunications, Inc. Director, Hillenbrand Industries. Dir, Ransburg Corp. Mbr Natl Acad of Engrg, Eta Kappa Nu (Natl Pres 1969-70), ASEE - President - June 1983 to June 1984; Lamme Award - 1980; Fellow - 1973. Fellow IEEE. AAAS - Fellow - 1981. Dir, Member, National Science Board, 1986-present. Author: 'An Introduction to the Principles of Communication Theory', 'Signal Detection Theory' and 'An Introduction to Electrical Design'. *Society Aff:* NAE, IEEE, ASEE, AAAS, ΤΒΠ, ΣΞ, HKN.

Hancuff, William R
Business: 11800 Sunrise Valley Drive, Reston, VA 22091
Position: VP *Employer:* James M. Montgomery, Consulting Engineer, Inc. *Education:* PhD/Environ Health Engrg/Univ of TX; Engr Deg/Sanitary Engrg/Stanford Univ; MS/Engrg Econ Plan/Stanford Univ; BS/CE/Loyola Univ *Born:* 4/28/43 Presently the manager of James M. Montgomery's Mid-Atlantic region operations, and has managerial responsiblity for all work performed in the five- state area and the District of Columbia. He is also responsible for technical supervision of all planning, studies, reports and design performed in this area. Dr. Hancuff has extensive experience in environmental engrg covering the complete range of water treatment, waste treatment, municipal waste treatment, pretreatment, design and construction, landfill studies, water treatment, feasibility studies, laboratory investigations, pilot plant operations, hydrology and storm drain design. The greatest portion of Dr. Hancuff's experience has been in the areas of water and wastewater engrg. *Society Aff:* APWA, ASCE, AWWA, CWPCF, WPCF, VSPE, VWPCA

Hand, John W
Business: 650 S Cherry St, Suite 400, Denver, CO 80222
Position: Vice President. *Employer:* PAce Co Consultants & Engrs, Inc. *Education:* BS/Chem Engg/Auburn Univ. *Born:* November 9, 1922 in Mobile, Alabama. BS in Chemical Engrg from Auburn University 1943. Chemical Warfare Service 1943-44; USNR Communications Officer 1944-46; Ideal Cement Company, Chemist, Research Engr, Executive Assistant, Director of Business Research, Vice President Administration, 1946-65; Cameron Engrs, Inc, Vice President 1965-present. Mbr AIChE, AAAS, IAEE and Colorado Mining Association, Director of Power Resources Corporation and Cameron Engrs Inc. Holds several patents in cement manufacturing, vertical kiln technology and coal gasification. Reg Prof Engr in Colorado. *Society Aff:* AIChE, AAS, IAEE, CMA.

Handelman, Robert A
Business: First Natl Tower, Akron, OH 44308
Position: Pres. *Employer:* Chemstress Consultant Co. *Education:* MS/ChE/Univ of Akron; BS/CE/IA State Univ. *Born:* 10/21/40. Reg PE in several states. Native of Chicago, IL. Married, with five children. Technical supervisor for Goodrich Gulf Chems until 1963. Proj coordinator & process engr for chem industry until 1965. Founded Chemstress Consultant Co, a full service proj mgt, process design facilities consultant firm. Served as its pres since 1965. Chrmn of Akron AIChE in 1977, first annual Akron area AIChE chem engr of the yr, 1978. Advisor to the University of Akron Engineering College. Supports civic, education & scientific progs. Enjoys tennis, golf & music. *Society Aff:* AIChE, NTSA.

Handwerker, Richard B
Home: 320 Del Zingro Dr, Davison, MI 48423
Position: Continuing Education Department Administrator. *Employer:* GMI Engineering & Management Institute. *Education:* BIE/Mfg Process/Gen Mot Inst. *Born:* June 16, 1924 Dayton Ohio. GMI 49 BIE. Served as 2nd Lt Pilot WWII. Jr Process Engr Delco Moraine Division of GMC 1946-53. Transferred to General Motors Institute Oct 1953, teach Industrial Hydraulic Control Systems and Manufacturing Process Courses. Retired July 1, 1981. Founding mbr of the Fluid Power Society and Charter Mbr and Past President of Chapter 1. Major Michigan Air National Guard and Air Force Reserve (Retired). April 1983 - Administrator for Continuing Education Dept. GMI Engineering & Management Institute. Responsible for all off-site and special training courses offered. *Society Aff:* FPS, ROA, ASTD.

Handy, Lyman L
Business: Dept of Pet Engr, Univ of So Calif, Los Angeles, CA 90089-1211
Position: Professor and Chmn, Pet Engrg. *Employer:* University of Southern California. *Education:* PhD/Chemistry/Univ of WA; BS/Chemistry/Univ of WA. *Born:* August 4, 1919 in Payette Idaho and reared in Lynden Washington. Served in US Navy 1942-46. Employed by Chevron Oil Field Research Company from 1951-66, highest poistion Sr Research Associate. Appointed Professor of Petroleum & Chemical Engrg at University of Southern California 1966- . Chmn, Pet Engrg 1966-86, & Chmn Chemcial Engrg 1969-76. Chmn, Orange County Section ACS 1969; Chmn LA Basin Section, Soc of Petroleum Engrs 1974-75. Distinguished Lecturer for Society of Petroleum Engg 1975-76. Bd of Drs, Soc of Petroleum Engrs 1978-81. Distinguished Member and Distinguished Service Award Soc. of Petroleum Engrs. 1984. *Society Aff:* SPE, ACS, AIChE, AIC, AAAS, ASEE.

Haneman, Vincent S, Jr
Business: 539 Duckering Bldg, Fairbanks, AK 99775
Position: Dean of Engineering *Employer:* University of Alaska *Education:* PhD/Aeronautical Engg/Univ of MI; MSE/Aeronautical Engg/Univ of MI; SB/Aeronautical Engg/MA Inst of Tech. *Born:* February 1924. Previous positions: Dean of Engg Auburn Univ: 1972-79 Assoc Dean of Engrg, Dir of Engrg Research, Professor Mech Engrg, Oklahoma State Univ; 1966-72 Senior Associate and President of a consulting firm, Haneman Associates Inc, 1960-66. Army Scientific Advisory Panel; AIAA Associate Fellow; Sr Mbr AAS; Mbr IEEE; Tau Beta Pi, Sigma Xi. Registered Professional Engr, Alabama, Ohio, Oklahoma, TX; Pres 1980-81. First-Vice President, Mbr, Bd/Dir and Chmn of the Engrg Res Council for American Society for Engrg Education, Mbr Bd of Dirs 1979-83; Chmn, Engineers' Week 1977-78 NSPE. Major General, USAFR Mbr Asst DCS R&D. Retired 1980. *Society Aff:* ASEE, NSPE, AIAA.

Hanes, N Bruce
Business: Dept Civil Engrg, Medford, MA 02155
Position: Professor/Chmn Civil Engineering. *Employer:* Tufts University. *Education:* PhD/Evironmental Eng/Univ of WI; MS/Environmental Eng/Univ of WI; BS/Civil Engg/North Dakota State Univ. *Born:* 1/24/33. Taught Univ of WI 1955-57, MT State Univ 1957-59 and Tufts Univ 1961- present. Worked as an Engr for the City of Milwaukee and the MT State Bd of Health. Elected Park Commissioner and Mbr of Bd of Health, town of Winchester, MA. One of the founders, elected chmn of the Bd of Dirs, 1973-74 and currently serving on the Bd of the New England Consortium on Environmental Protection. Have served as a consultant to the US Public Health Service, Environmental Protection Agency and the Natl Sci Foundation. PPres of the Assoc of Environmental Engg Profs. Chrmn Amer Acad of Environ Engrg Education Ctte. Interested in shaping Natl Policy with respect to the protection of the environment. *Society Aff:* AEEP, AAEE, AWWA, WPCF, ASEE.

Hanesian, Deran
Business: Dept of Chem Eng, Chemistry, and Environmental Science, New Jersey Institute of Tech, Newark, NJ 07102
Position: Prof/Chmn, Dept Chem Engrg & Chem. *Employer:* NJ Inst of Tech *Education:* BChE/Chem Engr/Cornell Univ; PhD/Chem Engr/Cornell Univ *Born:* 9/26/27. Niagara Falls NY. s. Vahan and Anna (Kabasakallian) H; With E I duPont de Nemours 1952-57, 60-63, production, dev, and res engrg Niagara Falls NY, 1952- 1957; res engr, Wilmington DE 1960-63; Prof and Chrmn Dept of Chem Engrg and Chemistry, NJIT 1963- ; served with AUS 1945-46. Mbr Amer Soc Engrg Edn, Amer Inst Chem Engrs, Amer Chem Soc, Natl Assoc Armenian Studies and Res (Dir 1972), Sigma Xi, Omega Chi Epsilon, Alpha Chi Sigma, Omicron Delta Kappa, Tau Beta Pi, Order of the Engineer. Past Chrmn Amern Inst Chem Engrg Prof Dev Ctte. P Chrmn, Amern Inst Chem Engrs, Ed Projs Ctte. Contributed articles to sci, tech journals. Mbr Adv Bd, Intl Chem Engrg (1973-79), Elected Fellow, Amer Inst Chem Engrg (1981), P Chrmn, Am Soc Eng Educ, Chem Eng Div and Div of Experimentation and Lab Oriented Studies, Reg PE NY and NJ. BChE, Cornell Univ, 1952, PhD Cornell Univ, 1961. Honors: Robert Van Houten Award for Teaching Excellence, 1977; Senior Fulbright Hays Grant to Soviet Armenia (USSR), 1982; American Society of Eng Educ. (ASEE) Mid Atlantic, AT&T Foundation Award for Excellence in Instruction of Eng. Students, 1986; NSF grant to Develop Process Dynamic Lab; NSF grant to Develop Chemical Reactor Engineering Laboratory; Deutscher Akademischer Austauschdienst (DAAD) Grant 1981; Visiting Professor, Univ of Edinburgh Scotland 1971; Visiting Professor, Algerian Petroleum Inst, Boumerdes Algeria 1978. *Society Aff:* AIChE, ACS, ASEE, ΩΧΕ, ΑΧΣ, ΣΞ, ΟΔΚ, ΤΒΠ, ORDER OF THE ENGINEER.

Haney, Paul D
Business: 1500 Meadow Lake Pkwy, Kansas City, MO 64114
Position: Consultant (Retired Partner) *Employer:* self *Education:* MS/Sanitary Engg/Harvard Univ; BS/Chem Engg/KS Univ. *Born:* 2/5/11. Mbr, National Academy of Engrg; Mbr, Sig Xi; Diplomate, American Academy of Environmental Engrs. Honorary Mbr, American Water Works Association. President (1968-69), (Honorary mbr 1977) Water Pollution Control Federation. Mbr, American Chemical Society. Fellow, American Society of Civil Engrs. Mbr, Amer Inst of Chem Engrs. Fellow, Amer Assoc for the Adv of Sci. Mbr, Nat'l Soc of PE. Fellow, Am. Public Health Assoc. Mbr, Cosmos Club of Washington, D.C., Distinguished Engrg Service Award, Univ. of Kansas, 1983. Registered professional engr. Professional Experience: state sanitary engrg activities, Kansas Department of Health & Environment; Associate Professor, University of Kansas and University of North Carolina. Commissioned Corps, US Public Health Service. Since 1954 with Black & Veatch, Engineers-Architects (retired partner; consultant since 1971). *Society Aff:* NAE, ASCE, AIChE, AAEE, NSPE, ACS, AAAS, APHA, AWWA, WPCF.

Hang, Daniel F
Home: 2012 Boudreau Dr, Urbana, IL 61801
Position: Pres *Employer:* HTH Assoc, Inc *Education:* MS/EE/Univ of IL, Urbana; BS/EE/Univ of IL, Urbana *Born:* 07/17/18 Proj engr GE, Schenectady, NY 1941-47. Prof (Emeritus) of Elec Engrg and of Nuclear Engrg, Univ of IL 1947-84 with continued research in engrg economy as applied to nuclear fuel mgmt and elec power. Secy, IL PE Examining Ctte 1970-. Co-founder-pres of HTH Assocs Inc 1978-, providing software and service in nuclear fuel mgmt to elec utilities. Engrg conslt to Argonne Natl Lab 1959-81, Army Corps of Engrs 1971-5, and Natl Sci Fdn 1974-6. Enjoy most participating sports and Faculty advisor to Tau Beta Pi, IL Alpha chapter. *Society Aff:* ANS, IEEE, ASEE, NSPE-ISPE, NCEE

Hanhart, Ernest H
Home: 9 McKim Avenue, Baltimore, MD 21212
Position: Consulting Mechanical Engineer. *Employer:* Self Employed. *Education:* BE/Mech/Johns Hopkins Univ. *Born:* 3/10/10. Mechanical Engr Bethlehem Steel Company 1933-45. Consulting Engr 1945 to present. JHU engrg faculty 1947-62. Special acheivements: established day world record rebuilding blast furnace; discovered fresh water production by blast furnaces; design and construct portable hydrogen liquefier and vacuum sealed valve for liquid hydrogen transfer; design welded composing frame; lighting fixture for reduced spectral reflection; design composing type casting machine for expanded capacity and mobile tilt self-balancing imposing cable; patent fabric washing machine; design infinitely variable/hydro cross-wave amplitude generator machine Life Fellow ASME, Vice President ASME, Chmn Organization Committee ASME. 1958-63; Chmn. Natl. Nominating Comm. ASME: 1957. Mbr Maryland Board of Registration for Professional Engrs and Land Surveyors; 1960-70; Pres MSPE. Life Member; MSPE Meritorius awards: Pi Tau Sig (Honorary); ASME, MSPE; NSPE, ESB; Founders' Award-Engrg Society of Baltimore; ASME Centennial Medallion *Society Aff:* ASME, NSPE, MSPE, ESB.

Hanink, Dean K
Home: Hannink Consulting, 145 Maple Crest Dr, Carmel, IN 46032
Position: Metallurgical Consultant *Employer:* Allison Division General Motors Corp. *Education:* BS/Met E/U of MI; Metallurgical Engrg/-/University of Michigan. *Born:* March 8, 1921 Triplett Mo. BS Metallurgical Engrg 1942 Univ of Mich. Reg Prof Engr Mich. Started career with Dodge Chicago Plant, Chrysler Corp. 1945 joined Res Labs Div, G M Corp as res engr. Transferred 1955 to Allison Div as ch metallurgist. Retired from position of Mgr. Engr. operations to become a consultant. Numerous pats granted & papers publ in SAE Transactions, ASTM Bulletin, Metal Progress & Prod Engrg Design Digest. Honored as an Amer Soc for Metals Fellow 1971, P Pres & Trustee 1975-76. Member SAE Aerospace Council. Member Panel for Redesign of Space Shuttle Solid Propellant Booster, Nat'l. Acad. of Sciences. *Society Aff:* AME, SAE, ASM.

Hankins, Richard P
Business: 1604 Santa Rosa Rd, PO Box K90, Richmond, VA 23288
Position: Pres. *Employer:* Hankins and Anderson, Inc. *Education:* BS/Elec Engg/VPI. *Born:* 3/25/12. Native of Richmond, VA. Mech engr for several firms prior to WWII. Served with Army Air Corps 1942-1946. Commenced private practice as consulting mech, elec and civ engr 1947. Inc 1973. Presenty chrmn, pres and treas of the corp. Registered Professional Engineer in fifteen states and the United Kingdom.. *Society Aff:* ACEC, NSPE, ASME, ASHRAE, IME, (UK).

Hanks, Richard W
Business: Dept of Chem Engg, Provo, UT 84602
Position: Prof & Chrmn. *Employer:* Brigham Young Univ. *Education:* PhD/ChE/Univ of UT; BS/ChE/Yale Univ. *Born:* 10/16/35 Res Engr, Union Carbide Oak Ridge Gaseous Diffusion Plant, 1960-1963. ChE prof, BYU, 1963-present. Pres. Richard W. Hanks Assoc. Inc., 1979-present. Res and publications in areas of non-Newtonian fluid mechanics, slurry pipeline hydraulics, transitional and turbulent flows in various types of ducts, thermodynamics of nonideal solutions. Over 20 publications, 4 books. Reg PE, UT No 3143. Consultant on slurry pipeline hydraulics, non-Newtonian fluid mechanics, transitional and turbulent flow, slurry rheology. *Society Aff:* AIChE, ASEE, SR, STA.

Hanley, Thomas R
Home: 914 Cambridge Ct, Terre Haute, IN 47802
Position: Associate Prof *Employer:* Rose-Hulman Institute of Technology *Education:* PhD/Chem Engg/VA Tech; MS/Chem Engg/VA Tech; MBA/Mgt/Wright State Univ; BS/Chem Engg/VA Tech. *Born:* 7/26/45. Native of Logan, WV. Married, Norma Kathryn Decker. Mfg engr and proj leader at Electro-Tec, Blacksburg, VA, during grad school. Served in USAF (1972- 1975) as a dev engr doing res with high temperature polymers and space system lubricants, WPAFB, Dayton, OH. Asst Prof of Chem Engg at Tulane Univ (75-79), New Orleans, LA. Asso Prof of Chem Engg at Rose-Hulman Inst of Tech (79-present) AIChE (New Orleans Sec Chrmn, 79-80), Terre Hante Sec Chrmn, 81-82, ASEE, NABC, SPE, ISOTT. Tau Beta Pi, Phi Kappa Phi, Lambda Upsilon, Omega Chi Epsilon, Sigma Xi. Consultant, Air Force Reservist, PE (LA, VA, IN). SAME award (1966 and 1967), American Legion award (1967), Distinguished Military Grad (1967), AIChE Prof Dev Award (1980), AIChE Outstanding Student Chapter Advisor (1979) *Society Aff:* AIChE, NABC, ASEE, SPE, ISOTT

Hann, Roy W, Jr
Home: 1300 Walton Drive, College Station, TX 77840
Position: Prof & Res Engr. *Employer:* Texas A&M University. *Education:* BSCE/Civ Engrg/Univ of OK; MCE/Civ Engrg/Univ of OK; PhD/Engr Sci/Univ of OK. *Born:* 3/21/34. in Oklahoma City, OK. BS, MCE & PhD from Univ of OK. USPHS Dallas 1957-59; Civil Engr C H Guernsey & Co 1959-60; Cons Civil Engr 1960-62; Asst Prof USC, Columbia 1962-65; Asst, Assoc Prof Env Engr, TX A&M Univ 1965-71; Sea Grant Dir 1976-77 Prof TX A&M Univ 1971- , Hd Environmental Engr 1970-75, 1980-86. Pres Civil Engr Sys Inc 1968- , Realtor-Owner Spring Valley Ranches. Pres Intl Spill Tech Corp 1979- Pres Hann Interprises 1963- . 1969 Outstanding Young Engr, TSPE; 1969 Paper Award, AWWA; 1970-71-72 Paper Award, TX Sect ASCE; 1972 Sea Grant Merit Award. Received 1983 Audubon Soc AAES Palladium Medal; Mbr ASCE, ASEE, AWWA, TSPE, NSPE, TWPCF, WPCF, BCS Bd of Realtors, Amer Brahma Assoc, Sigma Xi, Chi Epsilon, Omicron Delta Kappa, Sigma Tau, Tau Beta Pi, Sierra Club & Sigma Chi. Served on Intl (UNEP, IMCO, WHO, US Coast Guard) & natl (Natl Acad of Scis, Gov't & Industry) consltg assignments; contributions of over 150 publs. Specializing in Oil Pollution Control, Dir of Environmental Studies and Water Pollution Control Res. *Society Aff:* ASCE, AWWA, TSPE, WPCF.

Hanna, George P, Jr
Home: 7389 N Bond Ave, Fresno, CA 93710
Position: Prof, Civil Engg. & Dir., Center for Applied Engrg. Research *Employer:* CA State Univ, Fresno. *Education:* PhD/Civil Engg/Univ of Cincinnati; MS/Civil Engg/NY Univ; BS/Civil Engg/IL Inst of Technology. *Born:* 3/25/18. PhD/Civil Engrg/Univ of Cincinnati; MS/Civil Engg/NY U; BS/Civil Engrg/IL Inst of Tech. Native of Western NE. Served in Sanitary Corps, AUS 1943-46. Design engr Holmes, O'Brien & Gere, Cons Engrs Syracuse NY 1946-48. Dir Bureau of Sanitation, Syracuse Dept of health 1948-50. Asst Prof Civil Engr, Syracuse Univ 1950-52. Chf Engr Matl Handling Products Corp, Syracuse NY 1952-53. Engr Creole Petroleum Corp, Venezuela 1954-59. Prof Civil Engr & Dir Water Resources Ctr, OH State Univ 1959-69. Prof and Chmn Dept of Civil Engg, Univ of NB 1969- 72. Dean College of Engg & Technology, Univ of NB 1972-79. Prof of Civil Engg, CA State Univ Fresno 1979- . Author of articles on water quality and wastewater res. Mbr, ASCE (Fellow); AAEE, NSPE, AWWA, WPCF, ASEE, APWA, AEEP. Reg Prof Engr NY, OH, NB, CA *Society Aff:* ASCE, ASEE, NSPE, AWWA, WPCF, AAEE, AEEP, APWA

Hanna, Martin Jay, III,
Home: 10523 Liberty Rd, Randallstown, MD 21133
Position: Pres. *Employer:* Fire Engrg Corp. *Education:* BS/Fire Protection/Univ of MD. *Born:* 7/2/36. Pres and Principal Engr of Hanna Fire Engineering Corp, a consulting engg firm specializing in the fields of Fire Protection, Safety, Construction and Codes. Formerly, Fire Protection Engr with the MD Dept of Public Safety and Correctional Services, Office of the Fire Marshal, with responsibility for Medicare-Medicaid (USDHEW) Prog for Life Safety Surveys. Chief Fire Protection Engr with Baltimore County, MD attached to the Office of the Fire Chief. Have served as an Interim Instr with the Dept of Fire Protection Engg, Univ of MD, as a Field Instr with the MD Fire & Rescue Inst, and adjunct faculty, U.S. Fire Academy, FEMA. *Society Aff:* ASHRAE, NSPE, SFPE, ASPE, ASSE.

Hanna, Owen T
Business: Dept of Chem & Nucl Engr, Santa Barbara, CA 93106 *Employer:* Univ of CA. *Education:* PhD/Chem Engr/Purdue Univ; BS/Chem Engr/Purdue Univ. *Born:* 6/8/35. Native of Chicago, attended school in Terre haute, IN. Served with US Army at Caltech Jet Propulsion Lab 1961-62; Asst Prof ChE Rensselaer Poly Inst 1962- 65; Sr Res Engr, Boeing Co 1965-67; Currently with Univ CA Santa Barbara, Assoc Prof 1967-74, Dept Chrmn 1971-73, 1981-84; Prof 1974-Pres; Consultant to

Hanna, Owen T (Continued)
EG&G, US Navy, Lawrence Livermore Natl Lab. Los Alamos Natl Lab. Main technical interests include methods for solution of problems in transport phenomena and chemical reaction systems; applications of math in chem engg including math modeling, numerical and asymptotic methods, optimization and statistical methods. *Society Aff:* AIChE.

Hanna, Steven J
Business: School of Engrg, Southern Ill. Univ, Edwardsville, IL 62026-1275
Position: Prof & Assist Dean *Employer:* Southern Illinois Univ at Edwardsville *Education:* PhD/CE/Purdue Univ; MS/CE/Purdue Univ; BS/CE/Purdue Univ *Born:* 12/24/37 Native of Oaklandon, IN. Educated at Purdue Univ concentrating in the civil engineering area of transportation and materials. Undergraduate cooperative education student with the Indiana State Hwy Commission. Dept of Civil and Architectural Engrg faculty member at the Univ of WY 1967-1973. Returned to the midwest in 1973 as a member of the Department of Engrg and Technology faculty at Southern Illinois Univ at Edwardsville. Chairman of the Department of Engineering & Technology 1981-1983; Asst Dean (Acting) of the Sch of Engrg and Chrmn (Acting) of the Dept of Civil Engrg 1983-1984. Assumed duties as Prof and Assist Dean of the Sch of Engrg in 1984. Vice Chairman, Rocky Mountains District ASTM 1972-73. President Wyoming Section ASCE 1972-73. President St Clair Chapter of The IL Soc of PEs 1986-87. Past Pres and Consistory member, United Church of Christ of Highland, IL. Enjoy gardening, hunting and fishing. *Society Aff:* ASCE, ASEE, ASTM, NSPE.

Hanna, William J
Home: 27 Silver Spruce-NSR, Boulder, CO 80302
Position: Prof of EE. *Employer:* Univ of CO. *Education:* Profl/EE/Univ of CO; MS/EE/Univ of CO; BS/EE/Univ of CO. *Born:* 2/7/22. Native of Berthoud, CO. First Lt Sig C, US Army-Signal Corps Res Labs, Belmar, NJ, 1943-46. Faculty mbr, Univ of CO since June 1946. Prof of EE since 1962. Industrial experience-summers with Public Service Co of CO. Consultant, Los Alamos Scientific Labs 1965-1986. Chrmn, Denver Sec AIEE 1961-62; Pres Prof Engg of CO 1967-68; Pres Natl Council of Engg Examiners 1977-78. Mbr, CO State Bd of Engg Reg 1973 to 1984. PEC Alfred J Ryan Award, 1978; CA Soc of Prof Engg Archimedes Award, 1978; NCEE Distinguished Service Certificate, 1979. PEC Engineer of the Year 1968; Univ of Colo. Outstanding Alumni Award 1983; Colo. Eng. Reg. Board Appreciation Award 1987. Research in Electric Power Engineering. *Society Aff:* IEEE, ASEE, NSPE.

Hannabach, John
Home: 100 Catina Ct, Atlanta, GA 30328
Position: Pres *Employer:* Hancon *Education:* BS/EE/Purdue University *Born:* 07/17/32 Mr. Hannabach established HANCON, College/Industry Relations Consultants, in 1987. Prior to that, he was employed by General Electric for 30 years. He held numerous engrg and engrg mgmt positions from 1958-83 directing the efforts of engrs and scientists involved in development, analyses & design of electronic, structural & materials systems for aerospace & defense applications. From 1983-87, he was the Southern Regional Mgr for GE's Professional Recruiting and Univ Relations. He is an active commissioner with the Engrg Manpower Commission and serves on the Industrial Advisory Council of AAES. He is a bd mbr of Relations With Industry and served on the Advisory Bd for Quality of Engrg Education Project for ASEE. *Society Aff:* AAES, ASEE

Hannay, N Bruce
Business: 600 Mountain Avenue, Murray Hill, NJ 07974
Position: Vice President-Res . *Employer:* Bell Laboratories. *Education:* PhD/Phys Chem/Princeton Univ; MA/Physics/Princeton Univ; BA/Chemistry/Swarthmore College; PhD/Hon/Tel Aviv Univ; D.SC. Hon/Swarthmore College!Hon/Swarthmore College DSC/Hon/Polytechnic Institute of New York/Hon/Swarthmore College. *Born:* 2/9/21. in Mt Vernon, WA. Joined Bell Labs 1944; Chem Dir 1960-67, Exec Dir Res- Matls Sci & Engg Dir 1967-73, Vice Pres Res & 1973- . Mbr: NAS; NAE Foreign (secy); AAAS (Fellow); APS (Fellow); ECSPPres); ACS, IRI (PPres); Dirs of Indus Res P Chrmn. Adv Cttes: Univ of CA, Berkeley (Chrmn); Univ of CA San Diego; Stanford; Piny Princeton; Duke Brown; Lehigh; Clarkson College (Trustee); Harvard Univ (Chrmn); Council on Sci & Technology for Dev; Intl Inst for Applied Sys Analysis (Council mbr); Dept of State; Office of Sci & Technology Policy; Center for Sci & Technology Policy. Plenum Pub Corp Bd of Dir. Cons: Alexander von Humboldt Fnd; Cortexa International Fund (Banque de Paris et des Pays-Bas). *Society Aff:* NAE, NAS, APS, ACS, ECS, IRI, AAAS, DIR.

Hanneman, Rodney E
Business: PO Box 8 GE R&D Center, Schenectady, NY 12301
Position: Mgr-Matls Programs Operation *Employer:* General Electric Company. *Education:* PhD/Metallurgy/MA Inst of Technology; MS/Metallurgy/MA Inst of Technology; BS/Physical Metallurgy/WA State Univ. *Born:* 3/14/36. Native of Spokane, WA. Metallurgist GE's Hanford Labs, 1959. Res Staff Assoc with MIT Lincoln Labs 1959-63 in fields of electronic & optical matls & superpressure studies. Res Scientist at GE's Corp R&D Center 1963-68 in fields of high temperature thermodynamic & kinetic phenomena, diamonds & superhard matls, ceramics, interfacial henomena & physical metallurgy. Mgr of Inorganic Structures Branch 1968-74 & Inorganic Matls & Reactions Branch 1974-75 Mgr BWR Structural Matls Program 1975-77; Manager-Matls Char. Laboratory 1977-80; present position since 1980; currently respon for R&D progs in natural resoures, overall materials R&D strategic planning and aiding technology transitions. Recipient of ASM Geisler Award, ACS Chem Innovator recog, ASM Engg Matls Achievement Award, Edison Medallion. *Society Aff:* ASM, AIMS, ACS.

Hannigan, Eugene J
Home: 4434 Marywood Drive, Monroeville, PA 15146
Position: President *Employer:* Hannigan Engineering Associates Inc. *Education:* MSCE/Geotechnical Engg/Purdue Univ; BSCE/Civil Engg/Drexel Univ. *Born:* May 1939. Research on basic properties of clay structure. Has been practicing Geotechnical Consulting Engr in Pittsburgh since 1964. In 1973, founded Eugene Hannigan Consulting Engrs, which became Hannigan Engineering Assoc. Inc. in 1986. Registered Professional in four states and an Adjunct Assistant Professor of Civil Engrg at the University of Pittsburgh. Served ASCE as Pres of Pittsburgh Section; Secretary-Treasurer of District 4 Council; Chrmn of Natl Committee on Student Services, mbr of Steering Committee, 1978 Pittsburgh National Convention, & Mbr of Natl Committee on Soc Fellowships, Scholarships, Grants & Bequests, & Mbr of Exec Ctte of Ed Div. *Society Aff:* ASCE, NSPE, ASTM.

Hano, Ichiro
Home: 1-9 Ehara 1-chome Nakano-ku, Tokyo, Japan 165
Position: Professor Emeritus 1969. *Employer:* Waseda University. *Education:* Dr Engrg/Engg/Waseda Univ. *Born:* October 2 1898 Toyama Prefecture, Japan. Fellow: IEEE 1972; Honorary Mbr, Inst of Electrical Engrs of Japan 1968; IEEE Fellow Awarded 'for contributions to power systems engrg and in the development of power system stabilizing method.' Awards in IEE of Japan: Power Engrg Prize 1955, Achievement Prize 1965, Authorship Prize 1970. Founder's Memorial Academic Award: Waseda University 1965. National Award: Third Rising-Sun Prize, Japanese Government 1969.

Hanpeter, Robert W
Home: 8820 Hemingway Dr, St. Louis, MO 63126
Position: Sec-Tr & Ch Engr *Employer:* Engineering Sales Assocs, Inc. *Education:* BS/ME/MIT *Born:* 10/8/26 Native of St. Louis, MO. Married, 4 children. Spent 1 1/2 yrs in US Air Force in Air Traffic Control (1945-46), 4 yrs at MIT Cambridge, MA (1944-48). Business career in St. Louis with Wagner Electric Corp. (1948-52) as air brake design & production engr. Sunnen Products Co. (1952-59) as service Engr, Moehlenpah Engrg (1959-64) as Sales Engr. Founded present company (1964), Engi-

Hanpeter, Robert W (Continued)
neered Sales, a distributor of hydraulic-pneumatic products. *Society Aff:* FPS, SME, AISE, NDPA, St L Engrs Club

Hanratty, Thomas J
Business: Dept of Chem Eng, 205 RAL, 1209 W California, Urbana, IL 61801
Position: Prof Chem Engg. *Employer:* Univ of IL. *Education:* PhD/Chem Engg/Princeton; MS/Chem Engg/OH State; BChE/Chem Engg/Villanova. *Born:* 11/9/26. After receiving his PhD degree from Princeton Univ in 1953 Prof Hanratty joined the teaching staff of the Chem Engg Dept of the Univ of IL. Since then he has developed a broadly based res prog covering the areas of turbulence, boundary-layers, gas-liquid flows, convective heat and mass transfer and fluid- solid flows. The work has been recognized by three awards from the AIChE (Colburn, Walker and Profl Progress Awards) two awards of the ASEE (Curtis McGraw and Sr. Research Award), by election to the Natl Acad of Engrg by his receipt of an honorary doctorate degree from Villanova Univ and a Dist Engrg Alumnus Award from OH St Univ. Was appointed a Shell Dist Prof in 1981. *Society Aff:* AIChE, ACS, APS, ASEE, AAM.

Hansberry, John W
Home: 42 Bush St, S. Dartmouth, MA 02748
Position: Prof *Employer:* Southeastern Mass. Univ *Education:* PhD/Engrg/Brown Univ; ScM/Engrg/Brown Univ; BSME/Engrg/Rice Univ; BA/-/Rice Univ. *Born:* 9/18/43 School and college years spent in Houston, TX; Graduate of Rice Univ. PhD in Applied Mechanics and Engrg Science from Brown Univ. Prof of Mechanical Engrg at Southeastern MA Univ and Consulting Eng in Mech Design and Analysis of Mechanical and Power Plant Components. Active in the field of product liability; services available to the legal profession. Strongly active in local ASME section acvtities. Leisure activities include marathon running in which performances are regionally competitive in masters category. *Society Aff:* ASME.

Hansel, Paul G
Home: 1374 Monterey Blvd NE, St Petersburg, FL 33704
Position: Consultant, Sci & Technology (Self-employed) *Education:* BS/Engrg-Physics/Univ of KS. *Born:* 6/22/17. in Grand Island, NE. Res Engr (Radio Direction Finding) Signal Corps Engrg Labs, Ft Monmouth NJ 1941-47; chf Radio Engr Servo Corp of America, Hicksville NY, 1947-61; VP Electronic Communications Inc, St Petersburgh, FL 1961-79; N Numeous Pats, Inventor of Doppler Direction Finder & Doppler Omirange. Fellow IEEE, Recipient IEEE Pioneer Award 1970 and Univ of FL (Coll of Engrg) Disting Service Award 1976. Sci & Tech Conslt 1979-present. *Society Aff:* IEEE, AAAS, NYAS.

Hansen, Arthur G
Home: 815 Sugarbush Ridge, Zimsville, IN 46077
Position: Educational Consultant (Retired). *Education:* PhD/Mathematics/Case Inst of Tech; MS/Mathematics/Purdue Univ; BS/EE/Pudue Univ. *Born:* 2/28/25. Dean, Coll of Engrg 1966-69; Pres 1969-71 - GA Inst of Technology; Pres - Purdue Univ 1971-1982. Chancellor Texas A&M System 1982-1986. (Retired). Mbr: Natl Acad of Engrg, mbr. Amer Association for the Advancement of Sci, Chrmn Indiana Endowment for Educational Excellence, Board mbr American Electric Pwr., Navistar, Int. Paper Co. and Interlake Corp. Indiana Engineer the year, 1979. Past mbr, Genl Advisory Councils of Electric Power Res Inst & Gas Res Inst and Past Chrmn Natl Research Council Ctte on Minorities in Engrg. P Chrmn, Texas Science and Tech Council; P Mbr, Research Advisory Bd (DOE) Enjoys photography and fishing. *Society Aff:* NAE.

Hansen, Christian A
Business: LCP Chemicals & Plastics, Inc, CN 3106, Edison, NJ 08818
Position: Chmn & Chief Exec Officer *Employer:* LCP Chemicals & Plastics, Inc. *Education:* BS/Chem Engrg/Rice Univ. *Born:* 9/12/26. in New Braunfels, TX. US Navy; Exxon: Houston, Baytown, Linden, NJ 1948- 68; Dir Mfg Chemicals Div, GAF Corp NY City 1969-71; Founder, Pres, Chrmn, LCP Chemicals & Plastics, Inc. Edison NJ 1972-present. Mbr American Inst Chemical Engrs; Former Mbr Chemical Industry Council of NJ, 1965-1981 Chrmn, 1977-81; Councilman Bayton TX; United Fund, Union County, Man of Yr 1970; Mbr Baltusrol Golf Club Springfield, NJ: Mbr Navesink Country Club. Red Bank, NJ. Patentee in field; Former Chrmn and Current Mbr of the Bd of Dirs of The Chlorine Inst; Mbr of Chemical Manufacturers Assoc. *Society Aff:* AIChE, SCI.

Hansen, Edwin L
Home: R.4, Mahomet, IL 61853
Position: Owner (Self-employed) *Education:* MS/CE/Univ of IL; BS/Agri/IA State College. *Born:* February 2, 1911. Agricultural Engrg Portland Cement Assn 1941-46; Mbr Agri Engrg Ctte to China 1946-48; Partner, Hansen Bros Agri Engrg Service 1948- 53; Mbr Overseas Cons Inc, Iran 1948; Owner, E L Hansen & Co 1953-56; Prof, Univ of Ill 1956-75. Fellow ASAE, V Pres ASAE; P Pres Urbana Exchange Club. *Society Aff:* ASAE.

Hansen, Ethlyn Ann
Home: 77 Pikes Peak Dr, San Rafael, CA 94903
Position: Deputy Dist Director Operations & Toll Bridges. *Employer:* Calif Dept of Transportation *Education:* MS/CE/Trans Plng/Univ of CA, Berkeley; BS/CE/Univ of Ut *Born:* 1/26/31 Native to Salt Lake City, UT area. With the California Dept of Transportation, District 4, as an engr and planner since 1951 in the planning, design, construction, traffic and environmental planning branches. Is now Deputy District Director, Operations & Toll Bridges, in SF area, Caltrans. Responsible for all operational aspects of approx. 1500 mi of the most heavily travelled metropolitan freeways and toll bridges in the State of Ca. Licensed in CA as a PE in the civil and traffic disciplines. Officer and/or committee chairperson for ITE, SWE, San Francisco Bay Area Engrg Council. Active in may other civic and professional organizations. *Society Aff:* ITE, SWE, APWA, XE

Hansen, Grant L
Home: 10737 Fuerte Drive, La Mesa, CA 92041 *Employer:* Retired-Self Employed Consultant *Education:* D.Sc./Aero./National Univ; MA/Bus./American Univ; BS/EE/IL Inst of Techn. *Born:* 11/5/21 U. S. Navy World War II. Missile engrg and test, Douglas Aircraft 1947- 1960. VP-Launch Vehicles Convair 1960-1969. Assistant Secretary (R&D) U. S. Air Force 1969-1973. U. S. National Delegate NATO AGARD. VP, General Dynamics Corp and General Mgr Convair Division 1973-78. Pres, Systems Group, System Development Corp 1978-86. Retired Engineering and Management Consultant 1986-present. Pres AIAA 1975. IIT Alumni Hall of Fame 1984. NASA Distinguished Public Service Award 1975. USAF Exceptional Civilian Service Medal 1973 and 1983. Member Air Force Scientific Advisory Bd. Member National Academy of Engrg. Wife-Iris. Five children. *Society Aff:* AAAS, AIAA, AAS, IAA, DGLR

Hansen, Harold V
Home: 23823 178th Ave N, Cordova, IL 61242
Position: President. *Employer:* Harold V Hansen Inc - Cons Engr. *Education:* BS/AGR Engg/IA State College. *Born:* February 1915; native of Cordova Ill. Started with John Deere Planter Works 1936, advanced to Manager Product Engrg before retiring in 1974. 46 patents majority on corn planters. Cons Engr since 1974. Honorary awards; Fellow mbr ASAE; Gamma Sigma Delta, the honor society of Agriculture; Doctor of Business Council for Columbia S A. *Society Aff:* ASAE.

Hansen, Hugh J
Home: 3018 NW Lisa Pl, Corvallis, OR 97330
Position: Extension Agricultural Engineer *Employer:* Oregon State University. *Education:* MSAE/Agri Engrg/Cornell Univ; BSAE/Agri Engrg/ND State Univ. *Born:* March 30, 1923 Thief River Falls, Minnesota. Native of Kintyre, N D. Served in USMCR as fighter/test pilot 1942-46. Assistant Professor Purdue University 1952-54; Editor 1955-61, Editorial Director 1961-63, Publisher 1963- 74 of 'Electricity on

Hansen, Hugh J (Continued)
the Farm'; Mgr Western Regional Agri Engrg Service 1974- 78; Extension Agricultural Engineer 1979-present. Acting Agri Engr Dept Head 1986. Member American Society of Agri Engrs since 1952, director 1963-65, president 1971-72, No At Region Chmn 1963-64; fellow 1978; Engrs Joint Council Board Mbr 1972-74. National Safety Council Meritorious Service Award 1971. Executive Mgr National Farm Electrification Council 1972-74. Amer Soc of Agri Engrs George W. Kable Electrification Award 1980. *Society Aff:* ASAE, ASEE.

Hansen, James C
Business: P.O. Box 769, 78 River St, Springfield, VT 05156
Position: Pres *Employer:* James Hansen and Assocs *Education:* BS/EE/Univ of NM *Born:* 8/12/33 Native of Appleton, WI. Design engr for five years with WI MI Power Co., power supply engr for five years with CA Dept of Water Resources, and consulting engr for eight years with R. W. Beck and Assocs in Seattle and Boston specializing in feasibility and financing of power supply for electric utilities. Entered private practice as a consulting engr in 1976. Pres and principal owner of James Hansen and Assocs, consulting engrs, specializing in power supply and hydroelectric developement. Also Pres of American Hydroelectric Developement Corp, San Jose, CA. *Society Aff:* IEEE, NSPE.

Hansen, Kenneth D
Home: 5505 So. Emporia Circle, Englewood, CO 80111
Position: Senior Water Resources Engineer *Employer:* Portland Cement Association *Education:* MS/CE/Univ of CO; BS/Arch Eng/Univ of NM *Born:* 11/19/31 Structural Engr for Boeing Airplane Co. -Seattle and Ketchum & Konkel- Denver before joining Portland Cement Association in 1960 in Albuquerque. Assumed current responsibility as Senior Water Resources Engr in Denver 1971. Author of technical papers on soil-cement slope protection, roller compacted concrete and mass concrete for dams. Author ICOLD Bulletin "Soil- Cement for Embankment Dams." District 16 Director ASCE-three years starting 1981. Chrmn ASCE Professional Activities Ctte, 1984. Presidential Citation while president CO Section ASCE 1980. President Albuquerque Branch, ASCE 1967, Albuquerque Chapter NSPE 1971, and Albuquerque Chapter CSI 1966. *Society Aff:* ASCE, ACI, USCOLD

Hansen, Kent F
Home: Baker Bridge Rd, Lincoln, MA 01773
Position: Professor Nuclear Engineering. *Employer:* Massachusetts Inst of Technology. *Education:* ScD/Nuc Engrg/MIT; SB/Physics/MIT *Born:* August 10, 1931 in Chicago, Illinois. Senior Engr, Sylvania Electric Products, 1958-59. Ford Postdoctoral Fellow 1960-61, Assistant Professor 1961- 65, Associate Professor 1965-69, Professor 1969- , Nuclear Engrg Department MIT. Executive Officer 1972-74, Acting Department Head January - July 1975. Chmn of the Math & Computing Div of Amer Nuclear Soc (ANS) 1968-69; Bd/Dir of ANS 1971- 74; elected Fellow of ANS 1973. Professional interest in nuclear reactor theory, numerical methods & computer simulation. Assoc Dean of Engg, MIT, Jan 1979-1981; Arthur Holly Compton Award, Amer Nuclear Soc, June 1978. Director EG & G, Inc, 1979-. Mbr Natl Acad of Engrg - 1982, Assoc Dir, MIT Energy Lab - 1984 . *Society Aff:* ANG, ASEE, NAE

Hansen, Ralph W
Business: Dept Agri & Chem Engrg, Fort Collins, CO 80523
Position: Assoc Prof/Extension Agri Engr. *Employer:* Colorado State University. *Education:* MS/Agri Engrg/ND State Univ; BS/Agri Engrg/ND State Univ. *Born:* July 30, 1926. Extension Agricultural Engr and Associate Professor of Agricultural & Chemical Engrg. Agricultural Engrg Degrees from North Dakota State University. On Colorado State University faculty for 25 yrs. Extension Agricultural Engr for Iowa State University from 1953-56. Fulbright Lecturer at the University of Helsinki, Finland 1958-59; a Junior Design and Experimental Engr, New Idea Division of Avco Corp, Coldwater Ohio 1952-53; Assistant Utilization advisor, North Dakota Rural Electric Assoc, Grand Forks, North Dakota summer of 1950. Registered Professional Engr in the State of Colorado. Mbr of the American Society of Agri Engrs, served as Director, Chmn and Secretary of Rocky Mt Region; the American Society of Engrg Education; Phi Kappa Phi; Gamma Sigma Delta; Tau Beta Pi and listed in Who's Who in the West and American Men of Science. Major areas of research have been in agricultural pollution control and solar energy applications.. *Society Aff:* ASAE, GSD, ESP, ΣΞ

Hansen, Robert C
Home: 18651 Wells Drive, Tarzana, CA 91356
Position: President. *Employer:* R C Hansen Inc. *Education:* PhD/EE/Univ of IL; D Eng (Hon)/-/Univ of MO-Rolla; MS/EE/Univ of IL; BS/EE/MO School of Mines. *Born:* 8/3/26. Univ of IL 1950-55; sr staff engr microwave lab. Hughes Aircraft Co, 1955- 60; sr staff engr. Space Tech Labs, Inc, 1960-61; assoc dir satelite control, dir. Test Mission Analysis, MOL ops groups dir. Aerospace Corp, 1961-67; VP KMS Industries Tech Ctr Div KMS Industries Inc, 1967-71 Pres R C Hansen, Inc, 1971- . Served with USNR 1945-46. Reg PE, CA Fellow Inst Elec Engrs (London) and IEEE. IEEE group antennas & propagation chrmn 1963-64 and 1980. Mbr Am Phys Soc Intl Sci Radio Union (chrmn US comm VI 1967-69), Sigma Xi, Tau Beta Pi, Eta Kappa Nu, Phi Kappa Phi.

Hansen, Robert J
Home: 25 Cambridge Street, Winchester, MA 01890
Position: Principal. *Employer:* Hansen, Holley and Biggs Inc. *Education:* ScD/Civil Engg/MA Inst of Tech; BS/Civil Engg/Univ of WA. *Born:* May 1918 Tacoma, Washington. Research Engr 1940-45 for National Research Council, National Defense Res Committee of Office of Scientific Res and Dev. Research Associate to Prof of Civil Engrg MIT 1947-75. Prof Emeritus 1975- . Partner, Hansen, Holley and Biggs, Consulting Engrs 1955- . Principal, Newmark, Hansen and Associates 1958-68. Principal, Hansen, Holley and Biggs Inc 1975- . Recipient Army-Navy Certificate of Appreciation 1947; Moisseiff Award ASCE 1974; Reese Research Prize, ASCE 1975; Citation for Distinguished Service 1970, Dept of Defense. Mbr, Security Resources Panel, Executive Office of the President 1957; Mbr Sr Adv panel Air Force Ballistics Div USAF 195860; mbr Adv Com Civil Defense, Nat Acad Science 1959-69. *Society Aff:* ASCE.

Hansen, Torben C
Business: Lab. for Building Materials, Building 118, Technical University of Denmark, 2800 Lyngby Denmark
Position: Professor. *Employer:* Technical University of Denmark. *Education:* PhD/Concrete Technology/Royal Inst of Tech, Stockholm Sweden; MSc/Civil Engg/Technical Univ of Denmark. *Born:* 5/30/33. Professor of building materials at the Technical University of Denmark in Copenhagen. Formerly research engineer at the Swedish Cement and Concrete Research Inst, the Royal Inst of Technology in Stockholm, where he obtained a PhD degree in 1963. In the United States held positions as development engr at the Portland Cement Association, Skokie, Ill, and later as lecturer in engrg materials at University of California, Berkeley. 1965-67 was NSF Senior Foreign Scientist Fellow and visiting professor at Stanford University, California. 1956- , engaged in concrete res. ACI Wason Medal 1966. 1976-81 UNIDO Cons in concrete technology to the Indonesian Government 1976 & 1979. Chrmn of Rilem Tech Cttee 37-DRC on "Demolition and Reuse of Concrete–, Chrmn of Rilem Tech Cttee 94-CHC on "Concrete in Hot Climates–. *Society Aff:* ACI, RiLEM.

Hansen, Vaughn E
Business: 5620 S. 1475 E Bldg A, Salt Lake City, UT 84121
Position: Consulting Engineer-Owner. *Employer:* Vaughn Hansen Associates. *Education:* BS/Civil Engg/UT State Univ; MS/Civil Engg/UT State Univ; PhD/Mechs & Hydraulics/State Univ of IA. *Born:* 7/26/21. Registered PE. Recipient of Collingswood prize, and President Intermountain Section American Society of Civil Engrs. Mbr of honor societies of Phi Kappa Phi and Sigma Tau. Former Dir of Engrg Experiment Station and Utah Water Research Lab and Professor of Civil and

Hansen, Vaughn E (Continued)
Irrigation Engrg at Utah State University. Directed establishment of Inter-Amer Center for Land and Water Resource Dev for organization of Amer States. Directed irrigation res in 17 Western States for USDA. Pub more than 50 tech papers & author of revised edition of 'Irrigation, Principles and Practice'. *Society Aff:* ACEC, AWRA

Hansen, William P
Business: c/o Ridgecrest Group, Inc, 1425 Ridgecrest Circle, Denton, Texas 76205
Position: Management Consultant *Employer:* The Ridgecrest Group, Inc. *Education:* /-//-/Univ Houston; /-//-/UT Austin *Born:* Native of Texas: Career directed toward Project Management: Successful projects and major achievements in total spectrum of project development, conceptual design, detailed design, procurement, construction has been achieved. Director of integrated project programs including largest energy conservation program ever embarked upon--the Saudi Arabian Gas Gathering Project executed during years 1974 through 1983 utilizing over 3000 engineers and over 25000 field construction workers, and the 8 billion $ Esso Resources Canada, Ltd. Northern Alberta Refining. Proven leadership in project management, systems, motivating individuals and multi-engineering construction teams achieving significant objectives. *Society Aff:* AIChE, PMI

Hanson, Clayton L
Business: 1175 S Orchard - Suite 116, Boise, ID 83705
Position: Agri Engr. *Employer:* USDA, SEA-AR. *Education:* PhD/CE/UT State Univ; MS/CE/Univ of ID; BS/CE/ND State Univ. *Born:* 5/15/37. A native of Belfield, NC. Attended Meteorology Sch at the Univ of WA after completing BS Degree, and then spent two yrs as a US AF Forecaster, 1960-1962. Ari Engr in Rangeland Hydrologic Res prog in western SD with the USDA, ARS, 1963-1973. Presently Agri Engr with the Northwest Watershed Res Ctr, USDA, SEA- AR, doing stochastic and deterministic modeling of rangeland hydrologic systems with a special interest in modeling the mountainous precipitation regime. *Society Aff:* ASAE, ASCE, AGU, SRM.

Hanson, Dudley M
Home: 3812 N Thornwood, Davenport, IA 52806
Position: Supervisory Civil Engineer. *Employer:* US Army Engr District, Rock Island. *Education:* MPW/Public Works Admin/Univ of Pittsburgh; MS/Structures & Fnds/Univ of IA; BSCE/StrucEngr/Univ of IA. *Born:* 1/20/41. in LaCrosse, WI. With Rock Island Dist Corps of Engrs since Jun 1964 as Civil Engr, Struct Engr & Supervisory Civil Engr. Since 1986 Chief, Plng Div. Respon for mgmt and supervision of multi-disciplinary civil works planning and organization. Pres Tri-City Section, ASCE 1974-75; Chrmn Quint-Cities Joint Engrg Council 1972-73. Mbr Chi Epsilon, Pres Rock Island Post, Soc Amer Military Engrs 1979 Pres Port of Quad-Cities. Prop Club of US 1979-80; Mbr PIANC. Enjoy golf, bridge, reading, bicycle riding. *Society Aff:* ASCE, SAME, PIANC.

Hanson, John M
Business: 330 Pfingsten Road, Northbrook, IL 60062
Position: President. *Employer:* Wiss, Janney, Elstner Assocs. Inc. *Education:* PhD/Civil/Lehigh Univ; MS/Structural/IA State Univ; BS/Civil/SD State Univ. *Born:* 11/16/32. Lt USAF from 1953-55. Previously employed by Sverdrup & Parcel, J T Banner, Phillips-Carter-Osborn and PCA. With WJE since 1972. Over 30 publications. Mbr of PCI, IABSE, RILEM, & EERI. Fellow of ASCE & ACI. Active on prof cttes. P Chmn of ACI 215 Fatigue, PCI TAC, Segmental Const. & Research. Current Chmn of TRB A2CO0 Bridge. P Member Ill Structural Engineers' Examining Comm. Prof Engr in CO, OR, MN & MI and Struct Engr in IL. Corecipient of 1974 ASCE State-of-the-Art Award and 1976 Raymond C Reese Res Prize. Received ACI Bloem Disting Service Award 1976. Corecipient of 1978 PCI Martin P Korn Award and 1979 ASCE T Y Lin Award, Named Distinguished Engr of SDSU in 1979. Received Professional Achievement Citation in Engineering from Iowa St. U in 1981 *Society Aff:* ASCE, ACI, PCI, EERI

Hanson, Robert D
Business: Dept of Civ Engrg, Ann Arbor, MI 48109-2125
Position: Prof *Employer:* Univ of MI *Education:* PhD/CE/CA Inst of Tech; MS/CE/Univ of MN; BS/CE/Univ of MN *Born:* 7/27/35 Native of Albert Lea, MN. Engr with Pittsburg-Des Moines Steel Co., Asst. Prof for Univ of ND and Univ of CA at Davis before joining the Univ of MI in 1966. Served as CE Dept Chrmn from 1976-84. Active res and teaching program in earthquake engrg, structural design, supplemental mechanical damping of buildings, and structural repair and strengthening. Past Dir and VP of EERI, ASCE Reese Res Prize 1980. Mbr of NAE, ASCE, EERI and ACI. Currently technical coordinator for US/Japan Cooperative Large-Scale Earthquake Engrg Res Program and Member NAE Cttee on Earthquake Engrg. *Society Aff:* ASCE, EERI, ACI, NAE

Hanson, Roland S
Home: 10470 NW Lee, Portland, OR 97229
Position: President *Employer:* RS Hanson & Assoc. *Education:* MBA/Bus Mgmt/Fairleigh Dickinson; BS/Engg for Ind Mgmt/Fairleigh Dickinson. *Born:* July 1932. Native of Lyndhurst, NJ. and Senior Industrial engr with Textronix Inc Senior Consultant for Oregon Association of Hospitals-Systems Program. Industrial Engr & consultant with St Vincent Hospital & Medical Center, A T Kearney Inc, Beckman Instruments Inc, IBM, M & M's Candies, the Okonite Co. Registered PE (Industrial) Oregon & California Board 1967-69. Taught at Oregon State University & San Francisco State University. Written numerous articles and frequent lecturer. Vice President IIE 1973-76. Mbr IIE, NSPE. Fellow-AAAS. Industrial Engr of Year, San Francisco Bay Area 1970, Oregon 1972. Enjoy golf, home computing and furniture making. 80-81 Pres Engrs Coord Council of OR and Mbr Washington County, OR Public Works Advisory Committee. *Society Aff:* IIE, NSPE, ORSA, TIMS, AAAS.

Hanson, Walter E
Business: 1525 S 6th Street, Springfield, IL 62703
Position: Founder & Special Consultant. *Employer:* Hanson Engineers Incorporated. *Education:* MS/Civil Eng/Univ of IL; BS/Civil Eng/KS State Univ. *Born:* July 15, 1916. Navy Radar/Communications - Princeton, MIT 1939-44; MS University of Illinois 1947. Three years as US Navy officer. Taught six yrs, Instructor to Associate Prof, University of Illinois, in structural and geotechnical engrg; Three yrs as Engr of Bridge and Traffic Structures, Illinois Division of Highways; Formed predecessor to Hanson Engrs Inc 1954; ASCE- Cincinnati Section Award, 'Outstanding Contributions to Soil Mechanics and Foundation Engrg,' 1968; University of Illinois-Alumni Honor Award for Distinguished Service in Engrg, 1973; Who's Who in the World; Author and co- author of technical papers and textbook, 'Foundation Engrg' 1973. *Society Aff:* ASCE, NSPE, ASEE, ACEC, ACI, ASTM, AREA.

Hansson, Carolyn M
Business: Park Alle 345, DK-2605 Brondby Denmark
Position: Dept Head *Employer:* The Danish Corrosion Center *Education:* PhD/Metallurgy/Imperial College, London; BSc/Metallurgy/Imperial College, London. *Born:* Res Scientist Martin Marietta Labs 1966-70; Asst Prof Columbia Univ 1970- 71; Asst Prof 197173, Assoc Prof 1973-76 State Univ of N Y at Stony Brook; Mbr Tech Staff Bell Labs 1976-1980. Member, research staff, The Danish Corrosion Centre, 1980- Hardy Gold Medal of Met Soc of AIME 1970; Churchill Overseas Fellowship Cambridge 1977. Mbr Solid State Sci Ctte of NRC. Mbr Engrg Manpower Comm EJC 1975-77. Mbr Long Range Planning Ctte TMS/AIME 1975-78; Mbr Ctte on Met Profession TMS/AIME 1973-77. Guggenheim Fellowship 1977. Achievement Award of Society of Women Engineers, 1980. Mbr Danish Academy of Tech Sciences 1987- . *Society Aff:* IOM, AIME/TMS, ASM, ASTM, DMS, SEMS.

Hanway, John E, Jr
Business: PO Box 18, Naperville, IL 60540
Position: Consulting Engineer (Self-employed) *Education:* MS/ChE/WVA Univ; BS/

Hanway, John E, Jr (Continued)
ChE/WVA Univ. *Born:* May 1922. Post graduate studies Ohio State Univ 1950-54. Foreign service in USNR 1942-46. Div Ch, Batelle Inst, Cols OH 1961-65. Vice Pres Engrg, Copeland Process Corp, Oak Brook IL 1965-68. Asst Dir Res, Chicago Bridge & Iron Co, Oak Brook IL 1968-74. Independent cons engr 1974- . Reg P E in OH, IL & WI. Mbr AIChE, AIME, Sigma Xi. Chmn, AIChE National Mbrship Ctte 1965-67. 30 publications and 15 patents in fields of fluidized-bed technology and waste treatment. Hobbies: fishing, camping and gardening. *Society Aff:* AIChE, AIME, NSPE, ΣΞ.

Hanzel, Richard W
Business: 2001 South York Road, Oak Brook, IL 60521
Position: V Pres Engrg & Dir Tech Services. *Employer:* Sunbeam Corp, Internatl Group. *Education:* Professional Engg/Met Engg/MI Technological Univ; MS/Met Engg/MI Technological Univ; BS/Met Engg/MI Technological Univ; MBA/Bus/Univ of Chicago. *Born:* 11/7/24. in Cicero, IL. Assoc Res Metallurgist - Armour Research Fnd, Chicago 1951- 54; Ch Metallurgist R&D, Sunbeam Corp Chicago IL 1954-56; Mgr Res & Engg, Matls & Processes Div 1956-72; VP Engg & Dir Tech Services, Intl Group 1972- . Who's Who in the Midwest 1970. Elected Bd/Dir-Bell Federal Savings, Chicago 1975. Served with US Army 1943-46. Mbr ASM, Chmn Chicago-Western Chapter 1958-59; Chmn National Chapter Advisory council 1962; Chmn Editorial Committee Metal Progress 1967. Mbr of Public Service Committee of American Society for Metals, 1972-75. Contributor of articles to technical publications. Patentee in field. *Society Aff:* ASM.

Hapeman, Martin Jay
Business: 2901 E Lake Rd, Erie, PA 16531
Position: Mgr, Advance Product Concepts *Employer:* GE Co. *Education:* BME/ME/RPI. *Born:* 12/10/36. Marty received his BME Degree from RPI while working as Asst Mech Engr for NY State. After grad he served as Mech Engr (Ordinance) for Dept of the Army. In 1964, he joined GE Co as Mech Design Engr. He is a grad of the Co's Advanced Engg Prog. He had held management assignments in Mech Design, Mgt of Mining Ind Locomotive Engg, Propulsion Equip Engg and Diesel Engine Engg. He is currently Mgr. of Advanced Product Concepts. *Society Aff:* ASME, SAE.

Happ, H Heinz
Home: 1 River Road, Schenectady, NY 12345
Position: Consultant *Employer:* General Electric Company. *Education:* BSEE/EE/IL Inst of Tech; MSEE/EE/RPI; DSc/EE/Univ of Belgrade. *Born:* 6/27/28. HAPP, HARVEY HEINZ, electrical engineer, educator; b. Berlin, June 27, 1928; came to U.S., 1947, naturalized, 1953; s. Harry and Hertha (Friedmann) H.; m. Ruth Hollander, Nov. 17, 1951; children: Deborah Ann, Sandra Eva. B.S. in Elec. Engring, Ill. Inst. Tech., 1953; M.E.E., Rensselaer Poly Inst., Troy, N.Y., 1958; D.Sc., U. Belgrade, Yugoslavia, 1962. Registered profl. engr., N.Y. With Gen. Electric Co., 1954--; sr. application engr. Gen. Electric Co., Schenectady, 1968-72; mgr analytical engring. services Gen. Electric Co., 1972-77, mgr. advanced system tech., 1977-82, mgr. system analysis, 1982-1987, Consultant 1987--. also mem. faculty power system engring. course; lectr. colls. Author: Diakoptics and Networks (translated into Russian and Romanian), 1971, Piecewise Methods and Applications to Power Systems (translated into Chinese), 1980; editor: Gabriel Kron and Systems Theory, 1973; mem. editorial bd. Procs. IEEE; contbr. 140 articles and book revs. to profl. jours., chpts. to tech. books. Fellow IEEE (Prize Paper award region 5 1962, power systems engring. com. 1977, region 1 award 1980); mem. Tensor Soc. Gt. Brit. (v.p. 1972--), Conf. International des Grands Reseaux Electrique a Haute Tension, Internat. Power Systems Computations Conf. (co-founder 1962), Gen. Electric Co. Engrs. and Scientists Assn. (chmn. policy com. 1968-70), Ill. Inst. Tech. Alumni Assn., Sigma Xi, Tau Beta Pi, Eta Kappa Nu. *Society Aff:* IEEE, CIGRE, PSCC, TSGB.

Happel, John
Home: 69 Tompkins Ave, Hastings-on-Hudson, NY 10706
Position: President. *Employer:* Catalysis Research Corp. *Education:* DChE/Chem Engg/Poly Inst of Brooklyn; MS/Chem Engg/MIT; BS/Chem Engg/MIT. *Born:* April 1908. Native of New York City. Chemical Engr with Mobil Oil Co. Supervisor of 50 technical staff prior to WWII. Served on Neches Technical Ctte 194244, responsible for design and initial operation of world's largest butadiene plant for synthetic rubber. Chmn of Chemical Engrg Dept of New York University 1949-73. Organized Catalysis Research Corporation 1973 concerned with sponsored research in areas of energy conservation, alleviation of pollution and petrochemicals. Adj Prof at Columbia University since 1973. Certificate of Distinction PIB, Honor Scroll ACS, Tyler Award N Y Section AIChE, Fellow AIChE, & N Y Acad of Scis, Prof Engr in N Y state, Member of National Academy of Engineering. *Society Aff:* AIChE, ACS, ΣΞ, ΑΧΣ.

Harada, Tatsuya
Home: 1-21-2 Aobadai, Midori-ku, Yokohama-shi, Kanagawa-Ken, 227 Japan
Position: Assoc Dir *Employer:* Cent Res Institute of Electric Power Industry *Education:* PhD/EE/Tokyo Inst of Tech; BS/EE/Tokyo Inst of Tech *Born:* 10/5/24 Native of Yamanashi Prefecture, Japan. Sr. Research Eng for Central Research Institute of Electric Power Industry (CRIEPI) 1952-1963, specializing in impulse voltage tests of electric power apparatus. Fellow Research Engineer, Supervisor of High Voltage Section since 1964, Associate Director since 1972 in CRIEPI, doing researches on impulse voltage measurment, high current technique and application of opto-electronics to high voltage techniques. Editing Director, IEEJ 1976-1978. IEEJ'S Paper Prize 1966, Progress Awards 1968, 1971, 1973, Electronic Power Award 1975. Fellow IEEE 1981. *Society Aff:* IEEE, IEEJ, LSJ, JRED

Haralampu, George S
Business: 25 Res Dr, Westborough, MA 01581
Position: Dir of Engrg *Employer:* New England Electric System *Education:* MS/Elec Power Engg/Northeastern Univ; BS/EE/Tufts Univ. *Born:* 3/20/25. Born in Lynchburg, VA. Lived in NYC, & Greece before settling in the Boston area. Served with US Army 1943-1946. With New England Electric System since 1952 in system protection, planning, & computer applications. Field engr during construction of generating plant. Dir of Elec Engg 1975 to 1984; As Dir of Engrg, has responsibility in planning, computer engrg applications, HVDC interconnection between New England and Quebec, system protection, communications, distribution engrg, design of elec mech & struct facilities. Past Chrmn of Task Force on System Protection for the Northeast Power Coordinating Council, & Past Chrmn of the IEEE Surge Protective Devices Committee. Elected to grade of Fellow in IEEE in 1977. Received the Laurence F. Cleveland Award in 1987 by IEEE Boston Chapter. PES. *Society Aff:* IEEE.

Haralick, R M
Business: Elec Engg Dept, 148 Whittemore Hall, VPI & SU, Blacksburg, VA 24061
Position: Professor Electrical Engrg. *Employer:* VA Polytechnic Inst & State Univ. *Education:* PhD/EE/Univ of Kansas; MS/EE/Univ of Kansas; BS/EE/Univ of Kansas; BA/Math/Univ of Kansas *Born:* Sept 1943. Currently Prof of Electrical Engrg and Prof of Computer Science at VPI & SW. Research interests include pattern recognition, image processing, texture analysis, remote sensing, general systems, artificial intelligence, data compression, music, art, poetry. Sr mbr IEEE; 1975 ASEE Dow Award; 1975 Eta Kappa Nu Outstanding Young Engr Award, honorable mention; 1975 EIA Best Paper Award; NSF Faculty fellowship award 1977-79 to relate pattern recognition and diagnostic radiology; President of Douglas County Environmental Improvement Council 1976; and leader of KU Engrg School Explorer Post 1972-78. Dev KANDIDATS image processing software sys at KU & GIPSY image processing software at VPI & SU Assoc Editor of *IEEE Transactions on systems, Man, and Cybernetics*, Pattern Recognition, and *Computer Graphics and*

Haralick, R M (Continued)
Image Processing; Editorial Bd of *IEEE Transactions on Pattern Analysis and Machine Intelligence*. *Society Aff:* IEEE, ACM, PRS, SGSR.

Harbage, Robert P
Business: PO Box 32605, Charlotte, NC 28232
Position: Vice President & General Manager Actuator Div. *Employer:* Duff-Norton Company. *Education:* BAE/Power & Mech/OH State Univ; MS/Machine Design/ Purdue Univ. *Born:* February 1924. US Navy 1945-46. B Agri Engrg from Ohio State University; MS from Ppurdue Univeristy. Instructor, Purdue Univ. Stress Analyst, New Holland Machine Co. Development of agricultural harvesting equipment for New Idea Division, Avco Corp 1956-70; last ten years as Director of Engrg and mbr of Operating Committee. Vice President Engrg 1971-1985, VP & General Mgr. Actuator Div. of Duff Norton Co. 1985-present, manufacturer of hoisting and lifting equipment. Tau Beta Pi; Sigma Xi; Alpha Epsilon; Chmn Ohio Section ASAE, 1959; Chmn FIEI Agricultural Research Council, 1969; Celina School District Board of Education, 1968-70; Distinguished Alumnus of the College of Engrg, Ohio State University, 1970. *Society Aff:* ASAE, SAE.

Harbin, William T
Home: 3922 Gail Drive, Oakwood, GA 30566
Position: Ret. (Consulting Engr) *Education:* BS/ChE/GA Inst Tech. *Born:* 7/6/08. E R Squibb & Sons to 1946, res on antibiotics. Dept Hd, sulfer drug production, Celanese Corp of Amer. Plant Mgr plasticizer plant to 1952. Kolker Chem Co Newark NJ, Genl Chem Engr Prod Mgr. Thompson Chem Co (Teknor Apex Co), Pawtucket, RI. Chems & plastics mfg, Plant Mgr to 1973. Mbr ACS, AAAS, AIChe Fellow. Chairman RI Sect Mbr, Natl Adm Ctte. Elected AIChE Fellow 1972. AIChE (1972). *Society Aff:* ACS, AIChE, AAAS.

Harbour, William A
Business: 24 Aviation Rd, P.O. Box 5269, Albany, NY 12205
Position: Partner-in-Charge *Employer:* Clough, Harbour & Assocs *Education:* BCE/Transp/RPI; Assoc Applied Sci/Hgwy Tech/Hudson Valley Community Coll *Born:* 11/1/38 Mr. Harbour began his career with, what was then, John Clarkeson, Consulting Engr, in 1960. In 1971, he became Managing Partner and assumed responsibility for engrg and administration of a wide variety of projects involving the planning, design, and construction supervision of numerous civil engrg projects throughout NY, New England, and FL. the scope of Mr. Harbour's experience includes transportation design and planning with a heavy emphasis on the interstate networks, urban transportation and intercity mass transportation systems. He also has extensive experience in municipal engrg including the planning and design of several water distribution systems, elevated water storage tanks, zoning and subdivision regulations, and land use planning. Approximate Construction Value - 1 Billion Dollars. *Society Aff:* NYPF, NSPE, NYSPE, ITE, WPCF

Harbourt, Cyrus O
Business: Dept Elec Engrg, Columbia, MI 65201
Position: Prof. *Employer:* University of Missouri-Columbia. *Education:* PhD/Elec Engrg/Syracuse Univ; SM/Elec Engrg/MIT; BS/Elec Engrg/LA State Univ. *Born:* June 1931. Married Mary J Heuvel in 1952. Children: Ellen, Joan, Cyrus, Anna and Mary Alice. Served in US Army Ordnance Corps, Aberdeen Md 1955-57. Teaching experience at MIT, Univ Delaware, Syracuse, Univ Texas-Austin, and Univ Missouri-Columbia. Consultant to TRACOR, Humble Oil, Amer Council on Education. Technical interest in electric circuits, device modeling and simulation, power sys. Winner Teaching Excellence Award, Univ Texas 1963. ECPD Accreditation Visitor (IEEE), 1971-75. Chmn of EE at UMC 1967-77. Elec Engrg, Bonneville Power Admin, Portland, OR, 1977-78. Dir, Engg Extension, UMC, 1982-86 and 1987-present. Dean of Engineering, UMC, 1986-87. *Society Aff:* IEEE, NSPE.

Harchar, Joseph J
Home: 401 Payne Hill Rd, Clairton, PA 15025
Position: Consultant *Employer:* U.S. Army Corps of Engrs (Ret) *Education:* BS/CE/PA St Univ *Born:* 5/4/23 Native of the Pittsburgh area. Served with U.S. Navy 1942-44. Graduated coll in 1949. Registered in PA in 1954. Initially employed by American Bridge Co and Koppers, Inc. was with U.S. Army Corps of Engrs after 1958. Duties included assignment as Project Mgr on reservoir projects and was basically responsible for the planning of water resource projects to address flood control and streambank erosion problems. Retired from the Corps of Engrs in 1983 and currently operating as a conslt in the water resource mgmt field. Hobbies include sports, popular music and outdoor activities like gardening, hiking, fishing, etc. *Society Aff:* SAME, USCOLD, ΤΒΠ, ΧΕ

Harden, Charles A
Home: 809 S Willow Ave, Tampa, FL 33606
Position: Marine Surveyor & Consultant (Self-employed) *Education:* BS/ME/Univ of SC. *Born:* 12/5/25. Native of Norfolk, VA. US Naval Reserve 27 yrs-retired as LtCdr. Active duty 1943-1947 & 1950-1951. Marine Engr - US Army Corps of Engrs - 9 yrs; estimator & surveyor - MD Drydock Co, 4 1/2 yrs; Proj mgr Gibbs/ Aerojet Gen Shipyard - 5 yrs. Dist Sales mgr Bird-Johnson (CPP Propellers), 4 yrs; VP Arthur D Darden, Inc (Naval Arch & Marine Engrs) 2 yrs. Chief Engr - Tampa Ship Repair & D D Co - 2 1/2 yrs. Pres, Tampa Barge Service, 4 1/2 yrs. Consultant since early 1979. *Society Aff:* SNAME, SAME.

Harder, Edwin L
Home: 1204 Milton Avenue, Pittsburgh, PA 15218
Position: Senior Consultant. *Employer:* Westinghouse Electric Corp. - Retired. *Education:* PhD/Mathematics & Engineering/Univ of Pittsburgh; MS/Elec Engg/ Univ of Pittsburgh; BS/Elec Engg/Cornell Univ. *Born:* April 1905. Native of Buffalo, NY. First half of career in Railroad Electrification and Electric Power Engrg. Second half in Computers and Advanced System Engrg--both involving power system problems. After retirement writing on Energy--Resources, Processes, Basic Fundamentals. (Fundamentals of Energy Production, John Wiley & Sons, 1981) 65 US Patents, 150 Technical Papers on Control and Computers. IEEE Lamme Award for development in electrical machinery. AFIPS Distinguished Service Award for contributions as President of the US Computer Federation (AFIPS) and as US Representative to the International Federation (IFIP). National Academy of Engrg. Centennial Medal-IEEE. *Society Aff:* IEEE, AMS, NAE, SCS.

Harder, John E
Business: PO Box 341, Bloomington, IN 47402
Position: Advisory Engineer. *Employer:* Westinghouse Electric Corporation. *Education:* BS/Mech Engg/Carnegie Mellon; MBA/Bus/Indiana Univ; MS/Physics/ Indiana Univ. *Born:* 1930. Graduate study, Univ of Pittsburgh 1953-58. Employed by Westinghouse Elec Corp since 1952 as design & applications consultant for the Distribution Apparatus Div, Bloomington Ind. Principal products include power capacitors & equipment; surge arresters; circuit breakers, switches & fuses; coupling capacitors & line traps. Reg P E. Fellow IEEE; Mbr CIGRE, ASME. Vice Pres Region VI ASME 1974-77. *Society Aff:* IEEE, ASME.

Hardin, Bobby O
Home: Lexington, KY 40506-0046
Position: Prof of Civil Engrg *Employer:* Univ of KY *Education:* PhD/CE/Univ of FL; MS/CE/Univ of Ky; BS/CE/Univ of KY *Born:* 9/9/35 Served four yr term as Chairman of the Department of CE, 1973-1977. Awarded five national research prizes or medals Alfred Noble Prize, Walter Hubor Research Prize, Norman Medal, C.A. Hogenloger Award, and the Middlebrooks Award. Awarded three patents. Soil testing equipment developed by Professor Hardin is used throughout the USA and several foreign countries in about 40 laboratories. Besides 20 yrs of teaching, Dr. Hardin has worked for TVA on construction of Hickajack Dam and has consulted with numerous engineers on soil testing, constitutive equations for soils, and vibration of soils. *Society Aff:* ASCE, ASTM

Hardin, Clyde D
Home: 1 Sorrelwood Cross, Savannah, GA 31411 *Education:* BS/PHYSICS, MATH/ WAKE FOREST. *Born:* 5/26/25. Mr Hardin is a consultant in defense electronics. Served as a technical dir. US Army Electronic Res & Dev Command (Eradcom) 1980-81; assoc tech dir 1979-80; dir, Army Electronic Warfare Lab 1973-79; Dir Defense RDTE Group Korea 1971-73; spec asst to asst secy of the Army (R&D) 1969-71; consultant to the sci advisor, US military assistane command, Vietnam 1967-68; lab chief, Harry Diamond Labs, 1958-67; physicist/engr Natl Bureau of Stds/DOFL 1948-58. Fellow, IEEE. Defense meritorious civilian service medal, Army meritorious civilian service medal, Army R&D achievement award, ROK presidential natl security medal, AOC electronic warfare mgt medal. Veteran, WWII (Navy, aviation radar). *Society Aff:* IEEE, AFCEA, AOC, ADPA.

Hardin, Edwin M
Business: PO Box 101, Birmingham, AL 35201
Position: VP-Engr & Technology *Employer:* Rust Intl Corp *Education:* BS/CE/Univ of AL. *Born:* 6/11/26. Native of Birmingham, AL. Married to Edith Loraine Williams. Children Judith and Jenny. 1956-58 Jr Structural Engr & Technology, Rust Engg Co. 1958-63 Office Engr, Walter Schoel Engg Co 1963-68 Sr Structural Engr, Rust Engg Co 1968-70 Chief Structural Engr, Rust Engg Co 1970-74 Chief Design Engr, Rust Engg Co 1974-79 Chief Engr, Rust Intl Corp 1979 VP - Engrg & Tech, Rust Intl Corp. Tau Beta Pi, Chi Epsilon, Theta Tau, Phi Eta Sigma. Engg Council of Birmingham Engr of the Year - 1978, 1984. Woodman-Keith Award -1985. AL Soc of PE Engr of the Year - 1978-79. Baptist Mason (Shriner). *Society Aff:* ASCE, NSPE, SAME, NCEE, ASEE.

Harding, Richard S
Business: 7655 Redwood Blvd, Novato, CA 94947
Position: President. *Employer:* Harding-Lawson Associates. *Education:* BSCE/-/Univ of TX-Austin. *Born:* May 27, 1923. BS Civil Engrg Univ of Texas. Served in US Army Corps of Engrs (Europe, WW II; Korea, 1951). More then twenty-five yrs' experience in consulting soils and foundation engrg. Mbr of Seismic Investigation and Hazard Survey Advisory Committee for City and County of San Francisco; and American Nuclear Society Committee 2.11, providing geotech guidelines for nuclear power reactor sites. Mbr Amer Soc of Civil Engrs, Consulting Engrs Association of California, Structural Engrs Association of Northern California, Chi Epsilon, Tau Beta Pi. *Society Aff:* TBΠ, XE, CEAC, SEONC, ASCE.

Hardinge, Harlowe
Home: 556 Country Club Rd, York, PA 17403
Position: Chairman of the Board. *Employer:* Hardinge Company, Incorporated. *Education:* ME/-/Cornell Univ. *Born:* March 17, 1894 Denver Colorado. Sigma Xi Honorary Society. Served Overseas World War I, Cptn Signal Corps 1917-19. V Pres & General Manager Hardinge Co Inc 1922-39 York Penna. Pres & Chmn/Bd Hardinge Co Inc 1940- . Mbr M & M Soc of Amer 1950. Sold assets, incl many US & foreign pats, but not corp struct of Hardinge Co Inc, to Koppers Co in 1965. Legion of Honor, Mbr AIME Class of 1917. Robert H Richard Award recipient 1972. Class of 1975 SME-AIME Disting Mbr. Contrib author 'Taggart's Handbook' of Mineral Dressing. Past President Manufacturers' Association of York, Pa. Director Pennsylvania Chamber of Commerce. Former Director Natl Central Bank, York, Pa. Former mbr Administrative Bd Cornell Univ. Reg Prof Engr, Penna. Distinguished Pennsylvanian for 1981 *Society Aff:* AIME, M&MS, ΣΞ.

Hardison, Leslie C
Business: 600 North First Bank Dr, Palatine, IL 60067
Position: President and CEO. *Employer:* ARI Technologies Inc *Education:* BS/ME/IIT. *Born:* 2/16/29. in Chicago, IL. Employed by ITT & IITRI after graduation & subsequently by Univ Oil Products in design of refining processes in charge of pilot plant design group and process dev 1958-63. Dir of R&D, Catalytic Combustion Co 1963- 66. Technical Dir UOP Air Correction Division 1966-70. Cofounder & VP of Air Resources Inc, Chicago based engr-contractor firm spec in energy and environmental control sys 1970-78. Elected Pres and CEO of ARI Technologies, Inc Sept 1978. Mbr ASC, AIChE, ASME, APCA. Reg PE in IL, CT, CO & PA. *Society Aff:* ACS, AIChE, ASME, APCA.

Hardy, Robert M
Home: 11615 Edinboro Rd, Edmonton Alberta, Canada T6G 1Z7
Position: Chrmn of Bd and Prin *Employer:* Self Employed *Education:* DSc/Geotechnical Engrg/U of Manitoba; MSc/Geotechnical Engrg/McGill Univ; BSc/Geotechnical Engrg/U of Manitoba *Born:* Sept 1906. Native of Winnipeg Manitoba. Academic Staff U of Alberta 1931-59 & 1963-71; Dean of Fac of Engrg 1946-59 & 1963-71; Prof Emer CE 1971. Keefer Medal from EIC 1947; Centennial Medal of Canada 1967; Centennial Award of APEGG of Alberta 1968; Leggett Award of Canada Geotech Soc 1971; Gold Medal for 1973 of Can Council of PE's; Officer Order of Canada 1974; Achieve Award of Gov of Alberta 1974. Fellow Royal Soc of Canada; Fellow EIC; Life Mbr ASCE; Mbr ACEC. Current practice principally assoc with pipelines, mining of tar sand & tailings disposal sys. Founded R M Hardy 1951. *Society Aff:* EIC, ASCE, CGS, RSC.

Hardy, William C
Home: 1202 Glen Cove, Richardson, TX 75080
Position: Engrg Consit *Employer:* Sun Exploration & Prod Co. *Education:* MS/Petro Engr/Univ of OK; MBA/Bus/Univ of OK *Born:* 2/21/28 Served in US Navy 1945-1948. Attended Univ of OK and awarded BS and MS Degrees in Petrol Engrg in 1952 and 1953. Subsequently accepted position as Asst Prof in Petrol Engrg Dept of Univ of OK. Joined Sun Prod Co Res Organization in Dallas, TX in 1954. During 1960's studied math at SMU. During 1970's attended Univ of Dallas, earning MBA in 1974. Received many res assignments from 1954- 1972, the maj one in thermal methods of oil recovery. Awarded 20 US and 15 foreign patents. Assigned to Exploration Dept of Sun from 1972-1975 as Mgr of Heavy Oil Task Force; respons for locating, appraising, leasing, and developing heavy oils in US. From 1976 to 1983 was assigned to Bus Planning and Analysis Dept, with title of Planning Conslt for External Affairs; respons for making long-range forecasts of economic parameters critical to Sun's operations. Assigned to Res Dept 1983 to present with title Engrg Conslt, with respons in the area of thermal methods of oil recovery. Past Chrmn of Dallas Sect of SPE. Past Chrmn of Hydrocarbon Economic and Evaluation Symposium in Dallas. Past Chrmn of Petroleum Data Sys, a very large data base sponsored by the Univ of OK. Fellow of Amer Acad of Sci on Oil Producibility Ctte. Petroleum Engr of the Yr in Dallas 1982. Recipient of SPE Regional Serv Awd 1984. Publ 6 articles in the area of enhanced oil recovery tech and 10 articles on the economic impact of govt regulations on the petroleum indus. *Society Aff:* AIME

Harer, Kathleen F
Home: PO Box 21142, KSC, FL 32815
Position: Chief, Industrial Safety *Employer:* NASA Kennedy Space Center *Education:* BS/Ind Engg/Univ of WA; BS/Aero&Astron/Univ of WA; MBA/-/Univ TN. *Born:* 5/17/47. Ms Harer is currently chief indust. safety for NASA Kennedy Space Center where she has worked since 1983. Prior to that among several pos. held, Sr Safety Engg with JRB Assoc, Inc, a safety and health consulting firm, three yrs as a Safety Engg with the Occupational Safety and Health Administration (OSHA) and three yrs with the US Dept of Energy (formerly AEC and ERDA). Immediately prior to joining NASA, was proj mgr for a Dept Energy environm. cleanup proj. She has served in several officer positions and chaired committees in her tech societies, including the following: SWE 1987-88 Natn'l Pres, Pres-elect, VP, Treas, Exec Cttee Dir, Regional Dir - Smoky Mountain Section Pres and Sect Rep, Baltimore/Washington Sect Treasurer, National Mbrship and Statistics Committees Chair, and Reg Student Activities Coordinator; ASSE- East TM Chapter Pres and Treasurer. *Society Aff:* SWE, ASSE, Bd Certified Safety Professionals.

Harford, James J
Business: 1290 Ave Americas, New York, NY 10104
Position: Executive Secretary. *Employer:* Amer Inst Aero & Astronautics. *Education:* BE/ME/Yale Univ. *Born:* 8/19/24. Served as Navy engrg officer during WWII. Worked as an estimating engr for Worthington Corp and was an Assoc Editor of Modern Industry Magazine. Spent two years in Europe in 1952-53 writing articles under contract to the U S Mutual Security Agency on French, British and Italian industries. Became Executive Secretary of the American Rocket Society in 1953. When ARS merged with the Inst of Aerospace Sciences in 1963 to become the American Inst of Aeronautics and Astronautics, became Depty Exec Secy. Appointed Exec Secy 1964. Fellow AIAA & Fellow British Interplanetary Soc & AAAS; Assoc Fellow Royal Aeronautical Soc. *Society Aff:* BIS, AAAS, RAES, IEEE.

Hargens, Charles W, III
Home: 1006 Preston Rd, Philadelphia, PA 19118
Position: Staff Engineer. *Employer:* Franklin Res Center. *Education:* SB/EE/MA Inst of Technology. *Born:* Oct 21, 1918 Philadelphia. With Lockheed Aircraft Corp, 1941-42; Mass Inst of Tech 1942-45 (Radiation Lab) & Gilfillan (ITT) Calif; with RCA 1945-47; with Franklin Inst Res Labs 1947- , Tech Dir Elec Engrg Dept 1960-68, Fellow 1968- ; Assoc Prof Temple U 1976-77; Assoc Prof Drexel U 1978- ; mbr Elec Engrg Curriculum Comm Spring Garden College; Reg Prof Engr Pa.; Elected Fellow IEEE; Mbr Amer Soc Testing & Matls, Acoustical Soc Amer, Sigma Xi; Mass Inst of Tech Club P Pres. Patentee in field. Contrib to prof journals. *Society Aff:* ASA, IEEE, ASTM, ΣΞ

Harger, Robert O
Home: 23 Watchwater Way, Rockville, MD 20850
Position: Prof *Employer:* Univ of MD. *Education:* PhD/EE/Univ of MI; MSE/EE/Univ of MI; BSE/Math/Univ of MI. *Born:* 9/15/32. Line Officer, USN (1955-1958). Res Engr, Inst of Sci & Tech (1960-1968) and Asst Prof, EE (1963-1965), Univ of MI, Ann Arbor. Assoc Prof (1968-1975), Prof & Chrmn (1975-1980), Univ of MD. Fellow, IEEE (1979). Carlton Award, Best Paper (1977), IEEE Aerosp & Electronic Sys Soc. Consultant: Indus, Governmental, Nonprofit, & Public Interest Organizations, 1968-present. Author: Synthetic Aperture Radar Systems, Academic Press (1970). Editor: Optical Communication Theory; Dowden, Hutchinson, Ross (1977). *Society Aff:* IEEE, AAAS.

Hargis, John C
Business: 8110 East 46th Street South, Tulsa, OK 74145
Position: VP *Employer:* Netherton & Associates *Education:* BS/Arch.Eng/OK State Univ *Born:* 8/20/30 Native, OK. Served USCE 1951-1953, Company Commander Engrg Construction Company. General contractor 1953-1957, joined Netherton & Associates 1961, Responsibility of Vice President since 1973. Major responsibility is Director of Industrial Projects. Also serve on Advisory Board to OK State Tech. Have patented design athletic training device. Enjoy working wood and metals. *Society Aff:* AISC, AWS, PCI, ACEC

Hargreaves, George H
Home: 1660 East 1220 North, Logan, UT 84321
Position: Research Professor Emeritus *Employer:* Utah State University, Logan. *Education:* BS/CE/Univ of WY; BS/Soils/Univ of CA. *Born:* Apr 2, 1916. Delegate to U N ECAFE Water Resources Conference, 1953; Chmn Rocky Mountain Section ASAE, 1974; Chmn Surface Water Committee ASCE, 1974- 75; author or co-author of pre-investment studies, a soil survey report, river basin planning studies and numerous published papers on irrigation and water resources; active in consulting for international organizations & Utah State University. Prior employment with Corps of Engrs, State Department, Bureau of Reclamation, Navy and Department of Agriculture. Languages include Portuguese, Spanish & reading knowledge of French. *Society Aff:* ASAE, ASCE, ICID

Hargreaves, William J
4110 Arbor Dr, Midland, MI 48640
Position: VP & Dir *Employer:* Retired *Education:* BA/Chem Sci Engr/MI State Univ; Advanced Mgmt Program/Harvard Univ; Middle Mgmt Program/Case Inst of Tech. *Born:* 4/1/21. Graduated from Michigan State University with a Bachelors in Chemical Engrg in 1946. Attended Harvard's Advanced Management program in 1966. Entire career with Dow Corning, holding various positions in engrg & manufacturing; Plant Mgr in Elizabethtown Ky, Dir of Mfg. Became General Mgr of Dow Corning Europe in 1971. Returned to U S Jan 1, 1975 as Group VP for Operations. Named, Exec VP in 1977. Became mbr of Bd of Dirs in 1978. May 1981, Sr Mgmt Conslt. 1986 May Retired.

Harkel, Robert J
Business: Box 248, Chesterton, IN 46304
Position: Assistant Chief Metallurgist. *Employer:* Bethlehem Steel Corporation. *Education:* BS/Met E/Lafayette College. *Born:* April 1928. Native Bethlehem, Pa. Loop Course 1949 - Sparrows Point Plant to 1964 - Metallurgical and Quality Control functions - Metallurgist Supervisor Iron and Steelmaking - 1956-64, Assistant Chief Metallurgist - Burns Harbor Plant 1964 to present. Various ASM offices including Balt Chap Chmn 1960- 61. National Chapter Advisory Ctte and Calumet Chapter Board. Mbr AIME. Hobbies - golf and fishing. *Society Aff:* ASM, AIME.

Harleman, Donald R F
Business: MIT Rm 48-311, 77 Mass Ave, Cambridge, MA 02139
Position: Ford Prof of Eng. *Employer:* MIT. *Education:* ScD/Civil Eng/MIT; SM/Civil Eng/MIT; BS/Civil Eng/PA State Univ. *Born:* Dec 1922. Mbr of Civil Engrg faculty at MIT since 1950. Visiting Prof CIT 196263. Guggenheim Fellowship, Cambridge Univ 1969. Currently Ford Prof of Engrg. Teaching and research in fluid mechanics and water quality control in lakes, reservoirs, rivers, estuaries and coastal waters. Res and consulting on waste heat management associated with electric power generation. Mbr, National Academy of Engrg. Received Res Prize (1955), Hilgard Prizes (1971, 73) & Stevens Award (1973) from ASCE. Co-author of textbook: Fluid Dynamics. *Society Aff:* NAE, ASCE, WPCF, ASLO, AGU.

Harley, Theodore H
Business: 3636 Blvd of the Allies, Pittsburgh, PA 15213
Position: Pres. *Employer:* Vesuvius Crucible Co. *Education:* BS/Ceramic Engg/OH State Univ. *Born:* 3/29/21. in Columbus, OH. Received degree in Ceramic Engg at OH State Univ in 1943. Employed in Res at Harbison-Walker Refractories Co, Pgh, PA, 1943. Mbr US Merchant Marine 1944-1946. Employed as Refractories Engr, Johnstown Plant, Bethlehem Steel Co 1947-1951. Employed as Asst Masonry Supt, Steel Div of Ford Motor Co Dearborne, MI 1951-1954. Employed as Ceramic Engr Haws Refractory, Johnstown, PA 1954-1956. Joined Vesuvius Crucible Co 1956 as Asst Sales mgr. Elected as Pres Vesuvius Crucible Co 1958 to present. Married Lois D Long, Johnstown, PA, 1948. Three daughters Hope, Heather & Hilary. *Society Aff:* AIME.

Harling, Hugh W, Jr
Business: P.O. Box 1986, Orlando, FL 32802
Position: Pres *Employer:* Harling, Locklin & Assoc., Inc. *Education:* BS/Civil Engrg/Univ of FL; MBA/Bus Admin/FL State Univ *Born:* 2/20/43 Former Mayor City of Altamonte Springs, FL. Pres, Harling, Locklin & Assocs, Inc. Regional Mgr, Post, Buckley, Schuh & Jernigan, Inc. Pres, Innovation, Technology & Implementation, Inc. Dir of Utilities, City of Titusville, Titusville, FL. Utilities Engr, Utilities Dept, City of Titusville, FL. Project Engr, Black, Crow & Eidsness, Inc, Gainesville, FL. Design Engr, A. E. O'Neall & Assocs, Orlando, FL. *Society Aff:* ACEC, FICE, NSPE, ASCE, APWA, WPCF

Harllee, John W, Jr
Business: 3901 NW 29th Ave, Miami, FL 33142
Position: Vice President. *Employer:* PTL Inspectorate, Inc. *Education:* BSCE/Structural/Univ of FL. *Born:* 1928 Tampa. Resident Miami Fla 1929- . Fla

Harllee, John W, Jr (Continued)
Reg PE. Asst Proj Engr Fla SRD 1950-51. Asst Mgr Pittsburgh Testing Lab Jacksonville 1951-53. Vice Pres & Mgr Pittsburgh Testing Lab Miami 1953-87. Presently VP FL Region. Mbr Dade School Bd. Mbr Fla Public School Council. Trustee Baptist Bible Inst. Mbr NSPE, AWS, ASTM, ASNT, CSI ACI- Dir South Fla Chap. Fla Engrg Soc P. Pres State Society - Pres, Dir, Secy & Publications Dir Miami Chap. Fla Inst Cons Engrs - Dir/Chmn Engrg Lab Forum. Kiwanian-Lt. Gov 16 Div. BSA Ctteman. U S Optimist Dinghy Assn - Pres Dir & Secy/Treas (Youth Sailing). Fla Internatl Univ Engrg Advisory Comm. Miami Yacht Club. Who's Who in S & SW So. Fla Historial Assoc. Mbr.. *Society Aff:* NSPE, ASCE, ASTM, ACI, AWS.

Harlow, H Gilbert
Home: 17 Front St, Schenectady, NY 12305
Position: Prof *Employer:* Union Coll *Education:* MS/CE/Harvard Univ; BS/CE/Tufts Univ *Born:* 4/27/14 Chrmn of Civil Engrg at Union Coll for 29 years from 1950 to 1979. Presently, Prof of Civil Engrg. Consultant to Army, Navy, many contractors, consulting engrs, and lawyers primarily in the soil mechanics, foundations and concrete areas. Hybridist and originator of a strain of begonias widely grown in the eastern US. *Society Aff:* ASCE, ASEE, ΣΞ, ΤΒΠ

Harman, Charles M
1701 Tisdale Street, Durham, NC 27705
Position: Prof. *Employer:* Duke Univ. *Education:* PhD/ME/Univ of WI; MSME/ME/Univ of ND; BS/ME/Univ of MD. *Born:* 7/25/29. Duke Univ faculty mbr from 1961 to 1981. Served as Engg Sch Res Coordinator, Acting Chrmn of the Dept of Mech Engg and from 1970 to 1981 as Assoc Dean of the Grad School. Served in USN from 1949-51; Ford Fdn Fellow from 1960-61; consultant, Douglas Aircraft Co from 1961-64; advisor, Army Res Office from 1964-1975; Assoc Prog Dir, Natl Sci Fdn from 1967-68; Editor, Journal of Advanced Transportation from 1976 to present. Reg PE in WI and NC. *Society Aff:* ATRA, ASME, ASHRAE

Harman, Willis W
Business: 333 Ravenswood Ave, Menlo Park, CA 94025
Position: Prof. *Employer:* SRI Intl. *Education:* PhD/Elec Engg/Stanford Univ; MS/Phys/Stanford Univ; BS/Elec Engg/Univ of WA. *Born:* 8/16/18. Serves as Senior Social Scientist at SRI Intl, while holding a part-time professorship in the Dept of Engg-Economic Systems at Stanford Univ. Recently published a book entitled *An Incomplete Guide to the Future* and is currently working on projs relating to this work. Specialized profl competence includes work on policy analysis related to alternative future histories and work regarding maj societal problems.

Harmon, Don M
Business: Box 279, Garrison, ND 58540
Position: Div Mgr *Employer:* Otter Tail Power Co *Education:* BS/EE/Univ of ND *Born:* 8/10/29 Scobey, MT. US Army 1951-54, Infantry & Signal Corps, 2 years Far East. RCA Service Co 1956-1963. Univ of ND 1963-67. Westinghouse 1967-68. Otter Tail Power Co 1968 to present. Varied experience in design, installation & maintenance of: voice & data communications, manual telephone, HF point to point, VHF & UHF ground to air, high power search & heightfinder, 3D radar, power distribution and subtransmission. Limited experience torpedo & deep sea sonar. Active in civic organizations. Currently, Pres NDSPE. Enjoys hunting & fishing. *Society Aff:* IEEE, NSPE

Harmon, George W
Home: 1068 George Rd, Meadowbrook, PA 19046
Position: Retired *Employer:* Rohm & Haas Company *Education:* PhD/ChE/Univ of MD; BS/ChE/Univ of MD *Born:* 9/9/21. Native of Silver Spring, MD. Served with USN as Electronics Technician. Was Research Section Manager for Rohm & Haas Co in Process Res with responsibility for process hazards evaluation, mtls of construction, physical properties measurement for process scalings, & corporate procedure for safe handling of monomers. Elected to Fellow Grade of AIChE. Served as Chrmn & Treas of both the Phila Area & DE Valley Section of AIChE. Interested in golf, amateur radio, woodworking, & travel. Retired 1984.

Harmon, Kermit S
Home: 12103 Sea Shore, Houston, TX 77072
Position: Pres and CEO *Employer:* TX Energy Engrs, Inc *Born:* 2/26/36 A specialist in energy conservation and sys design, an inventor, a registered PE, and a lecturer and author of publications dealing with energy conservation and air-conditioning sys design. VP of Engrg and Dir of Mech and Elec Design for 3D/Intl in Houston from June, 1978 until July, 1980. As a Charter Mbr of the AEE, 1983 Natl Pres and founder of its Houston Chap, has served as Pres and Dir of the Houston Chap, Natl Elections Comm Chrmn, Pres of AEE Scholarship Foundation, and Treas of Engrs' Council of Houston and Chrmn of the Subcomm on Energy. Recipient of the 1980 "Energy Engr of-the-Year Award" presented bythe Houston AEE Chap. Recipient of the Distinguished Service Award presented by AEE in 1984 *Society Aff:* AEE, ASHRAE, CSI, NSPE

Harmon, Robert W
Home: R4 Box 135, Centralia, MO 65240
Position: Dir of Engg. *Employer:* A B Chance Co. *Education:* MSEE/Power Engg/Purdue Univ; BSEE/High Voltage Engg/Purdue Univ. *Born:* 10/22/29. Born, raised, & educated in IN. Grad teaching asstship while at Purdue Univ. Engg career in power utility related industries. Aro, Inc as engr in power systems for large wind tunnels. OH Brass Co as dev engr & dir, new product dev. A B Chance Co as chief engr & dir of engg. Hold 30plus US patents. Fellow IEEE. Numerous technical articles & presentations. Developed the fundamental engg principle for design of ultra high voltage insulation systems. Married, four sons. Active in whitewater canoeing, wilderness camping. *Society Aff:* IEEE, ΣΞ, ΤΒΠ, HKN.

Harmond, Jesse E
Home: 925 Witham Drive, Corvallis, OR 97330
Position: President. *Employer:* Progressive Golf, Inc. *Education:* BS/EE/MI State Univ. *Born:* Dec 10, 1906 Columbus, Miss. Diesel Elec Engrg, Kolola Springs Gravel Co 1932-33. Diesel Elec Engr US Army Engr's 1933-39. Res Project Ldr Cotton Packaging USDA, Stoneville Miss 1939-41. Asst Res Dir USDA Cotton Gin Lab 1941- 44. Dir USDA Dehydrated Food Compression Lab, Beltsville, MD 1944-45. Invest Head Fiber Flax Harvesting and Processing Res and Dev 1945-53. Section Head USDA Seed Harvesting and Processing 1953-69. Cons Engr 1969-present. FAO & Rockefeller Foundation Chile, Mexico, Equador, Colombia & Argentina 1969-75. Inventor and President of Progressive Golf. IndoorOutdoor Golf Game. 1969--A.S. A.E. Fellow, Sigma Xi, author, inventor. Hobbies: golfing and fishing. *Society Aff:* ASAE, OPEDA, ASAPC.

Harms, John E, Jr
Business: PO Box 5, Pasadena, MD 21122
Position: President. *Employer:* John E Harms Jr & Assocs. *Born:* Nov 12, 1921 Hagerstown, Md. Attended Gettysburg College and The Johns Hopkins University. BS in CE from The Johns Hopkins University. US Army Corps of Engrs 1943-46 - Major, retired. Employed Whitman Requardt & Assocs, Baltimore Md 1946-55. Pres of John E Harms Jr & Assocs Inc, Consulting Engrs 1955-present. Reg Prof Engr in Md, Delaware, D C. Past Treasurer, Cons Engrs Council of Md Inc. Active in many professional civic & religious organizations. Spare time spent in farming, boating, flying and general travel.

Harms, William O
Home: 4029 Hiawatha Dr, Knoxville, TN 37919 *Employer:* Retired. *Education:* PhD/Phys Metallurgy/UNiv of MN; MS/Phys Metallurgy/Univ of MN; BS/Metallurgical Engg/Wayne Univ. *Born:* Sept 1923, native of Alton, Ill. Served with Army Air Corps 1942-44. Taught Matls Course, Oak Ridge Sch of Reactor Technology 1955-56; Assoc Prof of Met Engrg, Univ of Tenn 1955-60. With Oak Ridge

Harms, William O (Continued)
Natl Lab in various capacities in Metals and Ceramics Div 1953-55 and 1960-72 including: Hd, Ceramics Lab; Dir, High Temp Matls Program & Isotopic Fuel Matls Prog; Sect Chf for Res and Dev, and Mgr ORNL LMFBR Prg; Dir, Lf Mbr, Prg 1972-78. Assumed current respon as Dir Nuclear Reactor Tech Progs, 1978-1985. Retired 1985. *Society Aff:* ASM, ANS, ΣΞ, ΤΒΠ, AIME.

Harnden, John D
Business: P O Box 43, Bldg 37, Schenectady, NY 12301
Position: Mgr, Technology, Advanced Elec Prod Oper. *Employer:* General Electric Company. *Born:* 9/20/28. Graduated from Union College in 1950 with a BEE. Mr Harnden is a recognized authority on the application of semi-conductor components to electric power systems. Mr Harnden joined the General Electric Company in 1950. Mr Harnden has written more than 37 technical papers and articles in the field of solid state electronics and has coauthored four books. He has been granted 86 U S patents and has had a total of 115 patent applications filed. He is a Fellow of the Inst of Electrical and Electronics Engineers. An active mbr of the Magnetics Soc. Member of the IEEE Spectrum Editorial Board. Member of the IEEE Spectrum Editorial Board.. He is also a mbr of the IEEE Admission & Advancement Ctte & Technical Ctte Chrmn, Control & Power Processing of the Magnetics Soc. *Society Aff:* IEEE, ASI, SSPI, IMS, IAS

Harness, Arminta J
Home: 1633 Nelson Ave, Manhattan Beach, CA 90266
Position: Retired *Education:* BE (Aero Engr)/Aero Engr/USC. *Born:* March 1928. BE (Aero Engrg) from Univ of Southern California. First woman engr to join US Air Force. Service from 1950-74. Retired Lt Colonel. Pioneering USAF assignments in engrg and R&D management. First woman Staff Development Engr and first woman awarded Master Missileman Badge. Employed by Westinghouse Hanford Company as Technical Assistant to the President 1974-1976 and as Manager, Laboratory Planning Dept 1976-79. Fellow Life Mbr and 1967-78 National President of the Society of Women Engrs. Fellow of the Institute for the Advancement of Engrg. IAE Engrg Achievement Merit Award 1971. *Society Aff:* SWE, AFA, DAV.

Harold, Robert A
Home: 1856 Cherrylawn Dr, Toledo, OH 43614
Position: Associate *Employer:* Samborn, Steketee, Otis & Evans *Born:* 10/11/39 Native of Lorain, OH. Attended OH Univ 1957-1962. Served as surveyor and photographer with Army 1963-65. Was Chief Engr of Clarke H. Joy Co before joining Samborn, Steketee, Otis & Evans in 1974. Elected an Associate in 1979. Current responsibilities include instrument and control standards and specifications co-ordination as well as project management. Sr Member of ISA. Participant of Intl Purdue Workshop on Industrial Computer Systems, Interfaces and Data Transmission Committee. Editor, TSPE Newsletter, *The Toledo Engr* 1980-present. Papers published on tone encoding. Currently working on computer room RFI shielding techniques. Hobbies include amateur radio, photography and fishing. Registered PE in the State of OH. *Society Aff:* NSPE, OSPE, TSPE, ISA

Harper, John D
Business: 1501 Alcoa Building, Pittsburgh, PA 15219
Position: Dir & Consultant. *Employer:* Aluminum Company of America. *Education:* BS/Elec Engg/Univ of TN. *Born:* 4/6/10. With Alcoa since 1933, from Elec Design Engr, Power Mgr to American Soc for Metals; Life Mbr, Inst of Elec & Electronics Engrs Inc; Natl Acad of Engg; Nathan W Dougherty Award, Univ of TN; PA Soc of Prof Engrs. Chmn of Communications Sattellite Corp, Washington, DC; Chmn of Coke Investors, Inc, Ft Worth, TX. *Society Aff:* AIEE, ASME, ASM, IEEE, ΤΒΠ, HKN, ΒΓΣ.

Harper, Judson M
Business: Office of VP for Research, Ft Collins, CO 80523
Position: VP for Research. *Employer:* CO State Univ. *Education:* PhD/Food Tech/IA State Univ; MS/Food Tech/IA State Univ; BS/ChE/IA State Univ. *Born:* 8/25/36 Native of Des Moines, IA. Taught at IA State Univ 1958-1964. Res engr & dept hd with Gen Mills, Inc, Minneapolis, MN 1964-1970. Assumed current position as VP for Research, CO CO State Univ 1982. Previously head, Dept of Agricultural & Chem Engr, CO State Univ 1970-82. Research interests include food extrusion and the analysis of food processing systems. Author of 75 journal publications, numerous reports and a book on food extrusion. Outstanding Alumnus ISU, 1970, Distinguished Service Award CSU, 1977, Food Engineering Award of Dairy & Food Ind Assoc & ASAE, 1983, and Professional Achievement Citation in Engineering, ISU, 1986. Enjoys sports & outside activities. *Society Aff:* AIChE, ASAE, AAAS, ASEE, IFT, ACS.

Harrawood, Paul
Business: Box 1607 Station B, Nashville, TN 37235
Position: Professor of Civil Engineering *Employer:* Vanderbilt Univ. *Education:* PhD/Civ Engg/NC State Univ; MS/Civ Engg/Univ of MO; BS/Civ Engg/Univ of MO. *Born:* 8/28/28. Native of Southern IL. Active duty in US Navy, 1951-54. On Engg faculty at Duke Univ 1956-67 as Instr., Asst. Prof. and Asst. Dean. Since 1967 at Vanderbilt Univ. as Assoc. Prof., Prof., Assoc. Dean, Acting Dean and Dean in Sch. of Engg. Professional experience with McDonnell Aircraft Corp. as Test Engr, Experimental Stress Analysis, and with US Army Corps of Engrs as Construction Mgt. Engr. 1986 Engr of the Yr, by Mid TN Soc of Professional Engrs. *Society Aff:* ASCE, SAME, ASEE, AAAS, ΣΞ, XE, ΤΒΠ.

Harrington, Burnett W, Jr
Business: 1401 Peachtree St, Suite 120, Atlanta, GA 30309
Position: Pres *Employer:* Harrington, George & Dunn, PC *Education:* MS/CE/Univ of So CA; BS/CE/Howard Univ *Born:* 8/3/40 Formally employed as a structural engr with the City of Los Angeles. Held the position of Chief Structural Engr for the Metropolitan Atlanta Rapid Transit Authority, where I assisted consultant in developing Design Criteria and Standard Drawings for structural elements of the system. Joined consulting firm of Harrington, George and Dunn in 1976 as Pres. Established both co and design policy and supervised firms structural design efforts. Authored a paper for US Dept of Transportation, *Investigation of Design Standards for Urban Rail Transit Elevated Structures*. Chrmn of Committee on Minority Programs for ASCE. *Society Aff:* ASCE, ACI, PTI

Harrington, Dean B
Home: 22 Via Maria Dr, Scotia, NY 12302
Position: Retired *Employer:* General Electric. *Education:* BS/EE/MIT. *Born:* Native of Schenectady NY. Engr with Genl Elec in Schenectady 1944-83. Engr in large steam-turbine-driven generators since 1948, respon for spec & advance elec engrg. Mgr Generator Advance Engrg 1967-83. Chmn IEEE Synchronous Machinery Subctte 1964-66, Chmn IEEE Rotating Machinery Ctte 1969-70, Fellow IEEE 1973- . Was U S rep to IEC Ctte & Working Group meetings in Moscow 1963, Tokyo 1965, Rome 1967, Washington 1970, London 1973, Brighton 1974, The Hague 1975 & London 1977, Stockholm 1980. Recieved 1981 IEEE Tesla Award. Elected to National Academy of Engineering in 1981. Enjoy classical music, hiking, carpentry. Awarded General Electric's Charles P Steinmetz Award 1977. Chmn IEC subcomm for Evaluation & classification of Rotating Machinery Insulation (SC 2 J) 1981 to present. Licensed Prof Engr in New York state. *Society Aff:* IEEE.

Harrington, Ferris T
Home: 801W Long Lake Rd F-2, Bloomfield Hills, MI 48013
Position: Vice President-Sales. *Employer:* Sperry Rand Corp, Vickers Div. *Education:* ME/-/Univ of Akron. *Born:* Mar 1903. Engr & Ch Engr Whitman & Barnes 1924-32; Field Engr & Sales Engrg Mgr, Vickers 1932-39; V P Sales 1939 retirement. Cons Foote Burt Co, Cleveland Oh; Pioneer Engrg & Mfg Co, Warren Mich. Expert witness on hydraulic sys for sev firms in courts. Inventor - 54 U S pats in hydraulics & machinery fields. Fellow & Life Mbr ASME. Mbr Engrg Soc of Detroit, SAE, AIAA. Chmn Aerospace Div ASME 1958. Mbr Theta Chi, Sigma

Harrington, Ferris T (Continued)
Tau, Omicron Delta Kappa, Plum Hollow Golf Club. *Society Aff:* AIAA, SAE, ASME

Harrington, H Richard
Home: 10 Valley Green Dr, Doylestown, PA 18901
Position: VP. *Employer:* Harrington-Robb Co. *Education:* BS/CE/Drexel Univ. *Born:* 10/19/28. Grad Drexel Evening Coll, 1955. Worked for Sharples Corp in 1950-58 as Process Dev Engr. Plant Engr for Anglo-Am Clay Corp 1958-1962. Application Engr for Mixing Equip Co 1962-1964. Dir of Application Engg Dept for Mixing Equip Co 1964 to 1970. Started own Sales rep Co with partner in 1970. Presently employ 9 engrs representing several process equip cos serving CPI. Married Doris Thompson Harrington. Three daughters, Dawn, Nancy, Rebecca. Patent held and co-authored technical papers. Active in church and civic activities. *Society Aff:* AIChE, WPCF.

Harrington, John J
Home: 2802 Jo Alyce Dr, Allison Park, PA 15101
Position: Principal - VP, Treas. *Employer:* KLH Engineers, Inc. *Education:* BS/CE/Univ of KY *Born:* 9/20/22 Employed by PA Railroad, prior to World War II. With US Army 95th Inf. Div., 320th Engr Combat Battalion 1942-45. BSCE (1950), Univ of KY. Morris Knowles Inc, responsible for structural and hydraulic design and construction surveillance for Philadelphia Water Dept - Queen Lane Pump Station and Water Plant, Samuel Baxter Water Plant and Belmont Water Plant facilities. Designed and developed several municipal water supply and wastewater projects in OH, NY, WV and PA. With BCM Inc, responsible for numerous industrial and municipal water supply and wastewater projects. Trustee, PA Sect. AWWA. Formed firm of Keith-Langford- Harrington Conslt Engrs, Inc (Now KLH Engineers Inc.) with George W. Keith and Leon E. Langford in 1982. Providing environ engrg services to municipalities and indust. *Society Aff:* AWWA, AISE, NSPE.

Harrington, John M
Business: P.O. Box 3333, 916 So 17th St, Wilmington, NC 28406
Position: VP/Structures *Employer:* Talbert, Cox & Assoc, Inc *Education:* Doctoral Studies/Engrg Sci/Univ of FL; MS/CE/GA Inst of Tech; BS/CE/GA Inst of Tech *Born:* 8/5/38 Experience includes the structural design of bldgs, bridges, airport pavements, water and waste treatment facilities, site development functions including roadways, drainage, sewage collection, and water distribution systems; construction management, cost estimating, soil conditions, land subdivision and development, and municipal engrg. He is a certified diver. Mr. Harrington has written programs to assist in roadway and structural design and structural analysis. Hobby: sailing. *Society Aff:* ASCE, NSPE

Harrington, John V
Home: 1048 San Mateo Drive, Punta Gorda, FL 33950
Position: Sr VP - R&D and Dir, COMSAT Labs. Retired *Employer:* Communications Satellite Corp. *Education:* ScD/EE/MIT; MEE/EE/Poly Inst of Brooklyn; BEE/EE/Cooper Union Inst of Tech. *Born:* 5/9/19. 1946-51 Res Engr, Air Force Cambridge Res Lab; 1951-73 Dir various maj res progs, Prof, EE & A&A, MIT, and First Dir of MIT's Ctr for Space Res. Joined COMSAT in 1973; Retired 1984. Fellow of IEEE, AIAA, AAAS 1977-81. Served on number of corporate Bds and Govt Advisory Committees. Awarded the USAF Medal for Exceptional Civilian Service, and Cooper Union Citation for Profl Achievement and Gano Duun Award. *Society Aff:* IEEE, AIAA, AAAS, ΤΒΠ, ΣΞ.

Harrington, Roger F
Business: ECE Dept, 111 Link Hall, Syracuse, NY 13210
Position: Prof Elec & Computer Engrg. *Employer:* Syracuse University. *Education:* BS/EE/Syracuse Univ; MS/EE/Syracuse Univ; PhD/EE/OH State Univ. *Born:* Dec 24, 1925. Prof Syracuse Univ 1952- . Visiting Prof Univ of Ill 1959, Univ of Calif Berkeley 1964, Tech Univ of Denmark 1969. IEEE Fellow award 1968. Best Paper awards IEEE AP Soc 1968, 1971, Syracuse Sect 1967, 1973, 1974. Disting Alumni Award Ohio State Univ 1970. Fulbright Award Denmark 1969, Sigma Xi Res Award 1971, Visiting Scientist Yugoslavia 1972, Natl Lectr IEEE AP Soc 1973-75. Books: 'Time-Harmonic ElectroMagnetic Fields', McGraw-Hill 1961, 'Field Computation by Moment Methods', MacMillan Co 1968. *Society Aff:* AAUP, IEEE, URSI.

Harrington, Roy L
528 Kerry Lake Dr, Newport News, VA 23602
Position: Naval Arch/Tech Mgr *Employer:* Newport News Shipbuilding. *Education:* MBA/Bus Adm/William & Mary; MS/NA & ME/U of MI; BSME/Mech Egr/NC State U. *Born:* Feb 1934. Served in US Army 1953-56. Joined Newport News Shipbuilding as engr. Received SNAME scholarship for graduate work. MS NAME from Univ of Michigan in 1961. Served on various SNAME Technical and Research Panels and Committees. Authored a number of technical papers. Held a succession of engrg mgmt positions at Newport News Shipbuilding and was promoted to Naval Architect and Tech Mgr in 1973. Served as tech editor of the SNAME text 'Marine Engrg' published in 1971. MBA College of William and Mary 1972. Received Vice Adm E L Cochrane award for the best SNAME section paper in 1967, 1973, and 1981. Elected mbr of SNAME Council 1972-74, 1975-77, and 1980-82; Mbr of SNAME Exec Ctte 1983-88. Elected SNAME VP 1986-88. Elected to the SNAME grade of Life Fellow in 1983. *Society Aff:* SNAME, ASNE.

Harriott, Bill L
Business: Box 2208, Sunnyvale, CA 94087
Position: Self-employed *Employer:* R&D Consultant *Education:* BS/Agri Engg/IA State Univ; MS/Agri Engg/Univ of AZ. *Born:* 12/15/30. Lake City, Iowa. International Consultant to manufacturers, govt. agencies, universities and trade assoc. Specializes in agricultural equipment research and development planning and evaluation. Formerly Research Consultant FMC Corp. Prof of Agr. Engr Univ of Arizona - Design Engr for Deere & Co. Served on Pres. Reagan's private section survey in Fed. Govt. Reg Prof Engr Calif, Arizona, Ill. Hon Soc Mbrships: Tau Beta Pi, Phi Kappa Phi, Gamma Sigma Delta, Alpha Zeta. Past President ASAE. Past Regl VP ASAE. Past Director, Council for Agricultural Science and Technology. Past Governor ARES. Twenty professional publications. Five patents. *Society Aff:* ASAE, NSPE, SAE, CAST, ESA, CSAE, IAgrE, AESA.

Harris, Aubrey L, III
Business: PO Box 7504, Little Rock, AR 72217
Position: Proj Engr. *Employer:* Harris Engg, Inc. *Education:* BSAE/Engrg/AR State Univ. *Born:* 6/1/50. Native of Forrest City, AR, grad from AR State Univ in 1973. I began working for Shreeve Engg, Ltd as a field Engr responsible for inspections of new construction and surveying. In 1974 advance to a Staff Engr in the Structural Design Dept responsible for industrial bldg and material handling structures. In 1975 became a Proj Engr responsible for complete plant design for grain and raw material mfg plants, and presently I have the increase position of investigating and testifying as a expert witness in structural failures. In 1983 opened Harris Engg Inc, a consltg engrg firm, designing indus bldg & material handling sys for industry in AR & MS. Enjoy hunting, fishing and wild life photography. *Society Aff:* ACEC.

Harris, Benjamin L
Home: 11323 Glenarm Rd, Glenarm, MD 21057
Position: Private Consultant *Education:* BE/Gas Engrg/Johns Hopkins U; PhD/ChE/Johns Hopkins U; Diploma/National Security/Industrial College of the Armed Forces (ICAF). *Born:* 8/1/17. Technical Service in Army RD&E 1941-6. Asst Prof of Chem Engg, Johns Hopkins Univ 1946-53. Pres small consulting firm 1946-63. Chief, Detection & Plants Branches of R&D Command, US Army Chem Corps 1952-55; Asst to CG; Asst to Deputy Commander of R&D Lab; Chief Operations Res Div 1955-66. Deputy Asst Dir of Defense Res & Engg, the Pentagon, 1966-70. Technical Dir Edgewood Arsenal 1970-77. Technical Dir, Chem Systems Lab 1977-1981. Sr Exec Service, US Govt. Pres, Engrg Res Co of Glenarm MD, Inc, 1981-85; private consultant 1985-date. Areas of responsibility included supervision of engrs

Harris, Benjamin L (Continued)
& scientists in mtl R&D, policy, budget & execution of prog involving 650-800 technical personnel & $100 million. Member, Maryland Scientific Council, Maryland Academy of Sciences, 1976-84 Member, Executive Board and Commissioner Baltimore Area Council and Member, National Committee, Boy Scouts of America. Chmn, MD State Bd of Reg for Prof Engrs. *Society Aff:* ACS, AAAS (FELLOW), AIChE (FELLOW), ΣΞ

Harris, Colin C
Business: Henry Krumb School of Mines, 907 Engrg Ctr, Columbia University, New York, NY 10027
Position: Prof (Mineral Engrg) *Employer:* Henry Krumb School of Mines, Columbia Univ *Education:* PhD/Coal Prep & Mineral Processing/Leeds Univ, England; BSc/Math-Physics/London Univ, England *Born:* 1/9/28 in Leeds, England. Served in British Armed Forces 1946-49. Taught and researched coal preparation and mineral processing, Leeds Univ, England 1953- 1960; Columbia Univ 1960-present. Advisor/consultant to industry, government funding agencies, universities, intl publishers. Member of organizing committee for several intl conferences. Author of over 100 publications in mineral processing, coal preparation, fine particle tech, flow and retention in porous media, flotation hydrodynamics and kinetics, design and scale-up of processing machinery. Member of Institution of Mining and Metallurgy (London), Operational Research Society (Great Britain). Chartered Engr (Great Britain). Editorial Board member of several Intl Journals of Mineral Processing. Publications committee, Awards Cttee AIME. Listed in "Who's Who in America" & "Who's Who in The World–, "Who's Who in Technology Today–. etc. *Society Aff:* AIME, IMM, ORS

Harris, Cyril M
Business: Mudd Bldg, New York, NY 10021
Position: C Batchelor Prof of EE & Prof of Arch *Employer:* Columbia Univ. *Education:* BA/Math/UCLA; MA/Physics/UCLA; PhD/Physics/MIT *Born:* 6/20/17. American Institute of Architects Medal (1980); Wallace Clement Sabine Medal, Acoustical Soc of Am (1979); Franklin Medal, Franklin Inst (1977); Honorary Award, US Inst for Theatre Tech (1977). Distinguished Service Award, UT Symphony Orchestra (1979), Natl Acad of Sci; Natl Acad of Engrg. Acoustical Consultant for: John F Kennedy Ctr, Wash DC; Orchestra Hall, Minneapolis; Krannert Ctr, Urbana; Symphony Hall, Salt Lake City; Metrolpolitan Opera House, new Avery Fisher Hall, Lincoln Ctr; Powell Symphony Hall, St Louis; Natl Ctr for the Performing Arts, Bombay. Publications: *Handbook of Noise Control*, 2nd Ed 1979. *Historic Architecture Sourcebook*, 1977. *Dictionary of Architecture and Construction*, 1975. *Shock and Vibration Handbook*, 2nd Ed 1976 (with C E Crede). *Acoustical Designing in Architecture*, Revised Ed. 1980. (with V O Knudsen). *Society Aff:* IEEE, ASA, AES

Harris, Donald R
Business: Dept of Nuclear Engrg, RPI, Troy, NY 12181
Position: Dir, Critical Facility, Assoc Prof *Employer:* RPI *Education:* PhD/Nuclear Engrg and Sci/RPI; MA/Phys/Princeton Univ; MS/Math/Carnegie- Mellon Univ; BS/Phys/Carnegie-Mellon Univ *Born:* 11/29/25 On staff of Rensselaer Polytechnic since 1975; previously Los Alamos 1968- 75; Bettis Atomic Power Lab 1960-68; Mbr ANS, active in seven divs/cttes; mbr Societe Francaise d'Energie Nucleaire, European Nuclear Soc, Amer Phys Soc, AAAS, Amer Anthropological Assoc, Soc for Amer Archaeology, Econ Hist Soc. Extensive academic experience; Current conslt, Yankee Atomic Power Co, Gen Elec, Exxon, Transnuclear; past conslt New England Nuclear, State of NY, Sci Applications Inc, Con Ed NY, Oak Ridge Natl Lab. Energy Res. Awards: Westinghouse Scholar, NY Acad of Sci, Alpha Nu Sigma, Tau Beta Pi, Sigma Xi. Author, over 450 tech reports and papers. *Society Aff:* ANS, SFEN, ENS, APS, AAA, SAA

Harris, Forest K
Home: 9905 Wildwood Rd, Kensington, MD 20895
Position: Consultant. *Employer:* National Bureau of Standards. *Education:* PhD/Physics/Johns Hopkins; MS/Physics/OK Univ; BA/Physics/OK Univ. *Born:* August 26, 1902 (Gibson County, Indiana). Physicist, National Bureau of Standards (Electricity Div) since 1925, Ch of Absolute Measurement Section 1958- 70, Cons to Electricity Div 1970- . Adjunct Prof of Engrg, George Washington University 1940-75. Assoc Ed 'Review of Scientific Instruments' 1938-74. Author of text 'Electrical Measurements' 1952. Dept of Commerce Silver Medal 1957, Edward Bennett Rosa Award 1967, Morris E Leeds Award (IEEE) 1972, Si Fluor Award (ISA) 1973. William Wildhack Award (CPEM) 1981 Life Fellow IEEE; Fellow Washington Acad of Sciences. *Society Aff:* IEEE, ISA.

Harris, Jay H
Business: College of Engrg, San Diego, CA 92182
Position: Dean *Employer:* San Diego State Univ *Education:* PhD/EE/UCLA; MS/EE/CA Inst of Tech; B/EE/Polytech Inst *Born:* 6/3/36 Born in Newark, NJ. Following his BEE degree (summa cum laude) he served eight yrs in antenne res at the Hughes Aircraft Co. During this period he was a Hughes Masters and Doctoral Fellow and a Fullbright-Hayes Fellow at the Univ of Paris (1965). In 1966 he joined the Elec Engrg Dept of the Univ of WA where he rose to Full Prof. His principal res interests were in integrated optics and optical communications and Biomedical Engrg. In 1974 he joined the Natl Sci Fdn where he served as Dir of Elec & Quantum Elec Programs and initiated the Fdn's Microelec effort. In 1980 he joined San Diego State Univ as Dean of the Coll of Engrg. He was made a Fellow of the IEEE for his efforts in integrated optics, and is a former Traveling Lecturer of the Optical Soc of America and a mbr of Tau Beta Pi and Sigma Xi. *Society Aff:* IEEE, OSA, RESNA, AAAS, ASEE

Harris, LeRoy S
Home: 1715 Queen Anne Gate, Westlake, OH 44145
Position: Dir *Employer:* MK-Ferguson Co *Education:* M.S./Mech. Eng'g/Penna State University; B.S./Mech. Eng'g/Penna State University *Born:* 04/05/24 Born in Philadelphia, Pa. Parents Philip and Bella Harris, married Florence Schwab, two children. Asst. Professor Villanova Univ. 1954-56; Manager R&D Schutte & Koerting Co. 1956-68; Asst. Pres. K-G Industries 1968-73; VP American Tara Corp. 1974-76; Manager Mech. Engrg. and Dir. Claims Consulting MK-Ferguson Company 1976-present. As Dir responsible for consulting engrg serving insurance industry. ASME; Fellow, VP 1980-82, Chmn Process Industries Div 1966-70, Chmn Environmental Affairs 1977-80, VChmn Bd on Research 1987. ASME Centennial Medal. Granted 30 patents on desuperheaters, flow meters, briquetting machines. Authored 33 papers on chemical process equipment. PE license 8 states. *Society Aff:* ASME

Harris, Michael E
Business: 877 S Alvernon Way, Tucson, AZ 85711
Position: Chief Structural Engineer. *Employer:* Rod Gomez & Associates Cons Engrs. *Born:* September 18, 1942. BSCE from Univ of Arizona, Tau Beta Pi. Reg Prof Engr in States of Arizona, California, New Mexico and Colorado. Bridge Design Engr with Orange County, California Highway Department. With Rod Gomez & Associates since 1965. Assumed current responsibilities as Chief Structural Engr and Director of Structural Engrg Design in 1969. Young Engr of the Year Arizona Chapter of NSPE 1972. Vice President of Structural Engrs Association of Arizona 1972-75; Secretary of Western States Conference of Structural Engrs 1974; Mbr of Arizona State Seismological Committee; Two time recipient of PCA Outstanding Structures Award in Az.

Harris, Norman H
Home: P.O. Box 520, Saugus, CA 91350
Position: Tenured Instructor *Employer:* Los Angeles Trade-Technical Coll *Education:* Ph.D./Ceramic Engrg/Univ of IL; MS/Ceramic Engrs/Univ of IL; BA/Chem/Whittier Coll *Born:* 9/26/41 Indigenous to Southern CA. Became interested in ceramics while working at Thatcher Glass Manufacturing Co. and Pacific Clay

Harris, Norman H (Continued)
Products Research Lab. After grad school obtained patent on electronic ceramic processing method at McDonnell Douglas Corp. Initiated and taught only ceramic engrg technology two-year A.S. program in Western U.S. at Los Angeles Trade-Technical Coll since 1976. General Partner in Ceramic Correspondence Inst since 1972 and Pres of Cer Tek Enterprises Inc. since 1968. Chrmn, Southern CA Section of the American Ceramic Society, 1981; Secretary/Treasurer, National Inst of Ceramic Engrs, 1981. Enjoy making music, hiking and prospecting with my wife, Cynthia. *Society Aff:* NICE, ACerS

Harris, Robert L
Business: Dept Envir Sci & Engrg, 201 H, Chapel Hill, NC 27514
Position: Prof *Employer:* Univ of NC *Education:* PhD/Envir Sci & Engrg/Univ of NC; MS/Envir Engrg/Harvard Univ; BS/ChE/Univ of AR *Born:* 7/18/24 Prof of Environmental Engrg. Dir, Occupational Health Studies Group, School of Public Health, Univ of NC. Certified industrial hygienist (Engrg). Dir, Bureau of Abatement and Control (1968-1970) and Chief, Field Investigations, Abatement Program (1964-68) of the Natl Air Pollution Control Administration. Chief, Engrg, Div of Occupational Health, Public Health Service (1960-64). Member Tau Beta Pi Assoc, Sigma Xi, Omicron Delta Kappa Society. Numerous publications in the field of environmental health. Current Member, Board of Scientific Counselors, National Institute for Occupational Safety and Health. *Society Aff:* AAIH, ABIH, AIHA, AAAS, SOEH

Harris, Robert S
Business: One North Krome Ave, Homestead, FL 33030
Position: Regional Manager. *Employer:* Post Buckley Schuh & Jernigan Inc. *Education:* N-A/Bus Admin/Univ of Miami; N-A/Civil Engg/Int Corresponence Sch; N-A/Land Surveying/Dade County Adult Education; N-A/Modern Bus/Alexander Hamilton Inst. *Born:* June 1938. Attd Univ Miami, ICS Civil Course. Native Miami, FL. Served with the Army Corps of Engrs, active & reserve 1958-64. Progressively Field Surveyor, Head of Survey Department, Principal Assoc & Reg Office Mgr with Post, Buckley, Schuh & Jernigan Inc, Miami based cons firm since 1962. Assumed current assign as regional office mgr for the firm's Homestead/FL Keys Office in 1973. Serves as County Engr for Monroe County under cons contract as well as project mgr for land development projects in So Dade County. Pres, Fla Soc of Prof Land Surveyors 1976-77; Florida Surveyor of the Year 1976, Fellow mbr ACSM, Listed in Who's Who in the South and Southwest 1980-81 and subsequent editions; appointed by former Florida Governor Askew as First Chrmn of Florida Public Land Survey Advisory Board 1977-79. *Society Aff:* ACSM, SAME, ASPRS, NHS, NSPS, NACE, CMLS, USNI

Harris, Ronald D
Business: Kraft, Inc, 801 Waukegan Road, Glenview, IL 60025
Position: VP, Product Development, Technology. *Employer:* Kraft, Inc. *Education:* MBA/Management/Univ of Cinn; MSc/ChE/OH State Univ; B/ChE/OH State Univ *Born:* 4/9/38 Currently VP of Product Development, Technology for Kraft, previously VP of Research & Development for Anderson Clayton Foods, and before that was Dir of Research & Development for the Clorox Co, and before that was a Group Leader for Foods, Procter & Gamble. Principal expertise is in the area of consumer product development, including fat and oil-based products, dairy products and their analogs, and household chemical specialties. Born in Norman, OK, but grew up in Columbus, OH. Married Judith Wright on 7/28/62; Children: Todd David, Scott Howard and Susanna Katherine. Served as Lt Intelligence, US Army. Honors, Distinguished Military Graduate, Phi Eta Sigma, Tau Beta Pi (Engrg), Phi Lambda Upsilon (Chemistry), and Delta Mu Delta (Business). *Society Aff:* ACS, IFT, AOCS

Harris, Roy H
Home: 2025 Nottingham Ln, Burlington, NC 27215
Position: Advanced Tech Systems V. Pres. *Employer:* AT&T Technologies *Education:* MEE/EE/Poly Inst of NY; BSEE/EE/GA Inst of Tech. *Born:* 12/17/28. Native of Madison, GA. Design engr in airborne radar for Hazeltine Electronics Corp. With Bell System since 1956. Mbr of Technical Staff with Bell Tel Labs with microwave design responsibilities for air defense systems. Since 1964 have worked for Western Elec in Military Systems design. Assumed current position as Adv. Tech. Sys. VP in 1986. Bd of Dirs IEEE 1978-79. SOUTHCON Bd of Dir 1980-83, IEEE Fellow 1983. 1984 Engrg Mgr of Year by Engr. Mgt Soc of IEEE. Bd of Trustees N.C. A&T State Univ. 1985-. Enjoy golf, tennis & skiing. *Society Aff:* IEEE, NSIA, ODK, ΤΒΠ, ΦΚΦ

Harris, Stephen E
Business: E L Ginzton Lab, Stanford, CA 94305
Position: Prof Elec Engrg/Applied Physics. *Employer:* Stanford University. *Education:* PhD/EE/Stanford Univ; MS/EE/Stanford Univ; BS/EE/Rensselaer Polytechnic. *Born:* 11/29/36. During 1960 was employed at Bell Telephone Labs. Since 1963 has been on the faculty of Stanford Univ where he is now a Prof of Elec Engrg & Applied Physics. Res has been in the areas of lasers and quantum electronics. Areas of contribution include optical communications, tunable lasers, the invention of the tunable acoustooptic filter, work on optical frequency conversion and harmonic generation techniques, and most recently contributions in the area of laser induced collisions. Has about 90 publications and 11 patents. Has received the Alfred Noble Prize in 1965, the Curtis W McGraw Res Award in 1973, a Guggenheim Fellowship in 1976, and the David Sarnoff Award in 1978. Is a Fellow of the IEEE, the American Physical Soc, Optical Soc of America, is a Mbr on the Natl Acad of Engg and the Natl Acad of Scis Received the RPI Davies Medal for Engineering Achievement (1984).. *Society Aff:* IEEE, APS, OSA, NAE, NAS.

Harris, Tedric A
Business: Postbus 50, 3430AB Nieuwegein, Netherland
Position: Managing Director *Employer:* SKF Engineering & Research Center BV *Born:* Feb 25, 1932. BSME & MSME Penn State 1953 & 1954: Tau Beta Pi, Pi Tau Sigma, Sigma Tau. Dev Test Engr Hamilton Std Div, United Aircraft Corp 1953-54; Analytic Des Engr Bettis Atomic Power Lab, Westinghouse Elec Corp 1955-56; Supr Bearing Tech, SKF Indus Inc 1960-68; Mbr Analytic Services 1968-71; Dir Corp Data Sys 1971-73; V P Info Sys, Services 1968-71; AB SKF Group HQ 1973-76. Pres, Specialty Bearings Div, SKF Indvs Inc 1977-79; Managing Dir, SKF Eng & Res Center BV. ASME Lub Div Exec Ctte 1971-72; Res Ctte Lub 1968-72; Fellow Mbr 1973. ASLE Deutsch Award 1965, Hodson Award 1968. Author 38 tech publs, incl book 'Rolling Bearing Analysis'. Holder 3 US pats.

Harris, Thomas R
Business: Box 1724, Station B, Nashville, TN 37211
Position: Dir of Biomedical Engrg *Employer:* Vanderbilt Univ *Education:* MD/Medicine/Vanderbilt Univ; PhD/ChE/Tulane Univ; MS/ChE/TX A&M Univ; BS/ChE/TX A&M Univ *Born:* 2/19/37 Born at San Angelo, TX. Educated in public schools of Dallas, TX. Ind experience with Standard Oil of CA (1958-60). Served in US Army as 2nd Lt (1958- 59). Joined faculty of Vanderbilt Univ as Asst Prof of Chem Engrg. Currently Prof of Biomed Engrg & Medicine. Dir of the Biomedical Engrg Program, Sch of Engrg and Dir of the Biomedical Engrg Service, Sch of Medicine since 1975. Res specialization in biological transport phenomena, cardiopulmonary physiology and biomathematics. *Society Aff:* AIChE, ASEE, BMES, APS, MS, IEEE-EMBS.

Harris, Walter R
Home: 409 Whitman Rd, SE, Winter Haven, FL 33880
Position: Retired *Education:* MS/EE/U of Pittsburgh; BS/EE W VA Univ. *Born:* June 1915. Tau Beta Pi and Eta Kappa Nu. Entire career with Westinghouse Elec Corp in Heavy Indus Drive & Control Sys Engrg. 20 pats. Over 50 tech papers & articles. Reg P E in Pa. Mbr AISE; Fellow IEEE, P Chmn Pgh Sect IEEE, P Pres

Harris, Walter R (Continued)
IEEE Indus Applications Soc; recipient of IEEE-IAS Outstanding Achievement Award 1975. Retired. *Society Aff:* AISE, IEEE.

Harris, Warren S
Business: 101 Agri Engr Bldg, Fayetteville, AR 72701
Position: Associate Professor. *Employer:* University of Arkansas. *Education:* MS/Agri Engg/Univ of IL; BS/Agri Engg/Univ of IL. *Born:* August 1923. Native of Bardolph, Ill. Served USNR 1944-46. Agri Engrg with USDA SCS, 2 yrs. Self employed 6 yrs. With Univ of Arkansas since 1958. Res in drainage, irrigation and erosion control. Teacher. Station Engr for Agri Exp Station. State, Regional & Natl ASAE cttes and/or offices. Chmn SW Region ASAE, 1976-77. Regional Dir (ASAE) 1978-80. Lions Club. *Society Aff:* ASAE, GSD, AE.

Harris, Weldon C
Business: 303 W Swamp Rd, Doylestown, PA 18901
Position: Owner. *Employer:* Weldon C Harris & Assocs. *Education:* MSCE/Civil-Sanitary Engr/Tulane Univ; BSCE/Civil Engr/GA Tech. *Born:* 4/30/30. Native of Atlanta, GA and attended GA Tech as a co-op student 1947-1952. Commissioned Ensign in US Navy Civil Engr Corps in 1953 and served four years as a project/operations officer with Atlantic Fleet Sea Bee Battalion. Established a consulting engg business in 1968 in environmental engg with 22 years experience in design, operations and construction mgt. Experience in water supply and treatment, waste water treatment and disposal and fire protection systems design. *Society Aff:* NSPE, AWWA, WPCF.

Harris, Wesley L
Home: 10 Janet Lane, Vernon, CT 06066
Position: Dean, School of Engrg Professor of Mechanical Engrg. *Employer:* Univ of Connecticut. *Education:* BS/Aero Engr/Univ of VA; MS/Aero Engr/Princeton Univ; PhD/Aero Engr/Princeton Univ *Born:* 10/29/41. in Richmond, VA. Taught on faculty at Univ of VA 1968-1972; MIT 1972-85. Consltg in areas of Isotope Separation in gas mixtures, solar energy sys, acoustics, and computational fluid dynamics. Current res in areas of external unsteady transonic flow, helicopter rotor acoustics, and wind turbine aerodynamics. More than 90 tech publs. Mbr and conslt to US Army Sci Bd 1979-86. Mgr of Computational Methods, NASA Hdqtrs, Washington, DC, 1979-1980. Mbr Tau Beta Pi and Sigma Xi. *Society Aff:* AIAA, AHS, ΤΒΠ, ΣΞ

Harris, William B
Business: Chemical Engrg Dept, College Station, TX 77843
Position: Prof *Employer:* TX A&M Univ *Education:* PhD/Agric Engrg/CO State Univ; MS/ChE/TX A&M Univ; BS/ChE/Univ of CO *Born:* 9/26/19 Native of CO. Worked for Allied Chem & Dye during WWII developing "Atabrine–, an anti-material drug and catalytic aniline process (1941-46). Worked at Apache Powder Co (explosives) & promoted to asst Technical Dir (1946- 1953). Returned to school and joined staff at TX A&M. Tenured Prof, active in Solar Energy and Alternative fuels. Fellow of AIChE. *Society Aff:* AIChE, ISEC, ΦΚΦ, ΤΒΠ, ΦΛΥ

Harris, William J, Jr
Business: 1920 L St NW Suite 620, Washington, DC 20036
Position: Vice President, Res/Test Dept. *Employer:* Association of American Railroads. *Education:* ScD/Metallurgy/MA Inst of Tech; MS/Engg/Purdue Univ; BS/Chem Engg/Purdue Univ. *Born:* June 1918. Native of South Bend Ind. Resident of Arlington, VA. Head, Aircraft Armor Sec, Bureau of Aeronautics, US Navy 1941-45. Head, Ferrous Alloys, Naval Res Lab 1947-51. Exec Secy, Materials Adv Board 1951-54, Exec Dir 1957-60. Asst Exec for Planning Secy, Engrg Div, Natl Res Council 1960-62. Asst Dir, Columbus Lab of Battelle Memorial Inst; Hd, Washington Office of Battelle 1954-57; 1962-70. Vice Pres, Res & Test, Assn Amer Railroads 1970- Fellow TMS: Fellow, ASM: Mathewson Medal AIME; Disting Alumnus Purdue; Railroad Man of Yr 1976 by Modern Railroads. Pres, Met Soc; Pres, EJC; Chmn Natl Matls Adv Bd. Numerous prof soc's - gov't & indus adv cttes. More than 40 publs. Elected to NAT Acad Engg 1977. Recip TRB Disting Serv Award, 1977. Hon Doctor of Eng Purdue 1978. *Society Aff:* AIME, TMS, ASM, ASME

Harrisberger, Lee
Business: University of Alabama, Drawer ME, Tuscaloosa, AL 35487
Position: Prof and Hd *Employer:* Univ of AL. *Education:* PhD/ME/Purdue; MS/ME/Univ of CO; BS/ME/Univ of OK. *Born:* 9/24/24. Native of CO, USNR 1943-46, Instr, Murray State Sch of Agri (1946-49). Prof at the Univ of UT (1950-54), NC State (1954-60), Purdue (1960-63). Served as Prof and Hd of Mech Engg at OK State Univ (1963-71), Dean of the Coll of Sci and Engg at the Univ of TX of the Permian Basin (1971-75). Pres of the American Soc for Engg Education (1975-76) and Andrew Mellon Visiting Prof of Engg at the Cooper Union (1976-77). Author of two texts, chapters in 5 books and over 70 publications, lecturer on educational innovations, creator of graphic art, and furniture. Currently Hd of Mech Engg, Dir of the Mech Engrg Design Clinic. ASPE Engr of the year 1984, Recipient of ASEE Fred Merryfield Design Award for Excellence in Engr Design Educ 1984, Recipient of ASEE Chester F. Carlsen Award for contributions to engg educ 1986. Honorary Life Mbr ASEE 1980, Fellow, ASEE 1983, Fellow, ASME 1985. *Society Aff:* ASEE, ASME, NSPE, WFS, APT.

Harrison, Blaine L
Business: 543 Byron St, Palo Alto, CA 94301
Position: Vice President *Employer:* Kennedy/Jenks/Chilton, Inc. *Education:* BS/Civil Engg/Univ of CA, Berkeley. *Born:* August 1932. Native Californian. Engaged in municipal engrg from 1955-63 at which time joined predecessor cons firm (founded 1934). Appointed to Bd of Dir 1979. Respon for design of extensive regional wastewater transport and disposal sys. Direct administrative policy and responsible for engrg/admin and project control. Became Chief Administrative Officer in firm 1984. Elected Fellow ASCE, 1972. Avocations: Audiophile; boating. *Society Aff:* ASCE, APWA, AWWA, ACEC

Harrison, Carter H, Jr
Home: 14960 NW Ridgetop Ct, Beaverton, OR 97006
Position: Chief Utilities Planning Engr. *Employer:* Dept. of Public Works, City of Salem. *Education:* Engr/Eng-Econ Plan/Stanford Univ; MS/Civil/Water Resources Plan/Stanford Univ; BS/Petro/Stanford Univ *Born:* 7/26/36 Montclair, NJ. US Army 1955-57, Antioch Coll 1957-59, Stanford Univ 1959- 1966. Acting Research Mgr - ASCE; Asst Prof - Auburn Univ; Sr Engr - Stevens, Thompson & Runyan, Inc (now CRS Engrs, Inc). Specializes in long-range water- supply and distribution planning, flood-plain analysis and computer applications to hydraulics. Pres, OR Section ASCE - 1981; Pres Engrg Coordinating Council of OR - 1979; US Deligate to No American - Australian Conference on Metrication - 1975. Hobbies: Computers, amateur radio, cycling, canoeing and skiing. *Society Aff:* ASCE, AWWA

Harrison, Charles W, Jr
Home: 2808 Alcazar St, NE, Albuquerque, NM 87110
Position: Consulting Engineer *Education:* PhD/EM Theory/Harvard Univ; ME/EM Theory/Harvard Univ; SM/Communication Engg/Harvard Univ; EE/Elec Engg/Univ of VA; BSE/Genl Eng/Univ of VA. *Born:* Sept 1913. Ensign USN 1939, CDR USN 1948. Retired USN 1957. Res in electromagnetics Sandia National Labs, Albuquerque 1957-73. Dir General ElectroMagnetics 1973- . Contrib numerous articles to prof journals. Co-author 'Antennas & Waves: A Modern Approach' MIT Press 1969. Reg P E: Va, N M, Mass. Lecturer, Harvard and Princeton Universities 1942-1944. Visiting Prof Christian Heritage Coll, El Cajon Calif 1976. Fellow IEEE, Mbr URSI, Sigma Xi. Electronics Achievment Award, IEEE 1966. Best paper award in IEEE, Elec Mag Compatibility Trans 1972. Hobbies: Hi Fi, Amateur Radio, Photography. BSE, EE U. of Va 1939-1940. SM, ME, PhD (Appl. Physics) Harvard Univ. 1942, 1952, 1954. *Society Aff:* ΣΞ, IEEE, URSI

Harrison, Douglas P
Business: Dept of Chem Engr, Baton Rouge, LA 70803
Position: Prof. *Employer:* LA State Univ. *Education:* PhD/ChE/Univ of TX; BS/ChE/Univ of TX. *Born:* 11/19/37. Native of Frost, TX. Joined LSU in 1969 as Asst Prof of Chem Engg. Prof since 1978. Dept Chrmn 1976-1979. Additional profl experience with Monsanto and US Enviromental Protection Agency. Res interests include chem reaction engg, catalysis, kinetics, coal processing, and air pollution control tech. Enjoys tennis and woodworking. *Society Aff:* AIChE, ASEE, APCA

Harrison, Gordon R
Home: 2915 Greenrock Trail, Doraville, GA 30340
Position: V P *Employer:* Electromagnetic Sci, Inc *Education:* PhD/Physics/Vanderbilt Univ; MS/Physics/Vanderbilt Univ; BS/Physics/Univ of Central Arkansas. *Born:* Dec 14, 1931. Native of Brinkley Arkansas. Employed by ORNL & Convair summers of 1953, 1954 & 1955. Joined Sperry Microwave, Clearwater Fla 1957. Performed res in materials for microwave applications. Progressed to Engrg Mgr respon for applied res. Joined the staff of EES, Ga Tech 1971. Assumed position as Lab Dir responsible for mgmt & leadership of a staff performing basic & applied res in energy, material, physical, environmental, and nuclear sciences 1972. Joined Electromagnetic Sci in June 1983 as V P and mgr, materials and new tech programs. Responsibilities for the fabrications of ferrimagnetic materials and their applications in microwave and millimeterwave components and subsystems. Mbr Sigma Pi Sigma, Sigma Xi, IEEE (Fellow 1975), S-MTT & S-MAG. *Society Aff:* IEEE, MTT-IEEE, MAG-IEEE, ΣΞ, SPIE.

Harrison, Michael A
Business: Computer Science Div, Berkeley, CA 94720
Position: Prof Computer Science *Employer:* Univ of CA, Berkeley *Education:* PhD/Comm Sci/U of MI; MSEE/Elec Engr/Case-Western; BS/Elec Engr/Case- Western Reserve. *Born:* 4/11/36. Prof of Computer Sci at Univ of CA at Berkeley. Cons Editor in Computer Sci for Addison Wesley Publ Co. Cons for various companies. Visiting Prof at MIT, Hebrew Univ in Jerusalem. Frankfurt and Stanford Guggenheim Fellow. Chrmn, ACM Special Interest Group in Automata & Computability Theory. Editor of Information Processing Letters, JACM, JCSS, TCS, DM and DAM. Chrmn ACM Sig Board; Mbr ACM Council 1978-. Vice President, ACM, 1980-1982. Ex Comm CSNet, Dir, Amer Federation of Information Processing Societies. Trustee, Charles Babbage Institute. Chrmn, Natl Res Council Panel on International Dev in Microelectronics and Computer Science. *Society Aff:* ACM, IEEE, AAAS.

Harrison, Otto R
Home: PO Box 296, New Canaan, CT 06840
Position: Gen Ops Manager *Employer:* Esso Libya *Education:* BS/Petrol Engr/Univ of TX; Assoc of Sci/Engr/Schreim Inst. *Born:* 7/18/35. Grad Univ of TX in 1959. Joined Exxon in 1959. Served in various engg assignments in TX, LA, (reg Engr in both states), Wash, DC & NY. Prior work includes Dist Mgr, off shore dist, Div Mgr, Mid Continent Div, Operations Mgr. Southeastern Div, Coordinator Engg & Planning & current assignment of Deputy Prod Mgr, Producing Div, Exxon Corp. Holder of US patents in drilling & production & developed CPC (Computer Production Control) for oil fields. Served on Pres's Exec Exchange prog. Currently Gen. Ops. Mgr of Esso Libya. *Society Aff:* AIME, API, ΤΒΠ, ΣΓΕ, NSPE.

Harrison, Stanley E
Business: 7915 Jones Branch Dr, McLean, VA 22102
Position: Pres & COO *Employer:* The BDM Corp *Education:* MS/EE/Univ of NM; BS/EE/OH State Univ *Born:* 11/19/30 Native, Northup, OH. Professional Services Firm Corporate Exec. Served with US Air Force, 1948-1952. Res Staff, Sandia Lab, NM, 1958-1963; Program Mgr, Nuclear Div, Martin Marietta, MD, 1963-68; The BDM Corp since 1968 (Dir, Albuquerque Operations, 1968-1970; Dir, Western Operation, 1970-72; VP, Operations, 1972-74; Exec VP and Chief Operating Officer 1974-83; Dir, 1978- Present; Assumed current responsibility as Pres and Chief Operating Officer in 1983). Also, Dir and Pres and COO, BDM Services Co, 1971-present. Dir, Pres and COO, Zapex Corp, 1975-present. V Chrmn-Operations and Dir, BDM Mgmt Services Co, 1975-present. Exec VP and Dir, BDM Intl, Inc 1979-present. *Society Aff:* IEEE, AFCEA, ADPA.

Harstad, Andrew E
Home: 5311 Frances Ave NE, Tacoma, WA 98422
Position: Pres. *Employer:* Incon Inc. *Education:* BS/CE/Univ of WA. *Born:* 2/1/24. Native of the Pacific Northwest. Served in Navy in WWII. Experience in engg with pulp & paper & AEC Proj at Hanford, WA. Mgd WA State's first Air Pollution Prog at City of Tacoma progressed from Technical Service to Sales Mgr with Pennwalt Ind Chems. Having been active in AIChE, AWWA & WPCF for yrs, formed INCON & serve as pres. Enjoy golfing & fishing. *Society Aff:* AIChE.

Harstead, Gunnar A
Business: 169 Kinderkamack Rd, Park Ridge, NJ 07656
Position: Pres *Employer:* Harstead Engrg Assocs, Inc *Education:* PhD/Structural Mech/NYU; MS/CE/Columbia Univ; BS/CE/Columbia Univ *Born:* 9/12/32 in Norway. Served as a commissioned officer in the US Navy, Pacific Fleet, during the Korean War. Employed as a design engr of bridges, bldg and special structures by Howard Needles and Severud Assocs. Entered nuclear power field with Westinghouse in 1965 as senior engr. Later employed as Supervising Civil engr and Assistant Chief Structural Engr with Burns & Roe Inc and Stone & Webster Engrg Corp, respectively. In 1975 became VP and part owner of Soot & Harstead Assocs. On July 1, 1979 became pres and major stockholder of Harstead Engrg Assocs, Inc. Had made presentations at national conferences and published papers in technical journals. *Society Aff:* ASCE, NSPE, ACEC

Hart, Alan S
Home: 320-41 Vallejo Drive, Millbrae, CA 94030
Position: District Engineer - retired. *Employer:* State of California. *Education:* BS/Civil Engrg/Univ of CA, Berkeley. *Born:* Dec 7, 1907. 44 yrs with Calif Div of Highways; last 21 yrs were as District Engr. Headed Highway Districts at Bishop, Eureka, Marysville & San Francisco. Engineered 1st freeway routings through Calif redwoods. Charge of const of I 80 freeway over Donner Summit (ASCE award winning proj), which has been officially named by the California legislature "Alan S. Hart Freeway." Located freeway routing through City of Sacramento. Charge of const of freeways in San Francisco Bay area. Pres S F Sect ASCE 1972-73; Mbr Natl Ctte ASCE. Mason, Shriner, Rotarian, Elk. Golf, bridge, photography. Married and a grandfather. *Society Aff:* ASCE.

Hart, David R
Home: 2630 Greenmont Drive, Birmingham, AL 35226
Position: Proj Mgr *Employer:* Rust Intl Corp *Education:* PhD/Chem Engg/Univ of AL; MS/Chem Engg/Univ of AL; BS/Chem/Auburn Univ. *Born:* 7/26/26. in PA. Served in US Navy 1944-46. Married, two children. Worked in industrial res and engg from initial degree, patents, publications. Adjunct teacher at Auburn Univ and at Univ of AL. Joined Rust in Birminghan im 1977; trasferred to OR in 1978 as General Manager of a process dev unit for study of the conversion of biomass to liquid fuels. Transferred to Allentown, PA in 1980 to work in Coal Liquefaction (SRC-I) and returned to Birmingham in 1981 to work in the project mgt field. Prin hobby is tournament contract bridge; Life Master of American Contract Bridge League. *Society Aff:* AIChE, ACS, ΣΞ, PMI.

Hart, Don A
Home: 44020 Galion Avenue, Lancaster, CA 93534
Position: Director AF Rocket Propulsion Lab *Employer:* US Air Force. *Education:* BSChE/Univ of TX-Austin. *Born:* Aug 1933. Chemical engr at DuPont 1956. Active duty (Lt) USAF 1956-59. Proj Officer Power Plant Lab, WrightPatterson AFB Ohio. Various supervisory/mgmt postions in solid & liquid rocketry at Air Force Rocket Propulsion Lab, Edwards AFB Calif 1959-70. Deputy Dir of Space Propulsion and

Hart, Don A (Continued)
Power Div, NASA Hq 1970- 72. Returned to Air Force Propulsion Lab in 1972; Dir of Plans & then Deputy Director. Appointed Director 1981. Assoc Fellow, AIAA. Air Force Meritorious Award Civilian Service Award 1970. *Society Aff:* AIAA.

Hart, Franklin D
Business: Ctr for Acoustical Studies, 2149 Burlington Laboratories, NC 27650
Position: Assoc. Dean of Engr. *Employer:* NC State Univ. *Education:* PhD/ME/NC State Univ; MS/ME/NC State Univ; BS/ME/NC State Univ. *Born:* 10/15/36. Dr F D Hart is Prof of Mech & Aerospace Engg & Assoc. Dean of Engr. for Research Programs NC State Univ, Raleigh, NC. Dr. Hart is in charge of all research programs in the School of Engr. This involves seven academic departments and four dedicated research laboratories.Dr Hart has specialized in noise & vibration control, especially in machinery design; has directed numerous res projs in noise control including res in the psychological & physiological effects of noise on man. He has published over 60 papers & res reports. *Society Aff:* ASA.

Hart, Jack B
Home: 5560 Broadmoor Pl, Indianapolis, IN 46208
Position: Assoc Prof of ME. *Employer:* IN Univ. *Education:* PhD/Bio-ME/Univ of MO; MS/ME/Univ of MO; BS/ME/Univ of MO. *Born:* 3/2/36. Native of Nora Springs, IA. Educated in MO. Served in a heavy ordinancy battalion in the US Army from 1956 to 1958. Taught engg courses at the Univ of MO, Purdue Univ at Indianapolis and currently teach at IN Univ - Purdue Univ at Indianapolis. Current res interests are associated with various bioengg projs. *Society Aff:* ASME, ASEE, BES.

Hart, Lyman H
Home: 2665 E Ave De Pueblo, Tucson, AZ 85718
Position: Consultant-Geology. *Employer:* Straus Minerals. *Education:* MS/Geology/Univ of WI; BS/Chemical Engg/Univ of WI. *Born:* Dec 1900; native of Madison Wis.. Mining geologist 16 yrs Anaconda; 6 yrs US Smelting Refining & Mining Co. Asst Genl Mgr W Minn Dept, Asarco; later Ch Geol N Y. V P Guggenheim Explor Co N Y; V P Straus Minerals N Y. Present: Cons Straus Minerals AIME Jackling Medal 1973; Distinguished Mbr Soc of Min Eng, AIME with Behre Dolbear & Co (Partime) '74-'78 Continuing as Conslt, Straus Minerals, N.Y. to Present *Society Aff:* AIME, SEG, GSA, AAAS, MMS, AIPG.

Hart, Patricia H
Home: 2355 NE Ocean Blvd Apt 28B, Stuart, FL 33494
Position: Asst Proj Engr. *Employer:* Pratt & Whitney Aircraft. *Education:* ME/Systems Engg/Univ of FL; BS/Chemistry/Univ of Cincinnati; BS/Math/Southern Methodist. *Born:* 2/18/27 Native of Cincinnati, OH. In 1956 started with Chance Vought Aircraft in Analog Simulations. With Pratt & Whitney Aircraft since 1959 in digital simulations and controls serving as prog mgr and asst proj engr. She is the p pres of the FL Sec of the Soc of Women Engrs, 1980 chrmn of the Palm Beach Sec of IEEE. She has a BS with honors in chemistry, BS in math, ME in engg and is a mbr of Iota Sigma Pi & Phi Mu Epsilon. *Society Aff:* SWE, IEEE.

Hart, Peter R
Home: 522 Pape Ave Apt 2, Toronto Ontario M4K 3R4 Canada
Position: Pres. *Employer:* Peter R Hart Inc. *Education:* BSc/Civ & Municipal/Univ College, London. *Born:* 6/11/33. Appointed Asst County Engr, Grey County, in 1958 and, in 1960, County Engr, Brant County. Entered consulting engg in 1964 by joining Damas & Smith Limited becoming Dept Dir, Road Design. From 1969-74 as Proj Mgr and Associate of M M Dillon Limited responsible for maj works. Principal of M H Kilpatrick Assoc Limited from 1974 until establishing his own practice in 1978. Chrmn, Engg Tech Div, Centennial Coll of Applied Arts and Tech. Pres CSCE 1978-79. VP EIC 1978- 79. Canadian Silver Jubilee Medal 1977. Mbr, Royal Ontario Museum. Blood Donor (70plus donations); Master Reader, Books for the Blind CNIB; Amateur Theatre, Actor and Dir, (summer stock, CBC-TV). Enjoys music. *Society Aff:* EIC, CSCE, ICE.

Hart, Robert N
Business: 600 Commerce Square, Charleston, WV 25301
Position: VP - Gen Mgr *Employer:* Bow Valley Petro, Inc *Education:* BS/Petro Engrg/WV Univ *Born:* 12/7/42 Native of Charleston, WV. Began my career as an underground gas storage engr with The Columbia Gas Transmission Corp in 1966. Advanced to the position of Mgr of Columbia's Storage Engrg Dept in 1978. Became an oil and gas consultant in 1980. Assumed current responsibility as VP - Gen Mgr for Bow Valley Exploration, Inc for exploration and production of oil and gas. Appalachian Petro Section SPE Chrmn - 1976. Member SPE Monograph Committee. Active in local children's museum/art gallery. Enjoy golf, tennis and photography. *Society Aff:* AIME

Hartenberg, Richard S
Home: 726 Laurel Avenue, Wilmette, IL 60091
Position: Prof of Mech Engrg, Emeritus. *Employer:* Northwestern University. *Education:* PhD/Engg Mechs/Univ of WI; MS/Mech Engg/Univ of WI; BS/Mech Engg/Univ of WI. *Born:* Feb 1907. Native of Chicago, Ill. Instructor at Wisc before going to Northwestern in 1941. Sr Engr, Forest Products Lab 1942-44. Ch Engr NDRC and BuAer projects during WWII. Teaching interests: machine design, kinematics, history of engrg. Serve on editorial bd of 'Mechanism & Machine Theory'. Author (With J Denavit) of *Kinematic Synthesis of Linkages.* Contrib to Dictionary of Scientific Biography, Encyclopaedia Britannica. Editor Mbr ASME History & Heritage Ctte. Western Electric Fund Teaching Award 1964; ASME Mechanism Ctte award 1974. ASME Centennial Medal 1980. Reg prof engr, Ill & Wisc; chartered mechanical engr, Great Britain. Fellow ASME & Inst of Mechanical Engrs. *Society Aff:* ASME, SHOT, ASEE, IMechE, VDI.

Harter, Marion M
Home: 8900 Jarboe, Kansas City, MO 64114
Position: Chief, Design Branch. *Employer:* Kansas City District, Corps of Engrs. *Education:* MS/Civil Engg/Univ of KS; BS/Aeronautical Engg/Univ of KS. *Born:* June 1926. Reg PE in Mo and Kansas. Native of St Joseph Mo. US Navy 1944-46. Corps of Engrs 1949- . Assumed present position as Ch, Design Branch of Engrg Div Mar 1976. Supervices 70 engrs, technicians, & respon for arch, mech, elec & struct design of civil and military work assigned to Kansas City District. Also respon for preparation of specifications and estimates of all work. Chairman ASTM Committee C-27, 1980-84 Chmn, ASTM Subcommittee C-27.3, 1975-79; Pres Kansas City Sec, ASCE 1974-75; Chmn District Council ASCE 1971-72; Prof & Scientific Award from Greater Kansas City Federal Exec Bd 1968. (Wife: Sue Rose Mounce Harter; Daughter: Rosina). *Society Aff:* ASCE, ASTM.

Hartig, Elmer O
Business: 1210 Massillon Road, Akron, OH 44315
Position: VP, Engineering & Research. *Employer:* Loral Systems Group. *Education:* PhD/Applied Physics/Harvard Univ; MS/Applied Physics/Harvard Univ; BS/Elect Engr/Univ of NH. *Born:* 1/28/23. Native of Evansville, Ind. Served with US Army 1944-46 (assoc with Manhatten Project at Columbia Univ & Los Alamos Sci Lab). Joined Goodyear Aerospace Corp in 1950. Helped pioneer efforts in dev of coherent synthetic aperture side-looking radar, microwave antennas & radomes, & missile terminal guidance sys. Other fields include radar data processors, laser scanner & recorders, and wide-band data transmission sys. A Fellow in the IEEE & mbr of Sigma Xi. In 1960 received Achievement Award from Phoenix Section of IEEE for outstanding contributions in field of radar. Mbr of Divisional Advisory Group for USAF Foreign Technology Div 1972-76 & mbr of USAF Scientific Advisory Bd, Guidance & Control Panel 1964-76. Served as mbr of AZ Atomic Energy Commission 1970-74. USAF Army Science Board member and consultant 1980-1984. *Society Aff:* IEEE, AIAA.

Hartigan, Michael J
Business: 224 S Michigan Ave, Chicago, IL 60604
Position: Vice President. *Employer:* McDonough Assoc Inc. *Education:* Cert Hwy Traffic/Transport/Yale Univ; BS/Civil Engg/Univ of Notre Dame. *Born:* Sept 1927. BSCE Univ of Notre Dame 1950; Certificate Highway Traffic, Yale Univ 1954. Reg Prof Engr in 4 states. Formerly Ch Engr for Ill State Toll Highway Authority, encompassing engrg mgmt of 187 mile system of toll highways in northern Ill. Was also Deputy Regional Transportation Engr in Chicago Metropolitan Area for Ill Dept of Transportation. Since 1973, Vice Pres of McDonough Associates Inc., assisting in general administration & mgmt of firm and directing transportation related activities. Fellow Inst of Transportation Engrs & President of Ill Section 1970. Arthur J Schmitt Foundation Scholarship, Notre Dame; RandMcNally Fellowship, Yale. *Society Aff:* ITE, ACEC, TRB, APWA.

Harting, Darrell R
Home: 16810-12th Ave SW, Seattle, WA 98166
Position: Technical Representative *Employer:* Boeing Aerospace Company. *Education:* BS/EE/Univ of WA. *Born:* October 1926. Native of Seattle, Washington. With Boeing Aerospace Company since 1950. Holds patents on -S/N- Fatigue Life Gage & a strain gage having a strainresistant electrical connection and is co-inventor of nondestructive testing technique for detecting & analyzing incipient failures in rotating components and co-inventor of a High Temperature Capacitive Strain Gage. Reg Prof Engr, States of WA & CA. Mbr of ISA (Dir, Test Measurement Div 1973-75, VPres District 9 1975-77, Pres 1980-81, and SESA (V Pres 1975-77, president 1978). *Society Aff:* SESA, ISA, ADPA, BSSM.

Hartley, Boyd A
Business: Fire Protection & Safety Engrg Dept, 10 W 32nd St, Chicago, IL 60616
Position: Chrmn *Employer:* IL Inst of Tech *Education:* BS/CE/Univ of PA *Born:* 4/23/25 USNR, 1943-45. Natl Bd of Fire Underwriters, 1948-1966. Assoc Prof, IL Inst of Tech, Fire Protection and Safety Engrg Dept, 1966-present; Chrmn, 1975-present. Honorary fraternities: Tau Beta Pi, Salamander, Compass and Chain. Member, Natl Fire Protection Assoc, Society Fire Protection Engrs (P Pres, Chicago Chapter), American Society Safety Engrs, Intl Assoc Fire Chiefs; served many committees. Corporate member, Underwriters Labs, Electrical Council (1956- present); Fire Council (1975-present). Trustee: IAFC Foundation, Greater Chicago Safety Council. Pres, IL Safety Council, 1984-85. FP Committee, Chicago, Assoc Commerce and Industry. PE Registration NY and IL. *Society Aff:* SFPE, ASSE, NFPA, IAFC

Hartley, Fred L
Business: PO Box 7600, Los Angeles, CA 90051
Position: Chief Executive Officer. *Employer:* Unocal Corporation. *Education:* Bachelor of Appl Sci/ChE/Univ of British Columbia. *Born:* 1/16/17. I am Chrmn of the Bd of Dirs, and Chief Executive Officer of Unocal Corporation. I joined the Co in 1939 as an engg trainee shortly after graduation from the Univ of British Columbia with a degree in chemical engg. My work includes process design, operations, supervision, and technological sales. I was elected Sr VP in 1960 in charge of Mktg, and in 1962 assumed charge of the Refining and Marketing Div, in Nov 1963 I was elected Exec VP, and Pres and CEO in 1964. I was elected Chrmn of the Bd of April 1974 (my present assignment). *Society Aff:* ACS, AIChE, AIC, IAE, NYAS, SAE, ΠΕΤ, SPE OF AIME.

Hartman, David E
Business: Northern Arizona University, P.O. Box 15000, Flagstaff, AZ 86011
Education: PhD/Materials Science/Rice Univ; MS/Metallurgical Engg/Rice Univ; BS/Mech Engg/Rice Univ; BA/Sci-Engrg/Rice Univ *Born:* 2/28/35. Born in Panama Canal Zone. Carrier pilot aboard USS Forrestal 1958-59. Sr Engr with Houston Res Inst, 1965-67. Part time instructor in Mech Engg, Univ of Houston, 1966-71. Sr Engr, Shell Pipe Line Corp Res and Dev Lab, 1967-72; failure analysis, LNG Vapor explosion res. Exec Officer, Naval Air Reserve squadron RTU-94, Naval Air Station, New Orleans, 1970-72. Owner, Oceanus Enterprises, scuba products and ocean related materials, 1972-74. Prof of mech and welding engg, LeTourneau College 1975-1985. Prof of mech engg at Northern Arizona University since 1985. Married Judith Ann O'Brien, 1968; Adopted Michael David Hartman, Chiapas, Mexico,
e 1977. Reg PE, TX, 1971. *Society Aff:* ASNT, ASTM, ASEE, ASME, NSPE, TSPE.

Hartman, John Paul
Business: College of Engineering, Univ. of Central Florida, 4000 Central FL Blvd, Orlando, FL 32816-0450
Position: Asst Dean, Prof of Engrg *Employer:* Univ of Central FL *Education:* PhD/CE/Univ of FL; SM/Soil Mech/Harvard Univ; BSCE/CE/WA Univ (St. Louis); BS/Math & Geol/Principia Coll *Born:* 8/15/36 Raised in St. Louis, MO. Instructor in US Naval Nuclear Power Program (Mare Island Naval Shipyard) from 1960-65. Soils Engr for Ardaman and Assocs in 1965. Sr Engr for Martin-Marietta Aerospace 1965-68. Had NSF Science Faculty Fellowship 1970-71. At the Univ of Central FL has been Chrmn of Civil Engrg and Environmental Sciences (1975-78) and Acting Chrmn of Mechanical Engrg and Aerospace Sciences (1978-79). Awarded Southeastern Section ASEE Western Elec Award for excellence in engrg education (1981). On ASEE Bd of Dirs (1981-1983) and member of ASME Natl History and Heritage Committee (1975-present). Elected Fellow of ASME (1986), and Fellow of FES (1987). *Society Aff:* ASCE, ASME, NSPE, ASEE, AREA, FES, SHOT, SIA

Hartman, Walter M
Business: 1605 Mutual Building, Detroit, MI 48226
Position: Engr Mgr *Employer:* PSG/The Hinchman Co *Education:* BS/Geol Engrg/Univ of Pittsburgh *Born:* 8/26/25 Native of McKeesport, PA. WWII service with US Navy, 1943-46. Worked in electrical testing with Westinghouse and seismic geophysics with Conoco. Corrosion consulting with The Hinchman Co and A.V. Smith Engrg Co (Divs of Professional Services Group, Inc) since 1960. Field engr, Philadelphia, PA, 1960-66. Established and managed Charlotte, NC, regional office 1966-1977. 1977 to 1984, primary responsibility as corrosion consultant and project mgr for new rapid transit systems in Washington, DC and Atlanta, GA. Presently Engrg Mgr Central Region Office, Detroit. Former Officer Atlanta Section, NACE. Present Officer Detroit Section, NACE. *Society Aff:* NSPE, NACE, SEGA

Hartmanis, Juris
Home: 324 Brookfield Rd, Ithaca, NY 14850
Position: Prof. *Employer:* Cornell Univ. *Education:* PhD/Math/CA Inst of Tech; MA/Math/Univ of KS City; CandPhil/Phys/Univ of Marburg. *Born:* 7/5/28. Juris Hartmanis, born in Riga Latvia, studied phys at the Univ of Marburg in Germany and math in this country, earning his PhD at the CA Inst of Tech in 1955. He has taught math at Cornell & OH State Univ and from 1968-65 was Res Scientist at GE Res Lab in Schenectady. From 1965-1971, he served as the first chrmn of the new Comp Sci Dept at Cornell Univ and resumed the chairmanship 1977-82. His main res interest are in the theory of computational complexity and theory of computing. Currently, he is Editor for Springer-Verlaj Lecture Notes in Comp Sci, Siam Journal for Computing, Journal of Comp & System Sciences & Assoc Editor for Math Systems Theory. *Society Aff:* ACM, AMS, MAA, SIAM, AAAS.

Hartmann, Richard J
Home: 5533 Warden Avenue, Edina, MN 55436
Position: President. *Employer:* Arjay Sales Inc. *Education:* BS/EE/Univ of MN. *Born:* 6/26/30. BSEE Univ MN 1958. Lifetime resident Minneapolis. US Navy 1948-52 Electronic Tech. Prod Engr Honeywell Aero on floated gyro's 1955-57. Control Data Corp 1957-65, Prod Engr, Design Engg Sec Hd & Reg Sales Mgr for servo components. Electro-Craft Corp 1965-67 - Reg Salels Mgr motors & speed controls. Formed Arjay Sales Inc, manufacturer's rep co in 1967. Served as Treas & Pres Twin City Sec Instrument Soc of Amer (ISA). Served as ISA Sec Delegate in District 6, 4 yrs. ISA District 6 VP 1976. Recipient: ISA Distinguished Service Award 1984. Hobbies: Camping, fishing & golf. *Society Aff:* ISA.

Hartsaw, William O
Home: 1407 Green Meadow Rd, Evansville, IN 47715
Position: Prof & Hd ME. *Employer:* Univ of Evansville. *Education:* PhD/Theo & Appl Mechanics/Univ of IL; MS/Engg Mechanics/Purdue Univ; BSME/ME/Purdue Univ. *Born:* 10/17/21 Born in Tell City, IN. Married Delma Stuckey, son Mark Alan. Became Instr of Engg at Evansville College 1946. Moved though the various ranks & became Prof of Engg in 1963. Engg Dept Hd 1958-61, Dir Sch of Engg 1961-1968. univ of Evansville's first Dean of Engg 1968-1975. Obtained ECPD accreditation of EE & ME progs in 1971. Mech Engg Dept Hd 1977-1986. Evansville Environmental Protection Agency Board (Vice Chairman) 1980- Received the ASME Centennial Medallion 1980; Vice Chariman Evansville Section, ASME-1980; Chairman, Evansville Ssection ASME 1981 Received the Tri-State Council for Engg & Sci 1979 Technical Achievement Award. NSF Faculty Fellow 1961-62. Active in local, regional & natl ASME. Advisory Council Boy Scouts, Evansville Urban Transportation Advisory Committee. 1983-VP ASME Region VI. 1983-VP ASME Region VI. Distinguished Prof of Mech Engg 1986- . *Society Aff:* AAAS, ASHRAE, ASME, ASEE, ASTM, AAUP.

Hartung, Herbert O
Business: 8390 Delmar Blvd, St. Louis, MO 63124
Position: Pres - Retired. *Employer:* St Louis Cty Water Co. *Education:* BSCE/Sanitary/KS Univ. *Born:* January 1909. Native of Kansas City Mo. Burns & McDonnell Cons Engrs Kansas City 1930-31. St Louis County Water Co 1930- , Pres 1973- . Retired Pres Mo Water Co, Independence 1973- . Retired VP Northern Ill Water Corp, Champaign 1959- . Diplomate Amer Acad San Engrs, AWWA: Life Mbr, Honorary Mbr, Fuller Award, Goodell Prize, Chmn Water Quality Div, Standards Council, Tech & Prof Council, numerous cttes. EPA Adv Ctte on Drinking Water Stds 1972-74. Retired Pres Missouri River Public Water Supplies Assn. *Society Aff:* AWWA.

Hartung, Roderick L
Business: PO Box 1300, Pascagoula, MS 39567
Position: Gen Mgr. *Employer:* Chevron USA, Inc. *Education:* MS/CE/Univ of MI; BS/CE/Univ of MI; BA/Pre Engg/Arbion College. *Born:* 7/25/35. Born & raised in MI. Attended Albion College on dual degree prog with Univ of MI. Joined Std Oil Co of CA in 1959 & have held increasingly responsible positions in engg (was asst chief engr-corp, 1974-1976) & mfg. Spent three yrs in Lima, Peru as Dir of Refineria Conchan, 1969-1972. Appointed to present position 12/76. Active in civ affairs including pres elect of local rotary and chamber of commerce. Married with three children. Enjoy boating, fishing, skiing, etc. *Society Aff:* AIChE.

Hartwig, Thomas L
Home: 9131 RidgeFel Ave, Pittsburgh, PA 15237
Position: Sr VP and Mgr Mun Environ Engrg *Employer:* Duncan, Lagnese and Assoc Inc *Education:* BS/CE/Univ of Notre Dame *Born:* 6/16/52 I graduated from the Univ of Notre Dame, with honors, in 1974 with a BS degree in Civil Engrg. Prior to graduation, I worked summers for surveying and engrg firms in Butler, PA. I have been with Duncan, Lagnese and Assocs since June 1974. Presently I am Sr VP and Mgr of Mun Environ Engrg. While at Notre Dame, I served at Pres of the Student Chapter ASCE. I am a registered Professional Engr (PE) and registered Professional Land Surveyor (PLS) in the state of PA. Also, I am a licensed sewage treatment and water treatment plant operator. My biography also appears in "Who's Who in the East–, "Who's Who in Amer–, and "Who's Who in the World–. *Society Aff:* NSPE, WPCF, ASCE, AWWA, WPCAP.

Hartz, Nelson W
Home: 8 Holland Rd, Pittsburgh, PA 15235
Position: Consultant #EMP# *Born:* June 1913. Native of Pittsburgh. BS Carnegie Inst Tech 1934. Reg Prof Engr Penna. Cert Safety Prof. Employed by Mine Safety Appliances Pittsburgh since 1934 except for 4 yrs in Air Force during WW II. Treas Instrument Soc of Amer 1968-70 & Pres, V P & Bd Mbr Amer Soc for Safety Res. Major experience has been with instrumentation for measuring toxic & explosive gases & analysis of process streams. *Society Aff:* ISA, ASSE.

Harvey, Francis S
Business: 143 Dewey St, Worcester, MA 01610
Position: Pres-Treas. *Employer:* Harvey & Tracy Assoc, Inc. *Education:* BS/Civ Engr/Worcester Polytechnic Inst; Dr (Hon)/Engr/Worcester Polytechnic Inst. *Born:* 12/20/14. Native of Worcester, MA, Cadet Engr-Ebasco Services 1937-40 Design Engr-AJD Co, 1940-46. Private Practice Consulting Structural Engr 1946-60 Pres-Treas, Harvey & Tracy Assoc 1960 to date. Practice in structural engg & arch. Reg Arch & Engr. Co-author of 1970 Boston Bldg Code. Mbr of MA Bd of Schoolhouse Stds, 1965-75. Mbr of MA State Bldg Code Commission 1979-81; Chrmn of Bd Bay State Savings Bank 1977-; Trustee Worcester Poly Inst. Mbr & 8th Recipient of the Ralph W Horne Fund Award, BSCE. Trustee St Vincent Hospital 1970- : Fellow A. S.C.E. - Life. *Society Aff:* ASCE, NSPE, CSI, ΤΒΠ.

Harvey, William B
Business: 1101 State Rd, Bldg N PO Box 623, Princeton, NJ 08540
Position: Pres. *Employer:* Van Note-Harvey Associates *Education:* BS/Civ Engg/Lafayette College. *Born:* 2/18/24. Reg PE: NJ, NY, PA, VA & FL. VP, Lanning Engg Cos, 1952-60, in charge of maj municipal sewerage projs. 1960 to present: Pres & Bd Chrmn, Van Note-Harvey Associates. Presently serve as Bd Chrmn, Nassau Surveying Co, Inc. Bd Chrmn, Penn State Engg, Inc (wholly owned subsidiary) since 1970. Bd Chrmn Vn Note- Harvey Assoc, Inc (VA corp) 1970 to 1980. Am responsible for all phases of operation for all of the above cos totaling 125 employees involved in gen municipal work, wastewater work (both public & private) & environmental work. *Society Aff:* ASCE, NSPE, WPCF.

Harvill, Lawrence R
Business: Univ of Redlands, Redlands, CA 92374
Position: Chrmn, Dept. of Engg. *Employer:* Univ of Redlands. *Education:* PhD/Applied Mech/UCLA; MS/Mech Engrg/UCLA; BS/Mech Engrg/UCLA *Born:* 01/27/35 Dr Harvill is a native of Southern CA. He has been a mbr of the faculty at the Univ of Redlands since 1964. At present he is chrmn of the Engrg and Computer Sci Dept. He also served the Univ as VP of Academic Affairs from 1970 thru 1973. Dr Harvill has authored several articles and two texts. Since 1985 Dr Harvill has served as a consultant to the McCrometer Div of Ametek located in Hemet, CA. *Society Aff:* ASME, ASEE, ΤΒΠ, ΣΞ

Harwood, Julius J
Home: 2258 Shorehill Dr, W Bloomfield, MI 48033
Position: Dir, Materials Sciences Lab. *Employer:* Engrg & Res Staff, Ford Motor Co. *Education:* MS/Metallurgy/Univ of MD; BS/Chemistry/City College of NY. *Born:* Dec 1918. Naval Gun Factory Washington, DC 1940-46; Office Naval Res, Hd Met Branch 1946-68; Ford Motor Co, 1960-present, current position 1974. Adjunct Prof Wayne State Univ; Bd of Control MI Tech Univ; Pres AIME 1976; Pres The Met Soc 1973; Chrmn Inst of Metals Div 1968. Fellow ASM & Fellow TMS. Fellow AAAS & Fellow Eng. Soc of Detroit CESD Cons to Natl Sci Fdn & OTA. Chrmn/Bd of Visitors, Engg Coll RPI. Editorial Adv Bd of "Mtls Sci & Engg" & "Treatises on Mtls Sci & Tech–. Governing Bd of Acta Met. Publ over 70 articles and edited 5 books in field of mtls. Mbr AAAS, MI Acad of Sci, Sigma Xi, ASM, AIME. Ford Citizen of Year 1968. Reg PE CA. Elected to Natl Acad of Engg 1976, Chrmn Natl Mtls Advisory Bd 1976-. *Society Aff:* ASM, AIME, AAAS, ESD, RESA, ΣΞ.

Harza, Richard D
Business: 150 South Wacker Dr, Chicago, IL 60606
Position: Chrmn *Employer:* Harza Engrg Co *Education:* MSc/Civ Engrg/Northwestern Univ; BSc/Mech Engrg/Northwestern Univ; DSc/(Hon)/Univ of WI. *Born:* 10/07/23 MS in CE - 1947, Northwestern Univ; BS in Mech Engrg - 1944, Northwestern Univ. Instructor Civ Engrg Dept, Northwestern Univ, 1946-47. With Harza Engrg Co since 1947. VP, 1953; Dir, Admin, Financial & Corp Operations,

Harza, Richard D (Continued)
1976-1980; Dir, Mgmt Group II, USA Resources, 1970-1976. Chrmn and Pres 1980-1983; Pres 1977-1986, Chrmn 1986- . Reg PE-AK, AZ, DC, IL, IN, IA, MI, MS, OH, PA, VA and WA. Mbr - AICE, AWRA (P Pres), CECI, ISPE, NSPE, USCOLD (P Chrmn; P Chrmn of USCOLD Environ Effects Ctte), ICOLD, ASME. Has written and presented numerous tech papers and articles on consltg engrg; has lectured to various professional socs and trade organizations. Is currently Proj Dir of Chicago's Tunnel and Reservoir Plan (TARP), and the Strontia Springs Diversion Dam in Colorado. *Society Aff:* ASCE, AWRA, USCOLD, NSPE, ICOLD

Hasan, A Rashid
Business: Chem Engrg Dept, Box 8101, Grand Forks, ND 58202
Position: Assoc Prof *Employer:* Univ of ND *Education:* PhD/ChE/Univ of Waterloo; MASc/ChE/Univ of Waterloo; BSc Engrg/ChE/Buet, Dacca *Born:* and brought up in Bangladesh. Came to Canada on Commonwealth Scholarship in 1973. PhD research was in the area of heat & momentum transfer in multiphase flow. Recent interests include multiphase flow metering, slurry rheology and heat transfer and pressure transient analysis for oil-wells. I am an avid bridge player and also like ping-pong. *Society Aff:* AIChE, SPE, ΣΞ.

Hasegawa, Akira
Business: 600 Mtn Ave, Murray Hill, NJ 07974
Position: Disting Mbr of Tech Staff & Adjunct Prof *Employer:* AT&T Bell Labs. *Education:* Dr Sci/Phys/Nagoya Univ; PhD/Elect Eng/Univ of CA; M Eng/Elect Eng/Osaka Univ; B Eng/Elect Eng/Osaka Univ *Born:* 6/17/34. Native of Japan, Came to US as a Fulbright student, obtained PhD in EE in 1964 at the Uiv of CA, Berkeley. Worked as an Assoc Prof a Osaka Univ until 1968 when joined Bell Labs. The fields of interest are plasma phys, space physics, non-linear optics & fluid dynamics. Published approximately 160 papers & two books. A fellow of Am Physical Soc, and of IEEE, mbr of Am Geophysical Union, Physical Soc of Japan & Sigma Xi. Since 1971 is an Adjunct Prof of Columbi Univ. *Society Aff:* APS, IEEE, AGU, ΣΞ, PSJ.

Hasfurther, Victor R
Home: 1710 Bill Nye, Laramie, WY 82070
Position: Prof *Employer:* Univ of WY. *Education:* PhD/Hydraulics & Fluid Mechanics/UT State Univ; BS/Civ Engg/UT State Univ. *Born:* 1/27/43. Prof in 1980. Raised in the UT-Southern ID Area. Became an Asst Prof of Civ Engrg at the Univ of WY in Sept 1969. Received the PhD degree from UT State Univ in 1971. Was promoted to Assoc Prof in 1975 and Full Prof in 1980. Awarded the John P Ellbogen Distinguished Classroom Teaching Award by the Univ of WY in 1979. Pres of WY Sec of ASCE 1978-79. Res areas have been in Evapotranspiration Waste Disposal Units & Surface & Groundwater effects of Coal Strip Mine Operations. *Society Aff:* ASCE, ASEE.

Hash, Lester J
Business: 224 W 9th St, Sioux Falls, SD 57102
Position: Manager *Employer:* City of Sioux Falls Water Dept. *Born:* Apr 1928. Native of Cornwith Iowa. After high school, two-year hitch in Marines. Graduate Sioux Falls Coll 1953 BS Math. Genl Sci teacher, coach Washington H S. BS Civil Engrg Sch of Mines & Tech, Rapid City S Dak 1958. Design Engr Minn Highway Dept 1958. 1959-61 Highway Construction Proj Engr & Estimator, Al Johnson Construction Co. 1961present Genl Supt Water Dept City of Sioux Falls. South Dakota Waterworks Operator 1969, Water Utility Man of the Yr 1971, State of S D. Natl Dir 1970-73 AWWA, AWWA Diamond Pin Award 1973, AWWA Ambassador Award 1975. Outstanding Alumni Award-Sioux Falls Coll 1980, AWWA Presidential Award 1981, AWWA Honorary Mbrship Award 1981.

Hashin, Zvi
Business: Faculty of Engineering, Telaviv Univ, Tel Aviv, Israel 69978
Position: Prof, Chair, Mechs. of Solids *Employer:* Tel Aviv Univ *Education:* Doctor of Science/Mechanics/Univ. of Paris (Sorbonne); M.Sc./Mechanics/Technion - Israel Inst. Technology; Civil Engineer/Civ. Eng./Technion - Israel Inst. Technology; B.Sc./Civ. Eng./Technion - Israel Inst. Technology *Born:* 06/24/29 After Doctorate: Lecturer, Senior Lecturer, 1957-59, Technion, Israel. Research Fellow, Harvard, 1959-60. Assoc Prof, Appl Mechs, 1960-65; Prof, 1965-71, Univ of PA. Prof, Materials Engrg, Technion, Israel, 1971-73. Prof, Tel Aviv Univ, Israel, 1973-Present, and founding chairman, Dept. of Solid Mechanics, Materials and Structures, 1973-77, 1979-81. Nathan Cummings Chair, Mechanics of Solids, 1980-Present. Visiting Prof, Univ of PA, 1977-79. Pres, Israel Society Theoretical & Applied Mechanics and Mbr, General Assembly IUTAM, 1976-Present Landau Prize, 1973. Founding medal of excellence in composite materials, 1984. Fellow ASME, 1979. Scientific Advisor, Materials Sciences Corp. Spring House, PA. 1970-Present. Co-chairman, IUTAM Symposia 1982, 1985. *Society Aff:* ASME, SES, AAM, ISTAM

Haskell, Arthur J
Home: 287 Sheridan Road, Oakland, CA 94618
Position: Sr VP *Employer:* Matson Navigation Co *Education:* Nav E/Nav Arc & Marine Engrg/MIT; BS/Engrg/US Naval Academy *Born:* 4/16/26 Native of Newark, NJ. Commissioned Ensign, US Navy, 1947; advanced through grades to Commander, 1966; retired 1970. Sr Procurement Engr for Natl Bulk Carriers, NY, 1956-1962; Asst Plant Mgr, Western Gear Corp, Belmont, CA, 1962- 63. Matson Navigation Co, 1964-present. Positions held: Project Engr, 1964; Mgr, Contract & Administration, July, 1965; Asst Mgr, Engrg & Maintenance, February, 1966; VP, December, 1970; Sr VP, April, 1973 to present. Presently responsible for design and construction of Matson's vessels and shoreside material handling equipment, operation, maintenance and repair of co's vessels and all purchasing activities. Chrmn, No. CA SNAME, 1969-1970; Natl VP, SNAME, 1973-present; SNAME Exec Committee, 1977-1980 & 1983-present; SNAME, Nominating Committee, 1981-83. Mbr Marine Bd, Natl Res Council 1981-85. American Bureau of Shipping, Bd of Mgrs 1987-present. *Society Aff:* SNAME

Haskell, Barry G
Business: 4C-538, Holmdel, NJ 07733
Position: Technical Staff *Employer:* AT&T Bell Labs *Education:* PhD/Elec Engin/Univ. California, Berkeley; MS/Elec Engin/Univ. California, Berkeley; BS/Elec Engin/Univ. California, Berkeley; AA/Engin/Pasadena City Col, Pasadena, CA *Born:* 09/01/41 From 1964 to 1968 he was a Research Assistant in the Univ of CA Electronics Research Lab, with one summer being spent at the Lawrence Livermore Lab. Since 1968 he has been at AT&T Bell Labs, Holmdel, NJ and is currently Head of the Visual Communications Research Dept. He has also taught graduate courses at Rutgers Univ, City College of NY and Columbia Univ. His research interests include digital transmission and coding of images, videotelephone, satellite television transmission, medical imaging as well as most other applications of digital image processing. *Society Aff:* IEEE, ΣXI, ΦBK

Hassani, Jay J
Business: 32 West Rd, Towson, MD 21204
Position: VP. *Employer:* Century Engg, Inc. *Education:* DIC/Reinforced Concrete/Imperial College, London Univ; BSc/Civ Engg/Tehran Univ. *Born:* 12/22/20. Native of Iran. Schooling in Italy, Austria, Iran and England. After completion of education, worked 5 yrs for Oil Co in Abadan, Iran, in various engg capacities. Immigrated to US in Jan 1955, joined Baltimore engg firm of Whitman, Requardt and Assoc as Structural Engr and held various positions during 20 yrs of association with this firm. Formed own business of consulting engg to shipyards in 1955, accepted position of proj dir of new shipyard in the Persian Gulf, from mid 1955 to end of 1978. Joined Century Engg, Inc as VP in charge of Shipyard/Marine Div in May 1979. *Society Aff:* ASCE, NSPE, SNAME.

Hassebroek, Lyle G
Business: 1500 114th Ave, SE, Bellevue, WA 98004
Position: VP and NW District Mgr *Employer:* CH2M HILL *Education:* BS/CE/Univ of WI *Born:* 9/11/41 Born and raised in Hudson, WI. Graduated from Univ of WI with BS CE in 1963. Joined CH2M HILL that year in Corvallis, OR in Sanitary engrg dept. Transferred to Seattle Regional Office in 1965. Have held increasingly responsible positions until this time where am Seattle Regional Mgr and Northwest District Mgr - responsible for CH2M HILL's operations in OR, WA, HI, AK and Western Canada. Have served on CH2M HILL, Inc Bd for 3 years and currently serve on CH2M HILL Canada Ltd Bd. Enjoy reading, hiking and landscape architecture. *Society Aff:* WPCF, APWA, ACEC

Hassialis, Menelaos D
Home: 122 Phelps Road, Ridgewood, NJ 07450
Position: Krumb Prof of Mining, Emeritus. *Employer:* Columbia Univ - Krumb Sch of Mines. *Education:* 5BA/Chem-Physics/Columbia College; PhD/Not Complete/Univ Apptment; DSc/- /Bard College. *Born:* Dec 1909. Graduate work physics-chem 1931-34. Appointed faculty Sch Engrg 1934, promoted thru ranks to Krumb Prof Mining. Chmn Krumb Sch of Mines 1957-67. Admin Bd Lamont-Doherty Geological Observatory 1967-78. Dir USAEC Lab 1951-58. US Representative UNESCO 1957. Amer Delegation Geneva Confs 1955 & 1958. Hd UN Missions Turkey, Spain 1964-65. Pres Pacific Uranium Mining Corp 1959-62. Bd Chmn: and CEO Sandvik Inc 1973-87, Cons Bd. Chmn Strata Bit Corp. 1983-date Mobil, Exxon, Kerr-McGee. Honors: DSc (hon) Bard Coll. Bicentennial Medal, Freiburg Acad 1965, Bicentennial Scroll Tech Univ of West Berlin 1968. Exec VP Sammine Exploration Co, Canada. Knight of Malta - OSJ 1985. *Society Aff:* ACS, AIME, MMSA.

Hassler, Francis J
Business: Box 7625, Raleigh, NC 27695
Position: Hd, Biological & Agri Engrg. *Employer:* N. C. State University. *Education:* PhD/Agri Engrg/MI State Univ; MS/Agri Engrg/MI State Univ; BS/Agri Engrg/Univ of MO. *Born:* 8/2/21. Native of rural Mo. Army Field Artillery 1943-46; discharged as 1st Lt. Joined NC State Univ 1950 as Asst Prof. Wm Neal Reynolds Prof & Head Biological & Agri Engrg NC State Univ. Elected Fellow ASAE 1975. Author Fellow AAAS 1983, Outstanding Engrg Aichievement Award 1983, By NC Soc of Engrs 33 publs; 3 pats in tobacco processing. Dir 1964-71 Ford Foundation Grant for Strengthening Post- Grad Training & Res in Agri Engrg at IIT, Kharagpus India. Interim Exec Dir 1964-66, Water Resources Res Inst NCSU. Spec int in academic challenges of phys- biol unification for bioengrg. Member: ASAE, ASEE, AAAS, Sigma Xi, Phi Kappa Phi, Tau Beta Pi, Gamma Sigma Delta, NC Soc of Engrs. *Society Aff:* ASAE, ASEE, AAAS, ΦΚΦ, ΣΞ.

Hassler, Paul C, Jr
Business: Cvl Engrg Dept, El Paso, TX 79968
Position: Prof of Civil Engrg *Employer:* The Univ of Texas at El Paso. *Education:* MS/CE/Univ of NM; BS/Engr/Grove City Coll. *Born:* Sept 7, 1922, Grove City PA. Reg Prof Engr TX; Public Surveyor TX. Exec Secy TX Sec ASCE 1961-67. V Pres TX Sec ASCE 1968-69. Pres TX Sec ASCE 1969-70. Dir ASCE, District 15 1971-74. Instr: TX Col of Mines 1948-51. Asst Pres, Prof TX Western Col 1954-56. Assoc Prof TX Western Col 1956-61. Present, Prof TX Western Col & Univ TX El Paso 1961-present Asst Dean of Engg. Univ of TX El Paso 1979-81, Pres El Paso Chpt TX Soc of Prof Engrs 1978-79. *Society Aff:* ASCE, SAME, TSPE.

Hatch, George W
Home: 4239 Jupiter Drive, Salt Lake City, UT 84117
Position: District Manager *Employer:* Mountain Bell. *Education:* BES/EE/Brigham Young Univ; MA/Organization Communications/Brigham Young Univ *Born:* February 22, 1928, Oxford Idaho. 1958 Bell Sys sponsored schools incl: Regional Communications Engrg Sch at U of Colo & Bell Sys Data Communication Sch at Coopertown N Y. Transmission Engr, Mountain Bell Salt Lake City Utah 1958-63. Proj Engr, Transmission & Microwave Radio, Mountain Bell Denver Colo 1963-66. Asst Engrg Mgr, Transmission Sys AT&T NYC 1967-70. Engrg Superintendent, Interoffice Facils, Mountain Bell Salt Lake City Utah 1970-76. USPE/NSPE, Pres Salt Lake City Chap USPE 1974. Presented paper on 'Communications for Large Aperture Seismic Aray' 1966, IEEE Internatl Communications Conf Philadelphia Pa. Assoc Gen Contractors Scholarship Award BYU 1958. Enjoy classical music, drama, gardening, fishing & golf. Pres USPE, 1980; District Mgr, Network 1976-81. *Society Aff:* NSPE

Hatch, Henry J
Home: 10 Palm Cir Dr, Honolulu, HI 96819
Position: Div Engr. *Employer:* US Army Engr Div - Pacific Ocean. *Education:* Masters/Geodetic Sci/OH State Univ. *Born:* 8/31/35. BrigGen US Army. B 8/35 in Pensacola. H 10 Palm Cir Ft Shafter HI 96819. Master OH State Univ grad of West Point 1957. Has commanded troops to brigade level & served in other assignments as builder, planner & researcher; now Div Engr, Pacific Ocean Div HI, respon for military & civ works planning, design, engg & const services within assigned areas in Pacific Basin. *Society Aff:* SAME.

Hatch, John E
Home: 409 Austin Avenue, Pittsburgh, PA 15243
Position: Mgr Tech Projs - Ingot & Fabricating. *Employer:* Alcoa-retired. *Education:* MSc/Metallurgy/Carnegie Mellon; EMet/Metallurgy/CO Mines. *Born:* 1/3/15. Grad Colo Mines 1936, Metallurgical Engrg. Employed at Youngstown Sheet & Tube as metallurgist in open hearth, blooming mill, sheet mill, seamless tube to 1940. Joined Alcoa 1940, metallurgist in aluminum & magnesium ingot, extrusion, & sheet. Installed & managed Univac I computer sect at Davenport works. Mgr of Met & Qual Assurance at Warrick & Davenport Works. Corporate Mgr Metallurgy & Quality Assurance for Ingot & Powder 1969-75. Mgr Special Projs to July 1976. Retired July 1976. Chmn Tri City (Rock Island, Moline, Davenport) Amer Soc for Metals 1960. Mbr of AIME Metals Div, Theta Tau, Tau Beta Pi, Sigma Xi engrg & honorary frats. Teaching Metallurgy Part Time Carnegie Mellon Univ, consulting in metallurgy of light metals, volunteer consultant with Internatl Exec Service Corps. *Society Aff:* AIME, ASM, IESC, TMS.

Hatch, Marvin D
619 Jefferson, Sedgwick, KS 67135
Position: Process Engr. *Employer:* Beech Aircraft Corp. *Education:* MS/Physical Science/KS State Teachers College, Now: Emporia State Univ; BS/Physical Science/KS State Teachers College, Now: Emporia State Univ. *Born:* 1/12/43. Raised on a farm 8 miles west of Burlington, Kansas. Taught in Kansas public schools 1964-69. With Cessan Aircraft Company since 1969. Prime respon is adhesives, sealants & chem processing. Pres Wichita Chap of SAMPE 1973. Presented paper on Hard Anodizing to SAE meeting 1971. Also serve on Bd/Dir of Satan Electronics Inc, Salina Kansas, with Beech Aircraft in 1985. Enjoy hunting, fishing & amateur radio. *Society Aff:* SAMPE.

Hatch, Philip H
Home: 70 Gibson Street, North East, PA 16428
Position: Ch Mech Officer - ret. *Employer:* Long Island R R. *Education:* SB/Elec Engrg/MIT. *Born:* May 1899 Albany, N.Y. Reg prof engr, Conn. USNR WWI. Two principal positions were Gen Mech Supt, New Haven RR & Ch Mech Officer, Long Island RR; in both was respon for design, application, maintenance of locomotives & cars & supporting facils. Had active role in dev & application to RR service of the diesel elec locomotive. Fellow & Life Mbr IEEE, former Mbr Amer RR Engrg Assn, Locomotive Maintenance Officers Assn, Bd/Examiners AIEE, Land Transp Ctte IEEE, var cttes AAR. Author, cons. Hobbies: fishing, gardening & golf. *Society Aff:* IEEE, IAS.

Hatch, Randolph T
Home: 12914 Autumn Dr, Silver Spring, MD 20904
Position: Assoc Prof. *Employer:* Univ of MD. *Education:* BS/ChE/Univ of CA; MS/Biochem Engg/MIT; PhD/Biochem Engg/MIT. *Born:* 4/27/45. Joined the Dept of ChE, Univ of MD in 1972 as an Asst Prof. Areas of res specialization are biochem

Hatch, Randolph T (Continued)
engg & fermentation. Promoted to Assoc Prof with tenure 1977. Lead res in computer control of fermentation & membrane separations. Also maintain res activity in semi-solid fermentation & in design of fermentors. Took one yr leave of absence in 1977-1978 to be Prog Dir for chem processes in the Engg Div, Natl Sci Fdn. *Society Aff:* AIChE, ACS, ASM, AAAS.

Hatcher, Stanley R
Business: 275 Slater Street, 17th Floor, Ottawa, Ontario, Canada K1A 1E5
Position: President *Employer:* Atomic Energy of Canada Limited *Education:* Ph.D./Chemical Eng./University of Toronto (Canada); M.Sc./Chemical Eng./ University of Birmingham (England); B.Sc./Chemical Eng./University of Birmingham (England) *Born:* 08/20/32 Born near Salisbury, England. Obtained B.Sc. 1953, and M.Sc. 1954, degrees in Chem Engrg from Univ of Birmingham. Received Ph.D. in Chem Engrg from the Univ of Toronto, 1958. Joined Atomic Energy of Canada Limited (AECL) in 1958, working in various positions at Chalk River Nuclear Research Labs (CRNL) and Whiteshell Nuclear Research Establishment (WNRE). Served as AECL liaison officer at the United Kingdom Atomic Energy Authority from 1968-69. Became V-P/G.M. of WNRE in 1978, joined AECL CANDU Operations as VP, Marketing/Sales in 1981. Appointed Pres, AECL Research Co in 1986. Appointed to the American Nuclear Soc BOD in November 1986. Fellow of Chem Institute of Canada, member of Canadian Nuclear Assoc, Canadian Nuclear Soc, Canadian Soc for Chem Engrg and Assoc of Prof Engrs of Ontario. *Society Aff:* ANS, CIC, CNA, CSCE

Hatcher, William J, Jr
Business: PO Box 6373, Tuscaloosa, AL 35487
Position: Research Professor, Dept of Chem Engrg *Employer:* The University of Alabama. *Education:* PhD/ChE/LA State Univ; MS/ChE/LA State Univ; BChE/ChE/ GA Inst Tech. *Born:* July 1935. USMC 1957-60. Res Engr for Exxon Res Labs 1960-69. Univ of Alabama fac since 1969, hd of the Dept of Chem & Met Engrg 1973-1983. Acting Dean, Coll of Engrg 1981-1983. Reg Prof Engr in Alabama. Fellow Amer Inst of Chem Engrs, Amer Soc for Engrg Ed. Fellow, Amer. Inst. of Chemists. *Society Aff:* AIChE, ASEE, AIC.

Hatfield, Kent E
Business: 375 Chipeta Wy, Bx 8009, Salt Lake City, UT 84108
Position: Exec VP. *Employer:* Ford, Bacon & Davis Utah Inc. *Education:* BChES/Chem Engg/Brigham Young Univ. *Born:* February 1934. Native of Utah. Lic Prof Engr, Mbr AIChE, SME & NSPE. Proj Engr for PPG Indus, specializing in heavy chem design & construction. Proj Mgr for NL Indus specializing in mineral extraction design & construction. Joined FB&D 1972 as its first employee in Utah & dev co into its current size & reputation. Executive V Pres & Genl Mgr, and director respon for co's nationwide design & const activities in mining, chem processing & nuclear engrg. Executive Vice President and Chief operating officer for Enercor a major developer of Tor Sand resources in the Western United States, responsible for all activities associated with Enercor's commercial Tor Sand development.. *Society Aff:* AIChE, SME, NSPE.

Hatheway, Allen W
Business: Dept of Geological Engg, 129 McNutt Hall, Univ of Missouri-Rolla, Rolla, MO 65401-0249
Position: Prof of Geological Eng *Employer:* Univ of MO - Rolla *Education:* AB/Geo/UCLA; MS/Geo Engr/Univ AZ; PhD/Geo Engr/Univ AZ; Prof Degree/Geo Engr/Univ AZ; Dip/Military Sci/Army War College. *Born:* 9/30/37. in Los Angeles. US Army 1961-63. Exploration Geologist (copper) & Res Assoc during grad studies 1964-68; Geotech Engr (staff engr to proj mgr) US Forest Serv, Woodward-Clyde & other cons firms 1969-75; Adjunct Asst Prof, Civil Engg, Univ So CA 1971-74; Proj Geologist, Shannon & Wilson 1975-76; Chief Geologist Haley & Aldrich 1976-1981; Adjunct Assoc Prof Geology, Boston Univ, 1979-1981 Professor of Geological Engg, Univ Missouri-ROLLA, 1981-present Col, Corps Engrs USAR 1963- ; 70 prof papers; co-editor *Geology in Siting of Nuclear Power Plants* 1979; Co-author of "Geology and Engineering" 3rd. Edition, 1988. CA Outstanding Young Civil Eng ASCE 1972; Daniel Mead Prize, ASCE 1975; Mbr Board on Earth Sciences, U.S. National Research Council (1987-1990). E. B. Burwell Award, GSA 1981 Mbr ASCE, AIME, AGU, GSA, SAME, AEG. Chmn Engg Geology Div, GSA, 1980; Pres, AEG, 1984. PE (Civil) CA, MA, (Geol) AZ; Reg Geologist CA, ME; Cert Engr Geologist CA. *Society Aff:* ASCE, AIME, SME, GSA, AEG, SAME, EERI

Hatheway, Alson E
Business: Suite 400, 595 E. Colorado Blvd, Pasadena, CA 91101
Position: Pres. *Employer:* Alson E Hatheway, Inc. *Education:* BS/Mech Engr/Univ of CA. *Born:* 11/15/35. A native of Southern California, Manager of Design Engineering at Xeros-EOS, Hughes Aircraft and Gould Inc. Founded Alson E Hatheway, Inc to promote sound methods in the development of new products. The firm offers optical and optomechanical system engineering including design and analysis; end-to-end integrated analysis including optics, structures, heat transfer, control systems, acoustics, fluid mechanics and other disciplines; prototyping and evaluation testing of system and technology concepts; reports, design reviews, seminars, tutorials and technical training sessions for customers' technologists. *Society Aff:* ASME, NSPE, OSA, AIAA, AOC, SPIE.

Hattersley, Robert S
Home: 8722 Long Lane, Cincinnati, OH 45231
Position: Section Head *Employer:* Procter & Gamble Co. *Education:* MS/Mech. Engr./Ohio State University; BS in ME/Mech. Engr./New Jersey Institute of Tech. *Born:* 09/04/31 U.S. Marine Corps Engineering Officer, 1954-57. Positions of increasing responsibility with Procter & Gamble: process design engineer 1957-61; production supervisor 1961-63; Engineering group leader 1963-65; plant engineer, Toronto, Canada 1965-67; plant engineer, Quincy, Mass. 1967-73; Production Operations Manager, Mexico City, Mexico 1973-76; Section head international engineering 1976-present. Specialized in chemical process design and plant startups. Pres, Cincinnati Chaper NSPE 1986-87. VP elect General Engineering ASME 1987-present. Registered in OH & SC. Amature cabinetmaker. *Society Aff:* ASME, NSPE

Hattery, Donald P
Business: 800 1st St, NW, Cedar Rapids, IA 52406
Position: Partner. *Employer:* Shive-Hattery Engrs *Education:* BS/Civil Engr/IA State Univ. *Born:* 3/11/30. in Iowa. Civil Engr with Penn RR 1952-53. US Army Corps of Engrs 1953-55. Instructor, Iowa State Univ 1955-56. Civil Engr & Materials Handling Supr, Link Belt Speeder Corp 1956-61. Cons Civil Engr as Partner in Shive-Hattery Engrs 1961- . Pres Cons Engrs Council of Iowa 1968. Pres Iowa Engrg Soc 1971. Fellow Amer cons Engrs Council. Pres Hawkeye Area Council Boy Scouts of Amer 1973-75. Mbr Linn Cty Regional Planning Comm since its beginning. Pres Cedar Rapids Chamber of Commerce 1979. Trustee, Mount Mercy Coll. Reg Prof Engr & Land Surveyor in Iowa. *Society Aff:* NSPE, ACSM, ACEC.

Hauck, Charles F
Home: Box 144, Bradfordwoods, PA 15015
Position: Director-Consultant *Education:* BS/Chem Math/John Carroll Univ; MS/-/ Case Western Reserve University *Born:* 9/26/12. Plant Engr Atlas Powder 1941-45, Sr Staff Design Engr Hagan Chem & Control 1945-51, Mgr V P Sales Chem Plants Div Blaw-Knox Co, Pgh 1952-64, V Pres Genl Mgr Chem Plants Div 1964-67, Sr V Pres 1967, Pres 1967-81. Dir White Consolidated Indus 1968-82. Mbr Amer Inst Chem Engrs (Exec Council P Chmn, Pgh) Amer Iron & Steel Inst, Natl Soc Prof Engrs, P Pres Western Pennsylvania, Boy Scouts, Dir Pittsburgh Des Moines Co Metcalfe Award ESWP. *Society Aff:* NSPE, AIChE, ESWP, PSPE.

Hauck, Frederick A
Business: 2501 Carew Tower, Cincinnati, OH 45202
Position: Pres. *Employer:* Hauck's Exploration Co. *Education:* Dr Sci; or Sci; Dr SCI; Dr Law; Humanities; Letters/Mining Engg/Univ of Munich, Germany *Born:*

Hauck, Frederick A (Continued)
12-28-1894 After grad from a preparatory school here in Cincinnati I spent a few yrs at the Univ of Munich, Germany. These yrs were from 1912-14. I became very interested in mining during the advent of the Atomic Energy Program at which time, through the help of the US Government, mined Cape Canaveral, Indian river and Banana River and had leases within the entire East Coast from Titusville to Fort Pierce. These were the yrs that I was a Board mbr of the Carborundum Co and through my operations at Nielbourne, FL, produced the first zirconium reactor grade metal, the contract of which was given the Carborundum Co at the Argonne Labs. During the past 50 yrs I have been associated with a group in Sonora, Mexico, mining mostly silver and copper and have been successful in my enterprises in that state. Am bd mbr Univ of Florida and Univ of Cinn benefactor Rollins College. Xavier Univ Cinn, visiting Prof Institute advance studies Princeton. Received Doctor of Sci, Hd University, Fla.

Hauck, George F W
Business: 600 W Mechanic St, Independence, MO 64050
Position: Associate Professor. *Employer:* University of Missouri. *Education:* PhD/Struct Engrg/Northwestern Univ; MArchE/Soil Mech/OK State Univ; BArchE/ Struct Engrg/OK State Univ. *Born:* Sept 1932 at Kassel (Germany). US cit since 1954. US Army (enlisted) & US Naval Reserve (Cmdr CEC). Early experience as structural designer (US Steel Corp) & res asst (OSU & NU). Chmn of Dept of Civil Engrg Tri-State Univ in Ind 1963-75. During that time also Visiting Prof at Tech Univ Munich & Bd/Chmn Tri- State Engrg, Inc. Mbr of ASEE (Chair, Metric Coord Com) & Fellow of ASCE (offices incl editor & Pres Ind Sect, Chair District 9 Council et al). Mbr Sigma Xi, Phi Kappa Phi, Tau Beta Pi, Sigma Tau & Chi Epsilon. Since 1975 Assoc Prof Civil Engrg Univ of Mo--Columbia, Coordinator of the Civil Engrg Program in Kansas City. Publs on res in structures Harry S Truman Award, 1986, civil engrg history & ed admin. Columnist, "Moving Metric," engineering education (1977-81) Co-author, Davis, Troxell, Hauck: Testing of Engineering Materials, 4/e, McGraw-Hill, NY, 1982. Author, *The Aqueduct of Nemausus*, McFarland, Jefferson NC, 1988. *Society Aff:* ASCE, ASEE, USMA, ASTM, SHOT.

Hauenstein, Henry W
Home: 5240 Carlingfort Drive, Toledo, OH 43623
Position: Partner. *Employer:* Finkbeiner, Pettis & Strout, Ltd. *Education:* BS/CE/Univ of Cincinnati. *Born:* April 1924. Native of Cincinnati Ohio. US Army 1943-46. Asst Engr City of Cincinnati 1948-52. With Finkbeiner, Pettis & Strout, Ltd since 1952; Partner since 1958; Chmn 1984-1986 Mbr Adv Bd of Examiners Ohio Water & Wastewater Operators. Diplomate Amer Acad of Environ Engrs. Fellow ASCE. Rotarian. Mbr WPCF, AWWA, NSPE, SAME. Reg Prof Engr Ohio, Mich, N. Carolina, Florida. *Society Aff:* AWWA, NSPE, WPCF, ASCE, AAEE, SAME.

Haug, Edward J, Jr
Home: 1407 Eastview Dr, Coralville, IA 52241
Position: Prof. *Employer:* Univ of IA. *Education:* PhD/Appl Mech/KS State Univ; MS/Appl Mech/KS State Univ; BS/Mech Engg/Univ of MO. *Born:* 9/15/40. Served as mech res engr while a Capt, US Army, at Rock Island Arsenal 1966- 68 and as a civilian from 1968 to 1970. From 1970 to 1972 served as chief of Army Weapon System Analysis and from 1972 to 1974 as chief of Army Weapon Systems Res, Weapons Command, Rock Island, IL. From 1974 to 1976 served as Chief of Concepts and Tech, US Army Armament Command. Prof of Mech Engg, Univ 1976 to present (adjunct Asst, Assoc, & Full Prof 1969-1976). Director of Univ of Iowa Center for Computer Aided Design 1981 to present. Consult with Army agencies, Battelle, Ford Motor Co, and Intertech Corp. Director, CADSI of Iowa City IA and Alphasel Inc. of Camarillo, CA. *Society Aff:* ASME, SES.

Haun, J William
Business: PO Box 1113, Minneapolis, MN 55440
Position: Vice President, Engrg Policy. *Employer:* General Mills, Inc. *Education:* PhD/Chem Engg/Univ of TX, Austin; MS/Chem Engg/Univ of TX, Austin; BS/ Chem Engg/Univ of TX, Austin. *Born:* Sept 1924. Although born in Birmingham, AL, consider Austin TX my home town. Marine Corps 1942-45 - mostly in sch at Univ of CO. First full-time job, Instructor in Chem engg at TX Univ. Monsanto Chem Co Plastics Div 1950-56, process dev & plant tech work on polymers, Group Ldr for High Pressure Polymerization. In res at Genl Mills Inc 1956-63, apptd a VP in 1963. Am functionally respon for engg at the corp level worldwide with emphasis on environ control & energy conservation. Fellow AIChE. Dist Engg Grad Univ of Texas, Mbr Univ of MN; Indus Adv Council, 1970-77; Chmn Envioon Qual Ctte NAM 1972-79; Chmn USA/BIAC Subctte on Environ; Chrmn of the Bd, Intl Center for Industry & the Environ. Mbr: Natl Task Force on Technology and Soc; US Chamber of commerce Committee on the Envioon Food Industry advisory Committee to the US Dept of Energy VChrmn. *Society Aff:* AIChE.

Haus, Hermann A
Business: 36-351 77 Mass Ave, Cambridge, MA 02139
Position: Institute Professor. *Employer:* MIT. *Education:* ScD/EE/MIT; MS/EE/RPI; BSc/EE/Union College. *Born:* 1925 Ljubljana, Yugoslavia. Attended the Technische Hochschule, Graz, and the Technische Hochschule, Wien Austria & received BS from Union Col, Schenectady N Y 1949, MS Rensselaer Polytechnic Inst 1951, ScD Mass Inst of Tech 1954. Joined fac of Elec Engrg at MIT. Engaged in res in electromagnetic theory & lasers in MIT's Res Lab of Electronics. Former Bd/Mbr editorial bds of the Journal of Applied Physics & of Electronics Letters, & Internatl Journal of Electronics. Visiting Mackay Prof Univ of Calif Berkeley summer 1968. Author or coauthor of four books & over 100 journal articles. Mbr of Sigma Xi, Eta Kappa Nu, Tau Beta Pi, the Natl Acad of Engrg, the Amer Physical Soc, Fellow IEEE, American Academy of Arts and Sciences. Received the Western Elec Fund Award of the Amer Soc of Engrg Educ in 1971, the 1984 Award of the Quantum Electronics and Applications Soc of the IEEE, the 1987 Charles Hard Townes Award of the Optical Society of America. *Society Aff:* IEEE, APS, NAE, AAAS, NAS, OSA.

Hauser, Consuelo M
Business: Mail S-0488 PO Box 179, Denver, CO 80201
Position: Senior Engineer *Employer:* Martin-Marietta Corp, Aerospace Div. *Education:* Postgrad/Civil Engrg/Univ of CO; M Eng/Structural Eng/Yale Univ; BSCE/Civil Eng/Univ of IL. *Born:* 10/5/29. Civil Engr with Henry A Pfisterer, Cons Engr 1952-53. Structural Engr with Ordnance Corps, Aberdeen Proving Ground MD 1953-54. Design Engr with Sanitary Engrg Sect, Dept of Public Works, City of County of Denver 1959-61. Partner Hauser Labs 1961-79; chief respon. Mbr ASCE, AIAA, CO Soc of Engrs, Soc of Women Engrs (Chmn 1976 Natl Convention), & Chi Epsilon. Reg Prof Engr CO. *Society Aff:* ASCE, AIAA, SWE, Co soc of Engrs.

Hauser, H Alan
Business: 400 Main St, E Hartford, CT 06108
Position: Mgr Mtls engg and Res. *Employer:* Pratt & Whitney Aircraft. *Education:* BS/Met/MIT. *Born:* 10/13/34. Received BS degree in Met from MIT in 1955 and joined Alcoa as a plant metallurgist. Went on active duty in US Army; discharged in 1958 as first Lt. Joined Pratt & Whitney Aircraft in 1958 and have served in various mtls positions to the present. Now Mgr, Mtls Engg and Res for Commerical Products Div. Total mtls responsibility (res, dev and application) for commercial gas turbine engines. Manage approximately 300 people (150 profls of all degree levels). *Society Aff:* ASM.

Hauser, John R
Business: Elec. & Comp. Engr. Dept, Raleigh, NC 27695
Position: Professor. *Employer:* North Carolina State University. *Education:* PhD/Elect Engr/Duke Univ; MS/Elect Engr/Duke Univ; BS/Elect Engr/NC State Univ. *Born:* Sept 1938. Native of Advance NC. Employed by Bell Telephone Labs in Winston-Salem NC from 1960-62. Res engr with Res Triangle Inst from 1963-66. Assumed present position with NCSU in 1966. Present teaching & res activities in

Hauser, John R (Continued)
solid state electronics with special emphasis on III-V materials & devices. Mbr of IEEE, APS, ASEE. Received ASEE award for excellence in engrg teaching in 1975. Received NCSU Res Award in 1978 for Excell in engrg res. *Society Aff:* IEEE, APS, ASEE.

Hauspurg, Arthur
Business: 4 Irving Place, New York, NY 10003
Position: President & Chief Operating Officer. *Employer:* Consolidated Edison Co. *Education:* MS/Elec Engg/Columbia Univ; BS/Elec Engg/Columbia Univ. *Born:* Aug 1925. Fellow IEEE. Served in US Navy 1943-46. Prior to joining Con Edison in 1969 was affiliated with Amer Elec Power, Ch Elec Engr 1966 & Asst V P 1968. Cons for NAS & in 1965-66 special cons to the FPC on analy of Northeast Blackout & preparation of Report to the Pres. & Dir of Empire St Elec Energy Res Corp; Mbr NAE; US Tech Ctte CIGRE; EEI-Power Engrg Ed Award; EEI-Panel of Judges; US USSR Working Grp on UHV Elec Trans; Columbia Univ Adv Council of Fac of Engrg & Applied Sci; Natl Acad of Engrg, Elec Engrg, Power Peer Grp; Exec Ctte of Northeast Power Coord Council. Dir. - Chancellor High Yield Fund, Inc.; Chancellor Tax-Exempt Daily Income Fund, Inc.; Chancellor High Yield Municipals; Chancellor New Decade Growth Fund, Inc.; Chamber of Commerce & Industry; Committee for Economic Dev; Economic Dev Council of NYC; Electric Power Res Inst; Reg Plan Assoc.

Hausz, Walter
Home: 4520 Via Vistosa, Santa Barbara, CA 93110
Position: Prog Mgr-Energy Conversion. *Employer:* GE-Ctr for Advanced Studies (TEMPO). *Education:* MS/EE/Columbia Univ; BS/EE/Columbia Univ. *Born:* May 1917. Gen Elec Co 1939-1981. Radar projs leader 1940-45 Schenectady; Mgr-Electronics Sys, Elec Lab Syracuse 1946-54. Mgr-Advanced Electronics Ctr Ithaca N Y 1954-56, leading a staff of 150 in radar, communications, computers, countermeasures. Gen Elec-TEMPO 1956-1981; Sect Mgr & Mgr-Engrg; led defense sys & nuclear weapons effects projs. Retired from GE 1981. Currently leading contract studies in advanced energy sys-solar energy, conservation as consultant (Eco-Energy Associates). Fellow IEEE & AAAS; Mbr Sigma Xi; GE Coffin Award; Sect Chmn IRE Los Angeles 1960; Chmn Engrg Foundation Res Confs Ctte 1967-70. *Society Aff:* IEEE, AAAS, ΣΞ, ASHRAE, NY ACAD OF SC.

Hauth, Willard E III
Business: 3901 S Saginaw Rd, Box 540, Midland, MI 48686
Position: Mgr, Ceramic Matls and Composites Dev *Employer:* Dow Corning Corp *Education:* MS/Ceramic Sci/SUNY-Coll of Ceramics at Alfred Univ; BS/Ceramic Engrg/SUNY- Coll of Ceramics at Alfred Univ *Born:* 05/20/48 Experienced in applied dev and engrg of matls, fabrication tech, and product forms in the areas of structural ceramics, composites, refractories, tech ceramics, specialty glasses, and glass ceramics. Prof activities include indust, natl lab, consltg, and contract res experience with several tech publs. Prof affiliations include active mbrshp in the Amer Ceramic Soc (Board of Trustees 1987-89) and the Natl Inst of Ceramic Engrs (Pres 1983-84, Exec Ctte 1980-89, Trustee 1987-89). AAES Bd of Govs (1984- 85). Received the 1983 Karl Schwartzwalder-Prof Achievement in Ceramic Engrg Awd as the outstanding young ceramic engr in the nation. *Society Aff:* NICE, ACerS, ASM International, ASTM, ASC

Havard, John F
Home: 18552 Augustine Rd, Nevada City, CA 95959
Position: Mining Consultant *Education:* EM/Mining Engg/Univ of WI; PhM/Geology/Univ of WI; BS/Mining Engg/Univ ov WI; PhB/Geology/Univ of WI. *Born:* Mar 1909. Attended Montana Sch of Mines. Advanced Mgmt Program Harvard Business Sch. Mining engr, geologist, mine supt, works mgr, ch engr of mines, US Gypsum Co 1935-52. Asst resident mgr, Potash Co of Amer 1952-53. V Pres Fibreboard Corp 1953-63. Mgr, Mineral Projs, V Pres, Sr V Pres, Sr Consultant Kaiser Engrs 1963-date. Reg prof geologist or engr in 4 states. Pres, Soc of Mining Engrs of AIME 1976, Henry Krumb lecturer 1979. Mbr US Natl Ctte on Geology 1975-79. AIME Hardinge Award 1982. Fellow AAAS. Mbr SEG, etc; AIME Hon Mem 1984; Univ of WI Distinguished Service Citation 1986. *Society Aff:* SME/AIME, SEG, CIM, AAAS.

Havens, William W, Jr
Business: American Physical Society, 335 East 45th Street, New York, NY 10017
Position: Prof of Appl Physics & Nuclear Engg. Emeritus; Exec Sec American Physical Society *Employer:* American Physical Society *Education:* PhD/Physics/Columbia Univ; MA/Physics/Columbia Univ; BS/Mathematics/City College of NY. *Born:* 3/31/20. Dr William W Havens, Jr is now Executive Secretary of the American Physical Society and Prof Emeritus of Appl Physics and Nuclear Engg at Columbia Univ. He received his BS from City College of NY in June 1939. He continued his studies as a grad student at Columbia Univ. In 1940 he was appointed Asst in Physics, at Columbia and has been assoc with Columbia Univ since that time. He received his MA in 1941 and his PhD in 1946. Dr Havens became a full Prof in 1955 and in 1961 was appointed Dir of the Div of Nuclear Science and Engg in the School of Engg and Appl Sci at Columbia Univ, holding this post until 1977. Dr Havens' specialty is Neutron Physics. He has authored or co-authored more than 70 articles in scientific journals of original res. Dr Havens is a fellow of The American Physical Society and has served as its Executive Sec since 1967. He has been a mbr of the Governing Bd of the American Inst of Physics since 1958 and a mbr of its Exec Comm since 1960. *Society Aff:* APS, ANS, AAPT, AAAS.

Havner, Kerry S
Business: Dept of Civ Engg, Box 7908, Raleigh, NC 27695
Position: Prof. *Employer:* NC State Univ. *Education:* PhD/Appl Mechanics/OK State Univ; MS/Structural Engr/OK State Univ; BS/Civ Engr/OK State Univ. *Born:* 2/20/34. Instr/Asst Prof Civ Engr, OK State Univ 1957-62; Sr Stress & Vibration Engr, AiRes Mfg Co, Phoenix, 1962-63; Sect Chief, Solid Mechanics Res, Douglas Aircraft Co Missile & Space Systems Div 1963-68; Lecturer (part-time), Civ Engr, Univ S CA 1965-68; Assoc Prof 1968-75, Prof (Mechanics) 1975, Dept of Civ Engr, NC State Univ. Visiting Fellow, Clare Hall and Senior Visitor, Dept. of Applied Math. & Theoretical Physics & Dept. of Engineering, Univ. of Cambridge, 1981. Alcoa Fnd Dist Engr Res Award, NC St Univ 1982. Alumni Assoc. Outstanding Research Award, NC St Univ 1987. Res in continuum mechanics of solids (in particular theory of metal plasticity, crystal plasticity). Over 60 res publications. *Society Aff:* ASCE, ASME, AAM, SES, SIAM.

Hawes, Richard D
Business: 1723 Harney Street, Omaha, NE 68102
Position: Manager, Water Operations. *Employer:* Metro Utils Dist of Omaha. *Education:* BSci/Mech Engr/Univ of NB. *Born:* Sept 10, 1925. Served in var engrg & supervisory capacities at MUD since 1950. In present position in charge of water treatment & pumping plnat operation & plant & distrib sys maintenance 1968- . Mbr Amer Soc of Mech Engrs, Amer Water Works Assn (Chmn Nebraska Sec 1972-73) (Director, 1983-1986). Water Utility Man of the Year Award 1973. Elder in Presbyterian Church. Active in Scouting. Reg PE State of Nebr. *Society Aff:* ASME, AWWA.

Hawk, G Wayne
Business: Box 470, NY 14052
Position: President *Employer:* G W Hawk, Inc *Education:* MS/Mech Engrg/Univ S CA; BS/Aero Engrg/Purdue Univ. *Born:* 2/21/28. Warren, OH. Bd/Dir: MHP Machines Inc, Buffalo; Computer Res Inc, Buffalo; ACME Elec Corp Olean; Learfan Ltd, Reno, Nev. Mbr Bd/Governors of Natl Conf on Fluid Power. Past Chrmn Natl Fluid Power Assoc. Bd/Dir: Buffalo Philharmonic Orchestra, Boy Scouts of Amer-The Greater Niagara Frontier Council; Pres Greater Buffalo Dev Fdn; Mbr Air Force Assn & P Pres 1974-75 of the Lawrence D Bell Chap of AFA; Mbr of the E Aurora Country Club E Aurora NY, Country Club of Buffalo, Crag Burn Club of E

Hawk, G Wayne (Continued)
Aurora, Buffalo Club; The Aero Club of Buffalo. Bd/Dir of Amer Red Cross, Buffalo Chap. BS Aerospace Engrg Purdue Univ 1951. MS Mech Engrg Univ of S CA 1955. Gamma Alpha Rho (Aero Engrg Hon). Chandelle Sq (Air Force Hon) Mbr: Dirs of Mfg-Hanover Trust Co- Buffalo NY; Greater Buffalo Dev Fdn, Buffalo, NY. Mbr Bd of Regents of Canisius Coll, Buffalo, NY. *Society Aff:* AIAA, ASME.

Hawk, Kenneth C
Business: PO Box 427, Binghamton, NY 13902
Position: President. *Employer:* Hawk Engineering, PC. *Education:* BSCE/Civil Eng/Univ of RI. *Born:* 7/7/22. Served in U.S. Army 1942-1946, European Theater of Operations. BSCE University of Rhode Island 1949. President of Hawk Engineering, p.c. founded 1955. Registered Professional Engineer and Land Surveyer, 8 states. Member of NYSAPLS, NYSSPE, CEC/NYS (Past President, Engineer-of-the-year (So. Tier Chapter); ASCM (Fellow); ASCE (Fellow, Past Pres. & Engineer-of-the-year, Ithaca Section); ACEC (Committee Chairman, Dir). Has served on local Planning Board, Community Charities (Dir); SUNY Foundation, BCC Foundation, Development Foundation (Dir); Chamber of Commerce (Past Pres.); International Rotary. Married Betty Jane Browning; 5 children. *Society Aff:* ACEC, ASCE, ACSM, NSPE

Hawkins, Neil M
Business: 201 More Hall, FX-10, University of Washington, Seattle, WA 98195
Position: Prof & Chrmn Civil Engrg-Adjunct Prof Arch *Employer:* University of Washington. *Education:* PhD/Civil Engrg/Univ of IL; MS/Civil Engrg/Univ of IL; BE/Civil Engrg/Univ of Sydney; BS/Math & Physics/Univ of Sydney. *Born:* 1/31/35. Born in Sydney, Australia. Taught struct engrg at Univ of Sydney 1961-68 and while on leave worked at Portland Cement Assn, Skokie 1966-67. With Univ of WA, Seattle since 1968, Appointed Chrmn of Civil Engrg, 1978. Natl Sci Foundation prin investigator since 1970. Edward Noyes Prize, Inst of Engrs Australia 1965; ACI Wason Medal 1969, ASCE State-of-Art Award 1974; ASCE Raymond Reese Award 1976; ACI Raymond Reese Award 1978; ACI Raymond Reese Award 1980; Chmn ASCE-ACI's Comm on Shear 1974-77. Chmn ASCE-ACI's Comm on Precast Concret 1979-82. Enjoys hiking & fishing. Dir, Amer Concrete Inst 1982-1985; Dir, Earthquake Engrg Research Inst, 1984-86; Mbr, Nat'l Science Advisory Ctte on Earthquake Engrg; 1983-1985. Mbr, Bd of Assessment, Center for Building Technology, NBS, 1986-88. *Society Aff:* ASCE, ACI, PTI, EERI.

Hawkins, Robert C
Business: General Electric Company, 1000 Western Avenue - Bldg. 174AC, Lynn, MA 01910
Position: General Manager, Advanced Technology Operations *Employer:* General Electric Company. *Education:* BSME/Engrg/VA Polytechnic Inst; Adv Mgmt Pgm/Harvard Bus Sch; Adv Mgmt Pgm/Williams Coll. *Born:* 03/20/27. A Magna Cum Laude grad of VA Polytechnic Inst with a BSME, Bob joined GE in 1957. His GE experience includes advanced design, Engrg Mgr of F101 Progs, Mgr of Evendale Commercial Engrg Oper, and Mgr Evendale Prod Engrg Oper. In his present position he is responsible for directing the Aircraft Engine Business Group's advanced technology developments in the areas of engine design, advanced materials, manufacturing and quality processes. *Society Aff:* AIAA, CES, HSF.

Hawkins, W Lincoln
Business: 26 High Street, Montclair, NJ 07042
Position: Consultant. *Employer:* Privately Employed. *Education:* PhD/Chemistry/McGill Univ; MS/Chemistry/Howard Univ; ChE/Chem Engg/Rensselaer Polytechnic Inst; LLD/Honorary/Montclair State; D Eng/Honorary/Stevens Inst of Technology; LLD/Honorary/Kean College; D. Sc./Honorary/Howard Univ. *Born:* Mar 21, 1911. Bell Labs: Asst Chem Dir 1974-76, Hd Dept of Applied Res 1972-74, Supr 1963-72, Mbr Tech Staff 1942-63; McGill, Lecturer 1938-41; Columbia Post Doctoral Fellow 1941-42. Cons for Bell Labs on ed programs and for Plastics Inst of Amer on polymer res. Mbr Natl Acad of Engrg, ACS. Fellow N Y Acad of Sci, Amer Inst of Chemists. Sigma Xi, Honor Scroll, Hon LLD Montclair State. Hon D. Eng Stevens Institute of Technology Approx 50 tech publs on stabilization of polymers, 14 pats, ed Polymer Stabilization. Hon LLD Kean College. Hon D. Sci. Howard Univ. Recipient of the International Award of the Society of Plastics Engineers. *Society Aff:* NAE, ACS, AIC, SPE.

Hawkins, Willis M
Home: 5239 Bubbling Well Lane, La Canaoa, CA 91011
Position: Sr Advisor *Employer:* Lockheed Corp *Education:* BSAE/Aeronautics/Univ of MI. *Born:* 12/1/13. Hon Dr Sc IL Col, Hon Dr Eng Univ of MI. Lockheed Aircraft since 1937; Ch Advanced Aircraft Design, Founder Mbr & VP-Asst Genl Mgr Missile & Space Div, VP-Genl Mgr Space Sys Div. Sr Corp VP-Sci & Eng (Ret 1974); Recalled 1976 Pres CA Co, 1979 SR VP Aircraft (ret 1980) Sr Advisor (Now). Asst Secy US Army (R&D) 1963-66 Dir, Wachenhut Corp, Avemco Corp. Mbr NAE, Hon Fellow AIAA, RAES. Mbr Naval Studies Bd, NRC. *Society Aff:* NAE, AIAA.

Hawks, Glenn C
Business: 2590 E. Main St 201, Ventura, CA 93003
Position: President *Employer:* Hawks & Associates *Education:* BS/CE/Univ of UT; MS/CE/CA State Long Beach. *Born:* July 1938. Reg Prof Engr in Calif & Utah. Natl Engrg Cert. US Navy 1961- 64. Design & field engrg with Orange County, Calif Flood Control Dist, 1964-68. PRC Toups Corp 1968 to 1981 Mgr of Ventura Regional Office 1970-1981; V Pres since 1972. Tech Dir for Toups Corp Ser in Natl Flood Insurance Study Program (FEMA). President of Hawks & Associates, May 1981. V Chmn Cons Engrs Council of Ventura County 1975. Pres of Santa Barbara Ventura Branch of Amer Soc of Civil Engrs 82-83 VP of Tri-Counties Civil Engrs & Land Surveyors 1984. Appointed to Ventura County Sewer Adv Bd 1971. Rotary, Chamber of Commerce & Boy Scouts of Amer. *Society Aff:* ASCE, APWA, ICIDFC, ASP.

Hawks, Roger J
Home: 218 North Darling St, Angola, IN 46703
Position: Dean of Engrg *Employer:* Tri-State Univ *Education:* PhD/Mech Engr/Univ of MD; MS/Aero & Astro/MIT; BS/Aero Engr/Univ of Cincinnati *Born:* 08/21/42 Native of Jeromesville, OH. Aerospace engr for NASA-Goddard Space Flight Center and Fairchild-Hiller Corp. Instructor in mech engrg at Univ of MD, 1967-72. Asst Prof of Mech Engrg, Clarkson Coll, 1972-77. Joined Tri-State Univ as Assoc Prof of Aero and Mech Engrg in 1977. Promoted to Prof in 1982. Chrmn of Aero and Mech Engrg Dept 1978-84. Became Dean of Engrg in 1983. SAE Ralph R Teetor Award in 1975. Have conducted res in automobile stability and aerodynamics, electrical transmission line galloping, and missile flight dynamics. *Society Aff:* AIAA, SAE, ASEE, ΣΞ

Hawley, Erwin T
Business: 7601 Rockville Rd, Indianapolis, IN 46206
Position: Mgr of Application Engineering. *Employer:* PT Components, Inc/Link-Belt Bearing Div *Education:* BS/Engr/Purdue Univ. *Born:* Dec 1929. Internatl Harvester Co 1951 & 1953. USMCR Cpt Communications Oct 1951-Oct 1953. Joined Link-Belt Co Jan 1954 Engrg Trainee. March 1955 Design Engr Bearing Div. Designed & placed into prod line of Spherical Roller Bearings. 1961 Supr Application Engr. 1965 Asst Ch Engr. 1970 Proj Mgr Cylindrical Roller Bearings coordinated engrg, tooling, capital equip & market res for FMC/Link- Belt. Proj resulted in new mfg facil-1973 Mgr Mkting Planning-1975 Mgr Internatl Business with direct marketing & mfg respon in five countries & prod respon for balance. Also Product Mgr for Roller Bearings - 1986 Mgr Application Engineering. *Society Aff:* SAE, ASAE.

Hawley, Frank E
Home: 21Goldenridge Ct, Santa Mateo, CA 94402
Position: Hgwy and Transp Conslt *Employer:* Self *Education:* BS/CE/Univ of NM *Born:* 1/20/24. In 1948 graduated from Univ of NM; is a reg engr; native of NM.

Hawley, Frank E (Continued)

He was a pilot in the European theater during WWII. In his 35 yrs with the Gov, worked successively as a Jr Engr, Res Engr, Area Engr, Planning Engr, Div Engr, Deputy Regional Admin & was apptd Reg Admin for the Fed Hwy Admin, San Francisco CA in July 1972. He was cited several times for outstanding serv & in 1969 received the Admin's Bronze Medal for Outstanding Serv. He left govt service in Dec 1980 and began practice as hgwy & transp conslt. Primary client is Western Hwy Inst, San Bruno, CA. He is a resident of San Mateo CA. *Society Aff:* ARTBA.

Haworth, Donald R

Business: 345 E 47th St 6th Fl, New York, NY 10017
Position: Assoc. Exec. Director. *Employer:* American Society of Mechanical Engrs. *Education:* PhD/Mech Engg/OK State Univ; MSME/MEch Engg/Purdue Univ; BSME/Mech Engg/Purdue Univ. Born: Jan 26, 1928. Children Merry, Melodie, Donald, Jared, Maria, Tobey. Designer Bell Helicopter Corp, Fort Worth 1952-54; Instr Purdue Univ 1954-56; Propulsion Engr Chance Vought Aircraft Co, Dallas 1956-58; Asst Prof Okla State Univ 1958-61, Assoc Prof 1962-66; Sr Scientist LTV Res, Dallas 1961-62; Prof & Chmn Dept of Mech Engrg, Univ of Neb Lincoln 1966-76; Mgng Dir, Education Amer Soc ME 1976-84, Assoc. Exec. Dir. 1984-. Reg P E Okla. Mbr Amer Soc Engrg Ed (ME Div Secy 1971-72, V Chmn 1972-73, Chmn 197374), Amer Soc ME (Chmn ME Dept Hds, Region VII 1968-70, Natl V Chmn Dept 1968-69, Chmn 1969-70, Mbr Pol Bd Ed 1969-75, Secy Region VII 1972-74). Engrg Council Prof Dev (Dir 1971-74), Sigma Xi, Pi Tau Sigma (Natl V P 1963-68, Natl Pres 1968-71), Tau Beta Pi, Phi Kappa Phi & Sigma Tau. *Society Aff:* ASME, ASEE, ASAE, CESSE.

Hawthorne, George B, Jr

Home: 505 Kimberly Rd, Warner Robins, GA 31093
Position: Electronics Engineer *Employer:* USAF, WR-ALC/MMECV *Education:* PhD/EE/GA Tech; MSEE/EE/GA Tech; BEE/EE/GA Tech. *Born:* 8/15/28. Bainbridge GA. Teaching & res at Ga Tech to 1963. Thereafter, analy/design of large computer-based air defense & spaceflight control sys with MITRE Corp. (Subdept Hd) for USAF & NASA. Since 1972 long-range planning & design of airports funded by FAA, with cons engrg firm in Columbus GA. Since 1977, Electronics Engr with USAF at Robins AFB, GA. Mbr IEEE, Sigma Xi & New York Acad of Sci. Reg Prof Engr in Mass & Ga. Enjoy music, writing & photography. Active in hobby club, Mensa & political process. Married, two children. *Society Aff:* IEEE.

Hawthorne, John W

Business: Suite 1100, 11 East Adams Street, Chicago, IL 60603
Position: VP *Employer:* Camp, Dresser & McKee *Education:* BS/CE/Tufts Univ; MS/Environ Engrg/Univ of Cinn; MBA/Business/VPI *Born:* 9/15/41 Has had a wide range of environmental engrg experience both in industry as well as consulting engrg. His consulting experience has covered studies, designs, construction administration and startup water, wastewater and solid waste facilities. He has served as an expert witness as well as Federal District Court's appointed steering committee responsible to their Chief Justice in conducting a full plant evaluation of the Detroit Wastewater Treatment Facilities. As VP for Camp, Dresser & McKee (the nation's largest environmental engrg firm), he serves as Asst Regional Mgr and Mgr of Operations. *Society Aff:* WPCF, AWWA, AAEE, ASCE, AMA

Hawthorne, V Terrey

Business: 2 Main St, Depew, NY 14043
Position: Dir Engr & Quality Assurance. *Employer:* Dresser Indus Inc-Transp Equip Div. *Born:* Oct 1934. BEE 1956 NC State Col, grad credits at Syracuse Univ & Univ of Chicago, Harvard Business Col Program for Mgmt Dev. 1956-60 Jr Engr & supervisory positions Pa RR. 1960-65 Engr GE Television Receiver Dept. 1965-74 Dir of Engrg to V Pres Keystone Railway Equip Co. 1975-present Dir of Engrg & Quality Assurance Dresser Transportation Equip Div. Prof Engrg license in Pa, New York & Ill. Computer programming experience FORTRAN & SPS. Specialty railroad freight car component design & testing.

Hay, D Robert

Home: R R 6, Lachute, Quebec, Canada J8H 3W8
Position: Pres *Employer:* Tektrend International, Inc. *Education:* B Eng/Metall Eng/McGill Univ; MS/Eng Phys/Cornell Univ; PhD/Matls Sci/Cornell Univ. *Born:* 4/1/39. Asst & Assoc Prof Drexel Univ, Phila 1966-72; Invited Prof Ecole Polytechnique, Montreal 1972-75. Dir, Technology Dev Center, Ecole Polytechnique, 1975-79. Pres, Tektrend Interntl Inc, Lachute, Quebec, 1973- present. Tech int: fracture mechanics, nondestructive testing, explosive welding, failure analysis, Artificial Intelligence, Pattern Recognition. Prof affiliation: ASM, ASNT, EIC, Tau Beta Pi, Sigma Xi, AMA, CIMM. 50 publs. Dir, Indus Innovation Ctr - Montreal, 1979-81; Tech Dir, Hydrogen Indus Coun 1982-present; Dir, The Tektrend Group, 1983-present. *Society Aff:* ASM, ASTM, ТВП, ΣΞ, AMA, ASNT.

Hay, Ralph C

Home: 2108 Lynwood, Champaign, IL 61820
Position: Consultant Prof Emeritus (Self-employed) *Education:* MS/Agri Engr/MI State Univ; Grad Study/Agri Engr/Univ of IL; Grad Study/Agri Engr/IA State Univ; BS/Agri Engr/KS State Univ. *Born:* 7/21/07. Reared on Kansas farm which I still own. 40 yrs service on Univ of IL staff - asst to prof-teaching, res, ext, & admin. Res in soil conservation & drainage. Secy State Soil Conservation Bd, Chmn State Water Survey. Executive Secy IL Land Improvement Contractors. Intl Campus Dir & Adviser in two new agricultural engg depts in India. ASAE Bd mbr & Chrmn of Soil & Water Div & Intl relations, Life Fellow & Intl Kishida Award. SCSA Fellow. Univ YMCA, Physical Fitness, Farm Mgt Phi Kappa Phi, Sigma Tau, Alpha Zeta, & Sigma Phi Epsilon. *Society Aff:* ASAE, SCSA, ΦΚΦ, ТВП.

Hay, Richard N

Home: 344 E Groveland Dr, Oak Creek, WI 53154
Position: Partner, Civ Asst. Div Mgr. *Employer:* Graef-Anhalt-Schloemer. *Education:* BS/Civ Engg/Univ of WI. *Born:* 6/29/27. Native, Racine, WI, US Marine Corps, 1944-45, 1951 Univ of WI grad; Partner, Civ Asst. Div Mgr & VP Graef-Anhalt-Schloemer, consulting firm, Milwaukee; former Oak Creek City Engr, initiated dept, established planning & engrg precedence; 1979-80 Pres & Founder of Egn Fdn of WI; most state & all local officerships; WI Soc of PE; 1985 Wis Eng-of-the-Year; 1974 Engr-of-the-Yr in Private Practice Sec; 1983 WSPE Pres Award; Rotary Club of Mitchell Field, all officerships, Rotarian of the Yr 1969; 27 yrs perfect weekly attendance; Paul Harris Fellow; active participant & organizer, local civic groups, city of Oak Creek, 1984 Citizen-of-the-Year; life mbr WI Alumni Assoc; enjoy sports; family life, good friends & soc, civ & fraternal involvement. *Society Aff:* NSPE, AWWA, APWA, ACSM.

Hay, William W

Home: 404 Engineering Hall, 1308 W. Green St, Urbana, IL 61801
Position: Prof Emeritus. *Employer:* Univ of IL. *Education:* PhD/CE/Univ of IL; MS/CE/Univ of IL; MgtE/Mgt Engg/Carnegie-Mellon Univ; BS/Mgt Engg/Carnegie-Mellon Univ. *Born:* 12/10/08. Bay City, MI. Early home Warren PA. Engg and Maintenance various railroads 1934-1943; 1st Lt Col Military Railway Service; Chief Engr Korean Railway for US Military Govt in Korea 1945-46. Asst to full prof Railway Civ Engg, Univ of IL 1947-1977. Railway consultant to firms, cities govts in US, Canada, Southern Africa, Venezuela. Author: Railroad Engg 1953 (revised 1982), An Introduction to Transportation Engg 1961 (revised 1977), Chapter: Railway Engg, Encyclopedia of Phys Sci and Tech, Vol. 11, Academic Press, 1987. TM5-370 Railway Construction in the Theater of Operations, Section 6 Railroad Engg Abbett's American Civ Engg Practice; Dir - AREA and Roadmasters; Honorary Mbr AREA 1979; Honorary Member Roadmasters 1985; Alumni Merit Award, Carnegie - Mellon Univ. 1979. *Society Aff:* AREA, RTMWA.

Hayakawa, T George

Business: Hayakawa Associates, 1180 So. Beverly Dr, Los Angeles, CA 90035
Position: President. *Employer:* Hayakawa Assocs Cons Engrs. *Education:* BE/Mech

Hayakawa, T George (Continued)

Engg/Univ of So CA. *Born:* 7/4/24. Native of Los Angeles. US Army - WW II. Prof reg in 22 states & the National Council of Engrg Examiners & Amer Cons Engrs Council. Fellow, Amer Soc of Heating, Refrigeration & AirConditioning Engrs, Mbr Cons Engrs Assn of Calif, Natl Soc of Prof Engrs. Directs the activ of Hayakawa Assocs, a cons engrg firm founded in 1959 specializing in mech & elec engrg for buildings. Hayakawa Assocs has been respon for the design of approximately 50 million square feet of projects throughout the US as well as in & Asia. Twice invited to Japan as lecturer & cons. Represented the US at the First Intl Symposium On Energy in Tokyo, Japan. *Society Aff:* ASHRAE, ACEC, NSPE, CEAC.

Hayashi, Izuo

Home: 1-34-15 Tsukushino, Machida, Tokyo Japan 194
Position: Director. *Employer:* Optoelectronics Technology Research Co Ltd. *Education:* PhD/Physics/Tokyo Univ; BD/Physics/Tokyo Univ. *Born:* 1922 in Tokyo. Was Assoc Prof in Tokyo Univ until joining the Bell Telephone Labs in 1964 where he engaged in res on semiconductor lasers & achieved the first cw operation at room temperature. From 1971 to 1982 was in the Central Res Labs Nippon Elec Co Ltd, 1982-87 Optoelectronics Joint Research Labs, 1987 Optoelectronics Technology Research Lab. & in charge of R&D of opto-electronic devices and matls. Received three awards in Japan & promoted to the Fellow grade in IEEE in 1976, GaAs Symposium Award & IEEE-ED Award 1984. *Society Aff:* APS, IEEE.

Haycock, Obed C

Home: 3390 S. 2770 East, Salt Lake City, UT 84109
Position: Prof Emeritus Elec Engrg. *Employer:* Retired. *Born:* Oct 5, 1901 Panguitch, Utah. s. George A & Elida (Crosby) Haycock; student Utah State Univ 1921-23; BSEE Univ Utah 1925; MS Purdue Univ 1931; m. Mary Harding Aug 12, 1926; children-Jean (Mrs Garl Gardiner), Don, Ralph, Richard, Lois (Mrs Roland Porter); Instr to Prof EE Univ Utah, Salt Lake City 1926-69; Res Engr Rutgers Univ 1944-45; owner, mgr radio sta KLGN, Logan 1954- 59; organized Salt Lake IRE; organized student branch Univ of Utah AIEE & IRE; Dir Upper Air Res Labs 1957-69; res in outer space conducted using rockets & satellite as vehicle; Mbr Amer Inst Elec Engrs, Sigma Xi, Tau Beta Pi, Outstanding Engr Utah Engrg Council; Fellow IEEE.

Hayden, Ralph L

Business: Riley, Park, Hayden & Assoc, 136 Marietta St. - NW Suite 310, Atlanta, GA 30303
Position: VP & Gen Mgr. *Employer:* Riley, Park, Hayden & Assoc, Inc. *Education:* 65 hours/Engg/Univ of Louisville. *Born:* 1/4/31. Grew up in Monticello, KY, where career started in resident engg type work for the KY Hgwy Dept for five yrs. After military service engg resumed by doing railroad location, design and construction coordination for consultant for 10 yrs. Gen Mgr and co-founder in 1967 of gen civ and surveying firm with emphasis on railroad engg. Our 200 man firm, with about 600 projs, works throughout the entire southeastern US. *Society Aff:* ASCE, AREA, ACSM, AR/WA, ASP.

Hayden, Richard E

Home: 34 Brooks Road, Sudbury, MA 01776
Position: Pres *Employer:* Tech Integration and Dev Group, Inc *Education:* MS/Mech Engr/Purdue Univ; BS/Mech Engr/Norwich Univ. *Born:* 3/23/46. Founded Technology Integration and Development Group, Inc (TIDG) in 1984. TIDG provides mult-disciplinary and R/D services for government labs and manufacturers of a variety of devices. Previously Assoc Div Dir at Bolt Beranek and Newman Inc (BBN). Joined BBN in 1969. Conducted res in accoustics of flow/surface interactions. Received SAE Wright Bros Award in 1974 for res paper on propulsion lift noise reduction. Managed depts respon for aeroacoustic res, wind engineering, and instrumentation transducer development. Also interested in engrg ed, and education philosophy. Former Pres Norwich Univ Alumni Assn, Trustee Norwich Univ; Pres, Sudbury (MA) Youth Soccer Association. *Society Aff:* AIAA, APCA, SAE, ASME.

Hayduk, Stephen G

Business: One Huntington Quadrangle, Melville, NY 11747
Position: Assoc *Employer:* Bienstock, Lucchesi, P.C. *Education:* B of Tech/CE Tech/Rochester Inst of Tech *Born:* 2/20/50 Stephen G. Hayduk, PE is an Assoc in the consulting engrg firm of Bienstock & Lucchesi, P.C., specializing in civil and environmental engrg. Mr. Hayduk is also and adjunct faculty member (since 1977) in Engrg Science and Technology at Suffolk County Community coll. He is a licensed PE in the states of NJ, PA, and VT, and is especially active in NSPE, where he holds the following positions on the NY State Society Level: Chrmn - Registration Committee. Vice Chrmn - PE in Private Practice. Mr. Hayduk was in Army Aviation, in the Republic of Vietnam, and is a dedicated husband, father of two, and fly fisherman. *Society Aff:* ASCE, NSPE, WPCF

Hayek, Sabih I

Business: 230 Hammond Bldg, Universty Park, PA 16802
Position: Professor of Engineering Mechanics. *Employer:* Pennsylvania State University. *Education:* D Eng Sc/Engrg Mechs/Columbia Univ; MSc/Civil Engrg/Roberts Coll; BSc/Civ Engrg/Roberts Coll. *Born:* 3/25/38. Asst Prof 1965-70, Assoc Prof 1970-76, Prof of Engrg Mech since 1976. Res in radiated noise, stress waves, noise barriers & acoustical diffraction. Author of 40 publs, 36 reports, 700-pg monograph entitled *Mathematical Methods in Engrg* & a 130-pg monograph entitled *Acoustics of Marine Structures*. Hd of vibration & radiation group in the Applied Res Lab at Penn State 1970-83. Asst Sec of the Amer Acad of Mech 1972-. Served as the Hydrodynamics Noise Monitor for NAVSEA Hydromech Adv Ctte 1975-77, Visiting Res Fellow at the Inst of Sound and Vibration Research, Southampton Univ, Southampton Eng 1972. Visiting Senior Res Scientist, Naval Ocean Sys Ctr, San Diego, CA, 1980-81. Elected Fellow of the Acoustical Soc of Amer. Chrmn of Structural Acoustics and Vibration Div of ASA. Pres of Central PA chap of the Soc of Amer 1970. Awarded the PSU College of Engrg Excellence in Res Award 1982. Mgr of Amer Natl Standards Ctte on "Outdoor Noise Barriers" and on "Damping Characterization-. *Society Aff:* ASME, ASA, INCE, AAM.

Hayes, Albert E, Jr

Business: PO Box 2946, Fullerton, CA 92633
Position: President. *Employer:* Albert Hayes & Assocs. *Education:* PhD/Engg/CA Western Univ. *Born:* Dec 1920 San Francisco. Mbr MIT Radiation Lab during WW II. Engr wth Hazeltine Corp, Bendix Radio Div, Emerson Radio, Hoffman Electronics, North Amer Aviation, Ampex Corp & Lockheed Aircraft Corp. Served as special cons to Natl Security Agency. Co-founder & Ch Engr of Western Radio Lab 1970-78, at which time restructured as Albert Hayes & Assocs, Reg Professional Engrs. Reg Prof (Elec) Engr in Ca & Nev. Sr Mbr IEEE. Mbr Conn Wireless Assn. Mbr. Order of Quiet Birdmen.. Mbr, Soc of Reliability Engg. First Pres & Charter Mbr of Braille Tech Press, Inc. Five issued US Pats. Numerous publs in field. *Society Aff:* IEEE, SRE.

Hayes, John C

Business: 2401 Hillsboro Rd, Nashville, TN 37212
Position: Pres *Employer:* John Coleman Hayes & Assoc., Inc. *Education:* Dr of Jurisprudence/Law/YMCA Night Law Sch; BS/CE/Vanderbilt Univ *Born:* 4/23/37 Probably the youngest engr to begin private practice in State of TN. Upon grad from Vanderbilt Univ in 1959, Mr. Hayes worked for the TN Aeronautics Commission where he held the positions of Asst and State Airport Engr. During his tenure with the Dept of Aeronautics he spent his evenings studying law. He received his Dr of Jurisprudence from YMCA Night Law Sch in 1965. John went into private practice at the age of 27 in 1964. His firm, John Coleman Hayes & Assoc., Inc. performs not only engrg, but architectural and planning services for clients in the southeastern US. *Society Aff:* ASCE, NSPE, SAME

Hayes, John M
Home: 312 Highland Drive, West Lafayette, IN 47906
Position: Prof Emeritus & Cons Structural Engr. *Employer:* Purdue University & Cons. *Education:* BSCE/CE/Purdue Univ; MS/Engr Mechs/Univ of TN; CE/CE/Purdue Univ. *Born:* May 18, 1909 Wingate Ind. 1931-48 practice of civil engrg. 1948-75 Teaching structural engrg Purdue Univ. 1975-present Prof Emeritus & Cons Structural Engr. 1962 Pres Ind Sec ASCE. 1963-66 Natl Dir & 1968-70 Natl V Pres ASCE. 1970-74 Chmn Exec Ctte Ed Div ASCE. 1971 Special Citation Award, Amer Inst of Steel Construction Inc. 1975 Outstanding Engrg Alumnus Award The Univ of Tenn. April 1980 elected Honorary Member, A.S.C.E. *Society Aff:* ASCE, AWS, SESA, NSPE, ACI, AISC, AREA, AAAS, ASTM

Hayes, Robert H
Home: 723 Davidson Street, Raleigh, NC 27609
Position: Consulting Engr (Self-employed) *Education:* BSCE/Civil Engg/Univ of NC-Chapel Hill. *Born:* June 1908. Two yrs sub-prof work. 1931-71 civilian engr with Corps of Engrs US Army on water resources dev & military construction. Was Ch of Engrg for 25 yrs respon for planning & design in all Corps offices where employed. Special assignment 1954-58 as Ch Engr US Sect, St Lawrence River Jt Bd of Engrs, for monitoring design & construction of St Lawrence River Power Proj. Since retirement Jan 1972 part-time cons engrg. Awarded George W Goethals Medal by SAME 1973. *Society Aff:* ASCE, SAME, USCOLD.

Hayes, Thomas B
Business: 1600 SW Blvd, Corvallis, OR 97330
Position: Consulting Engineer *Employer:* CH2M HILL, Inc. *Education:* SM/EE/MIT; BS/EE/OR State Univ. *Born:* Aug 1912. Native of Oregon. WW II - US Navy. Founding partner of CH2M HILL Inc. Consultant to Indus & Energy Sys Dept. Invented & dev the FLOmatcher scheme for controlling the discharge of electrically driven pumps. Mbr IEEE (Fellow), ASME, (Fellow) NSPE, Sigma Tau & Eta Kappa Nu. Member and Past President of Oregon State Bd of Engrg Examiners. Member (Oregon) Governors Alternate Energy Commission, Reg P E in Oregon, Washington, Idaho, Alaska, Calif, Montana, Nebraska & Va. Enjoys skiing, boating & photography. *Society Aff:* IEEE, ASME, NSPE.

Hayes, Thomas J, III
Home: 2646 Chestnut St, San Francisco, CA 94123
Position: Engrg and Mgmt Conslt; Dir *Employer:* Various *Education:* BS/-/US Military Acad, West Point NY; MS/CE/MA Inst of Technology, Cambridge; Dipl/Indl/Industrial College of the Armed Fooces, Ft McNair. *Born:* 8/26/14. US Army Corps of Engrs 1936-69, 2nd Lt to maj Gen. Major assignments incl Asst Military Attache, London; Asst Engr Commissioner, Washington DC; District Engr, Alaska, Little Rock & Omaha. From 1958-69 sucessively headed four Corps of Engrs organizations respon for ICBM Const Program, design & construction of NASA Space Ctrs, worldwide engr troop & mapping operations & South Atlantic Div. Retired 1969 Served as Pres and COE 1969-78 and Chrmn & Chief Exec Officer 1979- 80 of Intl Engrg Co, Inc. Retired Jan 31, 1980. Conslt and Dir (1979-). Numerous military awards: Goethals Award from SAME 1962. Life Mbr Natl Acad of Engg. Fellow ASCE and SAME; Mbr AUSA. Reg PE, AK and Dist of Columbia. *Society Aff:* AUSA, ASCE, SAME

Hayet, Leonard
Business: The Smith Karach Hayet Mayric Partnership, 175 Fontainebleau Blvd, Miami, FL 33172
Position: Sr Partner *Employer:* The Smith Korach Hayet Haynie Partnership *Education:* MSIE/Indust Engrg/Purdue Univ; BSEE/EE/Purdue Univ *Born:* 06/10/32 Native of Passaic NJ. Civ Engr Copr US Navy 1955-58. Employed by Radar & Assocs on electrification of railroads in Chile, South America. Performed original design for Cape Kennedy Launch Complexes. Sr Partner The Smith Korach Hayet Haynie Partnership for last 14 yrs - hds engrg depts, incl Elect, mech, Air Cond, Plumbing & Civ. Mbr Amer Inst of Elec Engrs, FL Engg Soc, Fl Inst of Cons Engrs, serves on the panel of Construction Arbitrators of the Amer Arbitration Assn. Mbr of Soc of Fire Protection Engg, Prof Engr FL., Reg. Prof. Planner, N.J., Trustee Emeritus-Peporicolon Cancer Research Inst, Firm member-American Hospital Assn. *Society Aff:* AIEE, FES, SFPE, FICE, AAA

Haynes, Frederick L
Home: 3806 Fort Hill Drive, Alexandria, VA 22310
Position: Dir, Cooperative Technology. *Employer:* US Dept of Commerce. *Education:* MBA/R&D Mgmt/Amer Univ; BSIE/Ind Engg/Northeastern Univ. *Born:* 10/27/34. Native of Portsmouth New Hamp. Served in both the US Navy 1955-57 and US Army 1960-62. Received the Army Commendation Medal 1962. Employed as Indus Engr with Sylvania Elec Prods & the Aluminum Co of Amer specializing in mfg process design and new prod dev. With the U S Gov since 1964 as a res and dev engr, AF Sys Command, Sr Indus Engr Hdquarters USAF, Asst Dir Logistics and Communications Div, US Gen Accounting Office respon for the review of gov indus activities and the application of advanced tech to improving mfg productivity, Dir, Cooperative Technology, US Dept of Commerce respon for the established of joint industry, univ & govmt cooperative technology center to spur the dev of critical technologies needed to enhance balance industrial growth. Assoc Dir, Natl Tech Info Serv, US Dept of Commerce responsible for Marketing and Customer Serv. Currently, Assoc for R & D Limited Partnership Development, Office of the Asst Secy Productivity, Tech & Innovation. US Dept of Commerce. Fellow IIE, Reg PE Mass. *Society Aff:* IIE, SME, CASA.

Haynes, John J
Business: Civil Engrg Dept, Box 19308, Arlington, TX 76019
Position: Prof of Civil Engrg *Employer:* Univ of TX at Arlington *Education:* PhD/CE/TX A&M Univ; M/CE/TX A&M Coll; BS/CE/TX Tech Coll *Born:* 9/13/25 in Dallas and attended public schools in McKinney, TX. Served in US Navy in WWII and served in Naval Reserve as LCDR in Civil Engr Corp until 1968. Design engr for TX Highway Dept in Wichita Falls and Fort Worth, 1949-1954. Assoc Prof at Arlington State Coll, 1954-1960. Prof and Dept Chrmn of Civil Engrg at The Univ of TX at Arlington, 1961-1972. Prof of Civil Engrg at Univ of TX, Arlington 1972-present. Outstanding Teacher in Coll of Engrg in 1978. Pres of the TX Section of ASCE in 1979-1980. Engr of the Year of Mid-Cities Chapter of TSPE in 1980. Active fisherman and golfer. *Society Aff:* ASCE, NSPE, TSPE, ITE, ASEE

Haynes, Munro K
Home: 3311 E. Terra Alta Blvd, Tucson, AZ 85716
Position: Senior Engineer *Employer:* IBM Corp. *Education:* Ph.D./Elec. Eng./Univ. of Illinois; M.S./Elec. Eng./Univ. of Illinois; B.S./Physics/Univ. of Rochester *Born:* 12/10/23 Ph.D. Thesis on magnetic cores for digital computer systems. Joined IBM in 1950, and initiated research that led to widespread use of magnetic core memory systems in computers. Since 1971 has worked in magnetic recording, on channel characterization, media evaluation, recording physics, and buried servo systems. Now engaged in magneto-optic storage research. A Reg PE in CO, has published a variety of papers, and holds more than thirty issued U.S. Patents, for which received IBM Eighth-Level Invention Award and IBM Outstanding Innovation Award. Elected IEEE Fellow in 1987. Private Pilot, Instrument Airplane. *Society Aff:* IEEE, ΣΞ

Haynes, Patricia Griffith
Home: 3806 Fort Hill Dr, Alexandria, VA 22310
Position: Prog Mgr *Employer:* DOT-FAA *Education:* BS/Ind Mgt/Am Univ; Studied-5 yrs/Ind Engg/Northeastern Univ; Studied-2 yrs/Ind Engr & Mgt/Fairleigh Dickinson Univ; Grad Study/Systems Mgt/G W Univ. *Born:* 5/22/33. BSIM, Am Univ; Grad Study, Systems Mgt, Geo Wash Univ; Studied Ind Engg (7 yrs) - Northeastern Univ & Fairleigh Dickinson Univ; Native of Malden, MA. Employed as Asst Engr, Raytheon Mfg Co, specializing in Quality/Reliability Control; Cost Analyst, Operations Evaluation Group, MA Inst of Tech/Chief of Naval Operations,

Haynes, Patricia Griffith (Continued)
Pentagon; Sr Methods Analyst & Acting Div Mgr, NASA Scientific & Technical Info. Facility, resp for progs assoc with mgt info control system & resource utilization; Ind Engg & Mgt consultant, Principal, focusing on needs of small mfg & Health Care Services; Special Asst to the Dir Dept of Licenses, Investigations & Inspections, D.C. Govt; Author et al *Productivity Improvement Handbook for State and Locl Government*, John Wiley & Sons, Inc NY, 1979. Natl VP Am Inst of Ind engrs, Bd/Trustees, IIE; Reg PE, CA; Sr Mbr Am Soc for Quality Control, Soc of Women Engrs, WA Soc of Engrs; Mbr, Natl Soc of PE. *Society Aff:* IIE, SWE, ASQC, NSPE, WSE.

Hays, Donald F
Home: 639 Alpine Ct, Rochester, MI 48063
Position: Dept Hd; Fluid Mech Dept *Employer:* Gen Mtrs Res Labs *Education:* MSME/ME/OR State Univ; BSME/ME/OR State Univ. *Born:* March 1929 in Portland, OR. Married 1955. US Army Corp Engrs 1954-56 with assignment in Japan. Joined the Gen Mtrs Res Labs in 1952 & is currently Dept Hd, Fluid Mech Dept. Has authored numerous papers in the field of hydrodynamic lubrication, variational techniques in non-equilibrium thermodynamics & fluid flow & edited book 'The Physics of Tire Traction'. Admin respon encompass the areas of seal dev. Tribology, surface characterization, vehicle aerodynamics, engine flow and combustion res. Fellow ASME; ASME Pi Tau Siga Gold Medal 1959; Chmn ASME Res Ctte on Lubrication 1974-76; Tech Editor ASME Journal of Lubrication Tech 1975-81. 1980 ASME Centenial Medallion; 1983 Mayo D. Hersey awd (ASME). Elder, Presbyterian Chruch, Mbr Phi Kappa Phi, Pi Tau Sigma, Sigma Tau, Pi Mu Epsilon, Sigma Xi, AAAS. Outside int incl music, photography, tennis & woodworking. *Society Aff:* ASLE, AAAS, ASME, SAE.

Hays, George E
Home: 1909 Moonlight Dr, Bartlesville, OK 74003
Position: Sr Engr Assoc. *Employer:* Phillips Petrol Co. *Education:* BS/CE/State Univ of IA. *Born:* 4/25/21. Native of Alton, IL. With Phillips Petrol Co since 1943 in R&D. Current position as Sr Engg Assoc since 1968. Responsible for consulting throughout Phillips on data, processes, & new tech; for setting up experimental progs integrating pilot plant work with process design needs; for computerization, totally automated control & data-taking system for entire pilot plants; mgt info system dev. Mbr ACS Petrol Res Fund Advisory Bd 1971-73. *Society Aff:* AIChE, ACS.

Hays, James E
Home: 580 Hamilton Rd, Birmingham, MI 48010
Position: VP *Employer:* Smith, Hinchman, & Grylls *Education:* MS/Civ Engr/Univ of IL; BS/Engr & Sci/US Military Acad. *Born:* 7/26/28. BS from US Military Acad. MS Univ of IL. Native of Columbus OH. Combat engr, Commander & Const Engr commander at platoon, co & battalion level in US, Korea, Germany, & Vietnam 1954-70. Deputy & Dist Engr, US Lake Survey 1963-65. Dist Engr, Detroit Dist 1973-76. Cmdr & Dir Construction Engg Res Lab. Pres SAME, Detroit 1975. Pres SAME, Chrmn 1978-79. Dir, Installations & Services, Army Matl Dev & Readiness Command, 1979-1980. 1980 to present, VP and Dir Civil/Trans Engrg Div, SH&G. *Society Aff:* SAME, NSPE, ASCE, ESD, MSPE, BSA.

Hays, Steve M
Business: 315 Union St, Suite 606, Nashville, TN 37201
Position: Prin Engr *Employer:* Gobbell, Hays & Assoc., Inc. *Education:* BE/ChE/Vanderbilt Univ *Born:* 3/2/51 Prin Engr for Gobbell, Hays & Assoc., Inc. Joined firm in 1979, became partner in 1980. Responsible for conceptual design of all engrg aspects of all the Co's projects. Prior to this time worked for E. I. DuPont in Martinsville, VA and Old Hickory, TN. Grad cum laude from Vanderbilt Univ in Chem Engrg. Active mbr of Natl & TN Societies of PE, Assoc of Energy Engrs, Consulting Engrs of TN, American Consulting engrs Council. Chrmn of TSPE Public Relations Ctte. Named TN's Young Engr of the Year for 1980 by TSPE. *Society Aff:* NSPE, ACEC, AEE

Hayt, William H, Jr
Home: 705 Sugar Hill Dr, W Lafayette, IN 47906
Position: Prof of Elec Engrg. *Employer:* Purdue Univ. *Education:* BSEE/EE/Purdue; MSEE/EE/Purdue; PhD/EE/Univ of IL. *Born:* July 1, 1920 Wilmette Ill. Field Serv Engr Sperry Gyroscope Co 1942-46. With EE Sch, Purdue U 1946- ; Hd EE Sch 1962-65. ASEE Western Elec Fund Award 1973. Author *Engrg Electromagnetics*, 4d ed 1981; (with J E Kemmerly) *Engrg Circuit Analysis*, 4th ed 1986; (with G E Hughes) *Intro to Elec Engrg* 1968; (with G W Neudeck) *Electronic Circuit Analysis & Design* 2d ed 1984. Dir of Duncan Elec Co 1962-76. Sagamore of the Wabash, State of Ind. Fellow IEEE. Mbr State Univs Telecommunications Coord Coun 1967-77; Chrm 1967-71 *Society Aff:* IEEE, ASEE, ΣΞ.

Hayter, R Reeves
Home: 3425 Pine Bluff, Paris, TX 75460
Position: VP. *Employer:* Hayter Engr, Inc *Education:* BSE/Civil/Duke Univ. *Born:* 5/25/49. Native of Paris, TX. Following grad, associated with City of St Joseph, MO, and Booker Assocs, St Louis. Secy-Treas and Pres-elect, NW Chapt MSPE. Chapt outstanding young engr. Assumed position with Hayter Engg in 1975 as Principal- in-Charge for projs serving some 50 municipal and industrial clients throughout TX and OK. 1977-78 chapt pres and outstanding Young Engr. TX Soc of PE. 1981 TSPE State Dir. Mbr Rotary, Chi Epsilon, WPCF, Episcopal Church. Licensed PE TX, OK, MO, AR, LA, NM. *Society Aff:* ASCE, NSPE.

Hayter, Richard B
Business: Ward Hall, Manhattan, KS 66506
Position: Dir, Engrg Extension *Employer:* Kansas State Univ *Education:* PhD/Mech Engg/KS State Univ; MS/Mech Engg/KS State Univ; BS/Mech Engg/SD State Univ. *Born:* March 1943. Air flow instrumentation engr for Pratt & Whitney Aircraft prior to entering the AF. Aircraft Weapons Sys Engr & Base Civil Engrg officer USAF 1966-70. NIH Bio environ Engr Trainee, Kansas State Univ 1970-74. Mech Engrg Asst Prof & Res Assoc with the Inst for Environ Res, Energy Conservation Adv, Kansas State Univ 1975-77. Genl Mgr & Exec VP Energy Mgmt & Control Corp 1977-80. Dir, Engrg Extension, Kansas State Univ. Recipient of the Dow-ASEE Outstanding Young Fac Award 1976, Registered Prof Engr. Enjoys backpacking, photography, woodworking & choral music. *Society Aff:* ASHRAE, NSPE, ASEE.

Haythornthwaite, Robert
Business: College of Engg & Arch, Philadelphia, PA 19122
Position: Prof Engrg Science *Employer:* Temple Univ. *Education:* PhD/Engg/London Univ, UK; MS/Appl Math/Brown Univ; BSc/Civil Engg/Durham Univ, UK; BSc/Elec Engg & Mech Engg/London Univ, UK. *Born:* 5/5/22. Native of Whitley Bay, England; US Citizen 1964. Scientific Officer at Bldg Res Sta Watford England 1942-47. Taught at Sheffield Univ 1947-53, Brown Univ 1953-59, Univ of Mich 1959-67. The PA State Univ 1967-79. Hd, Dept Engrg Mechs 1967-74. Temple Univ 1979- . Dean, Coll Engrg Tech 1979-81. Active in ASCE (P Chmn Engrg Mechs Div), ASEE (P Chmn Mechs Div), Amer Acad Mechs (Pres 1969-71). P Editor *Proc US Natl Congr Appl Mechs, Journal Engrg Mechs Div* ASCE, *Mechs* AAM. Commonwealth Fund Fellow 1950; Walter L Huber Cvl Engrg Res Prize 1963. Res ints in mechs of solids & soil mechs. *Society Aff:* ASCE, ASME, ASEE, AAM

Hayton, Ben C
Business: 4800 Fournace Pl, Bellaire, TX 77401
Position: Pres *Employer:* Texaco Chemical Co *Education:* BS/Chemical Engr/Rice Univ. *Born:* 8/22/25. Born in Brenham, TX. Served as an officer in the US Navy During 1945 & 1946. Joined Jefferson Chemical Co (A Subsidiary of Texaco) in 1946 & served in various res & mkting position in Port Arthur, Houston & NY until 1966. Joined Texaco's petrochemical dept in 1966 as mgr, chemical div. Became Genl Mgr in 1970 and was elected VP in charge of petrochemicals in 1972.

Hayton, Ben C (Continued)
He was named President of Texaco Chemical Company in 1980 and also still serves as a Vice President of Texaco Inc. *Society Aff:* AIChE, API, NPRA, SPI, ACS, CMA.

Hayworth, Curtis B
Home: 5 Egbert Ave, Morristown, NJ 07960
Position: Ev Engg. *Employer:* US Treasury. *Education:* ScD/Chem Engg/NYU; MChE/Chem Engg/NYU; BChE/Chem Engg/CCNY. *Born:* 12/1/20. Vienna, Austria; US Citizen. Gran ScD (Chem-Engg) from NYU 1949. Joined General Chem Div, Allied Chem Corp 1944 Lab Asst, Lab Super 45-48, Res Chem Engg 48-53, Asst Mgr Dev Res 53-54, Asst Dir 54-58, Dir 58-61, Asst Tech Dir 61-63, Asst Dir Central Res Lab - Allied Chem Corp 63-67, Sr Res Tech Assoc Corp R&D 67-70, Numerous patents, Publs. Retired. Joined World Patent Dev Corp as VP 70- 71, Pres 71-75. Negotiated licenses with Soviet Bloc Nations. Us Treasury 75-present. Valuation Engg-. Enjoy Photography *Society Aff:* AIChE, ACS, ΣΞ.

Hazard, Herbert R
Business: 2770 North Star Road, Columbus, OH 43221
Position: Pres *Employer:* Herbert R Hazard Inc *Education:* BS/Mech Engg/PA State Univ. *Born:* Aug 1917. Analytical Engr for Babcock & Wilcox for 5 yrs; Asst Proj Engr for Bendix Marine Div 1 yr. At Battelle Columbus Labs since 1945 as Res Engr, Asst Div Ch, Div Ch Thermal Engrg, Staff Cons, Fellow. Pres, Herbert R. Hazard, Inc., Engrg conslt, since Feb 1984. Planned & directed over 70 R&D programs advancing nuclear reactors, gas turbines, automotive engines, steam boilers, deep-sea diving, advanced power sys & synthetic fuels. Author of 70 papers, 6 pats. P Chmn ASME Gas Turbine Div & many ASME cttes. Mbr Commerce Tech Adv Bd Panels on High-Speed Ground Transportation & Surface-Effect Ships. Fellow ASME, Reg Engr Ohio. *Society Aff:* ASME.

Hazeltine, Barrett
Business: Div of Engg, Providence, RI 02912
Position: Prof Assoc Dean. *Employer:* Brown Univ. *Education:* PhD/EE/Univ of MI; MSE/EE/Princeton; BSE/Basic Engg/Princeton. *Born:* 11/7/31. At Brown Univ, since 1959. In 1970 and 1976, at the Univ of Zambia. In 1980-81, and 1983-84 at the Univ of Malawi. In 1964-65, at Raytheon Corp as participant in Residencies in Engg Practice Prog. In 1971-72, Chrmn of Providence Sec IEEE, in 1977-78, Pres of Providence Engg Soc. In 1968, received Western Elec Award for Excellence in Instruction. Interested in digital circuits and computers, Tech mgt, appropriate tech. Technology courses for non engineers. *Society Aff:* IEEE, ASEE.

Hazen, David C
Business: Engrg Quad, Princeton, NJ 08540
Position: Professor Aeronautical Engrg. *Employer:* Princeton University. *Education:* MSE/Aero Engg/Princeton Univ; BSE/Aero Engg/Princeton Univ. *Born:* July 1927. BSE 1948 Magna Cum Laude, MSE 1949 from Princeton Univ, Phi Beta Kappa & Sigma Xi. Started with Princeton Univ as Instr in 1949, apptd Prof in 1963. Served as Assoc Dean of Fac 1966-69, Assoc Chmn Dept of Aerospace & Mechanical Sci 1969-74. Assoc Fellow AIAA, Chmn Ed Ctte 1965-68, V Pres Ed 1971- 72. Bd/ Trustees: Univ of Petroleum & Minerals, Saudi Arabia 1973- , Robert Coll Turkey 1969- , Engr's Council for Prof Dev 1971. Naval Res Adv Ctte 1971-78, V!. .!. Chmn 1974-75, Chmn 1975-77. Naval Studies Bd of Natl Res Committee 1978-. . *Society Aff:* AIAA, ASEE, ΣΞ, ΦBK.

Hazen, Richard
Business: 76 Oliphant Av, Dobbs Ferry, NY 10522 *Employer:* Retired *Education:* MS/Sanitary Engg/Harvard; BS/Civil Engg/Columbia; AB/History/Dartmouth. *Born:* 8/5/11. Dir ASCE & AWWA; Pres AICE. Mbr NAE, Hon Mbr AWWA. Hon Member ASCE. Practice confined to water supply & water pollution control planning, design, financial studies etc. Engagements with many large cities & metropolitan areas such as NY, Detroit, WA, Baltimore, Sao Paulo, Brazil, Port Salo, Egypt; & scores of smaller communities. Work primarily for public agencies, with some indus waste; cons to Natl Comm on Water Quality. *Society Aff:* ASCE, AWWA, NEWWUS, WPCF, NAE.

Bartholomew, George A
Riding Trail Lane, Pittsburgh, PA 15215
Position: Consulting Engineer *Education:* MS/Chem Engg/Univ of Pittsburgh; BS/ Chem Engg/Univ of Pittsburgh; AB/- /Haverford College. *Born:* Jun 21, 1924. Advanced studies: Chem Engrg & Met Engrg, Carnegie-Mellon University; Bus Admin, University of Pittsburgh. Pilot Plant Coordinator, Pittsburgh Consolidation Coal Co 1948-50; low temp carbonization & gasification. Res Fellow, Mellon Inst 1950-55; industrial fabric res in filtration & air pollution.Electrochemical Section Leader, US Steel, Applied Res Lab 1955-57; electrolytic, catephoretic, electrostatic & thermal coatings. Tech Dir, Raw Materials Sales, US Steel 1957-67; Dir tech, market & facility evaluations for proposed processes. Initiated advanced product applications. Vice President, Research & Development, Burrell Construction & Supply Co 1968-81 ; admin of res, new product dev, quality standards & control, raw material allocation, energy conservation. Consulting Engineer 1981- Served on tech committees of Natl Slag Assn, PA. Asphalt Pavement Assn; Dir: PA Ready Mixed Concrete Assn; Pittsburgh Chapter, ACI, Sigma Xi, Phi Lambda Upsilon, Sigma Tau, Sigma Beta Sigma, Beta Rho Sigma. Clubs: Longue Vue Club, Masonic Bodies, Wyndemere Country Club. *Society Aff:* AIChE, ACerSoc, AAAS, NYAS, NAWCC

Head, Julian E
Home: 916 W Woodlawn Ave, N Augusta, SC 29841
Position: Exec VP. *Employer:* Patchen, Mingledorff & Assoc. *Education:* BSCE/Civ Engr/GA Inst of Tech. *Born:* 2/19/25. Native of Oxford, AL. Served with Army Service Group throughout Europe during WWII (1943-45). Served two yrs with TVA in Structural Design group. Active as Squad Leader & Proj engr with Patchen & Zimmerman, performing structural & civ design for seven yrs. Exec VP & Chief Engr, Patchen Mingledorff & Assoc for 20 yrs. Responsible for multimillion dollar projs & directed up to 50 engrs & draftsmen. Enjoy music & fishing. *Society Aff:* ASCE, NSPE, GSPE.

Healy, Edward R
Business: 206 West White Street, Champaign, IL 61820
Position: President. *Employer:* Nothern Illinois Water Corp. *Education:* BS/Civil Engrg/Univ of IL. *Born:* Dec 1918 Kankakee IL. Served in US Marine Corps 1941-46. With present co (investor-owned water utility) since 1949. Supr Engr 1949-55, Mgr Champaign Div 1955-56, V Pres 1956-73, Dir 1970, Pres 1973. Dir AWWA 1969-72, Mgmt Div Award (AWWA) 1963, Fuller Award (AWWA) 1971. Pres Natl Assoc of Water Cos 1975-76. Selected for Hon Mbrship in American Work Assoc 1979. Enjoys golf & pop music. Selected for Honorary Membership National association of Water Companies, October 1983. Retired, January 1, 1984 from active employment with Northern Illinois Water Corp although still active as mbr of bd of dirs. Also active as part-time conslt in private practice.. *Society Aff:* ASCE, NSPE, AWWA, FSPE.

Healy, John M
Business: 1525 So 6th St, Springfield, IL 62703
Position: VP & Partner *Employer:* Hanson Engrs *Education:* MS/CE/Found Engrg/ Univ of IL; BS/Agri/Univ of IL; BS/CE/Univ of IL *Born:* 12/28/32 From 1960 to 1962, Mr. Healy was a Research Asst in the Dept of Civil Engrg at the Univ of IL. Joined Hanson Engrs in 1962. During his employment with the firm, he has gained experience in all 50 states on a variety of projects ranging from dams, grain storage elevators, microwave towers, industrial projects, and buried blast-resistant structures. As a VP and a Dir of the firm, he has operational management responsibilities for the foundation engrg consulting practice and the quality control material testing service. *Society Aff:* ASCE, ISSMFE, USCOLD, ISPE

Heaston, Robert J
Home: 8231 Stonewall Dr, Vienna, VA 22180
Position: Tech Mgr. *Employer:* US Army. *Education:* PhD/CE/OH State Univ; MS/ CE/Univ of AR; BS/CE/Univ of AR. *Born:* 3/3/31. Raised in AR. Went into USAF at Wright-Patterson AFB in OH. After military discharge, entered civ service & began progressive res & dev assignments with the AF, Advanced Res Projs Agency, & the Army. Have worked in various disciplines including missile & aircraft propulsion, combustion, explosives, energy conversion, & terminal homing of smart weapons. In 1976 became tech mgr for the next generation of Army missiles, aircraft, combat vehicles, & guns. Have given several presentations and published articles since 1977 on redefinition of the four fundamental forces. Enjoy theorizing, whether in philosophy, psychology, or phys. Take delight in my family, scouting, & church. *Society Aff:* AAAS, AIChE, ADPA.

Heath, James R
Business: Dept of Electrical Engrg, Univ. of KY, Lexington, KY 40506
Position: Assoc Prof of Electrical Engrg *Employer:* Univ of KY *Education:* PhD/EE/Auburn Univ; MS/EE/Auburn Univ; B/EE/Auburn Univ *Born:* 5/20/45 In 1973, he was with the Coll of Engrg, Univ of SC, Columbia, as an Asst Prof of Electrical Engrg. From 1974 to 1978, he was a private consultant and an Adjunct Asst Prof with the Dept of Electrical Engrg, Auburn Univ, Auburn, AL. In 1978, he joined the Coll of Engrg, Univ of KY, Lexington, where he held the positions of Asst. Prof and Assoc Prof of Electrical Engrg, Dir of University Computing, Assoc VP for Information Systems Planning & Policy and Dir of the Computing Center. He is currently an Assoc Prof of Electrical Engrg. His current research interests are in computer architecture, software engineering, algorithems, digital system simulation, performance analysis and microprocessor applications. *Society Aff:* NSPE, IEEE, ACM, HKN, TBΠ, ΣΞ

Hebeler, Henry K
Business: PO Box 3707 (8A-71), Seattle, WA 98124
Position: Div President. *Employer:* The Boeing Company. *Education:* SB/Aeronautical/MIT; Aero/Aeronautical/MIT; MBS/Bus/MIT. *Born:* Aug 12, 1933 St. Louis. SB Mass Inst Tech 11955, Aero Engrg 1956, MBA 1970. Stress analysis Boeing Co Seattle 1956-58, Minuteman Dev Mgr 1959-63 SRAM Dev Mgr 1964-68, Genl Mgr Res & Engrg Div 1970-71, Office Corporate Business Dev 1972-74, V Pres 1973- ; Chmn Boecon, Boeing Environ Prods Inc & Boeing Engrg & Const, Engrs. Recipient Mead prize Mass Inst Tech 1955; Sperry Gyroscope Fellow 1956, Boeing Sloan Fellow 1969. Mbr Sigma Xi, Chi Phi, Tau Beta Pi. Club: Meridan Valley Country.

Hecht, Barry M
Home: 26 McKinley Dr, Delmar, NY 12054
Position: Principal Transportation Analyst. *Employer:* NY State Dept of Transportation. *Education:* BS/Civil Engg/Cornell Univ; -/transp engg & reg planning/Yale Univ. *Born:* 11/13/43. Native of Brooklyn, NY. Currently directs urban transportation planning n upstate NY for NY State Dept of Transportation. 1974-79, dir urban transportation planning for Buffalo, NY area. BS Cornell Univ, attended Yale Univ joint program in transportation eng and urban/regional planning, 1965-67. Mbr, ASCE; Assoc Mbr, ITE; Pres, Upstate NY Section of ITE, 1977. *Society Aff:* ASCE, ITE.

Hecht, Harold
Business: 370 7th Ave, New York, NY 10001
Position: Senior Partner. *Employer:* Hecht Hartmann & Concessi. *Born:* Nov 1922. BS in Mech Engrg from The Cooper Union. In act practice since 1953. Prof Engrg Licenses in the States of New York, New Hampshire, Mass, Maryland, Florida, D C, New Jersey, Conn, Missouri & Texas. Extensive experience in Master Planning, Energy Analysis, Site Inspections & Report. Act engaged in design concepts of all projs. Fellow Brooklyn Engrs Club. Mbr of New York State Soc of Prof Engrs. Served as Adj Prof of Architecture & Building Tech, New York Inst of Tech.

Heckard, Calvin W
Business: 2305 Ocean Blvd Bx539, Coos Bay, OR 97420
Position: Gen Mgr & Chief Engrg. *Employer:* Coos Bay-North Bend Water Board. *Education:* BS/CE/WSU. *Born:* Jan 17, 1926. ICMA Techniques of Municipal Admin 1958. Positions held: US Dept of interior, Bureau of Reclamation 1950 (Const Engrg, Grand Coulee Dam Washington). Washington State Dept of Highways 1950-51. Coos Bay-North Bend Water Board 1951- present. Major achievements incl rebuilding of water dist sys, const of new office ser center complex, new water source dev in Coos Bay Sand Dunes aquifer, changing billing sys to modern data processing & putting into operation a rate schedule which provides for long range dev. P Pres-Prof Engrs of Oregon, Southwest Chap. Mbr-Natl Soc of Prof Engrs. P Chmn-AWWA, PNWS. Engr of Year-PEO Southwest Chap 1970. *Society Aff:* NSPE, AWWA.

Heckel, Richard W
Business: Michigan Technological Univ, Dept of Met Engrg, Houghton, MI 49931
Position: Prof, Metallurgical Engrg. *Employer:* Michigan Technological University. *Education:* Ph.D./Metallurgical Eng./Carnegie-Mellon Univ; M.S. /Metallurgical Eng. /Carnegie-Mellon Univ; B.S. /Metallurgical Eng./Carnegie-Mellon Univ. *Born:* Jan 1934. Native of Pittsburgh, Pa. Engrg Materials Lab of DuPont 1959- 63, Drexel Univ 1963-71, MMS Dept Hd, Carnegie-Mellon Univ 1971-76. Areas of res experience incl powder met, composite materials, coatings, oxidation and diffusional phenomena in solids. Author of 90 pubs & 3 pats. Mbr of TMS-AIME, ASEE, APMI, ASM. Lindback Foundation Award, Drexel Univ 1968; Bradley Stoughton Award, ASM 1969; Adams Memorial Membership, American Welding Society 1966; Metallurgical Education Achievement Award, Philadelphia Chapter of ASM 1967; Fellow of ASM 1983. Enjoy fishing & boating. *Society Aff:* TMS-AIME, ASM, ASEE, APMI.

Hecker, Siegfried S
Business: MS A100, LANL, Los Alamos, NM 87545
Position: Director. *Employer:* Los Alamos Natl Lab; Univ Ca *Education:* PhD/Metallurgy/Case Western Reserve Univ; MS/Metallurgy/Case Inst of Tech; BS/ Metallurgy/Case Inst of Tech. *Born:* Oct 2, 1943. Native of Austria. Mbr of Tau Beta Pi, Alpha Sigma Mu & soc of Sigma Xi. Recipient of W P Sykes Outstanding Met Award at Case Tech in 1965, ASM Marcus Grossman Young Author Award 1975, Department of Energy E.O. Lawrence Award 1984. Postdoctural apointee at the Los Alamos Scientific Lab 1968-70. Sr Res Met at GM Res Labs 1970-73. Worked on sheet metal forming & plastic instability. Am presently Director of Los Alamos Natl Lab. Areas of int incl plutonium & uranium met & materials for space nuclear power sys, high rate deformation, in addition to plasticity and instability. Mbr of ASM & AIME. Act in local chap of ASM PPres & several natl cttes and past Board Member of AIME. Enjoy outdoors activities, especially skiing. *Society Aff:* ASM, TMS/AIME.

Heckler, Alan J
Business: 703 Curtis Street, Middletown, OH 45043
Position: Mgr-Matls Sci Research. *Employer:* Armco Steel Corporation. *Education:* BS/Met Engg/IIT; MS/Met Engg/Purdue; PhD/Met Engg/Purdue. *Born:* Sept 1937. Has been with Armco Steel Corp, Res & Tech since 1965. Presently a Mgr & is repspon for the Matls Sci Sec. Principal area of respon has been the analysis crystallite orientation dist in metals. Mbr of ASM, AIME, IMS & Alpha Sigma Mu. Mbr of the International Council on Alloy Phase Diagrams of ASM. *Society Aff:* ASM, AIME, AMS, AΣM.

Hedden, William D
Business: 1105 Truman Rd-Bx15607, Kansas City, MO 64106
Position: Senior Vice President. *Employer:* Calvin Communications, Inc. *Born:* June 13, 1918. Eastman Kodak Co 1940-42. Color film processing US Navy 1942-46, Lt Cmdr. Photographic Officer Calvin Communications Inc 1946-present, incl Lab

Hedden, William D (Continued)
Supt, Sr V P - Installed color motion picture processing sys; supervised motion picture printing, processing, engrg operations. Soc Motion Picture & Television Engrs - Governor, V Pres, Exec VP, Pres, Past Pres.

Hedger, Harold E
Home: 448 Woodbury Road, Glendale, CA 91206
Position: Consulting Civil Engineer. *Employer:* Self-employed. *Education:* BSc/Civil Eng/Univ of CA. *Born:* Dec 17, 1898 Riverside, Calif. US Navy 1918-19. Employed by Los Angeles Cty Flood Control Dist 1919 & was its Ch Engr 1938-59 when it rose to internatl prominence in fields of flood control & water conservation. Corps of Engrs 1942- 45, attaining Col. Water resources cons in Calif, India, Jamaica, the Philippines, Cyprus & Costa Rica 1959-75. Hon Mbr ASCE 1972; Army Legion of Merit 1950; Army Distinguished Civilian Ser 1959; Golden Beaver Award 1958; Univ Ca Engrg Adv Council 1945-69. *Society Aff:* ASCE, SAME, AGU, LACES.

Hedges, Harry G
Business: Div. of Computer and Computation Research, National Science Foundation, Washington, DC 20550
Position: Program Director. *Employer:* Nat'l Science Foundation *Education:* PhD/Elec Engrg/MI State Univ; MS/Elec Engrg/Univ of MI; BS/Elec Engrg/MI State Univ. *Born:* Oct 7, 1923 Lansing, Mich. Electronics Engr, Wright Air Dev Center 1949- 51; Res Assoc, Willow Run Res Center Univ of Mich 1951-54; InstrAssoc Prof, elec engrg Mich State Univ 1954-68; Asst Dean, Col of Engrg 1968-70; Prof & Chmn, Dept of Computer Sci 1969-84 Sr Staff Assoc, Program Director, Nat'l Science Foundation 1984- . Sr Mbr, Inst of Elec & Electronics Engrs; Chmn, Southeastern Mich Sec IEEE 1974; Amer Soc for Engrg Ed, Chmn, North Central Sec 1968-69; Mbr, Bd/Dir Natl Engrg Consortium 1968-71, 1975-77; Chmn, Computer Sci Bd 1974-75 Mbr 1973-; Mbr Sigma Xi, Eta Kappa Nu. *Society Aff:* IEEE, ACM, ΣΞ.

Hedley, William J
Home: 824 N Biltmore Dr, Clayton, MO 63105
Position: Consultant. *Employer:* Self. *Education:* Civil Engineer (Prof. Degree)/Civil Engg/WA Univ; BS/Civil Engg/WA Univ. *Born:* Nov 6, 1902 St Louis Cty, Mo. Engrg Dept of Wabash RR 1925-63, Ch Engr 1957-63, Asst V P 1963-64; Asst V P Norfolk & Western Railway 1964-67, retired Nov 1967; Cons, US Dept of Transportation 1968-85; Cons, Sverdrup & Parcel & Assocs 1968-present. Reg PE MO, Ill. Pres Engr's Club of St Louis 1950, Hon Mbr 1962, Achievement Award Gold Medal 1967. Pres AREA 1956, Hon Mbr 1965. Pres ASCE 1965, Civil Gov't Award 1969. Engr of Year Award, St Louis Chap MSPE 1964. Chmn Interprof Council on Environ Design 1967. V P St Louis Acad of Sci 1972- 1980. Hoover Medal 1973 from Joint Bd of Award ASCE, AIMME, ASME & IEEE. Bd/Trustees of Washington Univ 1959-62, Alumni Citation 1966, Alumni Achievement Award 1976. Mbr of Tau Beta Pi, Sigma Xi, Chi Epsilon. Natl Pres of Theta Xi Fraternity 1961-64. Mayor City of Clayton 1963-67; Chmn, City Plan Comm 1957-63; Man of the Yr Award 1974. Pres St Louis Cty Municipal League 1966, Mbr St Louis Cty Planning Comm 1958-62. Bd of Dir East-West Gateway Coord Council 1966, Chmn Exec Advisory Comm 1967-72. Exec Comm Hwy Research Bd 1968-71. Bd of Dir, Natl Council of State Garden Clubs 1954-65. Hobbies: gardening, photography, bridge. *Society Aff:* AAAS, ASCE, ΣΞ, TBΠ, AREA

Hedman, Dale E
Home: 1068 Maryland Ave, Schenectady, NY 12308
Position: VP Treasurer. *Employer:* Power Technologies, Inc. *Education:* MS/EE/Univ of NB; BS/EE/Univ of NB. *Born:* 8/22/35. Earned BSEE and MSEE from the Univ of NE. Started engg career in GE Co completing the 3-yr Advanced Engg Prog. Was one of the founders of Power Tech, Inc, in 1969 serving as Principal Engr responsible for equip application specification & presently co VP & Treas. Has served on the staff of the Power Tech Course offered to maj eastern utilities as a 2-yr post grad study prog. Responsible for the design & construction of the advanced Transient Network Analyzer installed at McGraw-Edison and 2 TNA's in the Peoples Republic of China. Responsible for maj engg switching surge studies conducted for elec utilities throughout the world. A Fellow in IEEE, mbr of CIGRE, IEEE rep to ANSI C62. *Society Aff:* IEEE, CIGRE.

Hedrick, Ira G
Business: Bethpage, NY 11714
Position: Presidential Asst for Corp Tech *Employer:* Grumman Aerospace Corporation. *Education:* Grad/Civ Engrg/Princeton Univ; BS/Civ Engrg/Univ of AR. *Born:* Feb 1913. Designer for Waddell & Hardesty. Ch Struc Engr, Eritrean Proj for Johnson, Drake & Piper. Struc Engr, US Army Engrs. Joined Grumman Aerospace Corp 1943: Ch Tech Engr 1957; VP Engrg 1963; Sr V Pres 1970; Presidential Asst for Corp Tech 1975. Respon for optimum alignment of corp's tech skills & resources on current progs & new business dev. Reg P E NY. Mbr Natl Acad of Engrg. Fellow AIAA. AIAA Sylvanus Albert Reed Award 1971; ASME 'Spirit of St Louis Award' 1967. Enjoy flying & tennis. NASA Group Achievement Award 1981; NASA Distinguished Public Service Medal 1984; Dept of the Air Force Exceptional Civilian Service Award 1984. *Society Aff:* NAE, AIAA, AFA, NSPE.

Hedrick, Lewis W
Home: 20724 Quedo Dr, Woodland Hills, CA 91364
Position: Manager *Employer:* Sky Harbor Office Plaza *Education:* BA/Science/Stanford Univ. *Born:* 10/2/26. Manager of Sky Harbor Office Plaza Complex. Oct 1926. Mech Engr, ASTP Stanford Univ, BS 1949. Native of Springfield OH. US Army WWII Philippines & Japan. Built & owned service stations, commercial bldgs & multi-residential structs in CA & TX. Qual control engr for Titanium Metals Corp. Field engr Bechtel Corp on maj petro refining projs. Mgr cost engg, planning & scheduling, proj admin, legal, pubs & computers Bechtel LA Mgr Bechtel European Power Opers. Natl Pres of AACE. Fellow to Inst for Advancement of Engg. Natl Dir of Assoc Commuter Trans. Holder of half dozen pats. Enjoys sports & agri enterprises. *Society Aff:* AACE, IAE, PCEA, ACT, PMI.

Heenan, William A
Home: Rt 1, 11 Carol Ln, Riviera, TX 78379
Position: Prof. *Employer:* TX A&I Univ. *Education:* PhD/ChE/Univ of Detroit; MS/ChE/Univ of Detroit; BChE/ChE/Univ of Detroit. *Born:* 4/17/41. Native of Landsdale, PA, Process Engr for Monsanto, 1964-67, Consultant T Atomic Power Dev Inc 1967-79. Taught at Univ of Puerto Rico, 1969-75, Univ of Houston, 1975-77, TX A&I, 1977-present, Managerial Consultant T Cache Corp, 1975-present, Process Control Consultant T Celanese Chem Corp, 1977-present. Res area is Computer Process Control and gross error detection and reconciliation. Enjoy sailing, hunting and fishing. *Society Aff:* AIChE, ACS, TBΠ.

Heer, Ewald
Home: 5329 Crown Ave, LaCanada, CA 91011
Position: Pres *Employer:* Heer Assoc Inc Engrg Conslt Group *Education:* DrEngSc/Engg Mech-CE/Tech Univ Hannover; CE/Civ Engg/Columbia Univ; MS/Civ Engg/Columbia Univ; BS/Physics/City Univ NY. *Born:* 7/28/30. After receiving a Dr Eng Sc, conducted and managed res and advanced dev projs at McDonnell Douglas and GE's Space Sci Lab before joining JPL in 1966. On assignment to NASA headquarters, played leading role in developing NASA plans for future space teleoperator and robot tech. Prog Mgr for res and dev at JPL until 1984. Assumed current responsibility in 1984. Organized two intl confs for systems tech in 1972 and 1975. Editor for "Mechanisms and Machine Theory–. At USC, Adjunct Prof of Indus and Systems Engg and Dir of the Inst for Technoeconomic Systems. Reg Engr in NY & CA. *Society Aff:* AIAA, ASCE, ASME, IEEE.

Heermann, Dale F
Business: USDA-ARS, Ag Engr Res Center, CSU Foothills Campus, Fort Collins, CO 80523
Position: Agricultural Engineer (Research Leader). *Employer:* USDA-ARS. *Education:* PhD/Agri Engg/CO State Univ; MSAE/Agri Engg/CO State Univ; BSAE/Agri Engg/Univ of NB. *Born:* 3/2/37. Res Agri Engr USDA-ARS. ASAE Paper Award. 1985 IA "Man of the Year–. *Society Aff:* ASAE, IA, ASA, SSSA, USICID.

Heffelfinger, John B
Business: 9233 Ward Pkwy, Suite 285, Kansas City, MO 64114
Position: Owner (Self-employed) *Education:* MS/EE/OH State Univ; BS/EE/Univ of KS. *Born:* 7/27/17. in Arkansas City, KS, son of John Byers and Lucile W (Parmenter) Heffelfinger. Employed by Collins Radio co, 1940-45, as Broadcast Equipment Engr and Chief Field Engr. Taught as Asst Prof of Physics at Park College, 1946-47. Since 1947 has owned and operated Consulting Communications Engg firm, specializing in Broadcasting and Telecasting, Reg in MO & NB. Avocational interests include horticulture, photography. *Society Aff:* IEEE, NSPE.

Hefner, Jerry N
Home: 108 Kittywake Dr, Newport News, VA 23602
Position: Supervisory Aerospace Engrg *Employer:* NASA *Education:* MS/Aerospace Engrg/NC State Univ; BS/ME/Old Dominion Univ *Born:* 7/3/44 Native of Norfolk, VA. Employed at NASA Langley Research Ctr since 1966 as Aerospace Engr. Head Civil Aircraft Branch, Advanced Vehicles Division. Also Manager of Langley Transport Technology and Viscous Drag Reduction Programs. Author/co-author of 60 scientific publications/presentations. *Society Aff:* AIAA

Hefter, Harry O
Business: 180 N Wabash Ave, Chicago, IL 60601
Position: Pres *Employer:* HOH Engrs, Inc. *Education:* BS/CE/Univ of IL *Born:* 5/7/29 Structural engr, Harza Engrg Co, 1951-53; Ch structural engr, Williams Engrg Co, 1953-54; Structural proj capt, Childs & Smith, 1954-56; Hd structural design, steel plants & bldg dept, Maccabee, Campbell & Assoc, 1956-57; Ch structural engr, Frank Klein Co, 1957-59; Foundry and chrmn of Harry O. Hefter Assoc, Inc, Chicago, 1959-present, HOH Construction Mgrs, Inc, 1973-present, HOH Engrs, Inc, 1974-present, Hefter Industries, Inc, 1974-present. Recipient award for engrg Inland Steel's Continuous Slab Caster and Water Pollution Control Facility, chosen "One of the Ten Outstanding Engrg Achievements in the US" by NSPE. *Society Aff:* NSPE, WSE, WSBFCPA, AISE, AFS

Hegbar, Howard R
Home: 353 Deepwood Dr, Wadsworth, OH 44281
Position: V Pres, Res & Engrg. *Employer:* Goodyear Aerospace Corp. *Education:* PhD/Elec Engr/Univ of WI; MS/Elec Engr/Univ of WI; BS/Elec Engr/ND State Univ. *Born:* 2/22/15. Res engr RCA Labs. 1941-46 in dev of UHF power tubes & microwave magnetrons. Dev Engrg Mgr Goodyear Aerospace Corp 1946- . Appointed VP for Res & Engrg 1974. Retired 1977. *Society Aff:* ΣΞ, TBΠ, IEEE.

Hegedus, L Louis
Business: W. R. Grace & Co, 7379 Route 32, Columbia, MD 21044
Position: VP, Inorganic Research *Employer:* W.R. Grace & Co., Res Div *Education:* PhD/ChE/Univ of CA, Berkeley; MS/ChE/Tech Univ, Budapest, Hungary *Born:* 4/13/41 Native of Budapest, Hungary. Experience includes the Research Inst for the Organic Chemical Industry (Budapest, Hungary, 1964-65), the Daimler-Benz AG (Mannheim, West Germany, 1965-68), and the Gen Motors Research Labs (Warren, MI, 1972-1980) where he was in charge of Catalysis Research. Assumed position of Dir, Inorganic Research Dept, Research Div, W.R. Grace & Co, Columbia, MD, in October, 1980. Promoted to current position of VP, Inorganic Res, in March 1984. Oversees research in catalysis, electrochemistry, technical ceramics, and construction materials. Allan P. Colburn Lecturer, Univ of DE, 1975; Chemical Eng of the Year, AIChE Detroit Section, 1978; AIChE Professional Progress Award, 1980; Leo Friend Award, ACS, 1981; Chrmn, AIChE Detroit Section, 1979-1980; Member of the Editorial Bd, AIChE Journal 1978-83; and Consulting Editor, 1985- ; Director, ISCRE inc. (Internatl. Symporia on Chemical Reaction Engineering); Scientific Council, Maryland Academy of Sciences; Advisory Boards at Berkeley, Princeton, Northwestern and Wisconsin; about 55 publications and patents in reaction engrg and catalysis. *Society Aff:* AIChE, ACS, ΣΞ, MRS.

Heger, James J
Home: 3317 Crestview Drive, Bethel Park, PA 15102 *Education:* BS/Met Engg/Carnegie Inst Tech; MS/Met Engr/Carnegie Inst Tech. *Born:* Feb 1918. Res metallurgist Babcock & Wilcox Tube Co. 1955- , U S Steel Corp: res engr, Ch Staff Engr & Ch Res Engr US Steel Primary Forming. Respon for initiation, coord & plan of stainless steel & hot rolling res. Now retired and a part-time instructor at Carnegie-Mellon University. Mbr Sub-Ctte Power Plant Matls NACA. Former Chmn ASTM, ASME, MPC Joint Ctte on Effect of Temp of Properties of Metal. Twenty tech pubs; co-author of book on stainless steel. Reg Prof Engr. ASTM Award of Merit, Fellow ASM & ASTM. *Society Aff:* ASM, AIME, ASTM, NACE, MPC, AISE.

Hehemann, Robert F
Business: Dept Met & Matls Sci, Cleveland, OH 44106
Position: Professor. *Employer:* Case Western Reserve University. *Education:* PhD/Met/Case Inst Tech; MS/Met/Case Inst Tech; BS/Met/Univ of MI. *Born:* Feb 1921. US Army 1943-46. Mbr of faculty Case Western Reserve Univ since 1946. Res Assoc Argonne Natl Lab summer 1965. Acting Hd Dept of Met 1967- 69. Mbr ASM, AIME, NACE, Sigma Xi, Tau Beta Pi. Bradley Stoughton Award ASM 1958 Fellow ASM. *Society Aff:* ASM, AIME, NACE, WACEW, AAAS

Hehn, Anton H
Business: 8060 N. Lawn Dale Ave, Skokie, IL 60076
Position: President *Employer:* Hehn & Assoc *Education:* BS/Mech Engrg/IL Inst of Tech; MS/Mech Engrg/NW Univ. *Born:* Feb 1937. Proj Engr from AMF Inc involved in the design of tape controlled, hydraulically actuated versatile transfer machine. 1962-68 with IIT Res Inst as Mgr, Fluid Sys & Lubrication Sect; respon for numerous programs in the fluid sys, controls & mechanisms of failure area. With GARD Inc since 1968. Assumed current respon as program Mgr of the automated sys dept in 1972. Engaged in a wide variety of projs dealing with machine design, mech power & control sys dev & hydraulic/pnuematic sys & component design. Presented numerous seminars in the USA & Japan dealing with fluid power. Pres Fluid Power Soc; Bd/Dir Fluid Power Soc; Mbr Fluid Power Co-ordinating Council. Dir of Engg For Graymills Corp, Manufacturer of Pumps and Cleaning Sys. Presently, Pres of Hehn & Assoc, Conslts in Fluid Power Sys and Controls and Pres of Shin Pyung USA, Inc, Mfgs of Hydraulic and Pneumatic Cyclinders, Presses, and Sys. Enjoy golf & swimming. *Society Aff:* FPS, SME, SAE.

Heiberg, Elvin R, III
Business: Ohio River Div, Cincinnati, OH 45202
Position: Division Engineer. *Employer:* US Army Corps of Engrs. *Born:* Mar 1932. BS USMA 1953; MSCE MIT; MA Govmt & MS Admin GWU. Var Army assigns through Exec to Sec of Army 1973, New Orleans District Engr 1974, promotion to BGen & Hd, Ohio River Div Corps of Engrs 1975. Other assigns incl R&D 1958-61, West Point Faculty 1965-68, Exec Office of Pres 196970, 1971. Awards: Silver Star, 3 Legions of Merit. Regional V Pres Ohio Valley, Soc of Amer Military Engrs 1976. Handball, raquetball. Father ret BGen, headed Dept of Mechs West Point 1969. Wife Kitty from Leavenworth, Kansas. 4 children. Nickname: 'Vald'.

Heid, Charles C,
Business: Naval Facilities Engg Command, Norfolk, VA 23511
Position: Rear Admiral-Cdr Atlantic Div. *Employer:* Commander, Atlantic Div. *Education:* MSCE/Civil Engg/Rensselaer Polytechnic. *Born:* 11/3/25. in El Paso, TX. 2 yrs TX College of Mines; grad US Naval Acad June 1947; BSCE, MSCE Rensselaer Polytech Inst. Commanded Mobile Const Battalion ASEVEN beginning 1963. During command battalion was selected twice "Best of Type" of Atlantic Fleet battalion. Battalion awarded Peltier Award 1963 as Navy's outstandin const battalion. Other commands incl: C O Const Battalion Ctr Davisville RI, Cdr 21st Naval Const Regiement, CO No Div NAVFACENGCOM Philadelphia PA, current Cdr Atlantic

Heid, Charles C, (Continued)
Div NAVFACENG Com. Reg Prof Engr TX. Regional VP SAME, Mbr of Center of Engg Bd of Advisors for Widener Univ, Chester, PA. *Society Aff:* SAME, APWA.

Heidema, Peter B
Business: 33 Baldwin Hill Rd, Littleton, MA 01460
Position: Consulting Engr (Self-employed) *Education:* CE/CE/Tech Univ, Delft, The Netherlands *Born:* 3/21/98 1936-1939 Head of Soils Testing, Fort Peck Dam, MT for Army Engrs. 1940- 1943 Head of Soils Testing, Galveston District, Corps of engrs, U.S.A. Houston, TX. 1943-1950 Head of Soils Testing and Soils Design, Galveston, District, C. of E. U.S. Army, Galvaston, TX. 1950-1956 Head of Soils Testing and Soils Design, Ft Worth District, C. of E. U.S. Army, Ft. Worth, TX. 1956-1957 Assistant Chief Soils Design, Passaonaquoddy Survey for New England and Division Corps of Engrs, U.S. Army in Boston, MA. 1957-1958 Supervisor, Soils Testing for Passamaquoddy Survey at Eastport, ME for Corps of Engrs, U.S. Army. 1958-1959 Head of Soils Design, Passamaquoddy Survey in Boston, MA and Waltham, MA for Corps of Engrs, New England Division. 1959-1968 Head of Soils Section of Foundations and Materials Branch in New England Division, Corps of Engrs, U.S. Army at Waltham, MA. 1955-1956 VP, Fort Worth Chapter, TX Society of PE. 1968-1981 Registered PE and Professional Land Surveyor, specializing in design and construction of earth dams, dikes and levees, continued thru 1984. *Society Aff:* NSPE, ASCE, ASCM, USCOLD, ISSMFE.

Heidlage, Robert F
Home: PO Box 781, Claremore, OK 74017
Position: Area Engr. *Employer:* USDA Soil Conservation Serv. *Education:* BS/Agri Engr/OK State Univ. *Born:* 6/30/37. Native of Chickasha, OK. Worked in farm equip sales for two yrs after college. Employed by Soil Conservation Services since 1962. Responsible for engg phase of conservation practices installed with assistance from SCS in Northeast OK. *Society Aff:* ASAE.

Heikes, Russell G
Business: Ind & Systems Engg, Atlanta, GA 30332
Position: Assoc Prof. *Employer:* GA Inst of Tech. *Education:* PhD/IE/TX Tech Univ; MS/IE/Univ of NB; BS/ME/Univ of NB. *Born:* 8/29/46. Joined ISyE faculty at GA Tech in 1972. Promoted to current rank in 1977. Have taught grad and undergrad courses and directed student res in the areas of quality control, inventory, applied statistics and OR. Have done res projs for DOT and the US Army. Published in *IIE Transactions, Management Science, Simulation* Policy Sciences & *Industrial Engineering*. Dir of Quality Control and Reliability Engg Div of IIE, 1979-81 and have served in various other div offices of IIE. *Society Aff:* IIE, ASA, ASQC, TIMS.

Heil, Charles E
Business: 300 Oceangate, Long Beach, CA 90801-5617
Position: VP *Employer:* ARCO Marine, Inc *Education:* MBA/Management/Temple Univ; BS/Naval Arch & Marine Engrg/Univ of MI *Born:* 7/7/29 Native of Cleveland, OH. Served as engrg officer in US Navy after graduation from MI for two years. Subsequent to discharge employed as lead naval architect at Bethlehem Steel Co, Baltimore yard for three years. With ARCO Marine, Inc, since 1957 in various positions, including naval architect, marine superintendent, mgr Evaluation and Chartering, Mgr Engrg, Construction and Maintenance. Represented Atlantic Richfield Co on 1969 voyage on MANHATTAN through the Northwest Passage. Appointed VP January 1980. Past Chrmn Los Angeles Section SNAME. *Society Aff:* SNAME

Heil, Dick C
Home: 9605 Flower Avenue, Silver Spring, MD 20901
Position: President *Employer:* Kappe Associates, Inc. *Education:* MSE/Sanitary Engrg/Univ of MI; BSSE/Sanitary Engrg/PA State Univ. *Born:* 5/14/25. Native of Coalport, Pa. Employed by Pa Dept of Health 1947-62. Ser as Regional Sanitary Engr 1954-62 in Pittsburgh Region. Sanitary Engr with Kappe Assocs Inc 1962-date. Exec VP since 1973, Pres 1978. Pres PA WWOA 1964-65. Secy- Treas Chesapeake Sect AWWA 1964-80. Diplomate & Staff Mbr AAEE. Mbr: Fellow ASCE, NSPE, WPCF, WPCA of Pa. International Director '81-'84 AWWA, President 1982 Chesapeake WPCA. Harry J Krum Award of PWWOA 1976. George Warren Fuller Award of AWWA 1977. Patentee. Reserve Officer, USPHS. Hobbies: Fishing, hunting & golf. Dir, Men's Guild, Holy Cross Hospital 1974-date. Mbr, Elks, Congressional County Club; Kenwood Golf & Country Club. Rosebud Gun Club. *Society Aff:* ASCE, NSPE, WPCF, AWWA, AAEE.

Heil, Terrence J
Business: 600 Stewart Street Suite 303, Seattle, WA 98101
Position: President *Employer:* Wieland Lindgren & Assoc Inc *Education:* MBA/Business/University of Michigan; BS/ME/S. Dak. School of Mines & Tech. *Born:* 06/17/33 Native of South Dakota. Worked for the Rapid City Journal as clerk and driver while attending high school and college. Engineering career sequence: Results Engr, Montana Dakota Utility Co., Mandan, ND; Results Engr, New York State Electric & Gas Corp., Binghamton, NY; Project Engr, Cummins & Barnard Inc., Ann Arbor, MI.; Mgr of Mechl Engrg Facilities Div., Commonwealth Assoc., Jackson, MI; VP, Power Systems, Orr Schelen Mayeron, Minneapolis, MN; Dir Mech Engrg, John Graham Company, Seattle, WA; currently Pres Wieland Lindgren & Assoc., Seattle, Wa. *Society Aff:* ASME, TAPPI, NSPE, SAME, AIPE

Heilmeier, George H
Business: 1400 Wilson Blvd, Arlington, VA 22209
Position: Director. *Employer:* Defense Advanced Res Projs Agency. *Born:* May 1936 Philadelphia, Pa. BS Elec Engrg Univ of Pa 1958; MSE, MA, PhD Solid State Materials & Electrons Princeton 1960-62. 1968 Head of Solid State Device Res at RCA Labs. 1970 Selected by the Pres of the US for White House Fellowship. 1971 Appointed Asst Dir of Defense Res & Engrg (Electrons & Phys Sciences). Jan 1975 Named Director Defense Advanced Res Projs Agency. Awards: IR-100 1968-69; Eta Kappa Nu Outstand Young Elec Eng 1968; David Sarnoff Outstanding Team Award 1969; Eta Kappa Nu Outstanding Young Elec Engr in the USA 1969; IEEE David Sarnoff Award 1976.

Heim, George E
Business: 125 Windsor Dr, Ste 107, Oak Brook, IL 60521
Position: Principal-in-Charge *Employer:* Harding-Lawson Assocs *Education:* PhD/Geol/Univ of IL; MS/Geol/Univ of MI; BS/Geol/Univ of MI *Born:* 3/25/34 in Buffalo, NY. Investigation of ground water resources and soils for MO Geological Survey. Teaching and research at Univ of WI - Milwaukee including soil mechanics and engrg properties of soils. Investigations of foundations for dam sites and rock tunnels at Harza Engrg Co. Geology, ground water, seismology, and foundation engrg for fossil and nuclear power plants including professional testimony before regulatory agencies at Dames & Moore and Sargent & Lundy. Principal-in-Charge of Chicago office of Harding-Lawson Assocs, consultants in geotechnical engrg. Chrmn of two ANS standards writing group. *Society Aff:* ASCE, ANS, ISSMFE, ISRM, USCOLD

Heimbach, Clinton L
Business: Dept of Civ Engg, PO Box 7908, Raleigh, NC 27695-7908
Position: Prof and Assoc Hd, Dept of Civ Engr *Employer:* NC State Univ. *Education:* PhD/Civ Engg/Univ of MI; MSCE/Civ Engg/Purdue Univ; BSE/Civ Engg/Univ of MI. *Born:* 2/2/21. Native of Lorain, OH. Served with Army Corps Engrs 1942-47 and 1950-52. Field engr Santa Fe Railway 1947-50 and roadmaster, 1952-56, specializing in maintenance of way problems and supervision. Faculty mbr, depts of civ engg Univ MI and OK State Univ. With NC State Univ in civ engg since 1965 specializing in transportation. Reg PE in two states. US Dept. Trans Faculty Fellow, Wash, DC, 1981-82. Appointed Assoc Dept Hd, 1984. Enjoys model railroading. *Society Aff:* ASCE, ITE, ASEE, AREA, TRB.

Heimlich, Barry N
Home: 3210 N. 36 St, Hollywood, FL 33021
Position: Pres *Employer:* Hi Medics, Inc. *Education:* MSE/ChE/Univ of MI; BChE/ChE/Poly Inst of Brooklyn. *Born:* 1/14/41. New York City 1/14/41. Process Res Engr. Esso R&E 1962-66. Joined Pfizer, Inc in 1966 as Dev Engr. Named Mgr-Chem Process Dev 1968. Group Mgr-Mfg 1971-73 respon for all domestic pharmaceuticals mfg. Named Group Mgr-Prod Engrg in 1973 in charge of Mfg Engrg, Indus Engrg & Facilities Engrg. Joined Key Pharmaceuticals Inc in 1974 as V P - Prod. Corporate Officer. Promoted to Sr VP- Corp Planning in 1982. Founder of HiMedics, Inc in 1984, engaged in pharmaceutical development. 4 U S Pats. Author of Tech Articles. Dev controlled drug delivery systems, including Theo-Dur, the world's leading drug used to treat asthma and Nitro-dur, a leading transdermal nitroglycerine patch for angina. Co-inventor of drug dispensing devices. Dev more economical processes for mfg of several important food additives & pharmaceuticals. Directed expansion of Key's production and corporate facilities. Negotiated numerous licensing, distribution, and merger agreements. Directed strategic planning. Tau Beta Pi, Amer Inst Chem Engrs. Outstanding Young Men of Amer 1975. Who's Who in South & Southwest since 1978. *Society Aff:* AIChE, SAA

Heinemann, Edward H
Home: Box 1795, Rancho Santa Fe, CA 92067
Position: President. *Employer:* Heinemann Associates. *Born:* Mar 14, 1908. Dr of Sci Northrop Univ. Douglas Aircraft 31 yrs, draftsman to V Pres Engrg, combat aircraft. Supervised design of SBD Dauntless, R3D-1 (DC-5) DB-7 Boston, A-20 Havoc, A-24, A-26 Invador, XCG-7 & XCG-8 Gliders, SB2D-1 & BT-1 Destroyers, TB2D Devastators, D-558 Skystreak, D-558-2 Skyrocket, F3D Skynight, A2D Skyshark, A3D Skywarrior, F4D-1 Skyray, F5D-1 Skylancer, A4D Skyhawk. 10 yrs Genl Dynamics as VP Engrg. Hon Fellow Royal Aero Soc, Amer Astro Soc, Hon Fellow Soc of Weight Engrs, Hon Mbr Tau Beta Pi, Mbr Assn Francaise des Ingenieurs et Techniciens, Amer Ord Assn, Soc of Automotive Engrs, Amer Soc of Naval Engrs, Soc of Naval Architects & Marine Engrg, Naval Inst, Natl Acad of Engrg, Aviation Hall of Fame, Natl Aeronautical Assn, Benjamin Franklin Royal Soc of Arts.

Heinemann, Heinz
Business: Lawrence Berkeley Lab, Univ of CA, Berkeley, CA 94720
Position: Sr Scientist. *Employer:* Univ of CA. *Education:* PhD/Phys Chem/Univ of Basel, Switzerland. *Born:* Aug 21, 1913. Mbr, Natl Acad Engrg 1976; E V Murphree Award in Indus & Engrg Chem ACS 1971; Eugene J Houdry Award in applied catalysis 1975; Distinguished lecturer in catalysis, AIChE 1970-71; Mbr: ACS, AIChE, AICh, Royal Soc, many others. About 150 pubs & pats. Exec Editor Catalysis Reviews. Indus connections: 1938-40 Ch Res Chemist, Rodessa Oil & Refining Corp, Shreveport, La; 1940-41 Danciger Oil & Refining Pampa Texas; 1941-48 Attapulgus Clay Co Philadelphia PA; 1948-57 Houdry Process Corp Marcus Hook PA; 1957-69 Dir Chem & Engrg Res The M W Kellogg Co; 1969-78 Mgr Catalysis Res & Mgr Res Contracts Mobil Res & Dev Corp Princeton NJ; 1978-present Staff Sr Scientist, Lawrence Berkeley Lab and Lecturer, Chem Eng, Univ of CA Berkeley Taught 1 year at Carnegie Inst of Tech. P Pres 'The Internatl Congress on Catalysis' many offices in prof societies. *Society Aff:* NAE, ACS, AICHE, AIC, Cat Soc, AAAS

Heinen, James A
Business: 1515 W. Wisconsin Ave, Milwaukee, WI 53233
Position: Prof *Employer:* Marquette Univ *Education:* PhD/EE/Marquette Univ; MS/EE/Marquette Univ; B/EE/Marquette Univ *Born:* 6/23/43 Native of Milwaukee, WI. Joined the faculty of Marquette Univ in 1969. Currently Prof of Electrical Engrg and Computer Science. Served as Chrmn of the Dept of Electrical Engrg during the period 1973-1976. Received Eta Kappa Nu. C. Holmes MacDonald Award for outstanding electrical engrg teaching in 1974 and the IEEE Milwaukee Section Memorial Award in 1981. Is a registered PE in the state of WI. *Society Aff:* IEEE, ASEE

Heiner, Clyde M
Business: 141 E First S, PO Box 11865, Salt Lake City, UT 84147-0865
Position: Sr. V.P. Corporate Development *Employer:* Questar Corporation *Education:* MBA/Bus Admin/Stanford Univ; BS/Structural Engg/Columbia Univ; BA/Liberal Arts & Humanities/Columbia College. *Born:* 4/4/38. Native of Ogden, UT. Current positions include Sr V P Questar Corporation; Pres and C.E.O., Questar Development Corp, Interstate Land Corp, and Interstate Brick Co. Responsibilities encompass investigation of internal and external opportunities for corporate growth and development, as well as overall management responsibility of a brick manufacturing facility. Previously held position of V P, Engg at Questar Corp, responsible for engg functions, regulatory affairs, consumer information and research. Previously employed as Assist Admin, Univ of Utah Hospital, responsible for general services, facilities, planning, patient accounting system design and operation. Member of Pacific Coast Gas Assoc. P Pres of Salt Lake Chapter AIIE. Mbr of the Bd, Salt Lake Chapter of the American Red Cross. Mbr of the Bd, College of Engrg Advisory Committee. Enjoy camping, gardening and family activities. *Society Aff:* PCGA.

Heins, Conrad P
Business: Civ Engg Dept, College Park, MD 20742
Position: Prof. *Employer:* Univ of MD. *Education:* PhD/Structures/Univ of MD; MSCE/Structures/Lehigh Univ; BSCE/Civ Engr/Drexel Univ. *Born:* 9/13/37. Author of over 150 referred technical articles & author of four text books: 1. "Bending/Torsonal Design" - D C Heath 1975. 2. "Applied Plate Theory" - D C heath 1977. 3. "Design of Modern Steel Highway Bridges" - J Wiley 1979. 4. "Design/Analysis of Structures" - M Dekker 1980, is Editor-in-Chief of "Civ Engr Design Journal–. M Dekker received Wash Acad of Sci 1976 Scientific Achievement Award in Engg; Soc of Sigma Xi - Annual Award for Scientific Achievement - MD Chapter; is mbr of Sigma Xi, Tau Beta Pi, CHIE ps, lou, Phi Kapp Phi; was Visiting Prof 1979 Japan Soc for Promotion of Sciences & is a PE in MD, VA, NJ & PA. Chrmn of Flexoral Mbrs, Comm of Metals ASCE, Chrmn TRB Committee on Dynamics & Bridge Field Testing. Enjoys fishing, hunting. *Society Aff:* ASCE, ACI, IABSE, AREA, SESA.

Heins, Robert F
Business: TAMS Bldg, 655 Third Ave, New York, NY 10017
Position: Partner. *Employer:* Tippetts-Abbett-McCarthy-Stratton *Education:* MSCE/-/Harvard Univ; BCE/-/Cooper Union; Bus Admin/-/Alexander Hamilton Inst; -/IBM Computg for Civ Engrg Execs/-. *Born:* Feb 1925. BSCE Cooper Union 1948; MSCE Harvard Univ 1949; Alexander Hamilton Inst, Bus Admin 1962; IBM-Computing for Civil Engrg Execs 1969. Reg PE & LS in New York. Reg PE in Alaska & Kenya. Fellow, Zimbabwe Inst of Engrs, Fellow, Amer Soc of Civil Engrs. Mbr, Harvard Engrs & Scientists, Amer Cons Engrs Council, USCOLD 1949, Instructor, NE Univ. 1948-61 occupied pos in Turkey, Spain, USA & Quebec as Soils & Paving Engr, Design & Const Engr, Ch Engr & Civil Engr. 1961 joined TAMS in Columbia & subsequently directed various US & Latin America Projs. 1972 appointed Assoc Partner. 1973 became Partner, respon for transportation & water resources dev projs in Middle East, Africa & Latin America. *Society Aff:* ASCE, ACEC, USCOLD, NYACE.

Heinsohn, Robert J
Business: Dept Mech Engg, Univ Park, PA 16802
Position: Prof Mech Engg. *Employer:* Penn State Univ. *Education:* PhD/ME/Michigan State Univ; SM/ME/MIT; BS/ME/RPI. *Born:* 8/28/32. I have supervised approximately 35 masters and doctoral theses and published over 100 papers in profl journals, magazines, symposia, etc. on air pollution control, industrial ventilation and conbustion. I am actively engaged in securing res support from govt and industry and have been the Principal Investigation for approximately 30 grants or contracts and a collaborator on several more. I am also a reg engr (PA)

Heinsohn, Robert J (Continued)
and cofounder and partner in Envirotherm, a private consulting firm in central PA. *Society Aff:* ASME, ASHRAE, APCA, AES.

Heintzleman, Walter G
Business: 570 Beatty Rd, Monroeville, PA 15146
Position: VP *Employer:* GAI Consultants, Inc *Education:* M/PA/Univ of Pittsburgh; BS/CE/Carnegie-Mellon Univ *Born:* 7/26/35 VP and Mgr, Civil/Structures/ Geotechnical Group at GAI Consultants since 1980. Responsible for engrg management of transportation, site development, geotechnical, structural engrg programs. PE in PA and NJ. Formerly Chief Engr at Urban Redevelopment Authority of Pittsburgh, managed planning/design/construction of $40 million in site improvements. Later, Engrg Mgr for Port Authority of Allegheny County (PA), responsible for design and construction of $150 million busway program. Past President Pittsburgh Section ASCE and Pittsburgh Chapter NSPE. *Society Aff:* ASCE, NSPE, ESWP, PMI, ASHE.

Heinz, Winfield B
Home: 922 S Barrington, Los Angeles, CA 90049
Position: Owner. *Employer:* Self. *Education:* MS/ME/GE Adv Course in Engg. *Born:* 3/5/02. in MN. BS in EE Univ of WA 1926; Sigma Xi, Pres Tau Beta Pi; electrician Wash Water Pwr Co; GE Co 3 yr Adv Crse Engg equiv MS in ME; Asst to Ch Eng Aero & Marine Engg Dept; m 1930 Rachel Edwards Clarke, one dau; Am Cyan Co Calco Div hd proc & proc equip dev & of Instr Engg Sec, Asst Ch Engg; Johns Hopkins Univ oper Res Off & Joint Chfs of Staff Weapons Sys Eval Gr; Ind consul engg NJ, PA, VA Reg ME those states. Work princ instr & aut control; Ch eng Askania Reg Div Gen Prec Equip Corp Chicago; Reg ME IL; Mgr Engg Instr Products Div Ampex Corp Redwood City CA; Consul US COmm Dept Indian Indus Fair N Delhi; Mem Ford found internat team for study small-scale indus, Gov of India; Lecturer & res eng UCLA 63-69; Fulbright Scholar Univ Rijeka Yugoslavia; currently ENVI solar panel & sys design, mfr & sale as ind enterprise;. *Society Aff:* ASME, ISA, NSPE, CSPE, SEIA, IAE.

Heiple, Loren R
Home: 1492 Century Dr, Fayetteville, AR 72703
Position: Emeritus Professor of Civil Engg. *Employer:* University of Arkansas. *Education:* BSCE/Civil Engg/IA State Univ; MS/Sanitary Engg/Harvard Univ; CE/ Civil Engg/IA State Univ; PhD/Civil Engg/Stanford Univ. *Born:* April 1918. Native of Hinsdale, Ill. Prior to WWII Ser & Dev Engr with Infilco Chicago; Jr San Engr Ia Ord Plant & Instruc Iowa St Coll. Cpt in Sanit Corps 1942-46 & ret Col US Army reserves 1975. Asst Prof Iowa St Coll 1946-48; City Engr Boone IA 1948-49; Cons Engr Public Admin Serv Chicago 1949-50. Prof & Hd Civil Engrg Dept 1950-71 & Dean of Engrg 1971-79 at Univ of Arkansas. Retired to rank of Professor Emeritus, 1984. Reg Prof Engr Iowa & Arkansas. Mbr of Arkansas Bd of Reg for Engrs & Land Surveyors 1972-84. *Society Aff:* ASCE, WPCF, CE, ASEE.

Heirman, Donald N
Business: AT&T Information Systems, Crawfords Corner Rd, Holmdel, NJ 07733
Position: Supervisor, Electromagnetic Compatibility/Product Safety. *Employer:* AT&T Information Systems. *Education:* MS/EE/Purdue Univ; BS/EE/Purdue Univ *Born:* 8/16/40 Born in Mishawaka, Indiana; currently supervises the Electromagnetic Compatibility/Product Safety Group which has responsibility for regulatory compliance interpretations, testing, and associated mitigation consultation as well as providing appropriate liaison with product safety underwriters; his IEEE activities include: Fellow of the Institute, Life Member of the EMC Society, Chairman of the Society's Standards Committee, and electromagnetic Measurements Technical Committee; Past President and Vice President of the Society; chairs several committees of Accredited Standards Committee C63 of the American National Standards Institute; lectures widely on EMC RF environments, emission and immunity measurement techniques, and test facility planning, construction and improvements. *Society Aff:* IEEE

Heiser, Will M
Business: 1617 J F Kennedy Blvd, Suite 1760, Philadelphia, PA 19103
Position: VP. *Employer:* O'Brien & Gere Engrs, Inc. *Education:* Grad Courses/Struct Engg/Univ of CO; BCE/Civ Engg/Geo Wash Univ; AB/Engg Math/Oberlin College. *Born:* 7/31/17. Hamilton, OH. Corps of Engrs 1939-40 Cincinnati, OH. Bureau of Ships 1940- 48 Wash, DC. Bureau of Reclamation 1948-53 Denver, CO. AID 1953-55 New Delhi, India. Justin & Courtney 1956-61, Phila, Iran, Pakistan. Tipton & Kalmbach 1961- 64 Pakistan. Partner Justin & Courtney 1964-73, Phila, Iran. VP Justin and Courtney, Inc 1973-78, Phila, Korea. VP O'Brien & Gere Engrs, Inc 1978-. ASCE Phila Section Dir 1972-80, Pres 1976, Natl Committee on Employment Conditions 1979-83. ACEC Officer Phila Section & PA Council. ICID Mbr 12th congress Comm 1979-84. Mbr ACI, ASTM, NSPE, USCOLD, AWRA, WWA, WPCF, APWA, SAME, ASTM. Former Chrmn New Britain Township Planning Commission. Former Trustee & Chrmn Bldg Comm Chalfont Methodist Church. Reg PE. 1974 Engr of the Yr. Bucks County PSPE. 1979 Outstanding Civil Eng Phila Sect ASCE. Wife Audrey, children Marcia, David & Charles. *Society Aff:* ASCE, ASTM, NSPE, SAME, WPCF.

Heitlinger, Igor
Business: 28 Richard Place, Trumbull, CT 06611
Position: Principal *Employer:* Heitlinger & Assoc *Education:* MS/ME/Univ of PA; BS/ME/City Coll of Cuny *Born:* 10/14/46 Over fifteen years experience in the chem process, utility and manufacturing industries, with responsibilities for the design and specification of mechanical systems and equipment from conception to startup. Prior to starting his own firm, he served as senior mechanical engr and project engr for the following major coprs: Olin Corp, United Engrs & Constructors Inc and Westinghouse Electric Corp. He is a member of the American Society of Mechanical Engrs, the Instrument Society of America, Pi Tau Sigma and the National Society of PE. He is registered PE in CT and NJ. *Society Aff:* ASME, ISA, NSPE, AIPE

Heitz, Robert G
Home: 3413 El Monte Drive, Concord, CA 94519
Position: Consultant *Education:* BS/Chemistry/CA Inst Tech. *Born:* Oct 1915. Res Chemist for Dow Chem Co Midland, Mich 1936-39. R&D Engr Dow, Pittsburg & Calif 1940-52. Dev of thermal chlorination processes. Dir of Res Western Div Dow 1952-70. Tech Director 1970- . Dev of hollow-fiber membrane devices. Personal achievement award, McGraw-Hill 1972. First award for Chem Engrg Practice AIChE 1974. John Fritz Medalist 1978. Consultant, 1980.. *Society Aff:* ACS, AAAS, AIChE, Rotary.

Hekimian, Norris C
Business: 11004 Homeplace La, Potomac, MD 20854
Position: VP, Research. *Employer:* Hekimian Labs, Inc *Education:* PhD//Univ of MD; MS//Univ of MD; B/EE/Geo Washington Univ *Born:* 1/14/26 Washington, DC. Radio Engr, Nat'l Bureau of Standards 1949-54; Branch Chief, Nat'l Security Agency 1954-1961; Ass't Dir of R&D Page Communications Engrs, 1961-68; Founder and Pres, Hekimian Labs, Inc, 1968-1981; Chief Executive Officer, Hekimian Labs, Inc 1981-83, VP, Research 1983-87; Consultant, Boggs and Hekimian, 1953-1960. Served in USAAF 1944-45. Recipient Geo Washington Univ Alumni Achievement Award, 1976. Member of IRE/IEEE since 1948. Elected Fellow of IEEE, 1980. Patron Award, Washington Section IEEE, 1979. Contributed articles in prof journals. Holds two dozen patents. Registered PE (MD) since 1969. *Society Aff:* IEEE, ΣΤ, HKN

Helander, Linn
Home: 3 So 615 Circle Dr, Warrenville, IL 60555
Position: Ret. (Prof Emeritus, Kansas State Univ.) *Education:* BS/Mech Engg/Univ of IL. *Born:* Aug 1891. During 1920's & early 1930's designed thermal sys for power generation. Employed by Westinghouse, UGI Contracting Co & Champion Fibre Co. Thereafter, engaged in teaching & res: Univ of Pittsburgh 2 yrs; Prof Mech

Helander, Linn (Continued)
Engrg, Kansas State Univ 24 yrs; Hd of Dept 22 yrs; Instituto Tecnologico de Aeronautica, Brazil 5 yrs. Cons on ASHRAE-AMCA Ctte respon for 1974 test standards for rating fans. P V Pres of ASME; E K Campbell Award & Distinguished Serv Award both from ASHRAE. Honored by an inscribed plaque in heat-transfer lab of ITA. *Society Aff:* ASMe, ASHRAE, ASEE, AAAS.

Helberg, Paul W
Home: Route 1, Box 97, Billings, OK 74630
Position: Sole Proprietor. *Employer:* Self-employed. *Education:* MS/Agri Engg/OK State Univ; BS/Agri Engg/OK State Univ. *Born:* 11/5/48. I am a native of Breckinridge, OK. I was employed by Miller-Newell Engrs, Ltd, Newport, AR, as a Proj Entr from 1971 to 1978 and was named an Assoc in 1977. Since Jan 1979, I have been a self-employed Consulting Engr in Billings, OK. I am a PE reg in both OK and AR. I was chrmn of ASAE, AR Sec, 1978-79. I received the award for Outstanding Young Agricultural Engr in AR for 1974 and the 1978 Southwest Region award for Distinguished Young Agricultural Engr. *Society Aff:* ASAE, NSPE.

Heldman, Dennis R
Business: National Food Processors Assn, 1401 New York Ave, NW, Washington, DC 20005
Position: Executive Vice President - Scientific Affairs *Employer:* National Food Processors Assn. *Education:* PhD/Agri Engr/MI State Univ; MS/Dairy Tech/OH State Univ; BS/Dairy Tech/OH State Univ *Born:* 6/12/38. Native of Hancock County, OH. Conducted teaching and res in food engg as faculty mbr at MI State Univ from 1965-84. Served as Dept Chrmn in Agricultural engg Dept at MI State from 1975-79. ASAE Young Researcher Award, 1974. ACE Fellow in Academic Admin, 1974-75. Advisory Editorial Bd for AVI Publishing Co, 1972-present. Editor for Journal of Food Process Engg, 1977-present. Distinguished Alumni Award from the OH State Univ, 1978. DFSA-ASAE Food Engr Award, 1981. Employment with Campbell Inst for Res & Tech from June 1, 1984-86. Current position with National Food Processors Assn began in July, 1986. *Society Aff:* ASAE, AIChE, IFT, AACC.

Helfman, Howard N
Business: 1545 Pontius Ave, Los Angeles, CA 90025
Position: Pres CEO *Employer:* Helfman/Haloossim & Assoc. *Education:* BS/EE/USC *Born:* 12/13/20 Educated at UCLA & USC. Served with Army Signal Corp & Signal Corp Engrg Labs WWII. Chief engr of Air Conditioning Co, Glendale, CA 1946-1956. Founded Climate Conditioning Co 1956. Founded Helfman Air Conditioning Co dba Howard N. Helfman & Assoc (consulting mech & elect engrs) 1962. Registered in CA, WA, FL, TX, PA, AZ, MA, WA, OR, UT. Senior extension teacher at UCLA since 1952. Past Pres, MEAC. Fellow, ASHRAE. Bd of Governors Westside YMCA. Sec-treas, APTEC Corp. *Society Aff:* ASHRAE, NSPE, ASPE, MEAC, AEE.

Helland, George A, Jr
Business: PO Bx 1212, Houston, TX 77001
Position: Exec V President-Operations. *Employer:* Cameron Iron Works, Inc. *Born:* Nov 1937. MBA Harvard Business School; BSME Univ of Texas. Native of San Antonio, Texas. Joined Cameron Iron Works in June 1961. Assumed current respons as Exec V Pres, Operations in June 1975 & Pres, V Pres and/or Dir of various subsidiaries. Mbr of various business int & engrg related organizations, presently Pres of Petroleum Equip Suppliers Assn. Mbr Bd/Dir, The Briarwood School. Elder St Philip Presbyterian Church. Hobbies incl bird shooting, photography, gourmet cooking.

Hellawell, Angus
Business: Materials Dept, Milwaukee, WI 53201
Position: Prof Chrmn. *Employer:* Univ WI. *Education:* DPhil/Metallurgy/Oxford Univ; MA/Metallurgy/England; BA/Chemistry/England. *Born:* 6/21/30. First degree in chemistry at Oxford Univ, 1949-53; continued for advanced degree, 1953-56, working on phase equilibria of iron alloys. Remained at Oxford 1957-1977, as univ lecturer specializing in res concerned with solidification and microstructure. Spent a yr as visiting prof at Stanford Univ, 1965-66, and MI Tech Univ, 1973-74. Enjoys music, gardening and mountaineering. *Society Aff:* ASM, AIME.

Heller, Gerald S
Business: Box M, Providence, RI 02912
Position: Prof of Engrg Dir-Materials Research Lab *Employer:* Brown Univ *Education:* PhD/Physics/Brown Univ; SCM/Applied Math/Brown Univ; SCB/ Physics/Wayne Univ *Born:* 9/5/20 Detroit MI. Staff member, Radiation Lab, MA Inst of Tech., radar development, 1942-45. Australian Group of Radiation Lab 1943-44. Assistant Prof Physics, Brown Univ 1948-54. Group Leader, Resonance Physics Group, Lincoln Lab, MA Inst of Tech., 1954-62. Visiting Prof of electrical engrg, MA Inst of Technology, 1962-63. Prof of Engrg Brown Univ, 1963-present. Dir of Materials Research Lab, Brown Univ, 1969-present. Materials Research Lab supports interdisciplinary research in the science of materials in all branches of engrg and the depts of physics, applied math and chem. *Society Aff:* IEEE

Heller, Kenneth G
Home: 335 Palomar Drive, Redwood City, CA 94062
Position: Principal. *Employer:* Kenneth G Heller, Cons Engr. *Education:* MS/Mech Engg/Univ of CA, Berkeley; BS/Mech Engg/Univ of CA, Berkeley. *Born:* Nov 1926 Dresden, Germany. m. 3 children. Ed in Europe & US Schs; BS Univ Cal at Berkeley 1949, MS 1965. Fluent several languages. Diversified tech/mgmt experience in indus & private practice. Hold numerous pats. Engr Grove Regulator Co Oakland Calif 1949-53; Advanced Tech Div Amer Standard Mountain View Calif 1953-64; appointed Supervisory Engr 1959, Mgr Aerospace Engrg 1961, Mgr Engrg 1963. Principal, Kenneth G Heller, Cons Engr Redwood City Calif 1964- . Reg Prof Engr CA. Mbr Cons Engrs Assn of CA. Assoc Fellow, Amer Inst of Aeronautics & Astronautics. Act in community in ed & 4-H youth activities. *Society Aff:* CEAC, ACEC, AIAA, CEAC.

Heller, Steven
Home: 5694 Mark Dale Dr, Dayton, OH 45459
Position: Acct. Exec. *Employer:* Ordiorne Ind. Advertising, Inc. *Education:* SB/Chem Engrg/MIT. *Born:* 4/1/22. SB Chem Engrg MIT 1943. Native of Chicago, IL. Ser with Army Ord 1943-46. Res Engr White Cap Co 1947-52. A O Smith Corp 1953-62 finally as Dir of Engrg of Glascote Products Subsidiary. Chemineer, Inc 1963-1983 Retired. Mgr, Mktg Communications. Chmn Dayton Sect AIChE. Mbr of Organizing Ctte & First Chmn Affiliate Socs Council of Engrg & Sci Inst of Dayton Reg Engr OH 27255. Mbr, Engr's Club of Dayton. Fellow AIChE 5-74. 1983 to present, Acc Exec, Odiorne Indus Adv, Inc. *Society Aff:* AIChE.

Hellickson, Martin L
Home: 3120 N. W. Greenbriar Place, Corvallis, OR 97330
Position: Assoc Prof *Employer:* Agricultural Engrg Dept *Education:* PhD/Agric Engrg/Univ of MN; MS/Agric Engrg/SD State Univ; BS/Agric Engrg/ND State Univ *Born:* 4/6/45 Native of Medora, ND. Served as a Lieutenant in the US Army, was with the 173rd Airborne Bridge in Vietnam. Joined teaching and research faculty in Agricultural Engrg Dept at OR State Univ fall 1975. Specialization areas include; agricultural structures and environment, application of alternative energies and energy conservation related to agricultural situations and extension of the post harvest life of frech fruits and vegetables. Member of Alpha Epsilon, Gamma Sigma Delta and Sigma Xi. Registered PE (AE) OR. Enjoys hunting, fishing, backpacking, camping, sports. *Society Aff:* ASAE, NGS, ΓΣΔ, ΣΞ, AE.

Hellickson, Mylo A
Business: Dept of Agri Engg, Brookings, SD 57007
Position: Prof. & Head. *Employer:* SD State Univ. *Education:* PhD/Engr/WV Univ; MS/Agri Engr/ND State Univ; BS/Agri Engr/ND State Univ. *Born:* 7/17/42. Native of Medora, ND. Asst Prof of Agri Engg, SD State Univ 1969-73; Assoc Prof 1973-1978; Prof 1978-1982. Prof. & Head 1982-present. Primary responsibilities are ad-

Hellickson, Mylo A (Continued)
ministration, teaching and res in the areas of livestock structures and environment (ventilation) and solar energy for crop drying and livestock building heating. Chrmn of the North Central Region of the ASAE 1977; outstanding Young Men of America 1970, 1977, 1978; ASAE Outstanding Paper Award 1977; Lecturer at sev European Univs, recipient of numerous res grants, holder of two patents; chrmn of four ASAE Natl Committees; SDSU Intl Rep to the North Central Intercollegiate athletic conf. Mbr of ASAE, ASEE, ASHRAE, Sigma Xi, Alpha Epsilon and Gamma Sigma Delta. *Society Aff:* ASAE, ASEE, CAST, NSPE.

Hellier, Charles J PE
Business: Essex Plaza, PO Box 818, Essex, CT 06426
Position: Pres *Employer:* Hellier Associates Inc. *Education:* -/Mech Eng/Norwich Tech; -/Metallurgy/Temple Univ; -/Ind. Mgmt/Penn State. *Born:* 9/16/34. Actively involved with ASNT for over 20 years serving as Edu Chrmn and Chrmn for the local sec, then progressing through the officers' chairs in the Natl Educ Council. recently completed a term as Chrmn, Bd of Dir, after serving 4 yrs on the Natl Bd of Dir, and completing one year terms as Treas, Secy, VP and Pres. In 1977, elected ASNT Fellow. Holds a Level III and was previously certified as Examiner to various Navships documents. In addition, he holds registration as a PE. Presently responsible for the admin and mgnt of the company. In addition, he dev and conducts varius educ programs, serves as a consult to industry, and promotes the tech of nondestructive testing. Mr. Hellier has developed coll credit courses in Nondestructive Testing at the Thames Valley State Tech Coll, Norwich, Ct and has lectured extensively throughout the world. He has authored and published many technical papers and several books dealing with various aspects of nondestructive testing. Mr. Hellier has over 25 years of diversified experience in the fields of education, inspection, equip. *Society Aff:* ASNT, ASME, AWS, ASM, ASQC, NSPE.

Hellman, Albert A
Business: 4221 Wilshire Blvd, Los Angeles, CA 90010
Position: Pres *Employer:* Hellman & Lober, Inc *Education:* BS/Aero Engr/West Coast Univ *Born:* 3/27/22 Founded Hellman & Lober, Inc in 1958 with Allen Lober (deceased) to provide a broad range of mechanical engrg services in the bldg construction field. Specialized in design and engrg of mechanical systems for large commercial and institutional projects such as hotels, office bldgs, hospitals and lab facilities. Since 1976, have designed several solar heating and air conditioning systems. Developed several innovative mechanisms and systems for reduction of energy consumption. Prior professional background includes employment in engrg and management positions in heavy industry, petrochemical engrg, aerospace research and development, architectural and consulting mechanical engrg firms. Registered PE in multiple states. *Society Aff:* ASHRAE, CEAC, ACEC

Hellstrand, Eric
Business: S-611 82 Nykoping, Nykoping, Sweden
Position: Dept Head of Reactor Technology *Employer:* Studsvik Energitelcnik AB. *Education:* Doctor of Techn/-/Chalmers Inst of Techn. *Born:* 9/14/23. Educ at Inst of Tech Stockholm. Visit Sch Brookhaven Natl Lab 1957-58 & 1965-66. Hd Sect for Core Analysis 1968-78, 1978-1981 Hd Dept of Reactor Sys and Nuclear Safety, since 1981 Hd Dept of Reactor Technology. Swedish Mbr NEA Ctte on Reactor Physics OECD 1968-76. Fellow ANS. Nuclear interest: Reactor Physics and Nuclear Safety.

Hellums, Jesse D
Business: 6100 South Main St, Houston, TX 77005
Position: Dean of Engrg *Employer:* William Marsh Rice Univ *Education:* PhD/ChE/Univ of MI; MS/ChE/Univ of MI; BS/ChE/Univ of TX *Born:* 8/19/29 Native of Rotan, TX. Process Engr, Mobil Oil Co., Beaumont, TX 1950-54; Assit. Prof Ch.E., Rice Univ, Houston 1960-65; Assoc Prof, 1965-68; Prof, 1968- ; Dir Biomed. Engrg Lab, 1968-80; Chrmn Dept Ch.E. 1969-1975; Dean of Engrg 1980- ; Adj. Prof, Baylor Coll of Med., Houston, 1960- ; NSF Sci. Fac. Fellow, U. Cambridge (England) 1967-68; Visiting Prof Imperial Coll, London, England 1973-74; Adj. Prof U. of TX Med. School, 1977- . 1st Lt. US Air Force 1954-56. *Society Aff:* AIChE, ACS, ASAIO, MS.

Hellwarth, Robert W
Business: Physics Dept - 0484, Los Angeles, CA 90089
Position: George Pfleger Prof Elec Engrg & Physics. *Employer:* Univ of S CA *Education:* DPhil/Physics/Oxford Univ (England); BSCE/EE/Princeton. *Born:* Dec 1930. D Phil from Oxford Univ (England) 1955, Rhodes Scholar at St John's Col; BSE from Princeton Univ 1952. Native of Detroit, MI. Res Scientist & Dept Mgr at Hughes Res Labs 1955-70. Sr Res Fellow & Res Assoc at CA Inst of Tech 1966-70. Visiting Assoc Prof of Univ of Ill 1964-65. Prof of elec engrg & physics at USC since 1970. Fellow of IEEE 1971. Hon: Natl Acad of Engg 1977. Recipient of "1983 Charles Hard Townes Awd of the Optical Soc of Am." Recipient of "1985 Quantum Electronics Award" of the IEEE, Natl. Acad. of Sciences 1986. Assoc ed IEEE Journal of Quantum Electonics 1965-77. Inventor of giant-pulse laser. Author of numerous scientific articles on solid state physics & quantum electonics. *Society Aff:* APS, AAAS, IEEE, OSA.

Helmer, Frederick T, III
Business: 2404 Maile Way, Honolulu, HI 96822
Position: Associate Professor of Management, and President of F. Theodore Helmer and Associates, Inc. *Employer:* Univ of HI. *Education:* BcEng/Mech Eng/Stevens Inst of Technology; MS/Ind Eng/Univ of Pittsburgh; PhD/Ind Eng/Univ of Pittsburgh. *Born:* 10/4/37. Native of Denville, NJ. Served with the USAF from 1959-79 as maintenance officer, procurement-prod officer & instructor at AF Acad. Presently Assoc Prof, College of Bus Admin, Univ of HI. PPres, Aerospace Div of Amer Inst of Indus Engrs. Act involved in sys acquisition & hospital mgmt res. President of F. Theodore Helmer and Associates, Inc, management consultants in long range planning, project management, productivity, and hospital/clinic management. *Society Aff:* IIE

Helmich, Melvin J
Home: Club Dr PO Bx 493, Mt Vernon, OH 43050
Position: Mgr, Product Engineering. *Employer:* Cooper Energy Services. *Born:* Sept 1924. BSME Purdue Univ. Native of Richmond, Ind. Prod dev engr, Cooper-Bessemer Co 1949. Significant achievements: dev of two & fourstroke cycle high-output turbocharged engines; intro of jet-powered gas turbine for gas compression applications; design & dev of the world's largest integral gas engine/compressor; medium speed integral gas engine/ compressor of high specific output & extreme compactness. Chmn, Diesel & Gas Engine Power Div, ASME 1974. ASME Diesel & Gas Power Award 1970. ASME Fellow 1975. Mbr of Pi Tau Sigma & Tau Beta Pi. Mbr of SAE. Avocation: motor sports.

Helms, Bennett L
Business: PO Box 491, Spartanburg, SC 29304
Position: Sr Vice President. *Employer:* Lockwood Greene Engineers. *Education:* BS/Chem Engr/Univ of SC. *Born:* Dec 1930. Resident of North & South Carolina. Ser three yrs USAF. Early career began as applications engr from Honeywell Indus Prods Group. Later Ch Instrument Engr & Proj Mgr for Lockwood Greene. Assumed pos of V Pres & then Sr V Pres with overall respon for Corporate Business Dev & Corporate Facilities Planning. Mbr Bd/Dirs Lockwood Greene . *Society Aff:* AIChE, TAPPI.

Helms, Jon D
Home: 86 Todmorden Dr, Rose Valley, Wallingford, PA 19086
Position: VP, Chemicals *Employer:* Sun Refining & Marketing Co *Education:* MS/Ind Mgt/MIT; MBA/Ind Mgt/Univ of Toledo; MS/CE/OH State Univ; BSCE/CE/OH State Univ. *Born:* 1/9/35. Native of Toledo, OH. Joined Sun Oil Co full time in 1957, & was Mgr, Process Engg at their Toledo Refinery 1966-1969. Was MIT Sloan Fellow 1970-71. Assumed present position as VP, Chemicals for Sun Re-

Helms, Jon D (Continued)
fining & Marketing Co, the refining/marketing/transportation div of Sun, in 1984. Am responsible for chemical businesses and systems activities for Sun Refining & Marketing. Mbr of AIChE. Active in United Methodist Church. *Society Aff:* AIChE, TBΠ.

Helms, Ronald N
Home: 4681 Gordon Drive, Boulder, CO 80303
Position: Prof/Lighting Cons. *Employer:* University of Colorado. *Education:* BArch/Arch Eng/Univ of IL; MS/Arch Eng/Univ of IL; PhD/Biophysic/OH State Univ. *Born:* March 16, 1939 in Peoria, Ill. He is Prof & Hd of Arch Engrg at the Univ of Colo & Principal in the firm of Lum-I-neering Assocs Lighting Cons. He received his Bachelor of Architecture degree (engrg option) in Jan 1963 & his Master of Sci in Architectural Engrg in Aug 1963 from the Univ of Ill. In Aug 1971 he received his PhD from Ohio State Univ in biophysics with an emphasis in Physiological Optics. He has pub a number of articles on Illumination Engrg in tech journals. He is the author of a text/reference book entitled "Illumination Engineering for Energy Efficient Luminous Environments" published by Prentice- Hall, Inc. A Mbr of IES Bd/Dir Dr Helms is also P Chmn of its Ed Ctte, a mbr of its Testing Procedures Ctte & a Mbr of the Computer Sub-Ctte. He is a Reg Prof Engr. Dr Helms has acted as a lighting cons to architects & engrs as well as being an 'expert witness' in numerous legal cases in the Denver area. *Society Aff:* IES.

Helser, Fred D, Jr
Business: 1110 Third Ave, Seattle, WA 98101
Position: Principal *Employer:* John Graham and Co. *Education:* BSEE/Elec Engr/Univ of WA. *Born:* 6/14/24. in OK, resident of Auburn, WA since 1938. Served in the US Navy 1943-1946. Joined John Graham and Co, Arch, Engrs & Planners in 1952. In 1980 assumed current responsibility as Principal and Dir of Arch & Engr Operations at John Graham & Co. Reg PE in 12 states. Active in business and community affairs. Dedicated Christian. Enjoy sports. *Society Aff:* NFPA, NSPE, WSPE, IEEE, IES, SAME, AEE, NIBS

Helstrom, Carl W
Business: E.C.E., C-014, U.C.S.D, La Jolla, CA 92093
Position: Professor. *Employer:* University of California, San Diego. *Education:* PhD/Physics/CA Inst of Tech; MS/Physics/CA Inst of Tech; BS/Eng Physics/Lehigh Univ. *Born:* Feb 1925 in Easton, Pa. USNR 1944-46. Westinghouse Res Labs 1951-66. Lecturer Sch of Engrg UCLA 1963-64. Prof Univ of Calif-San Diego since 1966. Editor IEEE Transactions on Inf Theory 1967-71. Professeur associe, Universite de Paris XI, Orsay 1973-74 and 1987. Chmn Dept of Applied Physics & Info Sci UCSD 1971-73 & 74-77. Phi Beta Kappa. Fellow IEEE & Optical Soc of Amer. Author of 'Statistical Theory of Signal Detection' Pergamon 1968 & 'Quantum Detection & Estimation Theory' Acad Press 1976 & 'Probability and Stochastic Processes For Engineers' Macmillan 1984. Centennial Medal, IEEE, 1984. *Society Aff:* IEEE, OSA.

Heltman, James W
Business: 4301 CT Ave, N.W, Washington, DC 20008
Position: VP *Employer:* ALPHATEC, p.c. *Education:* BS/CE/CO State Univ *Born:* 7/13/39 Native of Cleveland, OH. Educated in FL and CO. Resident of the District of Columbia since 1966. Naval Facilities Engrg Command, 1966-69. Civil engr and planner for Washington office of Leo A. Daly, 1969-76, becoming Chief civil engr and Assoc in 1975. Joined ALPHATEC in 1976 as Chief civil engr. Assumed present position as VP for civil engrg operations in 1980. Serves as Dir, National Capital Section, ASCE, 1981-83. Active in community affairs. Enjoy classical music, theater and hiking. *Society Aff:* ASCE, NSPE, ASTM

Helweg, Otto J
Business: International Programs, Davis, CA 95616
Position: Assoc Prof. *Employer:* Texas A&M Univ *Education:* PhD/Civ Engg/CO State Univ; MS/Civ Engg/UCLA; MDiv/Theology/Fuller; BS/Engg/US Naval Academy. *Born:* 2/1/36. Native of Watervilet, MI. Served 4 1/2 yrs in the US Navy. Currently a Capt, USNR, and Visiting Prof, Texas A&M. Was Field Hydrogeologist in Iran and a special consultant for the Iranian Ministry of Water and Power (1968-73). Later consulted in the US and Established the firm of Helweg & Assocs. Established the Water Studies Center at King Faisal Univ, KSA. Has authored or co-authored over fifty technical papers and has written two books, a text *Water Resources Planning* and *Improving Well and Pump Efficiency*. Special consultant to India for the UN in 1979. Main interests are theology and family. *Society Aff:* ASCE, NWWA, AGU, AWWA, AWRA.

Hemdal, John F
Home: 2808 Yost Blvd, Ann Arbor, MI 48104
Position: Prin. *Employer:* Hearing & Noise Assocs. *Education:* PhD/Elec Engg/Purdue Univ; MSEE/Elec Engg/Purdue Univ; BSEE/Elec Engg/Purdue Univ. *Born:* 7/29/34. Taught Electrical Engg at Purdue 1959-64, Univ of MI 1967-73, and Univ of MI (Dearborn) 1973 to present. Assoc Engr at Appl Physics Lab, Johns Hopkins Univ 1957-58. Res in acoustics, seismics, automatic speech recognition at Willow Run Labs, Univ of MI, 1964-73. Hd, Acoustics and Seismics Lab 1973. Res in energy, acoustics, remote sensing at the Environmental Res Inst of MI, 1973-78. Formed Hearing & Noise Assocs, 1978 and currently in private practise consulting in acoustics, noise control, energy. Author, "The Energy Ctr" Ann Arbor Sci Publishers 1979. Exec Acoustic Consultant, Zinder Engg, Inc. *Society Aff:* IEEE, ASA, NSPE, INCE.

Hemenway, Henry H
Home: 48 Hancock Hill, Worcester, MA 01609
Position: Retired. *Education:* BS/US Naval Academy. *Born:* Feb 1913, Philippines. Grad courses at Brooklyn Polytechnic Inst & Stevens Inst of Tech. Foster Wheeler Corp 1936-59 was Asst Mgr, Steam Dept. With Graver Tank Div of Trans Union Corp 1959-64 as V Pres, Res & Engrg. With Riley Stoker Corp 1964-68 as V Pres, Engrg. Was V Pres & Gen Mgr, Western Precipitation Div of Joy Mfg 1968-71. Was successively Group V Pres, Environ Sys; V Pres Engrg; Cons to the Carborundum Co 1971-present. Two pats-Steam Generators. *Society Aff:* ASME, APCA, C of C.

Hemphill, Adley W
Business: 10 E Baltimore St, Baltimore, MD 21202
Position: Exec Vice President. *Employer:* W R Grace & Co - Davison Chem Div. *Born:* Dec 19, 1925. BS Grove City Col 1947; MS Pitt 1950 Chem Engrg. Pittsburgh Coke & Chem operator 1949 & 1950. Davison Chem Co Div W R Grace 1951 to date. Res engr, pilot plant supr, res coordinator, plant mgr, gen mgr adsorbents, prod v pres, exec v pres-Petroleum Chems. Principal prods, silica gels, molecular sieves, fluid cracking catalysts, specialty catalysts. Outside interests: golf, music, handyman.

Hemphill, Dick W
Business: 2100 W State St, Ft Wayne, IN 46801
Position: Mgr Mtls Engg. *Employer:* Dana Corp. *Education:* BS/Metallurgical Engg/Univ of Pittsburgh. *Born:* 3/30/23. Native of Pittsburgh, PA. Attended Penn State College prior to WWII. Army Air Corp 1942-1945. Grad from Univ of Pittsburgh in 1948. Employed with Dana Corp since 1948. Assumed position of Chief Metallurgist in 1956. Present position as Mgr of Mtls Engg since 1970. Responsible for mtl, process and heat treat specifications plus the metallurgical and chem controls required to produce rear and front driving axles used in light truck and recreational vehicles. Enjoys bridge, bowling and golf. *Society Aff:* ASM, ASTM.

Hemsworth, Martin C
Home: 8040 S Clippinger Drive, Cincinnati, OH 45243
Position: Mgr-Design Technology. *Employer:* General Electric Company. *Education:* BS/ME/Univ of NB. *Born:* June 3, 1918. With GE Aircraft Engine Group since grad. Test Engrg Program 1940-41, Test Facils & Design Engrg Lynn MA 1941-48.

Hemsworth, Martin C (Continued)
Mgr-Test Facilities & Operations Cincinnati, Ohio 1948-51. Mgmt design & dev, advanced engines 1951- 60. Mgr-Engrg & later Advanced Design Engrg, Small Aircraft Engine Dept, Lynn, Mass 1960-64. Returned to Cinn as Mgr Engrg for TF39 & CF6 High Bypass Fan Engine Design & Dev 1964-71. Ch Engr, Group Engrg Div 1971-75 & in 1975 Mgr, Design Tech 1977, Mgr Energy Efficient Engine Program since 1980, Chief Engineer Cincinnati, Ohio. Member AIAA, Fellow ASME and Fellow SAE. Member National Academy of Engineering *Society Aff:* ASME, SAE.

Hench, Larry L
Business: Dept Matls Sci/Engrg, Gainesville, FL 32611
Position: Professor. *Employer:* University of Florida. *Education:* PhD/Ceramic Engr/OH State Univ; BCerE/Ceramic Engr/OH State Univ. *Born:* Nov 1938. PhD & B Ceramic Engrg from Ohio State. Native of Shelby, Ohio. Came to Univ of Fla in 1964 as Asst Prof in Ceramic Engrg. Has been Prof & Hd of the Ceramics Div of the Dept of Matls Sci & Engrg since 1972. Appointed Co-Dir of the Center of Res on Human Prostheses in 1973 & Dir of Biomedical Engrg Programs at the Univ of Fla in 1974. Is a mbr of several natl & internatl socs Pres of Soc for biomat & has ser on a number of cttes. Has pub over 120 tech papers & 8 bks and 6 pats. Received Amer Soc of Engrg Ed Res Award, Dow ASEE Outstanding Young Faculty Award & several other engrg related awards incl Soc for Biomat Clemson Award for Basic Res. *Society Aff:* SFB, ACS, ASEE, ΣΞ, KER.

Hench, Robert R
Business: 401 N Wash St, Rockville, MD 20850
Position: VP & Gen Mgr Information Processing Technology *Employer:* GE Info Services Co. *Education:* MA/Math/Univ of PA; BS/Math-Chemistry/Shippensburg State College. *Born:* 1/9/34. Bob holds a Bachelor of Science degree from Shippensburg State College & a Masters degree from the Univ of PA. His 29 years with GE have included assignments with both Information Services in Rockville and Aerospace in Valley Forge. In 1974, he was named General Manager - Information Services Technology Department & was responsible for design and development of the MARK III product. In 1976, he was named General Manager of the Marketing Department & was responsible for all marketing activities within the Information Services Business Division. In 1978, he became General Manager of the Engg Dept and was responsible for the MARK III, worldwide network, & IBM development efforts. In 1985 he assumed responsibility for all development & operational activities for GEISCO's data processing product lines. Location: Rockville.

Hendee, William R
Business: 4200 E 9th Avenue, Denver, CO 80262
Position: Professor & Chrmn, Dept of Radiology. *Employer:* Univ of CO Health Sciences Center. *Education:* PhD/Radiological physics/Univ of TX; BS/Physics/Millsaps College. *Born:* 1/1/38. Asst Prof of Phyiscs; Assoc Prof 6 Chmn, Dept of Physics & Astronomy, Millsaps Col, Jackson, MI 1962-65. At the Univ of CO Medical Center, Denver, was Asst Prof of Radiology (Medical Physics), Assoc Prof & Prof 1965-present Prof & Chrmn, 1978-present. Certified in radiological physics by the Amer Bd of Radiology & certified in health physics by the Amer Bd of Health Physics. Was AEC Fellow in radiological physics, Gilber X-ray Fellow in radiation physics & radiation biology, NSF Res Fellow, AEC Res Fellow, Campus Assoc for Danforth Foundation & received the Elda E Anderson award of the Health Physics Soc. Mbr of: Health Physics Soc, Amer Assn of Physicists in Med (Pres 1976-77) Soc of Nuclear Med (Pres-Elect 1979) (Chmn, Program Ctte 1976-77) Amer Col of Radiology (Fellow 1979), Soc of Photo-Optical Instrumentation Engrs (Bd of Governors) and others. *Society Aff:* HPS, AAPM, SNM, ACR.

Henderson, Angus D
Business: 120 Express St, Plainview, NY 11803
Position: Consulting Engineer. *Employer:* Henderson & Bodwell. *Education:* Intl Corr Sch 1934. *Born:* July 1906. Internatl Corr Sch 1934. Born Fall River, Mass. 22 yrs with NYC Water Supply; 5 yrs Exec V Pres Hydrotechnic Corp in Steel Indus. Since 1955 Cons Engr in private practice. Foreign assignments in So Amer, Spain, Sweden, Holland, Switzerland, India & Somalia. Fellow ASCE, active in Bayside Yacht Club. *Society Aff:* ASCE, AWWA, WPCF, ACEC.

Henderson, George E
Home: 150 Valley Road, Athens, GA 30606
Position: Retired. *Employer:* *. *Education:* BS/Agri Engr/OH State Univ. *Born:* 9/10/06. 1929-32 Rural Engr, OH Edison Co; 1932-37 Rural Engr, Dayton Pwr & Light Co; 1937-48 was with TVA, Agri Engrg Div; first as Hd of the Ed Sect, then as the Asst Ch & later Ch to which was subsequently added the Food Processing Div; 1949-73 Prof of Agri Engrg Univ of GA & Coordinator, then Exec Dir of the Amer Assn for Vocational Instructional Matls; 1973 retired as Prof Emeritus. Fellow, ASAE ser on numerous tech & admin cttes, Mbr of the Bd/Dir for 3 yrs; Secy then Chmn of the Southeast Sect of ASAE, elected Fellow 1972; Mbr of ASEE, Gamma Sigma Delta & Phi Kappa Phi. Author of 12 teaching publs. Massey-Ferguson Ed Award (ASAE), Distinguished Alumni Award, OH State Univ, Agri Engr, 1975; *Who's Who in America. Society Aff:* ASAE.

Henderson, James B
Business: Bx 2463 1 Shell Plaza, Houston, TX 77001
Position: Exec VP *Employer:* Shell Oil Co *Education:* BS/ChE/Purdue; PhD/ChE/Purdue. *Born:* 2/26/26. With Shell Chem Co 1949-72 in var Mfg, R&D & Mktg Pos. Genl Mgr Synethetic Rubber Div 1965-66; Gen Mgr Indus Chem Div 1966-69; VP 1969-72. With Shell Oil Co 1972-77. VP Mktg 1974-75; VP Oil Prods 1975-77. He resigned from Shell Oil in 1977 to accept a position with Shell Internatl Petrolem Co, Ltd, London, as Mktg Coord - Oil. He rejoined Shell Oil in Feb 1979 and was elected Exec VP and Dir of Shell Oil Co as of Sept 1979. Mbr Ame Inst Chem Engrs, Amer Chem Soc & Chrmn NAM 1983-4. Dir. Natl Gypsum Co. *Society Aff:* AIChE, ACS.

Henderson, Kenneth W
Business: P.O. Box 751, 2 Corporate Park Dr, White Plains, NY 10602
Position: VP *Employer:* Malcolm Pirnie, Inc. *Education:* MS/Sanitary Engrg/Harvard Univ; BS/Civ Engrg/NW Univ. *Born:* Jan 1, 1926. Native of Haverhill, MA. Ser US Navy 1943-47. Joined Malcolm Pirnie Engrs 1954, Partner 1963-70, VP, Malcolm Pirnie Inc 1970-present. Active in ACEC: Contract Documents Ctte AWWA: Hon Mbr 1977: Dir 1971-74; Fuller Awd 1969; Mbr Bd of Trustees, Distribution Div 1968-73; Mbr Standards Coun, 1968-73, V Chmn 1971-73. Chmn 1973-75; Standards Ctte, A-21; Chmn NY Sect 1967- 68 Natl Assn of Water Co's: Ctte on Water Resources. Author: Supervisory control of water sys, gen & fire serv water rate structures, & standardization procedures. Enjoy hunting, fishing & boating. *Society Aff:* ASCE, AWWA, CSI, NAWC, WPCF, NEWWA, ASTM.

Henderson, Robert E
Business: Applied Res Lab, PO Box 30, State College, PA 16804
Position: Prof., Mechanical Eng. *Employer:* PA State Unv., Applied Research Laboratory *Education:* PhD/ME/Cambridge Univ; MS/Aero Engg/PA State Univ; BS/Aero Engg/PA State Univ. *Born:* 11/1/35. Native of Shinglehouse, PA. Aerodynamics engr McDonnell Aircraft 1958-59. Employed by Applied Res Lab, PA State since 1959 conducting res in hydrodynamics & marine propulsors. Technical Supervisor of Turbomachinery & Propulsor Res 1964-80. Associate Director, Garfield Thomas Water Tunnel 1980-1983. Head, Fluid Dynamics and Turbomachinery Dept, 1983-present. Efforts include basic & applied res in unsteady flows and design of underwater propulsors, pumps & fans. Has held a joint teaching appointment in ME since 1974 presenting courses in fluid mechanics & turbomachinery. Mbr ASME Turbomachinery Committee 1979-Present; Leader of Hydroacoustics for US/FRG/NL Cooperative Res Prog 1973-81; mbr AIAA Underwater Propulsion Committee 1967-69; mbr SNAME H-8 Panel on Propellers 1981-87. Fellow ASME. *Society Aff:* ASME, SNAME, AIAA.

Henderson, S Milton
Business: Dept of Agri Engrg, Davis, CA 95616
Position: Professor-Emeritus. *Employer:* University of California. *Education:* MS/Agri Engg/IA State Univ; BS/Agri Engg/IA State Univ; BA/Physics/Simpson College. *Born:* Oct 1909. Native of Taylor County, Iowa. Prof Agri Engrg teaching & res at Iowa State Univ, Univ of Georgia & Univ of Calif, since 1947. Life Fellow of ASAE; Life Mbr of ASEE. Reg Engr in Calif. Have been V Pres & Chmn of Divs & many cttes of ASAE. Recipient of four tech paper awards from ASAE. Had tech assignments in Ireland, Thailand & Guatemala. Selected as Visiting Scholar to VPI. Prof area, agri process & food engrg with over 90 tech publs. *Society Aff:* ASAE, ASEE, ΦΚΦ, ΣΞ.

Hendricks, Charles D
Home: 2817 Pardee Place, Livermore, CA 94550
Position: Prof *Employer:* Univ of IL & Lawrence Livermore National Lab *Education:* PhD/Physics/Univ of UT; MS/Physics/Univ of WI; BS/Physics/UT State Univ *Born:* 12/5/26 In 1955 I became a member of the Technical Staff, Lincoln Lab, MIT. In 1956 I took a position as Asst Prof at the Univ of IL. In 1963 I was appointed a Prof and in 1966 became Prof of Elec Engrg and of Nuclear Engrg as well as dir of the Charged Particle Research Lab. During 1967-68 I was Visiting Prof, Dept of Elec Engrg at MIT; in 1971 during a period of sabbatical leave from the Univ of IL I received a Senior Research Fellowship at the Dept of Elec Engrg, Southampton Univ, Southampton, England. From 1962 to 1968 I served as a research consultant for the Xerox Corp. From 1974 to 1980 I was on leave of absence at the Univ of CA, Lawrence Livermore Lab, as Leader of the Target Fabrication Group in the Laser Fusion Program. In August 1980 I became Assoc Prog Leader in the Laser Fusion Prog and continued as Head of the Fusion Target Fabrication Element. I am the author and co-author of over 200 publications including articles on laser target fabrication and have been awarded several US Patents. *Society Aff:* APS, AIAA, AAAS, IEEE

Hendricks, David W
Home: 2306 Tanglewood Dr, Ft Collins, CO 80525
Position: A Prof *Employer:* CO State Univ. *Education:* PhD/Sanitary Engr/Univ of IA; MS/Irrigation Engg/UT State Univ; BS/Civ Engr/Univ of CA. *Born:* 9/10/31. Born in Springfield, MO, Sept 10, 1931. Attended schools in CA. Received BS in civ engg, Univ of CA, 1954; MS in irrigation engg, UT State Univ, 1960; PhD in sanitary engg, Univ of IA, 1965. Married in 1959 to Betty Ann Omo. Three children: Bridgette (b 1961), Philip (b 1964), Sara (b 1967). PE in CO, UT, ID. Diplomate, Am Acad of Environmental Engrs. Presently Prof of Civ Engg, Environmental Engg Prog at CO State Univ. Have directed or co-directed about twenty res projs totaling over 1.5 million dollars in funding. Have taught courses in water chemistry, water & wastewater treatment plant design, environmental health, solid waste mgt, industrial wastes, hydrology, fluid mechanics. Consultant for private cos, consulting engrs, local, state, fed govts & intl organizations on equip design, stream, pollution, water reuse, water planning, environmental impact, water quality environmental impact of dams, water filtration. About 50 publ including 3 bks, eg editor, *Environmental Design for Public Projects*, 1975; co-author *Technology Assessment for New Water Supplies*, 1977; co-ed, *Oper of Complex Water Systems*, 1983. *Society Aff:* ASCE, AIChE, WPCF, AEEP, ACS, AAEE, AWWA, AWRA.

Hendricks, Thomas A
Home: 615 N. Walnut St, West Chester, PA 19380
Position: President *Employer:* T.A. Hendricks & Assoc *Education:* JD/Law/St Louis Univ; BS/Civil Engg/Univ of KS. *Born:* March 3, 1928. Five yrs US Army Corp of Engrs in Europe & Far East. 8 yrs Standard Oil of Ohio as Div Engr; 13 yrs with Monsanto as Dir of Engrg; 10 yrs V Pres of Engrg & Mfg for Coen Co & Mbr of Bd. Mbr of AIChE, Sigma Tau, Tau Beta Pi. Socomy Vacuum Scholarship recipient. Mbr Illinois & Missouri Bar. Reg P E, APICS Fellow. *Society Aff:* AIChE, APICS

Hendricksen, William L
Business: PO Bx 55565, Houston, TX 77055
Position: President. *Employer:* Hendricksen Company Inc. *Education:* BS/EE/Univ of Houston. *Born:* Sept 2, 1925. High sch grad Spencer, Iowa 1942. Attended Iowa State Col 1 1/2 yrs. US Navy 2 1/2 yrs as Electronic Technician. After service moved to Houston Tex & attended night sch at Univ of Houston while working as Instrumentation Engr with Honeywell; BSEE 1952. 3 yrs at Lummus Co as Design Engr, 2 yrs with Union Switch & Signal as Sales/Applications Engr, 10 yrs as Sales Engr, Paul Condit Co. Formed Hendricksen Co 1967; has been there since as Pres. Mbr: AIChE, ISA, ACS, APCA, FGBMFI, Gulf Coast Measurement Soc, Gulf Coast Engrg & Sci Soc. P Mbr Rotary & Toastmasters. *Society Aff:* AIChE, ISA, ACS, APCA.

Hendrickson, Alfred A
Business: Dept. of Met Engrg, Michigan Technological Univ, Houghton, MI 49931
Position: Professor. *Employer:* Michigan Technological University. *Education:* BS/Met Engrg/MI Tech Univ; MS/Met Engrg/Columbia Univ; PhD/Mat Sci/Northwestern Univ *Born:* 05/18/29 b MI. Metallurgist, FWC Corp 1951-52; Ampco Metal Inc 1954-56; Lectr Northwestern Univ 1954-58; Assoc Prof 1960-64, Prof 1964-70, Forging Indus Prof 1970- , MI Tech Univ; Hon Res Fellow Univ of Birmingham UK 1970-71. Mbr ASM, AIME, SME, Sigma Xi. Met Trans Review Bd 1962-64, 1974-77; Inst Lecture Ctte, Chrmn 1964; Medical Matls 1966-68; Howe Medal, Grossman Awd Ctte 1976- . Intl Editorial Bd, Reviews of High Temperature Matls 1971- . Numerous publs in plasticity, metal forming. Natl Res Coun Visiting Prof, Natl Sun Yat-Sen Univ, 1982 (Conslt in forging process design, powder consolidation and metal processing). Tennis, fishing & skiing. *Society Aff:* ASM, AIME, APMI, SME

Hendrickson, Tom A
Home: 526 Hillcrest Rd, Ridgewood, NJ 07450
Position: Pres and C.E.O. *Employer:* Kollmorgen Corp. *Education:* MS/Physics/Georgetown Univ; AB/Physics/Harvard Univ. *Born:* 12/25/35. Born & raised on West Coast. Served as Officer, US Navy 1957-61; Engr in US Navy Nuclear Propulsion Prog (a joint US Navy & US Atomic Energy Commission Program) 1957-72. Dir Submarine Sys Div, US Naval Ships Sys Cmd & Div of Reactor Dev, USAEC 1966-72. With Burns & Roe, Inc 1972-1985. Served as Ch Nuclear Engr 1972-73. Deputy Dir-Engrg 1973-78. Asst to the Pres 1978-80. VP 1980-1985. VP, Corp Coordination and Planning 1983-1985. Dir, General Physics Corp. 1976 to 1985. Dir, Power Mgmt Assoc 1984 to 1985. President and C.E.O., Proto-Technology Corporation, 1985 to present. Director, Swuco, Inc. and Sisco, Inc 1986-present. Mbr: Amer Physical Soc. Amer Nuclear Soc & Amer Soc & Amer Soc of Mech Engrs. Lic Prof Engr in NY & NJ. *Society Aff:* APS, ASME, ANS.

Hendrie, Joseph M
Business: Brookhaven Lab Bldg 197C, Upton, NY 11973
Position: Sr Scientist *Employer:* Brookhaven Natn'l Lab. *Education:* PhD/Physics/Columbia Univ; BS/Physics/Case Inst of Tech. *Born:* 3/18/25. PPres Amer Nuclear Soc; PChmn US Nuclear Regulatory Comm; Mbr Natnl Acad Engg; Mbr Energy Engg Bd; Natnl Resch Council; Consult, Nuclear Power Plants for various utilities; Dir Houston Ind & Houston Lighting & Power Co.; Dir System Energy Resources, Inc. *Society Aff:* ANS, APS, ASME, IEEE, ACI, NSPE.

Hendron, Alfred J, Jr
Business: 2230 NCEL, 208 N Romine St, Urbana, IL 61801
Position: Professor Civil Engineering. *Employer:* University of Illinois. *Education:* PhD/Civil Engg/Univ of IL; MS/Civil Engg/Univ of IL; BS/Civil Engg/Univ of IL. *Born:* Oct 1937. Native of Clifton, Ill. Res Assoc Univ of Ill 1961-63. Ser in Army Corps of Engrs 1963-65 at Waterways Experiment Sta; res in soil & rock dynamics. Since 1965, Prof of Civil Engrg, Univ of Ill. Teaching & res in rock engrg, soil dynamics & tunneling. Cons to private indus & gov't on reactor foundations, slope

Hendron, Alfred J, Jr (Continued)
stability, blasting, dams, tunnels & chambers in rock. Received ASCE Walter L Huber Res Prize 1974, Mbr, Natl Acad of Engg. *Society Aff:* ASCE, ISRM, ISSMFE, EERI.

Henebry, William M
Business: Normandeau Associates, Inc, 25 Nashua Road, Bedford, NH 03102
Position: VP. *Employer:* Normandeau Associates, Inc. *Education:* BS/EE/Univ of IL. *Born:* 3/7/36. Native of Monticello, IL & grad of Univ of IL. Engr, Univ of IL, Dept of Phyds, 1960-63. Co-founder of consulting firm R-H Engg, 1963-66. Engg Mgr, Nucl Instrumentation Div EG&G, 1966-68, Gen Mgr, 1968-70; Gen Mgr, Data Products Group EG&G, 1970. Pres & Sr Consultant in marketing of technicl products & services, Sensoresearch Corp, 1971-75. Environmental Res & Tech, Inc (ERT) since 1975; Natl Market Mgr - Petrol, Chems, & Mining, 1975-77, VP Marketing 1977-79, VP, Planning & Commercial Dev. 1979-1981, Presently VP Commercial Development, Normandeau Associates, Inc. Author of various articles & papers, & holder of several patents. Involved in stds activities in several fields. *Society Aff:* AIChE, APS, IEEE, ISA.

Heneghan, John J
Home: 1789 Los Gatos-Almaden Rd, San Jose, CA 95124
Position: VP *Employer:* Wahler Assocs *Education:* BS/CE/IL Inst of Tech *Born:* 10/11/40 in Warrington, England. Registered Engr in six states. Married with three children. Professional experience includes five years with the CA Division of Safety of Dams, and a total of seventeen years of progressively increasing responsibility with private consulting firms. Experience has included extensive work in CA, HI, WA, and Western Canada. Approximately four years work experience in British Columbia and Yukon Territory. Joined Wahler Assocs in 1976 and presently Vice President of Foundation Engrg. On Bd of Dirs. Dir since 1977 and presently Pres of Soil and Foundation Engrs Assoc for 1981-82 year. Enjoys photography and travel. *Society Aff:* ASCE, USCOLD, EIC, CGS, ISSMFE.

Heney, Joseph E
Business: Chairman of the Board, Camp Dresser & McKee Inc, One Center Plaza, Boston, MA 02108
Position: Chmn of the Brd. *Employer:* Camp Dresser & McKee Inc. *Education:* MS/Sanitary Engg/Harvard Univ; BS/Civil Engg/Northeastern Univ. *Born:* 2/22/27. Mr. Heney has served CDM as Chrmn of the Brd since 1982. He was Pres from 1978 through 1984 and Exec VP from 1970 to 1977. He is a Dir of Camp Scott Furphy Pty. Ltd, a CDM affiliate in Australia. Mr. Heney was named an Assoc in 1962 and a CDM Partner in 1964. He has served as project dir and partner in charge for many major water engg projects. He is a Diplomate of Amer Acd of Environ Engrs; Tau Beta Pi. *Society Aff:* AAEE, AWWA, WPCF, ACEC, ASCE.

Henke, George R
Home: 5445 Dry Creek Rd, Napa, CA 94558
Position: Sr Engineer *Employer:* Bechtel National M & QS *Education:* AA/Genl/Napa; AS/NDT/Contra Costa. *Born:* Dec 2, 1929; native of Calif. USAF 1951-55. Air Weather Service. At AEC Test Site, Mercury, Nev. 1956-81 Kaiser Steel, Napa CA. AS degree in Matl Evaluation. 30 yrs in nondestructive testing. St of Calif teaching credential in Matl Evaluation. Chmn of Golden Gate Sect Amer Soc Nondestructive Test 1969-70; ASNT Fellow 1976. Known for work in radiation safety, NDT training, matl evaluation, failure problems & design. *Society Aff:* ASNT.

Henke, Norman W
Business: 3370 Miralom Ave, Anaheim, CA 92803
Position: Dir Operations *Employer:* Rockwell Intl. *Education:* BS/IE/IA State Univ; -/Exec Program/UCLA. *Born:* Dec 1932. BS from Iowa State Univ in Gen Engrg. Field Engr for Square D Co prior to serving with USAF in 1955-57. Ser as corp staff indus engr with Libby Owens Ford Glass Co specializing in advanced mfg methods. Joined Collins Radio in 1963 & has held progressively more respon pos in Mfg Mgmt. Currently Dir of Anaheim Operations for Satellite and Space Electronics Div. Eminent Engr Tau Beta Pi, 1985, V Pres, mbr Bd/Trustees IIE, Fellow IIE 1979, Engr of Yr 1975, Region XI IIE. *Society Aff:* IIE.

Henke, Russell F
Business: 2000 Eastman Dr, Milford, OH 45150
Position: Pres. *Employer:* Structural Dynamics Res Corp. *Education:* PhD/Mech Engg/Univ of Cincinnati; MS/Mech Engg/Univ of Cincinnati; ME/Mech Engr/Univ of Cincinnati. *Born:* 12/16/40. Ten yrs at Cincinnati Milacron in training and eventually Res Supervisor in Optimal Machine Tool Control. PhD thesis under Dr R Sridhar of CA Inst of Tech. Initiated mech design analysis computer services business of SDRC in 1969, when SDRC was one yr old. Became VP/COO in 1972, Pres in 1977, growing business at 40% per yr. SDRC has significant impact on how engg is done in worldwide vehicle and industrial plant systems, improving machine and human productivity, developing computer software which puts mech engrs at threshold of designer's revolution in the 1980's. *Society Aff:* ASME, PA-AMA

Henke, Russell W
Business: Box 106, Elm Grove, WI 53122
Position: Consulting Engineer. *Employer:* Russ Henke Associates. *Education:* BS/Mech Engrg/Univ of WI; MS/Mech Engrg/Univ of WI; Prof Mech/Engrg/Univ of WI. *Born:* April 28, 1924. Native of Milwaukee, Wisc. Pilot, US Navy Air Corps 1942-46. Internatl Pres, Fluid Power Soc 1966-67. Exec V Pres FPS. Dir of Res, Racine Hydraulics also Dir of Engrg, Ch R&D for several indus firms. Cons Engrg specializing in fluid power; cons on fluid power ed. Awards: Master Designer, Grey Iron Founders, Steel Founders; Author several books, numerous papers, & articles on fluid power, etc. Founding Dir, Fluid Power Inst MSOE. Active in standards work; prof soc's; internationally active in continuing educ. *Society Aff:* FPS, SAE.

Henley, Ernest J
Home: 359 Westminster, Houston, TX 77024
Position: Prof. *Employer:* Univ of Houston. *Education:* BS/ChE/Univ of DE; MS/ChE/Columbia Univ; DrEngSci/Engr/Columbia Univ. *Born:* 9/30/26. Asst Prof Chem Engg, Columbia Univ, 1953-58; Prof of Chem & Chem Engg, Stevens Inst of Tech, 1958-1964; Chief of Party, AID Mission, Univ of Brazil, 1964-1966; Prof of Chem Engg, Univ of Houston 1966-present; Assoc Dean, 1966- 1976. Author of 8 books, 83 res papers. Dir, RAI Res Corp, 1954-present; Procedyne Corp, 1959-present, Henley International, Inc. 1977-present; Houston Glass Fabrication Co, 1968-86. Continuous Learning Corp., 1980-86. Trustee, Cache Corp, 1975-86, COMAP Corp, 1980-86, COADE Corp, 1983-86, INACOM Corp, 1983-86. *Society Aff:* AIChE, ASEE, ACS.

Hennessy, John F
Business: 11 W 42nd St, New York, NY 10036
Position: Chmn & CEO *Employer:* Syska & Hennessy. *Education:* BS/Mech Eng/MIT; BS/Physics/Georgetown Univ *Born:* 7/18/28. Mr Hennessy is a licensed engg in 27 states and is a mbr of the National Council of Engineering Examiners. Military Service: In 1951-52 served with the USAF as Proj Officer for the Public Works Program, Air Res & Dev Command. Prof Act & Affil: Dir: Catholic Interracial Council; New York Assoc of Consulting Engrs (PPres); New York Heart Assoc (Campaign Chmn, New York City, 1972-75); Trustee: Cardinal Farley Military Academy (former) Whitby School (former) Hall of Science - New York; Clark College - Atlanta, GA. Chairman and Pres, In the Pink, Inc. & WTB, Ltd., both located at: 11 West 42nd Street New York, N.Y. 10036. *Society Aff:* ASME, ACEC, SAME, NSPE

Hennessy, Robert L
Business: 810 So. Flower St, Los Angeles, CA 90017
Position: Mgr of Productivity Analysis *Employer:* Southern CA Gas Co *Education:* M/BA/Univ of So CA; BS/ME/Univ of So CA *Born:* 9/21/27 San Bernardino, CA.

Hennessy, Robert L (Continued)
Navy V-5 program WW II. Employed with Southern CA Gas Co 33 years - started as junior engr on pipeline construction. Variety of staff and line assignments. 1966 responsible for creating corp's first industrial engrg dept. Later expanded to Systems Analysis and Industrial Engrg. For past 12 years have been in jobs as advisor to senior management on co overall effectiveness and management processes. Active in PE societies for many years. Past-pres and Dir of L.A. Chapter IIE, Industrial Engrg and Management Council. Founder of National Management Division of IIE. Awarded "Fellow" status - 1972. *Society Aff:* IIE, AIMC

Hennigan, Robert D
Home: 3882 Highland Avenue, Skaneateles, NY 13152
Position: Prof Grad Prog Environ Sci; Sch of Env & Res Eng; CH. Environmental Science Faculty. *Employer:* SUNY Coll of Environ Sci & Forestry. *Education:* BCE/Civ (San) Engrg/Manhatton Coll; MA/Pol Sci/Syracuse Univ. *Born:* Sept 1925. US Army WWII 1943-46. Engr NYS Health Dept 1949-58; Ch Water Pollution Control Sect 1958-1960. Asst Comm & Dir Pure Waters Div, NYS DH 1965- 67. Dir State Univ Water Res Ctr 1967-70. Principle Engr NYS Office for Local Gov't 1960-65. Exec Dir Temp State Comm on Water Supply Needs of SE New York 1970-75. Prof & Dir Grad Sch of Environ Sci, SUNY Col of Environ Sci & Forestry 1970- . Conslt. Chmn, Onon County Environ Mgmt Council 1971-79 . Exec Secy, NY Water Pollution Control Assn 1969-79. Fellow ASCE. Reg Engr, NY State. Diplomate Amer Acad Engrg. Achievement Awd for Public Service Manhattan Col Alumni Soc 1967, among others. Pres, Pres Elect V-Pres - New York Water Pollution Central Assoc Inc. - 1984-85 & 86. Chair - Faculty Environmental Science 1986- ; SUNY ESF Chair - Onondaga County Water Quality Management Agency 1987- . *Society Aff:* WPCF, ASCE, AWWA, ASPA, AWRA.

Henning, Rudolf E
Home: 400 Ponce de Leon Blvd, Belleair, FL 33516
Position: Prof *Employer:* Univ of S Fla, Col of Engrg. *Education:* D. Eng. Sc/Elec Eng/-; MSEE/Elec Eng/New York, NY; BSEE/Elec Eng/Columbia Univ. *Born:* Aug 3, 1923. US Army 1944-46. Sperry Rand Corp 1947-70 (Engrg Sect Hd, Sperry Gyroscope Div 1954-57, headed engrg activities, Sperry Microwave Electronics Div, Clearwater, Fla 1957-70). With Univ of S Fla since 1970, except Hd, Engrg Sciences Dept, Naval Electronics Lab Ctr 1971. Mbr ASEE, FES, NSPE, IEEE (Fellow 1965), Tau Beta Pi, Sigma Xi. Chmn Fla West Coast IRE Sect 1962, IEEE GMTT AdCom 1968; Chair MTT International Symposium 1965 & 79. IEEE Centennial Medal, 1984; IEEE Outstanding Counsellor Award, 1984; Florida West Coast Engineer of the Year - 1986; IEEE/MTTS/ARFTG/ Career Award - 1986. Pub & holds pats in Microwave & Instrumentation fields. Registered PE - Florida. Consultant. Active in church & community. *Society Aff:* IEEE, ASEE, NSPE, FES, ΣΞ, ΤΒΠ.

Henninger, G Ross
Home: C/O B. H. O'Brien, 121 Hughes Place North, Syracuse, NY 13210
Position: Professor Emeritus. *Employer:* Oregon Institute of Technology. *Education:* BSEE/Elec Engg/Univ of So CA. *Born:* 1898. Fellow IEEE. Life Mbr IEEE, ASEE, NSPE. Utility engrg 1922-24. Engrg Pubs 1924-50. Engrg Ed 1950-58. Engrg Tech Ed 1958-67. Dir, ASEE/Carnegie 'Natl Survey of Tech Inst Ed' 1956-58. US Delegate OECD 'Conf on Tech Ed & Indus', Germany 1961. US Corresp rep EJC & ECPD in FEANI-EUSEC internatl survey of engrg technician ed sponsored by OECD, Paris 1963. ECPD Engrg Tech Ctte 1956- 65; Natl Chmn 1963-65. Lt Col, Col Air Corps AUS 1942-45. Pres Ohio Col of Applied Sci/Ohio Mechs Inst, Cincinnati 1958-62. Cons, India Ministry of Ed, New Delhi 1970. ASEE Arthur L Williston Award 1964; James H McGraw Award 1973. Eta Kappa Nu 1953. Editor AIEE 1932-48. *Society Aff:* IEEE, ASEE, NSPE.

Henrie, Thomas A
Business: 2401 E St NW, Washington, DC 20241
Position: Chief Scientist. *Employer:* US Dept Interior, Bureau Mines. *Education:* PhD/Metallurgy/Univ of UT; BS/Chemistry/Brigham Young Univ. *Born:* 2/6/23. in UT. Chemist, Union Carbide 1955-58; Bureau of Mines, Boulder City Nev 1958-60; Res Dir, Bureau of Mines, Reno NV 1960-70; Deputy Dir & Assoc Dir, MMRD 1970- . Bureau of Mines scientist & administrator & formerly Deputy Dir 1970-74; respon for R&D to improve techniques for extraction processong minerals, incl coal, at decreasing econo, social & environ costs. Many pats in extractive metallurgy. Mbr AIME, the Mining & Met Soc of Amer, AAAS & UT Chap Sigma Xi. P Chmn AIME Extractive Met Div; PPres The Met Soc & VPres AIME, Fellow TMS (AIME). *Society Aff:* AIME, AAAS, A&MS, ΣΞ.

Henry, Donald J
Home: 233 Frankling Ave, Worthington, OH 43085
Position: Tech Dir, Matls Sciences-ret. *Employer:* General Motors Res Labs. *Education:* BS/Chemistry/Otterbein College; MS/Metallurgy/OH State Univ. *Born:* Sept 27, 1910 Columbus, Ohio. m Patricia Laird Mann 1951. Reg Prof Engr State of Mich 1951 present. Teacher, physics & chem, Westerville, Ohio public schs 1933-36. Fellow, Battelle Memorial Inst 1936-37. Res engr, GM Res Labs 1937-62; Dept Hd, Met Engrg 1962-69; Tech Dir, Matl Sci 1969-73. Ret 1973. Cons, Met Engrg 1973- . Mbr Sigma Xi & Alpha Sigma Mu. Patentee & author in field. Natl Trustee ASM 1969-71. Life Mbr ASM & American Foundrymen's Soc. Listed, Amer Men of Sci (12th ed), Who's Who Among Automotive Execs. *Society Aff:* ASM, AFS.

Henry, Edwin B, Jr
Home: 264 Arden Rd, Pittsburgh, PA 15216
Position: NDI Consultant *Employer:* Self *Education:* BS/Elec Engg/Univ of Pittsburgh. *Born:* 6/25/21. in Pittsburgh, PA. Radar maintenance & Repair Officer US Army 1943-46, retired Maj USAR. With US Steel since 1948. Lecturer in Elec engg 1950-55 (Evening Sch). Holder of several pats in nondestructive inspection & related machines. Basic patent on complementary ultrasonic/fluorographic inspection sys. Respon for dev of in-line fluoroscopic weld inspection sys using TV image presentation for line pipe. Mbr ISA. Fellow ASNT. Chmn, Indust Div Technical Council ASNT 1978-82. Mbr Soncis & Elec & Magnetic Methods Ctte, Steel Producers Committee. Mbr ASTM, Secr Real-Time Imaging Committee (E7.01.02) ASNT. Author of several papers on nondestructive inspection, and Chapter 51 - "Nondestructive Inspection–, *Making, Shaping & Treating of Steel,* 10th Edition. *Society Aff:* ASNT, ASTM, ISA.

Henry, H Clarke
Home: 15 Roman Rd, Thornhill, Ontario, Canada L3T 4J8
Position: Planning Assoc *Employer:* Imperial Oil Ltd *Education:* PhD/ChE/Univ of British Columbia; BSc/ChE/Queen's Univ *Born:* 12/17/43 Dr. Henry was born in Brockville, Ontario, Canada. He obtained an honours BSc in Chem Engrg from Queen's Univ in 1965 and a PhD in Chem Engrg form the Univ of British Columbia in 1969. Since 1969 he has held several positions of increasing responsibility within the Imperial Oil Res, Refining and Supply Depts. Dr Henry is an active mbr of the PE of Ontario and the Canadian Soc for Chem Engrg. From 1978 to 1981 he was a Natl Dir of the CSChE. Dr. Henry holds seven patents in the area of lubricating oil production. *Society Aff:* APEO, CSChE

Henry, Harold R
Business: Civil Engg Dept, PO Box 1468, Tuscaloosa, AL 35487
Position: Prof *Employer:* Univ of AL. *Education:* PhD/Civ Engrg/Columbia Univ; MS/Mech & Hydraulics/Univ of IA; BCE/Civ Engrg/GA Inst of Tech. *Born:* 6/2/28. in Henry County, GA. Taught at GA Inst of Tech, Columbia Univ, MI State Univ and the Univ of AL. Conslt to Ebasco Services, NASA, Army Missile Command, US Army Corps of Engrs, US Geological Survey, AL Civil Defense Agency, and other private firms. Civil Engg Dept hd at the Univ of AL since 1969-1984. Prof 1984- present. Prin Investigator of over $1,000,000 res contracts. Mbr of Natl Environmental. Health Scis Coun of NIH 1974-1978. ASCE Stevens Award 1952. Reg in AL, MI, NY. Mbr Tech Advisory Bd; Inst for Creation Res; Accreditation Visitor

Henry, Harold R (Continued)
for ABET (Accreditation Bd for Engrg and Tech). *Society Aff:* ASCE, AGU, ASEE, ΣΞ, ASPE.

Henry, Herman L
Business: Box 10348 Tech Station, Ruston, LA 71272
Position: Professor of Indus Engrg *Employer:* Louisiana Tech University. *Education:* BS/ME-EE/LA Poly Inst; MS/ME/IL Inst of Tech. *Born:* March 1918. Additional grad study Univ of Arkansas. Native of Jonesboro La. Taught at Ill Inst of Tech 1942-46. Taught at Louisiana Tech Univ 194651; 1955-present. Proj engr, Dow Chem Co 1951-55, specializing in indus steam-elec power plants. Assumed pos of Hd of Gen Engrg Dept, Louisiana Tech in 1956 dept name changed to Indus Engrg in 1963; dept name changed to Indus Engrg & Computer Sci in 1970. Mbr of NSPE, Mbr NCEE Sr Mbr IIE, Mbr ASEE, Mbr LA Engrg Soc; Mbr Tau Beta Pi, Alpha Pi Mu. Pres Monore Section LA Engrg Soc 1962; Chmn Prof Engrs in Educ, La Engrg Soc 1968. V Pres Shreveport Sect IIE 1966. Who's Who in Amer 1969. Cert Lay Speaker United Meth Church. Reg P E LA and TX. Visiting Ctte Mbr So Assn Coll's & Sch's. Outstanding Educator in Amer 1972. Personalities of South 1974. *Society Aff:* ASEE, IIE, NSPE, NCEE.

Henry, James J
Business: 2 World Trade Ctr 9528, New York, NY 10048
Position: Chairman of the Board. *Employer:* J J Henry Co, Inc. *Education:* BS/NA & ME/Webb Inst. *Born:* 1913. Grad Webb Inst of Naval Arch. 1935, received Amer Bureau of Shipping prize for Scholarship. Joined Tech Div of US Bureau of Marine Inspection (now the Office of Merchant Marine Safety of USCG. Later, joined the Ship Design Div of US Maritime Comm. In 1941, joined Consolidated Steel Corp, Wilmington, Calif as Naval Arch & Supt of Const. Formed own co in 1947 which is now one the leading Naval Arch firms in the world. Mbr, Bd/Mgrs of ABS; Mbr, Exec Ctte & P Pres of Soc of Naval Arch & Marine Engrs; P Chmn & currently V Chmn of Bd/Trustees of Webb Inst of Naval Arch; Former Mbr, Acad Adv Ctte of USCG Acad. *Society Aff:* SNAME, ASNE.

Henry, John B
Business: 367 W 2nd St, Dayton, OH 45402
Position: V P & Mgr *Employer:* Price Brothers Company. *Education:* MSCE/Civil Engr-Struct/OH State Univ; BCE/Civil Engr/OH State Univ. *Born:* Jan 1921 in Weston, West Virginia. Worked for US Army Corps of Engrs, Ohio River Div prior to WW II. Ser in Army Corps of Engrs in European & Pacific areas 1942-46. Asst Prof of Civil Engrg at Ohio State Univ teaching structural design 1948-50. With Price Brothers Co since 1950, advancing from Ch Engr to present mgmt pos with particular respon for the Engrg & Marketing functions- involved in the designing, mfg, selling & const of precast concrete bldg sys. Dir PCI, Mbr ASCE, ACI, SME. Reg P E 9 states. *Society Aff:* PCI, ASCE, ACI.

Henry, John W, IV
Home: 525 Ridge Road, Rt 11, Annapolis, MD 21401
Position: Senior Project Engineer. *Employer:* David W Taylor Naval Ship R&D Ctr. *Education:* BES/Mech Engr/Johns Hopkins Univ. *Born:* July 1939; raised on farm near Easton, Md. Earned BS in mech engrg at Johns Hopkins Univ 1961. Employed by US Navy working in field of machinery noise reduction since 1962. Furthered educ in acoustics at George Washington Univ & Catholic Univ. Authored a number of tech reports on Navy pump silencing. Received George E Melville Award 1969, US Pat Grant 1970, ASNE Solberg Award 1972 & Meritorious Civilian Service Award 1973. Currently Sr Proj Engr in pump silencing at David W Taylor Naval Ship R&D Ctr, Annapolis. Married; 3 children. Enjoy woodworking & gardening. *Society Aff:* ASNE.

Henry, Joseph D, Jr
Business: 425 Engrg Sciences Bldg, Morgantown, WV 26506
Position: Chrmn, ChE Dept *Employer:* WV Univ *Education:* BS/ChE/WV Univ; MSE/ChE/Univ of MI; PhD/ChE/Univ of MI *Born:* 3/29/41 The primary thrust of Joseph Henry's career has been in the area of research and development and separation processes and teaching of related subject matter. He organized and led a research group on novel separation processes in the Research Dept of Continental Oil Co from 1968 to 1972. He then joined the faculty of the Dept of Chem Engrg at WV Univ and developed a broad research program on separation processes. His research on multiple functional separation processes has led to the development of several new methods. He was Chrmn of the 1981 Gordon Research Conference on Separation and Purification; and is a member of the editorial bd of Separatoin Science and Tech and co-editor of *Separation and Purification Methods* effective January 1982. *Society Aff:* AIChE, ACS, AAAS

Henry, Laurence O (Pat)
Home: 4302 East 14th St, Tucson, AZ 85711
Position: Owner (Self-employed) *Education:* BS/Civil Engg/ ND State Univ. *Born:* 8/24/16. Native of Alexander, ND. Construction engg on Pentagon Building, 1942. Served as a Capt in the Army Corps of Engg, 1942-46. Field engg for Standard Oil Co. (Indian), 1946. Taught Civil Engg at NDSU, 1946-51. Chief engg for Sanitary Dist No 1 of Pima County, AZ, 1952-60. From 1960-present, consulting engg in Tucson, AZ. Recipient of the Arthur Sidney Bedell Award from WPCF, 1974. Arizona Dir of WPCF, 1970-73. Recepient of George Warren Fuller Award from AWWA, 1979. Dir of AWWA, 1977-80. Enjoy photography, and recipient of Merit Kodak International Newspaper Snapshost Award Contest, 1970. Reg PE in ND &AZ. Pres of AZ Water & Pollution Control Assoc, 1966-67. VP of AZ Soc of PE 1968. *Society Aff:* ASCE, NSPE, AWWA, WPCF.

Henry, Paul A
Business: Bx 272, Rt 236, Kittery, ME 03904
Position: President. *Employer:* Moulton Engrg Co. *Born:* Jan 4, 1931 Brookline, Mass. Studied music until 1952. US Army Bandsman through 1954. BS in Cvl Engrg Northeastern Univ in 1959. Portsmouth Naval Shipyard 13 yrs, 6 as Sect Hd of struct design group. Assumed Pres'y of Moulton Engrg Co in March 1972. Reg Prof Engr in Maine, New Hampshire & Mass. Mbr ASCE, NSPE, CEC.

Henry, Richard J
Business: PO Box 151, Latrobe, PA 15650
Position: Dir-Prod Res *Employer:* Teledyne Vasco. *Education:* MS/Met/Rensselaer Poly Inst; BS/Met/Carnegie Mellon Univ. *Born:* 1/28/42. Native of Worthington, PA. Dev Engr with Pratt & Whitney Aircraft (1963- 1965), working on directional solidification of superalloys at Experimental Foundry. With Teledyne Vasco since 1965, Dir-Prod Res since 1981. Responsible for dev of ultrahigh strength steels, received patent on unique wear resistant tool steel. Reg PE in PA. Taught undergrad met lab course - CMU 1968/1969. Chrmn Pittsburgh Chapter ASM, mbr of natl mbrship committee. *Society Aff:* ASM, ISS-AIME, APMI, SME.

Henry, Robert J
Business: Homer Res Lab, Bethlehem, PA 18016
Position: Engr. *Employer:* Bethlehem Steel Corp. *Education:* BS/Phys/Muhlenberg College. *Born:* 3/27/35. Engr in the Product Met Section with emphasis on res & dev of high strength rail steels. Patent & publications concerning rails & deformation & behavior of pearlitic steels. Some work involves failure analysis by use of optical and electron optical microscopes as well as electron microprobe. *Society Aff:* ASM, MAS.

Henske, John M
Business: 120 Long Ridge Rd, Stamford, CT 06904
Position: Chairman and Chief Executive Officer *Employer:* Olin Corporation. *Education:* BE/Chem Engg/Yale Univ. *Born:* 6/3/23. Army Corps of Engrs 1943-46. With Dow Chem Co from 1948-69 serving in Res, Mfg, Genl Mgmt. Dir, Dow Chem in 1965. Joined Olin Corp in 1969 as Group Pres, Chems. Elected Pres of Olin Corp in 1973, Chairman and CEO in 1980. Married Maryanne Decker 1944 &

Henske, John M (Continued)
have 4 children. Enjoy outdoor sports (sailing, golf, hunting, fishing), drama, music. *Society Aff:* AIChE, ACS, AMA, ΤΒΠ, ΣΞ.

Hensley, Floyd E
Business: Hensley Associates, 10305 Cricket Canyon, Oklahoma City, OK 73162
Position: Proprietor *Employer:* Hensley Assoc *Education:* BS/Civil Engr/Univ of OK. *Born:* March 1929. Native of Oklahoma City, Okla. Bridge designer (cons firm) seven yrs. Regional engr for AISC since 1964-1980. Participated in dev of 7th and 8th edition of steel const manual (AISC) & intro lecture series manual. Held nearly all offices in student, local & state ASCE, incl Pres of Oklahoma State Sect & Dist 16 Council Rep. Mbr Tau Beta Pi & Sigma Tau. Enjoy outdoor cooking, woodworking & fishing. VP Robberson Steel Co 1980-1984. Owner - Hensley Associates 1984. Hensley Associates specializes in structural engrg. *Society Aff:* ASCE, ACEC.

Hensley, Sam P
Business: 1536 Cobb Industrial Dr, Marietta, GA 30066
Position: Pres & CEO *Employer:* Hensley Internatl Inc. *Born:* BSCE from GA Inst of Tech; LLB from Atlanta Law Sch. Native of Marietta, GA. Entered private practice as co-founder of Hansley & Assoc in 1957. Experienced in legal Res, legislative analysis, trans planning & environ law as well as all aspects of civ engrg. Presently in charge of Dev of Pesph move System, including Res, dev, mfg and installation.

Herakovich, Carl T
Business: Dept. of Civil Engineering, University of Virginia, Charlottesville, VA 22901
Position: Prof of Civil Engineering *Employer:* Univ of VA *Education:* PhD/Mechanics/IL Inst of Tech; MS/Mechanics/Univ of KS; BS/CE/Rose-Hulman Inst of Tech. *Born:* 8/6/37. After 20 years at Virginia Tech, Kerakovich joined the Civil Engr department at the Univ of Virginia in Sept, 1987. Main res interests are in the field of composite mtls. Work has included finite element stress analysis, mtl characterization and thermal effects. Conceived the idea of the NASA-VA Tech Composites Prog and serve as its co-dir for fourteen years. Will be a member of the Center for Advanced Studies at the University of Virginia. *Society Aff:* ASME, ASCE, SES, SEM, ASEE, SAMPE

Herbert, Robert N
Business: 88 First St, Rm, 501, San Francisco, CA 94105
Position: Chrmn *Employer:* Herbert Engineering Corp. *Education:* SB/NA & ME/MIT. *Born:* Jan 13, 1924 Cotati, Calif. Amer Bureau of Shipping Student Award. Lic Prof Engr, State of Washington. Fellow, Soc of Naval Arch & Marine Engrs; P Chmn Northern Calif Sect. With US Navy during WWII & Korea. Worked for Philip F Spaulding & Assocs, Seattle 1953-1963 advancing to Ch Naval Architect. Formed own practice 1963 in San Francisco; incorporated 1975; retired 1983. Awarded David W. Taylor Medal from SNAME in 1986. Director of CiServ San Francisco. *Society Aff:* SNAME.

Herbst, John A
Business: 2319 Foothill, Suite 200, Salt Lake City, UT 84109
Position: Prof & Pres. *Employer:* Univ of UT & Control Interntn'l *Education:* D Engrg/Metallurgy/Univ of CA-Berkeley; MS/Metallurgy/Univ of CA-Berkeley; BS/ChE/Northwestern Univ *Born:* 2/8/42 Born in Chicago, IL. Worked as research engr for Intl Minerals and Chems in Skokie, IL and Saskatchewan, Canada. Assumed position of assoc scientist with Kennecott Copper Corp in Salt Lake City in 1968. Joined Univ of UT as assistant prof of metallurgy in 1971. Received AIME Hardy Gold Medal in 1972. Chrmn of Engrg Foundation Conference on Particulate Materials 1978, Chrmn of AIME Crushing and Grinding Committee 1979, Chrmn of National Academy of Science Committee on Comminution 1979-81. AIME Krumb Lecturer for 1983, Recipient of UT Dist Res and Dist Teaching Awards. Currently, President of control International, Inc. and Prof of Metallurgy at UT and dir, of the Center for Process Simulation and Control and the Generic Mineral Tech Center for Comminution. *Society Aff:* SME of AIME, TMS of AIME, AIChE, ISA.

Herbst, Rolf
Business: 906 2nd Ave NW, P.O. Box 708, Mandan, ND 58554
Position: Pres *Employer:* Toman Engrg Co *Education:* BS/Engrg/School of Mines, Recklinghausen, Germany *Born:* 12/1/27 Extensive engrg experience in West German mining districts; immigrated to US 1956; Employed by Toman engrg Co 1957, RPE in 1962; RLS, 1982; 28 years design & project experience with special expertise in water & wastewater treatment, water collection, storage & distribution, hydrology, surveying, streets & roads, lighting, swimming pools, airports. Acquired controlling interest in Toman Engrg 1979. *Society Aff:* NDCEC, NSPE, NDSLS

Herbst, Wolfgang B
Home: 7Foehrenstr, 8077 Putzbrunn, West Germany
Position: Mgr Advanced Design & Technology. *Employer:* Messerschmitt-Boelkow-Blohm Gmbh. *Education:* Dr/Eng./TU Berlin. *Born:* March 13, 1935. PhD from Tech Univ Berlin. Res Sci & Lecturer 1960-63. Proj engr & later Mgr of advanced design at VFW at Bremen. Employed with McDonnell Aircraft Co 1976-70, prime respon in the design of the F-15 Aircraft. Since 1971 employed with MBB at Munich as Hd of Advanced Design & since 1974 in addition as Head of Tech Depts of the Military Aircraft Div. Mbr of SAWE of its European Chap. Active pilot. *Society Aff:* AIAA, SAWE, DGLR.

Herbstman, Donald
Home: 13400 S W 72nd Avenue, Miami, FL 33156
Position: Corp. Dir of Safety & Risk Mgmt. *Employer:* Burger King Corp. *Education:* MA/Industrical Safety/NY Univ; BCE/Civil Eng/City College of NY. *Born:* 5/30/34. BCE from CCNY; MA Indus Safety NY Univ. Supt Municipal Protection Dept, NY Fire Insurance Rating Org 1958-66. Sr Safety Engr, Port Authority of NY & NJ 1966-73. Supervising engr in charge of const safety & security at the World Trade Ctr 1968-73. Corp Dir of Safety at Burger King Corp in 1973. Assumed present Postion of Corp. Dir of Safety and Risk Management 1982. Respon for all aspects of risk, claims mgmt., insurance & loss prevention, incl design of new office structures, interiors, mech sys, fire protection & security sys, safety programming, compliance with govmt safety & health regulations. Chrmn, Dade County Bd of Fire Safety and Appeals. Adj Prof. of Ind Engrg. U of Miami. Enjoy tennis, photography, running & swimming. *Society Aff:* SFPE, ASSE, ASIS, RIMS.

Herkenhoff, Phillip G
Business: P.O. Box 1217, Albuquerque, NM 87103
Position: Exec Vice President. *Employer:* Gordon Herkenhoff & Associates. *Education:* MS/Civil Eng/Cornell Univ; BCE/Civil Eng/Cornell Univ. *Born:* Dec 1, 1938. Field Engr, Design Engr & Proj Mgr with Gordon Herkenhoff & Assoc 1963-72. V Pres 1972-present; overall respon for operation of 110-employee cons firm. Reg Prof Engr in New Mexico, Arizona, Colorado, Utah & Oklahoma. Chmn Rocky Mt Sect AWWA 1973, V Pres & Natl Dir CEC, New Mexico 1976. Mbr ACEC, APWA, ASCE, AWWA. Trustee, Village of Los Ranchos de Albuquerque 1969-76. Hobbies incl: flying, fishing, camping, skiing, whitewater rafting & photography. *Society Aff:* ASCE, ACEC, AWWA, APWA.

Hermach, Francis L
Home: 2415 Eccleston St, Silver Spring, MD 20902
Position: Consultant. *Education:* BEE/EE/Geo Washington Univ. *Born:* Jan 8, 1917. Elec Engr Natl Bureau of Standards to 1963, Ch Elect Instruments Sect to 1972, Cons 1972- . Dev & verified instruments for elec measurements & methods for mitigating hazards from static elec. Designed highly accurate ac-dc comparators for voltage & current. Fellow IEEE, ISA, Washington Acad of Sci, Washington Philosophical Soc. Former Chmn IEEE Group on Instrumentation & Measurement, Former Director ISA Metrology Div & Precision Measurements Association. Numerous cttes of engrg socs, ANSI & NFPA. Taught elec measurements GWU. IEEE Morris E Leeds Award, and Centennial Medal, Dept of Commerce Silver Medal.

Hermach, Francis L (Continued)
Reg Prof Engr, DC. GWU Engineer Alumni Achievement Award. *Society Aff:* IEEE, ISA, PMA.

Herman, Elvin E
Home: 1200 Lachman Lane, Pacific Palisades, CA 90272
Position: Independent Consultant *Employer:* Self *Education:* BS/Elec Engrg/State Univ of IA. *Born:* 3/17/21. With Naval Res Lab 1942-51, specializing in radar & counter-countermeasure equip dev. With Natnl Bureau of Standards, Corona Labs 1951-53, design & dev of missile seeker equip. With Hughes Aircraft Co from 1953 until retirement July 1983. As Dept & Lab Mgr, directed and contributed to the dev of pulse doppler & synthetic array radar systems, signal processing & display equips. Tech Dir from 1970 until retirement. Now independent Conslt. Navy Meritorious Civilian Service Award in 1946; Recipient of LA Hyland Pat Award 1971; Fellow IEEE 1975; 21 patents granted. *Society Aff:* RESA, HKN.

Herman, Herbert
Business: Matls Sci Dept-SUNY, Stony Brook, NY 11794
Position: Prof Dept of Matls Sci and Marine Sciences *Employer:* SUNY *Education:* PhD/Materials Sci/Northwestern; MS/Materials Sci/Northwestern; BS/Physics/DePaul. *Born:* 6/15/34. Fulbright Scholar, Univ of Paris 1961-62 followed by res associateship at Argonne Natl Lab. Fac posts at Univ of Pa & State Univ of NY at Stony Brook (since 1968). Chmn of Matls Sci Dept, 1974-80. Ford Foundation Prof of Indus 1967-68. Liaison Sci, US Office of Naval Res London Br 1975-76. Currently active in matls sci & engrg, ocean tech and protective coatings technology. Mbr AIME/TMS, Amer Soc Met, Marine Tech Soc, Amer Phys Soc, Amer Cer Soc, Nat Assoc of Corr Engrg. Mbr Committees on Ion Implantation Metllrgy, Thermal Spray, Drilling Tech; Editor, Trtise Matls Sci, and Tech; Editor-in-Chief, Materials Sci and Engrg, Interntl journal. *Society Aff:* ASM, TMS, ASEE, APS, ACS.

Herman, Marvin
Business: Bx 894-Dept 5827, T26A, Indianapolis, IN 46142
Position: Ch, Materials Res & Engrg. *Employer:* Detroit Diesel Allison Div, GM Corp. *Education:* BS/Met Engr/Drexel Univ; MS/Met Engr/Univ of PA; PhD/Met Engr/Univ of PA. *Born:* March 1927. PhD 1965, MS 1953 Met Engrg Univ Pa; BS 1951 Met Engrg Drexel Inst. Directs all matls related work assoc with the div's aircraft & indus gas turbines. Incl: Matls Res, Matls Dev, Design Support-Matls Selection, Component Lab-Joining & Foundry, Lab Services-Chem & Mech Testing. Prior to joining Allison in 1965, employed at Franklin Inst 10 yrs 7 yrs Mgr, Met Lab. Patentee. Mbr of ASM, & Sigma Xi. *Society Aff:* ASM, ΣΞ.

Herman, Robert
Business: Ernest Cockrell, Jr. Hall, ECJ 6.800, Austin, TX 78712
Position: Prof *Employer:* Univ of TX at Austin *Education:* PhD/Physics/Princeton; MS/Physics/Princeton; BS/Physics/City College of NY *Born:* 8/29/14 Native of NYC. B.S. cum laude, CCNY, 1935; M.A., Princeton U., 1940, Ph.D., 1940. Fellow physics dept CCNY 1935-36; research assoc Moore School of Elec Engrg Univ PA, 1940-41; instr physics City Coll NY, 1941-42; supr chem physics group, physicist, assist to dir. Applied Physics Lab, John Hopkins Univ, 1942-55; cons physicist Gen Motors Research Labs, 1956, asst chrmn basic sci group, 1956-59, dept head theoretical physics dept, 1959-72, traffic sci dept, 1972-79, prof physics Center for Studies in Statis Mech and prof transportation, civil engrg, Univ of TX at Austin 1979 - present; Pres ORSA 1980-81; Recipient numerous awards including Naval Ordnance Devel 1945, Lanchester prize Johns Hopkins Univ and Ops Res Soc 1959, medal Univ Libre de Brux, 1963, Townsend Harris medal City Coll NY, 1963, Magellanic Premium, APS, 1975, Prix Georges Vanderlinden, Belgian Royal Acad, 1975. Fellow APS, WA Acad Sci, Franklin Inst (John Price Wetherill gold medal 1980); George E. Kimball medal 1976, Phi Beta Kappa, Sigma Xi. 1978, Natl Acad of Engg; 1979, Fellow, Amer Acad of Arts and Sci; Hon Doctorate in Engrg from Karlsruhe Univ 1984. Research in vibration-rotation spectra and molecular structure, infrared spectroscopy, solid state physics, astrophysics and cosmology, theory of traffic flow, high energy electron scattering. *Society Aff:* APS, ORSA, AAAS, NAE.

Hermann, Paul J
Business: Town Engrg Bldg, Ames, IA 50011
Position: Assoc Prof Aerospace Engrg. *Employer:* IA State Univ *Education:* BS/Aeronautical Engr/IA State Univ; MS/Theo & Applied Mech/IA State Univ. *Born:* Sept 30, 1923 Sheldahl, Iowa. Two yrs in the navy during WWII as radio tech. Dev Engr, Goodyear Aircraft (now Aerospace) Corp 1951-60, specializing in dynamic systems analysis, analog computer applications. Many early pubs, talks, in the field. Instructor, Theoretical & Applied Mechs, ISU 1947-50. Returned, 1960 ISU Dept Aerospace Engrg. Reg Aeronautical Engr Ohio & Iowa. Assoc Fellow AIAA; Sr Mbr IEEE; Life Mbr Soc for Computer Simulation (SCi). Mbr Bd/Dir SCi 1963-68, V Pres 1964, Pres 1965-67. Mbr Amer Soc for Engg Ed (ASEE); Mbr Intl Assoc for Mathematics and Computers in Simultation (IMACS). Organizer, first chmn IOWA sec AIAA. Jogger, private pilot, genealogist, philatalist. *Society Aff:* AIAA, IEEE, ASEE, SCS, IMACS.

Hermsen, Richard J
Home: 1755 Sumner Avenue, Claremont, CA 91711
Position: Elec Engrg Prof *Employer:* California State Polytechnic Univ. *Education:* PhD/Nuclear Engg/Univ of WI; MS/Nuclear Engg/Univ of WI; BS/Elec Engg/Univ of WI. *Born:* Oct 1928. Sr Engr & Design Specialist for Gen Dynamics; Mbr of the Tech Staff for Aerospace Corp. Specialized in the analysis & design of control systems for Aerospace vehicles. Prof of Elec & Computer Engrg at California State Polytechnic Univ Pomona. Teaching areas of systems theory; control & communication systems. Distinguished Teacher & Outstanding Prof at Cal Poly Univ Pomona 1971. Sr Mbr IEEE; Mbr ASEE, Tau Beta Pi, Sigma Xi, Eta Kappa Nu. Sr Aircraft Observer & Col in USAF Reserves. *Society Aff:* IEEE, ASEE, ROA.

Hernandez, Andres
Business: 2510 S. Florida Ave, Lakeland, FL 33803
Position: Pres. *Employer:* A H Construction Co. *Education:* MS/CE/Stevens Inst of Tech; BS/CE/Villanova Univ. *Born:* 9/14/41. Started profession as process engr with Exxon. Designed petrochem projs and operated plants in Spain, Malaysia, Philippines & Pakistan. Joined Davy Powergas as Sr Engr, became an Engg Supervisor & later was promoted to the Sales Dept. From Sales Mgr became responsible for business dev throughout Latin Am for Davy's Engg & Construction services. Formed own construction co in 1972. Now I am Pres of three corps: A H Construction Co, A H realty, Inc & Active Investors Co. In this capacity I make maj decisions involving administration, financial, sales & technical matters. Member Bd of Trustees, Polk Museum of Art. *Society Aff:* AIChE, NSPE, NAHB, TBΠ, NAR.

Hernandez, John W
Business: Box 3196, Las Cruces, NM 88003
Position: Prof *Employer:* New Mexico State Univ *Education:* BS/Civil Engr/U of New Mexico; MS/Saritory Engr/Purdue Univ; PhD/Water Resources/Harvard *Born:* Aug 17, 1929 Albuquerque, N M. PhD Harvard 1965; MS Purdue 1959; BS Univ of New Mexico 1951. Wide & varied experience in the water resources area. Has been employed by State agencies dealing primarily with water resources dev, control & regulation. Has been a Prof of Civil Engrg at New Mexico State Univ for the past nineteen yrs & the Dean of the Col from July 1975 to July 1980. While on leave from NMSU, he was Deputy Administrator of the US Environmental Protection Agency from 1981 to 1983. Principal contribs have been the dev of the water quality standards for the interstate streams of New Mexico & other res studies dealing with water quality in New Mexico. A mbr of many natl cttes, incl the Natl Drinking Water Adv Council & the Secr of Agri Ctte on Multiple Use of the Forest. Tau Beta Pi, Sigma Xi & others. *Society Aff:* ASCE, NSPE, AWWA, WPCF

Hernqvist, Karl G
Home: 667 Lake Drive, Princeton, NJ 08540
Position: Doctor. *Employer:* David Sanhoff Research Center *Education:* PhD/Electronics/Royal Inst of Tech, Stokholm, Sweden. *Born:* Sept 1922. Received doctorate in 1959 from the Royal Inst of Tech, Stockholm, Sweden. Joined RCA Labs 1952 & became a Fellow of the Labs in 1969. Presently doing res on gas lasers & physical electronics. Received seven RCA Labs Achievement Awards & the David Sarnoff Awards in 1974 and 1982. Authored 55 papers & holds 43 US pats. Fellow IEEE & a mbr of the Amer Physical Soc & Sigma Xi. *Society Aff:* IEEE, APS, ΣΞ.

Herold, Edward W
Home: 332 Riverside Drive E, Princeton, NJ 08540
Position: Consultant. *Employer:* Self. *Education:* DSc/Hon/Polytech Inst of Brooklyn; MSc/Physics/Polytech Inst of Brooklyn; BSc/Physics/Univ of VA. *Born:* 10/15/07. B.Sc., M.Sc., D.Sc. RCA Corp 1930-59, electron device res, becoming Dir, Electronic Res Lab of RCA Labs in 1951. V Pres Res, Varian Assoc, 1959-64. Returned to RCA, 1965, as Dir Tech for the Corp. Cons 1972- & Chrmn/Bd, Palisades Inst. 1969-1984. 47 US Pats, author 50 pubs incl co-authorship of book 'Color Television Picture Tubes'. Reg Prof Engr NJ. Fellow, IEEE & awarded the 1976 IEEE Founders Medal. *Society Aff:* IEEE.

Herr, James C
Business: PO Bx 748, Fort Worth, TX 76101
Position: Chief of Process Control. *Employer:* General Dynamics. *Education:* MS/Eng Admin/SMU; BS/Met Engr/MI Tech Univ. *Born:* Aug 1928 Battle Creek, Mich. Process Engr Interchemical Corp. Production Met Oliver Corp. With Gen Dynamics since 1954. Assumed current respon for Process Control 1968. Gen Dynamics Corp Chmn for Nondestructive Testing. Active in ASM & presently mbr Tech Bd. Reg Prof Engr. Enjoy classical music & golf. *Society Aff:* ASM, ASNT, ASTM, TSPE.

Herr, Lester A
Home: 7040 Cindy Lane, Annandale, VA 22003
Position: Sr VP *Employer:* T. Y. Lin International *Education:* MS/CE/OH State Univ; BCE/CE/OH State Univ; BS/Education/West Chester Univ. *Born:* Jan 1919 Parkesburg, Pa. Const inspector Pa Hwy Dept 1940-41; teacher public schs Pt Pleasant Beach, NJ 1941-42, 1945-46; Instructor Engrg Ohio State Univ 1947-50; Bridge & Hydraulic Engr, US Bur Public Res 1950-59; Ch Hydraulic Br Fed Hwy Admin 1959-69, Dep Ch Bridge Div 1969-73; Div Natl Hwy Inst 1973-74; Ch Bridge Engr 1974-80; Sr. V. Pres, T. Y. Lin International 1980- . Scoutmaster, Boy Scouts of Amer 1968-71, Comm 1971-75 . Served with AUS 1942-45. Recipient S c Trans Dept Silver Medal for meritorious serv 1969. Reg PE. Mbr ASCE, NSPE, AAAS & several hon societies. Received 1976 Dist Alumnus Award Ohio State Univ. *Society Aff:* ASCE, NSPE, AAAS.

Herrick, Robert A
Business: Herrick Engineering, Inc, 2004 Belvedere Drive, Toledo, OH 43614
Position: President *Employer:* Herrick Engineering, Inc. *Education:* BS/ChE/IA State Univ. *Born:* Oct 1934 Clinton, Iowa. Air pollution equip res, US Public Health Ser 1957-59. Evaluation of air pollution & indus hygiene, Bethlehem Steel Corp 1959-67. Cons engrg pos 1967-73. Responsible for all air, water, waste and industrial hygiene engr programs, Owen, Corning Fiberglass 1973-86. Consulting envir engr since 1986. Diplomate AAEE & ABIH. Pubs. Dir APCA 1973-1976. *Society Aff:* AIHA, APCA, NAC, AAEE

Herring, H James
Business: 20 Nassau Street, Princeton, NJ 08540
Position: President. *Employer:* Dynalysis of Princeton. *Education:* PhD/Engg/Princeton Univ; MS/Engg/Princeton Univ; BA/Engg/Harvard Coll; BS/Engg/Harvard Coll. *Born:* Aug 3, 1939 Larchmont, NY. Mbr of the Res Staff of Princeton Univ 1967- 76, studying turbulent flows. In 1969 formed Dynalysis of Princeton, a res & cons firm. Res interests incl turbulence modelling, numerical simulation of flow fields, turbulent boundary layers & wall jets. *Society Aff:* ΣΞ

Herring, Oren L
Business: Charleston, SC 29409
Position: Prof & Head, EE Dept. *Employer:* Citadel *Education:* MS/Elec Engr/Univ of MI; BS/Elec Engr/Citadel, The. *Born:* 3/17/24. Native of Spartanburg, SC. Began teaching career at The Citadel in 1949. Was appointed Prof and Head of Electrical Engg Dept at The Citadel in 1966. Reg as a PE in SC. Eta Kappa Nu, Tau Beta Pi. *Society Aff:* IEEE, ASEE.

Herriott, Donald R
Home: 1237 Isabel Dr, Sanibel, FL 33957
Position: Sr Scientific Advisor *Employer:* Perkin-Elmer Corp *Education:* -/EE/Brooklyn Polytechnic; -/Optics U of Rochester; -/Physics/Duke Univ. *Born:* 2/4/28. Physics, Duke Univ; Optics, Univ of Rochester; EE, Brooklyn Poly. Optical res in thin films, glass measurement & lens evaluation at Bausch & Lomb from 1950-56. Retired Bell Telephone Labs from 1956-81. Res on lasers (first gas laser, HeNe) interferometry, optical patterning of integrated circuits, electron beam patterning & X-ray printing of circuits. Dept Hd respon for new lithographic systems 1969 to 1981. Bd/Dir, Chrmn Laser Group, Publs Ctte, Chrmn of the Editorial Bd, etc for Optical Soc of Amer. Quantum Electronics Council, Program Chrmn, Chrmn of CLEA Meeting. Retired from Bell Telephone Labs 1981. Received Cledo-Brunetti Awd from IEEE, 1980; Received Patent of the Yr Awd 1979; Elected VP, Pres Elect, Pres of Optical Soc of Amer 1981. Received Fraunhofer Awd, Optical Soc 1983. Elected Natl Acad of Engrg 1981. Int: sailing, skiing & gardening. *Society Aff:* OSA, NAE, IEEE.

Herrmann, George
Business: Div Applied Mechs, Stanford, CA 94305
Position: Prof & Chmn. *Employer:* Stanford Univ *Education:* Dr Sc/-/Swiss Fed Inst of Tech; CE/-/Swiss Fed Inst of Tech. *Born:* April 19, 1921. Dipl CE 1945, Dr Tech Sci 1949 Swiss Fed Inst Tech, Zurich. Ecole Polytechnic, Montreal 1949-50; Columbia 1951-62; Walter P Murphy Prof, Northwestern 1962-69; Liaison Officer, ONR, London 1961-62; Stanford Univ 1970-present. ASME Applied Mech Div, Activites Ctte 1964, Exec Ctte 1969-74, Ch Honors & Awards Ctte 1970, Program Ctte 1971; Pres ACDM 1974-76; Founder (in 1958) & Editor-in-Ch of the English translation of the Russian Journal of Applied Mathematics & Mechs (PMM); Founder (in 1965) & Editor-in-Ch of the Internatl Journal of Solids & Structures. Author or co-author of over 200 tech papers; ed of over 12 volumes of conf proceedings & translation texts. Fellow ASME; Assoc Fellow AIAA; Charter mbr Biomedical Engrg Soc; Mbr ASCE, ASEE, SESA, Internatl Assn of Bridge & Structural Engrg, Soc for Natural Philosophy, SEA, Sigma Xi, Tau. *Society Aff:* ASCE, ASME.

Herrmann, Leonard R
Business: Dept of Civil Engrg, Davis, CA 95616
Position: Prof of Civil Engrg *Employer:* Univ of CA, Davis Campus *Education:* PhD/Struct Mech/UC Berkeley; MS/Civ Engrg/UC Berkeley; BS/Civ Engrg/UC Berkeley. *Born:* 3/21/36 Senior research engr for Aerojet-General Corp 1962-1965. Served as consultant, in the fields of structural mechanics and finite element analysis, to numerous private and governmental engrg organizations. Joined faculty at the Univ of CA, Davis Campus in 1965; served as dept chrmn from 1972-1975. *Society Aff:* ASCE.

Herrmann, Thomas A
Business: 4333w Clayton Ave, St Louis, MO 63110
Position: VP/Civil & Environ Div. *Employer:* Zurheide-Herrmann, Inc. *Education:* BS/CE/MO Sch of Mines. *Born:* Oct 30, 1928 St Louis, Mo. Principal, Zurheide-Herrmann, Inc Cons Engrs & Archs. Respon charge of all civil & environ engrg activities of multi- disciplined engrg firm. Reg Prof Engr in 9 states. Reg Land Sur-

Herrmann, Thomas A (Continued)
veyor. Mbr NSPE (P Dir), MSPE (P.Pres), St Louis Chap MSPE (P Pres), ASCE, AWWA, WPCF, CECMO, MW&SC, APWA, SAME & ROA. Retired mbr of US Army Reserve (Corps of Engrs) 35 yrs. (Rank Col). *Society Aff:* NSPE, ASCE, AWWA, WPCF, CECMO, SAME, ROA, APWA.

Herrmann, William A
Business: 4333w Clayton Avenue, St Louis, MO 63110
Position: Vice President. *Employer:* Zurheide-Herrmann, Inc. *Born:* 1922 St Louis, Mo. BSCE 1944 Iowa State Univ. Reg Structural & Prof Engr. V Pres & Principal Officer in cons engrg-arch firm founded in 1954 with respon for corp mgmt of multi-discipline staff serv indus & gov't. Previously was Const Supt, Asst Ch Struct & Ch Civil Engr of large indus const & engrg firm. Attained rank of Cmdr during active & reserve duty Navy Seabees. Bldg Commissioner 18 yrs & Alderman for two elected terms of City of 5,000. State Pres, Cons Engrs Council of Mo; Natl Dir, Amer Cons Engr Council; Bd/Dir of Concrete Council & Pride organizations of St Louis; Mbr Amer Soc of Civil Engrs, Natl & Mo Societies of Prof Engrs, Prof Code Ctte, numerous other prof orgs & cttes.

Herron, John L
Business: 5405 East Schaaf Rd, Cleveland, OH 44131
Position: President. *Employer:* Herron Testing Labs, Inc. *Education:* MS/Metallurgical Engg/Univ of MI; BS/Mech Engg/Univ of MI; BS/Metallurgical Engg/Univ of MI. *Born:* Cleveland, Ohio. Joined the present firm 1968 as Proj Met, becoming V Pres & Dir in 1970 and Pres in 1975. As Ch Exec Officer active in admin coordination & as tech adv for the corp dev of Indus Serv Div. Current prof affiliations are held in Amer Soc for Metals & Amer Soc of Mech Engrs. Presently hold prof licenses in the States of Michigan & Ohio. *Society Aff:* ASM, ASME.

Herron, William J
Home: 10718 Pineaire Dr, Sun City, AZ 85351
Position: Consultant (Self-employed) *Education:* BS/CE/Univ CA-Berkeley *Born:* 4/30/15 I served for 31 years with the Corps of Engrs, including 3 years active duty during WW II, 9 years with the Coastal Engrg Research Center and 19 years with the Los Angeles District. There, in charge of coastal engrg, I was responsible for design of 10 recreational harbors, including Marina del Reay and Dana Harbors, and improvements to San Diego and Los Angeles-Long Beach Harbors. Since 1970 I have continued in private practice as a consultant on numerous coastal projects. I have published technical articles, lectured at UCLA, Long Beach State and the Coastal Engrg Research Center and have been active in the ASCE Technical Groups and the ASBPA. Honors include: "Commendation" by the Dept of Army, "Outstanding Engr Merit Award" by the Inst for Advancement of Engrg, and the "Intl Coastal Engrg Award for 1980" by the American Society of Civil Engrs. *Society Aff:* ASCE

Hersh, Michael S
Home: 1134 Turquoise St, San Diego, CA 92109
Position: Mgr. *Employer:* Gen Dynamics. *Education:* MBA/Mgt/San Diego State Univ; MS/Met Engg/Stevens Inst of Tech; BE/Gen Engg/Stevens Inst of Tech. *Born:* 1/22/39. Mgr of Tool Eng & Tool Mfg, responsible for design, control, manufacture and service of all types of aircraft and missile tooling. Twenty yrs of res, dev, production support, and prog and functional mgt, primarily in mfg, mfg systems, Met and Joining. Presentation and publication of over 20 technical papers, Reg PE (Met - CA) 1966. Patent - Method of Sealing High of Year" award, 1971. Diploma - Prog Mgt from the Defense Systems Mgt College, 1977. Currently on ASM Nat Long Range Planning Committee. *Society Aff:* ASM, SME, NMA, Sigma, Beta, Gamma, Sigma, Iota Epsilon

Hershberger, W Delmar
Business: Sch Engrg UCLA, Los Angeles, CA 90024
Position: Professor Emeritus. *Employer:* Consultant. *Education:* PhD/Elec Engg/Univ of PA; MA/Physics/George Washington Univ; BA/Math & Physics/Goshen College, IN. *Born:* 5/10/03. Res Sci, RCA Labs 1937-49. Radar, Microwave Spectroscopy. 50 issued pats. Prof of Engrg UCLA 1949-70. P Hd of Electromagnetics Div & Lab. Fellow IEEE, cited for pioneer work on radar & frequency stabilization with microwave absorption lines. Paper on ultrasonics, microwave propagation, frequency stabilization, paramagnetic resonance, plasmas in solids. *Society Aff:* APS, IEEE, AAAS.

Hershenov, Bernard
Home: 22 Raleigh Rd, Kendall Park, NJ 08824
Position: Director, Marketing Coordination *Employer:* David Sarnoff Research Center Inc. *Education:* PhD/Electronics/U of Mich; MS/Math/U of Mich; BS/Physics/U of Mich. *Born:* 9/22/27. Dev Engr, GE Co, Schenectady 1959-60. RCA Labs since 1960 as Mbr of Tech Staff & Group Hd in high-power microwave tubes, microwave integrated circuits & solid state devices. Dir of RCA Res Labs, Inc Toyko, Japan 1972-75, Corp Engrg Staff, 1976-77, Hd, Energy Sys Analysis Group 1977-79, Dir of Opt. Electronics Research Lab, 1979-1987 (at RCA Labs). Currently, Director of Marketing at David Sarnoff Research Center (a subsidiary of SRI International) Mbr Adv Ctte, Magnetics Soc of IEEE, 1968-72; Secr-Treas 1971-72. Ser on Adv Ctte of MMM Conf, editorial bd of IEEE Transactions on MTT & Tech Ctte MTT-6. Two RCA Outstanding Achievement Awards & Fellow of IEEE. *Society Aff:* IEEE.

Hershey, Robert L
Home: 1255 New Hampshire Ave NW, Washington, DC 20036
Position: Division Vice President. *Employer:* Science Management Corporation *Education:* PhD/Engineering/Catholic Univ of Am; MS/ME/MIT; BS/ME/Tufts Univ. *Born:* 12/18/41. Res in instrumentation at Eastman Kodak, Bell Labs, Weston Instruments Acoustics at Bolt Beranek & Newman 1962-71. Mgt consulting at Booz, Allen and Hamilton 1971-79. With Science Management Corp since 1979. Developed Nat Waste Heat Utilization Plan for DOE, 1978. Developed projections of sitings of gasification combined cycle powerplants for DOE, 1979. Analyzed research programs for Gas Research Institute, 1979-81. Prepared National Fuel Cell Plan for DOE, 1981. Analyzed industrial technologies for Electric Power Research Institute, 1981-86, developed plans for robotics and advanced manufacturing technologies for loudspeaker producers (1983-84), analyzed acid rain control technologies for DOE (1982-85). Secy, NSPE PE in Ind Div 1973-75; Chairman, DC Chap, Acoustical Soc Am 1981-83; Chrmn, DC Sect, ASME 1978-79; Secy, DC Prof Council 1976; Pres DC Council Engg & Arch Societies 1978-79; Pres, DC Soc PE 1975-76 National Dir 1980-86; Chrmn, Wash Sect Robotics International, SME 1986-87. Secy, DC Bd of Registration for Professional Engineers. Tempo Enhancement Device 1975. Book: *How to Think with Numbers*, William Kaufmann, Inc, 1982. *Society Aff:* NSPE, ASME, ASA, SME.

Hershey, Robert V
Business: P.O. Box 510, Cloquet, MN 55720
Position: VP Mfg. Cloquet Unit. *Employer:* Potlatch Corp Northwest Paper Division *Education:* BS/ChE/IA State Univ *Born:* 11/5/32 Elected to Tau Beta Phi. Began employment in paper industry in June 1954, with American Can Co. Joined Potlatch Corp, Northwest Paper Division, in 1964, in the Research and Development Dept, and in May of 1980 became Plant Mgr. Actively involved in TAPPI, after joining in 1960, serving as officer and chrmn of various committees. In 1979 received the Coating and Graphic Arts Division Award and the Charles W. Engelhard Medallion, and served as TAPPI Director 1980-83. Currently VP Mfg Cloquet Unit, Northwest Paper Div., Potlatch Corp. *Society Aff:* TAPPI

Herskowitz, Gerald J
Business: Dept of Elec Eng & Comp Science, Castle Point Station, Hoboken, NJ 07030
Position: Prof of Elec Engg. *Employer:* Stevens Inst of Tech. *Education:* SEngSc/EE/NY Univ; MS/EE/Rutgers Univ; BEE/EE/Polytechnic Inst of NY. *Born:* 2/20/36. From NY City. Employed at RCA and Bell Lab before joining Stevens Inst of Tech in 1965. Promoted to Prof in 1972 and appointed Dir of Stevens Assoc, and indus liaison program, in 1975. During his two assignments at Bell Lab, Dr Herskowitz was respon for the dev of a variety of solid state circuits for magnetic memory and telecommunication system. His current res activities ctr about fiber optic communication sys for trunking and data networks. He is a telecommunications conslt to indus and government organizations. He is the author of two books in computer-aided design and 35 publications. Co-Chrmn of the Program Ctte of the IEEE Intl Communications Conference in 1976. *Society Aff:* IEEE, ASEE, TBΠ, ΣΞ, HKN

Hersman, Ferd W
Home: 826 Carini Lane, Cincinnati, OH 45218
Position: Pres *Employer:* Fischer Industrial Equipment Co. *Education:* BS/Chem Engrg/Univ of Cincinnati *Born:* 04/27/22 After some experience in R&D, Chem plant design & proj. engrg., specialized in the application and sale of equip to the chem process indus. Subsequently managed and purchased this manufacturers' representative agency (FIECo, 1983). AIChE, P Chmn OH Valley Sect., Engineers & Scientists of Cincinnati, Manufacturers' Agents Natl Assoc (MANA) Reg. PE in OH & KY, 31st Infantry WW II, Bronze Star, Cluster. Lt., USNR, 1950-1967; Greenhills Community Church, Presbyterian. Tennis, golf, BSA. Society Affiliation AIChE. *Society Aff:* AIChE, MANA

Hert, Oral H
Home: 5021 Turtle Creek Ct 2, Indianapolis, IN 46227
Position: Director, Bureau of Engineering. *Employer:* Indiana State Board of Health. *Education:* BS/Civil Engg/Purdue Univ. *Born:* May 1920 Indiana. BS Civil Engrg (sanitary option) 1948, Purdue Univ. Elected to Chi Epsilon. Ser in USAF 1941-45, 1951-53 retired with rank of major. Employed by the Indiana State Bd of Health since 1948 with increasing respons to present pos of Dir, Bureau of Engrg (300 employees) & Tech Secr, Stream Pollution Control Bd. Reg as a Prof Engr in Indiana. Mbr in Amer Soc of Civil Engrg, Amer Water Works Fed & Water Pollution Control Assoc. Ser as Pres of the Indiana Water Pollution Control Association & Diplomate Amer Acad Env Engg Dir to the Federation. Received Arthur Sidney Bedell Award in 1967. Enjoy fishing & square dancing. *Society Aff:* ASCE, WPCF, AWWA, AAEE.

Hertz, David B
Business: 1700 Broadway, New York, NY 10019
Position: Director. *Employer:* Prime Time Communications Inc *Education:* JD;PhD/In Eng/Columbia; MS/Naval Engg/USN Postgrad Sch; BS/Ind Eng/Columbus; BA/Economics/Columbia. *Born:* 3/25/19. Yoakum, TX. Retired Cmdr USNR. Chrmn, Prime Time Communications, Inc. Was Dir, McKinsey & Co Inc respon for tech dev, specializing in sys analysis, energy & the environ. Dr Hertz was a principal at Arthur Andersen & Co. Former Chrmn & Pres, The Inst of Mgmt Sci; P Pres, Operations Res Soc of Amer; Chrmn, Natl Acad of Sci Evaluation Panel for Inst of Applied Tech, Natl Bureau of Standards; Former Pres Intl Fed of Operations Res Societies. Author of *New Power for Mgmt: Computer Sys & Mgmt Sci* (McGraw-Hill); *The Theory & Practice of Indus Res* (McGraw-Hill); Trustee Columbia Univ; Trustee Columbia Univ Press. *Society Aff:* ORSA, TIMS, ASME, AIEE.

Hertzberg, Abraham
Business: Director, Aerospace & Energetics, Research Program, FL-10, University of Washington, Seattle, WA 98195
Position: Prof & Dir, Aero & Ener Res Prog *Employer:* University of Washington. *Education:* MS/Aero Engrg/VA Poly; BS/Aero Engrg/Cornell Univ. *Born:* July 8, 1922. BS Virginia Polytechnic Inst 1943, MS Cornell Univ 1949. 1943-46 aerodynamicist with Curtis-Wright Corp & 1944, 46 flight test engr with US Army. 1949 joined the Cornell Aeronautical Lab & was promoted in 1957 to Asst Hd & in 1959 to Hd of the Aerodynamics Res Dept. In 1965 was appointed to present pos of Prof Dept of Aeronautics & Astronautics & Director of the Aerospace & Energetics Res Lab, Univ of Washington. Since that time has been engaged in res on lasers & energy. Numerous prof & governmental cttes; 1966-67, Chmn of the NASA Res & Tech Ctte on Res & 1966-78, a mbr of the USAF Scientific Adv Bd. Fellow of the Amer Inst of Aeronautics & Astronautics, the Amer Physical Soc, Sigma Xi, Outstanding Educators of Amer & the Natl Acad of Engrs, NSF Plasma Dynamics Review Panel, DIA. *Society Aff:* AIAA, APS, AAES.

Hertzberg, Richard W
Business: Dept Mat Sci & Eng 5, Bethlehem, PA 18015
Position: Professor. *Employer:* Lehigh Univ. *Education:* BS/Mech Engrg/CCNY; MS/Metallurgy/MIT; PhD/Metallurgy/Lehigh. *Born:* Aug 1937 New York, NY. Fellow, ASM 1984. Eleanor and Joseph Libsch Outstanding Research Achievement Award, Lehigh University, 1983. William Woodside meml Lecturer, Detroit Chapter, ASM 1978. Outstanding Young Mbr, Lehigh Valley Chap, ASM 1973; Notable Achievement in Met Ed Award, Phila Chap ASM 1972; Alcoa Foundation Res Achievement Recognition Awds 1972 & 1973. Res Scientist United Aircraft Res Labs 1961-64; V Pres, Del Res Corp 1969-74; Visiting Prof, Ecole Polytechnique, Lausanne 1976. Currently NJ Zinc Prof. Prof Met & Dir Mech Behavior Lab, MRC. Authored 'Deformation & Fracture Mechs of Engrg Matls' 1976, 2nd ed. 1983; and co-authored "Fatigue of Engineering Plastics" 1980. Conducting res & teaching deformation, fracture & fatigue of metals & polymers. Cons to govt & indus. Enjoy travel, swimming, bicycling, tennis & cultural events. *Society Aff:* ASM, AIME, ΣΞ, NYAS.

Herum, Floyd L
Business: 590 Woody Hayes Drive, Columbus, OH 43210
Position: Prof *Employer:* OH State Univ *Education:* PhD/Agri Engrg/Purdue Univ; MS/Agri Engrg/IA State Univ; BS/Agri Engrg/IA State Univ *Born:* 03/02/28 Born near Estherville, IA. US Army enlisted 1946-48, 1950-51. Research assoc in grain storage and aeration, IA State Univ, 1954-55. Farm machinery specialist at Iraq Coll of Agriculture, 1955-57. Instructor in farmstead mechanization, Univ of IL, 1957-61. Assoc prof and prof, OH State Univ, 1964- pres, with initial two years in Brazil. Teaching, research, and graduate advising in physical properties of biological materials, processing of agricultural products, and electricity applications in agriculture. Causes and detection of grain quality deterioration in market. Consulting on process system malfunctions. Hobbies: woodworking, fishing, photography. *Society Aff:* ASAE, ASEE, IFT

Hervey, John H
Business: 2055 L St, N W, Washington, DC 20036
Position: General Manager-Distribution Services-WMA *Employer:* Chesapeake & Potomac Telephone Co. *Education:* Elec Engrg/Bliss Elec; Comm Engr/Univ of MO. *Born:* 04/28/23 Educated in the fields of electrical & communications engrg; has been involved in virtually all aspects of communications engrg and operations for over 30 yrs & continues as an active mbr of IEEE & the Washington Soc of Engrs. At the present time, is a General Mgr of The Chesapeake & Potomac Telephone Companies & is respon for Distribution Services in the Washington Metropolitan Area with an organization of over 3000 people. *Society Aff:* IEEE, WSE

Herwald, Seymour W
Business: 2282 Elmhill Rd, Pittsburgh, PA 15221
Position: Consultant. *Employer:* Self. *Born:* Jan 1917. PhD & MS from Univ of Pittsburgh; BS from Case Inst of Tech. Native of Cleveland, Ohio. Joined Westinghouse Elec Corp in 1939 at Pittsburgh. Work in automatic control culminated in design of airborne military electronics equip. 1952 Engrg Mgr; 1956 Div Mgr; 1959 V Pres Res; 1962 V Pres Electronic Components & Specialty Prods Group; 1968 V Pres Engrg; 1970 V Pres Engrg & Dev; 1975 V Pres Strategic Resources. Consultant IEEE, Fellow & Pres 1968; NEMA, Bd/Governors; NAE; ANSI served on Bd/Dir;

Herwald, Seymour W (Continued)
AF Decoration for Exceptional Civilian Serv 1974; Mbr AF Sci Adv Bd 1956-71. *Society Aff:* NAE, IEEE, ASME.

Herz, Eric
Business: 345 E 47th St, New York, NY 10017
Position: Exec Dir. *Employer:* Inst of Elec & Electronics Engrs. *Education:* BEE/Electronics/Poly Inst of Brooklyn. *Born:* 7/23/27. After grad, worked at Sperry Gyroscope Co on dev of LORAN -C. Moved to San Diego in 1957 to join General Dynamics. Supervised telemetry data processing engg, managed dev of a new military digital range measuring and communication system, participated in early space shuttle avionics devs, managed prog offices for cruise missile systems engg & support equip. Active as IEEE volunteer in local regional, and technical activities. Last volunteer office was VP-Technical Activities. Have been IEEE Dir since 1976. Moved to NY in 1979 to serve as Exec Dir & Gen Mgr of IEEE. *Society Aff:* IEEE.

Herzenberg, Aaron
Business: 7330 NW 12 St, Miami, FL 33126
Position: VP Engg. *Employer:* Alpha Consolidated Ind Inc. *Education:* BME/Power/NYU. *Born:* 11/8/19. Brooklyn, NY native. Reg PE. Patent for shop assembled solid fuel fired boiler granted 1970. Asst Chief Engr, Ind Div, Combustion Engg Inc 1945-1958. Chief Engr: Keeler Co 1958-1970. VP Engg at Intl Boiler Works 1970-1972. Mgr of Boiler Operations Superior Combustion 1972-1973. Mgr of Proposal Engg and Solid Waste Engg at Erie City Energy Div, 1973-1975. Proj Mgr at Metcalf & Eddy, 1975- 1976. Joined Alpha Consolidated Industries in 1976 as VP of Engg. Mbr of PA Boiler Advisory Bd from 1965-1975. Mbr of Williamsport Water & Sewer Authority, 1965-1970. Enjoy reading and music. *Society Aff:* ASME, ASME, NSPE.

Herzenberg, Caroline L
Business: Argonne National Lab, Bldg. 362, Argonne, IL 60439
Position: Physicist *Employer:* Argonne National Lab *Education:* PhD/Physics/Univ of Chicago; MS/Physics/Univ of Chicago; SB/Physics/MIT *Born:* 3/25/32 Nuclear physics researcher, Univ of Chicago (1958-9), Argonne Natl Lab (1959-61). Principal Investigator, NASA Apollo returned lunar sample analysis program, IIT Res Inst, 1967-71. Faculty member, IL Inst of Tech (1961-66); Univ of IL (1971-4); CA State Univ (1975-6). Developer of new fossil energy process control instrumentation and technology evaluation physicist, 1977-present, Argonne National Lab. Producer and host, television series on science and engrg, 1974-5. Chair, APS Committee on the Status of Women in Physics 1977-79; pres, AWIS Chicago Area Chapter, 1981; Treas, National AWIS, 1982-84; President-elect, national AWIS, 1985-87. General Chair, 1981 Symposium on Instrumentation and Control for Fossil Energy Processes. Private pilot. *Society Aff:* APS, AAAS, AWIS, ΣΞ

Herziger, William J
Home: N6OW5444 Edgewater Dr, Cedarburg, WI 53012
Position: Pres (Retired) *Employer:* Herziger & Assocs, Inc *Education:* BS/EE/Marquette Univ *Born:* 5/16/17 Native of Cedarburg, WI. Senior Project Engr, Switchgear and Control Division, Allis-Chalmers Manufacturing Co 1940-1951. District Mgr for Power Equipment Sales, Federal Pacific Electric Co 1951-1963. Pres of Herziger & Assocs, Inc, Consulting Electrical Engrs, Milwaukee, WI 1964-1982. Responsibilities included Corporate General Mgr and Engr In Charge of Reports, Studies and Chief Design Engr of Substations, Power Distribution Systems, Instrumentation and Control. Registered PE in WI and IL. Author of U. S. Patent No. 2,478,693 covering Electrical Control System and Apparatus. Pres of Electric Utility Commission, Cedarburg, WI. *Society Aff:* IEEE.

Herzog, Gerhard
Home: Via Collina 15, Viganello, Switzerland 6962
Position: Consultant. *Employer:* Retired. *Born:* 4-30-1904 PhD 1928, Mech Engrg 1925, Swiss Inst of Tech, Zurich: Prof Physics 193338. Res Assoc Birkbeck Col, London 1934, Res Assoc Univ of Chicago 1939-40. Fellow Amer Phys Soc. With Texaco: Physicist Res since 1940, advancing to Mgr Prod Res in Houston; from 1962-69 Mgr European Res in Brussels & Dir Texaco Belgium & Texaco Europe. Originator of many pats. Retired 1969. Since 1969 cons for indus in the field of dev & mgmt.

Hesler, Warren E
Business: 48 Leicester Sq, London WC2H 7LZ, England
Position: President. *Employer:* Catalytic International, Inc. *Born:* Oct 1923 Chicago, Ill. BS from Northwestern Univ. Grad studies NYU. Chem Engr & Sales Exec for Swenson Evaporator Co. Genl Sales Mgr Blaw Knox Co. Sales Mgr for Lummus Co. V Pres of Day & Zimmerman. Sales Mgr Catalytic Inc. Since 1973 Pres of Catalytic Internatl London England. Respon for Eastern Hemisphere design & const projs in the chem & oil refining fields. Fellow of AIChE. Mbr of TAPPI, IFT, SAME, Chemists Clubs', NYC. Author of numerous papers & holder of many pats in the field of evaporation, crystallisation & food processing.

Hess, Daniel E
76 Acco Drive, York, PA 17402
Position: Pres & C.E.O., Chain Products Group *Employer:* Acco Babcock Inc *Education:* BS/IE/PA State Univ. *Born:* Sept 1931. York Div Borg Warner Corp 1949-54, Co-Op 1954-58 Supervisor Standards & Methods Dev. Stephenson-Walsh & Assocs 1958-62, Mgmt & Engrg Cons York PA & Paris France. York Shipley Inc 1962-67 Ch Indus Engr. ACCO Babcock Inc.1967-present, Fellow, Amer Inst of Indus Engrs; other pos V P Prof Relations 1968-70; Honors & Awards Bd; Dir Internatl Relations; numerous Task Forces. Natl Soc of Prof Engrs, Assoc of Prof Engrs-Province of Ontario, Amer Welding Soc, Amer Soc of Testing & Matls, Amer Natl Standards Inst, Chmn US Delegation Tlll to Internatl Standards Orgn, Chamber of Commerce, Manufacturers Assn, Pres's Club of York Coll. Pres-Natl Assoc of chain mfgs. Bd of Dirs: Babcock Intl Inc, Fairfield CT; Acco Babcock Inc, Fairfield CT; Dominion Chain Inc, Stratford, Ontario (Pres & Chrmn of the Bd); Drovers & Mechanics Bank, York, PA; AAA Motor Club of PA, York, PA; Pace Resources, Inc, York, PA. *Society Aff:* ANSI, NSPE, IIE, ASTM, ISO.

Hess, Karl J
Business: Coordinated Science Lab, 1101 W. Springfield Ave, Urbana, IL 61801
Position: Research Prof *Employer:* Univ of IL *Education:* PhD/Physics & Math/ Univ of Vienna *Born:* 6/20/45 in Trumau, Austria. Assistant Prof 1970-1977 in Vienna, Dozent in 1977, came to the US permanently in 1977. Now Prof of Electrical Engrg at the Univ of IL, Urbana. Primary activities are res in semiconductor materials and theory of semiconductor devices. Married to Sylvia Horvath, children Karl and Ursula. *Society Aff:* IEEE

Hess, Wilmot N
Business: P.O. Box 3000, Boulder, CO 80307
Position: Director, National Center for Atmospheric Research *Employer:* National Center for Atmospheric Research *Education:* ScD/Hon/Oberlin College, OH; PhD/ Physics/Univ of Cal Berkeley; MA/Physics/Oberlin College; BS/EE/Columbia Univ. *Born:* 10/16/26. Lawrence Radiation Lab, incl work on nuclear bomb dev & testing; Dir, Plowshare Div, respon for developing schemes for peaceful bomb use; worked on nuclear experiments using cyclartron & bevatron. 1961-66 Dir, Lab for Theoretical Studies, NASA, GSFC, helped dev quantitative understanding of the Van Allen Radiation Belt of the earth. 1966-69: Dir, Sci & Applications, NASA- Houston, where I directed Apollo Sci program. 1970: Dir, Environ Res Labs, NOAA. Arthur S Fleming Award 1965; NASA Group Achievement Award 1969; AIAA G Edward Pendray Award 1969; Mbr, Natl Acad of Engrg 1976; 1980: Dir, Natl Center for Atmospheric Res. *Society Aff:* APS, AAAS, AMS, AGU.

Hesse, M Harry
Business: Rensselaer Polytechnic Inst, JEC 5008, Troy, NY 12181
Position: Prof. *Employer:* RPI. *Education:* DrIng/Elec Power Engg/Technical Univ

Hesse, M Harry (Continued)
Aachen; MSEE/Elec Power Engg/IL Inst of Tech; BEE/Elec Power Engg/Marquette Univ. *Born:* 3/20/27. in Milwaukee, WI. Served with US Navy, AETM 3/c, 1945-1946. With the GE Co, Pittsfield, 1956-1957 as transformer design engr; 1958-1970 as sr analytical engr in Schenectady. With RPI since 1970 as Prof of Elec Power Engg. Retained as consulting engrs by GE (LSTG) since 1971. Is a mbr of several IEEE committees, Fellow of the IEE (London), on panel of experts US CIGRE. Also a mbr of Tau Beta Pi, Eta Kappa Nu, Pi Mu Epsilon, Sigma Xi, and recipient of Westinghouse Elec Power Systems Fellowship & Fulbright Scholarship & 1980 EEI Engineering Educator Award. *Society Aff:* IEEE, IEE, ASEE, CIGRE.

Hessel, Alexander
2128 E14, Brooklyn, NY 11229
Position: Prof *Employer:* Polytechnic Univ *Education:* MSc/Physics/Hebrew Univ Jerusalem, Israel; DEE/Elec Engrg/Poly Inst of Brooklyn. *Born:* 10/19/16. Prof of Electrophysics Polytechnic Univ specializing in Electromagnetics, Open Periodic Structures & Phased Arrays. Prof Exper includes: Res Engr with the Israeli Ministry of Defense 1948-51 & 1953-56. Lecturer at Technion Haifa, Israel 1955-56. Res Assoc Microwave Res Inst 1958-60. Res Asst Prof 1960-62. Polytechnic Univ Brooklyn Assoc Prof 1962-67, Prof, 1967-. Fellow IEEE; mbr US Comm B of URSI, mbr GAP and MTT, mbr of Sigma Xi. Has authored & pub about 50 papers. *Society Aff:* IEEE, URSI

Hestekin, Walter E
Home: Walt Hestekin & Assoc Ltd, 704 S. Barstow St, Eau Claire, WI 54701
Position: President *Employer:* Walt Hestekin & Assoc, Ltd. *Education:* BSME/ME/Univ of WI-Madison. *Born:* US Navy 1942-45. Employed by Hovland Sheet Metal 1947-71. Corp Pres 1965- 71. NCEE Certificate. Prof Affiliate AIA. VP Larson Hestekson, Ayres Ltd - 1971- 1980 Present Position since April 1980. *Society Aff:* ASHRAE, NSPE, NCEE, AEE, EFP.

Hester, J Charles
Home: 4901 Wareham Dr, Arlington, TX 76017
Position: Program Mgr, High Temperature Power Systems *Employer:* LTV Aerospace & Defense Corp *Education:* PhD/ME/OK State Univ; MS/ME/OK State Univ; BS/ME/Arlington State Univ *Born:* 12/14/38 Was Asst prof of Mech Engrg at the Univ of TX-Austin (Aug 1966-67). Engrg Specialist, LTV Aerospace Corp, Missiles and Space Div (Aug 1967-Jan 1970). Assoc Prof of Mech Engrg at Clemson Univ (Jan 1970-Jan 1971). Assoc Prof and Dept Hd, Mech Engrg Dept, Clemson Univ (Feb 1971-Jan 1974). Assoc Dean, Coll of Engrg, Clemson Univ (Jan 1974-Jan 1977). Prof (1977-85). Program Manager LTV (1985-Present). Consultant to ind in energy use. Responsible for advanced energy systems and advanced air breathing missile systems planning, analysis and design. Served as Energy Advisor to Governors Edwards and Riley of SC. Presently Program Mgr of DARPA $45 Million High Temperature Gas Turbine Development Program. *Society Aff:* ASME, ASHRAE, ASEE, TSPE, NSPE

Hetnarski, Richard B
Business: Department of Mechanical Engineering/Rochester Institute of Technology Rochester, NY 14623
Position: Professor of Mechanical Engineering *Employer:* Rochester Institute of Technology *Education:* Dr. of Technical Sci. (PhD)/Mechanics/Polish Academy of Sciences, Warsaw; MS/Mathematics/University of Warsaw; MS/Mechanical Engineering/Technical University of Gdansk *Born:* 05/31/28 Designer, Design Bureau of Diesel Engines, Warsaw, 1952-4; Senior Engineer, Institute of Aircraft Research, Warsaw, 1955-59; Research Scientist, Polish Academy of Sciences, Warsaw, 1959-69; Vis. Assoc. Prof, Cornell Univ, 1969-70; Distinguished Vis. Prof, Rochester Inst of Tech, 1970-71; Prof of Mech Engrg, Rochester Inst of Tech, since 1971. Postdoctoral Fellowship, Columbia Univ, 1964-65; Northwestern Univ., 1965; Summer Fellowship and Sabbatical Leave: NASA Lewis Research Center, Cleveland, 1979-80; Visiting Prof, Univ of Paderborn, Germany, 1980. Editor-in-Chief, Journal of Thermal Stresses: An International Quarterly, since 1978; Editor, Handbook of Thermal Stresses, vol. 1-3, North-Holland, Amsterdam; Editor of: Dynamic Problems of Thermoelasticity, by W. Nowacki, Northoff, 1975; Dynamic Fracture Mechanics, vol. 1, by V. Z. Parton and V. G. Boriskovskii; Mechanics of Elastic-Plastic Fracture, by V. Z. Parton and E. M. Morozov. Author of numerous papers in Mechanics and Mathematics. *Society Aff:* ASME, AAM, ASEE

Hetrick, David L
Business: Dept of Nuclear and Energy Engrg, Tucson, AZ 85721
Position: Prof *Employer:* Univ of AZ *Education:* PhD/Phys/UCLA; MS/Phys/RPI; BS/Phys/RPI *Born:* 01/26/27 Born in Scranton, PA. Experience includes 8 years at Rockwell Intl, consltg for industries and natl labs, teaching phys at RPI and CA State Univ at Northridge, and teaching nuclear engrg at the Univ of AZ (since 1963). Part time admin Judge, US Nuclear Regulatory Comm. Res Interests: nuclear reactor dynamics, nuclear safety, applied math, and continuous sys simulation. *Society Aff:* ANS, APS, SIAM, SCS, AAAS, ASEE, IMACS

Hetzer, Hugh W
Home: 461 Sproul Road, Kirkwood, PA 17536
Position: retired *Education:* BS/Civil Engg/Carnegie Inst of Tech. *Born:* 12/13/13. Native of Moundsville, W Va. Distribution Engr W Va Water Ser Co Charleston, W Va 1936-42. Application Engr & Ser Mgr, Westco Pump Div, Joshua Hendy Iron Works St Louis, Mo 1942-44. Real Estate & Right of Way Engr W Va Water Ser Co Charleston, W Va 1944-57. Group Leader & Proj Mgr Union Carbide Co South Charleston, W Va 1957-62. Supt of Prod Chester Water Authority Chester, Pa 1962-78. Retired 12/31/78. Fellow ASCE, Fuller Awardee, AWWA, P Chmn Pa Sect AWWA. Life Mbr AWWA, Life Mbr ASCE. Reg Engr in W Va. 1978-1982 Sr Engg, Kelley, Gidley, Blair & Wolfe, Inc, Charleston, W Va; 1983-present Self-employed Consultant Water treatment and related fields. Received the Gold Water Drop Award from AWWA in 1987. (50 years of membership) Delivered a paper at both the W. Va. Section Annual Meeting and the Pa. Section Annual Meeting in 1987. Subject: Continuous Upflow Filtration. *Society Aff:* ASCE, AWWA.

Heuser, Henry V, Sr
Business: 1000 West Ormsby Street (P. O. Box 1918), Louisville, KY 40210 (40201)
Position: Chairman of the Board *Employer:* Henry Vogt Machine Co. *Education:* Doctor of Science (Honorary)/University of Louisville; BSME/Purdue University *Born:* 06/14/14 Born in Louisville and educated in Louisville public schools. After graduating from Purdue University, was employed by Henry Vogt Machine Co. in 1936. Continued in employment and served as President and CEO from 1953 to 1984. After retirement have remained active in the affairs of the Company, particularly in the field of research and development. *Society Aff:* ASME

Hewitt, Hudy C, Jr
Business: Box 5161, Cookeville, TN 38505
Position: Professor and Chairman, Mechanical Engrg Dept *Employer:* Tennessee Technological University. *Education:* PhD/Mech Engg/OK State Univ; MSc/Mech Engg/OH State Univ; BSc/Mech Engg/OK Univ. *Born:* 4/9/37. Served in the US Navy 1961-63 at the USN Postgrad Sch. Prof of Mech Engrg at Tennessee Tech Univ 1966-present. Mbr Phi Eta Sigma, Sigma Tau, Tau Beta Pi, Pi Tau Sigma. Natl Secy-Treas Pi Tau Sigma 1971-1983. President Pi Tau Sigma 1983-86. Editor-in-Chief Mechanical Engineering News 1980-84. Authored Textbook *Scope of Experimental Analysis.* Reg PE in Tennessee. Major int incl measurements, solar energy, heat transfer, ranching & fishing. *Society Aff:* ΠΤΣ, ΤΒΠ, ΦΗΣ, ΣΤ, ΟΔΚ, ASME, ASEE.

Hewitt, William A
Business: John Deere Rd, Moline, IL 61201
Position: Chrmn & CEO *Employer:* Deere & Co *Education:* AB/Econ/Univ of CA *Born:* 8/9/14 Native of San Francisco. Honorary Doctor of Laws Degrees, Augustana Coll, Rock Island, IL and Knox Coll, Galesburg, IL. Bd of Trustees, CA Inst

Hewitt, William A (Continued)
of Tech. Visiting Committee Harvard Univ School of Design and Harvard Business School. Stanford Res Inst Council, 1971-1979. Fellow, American Society of Agricultural Engrs, 1981. Joined Deere & Co 1948, Exec VP, 1954; Pres and Chief Executive Officer, 1955; currently Chrmn and Chief Executive Officer. Responsible for introduction of modern agricultural equipment including four wheel drive tractors and machines embodying advanced human factors relationships for operators. Intense interest in architecture and art. *Society Aff:* ASAE, SAE, AIA

Hewlett, Richard G
Home: 7909 Deepwell Dr, Bethesda, MD 20034
Position: Vice President, Senior Associate *Employer:* History Associates Incorporated *Education:* PhD/History/Univ of Chicago; MA/History/Univ of Chicago. *Born:* 1923 Toledo, Ohio. Intelligence specialist, Dept of the Air Force, Washington 1951-52. Program Analyst, US Atomic Energy Comm 1952-57; Ch Historian 1957-75. Author of 3 books on the history of atomic energy: The New World, 1939- 46; Atomic Shield, 1947-52; Nuclear Navy, 1939-62. Ch Historian, US Energy R&D Admin 1975-77. Ch Historian, US Dept of Energy, 1977-1980; vice president, senior associate, History Associates Incorporated, 1980-. 1969-70 David D Lloyd Prize in history; AEC Distinguished Ser Award 1973. Mbr, Exec Council, Soc for the History of Tech 1977-1980.. *Society Aff:* AHA, OAH, SHOT, SHFG, NCPH.

Hewlett, William R
Business: 1501 Page Mill Road, Palo Alto, CA 94304
Position: Vice Chmn of the Bd *Employer:* Hewlett-Packard Co. *Education:* MS/Elec Engg/MIT; EE/Elec Engrg/Stanford U; LLD/Hon-Law/U of CA, Berkeley 1966; LLD/Hon-Law/Yale Univ 1976; DSc/Hon/Kenyon Coll; EngD/Hon/Notre Dame & UT State Dartmouth Coll; LLD/Hon Law/Mills Coll. *Born:* 5/20/13. Ann Arbor, Michigan. Awards: Hon Life Mbr, IEEE, Founders Medal (with David Packard) 1973; Vermilye Medal (with David Packard) by the Franklin Inst, Philadelphia 1976; Medal of Achievement, Western Electronic Manufacturers Assoc. 1971; SAMA Scientific Apparatus Makers Ass. jointly with David Packard 1976. Honorary Degrees - University of Notre Dame - Eng.D 5/18/80; Utah State University, Eng.D 5/31/80. Member-National Academy of Sciences; Honorary Lifetime Membership - ISA; Member, American Philosophical Society. *Society Aff:* NAE, NAS, IEEE. ISA.

Heyborne, Robert L
Business: 3601 Pacific Avenue, Stockton, CA 95211
Position: Dean, School of Engineering. *Employer:* University of the Pacific. *Education:* PhD/Elec Engg/Stanford Univ; MS/Elec Engg/UT State Univ; BS/Elec Engg/UT State Univ. *Born:* 4/17/23. Ph.D./Elec. Engrg Stanford Univ; MS/Elec Engrg/UT State Univ; BS/Elec Engrg/UT State Univ. Instructor 1942-46, 1951-52 Naval Elec Sch, Del Monte, Ca & Memphis, Tenn. Prof Utah State Univ 1957-69, Dean of Engrg. Univ of the Pacific 1969-present. Indus conslt Publs, Journal of Geophysical Res, Engrg Ed. ASEE; Chrmn Rocky Mountain Section 1969; Chrmn Pacific Southwest Sect 1975 Chrmn, Cooperative Ed Div 1977, Relations with Indus Bd 1974-77, Chrmn Council of Sections zone IV 1982-1984. Chrmn, Calif State Engrg Liaison Ctte 1974. Awards for distinguished teaching, Honor Award, Consulting Engineers Assoc of California, 1985 & 1987; Amer Soc for Engg Educ Natl Clement J. Freund Award, 1986 & Alvah K. Borman Award, 1983; Distinguished Alumni Award, Utah State University 1979. Engr of the Year Award, San Joaquin Prof Engrs Council 1972. Mbr, IEEE, AGU, ASEE, URSI, Sigma Xi, Tau Beta Pi, Phi Kappa Phi, Eta Kappa Nu. *Society Aff:* AGU, IEEE, ASEE, URSI, ΣΞ, ΦΚΦ, ΤΒΠ, HKN.

Heydt, Gerald B
Business: 101 W Bern Street, Reading, PA 19601
Position: Mgr, Process Laboratory. *Employer:* Carpenter Technology Corporation. *Education:* BS/Metallurgy/Penn State Univ. *Born:* 11/18/26. Pub 5 articles. Hold 5 US Pats. Mbr: ASM, Welding Res Council (Amer Welding Soc), ASTM, AUS, APMI.

Heyer, Edwin F
Business: 2901 Ponce de Leon Blvd, Coral Gables, FL 33134
Position: President. *Employer:* Brill/Heyer Associates. *Education:* BSCE/Civil Eng/Univ of Miami. *Born:* Dec 1922. Tau Beta Pi; Natl Dir NSPE; Fellow & P V Pres Fla Engr Soc; Fellow Amer Soc Civil Engrs; ACI Mbr; South Fla Engr of Year 1969. Recipient J F Lincoln Arc Welding Foundation Design Award 1966. Cons Struc Engr specializing in Blast Design & Hurricane Wind Design. Formerly V Pres & Ch Structural Engr of 200 man Architect-Engr firm & instructor of Civil Engr at Univ of Miami. Author of Tech Papers in *Welding Engr & Fes Journal.* P Pres South Fla Interprofessional Council; Mbr Chamber of Commerce. *Society Aff:* NSPE, ASCE, ACI.

Heywang, Walter
Business: Otto-Hahn-Ring 6, Munich 83 W Germany 8000
Position: VP. *Employer:* Siemens AG. *Education:* Prof/Phys/Tech Univ Munchen; Dr/rer-nat/Univ Wurzburg; DiplPhys/Phys/Univ Wurzburg. *Born:* 10/1/23. After my studies of phys at the Univ of Wurzburg I have joined Siemens in 1950, where I am now in charge of the Corporate R&D labs. Additional activities: Mbr of supervisory bds for Vacuumschmelze AG & BESSY (Berlin Electro Synchrotron). Mbr of steering committees for the Fraunhofer Inst for Applied Solid State Phys & for Solid State Tech. Coeditor of the series "Solid State Electronics–, Springer, Berlin - Heidelberg - NY. *Society Aff:* IEEE, DPG, NTG.

Hiatt, Ralph E
1869 Field Road, Charlottesville, VA 22903
Position: Prof of Electrical Engineering. *Employer:* University of Michigan. *Education:* AB/Physics/IN Central Univ (Univ. of Indianapolis); MA/Physics/IN Univ. *Born:* 4/12/10. Native of Portland, Ind. Taught in public schs of Portland & Richmond, Ind prior to WWII. Staff Mbr, MIT Radiation Lab 1942-45, Hd of Antenna Field Sta 1944-45. Air Force Cambridge Res Lab 1945-58, Ch of Antenna Lab 1956-58. Univ of Mich 1958-1980, Prof of Elec Engrg, Dir Radiation Lab 1960-75; Res elecromagnetic scattering & antennas. Mbr IEEE Admin Ctte on Antennas Propagation 1962-74; V Chmn & Chmn 1969-70, News Letter Editor 1972-74. Fellow IEEE & AAAS, Cons to Gov't & Indus. Hobbies: photography & travel; married 1940, 3 children *Society Aff:* ΤΒΦ, HKN, ΣΞ.

Hibbard, Walter R
Home: 1403 Highland Circle, Blacksburg, VA 24060
Position: Univ Disting Prof of Engrg. Director, Virginia Center for Coal and Energy Research *Employer:* Va Polytechnic Inst & State Univ. *Education:* DEng/Metallurgy/Yale; AB/Chemistry/Wesleyan. *Born:* Jan 1918. Navy (Buships) 1942-46. Taught at Yale 1946-51. Managed met & ceramic res at GE Co 1951-65. Dir US Bureau of Mines 1965-68. V Pres, Res & later Tech Services, Owens Corning Fiberglas 1968-74. Energy Res & Dev Office, Federal Energy Office 1974. Taught at VPI&SU since 1974. Pres, AIME 1967. James Douglas Gold Medal 1968. Mineral Economics Award 1983. Fellow, Met Soc & AAAS; Chmn, Engrg Sect 1976. Fellow, Amer Soc for Metals & Amer Ceramic Soc. Mbr, Natl Acad of Engrg, Amer Acad of Arts & Sciences. Registered engineer in Virginia, Ohio and Connecticut. *Society Aff:* AIME, ASM, ACerS, ASEE, AAAS, NICE.

Hibbeler, Russell C
Business: Civil Engineering Dept, Schenectady, NY 12308
Position: Assistant Professor. *Employer:* Union College. *Born:* Jan 18, 1944. PhD from Northwestern Univ, BS & MS from the Univ of Ill, PE. Worked as Struc Engr for the City of Chicago & Chicago Bridge & Iron Co until 1967. Taught at Youngstown State Univ 1968-71. Performed PostDoctoral Res at Argonne Natl Lab & taught at Ill Inst of Tech 1972-74. Worked as a Staff Cons at Sargent & Lundy 1975; currently involved in teaching, res & textbook writing at Union Coll. Hold Mbr status in ASCE, ASME, ACI & ASEE. Also a mbr of Tau Beta Pi & Sigma Xi.

Hibbeln, Raymond J
Home: 4812 Wallbank, Downers Grove, IL 60515 *Employer:* Retired. *Education:* BS/Metallurgical Engrg/MI Technical Univ; MS/Metallurgical Engrg/Univ of MI. *Born:* 11/26/25. Ontonagon, MI. Alpha Sigma Mu. USMC. With Western Elec since 1951. Retired 1987. Dept Ch Engr of Mfg 1956 heading several dev & production engrg depts. Operations included heat treating, met labs, foundry, die casting, parts mfg & assembly, sheet metal, brass & magnetic matl rolling mill. Assumed Sr Engr, Matls Dev in 1975. 37 yr mbr ASM. P Chmn Chicago Chap, Mbr ASTM in A-6, B-1, B-2 & B-5 (Chrmn B-5 1982-85); (Sec. B-5 1987-?); ANSI in B-32; Microbeam Analysis Soc; USA-TAG on Copper and Copper Alloys; ISO TC/26 on Copper; Recipient of 1984 Award of Merit and elected Fellow of the Soc. Co-editor of ASTM STP-831 *Sampling and analysis of Copper Catholic* (1984). Enjoy weaving, travel, outdoors, & Scouts (four sons are Eagles). *Society Aff:* ASM, ASTM, ISO-TC/26, USA-TAG.

Hickling, Robert
Home: 8306 Huntington Road, Huntington Woods, MI 48070
Position: Sr. Staff Res. Engineer *Employer:* GM Research Labs *Education:* Ph.D./Engineering Science/Calif Inst. of Technology; M.A./Math/St. Andrews University, Scotland *Born:* 10/28/31 Born Bologna, Italy; U.S. Citizen; divorced, 3 children. Education: M.A. Univ of St. Andrews 1956, Ph.D. Caltech 1963. Professional Experience: Scientific Officer, British Royal Naval Scientific Service 1954-57; General Motor Research Labs 1963-present. Sr. Res. Eng 1963-70, Supervisory Research Engr 1970-76, Dept. Res. Engr 1976-80, Sr. Staff Res. Engr. 1980-present. Concurrent Positions: part-time lecturer Wayne State Univ 1963-64, 1983-85; Adjunct professor mech engrg Wayne State Univ 1987-. Technical interests: ultrasonic sensing for inspection and object recognition in manufacturing; combustion enhancement. *Society Aff:* ASME, ASA, INCE

Hickman, Roy S
Business: Santa Barbara, CA 91006
Position: Prof *Employer:* Univ of CA *Education:* PhD/ME/Univ of CA-Berkeley; BS/ME/Univ of CA-Berkeley *Born:* 4/13/34 Cryogenics engr at Lawrence Radiation Lab (1957-1961). Research Engr at Jet Propulsion Lab (1961-62). Senior Research Engr at Heliodyne (1962-1964). Asst Prof and Assoc Prof at Univ of Southern CA (1964-1968). Assoc Prof (1968-1975) and Prof of Mechanical and Environmental Engrg 1975 - . Dept Chrmn (1977-1981) of Mechanical and Environmental Engrg. Major fields of interest: Diagnostic fluid mechanics measurements, laser probes, Raman spectrometry of flames, electron beams, supersonic aerodynamics. *Society Aff:* ASME, AIAA, ASEE

Hickok, Robert L, Jr
Business: ECSE Dept, Plasma Dynamics Lab, Troy, NY 12181
Position: Prof *Employer:* RPI *Education:* PhD/Physics/RPI; MA/Physics/Dartmouth; BS/Physics/RPI *Born:* 2/25/29 Native of Schenectady, NY. Post Doc 1956-58 in Linear Electron Acceleratory Lab, Yale Univ. 1958-1971 employed at Mobil Research and Development Center, Princeton, NJ, where I was responsible for development of the fusion research program. Joined the faculty of the ECSE Dept, RPI in 1971 and is the dir of the Plasma Dynamics lab. I have also had primary responsibility for research projects at Princeton Plasma Lab, ORNL, LLNL and United Tech Research Center. Consultant for Gen Elec Co. Married and have 3 children. *Society Aff:* IEEE, NPSS, APS

Hicks, Earl J
Home: 2001 Skyline Place, Bartlesville, OK 74006 *Employer:* Retired from Phillips Petroleum Co. *Education:* BS/ME (Aero)/University of Utah; *Born:* 12/02/19 Earl J. Hicks' career spans 41 years, the last 37 of which were spent with Phillips Petroleum Company. For 34 of those years he worked in technical engrg developing an expertise and becoming an authority in vibration control, pulsation control (for pumps and compressors), dynamic considerations for piping systems, effects of pulsation on compressor performance and metering, noise, torsional vibration, vibration isolation, experimental stress analysis and special hydraulic surge considerations. His expertise has been recognized by the Oklahoma Society of Prof Engrs awarding him the "1982 Outstanding Engineer" award for the State of Oklahoma and in 1984 being elected a "Fellow–, ASME. The last three years of his career he served in management positions, first as the Dir of the Materials Engrg Lab and then as Branch Manager of Standards and Regulations. *Society Aff:* ASME, NSPE, OSPE, SEM

Hicks, Frank B, Jr
Business: 4560 Old Pinevlle Rd, Charlotte, NC 28210
Position: President. *Employer:* Frank B Hicks Associates, Inc. *Education:* BS/Civil Eng/NC State Univ. *Born:* Jan 18, 1931 NC. Established Frank B Hicks Assocs in 1966 to practice Civil, Structural & Sanitary Engrg, Surveying & Planning. Reg Engr in NC, SC, Tenn, Ga, Fla, W Va, Ky, Ohio, Md, Va. Reg Land Surveyor NC & SC. Fellow ASCE, PPres NC Sect ASCE. Mbr: NSPE, Prof Engr of NC, Cons Engr Council of NC, NC Soc of Surveyors, ACI. *Society Aff:* ASCE, PENC, NSPE, NCSS, ACEC, ACI.

Hicks, Jesse L
Home: 7300 Homestead Drive, Knoxville, TN 37918
Position: Project Manager, Senior Civil Engineer *Employer:* IT Corporation *Education:* MPA/Pub Adm/NC State Univ; BS/Bus Adm/Franklin Univ; BS/Agri Eng/Univ of TN. *Born:* 1922. Native of Paris, Tenn. Served with Army Corps of Engrs 1943-45; Field Engr with USDA, SCS 1949-60 in Tenn; State Engr in New Hamsphire 1960-65 with USDA, SCS; 1965-70 Asst State Dir, Ohio; 1970-71 Deputy State Dir, Ohio; 1971-81 served as State Dir of SCS in North Carolina. In charge of admin of all SCS programs incl watershed planning & const, river basin planning. State Chmn of NC Chap of ASAE in 1974. Received Outstanding performance ratings in 1969 & 1974; Governors Award, NC Wildlife Federation 1976; President Intl Soil Conservation Soc of Amer 1981, Consultant to Ecuadorian Govt on Soil Conservation in 1979; Elected Fellow in Amer Soc of Agri Engrs 1979. Regional Environmental Eng, Tenn. Div. of Health & Environment in 1983. Employed by IT Corp 1983-present as Senior Civic Eng. *Society Aff:* ASAE, SCSA, ASCE, ROA

Hicks, Philip E
Business: 10022 Bridlewood Ave, Orlando, FL 32817
Position: Pres *Employer:* Hicks & Assocs *Education:* PhD/IE/GA Tech; MSE/IE/Univ of FL; B/IE/Univ of FL *Born:* 3/16/33 Dr. Hicks is Pres of Hicks & Assocs, Consulting Industrial Engrs, Orlando, FL and Inst Dir, Industrial Inst, Orlando, FL. Dr. Hicks was chrmn of Industrial Engrg Depts at NC A&T State Univ and NM State Univ. Dr. Hicks has been active in teaching, research and consulting for the past twenty four years. Dr. Hicks is author of *Introduction to Industrial Engrg and Management Science*, McGraw- Hill, 1977, and co-author of *Orientation to Professional Practice*, McGraw- Hill, 1981. *Society Aff:* NSPE, IIE, ACEC

Hidzick, George M
Home: Box 3456 RFD, Long Grove, IL 60047
Position: VP & HPR Engg Officer. *Employer:* Kemper Group *Education:* BS/Engineering/Univ of CA, Berkeley. *Born:* 8/18/25. Native San Francisco, CA. Served in 342nd Infantry Regiment, 86th Div in Europe and Phillipines in WWII. BS degree Industrial Engg, Univ CA Berkeley Jan '50. PE Fire Protection - CA. Fire Protection Engr with Industrial Risk Insurers, Chief Engr, Pacific Regional Office in 1965. Since 1966 with HPR Dept of Kemper Group holding position as HPR Engg Mgr Western Div and presently HPR Engg Officer at Long Grove, IL world headquarters. Currenlty VP of American Protection Insurance Co and HPR Engg Officer of Lumbermens Mutual Casualty Co. and Kemper Intl Insurance Co. Corporate Member of & serve on Underwrites laboratories Fire Council and past mbr Bd of Dirs of SFPE and Bd. of Gov of S.F.P.E. Scientific and Research Foundation. Serve on various NFPA technical committees. Also serves on bd of dir of Natl Inst of

Hidzick, George M (Continued)
Bldg Sci and mbr of SFPE, ASME, ANS and NFPA. *Society Aff:* SFPE, ASME, ANS, NFPA.

Hienton, Truman E
Home: 6203 Carrollton Terrace, Hyattsville, MD 20781
Position: Retired. *Employer:* Self-Consulting. *Education:* MS/Agri Engg/IA State College; BS/Agri Engg/IA State College; BS/Agri Engg/OH State Univ; DSc/Hon/Purdue Univ; AE/-/OH State Univ. *Born:* May 1898 near Cleveland, Ohio. US Army Sept-Nov 1918. US Army Ordnance 1941-46, Cpt to Col USAR. Extension work in agri engrg at Univ of Nebraska & Purdue Univ 1921-24. Leader farm electrification proj at Purdue 1925-41. Farm Electrification Branch, Agri Engrg Res Div USDA 1946-68. Special attention to elec methods & equip for insect control. Golden Plate Award Amer Acad Achievement 1968; Cyrus Hall McCormick Medal 1967 *Society Aff:* ASAE, IEEE, AAAS

Hiett, Louis A
230 Windover Grove, Memphis, TN 38111
Position: Vice President. *Employer:* Buckeye Cellulose Corp. *Education:* BChE/Chem Engrg/GA Inst of Technology. *Born:* 10/2/25. U.S. Army 1943-45; married Virginia Scott Wingfield 1948; 3 children. Employed by The Buckeye Cellulose Corp., a subsidiary of The Procter & Gamble Co. since 1949. Tech Div 1949-61, 64-83; Plant Mgr 1961-64; Mgr Tech Div 1972- 83; Special Assignment 1983- . Prof activities include Cellulose Pulp, Cellulose Derivatives, & Nonwovens Dev, & Tech Exchange. Res Bd of Visitors, Memphis State Univ; Pulp & Paper Indus Advisory Ctte, Georgia Inst of Tech; Bd/Dir INDA. Served on Bds/Dirs of Jr Achievement (Pres), Shelby United Neighbors, & Memphis Metropolitan YMCA, Presbyterian Elder, Member AIChE, TAPPI. *Society Aff:* AIChE, TAPPI.

Higbie, Kenneth B
Home: 9311 Singleton Drive, Bethesda, MD 20034
Position: Deputy Director, Research Center Operations *Employer:* Bureau of Mines, US Dept of Interior. *Education:* BS/Chem Engg/OR State Univ. *Born:* Nov 17, 1924 Jamaica, New York. Chem Engr with Bureau of Mines at Albany, Oregon & Washington, DC 1948-57. Assoc Dir of Res for the Beryllium Corp 1957-61 respon for developing beryllium powder met & chem res activities. Rejoined the Bureau of Mines in 1961 as Aluminum & Bauxite Commodity Specialist. Sr Staff Met Adv to Asst DirMet, Bureau of Mines 1963-73. Named Ch of Div of Solid Wastes in 1973, respon for Bureau's broad program in mining, met, indus & urban wastes. Named Dept Dir off Res Center Operations in 1979, Respon for all Bureau's Res Facilities and the activities there of. Enjoys sports & travel. *Society Aff:* AIME, AAAS, ACS, ΣΞ.

Higdon, Archie
Home: 654 Rancho Dr, San Luis Obispo, CA 93401
Position: Retired. *Education:* PhD/Applied Math/IA State Univ; MS/Applied Math/IA State Univ; BS/Math Speech/SD State Univ. *Born:* 10/22/05. Recalled by USAF from Prof of Mechanics in Engg at IA State in 1951. Was Prof & Hd of Mechanics & Chrmn of Engg Sci Div at USAF Acad for many yrs. Also served as Prof & Hd of Math, Prof & Hd Phys, & as Assoc Dean of Faculty for Engg & Basic Sci. Retired from Permanent Prof USAF in 1967 to become Dean of Engg & Tech at CA Poly State Univ. Retired from Cal Poly State Univ in 1972 at age 67. *Society Aff:* ASEE, NSPE.

Higgins, Wayne R
Business: 14235 Winconsin Ave, Elm Grove, WI 53122
Position: Pres *Employer:* Traffic Engg Services, Inc *Education:* Prof Dev in Engg/Transportation/Univ of WI; BSCE/Transportation/Univ of WI. *Born:* 2/2/44. Temporary asst to Traffic Engr in Kenosha WI 1966-68. Traffic Signal & Lighting Engr City of Madison, WI 1969-75 and Traffic Operations Engr at Madison 1976-79. VP of Tucker Co Inc Jan 1979-present. Pres of Signal & Ltg Engr Sys, Inc 1981-85. Pres of Traffic Engineering Services, Inc 1985-present. Winner Edwin F Guth Lighting Design Award of Merit from IES for design of lighting system on John Nolen Causeway in Madison. PPres of Badger Section IES and WI Section ITE. Reg VP-Nort Central-IES 1978-80, Mbr IESNA Membership Committee 1979-Present-Chairman 1984-85, Registered Professional Engr in Wisconsin. *Society Aff:* ITE, IESNA, ASCE.

Higgins, William H C
Home: 20214 126th Ave, Sun City West, AZ 85375
Position: Retired. *Education:* Dr Eng/Hon/Purdue (1975); EE/EE/Purdue (1934); BS/EE/EE/Purdue (1929). *Born:* 4/26/08. Grammar & High Sch, LaPorte, Ind; 1929-34 AT&T Co, NYC carrier & program transmission; 1934-40 Bell Tel Lab, NYC fixed & mobile radio dev; 1940-61 Bell Tel Lab, Whippany, NJ radar & fire control sys, Nike guided missile sys, satellite & ICBM guidance sys, submarine dtection, radars for distant early warning line - Exec Dir 1952; 1961-66 Bell Tel Lab, Holmdel, NJ - Exec Dir Electronic Switching Sys for Bell Sys use; 1966-73 Bell Tel Lab, Holmdel, NJ - V P Switching Sys Dev - electronic & electro-mech switching syst 1973-77, North Elec Co, DE, OH Group VP Res & Dev-dev of advanced switching sys for telephony. 1977-78 Telecommunications Consultant, United Telecommunications, Inc. KS City, MO. IEEE Comm Society Edwin H Armstrong Achievement Awd 1977. Hon Dir of Engg, Purdue Univ 1978. Fellow IEEE 1960. *Society Aff:* IEEE

Higginson, R Keith
Business: 2520 South State St 214, Salt Lake City, UT 84115
Position: Pres *Employer:* Higginson and Assocs, Inc *Education:* BS/CE/UT State Univ; AA/Engrg/Boise State Univ *Born:* 5/20/30 Commissioner, US Bureau of Reclamation, 1977-1981, Dir, ID Dept of Water Resources, 1966-1977. Chief Engr, Water Rights Branch, UT State Engr's Office 1957-1965. Formerly Pres, Southern ID Section ASCE, Bear River Compact Commission, Columbia River Compact Commission, Western State Water Council, Pacific Northwest River Basins Commission, Interstate Conference on Water Problems, Active in church and community affairs, Now Pres of Higginson and Assocs a consulting firm specializing in water resources engrg, hydropower, and water rights. *Society Aff:* ASCE

Higgs, Wallace N
Business: 6022 Fairdale, Houston, TX 77057
Position: Exec VP. *Employer:* CGS Engg. *Education:* BCE/CE/Univ of FL; MBA/Mgt/GA State Univ. *Born:* 6/8/36. Native of Ft Pierce, FL. Assumed current position of Exec VP of CGS Engg Inc in Aug 1979 to operate the firm. Prior to joining CGS, was Asst VP of Atlanta, GA office of Henningson, Durham & Richardson, Inc. Served as Treas for GA State PEPP-1973, Cobb Chapter of GSPE-1970, & GA Engr Week-1970; Chrmn, GA Engr Week-1971-74. Mason & Shriner. *Society Aff:* NSPE, GSPE, APWA, AWWA, WPCF.

Hightower, George B
Business: 708 Antone St, NW, Atlanta, GA 30318
Position: President. *Employer:* Self employed - George B Hightower PE HVAC Consultant *Education:* BS/Elec Engg/VA Military Inst. *Born:* March 1911 Atlanta, Georgia. VMI-33-BSEE Application Engr Air Conditioning Contractors 1934-37; Engr & Dist Mgr Genl Elec Air Cond Dept 1937-42, 46-47; Pres Conditioned Air Engrs 1947-1982; Consultant 1982 to present, WWII Lt Col China 14th AAF Flying Tigers, Wing Exec Officer & Base Cmdr 1942-46, Legion of Merit & 2 Bronze Stars, ASHRAE Author, P Pres GSPE, AMCA, Atlanta Bldrs Exchange GA Engg Fnd; Chmn Atlanta & Ga HVAC Adv Comm GA Energy Comm, Natl Environ Balancing Bureau; Engr of the Yr GSPE 1971, ASHRAE/ALCO Award, ASHRAE Fellow & Life Mbr. Distinguished Service Award from NSPE, ASHRAE & AMICA. Married Emily Anderson; 2 daughters. Chmn 1972-73 NSPE/PEC, Exec Ctte 1972-74 NSPE; Natl Dir 1971-76 NSPE; Kiwanis, ABET Director on Alternates 1981-87. *Society Aff:* NSPE, ASHRAE, MCAA, NEBB, NAFE

Hightower, Joe W
Business: Dept of Chem Engrg, Houston, TX 77251-1892
Position: Prof *Employer:* Rice Univ *Education:* PhD/Physical Chem/Johns Hopkins Univ; MS/Physical Chem/Johns Hopkins Univ; BS/Chem & Math/Harding College *Born:* 9/14/36 Is a specialist in heterogeneous catalysis. His research has concentrated on the application of isotopic tracers to study kinetics, networks, mechanisms, and adsorption in catalytic systems of industrial significance. Being quite active in professional societies, Dr. Hightower is a former Chrmn of the Bd of Trustees of the Gordon Research Conferences, Chrmn of the ACS Petroleum Research Fund Advisory Bd, and a member of various AIChE committees. As a part of his consulting activities, Dr. Hightower has taught over 25 Short Courses on Catalysis in the US and in several foreign countries. He is the co-founder and Pres of the Human Resources Development Foundation, a benevolent organization in Houston. *Society Aff:* ACS, AIChE, ΣΞ, ΦΛΥ, ASTM, CSNA

Higinbotham, William A
Home: 11 N Howell's Point Rd, Bellport, NY 11713
Position: Retired *Employer:* Self *Education:* BA/Physics/Williams College. *Born:* 10/25/10. Bridgeport, Conn. Hon DSc. Williams Coll 1963. Cornell Grad Sch, physics 1932-40. MIT Radar Lab 1941-43. Manhattan District Proj at Los Alamos, NM 1944- 45; Hd Electronics Group 1945. Chairman, Federation of American Scientists, 1946; Executive-Secretary Federation of American Scientists, 1947. Assoc Hd Instrumentation Div, Brookhaven Natl Lab 1948-51; Hd Instrumentation Div 1952-68. Sr Scientist in Brookhaven Natl Lab Tech Support Group on nuclear matls safeguards 1969-84. Inst Radio Engr, Chmn of Nuclear Sci Group 1962; V Chmn 1957, 61. Fellow: IEEE, Amer Physical Soc, Amer Nuclear Soc, Amer Assn for Advancement of Sci. and Inst. Nuclear Materials Management. First Annual Award for Contributions to Nuclear Instrumentation, IEEE Nuclear Science Group, 1972; First Annual Distinguished Service Award, Institute of Nuclear Material Management, 1979. Currently consultant for Brookhaven et. al. *Society Aff:* IEEE, APS, ANS, AAAS, INMM.

Hilbers, Gerard H
Business: 30007 Van Dyke, Warren, MI 48090
Position: Mgr, Mfg Planning Dept. *Employer:* Chevrolet Motor Div, GMC. *Education:* BSE/Metallurgical/Univ of MI. *Born:* 3/9/28. Various pos in Met Dept of Chevrolet Div in forging, axle & aircraft engine plants. Became Ch Met of Chevrolet Axle Plant in 1957. Transferred to Mfg in 1961 as Supt, Indus Engrg. Various assignments in transmission & engine plants to Plant Mgr in 1970. Currently respon for Mfg Planning for entire Chevrolet Div incl proj control, make or buy allocation financial analysis, etc. P Chmn, Buffalo Chap ASM; currenlty Mbr 3 Natl Committees, ASM. *Society Aff:* ASM.

Hildebrand, Neal J
Business: P. O. Box 36487, Houston, TX 77036
Position: Mgr *Employer:* Chevron Geosciences Co *Education:* MS/Petro Engrg/Univ of CA-Berkeley; BS/Petro Engrg/Univ of CA-Berkeley *Born:* 10/4/30 Twenty three years experience as a Petroleum Engr. Various assignments in Production, Drilling, Reservoir, Evaluation, Organization, Computer, and Corp consulting. Currently Mgr of Production Services Division with 35 PE and physical scientists. Member of Society of Petroleum Engrs, past chrmn of the Golden Gate section and currently member of the National Continuing Education Committee. *Society Aff:* SPE

Hildebrandt, Peter W
Home: PV-11, Olympia, WA 98504
Position: Asst Dir for Air Programs. *Employer:* Washington State Dept of Ecology. *Education:* MS/Civil Engr/Univ of WA; BS/Civil Engr/Univ of WA. *Born:* 2/11/33. PChmn & former Dir of the PNWI Sect of APCA & presently serves as their meeting Site Selection Chmn. Also served for APCA on the Radioactive Substance Subctte of the Intersoc Ctte on Methods of Sampling & Analysis. V Chmn of EPA's Air Pollution Manpower Dev Adv Ctte. Former VP APCA. State representative on EPA's Standing Air Monitoring Work Group.

Hildyard, Benjamin G
Home: 5550 Calmuet Ave, La Jolla, CA 92037
Position: President *Employer:* BGH Engineering, Inc. *Education:* BS/Civil Engg/KS State Univ. *Born:* 8/21/16. Kansas. Over 40 yrs of civil engrg experience in water resources dev & utilization for domestic & overseas projs. Extensive background in tech & admin proj operation encompassing water supply & treatment, irrigation, hydroelectic power generation, wastewater treatment, drainage, flood control & harbor projs. Retired and presently owner of own consulting firm. Served with Army Corps of Engrs 1942-46. *Society Aff:* ASCE, ICID, WPCF, CWPCA, APWA, AWWA, SAME, ACEC, AAA.

Hileman, Andrew R
Business: 777 Penn Center Blvd, Pittsburgh, PA 15235
Position: Advisory Engineer. *Employer:* Westinghouse Electric Corporation. *Education:* MSEE/Elec Engr/Univ of Pittsburgh; BSEE/Elec Engr/Lehigh Univ. *Born:* Aug 1926. BS Lehigh Univ, MS Univ of Pittsburgh, Lamme Scholarship (Tech) Univ of Berlin, Germany. Fellow IEEE, Chmn ANS C92 Insulation Coordination, Mbr CIGRE SC33 on Insulation Coordination, P Chmn IEEE PES Surge Protective Device Ctte. Author of over 35 IEEE & CIGRE papers, three of which received the Power Engrg Soc's Prize Paper Award. Lec for Penn State Univ & Carnegie-Mellon Univ. With Advanced Sys Tech of Westinghouse Corp since 1951. Major fields are insulation coordination, overvoltage protection, lightning, switching surges & elec design of EHV & UHV transmission lines. *Society Aff:* IEEE, CIGRE, ASC C92

Hiler, Edward A
Business: Dept of Agri Engrg, College Station, TX 77843
Position: Professor & Head. *Employer:* Texas A&M University. *Education:* PhD/Agri Engg/OH State Univ; MS/Agri Engg/OH State Univ; BS/Agri Engg/OH State Univ. *Born:* May 14, 1939 Hamilton, Ohio. Served in the Dept of Agri Engrg at Texas A&M Univ as Asst Prof, 1966-69; Assoc Prof 1969-73; Prof 1973-74 & Prof & Hd 1974- . Respon for teaching & res as well as admin duties as Dept Hd. Reg PE in Texas 29140. Chmn Texas Sect ASAE 1974-75. Chmn Agri Engrg Div ASEE 1976-77. Elected for membership in the National Academy of Engineering in 1987. Recipient of Disting Young Agri Engr Award SW Region ASAE 1975. Rec'd 2 Paper Awards from ASAE in 1972, 1 in 1974. Recipient of Faculty Disting Achievement for Res TAMU 1973 & Disting Mbr for Res Award, TAMU Chap of Sigma Xi 1975. Recipient of ASAE Young Res Award, 1977; Distinguished Alumnus of College of Engg, The OH State Univ, 1978. *Society Aff:* ASAE, ASEE, NSPE, AAAS, AGU, NAE.

Hilf, Jack W
Home: 8505 E Temple Dr 458, Denver, CO 80237
Position: Consulting Engr (Self-employed) *Education:* PhD/CE/Univ of CO; MS/CE/Univ of CO; BCE/CE/NYU; BS/Chem & Bio/Coll of NY. *Born:* 5/4/12 Jr. Engr Corps of Engrs and Bd of Water Supply N.Y.C. 1941-1942. Military service private to major, CE 1942-1946. Design Engr Earth Dams U.S.B.R. Denver, CO. 1946-62. Chief of Information Retrieval USBR 1962-1965. Senior Water Research Scientist, Dept of Interior, Wash., DC 1965-66. Asst. Chief and Chief Division of Design USBR, Denver, 1966 to retirement in April 1975. Consulting engr on dams in 14 foreign countries and USA 1975. Prof of CE (adjoint) U. of CO 1976-81. Eight books, 21 papers and discussions. Colonel, Army CE (retired). Dept of Interior Gold Medal for Distinguished Service 1965. Enjoys tennis. *Society Aff:* ASCE, NSPE, USCOLD, ISSMFE.

Hilibrand, Jack
Business: Route 38, Bldg 206-1, Cherry Hill, NJ 08358
Position: Principal Staff Scientist *Employer:* GE/RCA Aerospace and Defense *Education:* ScD/EE/MIT; B/EE/City Coll of NY *Born:* 9/15/30 B. in NY City. After

Hilibrand, Jack (Continued)

Doctorate went to RCA Labs. Did basic research in Tunnel Diodes, Varactor Diodes. At RCA Solid State Division in 1960s managed Gallium Arsenide Device Research, Power Transistor Development, Complementary MOS Integrated Circuits, and Integrated Circuit Tech. At RCA Aerospace and Defense in 1970s and 1980's coordinated development of ULSI technology, design tools, and system application. Currently responsible for microelectronics planning at RCA Aerospace & Defense. Fellow, IEEE 1980; Mbr, IEEE Fellow Ctte 1983-86; Chmn, IEEE Fellow Ctte in 1987, 1988. *Society Aff:* IEEE, APS, ΣΞ

Hill, Alan T

Business: 310 K St, Suite 602, Anchorage, AK 99501
Position: Vice President/Regional Manager. *Employer:* CH2M Hill. *Education:* BS/Civil Engr/Univ of NV. *Born:* 6/5/38. Sacramento, CA. Mbr CA Council of Civil Engrs & Land Surveyors prof societies. Served as 1976 Chmn of ASCE Engg Surveys ctte. 1972 Pres of Western Assoc of Eng & Land Surveyors. Mbr of Rotary & Elks & Chamber of Commerce, Dir & organizer of North Valley Bank, Redding, CA. Prof registration in CA, ORE, NEV & AK. Joined Clair Hill & Assoc in 1961 as Staff Engr. Was Div Mgr in 1970 when Clair A Hill & Assoc & CH2M merged to form CH2M Hill. Married wife, Beverly, 4 children. Enjoy hunting, fishing, skiing & sports *Society Aff:* ASCE, ACSM, NSPE.

Hill, David A

Home: 1850 Kohler Dr, Boulder, CO 80303
Position: Electronics Engr. *Employer:* U.S. Dept. of Commerce *Education:* BS/EE/Ohio University; MS/EE/Ohio University; Ph.D./EE/Ohio State University *Born:* 04/21/42 Native of Cleveland, OH. Research scientist in Federal labs since 1970 working on electromagnetics, antennas, and propagation. Joined the National Bureau of Standards in 1982 to do research in electromagnetic interference problems and is also an adjoint professor in electrical engrg at the Univ of CO. Published more than 100 journal articles in electromagnetics. Has served as a technical editor for the IEEE Trans. on Geosci. & Rem. Sens., and is now an associate editor for the IEEE Trans. on Antennas & Propagation. Fellow of IEEE; mbr of the International Union of Radio Science. *Society Aff:* IEEE

Hill, Dexter

Business: Entex Bldg 3428, Houston, TX 77002
Position: Vice President. *Employer:* Matthew Hall Inc. *Education:* ME/ChE/TX A&M Univ; BS/ChE/Univ of TX. *Born:* Dec 1936 Texas. ME Texas A&M 1967. BS Univ of Texas 1960. Chmn Coastal Bend Sect AIChE 1966. Process Engr incl supr pos for Celanese Chem Coll 1960-67. Principal in cons firm of Purvin & Gertz, Inc 1967-76. Specialized in economic feasibility & proj dev. Assumed present pos with British based internatl engrg contractor early 1976. Respon for liaison with US based clients. *Society Aff:* AIChE.

Hill, Frederick J

Business: Dept of Elec Engg, Tucson, AZ 85721
Position: Prof. *Employer:* Univ of AZ. *Education:* PhD/EE/Univ of UT; MS/EE/Univ of UT; BS/EE/Univ of UT. *Born:* 9/2/36. Primary areas of interest are digital logic design, hardware description languages and test sequence generation. He developed the language AHPL and was responsible for its implementation in the form of a hardware compiler and function level simulator. He is co-author with L R Peterson of the books "Introduction to Switching Theory and Logical Design", (1968, 1974), "Digital Systems: Hardware Organization and Design" (1973, 1978) and "Digital Logic and Microprocessors" (1984). *Society Aff:* IEEE.

Hill, George R

Business: Biomaterials Profiling Center, 321 Chipeta Way, Research Park, Univ of Utah Salt Lake City, UT 84112
Position: EIMCO Prof *Employer:* Univ of UT *Education:* PhD/Chem/Cornell Univ; AB/Chem/Brigham Young Univ *Born:* Ogden, UT; m. Melba Parker, 8/25/41; pt-time instr Cornell, 1942-46; prof chem, 1950-72; chrmn Fuels Engrg, 1951-65; Dean Coll Mines and Mineral Ind, 1966-72; Dir US Office of Coal Research, 1972-73; Dept Dir Fossil Fuels electric Power Research Inst, Palo Alto, CA, 1973-77; Envirotech endowed prof, Univ UT, 1977- ; mem NRC comm Mineral and Energy Resources, 1976-81; mem FEAC, Dept Energy, 1977- ; mem UT Energy Council, 1978- ; Distinguished Service award, UT Petroleum Council, 1968; Outstanding PE award, UT Engrg Council, 1970; Henry H. Storch award, ACS, 1971; Honorary Doctorate of Science, BYU, 1980; Publications: coal conversion, oil shale, corrosion, catalysis. National Coal Council 1986- . *Society Aff:* ACS, AIChE, ΣΞ, ΦΚΦ

Hill, Jack M

Business: 229 West Bute St, Norfolk, VA 23510
Position: Sr VP *Employer:* MMM Design Group *Education:* BS/CE/VA Polytechnic Inst *Born:* 9/24/26 Native of Norfolk, VA. Served in US Army 1944-1946. Artillery and tanks instructor. Post graduate work in structures at VA Tech. Taught statics and strength Norfolk Extension VA Tech. Hull Engr Welding Shipyards 1949-1950. PE and field engr John B. McGanghy & Assoc. Assoc and advance to principal of 350 man consulting firm. Head Norfolk office with 90 professionals and technicians. Responsible for MMM Design Group as Dir of Design. Past Pres ASCE, VA Section 1980-1981, Civil Engr of Year Award received from VA Section 1981. Served as National Dir ASCM. Tennis, golf, swimming and fishing. *Society Aff:* ASCE, NSPE, SAME, ACSM, ACEC

Hill, James E

Home: 3601 Via La Selva, Palos Verdes Estates, CA 90274
Position: retired *Education:* BS/Chem/CA Inst of Tech *Born:* 11/25/96 Chief Fire Protection Engr with major oil co for 20 years followed by 20 years as independent consulting engr, citations from American Petroleum Inst, The Society for Advancement of Engrg and Society of Fire Protection Engrs in recognition of significant accomplishments and stature in engrg & valuable service to fire protection in industry & achievements in other professional societies. Particularly innovative in fire prevention techniques & tactics and strategy in control & extinguishment of major petroleum fires & emergencies. *Society Aff:* SFPE

Hill, James Stewart

Home: 263 N. Main Street, Hudson, OH 44236
Position: Principal Engr. *Employer:* EMXX Corp. *Education:* BSEE/EE/Case Western Reserve Univ. *Born:* 12/2/12. Wash, DC, Resident Cleveland/Hudson, OH 1919-1962. United Broadcasting Co, staff engr/chief engr, 1934-1953. Conducted propagation surveys & radio interference studies with Smith Electronics, Jansky & Bailey, Genistron, & RCA. Currently pres & principal engr, EMXX Corp, specializing in consulting services on electromagnetic compatibility. reg PE, OH. Active in IEEE Electromagnetic Compatibility Soc (EMCS), on Bd of Dir, Chrmn of Intl Affairs Committee. Recipient IEEE Centennial Medal and EMCS Certificate of Appreciation & Citation, 1977 Montreux EMC Symposium "for outstanding leadership promoting the intl exchange of EMC tech-. Life Fellow, IEEE. Assoc Editor, Electromagnetic Compatibility Newsletter. Author, EMC Handbook. Coauthor "A Guide to FCC Equipment Authorization-. Classic Car Buff. Listed in "Who's Who in the World" and "Who's Who in Finance and Industry-. *Society Aff:* IEEE, EMCS.

Hill, John S

Business: Box 169, Joplin, MO 64802
Position: Pres. *Employer:* John S Hill & Assoc Inc. *Education:* BS/Chemistry/TX Wesleyan College. *Born:* 8/9/24. 30 yrs in Ind and Private Engg in the fields of Chemistry, Chem Engg, Soils Mechanics and Engg Testing. Pres of John S Hill & Assoc, Inc, since 1974. Owner/Mgr of the Bruce Williams Labs, since 1960. Owner/Mgr of Bruce Williams' Consulting Engrs & Assoc. John S Hill & Assoc, Inc, is a team of Scientific Investigators specializing in Causes of Fires and Explosions and Accident Reconstruction involving trucks. The Bruce Williams Labs is a Commer-

Hill, John S (Continued)

cial Testing Co. Bruce Williams' Consulting Engrs & Assoc is a Geotechnical Engg Co. *Society Aff:* AIChE, ACS, ASTM, NSPE, AIME.

Hill, Luther H, Jr

Business: PO Bx 34306, Dallas, TX 75234
Position: President. *Employer:* Luther Hill & Assoc, Inc. *Education:* BS/Civil Eng/Rice Univ; BA/-/Rice Univ. *Born:* 5/15/34. BS Civil Engg from Rice Univ. Served with the US Army Corp of Engrs. Proj Engr & Proj Mgr with Henry C Beck Co. VP of Commerical Const with Chaney & James Const Co. Presently, Pres of Luther Hill & Assoc Inc. P Pres of the Dallas Chap of Associated Genl Contractors (ACG). Past Natl Bd/Dir of Amer Inst of Constructors (AIC). P. Chmn of the Adv Council of the Const Res Ctr-Univ of TX at Arlington. Past Chmn of the Adv Council-Const Degree Program at E TX State Univ. Past Chmn-US Natl Ctte W-65, Org & Mgmt of Const. Enjoy hunting & fishing. *Society Aff:* AIC.

Hill, Ole A, Jr

Home: 474 South York Street, Elmhurst, IL 60126
Position: Staff Assistant (retired). *Employer:* Commonwealth Edison, Co. *Education:* MBA/Bus Admin/Univ of Chicago; BS/Elec Engg/IA State Univ. *Born:* March 1907 Council Bluffs, Iowa. Delta Tau Delta, Eta Kappa Nu. Illuminating Engrg Soc, Fellow; Distinguished Ser Award 1970; Chmn Rural Lighting, Lighting & Air Conditioning, Fluid Milk Indus Lighting Cttes; Exec Chmn, Natl Tech Conf 1971. Pres, Chicago Lighting Inst 1956-63. Exec Dir, Elec Inst of Chicago 1963-72. Public Ser Co of Northern Ill 1928-54; Design & application of commercial, indus & street lighting. US Army 1942-46; respon for replacement training ctr instruction in map reading, scouting, aerial photograph interpretation. Ill Civil Defense Agency 1952-53; respon for dev & coordination of state's utilities emergency procedures. Commonwealth Edison Co 1954-72; Training & dev of engrg personnel in marketing activities; Field res in high frequency fluorescent lighting sys & lighting & air conditioning. *Society Aff:* IES.

Hill, Percy H

Business: Medford, MA 02115
Position: Prof Emeritus *Employer:* Tufts University. *Education:* SM/ME/Harvard Univ; BME/ME/RPI. *Born:* 2/19/23. Norfolk, Va. Reg Prof Engr Mass & New Hampshire. Lt jg USNR, Communications Dept; Hd USS Savo Island (CVE) Asiatic Theater, WWII 1944-46. Prof. Emeritus, Dept of Engrg Design, Tufts Univ, Medford, Mass 1948-1983 Pres, Stratford Labs, Englewood Cliffs, N.J. Chmn Elect, Engrg Design Graphics Div of ASEE 1971-72. Chrmn Elec Design in Engrg Ed. Div. of A.S.C.E. 1982-present. Author of nine textbooks & over 50 tech papers. Active cons & researcher. Co-author of two US Pats. Mbr of Tau Beta Pi, soc of Sigma Xi, ASME, ASEE, Human Factors Soc. Enjoys golf, tennis, sailing, fishing, gardening & model building. *Society Aff:* ΤΒΠ, ΣΞ, ASEE, ASME, HFS.

Hill, Richard F

Home: 16Rock Ridge Road, Newtown, CT 06470
Position: Dean of Science and Engineering *Employer:* University of Bridgeport *Education:* PhD/EE/Univ of WI; MS/EE/Univ of RI; BS/EE/Univ of RI. *Born:* Sept 1933. Native of Ashaway, RI. Assoc Prof of Elec Engrg; Prof of Ocean Engrg; Dir of Inst of Ocean Tech, URI 1960-68. Pres, Nereus Corp 1968-70. Dir of Environ Sciences & Mgmt, Ctr for Environ & Man, Inc 1970-71. Advisor on Environ Quality, Deputy Ch of Bureau of Power, Dir of Office of Energy Sys & Ch Engr for the Federal Power Comm 1971-76. Executive Manager, Engineering Societies Commission on Energy, Inc. 1977-80. Program Chmn for annual Energy Tech Confs. Sr Mbr IEEE. PE. *Society Aff:* IEEE, ASEE, AAAS, AGU, ASCE, NSPE.

Hill, Richard L

Business: P.O. Box 5722, Kingsport, TN 37663
Position: President *Employer:* Hill Associates *Education:* ME/Ind Engr/Univ of S FL; BS/IM/Univ of Tenn. *Born:* 2/25/35. Nashville, Tenn. Served with Army Adjutant Genl Corps 1954-56. Indus Engrg Staff: Western Elec Co; Eaton, Yale & Towne; Sperry Rand Corp; Internatl Minerals & Chem Corp; IRC, Div of TRW, Inc; Kingsport Press, Div of Arcata Corp. Personnel Dev Mgr, IMC. Mgr, Indus Engrg at: IRC; Magnavox; Kingsport Press. Assistant professor/Extension Specialist, VPI. Instructor, Mgmt Tech, IE Tech, WSCC. CEO, own firm, since 1981. Inst Dir, Graphic Arts Div, 1974-75 VP Region III, 1977-79, IIE. Regional Outstanding Indus Engr, IIE 1973. Epsilon Sigma Phi. Unit Superior Service Award, USDA 1978. Enjoy bridge, gardening, reading. *Society Aff:* IIE.

Hill, Robert B

Business: 1000 S Fremont Avenue, Alhambra, CA 91802
Position: Exec Vice President. *Employer:* C F Braun & Co. *Education:* MSChE/Chem/IL Inst of Technology; BSChE/Chem/SD Sch of Mines. *Born:* Sept 22, 1919 Huron, South Dakota. Chem Engr for Esso Standard Oil Co from July 1943-47. Joined C F Braun & Co in May 1947 as a Chem Engr. Has held pos of Sr Process Engr, Sales Coordinator, Hd of Process Engrg Dept, Hd of Engrg Div & V Pres respon for Proj Mgmt, Proj Controls & Quality Assurance & Exec VP. Mbr of the Amer Inst of Chem Engrs & the Proj Mgmt Inst. *Society Aff:* PMI, AIChE.

Hill, William P

Home: 1818 Foxcroft Lane No. 806, Allison Park, PA 15101
Position: Ret Sr V President. *Employer:* National Steel Corporation. *Education:* BS/Mech Engr/Univ of DE. *Born:* March 30, 1908 Rehobeth Beach, Del. Post grad Johns Hopkins 1937-52; Lehigh Univ 1954. With Bethlehem Steel Co, Sparrows Pt, Md 1933-53; Ch Engr 1947-48; Asst Genl Mgr 1948-53; Asst to V Pres Operations Bethlehem, Pa 1953-58; V Pres Engrg Natl Steel Corp, Pittsburgh 1958-71. Senior VP Natl Steel Corp 1971-1973. Cons WPB WWII. Pres Bethlehem Council Boy Scouts of Amer 1958. Recipient Silver Beaver award Boy Scouts of Amer 1970; Great Insignia in Gold Fed Pres, Austria 1972. Fellow ASME; Mbr AISI, AISE, AOA, ASME. Reg PE in Md, Ind, & Pa. President VP Services Inc., Executive VP. Chemical Resources Recovery Inc. *Society Aff:* ASME, AISI, AOA.

Hill, William W

Business: Angola, IN 46703
Position: Prof of Engrg *Employer:* Tri-State Univ *Education:* PhD/Fluid Mech/CO State Univ; MS/Aerodynamics/Purdue Univ; BS/ME/GA Inst of Tech *Born:* 9/10/29 Following baccalaureaute, served as Navy All-Weather Flight Instructor. Upon completing Masters program, served as Instructor of Aeronautical Chrmn. Continued at Tri-State on Civ Engrg faculty. Appointed Engrg Dean in 1974 and initiated Tri-State Engrg and Res Ctr. Appointed Sr VP in 1978; in 1981 accepted appointment to the Laurence L. Dresser Engrg Chair. Reg PE in IN, MI, OH. Served as consultant to numerous engrg firms and City of Angola. Active in NSPE, currently serving as Pres-Elect of the IN Soc. *Society Aff:* ASEE, ASCE, NSPE, ASEM

Hillberry, Ben M

Business: Sch of Mech Engg, W Lafayette, IN 47906
Position: Prof. *Employer:* Purdue Univ. *Education:* PhD/ME/IA State Univ; MS/ME/IA State Univ; BS/ME/IA State Univ. *Born:* 12/17/37. Chrmn of Design, Sch of Mech Engg, Purdue Univ. Actively involved in teaching and res. He is the author of numerous technical articles and is widely known for his res on fatigue behavior of mtls. In addition, he holds several patents in the area of mech design, is a consultant to industry, mbr of the bd of dirs of several cos, and is also noted for his res in bioengg. *Society Aff:* ASME, ASTM, ΣΞ.

Hillert, Mats H

Business: 10044 Stockholm, Sweden
Position: Prof. *Employer:* Royal Inst Tech. *Education:* Dr of Sc/Met/MIT; Civilingenjor/CE/Chalmers Inst Tech. *Born:* 11/28/24. Grad as chem engr from Chalmers Inst Tech in 1947. Employed at Swedish Inst for Metal Res 1948-1961. Leave of absence 1953-1956 for grad studies at MIT. Dr of Sci from MIT in 1956 with thesis on spinodal decomposition. Prof of Physical Met at the Royal Inst of Tech in Stock-

Hillert, Mats H (Continued)
holm from 1961. Vice rector 1974-1977. Res activities in thermodynamics of alloys, phase transformations in alloys & solidification of alloys. *Society Aff:* ASM, AIME, MS, KVA, IVA.

Hilliard, John K
Home: 1511 Clear View Ln, Santa Ana, CA 92705
Position: President. *Employer:* Hilliard & Bricken, Inc. *Born:* 10/22/01. BS Hamline Univ; Hon DS Hollywood Univ; Grad work Univ of MN in Communication Engg. Acoustical engr-United Artists studio 1928-33. MGM studio's asst to recording dir 1933-42. Radiation Lab 1942-43. Altec-Lansing VP 1943-50. LTV Res Dir 1960-70. Hilliard & Bricken, Inc, 1976- . Fellow: Acoustical Soc of Amer, IEEE, Audio Soc, SMPTE. Mbr, Natl Sci Fnd, CHABA group. Reg Prof Engr, CA E385. Initial mbr, Inst of Noise Control Engrs. Pub 70 articles & books on sound & acoustics.

Hillier, James
Home: 22 Arreton Rd, Princeton, NJ 08540
Position: Exec VP & Sr Scientist - Retired. *Employer:* RCA Corporation. *Education:* PhD/Phyiscs/Univ of Toronto; MA/Physics/Univ of Toronto; BA/Physics/Univ of Toronto. *Born:* 8/22/15. Pioneer work on electron microscopy as a grad student and at RCA 1937-53. Developed electron probe microanalayzer. In res & engrg mgmt, RCA Corp 1954-78. Ch Engr, Commercial Electronic Prods 1955-57. VP, RCA Labs 1958-68. Exec VP, Res & Engrg 1967-76. Exec V Pres & Sr Scientist 1976-78. Fellow APS, IEEE, AAAS. Eminent Mbr HKN. Mbr NAE (Council 1971). Mbr Indus Res Inst (Pres 1963). Albert Lasker Award 1960. IEEE David Sarnoff Medal 1967, Founders Medal 1981. Indus Res Inst Medal 1975 DSc (Hon) Univ of Toronto 1978, New Jersey Institute of Technology 1981. National Inventors Hall of Fame 1980. *Society Aff:* APS, NAE, IEEE, AAAS, EMSA.

Hilt, George H
Business: 3125 Carlisle Blvd NE, Albuquerque, NM 87110
Position: VP, Office Mgr *Employer:* Bovay Engrs, Inc *Education:* PhD/CE/IA St Univ; MS/CE/IA St Univ; BS/Gen. Engrg/U.S. Military Academy- West Point; AA/Drama/Phoenix Coll *Born:* 10/2/30 Native of Phoenix, AZ. Corps of Engrs, U.S. Army 1954-1979. Dir of Engrg U. S. Army Engr. Cmd. Vietnam, 1970-1971. Dir and Commander Waterways Experiment Station 1973-1976. Dir of Facilities Management DOD 1976-1978. Joined Bovay Engrs, Inc. (ENR TOP 100) 1979 as VP and Mgr, Albuquerque Office, a full-service consulting engrg office with Civil, Structural, Mechanical, and Electrical Engrg Design. NSPE Professional Development Award 1975. Registered PE seven states. *Society Aff:* NSPE, SAME, PIANC, APWA, ASCE

Himmel, Leon
Home: 39 Club Road, Upper Montclair, NJ 07043
Position: Director Information Systems *Employer:* ITT Corp *Education:* BSE/Elec Eng/CCNY. *Born:* Jan 1924. Pres ITT Avionics Div specialized in navigation, electronic countermeasures; Asst to the Pres ITT Corp, Dir of Special Projs; Cons; V Chmn Northern NJ AES; V Pres ASZD; Tau Beta Pi; Fellow IEEE. *Society Aff:* TBΠ, IEEE

Himmel, Seymour C
Home: 12700 Lake Ave, 1501, Lakewood, OH 44107
Position: Consult. *Employer:* Self. *Education:* PhD/Aero Engg/Case Western Res Univ; MSME/Mech Engg/Case Western Res Univ; BME/Mech Engg/City College on NY. *Born:* 10/24/24. in NY City. Served in WWII as Capt, Infantry. Taught mech engg courses in CCNY after grad. Joined NACA Aircraft Engine Res Lab in 1948 as Aeronautical Res Scientist. Conducted and Managed res in aircraft and space propulsion systems. Managed NASA Agena launch vehicle project. Respon for launches of Rangers VI-IX, Mariner Mars 64, OGOs, etc. Subsequently became Asst Dir for Launch Vehicles respon for dev and operation of Centaur and Agena launch vehicles. As Dir of Rockets and Vehicles was respon for all research in chem rocket propulsion, electric propulsion and communication as well as SERT II spacecraft. Served tour as Deputy Assoc Administrator-Tech in NASA Hdquarters. Dir of Aeronautics at Lewis Res Center. Assoc Dir of Lewis respon for internal operations of Center. Concurrently, served as Acting Director of Science & Technology responsible for all reasearch in basic disciplines & technologies for propulsion energy conversion systems. Recipient of NASA Distinguished Service Medal, NASA Medal for Outstanding Leadership. Exceptional Service Medal. Member, Aerospace Safety Advisory Panel. Now a consult in his fields of expertise. *Society Aff:* AIAA.

Himmelblau, David M
Home: 4609 Ridge Oak Drive, Austin, TX 78731
Position: Professor *Employer:* Dept of Chem Engrg Univ of Texas. *Education:* PhD/Chem Engg/Univ of WA; MS/Chem Engg/Univ of WA; BS/Chem Engg/MIT. *Born:* Aug 1923. Native of Chicago, Ill. Indus experience with Internatl Harvester Co, Chicago; Simpson Logging Co & Excell Battery Co Seattle. Asst Prof, Assoc Prof, Prof at Univ of Texas at Austin 1957- . Chmn, Chem Engrg Dept 1973-77. Res int in modeling, analysis & simulation. Author of 10 books & over 150 articles. Act in Amer Inst of Chem Engrs since 1957, Dir 1974-76. Editorial Bd, Indus Engrg Chemistry Process Design 1973-76. Pres CAChE Corp 1978-80. *Society Aff:* AIChE, ACS, SIAM, AMS.

Himmelman, Gerald L
Home: 26 Chapel Hill Dr, Nashua, NH 03063
Position: Systems Mgr. *Employer:* Robt Abel & Co. *Education:* Diploma/Arch Design/Hart Tech. *Born:* 2/14/34. I have spent 34 yrs in the field of Mtls Handling. During these yrs I have designed & sold numerous systems or supervised others in the design & sales of their systems. Having spent 20 yrs with a mfg, I am now with a distributor as Div Mgr to expand the co's sales of conveyors and related hardware. *Society Aff:* IMMS.

Hindle, Brooke
Home: 5114 Dalecarlia Dr, Washington, DC 20016
Position: Historian Emeritus Natl Museum American History, Smithsonian Institution *Employer:* Retired *Education:* AB/History/Brown Univ; MA/History/Univ of PA; PhD/History/Univ of PA. *Born:* 9/28/18. Studied at MIT & Brown Univ. Taught engrg & served as radar maintenance officer on CUE during the War. At NY Univ 1950-74 as Historian, Dept Hd & Col Dean; Dir of the Natl Museum of History & Tech., 1978-date, Sr Historian; 1974-78; Guggenheim Fellow, 1964-65; Visiting Prof at MIT 1971-72. Books: "The Pursuit of Sci in Revolutionary Amer," –David Rittenhouse," –Tech in Early Amer" and "Emulation and Invention;" co-author, "Engines of Changes: the American Industrial Revolution–; edited "Amer's Wooden Age," –Material Culture of the Wooden Age–; co-editor, "Bridge to the Future." *Society Aff:* SHOT, HSS, SIA, AHA.

Hinds, John A
Business: 570 Lexington Ave, NY, NY 10022
Position: Gen Mgr. *Employer:* GE. *Education:* MS/ME/Univ of Santa Clara; BA/Pre-Engg/Pomona College. *Born:* 7/1/36. *Society Aff:* ANS.

Hindson, Ralph D
Home: RR 1, Chelsea, Quebec, Joxino, Canada
Position: Consultant to the Iron and Steel Industry *Employer:* Self Employed *Born:* Grad McMaster Univ Hamilton, Ontario 1938 Met Engr. Mbr of Assn of Prof Engrs of Ontario. Twenty-five yrs with Steel Co of Canada Ltd in Sr Operating & Staff pos. Eleven yrs Sr Exec Officer with Gov't of Canada, Ottawa four years Managing Dir of Cansteel Corp & now a private consultant to the iron & steel industry. Fellow of ASM, Mbr & P V Pres of CIM & AISE; Mbr & Founding Chmn of Mech Working Ctte of AIME. Recipient of AISE First Kelly Award; CIM Conf of Mets Noranda-Airey Award; CIM Distinguished Lecs Award; the ASM Canadian Council Lec Award

Hines, Anthony L
Business: College of Engrg, Columbia, MO 65211
Position: Dean and Prof *Employer:* Univ of Missouri-Columbia *Education:* PhD/ME/Univ of TX; MS/ChE/OK State Univ; BS/ChE/Univ of OK. *Born:* 9/19/41. Native of Altus, OK. Process engr for Gulf-Gas 1967-68. Sr Assoc chem engr IBM 1969-72, specializing in polymers. Taught at Univ of TX, Austin 1972-73. Taught chem engg at GA Tech 1973-75. Taught chemical engineering at Colorado School of Mines 1975-1980. Joined University of Wyoming in 1980 as Professor of Chemical Engineering and Department Head. Joined OK State Univ in 1983 as Prof of Chem Engrg and Assoc Dean of Engrg Res. Became Dean of Engg at Univ of Missouri-Columbia in 1987. Teach graduate and undergraduate courses in microscopic and macroscopic mass transfer and fluid mechanics. Res areas deal with mass transfer studies and synthetic fuels (oil shale & tar sands). Chrmn of several technical sesions on synthetic fuels at AIChE meeting, 1977 and 1978. Chrmn of Fuels Div subcommittee of AIChE for Oil Shale and Tar Sands. Assoc Editor of journal In Situ. Textbook publ by Prentice-Hall, "Mass Transfer Fundamentals and Applications–. Enjoy hunting, fishing, & skiing. *Society Aff:* AIChE, AAAS.

Hines, Marion E
Business: South Avenue, Burlington, MA 01803
Position: V President, Chief Scientist *Employer:* Microwave Associates, Inc. *Education:* MS/Elec Engg/CA Inst Tech; BS/Meteorology/CA Inst Tech; BS/Appl Physics/CA Inst Tech. *Born:* Nov 30, 1918. Weather Officer US Army Air Force in WWII. Bell Tel Labs Tech Staff 1946-60; Dev of Microwave Electron Tubes, Storage Tubes, Parametric Amplifiers, Tunnel diode devices & digital communications techniques. Since 1960 at M/A-COM Microwave Assocs, Inc; now V Pres, Chief Scientist. Res & Dev of varactor harmonic generators, high power diode switches & phase shifters, diode oscillators & amplifiers, ferrite devices & antennas. Fellow, IEEE 1968 Life Fellow 1984. IEEE-S-MTT Microwave Prizes 1972 & 1977 and Microwave Career Award 1983, J J Ebers Award IEEE-S-ED, 1975. IEEE Lamme Medal 1983, IEEE Centennial Medal 1984 Numerous IEEE publs & US Pats in Microwave Device field. *Society Aff:* IEEE.

Hingorani, Narain G
Business: 3412 Hillview Ave, EPRI, Palo Alto, CA 94304
Position: VP & Dir Elec Syst Div. *Employer:* Electric Power Research Institute. *Education:* PhD/Elec Engrg/Manchester Univ; MSc/Elec Engrg/Manchester Univ; BE/Elec Engrg/Baroda Univ. *Born:* 6/15/31. Specialized in HVDC transmission through MSc & PhD, co-authored book on subject. For seven yrs taught at British univs. In 1968 joined BPA, Portland, Oregon with principal respon for Pacific DC Intertie. In 1974 joined EPRI as Program Mgr wth respon for AC & DC Station Equip. In 1984 became Dir of Transmission Dept with respon for transmission substations, lines and cables. In 1986 became VP & Dir-Elec Systems Div with respon for transmission, distribution, pwr plant elec. eqpt & System planning R&D. Authored over 100 papers. Fellow IEEE, Mbr IEEE Substations Ctte, Chrmn IEC Group on HVDC Performance specs., Mbr IEEE Pwr Engrg Soc Bd. Liaison member IEEE PES to Power Electronics Council and ANSI C34, Chrmn CIGRE Study Ctte 14 DC Links. Recipient PES Uno Lamm HVDC Award. *Society Aff:* CIGRE, PES.

Hinkle, Charles N
512 Parkridge Dr, West Lafayette, IN 47906
Position: Retired. *Education:* MSA/ME/Purdue Univ; BSAE/Engr/KS State Univ Formerly (KSAC). *Born:* 1/5/05. After 6 months of business college, four years at KSAC for a degree in Agricultural Engineering, seven years at Purdue, teaching and obtaining a MS degree Became Agri Engr in Sales Tech Ser Dept Standard Oil Co (Inc) 1936. Worked with SOCO res & Farm Equip Engrs in dev & testing Petroleum Prods with State Ext Agri Engrs, Farmers & Co reps in solving farm equip problems. During WWII personally made three colorsound movies, authored a booklet on tractors, which have been widely used in the mid-west by schs & at farm meetings. Have written over 200 articles for Co & Farm Publs. Lec, Photographer, appeared on Radio & TV. Dev 4H Tractor Program & Ldr Training Program. Received three Natl 4H Ctte Awards, Mbr API Agri Ctte SAE, ASAE-voted Life Fellow Mbr 1966, Life Deacon 1st Cong'l Church Maywood, Ill 1971. Retired 1966. Member-Optimist Club of Lafayette, Ind. *Society Aff:* ASAE, SAE, LOC.

Hinson, Tony D
Business: 2144 Melbourne St, PO Box 10068, Charleston, SC 29411
Position: Team Leader, Planning Services Br *Employer:* Southern Division Naval Facilities Engrg Command *Education:* Dip/Defense Planning/Naval War Coll; BS/CE/NC St Univ *Born:* 1/14/45 Native of High Point, NC. Completed NAVFACENGCOM Professional Development Center's Training program 1970. Field engr with projects in facilities design, construction maintenance and repair 1967-1972, while assigned to Gulf Division, NAVFACENGCOM, New Orleans, LA (1967) and Southern Division, NAVFACENGCOM, Charleston, SC (1970). Specialized in regional/urban/facilities planning in 1972. Assumed responsibility as Team Leader for preparation of site and facility development plans and feasibility studies in 1974. Pres, SC Section A.S.C.E. 1979-1980. Secretary, SC Council of Engrg Societies 1980-1981. SOUTHNAVFACENGCOM Employee of the Year 1973. Enjoy computer applications and photography. Working on MBA/Bus Mgmt. at The Citadel. *Society Aff:* ASCE, XE, SAME

Hipchen, Donald E
Business: 10301 9th St N, St Petersburg, FL 33716
Position: Pres. *Employer:* Jim Walter Res Corp. *Education:* BS/ChE/Univ of Pittsburgh. *Born:* 6/30/32. Native of Bradford, PA, married (wife Martha) with 3 children. Over thirty yrs of experience in res, dev, production & marketing of plastic products & processes, including flexible & rigid polyurethane, polyisocyanurate, polyvinylchloride, polyethylene, etc, with Union Carbide, Armstrong Cork, Allied Chem, & Jim Walter Corp. Assumed position as Pres, Jim Walter Res Corp in 1979, responsible for R&D in the areas of bldg & construction mtls, ind products, & specialty chems. Hold many patents & have authored numerous technical publications on cellular plastic products & tech. Hobbies are hunting, fishing, diving, photography, skiing. *Society Aff:* ACS, SPE, SPI, IRI.

Hira, Gulab G
Home: 139 South Rd, Bedford, MA 01730
Position: Sr. Proj Engr. *Employer:* The Gillette Co. *Education:* MBA/Bus Admin/Northeastern Univ; BE/Mech Elec Engg/Univ of Bombay (India); Int Sci/Phy-Chem/Univ of Bombay *Born:* 12/20/25. Born and educated in India and USA. Trained in Europe and UK. Postgrad work in USA. Started as Sales and Application Engr in 1948. Moved into Plant Engg in 1963. Certified Plant Engr. Certified Energy Mgr. Certified Cogeneration Professional. Reg PE. Chartered Engr. Built production facilities for Hoist-O-Mech Ltd 1963-68. Plant Engr for Ceat Tyres 1968. Facilities Sr. Proj Engr for Gillette, South Boston complex since 1968, responsible for utilities and pollution control activities. New England Plant Engr of the Yr Award 1976 & 1983. Pres Plant engrs Club AIPE, Boston chapter 1976-77. AIPE N.E. Group Director 1985 & 1986. AIPE President's Club Award 1980 & 1981. Pres ESNE 1981/82. Pres Met Chapter MSPE 1984-85. Pres MSPE 1986-87. Chrmn MA Engrs Week 1983. Chrmn N. E. Plant Engrs Conf 1983. *Society Aff:* AEE, NSPE, MSPE, AIPE, ASME, ESNE.

Hirai, Wallace A
Business: 109 Holomua St, Hilo, HI 96720
Position: Pres *Employer:* W. A. Hirai & Assoc Inc *Education:* MBA/Bus/Univ of HI; BS/ME/IA St Univ *Born:* 1/1/29 Wallace A. Hirai, a registered professional mechanical engr in the State of HI and the pres elect for the HI Society of PE 1982-83, is pres of a consulting engrg firm in Hilo, HI. He has specialized in the field of Energy as a consulting engr since leaving the firm of C. Brewer and Co as Chief Design Engr to form his own co. In recent years his firm has been very active in the field of

Hirai, Wallace A (Continued)
alternate energy technology and has worked on geothermal, wind, hydroelectric and biomass projects throughout the State of HI. *Society Aff:* NSPE, ASME, AWWA

Hirata, Edward Y
Home: 46-255 IkiiKi St, Kaneohe, HI 96744
Position: Director, Dept. of Transportation *Employer:* State of Hawaii *Education:* BS/Civil Engg/Univ of Hawaii. *Born:* May 6, 1933. Served in Army Corps of Engrs 1957-59. Proj Engr with various cons engrg firms 1959-62. Pres of Hirata, Shimakuro & Assoc, Inc Cons Engrs 1962-69. With City & County of Honolulu 1978-80, as Dir of Bldg Dept 1969- 71, Public Works Dir 1971-74. Mgr & Ch Engr Honolulu Bd of Water Supply. 1974- 78, managing dir, city & county of Honolulu. 1978-80, Pres Honolulu Woodtreating Co 1980-86, VP, Transmission & Distribution, Hawaiian Electric Co, Inc, currently director, Dept. of Transportation, State of Hawaii. *Society Aff:* NSPE, ASCE, AWWA, SAME.

Hird, Lyle F
Home: 5822 Barton Rd, Madison, WI 53711
Position: Sr Consultant *Employer:* Strand Assocs, Inc *Education:* BS/CE/Univ of WI-Madison *Born:* 7/28/29 in Cuba City, WI. Attended UW Platteville majoring in mining engrg prior to UW-Madison. Spent two years in US Army with duty in Korea surveying air fields. With Strand Assocs since 1955. In charge of wastewater treatment work until becoming pres of firm in 1970. Past pres of Madison chapter ASCE. Past pres of Consulting Engrs Council of WI. Member of ACEC Environmental Committee. Past member of NSPE Environmental Committee. Retired as Pres Strand Assocs 1984. Presently Sr Consultant to firm. *Society Aff:* ASCE, NSPE, AWWA, WPCF, ACEC

Hiroshi, Sakai M
Home: No. 5-38-3, Narita Higashi, Suginami, Tokyo, Japan 166
Position: Prof Emeritus *Employer:* Consulting Engineer *Education:* PhD/EE/Kyoto Univ; BS/EE/Port Arther Tech Univ *Born:* 11/9/21 Native of Japan. Appointed Engrg officer at Navy as Sublieutenant until WWII finished. Employed as Officer with the Minister of Railway and Transportation, 1945-49. Employed as Chief Engr of Planning & Designing Section of Toho Tele-Communication Co, Ltd., 1949-58. Appointed Assoc Prof of Electric Engrg Dept of Tokyo Metropolitan Univ, 1958-63. Appointed Prof Emeritus Sophia Univ in Tokyo, since 1963, performed as Head of Graduate Course, 1975-76, and performed as Head of Dept of same, 1976-80. Master Calligrapher in Japan and enjoy private piloting, classical music and fishing. *Society Aff:* IEEE, JIEE, JIEC

Hirota, Sam O
Business: 864 S Beretania St, Honolulu, HI 96813
Position: Pres. *Employer:* Sam O Hirota, Inc. *Education:* BS/CE/Univ of HI. *Born:* 5/23/12. After grad from Univ of HI worked for: priv consulting engrg and surveying firms, City & County of Honolulu Dept of Public Works, Territory of HI Hgwy Dept, and Deputy-Dir of HI State Dept of Trans. *Membership* in other than profl societies include: Rotarty Intl (Ala Moana Club), Honolulu Acad of Arts and Mid-Pacific Country Club. *Honors and achievements* are: Certificate oFf Merit from DV, Men & Women of HI, Honorary Mbr of CA Council of Civ Engrs & Land Surveyors and Distinguished Service award from Engrg Assn of HI. Received Engr of the Year Award 1981 from Hawaii Soc of PE. Dist Alumnus Univ of HA. *Hobbies* are: Oriental art and ceramics, Japanese swords, golfing. *Society Aff:* NSPE, ASCE, ACSM, ACEC.

Hirsch, Carl M
Home: 114 Hedgewood Dr, Greenbelt, MD 20770
Position: Manager of Professional Services *Employer:* American Society of Civil Engineers *Education:* BS/Civil Engrg./Virginia Military Institute *Born:* 01/27/40 Bronx H.S. of Sci NYC and V.M.I. graduate. Served in U.S. Army Infantry and Engineers 1961-68; twice decorated for valor in Vietnam War. Project Engineer for Amoco Oil Co. 1968-73; various construction mgmt and project mgmt positions 1973-79. Codes and Standards Engr for American Iron and Steel Inst (AISI) 1978-83. Chief Engr & Exec Dir for National Corrugated Steel Pipe Assoc 1983-85. Mgr of Prof Services for ASCE since 1986. Co-editor of A.I.S.I. "Handbook of Corrosion Protection for Steel Pile Structures in Marine Environments–. ASCE Fellow. U.S. Army Reserve Officer. *Society Aff:* ASCE, NSPE, ASTM, NFPA, NACE, TRB

Hirsch, Ephraim G
Business: Pier 1 1/2 Embarcadero, San Francisco, CA 94111
Position: Principal, E G Hirsch & Associates. *Employer:* Self. *Education:* Posgrad Studies/Structures/Univ of Rome; MS/Civil Engg/Univ of CA; BS/Civil Engg/Univ of CA. *Born:* June 21, 1931. Fulbright Scholar, Univ of Rome, Italy 1959-60. Lec Univ of Calif, Col of Environ Design, Dept of Arch 1962-70. Own practice, either individually or on partnership, as a cons struc engr since 1964. Specializing in close collaboration with archs to produce struc solutions to architectural design problems. Prof regs in Calif, Ore, Colo, Nev, Wash, Ohio, New York. Mbr ASCE, SEAONC, CEAC, ACI, PCI. Int: Classical music & opera, backpacking, mt climbing, skiing & travel. *Society Aff:* ASCE, ACI, PCI, SEANC, CEAC.

Hirsch, Lawrence
Business: 4420 Rainier Ave, San Diego, CA 92120
Position: President. *Employer:* Hirsch & Co. *Education:* MSCE/CE-SAN/Univ of TX; BSCE/CE/Syracuse Univ. *Born:* Sept 20, 1929. Served with AF during Korean War. Achieved rank of Capt. Pres of Hirsch & Co from 1966-present. Reg in Texas, New Jersey, Calif & Arizona. Dir San Diego Cty Water Authority & Calif Water Resources Assn. Articles pub in several tech mags. Firm recipient of awards of excellence by ASCE & CEAC. Outstanding Young Civil Engr, San Diego Sect ASCE 1966; US Delegate Internatl Conf, Water for Peace 1967. *Society Aff:* AWWA, WPCF, SAME, ASCE.

Hirsch, Robert L
Business: 2300 N Plano Pkwy, PRC A104, Plano, TX 75075
Position: VP Research & Technical Services *Employer:* ARCO Oil & Gas Co *Education:* PhD/Nuc Engr/Phy/Univ of IL; MS/Nuc Engr/Univ of MI; BS/Mech Engr/Univ of IL *Born:* 3/6/35 Native of Wilmette, IL. Married with 3 children. Approximately 20 years of professional experience. Served as Nuclear Engr, Atomics Intl; Dept Head, ITT Industrial Labs; Staff Member, U.S. A.E.C.; Dir of Controlled Thermonuclear Research, U.S. A.E.C.; Asst. Administrator for Solar, Geothermal and Advanced Energy Systems, U.S. ERDA; Deputy Mgr, Science and Technology Dept, Exxon Corp; General Mgr, Exploratory Research, Exxon Research and Engrg Co, Mgr, Baytown Research and Development Division, Exxon Research and Engrg Co, Currently VP Research & Technical Svcs, ARCO Oil and Gas Co. Consultant with DOE, NASA, ORNL, and Princeton Univ. Member of AAAS, ANS, API, and APS. Outside activities include golf, tennis, handball and swimming. *Society Aff:* AAAS, API, APS, SPE, FPA, NYAS, AAPG, IRI.

Hirsch, Sylvan R
Home: 63 Old Short Hills Rd, West Orange, NJ 07052
Position: Consulting Engineer. *Employer:* Self-employed. *Education:* ME/-/Cornell Univ. *Born:* July 1902 Savannah, Ga. PE: NY, NJ, Mass. Asst Cvl Engr Refrigeration Div Worthington Pump 1930-'37. Cvl Engr Brunner Mfg Co 1937-47. Mgr Engrg Worthington Pump Holyoke, Mass 1947-57. Asst to V Pres Engrg Worthington Corp 1957-61. Ecs Div Pall Corp LIC 1968. Blazer Corp, Dir Engrg Development & Research 1974. Mgr Engrg Heat Recovery Corp 1976. Chmn Natl Standards Ctte ASRE 1957. Chmn Central NY Sect ASRE 1945; Chmn Conn Sect ASRE 1953; Fellow ASHRAE. Life Mbr ASME. Dev method fill Aresol cans at atmospheric pressure. Design Refrigeration Unit to cool Apollo Space Unit prior to take off. Pats on control of compressors. Specialist on Energy Conservation Systems. Member N.Y. Sect. Program Com. ASME; Member of Gen. Com. ASME Machine Design Div.; Chairman No. Jersey Sec. ASME; Writer Technical Journals. *Society Aff:* ASHRAE, ASME.

Hirschberg, Erwin E
Home: 881 Parker Woods Dr, Rockford, IL 61102
Position: Consultant *Employer:* Self-employed Conslt Engr *Education:* ChE-Professional/Chem Engrg/Univ of Cincinnati. *Born:* 8/25/22. in Park Ridge, IL. US Army, OSS 1942-46. Victor Chem Div, Stauffer Chem 1950-53. Joined Eclipse, Inc 1953; Sales Engr, Sales Mgr, Engrg Mgr, VP Engrg 1969-78, VP Tech Services 1978-83. Hirschberg Forensic Engrg & Consltg Services 1983-present. Mbr IL P E Examining Ctte 1972-76. Mbr AIChE, APCA, Combustion Inst, ANSI Stds Subcttes (4) AGA Hall of Fame, Reg PE OH, IL, NJ, AR, NSPE, ISPE, Rockford Kiwanis. Leisure acts: Boating & fishing. *Society Aff:* NSPE, ISPE, AICHE.

Hirschfeld, Ronald C
Business: 1017 Main Street, Winchester, MA 01890
Position: Principal *Employer:* Geotechnical Engineers, Inc. *Education:* PhD/Soil Mech/Harvard Univ; SM/Soil Mech/Harvard Univ; BSCE/Civil Engrg/Union College. *Born:* 11/23/30. Instructor, Harvard Univ 1958-60, Asst Prof 1960-64; Assoc Prof, Mass Inst of Tech 1964-72; Principal, Geotechnical Engrs, Inc 1970-78, Pres 1974-1978, 1982-86; Cons, Comision Federal de Electricidad, Mexico 1973-75. Natl Dir Amer Society of Civil Engrgs, 1981-1984. Pres, Mass Sect Amer Soc of Civil Engrs 1973-74; Pres, Amer Consulting Engrs Council of New England 1981-1982 Nat'l Dir, Amer Consltg Engrs Council, 1982-1983; Chmn, New England Sec Assn of Engrg Geologists 1971-73, Mbr Natl Bd/Dir 1972-73. Editor (with S J Poulos), Embankment-Dam Engrg, 1973. Member Dean's Advisory Council, School of Engineering, University of Massachusetts. Mbr, Uranium Mill Tailings Study Panel, Nat'l Academy of Science. *Society Aff:* ASCE, GSA, AEG, USCOLD, ACEC, ISRM, ISSMFE.

Hirschfelder, Joseph D
Business: Dept. Chem, Madison, WI 53706
Position: Emeritus Prof *Employer:* Univ Wisc *Education:* (Hon)D.Sc./Theor. Chem & Physics/Univ. Southern Cal & Marquette Univ.; Ph. Theor. Chem & Physics/Princeton; B.S./Chem/Yale *Born:* 03/27/11 Born 1911, Univ, Wisc Chem Dept 1937-present. During WWII, Theory of interior ballistics for guns & rockets (Nat. Defense Res. Council), Los Alamos Atom Bomb (Leader for group working on effects of atom bomb), Cross-roads Bikini bomb test 1946 was "Chief Phenomenologist." After war made contributions to Theory of Flames & Detonations (Edgerton Award of International Combustion Soc.); transport properties gases & liquids (Hon. Life Member ASME); Intermolecular Forces (Debye Award Am. Chem Soc); National Medal of Science (1975, President Ford). *Society Aff:* ASME, BRSC, NRS, NAS, AAAS, AAAS

Hirschhorn, Isidor S
Home: 56 Greenwood Ave, W Orange, NJ 07052
Position: Pres *Employer:* Lanthanide Research Corp *Education:* MA/Physical Chemistry/Columbia Univ; BA/Sciences/Montclair State College. *Born:* 10/25/15. Pres, Lanthanide Research Corp since 1984. VP Ronson Metals Corp 1959-1983. Gen Mgr Ronson Metals Corp 1950-1958. Technical Dir Ronson Metals Corp 1949. Previously - Chem Engr, Hercules Powder Works, Baraboo, WI; Instr in Electronics - USAF; Instr in Chemistry - Drew Univ & Madison (NJ) High Sch. *Society Aff:* AIME, ASM, ACS, ECS.

Hirschhorn, Joel S
Home: 10036 Pratt Place, Silver Spring, MD 20910
Position: Senior Associate *Employer:* Office of Technology Assessment US, Congress *Education:* PhD/Matls Engg/Rensselaer Poly Inst; MS/Met E/Poly Inst Brooklyn; BMetE/Met E/Poly Inst Brooklyn. *Born:* 9/18/39. At OTA since 1978, director of assessments on steel industry hazardous waste, superfund, and waste reduction. Was a Prof at Univ of Wi 1978-65, initiated major teaching & res program on powder met. Also began teaching of matls sci for non- engrs. Other int areas: hazardous waste, environmental protection policy, impact of tech on soc, industrial competitiveness, innovation, productivity. Author of: *Introduction to Powder Met*, Matls Sci: For Strangers in a Land', ASM-MEI Course *Powder Met, Holes & Wholes.* Co-editor *Advanced Experimental Techs in Powder Met.* Received ASEE Dow Outstanding Young Faculty Award in 1973. Intl Travel Award from NSF and German Marshall fund of US. U.S. Paper of Excellence 1987 Int. Congress on Hazardous Materials Management. *Society Aff:* AAAS.

Hirschorn, Martin
Business: 1160 Commerce Avenue, Bronx, NY 10462
Position: President. *Employer:* Industrial Acoustics Co, Inc. *Education:* BSc/Engg/Univ of London. *Born:* 1/23/21. Postgrad work at Columbia & New York Universities. Native of Berlin, Germany; came to US in 1947, naturalized 1952. Engr, Burt, Boulton & Haywood, Ltd London 1939-45, Hd Wrightson Processes, Ltd London, Nuswift, Ltd Elland, Yorkshire, Enhg 1945-47; Engr M W Kellogg Co, NYC 1947-49; Pres, Chmn/Bd Indus Acoustics Co, NYC 1949- ; Pres Chmn/Bd Indus Acoustics Co, Ltd England; Intra Acoustics Co Ltd., Montreal, Quebec; Pres & Dir Natl Noise Abatement Council 1956-59. Holder US & foreign pats incl noise suppressor for jet aircraft. Contrib many articles prof journals. Mbr Acoustical Soc of Amer, Inst of Acoustics in London, Inst of Noise Control Engrs & ASHRAE. *Society Aff:* ASA, INCE, ASHRAE.

Hirshfield, Jay L
Business: Mason Lab, 9 Hillhouse Ave, New Haven, CT 06520
Position: Prof. *Employer:* Yale Univ. *Education:* PhD/Phys/MA Inst of Tech; MS/Phys/OH State Univ; MS/EE/USAF Inst of Tech; BS/EE/Univ of MD. *Born:* 10/24/31. Native of Wash, DC. Served with USAF 1952-57. Postdoctoral position at MA Inst of Tech 1957-61. NATO Postdoctoral Fellow FOM Inst Voor Plasma-Fysica, Jutphaas, the Netherlands 1961-62. John Simon Guggenheim Fellow, Lab Gas Ionizzati, Frascati, Italy 1968. Visiting Prof of Phys, Racah Inst of Phys, Hebrew Univ, Jerusalem 1972, 1977-78. Instr & Asst Prof Phys Dept Yale Univ 1961-63. Asst Prof - present position Prof of Appl Sci engaged in plasma phys res Engg & Appl Sci Dept Yale Univ 1961-. *Society Aff:* ΣΞ, ΣΠΣ, AIP.

Hirt, Charles R
Business: 7162 Reading Rd, Cincinnati, OH 45222
Position: Assoc Dir Engrg Dir's Financial Plan & Forecast *Employer:* Procter & Gamble Co. *Education:* BME/Mech Engg/Univ of Dayton. *Born:* April 9, 1919. 2nd Lt to Lt Col with USAF 1942-47. With The Procter & Gamble Co from 1947-date; assumed current respon as Assoc Dir for Engrg Div's Financial Planning & Forecasting 1978. Winner Mech Engrg Award of Excellence Univ of Dayton 1941. Founding Mbr & Fellow of the Amer Assn of Cost Engrs. Offices held in AACE: Dir 1957, 1962-63; V Pres 1958; Pres 1960-61. Helped org Southwestern Ohio Regional Sect AACE; ser as Dir 1959-60. Received AACE Life Mbr Award 1966 & AACE Award of Merit 1974. *Society Aff:* AACE

Hirt, George J
Business: P.O. Box 8405, Kansas City, MO 64114
Position: Partner *Employer:* Black & Veatch Consulting Engrs *Education:* BS/ME/Finlay Engrg Coll *Born:* 7/24/33 Native of Independence, MO. Began Engrg career with the US Army, Ordnance Corps as an Automotive Test Engr during the period 1953 thru 1955. Joined Black & Veatch Consulting Engrs in 1955. In 1956 began a twelve year assignment as a construction engr and construction project mgr on large fossil fuel power plants and transmission facilities at various locations throughout the US. Between 1969 and the present, has specialized in Military Facility Engrg and has advanced three such positions as mechanical designer, project engr, project mgr, chief mechanical engr and in 1979 became a partner of Black & Veatch. *Society Aff:* NSPE, SAME

Hirth, John P
Business: 116 W 19th Ave, Columbus, OH 43210
Position: Prof *Employer:* The OH State Univ *Education:* B Met E/Met E/OH State Univ; MSc/Met E/OH State Univ; PhD/Met E/Carnegie Inst of Tech. *Born:* Dec 16,

Hirth, John P (Continued)
1930. Fulbright Res Fellow, Bristol Univ, England 1957-58; Carnegie 1958-61. At Ohio State since July 1961. Ser as Visiting Prof at Stanford Univ 1968-69. Awds: Hardy Gold Medal, AIME 1960; ASM Bradley Stoughton Awd 1964; Curtis McGraw Awd, ASEE 1967; Fellow ASM 1971; Campbell Lectr ASM 1972; Fellow TMS-AIME 1974. R F Mehl Medal TMS-AIME 1980, C H Mathewson Gold Medal TMS-AIME, 1982. Chmn Phys Met Gordon Conf 1967; Chmn, Chem & Physics of Metals Ctte of AIME 1967; Chmn, ASM Transactions Ctte 1969; ARPA Matls Res Council 1968- ; Bd of Overseers, Acad for Contemporary Problems 1971-77; Natl Acad of Engrg 1974-present. Author or co-author of over 290 articles in the areas of nucleation & growth processes, dislocation theory & physical met. Editor, Scripta Metallurgica 1975-. *Society Aff:* ASM, TMS, ASEE, ΣΞ.

Hisao, Kimura
Home: 1-13-9 Higashitamagawa, Setagaya-Ku, Tokyo 158 Japan
Position: Professor. *Employer:* Meisei Univ. *Education:* PhD/Elec Engg/Tokyo Univ; BE/Elec Engg/Tokyo Univ. *Born:* Nov 1911 Tokyo, Japan. Joined Mitsubishi Elec Corp as Transformer Designer 1934-44, dev surge-proof transformers. With Central Res Lab 1945-49, studied abnormal voltages of power circuits. Mgr of Elec Power Engrg Div 1949- 66, analyzed power network problems with computer. Joined Seikei Univ as Prof 1966-77. Joined Meisei Univ as Prof 1977-present. Elec Power Prize from IEE of Japan 1958. Fellow of IEEE 1975. *Society Aff:* IEEE, IEE.

Histand, Michael B
Business: Dept ME, Colorado State, Fort Collins, CO 80523
Position: Prof *Employer:* CO State Univ *Education:* PhD/Bioeng/Stanford Univ *Born:* 10/31/42 Main research interests involve the application of ultrasonics to medical diagnosis, Biological signal processing, orthopedics, and expert systems. A visiting appointment at Kyoto Univ in 1976 and ETH, Zurich 1986 expanded interests in clinical applications of engrg measurement techniques. As a prof of Mechanical Engrg and Physiology his teaching responsibilities comprise bioengrg, instrumentation, cardiovascular physiology and robotics. *Society Aff:* ASME

Hitchcock, Leon W
Home: 73 Madbury Road, Durham, NH 03824
Position: Prof Emeritus Elec Engrg. *Employer:* University of New Hampshire. *Education:* BS/Elec Engg/Worcester Polytechnic Inst. *Born:* June 21, 1886. Educ: Medway Mass Public Schs; WPI 1908. Exp: Design & Test D & W Fuse Co; 1909 Inspector Lines & Bonding, B & N St Ry Co; 1910 Drafting, Testing, Inventory, N J Neal Cons Engr. Teaching: UNH 1910-56; Chmn Dept Elec Engrg 1921-53; Act Dean, Col of Tech 1940-45; Institutional Rep ESMWT Program 1940-45. P Mbr: AIEE; IEEE; NH Acad Sci; ASEE; NHSPE; TriCounty Elec Assn. Community Ser: Church Warden; Auditor; Chmn, Red Cross Fund Campaign; Made Map of Durham. Wrote History of Elec Engrg Dept Univ New Hampshire. Honors: Leon W Hitchcock Scholarship Award by Tri-County Elec Assn; 1946 Life Mbr & 1950 Fellow AIEE; 1960 Dedication of Hitchcock Hall; 1967 Robert H Goddard Award for Outstanding Prof Achievements presented by WPI. Listed in Amer Men of Sci; Who's Who in the East; Who's Who in Amer Educ. 1978 "PROFILE OF SERVICE AWARD for valuable contribution to elec engg education at the Univ of NH for 46 yrs including 32 yrs as chrmn of an expanding dept." *Society Aff:* IEEE.

Hitomi, Katsundo
Home: 34-32 Yoshida-nakaoji-cho, Sakyo-ku, Kyoto, Japan 606
Position: Professor *Employer:* Kyoto University *Education:* DEng/Mech. Eng./Kyoto University, Japan; MS/Mech. Eng./Kyoto University, Japan; BS/Mech. Eng./Kyoto University, Japan *Born:* 01/16/32 Native of Osaka, Japan. Research Asst, Penn State Univ, 1959-61. Assistant Prof, Kyoto Univ, 1961-65. Assoc Prof, Tokyo Inst of Technology, 1965-71. Alexander-von-Humboldt Fellow & Visiting Prof, Aachen Technical Univ, 1966-67. Coordinator & Lecturer, Japan Productivity Center, 1968-Present. Professor, Osaka Univ, 1971-80. Visiting Prof, Penn State Univ, 1975. Technical Advisor, Asian Productivity Organization, 1979. Prof, Kyoto Univ, 1979-Present. Advisory Prof, Beijing Inst of Technology, 1983-Present. Author of 12 books and 250 articles. Research Fields: Manufacturing Systems Engineering, Production Management, Computer-Integrated Manufacturing. *Society Aff:* ASME, IIE, JSME, JIMA, KSME

Hittinger, William C
Home: 149 Bellevue Avenue, Summit, NJ 07901
Position: Retired *Education:* BS/Metallurgy/Lehigh Univ; DrEng/-/Lehigh Univ. *Born:* Nov 1922. Served with Army Ordnance Dept 1943-46. Matls Engrs, Western Elec Co 1946-52. Prod Mgr, Natl Union Radio Corp 1952-54. Various assignments with Bell Telephone Labs from 1954-66 incl Exec Dir, Semiconductor Device & Electon Tube Div from 196266. Pres, Bellcomm Inc an AT&T subsidiary from 1966- 68. Pres, Genl Instrument Corp 1968-70. RCA Corp 1970-85. Assuming current assignment as Exec V Pres, Res & Engrg in 1976. Fellow IEEE, Mbr NAE, Trustee Lehigh Univ. Retired, 1986. *Society Aff:* IEEE, NAE

Hittman, Fred
Business: 9190 Red Branch Rd, Columbia, MD 21045
Position: President & Chairman of the Board. *Employer:* Hittman Materials & Medical Components, Inc *Education:* BS/Chem & Metallurgical Engg/Univ of MI. *Born:* May 13, 1929. BSC University of Michigan 1951 Grad studies in Nuclear Engrg & Physics Univ of Idaho (1952) & Drexel Inst of Tech (1956). Shift Supr Gen Elec's Hanford Nuclear Works 1951-53, Res Engr in peaceful uses of atomic energy at Brookhaven Natl Lab (1953-55). Various program mgmt pos ending in post of principal scientist at Glen L Martin Co Nuclear Div (1955-62). Founded Hittman Corp (predecessor of HMMC) (nuclear & medical services & prods firm) in 1962. Reg P E, Md. Prof Affiliation ANS, AAMI, AIChE. *Society Aff:* ANS, AAMI, AIChE.

Hix, Charles F, Jr
Business: 1285 Boston Ave, Bldg 22 DE, Bridgeport, CT 06602
Position: Genl Manager - Mfg Dept. *Employer:* General Electric Company. *Education:* BSME/Design/Univ of CO; BSEE/Electronics/Univ of CO. *Born:* 1926 Longmont Colo. Genl Manager-Mfg Dept, Housewares & Audio bus div 1977 & Mfg & Engrg Opereration, Audio Electronics Prods Dept 1975. Mgr Engrg for AEPD 1970. During 1969 ser as Mgr, Special Sys Section, Info Sys Equip Div, respon for the TRADAR Point-of-Sale Terminal & IRS Regional Ctr Data Reduction Programs. Mgmt assignments with Missile & Space Div as Mgr Engrg, Manned Orbiting Lab 1965; Mgr, Advanced Manned Space Engrg & Mgr, Advanced Missile Engrg 1962. In 1956 was Mgr, Component Engrg, Special Defense Programs with Missile & Ordnance Sys Dept. Joined GE on the Test Program in 1949 & ser as Supr of Creative Engrg Program. BSEE & BSME. Served as Supply Corps Officer US Navy. *Society Aff:* IEEE, AIAA.

Hixson, A Norman
Business: 306 Towne Bldg, Philadelphia, PA 19104
Position: Professor of Chemical Engineering. *Employer:* University of Pennsylvania. *Education:* PhD/Chem Engg/Columbia Univ; ChE/Chem Engg/Columbia Univ; BS/Chem Engg/Columbia Univ; BA/Engg/Columbia Univ. *Born:* July 1909 Iowa City, Iowa. Chem Engr on catalyst dev Intermetals Corp 1933-36. DuPont Co 1936-38 one of the first 3 engrs on Nylon dev. Univ of Pa 1938-1980. Prof Emeritus 1980-date Asst V Pres Grad Engrg Affairs 1954-75 in charge of all engrg grad programs. Cons Raw Matls Div, Bethlehem Steel Co 1944- 60 & 1964-75. Shell Visiting Prof Univ Col, London 1960-61, Visiting Prof Univ of Sydney 1975. Fellow AIChE 1974. Res on separation processes, ore beneficiation & SO_2 recovery. Hobbies: golf, travel & bridge. *Society Aff:* ACS, AIChE, ASEE.

Hixson, Elmer L
Home: 3103 White Rock Drive, Austin, TX 78731
Position: Professor Electrical Engrg. *Employer:* The University of Texas at Austin. *Education:* PhD/EE/Univ of TX; MSEE/EE/Univ of TX; BSEE/EE/Univ of TX. *Born:* 9/29/24. Served in US Navy worked at Navy Electronics Lab 1948-54; taught

Hixson, Elmer L (Continued)
at Univ of Texas since 1954. Res & teaching specialiy in acoustic & noise control. Sr Mbr IEEE, Mbr ASEE, Fellow Ascoustical Soc Amer, Sigma Xi, HKN, INCE. *Society Aff:* IEEE, ASEE, ASA, INCE.

Hixson, Thomas D
Business: 3666 Government St, P.O. Box 5444, Alexandria, LA 71301
Position: VP *Employer:* Meyer, Meyer, La Croix & Hixson *Education:* BS/CE/Univ of SC *Born:* 7/8/38 Atlanta, GA. Married, 3 children. Served US Air Force in Europe 1961. Civil Engr with US Forest Service 1962-1971. Various assignments in S.E. US including regional bridge engr. Joined firm of Meyer, Meyer, La Croix & Hixson, Inc in 1971. VP in charge of utility division. Past President of LA Engrg Society. Active in community affairs including Chamber of Commerce, church, Lions Club, United Way and local governmental advisory groups. Vice Chairman S.W. Region NSPE/DEPP. Private pilot. Enjoys hunting, fishing & golf. Registered engr in LA and MI. Land surveyor in LA. *Society Aff:* NSPE, ASCE, ASTM, ACEC, AAEE

Hjersted, Norman B
Business: Suite 2406, 106 W 14th St, Kansas City, MO 64105
Position: President. *Employer:* Conservation Chemical Company. *Born:* Aug 27, 1922 Chicago. Received a Chem Engrg degree from Rice Univ. Founded first co (1960) in US specifically provide environ services. Dev number of waste recycle & detoxification processes.

Hladky, Wallace F
Home: 15 Decision Way E, Washington Crossing, PA 18977
Position: Dir of Operations and Engrg *Employer:* Thiokol-Chem Div. *Education:* BS/Chem Engg/IA State Univ. *Born:* 10/3/27. Native of IA. Communications Officer, Chemical Mortar Batallion. With Salsbury Laboratories as Engineering Manager over twenty yrs. Joined southwest specialty chemicals in 1975 and appointed VP/Mfg in 1977. Merged with Thiokol Corp and appointed Dir of Operations with responsibility for Chemical Manufacturing at three plants in 1979. Received IA AIChE state award. Served as National Chairman of Career Guidance for AIChE three yrs after term as VChmn. Served in vaious civic, religous and political capacities at municipal level. *Society Aff:* AIChE.

Ho, Yu-Chi
Business: Pierce Hall, Harvard University, Cambridge, MA 02138
Position: Gordon MdKay Prof Engg Applied Math. *Employer:* Harvard University. *Education:* PhD/Applied Mathematics/Harvard Univ; SMEE/EE/MIT; SBEE/EE/MIT. *Born:* March 1, 1934. Naturalized US Citizen. 1955-58 Bendix Corp, Sr Engr co- inventor of four pats on various aspects of numerical control. 1961-present Harvard Univ, Prof of Engrg & Applied Mathematics. Cons to various gov't, nonprofit orgns & indus in decision & control & Manufacturing Automation problems. Author of two books & over 80 tech papers. Chmn & organizer of prof meetings & symposiums. Member of the National Academy of Engineering. *Society Aff:* IEEE, TIMS, SIAM, NAE.

Hoad, John G
Business: Post Office Box 1656, Southern Pines, NC 28387
Position: Consulting Engr *Employer:* Self-employed. *Education:* BS/CE/Univ of MI. *Born:* Sept 1909. Worked with Public Admin Serv, Mich State Hwy Dept, Edison Co, Army Air Forces WWII, Lt Col. Formed Hd Engrs, Inc in 1953 & was Pres & CEO until 1979. Domestic & foreign engagements in process & mfg indus & investorowned utilities. Principal work in environ pollution control; pulp, paper & cement mills; chemicals, power plants, mfg plants & foundries. P Pres AICE. V Pres & Dir 1970 EJC. Councillor 1970 AICE. *Society Aff:* ASCE, ASME, ACEC.

Hoadley, Peter G
Business: Box 1602, Sta B, Dept of Civ Engr, Nashville, TN 37235
Position: Prof of Civil Engr. *Employer:* Vanderbilt Univ. *Education:* PhD/Structural Engr/Univ of IL; MS/Structural Engr/Univ of IL; BSCE/Civ Engr/Duke Univ. *Born:* 12/30/34. Mbr of the Vanderbilt Univ Engg Faculty since 1961. Currently Prof of Civ Engg. Former Pres Nashville Sec of ASCE. Mbr of Tau Beta Pi and Phi Beta Kappa. Former Chrmn of ASEE-CE Div. Former Pres of Univ Club of Nashville. Finalist in White House Fellow Prog. Active on several ASCE and ASEE natl committees. PE in TN. Co-author of *Structural Engg* textbook. Former Editor of ASEE's *Civil Engg Ed.* Member ABET/EAC. *Society Aff:* ASCE, ASEE.

Hoag, David G
Home: 116 Winthrop, Medway, MA 02053
Position: Senior Technical Advisor *Employer:* Charles Stark Draper Lab, Inc. *Education:* SM/Instrumentation/MIT; SB/Elec/MIT. *Born:* Oct 1925. Employed by Charles Stark Draper Lab, Inc (formerly MIT Instrumentation Lab) since 1946. Tech Dir Polaris Missile Guidance Sys 1957-61. Tech Dir 1961-66 & Program Mgr 1966-72 for the design of navigation, guidance & control sys for the Apollo lunar landing mission spacecrafts. Assoc Editor ACTA Astronautica. Mbr Internatl Acad of Astronuatics. Pres Inst of Navigation. Fellow AIAA. National Academy of Engineering. NASA Public Serv Award. ION Thurlow Award. Special Award British Royal Inst of Navigation. AIAA Louis W Hill Space Transportation Award. *Society Aff:* AIAA, ION, NAE.

Hoag, James D
Business: 1901 Gratiot St Box 149, St Louis, MO 63166
Position: Manager, Safety & Health. *Employer:* Union Elec Co *Education:* BS/Ind Psy & Eng/Purdue Univ. *Born:* 7/14/24. in Downers Grove, IL. Served as Flight Engr with US Army AF in WWII. Safety Engr with Dept of the Army, Navy & Kaiser Aluminum, specializing in Ordnance & Chem Safety. With Union Elec Co since 1958 has been respon for dev, coordination of safety policy & mgmt programs. Past Pres, Amer Soc of Safety Engrs. Mbr & P Chmn Edison Elec Inst Codes & Stds Ctte & Mbr P Chmn of Safety & Indus Health Ctte. Since 1962, VP & Dir of Greater St Louis Safety Coun & MO Safety Coun. Enjoys Little Theatre work & photography. A Certified Safety Prof & Reg PE, Safety; State of CA. Lecturer-WA Univ-Industrial Mgmt; Fellow Amer Soc Safety Engrs, Mbr Human Factors Soc. *Society Aff:* ASSE, HFS.

Hoag, LaVerne L
Business: 202 W Boyd, Room 124, Norman, OK 73019
Position: Assoc Prof. *Employer:* Univ of OK. *Education:* PhD/IE/Univ of MI; MS/IE/Univ of MI; BS/IE/Univ of MI. *Born:* 8/11/41. Native of Ann Arbor, MI. Worked for Community Systems Fdn and the Univ of MI before completing a doctoral prog. Joined the faculty of the School of Industrial Engg at the Univ of OK in Sept, 1969. Has conducted res and written papers in the areas of work physiology, inspector performance, quality control sampling procedures, public transportation and hand tool design. Chrmn of the Ergonomics Group of AIHA for 1979-80. Is an amateur radio operator and a stained glass craftsman. *Society Aff:* IIE, HFS, ES, AIHA, ASEE.

Hoagland, Jack C
Home: 12452 Ranchview SW (Orange Co), Santa Ana, CA 92705
Position: Mgr, Adv. Programs *Employer:* Rockwell Intl *Education:* BS/Engr/CA Inst of Tech; Adv. Studies/EE-SYS/UCLA-UCI *Born:* 11/14/18 Research Engr, North American Aviation 1949-53; Systems Engr, Lockheed Missile & Space 1953-57; Mgr Research, The Ralph M Parson Co 1057-60; Consultant on Communication/Electronic Systems 1960-62; Mgr Identification, McDonnell Douglas 1962-70; Owner, Hoagland Enterprises (Electronic System) 1970-present; Mgr Adv Systems and Spec Prog, Rockwell Intl 1974-present. Served USNR 1944-51 as Research Officer for Missles/Satellite Development. Registered PE, EE/ME-CA. Published over 30 papers on space communication/electronic systems. Fellow IAE, Fellow IEEE; Chrmn Los Angeles Council/So. Area Region Six 1979-80. M Barry Carlton, IEEE, Award 1967, plus numerous outstanding leadership, achievement and engrg awards/

Hoagland, Jack C (Continued)
citations by IEEE, Rockwell, Douglas Aircraft, McDonnell-Douglas, 1966-1982, Rockwell-Engr of the Year 1982. Listed in Amer Men & Women in Sci, Who's Who in Amer. Mbr Old Crows, AOPA. 1986 NSA Meritious Service Award *Society Aff:* IEEE, IAE

Hobson, Kenneth H
Business: Univ Ave, Fairbanks, AK 99701
Position: Asst Prof, School of Engg. *Employer:* Univ of AK. *Education:* Masters/Engg Mgt/Univ of AK. *Born:* 8/14/20. in England. Received early educ & profl training there. Emigrated to Canada in 1953 & AK in 1965. Profl experience as follows: Hawker Aircraft, Kingston on Thames, Engg, Aircraft Tool Design 1941-45; CVA Jigs Molds & Tools, Shoreham, Sx England, Motor Vehicle Tood Design 1946; Ministry of Supply Armament Design Dept Langhurst, Sussex, Engg, Armored Vehicle Design 1946-53; leland Motors, Longeveil PQ, Canada, Motor Vehicle Design 1953-54; Canadair Montreal, PQ Canada, Design of Aircraft Tooling & test fixtures 1954-61/Vehicle design & testing 1961-65; Univ of AK, Educ, 1965 to present time. *Society Aff:* ASCE, SAME, NSPE, ASEE.

Hochstein, Samuel
Business: 28-11 Bridge Plaza N, Long Island City, NY 11101
Position: Deputy Commissioner, Chief Engr. *Employer:* New York City Dept of Traffic. *Education:* BCE/CE/CCNY. *Born:* Nov 1925 New York City. BSCE from CCNY 1949. Entered Dept of Traffic in 1950 & appointed to Deputy Commissioner-Ch Engr in 1969. Fellow of Inst of Trans Engrs. Mbr of Trans Res Bd. Mbr of Natl Advisory Ctte on Uniform Traffic Control Devices. P Pres ITE Metropolitan Sec. P Dir of Inst of Trans Engrs. *Society Aff:* ITE, TRB, NAC.

Hockenbury, Robert W
Business: JEC 5046, Dept Nucl Engg, Troy, NY 12181
Position: Assoc Prof. *Employer:* RPI. *Education:* PhD/Nucl Sci/RPI; MS/Phys/RPI; BS/Phys/Union College. *Born:* 7/31/28. Served in US Army, 1946-1947. With Knolls Atomic Power Lab from 1953-1958. At RPI since 1958. Performed res in neutron cross section measurements to 1976. Taught courses in neutron phys from 1970-1974. Currently teach Nuclear Reactor Analysis & Nuclear Reactor Reliability & Safety. Doing res (since 1975) in reliability methods emphasizing practical applications. Served on Executive Council of Society for Risk Analysis. Also doing experimental research on radiation-induced polymerization. *Society Aff:* Soc. for Risk Analysis

Hockett, John E
Home: 1765 38th St, Los Alamos, NM 87544
Position: Construction Project Manager *Employer:* Los Alamos Nat'l Lab, Univ of Calif. *Education:* DEng/Metallurgy/Univ of CA, Berkeley; Ms/Metallurgy/Univ of CA, Berkeley; BS/Met Engg/MI Tech Univ. *Born:* Apr 1919 Detroit. Served Corps of Engrs until transfer to Air Corps pilot, WWII. Air Medal and DFC with clusters, Asia-Pacific Process Engr, Douglas Aircraft, Sr Met, Reynolds Metals; Joined LANL 1952. Res in matls deformation. Currently respon for group doing selection & advanced dev of matls for nuclear weapons. Also const pro mgr-New Trition facility at Los Almos Natl Lab. Tau Beta Pi; Alpha Sigma Mu; Sigma Xi; Fellow Amer Inst of Chems; all Chap offices & Fellow, ASM; Reg Engr New Mexico. I jog, dirt motorcycle, woodwork, fly airplanes. *Society Aff:* AIC, AIME, ASM.

Hocott, Claude R
Home: 4538 Ivanhoe St, Houston, TX 77027
Position: Prof Emeritus. *Employer:* The University of Texas. *Education:* PhD/Chem Engg/Univ of TX; MS/Chem Engg/Univ of TX; BS/Chem Engg/Univ of TX; AA/Chemistry/Pan American Univ. *Born:* Nov 1909. Distinguished Engrg Grad 1971. Res Engr, Humble Oil & Refining Co; V Pres Esso Prod Res, Exec V Pres Exxon Prod Res 1937-74. Prof, The Univ of Texas 1974-79. Chmn, Petroleum Branch AIME; V Pres AIME 1970; Hon Mbr 1971; Mbr, Natl Acad of Engrg. *Society Aff:* AIME, AIChE, AAPG.

Hoddy, George W
Home: 508 W. Williams St, Owosso, MI 48867
Position: V Chrmn of Bd/Founder *Employer:* Universal Electric Co *Education:* PhD/Mod Ind/Sunshine Univ; MS/EE/OH State Univ; BS/EE/OH State Univ *Born:* 3/7/05 Native of Columbus, OH. Electrical engr, Day-Fan Elec Co, Dayton, OH, 1926- 29; Robbins & Myers, Inc, Springfield, OH, 1929-31; chief engr, Pioneer Div, Master Electric Co, Dayton, 1932-34; VP, general mgr, Redmond Co, Inc, Owosso, MI, 1934-43; pres, gen mgr, chief exec officer, Universal Electric, Owosso, 1942-71, chrmn bd, 1971-79, vice chrmn, 1979-; chrmn bd, American Universal Electric (India) Ltd, 1962-; Universal Electric Export, 1973-, Universal Electric Ltd, Gainsborough, England, 1974-; chrmn, Intertherm, Inc, 1980-; pres, Fiji Marina, Los Angeles, 1968-76; and many others. Member OH State Univ. Alumni Assn, Newcomen Soc. Red Cross of Constantine 1982; Citizen of Yr 1984 (Owosso- Corunna Area Ch of Comm) Mbr AAAS, AIEE. *Society Aff:* ΣΞ, ΤΒΠ, ΗΚΝ, ΠΜΕ, ΛΧΑ

Hodge, Jack S
Business: PO Bx 1070, Suffolk, VA 23434
Position: District Engineer. *Employer:* Va Dept of Highways & Transportation. *Education:* BS/CE/Univ of VA. *Born:* Feb 14, 1933. Presently the District Engr for the Suffolk District VA Dept of Hyws & Trans after having been Asst Resident Engr, Resident Engr & Asst District Engr in var locations in VA. Helped organize Blue Ridge Branch of VA Sect ASCE & first Pres of the Branch. I then became 3nd, 2nd, 1st V Pres & then Pres of A Sect ASCE. Currently Past President. *Society Aff:* ASCE.

Hodge, Kenneth E
Home: 42846 Cinema Avenue, Lancaster, CA 93534
Position: Dir of Aeronautical Projects Office *Employer:* NASA Dryden Flight Research Facility *Education:* M Mech E/Dyn & Controls/Brooklyn Poly; B Aero E/Aerodynamics/Rensselaer Poly *Born:* 1/25/28. Native of Claremont, NH. Aerospace industry R&D Experience with Grumman and Lockheed totalling 22 yrs, specializing in flight test. Grad of Stanford exec prog in 1969. Entered fed service in 1972 with Natl Aero & Space Council, exec office of the Pres. Joined NASA Headquarters in 1973 with mgt responsibility for res progs in aircraft operations & safety tech. In 1980 joined NASA Dryden Flight Research Center as Dir of Engrg. Became Dir, Aeronautical Projects Office in 1981. Reg PE, CA. *Society Aff:* AIAA, NAA, AOPA, ARRL, AWA, SEA, SSAE.

Hodge, Philip G, Jr
Business: University of Minnesota, 107 Akerman Hall, Minneapolis, MN 55455
Position: Professor of Mechs. *Employer:* U of Minnesota. *Education:* PhD/Applied Math/Brown Univ; AB/Math/Antioch College. *Born:* Nov 9 1920. Perm Teach Positions: UCLA, Poly Inst Brooklyn, Ill Inst Tech, U of Minnesota; Visit appoints: Brown, Stanford (NSF Postdoc Fellow), California (Russell Severance Springer Visit Prof); Books: Theory of Perfectly Plastic Solids (with W Prager), Elasticity-Plasticity (with J N Goodier), Plastic Analysis of Structures, Limit Analysis of Rotationally Symetric Plates & Shells, Continuum Mechanics; over 100 papers. ASME: Hon Mbr, P Chmn Applied Mechs Div, Natl Nom Ctte; Worcester Reed Warner Medal 1975. Elected National Academy of Engineering 1973. Am. Acad. Mech: Dist Service award 1984 ASCE: Theodore von Karman Medal 1985. Hobbies: mountain hiking, vegetable gardening, photography, running. *Society Aff:* ASME, AAUP, ΣΞ.

Hodge, Raymond J
Business: 1101 15th St NW, Washington, DC 20005
Position: Partner. *Employer:* Tippetts-Abbett-McCarthy-Stratton. *Education:* MCE/Civ Engrg/Cornell Univ; BCE/Civ Engrg/Manhattan Coll. *Born:* 5/15/22. Native of New York. US Naval Cvl Engrg Corps Officer; three yrs active duty Pacific WWII, two yrs Korean War. Asst Prof Civil Engrg Cornell Univ; Res Assoc Johns Hopkins Univ & Sloane Sch of Physics, Yale Univ. Partner, Tippetts-Abbett-

Hodge, Raymond J (Continued)
McCarthy-Stratton, Cons Engrs & Archs since 1968. Author "Dallas/Fort Worth Airport, A Major International Air Center–, "What are the Requisites for Expanding or Building New Airports?–, "The Economic Impact of International Airport Development–, "New Slants on Airports, Profile of the Airport Development Crisis–. Reg as Prof Land Surveyor & Engr in eleven states. *Society Aff:* ASCE, ACEC, NAE, SAME, AAEI, NSPE, ΤΒΠ, ΧΕ.

Hodges, David A
Business: Dept of Elec Engg & Comp Sci, Berkeley, CA 94720
Position: Prof. *Employer:* Univ of CA. *Education:* PhD/Elect Eng/UCA Berkeley; MS/Elect Eng/UCA Berkeley; BEE/Elect Eng/Cornell U *Born:* 8/25/37. Attended public schools in NY State. Joined components area of Bell Labs, Murray Hill, in 1966. Served as Hd of System Elements Res Dept at Bell Labs, Holmdel, in 1969-70. Mbr of the faculty at UC Berkeley since 1970. Res interests in analog & digital integrated circuits, ULSI for telecommunications applications, and mfg infor sys. *Society Aff:* IEEE

Hodges, George H
Home: Box 2661, Palm Beach, FL 33480
Position: Mgr Marine Dept *Employer:* Babcock & Wilcox Co *Education:* BS/Engrg/Yale Univ *Born:* 1/24/08 Graduated in 1930-Yale Univ-Sheffield Scientific School-Engrg. Joined Babcock & Wilcox Co in 1930-Engr Marine Dept 1930-1948-assistant to mgr 1948. Appointed mgr, Marine Dept in 1949 and served until retirement in 1968. Dept had 100 people. 1940-1946 active duty in Corps of Engrs, Army of The US-Grade of Capt thru Colonel. Last command was 2500 mem-Combat Group, XIX Corps. *Society Aff:* SNAME, SAME

Hodges, Joseph T
Home: 4200 E. 135th Street, Grandview, MO 64030
Position: VP *Employer:* Peterson Mfg Co *Education:* MBA/Bus Management/Univ of MO; B/ME/Syracuse Univ *Born:* 9/20/17 Native of Cornwall-on-Hudson, NY. Worked as Cadet Engr for Jones & Lawson Machine, for Glenn L. Martin Co in Baltimore in tool design, Production Engrg and as Controls Designer. Moved to Kansas City in 1949 with Bendix Corp, as Sr Engr, working on case design and fuzing systems for the atomic weapon program. In 1966 went with Peterson Manufacturing Co as VP of Engrg with responsibilities for Product Design, Development, Quality Control, and Quality Assurance. Hobbies - Sport Car Rallying, Raquetball, Reading, and CA Wines. *Society Aff:* SAE

Hodges, Lawrence H
Business: Technical Affairs Consultant, Ltd, 840 Lake Avenue, Lake Forum Building, PO Box 307, Racine, WI 53401-0307
Position: Consultant (Self-employed) *Education:* BS/Mech Engrg/Univ of WI; BS/Agri Engrg/TX A & M *Born:* July 1, 1920. BS in Agri Engrg, Texas A&M; BS in Mech Engrg, Univ of Wisc-Madison. Served US Army Artillery; ret from Army Reserves in 1962 with Rank of Lt Col. Asst Prof of Agri Engrg Univ of Wiscon-Madison. Former V Pres of T ech Affairs at J I Case Co, with respon for the Legal, Political & Social impacts of Tech. Currently a private consit specializing in Product Safety, Product Liability expert and Technical barriers to Intl Trade. ASAE P Pres 1973-74. ASAE Wisc Engr of the Yr in 1975. Univ of Wisc. Distinguished Service Citation in 1976. FIEI Engrg Merit Award - 1977. SAE's Arch T Colwell Coop Engrg Medal - 1982. TAMU Engrg Alumni Honor Award - 1984. National Academy of Engineering-1985. Active in num prof socs & trade assns. Author num tech papers & articles for trade mags. *Society Aff:* ASAE, SAE.

Hodges, R Dale
Business: 2708 N Acadian Thruwy W, Baton Rouge, LA 70805
Position: President. *Employer:* Pan American Engineers, Inc. *Education:* BS/Civil Engrg/MS State Univ. *Born:* Feb 1928. Engr in US Corps of Engrs 1949-55 N O Dist & Morocco, public works & airbase design; with Pan Amer Engrs 1955- . Partner 10 yrs. President since Jan 1976 when inc. Respon for firm policy & large proj dev. Mbr: ASCE, La ES, Miss ES; Pres La ACEC in 1974, Natl Dir ACEC 1976 American Arbitration Assoc. Enjoy sports. Baptist: Deacon, Minister of Music, Lic to Minister, Pres Louisiana Gas Assoc 1985. ACEC Fellow. Pres of Mississippi Gas Corp, a utility and owner of System Operators, a utility operating company. *Society Aff:* ASCE, ACEC, LAES, MSES.

Hodges, Raymond D
Business: 2708 N Acadian Thruway W, Baton Rouge, LA 70805
Position: Pres. *Employer:* Pan American Engg, Inc. *Education:* BS/Civil Engr/MS State Univ. *Born:* 2/12/28. Employed with US Army Corps of Engrs, N O Dist 1949 - 1951. Civil Projs, Mediterranean Div 1951 - 55, Air Base Engr. With Pan American Engrs, since 1955 as engr, partner and now Pres of corp. Responsible for engg proj of all types, water, sewer, highway, bridge and other, as well as company admin. CEC/L State Pres 1974-75. ACEC Nat'l Dir 1976-76. *Society Aff:* ASCE, NSPE, ACEC.

Hodges, Teddy O
Business: Coll of Engrg, Manhattan, KS 66506
Position: Assoc Dean & Dir of Engrg. *Employer:* Kansas State University. *Education:* BS/Agri Engrg/TX A&M Univ; MS/Agri Engg/IA State Univ; PhD/Agri Engg/MI State Univ. *Born:* 10/17/22. Native of Ravenna, Texas. Taught & res in Agri Engrg at Iowa State Univ, Univ of Arkansas, Univ of Missouri & Kansas State Univ between 1951-74. Assoc Dean of Engrg & Dir of Engrg Res Jan 1974-1981. Dir of ASAE 1972-74. Elected Fellow of ASAE 1976. Professor of Architectural Engineering & Construction Science, 1981-. *Society Aff:* ASAE, NSPE.

Hodges, William J
Home: C-12 Kimbrook, Greenwood, SC 29646
Position: Sr. Civil Engr. *Employer:* W E Gilbert & Assoc, Inc. *Education:* BCE/Civ Engg/Clemson Univ. *Born:* 8/4/27. in Ware Shoals, SC. Married former Ann Holley. Four daughters; Cheryl, Holley, Amye, Julie. Served with AFWESPAC 1946-47. Grad 1953 with high honor Clemson Univ. Honor fraternities: Tau Beta Pi, Phi Kappa Phi, Phi Eta Sigma. Biographee: Who's Who in the South & Southwest, Who's Who in SC, South Carolina Lives. Profl reg in SC, NC & GA. Joined Hearst-Coleman & Assoc in 1953 as field engr, VP in 1965, Pres & Chrmn in 1970. Assumed current responsibility as Senior Civil Engr upon merger with W E Gilbert & Assoc in 1978. Presently directs research & design incidental to unique structural problems & applications. *Society Aff:* NSPE.

Hodgkins, Franklin
Business: PO Bx 1498, Reading, PA 19603
Position: Asst to VP-Business Dev. *Employer:* Gilbert Assocs Inc. *Born:* Nov 1927 Reading PA. BS Engrg USMA West Point NY. Comm USAF Spec Weapons 1951-55. With Gilbert Assocs Inc 1955- . P E in PA, WI, NY. Respon for natl indus accounts related to services provided to the steel indus, food & beverages and select U S Govmt projects. Active in local gvmt acts.

Hodgson, Richard E
Home: 4201 Pierson Dr, Huntington Beach, CA 92649
Position: Sr. Project Manager *Employer:* Pacific Lighting Energy Systems *Education:* BS/Chem Engrg/MIT. *Born:* 1/16/28. Native of Stoneham, MA. Long time So CA resident. US Army 1946-48. Process Engr for Fluor Copr specializing in the design of Petroleum Refineries & Petrochemical Plants. Chem Engr for Exxon in Aruba, specialized in refinery troubleshooting. VP for Ben Holt Co in chrg of all engrg & const projs. VP Engrg for E & L Assocs Inc. VP operations, Reserve Synthetic Fuels. Energy Consultant, specializing in Process Design & Project Mgmt. Principle Process Engrg CF Braun & Co, Process Mgr for 200 MB/SD Kuwait Oil Refinery. Sr. Project Manager, Pacific Lighting Alt. Energy Plants. Thirty-six yrs experience in Process Plant Design & Management. *Society Aff:* AIChE.

Hodgson, Thom J
Business: Ind Systems Engg, Gainesville, FL 32605
Position: Prof. *Employer:* Univ of FL. *Education:* PhD/IE/Univ of MI; MBA/Quant Methods/Univ of MI; BSE/Sci/Univ of MI. *Born:* 3/6/38. *Society Aff:* IIE, ORSA, TIMS.

Hodsdon, Albert E, III
Business: 10 Common St, Waterville, ME 04901
Position: Principal *Employer:* A. E. Hodsdon Engrs *Education:* BS/Mech Eng/Univ of ME '69; ME/Interdiciplinary/Univ of ME '75 *Born:* 10/9/47 Native of Rumford, ME. Worked as a graduate assistant during graduate school at Univ of ME. Worked part time for Klienschmidt and Dutting Engrs, Pittsfield, ME 1970-1971. Worked as project mgr for Wright and Pierce Engrs from 1971-1974. Set up private practice in 1974 in the City of Waterville, ME. Have been specializing in water utility design and energy conservation design. Pres of ME Section-ASCE 1979-1980. Treasurer of CEM -1978-1984. Holds registrations in four states ME, NH, VT, and MA. *Society Aff:* ASCE, NSPE, CSI, AWWA, ASHRAE, CEM

Hoefle, Ronald A
Home: 145 S Edgelawn, Aurora, IL 60506
Position: VP/Engr. *Employer:* W E Deuchler Assoc. *Education:* BSCE/Civ Engr/Univ of IL. *Born:* 4/24/29. Native of Freeport, IL. Served with the USAF from 1951-1953. Began career with Dravo Corp, Pittsburgh, PA. Joined W E Deuchler Assoc in June, 1955. Became principal in Mar 1977. Assumed current responsibility as VP for Engg in 1978. Reg in IL as a PE and as a Structural Engr. Pres ISPE 1986-87. *Society Aff:* ASCE, NSPE, ACEE, WPCF, AWWA, WSE, ISPE.

Hoegfeldt, Jan M
Home: 5326 Minnetoga Terr, Minnetonka, MN 55343
Position: Sr Prin Metallurgist, Materls and Processes *Employer:* Honeywell Inc. *Education:* SM/Metallurgical Engrg/MIT; SB/Metallurgy/MIT. *Born:* July 1926 Valparaiso, Chile; now US citizen. PE in OH & MN. Dev & controlled quality of superalloys with Haynes Stellite. Improved tantalum & niobium powders for capacitors, Kemet 1962-65. Ran USAF superalloy extrusion prog for TRW Inc. Joined Honeywell 1968; jack-of-all-facits. Author and co- author of numerous papers ranging in topics from surface textures evaluations through pitfalls in tech transfers, but mainly concerned with the effects of high strain rates on the deformation and fracture of metals. Winner of sixteen consequtive Honeywell Engrg Excellence awds. Served as ASM Chmn Purdue Chapt & on Natl Dev Ctte. Mbr ASM Natl Nominating Ctte 1977; Elected FELLOW ASM 1979 Mbr ASM Advisory Tech Awareness Coun 1981-1984. Mbr ASMI Adwards Policy Cttee 1984-1987. Previous Dir Mid-Indiana Chap, ASNT. Past Dir of Intl Metallographic Soc; Genl Chmn 1975 Intl Metallographic Convention. Enjoys family, bridge, sports & MN. *Society Aff:* ASMI, ACBL, TBΠ, TMS OF AIME.

Hoekstra, Harold D
Home: 253 N Columbus St, Arlington, VA 22203
Position: Consulting Engr (Self-employed) *Education:* BS/Aero Engr/Univ of MI. *Born:* 8/18/02. Born in Chicago, IL & grew up in Battle Creek, MI. After college was Chief Engr, Crosley Aircraft Co Cincinnati, OH; proj engr and/or designer with Ford Motor Co Dearborn, MI, Curtiss Co Buffalo, NY & Stinson Aircraft Co Wayne, MI. With Fed Aviation Administration 1937-70 on aircraft certification & the dev of airworthiness regulations, including turbine & supersonic transports; from 1961- 70 was Chief, Engg & Safety Div. VP-Engg, Flight Safety Fdn, 1970-72. Consulting Engr, 1972-date. Author, Safety in Gen Aviation 1970. Inventor, 9 patents. Commercial pilot. *Society Aff:* RAeS, AIAA, SAE, Fellow of all three.

Hoel, Lester A
Business: Dept of Civil Engrg, Charlottesville, VA 22901
Position: Professor & Chairman. *Employer:* University of Virginia. *Education:* D Eng/Civil Engrg/Univ of CA (Berkeley); MCE/Civil Engrg/Polytechnic Inst of Brooklyn; BCE/Civil Engrg/City of NY. *Born:* 2/26/35. Presently Prof & Chmn, Dept of Civil Engrg Univ of Va, Charlottesville. Formerly Prof & Assoc Dir, Trans Res Inst Carnegie-Mellon U 1966-74. Alumni Award 1957; Fulbright Res Scholar 1964-65; ASCE Res Award 1976. TRB Best Paper Award 1977. Author of over 100 articles and reports in trans. Mbr of Chi Epsilon, Tau Beta Pi, Sigma Xi, ITE, ASCE, TRB & ASEE. Reg Prof Engr in California, Pennsylvania & Virginia

Hoelzeman, Ronald G
Business: Pittsburgh, PA 15261
Position: Assoc Chairman *Employer:* University of Pittsburgh. *Education:* PhD/EE/Pittsburgh; MS/EE/Pittsburgh; BS/EE/Pittsburgh. *Born:* 10/6/40. Sys Engr for Westinghouse Elec Corp specializing in Process Control Computer Sys. Taught at the Amer Univ of Beirut 1969-70. With Pitt since 1970. Currently Assoc Prof and Assoc Chrmn of the Elec Engrg Dept. IEEE Vice President for Educational Activities. *Society Aff:* IEEE, ASEE.

Hoeppner, David W
Business: Chairman and Professor, Dept. of Mech. Engr, University of Utah, Salt Lake City, UT 84112
Position: Professor Chairman - MIE Dept. *Employer:* U. of Utah. *Education:* PhD/Met Eng/Univ of WI-Madison; MS/Met Eng/Univ of WI-Madison; BME/Mech Engr/Marquette. *Born:* Dec 17, 1935. BME Marquette Univ, MS & PhD Univ of Wisc-Madison. Listed in Outstanding Young Men of Amer, Outstanding Educators of Amer; Mbr of ASM, ASEE, ASME, AIAA, ASTM, AIME, Sigma Xi, Jaycee Internatl Senator. Mbr of ASTM Ctte E9 on Fatigue (Exec Ctte & Subctte Chmn). Advisor for ASME to Office of Tech Assessment (US Congress), Cockburn Prof of Engg Design, Conslt to US Navy, US Air Force, Canadian Forces, NASA, NATO/AGARD/SMP, Rolls Royce Aeroengine Div, US FAA, Numerous Industries. Developed, and Principal lecturer in Course, on Aircraft Fatigue Course for U.S. Federal Aviation Adminstration. *Society Aff:* ASME, ASM, AIAA, ASEE, ASTM, SAE, ΣΞ.

Hoeppner, Steven A
Home: 17 Van Buren Rd, Pittsford, NY 14534
Position: Dist Sales Mgr. *Employer:* Olin Corp. *Education:* BS/Mgt/Univ of IL. *Born:* 11/10/38. in Minneapolis, MN. Grad from the Univ of IL in 1964. Joined the Brass Group - Olin Corp in Feb 1964 & held various positions prior to being appointed Dist Sales Mgr in 1972. Chrmn Rochester Chapter ASM & mbr of ASM Natl Chapter Advisory Committee. Enjoy woodworking, cooking & classical music. *Society Aff:* ASM.

Hoerner, George M, Jr
Business: Chem Engg Dept, Easton, PA 18042
Position: Professor, Dept Head *Employer:* Lafayette College. *Education:* PhD/ChE/Lehigh Univ; MEd/Educ/Univ of Rochester; BS/ChE/Lafayette College. *Born:* 2/12/29. *Society Aff:* AIChE, NSPE ASEE, ISA.

Hoff, Nicholas J
Home: 782 Esplanada Way, Stanford, CA 94305
Position: Professor Emeritus *Employer:* Stanford Univ *Education:* PhD/Engg Mech/Stanford Univ; Dipl Ing/Mech Engg/Polytech Inst Zurich. *Born:* Jan 1906. Diploma Swiss Federal Polytech Inst Zurich. Native of Hungary. Airplane designer, Hungary 1929-38. Taught at Polytech Inst Brooklyn 1940-57. Hd Dept Aeronautics/Astronautics Stanford Univ 1957-71. Prof Rensselaer Polytech Inst 1976-79. Active & held off in AIAA, ASME, ASCE, IUTAM; advisor/cons to NACA, NASA, USAF, USN, NATO & several aerospace co's. Mbr: Natl Acad of Engrg since 1965; corresponding Mbr Internatl Acad Astronautics since 1976. Worcester Reed Warner Medal ASME 1967; Pendray Award AIAA 1971; Structs, Struc Dynamics & Materials Award AIAA 1971; Karman Medal ASCE 1972; ASME Medal 1974; DSc (hon) Technion, Haifa 1980; Monie A. Ferst Medal Sigma Xi Soc 1982; I B Laskowitz Award NY Acad Sci 1983; Daniel Guggenheim Medal 1983; Mbr (hon) Hungarian Acad of Sci since 1986. *Society Aff:* ASME, AIAA, ASCE, RAeS.

Hoffman, Charles H
Home: 510 Crescent Parkway, Sea Girt, NY 08750 *Education:* MS/Elec Engg/MIT; BS/Elec Engg/Lehigh. *Born:* 9/5/17. In US Naval Reserves on active duty 1942-45. Joined Public Service Electric & Gas Co 1940, receiving assignments in sys operation, computer applications & power sys planning. Involved in interconnection planning & operations for many yres. Elected VP 1968. Assumed respon as Sr VP-Sys Planning & Interconnection in 1977. Retired in 1980. Author of numerous tech papers on power sys planning & pooling pubs between 1952-68. Fellow IEEE. Mbr CIGRE. Prof Engr in NJ. Mbr Phi Beta Kappa, Tau Beta Pi, Eta Kappa Nu, & Sigma Xi. Enjoy square dancing, travel & sailing. Married Louise Williams & has three children and two grandchildren. *Society Aff:* IEEE.

Hoffman, Conrad P
Business: 88 New Turnpike Rd, Troy, NY 12182
Position: Owner *Employer:* Hoffman Engrs & Surveyors Hoffman Assocs *Education:* BCE//RPI *Born:* 7/8/36 After graduation from R.P.I., served 4 years with the Army Corps. of Engrs in Germany and Vietnam. For past 10 years, owner of Hoffman Assocs and Hoffman engrs & Surveyors, civil engrg, architecture and land surveying firms with offices in Troy, NY and Pittsburg. Pres Rensselaer Cty. Chapter N.Y.S. PE Society (2 terms). City Engr for Mechanicville, NY, past 12 years. Member, Bd of Dirs NY State Easter Seals Society. Have lectured extensively on the physical accessibility needs of the handicapped. Married, 2 children. *Society Aff:* NSPE

Hoffman, Dwight S
Home: 1414 Alpowa Way, Moscow, ID 83843-2402
Position: Consultant. *Employer:* Self. *Education:* MS/Chemistry/Univ of ID; BS/Chem Eng/Univ of ID. *Born:* Aug 1916. Native of Leland, Idaho. Taught at the Univ of Idaho for 32 yrs. Assoc Dean of Engrg 1962-70. Dept Chmn 1974-76. Ret as Prof of Chem Engrg & Chmn Emeritus. Major academic int is thermodynamics & phase equilibria. Tech Cons for Phillips Petroleum Co. 1952-66. Visiting Prof U of Khartoum Sudan 1966 to establish curriculum in Chem Engrg. Had Fulbright grant 1976 and wrote chem engrg curriculum for Univ of Dar es Salaam, Tanzania. Western Elec Fund Award 1965. Consult for Educ program at Idaho Engrg Lab, 1976-78. Visiting Prof Chem Engr Wash State Univ 1977-80. Taught (part-time) courses Univ S. Calif in Applied Systems Dept, 1980-1985.Currently retired. *Society Aff:* AIChE, ACS, NSPE, ASEE, ΣΞ.

Hoffman, Harold L
Business: PO Box 2608, Houston, TX 77001
Position: Editor *Employer:* Hydrocarbon Processing. *Education:* BS/ChE/Rice Univ. *Born:* 7/30/26 Taught radar. Tech service, prod quality & process design for Amoco Oil Co Editor for Gulf Publishing's 'Petroleum Refiner' (now 'Hydrocarbon Processing') 1955-. Major contrib 1956 Best Issue Awd, Indus Mktg. Contrib to Kirk-Othmer's Encyclopedia of Chem Tech, McGraw-Hill Encyclopedia of Sci Tech, Liptak's Instrument Engrs Handbook & Reigel's Handbook of Indus Chem. Initiated computer program for Worldwide HPI Const Boxscore. Admin Bd of St Philips United Methodist Church. Dir of AIChE 1974-76. Mbr Tau Beta Pi, Sigma Xi, Res Dept, Houston Chamber of Commerce. Able Toastmaster level in Toastmasters Internatl *Society Aff:* AIChE, TBΠ, ΣΞ

Hoffman, Joe D
Business: Sch of Mech Engg, W Lafayette, IN 47907
Position: Prof. *Employer:* Purdue Univ. *Education:* PhD/ME/Purdue Univ; MS/ME/TX A&M Univ; BS/ME/TX A&M Univ. *Born:* 8/9/34. Profl interests include propulsion, gas dynamics, and computational fluid dynamics. Areas of specialization include performance prediction for aircraft inlets, solid rocket motors, & propulsive nozzles of all types. With Purdue Univ since 1960. Asst Prof 1963-73, Prof since 1973. Design Engr, Aerojet Solid Rocket Co, 1961-64. Res Aerospace Engr, Propulsion Lab, Army Missile Command, 1966-67. Approx 30 publications in refereed archival journals. Co-author (with M J Zucrow) of *Gas Dynamics*, two volumes, John Wiley, 1976. Consultant (approx 40 days per yr) for the propulsion ind, Dept of Defense Lab, & NASA. Native of USA. *Society Aff:* AIAA, ASME.

Hoffman, Myron
Business: Bainer Hall, Davis, CA 95616
Position: Prof. *Employer:* Univ of CA. *Education:* ScD/Instrum & Control/MIT; SM/Aero Engg/MIT; SB/Aero Engg/MIT. *Born:* 11/15/30. Taught in Aero & Astro Dept at MIT from 1955-56 and 1959-68. Lt in USAF from 1956-59; stationed at W-PAFB, Dayton, OH, in Air Res and Dev Command. Joined Mech Engg Dept of Univ of CA in 1968. Also consultant to Lawrence Livermore Lab on fusion reactor studies. Hobbies: skiing, fishing, Italian. *Society Aff:* ANS.

Hoffman, Nathan N
Home: 8100 Balcones Dr, Apt 164, Austin, TX 78759
Position: Consultant (Volunteer) School of Architecture *Employer:* University of Texas Austin *Education:* BS/Struct Design/Fenn College (Cleveland State Univ). *Born:* June 10, 1915 New York. Lt Cmdr, Civil Engrg Corps USNR, Exec Officer of Const Battalion Detachment in South Pacific, WWII. Received two letters of commendation. With present firm since 1946, engaged in Cons Struc Engrg. Received Gold Medal, Cleveland Engrg Soc. P Pres, Cons Engrs Council of Ohio. Engr of yr 1980-Cleveland Tech Soc Council Distinguished Consult of yr 1980-OH Assoc fo Consltg Engrs. *Society Aff:* NSPE, ASME, ACEC.

Hoffman, Paul W
Hoffman Process, Inc, P. O. Box 345, Sewickley, PA 15143
Position: President. *Employer:* Hoffman Engrg Inc. *Education:* LLB/Law/Wm McKinley Sch of Law; BSEE/Elec Engg/OH Univ. *Born:* 10/29/25. Founder & Ch Exec Officer of Hoffman Engrg Inc and subsidiary; Hoffman Process Mfg Co a manufae of process equip. BSEE Ohio Univ; LLB Wm McKinley Sch of Law & has been admitted to practice in the state of Ohio. Exec training programs: Tuck Sch of Business Admin, Dartmouth Col; Columbia Univ Grad Sch of Business Admin; Univ of Pittsburgh Grad School of Business Admin. Job scope: Respon for engrg in process field dealing with design, engrg & mfg of heat transfer equipment, distillation equipment & shop assembled process systems combining both heat transfer & distillation. Through Hoffman Process Mfg Co has developed machinery and specialty helical finned tubing for high temperature and corrosion resistant applications. Active in local community, church affairs, Rotary and Masonic work. Mbrships: Amer Soc of Chem Engrs, Natl Soc of Prof Engrs

Hoffman, Phillip R
Home: 3938 West Point Dr, Los Angeles, CA 90065
Position: Consulting Engr (Self-employed) *Education:* AB/Chem/UCLA; BE/CE/Univ of CA; MS/CE/Univ of So CA *Born:* 2/5/14 in Two Rivers, WI, 1914. Employed by the Dept of Water & Power, Los Angeles. 36 years, mostly in design of hydroelectric facilities. As Resource Research Engr developed the Castaic Power Project, 1,250,000 Kw, then the world's largest pumped storage project. After retirement in 1974, became consulting engr to irrigation districts, A/E firms, municipalities and the Corps of Engrs. Papers published include an innovative scheme "Hydroelectric Development, - Without Dams, Reservoirs and Penstocks–. "Energy Independence for the Island of Tahiti, by Conversion from Diesel to Hydro–, presented at "WATERPOWER '81–. Have sailboat and enjoyed sailing, - when there was time. Selected to receive the Rickey Medal award for 1981, for outstanding contribution to the science of hydroelectric engrg. *Society Aff:* ASCE, ACEC

Hoffman, Russell
2101 Gandin Rd, Cincinnati, OH 45208
Position: Conslt *Employer:* Self employed, R Hoffman & Associates *Education:* BS/Met Eng/Univ of Cincinnati. *Born:* 2/22/22. Native of Cincinnati. Co-oped as met. Chemist for Eagle Picher, later at Armco as Met Checker. Upon graduation, employed by Tool Steel Gear & Pinion Co in various metallurgical positions. Chief Metallurgist in 1955. Subsequently served in mgmt poisition incuding VP of Engg and Exec VP of sevl subsidiaries. In 1974, Grp VP of Xtek. Was also Exec VP of

Hoffman, Russell (Continued)
Xtek Canada & Xtek Coated Products. President XTEK Piping Systems Inc & VP XTEK. Retired from Xtek, Jan 1, 1983. Past Chrmn, Cincinnati Chapter ASM. Past Chrmn, Metallurgy & Mtls Committee, AGMA. Technical Div Exec Award 1972, AGMA. Disting Alumni Award 1969 Coll of Engg, Univ of Cincinnati. Dir of MMF, Inc, Badall Co, Inc, XTEK Piping Systems Inc. *Society Aff:* ASM.

Hoffman, Terrence W
Business: Principal Chemical Engr, Polysar Ltd, Vidal St. S, Sarnia, Ontario Canada N7T 7M2
Position: Principal Chem Engr. *Employer:* Polysar Ltd. *Education:* PhD/Chem Engr/McGill Univ; MSc/Chem Engr/Queen's Univ; BSc/Chem Engr/Queen's Univ. *Born:* Jan 3 1931 Kitchener Ontario. Lecturer Royal Milit Coll 1953-55; McGill U 1955-58. Prof positions McMaster U 1958-82, in Chem Engrg Dept, Dept Chmn 1964-70. Awarded Ford Found Residency in Engrg Practice at Hercules Inc 1966-67. CSChE ERCO Award 1970; Fellow of Chem Inst of Canada 1971, NSERC Sr Indust Fellowship at Polysar Ltd, 1979-80; Principal Chem Engr, Polysar Ltd. 1982-. *Society Aff:* CSChE, AIChE, APEO.

Hoffmann, Jean A
Home: 213, avenue Louise, Boite 8, 13-1050 Brussells, Belguim
Position: Professor & Elec Mech Engr. *Employer:* Brussels University. *Education:* -/Engr/Brussels Univ; -/Emeritus/-. *Born:* Oct 16, 1902 Marcinelle, Belgium. Degree EE, ME 1927 from Brussels Univ. Engr for 'Societe d'Electricite et de Mecanique' 1928-35. Asst for Elec Engrg for Faculty of Engrg of Brussels Univ 1928-45, Political Prisoner in Germany 1942-45. Prof of Elec Engrg at Brussels Univ 194573. Emeritus, 1973. Founder: teaching of Indus Electronics Y Live Fellow 1973 & Automatic Control at the Faculty of Elec Engrg. Pres, Internatl Assn for Analog Computation 1955-70. Live Fellow 1973. C027561 02bP.O. Box 20348 C027561 03Birmingham Fellow IEEE 1966, for contrib to Engrg, teaching and analog computation in Belgium. Other awards: 'Doyen d'Honneur du Travail scientifique' 1950; Certificate of Ser Montgommery. Enjoy Photography & Astronomy. *Society Aff:* ULB, SITEL, SRBE, SBA, IBRA, IEEE.

Hoffmann, Lewis E
Business: P.O. Box 20348, Birmingham, AL 35216
Position: President. *Employer:* Harry Hendon & Associates, Inc. *Born:* Dec 1914 Philadelphia PA. BSCE Drexel U 1937. Prior to WWII employed by US Coast & Geodetic Survey, Murray & Flood NY, US Army Engrs. US Army Corps of Engrs 1941-46. Staff Engr Polk Powell & Hendon Birmingham AL 1946. Principal of firm when reorganized as Harry Hendon & Assocs 1956. Pres of firm 1973- . Fellow ASCE & ACEC. Past Pres & Natl Dir CEC/AL. Vice President, ACEC 1981-82 Chmn ASCE-Pipeline Div 1963-64. Mbr Rotary Club. Enjoys rose growing & bridge.

Hoffmann, Ludwig C
Home: 6618 Malta Lane, McLean, VA 22101
Position: Marine Consultant *Employer:* Private Practice *Education:* BS/Aero Engg/MA Inst of Technology. *Born:* 9/8/06. Marine Engr for Bethlehem Steel Corp, US Navy & US Maritime Admin (Mar Ad). Ch Mar Ad Inspector at Newport News during Const of SS US & Mariner ships 1949; Ch, Office of Ship Const (MarAd) 1956; Asst Maritime Admin for Operations (Mar Ad) respon for Ship design const 8 operation, intermodal sys, environ act & subsidies 1968. Maritime cons 1973. Mbr Tech Ctte of Amer Bureau of Shipping & Amer Soc of Naval Engrs. Fellow SNAME; Dept of Commerce Gold & Silver Medals for maritime achievements; SNAME David Taylor Gold Medal. *Society Aff:* SNAME, ASNE.

Hoffmann, Michael R
Business: 138-78 Env Engr Sci, Pasadena, CA 91125
Position: Assoc Prof *Employer:* CA Inst of Tech *Education:* BA/Chem/Northwestern Univ; PhD/Chem/Brown Univ *Born:* 11/13/46 Dr Hoffmann was appointed a Research Fellow at the CA Inst of Tech under the NIEHS post-doctoral training program. While at Caltech Dr Hoffmann received his training in Environmental Engrg with a special emphasis on aquatic chem. His research has been focused on applied environmental kinetics in the areas of chem catalysis, oxidation of reduced sulfur compounds, cloud and fogwater chemistry, microbial catalysis, ligand substitution and photo-assisted catalysis. From 1975 to 1979, Dr Hoffmann was an Assistant Prof of Environmental Engrg in the Dept of Civil and Mineral Engrg at the Univ of MN where he taught basic environmental engrg courses in air and water pollution control. He was promoted to Assoc Prof with tenure in 1979. In the summer of 1980, Dr Hoffman moved to the CA Inst of Tech where he was appointed an Assoc Prof of Environmental Engrg Science. Dr Hoffmann is an assoc editor of the Journal of Geophysical Res and a mbr of the editorial bd of Environ Sci *Society Aff:* ACS, AGU, ASLO

Hoffmanner, Albert L
Business: 505 King Avenue, Columbus, OH 43201
Position: Assoc Mgr, Metalworking Section. *Employer:* Battelle Memorial Institute. *Education:* PhD/Physical Metallurghy/Carnegie-Mellon Univ; MS/Physical Metallurghy/Carnegie-Mellon Univ; BS/Metallurghy/PA State Univ. *Born:* July 1936. Native of West Chester PA. Principal Engr with TRW Inc - Equip Group 1963-74, with respon for metalworking & metal removal R&D in support of mfg divs & gov't sponsored res on metalworking. In Jan 1974 assumed current pos involved with res & dev on mfg sys & metalworking processes. Edited 'Metal Forming; Interrelating Between Theory & Practice'. P Chmn Forging Ctte, ASM Mech Working Div; Chmn 1974-76, AIME Shaping & Forming Ctte; Ctte Mbr, Natl Matls Adv Bd. *Society Aff:* ASM, AIME, SME.

Hofmann, Frederick J
Home: Box 169B Lyons Road, Basking Ridge, NJ 07920
Position: Sr Assoc *Employer:* T & M Assoc *Education:* BS/CE/NJ Inst of Tech. *Born:* 1/28/37 Lives in Liberty Corner, NJ with wife and four children. Past Lector and Mbr adult folk group at St James RC - Church, mbr of Bernards Township recreation ctte, Bernardsville YMCA Bd of Dirs, Boy Scout Troop Ctte. Joined T & M Assoc in 1984 after 22 yrs with Edwards and Kelcey, Inc. and three yrs in Air Force. Worked on hgwy and toll road feasibility studies and design and construction projects in MN, Brazil and NJ. Currently Asst. Mgr of Transportation Group; also past-Pres of NJ Section of ASCE, past-chrmn of ASCE's Natl Ctte on Geometric Design of Hgwys, vice-chmn of ASCE Hgwy Div Exec Ctte & currently Dir District 1 on ASCE's Natl Bd of Direction. Enjoys do-it-yourself projects. *Society Aff:* ASCE, NSPE, IBTTA

Hoft, Richard G
Business: 223 Elec Engg, Columbia, MO 65211
Position: Prof. *Employer:* Univ of MO. *Education:* PhD/Elec Engrg/IA State Univ. *Born:* 12/4/26. Dr Hoft was born in Wall Lake, IA, & joined GE in Schenectady as a test engr in 1948. From 1949-1956 he was an automatic control dev engr in the GE Corp R&D Ctr; from 1956-1960 he was mgr-elec control engg; from 1960-1963, mgr- converter circuits engg in the GE Corporate R&D Ctr. He joined the Univ of MO, Columbia, in 1965 & is currently a prof of elec engg-teaching, conducting res & performing consulting work. Dr Hoft is the author of two books and numerous technical papers on thyristor circuits & automatic control systems. *Society Aff:* IEEE, NSPE, ASEE.

Hogan, Brian R
Home: 18 Windsor Rd, Dover, MA 02030
Position: Pres *Employer:* Resource Tech Corp *Education:* LLB/Law/Catholic Univ; BECE/Civil Engrg/Vanderbilt Univ. *Born:* Aug 1936. 1958-62: served as Public Works officer in Navy Civil Engr Corps. 1962-71: Proj Mgr with Metcalf & Eddy Engrs, working in all areas of solid waste mgmt & resource recovery. 1971-73: Pres of Sanitas Tech & Dev Corp a wholly-owned subsidiary of natl, publicly held corp, specializing in solid waste collection & disposal. 1973-present Owner & Pres of Re-

Hogan, Brian R (Continued)
source Technology corp a firm specializing in solid waste control. P Pres of local section, ASCE; P Mbr of ASCE Natl Ctte on Public Affairs. Mbr of ASCE Natl Ctte on Resource Recovery *Society Aff:* ASCE, NSWMA

Hogan, C Lester
Business: 464 Ellis St MS 20-2234, Mountain View, CA 94042
Position: Dir & Conslt *Employer:* Fairchild Camera & Instrument Corp. *Education:* PhD/Physics/Lehigh Univ; MS/Physics/Lehigh Univ; BS/Chem Engrg/MT State Univ. *Born:* Feb 8, 1920 Great Falls MT. Instr Phys Lehigh Univ 1946-50; tech staff Bell Tele Labs 1950-52; Sub-Dept Hd 1952-53; Assoc Prof Harvard 1953-57, Gordon McKay Prof Applied Phys 1957-58; Gen Mgr Semiconductor Prods Div Motorola Inc Phoenix 1958-60, became V Pres 1960, Exec VP Dir until 1968; Pres Ch Exec Officer Fairchild Camera & Instrument Corp 1968-74, V Chmn of Bd 1974-79, Dir 1979-present; Dir First Interstate Bank; Dir Rolm Corp 1974- ; Dir Tab Products Co 1975- ; Dir Varian Assocs; Dir Timeplex Inc; Bd of Trustees, Lehigh Univ; Gen Chmn Internatl Conf on Magnetism & Magnetic Materials 1959-60; Adv Counc Dept Elec Engrg Princeton 1957-68; Adv Bd Coll of Engrg Univ of CA-Berkeley 1974- ; Visiting Comm Dept of Elec Engrg & Computer Sci MIT 1975- ; Mbr Visiting Comm Dept of Phys Lehigh U 1965-71, Natl Acad of Engrg 1977; Exec VP, IEEE 1978, VP Tech Activities 1979, Fellow 1961, Frederick Philips Gold Medal 1975; Awd of Merit, Amer Electr Assoc, 1978; Fellow, AAAS 1978; Hon *Society Aff:* ΣΞ, TBΠ, ΦΒΦ.

Hogan, Mervin B
Home: 921 Greenwood Terrace, Douglas Park, Salt Lake City, UT 84105
Position: Prof. Emeritus of M.E. *Employer:* Univ. of Utah *Education:* Ph.D./M.E./Univ. of Michigan; M.E./M.E./Univ. of Utah; M.S./M.E./Univ. of Pittsburgh; B.S.M.E./M.E./Univ. of Utah *Born:* 07/21/06 Have spent career about evenly between academic teaching and research, and industrial design and management. Includes head of Dept. of Mech. Engg. at Univ. of Utah; mech. design at Westinghouse Electric, Chicago Bridge & Iron, & General Electric. *Society Aff:* ASME, I Mech E, IEEE

Hoge, A Wesley
Home: 4050 Maulfair Drive, Allentown, Pa 18103
Position: Retired *Employer:* Air Products & Chemicals, Inc. *Education:* BMet/Met Engg/OH State Univ; AMP/Bus Mgt/Harvard Bus Sch. *Born:* Sep 1915. Attended Wittenberg Univ., B. Met. E. Ohio State Univ, 1937, Advanced Mgmt Program, Harvard Business School 1965. Native of Canton Ohio. Process Engr for UGI Co 1938-43. In 1943 joined Houdry Process Corp which was merged into Air Prods & Chems, Inc 1962. Houdry lic process tech to the chem & oil refining indus, mfs & sells catalysts & specialty chems. Supr Tech Service 1948-52; Mgr Engrg 1953-64; Vice Pres Engrg & Mfg of Houdry Div 1964-67; Pres Houdry Div of APCI 1967- . Named Vice Pres of Air Prods & Chems Inc 1971-80.Participated in local govmt & served on the bds/community & charitable organizations. Enjoy classical music, golf. *Society Aff:* AIChE, ACS, ΣΞ.

Hogg, Allan D
Home: Box 401, Kleinburg Ontario L0J 1C0 Canada
Position: Retired (Semi) *Employer:* *. *Education:* PhD/Mech Engg/Univ of Toronto; MASc/Mech Engg/Univ of Toronto; BE/Mech Engg/Univ of Saskatchewan. *Born:* Aug 1913. Fellow ASME & EIC; Mbr Assn of PEs of Ontario; Served on Exec Bd/Ontario Sect ASME incl Chmn; on cttes Toronto Branch EIC; APEO Specialist Designation Prog. Most of prof career in charge of group of PEs working on stress analysis, vibration, fluid mechs & rock mech problems in elec power indus. *Society Aff:* ASME, EIC.

Hogg, David C
Home: 4978 Carter Ct, Boulder, CO 80301
Position: Senior Scientist *Employer:* CIRES, Univ. CO *Education:* PhD/Physics/McGill Univ.; BSc/Radio Physics/Univ of Western Ontario; MSc/Physics/McGill Univ *Born:* Sep 1921 Vanguard Saskatchewan Canada. Canadian Army Radar 1940-45. Res in microwave diffraction & propagation, tropospheric scatter, millimeter-wave absorption by oxygen & water vapor, microwave noise temperature of the troposphere, low noise antennas, satellite projs: Echo, Telstar, Comstar, influence of rain on design of terrestrial & satellite radio-relay sys, optical & infra-red propagation; remote sensing of the winds, temperature, humidity, and liquid water in the tropospher. Fellow IEEE; P Chmn URSI-US Comm F; Distinguished Lectr IEEE on antennas & propagation; Mbr AAAS; Mbr Natl Acad of Engg; 1983 Silver Medal Award, US Dept of Commerce; 1984 Meritorious Achievement Award, Geoscience and Remote Sensing Soc of IEEE. Adjoint Prof, Univ. CO. *Society Aff:* IEEE, AAAS, NAE, URSI.

Hoggatt, John T
Business: P.O. Box 3999 Mail Stop 8Y-70, Seattle, WA 98124
Position: Tech Mgr-Parts, Mat & Processes *Employer:* The Boeing Aerospace Company *Education:* MS/Mech Engr/Seattle Univ; BS/Chem Engr/MI State Univ. *Born:* 9/29/37. Joined The Boeing Co 1959 as matls & Process Engr, specializing in nonmetallic matls; worked for Thiokol Chem Co as Lead Engr respon for R/D of filament wound composites & motor cases 1962-63; rejoined Boeing 1963. Currently tech mgr of parts, materials and processes, respon for all materials and processing personnel and activity within The Boeing Aerospace Co. Has been Prog mgr on over 40 govt contracts in the Material/Process Field. Mbr SAMPE & Bd of Dirs SAMPE; Pres. SAMPE; SAMPE Fellow; P Mbr ASTM, SPI. Enjoys sports & fishing *Society Aff:* SAMPE

Hognestad, Eivind
Business: 5420 Old Orchard Rd, Skokie, IL 60077
Position: Principal Consultant. *Employer:* Construction Technology Laboratories, Inc. (CTL). *Education:* DSc/Civil Engg/Norwegian Inst of Tech; MS/Civil Engg/Norwegian Inst of Tech; MS/Theor & Applied Mechs/Univ of IL, Urbana. *Born:* July 1921 Time Norway. Royal Norwegian Navy 1944-46. Faculty Mbr Univ of Ill-Urbana 1947-53; with Portland Cement Assn 1953-86, assumed current respon in CTL Laboratories 1987. Mbr NAE & Royal Norwegian Acad of Sci. Fellow, Res Award, Boase Award & Chicago Civil Engg of the Yr from ASCE. Hon Mbr, Wason Medal, Bloem, Lindau & Kennedy Awards from ACI. Enjoys classical music & golf. *Society Aff:* ASCE, ACI, NAE.

Hohmann, Edward C, Jr
Home: 918 E. Comstock Ave, Glendora, CA 91740
Position: Dean of Engrg *Employer:* CA State Polytechnic Univ, Pomona *Education:* PhD/ChE/Univ of S CA; MS/ChE/MI State Univ; BS/ChE/Univ of S CA *Born:* 11/21/44 Pasadena, CA. Studied Chem Engrg. Prof & Chrmn Chem & Materials Engrg CA Poly Univ. Frequent consultant to industry and government. Specializing in process design and process synthesis. *Society Aff:* AIChE, ASEE, ACS, TBΠ, ΣΞ

Hohns, H Murray
Home: 512 Pu'u'ikena Dr, Honolulu, HI 96821 *Employer:* Self-Employed *Education:* MSc/CE/Polytech Inst of Brooklyn; MA/Theology/Fuller Theological Seminary; BSc/CE/Tufts Univ; Attended/Bible/Phila Coll of Bible *Born:* 04/26/31 Native of Jamaica, NY. Began engrg career in 1951. Retired in 1986 after 20 yrs as Pres of Wagner-Hohns-Inglis-Inc, one of country's largest 250 engrg firms. Practice specialty has been in forensic engrg covering several thousand major problems throughout the world. Respon for 200,000,000 of in place construction. Authored or co-authored six books on construction scheduling and disputes. Now in private practice as arbitrator-mediator-consultant in Honolulu, HI. *Society Aff:* NSPE, ASCE

Hoisington, David B
Business: Code 73, Monterey, CA 93940
Position: Prof. *Employer:* Naval Postgrad School. *Education:* MS/EE/Univ of PA;

Hoisington, David B (Continued)
SB/EE/MA Inst of Tech. *Born:* 2/11/20. During WWII a Dev Engr for Hazeltine Electronics Corp, Little Neck, NY, on IFF systems. Following WWII a Proj Engr for Sperry Gyroscope Co, Lake Success, NY, in Armament Radar Div. Since 1947 with the Naval Postgrad School Currently Prof of Electronics and Chrmn of the Electronic Warfare Group, responsible for the interdisciplinary curriculum in Electronic Warfare. Presents courses in Electronic Warfare both at the Postgrad School and continuing education courses throughout the country. Res is largely in the area of electronic countermeasures and in the vulnerability of US Navy systems to foreign countermeasures. *Society Aff:* IEEE, AOC, ΣΞ.

Hoke, George C
Business: PO Drawer 971, Durham, NC 27702
Position: Vice President. *Employer:* W M Piatt & Company. *Education:* BS/CE/Duke Univ. *Born:* Jan 1921 Claremont N C. US Navy WWII. Upon graduation entered const of homes & light indus bldgs. Joined Piatt & Davis Cons Engrs 1952; firm later reorganized & inc as W M Piatt & Co; became Principal & V Pres upon reorganization. Mbr: ASCE, Pres N C Sect 1974; ACEC; Water Pollution Control Fed; N C Soc of Engrs; Amer Arbitration Assn, Panel of Arbitrators; Durham Engrs Club Pres 1980; Boy Scouts of Amer; Gov's Award & Silver Beaver Award. Lions Internatl. *Society Aff:* ASCE, ACEC, AAA, Lions.

Hoke, John H
Business: 209 Steidle Bldg, University Park, PA 16802
Position: Professor *Employer:* The Pennsylvania State University. *Education:* DEngr/ME/Johns Hopkins; MS/Metallurgy/Penn State; BS/Metallurgy/Penn State. *Born:* July 24, 1922 Greencastle Pa. Prof Engr. Served USAF 1943-46. Res Asst with Armco Steel 1948-51; Res Met Babcock & Wilcox 1954-57 specializing in stainless steels & high temperature materials; Supr stainless steel res sect Crucible Steel 1957-60; with Met Sect Penn St 1960- , teaching undergrad & grad courses & dir res on stainless steels, corrosion & mech behavior. Secy-Treas 1967-73, Pres 1964-65 & 1974-75 Penn St Chapt ASM. Enjoys outdoor activities particularly gardening, hunting & fishing. *Society Aff:* ASM, AIME, NACE, ASTM, ASEE.

Holbrook, George E
Home: Box 606 Cokesbury Village, Hockessin, DE 19707
Position: Retired. *Employer:* DuPont Co Inc. *Education:* DSc/Hon/Univ of MI; PhD/Chem Engg/Univ of MI; MS/Chem Engg/Univ of MI; BS/Chem Engg/Univ of MI. *Born:* Mar 4, 1909 St Louis. E I DuPont de Nemours & Co Inc Wilmington DE 1933- 76; Asst Genl Mgr Organic Chems Dept 1955-56, Genl Mgr Elastomer Chems Dept 1957-58, V Pres, Dir, Mbr Exec Ctte 1958-69, Dir Mbr Finance Bonus & Salary Cttes 1970-76. Fellow AICE (Pres 1958, Treas 1963-69); Prof Progress Award 1953; Founders' Award 1961 F. J. VanAntwerpen Award, 1980; Mbr EJC (V Pres, Dir, Mbr Exec Ctte 1960-61); Inst Chem Engrs London (Hon); NAE (Charter Mbr, Exec Ctte 1964), AIChE Eminent Chemical Engineers, 1983. *Society Aff:* AIChE, ACS, APS, AAAS, NAE, NYAS.

Holcomb, William F
Home: 19377 Keymar Way, Gaithersburg, MD 20879
Position: Engr Dir (Chem) *Employer:* US Pub Hlth Ser *Education:* MS/Met Engr/Univ of ID; BS/ChE/NM State Univ. *Born:* 12/14/35. Argonne Natl Lab, 1960-67; Aerojet Gen Corp, 1967-69; Fenix & Scisson consulting firm, 1969-70; Idaho Nuclear Corp, 1970-71; Allied Chem Corp, 1971- 74; specializing in radioactive waste mgt, nuclear fuel reprocessing and nuclear fuel fabrication. Since 1974, a US Public Health Service Commissioned Officer; 1974-1980 and 1983-present; detailed to US Environmental Protection Agency specializing in radioactive waste mgt and setting waste disposal stds; 1980-1983 assigned the National Institutes of Health as a health physicist. Reg Chem Engr, ID. AIChE Natl Public Relations Committee Chrmn 1978-1979, and received five AIChE Public Relations Natl Awards. Listed in Who's Who in the West, 13th Edition; Who's Who in American Education Leaders 1967-69; Who's Who in the East, 18th Edition. Over 40 publications. Engg College Instr, 1968-69. *Society Aff:* AIChE, COA-USPHS, ΣΞ.

Holden, Donald A
Home: 201 Montvue Dr, Charlottesville, VA 22901
Position: Retired *Education:* MS/Genl Engg/MA Inst of Tech; BS/Civil Engg/MA Inst of Tech. *Born:* Apr 7, 1910 Reading Mass. Parents Archer H & Alice W Holden. Married Eleanor W Watson Sept 4, 1937; 1 son Hugh Warren Holden. Employed Newport News Shipbuilding & Dry Dock Co 1934-70: Prod Mgr 1957-59, V Pres Prod Mgr 1959-60, Exec V Pres 1960-64, Pres 1964-69, Chmn 1965-70. Exec Dir Council of Independent Colleges 1971-83. Alumni Mbr MIT Corp 1967-72. V Pres & Trustee Mariners Museum 1965- . Fellow, Mbr & P Pres (1967-68) SNAME; Fellow ASME; Mbr ASNE; P Trustee NSIA; Mbr Navy League, Propeller Club. Phi Epsilon, Pi Tau Sigma. Farmington C C. *Society Aff:* SNAME, ASME, ASNE.

Holden, Frank C
Business: 505 King Ave, Columbus, OH 43201
Position: Manager Materials for Energy systems *Employer:* Columbus Labs-Battelle Memorial Inst. *Born:* Nov 1921 Bangor Maine. MES & MS Harvard Univ; BS Univ of Maine. Served Army Corps of Engrs Manhattan Dist 1942-46. Instr Univ of Maine 1949-51, With Battelle 1951- , conducting res in titanium alloy dev, refractory metals & mech met. Assumed current respon for mgmt of met res 1973. Mbr: Tau Beta Pi, Sigma Xi, Pi Mu Epsilon honorary socs; ASM & AIME tech socs.

Holder, Sidney G, Jr
Business: Quality & Safety, 1232 37th Place, N, Birmingham, AL 35234
Position: Mgr Engg Quality & Safety *Employer:* Fontaine Truck Equip Co. *Education:* MSc/Metallurgy/Univ of AL; MSc/Ind Mgr/Univ of TN; BSc/Metallurgy/Univ of AL. *Born:* 2/19/29. Pres: AL Junior Academy of Science - 1946 - High School Valedictorian: Senior Class-/Shades Cahoba High School - 1946. CoCommander: Pershing Rifles - Univ of ALA 1949. Pres Freshman & Senior Classes: - School of Chem, Metallurgy, & Ceramics - Univ of Al-1947 & 1950. American Soc for Metals - office currently & 31 yr mbr. ASSE Pres 1981-82. Greater Birmingham Safety League Pres 1975-77. *Society Aff:* ASM, ASSE, AWS.

Holderby, George D
Business: PO Box 516, St Albans, WV 25177
Position: Pres. *Employer:* Holderby Engg, Inc. *Education:* 2 yrs/Elec/AR Polytech; 1 yr/Air Conditioning/Milwaukee Sch of Engg. *Born:* 11/22/19. Native of Newark, AR, US Navy 1942-45; Sales Engineer, Trane Co. 1947-49; Mgr Heating Dept., Bailey-Ferrell Co. 1949-54; Chief Engr, Henry Elden and Assoc 1959-71; Pres, Chief Engr, Holderby Engrg, Inc. 1971-present; Reg Engr - West Virginia, Virginia, Ohio, Kentucky. *Society Aff:* ASHRAE.

Holdredge, Russell M
Business: UMC 41, Logan, UT 84322
Position: Assoc Dean *Employer:* UT State Univ *Education:* PhD/ME/Purdue Univ; MS/ME/Univ of CO; BS/ME/Univ of CO; BS/Bus Admin/Univ of CO *Born:* 7/22/33 Has been Assoc Dean of Engrg at UT State Univ since 1976. From 1970-76 he was Prof and Head of the Mechanical Engrg Dept at UT State Univ. He has been active in professional societies having served on several regional ASME committees and as chrmn of the UT Section. He has published several articles in the area of heat transfer. *Society Aff:* ASEE, ASME, ISES, ASHRAE

Hole, William G
Home: 76 Balfour Ave, Montreal, Canada H3P 1L6
Position: Vice Pres, Genl Mgr - ret. *Employer:* Amer Air Filter of Canada Ltd. *Education:* BSc/Civil Engg/Univ of Alberta. *Born:* Dec 1910 Canada. Post Grad studies Univ of London England. Cons Engr; V Pres & Genl Mgr, Chmn of Board American Air Filter of Canada Ltd. Canadian Exec Services Overseas 1973. Mbr ASHRAE 1942, Natl Pres 1970. Mbr Order of Engrs Quebec, Engrg Inst of Canada. *Society Aff:* ASHRAE.

Holfinger, Robert R
Business: 999 Pine Ave, S.E, Warren, OH 44481
Position: Asst Chief Engr *Employer:* Republic Steel Corp *Education:* B/ME/OH State Univ *Born:* 3/12/30 Native of Canton, OH. Assoc with Republic Steel Corp since 1947 holding various engrg positions in Canton, Youngstown, and Warren, OH. Currently Assistant Chief Engr of Mahoning Valley District. National Dir AISE 1971. Attended OH State Univ as a Sloan Scholar graduating 1960 with BME (cum laude). Co-founder, Dir, VP & Treasurer of Industrial Fabricators, Inc. Partner in Fabco Development Co. *Society Aff:* ASME, AIPE, AISE

Holl, Barton S
Business: 201 E Bowen St, Logan, OH 43138
Position: V President Production. *Employer:* The Logan Clay Products Co. *Education:* MS/Ceramic Engr/OH State; BCerEng/Ceramic Engr/OH State. *Born:* Feb 1923 Logan Ohio. MS Ceramic Engrg Ohio St Univ; B Cer Engrg OSU. With USA 1943-45. With Logan Clay Prods 1950- . Respons in engrg, maintenance & production. Trustee & P Pres Ohio Ceramic Indus Assn. Trustee & former Chmn of Board of Hocking Tech Coll. V Pres Logan Clay 1962-79 ; also corporate energy coordinator. 1979 chrmn of Bd CEO Logan Clay. *Society Aff:* ACS, NICE.

Holl, J William
Business: Appl Res Lab PO Box 30, State College, PA 16801
Position: Prof of Aerospace Engrg. *Employer:* Pennsylvania State University. *Education:* PhD/Mech Engg/Penn State Univ; MS/Mech Engg/Univ of IL; BS/Mech Engg/Univ of IL. *Born:* Feb 20, 1928 Danville Ill. Res Asst Univ of Ill 1949-51; Res Assoc Penn St Univ 1951-55, 1956-58; Asst Prof of Engrg Res Penn St Univ 1958-59; Assoc Prof of ME Univ of Nebr 1959-63; Assoc Prof of Aerospace Engrg Penn St Univ 1963-67, Prof 1967- . R T Knapp Award ASME 1970 (with A Kornhauser), Melville Medal ASME 1970 (with A Kornhauser); Lindbach Teaching Award Penn St Univ 1973; Outstanding Mech Engr Award Central Pa Sect ASME 1974; Outstanding Performance Award of the Applied Res Lab, Penn State Univ, 1977. Fellow ASME; Assoc Fellow AIAA, ASME Centennial Medallion 1980, VP ASME Basic Engrg Dept 1980-84. *Society Aff:* ASME, AIAA, ΣΞ.

Holladay, James F
Business: PO Box 1000, Carrollton, GA 30117
Position: Senior VP Corporate Engineering & Energy *Employer:* Southwire Company. *Education:* BSEE/Elec Engrg/GA Tech. *Born:* 4/5/22. Birmingham Ala. Served in WWII 1943-47. Wife: former Anne Wedsworth. Children: Jim - 29, David - 26, Cynthia - 23. With Southwire Co 1952- , Board of Dir 1958, VP of Plant Engrg 1963-Senior VP 1980. Corporate Chairman, Energy Board. Engr of the Year in Indus GSPE 1970; Mr Indus Award 1972. AIPE Fellow 1974. Who's Who in Ga 1975-76. Enjoys photography & gardening CPE-Cert Plant Eng. PE, CEM Certified Energy Mbr MF002682 CA 1978-1983. Nat'l Energy Mgmt Exec Award 1983 (By Assoc of Energy Engrs). *Society Aff:* AIPE, NSPE, AEE.

Holladay, William L
Home: 2173 Mar Vista Ave, Altadena, CA 91001
Position: Consulting Engineer *Employer:* Self-employed. *Education:* BS/EE/CA Inst Tech. *Born:* Dec 9, 1901 Fayette Mo. Reg ME & EE Calif. Pres Caltech Alumni Assn 1963; Pres Pasadena Lung Assn 1970; Pres ASHRAE 1968-69. Wolverine Diamond Key Award ASRE 1950. Mbr: IEEE, Tau Beta Pi. Fellow: ASHRAE, Inst for the Advancement of Engrg. Author of papers on weather data, heat transfer, low temperature refrigeration. Cons Engr 1952- , with Holladay & Westcott, Holladay Eggett & Helin; self-empl. 1972. *Society Aff:* AEE.

Holland, Charles D
Business: Dept of Chem Engrg, College Station, TX 77843
Position: Prof & Head Dept of Chem Engrg. *Employer:* Texas A&M Univ. *Education:* BChE/ChE/NC State; MS/ChE/Texas A&M Univ; PhD/ChE/Texas A&M Univ. *Born:* Oct 9, 1921. Served at Lt NSNR 1943-46. Engr Burlington Mills Corp 1947- 48. Grad Student 1948-53 Texas A&M. Joined Dept of Engrg Texas A&M 1952. Appointed Hd Dept Chem Engrg 1964. Cons for Monsanto 1970- . Outstanding Prof of Coll of Engrg 1962, 1984. Author of 6 books: 'Multicomponent Distillation', 'Unsteady State Processes with Applications in Multicomponent Distillation', 'Fundamentals & Modeling of Separation Processes: Absorption, Distillation, Evaporation & Extraction', 'Fundamentals of Chemical Reaction Engineering', 'Fundamentals of Multicomponent Distillation', 'Computer Methods for Solving Dynamic Separation Problems' & numerous articles in prof journals. President of Texas Institute for Advancement of Chemical Technology. *Society Aff:* AIChE, ACS, ASEE, AIC, ΣΞ, ΦΚΦ, ΦLAMDAEPISOLON, ΤΒΦ

Holland, Eugene P
Business: 500 Green Bay Rd, Kenilworth, IL 60043
Position: Pres. *Employer:* Coder Taylor Assoc, Inc. *Education:* BSCE/Structural Engr/Valparaiso Univ. *Born:* 4/3/35. Niagara Falls, NY, Apr 3, 1935; postgrad Univ CO, IIT. Structural Engr Ketchum & Konkel, Denver, 1957-61; prin engr advanced engr group Portland Cement Assn, Chg, 1961-64; pres Wiesinger-Holland Ltd, structural engrs, Chg 1964-78; Pres, Coder Taylor Assoc, Inc, 1978-present; adj prof Univ IL, Chg Circle. Reg Structural engr IL, PE 10 other states. Fellow Am Concrete Inst (chrmn ACI 318 bldg code com 1975-79, Bd of Dirs 1976-1981); mbr Prestressed Concrete Inst (pres 1976), Structural Engrs Assn IL (Treas, Dir 1972-77; chrmn code com 1975- 78). Deans Club (Valparaiso Univ). Co-author: Housing, 1976-John Wiley & Sons, Contbr articles to profl jours. Rec'd AI Del R Bloom Award, 1978. Biography in Who's Who in Am. *Society Aff:* ACI, PCI, NSPE, ISPE, SEAOI, ASCE.

Holland, Francis E
Business: PO Box 1467, E Willowbrook Plaza, Mason City, IA 50401
Position: Pres *Employer:* Wallace Holland Kastler Schmitz & Co. *Education:* BS/Civil Engg/IA State Univ. *Born:* 8/14/23. Consulting Civil Engr & Land Surveyor; Reg IA & MN. Army Air Corps 1943-46. BSCE ISU 1949. US Bureau of Reclamation 1950-52. Wallace Holland Kastler Schmitz & Co 1952 to present, Pres. American Soc of Civil Engrs, Fellow. IA Engg Soc Pres 1967. NSPE, Dir 1969-71. IA Consulting Engrs Council, Pres 1972-73. IA State Bd of Engg Examiners, 1975-present, chrmn 1979. Rotary Intl Pres Mason City Club 1977-78. Winnebago Council BSA, Pres 1971, 1974 and 1975. *Society Aff:* ASCE, NSPE, ACSM, IES.

Holland, Joe E, Jr
Home: 102 Leota Dr, Hendersonville, TN 37075
Position: Chief, Nashville Basin. *Employer:* TN Div of Water Quality Control. *Education:* BS/Agri Engr/Univ of TN. *Born:* 12/26/49. Reared on small Macon County, TN farm. Prior to working with Water Quality Control was employed by Security Mills Inc. In 1973 began current work as an environmental engr. Now in charge of a field office with a staff consisiting of engrs, chemists, biologists and other support personnel. Office administers water pollution control program in 41 middle TN counties. Chrmn, ASAE - TN Sec 1977. Active rock climber, canoeist and bicyclist, on local softball and soccer teams. *Society Aff:* ASAE.

Holland, John G
Home: 2035 E. Skyview Dr, Altadena, CA 91001
Position: Dir of Engr *Employer:* Jacobs Architects *Education:* M/BA/Univ of WI-Madison; BS/ME/Univ of WI-Madison *Born:* 6/13/48 Native of Wauwatosa, WI. Became involved in HVAC consulting business through father's firm of Holland & Kurtz, Inc. Worked for Flad & Assoc, the largest arch/engr firm in WI, and then Broyles & Broyles, Inc-a Ft Worth, TX based major Design/Construct mechanical contractor. Left Broyles to help start Mechanical Economics Co of America, Inc, also in TX. Has served as Dir of Engrg for Jacobs Architects, a member of the Jacobs Engrg Group in Pasadena, CA. Currently employed by Robert M. Young and Associates, a consulting mech engrg firm, as an Associate. Registered PE in six states. Enjoy travel, tennis and photography. *Society Aff:* NSPE, ASHRAE

Holland, Leslie A
Home: –Hazelwood– Balcombe Rd Pound Hill, Crawley Sussex, RH103NZ England *Position:* Prof *Employer:* Univ of Sussex *Education:* DSchc/Physics/Rouen Univ; D Tech/Applied Physics/Brunel Univ; CEng/Chartered Engineer/-; CPhys/Chartered Physicist/-. *Born:* 6/14/21 Commenced R&D studies in Vacuum, Surface and Thin Film fields in Edwards High Vacuum Ltd, in 1944 becoming Dir of Central Res Lab 1966-73. Assoc Reader, Physics Dept, Brunel Univ 1962-1976; Visiting Lecturer, Cranfield Coll Technology 1970-1974. Moved (1974) to Univ of Sussex to form a lab for Plasma Materials Processing. Author of papers, patents and reference works on foregoing subjects. Inventor of bent electron beam vapour source and rf plasma process for hard carbon deposition. Non-executive Dir of Sira Inst (1973-81). Consultant for plasma and thin film processes to industrial companies. Silver Medallist Plastics Inst 1955. Welch Medallist Amer Vacuum Soc 1976. Whitworth Fellow and Medal 1975. Pres Intl Union for Vacuum Science, Tech and Applications 1977-80. Mbr British Vacuum Council (current). Author of "Vacuum Deposition of Thin Films" & "The Properties of Glass Surfaces–, both Chapman & Hall, Pub., London. Co-Author & Editor of "Thin Film Microelectronics" & "Vacuum Manual–. *Society Aff:* FIEE, FIP, FSGT, OSA, AVS, IVS

Hollander, Gerald M
Home: 9612 Old Spring Rd, Kensington, MD 20895 *Position:* Consultant *Employer:* Self *Education:* BSME/Mech Eng/Univ of So Calif; MA/Govt/George Washington Univ; PhD/Engr Admin/Walden Univ of Adv Stud. *Born:* Oct 1921. Sr ME Pub Wks State Calif 1950-55; Supr Engrg Span B ases 1955-56; Mgr Italian Branch Lublin McGaughy 1956-60; Ch MechElec Engr McGaughy, Marshall McMillan Norfolk, Prof George Washington Univ 1979-1986; Dir, Mech-Elec Engr Vet Admin 1962-68; Dir Arch & Engrg Vet. Admin 1968-74; Dir Plan & Devlmnt Vet Admin 1974-83; Lecturer L A City Coll & D A Grad Sch. US Army 1943- 45. P Pres DC Soc of P E's; VP NSPE; Mbr ASHRAE; Mbr Bd of Reg P E's DC; Publs: 'An Old Span Custom & Problems Created for Amer Const in Spain'; 'Mod Design, Sys Approach for Res Bldgs' ; 'Value Engrg in Vet Admin'; 'The Art of Estimating'; 'Ingredients of Accurate Const Cost Estimating'; 'Life Cycle Cost A Concept in Need of Understanding', 'Life Cycle Cost Analysis the Key to Engineering'. Wife: Elaine K; children: Lois & Bret *Society Aff:* DCSPE, NSPE, ASHRAE.

Hollander, Gerhard L
Business: PO Bx 2276, Fullerton, CA 92633-0276 *Position:* Pres & Tech Dir *Employer:* Hollander Assocs *Education:* EE/EE/MIT; MS/EE/WA Univ; BSEE/EE/IL Inst of Tech. *Born:* Feb 1922 Berlin, Germany. US Cit 1943. Radio buyer & tech adv, Spiegel Inc 1940-42. US Navy 1943-45 awarded Navy Commendation for dev of firecontrol training devices. Staff & Prof at Ill Inst Tech & St Louis Univ. Staff, MIT Servo Lab 1952-54. Sr Engr & Engrg Mgr, Clevite Res Ctr, Philco Corp & Hughes Aircraft Co 1954-61 respon for business, sci & commercial computer sys dev. Pres & Tech Dir of Hollander Assoc Fullerton, Ca 1961-present. Primary projs in sys arch. Over 50 papers & pats incl original contribs on packet-switching, hierarchical memories, computer arch & profit control sys. IEEE Fellow award for computing-equip design & evaluation methods for large-scale sys. Mbr IEEE, ACM, ORSA, Sigma Xi. Chmn of many AIEE, IRE, IEEE, AFIPS tech cttes in computer sys. V Chmn & Chmn IEEE Co; Fellow, NCMA. IEEE Centennial Medal. *Society Aff:* IEEE, ACM, ORSA, ΣΞ, NCMA, PMI.

Hollander, Lawrence J
3 Stonehenge Lane, Albany, NY 12203 *Position:* Visiting Professor of Electrical Engineering and Computer Science *Employer:* Union College *Education:* Bachelor/Elec Engrg/NY Univ; Master/Elec Engrg/NY Univ. *Born:* 11/5/26. Cons in elec power generation, tranmission & distribution. Taught elec engrg at NY Univ, coll administration, & Dir of Evening Div of Sch of Engrg & Sci 1953-67. With the Amer Gas Assn 1967-69. Exec Secy, NY State Bd for Engrg & Land Surveying 1969-78. Mbr of Exec Council of Tau Beta Pi 1975-78; Sr Mbr of IEEE; P Pres of NY Chap, NY State Soc of Prof Engrs. Dir of Fellowships of Tau Beta Pi 1979-present; secretary, zone 1, Amer Soc for Engrg Education 1979-81; Associate Dean, The Cooper Union School of Engineering 1979-1986. *Society Aff:* IEEE, ASEE, ΤΒΠ, NSPE, HKN, ΠΤΣ.

Hollander, Milton B
Home: Technology Management, Inc., 1 Largo Dr. Box 4819, Stamford, CT 06907 *Position:* Executive Vice President *Employer:* Technology Management, Inc. *Education:* PhD/Engrg/Columbia; MS/Engrg/MIT; BS/Mech Engrg/Purdue. *Born:* 11/29/28. Native of Bayonne, NJ. US Corps of Engineers, Korea, 1946-48. Registered P.E. in NY, NJ, CT, PA & FL. With Technology Management, Inc. since retiring from Gulf & Western in 1985. Director & Technical Consultant, Omega Engineering, Inc. Trustees Executive Committee, Univ. of Bridgeport. Executive Committee, Connecticut Technology Institute, Univ. of Bridgeport. Member Industrial Research Institute and Directors of Industrial Research. Published over 25 technical papers e.g. temperature measurement, instrumentation, metal cutting. Granted over 50 US patents. Inventor of friction welding. Honors: Outstanding Alumnus Award, Purdue Univ. 1972; Outstanding Young Man in America 1965; DuPont Research Fellow, Columbia Univ. 1955-57; Research Fellow Am. Society Tool & Manufacturing Engineers 1954-55; Research Fellow M.I.T. 1952-53; Sigma Xi. *Society Aff:* ASME, ISA, SME, AWS.

Holle, Charles G
Home: 2540 Mass Ave NW, Washington, DC 20008 *Position:* Major General. *Employer:* Retired-US Army. *Education:* BS/US Military Acad; CE/Civil Engg/Rensselaer Poly In. *Born:* Oct 1898 Cincinnati, Ohio. Army Engr Sch; Command & Genl Staff Coll. Commissioned Corps of Engrs 1920. Duties: Engr Troops, Texas & Canal Zone; Civil Works, Greenville, Vicksburg, New Orleans, Atlanta; Amer Battle Monuments Comm Paris France; Panama Canal; Deputy Ch of Engrs. Bds: Chmn, River & Harbor; Pres, Beach Erosion; St Lawrence River Joint Canada & US. Ret Major Genl 1958. Decor: DSM, LM (OCL). Mbr: Amer Soc of Civil Engrs (Fellow); Soc of Amer Military Engrs (Charter, Fellow, Gold Medal 1962, Treas 1967-73); Amer Shore & Beach Preservation Assn (Pres 1959-72); Permanent Internatl Assn of Navigation Congresses (Secy Genl XXth Congress & Life Mbr). *Society Aff:* ASCE, SAME, ASBPA, PIANC

Hollenbeck, Leslie G
Home: 15 Robinson Dr, Bedford, MA 01730 *Position:* Senior Staff Engineer. *Employer:* AVCO Systems Division. *Education:* BS/Civil Engg/Northeastern Univ. *Born:* 6/24/22. Served with US Navy 1940-46. Weight Control Engr with Pratt & Whitney & Chance Vought prior to joining Avco in 1961. Currently specializing in Mass Properties Control of Ballistic Reentry Vehicles & Penetration Aids. Sr Mbr Soc of Allied Weight Engrs (SAWE). Currently holds pos of Dir SAWE. Enjoy tennis & hiking *Society Aff:* SAWE.

Hollett, Grant T, Jr
Home: 440 Prospect Ave, Elmhurst, IL 60126 *Position:* Exec VP *Employer:* Cherry Electrical Products Corp *Education:* BS/Mech Engg/Duke Univ. *Born:* 5/8/42. Nuclear Power Postgraduate School, US Navy 1965. Navy active duty 1964-69 included 4 yrs Nuclear Power and Auxiliary Engg assignments USS Enterprise. Youngest qualifier as Engg Officer of Watch. 1969-78 advanced through engg and manufacturing responsibilities for food and health care products of Procter and Gamble. 1978-81 headed Energy and Pollution Controls, Inc, subsidiary of Flick-Reedy, directing all operations of EPC, manufacturer of air pollution control equipment for industrial fumes and coal-fired systems. 1981-present with Cherry Electrical Products Corp, first as VP Production directing manufacturing operations for electromechanical minature switches, automotive switches & electronic assemblies. Fabrication responsibility for large injection moving & metal stamping operations. In 1986 promoted to exec VP heading all operations/sales for 3 company divisions. Patents for totally dry sulfur dioxide pollution con-

Hollett, Grant T, Jr (Continued)
trol system. Mbr ASME, APCA, SME, AES. Reg PE OH, IL. Pi Tau Sigma, Tau Beta Pi. *Society Aff:* ASME, APCA, SME, AES

Holley, Charles H
Business: 1 River Rd-Bldg 2-600, Schenectady, NY 12345 *Position:* Genl Mgr-Elec Utility Sys Engrg Dept. *Employer:* General Electric Company. *Education:* BS/EE/Duke Univ. *Born:* 4/15/19. Grad from Ford City, Pa High Sch & Duke Univ with and elec engrg degree & joined G E in 1941. In 1945 moved to Schenectady & held various pos in Gas Turbine & Large Steam Turbine-Generator engrg areas. Mgr of Generator Engrg for Large Steam Turbine-Generator Dept for 12 yrs. In 1974 was appointed Genl Mgr of the Elec Utility Sys Engrg Dept in the Energy Sys/Technology Div. In March 1980, he was appointed Mgr of Turbine Tech Assessment Operation in Gen Elec Turbine Business Group. He is responsible for the development and assurance of the quality of the Group's technical work and the technical performance of the Group's products. He is a reg NY State PE, a Fellow of the IEEE and a Mbr of the Natl Acad of Engrg. *Society Aff:* IEEE, NAE, CIGRE.

Holley, Edward R
Business: Dept of Civil Engrg, Univ. of Texas, Austin, TX 78712 *Position:* Professor. *Employer:* Univ of TX. *Education:* ScD/Civil Engg/MA Inst of Tech; MSCE/Civil Engg/GA Inst of Tech; BCE/Civil Engg/GA Inst of Tech. *Born:* Aug 1936. ScD MIT 1965; BSCE, MSCE Ga Tech 1960. Native of Atlanta, Ga. Joined fac of Univ of Ill-Urbana-Champaign in 1964 & Univ of TX-Austin in 1979. Spent one-yr leaves of absence at Bureau of Reclamation in Denver, Delft Hydraulics Lab in the Netherlands & Univ of Queensland in Brisbane, Australia. Primary prof & res ints currently environ fluid mechs incl transport of pollutants in rivers & coastal waters, outfalls & diffusers, thermal discharges, numerical simulations, & physical model studies. ASCE Hilgard Hydraulic Prize 1971 & 1980. Christian. *Society Aff:* ASCE, AGU, IAHR.

Holley, James R
Business: 20 West Market St, P.O. Box 43, York, PA 17405 *Position:* Pres *Employer:* James R. Holley & Assoc, Inc *Education:* Geol Engr/Geol/CO School of Mines *Born:* 4/4/42 Engrg education obtained at CO School of Mines, Golden, CO. Served three years in the US Marine Corps as combat engr, from 1964 to 1967. Viet Nam veteran. Employed by Buchart-Horn, Inc from 1967 till 1977. Experience covers general engrg, water supply and treatment, sanitary and industrial waste treatment, storm water management and construction management. Started James R. Holley & Assoc, Inc in 1977. *Society Aff:* NSPE, WPCF

Holliday, Frank J
Home: 10737 Livingston Dr, Northglenn, CO 80234 *Position:* Principal *Employer:* Woodward-Clyde Consultants *Education:* M/CE/MI State Univ; B/CE/Univ of WY *Born:* 10/24/38 Native of Laramie, WY. Served with the US Navy 1956-58. Completed Masters degree in December 1963 and started with Woodward-Clyde Consultants (WCC) as a junior engr in January 1964. Assumed increasing responsibilities. Am now a Principal and VP of WCC and as the mgr of their Denver office, responsible for both the technical and business aspects of the Denver practice. *Society Aff:* ASCE

Holliday, George H
Business: PO Box 576, Houston, TX 77001 *Position:* Sr Staff Environ Engr *Employer:* Shell Oil Company. *Education:* PhD/CE/U of Houston; EME/ME/USC; MS/CE/USC; BSME/ME/UC Berkeley. *Born:* 10/26/22. Served with Army Engrs 1942-46. Drilling & Field Facilities Engr for Shell Oil 1948-65. Sr Staff Mech Engr for Shell Dev Co 1965-70 specializing in offshore res. Sr Staff Environ Engr Shell Oil commencing 1970. Chmn API ctte on Standardization of Tubular Goods 1977-1980. Mbr ASME Environ Subctte. Secr S CA Oil Pollution Control Orgn, Engrg Ctte. Mbr Gulf Alaska Oil Clean-Up Engrg & Cook Inlet Mech Cttes. Chrmn TX Mid Continent Oil and Gas Assoc Prod subctte 1972-1982. Fellow ASME mbr of US delegation to Intl Standardization Tech Committee 67 (Tubular Goods). Distinguished Lecturer SPE 1980-1981 *Society Aff:* ASME.

Hollinger, Henry C
Home: 710 Loretta Pl, Lima, OH 45805 *Position:* Partner. *Employer:* Kohli & Kaliher Assoc Ltd. *Education:* BS/Civ Engg/OH Northern Univ. *Born:* 2/21/23. W.W. II - served on bomber crew Eighth Air Force. After college became proj engr supervising hgwy and bridge construction OH Dept of Hgwys and then became resident engr supevising construction of sewage treatment plants. Appointed City Engr (1955) then Dir Public Works, City of Lima, with responsibility directing municipal engr prog and maintenance of St and Sewer facilities. Initiated maj sewer imporvement prog. Became Partner (1965) then Managing Partner Kohli and Kaliher Assoc, Limited, consulting engrs and surveyors. Chapter Pres NSPE (1966), Pres Consulting Engrs of OH (1976). Hon Degree (Dr of Engr), Ohio Northern University (1982). Active civic affairs. *Society Aff:* ASCE, NSPE, WPCE, AWWA.

Hollingsworth, Dwight F
Home: 210 Buck Toe Hills Road, Kennett Square, PA 19348 *Position:* Retired. *Employer:* E I DuPont de Nemours & Co Inc. *Education:* MSE/Heat & Vent/Purdue Univ; BSME/Refrig/Purdue Univ. *Born:* 1/28/10. Ch Engr Earlham Coll 1932-35; E I DuPont 1936-71: Indus Engrg, Lubrication Cons, Cons Mgr, Special Asst Mgmt, Mgr Engrg Standards-Chmn Engrg Standards. Reg Prof Engr Delaware. Life Fellow Amer Soc of Mech Engrs; assisted organization of Amer Soc of Lubrication Engrs, charter member; Natl P Pres, William P Kuebler Awd 1960; EJC, Engrs of Distinction 1970; Chem Mfgs Assn, organized Engr. Adiv. Cttee, Past Chrmn 1955-71; Amer Natl Standards Inst, Organized and first Chrmn of Piping and Process Equipment Mgmt Standards Bd, Mbr-Bd/Dir, Consumer Coun, Metric Adv Ctte. Present activities: Community Affairs, Buck Toe Hills Assn Trustee. *Society Aff:* ASME.

Hollis, Mark D
Home: 411 Lone Palm Drive, Lakeland, FL 33801 *Position:* Cons, Environmental Engineering. *Employer:* Self. *Education:* DSc/Env Sci/Univ of FL; CE/San Eng/Univ of GA; BSCE/San Eng/Univ of GA. *Born:* Sept 1908 Georgia. Received Distinguished Ser Medal & ret USPHS 1961 with 30 yrs ser (15 as Asst Surgeon Genl, rank of Maj Genl). Central figure estab CDC, Atlanta 1946, served as 1st Dir. Administered natl progs Air-Water- Land pollution 1947-61. Dir Environ Health, World Health Org, Geneva Switzerland 1961-72 incl PAHO/WHO, Western Hemisphere. Natl Acad of Engrg since 1967. Pres, WPCF 1959-60. Life Mbr ASCE, APWA, AIDIS. Diplomate AAEE. Emerson Medal & Bedell Award 1964. Cosmos Club. Rotarian. *Society Aff:* NAE, AAEE, ASCE, APWA, WPCF.

Hollmann, Harold R
Business: 600 Grant St Rm 1844, Pittsburgh, PA 15230 *Position:* Gen Mgr. *Employer:* US Steel Corp. *Education:* BS/Met Engr/Univ of MO; -/Bus Admin/Univ of VA. *Born:* 3/11/27. Native of St Louis, MO. Received a BS degree in metallurgical engg from the Univ of MO & studied business administration at the Univ of VA. Joining US Steel as a met administration at the Univ of VA. Joining US Steel as a met trainee at the Gary Sheet & Tin Mill in 1949, he moved through a succession of assignments as a service met & resident met in US Steel's sales offices in Chicago, IL & Cincinnati. In 1957, transferred to Pittsburgh as a met engr with the sheet & strip met staff, moving up to asst mgr, and, in 1969, to mgr of the unit. Promoted to gen mgr product met in 1976, and to gen mgr, commercial, sheet products on Sept 1, 1979, where he assumed his present responsibilities. *Society Aff:* ASM, AISI, SAE, ANMC, AIMMPE.

Hollomon, J Herbert
Business: Center for Tech & Policy, 197 Bay State Rd, Boston, MA 02215
Position: Dir, Ctr for Tech and Policy *Employer:* Boston University *Education:* ScD/Metallurgy/MA Inst of Technology; BS/Physics/MIT. *Born:* Mar 1919. Resides in Brookline Mass. 1942-46 US Army (Maj). 1946-62 affiliated with G E. 1960 became Genl Mgr of Genl Engrg Lab, Schenectady. 1962- 68 Asst Secy of Commerce for Sci & Tech. 1968-70 Pres of Univ of Oklahoma. 1970- 72 cons to Pres & to the Provost of MIT. In 1972 Dir of Ctr for Policy Alternatives & Prof of Engrg. Founding Mbr of NAE. Besides sci writings, has written over 200 articles related to public policy, safety, consumer affairs & ed American Inst of Chemists Metallurgical Soc of Am Inst of Mines Metallurgical & Petroleum Eng (Fellow) Sigma Xi AAAS - Am Academy for Arts & Sciences (Fellow) Royal Swedish Academy of Engineering Sciences. *Society Aff:* ASM, AAAS, NAE.

Holloway, Frederic A L
Home: 7375 Boyce Dr, Baton Rouge, LA 70809
Position: Retired from Exxon Corp. *Education:* ScD/ChE/MA Inst of Tech; BS/ChE/GA Inst of Tech. *Born:* 11/8/14. ScD MIT "Perf Commercial Absorption Tower Packings" Instr 1938-39. Esso Std Oil Co, Process Engr Rubber & Chem Plants 1937-47; Hd Tech Depts 1948-52; Genl Supt & Asst Genl Mgr 1953-55; Mfg Dept NY; Genl Mgr East Coast Div & Asst Genl Mgr Mft 1956-60 Humble Oil & Refining Co 1961-62, VP Mfg Planning. Std Oil Co (NJ) Depty Coord Refining 1962-64. Esso Res & Engrg Co Pres & Dir 1964-68. Std Oil Co (named changed to Exxon 1972) Coord Corp Planning 1968-70. VP Corp Planning 1970-73. VP Sci Tech 1973-1978. Fellow AIChE. Council Treasurer and Exec Ctte NAE *Society Aff:* AIChE, NCE.

Holloway, Gale A
Business: 7 S 600 County Line Rd, Hinsdale, IL 60521
Position: Project Engineer. *Employer:* International Harvester Company. *Born:* March 1944 Joliet, Ill. BS 1968, MS 1969 Agri Engrg, Colorado State Univ. Mbrship in Alpha Epsilon 1969. Joined Internatl Harvester FEREC 1969. Design Engr specializing in electro-hydraulic controls & hydraulic components 1969-74. Current position Project Engr, hydraulic development of lab, responsible for engineering phases of operation. 'Who's Who in the Midwest'. Reg P E Ill, Mbr NSPE/ISPE. Mbr ASAE, Chicago Sect officer 1970-76, Chmn 1975; ASAE Natl Winter Meeting Arrangements Ctte 1970-75, Chmn 1975. Mbr FPS, Speaker/Author with publ in 1974 FPS Testing Symposium proceedings. IH participant on NFPA Cttes.

Holloway, William J
Home: Sunridge Road, Rindge, NH 03461
Position: President. *Employer:* Wiltec, Inc. *Education:* BS/ChE/Case Inst of Tech. *Born:* Jan 1926. Native of Salem Ohio. Served 10 yrs B F Goodrich Chem Co as a chem engr in polymers. With Borden Chem Co 15 yrs as Plant Mgr, then Operating Mgr of polymer prod. Last 14 yrs as Pres of Wiltec Inc a co fabricating finished prods from polymers. Pres Leominster Chamber of Commerce 1970. Enjoy sailing, skiing & bridge. *Society Aff:* AIChE, SPE, ACS.

Hollrah, Ronald L
Business: PO Box 8405, Kansas City, MO 64114
Position: Project Mgr *Employer:* Black & Veatch *Education:* PhD/CE/Univ of MO-Columbia; MS/CE/Univ of MO-Columbia; BS/CE/Univ of MO- Columbia *Born:* 9/7/42 Native of St. Charles, MO. National Science Foundation Tranineee and Instructor while in Graduate School. Served as an Instructor and Operations Officer with the U.S. Army in Korea. Honorably discharged with the rank of Captain in 1970. With Black & Veatch since 1971. Selected as the Young Engr of the Year by MSPE in 1975. Serves as the Project Mgr on several coal power generation facilities. Vice Chmn of ACI Specification Committee. Registered PE in five states. *Society Aff:* ACI, NSPE

Hollywood, John M
Home: 67 Halsey Drive, Old Greenwich, CT 06870
Position: Consultant. *Employer:* Goldmark Communications Corp. *Education:* MS/EE/MIT; BS/Communications/MIT. *Born:* Feb 1910 Red Bank N J. Reg P E Conn. Life Fellow IEEE for contribs to electronic countermeasures & color television. Joint Army-Navy Cert of Appreciation. 14 pats, many articles in Proc of the IEEE & elsewhere, chap of TV Engrg Handbook. Bd of Eds, Audio Engrg Soc. CBS Labs 1936-43 & 1949-72 as Sci Adv to Dr Peter C Goldmark, Pres, Branch in England of RRL of Harvard Univ, NDRC, OSRD 1943-45. Naval Res Lab 1945-46, Airborne Instruments Lab 1946-49, Cons. Interests are amateur radio WISK, hiking, music. *Society Aff:* IEEE, AES, NSPE.

Holman, Jack P
Home: 11407 Crest Brook, Dallas, TX 75230
Position: Professor *Employer:* So. Meth. Univ. *Education:* Ph.D./Mech. Engr./Oklahoma State Univ; MSME/Mech. Engr./Sou. Methodist Univ; BSME/Mech. Engr./Sou. Methodist University *Born:* 07/11/34 Native of Dallas Texas. Task Scientist, Aerospace Labs, Wright Field, 1958-60. Southern Methodist Univ: Assoc. Prof. 1960-66, Prof. and Dir Thermal and Fluid Sciences Center 1966-73, Chairman Civil and Mech Engrg 1973-78, Asst. Provost 1978-81. Chairman ASME Region X Mech Engrg Dept. Heads 1977-79, Chairman National Mech. Engr. Dividion of ASEE 1971. ASEE George Westinghouse Award 1972, ASME James Harry Potter Gold Medal 1986, ASME Worcester Reed Warner Gold Medal 1987. Research in several fields of heat transfer. Outstanding Prof. SMU eight times. Fellow ASME. Reg P.E. TX. *Society Aff:* ASME, ASEE

Holmboe, John A
Home: 5845 Doverwood Dr - 301, Culver City, CA 90230
Position: VP of Construction & Planning *Employer:* Hartfield-Zodys Inc *Education:* Bachelor/Civil Eng/Univ of MN; Jr Cert/Bus/Univ of MN. *Born:* 5/20/27. Mbr Chi Epsilon & Tau Beta Pi. Sewer Engr City of St Paul 1952-59. Genl Superintendent & VP Aagard Bldg Co 1959-67. Joined Dayton Hudson Properties as Budget Mgr in 1967 & progressed to pos of Dir of Construction in 1977. Joined Harfield-Zodys Inc in Jan 1980 as VP of Construction and Planning. Have held offices of Pres, VP, Secy & Treas of So Capitol Chap of Minn Soc of Prof Engrs. Have also held following pos in Minn Soc of PE state orgn: Pres, VP, Dir, State Mbrship Chrmn. Also am a mbr of Triangle Fraternity. *Society Aff:* NSPE.

Holmes, D Brainerd
Business: Bay Colony Corporate Center, 950 Winter Street, Suite 4350, Waltham, MA 02154
Position: Retired President of Raytheon Company. *Education:* BS/Elec Engg/Cornell. *Born:* 5/24/21. in NY. Design Engr Western Elec, also mbr Tech. Staff Bel Labs 1945-53. Initiated, dev first precision res transmission measuring set, other test equipment; participated dev long distance coaxial telephone and TV sys RCA 1953- 61. Genl mgr major defense sys div 1961; Proj mgr Navy Talos land based missile sys dev 1954-57; Air Force Atlas launch control and checkout equip dev 1957; USAF Ballistic Missile Early Warning Sys (BMEWS) 1958-61; First Dir of Manned Space Flight NASA 1961-63; Sr VP & Dir Raytheon Co, Lexington, MA 1963-69; Exec VP & Dir 1969-75; Pres & Dir 1975- . *Society Aff:* NAE, AIAA, ADPA, IEEE.

Holmes, Elwyn S
Business: Rm 233 Seaton Hall, Manhattan, KS 66506
Position: Ext Ag Engr *Employer:* KS State Univ *Education:* MS/Ag Engr/TX A&M; BS/Ag Engr/TX A&M *Born:* 4/27/22 Jewett, TX. Res assist TX A&M 1943-1946. Extension Agricultural Engrg, Univ of KY 1946-1956. Res and Extension Agricultural Engrg, Univ of FL 1956-1966. Extension Agricultural Engrg, KS State Univ 1966-current. Reg Engr KS 5611, FL 6545. Primary responsibilities for all electric power and processing in farm engrg applications. *Society Aff:* ASAE, NSPE

Holmes, Lyle A
Home: 437 Elm Rd, Barrington, IL 60010
Position: Consultant *Employer:* International Minerals & Chemical Corp. *Education:* BS/Ceramic Engrg/Univ of Saskatchewan. *Born:* Jan 1921 Minnesota. Five yrs Royal Canadian Engrs. Gen whiteware ceramic engrg duties three yrs. Salesrelated activities in the U S & Canadian ceramic & glass indus fifteen yrs. Pres & Dir, Amer Nepheline Corp. With Internatl Minerals & Chem Corp & affiliates in sales & genl mgmt since 1967. V Pres & Genl Mgr, IMCORE Div. Mbr NICE, Ceramics; PPres Amer Ceramic Soc. Interests in music, photography, tennis & trout fishing. *Society Aff:* ACS, AIME.

Holmes, William C
Home: 3689 Suzanne Way, Redding, CA 96002
Position: Owner. *Employer:* Holmes Engg. *Education:* MS/Aero Engrg/MIT; MS/Mech Eng/Stanford Univ; BA/Mech Eng/Stanford Univ. *Born:* 12/10/17. Born 12/10/17; Education: MS/Aero Eng/MIT, MS/Mech Eng/ Stanford Univ, BS/Mech Eng/Stanford Univ; Engg & Mgt resp. including Flight Test Officer Naval Air Test Cntr; Proj Aerodymanicist, Supvsr Prelim Design, Chief Design Engr (Northrop); Dir Prog Mgt Space Systems (LockheedP; VP/Gen Mgr West Coast (Radiation); Asst to Pres (IMI); Plant Mgr (UTC); Dir Operations (Teledyne); Pres & CEO (Filt-Aire); Sr Consultant (Englert & Co); Managing Ptnr (Xikon Vent Cap Group); Principal/Owner Holmes Engineering; Instr Univ Santa Clara Grad School & Shasta College; Dir Amer Inst Building Design; Pres. Econ. Devel. Corp Shasta County. *Society Aff:* AIAA, IEEE, ASHRAE, AIBD, AEE, NSPE, CSI

Holmes, William D
Business: 3 Canal Plaza, Portland, ME 04112
Position: Exec. V.P. *Employer:* Consumers Water Co *Education:* BS/Civ Engg/Univ of IL. *Born:* 4/13/29. Native of Downers Grove, IL. DeLeuw, Cather Engrs, 1951-1956 except 2 yrs in US Army. Kankakee Water Co, 1956-84; Asst Mgr 1956-1959, Exec VP 1960- 1965, Pres 1965-84. Exec V.P. Consumers Water Co 1984-present. Dir, Kankakee Federal Savings & Loan 1965-85; V.P. Natl Assn of Water Cos. Past: Pres, Univ of IL Alumni Assn; Chrmn, IL Section AWWA; Pres, Kankakee Chamber of Commerce; Pres, Rotary; Chrmn, Cancer Crusade; Dir, IL Heart Assn. Mbr: Phi Eta Sigma, Chi Epsilon and Sigma Tau Honorary Fraternities. *Society Aff:* ISPE, NSPE, AWWA.

Holonyak, Nick, Jr
Business: Dept of Elec and Computer Engrg, 1406 W. Green St, Urbana, IL 61801
Position: Prof (Elec Engr & Ctr for Advanced Study) *Employer:* Univ of IL *Education:* BS/EE/U of Ill. (Urbana); MS/EE/U of Ill. (Urbana); PhD/EE/U of Ill. (Urbana). *Born:* 1928. BS 1950; MS 1951; PhD 1954 Univ of IL. Texas Instruments Fellow 1953-54. Bell Labs 1954-55. US Army 1955-57. Genl Elec Syracuse 1957-63 (p-n-p-n devices, tunnel diodes & visible diode lasers & LED's). Co-author of 'Semiconductor Controlled Rectifiers' (Prentice-Hall Inc 1964). Editor Prentice- Hall series 'Solid State Physical Electronics'. Editorial Bd Proceedings IEEE 1966-74, Solid State Electronics 1970- , Semiconductors & Insulators 1976- . Cordiner Award GE 1962, IEEE Morris N Liebmann Award 1973, John Scott Medal 1975 for first (1962) visible diode lasers & Led's, Gallium Arsenide Conference Award with Welker Medal (1976), IEEE Jack A. Morton Award 1981, Electrochem Soc Solid State Sci & Tech Awd 1983. Fellow IEEE & APS. Mbr Natl Acad Engrg, Natl Acad Scis, American Acad of Arts and Scis, AAAS, Mathematical Assn Amer, Electrochemical Soc. *Society Aff:* IEEE, APS, ECS, MAA, AAAS.

Holt, Arthur W
Business: Tech A217, Washington, DC 20234
Position: President Arthur Holt Inc. *Employer:* Natl Bur Stds. *Education:* MA/Physics/Williams College; BA/Physics/Williams College. *Born:* May 15, 1921. SEAC Computer Staff NBS 1949-55. Commerce Dept Silver Medal 1954 for patent basic to dynamic RAM; 30 patents including early claims on helical scan TV magnetic tape recorders (VTRs). Chief Electronic Engr Rabinow Co & Rabinow Div Control Data 1955-69 developing early Optical Character Recognition Machines. V P Recognition Terminals 1969-71. Cons Recognition Equipment 1971-72. Founded Arthur Holt Inc 1972, a company which invents & develops new hardware for computer related applications. Fellow IEEE 1968. Most recent publ: Guest Editor, Pattern Recognition Society Vol 8 1976. *Society Aff:* IEEE.

Holt, Ben
Business: 201 S Lake Ave, Pasadena, CA 91101
Position: President. *Employer:* The Ben Holt Co. *Education:* BA/Chem/Stanford; MS/Chem Eng/MIT. *Born:* 1914 Caldwell Idaho. Engrg & mgmt positions Shell Dev Co, Union Oil Co of Calif, The Ralph Parsons Co & Amer Potash & Chemical Corp 1937-60. Founder & Pres The Ben Holt Co 1960- , engrs & constructors specializing in process plant design & const including geothermal installations. Fellow AIChE, Dir Geothermal Resources Council, Past Dir Fuels & Petrochemicals Div. Mbr Advisory Ctte on Geothermal Energy, Energy R/D Admin. Publs in geothermal & oil shale tech. Hobbies include ocean racing & tennis. *Society Aff:* AIChE.

Holt, Randolph E
Business: 7920 Mountain Rd, NE, Albuquerque, NM 87110
Position: President. *Employer:* Randy Holt & Associates Inc. *Education:* MS/Civil Engg/Univ of NM; BS/Civil Engg/Univ of NM. *Born:* Apr 1930 Albuquerque N M. USN 1953-56. Struct Engr for Eugene Zwoyer & Associates 1959-64. Asst Dir Cvl Engrg Res Facility 1961-70. With Randy Holt & Associates 1964- ; became Pres 1971, specializing in all types of struct design. Reg P E NM, Texas, Ariz & Colo. Pres NM ASCE 1972 Pres NM CEC 1980; Fellow ASCE & ACEC Mbr: NSPE, ACI, PCI, PTI, ICBO, AISC, CRSI & AWS. Enjoys photography, golf, tennis & cycling. *Society Aff:* ASCE, ACEC, PCI, PTI, ICBO, AISC, CRSI, AWS.

Holt, Sherwood G
Business: Consolidated Papers Inc, Wisconsin Rapids, WI 54494
Position: Dir Res & Dev. *Employer:* Consolidated Papers Inc. *Education:* PhD/Mech Engg/Cornell Univ; MME/Mechs/Cornell Univ; BME/Mech Engg/Cornell Univ. *Born:* 1921 Warsaw N Y. US Navy 1943-45; Assigned to Naval Res Lab Mechanical Dev Sect; proj related to study of ship board shock & vibration & dev of aircraft antenna hardware. Instr & Res Associate Cornell Univ 1944-51; E I duPont Proj Engr 1951-54, spinning methods & economic evaluations of processes for 'Dacron' manufac; Scott Paper Co Engrg Specialist, papermaking equipment dev 1955-60; with Consolidated Papers Inc, Dir of Res & Dev. Pres TAPPI;; TAPPI Fellow 1966. Hobbies: skiing & sailing. *Society Aff:* TAPPI.

Holt, Stephen
Home: 2777 E. Jamison Ave, Littleton, CO 80122
Position: Pres *Employer:* URS Co *Education:* BS/CE/CO Univ *Born:* 7/11/35 *Society Aff:* ASCE, CEC, NSPE, CTA, ARTBA

Holthaus, Thomas J
Home: Box 1505, Taif, Saudi Arabia
Position: Prog Dir. *Employer:* McDonnell-Douglas Corp. *Education:* BS/EE/St Louis Univ. *Born:* 3/8/33. Served with the USAF as an electronics officer 1955-59. Field engr for Douglas Aircraft specializing in missile guidance systems 1959-62. Proj engr and later branch-chief Thor missile advanced design. Mgr-Avionics Intergration for McDonnell-Douglas 1967-69. VP, & later Pres of 3R Coating Co, Inc, 1970-75. Section-Hd, air launched missile trainers for the Dept of Defense 1975-78. McDonnell-Douglas Astro-TVL as prog mgr, 1978-80, Mgr, Mktg 1980-81. Present position Mgr of Base Fiscal and Admin in Saudi Arabia with McDonnell-Douglas Services, Inc. Enjoy swimming & classical music. *Society Aff:* AIAA, AMA, TESS.

Holton, John H
Business: 466 Prospect Ave, Little Silver, NJ 07739
Position: President. *Employer:* Holton Industries Inc. *Education:* BSChE/-/Yale Univ. *Born:* Nov 30, 1920 Boston Mass. Lt USNR 1943-46. Plant Oper Mgr Niaga-

Holton, John H (Continued)
ra Chem Div FMC 1946-53; Genl Mgr Gallouhur Chem 1953-56; Tech Dir to Exec V P Natl Sugar Refining Co 1957-69; Treas Sugar Indus Technologists, Process Engrg Cons 1969-71; Pres Holton Indus serving biol, indus & res communities. Mbr AIChE, ACS & Food Indus Technologists. Enjoys boating & tennis. *Society Aff:* AIChE, ACS, IFT, AALAS.

Holtz, Wesley G
Home: 3250 Moore St, Wheat Ridge, CO 80033
Position: Consulting Engr *Employer:* Self *Education:* BS/CE/Univ of CA-Berkeley *Born:* 6/5/11 Formerly Chief of Soils Engrg and Deputy Chief of Res, US Bureau of Reclamation, Retired. Private Consulting Engr. Author of over 50 technical articles and papers on Geotechnical Engrg subjects. Former Dir and Honory Mbr ASTM, Former Secy and Chrmn ASTM Committee D-18 on Soils and Rocks for Engrg Purposes, Former Chrmn ASCE Div of Soil Mechanics and Foundation Engrg, Dept of the Interior Distinguished Service Award, Society of Sigma Xi, ASCE Wellington Prize, Vice Chrmn and Chrmn ASCE Research Council on Expansive Soils. *Society Aff:* ASCE, ISSMFE, USCOLD, ASTM, ASCE.

Holtzman, Arnold H
Home: 208 Stone Crop Rd, Wilmington, DE 19810
Position: Mgr New Business Programs. *Employer:* E I duPont de Nemours & Co Inc. *Education:* PhD/Metal Engrg/Lehigh Univ; MS/Metal Engrg/Lehigh Univ; BS/Metal Engrg/Drexel *Born:* May 1932. Attended public sch Phila Pa; Work experience includes magnetic materials, experimental heat treating, stainless steel mill fabrication, dissimilar metal bonding & processing, & explosives. Joined duPont 1957; R&D Mgr in lab & business roles; also as Regional Sales Mgr; currently Dir Dev Div-Central Res & Development Dept. Awarded Wetherill Medal 1969; named ASM Fellow 1972. Granted numerous patents. Married; 3 children. Hobbies are sports & piano. *Society Aff:* ASM.

Holway, Donald K
Business: 5300 South Yale Avenue, Tulsa, OK 74135
Position: V Pres & Principal Elec Engr. *Employer:* W R Holway & Associates. *Education:* MS/Elec Engg/MIT; BS/Physics/Univ of Chicago. *Born:* 1917 Rhode Island. Grad work in Theoretical Phys Univ of Chicago 1937 & 38, Elec & Hydroelec Engrg MIT 1938-39 & 1941-42; MSEE MIT 1947. Has served as Principal in Holway Companies 1942- ; presently Sr VP of W R Holway & Associates Tulsa, Div of Benham GroupInc. Also Sr VP Benham Inc. Mbr: IEEE, NSPE, Okla Soc of Prof Engrs, ACEC & Sigma Xi. Past Chmn Tulsa City-County Elec Examining & Appeals Board. Reg Prof Engr: Okla, Kan, Mo, MA. Resident of Tulsa Okla since 1918. Major proj respon: Markham Ferry Hydroelec & Salina Pumped-Storage Proj for Grand River Dam Authority; Lawrence, MA, re-dev of Great Stone Dam. *Society Aff:* IEEE, NSPE, ACEC, ΣΞ, AAAS.

Holway, William N
Business: 5314 South Yale Avenue, Tulsa, OK 74135
Position: CEO W R Holway & Associates. *Employer:* W R Holway & Associates. *Education:* BS/Civil/MIT. *Born:* Nov 1920. Naval Architecture Certificate Univ of Mich 1944. Exec V Pres, Dir of The Benham Group C.E.O., W R Holway & Associates subsidiary of The Benham Group. Principal of Holway co since 1946. Mbr ASCE; AWWA; ACI; Natl & Okla Soc of Prof Engrs; Cons Engrs Council of Okla, Pres 1963-64; CEC, Natl Dir 1964-65, Natl Secy-Treas 1968-69, Natl Pres Elect 1972-73 & Natl Pres 1973-74; Amer Inst of Cons Engrs. Reg Prof Engr: Okla, Ark. Fellow of ACEC. *Society Aff:* ACEC, NSPE, AWWA, ACI.

Holz, Harold A
Business: 1 University Plaze, Hackensack, NJ 07601
Position: Acct Exec - Polyolefins Div. *Employer:* Union Carbide Corp. *Education:* BS/ME/Stevens Tech *Born:* June 26, 1925 N Y City. Employed by Plastics Div Union Carbide Feb 1947; sales, engrg & mgmt assignments in N Y, Hartford, St Louis & Chgo; in present position N Y Office 1965- . Various local & internatl offices with SPE incl Internatl Pres 1975-76; Distinguished Mbr SPE 1976. Listed in 39th Edition Who's Who in Amer. USNR (V-12) 1943-46 Ltjg Mbr-Plastics Pioners Association-1977. *Society Aff:* SPE.

Holzmacher, Robert G
575 Broad Hollow Rd, Melville, NY 11747
Position: Pres *Employer:* Holzmacher, McLendon & Murrell, P.C. *Education:* M/CE/Polytechnic Inst of Brooklyn; B/CE/Polytechnic Inst of Brooklyn *Born:* 8/18/27 A licensed PE, a licensed professional planner, a licensed land surveyor, a fellow of the American Society of Civil Engrs and a diplomate of the American Academy of Environmental Engrs. A mbr of Chi Epsilon Honorary Civil Engrg Fraternity. A past state pres of ACEC/NY State, a past national dir of ACEC; a past pres of the Long Island Branch, Metropolitan Section, ASCE, and its 1980 "Engr of the Year–; past chrmn of the Long Island Assoc of Commerce and Industry Environmental Task Force; 1970 "Engr of the Year–, Suffok County Chapter, PE. *Society Aff:* ASCE, ACEC, AAEE, AWWA, WPCF.

Holzman, Albert G
Business: Industrial Engr Dept, 1048 Benedum Hall, Pittsburgh, PA 15261
Position: Professor & Chairman. *Employer:* University of Pittsburgh. *Education:* BS/Indus Engr/Univ of Pittsburgh; MS/Indus Engr/Univ of Pittsburgh; PhD/Economics/Univ of Pittsburgh. *Born:* 10/28/21. Chmn of Dept of Indus Engrg, Engrg Mgmt & Opers Res 1966- ; formerly Dir of Engrg Opers NASA Space & Tech Transfer Prog PA region. Elected Mbr Natl Acad of Engrg. Elected Fellow IIE. ED of Encyclopedia of Computer Sci & Tech (16 volumes); Ed of Operations Res Support Methodology; ED of Mathemetical Programming for Operations Researchers and Computer Scientists; res & publs in multi-criteria decision making, engr mgt, educ sys, nonlinear programming, optimization methods & information sys. Reg Prof Engr. Cons for major corps. Mbr: ORSA, TIMS, IIE & ASEE *Society Aff:* ASEE, IIE, ORSA, TIMS, NAE.

Homer, Eugene D
Business: Computer Science Dept, CW Post College of L I U, Greenvale, NY 11548
Position: Prof & Chrmn. *Employer:* Long Island Univ. *Education:* PhD/Industrial Engr/NY Univ; MIndE/Industrial Engr/NY Univ; BIndE/Industrial Engr/NY Univ. *Born:* 8/3/21. 1938-1942: Various jobs in factory admin. 1942-1946: Army Air corps. 1946- 1951: Production Control, American Lead Pencil Co, Inc. 1951-1952: Asst Govt Contracts Mgr, Lightolier, Inc. 1952-1957: Asst Head, Systems & Procedures Dept, Emerson Radio & Phonograph Corp. 1957-1969: NY Univ; instructor, asst prof, sr res scientist, assoc prof. 1969-1980: C W Post College of L I U; prof & chrmn of Mgt Sci & Engr Dept. 1980-present: C. W. Post College of LIU; prof & Chrmn of Computer Science Dept. *Society Aff:* IIE, ACM, AAAS, NYAS, TIMS, ASEE.

Homewood, Richard H
Home: 104 Haggetts Pond Rd, Andover, MA 01810 *Employer:* Retired *Education:* MSCE/Civil Eng/CA Inst of Tech; BSCE/Civil Eng/Univ of MA. *Born:* Sept 1927 Fitchburg Mass. Stress Analyst with Republic Aviation 1951-52; Struc Designer for AE firms 1953-56; with Avco Sys Div 1956-73, held supr positions in struc test, pyrotechnic & mech design, instrument maintenance, & environ test; Proj Mgr for dev of high power laser components; with Stone & Webster 1973-79, respon for engrg mech effort on nuclear power plant proj. Treas SESA 1970-81. With Avco Systems Div 1979-87, responsible for Mech Design of Reentry Systems. Enjoys skiing, golf & flying. *Society Aff:* SEM.

Honey, Richard C
Business: 333 Ravenswood Ave, Menlo Park, CA 94025
Position: Sr Staff Scientist. *Employer:* SRI Intl. *Education:* PhD/EE/Stanford; EE/EE/Stanford; BS/Phys/Cal Tech. *Born:* Mar 1924 Portland Ore. USN 1943-46. Res Asst Stanford Univ, wideband electron deflection sys; with Stanford Res Inst 1952-

Honey, Richard C (Continued)
, microwave & millimeter wave components & antenna sys incl wideband traveling-wave parametric amplifiers, instantaneous wideband direction-finding antennas, radar scanning antennas, phased arrays, & leaky-wave antennas. 1962- laser application progs, development of laser radar sys; laser eye & skin damage; instrumentation for treating ocular diseases; optical propagation in seawater. Fellow IEEE 1968; Fellow OSA 1977; Chmn Working Group on Lasers, AGED, DDR&E 1970-75; ANSI Z-136 Ctte on Laser Safety. Army Sci Bd 1978-84. *Society Aff:* OSA, IEEE, SPIE, LIA.

Honeycutt, Baxter D
Business: P.O. Box 1346, Houston, TX 77001
Position: Engr Mgr *Employer:* ARCO Oil & Gas Co *Education:* MS/Pet Engr/TX A&M Univ; BS/Pet Engr/TX A&M Univ *Born:* 9/12/81 Engrg Mgr for Exploration Support, Arco Oil & Gas Co. Married - Nell Treadaway, 1953; two children, Michael born 1958 and Gloria born 1963. Extended involvement in professional society; Education and Accreditation Committees, industrial and professional political action groups, minority and female professional development, civic interest center on education. Exploration and exploitation experience in Gulf of Mexico, East TX basin, Colombia SA, and currently world wide, all frontier areas. Military service - Korea, 1955, as Petroleum Testing Officer. Research assignments; well perforating, tertiary recovery and advanced drilling tech. *Society Aff:* SPE, API

Honeycutt, Kenneth E
Business: 312 Directors Drive, Knoxville, TN 37923
Position: VP, Special Projects *Employer:* IT Corporation *Education:* BS/ChE/TX A&M. *Born:* Sept 1928. Grew up in New London Texas. Associated prior to 1980 with Dow for entire career in Texas, La, Mich & Tenn; core experience has been designing & building plants, including assignments as Chief Proj Engr & Mgr of Process Engrg; also had genl exec & mgmt roles in operations, staff & business dev; most recent endeavor has been in mgmt of cons service to design environ control processes. Knoxville operation was acquired in 1980 by IT Corporation. Reg Prof Engr 7 states. Mbr Tenn Soc of Prof Engrs; Mbr AIChE. Enjoys classical music & hiking. *Society Aff:* AIChE, NSPE.

Hong, Franklin L
Home: 175 Tillman St, Staten Island, NY 10314
Position: Pres *Employer:* Lorrison, Inc *Education:* MSCE/Structural Engrg/Purdue Univ; BSCE/Structural Engrg/Purdue Univ *Born:* 1/9/28 I have been engrg since graduation from Purdue Univ. I have worked mostly with consulting engrg firms starting at Tippetts, Abbott, McCarthy and Stratton as Junior engr; Hardesty and Hanover as a designing engr; King & Gavaris as a Project Engr; Amman & Whitney as a Supervising Engr; as partner in charge of consulting engrg projects at Omnidata, Inc at E W Finley, PC as Chief Engr for the past 4 years with a firm of 120 people and now Principle of Lorrison, Inc, Olmos Building Supplies, Inc, Orco, Inc and Apachi Redi Mix. *Society Aff:* NSPE

Honnell, Martial A
Home: 1055 N College St, Auburn, AL 36830
Position: Prof. *Employer:* Auburn Univ. *Education:* EE/-/GA Inst of Tech; MS/EE/GA Inst of Tech; BSc/EE/GA Inst of Tech. *Born:* 10/23/10. Native of Lyons, France. With Radiomarine Corp of Am, New Orleans, 1928-30, Sou Broadcasting Stations, Inc, Atlanta, 1930-1936. Radio Div Pan Am Airways, Miami, 1936-37. Prof in charge communications & electronics courses, GA Tech 1937-53. VP & Chief Engr Measurements Corp, Boonton, NJ, 1953-58. Prof of EE Auburn Univ since 1958. Proj leader for numerous res Projs for NASA & US Army Missile Command, Huntsville, AL. First Prize Sigma Xi Res Award, GA Tech, 1951. Distinguished Grad Faculty Lectr, Auburn Univ 1976-77. Nine NASA Certificates of Recognition for technical contributions. Patents & numerous publications. *Society Aff:* IEEE, NSPE, ΣΞ, ΦΚΦ, ASEE.

Honsinger, Leroy V,
Home: 261 E Heritage Village, Southbury, CT 06488 *Employer:* Retired *Education:* MS/Naval Arch/MA Inst Technology; BS/Marine Engg/US Naval Acad; AMP/Advanced Mgmt/Harvard Bus Sch. *Born:* 1905. 39 yrs Naval service mostly in ship design & shipbuilding; Deputy Chief Bureau of Ships & Cmdr Naval Shipyards Long Beach & Mare Island. 11 yrs Todd Shipyards. Pres Amer Soc of Naval Engrs 1958; Pres SNAME 1975-76. Past V Chmn Board World Educ Inc. Former Chrmn Elmer A Sperry Bd of Award (under ASME). Retired. *Society Aff:* ASNE, SNAME.

Honstead, William H
Business: Fairchild Hall, Manhattan, KS 66506
Position: Exec VP *Employer:* KS State Univ Res Fdn *Education:* PhD/ChE/IA State Univ; MS/ChE/KS State Coll; BS/ChE/KS State Coll *Born:* 5/21/16 Native of Waterville, KS. Plant engr, Natl Aniline Div, Allied Chem and Dye Corp, Buffalo, NY, 1939-43, Married 1940, three children. Instructor to Assoc Prof, Chem Engg, KS State Univ, 1943-56. Full Prof, 1956. Hd of Dept 1960-68. KS Ind Extension Service, 1968-81, Dir, 1970-81. Exec VP, KSU Res Fdn, 1972- present. Res on dehydration of vegetative mtls, industrial utilization of sorghum gramns, heat transfer, fertilizer tech. Mbr Tau Beta Pi, Sigma Xi, Gamma Sigma Delta, Phi Kappa Phi. *Society Aff:* AIChE, NSPE, ACS, ASEE

Hood, A Craig
Home: 2 Kathwood La, Wayne, PA 19087
Position: Pres *Employer:* ACH Tech (Mgmt and Engrg Consltg Firm) *Education:* SB/Metallurgy/MIT. *Born:* Nov 25, 1928 Phila PA. With Standard Pressed Steel Co 1960- ; appointed Genl Mgr SPS Special Products Div 1973. With ACH Technologies 1983-Present. Mbr ASM, Chmn Phila Chapter 1963-64; Natl Advisory Ctte 1964-65; Fellow ASM 1975. Mbr Inst of Directors London. Lectr Ministry of Aviation Moscow 1973. Metal Properties Council, Natl Engrg Found 1965-70. Pat 3, 897,222 Protective Coatings for Ferrous Metals; other patent applications & tech papers in fields of stress corrosion, fastener, coatings & bolted connections. Honor: Mbr, Fastener Hall of Fame. *Society Aff:* ASM, AIME, SAE, SAMPE, APMS, ASTM.

Hood, John L, Jr
Business: PO Box 2641, Birmingham, AL 35291
Position: Field Supervisor-Commercial Services *Employer:* Georgia Power Company *Born:* Oct 1946 Birmingham Ala. BS Auburn Univ; Grad Sch UAB. Began employment with Ala Power Co 1969 as Commercial Lighting Specialist; transferred to Georgia Power 1981 current respon for Commercial Service to include energy conservation, chain accounts public relations & training.Mbr Birmingham Area Chamber of Commerce Pres Ctte & Diplomatic Corps, Birmingham Jaycees; Jefferson Cty Health Dept Ctte; Illuminating Engrg Soc, Regional V Pres S Central Region IES 1974-76; V Pres, Regional Activities IES 1976-78. Cpt Army Res. Enjoys golf & tennis. *Society Aff:* IES.

Hood, Russell
Business: 1479 Buffalo Place, Winnipeg Manitoba, Canada R3T 1L7
Position: V Pres & Mgr Manitoba & NW Ontario. *Employer:* Underwood McLellan & Associates Ltd. *Education:* BE/Civil/Univ of New South Wales; Assoc/Civil/Sydney Tech College. *Born:* 1932 Sydney Australia. Employed by New South Wales Dept of Railway & New South Wales Electricity Comm 1953-58. Joined Underwood McLellan & Assoc, Canada 1958; became Mgr Manitoba & NW Ontario 1966, became a V Pres of the co 1971. Prof activities include: Assoc of Prof Engrs of Manitoba, Pres 1971, Mbr of Council 1968, 69 & 70; Canadian Council of Prof Engrs, Pres 1973-74, V P 1972- 73, Mbr of Exec Ctte 1971-75, Dir 1971; Engg Inst of Canada, Pres 1978-79; Mbr Exec Committee 1976-80; RTAC; Hwys Res Board; Canadian Soc for Cvl Engrg. *Society Aff:* APEM, EIC, CSCE, RTAC.

Hood, Thomas B
Home: 506 Lyme Rock Rd, Bridgewater, NJ 08807
Position: Director of Industrial Engineering. *Employer:* ETHICON Inc Div of Johnson & Johnson. *Education:* BS/Engrg/Fairleigh Dickinson Univ; MBA/Ind Mgt/Fairleigh Dickinson Univ. *Born:* Native of Arlington N J. P E Calif. Regular Army 1950-51. Started engrg career with Weston Elec Instrument Corp Newark N J 1952; joined Johnson & Johnson New Brunswick N J as Sr Indus Engr 1957, promoted to Group Leader ETHICON, Inc Div of J&J 1965, Indus Engr Mgr 1968, Ch Indus Engr 1973, Dir of Indus Engrg 1976. Mbr ASPE, P Pres of Raritan Valley Chapt 168; IIE, P Natl Dir of Work Measurement & Methods Engrg Div 1974-75. Chrmn Bd Trustees Raritan Valley Chapt 168 IIE 1979-82 Adjunct Prof Grad Sch of Eng, Rutgers Univ. *Society Aff:* IIE, ASPE.

Hook, Melvin E
Home: 40 Oak Spring Dr, Pittsburgh, PA 15238
Position: Manager & Board Secy. *Employer:* Fox Chapel Authority (Water). *Education:* MS/Civil Engg/Univ of Pittsburgh; BS/Civil Engg/Univ of Pittsburgh. *Born:* Mar 1927 Pennsylvania. Reg Prof Sanitary Engr and land surveyor in PA. Served with Air Force Engrs in SW Pacific during WWII. With the Fox Chapel Authority (water) 1952-present, in complete charge of admin, water treatment & distribution. Chmn Pa Sect AWWA 1975. Pa Municipal Authorities Assn Dir 1975-78; Pa Municipal Authorities Assn Sahli Award 1974. Cubmaster 1974-76. Enjoys hunting & fishing. District Commissioner, Allegheny Trails Council, Boy Scouts of America, (1977- 80). PA Section AWWA Fuller Award 1978. AWWA Diamond Pin Club Award 1979. Mbr, PA State Bd for Cert of Sewage Treatment Plant & Waterworks Operators 1977- present. PA Municipal Authorities State Mbrship Chm 1977-78. PA Sect AWWA State Mbrship Chm 1978-1983. AWWA Ambassador Award-1980, Diplomate-Amer Acad of Environ Engrs (1979-present), Mbr Western PA Corrosion Coordinating Ctte; Mbr Water Utility Ctte. OH River Sanitation Commission; Mbr PA Allegheny River Protective Assn (1981-present); PA Class A Certified Waterworks Operator; Mbr Water Works Operators Assn of PA; Intl Dir AWWA (1980-1983); AWWA Honorary Mbr 1984, AWWA Platinum Presidential Award 1984, AWWA Diamond Pin Club Pres. 1984- 85, PA Municipal Authorities Assn Pres 1985-86. AWWA Natl Mbrshp Chrmn 1983-86. Reg. Prof Land Surveyor in PA. Trustee AWWA research foundation (1985-present). Commissioner-Ohio River Valley Water Sanitation Commission. Member AWWA General Policy Council. Life member AWWA. *Society Aff:* AWWA, AAEE, NSPE, SAME

Hook, Raymond
Home: 111 Long Meadow Drive, Clarksville, VA 23927
Position: Retired. *Employer:* Naval Air Systems Command. *Education:* BS/Mech Engg/KS State Univ. *Born:* Mar 1916 Osborne Kan. Aero Weight Engr Glenn L Martin Co Balto & Omaha prior to WW II. With Navy in Pacific as Cargo Officer 1944-46. Entered civil service with Navy BuAer 1946; served in projs & progs on aircraft & missile dev, specification coord, contract compliance, weight prediction methods & proposal evals; Weight Control Branch Head 1971, respon for all NavAir weight proposal evaluations. Retired 1975. Mbr Soc Allied Weight Engrs 40 yrs, serving 15 yrs on Govt/Indus coordinating panels. Elected SAWE Fellow 1975. Enjoys golf, tennis & travel. *Society Aff:* SAWE.

Hook, Richard W
Home: 205 Valley St, Horicon, WI 53032
Position: Div Engr. *Employer:* John Deeve Horicon Works. *Education:* BS/Agri Engg/Univ of IL. *Born:* 1/26/41. Native of Thomson, IL. Employed by John Deeve 1964-78 in various product design, engg mgt and energy conservation positions. Proj assignments included the responsibility for dev of machines such as cotton harvesters, rotary hoes, chisel plow, field cultivators and beet harvesters. Listed as inventor on 16 patents. Assumed current responsibility in 1978 as Div Engr for Grounds Care Equip including lawn and garden tractors, lawn tractors, riding mowers, skid steer loaders and attachments for all the same. Runner-up for ASAE Engg Concept of the Yr Award 1975. ASAE Young Designer Award 1977. *Society Aff:* ASAE, ASME, SAE.

Hook, Rollin E
Business: Res Ctr, 703 Curtis St, Middletown, OH 45043
Position: Principal Research Metallurgist *Employer:* Armco. *Education:* PhD/Met Engr/OH State Univ; MS/Met Engr/OH State Univ; BS/Met Engr/Purdue Univ. *Born:* 11/26/34. Native of Highland, IN. Served as officer in the USAF 1957-60 and as civilian AF employee 1960-67, assigned to Wright-Patterson AFB, OH conducting basic res on deformation & fracture of metals. With Armco, Res & Tech since 1967. Conduct res & dev on low carbon & high strength steels, metal forming, and processing. Held all Dayton Chapter ASM offices, 1972-76. Fellow, ASM, 1979. *Society Aff:* ASM.

Hooper, E Dale
Business: PO Box 101, Florham Park, NJ 07932
Position: Engg Advisor. *Employer:* Exxon Res & Engg Co. *Education:* MSChE/-/Newark College of Engg; BSChE/-/Univ of Pittsburgh. *Born:* 1/8/28. Native of Pittsburgh, PA. US Army 1946-47. With Exxon Res & Engg Co since grad in 1951, specializing in dev of maj processes for petrol refining & petrochems mfg. Contributed to dev of transfer line reactors for fluid catalytic cracking units, steam cracking of vacuum gas oils bimetallic catalysts for naphtha reforming & process for deep hydroconversion of heavy oils. Technical service for European affiliates 1961-63 & 1968-70. Section supervisor 1964-86. Reg engr in NJ. *Society Aff:* AIChE.

Hooper, W Euan
Business: PO Box 16858, Philadelphia, PA 19142
Position: Director Vehicle Technology. *Employer:* Boeing Vertol Company. *Born:* Mar 26, 1931 Newquay England. Master Degree in Mech Engrg Cambridge Univ 1954. Worked 7 yrs Bristol Helicopter Div (now Westland Helicopters) as a Dynamicist, Personal Asst to Chief Engr & then Chief Dynamics Engr for the Sycamore, 173 & Belvedere helicopters. He joined Vertol Div of Boeing Co 1962 as Dynamicist working on Model 107, CH-46 & CH-47 helicopters; in 1965 became Dynamics Unit Chief for Tilt Rotors, Tech Mgr for Model 347 Advanced Tech Prog 1968; in 1972 was appointed to present position in which he has been closely involved with all Boeing-Vertol helicopters

Hoover, John W
Home: 2107 N W Fourth Place, Gainesville, FL 32603
Position: Prof Emeritus of Aerospace Engrg. *Employer:* Univ of Florida. (Ret.) *Education:* MS/Aeronautical Engg/GA Inst of Techn; BS/Aeronautical Engg/Auburn Univ. *Born:* Feb 1916. Native of Demopolis Ala. Engrg with Alabama Hwy Dept prior to WWII; taught & administered ground school for USAF pilot school 1941-44; Design Engr Taylorcraft & Globe aircraft co's 1944-46; taught aircraft structs & design Univ of Ala 1946-51 & Univ of Fla 1951-85; Dept Hd 1956-59; Prof Emeritus 1985-present; summer work with Boeing, Pratt & Whitney, Piper & US Navy. Part-time cons. NSPE Bd of Dirs 1974- 78. Pres Florida Engrg Soc 1975-76. Assoc Dir & Acting Dir, NASA/FL STAC 1977- 80. Enjoys music & shopwork. *Society Aff:* ASEE, AIAA, NSPE.

Hoover, Mark D
Home: 11013 Phoenix NE, Albuquerque, NM 87112-1671
Position: Aerosol Scientist *Employer:* Lovelace Inhalation Toxicology Res Inst *Education:* PhD/Nuclear Engr/Univ of NM; MS/Nuclear Engr/Univ of NM; BS/Math- Eng/Carnegie-Mellon Univ *Born:* 08/09/48 Native of Faribault, MN. Served in the US Army Security Agency 1970-1974. Guest Scientist at the Fraunhofer Inst fur Toxikologie und Aerosolforschung in Grafschaft, West Germany, in 1977. Res scientist at the Lovelace Inhalation Toxicology Res Inst, Albuquerque, NM, 1977 to present. Open literature publs on aerosols from fusion and space nuclear energy sys; aerosols from fabrication of mixed-oxide reactor fuels; instrumentation for characterizing airborne particles; and on the inhalation hazards of plutonium recycle, use

Hoover, Mark D (Continued)
of thorium- uranium fuels, use of beryllium in reactor sys, and decommissioning of radioactively contaminated facilities. Chrmn, NM ANS Section 1985-87. Enjoys woodworking, music, and camping. *Society Aff:* ANS, TBΠ, AAAR

Hoovestol, Richard A
Home: 4026 Craig Dr, Duluth, GA 30136
Position: County Engineer *Employer:* Forsyth County. *Education:* PhB/Engg/Univ of ND; BS/Civ Engg/Univ of ND; Associate of Arts/Engg/Bismarck Jr College. *Born:* in New Salem, ND. Attended Almont High School. Attended Jamestown College, Bismarck Jr College, Univ of ND and Univ of S CA. Mbr of Delta Tau Delta, Social Fraternity, and Sigma Tau, Engg Scholastic Fraternity. Mbr - Order of the Engr. Worked 17 yrs in Aerospace Industry. Recipient of cost savings award at Lockheed - 1968. Surveying, soils mechanics and Civ Engg work total 17 yrs. Employed by Public Works Department, Engg Div. of Forsyth County. Received Engr of Yr in Govt- Metro Atlanta 1978. Received Engr of Year Award - GA 1981-82. Pres, DeKalb Chapter GSPE 1981-82. Married to Rachel A Rundle, four sons, Brent, David, Jon and Steven. Enjoy boating. *Society Aff:* ASCE, NSPE.

Hopcroft, John E
Business: 4130 Upson Hall, Ithaca, NY 14853
Position: Chairman/Computer Sci *Employer:* Cornell University *Education:* Ph.D./Electr. Eng./Stanford University; M.S./EE/Stanford University; B.S./EE/Seattle University *Born:* 10/07/87 John E. Hopcroft is professor and chairman of the Computer Science Dept at Cornell Univ, who received a PH.D. in Electrical Engrg from Stanford Univ in 1964, has written more than eighty articles and books on data structures, algorithms, and robotics. He serves as editor for numerous journals, consultant to major corporations, and advisor to various federal agencies. In addition, he directs the Computer Science's Robotics Project at Cornell which focuses on research issues in robotics. solid modeling, graphics, and automated design. He is a Fellow in the IEEE Amer Assoc of the Advancement of Sciences, and Amer Acad of Art & Sciences. *Society Aff:* IEEE, SIAM, ACM

Hopkins, John M
Business: 461 S Boylston Street, Los Angeles, CA 90017
Position: Pres, Union Energy Mining Div. *Employer:* Union Oil Co of Calif. *Education:* BS/Chem Engg/Univ of CO. *Born:* 5/30/20. Native of AZ. Received Chem Engg degree form Univ of CO. Joined present co in 1942 as engg trainee at San Francisco CA Refinery. In July, 1979, named Pres of Union Oil's newly formed div, Union Energy Mining Div with respon for the dev and operation of Union's oil shale and uranium resources. Prior to this, he was Sr VP of the 76 Div. Mbr AIChE, API & 25 Yr Club of Petroleum Indus. Enjoy golf, traveling & fishing. *Society Aff:* AICE, API.

Hopkins, Paul F
Business: 312 Bray Hall/Forest Engineering, Suny-College of Env. Sci & Forestry, Syracuse, NY 13210
Position: Assist Prof, Dept of Forest Engrg *Employer:* SUNY, Coll of Env Sci & Forestry *Education:* MS/Enviro and Resource Engrg/SUNY-Coll of Env Sci & Forestry; BS/Forestry/Univ of ME-Orono *Born:* 5/26/55 Native of Terryville, CT. Academic appointment SUNY, Col of Env Sci and Forestry, Dept of Forest Engrg, 1979-. Instruct courses in surveying, photogrammetry, and remote sensing. Conduct res in remote sensing and photogrammetry for US Forest Service, Defense Mapping Agency, state and local gov't agencies, and private industry. Honors include two scholarships at Univ of ME-Orono, Graduate Fellowship at Univ of WI- Madison, Attenhofen Memorial Scholarship and Western Great Lakes Region Student of the Year from ASPRS, and membership in Phi Kappa Phi and Xi Sigma Pi. Elected VP (1981), Pres (1982, 1983), and Natl Dir (1987-1990) of Central NY Region of ASPRS. *Society Aff:* ASPRS, ACSM, SAF.

Hopkins, Richard J
Home: 19 Lynnfield Drive, Glens Falls, NY 12801
Position: Retired *Education:* BS/Elec Engg/WA State Univ. *Born:* July 25, 1912. Native of Hoquiam Wash. Joined G E 1934; entire career in dev, design & application of power capacitors with General Electric Co. Before retirement was Mgr Engrg-Power Capacitor Prod Sect; since retirement respon for tech exchange with foreign licensees. From 1974 to 1982. Elected Fellow IEEE 1971. Mbr NEMA Capacitor Tech Comm 1956-1975. Chmn NEMA Capacitor Sect 1964-67. IEEE Capacitor Sub-comm 1957-75. US Rep CIGRE Study Comm 18 1960-67. US Delegate to CIGRE Study Comm 15-05 1970 & 1974. US Delegate IEC Tech Comm 33 1972 & 1974. Author 4 IEEE tech papers & 9 articles in the tech press. 1 pat. Mbr Rotary Internatl. Hobbies: woodworking, rifle & pistol competition. *Society Aff:* IEEE, CIGRE.

Hopkins, Robert E
Business: Lab for Laser Energetics, Rochester, NY 14627
Position: Sr Engr. *Employer:* Univ of Rochester. *Education:* PhD/Optics & Phys/Univ of Rochester; BS/Engg/MIT. *Born:* 6/30/15. Prof Inst of Optics Univ of Rochester 1950-1968. Pres of Optical Soc of Am. Mbr of Army Scientific Advisory Panel. Founder of Tropel Inc Fairport NY. Pres of Tropel 1969-1975. Sr Engr, Lab for Laser Energetics Univ of Rochester ('75 to present). Ives Medal Optical Soc of Am. Bd of Dirs SPIE. Optical System Design & Evaluation as a specialty. *Society Aff:* OSA, SPIE.

Hopkins, William E
Home: 11 Willard Street, Newton, MA 02158
Position: V Pres & Mgr of Projects. *Employer:* Stone & Webster Engrg Corp. *Education:* BE/Mech Engg/Johns Hopkins Univ. *Born:* 11/29/03. Native of Baltimore Md. Mech Engr for Potomac Edison Co, Allis-Chalmers Mgr Co & Stone & Webster Engrg Corp; subsequently Sr Cons Engr, Mgr of Projs & V Pres of Stone & Webster Engrg Corp. Pres Metropolitan Sect Mass Soc of P E's. Reg P E New York, Pa, Conn, Florida, Maine, Mass. 2nd Lt Engr Corps USA. Received certificates awarded by the ASME as Fellow, Chmn Power Div Exec Ctte & Mbr Medals Ctte. Alderman-at-Large City of Newton MA. Received ASME James N Landis Award 1978. *Society Aff:* ASME.

Hopkinson, Philip J
Business: 1701 College St, Ft Wayne, IN 46804
Position: Mgr-Engg. *Employer:* GE Co. *Education:* MS/EE/Poly Inst of NY; BS/EE/Worcester Poly Inst. *Born:* 4/11/44. in Burlington, VT. A grad of GE Doctoral Advanced Engg Course in 1970. Licensed PE AGE Employee since 1966 with maj background in Distribution Transformers, Dev & Design. Assumed current responsibility as Mgr of Engg for the Specialty Transformer Bus Dept in 1978, with product scope for Dry Type and Specialty Transformers as well as Electronic Power Supplies. Active Ind Stds participant with voting mbrship in NEMA St-8, ANSI C57 12.5, & High Voltage Apparatus Coordinating Committee. Advisor to Underwriters' Labs & co-ordinator for Product Safety. Holder of 2 US Patents & publisher of Technical Paper on Oil Inhibitors. *Society Aff:* IEEE.

Hoppel, Susan K
Home: 5715 Glade Circle, Lincoln, NB 68506
Position: Environmental Engr *Employer:* NB Dept. of Environmental Control *Education:* MS/CE/Univ of NB; BA/Math/DOANE Coll *Born:* 3/23/47 Worked for NB Natural Resources Commission 1971-1980 in water quality planning. Responsible for Salt Creek Basin-Lincoln Metropolitan Area Water Quality Management Plan, water quality plans for the 13 river basins in NB, and the statewide Section 208 water quality plan. Consulting engr 1980-1983, doing water quality-related work. With NB Dept. of Environmental Control, Wastewater Facilities Section, as review engineer since 1983. Registered engr in NB and IA. Certified wastewater treatment plant operator. NSPE state Outstanding Young Engr 1978. *Society Aff:* NSPE, ASCE, WPCF

Hoppenjans, J Richard
Business: 122 S St Clair St, PO Box 838, Toledo, OH 43696
Position: V.P. and Chief District Engineer *Employer:* Bowser-Morner Associates, Inc. *Education:* MSCE/Soil Mech/Univ of Dayton; BSCE/Soil Mech/Univ of Dayton *Born:* 12/10/45 From 1969 to 1972 - Conducted basic & applied res on USAF contracts at the Univ of Dayton in areas of tire/soil interaction. Since 1972, has been staff geotechnical engr & technical dir (1974) for Bowser-Morner, Inc, in Dayton, OH, & Toledo, OH. Since 1985, has been VP of Bowser-Morner Associates, Inc. Has held several positions as officer in the local sections of both ASCE & NSPE. Active in church & civic affairs. Hobbies include woodworking & outdoor activities. Currently V.P. and Chief District Engr in Toledo, OH office. *Society Aff:* NSPE, ASCE, ASFE, DFI, ACEC

Hopper, Grace M
Home: 1400 S Joyce St A1614, Arlington, VA 22202
Position: Como USNR/NAVDAC Code OOH. *Employer:* Dept of the Navy. *Education:* PhD/Math/Yale Univ; MA/Math/Yale Univ; BA/Math/Vassar Coll *Born:* 12/09/06 Prof Activities: 1931-43 Instructor to Assoc Prof Dept of Mathematics Vassar College; 1943 Asst Prof of Mathematics Barnard College; 1944-46 Mathematical Officer US Navy Bureau of Ordnance; 1946-49 Res Fellow in Engrg Scis & Applied Physics, Computation Lab Harvard Univ; 1949-52 Sr Mathematician Eckert-Mauchly Computer Corp; 1952-64 Sys Engr, Dir of Automatic Programming Dev, UNIVAC Div of Sperry Rand Corp; 1959- , Visiting Lectr to Adj Prof Moore School of Elec Engrg Univ of Penna; 1964-71 Staff Scientist Sys Programming, UNIVAC Div of Sperry Rand Corp (on military leave 1967-71) - ret 1971; 1967 - , active duty US Navy - Naval Data Automation Command Code OOH; Natl Acad of Engg, 1973; 1971-78 Distinguished Fellow British Computer Soc 1973; McDowell Award IEEE 1979, Legion of Merit 1973; Prof Lecturer in Mgnt Scis George Wash Univ. Has received over 20 Hon Doctorates, including South Coll, Univ PA, Clarkson Univ; Author of numerous articles on computer sci. *Society Aff:* IEEE, DPMA, ACM, AAAS

Hopper, Jack R
Home: 3590 Crestwood, Beaumont, TX 77706
Position: Prof & Hd Chem Engr Dept. *Employer:* Lamar Univ. *Education:* PhD/Chem Engg/LA State Univ; MChE/Chem Engr/Univ of DE; BS/Chem Engr/TX A&M Univ. *Born:* 5/12/37. Jack R Hopper is Prof & Hd of the Chem Engr Dept at Lamar Univ. He is past (1978-80) Chrmn of Group I of the Natl Programming Committee (NPC) of the Am Inst of Chem Engrs (AIChE) and also served as VChrmn of Group I (1976-78) & Chrmn (1974-76) & VChrmn of Area 1b of the NPC of AIChE & Chrmn of the 1986 National AIChE & Petro Chem Mtg in New Orleans. He served as Jr Res Engr (1959-60) & Asst Res Engr (1960-61) with Humble Oil & Refining Co R&D; & Res Engr with Exxon Res & Engr Co from 1964-1967. He came to Lamar Univ in 1969 as an Asst Prof & served as Assoc Prof (1972-75); Prof (1975-) and Acting Hd of the Chem Engg Dept (1974) before assuming the position as Hd in 1975. He holds four U.S. patents and some twenty-eight publications. He received the ASEE Gulf- Southwest Section Dow Outstanding Young Faculty Award in 1971 and was named a Fellow of AIChE in 1985. His primary res interests are in catalysis, kinetics, reactor modeling and hazardous materials research. *Society Aff:* AIChE, ACS, APCA, Catalysis Society, ASEE

Hoppmann, William H, II
Business: Univ of S Carolina, Columbia, SC 29208
Position: Professor Emeritus. *Employer:* Cons US Army. *Education:* PhD/Applied Mechs/Columbia Univ; MA/Math/George Wash Univ; BS/Physics & Math/College of Charleston, SC. *Born:* Sept 1908. Native South Carolinian. ASME Fellow. Prof of mech & engrg at JHU, RPI & USC; dev vibration & fluid mech labs at JHU & RPI; same for bioengrg at USC. 7 pats on res equip. Pub over 100 papers in sci & engrg journals. Presented invited lectures at major univs & prof socs. Patent cons in indus. Reviewer for tech journals & government sci contract proposals. Presently cons for US Army Aberdeen Proving Ground. Dev comprehensive tests on generalized modeling 1929-47: Model tester, naval architect, matls engr, mech engr & consultant in applied mechs for US Navy. *Society Aff:* ASME, IITΣ.

Hora, Michael E
673 Banbury, Bolingbrook, IL 60439
Position: VP *Employer:* A T Kearney. *Education:* MBA/Finance/Loyola Univ; BS/Engrg/Univ of IL. *Born:* 5/14/43. Oak Park, IL. Served in line mgt positions in the foundry industry, machining industry, chem processing and high-speed packaging. Was Gen Supervisor of Technical Services with Armour-Dial, Inc before joining A T Kearney. Responsibilities include establishing policy for the Plant Engg and Maintenance Mgt Consulting Group. Active involvement in gen Mfg Services Consulting, managing and directing consultants in client situations and developing business for the Firm. Maj responsibility includes maintaining high stds of excellence and professionalism. Is an avid jogger and backpacker. *Society Aff:* ASME.

Hord, William E
Business: SIU-Edwardsville Box 65, Edwardsville, IL 62026
Position: Chairman & Professor. *Employer:* Southern Ill Univ at Edwardsville. *Education:* PhD/Elec Eng/Univ of MO-Rolla; MS/Elec Engr/Univ of MO-Rolla; BS/Elec Engr/Univ of MO-Rolla. *Born:* Oct 17, 1938. Native of Warsaw Illinois. Engr for Sperry Gyroscope Co Great Neck N Y, involved in design & test of high power klystrons; Sr Engrg Specialist for Emerson Electric Co St Louis Mo, involved in design of antennas for airborne phased array radar; with Southern Illinois Univ at Edwardsville since 1969; respon for the direction of engrg & tech progs since 1974. Has served as cons to Emerson Electric Co & Monsanto Res Corp Dayton Ohio. *Society Aff:* IEEE, ASEE, ISPE.

Horie, Yasuyuki
Business: Dept of Civil Engrg, P.O. Box 5993, Raleigh, NC 27650
Position: Prof *Employer:* NC State Univ *Education:* PhD/Physics/WA State Univ; MS/EE/Yale Univ; BA/Physics/Intl Christian Univ *Born:* 7/6/37 Native of Tokyo, Japan. After the completion of graduate degrees in the US, Spent two years in Scotland and England as a visiting lecturer in math depts. Joined the NCSU in '69. Worked for Physics Intl Stanford Research Inst. Physical scientist with the US Army Research Office for '79 and '80, evaluating R&D policy and program in the div of engrg science. Specialize in Non-linear wave propagation, thermodynamics of shock propergation, defects in solids, and material synethesis under high-pressures. *Society Aff:* APS

Horlock, John
Business: Walton Hall, Walton, Milton Keynes, MK7 6AA UK
Position: Vice-Chancellor *Employer:* Open University *Education:* ScD 1976/Cambridge University, U.K.; PhD 1955; MA 1955; BA 1st class honours/Mechanical Sciences/Tripos; *Born:* 04/19/28 Dr. Horlock is Vice-Chancellor of the UK Open Univ. After graduating in Mech. Engrg. at Cambridge, he worked with Rolls Royce, returning to Cambridge as a Research Fellow and Lecturer. He subsequently held Engineering Chairs at Liverpool and Cambridge Universities, and Visiting Chairs at MIT and Pennsylvania State. He holds a number of Fellowships and has served on government committees. He is a Fellow and former V.P. of the Royal Society, he was formerly Chairman of the Aeronautical Research Cttee. and is currently Chairman of the Advisory Cttee. on the Safety of Nuclear Installations. He became Vice-Chancellor of Salford Univ. in 1974, moving to the Open Univ. in 1981. *Society Aff:* RS, Eng, I Mech E, R Ae S, ASME

Horn, Alvah J
Home: 34 Lloyden Dr, Atherton, CA 94025
Position: Petro Cons & Cons Prof. *Employer:* Self. *Education:* MSE '43/Aerological Engrg/US Naval Acad Annapolis, MD; AB '39/Chem/Stanford Univ. *Born:* Aug 1917 New Orleans, raised in CA oilfields. Worked for Standard Oil Co of CA; comm in Navy 1942. Ret to Standard 1946; Asst Div Engr 1952 spec in drilling &

Horn, Alvah J (Continued)
prod problems; Ch Div Engr 1962, then Ch Engr; in 1965 appted Asst Gen Mgr; Chmn, Training Comm, Chmn Mgmt Comm, Labor Contract; managed Standard's drilling & prod activities Alaska; retired Jan 1973 - presently cons in petro engrg, drilling & property mgmt. Mbr Bd of Dir: TRANSCON Inc 1973- . Midway Premier Oil Co 1980- , Murphy Energy Corp 1981- , Consltg Prof Sch of Earth Scis Stanford Univ 1975- . Hon Mbr 1968 PET, Hon Chrmn LA Reg Mtg 1968 SPE/AIME; Tau Beta Pi; Phi Lambda Upsilon; Life Mbr, Sch of Earth Sci's Stanford. Dir Reg I 1967-70 SPE; Dir 1971-72 AIME, V P 1973-74 AIME. Disting Serv Awd 1973 SPE/AIME. Disting Lecturer 1973-74 SPE/AIME. Managing Partner, AS Horn, Operator, Oil Producer 1983- . Distinguished Mbr, SPE, 1983- . *Society Aff:* SPE, AAPG, NCSEA.

Horn, John W
Home: 3612 Anclote Pl, Raleigh, NC 27607
Position: Prof of Engrg *Employer:* NC State Univ *Education:* MS/CE/MIT; BS/CE/WV Univ *Born:* 8/6/29 Native of Harpers Ferry, WV. Engr with AK Railroad. 1st Lt US Air Force 1952-54. Instructor at MIT. Assumed current position in 1956. Consulting engr 1956-66. Principal and chrmn of Bd with Kimley-Horn & Assocs Inc, consulting engrs, US, 1966 to 1982. Principal and Sr Consult 1982-date. Pres NC Section ASCE 1962-64. Dir Dist 6 ASCE 1964-66. Author of 20 res publ in transp engr. *Society Aff:* ASCE, NSPE, APWA, ITE, ASEE, ΣΞ

Horn, Russell E
Business: 40 S Richland Ave, PO Box M-55, York, PA 17405
Position: Chairman of the Board. *Employer:* Buchart-Horn Inc. *Education:* BS/CE/PA State Univ. *Born:* May 1912. Served in WWII with US Army Corps of Engrs 1941-45; retired as Col 1963. Co-founder Buchart Engrg 1945; Pres Buchart Engrg 1959-61; Pres Buchart-Horn Inc 1961-72; Chmn of the Bd since 1972. Pres PACE Resources Inc since 1970. Chmn Yorktowne Assoc Contractors Inc. Previously Natl Pres ASHE. Pres C E C of Penna & Pres Penn State Alumni Club York County. Currently Dir White Rose Motor Club AAA; Past Chmn and presently Dir of York County Chap Amer Natl Red Cross; Dir of Auto Club of Southern Pennsylvania. *Society Aff:* NSPE, ACEC, ASHE, ASCE.

Hornbeck, John A
Business: 111 Driftwood Pl, St Simons Island, GA 31522
Position: VP - Retired. *Employer:* Bell Labs. *Education:* PhD/Physics/MA Inst of Tech; AB/Physics/Oberlin. *Born:* 11/4/18. Native of Northfield Minn. Began career 1946 with Bell Labs as res physicist; 1962 named Pres & Dir of Bellcomm, Inc. formed to aid NASA in Apollo manned lunar-landing prog; 1966 elected V Pres of Western Electric Co & Pres & Dir of Sandia Corp, a subsidiary of Western Elec Co., operating labs in New Mexico and California for the Energy R/D Admin. 1972 elected VP of BTL. Fellow IEEE, Fellow Amer Physical Soc, Mbr Natl Acad of Engrg & Amer Assn for Advancement of Sci, former mbr Naval Studies Bd of the Natl Acad of Sci, also, Aerospace Safety Advisory Bd (NASA); cons to Energy R/D Admin. Retired, Aug 1979. *Society Aff:* APS, IEEE, ΦBK, ΣΞ, ΔΣP, AAAS.

Hornberger, Walter H
Home: 1804 Platt St, Niles, MI 49120
Position: Quality Control Eng. *Employer:* Wells Electronics, Inc. *Education:* BS/Met Engg/IIT. *Born:* 12/13/39. Originally from Chicago, IL. Attended Univ of IL, grad from IIT. Worked in Chicago area steel mills as a technician prior to grad & subsequently Mill Met at WI Steels Basic Oxygen & Strand Cast Dept Intl Harvester. Joined Natl-Std in 1967 as an R&D Met. Became heat treat mgr-met at Rockwell Intls, Automotive Products Div, in Allegan, MI in 1973. Returned to Natl-Std in 1974. Received PE License - MI 1974. Served as chapter chrmn - Notre Dame Chapter ASM 1978/79. Pres of Optimist Club of Niles 1978-79. Pres of MSPE. Blossomland Chpt 1983-84. Serving as MSPE State Dir Rep Blossomland Chpt 1984-85. Pres N.S. Credit Union 1981-87. Served on Salvation Army Advisory Board 1980-87. *Society Aff:* ASM Int'l, NSPE, MSPE.

Hornbogen, Erhard
Business: Institut fuer Werkstoffe, Ruhr University, D-4630 Bochum 1, Germany
Position: Prof Dr - Ing. *Employer:* Ruhr-Univ Bochum Inst fur Werkstoffe. *Education:* Dr/Phd Met/Univ Clausthal *Born:* Feb 2, 1930 Greiz (now GDR). Studied physical metallurgy at Univs Stuttgart & Clausthal; 1956 PhD on 'Shape memory effect due to martensitic transformation of beta-brass'. 1958-62 Res Engr US Steel Res Centre Monroeville PA; 1963-65 Max-Planck-Inst fur Metallforschung Stuttgart; 1965-68 Prof (metal physics) Univ of Gottingen; 1968- Prof (materials sci & engrg) Ruhr-Univ Bochum; 1971-72 Dean Dept of Mech & Civil Engrg. 1962 and 1963 Grossmann Award (ASM); 1964 Masing Preis (German Met Soc); 1976 Fellow (ASM); 1979 R F Mehl Award; Inst of metals Lecturer (AIME); 1984 Medaille Reaumur (Soc Franc de Met). *Society Aff:* ASM, DPG, DGM.

Hornburg, Charles D
Business: 1850 Northwest 69th Ave, Ft Lauderdale, FL 33313
Position: President. *Employer:* DSS Engrs Inc. *Education:* BSME/Heat Tranfer/Case Inst of Technology. *Born:* June 1931. Reg P E Florida & Ohio. Mbr ASME. 31 yrs exp in engrg, proj mgmt, design, const & opers; since 1968 Pres & majority stockholder of DSS Engrs Inc Ft Lauderdale Fla, a cons engrg frim spec in desalination, elec generation, geothermal resource utilization, solar energy, water supply & wastewater treatment. Previously Contract Mgr of Foster Wheeler Corps Condenser Dept & Proj Mgr for design & const of a 17 MGD Flash Desalting Test Module; Supr of Engrg at Maxim Div, AMF Inc & Proj Mgr on marine distilling plants of Griscom-Russell Co. *Society Aff:* ASME IDEA, WSIA.

Horne, Charles F, Jr
Business: PO Box 2507, Pomona, CA 91766
Position: Consultant. *Employer:* *. *Education:* MS/Electonics & Communications/Harvard Univ; BS/Electronics & Communications/US Naval Acad. *Born:* 1/3/06. in NY City. US Navy 1926-51, retired Rear Admiral. Administrator CAA 1951- 53. Div Mgr & Pres Genl Dynamics Pomona Div 1953-71; VP Genl Dynamics Corp 1961- 71. Present: Cons; Pres Industry-Education Councils of Amer; Mbr Economic Dev Council Los Angeles County; Chmn Tech Use Task Force, encouraging indus to use new tech, Los Angeles Area Chamber of Commerce; Mbr Manpower Adv Council Los Angeles County; Chrmn, Los Angeles County Private Industry Council. Awards: CAA Natl Regional Medal 1952, NMA Gold Knight 1962, EIA Medal of Honor 1964 (Hon Dir, P Pres), CACVE Governor's Award 1975. Fellow: IEEE, IAE. Assoc Fellow: AIAA. *Society Aff:* IEEE, AIAA, IAE, EIA.

Horner, John F
Home: 1332 Dartmouth Rd, Flossmoor, IL 60422
Position: VP. *Employer:* Amoco Oil Co. *Education:* PhD/CE/Purdue Univ; BS/CE/Purdue Univ. *Born:* 6/6/25. Married (Mary Rose). Has two daughters (Marian Christine Bliwas; Terry Lynne Horner). Began corporate career as Assoc Chem Engr in 1949 at the TX City refinery; rose progressively to current position (VP - Refining & Engg) held since Dec 1, 1971, in which position he also serves as a mbr of the Mgt Committee. Civic activities include mbr of the Bds of Dirs of Jr Achievement of Chicago; Jr Achievement Inc (natl organization); & IIT/IIT Res Inst. *Society Aff:* AIChE, API, NPRA.

Horning, John C
Business: 1 River Rd Bldg 36-229, Schenectady, NY 12345
Position: Mgr-Engrg RECO. *Employer:* General Electric Co. *Education:* BSME/Mech Engr Ht Power/Case Inst. *Born:* 12/19/24. Mech Engr from Case Inst of Tech; BSME 1948. Joined Genl Elec upon graduation & since then as had various functional & managerial assignments in engrg, mfg, marketing & plant const opers; was appointed to current assignment as Mgr-Engrg Real Estate & Const Oper 1972. P Dir of Engrg Soc of Cincinnati & currently a mbr of the Bldg Res Adv Bd and Federal Construction Council of the Natl Res Council - Natl Acad of Scis.

Hornsby, Clarence H, Jr
Business: PO Box 7, Catawba, SC 29704
Position: Pres & GM; Also: V. Pres & Gen Mgr Catawba Newprint Co. *Employer:* Bowater Carolina Co *Education:* BME/Mech Engg/Auburn Univ; -/Advan Mgmt Program/Harvard U Grad Sch of Bus. *Born:* 4/8/26. Native of Dothan, AL. Joined Int Paper in 1950 and held several engg and management position with IP before joining Bowater in 1961. Has been Chief Engr and held other management jobs including VP and Mill Mgr before becoming Pres & Gen Mgr in 1978. Also serves as VP & Gen Mgr of Catawba Newsprint Co, Dir of: First Union - Rock Hill, VP TAPPI, Vice Chrmn SC State Bd for Tech Educ. AL Reg 4503, AR Reg 1717. *Society Aff:* NSPE, TAPPI, SCSPE, CPPA.

Horowitz, Carl
Business: 2186 Mill Avenue, Brooklyn, NY 11234
Position: President. *Employer:* Polymer Res Corp of America. *Education:* BS/Chem Eng/Columbia Univ; MS/Chemistry/Polytechnic Inst of NY; PhD/Chemistry/Polytechnic Inst of NY. *Born:* 8/10/23. With Yardney Electric Corp 1951-63 advanced from Proj Engr to pos of V Pres; in 1963 started a successful co: Polymer Res Corp of Amer, has been its pres since the beginning; the co, now a public corp, spec in R&D & chem mfg. Received 2 Vander Vaaler Awards: one 1964 & the other 1972. Rec'd 1979 Award of MERIT from Pollution Eng Magazine. Enjoys classical music & physical activities *Society Aff:* AIChE, ACS, AATCC.

Horowitz, Ellis
Home: 1647 Campus Rd, Los Angeles, CA 90041
Position: Assoc Prof EE & Computer Sci. *Employer:* Univ of Southern Calif. *Born:* Feb 1944. PhD & MS Univ of Wisconsin Madison, BS Brooklyn College. Employed by IBM in summers of 1965/66/67; Instructor in computer sci at Univ Wisconsin 1969-70; Asst Prof in computer sci at Cornell Univ 1970-73; Assoc Prof of computer sci & elec engrg at Univ of Southern Calif 1973- . Mbr: IEEE, ACM, SIAM, AAAS & Sigmi Xi. Author of Fundamentals of Data Structs (Computer Sci Press) & Practical Strategies for Dev Large Software Sys (AddisonWesley) & numerous articles.

Horowitz, George F
Home: 4911 Lockhaven Ave, Los Angeles, CA 90041
Position: Principal Engr *Employer:* The Metropolitan Water District of S CA *Education:* MS/CE/Univ of So CA; BS/CE/Univ of CA-Berkeley *Born:* 5/19/24 in Vienna, Austria, spent 10 years in China before coming to the US in 1949. Formerly active in municipal engrg, the oil industry, and the private power industry, currently in charge of hydraulic engrg design and the safety of dams and reservoirs for the Metropolitan Water District of Southern CA. Author, or co-author of various technical papers and articles, pioneered the installation of mini-hydroelectric plants on domestic water supply distribution systems. Past chrmn, Hydraulics Technical Group, Professional Conduct Committee, and Professional Practice Committee of the Los Angeles Section, ASCE. Member, USCOLD Cttee on Earthquakes. *Society Aff:* ASCE, USCOLD, USCID, EERI, ΣΞ, AWWA

Horowitz, Isaac M
Business: CB 425, Univ of CO, Boulder, CO 80309
Position: Professor. *Employer:* Weizmann Inst of Science. *Education:* DEE/Systems/Polytech Inst of Brooklyn; MEE/Systems/Polytech Inst of Brooklyn; SB/Elec Engrg/MIT. *Born:* Dec 15, 1920 Safed Israel. Emigrated to Canada 1925, USA 1951, Israel 1969. BSc MIT 1948, MEE 1953, DEE 1956. Asst Prof 1956-58 Polytech Inst of Brooklyn; Volunteer Israel Army 1948-50; Hughes Res Labs 1958-64, Mbr of Sr Staff; Sr Scientist Guidance & Controls 1964-66; Prof CCNY 1966-67; Joint appointment at Univ of CO and Weizmann Inst (Cohen Chair) 1967-69. Winner 1956 Natl Electronics Conf Best Paper Award. m. Chana Shankman 1945; c. Chaya, Ruth, David, Dafna. *Society Aff:* IEEE.

Horowitz, Stanley H
Home: 17 Sherwood Dr, Plainview, NY 11803
Position: Asst Div Mgr-Elec Engrg Div *Employer:* Am Electric Power. *Education:* BEE/Elec Power/CCNY. *Born:* 8/5/25. Born & raised in NYC & a grad of the CCNY. Served as Navigator in USAF during WWII. Employed by Am Elec Power Service Corp as a relay engr, Hd of Relay Sec & presently Asst Div Mgr-Elec Engrgs Div. Author of over a dozen papers on system protection & control. Elected a Fellow of IEEE in 1979, past chrmn of IEEE Power System Relaying Committee & presently intl chrmn of CIGRE study committee on protection & control. *Society Aff:* IEEE, CIGRE.

Horrigan, Robert V
Home: 6941 Country Lakes Circle, Sarasota, FL 34243
Position: Consultant *Employer:* Transelco, Div of Ferre Corp. *Education:* BS/Chem Engg/MIT; MS/Chem Engg/MIT; PhD/Chem Engg/Yale Univ. *Born:* June 8, 1924. s. Thomas H & Viola M Horrigan. m. Marion F Phillips July 1, 1950. Children: Robert A, Stephen P, Paul M. Assoc Engr Brookhaven Natl Lab Upton N Y 1950-52; Res Investigator, TAM Div, N L Indus 1952-55, then Mgr, Dev Div 1955-60. Pres, Transelco Inc, Penn Yan N Y 1961-1976, producing specialty ceramic raw matls for electronic & surface finishing indus. USNR 1943-46 Pacific Theater. Prof Engr NY State & FL. Mbr AIChE, Amer Chem Soc, Amer Ceramic Soc; Sigma Xi, Tau Beta Pi. Mbr Yale Club of N Y City. Pats in field of applications of transition elements. Enjoy sailing & fishing. Retired as Genl Mgr, Transelco Div of Ferro Corp on July 1, 1979; now Consultant, Ferro Corp. *Society Aff:* ACS, AIChE.

Horsefield, David R
Home: 13554 No Lakewood Dr, Mequon, WI 53092
Position: Consulting Engineer. *Employer:* Self-employed *Education:* MSE/Sanitary Engg/Univ of MI; BS/Civil Engg/Univ of MA. *Born:* 1/24/31. 30 yrs exper in environ engg, in wastewater, water & hazardous waste mgmt. Joined CDM in 1960 as engr. Became VP in 1970, partner of the firm in 1972, Sr Vp & Mbr of the Bd of Dirs in 1972. Received the BSCE Desmond Fitzgerald Medal and Sanitary Section Award 1968-69; Univ of MA Engg Alumni Assn Award 1976. Dipl & Past Trustee-AAEE. Trustee at Large, The Amer Acad of Environmental Engrs 1976-1978; Bd of Control WPCF 1980-82; Mbr of the Bd of Dirs; Consulting Engrs Council of Wisconsin 1980-82. Self-employed Consulting Engr 1982-present. Bd of Dir Warzyn Engrg, Inc. 1984. Prof Engr in 19 states. *Society Aff:* WPCF, AWWA, ASCE, NSPE, ACEC, SAME

Horst, Neal A
Business: 55 S Richland Ave, PO Box M-55, York, PA 17405
Position: Vice President *Employer:* Buchart-Horn, Inc. *Education:* BS/CE/Lehigh Univ. *Born:* 12/20/41. Native of Lebanon, PA. Served as commissioned officer with the US Coast and Geodetic Survey, 1963-1965. Engr for Gannett Fleming Corddry & Carpenter, Inc in Dam Sec, Hydraulic Div. With Buchart-Horn, Inc since 1969 with assignments as Mgr, Harrisburg, PA office and Dir, Environmental Planning Div. Assumed responsibilities for the firm's business dev in 1977. Elected Vice President in 1980. Currently responsible for Environmental Group including Environmental, Chemistry & Earth Sciences, Construction Management, Rate Analysis/Valuation Divisions & B-H Laboratories. Pres, Central PA Sec ASCE, 1979-1980. Secy/Treas, Dist 4 Council ASCE, 1976-1979. *Society Aff:* ASCE, NSPE/PEPP, WPCF, AAEE, APWA, ACEC, SAME.

Horton, Billy M
Business: Univ Circle, Cleveland, OH 44106
Position: Prof *Employer:* Case Western Reserve Univ *Education:* MS/Physics/Univ of MD; BA/Physics/Univ of TX *Born:* 12/27/18 Native of Bartlett, TX, s. Walter H. and E. Loraine (Mitchusson) H., m. Hattie Grace Schultz 1941, c. Phillip E. and Stephen D., Instructor A.F. Technical School, Chanute Field, IL. 1941-42; Radar Officer, U.S. Army, 1942-46; Naval Research Lab, Wash., DC 1946-51; National Bureau of Standards; 1951-53; Harry Diamond Labs, 1953-74, Technical Dir 1961-74; Prof Mechanical and Aerospace Engrg, Case Western Reserve Univ, 1975-.

Horton, Billy M (Continued)
Awards: Arnold O. Beckman 1960; U.S. Army Exceptional Service 1965; John Scott 1966; Dept. of Defense Distinguished Service 1967; PTC Inst Inventor of the Year 1971; Control Systems, Noise, Friction, Foldability of Structures, Inventor of Fluid Amplification. *Society Aff:* NAE, IEEE, AAAS

Horton, Thomas E, Jr
Home: 209 St. Andrews Cr, Oxford, MS 38655
Position: Prof *Employer:* Univ of MS *Education:* PhD/Eng Mech/Univ of TX; MS/ME/Stanford Univ; BS/ME/Univ of TX *Born:* 1/12/35 Westinghouse Scholarship at Univ of TX and Caterpillar Research Fellowship at Stanford. Mechanical Engr developing offshore technology in 1958 & 59 at Shell Development Co. Teaching and research positions in fluid mechanics at the Univ of TX from 1960 to 63. Senior Research Engr engaged in planetary entry at the Jet Propulsion Lab of Cal Tech from 1963-66. Assoc Prof from 1966-71 and Prof of mechanical Engrg from 1971 to present at the Univ of MS also Research Engr investigating unsteady fluid mechanics, thermophysics, and high energy laser technology. First Dir of Laser Science Division of the U.S. Army High Energy Laser Lab 1975-76. Director of Reiton Corp. of Houston, Texas. *Society Aff:* ASME, AIAA, APS.

Horton, Thomas R
Home: 165 Soundview Ave, White Plains, NY 10606
Position: Pres & CEO *Employer:* American Management Assoc *Education:* PhD/Math/Univ of FL; MS/Math/Univ of FL; BS/Math/Stetson Univ; LLD/Pace Univ; DHL/Univ of Charleston; DHum/Stetson Univ. *Born:* 11/17/26. Native of Ft Pierce, FL. Taught Math at the Bolles Sch, Jacksonville; served US Army 1944-66. With IBM 1954-1982 as systems mgr, IBM dir of systems and application engg, divisional gen mgr and divisional vp. Pres & CEO, Amer Mgmt Assoc since 1982. Dir, Mastery Education Corp, Waterboro, MA & Perrigo Co, Allegan, MI; Visiting Committee to Linguistics and Philosophy, MIT; Editor and author. Edited *Traffic Control: Theory and Instrumentation*, Plenum Press (1965); author, *What Works for* Me, Random House (1986). *Society Aff:* AAHE, AAAS.

Horvath, Csaba
Business: Mason Lab, PO Box 2159, Yale Station, New Haven, CT 06520
Position: Prof in Chem Engg. *Employer:* Yale Univ. *Education:* PhD/Physical Chemistry/Univ of Frankfurt (M), W Germany; DiplIng/Chem Engg/Techn Univ, Budapest, Hungary. *Born:* 1/25/30. in Szolnok, Hungary. After graduation in 1952 served on the faculty of the Techn Univ, Budapest. From 1956 to 1960 with Farberke Hoechst AG in Frankfurt (M) W Germany. Received doctorate in 1963, immigrated to USA and was researcher at the medical schools of Harvard & Yale. Appointed to Assoc Prof at Yale in 1970. Currently Prof of Chemical Engineering at Yale. Teaches courses on biochemical engg and separation processes. Research interest: chromatography, biological reactors and biochemical separations. Consultant. Over 180 scientific publications and patents. *Society Aff:* AIChE, ACS, IFT, DECHEMA, AAAS, NYAS.

Horvay, Gabriel
Business: Amherst, MA 01003
Position: Professor Emeritus of Engineering *Employer:* University of Massachusetts *Education:* PhD/Physics/Columbia Univ; EE/EE/Columbia Univ; BS/EE/NY Univ. *Born:* 1908 Budapest, Hungary. Indus res: David Taylor Model Basin 1942-44, McDonnell Aircraft 194447, GE 1947-68 (Knolls Atomic Power Lab, R&D Ctr: stress analysis, heat transfer, materials science 47 applied math). Teaching: Talladega Coll, Ala 1935-39, Univ of Cincinnati 1940-42, Adj Prof Mechs RPI 1963-68; Prof of Engrg Univ of Mass since 1968; Emeritus 1978. Visiting Professor: National University of Mexico 1980; Fulbright Lecturer, Leningrad Polytechnic Institute 1981. Best paper awards: AIChE-ASME Heat Transfer Divs 1964; ASME Mechanism Div 1974. Fellow ASME; Mbr SIAM, AIME. Mbr PVRC Dynamic Analysis Sect. Cons for indus co's. *Society Aff:* ASME, AIME, SIAM.

Horzella, Theodore I
Home: 2404 Moreno Dr, Los Angeles, CA 90039
Position: VP, Dir *Employer:* Enviro Energy Corp - Burbank. *Education:* MS/Chem Engrg/IA State Univ; BS/Chemistry/Univ Santa Maria. *Born:* July 1927. Sigma Xi. Business Admin UCLA, A Hamilton Inst. Languages: Spanish, German, French. Mbr: AMA, AIChE, SAM. Papers: Agri Food Chem, Scientia, Chem Engrg Progress, Pulp & Paper, Chem Engrg. Tech & mktg mgmt US, Mexico, Europe & Australia. Business planning, marketing, merger analysis, training programs for mktg orgs. Estab & managed manufacturers' rep org. Negotiated foreign lic's & managed opers in Europe, Japan, Australia. Prod mgmt, P/L control, air pollution control sys, process heat exchanges, matl handling. Application engrg matls handling for coal preparation, fertilizer, chem & indus installations. *Society Aff:* AIChE, ΣΞ.

Horzewski, Jerome C
Home: 4969 S 15th St, Milwaukee, WI 53221
Position: General Supervisor, Met. Dept. *Employer:* Ladish Co - Cudahy Wisc. *Born:* 4/1/33. Native of Milwaukee. Served with Army Corps of Engrs 1953-55. With Ladish Co since 1951 in various pos incl Gen Supr of Process Control, Asst. Dir of Metallurgy Fittings Div, Gen Supvr, Metallurgical Dept and currently statistical process control specialist, quality dept. Ferrous Met Diploma from MEI 1969. Indus Mgmt from Lincoln Ext Inst in 1972. Attended The Inst for Productivity through Quality - Univ of TN, 1984. Active in ASM-Milwaukee Chap. Past Chmn, Natl MEI Ctte. Mbr Natl Eagle Scout Assn. Mbr ASQC. Enjoys camping, fishing & boating. *Society Aff:* ASM, ASQC, NESA.

Hoshide, Henry S
Business: 1150 S King St 800, Honolulu, HI 96814
Position: Director of Civil Engineering Dept. *Employer:* Wilson Okamoto & Associates, Inc. *Education:* MS/Civil Engg/Univ of CA, Berkeley; BS/Civil Engg/IA State Univ. *Born:* Dec 1932. Native of Lihue, Hawaii & served in US Army from 1953-55. With Fed Hwy Admin from 1958-66, leaving as Region of Hydraulics & Drainage Engr. Duties incl instruction, design review & special cons to const proj & problem areas. With Wilson Okamoto & Assocs 1968- . Respon for Civil Dept's & co's overall operations efficiency, incl preparation of proposals & client contact. Enjoy music, reading & church related work. *Society Aff:* ASCE, AWWA, ACEC.

Hoskins, Charles D
Business: 740 New Circle Road, Lexington, KY 40507
Position: Senior Staff Assistant. *Employer:* International Business Machines. *Born:* July 27, 1920 Oswego N Y. BS RPI 1948; Student Syracuse University 1950; Univ of Kentucky Grad Sch 1961-62. Asst Plant Metallurgist, Curtiss Wright Corp Caldwell N J 1948-50; Plant Metallurgist Ingersol-Rand Co Athens Pa 1950-51; Ch Metallurgist Chrysler Corp Newark Del 1951-56; Staff Metallurgist IBM 1956-66, Dev Metallurgist 1966-72, Sr Staff Asst 1972- . Reg P E Del, Ky. Teacher Chrysler Corp 1953-56, IBM 1957-58. Mbr SAE, ASM (35 yr mbr & served on various cttes), Amer Ordnance Assn & secondary sch rep of Rensselaer Polytechnic Inst. Mbr of the Amer Legion, Knights of Columbus 4th degree, BPO Elks & Hon Order of Kentucky Colonels. Received the Award of Merit given by ASM 1976.

Hoskins, Harry D, Jr
Home: 5428 St. Charles Ave, New Orleans, LA 70115
Position: Sr. VP *Employer:* Walk, Haydel & Assocs, Inc *Education:* EM/Mining/CO School of Mines *Born:* 4/8/14 Native of Denver, CO. Worked for the DuPont Corp prior to World War II. Integrated into the Regular Army and served from 1942 to 1963. Command assignments ranged from a Platoon to Joint Forces Command, Services Schools thru the Armed Forces Industrial Coll. Joined Gulf Oil Co's Central Engrg Office in 1963. Joined Walk, Haydel & Assocs, Inc in 1965. Currently serving as Senior VP of Long Range Planning which includes responsibility for corporate expansion, diversification and overseas operations. Enjoys racing sail boats and flying sail planes. *Society Aff:* AIME, AIME, NSPE, SAME

Hoskins, John R
Business: College of Mines, Dept Mining Engg & Met, Moscow, ID 83843
Position: Hd of Dept. *Employer:* Univ of ID. *Education:* PhD/Mining Engr/Univ of UT; BS/Mining Engr/Univ of ID. *Born:* 6/9/19. Eight yrs experience as mining engr and supervisor in surface and underground metal and non-metal operations in USA, Mexico & AK. Five yrs with US Bureau of Mines Res; Twenty-five yrs teaching mining engg. Consultant on production, operations and Health & Safety. Res interests are costs, production and operations. Past AIME AK Sec Chrmn, SME Education Bd Chrmn, SME Accreditations Committee Chrmn, SME/AIME Rep on ECPD Bd of Dirs, past EEIA Committee Mbr; past ASEE Minerals Div Chrmn. Trustee Northwest Mining Assn. *Society Aff:* AIME.

Hosler, Darrell M
Business: 8700 Indian Creek Pkwy, PO Box 25548, Overland Park, KS 66225
Position: VP, Admin *Employer:* Profl Services Div Armco Inc. *Education:* BS/ME/KS State Univ *Born:* 10/31/36 Born in Beloit, KS. With Burns & McDonnell in Kansas City since 1959. Presently assigned to parent co, Armco Inc, as the Profl Services Div's VP Admin, responsible for all administrative matters, and marketing and operating assistance. From 1959-1963 design engr for central heating and cooling systems and aircraft hydrant refueling sys. Proj mgr for large ind and aviation projs and the dev of new business, 1963-1979. Elected VP in 1979 responsible for directing domestic and intl marketing. Reg PE in five states. Chrmn, Kansas City Construction Ind Affairs Council 1977-1979. *Society Aff:* SMPS, SNPE.

Hosler, Ramon E
Home: 2071 Geronimo Trail, Maitland, FL 32751
Position: Prof *Employer:* Univ of Central FL *Education:* PhD/Chem Engr/Univ of IL; MS/Chem Engr/Univ of IL; BChE/Chem Engr/Univ of Dayton *Born:* 12/24/35 Born in Columbus, OH. Employed by Westinghouse Electric Corp (1961-1979) as sr engr, mgr of Lab Facilities and Tech Conslt to Naval Nuclear Powr Sch. Joined Univ of Central FL as Assoc Prof of Engrg in 1979, Prof 1985. Active in AIChE, Heat Transfer and Energy Conversion Div including Secy, Exec Ctte, Chrmn (1981), Awds Ctte Chrmn (1984). *Society Aff:* AIChE, ASME, ASEE

Hostetter, Gene H
Home: 8811 Gallant Dr, Huntington Beach, CA 92646
Position: Prof. *Employer:* Univ of CA, Irvine *Education:* PhD/Engg/Univ of CA; MS/Elec Engr/Univ of WA; BS/Elec Engr/Univ of WA. *Born:* 9/14/39. Positions with Stations DQDE, KAYO, KOL, KIRO-TV, Seattle, 1958-1967 while a lecturer at the Univ of WA. Prof of Elec Engg, CA State Univ, Long Beach 1968- 1981. Dept Chrmn 1975-1978, 1980-1981. Prof of Elec Engrg, Univ CA, Irvine since 1981. Dept Chrmn 1981-1984, Acting Dean of Engrg, 1984-present. Res interests and activity in the areas of microprocessor-based signal processing, control system design, electronic system design and electromagnetic theory. Sr mbr of IEEE, active in the Orange County Sec and in the Control Systems Soc. *Society Aff:* IEEE, ASEE, AAAS, HKN, ΣΞ.

Hotchkiss, Calvin M
Business: 1133 Ave of Americas, New York, NY 10036
Position: Coordinator of Engrg Services. *Employer:* Eastman Kodak Co. *Education:* Degre/Europ Culture/Univ of Paris; MA/Europ Lang/Univ of Rochester; BA/Pre-Med, Engrg & Humanities/Univ of Princeton. *Born:* March 25, 1921 Rochester, New York. The Hill Sch Pottstown, Pa 1939; Princeton Univ AB 1943; Univ of Rochester MA 1949; Univ of Paris 1950; N Y Inst of Finance 1965. Soc of Photographic Scientists & Engrs 1959- ; V Pres N Y Chap 1962-63. Soc of Motion Picture & TV Engrs; Chmn Nomenclature Ctte 1951-53, Program Chmn, 111th Tech Conf 1972, Asst Program Chmn 114th Tech Conf Oct 1973, Fellow 1973, Mgr N Y Sect 1972-73, SecyTreas N Y Sect 1974, Chmn N Y Sect 1975, P Chmn NY Sect 1976, Governor Natl Bd of Govs 1978-79, 1981-82, Chrmn Bd of Editors Journal of SMPTE 1980-81, 1981-82; Fdn of the Motion Picture Pioneers 1973- . Chmn Bd of Editors SMPTE Journal 1980-85. Program Chmn Natl Tech Conf SMPTE 10/84. *Society Aff:* SMPTE, SPSE, BKSYS.

Hottel, Hoyt C
Home: 27 Cambridge St, Winchester, MA 01890
Position: Prof Emer, Consultant (Self-employed) *Education:* MSc/CE/MIT; AB/Chemistry/IN Univ. *Born:* 1/15/03. On MIT faculty in Chem Engg, 1928-1968, latest as Carbon P Dubbs Prof of Chem Engg & Dir, MIT's Fuels Res Lab. WWII Hd of Fire Res Sec of Natl Defense Res Committee. Author of three books, chapters in 14 others, CA. 150 scientific and technical papers, primarily in areas of radiative transfer, combustion, ind furnaces, energy conversion. Mbr Natl Acad of Sci, Natl Acad of Engg, Am Acad of Arts & Scis. Recipient of US Medal for Merit, King's Medal (Gt Britain) 1980 Founders Award of NAE, & eight other US & foreign profl awards. Consultant in chem engg, especially fuels, furnaces, energy problems. *Society Aff:* AIChE, ACS.

Hougen, Joel O
Home: 1206 Falcon Ledge Dr, Austin, TX 78746
Position: Consultant (Self-employed) *Education:* PhD/ChE/Univ of MN; MS/ChE/Univ of MN; BS/ChE/Univ of WI *Born:* 2/26/14 Prof Emeritus, Chem Engr, Univ of TX. Actively engaged in consulting. Specialists in measurements, plant testing, process control system design. 120 publications and book, Measurements and Control Applications. ISA (1979). Born Tacoma, WA. Youngest son of Norwegian Lutheran clergyman. Practiced: Pan Am Refining, Union Oil, Monsanto. Educator: Univ of MN, UCLA, Univ of IL, Rensselaer Polytechnic Inst, St. Louis Univ, Univ of TX. Consultant: Chem and petrochemical industries Registered PE, state of TX. Pioneer in application of concepts of system engrg in the chem industry. *Society Aff:* ISA, AIChE, ASEE

Hougen, Oddvar
Business: Morrison Knudson Engrs Inc, 180 Howard St, San Francisco, CA 94105
Position: Chief Engr *Employer:* Morrison-Knudson Engrs Inc. *Education:* Diploma/Civil Engrg/Norwegian Inst of Tech *Born:* 8/10/26 in Norway. Served with the Norwegian Corps of Engrs 1950-1952. Consulting engr for industrial plants, etc, 1952-1956. Design engr at Snowy Mountains Hydroelectric Authority, Australia 1956-1959. Design and planning studies for US and overseas hydroelectric projects while employed by Harza Engrg Co, Chicago, 1959-1974 and has been Project Mgr for hydro projects in US and overseas with Intl Engrg Co since 1974. Appointed Dir and Chief Engr of IECO's Water Resources Division (later MKE) in 1981. Project Dir for the Piedra del Aquila Hydroelectric Project in Argentina 1985-87. Advisor to East China Hydroelectric Power and Design Institute at Hangzhoh, PRC, for the Shuikoh Hydroelectric Project. *Society Aff:* ASCE, USCOLD, PMI

Hougen, Olaf A
Home: 110 South Henry St, Apartment 509, Madison, WI 53703
Position: Ret. (Emeritus Prof of Chem Engrg) *Education:* PhD/Chem Engg/Tech Univ of Norway; PhD/Univ of Engg/Univ WI-Madison; ChE/Chem Engg/Univ WI-Madison; BS/Chem Engg/Univ of WA. *Born:* Oct 4, 1893 Manitowoc, Wisconsin. Hon PhD Tech Univ of Norway 1960. Instr-Prof Chem Engrg Univ of Wisc 1918-63. Interim Prof at Armour Inst of Tech 1937; UCLA 1950; Norway 1951; Japan 1957. Sci Attache Amer Embassy in Stockholm 1961-63. Co-author of 5 books, 7 bulletins & 180 papers. William H Walker Award 1944; Benjamin Smith Reynolds 1955; Fulbright Prof (Norway, Japan, Taiwan, India); Founders Award AIChE 1958; Esso Award 1961; Lamme Gold Medal 1961; Warren K Lewis Award 1964; Gold Cross of St Olav-Knight First Class 1968; Natl Acad of Engrs 1974; Plenary Lecturer AIChE 1976. WI Acad of Sci, Arts and LeMerr (1974 Hon); WI Alumnis Assoc (Disting Service, 1974). 1st non-Japanese to be elected honorary mbr of the Japanese Soc of Chem Engrs (1981). *Society Aff:* AIChE, ASEE, ACS, WASAL.

Hough, Carl H
Home: 6312 E Greenlake Way N, Seattle, WA 98103
Position: Mgr, Recruiting *Employer:* The Boeing Company. *Education:* BS/Ind

Hough, Carl H (Continued)
ED/CA State College CA, PA; MA/Ind Ed/Univ of MD, College Park MD. *Born:* Nov 1922 Brownsville Pa. Served with Army Air Corps as pilot 1942-45 & AF Reserve through 1958. Was an admin & teacher in secondary ed for eleven yrs. Joined The Boeing Co in Seattle in 1957. Presently, Mgr, Recruiting having held pos of Mgr Prof Placement, Admin of Sch Relations, Corp Training Mgr, Mgr of Prof Dev, Mgr of Ed & Training & Ch of Engrg Ed. V P for Finance, ASEE 1970-74; Chmn, RWI Div ASEE 1969. Pres, Northwest Placement Assn 1976. *Society Aff:* ASEE.

Hough, Eldred W
Business: Box 90, Carrollton, IL 62016
Position: Prof & Hd. Emeritus, Dept of Petr. Engr *Employer:* MS State Univ *Education:* PhD/Physics/CA Inst of Tech; MS/Physics/CA Inst of Tech; BS/Engr Physics/Univ of IL *Born:* 1/26/16 Senior Fellow Cal Tech 1946-1949; Senior Research Engr, Stanolind Oil and Gas Co. 1949-1952; Prof of Petr Engr Univ of TX, 1952-1961; Prof and Hd., Petr Engr, MS State Univ, 1961-1965; Prof of Engr and Asst. Dean of Technology So. IL Univ 1965-1969; Prof Chem Engr and Dean, Coll of Engrg and Sci, Univ of ME, 1969-1974; Prof and Hd., Petroleum Engrg Univ of Petr and Minerals (Dhahran, Saudi Arabia) 1974-1975. Prof, Chem Engrg Univ of ME, 1975-1976; Prof and Hd, Petroleum Engrg, MS State Univ, 1976-82. I was the first Prof of Petr Engrg in MS (1961) and in Saudi Arabia (1974) Prof Emeritus and Head Emeritus, Petroleum Engrg 1982- . *Society Aff:* AIME-SPE, AIChE, ASEE, ACS

Hough, James E
Business: 3398 West Galbraith Rd, Cincinnati, OH 45239
Position: Principal *Employer:* James E. Hough & Assocs *Education:* MS/Engrg Geol/Univ of KY; BS/Engrg Geol/Univ of KY; AS/Pre-Engrg/Paducah Jr Coll *Born:* 7/25/30 Native of Paducah, KY. Served with U. S. Army (CE) 1953-55. Married, two sons. Seven years broad experience in the public sector, state and federal levels. 24 years experience as consultant soil & foundation engr and geologist. Professionally registered in a number of states. Investigator of record on hundreds of geological and geotechnical engrg projects involving foundation analysis and terrain evaluation in connection with subsurface investigations for earthen and earth-supported structures. Lecturer at the Univ of Cincinnati. Writer of books and several papers on the subjects of landslides, engrg geology, and expert testimony. Avocation fishing. *Society Aff:* ACEC, ASCE, ASFE, AEG, GSA

Houpis, Constantine H
Home: 1125 Brittany Hills Dr, Dayton, OH 45459
Position: Prof Elec Engrg *Employer:* Wright Patterson Air Force Base *Education:* PhD/EE/Univ of WY; MS/EE/Univ of IL; BS/EE/Univ of IL *Born:* 6/16/22 Native of Lowell, MA. Served in US Army. Development Engr, Babcock & Wilcox, Alliance, OH. Instructor Elec Engr Wayne State Univ. Principal Elec Engr, Battelle Memorial Inst, Columbus, O. Prof of Elec Engrg, Air Force Inst of Tech since 1951. Guest Lecturer, National Technical Univ of Athens, Univ of Patras, Weizmann Inst of Science. Recipient Outstanding Engr award, Dayton area National Engr week 1962. Author (with J. J. D'Azzo) *Feedback Control System Analysis and Synthesis*, 1960, 2nd Ed 1966; *Linear Control System Analysis & Design*, 1975, 2nd Ed 1981, 3rd Ed 1988. Numerous articles in technical journals in the U.S. and Europe. *Society Aff:* IEEE, ASEE, TBΠ

Hourigan, Edward V
Business: 1220 Washington Avenue, Albany, NY 12232
Position: Deputy Ch Engr, Structures. *Employer:* N Y State Dept of Transportation; Retired. *Education:* BCE/Structures/Manhattan College. *Born:* March 1928 New York City. Served with US Army Corps of Engrs 1945-48. Grad studies RPI. Joined NYS DPW (now DOT) 1952 as Design Engr. 1961 appointed Exec Engr of NYS-NYC Metro Transportation Study & served with the initiation of the Tri-State Reg Planning Comm. Returned to Albany, N Y 1965 as Hd of Transportation Programming Sect & in 1967 assumed pos as Dir of Struct Design & Const. In addition, served as Proj Mgr for the West Side Hwy Proj in New York City 1971. to 1978 Mbr & P Pres Mohawk Hudson Sect ASCE, SAME, ITE, AASHTO & NAS. *Society Aff:* ASCE, SAME, ITE, AASHTO.

House, Hazen E
Home: 3529 Iskagna Drive, Knoxville, TN 37919
Position: Asst Ch Elec Engr (ret). *Employer:* Aluminum Co of America - Pgh. *Education:* SM/Elec Engg/MIT; SB/Elec Engg/MIT. *Born:* Sept 1906 Knoxville, Tenn. Began work with ALCOA summer 1928. Full time 1929 on Calderwood Hydroelectric Prof until completion 1931. 1937 Reg P E: Tenn. 1947-54 Assoc Prof EE Univ Tenn (pt-time). 1954-62 Ch, EE Div ALCOA Res Labs Massena N Y. 1962-64 Sr Sr Elec Engr ALCOA Pgh Pa. 1964-71 Asst Ch Elec Engr, ALCOA Pgh. IEEE-M30, SM45, F71, LF72. 1969-70 Chmn, Transmission & Distrib Ctte, Power Engrg Soc IEEE. ASME Mbr 1945. Sigma Xi mbr 1957, Mbr Emeritus 1974. Most important work: Elec & Thermal properties of ACSR & Stranded Aluminum Elec Conductors. Tech papers pub by AIEE (now IEEE) in 1958 used as a basis by the Aluminum Assn in their indus publs. *Society Aff:* IEEE, ASME, ΣΞ.

House, Robert W
Home: 1800 Kingsbury Dr, Nashville, TN 37215
Position: Orrin Henry Ingram Distinguished Prof *Employer:* Vanderbilt Univ. *Education:* PhD/EE-Comp Sci/PA State Univ; MS/Math/OH Univ; BS/Math/OH Univ; Mgt Dev Prog/-/Harvard Bus Sch. *Born:* 5/31/27. in Wellsville, OH. US Navy 1945-46. Presidential Exec with Agency for Intl Dev 1974-75. IEEE Bd of Dirs 1973-76. Pres of IEEE Systems, Man & Cybernetics Soc 1968. Mbr of Advisory Bd of Inst of Intl Law & Economic Dev 1973-78. Fellow of IEEE. Asst Prof of EE PA State 1959. Adjunct Prof of Math OH State Univ 1962- 68. Sr Fellow Battelle Columbus 1959-70. Mgr Social & Systems Sciences Dept Battelle Columbus 1970-74. Prof of Mgt Vanderbilt Univ 1975-79. Prof of Tech & Public Policy & Dir of the Prog, Sch of Engg Vanderbilt Univ 1975-81. Assoc Dean of the Grad. School of Vanderbilt Univ. 1981-82. Dean of the Grad Sch of Vanderbilt Univ, 1982-84. Orrin Henry Ingram Dist Prof and Dir of the Mgmt of Tech Program, Sch of Engrg Vanderbilt Univ 1984- . *Society Aff:* IEEE, AAAS, CS, SMC, EM.

Housner, George W
Business: Calif Inst of Tech, Pasadena, CA 91125
Position: C F Braun Prof of Engrg. *Employer:* California Inst of Technology. *Education:* PhD/Civ Engrg/CIT; BS/Civ Engrg/Univ of MI. *Born:* Dec 1910 Saginaw, Mich. Operations Analysis Sect, 15th AF 1943-45. On staff of Caltech since 1945. Specialty: earthquake engrg res. Cons on earthquake resistant design of major engrg projs. P Pres of Internatl Assn for Earthquake Engrg; P Pres of Earthquake Engrg Res Inst; P Pres of Seismological Soc Amer. Mbr of Natl Acad of Engrg & Natl Acad of Sci. Mbr Indian Natl Sci Acad. Author of three text books & 105 tech papers. US War Dept Distinguished Civilian Service Award 1945; ASEE Vincent Bendix Res Award 1967; ASCE Von Karman Medal 1974; ASCE Newmark Madal 1982. *Society Aff:* ASCE, SSA, EERI, ASEE, AGU.

Houston, Clyde E
Home: 1012 Miller Dr, Davis, CA 95616
Position: Head *Employer:* Clyde E. Houston & Assocs *Education:* BS/CE/Univ of AZ *Born:* 1/9/14 For 17 years was Irrigation Engr with US Dept of Agriculture in Western USA. Work consisted of planning small water structures and form irrigation system design and operation. Was Snow Survey Supervisor in NV. Left post of SCS State Conservation Engr for NV to join Univ of CA, Extension Service as Irrigation and Drainage Engr. Retired as Assistant State Dir in 1970 and joined FAO of United Nations in Rome, Italy, as Chief Water Service, for six years. Presently, Consultant in Irrigation and Drainage. Worked in over 50 different countries and authored over 150 publications on irrigation and drainage. *Society Aff:* ASCE

Houston, John P
Business: 7000 S W Adams, Peoria, IL 61641
Position: Dir, Industrial Relations - Retired. *Employer:* Keystone Steel & Wire Division. *Education:* BSME/Intern Comb Engines/Purdue Univ. *Born:* Dec 1924. BSME 1948 Purdue Univ. Reg PE Ind & Ill. Native of Pekin, Ill. Served in US Navy 1943-46. Design Engr for Falls Paper & Power & Scott Paper Co 1948-52. Started Indus Engrg Dept at Keystone Steel & Wire Co in 1952 & ser as Ch Indus Engr prior to assuming pos of Dir, Indus Relations Dept. Respon incl plant protection, medical ctr, insurance, salary admin, pensions, labor negotiations, etc. Mbr four yrs Ill PE Examining Ctte, Region VIII V Pres, IIE for three yrs, P Pres of Amer Inst of Indus Engrs. *Society Aff:* IIE, NSPE.

Houston, Joseph B, Jr
Business: 12150 Country Squire Lane, Saratoga, CA 95070
Position: Consultant *Employer:* Houston Research Associates *Education:* MS/Engrg Mgmt/Northeastern Univ; AB/Astronomy/Univ of TX. *Born:* 6/15/34. Native of Birmingham, Ala. Served in Army Corps Engrs as co comdr & army aviator 1956-61. Assoc Engr, Electro-Optics Div Perkin-Elmer Corp 1961-64; Sr Engr Itek Corp managing optical sys mfg programs 1964-68; Ch Optical Engr, Underwater Sys 1969-71. Staff Asst to Pres, E-O Div, Kollmorgen Corp 1971-73. Assumed respon for E-O Sys as V Pres, Applied Tech from Aug 1973 to Apr 1981. Formed Houston Res Assocs in April 1981. P Pres, New England Sect, OSA; Chmn OSA Tech Group; P Pres, Fellow and Life mbr SPIE; Assoc Editor, SPIE Journal of Optical Engrg. Editor, Opt Soc of America *Optical Workshop Notebook*. George W. Goddard Award, SPIE, 1982; Conslt, US Army Intelligence and Security Command, 1982-87; Outstanding Civilian Service Medal from Dept. of the Army. SPIE Representative to DOD - University Forum, 1983-present *Society Aff:* SPIE, OSA, AOC, AAAA, AUVS, NSL, IEEE, The Planetary Society

Houts, Ronald C
Business: Dept of Elec Engg, PO Box 6169, Tuscaloosa, AL 35487-6169
Position: Prof. *Employer:* Univ of AL. *Education:* PhD/Elec Engg/Univ of FL; MSEE/Elec Engg/Univ of FL; BSE/Elec Engg/Univ of FL. *Born:* 9/22/37. Served to Capt, US Army Signal Corps (1963-65). Instructor at Univ of FL (1961-63) and Adjunct Prof at Cochise Jr College (1964-65) and Univ of AZ (1965). Mbr of Engg Faculty Univ of AL-Tuscaloosa since 1966. Served as Army Material Command Visiting Res Prof to Army Missile Command (1974-1975). Conducted res projs for NASA (1966-74) and Army Res Office (1976-78). Consultant Battelle Memorial Labs (1979-82) and Universal Data Systems (1983-86). Visiting Prof of Elec Engrg, US Military Academy (1983-84). Awarded US Army Commendation Medal (1966) and Outstanding Civilian Service Medal (1984). Published over 50 technical papers and co-author of *Fundamentals of Analog and Digital Communication Systems*, Allyn & Bacon (1971). Technical Prog Chrmn, IEEE Natl Telecommunication Conf-1978. *Society Aff:* IEEE, ΣΞ, ΤΒΠ, ΦΚΦ, HKN, ΦΗΣ.

Hovey, Harry H
Business: 50 Wolf Road, Albany, NY 12233-3250
Position: Dir-Div of Air Resources *Employer:* N Y State Dept Environ Conservation. *Education:* MPH/Air Poll/Univ MN; BCE/Sanitary/Rensselaer Poly Inst. *Born:* Aug 1930. Lic Prof Engr since 1968. Diplomate, Amer Acad of Environ Engrs. US Army Medical Corps as Sanitary Engr 1953-55. Hydraulics Engrs, US Fish & Wildlife Service; Field Engr, Barker & Wheeler Cons. Air Pollution Control Engr N Y St 1958. Dir Air Resourcs 1976- to date. Natl Air Pollution Control Assn 1958 VP 1977-79; Chmn Mid-Atlantic States Sect 1968. Town of North Greensbush N Y Planning Bd 1967-72 & Zoning Bd of Appeals 1972- 1981. Ham radio operator KB2FC; Member EPA Clean Air Scientific Advisory Committee 1978-1981; IJC International Air Quality Advisory Board 1980-present. *Society Aff:* APCA, NSPE, AAEE.

Hovland, H John
Home: 781 Alvarado Rd, Berkeley, CA 94705
Position: Sr. Civil Engr *Employer:* Pacific Gas & Electric Co *Education:* PhD/Geot Engr/Univ CA-Berkeley; MS/Soil Mech(CE)/Brigham Young Univ; BES/CE/ Brigham Young Univ *Born:* 6/30/34 PhD in Geotechnical Engrg from U.C. Berkeley in 1970. Consulting Engrg for four years. Staff consultant in Geotechnical Engrg with Pacific Gas and Electric Co from 1972. Experience includes static and dynamic analysis of dams, three- dimensional finite element analysis of Helms underground powerhouse and Scott Dam, three-dimensional analysis of landslides, and various geotechnical concerns in the planning, siting and construction of highrise structures, dams, geothermal power plants, pipelines, tunnels, gas and electric transmission lines, substations and other industrial facilities. Emphasis of experience is with geotechnical aspects of constructing facilities in surface deposits and weak and problematic rock. 15 to 20 publications in national journals or proceedings of conferences. Developed a three-dimensional slope stability analysis method. *Society Aff:* ASCE, USCOLD, ICOLD.

Hovmand, Svend
Business: 9165 Rumsey Rd, Columbia, MD 21045
Position: Group VP *Employer:* Niro Atomizer Inc. *Education:* PhD/Chem Engg/ Univ of Cambridge, England; MS/Chem Engg/Tech Univ of Denmark. *Born:* Native of Denmark. Working with fluidization, particle, and ceramic technology. Since 1964, specializing in scale-up and desing of fluid beds, spray dryers and various powder agglomerating systems. R&D Mgr for A/S Niro Atomizer Copenhagen from 1971-77. VP for Niro Atomizer Inc, Columbia, MD, since 1977. Pres of Bowen Engrg NJ since 1981. VP and President of NIRO Ceramic Inc, Columbia from 1982, specializing in building turnkey ceramic plants, inclusive of Crossville Ceramics Company, Tenn. of which Niro Ceramic Inc is the managing general Partner. Director of Industrial Drying Course, Center for Professional Advancement, NJ since 1982. *Society Aff:* AIChE

Howard, G Michael
Business: Sch of Engg, UTC Bldg U-37, 191 Auditorium Road, Storrs, CT 06268
Position: Prof & Assoc Dean. *Employer:* Univ of CT. *Education:* PhD/ChE/Univ of CT; MEngr/ChE/Yale Univ; BS/ChE/Univ of Rochester. *Born:* 7/4/35. Have taught at CT since 1961, Assoc Dean of engg since 1974. Gen field of technical interest is process modeling and control which has led to recent res work in area of energy and energy conservation. Teaching interests include dev of courses in energy tech for both engg students and those with non-technical backgrounds. Won a univ teaching award. Admin responsibility for undergrad progs. Personal interest is athletics, enjoy playing tennis and basketball. *Society Aff:* AIChE, ACS, ASEE.

Howard, George C
Business: PO Box 591, Tulsa, OK 74102
Position: Lab Services Superintendent. *Employer:* Amoco Production Co Research Ctr. *Education:* BS/Petroleum & Mech Engg/Univ of OK. *Born:* 8/19/17. Employed by Stanolind Oil & Gas Co (predecessor to Amoco Prod Co) Feb 4, 1941; grad from Univ of Oklahoma in Jan 1941 with BS degree in Petroleum & Mech Engrg; 1947-72 respon for res & dev assoc with drilling & completing oil & gas wells, dev of procedures for subsea well completion sys, arctic & pollution control. Received award (Lester C Uren Award 1967) for dev of hydraulic fracturing, served as Distinguished Lecturer on cementing & fracturing & co- authored a monograph for the Soc of Petroleum Engrs. Served on the Petroleum Panel of Natl Security Indus Assn's Ocean Sci & Tech Ctte (NSIA/OSTAC) for the Pres's comm on Marine Sci, Engrg & Resources; authored more than 50 tech articles, holds 50 pats; Mgr Lab Services 1972-present. *Society Aff:* SPE.

Howard, George T
Business: 7046 Hollywood Blvd, Ste 600, Hollywood, CA 90028-6063
Position: Pres *Employer:* George Thomas Howard Assocs *Education:* MS/EE/MIT; BS/EE/MIT; BA/Physics/Reed Coll *Born:* 6/8/29 At an early age he decided his life's goal: to design theatres. Joined General Electric's Large Lamp Division in 1950. On

Howard, George T (Continued)
the team that developed the dichroic Filter and the 10K Fresnel spotlight for the motion picture industry. Also worked on the famous Nela Park Christmas lighting. Western regional mgr, Wakefield Lighting Co, before becoming VP/General Mgr of Kliegl Brothers Western Corp in 1965. From 1958, independent consultant for Presentation facilities. George Thomas Howard Assocs is now a world leader in the field of consulting for theatrical and presentation facilities; including all phases of electrical and mechanical systems, audio-visual, projection, and security systems for theatre, arena, recital & concert halls, exhibit area, radio and television production facilities. *Society Aff:* NSPE, SMPTE, IEEE, IES, ACEE

Howard, J Wendell
Business: PO Box 511 Bldg 54B, Kingsport, TN 37662
Position: Proj Mgr. *Employer:* TN Eastman Co. *Education:* BS/Civ Engr/Univ of TN MS/Engg Adm/Univ of TN. *Born:* 1/29/41. Employeed by TN Eastman Co since 1968. Positions held include Civ Engr, Sr Civ Engr, Engr in charge of Water Quality Engg & Proj Mgr. Engg responsibilities were the design of water & wastewater treatment facilities for TN, Carolina & AR Eastman Co plants. Currently proj mgr for seven divs at TN Eastman Co. Reg Engr in states of TN, SC & Ar. Former Pres (1976) of TN Valley Sec of ASCE. Authored & co-authored papers dealing with ind wastewater treatment processes. *Society Aff:* ASCE, NSPE, ΤΒΠ, XE.

Howard, John N
Business: AFGL/CA, Hanscom AFB, MA 01731
Position: Chief Scientist. *Employer:* Air Force Geophysics Laboratory. *Education:* PhD/Physics/OH State Univ; MSc/Physics/OH State Univ; BSc/Physics/Univ of FL. *Born:* 2/27/21. Infrared Molecular Physics. Since 1954 at AF Geophysics Lab (formerly named AF Cambridge Res Labs), Hd, Optics Lab 1961-64; Ch Sci 1964-present. Mbr, Bd/Dir Optical Soc of Amer 1962-present. Editor, Applied Optics 1962-present. Adv Ctte, Physics Today 1967-present (Chmn 1970-present). V Pres, Metric Assn 1965-72; Chmn, Metric Comm, Amer Geophys Union 1970-present. Mbr IEEE, Metric Ctte 1974- present. Specialties: Atmospheric physics, infrared optics, metrication. *Society Aff:* AGU, APS, OSA, AMS, SAS, RESA, SHS.

Howard L Hartman
Business: Department of Mineral Engineering, University of Alabama, P.O. Box 1468, Tuscaloosa, AL 35487-1468
Position: Prof and Drummond End Chr of Mining Engineering. *Employer:* Univ of AL. *Education:* PhD/Mining Eng/Univ of MN; MS/Mining Eng/Penn State Univ; BS/Mining Eng/Penn State Univ. *Born:* 8/7/24. Native of Indianapolis Ind. USNR 1944-46 - Ltjg. Employed as mining engr 1947-56 by Phelps Dodge Corp, Anaconda Co, Arizona Mine Inspector, Minn State Tax Comm. Taught at Minn 1950-54, Colo Sch of Mines 1954-57, Penn State 1947-48, 1957-67 (Instructor, Dept Hd, Assoc Dean), Calif State Univ at Sacramento 1967- 71 (Dean), Vanderbilt Univ 1971-80 (Dean), Ala 1980-(Prof and endowed chair). Tech specialties: mine ventilation & environ, rock drilling & excavation. Ed fields: mining engrg, socio-engrg. Publ over 100 articles & books. Chmn Fed Metal and Nonmetal Mine Safety Bd of Review 1971-75. AIME Mineral Indus Ed Award 1965, SME-AIME Distinguished Mbr 1982, SME-AIME Publication Bd. Award 1982. NUCEA Faculty Service Award, 1985. Int in opera & camping. *Society Aff:* SME-AIME, ASEE, ΤΒΠ, ΣΞ.

Howard P Emerson
Home: 1433 Lakewood Drive, Zachary, LA 70791-2746
Position: Retired *Education:* SB/Electrochemical Engrg/MIT; BS/Chemistry/ Dartmouth Coll *Born:* 12/21/01 After Dartmouth (1923) taught 3 yrs in Turkey at Robert Coll. After MIT (1928) worked for Western Electric, DuPont Ammonia Corp, and TVA. In 1947 was asked to start a new Dept of Indus Engrg at the Univ of TN, and design a curriculum. Early affiliation with the Inst of Indus Engrs, started in 1948. National Pres (1956-57), Fellow (in 1964), and Life Mbr. Reg PE. Participated in ASEE, EJC, ASQC and in Tau Beta Pi's objective of Liberal Learning. After retiring (1972) in FL, (now LA), my wife and I made a trip around the world. *Society Aff:* IIE, ΤΒΠ, CFIIE

Howard, Robert E
Home: 2015 St Clair, Brentwood, MO 63114
Position: Retired *Education:* PhD/Chem Engr/Purdue Univ; MS/Chem Engr/Purdue Univ; BS/Chem Engr/Purdue Univ. *Born:* Apr 1920. With Monsanto Corp since 1943. Served in various res & engrg pos, becoming Dir of Organic Div of Corporate Engrg Dept in 1965. Respon for Monsanto's activities in behalf of DOE's coal gasification demo plant program in 1976-1980. Fellow & participate in local & natl activities of AIChE. Active scouter & enjoy hiking, canoeing & music. *Society Aff:* AIChE, ACS.

Howard, Ronald A
Business: Dept of Engg-Economic Systems, Stanford, CA 94305
Position: Prof. *Employer:* Stanford Univ. *Education:* ScD/EE/MA Inst of Tech; EE/ EE/MA Inst of Tech; SM/EE/MA Inst of Tech; SB/EE-Economcs/MA Inst of Tech. *Born:* 8/27/34. Directs teaching and res in the Decision Analysis Prog, Dept of Engg- Economic Systems, Stanford Univ. He defined the profession of decision analysis in 1964 and has since supervised several doctoral dissertations in decision analysis every yr. He is the author of *Dynamic Programming and Markov Processes* (Wiley 1960); and *Dynamic Probabilistic Systems* (two volumes, Wiley, 1971). His current res interests are life-and-death decision making and creation of a coercion-free soc. He had been a mbr of the Natl Acad of Sci's Committee to Review the Wash DC Metropolitan Area Water Supply Study. He is currently a mbr of the Tech Adv Ctte, Office of Nuclear Waste Isolation, Battelle, Columbus, OH. *Society Aff:* ORSA, ORSUK, TIMS, IEEE, SRA, SSE.

Howard, William G, Jr
Home: 2080 E. Alameda Dr, Tempe, AZ 85282
Position: VP *Employer:* Motorola, Inc *Education:* PhD/EE/Univ of CA-Berkeley; MS/EE/Cornell Univ; B/EE/Cornell Univ *Born:* 11/6/41 Native of Newton, MA. Now residing in Tempe, AZ. Served as Assistant Prof of Electrical Engrg and Computer Sciences at the Univ of CA 1957-1969. Joined Motorola Inc in 1969. Assumed responsibility as Group Operations Mgr of Linear Integrated Circuits in 1973. Assumed responsibility as VP and Dir of Tech and Planning for the Semiconductor Sector in 1979. *Society Aff:* IEEE, AAAS

Howard, William J
Business: Albuquerque, NM 87185
Position: Executive Vice President. *Employer:* Sandia Laboratories. *Education:* BSME/NM State Univ. *Born:* 8/25/22. With Sandia Labs 1946-63. Asst to Secy of Defense (Atomic Energy) 1963-66. VP, Sandia Labs 1966-73; Exec VP 1973-present. US SALT Delegate, 1976. *Society Aff:* NAE.

Howarth, David S
Home: 2165 Chalet Dr, Rochester, MI 48063-3809
Position: Staff Res Engr *Employer:* GM Res Labs. *Education:* PhD/EE/Carnegie-Mellon Univ; MS/EE/Carnegie-Mellon Univ; BS/EE/Carnegie-Mellon Univ. *Born:* 5/29/43. New Kensington, PA. PhD thesis: semiconductor heterojunction devices. 1972- present: GMRL Electrical & Electronics Engg Dept specializing in electronic control, sensors, applications & theory, for automotive engines & electric vehicles. Awards: SAE Vincent Bendix Award for outstanding paper in automotive electronics, SAE Arch T Colwell Award for contributions to automotive engg, IEEE Vehicular Technology Soc Avant Garde Award for pioneering leadership contributions. Sr mbr IEEE. Profl activities: Reg PE; Automotive Tech Editor of IEEE Transactions on Vehicular Tech for 6 yrs. Patents: many on exhaust gas sensors for lean engine control. Publications: has written numerous of scientific & technical papers & made presentations at natl & intl confs. Mbr Rochester Area YMCA Bd of Directors. *Society Aff:* IEEE, ΣΞ, SAE, AVS, HKN.

Howarth, Elbert S
Home: 516 Keystone Dr, New Kensington, PA 15068
Position: Retired *Education:* MS/Mechs/Carnegie Inst of Technology; BS/Mech Engrg/Carnegie Inst of Technology. *Born:* 5/24/15. Native of Western PA. Entire prof career with Alcoa starting in 1936 as Res Engr & progressing ad Div Ch, Proj Mgr, Ch Engr, Asst Ch Const Engr, Asst Dir & Assoc Dir. Specializing in engrg mechs, facilities design & operation, supporting services incl shops, purchasing, computers fiscal planning. Respon for OSHA, security, tech infor & public relations. Reg Prof Engr, P Chrmn Pittsburgh Sect ASME, Fellow ASME. Mbr Sigma Xi, ASM, Engrg Soc of Western PA, Sci Res Soc of America, Bd. Citizens Genl Hospital, Dir Penn State Univ & Pa. Mbr New Kensington Area Chamber of Commerce, Chamber Foundation New Kensington Zoning Hearing Bd, City of New Kensington Industrial Dev Authority, Genl Advisory Committee of Northwest Westmoreland Area Vocational-Technical Sch, Mbr Hill Crest Country Club. *Society Aff:* ASME, ASM, ESWP, ΣΞ, RESA

Howe, Everett D
Home: 3112 Ptarmigan Drive 1, Walnut Creek, CA 94595
Position: Prof of Mech Engrg, Emeritus. *Employer:* University of Calif-Berkeley. *Education:* MS/Mech Engrg/Univ of CA; BS/Mech Engrg/Univ of CA. *Born:* Jan 10, 1903 Oakland, Calif. Indus experience: GE Co Lynn, Mass; Allis- Chalmers Co; Caterpillar Tractor Co; Hetch-Hetchy Proj, City of San Francisco. Served in the Univ of Calif 1928-present. Teaching of Mech Engrg courses in thermodynamics, heat transfer, fluid mechs, power generation. Res in combustion, heat transfer, desalination. Admin posts: Hd of the Mech Engrg Lab, Chmn of the Dept, Assoc Dean of the Coll of Engrg, Dir of the Sea Water Conversion Lab & coordinator of the Statewide res program in desalination. Mbrships: Fellow ASME; Mbr ASEE, Sigma Xi. Reg Mech Engr Calif. Mbr & Dir Internatl Solar Energy Soc. *Society Aff:* ASME, ASEE, ISES.

Howe, John P
Home: 5725 Waverly Ave, La Jolla, CA 92037
Position: Adjunct Prof. *Employer:* Retired - Self. *Education:* PhD/Phys-Chem/Brown Univ; BS/Chem/Hobart College. *Born:* 6/24/10. Native of Groton, NY. Taught physical chemistry Ohio State & Brown Univs, res photochemistry, spectroscopy, statistical mechanics, nucl mtls. Res assoc and Assoc Dir, Met Lab, Univ of Chicago. Organized & managed met at Knolls Atomic Power Lab, GE Co, Schenectady; res dept at Atomics Intl, & planned corporate res ctr, N Am Aviation Co. Ford Prof of Engg & Dir, Dept of Engg Phys, Cornell Univ. Chrmn dept of met, assoc dir of res & technical dir advanced energy systems, Gen Atomic. Co-edited "Journal of Nuclear Materials" & "Annals of Nuclear Energy–. Advisor on classification & mbr ACRS, US AEC. Co-chairman session Atoms for Peace 1955. Retired 1975. Courses books & publications on nucl energy, Univ of CA, San Diego. Conslt Mat Sci. *Society Aff:* ASM, APS, AIC, ACS, AAAS

Howell, Francis K
Home: Box 928, Harrisonburg, VA 22801
Position: Retired. *Born:* 1893. ME from Stevens Inst of Tech. Supt of Compressing Stations, Philadelphia Co Pittsburgh. Respon for transmission of 20-35 billion cf/per yr of natural gas from fields in Pa & W Va 1919-25. Sales Engr, Green Fuel Economizer Co Beacon, N Y. Application of auxiliary boiler plant equip 1926-32. Hd Dryer Dept Buell Engrg Co N Y C. Application 'Turbo' dryers to starch, glue, adhesives, resins 1935-37. Sales Engr, United Conveyor Corp, Chicago, pneumatic ashhandling 1938-39. Est engrg sales business in Richmond, Va incl ashhandling, flyash collectors & cooling towers 1940 until retirement 1961. Received Fellow status in ASME 1967.

Howell, Irvin N
Home: 2529 Comanche Drive, Birmingham, AL 35244
Position: Asst VP, Rates & Economics. *Employer:* South Central Bell Telephone Company. *Education:* BS/EE/The Citadel. *Born:* 01/03/22 Mine Disposal USCG 1940-46. Has held var pos with AT&T, Southern & South Central Bell since 1951 specializing in Transmission Engrg & Quality Control & Cost for Pricing. Current respon-Asst-VP-Rates & Economics. IEEE Fellow, IAS P Pres, Chrmn Policy & Planning, Chrmn Stds Bd mbr IEEE Bd of Dirs and Long Range Planning Ctte, ABET Visitor, Listed in Who's Who, Men of Achievement. Reg PE TN & MS. Enjoys Lopidery woodworking & breeding Chows. *Society Aff:* IAS, COM SOC.

Howell, John R
Home: 3200 Kerbey Lane, Austin, TX 78703
Position: Chairman *Employer:* Dept. of Mech. Engng. *Education:* PhD/Engineering/Case Institute of Technology; MS/Chem Engng/Case Institute of Technology; BS/Chem Engng/Case Institute of Technology *Born:* 06/13/36 Born Columbus, Ohio. Aerospace engineer, NASA Lewis Research Center, 1961-68; Associate Prof and Prof, ME Dept, U. of Houston, 1968-78; Prof, U. of Texas at Austin, 1978-83, E.C.H. Bantel Prof, 1983-present, Chairman, 1986-present; ASEE Ralph Coats Roe Award, 1987. Enjoy sailing, flying, and photography. My wife is Susan Conway, and I have three children, Reid, Keli, and David by a previous marriage. *Society Aff:* ASME, AIAA

Howell, John T
Business: US Coast Guard Academy, New London, CT 06320
Position: Head, Dept of Applied S&E *Employer:* USCG *Education:* MSE/Naval Arch, Marine Eng, Mech/Univ of MI; BS/E/USCG Academy *Born:* 10/22/36 Native of Seattle, WA. Last three tours have been Chief, Shipbuilding Branch, Executive Officer, Polar Class Icebreaker (Polar Sea), and Chief, Naval Engrg Branch, Thirteenth Coast Guard District. In addition to current Dept Head position, am Head Tennis Coach, teaching one engrg course, and pursuing a master's degree in management. Next goal: Command of a Polar Class Icebreaker. *Society Aff:* ASNE, ASEE

Hower, Glen L
Business: Washington St U, Pullman, WA 99164
Position: Prof, Dept of Elec Engrg. *Employer:* Washington State University. *Education:* PhD/EE/Stanford Univ; MS/EE/WA State Univ; BS/EE/WA State Univ. *Born:* Feb 1934 in Wenatchee, Washington. GE Co, Richland, Washington 1956-57. With Washington State Univ since 1957; Asst Dean Col of Engrg 1967-70; Alcoa Foundation Prof 1972-75; 1970-80,Prof & Chmn, Dept of Elec Engrg & Branch Mgr Elec Engrg Res; Res specialty in area of applied electromagnetics & wave propagation. ASEE, IEEE, AGU; Outside activity, backpacking. *Society Aff:* IEEE, AGU, ΣΞ, TBΠ, ASEE.

Howes, Fred S
Home: 3375 Ridgewood Ave 412, Montreal Quebec H3V 1B5 Canada
Position: Retired. *Education:* PhD/Elec Engg/Univ of London; MSc/Elec Engrg/McGill Univ; BSc/Elec Engrg/McGill Univ; DIC/Elec Engrg/Imperial Coll. *Born:* Jul 25 1896 Paris Ontario, Canada. BS, MS McGill Univ; DIC Imperial Coll; PhD U of London. McGill U 1924-65. Retired as Emerit Prof of EE. Hon Life Fellow IEEE; Hon Life Mbr: ASA, Sigma Xi, CAUT, MAUT. Hon. Life Mbr Corp Prof Engrs of Quebec. Publication since retirement - a book - a critical examination of the Synoptic Gospels - Matthew, Mark & Luke. Title - "This is The Prophet Jesus - An Evolutionary Approach to His Teaching–. available from the author. *Society Aff:* IEEE, ASA, ΣΞ, CAUT.

Howland, James C
Home: 2575 S. W. Whiteside Dr, Corvallis, OR 97333
Position: Sr Conslt *Employer:* CH2M Hill *Education:* MS/CE/MIT; BS/CE/OR State Univ *Born:* 6/2/16 Engr for Standard Oil Co of CA, 1939-1941. Officer US Army Corps of Engrs 1941-1946, Lt to Major. Service included heading the island design section for the pre-invasion planning and construction to make the island of Saipan into a major forward base. One of the founders (1946) and chief operating officer 1946 to 1974 of Cornell, Howland, Hayes and Merryfield, Consulting Engrs which

Howland, James C (Continued)
is now CH2M Hill a 2800 person organization with 40 offices. Currently Sr Conslt. Charter member and Pres (1969) Consulting Engrs Council of OR, member PE of OR, Honorary Member ASCE, Corvallis Planning Commission (1961-1974), trustee Linfield Coll (1978-), Tau Beta Pi, Phi Kappa Phi, Legion of Merit. *Society Aff:* ASCE, NSPE, TBΠ

Howland, Ray A
Business: 2951 Flowers Rd, So, Atlanta, GA 30341
Position: President. *Employer:* The Howland Company, Inc. *Education:* MS/EE/Cornell; BEE/EE/George Washington Univ. *Born:* Feb 1934. Reg Prof Engr Georgia. Native of Marshall, North Carolina. Served in US Army 1954-56. Engr for Jansky & Bailey specializing in electromagnetic compatibility high power radars & communication sys 1959-63. Sr Engr leading instrumentation design projs with Sci Atlanta up to 1973. Organized The Howland Co, Inc in 1975 to provide microwave instrumentation & sys engrg services. Mbr IEEE, ACEC, Rotary. Editor CPEM 1972 & 1976. President IEEE S/IM 1981. Enjoy hiking & camping. *Society Aff:* IEEE.

Howlett, Myles R
Home: 6826 Rosemont Drive, McLean, VA 22101
Position: Retired *Employer:* Forest Service-US Dept of Agri. *Education:* BSCE/Civil Engg/Syracuse Univ. *Born:* Aug 1921 Morrisville, N Y. Served with Corps of Engrs 1942-46. Joined Forest Service in Calif 1949. In 1956 was US Dept of Agri nominee for the William A Jump Award given annually for outstanding public service, for work on flood prevention projs following disastrous fires in Calif. In 1965-66 represented Forest Serv on Interagency Task Grp estab at request of Pres's Council on Recreation & Natural Beauty to formulate a prog for scenic rds & pkwys. Reg Prof Engr, an ASCE Fellow & P Natl Capital Sect Pres of ASCE. Dir of Engrg for USDA Forest Service from 1971-83. *Society Aff:* ASCE, NSPE, SAF, WSE, AFA.

Howse, Paul T Jr
Home: Rt. 7 Box 72B, Milton, FL 32570
Position: Sr Specialist. *Employer:* Monsanto Textiles Company. *Education:* BS/ME/Univ of AL. *Born:* Feb 1931 Birmgingham, Ala. Assoc Engr for Southern Res Inst 1959-62. Design Engr 'A' Hayes Internatl 1962-63. With Monsanto Textiles Co since 1963. Presently dev equip & processes for prod of Nylon Carpet Yarn. Pi Tau Sigma 1958, Tau Beta Pi 1959, Machinery Magazine 's award for 'Outstanding Excellence in Machine Design' 1959. 8 Pats.Reg PE Fla. V Pres for Region XI (Southeastern) ASME 1974-76. Enjoy golf, fishing, boating & traveling. *Society Aff:* ASME.

Howse, Roxy S
Business: 5750 Major Blvd 500, Orlando, FL 32805
Position: Pres, Major Design Corp;Ch Engr. *Employer:* Major Realty Corp. *Born:* Dec 1937. BSCE from Tenn Tech. Born & raised in Nashville, Tenn. Hwy Engr, State of Ill Hwy Dept 1959-61; Ch Co Engr, John L Burns Nashville, Tenn; Interstate Hwy Const 1961-64; Asst County Engr, Hillsborough Cty, Tampa, Fla 1964-67; City Engr, Melbourne, Fla 1967-69; Cty Engr & First Pub Wks Administrator Orange Cty, Orlando Fla 1969-72; Ch Engr of Major Realty Corp & Pres Major Design Corp, a cons engrg firm 1972- ; State (Fla) & Natl Chmn of Engrg Student Chap Ctte 1964-75; ASCE; FES - Pres, Central Fla Chap 1976-77; other var offices of FES 1964- ; FICE Career Guidance Ctte Chmn; Recipient 1976 FES Award for Distinguished Service.

Howson, Louis R
Home: 716 South Oak Street, Hinsdale, IL 60521
Position: Consultant *Education:* CE/-/Univ of WI; BSCE/-/Univ of WI. *Born:* April 19, 1887 Clinton, Iowa. Cert of Achievement 1949. Hon Chi Epsilon. Joined Alvord & Burdick 1908 Alvord, Burdick & Howson since 1922. Specializing in water resources, sanitary engrg & public utility valuation. Authored many papers. Ch Sanitary Engr for 6 lake studies in four U S Supreme Court cases re sewage treatment requirements at Chicago with varying diversions from Lake Michigan. 200 utility appraisals. Served Chicago, Detroit, Milwaukee, Louisville, DesMoines, Gary, Baton Rouge, St Louis, Roanoke, etc. 15 yrs mbr Ill State Bd of Conservation & Natural Resources. P Pres West Soc of Engrs, AWWA, ASCE - Dir & V Pres of EJC. *Society Aff:* ASCE, ASME, AWWA, APWA.

Hoyl, Alfred G
Business: Los Lagos Ranch, Rollinsville, CO 80474
Position: Pres *Employer:* Coal Fuels Corp *Education:* Engrg/Mining/CO School of Mines; Engrg/Geol-Geophysics/CO School of Mines *Born:* 9/27/13 CO native. Wife: Damaris. Children: Gregory, Damaris, Geoffrey. Registered PE, Land Surveyor, CO. US Coast & Geodetic Survey, 1934; Tela RR Co., Honduras, 1935-38; Climax Molybdenum Co., 1940; Phelps Dodge corp., Morenci, AZ, 1941; Corps of Engrs, AUS Combat Engr Line Officer; Engr Bd; Joint Army-Navy Engrg & Testing Bd, 1941-1946; Founder & Mgr, Concrete Masonry Corp, Elyria, OH, 1946- 52. Founder and Pres: Coal Fuels Corp, Coronado Silver Corp & Silver Ventures Corp: developing coal and precious metal properties in West 1952 to present. Dir, CO Mining Association. Rancher and Patentee in Field. Dist mbr, SME-AIME, 1980. Life mbr, CO Mining Assoc. Western Gov, Amer Mining Congress. Pres and Co- founder, Mineral Information Inst. *Society Aff:* SME-AIME, MMSA, AMC, CMA.

Hoyt, Jack W
Home: 4694 Lisann St, San Diego, CA 92117
Position: Prof of Mech Engg. *Employer:* San Diego State Univ *Education:* PhD/Engg/UCLA; MS/Engg/UCLA; BS/Engg/IL Inst of Technology. *Born:* Oct 1922. From 1944-46 res engr on gas turbines at the NACA Cleveland Lab. From 1948-79 a specialist in propulsion & hydrodynamics at the Naval Undersea Ctr San Diego, CA. Prof. of Mech. Engrg, Rutgers University, 1979-81. Prof. of Mech Engrg, San Diego State Univ, 1981-present. Began experimental studies with drag-reducing polymers in 1960 & have cont intensive effort since then. Many publ articles on the chem, rheological & hydrodynamics aspects of drag-reducing polymer solutions, & have obtained pats on special instruments for studying their properties. Freeman Scholar 1971 ASME. NUC Gilbert Curl Award 1975. *Society Aff:* ASME, SNAME, NYAS.

Hren, John J
Business: Dept of Mtls Sci & Engg, Gainesville, FL 32611
Position: Prof. *Employer:* Univ of FL. *Education:* PhD/Matls Sci/Stanford; MS/Matls Sci/U of IL; BS/Met Engr/U of WI *Born:* 12/3/33. Native of Milwaukee. Married (1957), five children. Law student & patent examiner (1958). Postdocs in mtls sci: Berkeley, Stuttgart, Cambridge (NSF Fellow 1962-64). Established electron microscopy lab, teaches & directs grad res at Univ of FL (1964-present), prof 1972. Chrmn, Mtls Sci Div ASM (1975-77), Dir EMSA (1977-79), President (1980), VP Mtls Consultants, Inc (1970-), Pres Metamics, Inc (1977-). Visiting scientist CSIRO-Tribophysics (1970), Oak Ridge Natl Lab (1979). Visiting Prof, Vanderbilt Univ (1975). Fulbright Fellow (1980). Coauthor four books & numerous papers on mtls sci, electron microscopy & field- ion microscopy. Hobbies: backpacking, canoeing, classical music, opera, and intl travel. Editorial bd Current Contents (1972-), Elec Micros Tech (1983-), dir Major Analytical Instrumentation Ctr U of FL (1982-), Chrmn, Adv Council Natl Ctr for Electron Microscopy Berkeley CA (1983-). *Society Aff:* ΣΞ, ΦΚΦ, TBΠ, AVS, APS, ASM, AAAS

Hribar, John A
Home: 610 Driftwood Dr, Pittsburgh, PA 15238
Position: Vice President. *Employer:* GAI Consultants Inc. *Education:* PhD/Civil Engg/Carnegie-Mellon Univ; MS/Civil Engg/Carnegie-Mellon Univ; BS/Civil Engg/Carnegie-Mellon Univ. *Born:* Jan 1934. Native of Pittsburgh, Penn. Asst Prof 1960-65, Assoc Prof 1965-69 & Lecturer 1969-70 in Dept of Civil Engrg at Carnegie-Mellon U. Specialized & published in Engrg Mechanics, Soil Mechanics, Found Engrg & Struct Mechanics. Joined GAI Consultants Inc & its predecessor, General

Hribar, John A (Continued)
Analytics Inc at its inception & became Exec V P in 1969. Respon for QA Prog & Engrg Mechanics Group which concentrates on projs in seismic analysis. Mbr ASCE, ASTM, ACI & ASEE. ASCE Collingwood Prize 1966. Pres-1980 Assoc of Soil & Fnd Engrs (ASFE). *Society Aff:* ASCE, ACI, ASTM ASEE.

Hrones, John A
Business: The Law School, Cleveland, OH 44106
Position: Provost Emeritus. *Employer:* Case Western Reserve Univ. *Education:* ScD/ME/MIT; SM/Mech Engrg/MIT; SB/Mech Engrg/MIT. *Born:* Sep 28, 1912. At MIT-instructor 1939, Asst Prof 1941, Assoc Prof 1945, Prof 1948. Head of Machine Design Div 1948-57. Dir Dynamic Analysis & Control Lab 1950-57. V P for Academic Affairs 1957, Provost 1964 of Case Inst of Tech. Provost Emeritus & Prof of Engrg 1976 Case Western Reserve Univ. Mbr of Tau Beta Pi, Sigma Xi, Natl Acad Engrg, Amer Acad Arts & Sciences. James Clayton Memorial Lecturer, Inst of Mech Engrs London 1960. Trustee of Cleveland Museum of Natural History, The Inst of Defense Analyses, The Asian Inst of Tech. Pres of The AIT Foundation Inc. *Society Aff:* ASME, ASEE, NAE.

Hrycak, Peter
Home: 19 Roselle Ave, Cranford, NJ 07016
Position: Prof of Mech Engg. *Employer:* NJ Inst of Tech. *Education:* PhD/ME/Univ of MN; MS/ME/Univ of MN; BS/Undesignated/Univ of MN; Dipl Volkswirt/Bus Adm/Univ of Tuebingen. *Born:* in Przemysl, Poland, Natl in 1956. Studied at Univ of Vienna (Austria) and Tuebingen. Came to US in 1949. Studied at Univ of MN; also tchg asst & instructor there. Mbr of tech staff, Bell Labs, Murray Hill, NJ 1960-65; sr proj engr, Curtiss-Wright Corp, 1965 from assoc prof to prof, NJIT, 1965-present. NASA grant 1967/68. NSF Grant 1982-85. One of the original Telstar designers. Published papers in natl & intl technical journals & took part in natl & intl technical confs. PE Licensed in NJ; pres, 1966/67 of Ukrainian Engrs's Soc of Am; Pi Tau Sigma, Tau Beta Pi, Sigma Xi. *Society Aff:* ASME, AIAA, ASEE, AGU, ACS, NYAS.

Hsiao, David K
Business: 2036 Neil Ave Mall, Columbus, OH 43210
Position: Prof of Computer & Infor Sci. *Employer:* Ohio State Univ. *Education:* PhD/CIS/Univ of PA; MS/Math/Miami Univ; BA/Math/Miami Univ. *Born:* 9/1/33. Previously, Visiting Prof at the Sloan Sch, MIT and Distinguished Visiting Scientist at Computer Corp of America. Now Prof at OH State Univ. Taught at Univ of PA, Faculty Assoc at IBM Res, San Jose, conducted res at Honeywell and consulted for government and industries. Pub numerous papers on database systems design and engg, his books - *Systems Programming - Concepts of Operation & Data Base* Systems, Addison Wesley, Monograph *Computer Security*, Co-authored, *ACM Monograph Series*, Academic Press, Paperback. *Collected Readings on a Database* Computer, edited, the OH State Univ, Session Chmn of various conferences & co-founder of the Intl Conference of Very Large Data Bases, Chmn of Planning Committee of VLDB, USA 1975, Belgium 1976, Jpan 1977, Germany 1978, Brazil 1979, Canada 1980. Currently serves as Editor-in-Chief of ACM Transactions on Database Systems (TODS) & Sr Mbr of IEEE. *Society Aff:* IEEE, ACM.

Hsiao, Mu-Yue
Home: 7 Fair Way, Poughkeepsie, NY 12603
Position: IBM Fellow *Employer:* IBM Corp. *Education:* PhD/EE/Univ of FL; MS/Math/Univ of IL; BS/EE/Taiwan Univ. *Born:* 7/17/33. Native of Hunan, China & came to US 1958. Served in Chinese Air Force, Taiwan 1956-58. With IBM since 1960 except from 1965-67 at Univ of Fla. Assumed current IBM Fellow respon of Laboratory Engrg respon Analysis Dept. Respon for computer hardware error detection & correction design, & system tech analysis. IEEE Fellow, Outstanding Invention Awards 1972 & 1979, 1983. Enjoy intelligent games & sports

Hsieh, Jui Sheng
Business: 323 High St, Newark, NJ 07102
Position: Prof of Mech Engg. *Employer:* NJ Inst of Tech. *Education:* PhD/ME/OH State Univ; MS/ME/Univ of KY; BE/ME/Wuhan Univ. *Born:* 3/5/21. Born Chungking, China; came to US 1948; natualized 1961; married 1961. PhD OH State Univ 1955; MS Univ KY 1950; BE Wuhan Univ 1943. Taught at Univ Bridgeport 1955-60; Newark College Engg 1960-65. Tenured full prof of mech engg at NJ Inst of Tech since 1965. Reg PE NJ. Author "Principles of Thermodynamics" (McGraw-Hill, 1975), Chinese edition of "Principles of Thermodynamics" (translated by Chinese Academy of Science, Published by People's Education Press, 1981), "Solar Energy Engrg" (Prentice-Hall, 1985). Listed Who's Who in Am. *Society Aff:* ASME, ASEE, ASES.

Hsiung, Andrew K
Home: 2115 NW Estaview Circle, Corvallis, OR 97330
Position: Retired; Consultant *Employer:* Self *Education:* PhD/Sanitary Engg/IA State Univ; MS/Sanitary Engg/Johns Hopkins Univ; BS/Civil Engg/Chiao-Tung Univ. *Born:* Jan 1920. Served for hwy, RR & municipalities in mainland China & Taiwan 1943-58. Sr Engr in charge of Taiwan Community Water Supply Prog 1959-64. 1967-84, wth Neptune MicroFLOC Inc as Sr Res Engr. 1985- , Self-Employed Consultant. Patentee in field of water & wastewater treatment. Spec in solids liquid separation. A Christian by faith. Rudolph Hering Medal 1970 ASCE. *Society Aff:* ASCE, AWWA.

Hsu, En Y
Business: Dept of Civil Engg, Stanford, CA 94305
Position: Prof. *Employer:* Stanford Univ. *Education:* PhD/Mechanics and Hydraulics/State Univ of IA; MS/Mechanics and Hydraulics/State Univ of IA; BE/Civil Engg/Natl Tsing Hua Univ, China. *Born:* 10/17/15. 1937-40 Res Assoc, Natl central Hydraulics Lab, China. 1940-43 Engr, Yunnan-Burma Hgwy Adm, China. 1944-50 Res Assoc, IA Inst of Hydraulic Res. 1950- 53 Res Engr, Hydrodynamic Lab, CA Inst of Tech. 1953-58 Physicist, Hydromechanics and Aerodynamics Lab, David Taylor Model Basin, Washington, DC, and Naval Ordinance Lab, MD. 1959-60 Supervisor, Hydrodynamics, Lockheed Missiles and Space Co, Sunnyvale, CA. 1961-date Faculty Mbr, Dept of Civil Engg, Stanford Univ. Part time Consultant at TRW. *Society Aff:* AGU.

Hsu, George C
Business: 6601 W Broad St, Richmond, VA 23261
Position: Mgr, Ind Stds. *Employer:* Reynolds Metals Co. *Education:* MS/Min Engg/Univ of ID; BS/Min Engg/Taipei Inst of Tech. *Born:* 4/21/39. & raised in China, immigrated to the US in 1962. Worked for ID Bureau of Mines & Geology before joined Reynolds Metal Co in 1967. Was a Sr Scientist in the Met Res Div specializing in alloy dev, sheet metal formability, heat treatment, direct reduction process, etc. Assumed current responsibility in 1974. Active in Aluminum Assn Technical Committee on Product Stds; ASTM Committees on Light Metals & Alloys, & on Metrication; ASME Boiler & Pressure Vessel Committee; ASM Handbook Committee; & Intl Standardization Organization TC 79 on Light Metals & their alloys. Past-Chrmn of Richmond Chapter of ASM. 1st Vice-Chrmn of Virginia Section of AIME. Also active in civic organizations & Chinese community. *Society Aff:* AIME, ASME, ASM, ASTM.

Hsu, Hsien-Wen
Business: Dept of Chem Engr, Knoxville, TN 37996-2200
Position: Prof. *Employer:* Univ of TN. *Education:* PhD/ChE/Univ of WI; MS/ChE/KS State Univ; BS/ChE/Natl Taiwan Univ. *Born:* 4/7/28. Native of Chia-yi, Taiwan. Naturalized US citizen. Taught in sch of Mech Engg, Purdue Univ 1959-1963. With the Univ of TN since 1964. Interested in transport phenomena of chem engg operations, zonal centrifugations, chem bioengg, & process optimization. Enjoy classical music, fishing, hiking, & orchid growing, etc. Like to travel & see the world if financially possible. *Society Aff:* AIChE, ACS, AAAS, SCE.

Hsu, Thomas TC
Business: Civil Engineering Dept, University of Houston, Houston, TX 77004
Position: Prof, Dept Civil Engrg. *Employer:* Univ of Houston *Education:* PhD/Struct Engg/Cornell Univ; MS/Struct Engg/Cornell Univ; BS/Arch Engg/Harbin Polytech. *Born:* Jul 28, 1933 in Swatow, China. Dev engr for Portland Cement Assn, Skokie, Ill 1962-68 doing res in struct concrete. Joined Univ of Miami as Assoc Prof 1968 & became Chmn of Civil Engrg Dept 1974. Joined Univ of Houston as Prof in 1980, chairman 1980-84. Received Wason Medal from ACI 1965, Res Award ASEE 1969 & Huber Civil Engrg Res Prize from ASCE 1974. Mbr ACI, ASCE, NSPE, PCI ASEE, HRB, AAUP. Active in dev of ACI Building Code. Prof Engr in Ill, Fla & Tx & involved in many cons works. *Society Aff:* ACI, ASCE, NSPE, PCI.

Hsu, Yih-Yun
Business: Dept. of Chem. & Nuc. Eng, Univ. of Maryland, College Park, MD 20742
Position: Prof *Employer:* Univ of MD *Education:* PhD/Chem Engrg/Univ of IL; MSc/Chem Engrg/Univ of IL; BSc/Chem Engrg/Taiwan Univ. *Born:* 7/10/30. Born in China; Postdoctoral training at the Northwestern Univ; at the NASA Lewis Res Ctr, as heat transfer res engr, 1959; also a visiting Prof at Taiwan Univ, 1964-1965. 1974-1981 in the Reactor Safety Res Div of AEC/NRC, he was responsible for thermal hydraulics and instrumentation. Since 1982, he has joined the faculty of Univ of MD as a Prof in the Dept of Chem & Nuc Engr. He co-authored a book on Boiling Two Phase Flow. He was chrmn of ANS Thermal-hydraulic Div 1983. AIChE HT and EC Div Chrmn, 1973 and received 1979 ASME HT Div Meml Award. He is a Fellow of both ANS & AIChE. His res interests are in the area of heat transfer & fluid flow especially those related to reactor safety & transients. He and his wfe, Shiao- Ying Chiang, have three children: Lewis, Ben and Cindy *Society Aff:* AIChE, ANS.

Hu, Hsun
Home: 3560 Mayer Dr, Murrysville, PA 15668
Position: Research Professor of Materials Science and Engineering *Employer:* University of Pittsburgh *Education:* PhD/Phys. Met/Notre Dame Univ.; MS/Met Engg/Lehigh Univ; BS/Met Engg/Chiaotung Univ. *Born:* Aug 17, 1927. Res Metallurgist in charge of X-Ray Diffraction Lab, Inst for Study of Metals, Univ of Chgo; Res Engr Westinghouse Elec Corp; Sr Scientist, Staff Scientist & Sec Head, E C Bain Lab for Fundamental Res; Sr. Res Consl, Res Lab U S Steel Corp; Research Professor, Univ of Pgh; Recipient Mathewson Gold Medal AIME; founder & editor Internatl Journal of Texture; mbr AIME; Mbr MRS; Fellow ASM. Over 100 Scientific & tech publs. *Society Aff:* TMS-AIME, ASM, MRS

Hu, Sung C
Business: San Francisco State Univ, 1600 Holloway, San Francisco, CA 94132
Position: Prof *Employer:* San Francisco State Univ *Education:* PhD/Elec Engr/OR State Univ; MS/Elec Engr/OR State Univ; BS/Electronics Engg/CA Poly State Univ. *Born:* 11/4/42. Joined San Francisco State Univ in 1980. With the Cleveland State Univ from 1970 to 1980. Teaching grad and undergrad courses in Elec Engg. Res work in simulation of distribution networks, cellular circuits, multi-valued logic, digital system reliability, and microprocessor applications. Author of more than 20 technical papers and a book on "Computer Logic Experiments." Very active in profl society activities. Mbr of several honorary societies and receiver of several merit certificates. *Society Aff:* IEEE, IEEE-CS, ASEE

Hu, Walter Wanwang
Business: Dept of Civil Engrg, Brookings, SD 57006
Position: Prof of Civil Engrg. *Employer:* South Dakota State Univ. *Education:* PhD/Hydraulics/MI State Univ; MS/Hydraulics/Univ of MN; BS/Hydraulics/Natl Taiwan Univ. *Born:* Jun 25, 1935. PhD Mich State Univ 1966; MS Univ of Minnesota 1959; BS Natl Taiwan Univ 1956. Taught Hydraulics in Bradley Univ Peoria, Ill for 12 yrs 1962-74. Ch Hydraulic Engr for Jones & Henry Engrg Ltd Toledo, Ohio 1974-75. Full Prof of Civil Engrg SDSU Brookings, SD since 1975. Respon for hydraulic prog & labs. Var consul services. *Society Aff:* ASCE.

Huang, C J
Business: Cullen Blvd, Houston, TX 77004
Position: Prof. *Employer:* Univ of Houston. *Education:* PhD/ChE/Univ of Toronto; MASc/ChE/Univ of Toronto; BSc/ChE/Natl Taiwan Univ. *Born:* 7/1/25. PhD (ChE), 1955, Univ of Toronto; Univ of Houston since 1955; Prof of Chem Engg, Dept Chrmn, 1962-1965; Assoc Dean, College of Engg,.1967-1969, Asst VP and Assoc Dean of Faculties, 1969-1973; Fulbright Visiting Prof, Natl Taiwan Univ, 1966; Special Chair Prof, Cheng-Kung Univ, 1968; Visiting Prof Univ of Tokyo, 1979; Dir, Pasadena Natl Bank (TX), Advisory Committee Mbr (population) to Secy of Health, Education & Welfare, 1973-1975; Honor Socs, Sigma Xi, Phi Kappa Phi, Tau Beta Pi; Mbr, AIChE; Best Publicatio Awards, AIChE TX Sec, 1959, 1964 & 1966; Profl Achievement Award, Chinese Inst of Engrs, NY, 1966; Fields of interest: Energy Conversion, Interphase Mass Transfer, Chem Process Economics & Design. *Society Aff:* AIChE.

Huang, Chi-Lung (Dominic)
Business: Dept of Mech Engrg, Manhattan, KS 66506
Position: Prof *Employer:* KS State Univ *Education:* BS/CE/Natl Taiwan Univ; MS/CE/Univ of IL; Dr of Engg/Engrg Sci/Yale Univ *Born:* 10/10/30 Native of Fukien, China. Engrg designer for Taiwan Power Co, China, between 1955-58, specializing in power plant structural design. Suspension bridge designer with Ammann and Whitney, NYC during 1960-61. Assumed current position as Prof in Mech Engrg at KS State Univ in 1964. Mbr of Sigma Xi, Phi Tau Phi, Tau Beta Pi, Phi Kappa Phi and Pi Tau Sigma. Seventy-two technical papers have been published. *Society Aff:* AIAA, ASME, AAM, SES

Huang, Ching-Rong
Business: 323 High St, Newark, NJ 07102
Position: Prof of ChE. *Employer:* NJ Inst of Tech. *Education:* PhD/CE/Univ of MI; MS/Math/Univ of MI; SM/CE/MIT; BS/CE/Natl Taiwan Univ. *Born:* 1/19/32. Joint NJIT, Dept of Chem Engg and Chemistry since 1966. Publications: 1 book, 30plus res publications and 20plus oral presentations. Res Grants: 3 multi-yr NSF grants and 3 grants from the State of NJ. Res Areas: Transport Phenomena, Thixotropic Fluids, Flow of non-Newtonian fluids. Biorheology, Math, Modeling of Chem Reactors, Tertiary Oil Recovery, Sedimentation Modeling. *Society Aff:* AIChE, ISBR.

Huang, Chin-Pao
Business: Department of Civil Engineering, University of Delaware, Newark, DE 19716
Position: Prof *Employer:* Univ of DE *Education:* BS/Civ Engrg/Natl Taiwan Univ; MS/Environ Engrg/Harvard Univ; PhD/Environ Engrg/Harvard Univ *Born:* 10/04/41 Born in the Island of Formosa, Dr Huang received his BS in Civ Engrg from the Natl Taiwan Univ in Taiwan. He came to this country in 1966 as a grad student while attending Harvard Univ. He completed his two grad degrees, MS and PhD from Harvard in 1967, and 1971, respectively then took a teaching position as an asst prof at Wayne State Univ in Detroit, MI right after his grad studies. Dr Huang joined the University of DE in 1974 and has been with DE since then. Dr Huang's res interests are applied interfacial chemistry, water and wastewater treatment and indust waste mgmt. Dr Huang has over 90 pubs and is currently Associate Ed for the Journal of Environ Engrg Div, Amer Soc of Civ Engrs. Dr Huang has been a constant recipient of res grants from the US Environ Protection Agency, Natl Sci Fdn, and The Dept of Interior on various res projs related to heavy metal chemistry. He has also extensive consltg service to private industries and governments. *Society Aff:* ASCE, AIChE, WPCF, ACS, IAWPCR

Huang, Eugene Y
Business: 400 Garnet, Houghton, MI 49931
Position: Consulting Engineer *Employer:* Self-Employed *Education:* ScD/CE/Univ of MI; MS/CE/Univ of UT *Born:* 11/28/17 Native of China. Naturalized US citizen. With Chinese National Highway Administration (from Assistant Engr to Assoc

Huang, Eugene Y (Continued)
Engr) 1941-48. Taught at Univ of IL, Urbana, (as Assistant Prof and Assoc Prof of Civil Engrg), 1954-63, and MI Tech Univ (as Prof of Transp Engrg), 1963-84. Consulting Engr on transportation systems planning, highway traffic studies and geometric design, and pavement materials and analysis. Author of *Manual of Current Practice for Design, Construction, and Maintenance of Soil-Aggregate Roads, An Overview of the American Transportation System*, and numerous articles in technical journals. Member of Tau Beta Pi, Chi Epsilon, Sigma Xi, and Phi Tau Phi. *Society Aff:* ASCE, ASTM, AAAS, AREA, TRB, TIMS, ASEE

Huang, Ju-Chang (Howard)
Business: Dept of Civ Engg, Rolla, MO 65401
Position: Prof & Dir of Environmental Res Ctr. *Employer:* Univ of MO. *Education:* PhD/Civ Engrg/Univ of TX-Austin; MS/Environ Health Engr/U of TX-Austin; BS/Civ Engrg/Natl Taiwan Univ *Born:* 1/3/41. Born in Taiwan, China & came to the US for grad studies in 1964. Received his doctoral degree in 1967 & has been teaching at Univ of MO - Rolla since then. Became full prof in 1975 & Dir of Environmental Res Ctr in 1979. Took a yr leave from Univ in 1972-1973 & worked for Austin, Smith & Assoc as Asst to the VP & Chief Sanitary Engr. Technical specialization & res interests are in water & wastewater treatment, hazardous waste management & disposal. Has more than eighty publications in books and journals in these fields. Served as a sanitary engrg consultant to several govt agencies & ind firms. Was selected as MO's 1976 Young Engr of the Yr. Received ASEE/Ford Fdn Resident Fellowship in 1972 & ASCE Walter L Huber Civ Engg Res Award in 1979. *Society Aff:* ASCE, AAEE, WPCF, ASEE, AEEP, OCEESA

Huang, T C
Business: Dept of Engrg Mechs, Madison, WI 53706
Position: Prof. *Employer:* Univ of Wisc. *Born:* Jan 1921 Shanghai, China. PhD & MS Univ of Ill 1952 & 1949; BS ChiaoTung Univ 1946. Res stress & vibration analyst 1952-55 at Internatl Harvester Co, Chicago. 1955-62 taught at Oregon State Univ, UCLA, Univ of Fla & Brown Univ. Prof of Engrg Mech Univ of Wisc - Madison since 1962. Author of 'Engineering Mechanics-Statics & Dynamics'. Res interests in dynamics & linear, nonlinear & random vibrations of discrete & continuous sys. ASME Fellow; Mbr AIAA, ASA, SESA, Engrg Science, ASEE, Acad of Mech, Sigma Xi & Phi Tau Phi.

Huang, Thomas S
Business: Coordinated Science Laboratory, Univ of Illinois, Urbana, IL 61801
Position: Prof. *Employer:* Univ. of Illinois *Education:* ScD/EE/MIT; MS/EE/MIT; BS/Elec Communication/Natl Taiwan Univ. *Born:* 6/26/36. Assoc Prof of EE, MIT, 1967-73. Prof of EE & Dir, Lab for Info & Signal Processing, Purdue Univ, 1973-80. Prof of EE, Univ of IL, Urbana - Champaign, 1980 present. Visiting Prof at ETH-Zurich, Switzerland; Technical Univ of Hannover, Germany; INRS-Telecommunications, Canada. Visiting Scientist, Rheinisches Landesmuseum, Bonn, Germany. Consultant to numerous ind firms & govt agencies, including MIT Lincoln Lab, Kodak, Natl Acad of Sci, IBM Res Lab- Zurich, & Hasler (Bern, Switzerland). An Editor of Springer book series in Info Sciences, & the Intl Journal Computer Graphics and Image Processing. Guggenheim Fellow, 1971-72. A V Humboldt Fdn US Sr Scientist Award, 1967-77. IEEE Fellow. OSA Fellow. *Society Aff:* IEEE, OSA, PRS.

Hubacker, Earl F, Jr
Business: P O Box 12910, Pensacola, FL 32521
Position: Dir of Community Services. *Employer:* City of Pensacola, Fla. *Born:* Jun 1929. BSE Mech Univ of Mich. Naval aviator active duty 1948-53. Active Reserve 1953- . Current rank Captain. Thermodynamic sys engr with McDonnell Aircraft Corp, St Louis. Res engr with Monsanto Corp, Pensacola, Fla. Employed by City of Pensacola since 1963. Respon include transportation, public works, parks, recreation & library activities. Pres Fla Engrg Soc. Reg Engr Fla & Ala.

Hubbard, Davis W
Business: Dept of Chem and Chem Engrg, Michigan Tech University, Houghton, MI 49931
Position: Prof *Employer:* MI Tech Univ *Education:* PhD/ChE/U of Wisconsin; MS/ChE/U of Wisconsin; BS/ChE/U of Wisconsin. *Born:* 11/26/35 Employed at Oak Ridge National Lab 1957-1958. Fulbright-Hayes fellow at the Univ of Padua 1964-1965. Employed by MI Tech Univ since 1965. Summer employment at Dow Chem Co, Midland, MI 1967. NATO Senior Fellow at Istituto per lo Studio della Dinamica Delle Grandi Masse in Venice, Italy 1976. Program Dir for Thermodynamics and Mass Transfer at the National Science Foundation, Washington, DC 1979-1980. NAVY-ASEE Fellow at Naval Res Lab, Washington, D.C. 1982. *Society Aff:* AIChE, ACS, IAGLR, ASEE.

Hubbard, Harvey H
Home: 23 Elm Ave, Newport News, VA 23601
Position: Sr Res Assoc *Employer:* College of William & Mary *Education:* BS/Elec Engg/Univ of VT. *Born:* 6/17/21. Native of Swanton VT, employed at Westinghouse Mfg Co - 1942. Served with US Army Signal Corps & US Army Air Corps, 1942-45. Res Engr for Natl Advisory Comm for Aeronautics, 1945-59, specializing in Aircraft Noise Control. Natl Aeronautics and Space admin, 1959-1980, res in aeroacoustics, propeller noise, jet and rocket engine noise, atmospheric propacation, sonic booms and effects of noise on structures and people. 1981-present, Coll of William & Mary, specializing in Large Wind Turbine Acoustics. ASA exec council, 1975-78 VP Elect, 1984. INCE Pres, 1979, AIAA Aeroacoustics Medal 1979, ASA Silver Medal 1978, Natl Aeronautics and Space Admin Gold Medal for Exceptional Scientific Achievement, 1969. *Society Aff:* ASA, INCE, AIAA.

Hubbell, Dean S
Home: 80 Hickory Ln, Tavares, FL 32778
Position: V P & Tech Dir - ret. *Employer:* H H Robertson Co, Pittsburgh Penn. *Education:* MS/Chem Eng/OH State Univ; BS/Chem Eng/OH State Univ; Chem Eng (hon)/Chem Eng/OH State Univ. *Born:* 3/9/04. Sr Indus Fellow, Mellon Inst 1923-58. V P & Tech Dir Architectural Products Div, H H Robertson Co 1957-69. Dev waterresistant form of magnesium oxychloride cement; dev concept of selfsanitizing floor surfacing; air-stable finely-divided cupreous powder for use on Mgocl cement, as the toxic ingredient in marine anti- fouling paint, as plant fungicide & the catalyst in synthesis of silicones. Contributed to tech of poreclain enameling of aluminum. P E Penn. Active mbr AIChE, ACS, Amer Ceramic Soc, Porcelain Enamel Inst, ASTM C-2, Chmn C-16 & C-22. Cosmos Club Wash, D C. Interested in Ecology. Dev 20 acres of 'pocket wildernes' & presented it to Fox Chapel Area School District in Pittsburgh, Penn. Dev a homestead with acreage in the NC mountains which we deeded intact to John C. Campbell Folk School, Brasstown, N.C. in 1980. Have over 60 U S & Foreign patents. *Society Aff:* ACS, ASTM.

Hubbell, John H
Business: Center for Radiation Res, National Bureau of Standards, Gaithersburg, MD 20899
Position: Physicist *Employer:* Natl Bureau of Standards *Education:* MS/Phys/Univ of MI, Ann Arbor; BSE/Engrg Phys/Univ of MI, Ann Arbor *Born:* 04/09/25 Born in Ann Arbor, boyhood in Manistee, MI. Saw action in WWII as machine- gunner, Army, Bronze Star Medal awarded. Scoutmaster, BSA Troop 17, Wash, DC 1953-60. With NBS since 1950, including Dir, X-Ray and Ionizing Radiation Data Center 1963-81. 68 publs on x-ray interactions with atoms and bulk material, including article "Radiation Gauging" in *Encycl Material Science & Engineering* (Pergamon 1986). One publ (1969) identified by Inst Sci Info as a "Citation Classic" (1982). Other honors: Elected Fellow ANS 1981; Faculty Medal, Tech Univ Prague 1982; SNM Aebersold Award 1985; ANS Radiation Indust Award 1985; elected Fellow HPS 1986. Enjoy photography, harmonica-playing, grandchildren, eclipse-chasing. *Society Aff:* ANS, APS, HPS, RRS, SNM, IRPS

Hubbert, M King
Home: 5208 Westwood Dr, Washington, DC 20016
Position: Consultant. *Employer:* Self employed. *Education:* PhD/Geology & Physics/Univ of Chicago; MS/Geology & Physics/Univ of Chicago; BS/Geology & Physics/Univ of Chicago; D. Sc./Honorary/Syracuse University; D. Sc./Honorary/Indiana State University. *Born:* 10/5/03. Native of TX. Professional career: Petroleum industry, univs, and IL and US Geological Surveys. Petroleum industry: 1927-28, Amerada Petroleum Corp; 1943-1964, exploration and production res, Shell cos, Houston. Univs: 1931-1940, Columbia, geology and geophysics; 1962-1968 (part time), Stanford, Prof of Geology and Geophysics; spring 1973 Univ of CA, Berkeley, Regents' Prof Geological Surveys: 1931-1937 (summers), IL and US; 1964-1976, US Geological Survey. Honors: Day Medal, geophysics; Penrose Medal, gen geology, GSA; Lucas Medal, petroleum engg, AIME; Rockefeller Public Service Award, Princton; William Smith Medal, hydrodynamic entrapment of petroleum, Geological Society of London; Cresson Medal, Franklin Institute; Vetlesen Prize & Medal, Columbia. *Society Aff:* NAS, AIME, GSA, AAPG, AGU, SExG, Canadian Soc. Pet. Geol., AAAS.

Huber, Robert J
Business: Dept. of Electrical Engineering, University of Utah, Salt Lake City, UT 84112
Position: Prof of Elec Engrg *Employer:* Univ of UT *Education:* PhD/Physics/Univ of UT; BS/Physics/Univ of UT *Born:* 7/10/35 Native of Payson, UT. Employed 1961-1966 by Argonne Natl Lab working in fast nuclear reactor kinetics. With General Instrument Corp 1967-1971 specializing in MOS integrated circuits and with the Univ of UT Div of Artificial Organs 1971-1976 working on a visual prosthesis. Joined the Univ of Utah Electrical Engrg faculty in 1977. Presently dir of the microelectronics lab in the Electrical Engrg Dept. Chrmn of the UT section of IEEE in 1976. Consultant in the field of silicon semiconductor device design and processing. Active in research in Solid State Chemical Sensors. *Society Aff:* IEEE, ΣΞ

Hubert, Marvin H
Business: Box 310, Old Westbury, NY 11568
Position: Pres. *Employer:* Hubert Technical Industries, Inc *Education:* BSIE/ME/Lehigh Univ. *Born:* 6/8/25. Cost and Design Consultant specializing in Mech & Elec fields. Provides Consulting Services to Owners such as IBM, Arch, Engrs, Contractors & Govt Agencies. Conducts Seminars throughout country, at the Univ of WI & the Univ of IL. Author of one of the most comprehensive & realistic Plumbing/Piping Manual, Natl Plumbing/Piping Estimator Manual, ever written for Contractors. Developed Computer Estimating Progs for the Mech & Electrical & General Contracting Industry. Consultant in Solar Energy, Energy Conservation and Balancing & Testing of Air & Water Systems. *Society Aff:* ASMR, NSPR, ASHRAE

Hucka, V Joseph
Business: Dept of Min Engrg, 107A WBB, Salt Lake City, UT 84112-1183
Position: Prof, Min Engrg *Employer:* Univ of UT *Education:* PhD/Mining Engrg/Tech Univ of Mines-Czech; MS/Mining Engrg/Tech Univ of Mines-Czech; BS/Mining Engrg/Tech Univ of Mines-Czech *Born:* 4/22/25 in Ostrava (Czechoslovakia). He studied at the Technical Univ of Mining and Metallurgy in Ostrava obtaining Min Engrg Diploma, and later (1966) also PhD degree. He worked in coal mines for six years in various underground operations. Later he joined the Scientific Research Coal Inst in Ostrava as the Head of Rock Mechanics Centre. For a time he worked in the Mining Research Inst in Essen, West Germany. After emigrating to Canada in 1968, he was appointed Prof in the Dept of Mines and Metallurgy at Laval Univ and in 1978 he joined the teaching staff at the Dept of Mining Engrg at the Univ of UT as Prof in Mining Engrg. He is a PE licensed in the State of UT. *Society Aff:* AIME, CIM.

Huckaba, Charles E
Home: 1500 Massachusetts Ave, NW, Apt 409, Washington, DC 20005 *Employer:* Pres., Charles Huckaba Associates, Inc *Education:* PhD/ChE/Univ of Cincinnati; MS/ChE/MIT; BE/ChE/Vanderbilt Univ. *Born:* 10/20/22. Native of Nashville, TN. Previously Dir, Engg Program Dev., Cooper Unon 1976-80; Sect Hd Natl Sci Fnd 1974-76; Prof of Rehabilitation Medicine & Mbr-at- Large Faculty of Engg, Columbia Univ 1967-74; Chmn of ChE Dept Drexel Univ 1963- 67; faculty Univ of FL 1955-63; faculty Lamar Univ 1952-55, DuPont Year-in- Industry participant 1961-62. Reg PE NY & PA. Spec interest in the social implications of sci & technology, in industry-government relations and the funding of entrepreneurial research. Operaphile & travel enthusiast. Fellow AIChE, Fellow AIC; Distinguisted Engg Alumnus Award Univ of Cincinnati 1971, Stephen L Tyler Award AIChE 1972 *Society Aff:* AIChE, ASEE.

Huckabay, Houston K
Business: Box 10348 LA Tech, Ruston, LA 71272
Position: Prof & Head of Chem Engg. *Employer:* Louisiana Tech Univ. *Education:* BS/Chem Engrg/LPI; MS/Chem Engrg/LSU; PhD/Chem Engrg/LSU. *Born:* Jul 1932 Shreveport, La. MS & PhD LSU; BS LPI. Weather Officer, Aerial Reconnaissance Weather Officer USAF 1954-57. Res Engr Crossett Co respon for paper pulping & bleaching R&D 1960-62; Sr Res Engr & Res Supr Forest Prods Div, Oiln Mathieson, respon for dev of protective paper coatings & FDA liaison 1962- 64. Joined La Tech Engrg faculty in 1964. Dir of Engrg Grad Studies 1967-78 respon for coordination & administration of master's & doctoral programs in engrg. Prof & Head of Ch Engrg with res & publ in engrg design & simulation. Cons to legal firms on fires & explosions to La State Revenue Dept. Secy-Treas, V Chmn, Chmn El Dorado Sect AIChE. Same offices in Ouachita Valley Section, ACS. Married with five children, Episcopalian, Republican. Hobbies are reading & fishing. Mbr, Bd of Direction, LA Engrg Soc, 1983-1987. Mbr, Engineers' Selection Bd for LA State Contracts, 1985. Mbr, NSPE. Registered PE Chemical in LA. *Society Aff:* AIChE, ASEE, ACS, LES, NSPE, NFPA.

Hucke, Edward E
Business: 2090 Dow Bldg, Ann Arbor, MI 48109
Position: Professor. *Employer:* Univ of Mich. *Education:* ScD/Metallurgy/MIT; SM/Metallurgy/MIT; SB/Metallurgy/MIT. *Born:* Sep 1930. Dir of Res for LFM Co 1953-55. Cons for over 50 companies. Bradley Staughton Award from ASM. Lilliequist Award from Steel Founders Soc. Outstanding Young Engr Award from Engrg Soc of Detroit. Author of 75 tech publ including patents in fields of electrochemistry; thermodynamics; materials processing; properties of metals, carbides & carbon. Fellow ASM, AIME, Inst of Metallurgists, Electrochemical Soc, Amer Carbon Soc, Inst of Chemists-Fellow. *Society Aff:* ASM, AIME, ECS, Inst of Chem, ACS, Inst of Met.

Huckelbridge, Arthur A
Business: Univ Circle, Cleveland, OH 44106
Position: Asst Prof of Civil Engr *Employer:* Case Western Reserve Univ *Education:* DEngr/Structural Engrg/UC Berkeley; MS/Structural Engrg/UC Berkeley; BS/CE/Northwestern Univ *Born:* 8/8/47 Born and raised on a farm in Downstate IL. Served with US Army Security Agency from 1970 to 1973 as linguist/transcriber. Doctoral studies, completed in 1977, were concerned with structural response to extreme seismic events. Current responsibilities include teaching and research in areas of structural design, structural dynamics and earthquake engrg. Actively consults in areas of dynamics, structural and earthquake engrg, including seismic review of nuclear facilities. Recipient of ASCE Moissieff Award, 1980. Enjoys sports, camping and outdoor activities. *Society Aff:* ASCE

Huckins, Harold A
Business: 2 Park Ave, New York, NY 10016
Position: Vice President. *Employer:* Halcon SD Group *Education:* BSChE/Chem Engg/Northeastern Univ. *Born:* Nov 1924. Grad studies Lowell Inst & Boston Univ. Process engr Monsanto (Boston) 1945-49; Koppers Co 1949-53. Joined Scientific Design Co (Halcon subsidiary) 1953 as Sr Process Engr. 1956 Proj Mgr for first

Huckins, Harold A (Continued)
commercialization of the Mid-Century terephthalic acid tech. 1960 promoted to Dir - Process Engrg/Proj Evaluation. From 1967-70 served as tech dir with Oxirane Corp, a joint venture for commercializing propylene oxide tech. Respon included plant start-up & managing plant tech staff. 1970 V P Oxirane Corp's Tech Operations. Returned to Halcon 1973 as V P-Tech Assessment Chmn of Central Jersey Subsection AIChE & has served as Chmn Continuing Education Cttes in NY & NJ, as Past Dir (1967-69) N Y Sect. Fellow 1976, Dir, Natl Matl Engg & Sci Div 1979- Dir Natl Management Div 1980. Initiator & Organizer of Natl Elec Car Symposium-AIChE & Princeton Univ 1978, "Advances in Materials Technology for Process Industries Needs" Conference in 1984 cosponsored by 26 technical societies. Dir Matl Technology Inst Columbus, OH 1977-present. Mbr AAAS, ACS, NACE & MENSIA. Holder of 7 patents. *Society Aff:* AICHE, AAAS, NACE, ACS.

Huddleston, John V
Home: 77 Depew Ave, Buffalo, NY 14214
Position: Tenured Prof *Employer:* SUNY at Buffalo *Education:* PhD/Applied Mech/Columbia Univ; MS/CE/Columbia Univ; BS/CE/Columbia Univ *Born:* 2/1/28 Elected to Tau Beta Pi, NY Alpha, 1951. Married Martha R. Hendry, 1952. Taught at Columbia Univ and Yale Univ before joining SUNY at Buffalo in 1967. Visiting Prof at Univ of Florence, 1974. Fulbright-Hays Lecturer in Malaysia, 1975-76. Fulbright Visiting Prof in Malaysia, 1984-85. Licensed PE in the State of NY. Founder and Pres of Exchange Computing Systems, Inc, Buffalo, NY. *Society Aff:* ASCE, ASME, TBΠ, XE

Huddleston, Robert L
Home: 301 Glenville Rd, Churchville, MD 21028
Position: Consultant; materials & nondestructive testing. *Employer:* Self. (Retired from Gov't. Svc) *Education:* MA/Liberal Arts/Johns Hopkins Univ; BS/Metallurgical Engg/VA Poly Inst. *Born:* Nov 15, 1928. Native of Covington, VA. Metallurgist, Detroit Induction Heating Co. Sr Metallurgist Koppers Co. Spec in ductile iron. US Army Ordnance 1954-56. Ch, Matls Eval Sect (Material Testing Directorate), 1956-1974. 1974-1981, Physical Test Branch respon for metrology, mech, metallurgical and chem analysis and nondestructive testing at Aberdeen Proving Ground. Group Leader, International Standardization, for US Army Test and Evaluation Command, with respons for all Army international testing agreements in the areas of materiel development and production testing, 1981-1985. Consultant, materials and NDT, 1985-present. Chmn Chesapeake Bay Sect ASNT 1964. Mbr ASNT Natl Tech and Educ Councils. Fellow ASNT 1976. ASM Silver Certificate 1973. ASM Handbook Ctte 1974. Mbr Washington Soc. of Engrs. Registered Profl Engr (Quality) in CA. *Society Aff:* ASM, ASNT, NSPE, WSE, TBΦ

Hudelson, George D
Home: 1946 Chard Rd, Cazenovia, NY 13035
Position: Retired *Education:* MS/Mech Engg/OH State Univ; BS/Mech Engg/Purdue Univ *Born:* 11/16/20 Bedford, IN. Worked in aircraft engine research at Wright Aero Corp and NACA Labs during World War II. Taught in Mechanical Engrg Dept OH State 1947- 1957. Joined Carrier Corp 1957, progressed from Chief Mechanical Engr 1965 to present position 1984. Responsible for overall corp research and for coordination with divisional engrg programs. Distinguished Engrg Alumnus award, Purdue, 1977 and OH State 1980. Corp representative to Industrial Research Inst, member Audit and Review Committee, Accreditation Bd for Engrg and Tech. Corp spokesman for governmental legislative and regulatory activities on energy- related matters. Retired Feb. 1, 1986. *Society Aff:* ASHRAE

Hudson, Carroll D
Business: Box 9000, Tyler, TX 75711
Position: VP *Employer:* ARCO Oil & Gas *Education:* BS/Petro Engrg/Univ of OK *Born:* 2/27/23 in Chattanooga, Comamche County, OK. Air Force - Flight Engr, B-26 - WWII. Univ of OK - BS Petroleum Engrg - 1951. 30 years with Atlantic Richfield Co as drilling and production engr, district engr, superintendent, operations mgr, VP & district mgr. Gulf Coast, Western Canada, East TX. *Society Aff:* AIME, SPE, TSPE

Hudson, Donald E
Business: 1201 E Calif Blvd, Pasadena, CA 91125
Position: Prof of Mech Engrg & Applied Mech. *Employer:* Calif Inst of Tech. *Education:* PhD/Mech Eng/CA Tech; MS/Mech Eng/CA Tech; BS/Mech Eng/CA Tech. *Born:* Feb 1916. Prof of Mech Engrg since 1955 & of ME & Applied Mech since 1963. Special fields: shock & vibration, dynamic instrumentation, earthquake engrg. Mbr Seismological Soc of Amer (Pres 1971-72, Soc for Experimental Stress Analy, Amer Geophysical Union, Earthquake Engrg Res Inst, Amer Soc for Engrg Education, Amer Soc of Mech Engrs (Fellow), Internatl Assn for Earthquake Engrg (Pres. 1980-), Natl Acad of Engrg. *Society Aff:* ASME, ASEE, SESA, SSA, AGU.

Hudson, John C, Jr
Business: 107 W Lee St, Sardis, MS 38666
Position: Pres. *Employer:* John C Hudson, Jr & Assoc Ltd. *Education:* BSME/ME/Univ of MS; -/ME/Northwest MS Jr College. *Born:* 11/13/35. Native of Olive Branch, MS. Worked in Memphis as Design Engr for Dover Corp, Alen & Hoshall & Hufft, Ragon & Valentine until 1970. Opened Consulting Engr Office in Sardis, MS 1970 & inc 1972. Corp grew from 4 employees to 9. Licensed to practice engg in 7 states. Past Pres of the Univ of MS Engg Alumni Chapter, Memphis Chapter of ASHRAE, NW Chapter MS Engg Soc & Panola Co Chapter the Univ of MS Alumni Assn. Past Dir of the Univ of MS Alumni Assn & Memphis Chapter NSPE. Mbr United Methodist Church & Masonic Lodge (32 deg). *Society Aff:* ASHRAE, NSPE.

Hudson, John L
Business: Dept Chemical Engineering, Thornton Hall, University of Virginia, Charlottesville, VA 22901
Position: Prof. *Employer:* Univ of Va. *Education:* PhD/Chem Engg/Northwestern Univ; MSE/Chem Engg/Princeton Univ; BS/Chem Engg/Univ of IL. *Born:* Jun 1937. Fulbright scholar Grenoble, France 1962-63. Tubingen, Germany 1982-83. On Faculty of Chem Engrg at Univ of Ill 1963-75. Mgr of Div of Air Pollution Control, Ill EPA 1974- 75 (on leave from Univ of Ill). Prof & Chmn of Chem Engrg Univ of Va 1975-85. Center Adv. Studies UVA 1985-86; Prof. 1986- . 60 publ in tech journals. *Society Aff:* AIChe, ACS, ASEE.

Hudson, Ralph A
Business: 2233 S. W. Canyon Rd, Portland, OR 97201
Position: VP *Employer:* MEI-Charlton, Inc *Education:* BT/Mech/OR Inst Tech; MS/Mech/U of Portland. *Born:* 2/21/47 Went through school as a technologist graduating with honors and recognition in Who's Who in American Colls. Registered mechanical engr in OR, CA, WA, and AK. A principal and VP of a small consulting engrg firm doing 1 million dollars of business per year. Tech assignments include fit-for-use engrg evaluation of industrial equipment, boilers, and paper mill facilities. Special projects of high temperature experimental stress analysis in pulp and paper recovery boilers have been especially challenging. Outside interest include exploring OR, motorcycles, sports cars, boating and flying. *Society Aff:* ASTM, NSPE, ACEC, TAPPI.

Hudson, Robert Y
Home: 4 Crestwood Dr, Vicksburg, MI 39180
Position: Consultant. *Employer:* Self. *Education:* BS/Math/TN Tech Univ; BS/CE/State Univ of IA; PE/CE/State Univ of IA. *Born:* 3/13/12. Native of Algood, Tenn. Reg P E 493 State of Miss. With US Army Engr Waterways Exper Station, CE, Vicksburg Miss 1937-72. Respon for conducting hydraulic models concerning breakwater stability & harbor wave action. Fellow ASCE 1959-71. Recipient War Dept Commendation for Meritorious Civilian Serv 1947, Dept of Army Decoration for Meritorious Civilian Service 1972 & George W Goethals Medal for 1971, SAME

Hudspeth, Elmer B, Jr
Home: 1613 55th St, Lubbock, TX 79412
Position: Ret. (Research Leader) *Education:* MS/Agric Engr/MI State Univ; BS/Agric Engr/TX A&M Univ. *Born:* 1/27/21. Native Texan. Served in Air Force 1942-45. Retired Air Force Res, present rank Lt Col. Instructor Agricultural Engrg Dept Texas A&M Univ 1946-50. Reg P E in State of TX 8842. Chmn Tex Sect ASAE 1968-69. Fellow ASAE. Progressive Farmer Man of the Year in Southwestern Agriculture 1971. Mbr Bd of trustees McMurry Col Abilene, TX 1971-74. With USDA SEA-AR since 1950. USDA Award for Superior Service 1955 & 1973. Enjoy fishing & golf. Elected Fellow ASAE 1957 Gerlad W Thomas outstanding Agriculturist Award from TX Tech Univ Lubbock, TX 1978. Retired 1980 after 30 years with USDA. *Society Aff:* ASAE.

Huemmer, Philip M
Home: 431 Orpheus Ave, Encinitas, CA 92024
Position: Retired (Engr Consultant). *Employer:* *. *Education:* BS/Chem Engg/Univ of WA-Seattle. *Born:* 6/7/13. Born in Chicago, IL. Married Louise E. Newcom, children Douglas B & Stephen P. Union Oil 1936-47 Process Supr - S F Refinery; Ehrhart & Assocs 1947-69 VP Engr; Procon Inc 1969-78 (subsidiary UOP), VP & Genl Mgr. Chmn LA Sect AIChE 1958; Mbr Pacific Energy Assn. Reg CA Chem & Mech; PE TX, AL, MS & LA. CA Contractors License "A-. *Society Aff:* AIChE.

Huey, Ben M
Home: 210 E. LaJolla, Tempe, AZ 85282
Position: Assoc Prof *Employer:* AZ State Univ *Education:* PhD/EE/Univ of AZ; MS/EE/Univ of AZ; BS/Math/Harding Univ *Born:* 10/10/45 Native of Marysville, OH. Assistant Prof, Univ of OK 1975-1978, Tech Program Chrmn, Microcomputer '77; Consultant, Telex Computer Products, 1977. Assoc Prof, AZ State Univ 1979-present. Pres, IEEE Computer Society - Central AZ Section 1981. Research activities are in computer architecture, design automation, high-level design languages, and computer design. *Society Aff:* IEEE, ACM, HKN

Huey, Stanton E
Business: 1309 Louisville Ave, Monroe, LA 71201
Position: Consulting Engr. *Employer:* S E Huey Co. *Education:* BS/CE/Washington Univ. *Born:* Jul 1898. Final Honors, Sigma Xi, Tau Beta Pi. Fellow Amer Consul Engrs Council. Life Mbr ASCE. Principal Asst to E C McGee Consulting Engr-2 years. Mbr Bd of State Engrs, La-6 years, Acting Ch State Engrs La - 1 yr, cons Engr 43 yrs. Democrat Presbyterian. *Society Aff:* ASCE, ACEC, ACSM, LES.

Huff, E Scott
Home: RR1, Box 255E, Hollis Center, ME 04042
Position: Prin/VP. *Employer:* BH2M, Engrs. Surveyors. Planners. *Education:* MS/Env Engg/OR State Univ; BS/CE/Univ of ME. *Born:* 9/29/50. Native of Bourne, MA. 1st position: engr, Camp Dresser & McKee Inc, Boston, MA, 1972-1973. Res engr, OR State Univ, 1974-1976. Environmental engr, Dale E, Caruthers Co, Gorham, ME, 1976-1978. Formed BH2M, Engrs, Gorham, ME in 1978. Present responsibilities include mgt of civ & environmental engg projs as well as client dev. Past Secy ME Sec, ASCE, 1977-79; currently President-Elect ME Sec, ASCE. *Society Aff:* ASCE, NSPE, CSI.

Huff, James E
Business: Process Engg, 566 Bldg, Midland, MI 48667
Position: Associate Process Consultant *Employer:* Dow Chem Co USA. *Education:* Dr of Engg/CE/Yale Univ; BS/CE/Univ of ID. *Born:* 4/7/28. Native of Moscow, ID. Employed by the Dow Chem Co since completion of grad work in 1955. Thirteen yrs in process res & pilot plant activities. Engaged in process design & troubleshooting work since 1968. Specialty areas include full- scale design from bench data, & analysis of runaway reaction hazards & emergency pressure relief system design. Chrmn of Midland Chapter Sigma Xi/RESA 1963; Chrmn of Mid-MI Sec AIChE 1971. Married, five chidren. Leisure activities include music, auto mechanics, & home improvement projs. 10 publications on chemical reactor design and overpressure protection. Visiting lecturer, U. Cal. at Santa Barbara, 1986-87. *Society Aff:* AIChE, Sigma Xi

Huffaker, Ray E
Business: 3321 E Slauson Ave, Los Angeles, CA 90058
Position: Asst V P. *Employer:* Ladish Pacific Div. *Education:* MEd/Education/Marquette Univ; BS Met E/Metallurgy/Purdue Univ. *Born:* Jan 1924. BS Met E Purdue Univ; M Education Marquette Univ. Native of Indiana. Taught in West Allis, Wisc Vocational School & at Univ of So Calif. Served with Army 1942-45, Officer in Corps of Engrs. Joined Ladish Co 1948 as met engr. Helped create Ladish Pacific Div 1955 as Ch Metallurgist. Presently Asst V P, Mgr of Operations. Fellow Amer Soc for Metals 1974. Served as Chmn of LA Chap ASM, Chmn of Westec 1965 & have been on sev natl ASM cttes. Enjoy jogging & working with Boy Scouts, YMCA & actively involved in local, district & confwide church programs.

Huffman, G David
Business: 611 N. Capitol, Indianapolis, IN 46204
Position: Dir, Energy Engrg & Research Div; Prof Math *Employer:* Indianapolis Ctr for Advanced Res (ICFAR). *Education:* Post-Doctoral/Aero Engr/Imperial College, Sci & Tech; PhD/Mech Engr/OH State Univ; MSc/Mech Engr/OH State Univ; BESc/Engr Sci/Marshall Univ. *Born:* May 1939. Post-doctoral studies at Imperial Coll of Sci & Tech London, England. Dir of Energy Engineering & Research Division, ICFAR since 1974. Respon for tech direction of Exploratory Res progs incl aerodynamic analyses, air quality studies & energy sys res. Also Prof of Math Sciences Purdue Univ School of Science at Indianapolis. Formerly res engr at WPAFB; Sect Ch - Res Dept, Prin Scientist Exper Fluid Mech Res, Ch - Aerothermodynamic Res & Ch - Mech Res at Detroit Diesel Allison Div Genl Motors, visiting Scientist, AF Academy- Colorado. Hobby-tennis. *Society Aff:* AIAA, ASHRAE, ΣΞ.

Huffman, John R
Business: 2835 Gilroy St, Los Angeles, CA 90039
Position: Chrmn of the Bd. *Employer:* Semco, Sweet & Mayers Inc. *Education:* BS/Mgmt Engg/Carnegie-Mellon Univ; MS/Industrial Engg/Univ of Pittsburgh; PhD/Eng/UCLA. *Born:* Oct 7, 1920. Wife-Alice E, son Leroy J. Dept of Indus Engrg Univ of So Calif 1947-54; General Tire & Rubber Corp 1957-68; Aerojet Gen Corp 1968-70; Semco, Sweet & Mayers Inc 1970-81. Intl Pres IMMS 1977-78; Pres LA Chap IIE 1976-77; Chmn Northern Ohio Chap Inst of Mgmt Sciences; Chmn Indus Engrg Div ASEE; Editorial Bd JOURNAL OF INDUSTRIAL ENGRG 1964-68. Mbr IIE, IMMS, TIMS; Hon Mbr Pi Tau Sigma; Assoc Mbr Sigma Xi; Prof Certified in Matl Handling & Matl Mgmt. Licensed Mech Engr (IN & CA) Licensed Industrial Eng (CA). Currently respon for dev of criteria & economic evaluation of warehousing & material handling systems and related data processing systems. *Society Aff:* IIE, IMMS, TIMS.

Hufnagel, Robert E
Business: 100 Wooster Hts Rd, Danbury, CT 06810
Position: Chief Scientist (1984-) *Employer:* Perkin-Elmer Corp *Education:* PhD/Eng Physics/Cornell Univ; BEP/Eng Physics/Cornell Univ *Born:* 5/24/32 Native of NJ. Joined Perkin-Elmer Corp in 1959. Dir of research 1967-1976. Tech dir of Optical Group 1976-84. Chief Scientist of Optical Group since 1984. Research, development, and publications in atmospheric optics, image formation, image quality, laser systems, stochastic systems analysis, and feedback control systems. *Society Aff:* OSA, IEEE

Huggins, William H
Business: 34th & Charles, Baltimore, MD 21218
Position: Prof of Elec Engrg. *Employer:* Johns Hopkins Univ. *Education:* ScD/EE/MIT; MS/EE/OR State Univ; BS/EE/OR State Univ. *Born:* Jan 11, 1919 Rupert, Idaho. Also studied precipitation-static radio interference while serving as

Huggins, William H (Continued)
an instructor there. Joined Radio Res Lab Harvard Univ 1944. 1946-54 civilian engr with Air Force Cambridge Res Ctr. Received Science Doctorate MIT 1953 with dissertation on 'A Theory of Hearing'. Joined the faculty of The Johns Hopkins Univ 1954 as Prof of EE. Was Chmn of Dept from July 1 1970-July 1, 1974. Chmn, Dept of Mech Sciences, June 1, 1978-June 30, 1981. Currently is interested in educational uses of computers & iconic communications. Recipient of Browder J Thompson Memorial Award 1948; 1955 Annual Award of Natl Electronics Conf; AF Decoration for Exceptional Civilian Service 1954; Christian R & Mary F Lindback Award for Disting Teaching 1961; Western Elec Fund Award for Excellence of Instruction of Engrg Students, Middle Atlantic Sect ASEE 1965; IEEE Education A. *Society Aff:* NAE, IEEE.

Hughen, James W
Business: 1020 Cromwell Bridge Rd, Baltimore, MD 21204
Position: VP *Employer:* Kidde Consultants, Inc *Education:* B/CE/Structures/Dartmouth Coll *Born:* 8/27/22 in Rockford, AL. Served with US Marine Corps for 4-1/2 years during World War II. Worked for J. E. Greiner Co, Baltimore, MD for 24 years on design of bridges; such as: Mystic River Bridge - MA, Delaware River Bridge - PA, Chesapeake Bay Bridge - MD, Cuyahoga River Bridge - OH, Susquehanna River Bridge - MD, Kanawha River Bridge - WV, Francis Scott Key Bridge - MD since 1972. With Kidde Consultants, Baltimore, MD since 1972. VP in charge of Structural and Transportation Divisions. *Society Aff:* ASCE, NSPE, MSPE, ESB, AISC, ACI

Hughes, Edwin L
Home: 447 Pauma Valley Way, Melbourne, FL 32940
Position: Conslt *Employer:* Self *Education:* MS/EE/Univ of IL; BS/EE/MO Sch of Mines *Born:* 08/11/24 Ed is currently a technology mgmt advisor to indust. He has done res in digital computers, optical storage and subscription television. He has managed large projs in inertial navigation, xerographic sys, high-speed printing and computer printers. Ed was Pres and CEO of FL Data Corp, VP of Engrg at Santec Corp, VP of Prod Dev (several titles) at Xerox Corp, and Technical Dir for Gen Motors. He is also listed in Who's Who in the East, Who's Who in the South, Who's Who in America, Who's Who in Finance and Industry, and has been honored as an outstanding alumnus by both the Univ of IL and MO School of Mines for his achievements in engrg mgmt and prod dev. He was honored at the 1985 NCC for his early contributions as a digital computer pioneer at the Univ of IL, 1949-1953. *Society Aff:* AAAS, NSPE, IEEE, TBΠ, HKN, ΦΚΦ

Hughes, Ian F
Business: 3210 Watling Street, E Chicago, IN 46312. (Mail Code 2-100)
Position: VP of Res & Qual *Employer:* Inland Steel Co. *Education:* PhD/Metallurgy/London Univ; BSc/Metallurgy/London Univ. *Born:* Apr 14, 1940 Isle-of-Man, England. Co-op prog Richard Thomas & Baldwin Steel Co 1958-62. BSc London 1962. PhD London 1966. Fellowship at Natl Physical Lab England 1965-67. Working on texture generation in sheet steels. Joined Inland Steel 1967; continued res in deep drawing & high strength steels. Became Gen Mgr of Res in 1984 and assumed current position of Vice Pres of Res & Qual in 1986. Mbr ASM; AIME; Past Pres American Deep Drawing Res Group. Interested in music, arts, travel & youth soccer. *Society Aff:* ASM, ADDRG, SAE, AIMMPE, MS, IEEE.

Hughes, Joseph Brian
Business: 1670 Kalakaua Ave 604, Honolulu, HI 96814
Position: President. *Employer:* J Brian Hughes & Associates, Inc. *Education:* BSc/Civil Engg/London Univ. *Born:* Feb 1931. BSc from London Univ. Chartered Engr UK; Reg Engr Hawaii & Guam. Royal Engrs, British Army, Middle East 1952-54. Design & Asst Resident Engr for Freeman Fox & Partner, London 1954-57. Engr with PWD British Solomon Islands 1957-59. Sr Engr W S Atkins & Partners, London 1959-61. Sr Engr PWD Queensland, Australia 1961-64. Sr Engr Alfred Yee & Assocs Honolulu 1964-66 & Peter Hsi Assocs 1966-67. Pres J Brian Hughes & Assocs since 1967. Respon for client relationships & overall design programs, for design of hi-rise, indus & commercial structures. *Society Aff:* ICE, CEE, SEAOH, ACI.

Hughes, Richard R
Business: 1415 Johnson Dr, Madison, WI 53706
Position: Prof of Chem Eng *Employer:* Univ of WI. *Education:* SB '42/Chem Engrg/MIT; SM '47/Chem Engrg/MIT; ScD '49/Chem Engrg/MIT. *Born:* 3/4/21. Raised in India, England, USA. Grad into US Army, - bomb disposal officer (42-44, Mediterranean), instr (44-46, W Point). MIT grad res with E R Gilliland. With Shell Dev: res engr (49-55); process engr (56); refinery technologist (57); supervisor (of group that dev CHEOPS) (57-62); Prof at ChE (68-), - five PhD's, seven MS', on comp methods & res, campus-wide computing activities, bldg planning. AIChE - fellow (71-), dir (69-71), cttes & divs. Cast Award for 1979, Founders Award 1980; VP 1981, Pres 1982, Editor, "Computers and Chem Engineering–. Married, three daughters, six grandchildren. Avocations: music, hiking, golf. *Society Aff:* AIChE, ACS, ACM, ORSA, TIMS, ASEE.

Hughes, Thomas J
Business: Div of Appl. Mechs, Durand Building Rm 252, Stanford, CA 94305
Position: Prof and Chrmn of the Div of Appl Mech *Employer:* Stanford University *Education:* PhD/Engrg Sci/UC Berkeley '74; MA/Math/UC Berkeley '74; MME/ME/Pratt Inst '67; BME/ME/Pratt Inst '65. *Born:* 8/3/43. Has held ind positions at Grumman Aircraft Engg and Gen Dynamics/Elec Boat Div. With Stanford Univ. since 1980. Presently Prof of Mech Engrg and Chrmn of the Div of Appl Mech. Main res interest is computer techniques for fluid, solid, and structural mechanics problems. A mbr of Phi Beta Kappa and Sigma Xi. Received Huber Res Prize, ASCE 1978, and Melville Medal, ASME 1979. Fellow, Amer Acad of Mech, Fellow ASME. Consultant to govt and private industry. *Society Aff:* ASME, ASCE, AIAA, SIAM.

Hughes, Thomas P
Business: Dept Hist & Soc of Sci, Philadelphia, PA 19174
Position: Prof History of Technolgy. *Employer:* University of Pennsylvania. *Education:* BME/Mech Eng/Univ of VA; PhD/History/Univ of VA. *Born:* 9/13/23. in Richmond, VA. Taught History of Tech at MIT & the Inst of Tech, SMU. In 1971, won the Dexter Prize of the Soc for the History of Tech & the Book Award of the TX Inst of Letters for Elmer Sperry (Johns Hopkins 1971). Chrmn of the Historical Adv Ctte of NASA; Mbr NASA Advisory Council; Mbr of the US Natl Ctte for the History & Philosophy of Sci. Pres of the Soc for the History of Technology. History of Technology Editor for Johns Hopkins Univ Press. Council of History of Sci Soc. *Society Aff:* SHOT, HSS, AHA.

Hughes, William L
Home: RR 1, Box 641, Stillwater, OK 74074
Position: President, In En Corporation *Employer:* Oklahoma State Univ. *Education:* PhD/EE/IA State Univ; MS/EE/IA State Univ; BS/EE/So Dak Sch of Mines & Tech. *Born:* 1926. US Navy WWII. Prof of EE Iowa State Univ until 1960. Hd Sch of EE Oklahoma St Univ until 1976; Instigator & Dir of Res programs in Color Television at ISU & Alternate Energy programs at OSU. Lic Amateur & Comm Radio Operator. Comm single & multi engine instrument rated pilot. Listed in Who's Who in America, Who's Who in Southwest etc. Fellow IEEE; active on several National IEEE Cttes, cons to many industries & a few foreign govmts. Hobbies: reading history of world religions, hunting, fishing & beekeeping. *Society Aff:* IEEE.

Hughto, Richard J
Business: 235 West Central St, Natick, MA 01760
Position: VP *Employer:* Rizzo Associates, Inc. *Education:* Ph.D./Water Res Eng./Cornell University; M.E./Env. Eng./Manhattan College; B.E./Civil Eng./Manhattan College *Born:* 10/28/50 Dr. Hughto is currently responsible for all hazardous waste site investigations and cleanup work undertaken by Rizzo Assoc. The work includes field investigations, remedial action designs and implementation. He was formerly employed by Camp Dresser & McKee in Boston, where he was Regional Mgr for a nationwide Superfund contract for the USEPA. He also completed numerous projects involving surface and groundwater modeling of the impacts of contaminants on receiving waters. Dr. Hughto also had experience at Cornell Univ as a researcher and instructor, and at the USEPA in the Region II office as a water quality modeling engineer. *Society Aff:* ASCE, WPCF, HMCRI, ACEC

Hugo N Halpert
Home: 16548 Wikiup Road, Ramona, CA 92065-4646
Position: President. *Employer:* Halpert Engineering Consultants, Inc. *Education:* MSE/Soil Mech/Univ of MI; BSE/Soil Mech/Univ of MI. *Born:* Jan 10, 1918. Worked for USED prior to WWII. Served with Army Corps of Engrs 1943-46. Soil & Fdn Engr for Veterans Admin 1949-54. Chief Soil Engr on Mass Turnpike for Howard, Needles, Tammen & Bergendoff 1954-57. Founded predecessor firm of Halpert Assocs in 1957 & have served as Chief Exec Officer since that date. VP, Nevada CEC 1963-64. Mbr ASCE, ASFE. Reg Prof Engrg in several states. *Society Aff:* ASCE, ASFE.

Hukill, Emory G
Business: 7013 Krick Road, Bedford, OH 44146
Position: President. *Employer:* Hukill Chemical Corporation. *Education:* BS/Chem Engg/Case Inst Tech. *Born:* 7/4/14. *Society Aff:* ACS, AIChE, ΑΧΣ, ΣΧ.

Hukill, William V
Home: 511 20th Street, Ames, IA 50010
Position: Retired. *Education:* BS/Mech Engg/OR Agri Col. *Born:* Oct 5, 1901 Corvallis, Ore. Married April 1, 1925 to Maybelle Coffman. Employed in res by US Dept Agri 1924-68. Retired Jan 1968. Fellow ASAE; Life Mbr ASHRAE; Mbr Sigma Xi, Mbr Gamma Sigma Delta. On Grad Fac of Iowa State Univ from about 1950. Received John Deere Gold Medal from ASAE in 1960. Received Golden Plate from Acad of Achievement in 1961. Res incl preservation of fruits & veges; also drying & storage of grains. In charge of joint res proj USDA & Iowa State Univ 1943-67. *Society Aff:* ASAE, ASHRAE.

Hulbert, Lewis E
Business: 505 King Ave, Columbus, OH 43201
Position: Research Leader *Employer:* Battelle Columbus Division *Education:* PhD/Eng. Mech/The Ohio State University; MS/Mathematics/Case Inst. of Technology; BS/Chemistry/Iowa State College *Born:* 11/15/24 Joined Battelle in 1952. Specialized in applications of the computer to solution of problems in heat transfer and diffusion theory, stresses in thermally & mechanically loaded plates, shells & solids. Pioneered in the application of the boundary point least squares method for stress analysis of bodies with complex shapes, 1962-72. Formed & managed a group for the development of advanced methods in structural analysis, 1967-73. Assumed current position as Research Leader providing technical & admin direction to large projects in 1973. Fellow of ASME. VP of Systems & Design Operating Group, 1985-89. Dedicated service award 1985. *Society Aff:* ASME, ΣΞ

Hulbert, Samuel F
Business: 5500 Wabash Ave, Terre Haute, IN 47803
Position: Pres. *Employer:* Rose-Hulman Inst of Tech. *Education:* PhD/Ceramic Science/Alfred Univ; BS/Ceramic Engineering/Alfred Univ. *Born:* 4/12/36. A native of Adams Center, NY, Dr Hulbert is an internationally-recognized authority on the engg aspects of developing artificial body parts and the use of carbons and ceramics as implant materials for the replacement of bone and teeth. He graduated from Alfred Univ where he received his BS in ceramic engg (1958), and the PhD in ceramic science (1964). Prior to becoming Pres of Rose-Hulman Inst of Tech, he had been dean of engg at Tulane Univ for three years. There he also held the dual appointment as prof of bioengineering and adjunct prof of biomaterials in the Dept of Surgery and the School of Medicine. His distinction as an engr, researcher and consultant has been augmented by the publication of more than 200 articles on his work. President of the Second World Congress on Biomaterials, Washington, D.C., 1984 *Society Aff:* ASEE, ASAIO, ASTM, AAMI, BMES, ACerS, ESB.

Hulbert, Thomas E
Business: Snell Engineering Ctr Rm 120, 360 Huntington Ave, Boston, MA 02115
Position: Dir, School of Engrg Technology and Assoc Dean, Engrg *Employer:* Northeastern Univ. *Education:* MS/Engg Mgt/Northeastern Univ; BMgtE/-/RPI. *Born:* 9/1/35. Native of Glens Falls, NY. Held positions as Ind Engr & Sr Ind Engr at Armstrong Cork Co & Raytheon Co from 1957-63. Joined faculty at Northeastern in 1963 as Instr. Promoted to Asst Prof then Assoc Prof in 1967. Served as Assoc Dir of Office of Educational Resources from 1967-69. Joined Dean's staff in 1969 as Asst Dean. Served on Governor's Mgt Task Force (MA) Sept-Dec 1975. Acting Dean May 1979 - July 1981. Appointed Dir of Lincoln Coll now School of Engineering Technology July 1982. Mbr Univ Delegation & Leader of an invited Delegation of Engrs to People's Republic of China in March and Sept 1980. Lecturer, Business Mgmt Seminar in PRC June 1984. Served as Conslt to Ind on long range planning, scheduling, methods, measurement, mgmt inventive plans, facilities design & productivity improvement. Past natl officer of IIE. Avid skier, volunteer ski patroller and CPR instr. *Society Aff:* IIE, ASEE, IIE.

Hulburt, Hugh M
Home: 2040 Thornwood Ave, Wilmette, IL 60091
Position: Prof of Chem Engg & Assoc Dean of Technol Inst *Employer:* Northwestern Univ. *Education:* PhD/Phys Chem/Univ of WI; MS/Phys Chem/Univ of WI; BA/Chemistry/Carroll College. *Born:* 10/27/17. Nashua, NH, Oct 27, 1917; s Clarence Hellings & Alice (McKinney) H; BA, Carroll Coll, Waukesha, WI, 1938; MS Univ WI, 1940, PhD, 1942; NRC fellow chemistry, Princeton, 1942-43. Sr res chemist Shell Oil Co, 1943-44; instr chem engg Cath Univ Am, 1946-51; supr engr, then dir res & dev Chem Constrn Co, NYC, 1951-56; with Am Cyanamid Co 1956-63, dir phys res, central res div, 1959-63; prof chem engg Northwestern Univ, 1963-, chrmn dept, 1964-70, assoc dean Grad Sch, 1975-1980; Assoc. Dean, Technological Inst, 1980- . Reilley Lectr Univ Notre Dame, 1967; vis prof Swiss Fed Tech Inst 1971; cons in field. Editor: IEC Process Design & Dev, 1962-. Contbr articles to profl jours. Home: 2040 Thornwood Av Wilmette IL 60091 Office: Dept. of Chem . Eng., Northwestern Univ Evanston IL 60201. *Society Aff:* AIChE, ACS, ASEE, APS, ECS.

Hull, David R
Home: 1900 S Ocean Blvd, Pompano Beach, FL 33062
Position: Cons-Electonics Mgmt. *Employer:* Self. *Education:* MS/Electronics/Harvard; BS/-/US Naval Acad. *Born:* Oct 29, 1903. Prof employment: US Navy 1921-48, final assignment - Bureau of Ships, Asst Ch for Electronics - retired Cpt; IT&T Corp, Tech Dir; Raytheon Co, V P; since 1960 Cons - Electronics Mgmt. Orgs: Inst Elec & Electronic Engrs, Life Fellow; Acoustical Soc of Amer, Fellow; Soc of Naval Engrs, Mbr Emeritus; Electronic Indus Assn, Pres 1958-60, Dir 1956-60, Hon Mbr. Awards: EIA Medal of Honor, US Navy Legion of Merit, Commendation Medal, 6 campaign medals. *Society Aff:* IEEE, ASA, ASNE, EIA.

Hull, Frederick C
Home: 109 Lavern St, Pittsburgh, PA 15235
Position: Consulting Metallurgist (Self-employed) *Education:* DSc/Met/Carnegie Mellon Univ; BSc/Met/Univ of MI. *Born:* 11/9/15. Consulting Metallurgist. With Westinghouse Res Lab from 1941 to 1981,. Mgr Met Sec 1954, Asst Mgr Met Dept 1957, Advisory Met 1957 to 1981. Alloy dev include precipitation hardened superalloys (DISCALOY), high strength weldable stainless steels (KROMARC58) and non-magnetic retaining rings. Other areas of interest are nucleation & growth of pearlite, failure analysis, grain size measurement & control, studies of sizes & distribution of grain shapes of metals, levitation melting, hot cracking tests, control of delta ferrite & sigma phase, & weldability. Awards: 1962 Dudley Medal ASTM),

Hull, Frederick C (Continued)
1974 Lincoln Medal AWS), 1978 elected Fellow of ASM. *Society Aff:* ASM, AIME, ASTM.

Hull, Maury L
Business: Dept of Mech Engg, Univ of Calif, Davis, CA 95616
Position: Assoc Prof *Employer:* Univ of CA, Davis. *Education:* BS/ME/Carnegie - Mellon Univ; MS/ME/Univ of CA, Berkeley; PhD/ME/Univ of CA, Berkeley *Born:* 7/18/47. Undergrad study was completed at Carnegie-Mellon Univ in 1969. Subsequent grad studies at the Univ of CA, Berkeley, culminated with a PhD degree in 1975. Following a year's postdoctoral work at Berkeley, he joined the faculty at Davis. His res interests ctr in the fields of instrumentation, experimental mechanics, biomechanics, mech properties of materials, and engg design. Prof Hull is also a consultant in litigation proceedings dealing with machine safety, snow skiing iccidents, and product safety in general. *Society Aff:* ASME, ASB, ISSS, ISBS.

Hull, William L
Business: 1206 W. Green St, Urbana, IL 61801
Position: Prof of Mech Engrg Emeritus *Employer:* University of Illinois. *Education:* BS & ME/ME/Univ of CO; M(AE)/Automotive/Chrysler Inst of Eng; MS/ME/Purdue. *Born:* April 1913. MS from Purdue Univ; BS & ME from Univ of Colorado; M (AE) from Chrysler Inst of Engrg. Engr with Chrysler Corp 1935-37. Taught at Purdue Univ 1937-40; Univ of Colorado 1940-47; Univ of Ill, Urbana since 1947. Prof of ME, Hd of Thermal Sys Div, Dir Automotive Engrg Lab. Res on fuel volatility & mixture distribution for Army Air Corps during WWII. Res on diesel engine combustion for Caterpillar Tractor Co 195662 & Internatl Harvester Co 1966-75. Chmn, Central Ill Sect SAE 1963-64. Mbr of Bd/Dir of SAE 1969-71. Fellow mbr of SAE. *Society Aff:* SAE.

Hulley, Clair M
Home: 11560 Deerfield Rd, Cincinnati, OH 45242
Position: Prof of Engg & Computer Graphic Sci. *Employer:* University of Cincinnati. *Education:* ME/Mech Engg/Univ of Cincinnati; BS/Engg/Univ of Cincinnati. *Born:* 3/6/25. Native Cincinnatian. Prof Industrial Engineering 1981. Prof Engg & Computer Graphic Sci 1977. Prof Engg Graphics-Coordinator Computer Graphics 1976, Prof in Charge Engg Graphics 1969-present. Teaching in Mech Engg. Engg Graphics & Engg Analysis Depts since 1947. Texts on Computer Graphics-Hulleytran Language 1970, Computer Plotting 1974, Coordinate Geom 1974, Contouring 1976. Matrix Methods in Computer Graphics 1977, Efficient Computer Programming 1980. Bits, Bytes and Binary Bumping 1981. Four articles on Computer Graphics in ASEE Journals. Cons (Computer Related) Louvers Inc - Cost Analy, Allis-Chalmers - Centrifugal Pump Patterns, Genl Elec - Tuned Boring Bars, Tool Steel Co - Welded Gear Box Design, Tech Dev, Airfoil Logic-Tensgridity Structures-NASA, 1977. Hobby: horticulture & oil painting. 1983 Life Member A S Me E; 1983 Award for Long Meritorius Dedication To Engrg Ed (One from faculty and one from ME And IE student body). Resolution of Appreciation from Alumni Organization for Obtaining Legislation to Promote Faculty Wellbeing. *Society Aff:* ASME, ASEE, AAUP.

Hulm, John K
Business: Bldg 801-3C54 - 1310 Beulah Rd, Pittsburgh, PA 15235
Position: Dir, Corp Research *Employer:* Westinghouse Corp *Education:* PhD/Physics/Cambridge Univ, England; MA/Physics/Cambridge Univ, England; BA/Physics/Cambridge Univ, England *Born:* 7/4/23 Thirty years experience in Materials and Solid State Tech, with special emphasis on Superconductors. Developed first Type II superconductors, first high-field magnets, superconducting generators, etc. Wetherill medal of Franklin Inst (1964), Intl Prize for New Materials, American Physical Society (1979). Elected National Academy of Engrg (1980). US Science Attache to UK (London) 1974-76. Assist Prof Physics. Univ of Chicago (1949-54). Westinghouse Electric Corp 1954-74, 76-present. Deputy head of Central R&D (current). *Society Aff:* ACS, NAE, AAAS

Hulsey, J Leroy
Business: 4915 Waters Edge Dr, Suite 170, Raleigh, NC 27606
Position: Pres *Employer:* Civil Engrg & Applied Research, Inc *Education:* PhD/Structural Engrg/Univ of MO-Rolla; Post Grad/Structural Engrg/Univ of IL; MS/CE/Univ of MO-Rolla; BS/CE/MO School of Mines and Metallurgy *Born:* 10/6/41 Summer 1964, Civil Engr, Soil Conservation Service; Oct. 1965-Sept. 1966, Design Engr, Daily & Assoc, Engrs, Inc, Champaign, IL; Sept. 1966-Aug. 1968, U. S. Army; Oct. 1968-Dec. 1971, Design Engr, Daily & Assoc, engrs, Inc; Jan. 1972-Dec. 1975, Grad. Student, Univ of MO-Rolla; Jan. 1976-Dec. 1979, Prof. of Civil Engrg, NC State Univ, Raleigh, NC; June 1979-Present, Pres of Civil Engrg & Applied Research, Inc, Raleigh, NC; June 1981-Present, Pres of Applied Computer Services, Inc, Raleigh, NC. Dr. Hulsey received outstanding teaching awards at the Univ of MO-Rola in 1972-1975. He also received an outstanding teaching award at NC State Univ in 1978-1979. Dr. Hulsey's work has been devoted to the development of state-of-art FEM analysis programs for soil-structure analysis, and bridge design and analysis. He currently serves on the ASCE Geotechnical analysis Committee, the TRB General Structures Committee, and is a licensed structural engr in IL. His past research activities include studies of semi- integral, end-bent, continuous composite girder bridges. Dr. Hulsey has also assisted many consultants and construction companies with special structural and soil problems in the building and specialized structural area. *Society Aff:* ASCE, ACI, ACEC, TRB, ΣΞ

Hulswitt, Charles E
2756 Tanglewood Drive, Wooster, Ohio 44691
Position: Vice President & Asst General Mgr. *Employer:* Astro Metallurgical Corporation. *Education:* BSChE/Chem Engg/Purdue Univ. *Born:* May 1928. BS Chem Engrg Purdue Univ 1956. Engrg Mgr Carbon Prods Div, Union Carbide Corp 1956-72. Also with D Q Kern Assocs specializing in heat transfer & thermal design 1962-71. Joined Astro Met Corp in 1972 as V Pres Engrg & subsequently named Asst Genl Mgr of the Corp. Prof specialties heat & mass transfer. Served with AUS 1952-54. Mbr Amer Inst of Chem Engrs, Amer Soc for Metals & Natl Assn of Corrosion Engrs. Author of sev articles on chem engrg processes. *Society Aff:* AIChE, ASM, NACE.

Hultberg, Dwain R
Home: 448H Rockdale Rd, Follansbee, WV 26037
Position: Sr Res Eng *Employer:* Natl Steel Res & Dev Labs. *Education:* MS/Chemistry/Carnegie Mellon Univ; BS/Chemistry/WVA Wesleyan Coll. *Born:* 10/7/46. Native of Follansbee, West Virginia Presently employed as a Supervisor - Tech Dev Dept - Weirton Steel Co. Respon for eval the surface quality of steel strip prods through all phases of processing operations. Involved in dev & refining those mill processes relating to the high speed cleaning & plating of steel strip. Experience in area of various types of filtration applied to steel processing. Amer Electroplaters Soc - Pgh Sect; Past Mbr. Inst of food Technologists Pgh Sect. Experience with high temperature porcelain enameling of steel through all process of customer use as well as steel mill production of product. *Society Aff:* SME, AES, NMDA.

Humber, Philip M
Business: 218 Hospital Dr N.E, Ft Walton Beach, FL 32548
Position: Pres *Employer:* Humber/Almond/Buythe, Inc *Education:* BS/EE/MS State Univ *Born:* 1/9/42 Tupelo, MS. Graduated 1958 Tupelo High School. Worked as field engr for Gulf Power Co Pensacola, FL from June 63 to Feb 68. Worked Mar 68-Feb 69 as facilities design engr at Ingalls Shipbuilding Corp (Div of Litton) in Pascagoula. Entered consulting practice in Ft Walton Beach, FL in February 1969. *Society Aff:* NSPE, ACEC

Humenik, Frank J
Business: P.O. Box 7625, Raleigh, NC 27695-7625
Position: Prof & A Dept Hd in Charge Ext Bio & Ag Engr Dept. *Employer:* NC State Univ. *Education:* PhD/Sanitary Engg/OH State Univ; MS/Sanitary Engg/OH State Univ; BSCE/Civ Engg/OH State Univ. *Born:* 5/26/37. Draftsman H K Ferguson Co, Cleveland, 1956-58; Instr, OH State Univ, Columbus, 1967-68; currently Prof and Associate Dept Hd in Charge Extension Biological and Agri Engg Dept, NC State Univ; NC Man of Yr in Agr, *Progressive Farmer*, 1975; Duggar Lecturer Auburn Univ, 1977; ASAE Gunlogson Countryside Engg Award, 1978; reg PE, OH & NC; author of technical publications; reviewer of grant proposals, technical publications and res reports in area of agri waste mgt, water quality and environmental engg. *Society Aff:* ASAE, ASCE, WPCF.

Humenik, Michael
Home: 17097 Cambridge, Allen Park, MI 48101
Position: Dir, Mfg Processes Lab. *Employer:* Ford Motor Company. *Education:* ScD/Ceramics/MIT; BS/Glass Tech/Alfred Univ. *Born:* Nov 1924. Native of Garfield N J. Served as pilot USAAF 1943-45. With Ford Motor Co since joining in 1952 as res scientist. Major prof activity: physico-chem aspects of cermet sys, dev of cermet matls & powder met. Held various pos in Res & currently Dir of Mfg Processes Lab, Res Staff. Mbr of ASM, AIME & Fellow Amer Cer Soc. *Society Aff:* ASM, AIME, ACS.

Hummel, David M, Jr
Business: P.O. Box 30615, Billings, MT 59107
Position: Area Construction Engineer Associate, Member Board of Directors *Employer:* Northern Engrg & Testing, Inc *Education:* MS/Civil Engrg/Stanford Univ; B Engr/Civil Engrg/Yale Univ *Born:* Nov 22, 1940. Native of Hamden Conn. Employed by Bechtel Corp 1963-69; Assoc with Empire Sand & Gravel, Billings 1969-76 - hwy, heavy & util const. COP Construction Co. 1976-1980 VP - Highway & Heavy Const. Pres of MT Sect ASCE 1974-75. Lic Prof Engr in MT & WY. Pacific NW Council, Chrmn 1978-79. Elected mbr to City of Billings Local Gov't Study Comm 1974-77. Outstanding Young Engr, Midland Empire Chap, Montana Soc of Engrs (NSPE) 1974. NSPE-PEC Mbr of Bd of Governors 1986- . Enjoy skiing, tennis & backpacking. *Society Aff:* ASCE, NSPE.

Hummell, John D
Business: 170 North High St, Columbus, OH 43215
Position: Managing Partner *Employer:* A. E. Stilson & Assocs *Education:* MS/ME/OH State Univ; BS/ME/OH State Univ *Born:* 3/28/24 A partner and mgr of a consulting engrg firm having a staff of approximately 200. Specializing in energy conservation, resource recovery systems, and air pollution control. Prior to entering the consulting field, had 16 years of experience in applied research in combustion, energy utilization, and performance of pollution control systems. Tech papers have been presented to professional assocs and patents have been issued on unique combustion systems. *Society Aff:* ASME, NSPE, APCA, NACE

Hummer, Robert H
Business: 1500 Planning Research Dr, McLean, VA 22102
Position: Dir., Corp. Facilities *Employer:* Planning Research Corp. An Emhart Subsidiary *Education:* B.S./B.C./Virginia Polytechnic Institute *Born:* 04/25/44 A native of Millwood, Va., Hummer received his BS degree from Virginia Polytechnic Inst and has earned graduate credits in Bus Admin. Corporate Plant Engrg staff of Boise Cascade 1968-72. Dir of Dev, Days Inns of America 1972-76. Commercial Dev 1976-80. With Planning Research Corp since 1981. Assumed current responsibilities in 1984. First became Mbr Amer Ins of Plant Engrs 1968, currently VP - Public and Professional Affairs and Bd of Dir. Active in the community, now Chairman - Health and Safety, National Capital Area Council, Boy Scouts of America. *Society Aff:* AIPE

Humpherys, Allan S
Business: Rt 1 Box 186, Kimberly, ID 83341
Position: Agricultural Engineer. *Employer:* USDA-Agricultural Res Service *Education:* MS/Civil Engg/UT State Univ; BS/Agri Engg/UT State Univ. *Born:* April 1926 Idaho Falls, Idaho. US Navy 1945-46. Grad studies Colo State Univ 1962-63. Res Asst Utah Agri Experiment Sta 1954-58, res in canal & reservoir linings, matls & techs for seepage control & water harvesting. Agri Res Serv USDA 1958- , spec in design, testing & dev of automated farm irrigation sys & farm water measurement. Author of tech & semi-popular publs. Chmn Pacific Northwest Region ASAE 1974; Mbr of ASAE, ASCE & ICID tech cttes. Mbr ASAE, ASCE, USICID; IA. Religion LDS. Enjoy camping, hiking & fishing. *Society Aff:* ASAE, ASCE, ICID, IA.

Humphrey, Albert S
Home: 4030 Charlotte St, Kansas City, MO 64110
Position: Chairman & Chief Exec Officer. *Employer:* Business Planning & Dev, Inc. *Education:* MS/Chem Engrg/MIT MBA/Finance-Mktg/Harvard Bus Sch; BS/Chem Engrg/Univ of IL. *Born:* 6/2/26. Non-exec Dir of AcquaMedia, Triad, Galley West Inc, Mannings, Delta Electronics, Intergrated Graphics, K F Beer GmbH Dusseldorf & Currently Director of: Tower Lysprodukter a/s Norway; Tower Lamp Co Ltd England. *Currently a Dir of: SANBROS Ltd London England, Tower Lamp Co Ltd Huntingdon, England Intl Computers Limited, Birmingham England Hart, Brown and Curtis Ltd London England. Amer Inst of Chem Engrs; Sci Res Soc of Amer, Sigma Tau, Sigma Xi, Tau Beta Pi, Dictionary of Intl Biographies, Leaders of the English Speaking World Royal Blue Book; Who's Who in the World, Who' Who in Amer Ed & Sci, Who's Who in CA, Who's Who in the West, The Natl Registry of Prominent Americans. Cons: NASA; Rolls Royce 1971 Ltd; W H Smith & Son Ltd London; Univ City Sci Ctr; Roche Prods Ltd; Bemrose Ltd; Leckenbee Steel; Anglia Canners Ltd; J Lyons/Tetley Tea; Mapleton Foods Ltd; Rist's Wires & Cables Ltd Caledonian Tractor & Equipment Co Ltd; Food Securities Group-British Foods; Seguros America Banamex SA. Mexico; Nordisk Aluminium a.s. Norway. Advanced Services plc (UK) and Thomas Waide and Sons Ltd (Leeds England).* *Society Aff:* AIChE, IM, ID, HAA, MITAA, UNIV OF IL AA, ΦΔΘ.

Humphrey, Arthur E
Home: RD 7 Quartermile Rd, Bethlehem, PA 18015
Position: T.L. Diamond Prof of Biochem Engg; Dir, Ctr for Molecular Biosci and Biotech. *Employer:* Lehigh University *Education:* PhD/ChE/Columbia; MS/Food Tech/MIT; MS/ChE/Univ ID; BS/ChE/Univ ID *Born:* 11/09/27 Native of Moscow, Idaho. Taught at the University of Pennsylvania for 27 yrs, serving as Dir of the School of Chem Engrg for 10 yrs and Dean of Engrg and Applied Sci for 8 yrs. VPres and Provost of Lehigh University for 6 years. Presently, T.L. Diamond Prof of Biochem Engg. Field of speciality-Biochemical Engrg/ Fermentation Tech. Author of three texts and over 200 professional publications. Mbr Natl Acad of Engr, AIChE Professional Progress Award 1972, AIChE Food, Bioengineering, and Pharmaceutical Award 1973, AIChE annual Lecture Award 1974, Fulbright, Univ. Tokyo, Japan 1964, Fulbright, Univ. New South Wales, 1972. Dir Cartech, Former Dir NBS, ABEC, Inc. *Society Aff:* AIChE, ASM, SIM, ASEE, ACS

Humphrey, James E, Jr
Home: 1357 Delong Road, Lexington, KY 40515
Position: Professional Engineer. *Employer:* The Webb Companies. *Education:* MSCE/Structures/Univ of KY; BSCE/Civil/Univ of KY. *Born:* 8/25/30. 2 yrs in US Army Corps of Engrs, Panama. Assoc of Kroboth Struct Engr 9 yrs, designing bridges & bldgs. V P of Young Engrs 3 yrs, V P of Consolidated Devs, Inc 7 yrs where respon for design & const of 16 shopping ctrs. Design & const mgr for Commercial Projs at White & Congleton 3 1/2 yrs. Pres. of Unity Structures, general contractors, 2 yrs. Project Manager at P.D.R. Consul. Engrs., 3 yrs. PPres of KY sect of ASCE; P. Natl. Dir. ASCE (KY ID & OH). PPres of Bluegrass Chap, KSPE; PPres of KSPE, received Distinguished Service Award in 1987. Active in civic & church affairs. Presently responsible construction management of projects for The Webb Companies. *Society Aff:* ASCE, NSPE.

Humphrey, Watts S
Home: Eight Hanson Rd, Darien, CT 06820
Position: Director, Technical Assignment *Employer:* IBM Corporation. *Education:*

Humphrey, Watts S (Continued)
MBA/Bus/Univ of Chicago; MS/Physics/IL Inst. of Tech; BS/Physics/Univ of Chicago. *Born:* 7/4/27. Computer Engr with Sylvania Elec Prods. Taught grad courses in switching circuits at Northeastern Univ. Author of book *Switching Circuits with Computer Applications.* With IBM since 1959. Various tech & mgmt pos incl, Dir of Programming; Dir of Endicott Dev Lab; Dev Div V P Tech Dev; Corp Dir, Policy Dev; Assumed current pos of Dir, Technical Assessment 1979. Holds 5 pats in computer design. P Chmn of the Boston Sect of the Computer Group at the IRE. Fellow of IEEE, Member ACM, Editorial Bd IEEE Spectrum *Society Aff:* IEEE.

Humphreys, Clark M
Home: 6805 Mayfield Rd, Apt 206, Mayfield Heights, OH 44124
Position: Retired. *Education:* BME/Mech Eng/OH State Univ; ME (Honorary)/-/OH State Univ. *Born:* Feb 9, 1902 Kenton, Ohio. Mech Engr in State Arch's Office Columbus, Ohio 1925-31. Asst Prof of Mech Engrg, Carnegie Inst of Tech 1931-46. Sr Engr & for last 4 yrs Asst Dir of ASHRAE Res Lab Cleveland, Ohio 1946-61. Mech Engr & for last 3 yrs Ch of the Engrg Branch, NIOSH, USPHS Cincinnati, Ohio 1961-74. Retired Dec 1974. Reg Prof Engr in Pa & Ohio. Fellow in ASHRAE. Author or co- author of approx forty tech articles. *Society Aff:* ASHRAE.

Humphreys, Jack B
Business: Dept of Civil Engg, The Univ. of Tennessee, Knoxville, TN 37916
Position: Prof. *Employer:* Univ of TN. *Education:* PhD/Civil Engg/TX A&M Univ; MS/Civil Engg/Univ of TN; BS/Civil Engg/Univ of TN. *Born:* 12/17/33. Native of Knoxville, TN. Officer in US Air Force, 1956-59. Bridge design engg with Mid-South Engg Co 1959-60. Asst Prof of Civil Engg, Univ of Southwestern LA 1962-64. Assoc Prof and Prof of Civil Engg, The Univ of TN 1966- present. Teaching and res in Traffic Engg and Highway Safety. Extensive consulting in traffic accident reconstruction and accident causation and in seminars on temporary traffic controls and governmental liability. Allstate Safety Crusade Certificate of Commendation, 1972. Hobbies: Waterskiing, Fishing. *Society Aff:* ITE, ASCE, TRB.

Humphreys, Kenneth K
Home: 305 Lebanon Avenue, Morgantown, WV 26505
Position: Executive Director *Employer:* American Assoc. of Cost Engineers *Education:* MS/Matls Sci Engr/WV Univ; BS/ChE/Carnegie Inst of Tech *Born:* Jan 1938. Native of Pittsburgh, Pa. Held various pos wth US Steel Corp; Applied Res Lab 1959-65. Specialized in ore reduction, blast furnace operations & coal carbonization. With Coal Res Bureau, W Va Univ (1965-81) dir major state & fed res programs. Also, Asst Dean & Prof, Coll of Mineral & Energy Resources (1965-81). Exec Dir, AACE since 1971. Author of co-author of over 200 tech papers, books & pats primarily related to coal utilization &/or cost engrg. Veteran leader 30 yrs Boy Scouts; numerous scouting awards incl the Silver Beaver, Award of Merit & the Woodbadge. Secy, Intl Cost Engg Council; Recipient C.T. Zimmerman Founders Award, AACE 1979; Recipient Distinction of Het Shaap Met Vijf Poten, Royal Netherlands Industries Fair & Dutch Assoc of Cost Engrs, 1977; Fellow British Association of Cost Engrs; Fellow AACE; Cert Cost Engr, AACE; Cert Cost Engr, Sociedad Mexicana de Ingeneria Economica y de Costos; Registered Prof Engr PA and W VA. Vice-President 1980-81, President-Elect 1981-82, President 1982-83, West Virginia Society of Professional Engineers. Dir 1971, Exec Dir 1971-present AACE. *Society Aff:* AIME, AACE, NSPE, SMIEC, ACE, ASAE, CESSE.

Humphries, Kenneth W
Business: College of Engg, Columbia, SC 29208
Position: Assoc Dean & Prof. *Employer:* Univ of SC. *Education:* PhD/Geotechnical Engg/NC State; MS/Soils Engg/Univ of SC; BS/Civil Engg/Univ of SC. *Born:* 7/29/31. Worked with Union Carbide Chem one yr and Kline Iron & Steel Co in Columbia, SC for 2 yrs. Taught at USC since 1962 except while working on doctorate. Founded Consulting Co "Foundation Engg Consultants, Inc" in 1973 - Serve as Pres and Prin. *Society Aff:* NSPE, ASCE, ASTM, EERI, ASEE.

Humphries, Stanley, Jr
Business: Dept. Electrical and Computer Eng, Univ. of New Mexico, Albuquerque, NM 87131
Position: Prof *Employer:* Univ of NM *Education:* PhD/Nuclear Engrg/U of CA, Berkeley; MS/Nuclear Engrg/U of CA, Berkeley; SB/Phys/MIT *Born:* 02/25/46 Employed as staff mbr at Lawrence Livermore Lab (1971), Los Almos Natl Lab (1972-73) and Sandia Natl Labs (1977-82). Cornell Univ: Asst Prof (1973-77). Currently Prof at Univ of NM and Dir of Inst for Accelerator and Plasma Beam Tech. Author of numerous scientific articles and textbook *Principles of Charged Particle* Acceleration (Wiley, 1985). Spouse: Sandra F, children: Colin J and Courtney E. *Society Aff:* IEEE, APS

Hundal, Mahendra S
Business: ME Dept, Votey Bldg, Burlington, VT 05405
Position: Prof, ME. *Employer:* Univ of VT. *Education:* PhD/ME/Univ of WI, Madison; MS/ME/Univ of WI, Madison; BE/ME/Osmania Univ, India. *Born:* 11/25/34. Design Engr, Tata Steel Co, Jamshedpur, India, 1954-60. Asst Prof, San Diego State Univ, San Diego, CA, 1964-67. Joined Univ of VT as Assoc Prof in 1967. Promoted to Prof in 1977. Author of over 70 published papers and technical reports. Consultant in vibrations, acoustics, noise control and stress analysis. Reg P E *Society Aff:* ASME, ΣΞ, INCE

Hung, Tin-Kan
Business: Dept of Civ Engg, Pittsburgh, PA 15261
Position: Res Prof. *Employer:* Univ of Pittsburgh. *Education:* PhD/Mechanics & Hydraulics/Univ of IA; MS/Civ Engg/Univ of IL; BS/Hydraulic Engg/Cheng-Kung Univ. *Born:* 6/12/36. Born in Nanking, China. Before joining Univ of Pittsburgh in 1975, was at IA Inst of Hydraulic Res & then Carnegie-Mellon Univ. Res areas in Computational Analysis of Nonlinear Viscous Flows & Biomechanics. A recipient of 1978 Walter L Huber civ Engg Res Prize. Served on editorial bd of Journal of Engg Mechanics Div, ASCE; editorial bd of Journal of Hydraulic Research, IAHR;Chrmn of Task Committee of Bioengg & Human Factors; & Chrmn of Fluids Committee in EMD, ASCE; Conf Co-chrmn of ASCE Specialty Conf in Human & Social Factors in Civ Engg Planning & Design in 1975 and 1979. Enjoy classical music & ice skating. *Society Aff:* ASCE, ASME, IAHR, ΣΞ, SOC OF NEURO. SCI.

Hunkin, Geoffrey G
Business: 4061 South Eliot, Englewood, CO 80110
Position: Pres. *Employer:* Hunkin Engineers Inc and Ground Water Sampling, Inc. *Education:* BS/Mining, Geology & Met Engg/Sch of Mines, Cornwall, U.K. *Born:* 8/2/23 Native of Cornwall, UK. Served with Royal Navy Air Service, Pilot, 1941-45. Supervisory and mgt positions in; ultra-deep gold mining to 10,000 ft, India: uranium mining, USA, Australia and Canada: lead-zinc, Ireland: gold, copper and base metals, USA, Africa. Developed first commercial uranium in-situ leaching, Stanrock Uranium ML, Canada, 1957-61. Managed six-company consortium for process dev of in-situ uranium recovery, USA 1972-74. Consulting engr in private practice 1974 onwards, operating mainly Western USA, Canada, Australia. AIME McConnell Award, 1977. Soc of Mining Engrs, Distinguished Mbr, 1977. The Inst of Mining and Metallurgy, Fellow, 1965. CIMM Registered P.E. CO, NM, TX, WY, MT. *Society Aff:* NSPE, AIME, SME, SPE, CECC, ASTM, CIMM

Hunley, William H
Home: 449 Argyle Dr, Alexandria, VA 22305
Position: Pres *Employer:* William H. Hunley, Inc. *Education:* BSE/Naval Architecture & Marine Eng/Univ of MI; BME/Mech Engrg/George Washington Univ *Born:* May 1, 1925 Mathews Va; s. Arthur Kimmel & Nannie Banks (Ashberry) H. Engrg Draftsman, Bureau Ships, USN Wash D C 1950-51; mech engr 1951-53; naval arch 1953- ; Hd Hull Mech Sys Design Grps 1954-63; Design Coordinator for Seplane Tenders 1958; Dept Hd Hull Mech Sys Design 1963-66; Acting Hd Hull Mech Sys 1966-67; Branch Hd Deck Mech Sys, Naval Ship Engrg Ctr, Hyattsville, Md 1967-70; Hd Naval Arch Subdiv 1970-73; Ch Naval Arch 1973-79, Tech Dir for Ship

Hunley, William H (Continued)
Design, 1979-80. Washington Manager, CDI Marine Co. 1980-84. President, CDI Marine Co. of VA. 1983-84. Dir of Naval Engrg, CASDE Corp., 1984-1987. President, William H. Hunley, Inc. 1987- . Guest Lectr Univ Calif Statewide Lec Series Ocean Engrg 1966; Bd/Dir Waterford Foundation 1975-77; Naval Sea Sys Sr Civilian Exec Planning Bd, Chmn 1976-77; Served with Infy, AUS 1943-46; European African Mideast Theatre. SNAME: Fellow, awarded Life Mbrshp, Chmn Chesapeake Sect 1974- 75, Mbr Council 1976-84, Exec Comm 1979-82, Mbrship Chrmn 1979- . Assn Sr Engrs: Pres 1968-69, Exec Dir 1976; Honorary Member 1981; Quarterdeck. ASNE: Mbr, Mbr Council 1978- Chmn ASTM Shipbuilding Standards Committee 1981-82. Member Propeller Club 1981- Member American Management Association, Editorial Board, ASTM Journal of Testing & Evaluation 1984-87. *Society Aff:* SNAME, ASNE, ASTM, ASE

Hunsaker, Barry
Business: 5599 San Felipe Rd, PO Box 4412, Houston, TX 77210
Position: Exec VP *Employer:* Southern Natural Gas Co. *Education:* BS/ME/Univ of AZ. *Born:* 5/20/26. Native of AZ. Served in US Army AF during WWII. Joined El Paso Natural Gas Co 1947; successively, Jr Engr, Engr, S Engr, Proj Mgr, Chief Engr, Dir of Engg, Asst VP, VP. Exec VP El Paso LNG Co; Pres and Dir El Paso LNG Co and El Paso Marine Co; Dir the El Paso Co. Retired 1980; Mbr of Council, SNAME; Mbr, Bd of Managers, Am Bureau of Shipping; Trustee, Univ of St Thomas; Formerly Chief operating officer intl LNG business and world's largest fleet of liquefied natural gas tankers. Now Exec Vice Pres-Gas Supply and Director Southern Natural Gas Co. *Society Aff:* ASME, SNAME.

Hunsinger, Bill J
Home: Urbana, IL 61801
Position: Assoc Prof. *Employer:* Univ of IL. *Education:* PhD/EE/Univ of IL; MSEE/EE/Bradley Univ; BSEE/EE/Univ of IL. *Born:* 2/23/39. Native of Roanoke, IL. Worked as a design Engr for GE, Bloomington, IL, 1961 to 1966. Developed surface Acoustic Wave Devices and Systems for Magnovox Co from 1966 to 1974. Assumed the present joint appointment as Prof & Res Principle Investigator at the Univ of IL. Present res is in the Surface acoustic wave device area. Presently a Sr Mbr of IEEE. *Society Aff:* IEEE.

Hunsperger, Robert G
Business: Dept of Elec Engrg, University of Delaware, Newark, DE 19716
Position: Prof *Employer:* Univ of DE *Education:* BSc/EE/Drexel U; MS/EE/Princeton; PhD/Applied Phys/Cornell U *Born:* 03/06/40 Dr Robert G Hunsperger has been a Prof of Elec Engrg at the Univ of DE since 1976. Prior to this he spent ten yers in semiconductor and optical device res as a member of the tech staff of Hughes Res Labs in Malibu, CA. He has taught at USC and at UCLA, and served as a Conslt in the fields of semiconductor and optical devices and sys. Organizations that have called upon Dr Hunsperger's expertise in this area include Hughes Aircraft Co, Martin Marietta Labs, El DuPont Co, and the US Army (ARRADCOM). He has over 60 publs and holds 14 patents. *Society Aff:* IEEE, OSA, SPIE, AAAS

Hunsucker, Robert D
Business: Geophysical Institute, University of Alaska, Fairbanks, AK 99775-0800
Position: Prof *Employer:* Univ of AK *Education:* PhD/EE/Univ of CO; MS/Physics/OR State Univ; BS/Physics/OR State Univ *Born:* 3/15/30 Born and raised in Portland, OR. Graduated from Benson Polytechnic High School. Served in US Navy from 1954-56 as Chief Engr of LST and Boat Group Commander on AKA in the Far East (Ens LTJG/Lt). Have been chapter officer for IEEE Group on Antennas & Propagation, and Pres of AK Sigma Xi Club. Approx 29 years experience in US Gov labs and Univ radio propagation research. Author or Co-author of one book and sixty technical papers. Have taught Undergraduate & Graduate courses in EE, Physics and Space Physics. Hold FCC 1st Class Radiotelephone & Advanced Amateur license KL7CYS & FAA Private Pilot Certif. *Society Aff:* AAAS, IEEE, ΣΞ, ΣΠΣ, HKN.

Hunt, Bobby R
Business: Electrical Engrg Dept, Tucson, AZ 85721
Position: Prof. *Employer:* Univ of AZ. *Education:* PhD/Systems Engg/Univ of AZ; MSc/Elec Engg/OK State Univ; BSc/Aero Engg/Wichita State Univ. *Born:* 8/24/41. After completion of PhD worked at Sandia Labs (Albuquerque, NM) for one yr doing Electrical Analysis of Strategic weapons. Went to Los Alamos Scientific Lab in 1968, worked in res & administration of digital image processing. Came to AZ in 1975. Currently Prof of Electrical Engg & Prof of Optical Sciences. Consultant to numerous ind/govt organizations. *Society Aff:* IEEE, OSA, ASP, SPIE.

Hunt, Charles A
Home: 706 Bloomfield Blvd, Jackson, MI 49203
Position: Exec Civil Engr *Employer:* Consumers Power Co *Education:* CE/Civil Engrg/Cornell Univ *Born:* 9/15/17 Native of Jackson, MI. Designer Frederick R. Harris, Inc. Project Engr Parsons, Brinckerhoff, Quade and Douglas. VP and Chief Engr Fargo Engrg Co. Engaged in designs and studies for drydocks, water supply, waterfront structures and hydroelectric dams; irrigation and hydroelectric schemes for Argentina and Colombia; conceptual designs pumped storage plants for major utilities including 2,000mW Ludington Project. Present position as consultant to management of power co on hydroelectric plants, Midland and Palisades nuclear plants and various fossil plants. Member Professional Panel-Science for Citizens Center Southwestern MI. Active in state and local ASCE. Former Chrmn of local Civil Service Bd. *Society Aff:* ASCE

Hunt, Charles H
Business: Space Technology Center, Box 8555, Philadelphia, PA 19101
Position: Mgr of Tech and Engrg *Employer:* Gereral Electric Co. *Education:* BSE/Marine Engg/ US Merchant Marine Academy; -/Fluid Dynamics/Advanced Engg Program-GE; -/Nuclear Engg/Stanford Univ. *Born:* 7/11/30. Native of Trenton, NJ. Worked for Electric Boat Co (Nautilus and Sea Wolf); served with US Navy (submarines) during the 1950s; with GE since 1955. Managed a 900 man multi-discipline engineering organization involved in major NASA and DOD space programs and new energy system engineering. He has lectured extensively with GE and at the Univ of MI and the Univ of PA. He is an Assoc Fellow of the AIAA and was a mbr of the Defense Sci Boards's Task Force on Shuttle Utilization and a consultant to DOD. He is presently Mgr of Tech & Engrg for the Aerospace Buoinena Group. *Society Aff:* AIAA, AFCEA.

Hunt, Donnell R
Business: 1304 W Pennsylvania, Urbana, IL 61801
Position: Prof of Agri Engg. *Employer:* Univ of IL, 338 AESB *Education:* PhD/Agri Engr-TAM/IA State Univ; MS/Agri Engr/IA State Univ; BS/Agri Engr/Purdue Univ. *Born:* 8/11/26. Reared and received schooling at Danville, IN. With US Army 1945-6. Served as full-time instr of farm machinery while receiving advanced degrees. Assoc Prof and Prof of Agri Engg at Univ of IL since 1960. Teaching and Res in farm field equip. Lengthy overseas assignments in Ireland (1969) and Sri Lanka (1977). Active in ASAE including technical editor for Transactions of ASAE and Chrmn of Power & Mech Div. Varied consulting work related to farm machinery design, marketing, and product reliability. Farm owner. Presbyterian Church Elder. *Society Aff:* ASAE, ASEE.

Hunt, Everett C
Home: 48 Cold Spring Hills Rd, Huntington, NY 11743
Position: Dir of Res, Prof of Marine Engrg *Employer:* Webb Institute of Naval Architecture *Education:* MS/Operations Res/Northeastern Univ; MS/Management Engrg/RPI; BS/Marine Engrg/US Merchant Marine Academy *Born:* 12/28/28 Served two years as an engr officer in US Navy before joining GE Co in 1954. During twenty three years with GE Co had many assignments with increasing engrg responsibility including steam design engr, turbine proposal engr, marine power plant proj-

Hunt, Everett C (Continued)
ect mgr, corp consulting staff, mgr engrg of heat transfer products, and quality control mgr for marine turbine and gears. In 1975 joined SUN SHIP as Dir with responsibility for new ship cost estimating, central scheduling, computer operations, and strategic planning. In 1975, accepted position as marine engrg prof at US Merchant Marine Academy. Appointed Head of engrg Dept at USMMA in July 1980. Registered PE (MA), Chartered Engr (United Kingdom), Chrmn of Inst of Marine Engrs (USA). In August 1984, joined Webb Institute of Naval Architecture as Dir of Res and Prof of Marine Engrg. *Society Aff:* IME, SNAME, IPEN

Hunt, Richard N
Home: 992 Oak Hills Way, Salt Lake City, UT 84108
Position: Mining Consultant. *Employer:* Semi-retired. *Education:* BS/-/Univ of WI. *Born:* April 16, 1893. BS & two yrs grad work at Univ of Wisc. Phi Beta Kappa. Sigma Xi. Ch Geologist US Smelting Refining & Mining Co 1925-29. Cons office Palo Alto, Calif 1930-39. Ch Geologist, V Pres & Dir US Smelting Refining & Mining Co 1938-58; also V Pres & Dir, Hecla Mining Co 195472. Hon Mbr & Legion of Honor, Amer Inst of Min, Met & Petro Engrs; Mbr Soc of Economic Geologists, Geol Soc of Amer, Amer Assn of Petro Geol's. *Society Aff:* AIME, SEG, GSA, AAPG.

Hunter, J Stuart
Home: 100 Bayard La, Princeton, NJ 08540
Position: Prof *Employer:* Princeton Univ *Education:* PhD/Exp Stat/Inst of Stat-NCSU; MS/Engrg Math/NC State Univ; BS/EE/NC State Univ *Born:* 6/3/23 US Army '42-'46. Statistical Techniques Research Group, Princeton Univ. '57-'59 Mathematics Research Ctr, Univ of WI '59-'61, Prof, School of Engrg, Princeton Univ, '61-present. Founding Editor, Technometrics, '59 Shewhart Medalist ASQC '71, Lecturer, 32 1/2 hour TV programs on Statistical Design of Experiments, '68; 10 on Statistics for Problem Solving and Decision Making, '71. Visiting lecturer: SBL Union of South Africa '75; KRSI, Korea '79; Dalian Inst of Tech '81. Chrmn Review Panel Applied Math Division NBS '79-80. Comm on National Statistics, Natl Acad of Science '75-'81. Frequent short course lecturer, three texts, forty papers, extensive consulting. *Society Aff:* AIChE, ASTM, ASQC, AAAS

Hunter, Joel
Home: 1122 N Ocean Blvd Gulf Stream, Delray Beach, FL 33444
Position: Retired. *Education:* BBA/Accounting/Emory Univ. *Born:* 9/11/05. Atlanta. Certified Public Accountant GA & NY. Practising in Atlanta then NY from 1942 as partner in Haskins & Sells. Mbr GA Bar VP. Finance of Crucible Steel Co 1951, then Exec/VP finally Pres and Chief Exec Officer. Retired 1967. Distinguished Life Mbr ASM 1957; Recipient Res Medal ASM 1968. *Society Aff:* ASM, AICPA.

Hunter, Lloyd P
Home: 10 Schoolhouse Lane, Rochester, NY 14618
Position: Prof Emeritus Electrical Engrg *Employer:* The University of Rochester. *Education:* DSc/Physics/Carnegie-Mellon Univ; MS/Physics/Carnegie-Mellon Univ; BS/Physics/MIT; BA/Physics/College of Wooster. *Born:* Feb 1916 Wooster, Ohio. Res Engr Westinghouse Elec Corp 1939-51. During WWII was assigned to the Univ of Calif Radiation Lab for isotope separation res. Later assigned to the Oak Ridge Natl Lab for power reactor dev. Organized solid state electron res & dev prog at IBM Corp 1951-63, Dir of Component Engrg 1960. Prof of Elec Engrg, Univ of Rochester since 1963 specializing in Solid State devices & Biomedical ultrasound. Retired 1981. Active in soaring. *Society Aff:* APS, IEEE.

Hunter, Robert K
Business: PO Box 4079, Gulfport, MS 39501
Position: Manager, Industrial & Commercial Sales. *Employer:* Mississippi Power Company. *Education:* BS/EE/MS State Univ. *Born:* Nov 11, 1922 Mantee, Mississippi. BSEE 1951 Mississippi State Univ. Ser with 3352nd Signal Serv BN, Army Signal Corps 1942-46 & with the Div Artillery, 31st Infantry Div during Korean conflict. Began employment with Mississippi Power Co 1953 & have held pos of Sales Engr, Div Sales Supr & Commercial Sales Mgr. P Pres, Mississippi Engrg Soc. Enjoy hunting & fishing. *Society Aff:* NSPE, ASHRAE, NFPA, IAEI.

Hunter, Robert Q
Home: 1228 Monte Vista Dr, Redlands, CA 92373
Position: Vice President. *Employer:* Saramco Inc. *Education:* ChEng/ChE/Pratt Inst; Dip/Ind Relations/UCLA; -/ChE/Univ of Pittsburgh. *Born:* Oct 1908. Grad of Pratt Inst in ChE followed by additional studies at Brooklyn Polytech, Univ of Pittsburgh & the Univ of Calif Los Angeles. Primary area of interest has been in the fields of synthetic resin & plasticizer dev & prod for such cos as Amer Cyanamid, Falk & Co & Allied Chemical. Mbr of AIChE, APCA & LASCT. Was Chmn of the Calif Manufac Assn Ctte which negotiated Rule 66 (for Solvent Control) with the Los Angeles County Air Pollution Control District resulting in Outstanding Citizenship Award by Los Angeles County & Calif Manufac Assn. *Society Aff:* AIChE, LASCT, APCA.

Huntington, William S
Home: 1650 Brookside Dr SE, Issaquah, WA 98027
Position: Dir Prod Dev Div. *Employer:* The Boeing Company. *Education:* MS/Aero Eng/MS State. *Born:* Nov 1928 Mississippi. BS 1949 & MS 1951 in Aeronautical Engrg at Mississippi State, employed by Boeing in Seattle that yr. First assignment involved in dev of Boeing owned transonic wind tunnel. He was then assigned to aerodynamic design & performance of first jet transport prototype (Dash 80), the KC-135 aerial re-fueling tanker & 707 commercial jet transports. After preliminary design studies of military & commercial derivatives of above airplanes, he was given respon for aerodynamic performance & design of the 737 twin jet & later became Ch of Tech for 737 program. In Jan 1971 he became Ch of Tech of the 707/727/737 Airplanes. His current assigment is Dir of Prod Dev for the 707/727/737 Div of the Boeing Commercial Airplane Co. He is a mbr of AIAA & Sigma Gamma Tau Natl Honor Soc in Aerospace Engrg. *Society Aff:* AIAA.

Hurd, Lewis R
Home: 80 Reeves Rd, Stockbridge, GA 30281
Position: Pres. *Employer:* Lewis Hurd Engrs. *Education:* BCE/Civ Engg/GA Tech. *Born:* 5/9/27. Native of GA. Employed by Southern Railway for Track Construction & maintenance 1944-58. Bridge Designer with the GA Hgwy Dept until 1965. Asst Hd of Civ & Structural Dept - Zimmerman, Evans & Leopold Engrs until 1966. Structural & Proj Engr - Simons-Eastern Co until 1973. Since 1973 serving as Founder & Pres of Lewis Hurd Engrs. For GA - Engr of the Yr in Private Practice 1974-75; Engr of the Yr 1977-78. Pres - GA Soc of PE - 1979-80 Chairman. Georgia PE in Private Practice. 1981-82, Chairman-Georgia Link of the Order of the Engineer-1978-79. *Society Aff:* NSPE, GSPE, ASCE.

Hurd, Walter L Jr
Home: 1840 Dalehurst Ave, Los Altos, CA 94022
Position: Principal *Employer:* W.L. Hurd Associates *Education:* MA/Guidance & Counseling/San Jose State Univ; BA/History & Polit Sci/Morningside Coll. Sioux City, IA. *Born:* 7/8/19. With Army A F 1941-46. Disting Flying Cross with 2 Oak Leaf Clusters & Air Medal with 3 Oak Leaf Clusters. Released from active duty as Lt Col. With Phillipine Airlines Inc 8 yrs as Ch Pilot, Flight Operations Mgr & VP Genl Operations Mgr. Quality Control Mgr for Natl Motor Bearing. Joined LMSC in 1958 as Reliability Engg Supr on Polaris Program. Served 3 yrs as Reliability and Quality Engg mgr of Missile Sys Div. From 1967-1977 as Prod Assurance Mgr, Space Sys Div. Promoted to Corp Dir of Quality Assurance & Product Safety, Lockheed Corp in 1977. Formed W.L. Hurd Associates in July 1986. Promoted Brig Gen AF Reserve 1962. ASQC: Mbr, Fellow 1963, VP 1969-72, Pres 1977-78, Chrmn of Bd of Dir 1978-79, Cert Qual Engr and Rel Engr, Edwards Medal 1986, Lancaster 1984, Grant 1974 & Lubelsky 1972 Awards. Dir EJC 1978-79. Reg prof Eng. CA & Exp Exam. Fellow BIS, Assoc Fellow AIAA. Mbr NSPE, IAQ, EOQC, SSS, NSS,

Hurd, Walter L Jr (Continued)
SSI, PSQC (Hon), AOQC (Hon) & NZOQA (Hon). *Society Aff:* ASQC, AIAA, NSPE, SSS, BIS, IAQ, EOQC, NSS, SSI, PSQC, AOQC, NZOQA.

Hurlbert, Don D
Home: 563 E 129th Terr, Kansas City, MO 64145
Position: City Engr. *Employer:* City, MO of KS City. *Education:* BSCE/Structural/Purdue Univ. *Born:* 10/5/26. Finished high sch at E Waterloo, IA. Served with Naval Air Corps 1944-46. Various engg positions with Frisco Ry Co, DuPont at Savannah River Plant, Raymond Intl in Thailand, City of Waterloo, IA, Water Dist No 1 of Johnson County, KS, & Riddle Engg. With KS City since 1965. Appointed City Engr in 1971. In charge of municipal improvements for Public Works. Past Pres of Hgwy Engrs Assn, Engrs Club of KS City, KS City Sec ASCE, and K. C. Metro Chapter of APWA Recipient of *Top Ten Public Works* Administrators Award 1979. Enjoy golf, fishing & bridge. *Society Aff:* ASCE, APWA.

Hurlburt, Harvey Z
Business: 8410 Manchester, Houston, TX 77012
Position: Lab Dir. *Employer:* Stauffer Chem. *Education:* ScD/ChE/MIT; MSChE/ChE/Univ of TX; BSChE/ChE/Univ of TX; BA/Chemistry/Univ of TX. *Born:* 9/2/21. in Kellogg, ID. Son of Harvey S and Vera Zeh Hurlburt. Married in 1943 to Gertrude M Lepick. Children: Geoffrey, Victoria, Veronica, Susan, Barbara, Claudia, Tobias, Octavia. Process Engr Uniroyal 1943-1947. Grad Asst at Univ of TX & MIT. Res Engr, Consolidated Chem Ind, Houston 1950-1953. Dir of Res, 1953- 1955. Dir of Peiser Labs, Stauffer 1956 to date. Patents on Inorganic Chem Processes (Mfg of sulfuric acid, liquid SO_2, liquid SO_3). Elected Fellow of AIChE in 1978. Reg PE in State of TX, 1954. *Society Aff:* AIChE, ΣΞ, ACS, ΩXE.

Hurlbut, Ronald L
Business: Director of Public Works, City of Fairfield, 1000 Webster St, Fairfield, CA 94533
Position: Dir of Public Works. *Employer:* City of Fairfield. *Education:* MS/Civil Engg/Univ of CA, Berkley; BS/Civil Engg/Sacramento State College. *Born:* April 1941. Taught variety of transport engrg courses at local colls incl Univ of Calif ext & Northwestern Univ. City of Oakland 1965-83. Deputy Dir of Public Works, Engg & Design. Director of Public Works, City of Fairfield 1983 to present. California registered Civil Engineer and Traffic Engineer; Pres Northern California Chapter APWA 1987; Pres District 6 Inst of Trans Engrs 1976; Blue Key Natl Hon Frat; Outstanding Sr Award SSC Alumni 1964. Enjoy woodworking, outside activities & Boy Scout Adult leadership. *Society Aff:* APWA.

Hurlebaus, Richard P
Home: 216 Madison Road, Huntingdon Valley, PA 19006
Position: Prod Res Engr *Employer:* The Budd Company. *Education:* BS/Met Engr/Univ of PA. *Born:* 2/15/25. Active in matls & welding res for The Budd Co since 1950. Contributor for Chapter Mat. For Welding Hbk. Mbr. of AWS. Author of tech papers on welding titanium & fatigue properties of welded steel joints. Inventor with 5 pats applied to welding. Am respon for welding res for the corp. Ints are outdoor sports, especially tennis. *Society Aff:* AWS.

Hurley, Daniel J
Home: 66 Bernhardt Dr, Buffalo, NY 14226
Position: Chrmn *Employer:* Nussbaumer & Clarke, Inc *Education:* CE//Cornell Univ *Born:* 5/3/10 Licensed PE in NY, MA and NJ. My professional experience includes five years with the American Bridge Co, five years with Jackson & Moreland of Boston in charge of structural design and resident field engr for the Mystic Power Station. For the past 38 years, I have been associated with Nussbaumer & Clarke, Inc of Buffalo in charge of structural design and co-ordination of resident field services. During the past 15 years I have been VP and Chrmn of the Bd of Nussbaumer & Clarke, with primarily administrative and financial responsibility. *Society Aff:* NYSPE, AWWA, APWA, WPCF, ACEC

Hurley, George F
Business: MS F631, P.O. Box 1663, Los Alamos, NM 87545
Position: Staff Mbr. *Employer:* Los Alamos Natl Lab *Education:* ScD/Phys Met/MIT; SM/Phys Met/MIT; SB/Met & Mtl Sci/MIT. *Born:* 11/15/40. A native of Newton, MA, was Sr Analytical Engr at Hamilton Std before joining Tyco Labs in 1966. Was responsible at Tyco for the conduct of R&D activities related to the application of EFG tech to production of shaped metal and ceramic single and duplex crystals. Tyco's Res Ctr later became the Mobil Tyco Solar Energy Corp. Here he was Principal Scientist, responsible for several progs in crystal growth and die dev. At Los Alamos, he has been engaged in development of ceramic and organic materials for fusion reactors. Currently he is Section Leader of the Ceramics and Powder Metallurgy Section of the Materials Processing Group. *Society Aff:* ASM.

Hurley, Ronald G
Home: 10123 Wolfriver Drive, Plymouth, MI 48170
Position: Research Scientist Senior. *Employer:* Ford Motor Company. *Education:* PhD/Solid State/PA State Univ; MS/Solid State/PA State Univ; BS/Chemistry/Marshall Univ. *Born:* 5/29/39. Native of Logan, West VA. Staff Mbr Los Alamos Sci Lab 1962-70. With Ford Motor Co's Engrg Res staff since 1973. Past respon incls scanning electron microscopy & high energy x-ray analysis of matls. Mbr Amer Foundryman Soc Amer Chem Soc, Microbeam Analysis Soc & Amer Soc for Metals. Past Chmn of ASM's Activity on Materialography & Microstructural Sci. Current respon include dev of energy mgmt program for metal melting activity, dev of sensors for monitoring hostile environments, and dev of CAD programs for high pressure aluminum die casting. *Society Aff:* ACS, MAS, ASM, AFS.

Hurst, William D
Home: 67 Kingsway, Winnipeg Manitoba, Canada R3M OG2
Position: Cons Civil Engr. *Employer:* Private Consultant. *Education:* BSc (C.E.)/Civil Engg/Univ of Manitoba; CE/Civil Engg/VA Polytechnic Inst & State Univ. *Born:* 3/15/08. Native of Winnipeg Can. BSc(C.E.) Civil Engrg Univ of Manitoba; C.E. Civil Engr Virginia Poly Inst and State Univ. Served as Teaching Fellow (1930-1) Served City of Winnipeg, respon charge 1934-72; Eg Engr of Water Wks, Deputy Engr, City Engr & Commissioner of Bldgs 1944-74; Commissioner of Public Works (1972) Chmn of Commissioners Greater Winnipeg Water & San Dists 1949-60; Chmn Winnipeg Parking Auth. Rivers & Streams Auth; Mbr many govt bds; Comm, Natl Cap Comm Ottawa Canada-Pres. Sr Res Assoc Amer Pub Works Assoc. Spec ints: Water supply and drainage urban drainage, environ aspects. Apptd to Order of Canada 1972. Honorary Mbr AWWA, Honorary Mbr APWA, Pres AWWA 1962-63, APWA 1958-59. V Chmn APWA Res Found. Honorary Fellow Institution of Water Engineers and Scientists (U.K.), Hon. Member Manitoba Assoc. of Architects. Author many prof papers. Reg. Prof Engr Manitoba, Minnesota. *Society Aff:* ASCE, AAEE, AWWA, APWA, EIC, IWES

Hurt, Kenn
Business: Caterpillar Inc, 600 W Washington Street, East Peoria, IL 61630
Position: Factory Engrg Mgr *Employer:* Caterpillar Inc *Education:* Grad/MFG/Caterpillar Inc 4-Year Machinist Apprentice; /Math/IL State Univ *Born:* 8/11/30 Native of Lincoln, IL. Attended IL State Univ. US Navy 1950-52. With Caterpillar Inc since 1949. Assumed current responsibility as Factory Engineering Mgr, 1979 with responsibility for Manufacturing Engrg and Planning Systems. Certified Manufacturing Engr. Dir SME 1979-1982. SME Award of Merit 1979. Olin C. Simpson award 1978. Enjoy antiques, gourmet cooking, and golf. *Society Aff:* SME

Hurt, Nathan H
Business: PO Box 628, Piketon, OH 45661
Position: Genl Mgr. *Employer:* Goodyear Atomic Corporation. *Education:* BSME/Mech Engrg/Univ of CO. *Born:* 6/7/21. Tau Beta Pi, Pi Sigma Tau. Native of Denver CO. Served in US Navy 1943-46. With Goodyear Tire & Rubber Co 1947- .

Hurt, Nathan H (Continued)
Served in var engrg & engrg mgmt assignments, domestic & foreign. Assumed current respon as Genl Mgr of Goodyear Atomic Corp, wholly owned subsidiary in 1977. As such am Exec Mgr of plant and as such am respon for all phases of plant operations. Mbr AIChE; & ASME; VP ASME; Reg Engr, OH. Enjoy skiing & requetball. *Society Aff:* ASME, AIChE, ASME.

Hurt, Ronald L
Business: P.O. Box 22738, 620 Euclid Ave, Lexington, KY 40522
Position: Sr VP *Employer:* Parrott, Ely and Hurt, Cons Engr *Education:* BS/CE/Univ of KY *Born:* 2/4/40 Native of Hazard, KY. Employed by Bell Engrs 1958-1968. Work included structural design for water and sewage facilities. A founding principal of Parrott, Ely and Hurt, Consulting Engrs, Inc in 1968. Directs the firm's Water Distribution and Treatment Division. Has designed and managed projects involving potable and waste water treatment, water distribution systems and water tanks and towers. Mr Hurt received the KY Young Engr of the Year award in 1973. He is Past Pres of the Bluegrass Chapter of KSPE and the Bluegrass Chapter of the Construction Specification Inst. He is also a licensed commercial pilot and cert flight instructor. *Society Aff:* NSPE, ASCE, NCSI, AWWA

Hurwitz, Henry, Jr
Home: 827 Jamaica Rd, Schenectady, NY 12309
Position: Pres *Employer:* Hurbits Associates. *Education:* PhD/Physics/Harvard; MA/Physics/Harvard; BA/Physics/Cornell. *Born:* Dec 1918. PhD Harvard Univ, BA Cornell. Physics instructor at Cornell for 3 yrs; worked at Los Alamos Sci Lab on fusion physics 1 & 1/2 yrs; came to General Electric in Schenectady 1946, joining the Knolls Atomic Power Lab there when it was 1st formed; 1957 joined the GE Corp R/D Center as Mgr of Nucleonics & Radiation. Present fields of interest incl classical & quantum theoretical physics, applied mathematics, computer sys applications & concepts & nuclear energy. 1961 recipient of US AEC EO Lawrence Memorial Award & 1975 recipient of GE/CR&D Coolidge Fellowship Award. Fellow of Amer Physical Soc, Amer Nuclear Soc, NY Acad of Scis & Amer Assn for the Advancement of Sci, Sr Mbr of IEEE, Pres, Hurbits Assoc. Schenectady, NY 12309. *Society Aff:* APS, ΣΞ, ΦΒΚ, ΤΒΠ, ANS, IEEE, AAAS.

Husa, V Lennie
E11020-18th Ave, Spokane, WA 99206
Position: Consltg Engr *Employer:* Self Employed *Education:* BS/Agr Engrg/WA State Univ. Specialization: Civil, Agricultural & Hydraulic Engr *Born:* July 1925 Belden N D. Reg P E Washington. Employed as consltg engr, retired from Soil Conservation Serv. Mbr Amer Soc of Agri Engrs & Washington Soc of Prof Engrs. P Chmn Spokane Chap Inland Empire Sect & Pacific Northwest Region ASAE. Past Pacific NW Regional Dir & Mbr Bd of Dirs ASAE. P Chmn Spokane Chap WSPE. Awarded Engr-of the-year Spokane Chap ASAE 1968 & Inland Empire Sect ASAE 1976. Active Boy Scouts of Amer. Awarded District Award of Merit 1975. *Society Aff:* NSPE, ASAE, WSPE.

Husby, Donald E
705 5th Avenue, P.O. Box 66, Madison, MN 56256
Position: Regional Mgr *Employer:* Cooper Lighting *Education:* BS/Engrg Physics/SD State Univ *Born:* 11/30/27 Native of Montevideo, MN. Wife-Beverly J. (Tilbury) married 38 years, Navy veteran, have 35 years experience in all phases of lighting industry, 23 years of design have yielded 21 US Patents, have written several papers in lighting, speaker at seminars and conferences, have served as VP for Nat Fortune 500 Corp, formed and served as pres of small ltg corp. Am National Dir of IESNA. Have served on many Natl Tech Comm. Responsible for recommended pratices in lighting for industry, roadways, sportslight, aviation and security. Have served as advisor to UL & ANSI. *Society Aff:* IESNA, NSPE

Huset, Elmer A
Home: 1766 Alameda, Saint Paul, MN 55113
Position: President *Employer:* Husett and Assoc *Education:* MPH/Public Health Engr/Univ of MN; B/CE/Univ of MN *Born:* 9/29/20 Native of Minneapolis/Saint Paul. Served with the Army Corps of Engrs, 1945-46; District engr - MN Dept of Health, 1947-53. Municipal Water Supply Section of MN Dept of Health, 1953-65, serving as Chief of Section 1955-65. With Saint Paul Water Utility 1965 to 1983. Organized Huset & Assoc in 1983. Dir - AWWA, 1979-81; Dir - AWWA Research Foundation, 1979-81. Fuller Award 1964. Enjoy golfing and traveling. *Society Aff:* SAME, AWWA.

Huskey, Harry D
Business: Information Science, Santa Cruz, CA 95064
Position: Professor. *Employer:* Univ of California. *Education:* PhD/Math/OH State Univ; MA/Math/OH State Univ; BS/Math/Univ of ID. *Born:* Jan 1916. Taught mathematics at Univ of Pennsylvania & worked on 1st electronic computer (ENIAC); designed & built automatic computer (SWAC) for Natl Bureau of Standards; designed Bendix G15 Computer; since 1954 Prof at Univ of Calif (Berkeley & Santa Cruz). Pres of Assn for Computing Machinery (1960-61) & edited IEEE Transactions on Computers (1965-70). Spent 2 yrs in India in tech transfer & managed a UC-UNESCO contract relative to Burma. Was Cons to UN on Computers for Dev US Sr Scientist awardee Fulbright-Von Humboldt Fdn 1974-75. Prof of Information Scis at UCSC Santa Cruz Calif. In 1978 received National Computer Conference Pioneer Award, 1982 received IEEE Computer Society's Pioneer Award and in 1984 received IEEE Centenial Award. *Society Aff:* IEEE, ACM, BCS, AMS, AAAS, ΣΞ.

Huss, Harry O
Home: Rt 1, Box 352, Uhrichsville, OH 44683
Position: Consulting Engg - Retired. *Employer:* Tech Support Dir-Edgewood Arsenal. *Education:* MS/Eng Math/Univ of Toledo; BS/Mech Engg/Drexel Univ. *Born:* March 1912. Grad work at Univ of Michigan & Johns Hopkins. Taught at Univ of Toledo & Internatl Correspondence Schools prior to WWII; Civilian Engr with US Army 1939-76; dev & staff engrg 1939-56; Dir of Prod Engrg & Asst to Ch Engr 1954-64; Ch Prod & Maintenance Engrg 1965-71; Ch Defense Div 1971-72; Deputy Dir of Tech Support Directorate 1972-76; lecturer on value engrg & mgmt. Reg P E Ohio & Maryland. Respon for providing all tech support (testing, design, tech graphics, facilities etc), 1981 ASME Legislative Fellow. *Society Aff:* NSPE, ASME.

Hussain, A K M F
Business: Dept. of Mechanical Engineering, Univ. of Houston, Houston, TX 77004
Position: Prof. *Employer:* Univ of Houston. *Education:* PhD/Mech Engg/Stanford Univ; MSME/Mech Engg/Stanford Univ; BSc/Mech Engg/E Pakistan Univ of Engg. *Born:* 1/20/43. Native of Bangladesh, where worked as newspaper correspondent and as a design engr. Editor, Inst of Engrs, Pakistan for 1963-65. Naturalized in 1977. Visiting Asst. Professor, Dept. of Mechanics, The Johns Hopkins Univ, 1969-71. Promoted to Prof of Mech Engg in 1976. Distinguished Univ Professor in 1985. Dir, aerodynamics & Turbulence Lab 1972-date. Served Univ of Houston in important committees including mission self-study (1973-75) and AD-HOC committee on Univ Finances (1975-78). Supervises res of doctoral and post-doc fellows in basic turbulence, aerodynamics, aeroacoustics, hydrodynamic stability, bio-fluid mech, chaos, funded by various federal agencies. Fulbright Scholar, Stanford Univ 1965-66. August Berner Honors Fellow, Stanford Univ 1966-67. Eckert Prize for Outsdtanding Ph.D. Thesis, Stanford Univ. (Awarded 1971). Research Excellence Award, Cullen College of Engr., Univ of Houston 1979. US-India Exchange Scholar, 1980. Research Excellence Award, Univ of Houston 1985. Freeman Scholar, ASME (1984). Enjoy tennis, table tennis, and yoga. *Society Aff:* ASME, AIAA, APS.

Hussain, Nihad A
Business: Dept of Mech Engg, San Diego, CA 92182
Position: Prof. of Mech Engrg and Assoc Dean of Engrg *Employer:* San Diego State Univ. *Education:* PhD/Heat Transfer/Univ of Notre Dame; MS/Mech Engrg/Purdue Univ; BS/Mech Engrg/Baghdad Univ *Born:* 1/9/41. Nihad Hussain was born in Iraq.

Hussain, Nihad A (Continued)
After his graduation from Baghdad Univ, he worked as an air conditioning engr. In 1963 he was awarded a fellowship to do grad work. He joined the faculty at San Diego State Univ in 1969. In 1973 he became a US citizen. He was a NASA-ASEE Res Fellow at Lewis Res Ctr in the summers of 1973 and 1974. During the summer of 1976 he was a USAF-ASEE Res fellow at Wright Patterson AFB. His res contributions are in the areas of Combined Convections in Internal flows; Radiation in Semitransparent media; and analysis of Two-phase Buoyant jets. He served as a tech conslt to a number of organizations, including Jac-Par Inc and Geoscience Inc both of Solana Beach, CA and to Burroughs Corp of San Diego. He is tech reviewer to the ASME Journal of Heat Transfer, the Intl Journal of Heat and Mass Transfer, the AIChE Journal and to the Nuclear Tech Journal. *Society Aff:* ASME, ASEE, ΣΞ, AAAS, ΤΒΠ.

Husseiny, Abdo A
Business: 125 S Third Sherman Pl, Ames, IA 50010
Position: Div Mgr. *Employer:* Science Applications, Inc. *Education:* PhD/Nucl Engr/Univ of WI; MS/Nucl Engr/Univ of WI; BS/EE/Univ of Alexandria; Dipl/Reactor Safety/MIT. *Born:* 7/7/36. Div Mgr of SAI since 1978. Chrmn of the Bd, TII, VA. Taught at IA State Univ, Carnegie-Melon Univ, Univ of WI, Univ of Alexandria, & Scared Heart College. Worked at Argonne Natl Lab, Brown Boverie (Vienna). Consulted for Los Alamos Scientific Lab & ITT. Author/editor of over 100 technicl papers, books on mathematical analysis, water resources, decision analysis, desalination journal. First Chrmn; IA/NB ANS & First Intl Conf on Iceberg Utilization. Listed in Who's Who in the Midwest, Book of Honors, Intl Who's Who in Community Service, Personalities of the West & Midwest, Notable Americans. Awarded ANS Certificate of Governance. Interviewed in hundreds of natl & intl newspapers & magazines including the London Financial Times, Los Angeles Times, Wall Street Journal, Times, & on radio & TV progs including Good Morning America. *Society Aff:* ANS, IEEE.

Hust, James L
Home: 4325 Henderson Circle, Jackson, MS 39206
Position: Cons Engrs. *Employer:* J L Hust & Assocs. *Born:* Sept 24, 1934 Meridian Mississippi. High School: Clinton High School Clinton Mississippi 1952; College: Mississippi St Univ, BS Aeronautical Engrg 1959; Grad: Mississippi St Univ MSCE 1971. Prof Socs: NSPE 1964, CEC 1970. Honors: Secretary-Treasurer Jackson Chap MSPE 1967-68, Asst Secr/Treas MSPE 1970-71, Secr/Treas MSPE 1971-72 & 1972-73, Secy/Treas CEC of Miss 1974-75, Pres Elect CEC of Miss 1975-76, Pres CEC of Miss 1976-77. Awards: Young Engr of Year MSPE 1968. Began practice of Struct Engrg in June of 1959 immediately after receiving BS Degree; served my term as Jr Engr, Assoc Engr with firm of Maxwell- Magonas & Spencer through 1964; whereupon the firm received me as a full partner; entire prof career has been assoc, in one way or another, with the original mbrs of that firm in the practice of design & preparation of contract documents for structs. *Society Aff:* NSPE, ACEC.

Hustead, Dennis D
Business: 531 E. Bethany Home Rd, Phoenix, AZ 85012
Position: Managing Engr. *Employer:* Boyle Engr Corp. *Education:* BS/Civ Engr/CA State Univ. *Born:* 9/14/40. Managing Engr of AZ operations for Boyle, specializing in water resource dev, and environmental engg, since 1972. Resident engr for irrigation proj in Sahara Desert Libya 1969-70. Proj mgr for various assignments at 8 Indian reservations in Southwest. Proj mgr for 3 irrigation dists importing water from CA Water Proj. Other experience includes domestic water & sewage system design, land dev & environmental studies. Envolved in bus dev activities in Middle East. Pres of Phx Chapter NSPE 1976-77. Pres of AZ Chapter NSPE 1981-82. Mbr of Scottsdale Bible Church & Phoenix Country Club. Hobby is golf. *Society Aff:* NSPE, ASCE, WPCF.

Husted, John E
Home: 2027 Eldorado Dr, Atlanta, GA 30345
Position: Prof Mineral Engr Emeritus *Employer:* Retired *Education:* PhD/Geol/FL State Univ; MA/Geol/Univ of VA; BS/Pre-Med/Hampden-Sydney Coll *Born:* 10/12/15 Taught high school biology, chem, and physics, 1938-40. Analytical chem and geologist with US Geological Survey, 1942-45. Plant chem, consol. Feldspar Corp, 1945-6. Resident geologist, VICC, 1951-55. Principal geologist, Battelle Memorial Inst, 1955-7. Taught Washington and Lee Univ (1946-8), Trinity Univ - TX (1949-51), Capitol Univ (1958). At GA Tech since 1958; Hd Mineral Engr Br, EES 1958-1971; Assoc Prof 1963; Prof, 1967; Prof, School of ChE, 1974; taught geology, mineralogy, mineral processing, mining, mineral economics, origin mineral resources, fossil fuels; Dir, GA Mining and Mineral Resources Inst, 1980-1984. Member Executive Reserve, US Emergency Minerals Administration. Retired March 31, 1984 from GA Tech. *Society Aff:* SME-AIME, GSA, AAPG, ΣΞ.

Huston, Norman E
Home: 4556 Winnequah Rd, Monona, WI 53716
Position: Prof Emeritus Retired. *Employer:* Univ of WI. *Education:* PhD/Phys/Univ of S CA; AB/Phys/Univ of CA. *Born:* 1/24/19. AB Univ of CA, PhD Univ of S CA. Born in Jefferson, IA. Res physicist Univ of CA Radiation Lab 1943-44; radar Officer and Observer, USNR, 1944-47; Res Engr to Dir, Radtn Tech and Instr Dept, Atomic Intl div, No American Aviation 1950-66; Prof Emeritus, Nuclear Engg 1984- ; Prof, Nuclear Engg and Dir, Instrumentation Systmes Ctr, Univ of WI-Madison, 1966-84. Also at Univ of WI, Dir, Ocean Engg Lab and Assoc Dir Marine Studies Ctr, 1967- 1970, and Dir Adv Ctr for Medical Tech and Systems 1971 to 1978. Mbr NAE-CIEBM Sub-Comm on Interaction with Ind and Task Grp on Ind Activity 1970-71. UNIDO Expert on Mission to Singapore 1972. Ldr four NSF and AID missions on scientific instrument projs to Egypt and Italy. Cons to NSF, AID, UNIDO and private industry. Fellow, AAAS, ISA and Inst MC. VP 1971-76, Pres 1979 ISA; VP & Treas NMA (NAA) 1978-79. *Society Aff:* AAAS, ISA, Inst MC, ANS, APS.

Huston, Ronald L
Business: Mail Loc 72, Cincinnati, OH 45221
Position: Prof Mech Dir Inst App Interd Res. *Employer:* Univ of Cincinnati. *Education:* PhD/Eng Mech/Univ of PA; MS/Eng Mech/Univ of PA; BS/Mech Eng/Univ of PA. *Born:* 8/5/37. Lic Engg: OH. Native of Johnstown, PA. Involved in Engg Educ at the Univ of Cincinnati since 1962; served as Hd of Dept of Engg Analysis 1969-75; currently hold pos of Prof of Mechanics & Dir of the Inst of Applied Interdisciplinary Res. Active in indus cons & govt sponsored res. Author of over 100 tech papers, Author of *Introduction to Finite Elements*, Marcel Dekker, 1984. Served as Div Dir, Div of Civil and Mech Engrg, Natl Sci Foundation, 1978-79. Served as Acting Sr VP and Provost, Univ of Cincinnati, 1982. *Society Aff:* ASME, ASEE, AIAA, SAE, HF, SIAM, SBC, SES.

Hutchinson, Charles A, Jr
Business: 23 Bendwood Dr, Sugar Land, TX 77478
Position: Consultant. *Employer:* Self. *Education:* MS/Chem Engg/Univ of CO; BS/Chem Engg/Univ of CO; BS/Elec Engg/Univ of CO. *Born:* June 1923. Native of Boulder Colorado. Taught at Univ of Colorado 1946- 48; res with Atlantic Richfield in reservoir engrg, reservoir mechs & proj evaluation 1949-59; Mgr Planning ARCO 1960; Mgr Foreign Exploration 1961-64; Mgr Exploration & Prod South Texas & Gulf of Mexico 1965-. Engr of Year 1967. SPE- AIME Pres 1976. Pub over 20 tech papers. SPE-AAPG Disting Lecturer 1968. Chamber of Commerce Bd & Exec Ctte 1967-72; United Fund Bd 1964-73, Pres 1973. Avid photographer, skiier, mountain climber, golfer & traveler. Spring Branch Bank Bd Houston 1972- VP ARCO Oil & Gas Co 1979 Now Mgmt Cons-Self Employed. *Society Aff:* SPE-AIME, AAPG, SEG.

Hutchinson, Charles E
Business: Thayer School of Engrg, Hanover, NH 03755
Position: Prof & Dean of Engrg *Employer:* Dartmouth Coll *Education:*

Hutchinson, Charles E (Continued)
PhD/EE/Stanford Univ; MSEE/EE/Stanford Univ; BSEE/EE/IL Inst of Tech *Born:* 12/18/35. in Parkersburg, WV. Served in the US Navy from 1957-1960. Res Specialist with the Autonetics Div of N Am Aviation (now Rockwell Intl) 1963-1965. Univ of MA, Amherst, 1965-1984, Prof & Hd of Dept of Elec & Comp Engrg. Joined Thayer School of Engrg, Dartmouth Coll in 1984 as Prof and Dean. *Society Aff:* IEEE, ASEE.

Hutchinson, Frank D, III
Business: 11 Penn Plaza, New York, NY 10001
Position: Pres *Employer:* Drano Energy Resources, Inc. *Education:* BCE/Civ Engg/ Cooper Union; Dipl/Nucl Engg/Intl Inst Nucl Sci & Engg; Dipl/Reactor Operations/ Duguesne Light Com-Reactor Oper. *Born:* 6/10/29. F D Hutchinson started with Gibbs & Hill in 1951 & is Pres. Frank & his wife, Diane Mary, have three children. Frank is a lic PE in 23 states & has Natl Certification from Engg Examiners (NCEE). Frank is a mbr of the honorary Civ Engg Fraternity, Chi Epsilon; a past chrmn of the NY Metropolitan Sec of ASME, & the present Chrmn of the Advisory Committee to the Nuclear Engg Dept of RPI. Frank received the ASME Metropolitan Sec Outstanding Leadership Award in 1976 & the Volmar Freis Lectr Award at RPI in 1978. *Society Aff:* ANS, ASCE, ASME, XE.

Hutchinson, Herbert A
Home: 5313 Bliss Court, Kettering, OH 45440
Position: System Dev Engrg Mgr (GS-15). *Employer:* Aeronautical Systems Div USAF. *Education:* MSAE/Aero Eng/GA Tech; BAE/Aero Eng/GA Tech. *Born:* March 1932. Native of New York City. US Army Officer (Korea) 1953-55; Aerodynamicist, Republic Aviation Corp (F-84 & F-105A) 1955-56; Wind tunnel Engr Georgia Tech Res Inst 1956-59; Aerodynamics Engr Convair/Fort Worth (B-58) 1959- 61; with USAF at Wright Patterson AFB Ohio since 1961. Conducted numerous design studies of flight sys concepts & R/D aeromechs contracts. Chaired Transonic- Supersonic Working Group of AACB for future major natl test facilities. Ch Sys Engr YF-16 & YF-17 progs 1971-75 & F-16 Air Combat Fighter 1975-76; Hd engrg efforts of High Energy laser prog 1976-79. Chf Sys Engr for TR-1 Prog 1980. Chf Sys Engr for air launched cruise missile 1981. Currently Chief, Flight Technology Division, ASD. *Society Aff:* AIAA.

Hutchinson, James R
Business: Dept of Civ Engg, Davis, CA 95616
Position: Prof. *Employer:* Univ of CA. *Education:* PhD/Engr Mechanics/Stanford; MLitt/Math/Univ of Pittsburgh; BS/Mech Engr/Stanford. *Born:* 6/1/32. Native of San Francisco. Engr for Westinghouse Atomic Power Div 1954-58. Res specialist for Lockheed Missiles and Space Co 1958-64. With Univ of CA, Davis since 1964. Consulted with Lockheed and Aerojet corps. Hobbies include singing Gilbert and Sullivan roles. *Society Aff:* ASME, AAM, ΣΞ

Hutchinson, John W
Home: 104 Tahoma Rd, Lexington, KY 40503
Position: Prof of Civil Engg. *Employer:* Univ of KY. *Education:* PhD/Civil Engg/ Univ of IL; MS/Civil Engg/Univ of IL; BS/Civil Engg/Univ of IL. *Born:* 12/1/27. Native of Williamstown, KY. Served with Army Corps of Engineers 1945-48 & 1951-53. Engg for Stephens-Adamson Mfg Co, specializing in bulk materials handling equipment. Taught 10 yrs at Univ of IL. With Univ of KY since 1964 in transportation education and res, directed the Univ of KY Multidisciplimary Accident Study team 1972-81. Univ of KY Res Foundation Res Award 1975, Alumni Assoc Great Teacher Award 1978, SAE Teetor Award 1979, Univ of IL Epstein Award 1958. R.E. Shaver Memorial Award for teaching exellence 1983, Sci Adv to Intl Ctr for Transp. Studies, Amalfi Italy. Enjoy fishing, gardening, ancient history, occult, anthropology and archery. *Society Aff:* TRB, ASCE, ITE, AAAM. ASTM.

Hutchinson, Richard C
Home: 716 Shady Lane, Lakeland, FL 33803
Position: Owner. *Employer:* Richard C Hutchinson P E. *Education:* MEA/Engrg Admin/Univ of So FL; BS/Indl Engg/MA Inst of Technology. *Born:* Dec 1916. Native of Franklin Mass. Married Mary Ellen Johnston 1940, 1 daughter Judith Hodges. Maintenance & Mgmt Cons since 1970 helping mgmt increase productivity of work force, dev preventive maintenance progs & improving admin procedures; additional 33 yrs of varied indus exper incl V Pres of Hapman Conveyors, Ch Indus Engr for Celanese, Crawford Mfg & American Wringer & other indus engrg positions. V Pres IIE 1974-76, Mbr IIE. Outstanding Indus Engr in Southeast 1964. Enjoy travel, crossword puzzles & bridge. Elected as a Fellow, in IIE-1980. *Society Aff:* AIIE.

Hutchison, Ira W
Business: Corporate Ctr, Midland, MI 48640
Position: Sr Mgmt Cons. *Employer:* Dow Corning Corp. *Education:* MChE/Chem Engg/Univ of Louisville; BChE/Chem Engg/Univ of Louisville. *Born:* 7/20/14. Louisville Kentucky. 1940-44 Dow Chemical Co Prod & Process Dev Engr; 1944- 1979 Dow Corning Corp various positions in mfg, sales, internatl opers, genl business mgmt. Important positions: V Pres Dir Internatl Opers, Pres Dow Corning Internatl, Officer &/or Dir Dow Corning foreign subsidiaries & assoc cos. Genl Mgr Medical Prods Business. Sr. Mgmt Cons. Retired 1979. Also V Pres for MI Molecular Inst Midland Mich, a fundamental res & advanced educ org, 1974-1982. Presently, Private Conslt *Society Aff:* AIChE.

Hutchison, Stuart M
Business: 1105 Williams, Great Bend, KS 67530
Position: Pres *Employer:* Evans-Bierly-Hutchison & Assoc *Education:* B/CE/KS State Univ *Born:* 3/7/28 Native of Wakeeney, KS. Served in US Navy 1948-1955. Naval Reserve until 1976. Retired with rank of Captain. With consulting firms in Topeka, KS 1957- 1968. Partner in Evans-Bierly-Hutchison & Assoc 1969 responsible for road and bridge section. Chrmn KS Consulting Engrs 1979, Golden Belt Chapter (NSPE) Pres 1969, 1981. KS Engrg Society State Dir 1970. *Society Aff:* ASCE, NSPE, ACEC

Hvorslev, Mikael Juul
Home: Highland Farms Health Center, Azalea Hall, Black Mountain, NC 28711
Position: Consultant Retired *Education:* Dr Tech/Geo Engg/Tech Univ of Vienna, Austria; Civ Engg/Hydraulics/Tech Univ of Denmark; (Danish –Civilingenior– corresponds to a masters degree). *Born:* 12/25/95. Native of Denmark, Civil Engg 1918, USA Citizen 1929. Activities 1918-32: Design and construction industrial structures, dams, hydo-electric and water- supply projs in Denmark, France, USA, Columbia SA. 1933-37 Res on strength of soils in Austria & Denmark; Dr tech 1936. 1938-49: fdn exploration & soil sampling sponsored by ASCE, Harvard Univ, USAE Waterways Experiment Station; comprehensive report 1949. Concurrently consultant on dams and spillways. 1950- 76: Resident geotechnical conslt USAE Waterway Experiment Station. Core boring in tropically weathered rocks and soils, Panama, & in ice and frozen stony till, AK and Greenland. Design of soil testing equip. Soil bearing capacity for inclined loads; failure of slopes & river banks. Time-lag in groundwater observations. Roads on moving glacier, greenland. Reports and papers. *Society Aff:* ASCE, USCOLD, AGU, ISSMFE, ATV.

Hwang, Charles C
Business: Mechanical Engrg Dept, Univ. of Pittsburgh, Pittsburgh, PA 15261
Position: Assoc Prof *Employer:* Univ of Pittsburgh *Education:* PhD/Engrg/Harvard Univ; MS/ME/KS State Univ; BS/ME/Nat Taiwan Univ *Born:* 4/7/30 Research engr at Carrier Corp, NY, specializing in refrigerant condensing heat transfer. Did work on ARC Plasmas in graduate study. Current research interests are: coal combustion, coke-oven gas combustion, heat transfer and fluid mechnics of fires, especially in mine entries. *Society Aff:* ASME, Combustion Institute.

Hwang, Ching-Lai
Business: Dept of Ind Engrg, Durland Hall, Manhattan, KS 66506
Position: Prof *Employer:* KS State Univ *Education:* PhD/ME/KS State Univ; MS/ ME/KS State Univ; BS/ME/Natl Taiwan Univ *Born:* 1/22/29 Dr. Hwang was born on Taiwan. Was a res engr and instructor at Tatung Engrg Co and Tatung Inst of Tech, Taipei, 1953-58. Has been an asst, assoc, and full prof of Ind engrg at KS State Univ, 1964-present. Has published four books: "Multiple Objective Decision Making–, "Multiple Attribute Decision Making–, "Group Decision Making under Multiple Criteria–, and "Optimization of System Reliability–; and 120 journal articles on Operations Research, Decision Theory, System Reliability, Expert Systems. *Society Aff:* IIE, ASME, JAACE

Hyde, John W
Home: 26 Winthrop Drive, Riverside, CT 06878
Position: Petroleum & Chemical Engineering Consultant *Employer:* Hyde Associates *Education:* BSc/Chem/London England; ChE/Chartered Chemical Engineer/ Ballersea Polytechnic England. *Born:* Jan 1910. Fellow Br Inst Chem Engrg (Battersea Poly), Chartered Engr, Fellow Inst Pet, AIChE, Holds 22 Pats in Pet Tech. Main career with The British Petroleum Co Ltd 1926-69 in fields of petrol refining. V Pres B P North America Inc (New York) 1963-69, Pace Co 1970-1979. V Pres & Dir Eastern Region. Expert in Petroleum Processing & Transportation incl Prod & Crude Oil Qualities. Considerable top mgmt experience in these fields. Has worked in European, Middle East & South American Countries. Considerable Experience as Expert Witness. Mbr of New York Yacht Club & United States Figure Skating Assn. Sailing & Ice Skating. *Society Aff:* Inst Pet, API, Inst Chem Engrg, IChE, GA.

Hyde, Roger L
Home: 1116 Belrose Road, Mayfield Heights, OH 44124
Position: Manager, Civil Engineering. *Employer:* Trygve Hoff-Roy F Weston. *Born:* Jul 26 1935 Conneaut OH. Married-5 children. BSCE Cleveland State U 1959. 1956-60 Merritt Chapman & Scott Corp Marine Const Engr. Since 1960 Trygve-Weston Engr to Dept Mgr. Reg Prof Engr OH, PA, WV, IL, KY, FL; Reg P S in OH. ASCE: Pres Cleveland Sect 1969, Chmn District 9 Council 1969, Pres OH Council 1970; Metro Cleveland Hgwy Users Fed Exec Ctte Chmn 1973-75; CSU Fenn Engrg Alumni Dir; Cleveland Engrg Soc Mbr; Inst Transportation Engrs Mbr; Soc Value Engrs Mbr.

Hyman, David S
Business: 2622 Maryland Ave, Baltimore, MD 21218
Position: Owner *Employer:* David S. Hyman & Assoc. *Education:* PhD/Geog & Env Engg/Johns Hopkins Univ; MS/Civil Engg/Johns Hopkins Univ; MLA/History of Ideas/Johns Hopkins Univ; BS/Civil Engg/Univ of MD. *Born:* 7/22/22. Native of Baltimore, MD. Served with US Army Combat engrs 1942-1946. Sr Engr Baltimore Bur Bldg Inspection 1949-1954. Consultant Baltimore Metro Study Commission 1963-1964. Partner, Hyman & Flower, engr and architects 1954-1964. Owner, David S Hyman & Assoc 1964-1978. Sr Partner Hyman & Macklin, engrs and architects 1979 to 1980 David S. Hyman & Assoc 1981-. Lecturer, Dept of Geography & Environmental Engg, Johns Hopkins Univ since 1975. Res in Precolumbian cement tech, engg and bldg construction since 1968 - sponsored by Natl Sci Fdn, FL St Museum & Natl Geographic Soc. *Society Aff:* AAAS, NSPE, AAA, AIA.

Hyman, Marshall L
Home: 11 High Hill Drive, Pittsford, NY 14534
Position: President. *Employer:* Nalge Co Subs Sybron Corp. *Education:* BSChE/Chem Engg/Univ of IL. *Born:* April 1924. Native of New York City. Served as Pilot in Army Air Force 1942-45. Chem Engr with NACA (now NASA) 1949-53 specializing in facilities design. With Sybron Corp & predecessor co's since 1953. Pos incl proj engr & sales mgmt at The Pfaudler Co. On loan to Oak Ridge Natl Lab (1955-57) as cons on fuel reprocessing processes. Inventor of Darex process for stainless steel fuels. Transferred to Nalge Co in 1969 as V Pres Marketing becoming Exec V Pres in 1972 & Pres in 1976. *Society Aff:* AIChE, ANS

Hymel, Norwood F
Business: 4747 Earhart Blvd, Suite 200, New Orleans, LA 70125
Position: Vice President. *Employer:* Fromherz Engrs Inc. *Education:* MS/Civil Engg/ Tulane Univ; BS/Civil Engg/Tulane Univ. *Born:* Jan 25, 1926. Native of New Orleans, La. Active duty with USNR 1944-47. Joined Fromherz Engrs in 1949. Promoted to Ch Engr in 1966 & V Pres in 1975. Respon for scheduling personnel assignments, decisions involving proj dev & engrg design, providing tech cons with proj engrs & supervising preparation of tech reports. Pres La Sect ASCE 1973. Cmdr USNR (retired). Reg Fallout Shelter Analyst 1961. *Society Aff:* ASCE, SAME, ROA.

Hyslop, Robert L
Business: 135 E Hancock St, Lansdale, PA 19446
Position: Exec V P & Asst Genl Mgr. *Employer:* Andale Company. *Education:* MS/ChE/Drexel Inst of Tech; BS/ChE/Drexel Inst of Tech. *Born:* Dec 1930. Native of State College, Pa. Employed by Andale Co (Designer & Mfg) since June 1956 starting as Sales Engr. Respon for evaluation of customer requirements & design modification of standard equip to accomplish required duty. Appointed Sales Mgr in Oct 1966, V Pres in Charge of Sales in Feb 1972 & Exec V Pres & Asst General Mgr in Feb 1974. Mbr AIChE & ASME. *Society Aff:* AIChE, ASME.

Hyzer, William G
Business: 136 South Garfield Ave, Janesville, WI 53545
Position: Consulting Engineer (Self-employed) *Education:* BEE/Elec Engg/Univ of MN; BS/Phyiscs/Univ of WI. *Born:* 3/25/25. Native of Janesville, WI. Served with US Navy 1944-47. Ch Physicist Parker Pen Co 1948-53. Cons Engr in Private Practice since 1953. Pres Photo Data Inst since 1968. Fellow SMPTE; VP 1966-69. DuPont Gold Medal 1969. Fellow SPIE; VP WSPE 1979- . Three pats. Author of *Engg & Scien High Speed Photography* 1962 & *Photographic Instrumentation Science & Engg* 1965. Editorial Bd 'Industrial R/D' & 'Photomethods' magazines. Author of regular column *Scientific Instrumentation* for 'Photomethods' Coleman Memorial Award 1980. Named 'Engineer of the Year' by WSPE 1981. *Society Aff:* SMPTE, SPIE, NSPE, WSPE, IES, AAFS, ASP

Iadavaia, Vincent A
Business: 299 Market St, Saddle Brook, NJ 07662
Position: Office Mgr. *Employer:* Havens & Emerson Inc. *Education:* BSCE/Environmental/Manhattan College. *Born:* 2/8/20. Native of NYC; grad from Manhattan College, BSCE Sanitary Option. Field Engr-M W Kellogg Inc-Baltimore-High Octane Gasoline Plant. Asst Dist Engr 1944- 46, NY Staee Health Dept, Kingston, NY. With Havens & Emerson since Jan 1946; Partner 1968; mbr Bd of Dirs 1969; Office Mgr Saddle Brook office, NJ, 1979. Specialize in water & wastewater treatment with associated projs in report, design & construction phase. *Society Aff:* ASCE, ASTM, NSPE, WPCF, AWWA.

Iams, Harley A
Home: 11933 Bernardo Ctr Ct, San Diego, CA 92128
Position: Consultant *Education:* BA/EE & ME/Stanford; Cert/PG Exec Prog/UCLA. *Born:* Mar 1905. Exec Program UCLA. Built movie scanner & 6 cathode ray TV receivers for home tests via KDKA, Westinghouse Elect 1928-29. At RCA 1931-40 contrib to TV pick-up tubes with sensitivity & picture qual adequate for commercial use. For this, joint Modern Pioneers Award, Natl Assn Mfgrs 1940. In WW II dev microwave radar devices & sys. Assoc Hd, Guided Missile Lab, Hughes Aircraft 1951-57; Assoc Dir Hughes Res Labs 1957-69. Conducted Engr Mgt seminars UCLA; also Amer Mgmt Assn. Mbr S Cal Selective Svc Adv Comm for Sci & Tech Personnel 1955-68. 15 pubs, 63 US patents. Reg Prof Engr Cal E-895. Life Fellow IEEE. *Society Aff:* ΦBK, IEEE.

Iberall, Arthur S
Home: 4675 Willis Ave, Sherman Oaks, CA 91403
Position: Visiting Scholar *Employer:* UCLA *Education:* BS/Physics/CCNY. *Born:*

Iberall, Arthur S (Continued)
1918, m. 1940, 4 daughters. Ed NYC schools; CCNY physics, mech engrg, BS 1940. Grad study-physics George Washington Univ 1942-45. Hon D Sc Ohio State Univ 1976. Physicist Natl Bureau of Standards 1941-53, instrumentation. Res dir- ARO 1953-54, aircraft accessories. Ch Physicist-Rand Dev 1954-65. Genl R&D-Genl Tech Serv Inc 1965-81, genl R&D & sys sci. Visiting scholar-UCLA 1981-, systems res. Mbr ASME 1952, Fellow 1970, chmn ACD-ASME 1972; Alza Disting Lec-Biomedical Engrg Soc 1975. Author 4 bks, ed 2 bks, 200 tech art, var patents & dev. *Society Aff:* ASME, APS, BES, ΣΞ, NYAS.

Ibison, James L
Home: 3107 Maroneal Blvd, Houston, TX 77025
Position: Dir of Facility Planning & Constr. *Employer:* Houston Independent School Dist. *Education:* BS/Mech Engg/Univ of Arkansas. *Born:* Sep 1912. Native of Greenwood Ark. Officer supervising constr with Army Corps Engrs 1940-46. Branch Mgr Westinghouse Elec Corp 1946-63. Pres Ibison Engrs 1963-67. Cons Engr & Dir of Facility Planning & Constr, Houston Independent Sch Dist 1967- . In charge of largest sch air cond proj in world, 226 existing schools, 58,000 tons of air cond costing 46 million dollars. Mbr ASHRAE since 1938 & elevated to grade of Fellow in 1973, Disting Serv Award 1976. Reg Prof Engr Texas since 1948. Ret Colonel USAF. Mbr Houston Principals' Assn. Who's Who in South & Southwest 1976-81 Life Mbr of ASHRAE. *Society Aff:* ASHRAE.

Ibuka, Masaru
Business: 7-35 Kitashinagawa-6, Shinagawa-ku, Tokyo, Japan 141
Position: Honorary Chairman. *Employer:* Sony Corporation. *Education:* Hon Doctorate in Engg/-/Sophia Univ; Dr of Science, honoris causa/-/Univ of Plano; Hon Dr of Sci/-/Waseda Univ; BS/Sch of Sci and Engg/Waseda Univ. *Born:* Apr 1908. Founded Tokyo Tsushin Kogyo K K (presently known as Sony Corp) 1946. Currently Hon Chmn of Sony. Other pos incl Pres of Japan Inst of Invention & Innovation, & Chmn of Early Dev Assn. Life Fellow IEEE, Foreign Mbr IVA (Royal Swe Acad of Engrg Sci), & Foreign Assoc Natl Acad of Engrg, Pres of Japan Audio Society. Vice-Chairman, Japan Associates for the International Exposition, Tsukuba, 1985. Chairman, The Railway Technical Research Institute (RTRI). Hon Doctorate Plano Univ 1974; Hon Doc Sophia Univ Tokyo 1976; Hon Dr of Sci, Waseda Univ 1979. Medal of Honor with Blue Ribbon from H M The Emperor of Japan. 1960; IEEE Founders Medal 1972. Decorated by H M The Emperor of Japan with the First Class order of the Sacred Treasure 1978. The Humanism and technology award from the Aspen Institute for Humanistic Studies 1981. Decorated by H.M. the Emperor of Japan with the First Class Order of the Rising Sun with the Grand Cordon 1986. Decorated by H.M. the King of Sweden with Commander First Class of the Royal Order of the Polar Star 1986. Author of bks & art on early child ed. Enjoy playing golf. *Society Aff:* IVA, IEEE, NAE.

Ide, John M
Home: 2320 Bowdoin Ave, Palo Alto, CA 94306
Position: Consultant. *Employer:* EPRI-Elec Power Res Inst. *Education:* DSc/Com Eng/Harvard;: MS/Com Eng/Harvard; BA/Physics/Pomona College. *Born:* 1907. Instructor at Harvard 1931-36. Geo-physical Engr Shell Oil Co 1936-41. US Naval Res Lab 1941-45. Ch Scientist USN Underwater Sound Lab, New London Conn 1945-59. In 1958 received Natl Civil Serv Career Award. Defense Admin Pentagon 1959-61. Dir SACLANT Res Ctr, La Spezia Italy 1961-64. (Under U S Defense Dept Contract Penn State Univ 1961-63. Estab as NATO Lab 1963). Dir Engrg Div Natl Sci Foun 1964-72. Hon PhD Pomona College 1960. Fellow IEEE 1971. Cons Elec Power Res Inst Palo Alto 1975-79. *Society Aff:* AIEE.

Idriss, I M
Business: 3 Embarcadero Ctr-700, San Francisco, CA 94111
Position: Principal & Vice Pres. *Employer:* Woodward-Clyde Consultants.
Education: PhD/Civil Engg/Univ of CA, Berkeley; MS/Civil Engg/Caltech; BCE/Civil Engg/Rensslaer Poly Inst. *Born:* Dec 1935. Has been engaged in practice of soil mech & foundation engrg. Has worked on foundation investigations for land dev, indus bldgs, power plants, earth dams & other structures. Has also been engaged in res, applications & consultations pertaining to the seismic response of soil masses & soil struct, eval of the failure potential of soils & soil struct interaction during earthquakes. As a result of this res, has formulated procedures & computer programs that are widely used in the geotech field of earthquake engrg. Has also been engaged in application of probabilistic procedures to the geotechnical engineering practice. Is a consulting Professor of civil engineering at Stanford University and is an invited Lecturer at sev universities & prof societies. Middlebrook Award 1971, J James R Croes Medal 1972, Walter Huber Res Prize 1975, Norman Medal, 1977 ASCE. *Society Aff:* ASCE, SSA, SEAONC, API, USCOLD, EERI.

Igdaloff, Harold B
Home: 11806 Bellagio Rd, Los Angeles, CA 90049
Position: Pres. *Employer:* Sungro Chem, Inc. *Education:* MSE/Chem Engg/Univ of MI; BSChEng/Chem Engg/Purdue Univ. *Born:* 7/26/25. in Toledo, OH. Served in AUS WWII. Joined Plaskon Div LOF Glass Co., 1948 in syn resin prod. Contd as Sr Chem Eng after acquistion by Allied Chemical Co from 1959-1961 Chief Eng Spec Resin, Inc. Became VP Operations, Silmar Chem Corp assigned to formation of subsidiary chem co. Later acquired by Vistron Corp in 1968. VP Operations, Roblen Resins Co. Div. Whittaker Corp. Assumed resp as Pres Sungro Chem, Inc. Mfr of fertilizers, herbicides,insecticides and spec chem packager in 1968. Part-time lecturer chem engg Toledo Univ. Charter member, PVP local section AICHE, Toledo. *Society Aff:* AIChE; ACS.

Igel, Clement Edward
Business: 100 North 12th R 712, St Louis, MO 63101
Position: Div. Staff Network Engg *Employer:* Southwestern Bell Tele Co. *Education:* BS/Engr/Washington Univ. *Born:* 5/1/24. Reg Prof Engr MO. US Army 1942-45. 33 yrs Southwestern Bell Tele Co, serving in var capacities in the Engg Dept. Instrumental in providing staff functions for network distribution services for the state of Missouri, including capital & maintenance budgets. *Society Aff:* MSPE, NSPE, IEEE, EBT.

Ignizio, James P
Business: Industrial Engineering, Univ of Houston, Houston, TX 77004
Position: Prof. & Chairman *Employer:* Univ of Houston *Education:* PhD/Operations Res/IE/VPI; MSE/Engr/AL; BSEE/EE/Univ of Akron. *Born:* 10/28/39. in Akron, OH. Married to former Cynthia Klausman of Manchester, OH. Two daughters: Karin and Laura. Following grad from the Univ of Akron, served as a Proj Engr and Mgr in the aerospace ind where -orponsible for the design and dev of the antenna systems for the Saturn/Apollo moon landing mission. Taught from 1968 through 1974 at the Univ of AL (Huntsville) in the area of Operations Res/Systems Engg. Taught from 1974 through 1986 at Penn State. Presently Prof & Chairman, Department of Industrial Engrg, at the Univ of Houston, author of 4 textbooks, several monographs and over 150 technical papers. *Society Aff:* IIE, ORSA, ASEE, TIMS, AAAI.

Imber, Murray
Home: 38 Woodland Ave, Glen Ridge, NJ 07028
Position: Prof. *Employer:* Poly Inst of NY. *Education:* EngScD/Mech Engg/Columbia Univ; MS/Mech Engg/Columbia Univ; BS/Mech Engg/Univ of IL. *Born:* 9/24/29. Native of NYC. Recipient of Union Carbridge Fellow. Taught at Columbia Univ, and WV Univ. Visiting Prof at Univ of Windsor, Canada, followed by Visiting Scientist at same univ. Currently, Prof of Mech Engg at Poly Inst of NY since 1965. Res includes non-linear heat transfer in solids, temperature measurement and inverse applications, Lift & Drag effects on golf balls, Mbr of Sigma Xi and Pi Tau Sigma. Currently, consulting mbr of the Ball and Implements Comm of US Golf Assoc. ASME comm mbr of Fundamental Heat Transfer. Lift & Drag Effects on

Imber, Murray (Continued)
golf balls. Ardent trout fisherman. Recipient AWU-DOE Sabratical Award. Fellow ASME. *Society Aff:* ASME, AIAA, ASEE, ASUP

Imhoff, John L
Business: Indus Engrg Dept, Fayetteville, AR 72701
Position: Disting Prof & Dir Prod Ctr - Dept of Indus Engrg *Employer:* Univ of Arkansas. *Education:* PhD/Indus Engr/OK State Univ; MS/Mech Engr/U of Minn; BS/Mech Engr/Duke Univ. *Born:* Feb 9, 1923. US Navy 1943-46. Facilities Design Engr Crosse & Blackwell; Engrg mgmt Armco. Asst Prof Minnesota. With Arkansas since 1952. Outstanding teacher award-1964. Cons to gov, military, & indus. AEC Fellow, CalifBerkeley; NSF Fellow Stanford. Prof engr Minn & Ark; Fellow IIE (past v p, dir, chmn Natl Council Acad Dept Hds); ASEE (past dir, reg chmn); APM (past natl pres); Mbr Sigma Xi, Tau Beta Pi, Phi Beta Kappa, Omicron Delta Kappa, Pi Tau Sigma, AAAS, ORSA. *Society Aff:* NSPE, ASEE, IIE, ΣΞ.

Imperato, Eugene G
Home: 26 Parkside Circle, Willingboro, NJ 08046
Position: Conslt. *Employer:* Self Employed *Education:* PhD/Solid St Sci/PA State Univ; BS/Metallurgical Engr/NM Inst of Mining & Tech; BS/Ceramic Engr/NM Inst of Mining & Tech *Born:* 8/6/47. B.S. degrees 1970 with Highest Honors. NDEA Title IV fellow 1970-72, Ph.D. 1976. 1975, NL Industries, Central Research Lab, specializing in secondary lead smelting and refining. 1976, FMC Corp, Chem Group Research Lab, specializing in industrial minerals and glass technology; Two patents. 1981, Mgr, Metallurgy Refinery Technical Group, Engelhard Industries Division, Engelhard Corp, specializing in the pyrometallurgical treatment of precious metal scrap. Since 1983 conslt on the processing of nonferrous and precious metals scrap and indust minerals. Regional Dir (Region I) TMS-AIME 1981-83. Past Chrmn, Philadelphia Section, AIME. Past Chrmn, Philadelphia Section, ACerS. Married; 5 children. *Society Aff:* TMS-AIME, ASM

Inaba, Yoshio
Home: 273 Waianuenue Ave, Hilo, HI 96720
Position: President. *Employer:* Inaba Engrg Inc. *Education:* BS/CE/Univ of HI. *Born:* Jan 1911 Kona, Hawaii. Reg Civil & Struct Engr & Land Surveyor. Appointed Hd of Bureau of Plans & Surveys, 2nd Asst Engr & Ch Engr 1953 to 63 of Dept of Public Works, Cty of Hawaii. Cons Engr 1963-65. Pres Inaba Engrg Inc since 1965. Appointed State Bd of Reg for Prof Engrs, Architects & Land Surveyors 1952 to 57. Asst Deputy Dir Cvl Defense, Cty of Hawaii 1956-63. Mbr Cons Engrs Council of Am & Hawaii, Natl Soc of Prof Engrs & Amer Congress on Surveying & Mapping. *Society Aff:* ACEC, NSPE, ACSM.

Inatome, Joseph T
Business: 10140 W Nine Mile Rd, Oak Park, MI 48237
Position: President. *Employer:* Inatome & Assoc Inc. *Education:* BSME/Mech Engg/Wayne State Univ. *Born:* June 3, 1925 San Francisco, Calif. P Pres Cons Engrs Council of Metro Detroit, Mech Engrg Dir CEC Michigan, Bd of Dir with Detroit CSI Chapter, Mbr ASHRAE, Mbr NSPE, & Mbr Internatl Solar Engrg Soc. Sept 1976 Founded Inacomp Computer Centers, Inc. The retail store operation has now expanded into (3) stores with (33) offices. Mr. Inatome serves as Chrmn of the Bd of Inocomp Computer Centers. *Society Aff:* ASHRAE, ACEC, NSPE, CSI.

Ince, A Nejat
Business: Back Ln, Melbourn, Cambridge, Herts, England
Position: Dir. *Employer:* PA Centre for Advanced Studies. *Born:* Nov 1928. PhD Cambridge Univ; B Sc Birmingham Univ Eng. Worked in transmission lab of Genl Elec Co UK 1955-56. With PTT, Turkey 1956-59 as Ch Engr for transmission sys. Worked for 1 yr with Turkish Broadcasting Corp as Ch Eng for planning of LF & MF stations. Joined SHAPE Tech Centre 1961, up to 1968 Hd, Radio Branch spec in air/ground & satellite communications. Principal planner/designer of NATO SATCOM Sys. Ch, Communications Div since 1968 engaged in dev & sys engrg of an automatically switched network, NICS for NATO. With Western Union Telg Co NJ, 1978-79 as Asst VP for Advanced Dev engaged in new developments in transmission, switching & terminal areas. Since Nov 1979, with PA Intl Manage. Cons as Dir of Centre for Advanced Studies, Cambridge, UK. Offering res and advisory skills in telematics, energy, economics and social & life sciences. Received 1979 IEEE award in Intl Communications for "contribution to satellite communications systems and the planning and design of automatically switched intl communication." Prof at Tech Univ of Istanbul and adviser to the Prime Minister. Fellow IEEE since 1972 & author of over 50 tech papers, books & reports. *Society Aff:* IEEE.

Incropera, Frank P
Business: School of Mech. Eng, W. Lafayette, IN 47907
Position: Prof. *Employer:* Purdue Univ. *Education:* SB/ME/MIT; MS/ME/Stanford; PhD/ME/Stanford *Born:* 5/12/39. Native of Lawrence, MA. Ind experience with Barry Controls Corp, Aerojet Gen Corp, Lockheed Missiles and Space Corp. Visiting Scientist with Natl Aero and Space Admin. Visiting Scholar at Univ of CA at Berkeley. Author of over one-hundred papers in the fields of heat transfer, bioengineering, and high temperature gas dynamics. Author of four textbooks on Molecular Structure and Thermodynamics and Heat Transfer. *Society Aff:* ASME, ASEE

Inculet, Ion I
Business: Univ. of Western Ontario, London, Ont Canada, N6A 5B9
Position: Prof. & Director, Applied Electronics Research Centre. *Employer:* Univ of Western Ontario. *Education:* MESc/Power/Laval Univ; Dipl Ing/Power (Elec)/Bucuresti Poly Inst. *Born:* 2/11/21. Prof Inculet, is the Director of the Applied Electrostatics Research Centre at the Univ of Western Ontario. Teaches electrostatics & electromech energy conversion. Has been directing several industry & govt (including NASA from USA) sponsored projs in electrostatic beneficiation of ores, air cleaning, electrostatic painting & coal desulfurization, aerial and ground electrostatic spraying of vegetation, etc. Previously employed by Canadian GE as Advance Dev Engr in motors & later as Mgr of Engg. He is Author of some 80 papers, 12 patents, a book & chapters in various books. Elected "Fellow" in IEEE. 1983 Recipient of the IEEE-IAS Outstanding Achievement Award. *Society Aff:* IEEE, ESA.

Ingalls, Larry W
Home: 203 Lake Shore Dr, Fredericksburg, VA 22405
Position: Partner. *Employer:* Sullivan, Donahoe & Ingalls. *Education:* BS/Civ Engg/VPI. *Born:* 12/17/41. Native of Fredericksburg, VA. Reg as PE in VA 1972. Joined firm of Sullivan, Donahoe & Ingalls in 1972 as full partner. Responsible for design projs, construction coordination, and business operations. Active in church work and civ clubs. Enjoy camping, fishing, and sports. P Pres Kiwanis Club of Stafford, Church Deacon, Member Stafford County Board of Zoning Apeals. *Society Aff:* NSPE, VSPE, ASCE.

Inger, George Roe
Business: Dept. of Aerospace Engineering Sciences, Boulder, CO 80319
Position: Chairman *Employer:* Univ. of Colorado *Education:* PhD/Aero Engg/Univ of MI; MS/Aero Engg/Wayne State Univ; BS/Aero Engg/Wayne State Univ. *Born:* 1/27/33. Detroit, MI. Res positions in Aerodyn & Heat Transfer with GM Res Labs, Bell & Douglas Aircraft Corp, 1956-1967; Chief, Fluid Phys Res Branch, Mc-Donnell-Douglas Astro West, 1967-1970; Visiting Prof & Heinneman Fellow, Von Karman Inst (Brussels), 1970-1971. Professor of Aero Engg, VPI & SU, 1971; Teaching of Aerodyn and Fluid Mechanics plus extensive res sponsored by NASA, NSF, ONR and AFOSR; Fulbright-Von Humboldt Fellow (Gottingen, W Germany), 1976- 1977. Phi Kappa Phi, Sigma Xi, Sigma Gamma Tau. Private pilot; handball and squash. *Society Aff:* AIAA, AAHS.

Ingerman, Peter Zilahy
Business: 40 Needlepoint Lane, Willingboro, NJ 08046-1997
Position: Sys Consultant (Self-employed) *Education:* MSE/EE/U of P; BA/Physics/U

Ingerman, Peter Zilahy (Continued)
of P *Born:* 12/9/34 BA physics 1958, MSE Elec Engrg 1963, Univ of PA. FLMI 1973, CLU 1975. CDP 1978 CCP 1979: Sr Mbr IEEE, Fellow Brit Computer Soc. Mbr NJ Acad Sci, AAAS, ACM, DPMA. 2 pats 1 book, num papers & lecturers. Holistic sys consulting, usually computer-oriented, with special concern for human-system interfaces. Life Mbr Sigma Xi (Scientific Research of North American). *Society Aff:* ACM, AAAS, DPMA, IEEE.

Ingerslev, Fritz H B
Home: 20 Vilvordevej, DK-2920 Charlottenlund, Denmark
Position: Professor. *Employer:* Tech Univ-2800 Lyngby, DK. *Education:* Dr Sc/EE/Tech Univ Denmark; MS/EE/Tech Univ Denmark *Born:* July 6, 1912, Denmark. Prof. The Acoustics lab, Tech Univ Denmark 1954-1982. Dir Acoustical Lab Acad of Tech Sci 1945-1981. Secy Comm on Acoustics, Int Union for Pure & Applied Physics 1960-69. Pres Danish Acoustical Soc 1955-63. Pres Scandinavian Acoustical Soc 1955-62. Pres 4th ICA Congress Copenhagen 1962. Pres INTER-NOISE Copenhagen 1973. Pres Int Inst of Noise Control Engrg 1974- . Pres Tech Comm TC 43 & TC 43/SCI, ISO 1968- . Fellow Amer Acoustical Soc. Fellow IEEE-USA. Honory Member Institute of Acoustics. UK Foreign assoc Natl Academy of Engrg USA 1982. *Society Aff:* DIF, ATV, DAS

Ingersoll, Alfred C
Home: 102 Estates Dr, Orinda, CA 94563
Position: Assoc Dean of Engrg-Cont Ed & Dir Engrg Extension. *Employer:* Univ of California L A. *Education:* PhD/CE/Univ of WI; MS/CE/Univ of WI; BS/CE/Univ of WI *Born:* 06/08/20 in Madison, WI. Lab of Linde Air Prod Co 1942-46; Dept of Civil Engrg CalTech 1950-60; Dean of Engrg Univ of S CA 1960-70; Assoc Dean Cont Ed UCLA Sch of Engrg & dir of Cont Ed in Engrg & Math, UCLA Ext 1970- . ASCE, Pres Los Angeles Sect 1970-71. ASEE, VP Sects West 1966-68, Chrmn Pacific Southwest Sect 1960-61. NSPE, Natl Dir 1969-77, Pres, NSPE Ed Found 1973-77. Reg Civil Engr CA. ASCE: Rudolph Hering Medal (Sanitary Engrg) 1957, Edmund Friedman Prof Recog Award 1969. SWE: Rodney Chipp Award 1971. SME: Region VII, Educator of Yr 1972. NSPE: VP Western Region 1977-79, Chrmn Publications Ctte, 1979-81. ASCE: Mbr, Exec Ctte, Educ Div, 1978- , Ctte on Continuing Profl Dev 1978-81. Sr Editor, Civil Engrg Series, Marcel Dekker, Inc, Pub. NSPE: Chrmn Engrs' Week Ctte 1982. Chrmn Continuing Profl Dev Ctte 1981- . Pres, CA Engrg Fdn 1979- . *Society Aff:* NSPE, ASCE, ASEE, AAEE, SHOT, IAE, TTS.

Ingram, Sydney B
Home: 4 Galloping Hill Rd, Basking Ridge, NJ 07920
Position: Retired. *Employer:* *. *Education:* PhD/Physics/CA Inst of Tech; BA/Math & Physics/Univ of British Columbia. *Born:* Oct 8, 1903 Bottineau Co N D; 1928-30 Natl Res Council Fellow, U of Mich, res in infra-red spectra; 1930-68 Bell Tele Labs, 1930-52 Dev Engr in charge of dev of electron tubes of var types for communications & military applications, 1952-68 Dir of Ed & Training & later Dir of Employment; 1968-71 Engrs Council for Prof Dev as Exec Secy & Secy. Retired 1971. *Society Aff:* IEEE, ASEE, APS.

Ingram, Troy L
Business: 600 - 18th St, Birmingham, AL 35291
Position: Chief Agri Engg. *Employer:* AL Power Co. *Education:* BS/Agri Engg/Auburn Univ. *Born:* 3/21/19. Native of Blount County, AL. Served as Electricians Mate 1st Class, US Navy Seabees in WWII. Employed by AL Power Co in 1936. Since 1948, have been involved in Co's rural electrification prog. Made Chief Agri Engr in 1962 and have been responsible for on-the-farm application of electricity and coordinating agriculturally related res with Auburn Univ and Univ of GA in areas of energy mgt and conservation. *Society Aff:* ASAE.

Inman, Byron N
Home: 8239 W Mercer Way, Mercer Island, WA 98040
Position: Consultant. *Education:* MS/ChE/CA Inst of Tech; BS/Chem/CA Inst of Tech. *Born:* 4/9/14. Native of Yakima, WA. Employed 1936 by DuPont Co as chemist. Became Plant Engr at El Monte, CA 1941. During WWII helped design & supervised construction of processes for KCN & Acrylonitrile. Dev work in Niagara Falls 1945-50, prepared conceptual designs for HCN & H2O2 processes. 1950-4 did basic design of these processes, which were then constructed numerous locations in US & abroad. Was Design & later Proj Engg Supt at Memphis. Elected Fellow AIChE 1974 for contributions to process deisgn. Retired 1977 to private development activity in micro fluidics. *Society Aff:* AIChE.

Inman, John W
Business: 118 Wilgart Way, Salinas, CA 93901
Position: Farm Advisor-Agri Engrg. *Employer:* Univ of California. *Education:* BS/Ag Engrg/OR State Univ; BS/Ag/OR State Univ *Born:* 07/25/40 Born and raised in OR. Undergrad work in Ag Engrg and Ag at OR State Univ, grad work in seed processing and testing 1963-65, OR State Univ. With Univ of CA in Salinas, CA since 1965 as Farm Advisor--Ag Engrg specializing in prod and harvesting equipment for vegetable crops including lettuce, celery, cauliflower, broccoli, Brussels sprouts, artichokes, and tomatoes. Other programs include integrated pest mgmt, safety and machinery for small farms. Specific interests in European farm machinery and its adaptability to US conditions. Have traveled to machinery shows in Italy and England and studied farm machinery there. Hobbies include travel and reading. Overseas consultant on vegetable mechanization, Romania 1971, USSR 1975, Mexico 1980-81. Northern CA--NV Sect ASAE, Chrmn 1971-72, Chrmn 1972-73, various cttes since then. PM-48 1968-72. Reg PE CA No. AG280. Grad CA Agri Leadership Prog Class VIII 1979. Agri Engrg conslt including legal work. *Society Aff:* ASAE, CAST

Inose, Hiroshi
Home: -39-9 Jingumae 5-Chome, Shibuya-Ku, Tokyo 150 Japan
Position: Director General *Employer:* National Center for Science Information Systems *Education:* Dr of Engrg/Elec Engrg/Univ of Tokyo. *Born:* 1/5/27. Hiroshi Inose served as a prof of electronic engrg at the Univ of Tokyo 1961-1987 & served as its dean of engrg and the dir of its Comp Ctr.Since 1987, he has been the d.g. of the National Center for Science Information System. For his works which have been mainly concerned with Digital Communications Tech & Rd Traffic Control, Dr Inose received a number of awards including the second Marconi Intl Fellowship & Japan Acad Prize. He is the Chrmn of Ctte for Scientific and Tech Policy at OECD since 1984. Dr Inose is a fellow of IEEE, a foreign assoc of the Natl Acad of Scis, and Natl Acad of Engrg, USA, & a foreign mbr of the Am Philosophical Soc. *Society Aff:* IEEE, NAS, APS, IPSJ, NAE, IEICEJ.

Inouye, Henry
Business: PO Box X, Oak Ridge, TN 37830
Position: Res Staff. *Employer:* Oak Ridge Natl Lab. *Education:* MS/Metallurgy/MA Inst of Tech; BS/Metallurgy/CO Sch of Mines. *Born:* 9/14/20. Native of CO. Served in US Army in Italy and France 1943-1946. Chemist- Metallurgist at: American Manganese Steel Co 1946-1947, Natl Bur Stds 1947, War Dept 1947-1950. Staff Metallurgist at Oak Ridge Natl Lab since 1952 specializing in high temperature alloy dev and gas-metal corrosion. Fellow ASM 1977, IR-100 Award 1979. Hobbies: fishing gardening. *Society Aff:* ASM.

Iorgulescu, Jorge
Business: 135 S LaSalle St, Chicago, IL 60603
Position: Sr VP Intl *Employer:* Liquid Carbonic Corp. *Education:* PMD/Bus Admin/Harvard Bus Sch; MS/CE/Argentina-La Plata Natl Univ. *Born:* 7/12/35. Born Buenos Aires, Argentina, July 12, 1935; s Nicolas & Ida (Mayer) I; student La Plata Indsl Sch (Argentin), 1948-53; CE, La Plata Univ (Argentina), 1959; prog mgt dev, Harvard Bus Sch, 1970; m Beatriz E Cobenas, July 24, 1964; chidren - Bernardo, Lionel, Andrew. Nylon plant engr du Pont de Nemours, Ducilo, Berazategui, Argentina, 1959-60, process cntrol chief, 1960; with Liquid Carbonic Argentina Sociedad

Iorgulescu, Jorge (Continued)
Anonima Indsl Coml, Buenos Aires, 1960-67, tech mgr, 1964-67; prodn & engg mgr. Liquid Carbonic Corp Intl div, Chgo, 1967-70, vp operations, 1970-. Bd dirs Liquid carbonic Mexico, Sociedad Anonima, Gases ind, Sociedad Anonima, Argentina, Liquid Carbonic Venezolana, Liquid Carbonic Argentina, Arfin S.A. Argentina, CAF S.A. Argentina; Mitsui Toatsu Liquid Carbonic, Inc, Japan Korea Liquid CAC Bontic, Inc; others. Mbr Am Mgt Assn, Am Inst Chem Engrg, AM Soc Mech Engrs, Harvard Bus Sch Alumni, Council of Ams, Assoc Harvard Alumni. BSA Scout Master Asst. Enjoys classical music, tennis, raquetball & squash. *Society Aff:* AIChE, ASME, AMA, AMBA.

Iosupovicz, Alexander
Home: 4637 Gesner St, San Diego, CA 92117
Position: Prof *Employer:* San Diego State Univ *Education:* PhD/Syst Info Sci/Syracuse Univ; MSc/EE/Technion, Israel; BSc/EE/Technion, Israel *Born:* 2/14/38 On the scientific staff of Bell-Northern Research, Canada, 1965-67, specializing in computer hardware design and testing. Asst Prof of systems engrg at RPI, 1970-75. On the faculty of CA State Univ, Northridge, 1975-78. On the faculty of San Diego State Univ since 1978, tenured full Prof since 1981. Current research and teaching in computer design, design automation of digital systems and fault tolerant computing. Have also been a consultant to industry in design automation, testing and simulation of digital systems, since 1976. *Society Aff:* IEEE

Iotti, Robert C
Business: 2 World Trade Ctr, New York City, NY 10048
Position: VP *Employer:* Ebasco Services Inc *Education:* PhD/Nuclear Engrg/KS State Univ; MS/Nuclear Engrg/KS State Univ; BS/Nuclear Engrg/KS State Univ *Born:* 11/25/41 Native of Karlsruhe (Germany). Taught nuclear engrg at KS State Univ prior to joining Ebasco in 1970. At Ebasco specialized in radiation analysis and design, fluid mech and engrg mech. Assumed current respon as VP Advanced Tech in 1984. Presently respon for planning and direction of all work in advanced structoral analysis, fluid mech, nuclear engrg and chem engrg. Also respon for on-line real time computer application. Author of numerous papers in several fields. Received best paper award from radiation shielding div of ANS. Mbr of ASME PUP NED Ctte and ANS Pipe Rupture Ctte.

Ipponmatsu, Tamaki
Business: 6-1 Ohtemachi 1-chome, Chiyoda-ku, Toyko, Japan 100
Position: Advisor *Employer:* Japan Atomic Power Co. *Education:* BS/Elec Engg/Kyoto Imperial Univ; PhD/Elec Engg/Osaka Imperial Univ. *Born:* 4/29/01. 1951 Managing Dir, currently Advisory, Kansai Elec Power Co. 1957 Exec VP, currently Exec Advisory, Japan Atomic Power Co. 1958 Counselor, currently Adivsor, Japan Atomic Energy Res Inst. 1958-59 Pres, Inst of Elec Engg of Japan. 1959 Exec Secy, currently Advisory, Japan Commitee for Econ Dev. 1964 Standing Dir, currently Vice Chrmn, Japan Atomic Industrial Forum. 1965-78 Counselor, Japan Atomic Energy Commission. 1965-67 Pres, Atomic Energy Soc of Japan. 1967 Counselor, Power Reactors and Nuclear Fuel Development Corp. 1972-78 Mbr, IAEA Scientific Advisory Committee. 1973 Fellow, IEEE; 1976-79 Dir, Atomic Industrial Forum (U.S.); 1978 Mbr, U.S. Natl Acad of Engg. 1959 Blue Ribbon Medal (Japan); 1971 Second Class Order of Rising Sun (Japan); 1977 Commander of the Order of the British Empire (UK). Enjoy "Go" and golf. *Society Aff:* NESJ, IEEJ, ANS, IEEE.

Irby, Raymond F
Home: P.O. Box 587, Oakton, VA 22124
Position: Owner. *Employer:* R F Irby Assoc. *Education:* M Engr/Elect Engr/U of Louisville; BEE/Elect Engr/U of Louisville. *Born:* 6/2/24. M Engr & BEE Univ of Louisville. Grad Bell Tele Labs Comm Dev Training Program. Served with Army Signal Corps 1942-46. Bell Tele Labs, Mbr Tech Staff, 1953-56. Melpar Inc, Proj Engr 1956-58. SCOPE Inc, Dir of Res & Dev 1958-62. AGA Corp, Prog Exec 1962-64. Private practice, elec engrg 1964- . Patents & papers in field of data trans. microwave sys. Reg P E VA, MD, W.VA. Pres VA Soc of Prof Engrs 1976- 77. NSPE Natl Dir 1973-75. Pres Nothern VA Chap of VSPE 1970-71. Sec/Treas WA Soc of Engrs, 1981-1985. Sr Mbr IEEE. Mbr Sigma Tau, Tau Beta Pi & Phi Kappa Phi. Soc Aff: NSPE, IEEE. *Society Aff:* NSPE, IEEE, WSE.

Ireland, Donald R
Business: 5383 Hollister Ave, Santa Barbara, CA 93111
Position: Mgr, DYNATUP Products. *Employer:* Effects Technology Inc. *Education:* BS/Physical Metallurgy/Univ of MN; MSE/Matls Engg/Univ of MI. *Born:* Mar 31, 1937. Exper in fracture analysis, materials dev, dynamic mech prop, failure analysis, radiation effects on reactor materials. Leading force in tech advancement & widespread acceptance of instrumented impact testing known as DYNATUP. Widely recog as auth on & innovator in dynamic testing tech & dev of equip & tech for instrumented impact testing. Mgr DYNATUP group of Effects Tech Inc. Pub more than 3 dozen papers. Mbr ASTM cttes, subcttes, and task groups. Also mbr ASM. *Society Aff:* ASTM, ASM, SPE

Ireland, Herbert O
Home: RRl Box 185C, Gilman, IL 60938
Position: Prof Emeritus. *Employer:* Univ of IL. *Education:* PhD/Civ Engg/Univ of IL; MS/Civ Engg/Univ of IL; BS/Gen Engg/Univ of IL. *Born:* 6/12/19. A native Illinoisian on active duty as a commissioned officer with the Army Corps of Engrs 1941-46. Academic staff, Univ of IL at Urbana-Champaign 1947-79 specializing in geotechnical engg, teaching introductory & grad level courses. Became Prof of Civ Engg Emeritus in 1979. Consultant on geotechnical problems and author of 33 published articles. Reg as both a Structural & PE in IL. Active in Methodist church. Enjoy hunting, fishing, horticulture, etc. *Society Aff:* ASCE, AREA, GSA, USCOLD.

Ireland, William, III
Home: 1456 Mullica Dr, North Brunswick, NJ 08902
Position: Asst State Conservation Engr. *Employer:* US Dept of Agri-Soil Conserv Serv. *Education:* Master/Pub Admin/Univ of Dayton; Bach/Agri Engg/OH State Univ. *Born:* Mar 1, 1939. Bach of Agri Engrg Ohio State Univ 1962. Master of Public Admin Univ of Dayton 1974. Reg Prof Engr. Mbr Amer Soc of Agri Engrs, Soil Conservation Soc of Amer, Natl Assn of Conservation Dist; Chmn Ohio Sect of Amer Soc of Agri Engrs 1974-75. US Army 1962-64; Soil Conservation Serv, Agri Engr Watershed Planning Party, Columbus Ohio 1964-65; Soil Conservation Serv, Agri Engr Area Office, Dayton Ohio 1965-71; Soil Conservation Serv, Area Engr, Dayton Ohio 1971-76; Soil Conservation Serv, Asst St Conservation Engr, Somerset NJ 1976- . *Society Aff:* ASAE, SCSA, NACD.

Ireson, W Grant
Home: 735 Alvarado Ct, Stanford, CA 94305
Position: Prof Emeritus *Employer:* Stanford Univ *Education:* MS/IE/VA Polytechnic Inst; BS/IE/VA Polytechnic Inst *Born:* 12/23/15 Industrial Engr Wayne Mfg Co, Waynesboro, VA, 1937-1940. Instructor to Acting Prof and acting Chrmn, I.E. Dept, VA Polytechnic Inst, 1940-48. Prof of IE, IL Inst of Tech, Chicago, 1948-1951; Prof of IE, Stanford Univ, 1951- present, Chrmn 1954-1975. Consultant to many companies, UNIDO, US Air Force and Navy and intl organizations. Special areas of expertise: Engrg Economics, Quality Control and Reliability, Educational Development in emerging nations, small business development. Author, co-author or editor of 5 books. Author of over 40 papers and invited speaker at over 40 conferences and symposia. Recipient of E.L. Grant Award (ASQC) Order of Civil Merit (Government of So Korea), Wellington Award (IIE Engrg Economy Div) Frank and Lillian Gilbreth Award (IIE), Tau Beta Pi Award for outstanding Undergraduate teaching, Stanford. Austin Bonis Award for contributions to reliability education (ASQC, Rel. Div.). *Society Aff:* ASEE, ASQC, IIE

Irish, Wilmot W
Home: 26 Dart Dr, Ithaca, NY 14850
Position: Assoc Professor. *Employer:* Cornell Univ. *Education:* MS/Agri Engg/Univ of IL; BA/Agri Eng/Univ of VT. *Born:* July 1928 Burlington, Vermont. BSAE 1950

Irish, Wilmot W (Continued)
Univ of Vermont; MS 1955 Univ of Illinois. US Infantry 1951-53, Korea, Bronze Star, First Lieutenant. Univ of Conn 1955-60, Extension Agri Engr & Asst Prof in farm struct planning. Since 1960 with Cornell Univ, Col of Agri & Life Sci, Dept of Agri Engrg with respon for adult ed programs & res about planning dairy sys & improving farm bldgs. Lic Prof Engr since 1966 NY. Natl Sci Foun Trainee 1966-67 Univ of Calif Davis. Secy-Treas 1972-74 North Atlantic Region ASAE. Chmn 1973-74 Struct & Environ Div, Amer Soc of Agri Engrs. Visiting Prof 1975 Univ of Mo, Columbia. Chmn 1977- 78 NAR-ASAE. *Society Aff:* ASAE, IAMFS.

Irvine, Leland K
Business: 1864 S State St Suite 270, Salt Lake City, UT 84115
Position: Consulting Engr. *Employer:* Acoustical Engrs, Inc. *Education:* BSEE/Elec Engr/Univ of UT. *Born:* 2/7/09. First acoustical proj broadcasting booth in LDS Taberanacle 1934, since then over 2200 projs in architecture acoustics, noise control, environmental surveys, sound systems etc in Western US, Nigeria and Canada. Civic activities have included chrmn C of C Winter Sports committee and airport committee. Bd mbr UT Symphony and SLC Public Library. Pres UT Mfg, Assn. Other interest - Dir bank & trust co. Reg PE, UT. Adjunct prof of Architecture - Grad Sch of Arch - Univ of UT 30 yrs. Past Natl Dir NSPE. Past Natl Pres Vermiculite Inst. Past chp Pres ASHRAE. Pres Elect Conslt Engrs Coun of UT. *Society Aff:* ACEC, NSPE, NCAC, INCE, ASTM

Irvine, Robert G
Home: 1393 Carthage Court, Claremont, CA 91711
Position: Professor *Employer:* CA State Polytechnic Univ, Pomona, CA *Education:* MS/EE/CA St Univ-LA; BS/EE/UT State Univ *Born:* 7/27/31 13 years of full time engrg design experience after bachelors degree. 12 years of full time teaching experience at univ level. 7 before masters degree, 5 after masters degree. 6 years of consulting experience (PE-CA) Author: Operational Amplifier Characteristics and Applications, 1981, Prentice-Hall Inc. Engrg text book at 4 year coll level. Past Section Chrmn, Foothill Section, IEEE. Member of WESCON committee. Member of committee; Measurement Science Conference. Program Ctte Chrmn, Mini-Micro West. Eta Kappa Nu, Tau Alpha Pi. *Society Aff:* IEEE, ASEE, PMA.

Irvine, Thomas F, Jr
Business: Dept of Mech Engrg, Stony Brook, NY 11794
Position: Prof of Engineering. *Employer:* State Univ of New York-Stony Brook. *Education:* PhD/ME/Univ of MN; MS/ME/Univ of MN; BS/EE/PA State Univ. *Born:* June 25, 1922 Northmont N J. Univ of Minn: Asst Prof 1956-58, Assoc Prof 1958-59. North Carolina State Univ: Prof 1959-61. State Univ NY-Stony Brook: Dean of Engrg 1961-72, Prof of Engrg 1972- . Fellow ASME, Tech Ed Jour Heat Trans 1960-63, Mbr Exec Comm Heat Trans Div 1964-70, Chmn 1968, Co-ed: 'Advances in Heat Trans', Pergamon United Engrg Series, (23 bks pub), 'Heat Transfer-Soviet Res', 'Heat Transfer Japanese Res', 'Previews in Heat & Mass Transfer'. Mbr Exec Ctte, Internatl Ctr of Heat & Mass Transfer 1968-84, Chmn 1972-74, 80-84 Author of 63 Journal papers. Fellow AAAS, Fellow Internatl Ctr of Heat and Mass Transfer. *Society Aff:* ASME, ΣΞ, AAAS.

Irving, Edward M
Business: 1754 Dana Avenue, Cincinnati, OH 45207
Position: V Pres-Genl Mgr. *Employer:* Inmont Corporation. *Born:* Sep 1928. BS Chem E New Jersey Inst of Tech (formerly NCE). Native of Stratford, Ontario, Canada. With Inmont Corp since 1947-V P/Genl Mgr 1973. Formerly V P/Mfg 1970. Mbr AIChE, Tau Beta Pi. Reg Engr in Province of Ontario. Mbr Royal Inst of Great Britain. Int: tennis, golf, skiing, riding, gardening.

Irwin, George R
Home: 7306 Edmonston Ave, College Park, MD 20740
Position: Prof of Mechanics. *Employer:* Dept of Mech Engrg, U of Md. *Education:* PhD/Physics/Univ of IL; MA/Physics/Univ of IL; AB/English/Knox College. *Born:* Feb 1907. Res scientist at Naval Res Lab 1937-67, Supt of Mech Div 1950- 67. Prof. of Mech., Lehigh Univ., 1967-72. US Navy Dist. Civ. Service Award 1946. Post-1946 res led to dev of fracture mech. ASTM Dudley Medal 1960, US Navy Conrad Award 1969, SEM Murray Lecturer 1973, ASTM Hon Mbrship 1974, ASM Sauveur Award 1974, Societe Francaise de Metallurgie Grande Medaille 1976. Mbr of Natl Acd of Eng 1977, ASME Nadai Award 1977, Lehigh Univ. Dir. Eng. (Hon) 1977, SEM Lazan Award 1977, Franklin Inst. Clamer Medal 1979. Techn. Uni. Vienna Tetmajer Award 1985, ASME Timoshenko Medal 1987. Current int are ed & applications aspects of fracture mech. *Society Aff:* SEM, NAE, AAAS, ASTM, ASME

Irwin, Kirk R
Home: 5L5 Highland Terrace, York, PA 17403
Position: VP Mfg *Employer:* The J.E. Baker Co *Education:* BS/ME/Lehigh Univ *Born:* 10/18/39 Native of Lansdowne, PA. Spent 11 years with Alpha Portland Cement Co, in operations and process engrg, including 3 years as plant mgr. Joined J. V. Warren, Inc to develop licensed oil & gas firing systems business for Rotary Kiln operations. Joined Gen Portland Inc for 2 years in operations prior to joining J.E. Baker Co, in 1975. Principal in firm. Current responsibilities include engrg, All Plant operations, & Rock Products Sales. Registered PE in PA, OH and TX. Married with 3 daughters. Enjoy golf, boating, home restoration and remodeling. *Society Aff:* ASME, NSPE, PSPE

Isaac, Maurice G
Business: 1285 Boston Rd, Bridgeport, CT 06602
Position: Prog Mgr *Employer:* GE. *Education:* MS/EE/Syracuse Univ. *Born:* 1/1/35. Maurice completed GE's Advanced Engg Prog & all PhD course work in System Sci at Poly Inst of Brooklyn. His early work addressed the response of linear and time variable systems to deterministics & stochastic stimuli. This work was reported in IEEE transactions on Electronic Computers in 1962. Subsequently he *Pioneered* the dev of Computer Aided Design techniques & the application of micro-electronics to microwave components. Following his assignment as the Mgr of R&D planning in the Aircraft Equip Div, he accepted responsibility from the Corporate Consulting function to develop and disseminate throughout GE, a full series of order-based training programs in Electronics Technology. Subsequently Maunce lead the team which completely redesigned Gen Electrics flagship entry level training program in Mfg Leadership. The result is a nine course distinguished curriculum which employs interactive video disk, personal computers, videotape, poin material & leader interview. *Society Aff:* IEEE.

Isaacs, Gerald W
Business: Agri Engrg Dept, Gainesville, FL 47907
Position: Prof & Chairman, Agri Engrg Dept. *Employer:* University of Fla. Gainsville *Education:* PhD/Agri Engrg/MI State Univ; BSEE/Elec Engrg/Purdue Univ; MSEE/Elec Engrg/Purdue Univ. *Born:* Sep 3, 1927 Crawfordsville, Ind. Farm reared in Parke Cty Ind. Enlisted ser in U S Navy as electronics tech 1945-46. BSEE Purdue Univ 1947 & MSEE in 1949. PhD with major in agri engrg Mich State Univ 1954. Mbr Agri Engrg fac Purdue Univ 1948-52 & 1954-81, & Hd of that dept 1964-81. Prof. and Chariman, Agr. Engr. Dept, Univ of Florida, Gainesville, 1981- . Private cons in area of grain drying & storage. Prof engr IN. Fellow & Past Pres ASAE. Mbr ASEE & NSPE. Mbr Tau Beta Pi, Gamma Sigma Delta, Pi Mu Epsilon, Phi Tau Sigma, Sigma Xi & Phi Kappa Phi hon. *Society Aff:* ASAE, ASEE, NSPE.

Isaacs, Jack L
Business: 1285 Boston Ave, Bridgeport, CT 06602
Position: Mgr-Res Engg. *Employer:* Gen Elec Co. *Education:* MS/Chem Engr/Northwestern Univ; BS/Chem Engr/Univ of KY. *Born:* 3/18/40. Native of Louisville, KY. With N American Aviation in nuclear testing, 1963-5. Joined Gen Elec Maj Appliance Labs, Louisville, KY, in 1965. Appointed Proj Mgr-Plastics Applications Ctr in 1975, specializing in developing plastics products in several Gen Elec

Isaacs, Jack L (Continued)
depts. Transferred to Bridgeport, CT, in 1978 as Mgr-Resident and Sourced Products Engg for the Gen Elec Housewares (Small Appliance) business with responsibility for engg at all domestic and off-shore plants. Intl Pres of the Soc of Plastics Engrs for 1978-9, and officer in numerous other organizations and societies. Married, four children. *Society Aff:* SPE, ΑΧΣ, ΣΞ.

Isaacson, Irwin
Business: 931 Canal St-Suite 600, New Orleans, LA 70112
Position: Chairman. *Employer:* Leo S Weil & Walter B Moses Inc. *Education:* BE/Mech Engr/Tulane. *Born:* Jul 18, 1925 New Orleans, La. Grad studies Elec Engrg 1965 LSU. Leo S Weil & Walter B Moses Inc, Cons Engrs, New Or leans, La 1947- , V P 1961-72, Pres 1972-1982. Chairman of Board 1982-1987. Written tech art for 'Elec Constr & Maintenance', 'Heating, Piping & Air Cond', 'Air Cond, Heating & Ventilating'. Mbr Tau Beta Pi, Pi Tau Sigma, IEEE (senior member), IES, IAEI. Pres Cons Engrs Council of New Orleans 1974-75. Mbr Amer Cons Engrs Council, NFPA, Bd of Elec Examiners-City of New Orleans. Served on Bds of Touro Synagogue & Jewish Community Ctr, New Orleans. Presently on Bd of Mgrs of Touro Infirmary, New Orleans. Regis: Elec & Mech-LA-1948, Prof Engr Miss 1971. *Society Aff:* IEEE, IES, IAEI, ΤΒΠ, ΠΤΣ.

Isaacson, LaVar K
Business: Dept of Mechanical & Ind Engrg, Salt Lake City, UT 84112
Position: Prof *Employer:* Univ of UT *Education:* PhD/ME/Univ of UT; BS/ME/Univ of UT *Born:* 7/15/34 in Provo, UT. Test engr for Marquardt Corp. Project Mgr and Program Mgr for Electric Power Research Inst. Sr Technical Specialist for Hercules, Inc, specializing in rocket nozzle boundary layer analysis techniques. Asst Prof, Assoc Prof, and Prof of Mechanical Engrg at The Univ of UT, with emphasis in thermodynamics, propulsion systems, turbulent flow and stability. *Society Aff:* AIAA, ASME

Isaak, Elmer B
Home: 79 W 12th St, New York, NY 10011
Position: Chairman of the Board *Employer:* URS Company Inc. *Education:* CE/Civil Engg/Cornell Univ; AB/Economics/Cornell Univ. *Born:* 9/6/12. Became Chairman of the Bd of URS Co Inc after 2 yrs as Pres of its predecessors, URS/Madigan-Praege & URS/Coverdale and Colpitts, and 1 1/2 yrs as Exec VP. Previously with predecessor firm Madigan-Hyland over 30 yrs, interrupted by 8 yrs in private practice as Elmer B Isaak, PE, PC. Specialities include transportation planning and feasiblity reports for large projs such as NY State Thruway and Verrazano-Narrows Bridge. Active in prof soc, & co-founder Joint Urban Manpower Prog (JUMP) for training disadvantaged youth in engg offices. Dir ASCE 1972-75, Asst Treas 1976-81. Mbr NY SSPE, ACEC, IBTTA. The Moles. Lic P E NY & FL. *Society Aff:* ASCE, NYSSPE, IBTTA, TRB, MOLES.

Isakoff, Sheldon E
Home: R.D. 1, Box 361, Chadds Ford, PA 19317
Position: Dir, Engg R&D. *Employer:* DuPont Co. *Education:* PhD/ChE/Columbia Univ; MS/ChE/Columbia Univ; BS/ChE/Columbia Univ. *Born:* 5/25/25. in Brooklyn, NY. Served in Navy, 1943-46; Engg officer. Grad Fellow, Brookhaven Natl Lab 1949-50. Joined DuPont in 1951 after completing PhD, Columbia. Wide variety of engg, R&D, & mgt assignment in chem engg, mtls tech, instrumentation, engg phys & automation. Assumed current position, in 1975, as dir, engg res & dev, responsible for new process & engg tech for DuPont's commercial operations. Fellow of AIChE; Dir 1977-79. Bd of Trustees, Federation of Mtls Socs 1976-79; Fellow of AAAS; Member of National Academy of Engineering. AIChE Founder's Award 1980; Mbr, Natl Materials Adv Bd 1980-82. AIChE Inst Lecturer 1984; Mat'ls Engg & Sci Div Award 1986. *Society Aff:* AIChE, ACS, AAAS, NAE.

Isenberg, Lionel
Home: 1205 Sunbird Ave, La Habra, CA 90631
Position: Pres. *Employer:* IR Assoc. *Education:* MBA/Bus Practice/Alexander Hamilton Inst; BS/Chem/Univ of CA; BS/EE/Univ of CA. *Born:* 2/26/25. Native of Los Angeles. Dir Tech Transfer, Rockwell Intl to 1971. Manager, Energy Projects, Jet Propulsion Labs.Pres, IR Assoc, 1971 to date. Specialty, Tech transfer from Aerospace Ind to Energy Systems Applicatins; Product Line Dev & Marketing. More than 20 maj publications incl 3 Books & 50US & Foreign Patents. Instr in Cryogenic & Mtls Engg. Current activities: Space Nuclear & Integrated Solar Thermal Power Systems, Synthetic Fuels processes and components, Magnetic Fusion key components. Honors: Sigma Xi, RESA, Presidential Award for Innovation, "Engineer of the Year–. Regional Bd Mbr, AIAA, IAF Intl Congress Steering Committee. Chrmn, Dir & Judging Chrmn Intl Sci & Engg Fair. Admin-Chman AIAA Annual MTS. Former mbr AIChE Natl Prog & Aerospace Committees & Editor AIChE Symposium Volumes. *Society Aff:* AIChE, AIAA, AIC, ASS, AAAS, AFA, ACS.

Isenberg, Martens H, Jr
Business: 250 Old Country Rd, Mineola, NY 11501
Position: District Manager *Employer:* Long Island Lighting Co. *Education:* -/Marine Engg/US Merchant Marine Acad. *Born:* Oct 1926. Native of Long Island NY. Joined M H Tredwell Co power plant constr. 1948 Long Island Lighting Co as Results Technician, Elec Prod. Adv to Plant Engr respon for oper procedures & training. Trans to Constr Dept. Respon for power plant equip & components. Joined Sales Dept as Lighting Engr 1956. Joined IES 1963. Conducted & taught lighting courses for 11 yrs. Founder Long Island Sect & V Pres Northeast Region 1973-75. District Manager for commercial & indus electrical and gas usage. Enjoys sailing. Patent for hydraulic mechanism for sailboats. Member of Air Conditioning Contractors of America. *Society Aff:* IES, ACCA, CPA.

Ishii, Thomas Koryu
Business: 1515 W Wisconsin Ave, Milwaukee, WI 53233
Position: Prof *Employer:* Marquette Univ *Education:* Dr of Engrg/EE/Nihon Univ-Japan; PhD/EE/Univ of WI; MS/EE/Univ of WI; BS/EE/Nihon Univ-Japan *Born:* 3/18/27 Born in Tokyo, Japan. From 1949 to 1956, he worked on research of microwave circuits and amplifiers, and instructed students at Nihon Univ. From 1956 to 1959, he worked on research of noise of microwave amplifiers at the Univ of WI. Since 1959 to date, he has been with Marquette Univ. At present, he is a Prof of Electrical Engrg. The research areas include millimeter waves and microwave. Quantum electronics is also in his area. Since 1949, he has published more than 300 technical papers in various Engrg and Scientific Journals. In 1969, he received IEEE Milwaukee Section Memorial Awards and in 1984, he received IEEE Centennial Medal Awards for his Contribution. *Society Aff:* IEEE, NSPE, ASEE, ΣΞ, HKN, ΤΒΠ, AAUP

Ishimaru, Akira
Business: Elec Engrg Dept, Seattle, WA 98195
Position: Prof of Elec Engineering. *Employer:* Univ of Washington. *Education:* PhD/Elec Engrg/Univ of WA; BS/Elec Engrg/Univ of Tokyo. *Born:* Mar 1928 Japan. Naturalized 1963. Employed at Bell Tele Labs. Visiting Assoc Prof Univ of Calif, Berkeley. Cons to Boeing Co & Jet Propulsion Lab on antennas & propagation. Currently Prof of Elec Engrg & Applied Mathematics at Univ of Washington. Mbr-atLrg of US Natl Ctte of URSI. Editor of Radio Sci. Fellow of IEEE. Ed Ctte chmn, Admin Ctte mbr, & 'Disting Lecturer' of IEEE Antennas & Propagation Soc. Ed Bd of IEEE Proceedings. 1968 Region 6 Achieve Award. Fellow of Optical Soc of Amer, IEEE Centennial Medal 1984, Chrmn, Commission B of US Natl Ctte of URSI, Assoc Editor of Journal of Optical Soc of Amer. *Society Aff:* IEEE, URSI, OSA.

Isidori, Alberto
Home: Via Monti Parioli 49a, Rome, Italy 00197
Position: Professor *Employer:* University of Rome *Education:* PhD/Autom. Control/University of Rome; MS/Electr. Engrg./University of Rome *Born:* 01/24/42 Native of Rapallo, Italy. Since 1975, tenured professor of Automatic Control in the Univ of

Isidori, Alberto (Continued)
Rome. Held visiting positions in Washington Univ, St. Louis, 1983, in Arizona State Univ, Tempe, 1986, in Univ of Ill, Urbana, 1987. Received the IEEE Control Systems Society's Outstanding Paper Award in 1981. Elected to Fellow Grade of IEEE in 1987. Since 1983, responsible for coordinating the research in control theory financed by the Ministry of Public Education in Italy. Research interests include systems and control theory. Enjoy classical music and sailing. *Society Aff:* IEEE

Israelsen, Orson Allen
Home: 15115 Interlachen Drive #526, Silver Spring, MD 20906
Position: Consultant to BSSC & NAS. *Employer:* Self employed. *Education:* MS/Civil Engg/Univ of UT; MA/International Affairs/George Washington Univ; BS/Civil Engg/UT State Univ. *Born:* Mar 1919. Asst Prof Construction Engrg San Jose State 1947-51. US Army & Air Force 1941-46 & 1951-67: Combat Pilot, Res & Dev, & Constr Engrg Officer, to Colonel USAF. Tech Dir & Genl Mgr, Environ Res Corp 1968-70. Exec Secy US Natl Ctte on Tunneling Tech, Natl Acad of Sci-Natl Acad of Engrg 1972-78. Manager, Washington Office Bovay Engrs, Inc 1978-80. Exec Secy Ctte on Nat. Disasters 1980-85, Exec Secy Ctte on Earthquake Engineering 1982-85, Consultant to NAS & BSSC 1985- , Reg PE Ohio, Colo, Vir. Legion of Merit for dir constr in Vietnam 1967. *Society Aff:* ASCE, EERI.

Israelson, Harold G
Business: 11000 West 78th Street, Eden Prairie, MN 55344
Position: President. *Employer:* Reese, Ellingson *Education:* BS/Civ Engr/Univ of MN. *Born:* Apr 9, 1928. Army Topographic Engrs 1946-48. Engr G M Orr Engrg 1953-56. Proj Engr-Highway Div-Ellerbe & Co 1956-59. Proj Engr Banister Engrg Co 1959-61. Proj Engr Northern Contracting Inc 1961-64. Pres Union Central Contracting Inc 1964-67. Proj Engr-Lindsay Engrg-Brauer & Assoc Inc 1967-71. Founded Israelson & Assoc Inc 1971. Currently Pres of this Cons Engrg Co spec in Civil Hydrology, Structural Engrg; Architecture, Planning & Land Surveying. Reg Engr Fla, Ill, Iowa, Minn, N Dak, S Dak, Wisc. Dir Minn Heart Assn. Chmn 1975 & 1976 Suburban Area Fund Drive. Hobbies: golf, skiing, hunting. *Society Aff:* ASCE, NSPE, WPCF, AWWA, APWA, AAA.

Ives, M Brian
Business: McMaster Univ, 1280 Main St W, Hamilton Ontario, Canada, L8S 4L7
Position: Assoc Dean of Engg. & Professor of Materials Science and Engineering. *Employer:* McMaster University. *Education:* PhD/Physics/Univ of Bristol; BSc/Physics/Univ of Bristol. *Born:* Sep 1934 Bournemouth, England; Res Met Engr, Carnegie Inst of Tech, Pittsburgh 1958-61; on fac of Mc Master University, Hamilton, Ontario, Canada, from 1961. Leaves at U of Milan, Italy 1967-68, Max Planck Institut fuer Eisenforschung, Germany 1975, Univ. Erlangen, Nurnberg, Germany, 1983. Teaches intro course on engrg materials employing a nonlecture format. Res Prog concerned with pitting corrosion. Mbr ASM, TMS-AIME, & NACE. President of ASM 1984-85. PEng; Fellow, ASM; Fellow, Royal Soc of Arts. *Society Aff:* TMS-AIME, ASM International, NACE.

Ives, Raymond H, Jr
Home: 5701 Aspen Ct, Virginia Beach, VA 23464
Position: Dir of Civ Engrg *Employer:* Navy Public Works Ctr *Education:* M/CE/VA Tech Univ; BS/CE/VA Tech Univ *Born:* 7/11/52 Native of Norfolk, VA. Civ Engr I for the City of VA Beach 1974-75 in charge of Site Plans and Subdiv. Joined the Consulting Engrg Firm of Glenn- Rollins and Assoc, Inc in 1976 specializing in all phases of civ-struct design. Appointed Dir of the Civ-Struct Div of the Navy Public works Ctr in 1980. Obtained profl reg in the state of VA in 1978. Community activities include: Red Cross instructor for Advanced Lifesaving at VA Beach YMCA, Youth and Adult Aquatics. Certified scuba diver and first-aid instructor. Enjoy tennis, sailing, electronics, and all water sports. *Society Aff:* ASCE, NSPE

Ivey, C Allen
Business: 5650 Peachtree Pkwy, Norcross (Atlanta), GA 30092
Position: Pres/Chrmn Bd *Employer:* Cerny & Ivey Engrs, Inc *Education:* B/ME/GA Inst of Tech *Born:* 10/8/34 Native of Atlanta, GA. Taught at GA Tech and was a Research Engr at the Engrg Experiment Station, 1955-1964. Chief Engr for GAB Services, Southeast Div, 1964-66. Founded Cerny & Ivey Engrs, Inc in 1967. In addition to administrative duties, is Principal Engr in failure analysis div of co. Has presented technical papers in Japan, England, and US. Serves as "expert witness" in technical cases in federal and state courts. Serves on technical committees of various societies promulgating standards and codes. Engr of Year in Private Practice, GSPE, 1980. Pres, Atlanta Chapter, GSPE 1980. Chrmn GSPE Ethics Committee, 1974-75. Registered Professional Engr in GA, AL, FL, MA, NC, SC, TN, KY, VA. *Society Aff:* ASM, ASTM, ASNT, NSPE, BSSM, SESA.

Iwan, Wilfred D
Business: Mail Code 104-44, Pasadena, CA 91125
Position: Prof of Appl Mechanics. *Employer:* CA Inst of Tech. *Education:* PhD/ME/CA Inst of Tech; MS/ME/CA Inst of Tech; BS/Engg/CA Inst of Tech. *Born:* 5/21/35. Served on the faculty of the USAF Acad from 1961 to 1964. Joined Caltech faculty in 1964. Maj fields of interest are vibration, dynamics & earthquake engg. Active in res & consulting. Chrmn of California Seismic Safety Commission, Chrmn of the Intl Strong Motion Army Council and Exec Secy of the Univs Council for Earthquake Engg Res. A native of CA. Married & has three sons. Heavily involved in church related activities. *Society Aff:* ASME, ΤΒΠ, ΣΞ.

Jablonsky, John L
Business: 85 John St, New York, NY 10038
Position: VP *Employer:* American Insurance Services Group, Inc *Education:* B/ME/NYU *Born:* 5/28/32 Joined the Natl Bd of Fire Underwriters in 1954 and continued with the American Insurance Assoc and Amer Insur Serv Group, Inc. Named Dir of Codes and Stds in 1966. Elected Asst VP in 1972 and VP of the Engrg & Safety Service in 1978. Is an active mbr of cttes of numerous natl and intl fire, safety, bldg, codes and stds organizations and has written and lectured extensively on fire and accident prevention and bldg construction. In 1970, appointed by Pres Nixon to the Natl Commission on Fire Prevention and Control. Elected Fellow, Soc of Fire Protection Engrs in Nov 1980. Serves on the Bd of Dir of the ANSI and Standards Council of the Natl Fire Protection Assoc. Resides in Port Wash, NY. *Society Aff:* SFPE, ASSE, ASTM

Jack, Robert L
Home: 4720 N 7th Rd, Arlington, VA 22203
Position: Consultant. *Employer:* Self employed. *Education:* BS/ME/Univ of TN. *Born:* 7/15/12. Design Engr for Combustion Engg. Joined Maritime Admin in 1939. Asst Chief Engr, Gulf Coast Regional Office at New Orleans during WWII 1942-45. Proj Engr Maritime Admin for preparation of design contract plans & specifications for Nuclear Ship Savannah. 20 yrs experience in trials, performance investigations, and guarantee surveys of merchant ships. Eight yrs Chrmn, Trial and Guarantee Survey Bds of MARAD. Retired to private consulting practice 1973. *Society Aff:* SNAME, ASME.

Jack, Sanford B, Jr
Home: 1129 Snowdon Dr, Knoxville, TN 37912
Position: Design Proj Manager *Employer:* TN Valley Authority. *Education:* BSCE Univ of TN. 1951 *Born:* 5/19/27 Native of Knoxville, TN. Married with three children. Served in US Marine Corps 1945-46. Civ Engr and engineering manager with TVA in steam plant design since 1951. Currently respon for structural mechanical & electrical design of facilities for TVA's fossil generating plants. Fellow ASCE. Pres of TN Valley Sec, ASCE, 1975. Active in First Baptist Church, Knoxville. Enjoy golf & supporting the Univ of TN athletics prog.

Jackins, George A, Jr
Business: 2066 Old Rocky Ridge Rd, Birmingham, AL 35216
Position: Pres *Employer:* Energy Management Consultants, Inc *Education:* B/IE/GA Inst of Tech *Born:* 10/21/34 Native of Baltimore, MD. Resident of Birmingham, AL since graduation from GA Tech in 1958. Active in the HVAC industry and engrg activities there. Formed Energy Management Consultants, Inc in 1975 to specifically address energy conservation and management in existing bldgs of all categories. Firm now has a broad client base through southeast. Active in local and natl ASHRAE activities. Member ACEC natl energy committee. *Society Aff:* ASHRAE, NSPE, ACEC, AEE

Jacks, Robert L
Home: 530 E 86th St, New York, NY 10028
Position: Corporate Mgr. *Employer:* Union Carbide Corp. *Education:* MBA/Mgt/CUNY; MS/ChE/MIT; BE/ChE/Tulane Univ. *Born:* June 1925. Currently Mgr-Admin & Planning, Health, Safety & Environ Affairs, for Union Carbide Corp. Previous pos include VP & Gen Mgr, Teller Environ Sys Inc; Mgr of Engg, Armour Agri Chem; Mgr of Projs and Mgr-Planning & Control, M W Kellogg; & Design Engr, Exxon Corp. Was also Asst Prof, ChE, LA State Univ 1946-50. Elected AIChE Fellow, 1974, & Natl Dir 1976. Mbr Tau Beta Pi & Sigma Xi. honored as "Eminent Engr" in NY by TBP in 1969. *Society Aff:* AIChE, ΤΒΠ, ΣΞ.

Jackson, Arthur J
Home: 174 Rolling Ridge Rd, Fairfield, CT 06430
Position: VP. *Employer:* Peabody Engg. *Education:* BS/Chem Engg/Worcester Polytechnic Inst. *Born:* 9/28/18. Native of Springfield MA. With Whitcock Mfg Co, West Hartford CT - Mfgs of heat transfer equip - from 1941 to 1969. Progressed through position of Sales Mgr and onto VP in 1965 with responsibility for sales & engg. 1970 began as Chief Engr at Peck Stow & Wilcox, Southington, CT - became Exec VP & Gen Mgr in 1971. Started Peabody Engg Corp - 1973; appointed VP in 1975. Peabody Mgfs Combustion Equip. *Society Aff:* AIChE.

Jackson, Clarence E
Home: 866 Mission Hill Ln, Worthington, OH 43085
Position: Prof Emeritus & Consultant. *Employer:* OH State Univ. *Education:* BS/Phys/Carleton College; -/Phys/Geo Wash Univ. *Born:* 9/4/06. 1930-37 - Dept of Met, Natl Bureau of Stds, Wash, DC. 1937-45 - US Navy - Naval Res Lab (Hd of Welding Sec), Wash, DC. 1945-64 - Union Carbide Linde Div, Niagara Falls, NY & Newark, NJ Res in Welding & Met. Assoc Mgr for Elec Welding Dev. 1964-77 - Prof, Dept of Welding Engg, OH State Univ, Columbus, OH. Teaching and grad advisor. 1977-date - Prof Emeritus, Dept of Welding Engg, OH State Univ. Consultant to govt and industry in Welding application. Over 50 domestic and foreign patents - Ext profl publications. Professional Engineer E-031063 Ohio State Board of Registration. *Society Aff:* AWS, AIME, ASM, IIW, BIW.

Jackson, David B
Business: Box 87, Berkeley Heights, NJ 07922
Position: Pres. *Employer:* Dave Jackson Homes, Inc. *Education:* BS/ChE/Lehigh Univ *Born:* 4/8/45. Wife, Suzanne, son David, Jr. B 1968. 1967-8 Res Engr with Hercules,Inc Wilmington DE, two patents plastics fabrication. 1968-71 technical sales rep, Hercules. Plastics Dept Cincinnati, OH. 1971-73 Market Dev GE Co. Chem div introduced PBT resins. 1973 to present, Pres Dave Jackson Homes, Inc, Berkeley Heights, NJ residential dev and bld co. Director and Past Pres Home Builders Assoc of Somerset/Morris, Secretary of NJ Builders Assoc, Dir Natl Assn of Home Builders, AIChE, Deacon of Congregational Church, Short Hills, NJ. Hobbies, golf, scuba diving, sailing. *Society Aff:* AIChE.

Jackson, David E
Business: 2901 Butterfield Rd, Oak Brook, IL 60532
Position: Pres, Katalco Corp. *Employer:* Nalco Chem Co. *Born:* Aug 1933. M Chem E Univ of OK, B Sc (Chem E) Queens Univ, Kingston, Ontario. Process Engr, Lago Oil, Aruba 1955-59. With Nalco Chem since 1960. Since 1970 as a bus area Group Mgr & since 1974 as Pres of Katalco Corp, a joint venture co with Imperial Chem Industries Ltd. Reg PE, TX, 1960.

Jackson, Harry A
Home: 17 Birch Ln, Groton, CT 06340
Position: Consultant (Self-employed) *Education:* BS/Naval Arch/Univ of MI; -/Nucl Power/GE Advanced Engg. *Born:* 12/7/16. A submarine designer of intl reputation who has cntributed significantly to the Navy's Nuclear Power, Polaris & submarine dev progs. At present, he is a self employed consultant in the field of Ocean Engg & is on the staff of the MA Inst of Tech as a Sr Guest Lectr. He is reg as a PE in the States of CT, WA & NH. He was Chrmn of the New England Sec of the Soc of Naval Architects & Marine Engrs for the 1977-1978 Season. ASNE Awarded him their Harold E. Saunders Award For lifetime contributions in May 1980. He is an ardent small boat sailor. *Society Aff:* SNAME, ASNE, RINA.

Jackson, Jesse B, Jr
Business: c/o Corco, Ponce, PR 00731
Position: Sr VP Operations. *Employer:* Commonwealth Oil Refining Co, Inc. *Born:* July 1923. MS Chem, BS Chem E Columbia Univ, BS Chemistry Univ of SC. Various positions in Creole Petrol Corp in Venezuela 1955-66. Dept Mgr-Refining Esso Argentina 1966-70. Mgr of Refining, Americana Hess Corp 1971-73. Assumed resent position Oct 1973. AIChE, Sigma Xi, Phi Lambda Upsilon. Enjoy hunting and woodworking.

Jackson, Melbourne L
Business: Chem Engg Dept, Univ. of Idaho, Moscow, ID 83843
Position: Prof & Dean of Engr, Emeritus. *Employer:* Univ of ID. *Education:* PhD/Chemical Engr/Univ of Minnesota; BsChE/Chemical Engr/Montana State Univ *Born:* Sept 27, 1915. BS Chem E MT State Univ 1941, PhD Univ of MN, 1948; reg engr, WA, ID; Fellow Am Inst of Chem Engrs; Hd, Chem Engr, 1953-1965; Dean of the Grad Sch & Coordinator of Res, 1965-1970; Dean of Engg 1973-74, Dean of Engg, 1978-80, 1983-, Univ of ID; mass transfer applied to air & water environmental control; 40plus papers in reviewed journals, presentations at natl & intl meetings; consultant to industry & govt, U.S. and foreign; ID Air Pollution Control Commission, 1959-72; Instr-Assoc Prof, Univ of MN, CO, MT State, 1942-1950; Hd, Process Dev, US Naval Ordnance Test Station 1950-53; maj prof for over 60 MS & PhD students; Doctor of Engineering (Honorary), 1980, Montana State University. *Society Aff:* AIChE, ACS.

Jackson, Ralph L
Home: 1001 Robin Rd, Muscatine, IA 52761
Position: Principal Elec Engr *Employer:* Stanley Consultants, Inc *Education:* BS/EE/Univ of IA *Born:* 4/8/25 Registered PE in IA in fields of Electrical and Mechanical engrg. Extensive experience in computer applications to engrg problems. Training in programming and operation of computers. Experience in all phases of fossile fueled central power station mechanical and electrical design. Married, 3 children. Active in community projects and church. On United Way Bd of Dirs. *Society Aff:* NSPE, IES, IEEE, SAME

Jackson, Robert B
Home: 1100 Vista Place, Apt. J, Wenatchee, WA 98801
Position: Consulting Engr (Self-employed) *Employer:* None *Education:* MS/CE/VA Polytech Inst; BS/CE/VA Polytech Inst *Born:* 1/21/11 Raised and educated in VA. Listed Who's Who in the West, 1980-81 Lt. Col Corps of Engrs 1941-1946; TVA hydro projects 1934-1941; Project Engr for Dravo on Morgantown Dam 1948-1950; General Construction Supt., Koppers Co., Baltimore 1950-1956; Resident Engr in charge of projects--Harza--on Priest Rapids and Wanapum Dams on Columbia River, WA State 1956-1961; Chief Construction Engr, Harza 1961-1965; Project Mgr, Mica Dam, B.C. Hydro 1965-1972; Project Supt., Ebasco on Keban Dam,

Jackson, Robert B (Continued)
Turkey 1973-1974; Consultant, Rock Island Dam Expansion 1975-1980. Fellow ASCE; Member USCOLD *Society Aff:* ASCE, USCOLD

Jackson, Roy
Business: Dept Chem Engg, Princeton University, Princeton, NJ 08544
Position: Prof. *Employer:* Princeton Univ *Education:* DSc/Chem Eng/Univ. of Edinburgh; MA/Physics/Cambridge Univ; BA/Physics/Cambridge Univ. *Born:* Oct 1931. D Sc from Univ of Edinburgh; MA from Cambridge Univ; Native of Manchester, England. Worked for Imperial Chem Industries 1955-61. Reader in Chem Engg, Univ of Edinburgh, 1961-68. Prof. Rice Univ 1968-77, nominated A J Hartsook Prof in 1973 & appointed Dept Chrmn in 1976. Prof. Univ. of Houston 1977-82. Prof. Princeton Univ. 1982-present. Fairchild Scholar. Cal. Inst. of Technology 1982-83. Main relaxation: sailing. *Society Aff:* AIChE.

Jackson, Stuart P
Home: 3732 Laurel Ridge Rd, Roanoke, VA 24017
Position: Pres *Employer:* Engrg and Marketing Corp of VA *Education:* PhD/Elec Eng/OH State Univ; MSc/Elec Eng/OH State Univ; BSE/Engg/Princeton. *Born:* 7/2/26. PhD OH State Univ. Also M Sc, BSE Princeton Univ. USN. Taught Lowell Technological, Lowell, MA. GE Co, North Elec, Solidstate Controls, Inc, Kollmorgen Corp-Inland Motor Div, Consulting-Power Conversion & minicomputer sys, delegate Intl Electrotechnical Commission, Alberta F Sperry Award 1973, mbr & held numerous offices in ISA, IEEE, & NEMA. Lic PE in MA, VA, OH, & LA. Thirteen patents, numerous papers. Married, 5 children, three grandchildren. Active in Presbyterian Church and Damascus Ministries. Enjoy scuba diving & fishing. *Society Aff:* ISA, IEEE.

Jackson, Warren, Jr
Home: 4871 Westbourne Rd, Lyndhurst, OH 44124
Position: Process Control Specialist-R&D. *Employer:* Std Oil Co. *Education:* MSEE/Electronics/Case Inst of Tech; BSEE/Electronics/Purdue Univ. *Born:* May 1922. MSEE Case Inst of Tech; BSEE Purdue Univ. Native of River Forest, IL. Police VHF radio dev River Forest 1941-42. US Army Signal Corps 1942-45, Ch Engr sev European AFN broadcasting stations. With Std Oil Co-OH since 1947, instrument R&D 7 yrs, process control & simulation in Process Engg Div 7 yrs, computer simulation & oper res in Mgt Sci Div 8 yrs, Supr of Instrumentation R&D 4 yrs & currently Sr Res Specialist R&D, respon for process control & lab instrumentation design, innovation & computer interfacing. 14 US pats issued & 7 tech papers pub. Cert of Merit awarded by Commanding Gen, European theater. Hon life mbr Soc for Computer Simulation (Simulation Councils, Inc) as charter bd mbr & chrmn of Midwestern Simulation Council. Mbr IEEE & AAAS. Hobbies: ham radio & electronic music. Reg Engr OH & FCC lic 1st class radiotelephone operator. *Society Aff:* IEEE, AAAS, SCS.

Jackson, William D
Business: P.O. Box 15128, Chevy Chase, MD 20815
Position: President *Employer:* HMJ Corp *Education:* Ph.D./Elec. Eng./University of Glasgow (Scotland); A.R.C.S.T./Elec. Eng./ University of Strathclyde, Scotland; B.Sc. (1st Class Hon.)/Elec. Eng./ University of Glasgow *Born:* 05/20/27 Native of Edinburgh, Scotland. Educated at Glasgow Univ and came to U.S. in 1955 as Fulbright United Kingdom Travel Scholar. U.S. Citizen. Career has included academia (M.I.T., Univ of Ill at Chicago Circle, Univ of TN Space Institute and The George Washington Univ), government (Dept of Energy and predecessor agencies) and industry. Currently Pres and Technical Dir of HMJ Corp, a metropolitan Washington based company specializing in advanced energy technology and energy system analysis. Consultant to numerous companies and organizations. Active in international cooperation, particularly magnetohydrodynamic (MHD) electrical power generation. *Society Aff:* IEEE, ASME, IEE(UK), AIAA, APS

Jaco, Charles M, Jr
Home: River Hills Plantation, Clover, SC 29710
Position: Partner. *Employer:* JCi Consultants *Education:* MChE/Thermodynamics/Univ of DE; BS/Sc & Engg/US Military Academy; -/Chem Engg/MS State Univ; Faculty/Engg/US Military Academy. *Born:* 1/28/24. Born Montgomery Co, MS, 1924. Business exec, engr & conslt, CMC (Certified Mgmt Conslt). Faculty mbr USMA 1956-59, co-author engr text. Military svc WWII, Korea; cadre, proj engr and div hd Missile Ctr, Huntsville, AL 1950-54; staff nuclear tests JTF VII Marshall In 1955-56; Mgr Corporate Dev, Dravo Corp, Pittsburgh, PA; Plant Mgr, Midland Ross Corp, Cleveland, OH; Pres & GM, Georgetown Ferreduction Corp, Georgetown, SC; Pres, Midrex Corp, Charlotte, NC. Partner, JCi Conslt and NHA Consult 1976-present. Holds patents. Assisted dev Midrex and other processes. Writer, Lecturer. *Society Aff:* AMA, IMC, AIChE, AISE

Jacob, Everett
Business: PO Box 3387, Houston, TX 77001
Position: Dir of Engg. *Employer:* Dow Chem USA. *Education:* MS/ME/OK State; BS/ME/OK State. *Born:* 6/9/26. Everett Jacob, dir of engg & gen mgr of Engg & Construction Services of Dow Chem USA, is a native of Yorktown, TX. He earned his BS & MS degrees in mech engg from OK State, where he grad in 1948 & 1949, respectively. He joined Dow in Dec, 1950, in the TX Div. He was named gen mgr of the LA Div in 1970 and then transferred to the TX Div as gen mgr in 1973. He has been in his present assignment since 1977. He is pres of the Houston Minority Purchasing Council, a mbr of the bd of dirs of the Texas Commerce Bank-Lakeside & is active in many civic organizations. *Society Aff:* NSPE, AIChE.

Jacobaeus, Christian
Home: Rattviksvagen 22, Bromma Sweden 161 42
Position: Sr VP. *Employer:* LM Ericsson Telephone Co. *Education:* DrElEngg/-/Royal Inst of Tech; BEE/-/Royal Inst of Tech Stockholm. *Born:* 6/15/11. Worked from 1935 in various positions in LM Ericsson Telephone Co, Stockholm. 1953-1976 as Exec VP Chief Technical Officer, from 1976 as consultant. Res in Telephone Traffic Theory Thesis "A Study on Congestion in Link Systems" 1950. Main Creator of Ericssons Crossbar switching systems. Various patents in telecommunications & electronics. Articles in technical press, papers to congresses, Intl Advisory council Intl Teletraffic Congress 1955-Doc Sc hc 1978. Univ of Lund, Sweden Maj gold Medal 1976 Swedish Acad of Engg Sciences. Alexander Graham Bell Medal 1979 IEEE Married 1) 1947-1972 Eva- Britta Widforss. 2) Irene Jonsson. Interests; travel, literature, golf. *Society Aff:* IEEE, AAAS, SAES, RSAS.

Jacobs, Donald H
Business: UNE 1001 Clover Dale Ave, Victoria, BC, Canada V8A 409
Position: Pres & Tech Dir. *Employer:* IEEE, APS, OSA, Comp Soc *Education:* BS/Physics/Rutgers Univ; MA/Physics/Duke Univ *Born:* 8/26/15. Electronics Eng, Arcturus Radio Tube Co, 1937-38. Optical Eng, Ilex Optical Co, 1938-40. Physicist Celanese Corp, 1940-42. Physicist, Nat'l Bur Stda, 1942- 43. Sr Physicist, US Naval Observatory, 1942-44. Physicist, Naval Ord Plant, Indpls, 1944-45. Res Project Supervisor, Univ of NM, 1945-46. Super Missile Guidance and Range Instrumentation, North American Aviation, 1946-48. Pres. & Tech Dir, Jacobs Instr Co, 1948-58. Pres & Tech Dir, Jacobs Instr Co, Ltd. 1958- present. Dev the highest velocity gun of its time. Directed the dev of the complete system of range instrumentation for guided missiles at Hollomon AFB. Conceived and dev the JAINCO aircraft navigational system. Developed the Mark 6 (AN/ARB-1) Bombing System. Conceived and built the first paralleled asynchronous digital computers (the JAINCOMP Family). Hold the basic patents on parallel digital computers. Invented & patented magnetic core random access computer storage. Inventor of the Digital Pulse Delay Generator. Author of *Fundamentals of Optical Engineering.* (McGraw-Hill, 1943, Editions Retz 1981). Inventor of the "JAINCO ZERO-Drift Chronometer, Jacobs-E supercomputer, and the Jacobs automatic pistol. Fellow, IEEE. *Society Aff:* IEEE, APS, OSA.

Jacobs, George
Business: US Bd for Intl Broadcasting, Suite 430, 1030 15th St NW, Wash, DC 20005
Position: Dir of Engg. *Employer:* US Govt. *Education:* MS/EE-Communications/Univ of MD; BSEE/Electronics-Comm/Pratt Inst. *Born:* 7/16/24. Completed 34th yr in governmental mgt and engg in the field of telecommunicatons. Maj accomplishments have been the dev of worldwide intl broadcasting systems, dev of satellite concepts and communication systesm, formulating maj natl & intl policies in the field of telecommunications, acting as a US govt consultant to a number of foreign countries in the dev of communication systems, and participating in more than a dozen maj intl telecommunication confs responsible for draftiong intl stds and regulations. Am author of more than 500 published technical articles on all aspects of telecommunications engg and mgt, and a co-author of the "Shortwave Radio Propagation Handbook-. Fellow in the IEEE and in the Radio Club of Am. Received Marconi Gold Medal Award for Supior Engg Accomplishments, 1977, a US govt Superior Honor ward in 1976, and an Outstanding Performance Award in 1980. Registered Professional Engineer in the District of Columbia and the state of Maryland. *Society Aff:* IEEE, RCA.

Jacobs, Harold R
Business: 208 Mech Engrg Bldg, Univ Park, PA 16802
Position: Hd of Mech Engrg & Prof *Employer:* PA State Univ *Education:* PhD/ME/OH State Univ; MS/ME/WA State Univ; BS/Gen Engrg/Univ of Portland *Born:* 11/19/36 Born in Portland, OR. He grad cum laude from the Univ of Portland in 1958. After working for GE Co he continued his education at WA State Univ and the OH State Univ, obtaining his MSME in 1961 and PhD in 1965. After working at the Aerospace Corp for three yrs he joined the faculty of the Univ of UT in 1967. During his sixteen yrs he was promoted to Prof in 1972. He has served as Chrmn of Civ Engrg as well as being Assoc Dean for Res of the Coll of Engrg. In 1984 he was appt Hd of Mech Engrg at the PA State Univ. He is the author of more than 80 technical publications in the fields of heat transfer, energy conversion and thermal effects. *Society Aff:* ASME, AIAA, ΣΞ

Jacobs, Ira
Business: AT&T Bell Laboratories, Room 3C-206, Crawfords Corner Road, Holmdel, NJ 07733
Position: Dir *Employer:* Bell Labs *Education:* PhD/Physics/Purdue Univ; MS/Physics/Purdue Univ; BS/Physics/City Coll of NY *Born:* 1/3/31 Native of Brooklyn, NY. Teaching Asst and Research Fellow at Purdue Univ. At Bell Labs since 1955. Worked in electromagnetic and communication theory, satellites, and space communications. Supervisor 1960, Dept Head 1962, Dir 1969. Lecturer in short courses at Purdue, MI, and PINY. Participant in Inst of Defense Analysis summer studies (1964, 1966) on satellite multiple access. 1974-1985, directed Bell Labs development on optical fiber telecommunications. IEEE Fellow for contributions and leadership in development of lightwave communication systems. Since 1985, Dir of Transmission Technology Lab with responsibility for digital transmission and signal processing technology. *Society Aff:* IEEE, APS, AAAS

Jacobs, J Donovan
Business: 500 Sansome St, San Francisco, CA 94111
Position: Chrmn *Employer:* Jacobs Assoc. *Education:* BS/CE/Univ of MN. *Born:* 12/24/08. Twenty yrs with Walsh Construction Co in construction engg and supervisory position. 1954 founded Jacobs Assoc to provide specialized technical services to the construction ind. Involved heavily in underground struct such as tunnels, shafts and chambers. Inventor of the Tunnel Sliding Floor, a device to speed the advance of underground excavation. Served as consultant to owners on public works projs including Oroville Dam, Snowy Mtns Scheme (Australia), Sewage Tunnels for Metropolitan Municipality of Greater Seattle; Melbourne (Australia) Underground Railway Loop Systems; Mbr 1969 NAE; Chrmn 1977 Natl Committee on Tunneling Tech. Golden Beaver Award for Engineering, The Beavers, 1980. Non- Member Award by the Moles, New York City, 1981. *Society Aff:* ASCE, AIME, NAE, IEA.

Jacobs, John D
Home: 115 Woodmont Boulevard, Apt #421, Nashville, TN 37205-2222
Position: Former: Prof of Mechanical Engineering *Employer:* Retired *Education:* MS/ChE/MIT; BE/ME/Vanderbilt Univ *Born:* 06/16/11 After 11 years of industrial experience in industrial engrg & personnel, joined faculty of Vanderbilt Univ School of Engrg in 1947 and attained the rank of Prof of Mech Engrg. Developed a number of new courses and served for about ten years as Chairman of the Curriculum Cttee of the School. Publications include *10SU,2* Years, a history of the Vanderbilt School of Engrg and, for the ASME Centennial, an *Inventory of Early Industrial Sites of Middle Tennessee.* Served a term as Chairman of the (then) Nashville Section of ASME. Am a Life Fellow of ASME. *Society Aff:* ASME,ΣΞ,TAU BETΠ,ΠTΣ

Jacobs, John E
Home: 631 Milburn, Evanston, IL 60201
Position: Walter P Murphy Prof; Dir. *Employer:* Northwestern Univ. *Born:* June 1920 Kansas City MO. BS, MS, PhD from Northwestern Univ. GE X-ray Dept Kansas City MO prior to WWII. USN Deck officer 1942-46. GE X-ray Dept Milwaukee WI & Res Sci GE Schenectady NY 1950-60. With Northwestern Univ since 1960. Appoint Dir of Biomed Engg Ctr 1961, Walter P Murphy Dist Prof 1969. Mbr NAE 1969, DSc (Hon) Univ of Strathclyde Glasgow Scotland 1972. 36 issued pats, 80 pubs. Maj res areas: ultrasonic imaging, dev of instrumentation for health care del. Outside interests: sailing, collecting & renovating antique clocks.

Jacobs, Joseph J
Business: 251 S Lake Ave, Pasadena, CA 91101
Position: Chrmn and CEO. *Employer:* Jacobs Engg Group Inc. *Education:* PhD/Chem Engg/Poly Inst of Brooklyn; MS/Chem Engg/Poly Inst of Brooklyn; BS/Chem Engg/Poly Inst of Brooklyn. *Born:* 6/13/16. Chem Engr, Autoxygen, Inc, NY 1939-42; Sr Chem Engr, Merck & Co, Rahway, NJ 1942-44; VP & Technical Dir, Chemurgic Corp, Richmond, CA 1944-47; Pres, Jacobs Engg Co 1947-74; Chrmn & Chief Exec Officer, Jacobs Engg Group Inc 1974 to date. Fellow of American Inst of Chem Engrs, American Inst of Chemists, and Inst for the Advancement of Engg. Contributor of numerous articles in technical journals and holder of various patents. Married: Violet (Jabara). Children: Margaret, Linda, Valerie. Clubs: Altadena Town & Country Club, CA Club, Annandale Golf Club, Pauma Valley Country Club, Union League Club (NY). *Society Aff:* ACS, AAAS, ΦΛΥ, ΣΞ.

Jacobs, Louis S
Business: 2605 W. Pratt Blvd, Chicago, IL 60645
Position: Pres *Employer:* Louis S. Jacobs & Assoc. *Education:* ScD/Safety/Indiana North Univ; PhD/Human Factors/Indiana North Univ; MS/Prof Mgmt/Ind North Univ; MS/Ind Engrg/Armour Inst of Tech; BS/Arch/IL Inst of Tech *Born:* 6/11/17 , reared, educated primarily in Chicago Educational Insts. Served with US Navy as Lieutenant over over three and one-half yrs. Asiatic-Pacific and American Theatres. Registered PE IL, DE, Registered CA as Industrial Engr, Safety Engr, Manuf Engr; Registered IL as Architect. Taught Industrial Engrg-IL Inst of Tech; Taught Architecture Univ of IL; Architecture and Engrg Loop Coll Chicago; VP, Chicago Chapter, IL Society PE; 1978-1983 1978-1983 Pres, IL Society of Architects; 1980-1982 VP, Construction Safety Assoc of America; Pres and Fellow, Systems Safety Society, NW Chapter 1978-1984; Diplomate, American Academy Environmental Engr; Pres, Louis S. Jacobs Assoc; Fellow SARA. BS Arch., Armour Inst Engrg Ind North Univ 1972; PhD Human Factors Engrg, Ind North Univ 1974; MS Prof Mgmt Ind North Prof School Mgmt. *Society Aff:* ASCE, NSPE, ISPE, ISD, NSFP, AIA.

Jacobson, David H
Business: Altron/Altech Group, PO Box 286, Boksburg, 1460, South Africa
Position: Exec Dir *Employer:* Allied Electronics Corp (Altron) *Education:* PhD/Auto Control/Imperial Coll, Univ of London; DIC/Engrg/Imperial Coll, Univ of London; BSc/EE/Univ of Witwatersrand, South Africa *Born:* 2/23/43 Born in Johannesburg,

Jacobson, David H (Continued)
South Africa. Post-doctoral Fellow, Asst Prof (1968- 71) and Assoc Prof (1971-72) at Harvard Univ and Visiting Research Assoc at the Univ of CA, Berkeley (1971). Prof in the Field of Mathematics at the Univ of Witwatersrand (1972-74). Dir of the Natl Research Inst for Mathematical Sciences of the CSIR (1975-1980). VP of the CSIR (1980-82) and Deputy Pres of the CSIR (1983-85). Exec Dir of the Altron Electronics Group since 1985. Authored or co-authored over 100 research articles and 4 monographs. Fellow of the IEEE, Fellow of the SAIEE. In 1974, was a recipient of the Jaycee's Four Outstanding Young South Africans Award for contributions to applied mathematics. Registered PE and reg natural scientist in South Africa. *Society Aff:* SAIEE, IEEE, ORSSA, SAMS, I Mkt M

Jacobson, E Paul
218 Dearborn, Sandpoint, ID 83864
Position: Drainage & Soil Cons Specialist. *Employer:* Harza Engg Co. *Education:* BS/Agr Engr/IA State Univ. *Born:* 10/23/09. Native of Harcourt, IA, Employed for 30 yrs by USDA Soil Conservation service, lastly as State Conservation Engr for IA. Developed system of parallel bench terraces with tile outlets. Presently on a semi retired basis working for Harza Engg Co on special projs. Author (with C B Richey, Carl W Hall) Agri Engrs' Handbook. Presently manage my farm which has complete erosion control system designed for optimum machinery operational efficiency. This farm is visited annually by technicians from all parts of the world. 1973 received ASAE Hancor Sail and Water Engg Award. 1981 received ASAE Doerfer Engineering Concept of the Year award. *Society Aff:* ASAE, SCSA, IES.

Jacobson, Nathan L
Business: 86 Main St, Chester, CT 06412
Position: Pres *Employer:* Nathan L Jacobson & Assocs., Inc. *Education:* BCE/Civ Engrg/Cornell Univ. *Born:* Jan 1929. Native of Chester CT. BCE Cornell Univ. Chi Epsilon. US Army 1947-48. A S Wikstrom Inc 1953-57 Const Proj Engr; Goodkind & O'Dea Cons Engrs 1957-72; Design Engr 1957-62, Ch Engr CT Office 1962-65, Mgr Water Res Div 1965- 72. Formed own cons engrg firm 1972. Mbr: ASCE, NSPE, PEPP, CEPP CSI CSI, WPCF, NEWPCA, NYWPCA, AWWA, NEWWA, ACEC, AAA. Elect Dir CEPP 1976. Spec in plan & design of Water Pollution Control & Water Trtmt facils. Served as Mbr of local Dev Comm, Bd/Finance, Reg Plan Agency; Chrmn Bd/Assessors, Plan Comm. Enjoy reading, boating, fishing, traveling. *Society Aff:* ASCE, NSPE.

Jacobus, David D
Home: 24 Acad Ln, Bellport, NY 11713
Position: Sr Mech Engr. *Employer:* Brookhaven Natl Lab. *Education:* ScD/-/MIT; ME/-/Stevens Inst of Tech. *Born:* 2/16/00. in Jersey City, NJ. ME Stevens Inst of Tech, 1921; ScD MIT, 1930; Married Margaret Penman on 2/24/26; children, David Penman, John Henry; Married 2nd, Elinor Hughes, 7/14/57; Stone & Webster Engg Corp, 1922-26; MIT, 1927-30; Stevens Inst of Tech 1931-40, Prof 1946-48; Keoffel & Esser Co, 1940-42; Radiation Lab, MIT, 1942-46. Recipient War & Navy Dept Award; Cambridge Electron Accelerator, Harvard Univ, 1957-66; Brookhaven Natl Lab; Hd ME Dept, 1949-57. Consulting engr 6G. Retired 1975. *Society Aff:* ASME, APS, AAAS.

Jacoby, George V
Business: 3333 Scott Blvd, Santa Clara, CA 95051
Position: Senior Professional Consultant *Employer:* ISS/Sperry Univac *Education:* Dipl. Ing/EE & Engr Econ/Royal Hungarian Tech Univ *Born:* 2/26/18 Native of Esztergom, Hungary, came to U.S. in 1950. 1953-58: Honeywell, R&D work on optimizing control and servos. 1958-71: RCA, Development of servo systems and signal equalization for magnetic recording. Invented and product developed the Delay Modulation Code, doubling the bit density of previous systems. This code became the standard of the digital recording industry, as the MFM code. In 1971 joined Sperry/Univac as Mgr, Advanced Recording Techniques. Invented the 3PM code, developed high density disk drives with further 50% density increase over the MFM code. In 1978 received the co's Outstanding Contributor's Award. Holds 20 patents. Fellow, IEEE 1981. Prof Engr of PA. *Society Aff:* IEEE

Jaffe, William J
Home: 1175 York Ave, Apt 9E, New York, NY 10021
Position: Distinguished Prof Emeritus *Employer:* NJ Inst of Tech. *Education:* EngrScD/Indus Engg/NYU; MS/Indus Engg/Columbia Univ; MA/Math/Columbia Univ; BS/Math/NYU. *Born:* 3/22/10. Naval Architect US Navy Dept (1941-5). From Instr to Distinguished Prof Newark College Engg (NJIT) (1946-75): Establish, administer, teach undergrad-grad courses, progs, res. Author books, audio, professional articles, Headed IE Std Terminology (ASME-IIE-ANSI). Work: Biomedical Engg, NY Academy of Medicine (Assoc Fellow). Intl Bodies: Inst Business Admin Mgt (Japan), Committee Manpower (Israel.). ANSI: Board of Standards Review. ASME: Standardization Bd. Listed among "Landmark Authors" ASME Mgt Centennial. PE CA. Fellow: ASME, AAAS, IIE, SAM. Omicron Delta Kappa, Pi Mu Epsilon, Alpha Pi Mu, Pi Delta Epsilon. Clubs: Chemists (NY). Press (San Francisco). *Society Aff:* ASME, IIE, AAAS, SAM, NYAcadMedicine, AMS.

Jaffee, Robert I
Business: 3412 Hillview Ave, Palo Alto, CA 94304
Position: Senior Technical Advisor *Employer:* Elec Power Res Inst. *Education:* PhD/Chem Engr/Univ MD; SM/Metallurgy/Harvard Univ; BS/Chem Engr/IL Inst Tech. *Born:* 7/11/17. After working as a physical metallurgist at Leeds and Northrup and the Univ of CA at Berkeley, I spent 32 years at Battelle's Columbus Laboratories, ending up as Chief Materials Scientist. I am currently head of materials research at the Elec Power Res Inst in Palo Alto. I am also consulting Prof in Materials Science at Stanford Univ. I was elected to NAE and am Honorary Mbr of ASM, and am a fellow of TMS, ASM, and the Inst of Metallurgists (UK). My research was mainly concerned with metallurgy of titanium, refractory metals, and power plant materials. Publications include 200 papers, 20 books, and 40 US patents. I was Pres of the Metallurgical Society of AIME in 1978-1979. I presented the 1976 H W Coillett Mem Lecture of ASTM, the 1977, Edward DeMille Campbell Mem. Lecture of ASM, and the 1985 Distinguished Joint Lectureship on Materials and Society of TMS-AIME and ASM. I received the 1984 James Douglas Gold Medal of AIME for distinguished achievement in nonferrous metallurgy. *Society Aff:* TMS-AIME, ASM.

Jager, Richard Henry Murray
Business: 491 Ridge Road, Overport, 4001, Durban, Natal, Republic of South Africa
Position: Sr Partner (Self-employed) *Education:* BSc/E/Mechanical/Univ of Natal, Diocesan Coll (Bishops) - Cape Town *Born:* 3/28/36 1959 - Asst Combustion Engr at Congella Power Station. 1960-61 - pupil engr with British Colonial Civil Service in Northern Rhodesia. 1962-63 - Chief Engr for Zambesi Sawmills in Northern Rhodesia. 1964-68 - Shell & BP Refineries South Africa. 1964 - Project Engr, Mechanical. 1965-66 - Project Engr special projects. 1967 - Seconded to Shell Intl Management Team for Base Oil Refinery. 1968 - Graduate Asst to Head of Operations. 1969-1971 - M.S. Zakrzewski & Partners-Sr Project Engr. 1972 - established private practice - Jager & Assocs, managing dir. *Society Aff:* AMA, AIChE, IIE, IChE, MSAIMech.E, SAIM, EASA

Jahn, Robert G
Business: D334 MAE Dept, SEAS, Princeton University, Princeton, NJ 08544
Position: Dean Emeritus, Professor of Aerospace Science *Employer:* Princeton Univ. *Education:* PhD/Phys/Princeton Univ; MA/Phys/Princeton Univ; BSc/ME-Phys Option/Princeton Univ. *Born:* 4/1/30. Dean Emeritus, Sch of Engg/Appl Sci, Princeton Univ. Teaching and res in magnetoplasmadynamics. Author, "Physics of Electric Propulsion." Chrmn of the Bd, Associated Univs, Inc. Director, Hercules, Inc. Mbr, Space Systems and Tech Advisory Committee, NASA Res and Advisory Council. Fellow, American Physical Soc and Am Inst of Aero & Astro. Recipient Shuichi Kusaki Prize in Phys, 1951 and Curtis W McGraw Res Award, Am Soc for

Jahn, Robert G (Continued)
Engg Ed, 1969. Bd of Trustees, Drexel Univ. *Society Aff:* APS, AIAA, ASEE, ΦBK, TBΠ, ΣΞ.

Jain, Anil K
Home: 517 Hubble Street, Davis, CA 95616
Position: Prof. *Employer:* Univ of CA, Davis *Education:* PhD/EE/Univ of Rochester; MS/EE/Univ of Rochester; B Tech/EE/IIT, Kharagpur *Born:* My specific area of expertise is: Digital Signal & Image Processing. Tech Honors and Awards: 1977-2 Achievement Awards, Interl Picture Coding Symp, Tokly. 1968-70 - System Science Fellowship, Univ of Rochester. My present tech involvement is in Electrical Engineering and Computer Sci. Senior Mbr of IEEE; Chmn, Inform Theory Group, Buffalo Sect, 1977-78. Employed at State Univ of NY as Assoc Prof, Buffalo, NY, from 1974-79, and at University of Southern California, as Assistant Prof., 1970-74. I had 10 yrs prof exper and 5 yrs mngmt exper. 1979-present-Prof, Dept of Elect and Computer Engrg, Univ of CA, Davis, CA 95616, 1983-IEEE Donald G Fink Prize, 1983-Topical Editor for Image and Signal Processing, Journal of the Optical Soc of America. (JOSA-A). *Society Aff:* IEEE, OSA

Jain, Kris K
Home: 4439 Lakewood Blvd, Naples, FL 33942
Position: Pres. *Employer:* Kris Jain & Assoc Inc. *Education:* MS/Structural/Univ of IL; BS/Civil Engg/UNiv of Roorkee, IN; BS/Phyiscs/Univ of Delhi, IN. *Born:* 5/31/35. A naturalized US citizen, immigrated from India in 1962. Worked as Asst Exec Engr with Govt of India from 1959-62. After masters took additional doctoral courses in CE. Reg PE in NC, NY, MA & FL; certified by NCEE. Worked with consulting firms in OH, NC, MA & FL, designing multistory bldgs and bridge structures. Chf structural engr with CE Maguire A/E, Waltham, MA 1972-74, at present, Pres of Kris Jain & Assoc, Consulting Engrs in Naples, FL. PPres of Naples East Rotary; PPres of South FL Section ASCE 1978-79; Recipient of outstanding Civil Engr award from South FL Section ASCE 1978-79. *Society Aff:* ASCE, FES, NSPE, ACI, PCI, PTI.

Jain, Ravinder K
Home: 1202 Devonshire Dr, Champaign, IL 61821
Position: Chief, Enviromental Div. *Employer:* US Army Corps of Engrs. *Education:* PhD/Civ Engr/TX Tech Univ; MS/Civ Engr/CA State Univ; BS/Civ Engr/CA State Univ; MPA/Management/Harvard Univ (1980). *Born:* 10/12/35. Presently Chief, Enviromental Div, US Army Corps of Engrs, Construction Engg Res Lab (CERL) and an Adjunct Prof at the Univ of IL, Urbana-Champaign and Research Affiliate Massachusetts Institute of Technology. Academic training includes a PhD in Civil Engrg. Gained diversified expertise in enviromental and water resource engg and systems analysis through a career in profl practice, res and teaching. Worked for CA Dept of Water Resources 1964-68; and a municipal engg consulting firm 1961-64. Recipient US Army R & D Achievement Award 1976 and Exec Dev Fellowship to attend Harvard Univ for 79 academic yr and was awarded the MPA degree from Harvard in 1980. Author of the book: Enviromental Impact Analysis - A New Dimension in Decision Making, 1980 (van Nostrand Reinhold), Chrmn ASCE Enviromental Engg Res Council 1975-77. Licensed Civ Engr CA; diplomate, Amer Acad of Environ engrs., Elected Fellow Churchill College, Cambridge University, 1986. *Society Aff:* ASCE, SAME, ASEE, IAWPRG, WPCF.

Jain, Subhash C
Business: Inst of Hydraulic Res, Iowa City, IA 52242
Position: Prof *Employer:* Univ of IA *Education:* PhD/Mech & Hydr/Univ of IA; MS/CE/Univ of Roorkee-India; BS/CE/Univ of Roorkee-India; BS/Sci/Univ of Agra-India *Born:* 4/6/38 Native of India. Taught in a Regional Engrg Coll, India prior to arrival in USA in 1967. With Univ of IA since 1967. Assumed position of Assoc Prof & Sr Res scientist in 1977 and Prof in 1982. Author of more than 100 technical papers and reports, and a textbook in Fluid Mechanics. Maj res interests: Waste-heat mgmt, river hydraulics, and hydraulic structures. Enjoy reading. *Society Aff:* ASCE, IAHR, AGU, ΣΞ

Jakes, William C, Jr
Home: 58 Wild Rose Dr, Andover, MA 01810
Position: Dir, Radio Transmission Lab. *Employer:* Bell Tel Labs, Inc. *Education:* PhD/EE/Northwestern Univ; MS/EE/Northwestern Univ; BS/EE/Northwestern Univ. *Born:* 5/15/22. Born in Milwaukee, WI. USN 1944-46, Lt in aircraft radar. Joined Bell labs in 1949 specializing in microwave antenna and propagation res. Proj Engr for Echo balloon satellite experiment 1959-61. Hon PhD IA Wesleyan 1961, Northwestern Univ Alumni Distinguished Award 1962. Hd, Radio Transmission Res 1963-71 for microwave mobile communication studies. Present position since 1971, responsible for Bell System analog and digital microwave radio systems design and dev. Fellow IEEE, Annual Paper Award IEEE vc 1971. Book "Microwave Mobile Communications–, 1974, Co-recipient Alexander Graham Bell Medal of IEEE, 1987. *Society Aff:* IEEE.

Jakeway, Lee A
Business: P.O. Box 1057, Aiea, HI 96701
Position: Agric Engr, Project Leader *Employer:* Hawaiian Sugar Planter's Assoc *Education:* MS/Agric Engrg/Univ of HI; BS/Agric Engrg/MI State Univ *Born:* 8/3/53 Native of Grand Rapids, MI. Moved to HI in 1975. After completion of degree, was a Research Assoc on the research staff of the Agricultural Engrg Dept, Univ of HI, 1977-79. Developed mechanical harvest systems for taro, a tropical root crop. In 1979, was employed by the East-West Ctr, Resource Systems Inst as a Research Fellow. Introduced nitrogen fixing tech to HI utilizing renewable energy sources. Presently employed by the Hawaiian Sugar Industry as scientific research staff Project Leader in Agricultural Engrg. Author of several publ in transactions of A.S.A.E and Hawaiian Sugar Technologists Reports. Reg PE in the state of HI since 1983. Past Chrmn of HI State Chapter of ASAE. Enjoy year round outdoor activities in Hawaii. Swimming, running, volleyball, and golf. *Society Aff:* ASAE, RI/SME, HSPE

Jalan, Vinod M
Home: 158 Hill St, Concord, MA 01742
Position: Pres *Employer:* ElectroChem, Inc. *Education:* PhD/Chem Engrg/Univ of FL; BS/Chem Engrg/Bombay Univ (India) *Born:* 05/02/43 My interests are electrochemical technology with emphasis on high surface-area materials and ionic membranes. This work has led to the invention of highly dispersed platinum alloy catalysts which have become a new generation of electrocatalysts. I have pioneered the use of copper based mixed oxides for high temperature desulfurization processes. Also, I have done R & D on electrode struct, redox batteries, methanol reforming, and coal gasifier hot gas cleanup. My publs number over 60, and I hold nine US Patents in the area of electrocatalysts, fuel cells, batteries, desulfurization, and coal gas cleanup. *Society Aff:* AIChE, ACS, ECS, MRS, Catalysis Society.

Jambor, George
Home: 7339 Univ Dr, Shreveport, LA 71105
Position: Owner *Employer:* George Jambor P.E. & Assoc. Consulting Engineer *Education:* BSME/Mech/Inst of Tech.; -/-/Budapest Institute of Technology *Born:* Dec 8, 1920 Budapest, Hungary. Elementary & Middle Sch Education received in Prague, Czechoslovakia. ME Sci degree from Inst of Tech, Budapest, Hungary in 1941. Employed by WM Co & Automobile, Tractor & Airplane Co of Budapest. After war, employed by UNRRA in Germany & Italy. Came to US in 1950. Naturalized in 1956. Employed by Paul O Rottmann, Consulting Engr since Apr 1950. Firm changed name in 1958 to Rottmann-Harriss-Jambor & Assocs, in 1968 to Harriss-Jambor & Assocs. Have been in charge of all Mech & Sanitary Engg since 1955. Reg Engg in LA, TX, AR, MS, AL, & TN. Mbr of ASHRAE, ACEC, NSPE, LES, A.E. E. Received all Elec Sys Design award 1969 & Energy Conservation Award 1974 from Bldg Sys Design for High Rise Mid-South Towers of Shreveport, LA. P Pres of Shreveport Chapt, Region VIII, ASHRAE. Firm name changed in

Jambor, George (Continued)
1987 to George Jambor, P.E. & Associates Consulting Engineers *Society Aff:* ASHRAE, ACEC, NSPE, LES, AEE.

James, Alexander
Business: PO Box 5882, Greenville, SC 29606
Position: Pres. *Employer:* Piedmont Crescent Eng Mgt Co *Education:* BS Chem/Chem/MS Coll; BChE/Ch Engrg/Univ AL. *Born:* 4/24/24. World authority on man-made fibers. Technical editor of International Fiber Journal. Pres of wholly owned consulting co catering to synthetic fiber & polymer industries throughout the world. Thirty Two yrs experience in design, operation & construction of man-made fiber plants. Presently operating a development laboratory catering to the man-made fiber industry. Registered professional Engineer in NC, SC, GA & FL. Married Louise Moore James, son Daniel is grad of Auburn Univ, daughter Amy is a grad of Wake Forest Univ and Univ Tennessee Grad Sch. *Society Aff:* NSPE, ACEC, AIChE.

James, Arthur M
Business: 319 SW Washington St, No. 614, Portland, OR 97204
Position: Pres *Employer:* Arthur M. James Engrs, Inc. *Education:* BS/CE/Coll of the City of NY *Born:* in NYC. Commissioned at Ft Belvoir, served in USAF Aviation engrs 1943-46. Employed by Hardesty & Hanover Bridge Engrs prior to moving to Oregon in 1947. Served as Bridge Engr, City of Portland, 1947-48. Employed as Structural Engr by several local architect-engrs prior to starting own consulting firm in 1951. Pres - Structural Engrs of OR 1958-59. Firm has done ind-commercial work as Prime for many yrs. Several patents including prestressed tanks, climbing cranes, long-span truss erection, "Lubritube" and "Sanigrate–. Six children: (M) 1 - in construction, (M) 1 - lawyer, (M) 1- doctor, (M) 1- artist, (F) 1 - Reporter, (M) - 17 yrs old (possible engr). *Society Aff:* ASCE, ACI, ACEC

James, Charles F, Jr
Business: Coll of Engrg & Appl Sci, UW-MILW, PO Box 784, Milwaukee, WI 53201
Position: Dean, Coll of Engrg & Appl Sci *Employer:* Univ of WI-Milwaukee *Education:* PhD/Ind Engr/Purdue Univ; MS/Ind Engr/Purdue Univ; BS/Mech Engr/Purdue Univ. *Born:* 7/16/31. Dean of Engrg & Appl Sci at UWM since Jan 1984, Chrmn of Indus Engrg at URI 1967-82. Sr Engr for McDonnell, 1963. Former mbr of faculties at Purdue and Univ MA. Employed in fifties as production engr by Shampaine Co, Colgate-Palmolive, and Kroger Co. Consultant to several industries and labor unions. Arbitrator listed by AAA and Fed. Mediation and Conciliation Service. Consultant in India several times. Res experience includes mfg processes, health systems, occupational safety, traffic safety and ROBOTICS. Held faculty appointment in New Zealand, 1979. Served as C Paul Stocker Dist Hd Prof at OH Univ, 1982-83. Outstanding Educators of America, 1973; Eminent Engr Award, Tau Beta Pi, 1978. Held several natl and State offices in IIE including pres of State Chapter. Member, Bd of Dir, Badger Meter Co., Milwaukee, WI; Registered P.E. WI. *Society Aff:* IIE, ASME, SME, ASEE, AFS, AAA, NSPE, WSPE

James, Ernest C
Home: 3009 Leta Ln, Sacramento, CA 95821
Position: Principal Engr; Chief-Civ Design. *Employer:* State of CA. *Born:* Apr 23, 1920, Ord, NB. BSCE Univ of CA Berkeley. Army Corps of Engrs 1940-46. Ret Reserve Rank of Maj. Currently Principal Engr; Chief, Civ Design, State of CA, Dept of Water Resources. responsible for Civ Engg Design of State Water Proj incl power plants, pumping plants, aqueducts, dams, & all other features. Fellow ASCE. Natl Dir NSPE; Mbr Natl Bd of Governors NSPE-PEG, Chrmn CSPE-PEG. Reg PE CA (Civ).

James, Jack N
Home: 1345 El Vago St, La Canada, CA 91001
Position: Staff Asst (Retired) *Employer:* CalTech/Jet Propulsion Lab. *Born:* Nov 1920, Dallas, TX. BS from Southern Methodist Univ, MS from Union Coll. Married, 4 children. Was a Lt in the Navy from 1943-46. Worked for GE, Schenectady NY, Res Engr from 1945-49. Worked for Radio Corp of Am, Camden, NJ, Res Engr from 1949-50. With Jet Propulsion Lab since 1950. Became Asst Lab Dir for Defense and Civil Programs June 1980. NASA Exceptional Scientific Achievement Medal for managing first spaceflights to Venus in 1962 & Mars in 1964. Louis W Hill Space Transport Award, Ballantine Medal from Franklin Inst of Phila, Distinguished Alumni, S.M.U.

James, John W
Business: 5330 MacCorkle Ave, South Charleston, WV 25309
Position: VP *Employer:* Triad Engrg Consultants *Education:* BS/CE/WV Inst of Tech *Born:* 7/8/46 With Ackenheil & Assocs 1968-1973, Project Mgr to Charleston Mgr. Had own consulting engrg firm 1973-1979 when merged with Triad. Has done geotechnical studies for many varied types of projects, primarily in WV, including dams, roads, bridges, large and small foundations, hazardous waste studies and landfills, environmental projects, and surface mining. Pres of WV Section of ASCE, 1981, Past Pres Charleston Branch of ASCE, Past Pres Charleston Chapter of WVSPE (NSPE). Married, four kids. Enjoys skiing, tennis, hunting, fishing and canoeing. *Society Aff:* ASCE, NSPE, ASTM, ACEC, AWWA

James, Lee A
Home: 1274 Old Meadow Rd, Upper St. Clair, PA 15241
Position: Advisory Engineer *Employer:* Westinghouse Electric Corp. *Education:* MSME/Mech. Engng./University of Washington; BSME/Mech. Engng/University of Washington *Born:* 09/10/37 Native of Dexter, Iowa. Following service as an Ordnance Officer in the US Army, worked as a stress analyst/designer at The Boeing Co in Seattle for nine years. Employed at the Dept. of Energy Hanford site (3 years with Battelle-Northwest, 17 years with Westinghouse Hanford Co.), rising to the level of Fellow Engr. Presently an Advisory Engr. with Westinghouse in West Mifflin, PA. Have specialized in fracture mechanics the last 20 years, publishing about 90 journal articles on that subject. Elected a Fellow of the ASME in 1986. *Society Aff:* ASME, ASTM, ASM, ΣΞ

James, Ralph K
Home: 227 Providence Rd, Annapolis, MD 21401
Position: . *Education:* MS/Naval Architect/MIT. *Born:* 5/21/06. USN 1928-1963, RAdmiral (ret), last duty Chief of Bureau of Ships, 1959- 1963. Pres ASNE - 1962-1963. Pres SNAME 1963-1965; Exec Dir CASL 1963-1969; Consultant, Bell Aero 1970-1978; Now retired. *Society Aff:* SNAME, ASNE.

James, Robert G
Business: 32 West Rd, Towson, MD 21204
Position: VP *Employer:* Century Engrg, Inc *Education:* MS/CE/Univ of MI; BS/CE/Wayne Univ *Born:* 1/18/33 Native of Detroit, MI. Served with Guided Missile Battalion, US Army 1955- 57. Bridge Engr with Detroit DPW prior to joining Green Assocs, Baltimore, MD, 1968 as chief structural engr, bridges and bldgs. Design responsibility for over 50 bridges. Partner with Century Engrg, Inc since 1974. Assumed current responsibility as VP, Environmental and mass transit projects, 1975. Mgr for major subway and transit design projects in Baltimore and Washington, DC. Responsible for numerous environmental projects throughout MD and VA. Including solid waste disposal and waste water treatment. *Society Aff:* ASCE, ITE

Jamshidi, Mohammad
Business: Dept of Electrical and Computer Engrg, Albuquerque, NM 87131
Position: Prof *Employer:* Univ of NM *Education:* PhD/EE/Univ of IL, Champaign-Urbana; MS/EE/Univ of IL, Champaign-Urbana; BS/EE/OR State Univ *Born:* 5/10/44 in Shiraz, Iran. He has taught and done research at Shiraz Univ (1971-79) and, IBM Watson Research Ctr, Yorktown Heights, NY (1975-77). He has been visiting Prof at Tech Univ of Denmark (Lyngby, DK) and Univ of Stuttgart (Stuttgart, FRG); Adv Engr IBM Info Prods Div (CO); visiting prof, Gen Motors Res Labs (MI); invited lecturer at Indian Inst of Tech at New Delhi and Bombai, Univ of

Jamshidi, Mohammad (Continued)
Baghdad (Baghdad, Iraq), Federal Inst of Tech (Zurich, SW), Middle East Tech Univ (Ankara, Turkey), Univ Bochum (Bochum, FRG), Twente Univ (Enchede, NL), Intl Inst of Appl Systems Analysis - IIASA (Luxenburg, Aus) and American Univ of Cairo (Cairo, Egypt); active participant in numerous natl and intl conferences; sr member of IEEE, founding editor of *IEEE Control Systems Magazine*; Assoc Editor of *IFAC J Automatica* (Pergamon Press) and *Large Scale Systems* (North-Holland). He is the author of 2 books, *Analog Simulation of Dynamic Systems*, Univ of IL 1971 and *Large Scale Systems-Modeling and Control*, Elsevier North Holland 1981; coauthor or author of 100 technical papers. Exec Edit Bd, *Encyclopedia for Sci & Tech*, Acad Press, 1986. *Society Aff:* IEEE

Jan, Hsien Y
Business: 65 Broadway, New York, NY 10006
Position: Asst to VP *Employer:* American Bureau of Shipping *Education:* Dr of Engrg/Naval Arch/Univ of CA-Berkeley; M Engrg/Naval Arch/Univ of CA- Berkeley; BS/Naval Arch/Chinese Naval Coll of Tech *Born:* 10/20/27 Dr. Jan holds the position of Asst to VP in the Ocean Engrg Div, American Bureau of Shipping, gaining 32 years of engrg experience in the field of ship design, structural analysis, applied research and development of strength standards. He actively participated in the activities of SNAME, SSC and other industrial communities. Presently, he serves as a member of 2 SNAME technical panels, and also a technical committee of ISSC 1985. He is the author of several technical papers and a recipient of the SNAME Capt Joseph E. Linnard Prize. *Society Aff:* SNAME

Janairo, Max R, Jr
Business: 1010 Brodhead Rd, Coraopolis, PA 15108
Position: CEO *Employer:* BIRO Tech, Inc *Education:* MS/CE/Struc Dynamics/Univ of IL; BS/Gen Sci/USMA, West Point *Born:* 2/10/33 in the Philippines. Served with Army Corps of Engrs from 1954 to 1978. Project engr NIKE missile construction program. Various military engrg and construction assignments in Europe and Far East. Engr Staff Asst for construction and engrg activities in Office of Secty of Defense. District Engr for Corps' Pittsburgh District. With Michael Baker, Jr., Inc 1978-1983 as, VP of Transportation. Exec VP Green Intl, Inc, 1983-1984. Chief Exec Officer of BIRO Tech Inc natl Dir, SAME, 1982-1985. *Society Aff:* ASCE, NSPE, SAME

Janes, Henry W
Home: 541 Park Ave, Towson, MD 21204
Position: Proj Engr. *Employer:* Whitman, Requardt & Assoc. *Education:* BS/CE/Univ of MD. *Born:* 8/11/17. Civ Service with C of E, soils, construction Wash Natl Airport, DC, Kindley AFB, Bermuda, Andrews AFB, MD 1939-43. USN (WWII) Officer in Charge of LCT- Pacific '43-'45. Corps of Engr-Wash, DC Dist, Materials Lab and Airfield additions 1945-50. Engr Res & Dev labs, Ft Belvoir, VA 1950-53. E Ocean Div, Geol, Soil & Materials, Airfield extensions Atlantic bases 1953-54. Whitman, Requardt & Assocs, Baltimore, MD 1954 to 82: Proj Engr in charge of Geotechnical Group, Design and Contr inspection, earth dams, graving docks, fdns and controlled fills. Reg Engr, MD, VA, DE, NJ, MS. Retired from Whitman, Requardt & Assocs August 31, 1982. *Society Aff:* ASCE, NSPE, ISSMFE.

Janes, Robert L
Home: Rt 11 Box 201, Brainerd, MN 56401 *Education:* PhD/Civ Engg/IIT; MS/Civ Engg/CA Inst of Tech; BS/Civ Engg/CA Inst of Tech. *Born:* 12/11/14. in St Paul, MN. Worked at Caltech on aerial torpedo dev during WWII. Struct & Mechanics Res at Armour Res, Chicago, then Pavement Dev for Portland Cement Assn until 1963. Teaching & res at OK State Univ since that date. Construction & Transportation courses. Active n ASCE -- state pres in 1977. Retired 1980. *Society Aff:* ASCE, NSPE.

Jang, Roland
Home: 19902 Via Escuela Dr, Saratoga, CA 95070
Position: Pres. *Employer:* Intl Diagnostic Tech. *Education:* MS/CE/Univ of CA; BS/Chemistry/Univ of CA. *Born:* 11/5/22. Current pres IDT, engaged in diagnostic instrument & reagents. Was founder. Prior founded Intl Medical Tech engaged in X-ray intensifying screens. Prior was VP of Memorex. An electronic co. Had positions in various industires. Health care, electronics, food, fertilizer. Participated worldwide. 56 yrs old. *Society Aff:* ACS, AIChE.

Janisse, Norman J
Home: 507 E Michigan St, Milwaukee, WI 53202
Position: Staff Consultant *Employer:* Johnson Controls, Inc. *Education:* BSME/-/Univ of WI. *Born:* 7/22/24. USN 1942-1946. College grad 1948. Immediately started profl career in the area of controls, their design, & application for comfort & energy conservation with Johnson Controls. Authored over 30 papers & articles on these subjs & made over 100 appearances as guest speaker to technical socs & colleges. Have managed the appliction & technical education activities of the co, with additional responsibility of mgr of energy mgt systems. Appreciable service on Natl Technical Soc Committees & as local chapter officer.Reg PE. Awarded fellow in ASHRAE in 1978. *Society Aff:* ASHRAE, AEE.

Janney, Jack R
Business: 330 Pfingsten Rd, Northbrook, IL 60062
Position: Chrmn/Bd. *Employer:* Wiss, Janney, Elstner & Assocs Inc. *Born:* June 17 1924. Native of CO. BS Arch Engg 1949, MS Struc Engg Univ of CO 1950. 6 yrs in R&D Labs of Portland Cement Assn. Originated Wiss, Janney, Elstner & Assocs Inc 1956. Prof experience in experimental stress analysis, chiefly in areas related to struc problems. Exper in load test & eval results of tests conducted on completed structs, incl containment vessels built for nuclear power stations. Mbr: ASCE, NSPE, ASTM, ACI (Mbr Bd/Dir), PCI (Mbr Bd/Dir). Has been on many cttes within these organizations. Reg Struc Engr IL & Reg PE IL, WI, CO, MA.

Jansen, Carl B
Home: Gateway Towers, apt. 9L, Pittsburgh, PA 15222
Position: Ret-Honorary Chrmn. *Employer:* Dravo Corp. *Education:* BS/CE/Union College. *Born:* 5/31/00. Received BS in Civ Engg degree from Union College 1922; Honorary Dr of Engg Union College 1949 and Carnegie Inst of Tech 1967. Employed 1922 by Dravo Contracting Co (later Dravo Corp) successively as field engr, constr superintendent, dept mgr, div gen mgr, 1946 pres and 1959 chrmn and ceo Dravo Corp; retired 1966. Dir Dravo Corp 1934-1966; Allegheny Ludlum Steel Corp 1960- 1971. Past pres & chrmn & continuing mbr of Exec Comm of Allegheny Conf on Community Dev; mbr of Airport Advisory Comm of Allegheny Co. Hon Mbr ASCE 1971; Hon Mbr Permanent Intl Assn of Navigation Congresses (PIANC) 1967. Moles Award for outstanding Achievement in Constr 1955. Metcalf Award for Outstanding Engg Achievement Engrs' Soc W PA 1968. Reg PE PA. Dir ASCE 1965-68; Dir EJC 1967-68. *Society Aff:* ASCE, SAME, PIANC.

Jansen, Robert B
Home: Route One, Box 297, Mt. Spokane Park Drive, Mead, WA 99021
Position: Consulting Civil Engineer *Education:* MSCE/Civ Engg/Univ of S CA; BSCE/Civ Engg/Univ of Denver. *Born:* 12/14/22. Native of Spokane, WA. Chrmn, Engg Bd of Inquiry, Baldwin Hills Dam Failure, 1964. Chief, CA Div of Dam Safety, 1965-68. Engr on CA Water Proj, Dept of Water Resources, leving as Chief, Design & Construction in 1977. Asst Commissioner for Engg & Res, US Bureau of Reclamation, 1977-1980. Exec Dir, Independent Panel to Review Cause of Teton Dam Failure, 1976-1977. Chrmn, US Committee on Large Dams, 1979-1981. Consulting Civil Engineer since 1980. Author of various publications on dam safety & govt mgt, including Dams and Public Safety. *Society Aff:* USCOLD, ASCE.

Janson, Lars-Eric
Home: Boxergrand 47, S-12362 Farsta, Sweden
Position: Hd of R&D. *Employer:* VBB Consulting Engg. *Education:* DrSc/Civil Engg/Royal Inst of Tech, Stockholm; MSc/Civil Engg/Royal Inst of Tech, Stock-

Janson, Lars-Eric (Continued)
holm. *Born:* 2/9/29. 1954 Design Engg, VBB Consulting Engineers and Architects, Stockholm. 1957 Res Engg, Dept of Hydraulics, Royal Inst of Tech, Stockholm. 1964 Assoc Prof, Dept of Hydraulics, Royal Inst Tech. 1972 Hd of Res and Dev VBB. 1973 Prof, Div of Water Resources and Environmnetal Control, Royal Inst of Tech, Stockholm. 1985-date fellow member of the Royal Swedish Academy of Engineering Sciences (IVA). Won, jointly with Stein Bendixen and Anders Harlaut, the 1978 Wesley W Horner Award, instituted in 1968 by the Sanitary Engineering Div of ASCE. A vast number of pub mainly in the fields of environmental engg, plastic piping, submarine piping, frost penetration depth in pipe trenches. *Society Aff:* ASCE, AGU, SKIF, IVA.

Janssen, Gail E
Home: 1612 Oakridge Ave, Kaukauna, WI 54130
Position: Pres & Chrmn of the Bd *Employer:* F & M Bank-Kaukauna *Education:* BS/Mech Eng/Univ of WI; BS/Agri/Univ of WI *Born:* 12/11/30 Operated a dairy farm at Oconto prior to entering college. Outstanding Young Farmer-Oconto county in 1955. Employed as Asst. Chief Engineer-Gehl Co; Chief Eng., VP Engineering, and President-Badger Northland, Inc. Currently President and Chairman of the Board F & M Bank and F & M Bancorporation, Inc. Engineer of the Year 1973, Wis. ASAE. Member Alpha Epsilon Honor Society. Registered Prof. Engineer, and Director of Wisconsin 4-H Foundation, and President of Wisconsin 4-H Foundation for 1981. Director of Finance ASAE 1981. *Society Aff:* ASAE, WSPE, NSPE

Janssen, James S
Home: 225 Bellaire Dr, New Orleans, LA 70124
Position: Fuels Consultant. *Employer:* Waldemar S Nelson & Co. *Education:* BS/CE/Tulane Univ. *Born:* 10/01/07. Born 1907. BS in Civ Engg Tulane Univ 1931, grad work in fluids and fuels. Most of career in natural gas engg at New Orleans Public Services Inc. Now Fuels Consultant with Waldemar S Nelson & Co, Engrs & Archs. Reg in LA (C & M). Active in LA Engg Soc, various committees. Vd Mbr, Pres in 1962. Active in Natl Soc Prof Engrs, various committes, PEI State Chrmn (1970-71), SW Regional PEI VChrmn (1972-73), formerly Natl Dir from LA. Mbr Amer Gas Assn, Sou Gas Assn, Amer Pub Works Assn (Pres N O Chapter, 1950-51), Engrs Club of N O (Pres, 1967), Natural Gas Men of N O (pres, 1972). New Orleans City Park Bd (Pres, 1968-69) and various civic and alumni groups. *Society Aff:* NSPE, AGA.

Janzen, Jerry L
Business: PO Box 599, Dallas, TX 75221
Position: Partner *Employer:* Arthur Young & Co *Education:* MS/IE/OK State Univ; BS/IE/OK State Univ *Born:* 10/28/36 in Enid, OK. He attended school at Phillips Univ and OK State Univ. His work experience includes engrg positions with the Western Elec Co, the Federal Aviation Administration, and 1 year as a visiting asst prof at the Univ of OK. He is presently a partner with Arthur Young & Co in charge of the Southwest Region industrial engrg practice. Mr. Janzen is a Registered PE, a Certified Manufacturing Engr, and a Certified Management Consultant. *Society Aff:* IIE, NSPE, IMC

Jaron, Dov
Business: Biomedical Eng & Sci Inst, Drexel University, 32 & Chestnut, Philadelphia, PA 19104
Position: Prof & Dir, BMES Inst *Employer:* Drexel Univ. *Education:* PhD/Biomedical Engg/Univ of PA; BS/EE/Univ of Denver. *Born:* 10/29/35. in Israel. Post secondary education in Israel & US. Sr Res Assoc in BME at Maimonides Hospital, NY, 1967-1970. Dir, Surgical Res lab, Sinai Hospital, Detroit, 1970-1973. Assoc Prof (1973-1977) and Prof (1977-1979) EE Dept, and Coordinator of Biomedical Engg Prog (1973-1979) Univ of RI. Assumed present position as Prof of Biomedical Engg & Prof Elect & Comp Engg and Dir of the Biomedical Engg and Sci Inst of Drexel Univ in 1980. Res activities: Control & optimization of heart assist devices, biomedical instrumentation, computer applications to health care, cardiovascular modeling. Over 120 scientific publications. Tau Beta Pi, Eta Kappa Nu, and Sigma Xi honor socs. Fellow IEEE. *Society Aff:* IEEE, ASAIO, AAMI, BMES, AAAS, ASEE, EMBS

Jarrett, Noel
Home: 149 Jefferson Ave, Lower Burrell, PA 15068
Position: Principal *Employer:* Noel Jarrett Associates *Education:* MS/Chem Engr/Univ of Michigan; BS/Chem Engr/Univ of Pittsburgh *Born:* 11/17/21. Joined Alcoa Labs as a res engr, Process Met Div Sept 1951. Advanced to sec hd, that div, 1956, asst chief in 1959, chief 1969, 1973, asst dir, Metal Prod Labs. May 1981-82 Tech Dir, Smelting R&D 1982 Tech Dir Chem Engr Greater part of career spent in field of smelting & purification of aluminum, & environ problems pertaining to aluminum smelting. One of those most active in dev of Alcoa Smelting Process. Holds 13 patents in fields of smelting & melting of aluminum. Prior to retirement 1987 was responsible for Chem Engrg R&D for Alcoa Labs. Currently principal in "Noel Jarrett Associates" a consulting firm specializing in electrochemistry, extractive metallurgy and research management. *Society Aff:* ASM, AIChE, AIME-TMS, Electrochem, NAE.

Jarvis, John J
Business: School of Industrial & Systems Engrg, Atlanta, GA 30338
Position: Prof *Employer:* GA Tech *Education:* PhD/Oper Res/Johns Hopkins Univ; MS/IE/Univ of AL; BS/IE/Univ of AL *Born:* 8/7/41 Joined the faculty at GA Tech in Industrial & Systems Engrg in February, 1968. Advanced to the rank of tenured Prof with area concentration in the field of Operations Research. Authored and co-authored a book and numerous articles in the area of Linear Programming and Network Flows. Presented several dozen seminars and lectures at natl scientific society meetings and major Univs across the US. Registered PE in the State of GA. Consultant to several dozen public and private organizations in the areas of transportation, distribution, production, inventory, scheduling and network analysis. *Society Aff:* IIE, ORSA, TIMS

Jaske, Robert T
Home: 7908 Chelton Rd, Bethesda, MD 20814 *Employer:* Federal Emergency Management Agency *Education:* BS/ChE/Northwestern *Born:* 4/13/23 Native of Chicago; 42 yrs with nuclear energy starting wth Manhattan District; 18 yrs at Hanford with GE Co; 9 yrs with Battelle Northwest specializing in water resources sys. 6 yrs with US Nuclear Reg Comm as environmental cons & spec proj mgr; with FEMA, Proj Officer, multi CPU decision support sys; Fellow AIChE, Fellow ASCE Dip and Former Trustee of Amer Acad of Envr Engrs, Past Chmn of AAES Coord Comm on Energy and present AIChE repr to same, Founder memb Envirn Div of AICHE; 55 pubs in water resources sys and energy tech; water res cons to National Water Comm; NAE Committee on Power Plant Siting. Enjoys public service projects, marksmanship and photography. Recd Public Service Award of AICHE in 1976. Reg PE WA & OR, & VA. *Society Aff:* AIChE, ASCE, AAEE

Jasper, Norman H
Home: 100 Cherry St, Panama City, FL 32401
Position: Pres. *Employer:* Lagoon Investment Co. *Education:* DrEngg/Aero Engg/Catholic Univ of Am; MS/Engg/Univ of MD; BSME/ME/CCNY. *Born:* 5/10/18. Naval Architect US Navy 1941-46. Naval ship R & D Ctr 1946-61 specializing in vibrations, acoustics, ship dynamics. Technical Dir Navy R & D Lab in Panama City, FL 1961-71. R & D in Mine & Torpedo Defense & Underwater Swimmer Tech. Sci Advisor Comphibpac 1972 & Comoptevfor, 1973. Pres Lagoon Investment Co 1974-. Published numerous technical papers. Civilian Distinguished Service Medals of Navy and of Defense Depts. Life Fellow ASME. Active in technical committees of Natl Acad Sci, ASME & SNAME. *Society Aff:* ASME, SNAME.

Jatczak, Chester F
Business: 1835 Dueber Ave, SW, Canton, OH 44706
Position: Res Scientist. *Employer:* Timken Co. *Education:* BS/Met Engg/Purdue Univ; Career Adv/Met Engg/MIT. *Born:* 5/4/22. BS Purdue, 1948; Sigma Gamma Epsilon Met Hon; MIT, 1965; AF Pilot WWII, Squadron Commander & retired Lt/Col, Air Natl Guard; employed by Timken since 1948, formerly mgr of Phys Met Res until raised to present position in 1974; recognized authority in areas of bearing mfg, alloy & tool steel dev, steel heat treating & processing, mech met & x-ray tech; publisher of over 50 technical papers, & two technical manuals each, SAE & ASM; holder of 10 patents; consultant to NMAB/NAS, NASA & DOT; elected fellow of ASM, 1976; invited foreign lectr at Int Conf on Mech Behaviour of Mtls, Kyoto, Japan, 1971; reg PE, OH & IN. *Society Aff:* ASM, SAE, NGA, NSPE.

Jeanes, Joe W
Business: 222 Cavalcade, P.O. Box 8768, Houston, TX 77249
Position: Chief Exec Officer and Bd Chmn *Employer:* Southwestern Labs *Education:* BS/CE/Univ of Houston *Born:* 7/9/26 Native Houston, TX. Served Navy WWII South Pacific. Since 1950 with Southwestern Labs, Inc, a Materials Engrg and Geotechnical Consultant firm having 15 offices in TX, & LA. Worked up through ranks assuming current responsibility as C.E.O. and Chmn in 1987. Licensed Engr TX and LA. Pres Houston Engrs Club 1976. Pres TSPE's San Jacinto Chapter 1972. TSPE Bd of Dirs 1980 and currently. TCEL Secty/Treasurer 1980-81, Pres 1984-85; Natl Chrmn ACIL's Construction Materials Testing Committee 1976-1981. Received ACIL's highest award 1977. Received Chapter Honor Membership Chi Epsilon 1980. Bd of Dirs First City Natl Bank-Northline. Bd of Dirs Brae-Burn Country Club 1980-83. Pres Brae-Burn CC 1982-83. Enjoy golf and hunting. Mbr ASTM, ACI, ASCE. Currently serving on Exec Cttee ACIL, and Bd of Dirs. Currently serving on NSPE Bd of Dirs. *Society Aff:* NSPE, ASCE, ACIL, TCEL, ACI, ASTM, ASFE, TSPE.

Jednoralski, J Neil
Business: 2216 Brookwood Ln, Salina, KS 67401
Position: Owner/Engr *Employer:* Jednoralski Engg. *Education:* BS/Agri Engr/Univ of IL; BS/Agri/Univ of IL. *Born:* 8/19/46. in Chicago, IL. Commissioned in Army Corps of Engrs 1970-71, Asst S-3, 36th Engr Bn (Const), Vinh Long, S Vietnam. Hydraulic Engr for IL Div of Waterways, in charge of flood surveillance and forecasting, and review of Lake MI shoreline permits. Water Resources Engr for Metcalf & Eddy/Alstot, March, & Guillou, Trenkle Slough drainage for IL Power Co nulcear power station at Clinton. Sr Proj Engr for Gannett Fleming Corddry & Carpenter; Sr Engr for Clyde E William & Assoc; Engr for Bucher, Willis & Ratcliff - on FIA flood insurance studies in 3 states. Smoky Valley Chapter of the KS Engg Society - Outstanding Young Engr - 1979. PE in IL, PA, CO, IN, & KS. *Society Aff:* ASAE, ASCE, IWRA, NSPE/PEPP, KES/PEPP, SAME, USCID, KCE, ISPF.

Jefferson, Thomas B
Business: Dept of Thermal & Environmental Engg, Dept. of Mech Engrg & Energy Processes, Carbondale, IL 62901
Position: Prof. *Employer:* Southern IL Univ. *Education:* PhD/ME, Heat Transf/Purdue Univ; MSME/ME, Heat Transf/Univ of NB; BSME/ME/KS State Univ. *Born:* 11/25/24. Native of Urich, MO. Served in Army Air Corps, WWII. Married Carolyn Chelf, Denver, 1946. Three children: Thomas C, Richard K, Terry Anne. Univ of NB, 1950- 52, Instr ME Purdue, 1952-58, Instr/Asst Prof ME Univ of AR, 1958-68, Prof & Hd, ME Dept; 1968-69, Assoc Dean of Engg. Southern IL Univ, Carbondale, 1969-78, Dean of Engg and Tech; 1978-84, Prof, Thermal & Environmental Engg, 1984- Present, Prof, Mech Engg. Ten summers at Martin-Marietta, Denver--Heat Transfer and Thermal Control Analysis on Titan I, II, & III, Skylab, and other projs. Reg PE: IN, AR, IL. *Society Aff:* ASME, ASEE, NSPE.

Jeffries, Neal P
Home: 3112 Cooper Rd, Cincinnati, OH 45241
Position: Exec Dir *Employer:* Ctr for Manufacturing Tech *Education:* BS/Mech Eng/Purdue; MS/Mech Eng/MIT; Engr/Mech Eng/Stanford; PhD/Mech Eng/U of Cincinnati. *Born:* 8/25/35 Founder and Exec Dir - Ctr for Manufacturing Tech (non-profit education, research and consulting inst). Dir of Academy for Professional Education (non- profit continuing education ctr). Awarded 1981 ASME Medal for continuing education. Member SME Intl Bd of Dirs, ASME Natl Professional Development Council. Active in natl and local activities of ASEE, ASTD and Engrg Society of Cincinnati. Served as Asst Prof and Research Assoc, Univ of Cinn (1967-1975); Mgr of Education, SDRC (Consulting Firm) (1975-78); Research Program Mgr, Gen Elec (1963-67); Engrg Project Officer, US Air Force (1958-1961). Listed in Who's Who in Midwest and American Men & Women of Science. *Society Aff:* SME, ASME.

Jeffries, Ronald F
Home: 1354 Cherokee Ave, Lynchburg, VA 24502
Position: Project Mgr *Employer:* Wiley & Wilson *Education:* BS/ME/GA Tech *Born:* 7/30/33 Worked 5 years for Babcock & Wilcox as Patent Engr and as Marketing Liaison Engr, both in nuclear engrg. Worked 1 year as a Bd Design Engr for Pratt & Whitney Aircraft, assigned to development of liquid hydrogen rocket engine. Worked approximately 18 years as consulting engr. Most of design experience is related to steam/power plants; however, am also experienced in other mechanical engrg disciplines. Managed Mechanical Design Dept of up to 28 people for 8 years. *Society Aff:* NSPE, ASME, APCA

Jeffries-Harris, Michael J
Home: 3 Waring Dr, Orpington, Kent, England BR66DN
Position: Staff Engr. *Employer:* Amoco Europe Inc. *Education:* PhD/Oil Tech/Imperial College of Sci & Tech; BS/Oil Tech/Imperial College of Sci & Tech. *Born:* 1/13/32. Native of London, England. Served with Royal Engrs 1950-51. Petrol Engr with Chevron Canada 1954-58 active in drilling, completion and production operations. 1961-70 sr res engr with Chevron Oil Field Res Co specializing in well completion & stimulation. Since 1970 Staff Petrol Engr with Amoco (UK) Exploration Co & Amoco Europ Inc active in reservoir engg & operational problems. AIME Ferguson Award 1970. Authored & co-authored five technical publications. Enjoy art, music, drama, golf & sailing. *Society Aff:* AIME.

Jelen, Frederic C
Home: 6170 Pansy Dr, Beaumont, TX 77706
Position: Conslt (Self-employed) *Education:* PhD/Physical Chemistry/Harvard Univ; MA/Physical Chemistry/Harvard Univ; SM/CE/MIT; SB/CE/MIT. *Born:* 1/17/10. Award of Merit, Am Assn of Cost Engrs 1970. Life Mbrship, Am Assn of Cost Engrs, Fellow AACE 1980. Author book "Cost and Optimization Engineering" McGraw- Hill 2nd ed., 1983. Editor "Cost Engineers Skills Course" handbook, AACE 1979. Conducted seminars on cost engg in North & S Am, & Europe. Consultant to Am Economic Fdn, Mobil Oil, Mobil Chem, DuPont, PPG Ind, INCO. Over 50 publications & 7 patents. Speakers' Bureau AIChE & AACE. Awards for presentation from both organizations. Listed Intl Who's Who in Engrg. Diamond Quality Award 1985 Assoc for Pushing Gravity Research. *Society Aff:* AACE

Jelinek, Robert V
Home: 6332 Ledgewood Dr, Jamesville, NY 13078
Position: Prof Paper Sci & Engrg *Employer:* SUNY Coll of Environ Sci & Forestry. *Education:* PhD/Chem Eng/Columbia; MS/Chem Eng/Columbia; BS/Chem Eng/Columbia. *Born:* Mar 5, 1926 NYC. Married, 3 sons. BS, MS, PhD Columbia Univ. Reg engr NY. Since 1972, Prof & Dean (1972-76) Sch of Environ & Resource Engg, SUNY Coll of Environ Sci & Forestry, Syracuse, NY. Prog Dir for Engg Chem, Natl Sci Fdn 1971-72. Taught chem engg at Syracuse Univ 1954-71 & Columbia Univ 1949-51, 1953-54. Chem engr Std Oil Dev Co 1951-53. Pvt US Army Engrs 1945-46, 2nd Lt to Capt US Army Reserve 1948-63. Mbr AIChE, ACS, NACE, TAPPI, NY Acad of Sci, Tau Beta Pi, Sigma Xi, Phi Lambda Upsilon, Alpha Chi Sigma. Specialties: computer simulation, process design, corrosion. *Society Aff:* AIChE, ACS, NACE, TAPPI.

Jenett, Eric
Business: PO Box 3, Houston, TX 77001
Position: VP. *Employer:* Brown & Root, Inc *Education:* MS/ChE/Columbia Univ; BS/ChE/Columbia Univ. *Born:* 8/11/23. Presently Proj Gen Mgr on $800mm Iraq-Saudi pipeline involving worldwide procurement, contracting, design and finance. Immediate prior assignment. Proj Gen Mgr on $350 million ethylene/polyethylene with utilities project; functions involved coordination and mgmt of cost, estimating, schedule and materials procurement for 4 engrs with Worldwide and worldscale construction co at site. Prior assignment was a Proj Mgr for design & construction of $600 million ethylene plant. Experience in engg, design, and proj mgr & in direction of comp- based tech applications primarily for the petrochem & petrol industries. Assignments involving Engg Div Mgt, Process Dept Mgt/Operation Dev, coordination & evaluation of comp-based tech, tech design engg, design proj mgr, foreign site supervision of engg & construction, process & utility design, pilot plant design, cost estimates & feasibility/economic studies of processes & utility schemes. *Society Aff:* PLU, AIChE, PMI, NSPE.

Jeng, Duen-Ren
Business: 2801 W Bancroft St, Toledo, OH 43606
Position: Prof. *Employer:* Univ of Toledo. *Education:* PhD/Mech Engg/Univ of IL; MS/Mech Engg/Univ of IL; BSME/Mech Engg/Natl Taiwan Univ. *Born:* 3/1/32. Native of Taipei, Taiwan. Taught in Univ of AL in Huntsville AL as an asst prof before joined the Univ of Toledo in 1967. Currently serve as a prof in Mech Engg specialized in the areas of Heat Transfer. Fluid Mechanics and Thermodynamics. Duties including teaching undergrad and grad courses in the above area and directing res in these area. Reg PE in state of OH. Chrmn of Engg Mechanics PhD prog at Univ of Toledo, 1976-1979. *Society Aff:* ASME, OSPE, ASEE, ΦΚΦ, ΣΞ.

Jenike, Andrew W
3 New Castle Drive, Nashua, NH 03060
Position: Consultant. *Employer:* Self-employed *Education:* PhD/Struct Engr/Kings Col, U of London; Dipl Engr/Mech Engr/Warsaw Tech Univ *Born:* 4/16/14. Andrew W Jenike was born in Poland, entered the USA in 1951. Following four yrs in the mining industry, he set up the Bulk Solids Flow Lab at the Univ of UT. He developed a theory and a quantitative method of design for flow of particulate solids in storage & reactor vessels. In consulting practice since 1954, he organized Jenike & Johanson, Inc, Consultants on Storage & Flow of Solids. He served as pres between 1966 & his retirement in 1979. He continues as an independent conslt. He published some fifty papers in the field & received the German Federal Republic Senior U.S. Scientist Von Humboldt Award as well as an Honorary Doctor of Tech from the Univ of Bradford, England. *Society Aff:* ASME, AIME, AIChE

Jenkins, Franklin H
Business: 2000 N 7th-Box 728, W Monroe, LA 71291
Position: Principal. *Employer:* Jenkins, Madden, Lazenby & Assoc. *Born:* Sept 1942. BSCE from LA Tech 1965. Started own consulting engg firm in 1969. Now employ thirty. Current Pres of W Monroe Chamber of Commerce. Currently hold offices in LA Engg Soc & Natl Soc of Sanitary Engrs. Am responsibe for mgt & operation of total co incl policy making. Enjoy fishing and hunting.

Jenkins, George S
Business: 2301 Connecticut Ave, NW 3B, Washington, DC 20008
Position: Engrg Consultant *Employer:* Self *Education:* MS/CE/The Johns Hopkins Univ; BE/CE/The Johns Hopkins Univ *Born:* 7/16/22 Baltimore, MD. Graduated The Johns Hopkins Univ 1943. Served as Engrg Officer US Navy WWII. Mechanical Engr 1947, Asst Prof Engrg 1947-51. Joined J.E. Greiner Co 1947, became Partner 1961, Pres 1970-75. Notable projects included Chesapeake Bay Bridges, Tampa Intl Airport, Space Shuttle Landing Facility, Section of Washington, DC Metro System, Key Bridge, Baltimore. Presently Special Engrg Consultant to Agency for Intl Development US State Dept. 1967-present, serve as Arbitrator to AAA. 1980 Member of Engrg Advisory Council - Whiting School of Engrg, Johns Hopkins Univ. 1970-present, member of Bd of Dirs, Easco Corp. *Society Aff:* ASCE, AAA, TRB

Jenkins, Ivor
Home: Drift Cottage, Worthing, NR Dereham, Norfolk UK England
Position: Consultant (Self-employed) *Education:* DSc/-/Univ of Wales; MSc/-/Univ of Wales; BSc/-/Univ of Wales. *Born:* 7/25/13. Mbr scientific staff GE Co Ltd, Res Labs, Wembley 1934-1944; Chief Met, Whitehead Iron & Steel Co Ltd, Newport, 1944-1946; Hd of Met Dept & Chief Met, GE Co Ltd, Wembley 1946-1961; Dir of Res, Manganese Bronze Holdings 1961-1969; Group Dir of Res, Delta metal Co Ltd & Managing Dir, Delta Mtls Res 1969-1978; viiting Prof, Dept of Met, Univ of Surrey, 1978-; Consultant 1978-. Fellow Inst of Met (Pres 1965-66); Pres Inst of Metals 1968-69; Fellow Am Soc Metals 1974; Fellow Metals Soc 1970; Fellow of Engrg 1978. Williams Prize Iron & Steel Inst 1946. Platinum Medal, Metals Soc 1978. Commander of the British Empire (CBE), 1970. *Society Aff:* IOM, ASM, MS, APMI.

Jenkins, Larry L
Home: 1635 North 106th St, Omaha, NB 68114
Position: VP *Employer:* Henningson, Durham and Richardson, Inc. *Education:* BS/ME/Univ of NB *Born:* 4/21/35 in Sioux City, IA. Registered P.E. With experience in HVAC, plumbing, piping, energy conservation, solar, lighting and power systems. Current responsibility is V. pres and Mechanical Section Manager and specializes in mechanical systems for buildings with emphasis on energy and economics. Past chrmn, NB Chapter ASME; past Pres, NB Chapter ASHRAE, past regional vice-chrmn, Education, Region IX, ASHRAE; Current Society Director and chairman, region IX, ASHRAE; past member TG/Ice maker heat pumps, ASHRAE; current mbr TG 4. BES, Building Envelope Systems, ASHRAE, Current Mbr TC 10.2 Automatic ice making plants, ASHRAE, Reg Award of Merit, Region IX, ASHRAE award winner in National CSI. Specification competition; author of Trade Publication Articles. Canoeing and outdoor sports are hobbies. *Society Aff:* ASHRAE, ASME

Jenkins, Leo B, Jr
Business: Speed Scientific Sch, Louisville, KY 40292
Position: Assoc Dean for Plan and Devel *Employer:* Univ of Louisville. *Education:* PhD/EE/Purdue Univ; MEngr/EE/Univ of Louisville; MEE/EE/Univ of Louisville; BEE/EE/Univ of Louisville. *Born:* 8/16/27. Assoc Dean, 1982; Asst Exec VP for Admin, Office of the Pres, 1979-82; Faculty Fellow, 1978-1979; Faculty mbr, Univ of Louisville since 1952; Chrmn, Dept of Elec Engr, 1970-1978. Summer employment: Westinghouse Elec, 1952; Allis- Chalmbers Co, 1956; North American Rockwell, 1969; Frequent consultant in area of specialization. NSF Sci Faculty Fellowship, 1960-61; Chrmn, Southeastern Assn of Elec engg Dept Hds, 1972-74; Dir, Southeastern Ctr for EE Education, 1974-76; Chrmn, Louisville Sec, IEEE, 1962; Pres, Univ Louisville Chapter, AAUP, 1974-75; Reg PE, KY. Sr Mbr, IEEE; Mbr, ASEE, AAUP, NSPE. 5 Honor Societies. Principal or co-Investigator, 11 contracts and grants. M Margaret Jean Woods, one child. *Society Aff:* IEEE, NSPE, ASEE, AAUP.

Jenkins, Marie H
Home: 2104 Kramer Lane, Austin, TX 78758
Position: Pres., Ch. of Bd. *Employer:* NAPP, Inc & LACE Engg. *Education:* BS/ChE/Univ of WA. *Born:* 4/22/29. Presently chrmn of the bd and pres NAPP, Inc, mfg of stationary source air sampling equip. Pres LACE Engg, wholly owned subsidiary of NAPP, Inc, providing engg services in air, water, noise pollution measurement and operating commercial chemical analytical laboratory. Native of Alexandria, LA, grad BS in chem Engg, Univ of WA. Also attended LA Poly Inst, Univ of TX, UCLA. First woman chrmn of an AIChE section (Mojave Desert, 1963). Reg PE CA, TX. Co-founder (founding Secy-Treas) Desert Empire Chap, CA Soc of PE.

Jenkins, Marie H (Continued)
Certified Prof Property Specialist. *Society Aff:* AIChE, NSPE, TSPE, NCMA, NPMA, APCA.

Jenkins, Peter E
Business: Mechanical Engineering Dept., 255 WSEC, Lincoln, NB 68588
Position: Chairman & Prof. *Employer:* Univ of Nebraska *Education:* Ph.D./M.E./Purdue University; M.S./M.E./S.M.U.; M.B.A./Business/Pepperdine University; B.S.M.E./M.E./Univ. of Kansas *Born:* 05/07/40 Extensive experience in managing and conducting basic and applied research programs. Directed industrial and government sponsored programs on all aspects of turbomachinery performance for pumps, compressors, and turbines. Responsible for industrial sponsored research consortium programs on centrifugal compressor surge, pump cavitation, and rotating stall for radial and axial compressors. Served as Director of the Turbomachinery Symposium and the International Pump Symposium held in Houston, Texas, for six years, and was the initial Director of the Air Force sponsored AFRAPT program at Texas A&M University. Served as Executive Vice-President and Director of Engineering of the Engine Corporation of America in Long Beach, California, directing R&D efforts in dual and gas turbine technology. *Society Aff:* ASME, ASHRAE, AIAA, SAE, NSPE

Jennett, J Charles
Business: Dean of Engrg, 109 Riggs Hall, Clemson, SC 29631
Position: Dean - Coll of Engrg & Prof Envir Sys Engrg *Employer:* Clemson University *Education:* PhD/CE/Univ of NM; MS/CE/Southern Methodist Univ; BS/CE/ Southern Methodist Univ. *Born:* 6/11/40. Engr, US Corps of Engrs, Southwestern Design Branch, 1962-1963. State of CA Dept of Water Resources, 1963-1964. Pitometer Assoc, 1964-65, 1969. Asst/Assoc Prof, Univ of MO-Rolla, 1969-1975. Prof and Chrmn, Civ Engrg Dept, Syracuse Univ, 1975-1981. Dean and Prof of Environ Sys Engrg, Clemson Univ, 1981-Pres. Reg P E states of MO, NY, and SC. Outstanding Young Engr of MO, 1974. Outstanding Teacher, Univ of MO-Rolla, 1969-70, 1974-75. Diplomate of the Amer Acad of Environ Engrs (by invitation). Outstanding Educator - SCSPE (Piedmont Chapter). Mbr Sigma Xi, Chi Epsilon, Phi Kappa Phi, and Sigma Tau Hon Fraternities. Mbr, Publication Bd of CRC Press, Inc. Editor (Eastern US) Minerals and the Environment Journal. ASCE Rep of the Engrg Council on Prof Dev, Env Engrg Div - Engrg Res Council, Water Supply and Resource Mgt Comm SCS P E - Educ Comm. Author of more than 100 papers, reports, book chapters in area of indus wastes, toxic metals, and trace organics in the environ. *Society Aff:* WPCF, ASCE, NSPE, SEGH, IAWPRC.

Jenniches, F Suzanne
Business: PO Box 746, MS124, Baltimore, MD 21203
Position: Mgr. *Employer:* Westinghouse Elec Corp. *Education:* MS/Env Eng/Johns Hopkins; BS/Biology/Clarion State *Born:* 2/28/48. Jan, 1970-June, 1974 Taught advanced biology to sr high sch students. July, 1974-Jan, 1975 Product Evaluation Engr - Automated Digital Test. Jan, 1975-Aug, 1976 Proj Engr - High-speed Memory Test Systems. Aug, 1976-May, 1977 Sr Product Evaluation Engr - Capital Equip Procurement; Long Range Planning. May, 1977-Feb, 1978 Test Supervisor - Automated Test Facilities. Feb, 1978-Feb 1980 Area Mgr, Printed Circuit Assembly - Mfg Operations of PC Assembly on two maj progs; all mechanized & automated PC Assembly Areas. 1979-1981 Baltimore/WA Soc of Women Engr Sec Pres Feb 1980-present Engineering Mgr, Mfg. R & D - Automated workcenters for P.C. Assembly, Cable & Harness, material handling utilizing sensored robots with in-process test & inspection; 1982 to present-Advanced Radar Operations Mgr with total mfg responsibility for B-18 offensive radar system. *Society Aff:* SWE, IES

Jennings, Burgess H
Home: 1500 Sheridan Rd, Wilmette, IL 60091
Position: Consulting Engr. *Employer:* Self-employed. *Education:* MS/ME/Lehigh Univ; MA/Math/Lehigh Univ; BE/ME/Johns Hopkins Univ. *Born:* 9/12/03. Educator, author and consultant. On faculty of Lehigh Univ 1926-40, at Northwestern Univ 1940-72 serving as prof, dept chrmn and Assoc Dean, currently Prof Emeritus of Mech Engg. Consultant to numerous cos. Active in profl societies: in ASHRAE Treas and Natl Pres (1949); in ASME committee work and sec activities; in ASLE secy and VP 1944-49. Author: nine engg texts covering environmental control and power; over 100 res papers and articles. Awards include: Richards Meml Award and Medal (1950 ASME), Worcester Reed Warner Medal (1972 ASME); Service and Achievement Award (1961 ASHRAE); elected to Natl Academy of Engg (1977); named Honorary member ASME (1978); F. Paul Anderson Medal (1981 ASHRAE). *Society Aff:* ASME, ASHRAE, ASEE, ASLE.

Jennings, Frank L
Business: 1222 Arterial Hgwy, Binghamton, NY 13901
Position: Pres. *Employer:* Jennings Engg Co. *Education:* MSME/ME/Univ of PA; BA & BSME/ME/Lehigh Univ. *Born:* 7/24/28. Reg PE in PA & NY. Reside in Binghamton, NY Dev Engr for Naval Air Mtl Ctr, Phila, PA, 1951-55. Officer in the Naval Reserve 1955-58. Mech design engr, GE Co for aerospace projs 1958-65. Proj Mgr & Asst Dir of R&D Link Group, Gen Precisio Systems, 1965-68 Engr Unit Mgr GE Co 1968-70, Mgr, Engg Co since 1970. Architectural mech & elec. engg consultants; Enjoys hunting, skeet, fishing, sailing & skiing. *Society Aff:* ASME, ACEC, ASHRAE, NSPE.

Jennings, Marshall E
Business: Gulf Coast Hydroscience Center, NSTL Station, MS 39529
Position: Project Chief *Employer:* U. S. Geological Survey *Education:* MS/CE/CO St Univ; BS/CE/Univ of TX. *Born:* 03/25/38 Native of San Antonio, TX. Began career as field engr with Geological Survey at various locations in TX. Headquarters assignments with USGS in Washington D. C. 1967-72, specializing in probability theory applications to natl streamflow gaging programs. Project Chief, Gulf Coast Hydroscience Ctr, 1972-present, specializing in deterministic modeling applications of hydraulics, watershed and streamflow, water-quality and sediment processes for aquatic environmental analyses. Led team effort for dam-break hazard analyses following Mt. St. Helen's eruption in 1980. Pres., MS Section, ASCE, 1981; Phi Kappa Phi, 1967. Enjoys reading history, classical music and working with Boy Scout groups. *Society Aff:* ASCE, AGU, AWRA, AIH

Jennings, Nathan C
Business: 1502 Augusta Dr, Houston, TX 77057
Position: VP-Refining. *Employer:* Independent Refining Corp. *Education:* BS/ChE/Univ of TX. *Born:* 5/13/40. Thirteen yrs experience in refining operations with major oil company specializing in catalytic cracking, alkylation, catalytic reforming, hydrocracking, aromatics extraction, hydrotreating, and chem treating operations and process tech. Assumed present position in Dec 1977. Respon for all phases of refining operations, Crude Evaluation Dept, and Operations Planning Dept. Reg PE in TX. *Society Aff:* AIChE.

Jennings, Paul C
Business: Mail Code 104-44, Pasadena, CA 91125
Position: Chmn, Div of Engg & Appl Sci. *Employer:* CA Inst of Tech. *Education:* PhD/Civ Engrg/CA Inst of Tech; MS/Civ Engrg/CA Inst of Tech; BS/Civ Engrg/ Colorado State Univ. *Born:* May 1936. PhD Caltech 1963; MS Caltech 1960; BS 1958 from CO State Univ; Instr USAF Acad 1963-65; Asst, Assoc & Prof Caltech 1965-; Exec Officer Appl Mech & Civ Engg 1975-; Chrmn, Div. of Engrg and Applied Science 1985- ; Mbr, Bd/Dir Appl Tech Council 1972-75; Mbr, Bd/Dir Earthquake Engg Res Inst 1967-70, Pres. 1981-83; Mbr, Bd/Dir SSA 1976- , Pres 1981; Mbr, Exec Ctte ASCE 1975-81; Author of numerous articles in earthquake engg & structural dynamics. Consultant on earthquake engg problems of tall bldgs, offshore drilling platforms, nucl power plants & other maj projs. *Society Aff:* ASCE, SSA, EERI, AGU.

Jennings, Stephen L
Business: 609 W North St, Salina, KS 67401
Position: Partner. *Employer:* Bucher & Willis, Cons Engrs, Planners & Arch. *Education:* MS/Environmental Engg/Purdue Univ; BS/Civil Engg/Univ of KS. *Born:* 9/11/34. Native of KS. Specialist in water pollution control. Prior to assuming present position, served as proj engr for maj water and wastewater treatment projs. Also responsible for maj structural engg projs. Received ACEC excellence award in 1977 for innovative water treatment plant design. Since assuming present position in 1975, direct firm's activities in municipal and industrial water and wastewater engg and in energy conservation and alternate energy systems utilization. Enjoy sports and furniture bldg, repair and refinishing. *Society Aff:* ACEC, ASCE, WPCF, AWWA, NSPE.

Jenny, Hans K
Business: 204-2, Cherry Hill, NJ 08358
Position: Mgr RCA Engr. *Employer:* RCA Corp. *Education:* MS/EE/Swiss Fed Inst of Tech. *Born:* 9/14/19. Native of Switzerland. Served as officer in Swiss Army (1939-45). Asst Prof Swiss Fed Inst of Tech - R & D Engr - Microwaves RCA 1946-1948 - engr leader 1948-49 - engg mgr, 1950-58 - Chief Engr - Microwave Ops 1959-66. Operations Mgr 1967-73. Corporate Res & Engr. Staff Exec, 1974-date. Led teams on microwave defense systems, commercial weather radar & radar for lunar landings. Developing progs to support viability of RCA's tech profs. *Society Aff:* IEEE.

Jens, Wayne H
Home: 1246 Balfour Rd, Grosse Pointe Park, MI 48230
Position: VP Nuclear Operations *Employer:* Detroit Edison. *Education:* PhD/ME/Purdue Univ; MS/ME/Purdue Univ; BS/ME/Univ of WI. *Born:* 12/20/21. Worked as Hd of Engg Analysis Group at Argonne Natl Lab. Employed at Nucl Dev Corp of Am, serving as Nucl Consultant to Dow Chem - Detroit Edison Proj; also held position of Proj Mgr of Belgian Test Reactor. Held various managerial and technical positions at Atomic Power Dev Assoc, Inc. Since joining Detroit Edison, served as Proj Mgr-Enrico Fermi Unit 2, Asst Mgr-Engg & Construction, Mgr-Engg & Construction Asst VP-Engg & Construction, prior to being appointed VP Nuclear Operations. *Society Aff:* AIF, ANS, ASME.

Jensen, Arthur S
Business: Westinghouse Adv Tech Div, Box 1521, Mail Stop 3531, Baltimore, MD 21203
Position: Sr Advisory Physicist. *Employer:* Westinghouse Adv Tech Div *Education:* PhD/Phys/U of PA 1941; MS/Phys/U of PA 1939; BS/Phys/U of PA 1938; Dipl/Computer Sci/Westinghouse Sch of Appl Engrg Sci 1977; Dipl/Adv Engrg/Westinghouse Sch of Appl Engrg Sci 1972 *Born:* 12/24/17. Sr Advisory Physicist, Westinghouse Adv Tech Div since 1965, sensor systems dev integrated circuits, CCDs charge coupled devices; Mgr Special Electron Devices 1957-65, storage and infrared camera tubes. Res physicist, RCA Labs, 1946-57. Officer-instr, Dept of Elec Engg, US Naval Academy 1941-45. Capt, USNR (ret), 5 yrs active duty WWII. Inventor: Radechon barrier grid signal storage tube; high contrast direct view storage tube; grating storage camera tube; fill/spill low noise CCD input; interconnecting integrated circuits. 25 issued US patents. PE 5017 MD; MD Bd of Reg for PE (Vice-chrmn) 1979-86; ABET/IEEE Ad Hoc Bd of Visitors. NCEE examination consultant 1986. Engineers Council of Maryland Outstanding Service Award 1986. Maryland State Governor's Citation 1986, (for service in Md St Bd of Reg's of P.E.). Life Fellow IEEE 1982. SID Award 1975; Westinghouse Special Corporate Patent Award 1972; Westinghouse Patent Awards 1965, 1977, 1982; Naval Reserve Medal 1951; Armed Forces Reserve Medal 1961; Fellow of Wash Acad of Sci. *Society Aff:* IEEE, APS, AAAS, SPIE

Jensen, Eugene T
Home: 3910 Parlington Rd, Topeka, KS 66610
Position: Dir Bureau Water Quality *Employer:* KS Dept Health & Environ *Education:* MPH/Public Admin/Univ of MI; BS/CE/CO State Univ *Born:* 3/12/21 Native of IL. Served as pilot in WWII and subsequently as Officer in US Public Health Service. Directed PHS marine health science programs, and justified, planned & constructed marine research labs in RI, AL & WA. Directed Natl estuarine study for Dept of Interior. Served as Regional Administrator for Federal Water Pollution Control Agency and as Deputy Asst Administrator for USEPA. Served as Exec Secty for VA Water Control Board. Commissioner, Interstate Commission on Potomac River, (Federal & VA). *Society Aff:* WPCF, AAEE, SCS

Jensen, Harold M
Business: 113 Main St, Box 187, Cedar Falls, IA 50613
Position: Pres *Employer:* Jensen Consulting Engineers Inc *Education:* MS/Environmental Engg/IA State Univ; BS/Civ Engg/IA State Univ. *Born:* 5/11/39. Native of Madrid, IA. Employed IA State Hgwy Commission 1961, US Army Engr Officer, 1962-64, Asst Fayette County Engr 1964, Asst Dir Water, Ames, IA 1966-68, City Engr & Dir of Public Works, Cedar Falls, IA 1969-73, VP Bert B Hanson & Assoc, Inc consulting engr 1973-80 Pres. Jensen Consulting Engineers Inc, since 1980.Pres CEC/IA 1979; IES Dunlap-Woodward Award 1973; Ctr for Ind Res & Service; ISU Advisory Council; Bd of Dirs for Metropolitan Transit Authority; Chrmn, Black Hawk County Bd of Health 1976-77; Chrmn, Civ Tech Bd, Hawkeye Inst of Tech. *Society Aff:* CEC/IA & ACEC, IES & NSPE, AAEE, ASCE, AWWA.

Jensen, Harvey M
Business: Appl Res Lab, PO Box 30, State College, PA 16804
Position: Assoc Prof of Engg Res. *Employer:* Penn State Univ. *Education:* MS/Elec Engg/Penn State Univ; BS/Elec Engg/IA State Univ. *Born:* 7/29/25. Born and raised on IA farm in Shelby County. Served with USAAF 1944-45. Received AF reserve commission from ROTC at IA State Univ. Grad with highest scholastic record of elec engg class. Test engr for Commonwealth Edison Co 1949- 55. Mbr of faculty at Penn State since 1955. Responsible for system analysis and simulation studies in the guidance & control department of the Appl Res Lab. Married, with three sons, all grad of Penn State. *Society Aff:* IEEE.

Jensen, Marvin E
Business: 2625 Redwing Road, Suite 325, Fort Collins, CO 80526
Position: National Program Leader, Water Mgmt. *Employer:* USDA-ARS. *Education:* PhD/Civil Engg/CO State Univ; MS/Agri Engg/ND State Univ; BS/Agri Engg/ND State Univ. *Born:* 12/23/26. Native of MN; Agri Engg Dept, ND State Univ, 1952-55; US Dept of Agri, 1955-present, Amerillo, TX, Fort Collins, CO, Kimberly, ID, Beltsville, MD and Fort Collins, Co. Currently National Program Leader, water management. Chmn, Executive Committee of the Irrigation and Drainage Div, ASCE, 1976-77; Chmn, Steering Committee, Soil and Water Div, ASAE, 1973-74; Director, Soil and Water, ASAE, 1980-85; V P, Tech Council, 1983-86; and Pres, International Commission on Irrigation and Drainage, 1984-87. *Society Aff:* ASAE, ASCE, SSSA, AAAS.

Jensen, Roger N
Home: 10759 Oro Vista Ave, Sunland, CA 91040
Position: Design Specialist, Sr. *Employer:* Lockheed-CA Co. *Born:* May 1928. BS in Civ Engg from Northwestern Univ, Evanston, IL. While attending sch, worked for the Dept of Subways & Superhgwys, City of Chicago. With Lockheed Aircraft Corp since 1951, specializing in weight & configuration control. Participated in the design of the YF-12A/SR-71, C-141A, C-5A & L-1011. Elected Exec VP of the Soc of Allied Weight Engrs 1975, 1976. Celebrated 25th wedding anniversary with Florence in June 1976 (& parents' 50th). Enjoy sailing, swimming, backpacking, bicycling.

Jensen, Roland J
Business: 414 Nicollet Mall, Minneapolis, MN 55401
Position: VP Eng & Const *Employer:* Northern States Power. *Education:* MS/Ind Mgt/Univ of MN; BS/ME/SD State Univ. *Born:* 10/19/29. Native of Lake Norden, SD. Adjunct Prof of Engg, ND State Univ; Chief Nuclear Engg 1968-1972, Mgr of Nuclear Plant Engg & Construction, 1972-1975. In charge of Corp Strategy & Planning 1975-1981, VP, Engrg & Construction. Northern States Power Co; Mbr of Bd of Dirs and Exec Committee of Bd, 1970-73. Hd of Power Div 1969. American Nuclear Soc; wine maker, photographer, runner, reader. *Society Aff:* ANS.

Jensen, Rolf H
Business: 1751 Lake Cook Rd, Deerfield, IL 60015
Position: Pres *Employer:* Rolf Jensen & Assoc. Inc. *Education:* Bachelor's/Fire Prot. Engrg./Illinois Institute of Technology *Born:* 01/16/29 Mr. Jensen is President of a fire protection engineering consulting company which he founded in 1969. He has been active in NFPA throughout his career and a member of the Automatic Sprinkler Cttee since its inception. He has also served SFPE continuously, been awarded its Grade of Fellow, served as pres of the Chicago Chapter and is currently pres of the national Society. He worked at Underwriters Labs for 12 years and now serves on its Fire Council. He was Prof and Chairman of the Fire Protection Engrg Dept at Ill Inst of Tech for 10 years. *Society Aff:* SFPE, NFPA, AAES, NSPE

Jentzen, Carl A
Business: 31 Wendever Dr, Suffield, CT 06078
Position: Consultant *Employer:* Jentzen Assoc, Inc *Education:* BChE/ChE/RPI; MS/Pulp & Paper/Inst of Paper Chem; PhD/Publ & Paper/Inst of Paper Chem *Born:* Dec 1937 New Haven, CT. MS & PhD Inst of Paper Chemistry; BS Chem E RPI. Appointed mill mgr of Westvaco's 1350 ton/day bleached board installation at Covington, VA 1972. Initially joined Westvaco 1963 & had attained pos of asst mill mgr-production when resigned in 1971 to assume active ownership role in a family business in WI. Advanced Mgt Prog, Harvard Bus Sch 1974. Mbr TAPPI Enjoy golf, tennis & skiing. President Lincoln Pulp & Paper Co 1978-1981; Started own consulting business June, 1981. Pres Jentzen Assoc, Inc (consulting) & Custom Corp (specialty chemical sales). *Society Aff:* TAPPI

Jeppson, Roland W
Business: Civ-Environ Engg Dept, Logan, UT 84322
Position: Prof. Dept Civ & Environ Engg. *Employer:* UT State Univ. *Education:* PhD/CEE/Stanford; MS/CEE/UT State U; BS/CEE/UT State U. *Born:* Aug 30, 1933. Prof of Civ & Environ Engg at UT State Univ, Logan, UT & a researcher at the UT Water Res Lab there. He received his BS in 1958 and MS in 1960 both in Civ Engg at UT State Univ, & his PhD in 1967 with a speciality in Fluid Mechanics from Stanford Univ. He is author of more than 50 res articles & a book & has received the J C Stevens Award for Meritorious Discussion by the Am Soc of Civ Engrs and a Horton award from American Geophysical Union. Other honors include a past NSF Sci Faculty Fellowship, & an appointment under the NORCUS prog of the Atomic Energy Commission. Prior to receiving his PhD he taught at Humboldt Sta Coll, CA 1960-64 & has taught since 1966 in Civ Engg & engaged in porous media, potential flow, hydraulic & hydrologic res at UT State Univ, UT. *Society Aff:* ASCE, ASEE, AGU.

Jepsen, John C
Home: 18315 Theiss Mail Rd, Spring, TX 77379
Position: Mgr Plans & Analysis. *Employer:* Shell Chem Co. *Education:* PhD/CE/Univ of OK; BS/Chem Tech/Univ of NV. *Born:* 1/25/34. Native of Minden, NV. Upon grad Univ of NV worked at Lawrence Radiation Lab, Livermore CA. Served two yrs in Chem Cor, US Army. Joined Shell Dev Co in 1962. Spent 5 yrs in R&D, transferred to Shell Chem Co in 1968 with subsequent assignments in plant technical & operations mgt. Corporate economics and planning, corporate R&O - Polymers mgt. Corporate Polymers Technical mgt and Bus Mgr - Polymers. Current assignment mgr Plans & Analysis - Elastomers and Plastics. Chem Mfg. Chrmn Student Chapter Committee AIChE 1979. Hobbies hunting, fishing, golf & skiing. Teach flying on part time basis. *Society Aff:* AIChE, NASCP.

Jerger, Edward W
Business: Coll of Engrg, Mech Engrg Dept, Notre Dame, South Bend, IN 46556
Position: Prof Mech Engrg *Employer:* Univ of Notre Dame. *Education:* BSME/ME/Marquette Univ; MS/ME/Univ of WI; PhD/T&AM/IA State Univ. *Born:* 3/13/22. Process Engr, WI Malting Co, Manitowoc, WI (46-48), Asst Prof, Mech Engr, IA State Univ (48-55), Assoc Prof Mech Engr, Univ Notre Dame (55-61), Prof & Hd Mech Engr, Notre Dame (61-68), Assoc Dean Engr, Notre Dame (68-82), Prof Mech Engrg (82-present). Reg Prof Engr, IN, IA. Pres, Pi Tau Sigma 1974-78. Consultant, 1965-68 Univ Madre y Maestra, Santiago, Dominican Republic. *Society Aff:* ΠΤΣ, ΤΒΠ, ΣΞ, ΦΚΦ, ASME, ASEE, NFPA, IAAI.

Jericho, Jack F
Home: 3630 Madrid Cir, Norcross, GA 30092
Position: Ex Dir Emeritus *Employer:* Inst of Ind Engrs *Born:* 11/08/12 During career, lived in Chicago, San Francisco, Denver, New York and Altanta. Served in key management functions of Industrial Engrg at United Airlines from 1947-63; received Award of Merit in 1963. Active in IIE from 1956 to 1976. Achieved natl Presidency in 1961-62 & became Exec. Dir. in 1963. Retired in 1976. Received IIE's Distinguished Service Award in 1963 & elected Fellow in 1969. Lectured at I.E. Conferences & Meetings for 20 years (500) & published at least 10 articles on Ind. Eng. Honorary Mbr of Alpha P. MU & registered in CA. *Society Aff:* IIE, ASEE

Jeris, John S
Business: Environmental Engg Manhattan College, Bronx, NY 10471
Position: Prof. & Dir of Environmental Engrg *Employer:* Manhattan College. *Education:* ScD/Sanitary Eng/MIT; MS/Civ Engg/MIT; BS/Civ Engg/MIT. *Born:* 6/6/30. Resides in Yonkers, NY. Served with Army Corps Engrs 1954-56. With Stearns and Wheler consultants 1956-59. Prof Manhattan College 1962 to present. Dir, Environmental Engg Grad Prog 1966-78; 1986- . VP res and dev, Ecolotrol, Inc since 1970. Res specialties, biological fluidized bed for water and waste water treatment, anaerobic processes and toxic materials biological treatment, biogas production from indus wastes. PE in NY and NJ. Outstanding Educator of America 1971; Allen Award, NYWPCF 1975; Thomas Camp Award from WPCF 1979. Over 30 technical publications. Holds several patents. *Society Aff:* WPCF, AWWA, IAWPR, AEEP.

Jerner, R Craig
Business: 3503 Charleston Rd, Norman, OK 73069
Position: President *Employer:* EMTEC Corp. *Education:* PhD/Met/Univ of Denver; MS/Met Engg/WA Univ; BS/Met Engg/WA Univ. *Born:* 10/12/38. Assoc Prof, Met Engg, OK Univ 1965-1978. Asst Dean, Grad College, OK Univ, 1971-72. Current activities: teaching, consultant to attorneys & various cos involving met failures, quality control and materials testing. Short course lctr, OK Ctr for Continuing Education. Associate Staff Mbr, US Dept of Transportation, Transportation Safety Inst (OK City). Std Oil Teaching Award for excellence in undergrad instruction; also awarded numerous res grants from various agencies including NSF, NASA & USAF. 1976-Present - Pres, Chrmn of the Bd of Emtec Corp and Sr Professional staff mbr. *Society Aff:* AIME, ASM, AAFS, ASTM

Jerome, Joseph J
Home: 20106 Balfour Rd, Harper Woods, MI 48225
Position: Technical Service Engr. *Employer:* Huron Cement Co. *Born:* Feb 1927. BSCE Univ of Detroit 1952. Twenty-four yrs experience in construction, fourteen present employer. Provide tech asst to indus. Mgr of ACI, MSPE, CSI, CIB of Detroit. Presently pres of MI Chap of ACI & Dir of CIB of Detroit. Enjoy flying as a hobby.

Jessel, Joseph J A
Home: 4911 35th Rd, N, Arlington, VA 22207
Position: Retired. *Education:* ScD/EE/Harvard Univ; MS/EE/Harvard Univ; BS/EE/Harvard Univ. *Born:* 12/20/07. in Lawrence, MA. Attended public sch in Methuen,

Jessel, Joseph J A (Continued)
MA. Grad Magna Cum Laude from Harvard with BS in EE in 1931. Earned ScD in 1934 & 1936, Asst EE in 1936, EE in 1945, Hd Sec of Transmission and Coordination in 1960, Chief Div of Elec Resources & Requirements in 1962, and Asst Chief Brueau of Power in 1965, retiring in 1972. IEEE Fellow award in 1961. Federal Power Commission Meritorious Service (1967), Special (1969) and Distinguished Service (1972) Awards. Expert witness in many legal-engg proceedings. Faculty mbr of grad sch, US Dept of Agri 1942-76. *Society Aff:* IEEE, NSPE, AAAS, WSE, HES, HGS.

Jesser, Benn W
Home: 8 Winant Rd, Princeton, NJ 08540
Position: Consultant *Education:* BS/CE/Princeton; MS/CE/Princeton. *Born:* 6/10/15. 2 1/2 yrs as control engr for DuPont. Taught chem engg at Princeton '38 to '42. '72 to '71 with M W Kellogg in design & construction; VP-Eastern Hemisphere Operations in London (1957-60), VP-Western Hemisphere Operations. Pres Hoechst-Uhde Corp (1971 to 1981). US subsidiary of Uhde GmbH & Hoechst AG. Retired and Consultant (1980-date). Mbr AIChE, ASME, Committee on Engg Law, Pres-Princeton Engg Assn, one of recipients for the Chem Engg Process Achievement Award for Kellogg ammonia process. *Society Aff:* AIChE, ASME.

Jesser, Roger F
Home: 10390 W 74th Pl, Arvada, CO 80005
Position: VP - Engrg & Constr *Employer:* Adolph Coors Co *Education:* BSCE/Civ-Struct Engrg/CO State Univ *Born:* 03/08/26 Native of CO. Served in US Army Air Corps 1944-46. Employed at Adolph Coors Co, Golden, CO. Involved in design and construction of malting, brewing, packaging facilities, water and waste water treatment, cogeneration, 2 piece aluminum can and glass bottle manufacture. VP of Construction 1976, present position since 1980. Pres of Professional Engrs of CO 1984-85. Mbr of Lutheran Med Ctr Bd of Dirs. Listed in Marques' Who's Who in Amer 1984-85. Reg PE in CO, LA, VA. *Society Aff:* NSPE

Jester, Guy E
Business: 2150 Kienlen Ave, St Louis, MO 63121
Position: VP. *Employer:* J S Alberici Construction Co, Inc. *Education:* PhD/Civ Engg/Univ of IL; MS/Civ Engg/Univ of IL; BS/Military Engr/US Military Acad; -/Adv Mgt for Executives/Univ of Pittsburg; -/-/Army War College; Command and General Staff College Engineer School *Born:* 10/20/29. Commissioned Army Corps of Engrs. Retired as Col. Engr Unit Commander - Korea & Vietnam. Asst Prof USMA, Deputy Dir Waterways Experiment Station. Div Engr Vietnam, Asst to chief of R&D, U.S. Army Chief Info Systems & Army Research Office U.S. Army. Dist Engr, St Louis: currently VP, J S Alberici Construction Co, Reg PE. past Pres, St Louis Section ASCE & SAME. Chairman, Bldg & Ind Dev Commission. Chairman, Bldg Code Review Committee. Past National Dir SAME. Pres Assn for Improvement for MS River. VChrmn, Prof Code Committee. Chrmn Bi-State Legislative Committee. Past Dir., Regional Commerce & Growth Asso. Dir Univ of IL Alumni Assoc., Director, Engr. Club of St. Louis, ASCE Visitor Com. for Accreditation Bd for Engrg Const. Man of Year 1980 *Society Aff:* ASCE, SAME, ΣΞ, ΦΚΦ.

Jester, Marvin R
Business: 2900 Vernon Pl, Cincinnati, OH 45219
Position: Vice Chmn. *Employer:* Kintech Services Inc. *Education:* ChE/Chem Engg/Univ of Cincinnati. *Born:* 8/23/25. Native Cov, KY. Served with USAF 1943-1946. With Kintech Services since 1963. Assumed presidency in 1978, V. Chmn. 1986. Spent ten yrs with Air Products & Chemicls Inc including 4 yrs in England starting intl operations local sec chrmn AIChE 1970. Fellow Grade-AIChE. Reg PE-OH. Enjoys classical music, golf. *Society Aff:* AIChE.

Jester, William A
Business: The Pennsylvania State University, University Park, PA 16802
Position: Assoc Prof of Nuc Engrg *Employer:* The PA State Univ *Education:* PhD/Chem Engrg/The PA State Univ; MS/Nuclear Engrg/The PA State Univ; BS/Chem Engrg/Drexel Inst of Tech *Born:* 06/16/34 Upon receiving his PhD, Dr Jester joined the faculty of the Penn State Nuclear Engrg Dept. His area of res includes the applications of ionizing radiation and radioactive materials in the solution of tech problems. He has also done extensive work in the dev and evaluation of highly sensitive radiation monitors. He has authored over 80 tech papers and has received over two million dollars in res and educational grants since 1967. He has been especially active in the Amer Nuclear Soc serving on the Isotope and Radiation Div, Education Div, and NEED Ctte; exec cttes and two terms as Secy for the Isotope and Radiation Div. Dr Jester was the 1985 recipient of the Natl Audubon Soc and the AAES Joan Hodes Queneau Palladium medal. *Society Aff:* ANS, ACS, ASEE, AAAS, AIC, ANSTA

Jewell, James Earl
Home: 77 Beale St., Rm # 2559, San Francisco, CA 94106
Position: Consultant in Lighting *Employer:* Self *Education:* MFA/Lighting/Yale Univ Sch of Drama; AB/Theatre/Univ of the Pacific. *Born:* 7/26/29. Native of Los Angeles, CA. Mbr of the technical theatre faculties at Smith College and Univ of CA at Berkeley. Consultant for the theatre facilities at numeous theatre and performing arts structures in CA and elsewhere. Lighting Designer for many theatrical productions. Lighting designer for many private residences, for commercial institutions and museums. Hd, Engg Div, Holzmueller Corp, 1957-1967; Sr Consultant, Bolt, Beranek and Newman, 1967-1968; Joined Pacific Gas and Electric in 1968 and assumed current reponibilities as company's chief lighting designer in 1969. Retired, March 1987. Fellow, American Theatre Association. Dir, Illuminating Engg Soc, 1979-81 VP, Illuminating Engg Soc., 1981-84, Pres 1984-85, Exec Comm, US Natl Comm for the Intl Commission on Illumination, 1978- 82. Treas, Intl Comm on Illumination 1983-87. *Society Aff:* IES, ATA, NTHP.

Jewell, Thomas K
Business: Dept of Civ Engg, Schenectady, NY 12308
Position: Assoc. Prof *Employer:* Union College. *Education:* PhD/Civil Engg/Univ of MA; MS/Environmental Engg/Univ of MA; BS/-/USMA *Born:* 2/18/46. Served in Army as helicopter pilot, small unit commander, & battalion & brigade staff officer from 1968-1973. With Dept of Civ Engg, Union College, Schenectady, NY since 1978, Chairman since 1986. Conducted complete curriculum review and update of Civ Engg prog at Union College. Served on the Academic Computing Advisory Bd. and the Faculty Review Board. Res interests are in the areas of development of educ computer application packages and application of systems analysis to Civ Engg design problems. Received Wesley W Horner Award from ASCE in 1979. Received Engineering Science/Association of Environmental Engineering Professors Doctoral Thesis Award in 1980. Author of 21 technical papers and publications. and a textbook, A Systems Approach to Civil Engineering Planning and Design, published by Harper and Row. Enjoy classical and country music, gardening, cross country and downhill skiing, and hiking/camping. Active as volunteer fireman and scout leader.. *Society Aff:* ASCE, ASEE, AWRA, NSPE.

Jewell, William J
Business: 202 Riley-Robb Hall, Ithaca, NY 14853
Position: Professor *Employer:* Dept Agri Engg, Cornell Univ. *Born:* May 5, 1941. PhD from Stanford Univ; Postdoctoral fellow Univ of London; ME Manhattan College; BS Univ of ME. Native of Fairfield, ME. Taught at Univ of TX, Austin & Univ of VT before assuming present position in 1973. Consultant to ind, municipalities and consulting firms. Elected mbr Bd/Dirs of Assn of Environmental Engg Profs. Mbr of: Am Soc of Civ Engrs, Res Prog Ctte for the Water Pollution Control Federation & Res Ctte Am Water Works Assn. Maj responsibility for agricultural waste mgt & rural environmental engg at Cornell Univ. Enjoy photography, rock gardening & sports cars.

Jeynes, Paul H
Home: 61 Undercliff Rd, Millburn, NJ 07041
Position: Retired. *Education:* ME/Energy/Yale. *Born:* 6/17/98. Ansonia, CT. Lab Asst, Yale, 1918-20; taught and conducted lab classes in Mech Engg. Married Margaret Elizabeth Owen, Seymour, CT (teacher) in 1925. Two sons; Paul H, Jr a sculptor; Dean Owen, with Procter and Gamble. With Public Service E & G Co, (Newark, NJ) 1920-63 in various capacities; engg economist at retirement. Originated Minimum Revenue Requirements Discipline; techniques for establishing service life and retirement patterns of depreciable plant; identifying and evaluating factors affecting cost of capital. Developed and taught courses in engg economy. Presently actively engaged in air pollution research. 50plus published technical papers; "Profitability and Economic Choice," IA State Univ Press, 1968. *Society Aff:* IEEE, ΣΞ.

Jimenez, Rudolf A
Business: Civ Engg Dept, Univ. of Arizona, Tucson, AZ 85721
Position: Prof. *Employer:* Univ of AZ. *Education:* PhD/CE/TX A&M Univ; MS/CE/Univ of AZ; BSCE/CE/Univ of AZ. *Born:* 5/13/26. in Eureka, IL. Served in Navy from 1944 to 1946. Worked for MacDonald Const Co in Panama Cana Zone 1946-1947. Profl work has been teaching at TX A&M Univ and Univ of AZ along with res in characterization of bituminous paving mixtures and flexible pavement design offices held include, chrmn HRB ctte, pres AZ Sec of ASCE, and pres of AAPT, Chrmn C.E. Ctte of NCEE, Chrmn Ariz St-Bd of Tech Registration, PEE Governor ASPE. *Society Aff:* AAPT, ASCE, NSPE, ΣΞ.

Jimeson, Robert M
Business: 1501 Gingerwood Ct, Vienna, VA 22180
Position: Consultant. *Employer:* RMJ Assoc. *Education:* MS/Engg Admin/Geo Wash Univ; BS/ChE/PA State Univ. *Born:* 1/29/21. Consultant in govt relations, energy and fuels, environmental mgt, chem engg, intl activities, and systems analysis. Served as Mgr, Fossil Tech Overview, Dept of Energy; profl staff, Office of Tech Assesment, Congress of the US; Asst Dir of Fuel Resources & Adviser on Environmental Quality, Fed Powr Commission; Chief, Fuel Policy & Prog Planning, Natl Air Pollution Control Admin, Dept Health Education & Welfare; res coordinator & chem engr, Bureau of Mines, Dept of Interior; chem sales, Union Carbide Corp; Organic Synthesis Fellow, Mellon Inst; aircraft design, Martin Marietta; taught at PA State Univ and Grad sch at Geo Wash Univ; reg PE in PA; published a book & over 50 technical papers. *Society Aff:* NSPE, AIChE, ACS.

Jischke, Martin C
Business: Chancellor, University of Missouri - Rolla, 206 Parker Hall, Rolla, MO 65401-0249
Position: Chancellor *Employer:* Univ of MO - Rolla *Education:* PhD/Aeronautics & Astronautics/MA Inst of Tech; SM/Aero & Astro/MA Inst of Tech; BS/Phys/IL Inst of Tech. *Born:* 8/7/41. in Chicago, IL. Joined faculty at Univ OK in 1968 as Asst Prof of Aerospace and Mech Engg, Assoc Prof in 1972, Prof in 1975, Dir AMNE 1977-1981, Dean of Engineering since 1981. In 1986 became Chancellor of the Univ of MO - Rolla. Mbr, Bd of Dir, MO Corp for Sci & Tech (1986). Listed in Who's Who in the South and Southwest. Summer work with NASA (1973), D Douglas Lab (1971), Battelle (1970), RAND Corp (1965). Consultant to several organizations. Received Regents Award for Superior Teaching in 1975, Ralph Teetor Award in 1971, White House Fellowship in 1975. Listed in Who's Who in South and Southwest, American Men of Sci, Outstanding Educators of America, & Intl Men of Achievement. Author of over forty reports and papers on fluid dynamics, heat transfer, and aerodynamics. Hobbies include gardening and hiking. *Society Aff:* AIAA, ASME, OSPE, AAUP, AAAS, ASEE, APS, NSPE.

Jobusch, Wallace E
Home: 5700 Bunker Hill Street, Apartment 2208, Pittsburgh, PA 15206
Position: Director, Physical Plant Division *Employer:* University of Pittsburgh *Education:* BS/Arch Engg/Univ of IL, Urbana. *Born:* 4/8/23. Native of Collinsville, IL. Structural Engg, Tucson, AZ, 1949-50. Staff of Purdue Univ, 1950-72, last 7 yrs as Dir of Fac Planning. From 1972-1980 staff of Panama Canal Co (Commission); 1972-78, As Chief Arch; 1978 to 1980 as Chief of Maintenance Div. 10/80 to present, Physical Plant Division, University of Pittsburgh. Mbr of Canal Zone Bd of Reg for Arch and Prof Engrs--7/72-10/79 serving as Secretary, VChmn & Chmn. Canal Zone Soc of PE: VP, 1977-78; Pres Elect, 1978-79; Pres, 1979-80. Reg PE, IN, PA; Reg Arch, IN, PA. Certified by National Council of Arch Reg Boards. *Society Aff:* NSPE, AIA.

Jodry, Charles L
Home: 13 Draycoach Dr, Chelmsford, MA 01824
Position: Sr Engr. *Employer:* Raytheon-Missile Systems Div. *Education:* BS/Ind Tech/Northeastern Univ. *Born:* 3/20/36. Native of Dayton, OH. Began career in Weights Engg with GE Co, Evendale, OH, in 1956. Since that time have held responsible Weights Engg positions with Avco, Itek and Raytheon in the Boston area. Currently in charge of Weight and Balance for the Advanced Medium Range Air-to-Air Missile (AMRAAM) validation prog at Raytheon Co. Charter mbr of SAWE, Boston Chapter. Presented a paper on "Antenna-Gimbal System Balancing" at the SAWE Intl Conf in May 1979 at NYC. *Society Aff:* SAWE.

Joel, Amos E, Jr
Home: 131 N Wyoming Ave, S Orange, NJ 07079
Position: Switching Consultant. Retired *Employer:* Bell Telephone Labs. *Education:* MS/EE/MIT; BS/EE/MIT *Born:* 3/12/18. Amos E. Joel, Jr, is a switching consultant. Retired from AT&T Bell Labs (1983). He joined AT&T Bell Labs in 1940. He was engaged in the design of early digital computers, and cryptanalysis machines, as well as switching systems of all types. He has been dept hd for dev planning of the Bell System's First ESS. He was also a major innovator in the development of the traffic service position system (TSPS) used throughout the US for operator service. In 1976 he was co- recipient of the Alexander Graham Bell Medal, and in 1981 the Franklin Institute Stuart Ballantine Medal for his work in electronic switching systems. He is the author of many works on switching, including the IEEE Press Books - Electronic Switching: Central Office Systems of the World, and Electronic Switching: Digital Central Office Systems of the World. He is also the principal author of History of Science in Engrg in the Bell Sys Sfjajing Tech (1983). Recepient of the 2nd Intern'l Telecomm Union - Plenery Award 1983 & the 1984 Columbian Award from the city of Genoa, Italy. He has also taught many courses on switching. He holds more than 65 patents in the field of telecommunications. He is a member of the National Academy of Engineering, a fellow of the IEEE and a licensed PE in the state of NY. *Society Aff:* ACM, IEEE, NAE, AAAS.

Joffe, Joseph
Home: 77 Parker Ave, Maplewood, NJ 07040
Position: Prof Emeritus. *Employer:* NJ Inst of Tech. *Education:* PhD/Chemistry/Columbia Univ; MA/Physics/Columbia Univ; BS/Chem Engg/Columbia Sch of Engg; BA/Chem, Phys/Columbia College. *Born:* 10/14/09. Grad of NYC public schools, Columbia College and Columbia Univ. Joined NJ Inst of Tech in 1932 as instr and rose to the rank of Prof of Chem Engg in 1940. In charge of sponsored res 1951-61. Chrmn of Dept of Chem Engg 1963-66. Chrmn of Dept of Chem Engg and Chemistry 1966-75. Distinguished Prof of Chem Engg 1965- 75; Emeritus 1975-. Summer employment with Exxon Res and Engg Co 1948-1977. Consulting chem engr 1975-. Fellow AIChE. *Society Aff:* AIChE, ACS, ASEE.

Johannes, Virgil I
Home: Two Cardinal Road, Holmdel, NJ 07733
Position: Dept Head *Employer:* AT&T Bell Labs *Education:* Eng.Sc.D/Electronics/Columbia University; M.Sc./Elec. Engrg./Columbia University; B.S./Elec. Engrg./City College of New York *Born:* 02/07/30 Native of Omaha, Nebraska. Taught at City College of New York, Fairleigh Dickinson Univ (Chairman of EE Dept.), and Columbia Univ. Joined Bell Labs in 1963, specializing in multiplexes, terminals, and repeaters for digital telecommunications on wire, coax, fiber,

Johannes, Virgil I (Continued)
radio, and satellite. Since 1983, has headed a department developing terminals for undersea fiber optic digital systems. Vice-chairman of CCITT Study Group XVIII "Digital Networks" 1979 to date. Licensed PE, NJ. Enjoys hiking, scuba diving, and good but inexpensive wines. *Society Aff:* IEEE

Johansen, Bruce E
Business: Ada, OH 45810
Position: Prof & Chrmn Elec. *Employer:* OH Northern Univ. *Education:* PhD/Elec Engg/Worcester Poly; MS/Phys/Univ of Pittsburgh; BES/Engg Sci/Cleveland State Univ. *Born:* 10/15/39. Education and early engg was in nuclear phys and mechanics. PhD was in the area of high voltage engg. Engg education has been main interest in recent yr with emphasis on undergrad engg. *Society Aff:* IEEE, ΤΒΠ, ΣΞ.

Johansen, Craig E
Home: P. O. Box 487, Castle Dale, UT 84513
Position: Pres. *Employer:* Johansen & Tuttle Engrg *Education:* BS/CE/UT St Univ. *Born:* 01/31/42 Special Engineering Consultant to Energy impacted Communities-creating financing vehicles and desigining long & short range improvements; Registered Professional Engr & Land Surveyor-UT. Principal expertise: The treatment, conveyance, distribution, collection, impounding, & diverting of water. Designing of highway systems & soil analysis. Construction contract specialist. Also experienced in Geology, and Agriculture Sciences. b. Provo, UT; Married June 22,1962, Jonnie Ward; Children: LaDawn, Peter, Gwen, Jennifer, Gracelyn, Merrial, Jonathan. *Society Aff:* ACSM, ASCS, ACEC.

Johansen, Nils I
Home: PO Box 82018, College, AK 99708
Position: Assoc Prof. *Employer:* Univ of AK. *Education:* PhD/CE-Engr Geol/Purdue Univ; MSCE/CE-Engr Geol/Purdue Univ; BSCE/CE/Purdue Univ. *Born:* 12/25/41. in Oslo, Norway. Hwy Engr I with the IN State Hgwy Comm 1967-71. Univ of AK since 1971. Acting Head, Dept of Mineral Engg 1974-76, 1982, Coordinator, Geological Engg Prog 1976 to present. Research through the Mineral Industry Research Lab on Permafrost and Arctic and Northern Dev. Author of papers and reports in these areas. Private consulting in Geotechnical Engineering. PE Alaska, Indiana. *Society Aff:* ASCE, SME-AIME.

Johanson, Jerry R
Business: 3485 Empresa Dr, San Luis Obispo, CA 93401
Position: Pres. *Employer:* Jenike & Johanson Inc. *Education:* PhD/ME/Univ of UT; BS/ME/Univ of UT. *Born:* 8/29/37. Received PhD in mech Engg from the Univ of UT in 1962 for theoretical work in flow of bulk solids. After four yrs in applied res at US Steel, he co-founded Jenike & Johanson, Inc, consultants in the storage & flow of bulk solids, North Billerica, MA and has been Pres since 1979. Has over 60 technical papers including articles on the flow of bulk solids, bin design, and compaction of bulk solids indicating his continuing theoretical work & application of theory to ind problems. Chrmn ASME Exec Comm Mtls Handling Engg Div 1973 ASME Henry Hess Award 1966. *Society Aff:* ASME, AIChE.

Johanson, K Arvid, Jr
Home: 201 Wearden Dr, Victoria, TX 77904
Position: Group Leader, Systems Engrg *Employer:* Union Carbide Corp, Unipol Systems Dept *Education:* MS/ChE/Univ of Hou; BS/EE/TX A&I. *Born:* 02/27/36 St Augustine, FL. m. Eleanor Friesen. Reg in CA, LA & TX. Served in USA Signal Corps 1959-61. Past Mbr World Wide Process Control Team. Mbr Control Systems Operating Ctte, UCC. Manages measurement, automation, communication & power engg group. Current VP & mbr Exec Bd responsible for Education, Instrument Soc of Am. Past Member President Society Advisory Committee, Sect Pres, District Officer, Div Officer, author, and session developer for ISA. Author for AIChE. Honors: Alpha Chi, Distinguished Military Grad, & AFCEA Award, TX A & I; Grad with distinction, the Signal Sch, USA; Special Service Award, New Orleans Sect ISA. Distinguished Service Award- Cresent Bend Section, ISA. R/Past Associational S S Dir, New Orleans and Guadalupe Baptist Assns. Former Dir Church Leader Tng, Baptist Temple. BSEE-TX *AEI, 1959; M.S. ChE-U.H-V.C., 1978.* *Society Aff:* ISA.

Johari, Om
Business: 10 W 35th St, Chicago, IL 60616
Position: Met Adv & Dir Annual SEM Symposia. *Employer:* IIT Res Inst-Metals Div. *Born:* Aug 1940. PhD & MS from Univ of CA, Berkeley; B Tech (Honors) from Indian Inst of Tech, Kharagpur. Post doctoral fellow & lectr Univ of CA, Berkeley before joining Drexel Univ as Asst Prof (1965-66). At IITRI since 1966, respon fr characterization of mtls in many applied & basic res problems using electron microscopy, micro-analysis, & related methods. Spec int in failure prevention through failure analysis. Author of nearly 50 pubs. Ed of the Proceedings & Dir of Annual Scanning Electron Microscope Symposia since 1968. Chrmn Mtls Characterization Activity ASM 1972-75; Marcus A Grossman Award 1966 ASM.

John, James E A
Business: Coll of Engrg, Amherst, MA 01003
Position: Dean of Engrg *Employer:* Univ of MA *Education:* PhD/Mech Engrg/Univ of MD; MS/Aero Engrg/Princeton Univ; BS/Aero Engrg/Princeton Univ *Born:* 11/06/33 Served as res engr at Airco, 1956-59, working on rocket fuels and oxidizers. Taught in Dept of Mech Engrg at the Univ of MD, 1959-1971; Exec Dir of Ctte on Motor Vehicle Emissions, Natl Acad of Scis, 1971-72. Prof and Chrmn, Dept of Mech Engrg, Univ of Toledo, 1972-77; Prof and Chrmn, Dept of Mech Engrg, OH State Univ, 1977-1982. Assumed current respon as Dean of Engrg at Univ of MA, 1983. Author of textbook, *Gas Dynamics*, (1969, 1984); Co-author of textbooks, *Introduction to Fluid Mechanics*, (1971, 1980); *Engg Thermodynamics*, (1980). *Society Aff:* ASME, SAE, ASEE

Johns, Roy F, Jr
Home: 1460 Coraopolis Heights Rd, Coraopolis, PA 15108
Position: Pres. *Employer:* Roy F Johns, Jr, Assoc. *Education:* BS/Civ Engr/Carnegie Mellon Univ. *Born:* 4/14/34. Native of Pittsburgh, PA. Served with US Army Corps of Engrs 1956-57. Design Engr for Richardson Gorden and Assoc 1955-1960, Pres of Roy F Johns, Jr, Assoc-Engrs 1960 to present. Roy F Johns, Jr, Associates specializes in land planning and site preparation for large scale communities with design population's ranging from 5,000 to 150,000. Pres & Dir of real estate dev and mining firm. Past Dir and Pres of PSPE Pittsburgh Chap. Past Dir ASCE Pittsburgh Chapter. Enjoy classical music and travel. Dir of Inter-Solar Ltd, London England - A Solar Energy Dev Firm. *Society Aff:* ASCE, SAME, NSPE.

Johns, Thomas G
Business: 2223 West Loop South, Suite 320, Houston, TX 77027
Position: General Mgr. *Employer:* Battelle Columbus Labs Houston Operations *Education:* Ph.D/Eng Mech/OH St Univ; MS/Eng Mech/OH St Univ; BE/Mech Eng/Youngstwon St Univ. *Born:* 05/20/43 Dr. Johns has a broad background in applied solid mechanics. Since joining Battelle's Columbus Labs in 1971, he has researched various problems in shell structures, pipelines, contact stresses, finite-element technology and inelastic buckling and stability, and is published in these areas. Dr. Johns has managed numerous research programs to advance the state of the art of various facets of offshore technology. Currently, Dr. Johns is General Mgr of Battelle Houston Operations, established in 1978 to work closely with the oil and gas industry in solution of porblems in a broad arena. He is currently a mbr and Technical Editor for the Offshore Technology of SPE; a mbr and Chrmn of Emerging Energy Technologies of the Petroleum Div. of ASME; a mbr of Sigma Xi, the Column Research Council and is registered Prof Engr in the State of OH. *Society Aff:* ASME, SPE

Johnson, A Franklin
Home: 222 Wildbrier Dr, Ballwin, MO 63011
Position: Asst Dean. *Employer:* WA Univ. *Education:* PhD/Phys/Univ of Toronto; MA/Phys/Univ of Toronto; BSc/Math/Univ of Alberta. *Born:* 10/8/17. in Canada. Capt, Royal Canadian Engrs 1939-45. Naturalized US citizen 1955. Res Engr, US Rubber Co, 1949-52. Res Engr Honeywell Inc, 1952-55, Magnetic mtls. Sr Res Physicist and Res Mgr 3M Co 1955-1964. Prof of Physics Monmouth College 1964-1978. Asst Dean, Sch of Engg and Appl Sci, WA Univ, St Louis, MO 1978-. *Society Aff:* APS, AAPT, AAAS, ΣΞ.

Johnson, Aldie E, Jr
Business: 130 Second Ave, Waltham, MA 02254
Position: VP. *Employer:* Teledyne Engrg Serv *Education:* BS/Aero Engr/Iowa State Univ. *Born:* Apr 8, 1925; BS Aero Engg 1947, IA State Univ; grad study Univ of VA 1948; Aero Res Scientist, Structures Res, NASA Langley Field, 1948-57; Sec Chief Applied Mechanics & Mgr, Structures Dept, Avco Sys Div, 1957-68; Mgr Engg & VP, Teledyne Engrg Serv 1968-; AIAA, Assoc Fellow, Mbr Structures Technical Ctte, Technical Prog Chrmn 12th Structures, Dynamics, Mtls Conf; SESA Fellow; Treas 1967-69, VP 1968-71, Pres 1971-72; EJC, Bd/Dir 1973-74, Unity Task Force; Tau Beta Pi; Phi Kappa Phi; Reg PE, MA. *Society Aff:* SESA, AIAA.

Johnson, Andrew B
Home: 302 Sycamore Ave, Morton, IL 61550
Position: Chief Mtls Engr. *Employer:* Caterpiller Tractor Co. *Education:* BS/Met Engr/Univ of WI. *Born:* 12/5/22. Employed by Caterpillar Tractor Co since 1947 in various met and engg positions. Presently Chief Mtls Engr in Engg Gen Offices responsible for Corp Eggg Mtls dev and application prog. Chrmn of the Corp Met Steering Committee. Mbr of Soc of Automotive Engrs "General Materials Council–. *Society Aff:* ASM.

Johnson, Anthony M
Business: 3404 N Orange Blossom Trail, Orlando, FL 32804
Position: Exec VP. *Employer:* Intl Laser Systems. *Education:* BS/CE/Univ of MD. *Born:* 7/12/26. in Baltimore, MD. Attended Univ of MD. BS in CE 1950. Held various positions with Bechman Instruments, Inc until 1962, joined Korad, subsidiary of Union Carbide Corp to start co in laser field. Served as Pres of T J Assoc from 1966 to 1972, a consulting firm when joined Intl Laser Systems where current position is Exec VP. *Society Aff:* AIChE, LIA, SPIE.

Johnson, Arthur T
Home: RD 2 Box 32 Castleton Rd, Darlington, MD 21034
Position: Prof *Employer:* Univ of MD. *Education:* PhD/Bioeng/Cornell Univ; MS/AgrEng/Cornell Univ; BAE/AgrEng/Cornell Univ. *Born:* 2/21/41. in East Meadow, NY. Lived in Newfield, NY. Attended Cornell Univ and attained PhD 1969 working on thermoregulation in poultry. Served in Viet Nam 1970-71 as Capt, USArmy. Employed as res engg at Edgewood Arsenal, MD 1971-74, working on thermal and respiratory stresses from CB gear. Asst Prof, Agri Eng and Phy Ed, Univ of MD 1974-1980. Associate Prof in 1980-1986. Professor 1986-present. Interests are agri-bioeng and instrumentation. Tres AEMB 1979-1982 VP AEMB 1982-84, AEMB Pres 1984-present; consultant for NBS-OERI, PE in MD, Chmn, Bioeng Comm, ASAE, 1976. Part-time farmer. Chrmn, Wash, DC-MD Section, ASAE, 1983-84; Chrmn, Bioengr & Physical of Resp Wear Comm, ISRP 1983-85. Sec-Treas ASEE AED 1985-86, V. Chm ASEE AED 1986-87, Chm ASEE AED 1987-88. *Society Aff:* ASAE, AIHA, IEEE-EMBS, AEMB, ACGIH, ASEE, MDS ΣΞ, ISRP

Johnson, Barbara C
Home: 210 N Wycliff St, San Pedro, CA 90732
Position: Engg Mgr. *Employer:* Rockwell Intl-Space Div. *Education:* BS/Gen Eng/Univ of IL *Born:* BS Univ of IL 1946 Gen Engg. Last assignment 1973 Mgr-Mission Requirements & Integration for the Space Shuttle Prog. Direct the flight performance & mission analysis for Rockwell Space Div. Previous assignments: 1968 Mgr-Sys Engg for Command & Service Module Progs for Rockwell-Space Div (Apollo, Skylab, ASTP); 1961 Engg Supervisor-Entry Performance Analysis; Pre-1961 System Analysis-Hound-Dog/Navaho Missile/Advanced Sys Prog. Assoc Fellow, AIAA; Sr Mbr, SWE; Outstanding Engr Merit Award for IAE 1976; Distinguished Alumni Award, Univ of IL 1975; SWE Achievement Award 1974; Univ of IL Medallion of Honor Award 1971. Retired 3/1/84. *Society Aff:* AIAA, IAE, SWE.

Johnson, Bernard T
Home: 154 Summit House, W Chester, PA 19380
Position: Prog Mgr UH-61A Uttas Helicopter. *Employer:* Boeing Vertol Co. *Born:* Jan 1921. BS in Aero Engg from the Univ of OK. Served in Army Air Corps 1942-46. Employed by The Boeing Co since 1946 as Flight Test Engr on B-50 & B-47 through 1954, Chief of Flight Test Operations in Seattle from 1954 to 1959, Mgr Minuteman Missile Test until 1963 including 1 1/2 yrs at Vandenburg AFB. Since 1963, have been testing helicopters at the Boeing Vertol Co in Phila, PA. Present position is Prog Mgr for the US Amry UH-61A helicopter. Mbr of AIAA & AHS.

Johnson, Charles C, Jr
Business: 11510 Georgia Ave, Suite 220, Silver Spring, MD 20902
Position: Pres *Employer:* C.C. Johnson & Assocs, Inc *Education:* MS/CE/Purdue Univ; BS/CE/Purdue Univ *Born:* 9/6/21 Native Des Moines, IA. US Public Health Service officer 1947-1971, attained rank of Asst Surgeon General 1968. Asst Commissioner Health, NY City, 1967-68. Administrator Environmental Health Service, Dept HEW, 1968-1970. Assoc Exec Dir, American Public Health Assoc, 1971-72. VP for Research & Development, WA Technical Inst 1972-74. VP Malcolm Pirnie Inc, Consulting Environmental Engrs 1974-79. Pres, C.C. Johnson & Assocs, Inc, Environmental Engrs 1979-present. Chrmn, Natl Drinking Water Advisory Council, 1975-1981. Walter F. Snyder Award by NEHA/NSF, 1977. Commissioner, Natl Capital Planning Commission, 1971-74. Distinguished Engrg Alumnus, Purdue Univ, 1967. *Society Aff:* AWWA, WPCF, APHA, NSPE, AAEE

Johnson, Charles E
Business: 1E-227A, AT&T Bell Laboretories, 1 Whippang Rd, Whippang, NH 07981
Position: Asst Prof *Employer:* Duke Univ & Med Ctr *Education:* PhD/ME/Duke Univ; BS/Gen Engrg/US Naval Acad *Born:* 7/19/44 Asst Prof, Mech Engrg; Asst Res Prof, Anesthesiology; Engrg Safety Officer, Hall (Hyperbaric) Lab. Instructor: grad courses in heat and mass transfer; undergrad courses in thermal sciences. Researcher: modelling of heat and mass transfer in condensers and from biomedical systems. Designer: biomedical instrumentation and equip, HVAC systems for hyperbaric chambers, and pressure vessel components. Licensed PE, NC. Principle editor, *Hyperbaric Diving Systems and Thermal Protection*, 1978. ASME: two term Chrmn, Ocean Engrg Div; chrmn technical sessions, and author of technical papers. Mbr Adv Bd, Southeastern Consortium for Undersea Res; engrg consultant for SE Undersea Res Facility. Engrg Duty Officer: Commander, USNR-R; Active duty 1966-72, SEA LAB III. Married: Nancy Margaret Sleeth, 4 children. *Society Aff:* ASME, ASHRAE, ΣΞ, UMS

Johnson, Charles L
Business: Box 40195, Columbus, IN 47201
Position: Packaging Specialist. *Employer:* Cummins Engine Co, Inc. *Born:* Mbr SPHE, IMMS, IIE. Present Natl Chrmn SPHE Loss & Damage Control Ctte. Chapt Rep to SPHE Natl Bd 1971-76. Annual entries SPHE Natl Packaging & Handling Competition 1967-75 - eleven entries with six awards including first in plastic non-cushioning class in 1975. Served as Judge, Natl Competition 1968, 69, 70, 73 & 74. Chapt Rep to SPHE Natl Bd Meetings 1971-76. Diversified experience in Ind Packaging, Ind Engg & Mtl Handling. Engineered reusable steel shipping skids for large diesel engines. Mbr Indus Adv Bd, IN State Univ - Packaging Sch, Terra Haute, IN. Enjoys wood carving. BA Univ of IL.

Johnson, Charles M
Business: 1701 N Ft Myer Dr, Arlington, VA 22209
Position: Mgr, Advanced Sys & Analysis. *Employer:* Fed Sys Div, IBM. *Education:* PhD/Phys/Duke Univ; BE/CE/Vanderbilt Univ. *Born:* May 1923. PhD from Duke Univ; BE from Vanderbilt Univ. Native of Nashville, TN. Res Scientist, Johns Hopkins Univ 1951-56. Mgr Microwave-Phys, Electronic Communications Inc 1956-60. Dir Adv Sys & Microwave Depts, Emerson Res Labs 1960-61. Mgr Applied Res Dept Fed Sys Div, IBM 1961-67. Deputy Safeguard Sys Mgr, Dept of Army (leave of absence from IBM) 1967-73. Mgr Adv Sys & WWMCCS Architecture Dev, IBM, 1973-. Principal fields of endeavor are: phased array radars, weapon systems, millimeter wave generation, microwave spectroscopy, Lasers, solid state devices, communications, data processing. Published 35 unclassified technical papers & handbooks. Sr Mbr of IEEE, Mbr of APS & Listed in Am Men & Women of Sci. Awarded Dept of Army medal for Exceptional Civilian Service. *Society Aff:* IEEE, APS.

Johnson, Charles P
Business: 1620 Elton Rd, Silver Springs, MD 20903
Position: VP-Secy. *Employer:* Johnson, McCordic & THompson, P A. *Born:* Sept 1924 Wash DC. Served as navigator in US Army Air Corps 1942-45. BSCE Univ of MD 1950. Reg PE in MD & VA. Formerly Exec VP, Greenhorne & O'Mara 1957-70. Formed Johnson, McCordic & Thompson P A in 1971. Present firm specializes in land planning & civ engg for large private devs. Dir ACEC Metropolitan Wash; Mbr ASCE; Past Pres, Suburban MD Engrs Soc.

Johnson, Claude W
Business: 7655 Old Springhouse Rd, McLean, VA 22102
Position: VP *Employer:* Pailen-Johnson Assocs, Inc *Education:* BS/EE/Howard Univ *Born:* 7/17/36 Native of Cincinnati, OH. Research engr for Lockheed Missiles & Space 1963- 68. Held a number of technical management positions with IBM Corp, including Divisional Systems Support Mgr, 1968-1980. Founded Pailen-Johnson Assocs in July, 1980. Firm provides high tech systems engrg services to US Dept of Defense. Disciplines include ASW, ADP, ECM and Solid State Physics. *Society Aff:* IEEE, NSPE

Johnson, David, L
Business: P. O. Box 1803, 800 First St., NW, Cedar Rapids, IA 52406
Position: VP *Employer:* Shive-Hattery Engineers & Architects Inc. *Education:* BS/CE/IA St Univ *Born:* 07/14/31 A native of rural Ames, IA, farmed for six years and graduated from IA State Univ. with a BSCE in Civil Engr in 1960. Served two years with Stanley Consultants and three and one-half years with Kruse Engrg Service. Joined Shive- Hattery & Assocs in 1966, assuming his current responsibility as VP in 1978. Mbr of IA Engrg Society, Pres 1981-82. Registered prof engr and land surveyor in IA and TX with experience in municpal engrg, industrial site development, storm water management, administration and water supply and distribution. *Society Aff:* NSPE, ASCE.

Johnson, Dean E
Business: 300 S Northwest Hwy, Park Ridge, IL 60068
Position: Pres & Chief Exec Officer *Employer:* Protection Mutual Ins Co *Education:* BS/Fire Prot & Safety Engrg/IL Inst of Tech *Born:* 8/16/31 in Kiester, MN. Received his CPCU designation in 1967. Married in 1951 to Margaret Mentink and has 2 children. Now resides in Lincolnshire, IL. 1953-1966 was employed by the IA Inspection Bureau in Des Moines, IA as Inspection Engr and was promoted to position of Asst Mgr. In 1966, joined Protection Mutual Ins Co as a field representative, was promoted to Regional VP in 1969. One year later, transferred to Home Office as VP and Dir of Underwriting; moved on to position Chief Operating Officer in 1974, Exec VP in 1976, and Pres and Chief Exec Officer in February of 1978. Remains in that position at present. Holds additional position of Pres of Park PM Corp of Park Ridge, IL, as well as other directorships. *Society Aff:* NFPA, SFPE, CPCU

Johnson, Don H
Business: P.O. Box 1892, Dept of Elec Engrg, Houston, TX 77251
Position: Prof *Employer:* Rice Univ *Education:* PhD/EE/MIT; EE/EE/MIT; SM/EE/MIT; SB/EE/MIT *Born:* 7/9/46 Native of Mt. Pleasant, TX. Received formal education at MIT; bestowed PhD in 1974. Staff mbr of MIT Lincoln Lab from 1974-77, specializing in Speech Signal Processing. Joined the faculty of Rice Univ in the Dept of Elec Engrg in 1977. Currently a Prof performing res in Sonar Signal Processing, Theory of Point Processes, and Neural Modeling. *Society Aff:* IEEE, ASA, ARO, AAAS

Johnson, Douglas E
Business: 2014 NE Sandy Blvd, Portland, OR 97232
Position: PE. *Employer:* Talbott, Wong & Assocs Inc. *Born:* Sept 18, 1935. BS, MS, ME from OR State Univ. Native of Portland, OR. Served in US Army 1955-57. Res engr, Chevron Res Co, 3 yrs. Fuels & lubricants res. Taught 3 yrs (ME) at Univ of ID. Proj & res engr, Freightliner Corp, (truck component design, ride quality & vehicle safety stds) 6 yrs. Talbott, Wong & Assoc since 1974 (Cons Engr). Soc of Automotive Engrs Teetor Award 1965. Chrmn, OR Sec, Soc of Automotive Engrs, 1976-77.

Johnson, Edward O
Business: 971-2 Aza 4-go, Zushi-machi, Machida City, Tokyo Japan 194-02
Position: Dir Res. *Employer:* RCA Res Labs, Inc. *Education:* BS/EE/Pratt Inst. *Born:* 12/21/19. 1941-1945 US Navy Electronics. 1948 BSEE Pratt Inst Brooklyn. 1949-1955 Grad courses phys & engg at Princeton Univ & Swiss Fed Inst of Tech. 1953 IEEE Editor's Award. 1960 Eta Kappa Nu Recognition. 1966 Mbr US Dept Commerce Panel on Elec Automobiles. RCA Res Labs gaseous electronics 1948-54; semiconductor surfaces and devices 1955-59; Group head. Chief Engr Solid State Div 1960-64; Mgr Dev & Tech. Programs Components Div 1965-69; Mgr Optoelectronics 1970-72; Mgr Tech Licensing 1973-75; Dir Research Tokyo Lab 1975-pres. Approx 30 pats and 25 technical papers *Society Aff:* IEEE.

Johnson, Edward W
Home: 4364 Lehigh Drive, Troy, Michigan 48098
Position: Retired *Education:* BS/ARSC/IIT; BS/ME/ITT *Born:* 05/03/09 Extensive Industrial and Government experience in analysis, design, labs/field testing of complex engrg programs in automotive, aeronautical, construction equipment, environmental, mechanical fields as designer, manager, and consultant. Involved in management/engrg studies for private firms on the advisability of new fields of product development, manufacturing site acquisitions, construction/repair of shipdocks, harbor facilities, marinas, embankments, and bridges. Engaged in engrg studies on recycling waste water under various climatic environments, desalination, solar and geothermal industrial power sources, and design criteria for sanitary landfills. Registered Prof Engr. Retired LCdr., CEC, USNR. *Society Aff:* NSPE, ASME, SAE, AIAA.

Johnson, Elliott B
Business: Div of Engg, Chico, CA 95927
Position: Prof of Civ Engr. *Employer:* CA State Univ. *Education:* MS/Municipal Engr/IA State Univ; BS/Civ Engg/SD State Univ. *Born:* 6/15/24. Asst City Engr Sioux City, IA; City Engr Yankton, SD Principal Engr Chula Vista, CA. Twenty-three yrs univ teaching experience and eight yrs municipal engg experience. *Society Aff:* ASCE, ASEE.

Johnson, Emory E
Home: 515 13th Ave, Brookings, SD 57006
Position: Prof of Civ Engr Emeritus. *Employer:* SD State Univ. *Education:* MS/Struct Mech/Univ of MI; BS/Civ Engr/Univ of NB. *Born:* 5/3/14. Native of NB with first degree from Univ of NB. Worked one yr with Dept of Hgwys and have taught at colleges and univs since 1937. This was MO School of Mines and Metal-

Johnson, Emory E (Continued)
lurgy 1937-41; SD State College 1941-1943; Univ of KS 1943-1946; CO State Univ 1946; SD State Univ Hd Civ Engr Dept 1947-1979. Prof Emeritus 1979-. Dir, American Soc of Civ Engrs 1967-1970. Exec Secy, SD Engg Soc 1977- 1982 Natl Pres Chi Epsilon, 1978-80; Dist Dir, Tau Beta Pi, 1976-1980. Current practice as forensic engineer in area of structural failures, and accident reconstruction. *Society Aff:* ASCE, ACI, NSPE, ASEE, ACSM.

Johnson, Ernest F
Business: Dept of Chem Engg, Princeton, NJ 08544
Position: Prof. Emeritus *Employer:* Princeton Univ. *Education:* BS/Chem Engrg/Lehigh Univ; PhD/Chem Engrg/Univ of PA *Born:* 4/4/18. Jamestown, NY. Res engr, area technical supervisor, synthetic organic chemicals manufacture, Allied Chem 1940-6. Thermodynamics Lab, Univ of PA 1946- 8. Since 1948 faculty mbr at Princeton Univ, Assoc Dean of the Faculty 1962-6, Chrmn Dept of Chem Engg 1977-8, associated with Plasma Phys Lab 1955-. Trustee Associated Univs, Inc 1962-8, Bd Chrmn 1966-8. Ind consultant 1949-. Dir Autodynamics, Inc 1969-85. Exec committees ACS Div Ind & Engg Chemistry (1965-7) AIChE Central Jersey Sec (1973-). ACS Council 1976-8. Author *Automatic Process Control*, McGraw Hill, & eighty articles in scientific & engg journals. Tau Beta Pi and Sigma Xi. *Society Aff:* ACS, AIChE, AAAS.

Johnson, Fielding H
Home: 961 Tifton, Baton Rouge, LA 70815
Position: President *Employer:* Barber & Johnson, Inc. *Education:* BS/Chem Eng/LA State Univ *Born:* 12/15/34. Native of Baton Rouge, LA. Attended Louisiana State Univ. in Baton Rouge. Served in USAF (SAC) 1957-59. Div Engr with E I Dupont de Nemours & Co, Inc from 1956-66. Joined Kaiser Aluminum & Chem Corp Co, Inc (1966-72) various assignments & Construction Mgr. Formed Barber & Johnson, Inc (1972-present) Consulting Engineering Council of LA; National Director 1980-81 State President 1979-80; Fellow ACEC; Mbr AIChE, NSPE; LA Engg Soc; Reg Engr LA, MS & TX. *Society Aff:* AIChE, NSPE, ACEC, LES.

Johnson, Gearold R
Business: Mech Engrg Department & the Center for Computer-Assisted Engrg, Ft Collins, CO 80523
Position: George T. Abell Prof of Engrg *Employer:* CO State Univ. *Education:* PhD/ME/Purdue Univ; MS/Engr/Purdue Univ; BS/Aero Engr/Purdue Univ. *Born:* 1/11/40. Raised in Des Moines, IA. Res engr for Boeing Co, Aerospace Div in orbital and flight mechanics, 1962-1965. NATO post-doctoral fellow, von Karman Inst for Fluid Dynamics in Belgium 1970-1971. With CO State Univ since 1971. Currently George T. Abell Professor of Engineering; Prof of Mech Engr and also Comp Sci. Dir, Center for Computer Assisted Engineering. Visiting Prof, Univ of Kent, Canterbury, England, 1978; Visiting Prof CA Inst of Tech, Pasadena, CA, 1984; Visiting scientist Shape Data Ltd, Cambridge, England 1986; IBM Summer Fellow, Boulder, CO, 1987. Responsible for dev of Computing Engg prog at CSU. Over 80 res publications. Enjoy hiking, bicycling and photography. *Society Aff:* IEEE, ACM

Johnson, George W
Home: 727 Fairwood Rd, Akron, OH 44319
Position: Mgr, Quality Assurance. *Employer:* Goodyear Aerospace Corp. *Born:* Native of Akron, OH. BSME from Univ of Akron, 1950. Grad courses in Statistics, Economics, Quality Control Mgt, Univ of IL. Proj Engr specializing in Aircraft Wheel & Brake Antiskid Design. Sec Hd 1954-57. Production Mgr of Aviation Products Plant in England. Became Qual Assurance Mgr - Aircraft Wheel and Brake Div 1965 at home plant. Assigned Worldwide Qual respon for all aircraft wheel & brake sys mfg or maintained by Goodyear Plants or Repair Stations in 1973. In addition, respon for Goodyear Ind Brake Plant Quality, 1975. Special interest in improving bus communications with the public & educational fields. Received following awards from the Am Soc for Quality Control: Saddoris 1967, Fellow 1973, Testimonial for Service on Bd, Regional Dir 1967-74 & Dir-at-Large 1974-76. Enjoy designing & bldg boats, studying ancient and modern history.

Johnson, Gerald N
Home: Rural Route One, Ames, IA 50010
Position: Owner *Employer:* Gerald N Johnson, PE. *Education:* PhDEE/EE/IA State Univ; MSEE/EE/IA State Univ; BSEE/EE/WA Univ (STL) *Born:* 2/18/42. St Louis, MO. During sr yr at WA Univ built and installed mobile remote broadcast system for a broadcast station. Joined high power transmitter dept at Collins Radio, Cedar Rapids and Dallas. Began grad work in 1966. Served in US Army as res engr during Vietnam. Returned to grad school doing work in network and electric power system analysis by computer. Formed consulting firm which works in acoustics, electrical and electronic design and safety analysis, and services to litigants. Play Tuba and double bass in local musical productions, garden, and experiment with solar energy and UHF comm. *Society Aff:* NSPE, NFPA, IES.

Johnson, Glen E
Business: Box 8, Station B, Nashville, TN 37235
Position: Assoc Prof *Employer:* Vanderbilt Univ *Education:* PhD/ME/Vanderbilt Univ; MS/ME/GA Inst of Tech; BS/ME/Worcester Polytechnic Inst *Born:* 5/29/51 Rochester, NY. Pres's Fellow at GA Tech. Harold Stirling Vanderbilt Graduate Scholar at Vanderbilt. Recip of 1984 SAE Ralph R. Teetor Educational Award. Formerly Mechanical Engr in Machine Design at TN Eastman Co, and Asst Prof of Mechanical Engrg at Univ of VA. Principal Investigator on awards from Natl Science Foundation, U.S. Air Force, and US Dept of Transportation. Chrmn of ASME Design Automation Committee. Former Assoc Editor in Design Automation area for ASME Journals. Mbr of ASME Computer Aided Design Committee. Author of numerous technical papers, reports, and articles in the areas of optimal design, mechanical vibration, and noise control. PE in TN & VA. *Society Aff:* ASME, ASA, MPS, SAE, AGMA

Johnson, H Richard
Business: 3333 Hillview Ave, Palo Alto, CA 94304
Position: Prof. *Employer:* Watkins-Johnson Co. *Born:* Apr 26, 1926 Jersey City NJ. BEE with Distinction, Cornell Univ 1946; PhD Phys MIT 1952. Prior to being co-founder Watkins-Johnson Co was with Hughes Aircraft Co. Lect in EE at UCLA 1956-57 & at Stanford 1958-68. Hon mention in HKN "Outstanding Young Elec Engr" awards for 1956. Fellow IEEE, Mbr natl Acad of Egg, Am Phys Soc, Tau Beta Pi, Eta Kappa Nu, Phi Kappa Phi, Sigma Xi & RESA. Author 19 pubs in tech journals, & 7 patents.

Johnson, Horace A
Business: 6247 Navigation, Houston, TX 77011
Position: Engg Mgr. *Employer:* BJ-Hughes Inc. *Born:* Mar 1924. BS Case inst of Tech. Grad cert in BA UCLA. Served in US Army 1942-45. Res Engr Crane Co specializing in wear & erosion of metals. With BJ- Hughes since 1951. Was Met, Chief Met prior to present position. Assumed current responsibility as Engg Mgr Machinery Div in 1973. Mbr ASM, P Chrmn TX Chap. Mbr Natl Nominating Ctte 1970.

Johnson, Howard K
Home: 4646 S. Landings Dr, Ft. Myers, FL 33907
Position: Retired. *Education:* MS/Agri Engg/Purdue Univ; BS/Agri Engg/Purdue Univ. *Born:* 9/13/23. Pilot, Army Air Corps, 1943-45. Teaching & Res, Purdue, 1951-54. Design of grain storage and grain processing equip, 1954-62. Dir of Engg, A O Smith Harvestore Products, 1962-68. VP, Engg and Mfg, Farm Fans, Inc, 1968-78. Pres, Farm Fans, Inc, 1979-81. Retired 1981. *Society Aff:* ASAE.

Johnson, Howard P
Home: 1219 Wisconsin, Ames, IA 50010
Position: Prof and Head *Employer:* IA State Univ. *Education:* PhD/AgE-CE Engg/IA State Univ; MS/Mech & Hydraulics/State Univ of IA; MS/Ag Engg/IA State Univ;

Johnson, Howard P (Continued)
BS/Ag Engg/IA State Univ *Born:* 1/27/23. Native of IA. Served in WWII in Armed Force & Artillery, 1943-1946. After 6 months field experience, joined IA State in 1950 as instr & res assoc in land and water resources teaching & res. Appointed area leader in 1956; Promoted to prof in 1963. Appointed Head of Department in 1981. Res emphasis is Hydrology & non- point pollution control, Chrmn, Soil & Water Div, ASAE, 1970; ASAE Administrative Council, 1974-76. ASAE Fellow, 1976. Hancor Award, 1978. Engr of the Year, Mid-Central Section, ASAE, 1982. Award of Merit, Gamma Sigma Delta, 1984. Anson Marston Distinguished Professor of Engineering, 1986. Enjoy natural history, fishing, & classical music. *Society Aff:* ASAE, ASEE, AAAS, NSPE, IAS, SCSA.

Johnson, Hubert O, Jr
Business: PO Box 2246, Universal City, TX 78148
Position: Engrg Consultant (Self-employed) *Education:* ME/Civ Engr/A&M College of TX; BS/Engr Admn/A&M College of TX. *Born:* 5/2/21. Reg PE State of TX. Fellow of ASCE & SAME. Retired from USAF while serving as Dir of Facility Maintenance, Office of Secy of Defense. Served as Deputy Chief of Staff for Civ Engg, USAF Europe (1970-71) & AF Logistics Command (1971- 72). Past Regional VP SAME for OH Valley Region & the TX Region. Natl Pres SAME 1973-74. Subsequent to separation from military service 1974 self employed as engg & construction consultant, & engr principal of Rehler, Reitzer and Johnson - an architect-engr firm engaged in design & profl services in Saudi Arabia (1976-79). *Society Aff:* ASCE, NSPE, SAME.

Johnson, Ingolf Birger
Home: 1508 Barclay Pl, Schenectady, NY 12309
Position: Engg Cons. *Employer:* Self. *Education:* BEE/Elec Engg/Polytechnic Univ; MEE/Elec Engg/Polytechnic Univ. *Born:* 9/29/13. Native of Brooklyn, NY. Grad Fellow, Polytechnic Univ 1937-39. Engr with GE Co in various technical and managerial assignments in power transmission from 1939-78. Activities included economics; lightning, arc and high voltage phenomena; surges and surge control and protection of lines, stations and equip; EHV series capacitor system design and application; Power Sys Engrg Course. Assumed current status of Engg Cons since retirement from GE Co in 1978. Chrmn IEEE Trans and Dist Comm (1957-59) and Surge Prot Dev Comm (1974-76). IEEE: William Martin Habirshaw medal and Award 1966; GE Co Steinmetz Award 1975. IEEE: Centennial Medal Award, 1984, IEEE: The Lamme Medal 1986, IEEE: Fellow 1958, Life-Fellow 1978. Has authored or co- authored over 70 technical papers. In 1980 elected to membership in National Academy of Engineering. Dist Alumnus Award, Polytechnic Univ, 1983. Reg PE Engr NY. *Society Aff:* IEEE-PES, CIGRE, NAE.

Johnson, J Stuart
Home: 1000 Vista Dr, Rolla, MO 65401
Position: Dean Emeritus (Engg), Prof Emeritus (EE). *Employer:* Univ of MO-Rolla *Education:* PhD/EE/IA State College; MS/EE/Univ of MO; BS/EE/Univ of MO; ScD/Honorary/Lawrence Inst of Tech. *Born:* 5/8/12. Gower, MO 1912; Married Lucille Woodson, 1934; Instr IA State 1936-7, MO Sch of Mines (EE) 1937-9; Asst Prof 1939-44; Lt USN 1944-46; Assoc Res Engr Univ of FL 1946-47. Prof EE & Asst Dean Engg Univ of FL 1947-54. Hd EE Purdue Univ 1954-7; Dean Engg Wayne State Univ 1957-67. Dean Engg Univ of MO-Rolla 1967-77. Retired 1977. *Society Aff:* IEEE, ASEE.

Johnson, J Wallace
Home: 5110 Park Ave, Richmond, VA 23226
Position: Owner (Self-employed) *Education:* BS/CE/VPI. *Born:* 10/9/27. Providing structural & civil engg services to other engrs, arch & gen contractors for the past fourteen yrs in the state of VA & FL. Employee of consltg engg from 1959-1972 in both the civ & struct divs. Ctte mbr for engg education, Consltg Engrs Council of VA. *Society Aff:* ASCE, NSPE.

Johnson, James A
Home: 11000 Henderson Rd, Fairfax Station, VA 22039
Position: Engg. Consultant *Employer:* self *Education:* MS/IE/Stanford Univ; BS/Gen Engrg/USMA, West Point, NY *Born:* 6/2/24 Stoughton, WI; enlisted US Army 1943; graduated West Point 1947; Commissioned 2nd Lt Corps of Engrs; advanced through grades to Major General; Service in US and France, Korea and Vietnam; district engr Philadelphia (1968- 1970); Commanding General, Engr Command, Vietnam and Director of Const, MACV (1970-72); Commanding General Ft. Belvoir and The Engr School (1974-77); Div Engr North Atlantic Div (1977-79); Deputy Chief of Engrs, US Army (1979-1980); retired August 1, 1980. Sr VP, The Kuljian Corp (1980-1983); Ex VP, Eastern Indemnity Corp (1984-) Assoc, Mgr, Defense Science Board (1985); Engr Consultant (1986-). Decorated 2 DSM, Silver Star, 2 Legion of Merit, 2 Bronze Star, Air Medal JT SVCS Commendation Medal; Army Commendation Medal; 2 Purple Heart; Combat Inf Badge; Parachute Badge. Pres, Philadelphia post SAME, (1968-69); Pres NY City Post (1978-79); SAME Bd of Dirs (1973-74); Registered PE, Civil, VT. Married 1949 Kathleen Clare Smith, children, Pamela Marie, James Allen, Jr., Mark Thomas and Stephen Victor. *Society Aff:* SAME, AACE

Johnson, James K
Business: 104 Rhodes Hall, Clemson, SC 29631
Position: Prof, ME & Dir of Cont. Eng Ed *Employer:* Clemson Univ. *Education:* MS/Mech Engr/GA Inst of Tech; MS/Mech Engr/Clemson Univ; BS/Mech Engr/Clemson College. *Born:* 2/5/28. Mech Engr and educator; engr, Clinton Textiles, 1950-1955; Mech Engg Faculty, Clemson Univ, 1955-present; Sci Faculty Fellow at GA Tech, 1963-1964; Prof and Dir, Continuing Engg Educ, Clemson Univ, 1972-present; Mbr, Bd of Dir, State Certification Bd of Environmental System Operators, 1972-1974; "Engg Educator of the Yr" - 1975, SCSPE, Piedmont Chapter; Elected as Honorary Life- Mbr, Intl Soc for Biomaterials, 1975; Mbr Bd of Dir, SC Soc Professional Engrs 1975-1979; Advisory Bd, Tri-County Technical College, 1970-1976; VP and Mbr of Council, American Soc of Mechanical Engrs, 1979-1981. *Society Aff:* ASME, ISB, ASEE, NSPE, SCSPE.

Johnson, James R
Home: Rt 1, Box 231B, River Falls, WI 54022
Position: Consultant. *Employer:* 3M Co, Retired *Education:* PhD/Ceramic Engr/OH State Univ; MS/Ceramic Engr/OH State Univ; BS/Ceramic Engr/OH State Univ. *Born:* 1/2/23. Mtls Sci and Engg. Phys Sci. Energy Systems. 25 US patents and 60 publications. Natl Academy of Engg. Distinguished Alumnus, OSU, Carlton Soc, 3M Asst Prof Univ TX 1950-51 Technical Advisor, Ceramics, Oak Ridge Natl Lab, 1951- 56. 3M Central Res Labs 1956-79. Dir, Physical Sci, 1962-72. Dir, Advanced Res, and executive scientist 1972-79. Educator, Univ MN; Univ WI, Stout; and Consultant, 3M Co 1979-. *Society Aff:* ACerS, AAAS, NICE.

Johnson, James W
Business: Chem Engg Dept, Univ. of Mo-Rolla, Rolla, MO 65401
Position: Chrmn, ChE Dept. *Employer:* Univ of MO.-Rolla *Education:* PhD/ChE/Univ of MO; MS/ChE/MO Sch of Mines & Met; BS/ChE/MO Sch of Mines & Met; BS, Educ/Math-Chemistry/Southeastern OK State College. *Born:* 5/25/30. Nativ of Nashoba, OK. Married to Vera Hamman, sons Christopher (age 26) and Victor (age 21). Served with US Army (1953-55), US Army Reserve (1955-67). Post- doctoral fellow, Electrochemistry Lab, Univ of PA (1962-63). Mbr Christian Church, Rotary. Faculty mbr in Chem Engg Dept, Univ of MO-Rolla, since 1958. Assumed present position as Dept Chrmn in 1979. Res Assoc in UMR Grad Ctr for Mtls Res since 1965. Hobbies livestock, antiques, real estate. *Society Aff:* AIChE, ASEE

Johnson, Joe W
Home: 266 Lake Dr, Berkeley, CA 94708
Position: Prof Emeritus. *Education:* BS/CE/Univ of CA; MS/CE/Univ of CA. *Born:* 7/19/08. Native of Pittsburg, KS. Hydraulic Res at Waterways Exp Sta, Vicksburg MS 1934-35; Res on Sediment Transport, Soil Cons Service, Wash, DC 1935-1942. Taught hydraulic engg and coastal engg at Univ of CA, Berkeley, CA from 1942-1975. Now consulting coastal engr, and editor of *Shore & Beach*, Journal of The American Shore & Beach Preservation Assoc. *Society Aff:* ASCE, ASBPA.

Johnson, Joseph E
Home: 33 Sycamore Hill Rd, Bernardsville, NJ 07924
Position: Fire Protection Consultant *Employer:* Self Employed *Education:* BS/ME/OR State College. *Born:* 12/12/14. Native of Dallas, TX. Chief Engr TX Automatic Sprinkler Co 1934-1947. Served with Army Corps Engrs 1943-1946. Pres Viking Fire Protection Co 1947- 1963. Pres Pyrotronics, Inc 1964-1979. Natl Pres SFPE 1965-1967. Treas & Bd Mbr Natl Fire Protection Assn. Mbr, Bd of Governors, Natl Elec Mfg Assn. Treas, Fire Detection Inst. *Society Aff:* NSPE, ASME, SFPE.

Johnson, Leland R
Home: 7010 Plantation Dr, Hermitage, TN 37076
Position: Dir. *Employer:* Clio Res Co. *Education:* PhD/History/Vanderbilt Univ; MA/History/Vanderbilt Univ; MA/History/St Louis Univ; BS/History/Murray State Univ. *Born:* 11/1/37. Author of histories of Army Engr Districts and scholarly - technical journal articles. Directs surveys and navigability studies, disaster asst reports. Prepares technical and public relations mtls for publications. SAME Toulmin Medal 1973, 1976, and 1979. *Society Aff:* SAME.

Johnson, Lloyd H
Home: 28 Sutherland Dr, Scotia, NY 12302
Position: Mgr-Tech. Req. Eng. *Employer:* Gen. Elec. Co. *Education:* B.M.E./Mech. Eng./Univ. of Minn. *Born:* 07/10/27 37 years experience in the design of steam turbines and their controls. Design of buckets, piping, expansion joints, exhaust hoods, admission valves & their controls. In 1953 first solved piping expansion problems on a computer. Major contributor to ASME publication TDP-2-1985 and TWDPS-1 "Recommended Practices for the Prevention of Water Damage to Steam Turbines Used for Electric Power Production" Fellow-ASME *Society Aff:* ASME

Johnson, Loering M
Business: PO Box 372, Tariffville, CT 06081-0372
Position: sole proprietor (consulting engineer) *Employer:* LMJ Enterprises (self-employed) *Education:* MSEngSci/Electrical/RPI; BSEE/Control Sys/Univ of ND. *Born:* 9/22/26. Principal consulting engineer, LMJ Enterprises. Adjunct Instructor (automatic control design), Univ. of Hartford (CT). Lic. PE (CT). Employed by du Pont 1952-55. With Combustion Engg 1955-1985. Involved with design of instrumentation, controls and electrical systems for nuclear-power production, propulsion & electric-generating plants. Developed, or initiated use of, new instrument systems for radioactivity scanning and monitoring. Co-inventor of an on-line boron concentration monitor. Author of the Nuclear-Fueled Steam Generation section of the Standard Handbook for Electrical Engineers, 12th ed., McGraw-Hill, 1987. Fellow of the IEEE. Member of the IEEE Stds. Bd. - 1978-81. Chrmn., IEEE Env. Qual. Cmte. - 1983-84. NSPE/CSPE Natl. Dir. - 1981-84. Awarded an IEEE Stds. Medallion - 1975; Eng. of the Year, Industry, CSPE - 1983; Toastmaster of the Year, TI - 1983. *Society Aff:* NSPE, IEEE, TI, ΣΞ.

Johnson, Lynwood A
Business: School of Industrial & Systems Engrg, Georgia Inst of Tech, Atlanta, GA 30332
Position: Professor *Employer:* GA Inst of Tech *Education:* PhD/IE/GA Inst of Tech; MS/IE/GA Inst of Tech; B/IE/GA Inst of Tech *Born:* 10/4/33 Macon, GA. Industrial engr, E.I. DuPont, 1955-57. Callaway Fellow, GA Tech, 1957-58. School of industrial engrg, GA Tech: Instructor 1958-1960, Asst Prof, 1960-64. Supervisor, Operations Research Div, Kurt Salmon Assocs, 1964-66. School of Industrial & Systems Engrg, GA Tech: Assoc Prof 1966-68; Prof 1968 -. Visiting Prof, Thayer School of Engrg, Dartmouth Coll, 1967. Visiting Prof, Dept of Systems and Industrial Engrg, Univ of AZ, 1981-82. Consultant in industrial engrg, operations research and applied statistics. Visiting Professor, Dept. of Mech. Engrg., Univ Wash, Seattle, 1985. *Society Aff:* IIE, APICS, ORSA, TIMS

Johnson, Major A
Business: Ct St Plant 4-58, Syracuse, NY 13221
Position: Mgr-Div Tech Plng *Employer:* GE Co. *Born:* BSEE, Purdue, 1948. MEE, Syracuse, 1963. Mbr Eta Kappa Nu, Tau Beta Pi, Sigma Xi. American Defense Preparedness Assn., U.S. Naval Institute, and Radar Systems Panel of IEEE Aerospace Electronics Sys. Soc. Employed by GE Co. since 1948 & grad of GE Adv. Engg Prog. Currently Mgr., Div. Technology Planning, Electronic Systems Division. Mr. Johnson is also a member of the GE Aerospace Group Advanced Development Council and is a 1981 winner of the GE Steinmetz Award for technical achievement.

Johnson, Marvin M
Business: 175 Nebraska Hall, UNL, Lincoln, NB 68588 - 0518
Position: Prof. *Employer:* Univ of Nebraska *Education:* PhD/IEL Univ of IA; MS/IEL Univ of IA; BSME/Prod Mgt Option/Purdue Univ *Born:* 04/21/25 Native of Neligh, Nebraska. Served with Army 1473rd Engrs, WWII. Employed by Houdaille-Hershey 1949-52, Bell and Howell 1952-54, Bendix 1954-1964. Subsequently Lecturer for Univ of IA, Prof of Industrial and Management Systems Engrg, Univ of NB, Lincoln, NE. Consultant to USAID/Afghanistan to establish Management Engr., and Processing Engr. 1975-76. Also participated in Egypt. Visiting Prof in Indust Engrg, RUM Univ of PR, 1982-83. Editor for Simulation, IIE Management and Quality Control and Reliability Engrg Newsletter, and for American Journal of Mathematical and Management Sciences. Fellow of IIE, Director of OC&RE IIE Division, officer of local IIE Chapters, VP of ASA local chapter, Advisor for UNL Toastmasters Club, author of many technical publications and co-author of three books. *Society Aff:* IIE, ASME, ASEE, ASA, ORSA, TIMS.

Johnson, Maurice V, Jr
Business: Sunkit Res Ctr, 760 E Sunkist St, Ontario, CA 91761
Position: VP, Research & Development *Employer:* Sunkist Growers, Inc. *Education:* BS/Ag Eng/Univ of CA, Davis *Born:* 9/13/25. BS with honors, Univ of CA, Berkeley. Attended Univ of CA, Davis. ASAE Fellow, Dir ASAE 1969-70. Sr mbr IIE, mbr ASME. Reg PE. Mbr Univ of CA Pres's Engg Advisory Council. VP, CCQC & VChrmn of the Bd, CCQC. Served US Army (Sgt) European Theater, Prisoner of War, awarded Purple Heart, three Battle Stars. VP - Research & Development for Sunkist Growers. Holds 15 patents on specialized equip for handling, packing & transporting fresh citrus. Formerly designed farm machinery for Harry Ferguson, Inc. Served on staff of Univ of CA, Davis. Hobbies - golf & swimming. Mbr of Bd of Visitors, College of Eng, Univ of CA-Davis. *Society Aff:* ASAE, ASME, AAAS, IIE

Johnson, Milton R, Jr
Home: 810 Robert St, Ruston, LA 71270
Position: Prof Emeritus *Employer:* LA Tech Univ. *Education:* PhD/Elec Engg/TX A & M Univ; MS/Elec Engg/OK State Univ; BS/Mech Elec Engg/LA Tech Univ. *Born:* 11/5/19. Native of Shreveport, LA; Prof Emeritus 1986-present; Head, Dept. of Elec Engg. 1980-1985; Prof of Elec Engg, LA Tech Univ, Ruston, LA 1955-1980, 1985-1986; Assoc Prof, 1954-55; Asst Prof, 1947-54; Maj teaching in elec power sys and electro-mech energy conversion; Active in dept admin, curriculum and course dev; Dir of Electro-mech Instrumentation and Solid-state Electronic Device seminars; Design Engr, Gen Elec Co, Aircraft magnetoes, dc motors, and single-phase induction motors, Ft Wayne, IN, 1941-47; Design consultant for campus and REA Co-op power distribution systems, 1954-73: NSF Sci Faculty Fellow, 1959-60; Mbr of Tau Beta Pi and Eta Kappa Nu; Baptist. *Society Aff:* IEEE, ASEE, LES.

Johnson, Morris V
Home: 50720 Woodbury Way, Granger, IN 46530
Position: Div. Gen Mgr. *Employer:* Natl. Std. Co (Forged Products Div.) *Education:* MS/ME/Univ of KY; BS/ME/Univ of KY. *Born:* 3/13/25. in Shelbyville, KY. Married Mary Frances Jackson, 6-30-51. Children - Morris V, Jeffrey Lee; Engr So Bell Tel & Tel Co, Louisville, 1950-51. Design Engr B F Goodrich Chem Co, Louisville, KY, 1951-54. Chem Design Engr, 1956-62. Maint Supt, 1962-64 Prod Engr Univ of KY. Wennergren Aero Res Lab Lexington, KY, 1954-56. Pct Mgr Delmonico Foods Co, Louisville, 1964-66; VP Mfg/Engr Martin Sheets Co, Louisville, 1966-69; Gen Mgr Machine Div Reynolds Metals Co, 1969-77; Gen. Mgr. Rawls Div. Natl. Std Co., 1978-81; Gen Mgr Forged Products Div Natl Std Co 1981-. Served in US Army 1945-46; Reg PE, KY; Mbr Tau Beta Pi, Pi Tau Sigma, Sigma Xi, Triangle, Knollwood Country Club. *Society Aff:* ASME, AMA.

Johnson, Nolan L, Jr
Business: 1000 Crescent Ave NE, Atlanta, GA 30309
Position: Proj Engr. *Employer:* Souther Engg Co of GA. *Education:* MS/Sanitary Engg/GA Inst of Tech; MBA/Mgt/GA State Univ; BCE/Civ Engg/GA Inst of Tech. *Born:* 12/17/46. Construction Officer in US Army Corps of Engr 1971-1973 bldg roads & bridges. With City of Atlanta was responsible for design of maj water & sewage treatment capital improvement projs. Directed study making recommendations to reduce damage caused by flooding of maj streams in Atlanta, GA. Design Engr for Keck & Wood, Inc performing hydraulic analyses & pumping station design for sewage treatment facilities. With Souther Engg Co of GA presently serving as proj mgr for several 201 Facilities Plans & Rural Water Systems. Chi Epsilon honorary fraternity; 1978 ASCE Collingwood Prize recipient for ASCE publication. *Society Aff:* WPCF, ASCE.

Johnson, Orwic A
Business: PO Box 1171, 3640 Commerce Dr, Columbus, IN 47201
Position: Owner. *Employer:* Columbus Surveying & Engr. *Education:* Self taught. *Born:* 7/9/39. Native of Columbus IN. Employed by several local consultants until Jan 1, 1975 when I formed Columbus Surveying & Engineering. Licensed as a PE and Land Surveyor in IN, OH, KY, and IL. Past Pres of James B Eads Chapter of IN Soc of Professional Engrs and past pres of IN Soc of Professional Land Surveyors an affiliate of ACSM. Enjoy woodcraft, fishing, and boating. *Society Aff:* NSPE, ACSM, American R/W Assoc.

Johnson, Paul C
Home: 135 Nichols Rd, Cohasset, MA 02025
Position: President *Employer:* Paul Johnson Assoc *Education:* BS Mech Engg, IA State Univ, Ames *Born:* -/-/39 With Chicago Bridge & Iron for 10 yrs, specializing in design, estimating, sales & start-up of low temperature and cryogenic liquefaction & refrigerated storage facilities. With Distrigas for 3 yrs. As chief engr, respon for technical activities associated with design, construction, start-up and operation of LNG facilities. Established own consulting engrg firm in 1974. Firm provides consulting engineering and other technical services for faciltiies which process, store and/or transport refrigerated liquefied gasses (LNG, LPG, etc). Presented numerous technical papers. PE in IL, MA & NY. Mbr AGA, LNG committee. Hobbies include bridge and sailing. *Society Aff:* ASME, AGA, NEGA, CEC

Johnson, R Barry
Home: 12108 Radium, San Antonio, TX 78216
Position: Asst to the Pres. *Employer:* Texas Medical Instruments, Inc. *Education:* MS/Optics/Univ of Rochester; BS/Phys/GA Inst of Tech. *Born:* 4/8/46. Consultant, 1975-. Adjunct Asst Prof, Univ of AL in Huntsville, 1973-1978. Lectr, Southeastern Inst of Tech, 1976 Dir, Mgr & Asst to the Pres, Science Applications, Inc, 1978-1981; Mgr Optics, Dynetics, 1976-1978; Sr Engg, TX Instruments Inc, 1968-1972; Principal Engg, Teledyne Brown in infrared & optical systems 1972-1976. Awarded several US & foreign patents. Published numerous technical papers. Huntsville Section, OSA, Pres, 1975; VP, 1974. Mbr SPIE, OSA, SPIE Natl Technical Council, Assoc Editor, 1976-, Forum Editor & Columnist, 1975-, Guest Editor, Optical Engg, SPIE service citation. *Society Aff:* SPIE, OSA.

Johnson, Ray C
Business: Worcester Polytechnic Inst, Worcester, MA 01609
Position: Gleason Prof. of Mech. Engg. *Employer:* Rochester Inst. of Tech. *Education:* Ph.D./Mech & Aero Sci/Univ. of Rochester; MS/ME/Univ of Rochester; BS/ME/Univ of Rochester. *Born:* 8/26/27. Author of many tech articles and two engg textbooks: (1) "Optimum Design of Mechanical Elements–, John Wiley & Sons, Inc, 1961, second edition 1980, and (2) "Mechanical Design Synthesis–, Van Nostrand-Reinhold Co, 1971, second edition by R E Krieger Co in 1978. Consulting engr in mech design to many US cos. Patentee in field of mech design. Mbr of Natl Acad of Engg in Mexico. Speaker at many seminars on mech engg design in the USA, Mexico, and England. Reg PE in six states. *Society Aff:* ASME, NSPE, ASEE, SESA, AAUP.

Johnson, Raymond C
Business: 2800 SW Archer Rd, PO Box 1253, Gainesville, FL 32602
Position: Prof of EE. *Employer:* Univ of FL. *Education:* MS/EE/Univ of FL; BS/EE/TX A&M. *Born:* 9/29/22. in Galveston, TX. Served in Army Signal Corps 1943-46. Joined Univ of FL faculty as Asst Prof of Elec Engg in 1946. Presently tenured, Prof of EE, & Dir of Electronic Communications Lab. Current interest is in communications theory & res related to electronic distance measuring devices.

Johnson, Richard C
Business: Micro-J, Inc, 467 Tenth Street, NW, Atlanta, GA 30318
Position: Pres. *Employer:* Micro-J, Inc. *Education:* PhD/Phys/GA Inst of Tech; MS/Phys/GA Inst of Tech; BS/Phys/GA Inst of Tech. *Born:* 5/9/30. Served in US Navy (1953-55) as Electronics Officer aboard a destroyer. 1956-87 was actively engaged in appl res and dev work at GA Tech Research Institute. Hd, Radar Branch (1963-68); Chief, Electronics Div (1968-72); Mgr, Systems and Techniques Dept (1972-75); Assoc Dir of EES (1975-79). 1987- Pres of Micro-J, Inc. Reg PE (Elec) in State of GA; Fellow of IEEE. Active in IEEE Antennas and Propogation Soc (Newsletter Editor 1975-78, Distinguished Lecturer 1977-78, VP 1979, Pres 1980). Technical interests include microwave antennas and measurements, radar, microwave components, and applied physics. *Society Aff:* AOC, APS/IEEE, ΣΞ.

Johnson, Robert D
Business: 20 E Worcester St, Worcester, MA 01604
Position: Traffic Engr. *Employer:* City of Worcester-Dept. Traf Engrg *Born:* Dec 1933. CE degree Northeastern Univ. Traffic Engr of Worcester from 1966; Exec Dir Off-Street Parking Bd; Chrmn of Worcester Regional Transit Auth; Pres NE Sec Inst of Transportation Engrs 1975; VChrmn Dist 1 of ITE 1976; Reg PE MA, Reg Land Surveyor MA.

Johnson, Robert E
Home: 210 Outlook Dr, Houston, TX 77034
Position: Asst-Payloads, Struct & Mech Div. *Employer:* NASA Johnson Space Ctr. *Born:* Aug 1933. BA & BS from Rice Univ, Houston, TX. Native of Houston, TX. Design Engr & Sr Mtls Engr with Gen Dynamics & Ling-Temco-Vought prior to joining NASA in 1962. Current areas of responsibility include structural & mech design & analysis necessary to define payloads to be flown in NASA Space Shuttle Prog. Also responsible for specific areas of res associated with advanced spacecraft design. Mbr of ASME & ASM Technical Cttes. Recipient of NASA Cert of Commendation for mtls dev.

Johnson, Robert L
Business: 5301 Bolsa Ave, Huntington Beach, CA 92647
Position: Pres. *Employer:* McDonnell Douglas Corp. *Education:* MS/-/Univ of CA; BS/-/Univ of CA. *Born:* 5/-/20. Native of Winslow, AZ. 1942-46 US Navy, Aircraft Instruments Officer Lt. 1946-69 Douglas Aircraft Co, last position as VP, Manned

Johnson, Robert L (Continued)
Orbiting Lab. 1969-73 served as Asst Secy of the Army (R&D). 1973-75 Corporate VP of Engg & Res, McDonnell Douglas Corp. In 1975 assumed current responsibilities as Pres of MDAC. Additionally, on 8 Dec. 1980, named Corporate vice president-group executive, MDC assuming responsibility for West Coast aerospace operations. Mbr of the Univ of CA Engg Advisory Council; AIAA-Fellow; Natl Academy of Engg; recipient of the Am Rocket Soc's James H Wyld Meml Award.

Johnson, Robert M
Business: Mech Engg Dept, Arlington, TX 76015
Position: Prof and Associate Dean of the Graduate School. *Employer:* Univ of TX. *Education:* PhD/Engg Sci/Univ of OK; MMetEngg/Met Engg/Univ of OK; BS/Met Engg/Univ of OK. *Born:* 3/28/39. Born in OK City & attended OK City Schools & the Univ of OK. Received 3 yr NASA Traineeship as grad student. Upon grad with PhD, took position of Asst Prof at the Univ of TX at Arlington in 1967. Promoted to Assoc Prof in 1970. In 1976, an NSF summer faculty res participant at Vought Corp Advanced Tech Ctr in Dallas 1977-78, Sr Scientist at Vought ATC. Promoted to Prof of Mtls Sci at UT Arlington, 1979. Named Associate Dean of the Graduate School, at the University of Texas at Arlington, 1980 Chrmn of N TX Chapter ASM 1974-75. Married, Mary. Daughter, Dana. Enjoy classical music, reading & travel. *Society Aff:* ASM, ASEE, ΣΞ, TBK, ΠΤΣ.

Johnson, Robert R
Business: 1675 Maple Rd, Troy, MI 48084
Position: Sr VP Engrg & Info Sys *Employer:* Energy Conversion Devices Inc. *Education:* PhD/EE/CalTech; MEngr/EE/Yale; BS/EE/Univ WI. *Born:* 6/20/28. Worked with GE's Transient Analyser in 1950; did logic design of first MX1179 airborne firecontrol & navigational digital computer for Hughes Aircraft; at GE-Computer Dept was responsible for Bank of American - Erma proj, & mgr of engg in Phoenix. Joined Burroughs in 1964 & was in charge of all developmental engr from 1968-1979 as VP-Engr & VP-Advanced Tech 1980-81. Currently Sr VP Engr & Info Sys for ECD. *Society Aff:* IEEE, ACM.

Johnson, Ronald A
Business: 323 Duckering, Univ of Alaska, Fairbanks, AK 99775-0660
Position: Prof *Employer:* Univ of AK *Education:* PhD/Aerospace/Cornell Univ; MS/Aerospace/Cornell Univ; ScB/Aerospace/Brown Univ *Born:* 12/20/43 After graduating from Cornell Univ in 1969, Dr. Johnson worked for 7 years as a staff scientist at Avco Systems Div in Wilmington, MA. There, he worked on a variety of environmental engrg and re-entry physics problems. Since 1976, he has taught courses and performed research in both mechanical and environmental quality engrg at the Univ of AK in Fairbanks. His current research deals with air pollution and remote energy production. *Society Aff:* WPCF, ASME

Johnson, Stanley E
Home: 6983 Ardelle Dr, Reynoldsburg, OH 43068
Position: Chief Engr. *Employer:* Burgess & Niple, Limited. *Education:* BCE/Civ Engrg/OH State Univ; MS/Wastewater & Water Resources/OH State Univ; MBA/Finance/OH State Univ. *Born:* 6/14/46. Native of Columbus, OH. With Burgess & Niple, Limited since 1970. Assigned to Burgess & Niple, Limited's WV office in charge of Hydraulics & Hydrology Sec 1976-78. Assumed current responsibility as Chief Engr in the Hydraulics and Hydrology Sec in Burgess & Niple, Limited's home office in Columbus, OH 1978. Pres, Central OH Sec of ASCE 1978. Boy Scouts of Am, Eagle Scout Award 1960. Mbr, Natl Civ Engg Honorary Fraternity, Chi Epsilon. Pres, Columbus Breakfast Sertoma Club, 1974. Sertoma Intl Gem Award 1970, & Silver Honor Club Award 1974. Judge for the 35th Intl Sci and Engrg Fair, May 7-13, 1984. Delegate to the OH Council of Local Sections ASCE 1984-87. Pres Northwest Sertuma Club 1983. *Society Aff:* ASCE, NSPE, PLSO, AMBA, AWWA.

Johnson, Ted D
Home: 2930 S Jay St, Denver, CO 80227
Position: VP & Denver Office Manager *Employer:* Woodward-Clyde Consultants *Education:* MS/CE/Univ of IL; BS/CE/Univ of IL *Born:* 9/19/39 Native of Galesburg, IL. Staff engr with Clark, Dietz and Assocs in Urbana, IL 1963-67. Moved to Denver, CO in 1967. Staff civil engr with Terrametrics, Inc 1967-68. Joined Woodward-Clyde Consultants in 1968, became VP in 1980. Currently, a Sr Assoc and Denver Office Mgr in Woodward-Clyde Consultants. Pres of CO Section of American Society of Civil Engrs in 1984-85. Enjoys skiing, swimming, hiking and back packing. *Society Aff:* ASCE, USCOLD, ISSMFE

Johnson, Vern R
Home: 1030 W San Miguel Cir, Tucson, AZ 85704
Position: Assoc Dean *Employer:* Coll of Engrg, Univ of AZ *Education:* PhD/EE/Univ of UT; BS/EE/Univ of UT *Born:* 02/25/37 Native of Salt Lake City, UT. Res engr at Microwave Elec Corp, Palo Alto, CA specializing in microwave acoustics and lasers, 1964-67. Univ of AZ: EE Faculty, 1967-79; Assoc Dean for Acad Affairs, Coll of Engrg, 1979-present. Visiting fellow in Acad Admin, Provost's Office, Cornell Univ, 1982-83. Region 6 IEEE Achievement Award, 1971. Chrmn, Tuscon Sect IEEE, 1970-71. Region 6 IEEE Exec Ctte, 1980-present. Mbr of the Bd of Dirs, Natl Consortium for Grad Degrees for Minorities in Engrg, Inc (GEM), 1982-present. Chrmn, Pacific Southwest Sect ASEE, 1984-85. *Society Aff:* IEEE, ASEE

Johnson, Victor J
Business: 1380 55th St, Boulder, CO 80303
Position: Consulting Engr (Self-employed) *Education:* MS/Mech Engrg/Univ of MD; BS/Mech Engrg/Tri-state Univ *Born:* Apr 1916. BS Tri-State Univ. MS Univ of MD. Born in NB & reared in SD. Valedictorian of 1939 grad class Tri-State Coll. First employed as Jr Des Engr Bucyrus Erie Co. Served as Ind Engr with the Panama Canal during WWII. Hd, Facilities Engrg Br, Naval Res Lab 1946-50. Proj Engr & Ch, Liquefier Oper Sect NBS 1950-57. Operations Mgr Stearns Roger Corp 1957-58 for construction & operation large USAF hydrogen liquefier plant. Dir, Cryogenic Data Ctr, NBS 1958-72. Visiting Lectr, Univ of CO for grad level engrg courses 1954-55, 1972, 1980. Dept of Commerce Gold Medal award for exceptional service 1953. Dist Serv Award 1966 ASHRAE, Dist Alumni Award TSC, Fellow 1973 ASHRAE. Chrmn, Honors & Awards Ctte 1970-71 ASHRAE, Journal Ctte 1976-77, Historical Ctte 1981-82. ASME Legislative Fellow 1984-5. *Society Aff:* ASME, ASHRAE, AIChE, IIR, RESA

Johnson, Virgil A
Business: PO Box 3986, Odessa, TX 79760
Position: VP. *Employer:* El Paso Products Co. *Education:* BSc/ChE/Univ of MO *Born:* 2/25/20. Native of Kansas City, MO. Served with Army Corps of Engrs 1940-46. Process Design Engr, Sinclair Res Lab, 1947-50. Technical Service Engr & Area Foreman, Sinclair Rubber, 1950-55. Petro-Tex Chem, 1955-57. Joined El Paso Products Co 1957 as Process Engr, held positions as Asst Chief Engr Chief Engr, Mgr of Engg, Dir of Corporate Planning, & assumed position as VP & Technical Dir in 1972. VP of Planning & Business Development since June, 1979. *Society Aff:* AIChE, NSPE, TSPE.

Johnson, Walter C
Home: 20 McCosh Cir, Princeton, NJ 08540
Position: Arthur LeGrand Doty Prof of Elec Eng, Emeritus. *Employer:* Princeton Univ. *Education:* EE/EE/Penn State Univ; BS/EE/Penn State Univ. *Born:* 1/6/13. Engr, GE Co, 1934-37. Mbr of the faculty, Princeton Univ, 1937-81. Chrmn, Dept of Elec Engg, 1950-65. Consultant for various cos; resident visitor Bell Telephone Labs 1968. Fellow IEEE; mbr American Physical Soc, Sigma Xi. Natl Award, Best Initial Paper, AIEE, 1939; Western Electric award for excellence in engg education, American Soc for Engg Education, 1967. Author of three books. Res and publications in electronic mtls and devices. *Society Aff:* HKN, TBΠ.

Johnson, Walter E
Business: Appliance Pk/35-1115, Louisville, KY 40225
Position: Mgr, Met Lab. *Employer:* GE Co. *Born:* Nov 10, 1927 Maynard, MA. After Army service, grad from MIT 1951 with BS in Met Engg. Worked for Reynolds Metals & US Naval Ordnance Plant before joining GE in 1960. Worked in Met & Ceramics Lab of the Maj Appliance Labs in the areas of sheetmetal forming & steel applications until became Mgr, Met Lab. Tr of Am Soc for Metals, former Chrmn of Am Deep Drawing Res Group, & known for analysis of forming of sheet steel.

Johnson, Walter K
Home: 5321 - 29th Ave So, Minneapolis, MN 55417
Position: Dir of Quality Control. *Employer:* Metropolitan Waste Control Commission. *Education:* PhD/Civ Engg/Univ of MN; MSCE/Civ Engg/Univ of MN; BSCE/Civ Engg/Univ of MN. *Born:* 8/28/23. Native of Minneapolis, MN. Served as Meterologist, USAAF, 1943-46, Capt. With Greeley & Hansen, Chicago, 1948-49; Infilco, Inc Tucson, 1951-52; Toltz, King, Duvall, Anderson & Assoc, St Paul, 1952-55; faculty at the Univ of MN, 1955-75, Assoc Prof of Civ Engg, 1965-74, Prof 1974-75, Adjunct Prof, 1975-; Dir of Planning, Metropolitan Waste Control Commission, St Paul, 1975-. Mpls Environmental Protection Agency Res fellow, Brit Water Pollution Res Lab, Stevenage, England, 1971. Diplmate, Am Acad of Env Engrs; Reg PE, MN; Patentee wastewater sampler. Res on wastewater treatment, nitrogen removal by dentrification. *Society Aff:* ASCE, AWWA, IAWPR, CSWPCA.

Johnson, Walter R
Business: 35 Norseman Ave, Gloucester, MA 01930
Position: Consultant (Self-employed) *Education:* Bachelor/Met Engg/MA Inst Tech. *Born:* 2/10/27. Ten yrs as Instr, Met Dept, MIT, followed by eleven yrs as Met, Raytheon Co, Missile Systems Div. Currently in eighteenth yr of private practice as an independent Met Consultant to Gen Industry. Pres, Carbide Coating Tech, Inc, Mbr of Exec Committee of Boston Chapter, ASM for seventeen yrs, & Chrmn, 1975-76. Specializes in generation of new business through creative utilization of updated material and process technologies to develop new products. Qube Resources Associate. *Society Aff:* ASM.

Johnson, Wendell E
Home: 1524 Woodacre Dr, McLean, VA 22101
Position: Consulting Engr (Self-emp) *Born:* Sept 1910. BS cum laude Univ of MN 1931. Native of Minneapolis. MN Hgwy Dept, Corps of Engrs & Panama Canal prior to WWII. Served with Corps of Engrs in ETO during WWII; from 1945 to 1970 served with Corps on water resources and military construction. Retired in 1970 from position as Ch of Engg Div for Civ Works in Wash DC. Since 1970 Consultant on water resources in US, Canada, Greece, Middle East, Asia & S Am. Mbr NAE, Hon Mbr ASCE, VP 1970-73 ICOLD, Fellow SAME; Mbr USCOLD, NSPE, Chi Epsilon, Tau Beta Pi. Outstanding Achievement Award Univ of MN 1968.

Johnson, William B
Home: 4219 Stern Ave, Sherman Oaks, CA 91432
Position: CAD/CAM/CIM & Automation Cons. *Employer:* Rockwell Intl. *Education:* BSME/Mech Aero/Aero Indus Tech Inst. *Born:* 5/20/22. May 1922. BSME Aero Ind Technical Inst-PE Mfg Engg, C Mfg E. Taught advanced automation classes Univ of MI 1961-68. Pioneer numerical control tech. Author 98 technical papers-5 books. Serves as CAD/CAM/CIM & Automation Consultant to Rockwell Intl, DOD, NASA, Army, & AF. Fellow Inst Production Engrs - Great Britain; Fellow Inst for the Advancement of Engg; Intl Pres Soc of Mfg Engrs, 1975-76. SME Intl Res Medal, 1964; Engr of the Yr, 1964; DOD Citation Award, 1965; NASA Achievement Award, 1970; Mgt Res Inst Distinguished Public Service Citation, 1975; CO Centennial Award, 1976. Engr of the Year - 1979; William B Johnson CAD/CAM Medal - 1981; Brigham Young Univ; Pres Inst for the Adv of Engrg 1982-83, William B Johnson Dist Contrib Award - 1982 SFVEC; Dist Engrg Achievement Award - 1983 IAE; Dist Engrg Interprofessional Award - 1983 NSPE/CSPE; Intl Joseph A Siegel Award - 1984 SME; Intl Gold Medal - 1984 IPRODE - Great Britain. Fellow - SME - 1986. IAE Internatl Frank E. Reeves Distinguished Achievements Memorial Award 1986. *Society Aff:* IAE, ASM, NSPE, SME, NCS, CASA.

Johnson, William C
Home: 4836 Carey Dr, Manlius, NY 13104
Position: VP Engr. *Employer:* Otisca Ind Ltd. *Education:* BS/CE/MI Tech Univ. *Born:* 6/12/21. Marquette, MI. Capt Army Corps Engrs 1942-1946 decorated. Engr with Toni Co & Gillette 1946-1948. Production & Intl Exec Engr Upjohn Co 1948-1973. Chef Engr Intl Div 1973-1975. Upjohn Award 1956. PE MI 1959. Dir Engr Bristol-Myers 1975- 1978. Chief Engr & VP Engr Otisca Ind Ltd 1978-present. Responsible for design & construction energy related process plants employing new tech. Hobbies include numismatics & gardening. Married, six children. Past mbr MSPE, NSPE, AACE, APCA. *Society Aff:* AIChE.

Johnson, William H
Home: 2025 Blue Hills Rd, Manhattan, KS 66502
Position: Prof Director. *Employer:* KS State Univ. *Education:* PhD/Agr'l Engg/MI State Univ; MS/Agr'l Engg/OH State Univ; BS/Agr'l Engg/OH State Univ. *Born:* 9/3/22. Native of Sidney, OH. Served with Army Corps Engrs 1943-46. Instr through Prof with the OH Agri Experiment Station, later called the OH Agri Res & Dev Ctr, 1948-69. Assoc Chrmn Agri Engg 1959-68. Acting Chrmn Agri Engg, OH State 1968-69. Visiting prof with TX A&M Univ 1969-70. Prof & Hd, Agri Engg, KS State Univ 1970-81. Prof & Director, Engineering Experiment Station, KS State Univ. 1981-present. Distinguished Alumnus, Coll of Engg OH State Univ. VP ASAE 1977-80, Pres 1986-87. KES Pres 1985-86, Engr of year 1987. Technical interests: harvesting and energy in agriculture. *Society Aff:* ASAE, NSPE, KES.

Johnson, William R
Business: 77 Beale St, Rm 1836, San Francisco, CA 94106
Position: Consulting Engr (Self-employed) *Education:* EE/EE/Stanford Univ; BS/EE/GA Tech. *Born:* 10/23/14. Native of OK. Engg education at GA Tech and Stanford. Profl career with Pacific Gas & Elec Co: Chief Elec Generation and Transmission Engr, 1951; Chief Elec Engr, 1976. Chrmn of Pacific Intertie Systems technical Studies Task Force (1964-67); Technical Prog Committee, 1972 Summer Meeting PES, IEEE; EPRI Task Force on HVDC Transmission Res (1973-75); EPRI Elec Systems Div Committee (1976- 78). Mbr of: CIGRE and US Rep Committee No 41 (1976-78). AEIC Committee on Power Distribution. US/USSR Technical Interchange Committee on HV Transmission. EPRI Res Advisory Committee 1976-78; Habirshaw Award, IEEE -1981. *Society Aff:* IEEE, CIGRE.

Johnson, Woodrow E
Business: 2001 Lebanon Rd, W Mifflin, PA 15122
Position: VP - Gen Mgr Trans Div. *Employer:* Westinghouse Elec Corp. *Born:* BS Hamline Univ, MS & PhD Brown Univ. Named VP & Gen Mgr of the Westinghouse Trans Div in 1971. Prior to this apptment was VP & Gen Mgr of the Astronucl/Underseas Div. Joined Westinghouse in 1949 as Sect Mgr in solid-state phys res at the Bettis Atomic Pwr Lab. Later assigned full respon for tech oper during the const & oper of the USS Nautilus prototype reactor facil at the ID test site. Prior to accepting the pos at the Astronucl/Underseas Div, spent 12 yrs in Westinghouse commercial nucl pwr activities. Awarded Westinghouse Order of Merit & authored & co-authored over 20 pubs dealing with nucl & aero-space subjects. Mbr: AIF, NSIA, AIAA, ANS, IRT, ATA, AIP & NAE.

Johnson, Zane Q
Business: 439 Seventh Ave, PO Box 1166, Pittsburgh, PA 15230
Position: Pres. *Employer:* Gulf Sci & Tech Co. *Education:* BS/ChE/Univ of OK. *Born:* 3/5/24. Joined Gulf Oil Corp at Port Arthur Refinery 1947. Dir-Planning, Refining Dept, Pgh 1958; Gen Mgr Gulf Oil Raffinaderij, Netherlands 1962; Mgr-Refining Gulf Eastern, London, 1964; Mgr-Distribution and Engg, Marketing, Houston 1967; Exec VP Gulf Oil-US, Houston, 1968; Pres Gulf Gen Atomic Co,

Johnson, Zane Q (Continued)
CA, 1969; Exec VP, Gulf Oil Corp, Pittsburgh 1970; Pres Gulf Sci and Tech Co, Pittsburgh, 1975. Dir Boy Scouts of Am, Duquesne Univ, Natl Assn of Mfg, Sch of Chem Engg and Mtls Sci of the Univ of OK. Trustee Shadyside Hospital. Mbr Am Inst of Chem Engrs, the Am Petroleum Inst. *Society Aff:* AIChE.

Johnston, Bruce G
Home: 5025 E Calle Barril, Tucson, AZ 85718
Position: Prof Emeritus. *Employer:* Retired. *Education:* PhD/CE/Columbia Univ; MSCE/CE/Lehigh Univ; BSCE/CE/Univ of IL. *Born:* 10/13/05. Early experience, concrete mix control at Coolidge Dam. Instr, Columbia Univ, 1934-38. In charge of structural res at Fritz Engg Lab, Lehigh Univ, 1938 to 1950, except for war yrs leave. Dir, 1947-1950. Initiated Lehigh res on plastic design and steel column behavior. Cofounder, Column Res Council. Prof of Structural Engg, Univ of MI, 1950-1968. Consultant on structural steel res and design specifications. Editor, through 3 editions, "Guide to Stability Design Criteria for Metal Structures." Coauthor, "Basic Steel Design" (3rd ed. 1986). Author or coauthor of 75 papers on steel structures res. Hon M ASCE, Mbr NAE. *Society Aff:* ASCE, NAE.

Johnston, Burleigh Clay, III
Business: 1095 Harbor Ave, Memphis, TN 38113
Position: Plant Mgr *Employer:* Great Dane Trailers *Education:* BS/IE/GA Inst of Tech; Grad Studies/IE/Memphis State Univ *Born:* 12/27/46 in Ft Smith, Ark. 1968-72 Ind Engrg working in Plant Engrg Group; 1972-79 Plant Engrg for Arrow Div; 1979-83, Asst Plant Mgr; 1984 Mgr of the Platform Trailer Plant, all with Great Dane Trailers. 1980-1983, IIE VP for Region VII (AR, LA, MS, AL, TN); 1974-78, Dir of the Memphis, TN chapter; 1972, Pres of the Savannah, GA chapter; 1969, Treas for Region IV Conference. Currently Chrmn of the Top Mgmt Adv Bd to the Memphis Chapter of IMC, 1979, Secy/Treas for Region IV (west of MS River); 1978 Pres of the Memphis Chapter. 1980-83 Exec Dir of Memphis Marriage Encounter. *Society Aff:* IIE, IMC

Johnston, E Russell, Jr
Business: Dept of Civ Engg U-37, Storrs, CT 06268
Position: Prof. *Employer:* Univ of CT. *Education:* ScD/Structural Engr/MIT; MS/Civ Engg/MIT; BCE/Civ Engg/Univ of DE. *Born:* 12/26/25. Engg educator and author; born Philadelphia, PA. Structural designer, Boston, late 1940's. Consultant and expert witness on structural analysis and design for various agencies and clients. Author (with F P Beer) of engg textbooks including: *Mechanics for Engineers*, 1956 (3ed, 1976); *Vector Mechanics for Engineers, Statics and Dynamics*, 1962 (4ed, 1984); *Mechanics of Materials*, (1981). Chrmn, New England ASEE, 1972-73; Hd, Dept of Civ Engg, Univ of CT, 1972-77; Western Elec Fund Award for Excellence in Engg Education, ASEE, 1968. Academic Guest: ETH, Zurich, 1970,1978; Imperial College, London, 1977; Alumni Award for Excellence in Teaching, 1984. *Society Aff:* ASCE, ASEE, IABSE.

Johnston, James P
Business: Dept of Mech Engg, Stanford, CA 94305
Position: Full Prof. *Employer:* Stanford Univ. *Education:* ScD/Mech Engrg/MIT; MS/Mech Engrg/MIT; BS/Mech Engrg/MIT. *Born:* 5/11/31. Involved for over 25 yrs with basic and applied res in fluid mechanics, turbulent shear flows and turbomachinery. At Stanford Univ since 1961 following three yrs of ind res. Current activities include teaching grad and undergrad courses in fluid mechanics, supervision of PhD student res, admin of contracts from govt (NASA, NSF, ONR, AFOSR), consulting for industry. Author numerous papers in archival journals. Assoc Editor *Journal of Fluids Engineering* (1977- 1980). ASME Freeman Fellowship in 1967/68 while Visiting Scientist at Natl Physical Lab. Teddington, England. ASME R T Knapp award for res accomplishment, 1975. Fellow ASME. *Society Aff:* ASME, AIAA, AAAS, ΣΞ.

Johnston, Roy G
Business: 1660 W 3rd St, Los Angeles, CA 90017
Position: VP. *Employer:* Brandow & Johnston Assoc. *Education:* BSCE/Civ Engg/Univ of S CA. *Born:* 1/7/14. Founding partner VP - Consulting Structural Engg firm Brandow/Johnston Assoc specializing in high rise bldg structures & earthquake safety engg. Past Pres CA State Bd of Reg PE. Past Pres SEAOC. Past Dir EERI, Past BD Chrmn BSSC, Mbr of Veterans Admin Advisory Committee on Structural Safety. Bd Chrmn Westmont College. Distinguished Civ Engg Alumni Award '82. Outstanding Engr Merit Award Los Angeles '76. National Academy of Engineers. George Washington Award - Advancement of Engineering '88 - Mbr State of Calif. Building Standards Commission. *Society Aff:* ASCE, EERI, SEAOC, ACI, BSSC.

Johnston, Wallace O
Business: 1112 Kenyon Ave, Plainfield, NJ 07060
Position: Consulting Engr (Self-employed) *Employer:* Wallace Johnston Engrs *Education:* B/ME/City Coll of NY *Born:* 11/8/29 Native of NY. Served in US Air Force 1950-54. Honorable mention award, 1961, Lincoln Welding Inst Engrg students competition. Obtained Professional Engrs License, 1967, State of NY. Now licensed in 14 other States. Served 2 year's Bd Dir of NYACE in 1969 and 1970, Johnston designed an unique ventilation system for American Airlines, LaGuardia Airport, NY, that solved a major jet fumes pollution problem in their baggage area. Technical paper published in Bldg Systems Design, 1976. Private practice since May, 1968. Now Principal of Wallace Johnston Engineers, Consulting Electrical/Mechanical Engrs. *Society Aff:* ASME, NSPE, ASHRAE, AACE, NYACE, MAA.

Johnston, Walter E
Home: 4000 Castle Ct, Raleigh, NC 27612
Position: System Energy Applications Engineer *Employer:* Carolina Power & Light Co. *Education:* MSME/Steam/Virginia Poly-Technic Institute & State U. *Born:* 03/04/34 National VChmn, Cogeneration Inst; mbr of Editorial Bd, "Cogeneration Journal–; mbr, ASME Cttee PTC 3.9 "Performance Testing of Steam Reaps" (code); Reg PE NC & SC; Certified Plant Engr.; Certified Energy Mgr; Certified Cogeneration Professional. Mbr Certification Bd for Cogeneration Professionals; NC 1987 Engr of Yr; Pres, NC Ch AEE 1987; ppres Raleigh Ch AIPE; current Area Dir (5.1) AIPE. Authored four energy papers presented at natn'l conventions and published in technical publications. Past instructor, energy mgmt, continuing Ed Dept, UPI & SU. Energy work Japan, Australia, Italy, Austria, England, Canada, USA. *Society Aff:* ASME, AEE, APEM, ASHRAE, AIPE, NSPE

Johnston, Waymon L
Business: Industrial Engrg Dept, College Station, TX 77843
Position: Asso Prof Safety Engr Program Head *Employer:* TX A&M Univ *Education:* PhD/IE/TX Tech Univ; MS/IE/Univ of MO-Columbia; BS/ME/Univ of MO-Rolla *Born:* 10/3/35 Native of Greenbrier, AR. Field engr for Dresser-Atlas Corp specializing in electronic oil well surveying 1958-1967. With McDonnell-Douglas (1965-66) as Planning Engr on Project Gemini. With TX A&M Univ since 1969. Directed Safety Engrg Graduate Intern Program under contract with US Army at Red River Army Depot from 1969-1978. Assumed Asst Dept Head position in 1980, head of safety engr program since 1984, Bd of Dirs, System Safety Society 1982; Pres of IIE (local chapter, 1974); Bd of Divs, TX Safety Assoc, 1984; Active Conslt in Forensic Engr, Human Factors Engr, and Product Safety Engr. *Society Aff:* HFS, ASSE, NSPE, SSS, IIE.

Johnston, William N
Business: 65 Broadway, New York, NY 10006
Position: Chrmn & Pres. *Employer:* American Bureau of Shipping. *Education:* BS/Naval Architecture & Marine Engg/MA Inst of Tech; BS/Mech Engg/Auburn Univ. *Born:* 7/11/22. Native of Mobile, AL. Elected to Tau Beta Pi, honorary engg fraternity, and to Pi Tau Sigma, honorary mech engg fraternity. Served as 2nd and 1st Lt with the US Army Corps of Engrs 1943-1946. Joined American Bureau of Shipping (intl ship classification society) in 1951. From 1951 to 1972 - Surveyor with promotions to Area Principal Surveyor both Field and Technical in the US

Johnston, William N (Continued)
and Europe. 1972-1979 Asst to Chrmn, VP, Sr VP, Pres NY hdquarters. Assumed current responsibilities as Chrmn & Pres May 1979. Also Chrmn and Pres of ABS Group of Cos, Inc, Chrmn of ABS Worldwide Technical Services, Inc, Pres of ABS Computers, Inc, Pres of EXAM Co and Pres of ABS Properties, Inc. Mbr of the Bd of Trustees of the Webb Inst of Naval Architecture. Enjoy classical music, classical ballet and opera. *Society Aff:* SNAME, AWS, ASNE, RINA, IME.

Johnstone, Edward L
Business: 2404 PA Ave, Evansville, IN 47721
Position: Senior Vice President *Employer:* Mead Johnson & Co. *Education:* BS/CE/Purdue Univ. *Born:* 8/14/21. Native of Evansville, IN. Worked for Mobil Oil Company in Tech Service Division & Sone & Fleming Refinery prior to WWII. Served with USN 1944-46. With Mead Johnson & Co. since 1946. Became Vice President, Engineering in 1968. Became VP Operations in 1973, responsible for Engrg; Maintenance & Services; Production; Materials Management; & Quality Assurance. Assumed current position, Senior Vice President in 1980 -- responsible for Operations, Personnel and Quality Assurance. Past Chairman-Board of Directors-Infant Formula Council; Chairman-Steering Committee-Pharmaceutical Mfg. Assn; Dir-IN Mfg. Association. *Society Aff:* AIChE.

Jolles, Mitchell I
Business: Dept of Mech & Aerospace Engg, Columbia, MO 65211
Position: Assoc Prof. *Employer:* Univ of MO. *Education:* PhD/Engg Mech/VPI & SU; MS/Applied Mechanics/Poly Inst of Brooklyn; BS/Aerospace Engg/Poly Inst of Brooklyn. *Born:* 2/10/53. Holds joint appointment as Assoc Prof in Mech & Aerospace Engg & Nuclear Engg at Univ of MO-Columbia. Res in stress analysis of cracks, effects of irradiation & corrosive environments, fracture & fatigue crack growth prediction has led to numerous published articles & technical presentations. Serves on fracture, fatigue & nuclear tech committees. Reviews papers for journals & special publications. Teaches courses in fracture, nuclear mtls, fatigue, mechanics. Recipient of 1979 Ralph R Teetor Education Award by SAE, 1979 Dow Outstanding Young Faculty Award by ASEE. Mbr of Sigma Xi. Previously on faculty of Notre Dame & VPI. *Society Aff:* SESA, ASTM, ANS, SES, ASEE

Jolliff, James V
Home: 191 Cardamon Dr, Edgewater, MD 21037
Position: VP *Employer:* Jolliff Enterprises Inc *Education:* PhD/Ocean Engrg/Catholic Univ of America; MS/Financial Mgmt/George Washington Univ; MS/Naval Arch/Webb Inst of Naval Arch; BS/Marine Engrg/Webb Inst of Naval Arch; BS/Gen Engrg/US Naval Acad. *Born:* 6/14/32 Dr Jolliff culminated his education with a Doctorate in Ocean Engrg in 1972. He has served in many shipyard and headquarters naval engrg assignments during the past 30 years including Head, Ship Survivability Office, Naval Ship Engrg Ctr and Commanding Officer, Naval Coastal System Ctr. He served as Chrmn of the ASNE Journal Committee for 8 years and has received four Pres's Awards, the Frank G. Law - Award, and the Jimmie Hamilton Award from this society. He has published in excess of 30 technical papers, is the winner of 4 Valley Forge Freedom Medals and wears the Legion of Merit. Currently VP, Jolliff Enterprises Inc. *Society Aff:* ASNE, SNAME, NAJA, ISA, ASA, ΣΞ, GAGB, FCGA, GIAAA

Jolly, Joe D
Home: 302 Carlisle Ave, Winnsboro, SC 29180
Position: Mgr-Plt Engg/Purchasing/Warehousing *Employer:* Uniroyal Goodrich Tire Co. *Education:* BS/Mech Engr/Clemson Univ *Born:* 12/3/24. Native of Gaffney, SC. Served in Army Air Force 1943-1946. Engr with M Lowenstien and Sons 1952-1957. With Uniroyal, Inc since 1957. Assumed current responsibility as mgr, Plant Engg in 1964. Reg PE in SC, GA. & N.C. Certified Plant Engr by AIPE. Chapter Pres of AIPE in 1972. Responsible for Engg, Environmental Control, and Energy Conservation at Complany complex in Winnsboro SC which includes Div, Res and Dev, and Div Headquarters. Serve in leadership roll in community and church affairs. Enjoy fishing and boating. Member South Carolina Society of Engineers; Listed in Who's Who in Technology Today - 1981. Who's Who in South & Southwest - 1982-83. *Society Aff:* AIPE, NSPE

Jonas, John J
Business: Dept. of Metallurgical Engineering, McGill University, 3450 University St, Montreal, Canada H3A 2A7
Position: CSIRA-NSERC Research Professor of Steel Processing *Employer:* McGill Univ. *Education:* PhD/Mech Sci/Cambridge Univ; BEngg/Met Engg/McGill Univ. *Born:* 12/8/32. Native of Montreal, Canada. Holds teaching post at McGill Univ since 1960. Active in res relating to metal forming, the high temperature deformation of metals, plastic instability & flow localization, the processing of HSLA steels, and the development of deformation textures. *Society Aff:* ASM, AIME, CIMM, ISIJ.

Jones, Alfred W
Home: 5369 Hickory Bend, Bloomfield Hills, MI 48013
Position: Prof. *Employer:* Wayne State Univ *Education:* Ph.D/Math/Columbia Univ; MA/Math/Columbia Univ; BA/Philosophy/Columbia Coll *Born:* 07/06/15 in New York City. Taught mathematics at Yale prior to World War II. Served as operations analyst for the Far East Air Forces. Assoc Prof of mathematics at RPI 1947-1957. Systems Engr at Bell Telephone Labs until 1964 when joined the Inst for Defense Analysis in Washington, DC. First chrmn Dept of Industrial Engrg and Operations Research at Wayne State. Fellow AAAS and former mbr Inst for Advanced Study at Princeton. Consultant to government and industry on Systems Modeling and Simulation. *Society Aff:* IIE, ORSA, ACM, TIMS, HFS

Jones, Andrew D
Home: 6600 Burciaga, El Paso, TX 79912
Position: Prof & Chrmn Civil Engrg Dept *Employer:* Univ of TX at El Paso *Education:* PhD/Trans/Purdue Univ; MS/CE/Univ of TX - Austin; BS/CE/Univ of Houston *Born:* 2/23/30 Registered PE in CA and TX. Dept Head Civil Engrg, CA Polytechnic State Univ, San Luis Obispo, CA, 1972-79; District Construction Engr, El Paso, TX Highway Dept, 1965-1970; Supervising Planning Engr, Austin, TX Highway Dept 1961-65; Sr Design Engr, Houston, TX Highway Dept 1956-1961; Design Engr, Houston, TX Highway Dept, 1953-1961. Member Tau Beta Pi, Chi Epsilon and Sigma Xi. Private consultant in TX and CA including subdivision development, traffic circulation, noise and air quality studies. Member of El Paso Regional Transportation Steering Committee, two Transportation Bd Committees and numerous El Paso City Committees. *Society Aff:* ASCE, NSPE, ASEE, ITE

Jones, Barclay G
Home: 310 E Holmes St, Urbana, IL 61801
Position: Prof ME & Nuc Engg; head, Dept Nucl Engg. *Employer:* Univ of IL. *Education:* PhD/Nuclear Engg/Univ of IL; MS/Nuclear Engg/Univ of IL; BE/ME/Univ of Sask. *Born:* 5/6/31. Native of Sask, Canada. Recipient PE prize, Univ of Sask, 1954. Athlone Fellow 1954-56. Atomic power ind experience: AERE, Harwell, England; Canadair Ltd, Montreal, Canada; Westinghouse APD, Pittsburgh, PA, 1956-58. Academic experience: Univ of IL 1963-, Prof since 1972. Res interests: fluid dynamics, heat transfer, turbulence, aerodynamic noise, two-phase systems, and reactor system simulation. Consultant to industry and natl labs. Author of more than 100 published technical articles. Elder and deacon, First United Presbyterian, Urbana, IL. Bd of dirs, Developmental Services Ctr, Champaign 1975-83, vp 1975-76, pres 1976-80. Bd of dirs: Champaign County Mental Health Bd. 1983-. Served RCAF 1950-1954. Confirmed jogger. Married Rebekah I. Scolnick August 15, 1959. Three children, Deborah born 1962, Allison born 1963 and Catherine born 1969. *Society Aff:* ANS, ASME, AIAA, CSME, AAAS, ΣΞ

Jones, Benjamin A, Jr
Business: Agri Experiment Station, 1301 W Gregory, Urbana, IL 61801
Position: Assoc Dir. *Employer:* Univ of IL. *Education:* PhD/CE/Univ of IL; MS/Agri

Jones, Benjamin A, Jr (Continued)
Engr/Univ of IL; BS/Agri Engr/Univ of IL. *Born:* 4/16/26. Received BS & MS degrees in Agri Engg in 1949 and 1950 & PhD (Civ Engg) in 1958 from the Univ of IL, Urbana. Asst in Agri Engg Univ of IL 1949-50. Asst Prof & Asst Ext Agri Engg, Univ of VT 1950-52. Returned to the Agri Engg Dept, Univ of IL in 1952 as Instr. Promoted through profl ranks to Prof in 1964. Appointed Assoc Dir of the Agri Experiment Station in 1973. Mbr grad faculty since 1958. Special interest in land drainage, irrigation, and engg aspects of conservation. *Society Aff:* ASAE, ASEE, SCSA, ΣΞ, AE, ΓΣΔ, CAST.

Jones, Charles E
Home: 9200 Idlewood, Mentor, OH 44060
Position: Retired *Education:* PhD/ME/Cornell Univ; MS/ME/TX A&M; BS/ME/CCNY. *Born:* 4/20/20. In 1920 - Joined Babcock & Wilcox in 1954 as a res engr at the firm's Alliance, OH Res Ctr. Held various posts there before being named Dir in 1968. Names Pres of Bailey Meter Co in 1971 and joined the Group VP's staff in 1975. Retired 1980. NYC native, a 1947 engg grad of CCNY, earned masters from TX A&M in 1951 and received doctorate from Cornell Univ in 1957. 1972 winner of ASME's Richards Meml Award, served as ASME vp and on numerous other Bds and Committees. Pres ASME 1980-81. *Society Aff:* ASME.

Jones, Clarence R
Home: 3415 Walton Way, Augusta, GA 30909
Position: Pres Bd Chairman, Clarence R. Jones, Consultant, Ltd. *Employer:* Clarence R Jones, Consultant, Ltd *Education:* MS/ME/Clemson University;BS/ME/Clemson University *Born:* 11/7/23. Born Ashton, SC. Wife: Eunice Polk Jones, Past Chrmn Ladies Aux NSPE. Daughters: Susan 34 & Debbie 32. Past VP NSPE, P VChrmn NSPE-PEPP, Past Chrmn NAELC. Past Chrmn CSRA-ASME. GA Engr of Yr 1974. Mil Serv as officer WWII with overseas service in the Pacific Theatre of Operations. BS and MS degrees in mechanical engineering Clemson University. Since 1950 to date, service as principal partner, president, board chairman and owner of consulting firms providing engineering and architectural services for major projects in the public and private sectors throughout the USA, the Middle East and in other foreign countries. *Society Aff:* NSPE, GSPE, ASME, AEE, ASHRAE, SAME.

Jones, Edward M T
Home: 99 Stonegate Rd, Portola Valley, CA 94025
Position: Exec VP. *Employer:* Tech for Communications Intern. *Born:* Aug 1924. MS & PhD from Stanford; BS from Swarthmore. Served as Navy Radar Officer 1944-46. Res Engr Stanford Res Inst 1950-58, Hd Microwave Group 1958-61. Dir of Engg TRG-West 1961-63. VP Engg at Tech for Communications Intl 1968-75, now Exec VP. Interests: antennas & microwave components. Fellow IEEE, co-author "Microwave Filters, Impedance Matching Networks & Coupling Structures-".

Jones, Edward O, Jr
Business: Sch of Engg, Auburn, AL 36849
Position: Asst Dean of Engg. *Employer:* Auburn Univ. *Born:* June 1922. MSME Univ of IL; BSME & BSEE Auburn Univ. Native of Dothan, AL. Taught Mech Engg, Auburn Univ; several yrs as Asst Dept Hd. Tooling Engr and Inter-Plant Coordinator, Consolidated Vultee Aircraft Corp. Consultant involving thin-shell pressure vessel stress analysis, pump evaluation, i c engine analysis & weld analysis. Thin-shell pressure vessel stress analysis and fracture mechanics publications. Mbr or past mbr Tau Beta Pi, Sigma Xi, Eta Kappa Nu, Pi Tau Sigma, ASEE, AL Acad of Sci, Who's Who in Engg Education, Am Men of Sci, Who's Who in South & Southwest, NSPE, ASPE, SAE, SESA. Reg PE, State of AL.

Jones, Edwin
Business: 345 E. 47th St, New York, NY 10017
Position: Director, Codes and Standards *Employer:* American Society of Civil Engineers *Education:* BSE/Civil/Drexel Univ *Born:* 05/22/30 Native of Trenton, NJ. Currently Director, Codes and Standards, American Society of Civil Engrs, since April, 1982. Responsible for mgmt of ASCE's voluntary consensus standards program. Retired in 1980 as design bureau chief, NJ Dept of Transportation after 32 years service in various technician and engrg grades. Was responsible for contract, plans and specifications, right-of-way engrg, and contracts and agreements for highway and transit projects. Fellow of ASCE. Licensed Profl Engr in NJ. Pres, NJ Society of Profl Engrs 1981-82. Committee chrmn ASCE Committee on Specifications and ASTM Precast Concrete Standards Committee. *Society Aff:* ASCE, NSPE, ASTM, ACI, NIBS, BOCA, ICBO, SBCCI.

Jones, Elmer
Business: 2036 Wooddale Blvd, Suite P, Baton Rouge, LA 70806
Position: Pres *Employer:* E. Jones & Assocs, Inc *Education:* BS/ME/Southern Univ *Born:* 12/24/46 I was born and reared near Winnsboro, LA, the son of a sharecropper. I attended local schools and graduated valedictorian of my high school class. I worked my way through Southern Univ, Baton Rouge, graduating with a BS in Mechanical Engrg. I was offered a position with the Chevrolet Motor Div, Warren, MI. I left there in 1973 to return to LA to work with Kaiser Chemical. In 1976, I was called to the ministry. In 1977, I began my own consulting engrg firm, Elite Engrg, Inc. and in 1979 changed the name to E. Jones & Assocs, Inc. I am Pres of the firm employing 20plus people with gross receipts of $600-750,000 for the past year. I am 33 years old, married to Mary E. Todd, and have 2 children. *Society Aff:* ASME, ACEC, LES

Jones, Everett Bruce
Home: 1209 Robertson, Fort Collins, CO 80524
Position: Pres *Employer:* Resource Consultants, Inc *Education:* PhD/Watershed Mgt/CO State Univ; MS/Meteorology/PA State Univ; BS/CE/Univ of WY *Born:* 9/23/33 Served 2 years as officer on active duty with US Army Corps of Engrs. Professional background includes work in state government as WY Chief of Water Development, in the academic sector as an Asst and then Assoc Prof, in research as Asst Dir of the PA State Univ Inst of Research on Land and Water Resources, as well as various positions in the private sector. Assumed presidency of Resource Consultants, Inc (formerly M.W. Bittinger and Assocs) in 1977. Active in Rotary, church and political organizations. *Society Aff:* ASCE, ASAE, AGU, AMS

Jones, Francis T
Business: Dept of Chem & Chem Engrg, Stevens Inst. of Technology, Hoboken, NJ 07030
Position: Dept Head *Employer:* Stevens Inst of Tech *Education:* PhD/Physical Chem/Polytechnic Inst of Brooklyn; ME/Engrg/Stevens Inst of Tech; BS/Chem/PA State Univ *Born:* 10/19/33 Worked summers at Sarnoff Research Ctr, RCA Labs (Princeton, NJ); GEC Ltd. Fellow of Cookridge High Energy Radiation Ctr, Leeds, England. Industrial employment at Union Carbide Nuclear Div, Tuxedo Park, NY - patent in chemonuclear processing. Visiting faculty at Oak Ridge Natl Lab. Faculty member at Stevens Inst of Tech, becoming Head of Chemistry and Chemical Engrg in 1979. Architectural design of lab bldgs in US and Algeria. *Society Aff:* ACS, AIC, AIChE.

Jones, Gary D
Business: Bldg 3-Rm 236, Louisville, KY 40225
Position: Gen Mgr-D&D Engg Dept. *Employer:* GE Co. *Born:* Oct 21, 1932, at Hastings, NB. He attended the Univ of NB & grad with a BS degree in ME in 1955. He had a brief assignment with GE & then spent the next three yrs in the USN as a Destroyer Gunnery Officer. He returned to GE in 1958 & entered the Creative Engg Prog. Upon completion of the prog in 1960, he joined the Household Refrig Div & had numerous design & managerial assignments in the Refrig Engg Dept. In 1975, he was appointed Gen Mgr-Dishwasher & Disposal Engg Dept.

Jones, Howard St C, Jr
Home: 6200 Sligo Mill Rd, NE, Wash, DC 20011
Position: Technical Consultant *Employer:* Harry Diamond Labs. *Education:* DSc/Sci/VA Union Univ; MSEE/EE/Bucknell Univ; BS/Math-Phys/VA Union Univ; Cert/Engrg/Howard Univ. *Born:* 8/18/21. Native of Richmond, VA. Employed with the Natl Bureau of Stds (NBS) as Asst Electro-Mech Engr, 1944. Served in US Army 1944-46 as Instr in Mech Engg. Electronic Physicist at NBS 1946-53. Electronic Scientist at Harry Diamond Labs (HDL) 1953-59. Supervisory Electronic Engr HDL, 1959-68. Supervisory Physical Scientist HDL from 1968-1980. Technical consultant 1980 to present. Adj Prof Howard Univ Sch of Engrg 1982-83. Direct, plan, coordinate res & dev progs involving microwave systems. Asst Prof Sch of Engg, Howard Univ, 1959-63. Five maj awards Harry Diamond Labs & three from Dept of Army. Thirty-five technical publications & 31 US patents. Reg PE, DC. *Society Aff:* AAAS, IEEE, WAS, APS, MTTS.

Jones, Howard T
Business: 1701 College, Ft Wayne, IN 46804
Position: Mgr-Engg. *Employer:* GE Co. *Born:* May 1923. PhD from Clemson Univ in 1972. BS from Western Carolina in 1966; PE IN & NC; hold 10 patents on insulation & HID ballasts. Served 2 yrs with the Signal Corps & 12 yrs as electronic & TV technician. With GE Co since 1956 as Design Engr, Subsection Mgr; & Section Mgr on HID ballasts, airport lighting & power transformers. Hobbies are macro-photograph & travel.

Jones, Irving W
Business: Dept of Civ Engg, Washington, DC 20059
Position: Prof, Dept of Civ Engg. *Employer:* Howard Univ. *Born:* Nov 1930. PhD from Poly Inst of Brooklyn, MS from Columbia Univ, BSCE from Howard Univ. Taught mechanics as AF Officer 1954-56, later entered aerospace structures field as res structural engr, Grumman Aerospace; was co- founder & Assoc for Appl Tech Assoc Inc, a consulting firm 1963-69; became full- time academician in 1969 as Assoc Prof at Howard Univ where maj task was dev of grad progs. Was Chrmn of CE Dept 1971-81; was engrg specialist at The Aerospace Corp 1981-83 while on leave. Areas of res include computer-aided structural design & analysis, dynamic analysis of complex structures.

Jones, James A
Business: 411 Fayetteville St, Raleigh, NC 27602
Position: Vice Chairman *Employer:* Carolina Power & Light Co. *Education:* BS/ME/NC State Univ. *Born:* 9/18/17. in Anderson, SC. With Carolina Power & Light Co since 1951. Elected vp 1969; sr vp 1970; dir 1971; exec vp 1973; chief operating officer 1976; sr exec vp 1979; Vice Chairman 1981. Past pres, NCSU Engg Fdn. Named Distinguished Engg Alumnus, NCSU, 1974; Outstanding Engr, NC Soc of Engrs, 1974; Engr of the yr by the Raleigh Engrs Club, 1976. Mbr of Atomic Industrial Forum, Pi Tau Sigma and Phi Kappa Phi. *Society Aff:* ASME, ASNE, PENC, NSPE, NCSE.

Jones, James R
Business: PO Box 235, St. Louis, MO 63166
Position: VP Environmental Affairs *Employer:* Peabody Coal Co *Education:* BS/ME/Purdue Univ *Born:* 1/19/21 Native of Michigan City, IN. Design engr, Goodyear Aircraft Corp 1942-46. Coal Combustion Engr & Chief Technical Services Peabody Coal Co and predecessors 1946-1970. Dir, Environmental Quality 1970-79. Assumed responsibility as VP, Environmental Affairs, Peabody mining and sales in 1979 directing activities in fields of air and water pollution control, land reclamation and environmental assessments of coal mining and use. Member KY Air Pollution Control Commission 1966 to 1972. Member two Natl Academy of Sciences studies. Mbr Bd of Governors of ASME 1983 to date. *Society Aff:* ASME

Jones, Joe P
Business: 811 Lamar St, Fort Worth, TX 76102
Position: VP *Employer:* Freese & Nichols *Education:* BS/Arch Engrg/Univ of TX *Born:* 10/6/28 Worked 6 years for the Corps of Engrs. Joined Freese and Nichols 1955. Became a principal in 1971. Now VP. Served TX Society of Professional Engrs in many offices including Pres, 1981-82. Chosen TX Engr of the Year in 1979. Club Pres and Lt Governer in Kiwanis. Scoutmaster and Bd Member in BSA Council. Received Silver Beaver and Vigil Honor from BSA. Served 4 years as Chrmn, Fort Worth Park and Recreation Bd. Treas NSPE 1984-. Fellow, SAME; Mbr Chi Epsilon Honor Soc. Fort Worth native, married, has 4 children. *Society Aff:* NSPE, ASCE, SAME, ASAE.

Jones, John D
Business: 2162 Front St, Cuyahoga Falls, OH 44221
Position: Chrmn of the Bd *Employer:* John David Jones & Assoc, Inc. *Education:* BS/CE/Univ of Arkon *Born:* 10/4/23 in PA. Upon grad, served as field engr for American Telephone and Telegraph Co, plan examiner for City of Akron and ch structural engr for Firestone Tire and Rubber Co. Married to Rose Capriola and has engr son, James, and daughters, Lynn and Stephanie. At present time is Chrmn of the Bd of the Co that he founded in 1958. The Co is now one of the top 500 design firms in America performing work in many states. *Society Aff:* ASCE, NSPE, ACSM, OACE

Jones, Joseph K
Business: PO Box 30067, Raleigh, NC 27612
Position: VP & Dir. *Employer:* Cotton Inc. *Education:* BS/Agri Engg/MS State Univ. *Born:* 6/22/22. Native of MS. US Marine Corps during WWII and Korean conflict. Worked in area of cotton mechanization res and product dev with USDA and private industry for ten yrs. Since 1957 planning and supervising contract res and product dev in the production and processing of cotton. Current position of VP and Dir of Processing and Handling res, Cotton Inc, since 1977, retired - 1984; Dir of ASAE, 1966-68; Man of the Yr award in Southern Agri, 1974; elected to fellow ASAE, 1977. Retired - 1984. *Society Aff:* ASAE.

Jones, Lee S
Business: 5741 Rostrata, Buena Park, CA 90621
Position: Director of Merchandising *Employer:* AMF Voit Inc. *Education:* BS/Chem Engr/IA State Univ; MBA/Bus/Pepperdine Univ *Born:* 2/9/37. Born and raised in SD. Process Engr with ADM Co MN 1959-1963; Sr Proj Engr, Gen Mills 1963-1968; Dir of Planning, Slim Jim Inc, 1968; VP Mfg, Model Products Corp 1968-1971; Mgr of Mfg Engr, AMF Voit 1971-73; VP Operations, Peterson Baby Products 1973-1975; Pres A&E Doron Plastics 1975-78; Self-employed 1978-79; Dir of Prod Dev AMF Voit 1979-1981 Director of Merchandising, AMF Voit, 1981-date. Holds three patents. Specializes in consumer products. Hobbies: youth soccer and photography. *Society Aff:* AIChE.

Jones, Marvin R
Home: 414 Flintdale Rd, Houston, TX 77024
Position: Consulting Engr (Self-employed) *Employer:* Self *Education:* 90 hrs/Mech Engr/Univ of OK. *Born:* 11/3/14. Bristow, OK. Worked as Product Dev Engr with American Iron and Machine Works, Hughes Tool Co, Oil Ctr Tool Co, Petromec, Cameron Iron Works, Inc. and Koomey Inc. More than 60 US Patents, more than 120 foreign patents. Published a number of engr papers on equipment for oil drilling and producing. Reg Prof Eng TX 1942, Mem ASME 1947, Fellow ASME 1969, Life Fellow ASME 1985. Served with Corps of Engrs WWII. Elected Mem Pi Tau Sigma, 1973, Honorary Status. Has spoken on eng. subjects in England, Germany, Austria, Hungary, Japan, Taiwan, Norway, Russia, Italy, Holland, Indonesia, Singapore, Jugo Slavjia, & Romania. *Society Aff:* ASME, SPE, API, ΠΤΣ, EAA

Jones, Mary V
Business: 7511 Wellington Rd, Gainesville, VA 22065
Position: Manager Design Department *Employer:* Atlantic Res Corp. *Education:*

Jones, Mary V (Continued)
MS/ME/Geo Wash Univ; BS/ME/VPI; Adv Grad Studies/Acoustics/Geo Wash Univ. *Born:* 9/19/40. Native of Blacksburg, VA. Daughter of J B & Evangeline Jones (J B in Who's Who Engg). Atlantic Res Corp 1962-present. Design of Solid Propellant Rockets. Manager Design Department 1986-present, Chief Mechanical Design Group 1982-86. Chief Engineer Multiple Launch Rocket Motor (MLRS) 1980-82, Design Team Leader for Stinger Rocket Motor and Tomahawk Booster 1973-80. Mbr, ASME; Sr Mbr, Soc of Women Engrs, President Baltimore/Wash. Section 1982-84; Mbr, WA Soc of Engrgs; PE VA; First native VA Women Reg. Tau Beta Pi Woman's Badge 1961, Tau Beta Pi 1969, Pi Tau Sigma 1960, Phi Kappa Phi 1961, Omicon Delta Kappa 1975, Outstanding Young Women in America 1971, Who's Who American Colleges and Univ 1962. Who's Who of American Women, Who's Who in the South and Southeast. Member of State Board of Architects Professional Engineers Land Surveyors and Landscape Architects, Commonwealth of VA 1982-present, President 1985; Member Board of Visitors Virginia Polytechnic Institute; Member National Council of Engineering Examiners. *Society Aff:* ASME, SWE, AIAA

Jones, Owen C, Jr
Business: NES Bldg, Room I-22, Troy, NY 12181
Position: Prof of Nuclear Engrg & Director-Center for Multiphase Research *Employer:* Rensselaer Polytechnic Inst *Education:* PhD/ME/Rensselaer Polytech Inst; MS/ME/Rensselaer Polytech Inst; BS/ME/Univ of MA *Born:* 10/18/36 In the twenty-five yrs of his profl career, Dr. Jones has made many accomplishments in areas of res dealing with heat transfer and fluid mech of single- and two-phase flows, having authored or coauthored over 250 reports, publications, and books dealing with both experimental and analytical areas. He has contributed much to the knowledge of heat transfer and fluid mechanics of multiphase flows while working at three natl labs Knolls Atomic Power Lab (1962-74), Argonne Natl lab (1974-76), Brookhaven Natl Lab (1976-81), and as a Professor of Nuclear Engineering and Director of the Center for Multiphase Research at Rensselaer Polytechnic Institute. As Fellow of the ASME, current mbr of ANS, AIAA, and ICHMT, and past mbr of the AIChE, has held many ctte positions, & organized tech sessions, courses & symposia. Mbr Tau Beta Pi, Pi Kappa Phi, Phi Eta Sigma, & Sigma Xi. *Society Aff:* ASME, ICHMT, ANS, AIAA

Jones, Phillip R
Business: 1419 N Palafox St, Pensacola, FL 32501
Position: Gen Mgr. *Employer:* Phillip R Jones & Assoc. *Education:* BS/Architectural Engg/Univ of KS. *Born:* 6/25/28 Active in civ, structural and architectural engg in SE for past 31 yrs. I specialize in engg for military construction and energy conservation projs. I served in Civ Engr Corps of US Navy, 1951 to 1953, prior to accepting position with firm in NM. Presently gen mgr for multi-disciplined consulting engg firm. I am a fellow mbr of FL Engg Soc. *Society Aff:* NSPE, ACEC.

Jones, Raymond M
Business: 6406 Georgia Ave, NW, Wash, DC 20012
Position: Pres (Self-employed) *Employer:* RM Jones & Associates *Education:* MS/Sanitary Engg/Univ MI; BS/CE/Howard Univ. *Born:* 11/29/22. Originally from St Louis, MO. Obtained BS Degree in Civ Engg from Howard Univ where has been a Prof of Civ Engg since 1959. Founder and Pres of R M Jones & Assoc since 1965. Responsible for marketing and administration. Twenty-five yrs of experience in Civ, Sanitary, Struct Engg and Surveying. Prior to 1965 summers were spent working in the capacity of Hydraulic Engr (US Bureau of Stds), Naval Architect (US Navy), and as a Sanitary Engr (US Public Health Service). He has directed several training progs for Foreign Students of Engg and Architecture at Howard Univ. *Society Aff:* NTA, ASCE, APHA, ASEE, NABCE.

Jones, Richard A
Home: 957 Tanglebriar, Fayetteville, AR 72701
Position: Assoc Prof *Employer:* Univ of AR. *Education:* BSEE/EE/Univ of AR; MSEE/EE/Univ of AR; EE/EE/Southern Methodist Univ; PhD/EE/Southern Methodist Univ. *Born:* 9/23/37. After receiving the PhD, Dr Richard A Jones joined the Boeing Co, Seattle, WA where he was prog mgr on several projs within the communication & signal processing area. His res interest included Rate Distortion Theoretic Studies on multidimensional signals such as images. Dr Jones joined the faculty at the Univ of AR in Aug 1977. He has received multi-yr funding from the AF Office of Scientific Res to pursue his area of res. Dr Jones has 34 publications, presentations, & reports. He has served as a session chrmn on two intl confs, as a paper referee for the IEEE journals and has served on the exec committee for the 1973 Intl Conf on Communications. He has been an invited speaker at several symposia. He is currently a Congressional Advisor on Sci & Tech & a consultant to the government and the oil industry on signal processing, telemetry, and communications. *Society Aff:* IEEE, SPIE.

Jones, Richard C
Home: 2705 Shadowbrook, SE, Grand Rapids, MI 49506
Position: Owner *Employer:* Richard Jones & Assocs *Education:* BS/CE/Univ of WI *Born:* 1/14/42 16 years experience in planning, design, and contract administration of private and governmental projects. He has been responsible for managing water, waste water, highway, bridge, commercial bldgs, industrial bldgs, solid waste and marina projects. Prior to forming his own consulting practice, Mr. Jones held responsible positions with several major consulting firms. In addition to involvement in civic organizations, Mr. Jones has been active in natl professional societies of which he has held various offices. Mr. Jones presented a paper entitled *Alternative Waste Water Systems for Small Communities* at the sixth natl conference on Individual Onsite Waste Water Systems sponsored by the Natl Sanitation Foundation. Paper published in proceedings of the same. *Society Aff:* ASCE, NSPE

Jones, Richard T
Home: 713 Winchester Rd, Broomall, PA 19008
Position: Engr-in-charge, Control Eng. *Employer:* Phila Electric Co. *Education:* BS/Elec Engg/Drexel Univ. *Born:* 10/12/24. Native of Phila, PA. Served in Army Air Corps 1943-46. Employed by Phila Elec Co since 1942. Responsible for startup of control systems in power generating facilities as a Field Engr. Resonsible for instrumentation and control systems design and implementation while working in Mech Engg Div 1962- 1972. Presently Eng-in-charge Control Engrg Section of Elec Engg Div Responsible for control and safety systems in nuclear and fossil generating plants. Past Dir - Power Div - ISA. Past chrmn - Nuclear Power Plant Stds Committee - ISA. Present Director - Standards & Practices Board - ISA, Active in civic affairs. Enjoy Amateur Radio, photography and music. *Society Aff:* IEEE, ISA.

Jones, Robert A
Home: 48 High Pastures Ct, Ridgefield, CT 06877
Position: Senior Saff Engr *Employer:* Perkin-Elmer Corp. *Education:* MS/Physics/Syracuse Univ; BS/Physics/Union Coll *Born:* 02/25/38 From 1960 to 1964, Mr. Jones was an optical physicist at Rome Air Development Ctr. He then joined the Perkin-Elmer Corp where he is presently responsible for Computer Controlled Polishing Operations. Mr. Jones's past research has covered a wide spectrum including image processing, image evaluation, and electro-optical systems. His current research involves automated optical fabrication techniques. Mr. Jones has authored numerous technical presentations and journal papers. He has been elected a fellow of the Optical Society of America and a mbr of Sigma Xi. He was also paper chrmn for the "Photo-Electronic Imaging Symposium" and a section editor for APSE. *Society Aff:* OSA, SPIE, ΣΞ

Jones, Robert R
Home: One Bay Tree Ln, Bethesda, MD 20816
Position: Engr. *Employer:* Natl Bureau of Stds. *Education:* BEng/Naval Arch/Univ of MI *Born:* 2/26/22. Kearney, NB. US Coast Guard to June 1943. Tau Beta Pi, Univ of MI. Marine HVAC at MD Drydock Co, Baltimore, 1943-46. Served with Public

Jones, Robert R (Continued)
Bldg Service, GSA, Wash, DC in Design & Construction from 1946-53 & 1958-67; chief mech & elec engr. Assoc, Gen Engr Assoc, Wash, DC, 1953-58; responsible charge for three US Embassies. Syska & Hennessy, Consulting Engrs, NYC, DC, & LA, 1967-74; Principle 1969-74. Staff engr, Federal Energy Administration, 1974-5. Prog mgr, Natl Bureau of Stds, 1975-present. Fellow, ASHRAE; Engr of the Yr, MW Consulting Engrs Council; Silver Beaver, Boy Scouts of Am; Measurement Services Awd, Natl Bureau of Standards; Bronze Medal, Dept of Comm. *Society Aff:* ASHRAE, ASTM

Jones, Roger C
Home: 5809 E Third St, Tucson, AZ 85711
Position: Prof. *Employer:* Univ of AZ. *Education:* PhD/EE/Univ of MD; MS/EE/Univ of MD; BS/EE/Univ of NB. *Born:* 8/17/19. A native of SD, serving in the AUS, Western Pacific Campaigns, 1942-45. From 1949-57, Electronic Scientist, Naval Res Lab, Wash, DC, specializing in Nucl Weapons, Plasma Physics & Radar Techniques. From 1957-64, Staff Sr Engr, Consulting Proj Engr, Sec Hd of Physics & Chief Scientist for Phys, Melpar, Inc, Falls Church, VA. From 1964-present, Prof of EE (1978-1986) Adjunct Prof of Radiology and (1986-present) Adjunct Professor of Radiation-Oncology, Univ of AZ, Tucson, AZ 85721, engaging in Lightning Res, Infrared Engg, Laser Engrg, Quantum Electronics & Bioengg. *Society Aff:* IEEE, APS, OSOA, AAAS, BES, NSPE, ACSM.

Jones, Ronald A
Business: 2720 Nolensville Rd, Nashville, TN 37211
Position: Secty/Treasurer *Employer:* Geotek Engrg Co, Inc *Education:* PhD/Geotech Engrg/Univ of IL; MS/CE/Vanderbilt Univ; BS/CE/Vanderbilt Univ *Born:* 10/10/43 Am a Licensed PE in six states and Consulting Engr in Geotechnical and Materials Engrg. Also have been Civil Engrg Prof for TN State Univ and a Civil Engr with the US Corps of Engrs, Nasville District. Am presently principal and Secty/Treasurer of Geotek Engrg Co, Inc, a position held since the co was founded in 1974. Hobbies include handball, scuba diving and underwater photography. *Society Aff:* ASCE, SAME, NSPE, ASME, ASTM

Jones, Russel C
Business: 132 Hullihen Hall, Univ of Delaware, Newark, DE 19716
Position: President *Employer:* Univ of Delaware *Born:* Oct 1935. PhD, MS & BS Carnegie Inst of Tech. Structural Engr with Hunting, Larsen & Dunnells 1957-59; Asst & Assoc Prof of Civ Engrg at MIT 1963- 71; Prof & Chrmn, Dept of Civ Engrg, OH State Univ, 1971-76; Dean, Sch of Engrg, Univ of MA 1977-81; VP for Academic Affairs, Boston Univ, 1981-87. Pres, Univ of Delaware, 1987-. Teaching & res interests: structural mtls, composite mtls, construction sys, bldg construction, professionalism & ethics, & engrg education. ASCE Collingwood Prize 1966. ASCE Edmund Friedman Professional Recognition Award 1981. ASCE Dir 1969-71 & 1972-75. ASCE VP 1976-77. ASEE, NSPE IEEE, & AAAS.

Jones, Thomas J
Business: 715 Stadium Dr, San Antonio, TX 78284
Position: Assoc Prof. *Employer:* Trinity Univ. *Education:* ScD/EE/NM State Univ; BS/EE/Univ of TX. *Born:* 7/2/29. A native of TX. Held engg design positions with the GE Co and Westinghouse Aerospace. Presently teaching electronics and computer hardware courses at Trinity Univ. Maintains active consulting service. *Society Aff:* IEEE, ΣΞ.

Jones, Trevor O
Home: 18400 Shelburne, Shaker Heights, OH 44118
Position: Group Vice President *Employer:* TRW Inc Automotive Worldwide. *Education:* HNC/Elec/Aston Tech College; ONC/Mech/Liverpool Tech College *Born:* 11/3/30. Raised and educated in England and was employed by GEC from 1950 to 1957. Proj Mgr Nuclear Ship Savannah at Allis-Chalmers 1958-1959. Nineteen yrs with Gen Motors Corp. From 1959 to 1970 engaged in Aerospace Space Progs including B- 52 bombing navigation system and Apollo computers. From 1970 to 1978 engaged in automative engg including Dir Electronic Control Systems, Dir Advance Product Engg and Dir GM Proving Grounds. Joined TRW in 1978 as VP Engineering, 1979 Group VP Electronics Group, 1985 Group VP Sales, Marketing, Strategic Planning and Business Development. Member, NAE. Fellow of IEEE and IEE. SAE Arch T Colwell and Vincent J Bendix Awards. *Society Aff:* IEEE, IEE, SAE.

Jones, Walter V
Business: 3100 Pleasanton Ave, Boise, ID 83704
Position: Ch Engr: Soil Mech & Fdns. *Employer:* Northern Testing Labs. *Born:* Feb 26, 1934. BS & MS Univ of ID. Taught at Univ of ID & conducted extension & short courses at other locations in ID. Geotechnical Engg engagements in CA (2 yrs), ID (9 yrs). In current assignment & principal in Northern Testing Labs since 1972. Published several articles on Geotechnical Engg subjects. Served on ASCE, Southern ID Sec, Bd/Dir, for 5 yrs (Pres 1975). Active in Mormon Church & Boy Scouts of Am.

Jones, Wesley N
Business: P. O. Box 1037, Severna Park, MD 21146
Position: Pres. *Employer:* WESGIN Enterprises, Inc. *Education:* BEE/Elect/NC State Univ *Born:* 11/22/23 Currently Pres, WESGIN Enterprises, Inc; co supplies high tech services to the elect indus. Areas of expertise include performance prediction for complex sys, reliability/failure analysis of semiconductor and microwave circuits, analysis of circuits using SPICE and selection/design/installation of specialized computer sys. From 1974 to 1978, Mgr, LSI Design Ctr, Bendix Aerospace Elect Group. Previously, Fellow Engr, Westinghouse Defense and Elect Sys Ctr; specialty Large Scale Integrated Circuits and computer simulations to predict performance. Eight patents, author of several papers; co-author of book "Integrated Electronic Systems" (Prentice-Hall). *Society Aff:* NSPE, IEEE, MSPE

Jones, Wilfred
Home: 605 Maple Ln, Edgeworth, Sewickley, PA 15143
Position: Pres, CEO. *Employer:* Pittsburgh-Des Moines Corporation. *Education:* CEng & Higher Dipl/ME/Enfield College of Tech. *Born:* 5/11/22. in England. Field engr to Mgr of Projs 1946-68 in design & construction of oilfield, refinery, gas, chem, synthetic fiber & nucl projs for British Petrol; Monsanto; Allied Chem; & Humphreys & Glasgow, including seven yrs in the Mid- East. Group VP, Daniel Intl Corp, 1968-76, responsible for power, chem & ind plant construction. Exec VP, Austin Idustries, 1976-78, administering subsidiary cos constructing power, oil, chem, ind projs. Currently Pres & CEO, Pittsburgh- Des Moines Corporation, intl engg, fabrication & construction co. Received ASME Outstanding Leadership Award 1975. *Society Aff:* ASME, AIChE, IMechE, InstPet.

Jones, William B, Jr
Home: 2612 Melba Cir, Bryan, TX 77801
Position: Prof *Employer:* TX A&M Univ. *Education:* PhD/EE/GA Inst of Tech; MS/EE/GA Inst of Tech; BS/EE/GA Inst of Tech. *Born:* 9/17/24. Native of Fairburn, GA. Served in USNR 1943-46. Mbr of Technical Staff, Hughes Aircraft Co, 1954-58. Prof of Elec Engg, GA Tech, 1958-67. Prof & Hd, Elec Engg Dept, TX A&M Univ, 1967-84. Visiting Prof., Univ. of FL, 1984-85. Prof of Elec Engg, Texas A&M Univ., 1985-. Chrmn, Atlanta Sec IEEE, 1961-62; Editor, IEEE Transactions on Communications Systems, 1960-61; Chrmn IEEE Communication Tech Group, 1966-68; VP, IEEE Communications Soc, 1971-73; Mbr IEEE Technical Activities Bd, 1963-70. Author: "Introduction to Optical Fiber Communications Systems," Holt, Rinehart and Winston, 1987. *Society Aff:* IEEE, OSA

Jones, William F
Home: 1156 Snowberry Ct, Sunnyvale, CA 94087
Position: Pres. *Employer:* William F Jones, Inc. *Education:* MS/Civ Engg/CA Inst of Tech; BSc/Civ Engg/Durham Univ. *Born:* 2/9/28. Native of South Shield, England.

Jones, William F (Continued)
Rotary Intl Fdn Fellow 1949-50. With Waterhouse and Rounthwaite (England) 1950-52; CA Inst of Tech and Converse Fdn Engg Co 1952-55; Skidmore, Owings and Merrill 1955-58; Testing and Controls, Inc and Gribaldo, Jacobs, Jones and Assoc 1958-73; William F Jones, Inc, 1973 to date. Since 1967 have served as VChrmn. NSPE/PEPP; Pres, CA Soc of PE and VP NSPE. Co-author of a number of geotech papers. Private pilot. *Society Aff:* NSPE, SSA, CEC, NAFE.

Jong, Mark M T
Business: 1845 Fairmount, Campus Box 44, Wichita, KS 67208
Position: Prof *Employer:* Wichita State Univ *Education:* PhD/EE/Univ of MO-Columbia; MS/EE/SD School of Mines & Tech; BS/EE/Natl Taiwan Univ *Born:* 10/20/37 Native of Taiwan. Taught at Natl Taiwan Univ and Univ of MO-Columbia prior to completing PhD in 1967. With Wichita State Univ since then. Is Prof and Graduate Coordinator in Electrical Engrg Dept. Main interests in circuit and system theory, signal processing and energy research with over 30 publications and technical reports. Author of textbook, *Methods of Discrete Signal and System Analysis*, McGraw-Hill 1982 copyright. Sr member of IEEE; served as officers in IEEE Wichita Section. *Society Aff:* IEEE, ASEE

Jonke, Albert A
Business: 9700 S Cass Ave, Argonne, IL 60439
Position: Prog Dir. *Employer:* Argonne Natl Lab. *Education:* MS/ChE/IIT; BChE/ChE/Cleveland State Univ. *Born:* 11/27/20. With Allied Chem Corp in 1943 in production of chems. Joined Argonne Natl Lab 1947, engaged in chem process dev in nucl & fossil energy fields. Became Group Leader 1948, Sec Hd 1963, Assoc Div Dir 1975, Prog Dir 1979. Specialist in fluidization processes in nucl fuel cycle, fluidized bed combustion of coal, coal tech. Author over 100 papers & reports, holder of 15 patents. Mbr several profl societies. Fellow Grade 1970 ANS; Fellow Grade 1978 AIChE. Enjoy camping, gardening, traveling. *Society Aff:* AIChE.

Jonsson, Jens J
Home: 1710 N. Lambert Lane, Provo, UT 84604
Position: Prof of EE *Employer:* Brigham Young Univ *Education:* PhD/Purdue Univ; MS/ME/Purdue Univ; BS/EE/Univ of UT; BS/GE/Univ of UT *Born:* 4/4/22 Industrial experience with North American Aviation, Gen Elec Co, Convair Astronautics and Bell Telephone Labs. Consultant for UNESCO at Middle East Technical Univ, Ankara, Turkey and the Polytechnic Inst of Bucharest, Romania. Professional and education projects at Stanford, MI State Univ and Univ of MI. Prof of Elec Engrg at Brigham Young Univ from 1953 to present. Dir of Engrg Analysis Computer Ctr, 1964-1972. Two books plus 20 publications in professional journals. Recepient of award for teaching excellence, Western Elec Fund; Certificate of Recognition for Contribution to engrg, UT Engrg Council; Community Service Award, UT Section of IEEE. Registered PE, UT. Mbr of IEEE, ASEE, Sigma Xi, and Eta Kappa Nu. *Society Aff:* IEEE, ASEE, ΣΞ, HKN

Jonsson, John Erik
Business: 3300 Republic Natl Bank Tower, Dallas, TX 75201
Position: Honorary Dir. *Employer:* TX Instruments Inc. *Education:* BSME/ME/RPI. *Born:* 9/6/01. In 1930 joined Geophyscial Service Inc from which the technologically based, world wide TX Instruments Inc evolved. A key participant in TI's growth & dev, Jonsson served as Pres, 1951-58; Bd Chrmn, 1958-66; & Hon Chrmn, 1966-77. His avocation has been civic & educational service (from Mayor of Dallas, 1964- 71 to innumerable Trustee positions). His honors include ten hon degrees; the Natl Acad of Engrg Founders Medal; the Gantt & Hoover Medals; & Hall of Fame for Bus Leadership; John Ericsson Medal, American Society of Swedish Engrs, 1980; Swedish-Amer of the Year, 1983. *Society Aff:* NAE, AMA, SEG

Joplin, John F
Business: P.O. Box 36468, 8601 Boone Rd, Houston, TX 77036
Position: Chrmn *Employer:* Hutchison Hayes Intl, Inc *Education:* BS/ME/Rice Univ *Born:* 8/20/24 Pres - Oil & Gas Cryogenics, Inc; Pres - Turbines Hispana Oil & Gas Co; Executive VP - Oil & Gas Supply Co; Exec VP - Oil & Gas Manufacturing Co.

Jordan, Angel G
Business: Carnegie Mellon Univ, 5000 Forbes Ave, Pittsburgh, PA 15213
Position: Provost *Employer:* Carnegie-Mellon Univ. *Education:* PhD/EE/Carnegie-Mellon Univ; MS/EE/Carnegie-Mellon Univ; MS/Phys/Univ of Zaragoza, Spain. *Born:* Sept 1930. MS & PhD from Carnegie-Mellon Univ, MS Univ of Zaragoza, Spain. Native of Pamplona, Spain. Naturalized in 1966. Worked & taught in Ordnance Navy Labs in Madrid 1952-56; worked at Mellon Inst in Pittsburgh in 1958; with Carnegie Mellon Univ since 1959. Assumed headship of EE in 1969, Deanship of Engineering in 1979; Provost of CMU in 1983. Fellow IEEE; member APS & ASEE; Fellow AAAS; member National Academy of Engineering. Recipient of NATO Sr Sci Award 1976. Dir Keithley Instruments Inc, Cleveland OH; Dir, Command Systems Inc, Pittsburgh PA; Dir, California Micro Devices Inc., Milpitas Cal; Dir, Allegheny Heart Inst, Pittsburgh PA; Dir, Allegheny Singer Res Inst, Pittsburgh PA; Chrmn of the Bd, MPC Corp, Pittsburgh PA; Dir, Pittsburgh High Tech Council; Trustee Lord Corp Foundation, Erie PA. *Society Aff:* IEEE, APS, ASEE, AAAS, NAE.

Jordan, Bernard C, Jr
Business: 279 Lower Woodville Rd, PO Box 1267, Natchez, MS 39120
Position: Partner. *Employer:* Jordan, Kaiser & Sessions. *Education:* BS/Civ Engg/Auburn Univ. *Born:* 7/20/15. Native of Natchez, MS. Grad of Auburn Univ 1938. 1939-41: Plant Engr, The TX Co, Bayonne, NJ 1941-46: US Army Corps of Engrs, Maj 1946-49: Dir of Public Works and City Engr, Natchez, MS 1949-51: Ind Engr, Intl Paper Co, Plant Engr 1951 to present: Private practice of Civ Engg. Design and construction of rds, bridges, ports, drainage systems, water and sewer systems and waste disposal, etc. Reg MS, LA & AL. *Society Aff:* ASCE, NSPE, NACE.

Jordan, Edward C
Business: Dept of Electrical Engrg, 1406 W Green, Urbana, IL 61801
Position: Prof Emeritus *Employer:* Univ of IL *Education:* PhD/EE/OH State Univ; MS/EE/Univ of Alberta; BS/EE/Univ of Alberta *Born:* 12/31/10 in Edmonton, Alberta, Canada. Control Operator, Radio Station CKUA, 7 years. Electrical Engr, Intl Nickel Co, 2 years. Taught electrical engrg at Worcester Polytechnic Inst, OH State Univ, and Univ of IL. Research on antennas and radio direction finding 1938-1960. Head of EE Dept, Univ of IL, 1954-1979. Fellow IEEE. Chrmn, Antennas and Propagation Society 1960. Chrmn, USNC/URSI 1966-70. IEEE Education Medal 1968. OSU Centennial Award, 1970. Eminent Member Eta Kappa Nu, 1974. Mbr, NAE. *Society Aff:* NAE, IEEE, HKN, TBPI

Jordan, Frederick E
Business: 111 New Montgomery St, San Francisco, CA 94105
Position: President *Employer:* Jordan Associates, Inc *Education:* MS/CE/Stanford Univ; BS/CE/Howard Univ. *Born:* Apr 27, 1937. Native Wash DC. BSCE Howard Univ; MSCE Stanford Univ. Reg Civ & PE in CA & 15 other states. Since 1968 directed consulting engg firm of Jordan/Avent & Assoc Inc with offices in CA, WA, MO, IL, on bridges, bldgs, hgwys, utilities, rapid transit, wastewater treatment & environmental projs. Worked for 6 organizations prior to private practice. Mbr 8 societies. Past Secy & Dir of San Francisco Sect ASCE, Founding Mbr & Pres of Engg Soc Ctte for Manpower Training Inc, Northern CA Council of Black PE, Western Assn of Minority Cons Engrs & Natl Council of Minority Cons Engrs. Numerous awards & articles.

Jordan, James J
Home: 5236 Overbrook Wy, Sacramento, CA 95841
Position: Owner *Employer:* J.J. Jordan Arch-Engr *Education:* BA/Architecture/OK State Univ; BS/Arch. Engg/OK State Univ. *Born:* Jan 7, 1926. Post-grad OK Univ 1959 & 1962, CA State Univ Sacto 1968-69, Air Univ Dayton 1974 & 1975. Navy 1944-46; Arch Engr, Ch of Design Tinker AFB OK 1952-65; Private A-E Practice

Jordan, James J (Continued)
OK City 1957-65. Pres Tinker Soc PE 1962; Rec'd USAF Civ Engg Meritorious Achievement Award for Prof Excellence Wash DC 1964; State PEG organizing Chrmn OK Soc PE 1965; Sacto Soc Mil Engrs Bd of Governors 1966; Pres McClellan Soc PE 1967; Sacto Valley Jt Engr Council Bd of Gov 1968; Reg PE OK & Mbr NSPE 1955 to 81; Licensed Architect OK 1957 to 81 CA since 1975. Articles published in many publications regarding computer Aided Design since 1981. Copywrite owner at 6 computer programs.

Jordan, Kenneth L, Jr
Home: 8308 Still Spring Ct, Bethesda, MD 20817
Position: VP *Employer:* Science Applications, Intl Corp. *Education:* ScD/EE/MIT; SM/EE/MIT; BEE/EE/RPI. *Born:* 5/10/33. Native of Portland, Maine. Attended Univ of Paris as Fulbright Scholar, 1956-57. Worked at MIT Lincoln Lab from 1960-73. Became Group Leader in 1968. Specialized in Communications and space systems. Worked as Dir, Strategic and space systems in the office of Secretary of Defense (C3I) from 1973-74 and Asst Dir (Systems) from 1974-76. Became Prin Deputy Asst Secy of Air Force (R&D) from 1976-79. Now works as VP (C3I) for Science Applications Intl Corp. *Society Aff:* IEEE, AIAA.

Jordan, Mark H
Business: 256 Broadway, 3d Floor, Troy, NY 12180
Position: Prin. *Employer:* self. *Education:* PhD/Mgt/Rensselaer Poly Inst; MCE/Structures/Rensselaer; BCE/Sanitary Engg/Rensselaer; BS/Naval Sci & Engg/US Naval Academy. *Born:* 4/10/15. Lawrence, MA. Educated USNA, Rensselaer, Cornell, George Washington. Commissioned ensign USN 1937, advanced to capt (CEC) 1955. Commanded 6th Construction Battalion 1943-44, 103rd Construction Battalion 1951-52, Civ Engr Corps Sch 1960-63. Div dir BuDocks (now NAVFAC) 1944-46, 1958-59. Directed design, construction Atlantic overseas areas 1952-56, Washington (DC) Naval Dist 1959-60. Directed construction, public works Boston Shipyard 1946-48, Bayonne Naval Base 1949-51. Assoc prof Univ of MO 1966-67. Prof Rensselaer 1967-77; dean continuing studies 1967-72; chrmn civ engg 1972-73; prof emeritus 1977. consulting engr, Albany, NY, 1976-85; Troy, NY, 1985- . Construction arbitrator. Author numerous papers. Mbr natl engg societies, natl committees. *Society Aff:* ASCE, NSPE, APWA, SAME, ACEC, CSI, AAA

Jordan, Michael A
Business: 3666 Grand Ave, Oakland, CA 94610
Position: Partner-Structural Engr. *Employer:* Liftech Consultants Inc *Education:* MS/Struct Mech/UC Berkeley; BS/CE/UC Berkeley. *Born:* Nov 2, 1934. BS Civ Engg, Univ of CA, Berkeley 1956; MS Struct Mech, Univ of CA, Berkeley 1961. Spec in complex struct design in analysis. Designs incl crane & rigging struct, seagoing struct, roll-on, roll-off facil, container terminals, & tramways, utilizing computer tech & appl math. Dockside cranes (up to 1400-ton capacity) are located in N Am, S Am, Europe, Asia, Africa, Australia, & the British Isles. Cited by "Engg News Record" as one of the "Men Who Made Marks" in the constr indus in 1971. Designed the container cranes & roll-on, rool-off struct on the Pt of Anchorage dock facil. The designer of the facil received the Grand Conceptor Award in 1975 by CEC. Mbr Bd/Dir Oakland Symphony Assn. Enjoys skiing, backpacking, & sailing. *Society Aff:* AISC, AITC, ACI, SEAOC.

Jordan, Richard C
Home: 1586 Burton St, St Paul, MN 55108
Position: Assoc Dean. *Employer:* Univ of MN. *Education:* PhD/Mech Engg/Univ of MN; MS/Mech Engg/Univ of MN; BAeroE/Aero Engg/Univ of MN. *Born:* 4/16/09. Instr, Petrol Engg, Univ of Tulsa, 1936-7; Instr, Asst Prof, Assoc Prof, Prof, Univ of MN 1937-; Hd, Dept Mech Engg, Sch Mech & Aerospace Engg, 1950-77; Assoc Dean, Inst of Tech, 1977-; Chrmn, NSF, Engg Sci Advisory Panel, 1954-7; Mbr and Chrmn, US-Brazil Committee on Ind Res, 1967-9; VP, Scientific Council and Exec Committee, IIR, 1967-71; Honorary Mbr, IIR; Recipient, F Paul Anderson Medal, Campbell Award, Outstanding Publications Key Award; Fellow and Presidential Mbr, ASHRAE, Fellow, ASME and AAAS; Natl Acad of Engg; Engr of Yr, MN, 1972; Outstanding Achievement Award, Univ of MN, 1979; Author 215 technical publications and books. *Society Aff:* AAAS, ASHRAE, ASME, NSPE, MSPE.

Jordan, William D
Home: 1501 High Forest Dr N, Tuscaloosa, AL 35406
Position: Prof. *Employer:* Univ of AL. *Education:* PhD/Theor & Appl Mech/Univ of IL; MS/Civ Engg/Univ of AL; BS/ME/Univ of AL. *Born:* 2/5/22. Native of Pickens Co, AL. Served with US Army Corps of Engrs & Ordnance Dept 1942-46. Retired as Col, US Army Reserve 1973. Taught Engg Mechanics at the Univ of AL 1946-86. Assumed responsibility as dept hd in 1961. Retired in 1986. *Society Aff:* SES, ASEE, AAM, NSPE, ASME.

Jorden, James R
Business: P.O. Box 2463, Houston, TX 77001
Position: Manager, Petroleum Engineering Research Dept. *Employer:* Shell Development Co. *Education:* BS/Petro Engrg/Univ of Tulsa *Born:* 4/16/34 Manager, Petroleum Engineering Research Dept, Shell Development Co. Joined Shell in 1957. Served with USAF 1957-1960. Held engrg positions, both technical and supervisory, of increasing responsibility in both research and operations organizations of Shell. Operations assignments were in petroleum E&P activities in LA, Gulf of Mexico, Rocky Mountains, CA, AK. Petrophysical Engr. Advisor in corporate Head Office of Shell Oil Co. Member of SPE. Co-author SPE Well Logging Monograph. SPE Annual Meeting general chrmn 1981. Member SPE Bd of Dirs, 1979-1982. SPE Pres, 1984. Ferguson Medal Committee Chrmn 1974. SPE Well Logging Technical Program Committee Chrmn 1977. Holds two patents. Author of several publications in SPE technical literature. Member of industry advisory committees for petr engr departments of U. of Tulsa and Colorado School of Mines. 1985 inductee into U. of Tulsa College of Engr Hall of Fame. *Society Aff:* SPE, SPWLA

Jorgensen, Gordon D
Business: 3003 N Central 1507, Phoenix, AZ 85012
Position: Partner. *Employer:* R W Beck & Assoc. *Education:* BSEE/Power/Univ of WA; Grad Studies/Bus & Mgt/Univ of WA. *Born:* 4/29/21. Grad studies Bus-Mgt Univ WA. Experienced in utility operations, including organization and mgt, fiscal matters, elec transmission and distribution, power supply, system planning and design, feasibility studies, valuations and rates. Partner in the Firm and mgr of its Phoenix office; also VP for intl operations. Sr Mbr Inst of Elec & Electronics Enrs (p Secy-Chrmn), also Natl Society PE, AZ Consulting Engrs Assn. Reg PE 10 states. Author numerous papers power utility supply and operations. Active in tennis- natl office US Tennis Assn. Cited by Gov of Honduras - reorganization of power complex. *Society Aff:* IEEE, NSPE, ACEA, ISES.

Jorgensen, Ib Falk
Business: 3 Park Central, Suite 625, 1515Arapahoe, Denver, CO 80202
Position: Chrmn of the Board *Employer:* Jorgensen and Hendrickson Engrs, Inc. *Education:* MS/Structural Eng/Tech Univ of Denmark; MS/CE/Tech Univ of Denmark *Born:* April 1922 Aalborg, Denmark. Immigrated to the USA in 1951, founded consulting engrg business, whose activities are: Structural design, specifications and supervisions for all types of structures including commercial bldgs, hospitals, schools, churches, public bldgs, towers, parking garages and special bldgs requiring complex structural analysis such as shells and aircraft hangars. Cvl engrg, bridges and water storage reservoirs. Industrial engrg including material handling, structures and facilities. Engrg evaluations. Investigations, reports and court testimony. Computer facilities. Consultants to private clientele, architects, engrs, contractors, municipalities, and government agencies. Licensed in several states. Past Chrmn of ACI Committee 422, Lateral Loads on Multistory Structures. Past Chrmn of ACI Committee 344, Prestressed Circular Structures, ACI; Past Natl Dir, ACI; Past President of Rocky Mountain Chapter of American Concrete Institute; Past Chairman of Rocky Mountain section of American Society for Testing and Materials. Past President of Structural Engrs Association of Colorado; enjoys skiing, sail boating and tennis. *Society Aff:* ACI, ASCE, ACEC, NSPE, AWWA

Jorgenson, James L
Home: 2927 Edgemont St, Fargo, ND 58102
Position: Prof of Civ Engg *Employer:* ND State Univ. *Born:* 5/30/36 BSCE Univ of ND 1958, MSCE Northwestern Univ 1959, PhD CE Purdue Univ 1966. Native of ND. Practiced structural engg with a consulting firm in Minneapolis in 1959-60 & with Kaiser Engrs in Oakland CA in 1969-70. Taught structural engg & fdn engg courses at NDSU from 1960-63 & again from 1965-69. Concurrently conducted res & wrote publications on: subgrade strength, subrage compaction, and statistical aspects of construction specifications. 1970 to present, Prof. of Civ Engg at NDSU. Also conducting res on pavement performance, strength of light gage steel bldgs & collapse load on a reinforced concrete bridge. Currently active as a structural & fdn engg consultant.

Jorne, Jacob
Business: Dept of Chemical Engrg, Univ Rochester, Rochester, NY 14627
Position: Prof *Employer:* Univ of Rochester *Education:* PhD/ChE/Univ of CA-Berkeley; MSc/ChE/Technion, Israel; BSc/ChE/Technion, Israel *Born:* 7/24/41 in Israel. Served as a Lieutenant in the Armor Corps of the Israel Defense Forces, 1963-66. Research Asst, Lawrence Berkeley Lab, 1967-1972. Joined Wayne State Univ in 1972 as an Asst Prof, and became Full Prof in 1981. Joined Univ of Rochester as a Full Prof in 1982. Major fields of research are electrochemical engrg, microelectronics processing, high energy batteries and fuel cells, oscillating chemical reactions, and the stability of ecosystems. Serves as a consultant in the area of energy conversion and storage. Chrmn of the Detroit Section of the American Inst of Chemical Engrs 1980-81. Recipient of the Chemical Engr of The Year Award, AIChE Detroit Section, 1979. *Society Aff:* AIChE, ECS

Joseph, Thomas L
Home: 1583 Northrup St, St. Paul, MN 55108
Position: Consulting Metallurgist (Self-employed) *Born:* Beaver, Utah & received Bachelor & Masters degrees from the University of Utah. During one half century with the U S Bureau of Mines, The University of Minnesota and self-employment, attention was focused upon correcting the ineffective gas-solid contact, associated with inadequate size preparation of the burden. High top pressure, adequate crushing, agglomeration by sintering and/or pelletizing were recognized the world over as ways to decrease labor, capital and transportation costs by increasing productivity. The benefits of high top pressures, high blast temperatures, sized burdens, fuel injection and conservation were recognized by foreign and domestic lecture tour and by a series of Honors and Awards.

Joshi, Aravind K
Business: , Philadelphia, PA 19104
Position: Prof & Chrmn *Employer:* Univ of PA *Education:* PhD/EE/Univ of PA; MS/EE/Univ of PA; BE/ME & EE/Univ of Poona-India *Born:* 8/5/29 Dr. Joshi is a Prof and the Chrmn of the Dept of Comp and Info Sci; he is also a co-dir of the Cognitive Sci Program. His maj res areas are natural language processing, artifical intelligence, and theory of computation. He was a Guggenheim Fellow in 1972, a mbr of the Inst for Advanced Study, Princeton in 1972, elected a Fellow of the Inst of Elec and Electronic (IEEE) in 1975, and elected as the Pres of the Assoc for Computational Linguistics (ACL) for 1974- 75. In June 1983, he was appointed as the Henry Salvaton Prof of Computer and Cognitive Sciences. *Society Aff:* IEEE, ACM, LSA, CSS

Joslyn, John A
Business: 50 Fordham Rd, Wilmington, MA 01887
Position: Mgr-Engg. *Employer:* GE Co. *Born:* Sept 1931 St Louis MO. Received a BEE from Yale Univ in 1953 & an MEE from Syracuse Univ in 1958. Served as a commissioned officer in the USN 1953-56. With GE since 1953 initially specializing in electronic design for Navy missile & gun control systems. First engr mgt assignment occured in 1967. Assumed current responsibility as Mgr, Engg for design of aircraft instruments in early 1973. Sr Mbr IEEE since 1970. Enjoy skiing, golf, & tennis.

Jost, H Peter
Business: Angel Lodge Labs, Nobel Rd, Edmonton, London England N18 3DB
Position: Chairman & Managing Director. *Employer:* K. S. Paul Products Limited. *Education:* Hon DSc/-/Univ of Salford; Hon D Tech/Major/Council for National Academic Awards (CNAA) (Liverpool Polytechic) *Born:* 1/25/21. Appr Associated Metal Wks, Glasgow. D Napier & Son Liverpool. College of Technology Manchester. 1943-44 Methods Engr K & L Steelfounders & Engrs; 1945-49 Ch Planning Engr, Datim Machine Tool Co Ltd, London; 1949-55 Genl Mgr, 1951 Dir, Trier Bros Ltd, London; from 1955 Managing Dir, 1973-77 Chmn, Centralube Ltd, London; from 1955 Managing Dir, 1974 Chmn, K S Paul Products Ltd, London; from 1971 Dir, Stothert & Pitt Plc Bath; 1965 Chmn Lubrication Working Group, Dept Ed & Sci, UK; 1965 Chmn Lubrication Working Group, Dept Ed & Sci, UK; "Jost" Report 1965; 1966-72 Chmn Ctte on Tribology Ministry of Tech, Dept of trade & Indus; 1972-74 Dept Chmn Ctte for Incus Tech, Dept of Trade & Indus; Mbr Ct 1970- & Council 1974-84, Univ of Salford; Vice President 1987; Mbr Inst of Mech Engr of Council 1974, Pres 1977-78 Inst of Prod Engrs; Mbr of Council 1974 Mbr of Finance Bd; Mbr of Disciplinary Board; Inst of Mech Engr; Member of Board 1977-83. Elected Member of Executive 1979-83. Chrmn of Home Affairs Committee 1980-83; Council of Engrg Inst; Pres intl Tribiology Council 1973- . Chrmn of Engrg & General Equip Ltd 1977- . Chrmn of Associated Tech Group Ltd 1976- . *Society Aff:* IME, !PE, ASME, IM

Joy, Edward B
Business: School of Electrical Engrg, Georgia Inst of Tech, Atlanta, GA 30332-0250
Position: Tenured Prof *Employer:* GA Inst of Tech *Education:* PhD inEE/Elect Engr/GA Inst of Tech; MSEE/Elect Engr/GA Inst of Tech; BEE/Elect Engr/GA Inst of Tech. *Born:* 11/15/41 in Troy, NY. Served in the US Navy as electrical division officer aboard an aircraft carrier after receiving Bachelor degree. Was Schlumberger Fellow, NASA Fellow and Scientific Atlanta Fellow in graduate school. Since 1970, has been conducting research and teaching in electromagnetics at GA Tech. He is author of more than 130 technical journal papers and major research reports in the areas of near-field antenna measurements, radome analysis and grounding systems. He has been the principal investigator and project dir on over 35 sponsored research programs. He is a consultant to several US and European companies in the areas of antenna measurements and radome design. He is a Sr member of the IEEE and Technical Coordinator of the Antenna Measurement Techniques Association. *Society Aff:* IEEE APS, AMTA

Ju, Frederick D
Business: Mech Eng Dept, UNM, Albuquerque, NM 87131
Position: Presidential Prof. *Employer:* Univ. of New Mexico *Education:* PhD/TAM/Univ. Illinois; MS/Mech Eng/Univ. Illinois; BS/ME/Univ. Houston *Born:* 09/21/29 Born in Shanghai, Republic of China. Attended Univ. of Amoy & Univ. of Taiwan prior to immigration to U.S. Chn, Mech Eng. Dept, UNM, 1973-76. Presidential Prof, UNM, 1985 to present. Fellow, ASME, 1987. Consultant, Los Alamos Nat'l Lab. Specialty in Solid Mechanics and Dynamics. Current researches in damage mechanics: fracture damage diagnosis, thermomechanical cracking in tribological environment, and solid propellant hazard. Enjoys classical music. *Society Aff:* ASME, SEM, ΣΞ, SES

Judd, Gary
Business: Pittsburgh Bldg, Troy, NY 12181
Position: VProvost, Dean of Grad Sch. *Employer:* RPI. *Education:* PhD/Mtls Engg/RPI; BS/Mtls Engg/RPI. *Born:* 9/24/42. Became a faculty mbr in Mtls Engg at RPI in 1967, an assoc prof in 1972, and a prof in 1976. Res areas are physical met, elec-

Judd, Gary (Continued)
tron optics instrumentation techniques, & biomtls. Became Acting Chrmn of Mtls Engg in 1974 & VProvost for Plans & Resources in 1975, Acting Provost 1983 and 1985. Assumed current position of VProvost for Academic Affairs & Dean of the Grad Sch in 1979. Am responsible for the Registrar, Academic Support services, Grad Admissions, Grad Sch progs and financial aid. Tau Beta Pi; Sigma Xi; Alpha Sigma Mu; Phi Lambda Upsilon; Geisler Award, Eastern NY Chapter, ASM. *Society Aff:* ASM, AIME, MAS, ASEE, AAAS

Judd, William R
Home: 200 Quincy St, W Lafayette, IN 47906
Position: Prof Rock Mechanics. *Employer:* Purdue Univ. *Education:* AB/Geology/U of CO; None/Civil Engineering/U of Co Graduate School *Born:* 8/16/17. Prof of rock mechanics & legal aspects of engrg, Purdue Univ & Former Hd, Geotechnical Engrg Dept; former hd, Basing Tech, The RAND Corp; former liaison engrg geologist, US Bureau of Reclamation. Geotechnical conslt to domestic & foreign agencies & cos since 1950. Editor-in-Chief, "Engineering Geology" since 1972. Former Chairman, NRC US Natl Ctte on Rock Mechanics. Former Exec Council, US Ctte on Large Dams; Natl Res Council Ctte on Safety of Dams; Panel of Arbitrators, American Arbitration Assoc; Geologic Review Group, Office of Nuclear Waste Isolation. Honors: Alex du Toit Memorial Lecturer (S Africa & Rhodesia), Sigma Xi, US Bureau of Reclamation Merit Awd, US Natl ctte on Rock Mechanics Special Awd. Reg Engr, Certified Engrg Geologist. *Society Aff:* ASCE, GSA, SAIMM, AEG, IAEG, ISRM, ISEG.

Juergens, Robert B
Business: 4405 Talmadge Rd, Toledo, OH 43623
Position: Proj Coordinator. *Employer:* Finkbeiner, Pettis & Strout Ltd. *Born:* Feb 1923. Attended Univ of Toledo; BS Marquette Univ. Native of Toledo OH. Served with Navy Civ Engr Corps (Seabees) 1943-46 & 1951-52. With Finkbeiner Pettis & Strout Ltd since 1947. Assoc Engr 1958. Partner 1971. Reg PE OH, MI, IN, WV, KY, IA. Fellow ASCE Secy-Treas 1959-60; Mbr Natl, OH & Toledo Socs of PE, Am Concrete Inst, Am Water Works Assn, Water Pollution Control Fed, Am Soc for Testing Mtls & Cons Engrs of OH. Served on var Toledo YMCA Bds the past 15 yrs.

Jumikis, Alfreds R
Home: 817 Hoes Lane, Piscataway, NJ 08854
Position: Prof Emeritus of Rutgers Univ *Employer:* Retired *Education:* Dr Eng Sc/Soil Mech/U of Latvia; Dr Tech/Foundation Engrg/Tech U of Vienna, Austria; Dr Ing/Thermal Soil Mech/U Stuttgart, West Germany. *Born:* 12/7/07 Since 1952, teaching at Rutgers Univ. Retired in 1978 as a Distinguished Prof, to become a Prof Emeritus of Rutgers Univ. Dir of the Joint Highway Research Project between Rutgers Univ, NJ State Highway Dept, and the Bureau of Public Roads in Washington, DC, 1952-55. Dir of the Exec Committee of the Foundations and Soil Mechanics Div of the Metropolitan Section (NY) of the ASCE (1962-65). Licensed PE in the states of DE, NJ and NY. Authored 115 publications on Soil Mechanics, Rock Mechanics, Thermal Soil Mechanics and Foundation Engrg. Recipient of 3 Natl Science Foundation research grants. A two-fold Laureate of Latvian Culture Fund in the Western Hemisphere. *Society Aff:* ASCE, NSPE, ISRM, NYAS, ASEE, USCOLD, AAUP, ΣΞ.

Jumper, Billy S
Business: 1111 W Loop South, Houston, TX 77027
Position: VP. *Employer:* Caudill Rowlett Scott. *Born:* Sept 1936 Lawton, OK. BSCE Univ of NM 1960. Field Engr US Bureau of Reclamation Glen Canyon Unit 1960-61. Field Engr Robet E McKee Gen Contr Inc 1961-64. Struct Engr & Proj Engr with Bovay Engrs Inc industrial, institutional, commercial, govt & airport facilities 1964-67. Private Cons Engr to architects & industry 1967-70. Struct Engr, Caudill Rowlett scott 1970-, educ, medical & commercial facil. Assumed present duties as VP & Mgr of Struct Engg 1973. Lic Engr 12 states. Practice has included considerable intl work. Mbr NSPE, TSPE, ACI, AISC. Prog Committeeman Northwest Houston TSPE 1973. Prog Chrmn & VP Houston Chapter ACI 1975. ACI Chapter Pres 1976.

Jung, John B
Business: 55 S Richland Ave, PO Box M-55, York, PA 17405
Position: Pres-Treas. *Employer:* Buchart-Horn, Inc. *Education:* BS/Civ Engrg/U of IL; (Undergraduate)/Civ Engrg/U of AZ *Born:* 3/28/26. Born and raised in Waterloo, IL. Mbr of college varsity baseball team and Kappa Sigma fraternity. Experience includes: bridge construction engr, Modjeski and Masters (1949-51); guided missile proj engr, US Army (1952-53); Design engr and sr proj engr, Michael Baker, Jr, Inc (1954-66). Joined Buchart-Horn in 1966 as asst vp, promoted to vp in 1970 exec vp in 1971. Elected pres and treas in 1976; responsible to the bd of dirs for all operations and profit. Also pres of Lisiecki, Dorsey, Kohler & Purdy, Inc. (Baltimore, MD), affiliated co. Reg PE in 12 states and DC. Also, president of Buchart-Horn, Inc. of Virginia (affiliated company), member of Bd of Dirs of Basco Assoc, Inc. (affiliated company), mbr of Bd of Dirs of PACE Resources, Inc. (parent company), mbr of Bd of Dirs of B-H Computer Systems, Inc. (affiliated company). *Society Aff:* ARTBA, ASCE, ASHE, ACEC, NSPE

Junger, Miguel C
Home: 90 Fletcher Rd, Belmont, MA 02178
Position: Pres/Principal Scientist *Employer:* Cambridge Acoustical Assoc Inc. *Education:* ScD/Appl Mech/Harvard Univ; SM/ME/MIT; BS/ME/MIT *Born:* in Germany in 1923, brought up in Spain & France, came to US 1941. Grad MIT in 1944 ME, 1946 MS Worked 3 yrs with C E Crede on vibration and shock isolation. Gordon McKay Scholar at Harvard (1950-51), ScD (Appl Mech 1951). Worked 4 yrs in underwater acoustics with F V Hunt as Postdoctoral Res Fellow, Acoustics Res lab, Harvard. 1955, founder ptnr, Cambridge Acoustical Assocs, consulting firm primarily active in underwater acoustics & noise control. Pres since 1959 when firm was incorporated. Part-time position: Sr Visiting Lecturer, Dept Ocean Engg, MIT, 1968-78; Office of Naval Res expert to evaluate acoustics res in Europe 1975; visiting prof, Universite de Technologie de Compiegne, France, 1975-77. Res: Vibrations & sound radiation & scattering by submerged structures; underwater sound sources; waveguides. Developed sound absorptive concrete block (–Soundblox–) Published in J Acoust Soc Americ, J Appl Mech, J Noise Control Engg, USN J Underwater Acoustics, ASME Colloquium proceed. Books: *Sound, Structures, and Their Interaction*, 2nd. ed., with D Feit (MIT Press, 1986), & *Elements d'Acoustique Physique*, with M Perulli (Maloine, Paris, France, 1978) *Society Aff:* ASA, AAM, ASME

Juran, Joseph M
Business: 866 United Nations Plaza, New York, NY 10017
Position: President *Employer:* Juran Enterprises, Inc. *Education:* BS/EE/Univ of MN; JD/Law/Loyola Univ. *Born:* 12/24/04. Sometime engr, ind exec, corporate dir, govt admin, univ prof. Since 1945, consultant to ind & govt insts. Author 10 books & numerous papers, principally in quality control, widely used as references & texts. Conducted training courses in quality control for over 20,000 mgr in over 30 countries. Numerous honors, awards. *Society Aff:* ASME, IIE, ASQC, AAAS.

Jury, Eliahu I
Business: Dept of EE and Comp Equip, Coral Gables, FL 94720
Position: Prof Elec Engg. *Employer:* University of Miami *Born:* May 1923 Iraq. ScD Columbia Univ, MS Harvard Univ, EE Israel Inst of Tech, Haifa Israel. Mbr of faculty at Univ of CA Berkeley since 1953. Res Engr Columbia Univ 1953-54. Expert Examiner CA State Bd of Reg of PE 1954-55. NSF Sr Postdoctoral Fellow 1965-66. Fellow IEEE 1968. Mbr NY Acad of Sci 1974. Listed in Am Men of Sci. Selected by the Natl Ctte of the USSR on Automatic Control to be listed in the Guide book "Who's Who in Automatic Control–. VP Sigma Xi Soc 1976. President (Berkeley Chapter) 1977. Senior Fulbright-Hays Fellow, 1979. ASME Centennial Medalist,

Jury, Eliahu I (Continued)
Hon mbr Sigma Xi, Hon Doctor of Tech Sciences, 1982, Swiss Fed Inst of Tech (ETH) Zurich, Research Prof Univ of Miami, since 1982. The First Recipient of IEEE Circuits and Systems Educ Award (1985). Rufus Oldenburger Medalist for the ASME, (1986).

Jury, Stanley H
Home: 6008 Kaywood Dr, Knoxville, TN 37920
Position: Prof Emeritus. *Employer:* Univ of TN. *Education:* PhD/ChE/Univ of Cinn; MS/ChE/Univ of Louisville; BChE/ChE/Univ of Cinn. *Born:* 3/3/16. Ordnance officer, rank of Maj, four & one half yrs European Theatre, WWII, bronze star. Engr & officer of Polacoat, Inc. Taught chem engg Univ of TN (1949- 1978). Dir NED of AIChE. Active in local section activities. Elected Fellow of AIChE. Consultant for Oak Ridge Natl Lab, Wright Patterson Air Base, Aero Corp, Honeywell, Dorr Oliver Co, Cherokee Explosives, GE. Guest lectr Oak Ridge Inst for Nuclear Studies, Gulf Oil, Bacteriological Warfare Ctr among others. *Society Aff:* AIChE, AAUP, ASEE, ΣΞ.

Justin, Joel B
Home: 2401 Pennsylvania Ave, Philadelphia, PA 19130
Position: Consulting Engr (Self-employed) *Education:* BS/CE/Cornell Univ. *Born:* 11/23/07 Various construction projects as Field Engr prior to 1932. TVA as Field Engr on large Hydro projects - 1933-1938. Ten years Mgr, Hydro Plants for Appalachian Electric Power Co. Partner Justin-Courtney Consulting Engrs - 1948. Pres of Justin-Courtney Consulting Engrs - 1973. Independent Consultant since 1978. Fifty years experience in development, design construction and management of Water Resource projects, domestic and foreign. These have involved multi- purpose projects for water supply, irrigation, flood control, hydroelectric power and pumped-storage projects in USA, Iran, India, Pakistan, Angola, Korea and Dominican Republic. Registered Profl Engr in several states, contributor for papers ASME and ASCE. *Society Aff:* ASCE, ASME, USCOLD, USCIDF.

Justus, Edgar J
Business: One St Lawrence Ave, Beloit, WI 53511
Position: VP of Technology (Retired) *Employer:* Beloit Corp. *Education:* BME/-/GA Tech. *Born:* 5/23/17. Accumulated 32 yrs profl, and 20 yrs managerial experience as Dir of Res and Chief Engr at VP level. Holds 100 US patents on paper machines and automatic controls. Served as consultant relating to Patent disclosures; has experience in evaluation of Patent Office actions. Has given expert and tech testimony in patent/trade secret matters; has consulted on licensing and acquisition of patents. Author of 40 tech articles. Principle expertise involves: Paper machines - auto controls, hydraulics. Received TAPPI Fellow Award Feb 25, 1975. Responsible for worldwide R&D activities of Beloit Corp, Registered Professional Engr, Wicsonsin, Tappi Engrg Award 1980. *Society Aff:* ASME, SAE, TAPPI, NSPE.

Kabel, Robert L
Business: 164 Fenske Lab, University Park, PA 16802
Position: Prof of Chem Engrg. *Employer:* The Pennsylvania State University. *Education:* PhD/Chem Engr/Univ WA; BS/Chem Engr/Univ IL. *Born:* Apr 3, 1932. Married, two children. With Penn State since 1963. Fellow Amer Inst Chem Engrs (Chmn Ed Projs Ctte 1975-77, AIChE J Editorial Board 1980-85). Mbr Amer Chem Soc (Chmn Central Pa Sect 1970). Cons. AF Commendation Medal 1963. Fellow, Royal Norwegian Council for Sci & Indus Res 1971-72. Fellowship, NATO Advanced Study Inst 'Analysis of Fluid-Solid Catalytic Reactors' Belgium 1974. Visiting Lectureship, Pahlavi Univ, Shiraz, Iran, Apr 20 - May 2, 1978. Co-editor, Author: Scaleup of Chemical Processes, 1985. Outstanding Teaching Awd, Coll of Engrg, The PA State Univ, 1982. Amoco Fdn Outstanding Teaching Awd, The PA State Univ, 1983. Western Elec Fund Awd for Excellence in Instruction, Middle Atlantic Section, Amer Soc for Engrg Educ, 1983. CMA Natl Catalyst Awd for Excellence in Chem Teaching, Chem Mfgs Assoc, 1984. Premier Teaching Awd, Coll of Engrg, The PA State Univ, 1984. Distinguished Achievement Award, Arizona State University, 1985. Outstanding Advising Award, College of Engineering, The PA State University, 1986. Res int: Reaction kinetics, Adsorption, Heterogeneous Catalysis, Process dynamics, Mathematical modeling of natural processes, scaleup. Hon Soc Mbrship: Sigma Xi, Tau Beta Pi, Phi Lambda Upsilon. *Society Aff:* AIChE, ACS, ASEE, AAUP.

Kablach, Thomas L
Home: 294 E Deer Park Dr, Clairton, PA 15045
Position: Mgr of Quality Control. *Employer:* Bucyrus-Erie Co, Glassport Plant. *Born:* July 1925. BS Met Engrg Carnegie-Mellon 1950. Served with USAF 194445 in Europe. Prior tech pos: Met Engr Mesta Machine Co; Ch Metallurgist Titusville Forge Co; Tech Dir Roll Mgr Inst; Tech Dir Erie Forge & Steel Co. Presently Mgr of Quality Control Bucyrus-Erie, Glassport Plant. Respon for all tech activities. Formerly Chmn, Steel Res Comm, natl Amer Foundrymen's Soc. Formerly on natl Ed Comm of ASM. Formerly Chmn, NW Pa Chap of AFS. Served as chmn of many subcttes of local AFS & ASM. Presently Chmn of Prog Comm, Steel Div, natl AFS. Active with several ASTM cttes.

Kadaba, Prasad K
Home: 3411 Brookhaven Dr, Lexington, KY 40502
Position: Prof. *Employer:* Univ of KY. *Education:* PhD/Phys/UCLA; MS/EE/Caltech; MSc/Phys/Univ of Mysore; BSc/Phys/Univ of Mysore. *Born:* 2/14/24. Native of Bangalore, India. Res Scientist at Natl Physical Lab, New Delhi 1950-52. Asst Superintendent of Dev, Defense Res Lab, India 1952-53. Taught at Newark College of Engg 1957-59. Taught at Memphis State Univ 1967-68. With Univ of KY 1954-57, 1959-66 & since 1968. Intl Res & Exchanges Bd Fellow 1973. Sr Fullbright-Hays Res Scholar to Yugoslavia 1974. Visiting Scientist to NASA, Argonne Natl Lab, OARU & WPAFB. Author of several res publications in the areas of Dielectrics, Magnetic Resonance Conductive Polymers & Pollution. Principal Investigator of Res grants from various Fed & State agencies Government of India Merit Scholarship to the USA 1945-47. Outstanding Teacher Award, University of Kentucky 1979 Vis Prof under the Air Force Systems Command Univ Resident Res Prog 1980-81 and 1981-82. Visiting Scientist to Ft Belvoir Army Res Ctr. *Society Aff:* IEEE, HKN, ΣΞ, TBΠ.

Kadaba, Prasanna V
Business: School of Mech. Engg, Georgia Inst of Tech, Atlanta, GA 30332-0405
Position: Assoc Prof *Employer:* GA Inst of Tech *Education:* PhD/ME/IL Inst of Tech; MS/ME/Univ of KY; BS/ME/Univ of Mysore, India; BS/EE/Univ of Mysore, India *Born:* 7/4/31 Originally from India. Naturalized citizen of US since 1972. Taught at IL Tech during 1956 to 1963. Worked in Research and Development centers of Borg Warner and Westinghouse from 1963 to 1969. Since 1969 teaching in School of Mechanical Engrg of GA Inst of Tech. Specilizing in Thermal Sciences with emphasis in Thermal Component and System Modeling Fluidized beds, Refrigeration systems, solar assisted heat pumps, Airconditioning load analysis and system simulation, Energy conservation and Second law analysis. Held number of short time significant assignments at NBS (National Bureau of Standards), NASA, LBL (Lawrence Berkeley Laboratory), Copeland Corporation and Visiting Professor at Univ. of Carabobo, Valencia, Venezuela. Member of Energy Action Committee of GA Society of PE and Energy Engrg Multi-disciplinary program. Listed in Intl Who's Who of Intellectuals, American Men and Women of Sciences and several other biographical listings. Enjoy photography. *Society Aff:* ASME, ASHRAE, ASEE, ASTM, SAME, ΠΤΣ, TBΠ, ΣΞ.

Kaden, Richard A
Business: Bldg 602, City-County Airport, Walla Walla, WA 99362
Position: Civ Engr. *Employer:* US Army Corps of Engrs. *Education:* MS/Engrg/U of CA, Berkeley; BS/CE/WA State Univ. *Born:* 5/10/39. Spent more than 26 yrs in the design & construction of water resources projs. Field experience include areas on concrete tech; Geotech Engg, rock & soil mechs; & instrumentation & Measure-

Kaden, Richard A (Continued)
ments for heavy construction. Served 9 yrs as Chief, Mtls Sec, Fdns & Mtls Branch as the Dist's authority in the field of concrete tech & mtls. Served 2 yrs as Asst to chief, Engg Div & 6 months within the Office of Policy, HQDA Office, Chief of Engrs, Wash, DC. Past Pres, Columbia Sec, ASCE; VP Walla Walla Post, SAME; and Mbr of ACI Ctte 210, 506 & 117; R&D Award in 1976; SAME Service Award 1984. *Society Aff:* ASCE, ACI, SAME.

Kadey, Frederic L, Jr
Business: 7653 South Rosemary Cir, Englewood, CO 80112
Position: Conslt-Indus Minerals *Employer:* Self Employed *Education:* MA/Geology/Harvard Univ; BSc/Geology/Rutgers Univ. *Born:* 6/21/18. Raised in Buffalo, NY. Served in US Army, European-African theatre 1941- 1945. Decorated with Croix de Guerre. Teaching Fellow, Harvard Univ 1946-1947. Geological Field Asst, Sinclair Oil Co, 1946. Married Brenda Boocock, 1950; two children: Brenda Catherine (King) and Frederic Lionel III. US Steel Corp Res Ctr, 1947-1951. With Johns-Manville from 1951- 1983. Responsible from 1971-1983 for Ind Mineral Exploration. Natl Defense Exec Reservist. Metals and Minerals Branch, US Dept of Interior. Chrmn, Industrial Minerals Div of SME, 1977. Dir SME. Pres of SME 1984. Pres NY State Sec AIPG 1967-1968. Patentee perlite processing. Contributor of numerous articles on perlite and diatomite. Republican. Episcopalian. *Society Aff:* SME, AAAS, MSA, AIPG, ΣΞ, ΑΣΦ.

Kadota, T Theodore
Business: AT&T Bell Labs (2C356), b600 Mountain Ave, Murray Hill, NJ 07974
Position: Member of Tech Staff *Employer:* AT&T Bell Labs *Education:* PhD/EE/Univ of CA-Berkeley; MS/EE/Univ of CA-Berkeley; BS/EE/Yokohama Nat'l Univ-Japan *Born:* 11/14/30 After completion of the PhD program at Univ of CA, he has worked at AT&T Bell Labs: in Military Research Dept in Whippany (1960-1966) and in Math Sciences Research Center in Murray Hill (1966-present). During this period, he was a Visiting Prof at Stanford Univ, Univ of CA (Berkeley) and Univ of HI in 1974, 1975 and 1978 respectively. His research interest has been theory of stochastic processes and its application to signal detection, estimation and information theory and modelling of random phenomena in underwater acoustics and infrared optics. He is a Fellow of IEEE, a former Assoc Editor of IEEE Trans Information theory. *Society Aff:* IEEE

Kaenel, Reg A
Home: PO Box 2495, Humble, TX 77338
Position: Manager, Electrical Engg. *Employer:* NL Petroleum Services. *Education:* BSEE/-/Swiss Fed Inst of Technology; MSEE/-/Swiss Fed Inst of Technology; ScD/-/ Swiss Federal Inst of Technology. *Born:* 10/22/29. in Berne, Switzerland from Swiss parents; US Citizen. Asst at Inst of Telecommunications of ETH 1955-58. With Bell Tele Labs 1958-71; respon for efforts in sys engg & device dev. Also, Hd, Feasibility Studies with space prog. With AMF, Inc 1971-78, Dir of Elec Engg. Also, Ch Engr of automatic bowling scorer equip & cons to bus units. With NL Petroleum Services, Inc, 1978-1980, Mngr. of Electrical Engineering for Drilling Systems. With RAK Enterprises, Inc. , 1980- , COE; micro-computer-based systems. Pres IEEE-ASSP Soc 1970-72, Chmn IEEE Cttes. Pub in archieval journals, hold 35 pats. Fellow IEEE. Enjoy skiing, tennis, boating & flying. *Society Aff:* IEEE, IEEE, IEEEAESP, ACM.

Kage, Arthur V
Business: 3100 Broadway, Kansas City, MO 64111
Position: VP & Dist Mgr *Employer:* The Austin Co *Education:* BS/CE/KS State Univ *Born:* 4/13/23 After completing military service in WWII, I completed my formal education and joined The Austin Co as a designer in Civ Engrg. During the next 30 yrs, I moved through the various positions of Field Engr, Estimating, Purchasing, Proj Engr, Proj Mgr, Asst Dist Mgr, and presently hold the position of VP and Dist Mgr for The Austin Co Central Dist which involves 14 states. I am past Dir of City Fed Savings and Loan Assoc in NJ, Past Vestryman of St. Luke's Episcopal Church, and past Pres of the Clark NJ Municipal Pool Assoc, and have served actively in Scouting for seven yrs. *Society Aff:* NCEE, ASCE

Kaghan, Walter S
Business: Walter S Kaghan, Ph.D, Inc, Sarasota, FL 33583
Position: Pres. *Employer:* Walter S Kaghan, PhD, Inc. *Education:* PhD/ChE/Purdue Univ; MChE/ChE/NYU; BChE/ChE/CCNY. *Born:* 4/16/19. Active as a consulting chem engr serving the polymer, resins, films & packaging inds since 1975. Prior to that, employed for twenty-five yrs in Olin Corp Film Div as dir of res dev & engg. Earlier experience included employment as chem engr with E I DuPont, M W Kellogg & St Regis Paper Co & three yrs as asst prof of chem engg at Rose-Hulman Inst of Tech. *Society Aff:* AAAS, AIChE, SPE, ACS, TAPPI.

Kahan, William M
Business: Elect Engg & Comp Sci Dept, Evans Hall, Berkley, CA 94720
Position: Prof. *Employer:* Univ of CA. *Education:* PhD/Mathematics/Univ of Toronto, Canada; MA/Mathematics/Univ of Toronto, Canada; BA/Mathematics/Univ of Toronto, Canada. *Born:* 6/5/33. Has worked on computers since 1953 especially on design, error-analysis, validation & automatic diagnosis of numerical algorithms for scientific, engg and financial calculations. Teaching & research at univsities of Toronto, Cambridge England, Stanford, CA at Berkeley. Consultant on floating point hardware and support software for IBM, Hewlett-Packard calculators, INTEL, IEEE- CS standards. *Society Aff:* ACM, SIAM, AMS.

Kahl, William R
Home: PO Box 72, 4900 Kemp Rd, Reisterstown, MD 21136
Position: Retired *Education:* BE/Civil/JHU; MCE/Civil/Poly Inst Brooklyn. *Born:* 5/6/11. Inspector MD State Rds Comm 1929, 1930, 1934-35. C&GS MD 1933-34. Engr B&O RR 1935-40. Struct Engr Portland Cement Assn 1940-46. Ptnr, Rummel, Klepper & Kahl 1947-81. Pres MD Sect of ASCE 1954; Pres Balto Post SAME 1952; Pres Cons Engrs Council of MD 1957; Mbr ECPD Accreditation Team 1966-67; EJC of MD 1957- 66; Exec Reserve, Bureau Public Roads 1961-75; MD Bd of Reg of PE 1961-66; Review Bd Civil Service Comm Balto & State of MD 1956-63. Reg PE in 17 states. Episcopal Church, Mason, Shriner. Hobby, Antique Cars. *Society Aff:* ASCE, SAME, CEC.

Kahles, John F
Business: Metcut Research Associates Inc, 3980 Rosslyn Drive, Cincinnati, OH 45209
Position: Sr VP *Employer:* Metcut Research Associates Inc. *Education:* PD/Met Engrg/Univ of Cincinnati; BS/Ch Engrg/Armour Inst. *Born:* Sept 1914. Sr VP Metcut Res Assocs Inc, Cincinnati. Adjunct Prof, Met Engrg (Cincinnati). Mbr Tau Beta Pi, Sigma Xi, Phi Lambda Upsilon. Mbr numerous tech societies & active as officer, ctte chmn & ctte mbr at local & natl levels. Over 90 publs in fields of Met, Machinability, Surface Integrity & Info Tech. Reg PE Ohio. Fellow ASM; 1971 'Engr of Year', Cincinnati; Joseph Whitworth Prize 1968, Inst of Mech Engrs, England; 1972 Distinguished Alumnus Award (Cincinnati); 1973 Res Medal, SME, Natl Acad Engrg, 1984, Mbr, Intl Inst for Production Engrg Res., Fellow SME. *Society Aff:* SME, ASM, ASTM, ASIS, TMS, AFS, AWS, ASNT.

Kahn, Lawrence F
Business: School of Civil Engrg, Georgia Institute of Technology, Atlanta, GA 30332
Position: Assoc Prof *Employer:* GA Inst of Tech *Education:* PhD/CE/Univ of MI; MS/CE/Univ of IL; BS/CE/Stanford Univ *Born:* 1/26/45 Native of San Francisco, CA. Structural engr for 4 years at the US Naval Civil Engrg Lab specializing in the development of undersea concrete structures. Structural engr for one year with Bechtel Power Corp. Experimental research since 1972 concentrating on the repair and strengthening of structures. Current dir of CAE/CAD res and dev for GTSTRUDL, GTICES Sys Lab. Pres Atlanta Chapter ACI, 1982. ASCE Raymond

Kahn, Lawrence F (Continued)
C. Reese Research Prize, 1980. Chrmn ACI Tech ctte, Rehabilitation. Forensics conslt; Reg PE. *Society Aff:* ASCE, ACI, TMS

Kahn, Leonard R
Home: 137 E 36 St, New York, NY 10016
Position: President. *Employer:* Kahn Communications, Inc. *Education:* BEE/BEE/Polytech Inst of Brooklyn. *Born:* June 16, 1926. Adj Prof of Elec Engrg, US Army, RCA, Crosby Labs, Kahn Res Labs. Cons major firms & US Gov't. Patentee of communications & broadcasting sys. Pres Kahn Communications Inc: developer, manufacturer SSB equip, diversity equip, Symmetra-Peak, Voice-Line, Proline, CSSB, ratio sq equip & AM Stereo. Author over 20 papers. 1980 Armstrong Medal Fellow, p mbr Bd/Dir Radio Club of Amer. Pat Agent, Reg Prof Engr. Mbr Tau Beta Pi, Eta Kappa Nu, Sigma Xi. Fellow IEEE. Has been awarded over 70 U.S. Patents.

Kahn, Robert E
Home: 4819 Reservoir Rd, NW, Washington, DC 20007
Position: Pres *Employer:* Corp. for National Research Initiatives *Education:* PhD/EE/Princeton Univ; MA/EE/Princeton Univ; B/EE/City Coll of NY *Born:* 12/23/38 Was an assistant prof of Electrical Engrg at MIT prior to joining Bolt Beranek and Newman where he was responsible for the system design of the ARPANET, the world's first packet switching network and conducted the first public demonstration of the network. In 1972 he joined DARPA where he became Dir, Information Processing Techniques Office. In 1986, he became Pres of Corp for National Research Initiatives, A Non-profit Organization to provide Leadership & Funding for Dev of Information Infrastructure. Dr Kahn has lectured widely and has written numerous technical papers. He is a fellow of the IEEE and a member of the National Academy of Engg. *Society Aff:* IEEE, ACM, MAA, ΣΞ, HKN, ΤΒΠ

Kahn, Walter K
Business: Dept of Elect Engrg & Comp Sci, Washington, DC 20052
Position: Prof *Employer:* George Washington Univ *Education:* DEE/EE/Polytech University NY; MEE/EE/Polytech University NY; BEE/EE/Cooper Union *Born:* 3/24/29 1951, employed by The Wheeler Labs, NY (Division Hazeltine Corp) in monopulse radar development. 1954-69 Polytechnic Inst of (Brooklyn) NY, Prof of electrophysics and assistant to Dir, Microwave Research Inst. 1967-68 Civilian Liaison Scientist, the Office of Naval Research, London, UK. Over 50 papers in electromagnetics and optics. Consultant to US Government and major electronics corps. Fellow IEEE. 1977-80 Editor, IEEE TRANSACTIONS on Antennas and Propagation. Prof of Engrg and Applied Science, Dept of Electrical Engrg and Computer Science (Chrmn 1970-74) from 1969 to date. Dir, ANRO Engrg Consultants, Inc., Sarasota, FL. *Society Aff:* IEEE, OSA, AAAS, ΤΒΠ, ΣΧSI, HKNU.

Kahne, Stephen J
Home: 462 State St, Brooklyn, NY 11217
Position: Prof *Employer:* Polytechnic Inst of NY *Education:* PhD/EE/U of IL; MS/EE/U of IL; BEE/EE/Cornell U *Born:* 04/05/37 Formerly on EE faculty, Univ of MN (1966-1976), Prof & Chrmn of Dept of Sys Engrg, Case Western Reserve U (1976-1983), Dir NSF Division of Electrical, Computer & Sys Engr (1980-1982), Dir & Consltg Partner of InterDesign Inc, Minneapolis (1967-1976), Dean of Engr, Polytechnic Inst of NY 1983-4. Currently Officer, Intl Fed of Automatic Control. Conslt to 25 companies in past 20 years. Author of 60 papers. Formerly Editor, IEEE Transactions of Automatic Control, VP-Tech Activities IEEE (1984-1985). *Society Aff:* IEEE, AAAS, ASEE

Kahng, Dawon
Business: PO Box 550, Martinsville, NJ 08836 *Employer:* Independent Consultant *Education:* PhD/EE/Ohio State U.; M. Sc./EE/Ohio State U.; BSc/Physics/Seoul Nat'l U. *Born:* May 4, 1931. PhD Ohio State Univ; BSc Seoul Univ. Native of Seoul, Korea. With Bell Labs since 1959. Supr since 1964. Fellow since 1984. Ret Independent Consultant since 1987. Specialties include semiconductor device physics, especially MOS transitors and ICs, Schottky diodes, thin-film electroluminescence, CCDs & nonvolatile memories; Stuart Ballantine Medal 1975. Ohio State Univ. Distinguished Alumnms Award 1986. *Society Aff:* Fellow, IEEE

Kahng, Seun K
Business: 202 W Boyd St, CEC 219, Elec Engr & Comp Sci, Norman, OK 73019
Position: Prof & Dir *Employer:* Univ of OK *Education:* BSEE/EE/Seoul Nat Univ; MEE/EE/Univ of VA; DSc/EE/Univ of VA *Born:* 1/16/36 Native of Seoul, Korea. Dir of the School of Electrical Engrg and Computer Sci at The Univ of OK in 1979, Prof in 1978, joined faculty of Univ of OK in 1968 as an Assistant Prof. Research work with NASA 1975-76, summer work with NASA (1974). Responsible for solid state device lab at Univ of OK. *Society Aff:* IEEE

Kahrilas, Peter J
Home: 124 San Clemente La, Placentia, CA 92670
Position: Radar Systems Laboratory Mgr *Employer:* Hughes Aircraft Co *Education:* M/EE/Poly Inst of Bklyn; B/EE/City Coll of NY *Born:* 7/3/22 Mr Kahrilas is currently mgr, Radar Systems Lab engaged in tech dev of advanced radar systsms and mgmt of production systems including medium and long range radars, mortar and artillery locators. He is involved with dev of new generations of radar sensor systems which integrate new technologies with modern system architectures. Mr Kahrilas has lectured at Columbia Univ, Univ of Southern CA, UCLA, Northeastern Univ, George Washington U, the Indian Inst of Tech, Kanpur, India, and various national and international symposia. Mr Kahrilas holds several patents in Electronic Scanning Radar (ESR) tech, has authored over 40 tech papers and the Electronic Scanning Radar Systems (ESRS) Design Handbook (Artech House 1976). He is a Fellow of the IEEE and holds PE Licenses in NY and CA. *Society Aff:* IEEE

Kailath, Thomas
Home: Dept of Elec Engrg, Stanford, CA 94305
Position: Prof of Elec Engrg. *Employer:* Stanford University. *Education:* ScD/Elec Engg/MA Inst of Tech; SM/Elec Engg/MA Inst of Tech; BE/Telecommunications/ Univ of Poona, India. *Born:* 6/7/35. Jet Propulsion Labs & Caltech 1961-63. Then at Stanford Univ, Dept. of Elec. Eng. since 1968, Dir of Info Sys Lab from 1971-81, and Associate Chairman since Jan 1981. Briefer periods at Lincoln Labs, Bell Labs, UC Berkeley, Indian Statistical Inst, Indian Inst of Sci, Imperial Col, Cambridge Univ., Weizmann Inst. of Sci., Pres IEEE Goup on Info Theory 1975. Bd/ Governors, Info Theory Grp 1971-76; Admin Ctte, IEEE Control Sys Soc 1971-77; Assoc Editor IEEE Transactions on Info Theory 1971-75; Guest Editor, IEEE Transactions on Automatic Control Dec 1974. Editor, Info & Sys Sci Series, Prentice-Hall; Edit Bd of the IEEE Press & of several journals. Res activities in info theory, detection & estimation theory, communication & control sys, mathematical statistics, functional analysis, numerical analysis & linear sys theory. Author, Linear Systems, Prentice-Hall, Inc., 1980. Guggenheim Fellow 1969-70. Churchill Fellow, 1977. Michael Fellow, 1984. Fellow IEEE 1970; Fellow IMS, 1975, Natl Acad of Engg, 1984, Educ Award of Amer Automatic Control Council, 1986. Several outstanding paper awards. *Society Aff:* IEEE, SIAM, AMS, SEG, MAA, ACM.

Kain, Charles F
Business: 8733 W Chester Pike, Upper Darby, PA 19082
Position: Sole Owner. *Employer:* Zenith Engrs, Inc. *Education:* BS/Civ Engg/ Villanova Univ. *Born:* 11/7/16. Resident of Havertown, PA. Worked as draftsman and structural designer, various firms. Served in Army Air Force in WWII. Organized own Consulting firm in 1948, practiced under own name, then Kain & Hooven, and currently since 1970, Zenith Engrs, Inc. As individual and chief officer of firm, designed commercial, institutional, and municipal bldgs, bridges, and other civ projs. Active in PSPE & NSPE, DE County Chapter Pres, State Chrmn of PE in Private Practice, State Pres (1974-75) of PSPE, Natl Dir (1974-77) of NSPE. Honors: Engr of the Yr, Delaware County, 1976; Recipient of Morehouse Award, Villanova Univ, 1975. Registered professional engineer in Pennsylvania, New

Kain, Charles F (Continued)
Jersey, New York, Delaware, Maryland, Virginia. Registered Landscape Architect in Pennsylvania. *Society Aff:* ASCE, NSPE, ACI.

Kain, Richard Y
Business: 123 Church St SE, Minneapolis, MN 55455
Position: Prof. *Employer:* Univ of MN. *Education:* ScD/EE/MIT; MS/EE/MIT; BS/EE/MIT. *Born:* 1/20/36 Raised in Louisville, KY. Taught at MIT before moving to the Univ of MN in 1966. Consultant on computer system design, especially secure computer systems. Enjoys photography & classical music. *Society Aff:* IEEE, AAAS, ACM, ΣΞ, HKN.

Kaisel, Stanley F
Business: 600 Albion Avenue, Woodside, CA 94062
Position: Consultant. *Employer:* Self-employed. *Education:* PhD/Elec Engg/Stanford Univ; MA/Elec Engg/Stanford Univ; BSc/Elec Engg/WA Univ. *Born:* Aug 1922. Native of St Louis, Mo. Special Res Assoc, Radio Res Lab Harvard 1943-45. Res Engr RCA Labs; Res Assoc, Electronics Res Lab Stanford; Mgr of Engrg, Litton Indus Electron Tube Div; Founder & Pres, Microwave Electronics Corp 1959-69 (now Teledyne MEC). Cons since 1969. Fellow IEEE. Bd/Dir, IEEE. Specialty microwave tubes, electronics warfare sys. *Society Aff:* IEEE.

Kaiser, C Hayden, Jr
Business: 279 Lower Woodville Rd, PO Box 1267, Natchez, MS 39120
Position: Partner. *Employer:* Jordan, Kaiser & Sessions. *Education:* BS/CE/LA State Univ. *Born:* 5/14/28. Native of Natchez, MS. Served as commissioned officer for 2 yrs in US Corps of Engrs after grad from college. Partner in firm of Jordan, Kaiser & Sessions, Consulting Civ Engrs since 1954. Natl Dir 1969-72 CEC, mbr & pres 4 years of MS St Bd of Reg for PE and LS (1971-1982). *Society Aff:* ACEC, ASCE, SAME, CEC/MISS, MES.

Kaiser, Edgar F
Business: 300 Lakeside Drive, Oakland, CA 94666
Position: Chrmn Emeritus. *Employer:* Kaiser Steel Corp. *Education:* -/Economics/Univ of CA, Berkeley. *Born:* 7/29/08. Career: started as supt of pipeline proj in Midwest; shovel foreman & later supt of canyon excavation, Hoover Dam; Admin Mgr, main spillway, Bonneville Dam; Proj Mgr Grand Coulee Dam; Mgr 3 shipyards in Portland Ore 1942-45; Gen Mgr Kaiser-Fraser Corp 1945 - subsequently appointed Pres; Pres Kaiser Indus Corp 1956-67; Chmn/Bd Kaiser Aluminum & Chem Corp, Kaiser Cement Corp 1967-78; Chrmn/Bd, Kaiser Steel Corp, 1967-1980 Engrg News-Record's Const Man of Yr 1968; Hoover Medal 1969; Golden Beaver Award 1975.

Kaiser, Hugh W
Home: 1977 Halekoa Dr, Honolulu, HI 96821
Position: Supv Gen Engr. *Employer:* Pearl Harbor Naval Shipyard. *Education:* BArchitecturalEngg/Architectural Engg/WA State Univ. *Born:* 9/26/42. With Pearl Harbor Naval Shipyard since 1966: Naval Architect 1966-68; producton troubleshooter 1968-74; developed and implemented computerized Ships Force Overhaul Mgt System 1974-75; hd of Waterfront Engg Support Team since 1975; Kepner-Tregoe mgt process Instr; K-T process mgt consultant. HI Sec SNAME: Chrmn 1977, VChrmn 1976, Secy-Treas 1974, Mbrship Chrmn three yrs, 1982 Spring Meeting Steering Committee Chrmn. Enjoy Scuba diving and underwater photography, have done underwater hull photography/survey for local ship repair facilities. Church activities include Moderator, Vice Moderator, Stewardship Chrmn, Lay Reader, and a Bible study group at work. *Society Aff:* SNAME.

Kaiser, Wolfgang A
Business: Breitscheidstr 2, Stuttgart 1, Germany D-7000
Position: Full Prof and Dir *Employer:* Univ of Stuttgart *Education:* Dr-Ing/EE/Univ Stuttgart; Dipl-Ing/EE/Univ of Stuttgart *Born:* in Schoental, Germany. Studied electrical engrg at the Univ of Stuttgart 1947 - 1954. Research engr, then lab head and finally dir of R & D in data division of Standard Elektrik Lorenz AG (SEL, a subsidiary of ITT) in the years 1954 - 1967. Since then full prof and dir of Inst for Telecommunications of Univ Stuttgart. Fellow IEEE. Head of research council of Muenchner Kreis. Bd member of SEL. Mbr of Heidelberger Academy of Sciences. Honorary degree: Dr.-Ing. E.U. *Society Aff:* IEEE, NTG, VDE, MK.

Kakretz, Albert E
Home: 2166 Morrow Ave, Schenectady, NY 12309
Position: General Manager *Employer:* General Electric Co *Education:* Bach. Engineering/Marine/SUNY Maritime College *Born:* 07/26/29 Navy veteran and graduate of General Electric's three-year Advanced Engrg Program with extensive experience in application of heat transfer and fluid flow principles to the design of nuclear power plants. Has managed naval nuclear prototype power plant test facilities. For the past 10 years has been General Manager of the Knolls Atomic Power Lab, directing nearly 2,000 engineers and scientists and 1,500 support personnel in the design and support of naval power plants for cruisers, attack submarines, and fleet ballistic missile submarines. Is an ASME Fellow. *Society Aff:* ASME

Kalajian, Edward H
Business: Melbourne, FL 32901
Position: Dept Head *Employer:* FL Inst of Tech *Education:* PhD/Ocean Engrg/Univ of MA; MS/CE/Univ of MA; BS/CE/Univ of MD *Born:* 4/19/41 Native of Bergen County, NJ. After coll he was a Civil Engr with the US Naval Facilities Engrg Command from 1962 to 1968. Completed graduate education and joined the faculty of FL Inst of Tech in 1971. Presently is Prof and Head Civil Engrg Dept at FIT. Registered PE in FL. Enjoys Family. *Society Aff:* ASCE, ASEE

Kalbach, John F
Business: 920 Alta Pine Drive, Altadena, CA 91001
Position: Engg Consultant. *Employer:* Independent Contractor. *Education:* BSEE/EE/Univ of WA. *Born:* 1/2/14. GE Co 1937-47; Advanced Engg course, high speed turbine generators. Univ of CA 1947-51; lectr participated in design of 14 million volt Van de Graaff at Los Alamos. Wm Miller Instruments 1951-55; Sr Designer & Engg Mgr for oscillographs, transducers, line transient recorders, analog computers. ElectroData/Burroughs 1955-79; managed Quality Assurance, Ed, Field Engg, Reliability & Advanced Dev activitites; Corp Staff Cons, 1979-Independent Consultant. Reg PE CA. IAE (Fellow), IEEE (Fellow), ACM, ESA, Electronic Club (Los Angeles). Hobbies: boating & flying. *Society Aff:* IEEE, IAE, ACM, ESA, AOPA.

Kalbfell, David C
Home: 941 Rosecrans St, San Diego, CA 92106
Position: Chrmn. *Employer:* Instruments, Inc. *Education:* PhD/Physics/Univ of CA, Berkeley; MA/Physics/Univ of CA, Berkeley; AB/Physics/Univ of CA, Los Angeles. *Born:* Aug 1914. Physicist for Standard Oil of Calif 1939-41. Univ of Calif, Div of War Res from 1941-46. Prof at San Diego State Univ from 1948-72. Pres Kalbfell Labs from 1944-53. Pres Kalbfell Electronix from 1953-present. Chmn Instruments Inc from 1958-present. Dev electronic instruments & power amplifiers. Holds 14 pats. Cons to Gen Dynamics, Lockheed, Beckman Instruments, Rohr Corp, Cubic Corp & others. Fellow IEEE, Reg Elec Engr in Calif, Phi Beta Kappa, Sigma Xi, Pi Mu Epsilon. *Society Aff:* IEEE.

Kalfus, Stanley
Business: 309 Fifth Ave, NY, NY 10016
Position: Pres. *Employer:* Stanley Kalfus, PC. *Education:* BSME//Purdue Univ; BA//NYU. *Born:* 1/21/25. A native of NYC. Served in the Army 1943-45. Engr with mech contractor and consulting engrs before forming own consulting firm in 1957. Inc in 1974. Pres of Stanley Kalfus, PC. A profl engr licensed in NY, NJ, CT, MA, PA, MD, WI, TX, FL, DC. Listed as a biographee in Who's Who in the East. *Society Aff:* NSPE, ASHRAE.

Kalin, Thomas E
Business: 1500 Meadow Lake Pkwy, Kansas City, MO 64114
Position: Partner *Employer:* Black & Veatch Conslt Engrs *Education:* BS/ME/Univ of MO-Rolla. *Born:* 08/28/35 Native of St. Joseph, MO and the son of Swiss immigrants. Assigned to central power station design, construction, and startup since 1957, with the firm of Black & Veatch Consulting Engrs. Was appointed partner on January 1, 1981. Currently project mgr on three fossil fueled electric generating stations. Responsible for directing, scheduling, and cost program development for major power stations, as well as new business development. Have traveled extensively, including U.S. and overseas. Have participated in numerous high school career guidance counselor seminars. Enjoy restoring classic Mercedes-Benz automobiles and horticulture. Married and has three children. Registered PE in MO, NB, MI, and KS, VT, SD, WY. *Society Aff:* NSPE, ASME, AWS, PEPP.

Kalish, Herbert S
Home: 65 Falmouth St, Short Hills, NJ 07078
Position: VP. *Employer:* Adamas Carbide Corp. *Education:* MetE/Met Engg/Univ of MO; MS/Met/Univ of PA; BS/Met Engg/Univ of MO. *Born:* 8/11/22. Native of NY. VP, Technical Dir, Adamas Carbide Corp. With Third Army during WWII. Formerly, Commercial Fuel Mgr, United Nuclear Corp; Mgr of Mtls Res Olin Mathiseon Chem Corp; Engg Mgr, Sylvania Elec Products, Inc. Known for work in cemented carbide, powder met, alloy dev & mtls for nuclear applications. Over thirty technical papers & patents. First Chrmn Long Island Chapter, Past Chrmn NJ Chapter ASM. Has been on Natl Nominating Committee. and now Trustee of ASM. Past Chrmn of NY Sec of APMI. Chrmn, Demands for Tungsten on Natl Mtls Advisory Bd. Represents USA at Intl Organization for Standardization meetings on hardmetals, & Tooling Committee of Intl Cold Forging Group. *Society Aff:* ASM, AIME, APMI, ASTM, SME

Kallen, Howard P
Business: 1271 Ave the Americas, New York, NY 10020
Position: Partner. *Employer:* Kallen & Lemelson, Consulting Engineers *Education:* MME/Mech Engg/Polytechnic Inst of Brooklyn; BME/Mech Engg/Cornell Univ. *Born:* Apr 5, 1924. Professional Engineers License in 15 states. Mbr of Tau Beta Pi and Pi Tau Sigma, Engrg Hon Soc & Phi Kappa Phi, Sci Hon Soc. Sr Partner in firm of Kallen & Lemelson, Cons Engrs specializing in Mech & Elec Engrg for the Bldg Const Indus since 1957. Former mbr of Bd/Dir, New York Assn of Cons Engrs. Firm was recently presented with Natl Honor Award of the Amer Cons Engrs Council, for IBM Mfg Facilities Design in Manassas Va, and first prize in Mechanical Engineering by the N.Y. Assn. of Cons Engrs for their work at St. Vincents Medical Center, Bridgeport, CT. *Society Aff:* ASME, ASHRAE, NSPE, ACEC.

Kallis, Stephen A, Jr
Business: 200 Bahav Ave, MS CF01-1/M18, W Concord, MA 01742
Position: Sr Tech Communication Specialist. *Employer:* Digital Equipment Corporation. *Education:* AB/Phil/Columbia Univ *Born:* April 24, 1937. US Navy (Ens-Ltjg) 1959-61. Misc assignments Chrysler Space Div Huntsville, Ala 1962-66. Northrop Space Div 1966. Raytheon Space & Info Sys 1967-68. Digital Equip Corp 1968-present. Dev several software packages, incl Cinetimer-8 for prod of timing control tapes for motion picture labs. Soc affiliations: SMPTE, AAAS. P Chmn Boston Chap SMPTE 1972-74. Previously on Bd/Edit SMPTE Journal. Respon for config & programming computer sys Three Days of Condor film. Lic pilot. Freelance Author. *Society Aff:* SMPTE, AAAS, AOPA, BCI

Kallsen, Henry A
Box 6316, University of AL, Tuscaloosa, AL 35486
Position: Prof of IE. *Employer:* Univ of AL. *Education:* PhD/CE/Univ of WI; MS/CE/Univ of WI; BS/CE/IA State Univ. *Born:* 3/25/26. Raised in Jasper, MN. Served in US Navy during WWII. First job out of college was in Maintenance of Way with Wabash RR. Also taught at Univ of WI and LA Tech Univ (Acting Dept Hd of CE). Asst Exec Sec of ASEE just before it moved to Wash, DC. Asst Dean of Engg at Univ of AL for seven yrs. Acting IE Dept. Hd 1985-86. Mbr of delegation to study engg education in northern Europe - 1971. Co-author of "A Beginning Course in Computer Science Using UBASIC" Kendall/Hunt, 1979. Past Lt-Gov of Kiwanis Intl. *Society Aff:* ASCE, IIE.

Kalmbach, Donald A
Home: 2835 Aspen Lane, Bloomfield Hills, MI 48013
Position: Asst V Pres-Engineering. *Employer:* Michigan Bell Telephone Co. *Born:* Arch Engrg I C S . Native of Detroit, Michigan. Employed in auto and airplane indus prior to joining MBT. Served as Asst Supt for Western Elec in 1957 & as Engrg Mgr - Bldgs for AT&T Co 1969-70. Assumed current respon as Asst V Pres-Engrg for MBT Co in 1970. Am also respon for MBT's efforts towards natl conservation of energy & modernization of telecommunications equip. Mbr of the Greater Detroit Economic Dev Corp & the Metropolitan Const Users Council, Exec Ctte. P Dir of the Bldg Res Inst, Natl Res Council. Hobbies: golf & furniture finishing.

Kalpakjian, Serope
Business: 10 W 32nd St, Chicago, IL 60616
Position: Prof of Mech Engrg. *Employer:* Illinois Institute of Technology. *Education:* MS/Mech Eng/MIT; MS/Mech Eng/Harvard; BS/Mech Eng/Robert College. *Born:* 5/6/28. Did res in grinding, friction & wear. Res Supr Cincinnati Milacron, Inc specializing in metal forming processes. Since 1963 with IIT teaching & res in Mfg processes; also fac advisor to student orgs. Winner of 'Excellence in Teaching Award', 'Centennial Medallion' of ASME & 'M.E. Merchant Award' for textbook 'Manufacturing Processes for Engineering Materials' (Addison, Wesley, 1984) & author of three books, contrib to books, handbooks, encyclopedias & prof journals. Research in mfg processes & equip, fracture of abrasive disks, friction & wear. Active in prof soc's & cons to indus. *Society Aff:* ASME, NAMRI., ASM INT, ΣΞ, CIRP, SME.

Kalteyer, Charles F
Home: 1501 N. Garfield, Midland, TX 79701
Position: Chief Proration Engr *Employer:* Gulf Oil Corporation *Education:* BS/ME/Univ of TX-Austin *Born:* 11/07/24 A native of San Antonio TX, Commissioned an Ensign - US Naval Reserve 1945 at Notre Dame. Employed by Gulf Oil Corp. Production Dept since 1946 starting as roustabout and roughneck in South Louisiana bay area, Field Engr in East TX, District Reservoir and District Production Engineer in Gulf Coast, Area Engr in Permian Basin at Hobbs, N.M. and Chief Proration Engr in Midland, TX. Chrmn Gulf Coast Section SPE 1965. Chrmn Various Natl AIME and SPE Councils and Committees 1966 and 1969, Dir SPE 1979-81, Midland Coll Petroleum Tech Advisory Committee, Dir Permian Basin Graduate Ctr, SPE Section Service Award 1981. *Society Aff:* AIME

Kalvinskas, John J
Home: 316 Pasadena Ave, S Pasadena, CA 91030
Position: Proj Mgr. *Employer:* Jet Propulsion Lab. *Education:* PhD/ChE/CA Inst of Tech; MS/ChE/MIT; BS/ChE/MIT. *Born:* 1/14/27. Native of Vineland, NJ. Service in USN, 1944-46. Chem engr at DuPont in dev of new organic chem processes (1952-55, 59-60). Rockwell Intl (1960-70), variety of engg mgt positions including: Supervisor-Basic Studies of advanced propulsion & new tech; Mgr-Propellant Engg; Mgr/Dir, Life Sciences Operations. Pres & Dir, Resource Dynamics Corp, an environmental engg co (1970-1974). Other affiliations, Corporate Res Dir, Monogram Industries, Inc, Proj Mgr, Holmes and Narver, Inc. Current affiliation, Jet Propulsion Lab, CA Inst of Tech, Proj Mgr, Energy & Tech Application Progs including coal & biomass & toxic-hazardous waste disposal. Hobbies include golf, fishing. *Society Aff:* AIChE, ACS, NYAS, ΣΞ.

Kamal, Aditya K
Business: PO Box 208, Beford, MA 01730
Position: Div Staff, Communication Div *Employer:* The Mitre Corp. *Education:* Doctor of Engineering/Electrical/Sorbonne Paris (France); DIISc/Electronics/Indian Institute of Science Bangalore (India); MSc/Physics/Ailahabad Univ (India); BSc/

Kamal, Aditya K (Continued)
Physics/B.H.U. (India). *Born:* 05/05/27 MS 1948; DIISc 1951; Dr Engrg 1957 (Paris), course on Indus Mgmt 1952-53 (London), special summer course on "Optical Masers" 1962 (USA). Exper: Lectr Delhi Polytech 1951-52. Engr, Marconi's Wireless Telegraph Co UK 1952-54. Sr Engr CSF Paris 1954-57. Hd of Microwave Div, Gov of India CEERI Pilani 1957-58. Sr Proj Engr Ligne Telegraphique et Telephonique, Paris 1958-59. Dir Millimeter Wave and Quantum Electronics Labs. Purdue Univ USA 1959-64. Visiting Prof SC Univ USA 1965-66. Prof & Hd Electronics & Commun Engrg Dept, Univ of Roorkee India. Visiting Prof, Ecole Nationale Superieure D'Electronique et Radioelectricite Grenoble France, Apr 1978 - Sept 1978. Vice Prof Boston Univ Sept 1978 - Aug 1980. Visiting Scientist, MIT 1979-80. Presently, Div Staff, Communication Div, MITRE Corp Bedford, MA 01730. Fellow IEEE (USA), Fellow ASc (India), Fellow I E (India); Mbr IEE (London), Sigma Xi (USA), IETE (India). Visited USA, France, UK, Canada, Japan sev times.

Kamal, Medhat M
Business: PO Box 591, Tulsa, OK 74102
Position: Tech Group Leader *Employer:* Amoco Production Co. *Education:* PhD/Petrol Engg/Stanford Univ; MS/Petrol Engg/Stanford Univ; MSc/Engg/Cairo Univ; BSc/Petrol Engg/Cairo Univ. *Born:* 1/11/45. in Cairo, Egypt. Taught Petrol Engg at Cairo Univ while working on his Masters degree in enhanced oil recovery (1965-1969). Worked in the areas of reservoir mechanics and pressure transient theory at Stanford Univ. Joined Amoco Production Res in 1974 to work in the Reservoir Performance and Evaluation Sec where he is currently a Tech Group Leader. Published 9 papers so far in the Petrol Engg literature. Awarded the SPE Cedric K Ferguson Medal in 1977 for the best paper published by an author under 33. Director for the Mid- Continent Sec. of SPE. Past-Chrmn of Continuing Education for Mid-Continent Sec. Chrmn of the Natl Text Book Ctte of SPE. Enjoys traveling, soccer, skiing, and tennis. Married to Gina 1978. One child, Sally 1982. *Society Aff:* SPE, CIM.

Kamal, Musa R
Business: Chem Eng Dept, 3480 University St, Montreal Canada H3A 2A7
Position: Professor & Chrmn Dept of Chemical Engrg; Dir, Brace Research Inst. *Employer:* McGill University. *Education:* PhD/Chem Engg/Carnegie-Mellon Univ; M Eng/Chem Engg/Carnegie-Mellon Univ; BS/Chem Engg/Univ of IL, Urbana. *Born:* 12/08/34. Worked at Amer Cyanamid Co Central Res Labs & became res group leader in Plastics & Resins Div; joined McGill Univ 1967. Chrmn, Dept of Chemical Engg, 1983- ; Dir, Brace Research Inst, 1987- . Dir, Microeconomics & sectoral studies section, Morocco Five-year Indust Dev Plan, Dar Al-Handasah Consultants, Rabat, Morocco. Pres, Tulkarm Enterprises Ltd, 1976- . Spec plastices processing, properties & polymer blends and composites. Program Chmn, 2nd World Congress of Chem Engg, Montreal, 1981 VP, World Congress of Chemical Engineering III Tokyo, Japa, 1986; General Chrmn, 2nd Annual Meeting, Polymer Processing Society, Montreal, 1986; Chrmn, International Polymer Processing Group (IPPG), 1982- . Ed book on plastics weathering & wrote many articles. Holds several patents. Treas Canadian Soc for Chem engrg 1969-72; Dir Chemical Inst of Canada Plastics Educ Foundation & various div of Soc of Plastics Engrs & Chem Inst of Canada. Mbr: AIChE, SPE, ACE, AAAS, Soc of Rheology, Quebec Order of Engrs, Polymer Processing Society, American Academy of Mechanics. Intl Education Award, Soc of Plastics Engrs (1984), Kuwait Prize for Sci and Tech (1983). Canplast Award, Soc of Plastics Industry (SPI), Canada (1985). Ed Board Mbr: Polymer Engrg and Sci, Polymer Engrg Reviews, Advances in Polymer Tech, Polymer Titles. *Society Aff:* AIChE, SPE, OEQ, ACS, CSChE, CIC, AAAS, NYAS, SocRheo, Acad Mech, AIP, PPS.

Kamat, Satish J
Business: Indus Engrg Box 4230, Las Cruces, NM 88003
Position: Prof Indus Engrg. *Employer:* New Mexico State University. *Education:* PhD/IE/TX Tech Univ; MS/IE/IIT; BE/ME/Univ of Bombay. *Born:* 8/7/44. Entered USA 1966. Worked summer 1967 with EVCO & Assocs Inc & summer 1968 with STEBCO Prods Corp; worked with Lubbock Mfg Co as Systems Analyst & Programmer 1970; joined N M St Univ as Asst Prof of Indus Engrg 1970, promoted to Assoc Prof 1975, promoted to Professor in 1980. Also held apptment as Expert with OMEW (ECOM), US Army 1976-78. Mbr: IIE, ORSA, ASEE, Alpha Pi Mu, Phi Kappa Phi & Sigma Xi. Enjoys golf & fishing. Took sabbatical eave at Storage Technology Corporation as Sr IE during 1979-80. *Society Aff:* IIE, ASEE, SME.

Kamel, Hussein A
Business: AME Dept, Bldg 16, Rm 210-B, Tucson, AZ 85721
Position: Prof. *Employer:* Univ of AZ. *Education:* PhD/Aircraft Structures, Engg/Imperial College of Sci & Tech; BSc/Aero/Cairo Univ. *Born:* 06/8/34. Dr Hussein A Kamel is a Prof of Aerospace & Mech Engg at the Univ of AZ. Dr Kamel holds a BSc in Aero Engg from the Univ of Cairo, & a PhD in Aero Engg from the Imperial College of Sci & Tech in London, England. He has taught at the Tech Hochschole, Stuttgart, Germany. His fields of teaching & residence encompass theory of structures, continuum mechanics, numerical analysis, software engg, & comp graphics. He was involved in the design & implementation of such codes as ASKA, DAISY, & GIFTS. *Society Aff:* AIAA, ASME, SNAME.

Kaminow, Ivan P
Business: AT & T Bell Labs, Box 400-Room HOH R219, Holmdel, NJ 07733
Position: Department Head. *Employer:* AT&T Bell Laboratories. *Education:* PhD/Appl Physics/Harvard; AM/Appl Physics/Harvard; MSE/Elec Engg/UCLA; BSEE/Elec Engg/Union College, NY. *Born:* 3/3/30. 1952-54 Hughes Aircraft Co. Culver City CA; Hughes Fellow; res on microwave antennas. 1956- Bell Labs Holmdel NJ; res on ferrites, ferroelectrics, nonlinear optics, raman scattering, electrooptic devices, optical fibers, optical communications, semiconductor lasers, lightwave networks. 1986 Adjunct Prof. Columbia Univ., NYC. 1977 visiting Lectr Univ Ca, Berkeley. 1968 Visit Lectr Princeton. Author, *Introduction to Electrooptic Devices*; co- Editor, *Optical Fiber Telecommunications II, Laser Devices & Applications*. Fellow: Inst Elec & Electronic Engrs, Amer Physical Soc & Optical Soc of Amer. Mbr, Natl Acad of Engrg. *Society Aff:* IEEE, APS, OSA, NAE.

Kaminski, Donald W
Business: 11955 Shaker Blvd 200, Cleveland, OH 44120
Position: President. *Employer:* Euthenics, Inc. *Education:* BS/Civil Engg/Univ of IL. *Born:* 1/16/29. Hwy Design Engr for Howard, Needles, Tammen & Bergendoff, Kansas City Mo 1949-51. Officer Navy Civil Engr Corps Seattle Wash 1952-55, Sect Ch & Proj Mgr HNTB's Cleveland Office 1955-72 - respon charge of num projs incl design of about 50 mi urban interstate freeway. Named Ch Engr of Euthenics Inc Spring 1972, elected Pres Fall 1972; firm's major work has incl design of hwys, bridges, sewer tunnels & wastewater treatment plants. Ints incl sports, chess & history. *Society Aff:* ASCE, ITE, NSPE, PEPP.

Kaminsky, Frank C
Business: 112 Marston, Amherst, MA 01003
Position: Prof. *Employer:* Univ of MA. *Education:* PhD/Operations Res/Northwestern Univ; MS/Operations Res/Northwestern Univ; BS/Mech Engg/Univ of CT. *Born:* 5/16/36. Since 1965, as a faculty mbr at the Univ of MA, I have been involved in teaching & res with a maj emphasis on the dev of planning tech for the treatment of societal problems. In particular I have developed models in the areas of regional wastewater treatment; solid waste mgt; health care systems; and wind energy systems. My current res is directed toward the dev & use of math-computer models to identify areas of high wind potential to assist in the dev & implementation of wind energy conversion systems for the production of elec energy. *Society Aff:* ISES.

Kammash, Terry
Business: Dept of Nuclear Engrg, Ann Arbor, MI 48109
Position: Stephen S Attwood Distinguished Prof. *Employer:* University of Michigan. *Education:* PhD/Nuclear Engg/Univ of MI; MS/Aero Engg/Penn State Univ; BS/Aero Engg/Penn State Univ. *Born:* Jan 1927. Taught Applied Mechanics, Aero Engrg & Nuclear Engrg at Penn St & Univ of Mich 1952- . Field of interest & expertise is power generation from Controlled Fusion reactors. Author of 1st book on Fusion Reactor Physics & Tech (Ann Arbor Sci Publishers Inc 1975). Respon, formation of Controlled Fusion Div of Amer Nuclear Soc & served as its 1st chmn 1970-72. Cons on fusion physics & engrg at Lawrence Livermore Lab, Oak Ridge Natl Lab, Argonne Natl Lab & Battelle Pacific NW Lab 1962- . Served on many advisory cttes for the Magnetic Fusion Div of ERDA. Fellow: ANS & APS. Mbr: ASEE, Sigma Xi, Phi Kappa Phi, Tau Beta Pi, Sigma Tau. Who's Who in the Midwest, Who's Who in Atoms & Amer Men of Sci. Recipient of the Arthur H Compton Award, Amer Nuclear Soc (ANS), 1977, and the outstanding Achievement Award of the Controlled Fusion Div, ANS, 1977. *Society Aff:* APS, ANS, ASEE.

Kammermeyer, Karl
Home: 404 1/2 E Bloomington, Iowa City, IA 52240
Position: Professor Emeritus. *Employer:* University of Iowa. *Education:* DSc/ChE/Univ of MI; MSE/ChE/Univ of MI; BSE/ChE/Univ of MI; -/Math/Univ of MI. *Born:* 1904. Married Cordelia G Meyers 1930; son, John Karl. Dev Engr Standard Oil Co of Ind 1933-36; Refinery Chief, Chemist & Chem Engr Pure Oil Co 1936-39; Asst Prof of Chem Engrg Drexel Inst of Tech 1939-42; Dir of Res Publicker Indus Inc 1942-47; Mgr R/D Chems Div Glenn L Martin Co 1947-49; Prof & Head Dept of Chem Engrg Univ of Iowa 1949-73, Emeritus 1974. Author of 70 publs & 6 books. Mbr: AAAS, Amer Chem Soc, ASEE, Sigma Xi, Tau Beta Pi; Fellow AIChE. Club: Triangle. Off: 121 Chem Bldg, U of Iowa, Iowa City. *Society Aff:* AIChE, ACS, AAAS.

Kamperman, George W
Business: 26100 Woodland Trail, Kansasville, WI 53139
Position: President. *Employer:* Kamperman Associates Inc. *Education:* BS/Physics/ALMA COllege; -/Acoustics/MIT Grad Sch. *Born:* May 1926. Graduate study (acoustics) MIT. Bolt Beranek & Newman Inc 1952-72; Pres Kamperman Assocs Inc 1972- . Fellow: Acoustical Soc of Amer, Mbr: Amer Indus Hygiene Assn, Inst of Noise Control Engr., Am Soc. Heating & Refrig. and Air Cond. Eng. Reg Prof Engr. For several yrs in charge of Instrumentaion Lab at Bolt Beranek & Newman Inc where respon for dev new equipment & procedures for measurement of sound & vibration. Lectr at many courses & seminars. Has had considerable experience in noise & vibration control of machinery & indus plants. Proj respon have included noise & vibration measurement & control in a wide variety of machinery. Has had a major role in writing over 12 Regulations, Standards & Test Codes plus many tech publs. *Society Aff:* ASA, INCE, AIHA, ASHREA.

Kampmeyer, John E,
Home: 269 Butler Rd, Springfield, PA 19064
Position: VP. *Employer:* Maida Engrg Inc. *Education:* MSME, Mech Engr/Drexel Univ; BSME/Mech Engr/Drexel Univ *Born:* 5/24/38. VP, Maida Eng Inc responsible for all Mech and Fire Safety Engrg and marketing of Governmental and Commercial projects. Conslt to numerous bldg owners in Phila on projects involving fire protection, life safety, code interpretation and HVAC. Formerly, Assoc and Supervising Engr, Fire Safety Dept, for the Kling Partnership, responsible for fire and life safety programming, code cordination, engg and design, and marketing of fire safety engg and design services. Also Kling Proj Mgr, mech and elec for the 2,500,000 sq ft AT&T hdquarters, Basking Ridge, NJ PA SPE Representative to and Secy of Construction Regulatory Coalition, a business & profe organization devoted to developing a construction code system for PA. Dir of Del Valley Chpt, SFPE. Course Dev and Instructor of courses on fire protection and codes for the city of Phila and the Phila Engrs Club. Member of NFPA Subcommittee on Smoke Control Systems Document. *Society Aff:* NSPE, SFPE, ASHRAE, ASME, NFPA, BOCA.

Kanal, Laveen N
Business: Dept of Computer Sci, University of Maryland, College Park, MD 20742
Position: Professor of Computer Science. *Employer:* Univ of Md, College Park. *Education:* PhD/Elec Engg/Univ of Penn Phila, PA; MSEE/Elec Engg/Univ of WA, Seattle; BSEE/Elec Engg/Univ of WA, Seattle. *Born:* 9/29/31. In indus worked as Communication Sys Design Engr at Canadian G E; Res Engr Moore Sch of EE Univ of Pa; Mgr Machine Intelligence Lab Genl Dynamics/Electronics; Mgr Advanced Engrg & Res Philco-Ford Corp. Now Managing Dir LNK Corp, a mgmt cons & R&D firm. In educ has been Adjunct Prof and visiting Prof of Oper Res, Statistics, Regional Sci at Wharton Grad Sch of Business Univ of Pa; Adjunct Prof of EE at Lehigh Univ; now Prof of Computer Sci & Dir of Lab for Machine Intelligence & Pattern Analysis at Univ of Md. Elected Fellow IEEE & Amer Assn Advancement of Sci. Ed books on pattern recognition & holds pats in pattern recognition *Society Aff:* IEEE, ACM, ASA, ΣΞ, AAAS, SME, AAAI.

Kanat, Walter
Business: 1391 S 49th St, Richmond, CA 94804
Position: Chief Engr. *Employer:* Stauffer Chem Co. *Education:* BS/CE/Univ of S CA. *Born:* 2/4/24. Entire prof career with Stauffer Chem Co. The earlier yrs with the Ind & Agri production depts in various assignments ranging from plant engg to plnt mgr. Specific production operation included sulfuric acid, chlorine/caustic, sulfur recovery plant, inorganic fertilizer, insecticides & miscellaneous others. Fourteen yrs with the corporate engg dept has been as proj engr mgr for a variety of large capital projs, such as soda ash refinery, sulfuric acid plants, chlorine/caustic, herbicides & insecticide facilities, environmental projs & others. The last five yrs, has managed an office of 100 to 115 people as Chief Engr & has broadly directed studies, evaluations, design & construction for a significant part of the co's capital prog. *Society Aff:* AIChE.

Kandoian, Armig G
Home: 195 Orchard Pl, Ridgewood, NJ 07450
Position: Telecommunications Engrg Cons (Self-employed) *Education:* BS/-/Harvard; MS/-/Harvard; DrSci/-/Newark College of Engg-NY Inst of Tech. *Born:* Nov 28, 1911 Van Armenia. Came to USA 1922. Educ: Public Sch Springfield Mass; BS & MS Harvard Univ 1930-35; Dr of Engrg (Hon) Newark Coll of Engrg 1967. With ITT 1935-65 to V Pres & Genl Mgr of ITT Labs; Gen Mgr & later Pres Communication Sys Inc 1965-68; Dir Office of Telecommunication US Dept of Commerce Wash DC 1970-73. Has done extensive work on antennas, navigation radar, broadcast, cable TV & all aspects of telecommunication. Many pats & publs. IEEE Achievement Award, Eta Kappa Nu Award. Chmn Ed Board of Reference Data for Radio Engrs. 1980 IEEE Internatl Communications Award- . Chrmn, IEEE Ctte on Telecommunication Policy 1980-81. *Society Aff:* IEEE, IEEE.

Kane, George E
Business: Packard Lab 19, Bethlehem, PA 18015
Position: Chairman Dept of Indus Engrg. *Employer:* Indus Engrg Dept-Lehigh Univ. *Education:* MS/IE/Lehigh Univ; BS/IE/PA State Univ. *Born:* Sept 1925, native of York Pa. Prof Engr Pa. US Navy 1943-46, 1953-54. Indus Engr Western Elec Co 1948-50; With Lehigh Univ 1950- . Principal interests in machining of metal parts & mfg methods. Author of over 50 tech papers, holder of 1 patent, contributor to Production Handbook. Dir of numerous res projs in machinability, tool testing, tool materials. Cons to private indus & gov't for 25 yrs. Mbr: IIE, ASME, ASEE, SME, Soc of Carbide Engrs, ASM, Sigma Xi, Tau Beta Pi, Alpha Pi Mu. *Society Aff:* ASME, ASEE, IIE, SME, ASM.

Kane, Harrison
Business: Department of Civil and Environmental Engineering, Univ of IA, Iowa City, IA 52242
Position: Prof. *Employer:* Univ of IA. *Education:* PhD/Civil Engg/Univ of IL; MS/

Kane, Harrison (Continued)
Civil Engg/Columbia Univ; BCE/Civil Engg/City College of NY. *Born:* 1/2/25. in Brooklyn, NY. Served in US Army 1943-46, 1st Lt Armed Force. Structural design engr for Parsons, Brinckerholf, Hall and MacDonald 1945-51 and 1953-56. Planning Engg for Dept of Army in Germany, 1951-53. Taught at City College of NY, PA State Univ, and Univ of IL. On faculty of Univ of IA since 1965, Chrmn of Civil Engg Dept 1971-85, Reg PE in IA. Bd Mbr of IA State Bd of Engg Examiners 1975-82. *Society Aff:* ASCE, ASEE, AAUP.

Kane, John M
Home: 5614 Coach Gate Wynde, Louisville, KY 40207
Position: Industrial Consultant. *Employer:* Self. *Education:* BS/Mech Engr/Univ of KY *Born:* July 31, 1908 Scotia N Y. Amer Air Filter Co 1933-67, V Pres 1962-67; Trustee Foundry Educ Foundation 1963-65; Pres Occupational Health Inst of Ky 1957-60; Pres Natl Castings Council 1966; Pres Foundry Equip Manufac Assn 1965- 66; V Pres Louisville Safety Council 1967; Ky St Air Pollution Comm 1966-70; Natl Alumni Representative Tau Beta Pi 1966-71. Fellow ASHRAE; Dir AIHA; Mbr AFS. Author of over 25 tech papers on Indus Ventilation, Air Pollution Control; co-author 'Design of Indus Ventilation Sys' Alden & Kane 5th Ed, Indus Press 1981. Lic Prof Engr (KY). *Society Aff:* AFS, AIHA, ASHRAE

Kane, Thomas R
Home: 817 Lathrop Dr, Stanford, CA 94305
Position: Prof. *Employer:* Stanford Univ. *Education:* PhD/Appl Mechanics/Columbia Univ; MS/CE/Columbia Univ; BS/CE/Columbia Univ; BS/Math/Columbia Univ. *Born:* 3/23/24. Native of Vienna, Austria. Came to the US in 1938. Served in the Army as combat photographer during WWII. Taught at the Univ of PA from 1953 until 1961, has been at Stanford Univ since 1961. Fulbright lectr at UMIST, England, 1959. US Acad of Sci, Soviet Acad of Sci Exchange Scholar, 1968. Visiting Prof, Fed Univ of Rio de Janeiro, 1971. Fellow, ASME. Fellow, AAS. Rec Dirk Brouwer Award, AAS. Visiting Prof, Shanghai Jiao Tong Univ, 1981. Author of five books, more than 100 technical papers. *Society Aff:* ASME, AAS, ΣΞ, ΤΒΠ.

Kaneko, Hisashi
Home: 3 - 22 - 16 Daizawa, Setagaya, Tokyo 155, Japan
Position: VP *Employer:* NEC Corp *Education:* BS/EE/Univ of Tokyo; MS/EE/Univ of CA; Dr Eng/EE/Univ of Tokyo. *Born:* 11/19/33 Dr Kaneko joined the Central Res Labs of Nippon Elec Co 1956, and has devoted himself to the res and dev of digital communication sys and equipment. Meanwhile, from 1960-62 he worked at the Electronics Res Labs of Univ of CA, Berkeley, and 1968-70 he worked at the Bell Telephone Labs, Holmdel. He is currently VP in charge of Transmission and Terminals Group, NEC Corp respon for the dev and mfg of transmission equipment. He is a Fellow of IEEE. *Society Aff:* IEEE, IEICE.

Kangas, Ralph A
Business: 1707 Linda Vista Ln, Boise, ID 83704
Position: President. *Employer:* Smith & Kangas Engrs Inc. *Born:* Apr 12, 1939 Boise Idaho. BSCE Utah St Univ 1962. Theta Tau, Sigma Tau, Student Chapt ASCE, Engrs Club. Reliability Engr The Boeing Co 1962-65; Instr Highline Coll Kent Wash 1965-67; Cons Engr 1965-67; with Smith & Kangas Engrs Inc 1967- , Pres 1974- . S Idaho Sect ASCE: Secy 1970, 2nd VP 1971, 1st VP 1972, Pres 1973, PNC Delegate 1974-75. ASCE Ctte on Public Affairs & Legislation 1975- 77. ISPE Legislative Ctte 1973-76. ITAC 1975-77.

Kant, Edward J
Business: 1074 Industrial Blvd, Naples, FL 33942
Position: Prof Assoc *Employer:* Project Mgmt Assocs, Inc *Education:* BS/Civil Engrg/Carnegie-Mellon Univ. *Born:* 7/21/39. Grad study Yale Univ 1963-65; grad study Univ of CT 1975-79. Const Coordinator C W Blakeslee & Sons Inc 1965-68; proj Engr Cahn Engrs Inc 1969; Cost Engr & Prof Engr L G Defelice Inc 1970; Chief Engr HUB Copr 1971; private practice, Edward J Kant & Assocs Inc 1971-77. Pres & Prin Engr; Loureiro Engrg Assoc, 1977-79, Sr Proj Mgr; Craven Thomspon & Assoc, Inc, Exec VP 1979-1980. Collier Enterprises, 1980-1981, Dir of Eng; Dev Mgmt Corp, 1981-1983, Pres; Listed in Who's Who In The East 1975- ; mbr var prof organizations & socs; panelist, AmerArbitration Assn. Community service included Sewer Comm, Econ Dev Comm, Bus Advisory Council. Naples Airport Authority; Collier County Water Mgmt Bd. *Society Aff:* ASCE, NSPE, FES, AAA.

Kanz, Anthony C
Home: 816 Woodley Dr, Rockville, MD 20850
Position: Deputy Dir Dept of Transportation. *Employer:* Montgomery County Government. *Education:* BSCE/Civil Engg/Univ of WI; Cert/Highway Traffic Engrg/Yale Univ-Graduate Sch. *Born:* Feb 2, 1927, native of Milwaukee Wis. Certificate in Hwy Traffic Yale Univ 1954. City of Milwaukee Staff Traffic Engr 1952-55; City of Phila Traffic Planning & Survey Engr 1955-57; City of Rockford Ill City Traffic Engr 1957-64; Montgomery Co Md Chief Bureau of Traffic Engrg 1964-73, Deputy Dir Dept of Transportation 1973- . Fellow ITE; Pres Wash D C Sect ITE 1971 & 1972. Director, International ITE 1980, 1981 & 1982; Mbr: APWA, NSPE, Maryland Society of Professional Engineers, Professsional Engineer, State of California, Cty Engrs Assn of Md & Natl Ctte on MUTCD; Delegate to Constitutional Convention for restructuring ITE Constitution 1971. Recipient Citation & Commendation for Achievement from City-County Planning Comm, Rockford-Winnebago County Ill 1960 Member, City of Rockville, Traffic & Transportation Commission 1972 to Present. *Society Aff:* ITE, APWA, TRB.

Kao, Charles K
Business: 7635 Plantation Rd, Roanoke, VA 24019
Position: VP. *Employer:* ITT EOPD. *Education:* PhD/EE/Univ of London; BSc/EE/Univ of London. *Born:* 11/4/33. Born in Shaghai, China, & educated in Great Britain. Pioneered the field of optical fiber communications (results published in 1966). Experience includes optical fiber communication systems & quasioptical techniques applicable to microwaves. Chrmn & Prof of the Electronics Dept of the Chinese Univ of Hong Kong (1970-1974). Awarded the 1976 Morey Award (Am Ceramic Soc), Stewart Ballantine Medal (Franklin Inst), 1978 Rank Prize (Rank Trust Funds of England), Morris N Liebmann Meml Award (IEEE), & 1979 L M Ericsson Prize Award (Sweden). Enjoys tennis & pottery making with his wife & two children. *Society Aff:* IEEE, IEE.

Kao, Timothy W
Business: Dept of Civ Engg, Washington, DC 20064
Position: Prof. *Employer:* Catholic Univ of America. *Education:* PhD/Engg Mechanics/Univ of MI; MSE/Engg Mech/Univ of MI; BSc/Civ Engg/Univ of Hong Kong. *Born:* 7/20/37. Born in Shanghai, China. Received secondary school and college education in Hong Kong. Came to US in 1959. Received grad education at the Univ of MI. Res Engr, Cities Service Res and Dev Co Tulsa, OK, 1961. Res Fellow, CA Inst of Tech, 1963-64. Joined the Catholic Univ of America in 1964 as Asst Prof, Assoc Prof, 1966 and Prof, 1970. Principal investigator of over a dozen res projs sponsored by NSF and ONR. Consultant to various cos and govt agencies. Visiting oceanographer at NASA/GSFC. Married May Lee, Two children: Michelle and Erika. Naturalized US Citizen. Assoc Dean, School of Engineering and Architecture, 1981. Chrmn, Dept of Civil Engg., 1981. *Society Aff:* ASCE, AMS.

Kapcsandy, Louis E
Business: 900 Poplar Place, S, Seattle, WA 98114
Position: President *Employer:* Baugh Construction Co. *Education:* BS/ChE/Tech Univ of Budapest. *Born:* June 1936. Dipl Ing (BSChE) Tech Univ of Budapest; attended Univ of San Francisco Law Sch & Univ of Calif - Berkeley. Native of Budapest Hungary; emigrated to U S 1957. USA Corps of Engrs 1959-61. Played prof football with San Diego Chargers. Norton Co 1964-72 specializing in packed tower design & marketing; Prod Mgr Koch Engrg 1972-74 respon for Chemical Process Equip, fractionation trays & static mixers; flow industries 1974-78 Vice Pres./G.M.

Kapcsandy, Louis E (Continued)
respon for high pressure intensifiers & waterjet cutting products. 1978-1984 Pres/CEO Fentron Industries respon for aluminum & glass curtain walls & window walls for high-rise buildings; President Baugh Construction respon for commercial building construction of 86th largest general contractor in U.S. *Society Aff:* AIChE, SAME, TAPPI, ACS.

Kapfer, William H
Business: 2247 Pineland Drive, Englewood, FL 33533
Position: Prof of Chem Engrg *Employer:* Polytechnic Univ *Education:* Engr ScD/ChE/NYU; M/ChE/NYU; B/ChE/NYU *Born:* 2/27/17 Chem engr at Hercules, Inc, 1941-1949, specializing in acid manufacture, plant startup, naval stores and cellulose products research and development. General Aniline Fellow at NY Univ, 1949-51. Prof of Chem Engrg at NY Univ 1951- 1973; Assistant Dean 1968-1973. Since 1973, Prof of Chem Engrg at Polytechnic Univ. Also served at Polytechnic Univ as Registrar, Assoc Dir of Special Programs, Dir of Placement, and Acting Dean of Students. Present tech interests are chem economics, chem plant design, and fluid mechanics. *Society Aff:* ACS, AIChE, AIC, AACE, ΣΞ

Kapila, Ved P
Business: 7439 Middlebelt Rd Ste 2, W Bloomfield, MI 48033
Position: Pres. *Employer:* Kapila & Assoc, Kapila Const Co Inc *Education:* MS/Civ-Structural/Univ of MI; MBA/Mgt/Wayne State Univ; Value Engg Design and Construction/Value Engg Seminar/Univ of WI; BS/Civ-Struct/Univ of MI; Dipl/Civ/Engg Sch. *Born:* 12/27/32. Born in India, naturalized US Citizen. Engg officer, Punjab State, India. Design Engr, Ayres, Lewis, Norris & May, Ann Arbor, MI (1964-65). Obenchain Corp, Dearborn, MI, (1965-65). VP of Engg, Chief Civ/Structural Engr, O Germany, Inc, Warren, MI (1966-76). Dir of Proj Services, Chief Planning & Scheduling, Chief Quality Assurance, Chief of Client Purchasing, Sr Proj Mgr, Hoad Engg, Inc, Ypsilanti, MI (1976-78). Pres, Kapila & Assoc, Engrs, Planners and Surveyors; Pres Kapila Construction Co, Inc, Developers, Designers & Custom Residential Builders & Kapila Contracting Co Inc Commercial, Industrial & Institutional-Construction Managers/General Contrators 7439 Middlebelt Rd, Suite 2, W Bloomfield, MI (1978-present). Reg PE - MI, VA, GA, Natl Engg Council. Reg Land Surveyor, MI. *Society Aff:* NSPE, ASCE, MSRLS, AISC, SAVE, ASAC, ACI, BASM, NAHB.

Kaplan, Herman
Home: Garden Road, Harrison, NY 10528
Position: Manager - Owner. *Employer:* Process Engineering Co. *Education:* MA/Chem/Brooklyn Coll; B. Sanitary Eng/Sanitary Eng/Univ of Ill; BS/Chem/Coll of City of NY. *Born:* 6/4/18. BS Chem CCNY 1938 (Phi Beta Kappa); MA Chem Brooklyn Coll 1940; B Cvl Engrg Univ of Ill 1949. Sanitary Engrg Corps US Army 1943-46, 1st Lt. Chief Engr Trubek Labs (now UOP Chem) 1950-65; Formed Elan Chem Co 1965 (owner 1965-1978), manufac of chems for Essential Oil Indus. Mbr ACS & AIChE. Prof Engr St of N Y, N J, Ill. Enjoys classical music, gardening & sculpture. *Society Aff:* ACS, AIChE.

Kaplan, Marshall H
Business: 233 Hammond Bldg, Univ Park, PA 16802
Position: Prof of Aerospace Engg. *Employer:* PA State Univ. *Education:* PhD/Aero & Astro/Stanford Univ; SM/Aero & Astro/MIT; BS/Aeronautical Engr/Wayne State Univ. *Born:* 11/5/39. Native of Detroit, MI. Attended Wayne State Univ, receiving a BS in Aero. Engg with "Distinction," in 1961. Received SM degree in 1962, followed by four yrs in the spacecraft industry at Hughes Aircraft Co and Ford Aerospace Corp. Received PhD in 1968. Currently Prof of Aerospace Engg and Dir of the Astronautical Res Lab at the PA State Univ. Author of two recent books: *Space Shuttle: America's Wings to the Future*, Aero Publishers, 1978, and *Modern Spacecraft Dynamics and Control*, John Wiley and Sons, 1976. Recipient of the 1978 Award for Outstanding Achievement in Res in the College of Engg at Penn State Univ. Is an instrument rated pilot and owner of a Beechcraft Bonanza in which he logs over 200 hours per yr. *Society Aff:* AIAA, AAS, AAM, ASEE.

Kaplow, Roy
Business: Rm 13-5106, 77 Massachusetts Ave, Cambridge, MA 02139
Position: Prof. *Employer:* MA Inst of Tech. *Education:* PhD/Metallurgy & Mtls Sci/MA Inst of Tech; SB/Physics/MA Inst of Tech. *Born:* 8/17/32. Native NYC; Education at MA Inst of Tech (SB Physics, 1954; ScD Mtls Sci and Engg, 1958); Profl career primarily at MIT in res and education -- presently Prof of Mtls Sci and Engg and of Education; Present res interests in rapidly quenched mtls, surface properties, solar energy (especially photovoltaics), software and hardware for interactive computer and video info and educational systems; Teaching responsibilities regarding structures and properties of mtls, dev and applications of computer systems; Consultant to govt and industry in areas of mtls, photovoltaic and magnetic devices, dev of computer hardware, and use of computers in teaching; Approximately 80 publications and 14 patents. *Society Aff:* APS, ACM, AAUP, MSES, AAS, NSPE

Kapp, J (Pat) P
Business: 15892 Pasadena Ave, Tustin, CA 92680
Position: Pres *Employer:* J. P. Kapp & Associates, Inc. *Education:* BS/CE/UCLA *Born:* 10/15/41 Registered Cvl Engr in CA. Started out as a draftsman and attended night school to obtain degree in cvl engrg and registration as an engr. At the age of 33 formed cvl engrg, land planning and surveying firm. With an eye toward the future J. P. Kapp purchased highly sophisticated computer equipment and software for use in the mapping, surveying and design areas. Within one year the system has been updated with additional analysis and design programs. J. P. Kapp has been a forerunner in the filed of computer tech. Enjoys opera, and classical music and bicycling. *Society Aff:* ASCE, NSPE, CSPE, CCCE&LS

Kappe, Stanley E
Travilah Rd Box 1036, Rockville, MD 20850
Position: Chrmn of the Bd & CEO *Employer:* Kappe Associates Inc. *Education:* BS/Civ Engrg/PA State Univ. *Born:* Aug 11, 1908. June 1930 Pa Bureau of Sanitary Engrg. Nov 1935 US Army Engrs, charge pollution study Del River & main tributaries. Sept 1936 Eastern Mgr Chicago Pump. Nov 1946- , Chrmn and CEO Kappe Assocs Inc environ application engrs & sci res org, Rockville Md. Reg Prof Engr Pa & D C. Special Cons: Hdq USAF, US Public Health Serv. Exec Dir Amer Acad Environ Engrs, editor Diplomate newsletter, Fair award, hon diplomate award 1981; Internatl Dir, Fuller Awardee & Hon Mbr AWWA. Mbr many prof socs. Lectr, author, patentee. *Society Aff:* ASCE, AWWA, WPCF, AAEE, NSPE.

Kappelt, George F
Home: 5138 Lower Mtn Rd, Lockport, NY 14094
Position: Retired *Employer:* Retired *Education:* BA/Chemistry/Univ of Buffalo. *Born:* Jan 1919. Served entire prof career with Bell Aerospace Co 1940- . Metallurgist, Group Leader, Chief, presently Dir of Engrg Labs & Test Dept. Active in SAE & ASM local & natl levels; Mbr & Chmn various cttes; served as Trustee ASM 1966-68. Also Mbr AIA Matls & Structs Ctte. Past Dir and Tres of EJC. Author of numerous papers & publs, most recent Prediction of Material Performance in Structures & Components, NAE, NMAB 1972. Fellow ASM 1970. *Society Aff:* SAE, ASM.

Kaprielian, Elmer F
Home: 39 Elmwood Ct, Walnut Creek, CA 94596
Position: Sr. Vice President, Retired. *Employer:* Self Employed (Consultant) *Education:* BE/EE/USC. *Born:* 1/12/22. Place of birth - Fresno, CA. Attended Fresno State College. Grad from Univ of S CA in 1942 with BE Degree in EE. Served in the US Naval Reserve in WWII attaining the rank of Lt. Sr grade - employed for 40plus yrs by the Pacific Gas and Elec Co with primary responsibilities in Generation (Conventional Steam Geothermal) Hydroelec & Elec Opera. Have made significant contributions to the CA Power Pool, Western Systems Coordinating Council, N Am Power Systems Interconnection Committee & the Natl Elec Re-

Kaprielian, Elmer F (Continued)
laibility Council - Fellow of IEEE. 1977, Tau Beta Pi Eminent Engr AW from Univ of CA, Berkeley, 1984 - Dist Engr, CA Poly San Luis Obispo. *Society Aff:* IEEE, ΤΒΠ.

Kapur, Kailash C
Business: Dept of Indust Engrg & Operations Res, Detroit, MI 48202
Position: Prof *Employer:* Wayne State Univ *Education:* PhD/Indus Engrg/Univ of CA, Berkeley; MS/Operations Research/Univ of CA, Berkeley; M Tech/IE-OR/IIT, India; BS/Mech Engrg/Delhi Univ, India *Born:* 8/17/41 Prof of Industrial Engrg & Operations Research, Wayne State Univ, Detroit, MI. Has worked as a Senior Research Engr with General Motors Research Labs, 69- 70. Visiting scholar, Ford Motor Co, 73. Consultant to Bendix, TACOM-US Army and 3M. Co-chrmn and seminar leader of the special seminars on Reliability in Product Design & Testing. Co-author of book *Reliability in Engrg Design* John Wiley, 1976. Published over 30 research papers in various scholarly and professional journals. Membership committee chrmn, ORSA, 1980-81; Program Chrmn, IIE, 1980-81, Assoc Editor, IIE, 1978-cont, presented over 20 papers at Professional Society Meetings. *Society Aff:* ASQC, IIE, ORSA, TIMS

Karadimos, Angelo S
Home: 4724 Bel Pre Rd, Rockville, MD 20853
Position: Supervisory Naval Architect. *Employer:* Navy Dept - Naval Sea Systems Command *Education:* Assoc/Aeronautical Engr/Embry Riddle Tech Inst; 2 yrs/ Aeronautical Engr/UCLA Exten. *Born:* 6/3/28. Degree in Aeronautical Engrg from Embry Riddle Tech Inst. 2 yrs in Army Engrs 1950-52. Mass Property Lead Engr Martin Marietta Corp on Titan Missile (Gemini) & many other advanced preliminary design concepts. Currently working as chief naval architect weight engr on Navy auxiliary & amphibious ships. SAWE Natl Officer, Pres, Sr VP, EVP, Tech Dir, Bd of Dir Mbr; received SAWE (highest award) Hon Fellow Award 1972. Enjoys modern music, golf & sports in genl, Married (Maurine) & has two daughters (Diane & Tina). *Society Aff:* SAWE, ASE.

Karady, George Gy
Home: 12 Douglas Dr, Princeton, NJ 08540
Position: Dir Elec Sys *Employer:* EBASCO Service. *Education:* MS/EE/Polytechnical Univ of Budapest; D Eng/EE/Poly Univ of Budapest *Born:* 8/17/30. Registered PE in NY and NJ, Dir Elec Sys of EBASCO, responsible consltg serv on elec system planning, distribution & transmission studies, advance engrg tech application. In addition, Adjunct Prof at the Poly Inst of NY. Formerly served as Chief Consltg Elec Engr, Chief Engr of Computer Tech and elec task leader of the Tokamak Fusion Test Reactor Proj (Princeton). Previously he was Prog Mgr in Hydro Quebec Inst of Res (CD). Worked as Assoc (Docent) Prof in the Tech Univ of Budapest (Hungary). Author of more than sixty tech papers & was elected IEEE fellow in 1978. Main field of interest: High voltage technique; transmission lines; power sys analyses, computer application, HVDC, elec indus electronics. *Society Aff:* IEEE, CIGRE

Karakash, John J
Business: Coll of Engrg/Phys Sci, Bethlehem, PA 18015
Position: Dean of Coll of Engrg & Phys Sci. *Employer:* Lehigh University. *Education:* Doctor of Eng'g/Eng/Lehigh; MS/EG/Univ of PA *Born:* June 14, 1914 in Turkey of Greek parents. Educ Dir 6th Serv Command Radar Sch 1940-42. Res Engr Computers & Radar, Moore Sch 1942-45; Asst Prof Lehigh Univ 1946 (res on microwave filters), Prof & Head EE Dept 1956, Disting Prof 1963, Dean of Coll 1966. Cons Bell Tel Labs 1950-56; Educ Cons Commonwealth of Puerto Rico 1973-75. Educ Cons to Engineering Colleges 1974- . Consultant IBM, 1980- . Life Fellow IEEE; Mbr: ASEE, Phi Beta Kappa, Sigma Xi, Ta Beta Pi, Omicron Delta Kappa. Robinson (1948) & Hillman (1963 & 1981) Awards for disting service Outstanding teaching & service student awards 1968 & 1975. IEEE Centennial Medalist, 1984. Mbr Bd of Trustees Wilkes College, 1980- . Mbr Bd of Dir, Komline-Sanderson Mfg Co. 1980- . *Society Aff:* IEEE, ASEE

Karara, Houssam M
Home: 1809 Coventry Dr, Champaign, IL 61821-5239
Position: Prof *Employer:* Univ of IL *Education:* DSc/Geodetic Sci/Swiss Fed Inst of Tech-Zurich; BSc/CE/Cairo Univ *Born:* 9/5/28 Native of Egypt, naturalized US citizen. Worked at Egyptian Ministry of Public Works, and at La Grande Dixence Power Project in Switzerland. Since 1957, teaching & research in civil engrg (Photogrammetry) at Univ of IL at Urbana- Champaign, as prof of civil & of Geography since 1966. Registered PE in IL. Pres (1972-76) of Commission V (Non-Topographic Photogrammetry) of the Intl Society for Photogrammetry. Editor-in-Chief, *Handbook of Non-Topographic Photogrammetry* (1979). Author-Editor of *Chapter XVI (Non-Topographic Photogrammetry) of the Manual of Photogrammetry*, 4th ed, 1980. Editor-in-Chief, *Handbook of Non-Topographic Photogrammetry*, Second Edition (1987). Author of some 50 technical papers on Photogrammetry and Photogrammetric engineering. *Society Aff:* ASPRS, ACSM, CISM, SBPTC, SFPT, DGPF.

Karasiewicz, Walter R
Business: P.O. Box 106, Columbia, SC 29202
Position: District Director *Employer:* Post, Buckley, Schuh & Jernigan, Inc *Education:* MSSE/Sanitary Engrg/GA Inst of Tech; BSCE/Civ Engrg/U of Miami, FL. *Born:* 12/23/42 1965-1969 Poznak & Assoc, Asbury Park, NJ; 1966-1967, Hensley Schmidt, Atlanta, GA,; 1967-1969, NJ State Dept of Health, Trenton, NJ; 1969-1971, Black, Crow & Eidsness, Atlanta, GA; 1971-1972, Brown and Root, Houston, TX; 1974-1977, Harwood Beebe Com, Colal, SC; 1977-1981, NcNair, Gordon, Johnson & Karasiewicz, Columbia, SC. Current position, Post, Buckely, Schuh & Jernigan, Inc Columbia, SC, District Director. *Society Aff:* WPCF, AWWA, NSPE, ASCE, AAEE, AAA.

Karassik, Igor J
Business: 233 Mt. Airy Rd, Basking Ridge, NJ 07920
Position: Chief Consulting Engr. *Employer:* Worthington Group, McGraw-Edison Co.. *Education:* MS/Mech Eng/Carnegie Tech; BS/Mech Eng/Carnegie Tech. *Born:* Dec 1911 in Russia. With Worthington 1934- , specialized in High Pressure Centrifugal Pumps 1936-59, Mgr of Planning Harrison Div 1959-63, Genl Mgr Advance Products Div 1963-70, VP & Chief Consulting Engr 1971-76, present position 1977- . Fellow ASME. P E in N J. Mbr Tau Beta Pi, Pi Tau Sigma, Sigma Xi. Has written numerous articles on pumps & steam power plants; author of books: 'Centrifugal Pumps - Selection, Operation & Maintenance', 'Engineer's Guide to Centrifugal Pumps' and 'Centrifugal Pump Clinie'; co-author of 'Pump Questions & Answers'; co-editor of 'Pump Handbook' publ by McGraw Hill Book Co. Recipient of the first ASME Henry R. Worthington Medal in 1980. *Society Aff:* ASME.

Karchner, George H
Business: Homer Research Labs, Bethlehem, PA 18015
Position: Research Engineer. *Employer:* Bethlehem Steel Corporation. *Education:* MS/Metallurgy/PA State Univ; BS/Metallurgy/PA State Univ. *Born:* Mar 1938. Employed as a Res Engr in Phys Metallurgy at Homer Research Labs, Bethlehem Research Engr in Physical Metallurgy at Homer Research Labs, Bethlehem Steel Corp 1961- ; working in alloy & process development. Received Marcus A Grossman Young Authors Award 1964. *Society Aff:* ASM.

Karcich, Matthew F
Home: 2410 Magellan, Colorado Springs, CO 80907
Position: VP Secy, Treas (Ret). *Employer:* Karcich & Weber Inc. *Education:* BS/Civ Engg/Univ of CO. *Born:* 5/27/15. Native of Trinidad, CO. Employed by Corps of Engrs Tulsa Dist, 1939-1941; USAF 1941-1945; USAFR Ret 1975; Boeing Aircraft Co 1951-1952 Redesign of parts for Ind Mobilization B-47 & B-52 aircraft; UT Const Co 1953-1954 Proj Engg. Housing; R Keith Hook & Assoc & Henningson, Durham & Richardson Municipal Projs. Paving-Water Treatment-Waste Water

Karcich, Matthew F (Continued)
Treatment, etc. Design & Field Engg 1954- 1965; Past Mbr & Chrmn Urban Drainage Bd City of CO Springs; Pres-Pikes Peak Chapter Reorganization-CO Springs Branch SCE; State Dir-NSPE: Principal and Partner K & W Inc CO Springs Apr 1963 - July 1979. Municipal Projs-Sold my 1/2 interest to Partner & Retired. Intend to pursue part time availability as consultant. *Society Aff:* ASCE, NSPE, CSI, ACI.

Kardos, John L
Business: Campus Box 1087, Materials Res. Lab, Washington University, St. Louis, MO 63130
Position: Prof. Director, Materials Research Laboratory *Employer:* WA Univ. *Education:* PhD/Polymer Engg/Case Western Reserve Univ; MS/Chem Engg/Univ of IL; BS/Chem Engg/Penn State Univ. *Born:* 4/19/39. Dr Kardos joined WA Univ in 1965 as Asst Prof and became Prof of Chem Engg in 1974. Since 1971 he has held the position of Dir of the Materials Res Lab and served during 1977-78 as Acting Chrmn of Chem Engg. His res efforts are in the area of composite and biomedical materials, including structure-property prediction, process modeling, rheology and characterization. He is currently active as a consultant to US industry in the area of reinforced plastics. He currently serves on the editorial bds of the Composites Sci and Tech and Polymer Composites. He received the 1981 Materials Engineering and Sciences Division Award of the American Institute of Chemical Engineers. *Society Aff:* AIChE, SPE, APS, ACS, SR, ASEE, SAMPE.

Kareem, Ahsan
Home: 7955 Dawnridge, Houston, TX 77071
Position: Assoc Prof & Independent Cons *Employer:* Univ of Houston *Education:* PhD/Struct Engr/CO State Univ; MSc/Struct Engr/Univ of HI/MIT; BSc/Struct Engr/Univ of Engrg & Tech-Pakistan *Born:* 9/29/47 Struct engr, Harza Engrg Co Intl, 1968-71; grad student 1971-77. Res Assoc, CO State Univ, 1977-78; Asst Prof, 1978-84; Assoc Prof, and Dir of Struct Aerodynamics and Ocean Sys Labs, Univ of Houston since 1984; recipient of the 1984 Presidential Young Investigator award; author of 60 tech papers; and independent conslt to indus spec in stochastic environ loads (i.e., wind, waves and earthquakes on land-based, aerospace and offshore struct). Listed in International Who's Who in Engrg, Who's Who in Technology Today, Who's Who in South and Southwest, Amer Men and Women of Science. Mbr, Natl Res Council Ctte on Natural Disasters. Enjoy playing racquetball, softball and tennis. *Society Aff:* ASCE, AIAA

Karger, Delmar W
Home: 506 Circle Dr, De Funiak Springs, FL 32433
Position: Consultant *Employer:* self *Education:* MS/Gen Engr/Univ of Pittsburgh; BS/EE/Valparaiso Univ. *Born:* 5/9/13 Reg Engr., Consultant, Expert Witness, Corp Dir & Investor. Over 80 man-years experience as a corporate director. Author: Engineered Work Measurement, 1986; Advanced Work Measurement, 1982; Managing Engrg & Res, 1980; Long Range Planning & Corp. Strategy, 1979; New Product Venture Management, 1972; Advanced Investment Decision Making, 1988; Engrg & Res Management; 1989. Over 6 other books and 100 articles. Educator: Dean Emeritus of School of Management, Ford Foundation Prof. Emeritus former Head of Indust Engrg, RPI, Troy, NY. 51 yrs. of professional exper, 25 in industry. Fellow of four professional societies. Biography in over 24 Who's Who publications. *Society Aff:* IIE, AAAS, SAM, MTM, IEEE

Karkalits, O Carroll
Business: Mcneese St Univ, Lake Charles, LA 70609
Position: Dean College of Engrg & Tech. *Employer:* McNeese State University. *Education:* BS/Chem Engg/Rice Univ; MS/Chem Engg/Univ of MI; PhD/Chem Engg/Univ of MI. *Born:* May 31, 1916 Pauls Valley Oklahoma. Engrg for Shell Oil 1938-40; Univ of Mich Teaching Fellow 1943-45, Instr 1945-47; Group Leader in Proc Dev at Amer Cyanamid 1948-56; at Petro-Tex Chem Corp, Mgr Res 1956-66, Asst Dir of Engrg 1966-72; Dean Col of Engrg & Tech 1972- . Chmn So Texas Sect AIChE 196970. Mbr: AIChE, ASEE, AAAS, Amer Assn of Cost Engrs, Amer Sci Affiliation, La Engrg Soc. Reg Prof Engr: La & Tex. (As of spring 1984, Fellow of AIChE, mbr also of NSPE). *Society Aff:* AIChE, ASEE, AAAS, AACE, NSPE.

Karlin, John E
Home: Wigwam Rd, Locust, NJ 07760
Position: Retired. *Education:* BA/U of Cape Town; MA/U of Cape Town; PhD/U of Chicago *Born:* Feb 1918. PhD Univ of Chicago 1942. Res Assoc at Psycho-Acoustic Lab Harvard Univ 1942-45; Bell Labs Mbr of Tech Staff 1945-52, Head of Human Factors Engrg Dept 1952-77. Fellow IEEE, Human Factors Soc, & APA Div of Engrg Psychologists. Interests: ecology, arts & tennis. *Society Aff:* APA, HFS, IEEE, ASA, CAS.

Karmel, Kenneth E
Home: 371 Overlook Dr, West Lafayette, IN 47906
Position: President *Employer:* Delta K Co *Education:* BChE/Chem Engg/Cornell Univ. *Born:* 2/6/32. NYC, son of Mabel & Charles. Employed Rohm & Haas Co, Phila 1954-1966, since 1966 by Great Lakes Chem Corp. VP, Great Lakes and Arkansas Chem since 1966. Dir of AR Chem sice 1976. Reg Prof Engg, Penna and AR. Founder, Delta K Co, 1981. *Society Aff:* AICHE.

Karmel, Paul R
Home: 26 Hopke Ave, Hastings-on-Hudson, NY 10706
Position: Prof *Employer:* City College of CUNY. *Education:* DEngrSc/EE/Columbia Univ; MS/EE/MIT; BE/EE/Cornell Univ. *Born:* 12/20/34. Teaching undergrad & grad elec engg at the CCNY since 1964, specializing in electromagnetics, microwaves and antennas. Acting Dean of Engg 1973-1975. Mbr of technical staff COMSAT Labs (1977-1979), Antenna Dept. BE, Cornell '56; MSc, MIT '57; D Engr Sc, Columbia '64. *Society Aff:* IEEE, ASEE, ΣΞ.

Karmele, Nicholas L
Home: 5808 Willow Glen Ct, Dayton, OH 45431
Position: Technical Dir. *Employer:* Foreign Tech Div, Wright-Patterson AFB. *Education:* MS/Public Admin/Geo Wash Univ; MS/ME/OH State Univ; BS/ME/ Northwestern Univ. *Born:* 8/2/33. Served a tour of duty as an AF officer with the Aerospace Technical Intelligence Ctr at Wright-Patterson AFB, OH. Upon completing military duty in 1958, accepted a Civ Service appointment. Assumed positions of increasing responsibility in scientific & technical intelligence analysis of foreign systems as an engg analyst, methods engr, group leader, branch chief, & currently as Technical Dir. Directs a 375 person organization in the processing and analysis of sensor data. Is a Sr Exec Service employee in the fed govt. Was selected by USAF to attend the Ind College of the Armed Forces in 1974-1975. Enjoys reading, woodworking & sports. *Society Aff:* ΣΞ, ΤΒΠ, ΠΤΣ.

Karn, Richard W
Business: 2551 Merced Street, San Leandro, CA 94577
Position: President/CEO *Employer:* Bissell & Karn, Inc. *Education:* B.S./Civil Engrg. /University of Calif., Berkeley *Born:* 07/19/27 Born in Oakland, CA. Served in U.S. Navy 1945-46. With Alameda County Flood Control & Water Conservation Dist 1950-66; Engineer-Manager 1962-66. Since 1966 with Bissell & Karn, Inc. as VP and Pres. Responsible for direction of six-office firm providing civil engrg services throughout N. CA with specialization in transportation, water resources, flood control, land planning & development, and solid waste management. Pres, ASCE 1984-85; Chairman, AAES 1987; Honorary Fellow, Institution of Civil Engineers (U.K.) 1985. Enjoys sailing, classical music, woodworking. *Society Aff:* ASCE, ACEC, APWA, NSPE, SAME

Karnaugh, Maurice
Business: PO Box 218, Yorktown Heights, NY 10598
Position: Research Staff Member. *Employer:* IBM Thomas J Watson Research Center. *Education:* PhD/Physics/Yale; MS/Physics/Yale; BS/Physics/CCNY. *Born:* Oct 1924. 1952-66 at Bell Telephone Labs Murray Hill N J working on digital tech

Karnaugh, Maurice (Continued)
in telephony; 1966-70 at Federal Sys Div IBM Gaithersburg Md; 1970- at Res Div IBM Yorktown Heights N Y. Fellow of IEEE, Secy Genl of Internatl Council for Computer Communication, Mbr of IEEE Communication & Computer Soc's, ACM SIGART, SIGCOMM, SIGOA. *Society Aff:* IEEE, ACM, ΦΒΚ, ΣΞ.

Karni, Shlomo
Business: Dept of EE & CE, UNM, Albuquerque, NM 87131
Position: Prof of Elec & Computer Engrg *Employer:* University of New Mexico. *Education:* PhD/EE/Univ of IL, Urbana; MEng/EE/Yale Univ; BS/EE/Technion, Israel. *Born:* June 1932. Visiting positions: Univ of Hawaii, Tel Aviv Univ (helped found sch of engrg, served as 1st Dean 1970-71). At Univ of N M 1961- . Industrial work: Palestine Power Co, Gulton Indus, Westinghouse, Los Alamos National Lab, AF Weapons Labs Doe. On editorial board IEEE Spectrum 1969-72; Assoc Ed IEEE Journal Circuits & Systems 1975-1981; Ed. 1981-86; on editorial board, IEEE Press, 1986- . Authored 5 textbooks (1 translated into Russian), num articles. Correspondent, Acad of Hebrew Language, Jerusalem 1970. Fellow IEEE, 1976. *Society Aff:* IEEE, AAUP, ASEE, ΤΒΠ.

Karpenko, Victor N
Business: P O Box 808 - East Ave, Livermore, CA 94550
Position: Proj Mgr. *Employer:* Lawrence Livermore Laboratory. *Education:* BS/Mech Engg/Univ of CA at Berkeley, CA. *Born:* 1/23/22. Currently Proj Mgr of Mirror Fusion Test Facility Lawrence Livermore Lab (LLL); worked in ordnance equip, utility sys & nuclear reactor design fields before joining LLL in 1957; during assoc with LLL had respon ranging from nuclear explosive deisgn, hardening of sys to nuclear weapons effects; and div leader respon for dev of engg tech necessary to conduct & contain nuclear underground tests; in current capacity respon for planning, engg & construction of Mirror Fusion Test Facility. Mbr of various Dept of Energy Review Cttees: Nuclear Test Containment Evaluation, Fusion Reactor Safety and Design of Toroidal Core Experiment (TFCX). Published papers on hardening of thermonuclear weapons, seismic safety, concept of undergrounding of nuclear power reactors, and Magnetic Fusion. *Society Aff:* ANS.

Karplus, Walter J
Business: 3732 Boelter Hall., UCLA, Los Angeles, CA 90024
Position: Prof. *Employer:* University of California at L A. *Education:* PhD/Engg/Univ of CA; MS/Elec Engg/Univ of CA; BS/Elec Engg/Cornell Univ. *Born:* Apr 1927, native of Vienna Austria. US Navy 1945-46. Field Party Chief Sun Oil Co; Res Engr Internatl Geophys Inc & Hughes Aircraft Co. Reg Prof Elec Engrg, Calif 1955. Mbr of Fac Sch of Engrg & Applied Sci UCLA 1955- ; Chmn Computer Sci Dept 1972-79. Res in On-Line Computation, Hybrid Computation, Math Modeling & Simulation, Digital Simulation Languages. Fulbright Fellow 1961; Guggenheim 1968; Fellow IEEE 1971. Author or co-author of 7 tech books & over 100 tech papers in EE & Computer Sci; 4 patents. *Society Aff:* SCS, ACM, IEEE, IMACS.

Karr, Andrew E
Business: Chem-pro Division, Otto N. York Co, 42 Intervale Rd, Parsippany, NJ 07054-0918
Position: Technical Mgr Karr Column *Employer:* Chem-Pro Division, Otto H. Yorklo *Education:* BChE/Chem Eng/Cooper Union; MChE/Chem Eng/Polytechnic Inst of Brooklyn; DChE/Chem Eng/Polytechnic Inst of Brooklyn. *Born:* Sept 1917. With Amer Cyanamid 1941-44 & 1946-47; US Navy 1944-46. With Hoffmann LaRoche Inc 1947-76; Mgr of Chem Engrg Process Dev & Design; in 1970 became Mgr of Process & Proj Engrg. Dev Karr Reciprocating Plate Extraction Column. Taught grad course in liquid-liquid extraction at Brooklyn Polytechnic. Fellow AIChE; Mbr ACS, Tau Beta Pi. Reg Prof Engr, N J. Author of 10 patents & 27 publs. Active in local & natl cttes of AIChE. Enjoys tennis. *Society Aff:* AIChE, ACS, AIME, ΣΞ.

Karren, Kenneth W
Home: 424 E 4750 N, Provo, UT 84601
Position: Consulting . *Employer:* Self-employed. *Education:* BS/Civil Engg/Univ of UT; MS/Civil Engg/Univ of UT; PhD/Civil Engg/Cornell Univ. *Born:* 5/20/32. USN Cvl Engr Corps 3 yrs; Phillips Petro 1 yr; Prestressed Concrete Div Otto Buehner & Co 4 yrs; Univ Prof BYU 17 yrs; Cons to Hercules Inc on finite element methods & struct mech 11 yrs; P Pres Utah Sect ASCE; Consultant. *Society Aff:* ASCE.

Karsky, Thomas J
Business: Ag Engrg Dept, Moscow, ID 83843
Position: Extension Safety Specialist *Employer:* Univ of ID *Education:* MS/Ag Mechanization/ND State Univ; BS/Ag Mechanization/ND State Univ *Born:* 12/2/50 Native of Langdon, ND. Worked in the field production dept for Green Giant Co, Montgomery, MN 1974-1977. With the Univ of ID Cooperative Extension Service since 1977. Responsible for promoting and conducting farm safety programs in ID. Also some responsibilities in farm machinery educational area. Inland Empire Section ASAE, Chrmn 1981, Chrmn Elect 1980, 1st Vice Chrmn 1979, Secy-Treasurer 1978. Enjoy camping, hunting and fishing. *Society Aff:* ASAE, NIFS

Karstens, Gerald A
Home: 2506 Hidden Glen, San Antonio, TX 78232
Position: Retired. *Education:* BSAE/Agri Engr/SD State Univ. *Born:* 1914 Elkton S D. Grad Study Purdue Univ. Employed by USDA 1941-46; Purdue Univ, Asst Prof 1946-53; Amer Trade Publ Co, V Pres 1953-65; Amer Feed Manufacs Assn, V Pres 1965-1979. Mbr ASAE (Fellow 1970), Board of Dir 1968-69 & 1973-75, Finance Ctte & 1968 Chmn, Grain & Feed Processing & Storage Ctte, Chmn of Ctte on processing standards, 4 yrs mbr of Tech Paper Awards Ctte; Natl Fire Protection Assn & mbr of Grain, Feed & Agri Standards Ctte; Natl Assn of Exposition Mgrs Certified Exposition Manager, 1977; Chicago Feed Club, P Pres, P Secy. Selected during Jr yr of coll for Farm Equipment Inst Summer Sch of Indus. Represents AFMA on Natl Safety Council; Mbr of Air Pollution Control Assn; Chicago Press Club; Life Fellow, ASAE (1979). *Society Aff:* ASAE.

Kartiganer, Herbert L
Business: 555 Rte 94/New Windsor, Newburgh, NY 12550
Position: President. *Employer:* Kartiganer Associates, Prof Corp. Consulting Engineers. *Education:* BCE/Civil Engg/Syracuse Univ. *Born:* Sept 1929. Engrg Officer USAF Korean War. Founder Kartiganer Assocs Cons Engrs 1957; Sanitary Sci & Labs Inc Newburgh N Y. N Y; St Lic Oper of Public Water & Wastewater Treatment Plants; Cert Value Engr; Reg Prof Engr/Surveyor: N Y, NJ, FL, Conn, Mass & Pa. Fellow ASCE; Mbr ASTM, AWWA, SAME, Natl Water Poll Control Fed, Amer Arbit Assn, Prof Serv Bus Mgmt Assn; Amer Cons Surveying & Mapping; NSPE; 1974-75 Pres CEC/NYS. Recipient: Civic Achievement Award 1971 Town of New Windsor, Cons Engrs Council NYS Grand Award of Engrg Excellence 1970. *Society Aff:* ASCE, ACEC, ASTM, AWWA, WPCA, AAA, NSPE.

Kashyap, Rangasami L
Business: School of Electrical Engrg, W. Lafayette, IN 47907
Position: Prof of EE *Employer:* Purdue Univ *Education:* PhD/Engrg/Harvard Univ; ME/Elec Engrg/Indian Inst of Science; BE/Elec Engrg/Indian Inst of Sci *Born:* 3/28/38 in Mysore, India; McKay prize fellow at Harvard. Joined Purdue in 1966; currently prof of Elec Engrg at Purdue; Author of numerous original research papers published in various IEEE transactions, author of many book chapters and books including *Stochastic dynamic models from empirical data* published by Academic Press; winner of the best original research award in the National Electronics Conference, 1966; Fellow of the IEEE, 1979. *Society Aff:* IEEE, ACM

Kasper, Arthur S
3194 Gulf Coast Dr, Spring Hill, FL 33526
Position: Technical Director, Automotive *Employer:* Reynolds Metals Company *Education:* BSE/Met Engg/Univ of MI. *Born:* 9/13/24. Technical Director - Reynolds Metals Company, Supervisor Chrysler Sheet Metal Engg 1970 - Sr Mtls Engg

Kasper, Arthur S (Continued)
1968 - Met Engg 1956 - Product Engg Hall Lamp Co 1955 - Process Engg Ford Motor Co 1955 - Staff Met Hoadaille 1952 - Res Met 1946. BSE Met Engg Univ of MI 1946 - Ensign US Naval Acad 1946. ASM elected Fellow 1978 - Authored and presented many technical papers at various technical and business meetings involving US & foreign participants in this country and abroad. IDDRG 1976 Technical Prog Chrmn - Exec Comm's ADDRG, ASM - SAE. Arch T Cowell Award outstanding paper 1976 - SAE, ISTC, DIU-32, Chrmn-SAE Gen Mtls Council, SAE Non-Ferrous Committee - Chairman, SAE-Wrought Aluminum Subcommittee - Chairman.. *Society Aff:* ASM, ASTM, ADDRG, SAE.

Kassam, Saleem A
Business: Dept of Elect Engrg, University of Pennsylvania, Philadelphia, PA 19104
Position: Prof *Employer:* Univ of PA *Education:* PhD/Elect Eng/Princeton U; BS/Engrg/Swarthmore Coll; MSE/Elect Eng/Princeton U; MA/Elect Eng/Princeton U. *Born:* 6/16/49 in Dar-es-Salaam, Tanzania. Resident in USA since 1970. Joined Univ of PA as Asst Prof in 1975, promoted to Assoc Prof in 1980, Full Prof in 1986. Active in sponsored basic res for AF & Navy. Consultant in Ind Res Programs. Married, three children. *Society Aff:* IEEE

Kastenberg, William E
Business: 401 Hilgard Ave, Los Angeles, CA 90024
Position: Prof *Employer:* UCLA *Education:* PhD/Nuclear Engrg/Univ of CA, Berkeley; MS/Nuclear Engrg/UCLA; BS/Engrg/UCLA *Born:* 06/25/39 Prof and Chrmn; Dept of Mech, Aero and Nuclear Engrg, UCLA, 1985-present. Assistant Dean of Graduate Studies, UCLA, 1981-1985. Res interests include nuclear reactor safety, probabilistic risk assessment, design and dev of advanced reactors (LMFBRS), fusion sys and space reactors. Has taught courses on reactor phys, safety and dynamics, environ engrg and applied mathematics. Conslt to US Nuclear Regulatory Comm, Natl Labs and Indus. *Society Aff:* ANS, AAAS

Kastor, Ross L
Home: 13514 Dripping Springs, Houston, TX 77083
Position: Engineering Advisor *Employer:* Shell Oil Company *Education:* B.S. and M. SC.,/Mechanical Engr./Ohio State Univ *Born:* 10/05/22 Joined Shell Oil Company in 1950. Had numerous assignments with Shell Production in Texas and Louisiana. Special assignments include two years in England as Design Assistant to Sir Frank Whittle, two years as Manager, Mech Eng Res, Shell Development Company, Houston, two years as Chief-Mech Engg in Houston Area/Region. Drilling Engineering Lecturer in Shell's Technical Training Dept for 16 years. Currently developing Drilling Computer Systems for Shell. Reg PE - Texas and Ohio. Member Tau Beta Pi; Pi Tau Sigma; Member ASME Executive Committee, South Texas Section ASME. Director OTC, Inc. Fellow member ASME, Member SPE *Society Aff:* ASME, SPE

Katell, Sidney
Home: 1464 Dogwood Ave, Morgantown, WV 26505
Position: Professor. *Employer:* West Virginia Univ. *Education:* -/Ind Eng/Pittsburgh; BChE/Chem Eng/NY Univ. *Born:* Feb 2, 1915 New York. Grad work Univ of Pgh. Presently Prof of Mineral Economics W Va Univ's Coll of Mineral & Energy Resources. Previously Chief of Process Evaluation US Bureau of Mines Morgantown W Va. Founding Mbr & P Pres of AACE & of the W Va Sect AACE; has received Award of Merit from AACE, Meritorious Service Award from Dept of Interior & Award of Merit from Oper Sect of Amer Gas Assn. *Society Aff:* AACE, AIChE, ACS, AGA.

Kates, Willard A
Home: 690 Lyons Circle, Highland Park, IL 60035
Position: Pres. *Employer:* W A Kates Co. *Education:* BS/EE/Univ of WI. *Born:* 6/21/01. 1921-23 testing, design of electrical machinery at GE and Westinghouse. 1923-29 Power plant, substation design, reports, appraisals, day and Timmermann and Unitet Engrs. 1929-44 Specialty glass business, Corning Glass Works. Managed Industrial, Pyrex Housewares, Consumers div. 1948 Pres, The W A Kates C - makers of automatic flow rate controllers. *Society Aff:* ISA, AIChE, IEEE, ΦΚΦ, ΠΒΤ, ΗΚΝ.

Kato, Walter Y
Business: Dept of Nuclear Energy, Upton, NY 11973
Position: Deputy Chairman *Employer:* Brookhaven National Laboratory. *Education:* PhD/Physics/PA State Univ; MS/Physics/Univ of IL; BS/Physics/Haverford College. *Born:* Aug 1924 Chicago Ill. Deputy Chairman 1980- Assoc Chmn for Reactor Safety (1975-80) & Sr Nuclear Engr, Dept of Applied Sci 1975-77, Dept of Nuclear Energy 1977- , Brookhaven Natl Lab 1975- ; Sr Physicist, Applied Phys Div 1969-75; Head, Fast Reactor Experiments Sec, Applied Phys Div 1963-70; Assoc Physicist, Reactor Engrg Div 1954-63; Argonne Natl Lab 1953-75; Visiting Prof, Nuclear Engrg Dept Univ of Mich 1974-75; Jr Res Assoc Brookhaven Natl Lab 1952- 53. Specialist in neutron & reactor physics, nuclear engrg & reactor safety. Fulbright Fellowship (Japan) 1958-59, Lectr Tokyo Univ & res Japan Atomic Energy Res Inst 1958-59. Argonne Univ Assn Distin Appointment Award 1974; Fellow & Mbr Board of Dirs, Amer Nuc Soc; Mbr APS, Sigma Xi & AAAS. Cons to Dir Off Nuc Reg Res, USNRC. *Society Aff:* ANS, APS, AAAS.

Katsanis, Theodore
Business: 21000 Brookpark Rd, Cleveland, OH 44135
Position: Aerospace Engineer. *Employer:* NASA, Lewis Research Center. *Education:* PhD/Math/Case Inst of Tech; MS/Math/Univ of WA; BS/Aero Engg/Parks College, St Louis Univ. *Born:* July 1925. PhD Case Inst of Tech; Designed logging machinery, Smith- Berger 1952-62; with NASA 1963- . Current work is in analysis of internal flow through turbomachinery. Several computer progs have been developed for this analysis & are widely used both in US & many other countries. 27 publs. SAE Manly Memorial Award 1974. *Society Aff:* ASME, AIAA.

Kattamis, Theo Z
Business: Dept of Met, U-136, Storrs, CT 06268
Position: Prof. *Employer:* Univ of CT. *Education:* ScD/Met/MA Inst of Tech; MS/Met/MA Inst of Tech; Dipl Engg in Mining/- /Univ of Liege, Belgium; Geol and Met/-/Univ of Liege, Belgium *Born:* 5/7/35. Taught Geology of Mineral Deposits at the Univ of Liege, Belgium (1960-62). Res Assoc in Met at MIT (1965-69). Joined the Dept of Met at the Univ of CT in 1969, where he teaches courses and directs res in metals casting and solidification, composite materials, metals joining, mtls processing in gen and also in the history of tech. Mbr of several technical committees, active in local chapters of technicl societies and in industrial consulting. *Society Aff:* ASM, AIME, AFS.

Kattus, J Robert
Business: Birmingham, AL 35203
Position: Consulting Metallurgist. *Employer:* Associated Metallurgical Consultants, Inc. *Education:* BS/Met Eng/Purdue Univ. *Born:* Aug 25, 1922 Cincinnati Oh. Ltjg USNR WW II. Metallurgist Aluminum Indus Inc Cincinnati Ohio 1946-48; Chief Metallurgist Anderson Elec Co Birmingham Ala 1948-52; Dir Metallurgy Res Southern Res Inst Birmingham Ala 1952-66; Cons Metallurgist 1966-79; Vice Pres Assoc Metallurgical Consultants, Inc. Natl Trustee ASM 1965-67; Chmn ASTM Ctte A-4 on Iron Castings 1965-69; Chmn SE Dist ASTM 1963-65; Award of Merit ASTM; Fellow ASM & ASTM. Publ 25 tech papers; co- author Aerospace Structural Metals Handbook. *Society Aff:* ASM, ASTM.

Katz, Donald L
Business: Dept of Chem Engrg, Ann Arbor, MI 48109
Position: Prof Emeritus of Chem Engg. *Employer:* Consulting Engr. *Education:* PhD/ChE/Univ of MI; MS/ChE/Univ of MI; BSE/ChE/Univ of MI. *Born:* Aug 1, 1907 Jackson County Mich. 1933-36 Phillips Petro Co Bartlesville Okla; Univ of Mich faculty mbr 1936-77, Chmn of Dept 1951-62, A H White Univ Prof. 1966-77, currently Emeritus Prof. Mbr: Amer Chem Soc, AIChE, SPE of AIME, ASME,

Katz, Donald L (Continued)
ANS, AAAS, ASEE, AGA, AAPG, NAE; Pres AIChE 1959; elected Mbr of NAE 1968. 285 tech publs incl some 15 books, such as 'Handbook of Natural Gas Engineering'. Recipient of 13 awards or honors incl the Founders, Lewis & Walker Award of AIChE. Received Natl Medal of Sci from Pres Reagan May 24, 1983. Cons engr to over 75 companies & govt organizations. *Society Aff:* AIChE, ASME, SPE, AIME, NAE, ASA, ACS, ANS, AAAS, ASEE, AGA, AAPG.

Katz, Erich A
Business: 344 Great Neck Rd, Great Neck, NY 11021
Position: Pres. *Employer:* E K Construction Co Inc. *Education:* MCE/Structural/NYU; BS/Civ Engg/Inst of Tech. *Born:* 3/25/26. PE (NY, NJ & FL). 1948-1960 Structural Engr for maj engg firm - design of industrial & petrochemical plants. 1960-present Pres of E K Construction Co Inc - Active in construction of public bldgs ($3-16,000,000 range), firehouses, schools, post-offices, libraries, community ctrs, maintenance shops for the US Army, medical & dental facilities for the US Navy, dormitories & similar facilities for fed, state & city govts. *Society Aff:* ASCE, NYPE.

Katz, I Norman
Business: Washington Univ, Box 1040, St. Louis, MO 63130
Position: Prof. *Employer:* WA Univ. *Education:* PhD/Mathematics/MA Inst of Tech; MS/Mathematics/Yeshiva Univ; BA/Mathematics/Yeshiva Univ. *Born:* 4/14/32. Senior Scientist at AVCO/Res and Advanced Dev, Wilmington, MA 1959-62; Sec Chief, Mathematical Analysis 1963-65; Mgr, Mathematics Dept 1966-67. Assoc Prof at School of Engg and Applied Sci, WA Univ 1967-1974; Prof of Applied Mathematics and Systems Sci, 1975-; Co-chrmn of the Undergrad Prog in Systems Sci and Mathematics. Professional interests: Finite Element Methods, Numerical Analysis, Algorithms for parallel computation, Optimal Facility Location. Visiting Consultant for SIAM, 1979. *Society Aff:* SIAM, ORSA, AMS, ACM (SIGNUM), MAA.

Katz, Isadore
Business: Applied Physics Lab, Laurel, MD 20810
Position: Physicist, Principal Staff. *Employer:* The Johns Hopkins University. *Education:* BS/Physics/Temple Univ. *Born:* Oct 1916 Philadelphia Pa. Grad work in phys Univ of Md; W S Parsons Fellow Johns Hopkins Univ 1968-69. Physicist MIT Radiation Lab 1942-46; Naval Res Lab 1946-52; Applied Phys Lab - Johns Hopkins Univ 1952- . Specialist in Radio Wave Propagation, Radar Meteorology, Turbulence, Rough Surface Scattering, Microwave sensing of the Atmosphere from Satellites. Head: Radar Atmospheric Phys Group APL. Fellow: IEEE. Mbr: Amer Meteorological Soc, Amer Geophys Union, Intl Sci Radio Union, Geophys Res Bd Natl Acad of Sci 1972-75. *Society Aff:* IEEE, AMS, AGU, URSI.

Katz, J Lawrence
Business: Department of Biomedical Engineering, Troy, NY 12181
Position: Prof. *Employer:* Rensselaer Poly Inst. *Education:* PhD/Phys/Poly Inst of Brooklyn; MS/Phys/Poly Inst of Brooklyn; BS/Phys/Poly Inst of Brooklyn. *Born:* 12/18/27. Born in Brooklyn, NY. Served in US Navy 1946-48. With RPI since 1956. NSF Sci Faculty Fellow & Honorary Res Asst in Crystallography, Univ College London 1959-60. Visiting Prof Biomedical Engg and Oral Biology, Univ of Miami 1969-70. VA Scientific Review & Evaluation Board for Rehabilitation Research & Development 1981-83. Pres, 1983-84, Member, Board of Directors, Biomedical Engineering Society 1982-85. Pres, Soc for Biomtls 1978-79; Clemson Award for Biomtls 1975. Chrmn, Biomedical engg Div, ASEE 1978-79. Guggenheim Fellow, Visiting Lecturer of Orthopaedics at Harvard Medical Sch & Visiting Biophysicist, Children's Hospital Medical Ctr 1978. Organizer and Dir of the Ctr for Biomedical Engg 1974-82. Chairman, Department of Biomedical Engineering 1982-85. NIH Senior International Fellow to Q.M.C., U. of London, 1985 and Lab. de Recherche Orthopediques, Paris, 1986. Res on the structure, function and properties of bone and skeletal biomtls. Active in politics, civic theater, stamp collecting & folk singing. *Society Aff:* ASME, IEEE, APS, SB, BMES, ASB, AIME, ORS, IADR, ASEE

Katz, Joseph L
Business: The Johns Hopkins University, Dept. of Chemical Engineering, Baltimore, MD 21218
Position: Prof of Chem Engg. *Employer:* Johns Hopkins Univ. *Education:* PhD/Chemistry/Univ of Chicago; BS/Chemistry/Univ of Chicago. *Born:* 8/4/38. in Colon, Panama. Amanuensis, Univ of Copenhagen 1963-64. From 1964-70 was a Mbr of Technical Staff at North American Aviation Sci Ctr. Joined faculty at Clarkson College of Tech; Assoc Prof of Chem Engg 1970-75, Prof of Chem Engrg 1975-79. Since 1979 has been at the Johns Hopkins Univ as Prof of Chem Engrg. Dept. Chrm 1981-84; Dir, Energy Research Institute 1981-83. Received the J W Graham Jr Res Prize, 1975, a John Simon Guggenheim Meml Fdn Fellowship in 1976 and the MD Chemist of the Year award, 1982. *Society Aff:* AIChE, ACS, APS, ΣΞ.

Katz, Marvin L
Business: Chief Petroleum Co, 14330 Midway Road, Suite 229, Dallas, TX 75244
Position: VP Engg *Employer:* Chief Petroleum Co. *Education:* PhD/Chem Engg/Univ of MI; MS/Chem Engg/Univ of MI; BS/Chem Engg/Univ of MI. *Born:* 12/12/35. Production res, Sinclair Oil & Gas, 1960-64. Mgr, Sinclair's Systems & Computing 1966-69. Mgr, Admin Dept, ARCO, Dallas, 1969-72. Mgr and VP, Res & Dev, ARCO, 1972-82. Responsible for petroleum production res, exploration res, and synfuels, coal, and minerals res. VP Planning & Eval, ARCO, Dallas, 1982-84. VP Engg, ARCO, Dallas, 1984-86. Chrmn, Tech Task Group of Natl Petroleum Council's Comm on Enhanced Recovery Techniques for Oil & Gas in the US, 1976. VP, Engg, Chief Petroleum Co., Dallas 1986-present. Pres SPE of AIME, 1980. *Society Aff:* SPE of AIME, AIChE

Katz, Peter
Home: 168 Carnavon Pkwy, Nashville, TN 37205
Position: Engrg Consultant *Employer:* self *Education:* M/ME/Polytechnic IN of Bklyn; B/ME/Cooper Union; BA/Business Adm/Hamilton Univ *Born:* 7/11/22 I was in executive positions in Engrg and Management at ROWE Mfg (vending machines), Varityper (Typewriters), New Brunswick Scientific (Instrumentation) and NASA - to develop new products and production machinery. Last-VP and Founder of Precision Tubular Heater Co. Designed and built all their machinery to automatically produce electric heating elements for major and minor appliances. Presently engrg consultant for Teledyne, Chromalox, and TENNSCO Corp to develop new products and automated production machinery. Hold several patents and product awards. *Society Aff:* NSPE

Katz, William E
Home: 11 Sunset Rd, Weston, MA 02193
Position: Exec V P *Employer:* Ionics, Inc. *Education:* SB/Chem Engg/MA Inst of Tech; SM/Chem Engg Pract/MA Inst of Tech. *Born:* 6/12/24. Honesdale, PA; graduated Honesdale High School, 1941. Entered MIT in 1941. Active duty US Army 1943-46. In charge of island communication station Tinian, Mariana Islands, 1945-46. Returned MIT 1946-49. Wrote music for 1948 and 1949 Tech Shows. Joined Ionics, Inc 1949, ad Chem Engg, later Treasurer. Currently Exec VP and Dir of Ionics, respon for Water Systems Div. Involved in the first US Municipal Electrodialysis Water Desalting Plants at Coalinga, CA, and Buckeye, AZ. Has supervised dev of the electrodialysis reversal (EDR) process for water desalting. *Society Aff:* AIChE, AWWA, NWSIA.

Katz, William J
Home: 220 W Cherokee Cir, Milwaukee, WI 53217
Position: Pres *Employer:* WJK Assocs Ltd *Education:* PhD/CH Engr/U WI Madison; MS/CH Engr/U WI Madison; BS/CH Engr/U of IL *Born:* Jan 1925. MS & PhD Univ of Wis; BS Univ of Ill. Res Assoc Univ of Wis 1951-53; R/D Rexnord Inc 1953-70; Div Mgr of Govt Contracts R/D Envirex Inc 1970-74, V Pres R/D 1974-76. 1976-81 Tech. Dir. Milw Metro Div Sew Dist. Div Pres Camp Dresser McKee.

Katz, William J (Continued)
Pres. WJK Environmental System Ltd present Professor, Env. Engr., M. U. present. WPCF Eddy Medal 1955; WPCF Gascoigne Medal 1970. Enjoys tennis & classical music. *Society Aff:* AIChE, ASCE, APWA, AAAS.

Katzen, Raphael
Business: 1050 Delta Ave, Cincinnati, OH 45208
Position: President *Employer:* Raphael Katzen Associates International *Education:* DChE/Chem Engg/Polytechnic Inst of Brooklyn; MChE/Chem Engg/Polytechnic Inst of Brooklyn; BChE/Chem Engg/Polytechnic Inst of Brooklyn. *Born:* July 28, 1915 Baltimore Md. Tech Supr Northwood Chem Co Phelps Wis 1937- 42; Tech Supr Northwood Proj Diamond Alkli Co Painesville Ohio 1942-44; Proj Engr, Proj Mgr, Mgr Engrg Div Vulcan Engrg Cincinnati Oh 1944-53; started cons practice Oct 1953. Managing Ptnr Raphael Katzen Assocs & Pres Raphael Katzen Assocs International Inc Cincinnati Ohio; Pres CP Assocs Ltd Montreal Quebec Canada. Fellow Amer Cons Engrs Council, AIChE, AIC. P E registration 17 states & D C. Mbr Natl Panel of Arbitrators, The Amer Arbitration Assn. *Society Aff:* AIChE, ACS, AIC, ACEC, TAPPI.

Katzenelson, Jacob
Business: Technion, Haifa Israel
Position: Riesman Prof of Elect Engrg *Employer:* Technion, Israel Inst of Tech *Education:* SCD/EE/MIT 1962; MSc/EE/Technion; BSc/EE/Technion *Born:* 01/14/35 Bell Labs, 1962-69. Proj MAC, MIT 1964-66. Work on nonlinear electronic circuits and their simulation. Technien 1966-. 1971-73 Visiting Prof, UC Berkeley, CA; 1979-81 Bell Labs, Holendel, NJ. Work on higher level programming languages and their application to computer aided design. *Society Aff:* ACM, IEEE, IAIP

Kauffman, Kenneth O
Home: 5393 Golf Course Dr, Morrison, CO 80465
Position: Chief-Water Resource Dir *Employer:* Intl Engrg Co, Inc *Education:* BS/CE/KS State Univ *Born:* 8/26/24 Served with Infantry Division in US and South Pacific 1943-1946. Water Resources Planning Engr, US Bureau of Reclamation 1950-1980. Project Planning Engr, McCook, NB 1950-1970. Chief, Western US Water Plan, Denver, CO 1970-1974. Assist Chief, Division of Tech Planning, Denver, 1974-1980. Meritorious Service Award, Dept of the Interior 1977. Tech Advisor to United Nations Delegation at World Water Conference, Mor del Plata, Argentina 1976. Advisor to Greece on Irrigation and Hydropower Development, 1978. Responsible for Intl Engrg Co's Water Resource Consulting Program, Rocky Mountain Area since 1980. Member ASCE's National Water Policy Committee since 1980. *Society Aff:* ASCE

Kauffmann, William M
Home: 2577 Nottingham Dr, Parma, OH 44134
Position: Consulting Engr. *Employer:* Self. *Education:* ME/IIT. *Born:* May 1904, native of Parma Ohio. Design Engr Superior Gas Engine Co Springfield Ohio 1926-30; Chief Designer Natl Transit Co Oil City Pa 1930-33; Field Engr Worthington Corp Buffalo N Y 1933-44; Asst Chief Engr Mack Truck Co Plainfield N J 1944-48; Mgr Res Worthington Corp 1948-52, Chief Engr 1952-60; Cons Engr 1960-65; Cons Worthington Service Corp Cleveland Ohio 1965-69; Private Practice 1969- . Holds 27 patents relating to internal combustion engines & compressor sys, energy & application. Author of over 150 engrg articles. Cons Ed 'Cryogenic & Indus Gasses'. Fellow ASME. Secy Diesel & Gas Engine Power Div ASME 1957-75. Admin Board Seven Hills United Meth Church. Hobbies: golf & reading. P E Ohio & N Y. *Society Aff:* ASME, NSPE.

Kaufman, Harold B, Jr
Home: 205 E 78th St Apt 12-J, New York, NY 10021
Position: Manager Process Engineering *Employer:* DCA Food Industries Inc. *Education:* BSChE/Chem Engg/Columbia Univ; AB/Chem Engg/Columbia Univ. *Born:* Sept 1923. US Army 1943-46. Positions in Chem Engrg 1948-58 with G E, H K Ferguson, Bechtel; since 1958 with DCA Food Indus Inc, currently Mgr Process Engrg, respon for application of sound Ch Engrg principles to solve a wide variety of complex problems of the food processing industry. Author of 28 US patents; initiated & dev 1st known application of microwave power to fermentation processes in food indus. Merit Award for Outstanding Personal Achievement in Chem Engrg. Chmn Standards Ctte AIChE's FPBE Div. Mbr review board of NSF. Fellow AIChE. *Society Aff:* AIChE, ACS.

Kaufman, Herbert L
Home: 188 W. Ramapo Ave, Mahwah, NJ 07430
Position: Partner *Employer:* Clinton Bogert Assocs *Education:* B/CE/Cooper Union Inst of Tech *Born:* 01/13/19 Jersey City, NJ. Served in U.S. Army in 1944-46. From 1946 to 1962 served as Engr with several consulting engrg firms, including 6 years as an Assoc with Parsons Brinkerhoff Quade & Douglas. Partner in Clinton Bogert Assocs since 1962. Directed development of advanced concepts in combined sewer overflow pollution abatement, for water and sewage treatment, for converting sewage sludge to a resource, and for complex hydrologic river analysis for water supply and power. Was responsible for engrg studies which resulted in the development of regional sewage and water systems. *Society Aff:* AWWA, AAEE, ASCE, FSWA, ACEC.

Kaufman, Howard
Business: Elec & Systems Engg Dept, Elec Systems and Computer Engg Dept, Troy, NY 12181
Position: Prof *Employer:* Rensselaer Poly Inst. *Education:* PhD/EE/RPI; MEE/EE/RPI; BEE/EE/RPI. *Born:* 4/24/40. Born in Saratoga Springs, NY. Res Engr with Cornell Aero Lab from 1965-1968 with emphasis on digital systems. Systems Engr with GE Res & Dev Lab in 1968. With RPI since Sept 1969. Responsibilities include coordinator of the Elec and Systems Engg Dept's Control Systems group as well as instructor of courses in modern control theory. Res interests include adaptive control, estimation theory, and computational methods. Held ASEE summer faculty fellowship at NASA Langley in 1972. IEEE Sr mbrship granted in 1975. Consultant for various ind and res facilities. *Society Aff:* IEEE.

Kaufman, Irving
Business: Electrical & Computer Engrg, Arizona State University, Tempe, AZ 85287
Position: Prof of Engrg *Employer:* Arizona State University. *Education:* PhD/Elec Engg/Univ of Ill; MS/Elec Engg/Univ of Ill; BE/ Elec Engg/Vanderbilt Univ. *Born:* 1925 Geinsheim Germany; s. Albert & Hedwig Kaufmann. m. Ruby Dordek 1950; ch: Eve, Sharon, Julie. RCA Victor 1945-48; Univ of Ill 1948-57; Head Microwave Res Ramo Wooldridge Corp, then TRW Space Tech Labs 1957-65; Prof of Engrg Ariz St Univ 1965- , Dir Solid State Res Lab 1970-78. Sr Fulbright Fellow (CNR Florence Italy) 1964-65, 1973-74; Visiting Prof Univ of Auckland New Zealand 1974. Liaison Scientist. U.S. Office of Naval Research, London, England, 1978-80. 16 US Patents; contribution to 'Methods of Exptl Phys' Vol 8; numerous arts on microwave & device electronics. IEEE Phoenix Sect Achievement Award 1968; Fellow IEEE. Distinguished Res Award, Arizona State Univ, 1986-87. *Society Aff:* IEEE, APS.

Kaufman, Jerome W
Business: RCA Bldg 17A-3-1, Camden, NJ 08102
Position: Metallurgical Engineer. *Employer:* RCA Corporation. *Education:* BS/Met E/Lehigh Univ. *Born:* May 18, 1917. Univ of Penn training towards MS in Met; Grad of course, Solid State Electronics 1960 Penn St Univ; Grad of course, Value Engrg 1958 RCA. Has prof engrg lic by examination in PA & N J; was Delegate for ASM to ETSCO; Chmn of JETS; Dir & Treas Phila Chapt ASM, received Chapt Chmn's Award 1974; elected Fellow ASM 1975. Chief Metallurgical Engrg for RCA; Sustaining Mbr Representative in ASM. Is authority on joining & materials utilization. *Society Aff:* ASM.

Kaufman, John E
Home: 1752 Newfield Ave, Stamford, CT 06903
Position: Technical Director. *Employer:* Illuminating Engineering Society. *Education:* BEE/Power/Cornell Univ. *Born:* Feb 1927. After serving with USN 1945-46 & upon grad from coll in 1949 joined Amer Elec Power Serv Corp as illuminating engr. Since 1951 with Illuminating Engrg Soc; in 1967 became Tech Dir. Ed of 'IES Lighting Handbook'. Past Mbr of Council of AAAS; Past Mbr Board of ESL. V P, US National Comm of CIE. Contrib on illumination in several reference books. Fellow IES & AAAS. IES Distinguished Service Award, 1986. *Society Aff:* IES, IEEE, AAAS, NSPE.

Kaufman, John Gilbert, Jr
Business: Alcoa Bldg, Pittsburgh, PA 15219
Position: Mgr. *Employer:* Alcoa (Aluminum Co of America). *Education:* Masters/Met/Carnegie Mellon Univ; Masters/CE/Carnegie Mellon Univ; BS/CE/Carnegie Mellon Univ. *Born:* 10/14/31. 25 yrs of experience in aluminum, including engg properties, fracture mechanics, fatigue, met and ollsy dev. Special interest in aero, automative and crygenic applications of aluminum. Author of more than 80 technical papers in the above mentioned fields, and several book chapters. Received Award of Merit and fellow grade mbrship from ASTM in 1973, and elected to fellow grad in ASM in 1978. Chrmn of ASTM Committee E-24 on fracture testing. *Society Aff:* ASTM, ASM.

Kaufman, Larry
Business: ManLabs, Inc, 21 Erie St, Cambridge, MA 02139
Position: President *Employer:* ManLabs Inc Cambridge Mass. *Education:* ScD/Metallurgy/MIT; BMetEng/Metallurgy/Brooklyn Polytechnic Inst. *Born:* June 1931. Sr Scientist Lincoln Lab 1955-58; joined ManLabs Inc 1958. Active researcher in thermochemistry & phase stability for 30 yrs. Publ 110 papers & several books in this field. Holds patents on unique high temp materials. Chmn of Thermodynamics Activity for Amer Soc for Metals. AIME Rosster Raymond Award 1964; Fellow of the American Socity for Metals, 1984; pioneered in application of computer techniques for calculating phase diagrams. Founded CALPHAD Proj 1976. Ed in Chief of CALPHAD journal; Pres of CALPHAD Inc. Dir of ManLabs Inc; Dir of Breton Corp. Managing Partner of Fort Washington Associates. *Society Aff:* ASM, AIME.

Kauss, Myron R
Home: 3208 So 113th St, Omaha, NE 68144
Position: Owner *Employer:* Mike Kauss & Assoc. *Education:* BS/Civil Engrg/Univ of Omaha *Born:* 6/6/35 Born Omaha, NE. Educated University of Nebraska Field Engineer-Northern Natural Gas Co 1957 to 1960. Land surveyor to manager field operations and eventually Vice President of Operations the Schemmer Associates Inc., 1960 to 1981. Currently owner of Land Survey Consulting business in Omaha NE. Served in US Navy "seabees–. Registered Land Surveyor in several Midwest, Southwest & Southeastern States PPres & COB American Soc. of Certified Engrg. Technicians. Mbr Omaha Chamber Commerce Professional surveyors assoc of NE, ACSM, ACEC, ASEE. Serves Boy Scouts of America Girl Scout Bds & various other civic & religious. *Society Aff:* ACSM, ACEC, ASCET, ASEE, PSAN

Kauzlarich, James J
Business: Dept. of Mech Engrg, Charlottesville, VA 22903
Position: Prof. *Employer:* Univ. of Virginia *Education:* PhD/Mech Eng./Northwestern U.; MSME/Mech. E./Columbia U.; BSME/Mech. E./U. of Iowa. *Born:* 09/27/27 Taught at Worcester Polytechnic Inst, 1958-61; Univ WA 1951-53; at Univ VA since 1963 as Chair, Mech Engrg. Worked for General Electric, Schenectady and Boeing Co, Seattle. Mbr Tau Beta Pi, Pi Tau Sigma and Sigma Xi. Published papers in rehabilitation engrg, tribology, heat transfer and fluid mechanics. Visiting Prof, Cambridge Univ, 1970-71, Suni Swansea 1984-85. Inventor of friction welding machine, hydraulic bearings, medical catheter and self-damping caster wheel. *Society Aff:* ASME, ASLE, ASEE

Kawahigashi, Ted S
Business: 745 Fort St Mall 900, Honolulu, HI 96813
Position: Corporate Secy & Asst Chief Engr. *Employer:* Austin, Tsutsumi & Assocs Inc. *Education:* MSCE/Civil Engrg/Univ of S CA; BSCE/Civil Engrg/Univ of HI. *Born:* Apr 1936. BSCE Univ of Hawaii 1957; MSCE USC 1968. Cvl Engr for Public Works Off at Pearl Harbor 14 ND 1957-61. USA Corps of Engrs 1958-60. Calif Div of Hwys (CalTrans) 1962-72; Dist Freeway Opers Engr, Dist 7 (L A) respon for sys opers & surveillance; with Austin, Tsutsumi & Assocs Inc 1972- , respon for hwys, traffic, drainage projs & for preparation of environ impact assessments. Mbr ASCE, APWA. Reg Engr Hawaii & Calif. *Society Aff:* ASCE, APWA.

Kay, William W
Home: The Highlands, P.O. Box 100, Drums, PA 18222
Position: Consultant *Employer:* W. W. Kay & Assoc *Education:* Engr of Mines//MO Sch of Mines; BS/Mining Engg/MO Sch of Mines. *Born:* 1/17/12. Mining Engr for large Anthracite Co. Spent many yrs in the explosives Dept of E I duPont de Nemours & Co in all types of mining, with accent on bituminous and anthracite coal. Instrumental in developing high speed tunnel equip and procedures. As a mining specialist - in charge of nationwide diamond tool prog. Also introduced resin roof bolts to the coal mining industry. Recognized as an explosives expert. Took early retirement from duPont to go into private practice as a Consulting mining engr. Became a mining consultant to Behre Dolbear & Co Inc of NYC. consultants and advisors to the minerals industry worldwide. Reg PE in PA and DE. Recently formed W. W. Kay & Assoc to do mining & civil engineering & consulting.. *Society Aff:* SME, NSPE, PSPE.

Kaya, Azmi
Home: 2365 Woodpark Rd, Akron, OH 44313
Position: Assoc Prof Mech Engg. *Employer:* Univ of Akron. *Education:* PhD/ME/Univ of MN; MS/EE/Univ of MN; MS/ME/Univ of WI; BS/EE/Technical College. *Born:* 2/1/33. Born in Turkey, married (Suna) two children. NATO visiting Expert; UN Energy consultant; Honeywell industrial experience 1963-70; Univ of Akron 1970- 81; Leave with B&W 1977/78. *Society Aff:* ASME, IEEE, ISA, ΣΞ, ASEE, TAPPI

Kayton, Myron
Home: 722 Adelaide Place, Santa Monica, CA 90402
Position: Pres *Employer:* Kayton Engrg Co *Education:* BS/ME/Cooper Union; MS/EE/Harvard; PhD/EE/MIT *Born:* 4/26/34 NY City; Bronx High School of Science. Designed earliest multisensor navigation sys at Litton Indus. Deputy Mgr of Lunar Module electronics project office at NASA. Developed digital avionics for several manned spacecraft, aircraft, and simulators for TRW, Inc and Kayton Engrg. Introduced multiplexing techniques to the power generation industry. Innovative contributions to R&D cost analysis and fault-tolerant computing. Taught "Power Plant Simulation" and "Multisensor Navigation" at UCLA. National Vice-President and conference Chmrn of IEEE. Founding Dir, Caltech-MIT Enterprise Forum - Director, Wincon Conference. Dir and past pres, Harvard Club of Southern CA; member of Alumni Council, The Cooper Union. Author of 30 papers and a standard reference book. Occasional writer on tech and economics subjects. Registered PE. *Society Aff:* IEEE, AAAS, ASME

Kazan, Benjamin
Business: Xerox Research Center, 3333 Coyote Hill Rd, Palo Alto, CA 94304
Position: Consultant - Display Research. *Employer:* Self-employed. *Education:* PhD/Physics/Tech Univ of Munich; MA/Physics/Columbia Univ; BS/Physics/CA Inst of Technology. *Born:* May 8, 1917 N Y C. Head Special Purpose Tube Sect, Signal Corps Engrg Labs 1940-50; Mbr Tech Staff RCA Labs 1951-58; Hd Solid-State Display Group Hughes Res Labs 1958-61; Chief Scientist Solid-State Div Electro-Optical Sys Inc 1962-68; Head Display Group IBM Res Labs 1968-74; Corp Cons on Display Res Xerox Res Labs 1974-1985; Consultant 1985- . Medal of Amer Roentgen-Ray Soc; Coolidge Award, G E Co; Fellow IEEE; Fellow SID; Mbr

Kazan, Benjamin (Continued)
APS. Committee mbr of sub-group of Advisory Group on Electron Devices 1972-1982. Assoc Ed IEEE Trans on Electron Devices 1975-1983; Assoc. Ed of Advances in Electronics and Electron Physics series; coauthor of books: 'Storage Tubes' & 'Electronic Image Storage'. *Society Aff:* IEEE, SID, APS.

Kazemi, Hossein
Business: P.O. Box 269, Littleton, CO 80160
Position: Manager, Reservoir Management *Employer:* Marathon Oil Co *Education:* PhD/Petro Engrg/Univ of TX; BS/Petro Engrg/Univ of TX *Born:* 3/11/38 Mgr, Reservoir Mgt., Prod Tech Group Marathon Oil Co, Exploration and Technology Ctr. Also, currently Adjunct Assoc Prof of Petro Engrg, CO Sch of Mines. Sr Res Engr, Atlantic Richfield R&D, Dallas, 1969; Sr Res Engr, Sinclair Res, Tulsa, 1963-69; Lecturer in Mathematics, Univ of Tulsa, 1967-69. 27 publications in Petro Reservoir Engrg, Pressure Transient Testing in Wells and Numerical Simulation. Received 1980 Denver Petro Section Henry Mattson Technical Service Award; Univ of TX Engrg Fellow, 1961. Reg PE, CO. *Society Aff:* SPE

Kealey, T Robert
Business: PO Box 2345, Harrisburg, PA 17105
Position: Consultant. *Employer:* Modjeski & Masters; Retired. *Education:* BS/Civil Engrg/Drexel Univ; MS/Civil Engrg/MA Inst of Tech. *Born:* 2/16/21. Army Corps Engrs 1942-45. CE Instr at Drexel 1946; with Modjeski & Masters 1947- . Design Engr, major bridge structs incl Greater New orleans Bridges 1 and 2 & Baton Rouge I-10 Bridge in La; Newburgh-Beacon Bridges in N Y; Theodore Roosevelt Bridge in Wash D C; Girard Point I-95 Bridge in Phila, Pa. Luling Bridge in La, MS River Bridge at Quincy, Ill. Invesigation of collapse of Silver Bridge in W Va, Yadkin River Bridge in N C & Calvin Bridge in Oklahoma; Became Ptnr & Chief Engr 1960; Managing Ptnr 1971 to 1986. Consultant 1986- . *Society Aff:* NSPE, ASCE, AREA, ACEC, ATC.

Keamy, Mitchell F, Jr
Business: PO Box 512, Milwaukee, WI 53201
Position: Group Exec & V P. *Employer:* Allis-Chalmers. *Education:* MBA/Mgmt/NY Univ; MChE/Chem Engg/NY Univ; SB/ChE/MIT. *Born:* May 1927. SBChE MIT, MChE & MBA NYU. Allis-Chalmers 1966- , Group Exec & V P 1974- . Exec respon for divs & subsidiaries comprising Minerals Processing Group engaged in providing processes & equipment & engrg/const mgmt for minerals, metals & paper processing facilities (i e Cement & Chems, Metals Processing, Mining Sys, Reduction Sys VoithAllis Inc) & subsidiary operating companies producing sponge iron. Reg Prof Engr: N Y, Wis, Mich, Minn, Mass, Ky, Maine. AIChE, ACS, AIME, Amer Mgmt Assn. *Society Aff:* AIChE, AIMME, ACS, AMC, AMA.

Keane, Barry P
Home: 204 Kings Way, Clemson, SC 29631
Position: Assoc Prof. *Employer:* Clemson Univ. *Education:* PhD/EE/Univ of FL; MS/EE/Univ of FL; BS/EE/Univ of FL. *Born:* 2/26/48. Grad study at Univ of FL. Topics: Beta activity in EEG, biological rhythms. Moore scholarship recipient, NDEA IV Fellowship. Joined Electrical and Computer Engg faculty of Clemson Univ 1975. Assoc Prof, tenure, 1979. VP, Inotec Inc, consultants, 1977 to present. 28 publications, guest editor Medical Instrumentation, 1979. Research interests: ambulatory medical instrumentation, arrhythmia detection, signal processing, EEG monitoring, epilepsy, machine intelligence. Developed biofeedback training tool for epileptics, inventor of "ICAM" method of arrhythmia monitoring, 1979. *Society Aff:* IEEE, AAMI, AEMB.

Keane, Robert G, Jr
Home: 8613 N Bali St, Ellicott City, MD 21043
Position: Hd, Hull Form Design & Performance Branch *Employer:* Naval Sea Sys Command (NAVSEA). *Education:* BES/Mech Engg/Johns Hopkins Univ; ME/Mech Engg/Stevens Inst of Tech; MSE/Naval Architect/Univ of MI. *Born:* 8/31/41. Native of Baltimore, MD. Mech engr with Westinghouse; marine engr with Ellicott Machine Corp; taugh mech engg at SIT; naval architect with Naval Ship R/D Center; with NAVSEA 1968- , former Hd of Hull Equipment Branch with over 20 naval architects - assumed current respon for hull from design and hydrodynamic performance of all naval ships with over 20 naval architects 1973. Mbr of Hydrodynamics Committee & Seakeeping Panel 1974-79, SNAME; ASNE, Natl Publicity Ctte 1976-77 & Papers Ctte 1978-79; Pres DCCEAS (DC Council of Engg & Arch Soc's) 1974-75, also VP, Secy & Treas. DCCEAS Bicentennial Award 1976. Enjoys gardening & working with youth. *Society Aff:* SNAME, ASNE, SAWE.

Kear, Bernard H
Business: Aircraft Rd, Middletown, CT 06457
Position: Sr Consulting Scientist. *Employer:* Pratt & Whitney Aircraft. *Education:* DSc/-/Univ of Birmingham, England; PhD/-/Univ of Birmingham; BSc/Univ of Birmingham, England. *Born:* July 1931; native of Port Talbot S Wales. BSc 1954, PhD 1957, DSc 1970 Birmingham Univ England. Emigrated to U S 1959; U S citizen Jan 1965. Since 1963 with gas-turbine engine division of Pratt & Whitney Aircraft. Over 10 yrs R/D work in high temp alloys field. Assumed present position as Sr Consulting Scientist 1977. Author of 105 tech publs, 20 issued patents; ed of 3 tech books. Mathewson Gold Medal AIME 1971; Howe Medal ASM 1970. Fellow ASM 1976. National Acad of Engg 1979. Enjoys tennis, swimming & sailing. *Society Aff:* TMS-AIME, ASM, EMSA.

Kear, Edward B
Business: Dean-Div. of Special Programs, Clarkson Univ, Potsdam, NY 13676
Position: Dean-Div of Special Programs *Employer:* Clarkson University. *Education:* PhD/Mech Engr/Cornell Univ; MS/Mech Engr/Cornell Univ; BS/Mech Engr/Clarkson Univ *Born:* 3/23/32. A native of Yorktown Heights, NY; worked for Hamilton Std and AVCO Corp prior to coming to the Mech Engg Dept at Clarkson Univ in 1956. For many yrs taught courses in control systems analysis. Chrmn of the Mech and Indust Engr Dept 1972-80. Has lead the dev of 2 major wind energy projs - one of the horizontal axis type and the second a vertical axis darrieus wind turbine. At the present time he is serving as the Dean of the Div of Special Programs. *Society Aff:* ASME, ASEE, NSPE

Kearns, John J
Business: Half Acre Rd, Cranbury, NJ 08512
Position: Vice President - Production. *Employer:* Carter-Wallace Inc. *Education:* MS/Chem Engg/Stevens; BS/Chem Eng/Columbia. *Born:* July 1928; native of Brooklyn N Y. MS Stevens Inst 1960; BSChE Columbia Sch of Engrg 1949. Lever Bros Co 1949-62 primarily in Food Process Dev; Carter- Wallace Inc 1962- , initially respon for low-calorie margarine dev & prod, became Corp Opers Engr 1967, V Pres Prod & Engrg Domestic Opers 1970, V Pres Prod - Corp 1975- . Adjunct Asst Prof Mgmt Sci Pace Univ 1973- . Holds 3 patents for Food Processes. AIChE Central Jersey Sect 1974; Mbr AIChE - EPIC Ctte 1975; Mbr Proj Mgmt Inst 1972. *Society Aff:* AIChE, PMI.

Keates, Richard J
Home: 3818 West LeMont Blvd, Mequon, WI 53092
Position: President. *Employer:* CMC Corp. *Education:* BS/ChE/Cooper Union. *Born:* July 1928; native of Port Washington N Y. Infantry & Chem Corps, Officer US Army Korean War. Const, startup & oper engr Goodyear Atomic Corp's gaseous diffusion plant; 15 yrs with G E Co's Plastics, Polymers & Med Equip Divs, Mgr of Manufac Opers in Polymers, Facilities & Planning Mgr for Med Opers. Kirkpatrick Award winner 1968. Pres CMC Corp 1971- . Reg Prof Engr. Enjoys participative sports: tennis, golf, skiing & sailing. *Society Aff:* NSPE, AIChE, WSPE.

Keaton, Clyde D
Home: 342 Silver Hill Rd, Concord, MA 01742
Position: Mgr-Engg. *Employer:* GE Co. *Education:* MS/ME/RPI; BS/ME/Villanova Univ. *Born:* 8/3/42. Clyde received a BSME from Villanova Univ & his MS from

Keaton, Clyde D (Continued)
RPI. He joined GE's large turbine dept in Schenectady, NY, in 1964. In 1967 he became a consulting engr for a number of operations including the aircraft engine group. In 1970 he was appointed mgr-mst-18 proj which was involved with a dev submarine propulsion system. In 1972 he was named mgr-mech marine engg for the marine turbine & gear products dept. He was appointed mgr-mfg engg for the medium steam turbine dept in Dec 1973 & joined the gen purpose motor dept as mgr-engg sec in June 1977. Present position, Nov 1978, Fitchburg, MA. *Society Aff:* ASME.

Keays, Keatinge
Business: MIT Branch PO Box 1, Cambridge, MS 02139
Position: Administrative Officer & Lecturer *Employer:* Dept of Ocean Engineering - MIT. *Education:* NAVE/Naval Engg/MIT; BS/-/US Naval Acad. *Born:* Sept 1926. Spec in submarine design & const USN until 1963 except for brief period in private practice (1957-59) San Francisco. 1963-68 Proj Mgr R/D Genl DynamicsElectric Boat; dir projs involving Large Object Salvage, Deep Diving Sys & Small Submersibles; dir preparation of USN Submarine Salvage Manual authoring chapter on mooring. Apptd to Academic Staff Dept of Ocean Engrg MIT 1968. Areas of spec interest incl Ocean Engrg Structs, Mooring & Deep Ocean Salvage. Mbr Council 1972-73 SNAME, Asst Secy-Treas 1963 ASNE. *Society Aff:* ASME, ASNE, SNAME, NSPE.

Kececioglu, Dimitri B
Home: 7340 N La Oesta, Tucson, AZ 85704
Position: Prof. *Employer:* Univ of AZ. *Education:* PhD/Engg Mechanics/Purdue; MS/Ind Engg/Purdue; BS/ME/Robert College. *Born:* 12/26/22. Prof of Aerospace & Mech Engg at the Univ of AZ since 1963, and Reliability and Maintainability Engg Consultant. In charge of a ten-course curriculum in Reliability Engg. Dir of the Allis-Chalmers Corp Reliability Prog 1961-1963. Fulbright Scholar. Over 112 scientific publications. Contributed to six books. Dir of the Annual Reliability Engg & Mgt, and of the Annual Reliability Testing Insts. Lectr worldwide over 230 seminars and short courses. Holds five patents. Received "Certificate of Excellence" from the Soc of Reliability Engrs; the Ralph R Teetor Award of the Soc of Automotive Engrs for "significant contributions to teaching, research and development during 1977;" the 1980 "Reliability Education Advancement Award" for "outstanding contributions in the development and presentation of meritorious reliability educational programs, the 1981 "Allen Chop Technical Achievement Award" for "outstanding contributions to the reliability science and technology," both from the American Society for Quality Control; the 1983 Anderson Prize of the College of Engrg, Univ of Arizona for "Engineering the Masters Degree Program in Reliability Engineering;" Honorary Member Golden Key National Honor Society, and Honorary Professor Shanghai Univ of Technology in 1984. *Society Aff:* ASME, IEEE, ASQC, AIAA, ASEE, SRE, SAE, ΣΞ, ΤΒΠ.

Keefner, Eugene F
Business: 2674 Dawson Ave, Signal Hill, CA 90806
Position: Pres/Owner. *Employer:* Keefner Engg Corp. *Education:* Certificate/N-A/Industrial College of the Armed Forces. *Born:* 7/14/21. Native of Detroit, MI, began engg career in USN 1939-45. In practice of experience Engg for 40 yrs in Industry as; Steam Power Plant Generation, Space Sci Test Engg, Institutional Engg, Facilities Engg Design and Construction. Since 1972 self-employed as Consultant to maj industrial firms in wastewater control. Since 1976 as designer-mfg under two patents a maj contribution in Engg has been to develop automation and state-of-the-art advancement to the Rendering Industry, world-wide. Present owner-operator of two Water Utility Co's an Import-Export firm, an Instrument Repair facility, OEM manufacturer of electronic control systems. *Society Aff:* NSPE, CSPE, AIPE, ISA.

Keegan, John J
Business: 1304 Buckley Rd, Syracuse, NY 13201
Position: Vice President. *Employer:* O'Brien & Gere Engineers Inc. *Education:* MSSE/Sanitary Engr/Manhattan Coll; BSCE/Civ Engr/Manhattan Coll *Born:* 1938; N Y C native. Chi Epsilon. Proj Engr Stearns & Wheler Cazenovia N Y 1959-65; O'Brien & Gere Engrs 1965- , Principal Engr 1967, V Pres 1971. P Pres Cent N Y Sect ASCE. Diplomate Amer Acad of Environ Engrs. Mbr ASCE, WPCF, AWWA. Respon for firm's Sanitary Engrg Div including design of facilities, special studies & investigations for domestic & industrial wastewater treatment facilities. Mbr Citizens Foundation, Lions Club. Reg Prof Engr: N Y, N J, VT, OH, DE, MD, PA, MA. Adjunct Prof Civil Engg Syracuse Univ. Mbr Natl Bd of Control WPCF. *Society Aff:* ASCE, AWWA, WPCF

Keeler, Stuart P
Home: 23145 W River Rd, Grosse Ile, MI 48138
Position: Senior Staff Scientist *Employer:* Budd Company Technical Center. *Education:* ScD/Mech Met/MIT; SB/Gen Eng/MIT; AB/Math-Physics/Ripon College. *Born:* Sept 1934. Joined Natl Steel Corp: 1964 Supr Tech Dev; 1967 Supr of Flat Rolled/Prods Applications 1978 Mgr Auto Res; 1983 Staff Applications Specialist. 1987 joined Budd Company Technical Center as Sr. Staff Scientist. Pioneered work with Forming Limit Diagrams & sheet metal formability. Became Pres of Amer Deep Drawing Group 1968-72; only North American to be elected Pres (1971-73) of 20 country Internatl Deep Drawing Res Group. Served on natl Tech Div Board & Long Range Planning of ASM. 1979 Fellow of ASM, Mbr AIME, SAE. Hobby is directing & producing amateur theatre plays, as well as boating and Coast Guard Aux. *Society Aff:* ASM, SAE, AIME.

Keenan, Arthur J
Home: 8982 Pebble Beach Cir, Westminster, CA 92683
Position: Mbr Engrg Tech Staff *Employer:* TRW System Inc *Education:* MS/Engrg/Univ of Beverly Hills; BS/Engrg/Univ of Beverly Hills; AA/Bus/Long Beach City College. *Born:* 8/4/30. Residence Westminster, CA. Reg PE, CA. Profl history: 1980-present, Mbr Engrg Tech Staff TRW System Inc; 1972-1980, Gen Supervisor Ind Engg Douglas Aircraft Co; 1971-72 Mgt Consultant, Alexander Proudfoot; 1970-71 Proj Engr, Time Zero Corp; 1967-70 Production Mgr, TRW System, Inc; 1976-present, Pres, Keenan, Young & Assoc - Mfg Consultants. IIE Inst Activities: 1979-present, Chrmn Region XII Mbrship Planning; 1978-79, Chrmn Aerospace Div: 1977-78 Div Dir, Aerospace Div; 1975-76, Pres Pacific View Chapter CA; 1978 Recipient, IIE Aerospace Dir Award, 1977 Recipient, Engr of the Yr Award. 1976 US Treasury - Commissioner's Citation. Speaker at Inst Confs. Chapter meetings & other soc meetings. *Society Aff:* IIE, PCS.

Keenan, John D
Business: 277 Towne Bldg, Philadelphia, PA 19104
Position: Assoc Prof *Employer:* Univ of PA *Education:* PhD/CE/Syracuse Univ; MS/CE/Syracuse Univ; BA/Bio/SUNY *Born:* 3/16/44 Mbr of PA faculty since 1972, & undergrad curriculum chrmn in civ engrg since 1978. Maj res area is the biological and health aspects of environmental pollution; principal foci are resource recovery and utilization of waste mtls, waste treatment, water supply and treatment, and biological and health effects. *Society Aff:* ASCE, WPCF, AWWA.

Keeney, Norwood H, Jr
Business: P.O. Box 695, Newport, NH 03773
Position: Prof-Dept Hd. *Employer:* Univ of Lowell. *Education:* PhD/-/Univ of Manchester; MS/CE/Univ of ME; BS/Chemistry/Trinity College. *Born:* 7/10/24. Native Hartford, CT, married Phyllis R Mottram, 1 son. Paper Chemist, Fram Corp, E Providence RI 1950-3, Faculty Univ Lowell 1953 to date, Dept Chrmn 1976- ; Served USAF 1943-5; Other mbrships Sigma Xi, Tech Assn Pulp & Paper Assn, NH Soc PE, VFW. Area of interest-pulp & paper, physics of paper. *Society Aff:* AIChE, NSPE.

Keepin, G Robert
Business: MS 550, Los Alamos, NM 87545
Position: Laboratory Fellow. *Employer:* Los Alamos National Laboratory. *Education:* PhD/Physics/Northwestern Univ; BS, MS/Physics/MIT; PhB/Physics/Univ of Chicago *Born:* Dec 1923; native of Oak Park Ill. PhD Northwestern Univ; BS, MS, PhB MIT. US AEC Postdoctoral Fellow Univ of Calif 1949-50. Cons to Argonne Natl Lab & Los Alamos Sci Lab & US delegate to 1st Geneva Conference (Atoms for Peace) 1955. 1963-65 headed Phys Div Internatl Atomic Energy Agency Vienna. (IAEA Tech Advisor to 3rd Geneva Conference 1964). Contributions to fission phys, nuclear engrg, reactor kinetics & control are well-known. Inventor of techs for nondestructive detection, identification & assay of fissionable materials & their applications to natl & internatl nuclear safeguards, inspection & surveillance under the Nuclear Nonproliferation Treaty. Fellow 1965 APS, Fellow 1967 ANS. Recipient of ANS 1973 Annual Award for Nuclear Materials Safeguards Technology. Fellow, Inst of Nuclear Mtls Mgmt, and recipient of INMM Distinguished Service Awd, 1983. Nuclear Safeguards Adv to Dir Gen, IAEA, Vienna, 1982-85. In 1985 named Fellow of the Los Alamos National Laboratory. *Society Aff:* ANS, APS, INMM

Keer, Leon M
Business: Department of Civil Engineering, Evanston, IL 60201
Position: Prof and Assoc Dean *Employer:* Northwestern Univ *Education:* PhD/Eng Mech/University of Minnesota; MS/ME/Caltech; BS/ME/Caltech *Born:* 9/13/34 Leon M. Kerr was born in Los Angeles, CA in 1934. He was educated at the CA Inst of Tech and the University of Minnesota. He was a NATO Fellow and a Guggenheim Fellow. In 1964 he joined the Engineering school at Northwestern Univ where he is currently Prof of Civil and Mech Engrg and Assoc Dean for Graduate Studies and Research. He is a Fellow of ASME & the American Academy of Mechanics. His research is in fracture mechanics, contact fatigue and dynamics of structures *Society Aff:* ASME, ASCE, ASA, ASEE

Keevil, James M
Home: 1100 Penny St. SE, N. Canton, OH 44720
Position: Sr. Engr. *Employer:* County of Summit Dept of Environmental Services *Education:* BS/Civ Engg/Tri-State Univ. *Born:* 3/6/41. Native of Portsmouth, OH. After grad from Tri-State Univ in 1964 was a proj engr for a consulting engr firm in Mansfield, OH. Was with Floyd Browne Assoc Limited from 1968-85. Began as a proj engr for Comprehensive Water & Sewer Studies for several OH counties. Was Dir of the Firm's Eastern Office 1977-85. Currently employed with County of Summit, Dept of Environmental Services as a Senior Engr. Tau Sigma Eta, an honorary engr fraternity at Tri-State Univ. Enjoy stamp collecting & home computer. *Society Aff:* ASCE, AWWA, WPCF.

Kegg, Richard L
Business: 4701 Marburg Ave, Cincinnati, OH 45209
Position: R&D Dir *Employer:* Cincinnati Milacron *Education:* PhD/Mech Eng/Univ of Cincinnati; MS/Mech Eng/Univ of Cincinnati; BS/Mech Eng/Univ of Cincinnati *Born:* 04/17/35 Res and dev employee of Cincinnati Milacron since 1958. Carried out and directed work on mfg process and vibration res, prototype testing, numerical and adaptive control devs, reliability and unattended machining. Directed the dev of a number of patented innovations in the design of machine tools and controls. Currently Dir of Advanced Mfg R&D. Author of a number of tech papers. Lectured on mfg tech around the world. Natl dir of SME. Reg PE in OH. Certified Reliability Engr (ASQC). Certified Robotics Engr (SME). *Society Aff:* SME, ASME, ASQC, ΣΞ, CIRP

Kehde, Howard
Business: Michigan Div, Bldg 633, Midland, MI 48640
Position: Eng Process Specialist. *Employer:* Dow Chemical USA. *Education:* DEng/ChE/NY Univ; MChE/ChE/NY Univ; BS/ChE/Columbia Univ. *Born:* 11/15/21. Native of St Louis, MO. Res Engr for Union Oil Co of CA and Foster Wheeler, respon for dev of processes using moving beds solids for gas separation by adsorption and for catalysis in reforming of hydrocarbons. With Dow Chemical USA since 1958 with many contributions to chemical refineries and unique styrene units. Chrmn, Publication, AIChE for yrs. Elected a Fellow of AIChE. Reg PE in MI, NY & CA. Also ordained Presbyterian Minister. *Society Aff:* AIChE.

Kehlbeck, Joseph H
Home: 7309 Shadwell Ln, Prospect, KY 40059
Position: General Manager-International Purchasing. *Employer:* G E - Corporate Sourcing. *Education:* MBA/Mgmt/Rutgers Univ; BS/Engg/Univ of IA. *Born:* 9/14/26. USA Paratrooper WII. Prod Engr Union Carbide & Carbon Corp 1950; joined G E 1951 as Time Study Engr. Worked for G E in Bloomfield NJ, Tyler TX, Balto MD, Trenton, NJ & Louisville, KY. Large num of line & staff positions leading to Mgr Mfg in Trenton 1969. Mgr Mfg Range Operation in Louisville 1972 & Mgr. Material Resource 1977 and Present Position in 1986. Pres AIIE 1977. Reg P E. Listed in Who's Who in Amer. Listed in Who's Who in Industry & Finance. 1978-IIE Rep to US-Gao Productivity Advisory Bd; 1980-IIE Rep to US GAO Quality Advisory Bd; 1978-Outstanding I E in KY. 1982- Honorary Mbr Tau Beta Phi, 1982 Fellow IIE. Enjoys tennis & golf. Married; 1 child. *Society Aff:* IIE.

Kehoe, Thomas J
Business: 2500 Harbor Blvd, Fullerton, CA 92634
Position: Mgr Tech Servs-Dept. *Employer:* Beckman Instruments Inc. *Education:* BS/Chemistry/Loyola Univ of Los Angeles. *Born:* June 16, 1919 Bisbee Ariz. Reg Prof Engr St of Calif: ChE 431 & CS Engr 24. BS Chem Loyola Univ Calif. R/D Lab Supr Amer Potash & Chem (Kerr McGee) 1942-49; Assoc with Pomeroy & Assocs, Cons - sanitation & waste treatment 1949- 54; Tech Servs Mgr Beckman Instruments Inc 1954- . Authored over 30 papers on on-line analyzers & process control plus chapters in books by Considine & Kolthoff & Elving; US Patent 3,976,104. Fellow AIC & Instrument Soc of Amer; Mbr ACS, ASTM (Dir 1964). Analysis Inst Div ISA, V Pres Tech Dept ISA, Natl Pres ISA 1970. Enjoys oil painting, sketching & stamp collecting. *Society Aff:* ISA, AIChE, ASTM, ACS.

Kehrl, Howard H
Business: 3044 W Grand Blvd, Detroit, MI 48202
Position: Vice Chairman *Employer:* General Motors Corporation. *Education:* MS/Industrial Mgmt/MIT; MS/Eng Mech/Univ of Notre Dame; BS/Eng Mech/IL Inst of Technology. *Born:* 2/2/23. Native of Detroit. Attended Notre Dame Midshipmen's Sch 1944 under Naval Officers Training Prog. Taught engrg at Notre Dame. Joined G M Res Labs 1948; in 1981 assumed current respons jurisdiction over the Corporation's Technical Staffs, Operating Staffs, and Public Affairs Groups. Mr. Kehrl serves on the boards of directors of the Detroit's Harper-Grace Hospitals and Metropolitan Detroit's United Foundation; elected Chariman of the Board of the United Negro College Fund. He is on the Board of Directors for the National Action Council for Minorities in Education and is a Director of the Chamber of Commerce of the US. Bd of the Motor Vehicle Manufacturers Assoc., the Economic Club of Detroit, the Soc of Manufacturing Engrg Educ Fdn., and is VChrmn of the GM Cancer Res Fdn. *Society Aff:* SAE, ESD, AOT,ΤΒΠ.

Keigher, Donald J
Home: 110 Rover Blvd, Los Alamos, NM 87544
Position: Fire Protection Conslt *Employer:* Los Alamos Natl. Lab. *Education:* BS/FP Engr/IL Inst of Tech; BS/Ind Engr/IL Inst of Tech *Born:* 12/26/20. Grad studies Univ of WA; Oak Ridge Sch Reactor Tech. US Navy 1943-46; Chf Engr USS KNAPP; Lt USNR (R). PE in IL, CA, & NM also CSP. Ins Engr 1946-49; Fire Prot & Safety USAEC 1949-69; Chf Indus Safety Branch RLO-AEC 1962-69; Chief Indus Safety Branch NASA Hq 1969-71. Active NFPA Tech Cttes; Computer Protection 1969-87; Atomic Energy 1977-87; Halon sys 1971-85, Chrmn (1975-80). SFPE Bd of Dirs 1976-82; Elected SFPE Fellow-1981. NFPA Bd of Dirs 1980-86.

Keigher, Donald J (Continued)
Awarded NFPA DSA (1986). Outstanding Young Man in Govt 1959; NASA Superior Performance 1970. *Society Aff:* SFPE, NFPA, NSPE, ASSE

Keil, Alfred A H
Business: 77 Mass Ave Rm 3-282, Cambridge, MA 02139
Position: Prof Emeritus, Senior Lecturer *Employer:* Massachusetts Inst of Technology. *Education:* ScD/Physics/Univ Breslau, Germany. *Born:* 5/1/13. Res Assoc with German Navy 1940-45; various positions of increasing respon with US Navy pioneering in many aspects of ship-related R/D 1947- ; 1st Tech Dir of Navy's David Taylor Model Basin 1963-66; joined MIT as Prof & Head Dept of Naval Architecture 1966. Led substantial broadening of educ prog toward Ocean Engrg. Dean of Sch of Engrg MIT 1971-77. 1977-78 Ford Prof of Eng 1978 Ford Prof of Eng, emeritus. Mbr NAE 1966- ; Mbr Naval Res Advisory Ctte 1971-75; US Navy Distinguished Civilian Service Award 1963; Mbr Natl Adv Committee for Oceans & Atmosphere 1977-79; Gibbs Bros Gold Medal Award 1967 NAS, Gold Medal Award 1965 ASNE, Lockheed Award Marine Sci & Eng, 1979, MTS. *Society Aff:* ASNE, MTS.

Keiser, Bernhard E
Business: Keiser Engg, Inc, 2046 Carrhill Rd, Vienna, VA 22180
Position: Pres *Employer:* Keiser Engrg, Inc *Education:* DSc/EE/Washington Univ St. Louis; MS/EE/Washington Univ St. Louis; BS/EE/Washington Univ St. Louis *Born:* 11/14/28 Bernhard E. Keiser, DScEE, Consulting Engr, Served in advanced engrg and management positions in several major corp prior to establishing own consulting engrg practice. Registered Profl Engr in VA, MD and the District of Columbia. Performs telecommunications system engrg studies dealing with satellite, microwave and cable transmission, also data communications. Has served as expert witness in US and Canada. Author of three books and 26 technical papers. Holds two US patents. Fellow of IEEE and Washington Academy of Sciences. Served as Chrmn of Northern Virginia Section of IEEE. *Society Aff:* IEEE, AFCEA, RCA.

Keiser, Edwin C
Business: Suite 329, Clark Bldg, Columbia, MD 21044
Position: Manager of Engineering *Employer:* Acres American Incorporated *Education:* MS/Civil Engg/Princeton Univ; BS/Genl/US Military Acad. *Born:* Aug 21, 1930 Windsor Colo. Grad US Army War Coll 1970. Served in var engrg troop units including commanding 1st Engr Battalion, 299th Engr Battalion & 24th Engr Group and 18th Engg Bde; const officer & Camp Cdr at Camp Tuto Greenland 196062; taught engrg mechs at US Military Acad 1962-65; Dept of Army Staff 1967-69; US Army Corps of Engrs Dist Engr Offices - N Y Deputy Dist Engr for West Point Projs 1971-73 & Savannah Dist Engr 1973-76; Asst Deputy Chf of Staff Engg, HQ US Army Europe 1978-79-1980; Manager of Engineering, Acres American Inc.; Dir of Const Opus & Fac Mgmt, Office of Secy of Defense, 1979- 1980; Manager of Engineering, Acres American Inc. *Society Aff:* SAME, AUSA.

Keister, William
Home: 66 Baltusrol Way, Short Hills, NJ 07078
Position: Consultant. *Employer:* Self. *Education:* BSEE/EE/Auburn Univ. *Born:* June 1907, Montgomery Ala. Joined Bell Labs 1930: Initial work on toll switching & signaling circuits; taught radar in Bell Labs Sch for War Training; after the war planned & taught courses in switching principles. Coauthor of The Design of Switching Circuits; in 1952 headed 1st exploratory dev of electronic switching sys; introduced concept of stored program control; appointed Dir of Switching Sys Studies Center 1958; retired 1972 as Dir of Computing Tech Center. Fellow IEEE; Ben S Gilmer Award Auburn Univ, co-recipient of the Alexander Graham Bell Medal 1976. *Society Aff:* IEEE, ΤΒΠ, HKN, ΦΚΦ.

Keith, Frederick S
Business: 10 Albany St, Cazenovia, NY 13035
Position: Partner *Employer:* Stearns & Wheler Engrs & Scientists *Education:* MBA/Bus Ad/Syracuse Univ; BCE/CE/Cornell Univ *Born:* 6/7/33 and raised in Newton, MA. Grad with BCE, Cornell Univ 1956. Served with US Army Base Topographic Co in 1957-58. Assoc Engr, B. K. Hough Ithaca, NY in soils & foundation engrg. City Engr, city of Syracuse, NY 1966-1968. Since 1968 partner with Stearns & Wheler, responsible for firm's activities in architecture, structures and Civ Engrg. Also serves as dir of personnel and general adm. Graduated with MBA from Syracuse Univ. Likes golf, sailing, reading & classical music. Has restored 1796 house. *Society Aff:* ASCE, APWA, NSPE, ACEC, PSMA.

Keith, George W
Business: 843 Fourth Ave, Coraopolis, PA 15108
Position: Chief Engineer - Environ Systems. *Employer:* The Chester Engineers. *Born:* July 18, 1926, native of Port Allegany Pa. BSCE Auburn Univ 1948. Resident of Upper St Clair Township. Reg Prof Engr: Pa, W V, Ohio, Md. Mbr WPCF, AWWA, AAA, NSPE, ASCE; Penna Federation Dir on Board of Control of WPCF (1975- 78). Bedell Award 1971. Respon for planning & design of numerous wastewater collection & treatment & water supply treatment & distrib facilities as cons engr to municipalities & other governmental agencies.

Keith, James M
Home: 130 Jessie St, San Francisco, CA 94105
Position: Consulting Engr. *Employer:* Self employed *Education:* BS/Civil Engg/Univ of IL. *Born:* Jan 29, 1922. Formerly Principal in charge of marine & waterfront engrg work for John A. Blume & Assoc. Engrs. Also extensive experience in struct response consultation in connection with blasting. ASCE; Cons Engrs Assn of CA. Joint author of *Rincon Offshore Island and Causeway*, ASCE Proceedings Paper 2170, and several other technical papers and articles. Received Amer Concrete Inst Construction Practice Award (1968). Inventor of US Pat 3,793,845 concerning anchoarage of submarine pipelines. Delivered two lectures at NATO sponsored Advanced Study Inst on Berthing and Mooring of Ships at Wallingford, England, in 1973. Developed seismic isolation system for a floating drydock fixed mooring system. *Society Aff:* ASCE.

Keith, Richard L
Business: 4701 Old Redwood Hgwy, Santa Rosa, CA 95401
Position: Pres (Self-employed) *Education:* BS/Civ Engg/IL Inst of Tech. *Born:* 4/11/30. Pres of multi-discipline Consulting Firm specializing in the design of wineries, with projs throughout the US as well as Canada, Mexico and Australia. Guest lecturer on winery design at the Univ of CA as well as seminars in San Francisco, PA & NY. Co-author of paper with Univ of CA entitled *Sample Costs for Construction of Table Wine Wineries in CA*. Listed 1978-79 edition of "Who's Who in the West–. Chrmn Sonoma County Planning Commission, Chrmn Sonoma County Air Pollution Control Bd, Past Pres, Local Chapter of the American Soc of Civil Engrs. *Society Aff:* NSPE, ASCE, AIP.

Keith, Theo G, Jr
Home: 3866 Laplante Rd, Monclova, OH 43542
Position: Chrmn & Prof *Employer:* The Univ of Toledo *Education:* PhD/Mech Engr/U of MD; MS/Mech Engr/U of MD; BME/Mech Engr/Fenn Coll. *Born:* 7/2/39 Native of Cleveland, OH. Mechanical engr for the Naval Ship Research Development Center Annapolis, MD 1964-71. With the Univ of Toledo since 1971. Assumed current responsibility as MECHANICAL ENGRG DEPT CHRMN in 1977. Promoted to Assoc Prof with tenure 1976. Promoted to Full Prof 1981. Employed by NASA Lewis Research Center summers 1972-1975. Won a Univ of Toledo Outstanding Teacher Award 1977. Ralph Teetor SAE Award 1978. Numerous journal articles, reports, conference presentations and reviews. *Society Aff:* ASME, AIAA, SAE, ASEE.

Keithley, Joseph F
Business: 28775 Aurora Rd, Cleveland, OH 44139
Position: Chmn of the Board. *Employer:* Keithley Instruments Inc. *Education:* SM/-/MIT; SB/-/MIT. *Born:* 1915. Mbr Tech Staff Bell Labs N Y; Scientist Naval Ordnance Lab working on underwater mine firing devices, received Disting Civilian Service Award from Navy 1945; Founded Keithley Instruments 1946, manufacs electronic test & measuring instrumentation sold throughout the world, presently Chmn of the Board. Fellow IEEE; served Cleveland Sect as Secy & Chmn; has been mbr Admin Ctte of Group on Indus Electronics & Control Instrumentation & Group on Instrumentation & Measurements of which he was Pres. Recipient IECI Achievement Award, Cecon Medal of Achievement Award 1976, and IEEE Instrumentation and Measurement Soc 1983 Soc Award. *Society Aff:* IEEE, AAAS.

Kell, James H
Business: JHK & Associates, Box 3727, San Francisco, CA 94119
Position: Chrmn *Employer:* JHK & Assocs *Education:* MSCE/Transp Engr/Purdue Univ; BSCE/Civ Engr/Purdue Univ *Born:* 2/25/30 Post Grad, Transp, Univ of CA-Berkeley. Began career as Research Assoc for the Joint Hgwy Research Project at Purdue Univ. Spent military service as Research engr for the Transportation Research and Development Command of the US Army at Ft Eustis, VA. Served as Lecturer in Transportation Engrg and Research Engr for the Inst of Transportation and Traffic Engrg, Univ of CA, Berkeley. Entered private consulting in 1964, first as Principal Traffic Engr for Traffic Research Corp, Principal for Peat, Marwick, Mitchell & Co, and, since 1971, as Pres then Chairman, of JHK & Assocs. Recipient of the 1981 Theodore M. Matson Award for outstanding contributions to the advancement of the science and profession of traffic engrg. *Society Aff:* ITE, ASCE, TRB

Kelleher, John J
Home: 3717 King Arthur Rd, Annandale, VA 22003
Position: Telecomm Conslt *Employer:* Self *Born:* 10/14/14 Major fields of interest and activity are space comms sys, and efficient utilization of orbit - spectrum resources. Military and civilian radio operator and technician, 1932/1940. Civilian employee, US Army Signal Corps, 1940/1962. Civilian employee, NASA Office of Space Applications, 1962/1969. Systematics Gen Corp, 1969/1980. Mbr of numerous US delegations of ITU and CCIR meetings, 1963 to 1983, including ITU Space WARCs in 1963, '71, '77, '79 and 83. Recipient DA Decoration for Meritorious Civilian Service (1960), NASA Exceptional Service Medal (1969) and IEEE Centennial Medal. Elected Fellow IEEE in 1980 "for contributions to intl radio regulations–. *Society Aff:* AIAA, IEEE, SOWP.

Kelleher, Matthew D
Business: Mech. Engr. Dept., NPS Code 69KK, Monterey, CA 93943
Position: Prof *Employer:* Naval Postgrad School *Education:* Ph.D./Mech. Engr/Univ. of Notre Dame; MS ME/Mech. Engr/Univ. of Notre Dame; BS/Engr. Sci/Univ of Notre Dame *Born:* 02/01/39 Native of New York City. Began teaching career as Asst Prof at Univ of Notre Dame. Ford Foundation Postdoctoral Fellow at Thayer School of Engrg at Dartmouth College. NASA/ASEE Summer Faculty Fellow at Stanford Univ and NASA Ames Research Center. Joined faculty of Naval Postgraduate School in 1967. Currently Prof and Assoc Chairman, Mech Engrg Dept. Fellow of ASME. Active in ASME Heat Transfer Div; served as Chairman, General Papers Cttee; Chairman, National Heat Transfer Conference Coordinating Cttee; member of Cttee on Aeronautical and Astronautical Heat Transfer. Served as Chairman of Advisory Cttee to Monterey County, CA Planning Commission. *Society Aff:* ASME

Kellenbarger, Richard W
Home: 2050 Northtowne Ct, Columbus, OH 43229
Position: Chief Metallurgist. *Employer:* The Columbus Auto Parts Co. *Born:* Aug 1931; Native Columbus Ohio. BS WPI. USAF - Korean Conflict. Plant Metallurgist Wyman-Gordon 1953-58; engaged in met res Battelle Memorial Inst 1958-65, spec in dev of pure iron, foamed metals & powder met; Chief Metallurgist Abex Corp, S K Wellman Div 1965-67; Columbus Auto Parts Co 1967- ; taught courses in met at Columbus Tech Inst. Chmn Colum Chap ASM 1969-70. Presented sev papers & articles on forging tech. Favorite hobbies are golf & table tennis.

Keller, Arthur C
Home: 125 White Plains Rd, Bronxville, NY 10708
Position: R&D Consultant. *Employer:* Self. *Born:* Aug 1901 NYC. BS, MS, EE. Attended Cooper Union, Yale Univ & Columbia Univ. Lic P E NY. Has written 36 papers & received 40 U S patents (150 worldwide) in the fields of electromech devices, sound recording & reproducing, Sonar, switching apparatus, electronic heating, sputtering, magnetic tape & complete telephone sys. Incl are the 1936 chaps on Recording in Pender's EE Handbook & the Encyclopedia Britannica articles on the 'Gramaphone' 1950 & 1957. Tech Dir at Bell Labs 1949-65. Now an independent Internatl R&D Cons. Has received 2 US Navy Citations 1943. First Prize paper in 1954 AIEE, The Emile Berliner award 1962 for the Stereophonic Disc Record & in 1966 3 awards from the Natl Assn of Relay Manufacturers. Cons to Munitions Bd 195053, Cons to Dept of Defense 1955-63. Has been on the Bd of Dir of sev co's. Mbr APS, IEEE (Fellow), Acoustical Soc of Am (Fellow), Audio Eng Soc; Yale Eng. Assoc. *Society Aff:* IEEE, ASA, APS, AES, YSE, ARRL.

Keller, Charles W
Business: 1500 Meadow Lake Pkwy, Kansas City, MO 64114
Position: Exec Ptnr-Head/Management Services Div. *Employer:* Black & Veatch Engineers-Architects *Education:* BS/Civil Eng/Univ of KS. *Born:* 11/27/25. With Black & Veatch 1946- , originally as cvl design engr; 1953- , spec in utility rate studies, property valuation, fesibility studies, depreciation studies & other economic matters; became Ptnr 1964 & Exec Ptnr & Head of Div 1974. Also Chrmn of Bd, Black & Veatch, Inc, Consulting Engrs, Asheboro NC, and Marshall and Brown, Architects-Engineers, Kansas City Mo. Director, The Pritchard Corporation, Kansas City MO. Pres CEC of MO 1968; Board of Dirs ACEC 1969. Reg 26 states. Officer in USN, Cvl Engr 1945-46 & 1951-53. Fellow ASCE, ACEC *Society Aff:* ASCE, AWWA, NSPE, WPCF, ACEC, KES

Keller, Edward L
Business: Dept of Elec Engg & Comp Sci, Berkeley, CA 94720
Position: Prof. *Employer:* Univ of CA. *Education:* PhD/Biomed Engg/Johns Hopkins Univ; BS/Engg Sci/US Naval Academy. *Born:* 3/6/39. in Rapid City, SD. Served with Navy as shipboard electronics officer 1961- 65. NASA and NIH predoctoral fellowships at Johns Hopkins Univ, 1965-71. Appointed asst prof, Dept of Elec Engg and Comp Sci, Univ of CA, Berkeley, 1972. Alexander von Humboldt, Sr US Scientist, Frankfurt, German Fed Republic, 1977- 78. Became prof, Berkeley, 1979. Assumed joint research position as Sr Scientist, Inst of Medical Sciences, Pacific Medical Ctr, San Francisco, CA, 1979. *Society Aff:* AAAS, IEEE.

Keller, George V
Home: 5 Dekker Dr, Golden, CO 80401
Position: Professor. *Employer:* Colorado School of Mines. *Born:* Dec 1927; native of New Kensington Pa. PhD, MSc & BSc Penn St Univ. US Navy 1944-45. Geophysicist with US Geol Survey 1952-64 spec in dev of elec prospecting methods; on faculty Colorado Sch of Mines 1964- , Head Dept of Geophysics 1974- . Pres Group 7 Inc, a company performing contract geophys exploration 1969- ; Pres Geotherm Exploration & Dev Co (Hawaii) 1974- ; Pres of Geophysics Fund Inc (non-profit) 1975- .

Keller, Jack W
Business: 12075 East 45th Ave, Suite 400, Denver, CO 80239
Position: Denver Mgr *Employer:* Black & Veatch Consulting Engrs *Education:* BS/CE/Univ of KS *Born:* 7/13/30 Born in Ottawa, KS. Attended Ottawa Univ then transferred to Univ of KS. Employed for short time for power plant contractor in Leavenworth, KS then took a position with Black & Veatch Consulting Engrs in Kansas City, MO. After 11 yrs in Kansas City, MO, transferred to a new branch office in Denver, CO where has been located for 17 yrs. Presently Mgr of Black & Veatch's Denver Office. *Society Aff:* ASCE, ASM, ACI, AWWA, WPCF

Keller, Joseph B
Business: Dept of Math, Stanford, CA 94305
Position: Prof. *Employer:* Stanford Univ. *Education:* PhD/Math/NYU; MS/Phys/NYU; BA/Phys/NYU. *Born:* 7/31/23. I was an instr in phys at Princeton Univ in 1943-44. Then I worked for the Columbia Univ Div of War Res in underwater acoustics from 1944-1946. I returned to NYU to get my PhD in math & I remained there until 1979. During that time, I worked in wave propagation, with special reference to radar applications, as well as in fluid dynamics & other branches of appl math. I also spent the yr 1953-54 at the Office of Naval Res, and other periods visiting various univs & inds. In addition, I served as a consultant to various ind & govt labs. In 1976 & 1977, I won the Lester R Ford Award of the Math Assn of Am, in 1977 I was chosen to be the Gibbs Lectr of the Am Math Soc, and the Hedrick Lectr of the Math Assn of Am. In 1979 I was awarded an Honorary Doctorate by the Technical Univ of Denmark. In 1979 I moved permanently to Stanford Univ, where I am Prof of Math & Mech Engg. In 1981 I was awarded the Eringen Medal of the Society of Engineering Science. In 1983 I was von Neumann Lecturer of SIAM; in 1984 I received the Timoshenko Medal of ASME; & in 1986 I was elected a Foreign Member of the Royal Society. *Society Aff:* SIAM, AMS, NAS, AAAS, USNCTAM, USNCURSI.

Keller, Kenneth H
Business: 202 Morrill Hall, 100 Church St SE, Minneapolis, MN 55455
Position: Pres *Employer:* Univ of MN. *Education:* PhD/ChE/Johns Hopkins Univ; MSE/ChE/Johns Hopkins Univ; BS/ChE/Columbia Univ; BA/Liberal Arts/Columbia Univ. *Born:* 10/19/34. Following his undergrad training at Columbia, Dr Keller spent four yrs with Admiral H G Rickover in the AEC's Naval Reactors Branch. He then earned his PhD at Johns Hopkins and, in 1964, joined the faculty at the Univ of MN where he is now the Pres. Previously he was Vice President for Academic Affairs and Professor of Chemical Engineering. His res is in the area of biomedical engg & has addressed blood transport problems, artificial organ design, & chemotaxis. He has been a Sigma Xi Natl Lectr for two yrs & Pres of the Am Soc for Artificial Internal Organs. *Society Aff:* AIChE, ASAIO, NYAS.

Keller, Walter D
Business: 305 Geology Bldg, Columbia, MO 65211
Position: Retired Prof. *Employer:* Univ of Missouri - Columbia *Education:* PhD/Geol/Univ of MO; AM/Geol/Harvard; AB/Geol/Univ of MO; BS/Ceramics/MO Sch of Mines & Met. *Born:* 3/13/00. Prof of geology, Univ of MO-Columbia, consultant, sometime employee A P Green Refractories Co, US Geological Survey. Main interests in geology, ceramic engg, clay mineralogy, weathering of rocks. Recipient Hardinge Award (AIME), recipient of Twenhofel Medal (Soc. Economic Paleontologists and Mineralogists), elected Distinguished Mbr (AIME), Distinguished Mbr (Clay Minerals Soc), Neil A Miner Award (Natl Assn Geol Teachers), Who's Who In America; Fellow, Geological Soc of America, Mineralogical Soc of America. Hobby, photography. *Society Aff:* AAAS, AIME, GSA, CMS, MSA, AAPG, SEPM, MSoc.

Kelley, Henry J
Business: VPI & SU, Blacksburg, VA 24061
Position: Chaired Professor. *Employer:* VPI & SU. *Education:* ScD/AeE/NYU; MS/Math/NYU; BAeE/AeE/NYU. *Born:* 2/8/26. Res Engr, Grumman Aerospace Corp, 1948-63; Asst Chief of Res, 1961-63. VP, Analytical Mechanics Associates, Inc, Jericho, NY, 1963-78. Prof of Aerospace Engg, VPI, 1978-present. Pres, Optimization Inc, 1979-present. Dr Kelley has published numerous papers on optimization theory and aerospace applications,. He has received three AIAA Natl Awards; the Mechanics & Control of Flight Award in 1973, the Pendray Award in 1979 and the Fellow Award in 1980. He is founder of AIAA's Journal of Guidance, Control & Dynamics. He became the first holder of the Christopher Kraft Chair of Aerospace Engrg at VPI & SU in Sept. 1984. *Society Aff:* AIAA, IEEE, AAS, SIAM.

Kelley, Richard B
Business: 2095 Hammond Drive, Schaumburg, IL 60195
Position: Vice President *Employer:* The Sievert Group *Education:* -/Chem Engg/IL Inst of Tech; -/Chem/Elmhurst College; -/Chem/Northern IL Univ. *Born:* 12/6/46. Chicago, Il. Native of Il area. Richards of Rockford, Inc. (division of Darling-Delaware, Inc.) from 1971-1980: Pres. from 1973 to 1980. Joined the Sievert Group in 1980 (Professional Engineers and Constructors) U.S. Patents 3, 771,724, 3,833, 173, 3,998,389, 968,439. Lincoln Design Award 1975. Outstanding Author, Pollution Engineering, 1977. ASHRAE Design Award, 1981. *Society Aff:* AIChE, ASME, ACS, ASHRAE

Kelley, Thomas N
Business: 695 Ohio St, Lockport, NY 14094
Position: VP, Tech. *Employer:* Guterl Sp Steel Corp. *Education:* BS/Met Engrg./Purdue Univ; MS/Met/Univ of Pittsburgh. *Born:* 2/1/31. Toledo, OH. Employed Universal-Cyclops Specialty Steel Div, 7 yrs; Howmet Corp, 5 yrs; and Stellite Div, Cabot Corp, 12 yrs in supervision of process metallurgy for the production of special purpose steels and superalloys. Chief Met at Stellite Div from 1967 to 1979. Assumed present position as VP, Tech for the Guterl Specialty Steel Corp in 1979. Elected ASM Fellow and Chrmn of the Central IN Chapter in 1978. Author and coauthor of published papers, articles and seminar presentations on electroslag remelting and the mill processing of special purpose alloy wrought products. *Society Aff:* ASM, AIME, AISI.

Kelley, Wendell J
Business: 500 S 27th St, Decatur, IL 62525
Position: Chairman and President. *Employer:* Illinois Power Co. *Education:* BS/Elect Engg/Univ of IL. *Born:* May 2 1926 Champaign Ill. Empl by Ill Power Co in Decatur Ill 1949- . 1st duties, engr in operating dept; became Operating Engr 1954; beginning 1955 worked in a variety of admin capacities; after serving as mgr of personnel, elected V Pres 1961; Elected to Board of Dir & Pres 1966; elected Chmn & Pres 1976. Awards: Alumni Honor Award, Coll of Engrg Award for Disting Service in Engrg Univ of Ill Apr 19, 1974 at Honors Award Convocation; Disting Alumnus Award, Elec Engrg Dept Univ of Ill, Elec Engrg Alumni Assn Oct 1973. Named Fellow IEEE Nov 1971. *Society Aff:* NSPE.

Kellogg, Herbert H
Home: Closter Rd, Palisades, NY 19064
Position: Prof. *Employer:* Columbia Univ. *Education:* MS/Met/Columbia Univ; BS/Met/Columbia Univ. *Born:* 2/24/20. Native of NYC. Grad of Sch of Mines, Columbia Univ. Asst Prof of Mineral Preparation, Penn State Univ, 1942-46. Joined Columbia faculty in 1946. Presently Stanley-Thompson Prof of Chem Metallurgy. Mbr, Bd of Dirs, the Metallurgical Soc (AIME) 1958-59, 1967-69. Fellow of the Metallurgical Soc, 1972. Recipient of James Douglas Gold Medal (AIME), 1973. Elected to Natl Acad of Engg, 1978. Author of numerous technical publications on thermochemistry of metal extraction and energy conservation in metal production. Consultant to industry. *Society Aff:* AIME, NAE, IMM.

Kellogg, Joseph C
Business: 5601 S Broadway 400, Littleton, CO 80121
Position: President. *Employer:* Kellogg Corporation. *Education:* BS/CE-Bus Admin/Univ of MN. *Born:* Apr 24, 1926. USNR. Al Johnson Const 1951-70, Field Engr, V Pres-Mgr Western Div; Kellogg Corp 1970- , Pres. Guest Lectr: Stanford, UC Berkeley, Univ of Minn, Univ of Colo, Cornell Univ, Colo St Univ. Recipient Outstanding Citizenship Award United Fund Hennepin County, Minneapolis 1967; Outstanding Const Engr Award; Const Mgmt Award 1974, Bureau Reclamation Task Ctte joint Ctte/ ASCE, ASEE, AGC; Amer Inst of Constructors, C of C, Beavers, Moles, Toastmasters Internatl. Author: publs on cost control sys, contract mgmt, systematic problem solving, production planning, environmental design tech, research and development, risk management. *Society Aff:* ASCE, USCOLD, MOLES, AIC.

Kellogg, Louis Henry
Home: RFD Burnett Rd, Mendham, NJ 07945
Position: President *Employer:* Kelloge Enterprises, Inc. *Education:* BEE/Communications Engg/OH State Univ. *Born:* 1916; native of Madison Oh. Grad study Ohio St U. Signal Corps officer, radar dev, testing WW II. Bell Tel Labs: 1945 Nike anti-aircraft sys, radar, computer dev; Dir Nike antiballistic missile sys testing (NM, Calif) 1960; Nike X Sys, Safeguard ABM Sys, computer software dev, testing, eval for Kwajalein MI 1968; Dir final real time software dev, testing, delivery for Safeguard, N D 1973; ret Bell Tel Labs 1975. Part time Consultant Sys Engg Dev Programs. Awarded Dept of Army Outstanding Civilian Serv Medal for Achieves in Ballistic Missile Def Progs 1974; Ohio St Disting Alumnus 1970. Sr Life Mbr IEEE. Enjoys sailing. *Society Aff:* IEEE.

Kelly, Alphonsus G
Home: 5 Ailesbury Road, Ballsbridge, Dublin 4, Ireland
Position: Engineering Director *Employer:* Electricity Supply Board *Education:* PhD/Hydraulics/National University of Ireland; B.E./Mechanical and Electrical/University College, Dublin *Born:* 06/01/26 Author of "How to Make Life Easier at Work" McGraw Hill, 1988 - an enlarged version of a 1973 book that was a best seller in 3 languages. Has contributed over 20 technical articles on engrg topics in the USA, UK, and France; best known is perhaps his proof that a syphon lifts water by more than 34 ft for which discovery he merits an entry in the Guinness Book of Records. Is Dir of Egrg for the national electric utility of Ireland and Chairman of the Bd of its separate entity for international consultancy. *Society Aff:* ASME, I Mech E, IEI

Kelly, Douglas V
Home: 19 Tolan Way, Lafayette, CA 94549
Position: Chief Mechanical & Nuclear Engr. *Employer:* Pacific Gas & Electric Co. *Education:* BS/Naval Arch & Marine Engg/MIT. *Born:* Oct 21, 1918 Berkeley Calif. US Navy 1942-46. 1946-48 Shipbldg Div Bethlehem Steel Co-Outfitting Supt; 1948-81, Pacific Gas & Elec Co's Engrg Dept involved in mech design of thermal power plants, appointed Chief Mech Engr 1963, since then respon for design of over 6, 000,000 kW of generating capacity incl fossil-fired, nuclear, geothermal, combined cycle, & combustion gas turbines. Mbr ANS, ASNE; Fellow ASME. Reg. Prof. Engr. (Mech., Nuclear) State of CA. Self-employed Consulting Engr. 1981 to present. *Society Aff:* ANS, ASNE, ASME.

Kelly, Frank J, Jr
Business: Rudolph & Sletten, Inc, 989 E. Hillsdale Blvd, Foster City, CA 94404
Position: VP Marketing *Employer:* Rudolph & Sletten, Inc *Education:* BS/EE/US Naval Acad. *Born:* 11/23/21. Native of Oakland, CA. BS US Naval Acad 1944. Served in submarines WWII 1943-46. Entered genl contracting bus 1947; chief estimator for Barrett Const Co San Francisco through 1966; joined Stone Marraccini & Patterson archs & planners as Chief Estimator 1967. Retired as Senior VP in Nov 1982, presently VP of Marketing for Rudolph and Sletten, Inc, construction mgrs. PPres of AACE 1977- 78. Avid golfer and tennis player. *Society Aff:* AACE.

Kelly, James L
Business: 90 Church St, New York, NY 10007
Position: Maj Gen USA/Engrg Div-N Atlantic. *Employer:* US Army Corps of Engineers. *Born:* July 30, 1929 Patchogue N Y. BS USMA 1950; MS Caltech 1955; MS Geo Washington Univ 1969. Served in US Army units in Korea 1951-52, Germany 1956-59, Vietnam 1966-67, Korea 1971-72; Dept of Army Staff 196768; Military Asst & Exec Officer to Secy of Army 1969-71; Deputy Dir Cvl Works, Office Chief of Engrs 1972-74; Div Engr N Atlantic Div 197477. Mbr SAME, Regional V Pres SAME. Attended Command & Genl Staff Coll 1962-63; Armed Forces Staff Coll 1965-66; Natl War Coll 1968-69.

Kelly, John W
Home: 27-6 Concord Greene, Concord, MA 01742
Position: Vice President Operations *Employer:* Badger Engineers Inc. *Education:* BS/Mech Engrg/Swarthmore Coll; MS/Mech Engrg/Stevens Inst of Tech *Born:* May 1924. BS Swarthmore; MS Stevens Inst of Tech; Mgmt Training MIT. Officer in US Navy. 1946-61 employed by Exxon Res & Engrg Co Florham Park N J, in various engrg & proj mgmt capacities both US & foreign; joined Badger 1962, Engrg Mgr of Paris office 1970-71, Mgr Proj Engrg Cambridge office 1972-74, 1975- , V Pres Operations respon for Engrg, Procurement, Const, Personnel & Planning. Reg Prof Engr: N J, Fla, Mass. Senior VP Operations 1981. Mbr AIChE, ASME, NSPE; served on exec ctte of AIChE, Engrg & Const Contracting Ctte. Enjoys music, tennis & golf. Bd of Dirs Natl Constructors Assoc; Mgmt Dev Program Northeastern Univ. *Society Aff:* ASME, AIChE, NSPE

Kelly, Joseph D
Business: 6624 Romaine St, Hollywood, CA 90038
Position: Senior Vice President. *Employer:* Glen Glenn Sound Co - Hollywood. *Born:* June 1921. Studied Electronic Engrg at UC Berkeley & UCLA. With Glen Glenn Sound Co 1948- , holding various positions including Chief Engr, Sound Dir & now Sr V P. Fellow Mbr SMPTE, Sr Mbr IEEE; Mbr AES, AMPAS, NATAS. Mbr of Seminar Faculty BYU 1969-71 on Audio Recording Techniques. Recipient of Motion Picture Academy Award 1974 (Oscar) for tech achievements which exhibited high level of engrg & tech merit which was important to the progress of the industry; recipient of SMPTE Samuel L Warner Memorial Award 1975 for outstanding contributions in the design & dev of new & improved methods & appar for sound- on-film motion pictures.

Kelly, Keith A
Home: 905 Jamieson Rd, Lutherville, MD 21093
Position: Owner *Employer:* Keith A. Kelly Cons Engrs *Education:* MS/Structural/Johns Hopkins Univ; BE/Civil/Johns Hopkins Univ. *Born:* Dec 1923; native of Balto Md. WWII Engr Combat Grp. Bridge Design Engr Md Roads Comm; 20 yrs private practice including Pres & Chief Engr Baker - Wibberley & Assoc (ENR top) & 8 yrs as Ptnr Stevenson & Kelly, Struct/Cvl Engrs; since 1971 has been Dir of State Office respon for land acquisition, design, const & maintenance of all state capital improvements. Fac mbr JHU Evening Coll 12 yrs. P E 11 states. Fellow ASCE; P Treas & St Dir MSPE; P Pres & Natl Dir CEC/Md; P Pres Md Sect ASCE; Mbr APWA. Since Jan, 1977, has been VP & Chf of Bldg Structures Div of Kidde Consultants, Inc (77 on ENR top 500). Established K A Kelly Cons Engrs in Mar 1981 specializing in structual engrg. Currently teaches Advanced Steel Design and Design of Masonry Timber and Aluminum Structures in the Whiting School of Engg, The Johns Hopkins Univ. *Society Aff:* ASCE.

Kelly, Neil W
Business: 525 Brush Creek Blvd, Kansas City, MO 64110
Position: Owner. *Employer:* Neil W Kelly & Associates. *Born:* Oct 1921 North Canton Ohio. BSChE Univ of Ks 1943. With Pritchard Prod Corp 1949-73 spec in water cooling towers, Mgr of Engrg 1956, Mgr of Mfg 1959; V Pres of Opers 1965, Pres 1971; currently acting as cons on all aspects of cooling towers & other heat rejection sys. Mbr ASME & AIChE; AIChE Continuing Educ lectr on Water Cooling Tower Tech; Mbr ASME cttes on Cooling Tower Testing & on Cooling Tower Res. Author & publ of Kelly's Handbook of Crossflow Cooling Tower Performance 1976.

Kelly, Robert E
Business: Univ of California, Boelter Hall 5731, Los Angeles, CA 90024
Position: Prof. *Employer:* Univ of CA. *Education:* ScD/Aero & Astro/MIT; MS/Engg/MIT; BS/Aero Engg/RPI; BA/Liberal Arts/Franklin & Marshall Coll. *Born:* 10/20/34. Postdoctoral res at Natl Physical Lab (UK), 1964-66, & Inst of Geophys & Planetary Phys, Univ CA, San Diego, 1966-67. Mbr of Dept. Mech, Aerospace and Nuclear Engrg, Sch of Engg & App Sciences, UCLA since 1967 (Full Prof since 1975). Consultant at Hughes Aircraft Co (Culver City, CA) 1976-83. Author of numerous res papers on thermal convection, flow instabilities, stratified flows and

Kelly, Robert E (Continued)
other aspects of fluid mechanics. Mbr of Exec Committee, Div of Fluid Dynamics, APS (1978-1982) and Chrmn (1980-81). *Society Aff:* ASME, APS, AIAA, SIAM

Kelly, Thomas J
Home: 19 Charleston Dr, Huntington, NY 11743
Position: VP Info Res Mgmt *Employer:* Grumman Aerospace Corp. *Education:* MS/ME/Columbia Univ; MS/Ind Mgt/MIT; BME/ME/Cornell. *Born:* 6/14/29. Kelly, Thomas J. h 19 Charleston Dr Huntington NY 11743. VP, Engg Grumman Aerospace Corp. B June 1929. BME Cornell 1951. MSME Columbia 1956. MSIE MIT 1970. Joined Grumman 1947. Summer student employee. Propulsion Engr Rigel Missile Prog 1951. Grp Ldr F11F iger & F11F-1F Super Tiger Prog 1953-1956. USAF 1st Lt WPAFB 1956-1958 as Performance Engr B58, F105 & Hound Dog Missile Prog Grp Ldr Rocket Propulsion Dev Engr Lockheed 1958. Return to Grumman 1959 as Asst Chief Propulsion. Engg Proj Ldr Apollo & LM studies. Proj Engr, Engr Mgr & Deputy Prog Mgr on LM Prog. Deputy Dir Space Shuttle Prog 1970-1972. Elected VP 1971, VP Engg 1976. VP IRM 1983 Licensed NYPE. Fellow AIAA & AAS, mbr ASME NASA Distinguished Public Service Medal for LM contributions. First Cornell Engg Award 1972. Honary Doctor of Science State Univ of NY at Farmingdale 1983. Tau Beta Pi, Pi Tau Sigma. Quill & Dagger. Natl Audubon Soc, Boy Scouts, Sailing. *Society Aff:* AIAA, ASME.

Kelly, Thomas R
Home: 2825 Monogram Ave, Long Beach, CA 90815
Position: Mgr-Engrg Tech Req'mnts Branch *Employer:* Rockwell Internatl - Space Transport & Sys Group *Education:* -/Mech Engg/Geo Pepperdine -/Admin Sciences/ Pepperdine Univ. *Born:* June 1925; Calif native. Reg Prof Engr MF121. Major interests & accomplishments are in the field of Standardization; both design, mfg & mgmt. With Rockwell Internatl Space Div (formerly N Amer Aviation) 1955- ; has served in various Design Standardization, Value Engrg & Producibility Engrg mgmt positions on following major high tech-dev Aero-Space contracts: Navaho, Hound Dog, 2nd Stage Saturn V, Apollo CSM & currently Shuttle Orbiter. Mbr of College of Fellows of Soc for Advancement of Engrg; Sr Mbr & Fellow Standard Engrs Soc, now serving as its immediate past president Mbr Chancellors Advisory Council, Pepperdine Univ. *Society Aff:* SES, NMA, SAE.

Kelsey, Chester C
235 E. Jericho Tpke, P.O. Box 109, Mineola, NY 11501
Position: Partner. *Employer:* Sidney B Bowne & Son. *Education:* BSCE/Civil Engg/ CCNY. *Born:* Apr 19, 1929 Floral Park N Y. P E NY, NJ, Penn, Ct, Maine, Vermont, Dist. of Columbia, Texas, Rhode Island, Massach., Virginia, Alaska, Fla; L S NY, Ct, Fla, Penna.; Prof Planner NJ. A-E Firm Sidney B Bowne & Son, Mineola N Y. ASCE Fellow; N Y St Soc of Prof Engrs, Past-Pres; Amer Arbit Assn; ASTM; Cons Engrs Council; N Y St Assn Professions Past-Pres.; Const Specifications Inst; Amer Pub Wks Assn; Past-Chmn Publ Ctte Natl Soc of Prof Engrs; N Y St Soc PE's Chap Engr of Yr Award 1975. NY St. Soc. PE's Engr. of Yr, Award 1977; NY St. Soc. PE's Maritorious Svc. 1986 and 1981; Mrt. Sec. ASCE Director 1979-1981. 1951-53 US Army, Civic: Boy Scouts; United Methodist Church; Natl Adv Council Amer. Red Cross. VP Long Island Old Car Club; 1987, Outstanding Prof Engr in Mgmt Award, NY State Soc of Prof Engrs-Nassau County Chapter. 1983, Long Island Disting Leadership Aware, Long Island Business. Chrmn of the Natl Soc of Prof Engrs (NSPE) Prof Engrs in Private Practice (PEPP) from July, 1986 to July, 1987. VP Natl Soc of Prof Engrs (NSPE) 1986-87. *Society Aff:* NSPE, APWA, NPA, AWWA, ASCE, ACEC.

Kelsey, Ronald A
Business: , Alcoa Ctr, PA 15069
Position: Sec Hd. *Employer:* Alcoa Labs. *Education:* MS/Civ Engg/Carnegie-Mellon Univ; BS/Civ Engg/Poly Inst of Brooklyn. *Born:* 3/29/23. BS Poly Inst of NY; MS Carnegie Mellon Inst; USAF 1942-1945. Joined Alcoa Labs 1949; maj activities incl mtl properties, stress analyses, and structural performance. Joined Gen Dynamics 1955, Electric Boat Div & Gen Atomic Div; maj activities incl structural design, mtl properties, & nuclear engg. Returned to Alcoa Labs 1960, responsible for fatigue & fracture mechanics, mech fasteners & welding, sheet metal formability, metal deformation. Inventor of forming processes & ballistic armors. Mbr of delegation of USA/USSR Sci and Tech Exchange Prog-Mtls for Cryogenic Applications, 1974 to present; Chrmn, Welding Res Council Aluminum Alloys Committee, 1968 to present; mbr AWS Subcommittee 9 Structural Welding Code-Aluminum; Chrmn, International Comm. for Aluminum Fatigue Data Exchange and Evaluation, 1980 to present; Member, Metal Properties Council Tech. Advisor Comm, 1980 to present. Fellow, ASM; Mbr of AWS, Sigma Xi. *Society Aff:* ASM, AWS, SESA, ΣΞ.

Kelsey, Stephen J
Business: P.O. Box 27407, Salt Lake City, UT 84127
Position: Mgr, Eng Svc. *Employer:* Thatcher Company. *Education:* PhD/Chem Engg/ Univ of UT; MES/Chem Engg/Univ of UT; BS/Chem Engg/Univ of Ut. *Born:* 1/15/40. After obtaining a BSChE in 1965 from the Univ of UT, Dr Kelsey worked for 18 months as a Process Engg for Celanese Chem in Pampa, TX. He then returned to the Univ of UT working as an Instructor and Res Asst in Chem Engg and for the US Bureau of Mines Exper Station in Salt Lake City while obtaining his advanced degrees. While there he became assoc with the Inst for Biomedical Engg headed by Dr W J Kolff conducting numerous studies on blood flows in artificial heart valves with Dr Harvey Greenfield. He held Sr Res Asst in Biomedical Engg at the Univ of UT concurrently from about 1969 until 1975. He then was Operations Mgr for Wasatch Chem Div of Entrada Industries until April of 1978 when he accepted his present position. He is the author of several tech articles published in AIChE J, Chem Engg Sci and International J of Numerial Methods in Eng as well as numerous presentations at Natl and Int'l Scientific Conventions. During the last ten years, he has been engaged in designing, building & running small chemical plants & terminals in Missoula, Montana; Carlin, Nevada and Henderson, Nevada; as well as Thatcher's main facility in Salt Lake City, Utah. *Society Aff:* AIChE.

Kemp, E L
429 Riley Street, Morgantown, WVA 26505
Position: Prof. *Employer:* WV Univ. *Education:* PhD/Struct Mechanics/Univ of IL; MS/Engg/Univ of London; DIC/CE/Imperial College of Sci & Tech; BS/CE/Univ of IL. *Born:* 10/1/31. Asst. Engr IL State Water Survey, 1952, Asst. Engr, US Army Corps of Engrs, 1953-1954, Fulbright Fellow, Imperial College, 1954-1955, structural engr, London, 1955-1959. Fellow, Univ of IL 1959-1962. Joined WVU in 1962 as Assoc Prof of Civ Engg, Prof 1965, Dept chrmn 1967-74. Fellow American Council of Learned Societies 1975-76. Prof of the History of Sci and Tech and of Civ Engg 1977-present. Regents' Fellow, Smithsonian Institution 1983-84. Regents' Fellow, Smithsonian Institution 1983-84. Active in res in both structural engg and the history of tech. Practicing ind archaeologist and preservation engr. Extensive publications, res and technical committee mbrships in history of tech, engg and related field. *Society Aff:* ICE, IStructE, ASCE, ACI, SIA.

Kemp, Harold S
Home: 20 Crestfield Rd, Wilmington, DE 19810 *Employer:* E I duPont de Nemours & Co Inc. - Retired. *Education:* PhD/Chem Engrg/Univ of MI; MS/Chem Engrg/ Univ of MI; BS/Chem Engrg/Univ of NM. *Born:* May 1917; native of Ishpeming Mich. Joined duPont's Engrg Dept in March 1943 as Res Engr at Engrg Res Lab; transferred to Engrg Service Div 1960 as Cons Mgr; respon for tech consultation within duPont on mass transfer, heat transfer, fluid flow & agitation & mixing. Prog co-chmn AIChE meetings in London & Kyoto. Fellow AIChE 1973; Chmn Exec Ctte of AIChE's Design Inst for Multiphase Processing 1973-78; Prog Chmn AIChE Meeting in Philadelphia; Founders' Award AIChE 1977; Chmn Admin Ctte of AIChE's Design Inst for Emergency Relief Sys 1977-84; Elected Director of AIChE for three yr term 1981-83; Elected VP of AIChE for 1985, Pres for 1986, and Past Pres for 1987. Chairman Ad Hoc Task Force, 1985, for AIChE's Center for Chemi-

Kemp, Harold S (Continued)
cal Process Safety; Member of CCPS Technical Committee and Advisory and Management Boards 1985-Present. *Society Aff:* AIChE, AAAS.

Kemp, Lebbeus C
Business: 12318 Huntingwick, Houston, TX 77024
Position: Retired. *Employer:* Self. *Education:* BS/ChE/Rice Univ. *Born:* 10/8/07. With Texaco Inc 1929-72; petro res, Dir of Res, Asst to V Pres, Asst to Sr V Pres, Genl Mgr Petrochem Dept 1955-57, V Pres 1957-59, V Pres Res (world-wide) 1959-68, V Pres spl assign 1968-71, Sr Officer Houston 1971-72. Now cons in genl mgmt. Past Dir Jefferson Chem Co Texas; US Chem Co; Neches Butane Prod Co. Reg Prof Engr: N Y, Tex, Va. Fellow AAS, AIChE; Mbr Amer Chem Soc, Rive Univ Assn (Life), Tau Beta Pi. Sigma Xi Clubs: Houston, Chemists (NYC). Dir AIChE 1959-61, Trustee United Engr Trustees 1960-63 *Society Aff:* AIChE, ACS, AIAA, AAAS.

Kemp, Robert M
Business: P.O. Box 1365, Alta Loma, CA 91701
Position: Managing Partner *Employer:* R. M. Kemp Group *Education:* PhD/Mgmt/Claremont Grad Sch; MS/Systems Engrg/UCCLA; MBA/Bus Mgmt/ Claremont Grad Sch; BS/Physics/Univ of Redlands. *Born:* 03/26/26 Native of San Bernardino, CA. Served in US Navy 1943-47, as Communications Officer. Structures Analyst and Designer for Consolidated Vultee Corp, 1951-58. Chief Engr of South Gate Aluminum Co 1958-59. Project Mgr for Convair Corp 1960- 70. VP-Program Dir, 1971-77 and VP Research and Engrg, 1977-81 for General Dynamics Corp. Currently R. M. Kemp Group-Management Consultants and PE. Mbr UCLA Engrg Deans Council and Delta Committee. Lecturer Univ of La Verne. Grad School of Bus. Author: *Effective Mngmt of High Tech Projs.* Claremont Univ Press. 1983. Avid runner and swimmer who breeds horses. *Society Aff:* ASCE, ADPA, ASNE.

Kemper, John D
Business: Dept of Mech Engrg, Univ. of California, Davis, CA 95616
Position: Professor of Mechanical Engineering *Employer:* University of California - Davis. *Education:* PhD/Structural Mechs/Univ of CO; MS/Engg/UCLA; BS/Engg/ UCLA. *Born:* 5/29/24. 1943-44 Lockheed Aircraft Corp; 1944-46 US Army Air Corps (Sgt); 1949-55 Telecomputing Corp; 1955-56 H A Wagner Co; 1956-62 Marchant Div SCM Corp - VP of Engrg 1960-62; 1962- , Coll of Engrg Univ of Calif, Davis, Calif - Assoc Prof to Prof, Dean of Engrg 1969-83. Author of "Engineers & Their Profession," 3rd ed, 1982; "Introduction to the Engineering Profession," 1985. Panel Chairman: "Engineering Graduate Education and Research," National Academy Press, 1985; "The Teacher of Engineering: Preparation and Practice, ASEE, 1986. Prof fields of int: shock & vibration, stress analysis, mech design. Fellow ASME, AAAS. Registered Prof Engineer, California. Personal hobbies & ints: western Americana, conservation, hiking, mountaineering *Society Aff:* ASME, ASEE, NSPE, AAAS.

Kempner, Joseph
Business: 333 Jay St, Brooklyn, NY 11201
Position: Professor. *Employer:* Polytechnic Univ. *Education:* PhD/Applied Mech/ Poly Inst of Brooklyn; MAeE/Aero Engrg/Poly Inst of Brooklyn; BAcE/Aero Engrg/ Poly Inst of Brooklyn. *Born:* Apr 1923 Brooklyn N Y. P E lic in N Y. Aeronaut Engr, Structs Res Langley Field NACA 1944-47. Continued res including creep buckling & shell theory at Polytech Inst of Brooklyn; Appointed Head of Dept of Aerospace Engrg & Applied Mech 1966-76; since 1957 principal investigator on sponsored res projs on aerospace & submersible vehicles. Publ large num of res papers; mbr Ed Adv Bd, Inter Jour Nonlinear Mech. Mbr Advisory Group II, Ship Res Ctte & Ctte on Basic Res, Advisory to Army Res Office. Laskowitz Gold Medal for Aerospace Res, N Y Acad of Scis. *Society Aff:* AIAA, ASME, ASEE, AAM.

Kendall, H Benne
Business: Dept. of Chemical Engineering, 184 Stocker Center, Ohio University, Athens, OH 45701
Position: Prof. *Employer:* OH Univ. *Education:* PhD/ChE/Case Inst of Tech; MS/ ChE/Case Inst of Tech; BS/ChE/Grove City Coll *Born:* 4/27/23. in Midland MI where father was a chem engr with Dow. Served in USAF 1943- 46. Married Patricia Emmons in Scranton, PA, Sept 1948. Four children, three grandchildren. Teaching experience in chem engg: Univ of MI: 1950; Case Inst of Tech; 1951-1960; OH Univ: 1960-present, Chrmn: 1961-67; Actg Chrmn 1971-72, 1982-83. Ind Experience: Dow Chem Co, Midland, MI: summers, 1941, 42, 47, 48, 50, 56; Shell Oil, Houston (Sabbatical from OH Univ) 1968-82. Principal interests: Appl kinetics, catalytic processing, history of tech. Reg football official (OH). *Society Aff:* AIChE, ACS, ΣΞ, ΤΒΠ, ΑΧΣ, ΟΔΚ.

Kendall, John M
Business: Head, Computer Science Dept, Univ of Calgary, Calgary Alberta, T2N 1N4 Canada
Position: Professor of Computer Sci & Head of Dept *Employer:* The University of Calgary. *Education:* PhD/Microelectronics/Univ of Birmingham; MSc/ Semiconductor/Univ of Birmingham; BSc/Physics/Univ of Birmingham; FELLOW Inst P/Physics/Univ of Birmingham; P Eng/Engineering/Univ of Birmingham *Born:* Dec 1936 Birmingham England. P Eng (Alberta), F Inst P (England). Taught at Birmingham, London & Lakehead Ontario; presently Prof of CPSC & Head of Dept. Author of 3 textbooks & numerous res papers; active cons to indus & govt; has always been active in upgrading Engrg & Phys syllabi at univ, provincial & natl levels. Winner of 1976 ASEE Western Elec Award for teaching excellence. Keen outdoorsman & very active in the Arts. Was Pres of Calgary Region Arts Fnd (1974-78) and Festival Calgary (1972-74). Recently returned to U of C after 3 1/2 yrs in Industry R & D Facility. *Society Aff:* F Inst P, P Eng

Kendall, Percy Raymond
Home: 900 N. 86th Way, Scottsdale, AZ 85257
Position: Mgr Energy Systems *Employer:* Motorola, Inc *Education:* Cert/Marketing and Managerial Econ/Univ of Chicago Bus School; Cert/Ind Cost Accounting and Labor Relations/Univ of IA; BS/EE/Case Western Reserve Univ *Born:* 7/10/14 40plus years in engrg, application systems, marketing and management. After several years as assistant mgr of final test dept, design engr, subcontract coordinating engr, mgr crystal dept and small order dept at Collins Radio, developed experience in tech marketing, application engrg, system engrg and marketing management at Motorola and Raytheon. Tech and management papers and conference leadership include: proposal preparation, compensation of engrs in marketing, design-to cost, acquisition management, energy conservation, generation and management. Assoc activities include: Chrmn of local chapters of IEEE, AEE, professional papers, national and local; National pres of AEE, 1982. Conference chrmn: IIE, AEE, IEEE, GCMA, FEE. Lecturer and adjunct prof: Babson Inst, Northeastern Univ. Pres: Tidewater Coll, 1965-1968; Profl Engr, OH FAA Licensed Pilot: Land, Sea, and Instrument ratings. FCC: 1st Class Radiotelephone license. *Society Aff:* AEE, AIEE

Kendrick, Junius L
Business: 244 Dexter Ave, Montgomery, AL 36104
Position: Agriculture Engineer. *Employer:* Alabama Power Company. *Education:* BS/Agricultural Eng/Auburn Univ. *Born:* June 1922. Served with 1st & 3rd Army in 5 major campaigns in WWII while assigned to 456 Mobil Anti-Air Craft Bn. Earned BS in Agri Engrg from Auburn Univ. Reg P E Ala 1959- . Respon for energy application on the farm in Central Ala, encouraging wise energy use & upgrading power facilities for rural dev projs. Recipient of Natl G B Gunlogson Engrg Award 1975. Has served as Chmn of Ala Sect ASAE; Chaired Ala's Top Rural Dev Ctte 1975; Hon Mbr Ala's 4-H Clubs, FFA Assn & FHA Assn. *Society Aff:* ASAE.

Kenison, Leon S
Home: 82 West St, Concord, NH 03301
Position: Director, Project Development Division. *Employer:* N H Dept of Trans-

Kenison, Leon S (Continued)
portation. *Education:* MBA/Admin/Univ of NH; BS/Civil Engg/Univ of NH. *Born:* Dec 1941; native of Jefferson N H. Married; 5 children. Const Proj Engr with Dept 1964-71; Asst Maintenance Engr 1971-87. Director Project Development 1987- . Reg Prof Engr in N H. Mbr NSPE, ASCE; Treas, Secy N E Council Delegate, VP, Pres. of NH Sect ASCE; Chmn NE Council ASCE 1975-76. Mbr Concord Historic District Commission & Planning Bd; Ward Moderator, certified basketball official, Little League & Babe Ruth League Umpire. Mbr Capital City Rotary Club. Hobbies include golf, basketball, squash, home remodeling, reading, gardening, hunting, fishing. *Society Aff:* ASCE, NSPE.

Kenji, Kakizaki
Home: 1332, Tsujido, Fujisawa-shi 251 Japan
Position: Exec VP *Employer:* Tokyo Elec Co, Ltd *Education:* PhD/EE/Waseda Univ; BS/EE/Waseda Univ. *Born:* 3/2/19. Exec VP, Tokyo Electric Co, Ltd. Fellow of IEEE. *Society Aff:* IEEE, IECE.

Kennedy, Alfred J
Business: 114 Cromwell Road, London SW7 UK
Position: Deputy Director *Employer:* The Technical Change Centre *Education:* DSc/Metal Phys/London Univ; PhD/Metal Phys/London Univ; BSc/Phys/London Univ. *Born:* 11/6/21. Lectr in phys, London Univ, 1947-50; Res Fellow, Royal Inst, London, 1950- 54; Hd of Metal Phys Gp, British Iron & Steel Res Assoc, 1954-7; Prof and Hd of Mtls Dept, College of Aero, Cranfield, 1957-66; Dir, British Non-Ferrous Metals Res Assoc, 1966-77; Director of Research, The Delta Metal Company 1978-81; Pres, Inst of Metals, 1970-71; Pres, Inst of Met, 1976-77; Platinum Medallist, the Metals Soc, 1976; Commander of the Order of the British Empire (CBE) 1979. Fellow of Engrg. Author of "Processes of Creep and Fatigue in Metals" (1962); "The Materials Background to Space Technology" (1964); Editor of Creep & Stress Relaxation in Metals (1965) & High Temperature Mtls (1968). Res papers & articles principally on deformation & fracture in crystalline mtls. *Society Aff:* IEE, Inst Met, IMM, Inst Phys.

Kennedy, Bruce A
Business: Suite 180, 1658 Cole Blvd, Golden, CO 80401
Position: Prin & VP. *Employer:* Golder Assoc. *Education:* BSc/Mining Engg/Imperial College-Royal Sch of Mines-Univ of London; ARSM/Mining Engg/Imperial College-Royal Sch of Mines-Univ of London. *Born:* 7/24/44. London, England. US Citizen 1972. Mining Engr the Anaconda Co 1965-1972. Prin and VP Golder Assoc 1972 to date responsible for co's surface mining and geological consulting worldwide. Dir of AIME and SME of AIME. Recipient of Robert Peele Meml Award 1971. Reg PE in WA and British Columbia. Distinquished member SME of AIME 1980. *Society Aff:* AIME, IMM, MMSA

Kennedy, Clyde M
Business: 2600 Century Pkwy NE, Atlanta, GA 30345
Position: President & Chief Executive Officer. *Employer:* Law Engineering Testing Company. *Born:* May 1931; native of Atlanta Ga. BSCE & MSCE Ga Inst of Tech. Lab Technician for Law Engrg Testing Co 1952-53. US Army 1953-55. Resumed employment with Law Engrg Testing Co: 1955-61 Soil Engr, 1961-62 Chief Soil Engr, 1962-66 Atlanta Branch Mgr, 1966-74 V Pres & Dir of Engrg, 1974-75 Exec V Pres, 1975- present Pres & Chief Exec Officer. Reg cvl engr: Ga, N C & Conn. Mbr ASCE, ACEC, Internatl Soc of Soil Mech & Foundation Engrg; Peachtree Chorus Atlanta Ga Chap.

Kennedy, D J Laurie
Business: Department of Civil Engineering, Edmonton Alberta Canada
Position: Professor *Employer:* University of Alberta *Education:* PhD/Civil Engg/Univ of IL; MS/Civil Engg/Univ of IL; BASc/Civil Engg/Univ of Toronto. *Born:* Feb 28, 1929. Prof Engr Ontario Quebec Alberta; Fellow Engrg Inst of Canada FCSCE. FASCE - Sigma Xi; Pres Can Soc for Cvl Engrg 1975- 76; Chmn Can Standards Assn Cttes', Steel Struct for Bldgs' 1968- ; Steel Fixed Offshore Structures 1984- ; Mbr Standing Ctte on Struct Design; Natl Bldg Code Canada 1969- . Ptnr & Dir Morrison Hershfield Burgess & Huggins Cons Engrs 1962-1982. Taught Univ of Toronto Cvl Engrg 1956-70; Chmn Div Solid Mech & Struct Engrg Carleton Univ 1970-73; Dean Faculty of Engrg Univ of Windsor 1973-79. Prof Univ of Alberta Cvl Engrg 1982-. *Society Aff:* EIC, CSCE, ASCE, ACI, AWS.

Kennedy, David D
Business: 657 Howard St, San Francisco, CA 94105
Position: Pres *Employer:* Kennedy/Jenks/Chilton *Education:* MS/Civ Engrg/Stanford U; BA/Chem/Whitman Coll. *Born:* 6/14/39 Native of San Francisco, CA. Served in US Public Health Service 1961-1963 on Great Lakes water quality survey. With Kennedy/Jenks Engrs since 1965 serving as project engr, project mgr, VP. Became pres in 1979. Worked on water resources project in Cambodia, water supply and sewerage projects in Argentina, Peru, Guam and American Samoa. Responsible charge of water quality, water supply & treatment and sewerage projects in western US. Pres of Consulting Engrs Assoc of CA 1986-87. Bd of Dir, Engrs club of San Francisco 1987- . *Society Aff:* ASCE, AWWA, CEAC, WPCF, AAEE, ECSF, USCOLD, SAME.

Kennedy, David P
Home: 2227 NW 16 Ave, Gainesville, FL 32605
Position: Pres. *Employer:* D P Kennedy & Assoc. *Born:* Nov 1923 Boston Mass. Radiation Lab MIT 1940-41; Raytheon Co Waltham Mass 1941-43. Served with Army Signal Corps WWII 1943-46 teaching electronics. After WWII Raytheon Co 1946-58; microwave communications, semiconductor devices; Instr Northeastern Univ Grad Sch of Engrg 1953-58, theory of semiconductor device oper; IBM Poughkeepsie N Y 1958-73, math analysis of semiconductor devices; Prof of EE Univ of FL Gainesville 1973-76; Pres D P Kennedy and Assoc 1976-Pres; presently Prof of EE Univ of Fla Gainesville. Has publ over 50 papers on theory of semiconductor device oper & contrib to 3 books on this subject. Fellow IEEE; Eminent Engr, Tau Beta Pi; Full Mbr Sigma Xi.

Kennedy, Donald D
Home: 104 Warrane Rd, Willoughby N S W, Australia 2068
Position: Retired. *Born:* Mar 28, 1908 Tumut NSW. Educ at Manaro Grammar Sch, Cooma NSW. Joined Western Elec as Sound Engr Jan 1929 - later apptd Mgr var branches in Australia. Served 1939-45 in war as Radar Officer RAAF, rank Squadron Leader. Joined Australian Govt as Recording Engr 1946, held positions during following yrs: Studio Mgr, Tech Services Officer, Spec Projs Officer; respon for planning & implementation of special multi screen 70mm film presentation Expo 70 Japan; carried out extensive res on Film Industry of Peoples Republic of China 1973 & 74 & presented paper at SMPTE Conference. Life Fellow of SMPTE. Mbr Soc Aust Cinema Pioneers. *Society Aff:* SMPTE, SACP

Kennedy, Edmund P
Business: 800 Independence Ave SW, Wash, DC 20591
Position: Div Chief. *Employer:* Federal Aviation Adm. *Education:* AMDP/Public Adm/Syracuse Univ; BEE/Electronics/Manhattan College. *Born:* 10/27/37. Native of NYC. Career with FAA began in 1959 & includes tours of duty in NY, Denver & Wash DC. Present position - Deputy Director, Airway Facilities Service FAA responsible the Mgt of Air Traffic Control Systems and facilities. One of twenty selected for FAA's first mgt prog (AMDP) at Syracuse Univ-1964. Married-three children. Enjoys coaching sports, coin & stamp collecting, hunting & fishing. Reg PE - state of CA.

Kennedy, Francis E, Jr
Home: 4 Haskins Road, Hanover, NH 03755
Position: Prof of Engrg *Employer:* Thayer Sch of Engrg-Dartmouth Coll. *Education:* PhD/Mechs/RPI; MS/Mech Engg/Stanford Univ; BS/Mech Engg/WPI. *Born:* 9/6/41. US Army Lt 1965-67. At Wyman-Gordon Co 1967-69 as dev engr in metal forming

Kennedy, Francis E, Jr (Continued)
opers. Post-doctoral assoc at RPI 1973-74. Joined faculty at Dartmouth 1974; teaches & supervises students in mech engrg & does experimental & analytical res in tribology & mech design. Cons to several indus & law firms. Author or co- author of a number of tech papers & reports. Recipient to 1st ASME Newkirk Tribology Award & SAE Teetor Award, both in 1976 *Society Aff:* ASME, ASEE, SAE, ASM, ASLE.

Kennedy, Jan B
Home: 25 Palmer Dr, South Windsor, CT 06074
Position: Research Engineer. *Employer:* United Technologies Research Center. *Education:* BSME/ME/Princeton; MSAE/AE/Princeton; diploma/AE/Von Karman; 2 years business graduate corses AE/Univ Conn *Born:* 6/22/41. BSME & MSAE Princeton Univ. Mbr Phi Beta Kappa & Tau Beta Pi. Two years business school at Univ of CT. 2 yrs instr Von Karman Inst in Brussels working in hypersonic cavity flow with transpiration cooling & blunt reentry vehicles. Joined United Technologies Research Center in 1968; dev prog for viscous- inviscid interaction in scramjets & laser cavity flow, conducted dev progs in integral rocket ramburners & recently dev and use instrumentation to design and characterize fuel injectors and sprays in gas turbine engines. Instrumentation includes laser velocitimetry, droplet sizing interferometry and light scattering. 2 term Chmn AIAA Conn Sect; ASME Chmn Princeton Student Chap; started European AIAA Chap at VKI. ASTM E29.04 mbr. 1976 Vir Tree Farmer of the Year. Tennis, skiing & woodworking are avocations. *Society Aff:* AIAA, ASTM, ΦBK.

Kennedy, Jerome D, Sr
Home: 3089 Country Club Ct, Palo Alto Hills, CA 94304
Position: Vice-President of Marketing. *Employer:* Advanced Electronics Design Inc. *Education:* BS/Electronics/Univ of IL. *Born:* Apr 6, 1929; native of Freeport Ill. Korea - US Army Counter-mortar team. Briefly in servo engrg at McDonnell Aircraft Co; 8 yrs wth Electronic Assocs Inc, selling analog & hybrid computers; Mgmt & Internatl Marketing of Applied Dynamics Inc 5 yrs; Operated J D Kennedy Corp representing Applied Dynamics, Data Genl Corp, Sycor, etc; Pres of Basic Computing Arts Inc, manufac of the Data Sentinel Computer Security Sys for IBM 360/370's from 1971-74; Marketing V Pres & Dir of Advanced Electronics Design Inc, manufacs of disk controllers and colorgraphics systems 1974- . Coinventor of Radar computer sys, patent . Various offices in IEEE, Simulation Council Inc & was Pres of Palo Alto Rotary Club 1971. Life Mbr Simulation Councils Inc. *Society Aff:* ACM.

Kennedy, John F
Business: Inst of Hydraulic Res, Iowa City, IA 52242
Position: Institute Director & Professor. *Employer:* The University of Iowa. *Education:* PhD/Civil Engg/CA Inst of Technology; MS/Civil Engg/CA Inst of Technology; BS/Civil Engg/Univ of Notre Dame. *Born:* 12/17/33. Res Fellow in Engrg Caltech 1960-61; Asst then Assoc Prof MIT 1961-66; Professor of Fluid Mechanics and Dir Iowa Inst of Hydraulic Res The University of Iowa 1966- ; Chairman, Div Energy Engrg 1974-76. Carver Distinguished Prof of Engrg 1981- . Principal tech interests incl river hydraulics, ice engrg, cooling tower tech & density stratified flows. Recipient of ASCE Stevens , Huber & Hilgard prizes. Elected to NAE 1973. Elected Pres of IAHR in 1979. *Society Aff:* IAHR, ASCE, NAE, ASME, ASEE.

Kennedy, Lawrence A
Business: Dept of Mech Engrg 206 W 18th Ave, Columbus, OH 43210
Position: Prof & Chairman *Employer:* The Ohio State Univ *Education:* Ph.D./Mech. Engr & Astronautical Sci/Northwestern Univ; M.S./Mech Engr/Northwestern Univ; B.S./Mech Engr/University of Detroit *Born:* 05/31/37 Prof Kennedy is Prof & Chairman of Mech Engrg at the Ohio State Univ. He was a faculty member at SUNY/Buffalo 1964-83, Visiting Prof at Univ of CA/San Diego (1968-69) and the von Karman Inst for fluid Dynamics/Belgium (1971-72), a Goebel Visiting Prof at the Univ of Mich (1980-81). He is a Fellow of ASME, an Associate Fellow of AIAA, a recipient of the SAE Ralph R. Teetor Award (1984) and the AT&T Foundation Award (1987). Dr. Kennedy was a NATO Fellow (1971-72), an AGARD Lecturer (1971-72), a NSF Faculty Fellow (1968-69) and a W.P. Murphy Fellow (1960-63). He is an editor of J. of Experimental Thermal & Fluid Sciences (1987-), an associate editor of Applied Mechanics Reviews (1985-). He is a contributor of numerous articles in engrg journals. *Society Aff:* AIAA, ASME, The Combustion Inst., SAE, APS, ASEE

Kennedy, Robert M
Business: 657 Howard St, San Francisco, CA 94105
Position: Chairman, Executive Vice President. *Employer:* Kennedy/Jenks/Chilton Inc. *Education:* MS/San Engr/Johns Hopkins Univ; AB/San Engr/Stanford Univ; -/Science/Whittier College. *Born:* Oct 1916. Born in Berkeley Calif. With USBR 1939-40; The Panama Canal 1940-42; PA Engr USPHS 1943-46; Ptnr Kennedy Engineers 1947-77; Ptnr Pacific Environ Lab 1947-72. Respon for dir of engrg opers for Kennedy/Jenks/Chilton in USA & abroad. P Pres CEAC; P V Pres & Dir ACEC; P Pres SF Sect ASCE; P Pres USA Sect AIDIS; Diplomate AAEE; Fellow ASCE, APHA, ACEC; Mbr AIDIS, AWWA, WPCF, APWA, USCOLD, NACE; Hon Mbr IWPC (England); Trustee, Whittier College. *Society Aff:* ASCE, ACEC, CEAC, AAEE, APHA, AWWA, WPCF, APWA, OSCOLD, AAEE, AIDIS, IWPC.

Kennedy, Robert S
Business: EECS Dept Rm 35209, Cambridge, MA 02139
Position: Professor of Elec Engrg. *Employer:* Mass Institute of Technology. *Education:* ScD/Elec Engr/MIT; SM/Elec Engr/MIT; BSEE/Elec Engr/Univ of KS *Born:* Dec 1933 Augusta Ks. Active duty USNR-AEC 1955-57. Staff Mbr Lincoln Lab 1963-64. Faculty Dept of EE & Computer Sci MIT 1964- . Res interests: communication in random channels 1963-68, optical detection & communication 1968-80. Local communications network 1980. Cons to govt & indus. Fellow IEEE; Mbr AIP, OSA, Sigma Xi, Tau Beta Pi, Omicron Delta Kappa, Eta Kappa Nu. Flight instr as an avocation. *Society Aff:* ΣΞΤΒΠ, IEEE, ODK

Kennedy, Thomas W
Business: 3208 Red River Center for Transportation Res, Austin, TX 78705
Position: Prof. *Employer:* Univ of Texas. *Education:* PhD/Civil Engg/Univ of IL; MS/Civil Engg/Univ of IL; BS/Civil Engg/Univ of IL. *Born:* 1/7/38. Taught three yrs at the Univ of IL - Urbana and since 1965 at the Univ of TX - Austin. Served as Dir, Council for Advanced Transportation Studies and Asst Univ VP for Res; Served as Assoc Dean for Res and Planning, 1979-1986 and Assoc Dean for Res and Facil 1986-8/31/87. September 1, 1987, returned to the Dept of Civil Engg to head a $6.5 million, 5 year *Strategic Highway Research Program* contract. Has been active in both teaching and res and has published more than 225 technical reports and papers. Presently serves on 14 natl tech comm of various professional societies. Primary areas of interest are materials, pavements, transportation and materials characterization. *Society Aff:* TRB, AAPT, ASTM, ASCE.

Kennel, William E
Business: 200 E Randolph Dr, Chicago, IL 60601
Position: Exec VP. *Employer:* Amoco Chemicals Corp. *Education:* ScD/CE/MIT; MS/CE/MIT; BS/CE/Univ of IL. *Born:* 8/11/17. Native of St Louis, MO. Employed as Process Engr at A E Staley Mfg prior to WWII. Served with US Army 1st Armored Div 1942-1946. Discharged with rank of Maj. Obtained MS & ScD degrees at MIT following WWII. Employed at Std Oil Co Res Dept in 1948. Assumed various positions in chem subsidiary, Amoco Chems Corp. Served as VP Res, VP Plastics Div, VP Marketing & assumed present position in 1975. *Society Aff:* AIChE, ACS.

Kennemer, Robert E
Home: 1461 Valle Vista-Unit 20, Pekin, IL 61554
Position: Retired. *Education:* BS/ME/Univ of ID. *Born:* Mar 14, 1920. Has been employed continuously by Caterpillar since 1942, with the exception of 1 1/2 yrs spent in US Navy. In 36 yr career with Caterpillar spent first 20 yrs in Res Dept doing

Kennemer, Robert E (Continued)

dev work on diesel engines; then in charge of product engrg on turbine engines for a period of 6 yrs; returned to field of diesel engines 1969 managing engr work on heavy duty diesel engines; in 1971 appointed Ch Engr for Indus Div. Retired 1978. Reg Prof Engr in Ill. Since becoming Mbr of SAE has served on many cttes & on Board of Dirs. Enjoys travel, photography, tennis, flying & is licensed pilot. *Society Aff:* SAE, ASME.

Kenner, Vernal H

Home: 5569 Millwheel Ct, Hilliard, OH 43026
Position: Assoc Prof *Employer:* OH State Univ *Education:* PhD/Engrg Sci/Univ of CA-Berkeley; MS/Applied Mech/Univ of CA-Berkeley; BS/ME/Univ of CA-Berkeley; AA/ME/City Coll of San Francisco *Born:* 6/12/41 Native of Washington, DC. San Francisco. CA Public Schools. Faculty, Mechanical Engrg Dept, Wayne State Univ (Detroit) 1971-78. Faculty, Graduate Aeronautical Labs, CA Inst of Tech, 1978-80. Assoc Prof, The OH State Univ since 1980. Registered PE, MI. Current research area: mechanical properties of polymeric materials. Publications in the areas of wave propagation in solids and fluids, experimental mechanics, biomechanics and material characterization. Consultant. *Society Aff:* ASME, SESA, AAM

Kensinger, Robert S

Home: 2619 65th Ave N, Minneapolis, MN 55430
Position: Proj Supervisor. *Employer:* Honeywell, Inc. *Education:* BAE/Aero Engg/Univ of MN. *Born:* 12/7/31. Native of St Paul, MN. Assoc Engr with Braniff Intl Airways prior to grad. Engr with N Am Rockwell Columbus Div in Aerodynamics & Airloads. Joined Honeywell Inc as a Res Engr in automatic flight controls in the Avionics Div. Currently a proj Supervisor in the Defense Systems Div responsible for aerodynamic R&D of air delivered weapons. Exec Bd mbr of the Ballistics & Vulnerability Div of ADPA. Chrmn of the 4th Vulnerability Symposium sponsored by ADPA. *Society Aff:* AIAA, ADPA.

Kent, James A

Business: 4001 W McNichols, Detroit, MI 48221
Position: Dean, Coll of Engg & Sci. *Employer:* Univ of Detroit. *Education:* PhD/Chem Engg/WV Univ; MS/Chem Engg/WV Univ; BS/Chem Engg/WV Univ. *Born:* 2/10/22. PhD, MS, BS ChE from WVU. Native of Morgantown, WV. Served with Army Corps Engr 1943-46. Res engr with Dow Chemical Co 1950-52. Res group leader with Monsanto Chem Co 1952-54. With WVU 1954-67 on chem engg faculty and from 1962-67 as Assoc Dean/Dir of the Engg Experiment Station. Dean of Engg, MI Tech Univ, 1967-78; Dean of Engg and Sci, Univ of Detroit, 1978-. Hazardous waste management. Advisory Comm: Office of Tech Assessment, 1975-77; Medical Devices Applications Program of NHLI 1969-74; MI Task Force Hazardous Wastes 1977-79. Chrm MI Hazardous Waste Management Planning Committee, 1980- . Music, golf. *Society Aff:* AIChE, ASEE.

Kentzer, Czeslaw P

Business: Aeronautics & Astronautics, Grissom H., Purdue U, W Lafayette, IN 47907
Position: Assoc Prof. *Employer:* Purdue Univ. *Education:* PhD/Aero Engr/Purdue Univ; MS/Aero Engr/Purdue Univ; BS/ME/San Diego State. *Born:* 6/29/25. Served with Polish, French and British Armies in WWII. Studied Aeronautical Engr in USA. Teaches aerodynamics at Purdue Univ since 1958. Special interests include gasdynamics, reactive flows, transonic flows, random wave motion applied to acoustics and to turbulence, computational fluid dynamics. *Society Aff:* AIAA, AMS.

Kenyon, Richard A

Home: 57 Old Forge Ln, Pittsford, NY 14534
Position: Dean, College of Engineering. *Employer:* Rochester Inst of Tech - N Y. *Education:* PhD/ME/Syracuse Univ; MS/ME/Cornell Univ; BME/ME/Clarkson College. *Born:* 4/8/33. Prof Engr N Y St. Additional study at Colo St Univ & MIT. Prof career includes employment with G E Co as gas turbine engr & service on faculties of Cornell Univ, Clarkson Coll of Tech & Rochester Inst of Tech. Appointed Prof & Head of Dept of ME 1970 & Dean of Coll of Engrg 1971 at Rochester Inst of Tech. Fellow ASME, ASME Region III VP 1978-82, ASME Sr VP 1982-84, ASME Bd of Governors 1984-86 ASEE, NSPE, NYSSPE, Tau Beta Pi, Sigma Xi, Pi Tau Sigma. Prof interest resource & energy recovery from solid waste Assoc of Engrg Colleges of NYS, Pres 1980-86; New York State Bd for Engrg and Land Surveying, 1981-86; NCEE, 1981-86 *Society Aff:* ASME, NSPE, ASEE, NCEE.

Kenyon, Wilfred

Home: 1339 Leisure World, Mesa, AZ 85206
Position: Consulting Engineer. *Employer:* Self Employed. *Education:* Grad/Aero Eng/London Univ; Grad/Mech Eng/London Univ. *Born:* Feb 1908 England. Grad in Aero & Mech Engrg London Univ England. Aero Engrg apprenticeship. Group Engr Vickers Armstrongs Ltd (now British Aircraft Corp). Over 22 yrs - Lockheed-Calif Co; Sr Design Specialist Military Escape Sys; Principal Investigator for Lockheed funded R/D Advanced Escape Sys Concept Studies; participated in various accident investigations. Reg Prof Aero Engr Ariz 10197 & Prof Mech Engr Ariz 10670. Assoc Fellow AIAA, Mbr SAFE. Presently self-employed as Cons Engr specializing in aero accident investigation. *Society Aff:* AIAA, SAFE.

Keopsell, Paul L

Business: Box 2219, Dept. of Civil Engrg, South Dakota State U, Brookings, SD 57007
Position: Prof of Civil Engrg. *Employer:* South Dakota State University. *Education:* BS/Civ Engrg/SD State Univ; MS/Struc Engrg/Univ of WA; PhD/Struct Engrg/OK State Univ. *Born:* June 17, 1930. Engrg specialty, numerical methods, mechanics, struct design, computerised design (earthwork, networks, simulation). Stress analyst Boeing Seattle 1952-57. Civil engrg staff SDSU 1957-65, Dir of Res & Computer Ctr 1965-1981, Prof of Cvl Engrg, 1965-. Pres S D Sect ASCE 1962; Pres S D Engrg Soc 1967, Natl Dir 1971-77, VP NSPE 1979-1981. Mbr ACI, ACM, ASCE, NSPE, Chi Epsilon, Sigma Xi, Phi Kappa Phi, Tau Beta Pi. Married Delores (Johnson); children Steven, Royal, Pamela. Reg Engr S D. Bd of Directors, Brookings Economic Development Center, 1986-; Secr-Treas SD Section ASCE, 1980-; Secr-Treas N.E. Chapter, SD Engrg Soc. 1987-. *Society Aff:* ASCE, NSPE, ACM, ACI.

Kephart, John T, Jr

Home: 700 Greenwood Road, Wilmington, DE 19807
Position: Retired *Education:* BSME/Mech. Engg./University of Utah 1948 *Born:* 02/17/25 Retired in 1985 as Principal Consultant, Engrg Dept, E. I. DuPont de Nemours & Company, Wilmington, DE. Life Fellow ASME. Chairman, ASME Fluid Transients Cttee, 1964-65. Ch, ASME Industrial Power Conference, 1982. Ch, ASME Industrial Operations Cttee (Power Div), 1983-85. Author of three papers (co-author in one case) and four discussions published by ASME in fields of fluid transients, nuclear power and coal-fired steam generation. Life Member NSPE. Pres, DE SPE, 1965-66. Dir, NSPE, 1967-71. Pres, Delaware Council of Engrg Societies, 1968-69. Named Delaware's "Outstanding Professional Engineer of the Year-, 1977. Member, Tau Beta Pi and Pi Tau Sigma honorary societies. *Society Aff:* ASME, NSPE

Keplinger, Henry F

Business: The Keplinger Companies, 3200 Entex Bldg, 1200 Milam, Houston, TX 77002
Position: Pres. *Employer:* Keplinger and Assoc. *Education:* MS/Petol Engg/Univ Tulsa; BS/Petrol Engg/Univ OK. *Born:* 8/25/37. H. F. Keplinger is Pres and Chrmn of the Bd of the Keplinger Companies. M Keplinger acts in a managerial, engrg and geological capacities to coordinate the various assignments and projects in the energy consulting profession undertaken by his firm, Keplinger and Assoc, Inc, Intl energy consultants. He has served as personal emissary and special advisor to clients, both domestic and foreign, in all phases of the indus, including finance; has conducted on- site inspections of energy oriented areas of interest throughout the world; and has expertise in energy and financial forecasts in exploration, dev, pro-

Keplinger, Henry F (Continued)

duction, mktg. He is an acknowledged energy expert and spokesman and is author of many articles and publications. *Society Aff:* AIME, SPE of AIME, API, SPWLA, AAPG.

Kepner, Robert A

Home: 630 Miller Dr, Davis, CA 95616
Position: Prof of Agricultural Engrg Emeritus, Univ of Calif. *Education:* BS/Agr Eng/Univ of CA. *Born:* May 1915 Southern Calif. Design & testing of aircraft heaters, Stewart- Warner Corp, Chicago & Indianapolis 1942-47. Agri Engrg Dept Univ of Calif, Davis 1937-42 & 1947-81. Orchard heating res 1937-42; Teaching & res in area of agri machinery 1947-81; major emphasis in res has been on equipment & methods for harvesting field crops & asparagus. Principal author of the engrg textbook 'Principles of Farm Machinery'. Mbr numerous ASAE natl cttes; Fellow ASAE 1970. *Society Aff:* ASAE, ASEE.

Keppler, William E

Home: 773 Midway Ln, Blue Bell, PA 19422
Position: Senior VP Technical Operations/Management Sys *Employer:* Schering-Plough Corporation. *Education:* MChE/Chem Eng/NY Univ; BChE/Chem Eng/Pratt Inst. *Born:* June 1922. Taught chem engrg at Cooper Union, NYU & Bucknell; 1945-50 Heyden Chem; 1950-71 Merck & Co Inc, last position V Pres Opers MSD Div; Squibb Corp 1971-72, V Pres; Mgmt Cons 1973-75; President Engel Industries, St. Louis - 1974-1975. Current Position, worldwide respon at corporate level for capital investments, Engineering and Computer Systems and Operations. Fellow AIChE. Enjoys tournament bridge & golf. *Society Aff:* AIChE.

Kerber, Ronald L

Home: E Lansing, MI 48824
Position: Prof & Asso Dean of Engr *Employer:* MI State Univ. *Education:* BS/Engr Sci/Purdue; MS/Engr Sci/Cal Tech; PhD/Engr Sci/Cal Tech *Born:* 7/2/43. Native of Lafayette, IN; joined the Mech Engg Dept at MI State Univ in 1969. Took a leave of absence to work for Aerospace Corp from 1971-1972, as a Mbr of the Technical Staff. Received joint appointment with Electrcl Engg and Mech Engg Dept at MI State Univ in 1976, and received the rank of prof in 1978. Assumed present position as Assoc Dean of Engg in 1980. Took a leave of absence to become a program manager at Defense Advanced Res Projects Agency (DARPA) from 1983-1984. Principal res areas include laser physics, chem kinetics and plasma chemistry. *Society Aff:* ASME, IEEE, ASEE, AIAA, SME

Kerkhoff, Harry Piehl van den

Business: 215 Vankerhoff Pl, North Augusta, SC 29841
Position: Self-employed *Born:* 08/23/08 I completed the fifth grade. My foster parents died and I went to work in the construction industry at the age of 12. Have worked 61 years. I took numerous correspondence courses, among which was structural engrg and architecture, also I employed 7 registered engrs as tutors over a period of 5 1/2 years. I am registered in 27 states and the District of Columbia. Over 50 years experience in the bldg industry. I have designed and built schools, gymnasiums, offices, churches, factories, warehouses, and done numerous renovations of classical architecture. I do analysis of structures, including structural failures, and preparation of reports showing findings in connection with said analyses. I feel that my work is indicative of my competence. *Society Aff:* ASCE, NSPE, AWS, ACI, NFPA

Kern, Roy F

Home: 818 E Euclid Ave, Peoria, IL 61614
Position: Owner. *Employer:* Kern Engg Co. *Education:* BS/Physics/Macalester College; BS/Chem Engg/Marquette Univ. *Born:* 10/25/18. Authority on steel and heat treatment selection, particularly in heavy off- the-road machinery. Also reknowned for the dev of Boron Steels and their uses. Book publ - 445 pgs - 1979. *Steel Selection A Guide for Improving Performance and Profits* John Wiley & Sons, Inc, Publishers. Wiley Interscience Div New York, Chichester, Brisbane, Toronto. Being published in serveral languages. Only book available to guide to selecting the lowest cost steel that will provide desired engineering qualities. *Society Aff:* ASM, SAE, AWS, SME, NAA.

Kerpan, Stephen J

Home: 1420 N Palm St, La Habra, CA 90631
Position: VP Operations & Gen Mgr. *Employer:* Thiem Industries Inc. *Education:* MBA/Gen Admin/Pepperdine Univ; BSME/ME/Bradley Univ. *Born:* 9/17/35. Mr Kerpan is a businessman-industrialist having over 30 years of successful experience with major consumer products and high technology corporations. Twenty of those years have been in sr mgmt positions. Mr Kerpan negotiated the purchase of product lines from a major aerospace firm and quickly molded them into a profitable and viable division for a medium sized corp. 1974-82, Mr Kerpan directed large divisions of Pet Inc, General Host Corp, Continental Grain as VP-Dir and mbr of the exec cttes. He also has been on the bd of dir of several companies. *Society Aff:* AIChE, ASME, NCMA.

Kerr, Arnold D

Business: Dept of Civ Engrg, DuPont Hall, Newark, DE 19716
Position: Prof *Employer:* Univ of DE *Education:* Dipl Ing/CE/Tech Univ of Munich, Germany; MS/Engrg Mech/Northwestern Univ; PhD/Mech & Math/Northwestern Univ *Born:* 03/09/28 During 1955, Bridge Design Engr, Hazelet & Erdal, Consltg Engrs. 1958-1959, Asst Res Scientist, CIMS, NYU. 1959-1961, Asst Prof, Aeronautics & Astronautics, NYU. 1961-1966, Assoc Prof, Aeronautics & Astronautics, NYU. 1966-1973, Prof, Aeronautics & Astronautics, NYU. 1967-1973, Dir, Lab of Mech of Solids, NYU. 1973-1978, Visiting Prof, Dept of Civ Engrg, Princeton Univ. 1978-present, Prof, Dept of Civ Engrg, Univ of DE. Author of numerous publs in analysis of structures, ice mechanics, and railway engrg. Editor of books *Railroad Track Mechanics and Technology*, Pergamon Press, 1978 and *Productivity in US Railroads*, Pergamon Press, 1980. Conslt to gov and indus. *Society Aff:* ASME, AREA, TRB

Kerr, Arthur J

Business: 5940 Baum Sq, Pittsburgh, PA 15206
Position: Charman of Board. *Employer:* Kerr Engineered Sales Co. *Born:* Nov 26, 1897. BSME Carnegie Tech. Fellow ASME. Contrib to dev of pipelining. While in charge of the SW Oper of Rockwell Mfg Co, worked on the prototype of the 1st remotely controlled automatic pump station. Secy, then Chmn Mid-Cont Sect ASME & headed estab of Petro Div ASME. Became a Regional V P of ASME & is founder & P Pres of Engrs Club of Tulsa. In 1944 moved to Pgh with Rockwell as V P. During 1951 & 1952 headed Genl Components Sect of Natl Prod Auth, U S Dept of Commerce. In 1952 left Rockwell to form Kerr Engrg Sales Co. NGAA Award 1958. In 1972 Elected Fellow in ASME.

Kerr, John R

Business: PO Box 4332, Downey, CA 90241
Position: Consulting Engineer. *Employer:* John R Kerr & Associates. *Education:* BSME/Mech Engr/U of AZ *Born:* Mar 1917 Buenos Aires Argentina. Post grad USC 1958. Asst Dist Ch Engr York Corp LA 1940-49; V P Heide & Cook Ltd, mech contr Honolulu 1950-55; mech engr Daniel, Mann, Johnson & Mendenhall LA 1956-57; Ch Mech Engr Victor Gruen Assocs LA 1958-60; Pres John Kerr & Assocs Inc 1961-69; cons mech engr private practice Downey, CA 1969- ; Instr dynamics Univ of Hawaii 1953; guest lectr USC Arch Coll 1959-61; Mbr air Cond Code Comm Hawaii 1952. Served to Lt USNR 1942-45 ETO, PTO; Reg Prof Engr Calif, Hawaii, Nev, Md, Colo, Conn, IL, WA, AZ & OR. Mbr CEAC, ASHRAE, ASPE, Theta Tau. *Society Aff:* CEAC, ASHRAE, ASPE

Kerr, William

Home: 2009 Hall, Ann Arbor, MI 48104
Position: Professor of Nuclear Engineering. *Employer:* University of Michigan.

Kerr, William (Continued)
Education: BS/EE/Univ of TN; MS/EE/Univ of TN; PhD/EE/Univ of MI. *Born:* Aug 1919. Taught at Univ of Tenn before WWII. Army Med Dept 1944-46. Taught EE & Nuclear Engrg Univ of MI; Chmn Dept of Nuclear Engrg Univ of Mich 1961-74; Dir Mich Memorial-Phoenix Proj Univ of Mich 1961- ; Phoenix Proj supports dev of peaceful uses of nuclear energy. Dir Office of Energy Res, Univ of MI 1978- . Mbr Advisory Ctte on Reactor Safeguards US Nuclear Regulatory Comm 1972- . Fellow Amer Nuclear Soc; Fellow American Assn for the Advancement of Sci; Amer Nuclear Soc A H Compton Award 1974. *Society Aff:* ANS, IEEE, AAAS, AAAS

Kerrebrock, Jack L
Home: 1220 N. Meade St 6, Arlington, VA 22209
Position: Assoc Adm, Office of A&S Tech *Employer:* NASA *Education:* PhD/Jet Propulsion/CA Inst of Tech; MS/ME/Yale Univ; BS/ME/OR State Coll *Born:* 2/6/28 Currently Assoc Adm, Office of Aeronautics and Space Tech, NASA. On leave as Richard Cockburn Maclaurin Prof and Head of Dept of Aeronautics and Astronautics, MA Inst of Tech. Taught and conducted research in energy conversion and propulsion since 1956. Early work was on nuclear rockets, space propulsion and power, magneto hydrodynamic generators; more recently, on fluid mechanics of turbomachinery for aircraft engines. Authored text *Aircraft Engines and Gas Turbines.* Consultant to government and industry. Served on many committees advisory to NASA and DOD. Member, National Academy of Engrg; Fellow, AIAA; ASME Gas Turbine Power award, 1971; 1980 AIAA Dryden Lecturer. *Society Aff:* AIAA, APS

Kersten, Robert D
Business: Univ of Central Florida, PO Box 25000, Orlando, FL 32816
Position: Dean, College of Engineering. *Employer:* Univ of Central FL (formerly FTU). *Education:* BS/Mathematics/OK State Univ; MS/Engr Mech/OK State Univ; PhD/Fluid Mech/Northwestern Univ. *Born:* 1/30/27. USNR 1945-47. Hydraulic Engr US Dept of Interior 1949-53; Okla St Univ Faculty 1953-56; Res Engr Jersey Prod Res Co 1956-57; Ariz St Univ Faculty 1957- 68, Chmn CE 1960-68; Dir Office of Univ Res Univ Central FL 1968-69 Prof of Engrg & Dean of Engrg Univ of Central FL 1968- . Frequent contributor to tech & prof literature. Listed in Who's Who Amer Men of Sci, Who's Who in Amer Educ. Special interests in Water Resources, fluid mech, engrg manpower & hist of engrg & tech. ABET Ad Hoc Visitor, Audit & Review Committee; Engrg Accreditation Commission Chair Elect 1987-88; Mbr of Bd of Trustees NSPE/ICET 1972-80. NSPE VP Chrmn Professional Engineers in Education 1984-85. Education Committee Trustees NSPE Educational Foundation; ASEE Accreditation Processes Comm; NCEE Liaison Committee; NCEE/ABET Relations Committee. Chrmn Engrg Education Park Force (NSPE, ASEE, EDC, NASULGC) Mbr NSPE, ASCE, ASEE, NCEE, Sigma Xi, Tau Beta Pi, Chi Epsilon, Pi Gamma Mu, Alpha Pi Mu, Eta Kappa Nu, Pi Tau Sigma, Tau Alpha Pi. *Society Aff:* ASCE, NSPE, ASEE, NCEE.

Kertz, Hubert L
Home: 575 Easton Ave Apt 12D, Somerset, NJ 08873
Position: President & Managing Director. *Employer:* American Bell International Inc. *Education:* EE/Radio/Stanford Univ; AB/Engg/Stanford Univ. *Born:* July 1910. BS, MS-EE Stanford Univ 1934-36. With Bell Sys 1926- , V Pres Opers Pacific Tel 1958, V Pres Amer Tel & Tel Co 1964-75. CE officer ABII, a wholly owned subsidiary of AT&T, Aug 1975- . Fellow AIEEE. Reg Engr Calif & N Y.

Keshavan, Krishnaswamiengar
Business: Dept of Civ Engg, Worcester, MA 01609
Position: Prof, Dept of Civ Engg. *Employer:* Worcester Poly Inst. *Education:* PhD/Environmental Engg/Cornell Univ; MS/Environmental Engg/State Univ of IA; BS/Civ Engg/Univ of Mysore; BS/Phsy, Chem & Math/Univ of Mysore. *Born:* 6/5/29. Naturalized US Citizen; Assoc Prof of Civ Engg at Univ of ME, Orono, ME from 1963-67; Assoc Prof of Civ Engg at WPI from 1967-72; Prof Civ Engg at WPI from 1976-present; UNESCO Sr Advisor at the Univ of the Philippines 1975-76; UNESCO Consultant to Missions in Philippines, Paris, Turkey, Venezuela, India from 1975-1978; Prin Investigator of Water Quality Res supported by NSF, Dept of Interior, Commonwealth of MA. More than 25 publications. *Society Aff:* ASCE, ASEE, WPCF, ΣΞ.

Keshian, Berg, Jr
Business: 12345 Alameda Pkwy, Lakewood, CO 80228
Position: Senior Geotechnical Engr *Employer:* Golder Assocs *Education:* MS/Geotech Engr/MIT; BS/CE/Univ of KS; AA/ Architectural Engr/Wentworth Inst *Born:* 08/06/48 Born and raised Boston, MA. Engr Wolsey Assocs specializing in pre and post blast monitoring. Geotechnical Engr Golder Assocs, specializing in foundation design and inspection. 1975-1981 Senior Project Engr responsible for the design and construction of several earth dams. Also performed state-of-the-art work in large strain consolidation of fine-grained wastes and published several papers on the subject. VP Ridge Branch Section ASCE. 1981 joined Golder Assoc., Denver, CO as Sr Geotechnical Engr responsible for analytical engrg and technical supervision for a variety of geotechnical and mining projects. Enjoy skiing, tennis, sailing, backpacking, bicycling. *Society Aff:* ASCE, AIME, NSPE, ISSMFE

Kesler, Clyde E
Business: Newmark Civil Engrg Laboratory, 208 North Romine Street, Urbana, IL 61801
Position: Emeritus Professor. *Employer:* Civil Engrg Dept Univ of Ill. *Education:* MS/Structures/Univ of IL; BS/Civil Engg/Univ of IL. *Born:* 5/7/22. Served with US Army Corps of Engrs 1943-46. Engrg Aid for IL Central Railrd 1946; Univ of IL faculty 1946- ; Cons in field of concrete. Over 70 publs. Hon Mbr & P Pres Amer Concrete Inst; Fellow ASCE. Sanford E Thompson Award 1958, Amer Society for Testing and Matls. Stanton Walker Lectr 1970 Univ of MD. Alfred E Lindau Award 1971 Amer Concrete Inst National Acad of Engg 1977. Reg P E in IL, Honor Mbr, Wire Reinforcement Inst 1983, Halliburton Award, Univ of IL 1982. *Society Aff:* ASCE, ACI, ASEE, NAE.

Kesler, Joel P
Home: 10504 Walnut Dr, Kansas City, MO 64114
Position: Retired. *Employer:* Black & Veatch Consulting Engineers. *Education:* BS/EE/KS State Univ. *Born:* 9/4/11. 1933 KS Hwy Comm; 1934 Westinghouse, Application Engr. 1947. Black & Veatch Cons Engrs Proj Elec Engr, Proj Engr, 1950 Hd Elec Dept, 1956 Ptnr in firm; 1966 Mgr Engg & Services Power div respon through the 7 dept hds for engg studies, design & services including forecasting div resource requirements, training & production of studies, plans & specifications, on time under quality control for all power projs including fossil nuclear power plants, substations, transmission & distribution sys. Mbr MSPE-NSPE, ASME, IEEE, ANS, Sigma Tau. Served in many offices & cttes, Sect & Natl IEEE & Chap & St MSPE. Retired Jan 1, 1979. *Society Aff:* NSPE-NSPE, ASME, IEEE, ANS.

Kessler, David P
Business: CMET Bldg, W Lafayette, IN 47907
Position: Head, Interdisciplinary Engrg & Prof of ChE *Employer:* Purdue Univ. *Education:* PhD/ChE/Univ of MI; MS/ChE/Univ of MI; BS/ChE/Purdue Univ. *Born:* 11/1/34. Hd, Interdisciplinary Engrg, Prof of Chem Engg, Purdue Univ. Reg PE, IN, OH. Mediator, fact finder, Arbitrator. Prior to academic career employed at Procter and Gamble and Dow chem. Co-inventor of wearable artificial kidney (US patent 731,826). Ind consultant. Ad-Hoc accreditor, AIChE. Hobbies tennis, squash, snooker. *Society Aff:* AIChE, ASEE, SPIDR.

Kessler, Frederick M
Home: 31 Shady Ln, Bound Brook, NJ 08805
Position: Pres *Employer:* FMK Tech, Inc *Education:* PhD/Elec Engg/Rutgers Univ; MS/Elec Engg/Rutgers Univ; BME/Mech Engg/CCNY. *Born:* 5/15/32. Native of NYC. Engg Duty Officer, USN 1955-1959. Submarine vibration res at David Taylor Model Basin, 1959-1961. Developed quiet products including whisperized air compressor for Ingersoll-Rand Co, 1961-1968. PhD candidate and instructor,

Kessler, Frederick M (Continued)
Rutgers Univ, 1968-1971. VP and mgr of engg, Lewis S Goodfriend & Assoc, 1971-1973. Managing partner of NY/NJ office and hd of Engg Acoustics for Dames & Moore 1973-1983. Presently Pres FMK Tech, Inc. Consultant to EPA, DOT, COE, and private sector mfg, mining, and utility clients. Adjunct Prof at Stevens Inst of Tech 1973-1980. VP Tech Affairs, INCE 1984-87. Mbr Environmental Commission, Bridgewater, NJ 1975-1981 Assoc Ed, JASA 1982-85. Fellow ASA. Pres-Elect INCE. *Society Aff:* INCE, ASA, IEEE.

Kessler, George W
Home: 720 Williams Dr, Winter Park, FL 32789
Position: Consultant, Power Generation Group. *Employer:* The Babcock & Wilcox Co. *Education:* BS/Mech Engg/Univ of IL. *Born:* Mar 1908. With Babcock & Wilcox Co as apprentice engr 1930-31; Sales Engr 1931-32; Analytical Engr 1932-33; Engr, Grp Leader, Head of Marine Design & Estimating Div 1933-46; Application Engr 194651; Asst Mgr Proposition Dept 1951- 53; Asst Chief Engr 1953-54; Chief Engr 1954-61; V Pres Engrg Power Generation Div 1961-70; V Pres Engrg & Tech & Group Staff 1970-73. Author of many tech articles on steam generation & holder of patent awards. Dir Metal Properties Council 1965- ; Fellow ASME 1964; Mbr NAE 1969. *Society Aff:* SNAME, ASNE, ASME, NAE.

Kessler, Harold D
Home: Howmet Turbine Components Corp, Muskegon, MI 44443
Position: Dir Tech & Mkt Planning, Titanian Alloy Operations *Employer:* Howmet Turbine Components Corp. *Education:* MS/Metallurgy Engg/IL Inst of Tech; BS/Metallurgy Engg/Case Inst of Technology. *Born:* Dec 28, 1921. Employed as Met Engr 1942-46 at NASA, Lewis Lab Cleveland & AF Matls Lab Wright Field. 1946-54 Supr Non-Ferrous Met Res IITRI Chgo. 1954- 64 Supr Met Engrg & Mgr Tech Lab TMCA Henderson Nev & Toronto Ohio. 1964- , Ch Metallurgist, Tech Dir & V P Matls Engrg RMI Co Niles Ohio, Dir Tech & Market Planning, Reactive Met Op, Howmet Turbine Components Corp. Has written about 75 tech papers; 7 pats. Ed of MEI Titanium & its Alloys Course ASM; Bd of Trustees ASM (current) held office on num of natl cttes of ASM; Fellow ASM; 'founder' Recog of Titanium Indus; Sigma Xi. *Society Aff:* ASM, AIME, APMI, ASTM.

Kessler, Thomas J, Sr
30 Harwood Lane, East Rochester, N.Y 14445
Position: Product Development Mgr. *Employer:* Xerox Corp. *Education:* PhD/ME/Rutgers Univ; MME/ME/NYU; BS/ME/Rutgers Univ. *Born:* 11/13/38. Born and raised in Monmouth County, NJ. Instr & post doctoral res at Rutgers Univ. PhD Thesis res on supersonics separted flows. Applied res/advanced dev at Bell Labs in the areas of reentry phys, unsteady shock waves, cooling of electronic equip, multilayer printed circuit bds, & electro chemistry. At Xerox area mgr responsible for advanced dev/design of subsystems for xerographic copiers & duplicators; area mgr for applied res on novel copiers & printers; Product Development Mgr. printers. 1979 Rutgers Engg Soc Award for Outstanding Achievement in Engg. Mbr of the Natl Ski Patrol. Hobbies: skiing, hunting & water skiing. *Society Aff:* ASME.

Kester, Granville, Jr
Business: PO Box 280, Beaver, PA 15009
Position: President-Northern Division *Employer:* Michael Baker Jr Inc-Cons Engrs. *Education:* BSCE/Civil Engrg/West VA Univ. *Born:* Apr 1923 St Mary's W Va. Served in US Navy during WWII. Empl by Michael Baker Jr Inc 1949- , as Designer, Engr, Proj Engr, Asst V Pres, V Pres & 1972- , Exec V Pres in charge of Northern Opers of Co. 1980 President, Northern Division Reg Prof Engr in 25 states & DC. Responsible for CE design of major sport stadiums, indus bldgs, water & wastewater, mapping & transportation projs. P Pres Pittsburgh Post SAME, Beaver County Chap PSPE (Engr of the Year 1964), Mbr Tau Beta Pi, ASCE & PEPP. *Society Aff:* ASCE, SAME, NSPE, TBΠ.

Kestin, Joseph
Business: Box D - Div of Engrg, Providence, RI 02912
Position: Professor of Engineering. *Employer:* Brown University. *Education:* DSc/Engrg/Univ of London; PhD/Engrg/Univ of London; Dipl Ing/Mech Engrg/Engrg Univ Warsaw (Poland); Drhc/Mechanics/Univ Claude Bernard, Lyon, France (honorary) *Born:* 1913 Warsaw Poland. Hd Mech Engrg Polish Univ Coll London. Prof Brown Univ 1952- . US Delegate Intern Steam Tables Confs. Pres Intl Assn Prop of Steam 1974-76. Mbr OSRD, Chmn OSRD Panel 1975-80; Mbr Natl Data Adv Bd. Fellow Inst Mech Engrs (Prize 1949) ASME. Mbr Exec Ctte Appl Mech 1968-73, Chmn 1972. Tech Ed Jour Appl Mech 1956-71. Mbr PBRes, Mbr Natl Nominating Ctte. Num papers in prof journals. Translator 4 tech books; author 3 tech books. Mbr ed bds of sev tech journals. Recipient of 1981 James Harry Potter Gold Medal (ASME) for thermodynamics Lecturer at 1979 Nobel Symposium in honor of the bicentennary of J. J. Berzelius. 1968 Special Advisor to Chancellor of Teheren Univ. 1972 Fulbright Lecturer at Inst. Sup. Tecnico, Lisbon, France 1966 Prof. Associe U. of Paris (Sarbonne); 1974 Prof. Associe U. Claude Bernard (Lyon); Lecturer Instituto Superior Technico, Lisbon; Lecturer Norges Tekniske Hoegskole, Trondheim, Norway; Advisor, Univ of Tehran, Iran; visiting prof, Imperial Coll, London. Currently visiting prof Univ of Maryland, College Park, Md. Mbr, Nat Acad of Engg. Fellow (1984-85) of Inst for Adv Study, Berlin. Recipient of Humboldt Prize of the Humboldt Foundation. *Society Aff:* ASME, ΣΞ, NAE.

Ketchledge, Raymond W
Business: Whippany Rd, Whippany, NJ 07981
Position: Exec Dir - Military Systems Div. *Employer:* Bell Laboratories. *Education:* BS/EE/MIT; MS/EE/MIT. *Born:* Dec 8, 1919. With Bell Labs since 1942; initial assignments in military sys, submarine cable, coaxial carrier & electron tube dev. In 1954 joined new electronic switching sys org; 1956 Asst Dir; 1959 Dir Electronic Switching Lab respon for ESS hardware; 1966 Exec Dir Electronic Switching Div; Oct 1975 Exec Dir Ocean Sys Div. Mbr Sigma Xi. 58 patents. 1965 Fellow IEEE; 1970 Mbr NAE. 1976 Alexander Graham Bell Medal. *Society Aff:* IEEE, NAE.

Ketchum, Gardner M
Home: 1307 Glenwood Blvd, Schenectady, NY 12308
Position: Professor Emeritus of Mechanical Engineering. *Employer:* Union College. *Education:* ScD/Mech Eng/MIT; SM/Mech Eng/MIT; SB/Mech Eng/MIT. *Born:* Oct 20, 1919 Phila Pa. Fellow ASME; Mbr ASEE, Sigma Xi, TBP. Instr at MIT, then dev engr G E Co 1949-53; Assoc Prof, then Prof at Union Coll 1953-85, Chmn of Mech Engrg Dept 1962-74; Co-dir Prog in Admin & Engrg Sys 1970-85. Has served as cons to various components of the G E Co, Alco Prods Inc, US Army & others. *Society Aff:* ASME, ASEE, ΣΞ, TBM.

Ketchum, Milo S
Home: 58 Willowbrook Rd, Storrs, CT 06268
Position: Professor of Civil Engineering. *Employer:* Univ of Connecticut. *Education:* BS/Civil/Univ of IL; MS/Civil/Univ of IL. *Born:* 3/8/10. Hon DSc Univ of CO 1976. Founded Ketchum, Konkel, Barrett, Nickel Austin, Denver CO 1945; cons to firm 1967- ; Prof Univ of CT 1967-78 Mbr of Juries - AISC Architectural Awards of Excellence 1975; Progressive Architecture Annual Design Awards 1958. Featured in "The Consulting Engr" magazine Feb 1958. Distinguished Alumnus Award, Univ of IL, Urbana, 1979; Hon Mbr, ASCE, 1978. Fellow: ACI, ACEC, Inst of Struct Engrs. Received Henry C Turner Medal ACI for pioneer work in thin shell structs 1968. Mbr Tech Activities Ctte & Bd/Dir ACI. Author "Handbook of Standard Structural Details for Bldgs" and many articles. Position: Professor Emeritus of Civil Engrg. "Architectural Record" Featured his firm and his philosophy in August 1975. Served on Jury for ASCE Outstanding Civil Engineering Achievements for both 1980 and 1981. Named Editor-in-Chief "Structural Engineering Practice–, published by Marcel Dekker, Inc 1981. Society Affiliations: ASCE, ACI, ACEC, ASEE, Institution of Structural Engineers (London), International Assoc. for Snell and Space Structures.. *Society Aff:* ASCE, ACI, ACEC, ASEE.

Kett, Irving
Business: 5151 State Univ Dr, Los Angeles, CA 90032
Position: Prof of Engr. *Employer:* CA State Univ. *Education:* DrSc in Engg/Transportation/NYU; MBA/Mgt/NYU; MCE/Structures/Brooklyn Poly Inst; BCE/Civ Engg/College of City of NY. *Born:* 1/13/23. My Profl career began as a hgwy engr with the NY State Dept of Public Works in 1946, after serving in the US Army during WWII. In 1967 I left the state of NY to accept a contract as Chief Design Engr for the Israel Hgwy Dept. In 1971 I accepted a profship at CA State Univ-LA. Presently I am on a three yr tour of active duty as a Col in the US Army. I was recalled from an active reserve status to work on the airbase bldg projs which the US Govt committed itself to in Israel. I am on leave from the Univ. *Society Aff:* ASCE, AAPT, ASME, ASEE.

Ketterer, Frederick D
Business: Univ. Pennsylvania, Moore School of EE, Philadelphia, PA 19104
Position: Prof *Employer:* Univ of PA *Education:* PhD/EE/MIT; MS/EE/Univ of PA; BA/Phys/Univ of PA *Born:* 11/27/32 Native of Philadelphia. With DuPont Co 1954-58 in Physical Test Methods Dev and as Instrument Design Engr. Designed Automatic Checkout Equip for a North American Aviation and GE 1959-60. Res at MIT 1961-65 in MUD and EHD. A Prof at the Moore Sch of EE at Univ of PA since 1965. Activities include organ preservation by freezing and microwave thawing, hyperthermia therapy under microcomputer control, and dev of microcomputer lab facility. Received MIT, Lindbach, and S. R. Warren Outstanding Teaching Awards.

Ketvirtis, Antanas
Fenco Consultants Ltd, 1 Yonge St, Toronto Ont, M5E 1E7 Canada
Position: Ch Elec Engr & Dir/Visual Environ. *Employer:* Fenco Consultants Ltd. *Education:* EE/Power Distr/Eng Coll of KS. *Born:* 1919. Fellow IES; Sr Mbr IEEE. Engrg Coll Kaunas & Univ of Kaunas Lithuania. Chief Elec Engr & Dir Visual Environment Fenco Cons Ltd Toronto. Credited with significant improvements in hwy lighting design techs & res related to the visual needs in vehicular transportation. Ctte activities: TRB, IES, CIE, RTAC & CSA. Author of textbook 'Highway Lighting Engrg' 1967 & a major res study 'Road Illumination & Traffic Safety' 1975; has written & presented 50 tech papers mostly on light apapplication in transportation. Lectr Ontario Traffic Sch & Ryerson Polytech Toronto. For Disting Engrg Achievement awarded Engrg Medal 1974 by Assn of Prof Engrs Ontario; Merit Award 1968 Assn of Cons Engrs Canada. *Society Aff:* FIES, SM, IEEE.

Kevorkian, Hugo
Business: 1414 Stanislaus, Fresno, CA 93706
Position: Principal Geotech Engr-Corp Secy. *Employer:* Braun, Skaggs, Kevorkian & Simons. *Born:* Oct 14, 1935, France. BS Calif St Univ Fresno. Lectr Soil Mechs-Foundations Calif St Univ Fresno 1968-75; Sr Geotech Engr for Braun, Skaggs, Kevorkian & Simons Engrs & Geols Inc. Mbr ASCE (Pres Fresno Chap 1970- 71); NSPE (Pres Fresno Chap 1973-74) & CSPE; Calif Council of Cvl Engrs & Land Surveyors; Cons Engrs Assn of Calif; Const Specification Inst; Seismological Soc of Amer. Licences: CE in Calif & Nev. Who's Who in the West & Who's Who in Amer.

Key, Carroll L, Jr
Business: PO Box 30, State College, PA 16801
Position: Assistant Director. *Employer:* Applied Research Lab, Penn St Univ. *Education:* MS/Physics/Univ of TX; BS/Math & Physics/Southwest TX State College. *Born:* Jan 1921 Nashville Tenn. US Navy 1942-46, Lt Cdr. Naval Res Lab 1946-48; Finch Telecommunications 1948-49; Applied Res Lab PSU 1949- : Asst Dir, Head of Ocean Tech Dept, Prof of Engrg Res. Primary respon prog of res in advanced acoustic torpedo weapon sys. Served on many Navy cttes & study groups incl internatl tech, advisory & study groups. Fellow ASA; Sr Mbr IEEE & Sigma Xi; Mbr ADPA. Has publ articles in USN JUA & has patents for various underwater devices. Assoc Editor for Sonar & Undersea Sys, IEEE Transactions on Aerospace & Electronic Sys. *Society Aff:* ASA, IEEE, ΣΞ.

Key Joe, W
Business: 2060 N Loop West, Houston, TX 77018
Position: Sr. VP *Employer:* Williams Brothers Engrg Co *Education:* MS/Structural Engrg/Rice Univ; BA/Civ Engrg & Architecture/Rice Univ *Born:* 11/22/34 Native of Pampa, TX. Served as Civil Engr Corps officer of the US Navy from 1960 through 1968. Awarded Joint Services Commendation medal and two Navy Commendation medals for outstanding performance. Founder of Ocean Resources Engrg, Inc in 1972, which was acquired by Williams Brothers Engrg Co in 1981. Active in civic affairs and a recognized leader in the development of complex, innovative offshore structures and equip. *Society Aff:* NSPE, TSPE, ASCE, SNAME

Keyes, Conrad G, Jr
Business: CAG Engrg Dept, Box 30001, Las Cruces, NM 88003
Position: Prof & Head Emeritus *Employer:* New Mexico State Univ. *Education:* ScD/Water Resources/NM St Univ; MSCE/Highways/NM St Univ; BSCE/Gen & Civ Engrg/NM A&MA. *Born:* 08/16/37 Have served as Dir, thru Past-Pres of the NM Section of ASCE. Was the General Conference Chrmn of the 1979 & 1985 Irrigation & Drainage Div Specialty Conferences and have served as chrmn of the Irrigation & Drainage Div of ASCE. Presently serving on Mgt. Group D of ASCE. Have served as Secy thru Pres of the Weather Modification Assoc. Presently the Exec Dir of the North American Interstate Weather Modification Council. Have held the rank of Instructor, Asst Prof, Assoc Prof, & Prof and as Head of Civil, Agricultural, & Geological Engrg at NM State Univ. *Society Aff:* ASCE, NSPE, AMS, AWRA, AIH.

Keyes, Fenton G
Business: 321 S Main St, Providence, RI 02903
Position: Sr Partner *Employer:* Keyes Associates. *Education:* BSCE/Civil/Northeastern Univ. *Born:* 8/24/16. 4 1/2 yrs experience US Army Corps of Engrs, 2 yrs with Seabees; 5 yrs with a private engrg firm culminating in formation of Fenton G Keyes Assocs 1951. Present position is Sr Ptnr, Keyes Assocs, which employs approximately 150 employees in 4 New England offices & also in Lagos & Benin City Nigeria. Disting Engr Award R I Sect NSPE 1957. Fellow ASCE; R I P Pres & Dir ASCE; Mbr, P Pres & Dir R I Chap NSPE; Dir R I Chap SAME. Mbr of several other engrg socs & civic orgs & Reg Prof Engr. Hobbies are golf & gardening 1982 Award - Outstanding Civil Engr, Northeastern Univ, 1983 Prov Engr Soc Award - Freeman Award As Outstanding Engr of RI. *Society Aff:* ASCE, ACEC, ACI, AISC, NSPE, USCOLD, WPCF, ASTM, SAME.

Keyes, Marion, A, IV
Home: 120 Riverstone Dr, Chagrin Falls, OH 44092
Position: President *Employer:* Bailey Controls Co *Education:* BS/ChE/Stanford Univ; MS/EE/Univ of IL; MBA/Baldwin-Wallace. *Born:* May 11, 1938 Bellingham Wash. Married 1962; 3 children. BS Chem Engrg, Stanford Univ 1960; MBA Baldwin-Wallace Coll 1981; Mbr of Bd of Dir Cleveland United Cerebral Palsy Assn, Euclid Gen Hospital, Assoc Indus of Cleveland; Pres NE Ohio Council, Boy Scouts of Amer; mbr, Cleveland World Trade Assn. Grad certificate EE/Control Sys, Univ of Ill 1968. Employed Ketchikan Pulp Co 1960- 63; Dir Engrg Control Sys Div Beloit Corp 1963-70; Genl Mgr Digital Sys Div Taylor Instrument Co 1970-75; VP Engrg Bailey Meter Co 1975; promoted to President, Bailey Controls August, 1980. Mbr Amer Automatic Control Council; Mbr: Automation Res Council, AIChE, TAPPI (P Chmn TAPPI Process Control Ctte), Amer Mgmt Assns, Amer Chem Soc, Instrument Soc of Amer. Holds 22 patents; author of over 50 tech papers & 1 book. Reg Prof Engr: Calif, N Y, Wis, Ill, Ohio. *Society Aff:* ARC, AIChE, TAPPI, ACS, ISA, IEEE, AACC.

Keyes, Robert W
Business: PO Box 218, Yorktown Heights, NY 10598
Position: Research Staff Member. *Employer:* International Business Machines Corp.

Keyes, Robert W (Continued)
Education: PhD/Phyiscs/Univ of Chicago; BS/Physics/Univ of Chicago. *Born:* 12/2/21. Native of Chicago. Employed Argonne Natl Lab 1946-50; Westinghouse 1953-60; IBM Res Lab 1960- : res in solid state phys & electronics & physical limits in information processing. Cons NAS Phys Survey 1970-72; NAS/NAE/NRC Evaluation Panel for NBS Inorganic Materials Div 1970-73; IEEE Intersociety Relations Ctte 1975-77; Chmn Gordon Conference on Microstructure Fabrication 1976; V Chmn APS Ctte on Applications of Phys 1976; Assoc Ed 'Review of Modern Physics'; Correspondent 'Comments on Solid State Phys'. IEEE W R G Baker Prize 1976 for paper 'Phys Limits in Digital Electronics'. Mbr NAE; Fellow IEEE, APS.

Keys, John W, III
Home: 8411 Crestwood Dr, Boise, ID 83704
Position: Asst Regional Dir *Employer:* US Bureau of Reclamation *Education:* MS/CE/Brigham Young Univ; B/CE/GA Inst of Tech *Born:* 3/25/42 He has worked with the Bureau of Reclamation since graduation from GA Tech in 1964. Specializing in hydrology, he worked on projects in UT and ND, was Regional Hydrologist in Billings, MT, and was chief Hydrologist at the Engrg and Research Center in Denver. From 1976 through 1979, he was Chief of the CO River Water Quality Office in Denver. After a year in Washington, DC, he was appointed to his present position, Assistant Regional Dir, Pacific Northwest Region, in Boise, ID. He is a registered PE in ND, MT, WY, and CO. *Society Aff:* ASCE, NSPE, NWRA, ICID, USCOLD

Kezios, Stothe P
Home: 1060 Winding Creek NW, Atlanta, GA 30328
Position: Prof & Dir-Sch of Mech Engrg. *Employer:* Georgia Institute of Technology. *Born:* April 21. Native of Chgo. PhD, MS & BS in ME ITT. Consultant. Taught grad & undergrad levels IIT 1942-67; fundamental res in heat transfer & thermodynamics. Commissioned CE Corps USNR Seabees WWII 1944-46. Dir of Heat Transfer Lab IIT 1955-67 & co-founder 1943; with Ga Tech 1967- . Cons to numerous indus & govt agencies; numerous invited lectures & talks; author of num tech articles & reports; completed Jakob's treatise 'Heat Transfer' Vol II. Founder, Cons Ed, Sr Tech Ed 'Journal of Heat Transfer' (ASME). Mbr ASME, ASEE, AAAS; Active ASME over 30 yrs; Chmn Exec Ctte HTD 1958, Chmn Chicago Sect 1959, V Pres & Mbr of Council (Communications) 1972- , Pres Elect 1976; internatl engrg & educ affairs: US representative to UPADI, Internatl Heat Transfer Confs, Secy (Boulder-London) 1961, Advisor (Chicago) 1966, Sci Ctte (Versailles) 1971, Tokyo 1974.

Khachaturlan, Narbey
Business: 1116 Newmark Laboratory, 208 North Romine, Urbana, IL 61801
Position: Prof, Assoc Hd of the Dept of Civil Engrg *Employer:* Univ of IL. *Education:* PhD/Struct Engg/Univ of IL; MS/Struct Engg/Univ of IL; BSCE/Civil Engg/Univ of IL. *Born:* 1/12/24. His area of specialization is structural engg with his res and teaching efforts principally in structural concrete and in structural optimization. He is the author of numerous publications and is the sr author of the textbook "Prestressed Concrete" published by McGraw-Hill. His contributions include dev of methods for discrete optimization of structures and studies on the physical significance of local minima. His practical work includes design of hgwy bridges, dev of design criteria for prestressed concrete structures and investigation of failures. A reg structural engr in IL. He has served two terms on the IL Structural Engr Examining Committee. *Society Aff:* ASCE, ISPE, NSPE, ACI, SEAOI, ΣΞ, ΦΚΦ, ΧΕ.

Khalil, Tarek M
Business: Dept of Industrial Engg, Coral Gables, FL 33124
Position: Prof and Chrmn of Indus Engrg *Employer:* Univ of Miami. *Education:* PhD/IE/TX Tech Univ; MSIE/IE/TX Tech Univ; BME/Mech E/Cairo Univ *Born:* 6/12/41. Prof and chairman of Industrial Engg, Prof of Biomedical Engg at the Univ of Miami, Dir of the Environmental Health and Safety Progs, Chrmn of the Grad Faculty of the School of Engg and Architecture UM. Prior to UM he was asst prof of ISE and Dir of the Human Performance Engg Lab, Univ of FL. Also taught at TX Tech and Cairo Univ. Industrial and consulting experience in Egypt, W Germany, USA. Reg PE state of FL and Egyptian Syndicate of Engrs. Pres Miami Chapter IIE 1976. Winner 1st Ergonomics Div Award IIE and the Human Factors Society's Jack A Kraft Award. *Society Aff:* IIE, HFS, AIHA, ASSE, SME, Int S B.

Khan, Fazlur R
Business: 33 W Monroe St, Chicago, IL 60603
Position: Partner. *Employer:* Skidmore, Owings & Merrill. *Education:* BE/Civil Engg/Univ of Dacca Bangladesh; MS/Civil Engg/Univ of IL; MS/Theoretical Applied Mechs/Univ of IL; PhD/Structural Engg/Univ of IL. *Born:* Apr 3, 1929 Dacca Bangladesh. Joined Skidmore, Owings & Merrill 1955, respon for engineering design of many major architectural projs including John Hancock Center & Sears Tower Chicago Ill. Haj Terminal King Abdul Aziz Int'l Airport, Seddah, Saudi Arbia Mbr (ACI Fellow), ASCE (Fellow), AWS, NAE. Publ over 100 tech papers relating to struct analysis & design. Wason Medal ACI 1971; Const's Man of Year Engrg-News Record 1972; Chicago Cvl Engr of the Year Ill Sect ASCE 1972; Alfred Lindau Award ACI 1973; Honorary Doctor of Science, Northwestern, 1973; J Lloyd Kimbrough Medal AISC 1973; Oscar Faber Medal Inst of Struct Eng, London Eng; Honorary Doctor of Engr Lehigh Univ 1980; Honorary Doctor of Technical Sciences, ETH, Zurich, 1980. *Society Aff:* NAE, ASCE, ACI, AWS.

Khan, Mohammad S
Business: 8325 NW 53rd St, Miami, FL 33166
Position: President *Employer:* Siddiq Khan & Assoc., Inc. *Education:* MSCE/Civil Engrg/Univ of Miami; BSCE/Civil Engrg/W. Pakistan Univ of Engrg & Tech Lahore, Parkistan *Born:* 5/13/41 Specializing in multistory structures, parking garages, warehouses, office buildings, precast & post tensioned structures, planning, site utilities & dev. VP — Crouse & Partners, Consulting Engrs. Miami, Fla 1974-78. Native of Pakistan, obtained BSCE in 1964 from W. Paksitan Univ of Engrg & Tech, Lahore. Prior to imigrating to USA was responsible for design and construction coordination of paper and board mill project for Packages Ltd, Lahore, Pakistan. Received MSCE Degree from Univ of Miami in 1969. Registered Prof Engr in the States of Florida, Michigan and Maryland. *Society Aff:* ASCE, NSPE, ACI

Khare, Ashok K
Business: Front St, Irvine, PA 16329
Position: Development Metallurgist *Employer:* Natl Forge Co *Education:* MS/Met Engrg/Stevens Inst of Tech; B Tech/Met Engrg/Indian Inst of Tech; BSC/Phys, Chem & math/Agra Univ. *Born:* 8/7/48 Born and educated in India. Came to USA in 1969. Received M.S. in Met Engrg in 1971 from Stevens Inst of Tech - Hoboken, N.J. *ASM* activities on local as well as natl level include - N.W. PA Chap - Chap Chrmn - 1978-79; Natl Mbrship Ctte 1978-82, two years Secy, two years Chrmn; Natl Chapter Mbrship Coun - Chrmn 1982-83; Engrg Assoc Achievement Award Selection Ctte - ASM Natl for 1984 Award - Chrmn; Natl Nominating Ctte - Mbr 1982; Natl Study Ctte on Mbr participation - mbr 1983; Process Control & Reliability Analysis Ctte - MTQC Tech Div - Chrmn 1982 to present. *ASME* Mbr Tech Subctte on High Pressure Tech - Pressure Vessel & Piping Div. Liaison Assignment: Mtls & Fabrication. Reviewer of proposals for Natl Sci Foundation, Wash. D.C. Edited tech book under ASM reference publications: "Ferritic Steels for High Temperature Applications–. Published in 1983 by ASM. *PATENTS*: Co-inventor of Indian Patent (1971): "Preparation of Cooper Coated Graphite Particles and *Society Aff:* ASM, ASME, AAAS.

Kibler, J B, Jr
Business: 500 E 8th Rm 1312, Kansas City, MO 64106
Position: VP-Customer Service. *Employer:* Southwestern Bell Telephone Co. *Education:* BS/EE/OK A&M College (New OK State Univ). *Born:* 6/8/22. BSEE OK State Univ, Stillwater OK. Retired Major USAF. With Southwestern Bell Tel Co 1947- ; assumed current respon as VP Customer Service 1978. In 1974 received

Kibler, J B, Jr (Continued)
MO Soc of Prof Engrs "Prof Engrs in Indus" Award for both Chap & State; 1976 State Dir MO Soc Prof Engrs. Enjoys fishing & golf. *Society Aff:* IEEE.

Kidd, George E
Home: 1362 Rodeo Dr, LaJolla, CA 92037
Position: Former Head - Genl Engrg Div. *Employer:* Gulf Genl Atomic San Diego (Retired). *Born:* Nov 1904 Germiston Transvaal S Africa. MSME Yale 1930; BS Ore St Coll. Fellow ASME; formerly Reg Engr: N Y, N J, Conn, Texas, La, Ill, Ind. Early experience: plant & facilities design, chem & petro-chem co's (du Pont, Magnolia Petro) 1930-46. Later experience: engrg mgmt, progs, budgets, personnel - Amer Cyanamid, Asst Ch Engr 1947-52; Air Reduction Co, Asst Ch Engr 1953-59; Gulf Genl Atomic, Div Hd 1959-69. War-time serv: Manhattan Proj Chicago. 2nd career: Newspaper Columnist; Volunteer, County Govt cttes & task forces.

Kidd, Keith H
Business: 121 Kennedy Ave, Toronto, Ontario, Canada M6S 2X8
Position: Sr Vice President. *Employer:* Leighton & Kidd Ltd. *Education:* MASc/Eng Physics/Univ of Tornoto; BASc/Eng Physics/Univ of Toronto. *Born:* Feb 16, 1921. Fellow IEEE 1966. Career: Ontario Hydro Res & Planning 1946-56; B C Hydro & Power Authority 1956-61, Engrg Exec; Private Cons Practice 1961- , power sys tech & econ planning, financing, rates. Ctte activities: IEEE, Transmission & Distrib, Sys Engrg, Protective Devices; IEEE-EEI-NEMA Insulation Co-ordination. *Society Aff:* IEEE.

Kidder, Allan H
Home: 118 Owen Ave, Lansdowne, PA 19050
Position: Cons-Elec Power Trans & Distrib. *Employer:* Self-employed. *Education:* SM/Elec Engg/MIT; BS/Elec Engg/MIT. *Born:* Dec 1899 Proctor Vt. MS MIT on completion of res for G E Co on Pressure Transmission Phenomena in Fluid Columns. Empl by Phila Elec Co 1923-66 serving successively as Supervisor Load Forecasts, Engr in Charge Elec Sys Planning, preparation of Condt Budget, Distrib Sys Cost Analyses; also respon for co's progs on Emergency Load Ratings, Preferred Voltage Ratings, Pipe Type Cable Sys Dev, Power Sys Parameters, Depreciation of Elec Gas & Steam Sys Plant & an Introduction to the EEI Underground Sys Reference Book. Organized & dir the co's Res Div 1956-66. Cond on Elec Power Transmission & Distrib problems 1941- and author *Standing-Wave Energy Dissipation From a Span of Overhead Conductor*, IEEE-PES Paper (A76 188-3) presented in Jan 1976. *Society Aff:* IEEE, NSPE.

Kidder, Ernest H
Home: 1709 Cahill Dr, East Lansing, MI 48823
Position: Professor Emeritus *Employer:* Retired. *Education:* BS/Agr/Engr/U of Minn; MS/Civ Engr/U of IL. *Born:* July 1912. 1935-43 & 1946-49 Soil Conservation Serv, Res Div; 1943-45 US Naval Res WW II EnsignLt; 1949 Mich St Univ - Teaching extension & res in soil & water conservation engrg. Res: Drainage of plastic till soils of NE Ill; sprinkler irrigation for frost protection. 1955 Sprinkler Irrigation Assn Man of the Year; Chmn Soil & Water Div ASAE; 1963-64 ASAE Board of Dirs; 1965 NSF Water Inst Utah St Univ; 1970 AID Cons Irrigation Argentina; 1972 Hancock Brick & Tile Award ASAE; 1976 Fellow ASAE 1979 Life Fellow ASAE. 1979-86 Irrigation Consultant. *Society Aff:* AZ, ΓΣΔ, ΣΞ, ASAE

Kidnay, Arthur J
Home: 6141 Van Gordon St, Arvada, CO 80004
Position: Prof and Dept Hd *Employer:* CO Sch of Mines. *Education:* PhD/CE/CO Sch of Mines; MS/CE/Univ of CO; BS/CE/CO Sch of Mines. *Born:* 4/4/34. Res engr with Chem Engrg Science Div of the Natl Bureau of Stds since 1958. Position has been part-time since joining the faculty of the Chem Engg Dept at the CO Sch of Mines in 1968. Promoted to Prof in 1976 and dept hd in 1983. 60 publications in the general field of thermodynamics. Teaching activities include thermodynamics, unit operations, & engg design. Active in the local section of AIChE, having held the offices of treas, secy, vchrmn, & chrmn. Mbr of the Natl Cryogenics Committee of AIChE. NATO Sr Sci Fellow at the Univ of Oxford, summer 1973. *Society Aff:* AIChE.

Kielhorn, William H
Business: Box 7001, Longview, TX 75607
Position: Assoc Prof Weld Engg *Employer:* LeTourneau Coll *Education:* MS/ME/Univ of WI; BS/Welding Engg/Le Tourneau Coll *Born:* 6/25/31 Taught Welding Engrg for 21 years. Prior to that was welding engr for Worthington Corp. Have been active in American Welding Society, espically East TX Section lectured at Dallas, Tulsa, Louisville, Shreveport, Longview sections. Taught several welding seminars consultation services with Sun Pipeline, Travelers Insur, Roberts Harbour Attorneys, Stemco & Trailmobile. TX State PE No. 46958, issued Jan 1980. Taught Metals Engrg Inst course on welding at Norfolk Naval Shipyard 1980; Also developed welding procedure for crane wheels there in 1979. *Society Aff:* AWS, ASM

Kiely, John R
Home: 206 Manzanita Way, Woodside, CA 94062
Position: Consultant. *Education:* BS/C.E./Univ of WA. *Born:* Nov 1906. Phi Beta Kappa, Sigma Xi, Tau Beta Pi. Identified many yrs with large engrg & const projs, especially nuclear power & mining plants. Mbr NAE; Fellow ASCE & ASME; Mbr AIME, USCOLD. P Pres & Dir EJC & Beavers. Honorary Chmn,Int'l Exec. Council of WEC. Past Chmn Energy Conservation Comm of World Energy Conference; P Chmn US Natl Ctte of WEC; Recipient of Golden Beaver Award for Mgmt in Engrg & Const. *Society Aff:* ASCE, ASME, AIME, USCOLD.

Kienow, Paul E
Home: 515 W Bulter Dr, Phoenix, AZ 85021
Position: VP. *Employer:* STRAAM Engrs, Inc. *Education:* BS/Engr-Physics/Univ of AZ; BS/Civil Engr/Univ of AZ. *Born:* 4/16/40. Reg PE (civil) in AZ. Served with US Navy Civil engr Corps 1962-1965, at Public Works Ctr, Guam. Field Engr for Southern Pacific Pipe Lines, Inc (Corrosion), 1966-1967. Industrial Engr for MN Mining and Mfg, 1968. Grad work under a Water Pollution Control Act Fellowship at the Univ of AZ 1969-1970. Chief Engr for Wheeler, Petterson and coffeen, Tucson, AZ, 1970-1971. Established Paul Kienow and Assoc, Inc, Tucson, AZ, 1972-1975. Dist Engr, Hydro Conduit Corp, Phoenix, AZ, 1976-1979. VP, STRAAM Engrs, Inc, Phoenix, AZ. Mgr of StRAAM's Phoenix, AZ office. Mbr of US Navy Reserve (Lt Commander) assigned to Public Eorks Ctr, Guam. Married, with two children. *Society Aff:* ASCE, NSPE, SAME, WPCF.

Kienzle, Gary M
Business: 611 Cascade West Pkwy SE, Grand Rapids, MI 49506
Position: Exec VP/Treas *Employer:* WW Engineering & Science *Education:* MBA/Business Administration/Western MI Univ; MS/Civil Struct Engrg/MI State Univ; BS/CE/MI State Univ. *Born:* 8/12/47 Native of Grand Rapids, MI. Research Engr with Exxon Production Research Co, Houston, TX 1970-72. Associated with William & Works, Consulting Engrg Firm, since 1972. Project Mgr in charge of Structural Engrg since 1975. Elected to Williams & Works Bd of Dirs, 1981. Pres of MI Society of PE/PE in Private Practice 1981-82. MI Society of PE Young Engr of the Year 1981. Enjoy sailing, racquetball, and cross-country skiing. Licensed private pilot. Elected Treas of Williams & Works, 1982. Elected Exec VP of WW Engineering & Science, holding company of Williams & Works, 1983. *Society Aff:* NSPE, ASCE

Kiersch, George A
Home: 4750 N. Camino Luz, Tucson, AZ 85718
Position: Conslt/Emeritus Prof Cornell U *Employer:* Kiersch Associates, GeoSciences & Resources Consultants. *Education:* Sr. Postdoctorate Fellow/Geomechanics/Tech Univ Vienna, Austria; Ph.D/Geol & Mining Engrg/Univ AZ; Geol Engr/Geo & Engrg/CO Sch of Mines. *Born:* 04/15/18 Internationally known specialist engrg geology. Ofcr Army Corps engrs 1942- 45. Supervising geologist dams/reservoirs US

Kiersch, George A (Continued)
Corps Engrs, CA and Intl Boundary and Water Commission USA-Mexico 1948-51. Dir, mineral resources survey Navajo- Hopi Indian Reservations, AZ-UT 1952-55. Exploration mgr, Southern Pacific Co western states 1956-60. Prof geological sciences, Cornell Univ since 1960, chrmn 1965-71; Emeritus Prof 1978-. Geologic conslt since 1960 for over 200 projects USA/foreign. First recipient Holdredge Award, AEG, 1965, Honorary Member, 1985. Chrmn, Div Engrg Geology, Geological Society America 1960-61. Mbr seven Natl Research Council Committees 1966-87. Chrmn, Ctte of Intl Assoc Engrg Geologists 1983-86, Vice-President, North America 1986-1990. Published over 65 papers and volumes; served as editor Geological Society America and Engrg Geology Journal. Editor, "The Heritage of Engineering Geology--The First Hundred Years-, 30-Contributing Authors for Geological Society of America, Boulder, Colorado (DNAG, CV-3). Chairman, Committee on "Environmental Geology-Hazards- & Geophysics" for International Lithosphere Program, 1986-1991. *Society Aff:* ASCE, GSA, SEG, USCOLD, IAEG, ISRM, AEG

Kieschnick, William F
Business: 515 S Flower St, Los Angeles, CA 90071
Position: President *Employer:* Intergroup (consulting and business development) *Education:* BS/Chem Eng/Rice Univ; Certific/Phys, Meterology/UCLA; Certific/Oceanography/Scripps-La Jolla *Born:* 1/5/23. Dallas Texas. Courses in Phys, Meteorology UCLA; Oceanography Scipps Inst. Prof exper in oil & gas prod res & dev; internatl opers; oil & gas exploration & prod opers; synthetic fuels from shale oil, tar sands & coal; petrochem mfg & mktg; corporate planning; petro refining & mkting; pipelines & tanker transportation of crude oil. Currently Director and retired CEO, ARCO; President Intergroup, development and consulting start-up companies. Outside Activities: Director First Interstate Bancorp, TRW, Atlantic Richfield, Pacific Mutual, Korn/Ferry International; Honorary Director, American Petroleum Institute; Trustee, California Institute of Technology; Chairman of the Board of Trustees, The Museum of Contemporary Art; Trustee, Carnegie Institution of Washington. *Society Aff:* API, AFAR, AIChE, AIME

Kiesling, Ernst W
Business: Box 4089, Lubbock, TX 79409-4089
Position: Prof & Chrmn *Employer:* TX Tech Univ *Education:* PhD/Mech/MI State Univ; MS/Mech/MI State Univ; BS/ME/TX Tech Coll *Born:* 4/8/34 NSF Science Faculty Fellow, 1963-64. Senior Research Engr, Southwest Research Inst, San Antonio, TX 1966-69. Prof and Chrmn, Civil Engrg, TX Tech, 1969-present. Research in structures, housing, solar applications to housing, earth sheltered housing. Married, Juanita Haseloff Kiesling. Three children, Carol, Chris, Max. *Society Aff:* NSPE, ASCE, ASEE, AUA, AS of ISES

Kight, Max H
Home: 9335 E Center Ave 7C, Denver, CO 80231
Position: Retired. *Employer:* *. *Education:* BS/Elec Engg/Univ of IL. *Born:* Oct 26, 1900, Bluffdale Texas. BSEE 1923 Univ of Ill. Was Asst Chief Designing Engr US Bureau of Reclamation Denver Colo at time of retirement. IEEE Fellow Award 'For contributions to the establishment of large scale hydro plants & associated power transmission' June 1959. Also Disting Serv Award US Dept of the Interior 1966. *Society Aff:* IEEE.

Kihn, Harry
Home: 30 Green Ave, Lawrenceville, NJ 08648
Position: Electronics Consultant. *Employer:* Kihn Assoc., Inc. *Education:* MSEE/Elec Circuits/Univ PA; BSEE/EE/Cooper Union Inst Tech; Teaching Certification/Educ/NYU; Teaching Certification/Educ/MT Teach Coll. *Born:* 1912 Austria. Educ studies NYU, Montclair St Coll. Res Engr Hygrade- Sylvania Ferris Instrument Corp; joined RCA 1939, Res Engr on circuits, TV, frequency modulation. During WWII dev radar, altimeters, anti-submarine devices. Post war res projs included: air navigation, color TV, 35 GHz radar, infrared sys. Headed Task Force on Wizard Missile Detection Prog. Joined Staff of Exec V Pres Res & Engg, respon for computers, solid state devices & defense projs. Initiated progs in integrated circuits, advanced computer tech. Appointed Tech Advisor Corporate Patents & Licensing. Pres, Kihn Associates Inc. 24 patents, 12 papers, Prof activities: Mtls Advisory Bd Natl Acad Sci 1960, TAB, IEEE, 1965- 68, Life Fellow IEEE; Mbr Sigma Xi; NY Acad Sci; AAAS: ADPA Prof Engr NJ. Active in civic affairs (School Bd, Boy Scouts, Rotary, etc.) Mbr Program Review Committee Nuclear Waste Isolation Project D.O.E Senior Editor, Governors Commission Science and Tech Recipient Centennial Medal. President, SPE, Mercer Chapter, Engineer of the Year. *Society Aff:* IEEE, ΣΞ, NYAS, NSPE, AAAS, ADPA, AMA

Kilby, Jack S
6600 LBJ Freeway, Suite 4155, Dallas, Texas 75240
Position: Consultant. *Employer:* Self-Employed. *Education:* BS/EE/Univ of IL; MS/EE/Univ of WI. *Born:* Nov 8, 1923. Served with OSS in WWII. Started work with Centralab Div of Globe-Union Inc as engr on ceramic based printed circuits. Joined Texas Instruments in 1958 & invented the monolithic integrated circuit. Held a number of positions in semiconductor R/D & mgmt with last title Asst V Pres & Dir of Engrg. Took leave of absence in 1970 to do independent dev of new products, consulting. Fellow IEEE. Mbr NAE. Recipient of Sarnoff Medal IEEE 1967, Ballantine Medal of the Franklin Institute 1967, Natl Medal of Sci 1969, Zworykin Medal of the NAE 1975, Brunetti Award IEEE 1978, IEEE Medal of Honor 1986, Natl Inventors Hall of Fame 1982, Dr Eng (Hon) U of Miami 1982, Rochester Institute of Technology, 1987. *Society Aff:* IEEE, NAE.

Killebrew, James R
Business: 600 Petrol Bldg, Wichita Falls, TX 76301
Position: Chrmn of Bd. *Employer:* Killebrew Rucker Assoc Inc. *Education:* BSArch Engg/Structural/Univ of TX. *Born:* 12/10/18. James R Killebrew was born & attended public schools in Okmulgee, OK. He served as a Naval Aviator in the Pacific Theatre during WWII & retired as capt, USNR. He is chrmn of the Bd, Killebrew/Rucker/Assoc, Inc. The firm designs hospitals, univs, schools, commercial & ind bldgs. The AC Spark Plug Plant, Div of GM is the largest precast concrete structure in this area. Jim was elevated to the coveted position of Fellowship, AIA 1975, & is past Pres, Wichita Falls Chapter AIA, & TX Soc PE. He is a mbr of Rotary, & serves as Elder, First Christian Church. *Society Aff:* ASHRAE, NSPE, AIA, ACI.

Killgore, Charles A
Home: 506 Hundred Oaks Dr, Ruston, LA 71270
Position: President. *Employer:* Killgore's Inc. *Education:* BS/Chem Engg/LA Tech Univ; MS/Chem Engg/LA Tech Univ. *Born:* Aug 1934; native of Lisbon La. Work toward PhD Okla St Univ. Chem Engr with Amoco prior to entering USAF 1956-59; faculty Coll of Engrg LTU 1959-62; Dir Nuclear Center LTU 1962-73; Assoc Dean Engrg 1973-76. Mbr AIChE, ASEE, ANS & La Engrg Soc. Reg in La. Outstanding Young Man 1968; Dow Chem Outstanding Young Fac Award; Andrew M Lockett Award LES: Dir Engrg Res LTU; Frost Foundation Award for excellence in teaching. Silver Beaver Award in Scouting. Univ representative to Natl Res Council. Author & Cons. *Society Aff:* AIChE, ASEE, ANS, HPS, AAPM, ASRT.

Killian, James R
Business: 77 Massachusetts Ave, Cambridge, MA 02139
Position: Retired Pres and Chairman of MIT *Employer:* MIT. *Education:* SB/Bus & Engrg Admin/MIT *Born:* 7/24/04. Pres MIT 1949-59; Chmn of the MIT Corp 1959-71; Honorary Chmn of the Corp 1971-79. Mbr of NAE. Special Asst to Pres Eisenhower for Sci & Tech 1957-59; Chmn of Pres's Sci Advisory Ctte 1957-59. Public Welfare Medal, Natl Acad Scis 1957; Hoover Medal 1963; Award of Merit AICE 1958; WA Award West Soc of Engrs 1959; Hon Mbr ASEE 1963; Sylvanus Thayer Award, US Military Acad 1978; Vannevar Bush Award of the National Science Board, 1980; Fellow Amer Acad of Arts & Scis; Dir G M Corp 1959-75; AT&T Co 1963-77; Polaroid Corp 1959- ; Ingersoll-Rand Co 1971-76; IBM, 1959-62; Cabot

Killian, James R (Continued)
Corporation, 1963-75. Trustee, Mitre Corp 1960-82 . *Society Aff:* NAE, AAAS, ASEE

Killian, Stanley C
Home: Lakeside Terr Rt 6, Mountain Home, AR 72653
Position: Vice Pres & Genl Mgr, Retired. *Employer:* Lapp Insulator Div - Interpace Corp. *Education:* BS/Physics/Univ of MI. *Born:* July 4, 1906. Chief Engr 1944-59, V Pres & Genl Mgr 1959-62 Delta Star Elec Div of H K Porter Co; V Pres of Engrg 196269, V Pres & Genl Mgr 1969-71 Lapp Insulator Div of the Interpace Corp. Fellow IEEE. Prize paper AIEE Great Lakes Dist 1951. Mbr of many IEEE, ASA & NEMA cttes. US Delegate to Internatl Electrotech Ctte in Stockholm Sweden 1958. Author many IEEE papers & trade journal articles. Holder of 17 patents. Main hobby: Archaeology. *Society Aff:* IEEE.

Killough, John E
Home: 1936 Carmel, Plano, TX 75075
Position: Research Director *Employer:* Arco Oil & Gas Co. *Education:* MCE/CE/Rice Univ; BA/CE/Rice Univ. *Born:* 6/19/47. Native of Baytown, TX. Res Engr with Exxon Production Res Co, Houston, TX specializing in reservoir simulation & numerical analysis, 1971-1976. Engr for Aramco, Dhahran, Saudi Arabia 1974-1976 in reservoir simulation & petrol computer applications. With Arco Oil & Gas Production Res Ctr since 1976 as director of math model dev. ASME Rossiter W Raymond Award, 1976. ASCE, AIME, IEEE, ASME & Western Soc of Engrs Alfred Noble Prize, 1977, for paper entitled "Reservoir Simulation with History-Dependent Saturation Functions–. *Society Aff:* SPE of AIME.

Kim, Donald, C W
Business: 677 Ala Moana Blvd, Suite 1016, Honolulu, HI 96813
Position: Pres & Chrmn of the Bd *Employer:* R M Towill Corp. *Education:* BS/Civ Engg/Univ of HI. *Born:* 12/14/28. Joined R M Towill Corp as Design Engr, 1957; Proj Engr, 1965; Mgr of Engg Dept, 1969; VP, 1971; Exec VP, 1972; Pres. , 1978; currently Pres and Chrmn of the Bd. Responsible for all depts in the corp; Planning & Land Dev Services, Surveying, Photogrammetry, Energy & Mechanical Engineering, Environmental & Civil Engg, Construction Mgt & Accounting. Also Chrmn of the Bd, since 1982, AMKOR A & E, Inc., complete architectural and engrg firm in Seoul, Korea. Served in Civ Engg Advisory Committe for Honolulu Community Coll and Engg Liaison Committe, Univ of HI. Past Natl Dir of NSPE, Past Pres, HI Soc of Professional Engrs, Past Dir, Honolulu Post of SAME, Past Chrmn, HI Council of Engg Societies, and mbr of Rotary Club of Honolulu, ASCE, AWWA, ASTM, AWPCF and Pres' Club of Univ of Hawaii. *Society Aff:* AWPCF, ASCE, ASTM, NSPE, SAME, AWWA,

Kim, Dong K
Home: 4194 Big Spruce Dr, Akron, OH 44316
Position: R & D Associate *Employer:* Goodyear Tire & Rubber Co. *Education:* PhD/Mtls Engg/NC State Univ; MS/Mtls Engg/NC State Univ; BS/Mining Engg/Seoul Natl Univ. *Born:* 3/18/43. Native of Korea. US citizen 1978. Post-doctorate Res Assoc at Lehigh Univ 1973-76. Res Scientist for Ctr for Surface & Coatings Res, Lehigh Univ 1976-78, concerning with adhesion & durability of protective coatings on metallic substrate. With Goodyear Tire & Rubber Co since 1978, responsible for res on mechanism of rubber-steel adhesion & dev of wire process and materials for tire reinforcing. SAE Arch T Colwell Merit Award 1979. *Society Aff:* ASM, AIME, ΣΞ

Kim, Hyough (Hugh) S
Business: 1999 Pennsylvania Ave, Hagerstown, MD 21740
Position: Chief Incoming Material & Met Lab. *Employer:* Mack Trucks Inc. *Education:* MMetE/Metallurgy/RPI; BS/Metallurgy/Lafayette College. *Born:* Feb 4, 1933, Seoul Korea, US Citizenship 1969. M Met Engrg RPI Troy N Y May 1963; BS Lafayette Coll Easton Pa. Res Asst for Material Engrg Dept RPI, spec dispersion strengthened SAP-type alloys by means of electron microscopy 1963-66; with Mack Trucks Inc 1966- , Metallurgist & Section Head, Met Lab. Current respon as Sect Mgr, Incoming Material & Met Lab; organizes, manages & develops the functions of Met Lab & Incoming Material Inspection Group. Ctte Mbr of Heat Treating Div ASM 1974- . Enjoys classical music & photography. *Society Aff:* ASM, KSEA.

Kim, Jai B
Business: Head, Department of Civil Engr, Bucknell University, Lewisburg, PA 17837
Position: Chrmn of Civ Engg Dept. *Employer:* Bucknell Univ. *Education:* PhD/Civ Engr/Univ of MD; MS/Civ Engr/OR State Univ; BS/Civ Engr/OR State Univ. *Born:* 5/17/34. Native of Korea, received all of collegiate educ in the US since 1955. Chief Hgwy Res Engr for the DC Govt 1964-1966. With Bucknell Univ since 1966, teaching, res and consulting in the areas of structural and fdn engg. Specialist in deep fdns (piles), precast concrete panel structures, and bridge engrg. Co-developed an arch-hanger-additional floor beam reinforcement sys for steel truss bridges. This sys is being applied to aging truss bridge which can be upgraded to carry modern traffic loads at minimal costs for another century. This system has been granted for a patent. *Society Aff:* ASCE.

Kim, Jongsol
Home: 1909 Polo Ct, San Mateo, CA 94403
Position: Principal and Owner *Employer:* Kim and Assoc *Education:* PhD/Soil Mech/Univ of WI-Madison; MS/Soil Mech/Univ of WI-Madison; BS/CE/Univ of WI-Madison *Born:* 8/21/30 in Yonan, Korea. Taught surveying and engrg mechanics at Univs of WI and Guam. With Trans-Asia Group (A-E firm), progressed from Civil Engr to Senior Soil Engr and Engrg Mgr. Individual practice as Soils Engrg Consultant in Guam and CA. With William F. Jones, Inc. (geotechnical consulting firm) in San Mateo, CA, starting as Sr Soil Engr in 1976 and managing the Co as Exec VP for six years. In 1985 started his own geotechnical consulting firm of Kim and Assoc in San Mateo, CA. Treasurer of PEPP Division of CSPE, 1981. Registered PE in four states, Guam and Korea. *Society Aff:* ASCE, NSPE, CEAC.

Kim, Kyekyoon (Kevin)
Business: Dept of Elec. & Computer Engrg, University of Illinois, 1406 W. Green St, Urbana, IL 61801
Position: Prof *Employer:* Univ of IL *Education:* PhD/Applied Physics/Cornell; MS/Nuclear Sci/Cornell; BS/Nuclear Engrg/Seoul Natl Univ Korea *Born:* 10/5/41 Native of Seoul, Korea. Came to US in 1966 to pursue graduate education. Research Assistant in Nuclear Engrg and Applied Physics (1966-71), and Postdoctoral Fellow in Applied Physics (1971-2) at Cornell. Research Assoc in Chem (1972-4) and Electrical Engrg (1974-6) at the Univ of IL. Became naturalized US citizen in 1977. Assistant Prof (1976-81) and Assoc Prof (1981-85) of Electrical Engrg and Nuclear Engrg and Prof (since 1985), also of Mech Engrg. Acting Dir (1976-80) and Dir (since 1980) of Fusion Tech and Charged Particle Research Lab at the Univ of IL. Member, NASA Science Working Group. Married in 1969. Two children. *Society Aff:* IEEE, APS, AVS

Kim, Myunghwan
Business: Phillips Hall, Ithaca, NY 14853
Position: Prof of Elec Engg. *Employer:* Cornell Univ. *Education:* PhD/EE/Yale Univ; MEngg/EE/Yale Univ; BS/EE/Univ of AL. *Born:* 2/8/32. Native of Seoul, Korea. Elec Engr for TN Valley Authority, 59-60; Natl Res Council Sr Res Assoc at Jet Propulsin Lab 68-69; visiting assoc for CA Inst of Tech, 1969; 62-present, Prof of Elec Engg, Cornell Univ. Consultant for IBM, Bristol Co, US Army Missile Command, & Natl Cancer Inst, Visiting Prof, Korea Advanced Inst of Tech & Korea Univ, 1982. *Society Aff:* IEEE, NYAS.

Kim, Rhyn H
Home: 2726 Wamath Dr, Charlotte, NC 28210
Position: Professor. *Employer:* Univ of NC, Charlotte. *Education:* PhD/Mech Engg/MI State Univ; MS/Mech Engg/MI State Univ; BS/Mech Engg/Seoul National Univ. *Born:* 2/4/36. Taught thermal sci and engg at the Univ of NC since 1965. As a NASA-ASEE Summer Res Faculty Fellow, analyzed a heat conduction problem of multi-layered reusable insulating surfaces for space shuttles and wrote a computer program. Was on a leave from the Univ in 1976-78. Worked for Office of Air Quality Planning Standards, Us Environ Protec Agcy (EPA); Engaged in energy strategies ad planning activ in assoc with totatl energe demand supply and pollution control technologies for the US in 1976-77. Participated in 1978 and continued the work for IERL, EPA as consultant. In 1981, as a member of Gas Research Institute - ASEE Summer Research Associateship, analyzed a simulation model for a phochemistry system in association with gas energy supplies from inorganic resources. As a NASA-ASEE Summer Res Faculty Fellow, developed software programs for a data acquisition system with VAX11/750 in association with the laser velocimetry used in a compressor of a turbine. As a consultant, participated in the environmental safety work of Duke Power Co. Am interested in res and dev of energy conversions and digital computer simulations of the energy conversion processes. Pub numerous art in the energy conversion processes. *Society Aff:* ASME, ASHRAE, ISA.

Kim, Thomas J
Business: 204 Wales Hall, Kingston, RI 02881
Position: Chrmn. *Employer:* Univ of RI. *Education:* PhD/Appl Mech/Univ of IL; MS/Math/Villanova Univ; MS/Mech Engg/Seoul Natl Univ; BS/Mech Engg/Seoul Natl Univ. *Born:* 10/13/39. Native of Seoul, Korea. Taught at the Univ of IL (Instr, 1966-67), Villanova Univ (Asst Prof, 1967-68), and the Univ of RI (Prof, 1968-). Appointed chrmn of dept of mech engg in 1979. Past and current res activities and interest are high strength ceramic processing, abrasive waterjet processing, and mechanics of mtls area. 1972 Tau Beta Pi Teaching Excellence Award. Published about 50 technical articles. *Society Aff:* ASME, ACS, ASEE, SME.

Kim, Young C
Business: Dept of Civ Engg, Los Angeles, CA 90032
Position: Prof & Chrmn. *Employer:* CA State Univ. *Education:* PhD/Civ Engg/Univ of S CA; MS/Civ Engg/CA Inst of Tech; BS/Civ Engg/Univ of S CA. *Born:* 5/25/36. He is currently Prof & Chrmn of Civ Engg at CA State Univ, Los Angeles. He was Visiting Scholar in Coastal Engg at Univ of CA, Berkeley in 1971, NATA Sr Fellow in Sci at Delft Tech Univ, Holland in 1975, Visiting Scientist, the US- Japan Cooperative Sci Prog at Osaka City Univ, Japan in 1976, & Resident Consultant for Sci Engg Assoc from 1967 to present. He is past Chrmn of the natl Committee on Res of the Waterway, Pt, Coastal & Ocean Div & presently he is Chrmn of Task Committee on Ocean Energy & Mbr of the natl Exec Committee on Technical Council on Res of the Am Soc of Civ Engrs. *Society Aff:* ASCE, AAAS, ASEE, IAHR.

Kimball, Charles N
Business: 425 Volker Blvd, Kansas City, MO 64110
Position: President Emeritus. *Employer:* Midwest Research Institute. *Education:* ScD/Comm Eng/Harvard; SM/Comm Eng/Harvard; BS/Elec Eng/Northeastern. *Born:* Apr 1911. Res in FM, TV, Radar RCA 1934-41; V Pres & Dir Aircraft Accessories Corp Kansas City 1941-46 (radar countermeasures & communications); V Pres C J Patterson Co Kansas City 1946-48; Dir Res Labs Bendix, Detroit 1948-50 (upper altitude meteorological res, communications phenomena); Pres Midwest Res Inst 1950-75 Chrm Bd of Trustees 1975-79 (non-profit multi-discipline inst). Fellow IEEE. Received various awards & honorary degrees. Named Mr Kansas City 1973. Mbr Board of Dirs TWA Inc & Hallmark Cards Inc. Trustee several educ insts Chrmn Tech Advisory Cttee Office of Tech Accessment, Wash DC; Chrmn greater KC Commity Foundation. *Society Aff:* ΤΒΠ, HKN, ΣΞ, IEEE.

Kimball, L Robert
Business: 615 W Highland Ave, Ebensburg, PA 15931
Position: President. *Employer:* Self-Employed. *Education:* Bachelors/Civil Engr/Penn State Univ. *Born:* Mar 6, 1923. USAF 1943-45 & USA Corps of Engrs 1950-52. Founded firm of L Robert Kimball Cons Engrs 1953 & now employees approximately 450 people & has branch offices in 5 cities throughout the US. Under his direction the firm has grown to include services in the fields of san & environ engrg, struct design, transportation, earth sci, airports & various other areas. Reg Prof Engr in Pa, Ala, Mass, NH, Neb & Va. Mbr ACEC, NSPE, Pa Soc of Prof Engrs in Private Practice, AAEE - Diplomate, & numerous other prof orgs. *Society Aff:* ACEC, NSPE, AAEE.

Kimball, William R
Business: 2160 South Clermont St, Denver, CO 80222
Position: Exec VP *Employer:* McFall, Konkel & Kimball Consulting Engrs Inc *Education:* BS/Mech Engr/Univ of Denver *Born:* 7/18/40 Registered engr in CO, and ten other states. A member of the consulting firm of McFall, Konkel & Kimball or predicessor firms for 23 years. Partner in charge of mechanical design of Denver Writers Square & Boulevard Plaza Complexs, Writers Sq, Aspen Post Office, Aspen Airport, Aspen Sport Obermeyer Office/Warehouse. The latter two projects have both received the Owens Corning Energy Conservation Award and Obermeyer has also received the 1981 ACEC Honor Award for Engrg Excellance. Region IX Vice Chrmn and Past Pres of Rocky Mountain Chapter of ASHRAE. Major design interests include heat reclaim, conservation as well as active passive solar energy. *Society Aff:* ASHRAE, NSPE, CECC, PEC, ACEC

Kimbark, Edward W
Home: 3233 N E Thompson St, Portland, OR 97212
Position: Consulting Engr. *Employer:* Bonneville Power Admin - Portland. *Education:* ScD/EE/MIT; SM/EE/MIT; EE/EE/Northwestern; BS/EE/Northwestern. *Born:* Sept 1902 Chicago. Studied elec engrg at Northwestern Univ 1920-25 & at MIT 1932-37, receiving ScD in 1937. Taught EE at Univ of Calif Berkeley 1927-29; MIT 1933-37 & 1942-44; Polytech Inst of Brooklyn 1937-39; Instituto Tecnologico de Aeronautica (Sao Jose dos Campos Brazil) 1950-55; Dean Sch of Engrg Seattle Univ 1955-62; Elec Engr Bonneville Power Admin 1962- . Author of *Power System Stability* (3 vols) 1948-56; *Electrical Transmission of Power & Signals* 1949; *Direct-Current Transmission* 1971. Contributed papers on symmetrical components, transient stability, single-pole switching & overvoltages caused by faults, to AIEE & IEEE. Fellow AIEE & IEEE 1948- . Won gold medal for disting service from US Dept of the Interior 1975 and Habirshaw Award from IEEE 1980. Elected to National Acad of Engg 1979. *Society Aff:* IEEE, NAE.

Kimberling, Charles L
Home: 6506 E 27th Pl, Tulsa, OK 74129
Position: Manager of Engineering. *Employer:* City of Tulsa-Water & Sewer Dept. *Education:* Master/Civil Engg/OK State Univ; BS/Civil Engg/Univ of AK; AS/Engg/AK Polytechnic College. *Born:* Feb 7, 1928. Assoc of Sci Ark Polytech Coll. Native of Paris Ark. Resident Engr on enlargements to Mohawk Water Treatment Plant, City of Tulsa 1952-54; Design Engr & Plant Engr Lee C Moore 1954-55; Stress & Dynamics Group at Douglas Aircraft 1955-56; With City of Tulsa Water & Sewer Dept 1956- . Assumed present respon of Asst Supt & Mgr of Engrg 1969. Dir NSPE 1971-75; Pres OSPE 1974-75; Chmn SW Sect AWWA 1975-76; Mbr St Board of Reg for Prof Engrs & Land Surveyors, St of Okla 1974-79 & 1979-84; Natl Dir, AWWA 1979-81. *Society Aff:* ASCE, NSPE, AWWA, WPCF.

Kimberly, A Elliott
Home: 711 Abbey Rd, Birmingham, MI 48008
Position: VP. *Employer:* Kimbro, Ltd *Education:* MS/Engg/OH State Univ; MS/Auto Engg/Chrysler Inst of Engg; BS/Mech Engg/OH State Univ. *Born:* Apr 1913 Columbus Ohio. With Chrysler Corp 1935- ; Dept Head Transmission Lab 1939-41; Dept Head Mech Labs 1942-52; Ch Engr DeSoto Div 1953- 59; Ch Engr Vehicle Reliability, Engrg & Prod Dev 1960-78 (Retired). Admin corp- wide Reliability & Warranty Prog. P Chmn: Ohio St Univ Alumni Advisory Board, Ctte of 100 for Engrg. OSU Awards: Engrg Disting Alumnus, Alumni Centennial, Disting Service. Mbr: Soc of

Kimberly, A Elliott (Continued)
Auto Engrs, Tau Beta Pi, Delta Tau Delta, US Power Squadrons, OSU Assn. *Society Aff:* SAE, TBII.

Kimbrough, Emmett A, Jr
Business: Box 5465, Miss State, MS 39762
Position: Agri Engr. *Employer:* MS State Univ. *Education:* MS/Agri Engg/TX A&M Univ; BS/Agri Engg/MS State Univ. *Born:* 10/26/21. Served in Air Corp - 1942-45; later AF Retired Lt Colonel Reserve-1971. Employed by MSU since 1948, leave of absence 1959-61, served with TX A&M in the Univ of Daeca, East Pakistan. Consultant to Brazilian Ministry of Agri 1966, new res station at Brasilia. Received "Engr of the yr" award MS Sec ASAE 1975. Chrmn, State Sec 1976. Developed "Faculty Awards Program" MS State Alumni Assn 1964 and is in its 16th yr. Present position: In charge of Engg Services for the MS Agri and Forestry Experiment Station, MSU which designs the res facilities for the system. *Society Aff:* ASAE.

Kimel, William R
Home: 900 Yale, Columbia, MO 65203
Position: Dean Emeritus, Coll of Engrg *Employer:* Univ of MO-Columbia *Education:* PhD/Engrg Mech/Univ of WI-Madison; MS/Mech Engrg/KS State Univ; BS/Mech Engrg/KS State Univ *Born:* May 1922. Instr, Asst Prof, Assoc Prof Mech Engrg 1946-58, Prof & Hd Nuc Engrg Ks St Univ 1958-68; Engr with Goodyear Tire & Rubber Co 194446; Boeing Airplane Co 1953; Westinghouse 1954; US Forest Products Lab 1955-56; Argonne Natl Lab 1958; Prof of Nuc Engrg & Dean Coll of Engrg Univ of Mo, Columbia 1968-86; Emeritus Dean, & Prof. UMC, 1986-. Mbr Board of Dirs ECPD 1971-77 (obs 1972-), Amer Nuclear Soc 1973-76, 1977-80, ASEE ECC Council 1975-77, Board of Directors, JETS 1976- , Columbia Chamber of Commerce 1968-84, Columbia Indus Dev Comm 1973-85, Columbia Rotary 1976- . PPres, ANS, JETS, MSPE. Col. Rot. V Pres, National Soc. Prof. Engrg., 1986-87. Disting. Service Award Ks St Univ 1972, Univ. of MO 1986. Fellow Amer Nuc Soc 1969. Fellow, Am Soc of Mech Engineers 1979. Faculty-Alumni Hon Award, Univ of MO-Columbia, 1979. Amer Soc of Military Engineers Bliss Award 1982. Univ of Wisconsin Coll of Engrg Distinguished Service Citation 1982. Engineers Club of St. Louis Award of Merit 1983. *Society Aff:* ANS, ASME, NSPE, JETS

Kimley, Robert J
Business: PO Box 33037, Raleigh, NC 27606
Position: Principal. *Employer:* Kimley-Horn & Assocs Inc. *Born:* Oct 1924, Buffalo N Y. MSCE & BSCE N C St Univ. With Marine Corps 1942- 46. Organized & dir Traffic Engrg Dept, City of Greensboro NC 195157; Advanced Planning Dept Head N C St Hwy Comm 1957-66; founded KimleyHorn & Assocs Inc 1966, serving as Pres to 1972, Chmn of Board of Dirs 1972-75 & presently serving as Principal. P Pres N C Sect ASCE; P Dir ACEC.

Kimm, James W
Business: 300 West Bank Bldg, 1601 22nd St, West Des Moines, IA 50265
Position: Pres *Employer:* Veenstra & Kimm, Inc. Engrg & Planners *Education:* BS/CE/Univ of IA *Born:* 09/26/25 Huron, SD; 399th Infantry - World War II; 6 years IA State Dept of Health - water pollution control; 5 years Stanley Engrg Co, Muscatine, IA; Veenstra & Kimm, Inc since 1961; Chi Epsilon; Tau Beta Pi; Presbyterian elder; Chamber of Commerce; Who's Who in Midwest; Engrs of Distinction by Engrs Joint Council. *Society Aff:* CEC/IA, IES, NSPE, IWPCA, WPCF, AWWA, APWA, CSI

Kinane, John M
Home: 1002 Foxwood Dr, Clifton Park, NY 12065
Position: Mgr, Indus & Sp Proj Opers *Employer:* Gen Elec Co *Education:* BS/Mech Engrg/Rutgers; MS/Mgmt Service/Stevens Inst *Born:* 3/26/41. Native of Teaneck, NJ. Served in the US Army between 1963 and 1966. Obtained commission through Army Artillery Officers Candidate Sch. Discharged as a second lt. Over 21 yrs experience in Proj Mgmt, Estimating and Cost Engrg for petrol chemicals, food and pharmaceutical industries. Assumed present position as Mgr of Ind & Sp Proj Opers with GE in 1984 with Opers responsibility for all of Gen Elec Ind & Sp Projs. *Society Aff:* AACE

Kind, Dieter H
Business: Bundesallee 100, Braunschweig Germany 3300
Position: Pres. *Employer:* Phys-Tech Bundesanstalt. *Education:* Dr-Ing/EE/Technical Univ of Munich; Dipl-Ing/EE/Technical Univ of Munich. *Born:* 10/5/29. Awarded Dr-Ing in 1957 with a thesis on the impulse breakdown of atmospheric air. Profl career at MeBwandler-Bau GmbH in Bamber, 1962 full Prof at the TU Braunschweig for HV-technique. Since 1975 Pres f Physikalisch- Technische Bundesanstalt (PTB) in Braunschweig. *Society Aff:* IEEE, VDE, DPG.

Kindl, Fred H
Business: P O Box 714, Schenectady, NY 12301
Position: President. *Employer:* Encotech Inc. *Education:* BSME/Mech Eng/Carnegie Inst of Tech *Born:* July 4, 1920. BSME Carnegie Inst Tech. Elected to Tau Beta Pi, Fellow Grade ASME; N Y St Prof Engr. 22 yrs experience in Power Generation field with G E Co: Thermal Power Sys Div 4 yrs - Gas Turbine cycle studies, component design, combustion dev; Steam Turbine Generator Business 14 yrs - control sys & component design, mgmt of lab activities, Mgr of Generator Mech Design; Gas Turbine Business 4 yrs - Mgr Engrg, Medium Gas Turbine Dept. President Encotech Inc cons engrg firm in Power Generation field 14 yrs. *Society Aff:* ASME, ASM, ASTM

Kindle, Kenneth W
Business: 3218 N. Fourth St, Longview, TX 75601
Position: Pres *Employer:* Kindle, Stone & Assoc, Inc *Education:* BS/CE/TX A & M Univ *Born:* 08/31/35 Native of Longview, TX. Presently Chrmn of the Bd and Pres of Kindle, Stone & Assocs, Inc - held since 1978. Prior, was Exec VP and Partner of Hart Engrg Co of Longview, TX. Experience prior to 1976 included City Engr for City of Longview, TX and Exec VP/Partner of B. L. Nelson Assocs, Consulting Engrs, Dallas, Tx. Presently serving as Pres of local chapter of TSPE and State Dir of the TX Section of the ASCE serving on ACEC/EPA Liason Committee and past chrmn of the Membership Committee, Consulting Engrs Council of TX. Also mbr of the Bd of Dir of CEC-TX. Presently serving as bd mbr of First Natl Bank of Longview and Chrmn of the Bd for the Leadership Longview Program of the Longview Chamber of Commerce. Served as Pres of NE-TX Chapter of ASCE in 1980-81. Enjoys golf, snow skiing and reading. *Society Aff:* ASCE, ASTM, NSPE, WPCF, ACEC

King, Albert I
Bioengrg Center, 418 Health Sciences Bldg, Detroit, MI 48202
Position: Prof & Dir *Employer:* Wayne State Univ *Education:* PhD/ME/Wayne State Univ; MS/Engrg Mech/Wayne State Univ; BS/CE/Univ of Hong Kong *Born:* 6/12/34 Joined the Wayne State Univ faculty in Sept 1966 as an Assistant Prof of Engrg Mechanics. From 1971-1976 he was the recipient of a Research Career Development Award from the National Inst of General Medical Sciences. The principal areas of Dr King's research are low back pain, human response to impact acceleration, vehicular safety, and clinical biomechanics. He is a member of the Intl Society for the Study of the Lumbar Spine, Sigma Xi, & Tau Beta Pi. In 1980 received the Charles Russ Richards Memorial Award from the ASME. In 1984 received the Volvo Award for res in low back pain and was elected to Fellow of ASME. He is principal investigator of a CDC Ctr Grant for Injury Prevention Res, one of five Ctrs in the U.S. *Society Aff:* ASME, SAE, ASEE, AAOS, ORS.

King, C Judson
Business: Dept of Chem Engrg, Berkeley, CA 94720
Position: Prof of Chem Engrg & Dean, College of Chemistry. *Employer:* University of California. *Education:* SM/Chem Engg/MIT; ScD/Chem Engg/MIT; BE/Chem Engg/Yale Univ. *Born:* Sept 1934. Asst Prof MIT 1959-63 & Dir Bayway Station Practice Sch; Asst Prof, Assoc Prof & Prof Univ of Calif, Berkeley 1963- ; V Chmn (1967-72) & Chmn (1972-81) Chem Engrg Dept; Dean, College of Chemistry (1981-). Interests in separation processes, food dehydration & concentration & process design & synthesis. Author of 'Separation Processes' McGraw-Hill 1971; 2nd ed, 1980 & 'Freeze-Drying of Foods' CRC Press 1971. AIChE Inst Lectr 1973; Food, Pharmaceutical & Bioengrg Div Award 1975; William H Walker Award 1976. ASEE George Westinghouse Award 1978, National Academy of Engineering, 1981. Scoutmaster, Troop 100 BSA Kensington Calif (1975-1986). Afficionado of Sierra Nevada mountains. *Society Aff:* AIChE, ASEE, ACS, AAAS.

King, Cecil N
Home: 415 Amberidge Trail NW, Atlanta, GA 30328
Position: Owner *Employer:* King Consulting Service *Education:* MS/Automotive Engr/Chrysler Inst of Engr; BSME/ME/Auburn Univ; BS in Eco/Accounting/ Samford Univ. *Born:* 4/26/19. and originally educated as accountant in Birmingham AL. Married - Active Christian - served USAAF, WWII as heavy bombardment maintenance officer attained zero losses of assigned aircraft during tour. Reeducated as engr while serving univ as accounting instructor. Tau Beta Pi. Managed Chrysler Foundries; began consulting and managing foreign foundries in 1960. Asst Prof of Mgt-Lawrence Inst of Tech between foreign assignments. Recently completed complete tooling for foreign Hi-Production Foundry. Presently consulting as independent. Specialties: organization; training; mgt systems; foundry tooling and processes. Avocations: studying, singing, camping, golf. Publications: Modern Casting-1957 - Foundry, 1969. *Society Aff:* SAE, AFS.

King, Donald D
Home: 920 Hardscrabble Rd, Chappaqua, NY 10514
Position: President. *Employer:* Philips Labs Div-N Amer Philips Corp. *Education:* PhD/Physics/Harvard; AM/Physics/Harvard; AB/Eng Sci/Harvard. *Born:* Aug 1919. Served as Asst Prof at Harvard & lectr in EE at Hopkins. Was Dir of Radiation Lab Hopkins 1955-56, V Pres Res of Electronic Communications 1956-64 & Dir of Electronics Res Lab Aerospace Corp 1964-67; has been Pres Philips Labs Div North Amer Philips Corp 1968- . Res publs have been in microwaves, antennas & electronic sys. Hon Life Mbr of the Soc of Microwave Theory & Techniques, IEEE & Dir Indus Res Inst. Current interests are in internatl res mgmt. *Society Aff:* IEEE, APS, AAAS, IRI.

King, Franklin G
Business: Professor and Chairman, Dept. of Chemical Engineering, North Carolina a & T State Univ, Greensboro, NC 27411
Position: Prof & Chrmn *Employer:* NC A&T State Univ. *Education:* BS/Chem Engg/ Penn State; MS/Chem Engg/KS State; DSc/Chem Engg/Stevens Inst of Tech; M Ed/ Education/Harvard Univ *Born:* 9/23/39 Native of Mahanoy City, PA. Taught at Howard Univ (13 years) and Lafayette Coll (6 years) before joining NC A&T State Univ. Industrial experience with Uniroyal and American Cyanamid. Consultant to Maxwell House, Gen Elec, Western Elec, Union Carbide, and National Institutes of Health. Author of papers on pharmacokinetics, spandex fibers, bacterial fermentation, self-paced instruction and learning modules. Interested in computer control, biotech for energy production and pollution abatement, kinetics of anti-cancer drugs and environmental toxicants, coal conversion, and the education of minority engrs. Enjoy tennis, camping, country music. *Society Aff:* AIChE, ASEE

King, James R
Home: Currier Rd, Tamworth, NH 03886
Position: Conslt *Employer:* Self *Education:* BS/Textile Engrg/Lowell Technological Inst Now U/Lowell (MA) *Born:* 10/17/23 Native of Lowell, MA. Employed in quality control and reliability engrg by Merrimack Mfg Co, Polaroid, CBS Elecs, Honeywell, Avco RAD, Rockwell Intl and Sprague Elec prior to establishing TEAM Conslts. He is the author of PROBABILITY CHARTS FOR DECISION MAKING and FRUGAL SAMPLING SCHEMES & papers applying statistical distribution theory to life testing and complex engrg problems. He has taught statistical subjects at LTI, Northeastern Univ and SUNH. He is currently a mbr of the editorial bds of the Amer Journal of Mathematical and Mgmt Scis, and Reliability Engineering. Reg PE NH. *Society Aff:* ASQC

King, James R, Jr
Business: 910 Collier St, Fort Worth, TX 76102
Position: Chairman and CEO *Employer:* Rady & Associates Inc. *Education:* BS/Civil Engg/Univ of TX. *Born:* Mar 1928 Moran Tex. Resident Engr Amarillo Texas for Freese & Nichols 1952-53; 1953- , Rady & Assocs Inc, design engr for water & sewerage facilities, Pres 1964- . P Pres of Cons Engrs Council of Texas; P Pres Texas Sect ASCE. Mbr ASCE, NSPE, TSPE, ACEC, Water Pollution Control Federation, Amer Soc of Military Engrs, Chi Epsilon, Tau Beta Pi, Special Hon Mbr Chi Epsilon, UT-Arlington. Mbr Rotary Club of Fort Worth, Fort Worth Club, Ridglea Country Club. Hobby: golf. *Society Aff:* ASCE, NSPE, ACEC, WPCF.

King, Joe H
Business: 12800 Lynn Townsend Dr, Detroit, MI 48288
Position: Mgr, Met Engg Dept. *Employer:* Chrysler Corp. *Education:* MBA/Bus/MI State Univ; MAE/Auto Engg/Chrysler Inst Engg; BSMetE/Met Engg/MI State Univ. *Born:* 9/25/27. Resident of MI. Entire career at Chrysler Corp, Engg Div. Mgr, Met Engg Dept since 1971. Responsible for Met Specifications, Failure Analysis, Corrosion, Electroplating and Sheet Metal Engg in the Corp Product Engg Function SAE - ISTC Exec Committee, Chapter Chrmn ASNT '70-'71. ASM - Advisory Tech Awareness Committee - '71-'79. MVMA - Engg Mtls Committee, Tau Beta Pi. *Society Aff:* ASM, SAE.

King, Joe J
Home: 3708 Chevy Chase, Houston, TX 77019
Position: Business Mgt/Engrg Cons (Self-emp) *Education:* BS/ME/Univ of TX. *Born:* July 15, 1901 Native of Waco, TX. AUS Office Ch of Engrs 1942-45, respon military pipelines overseas. Pioneer 1925 in design of high pressure welded gas & liquid pipelines. Pioneer 1940 in design of high pressure natural gas cycling installations. Pioneer 1946 in design of elec & gas turbine operated centrifugal compressor facilities for high pressure pipelines. Early participant in natl gas pipelines codes. Recipient Distinguished Engr Grad Award UT 1964. Recip ASME J Hall Taylor Award 1975. Ret Exec Tenneco Inc, Tenneco Chems & Petro Tex Chem Corp. *Society Aff:* ASME.

King, John E
Business: 422 W Sixth St, Lansdale, PA 19446
Position: Sales Manager. *Employer:* Precision Rebuilding Corp. *Education:* BS/Metallurgy/PA State Univ. *Born:* Sep 1919. Served as Lt USNR Ordnance on carrier aircraft. On US Army Ordnance list of Engrg Know How 1947-57. One of six engrs on dev of cold extrusion of steel in the US. Past Chap Chmn of Phila Chap Amer Soc for Metals; served on Natl Engrg Tech Ctte for Amer Soc for Metals. 1968-72 managed div & helped extend the inertia welding process using Caterpillar equip. Past Bd Pres of the North Penn YMCA. Enjoy bridge, tennis, golf & swimming. Reg Prof Engr of the State of Pa. *Society Aff:* ASME.

King, L Ellis
Business: UNC-Charlotte, Dept of Civil Engrg, Charlotte, NC 28223
Position: Prof & Dept Chrmn. *Employer:* Univ of NC at Charlotte *Education:* DrEngr/Transportation/Univ of CA, Berkeley; MS/Transp & Soil Mech/NC State Univ; BS/CE/NC State Univ. *Born:* 8/21/39. Native of Jamestown, NC. Was on faculty of WV Univ, Univ of CO, and Wayne State Univ before assuming present position as Prof & Chrmn, Dept of Civil Engrg, Univ of NC at Charlotte. Res experience has included projs in the areas of transportation and traffic engg, hgwy lighting and signing, human factors and transportation planning. Consulting experience has been in the same areas. Was awarded the Walter L Huber Civ Engg Res Prize of the

King, L Ellis (Continued)
Am Soc of civ Engrs in 1973. Enjoys classical music, reading, photography and classic automobiles. *Society Aff:* ASCE, ITE, NSPE, HFS, ASEE, ORSA.

King, Myron D
Business: 1009 MacDonald Ave, Richmond, CA 94801
Position: President *Employer:* Arlington Financial Corp; Arlington Engineers *Education:* AA/Mgt/Contra Costa College. *Born:* 5/1/26. Over twenty five yrs of Indus Engrg & Mgmt Consulting with both large & small organizations resulting in the formation of Arlington Engineers 15 yrs ago. Related endeavors include Myron D King Assoc, Mgmt Conslt, Principal; Arlington Financial Corp, Financial Conslt; Reg PE, State of CA. Past Pres, AM Inst of Ind Engrs, San Francisco - Oakland. Instructor in Indus Engrg and Indus Supervision, State of CA. *Society Aff:* NSPE, IIE, AEE, CSPE, IAFP.

King, Paul H
Business: College of Engineering, Northeastern University, Boston, MA 02115
Position: Dean, College of Engineering. *Employer:* Northeastern University *Education:* BS/Civil Engg/CA Inst of Technology; MS/Environ Engg/CA Inst of Technology; PhD/Environ Engg/Stanford Univ. *Born:* July 1936. Served as engr with US Public Health Service 1957-59. Sanitary Engr, Brown & Caldwell 1959-60. Previous fac experience at Univ of Ky 1961-63. Fac mbr at VPI & SU 1966-79. Admin Chmn of Eviron Engrg program at VPI. Active in res in physical-chemical water & wastewater treatment processes. Pres, AEEP 1976; ASCE Huber Res Prize 1976. Prof & Hd, Dept of Civil Engg, Univ of AZ 1979-86. Now Dean, College of Engineering, Northeastern University, Boston, MA. *Society Aff:* ASCE, WPCF, AWWA, AEEP, TBΠ, ΣΞ, ASEE.

King, Paul H
Business: Box 1631, Station B, Nashville, TN 37235
Position: Biomedical Engr *Employer:* Vanderbilt Univ *Education:* PhD/ME/Vanderbilt Univ; MS/Engrg/Case Inst of Tech; BS/Engrg Sci/Case Inst of Tech *Born:* 9/4/41 Twenty-three years teaching experience in Mechanical and Biomedical Engrg. Research experience in basic instrumentation, metabolism, cardiology, orthopedics, nuclear medicine and anesthesiology. Currently working on Computer assisted monitoring in the operating room (anesthesiology). One patent, 31 publ PE. *Society Aff:* AAMI, ASEE, Mensa.

King, Randolph W
Home: 1312 Kinloch Circle, Arnold, MD 21012
Position: Consultant on Public Service matters *Employer:* Self *Education:* Naval Engg/Nav Arch & Marine Engg/MIT; BS/-/US Naval Acad. *Born:* May 1923. Undergrad Stanford Univ. Served 35 yrs in US Navy. Extensive wartime experience at sea. Selected as engrg specialist. Design proj officer for hydrofoils & missile warships. Hd Engrg Dept, Naval Acad. Mg'd major res programs & dir large res complex. Operated naval shipyard for repair & conversion surface ships & submarines. Completed active duty 1975 grade of Rear Admiral. Cons in marine matters. Exec Officer, National Academy of Engineering. Currently engaged in extensive advisory public service activities. Active in Soc of Naval Archs & Marine Engrs. Pres, Amer Soc of Naval Engrs 1975. *Society Aff:* ASNE, SNAME, IPEN, USNI.

King, Reno C
Home: 303 Grosvenor Street, Douglaston, NY 11363
Position: Assoc Dean of Admin - Retired. *Employer:* Queensborough Community College - Retired. *Education:* DSc/Appl Mech/Stevens Inst of Tech; MME/Mechanical Engrg/NY Univ; BME/Mech Engrg/Cooper Union *Born:* Sept 1917. Native of Va; lives in NYC (since 1938). Mech Proj Engr, Ebasco Services Inc 1948-55; Prof of Mech Engrg NYU 1955-68; Prof Mech Engrg & Assoc Dean, Queenborough Community Col 1968-79. Editor, Piping Handbook; Mbr, Bd/Examiners of Prof Engrs & Land Surveyors 1959-69. *Society Aff:* ASME

King, Richard L
Business: 645 Semmes Street, Memphis, TN 38111
Position: President. *Employer:* King Engineering Cons, Inc. *Education:* BS/Civil Engg/Tufts College. *Born:* Jan 7, 1923. Native of Paris, Tennessee. Lt in USNR Civil Engr Corps. Worked for Allen & Hoshall, Cons Engrs in Memphis for 6 yrs. Est proprietorship in 1953 in Memphis. Formed King Engrg Cons, Inc in 1970. Reg Engr in 11 states. Specializes in municipal service. P Pres Cons Engrs of Memphis. Mbrship in ACEC, ASCE, NSPE, TSPE, TWWA. *Society Aff:* ACEC, ASCE, NSPE, AWWA.

King, Robert W
Business: 270 Park Ave, NY, NY 10017
Position: Pres Medical Products Div. *Employer:* Union Carbide Corp. *Education:* MS/ChE/MIT; BS/ChE/Univ of Denver. *Born:* 9/26/18. Bob King joined Union Carbide in 1942 as an engr at the Corp's chem facility in S Charleston, WV. He served in mgt positions there until 1960, when he moved to the NY Office as Dir of Engg for the then Chem Div. He remained with that Div in mgt positions until 1965 when he moved to the Linde Div as a VP & Gen Mgr of the Cryogenic Products Dept. In Jan 1974, he became a Sr VP of Linde. In Jan 1977, he became Dir of Corporate Dev, Union Carbide Corp. Mr King was appointed Pres of the newly-formed Medical Products Div in Apr 1978. *Society Aff:* AIChE.

King, Ronold W P
Home: 92 Hillcrest Pkwy, Winchester, MA 01890
Position: Prof, Consultant. *Employer:* Harvard Univ; self-employed. *Education:* PhD/Electrodynamics/Univ of WI; SM/Physics/Univ of Rochester; AM/Hon/Harvard Univ; AB/Physics/Univ of Rochester. *Born:* Sept 1905. Grad work at Univ of Munich Germany & Cornell Univ. Taught at Univ of Rochester, Univ of Wisc, Lafayette Col, Harvard Univ. On Harvard fac 1938-72 when became Prof Emeritus. Author of 10 tech books; articles in Encyclopaedia of Physics, Colliers Encyclopaedia, Dictionary of Physics Encyclopaedia of Science and Technology; over 200 tech & sci publs in journals. Guggenheim Fellow; Life Fellow IEEE; Fellow APS; Corres Mbr Bavarian Acad Sci; Mbr AAAS, Amer Acad Arts & Sci, URSI Commission B. Distinguished Service Award, Univ of Wisc Sch of Engrg; Eminent Mbr Award Eta Kappa Nu; IBM Distinguished Scholar, Northeastern Univ, 1985; Harold Pender Award, Univ of Penn Moore School, 1986. *Society Aff:* IEEE, APS, AAAS, URSI, ΦBK, ΣΞ

King, T A
Home: 1414 East West Hwy, Adelphi, MD 20783
Position: Chief Engineer, Navigation *Employer:* Dept of Navy (SSPO-2401). *Education:* MS/Opticts/Univ of Rochester; BS/Physics/Univ of Rochester. *Born:* March 1923. Univ of Rochester Res Asst 1949-51 Scintillation Counters, Infrared Sys; Commerce Dept (Natl Bureau of Standards) 1951-54 - Basic Characteristics of Optical Matls; Navy (SHIPS) 1954-56 Navigation Equip ; Navy Dept (POLARIS/POSEIDON/TRIDENT FBM Strategic Weapon Systems) 1957-present. Computer, Inertial, Electronic, Satellite, Sonar, Geophysics, Navigation Sys Dev, Acquisition, Logistics; Navy Superior Civilian Service Awards; Harvard Advanced Mgmt Prog 1969 (AMP 57); Senior Executive Service; Pres Stratford Mgmt Corp; Business/Financial Cons 1973-date; Civic org pres; transp cons (Md DOT); Wild Rivers Comm. (Md). *Society Aff:* AAAS, MTS.

King, Thomas B
Business: Rm 8-106, Cambridge, MA 02139
Position: Professor of Metallurgy. *Employer:* Massachusetts Inst of Technology. *Education:* PhD/Metallurgy/Glasgow Univ; BSe/Physical Chem/Glasgow Univ. *Born:* April 27, 1923 Motherwell, Scotland. BSc 1st Class Honors 1945; PhD 1950 U of Glasgow, Scotland. Lectr U of Strathclyde 1949-53. Asst Prof 1953-57, Assoc Prof 1957-61, Prof 1961- ; Hd of Dept, Met & Matls Sci 1962-72 Mass Inst of Tech. Naturalized 1959. Prof societies: Amer Inst of Mining & Met Engrs (AIME) Dir, Met Soc 1964-67; Amer Soc for Metals (Sauveur Lectr 1976); Honors: Fellow

King, Thomas B (Continued)
Amer Acad of Arts & Sci 1965; Fellow Met Soc of AIME 1975; Mbr Sigma Xi; Hon Mbr Alpha Sigma Mu 1964. Res ints & pubs: properties of silicates - diffusion; gas solubility, anionic structure, electrochemical kinetics at high temps; reaction; plasma-are meltings between gases, slags & metals; high-gradient magnetic separation; plasma-are melting. *Society Aff:* TMS-AIME, ASM.

Kingery, William D
Business: Rm 13-4090, Cambridge, MA 02139
Position: Kyocera Prof of Ceramics. *Employer:* Massachusetts Inst of Technology. *Education:* ScD/Ceramics/MIT *Born:* Joined the MIT fac in 1951 becoming full Prof in 1962. Apptd Kyocera Prof 1984. His work has had a major influence in ceramic ed & ceramic sci throughout the world, & he has made significant contribs to related fields such as ceramic tech, glaciology, met, matls sci & archaeology. Elected a mbr of the Natl Acad of Engrg in 1975, being cited for 'leadership in the sci & engrg of ceramic matls, spanning the whole spectrum of physical phenomena, struct property relationships, innovative processing & applications to modern techs'. *Society Aff:* ACerS, NAE, AAAS.

Kingsbury, Herbert B
Home: 500 Stamford Drive, Newark, DE 19711
Position: Prof Dept Mech Eng *Employer:* Univ of DE *Education:* PhD/Eng Mech/Univ of PA; MS/Mech Eng/Univ of PA; BS/Mech Eng/Univ of CT *Born:* 02/15/34 Prof Kingsbury has previously been employed by Gen Elec Co Missile and Space Div and the Dept of Aerospace Engrg at the PA State Univ. Since 1968, Prof Kingsbury has been a mbr ot the faculty of the Dept of Mech and Aerospace Engrg at the Univ of DE. He teaches undergrad and grad courses in Dynamics, Solid Mech, Structural Mech, and Bioengrg. His res interests include biomech, mech of porous fluid-filled solids, structural dynamics and struct damping. He is an author of fifty proceedings and journal publs and of more than sixty major tech reports. *Society Aff:* ASME, SES, AAM, ASB

Kingsbury, Howard F
Business: Dept of Arch Engrg, University Park, PA 16802
Position: Associate Professor. *Employer:* The Pennsylvania State University. *Education:* MS/Ceramic Sci/Penn State; BS/Glass Tech/Alfred Univ. *Born:* March 13, 1922. Reg Prof Engr Pa. Native of Elmira, N Y. Prior to joining Penn State in 1964 was with Pittsburgh-Corning Corp, last as mgr of Acoustics Lab. Present respons are teaching, res & cons in acoustics, primarily in bldgs; P Dir, Inst of Noise Control Engrg (INCE); Chmn, Ctte E-33 on Environ Acoustics of ASTM; Mbr & P Chmn of Tech Ctte TC 2.6, Sound & Vibration, ASHRAE; Exec Ctte, Engrg Acoustics Ctte, ASEE; Acoustical Soc of Amer. Outside ints incl local gov elected official, golf & hunting. *Society Aff:* ASTM, ASA, ASHRAE, NCAC, INCE.

Kingsbury, James E
Business: Code EA01, Marshall Space Ctr, AL 35812
Position: Dir, Sci & Engrg Directorate. *Employer:* NASA-George Marshall Flight Ctr. *Education:* BS/EE/Penn State. *Born:* 1928. Ed in Public Schs Wilkes-Barre, Pa. A Matls Dev Engr at Redstone Arsenal from 1951-60 specializing in high temp matls applications, welding & NDT dev. At NASA, Marshall Space Flight Ctr, a Dev Engr - Matls 1960-69 & Mech Sys Ch Engr, Skylab Program from 1969-73. Received NASA Exceptional Service Medal in 1969; NASA Distinguished Service Medal in 1973 and the AIAA Huntsville Sect Oberth Award in 1974. Presidential Award for Meritorious Service, 1980; AIAA Fellow 1981; NASA Distinguished Service Medal 1981; AAS Fellow, 1977. *Society Aff:* AIAA, AAS

Kingslake, Rudolf
Home: 56 Westland Ave, Rochester, NY 14618
Position: Emeritus Professor of Optics *Employer:* University of Rochester. *Education:* DSC/Tech Optics/Imperial College of Sci & Technology; MSc/Tech Optics/Imperial College of Sci & Technology; BSc/Tech Optics/London, England; Hon DSc/Optics/Univ of Rochester. *Born:* 1903. Attended the Dept of Tech Optics at the Imperial Col, London. Came to US in 1929 to help found the Inst of Optics at the Univ of Rochester. In 1937 joined Eastman Kodak Co as a lens designer & became hd of the lens design dept in 1939. Retired 1969. Received Progress Medal of SMPTE in 1964 & the Ives Medal of the Optical Soc of Amer in 1973 (Pres 1947-49). Gold Medal of SPIE in 1980 Author "Lenses in Photography–, "Lens Design Fundamentals" and "Optical System Design–. Editor 1965 'Applied Optics & Optical Engrg'(5 vols). Has written over seventy tech papers, mainly on lens design & optical engrg. *Society Aff:* OSA, SPIE, SPSE, SMPTE.

Kingston, Robert H
Business: MIT Lincoln Lab, Lexington, MA 02173
Position: Lincoln Lab Adj Prof, Dept. EECS *Employer:* MIT. *Education:* PhD/Physics/MIT; MS/Elec Engg/MIT; BS/Elec Engg/MIT. *Born:* Feb 13, 1928. Mbr Technical Staff, Bell Labs 1951-52. MIT Lincoln Lab 1952-present. Hd, Optics Div 196972. Ldr, Infrared Radar Group 1972-77. Sr Staff, 1977-present. Adjunct Prof Dept. EECs, M.I.T., 1986-pres. Visiting Assoc Prof EE Dept Stanford 1964-65. Specialties: Semiconductor physics, masers, lasers, optical & infrared detection. Editor, IEEE. Journal of Quantum Electronics 1965- 70. Chmn Special Group on Optical Masers DOD 1962-66. Lectr MIT EE Dept 1975. Fellow Amer Phys Soc; Fellow, IEEE. IEEE Centenial Medal of Honor, 1984. Fellow, Optical Society of America. Elected Town Meeting Mbr Lexington, Mass 1958-77. Former mbr, Appropriations Ctte, former mbr & chmn Capital Expenditures Ctte. Editor Semiconductor Surface Physics, Univ of PA Press, 1957 Author, Detection of Optical & Infrared Radiation, Springer-Verlag, 1978. *Society Aff:* IEEE, APS, OSA.

Kinnen, Edwin
Business: Dept of Electrical Engrg, Univ. of Rochester, Rochester, NY 14627
Position: Prof *Employer:* Univ of Rochester *Education:* PhD/EE/Purdue Univ; MS/EE/Yale Univ; BS/Engrg/Univ of Buffalo *Born:* 3/9/25 Industrial Positions: NY State Electric & Gas Corp, 49-50; Westinghouse Research Lab, 50-55; Consultant, 55-78. Other faculty positions: Univ of Pittsburgh, 52-55; Purdue Univ, 55-58; Univ of MN, 59-63. PE Registration: NY State. Over 70 tech papers & 15 tech reports in fields of automatic controls, nonlinear stability, biomedical engrg., computer aided design. Westinghouse Fellow, 56-57; Japan Society for promotion of science-visiting prof, Hokkaido Univ, 70; NIH Special Fellow, Univ of WA, 71; Netherlands Science Research (Zwo) Scientist, State Univ of Utrecht, 78. *Society Aff:* AAAS, IEEE, BME, ΣΞ, RESNA.

Kino, Gordon S
Business: Ginzton Lab, Stanford, CA 94305
Position: Professor, Electrical Engineering. *Employer:* Stanford University. *Born:* June 15, 1928 Melbourne, Australia. Prof, Elec Engrg Stanford Univ. BSc & MSc (mathematics) London Univ England; PhD (elec engrg) Stanford Univ. He has worked on microwave tubes, electron guns, plasmas, Gunn effect, microwave acoustics & acoustics imaging devices for medical instrumentation & nondestructive testing. Pub approx 150 papers. Guggenheim Fellow 1967, IEEE Fellow, Fellow Amer Phys Soc, Mbr Natl Acad of Engrg & mbr Matls Res Council DARPA.

Kinser, Donald L
Business: P.O. Box 1689-B, Nashville, TN 37235
Position: Prof *Employer:* Vanderbilt Univ *Education:* PhD/Mat Sci/Univ of FL; BS/Metallurgy/Univ of FL *Born:* 9/28/41 Native Tennessean. Joined Vanderbilt Univ in 1968, full professorship since 1976. Participant in USDOE-ORAU high level radioactive waste program; awarded NASA Certificate of Recognition in August, 1976 and again in April, 1980. Extensive consulting experience in failure analysis. Presently involved in DNR supported study of radiation effects on glasses. Appointed Fulbright Scholar to Syria, Middle East 11/81 - 1/82; NASA-LDEF Principal Investigator on Earth orbit radiation effects upon glass, 1980-84. Investigator on Free Election Laser Effects on Materials (NRL) Consultant to USIA. *Society Aff:* ACS, ASM, SGT(UK), SPIE.

Kinsey, Paul R
Business: Federal Paper Bd Co, Riegelwood, NC 28456
Position: Mgr of Manufacturing Services. *Employer:* Federal Paper Board Co, Inc. *Education:* MS/Pulp & Paper Tech/Univ of ME; BS/ChE/Purdue Univ. *Born:* June 22, 1929 native of Hammond, Ind. BS in Chem Engrg from Purdue Univ; MS in Pulp & Paper Tech from Univ of Maine. Joined Union Bag in 1951. Joined Federal Paper Board (formerly Riegel Paper Corp) in 1956. Current pos Mgr of Mfg Services. Have served in this capacity since April 26, 1965. Have served in many capacities in TAPPI, incl Bd/Dir. Active in Boy Scouts of Amer. Have been scoutmaster, ctte chmn, post advisor, Council Bd/Dir; currently District Scouting Chmn. Outside hobbies incl golf, camping, coin & stamp collecting. *Society Aff:* TAPPI.

Kinstler, John R
Business: 4000 Collins, Lansing, MI 48910
Position: Vice President Engineering *Employer:* Motor Wheel Corp. *Education:* BS/Phys/Case Inst of Tech. *Born:* 7/9/48. Native of Milwaukee, WI. Started with Goodyear tire & rubber in 1970 as design engr at metal products div. Transferred to Motor Wheel Corp in 1972 as staff product engr - truck wheels. 1974 - sr product engr; 1976 - mgr - truck rim & wheel engg; 1978 - mgr - truck rim, wheel & brake engg. 1979 - mgr advanced concepts. 1980 - Mgr Passenger Car Wheel and Brake Engineering. 1981 - Chief Engr-Product Engrg, 1983 - Dir of Mfg Engrg. Named VP Engrg 1985. Chmn Intnatl Standards SAE Subcttee; Chmn SAE Bolt Pattern Task Group. Mbr SAE truck wheel subcttee & Pass Car Wheel Subcttee. Holds several patents on wheels. Was 1986 Chmn Mid-Michigan Section SAE. *Society Aff:* SAE.

Kintigh, John K
Business: P.O. Box 8405, Kansas City, MO 64114
Position: Partner *Employer:* Black & Veatch Cons Engr *Education:* BS/ME/Univ of Rochester *Born:* 11/30/24 Associated with Black & Veatch Cons Engrs 35 yrs, partner in firm 27 yrs. As proj mgr in Power Div, have managed design and construction of a number of very large coal and gas fueled electric generation stations. Special assignments have included work in connection with power plant site studies and selection, electric and steam system studies, air pollution control studies, facilities design, and testing, feasibility studies, water supply investigations, dist steam sys studies, street lighting studies, and investigations including economic evaluation studies of nuclear power, solar power and geothermal power. Reg PE in six states. *Society Aff:* ANS, ASME, ISES, NSPE, MSPE

Kinyon, Gerald E
Business: 525 West Washington St, South Bend, IN 46601
Position: Chief Structural Engr *Employer:* Lawson-Fisher Assocs *Education:* MS/BA/IN Univ; BS/CE/Purdue Univ *Born:* 10/27/39 Native of South Bend, IN. Served in US Army Corps of Engrs 1963-1967, including one year as a Combat Engr co commander in the Republic of Viet Nam. Honorably Discharged in 1967 after attaining the rank of captain. Bridge and building structural design engr with Chas W. Cole & Son, Inc 1967-1973. Supvr Bridge Design Section 1973. Joined Lawson Assocs as project engr in 1974. Chief structural engr Lawson-Fisher Assocs in 1977. Pres of IN Section ASCE 1980-1981. Hobbies: running, gardening & carpentry. Interests: energy conservation & economics. *Society Aff:* ASCE, AISC, NSPE

Kinzel, Augustus B
Home: 1738 Castellana Rd, La Jolla, CA 92037
Position: Consultant. *Employer:* Self. *Education:* DSc/Chemistry/Univ of Nancy, France; DMetIng/Metallurgy/Univ of Nancy, France; BS/Genl Engg/MIT; AB/Mathematics/Columbia. *Born:* July 26, 1900. 9 hon degrees. Mbr NAS, Founding Pres NAE, Amer Philosophical Soc, P Pres AIME, EJC. Salk Inst, Union Carbide & Carbon Res Lab, past V P Res Union Carbide Corp. Washington Award. P E in N Y & Calif. James Douglas Medal & many others. Hon Mbr AIME & ASM. Howe Lectr & many others. Over 100 pubs; over 60 pats; author Alloys of Iron & Chromium. *Society Aff:* AIME, ASM, NAS, NAE.

Kipp, Raymond J
Business: 1515 W Wisconsin Ave, Milwaukee, WI 53233
Position: Dean, Coll of Engr. *Employer:* Marquette Univ. *Education:* PhD/Environmental Engr/Univ of WI; MS/Hydraulic Engr/Univ of WI; BS/Civ Engr/Marquette Univ. *Born:* 12/7/22. Native of Ossian, IA. US Navy in WWII, Air Medal with 2 clusters. Consultant in environmental. 1972 Engr of the Year Award from Engrs and Scientists of Milwaukee. Governor appointed mbr of Metropolitan Sewerage Dist. 31 yrs in education of engrs. *Society Aff:* ASCE, NSPE, WPCF, AWWA, AAEE, ASEE, ESM.

Kirby, Ralph C
Home: 116 Southwood Avenue, Silver Spring, MD 20901
Position: Chief Engineer *Employer:* US Bureau of Mines. *Education:* BChemE/Chem Engg/Catholic Univ of America. *Born:* July 21, 1925 Washington, DC. US Army 1943-45 incl serv with 1st Infantry Div in Belgium & Germany. Joined Bureau of Mines 1950 for extractive met R&D. Respon early 1960's for cost estimates on processes for producing met- grade alumina from US resources. Assigned 1965 to tech mgmt (staff) of Met activity of the Bureau of Mines. From 1972-76, respon for dir of major staff div that aided Asst Dir, Met in planning, dev & managing the progs & opers at eight field facilities as well as contract work concerned with R&D on processing & utilizing metals & minerals. Asst Dir, Met July 1976-79, Dir, Mineral Resources Technology, July 1979- 82, managed & directed fundamental Bureau of Mines R&D program covering entire mineral cycle (functionally both mining & metallurgy). Asst Dir, Planning and Budget, June 1982-84, responsible for Bureau of Mines program policy, planning system, and budget preparation; Chief Engr, July 1984- , top engrg counselor to Dir for planning, evaluating, and guiding Bureau of Mines engrg programs. Bd/Dir of The Met Soc, AIME 1976-79. Mbr AIChE, ASM, Sigma Xi, AAAS, Washington Society of Engineers. *Society Aff:* AIME, AIChE, AAAS, ΣΞ, ASM

Kirby, Robert E
Home: 336 Presidential Way, Guilderland, NY 12084
Position: Regional Fed Hwy Admin. *Employer:* US Government - Dept of Transp. *Education:* BSCE/Civil Engg/Northeastern Univ. *Born:* Jan 19, 1924 Milford, Mass. Ed: BSCE from Northeastern Univ Boston, Mass; Hwy Mgmt Inst, Univ of Mississippi. Pos: Respon for admin of hwy & other transp programs in Fed Hwy Admin, Region One (New England, New York, New Jersey, Puerto Rico, Virgin Islands). Mbrships: Amer Pub Wks Assn, Amer Assn of St Hwy & Transp Officials, NE Assn of St Hwy & Transp Officials (Bd/Dir), Fed Exec Assn of NE New York (Past Pres), TriState Regional Planning Comm (Commissioner). Two Fed Hwy Outstanding Performance Awards. Resident Guilderland, N Y. Military Awards: Air Force Distinguished Flying Cross, 5 Air Medals. *Society Aff:* AASHTO, NASHTO, APWA.

Kirchmayer, Leon K
Business: GE Co, 1 River Road, Schenectady, NY 12345
Position: Mgr, Advanced Sys Technology & Planning. *Employer:* General Electric Company. *Education:* PhD/Elec Engg/Univ of WI; MS/Elec Engg/Univ of Wi; BS/Elec Engg/Marquette Univ. *Born:* July 1924. Grad Sr Exec Program MIT. With GE as Mgr AST&P, respon for dev of advanced concepts of elec utility planning, operation & control. Books: Author of two, co-author of one, co-editor of three, in field of elec engrg. Author of 96 Tech Papers. Credited with several pats relating to computer control of power sys. Officer in IEEE, ASME, Natl Soc of Prof Engrs (named Engr of Year for local chap 1966), IFAC, AAC, CIGRE; Mbr of ORSA. Fellow in IEEE & ASME; Awarded Distinguished Service Citation by Univ of WI, Mbr, Natl Acad of Engg. *Society Aff:* IEEE, ASME, NSPE, ORSA, CIGRE.

Kirchner, Walter L
Business: PO Box 1663, MS K575, Los Alamos Natl Lab, Los Alamos, NM 87545
Position: Program Mgr *Employer:* Los Alamos Natl Lab *Education:* PhD/Nucl Engrg/MIT; MS/Nucl Engrg/MIT; BS/Marine Engrg/US Merchant Marine Acad *Born:* 09/21/47 Program Mgr, Defense Reactors, Los Alamos Natl Lab, Los Alamos,

Kirchner, Walter L (Continued)
NM. Tech mgr for innovative reactor design projects for special military power applications. Previously supervised groups working in advanced reactor design and diverse aspects of reactor safety res (numerical fluid dynamics, experiments, analysis, and advanced instrumentation). Conslt to the Kemeny Comm (The Pres' Comm on the Accident at Three-Mile Island). Respon for heat transfer analysis methods for initial TRAC codes. Formerly reactor operator/engrg officer on NS Savannah. Numerous publs on reactor power systems and thermal hydraulics of reactor safety. Mbr of the ASME and ANS (Thermal-Hydraulics Exec Ctte, 1982-85). *Society Aff:* ASME, ANS

Kirjassoff, Gordon L
Business: Edwards and Kelcey, Inc, 70 South Orange Ave, Livingston, NJ 07039
Position: Chmn Emeritus *Employer:* Edwards and Kelcey, Inc. *Education:* BS/Civil Engg/Drexel Univ. *Born:* Dec 5, 1922 Phila, Pa. s. Louis S & Belle (Gordon). m. Enid Newfield Oct 22, 1944 (dec Oct 31, 1963); children Kim M, David E; m. 2d Constance T Lovenstein; children Janet L, Linda A. Proj Engr Edwards & Kelcey Newark 1947- 53, Assoc 1953-58, Partner Boston 1958-78; Exec V Pres Edwards & Kelcey Inc 1958-78, respon for admin & proj dev in the Boston & Mpls (since 1976) offices in fields of planning, traffic engrg & major transp facilities. Chrmn of Bd and Pres Edwards and Kelcey, Inc, Livingston, NJ 1978-87. Chmn Emeritus Edwards and Kelcey, Inc. 1987- . Served with US Maritime Service 1944-46. Reg Prof Engr in 16 states. Mbr Natl Soc Prof Engrs, ASCE, ACEC, CECNJ, Pres '87-'88; SAME Home: Claridge House II, Apt 2CW, Verona, NJ 07044. *Society Aff:* ASCE, NSPE, ACEC, SAME, CECNJ.

Kirk, Dale E
Business: Agri Engrg Dept, Corvallis, OR 97331
Position: Professor. *Employer:* Oregon State Univ. *Education:* MS/Agri Engrg/Michigan State Univ; BS/Agri Engrg/Oregon State Univ. *Born:* July 1918. Native of Payette, Idaho. Supr farm surveying crews for AAA prior to WWII. Served as Ordnance Officer in US Navy 1944-46. Engaged in teaching & res in agri engrg at Oregon State Univ since 1946. Specialized in mech harvesting & processing of agri prods. Teaching courses in food engrg & the energetics of food prod. Acting Hd of Agric Engr Dept 1969-71, 1975-76, 1980-81. Chmn PNW Region ASAE 1966-67. Natl Bd/Dir ASAE 1971-73. Elected Fellow ASAE 1975. Enjoy activities involving preservation & prudent harvest of our forests & wildlife. *Society Aff:* ASAE, IFT, ΣΞ.

Kirk, Donald E
Business: Dept. of Elec. & Computer Engr, Naval Postgraduate School, Monterey, CA 93943
Position: Prof, ECE Department *Education:* PhD/EE/Univ of IL; MS/EE/Naval Postgrad Sch; BS/EE/Worcester Poly Inst. *Born:* 4/4/37. in Baltimore, MD. Faculty mbr at Naval Postgrad Sch since 1965; presently Prof of Elec & Comp Engrg Dept. Ind experience with M.I.T. Lincoln Laboratory, Grumman Aircraft and Sangamo Elec Co. Author of papers in control, estimation, signal processing & books *Optimal Control Theory: An Introduction*, Prentice-Hall, 1970. *First Principles of Discrete Systems & Digital Signal Processing*, Addison-Wesley, 1987. *Society Aff:* IEEE, ASEE, ΣΞ.

Kirk, Ivan W
Business: Pest Control Engineering Research, Room 231, Ag. Engineering Building, USDA, ARS, Texas A&M Univ, College Station, TX 77843
Position: Agricultural Engineer *Employer:* USDA, ARS *Education:* PhD/Agric Engrg/Auburn Univ; MS/Agric Engrg/Clemson Univ; BS/Agric Engrg/TX Tech Univ *Born:* 1/25/37 Native of Lark, TX. Agricultural Engr for US Dept of Agriculture, ARS, Lubbock, TX and Auburn, AL 1960-1971; Part-time Instructor and Assistant Prof, Dept of Agricultural Engrg, TX Tech, Lubbock 1963-65; Lab Dir, USDA-ARS, Mesilla Park, NM 1971-77; Assoc Center Dir, Southern Regional Research Center, New Orleans, LA 1977-1982; Dir, Southern Regional Research Center, New Orleans, LA 1982-1987; Agricultural Engr, USDA, ARS, College Station, TX 1987-Present. Registered PE, TX. Chrmn, LA Section ASAE 1981-82; Chrmn, ASAE Environmental Air Quality Committee 1980-81. National Cotton Council Fellowship, 1959; Distinguished Young Agricultural Engr Award, 1971, SW Region, ASAE; Outstanding Young Men of America Award, 1973; Authur S. Flemming Award, 1975; Alumni Citation Award, Abilene Chirstian Univ, 1978; New Orleans FBA Award, 1980. Past Pres Slidell LA Evening Lions Club. *Society Aff:* ASAE, AAAS, ΣΞ

Kirk, Thomas T
Business: P.O. Box 8674, South Charleston, WV 25303
Position: Planning and Development Engr *Employer:* Hobet Mining and Construction Co, Inc *Education:* B/CE/Univ of Cinn *Born:* 10/29/32 Mr Kirk's engrg career has spanned more than 25 years and has included positions in government, consulting and industry. He currently is registered in three states, and saw commissioned service in the Corps of Engrs. He has served as pres of the WV Section, ASCE, and was Chrmn of the Committee on Standards of Practice, ASCE. *Society Aff:* ASCE

Kirk, Walter B
Home: 115 Leslie Road, Monroeville, PA 15146
Position: Dir, Railway Equip Engrg (ret). *Employer:* Wabco Div, Amer Standard Corp. *Education:* BS/Mech Engrg/Purdue Univ. *Born:* Oct 1904 Brazil Ind. Test Engr, Field Engr, Engr of Tests, Engrg Mgr, Dir Railway Equip Engrg for Wabco July 1926-Nov 1969. Holder of 21 pats covering railroad brake equip 1939-71. Respon for many major devs in air brake equip used worldwide. Penna Prof Engr since 1946. Mbr of ASME since 1945, Fellow 1976. Retired from WABCO - 1969 Assisted South African Railways in their change over from vacuum brake to air pressure brake for freight trains by information and assistance to training operators. In 1970. Assisted ISCOR of South Africa Begin operation of air pressure braked ore hauling trains on their 535 mile Sishem to Saldanha Newly built railroad in 1976. *Society Aff:* ASME.

Kirk, Wilber W
Business: P.O. Box 656, Wrightsville Beach, NC 28480
Position: Pres of LaQue Ctr for Corrosion Tech. *Employer:* LaQue Center for Corrosion Technology, Inc. *Education:* MS/Met Engr/OH State Univ; BS/Math-Phys/Otterbein College. *Born:* 9/21/32. Engr with Westinghouse Bettis Atomic Power Lab 1958-62. Corrosion engr with Inco LaQue Ctr for Corrosion Tech 1962, becoming Res Supervisor in 1967 & Mgr, 1968. Appointed Dir in 1979 with responsibilities for intl projs & long-range scope of Inco corrosion res. Pres of LaQue Ctr for Corrosion Tech, Inc in 1982. Fellow of ASM 1977; chrmn, Offshore Tech Conf Exec Committee 1976-77. Member Bd of Trustees of Cape Fear Technical Inst. Pres, N.C. Educational, Historical & Scientific Foundation. Served with Army Security Agency two yrs following college. Rotarian; Methodist; enjoys golf, jogging & semi-classical music. *Society Aff:* TMS-AIME, ASM, NACE, ASTM

Kirkby, Maurice A
Business: 1750 Midland Bldg, Cleveland, OH 44115
Position: Senior VP, Oil & Gas *Employer:* The Standard Oil Co (OH) *Education:* MA/Mech Sci/Cambridge England; BA/Mech Sci/Cambridge England *Born:* 4/12/29 Native of Nottinghamshire, England. After service in the RAF, spent two years in the iron and steel industry in Scotland. Joined the British Petroleum Co in 1954, and served in various positions in drilling, petroleum engrg and production in the UK, Africa and the Middle East. Managed BP's North Sea operation 1974-76, and worldwide exploration and production 1976-80. Joined Sohio in 1980 and am in charge of their exploration and production activities. Dir, SPE of AIME 1979/80 and 1981-. *Society Aff:* AIME, F Eng, IMM, I Mech E

Kirkendall, Ernest
Home: Apt. 909, 5100 Fillmore Ave, Alexandria, VA 22311
Position: Retired. *Education:* DSc/Metallurical Engg/Univ of MI; MSE/Metallurgical Engg/Univ of MI; BS/Chemical Engg/Wayne State Univ. *Born:* July 6, 1914 East

Kirkendall, Ernest (Continued)
Jordan Mich. Instructor, then Asst Prof of Metallurgical Engrg at Wayne State Univ 1937-46; carried on metallurgical res leading to the discovery of the diffusion in metals phenomenon known as 'Kirkendall Effect'. Secy Metals Div AIME 1946-55; Genl Secy AIME 1955-63; Secy EF 1963-65; Secy & Genl Mgr UET 1963-65. Beginning 1965 employed by Amer Iron & Steel Inst & from 1968-1979 was V Pres Mfg & Res had staff respon for some 35 tech cttes dealing with opers, quality control & res. Retired 12/31/79 Dir 1967- 1980 Metal Properties Council; Tr 1971-80 EI Part-time prof, Univ of the Dct C, 1982-85. *Society Aff:* AIME, ASM Fellow, NSPE.

Kirkpatrick, Edward S
Home: PO Box 236, Berkeley Springs, WV 25411
Position: Senior Vice President. *Employer:* Pennsylvania Glass Sand Corp. *Education:* MS/CE/GA Inst of Tech; BS/CE/GA Inst of Tech. *Born:* Jan 1926. Native of Suffolk Virginia. Served in Army Air Corps 1943-46; Field Engr Infilco Inc 1952-54; District Engr St Louis County Health Dept 1954- 56; Exec Asst to the Secy ASCE 1956-62; with Pennsylvania Glass Sand Corp since 1962; Sr. V Pres since 1968-1980; respon for minerals exploration & corporate div. Chmn District 6 Council ASCE 1966; Pres West Virginia Sect ASCE 1967; Supreme Council Chi Epsilon 1970-78, Natl Marshall 1974-78; Mbr Visiting Ctte for the Coll of Engrg West Virginia Univ 1972- , Chmn 1975-80. Married Marion Morgan 1948; children: Douglas, Craig & Pamela. *Society Aff:* ASCE.

Kirkpatrick, Edward T
Business: 550 Huntington Ave, Boston, MA 02115
Position: President. *Employer:* Wentworth Inst of Tech. *Education:* PhD/Mech Engg/Carnegie Inst; MS/Mech Engg/Carnegie Inst; BASc/Mech Engg/Univ of British Columbia *Born:* Jan 1925. Native of Cranbrook British Columbia. Pres, Wentworth Inst of Tech since 1971. Recent activities incl Dir of ASEE & Commissioner of EMC & work with establishing educational facilities in Iran (Shiraz Tech Inst), Saudi Arabia (Univ of Petroleum & Minerals) & Algiers (INELEC). Served as Pres of Rochester Engrg Soc & of the New England Sect of ASEE. Listed in Whos Who in Education. Received Distin Civilian Award; US Army 1971 Penfield Prize & Engrg Inst of Canada Prize. Enjoys flying & const of experimental aircraft. *Society Aff:* ASEE, EMC, ASME, NSPE, AAES, JETS

Kirkwood, Roderick R
Home: 10820 23rd Ave NE, Seattle, WA 98125
Position: President *Employer:* John Graham Associates *Education:* Postgrad/-/Univ of WA; BS/-/Seattle Univ. *Born:* 1/11/20. in St Paul, MN. Educ: Univ of WA, Seattle Univ, BS 1954. US Army Sig Corps 1943-46; Mech Draftsman & Designer, Seattle WA 1941-47; Mech Engr, CA Pangborn, Seattle, WA 1947-52; John Graham and Co, Seattle WA: Dir of Engr 1952-6-, Dir of Operations 1969- , Partner 1973- . Reg PE 14 states. ASHRAE: dir and officer 1964-73, International Pres 1973-74, Fellow 1967. Names Engr of Yr. WSPE and Puget Sound Engg Council, 1963. Prin works: Office blgs: Ala Moana Honolulu; Wells Fargo, San Francisco CA; Lincoln Rochester, Rochester NY; Bank of CA Center, Seattle WA; State Office Bldg, Juneau AL; Fedl Office Bldg, Seattle WA. Shopping Centers: Lloyd Center, Portland OR; Ala Moana, Honolulu. Other projs: Space Needle, Seattle WA. Madigan Army Medical Center, Ft. Lewis, WA. *Society Aff:* ASHRAE, NSPE, TBΠ.

Kirkwood, Thomas C
Business: 8080 Ward Pkwy, Suite 200, Kansas City, MO 64114
Position: Pres *Employer:* A C Kirkwood & Assoc, PC *Education:* MS/Elec Engr/Stanford Univ; BS/Elec Engr/Univ of OK *Born:* 6/6/26 Raised in Kansas City, Mr Kirkwood came to work for the Firm in 1948 and is now Pres. He has had total proj mgmt responsibility for design and consultation of elec power sys, and has been involved in fed and state funded projs involving agencies such as EPA, DOE, DOD and others. He is active in many state and natl engrg societies, including NSPE and ACEC. *Society Aff:* NSPE, ACEC, IEEE, SAME, ULI, AWWA

Kirmser, Philip G
Business: Kansas State University, Manhattan, KS 66506
Position: Prof Engrg and Math *Employer:* KS State Univ *Education:* PhD/Math/Univ of MN; MS/Math/Univ of MN; B/ChE/Univ of MN *Born:* 12/17/19 Consulting engr and mathematician, licensed in the State of KS. Author or co-author of more than 50 papers in applied mathematics and engrg. Co-inventor for two patents, one a Chinese Typewriter. Sometime Fulbright Scholar to the Netherlands, Visiting Lecturer for the Soc for Ind and Applied Mathematics, Co- Chrmn of the 16th Midwestern Mech Conference and mbr of the Bd of Dirs, visiting scientist to Battelle, Switzerland 1970, visiting prof of Mathematics, Ecole Polytechnique Federale Lausanne, 1978. Consultant to Digital Equipment Company, 1985. Prof of Mathematics and Engrg. *Society Aff:* MAA, AMS, SIAM, ASEE, KIVI

Kiss, Ronald K
Home: 14233 Briarwood Terrace, Rockville, MD 20853
Position: Asst Deputy Commander for Surface Ships *Employer:* Naval Sea Sys Command (NAVSEA) *Education:* BS/Naval Arch & Mar Eng/Webb Inst; MS/Naval Arch/Univ of CA *Born:* 2/19/41. Attended Prog for Mgmt Dev at Harvard Business School. With NAVSEA 1982; previously served as Ch Preliminary Design, Ch, Div of Ship Design, Director, Office of Ship Construction and Asst Admin for Shipbuilding & Ship Operations at the Maritime Administration before assuming present position 1982; Advise and assist Deputy Commander (RADN-US Navy) on all aspects of acquisition of surface ships and craft including planning, programming and budgeting as well as design, test, evaluation and new construction trials. VP SNAME, Past VP ASNE, Royal Inst of Naval Architects. Author of numerous tech papers related to ship design & const. *Society Aff:* SNAME, ASNE, RINA, ASE

Kisslinger, Fred
Business: Dept of Met Engg, McNutt Hall, Rolla, MO 65401
Position: Prof of Met Engg. *Employer:* Univ of MO. *Education:* PhD/Met Engr/Univ of Cincinnati; MS/Met Engr/Univ of Cincinnati; BS/CE/Univ of MO. *Born:* 11/19/19. in St Louis, MO. Taught met engg at IL Inst of Tech 1947-64. Was partner in consulting firm of Gordon & Kisslinger doing failure analyses & accident investigations. Teaching at Univ of MO-Rolla since 1964. Am now teaching gen met, metallurgical thermodynamics (using Keller Plan & my own text), heat treating & controlled atmospheres. Partner in consulting firm of Askeland, Kisslinger & Wolf doing failure analyses, accident investigations and some met & gen engg work. Married Rhea Bond of Akron, OH and have four children. Enjoy photography & fishing. *Society Aff:* ASM, TMS, ASEE, AAUP

Kitlinski, Felix T
Business: 3608 N Progress Ave, Harrisburg, PA 17110
Position: President & Chief Engr. *Employer:* F T Kitlinski & Assocs Inc. *Born:* 12/26/24 Undergrad work Brigham Young Univ, BSCE Penn State Univ, pursued Master's work Columbia Univ. Native of Nanticoke Pa. Served with Army Infantry Div during WWII in Europe, Holder of Bronze & Silver Star medals. Soils Engr with Pa Turnpike Comm & Pa Genl State Authority 1949-53; Assoc as Soils Engr with Berger Assocs Harrisburg Pa & Abbott, Merkt & Co of NY 1953-59; private practice in Harrisburg 1959- in field of Geotech Engrg, serving as Pres & Ch Engr. PPres Harrisburgh Chap PSPE & Central Pa Sect ASCE.

Kitsuregawa, Takashi
Business: 2-2-3 Marunouchi, Chiyoda-ku, Tokyo, Japan 100
Position: Advisor *Employer:* Mitsubishi Electric Corp. *Education:* Dr/Engrg/Osaka Univ; BS/Physics/Osaka Univ. *Born:* May 1917. Native of Osaka Japan. With Mitsubishi Electric Corp (MELCO) since 1941; Managing Dir, Hdquarters Eng & Mfg, Hdquarters R&D 1979. Progress Award Japan Elec Manufacturer's Assn 1959; Achievement Award Inst of Elec & Communication Engrs of Japan (IECEJ) 1967; Citation AP-S of IEEE 1974; 12 Awards Invention Assn of Japan; Medal of Honor with Purple Ribbon by Japanese Govt 1971; Disting Services Award of IECEJ 1977 and JEMA 1983; Fellow IEEE 1976. Chaired TG-AP of IECEJ 1965-71, Tokyo

Kitsuregawa, Takashi (Continued)
Chap of AP-S, IEEE 1973-74, Kansai Regional Sects IECEJ 1965-71; VP of Soc Instrument and Control Engrs of Japan 1978-79; VP of IECEJ 1979-80 & others. *Society Aff:* IECEJ, SICEJ, IEEJ, JSAP, PSJ.

Kittel, J Howard
Business: 9700 S Cass Ave, Argonne, IL 60439
Position: Manager Office of Waste Management Programs *Employer:* Argonne National Lab. *Education:* BS/Met Eng/WA St Univ *Born:* 10/9/19. Native of Ritzville, WA. Met Res Engg at NASA Lewis Res Center Cleveland OH 1943-47, 1950; Sr Metallurgist at Argonne Natl Lab Argonne IL 1947-49, 1951- ; numerous publications on effects of irradiation on reactor matls; presently Mgr of Office of Waste Mgmt Programs. Served as delegate to UN Intl Confs on Peaceful Uses of Atomic Energy. Elected to Bd of Dirs Amer Nuclear Soc. Awarded Cert of Merit by Amer Nuclear Soc. Fellow of Amer Nuclear Soc. *Society Aff:* ANS, ΣΞ

Kitterman, Layton
Business: PO Box 226227, Dallas, TX 75266
Position: V Pres Process Engrg & Res. *Employer:* Glitsch Inc. *Education:* BS/Petroleum Refining/Tulsa Univ. *Born:* Oct 28, 1928. Native of Oklahoma. US Marine Corp 1946-48. Gas Engr with Sunray-Midcontinent Oil Co (now Sun Oil) 1953-56; joined Glitsch Inc 1956; assumed current respon as V Pres of Process Engrg & Res 1973. Coordinate engrg problems with 3 affiliates & 5 lics internationally. Have spent considerable time in the last 5 yrs designing trays & assoc equip for large scale heavy plants for nuclear power plants. Outdoor activities & boating. *Society Aff:* AIChE.

Kittner, Edwin H
Home: 604 East Ave, Blue Rapids, KS 66411
Position: Enginr & Maintenance Superintendent *Employer:* Georgia-Pacific Corp *Education:* BS/ME/KS St Univ; Assoc Degree/Pre-Engrg/Assoc Coll of Upper NY *Born:* 3/7/25 in Utica, NY area, served with Third Army, ETO, 1943-1946. With Certainteed Products Corp. as Plant Engr; Bestwall Gypsum Co as Project Engr in charge of building and relocating a plant to new site. With Georgia-Pacific as Engrg and Maintenance Superintendent. Presented twelve professional papers at national engrg conferences; had eighteen engrg articles published in trade journals; contributing author to four engrg handbooks; spoken at many engrg seminars regionally and at local Univ. Past Pres of local engrg society and served on many state engrg ctte. Very active in civic activities, both local and state level. *Society Aff:* NSPE, KES, TVES

Kittrell, James R
Business: PO Box 368, Amherst, MA 01004
Position: Pres *Employer:* KSE, Inc *Education:* BS/ChE/OK State Univ; MS/ChE/Univ of WI; PhD/ChE/Univ of WI *Born:* 10/28/40. in Arkansas City, KS. Worked in R&D & mfg for Kerr-McGee, Exxon, DuPont, Chevron, Std Oil of CA. Prof at Univ MA 1970-80. Consulted widely in reactor modeling & energy planning, both for domestic cos & governmental agencies & intl areas. Pres of process product R&D co. Has published over 70 articles & inventor/coinventor on 40 patents. *Society Aff:* ACS, AIChE

Kiviat, Philip J
Home: 11007 Old Coach Road, Potomac, MD 20854
Position: Vice-President, Business Operations. *Employer:* SAGE Federal Systems, Inc. *Education:* MIE/Opns Res/Cornell Univ; BME/Ind Eng/Cornell Univ. *Born:* Oct 1937. DBA coursework at Univ of So Calif. Reg P E in Calif. Opers Res U S Steel 1961-63; Staff RAND Corp 1963-69; Pres Simulation Assn 1969-71; Dir Mgr Sys Control 1971-72; Tech Dir at FEDSIM 1972-78; Client Mgr SEI Computer Svces 1978-1980; Vice Pres. CTEC, Inc. 1980-83. VP Federal Products Division SAGE Systems, Inc 1983-84. VP Business Operations, SAGE Federal Systems, Inc. 1985- . Chmn ACM SIGMETRICS 1973-77. ACM/SIGSIM Special Achievement Award 1974. Interagency Ctte on ADP Award of Excellence 1976. 1976 A. A. Michelson Award Winner. Pres. Computer Measurement Group 1978-1980. Distin Lectur of IIE. Chmn Federal Info Proc Stds Task Group 10 on Computer Performance Mgmt. Mbr: ACM, IEEE Computer Soc, Cornell Soc of Engrs, ACM, IEEE Computer Soc, Cornell Soc of Engrs. 1978 Chairman of President Carter's Data Processing Reorganization Project. *Society Aff:* ACM, IEEE, CMG.

Kivioja, Lassi A
Home: 60 Blackfoot Ct, Lafayette, IN 47905
Position: Prof. *Employer:* Purdue Univ. *Education:* PhD/Geodetic Sci/OH State Univ; MS/Physics/Univ of Helsinki, Finland. *Born:* 3/29/27. in Finland, Grad from Univ of Helsinki MS in 1951. Asst at the Finnish Geodetic Inst and at the International Isostatic Inst Part time 1949-51. Drafted into the Finnish Armed Forces 1951-52. High School Teacher in Mathematics & Physics in Finland 1952-55. Immigrated to USA in 1955. Res Associate and Instructor in Geodetic Sci at The OH State Univ 1955-63. PhD 1963, prof in Geodesy at Prudue Univ since 1964. Special Interests: Geodetic Instruments, Geometric, Physical & Celestial Geodesy. Inventor of US Patent No 4149321. "Mercury Leveling Instruments–. *Society Aff:* AGU, ACSM.

Klaiber, Wayne F
Business: Town Engr Bldg, Ames, IA 50011
Position: Prof. & Mgr of Bridge Engg Ctr *Employer:* IA State Univ. *Education:* PhD/Structural Engr/Purdue Univ; MS/Structural Engr/Purdue Univ; BS/Civil Engr/Purdue Univ. *Born:* 10/7/40. *Society Aff:* ASCE, ACI, AREA.

Klancko, Robert John
Business: 2 Orchard Rd, Woodbridge, CT 06525
Position: Mgr Special Proj *Employer:* United Illuminciting Co. *Education:* BSE/Chem E/Univ of CT; Grad/Chem/Southern CT State Univ. *Born:* 9/8/45. AIChE Professional Recognition Certificate, Certified Safety Profl, Certified Hazard Control Mgr - Master Level, 1981 Who's Who in Technology Today, 17th Ed Who's Who in the East. Reg PE CT. S CT Chapter ASM Young Mbrs Award 1978, Chapter Chrmn's Award 1980. Chrmn CT Met week 1977, 1978 - was founder of CT Met Week - 1977 ASM National Acad of metals and materials. Lecturer Special Prog Hartford Graduate Ctr 1971-76, Chief Lecturer - Coordinator Environmental Tech Prog Waterbury State Technical College 1982- present, Chrmn Woodbridge Inlands, Wetlands Agency. *Society Aff:* ASM, CFREP.

Klapper, Jacob
Business: 323 King Blvd, Newark, NJ 07102
Position: Prof & Chairman of Elec Engg. *Employer:* NJ Inst of Tech. *Education:* EngScD/EE/NYU; MS/EE/Columbia Univ; BEE/EE/CCNY. *Born:* 9/17/30. Joined NJ Inst of Tech in 1967; currently Prof of EE & Chairman of Dept. Previous positions were with CBS-Columbia (1952-56) Columbia Univ (1957-60), CCNY (1956-59), and RCA (1960-67). David Sarnoff Fellow (1962- 64). *Editor of Selected Papers on Frequency Modulation (Dover 1970), co-author of book Phase-Locked and Frequency- Feedback Systems (Academic Press 1972) and of over 40 tech papers and patents. Conslt to RFL, Microlab/FXR, ITT, Singer, Foxboro, Hazeltine, and others. Dir of short courses (Ctr for Profl Advancement). Dir of Tutorials for ICC and NTC. Lectured widely for IEEE in the US and overseas. Society Aff:* IEEE, ComSoc.

Klasing, Donald E
Business: 111 W. Locust St, PO Box 789, El Dorado, KS 67042
Position: Pres. *Employer:* Century Plastics, Inc. *Education:* MS/CE/Columbia Univ; MBA/Finance, Marketing/St Louis Univ; BA/Chemistry/MacMurray College. *Born:* 5/24/42. Born & raised in St Louis MO, educated in public schools. After liberal-arts MacMurray College, attended Columbia on combined plan. MS thesis was distillation comp prog. Process design, including phosphoric & sulpuric acid plants, while with Monsanto Co 1966-1972. MBA degree at night 1968-1973. From 1972 to 1975, sales engr for Sandvik Conveyor Inc. From 1975 to 1978, product mgr for

Klasing, Donald E (Continued)
tower packing at Glitsch Inc. 1978 to present at Glitsch subsidiary Century Plastics, an injection molder specializing in engg plastics, making parts for aircraft, recreation, & ind equip as well as tower packing for Glitsch. *Society Aff:* AIChE, SPE.

Klass, Philip J
Business: 404 N St SW, Washington, DC 20024
Position: Active Retirement, Self-Employed *Employer:* Contribut. Editor, High Technology magazine. Contribut. Avionics Editor, Av Week & Space Technology *Education:* BS/Elec Engr/IA State Univ *Born:* 11/8/19. 1941-52 Engr with Genl Elec in various phases of GE's avionics activies; 1952-1986. 1986- Tech. Journalist (self employed). Sr Avionics Editor, Aviation Week & Space Tech magazine - McGraw-Hill. Fellow IEEE; Mbr Amer Assn for Advancement of Sci; Natl Press Club; Aviation/Space Writers Assn. Author *Secret Sentries In Space* (1971 Random House), *UFOs Explained* (1975 Random House), *UFOs; The Public Deceived* (1983 Prometheus Books); Fellow; Committee for the Scientific Investigation of Claims of the Paranormal. *Society Aff:* IEEE, AAAS.

Klassen, Jacob
Home: 942 Dresden Crescent, Ottawa Ontario K2B 5J1 Canada
Position: Technical Consultant *Employer:* Energy Mines & Resources Canada *Education:* BE/Chem Engg/Univ of Saskatchewan. *Born:* 1921 USSR; arrived in Canada 1924. High School Regina Central Collegiate, Saskatchewan Univ, Univ of Saskatchewan BE 1943. Recipient of Governor Generals Gold Medal. 1943-46 Engr Div of Applied Biology Natl Res Council Ottawa; 1946-49 Ch Engr Engrg Installations Ltd, Mech Contractors Montreal; 1949-52 Mech Engr McDougall & Friedman Cons Engrs Montreal; 1952-74 Pres J Klassen & Assocs Ltd Cons Engrs Ottawa; 1974-1981 Dept of Public Works of Canada. Mbr: CEQ, APEO, ASHRAE; 1981- Sr Engr, Intl Energy Program, Energy Mines & Resources Canada, 1986- Technical Consultant, Energy Mines & Resources Canada, 1982- Fellow of ASHRAE. *Society Aff:* ASHRAE, CIE.

Klebanoff, Gregory, Jr
Business: PO Dr M-1033 N Hwy 427, Longwood, FL 32750
Position: Vice President, Genl Mgr. *Employer:* Plasti-Mac Div Standard Container Co. *Born:* Aug 1927. BSChE Pratt Inst. Served in USNR as electronics tech 194446. Service Engr for Permutit Co 1951-58 on ion exchange & indus water treatment equipment; Sales Engr for Bulkley Dunton Pulp Co 1958-61 on indus waste treatment equipment; various line & staff positions with Beloit Corp 1961-73 incl sales, advertising & promotion, market res, foreign engrg liaison & product mgr for plastic blow molding equipment. A principal officer in 1973 start of Plasti-Mac Inc for manufac blow molding machinery & sys, & continuing since Jan 1976 acquisition by Standard Container Co. Pats issued & pending. Enjoys soaring, tennis, snorkeling & chess.

Kleen, Werner J
Home: Denningerstr 36, 8000 Munich 80, West Germany
Position: Director of Research, Professor (retired). *Employer:* Siemens AG (ret) Techn Univ Munich. *Education:* Dr Phil Nat/Dr Habil Prof/Univ Heidelberg *Born:* 10/29/07 b Hamburg. Dr Phil Nat & Dr Habil Univ Heidelberg. Until 1946 Telefunken Berlin, dev of electron tubes; 1946-50 Comp Gen de Telegrphsie sans 1 Fil (CSF), Paris, 1950-52 & Instituto Nacional de Electronica Madrid; until 1967 Dir of Res Lab Siemens Munich; 1968-71 Dir of European Space Res & Tech Ctr Noordwijk/Netherlands & br Directorate of European Space Organization (ESRO now ESA); since 1955 Hon Prof Techn Univ Munich. Author & editor of several books on Electron Tubes & Lasers. 1955 GaussWeber Medal Univ Goettingen, 1957 Fellow IEEE 1961-62 Pres of Nachrichtechnische Gesellschaft (NTG W Germany), 1978 Ring of Honour Verband Deutscher Elektrotechniker (VDE W Germany), 1980 Microwave Career Awd, MTT, IEEE, 1982 Frederik Philips Awd, IEEE. *Society Aff:* DPG, VDE, NTE, IEEE

Klehn, Henry, Jr
Business: 1100 Glendon Ave, Los Angeles, CA 90024
Position: Partner. *Employer:* Dames & Moore. *Education:* MS/Engr Science/Univ of CA, Berkeley; BS/Geological Engg/Univ of CA, Berkeley. *Born:* July 1936 California. Entire prof career with Dames & Moore; joined firm 1960 in San Francisco, moving to Hawaii 1961 & Los Angeles 1963; has been Principal 1966- , Ptnr 1970- . Primary respon for geotech work at Walt Disney World in Fla & refinery projs in Puerto Rico & Ill; 1971-74 managed opers in Singapore & Indonesia on major petro & mining projs; since 1975 managed L A office, serving on Elec Ctte. Enjoys sports & gardening. *Society Aff:* ASCE, AIME, ΤΒΠ, ΣΞ.

Klein, Daniel
Home: 46 Foster Square, Pittsburgh, PA 15212
Position: Consultant *Employer:* Westinghouse Electric Corporation (Retired). *Education:* BA/Physics/Brooklyn College; PhD/Physics/Univ of Rochester. *Born:* 1/16/27. Served in US Navy WWII. Joined Westinghouse 1954 at Bettis Lab; worked on Reactor Phys in support of Nautilus & Shipping-Port Reactors; assumed Tech Dir position of Fusion Dept 1973. Fellow ANS, Mbr of Phi Beta Kappa & Sigma Xi. Retired Westinghouse 1986. Presently Consultant. *Society Aff:* ANS, APS, ΦΒΚ, ΣΞ.

Klein, H Joseph
Business: 1020 W Park Ave, Kokomo, IN 46901
Position: VP *Employer:* Haynes International Corp. *Education:* PhD/Met Engr/Univ of TN; MS/Met Engr/Univ of AL; BS/Met Engr/Purdue Univ. *Born:* 7/5/41. Joined Stellite as Sr Engr in 1969 in the Process Met Sec, appointed Group Leader 1973; Sec Mgr Process Met 1974 & as Dir of Tech 1978. Named Stellite Operations Mgr in 1980, Plant Manager 1982. Appointed Cabot Operations Manager 1984, VP Haynes International Inc. 1987. Responsible for advances in Process Met of High Performance Alloys & in particular the advances in electroslag remelting including ESR shapes & process modeling. Fellow of ASM. Received Distinguished Eng Alumnus Award from Purdue Univ in 1979 and Univ of Tenn 1984. Mbr of Natl Mtls Advisory Bd ad hoc Committee on Electroslag Remelting and Plasma Melting; Natl Acad of Sciences Committee for the "Review of the US/USSR Agreement on Cooperation in the Field of Science and Technology." Presently on the committee for Joint Cooperation in Electro Met between USA and USSR. Trustee of ASM International, Past Chmn of the Technical Division Board, & Energy Div of ASM, & Past Chrmn Vacuum Met Div of the Am Vacuum Soc, VP of Purdue Engr Assoc. *Society Aff:* ASM, AIME, AVS.

Klein, Imrich
Business: 67 Veronica Ave, Somerset, NJ 08873
Position: President. *Employer:* Scientific Process & Research Inc. *Education:* PhD/Chem Eng/Case Inst of Tech; MSc/Chem Eng/Case Inst of Tech; BSc/Chem Eng/Technion, Israel Inst of Tech *Born:* 9/14/28. in Kosice Czechoslovakia. Assumed present post of Pres of Sci Process & Res Inc Oct 1968; previously Sr Res Engr with Engg Res Center Western Elec Co, Sr Engr with Exxon Res & Engg Co, Res Engr with E I du Pont de Nemours & Co. Co- authored 2 books: *Engg Principles of Plasticatings Extrusion & Computer Progs for Plastics Engrs.* Author of 100 tech papers & books chaps. Writes monthly column *Processors 'Corner'* in Plastics World Magazine. author of sevl patents. Married, 2 sons. *Society Aff:* SPE

Klein, Joseph P, III
Home: 1941 Branch View Dr, Marietta, GA 30062
Position: Sr Engr. *Employer:* Law Engg Testing Co. *Education:* MS/Civ Engg/George Wash Univ; BE/Civ Engg/Villanova Univ. *Born:* 6/7/47. in Ithaca, NY. Began career with Potomac Elec Power Co. Joined Law Engg in 1972. Geotechnical Engg Mgr in WA Office until 1977 assignments as Gen Mgr in Riyadh, Saudi Arabia. In 1979 assigned to Atlanta. Responsible for marketing and execution of engg projs in Saudi Arabia with US clients. Served as Committee Chrmn and officer in Natl Capitol Sec, ASCE and N VA Chap VSPE. Served on engg faculty of N VA Community

Klein, Joseph P, III (Continued)
College; Univ of VA Continuing Ed Prog, and Asst Prof at Howard Univ. Received ASCE Edmund Feedman Young Engr Award for Profl Achievement in 1977. *Society Aff:* ASCE, ISSMFE, NSPE.

Klein, Lisa C
Business: Ceramics Dept PO Box 909, Piscataway, NJ 08855-0909
Position: Professor *Employer:* Rutgers Univ. *Education:* PhD/Ceramics/MA Inst of Tech; BS/Met/MA Inst of Tech. *Born:* 12/7/51. Since Spring 1977, prof in the Ceramics Dept, Rutgers Univ, divides time between teaching and res; res interests include studies of sol-gel processing, flow behavior and transformation kinetics, leading to new uses of glass and ceramics for thin films, membranes and substrates; faculty advisor for Rutgers Student Sec of the Society of Women Engrs; native of Wilmington, DE. *Society Aff:* ACerS, ASM, SWE, ΣΞ.

Klein, Morton
Business: 301 A SW Mudd Bldg, NY, NY 10027
Position: Prof. *Employer:* Columbia Univ. *Education:* Eng ScD/Ind Eng/Columbia; MS/Mech Eng/Columbia; BS/Mech Eng/Duke *Born:* 8/9/25. Native of NYC. Res, dev, design and production planning work for private firms and govt, 1947-1954. At Columbia Univ since 1956, teaching Industrial Engg and Operations Res. Chrmn, I. E.-O.R. Dept 1982-85. Participant in govt res projs on surveillance procedures related to production, quality control and inventory mgt 1957-1974. Editor: *Management Science*, 1960-1977. Author: (with Cyrus Derman) *Probability & Statistical Inference for Engineers*, 1959. Research and publications on production planning and scheduling, early cancer detection examination scheduling, network flows and operations research. *Society Aff:* IIE, ORSA, TIMS

Klein, Philip H
Business: 3330 West Friendly Avenue, Greensboro, NC 27410
Position: Dir-Corp. Eng'g *Employer:* Burlington Ind, Inc. *Education:* BS/Ch.E./Purdue University *Born:* 06/22/33 32 year career in engrg in chemical, fiber, glass, and textile industries. Chief Engr of Owens Corning Fiberglas Corp from 1968-76. Director of Corp Engrg at Burlington Industries from 1976. Patent awards include chemical reclaim and chemical mixing apparatus and process control systems. Developed engrg productivity measurement and project control systems for industry engrg staff depts. Guest lectures at midwest and southeastern univs. Technical venture business development in energy systems, roofing, and precast concrete products. *Society Aff:* AIChE, AAES

Klein, Ronald L
Home: 124 Poplar Dr, Bakers Ridge, Morgantown, WV 26505
Position: Prof & Chrmn of Elec Engr. *Employer:* WV Univ. *Education:* PhD/Elec Engr/Univ of IA; MSEE/Elec Engr/Univ of IL; BSEE/Elec Engr/Univ of IL. *Born:* 4/26/39. in Bloomington, IL. Student Univ of IL 1958-1963. Mbr of Tech staff, Bell Telephone Labs 1963-1967. Grad work in EE at Univ of IA, IA City 1967-1969. PhD 1969 with research topic: sensitivity and estimation in stochastic systems. Asst Prof EE Univ of KS 1969. Assoc Prof, EE, KU, 1974. Prof EE, KU, 1978. Res publications primarily in Nonlinear Estimation and Systems Identification. Chrmn and Prof of EE Dept, WV Univ, 1979. *Society Aff:* IEEE, ASEE, ΣΞ.

Klein, Stanley J
Home: 2 Stoney Clover La, Pittsford, NY 14534
Position: Consultant, Engr (Self-employed) *Education:* BSME/Mech Engrg/Univ of Rochester *Born:* 11/17/16 Mr Klein is a Consulting Safety Engr. He holds a NYSPE License and accreditations from the Bd of Certified Safety Professionals and the National Council of Engrg Examiners. He has appeared before the National Commission on Product Safety and Monroe County Trial Lawyers Assoc and the KS State Legislature. Mr Klein was appointed Engrg Consultant to the United Nations Industrial Development Organization. His practice covers the US including Virgin Islands, Puerto Rico, Central America and Africa, serving private industry, the legal profession and insurance companies. Mr Klein is the author of the book, "How to Avoid Products Liability: A Management Guide–, published by IBP Division of Prentice Hall. *Society Aff:* NSPE, NYSSPE, SAE, ASSE, NSC, WSO

Klein, Virginia L
Home: RR6, Iowa City, IA 52240
Position: Coordinator. *Employer:* Univ of IA Hospitals & Clinics. *Education:* DEng/Systems Engg/RPI; MSEE/EE/NYU; BEE/EE/CCNY. *Born:* 2/17/43. Educated in NY. Am a reg engr. Was EE in Design & Dev in areas of Communications, Computers, Inertial Guidance & Nucl Power. Since 1970 have been applying Systems Engg methods to biomedical problems. This currently includes integrating patient monitoring applications into overall hospital info system & planning for future mgt of patient data. *Society Aff:* IEEE, SWE.

Kleiner, David E
Home: 2120 Hickory La, Schaumburg, IL 60195
Position: President *Employer:* Environmental Services *Education:* MS/CE/Northwestern Univ; BS/CE/Valparaiso Univ *Born:* 6/12/36 With Harza Engrg Co since 1959. Presently, Pres Harza Environmental Services, Inc. consulting engrg firm of about 35 engrs and geologists. Firm performs studies, investigations, design and consultation for projects in water supply, treatment and distribution, hazardous waste mgmt, stormwater mgmt, groundwater assessment and remediation, wastewater collection and treatment and solid waste mgmt. Selected as Young Engr of the Year by Chicago Chapter of the IL Society of PE. Author of approximately 20 technical papers. *Society Aff:* NSPE, ASCE, USCOLD, SAME, EERI.

Kleiner, Fredric
Home: 11 Lexington Rd, New City, NY 10956
Position: Lab Mgr. *Employer:* Gen Foods Corp. *Education:* PhD/ChE/PA State Univ; MS/ChE/Columbia Univ; BS/ChE/Univ of PA. *Born:* 10/4/38. Native of Atlantic City, NJ. Began career in res & dev with Gen Foods Corp in 1967. For next 6 yrs, specialized in dehydration methods, concentration and separation techniques, crystallization & electronic heating methods. Have patents & publications in dielectric & microwave food processing. Managed physical chemistry sec for 3 yrs before assuming current responsibility as lab mgr of engg res in 1976. 1972 author & chrmn of AIChE natl student contest problem. 1973-1976 membership chrmn 1979 Div Chrmn of food, pharmaceutical & bioengg div 1978 Program Chrmn. Married, 2 children. Hobbies: philately, chess, golf, softball. *Society Aff:* AIChE.

Kleinfelder, James H
Business: 1501 N Broadway, Suite 308, Walnut Creek, CA 94596
Position: President. *Employer:* J H Kleinfelder & Assocs. *Education:* MS/CE/UC Berkeley; BS/CE/UC Berkeley. *Born:* 11/12/34. MS & BS Univ of CA, Berkeley. Pres J H Kleinfelder & Assocs with offices in CA, NV & Saudi Arabia; Ptnr of FAS Trading & Construction HQ in Rigadh, Saudi Arabia. Served on "Methodology for Delinating Mudslide Hazards" Ctte Natl Acad of Sci; Mbr ASCE, NSPE, ACEC, CEAC, SEOA, ASEE. *Society Aff:* ASFE, ASCE, ACEC.

Kleinman, Irving
Business: 90 New York Ave, Massapequa, NY 11758
Position: Pres *Employer:* Irving Kleinman & Assoc P.C. *Education:* BME/ME/City Coll of NY *Born:* 4/9/29 Native of NY City. Received education in the NY City public school system (including Bronx High School of Science and City Coll). Elected to Pi Tau Sigma and Tau Beta Pi honorary societies. After 10 years of experience in contracting and consulting, formed (in 1958) the engrg firm of which I am now pres. Hold multiple state licences to practice engrg; am a Certified Fallout Shelter Analyst; a member of the National Panel of Arbitrators of the American Arbitration Association. Past pres of the Consulting Engrs Council of NY State and a National Dir of the American Consulting Engrs Council. *Society Aff:* ASPE, ACEC, AEE, ASHRAE, ASME, IES, NFPA, NSPE

Kleinrock, Leonard
Business: Boelter Hall 3732, UCLA, Los Angeles, CA 90024-1600
Position: Prof *Employer:* Univ of CA *Education:* PhD/EE/MIT; MS/EE/MIT; BS/EE/City Coll of NY *Born:* 6/13/34 In 1963 he joined the faculty of the School of Engrg and Applied Science, Univ of CA, Los Angeles, where he is now Prof of Computer Science. His research spans the fields of computer networks, computer systems modeling and analysis, queueing theory, and resource sharing and allocation in general. At UCLA, he directs a group in advanced teleprocessing systems and computer networks. He is a Guggenheim Fellow, an IEEE Fellow, and has received various outstanding teacher and best paper awards, including the 1976 Lanchester prize for the outstanding paper in operations research, and the ICC '78 prize-winning paper award. He is a member of the National Academy of Engrg. In July 1986, he was awarded the 12th International Marconi Fellowship; In 1982, selected to receive the CCNY Townsend Harris Medal and co-winner of the L.M. Ericsson Prize. He is a member of the IBM Advisory Cttee since 1986. *Society Aff:* NAE, IEEE, ACM, ORSA, IFIP

Kleis, Robert W
Home: 6520 Sumner, Lincoln, NE 68506
Position: Executive Dean, International Affairs. *Employer:* University of Nebraska. *Education:* BS/Agri Engrg/MI State College; MS/Agri Engrg/MI State Univ; PhD/Agri Engrg/MI State Univ. *Born:* Nov 1925 Martin Mich. Agri Engrg teaching & res, Mich St Univ & Univ of Ill 1949-57; Head Agri Engrg Depts, Univs of Mass & Neb 1957-68; Assoc Dean & Dir Agri Experiment Station, Univ of Neb 1967-76; Dean of Internatl Agri Progs 1976-1984; Exec Dean, Intl Affairs. Executive Director, Board for International Food and Agricultural Development, U.S. Dept. of State, 1985-87. Approx 60 tech publs & 100 extension writings. Numerous natl, regional & state cttes & offices in prof orgs. Reg Engr, Private Pilot, Licensed Realtor, Cons, Rotarian, Shriner, mbr of several hon socs & active in civic & political affairs. *Society Aff:* ASAE, NSPE.

Klema, Ernest D
Home: 53 Adams St, Medford, MA 02155
Position: Prof Emeritus Dean Emeritus *Employer:* Coll of Engrg - Tufts Univ. *Education:* PhD/Physics/Rice Univ; MA/Physics/Univ of KS; AB/Chemistry/Univ of KS. *Born:* Oct 1920. Prof Engrg Sci, Adj Prof Internatl Politics, Fletcher Sch of Law & Diplomacy Tufts Univ; Dean, Coll of Engrg Tufts Univ 1968-73; Chmn Dept of Engrg Sci, Northwestern Univ 1960-66. Fellow APS, ANS; Sr Mbr IEEE. Chmn Subctte on Neutron Measurements & Standards of the Ctte on Nuclear Sci of Natl Res Council 1958-63, Ctte on Energy Conversion ASEE 1965-66, Student Prog Ctte, Winter 1967 meeting of ANS. Dev of high-voltage high-resistivity silicon surface-barrier detectors as reaction proton spectrometers & of low-resistivity silicon detectors for heavy ions. Dean of Engineering, Professor of Engineering, Emeritus, Feb., 1987. *Society Aff:* APS, ANS, IEEE.

Klement, William, Jr
Business: P.O. Box 5153, Vancouver, WA 98668-5153
Position: Assoc Prof *Employer:* Univ of CA *Education:* BS/Physics/Caltech; PhD/Engrg Sci/Caltech *Born:* 9/30/37 Chicago. Caltech, BS Physics, 1958. PhD Engrg Sci, 1962. Cornell, Physics, GE Fellow 1958-9. UCLA, Geophysics, Post Doc, 1962-4. KTH, Stockholm, NATO Fellow, 1963. Univ CA Berkeley, Physics, Miller Fellow, 1964-6. UCLA Engrg, Faculty, 1966-. ANU, Canberra, Guggenheim Fellow, 1968-9. Univ Chile, Santiago, Ford Foundation Exchange, 1973. NPRL-CSIR, Pretoria, 1974-6. Univ Oxford, UK, Sabbatical, 1979. Pioneering work in rapid quenching from melt-metallic glasses, etc-co-recipient 1980 American Physical Society Intl Prize for New Materials. Systematic work on high pressure phase transitions. Have worked professionally on five continents, travelled in half of countries of world. *Society Aff:* TBΠ, ΣΞ

Kletsky, Earl J
Business: College of Engrg, 203 Link Hall, NY 13244
Position: Prof *Employer:* Syracuse Univ. *Education:* PhD/Elec Engrg/Syracuse Univ; MS/Elec Engrg/MIT; BS/Elec Engrg/MIT. *Born:* 7/22/30. Native of Springfield, MA. Served as Elec Engr with USAF 1953-55. Awarded the Torchiana Fellowship in 1955 by the Dutch Govt and spent two yrs at the Technische Hogeschool in Delft, Netherlands. Inst for Sensory Res at Syracuse Univ since 1964-1980. Admin Dir 1974-80. Prog Coordinator of the Undergrad Prog in Bioengg in the College of Engg since 1973-1983. Asst Dean for Undergraduate Affairs in the Coll of Engrgs since 1982. Res involves real-time analog modeling and digital computer simulation of the peripheral auditory system of mammals and man. Active in amateur radio and a grower of orchids. *Society Aff:* ASEE, IEEE, NYAS.

Klieger, Paul
Home: Consultant, 2050 Valencia Dr, Northbrook, IL 60062
Position: Conslt *Employer:* Self *Education:* BS/Civil Engg/Univ of WI. *Born:* Oct 1916; native of Chicago Ill area. Joined Portland Cement Assn Res Labs 1941; assisted in Long-Time Study of Cement Performance in Concrete; appointed Consultant to R&D/CTL 1982. Suprs & admins res projs, both dues- supported & contract, in concrete tech field. Mbr Amer Concrete Inst (Fellow), Amer Soc for Testing & Materials (Fellow), Prestressed Concrete Inst, & Transportation Res Board. Active on tech cttes of ACI, ASTM, PCI & TRB. Retired in Oct. 1986. Currently Consultant, Concrete and Concrete Materials. *Society Aff:* ACI, ASTM, PCI, TRB.

Klimek, Edmund J
Business: Wolf & Algonquin Rds, Des Plaines, IL 60018
Position: Materials Consultant *Employer:* Borg Warner Auto Res Ctr *Education:* BS/Met E/Univ of IL. *Born:* 8/15/25. Quality, failure analysis, heat treat & welding metallurgist for Studebaker & Intl Harvester Co. Ch Metallographer Armour Res Foundation (now IITRI). With Borg- Warner 1965- ; materials & manufac cons to oper divs engaged in aircraft, automotive, farm equipment, oil field, air conditioning & nuclear component indus; R/D in areas of surface treatments, casting, heat treatment & mfg processes in metals. Dev of specifications & quality assurance procedures for cemented carbides, ceramics and mfg. carbons. Part-time instr at univ & indus levels 1956- , in met, heat treatment & mfg. Chmn Chicago-Western Chap ASM 1976.

Klimovich, John
Business: P.O. Box 06059, Portland, OR 97206
Position: President. *Employer:* KSA Inc. *Education:* BASc/CE/Univ of British Columbia. *Born:* 3/22/27. Upon grad entered cons field as design engr for Sandwell & Co Ltd in Vancouver B C; following 5 yrs as Hd of their Cvl & Struct Design Group served as Resident Engr during const of 2 successive projs (a 250 TPD Kraft Paper & Linerboard Mill at Toledo Ore & a 120 TPD Newsprint Mill in Khulna E Pakistan); subsequently was appointed Ch Engr of Sandwell Internatl Inc Portland Ore 1963 & Ch Engr of Sandwell Vancouver 1965; in 1971 was appointed V Pres & Dir of US & Latin Amer Opers for Sandwell Internatl Inc Portland Ore; in 1972 elected to Board of Dirs & in 1974 elected Pres. In 1979 resigned to form and head KSA Inc, Consulting Engrs, as Pres. Reg Prof Engr in B C, Ore, Pakistan, Wash, Tenn. Mbr Paper Indus Mgmt Assn; Tech Sect Assn of Pulp & Paper Indus; Cons Engrs of Ore & U S; Amer Mgmt Assn. *Society Aff:* PIMA, TAPPI, CPPA, NSPE.

Kline, Donald H
Home: 413 Ramblewood Dr, Raleigh, NC 27609
Position: VP-Engineering *Employer:* Kimley-Horn & Assoc, Inc. *Education:* MSCE/Structural Engg/Univ of IL; BCE/Civ Engg/NC State Univ. *Born:* 10/21/33. Served two years in Corps of Engrs in hydrology and construction. Conslt practice has included unique design and investigations, principally of bldgs. Current responsibility is for structural engrg practice and firmwide liability management of regional firm specializing in transportation engrg, parking facilities design, bridge design, other structures, and general civil engrg. Have served on numerous statewide bodies for bldg code revision, certification of testing labs, improvement of contracting procedures, & initiation of technical educ groups for structl & fnd engrg. Have served on & chaired natl ctte for Peer Review of conslt engrg firms. Principal profl activities have been with ASCE, Amer Consltg Engrs Council, Prof Engrs of North Carolina (Engr of the Year, 1984). *Society Aff:* ASCE, NSPE, ACI, AISC.

Kline, Jacob
Business: PO Box 248294, Coral Gables, FL 33124
Position: Prof & Chrmn. *Employer:* Univ of Miami. *Education:* PhD/Biomedical & Elec Engg/IA State Univ of Sci & Tech; MS/EE/MA Inst of Tech; BS/EE/MA Inst of Tech. *Born:* 8/3/17. Native of Boston, MA, Electronics Engr, IT&T, 1942-46; Video Sec Chief, Boson Univ Optical Res Lab, 1946-48; Asst Prof, Assoc Prof, EE, Coordinator of Biomedical Engg Prog of Univ of Rhode Island, Research Engineer MIT 1951-52; 1952-66; NSF Sci Faculty Fellow, 1960-62; NASA-ASEE Fellow Ames Res Lab, Stanford Univ, Summers 1965, 1966; Prof & Chrmn, Dept of Biomedical Engg, Univ of Miami, 1966-present; Certified Clinical Engr; Consultant, Forensic Biomedical Engg; Fellow of the Am Academy of Dental Electrosurgery; Fellow AAAS; mbr Ad Com, G-EMB of IEEE, 1968-74; Bd of Dirs, Assn for the Advancement of Medical Instrumentation, 1976-79; Chairman of AMMI Foundation Board of Trustees 1978-1985 and Awards Committee; author. *Society Aff:* AAMI, ASAIO, IEEE, American Academy of Dental Electrosurgery.

Kline, Stephen J
Business: Dept Mech Engrg, Stanford Univ, CA 94305
Position: Prof ME & Prof Values, Tech & Soc. *Employer:* Stanford University. *Education:* ScD/ME/MA Inst of Tech; MS/ME/Stanford; BA/ME/Stanford. *Born:* Feb 1922 Los Angeles. Office Ch of Ordnance US Army 1943-46. Res Analyst Rocket Propulsion Sys N Amer Aviation 1946-48. Grad work Stanford & MIT 1949-52. Stanford Faculty 1952- ; Dir Thermosci Div 1961-73, Chmn ME; Dept 1966. P Chmn Fluid Mech Ctte & Fluids Engrg Div ASME. Res Areas: internal flows, diffusers, production of turbulence; turbulent shear flows; separation; separated flows; uncertainty analysis & mensuration; flow visualization; dimensional analysis, modeling & similitude; continuum & variational thermodynamics. Awards & Medals: Melville Medal ASME; George Stephenson Prize Inst Mech Engrg; Golden Eagle Cine Award; Bucraino Prize (Italy); Fluids Engrg Award ASME. ASME Centennial Award. Mbr Natl Acad in Engrg. *Society Aff:* ASME, TBΠ, AIAA, NAE.

Klingensmith, Russell S
Business: PO Box 1963, Harrisburg, PA 17105
Position: Chief, Mine Drainage Control Group & Ptnr. *Employer:* Gannett Fleming Corddry & Carpenter, Inc. *Education:* MS/Sanitary Engg/Univ of NC; BS/Civil Engg/Carnegie-Mellon Univ. *Born:* July 1920; native of Ford City Pa. Served in Army Air Corps 1942-46. Field Engr, Office Mgr & Head of Pa's Mine Drainage Control Prog 1949-66. With GFC&C Inc 1966- , Ptnr 1973- . Respon for water pollution control activities related to coal & mineral mining - investigations, reports, const plans & specifications, supervision of const & protection to public water supplies. Enjoys the outdoors & hunting. *Society Aff:* WPCF.

Klinger, Allen
Business: 3531-C BH, Los Angeles, CA 90024
Position: Prof. *Employer:* UCLA. *Education:* PhD/EE/Univ of CA, Berkeley; MS/EE/CA Inst of Tech; BEE/EE/Cooper Union. *Born:* 4/2/37. BEE 1957; MS 1978: PhD 1966. Fellow IEEE for contributions to image analysis by means of computers. Tech Prog Chmn, Third International Joint Conference on Pattern Recognition, 1976. Chmn conference on data structures pattern recognition and computer graphics, 1975 and Co-Editor two books, Academic Press. Pub papers in IEEE Transactions (Control, Computers, Systems Man & Cybernetics, Pattern Analysis and Machine Intelligence) and other journals and books. Prof, UCLA Computer Sci Dept. Employment/consltg with Aerospace Rand, SDC, JPL, and World Bank. Research on artificial intelligence, pattern analysis, data structures for computer vision. *Society Aff:* IEEE, IEEE Computer Society, AAAI.

Klinges, David H
Business: Room 1018 Martin Tower, Bethlehem, PA 18016
Position: VP *Employer:* Bethlehem Steel Corp *Education:* LLB/Law/Yale Univ; AB/Govt/Franklin & Marshall *Born:* 7/22/28 Served with US Navy from 1953 to 1956, retired from the Naval Reserve with the rank of Lieutenant Commander. In 1956, joined the law firm of Haight, Gardner, Poor & Havens in NY City. In 1961, became associated with Bethlehem Steel Corp as a maritime attorney. Promoted to senior maritime attorney in 1965 and took on additional responsibilities of assistant secy in 1970. Named general mgr of sales of shipbuilding in 1972. Elected assistant VP, shipbuilding, in 1975. In 1977, attended Harvard's Advanced Management Program. Elected VP, shipbuilding, Feb 1, 1978. *Society Aff:* SNAME

Klink, Arthur E
Business: R50D-2 Scott Ave, Rahway, NJ 07065
Position: Research Fellow. *Employer:* Merck & Co Inc. *Born:* June 1943. PhD, MS, BS, BA Columbia Univ. Taught 2 yrs at Columbia prior to joining Merck & Co Inc Rahway N J. 10 yrs experience with Merck in the area of Process Dev starting from bench scale chem & biology through the Pilot Plant & on to full-scale design, start-up & long term optimization & consultation. Chmn Faculty Advisory Ctte, Columbia Chem Engrg Dept; Chmn Prof Dev Ctte & Mbr Exec Ctte N J Sect AIChE. Enjoys skiing, sailing & tennis.

Klinker, Richard L
Home: 2355 Davidsonville Rd, Gambrills, MD 21054
Position: Chief, Engrg Appl Branch *Employer:* General Services Admin *Education:* BS/Fire Protection Engrg/Univ of MD *Born:* 5/22/41 Native of Washington DC. Engr for MD Fire Underwriters Rating Bureau from 1963-1966. Employed by Chesapeake Division, Naval Facilities Engrg Command from 1966-1975. With GSA since 1975 as head of the Engrg Application's Branch, Accident and Fire Prevention Division. Am responsible for national fire protection engrg policy for new and existing Government owned and leased bldgs and the application of that policy. *Society Aff:* SFPE

Klinzing, George E
Business: 1235 Benedum Hall, Pittsburgh, PA 15261
Position: Prof *Employer:* Univ Pittsburgh *Education:* BS/ChE/Univ of Pittsburgh; MS/ChE/CMU; PhD/ChE/CMU *Born:* 3/22/38 An educator since 1963. Three years on assignment at Central Univ, Quito Ecuadar, Hon Prof Central Univ. Extensive teaching & educational consulting in Latin America. Research in gas-solid transport and mass transfer. Book - *Gas- Solid Transport* McGraw-Hill special interest in electrostatics in Pneumatic Transport, Mass Transfer in Liquid-Liquid Systems, SO2 and NOx Scrubbing Synfuels comparative studies, particle characterization filter cake analysis, Western Electric Teaching Award. *Society Aff:* AIChE, ASEE, ΣΞ

Klion, Daniel E
Home: 23 Jill Dr, W Nyack, NY 10994
Position: Chief Proj Engr. *Employer:* Ebasco Services, Inc. *Education:* BSME/Gen/Univ of IL. *Born:* 11/30/25. b NYC, Nov 30, 1925; s Harry & Jeannette Hilda (Stern) K; BS in ME, Univ of IL, 1950; m Helga Eichenwald, Sept 19, 1954; children-Pamela Diane, Roger Jay, Douglas Alan, Andrea Beth. With Ebasco Services, Inc, NYC 1950-55, 67-73 chief proj engr 1978-; with Burns & Roe, Inc, Oradell, NJ, 1956-63, Energy Corp of Am, NYC 1963-67, Stone & Webster Engg Corp, NYC 1973-78; mbr faculty NYU, NYC, 1965. Drainage Commr Town of Clarkstown (NY), 1968-69. Served US Army 1944-46. Reg PE, NY, PA, FL, TX, LA, GA, VA. Mbr ASME. Author: Handbook of Elec Engrs, 1978; patent application for film flow over plate surface. Home: 23 Jill Dr. W Myack NY 10994. Office: Two World Trade Ctr NYC, NY 10006. *Society Aff:* ASME, PMI, AIF.

Klipp, Verome M
Home: 2535 Navarra Dr, Carlsbad, CA 92008
Position: Pres *Employer:* GK2C Inc *Education:* BS/ME/IL Inst of Tech; BS/EE/IL Inst of Tech *Born:* 3/3/22 Subsequent to graduation in 1943, took advanced courses in fluid flow, heat transfer. 1943 to 1945 - Marine diesel engr for US Maritime Comm. 1945 to 1946 - US Army-Tech Sgt assigned to Los Alamos Project. 1946 to 1952 - Lundstum & Skubic Inc, chief engr for design & installation of bulk oil terminals, pumping stations, propane gas plants for Standard Oil, Shell Oil, Humble Oil. 1952 to 1963 - Abramson & Klipp-consulting engrs in private practice continuing as Jerome M. Klipp & Assoc from 1963 to 1981. Design of plumbing, HVAC, process piping & electrical systems in commercial, institutional, industrial and large residential bldgs. Excess of $20,000,000 per annum. Licensed as PE in 24 states throughout career. *Society Aff:* ASHRAE, CEC

Klippstein, Karl H
Business: Res Lab, 125 Jamison Ln, Monroeville, PA 15146
Position: Assoc Res Consultant. *Employer:* US Steel Corp. *Education:* MS/Civ/Rose Poly Inst; BS/Structural/Staatl Ing-Schule Dortmund; Technical Draftsman/Struc-Mech/Berufschule Dortmund. *Born:* Active in design & product dev (1956-1963) of ind specialty cranes, prefabricated metal bldgs; Gen Mgr (1963-1967) of the Jucho Co, Portland, OR; design, fabrication, & erection of pre-engineered steel systems; bridges; gen contracting. With US Steel Res since 1967; now heading Structures Res Group, conducting structural-mech-thermal res on bridges, blast furnaces, high-rise structures, equip. Papers & lectures mainly on fatigue of connections, cold- formed steel. Chrmn of AISI Advisory Group on Specifications for Design of Cold- Formed Steel Structural Mbr since 1977. Co-winner of ASCE 1979 Arthur M Wellington Prize. Mbr Natl Soc of PE (NSPE). *Society Aff:* NSPE.

Klipsch, Paul W
Business: PO Box 688, Hope, AR 71801
Position: Chairman of the Board. *Employer:* Klipsch & Associates Inc. *Education:* Dr Laws, Hon/NM State Univ; Engr/EE/Stanford; BS/EE/NM State Univ *Born:* 3/9/04. Elec Locomotive maintenance, Tocopilla Chile 1928-31; Geophysics 1934-41, Army Ordnance 1941-45; Audio Engrg (loudspeaker designs & manufac) Klipsch & Assocs 1946- . Num papers & pats in geophysics, firearms, ordnance & audio. Fellow Life, IEEE, Audio Engrg Soc; Mbr Acoustic Soc of Amer, Tau Beta Pi, Sigma Xi, Eta Kappa Nu. Episcopalian, Shriner, NRA (Life Mbr), listed Who's Who in Engrg (Lewis) 1938- . Silver Medal Audio Eng Soc 1978. Dr Laws Hons Causa 1981 NMSU. IEEE Centenial medal 1984. *Society Aff:* IEEE, AES, ASA

Klock, Neil H, Jr
Home: 4711 Whitehall Blvd, Alexandria, LA 71301
Position: Senior Engr *Employer:* Meyer, Meyer, Lacroix & Hixson, Inc *Education:* BS/Petroleum Engrg/LA State Univ *Born:* 2/5/37 Recently completed serving as Pres of the Alexandria, LA Chapter of the LA Engrg Society for 1980-81. He is a registered PE in civil and petroleum engrg and has worked as a senior design engr with Meyer, Meyer, LaCroix & Hixson, Inc, Alexandria, LA, since 1974. Mr Klock is a member of the Kiwanis Club, served on the Alexandria, LA Club's Bd of Direction for 1983-84, and was a Delegate to Kiwanis Intl Convention in Phoenix, AZ in June, 1984. He recently completed serving as State Dir, LA Engrg Society for 2 years from 1982-83 & 1983-84. Mr Klock is currently serving on the vestry of St Timothy's Episcopal Church, Alexandria, LA for a 3 yr term from 1983-86 and is currently serving on the Bd of Direction of LA State Univ's College of Engrg Alumni Assoc. *Society Aff:* NSPE

Kloeker, Delmar L
Home: 1416 Robin Hood Dr, Seymour, IN 47274
Position: President *Employer:* Traffic Engrg Studies, Inc *Education:* BS/Civil Engr/ Purdue Univ. *Born:* June 14, 1937; native of Seymour Ind. Asst Traffic Engr in Seymour, Dist of Ind St Hwy 4 yrs; Dist Traffic Engr of Ind St Hwy 4 yrs; Traffic Engr with R H Ludwig & Co 3 1/2 yrs; Traffic Engr with SIECO, Inc, 1973-83. President/Owner of Traffic Engr Studies, Inc 1983 to present. ITE - Fellow; Mbr ASCE, NSPE; P Pres & Newsletter Ed Ind Sect ITE. Respon with Traffic Engrg Studies Inc include design of traffic control devices, geometric design proj dev & client contact. Reg P E in Ind, Ohio, KY. Short course work at Purdue Univ, Univ of Md & Ga Tech, Univ of Tennessee, Univ of Wisconsin. Enjoys golf & travelling. April 1, 1983 formed Traffic Engrg Studies, Inc. Provides Traffic Engrg Consulting rental of Traffic Engrg Study Equipment. *Society Aff:* ITE, ASCE, NSPE.

Klopp, Richard P
Business: 1500 Market St, Philadelphia, PA 19102
Position: Chmn of the Bd *Employer:* Catalytic Inc. *Education:* BSCHE/Chem Engr/ Cornell Univ *Born:* Mar 1921. Joined Catalytic Inc 1952. Prior to becoming Pres in 1966 was successively Gulf Coast Area Sales Mgr, Mgr of Commercial Sales, Genl Sales Mgr, V Pres of Engrg & Sales & Exec V Pres. Previously had assignments in process & proj engrg & proj mgmt. Appointed Catalytic's Ch Exec Officer in 1968 & Chmn of the Board in 1971. Also is Mbr of AIChE, API, NPRA, Natl Soc of Engrs & a reg P E. *Society Aff:* AICHE, NSPE, API, NPRA

Klossner, Herbert O
Business: 320 Washington Ave S, Hopkins, MN 55343
Position: Dir. *Employer:* Hennepin Cty Dept of Transportation. *Education:* BSCE/Civil Engrg/Univ of MN. *Born:* 5/7/24. in Rice Lake, WI. Navigator 15th AF 1942-45. Various Engrg pos Milwaukee & W Allis 1949-65. Ch-Bureau of Engrg & Parking, Montgomery County MD 1965-68; managed county hwy sys & 1 of country's largest public parking sys. Ch Engr Hennepin County MN 1968-77. Dir DOT Henn Co 1977- . Engrg & mgmt control of hwy sys incl major freeway div. PPres Minn Sect ASCE; Past Pres NACE; Bd of Dirs ARTBA, NACO, & MGRI; PPres ARTBA Transportation Officials Div; NACE Res Chmn respon for prod of 17 volume action guides for County Engrs-Mgrs; 1974 selected 1 of Top 10 Public Works Officials Pres NACE 1979-80. NSPE Gov Advisory Group, Univ of Minn Civil & Mineral Engrg Adv Group. Recipient ARTBA's Ralph R Bartelsmeyer Award. ASCE Fellow. *Society Aff:* ASCE, NSPE, NACE, ARTBA, APWA, NACO, TRB.

Klotz, Bill W
Business: 1155 Dairy Ashford, Suite 705, Houston, TX 77029
Position: President. *Employer:* Klotz/Associates, Inc. *Education:* BSCE/Civil Engg/ TX A&M Univ. *Born:* 10/9/25. 1943-46 USAF Aerial Navigator -2nd Lt. Reg P E: Texas, Colorado, La, Alaska, Miss, Okla Mbrships: ASCE (Fellow); ACEC (Fellow) & NSPE, NCEE; PPres & Dir Cons Engrs Council, Texas; Texas State Board of Registration for Professional Engineers; Texas A&M Univ. Civil Engineering Advisory Council. Houston Engrg & Sci Soc; Tau Beta Pi; Texas Water Conservation Assn; Houston Chamber of Commerce; Past Chmn, Transp comm P Pres Victoria Chap TSPE; Victoria Chamber of Commerce; Jr Achievement of Victoria. Associated with Lockwood, Andrews & Newman, Inc 1948-1983, President 1974-1983; Director, Public Works, City of Houston, 1983; President, Klotz/Associates, Inc. - 1985 to date. *Society Aff:* ACEC, ASCE, NSPE, NCEE.

Klovning, John F
Business: 604 Wilson Ave, Menomonie, WI 54751
Position: Pres. *Employer:* Cedar Corp. *Education:* BSCE/Civ Engg/Univ of WI. *Born:* 9/12/44. & educated in WI. Army Intelligence Service 1965-67. Founded Cedar Corp (Menomonie Engg) 1975. Hold Admin & Design Supervision Responsibility for current staff of 28. Concurrently Reg PE E-13919, Architect A-4595, Land Surveyor S-1085, & Certified Soil Tester C595--State of WI. Serve as advisor to Village of Woodville & City of Menomonie Plan Commission. Prior to assuming Presidency of Cedar Corp, was employed by Owen Ayres & Assoc. Established & managed Comp Div servicing A/E disciplines plus accounting. Initiated original progs for Road Design, Water Distributicn Modeling & Computer Drafting. Free time enjoy cultivating cacti & succulents. *Society Aff:* ASCE, NSPE, CSI, AMA.

Klueter, Herschel H
Business: Bldg - 303 BARC - E, Beltsville, MD 20705
Position: Agri Engr. *Employer:* USDA-ARS Sensing Systems *Education:* PhD/Agri Engr/Purdue Univ; MS/Agri Engr/Univ of IL; BS/Agri Engr/Univ of IL. *Born:* 8/14/32. Raised on dairy farm in Edwardsville, IL. US Army 1953-1955. With USDA since 1959. Worked on res instrumentation and controls for animal and plant environment until 1976. 1976-1980 principal investigator for wind energy applications for storage and processing of agricultural products. Currently Agri Engr working on res instrumentation on Integrated Ctrls for Greenhouses. Also taught classes at Univ of IL and Univ of MD. Enjoy sports and square dancing. Reg PE in St of MD 9525. *Society Aff:* ASAE.

Klus, John P
Business: 432 N Lake St, Madison, WI 53706
Position: Professor Engr. Prof. Dev. *Employer:* Univ of Wisc-Madison. *Education:* PhD/Civil Engr/Univ of WI; MS/Civil Engr/MI Technological Univ; BS/Civil Engr/ MI Technological Univ. *Born:* June 1935. Prof of Civil Engrg. Taught at Mich Tech & Univ Wisc. At Wisc created concept of the Prof Dev Degree in Engrg & instrumental in dev other new continuing educ progs. Chmn 1972-present UNESCO Working Group on Continuing Educ. Numerous cttes: natl & local prof societies, univ & state. Co-author of engrg review text, Administration of Continuing Education text, & author of many tech & continuing ed publs. Awards incl Fulbright Scholar in Finland in 1966-67 & in 1985 & Distinguished Service Award ASEE/CES 1976. *Society Aff:* NSPE, ASCE, ASEE, AAAS.

Knaebel, John B
Home: PO Box 1329, Winston, OR 97496
Position: Mining Cousultant (Self-employed) *Education:* EM/Min & Geo/Stanford Univ; BS/Engrg/Stanford Univ. *Born:* 1/1/06 US Bur Mines (mining methods & costs) 1932-34. Developed, equipped, operated profitably a gold mine, Mindanao, PI 1934-37. Exploration & development, Western US, Mexico, Central America, BC 1937-40. Exploration, development, plant const, operation, lead-zinc and arsenic mines, NM & UT for US Smelting Refining & Mining, 1940-46. Anaconda Co, 1946-71, in charge expl & devel in Brit Guiana & Brasil (gold), uranium in NM (discovered, developed & operated Jackpile mine, largest freeworld producer) 1951-59. Iron ore, Ontario, copper in BC, N.B. Mex., and AZ 1959-68. Became VP New Mines 1964-71. Pres & VP various subsidiaries. Developed and operated Twin Buttes mine. Various Canadian properties. Mining man of year (Mining World magazine 1957). Recipient Saunders Medal (distinguished achievement in mining). AIME 1959; Jackling Award Soc Mining Engrs 1972; Distinguished member, AIME 1975. Mining consultant 1971 to date. *Society Aff:* AIME, SPE (Ontario & BC), AOPA

Knapp, Charles A
Business: PO Box 478, West Chatham, MA 02669
Position: Conslt *Employer:* Self *Education:* BCE/Environ Engr/NY Univ. *Born:* August 1916. Certificate of Value Engrg 1976. Jr Engr Army CE 1941-42. Proj & Field Eng Plant Engrg Div Dorr Co 1942-44. Sanitary Engr, Eastern Mgr & Natl Mgr, Sanitary Engrg Div Dorr-Oliver Inc manufacturing equip water/wastewater treatment 1944-63. Sr Assoc then V P Dir of Business Dev Metcalf & Eddy Inc Environ Engrs. 1963-82 Reg PE Mass, Conn, N Y. Mbr New England Water Pollution Control Assn (Pres 1974) Bd/Control Water Pollution Control Fed; Amer Public Works Assn (New England Chapter Sec/Treas 1979-85 & Man of Yr 1972); diplomate Amer Acad of Environ Engrs; Mbr New Eng Water Works Assn; Fellow Amer Soc of Civil Engrs; patentee sedimentavion equip, prefabricated sewage treatment plant. *Society Aff:* ASCE, WPCF, APWA.

Knapp, Roy M
Home: 3910 Northridge, Norman, OK 73072
Position: Prof & Dir *Employer:* Univ of OK. *Education:* DE/-/KS Univ; MS/Pet E/KS Univ; BS/Mech E/KS Univ. *Born:* 5/20/40. Native of Gridley, KS. Engg in Gas Supply, Employee Relations, and corporate Operations Res Northern Natural Gas Omaha, NB 1964-70. Returned to school for doctorate and a hydrological study of the Little Arkansas River basin. Taught petroleum engr at the Univ of TX 73-78. Currently Prof/Dir of Petr and Geological Engg at the University of Oklahoma. Special interests, microbial EOR, reservoir engr and simulation, and optimization methods. SPE: Prog, E and A & Manpower Committees, Distinguished Lecturer & PETE Faculty Award. Honoraries Tau Beta Pi, Sigma Tau, Pi Tau Sigma, Pi Epsilon Tau. Enjoy classical music, reading, handball, and canoeing. *Society Aff:* SPE, ASEE, IMACS

Knauss, Wolfgang G
Business: Caltech 105-50, Pasadena, CA 91125
Position: Prof of Aero. *Employer:* CA Inst of Tech. *Education:* PhD/Aero/CA Inst of Tech; MS/Aero/CA Inst of Tech; BS/Engg/CA Inst of Tech. *Born:* 12/12/33. Native of Germany, US college educ; Currently Prof of Aero and Appl. Mech. at Caltech. Res interests in linear & nonlinear viscoelasticity, mechanics of fracture in polymers & adhesion mechanics with emphasis on time-dependent quasistatic & dynamic effects; strong experimental interest. Consultant to tire & rubber industry, aerospace industry, & govt organizations. *Society Aff:* SR, AAM, SAE, SEM.

Knerr, Reinhard H
Business: AT&T, 555 Union Blvd, Allentown, PA 18103
Position: Supvr *Employer:* AT&T Bell Labs *Education:* PhD/EE/Lehigh Univ; MS/ EE/Lehigh Univ; Dipl Ing/EE/Ecole Nationale Superieure; Cand. Ing./EE/-. *Born:* 2/18/39 Native of Pirmasens, W Germany. Came to the US as a NATO scholar in 1962. Joined Bell Labs as a Member of Tech Staff in 1968. Was involved in research on microwave circulators, IMPATT power amplifiers, low noise and power GaAsFET amplifiers and satellite receivers. Published extensively in the field and holds 6 patents. He was then responsible for passive lightwave components & integrated optics dev. Presently, supervises work in lightwave local area network tech. Fellow IEEE, Member of IEEE-MTTS ADCOM, Editor of the IEEE Transactions on Microwave Theory and Techniques 1980, 81, 82. Pres MTTS 1986. C21576720AT&T Bell Labs *Society Aff:* MTTS, CS.

Knief, Ronald A
Business: PO Box 480, Middletown, PA 17057
Position: Mgr *Employer:* GPU Nucl Corp *Education:* PhD/Nucl Engr/Univ Illinois at Urbana-Champaign; BA/Phys Math Econ/Albion (Michigan) Coll *Born:* 10/08/44 Native of Dearborn, MI. Nucl Fuel Mgmt physicist for combustion engr 1972- 74. Assoc Prof of Chemical and Nuclear Engr at Univ NM 1974-80 specializing in nuclear reactor and fuel cycle safety and safeguards. Joined GPU Nuclear Corp in 1980 as mgr of training for Three Mile Island Nuclear Station; assigned as co-chrmn TMI-2 Programmatic Safety Overview Ctte in 1984-85. Author of books: *Nuclear Energy Technology* and *Nuclear Criticality Safety.* Chrmn ANS NM chapter, 1977, central PA chapter, 1984, and Nuclear Criticality Safety Division, 1979. Director nuclear criticality safety short courses since 1975. *Society Aff:* ANS, AAAS, INMM, ΣΞ

Knight, C Raymond
Home: 362 Overlook Trail, Annapolis, MD 21401
Position: Exec VP-Gen Mgr (Retired) *Employer:* ARINC Research Corporation. *Education:* MS/Physics/George Washington Univ; BS/Elect Engrg/Univ of UT *Born:* 9/25/18. Employed by Genl Electric Co 1940 as student engr. Worked in Electronics Dept of GE until 1951, advancing to pos of Mgr Electron Tube Application Engrg; later to Mgr Advanced Dev (tubes). Joined Aeronautical Radio Div as Asst Dir Reliability Res Dept 1951. After formation of ARINC Res Corp subsidiary, in 1958 became VP-Genl Mgr, Executive VP-Genl Mgr 1976, and retired in 1979. A Lic Prof Engr in D C & MD. Life-Fellow Mbr of the IEEE & Pres of the Reliability Group 1973- 75. Recipient of the Grp Annual Reliability Award 1976 and the IEEE Centennial Medal 1984. Hobbies: golf, cabinet making. *Society Aff:* IEEE(R)

Knight, Doyle D
Business: Dept of Mechanical and Aerospace Engrg, PO Box 909, Piscataway, NJ 08854
Position: Assoc Prof *Employer:* Rutgers Univ *Education:* PhD/Aeronautics/CA Inst of Tech; MS/Aeronautics/CA Inst of Tech; BS/Engrg/CA Inst of Tech; BA/Physics/Occidental Coll *Born:* 3/13/49 My research interests are broadly based in the area of theoretical fluid mechanics, with particular emphasis on solution of engrg problems. My work has involved both analytical and computational efforts in a variety of areas including the generation of water waves by wind, high speed and low speed boundary layer flows, high speed aircraft inlet flowfields with strong shockboundary layer interaction, decay of turbulent vortices, and the dynamics of free turbulent mixing flows. *Society Aff:* AIAA, APS, ASME

Knight, F James
Business: PO Box 1963, Harrisburg, PA 17105
Position: VP and Dir *Employer:* Gannett Fleming Geotechnical Engrs, Inc *Education:* MS/Geological Engrg/SD Sch of Mines & Tech; BS/Geological Engrg/MI Tech Univ *Born:* 05/14/34 Native of MI. Worked in mineral exploration and mine property evaluation for CF&I Corp in CO, AZ and other western states prior to completing MS. Was a graduate teaching assistant and wrote a thesis on complex diabase intunsions in North Central AZ. Joined GFC&C in 1963. Was Senior Engrg Geologist prior to assuming present position in 1979. Supervises staff of geologists and geotechnical engrs who provide consulting service in a wide range of engrg, architectural and mining applications. Reg PE/Geologist in PA, DE, AZ, IN & VA. Has held offices in the Society of PE including Pres of Harrisburg Chapter and PA State Membership Chrmn. Enjoys sailing. *Society Aff:* NSPE, ASCE, AEG

Knight, Herman E
Home: 6 Doral La, Columbine Valley, Littleton, CO 80123
Position: VP *Employer:* Gulf Mineral Resources Co (Gulf Oil) *Education:* Master/Mining Engrg/CO School of Mines *Born:* 10/16/16 Native of KY. Have held positions of Mining Engr, Chief Engr, Division Mgr for two major companies. Presently VP of large oil and mining co after several positions of Chief Engr, VP operations and VP engrg. Pres of Southern IL section of AIME. Member Governors Advisory Bd state of KY 1956-64. Have worked on mining projects in many Provinces of Canada, states of Mexico, Ethiopia, Tunisia, Spanish Sahara, Morroco, Spain and Perv. Army officer-World War II Europian Theatre. Registered PE: KY, MO, NM, and CO. *Society Aff:* AIME, SAIMM

Knight, Kenneth T
Home: 2620 Wells Ave, Raleigh, NC 27608
Position: Retired *Education:* BSEE/EE/Duke Univ. *Born:* Feb 27, 1912 Durham, N C. Engr CP&L Co 1933-43. Power Plant Operating Engr Camp LeJeune-Cherry Point 1943-47. Dir of Utilities Rocky Mt, N C 1947-50. PE North Carolina. Wm C Olsen & Assocs since 1950: Partner 1962, V Pres & Treas 1974. IEEE M 34, SM-48, State Chmn 1955-56. Duke Univ Coll of Engrg 'Distinguished Alumni Award' 1961. N C Assn for Retarded Children, Pres 1963-65. N C Soc of Engrs, Pres 1961. ASME, Mbr 1952, Chmn Prof Practice Ctte 1973-74, V Pres Region IV 1975-77. Pres Episcopal Laymen of Diocese of N C 1967-69. Enjoy family, church work, civic work & sports. *Society Aff:* IEEE, ASME, NSPE.

Knight, Ralph M
Home: 9638 E Mariposa Grande St, Scottsdale, AR 85255
Position: Retired. *Education:* BSChE/Chem Engg/Newark College of Engg. *Born:* Apr 19, 1915 Limestone Maine. m. Nancy Green Aug 8, 1964; children: Marie Elizabeth, Deborah Angelin, Calvin Lewis. Ed: High school Limestone, Maine 1933; Cum Laude Indus Cooperative Work from Newark, N J Col of Engrg 1939; post grad studies in Chem Engrg at Univ of Delaware, Newark, Del 1940-41. Major org affiliation: Amer Inst Chem Engrs, Fellow 1973; Mfg Chemists Assn, Dir 1975. Dev Chem Engr to Asst Div Supt, DuPont, Wilmington, Del 1939-52. Res Assoc, Standard Oil of Indiana, Whiting Res Labs 1952-53. V Pres Polymer Dev, Natl Distillers & Chem Corp 1953-60. Dart Indus Inc Los Angeles, Calif 1960-present; Exec V Pres & Dir of Dart Indus Inc; Pres & Ch Exec Officer Chem Group; V Pres C T Film Corp; Chmn Bd/Dir Dart Environ & Services Corp; Mbr Bd/Dir Estireno del Zulia (Venezuela) 1984 - semi-retired- Conslt-Mbr Bd/Dir Zimpro. Inc. *Society Aff:* AIChE, Tau Beta Pi.

Knoebel, David H
Business: Dept of Engg, Lake Charles, LA 70609
Position: Assoc Prof. *Employer:* McNeese State Univ. *Education:* PhD/Chem Engg/OK State Univ; MS/Chem Engg/OK State Univ; BS/Chem Engg/MI Tech Univ. *Born:* 1/7/40. Held positions of sr engr and sr res supervisor with DuPont, Savannah River lab, 1966-1974, nuclear reactor heat transfer and fluid mechanics res. Held a Visiting Prof position at Natl Tsing Hua Univ, Hsinchu, Taiwan from 1975-1977 in Nuclear Engg Dept. Am presently an Assoc Prof in Chem Engg at McNeese State Univ. Reg PE in LA. *Society Aff:* AIChE, NSPE, LES.

Knoedler, Elmer L
Business: 31 Light St, Baltimore, MD 21202
Position: 1982 Staff Consultant *Employer:* Sheppard T Powell Associates. *Education:* PhD/Chem/Columbia Univ; MChE/Chem/Columbia Univ; ME/Mech/Cornell Univ; Adv/Chem/Johns Hopkins. *Born:* Feb 1912. Engr Atlantic Refining Co 1934-35. Asst Supt Charge R&D, Davis Emergency Equip Co 1936-37. Charge Dev Metal Powder Proc, Metals Disintegrating Co 1939-41. Cons Engr, Sheppard T Powell 1941-58; Partner, 1958-present. Past Mbr, Bd/Reg for Prof Engrs & Land Surveyors, Md; Reg P E 22 states. Mbr AIChE; Fellow ASME, Fellow Amer Inst of Chemists; Mbr ACS; Fellow ACEC; Mbr ASTM; Engrs Club of Balto; S Sigma Xi; Phi Lambda Upsilon; Cornell Club of Md; Author num articles in tech & prof journals. *Society Aff:* ASME, ACS, AIChE, AIC, ACEC.

Knoell, William H
Business: 650 Washington Rd, Pittsburgh, PA 15228
Position: Pres & Ch Exec Officer (Mbr Bd Dirs). *Employer:* Cyclops Corp. *Education:* JD/-/Univ of Pittsburgh Law Sch; BS/Mech Eng/Carnegie Inst of Technology. *Born:* Aug 1, 1924 Pittsburgh. Bus career: practiced law 1949-50 Shoemaker & Knoell; Asst to Exec V Pres 1950-55 Pgh Corning Corp; Asst Secy, Secy, V Pres 1955-67, Crucible Steel Co of Amer; V Pres & Asst to Pres, a Dir, Exec V Pres, Pres, CEO Cyclops Corp 1967- . Pi Tau Sigma, Theta Tau, Beta Theta Pi, Phi Alpha Delta. Clubs: Univ, Duquesne, St Clair Country, Rolling Rock, Laurel Valley Golf. Mbr Bd of Dirs: Amer Sterilizer Co, Mbr Bd of Dirs: Duquesne Light Co Koppers Co Amer Iron & Steel Inst, Fed Reserve Bank of Cleveland, St Clair Mem Hospital, United Way of Allegheny County, Trustee (& mbr Exec Ctte) Carnegie-Mellon Univ, Mid-Atlantic Legal Fdn, Duquesne Club Chrmn, Adv Council of Jr Achievement of Penna.

Knoll, Glenn F
Business: 119 Cooley Lab, Ann Arbor, MI 48109
Position: Prof. & Chmn. *Employer:* Univ of Mich *Education:* B.S./Chem. Eng./Case Inst of Tech; M.S./Chem. Eng./Stanford; Ph. D./Nucl. Eng./Univ of Michigan *Born:* 08/03/35 Univ. of Mich Faculty mbr. 1962, Asst Prof; 1967, Assoc Prof; 1972, Professor, 1979 Dept Chairman. Nuclear engr, 1965-66, U.S. Scientist, Nuclear Research Center, Karlsruite, Germany. 1973 Senior Visiting Fellow, Univ. of Surrey. NSF Predoctoral Fellowship, Fulbright Travel Grant. Glenn Murphy Award, ASEE in 1979. Fellow of ANS and IEEE. Licensed P.E., Mich. Consultant to 15 industrial organizations. *Society Aff:* ANS, IEEE, ASEE

Knoop, Frederick R, Jr
Home: 219 Oak Forest Ave, Baltimore, MD 21228
Position: Partner. Retired *Employer:* Whitman, Requardt & Assoc. *Education:* BE/Civ Engg/Johns Hopkins Univ. *Born:* 5/4/20. With the exception of a five-yr period in gen contracting, was with Whitman, Requardt & Assoc, Consulting Engrs, 1946-85, in charge of various pojs including commercial and industrial bldgs, land

Knoop, Frederick R, Jr (Continued)
and utility master planning, water supply and sanitation. Partner in Whitman, Requardt & Assoc from 1964 until retirement in 1986. Served as officer in the US Army Corps of Engrs from 1941 to 1945. Final assignment Mapping Officer on the combined U.S.-British hdquarters staff of the Southeast Asia Command. Co-authored the section of the Std Handbook for Civ Engrs titled "Municipal and Regional Planning–, and written articles dealing with "The Role of Consulting Engrs in Plant Site Selection" and "Percentage Fees for Engineering Services." Has served as Dir of Dist 5 ASCE 1970-1973; Pres, MD Sec 1964; and on various other national committees of ASCE including chairmanship of The Committee on Profl Conduct. *Society Aff:* ASCE, AAEE.

Knop, Charles M
Business: 10500 W. 153rd St, Orland Park, IN 60462
Position: Chief Scientist, Dir. Antenna Res. *Employer:* Andrew Corp. *Education:* Ph.D.E.E/E.E./Ill. Institute of Technology; M.S.E.E./E.E./Ill. Institute of Technology; B.S.E.E./E.E./Ill. Institute of Technology *Born:* 02/18/31 Initial career spent soliciting Government R/D in technical areas of antennas and wave propagation at midwest and west coast R/D organizations; Since 1969 devoted R/D to commercial telecommunication antennas (Terrestrial and Satellite ESAs) at Andrew Corp; now investigating new areas for antennas/E.M. propagation applications as Chief Scientist at Andrew Corp. *Society Aff:* IEEE, APS, MTT, EMC

Knorr, David E
Business: Rt 2, Box 83, Humboldt, NE 68376
Position: Principal *Employer:* Knorr Engg *Education:* BS/Agri Engr/Univ of NE. *Born:* 4/14/48. Native of Waco, NE. Owner of Knorr Engineering since 1984, an Engg Consulting Company specializing in Agricultural Engg, Irrigation design, feasibility studies, quality control, pump & hydraulic design & expert witness. Experience with Westernland Roller Co. and Ingersoll-Rand Co. as product, test and sales engr as well as marketing. Professional activities include, past State Pres of ASAE and past state bd mbr & local pres of NSPE. Active in Rotary, Economic Development Board, family activities. *Society Aff:* ASAE. #TTL#Mr

Knorr, Jeffrey B
Business: Dept of Elec Engg, Monterey, CA 93940
Position: Prof *Employer:* Naval Postgrad School. *Education:* PhD/EE/Cornell; MS/EE/Penn State; BS/EE/Penn State *Born:* 5/8/40. LCDR USNR, active duty 1964-1967. Joined EE Faculty at Naval Postgrad School as Asst Prof in 1970; promoted to Assoc Prof in 1975, Prof in 1982. Appointed to Electronic Warfare Academic Group in 1976 and helped develop Navy's first MS degree prog in EW Systems. Appointed to Space Systems Academic Cttee in 1982 and helped develop Navy's MS program in Space Systems Engrg. Currently responsible for teaching and res in gen areas of electromagnetics, space systems and electronic warfare. Principal interest in microwave/millimeter waves and signal intelligence. Mbr IEEE Microwave Soc, Sigma Xi and Assoc of Old Crows. Married Suzanne Rumbaugh (1963). Daughters Kimberly (1966) and Tracy (1969). *Society Aff:* IEEE, AOC, ΣΞ

Knott, Albert W
Home: 1842 S Ivanhoe Street, Denver, CO 80224
Position: Dr *Employer:* Knott Laboratory, Inc. *Education:* PhD/Civil Engg/Stanford Univ; MS/Civil Engg/Univ of CA, Berkeley; BS/Arch Engg/Univ of CO, Boulder. *Born:* 4/26/31. Taught at Univ of Colorado, Penn State Univ, Univ of Oklahoma; engaged in private cons, struct, mech & arch testing. Principal in testing labs 1969-present; Reg PE. Primary ints: failure analysis of struc & mech sys; P Pres: Prof Engrs Colorado. Mbr NSPE, ASCE, SESA, ACI, AAAS, Sigma Xi, Metropolitan Sci Ctr, Pres, Knott Labortory, Inc, Denver Colorado. *Society Aff:* NSPE, ASCE, ACI, SESA.

Knott, Roger L
Business: 1150 W 3rd Street, Cleveland, OH 44113
Position: VP *Employer:* HWH Architects Engineers Planners Inc. *Education:* B.E.E./Electr. Engr./Fenn College (Now Part of CLeveland State Univ.) *Born:* 04/21/31 Native of Cleveland, Ohio. Electrical Engineer for the American Ship Building Co for six years. Chief electrical engineer for Adache Assoc., architects and engineers 1962-67. Joined HWH in 1967 and became V.P. 1975. Assumed present duties as Manager of Electrical Engrg 1973. Senior VP (Pres-Elect) of IESNA 1987-88. Reg engineer in OH, IL, TN, UT, CA, FL, and RI. Enjoy travel and photography. *Society Aff:* IESNA, IEEE, NFPA

Knotts, Burton R
Home: 7316 Dahlia Dr, Little Rock, AR 72209
Position: Electrical Engr *Employer:* Deitz Engineers Inc. *Education:* BS/EE/Univ of AR *Born:* 10/24/30 Native of Randolph Co, AR. Served with US Air Force 1949-52 as Radio Operator, in Japan and Korea 2 years. Worked in Plant Engrg for Pontiac Motor Division, in Electrical Design for 15 years in the Little Rock District, Corps of Engrs, Retired as Head, Specifications Dept, in December 1985, after 12 years of being responsible for all construction specifications prepared in the District. Currently is a Consulting Engineer with Deitz Engineers Inc., Little Rock. Past Pres, Little Rock Federal Employees Toastmasters Club. Enjoys music and genealogy. Has published 3 books on family genealogy. *Society Aff:* IEEE, SAME, CSI

Knowler, Lloyd A
Business: Division of Mathematical Sciences, The University of Iowa, Iowa City, IA 52242 *Employer:* The Univ of Iowa *Education:* Phd, MS, BA/-/Univ of IA *Born:* 01/30/08 A founder of American Society for Quality Control. Honors: Shewhart Medal, Oakley Awd, Fellow. Chair: Several important cttes, including Chair Prog Ctte of First Annual Meeting. Chair: Dept of Mathematics and Astronomy, The Univ of IA, 1946-59. Taught or Conducted: over 100 short courses in various univ. Several thousand representatives of indus. Consulted and/or taught in several major companies in US and Foreign. Listed in several biographical works, including American Men and Women in Science, Who's Who in America, Royal Blue Book, Who's Who in the World, Two Thousand Men of Achievement. Distinguished Alumni Achievement Award, Phi Beta Kappa, Society of Sigma Xi, Pi Mu Epsilon, Prof Engr, Certified Quality Engr, Certified Reliability Engr, Enrolled Actuary, Certified Pension Consultant. *Society Aff:* ASQC, ASPA, IAA, AAA

Knowles, Cyrus P
Business: 1401 Wilson Blvd, Suite 500, Arlington, VA 22209
Position: VP & Manager, Washington Operations. *Employer:* R&D Associates. *Education:* PhD/ME/Cornell Univ; MME/ME/Cornell Univ; BME/ME/Cornell Univ. *Born:* 8/19/37. Ford Fdn fellow, Cornell Univ 1962-65, Instr Cornell Univ 1966-67. Sr Staff Scientist, Avco Corp, Wilmington, MA 1967; Staff Mbr, Special Nucl Effects Lab, Gulf Gen Atomic, La Jolla CA 1967-68; Staff Scientist, Radiation Hydrodynamics Div Mgr, Assoc Mgr Theoretical Sci Div. Systems Sci & Software, La Jolla CA 1968-72; Sr Scientist, Assoc Div Mgr, Proj Mgr R&D Assoc, Marina Del Rey, CA 1972-1979; Asst to Deputy Dir for Sci & Tech, Defense Nucl Agency, WA DC, 1979-82. Assistant Deputy Undersecretary of Defense (Offensive & Space Systems), Office of the Secretary of Defense 1982-84. VP & Manager of Nuclear Weapons Division, R&D Associates, Marina Del Rey, Calif 1984-87; VP & Manager, Washington Operations R&D Associates 1987- . Res & publications on hydrodynamics, radiative transfer, nucl weapons effects, testing, weapon systems, & policy issues. Responsible for broad range of res & testing for Dept of Defense. *Society Aff:* APS, AIAA.

Knowles, James K
Business: 1201 E CA Blvd, Pasadena, CA 91125
Position: Prof. *Employer:* CA Inst of Tech. *Education:* PhD/Math/MIT; SB/Math/MIT; D. Sc./Honorary/National Univ. Ireland. *Born:* 4/14/31. Native of Cleveland, OH. Instructor in mathematics, MIT, 1957-58. Caltech faculty, 1958-present. Now Prof of Applied Mechanics. Primary interests: mechanics of solids, applied mathematics. *Society Aff:* ASME, SIAM, AAUP.

Knudsen, Clarence V
Home: 243 North 24th, Camp HIll, PA 17011
Position: Consultant Retired *Employer:* Michael Baker, Jr., Inc. *Education:* BS/CE/Univ of NB *Born:* 6/30/09 NE Bridge Dept-Designer Modjeski & Master-Project Eng, Michael Baker, Jr Inc Chief Bridge Eng, Office Mgr, Chief Eng, Consultant (Retired). Chief Engr on following projects: New River Gorge Bridge-Longest Arch in World. Betsy Ross Bridge over Del River. Three Ohio River Bridges. Michie Stadium-West Point Military Academy. Beaver Stadium-Penn State Univ. Stadium at Vanderbilt Univ. Retired Consultant on variety of assignments, including expert witness and value engineering. *Society Aff:* ASCE, NSPE

Knudsen, Dag I
Business: 270 Metro Sq. Bldg, St. Paul, MN 55101
Position: VP *Employer:* EMAinc-Computer System Consultant *Education:* BS/EE/OR State Univ *Born:* 9/9/40 Sr. Member of IEEE/Active in its Power Engrg Society. Sr Member of ISA-Dir (1979-1981) of its Water and Wastewater Industries Div. Registered PE (EE) MN. Registered Control System Eng-CA. Speaker on Automatic Control Systems, Computer Systems, and Instrumentation at National and Local Meetings of ISA, IEEE, UTC, WPCF, AWWA. 16 years of progressive experience as field sales engr, application specialist, branch mgr, and Systems Application Engr (with Leeds & Northrup Co), Project Mgr, and VP (with EMAinc-Computer System Consultants). *Society Aff:* IEEE, ISA, WPCF, AWWA, AMA

Knudsen, James G
Business: Oregon St U, Corvallis, OR 97331
Position: Prof *Employer:* Oregon St U *Education:* PhD/Chem Engg/Univ of MI; MS/Phys Chem/Univ of Alberta; BS/Chem Engg/Univ of Alberta. *Born:* March 27, 1920. Native of Youngstown Alberta, Canada. With Oregon State Univ since 1949, performing teaching & res in chem engrg. Assumed respon as Asst Dean of Engrg in charge of Engrg Experiment Station in 1959-1971 Assoc Dean in 1971-1981. Also is Dir (1974-80) of Oregon State Univ Office of Energy Res & Dev. Dir, Amer Inst of Chem Engrs 1974-76. VP 1979 President 1980 D.Q. Kern Award 1981, Founders Award 1977. es, cons & publ in heat transfer & fluid mechs. Hobbies incl stamp collecting, woodworking & gardening. *Society Aff:* AIChE, ACS

Knuth, Eldon L
Home: 18085 Boris Dr, Encino, CA 91316
Position: Prof. *Employer:* UCLA. *Education:* PhD/Aero/CA Inst of Tech; MS/AeroE/Purdue Univ; BS/AeroE/Purdue Univ. *Born:* 5/10/25. Aerothermodynamics Group Leader, Aerophysics Dev Corp, 1953-56; Assoc Res Engr, UCLA, 1956-58; Assoc Prof, UCLA, 1959-65; Prof, UCLA, 1965-. Gen Chrmn, Heat Transfer & Fluid Mechanics Inst, 1959; Hd, Molecular-Beam Lab, 1961-; Chrmn, Energy & Kinetics Dept, 1969-75; von Humboldt Awardee & Visiting Scientist, Max-Planck Institut fur Stromungsforschung, Gottingen, W Germany, 1975-76. Author, Introduction to Statistical Thermodynamics, 1966; patentee, Radial-Flow Turbomolecular Pump; author, numerous papers in field. Reg Chem Engr, Mech Engr, & Fire Protection Engr. Consultant to aerospace ind in area of rarefied gas dynamics; consultant to litigants in area of fires & explosions. *Society Aff:* AIAA, AIChE, APS, ASEE, NFPA, ΣΞ, AAUP.

Ko, Hon-Yim
Business: Dept of Civil Engrg, Boulder, CO 80309
Position: Professor of Civil Engrg. *Employer:* University of Colorado. *Education:* BS/Civil Engrg/Univ of Hong Kong; MS/Civil Engrg/CA Inst of Tech; PhD/Civil Engrg/CA Inst of Tech *Born:* Jan 18, 1940 Kong Kong. After working at the Jet Propulsion Lab, joined Univ of Colorado at Boulder in 1967 & became Prof of Civil Engrg in 1975 and Chrmn of Dept in 1983. Mbr ASCE, ASEE, Soc of Experimental Stress Analysis & Chi Epsilon. ASEE Dow Outstanding Young Faculty Mbr 1975 ASCE Huber Research Prize, 1979. Res specialties in constitutive properties of matls such as soils, rocks, concrete & coal and centrifugal modeling of geotechnical structures. Holds two U S Pats in multiaxial test devices. Cons on Surveyor & Viking projs & geotechnical engrg projs. *Society Aff:* ASCE, ASEE, SESA, XE.

Ko, Hsien C
Business: 2015 Neil Ave, Columbus, OH 43210
Position: Professor & Chrmn of Electrical Engg. *Employer:* Ohio State University. *Education:* PhD/EE/Oh State Univ; MSc/EE/OH State Univ; BSc/EE/Natl Taiwan Univ. *Born:* Apr 28, 1928 Formosa. Radio Wave Res Labs Formosa 1951-52. With Ohio St Univ 1952- . Asst Dir Radio Observatory 1955-65. Currently Prof and Chrmn of EE & Prof of Astronomy. Assoc Ed of Radio Science 1967-75. Fellow IEEE & Royal Astronomical Soc London. Res activities in electromagnetics & antennas, radio astronomy & astrophysics, plasma radiation, relativistic electrodynamics. *Society Aff:* IEEE, URSI, ASEE.

Ko, Wen H
Business: Rm-107, Bingham Bldg, 10900 Euclid Ave, Cleveland, OH 44106
Position: Prof of EE *Employer:* Case Western Reserve Univ. *Education:* -/Medical Engg/Stanford Univ; PhDEE/EE/Case Inst of Tech; MS/EE/Case Inst of Tech; BS/EE/Natl Amoy Univ. *Born:* 4/12/23. Wen Hsiung Ko was born in Fukien, China, 1923. He received the BS degree from the Chinese Natl Univ of Amoy in 1946, & the MS & PhD in EE from Case Inst of Tech, Cleveland, Ohio, USA, in 1956 & 1959, respectively. Since 1954 he has been with Case Inst & Case Western Reserve Univ as a grad asst, res assoc & faculty mbr. He is now a Prof of EE & a Prof of Biomedical Engg. Dr Ko was Dir of the Engg Design ctr, the Microelectronics Lab for Biomedical Sci, & the Biomedical Electronics Resource at Case Western Reserve Univ from 1970 to 1983. His interests are in microelectronics, control systems, micro-transducers design, & biomedical instrumentation. He is a Fellow of IEEE, a mbr of Sigma Xi, Eta Kappa Nu & ISHM. *Society Aff:* IEEE, ΣΞ, HKN, BME, ISHM.

Kobayashi, Albert S
Business: Dept of Mech Engrg, Seattle, WA 98195
Position: Professor. *Employer:* University of Washington. *Education:* BS/Precision Engg/Univ of Tokyo; MS/Mech Engg/Univ of WA; PhD/Mech Engg/IL Inst of Tech. *Born:* 12/9/24. Native of Chicago IL & Tokyo Japan. Tool engr in Konishiroku Photo Indus 1947-50. Design engr for IL Tool Works 1953-55. Res engr for Armour Res Foundation IIT spec in experimental mechanics 1955-58. with Univ of WA 1958- , starting from Asst Prof, Assoc Prof & currently Prof in ME. More than 225 publs in field of struct mechs, experimental mechs, fracture mechs, & biomechs. F T Tatnall Award, B. J. Lazan and R. E. Peterson Awards and W. M. Murray Medal from SESA in 1973, 1981, 1984 and 1984, respectively and SESA Fellow in 1977. Also ASME Fellow in 1981. Various mbrships' chmnships in SESA cttes. Ed of 5 books incl SESA Monograph Experimental Techniques in Fracture Mechs I & II and Manual an Experimental Analysis, 2nd and 3rd editions and Handbook of Experimental Mechanics, Assoc. ED. of Trans. of Japan Soc. for Composite Materials 1975- . *Society Aff:* SEM, ASME, ACS, NAE.

Kobayashi, Koji
Business: 33-1, Shiba 5-chome, Minato-ku, Tokyo 108 Japan
Position: Chmn of the Board & Ch Exec Officer. *Employer:* NEC Corp *Education:* Doctor/Engrg/Tokyo Imperial Univ; Doctor (hon)/Laws/Monmouth College; Doctor (hon)/Engrg/Polytechnic Inst; Doctor (hon)/-/Autonomous Univ. of Guadalajara, Mexico; Doctor (hon)/Science/Univ of Philippines; Membership (hon)/-/Nat'l Inst for Higher Education. *Born:* 2/17/07. Joined NEC Corp 1929, Dir 1949, Sr V Pres 1956-61, Exec V Pres 1961-62, Sr Exec V Pres 1962-64, Pres 1964-76, Chmn of the Board & Ch Exec Officer 1976- ; Governing Dir Japan Federation of Employers' Assoc 1966- ; Permanent Dir Japan Mgmt Assn 1953- ; Purple (1957)/Blue Ribbon Medals (1964), and 1st Class Order of Sacred Treasure, from His majesty the Emperor of Japan 1978; Frederik Philips Award (1977)/IEEE Founders medal (1984) from IEEE; Special United Nations Medal of Peace for International Law Enforcement Cooperation from International Narcotic Enforcement Officers Association of

Kobayashi, Koji (Continued)
the United Nations (1986). *Society Aff:* FEO, CIOS, IEEE, COR, NAE, IIC, IIASA, SATW.

Kobayashi, Riki
Business: Geo. R. Brown School of Engg, Dept. of Ch. E, Rice Univ., P.O. Box 1892, Houston, TX 77251
Position: Louis Calder Prof of Chem Engrg. *Employer:* Rice University. *Education:* PhD/ChE/Univ of MI; MSE/ChE/Univ of MI; BS/ChE/Rice Inst *Born:* May 13, 1924. Awarded Chair at Rice Univ: The Louis Calder Professorship Chair in Chem Engrg 1967. D L Katz Distin Lectr Apr 1975; AIChE Fellow 1975. Author of over 170 papers in Chem Engrg, Petro Tech, Physical Chem & Chem Phys: 1949- . Co-author: *Handbook of Natural Gas Engrg* McGraw-Hill 1959. Cons in Oil & Natural Gas Tech. Also: Fellow, Am Inst Chemist. *Society Aff:* AIChE, ABS, Am Inst Phys, AIME

Kobayashi, Shiro
Business: Dept of Mech Engg, Berkeley, CA 94720
Position: Prof. *Employer:* Univ of CA, Berkeley. *Education:* PhD/Mech Engg/Univ of CA Berkeley; MS/Mech Engg/Univ of CA Berkeley; BS/Precision Engg/Tokyo Univ Tokyo, Japan. *Born:* 2/21/24. Came to the US in 1956. A Fellow of ASME and of SME, and serves as an editorial bd mbr of several intl journals. Received the ASME Blackall Machine Tool and Gage Award and the Aida Engg Award from Japan. A Battelle Visiting Prof at the OH State Univ in 1967, E A Taylor Visiting Prof at the Univ of Birmingham, England, in 1970 and a Miller Res Prof at the Univ of CA, Berkeley, 1977. Co-author of the book "Mechanics of Plastic Deformation in Metal Processing–. About 150 technical papers. Natl Acad of Engrg, Mbr (1980). SME 1983 Gold Medal. *Society Aff:* ASM, ASME, SME, NAE.

Kobler, Helmut G
Home: 339 Fourth Ave, San Francisco, CA 94118
Position: V. P. General Manager *Employer:* Harrison Western Corp. *Education:* BSC/Mining-Constr/Univ of Pribram. *Born:* 1/18/28. Native of Austria. Grad 1949 from Mining Univ Pribram Czechoslovakia. Worked in coal mines in England, gold and basemetal mines in Canada. Held a number of engg, supervisory, and mgt positions on maj underground construction projs - the La Planicie and Ocumitos Hgwy Tunnels in Venezuela; Montreal Subway tunnels & stations; the Utardes No 3 Underground Power Pro in Quebec; the Yamuna Hydro Power Proj in India; the Columbia River Diversion Tunnels at Mica Creek, British Columbia; the 12th St Station and Tunnels of BART, Oakland, CA. In Chile he managed the construction and start-up operation of the Andina Rio Blanco 10, 000 TPD Underground Copper Mine. 1971 joined Jacobs Assoc as Resident Mgr of the Tapanti Hydro Proj in Costa Rica. 1973 became an Assoc of the firm and in 1979 was named VP & Gen Mgr of Jacobs Assoc, SA. March 1981 joined Harrison Western Corporation, an underground construction and mining company, as VP & General Manager of their Latin American Division.. *Society Aff:* ASCE, AIME, NSPE, SAME, APEO.

Kobs, Alfred W
Home: 7619 Windswept Ln, Houston, TX 77063
Position: President. *Employer:* Kobs Engineering Inc. *Education:* MSIE/Ind Engg/Univ of Houston; BSChE/Chemical Engg/Univ of TX-Austin; AA/- /Lon Morris College. *Born:* Sept 1919; native of Houston. Field Engr Eastern States Petro. US Navy 1944-46 & 1950-52. Production Engr Union Carbide Corp 1946-62. Pres Kobs Engrg Inc 1962- . Mbr AIChE, ISA, ASHRAE, TSPE & HESS. Married to former A Pidd Miller; 1 daughter. *Society Aff:* AIChE, TSPE, ISA, ASHRAE.

Kocaoglu, Dundar F
Business: 1178-B BEH, University of Pittsburgh, Pittsburgh, PA 15261
Position: Dir, Engrg Management Program *Employer:* Univ of Pittsburgh *Education:* PhD/Op Research/Univ of Pittsburgh; MS/IE/Univ of Pittsburgh; MS/CE/Lehigh Univ; BS/CE/Robert Coll-Turkey *Born:* 6/1/39 Design engr at Modjeski & Masters, 1962-63; structural engr at United Engrs, 1963-66 and 1969-71; partner in Tekser consulting firm in Turkey, 1966-69. With the Univ of Pittsburgh since 1971. Currently Assoc Prof of Industrial Engrg and Dir of Engrg Management Program. Principal Investigator for research projects totaling over $1 million. Founder and Chrmn of TIMS Coll on Engrg Management (1979-81), Chrmn of ASEE Engrg Management Division (1982-83), Dir of American Society for Engrg Management (ASEM) (1981-86), Ad Com Mbr, IEEE Engrg Mgmt-Soc (1982-Present), Pres, Omega Rho Intl Honor Soc (1984-86), Honorary Member, MIM-Society of Engineers, Scientists and Architects. Recipient of UN grants for tech management and tech transfer 1979-80, 87. Co-Author of *Engineering Management* (McGraw-Hill, 1981), Editor of *Management of R&D and Engineering* (Elsevier/North Holland, 1987) and *Handbook of Technology Management* (John Wiley and Sons, Inc. 1988). Author of over 30 papers on engrg and R & D, Editor of Wiley Series in Engrg Management. Editor-in-Chief of IEEE Transactions on Engineering Management. Listed in Who's Who in Science and Technology, Who's Who in the East, Who's Who in the World and Who's Who in Turkey. *Society Aff:* ASEM, ASCE, ASEE, IEEE-EMS, TIMS, Omega Rho, MIM

Koch, Leonard J
Business: 6 Post Dr, Decatur, IL 62521
Position: Conslt *Employer:* Self *Education:* MBA/Bus Admin/Univ of Chicago; BS/Mech Engrg/IL Inst of Technology. *Born:* 3/30/20. 1948-72 Argonne Natl Lab: Assoc Proj Engr for EBR-I, Proj Mgr of EBR-II, Dir of Reactor Engrg Div; 1972-83, IL Power Co VP. Respon for Nuclear Power Engineering and operations. Member National Academy of Engineering; Fellow of Amer Nuclear Soc; Recipient of Pi Tau Sigma "Richards Meml Award" 1965; US Delegate to 3 Internatl Atomic Energy Conferences; Adjunt Prof Univ of IL, 1983- . *Society Aff:* ANS.

Koch, William A
Business: 343 State St, Rochester, NY 14650
Position: VP, Gen Mgr, Motion Picture & Audiovisual Mkts Div *Employer:* Eastman Kodak Company. *Born:* July 19, 1926. Gettysburg Coll. AB Major in Physics 1949. Advanced from Chem Technician to Motion Picture Engrg Representative to Ch Engr in 3 regional offices of Motion Picture Film Dept of Eastman Kodak Co 194967. Became part of field mgmt for Motion Picture & Educ Markets Div & later Motion Picture & Audiovisual Markets Div 1967-74. Advanced to Sales Manager 1974. Assumed present position 1982. Served 2 terms as Gov of Soc of Motion Picture & TV Engrs & was elected Fellow Mbr of Soc 1972. *Society Aff:* SMPTE, BKSTS.

Koch, William H
Business: Natl Av Facils Exp Ctr, Atlantic City, NJ 08405
Position: Ch Simulation & Analysis Div. *Employer:* Federal Aviation Administration. *Born:* Aug 1933. BSEE Univ of Conn; grad work in phys & computer sci G W Univ & Univ of Va. R/D work Melpac Inc in battlefield surveillance sys, ECM & high speed computer circuits. With Federal Aviation Admin, Natl Aviation Facilities Experimental Ctr 1959- , in field of air traffic control automation. Mbr AIAA & ORSA, Charter Mbr Univ of Conn Engrg Alumni Assn.

Kocian, James J
Business: 5251 Westheimer/B 22029, Houston, TX 77027
Position: Sr Vice Pres. *Employer:* Kaneb Services Inc. *Education:* BS/ChE/TX A&M. *Born:* Sept 30, 1935; native of Shiner Texas. BSChE Texas A&M Univ 1956. Engr Union Carbide Corp serving in various capacities in Olefins mfg, business mgmt & long-range planning. Kaneb Services Inc Nov 1, 1973- ; V Pres Planning, Sr VP-coal oper, respon for long-range planning & coordination of all coal activities of the Co. Reg PE: Texas. *Society Aff:* AIChE.

Kocks, U Fred
Business: Center for Materials Science, Los Alamos Nat. Lab, Los Alamos, NM 87545

Kocks, U Fred (Continued)
Position: Fellow *Employer:* Los Alamos Nat. Lab. *Education:* PhD/Applied Physics/Harvard; MS/Physics/Gottingen; BS/Physics/Stuttgart. *Born:* 11/25/29. Studied physics at Univ of Stuttgart & Gottingen Germany. Immigrated 1955, US citizen 1962. Ph.D. Harvard 1959 Asst Prof Harvard to 1965. Senior Scientist, Argonne Natl Lab 1965-83, Visiting Prof Tech Univs of Munich 1964 & Aachen 1971-72. Visiting Prof. McMaster Univ Hamilton Ont 1978. Humboldt Awardee TH Aachen Germany 1979. Honorary Dr. Tech., Tampere Institute of Technology (Finland) 1982. Senior Scientist Award, Japan Soc. for the Promotion of Science 1985. Fellow, The Metallurgical Society of AIME 1987, ASM. *Society Aff:* TMS-AIME, ASM

Koczak, Michael J
Business: Drexel University, Dept of Materials Engr, Philadelphia, PA 19104
Position: Prof *Employer:* Drexel Univ *Education:* PhD/Metallurgy & Materials Sci/Univ of PA; MS/Metallurgy & Materials Sci/Univ of PA; BS/Metallurgical Engrg/Polytech Inst of NY *Born:* 4/29/44 Recipient of the prestigious Lindback Award for outstanding undergraduate teaching, he conducts courses in composite materials, powder metallurgy, ceramics and microscopy. Research interests include powder metallurgy, metal and ceramic matrix composites and facture analysis with over eighty publications in these areas. Research funding in these areas has been provided by NASA, ONR, AFOSR as well as other governmental and industrial sources. He received a senior visiting fellowship from the Science Research Council of England, while on faculty leave at the Univ of Surrey. He also serves as a consultant to industry and governmental agencies. He served as a liason sci for the Office of Naval Res Far East Tokyo Office. For service in this capacity he rec the meritorius civilian service Medal from the US Navy. In addition he rec Drexel Univ's res scholar award and the Dept of Mat Engrg teaching award. He is a mbr of AIME, ASM, ACS, SAMPE. Tau Beta Pi, Alpha Sigma Mu, and Sigma Xi. *Society Aff:* ASM, AIME, MPIF, ACS, SAMPE, ΣΞ, ΤΒΦ.

Kodali, V Prasad
Business: Dept Electronics, Lok Nayak Bhavan, New Delhi 110003, India
Position: Advisor, Govt of India. *Employer:* Government of India *Education:* PhD/EE/Univ of Leeds, England; MS/EE/Case Inst of Tech; BE/Telecom Engrg/Univ of Madras, India *Born:* 11/1/39 Taught at the Univ of New Hampshire (1963-64) USA. Worked with Microwave Assocs Limited (1967-68) England. Worked at Tata Inst of Fundamental Research, Bombay, India with primary responsibility of design and development of tracking radars for India's Space programmes. Published 25 research papers in circuit synthesis, computers, microwave semiconductors and radar electronics. Held position as Dir (Tech) in the Dept of Electronics of Government of India from 1973 to 1984 with responsibility for work in the areas of radars, navigational aids, sonars and Under-water electronics. Held (1984-86) position as Dir of Electronics & Instrumentation in the Defense Res and Dev organization with respon for Defense related R&D in the areas of electronics and opto electronics. Holding (since 1986) position as Adviser Dept of Electronics with responsibility for microwave and defense electronics areas. *Society Aff:* IEEE-Fellow, IE-Fellow, IETE-Fellow.

Koehler, Melvin L
Home: 1146 Hathaway Lane, Marion, OH 43302
Position: Consultant *Employer:* Self employed. *Education:* BSc/Civ Engg/Univ of NB. *Born:* 10/30/26. Native of Wausa, NB. Served in USNR, 1944-46. Grad, Univ of NB. Employed as field engr designer, US Bureau of Reclamation; design engr with Silas Mason Co & planning engr with Peter Kiewit Son's Co. With Floyd Browne Assoc, Inc from 1955 until retirement in 1986 in various capacities involving water & sewer projs; Served as Gen Mgr & Chrmn from 1977 until retirement; directed all operations of the regional consulting engineering firm and its laboratory and comp subsidiaries. Currently self employed as consultant. Pres Marion Chapter OSPE, 1970; Rotarian; Commander CEC USNR-R. *Society Aff:* ASCE, NSPE, WPCF, ASME.

Koehn, Enno
Business: Civil Engineering Dept, Lamar Univ., P.O. Box 10024, Beaumont, TX 77710
Position: Professor & Head of Civil Engineering. *Employer:* Lamar University *Education:* PhD/Civil Engr/Wayne State Univ; MS/Civil Engr/Columbia Univ; BCE/Civil Engr/City College, NY. *Born:* 4/29/36. Flushing NY; BCE CCNY; MS Columbia; MCE NYU; PhD Wayne St Univ. Res Engr N Am Aviation Columbus OH; Asst Prof Engg Sci LIU Brookville NY; Educ Specialist IBM Burlington VT; Prof Civ Engg OH Northern Univ Ada. Chrmn, Dept of Civil Engr, Lamar Univ.; NSF res grantee; Dept of Defense grantee; Summer Fellow MIT. Reg PE: NY, OH, IN, TX. Mbr Sigma Xi, Tau Beta Pi, Chi Epsilon; AACE; NSPE; ASEE; fellow ASCE (chrmn ctte on social & environ concerns in const), Past Pres Toledo and Indiana Sects; AAUP, Treas OH N Univ Chap. Outstanding Educator of Amer Award. Author: Syllabus for Computer Aided Design; Computer Progs for Civ Engg Students; Physical Testing for Civ Engrs; Pre Dsgn Cost Est Function for Bldgs; Social and Environ Costs in Const; Cost of Social & Environ Reg; Cost of Delays in Const; Using Practitioners to Teach-Observations during a one-yr experience; Cost of environ impact statements; Work Sampling and Unit Manhours; Benefits and Costs of EEO Reg in Const; Fuzzy Sets in Const Engrg; Geo Distribution of Work Environ; Costs and Benefits of MBE Reg in Const; Climate Effects on Const; Const Proj Environ-Perceptions of Mgmt; Internat Labor Producitivity Factors; Amer Dsgn & Const Tech, *Society Aff:* ASCE, ASEE, AACE, NSPE.

Koehne, Anthony J
Home: 285 Oak Ct, Severna Park, MD 21146
Position: Vice Pres & Chief Operating Officer. *Employer:* Poly-Seal Corporation. *Education:* MBA/-/Harvard Univ; BS/Mech Engr/Univ of WI. *Born:* Dec 1920. Capt Army Ordnance 1942-46. Maintenance Engr Exxon 1948-53, Plant Engr to Plant Mgr Allied Chem 1953-66; Genl Mgr Gordon Chem 1966-68; V Pres Poly Seal Corp 1966-present. Mbr AIChE, SPE. Married, 5 children. Enjoys fishing. *Society Aff:* AIChE, SPE.

Koelliker, James K
Business: Dept of Civ Engrg, Seaton Hall, Manhattan, KS 66506
Position: Prof *Employer:* KS State Univ *Education:* PhD/Agric Engrg/IA State Univ; MS/Water Resources/IA State Univ; BS/Agric Engrg/KS State Univ *Born:* 3/24/44 Native of White Cloud, KS. Civ engr for Soil Conservation Service, 1966-67. Civ engrg officer with USAF, 1971-73. Teaching and res experience in Agric Engrg Dept, KS State Univ, 1973-76; Agric Engrg Dept, OR State Univ, 1976-78. Assumed current responsibility in 1978 as Prof of Civ Engrg Dept for teaching and res in water resources and environmental engrg. Res work concentrated in applications of continuous computer simulation of the water budget and groundwater recharge. Reg PE, KS & CO. *Society Aff:* ASAE, ASCE, NSPE, ASEE.

Koelzer, Victor A
Home: 1604 Miramont Dr, Fort Collins, CO 80524
Position: Prof. *Employer:* Colorado State University. *Education:* MSc/Hydraulics/Univ of IA; BSc/Civil Engg/Univ of KS. *Born:* 5/3/14. Career in water resources dev, including 1938-40 US Geol Survey IA City IA; 1940-42 US Corps of Engrs Louisville, KY; 1942-46 Officer USA Navy Civil Engg Corps; 1946-56 US Bureau of Reclamation Denver & Loveland CO & Wash, DC; 1956-69 Harza Engg Co Chicago (Assoc 6 yrs, VP 1 yr); 1969-72 US Natl Water Comm Wash DC (Ch Engr); 1972-84 Independent Cons, Prof Civil Engg, CO State Univ & Pres Engg Farms, Inc. FT Collins CO. Awarded 1st Julian Hinds Award of ASCE for contributions to 'planning, dev & mgmt of water resources'. *Society Aff:* ASCE, ASEE, AWRA, ICID.

Koeneman, Robert M
Home: 1615 N 78th Ave, Elmwood Park, IL 60635
Position: VP & Gen Mgr. *Employer:* Remcor Products Co. *Education:* MBA/Finance/Univ of Chicago; Engg Chemistry/Engg/Christian Brothers College. *Born:* 12/22/42. Presently chief operating officer of Remcor Products Co, a Chicago based, firm specializing in the mfg of engineered refrig products. These range from ice makers & dispensers to specialty refrigeration packages. Several patents filed for ice dispenser devs. Previous experience includes proj mgt with responsibility for the design & construction of ind & cryogenic gas plants, including CO-H2, air separation, CO2, & medical gases. Education includes undergrad degree in Engg Chemistry, & MBA (finance) from the Univ of Chicago. Active for 15 yrs in the Natl & Chicago chapters of AIChE. Previously pres of student chapter of ACS. *Society Aff:* AIChE.

Koenig, Charles Louis
Home: 26890 Sherwood Forest, San Antonio, TX 78258
Position: President. *Employer:* Louis Koenig-Research (A Corp). *Education:* PhD/Phys Chem/NYU; BS/Chem/NYU. *Born:* Oct 11, 1911 Yonkers N Y. PhD NYU Physical Chem 1936. Chemist, Solvay Process Co, Syracuse; alkalies, heavy chems, salts, brines. Chemist & tech sales Lithaloys Corp N Y; lithium & calcium. Ch Research Branch AEC, N Y. Dept Chmn to V Pres of Southwest non-profit res insts Armour, Stanford, Southwest 1947-56. Pres & Principal Louis Koenig-Res 1956- , process design economics, planning in water & pollution control. Adv Board to Secys of Interior on Saline Water Prog, Advisory Board to Public Health Service on Advanced Waste Treatment Prog. Editor, Inorganic Indus Chems, Chem Abstracts. *Society Aff:* ACS, AIChE, AWWA, ΣΞ.

Koenig, George J
Business: 1320 Arch St, Philadelphia, PA 19107
Position: Engrg Mgr *Employer:* D'Ambly, Inc *Education:* MBA/Management/Drexel Univ; BS/ME/Drexel Univ *Born:* 1/13/42 in Philadelphia, PA. Worked as Mechanical Engr for Naval Facilities Engrg Command, 1965-1970. In Plant Engrg at RCA, 1970-1972. Have worked in consulting field since 1972, advancing from Project Engr to Project Mgr to Engrg Mgr. Responsible for technical content and quality control of designs for Philadelphia's Oldest Consulting Engrg Firm. Have special interest in heat recovery and energy conservation. *Society Aff:* ASHRAE, CSI

Koenig, Herbert A
Home: 75 Tumblebrook Drive, S Windsor, CT 06074
Position: Prof *Employer:* Univ of CT *Education:* PhD/Applied Mech/PA State Univ; MS/ME/Drexel Univ; BS/ME/City Coll of NY *Born:* 2/3/38 Native of NYC. Taught at Drexel Univ 1963-1965. Hold joint appointment as Prof of Mech Engrg and Orthodontics. Research in Biomechanics, Finite Element Analysis, Wave Propagation Thermal Stress Analysis and Compressible Fluid Flow. Consultant to Pratt and Whitney Aircraft (15 yrs), Burnoy Corporation (3 yrs). Industrial Experience as Development Engr with Hazeltine Electronics. RCA, General Electric (5 yrs). *Society Aff:* ASME

Koenig, John H
Home: Rt 1 Box 192, Delta, CO 81416
Position: Professor Emeritus. *Employer:* Rutgers University. *Education:* PhD/Cer Eng/OH State Univ; MSc/Aero Eng/OH State Univ; BChE/Chem Eng/OH State Univ; ChE/Chem Engg/OH State Univ. *Born:* Apr 1909. G E Co Lab 1931-35; O S U Engrg Exp Station - Res Fellow 1935- 38; Hall Co Res Dir 1938-42; Dept of Navy Electronics Div 1942-45; Rutgers The St Univ-Dir of Sch of Ceramics & N J Ceramic Res Station 1945-69, Prof Emeritus 1969- . Pres Natl Ceramic Educ Council 1955-56; Pres Keramos Natl Hon Ceramic Engrg Frat 1956-58; Dir Amer Soc for Testing & Materials 1957-60; Pres Amer Ceramic Soc 1962-63; Pres VI Internatl Congress on Glass 1962-65. Author & recipient of numerous awards. *Society Aff:* Amer Cer Soc, Can Cer Soc, ASTM.

Koepfinger, Joseph L
Home: 119 Windy Willows Dr, Coraopolis, PA 15108
Position: Dir System Studies and Research *Employer:* Duquesne Light Co. *Education:* MS/EE/Univ of Pgh; BS/EE/Univ of Pgh *Born:* 5/6/25. Native of Pittsburgh PA area. Employed since 1949 at Duquesne Light Co in the field of power sys protection, communications, & automation. Chmn of IEEE Standards Bd 1974 & 75, 78, 79, Past Chmn of Surge Protection Devices Ctte of the Power Engrg Soc of IEEE. Elected Fellow of IEEE 1975. Mbr Bd of Dirs of IEEE 1978-79, Chrmn ANSI C62 Committee on Surge Aressters, USNC Tech Advisor to IEC TC37. Currently responsible for corporate research activities. Hobbies: gardening & music. *Society Aff:* ANSI, IEEE

Koerner, Ernest L
Home: 12721 St Andrews Terrace, Oklahoma City, OK 73120
Position: President. *Employer:* TECHRAD Inc *Education:* PhD/Chem Engr/IA State Univ; MS/Chem Engr/IA State Univ; BChE/Chem Engr/Univ of Dayton. *Born:* Mar 17, 1931; native of Willoughby Ohio. Sect Leader for Union Carbide Metals Co 1957-59; Res Specialist Monsanto Co 1959-67; Sr Res Group Leader Kerr-McGee Corp 1967-70; founded Technology Research & Development 1970 as Div of Benham-Blair & Affiliates Inc. Purchased Tech Res & Dev from Benham-Blair in 1983. Renamed firm TECHRAD Inc. Holder of 17 pats on metal recovery processes & phosphoric acid extraction processes. Currently respon for environ & energy progs & indus waste treatment process design & devs for clients. Enjoys boating, fishing & sports. Reg P E Okla. *Society Aff:* AIChE, AIME, NSPE, OSPE, AIPE

Koerner, Robert M
Business: Dept of CE, Philadelphia, PA 19064
Position: Prof of Civil Engineering. *Employer:* Drexel University. *Born:* Dec 2, 1933. BSCE 1956, MSCE 1963 Drexel Univ; PhD 1968 Duke Univ. Prof Engr in Pa & Del. Indus exper in heavy const & geotech engrg cons. Current res interests in acoustic emission monitoring of soil, rock, coal, pipelines, concrete, etc; foundation techniques; blasting vibrations; fundamentals of particulate behavior. Over 60 tech publs. Former Pres Phila Section ASCE & Chmn of Mid Atlantic Section ASEE. Long distance runner.

Koff, Bernard L
Business: G E Co - Maildrop J-44, Cincinnati, OH 45215
Position: Chief Engineer, Engrg Dept. *Employer:* G E Co - Aircraft Engine Group. *Education:* BS/Mech Engg/Clarkson College of Technology; ME/Mech Engg/NY Univ. *Born:* 3/24/27. Mr Koff obtained a BSME in 1951 from Clarkson College of Technology & a MME in 1958 from NY Univ. He joined Genl Elec in 1958 & has been a pioneer in the design & dev of aircraft gas turbine engines incorporating numerous technology advances used throughout the industry. He has held mgmt positions in Turbomachinery Design, Preliminary Design, Advanced Engg, Genl Mgr-Dev Production Engg & Chief Engr. He holds major pats on numerous GE engines & has published numerous papers dealing with engine design & dev & continuing education for engineers. Present position, April 1975. Location: Cincinnati, OH. *Society Aff:* ASME, AIA, AIAA, SAE.

Kogos, Laurence
Home: 6 Calumet Rd, Westport, CT 06880
Position: Vice President, C.O.O. (80 to date) *Employer:* Roper Plastics, Inc. *Education:* BS/Chem Engr/Northeastern Univ. *Born:* July 1929; native of Boston Mass. Tech Dir H M Sawyer Co to 1958; Genl Mgr Farrington Texol to 1962; Genl Mgr various divs of Whittaker Corp, Pres Polymeric Fabricants Div Whittaker to 1973; Pres Hatco Plastics Div W R Grace & Co to 1979. Married Rae; 2 children. Mbr AIChE & SPE. *Society Aff:* SPE, AIChE.

Kogstad, Rolf E
Business: 800 Third Ave, NY, NY 10022
Position: Pres & Gen Mgr. *Employer:* Norsk Hydro Sales Corp. *Education:*

Kogstad, Rolf E (Continued)
BSc/CE/Univ of Glasgow. *Born:* 7/21/32. Immigrated from Norway in 1955 following education in Scotland & worked for oil cos in the US & abroad in agri chems. With Norsk Hydro since 1971 & from 1973 Pres & Gen Mgr of Norsk Hydro Sales Corp in NY. *Society Aff:* AIChE, API, SPI.

Koh, Severino L
Home: 801 Cottonwood St, Morgantown, WV 26505
Position: Prof & Chairman, Mech Engg & Mechs *Employer:* West Virginia Univ *Education:* PhD/Engg Sci/Purdue Univ; MS/Engg Mechanics/PA State Univ; BS/ME/Natl Univ; BS/Meteorology/NYU. *Born:* 1/8/27. Meteorologist, Philippine Weather Bureau, 1948-54. Res asst, Johns Hopkins, 1954-55. Instr, PA State, 1955-57. Instr, Purdue, 1957-59. Res assoc, Gen Tech, 1959-61. Mech Engr, GE, 1961-62. Visiting res assoc to full prof, Purdue, 1962- 1980; asst hd, Div of Interdisciplinary Engg Studies, Purdue, 1977-1980. Head, Dept of Engg, Purdue Calumet, 1980-81. Chairman, Dept of Mechanical Engg and Mechanics, West Virginia University, since 1981. Visiting prof, Technische Universitat Clausthal, 1968-69; Technische Universitat Karlsruhe, 1969; Universitat Bonn, 1974-75. AMOCO Outstanding Teaching Award, 1967; Sigma Gamma Tau Outstanding Prof, 1968; Humboldt Award (Germany), 1974; Balik Scientist (Philippines), 1976. Most Outstanding Filipino in the Midwest, U.S.A. 1979-80 in Science (Cavite Association of America), 1980.Past Secy, Past Dir, Editor of *The Engineering Science Perspective*, Soc of Engg Sci. Pres & Dir, Phil - American Acad of Sci & Engrg, 1980-81. *Society Aff:* ASME, ASEE, SES, SoR, GAMM, Acad of Machs, Phil-American Acad of Sci & Engg..

Kohl, Walter H
Home: 36 Woodleigh Rd, Watertown, MA 02172
Position: Materials Consultant. *Employer:* Self. *Education:* Dr Ing/Engg Physics/Dresden, Germany. *Born:* Jan 1905 Kitzingen Bavaria Germany. Dipl Ing/Dr Ing Engrg Phys Tech Univ Dresden Germany 1928 & 30. Process Dev for Electron Tubes at Rogers Electron Tubes Ltd Toronto Ontario 1930-45 (Proj Engr, Chief Engr, V Pres Dir). Special Lectr Univ Toronto (Dev Electron Microscope). Positions in USA: Collins Radio Co, Res Div Cedar Rapids Ia 1946-52; Stanford Univ 1952-58; Sylvania Elec Prod Mountain View 1958-62; NASA-ERC Cambridge Mass 1966-70; Independent Cons on materials, techniques & devices for vacuum devices, Van Nostrand-Reinhold N Y 1967. Fellow IEEE 1957, ACS 1967; Mbr APS, AAAS, Sigma Xi. 10 patents. *Society Aff:* APS, AAAS, IEEE.

Kohloss, Frederick H
Business: 345 Queen St 401, Honolulu, HI 96813
Position: President. *Employer:* Frederick H Kohloss & Assocs, Inc *Education:* JD/Law/George Washington Univ; MME/Engg/Univ of DE; BS/Mech Engg/Univ of MD. *Born:* 12/4/22. Texas. Army Corps Engrs 1943-46. Engrg teaching GWU & part-time Univ of Hawaii. Dev, standards & contracting experience. 1956- , cons engr; offices in Hawaii, California, Arizona, Colorado & Australia. Fellow ASHRAE 1973, Fellow ASME 1976; Sr. Mbr IEEE Mbr NSPE, IES, SAME, ASPE; Fellow ACEC. Reg P E (mech & elec), holder NEC Certificate. Chartered Engr Australia. Fellow Inst. of Engrs. Australia Member Prof Engr of NZ Active in building mechanical and electrial systems, air conditioning & energy sys design & analysis *Society Aff:* ASHRAE, ASME, IEEE, IES, SAME, ACEC, ASPE.

Kohne, Richard E
Business: 180 Howard St, San Francisco, CA 94105
Position: Pres. *Employer:* Morrison-Knudsen Engineers, Inc. *Education:* BSCE/Struct Engg/Univ of CA, Berkeley. *Born:* May 1924. Reg Engr Calif. Fellow ASCE. 1948-56 Struct Engr with Pacific Gas & Elec Co for hydroelectric projs; with Morrison-Knudson Engrs Inc Co 1956- , worked on major energy & mining related projs worldwide. Supr engrg for hydro dams from 300 MW to 12,000 MW, deep-water ports to 300, 000 DWT, heavy-haul railroads to 50, 000,000 tpy traffic. 1956-65 Mgr engrg offices Brazil, 1966-70 VP & Mgr Latin Amer, 1971-78 Exec VP Worldwide Opns, 1979 president. Mbr ASCE, SAME, ANS, USCOLD, AIME, AMA, ACEC, & CEAC. *Society Aff:* ASCE, SAME, ANS, USCOLD, AIME, AMA, ACEC, CEAC, ECSF, AUSA

Koide, Frank T
Business: PO Box 22786, Honolulu, HI 96822
Position: Prof. *Employer:* Univ of HI. *Education:* PhD/Physiology/Univ of IA; Masters/EE/Clarkson Coll of Tech; BS/EE/Univ of IL. *Born:* 12/25/35. in Honolulu, HI. Publications engr, then electronics engr for Collins Radio Co, specializing in propagation res, 1959-61. Adult education teacher in electronics, Cedar Rapids, IA. Predoctoral fellow 1962-66 at Univ of IA. Asst prof at IA State Univ 1966-68. Principal biomedical engr for Tech Inc, 1968-69, engaged in biomechanics res & instrumentation. Since 1969 assoc prof, then prof of elec engg & physiology, Univ of HI. Also, consultant & external examiner to Chines Univ of Hong Kong. NASA-ASEE Space Systems Inst Design Fellow, 1967. NSF Digital & Analogue Elec Inst Fellow, 1972. *Society Aff:* IEEE, AAMI, ΣΞ.

Koiter, Warner T
Business: Mekelweg 2, Delft Netherlands
Position: Prof Emeritus *Employer:* Delft University of Technology. *Education:* DSc/Mech Eng/Delft Univ Techn. *Born:* June 16, 1914. Mech Engr degree Delft 1936; PhD Eng Sci Delft 1945. Aeronautical Res Inst Amsterdam 1936-38; Govt Patent Office 1938-39; Dept of Cvl Aviation 1939-49; Prof Applied Mech Delft 1949-79. Mbr Netherlands Aircraft Dev Board 1946-79; Mbr Royal Netherlands Acad Sci, Amer Acad Arts & Sci, Deutsche Akademie der Noturforscher Leopoldina. Natl Acad Engg. Acad. des Sciences Paris, Istituto Lombardo Accademia di Science e Lettere Milano. Royal Soc London. Recipient Von Karman Medal ASCE 1965, Timoshenko Medal ASME 1968, Modesto Panetti Prize 1971. Hon degree Univ Leicester 1969. Univ Glasgow 1978, Ruhr Univ Bochum 1978, Univ Gent 1979 Univ Liege 1986. Hon. mbr. ASME 1980. Mbr Bureau 1956-60, Treas 1960-68, Pres 1968-72, V Pres 1972-76 of Internatl Union of Theoretical & Applied Mechanics (IUTAM). *Society Aff:* ASME

Kok, Hans G
Business: 1700 Broadway, New York, NY 10019
Position: VP *Employer:* Treadwell Corp *Education:* Dipl-Ing/CE/Technische Hochschule-Germany; Ing/Arch Eng/Hamburg Engrg Coll- Germany; Ing/CE/Suderburg Engrg Coll-Germany *Born:* 4/15/23 Arrived in the US in 1951 and was naturalized in 1959. Structural Design Engr with Lummus Co, NY City, 1951-1953. With Treadwell Corp since 1953. Design engr and Head of Structural Section, 1956-1962. Chief Structural Engr, 1962- 1963. Mgr, Plant Design, 1963-1969. Assistant VP, Engrg, 1969-1973. VP, Engrg, 1973 to present. Pres, Treadwell Corp of MI, Inc, 1973 to present. Dir, Bassett Miller Treadwell Pty Ltd, Sidney, Australia, since 1974. Chrmn, Executive Council on Engrg Laws, 1976. Recipient of First Award, James F. Lincoln Arc Welding Foundation, 1966. Chrmn, Material Handling Committee, Society of Mining Engrs of AIME, 1979/1980. PE, registered in NY and 26 other states. *Society Aff:* ASCE, NSPE, AIME, AMA, AMC

Kokjer, Kenneth J
Business: EE Dept, Univ. of Alaska, Fairbanks, AK 99775
Position: Assoc Prof *Employer:* Univ of AK *Education:* PhD/Biophysics/Univ of IL at Champaign-Urbana; MS/Biophysics/Univ of IL at Champaign-Urbana; BA/Physics/NB Wesleyan Univ *Born:* 2/27/41 Assoc Prof of Electrical Engrg and Biophysics. Joint appointment between EE and the Inst of Artic Biology (IAB). Teaching and research in EE and IAB, specializing in small computer applications to telecommunications and to on-line support for biomedical research laboratories. EE Dept Head from fall, 1977 to spring 1981. Consulting in remote data collection and instrumentation. Outside activities include flying, photography, and vocal performance. Registered PE (EE). *Society Aff:* IEEE.

Kolditz, Loren C
Business: Baldwin Rd, Rt 1, New Athens, IL 62264
Position: Pres. *Employer:* Loren C Kolditz Engg Co. *Education:* BS/CE/Univ of IL. *Born:* 8/12/30. My consulting engg experience involving natural gas pipeline & distribution facilities began in 1953 after grad from the Univ of IL & discharged from the USAF. This continued until 1965 when I formed a new firm to provide consulting services to small natural gas distribution systems. My current practice includes all phases of gas design, operation, mgt, & safety compliance. All of the work is closely coordinated with regulatory matters at the state & fed levels. *Society Aff:* AGA, ASCE, NSPE.

Kolflat, Tor D
Business: 55 E Monroe St, Chicago, IL 60603
Position: Partner. *Employer:* Sargent & Lundy. *Born:* Nov 1925; native of Chicago area. BSME & MSME Purdue. Air Corps 194445. With Commonwealth Edison 1951-54 at Waukegan Plant; joined Sargent & Lundy cons engrs 1954, became Ptnr 1966, respon for engrg design of mech features of power plants & admin control of mech personnel. Lic P E in Ill & other states. Hon Fellow ASME, Chmn Power Div ASME 1974-75; Mbr IEEE, ANS, WSE, AIF, ISPE, AWWA; Mbr Ill Energy Resources Comm; Mech Chmn Amer Power Conference 1965- . Submitted engrg testimony to US Senate Ctte & US Supreme Court. Over 28 tech papers publ.

Koller, Earl L
Business: Castle Point Station, Hoboken, NJ 07030
Position: Prof. *Employer:* Stevens Inst of Tech. *Education:* AB/Physics/Columbia College; AM/Physics/Columbia Univ; PhD/Physics/Columbia Univ *Born:* 12/8/31. in Brooklyn, NY and educated at Columbia College & Univ. Since obtaining the PhD in 1959, have been on the faculty of Stevens Inst, teaching scientists & engrs. My res activity has been in the area of particle physics first using nuclear emulsions techniques and during the past 17 yrs using first bare bubble chambers and more recently, hybrid systems. As part of a multi-univ international consortium, have helped to develop the software for this system as well as analysis procedures. Avid tennis player & skier. *Society Aff:* APS, AAUP, ΣΞ, ΦBK

Kollerbohm, Fred A
Business: SE Tower Prudential Ctr, Boston, MA 02199
Position: Proj Coordinator. *Employer:* C T Main Intl Inc. *Education:* BS/Civ Engg/Healds Engg College. *Born:* 9/23/36. Native of Nicaragua & raised in San Francisco. Grad from private schools and colleges. Twenty yrs plus of experience in design, supervision, quality control, inspection & mgt. Admin, in civ-struct engg with the largest cos in US & foreign. Eight yrs of experience overseas. Reg PE in US & foreign countries. Flight in Spanish, Portuguese & Italian. Reg as Gen Bldg Contractor in US & foreign countries. *Society Aff:* ASCE, NSPE, WSPE.

Kollitides, Ernest A
Business: 393 Seventh Ave, New York, NY 10001
Position: VP - Intl Operations *Employer:* Gibbs & Hill, Inc. *Education:* MS/EE/Polytech Inst of Brooklyn; B/EE/NYU; BA/Math & Phys/NYU *Born:* Mr. Kollitides has over 22 yrs of experience in the power energy field. In his present position, he manages a staff of dirs responsible for the worldwide Mktg and Bus Dev activities of Gibbs & Hill. From 1966 to Mar 1981, he was with Ebasco Services, Inc. and served progressively as Prin Engr, Supervising Engr, Proj Mgr/Proj Engr, Asst Ch Elec Engr, and from 1977-81, as the company's Corporate Ch Elec Engr responsible for directing the profl staff of the Elec Engrg and Design Depts in engrg and designing all the projects in the corporate hdquarters of the company and its regional and overseas offices. He also participated in several business dev and various presentation activities of the company. Prior to joining Ebasco, Mr. Kollitides occupied various engrg positions in the Elec Engrg Dept of Consolidated Edison Co of NY, Inc. Reg with the Natl Council of Engrg Examiners. Reg PE in 23 states. Authored several papers presented to organizations such as the Atomic Ind Forum, the American Power, Conference, IEEE, etc. *Society Aff:* IEEE, AEIC, NPEC, AIF, CNNP

Kolsky, Herbert
Business: Brown University, Div of Applied Math, Providence, RI 02912
Position: Prof. *Employer:* Brown Univ. *Education:* DSc/Phys/Univ of London; PhD/Appl Physical Chem/Imperial College; BSc/Phys/Imperial College. *Born:* 9/22/16. Hd of Phys Dept, ICI Ltd - 1946-1956. Visiting Prof of Engg, Brown Univ - 1956-1958. Principal Scientific Officer (Special Merit , Ministry of Supply - 1958-1960. Prof of Engg, Appl Math & Phys. Brown Univ - 1960-present. Russell Severance Springer Visiting Prof of Mechanics, Univ of CA, Berkeley 1978. Visiting Prof of Engg, ETH, Zurich - 1979. *Society Aff:* FInstP, FPhySoc, BSR, S of Rheo, SEC, AAM.

Komoriya, Hideko (Heidi)
Home: 420 Lexington Ave, NY, NY 10170
Position: Pres *Employer:* Intl Tech Serv, Inc *Education:* PhD/Nucl Eng/Univ of Del; MA/Math/Univ of Del; BA/Math/Cedar Crest Coll *Born:* 09/11/46 Nuclear Engr for Gen Elec Co, San Jose, CA 1975-1976. Nuclear Engr with Argonne Natl Lab, Argonne, Ill 1977-1984, specializing in nuclear power plant safety and review of detailed computer model dev of Light Water Reactor systems. Awarded citation from US Nuclear Regulatory Comm in recognition of her outstanding tech contribution to the Staff's review of Light Water Reactor safety concerns in 1983. Co-founder and Pres of Intl Tech Serv since 1980 providing expert consulting and training services in nuclear power plant safety analysis world-wide. Interests include classical music, skiing, historical architecture, horticulture and fishing. *Society Aff:* ANS

Koncel, Edward F, Jr
Business: Productivity Management Associates, 7229 Red Oak Dr, Crystal Lake, IL 60012
Position: Admn. *Employer:* Commonwealth Edison Co. *Education:* MS/EE/IIT; BS/EE/IIT. *Born:* 8/3/27. Fellow IEEE. Fortescue Fellow. Sigma Xi, Tau Beta Pi, Eta Kappa Nu. In early 1950's participated in feasibility & economic studies of developing nucl fueled power plants. Conducted res of lightning performance of transmission lines. Formerly Staff Asst to Treas, to VP of Engg & Operations, to VP of Finance. As Supervisor of Economic Res, directed wide range of studies, including nucl and fossil generating unit additions. Taught Engg Economics. Directed dev of computerized data bank system for nucl fuel accountability. Currently Administrator of developing & installing ind engg techniques for managing maintenance in fossil & nucl generating stations. *Society Aff:* IEEE, ANS.

Kondo, Yoshio
Home: 29, Higashi-Takagicho, Shimogamo, Sakyo-ku, Kyoto 606 Japan
Position: Prof Emeritus *Employer:* Kyoto University *Education:* Dr of Engrg/Metallurgy/Kyoto Univ; B Engrg/Metallurgy/Kyoto Univ *Born:* 02/17/24 1945Grad from Kyoto Imperial Univ 1951 Assoc Prof of Metallurgy, Kyoto Univ 1960 Awarded Dr of Engrg from Kyoto Univ; 1961 Prof of Metallurgy, Kyoto Univ 1962-1964 Res Assoc, MIT, Cambridge, Mass USA; 1977-1979 Mbr of Council Board, Kyoto Univ; 1983-1985 Dean, Faculty of Engrg, Kyoto Univ; 1971 Awarded EMD Science Award from AIME; 1971 Awarded Deming Prize from Deming Prize Awarding Ctte; 1976 Awarded E L Grant Award from AIME; 1981 Awarded Tanigawa-Harris Award from JIM. *Society Aff:* AIME, ASQC, ES, MMIJ, JIM, IAQ

Konrad, William N
Home: 1370 Fairhaven Blvd, Elm Grove, WI 53122
Position: VP Intl & Bus Dev. *Employer:* Envirex - A Rexnord Co. *Education:* BS/Civil Engg/Purdue Univ. *Born:* 9/5/23. in E Chicago, IN; Army Corps Engrs 1943-46. With Rexnord 1948- , with current respon as VP of Intl & Bus Dev respon for strategic dev progs, mktg of new prods, advertising & foreign opers. Assumed current respon 1978. Pres Water & Waste Water Equipment Manufac Assn; Pres Central States Water Pollution Control Assn; Dir Water Pollution Control Federa-

Konrad, William N (Continued)
tion. Mbr EPA Tech Advisory Group. Enjoys golf, tennis & gardening. *Society Aff:* WPCF, AWWA, NSPE, AAEE.

Konski, James L
Business: Old Engine House No. 2, 727 N Salina St, Syracuse, NY 13208
Position: Pres/Principal *Employer:* Konski Engrs, PC *Education:* MS/CE/U of MO; BS/CE/U of MO, Columbia, MO *Born:* 11/4/17 Engr for Bureau of Yards and Docks, US Navy; Structural Engr, Sanderson & Porter; Field Engr, Ebasco Services Inc. Organized branch office in Syracuse 1953 for consulting firm engaged in highway, bridge and other Civil Works. Founded Konski Engrs 1957, Senior Principal, responsibilities include highways, bridges, dams, parking facilities, feasibility studies, investigations of failures, planning, design and construction. Licensed PE in NY, KY, RI, KS, National Council of State Board of Engrg Examiners and licensed PLS in KY. Fellow and Past VP ASCE. Recipient of the University of Missouri 1986 Award for Distinguished Service in Engineering. *Society Aff:* ASCE, NSPE, SAME, ACEC, ASP

Konz, Stephan A
Business: Dept of Ind Engrg, Manhattan, KS 66506
Position: Prof *Employer:* KS State Univ *Education:* PhD/IE/Univ of IL; MS/IE/State Univ of IA; MBA/Bus Adm/Univ of MI; BS/IE/Univ of MI *Born:* 11/25/33 *Society Aff:* HFS, IIE, ASHRAE, ASQC, AIHA

Koo, Benjamin
Business: 2801 W Bancroft St, Toledo, OH 43606
Position: Prof. *Employer:* Univ of Toledo. *Education:* PhD/Structural Engrg/Cornell Univ; MS/Structural Engr/Cornell Univ; BSCE/Civil Engrg/St John's Univ *Born:* 4/4/20. Shanghai, China, US citizen. About 14 yrs consulting engg experience. With Civ Engg Dept; Univ of Toledo, since 1965. Winner of the outstanding Teacher's award, 1974. Approximately 75 technical and educational publications in periodicals. Res interests: structural engg, probabilitic approach to design, stability analyses, reliability and fire-safety. PE - OH & NY. Honor Soc mbr: Tau Beta Pi; Phi Kappa Phi; Sigma Xi. *Society Aff:* ASCE, ASEE

Koo, Jayoung
Business: Annandale, NJ 08801
Position: Staff Metallurgist *Employer:* Exxon Research and Engineering Company *Education:* PhD/Mat Sci & Engg/Univ of CA; MS/Mat Sci & Engg/Univ of CA; BS/Metallur Engg/Seoul Natl Univ. *Born:* 12/7/48. in Busan, Korea. Came to the USA in 1973. Worked for Lawrence Berkeley Lab, conducting res in the dev of structural steels for energy savings 1973-78. Asst. Prof at Rutgers Univ 1978-80, teaching electronmicroscopy, design of advanced engg alloys, etc. Current position at corporate research science Lab., Exxon Research and Engineering Company as staff metallurgist since 1980. Res progs on welding, catalysts, electronmicroscopy, and high strength structural steels. Hold US Patent No 4,067,756 "High Strength, High Ductility Low Carbon Steel–. First place award (1979) and other honorable awards (1976, 1978) at the Intl Metallographic Exhibit, jointly sponsored by ASM and IMS. Enjoy sports and fishing. *Society Aff:* ASM, EMSA, AIME.

Koonce, K T
Business: 1800 Avenue of Stars, Los Angeles, CA 90067
Position: Div Prod Mgr *Employer:* Exxon Co, USA *Education:* PhD/ChE/Rice Univ; BS/ChE/Rice Univ *Born:* 6/1/38 Native of Corpus Christi, TX. With Exxon Production Research Co and Exxon Co, USA since 1963. Six years in research and engrg mgmt. Prior to present assignment, had operational responsiblity for Exxon's AK North Slope activities. Presently based in Los Angeles and is responsible for Exxon's oil and gas production and onshore and offshore drilling operations in CA and AK. Chairman SPE GEM/TIC Committee 1980, presently SPE Director-at-Large. Enjoys tennis, resides in Chatsworth, CA. *Society Aff:* SPE OF AIME, TBΠ

Koonsman, George L
Business: PO Drawer 1307, Arvada, CO 80001
Position: Consultant *Employer:* Centennial Engrg Inc. *Education:* MSCE/Irrigation/CO State Univ; BSCE/Civil Eng/CO State Univ. *Born:* Nov 1922 Lamar Colo. B24 Pilot WWII. Instr Cvl Engrg 1947-51. Ideal Cement Co 1951-59 as Plant Engr, Asst Plant Mgr & Proj Mgr. 1960-62 Cement Engr for Amer Marietta. 1962-74 V Pres Ken R White Cons Engrs Denver. 1974- , Chairman of Centennial Engrg Inc. Mbr: ASCE (Fellow), NSPE, AIME, Pres 1980 Cons Engrs Council of Colo, Colo Mining Assn, Rocky Mt Coal Mining Inst, Pres 1980 Denver Kiwanis Club, Colo Ground Water Comm Mbr 1968-73. *Society Aff:* NSPE, ASCE, ACEC.

Koopman, Richard J W
Home: 2201 St Clair, Brentwood, MO 63144
Position: Samuel C Sachs Prof Emeritus of EE. *Employer:* Washington University. *Education:* PhD/EE/Univ of MO.; MS/EE/Yale Univ; BS/EE/Univ of WI. *Born:* June 1905 St Louis. Taught elec engrg at Yale, Mich Tech & Univ of Ks; Head Electro-Mech Section Cornell Aeronautical Lab 1943-46; at Wash Univ since 1946 as Assoc Prof, Prof & Dept Chmn 1949-65. Fellow IEEE, Chmn St Louis Section 1957; Chmn Natl Ctte on Aerospace Instrumentation 1959-60. Prize for Initial Paper (with H R Reed) 1937. Dir & Pres of Eta Kappa Nu 1964-65; Pres Engrs Club St Louis 1960. Listed in Who's Who in Amer for more than twenty years First Sammuel C. Sachs Prof of Elect Engrg, Washington Univ, Mbr History Ctte St Louis Electrical Bd.- . Active in prof socs & EE cons. *Society Aff:* IEEE, APS, ASEE, TBΠ, ΣΞ.

Kopczyk, Ronald J
Business: 110 Freeman Hall, Clemson, SC 29631
Position: Assoc Prof *Employer:* Clemson Univ *Education:* MS/EE/Clemson Univ; BS/EE/Univ of IL; G/ET/Coyne Technical Inst *Born:* 4/2/39 Native of LaSalle, IL. Assoc Prof of Engrg and Engrg Tech and Dir of Engrg Services at Clemson Univ. Special interests include Electronic Instrumentation, Programmable Logic Controllers for Education & Faculty Automation and Engrg Highspeed Cinematography and Data Acquisition. Enjoy hunting, fishing, photography, and mechanicing. Registered PE in SC. *Society Aff:* IEEE.

Kopetz, Marion J
Business: 1805 Grand, Kansas City, MO 64108
Position: Chief Engr, Spec. Projects *Employer:* Howard Needles Tammen & Bergendoff. *Education:* Dr of Engg/Civ/KS Univ; MS/Structural/KS Univ; BS/Civ/Univ of CO. *Born:* 9/21/25. Native of S IL. Served in 3rd Army in Europe WWII. Bridge designer OR Hgwy Dept 1951-2. With Howard Needles Tammen & Bergendoff since 1952. Hd of KS City office bridge sec 1964-76, chief structural engr since 1976. Adjunct Prof at KS Univ & Univ of MO at KS City. Designer of maj prize winning bridges throughout US. Past pres of ASCE KS City Sec & MSPE Western Chapter. Past Dist Governor Toastmasters Intl. *Society Aff:* ASCE, NSPE, ACI

Koplik, Bernard
Home: 74 Paerdegat 2 St, Brooklyn, NY 11236
Position: Chrmn and Prof. *Employer:* New Jersey Institute of Technology *Education:* PhD/Mech Engr/PINY; MS/Mech Engr/Columbia Univ; BME/Mech Engr/City College of NY. *Born:* 4/4/34. Faculty appointment at the City College of NY (1955-64), NSF Sci Faculty Fellow (1964-65), Asst Prof (1965-68) and Assoc Prof (1968-75) at the Poly Inst of NY, Chrmn and Prof at Manhattan College (1975-1981). Joined the faculty of New Jersey Institute of Technology in 1981 and presently Prof and Chrmn of the Mech Engr Dept. Industrial Consultant to Naval Appl Sci Lab, Gen Elec, Pitney Bowes and Brookhaven Natl Lab. Published more than 45 technical papers in areas of vibrations of plates and shells, dynamic response of nuclear reactors and biomechanics. Prin investigator for AFOSR and proj dir for several NSF-URP grants. Natl Secretary of the Mech Engr Dept Heads Comm of ASME (1979-1980), Exec Bd of ASME Region II (1983-present), Mbr of Bd of Trustees of Engrg Soc Library (1981-present), Study Comm on Nuclear Power Plant Structural Loads

Koplik, Bernard (Continued)
(1981-present), Prog Chrmn for Grad Div of ASEE (1985-86), Fellow of ASME (1985). *Society Aff:* ASME, ASEE, NSPE, ΠΤΣ, ΣΞ, TBΠ.

Koplowitz, Jack
Business: Dept of Elec & Comp Engg, Potsdam, NY 13676
Position: Assoc Prof. *Employer:* Clarkson Univ. *Education:* PhD/Elec Engg/Univ of CO; MEE/Elec Engg/Stanford Univ; BEE/Elec Engg/City College of NY. *Born:* 3/12/44. From 1968 to 1970 worked at Bell Labs in the data communications area. In 1973 received PhD in elec engg from the Univ of CO. Since that time have been with the dept of elec and comp engg at Clarkson Univ; presently an Assoc Prof. His res areas of interest include Communication theory, pattern recognition, image processing. *Society Aff:* IEEE, ΣΞ, HKN, TBΠ.

Koplowitz, Sol
Home: 560 Dale Ct, River Vale, NJ 07675
Position: Consulting Engineer *Education:* MChE/Chem Engg/Columbia; BS/Chem/CCNY. *Born:* Mar 1915 N Y City. San Engr J E Sirrine & Co & N J Pease & Co 1940-43; US Army Sanitary Corps 1943-46; with Havens & Emerson, Inc 1947-1980, became Partner 1971 in respon charge of planning & design of wastewater treatment & collection, water supply, treatment & distribution. Consulting Engineer 1980- . Participated in preparation of ASCE Manual of Sewage Treatment Design, Design & Construction of Gravity Sanitary Sewers, Glossary of Water & Wastewater Control Engrg, Pipeline Design Manual. P Chmn ASCE Ctte on Pipeline Planning; P Chmn ASCE Pipeline Div Exec Ctte, P Chmn ASCE Mgmt Group E. Reg P E: N Y, N J, ASCE Fellow; AAEE Diplomate; AWWA; WPCF; Sigma Xi. ASCE Rep. on Engineering Societies Library Board. *Society Aff:* ASCE, AAEE, AWWA, WPCF.

Koppel, Lowell B
Business: Setpoint, Inc, 950 Threadneedle, Suite 200, Houston, TX 77079
Position: Dir, Special Projects & Sr Consultant *Employer:* Setpoint, Inc. *Education:* PhD/ChE/Northwestern; MSE/ChE/MI; BS/ChE/Northwestern *Born:* 9/13/35. Dir, Special Projects for Setpoint, Inc., 1985-present. Chem Engg faculty mbr at Purdue 1961-85. Hd of Chem Engg, 1973-81. Author of textbooks, "Process Systems Analysis & Control" & "Introduction to Control Theory–; frequent contributor to chem engg literature; consultant to industry and govt. *Society Aff:* AIChE, ASEE, ISA, APICS.

Koppenaal, Theodore J
Home: 21095 Whitebark, Mission Viejo, CA 92692
Position: Owner *Employer:* Koppenaal & Associates *Education:* BS/Met Eng/Univ of WI; MS/Met Eng/Univ of IL; PhD/Metallurgy & Matls Sci/Northwestern Univ. *Born:* Dec 1931; native of Milwaukee Wis. 2 yrs active duty USN as Engrg Officer. 6 yrs with Atomic Energy Comm at Argonne Natl Lab as fundamental researcher examining strengthening mechanisms & radiation damage in solids. With Ford Aerospace & Communications Corp (formerly Philco-Ford Corp) 11 yrs; respon for materials dev on ordnance, armor, missile control, satellite & reentry sys. Aerojet Ordnance Co, 1979-1981; respon for process dev and manufacturing engrg on heavy metal projectiles. Koppenaal & Associates, 1981- ; consultants and fabricators of depleted uranium and hazardous waste disposal systems. 1973 WESTEC Conference Prog Chmn; ASM Chap Chmn 1974-75. Reg Prof Met Engr in Calif. *Society Aff:* ASM.

Koptionak, William
Home: 8045 Carlette St, La Mesa, CA 92041
Position: Vice Pres *Employer:* Neste, Brudin & Stone Inc. *Education:* BS/Civ Engg/Rutgers Univ. *Born:* 10/24/27. Served as Office/Resident Engr, Metcalf and Eddy, 1951-54, Sayreville, NJ; as Plant Superintendent, Middlesex County Sewerage Authority, 1954-57, Sayreville, NJ; as principal Engr, Proj Mgr, Kupper Engrs, 1957-63, Piscataway, NJ; as Sr Sanitary Engr, Proj Mgr, Chief of Operations, 1963-66, San Diego County, CA; as VP, Chief Engr, Hirsch and Koptionak, 1966-78. Recipient of Outstanding Engg Achievement Awards from American Soc of Civ Engrs and Consulting Engrs Assn of CA, 1973-74 for two water reclamation projs in CA - the San Diego Wild Animal Park and the Industrial Waste Treatment System at the Naval Air Station, North Island, San Diego, CA. Wife, Sally; children, Bruce and Lynn. *Society Aff:* ASCE, NSPE, SAME, WPCF, AAEE.

Korb, Lawrence J
Home: 251 Violet Ln, Orange, CA 92669
Position: Supr Metallurgy Group *Employer:* Rockwell Internatl Space Division. *Education:* Bach ChE/Chem Engg/Rensselaer Polytechnic Inst. *Born:* Apr 1930. Resident Orange Calif. Officer USN 1952-55. Sales Engr Alcoa 1955-59. With Rockwell Internatl Space Div 1959- , spec in spacecraft materials & processing; Engrg Supr of Met Group on Apollo CSM & Met and Ceramic Group on Space Shuttle Orbiter Progs. Reg Met & Corrosion Engr, CA. Mbr Natl ASM Cttes: Metal Progress - Edit Ctte 1968-72; Engrg Materials Achievement Award Ctte 1974- 75; Aerospace Group MS&DD 1971-75. Handbook Committee 1978-83. Chairman Publication Council 1984-85. Author or co-author of books & tech papers covering spacecraft materials, processes & designs. Mbr Met Adv Bd, Cerritos Coll 1970- 74. Mbr Composites Recast 1972; Lectr UCLA, Metal Matrix Composites 1968-69. USC-Spacecraft Mtls-Spacecraft Sys Design-1979. Enjoys jogging, sketching, golf, tennis. *Society Aff:* ASM.

Korb, Monte W
Business: 609 Beachview Dr, St Simons Island, GA 31522
Position: Engr. *Employer:* Korb Engg Co. *Education:* BME/ME/GA Tech. *Born:* 1/14/23. Performed engg duties for Bethlehem Steel Co, Thiokol Corp, Army Ballistic Missile agency, Hi-Shear Corp, GA Tech, and W R Grace & Co. Published several Technical papers and was granted three patents. Established the Korb Engg Co in 1965. *Society Aff:* ASHRAE, AWWA, NSPE, WPCF.

Korbitz, William E
Home: 8692 W 84th Circle, Arvada, CO 80005
Position: Mgr Engr *Employer:* Boyle Engrg Corp *Education:* MS/CE/Univ of WI; BS/CE/Univ of WI; Bach/Naval Sci/Univ of WI *Born:* 12/5/26 Mgr of Metropolitan Denver Sewage Disposal District; Public Works, Dir of City of Omaha, NB and City of Boulder CO; City Eng of Cities of Niles, MI, Cadillac MI and Big Rapids, MI; Dir of Graduate Center of Public Works Engrg and Administration at Univ of Pittsburgh (PA); Mgr Eng of Denver office of Boyle Engrg Corporation; Staff Officer of US Naval Mobile Construction Battalion One. *Society Aff:* AAEE, APWA, ASCE, NSPE, WPCF

Korchak, Ernest I
Business: Scientific Design Company, New York, NY 10016
Position: President. *Employer:* Scientific Design Company *Education:* ScD/Chem Eng/MIT; SM/Chem Eng/MIT; BCHEME/Chem Eng/Unv of Melbourne (Aust). *Born:* 2/15/34. Joined Halcon Intl Inc 1964 as chem engr; in 1972 apptd Genl Mgr of Halcon Catalyst Industries R&D Corp Div, respon for all mktg, res, dev & sales; elected VP 1974. Elected Pres of Halcon Catalyst IN and its parent Halcon R&D Corp. Mbr AIChE, ACS, Royal Australian Chem Inst. *Society Aff:* AIChE, ACS, RACI.

Korchynsky, Michael
Home: 2770 Milford Dr, Bethel Park, PA 15102
Position: Consultant in Metallurgy *Employer:* Korchynsky and Associates *Education:* Dipl Ing/Metals Tech/Tech Univ, Lviv (Lvov) Ukraine *Born:* Apr 11, 1918 Kiev Ukraine. Married; 3 children. Positions held: Ch Engr CE US Army Germany 1945-50, Res Metallurgist & Tech Supr Union Carbide Niagara Falls NY 1951-60, Res Supr & Dir Prod Res Jones & Laughlin Steel Corp Pgh Pa 1961-73, Dir Alloy Dev Union Carbide Corp Metals Div NY 1973-86. 1986-Principal: Korchynsky and Associates, Consultants in Metallurgy. Fields of res: steels, superalloys, refractory

Korchynsky, Michael (Continued)
metals, nuclear fuels, powder met. Expertise in indus res mgmt. Major contribution to HSLA steels dev. AISI Gary medalist 1965; ASM Carnegie lectr Pgh 1973, ASM Fellow 1976. Union Carbide Corp. Sr Corp Dev Fellow, 1979. AIME Hove Memorial lecturer, 1983. ASM WH Eisenman award, 1984. ASM E.C. Bain Award, 1986. Worldwide lecturing. Numerous tech publs & patents. *Society Aff:* AIME, ASM, SAE, VDEh

Korda, Peter E
Business: 4621 Reed Rd, Columbus, OH 43220
Position: President *Employer:* Korda/Nemeth Engineering, Inc. *Education:* PhD/Engg Mech/OH State Univ; DiplEngg/Struct/Univ of Budapest. *Born:* 12/5/31. From 1954 to 1960, struct designer in Budapest, London and Montreal. Starts grad sch in 1960. Upon receiving his doctorate in 1964, he stays on the faculty at OH State Univ as well as starting a consulting engg co. A full prof since 1967 he was in charge of the structural prog of the grad curriculum of the Sch of Architecture at OH State (Part-time) until his resignation in 1980. Pres of Korda/Nemeth Engineering, a consulting firm which provides structural, mechanical, electrical and civil engineering services for buildings, bridges and highways. *Society Aff:* ASCE, ACI, NSPE, AISC, IASS.

Kordish, Emil
Business: 1035 North Calvert St, Baltimore, MD 21202
Position: Partner *Employer:* Rummel, Klepper & Kahl *Education:* BS/CE/Bucknell Univ *Born:* 1/8/21 Native of NY City. Educated in public schools and Stuyvesant High School. Served in US Army 1942-46 during World War II at Aberdeen Proving Ground, MD. Structural designer and construction engr for Sandlass, Weiman & Assocs (1947), predecessor firm to Rummel, Klepper & Kahl, Baltimore, MD. Partner in RK&K 1956 to date. Responsible for hgwys, construction management and inspection. Pres - SAME, Baltimore Post 1965; ASCE, MD Section 1968; CEC MD, 1976. Advisory Bd Member to MD Legislature Purchasing and Procurement Policies Task Force 1977-78. Panel Member American Arbitration Association. Trustee Bucknell Univ and Dir Engrg Alumni Association. *Society Aff:* ASCE, NSPE, SAME, ASTM, ACEC

Korf, Victor W
Home: 2609-60th Court NW, Olympia, WA 98502
Position: Deputy Sec *Employer:* Wash St. Dept of Trans *Education:* BS/CE/WA State Univ *Born:* 11/25/31 Native of Portland, OR. Served with, among others, Federal Bureau of Rds, and the Army Corps of Engrs from 1953-55. With WA State Department of Transporation since 1956. Assumed current responsiblity as Deputy Secretary of Transportation with coordination and overview of day-to-day Department operations, in 1978. Am responsible for advising, coordinating, arbitrating, decisionmaking and evaluation situations in response to day-to-day problems. I exercise judgment regarding the interpretation of policy and establish Departmentwide procedures in operating within an established policy. President, Tacoma Section ASCE, 1981. *Society Aff:* ASCE, APWA

Korfhage, Robert R
Business: Dept of Computer Sci, Dallas, TX 75275
Position: Professor *Employer:* Southern Methodist University. *Education:* PhD/Math/Univ of Mich; MS/Math/Univ of Mich; BSE/Math/Univ of Mich. *Born:* Dec 1930. BSE, MS & PhD Univ of Mich. 2 yrs with United Aircraft Corp; Cons to various indus & govt labs; taught at Univ of Mich, N C St Univ, Purdue Univ & So Meth Univ (Dept Chmn SMU 1970-72). Mbr ACM, & Sigma Xi. Has held several natl positions in ACM incl Chmn of Special Interest Group on Information Retrieval 1973-75 & Regional Representative 1975-81. Prog Chmn 1977 Natl Computer Conference. Pres, Data Organization & Mgmt, Inc. *Society Aff:* ACM, ΣΞ.

Korkegi, Robert H
Home: 4418 Springdale St NW, Washington, DC 20016
Position: Consultant Exec Dir *Employer:* National Research Council *Education:* PhD/Aero & Math/CA Inst of Tech; MS/Aero Engg/CA Inst of Tech; BS/ME/Lehigh Univ. *Born:* 12/3/25. Educated in European & Am schools. Served with Army Corps of Engrs, 1944- 46. Res Assoc & Lectr, Univof S CA 1954-57. Consultant to DOD & AGARD (NATO) 1956. Technical Dir, von Karman Inst for Fluid Dynamics, Belgium, 1957-64. Dir, Hypersnnic Res Lab, ARL, USAF, 1964-75. Dir of NATO's Advisory Group for Aerospace Res & Dev (AGARD), Paris, 1976-79. Visiting Prof, Geo Wash Univ, 1979- 1981. Mbr of several natl & intl committees incl AIAA, NASA, NATO Sci Comm Ad Hoc Groups, AGARD, & the Bd of the von Karman Inst. Gen Chrmn of AIAA's 12th Aerospace Sciences Meeting. Author of 50 scientific publications. Exec Dir in Natl Res Council, 1981-present. Fellow AIAA. *Society Aff:* ΣΞ, ΤΒΠ, ΠΤΣ, ΟΔΚ, ΠΜΕ.

Korman, Nathaniel I
Home: 108 Yucca Lane, Placitas, NM 87043
Position: President. *Employer:* Ventures Res & Dev Group. *Education:* PhD/Physics/U of PA *Born:* Feb 1916. Mbr Sigma Xi. MIT, SM; Charles A Coffin Fellow. Univ of Pa, PhD. RCA Corp, various engrg positions: Ch Sys Engr, Dir Advanced Military Sys, Dir Med Plans & Progs; RCA Award of Merit; US Dept of Defense, Cons to R/D Board, Cons to ARPA. Pres Ventures R/D Group; dev of new business ventures, especially color separation for Graphic Arts. *Society Aff:* IEEE

Korn, Granino A
Business: Dept of EE, Tucson, AZ 85721
Position: Professor of EE. *Employer:* University of Arizona. *Education:* PhD/Physics/Brown Univ. *Born:* May 7, 1922. US Navy 1944-46. Proj Engr Sperry Gyroscope 1946-48; Staff Engr Lockheed Aircraft 1949-52; Head Analysis Group Curtiss-Wright 1948-49; Cons under own name 1952- . Sr Sci Simulation Award 1968; winner Humboldt Award 1975 (W Germany). Fellow, IEEE, 1979 Assoc Ed 'Simulation'. Sci Ctte IMACS; Sigma Xi, IEEE, SCS. Author. Co-Editor-inChief: 'Computer Handbook', 'Digital Computer User's Handbook'. *Society Aff:* IEEE, SCS, IMACS, ΣΞ

Kornei, Otto
Home: Del Mesa Carmel 265, Carmel, CA 93921
Position: Retired. *Employer:* *. *Education:* MSEE/EE/Techn-Hochschule, Vienna; MSEE/EE/Techn-Hochschule, Berlin. *Born:* 1903 Vienna Austria. ING (MS) EE Techn Hochschule, Vienna; DIPL ING EE Techn Hochschule, Berlin. IEEE, AES. Retired. Extensive work in facsimile, disc, film & magnetic recording with orgs in Europe & USA. As an independent cons in NYC instrumental in initial dev of Xerography in 1938. Subsequently important contributions to tech of magnetic recording, incl video, resulting in nine original papers & over 30 patents. Last employment as dept head with IBM San Jose Calif. Fellow IEEE; 1968; citation: "For fundamental and extensive contributions to magnetic recording technology for video and audio applications as well as creative contributions to the beginnings of xerography." Hon Mbr Audio Engrg Soc. *Society Aff:* IEEE, AES.

Kornhauser, Edward T
Business: Div of Engg., Brown University, Providence, RI 02912
Position: Prof. *Employer:* Brown Univ. *Education:* PhD/Applied Physics/Harvard Univ; MSc/Applied Physics/Harvard Univ; BEE/Elec Engg/Cornell Univ. *Born:* 6/30/25. Louisville, KY. Served in US Navy, 1943-46. Instructor, Harvard Univ, Dept of Applied Physics, 1949-51. Brown Univ, 1951-present, currently Prof of Engg. NATO Fellow, Univ of Bristol, 1959-60. Academic Visitor, Oxford Univ, 1966-67; Imperial College, London, 1974, Visiting Professor, Univ. of CA, San Diego, 1981-82. *Society Aff:* IEEE, ΣΞ, ΤΒΠ, HKN.

Kornsand, Norman J
Home: 112 Millbrae Ct, Walnut Creek, CA 94598
Position: Western Regional Mgr *Employer:* Schirmer Engrg Corp *Education:* BS/Fire Prot Engr/IL Inst of Tech; MBA/Personnel/De Paul Univ *Born:* 5/5/48 Rolf Jensen & Assoc, Inc: 1977-1982, Mgr - San Francisco Office; 1973-77, Assoc Ins Services Office: 1970-73, Public Protection Supervisor; 1970, Engr. PE: IL, CA, NV, AZ, OR, CO, WA. Private Investigator: CA. Schirmer Engrg Corp: 1982-present - Western Regional Office, Western Regional & Engrg Mgr. *Society Aff:* NFPA, SFPE, ICBO

Koros, Peter J
Home: 154 Maple Heights Rd, Pittsburgh, PA 15232
Position: Senior Research Consultant *Employer:* LTV Steel Company *Education:* ScD/Metallurgy/MIT; SM/Metallurgy/MIT; BS/Met Eng/Drexel Univ. *Born:* Res Eng J&L Steel 1958, Res Supr 1963, Ch Process Metallurgist Quality Control, 1965; Dir - Process Metallurgy 1975-79, Dir of Res - Special Proj 1979-80, Mgr, Process Dev 1980-82; Sr Res Assoc 1982-84, Sr Res Conslt 1984-. Responsible for new tech assessment, electric furnace technologies, tech exchanges, desulfurization of hot metal & steel, process analysis. 55 papers, 7 US pats. Honors: ISS-AIME Distinguished Mbr (Fellow grade), AISI Gold Medal 1978 & Reg Tech Award 1969, Intl Magnesium Assoc 1978, AIME's Toy 1962, Herty 1963; McKune 1963; J&L's Jalmet 1963. Prof Soc's: AIME, ASM, AISI & Eastern Staes Blast Furnace & Coke Oven Assn. Listed in Amer M/W of Sci. Past-Chrmn, Adv Council, US Bureau of Mines Generic Ctr for Pyrometallurgy, Program Chrmn 5th Intl Iron & Steel Congress (1986); formerly, Editorial Policy Chrmn for Transactions - ISS-AIME; Bd of Dirs AIME 1974, Ch Iron & Steel Div. In ASM, Govt Rules Comm, 1972-75. In 1969-72, Spearheaded move to restructure AIME by formation of Iron & Steel Soc. *Society Aff:* AIME, ASM, AISI.

Korpela, Seppo A
Business: 206 West 18th Ave, Columbus, OH 43210
Position: Assoc Prof *Employer:* OH State Univ *Education:* PhD/ME/Univ of MI; MS/ME/Univ of MI; BSE/ME/Brigham Young Univ *Born:* 4/8/43 Born in Finland. Immigrated to Canada in 1960, and then to the US in 1963. Naturalized American citizen in 1968. Worked in the Heat Transfer Laboratory of the Univ of MI as a graduate student where he won the Distinguished Achievement Award for the Outstanding Mechanical Engrg Graduate Student in 1971. Joined the Department of Mechanical Engrg at the OH State Univ in 1972 as an Assistant Prof. Promoted to Assoc Prof in 1977. Sabbatical leave at Princeton Univ in 1980. Visiting lecturer at Lappeenranta Technical Univ, Finland. *Society Aff:* ASME, APS.

Korte, George B, Jr
Business: 8401 Arlington Blvd, Fairfax, VA 22031
Position: Sr Assoc *Employer:* Dewberry & Davis *Education:* M/Bus Admin/Univ of VA; BS/Civ Engrg/George Wash Univ *Born:* 9/6/49 Native of Northern VA suburbs of Wash, DC. Civ Engr with Dewberry & Davis since receiving MBA in 1974. Has worked with the Natl Flood Ins Program since 1975. Became Asst Proj Mgr of 150-man proj in 1978. Became Mgr of Computer Aided Design and Drafting in 1982 and implemented his firm's first CADD system at that time. Became manager of Mapping and Computer Graphics Services in 1986. Named an Assoc in Dewberry & Davis in 1979, Sr Assoc in 1983. Pres of Northern VA Chapter, VA Soc of PE 1980-81. Chapter Young Engr of the Yr, 1980. VSPE Outstanding Service Award, 1981. Married the former Helen Spence of Falls Church, VA in 1973. Three children, Mary Ellen, 12, Christopher, 9, and Kathleen, 6. *Society Aff:* NSPE, NCGA

Korwek, Alexander D
Home: 34 Island Dr, Norwalk, CT 06855
Position: Sec & Gen Mgr (CEO) *Employer:* United Engrg Trustees, Inc *Education:* MBA/Bus Admn/Univ of Utah; BSBA/Bus Mgmnt/Washington Univ (St Louis, MO) *Born:* 02/20/32 Involved in the application of the electronic computer from its inception in the areas of engrg, science, mfg and bus. Was a part of a team doing state- of-the-art computer applications in the areas of mfg control, process control and operations analysis and res, as well as pioneering computer mgmt concepts. Did extensive work in the early use of the computer in cost estimation, developing Cost Estimating Relationships, and applying these concepts to practical use in the aerospace indus. Progressed into mgmt, applying methods and systems tech to the bus, financial, adm and mgmt problems as well as connecting these functions to the engrg and operational aspects of the firm. *Society Aff:* ASCE, CESSE, ASAE

Korynta, Earl D
Business: 2515 A Street, Anchorage, AK 99502
Position: Princ *Employer:* USKH, Inc *Education:* MS/EM/Univ of AK; BS/CE/Univ of ND *Born:* 5/3/43 Owner and chief engr of medium size Alaskan consulting firm, USKH, Inc offering engrg, architectural and surveying services. Past president of Anchorage Branch and Alaska Section American Society of Civil Engr (ASCE). Past Pres Alaska Prof Design Council. Named "Engr of the Year" in Alaska by fellow engr representing seven different National Societies. Married, three boys, commercial pilot. *Society Aff:* ASCE, NSPE, PEPP, ASTM, CECA.

Kosar, Halit M
Business: Univ Square, Erie, PA 16541
Position: Dean Science & Engineering Div. *Employer:* Gannon Univ *Education:* PhD/ME/Istanbul Tech Univ; MS/ME/Istanbul Tech Univ. *Born:* Nov 1924; native of Istanbul Turkey. PhD & MS Istanbul Tech Univ. 1 yr factory training in US 1950. Taught at Istanbul Tech Univ. Postdoc work at Case Inst. Asst Prof Gannon Coll. Summer work Boelkow GmbH Germany. Assoc Prof Robert Coll Istanbul & docent Inst Tech Univ 1967-69. Assoc Prof Gannon Coll & assumed current respon as Dean of Sci & Engrg Div 1973. Prof in Mech Engrg. Reg in Penn 1970. Mbr NSPE, ASEE, ASME, IEEE, SME, SAE, VDI, TMMO. Engr of the Year Award ASME 1976, PSPE 1980, IEEE 1983, Dedicated Service Award PSPE 1982. Enjoys reading, watch repair. *Society Aff:* NSPE, ASEE, ASME, IEEE, SME, SAE, VDI, TMMO

Koshar, Louis D
Home: 23 Serpentine La, Old Bethpage, NY 11804
Position: VP *Employer:* Pavlo Engrg Co *Education:* MCE/Structures/CCNY; BCE/Str Civil Soils/CCNY *Born:* 8/19/28 Currently VP of Pavlo Engrg Co in charge of design and planning of bridge and hgwy projects in various states such as WV, MA, NH and NJ including major bridges over the OH River at Parkersburg and East Huntington, WV. Presented papers on design of steel cable stayed bridges at SASHTO in 1979, ASCE at Univ of Cincinnati and at CCNY in 1980. Was Secretary of Nassau County Chapter NYSSPE from 1961 to 1966. Registered PE in NY, FL and WV. In charge of the design of the Harvard Bridge Reconstruction carrying Massachusetts Ave over the Charles River from Boston to Cambridge. Presented a paper on the design of the Harvard Bridge at MIT in 1984. *Society Aff:* ASCE, NSPE, ASHE, NYSSPE

Kosmahl, Henry G
Home: 7419 River Rd, Olmsted Falls, OH 44138
Position: Consultant on Electron Beam Devices *Employer:* Self *Education:* PhD/Electron Physics/Univ of Darmstadt *Born:* 12/14/19. Native of Germany. Instr of Phys at the Univ of Darmstadt, 1949-1952. Dissertation on high efficiency Klystrons, achieved 35% in 1949. Res Physicist at Telefunken Labs, Ulm, Germany 1952-1956. Presently Conslt to Nasa LeRes Ctr, Cleveland, OH. Conslt to Hughes AC, EDD in Torrance, CA. Works on high efficiency microwave Amplifiers, high efficiency noval Depressed collectors, beam reconditioning dynamic velocity tapers, multi-dimensional computer Progs, high efficiency space Tubes, advanced concepts for millimeter waves. Micron size field emission Amplifers for the 100Ghz-Tera Hertz Frequency Range. NASA Sci Achievement Medal 1974, IEEE/DOD Tech Advancement Awd 1977, IEEE Fellow 1979. NASA Major Invention Awds $25,000; 1980 and 1987. Published 20 patents, 10 Intl, 2 used in Space & DOD. Functions as Assoc Editor, Transactions on Electron Devices, IEEE. Enjoy classical music & swimming & walking. *Society Aff:* IEEE

Koss, John P
Home: 14 Mountain Gate Rd, Ashland, MA 01721
Position: VP, Operations *Employer:* General Cinema Beverages, Inc. (Div. of General Cinema Corp.) *Education:* MSc/IE/OH State Univ; BS/IE/OH State Univ *Born:* 5/31/24 Native Akron, OH. Army Air Force, Cyptrographer, WWII Pacific. Taught at OH State Univ & Kent State Univ, Ind Engrg. Developed IE in Construction with H.K. Ferguson Co 1952. New Physical distribution Concept with National Distribution Services 1972. With General Cinema Beverages, Inc since 1974 as VP, Operations: Industrial Engrg, Fleet Operations, Construction, Plant Engrg, Physical Distribution & Quality Control; largest Soft Drink bottler in USA. Pres, IIE Akron Chapter 1970; OSU Pres's Committee for Tomorrow. Registered PE: OH, MA, MI, CA, GA, MD, FL, IL. Enjoys flying & golf. *Society Aff:* IIE, AAAS, NSPE, FSPE, SSDT

Kossa, Miklos M
Business: 1760 Solano Ave, Berkeley, CA 94707
Position: Owner and Naval Architect. *Employer:* M M Kossa, Naval Arch-Marine Cons. *Education:* BSE/NA & Mar Eng/Univ of MI *Born:* Dec 1934, Budapest, Hungary. Attended Tech Univ of Budapest. Emigrated to US 1957. With Todd Shipyards Corp: Design Engr, Los Angeles Div, 1959-61; Naval Arch, San Francisco Div, 1961-71. Founded own consulting firm 1971. Chmn N Calif Sec, The Soc of Naval Archs and Marine Engrs, 1975-76. Mbr Advisory Ctte on continuing Maritime Educ, Calif Maritime Acad, Visiting Lecturer, Univ of CA, Berkeley, 1982.
Society Aff: SNAME

Koster, William P
Business: 3980 Rosslyn Dr, Cincinnati, OH 45209
Position: President. *Employer:* Metcut Research Assocs Inc. *Education:* PhD/Met Engg/Univ of Cincinnati; MSc/Met Engg/Univ of Cincinnati; BS/Mech Engg/Rutgers Univ. *Born:* April 1929. Native of NJ. Jr Engr, American Smelting & Refining Co Perth Amboy, NJ 1946-51 part time. Graduate Asst, Univ of Cincinnati 1951-53. With Metcut Research Assocs Inc since 1953: assumed position as Asst Dir of Metallurgical Engrg at Metcut in 1954; Dir of Metallurgical Engrg since 1958; elected V Pres in 1958 & assumed responsibility as Exec V Pres in 1975 and Pres 1978. Prime engrg work in recent years has been evaluation of surface machining & finishing variables to detrmine their influence on reliability and performance of machined components. Mbr of several professional socs: ASM International: Chmn of Cincinnati Chapter, 1961-62; elected Fellow ASM, 1976; elected Trustee 1985-88. SME: Awarded Gold Medal, 1978; elected Fellow SME, 1986. SAMPE: Chmn of Midwest Chapter, 1974-75; Nat'l Director, 1975-77; Nat'l President, 1979-80; elected Fellow SAMPE, 1982. *Society Aff:* SAMPE, ASM, SME, ESC

Kotfila, Ralph J
Home: 206 Monroe Mill Dr, Ballwin, MO 63011
Position: Lead Engr. *Employer:* McDonnell Aircraft Co. *Education:* BS/Met Engg/Purdue Univ. *Born:* 12/23/19. Associated with aerospace industry for 27 yrs. As Maj & Proj Engr in 1952, was responsible for early res in titanium met at the AF Mtls Lab. Published first technical literature relating to hydrogen embrittlement.& heat treatment of titanium alloys. Recognized as a "Titanium Founder" by others in the industry. As section chief at Aerojet in 1958, was responsible for res on the use & fabrication of refractory metals for rocket nozzle application. At MCAIR since 1971; working on titanium net shape mfg tech including superplastic forming, isothermal forging & powder met. Favorite sport, golf. *Society Aff:* ASM, AIME.

Kotnick, George
Home: 408 Glenwood Ave, Moylan, PA 19065
Position: Consultant *Employer:* Self *Education:* BSME/Power/OK State Univ. *Born:* Feb 1952. Native of Penn. Merchant Marine & Naval Officer prior to & during WWII. With Phila Elec Co '52 to '80. Project Engineering Manager, United Engineers & Constructors 1980 to 83. Currently power plant consultant. Supervision of design of elec generation systems, particularly environmental aspects. Reg PE; Diplomate Amer Acad of Environmental Engrs. Pres, Amer Soc of Mech Engrs 1984-1985. Sec-treasurer AAES, 1987. Hobbies include photography, fishing & ship modeling. *Society Aff:* ASME, ANS

Kottcamp, Edward H Jr
Business: SPS Technologies, Route 332, Newtown, PA 18940
Position: Group V.P. *Employer:* SPS Technologies *Education:* PhD/Met Engg/Lehigh Univ; MS/Metallurgy/Lehigh Univ; BS/Met & Mat. Science/Lehigh Univ. *Born:* 7/12/34. Presently Group VP, SPS Technologies (1987). Exec. VP Bethlehem Steel 1982-87 & VP Research, Bethlehem. Published in the areas of brittle fracture, cold extrusion of steels and pressure vessel design. Recipient of the American Welding Society's William Sparagan Award and US Army Munitions Command Award. Grad of the Prog for Mgt Dev, Harvard Business School 1973. Presently a mbr and trustee of ASM, mbr of AISI & IRI. *Society Aff:* ASM, AISI, IRI, WRC.

Kotval, Peshotan S
Home: 8 Verne Pl, Hartsdale, NY 10530
Position: Attending Radiologist *Employer:* New York Medical College *Education:* MD/Medicine/NY Med Coll; PhD/Physical Met/Sheffield Univ, England; M. Met/Physical Met/Sheffield Univ, England; MBA/Finance/Pace Univ NY; BS/Physics/Nagpur Univ India *Born:* 8/31/42. Pesho Kotval came to the US in 1966 after having received his PhD from the Univ of Sheffield, England. Directly upon his arrival, he joined Union Carbide Corp. His career at Union Carbide has spanned a broad involvement with Mtls Research and Engrg and its mgt. This work has included the design of high-temperature superalloys, powder-met composites, wearmtls, crystal-growth, solar-energy related mtls and medical tech. From his involvement with Medical Tech, in 1979, Dr Kotval joined NY Med Coll and received his MD degree in 1983. He became a Diplomate, Amer Bd of Radiology in 1987. He is presently an Attending Radiologist in the Dept of Radiology at NY Med Coll. Dr Kotval is also Prof of Mgt Sci at Pace Univ where he obtained his MBA degree in 1977. He has published some 40 tech papers in peer-review journals, has been awarded 8 US patents, and, in 1978, was elected a Fellow of the American Soc of Metals for his contributions to superalloy met. He is married (wife is a banker), has two children and resides in Hartsdale, NY. *Society Aff:* ASM, AMA

Koul, M Kishen
Business: 20 Lake Dr, E Windsor, NJ 08520
Position: Sr Research Scientist *Employer:* Johnson & Johnson Co *Education:* PhD *Born:* 09/10/41 Kashir India, BSc J&KU, Kashmir India 1959; BSc Met Engrg BHU, Varanasi India 1963. Joined Union Carbide India Ltd in 1963 as metallurgist. Came to US in 1965 joining MIT as Res Asst, Materials Sci Ctr; received PhD in Material Sci in 1968. Continued with MIT as mbr of DSR staff until joining Union Carbide 1969 as Res Scientist, Ferroalloy Div; became proj engr, new prod dev grp of Mining & Metals Div in 1970, respon for developing new prods & mkts for vanadium, tungsten, chromium, & manganese in the steel, aluminum & titanium industries. Joined Foote Mineral Co in 1974, and was made Mgr Steel Res & Dev. Respon for res & dev in the areas of HSLA steels, deoxidation, desulfurization & sulfide modification, & boron steels. Went to India in 1979 as Exec VP of Indian Metals and Ferro-Alloys Ltd. Returned to USA and joined Johnson & Johnson Co as a Sr Res Scientist in 1980. Mbr ASM, AIME. Listed in Amer Men & Women of Sci. *Society Aff:* ASM, AIME

Kouts, Herbert JC
Home: 249 South Country Road, Brookhaven, NY 11719
Position: Dept Chairman *Employer:* Brookhaven National Lab. *Education:* PhD/Physics/Princeton University; MS/Physics/Louisiana State University; BA/Mathematics/Louisiana State University *Born:* 12/18/19 Brookhaven National Lab 1950-73; 1976-. Research on reactor shielding, reactor physics, nuclear materials safeguards. 1973-76, Dir Safety Research, US Atomic Energy Comm, US Nuclear Regulatory Comm. Currently Chmn Nuclear Energy Dept, Brookhaven National

Kouts, Herbert JC (Continued)
Lab. 1962-67 USAEC Adv Cttee on Reactor Safeguards, Chman 1964. Mbr International Atomic Energy Agency's International Nuclear Safety Advisory Group (INSAG); E.O. Lawrence Award 1963; AEC Distinguished Service Award 1974; USNRC Distinguished Service Award 1976; ANS Theos Thompson Award 1982.
Society Aff: NAE, ANS

Kovaly, John J
Home: 3 Tubwreck Dr, Dover, MA 02030
Position: Consulting Engr *Employer:* Raytheon Co Missile Systems Division
Education: MS/Physics, Math/Univ of IL; BS/Physics, Math/Muskingum Coll *Born:* 6/12/28 In 1981 elected to Fellow grade IEEE for contributions to Synthetic Aperture Radar. Present position, Assistant to PATRIOT Program Mgr, Raytheon Co, Missile systems Division since 1976. PATRIOT is Army's Air Defense Missile System for 1980s and beyond. In 1974 named to position of Consulting Engr, highest professional scientific and engrg level obtainable at Raytheon Co. With Raytheon Co since 1965. Equipment Engrg Mgr B-1 Multimode Radar, Electronically Agile Radar, Hardsite Radar and Systems Engr for Synthetic Aperture Radar Guidance (SARG), Advanced Air-to-Surface Missile Terminal Guidance. Sylvania Electonic Products, Inc 1958 to 1965, radar systems design for Applications Tech Satellite Transportable Ground Station, Multifunction Array Radar, CODIPHASE phased array techniques program and Radar Signature Program. Naval Officer 1955 to 1958 with electronic designator specialty. Coordinated Science Lab, Univ of IL, 1951 to 1955, experimental work airborne coherent radars for ground and sea clutter characterization, submarine snorkel and periscope detection and synthetic aperture radar development. Pres, Boston section IEEE Group-Aerospace and Electronic Systems - 1972. Author of *Synthetic Aperture Radar*, Artech House Inc, 1976. Taught at UCLA and Northeastern Univ. Served on government advisory councils.
Society Aff: IEEE

Kovarik, James C
Business: Court St Plant 5-E7, Syracuse, NY 13201
Position: Manager-HMED Equipment Engineering. *Employer:* General Electric Company. *Born:* Native of Syracuse NY. BSEE 1954 St Louis Univ. Joined GE on the Advanced Tech prog. US Army 1954. Rejoined GE & completed ATC prog graduating from course 1959. Advanced dev engrg 1959-66. Became Mgr-Sys Engrg for Nike-X-Prog 1966. In 1967 appointed Mgr-Radar Receiver & Processor Engr & in 1968 named Subsection Mgr of Equipment Engrg. In 1974 the org elevated to Section level. Pat: Multiple IF Side-Lobe Cancellor 1962.

Kovats, Andre
Home: 13. Baker Rd, Livingston, NJ 07039
Position: Retired *Employer:* Foster Wheeler Energy Comp. *Education:* Pipl. Engineer, Mech. and Electric/Technical University of Budapest; High school-matura/Piarist Gymnasium of Budapest *Born:* 05/07/97 1916-18. Artillery-lieutenant in the Austro-Hungarian imperial army on the Russian-front. 3 medals of bravery; 1923-38, Chief-Engr., Turbo-machinery Dept. of the Ganz & Co. Budapest; 1938-45, Mgr, Diesel-train Dept. of the Ganz & Co. Budapest; 1948-54, Chief-Engr for pumps, compressors and Gas turbines, Rateau Soc. Paris, France; 1954-72, Chief-Engr. for Pumps, Foster-Wheeler Co. Livingston, New Jersey; 1972-82, Consultant. Designed the biggest volute-pump in the world for Egypt, and the three biggest concrete casing pumps in the U.S.A. Author of 2 books in French (translated also in German and Turkish): Pumps and Compressors; author, one book in English: Pumps and Compressors. 12 Papers published by the ASME. *Society Aff:* AAAS, ASME

Kowalonek, John M
Business: PO Box 6459, Bridgeport, CT 06606
Position: Dean of Engg. *Employer:* Bridgeport Engg Inst. & Sikorsky Aircraft Div OTC. *Education:* MSME/ME/Columbia Univ; BME/ME/Syracuse Univ. *Born:* 9/29/26. Dean of Engg at BEI for past 17 yrs. Began as instr in fluid mechanics, kinematics, graphics and machine design 33 yrs ago. Was elected "Fellow of the Inst-, highest BEI Honor given to persons contributing to the advancement of the Inst. He served as ME Dept VChrmn and later Chrmn prior to becoming Dean. Directed the Planning of BEI's Lab. He was instrumental in developing courses in drafting, biomedical engg, in addition to counseling sr students in proj work. Mr Kowalonek is employed as a Design Proj Engr, Propulsion Systems, at Sikorsky Div, United Technologies and has been there for over 24 yrs. Before joining Sikorsky in 1963, he was employed at AVCO Lycoming as a Dev Engr, Thermodynamics & Controls. In 1981, he was appointed a Designated Engineering Representative for the Federal Aviation Administration, specializing in Gas Turbine Power Plant Systems and Installations. In 1986, he was appointed Technical Manager-S-76 Engines at Sikorsky. *Society Aff:* ASME, ASEE.

Kowalski, Carl F
Business: 201 West Center Court, Schaumburg, IL 60196
Position: Bureau Chief of Traffic. *Employer:* Illinois Dept of Transportation. *Education:* BS/CE/U of Illinois. *Born:* Jan 1940. BSCE Univ of Ill. Directly respon for all traffic controls on over 8000 lane miles of urban arterial streets & 1200 lane miles of freeways carrying over 200,000 ADT in 6 county area which includes Chicago. With IDOT 1965- . Served as Pres of the Ill Sect of the Inst of Transportation Engrs 1974. Lectrs on Freeway Opers at Northwestern Univ Traffic Inst.

Kowalski, Philip L
Business: Ten Springborn Center, Enfield, CT 06082
Position: Pres *Employer:* Springborn Laboratories *Education:* M/BA/Univ of CT; BS/ChE/Rensselaer Polytechnic Inst. *Born:* 8/31/43 1965-67 Product/Process Engr Proctor and Gamble Co, Cincinnati, OH. 1967-74 American Cyanamid Co Res Engr, Stamford, CT. Production Engr, Wallingford, CT Polymer Plant. 1974-present Springborn Labs, Inc., Enfield, CT. Proj Mgr- business dev 1974. Gen Mgr-Contact Mfg 1975. VP-Res, Dev, and Engrg 1977. Pres- 1980. Outside interests golf, fishing, and classical music. *Society Aff:* AIChE, SPE, AMA

Kowalski, Sylvester Jos
Home: 1 Robin Lake Dr, Cherry Hill, NJ 08003
Position: VP, Nuc. Engr. *Employer:* Phila. Electric Co. *Education:* B.S./Mech. Engr./Drexel Univ. *Born:* 02/01/29 Resided in Philadelphia area for 45 years. Entire professional career with Philadelphia Electric Co. Progressed from junior engr to VP engrg and research. Currently, VP nuclear engrg for three operating plants and one plant under construction. For the past 16 years, involved in numerous positions associated with the nuclear power construction and engrg prog for Philadelphia Electric Co. *Society Aff:* ANS, ASME, PSPE

Kowel, Stephen T
Business: Dept of Elec & Comp Engg, University of California, Bainer Hall, Davis, CA 95616
Position: Prof of Elec Engg. *Employer:* Univ of CA, Davis *Education:* PhD/EE/Univ of PA; MSEE/EE/Poly Inst of Brooklyn; BSEE/EE/Univ of PA. *Born:* 11/20/42. Dr Kowel joined the faculty of Syracuse Univ, Syracuse, NY, as Asst Prof of Elec Engg in 1969, becoming Assoc Prof in 1974, & Prof in 1979. He was a visiting prof in the school of Elec Engg and in the National Research and Resource Facility for Submicron Structures at Cornell Univ, Ithaca, NY, 1982-83. He joined the faculty of the Univ of CA, Davis, as Prof of Elec and Computer Engg in 1984. Dr Kowel became Vice Chair of the ECE Dept in 1986. He has published approx fifty technical papers & has been awarded eight US patents. His current interests are in physical electronics and optics, principally on devices for imaging by direct electronic Fourier transformation, in electro-optics, optical processing, and adaptive optics. *Society Aff:* IEEE, AAAS, ASEE, AAUP.

Kowtna, Christopher C
Business: Medical Products Dept, Barley Mill Plaza P26-2122, Wilmington, DE 19898

Kowtna, Christopher C (Continued)
Position: Mgr, Regulatory Affairs *Employer:* E I Du Pont De Nemours & Co, Inc *Education:* MBA/Quantitative Methods/Widener Univ; BS/Chemistry/Widener Univ *Born:* 12/05/46 Native of Edison NJ. Held Quality Assurance positions in DoD prior to joining Merck, Sharp, and Dohme. Positions in Mfg, QC, and Govt Liaison. With Du Pont since 1978 working in medical device and pharmaceutical divs QA, Plant Tech, and Regulatory Affairs. Assumed current position Nov 1985 with worldwide respon. Regional Councilor ASQC 1982 to date. Vice Chmn sel section ASQC 1987-88. ASQC Cert Quality Engr and Sr Mbr. *Society Aff:* ASQC, RAPS

Kozik, Thomas J
Business: Mech Engrg Dept, College Station, TX 77843
Position: Prof. of Mech. Eng. *Employer:* Texas A&M Univ *Education:* Ph.D./Eng. Mechanics/The Ohio State University; M.S./ Eng. Mechanics/The Ohio State University; B.S./Aero. Eng./Rensselaer Polytechnic Inst. *Born:* 04/09/30 Thomas J. Kozik was born in Jersey City, N.J.. After receiving his B.S. degree. In 1952, he worked as a design engineer for Curtiss Wright Corp. in Caldwell, N.J. until drafted into the U.S. Army in 1954. While in the service Kozik was stationed at Fort Huachuca AZ, and was discharged in 1956 at which time he enrolled at Ohio State Univ to pursue graduate work. He received his M.S. in 1958 and his Ph.D in 1962. He also married Freda Fay Larson in 1958. Kozik came to Texas A&M Univ in 1963 and has been there since. In 1985 he was named Dietz Professor of Mechanical Engrg. He is a Fellow of the ASME, past Chairman of the Petroleum Div of the ASME, past chairman of the Energy Technology Conference & Exhibition and recipient of the General Dynamics Award for excellence in teaching and research.
e *Society Aff:* ASME, ASEE, AAAS, NYAS, AAUP

Kraehenbuehl, John O
Home: 1105 Ohio Ave, Canon City, CO 81212
Position: Retired. *Employer:* *. *Education:* EE/Elec Engg/Univ of IL; MS Eng/Mech Engg/Univ of TN; BSEE/Elec Engg/Univ of TN; BSME/Mech Engg/Univ of TN. *Born:* Oct 4, 1894 Knoxville Tenn. Educ Tenn, Ky public schs; m. Kathryn Irene Walter Aug 31, 1922; s. John David & Roger Gale, d. Mrs W G Dunbar. 1922-56 Prof Elec Engrg (Emiritus) Univ of Ill; 1956-62 Prof CE (Emeritus) Colo Coll; Teaching & Cons Chmn & mbr various natl & internatl comms on Illumination; Past Mbr Optical Soc of Amer, Phi Kappa Phi, Tau Beta Pi, Sigma Xi, Delta Epsilon, Illum Engrg Soc Emeritus; 1957 citation Gold Medal Illum Engrg Soc; citation Chicago Lighting Inst 1957. Author of Elec Engrg Circuits, Machines jointly; Illuminating; various papers on Elec Engrg & Illumination. Christian. *Society Aff:* IES.

Kraemer, M Scott
Business: 1200 Smith St. Suite 1900, 2 Allen Center, Houston, TX 77002
Position: VP & Region Mgr. *Employer:* Champlin Petroleum Co. *Education:* BS/Petroleum Engg/TX A&M Univ. *Born:* 5/25/21. Native of Victoria, TX. Upon graduating from TX A&M Univ, joined Amoco Production Co in 1943. After holding various engg supervisory positions, including chief engr from 1966 to 1973, & Regional Prod. Mgr 1973-1981-Assumed present position July 1,1981. A long time mbr of the Soc of Petroleum Engrs, serving on various comms as chrmn, chrmn of the exec comm of Offshore Tech Conf, and past pres of SPE and AIME. Recipient of the Soc of Petroleum Engrs' Distinguished Service Award and API's Citation for Service Award. Enjoy hunting, fishing and golf. *Society Aff:* AIME.

Kraft, Christopher C, Jr
Business: NASA Rd 1, Houston, TX 77058
Position: Director. *Employer:* NASA Johnson Space Center. *Education:* Hon Doctorate/Indus Inst of Tech; Hon Doctorate/St Louis Univ; Hon Doctorate/Villanova Univ; BS/Aero Eng/VPI. *Born:* 2/28/24. Native of Phoebus Va. Entered Federal Service Jan 1945; 1 of original mbrs of Space Task Group Oct 1958; named Dir NASA Johnson Space Cntr Jan 1972. Named 1 of 100 outstanding young leaders in the nation by eds of Life Magazine 1962; Arthur S Fleming Award 1963 (1 of 10 outstanding young men in govt career service); Spirit of St Louis Medal, NASA Outstanding Leadership Award from Pres of US; Disting Alumnus Citation VPI, Virginian of the Year; 2 NASA Disting Service Medals; Amer Astro Soc Space Flight Award; Louis W Hill Space Transportation Award; NCAA Theodore Roosevelt Award; VFW Natl Space Award; ASME Medal. Fellow AIAA & Amer Astronautical Soc; Mbr NAE. Enjoys golf. *Society Aff:* AIAA, AAS, NAE.

Kraft, Leland M, Jr
Business: 6100 Hillcroft, Houston, TX 77036
Position: Mgr, Spec Proj Grp. *Employer:* McClelland Engrs, Inc. *Education:* PhD/Geo Engg/OH State Univ; MSCE/Geo Eng/OH State Univ; BSCE/Geo Engg/OH State Univ. *Born:* 2/27/42. Worked for a time as an asst prof of civil engg at Auburn Univ. Mgr of the spec proj group, McClelland Engrs, Inc. Involved as a foundation engg on such projects as petroleum and chem process units, paper mills, multistory buildings and offshore structures. Respon for studies on the design and installation of foundations for offshore structures in various parts of the world, including large tower-type structures in the North Sea and on the Continental slope in the Gulf of Mexico, and for mudslide-resistant sturctures in the Mississippi Delta. *Society Aff:* ASCE, TRB, ISSMFE.

Kraft, R Wayne
Business: Dept of Materials Science & Engineering, Whitaker Lab, Building 5, Bethlehem, PA 18015
Position: Prof. *Employer:* Lehigh Univ. *Education:* PhD/Met/Univ of MI; MS/Met/Univ of MI; BS/Met/Lehigh Univ. *Born:* 1/14/25. *Society Aff:* ASM, AAAS, SGSR.

Kraft, Walter H
Business: Edwards and Kelcey, Inc, 70 S Orange Ave, Livingston, NJ 07039
Position: Sr. VP & Chief, Traffic & Transp. Div. *Employer:* Edwards & Kelcey, Inc. *Education:* DEngSc/CE/NJ Inst of Tech; MS/CE/Newark Coll of Engrg; BS/CE/Newark Coll of Engrg. *Born:* 12/31/38. Reg PE in CA (Traffic), KY, NJ, NY, PA, MN, DE, CT, TX & OH. Is Chief of Traffic Engrg & Trans Planning Div with Edwards & Kelcey, Inc since 1973. Became a VP in 1977 Senior VP 1984-present. Adj Prof at the NJ Inst of Tech & the Poly Inst of NY. Previously Adj Lectr at Carnegie-Mellon Univ & St John's Univ. Pres, ITE, Metropolitan Sec of NY & NJ, 1978; Chmn Urban Trans Div, ASCE, 1977-1979; Pres, NJ Branch, ASCE, 1970-71. Chmn, ITE Dist One during 1981 & 82; Section Dir, ITE Metropolitan Section of NY & NJ, 1980-81; Chmn, ASCE Natl Trans Policy Ctte, 1980-81; ASCE, Mgmt Group C, Contact Mbr, 1981-85; Chmn, TRB Ctte AIEO3 Intermodel Trans Facilities, 1980-86; ITE, Dist One, Intl Dir, 1984-86; and Prog Chmn, NY State Assoc of Traffic Safety Bds, 1980-81. Southern NJ Dev Coun, Trustee, 1984- ITE, International VP, 1986; & ITE, International Pres, 1987. MAUDEP, Director & Program Chairperson, 1978-85. NJIT, Member Civil and Environmental Advisory Committee to the Board of Trustees, 1982-present. ASCE Metropolitan Section Robert Ridgeway Award, 1962; Chi Epsilon; Tau Beta Pi; Special Service award, MAUDEP, 1975; Recipient of the Frank Masters Award, ASCE, 1982; ITE District One Distinguished Service Award, 1986; & ITE Ivor S. Wisepart Transportation Engineer Award, 1986. Enjoys folk dancing & bicycling. *Society Aff:* ITE, ASCE, TRB, SNJDC, ASHE, NSPE, NJIT, MAUDEP, SCS, TIEDC, AMA.

Krahl, Nat W
Business: 1111 West Loop South, Houston, TX 77027
Position: Sr VP & Dir of Structural Engrg *Employer:* Caudill Rowlett Scott, Inc *Education:* PhD/Civil Engg/Univ of IL; MS/Civil Engg/Univ of IL; BS/Civil Engg/Rice Inst; BA/Civil Engg/Rice Inst. *Born:* 9/30/21. Houston TX. m. Victoria Ferguson; 5 children. Line officer US Navy 1944-46. Struct engr Walter P Moore Cons Engr 1946-57; faculty mbr Rice Univ 1957-80; Prof Cvl Engrg & Architecture 1968-80; Chrmn of Cvl Engrg 1972-77; Ptnr Krahl & Gaddy Engrs 1969-76; Principal Nat Krahl & Assocs Cons Engrs 1976-80. Sr VP and Dir of Structural Engrg, Caudill Rowlett Scott Inc, Architects Planners Engrs, 1111 West Loop South, Houston, TX 77027. 1974 Design Award Lincoln Arc Welding Fdn. 1974 State-of-the-Art of Cvl Engrg Award for tech paper. Struct designer of over 500 bldgs & special structs. Hobby, photography. *Society Aff:* ASCE, IABSE, IASS, ACEC.

Kraimer, Frank B
Business: PO Box 336, Madison Heights, MI 48071
Position: Pres *Employer:* International Manufacturing Inc. *Education:* Assoc/Mech Engr/Lawrence Tech *Born:* 5/8/34 Presently live in Union Lake. Have been in the manufacturing business for 30 years. For 8 years I have been self-employed. Have taught special classes on machining procedures. Our company is involved in setting-up new companies as consultants and working with trouble companies in helping them get re- established. Have been active in SME as chairman of chapter and region. Also acting in selling of new and used machines, and rebuilding of used machines. I enjoy raquetball and golf. *Society Aff:* SME, SMR, NSPE, ESD

Krajcinovic, Dusan
Business: CEMM, MS 246, Univ. of IL, Box 4348, Chicago, IL 60680
Position: Prof of Structural Engineering. *Employer:* Univ of Ill at Chicago Circle. *Education:* PhD/Theo Appl Mech/Northwestern Univ; MSc/Structural Mech/Univ of Beograd, Yug; DE/Structures/Univ of Beograd. *Born:* Mar 1935. Taught at Univ of Beograd; mbr of Tech Staff at Ingersoll Rand Res Inc & Argonne Natl Lab; with Univ of Ill 1973- . Author of over 120 res papers & 4 books. Cons for Argonne Natl Lab, Sargent & Lundy Engrs, Ingersoll Rand Co, Fermilas, Gabinete da Area des Sines, World Bank, EPRI Calif & Energoprojekt Beograd. Charter Mbr Amer Acad of Mechs, Fellow ASME (mbr 3 AMD cttes), Struct Stability Res Council, Int Assoc. Struct Eng Reactor Techn. *Society Aff:* ASME, AAM, SSRC.

Kral, George J
Business: 111 Merchant St, Cincinnati, OH 45246
Position: Pres. *Employer:* KZF Inc. *Education:* CE/Civ Engrg/Univ of Cincinnati. *Born:* 7/30/20. in Schuyler, NB. Grad Schuyler HS 1938; Civil Engr Degree Advanced Studies WA Univ 1943-45; Univ of Cincinnati 1952-53. Disting Engr Alumnus Award Univ of Cincinnati 1970. Reg P E 14 states; Reg Land Surveyor 3 states. Founder 1956, now Pres of 110-man KZF Engrs-Architects-Planners-Surveyors. ASCE; Fellow; Natl Dir 1969-72; Chmn Mgmt Group 'A' TAC-ASCE; Engr-of-the-Yr Cincinnati 1978. Mbr: Amer Consulting Engrs Council; Pres Cincinnati Post Soc of Amer Military Engrs 1984-85; ASCE Edmund Friedman Prof Recognition Award 1983. *Society Aff:* ASCE, ACEC, SAME.

Kram, Harvey
Business: 59-25 Little Neck Parkway, Little Neck, NY 11362
Position: Group VP Operations *Employer:* Leviton Mfg Co, Inc. *Education:* BS/ME/MIT *Born:* 8/16/21 Naval Officer, WWII; formerly Chief Eng of Burchell Products, Inc., E.A. Laboratories, Inc. and VP, Operations of Revelon, currently Group VP, Operations for Leviton and President of Southern Devices, Inc. and Thyrotek, Inc. subsidiaries of Leviton. Married to Eleanore and have two children, Leonard Kram, MD of Malibu, California and Kathy Kram, Ph.D. of Boston, MA. Mbr of the Woodcrest Club of Syossett, New York and the Inverrary Country Club of FT Lauderdale. Former Officer of the MIT Club of NY and member of the Educational Council of MIT. Registered PE in MA, NY, FL and CA. Mbr of ASPE, ASME, SAE, SME, Beaver Key Soc of MIT. *Society Aff:* SAE, ASME, SME, ASPE

Kramer, Andrew W
Home: 667 Rockland Ave, Lake Bluff, IL 60044
Position: Consulting Editor Power Engrg. *Employer:* Technical Publishing Co. *Education:* BS/Elect Eng/Armour Inst of Tech. *Born:* Apr 1893 Chicago Ill. With Commonwealth Edison Co Chicago & G E Co Schenectady N Y 1915-17. Commissioned Ensign US Steam Engrg Sch, Stevens Inst. Served in US Navy WWI. With Western Elec Co 1918-20. Joined Tech Publ Co Chicago 1920 & served on Ed Staff of 'Power Engineering' as Elec Ed, Managing Ed & Ed until 1958. Appointed mbr Bikini Atomic Bomb Tests 1946. Mbr Advisory Ctte on Tech Info USAEC 1953-65. Chmn Nuclear Div Amer Power Conference 1962-72. Ed 'Atomics' 1958-65. Author Indus Electronics, Boiling Water Reactors, Nuclear Propulsion for Merchant Ships, Understanding the Nuclear Reactor, etc. Lifelong radio amateur & mbr ARRL 1924- . Mbr Amer Alpine Club & Life Mbr Alpine Club, Canada. Life Mbr AIEE & IRE before merger. Fellow IEEE 1971. *Society Aff:* IEEE.

Kramer, Daniel E
Home: 2009 Woodland Drive, Yardley, PA 19067
Position: Chf Engr *Employer:* Kramer Trenton Co *Education:* Juris Doctor/Law/Temple Univ School of Law; BS/CH.E. /Columbia Univ of School Engrg *Born:* 4/27/28 Mech Engr, pat. atty; b. NYC. s. Israel and Ella K.; BS, Columbia U., 1950 J.D Temple U., 1976; m. Doris Silk, Sept 10, 1950; children--David Bruce, Susan Lee, Matthew Alan. Res engr Columbia U., 1951-53; lab technician Kramer Trenton Co, 1953-58, natl svc mgr, 1958-64, ch engr since 1964-; admitted to NJ, PA bar, 1976. Fellow ASHRAE; chrmn ASHRAE Std Comm, and 4 testing Stds Comms, Chrmn Safety Comm; RSES Certif. Mbr Alco Medal; mbr Inst Environ Sci; Am Bar Assn, NJ Bar Assn, PA Bar Assn, Am Patent Law Assn, Patent Office Soc, Contbr tech articles to profl jours; 36 patents in Ind Refrig Sys and Controls. Office Box 820, Trenton, NJ, 08605. *Society Aff:* ASHRAE, ASTM, ASM, ABA, POS.

Kramer, Edward J
Business: Dept Mtls Sci & Engr, Bard Hall, Ithaca, NY 14853
Position: Prof. *Employer:* Cornell Univ. *Education:* PhD/Met & Mat Sci/Carnegie-Mellon; BChE/Chem Engrg/Cornell Univ *Born:* 08/05/39 NATO Postdoctoral Fellow, Dept of Met, Univ of Oxford, England 1966-67. Appointed Asst Prof, Dept of Mtls Sci & Engrg, Cornell Univ, 1967, promoted to Assoc Prof 1972 and to Prof 1979. Gauss Prof of the Akademie der Wissenschaften, Gottingen, W Germany, 1979; Visiting Prof, Ecole Polytechnique Federale de Lausanne, 1982. Fellow, Amer Phys Soc; Mbr, ACS, AAAS, SPE, Bohmische Physical Society. Awarded Amer. Phys. Soc. High Polymer Physics Prize 1985; Distinguished U.S. Senior Scientist Award of the Humboldt-Stiftung 1987; Guggenheim fellow 1987-88. Current res interests include mech properties, crazing, fracture and diffusion of polymers investigated using transmission electron microscopy, small angle X-ray scattering and Rutherford backscattering, as well as forward recoil, spectrometry. *Society Aff:* APS, ACS, AAAS, SPE, BPS.

Kramer, Irvin R
Business: Annapolis Lab, Annapolis, MD 21402
Position: Tech Advisor-Materials (Code 2802). *Employer:* USN David W Taylor Nav Ship R&D Ctr. *Education:* Dr Eng/Metallurgy/Johns Hopkins; MS/Chem Engg/Johns Hopkins; BS/Chem/Johns Hopkins. *Born:* Sept 18, 1912 Balto Md. m. 1935. Chemist Amer Radiator & Standard Sanitary Mfg Corp Md 1935-38; Metallurgist & Ch Spec Alloys Sect Naval Res Lab 1938-46; Metallurgist, Phys Sci & Head Mech & Materials Branch, Off Naval Res 1946-51; Asst to Pres Horizons Titanium Corp N Y 1951-53; V Pres Mercast Corp N Y 1953-55; Ch Materials Res Martin Co, Mgr IR&D Res 1955-74; Tech Adv USN DTNSRDC 1974-81. Res Scientist Univ of Maryland 1981- . Mbr Heat Resistant Materials Ctte, Natl Adv Ctte Aeronaut; Met Panel R/D, Board, Dept Natl Defense; Ctte Ship Steel, Natl Res Council 1944. Soc Metals; AIME, Board Materials & Engrg Sci; Board Mercast Corp; Fellow ASM, Merit Navy Civilian Award; Dr of Univ, Hopkins; Fellow Hopkins. Surface Effect in Flow & Fracture, Fatigue, Stress Corrosion, Hydrogen Embrittlement. *Society Aff:* ASM, AIME.

Kramer, Raymond E
Home: 441 Arbor Circle, Youngstown, OH 44505
Position: Prof E.E. *Employer:* Youngstown State Univ *Education:* MS/Electrical/Case-Western Reserve; BS/Physics/Heidelberg *Born:* 2/2/19 Prof Electrical Engrg Dept. Youngstown State Univ. 1950 to Date. Charter Chairman, Electrical Engrg 1950-1976 YSU, Nassa Fellowship Grant 1969, NDEA Fellowship

Kramer, Raymond E (Continued)
Grant 1970. Research in the area of Gravity Waves Plasmas & Solid State. Wife Maryann. Children: Michael J. Kramer, Philip T. Kramer, Kathryn M. Kramer. Captain, US AFR 1943-1953. Registered Engr State of OH. Consultant and Design Engr for Ohio Bell 1960, U.S. Steel 1951. Westinghouse 1954-55, ARC Research 1951-60. *Society Aff:* IEEE, AAAS, TBΠ

Kramer, Samuel
Home: 11709 Rosalinda Dr, Potomac, MD 20854
Position: Deputy Dir *Employer:* Natl Engg Lab. *Education:* BCE/Civ Engg/CCNY. *Born:* 6/2/28. Deputy Dir since 1980, Assoc Dir, 1978-1980 Natl Engg Lab, Natl Bureau of Stds;; Deputy Dir, Ctr for Bldg Tech, NBS, 1970-1978; Office of Mgt & Budget, Exec Office of the Pres, 1966-1970; Chief AF Design & Asst Chief Military Works, Engg Div, US Army Corps of Engrs, 1955-1966; Engr in Private Sector with maj engg & construction cos, 1950-1955; Military Service, Corps of Engrs, 1950-1953; Reg PE, (PE); Elected to Tau Beta Phi & Chi Epsilon, Natl Honorary Engg Organizations; Active in community & religious organizations; Served on many Exec Office & White House task forces as well as on sub-Cabinet-level & inter-agency committees. *Society Aff:* ASCE, NSPE, AAAS.

Kranich, Wilmer L
Business: Dept of Chem Engrg, Worcester, MA 01609
Position: Prof, Chem Engrg-Dean, Grad Studies. *Employer:* Worcester Polytechnic Institute. *Education:* PhD/Chem Engrg/Cornell Univ; BS/Chem Engrg/Univ of PA. *Born:* Nov 1919 Phila Pa. Instr Cornell 1941-44 USNR 1944-46 (rocket & guided missile res). Asst Prof Princeton Univ 1946-48; at Worcester Poly Inst 1948- 1985: Dept Head for 17 yrs, respon for planning of Goddard Hall of Chem Engrg & Chem, Dean of Grad Studies & Gordon Prof. Retired 1985. Staff cons at Arthur D Little Inc for 25 yrs. Research Scientist, Hungarian Academy of Sciences, Budapest 1983. Res publs primarily on kinetics & catalysis. Mbr AIChE Educ Projs Ctte. Hobbies include choral singing, painting & sculpture, hatha yoga & water skiing. *Society Aff:* AIChE, ACS, ASEE, ΣΞ, ΑΧΣ

Krapek, Anton
Home: 2229 W. Window Rock Dr, Tucson, AZ 85745
Position: Project Engineer & Mechanical Specification Writer *Employer:* Anderson DeBartolo, Pan, Inc *Education:* BSAeE/Aero Engg/Aeronautical Univ of Chicago. *Born:* Apr 1926 Moravia Czechoslovakia. US Citizen. BSAeE Aeronautical Univ Chicago. Reg Prof Engr: Mich, Arizona. Army Air Force 1944-46. With E Roger Hewitt Assocs Inc 1952-62 becoming V Pres 1957. Pioneered use of high temp hot water in plating processes. With architectural firms 1962-84. Vice President, Gazall, Krapek & Associates 1968-79. Present affiliation since 1985. Mbr, ASHRAE. Charter mbr MAP, Eastern Mich Chap ASHRAE. Author of over 20 articles on HPAC. 10 yrs on Board of Eds of Heating, Piping & Air Conditioning Magazine. ASHRAE Fellow Award 1975; Bronze Medallion, Hospitals, Journal of Amer Hosp Assn. PPres Eastern Mich Chap ASHRAE. Listed in Outstanding Amer, Who's Who in Michigan, Who's Who in the Midwest. *Society Aff:* ASHRAE

Krasavage, Kenneth W
Business: 2404 Pennsylvania Ave, Evansville, IN 47721
Position: Vice Pres *Employer:* Bristol-Myers Co *Education:* BS/Mech Eng/Univ of Wisconsin *Born:* 06/24/42 Native of Wisconsin Rapids, Wisconsin. Held various process control and facility engrg positions with US Steel Corp and Kimberly Clark Corp. With the Mead Johnson Div of Bristol-Myers Co since 1972 in several engrg and prod mgmt positions. VP, Engrg at Mead Johnson since 1980 with respon for all facility, process, indus, environ and pkg design engrg and maintenance. VP Production-Jan 1987. Registered Prof Engr. Mbr of the Pharmaceutical Mfrs' Assoc Engrg Ctte since 1982; PMA Engrg Ctte VChrmn in 1985. Enjoys golf, hunting, and fishing. *Society Aff:* PMA, ISPE.

Krase, Norman W
Home: Box 266 Crosslands, Kennett Square, PA 19348
Position: Retired. *Employer:* DuPont. *Education:* PhD/Chem Engr/Yale Univ; MS/Chem Engr/Amer Univ; BS/Chem Engr/Univ of IL. *Born:* Aug 1895. Chem Warfare Service 1917-18; Asst Prof Chem Engrg Univ of Ill 1926-36; Prof Chem Engrg Univ of Pa 1936-39; R/D DuPont Co 1939-45. Dir AIChE 1944-45, Sci Cons US Army 1945. Tech Investigator DuPont Co 1946-60. Hobbies: wood working, travel, photography. *Society Aff:* AIChE, ACS.

Krashes, David
Business: 241 W Boylston St, West Boylston, MA 01583
Position: President. *Employer:* Mass Materials Research Inc. *Education:* PhD/Metallurgy/RPI; MS/Metallurgy/RPI; BS/Physic/RPI. *Born:* Jan 1925; native of Long Island N Y. Attended Ohio St Univ before WWII. Army Infantry 99th Div Europe 1943-45. 2 yrs dev engr for Nuclear Metals Inc spec in heat resistant alloys; 5 yrs (1957-62) Assoc Prof of Met at Worcester Polytech Inst; also Lacrosse Coach. Founded Mass Materials Res Inc 1962 to perform met problem solving. Co owns Lehigh Test Labs Inc Wilmington Dela. Also owns Ct Metallurgical Inc, East Hartford CT & DK Pres of it also. Pres of Lehigh & Mass Materials; both labs perform prod res, problem solving, failure analysis & routine testing for east coast indus. During 1960's dev failure analysis techniques including electron fractography. Respon for corporate admin. 1981-82 Pres of Amer Soc for Metals. Treas & Trustee of Amer Soc for Metals 1972-74; Fellow ASM 1970. Mbr Bd of Dirs of publicly-owned Richard D Brew Co. Enjoys hiking, mountain climbing, skiing, athletics. Pres of Worcester Boys Club. *Society Aff:* ASM, AIME, ASTM.

Kratky, Vladimir J
Home: 634 Gaines Dr, Ottawa Ontario, Canada K1J 7W6
Position: Sr Res Officer. *Employer:* National Research Council of Canada. *Education:* PhD/Photogrammetry/Tech Univ Brno, Cz; MS/Geodesy/Tech Univ Brno, Cz. *Born:* May 1928. Assoc Prof at TECH Univ teaching & conducting res in photogrammery. Since 1968 with Natl Res Council of Canada working in analytical photogrammetry with emphasis on non-cartographic applications & satellite imageries. Was respon for dev of sys for geometric processing of images in Canadian LANDSAT Prog. Current involvement in on-line analytical photogrammetry. Publ over 60 tech papers. Recipient of ASP Abrams Grand Award 1975 and 1979. *Society Aff:* CIS, ASP, ISP.

Kraus, C Raymond
Business: 845 Mt Moro Rd, Villanova, PA 19085
Position: President. *Employer:* Cons Communications Engrs Inc. *Education:* BS/EE/PA State Univ *Born:* Phila. Transmission, radio & switching design for Bell Tel Co of Pa; Radio & Genl Transmission Engr, etc. Innovated sys design with economic benefits: pole mounted base stations for large mobile radio sys, customer dialing sys. Founded Consulting Communications Engrs Inc 1965: design of switching & satellite sys, nationwide quality of telephone service studies; design of new one-way (Voicegram) & one-way record message (Faxgram) sys for universal use. Patents on sys. Spokesman for the public for sophisticated telephone & related services. Fellow IEEE. Interested in music, opera, theatre. Eta Kappa Nu, Tau Beta Pi, Phi Kappa Phi, Sigma Xi. *Society Aff:* IEEE, HKN, TBΠ, ΦΚΦ, ΣΞ

Kraus, Harry
Business: 275 Windsor St, Hartford, CT 06120
Position: Dir of Engg Studies. *Employer:* Hartford Grad Ctr. *Education:* PhD/Mech Engr/Univ of Pittsburgh; MS/Mech Engr/Univ of Pittsburgh; BME/Mech Engr/NYU. *Born:* 7/23/32. From 1953 to 1954 Grad Student Training Prog at Westinghouse Electric Corp. Then in 1954 to 1956, Lt USAF. Returned to Westinghouse Electric Corp, Bettis Atomic Power Lab during 1956 to 1963. From 1963 to 1966 Pratt and Whitney Aircraft Group of United Technologies Corp. Went into teaching in 1966 at SUNY, Stony Brook. Took present position at Hartford Grad Ctr, in 1968. Duties have been both teaching and admin. Consultant for Exxon Res and Engg, Combustion Engg, Lenox Hill Hospital and other firms. Res interests involve

Kraus, Harry (Continued)
design and analysis of pressure vessels. Also ortho paedic Biomechanics. *Society Aff:* ASME.

Kraus, Robert A
Home: 21300 Claythorne Rd, Shaker Heights, OH 44122
Position: VP Tech Develop. *Employer:* Republic Steel Corp. *Education:* MS/ME/IL Inst of Tech; BS/ME/IL Inst of Tech. *Born:* 10/8/20. Began career with Republic Steel Co in 1939 in Chicago, IL as a student engr. Held a number of supervisory positions in Chicago and Warren, OH, prior to moving to corp hdquarters in Cleveland, OH in 1962. In present role, has full responsibility to monitor the dev and utilization of new steelmaking tech and processes on a worldwide basis. A native of Chicago, received BS & MS degrees in Mech Engg from the IL Inst of Tech. Married, has four children. *Society Aff:* AISE, OSPE, NSPE, ASEE, EJC.

Kraus, Wayne P
Business: P O Box 591, Tulsa, OK 74102
Position: Staff Research Engr *Employer:* Amoco Production Co. *Education:* PhD/ChE/LA State Univ; MS/ChE/LA St Univ; BS/ChE/LA State Univ *Born:* 10/19/42 in New Orleans, LA the third son of Frank and Mildred Kraus. Educated in the private schools of New Orleans, including Holy Cross High School. Received Academic honors in high school. An avid student of chemistry and physics from his youth. Went on to receive the BS MS PhD in Chemical Engrg from LA State Univ in Baton Rouge. Did process development research for the Ethyl Corp, 1968-70; received PhD in ChE, 1972; did reaction kinetics research, process development and reservoir simulation development for Phillips Petroleum Co., Bartlesville Research Center, 1972-76; did process simulation development and currently doing reservoir simulation development at Amoco Production Research, Tulsa, OK. Married to former Peggie Rose Weeks. *Society Aff:* AICHE, SPE

Krause, Henry M, Jr
Home: 158 Plantation Rd, Houston, TX 77024
Position: Petroleum Consultant (Self-employed) *Education:* BS/Petr Engg/Louisiana State Univ; Mgmt/Inst for Mgmt/Northwestern Univ. *Born:* 11/6/12. Native of Sicily Island, LA. Retired from Exxon Co. USA Feb 1977 after 40.5 yrs of service in Engrg and Mgmt assignments. Last 10 yrs spent in unitizing major Gulf Coast & East TX oil fields for secondary recovery projects. Since 1977, employed as Petroleum Consultant in oil & gas operations by Westland Oil Dev Corp, Houston, TX. Mbr AFME 1946; Chmn LS SPE 1951 & 53; SPE Dir 1959-61 & 1965-67; Pres Elec 1965; Pres 1966; P Pres 1967; AIME dir 1965-67. Gulf Coast Sect Service Award 1970; SPE Distinguished Service Award 1972. Enjoys outdoors, hunting & fishing. *Society Aff:* SPE.

Krauss, George
Business: Dept of Met Engrg, Golden, CO 80401
Position: AMAX Foundation Prof of Phys Met & Director Advanced Steel Processing and Products Research Center. *Employer:* Colorado School of Mines. *Education:* ScD/Metallurgy/MIT; MS/Metallurgy/MIT; BS Met E/Met Eng/Lehigh Univ. *Born:* 5/14/33. BS Lehigh Univ 1955, MS and ScD MIT 1958 and 1961. NSF Postdoctoral Fellow Max-Planck-Institut fur Eisenforschung, Dusseldorf Germany 1962-63. Held ranks of Asst, Assoc & Full Prof Lehigh Univ 1963-75. Reg Prof Engr in Pa & Colo. Chmn Lehigh Valley Chap ASM 1971; Chmn Ferrous Met Ctte AIME 1970; Mbr Heat Treatment Div Council ASM & Long Range Planning Ctte ASM; Key reviewer 'Metallurgical Transactions'. Interests include phase transformation, heat treatment, materials selection, failure analysis & microstruct characterization. Editor: of Heat Treating, Author of book *Principles of Heat Treatment of Steel*, ASM, 1980 Author of over 125 technical papers. Ed of book *Deformation Processing and Structure*, ASM, 1984, CoEditor of Fundamentals of Microalloying Forging Steels, TMS-AIME, 1987. *Society Aff:* ASM, TMS-AIME, IFHT, ISS-AIME.

Kraut, Seymour
Business: 165 Needham St, Newton Highlands, MA 02161
Position: Sr. VP & Gen Mgr. *Employer:* Honeywell Bull Inc. *Education:* BSCE/CE/Pratt Inst. *Born:* 5/1/32. Sy began his career as Dev Engr on the GE Chem-Met Training Prog in Syracuse, NY in 1953 & had various design & dev engg assignments. He later held mgt positions with GE in the fields of semiconductors, silicones & electronics until 1965, when he assumed the position of Mgr, Semiconductor Processing Operation in Semiconductor Products Dept. In 1969, he was promoted to the position of Mgr, Mfg for the Nucl Fuels Dept in San Jose, CA. Sy was transferred to the Comp bus in 1969, & in 1971, he was appointed Dir - Mfg for Phoenix Comp Operations. In 1975, he was appointed VP - Phoenix Comp Operations. In 1977, he was promoted to VP of Minicomp Systems & Terminals Operation in the Small/Medium Info Systems Div. In 1979, he was promoted to VP & G. Mgr of the Honeywell Customer Services Division, & in 1986 he was promoted to Senior VP & G Mgr of the Honeywell Bull Customer Services Division. *Society Aff:* ASME, AIChE, ECS, AFSM.

Kravitz, Lawrence C
Home: 7128 Wolftree Ln, Rockville, MD 20852
Position: VP & Director-Technology. *Employer:* Allied-Signal Inc. *Education:* PhD/Applied Physics/Harvard U; MSEE/EE/Air Force Inst of Tech; BS/EE/KS Univ *Born:* 7/27/32. Native of NY, NY; USAF, 1954-1958; GE Corp Res & Dev Ctr, 1963-1973; Dir of Electronics Res AF Office of Scientific Res (AFOSR), 1973-1977; Dir, AFOSR, 1977-1981; Dir of Res, Bendix Advanced Tech Center, 1981-1983, Dir-Tech, Bendix Aerospace Sector, 1983-86; VP and Director-Technology, Bendix Aerospace Sector 1986- . *Society Aff:* IEEE

Krawczyk, Theodore A
Home: 715 Sheffield Ct, Lake Forest, IL 60045
Position: Corporate Dir of Matls Mgmt. *Employer:* Hollister, Incorporated *Education:* BSIE/Indust Engg/Penn State; MBA/Indust Mgmnt/Univ of So CA. *Born:* 12/24/28. Division Mgr of Indus Engrg at Genl Foods Intl specializing in warehouse sizing & design, matl handing & cost controls. Dir of Matls Mgmt Columbus Coated Fabrics Subsidsary of Borden Chemical. Currently Corporate Dir of Mats Mgmt Hollister Incorporated. Lic PE: PA, NJ & CA. Pres, Intl Matl Mgmt Soc 1978-79; Currently Mbr Advisory Ctte IMMS 1980-81. Mbr Matl Advisory Committee-William Rainey Harper College. *Society Aff:* IMMS, IIE.

Krawinkler, Helmut
Business: Dept of Civ Engg, Stanford, CA 94305
Position: Prof *Employer:* Stanford Univ. *Education:* PhD/Structural Engg/Univ of CA; MS/Structural Engg/CA State Univ; Dipl Eng/Civ Engg/Tech Univ of Vienna. *Born:* 4/6/40. Native of Innsbruck, Austria. Worked as design engr in Vienna, Austria, 1964-65. Came to the US in 1965 on a Fulbright grant. Res engr, Univ of CA, at Berkeley, 1971-72, Lecturer of Civ Engg, CA State Univ, San Jose, 1972-73. At Stanford since 1973. Current responsibilities include teaching of structural design courses and res in structural and earthquake engg. Consultant to industrial organizations and mbr of several natl engg committees. Co-author of a book on structural design and of 60 proofl papers and reports. Enjoy classical music, good food and outdoor activities. *Society Aff:* ASCE, EERI.

Kreer, John B
Home: 1834 Pinecrest Dr, E Lansing, MI 48823
Position: Prof. *Employer:* MI State Univ. *Education:* PhD/Elec Engg/Univ of IL; MS/Elec Engg/Univ of IL; BS/Elec Engg/IA State Univ. *Born:* 9/25/27. Asst Prof at the Univ of IL 1957-59. Associate Prof/Prof at WV Univ 1959- 64. Assoc Prof/Prof at MI State Univ 1964-present. Traffic Res Engr with Traffic Systems Div, Office of Res, Fed Hgwy Admin 1974-75. Dept Chrmn 1976-87. *Society Aff:* IEEE, ASEE, NSPE.

Kreider, Jan F
Business: Joint Center for Energy Management, Univ. of Colorado, Boulder, CO 80309-0428
Position: Pres & Director *Employer:* Univ of Colorado *Education:* PhD/Chem.E/U of CO; MS/ME/U of CO; BS/ME/Case Inst of Tech *Born:* 12/11/42 Dr. Jan F. Kreider is a prof at the Univ of Co and director of its Joint Center for Energy Management. He specializes in the design and economic analysis of solar energy and energy conservation systems. He has written six books and several dozen technical papers. He is Editor-in-Chief of the McGraw-Hill Solar Energy Handbook. Dr. Kreider was the solar consultant for both the largest and the second-largest solar heating systems in the world as well as many other large solar projects. He has also assisted in the design of many solar systems for residences and has served as consultant to many US and foreign firms, government agencies and private clients. Two of his projects have been winners of the Owens-Corning Award. Dr. Kreider is a registered PE and member of several honorary and professional societies. *Society Aff:* ASME, ASHRAE, AAAS, ISES, ACEC, ASEE.

Kreider, Kenneth G
Home: 9232 Copenhaver Dr, Potomac, MD 20854
Position: Div Chief. *Employer:* National Bureau of Standards. *Education:* ScD/Mat Sci/MA Inst of Tech; SM/Metallurgy/MA Inst of Tech; SB/Metallurgy/MA Inst of Tech. *Born:* 5/21/37. Native of Lancaster, PA, studied metallurgy at MIT. SB and SM thesis on plasma spraying of metal coatings and ScD thesis on desalinating sea water. Married Carole Compton; children are Cynthia, Brett, and Kit. Military Service at Army Materials Res Agency. Dev processes for fabricating alumninum boron and other metal matrix composities at United Aircraft Res Lab (six patents). At National Bureau of Standards was Program Mgr for Industrial Energy Conservation before becoming Chief of the Thermal Processes Div in which measurement tech is developed for industrial harsh environments, combustion, heat transfer, moisture, and humidity. Presently heading prog in Chem Process Sensors. *Society Aff:* SEMI, ASM, ASTM, MRS.

Kreifeldt, John G
Business: Dept. Engineering Design, Tufts University, Medford, MA 02155
Position: Prof. *Employer:* Tufts Univ. *Education:* PhD/Human factors/Case Western Reserve; MS/EE/MIT; BS/EE/UCLA. *Born:* 10/7/34. Native of Manistee, MI. Prof of Engg Design, Tufts Univ, 1969-present. VP of Applied Ergonomics Corp, a product design consulting co. Published numerous papers and holds several patents. Initiated the first undergrad maj in Engg- Psychology (1971) in the nation. Consultant to business, legal firms, govt and hospitals. Conducts supported res in several human factors areas. *Society Aff:* HFS.

Kreisman, Herbert
Business: 2001 Estes Ave, Elk Grove Village, IL 60007
Position: Chrmn of the Bd *Employer:* Advance Mech Sys Inc *Education:* BS/ME/IL Inst of Tech *Born:* Oct 11, 1912. Mbr Tau Beta Pi Natl Engrg Honorary. Fellow ASHRAE, P Pres Ill Chap ASHRAE; P Pres Mech Contractors Chicago Assn; P Dir & recipient of Disting Serv Award Mech Contractors Assn of Amer. Engaged in mech contracting for 41 yrs as engr, V P, Pres & Chrmn of Bd of co. Mbr Natl Board of Arbitrators, Amer Arbitration Assn. Co-Chmn Const Indus Affairs Ctte of Chicago. Mbr Board of Trustees Associated Specialty Contractors. *Society Aff:* ASHRAE

Kreith, Frank
Home: 1485 Sierra Dr, Boulder, CO 80302
Position: Senior Research Fellow *Employer:* SERI *Education:* Doctorate/Science/Univ. of Paris; M.S./Engineering/U. of Calif., Los Angeles; B.S./Mech. Eng/U. of Calif., Berkeley *Born:* 12/15/22 Native of Vienna, Austria, came to U.S. in 1940; Research Engineer JPL, 1945-49; Guggenheim Fellow Princeton, 1949-51; Asst Prof of Mech Engrg at U.C. Berkeley, 1951-53; Assoc Prof, Lehigh, 1953-59; Prof, Univ of Colo, 1959-79; Chief Solar Thermal Conversion, 1977-85 and Senior Research Fellow, 1985-present, Solar Energy Research Inst. Published 9 books including Principles of Heat Transfer and 120 articles; co-edited 7 books, including Handbook of Solar Energy 1980, Direct Contact Heat Transfer 1987; 1978 Heat Transfer Memorial Award, 1983 Worcester Medal, 1985 Max Jakob Award; Fellow ASME and Editor Journal of Solar Energy Engrg. Directed construction of seven major research labs in solar and ocean thermal conversion. Enjoy music, skiing, and hiking. *Society Aff:* ASME, ISES, ASES, ΣΞ, AAAS

Kremidas, John P
Business: 8401 W 47th St, McCook, IL 60525
Position: Section Hd. *Employer:* Armak Res & Dev. *Education:* PhD/Met/Natl Tech Univ; MSc/Met/Univ of MN; BSc/Sci/Wheaton College. *Born:* 1/4/48. Born in Patras Greece Jan 4, 1948, educated in USA. 1973-75 Res Engr Asst at Univ of MN Mineral Res Ctr. 1975-1977 Sr Res Engr at Cities Service (Citgo Oil Co) Res Ctr, Cranbury, NJ. Involved in Applied Math Appl in Met. 1977-now Sec Hd at Armak Res Ctr, Chicago, IL. Involved in extractive met, res of uranium ferrous & base metals & ind minerals. Married, with two daughters and one son. *Society Aff:* AIME SME, AIME TMS, IMM, CIM.

Krendel, Ezra S
Home: 211 Cornell Ave, Swarthmore, PA 19081
Position: Professor Systems Engineering *Employer:* University of Pennsylvania. *Education:* AM/Soc Rel/Harvard; ScM/Physics/MIT; AB/Physics/Brooklyn College. *Born:* Mar 5, 1925. Chmn Board of Advisors 1969-70; Dir 1967-69, Mgmt & Behavioral Sci Ctr Wharton Sch Univ of Pa. Cons & Sr Advisor 1966; Tech Dir 1963-66; Lab Mgr 1955-63; Res Engr 1949-55 The Franklin Inst Res Labs Phila Pa. Fellow: IEEE, Amer Psychological Assn, Human Factors Soc & AAAS. Mbr of Opers Res Soc of Amer & P Chmn of its Mbrship Ctte. Mbr of The Ergonomics Soc. Has been mbr of many US Govt cttes including several NAS/NRC cttes & NASA Res Advisory Ctte on Control, Guidance & Navigation. Has been invited NATO Lectr in England, France, Italy, Greece & Turkey 1968 & Germany 1971. In 1960 together with D T McRuer was awarded Louis E Levy Gold Medal of The Franklin Inst. *Society Aff:* IEEE, APA, HFSA, AAAS, The Ergonomics Society, ORSA.

Krenkel, Peter A
Reno, NV 89506
Position: Dean, Coll of Engr *Employer:* Univ of NV *Education:* PhD/San Engr/Univ of CA; MS/CE/Univ of CA; BS/CE/Univ of CA; AA/Engr/City Coll of San Francisco *Born:* 01/03/30 Krenkel, Peter Ashton, engr, educator, univ dean; b San Francisco, Jan 3, 1930; AA, Coll, City San Francisco, 1952; BS, U CA-Berkeley, 1956, MS, 1958, PhD, 1960. Reg prof engr, GA, TN, NV, NC Instr U CA at Berkeley, 1958-60; founder Assoc Water & Air Resources Engrs, Inc, Nashville, 1968-; chmn, prof dept environ and water resources engrg. Vanderbilt U, Nashville, 1960-73; dir div environ planning TVA, 1974-78; exec dir Water Resources Ctr U NV Reno, 1978-82; dean Coll Engrg 1982-; disting lectr Am Inst Chem Engrgs; cons WHO, Intl Joint Commn on Great Lakes Water Quality; US EPA, US Dept Energy, Roy F Weston, Inc; chrmn thermal pollution panel Nat Water Commn, Washington, 1970-, TN Air Conservation Commn, 1971-. Author: (with V Novotny) Water Quality Mgmt, 1980; ed: (with F L Parker) Thermal Pollution, Biological Aspects, 1970, Thermal Pollution, Engrg Aspects, 1970, Water Quality Monitoring in Europe, 1972, Heavy Metals in the Aquatic Environ. Pres TN Lung Assn, 1974-75. Severed with AUS, 1953-55. Fellow USPHS, 1963; recipient awd outstanding res san engrg ASCE, 1963, Skill, Integrity. Respon awd Am Gen Contractors, 1984. Mbr Am Water Works Assn, Water Pollution Ctrl Fdn (bd control). Air Pollution Ctrl Assn, Am Public Health Assn, Am Inst Chem Engrs, ASCE. Intl Assn Water Pollution Res (governing bd), Am Acad Environ Engrg (diplomate). Sigma Xi, Tau Beta Pi, Chi Epsilon. *Society Aff:* AIChE, WPCF, ASCE

Krenzer, Bette A
Home: 11610 S 31st St, Omaha, NE 68123
Position: Technical Advisor *Employer:* USAF Foreign Technology Div. *Education:* BS/Engr Physics/Univ of KS. *Born:* 7/4/23. in KS City, MO. Now in 38th yr of Prof career for the USAF. Technical Laison to Hq Strategic Air Cond (Hq SAC) for the Cmdr of Hq Foreign Technology Div (Hq FTD) from Nov 1978. Awarded Meritorious Civilian Service Medal for engg accomplishment. Sr Mbr IEEE & SWE; served SWE as mbr of Council of Section Resp & of Exec Ctte. Was Natl Chmn of Career Guidance & Secy for 2 terms. *Society Aff:* IEEE, SWE, NSPE, AOC.

Krenzer, BK
Home: 420-E-64th Terr, Kansas City, MO 64131
Position: Consultant *Employer:* Self *Education:* Bachelor/Engr. Physics/Univ. of Kansas *Born:* 07/04/23 Served as pres of SWE during the 1986-87 period after a year as the Society's first president-elect. Employed as a civilian with the USAF for almost 40 years at increasing professional levels in electronic warfare. Was twice awarded the Meritorious Service Medal by the AF for that work, particularly in data storage, handling and processing, program management, and technical writing. Other recognitions include designation as Business Woman of the Year by the American Business Association and induction into the University of Kansas Women's Hall of Fame. I am a Senior member of IEEE and a Fellow of SWE. *Society Aff:* IEEE, SWE

Krenzer, Robert W
Home: 8426 Quay Dr, Arvada, CO 80003
Position: Program Manager *Employer:* Rockwell Intl. *Education:* PhD/Met/Univ of Denver; MS/Met/Univ of Denver; MSBA/BUS/Univ of Northern CO; BS/Met/CO Sch of Mines. *Born:* 8/17/40. Native of Springfield, IL. With Rockwell Intl since 1972. Published several articles in Metallurgy; Rockwell Engr of the Yr 1976; Chapter Chrmn, ASM 1978; Natl Cttee, ASM 1977-87; Leadership Denver Prog, Denver Chamber of Commerce 1979. ASM Fellow 1981; Tour guide, Ramses II Egyptian Exhibit, Denver Museum Natural History, 1987. Developed High Adventure Mountaineering Program, Boy Scouts, 1986-87. Enjy geology, hiking, Egyptian history, & volunteer work, Rockwell Speaker's Bureau, Denver Museum Speaker's Bureau, Boy Scouts, Tahosa Alumni Assoc. *Society Aff:* ASM, EES, IIE

Krepchin, David M
Business: 1007 5th Ave, San Diego, CA 92101
Position: V Pres, Mgr SD Area Branch *Employer:* M Rosenblatt & Son Inc. *Education:* MS/CE/UC Berkeley; BS/CE/CCNY. *Born:* 7/18/29. Naval Architect with N Y Naval Shipyard & Supr of Shipbldg N Y 1950-1953. US Army Corps of Engrs 1953-55. With MR&S since Oct 1955. Assumed present duties as Mgr & Ch Engr San Diego Area Branch 1972. Served as SNAME San Diego Section Chmn 1974-75; ASNE San Diego Sect Chmn 1976-77. *Society Aff:* SNAME, ASNE, ASCE, NSPE.

Kress, Ralph H
Home: 4444 Knoxville Ave, Peoria, IL 61614
Position: Ex V Pres *Employer:* Kress Corp *Education:* Applied Math/Lowell Inst; ME/MIT *Born:* 7/10/04 Retail Truck Salesman and Sales Mgr 1922-39; Engr and Tech Representative, Chevrolet Division, and Fleet Division, General Motors Corporation 1932-42 and 1946-50; Captain and Major AUS-TC 1943-46 (present title Major USAR-TC); Legion of Merit; Executive VP, Dart Truck Co., Kansas City 1950-55; Kress Automotive Engrg Cont 1955; Mgr, Truck Division, Letourneau-Westing-Denm Co., Peoria, IL 1956-62; Mgr, Truck Development Caterpillar Tractor Co., Peoria, IL 1962-69; Executive VP, Kress Corporation, Brimfield, IL since 1969. Received 26 patents for motor vehicles and motor truck. SAE Fellow 1978. *Society Aff:* SAR, Rotary, PE, AUSA, IMA, AL, YMCA.

Kressel, Henry
Business: 227 Park Ave, New York, NY 10172
Position: Staff VP, Solid State Research. *Employer:* RCA Laboratories. *Education:* PhD/Materials Sci/Univ of PA; MBA/Univ of PA; MA/Appl Phys/Harvard Univ; BA/Phys/Yeshiva Univ *Born:* 1/24/34. PhD & MBA Univ of Pa; MS Harvard Univ; BA Yeshiva Univ. With RCA 1959-67, Solid State Div (silicon transistor & microwave devices); RCA Labs 1967-84, respon for optoelectronic device res (semiconductor lasers & LEDs), silicon power devices CCD's and integrated circuits. With EM Warburg, Pincus and Co, 1984-. Recipient David Sarnoff Award (RCA) 1974. And IEEE, 1984. Fellow IEEE & APS. Mbr Natl Acad of Engrg. Past-President, IEEE quartrum Electronics and Applications Society. 32 US pats; author of over 130 publs in fields of optoelectronics, semi conductor devices, microwave & materials res. Co-author of book *Semi Conductor Lasers and the Terojunction LEDs.* Academic Pres, 1977 *Society Aff:* APS, IEEE.

Kresser, Jean V
Home: 823 Darien Way, San Francisco, CA 94127
Position: Power Systems Consultant. *Employer:* Self-employed. *Education:* BS/Elec Engg/MA Inst of Tech. *Born:* Sept 1904 Noumea New Caledonia. US Citizen since 1930. Engr Westinghouse Elec Corp until Oct 1969. Fellow Engr at retirement respon for application of elec equipment in indus in Pacific Coast Zone; design of control & protective sys. Power Sys Cons Oct 1969- . Worked as independent cons to elec indus & engrg firms on projs in Calif, Hawaii, Iran & S Amer. Author & co-author of numerous tech articles & papers. Lectr on elec sys at extension courses, IEEE & indus sponsored courses. Awarded Best Paper Prize by AIEE & Outstanding Engrg Performance by Westinghouse. Fellow & Life Mbr IEEE. *Society Aff:* IEEE.

Kretz, Anna S
Home: 16169 Silver Shore Dr, Linden, MI 48451
Position: Staff Engr *Employer:* Buick Motor Div., G.M.C. *Education:* MA/Bus Admin/Cen MI Univ; BS/Engrg/Oakland Univ *Born:* 9/24/51 Birthplace-Poland. Finished high school, undergraduate and graduate degrees in MI. With Buick since 1972-assignments including R & D and current production. Primary control systems oriented; all support hardware; mechanical design in instrument panels, supervisor of Body, A/C, Electrical Test Area. Assumed current responsibility June, 1981 of all electronic and electrical systems (less engine emission related). Served on various SAE committees locally and nationally; currently on the SAE BD of Dirs. Have recruited women and minorities on college campuses for Buick. Spoken at GMI and U of MI Flint, on the subject of Women in Engrg. Enjoy golf and skiing. *Society Aff:* SAE

Kretzmer, Ernest R
Business: 13 Blue Hills Dr, Holmdel, NJ 07733
Position: Retired *Education:* ScD/EE/MIT; MS/EE/MIT; BS/EE/Worcester Poly Inst. *Born:* 12/24/24. Currently Dir, Special Studies in the Legal & Patent Div. Joined Bel Labs in 1949 working in the TV Res Dept & from 1970 to 1978 was Dir of the Data Communications Tech & Applications Lab, having assumed present position in July 1978. Have had 26 patents issued on circuits mostly related to tv & data communication systems. Am a mbr of Sigma Xi & a Fellow of the IEEE. Enjoy classical music & outdoor activities. Retired from AT&T Information Systems Dec. 31, 1983. (Director, Customer Systems Research). *Society Aff:* IEEE.

Kretzschmar, John R
Business: 10850 Middleton Pike, Dunbridge, OH 43414
Position: Pres *Employer:* Blako Industries, Inc. *Education:* BS/ChE/Univ of MO *Born:* 6/11/33 St. Joseph, MO. Graduate Lafayette HS '51. Honor graduate. Univ of MO BS ChE '56. US Air Force 1957-60, 1st Navigator C-124 Aircraft. Spencer Chem Co 1956-57 trainee. KC, MO 1960-65 Ind & Plastic Prod Salesman. Rexene Polymers 1965-66 Salesman. Paramus, NJ 1966-67 Sales Supervisor. 1967-69 Mid-West Dist Mgr. 1969-70 Nat'l Sales Mgr. Blako Industries 1970-present Founder and President Extruder of Plastic Films. SPE Pres Toledo Section 1974. Nat'l Councilman 1978-81 Nat'l Ex. Comm 1981-present. SPE Treas 1982-83; SPE Sec 1983-

Kretzschmar, John R (Continued)
1984; SPE 2nd VP 1984-85. SPE 1st VP 1985-86, SPE Pres. 1986-87 SPE Pres 1987-88. *Society Aff:* SPE

Kretzschmar, Richard J
Home: 3312 Mayfair Rd, Baltimore, MD 21207
Position: Chief of Water Engineering Division *Employer:* City of Baltimore. *Education:* BS/Civil/Johns Hopkins Univ. *Born:* Feb 1928. Employed by the City of Baltimore, Bureau of Water Supply in Feb 1947. Appointed Ch of Water Div 1968. U S Army 2nd Armored Div 1950-52. Reg PE Md 1962. Chmn Chesapeake Sect AWWA 1971-72; Pres Md Sect ASCE 1974-75. George Warren Fuller Award AWWA 1975. Tau Beta Pi 1976. Greeley Award APWA 1985. *Society Aff:* AWWA, ASCE, APWA.

Kreuzer, Barton
Home: 3 Tall Timbers Dr, Princeton, NJ 08540
Position: Ret. *Employer:* *. *Education:* EE/Elec Engg/Polytechnic Inst of NY. *Born:* Feb 1909. Native of NY. Joined RCA R & D 1928, helped establish pioneer TV station W2XBS in NY 1928-29. Mgr RCA Film Recording Activities 1934 NY, and 1937-43 Hollywood. Various mgmt assignments in RCA Indus Prods in Camden NJ 1943-54. Dir RCA Corp Prod Planning 1954-53. One of founders, as Marketing Mgr RCA Astro-Electronic Div, which built & launched Tiros, the 1st meteorological satellite - followed by early communication, navigation & moon probe satellites. Div V P 1961 at RCA Space Ctr Princeton NJ. Returned to Camden 1967 to head Broadcast & Communication Equip Div of RCA; Corp V P 1968; Exec V P 1969. RCA Exec V P Consumer Electronics with world-wide general mgmt respons for all RCA TV & related business, headquartered in Indianapolis, Ind 1970-73. City-wide General Chmn Indianapolis United Fund 1973. RCA Exec V P Corp. *Society Aff:* SMPTE, AIEE.

Krezdorn, Roy R
Home: 1501 Hillmont, Austin, TX 78704
Position: Retired *Education:* PE/Engrg/TX A&M Univ; MS/EE/TX Univ, Austin; BS/EE/TX A&M Univ. *Born:* 1/30/10. Native of Seguin Texas. Served as engr, tech advisor, & cons to lower Colorado River Authority 1939-57. Ch Elec Engr on Buchanan Pump-Back installation, & 2 other hydro electric installations. Organized Texas Engrg Assocs in 1957, a cons firm serving the electric utilities in the state. Charter mbr of Power Distribution & Protective Relay Conf Planning Cttes, chairing PDC. Has taught EE since 1941 at U of Texas. Life Fellow in IEEE. Who's Who Texas, South, Southwest, Who's Who in Education, others. Retired teaching June 1977. Retired, pres TX Engrg Assocs in 1981. Enjoys flying & fishing. *Society Aff:* IEEE, NSPE

Kridner, Kenneth
Home: 5731 Arboles Dr, Houston, TX 77035
Position: Owner *Employer:* Tri Plek Productions *Born:* July 1921 Bay City Texas. BSChE Rice Univ, Houston Texas. Married Eloise. 4 children: Kirk, Michael, Mark, & Kathy. Served in U S Navy WWII as Electronic Technician 2c, Far East Theatre. Plant engr for Southern Acid & Sulphur Co; plant engr for Stanolind Oil & Gas Co - Gas Measurement/Test Engr; Gas Contract Engr. Sr Dev Engr for Colo Interstate Gas Co.; Measurement Superintendent for SARGO, SA on Argentine oil/gas pipeline proj; Ed of Petro/Chem Engr; Gas Ed of Pipe Line Indus;, Ed of Chilton's Gas Oil & Gas Energy; Ed of Gas Digest; & owner of Triplek Productions, publishers of Gas Digest. Past Pres of Amer Assn of Cost Engrs, Past Pres of Natl Gas Measurement Assn.

Krieg, Edwin H, Jr
Home: 108 Normandy Rd, Oak Ridge, TN 37830
Position: Engg Site Mgr Oak Ridge Natl Lab *Employer:* Martin Marietta Energy Systems *Education:* MS/Engg/Stevens Inst of Tech; BS/Mech Engg/Brown Univ. *Born:* Aug 1934. BS Brown Univ; MS Stevens Inst of Tech. Native of Ridgewood NJ. Ltjg US Navy 1957-60. Operations Analyst, Distribution, Materials, & Manufacturing Planning & Control Mgr for Linde Div Union Carbide 1960-68. With Nuclear Div Union Carbide 1969-84 as Sr Engrg Specialist, Quality Assurance Coordinator, Superintendent of Operations Planning, Maintenance Engrg, Field Maintenance, Barrier Manufacturing & Deputy Mgr Capacity Expansion Programs. Martin Marietta Energy Systems, 1984 to present Oak Ridge Natl Lab Engrg Site Mgr. Secy of Environmental Quality Advisory Bd, & Pres United Nations Ctte of Oak Ridge. Loaned executive United Fund. Bd/Dir East Tenn Res Corp, Oak Ridge Chamber of Commerce & Grove Dev Corp. Reg PE 4155 State of CA. Pres, Oak Ridge Chapter, NSPE. *Society Aff:* NSPE.

Kriegel, Monroe W
Okla. State Univ, 512 Engr North, Stillwater, OK 74078
Position: Director of Engrg Extension Emeritus. *Employer:* Oklahoma State University. *Education:* PhD/Chem Engr/Univ of TX-Austin; MS/Chem Engr/Univ of TX-Austin; BS/Chem Engr/Univ of TX-Austin. *Born:* 7/30/12. From 1940-64 worked for Carter Oil & Jersey Prod Res Co (affiliates of Exxon) as res engr & administrator in petroleum production. Joined Oklahoma State 1964. Since 1966, Dir of Engrg Extension with respon for all progs in Engrg, Tech, & Fire Service Training. Charter mbr of natl ctte on Continuing Engrg Educ for AIChE, AIME, & SPE. In ASEE served as Chmn of Continuing Engrg Studies Div & mbr of natl Bd/Dir. Represented US at internatl meetings on continuing Engrg Educ in Helsinki in 1972, Tunisia in 1975, & Aachen Germany in 1976. Enjoys driving tractor on his small cattle ranch. Retired from active administration and teaching June 1978. Currently Consultant in Continuing Engineering Education. Clients include United Nations Development Program and the Oklahoma State Regents for Higher Education. *Society Aff:* ASEE, SPE, AIChE.

Krieger, Charles H
Home: 1790 Forest View Ave, Hillsborough, CA 94010
Position: President & Chief Elec Engineer. *Employer:* Charles H Krieger & Associates Inc. *Born:* Feb 20, 1923. BSEE U C Berkeley 1947. Reg Elec Engr in Calif, Nev, & Texas. Signal Corps 1943-46. Formed own practice in Cons Elec Engrg 1957. Spec in preparation of plans/specifications & job visits, inspection reports; electrical surveys & reports, etc for indus, commercial & other types of const work associated with elec consulting services to Architects & Owners. Special interest in illumination, elec audio/visual controls, power distribution & uninterruptible power supply systems. P Pres IES Golden Gate Sect. Bd/Dir Cons Engrs Assn of Calif.

Kriesel, William G
Business: 2200 LaFontain St, PO Box 1708, Ft Wayne, IN 46801
Position: Pres. *Employer:* Masolite Concrete Products, Inc. *Education:* MS/Civ Engg/Purdue Univ; BS/Civ Engg/Purdue Univ. *Born:* 10/6/34. in Michigan City, IN. Served in US Army from 1954 to 1956. Entered Purdue Univ after Army service. Served as Chief Engr for Martin Marietta Corp, Lafayette, IN from 1961 to 1965. Formed Contech Architects & Engrs, Inc in 1965 & served as its Pres until June 1978 when he became Pres of Masolite Concrete Products, Inc. Is a past mbr of Prestressed Concrete Inst's Technical Activities Committee. Enjoys playing golf. *Society Aff:* NSPE, ASCE.

Krishna, Gopal TK
Home: 329 35th St, West Des Moines, IA 50265
Position: Pres *Employer:* Krishna Engrg Consultants, Inc *Education:* MBA/Bus/Drake Univ; MS/EE/Univ of KS; BE/EE/Osmania Univ; *Born:* 2/16/47 to Srinivas Achariar and Rajammal in Hyderabad, India. Attended Methodist Boys Multipurpose Higher Secondary School. Came to USA in 1969. Worked as design engr for Veenstra & Kimm, Inc, 1970-1980. Served as Secretary and VP of Central IA Chapter of IA Engrg Society (IES). Won Outstanding Young Engr award presented by Central IA Chapter of IES, 1979. Won John Dunlap-Sherman Woodward award presented by IES for outstanding technical or professional achievement before age thirty-five, 1979. At present, Pres of Krishna Engrg Consultants, Inc.

Krishna, Gopal TK (Continued)
Offers electrical, mechanical and environmental engrg services primarily to other consulting engrs and architects. Completed course work at IA St Univ for MS in Sanitary Engrg. *Society Aff:* IEEE, NSPE, WPCF, AWWA, NFPA

Krishnamurthy, Natarajan
Business: Cudworth Hall Univ Sta, Birmingham, AL 35294
Position: Prof & Chrmn of CE. *Employer:* Univ of AL. *Education:* PhD/Structures/Univ of CO; MS/CE-Structures/Univ of CO; BE/Civ Engg/Univ of Mysore; BSc/Phys, Chem, Math/Univ of Mysore. *Born:* 10/29/31. in Bangalore, India. Studied in Burma, India, and USA. Taught at Natl Inst of Engg (Univ of Mysore), Mysore, India, 1955-1959 and 1962-1967; at Auburn Univ, Auburn, AL, 1967-1975; and at Vanderbilt Univ, Nashville, TN, 1975-1979. Assumed current position in Aug, 1979. Industrial experience with Union Carbide Corp, summer 1968 and 1969, on finite element analysis of nuclear components. Consulting, on matrix methods of analysis and automated design of structures, and finite element analysis of structural components. Res and publications, in structural engg, with emphasis on computer applications and steel bolted connections. Hindu; married 1955, one daughter, born 1956, married 1979. *Society Aff:* ASCE, ASEE, ACI, NSPE, IE.

Kristiansen, Magne
Business: EE Dept, Lubbock, TX 79409-4439
Position: P W Horn Prof. *Employer:* TX Tech Univ. *Education:* PhD/EE/Univ of TX at Austin; BS/EE/Univ of TX at Austin. *Born:* 4/14/32. Born in Elverum, Norway. Nationalized US citizen in 1967. Consultant to industry & natl labs. Active in organizing IEEE, NPSS. Organizer of several natl & intl confs on Plasma Physics & Pulsed Power Tech. Fellow of IEEE. NATO Sr Fellow in Sci-1975. Fellow of Am Phys Soc, Japan Soc of Promotion of Sci Fellow- 1979. Enjoy skiing, swimming & jogging. Published 200 papers and one book which is translated to Russian, Japanese and Chinese. Reg PE (TX). USAF Meritorious Civilian Service Award. Member numerous Federal Advisory Committees. *Society Aff:* ASEE, IEEE, APS, AAAS.

Kristinsson, Gudmundur E
Home: 304-1363 Clyde Ave, West Vancouver, BC
Position: Naval Architect. *Employer:* MacMillan Bloedel Limited. *Education:* BSc/Ship Structure and Engr/Helsingor Skins Tech, Helsingor Denmark *Born:* 4/27/27. in Iceland. Educated in Denmark & emigrated to Canada 1951. Since that time has been in charge of numerous design & shipbuilding projs including 5 yrs in Japan. From 1960-74 was Pres of Commercial Marine Services Ltd Montreal. In Jan 1974 moved to Vancouver BC from 1974-76 employment with MacMillan Bloedel Ltd in theor Transp Research Dept in Jan 1977 Commenced present employment as pres Kris Krishinsson & Assocs, Naval Archs. *Society Aff:* SNAME

Kritikos, Haralambos N
Business: 200 S 33 St, Philadelphia, PA 19104
Position: Prof *Employer:* Univ of PA *Education:* PhD/EE/Univ of PA; MS/EE/Worcester Polytech Univ; BS/EE/Worcester Polytech Univ *Born:* 3/8/33 Born in Tripolis, Greece. Received BS & MS from Worcester Polytech Inst in 1954-56 respectively. PhD from Univ of PA 1961. He appointed Asst Prof in 1962 at the Moore Sch of Elec Engrg Univ of PA. He was a Res Fellow at the CA Inst of Tech during 1966-67. He returned to the Moore Sch as an Assoc Prof, and now holds the position of Prof. He is currently teaching and is conducting res in the area of electromagnetic Theory and Applications. *Society Aff:* IEEE

Krivsky, William A
Business: P O Box 860, Valley Forge, PA 19482
Position: Vice President. *Employer:* Certain-teed Corporation. *Education:* ScD/Metallurgy/MIT; SB/Metallurgy/MIT. *Born:* March 1927. ScD & SB from MIT. Native of Stafford Springs, Conn. USAAF 1944-47. Mgr Metals Res at Union Carbide Corp 1954-59. V P Brush Beryllium Co 1959-65. V P Continental Copper & Steel Indus 1965-69. Grp V P Chem Cable Corp 1969-71. Pres Crucible Specialty Metals Div, Colt Industries 1971-74. V P Group Exec Certain-teed Corp since 1974-77. SR VPres Cert Corp since 1977. Inventor of Linde AOD steelmaking process. Gold Medal Award 1959 AIMME. Extractive Met Lecturer AIMME 1971. Dir AIMME 1970-72. Clamer Medal, Franklin Inst 1977. *Society Aff:* AIMME, ASM, AMA.

Krizak, Eugene J
Home: 2114 Wyoming St, Baytown, TX 77520
Position: Manager of Engrg, Mech & Civil. *Employer:* Upjohn Co, Polymer Chemicals Div. *Born:* Oct 1931. BS U of Texas. Native of Baytown Texas. Specialized in proj engrg & proj mgmt for 20 years. Mbr TSPE. Enjoys flying, fishing, water skiing & camping.

Krizek, Raymond J
Home: 1366 Sanford Lane, Glenview, IL 60025
Position: Professor of Civil Engineering. *Employer:* Northwestern University. *Education:* PhD/Civil Engg/Northwestern Univ; MS/Civil Engg/Univ of MD; BE/Civil Engg/Johns Hopkins Univ. *Born:* 6/5/32. 1932 Baltimore Md. Served with U S Army Corps of Engrs 1955-57. Taught at U of Md, Catholic U of Amer, & Northwestern U. Res contribs in constitutive relations for soils, engrg assessment of dredging & disposal opers, soil- structure interaction of buried conduits, engrg properties of FGD scrubber sludge & soil improvment techniques. ASTM Hogentogler Award 1970; ASCE Huber Prize 1971; NSF Distinguished Visiting Scholar 1972. Stanley F. Pepper Chair 1987 Enjoys sports. *Society Aff:* ASCE, ISSMFE.

Kroeker, J Donald
Home: 6831 S E Brownlee Rd, Portland, OR 97222
Position: Retired. *Education:* Prof ME/Mechanical/OR State Univ; BSCE/Civil/OR State Univ. *Born:* April 1900 in Russia. Prof ME 1937 Oregon State Univ, mbr 4 honor societies. Publs Ed & res, AGA Testing Labs, Cleveland 1927-30; Dev Engr, Pa Furnace & Iron, Warren Pa 1930-32. HVAC Cons Engr, Portland, 1936-76, except in WWII, LtC US Army Engrg Dept, Alaska. Reg PE in Oregon, WA, AK, ID. Fellow ASHRAE & ASME; Mbr NSPE, CSI, PEO. Founder Mbr ACEC. 1st Pres Ore Chap ASHRAE & CECO. Tres ASHRAE 1952. Rotary Club since 1936. Served on many ASHRAE cttes. Contributed many tech papers to societies & trade journals. Designer of the 2 largest heat pump installations in the world. On WWII energy conservation, developed criteria applied across US by Dept of Interior Retired June 30, 1976. Currently active in energy conservation particularly in buildings and on 4 Committees on energy conservation. *Society Aff:* ΣΔΧ.

Kroemer, Herbert
Business: Dept of E&CE, Santa Barbara, CA 93106
Position: Professor. *Employer:* Univ of Calif-Santa Barbara. *Education:* PhD/Physics/Univ Gottingen *Born:* 1928. PhD in Physics U of Goettingen, Germany 1952. Since then worked in several labs in various areas of electronic device & materials res, such as high-freq transistors, heterojunctions, injection lasers, Gunn effect. Fellow IEEE & APS. Chmn 1972 Internatl Symp on GaAs & related compounds, & 1974 Device Res Conf. Recipient 1973 J J Ebers Award of the IEEE. Recipient 1982 Sr Res Award Am Soc Engrg Educ. Recipient 1982 GaAs Symposium Award & Heinrich Welker Medal of 10th Internat. Symp on GaAs & Related Compounds. Current work on heterojunctions & on molecular beam epitaxy. *Society Aff:* IEEE

Kroeze, Henry
Home: 303 N Holden, Greensboro, NC 27410
Position: Retired (Prof Emeritus) *Education:* Ph.D/Elec Engg/Univ of Delft; MS/Elec Engg/Univ of Delft; BS/Elec Engr/Univ of Delft; P.E. *Born:* 5/22/21. (Retired). Native of The Netherlands. 1st Lt Royal Netherlands Army, WWII; decorated by the British. 20 yrs in positions with industry (from proj engr to gen mgt). Full time

Kroeze, Henry (Continued)
with the Univ of WI Ctr System since 1967. Consultant to maj corps. "Engineer of the Year in Education, State of Wisconsin" award in 1972. First Rep of the engg profession on the US Metric Bd (presidential appointment, confirmed by the US Senate). Chrmn Dept of Engg & Comp Sci since 1971-1981 (UW Ctr System). Adviser to the Intl Special Olympics. Over 40 publication. in the fields of engg mechanics, engg economy, technical education and metrication. Past Mbr Bd of Dirs American Nat Metric Council. Past Pres, Piedmont Council of Engrg & Tech Soc; Score Counselor. *Society Aff:* IIE, ASEE, KIVI, AIPE, GEC

Krohn, Robert F
Business: 8404 Indian Hills Dr, Omaha, NB 68114
Position: Pres & CEO *Employer:* HDR, Inc *Education:* BS/Civ Engg/Univ of NB. *Born:* 4/9/33. Public engr, NB Health ept, 1960. 1962, civ engr, HDR, Omaha. Worked on Western Union Microwave System, hgwys, land devs. Promoted to Asst to the Pres, 1967. Named exec vp, 1969, transferred to DC as firm's liaison with fed govt, intl operations & Mid-Atlantic office mgt. Helped devlop TRIDENT Submarine Support Site Transit sections; US Embassy, Brazil; King Abdulaziz Univ Health Sciences Ctr, Jeddah, Saudi Arabia. HDR Pres, 1976. Responsible for firm's 24 offices, with projs completed or underway in 45 states, 28 foreign countries. NB Outstanding Young Engr, 1968. *Society Aff:* ACEC, ASCE, SAME, NSPE.

Krommenhoek, Daniel J, Jr
Business: Knolls Atomic Power Lab, Schenectady, NY 12301
Position: Mgr. *Employer:* General Electric. *Education:* MS/ME/Cornell Univ; BS/ME/Lafayette College. *Born:* Aug 1941; native of N Caldwell N J. BS Lafayette Coll; MS Cornell Univ. Mbr Tau Beta Pi, Pi Tau Sigma Hon Socs; Natl Sci Foundation Traineeship. Patent. With G E 1966- ; Engr, advanced design & advanced concepts; assumed lead engr respon at Knolls Atomic Power Lab for Plant & Safety Analysis (nuclear plants). Former Chmn NE N Y AIAA, Mgr, Proj Performance Analysis. *Society Aff:* AIAA, ASME.

Krone, Ray B
Business: College of Engg, Davis, CA 95616
Position: Assoc Dean/Prof. *Employer:* Univ of CA., Davis. *Education:* PhD/CE/Univ of CA, Berkeley; MS/CE/Univ of CA, Berkeley; BS/Soil Sci/Univ of CA, Berkeley. *Born:* 6/7/22. Photo reconnaissance pilot USAAF, 1944-5, Res Soil Scientist to Staff Sanitary Engr and Assoc Res Engr, Sanitry Engg Res Lab & Hydraulicss Engg Lab, UC Berkeley, 1950 to 1964. Assoc Prof of Civ Engg, UC Davis, 1964-70, Prof of CE, 1970-, Chair, Dept of CE, 1968-72, Assoc Dean for Res, College of Engg, UCD 1972-, Consultant on cohesive sediment transport & tidal hydraulics, Waterways Experiment Station, US Army Corps of Engrs, 1960-, USACE Committee on Tidal Hydraulic, 1975-, consultant to CA Attorney Gen 1969- and to many cnsulting firms, ports, and public agencies concerned with water quality mgt, sediment problems, marsh restoration, and harbor design. *Society Aff:* ASCE, WPCF, AGU, AAAS, Estuarine Research Federation.

Kroner, Klaus E
Business: College of Engrg, Dept of Industrial Engin, Amherst, MA 01003
Position: Assoc Prof (ret'd) *Employer:* Univ of MA *Education:* M/BA/American Inter Coll; B/EE/NYU; BA/Physics/Coll of Wooster *Born:* 7/19/26 Taught engrg graphics and related topics at NYU, U of ME, and U of MA. until 1969. Since then in Dept. of Industrial Engin. at U of MA with emphasis on industrial development, plant layout, engrg, economics, and computer applications. Also interested in metrication. Member of Engin. Design Graphics Div of ASEE since 1953, holding numerous positions. Active in town affairs and many civic efforts. *Society Aff:* ASEE

Krueger, Earl A
Business: 2320 Univ Ave, Madison, WI 53705
Position: CBD & Pres. *Employer:* Mead & Hunt, Inc. *Education:* BS/Civ Engg/Univ of WI. *Born:* 7/24/20. and educated in WI. Started in Consulting Engg in 1944. Joined Mead and Hunt, Inc in 1947. Became Pres in 1960 and elected Chrmn of the Bd in 1966. Firm is involved in gen practice with specialties in hydro-electric dev and dairy industry. A former Dir of ACEC and Pres of WI Chapter. Served as Dir of area Chapter of NSPE. Together with wife, Catherine, parent of two daughters and two sons. Enjoys golf, fishing and sail boating. *Society Aff:* NSPE, ACEC.

Kruger, Charles H, Jr
Business: Dept of Mech Engg, Stanford, CA 94305
Position: Prof. *Employer:* Stanford Univ. *Education:* PhD/Mech Engr/MIT; DIC/Mech Engr/Imperial College (London); SB/Mech Engr/MIT *Born:* 10/4/34. Native of Okla City, OK. Asst Prof, MIT, 1960. Res Scientist, Lockheed Res Labs, 1960-1962. Mech Engg Dept, Stanford Univ, since 1962; Prof since 1967; Chrmn since 1982. NSF Sr Postdoctoral Fellow, 1968-1969, at Inst for Plasma Physics, Munich and at Harvard. Assoc Editor AIAA Journal 1968-1971. Teaching and res interests include partially ionized plasmas, combustion, and air pollution. Mbr, Hearing Bd Bay Area Air Quality Mgt Dist 1970-1983 and Steering Com, Engg Aspects of MHD. Lecturer in air pollution, Norwegian Inst of Tech, 1979. Visiting prof, Princeton Univ, 1979-1980. AIAA Fluid and Plasmadynamics award, 1979. *Society Aff:* AIAA, ASME, APS, Comb Inst

Kruger, Fredrick C
Business: 145 Wildwood Way, Woodside, CA 94062
Position: Self. *Education:* PhD/Geology/Harvard Univ; MA/Geology/Univ of MN; BA/Geology/Univ of MN. *Born:* 4/1/12. Native of St Paul MN. Mining geologist with Cerro de Pasco Copper Corp 1941-49; Prof of economic geology at TN & cons mining geologist 1949-52; Asst Ch Geologist for Reynolds Aluminum 1952-57; Ch Geologist VP of Mining & Exploration Div of Intl Minerals & Chems 1957-66; AIME Krumb Lectr 1967-69, Centennial Lectr 1970, Hardinge Award 1972. Stanford Endowed Chair: Donald Steel Prof of Mining Geology 1971; Prof and Hd of Dept of Applied Earth Sciences, Assoc Dean for Res, Stanford Univ, Sch of Earth Scis, and private consult 1966-71; since then private consult mining geologist. *Society Aff:* AIME, SEG.

Kruger, Paul
Business: Civil Engr Dept, Stanford University, Stanford, CA 94305
Position: Prof. *Employer:* Stanford Univ. *Education:* PhD/Nucl Chem/Univ of Chicago; BS/Chem/MIT. *Born:* 6/7/25. Native of Jersey City, NJ. Served with USAF 1943-46. Res Engr for Gen Motors Corp 1953-4. VP & Mgr - Nuclear Sci & Engr Corp Pittsburgh, PA 1954-60; Hazleton Nucl Sci Corp Palo Alto, CA 1960-61. Appointed Prof of Nuclear Civ Engg at Stanford Univ since 1962. Prof Engr in Commonwealth PA since 1954; Chrmn of the ASCE Div. Environmental Engr Exec Comm. Visiting Prof at Univ CA Berkeley 1968-9 at MIT 1979-80. *Society Aff:* ANS, ASCE, ACS, AGU.

Krum, Glenn L
Business: P O Box 3020, Casper, WY 82602
Position: Supervisory Engr *Employer:* Exxon Minerals *Education:* M SC/Mining Eng/CO School of Mines; B SC/Mining Eng/CO School of Mines; BA/Geo/Univ of CO *Born:* 8/18/48 in Ft. Ord, CA. The son of Carl and Nelda N. (Lohopener) Krum. Married Anne Louise Hidding on April 22, 1977, two children Robert Alan and Cynthia Elaine. Following graduation from the Univ of CO served as a Division Officer in the US Navy seeing service in Vietnam. Returned to school in 1973 graduating in 1975 and 1977. Joined Exxon Minerals in 1977. Currently Supervisory Engr at The Highland Uranium Operations in WY. Publications: *Oil Shale Resource Assessment for in-site Retorting Phase 1: Picance Creek Basin, CO* with Dr. Gentry and Dr. Carpenter for Lawrence Livermore Laboratory. *Considerations of Inflation in Lead Mine Evaluation for APCOM Symposium (1979)-Robert Peele Award winning paper in 1980. Society Aff:* AIME-SME

Kruse, Cameron G
Business: 6800 S Cty Rd 18-35108, Minneapolis, MN 55435
Position: Sr. Vice President - Engineering and President. *Employer:* Braun Engrg Testing Co-Braun Envir Labs, Inc *Education:* MSCE/Civ Engrg/SD State Univ; BSCE/Civ Engrg/SD State Univ. *Born:* Dec 1941 Bryant S D. BSCE & MSCE South Dakota St Univ. Reg P E in Minn. With Braun Engrg Testing 1968- ; Sr. VP Engrg respon for engrg supr of field & lab testing & engrg staff. Also Pres Braun Envir Lab. Prof Org Activities: P St Pres MSPE; Natl Dir NSPE; V Pres. CEC/MN; Mbr MSPE, NSPE, PEPP, ASCE, CSI, ACIL, CEC/MN & Minn Geotech Soc; Minn Young Engr of Year 1973; guest lectr at schs & seminars; numerous publs. Community Activities, Chairman City Planning Commisssion: active in church educ, ushering & other activities; several positions in community athletic assn, auxiliary police; several positions in local political activities. *Society Aff:* ASCE, NSPE, ACEC, CSI, ACIL.

Krusling, James R
Business: Room 334 City Hall, Cincinnati, OH 45202
Position: Dir of Buildings and Inspections *Employer:* City of Cincinnati. *Education:* JD/Law/Chase College; MS/Structures/Univ of Cincinnati; CE/Civil Engg/Univ of Cincinnati. *Born:* 10/25/26. Native of Cincinnati OH. Career employee of City of Cincinnati starting as Co-op Student 1944; promoted to current position 1981 after almost 18 yrs as City Engr. Respon for Admin and Enforcement of Bldg Code and other codes and ordinances, has staff of 124 including 12 Prof Engrs and Archs. Lectr on Struct Engrg at Univ of Cincinnati Evening Coll 1959- . Title: Adjunct Prof of Civ Engg. Mbr: ASCE, P Pres Cincinnati Section; P Pres OH Council of Local Sections. APWA, P Pres OH Chap; Engrg Soc of Cincinnati; Tau Beta Pi. Hon mbr Alpha Sigma Lambda Hon Soc. *Society Aff:* ASCE, APWA, IME.

Kruty, Samuel
Home: 7051 N Oleander Ave, Chicago, IL 60631
Position: Vice President-Production Services. *Employer:* Dearborn Chemical (U S). *Education:* BSChE/Engg/Northwestern Tech Inst. *Born:* 13/11/21. BSChE Northwestern Tech Inst Evanston Ill, 1946. With USAAF as bombadier 1943-45. Dev Engr for Monsanto Chem Co Krummrich Plant 1947-50; with Dearborn Chem (US) 1951-79, assumed current resp as VP Prod Services (prod, purchasing, traffic) 1966. Served as Pres of local Indus Council; Mbr AIChE, ACS, AAAS. Serious avocational interests in palentology, prairie restoration & Maya Civilization. Retired as of April 1979. *Society Aff:* AIChE, ACS, AAAS.

Ksienski, Aharon A
Home: 1780 Lynnhaven Dr, Columbus, OH 43221
Position: Prof. *Employer:* OH State Univ. *Education:* PhD/EE/Univ S CA; MSc/EE/Univ S CA; BE/ME/Inst Mech Engr. *Born:* 6/23/24. Israeli AF 1948-51, Capt in charge of aircraft elec sch; Hughes Aircraft Co, 1958-1967. Hd of Res Staff of Antenna Dept. Res in antenna design & array optimization for angular resolution, signal processing antennas; OH State Univ, 1967-present. Prof of EE, Dir of Communication & Radar Systems, ElectroSci Lab. Res in radar target identification, signal processing antenna systems and satellite communication; Chrmn of US Commission 6 of the Intl Radio Sci Union 1972-75; Fellow of IEEE; Recipient of the Lord Brabazon Award 1967 & 1976 from the Inst of Electronic & Radio Engrs London, England; Commendation from the Gen Assembly of the State of OH 1977; R W Thompson Meritorious Achievement award from the OH State Univ, 1969. *Society Aff:* IEEE, URSI.

Ku, Yu H
Business: Moore School, Univ. of Penna, Philadelphia, PA 19104
Position: Emeritus Professor. *Employer:* University of Pennsylvania. *Education:* ScD/EE/MIT; SM/EE/MIT; SB/EE/MIT. *Born:* Dec 24, 1902 Wusih, Kiangsu China. Dean of Engrg Tsing Hua Univ; Pres & Dean of Engrg Central Univ; Visiting Prof MIT; Prof of EE Univ of Pa. Awards: AIEE Fellow 1945; IRE Fellow 1961 'for contributions to nonlinear circuit analysis'; IEEE Lamme Medal 1972 'in recognition of his outstanding contributions to analysis of a-c machines & systems'; Fellow IEE; Academia Sinica; Gold Medal Chinese IEE 1972; Personal Mbr of Genl Assembly IUTAM; Gold Medal Ministry of Educ ROC 1960; Gold Medal Pro Mundi Beneficio Brazilian Acad of Humanities 1975. Mbr ASEE, Sigma Xi, Eta Kappa Nu, Phi Tau Phi; US Natl Ctte of Theoretical & Applied Mechs. *Society Aff:* IEEE, IEE, ASEE, CIEE

Kube, Wayne R
Business: Box 8101 Univ Station, Grand Forks, ND 58202
Position: Prof, Chem Engr. *Employer:* Univ of ND. *Education:* MS/Chem Engg/MI Tech Univ; BS/Chem Engg/MI Tech Univ. *Born:* 05/03/22 Born in Mancelona, MI. Served with Army Corps engrs 1943-44. Worked on Manhattan Proj 1944-46. Taught MTU 1946-48. Chem Engr, US Bureau of Mines 1948- 52. Asst Prof to prof, Univ of ND 1952-80. Chem Engr (Fac), Grand Forks Energy Tech Ctr, 1952-80. (Retired) Served on Natl Comm AIChE, US mbr and chrmn. Co- chrmn for 12 Biennal Lignite Symposia. Specialized in res-dev on low-rank fuels. Reg Engr in ND. *Society Aff:* AAAS, AIChE, ACS, NDAC, EES.

Kubick, Raymond A, Jr
Home: 4913 Angeles Crest Cir, La Canada, CA 91011
Position: Chief Instrument Engineer. *Employer:* Ralph M Parsons Co Pasadena Calif. *Education:* ME/Genl Engg/Stevens Inst of Tech; MS/Industrial Mgmt/Stevens Inst of Tech. *Born:* Dec 1924. ME & MS Stevens Inst of Tech. Engr with Honeywell Inc, Curtiss Wright Corp, Genl Foods Corp, Air Reduction Co & C F Braun & Co; joined Ralph M Parsons Co 1966. Instrument Dept Head Parsons-Jurden Div N Y until 1972; Proj Engr LA Office until appointed co's Ch Instrument Engr 1973. Taught evening instrument courses Newark Coll of Engrg 196371. Author of Chap Instrumentation in Mining & Mineral Indus in handbook 'Instrumentation in the Process Industries'. Mbr ASME 1962; Fellow ISA 1971, ISA Disting Soc-Service Award 1975, ISA Dir 1971-74, V Pres-Elect ISA Educ Dept. P E Control Sys Engr Calif. *Society Aff:* ISA, ASME.

Kubitz, William J
Business: University of Illinois, 240 Digital Computer Laboratory, 1304 W. Springfield Ave, Urbana, IL 61801
Position: Prof and Assoc Head of Computer Science *Employer:* Univ of IL. *Education:* BS/Engr Physics/U of IL; MS/Physics/U of IL; PhD/EE/U of IL *Born:* 12/27/38. Native of Freeport, IL. Attended the Univ of IL 1957-62 and 1964-68. Dev engr for the GE Co 1962-64. 1968-70 res asst prof, Dept of Comp Sci, Univ of IL. 1970-74 asst prof. 1974-85 assoc prof. 1985 to present, professor and assoc head of the dept of Computer Science. Current res interests involve the the automated design of digital integrated circuits and integrated object-oriented graphics for networked, single user workstation envirionments. *Society Aff:* IEEE, ACM, SID, AAAS.

Kuch, Eugene R
Home: 2510 Oak St, Quincy, IL 62301
Position: Chief Metallurgist Mgr of Metallurgy/Quality Assurance Mgr *Employer:* Cooper Mach. Group - Cooper Ind. *Education:* BS/Chem Eng/Quincy Coll *Born:* Apr 13, 1923; native of Freeport Ill. With Army Air Corp 1943-46. Lubrication Engr with Gardner Denver; Metallurgist respon for design of plating & Heat Treat equipment & process; later in foundry area respon for casting process; assumed current respon as Ch Metallurgist respon for above & material application & testing. Dir AFS & mbr Handbook Ctte ASM. Dir of Ductile Iron Soc mbr ANSI ISO Ctte lubrication. Mbr ductile iron Res. Ctte. Mbr AFS Ctte compacted graphite iron. Mbr cast iron welding ctte AWS. *Society Aff:* DIS, AFS, AESF, ASNT, AWS, ASM, ASTM, ASLE, NACE, ASQC

Kuchinski, Frank Leonard
Home: 123 Hulda Hill Rd, Wilton, CT 06897
Position: Pres. *Employer:* Process Technology Consultants. *Education:* BS/ChE/VPI;

Kuchinski, Frank Leonard (Continued)
MS/ChE/Lehigh Univ. *Born:* Jan 1, 1928 Manville N J. Sales Exec for major engrg & const co. BS ChE VPI; MS ChE Lehigh Univ; Post Grad work in Polymer Chem. Served in Navy Air Corps. Prof Employment: E I DuPont de Nemours & Co as dev engr; Goodyear Tire & Rubber Co as dev & proj engr; as R/D, Mgr Sales & V Pres-Sales. Areas of Spec: Cons, polymer plant prod problems; sales-plant design & const. Tech publs. Mbr AIChE. Presently with Process Tech Consultants, Pres. *Society Aff:* AIChE.

Kuck, David J
Business: Center for Supercomputing Research & Development, University of Illinois at Urbana-Champaign, 104 S. Wright Street, Urbana, Illinois 61801
Position: Prof., Dir. Center for Supercomputing R&D *Employer:* Univ of IL. *Education:* PhD/Engrg/Northwestern Univ; MS/Engrg/Northwestern Univ; BSEE/Engrg/Univ of MI. *Born:* 10/3/37. Director, Center for Supercomputing Research & Development 1984- , and Professor of Computer Science & Electrical & Computer Engrg 1965- , Univ of Illinois at Urbana-Champaign; Ford Postdoctoral Fellow & Asst Prof of EE, MIT, 1963-65. Res interests in parallel supercomputing. Consultant to Alliant Computer Systems, National Science Foundation; Advisory Boards of Sequent Computer Systems, Scientific Computer Systems, Dana Group. A principal architect of Burroughs BSP, as well as several other comp systems. Patentee supercomputer, 1979. Pres. Kuck & Associates, Inc. a comp system and software consulting firm. Author of more than 80 technical papers & The Structure of Computers and Computations, Vol 1. Editor of 10 scientific & engrg journals. *Society Aff:* ACM, IEEE, AAAS.

Kudroff, Marvin J
Home: 508 Via Bodega, Palos Verdes Ests, CA 90274
Position: V P & Principal-Military/Indus Progs. *Employer:* Daniel Mann Johnson & Mendenhall. *Education:* BS/Arch-Engg/Penn State; Grad Work/Struct Analysis/USC & UCLA. *Born:* June 14, 1923. BS Penn St Arch Engrg; grad work USC & UCLA Struct Analysis. Awarded Outstanding Alumnus Award 1975 from Coll of Engrg Penn St; Merit Award from Inst for Advancement of Engrg (IAE); P Pres Struct Engrs Assn of S Calif & LA Post Soc of Amer Military Engrs (SAME) - Past Natl Dir SAME. Holds lic as Cvl Engr & Struct Engr in Calif & P E in Colo. With Daniel, Mann, Johnson & Mendenhall 1947- , presently spec in military & indus projs; Principal-in-Charge for joint venture serving Air Force in design of Aeropropulsion Sys Test Facility Tullahoma Tenn, Space Shuttle Orbiter processing Facility, Vandenberg Air Force Base & Satellite Center for Air Force. Fellow: ASCE, SAME & IAE. *Society Aff:* SAME, ASCE, AAA, IAE.

Kudryk, Val
Business: Central Research Dept, 901 Oak Tree Rd, South Plainfield, NJ 07080
Position: Mgr *Employer:* ASARCO Incorp *Education:* PhD/Met. Eng./Columbia Univ; MA Sc/Met. Eng./Univ of British Columbia; BA Sc/ChE/Univ of Alberta *Born:* 3/2/24 in Chipman, Alberta Canada. Awarded Shell Oil, Sherritt-Gordon and William Campbell Fellowships. Elected to Sigma Xi and licensed as a PE in the state of NY. Authored several technical papers and granted two patents. Prior positions included Asst VP at Nichols Engrg & Research Corp; VP, Accurate Specialities; Mgr of Metallurgical Center at Lummus Engrg. *Society Aff:* AIME, ASM, ASTM, IRI

Kuebler, Ronald C
Home: 6601 Eastbrook Rd, Columbia, SC 29206 *Employer:* Self *Education:* BS/Engrng (Met&Matl Sc)/Case Western Reserve. Univ *Born:* 3/31/48. (Specialties: Powder Metallurgy, Communication Disorders, Employment & Handicapped). Met at Ceromet, Inc in Ind, CA. Assoc Mfgr Engr at Burroughs Corp; worked in powder metal parts fabrication. As gen partner (Owner- Engr) started a powder metal parts mfg plant, called Pow Form Co. Principal Engr (Briquetting Operations) for a Precision Forged Products Div. Metallurgist for Wheel Trueing Tool Company; Mfg of Diamond Tools. Activity in Professional Societies as V Chrmn of Natl Young Mbrs committee of ASM & V Chrmn of the Detroit Sec of APMI. Organized and instructed an in-plant Basic Powder Metallurgy course. Hired and trained a retarded (Down's Syndrome) man to be a fully integrated crew mbr in a retail business. Active as Scout Coordinator in BSA & an avid handball player. Also enjoys tennis, jogging, camping & fishing. *Society Aff:* APMI, TASH

Kuehl, Hans H
Business: Electrical Eng Dept, Los Angeles, CA 90089-0271
Position: Prof *Employer:* Univ of Southern CA *Education:* PhD/EE/CA Inst of Tech; MS/EE/CA Inst of Tech; BS/EE/Princeton Univ *Born:* 3/16/33 Faculty member in electrical engrg at the Univ of Southern CA since 1960; Assist Prof 1960-63; Assoc Prof 1963-72, Prof 1972 to present. At USC, received Teaching Excellence Award (1964), Outstanding Electrical Engrg Faculty Award (1977), and Haliburton Award for Exceptional Service (1980). Elected Fellow of IEEE in 1980 for basic theory governing interaction of antennas in plasmas. During 1979-80, chairman of committee determining academic standards at USC. Presently, Assoc Chrmn of Elec Engrg - Electrophysics Dept. Principal Investigator on 8 research grants from National Science Foundation since 1965. *Society Aff:* IEEE, APS, URSI, ΦΒΚ, ΤΒΠ, HKN.

Kuehl, Neal R
Business: P.O. Box 458, Storm Lake, IA 50588
Position: Pres *Employer:* Kuehl & Payer, Ltd *Education:* MS/Environmental Health Engg/Univ of KS; BS/Civ Engg/IA State Univ. *Born:* 1/4/43. Spent early yrs on family farm in Northeast IA. Received officer's commission and spent tour of duty with Army Corp of Engrs including Post Engr assignments and duty in Vietnam. Engr with CA Dept of Water Resources and Burns and McDonnell Engg Co. Was with Otto & Culver, PC and Associated Engineers, Inc. From 1973-1983 serving as executive VP and Mgr of Prof Services. Formed Kuehl & Payer, Ltd in 1983. Currently am Pres of twelve men Civil/Environ Consltg Firm. Mbr of Natl Soc of Professional Engrs, consltg Engrs Council/Iowa Kiwanis Intl and United Methodist Church. *Society Aff:* ASCE, NSPE, IES, CEC/IA, ACEC.

Kuehner, Richard L
Home: 1106 W Busse Ave, Mt Prospect, IL 60056
Position: Consultant *Employer:* Independent *Education:* PhD/Microbiology/Yale Univ; AB/Biology/Allegheny Coll *Born:* Nov 1917. PhD Yale Univ; AB Allegheny Coll. Employed 1943-57 York Corp; 1957-1978 by Borg-Warner Central Office; 1978-Consultant in significant air qualities & qualities control. Holds about 30 domestic & foreign pats in field; about 40 publs in prof journals, handbooks, standards, encyclopedias & handbooks. Active in 8 prof socs; Fellow in ASHRAE & Gt Britain's Royal Soc of Health. Best known for pioneering odor control engrg, organizing natl conferences for many prof socs & approximately 35 yr participation in ASHRAE - particulates, odors, gases, sound, physiology & R&T cttes. *Society Aff:* ASHRAE, ΦΒΚ, ΣΞ

Kuers, Marvin M
Home: 1604 Bluebonnet Circle, College Station, TX 77840
Position: Sr. Lecturer *Employer:* Texas A&M University *Education:* MS/Ind Engr/SMU; BS/Mgmt Engr/TX A&M. *Born:* June 1923. Reg Prof Engr Texas. Military service 1943-46 (Infantry). Safety Engr with Employers Casualty Co 1949; Methods Engr Collins Radio 1954; Safety Dir & Ins Mgr J E Bush Co 1957; joined Bell Helicopter 1959, held positions of Area Indus Engr, Fab Area Indus Engrg Supv. Methods Engrg Supv & Chief, Area Indus Engrs. Sr. Proj Eng; 1980 Lecturer Texas A&M University; Panel mbr Amer Arbitration Assn; Regional V Pres IIE; Indus Representative Dr of Engrg Prog Texas A&M Univ. Sr Lecturer, Industrial Engg, Texas A&M Univ. Dir, Ind & Labor Relations Div, Inst of Industrial Engrs Panel Mbr Amer Arbitration Assoc. *Society Aff:* IIE.

Kuesel, Thomas R
Business: One Penn Plaza, 250 W 34th St, New York, NY 10119
Position: Sr VP, Partner, Dir *Employer:* Parsons, Brinckerhoff, Quade & Douglas, Inc *Education:* ME/CE/Yale Univ; BE/CE/Yale Univ *Born:* 07/30/26 Richmond Hill, NY. BE highest honors 1946, M Eng'g 1947 Yale Univ. With Parsons, Brinckerhoff, Quade & Douglas (NY/SF) since 1947-partner, Sr VP, Dir since 1968. Mbr US Natl Committee Tunneling Tech 1972-74. Fellow ASCE, ACEC. Mbr Natl Acad Engrg, Intl Assoc Bridge & Structural Engrg, British Tunneling Society, The Moles, Tau Beta Pi, Sigma Xi. Designer over 100 bridges, 60 tunnels, including Newport Suspension Bridge (RI), NORAD Combat Operations Ctr (CO), San Francisco BART Sys, Hampton Roads Bridge-Tunnel (VA), Ft McHenry Tunnel (MD), Hood Canal Floating Bridge (WA), subways Boston, NY, Baltimore, Washington, Atlanta, Caracas. *Society Aff:* NAE, ASCE, ACEC, IABSE, BTS, ΤΒΠ, ΣΞ

Kuh, Ernest S
Business: Dept. of EECS, Univ. of California, Berkeley, CA 94720
Position: Professor, Dept. of EECS *Employer:* University of California. *Education:* Ph.D./Elect Engg./Stanford Univ (52); M.S./Elect Engg. /Mit (1950); B.S./Elect Engg./Univ of Michigan ('49). *Born:* 10/02/28 in Peking, China. BS Univ of MI, SM MIT; PhD Stanford. Bell Tele Labs 1952- 56; U C Berkeley EE Faculty 1956- , Chmn EE & Computer Sci Dept 1968-72, Dean College of Engg 1973-80. Co-author: *Principles of Circuit Synthesis*; *Theory of Linear Active Networks*; *Basic Circuit Theory Linear & Nonlinear Circuits*. IEEE: Fellow; Pres, Circuits & Sys Soc 1972; Bd of Dirs 1976-77. Mbr Steering Committee of Evaluation Panel Natl Bureau of Stds. Mbr. NSF Ad. Comm. on Engineering and Appd. Sci. NSF Sr Post-doctoral Fellow 1962; Miller Res Professional Berkeley 1965; Natl Electronics Conference Award 1966; Disting Alumni Award Univ of MI 1970; IEEE Guillemen-Cauer Award 1973. Humboldt Fnd Senior Scientist Award, 1978; Hon Prof Shanghai Jiao Tong Univ 1979; Education Medal IEEE 1981; Lamme Medal ASEE 1981. Mbr NAE 1975, Mbr Academia Sinica 1976, Fellow AAAS. *Society Aff:* IEEE, ASEE, AAAS, NSF, NAE.

Kuhlke, William C
Business: One Shell Plaza, Houston, TX 77001
Position: Market Res Rep. *Employer:* Shell Chem Co. *Education:* BS/Chem Engg/Lehigh Univ. *Born:* 10/13/30. Mr Kuhlke has been with Shell since 1952 in a variety of marketing and corporate planning functions. He currently has responsibility for Thermoplastics Market Res. Mr Kuhlke is also President of the Soc of Plastic Engrs and past chrmn of its Intl Mbrship Committee. He is very active in Resins Statistics Committee of the Soc of the Plastic Industry Inc, previously being chrmn of the Polystyrene Subcommittee. He is in his eleventh yr of participation in Little League Baseball activities and is the father of four children - William Jr, Stephen, Susan and Patricia. He married Carol McDonald in 1956. He has also been chrmn of the Furniture Div of the Soc of the Plastic Indus Inc. *Society Aff:* CMRA, AIChE.

Kuhlman, John M
Business: Dept Mechanical & Aerospace Engg, West Virginia Univ, Morgantown, WV 26506
Position: Professor. *Employer:* West Virginia University. *Education:* PhD/Engg/Case Western Reserve Univ; MS/Engg/Case Western Reserve Univ; BS/Mech Engg/Case Western Reserve Univ. *Born:* 6/1/48. Mech Engg educator and researcher active in the areas of fluid mechanics and aerodynamics. Current res interests include: advanced panel methods for subsonic linear aerodynamic theory, optimum aerodynamic design methods for aircraft wings, propulsion induced aerodynamics for jet VTOL aircraft, turbulent jet mixing, turbulence measurements, buoyant thermal plume modelling, hydrodynamic stability theory, fluid mech of gas flow spark gaps, and use of winglets on low aspect ratio wings. *Society Aff:* AIAA, ASME, ASEE, ΤΒΠ, ΣΞ, ΠΤΣ.

Kuhn, Donald J
Business: 4441 Tonawanda Creek Rd N, PO Box 142, N Tonawanda, NY 14120
Position: Pres. *Employer:* Secured Landfill Contr Inc. *Education:* BS/Chemical Engineering/Tri-State College. *Born:* 2/7/43. Mr Kuhn has been a maj contributor to the dev of the centralized hazardous waste disposal facility concept. Since 1971 he has been responsible for maj disposal site dev & remedial actions both in the US & Canada. He also has developed accident prevention progs, performed economic evaluations for waste disposal alternatives, & operator training progs. He has been called as an expert witness in environmental hearings, & lectures on hazardous waste disposal. *Society Aff:* AIChE, WPCF, NWWA, ASTM

Kuhn, E Michael
Business: PO Box 7258, Philadelphia, PA 19101
Position: V Pres & Genl Mgr Ethylene Prods. *Employer:* ARCO/Polymers Inc. *Education:* BS/Chem Engg/Univ of Pittsburgh. *Born:* Apr 1921. BS ChE Univ of Pgh; Cornell Univ. Officers Training Sch; with US Navy 1943-46; Lt sr grade, commanding officer, amphibious ship. Harvard Business Sch 1965. Chem Engr synthetic fuels prog US Bureau of Mines 1946-52; 1952-64 Koppers Co, Mgr of Pilot Plants & Admin Asst to V Pres & Dir of Res; ARCO/Polymers Inc (formerly Sinclair-Koppers Co) 1965- , V Pres & Mgr Ethylene Prods. Hon socs: Sigma Tau & Phi Eta Sigma. Presbyterian Elder; former sch dir; music & sports. *Society Aff:* AIChE.

Kuhn, Howard A
Home: 5408 Peach Dr, Gibsonia, PA 15044
Position: Prof. *Employer:* University of Pittsburgh. *Education:* PhD/Mech Engg/Carnegie-melon Univ; MS/Mech Engg/Carnegie-Mellon Univ; BS/Mech Engg/Carnegie-Mellon Univ. *Born:* Dec 1940; native of Pittsburgh Pa. Thesis title 'The Minimum Gauge Problem in Thin Strip Rolling'. Asst Prof 1966-71, Assoc Prof 1971-74 Dept of Met Engrg Drexel Univ; Assoc Prof of Met & Materials Engrg & Assoc Prof of Mech Engrg 1975-77, Prof of Materials Science & Engg & Prof of Mech Engg 1977- Univ of Pgh. Teaching, res & cons in metal deformation processing, failure analysis & materials engrg design; spec in powder forging processes, workability & ductile fracture in forming. Fellow ASM, Mech Working & Forming Div, Powder Met Activity, Process Modeling; APMI, Editorial Board; ASME, Production Engrg; SME. *Society Aff:* ASME, ASM, APMI, SME.

Kuhn, Paul A
Business: 222 S Riverside Plaza, Chicago, IL 60606-5965
Position: Partner. *Employer:* Greeley and Hansen. *Education:* MSCE/Civil Engg/Univ of WI; BS/Sanitary Engg/Univ of IL. *Born:* 3/21/31. in Chicago, IL. Engr Pacific Flush Tank Co 1954-55; Res Asst Univ of WI 1955-56; Engr & Hd San Design Sect Stanley Engr Co 1956-61; Engr & Proj Mgr Greeley & Hansen 1961-68, Assoc 1968-71, Ptnr 1971. Fellow ASCE; VP Tri-City Sect 1961; Pres IL Sect 1970-71; Chmn Ctte on Sects & District Councils 1970-71; Exec Ctte Admin Div 1973-74; Vice President 1981-83. Professional Activities Ctte 1981-83; Budget Ctte 1981-84, Chrmn 1982-83; Exec Ctte 1981-83; Ctte on Soc Honors 1981-83; Task Ctte on Relocation of ASCE Headquarters, Chrmn 1982-83; Ctte on Constitution and Bylaws 1983-87, Chrmn 1983-84. Professional Activities Planning Ctte, 1985-90. IL Engg Council 1971-74. Washington Award Commission 1974-81. ISPE; NSPE-PEPP. ACEC; Chmn Environ Ctte 1976-79. Central States Water Pollution Control Assn V Pres 1973-75, Pres 1975- 76, Dir to WPCF 1978-81. WPCF: Arthur Sidney Bedell Award, 1978 Constitution and Bylaws Ctte 1979-85, Chrmn 1982-85, Treasurer 1985-88, Executive Ctte 1985-88. AWWA, APWA, CSI, Dipl, Amer Acd Environ Engrs, Amer Underground-Space Assn. ACEC Environmental Ctte 1974-81, Reg PE 15 states. Ira O Baker Univ of IL 1954; Harry E Jordan Scholarship AWWA 1955. Chicago Civil Engineer of the Year 1986. Hobbies: Photography, gardening & fishing. *Society Aff:* ASCE, ACEC, APWA, AWWA, CSI, NSPE, WPCF, AUA, AAEE

Kuivinen, Thomas O
Home: 970 E Foothills Dr, Tucson, AZ 85718
Position: Former Mgr of Tech Engrg (retired). *Employer:* Cooper-Bessemer Co/Div Cooper Indus. *Education:* Mech Engg/Design/OH State Univ; B ME/Design/OH State Univ. *Born:* 1929. Ohio State grad. Early work engrg & testing compressors.

Kuivinen, Thomas O (Continued)
Dev Appl Mechs Dept. Prof degree Ohio St 1937. Sev yrs in design of engines & compressors through stress analysis, vibration & matl selection. On Ch Engr's staff starting 1938 & Mgr Res of Tech Engrg 1954; respon for res in matls, appl mech, gas compression, tech writing & computer engrg. Presented many papers on design analysis, fatigue testing & prod design. Was mbr ASM cttes on cast iron design & engine valve matls; also ASME engine governor specifications; 14 yrs Exec Ctte Diesel & Gas Engine Div, Chmn 1962; ASME Fellow; ASME Meritorius Service Award 1970. Retired Dec 1967. Enjoys music, golf & landscape painting in oil. *Society Aff:* ASME.

Kulacki, Francis A
Business: College of Engineering, Colorado State Univ, Ft. Collins, CO 80523
Position: Dean of Engrg and Prof, Mech Engg. *Employer:* Colorado State Univ. *Education:* PhD/Mech Engrg/Univ of Minn; MS Gas Engrg/Gas Engrg/IL Inst Tech; BS ME/Mech Engrg/IL Inst Tech *Born:* 05/21/42 Native of Baltimore, MD and attended Poly Inst in "A" Course. Taught and conducted res at Ohio State Univ 1971-1980 in mech engrg; joined Univ of DE in 1980 as Chrmn of Mech & Aerospace Engrg; joined Colo State Univ, 1986, as Dean of Engg. Specialized in thermal sciences energy engrg, with emphasis on heat transfer (thermal convection, convection in porous media, and mixed convection), nuclear reactor safety, nuclear waste disposal, catalytic combustion, and electrohydrodynamics. Contributed fundamental engrg heat transfer correlations for fluids with volumetric heating and in natural convection in porous media. More than 100 tech papers and reports. Exec Cttes of ASEE Mech Engrg Div and ASME Heat Transfer Div. Bd of Visitors of Swarthmore Coll. American Men and Women of Sci, Who's Who in Midwest. *Society Aff:* ASME, ASEE, AAAS, ΣΞ, ΠΤΣ, ΤΒΠ.

Kulcinski, Gerald L
Business: 1500 Johnson Dr, Madison, WI 53706
Position: Prof. *Employer:* Univ of WI. *Education:* BS/Chem Engr/Univ of WI; MS/Nucl Engr/Univ of WI; PhD/Nucl Engr/Univ of WI *Born:* 10/27/39. Sr Res Scientist, Radiation Damage Effects to Mtls, Battelle Northwest Lab, 1965-1972; Assoc Prof Nucl Engg 1972-4, Prof Nucl Engg 1974-present; Technical Chrmn, 2nd Topical Meeting Nuclear Fusion Tech, Richland, WA, 1976; Fellow Am Nuclear Soc, 1978; Curtis J McGraw Award, 1978; Outstanding Achievement Award - AM. Nucl. Soc. 1980. Grainger Prof of Nuclear Engrg, 1984. Dir, Univ of WI Fusion Engg Prog 1973-4, 1979-1984, Dir Inst Fusion Tech, 1984-present. *Society Aff:* ANS

Kulhawy, Fred H
Business: Hollister Hall, School of Civil and Env. Engrg, Cornell University, Ithaca, NY 14853-3501
Position: Prof *Employer:* Cornell Univ *Education:* PhD/CE/Univ of CA-Berkeley; MS/CE/Newark Coll of Engrg; BS/CE/Newark Coll of Engrg *Born:* 9/8/43 in Kansas and raised in NJ. Teaching or research at Newark Coll of Engrg, CA (Berkeley), Syracuse and Cornell. Area representative for geotechnical engrg at Cornell; also graduate faculty in geology. Professional practice with Storch Engrs and Raamot Assocs. Private geotechnical consultant since 1969. Fellow or member of 11 major societies. Society positions have included: student chapter advisor, section pres, chairman of national committees, etc. Honors include Chi Epsilon, Sigma Xi, ASCE Edmund Friedman Young Engrs Award for Professional Achievement, ASCE Walter L. Huber Civil Engrg Res Prize, and Fulbright Scholar. Author of 145 publications. *Society Aff:* ASCE, ASEE, ASTM, AEG, GSA, IEEE, IAEG, ISRM, ISSMFE, TRB, USCOLD

Kulieke, Frederick C, Jr
Home: 305 Smith Dr, Tallmadge, OH 44278
Position: Senior Design Engr. *Employer:* Midland Ross Co Tech Center. *Education:* BSME/Engg/IL Inst of Technology. *Born:* 7/22/18. in Chicago, IL. Apprentice draftsman, asst ME & ME of Amer Steel Foundries, Coupler Sect 1936-68; served on Coupler Manufacs Engrs Ctte 1955-68. Ensign in buOrdnance USNR 1943-46. Elected Fellow in ASME 1970. Machine design engr Akron Std Mold 1968-72. Maintenance & engg St Thomas Hosp becoming Dir of Dept 1976. Resigned St Thomas Hospital Oct 1978. Sr Design Engr at Midland Ross Co, Tech Center. PE State of OHE-017802. *Society Aff:* ASME.

Kulisz, Andrzej
Home: 2367 Pinewood Circle, Minneapolis, MN 55432-6250
Position: Manager *Employer:* Eaton Corp. *Education:* MS/Administration & Management/Maritime Academy; BS/Navigation Systems & Instrument./Maritime Academy *Born:* 03/01/45 Native of Poland. Dir and Chief Engr of the inertial navigation system international dev prog. Staff engr for Medtronic, Inc. in the first successful dev of an implantable, programmable drug infusion pump. Assumed current responsibility as mgr of a new engrg venture of Eaton Corp. to develop electronic controls and guidance for hydraulically propelled off-highway vehicles. Recipient of two international awards for accomplishments in the field of applied engrg and mgt of industrial operations. Mbr of the Standard Cttee on Electronics (NFPA). Married to wife Ursula. Father of three sons, Tad, Thomas and Jonathan. Enjoy flying, reading and music. *Society Aff:* NFPA

Kulkarni, Kishor M
Home: 25142 Edgemont Dr, Richmond Heights, OH 44143
Position: Section Hd. *Employer:* SCM Glidden Metals. *Education:* PhD/Human Factors/Engr Production; Univ of Birmingham, England; MSc/Human Factors/Engr Production; Univ of Birmingham, England; Grad Dipl/Ergonimics & Work Design/Engr Production; Univ of Birmingham, England. *Born:* 12/4/43. Since 1977 respon for powder met tool steel R&D with considerable mktg & mfg interface. 1966-77 with IIT Res Inst, Chicago in contract res. As Mgr of Metalworking & Foundry managed progs on isothermal furgoing squeeze casting, techno-economic analysis, plastics processing etc. Currently Chrmn of ASM Mech Working & Forming Div & Mbr of Tech Divs Bd. Formerly Chrmn of Mfg Tech Ctte of ASME Gas Turbine Div over 25 Tech papers, many talks 6 numerous reports. PE in IL orgininally from India. Wife Filipino business grad. Interests are tennis, skiing, chess, impressionist paintings, real estate & stocks. *Society Aff:* ASME, SME, ASME, APMI.

Kullay, Emery
Home: 62 Ash St, Englewood Cliffs, NJ 07632
Position: Pres. *Employer:* EMK Construction Consultants Inc. *Education:* MME/ME/NYU. *Born:* 7/1/32. From 1960 to 1973 Chief Hvac Engr with Devenco Inc NYC. From 1973 to present Pres & Owner of EMK Construction Consultants Inc NYC. Received Progressive Archtl Design Awards for IBM office bldg complex Gaithersburg, MD 1964. Underground garage, City of Rochester, NY 1967. *Society Aff:* NSPE, ASME, ASHRAE, SAME.

Kumar, K S P
EE 254, 123 Church St SE, Minneapolis, MN 55455
Position: Prof. *Employer:* Univ of MN. *Education:* PhD/EE/Purdue; MSEE/EE/Purdue; DIISc/EE/Indian Inst of Sci. *Born:* 5/12/35. in India, now settled in USA. Summer work at Honeywell, Sylvania, Rias Baltimore. Consultant to ind on control problems. Interests in process control, transportation systems and adaptive control. Enjoy gardening, photography and classical Indian music. *Society Aff:* IEEE, ASEE.

Kummer, Wolfgang H
Business: 2830 Victory Pkwy, Santa Monica, CA 90405
Position: Chief Scientist Lab. *Employer:* KZF, Inc. *Education:* PhD/EE/UC Berkeley; MS/EE/UC Berkeley; BS/Engg/UC Berkeley. *Born:* 10/10/25. in Schuyler, NB. Grad Schuyler HS 1938; Univ of Cincinnati College of Engg, Civil Engg 1943; Advanced Studies WA Univ 1945-45; Univ of Cincinnati 1952-53. Disting Engr Alumnus Award Univ of Cincinnati 1970. Reg P E 14 state; Reg Land Surveyor 3 states. Founder 1956, now Pres of 110-men KZF Engrs-Architects- Planners-Surveyors. ASCE: Fellow; all elective offices Cincinnati Sec & Dist 9 Council; Natl Dir 1969-72; Assoc chmn Natl Convention 1974; Chmn Mgmt Group "A" TAC-ASCE; Engr-

Kummer, Wolfgang H (Continued)
of-the-Year--Cincinnati 1978. Mbr: Amer Consulting Engineers Council. *Society Aff:* ASCE, ACEC.

Kummler, Ralph H
Business: Dept ChE & Met Engrg, Detroit, MI 48202
Position: Chairman & Prof Chem & Met Engrg. *Employer:* Wayne State Univ. *Education:* PhD/Chem Engg/Johns Hopkins Univ; BS/Chem Engg/Rensselaer Polytechnic Inst *Born:* Nov 1940. Staff Engr Genl Electric Space Sci Lab 1965-69, Assoc Prof of Environmental Engrg. Wayne State Univ 1969-75; Chmn Chem & Metallurgical Engrg Wayne State Univ 1974- ; Prof of Chem Engrg 1975- . Author of numerous papers and patents on chem kinetics. Engrg Soc of Detroit Young Engr of the Year 1974. Aiche Detroit Section ChE of year, 1981 Cons Engr Urban Sci Applications, Inc 1978- , Mbr AIChE, APCA, ACS, Fellow of Engrg Soc of Detroit, Fellow of AIC. *Society Aff:* AIChE, ACS, APCA, AIC

Kundert, Warren R
Business: P O Box 315, Harvard, MA 01451
Position: Consultant *Employer:* Industrial Resources *Education:* MS/EE/Northeastern Univ; BS/EE/Northeastern Univ *Born:* 1/17/36 Consultant in noise control and instrumentation. Formerly VP, Engrg Genrad Enviromedics Div and Gen Mgr for Acoustics and Signal Analysis, Genrad. Extensive experience in the development of measuring instrumentation. Active in the development of National (ANSI) and International (IEC and ISO) Standards for Noise Measurement and Instrumentation. US technical Advisor to IEC TC-29 Electroacoustics. President-Elect, Former Secretary and Dir of Inst of Noise Control Engrg. Mbr of Editorial Bd of Noise Control Engrg and Contributing Editor, Sound and Vibration. Elected Fellow Acoustical Society of America. Winner of IEEE GA&E Award. Elected to HKN. Author of many published papers on Instrumental and Measurement. *Society Aff:* INCE, ASA, IEEE

Kunka, Peter
Business: 2333 West Northern Ave, Phoenix, AZ 85021
Position: VP *Employer:* Lowry-Sorensen-Willcoxson Engr *Born:* 09/03/46 Born in Austria, raised in Brooklyn, NY. Graduated Brooklyn Technical HS. Designer for Jaros, Baum & Bolles, NYC (1975-73) specializing in Hilton and Inter-Continental Hotels worldwide. Since 1973 with Lowry-Sorensen-Willcoxson. Registered mech engr in AZ, 1978; CA, 1979. Expertise in Class 100 clean rooms for semiconductor industry, dehumidification, deionized water, process piping, etc, and major institutional, commercial projects. Pres Phoenix ASHRAE (1982- 83); Regional Chrmn ASPE (1980-81); Pres Phoenix ASPE (1978-80). Married Luba 1966; two sons, Peter Jr. and Andy. Enjoy my family, travel and fine food. *Society Aff:* ASHRAE, ASME, ASTM, NSPE

Kunkel, Paul F
Home: 960 Evergreen St, Emmaus, PA 18049
Position: VP. *Employer:* McTish, Kunkel & Assoc. *Education:* BSME/ME/Lehigh Univ. *Born:* 4/23/26. Native of Kutztown, PA. Served with US Army in the Infantry 1944-46. Joined Lehigh Valley Dairy in Allentown, PA. Where responsible for plant engg & later served as dir of engg ('64-'70). Joined A L Wesenberger Assoc in Allentown 1971. Responsible for municipal reps, proj mgt, & client relations as administrative asst to the pres. Assumed present position of vp, McTish, Kunkel and Assoc 1979, responsible for municipal services div, public relations, & business mgt. Past Pres of Lehigh Valley Chapter, PSPE. Active in civic & fraternal organizations. *Society Aff:* NSPE, ASME.

Kuno, Hiromu John
Home: 28009 Seashell Way, Palos Verdes, CA 90274
Position: Asst Div Mgr *Employer:* Hughes Aircraft Co, Microwave Products Div *Education:* PhD/EE/Univ of CA; MS/Engr/Univ of CA; BS/Engr/Univ of CA. *Born:* 3/27/38. Born in Osaka, Japan. Joined Hughes in 1969 & has developed the millimeterwave product line. Currently Asst Div Mgr, & Mgr Torrance Res Ctr, Microwave Products Div. Prior, with RCA, Sarnoff Res Ctr, & NCR, Electronics Div. In 1979 awarded Fellow, IEEE for contribution to the field generation & modulation of millimeter-waves. Published or presented more than 100 technical papers, holds several patents, organized or chaired many technical conf sessions & seminars. Mbr of IEEE-MIT Admin Committee. ITTEE. *Society Aff:* IEEE, ΤΒΠ, ΣΞ.

Kunreuther, Frederick
Business: One North Broadway, White Plains, NY 10601
Position: President. *Employer:* F Kunreuther Assocs Inc. *Education:* BS/Bus & Engg/MIT *Born:* Sept 25, 1916 Germany, to USA 1934, citizen 1941. Mass Inst of Tech 1938/41 BSChE & Business Admin, Tau Beta Pi. Shell Oil Co 1941-70: Ch Res Technologist 1950-54 in Houston respon for R/D in cat cracking & all other refinery conversion processes, fluidized bed sys, separation processes etc; Mgr Refinery Tech Dept Houston 1954-56; Head Office Mfg Tech Dept 1956-65, Mgr Process Engrg 1965-70 respon for all refinery & petrochem plant designs, refinery trouble shooting etc; took early retirement when co moved from N Y to Houston. F Kunreuther Assocs Inc 1970- : cons to major chem & petroleum refining co's in process engrg, process economics, proj dev, conceptual design, design review etc. *Society Aff:* AIChE, ACS.

Kunze, Otto R
Home: 1002 Milner, College Station, TX 77840
Position: Professor. *Employer:* Agri Engr Dept, TAMU. *Education:* BS/Agri Engg/TX A&M Univ; MS/Agri Engg/IA State Univ; PhD/Agri Engg/MI State Univ. *Born:* 5/27/25. Native of Texas. High School Valedictorian. WWII veteran, European Theater Campaign Ribbon, 2 bronze stars. Recipient TAMU Distinguished Honor Award. Married, 4 children. Agri & Indus Engr 1951-56 Central Power & Light Co; Assoc Prof TAMU 1956-61 - 1964-69; Prof 1969- . NSF Sci Faculty Fellow 1962 MSU. Author & co- author of over 53 sci & tech articles. Primary author of three chapters in technical books. Rice process engrg cons India--Kharagpur 1975, Pantnagar 1985; Taiwan--Taichung 1985. ASAE Dir 1972-74, 41 ctte years serv to ASAE at natl level. Mbr of the TX Air Control Bd; Appted by the Governor of the State of TX - 1979-89. Elected to the grade of "Fellow" in the Amer Soc of Agri Engrs, 1978. Registered Professional Engineer in Texas, Registration No. 24709. *Society Aff:* NSPE, ASAE, ASEE, TSPE.

Kuo, Franklin F
Business: DTACCS/ADP, Washington, DC 20301
Position: Asst Dir ADP/Teleprocessing. *Employer:* Office of Secy of Defense (ODTACCS). *Born:* April 1934. BS, MS, PhD from Univ of Illinois. Currently Asst Dir (ADP/Teleprocessing) Office of the Dir Telecommunications & Command & Control Sys, Office of the Secy of Defense Washington D C; on leave from the Univ of Hawaii where has been Prof of Elec Engrg since 1966; from 1960-66 with Bell Labs Murray Hill N J. Author or co-author of 6 books, many journal papers. V Chmn ACM/SIGCOMM. Fellow IEEE.

Kupchik, George J
88 Central Park West, New York, NY 10023
Position: Prof Emeritus *Employer:* Hunter College *Education:* Eng ScD/Sanitary Engg/NY Univ; MCE/Sanitary Engg/NY Univ; MS/Microbiology/NY Univ; BS/Biology-Chem/Brooklyn College. *Born:* 7/28/14. New York City. Sanitary Corps US Army WWII; Sanitary Engr Health Depts Yonkers & N Y C 1952-62; Ch Engr Sanitation Dept N Y C 1968-70; with Amer Public Health Assn as Dir of Environ Health 1962-68 & Deputy Exec Dir (Env) 1970-71; Chmn School of Health Scis 1974-75; Acting Dean 1980-81; Dir & Prof Environ Health Sci 1971-84. Bd of Tr Amer Acad Environ Engrs 1968-75, NYC Environ Control Bd since 1982. Reg P E NY & NJ. Mbr APCA, AWWA, WPCF, APHA, AIDIS, IAHA, ASME (SWD). Kenneth Allen Memorial Award, N Y Pollution Control Assn 1962. Distinguished Service Award Am Public Health Assn 1986. Certified Asbestos Investigator, City of

Kupchik, George J (Continued)
New York, 1987. *Society Aff:* APHA, APCA, AWWA, WPCF, AIHA, AIDIS, ASME.

Kupelian, Vahey S
Home: 4806 Falstone Ave, Chevy Chase, MD 20815
Position: Deputy for BMD (S&TNF) *Employer:* Office of Sec. of Defense (S&TNF) *Education:* SB/Mech-AE/MIT. *Born:* 6/23/12. Pratt & Whitney Aircraft 1939 dev of aircraft engines; to Goodyear Aircraft 1941, Chief of Propulsion and Flight Test Engrg and responsible for dev of F2G which was fastest WWII US airplane & final winner of Thompson Trophy Air Classic. To Wash 1949: Chief Engr for Missile Systems at Naval Weapons Plant, Chief Mechanics Branch of Advanced Res Projs Agency, & Asst Dir of Army's Advanced Ballistic Missile Defense Agency, Assoc Deputy Asst Secy of Navy (RE&S) in 1977, Chief Scientist of Army BMDPO in 1983. Current position in Office of Sec. of Defense: Strategic & Theater Nuclear Forces---Deputy for BMD. Recipient of Defense Dept's Meritorious Civilian Service Medal, Army's Decoration for Exceptional Civilian Service, and AIAA Missile Systems Award. *Society Aff:* AIAA.

Kupper, Charles J, Jr
Business: 1034 Millstone Rd, Somerville, NJ 08876
Position: Pres. *Employer:* Self-employed. *Education:* MSCE/Sanitary/NJ Inst of Tech; BSCE/Civil/NJ Inst of Tech *Born:* 9/5/29. Involved in all phases of engg related to water and wastewater treatment facilities. Experience includes Proj Mgt supervision and Resident Engg, preparation of comprehensive water and sewerage master plans, feasibility studies, evaluation studies, rules and regulations for water and sewerage systems, preliminary and final design for dams, water supply and treatment facilities, wastewater collection and treatment facilities, construction mgt, solid waste and transportation. Construction claims expert witness Analysis, Evaluation and Reports. A licensed PE and Land Surveyor in NJ. Lic PE, NY, PA & FL and a licensed Profl Planner in NJ. American Arbitration Assn and other local and civic organizations. *Society Aff:* AWWA, NSPE, NJSPE, ASCE, AAEE, WPCF.

Kuranz, Joseph H
426 Westminster Dr, Waukesha, Wis 53186
Position: Dir-Municipal Advisory Services *Employer:* Donohue & Assoc, Inc. *Education:* PhD/Engrg/Tri-State Univ; BS/Mech Engrg/Tri-State Univ; MS/Pol Sci/Univ of WI. *Born:* 03/15/20 BSME Tri-State Univ, MS Political Sci Univ of WI. US Merchant Marine prior to WWII; served with US Navy as Engrg Officer 1942-46; retired as Commander USNR 1970; Design Engr Remington Rand Corp 1946-47; Chief Engr Waukesha Water Utility 1947-57; Genl Mgr Waukesha Water Utility 1957-85. Pres Amer Water Works Assn 1969-70; Pres WI Soc of PE 1966-67. Tri-State Univ & Univ of WI Distin Serv & Outstanding Alumni Awards; AWWA Fuller, Honorary Mbr & Best Paper Awards; contrib to numerous pubs & periodicals. Pres Natl Soc of PE 1986-87. WI Soc of PE Engrs, Engr of the Yr Award 1979. Diplomate of the Amer Acad of Environ Engrs. Reg PE LS and Marine Engr. Conslt World Health Org, Public Admin Service, and Abt Assoc Inc. Dir-Municipal Advisory Services Donohue & Assoc 1985- . Hon Dr of Engg degree Tri-State Univ. *Society Aff:* AWWA, NSPE, AAEE, EAA.

Kurokawa, Kaneyuki
1015 Kamikodanaka Nakahara, Kawasaki Japan 211
Position: Managing Director. *Employer:* Fujitsu Labs Ltd. *Education:* PhD/Elec Engg/Univ of Tokyo; BS/Elec Engg/Univ of Tokyo. *Born:* Aug 14, 1928. Native of Tokyo Japan. Asst Prof at Univ of Tokyo 1957-63; Bell Telephone Labs 1959-61 & 1963-75; Fujitsu Labs 1975- . IEEE Fellow 1974. Pubs: 1 book An Introduction to the Theory of Microwave Circuits Academic Press 1968 & more than 30 tech papers on Microwave Circuits, Balanced Amplifiers, Path Length Modulators & Oscillators. *Society Aff:* IEEE, ACM, IECE.

Kurth, Frank R
Business: 334 Wellington St N, Hamilton Ontario Canada L8N 3T1
Position: Gen Supt Canada Wks. *Employer:* Stelco Inc *Education:* BSc/Met Science/McMaster Univ; P.E.. *Born:* 12/11/25. in Germany. Immigrated to Canada in 1953. 1958 - with the Steel Co of Canada, Limited as Metallographer. 1959 - to Canadian Drawn Steel as Met - later Mgr. Since 1967 - present position - responsible for Specialty Wire Mill & Specialty Fastener Plant. Served as Chrmn Canadian Inst of Mining 1964-1965. Past Chrmn ASM Ontario Chapter. Presently: Chrmn of Canadian Council ASM. Chrmn Advisory Committee - Energy Systems - Mohawk College. *Society Aff:* ASM, WA, APEO.

Kurtz, Max
Home: 33-47 91st St, Flushing, NY 11372
Position: Consultant (Self-employed) *Education:* BBA/Accounting/CCNY. *Born:* 3/25/20. Consulting engr; PE license, NY State. Author of five engrg books published by McGraw-Hill, including "Handbook of Engineering Economics" and "Handbook of Applied Mathematics for Engineers and Scientists" (in preparation). Major contributing author, "Standard Handbook of Engineering Calculations" (McGraw- Hill); sole author of Civil Engrg & Engrg Economics sections. Has presented seminars in Engrg Economics, in the US and the Netherlands. Since 1961, instr in review courses for PE examinations, covering Mech Engrg, Civil Engrg, & the basic sciences, for ASME, NYU, NY Telephone, City of NY, & other major organizations. Recipient of Honor Award by Kings Cty Chpt, NYSSPE. *Society Aff:* NSPE.

Kurzweg, Ulrich H
Business: Dept of Engg Sci, Gainesville, FL 32611
Position: Prof. *Employer:* Univ of FL. *Education:* PhD/Phys/Princeton Univ; MA/Phys/Princeton Univ; BS/Phys/Univ of MD. *Born:* 9/16/36. Born in Jena, Germany. Came to US, 1947, naturalized 1952; s Hermann Herbert and Erna Herta (Michaelis) K; BS Univ MD, 1958; MA (Woodrow Wilson fellow), Princeton, 1959, PhD in Phys, 1961; m Sophia Speth, Dec 21, 1963; 1 du, Tina. Sr theoretical physicist United Aircraft Res Labs, East Hartford, CT, 1962-68; prof of engg sciences 1976-present Univ of Florida; assoc prof engg sci and mechanics Univ FL, Gainesville, 1968-75; adj asst prof Rensselaer Poly Inst Hartford Gra Ctr, 1964-67, adj assoc prof 1967-68. Fulbright grantee, 1961-62; recipient Sigma Tau - Tau Beta Pi award for excellence in undergrad engg teaching, 1970, 73. NASA certificate of recognition for the dev of large focal length parabolic mirrors. Mbr Am Phys Soc, NY Acad Scis, AAS, Sigma Xi, Sigma Tau. Contbr articles to sci jours and revs. Holds US patents on thermal pump and diffusional separation of gases by oscillations. *Society Aff:* AAAS, APS, NYAS.

Kurzynske, Frank
Business: , TN 37212
Position: President. *Employer:* Kurzynske & Assocs. *Born:* Nov 6, 1925 Fond du Lac Wisconsin. Texas A&M 1943 BSE, Tenn Tech 1949; MSE Kansas State 1950. Married Edna B Trent Nashville Tennessee, parents of 5 sons. Employed as a design engr with Nashville Bridge Co 1950-51; Principal In- House Designer, Plant Engrg Dept AVCO Corp 1951-55; Mbr E C Horn engrg firm 1955-56; established firm of Kurzynske & Assocs Nov 1, 1956 - principal office in Nashville Tennessee & in Aug 1972 opened branch office in Knoxville Tennessee. Mbr of NSPE, CEC, CSI & ASHRAE. Author or sailing stories and technical sailing papers that have been published in the USA and the United Kingdom. Also listed in Who's Who in America.

Kushner, Harold J
Div Appl Math, Providence, RI 02912
Position: Professor. *Employer:* Brown Univ. *Education:* PhD/Elec Engg/Univ of WI; MS/Elec Engg/Univ of WI; BEE/Elec Engg/CCNY. *Born:* 1933. Fellow IEEE. Spec in statistical problems in Control & Sys. Nonlinear filtering, numerical and approximation methods, optimization. Cons to Govt & Indus. Author 5 books & numerous articles. *Society Aff:* IEEE, SIAM, ORSA, IMS, IMS.

Kushner, Morton
Home: 16850 NE 25th St, Bellevue, WA 98008-2327
Position: Asst. Director Engrg Technology. *Employer:* Boeing Aerospace Co. *Education:* MA/Chemistry/Temple Univ; CE/Civ Engg/VPI; BA/Chemistry/Temple Univ. *Born:* 10/23/22. Technical Expertise includes development & management of large parts, materials & process organization for electrical/electronic & structural design. in metals & composites. 35 yrs experience in Aerospace military and NASA Design & Delivery of Hardware for Boeing Aerospace & Boeing Vertol Companies. Past National Pres of SAMPE & Member of Bd of Directors. SAMPE Fellow. Most recent experience in new light-weight aluminum alloys (Alhi) & high modulus organic composites (graphite/modified epoxy). *Society Aff:* SAMPE.

Kusuda, Tamami
Business: NBS (Natl Bureau of Stds), Gaithersburg, MD 20899
Position: Guest worker *Employer:* NBS (Natl Bureau of Stds). *Education:* PhD/ME/Univ of MN; MS/ME/Univ of WA; BS/Precision Engg/Tokyo Univ. *Born:* 6/24/25. Native of Seattle WA. Taught in Okayama Univ, Seton Hall Univ & Geo Wash Univ. Proj & staff engr for Worthington Air-Conditioning Co during 1955 through 1962 in air conditioner & compresser design. With Natl Bureau of Stds since 1962-86. Responsible for bldg heat transfer & energy analysis res. Assigned current position as Chief of Thermal Analysis Prog in 1974. Consultant to US Department of Commerce & Natl Technical Information Service on Japanese Technical Literature Act of 1986-Jan 1987. Reg PE in the state of NJ & MD. Recipients of US Dept Commerce Silver and Gold Medals Award. Wolverline award and Crosby Field award of ASHRAE for best technical papers of 1956 & 1976 respectively. *Society Aff:* ASME, ASHRAE, AAAS.

Kuzmanovic, Bogdan O
Business: Dept of Civil Engrg, Lawrence, KS 66045
Position: Prof *Employer:* Univ Of KS *Education:* Dr. Sci Tech/Structures/Serbian Academy of Sciences, Yugoslavia; MS/Structures/Univ of Beograd, Yugoslavia; Dipl Eng/CE/Univ of Beograd *Born:* 7/16/14 in Beograd, Yugoslavia. Worked in Ministry of Trans prior to WWII. Served with Army Corps Engrs (Yugoslavia) 1941-45. Sr design Engr with MOSTOGRADNJA, Beograd till 1951. Prof of Steel Structures, Univ of Sarajevo, Yug till 1958. Dean School of Engrg, Univ of KHARTUM, Sudan till 1965. Prof of Civ Engrg, Dept of Civ Engrg, Univ of KS, Lawrence KS till present. Consultant to Howard, Needles, Tauman and Bergendoff 1968-1971, to BEISWENGER, HOCH and Associates 1979 till present. Past Pres Kansas Section of ASCE. *Society Aff:* ASCE, IABSE

Kvalseth, Tarald O
111 Church St SE, Minneapolis, MN 55455
Position: Prof *Employer:* Univ of MN. *Education:* PhD/Ind Engg/Univ of CA; MS/Ind Engg/Univ of CA; BSc/ME/Univ of Durham. *Born:* 11/7/38. Native of Telemark, Norway. After receiving his PhD from the Univ of CA, Berkeley, 1971, he has been an Asst Prof of Ind and Systems Engg at GA Inst of Tech and a Sr Lecturer and hd of the Div of Ind Mgt at the Norwegian Inst of Tech. He joined the Univ of MN in 1979 where he is a Prof of Mech Engrg and the Dir of the Indus Engrg Div. His primary interests are in the area of human factors engrg (ergonomics). Council mbr of Nordic Ergonomics Soc 1977-1979 and Intl Ergonomics Assoc 1977-1979, VP of Intl Ergonomics Assoc 1982-85. *Society Aff:* IIE, IEEE, HFS, ES, AAAS, ΣΞ.

Kwatra, Subhash C
Home: 2845 Quail Run Dr, Toledo, OH 43615
Position: Assoc Prof *Employer:* Univ of Toledo *Education:* PhD/EE/Univ of South FL; MS/EE/Birla Inst of Tech & Sci-India; BS/EE/Birla Inst of Tech & Sci-India *Born:* 11/12/41 Native of India. Active in Satellite Communication Res. Supervising several grad students. Developing a communications program at the Univ of Toledo. Participate in Natl and Intl Conferences on Communications and Digital Signal processing. Have a number of publications. Enjoy all sports, especially tennis. *Society Aff:* IEEE

Kwiatkowski, Robert W
Home: 6 Sheffield Rd, Stoneham, MA 02180
Position: VP *Employer:* Chas T. Main, Inc *Education:* MS/CE/Northeastern Univ; BS/CE/Northeastern Univ *Born:* 06/22/32 in Chelsea, MA and settled in Stoneham, MA. Participated in several community activities, such as Scouting, Athletics and Parent-Teacher Leagues. Served in US Army 1955-57. With MAIN since 1954. Started as cvl designer and progressed to cvl engr, involved in cvl soils and structural design, hydraulics and hydrology, specifications and power studies for Hydro projects. Has been Project Engr for various water resources projects, including analyses, planning, studies and design. Served Project Mgr of several pumped storage developments, domestically and internationally. Was Mgr of Projects responsible for performance scheduling and staffing of all major projects. Currently VP of Hydro Power & Water Resources Div of MAIN, responsible for all div operations consisting of hydroelectric, water resources & agro-systems projects *Society Aff:* ASCE, BSCE, USCOLD, AWRA, USCID.

Kwok, Chin-Fun
Home: 5109 Philip Rd, Annandale, VA 22003
Position: Dir of Arch & Engg. *Employer:* Veterans Admin. *Education:* MSCE/CE/Univ of PA; BSCE/CE/St John's Univ; BSME/ME/Case Inst of Tech. *Born:* 2/14/26. Have been with Veterans Admin since 1962 & am in charge of working drawings and specifications for hospital & cementery construction. *Society Aff:* ASCE, ASHRAE.

Kyanka, George H
Business: S U N Y Coll Env Sci, Syracuse, NY 13210
Position: Prof - Dept Chrmn *Employer:* State Univ of New York.-ESF Syracusse *Education:* PhD/Mech Eng/Syracuse Univ; MS/Mech Engg/Syracuse Univ; BS/Mech Engg/Aero Eng/Syracuse Univ. *Born:* July 1941. Taught Mech Tech at Onondaga Community Coll; now teaching Wood Prods Engrg at SUNY Env Sci & Forestry; Cons on prods liability & wood prods; Dir of the SUNY Env Sci & Forestry prog in training minority & disadvantaged students in engrg & physical sci 1971-1979. Recipient of SUNY Chancellors Award for excellence in undergrad teaching in 1973. Selected by NATO 1975 as an outstanding res worker in wood sci & engrg. Outside activities incl a great variety of sports & handcraft work. *Society Aff:* ASME, ASTM, FPRS, SWST, SEM

Kydd, Paul H
Business: P O Box 6047, Lawrenceville, NJ 08648
Position: Vice President Technology. *Employer:* Hydrocarbon Res Inc. *Education:* PhD/Chemistry/Harvard; AB/Chemistry/Princeton. *Born:* Nov 25, 1930. Joined Hydrocarbon Res Inc as V Pres-Tech Dec 1, 1975; respon for the R/D Center of the co at Trenton N J; served as Mgr of the Chem Processes Branch at the Genl Elec Corp R/D Center in Schenectady New York 1966- 75; 1957-66 pursued individual res at Genl Elec in the field of combustion, gas turbines, heat exchanger dev, flame chemistry & physics, molecular beam chemistry & mass spectrometry. Author of 25 tech papers & 20 pats. Mbr of Phi Beta Kappa, Sigma Xi, AIChE & AAAS. Chairman Central Jersey section AIChE 1980- 81. *Society Aff:* AIChE, AAAS.

Kydonieus, Agis
Home: 1409 2nd Ave, New York, NY 10021
Position: Pres *Employer:* Hercon Labs Corp; a subsidiary of Health Chem Corp. *Education:* PhD/ChE/Univ of FL; BSCE/ChE/Univ of FL. *Born:* 4/22/38. Native of Samos, Greece; taught chem engg at Cooper Union 1968-1971; Asst Dir of Biomedical Engg with Baxter Labs in charge of Blood Bag & Intravenous Solution Bag Depts, corporate packaging, pilot plant facilities, polymer dev and special projs. In present position since 1974 in charge of Health Chems. Hercon Lab Subsidiary including R & D, production, QC, govt regulations, marketing. Editor of seven books & over 100 publications and patents. *Society Aff:* AIChE, SPE, CRS, AAPS.

Kydonieus, Agis F
Business: 1107 Broadway, New York, NY 10010
Position: Exec Vice Pres *Employer:* Herculite Products Co. *Education:* PhD/ChE/Univ of FL. *Born:* April 1938. PhD Univ of Florida. Native of Samos Greece. Chem Engr with Union Carbide & Union Camp before obtaining PhD & taught Chem Engrg at Cooper Union 1968-70; Pres & Principal of Chemtech Inc 1968-71 spec in polymer mfg & adhesive processing; Asst Dir of Corp BioMedical Engrg Baxter Labs 1973-74 in charge of corp R/D incl blood bags, intravenous solution bags, packaging, pilot plant facilities, polymer dev & processing, inhalation therapy & special projs; joined Health-Chem 1971; V Pres of the Hercon Div since 1974; respon for Div P & L, prod, quality control, R/D, new prod marketing, EPA regulations. Exec VP since 1979. *Society Aff:* AIChE, ACS, ESA, SPE.

Kyle, Benjamin G
Business: Durland Hall, Manhattan, KS 66506
Position: Prof *Employer:* KS State Univ *Education:* PhD/ChE/Univ of FL; MS/ChE/Univ of FL; BS/ChE/GA Tech *Born:* 12/4/27 Native of Atlanta, GA. Worked for Monsanto 1950-53. With KS State Univ since 1958. Worked summers for DuPont, Dow Chem Co and Phillips Petro Co. Enjoys gardening, fishing, hiking, reading, classical music and opera. *Society Aff:* AIChE, AAAS, ACS.

Kyle, William D, Jr
1425 West Dean Road, Milwaukee, WI 53217
Position: Ret. (Chrmn of the Bd & Pres) *Education:* BS AE EE/Electrical Eng/Cornell *Born:* May 18, 1915 Milwaukee; s. William Davidson & Margaret (Adams) K; m. Norma Timberman Feb 8, 1941; children: Susan, Nancy, Julie, Wendy, Robin. Grad Cornell 1936. Engr Wisc Elec Power Co Milw 1936-37; Engr Line Material Co South Milw Wisc 1937-39; Exec V Pres Milw 1947-49; Pres 1949-56 (div McGraw Elec Co 1949-56); Pres Kyle Co 1939-47 (merged with Line Material Co 1948); V Pres, Dir McGraw Elec Co Chicago resigned 1956; Pres Kyova Pipe Co Milw & Ironton O 1956- ; Chmn, Dir Kyle Co Milw; Chmn Bd, Ch Exec Officer Congoleum Corp (Wisc) Retired 1980; Pres, Dir K G Corp, Marietta Res & Investment Co; Dir Dumore Co Racine Wisc Retired, Wehr Corp (Retired), First Wisc Trust Co, First Wisc Corp (all Milw), First Wisc Natl Bank Retired, Water Pollution Control Corp Retired. *Society Aff:* AIEE

Laaspere, Thomas
Business: Thayer School of Engg, Hanover, NH 03755
Position: Prof. *Employer:* Dartmouth College. *Education:* PhD/Wave Propagation/Cornell Univ; MS/Wave Propagation/Cornell Univ; BS/EE/Univ of VT. *Born:* 3/17/27. Worked as a res assoc at Cornell Univ 1960-1961 on the Arecibo Radar Proj. Has taught engg at Dartmouth since 1961. In the 1960's and early 1970's participated in space res with VLF-LF receivers flown on the ogo series of spacecraft. Has also worked on res and dev projs in the area of elec load mgt. *Society Aff:* ΣΞ.

Labadie, John W
Business: Engrg Research Ctr, Foothills Campus, Colorado State University, Fort Collins, CO 80523
Position: Prof *Employer:* CO State Univ *Education:* PhD/Operations Research/Univ of CA-Berkeley; MS/Water Res Engrg/UCLA; BS/Engrg/UCLA *Born:* 11/29/42 Teaching experience in: water resource systems analysis, urban water management, optimization methods, and computer-aided engrg. Principal investigator for 35 research projects totaling over $1.5 million as funded by the World Bank, Natl Science Foundation, US Bureau of Reclamation, US Dept of Agriculture, the USDOE and other agencies. Consultant with U.S.A.I.D., US Army Corps of Engrs and the World Bank. Mgr of 30 lecture videotape course on: "Management of Water Resources: A Systems Approach," with training courses delivered in Bangladesh, Belgium,, Dominican Republic, Egypt, the Philippines, the United Kingdom, and the U.S. Author of over 130 journal articles, papers and research reports. Co-Winner of Amer Soc of Agric Engrs Paper Award. *Society Aff:* ΣΞ, AGU, ASCE, ORSA

Labate, Samuel
Business: 10 Moulton St, Cambridge, MA 02238
Position: Chmn of the Bd. *Employer:* Bolt Beranek & Newman Inc. *Education:* MS/Acoustics/MA Inst Tech; AB/Math/Lafayette College. *Born:* Dec 1918. Native of Easton Pa. Taught math 1940-41 at Univ of Penna. Worked for E I duPont 1941-42. US Army 1942-46. MS EE MIT 1948. Joined a co (Bolt Beranek & Newman Inc) after MIT as a cons engr in acoustics. Assumed a pos of Div Dir, Exec V P, Pres & in 1976 became Chmn of the Bd. *Society Aff:* ΦBK, ΣΞ.

Labbe, Leo P
Home: 45 Lafayette Lane, Basking Ridge, NJ 07920
Position: Engrg Mgr. *Employer:* AT&T. *Born:* June 1933. MSEE Northeastern Univ, BSEE Univ of New Hampshire. Air Force 1951-55. Mbr Tech Staff Bell Tele Labs 1959-72. Spec in Design of Solid State Broad Band Feedback Amplifiers, Long Haul Coaxial Cable Sys Line Repeaters & Line Regulators. Respon for Long Haul Coaxial Cable Sys Field Eval & the BTL Planning Function for Nationwide Sys Implementation. Assumed present respon as Engrg Mgr 1972. Respon for Std Design Methods & Transmission Objectives for the Bell Sys Switched Network & Private Line Services. Mbr Tau Beta Pi. Enjoy tennis, sailing, skiing.

LaBelle, Jack E
Home: 7 Elgin Pl, Apt 102, Dundin, FL 33528
Position: Chief Metallurgist - Retired. *Employer:* GM Corp - Detroit Diesel Allison Div. *Education:* BS/Chem Engg/MI State Univ. *Born:* Nov 1914 Frankfort Mich. Joined Detroit Diesel Engine Div G M 1937 as chemist, 1939 metallurgist. US Army Europe 1942-45. Rejoined Detroit Diesel as Asst Ch Metallurgist 1946, Ch Metallurgist 1950. From 1958-60 also chmn - G M Met Ctte. Prin engrg contrib involve Fastener Dev, Cast Iron Met & Devs for Fatigue Prevention. Current Respon Member-Technical Board, Kolene Corp. Received 1969 Natl Award of Merit ASM; elected Fellow ASM 1970. Hobbies: sailing, ornithology & history. *Society Aff:* ASM.

LaBoon, John F, Sr
Home: 60 Academy Ave, Pittsburgh, PA 15228
Position: Exec Dir & Chief Engr-ret. *Employer:* Allegheny County Sanitary Authority. *Education:* CE/Thesis-San Engg/Carnegie Inst of Tech; BSCE/Straight CE/Now Carnegie- Mellon Univ. *Born:* Native of Pittsburgh Pa. Partner Chester Engrs, Cons Engrs, Pgh. Dir of Works Dept Allegheny County, Pgh before WW II. Colonel, Military Governor WW II. Chmn of Bd & Ch Engr during planning & design phases, Exec Dir & Ch Engr during const & oper phases, Allegheny County Sanitary Auth. Cons San Engr since retirement. Hon Mbr ASCE. Life Mbr C-M Univ. *Society Aff:* ASCE, ESWP, APWA, WPCF, AWWA.

Labosky, John J
Business: 455 W. Fort St, Detroit, MI 48226
Position: Associate *Employer:* Smith, Hinchman & Grylls Assocs, Inc *Education:* Juris Doctor/Law/Univ of Detroit-Sch of Law; BSAE/Arch Engr/PA St Univ; Engrg Study/Engrg/John Carroll Univ *Born:* 08/24/48 Senior Engr and Assoc with AE firm of Smith, Hinchman & Grylls Assocs Inc. Currently Project Mgr for design of $100 million Defense Intelligence Analysis Ctr in Washington, DC. Formerly Mech Project Engr and Mech Discipline Head for H & G's Energy, Conservation & Research Div. BAE Penn State 1971; Juris Doctor, Univ of Detroit 1978. Completed summer law quarter in residence at Oxford Univ, 1977. Registered PE, mbr of Michigan and Federal Bar. Selected "Young Engr of Year" for 1980 by Detroit Chapter, MSPE. Selected Juror for 1978 OCF Energy Conservation Award Prog. Mbr of Bd and VP of Detroit Chapter, MSPE. *Society Aff:* NSPE, MSPE, TBΠ, ABA, ESD

Lach, Alexander A
Home: 20 Apache Ct, East Brunswick, NJ 08816
Position: Project Engr. *Employer:* Middlesex Cty Sewerage Auth. *Education:* BS/Civil Eng/Lafayette College. *Born:* March 23, 1932 South River N J. BSCE Lafayette Coll. US Army 1953-55. 1957-58 Office Engr with Metcalf & Eddy Inc

Lach, Alexander A (Continued)
during const of the Mddx Cty Sewerage Auth Treatment Plant (78 M G D) & Trunk Sys (27 miles). 195974 Plant Supt of Treatment Plant & Trunk Sys. 1974- , Proj Engr const of 120 MGD UNOX Activated Sludge Treatment Plant & Trunk Sys Expansion (27 miles of 60 inch-132 inch pipelines) & 3 pumping stations (340 MGD to 5 MGD). Lic P E NJ & Lic S-1 Treatment Plant Operator NJ. Mbr Natl Soc of Prof Engrs, Amer Soc of Civil Engrs, NJ Dir of the Water Pollution Control Fed, P Pres of the NJ Water Pollution Control Assn, NJ Bd of Examiners on Licensing of Supts of Water & Wastewater Treatment Plants & NJDEP Commissioner's Adv Ctte on Water & Wastewater Oper Training. *Society Aff:* NSPE, WPCF, ASCE.

Lachman, Walter L
Business: Biddeford Indus Pk, Biddeford, ME 04005
Position: President. *Employer:* Fiber Materials Inc. *Born:* June 1931. MS Matls Sci Univ of Calif; BS Univ of Pittsburg. US Army in Korean War. Prior to founding Fiber Materials Inc in 1969 worked at Inland Steel, North Amer Aviation, Ampex, HITCO, AVCO & Materials Inc. Dev novel 3- dimensional weaving tech for manufacture of quartz & graphite fiber reinforcements leading to estab of Fiber Materials Inc. Aluminum graphite patent. Pres of corp 1969- . Fellow AIAA & P Chmn & organizer of Boston Chap SAMPE.

Lackey, James F, Jr
Home: 130 Byron Dr, Pleasant Hill, CA 94523
Position: Chief Engineer Water Resources *Employer:* Morrison Knudsen Engineers Inc *Education:* BS/Civil Engg/OK State Univ. *Born:* 1/3/32. in Tatum, NM. Engr Morrison Knudsen Sedalia MO 1962-63; Asst Civil Engg Intl Engr Co Inc San Francisco 1963-64, Assoc Civil Engr 1964-67, Civil Engr 1967-69, Sr Civil Engr 1969-73, Principal Engr 1973-74, Ch Civil engr 1974-77, Chief Engr 1978-82, Design Mgr 83-86, Served to Capt USA 1952-59, decorated Commendation Medal with oak leaf cluster. Kavanaugh Community Bldg Award Scholar 1961. Reg P E CA, OK, TX, NM, AZ, NB, WA, UT, ID, CO. Mbr SAME, ASCE, ACI, AREA, USCOLD, Natl & CA Soc of Prof Engrs (P Dir, Chap Pres). Morrison-Knudsen Engineers Inc. Chief engr water resources 1987. *Society Aff:* ASCE, ACI, AREA, NSPE, SAME, USCOLD.

Lackey, William M
Home: 3713 W 30th Terr, Topeka, KS 66614
Position: Dir of Operations *Employer:* KS Dept of Transp *Education:* MS/Transp Engr/KSU; BS/Civil Engr/KSU; AA/Pre-Engr/Hutchinson Comm Coll *Born:* 11/26/39 Native of Hutchinson, KS. Worked for KS Dept of Transp since grad in 1963 as Project Engr, Resident Engr, Asst Engr of Constr, District Engr, Chief of Const & Maint and Dir of Operations since 1983. Have held all chapter offices and state offices including Pres in 1984-85 of KS Engrg Soc. Active in KSU athletic booster clubs, and officer in local credit union. Married with three daughters. Enjoy traveling and sports. National Dir, NSPE/KS. Mmbr, TRB Comm. *A3C01-Maintenance & Operations Mgmnt. Society Aff:* NSPE, ASCE

Lacy, Floyd P
Home: 4416 Thorp Rd, Knoxville, TN 37920
Position: Asst Mgr of Engg Design. *Employer:* TN Valley Authority. *Education:* BS/Civ Engg/Univ of KY. *Born:* 10/6/16. Native of Hopkinsville, KY. With Memphis Dist of US Army Engrs in 1939. On active duty in the US Army serving with the amphibious engrs in Europe and in the Pacific; discharged as Capt. With TN Valley Authority since 1939. Served as chief civ engr 1969-1973. Assumed present position of Asst Mgr of Engg Design in 1973, assisting in overall exec mgt of div of engg design & specifically responsible for coordinating activities of architectural design, civil, elec, & mech engg & design branches, & dam safety. Mbr of US Dept of Interior Teton Dam failure review group. Enjoy TN walking horses. *Society Aff:* USCOLD, ASCE.

Lacy, Robert R
1321-8th Ave No, Great Falls, MT 59401
Position: Owner. *Employer:* Lacy's Structural Engineering. *Born:* 1934. Raised in Montana where he received BS 1957 & MS 1964 in Civil Engrg from Montana State Univ. Has worked mostly for cons firms & in the field of structural engrg. In 1976, started own cons firm. Although a mbr of sev engrg orgs, principal organizational activity has been in ASCE, in which he has held sev offices, the highest of which is Sect Pres of the Montana Sect 1973-74.

Lacy, William J
Home: 9114 Cherry Tree Dr, Alexandria, VA 22309
Position: Conslt Industrial Pollution Control *Employer:* Self *Education:* BS/Chem/Univ of CT; DSc/Chem/Paul Sabatiev Univ. *Born:* May 1928. Completed course requirements for an Engrg PhD; post grad studies at NYU, Oak Ridge Inst. Nuclear Studies, Mich St Univ, Univ of Mich, Dr Sci (chemistry) Honary Paul Sabatiev U. Toulouse, France 1983. Reg PE Washington, D.C. over 153 pubs, contrib to twelve textbooks. Engr R&D Labs Corps of Engrs 1952-56; Oak Ridge Natl Lab. 1957-58; Exec Office of the Pres (OCD) 1958-62; Office of Secy of Defense (Pentagon R&D) 1962-67; FWQA, FWPCA, 1967-70 Environ Protection Agency 1970-84. Intl Lecturer. Dir Environ Div AIChE Fellow AIChE 1986; Diplomate Amer Acad of Environ Engrg; Bd of Dirs Intl Ozone Assoc, Mbr Tech Adv Bd Water & Wastewater Equip Mfg Assn; Policy and Planning Bd of the EPA Advance Environ Engrg Res Ctrs; EPA Bronze Medal 1983 US Govt Distinguished Service Medal 1984; Amer Chem Soc, Zero Pollution Discharge Symposium, Chrmn or co-chrmn of 31 conferences and symposia, conslt to France, Canada, Italy, Japan, Thailand, India, Poland, Egypt, Sweden, USSR and UNEP, OECD. Editorial Advisory Bds. "Industrial Water Engineering–, "Pollution Engineering" –Environmental Progress–, "Environmental Protection Technology" and "International Environmental Journal–. Dir EPA Indus R&D Div 1970-75, Principal Engr. Sci Adv EPA 1975-79, Dir EPA Water & Hazardous Material Div, 1979-84. Currently conslt to govt and indus on industrial and environ problems. *Society Aff:* AIChE, AAEE, IOA, WPCF, ASTM.

Lacz, John A
Business: 662 Goffle Rd, Hawthorne, NJ 07506
Position: Principal *Employer:* LAN Assocs, Inc *Education:* ME/Mech and Civil/Stevens Inst of Tech *Born:* 8/22/32 *Society Aff:* APA, ASPO, NSPE, CEC, ASTM

Ladd, Charles C
Business: Dept of Civ Engg., MIT, Cambridge, MA 02139
Position: Prof. Civil Engrg. *Employer:* MIT. *Education:* ScD/Geotechnical Engr/MIT; SM/Civ Engr/MIT; SB/Bldg Engr & Const/MIT; AB/Phys/Bowdoin College. *Born:* 11/23/32. Brooklyn, NY. MIT Dept of Civ Engr Faculty since 1961, specializing in Geotechnical Engr; effective teaching award, 1980. Dir, Ctr for Scientific Excellence in Offshore Engr. Fellow ASCE (Huber res prize, 1969; Croes Medal, 1973; Norman Medal, 1976 Terzaghi Lecture, 1986). Past pres BSCES. Gen Reporter, 9th ICSMFE, Tokyo, 1977; Co-Gen Reporter, 11th ICSMFE, San Francisco, 1985 Past Chrmn, Concord Dept Public Works. Consultant on soft ground construction projects in U.S. and abroad. Elected NAE 1983. Society Aff: ASCE, ASEE, ASTM, NSPE, NAE. *Society Aff:* ASCE, ASEE, ASTM, NSPE, NAE, CGS, BGS

Ladd, Conrad M
Business: 1780 S. Bellaire, Suite 809, Denver, CO 80222
Position: Chairman & CEO *Employer:* Senior Management Consultants Inc. *Education:* BSME/Steam Power/University of Michigan *Born:* 12/16/26 Native of Vermont and Michigan. Served in Navy in WWII. Design and development engineer on Nautilus submarine reactor system components and Fermi breeder reactor cove design and commercial prototype plant. Nuclear product manger for Brush Beryllium, subsequently marketing manager at Atomics International and Stone & Webster Engineering Corp. VP and Dir of Subsidiaries of Stone & Webster and Stearns-Roger Engrg Corp., where he served as Business Development Manager. Founder, Chairman & CEO, Sr Mgmt Consultants, Inc. with 13 engrg professionals. He is Chairman, ASME Energy Cttee, 1987-9; formerly Chairman, ASME Power

Ladd, Conrad M (Continued)
Div Exec Cttee and Task Groups on Acid Rain and on Clean Coal Tech. Enjoys sailboat racing, skiing, hunting. *Society Aff:* ASME

Laengrich, Arthur R
Business: Ortloff Engineers, Ltd, 2000 Wilco Building, Midland, TX 79701
Position: President & CEO *Employer:* Ortloff Engineers, Ltd. *Education:* BS/Chem Engg/OK State. *Born:* Feb 1933. BS Chem Engrg Oklahoma State Univ. Reg P E. Production Engr Union Carbide Chem Co through 1964. Var assignments incl Mgr of Proj's & Mgr of Engrg with Elcor Chem Corp through 1970. With The Ortloff Corp, an engrg & const firm, since 1970. Currently President & CEO of a subsidiary, Ortloff Engineers, LTD. Special expertise in technology licensing & the design & oper of natural gas treating, sulfur recovery, & pollution control sys. Active & held offices in AIChE Local Chap, other tech soc's. *Society Aff:* AIChE, APCA, NSPE

Lafferty, James F
Business: Univ of KY, Rose St, Lexington, KY 40506
Position: Prof & Director *Employer:* Wenner-Gren Research Lab *Education:* PhD/Nuclear Engr/Univ of MI; MS/Nuclear Engr/Univ of MI; MS/ME/Univ of So CA; BS/ME/Univ of KY *Born:* 12/23/27 Native of Bowling Green, KY. Served with 25th Infantry Div in Korea 1950- 51. Design Engr with Hughes Aircraft Co 1955-57. Faculty of Coll of Engrg, Univ of KY since 1958. Assumed current responsibility as Dir of Wenner-Gren Research Lab in 1969. Responsible for supervision of approximately 45 research personnel active in Biomedical Engrg Research. Primary research in musculoskeletal biodynamics. Responsible for Biomedical Engrg academic program. Consultant for protective sports equipment and vehicle crashworthiness. *Society Aff:* ASME, ASEE, ORS

Lafferty, James M
1202 Hedgewood Lane, Schenectady, NY 12309
Position: Consultant *Education:* PhD/Elec Engr/Univ of MI; MS/Physics/Univ of MI; BS/Engr Physics/Univ of MI. *Born:* Apr 1916 Battle Creek Mich. Joined Genl Elec Res Lab in 1942 as a res assoc. Worked in the fields of phys electronics, vacuum & plasma physics. Became Sect Mgr in 1956, Lab Mgr 1968 and consultant 1981. Pub more than 30 tech papers & have 66 pats. Inventor of the lanthanium boride cathode, hot-cathode magnetron ionization gauge & high power triggered vacuum gap. Recipient of Naval Ordnance Dev Award for work on VT proximity fuse 1946; Disting Alumnus Citation Univ of Mich 1953; IR-100 Award for Triggered Vacuum Gap 1968; 1979 IEEE Lamme Medal. Fellow IEEE, APS & AAAS. P Pres and honorary life member of the Amer Vacuum Soc. Lic PE NY. Member National Academy of Engineering 1981; Lamme Medal, IEEE (1979); President of International Union for Vacuum Science, Technique and Application (1980-83). *Society Aff:* APS, IEEE, AAAS, AVS.

Lafferty, Raymond E
Business: 499 Pomeroy Rd, P.O. Box 122, Parsippany, NJ 07054
Position: Sr VP *Employer:* Boonton Electronics *Born:* 7/12/18 Joined Boonton Electronics as Chief Engr in 1960, elected VP of Engrg in 1966, and Sr VP and Gen Mgr in 1981. Asst Chief Engr of the Daven Co from 1957 to 1960. Sr Development Engr with the Natl Broadcasting Co from 1948 to 1957. Earlier experience in broadcasting, teaching, and the manufacture of r.f. test equipment dating to 1940. Elected a Fellow of the IEEE in 1980 for contributions to loss measurements of reactive components. *Society Aff:* IEEE

Laflen, John M
Home: 222 So Hazel, Ames, IA 50010
Position: Supervisory Agricultural Engr. *Employer:* Agricultural Res Service USDA. *Education:* PhD/Agri Engg/IA State Univ; MS/Agri Engg/Univ of MO; BS/Agri Engg/Univ of MO. *Born:* Aug 31, 1936. Native of Vernon County, Missouri. Employment: 1960-67 Surface Drainage Res USDA-ARS Baton Rouge La. 1967- , res on soil & water loss from agri lands as affected by conservation structures & tillage sys. Chmn Iowa Sect ASAE 1974-75; Chmn ASAE Erosion Control Res Ctte 1974-76; V Chmn ASAE Erosion Control Group 1975-77; Chmn ASAE 1977 Erosion & Sedimentation Symposium. V Chmn ASAE Soil & Water Div 1979- . *Society Aff:* ASAE, AAAS, SCSA.

Lafranchi, Edward A
Business: P O Box 808, L-151, Livermore, CA 94550
Position: Dept Hd, Electronics Engrg Dept. *Employer:* Lawrence Livermore Lab. *Education:* BS/EE/Univ of Santa Clara *Born:* July 1928. US Air Force 1951-53. Employed by the Univ of Calif, Lawrence Labs 1950- . Have held pos of engr, Group Leader, Div Leader & appointed to current pos, Electronics Engrg Dept Hd in Sept 1973. This pos is respon for the overall engrg mgmt of a Dept comprised of 6 Div's & a staff of approx 1200 employees. The tech work of the Dept is very broad & covers such diverse specialities as pulse power, computer sys, microelectronics R&D, electronic materials R&D, instrumentation sys & mathematical sys R&D. Sr Mbr IEEE. *Society Aff:* IEEE

Lage, David A
Home: 8 Evangeline Lane, Woburn, MA 01801
Position: VP, Commerica Products *Employer:* General Electrodynamics *Education:* BS/Elec Engg/Northeastern Univ; AE/Elec Engg/NY Sch Indust Tech. *Born:* July 1929. Native of NYC. Tech Rep to US Air Force for 2 yrs, staff engr at Continental Electrolog Corp Brooklyn NY. Proj Engr BLH Electronics. Since 1970, product mgr for Proprietary Sys, Aircraft & Hwy Sys. Holder of patent on integral aircraft blown tire & low strut indicator. Active in Soc of Allied Wt Engrs, held all Chap office pos, & Internatl Bridge, Tunnel, & Turnpike Assn. Active on Engrg & Maintenance Cttes. Assn of Iron & Steel Engrs and Parental Drug Assn Inc. *Society Aff:* SAWE, IBTTA, PDA, AISE.

Lager, John A
c/o Metcalf & Eddy, 1029 Corporation Way, Palo Alto, CA 94303
Position: Vice President. *Employer:* Metcalf & Eddy Inc. *Education:* MS/Sanitary Engr/Harvard Univ; BS/Civil Engr/Northeastern Univ. *Born:* Nov 1934. Grad & extension courses mgmt, sys & computer applications. Reg P E 6 states; Diplomate AAEE; Mbr ASCE, WPCF, APWA Tau Beta Pi. Cons to EPA, NSF, & sev municipalities in the field of combined sewer overflows & urban stormwater controls. With Metcalf & Eddy 25 yrs spec both nationally & internationally in wastewater sys design, construction, operations, R&D & regional planning for point & nonpoint discharge impacts. Recipient 1976 ASCE State-of-the-Art of Civil Engrg Award. *Society Aff:* AAEE, ASCE, WPCF, APWA, SAME.

Lagerstrom, John E
Home: 2301 Jameson So, Lincoln, NE 68512
Position: Dir, Engrg Extension. *Employer:* Univ of Nebraska. *Born:* Dec 1922. BS, MS & PhD EE Iowa State. Mbr Iowa State engrg staff 1946-66 incl Acting Dir Iowa State Tech Inst, Acting Hd Dept of Arch, Assoc Dean of Engrg. Asesor del Rector, Natl Engrg Univ Lima Peru 1964-66. Dean of Engrg So Dakota State Univ 1966-71. Chmn Elec Engrg Univ of Nebraska 1971-73. Dir of Engrg Extension Univ of Nebraska 1973- . Reg P E Iowa, S Dak, Nebr. Cons on elec machine design Winpower Mfg Co 1952-56. Pres Eta Kappa Nu 1963-64. Dir NSPE 1962-64. Dir ASEE 1973-75.

Lagnese, Joseph F
Home: 3066 Woodland Rd, Allison Park, PA 15101
Position: Private Environmental *Employer:* Engrg Consultant *Education:* BS/Civil Engg/Univ of Pittsburgh; MS/Sanitary Engg/Johns Hopkins. *Born:* 5/20/29. Pgh PA. Penna Water Pollution Control Assn, Ed 1958-62; mbr Bd of Dir 1962- 69; Pres 1969-70. Water Pollution Control Fed, Dir 1968-69; Pres 1971-72. ASCE, Dir Pgh Sect 1968-70; Pres 1976. Penna Soc of PE's. Reg PE PA. Amer Acad of Environ Engrs, Diplomate and trustee 1983-1985, Duncan, Lagnese & Assocs Inc, Cons

Lagnese, Joseph F (Continued)
Engrs Pgh PA 1958-81. Adj Asst Prof CE Dept Univ of Pgh 1968- . Lecturer, Carnegie Mellon Univ 1979- and Univ of NC 1981-. Natl Reg of Prominent Americans & Intl Notables 1974-75. Chi Epsilon: Natl Cvl Engrg Frat. Who's Who in Amer. Bedell Award, Water Pollution Control Fed 1967. Engrs of Distinction. Citation for Serv to Const Indus "Engrg News Record" 1972. Award for Disting Serv in Support of Water Pollution Control, Natl Clay Pipe Inst 1973, Haseltine Award, PA Water Pollution Conslt Assoc., Gordon Maskew Fair Award. Am. Ac. of Env. Engrs, 1986. *Society Aff:* WPCF, ASCE, NSPE, AAEE.

LaGrone, Alfred H
Home: 3925 Sierra Dr, Austin, TX 78731
Position: Prof/Dir-Antennas & Propagation Lab. *Employer:* Univ of Texas at Austin. *Education:* PhD/Elec Engg/Univ of TX at Austin; MSEE/Elec Engg/Univ of TX at Austin; BSEE/Elec Engg/Univ of TX at Austin. *Born:* Sept 25, 1912. Reg P E. Engr for San Antonio Pub Serv Co San Antonio Tx 1938-47. US Navy WW II, 1942-46 Capt USNR. Res Engr Univ of Texas at Austin 1946-54. Assoc Prof 1954-60. Prof 1960- . Dir of Antennas & Propagation Lab 1966- . Cons for Amer Broadcasting Co, Collins Radio, Honeywell Inc, Texas Nuclear, others. Deputy Chmn 1973-76, Chmn 1976-79, Comm F, USNC/URSI. Scott- Helt Mem Award IRE 1960. Fellow IRE 1961. Chmn Natl Fellow Ctte IRE. Mbr Sigma Xi, Tau Beta Pi, & Eta Kappa Nu. *Society Aff:* IEEE, URSI, ΣΞ.

Laguros, Joakim G
Business: 202 W Boyd, Norman, OK 73019
Position: Prof. *Employer:* Univ of OK. *Education:* PhD/Civ Engg/IA State Univ; MS/Civ Engg/ISU; BS/Civ Engg/Robert College. *Born:* 2/4/24. Taught at Robert College, IA State Univ, OH Univ. Since 1963 with the Univ of OK specializing in geotechnical and hgwy materials engg. Res and consulting experience in the above areas and groundwater, too, with 45 publications. Mem Comm on Physicochemical Phenomena in Soils, TRB, 1973-1983; Secretary-Treasurer 1975-1983, OK Sec ASCE. *Society Aff:* ASCE, ASEE, TRB, CMS.

Lagvankar, Ashok L
Home: 1503 Darwin Ln, Wheaton, IL 60187
Position: VP *Employer:* Harza Environmental Services Inc. *Education:* PhD/Environ Eng/Northwestern Univ; MS/Public Health Eng/Maharaja Sayajirao Univ of Baroda; BS/Civil Engrg/MSUB *Born:* 10/09/36 Native of India. Immigrated to USA in 1963. Was Asst Prof at the Faculty of Engrg and Tech MSU, India, 1958-63. In 1968, after obtaining PhD, joined Roy F. Weston Inc as a Sr Environ Engr. Joined Harza Engrg Co in 1972. Assumed progressive responsibility and became chief environ engr of the firm in 1980. In addition assumed respon of the Head of Environ Engrg Dept in 1981. In 1986 became a VP in charge of engineering operations of Harza Environmental Services Inc, a Wholly owned subsidiary of Harza Engrg Co. Recipient of honors including Univ Grants Commission of India Sr Fellowship, (1960-62) Walter P Murphy Fellowship (1965) and AWWA's Academic Achievement Award for best res in water supply field, 1968. Author of numerous tech publs, and mbr of tech cttes. (Listed in Who's Who in Midwest, 1984). Enjoys tennis, golf, bridge and classical music. *Society Aff:* AAEE, ASCE, AWWA, WPCF

Lahey, Richard T, Jr
Business: Dept Nuclear Engrg, Troy, NY 12181
Position: Chmn Dept of Nuclear Engrg. *Employer:* Rensselaer Polytechnic Inst. *Education:* PhD/Mech Eng/Stanford Univ; ME/Engg Mechs/Columbia Univ; MS/Mech Engg/Rensselaer Polytechnic Inst; BS/Marine Eng/US Merchant Marine Acad. *Born:* 2/20/39. Laing Properties plcin St Petersburg, FL. Engr with KAPL 1961-64 engaged in design & analysis of naval nuclear submarines. Employed by G E 1966-75 in var tech & managerial pos incl Mgr Heat Transfer Mechanisms, Mgr Core Dev, Mgr Core & Safety Dev. Currently Chrmn Dept of Nuclear Engg RPI. Extensive publs in area of two-phase flow and heat transfer & LWR safety. Extensive gov & indus cons. Fellow ASME (Past Chrmn K-13), Fellow ANS (Past Chrmn Thermal-Hydraulics Div; Past Mbr, Bd of Dirs), Past ECPD Council rep of ASME, Tech Ed, Journal of Nuclear Engrg & Design. *Society Aff:* ANS, ASME, ΣΞ, ASEE.

Lahey, Robert W
Home: 180 Statesir Pl, Red Bank, NJ 07701
Position: Dir of Engg *Employer:* Jersey Shore Medical Center *Education:* Bachelor of Engg/Mech Engg/Stevens Inst of Tech. *Born:* 5/20/46. in E Orange, NJ; Edward William and Mary (Dykes) L. With Amer Smelting & Refining Co, Perth Amboy, NJ 1965-68 as a Dev Engr; With NJ Dept of Environmental Protection, Trenton, NJ 1969-72 as Environmental Engr, and then Sr Environmental Engr, developed NJDMV exhaust emission test for inspection stations; with Amerada Hess Corp, Woodbridge, NJ 1972-date, as Tech Service Engr, and then Mgr of Tech Services Dept, Dir of Engg with Bronx Lebanon Hospital Center, Bronx, NY 1983-86; Jersey Shore Medical Center, Neptune, NJ, 1986 to present as Dir of Engg. Served with USAF, 1965-67. Mbr Soc of Automotive Engrs, Assn of Energy Engrs (founder and two year Pres of NJ Chapter Natl awards ctte chrmn, Natl VP, mbr Natl Energy policy council), Amer Inst of Plant Engrs, Air Pollution Control Assn (previous mbr of Tech Ctte) GNYHES (Secretary), ASHE, NFPA, Chi Phi Fraternity; Publication *NJ Repair Project-Tune-Up at Idle* in the Journal of the APCA presented at Natl Convention; Editor *Energy Economics & Mgmt* Magazine. *Society Aff:* AEE, SAE, AIPE, APCA, ASHE, NFPA, GNYHES.

Lahti, Leslie E
2801 W Bancroft, Toledo, OH 43606
Position: Prof of Chem Engrg & Dean of Engineering *Employer:* Univ of Toledo. *Education:* PhD/Chem Engr/Carnegie Tech; MS/Chem Engr/MI State; BS/Chem Engr/Tri-State. *Born:* July 1932. Indus exper as glass technologist with Corning Glass & dev engr with Ren Plastics; summers and/or cons with Amoco Oil, Whirlpool Corp, Inland Chem & Great Lakes Chem. Acad exper at Tri-State & Purdue. Joined Univ of Toledo 1967 - Chmn 1972-80. Dean since 1980. Res ints in process design, nucleation & crystallization. Mbr Tau Beta Pi, Phi Kappa Phi, Sigma Xi. PE Ohio. *Society Aff:* AIChE, ACS, ASEE, NSPE.

Laing, Kirby (Sir)
Home: Flat 2, 47 Warrington Crescent, London, England, W91EJ *Employer:* Retired *Education:* MA/-/Cambridge. *Born:* July 21, 1916 Carlisle Eng. s. John & Beatrice (Harland) L. m. (1) Joan Dorothy Bratt July 8 1939; (2) Mary Isobel Lewis; 3 sons. Pupil John Laing Const Ltd London 1937-39; Dir 1939-46; Joint Managing Dir 1946-57; Chmn Laing Group of Co's 1957-76; Chm Laing Properties 1978-87. Served with Royal Engrs WW II; created Knight Bachelor 1968. Fellow Inst C E (P Pres); Hon Mbr Amer Soc C E; Hon Mbr Natl Fed Bldg Trades Employers (Pres 1965, 1967). Chmn Natl Joint Council for Bldg Indus 1968- 74. Club: Naval & Military (London). Contrib articles professionally. *Society Aff:* FEng, FICE, MASCE.

Laird, Campbell
Business: Dept of Mtls Sci & Engg, Phila, PA 19104
Position: Prof. *Employer:* Univ of PA. *Education:* PhD/Met/Univ of Cambridge; MA/Met/Univ of Cambridge; BA/Natural Sciences/Univ of Cambridge. *Born:* 6/17/36. Native of Scotland; Lt, Royal Engrs, 1954-56; Fellow of Christ's College, Cambridge, 1961-65; Ford Motor Co Scientific Lab, 1963-1968; Battelle Visiting Prof, OH State Univ, 1968-69; 1969-present, Dept of Mtls Sci, Univ of PA; Hd of Dept 1974-79; interested in precipitate morphology, electron optical tools, fatigue of metals; over 100 publications. *Society Aff:* TMS-AIME, ASM, RI.

Laird, Harry G
Home: 185 S Sea Ave, W Yarmouth, MA 02673
Position: Mgr Ind Engg. *Employer:* Polaroid Corp. *Education:* BS/Ind Engg/Penn State Univ. *Born:* 11/9/25. in Tyrome, PA served in USAF 1943-1945. Have held positions from Sr Engr to Corporate Mgr of IE with C H Marsland, Riegel Paper Corp, Mohasco Industries and currently Polaroid Corp. Areas of specialization include matl handling and packaging, facilities design, and physical distribution.

Laird, Harry G (Continued)
Served as Chapter Pres, Regional VP, Natl VP Finance and Dir of EJC Affairs for IIE. Currently chrmn of the Planning and Organization Commission for IIE. Elected a fellow of IIE in 1977. *Society Aff:* IIE.

Laird, William J, Jr
Home: 42755 Buckingham Dr, Sterling Hts, MI 48078
Position: V Pres, Mkting & Research. *Employer:* Upton Industries Inc. *Born:* Feb 1943. Assoc in Sci degree from Henry Ford Community College in metallurgy. Has completed undergrad work at U of Mich in Met & Chem Engrg. Joined Upton Indus Inc 1971 & Mgr Utek Met Div; promoted to Corp Sales Mgr 1972; promoted to present pos of V P Mkting & Res 1973. Elected to Bd of Upton Indus 1974. Is also Tech Cons to Sullivan Chem Co Inc. Pres of the Curriculum Adv Ctte, Tech Div, Henry Ford Community College & as a Detroit Chap Mbr; named Outstanding Young Mbr of the Amer Soc for Metals for contrib to Met & Matl Sci & to the Soc. P Chmn Young Mbrs Ctte & Secy Educ Ctte Natl ASM. Sky diver, SCUBA diving instructor & golfer.

Laithwaite, Eric R
Business: Dept of Elec Engrg, Imperial College, Exhibition Rd, London SW7 2BT England
Position: Prof of Heavy Elec Engrg. *Employer:* Imperial Coll of Sci & Tech, London. *Education:* DSc/EE/Univ of Manchester; PhD/EE/Univ of Manchester; MSc/EE/Univ of Manchester; BSc/EE/Univ of Manchester. *Born:* June 1921 Atherton Lancashire. RAF 1941-46 (RAE Farnborough 1943-46). Asst Lectr 1950-53, Lectr 1953-57, Sr Lectr 1957-64 Manchester Univ. Prof Heavy Elec Engrg Imperial Coll 1964-1986. Prof. Emeritus 1986-. External Prof of Applied Elec Royal Inst London 1967-76. Author 13 books & over 250 papers on linear motors, electromagnetic levitation & allied subjects. Chmn IEE Power Bd 1971-72. Dir Davy Linear Motors Ltd, Landspeed Ltd. Cotswold Research Ltd, Con to Brian Colquhoun & Ptnrs. Royal Soc S G Brown Award & Gold Medal 1966. IEEE Nikola Tesla Award 1986. Recreation: entomology. *Society Aff:* IEE, IEEE, RSA.

Laity, Ronald L
Business: 209 East Washington Ave, Jackson, MI 49201
Position: Staff Engr/Asst Treas *Employer:* Gilbert/Commonwealth; Commonwealth Assocs Inc *Education:* BSEE/Electrical Power/MI Tech Univ *Born:* 07/16/27 in Hancock and raised at Pontiac, MI. Served apprenticeship with General Motors Corp prior to and after serving with the US Air Force in West Germany. Various positions with Commonwealth Assocs Inc. (Gilbert/Commonwealth) since 1953 including mgmt of consltg, planning and design services for industrial, institutional, governmental and commercial facilities. Asst Treas - 1972. Primary field is industrial energy sys and facilities engrg. Registered PE in MI and 8 other states. Served on local and state Ethical Practices Committee of MI Society of PE and as ofcr of various committees and groups of Southeastern MI Section of IEEE. Deacon and Elder in First Presbyterian Church. *Society Aff:* NSPE, IEEE, ESD, PESD, TBΠ, HKN

Lake, James M
Business: 543 3rd Ave - Rm 206, Fairbanks, AK 99701
Position: President *Employer:* Lake & Boswell, Consulting Elect Engrs, Inc. *Education:* BS/Elec Engg/Univ of WI. *Born:* Aug 1918 Harvard Ill. Resident Fairbanks Alaska 1944- . US Army Signal Corps 1942-43. Tech Rep for Minneapolis Honeywell with US Army Air Corps 1943- 44. Asst to Resident Engr Corps of Engrs Ladd AFB 1948-53. Reg Elec Engr Alaska 1951- . Cons Elec Engr in private practice since 1954. Sr Mbr IEEE. Mbr Alaska St Bd of Registration for Engrs, Archs & Land Surveyors 1970-81. Pres Fairbanks Chap Alaska Soc of Prof Engrs 1974. Pres Alaska Soc of Prof Engrs 1977. *Society Aff:* NSPE, IEEE, IES.

Lake, Thomas D
Business: P O Box 6288, Anaheim, CA 92806
Position: Sr VP - Prin Engr. *Employer:* Converse Ward Davis Dixon. *Education:* BS/Civil Engg/Univ of IL. *Born:* 4/29/23. With Converse Ward Davis Dixon 1952- . Bd of Dir 1969- . Currently Sr VP & Sr Consultant to Anaheim office. Pres Soil & Fnd Engrs Assn of CA 1976-77. *Society Aff:* ASCE, NSPE.

Lakshminarayan, Mysore R
Home: 26751 Via Matador, Mission Viejo, CA 92691
Position: Tenured Prof *Employer:* Cal. Poly Univ *Education:* PhD/Elec/U of DE; MEE/Elec/U of DE; BE/Elec/Bangalore U; PE/Elec/California. *Born:* 4/24/44 Native of Bangalore, India. Taught for a year at Mysore Univ. Worked as a Post-Doctoral Research Fellow from 1977 to 1978. Joined California State Polytechnic Univ as a Faculty Member in 1978. Became tenured Prof in 1985. Published Papers in the areas of Solid State Devices, Microwave Circuits, Laser Modulation, Integrated Optics, and Fiber Optics. Recipient of a patent on microwave and optical systems, NASA tech brief award, and extraordinary service to Univ Award. *Society Aff:* HKN, IEEE.

Lakshminarayana, Budugur
Business: 233 Hammond Bldg, Dept of Aero Engg, Univ Park, PA 16802
Position: Evan Pugh Professor of Aerospace Engrg. *Employer:* PA State Univ. *Education:* D Eng/Mech Engr/The Univ of Liverpool; PhD/Mech Engr/The Univ of Liverpool; BE/Mech Engr/Mysore Univ *Born:* 2/15/35. Engaged in res and teaching in the field of aero propulsion, turbomachinery and fluid mechanics. Visiting assoc prof, MA Inst of Tech, 1971. He has published over 102 refereed papers and 46 reports and is the co-editor of two books. Assoc Editor of ASME *Journal of Fluids Engineering* (1980-83), and editorial bd mbr of Intl *Journal of Thermal Engineering*. He received the annual award (1977, 1983) for Outstanding Res at the PA State Univ and the Henry R Worthington award in 1977. Received the Premier Res Awd for 1984 at the PA State Univ. He directed the ASME PDP course on turbomachinery aerodynamics and has been the lecturer for the two week ASME/IA State course on turbomachinery. *Society Aff:* AIAA, ASME, ASEE

Lalas, Demetrius P
Home: 612 Woodland, Birmingham, MI 48009
Position: Prof *Employer:* Wayne State Univ *Education:* PhD/Aerospace Eng/Cornell Univ; MAeroE/Aeronautical Eng/Cornell Univ; BS/Physics/Hamilton Coll *Born:* 9/28/42 in Athens, Greece. Worked briefly for Curtiss-Wright Corpo. Since 1968, at Dept of Mechanical Engrg, Wayne State Univ. In 1973-74, visiting Fellow, Coop Instit. for Research in Environmental Sciences, U of Co. From 1976 to 1982, Dir Meteorolgical Institute, National Observatory of Athens Greece. From 1980, adjunct Prof, GA Inst of Tech. Since 1982, involved in solar and wind energy utilization at the European Economic Communities. Conslt to the Greek Govt and Public Power Corp on Air Pollution and Environ Matters. *Society Aff:* ASME, AMS, ESD

Lam, Tenny N
Business: Aldrin Corp, P.O. Box 741, Davis, CA 95617
Position: President *Employer:* Aldrin Corp. *Education:* D/ENG/CE/Univ of CA-Berkeley; ME/CE/Univ of CA-Berkeley; BS/CE/Univ of CA- Berkeley *Born:* 11/28/40 Native of Hong Kong, naturalized US citizen. Registered PE in CA and MI. Involved in teaching, research research, and consulting in transportation systems engineering. Assistant Prof, Univ of MO, Columbia 1966-68. Senior Research Eng, General Motors Research Laboratories, 1968-74. Prof. Univ. of CA; Davis 1974-87; Reader in Civil Engineering, Univ. of Hong Kong, 1987- . *Society Aff:* ASCE, ORSA

LaMantia, Charles R
Business: Acorn Park, Cambridge, MA 02140
Position: Pres & Chief Operating Officer, Director *Employer:* Arthur D Little, Inc. *Education:* ScD/ChE/Columbia Univ; MS/ChE/Columbia Univ; BS/ChE/Columbia Univ; BA/- /Columbia College. *Born:* 6/12/39. Dr Charles R LaMantia is Pres & Chief Operating Officer, Arthur D Little, Inc., (ADL). He joined the company in 1967 as a member of its engineering staff. He was elected VP in 1977 and from

LaMantia, Charles R (Continued)
1977-81 assumed responsibility for Chemical & Metallurgical Engg. In 1981, he left ADL to become Pres & Chief Executive Officer of Koch Process Systems, Inc. He rejoined ADL in September 1986 as Pres & Chief Operating Officer and as a mbr of its Board of Directors. *Society Aff:* AIChE.

Lamarre, Bernard C
Home: 4850 Cedar Crescent, Montreal, Quebec Canada H3W 2H9
Position: President & CEO. *Employer:* Lavalin Inc. *Education:* MS/Structure/Univ of London, England; DIC/Structure/Imperial College Eng; BSc/Civil/Ecole Polytechnique Montreal. *Born:* Aug 1931 Chicoutimi Quebec Canada. Athlone Scholar. Fellow of Engr Inst of Canada. Doctor Honoris Causa of the following Universities: St. Francis-Xavier, Nova Scotia; Waterloo, Ontario; Concordia, Montreal; U de M, Montreal; Sherbrooke, Quebec; UQUAC Chicoutimi, Quebec; and Queen's, Kingston, Ontairo. Fellow of Royal Canadian Soc of Architecture. With the Lavalin Group & its predecessor since 1955, first as struct designer, then successively as soil engr, pub work proj engr & ch engr. Principal Ptnr 1963 & finally with the incorporation of the Group in 1972 became shareholder & Pres & CEO of the Group - which employs over 7000 engrs, technicians & other permanent staff, & offers prof services in most fields of engrg, proj & const mgmt, Procurement & const. and in petrochemical and rapid transit equipment manufacturing. Officer of the Order of Canada and of the Order of Quebec. *Society Aff:* ASCE, EIC, CSCE, OIQ.

Lamarsh, John R
Home: 68 N Chatsworth Ave, Larchmont, NY 10538
Position: Hd, Dept of Nuclear Engg. *Employer:* Poly Inst of NY. *Education:* PhD/Physics/MIT; BS/Gen Sci/MIT. *Born:* 3/12/28. Post PhD profl work: 1952-53, Aircraft Nuclear Propulsion Proj, Pratt and Whitney Aircraft; 1956-57, Asst Prof of Physics, NYU; 1957-62. Asst Prof Engg Physics, Cornell Univ; 1962-66, Assoc Prof Nuclear Engg, NYU; 1966-73, Prof Nuclear Engg, Chrmn of Dept, NYU; 1973-present, Prof Nuclear Engg, Hd of Dept, Poly Inst of NY. Consult to Congressional Res Service, US Gen Accounting Office, Office of Tech Assessment. PE, NY State: Fellow, American Nuclear Soc. Author of two books on nuclear engg. Consultant to National Science Foundation. Administrative Judge, US Nuclear Regulatory Commission, 1981- Arthur Holly Compton Award, Am. Nuclear Soc., 1980 Member, Technical Advisory Committee on Radiation, City of New York. *Society Aff:* ANS, APS.

Lamb, Charles W
Business: USPO Box 33064, Wright-Patt AFB, OH 45433
Position: Deputy Chief of Staff, Engrg & Serv. *Employer:* US Air Force. *Born:* Feb 1931. BS US Naval Acad ('Amer Coll Student Leaders 1953'); BSCE & MSCE Univ of Oklahoma ('Outstanding Graduating Cvl Engrg Student'); Squadron Officers' School ('Disting Grad'); Air Command & Staff College. Native of Nashville Tenn. Continuous military serv as engrg officer since 1953, holding respon assignments at base, command & Hq USAF level. Assignments incl Base Cvl Engr at L G Hanscom Field Mass & Patrick AFB Fla & Hickam AFB Hawaii; Project Turnkey for const of Tuy Hoa AB, RVN; Directorate of Cvl Engrg Hq USAF, DCS Engrg & Services, Hq Air Training Command; DCS Engrg & Services AF Logistics Command. Legion of Merit, Bronze Star, Meritorious Service Medal, Commendation Medal & SAME Newman Medal 1976. Sigma Xi, ASCE, P Pres SA Chap SAME, APWA, PCI, Toastmasters, BSA.

Lamb, Donald R
Business: 910 Garfield St, Laramie, WY 82070
Position: Partner. *Employer:* Associated Engrs. *Education:* PhD/Civil/Purdue Univ; CE/Civil/Univ of WY; MSCE/Civil/Univ of WY; BS/Gen E Pt Option/Univ of WY; BA/Math Sci/Hastings College. *Born:* 5/6/23. PhD Purdue Univ; BA Hastings College; BS & MS Univ of WY. Native of Casper WY. Taught in pub schs in Western NB 1946-49. Army Air Corps 1942-49 as navigator. Taught C E at UW 1951- . In charge of Pub Relations & guidance at Engrg College 1975- . Partner Associates Engrs, a cons firm, since 1977. Past Chmn air Resources Council Laramie City Council, Chmn Environ Qual Council 1973- 75. Pres of Bd, Laramie Area Indus Dev Inc 1974-79. *Society Aff:* ASCE, NSPE, ASEE, ARTBA.

Lamb, Jamie Parker
Business: Dept of Aerospace Engineering, Univ of Texas, Austin, TX 78712
Position: Prof of Mech and Aerospace Engg *Employer:* Univ of TX. *Education:* PhD/Mech Engg/Univ of IL; MS/Mech Engg/Univ of IL; BS/Mech Engg/Auburn Univ. *Born:* 9/21/33. Served in USAF as proj engr at Wright Field, OH. Taught for two yrs at NC State Univ and at Univ of TX since 1963. Served six yrs as Chrmn of Mech Engrg Dept and as Assoc Dean of Engrg. Currently serving as Chrmn of Aerospace Engrg and Engrg Mechs Dept. Ernest Cockrell Memorial Prof; Author or co-author of more than 75 research papers and reports; Fellow ASME, Assoc Fellow AIAA. Served in ASME as Assoc Editor, Journal of Fluids Engrg, as Chrmn of Centrl TX Sec, and Chrmn of the Fluid Mechs Tech Cttee, Fluids Engrg Div. Served in ASEE as Chrmn of Mech Engrg Div and as Chrmn of Summer Faculty Progs Ctte. *Society Aff:* ASME, AIAA, ASEE, NSPE.

Lambe, T William
Home: Apt 5-4B 100 Memorial Dr, Cambridge, MA 02142
Position: Edmund K Turner Prof of Cvl Engrg. *Employer:* MIT. *Born:* Nov 28, 1920 Raleigh N C. m. Sept 13, 1947 to Catharine Canby Cadbury. BS N C State 1942; SM MIT 1944; ScD MIT 1948. Since 1945 on Acad Staff MIT; author 1 textbook & co-author 2 textbooks; author or co-author more than 80 tech papers. ASCE Collingswood Prize 1951; Desmond Fitzgerald Medal of BSCE 1954 & 1965; ASCE Wellington Prize 1961; ASCE Norman Medal 1964; NASA Cert of Appreciation 1969; ASCE Terzaghi Lecturer 1970; Rankine Lecturer 1973; Geotech Sect Prize of BSCE 1973; Keynote Speaker at Pan Amer, Australia-New Zealand, Asian, & S E Asian Confs; R P Davis Lecturer 1973; Genl Reporter VIII Soils Conf Moscow 1973; Terzaghi Mem Lecturer Istanbul 1973; Keynote Speaker V Brazilian Conf 1974; ASCE Karl Terzaghi Award 1975. Mbr NAE; Fellow ASCE; Fellow ICE.

Lamberson, Leonard R
Home: 21438 Meridian Rd, Grosse Ile, MI 48138
Position: Prof./Chairman *Employer:* Wayne State Univ. *Education:* PhD/Ind Engg/TX A&M Univ; MS/Ind Engg/NC State Univ; BME/ME/GM Inst. *Born:* 11/18/37. Native of MI. Ind experience with the GM Corp as a production foreman and work stds engr, 1957-1964. Has held previous profl positions at GM Inst & TX A&M Univ. With Wayne State Univ, College ofEngg, since 1970. Res and consulting in design reliability & reliability prog mgt. Also res interests in quality control & mfg. Publications include text book in reliability. Won outstanding teacher award at two univs. *Society Aff:* ASEE, IIE, ASQC.

Lambert, Jerry R
Business: Dept of Agric Engr, Clemson, SC 29631
Position: Prof. *Employer:* Clemson Univ. *Education:* PhD/Agric Engr/NC State Univ; MSE/Agric Engr/Univ of FL; BAgE/Agric Engr/Univ of FL. *Born:* 9/16/36. Native of Benton, IL & Auburndale, FL. Design Engr Trainee with USDA-Soil Conservation Service in FL and SC 1958-60. After grad school, joined faculty of agricultural engg at Clemson in 1964. Assumed prof position in 1972. Teach courses in computational systems, including personal microcomputers, soil & water resources engg, and instrumentation. Res in computer simulation of agricultural crop response to physical environment, with emphasis on water. Consulting in area of hardware & software for small computers for irrigation needs, water supply, and flood control. Enjoy farming and amateur radio. *Society Aff:* ASAE

Lambert, Wesley R
Home: 1568 Dapple Ave, Camarillo, CA 93010
Position: Partner *Employer:* N V Systems. *Born:* Mar 16, 1924 San Francisco. US Navy WW II. Managed Gov Photo-optical Engrg Orgs at major Dept of Def Test Ranges 1951-75. Currently Partner N V Sys of Camarillo Calif. Spec in design &

Lambert, Wesley R (Continued)
util of tracking telescopes & high speed instrumentation camera sys to record engrg test data of missile launches, flights & target intercepts. 13 yrs Rep of var military commands at the Optical Sys Group of the Range Commanders Council. 9 yrs Natl Gov of Soc of Photo- Optical Instrumentation Engrs & 8 yrs Assoc Ed of Optical Engrg/SPIE Journal.

Lambrakis, Konstantine C
Business: 300 Orange Ave, W Haven, CT 06516
Position: Dean of Engg. *Employer:* Univ of New Haven. *Education:* PhD/AE/Rensselaer Poly Inst; MS/ME/Univ of Bridgeport; BS/EE/Univ of Bridgeport. *Born:* 1/30/36. Joined the faculty of the Univ of New Haven on 1965 having worked in industry for about seven yrs as an electrical Design and Dev Engr and as a Sr Mech Engr. Served as chrmn of the Univ of New Haven Mech Engg Dept from 1973 to 1976 and as Dean of the Engg Sch from 1976 to the present. *Society Aff:* ASME, AAAS, ASEE, AIAA.

Lambur, Charles H
Business: 303 Fifth Ave, New York, NY 10016
Position: Pres. *Employer:* Schneider of Paris Inc. *Education:* BS/Mining/Univ of MO. *Born:* 10/30/10. Reg PE in Mining. BS in Mining Engg, MO Sch of Mines & Met. 40 yrs experience all phases of mining. Has been (1) Pres & principal owner - US Collieries, Inc, (2) Operating Pres of Warren Foundry & Pipe - Mt Hope Mine, (3) Pres, Tekara Intl (Turkish coal mine dev), (4) Dir of Operations - Simpson Creek Collieries of Std Ore & Alloy Corp & Sparta Coal Co. Authored mining articles for various mining journals. Chrmn of Bd of Schneider of Paris, Inc, USA. Assoc Intl Exec Service Corps, NY. Consultant, Corpozulia, Venezuela. Consulant Chachuah, BA Argentina. Mbr, Earth Resources Group, USA. *Society Aff:* AIME, NSPE, MSOPE.

Lamm, A Uno
Home: 365 Moseley Road, Hillsborough, CA 94010
Position: Eng Consultant (Self-employed) *Education:* Dr of Tech/-/Royal Inst of Tech, Stockholm, Sweden *Born:* 5/22/04 With ASEA Co, Sweden, 1928-1970, led successively development mercury arc rectifiers, HV ionic valves and system technology for HVDC transmission, HV Switchgear Dept, Atomic Energy Dept. Electrotechnical Dir in charge of all electrical departments and laboratories. Moved to USA 1965, Chairman Steering Committee ASEA-GE Joint venture for Pacific HVCD Intertie construction. Since 1969, Consultant. About 150 patented inventions. Gold Medal, Royal Academy of Engineering Sciences, Arnberg Prize, Royal Academy of Science, John Ericson Medal, American Society of Swedish Eng, Lamme Medal, IEEE, USA, Doctor Honoris Causa, Danish Inst of Tech, Honorary Fellow-Manchester University, England, French Ordre du Merite pour la Recherche et l'Invention. In 1981 Power Engrg Soc (Section of IEEE) instituted the Uno Lamm High Voltage Direct Current (HVDC) Award. *Society Aff:* NAE, KVA, KVS, IVA

Lammie, James L
Home: 1421 Epping Forest Dr, Atlanta, GA 30319
Position: Senior Vice President *Employer:* Parsons, Brinckerhoff/Tudor Engg Co. *Education:* MSE/CE/Purdue Univ; MSBA/Bus/Geo Wash; BS/Military Engg/USA Military Acad. *Born:* 9/19/31. Served in Army Corps of Engrs, 1953-74. Retired as Col as San Francisco Dist Engr. Currently SVP of Parsons Brinckerhoff and Proj Dir of Parsons Brinckerhoff/Tudor (PB/T) as Gen Engg Consultant to Metropolitan Atlanta Rapid Transit Authority. *Society Aff:* ASCE, SAME.

Lamont, Joseph, Jr
Home: 4233 N E 75th St, Seattle, WA 98115
Position: Senior Partner *Employer:* Dames & Moore. *Education:* LLD/Law-Hon/Geneva College; BS/Civil Engg/Univ of WA. *Born:* Apr 1926. Royal Canadian Air Force & US Army Air Force WW II. With Dames & Moore since 1950. Managed Seattle officee 1959-72. Spec in foundation studies for high-rise bldgs, heavy indus structures, hwys, airports, marine terminals. Elected Partner 1963. Regional Mgr N Amer offices 1972 to 1982 Senior geotechnical engineering consultant in Seattle office 1982 to present. Member Dames & Moore Board of Directors 1987. Elected Senior Partner 1984. Reg P E Wash, Alaska, Montana. P Pres & Dir Cons Engrs Council of Wash. 'Engr of Yr' award by CECW 1974. *Society Aff:* ACEC, ASCE, CECW.

Lamp, Benson J
Home: 1385 Lake Crescent Dr, Bloomfield Hills, MI 48013
Position: Business Planning Mgr, Ford Tractor Opera *Employer:* Ford Motor Co. *Education:* PhD/Agri Engg/MI State Univ; MS/Mech Engg/OH State Univ; BAE/Agri Engg/OH State Univ; BS/Agri/OH State Univ. *Born:* Oct 1925. Native of Columbus Ohio. Taught & conducted res at Ohio State until 1961. Product Planning Mgr for Massey Ferguson until 1966 & with Ford Motor since, serving in Planning & Mkting pos on worldwide basis. V P Tech Council ASAE 1969-71, Pres ASAE 1985-1986. *Society Aff:* ASAE.

Lampard, Douglas G
Business: Dept of Elec Engrg, Monash Univ, Melbourne Victoria, Australia
Position: Prof & Chmn, Dept of Elec Engrg. *Employer:* Monash Univ. *Education:* Ph.D/Mechanical Sciences/Cambridge University (England); M.Sc./Physics/Sydney University; B.Sc (1st Class Honours)/Physics/Sydney Univ. *Born:* May 4, 1927 Sydney Australia. BSc with 1st class honours Physics 1951; MSc Physics 1954. From 1951-62, res scientist with the Div of Electrotech of the Commonwealth Sci & Indus Res Org & in 1962 was appointed foundation Prof of Elec Engrg at Monash Univ. Fellow of the Inst of Elec & Electronics Engrs & served on the Bd/Dir 1970-71. Interests are in stochastic processes, electric circuit theory & in the application of elec engrg to physiological problems. *Society Aff:* AA, IE Aust, IEE, IEEE

Lampert, Murray A
Home: 84 Mason Dr, Princeton, NJ 08540
Position: Professor Emeritus. *Employer:* Princeton Univ. *Education:* BA, 1942/Mathematics/Harvard College; MA, 1945/Physics/Harvard University *Born:* Nov 29, 1921. BA '42 Harvard, MA '45 Harvard. Entered electronics during WWII. 1945-'49 studied physics at Univ. Calif. (Berkeley). 1949-'52 ITT Labs, Nutley, N.J., mostly in microwaves and gas discharges. 1952-'66 RCA Labs, Princeton, N.J., specializing in electronic physics of solids. Author or co- author about 50 papers, several review articles and book in this area. 1966-'74 Professor, EE, Princeton University; 1974 - 1986 inactive status due to medical disability; Emeritus as of Dec 1, 1986. Acting Dir RCA Lab, Tokyo, Japan 1962. Chm 1966 IEEE Solid State Device Res. Conf. Since 1974, main inter]- a]-aests are biophysics and the neurosciences. Since 1979 has authored or co- authored a dozen articles on electrolyte/polyelectrolyte theory. Fellow IEEE, APS and AAAS. *Society Aff:* APS, IEEE, AAAS, SFN, BS, ΣX.

Lampert, Seymour
Home: 5722 Oakley Terrace, Irvine, CA 92715
Position: Sr Res Assoc. *Employer:* Univ of S CA. *Education:* PhD/Engg/CA Inst of Tech; AE/Aero Engg/CA Inst of Tech; MS/Aero Engg/CA Inst of Tech; BS/AE/GA Inst of Tech. *Born:* 3/5/20. Native of Brooklyn, NY. Taught mathematics GA Tech 1943-1944. Served USNR AAL Det 1944-1946 - Aero Res Scientist. Ames Lab NACA 1944-1951; Res Engin Caltech JPL 1951-1954 Ch Engin Odin Assoc 1955-1957, Dept Mgr Appl Mechanics Aeronutronic, Div of Ford MC 1957-1962; Dir Advanced Systems Res, N American S & ID 1962-1967; VP and Div Dir, Systems Assoc Inc 1967-1971, since 1971 Staff Mbr USC - Conducting Res and Teaching Solar Energy Syst Design. Presently Board of Directors; Davato Corp - Past Mbr of Exec Council or County AIAA & Chrmn Spec Comm Orange & LA County AIAA - Formerly Editor-In-Chief "Journal of Solar Sciences–. *Society Aff:* AAUP, ΣΞ, TBΠ, ΦΗΣ.

Lampson, F Keith
Business: 16555 Saticoy St, Van Nuys, CA 91409
Position: Dir of Matls Engrg *Employer:* The Marquardt Co - Div of ISC Corp. *Education:* BS/Met Engg/Univ of IL. *Born:* 8/7/24. Native of Brookings S Dak. Metallurgist with NEPA, Oak Ridge Tenn; metallurgist with Allison Div GMC; Pacific Coast Tech Rep for Allegheny-Ludlum Steel Corp. Assumed dir of Matls Engg Dept Marquardt Co Sept 1965. Have added respon for dir of Configuration Mgmt & related activities as Engrg Library, Art & Photo Labs. Product line covers sophisticated propulsion sys such as bi- & mono-propellant rocket engines, integral rocket/ramjet propulsion sys, Ram air turbines & rotating accessories. Author sev tech articles. Mbr AMS Div of SAE, Matls & Structures Comm of AIA, Aerospace Activities Ctte of ASM, P Mbr Natl Finance Ctte ASM. Hobbies: reading & photo. Former mbr on ASM Bd of Trustees, Past Chrmn of San Fernando Valley Chapter of ASM; Chrmn of Western Metals Congress, 1972. Liaison Rep. between Ams Div. of SAE and E-24, Fracture Toughness Comm. of ASTM, Fellow of ASM, Member of the Metullurgical Society of AIME, Tech. Reg PE; mbr of Air Force Studies Bd Net Shape Tech Comm of Res Council. *Society Aff:* ASM, AIME, ASTM, SAMPE.

Lanc, John J
Home: 53 Vincent Dr, Middletown, NY 10940
Position: Pres. *Employer:* Lanc & Tully Engrg & Surveying, P.C. *Education:* Diplom Ingenieur/CE/Czech Technical Univ. *Born:* 9/6/41. Native of Prague, Czechoslovakia. Geodetic engr and civ engr-in charge on several maj ind construction sites in Czechoslovakia. With Eustance & Horowitz, PC 1969; as VP 1975-85; Pres of Lanc & Tully Engrg & Surveying, P.C. 1985-present. Consulting engr for several municipalities, design and construction inspection of public water, sewer, drainage and hwy systems, planning residential devs, large land surveys. From 1975 officer of Orange-Sullivan Chapter of NY State Soc of PE, president in 1981-82. Licensed in NY, NJ,PA,CT,TX. *Society Aff:* NSPE, ASCE, ACSM, NWWA.

Lancaster, Tom R
Business: 3040 Delta Hwy. North, Eugene, OR 97401
Position: Traffic Eng *Employer:* Lane County *Education:* MS/Transportation Eng/Univ of CA-Berkeley; BS/CE/OR State Univ *Born:* 7/22/48 and raised in OR, graduate of OR State Univ and Univ of CA-Berkeley. Worked in Traffic Section of OR State Hwy Div. 1970-76, presently Traffic Eng for Lane County Public Works Department. Pres of Inst of Transportation Eng District 6 1981-82. Registered PE in State of OR. *Society Aff:* ASCE, ITE

Lance, George M
Business: College of Engg, Iowa City, IA 52242
Position: Prof. *Employer:* Univ of IA. *Education:* MS/Instrumentation Engg/Case Inst of Tech; BS/Mech Engg/Case Inst of Tech. *Born:* 12/4/28. Born Youngstown, OH, 1928. Married, 5 children. US Navy, 1946-48. Profl and honorary societies ASME, IEEE, ASEE, Tau Beta Pi, Pi Tau Sigma, Sigma Xi. Reg OH, IA. Industrial experience-TRW, Res Engr 1954-56; Moog Servocontrols, Sr Systems Engr 1960-61; consulting and summer appointments include McDonnell- Douglas, Boeing, US Army Weapons Command, Harris-Intertype. Academic experience- WA Univ, Lectr 1956-60; Univ of IA 1961-present, Acting Chrmn Mech Engg 1972-74, Associate Dean 1974-79, currently Prof Mech Engrg. Principal areas of interest- control systems and optimization. *Society Aff:* ASME, IEEE, ASEE.

Land, Cecil E
Business: Sandia National Labs, Division 1112, Albuquerque, NM 87185
Position: Distinguished Member, Technical Staff, Ion Implantation Physics *Employer:* Sandia National Labs *Education:* DSc/(HON.)/OK Christian College; BS/Elec Engg/OK State Univ. *Born:* Jan 8, 1926 Lebanon Mo. US Army Infantry 1944-46. Power line carrier sys design engr with Westinghouse Electronics Div 1950-56. Joined Sandia Labs as a Staff Mbr in weapons sys engrg 1956. Engaged in applied res in ferroelectric ceramic matls & devices since 1961. Discovered electrooptic properties of ferroelectric ceramics. Chmn IEEE Committee on Ferroelectrics 1976-77; 1978- Awards: N Mex Soc of Prof Engrs, Albuquerque Chap, Engr of Yr 1973; IEEE Fellow 1974; Soc for Info Display, Frances Rice Darne Mem Award 1976; NM Acad of Sci, Disting Scientist Award 1976; DSc (Hon), OK Christian College, 1978; Soc for Infor Display Fellow 1980; Am Ceram Soc Fellow 1981; Japan Soc. for Promotion of Science Fellow 1983; Distinguished Member of the Technical Staff Award, Sandia National Laboratories 1983; IEEE UFFC-S Recognition Award 1986. *Society Aff:* IEEE, APS, Am. Ceram. Soc., OSA, SID, SPIE, MRS

Landau, Herbert B
Business: 345 E 47 St, NY, NY 10017
Position: Pres *Employer:* Engrg Info, Inc *Education:* MS/Info/Columbia Univ; BA/Chem/Hunter Coll *Born:* 10/24/40 Herbert B Landau is Pres and Chief Exec Officer of Engrg Info, Inc, a not- for-profit info serv founded in 1884. He has held various positions in the information industry including Mgr of Info Sys for the Solar Energy Res Institute, Dir of Scientific and Tech Info Serv for Mead Data Central Corp, Principal Mgr of Auerbach Assoc, Inc and Library Operations Suprv for Bell Telephone Lab. He is Past Pres of the Natl Fed of Abstracting and Info Serv, is a Past-Pres of the Amer Soc for Info Science and has been a Natl Lecturer for the Assoc for Computing Machinery. He has authored over 40 papers in various aspects of tech info handling. *Society Aff:* ASIS, NFAIS, SLA

Landau, Ralph
Business: 2 Park Ave, New York, NY 10016
Position: General Manager and Consulting Professor of Economics *Employer:* Listowel, Inc. & Stanford University *Education:* ScD/Chem Engg/MIT. *Born:* 5/19/16. Engr M W Kellogg Co; Exec VP & co-founded Scientific Design Co Inc 1946-63. Pres 1963-75 Halcon Intl; Chrmn 1975-81. Chrmn The Halcon SD Group Inc 1981-82. Life Corp Mbr MIT; Life Trustee Univ PA; Trustee, CA Inst Tech; Trustee, Cold Spring Harbor Lab; Vice Pres Natl Acad Engrg 1981-. Chmn, Amer Sect Soc Chem Ind (1979); Adj Prof of Tech, Mgmt & Soc Univ of PA. Consulting Prof Chem Eng & Economics, Stanford Univ, 1983-. Fellow of Faculty, Kennedy Schl., Harvard; Dir Alum Co of Amer., (1977-87) Chem Indus Medal SCI 1973; AIChE Petrochem & Petro Div Award 1972; Winthrop-Sears Award 1977; Perkin Medal 1981. Nat'l. Medal of Technology, 1985; John Fritz Medal, 1987. Author about 100 papers; many patents; 1 text on chem plant design; 3 volumes as co-editor *Society Aff:* AIChE, ACS, AEA, NAE, AAAS, AAAS

Landauer, Rolf
Business: P O Box 218, Yorktown Hts, NY 10598
Position: IBM Fellow. *Employer:* IBM T J Watson Research Ctr. *Education:* PhD/Physics/Harvard Univ; AM/Physics/Harvard Univ; SB/Physics/Harvard Univ. *Born:* Feb 4, 1927. Electronic Technician's Mate USNR 1945-46. Lewis Lab NACA 1950-52. With IBM since 1952. Dir Phys Sci 1962-66, IBM Asst Dir Res 1966-69, IBM Fellow 1969- . Had key role in initiating IBM work in injection laser & in large scale integration. LSI was name of IBM's project, later taken up by others. Personal res in: computing devices, electron transport theory, ferroelectricity, nonlinear electromagnetic wave prop, stat mech of the computational process. Mbr Natl Acad of Engg.

Landen, David
Home: 6224 N 29th St, Arlington, VA 22207
Position: Res Civ Engr. *Employer:* Retired. *Education:* BS/Geodetic and Cartographic Science/Geo Washington Univ; Washington, DC. *Born:* July 4, 1908. Attended MIT 1926-29; BS in Geodetic & Cartographic Sci GWU 1969. m. Jeanne Sybil Lesnick Oct 7, 1945; c. Deborah Dena & James Edward. Field topographic surveys USGS 1932-37; Ch, Aerial Mapping Prog, Maine, for US Dept of Agri 1937-40; Returned to USGS; Ch, Photo-Topography, Br of Spec Maps USGS 1941-49; detailed as Ch Instr Sch of Photo-Topography, Lowry Field Colo 1943. Office of Res & Tech Standards USGS 1949-1974. Retired from USGS 1974; Reg P E DC. Citations: U S Dept of Int, Aero Charting 1943; Ronne Antarctic Res Expedition - Feature named

Landen, David (Continued)
Mt David Landen 1948; US Air Force, Charting with Radar Photos 1950; Int Soc of Photogrammetry. Co-author 'Multilingual Dictionary Photogrammetry' 1962. Louis Struck Award ASP 1971. Presidential Citations ASP 1966 & 1976. Assoc Ed in Ch, Manual of Remote Sensing ASP. Assoc Prof Lect GWU 1966-1978 (part-time). Life Member; ASP, ACSM, WSE. *Society Aff:* APS, ACSM, NAGE, WSE.

Lander, Horace N
Business: 1 Greenwich Plaza, Greenwich, CT 06830
Position: Sr V P - Res & Dev. *Employer:* Molybdenum Div of AMAX Inc. *Education:* ScD/Metallurgy/MIT; BS/Metallurgy/MIT. *Born:* May 28, 1923 Cambridge Mass. Served with Metal Hydrides Inc, Beverly Mass, Met Group Leader; Jones & Laughlin Steel Corp, Supr Process Dev; Youngstown Sheet & Tube Co, Dir of Res. Joined Climax Molybdenum Co 1970 as V P - Res & Dev. Appointed Sr V P Res & Dev 1976. Respon for dir Div's res & internatl mkt dev activities. Mbr Indus Res Inst; Amer Iron & Steel Inst - & similar inst's in UK & Japan; Amer Inst of Mining, Met & Petro Engrs - awarded R W Hunt Award 1960; Amer Soc for Metals - Fellow 1970. *Society Aff:* ASM, AISI, IRI, AIME.

Landes, Spencer H
Business: 15731 Steamboat Ln, Houston, TX 77079
Position: Pres. *Employer:* S H Landes, Inc. *Education:* MS/ChE/OK State Univ; BS/ChE/Univ of IL. *Born:* 7/8/22. Native of IL. Service with field artillery WWII. Work history includes res in synthetic rubber polymerization, gas plant design, design of crude-oil & natural gas production systems. Formed engg-consulting firm in 1971 to furnished expertise to oil & gas producing cos. Work sphere is intl with branches in Europe & Malaysia. Reg PE TX, OK, LA & AK. *Society Aff:* AIChE.

Landgrebe, David A
Business: School of Elec Engg, W Lafayette, IN 47907
Position: Prof. *Employer:* Purdue Univ. *Education:* PhD/EE/Purdue; MSEE/EE/Purdue; BSEE/EE/Purdue *Born:* 4/12/34. Native of Huntingburg, IN. Held positions at Bell Tel Labs, Interstate Electronics Corp, and Douglas Aircraft. Teaching & Res on Signal Representation and Info Systems at Purdue. Presently Prof of Elec Engrg. Dir of Purdue's Lab for Applications of Remote Sensing (LARS) from 1969 to 1981. Associate Dean of Engineering and Director of the Engineering Experiment Station from 1981-1984. Lectured widely on remote sensing. Consultant to Douglas Aircraft, NASA & the Natl Acad of Sci. Co-author of text "Remote Sensing: the Quantitative Approach–. Assoc Editor of Journal - Remote Sensing of Environment. Mbr of Tau Beta Pi, Eta Kappa Nu & Sigma Xi honorary socs. Recipient of NASA Medal for Exceptional Scientific Achievement, 1973. Elected Fellow of IEEE, 1977. *Society Aff:* IEEE, ASEE, AAAS, ASPRS.

Landis, Fred
Business: PO Box 784, Milwaukee, WI 53201
Position: Prof of Mech Engrg *Employer:* Univ of Wisconsin-Milwaukee. *Education:* ScD/Mech Eng/MA Inst of Tech; SM/Mech Eng/MA Inst of Tech; B Eng/Mech Eng/McGill Univ. *Born:* Mar 1923 Munich Germany. Design engr Montreal 1945-47. Asst Prof Stanford Univ 1950-52, at Northrop Aircraft 1952-53. Asst Prof to Prof at New York Univ 1953-73, Dept Chmn Mech Engrg 1963-73. Dean of Intercampus Progs Poly Univ of New York 1973-74. Dean, Coll of Engrg & Appl Sci Univ of Wisconsin- Milwaukee 1974-83. Prof Mech Engrg 1984-. Served on Heat Transfer Div Exec Comm ASME 1966-74; currently on Pol Boards ASME. Chairman, Publications Committee ASME 1985- Vice President, Prop. Dev. ASME 1985-89. Author over 40 prof & sci publs. Extensive pub serv, incl sch bds & on N Y State Commissions. *Society Aff:* ASME, ASEE, AIAA.

Landis, James N
Home: 2701-5 Golden Raid Rd, Walnut Creek, CA 94595
Position: Consultant (Self-employed) *Education:* BS/Mech Engrg/Univ of MI. *Born:* 8/18/99. Math instr Univ of MI; engrg asst Publ Service Comm Indiana; var pos ME Dept and Mech Engr Brooklyn Edison Co 1923-38; Mgr Contract Control & Inspection Dept, Asst Mech Engr Consol Edison Co of NY 1938-48; Bechtel Corp San Fran, Ch Power Engr 1948-52, VP 1953-64, Exec Cons 1964-69, Consultant 1969-77. Cons steam elec power plants using fossil & nuclear fuels. P Pres Engrs Joint Council; P Chmn US Natl Comm World Energy Conf; Mbr Natl Acad Engrg; Hon Mbr Amer Soc ME (Fellow, P Dir-at-Lrg, P Pres); Fellow Amer Nuclear Soc; Sigma Xi; Tau Beta Pi; P E NY, CA, FL, MI. *Society Aff:* ASME, NAE.

Landsberg, Helmut E
Business: Inst of Phys Sci & Techology, College Park, MD 20742
Position: Prof Inst Phys Sci & Technology. *Employer:* Univ of Maryland. *Education:* PhD/Geophysics & Meteorology/Univ of Frankfurt *Born:* Feb 1906. Native of Frankfurt Germany. Received educ there. Had post-doc apptment there. Came to U S 1934. Teaching jobs at Penn St & Univ of Chicago. Cons & Oper Analyst USAAC during WW II. Res Exec jobs Res & Dev Bd, Air Force Cambridge Res Center; Dir of Climatology, Weather Bureau; Dir US Environ Data Service. 1964-76: Chmn Univ of Md Meteorology Prog - res & pubs on applied climatology, specially used in arch & engrg - air pollution problems - effects of urbanization on ATMOS environ. Mbr Comm on Environ of var natl orgs. Special award in Biometeorology 1961 AMS; Brooks Award 1972 AMS. Cleveland Globe Award AMS 1983. Pres 1968-70 Amer Geophys Union. Bowie Medal 1978. V P 1960-62 Amer Meteorological Soc; IMO prize World Meteorol Organization 1979. Alfred Wegener Medal, German Meteorological Soc, 1980. William F Peterson Fnd Gold Medal Award 1982. *Society Aff:* OPCA, ASTM, AMS, AGU

Landsburg, Alexander C
Home: 307 Williamsburg Dr, Silver Spring, MD 20901
Position: CAORF Liaison *Employer:* Maritime Admin, Dept of Transportation *Education:* MSE/Naval Architecture/Univ of MI; BSE/Naval Architecture/Univ of MI; PMD/Business Administration/Harvard Business School *Born:* 12/23/42. Worked for the Maritime Admin since grad involved with ship design and operations. Mgr Design Dev 1973 & Mgr Computer-Aided Ship Design 1976 in the Div of Naval Arch, Office of Ship Const. Chf of Environment Activities and Mgr Computer-Aided Cost Analysis in 1978 in the Office of Shipbuilding Costs. Since 1985, CAORF Liaison in Office of Advanced Ship Operations. Mbr US delegation to 1973 IMCO Marine Pollution Convention. Soc of Naval Archs & Marine Engrs (SNAME); chrm Chesapeake Section, chrmn of Economic Analysis of Transportation Systems Panel, mbr serving on controllability panel. SNAME Cochrane Award 1974 and 1976 for best Sect paper. Mbr Assn for Computing Machinery (ACM) & ACM Spec Interest Group for Computer Graphics (SIGGRAPH). Married, 4 children, plays classical piano, tennis, builds small boats & houses, teaches use of home computers. *Society Aff:* SNAME, ACM, SIGGGRAPH, ASNE.

Landweber, Louis
Business: Institute of Hydraulic Research, The Univ. of Iowa, Iowa City, IA 52242
Position: Prof. Emeritus *Employer:* Inst of Hydraulic Res. *Education:* PhD/Physics/Univ of MD; MA/Physics/George Washington Univ; BS/Math/City Collee of NY. *Born:* 1/8/12. Employed at David Taylor Model Basin, 1932-1954, as Hd of Hydrodynamics Div. Presently Prof Emeritus in Mech Engg Dept. of Engg College, and Res Eng, Inst of Hydraulic Res, The Univ of IA, since 1954. Conduct res, have published many papers, and contributed to several books in field of Ship Hydrodynamics. Received Ward and Kenyon medals from CCNY, and the Davidson Medal of SNAME in 1978. Was Fifth David Taylor Lecturer at David W Taylor Naval Ship Res & Dev Ctr in 1978. *Society Aff:* SNAME, ATMA, AAM, NAE.

Lane, Charles E III
Business: 210 South 28th St, Birmingham, AL 35233
Position: Pres *Employer:* Lane/Bishop/Hodnett, Inc. *Education:* B-Arch/Structure/Auburn Univ *Born:* 10/12/29 Native of Memphis, TN. Recipient of AIA Silver Medal upon graduation from Auburn Univ in 1953. Prior employers:

Lane, Charles E III (Continued)
Van Keuren, Davis & Co, the Ralph M. Parsons Co, Davis, Speake & Thrasher, and Davis, Speake & Assoc. Registered Architect with NCARB certificate. Associate, Structural Eng, with Davis, Speake & Associates, 1963- 75. In 1975 opened Lane/Hodnett, Stuctural Eng. By merger in 1978 formed Lane/Bishop/Hodnett, Inc. Structural engr for Samford Univ, Haley Ctr at Auburn Univ, Montclair Baptist Medical Ctr, Park Place Office Tower in Birmingham, and Dauphin Way Baptist Church in Mobile. Enjoy woodworking and sailing. *Society Aff:* ACI, AISC, ASCE, ACEC, CRSI

Lane, Golden E, Jr
Business: Campus P. O. Box 25, Albuquerque, NM 87131
Position: Mgr, Structural Mech Div *Employer:* Univ of NM; NM Engrg Research Inst *Education:* PhD/CE/Univ of NM; MS/CE/Univ of NM; BS/CE/Univ of NM *Born:* 02/16/39 in Ogden, UT. Lived in Albuquerque, NM since 1954. Research Engr at UNM/NMERI since 1970. Mgr of Structural Mech Div since 1973. Served as Principal Investigator on several research projects dealing with reinforced concrete mbr response and high intensity dynamic loading of reinforced concrete models of hardened strategic structures. In the NM Section of ASCE have served a mbr of Bd of Dirs of Albuquerque Branch, Chrmn of Structural Engrg and Engrg Mech Group and Section Pres-1980. Mbr of ACI committee 444, Models of Concrete Structures. *Society Aff:* ASCE, ACI, NSPE, ASTM

Lane, James W
Home: 939 Autumn Dr, Jackson, MS 39212
Position: Chief Elec Engr. *Employer:* Cooke, Douglass & Farr, Ltd. *Education:* MS/Elec Engr/MS State Univ; MBA/Bus/MS College; BS/Elec Engg/MS State Univ. *Born:* 9/2/38. Born Sumner, MS, Sept 2, 1938; parents, Fred C & Essie L (Tierce) Lane; m Joyce Ann Rains, May 27, 1959; children-Jim, Jeffrey Alan. BSEE, MSEE, MS S Univ, MBA MS Coll. Engr North, Beasley and Swayze, cons engrs; Jackson 1966-71; design engr Leigh Watkins III & Assocs, cons engrs, Jackson, MS, 1971-74; prin James W Lane Cons Engr; 1974-76, sr elec engr, Cooke, Douglass & Farr Ltd, Architects & Engrs, Jackson, MS, 1976-83; Prin: James W Lane Cons Engr 1983-85. Chief Elec. Engr. Cooke, Douglass & Farr, LTD. 1985-present. Active Boy Scouts of Am Recipient, Eagle Scout, Silver Explorer, God and Country awards Boy Scouts Am. Reg PE AL, MS, TX, LA, FL, WY, SC, VA, Oklahoma NCEE certificate. Lted in Marquis Who's Who in the South and Southwest, who's who in Technology. *Society Aff:* NSPE, MES.

Lane, Joseph R
Business: 2101 Constitution Ave, Washington, DC 20418
Position: Sr. Program Officer *Employer:* National Acad of Sciences. *Education:* ScD/Metallurgy/Mass Inst of Tech; BS/Met Engrg/U of IL. *Born:* 1917. Worked on uranium slug coatings at Univ of Chicago Met Lab (Manhattan Proj) before grad sch. Headed High-Temp Alloys Sect Naval Res Lab 1950-55. Since then, Staff Metallurgist & Program Officer, Natl Matls Adv Bd of the Natl Acad of Sci, respon for organizing cttes in response to tech problems of gov agencies & seeing that a balanced report ensues. Active in local chap (Chmn 1969-70) & Natl Cttes of Amer Soc of Metals. Mbr TMS, SAMPE & Soc Mfg Engrs. Sigma Xi. Fellow ASM. *Society Aff:* ASM, TMS, SME, SAMPE.

Lang, Edmund H
Business: PO Box 1609, Tustin, CA 92681
Position: Consultant (Self-employed) *Education:* MS/CE/PA State Univ; BS/CE/Carnegie Inst Tech *Born:* Jan 1912. Engr on hydraulic models, Panama, & on hydraulic studies. Taught civil works courses at The Engineer School. Dist Engr U S Lake Survey respon for hydraulic studies of St Lawrence Seaway & Niagara power projects. Dir Waterways Exper Station, respon for work of 200 res engr & sci professionals. Resident Mbr, Bd of Engrs for Rivers & Harbors - Ch Oper Exec respon for analysis of water resources proj's. Active Colonel, Corps of Engrs to 1967. Alternate Foreign Trades-Zone Bd. ASCE Staff Dir of Prof Services from 1970-77. Consultant from 1977. *Society Aff:* ASCE, ASEE, SAME, PIANC

Lang, Edward W
Home: 421 Canterbury Ln, Gulf Breeze, FL 32561
Position: Technical Assoc. *Employer:* St Regis Paper Co. *Education:* BS/ChE/Univ of MO; AB/Chemistry/Univ of MO. *Born:* 3/5/18. in MO, educated at Westminster College & Univ of MO. Spent three yrs in USN Lt (jg) during WWII. Work experience primarily in res & dev at TN Valley Authority (2 yr), Southern Res Inst (19 yr), & St Regis Paper Co (11 yr). Fields of experience include minerals processing, organic chems process dev, coal conversion, water desalination, water pollution control. Positions include: Hd, Chem Engrg Div, Southern Res Inst; & Proj Leader, Hydropyrolysis Dev, St Regis Paper Co. Publications: 20 technical papers. Honors: Tau Beta Pi & AIChE Fellow. *Society Aff:* AIChE.

Lang, Hans J
Home: 136 Westervelt Ave, Tenafly, NJ 07670
Position: Pres *Employer:* Lang Assoc, Inc *Education:* MS/ME/MIT; ME/ME/Stevens Inst of Tech. *Born:* 11/17/12 H. J. Lang, who introduced the Lang Factors for cost estimation, is Pres of Lang Assoc, Inc. He has been assoc with process engrg and construction organizations for most of his working career. Lang was educated at the Stevens Inst of Tech, ME 1934, Mass Inst of Tech, MS 1936, and was admitted to the California Bar in 1955. Lang has published articles on estimating, profitability, capital mgmt and other facets of cost engrg, and he has a long and current record in managing process engrg and construction co. This record includes 12 years with C F Braun & Co, 6 years with the Lummus Co as VP of North American Operations, 3 years with Procon as Pres and Chief Exec Ofcr, and 10 years with Pritchard, as Pres and Chief Exec Ofcr. *Society Aff:* AACE, ASME, AIChE, ABA, State Bar.

Lang, Harold H
Home: 9046 Maplewood Drive, Berrien Springs, MI 49103
Position: Chmn, Engrg Dept. *Employer:* Andrews Univ. *Education:* PhD/Mech Engg/Univ of MI; MSE/Mech Engg/Univ of MI; BSE/Mech Engg/Walla Walla College; BA/Physics/Walla Walla College. *Born:* Jan 1932. Res engr Rocketdyne Div, North Amer Aviation Inc 1954-56, did exper work with model rocket engines. US Army 1956-58. Dev engr Bettis Atomic Power Lab, Westinghouse Elec Corp 1958-63, dev testing techs & equip for inspection of nuclear reactor core components. Chmn Engrg Dept Andrews Univ 1963- . Reg PE State of MI. Dissertation Title-A study of the Characteristics of Automotive Hydraulic Dampers at High Stroking Frequencies. *Society Aff:* ASEE, ASME, NSPE, MSPE.

Lang, Martin
Business: 1 World Trade Ctr, Suite 2637, New York City, NY 10048
Position: VP, Mgmt Consulting Services *Employer:* Camp Dresser & McKee. *Education:* MS/Sanitary Engg/NYU; BS/CE/CCNY. *Born:* 1/28/15. Native of NYC. Successively promoted, over 40 yrs, to inspection, design, construction, res & administrative assignments in wastewater & water progs of NYC, culminating in appointment as Commissioner of Water Resources Sanitation, & Comr of Parks & Recreation. Pres of WPCF in 1978-79. Now VP of Mgt Consulting Services of Camp Dresser & McKee. Among other awards, Camp Medal of WPCF, Civ Govt Award ASCE, Nichols Medal, & Man of the Yr, APWA, Cleary Award AAEE, Mayor's Medal, NYC. *Society Aff:* WPCF, ASCE, AWWA, ACS, NSPE, AAEE, NAE.

Lang, William W
Home: 29 Hornbeck Ridge, Poughkeepsie, NY 12603
Position: Program Mgr, Acoustics Tech. *Employer:* IBM Corp. *Education:* PhD/Physics/IA State Univ; MS/Physics/MIT; BS/Physics/IA State Univ. *Born:* Aug 1926. Native of Boston Mass. Cons in noise control engrg 1949-57. Taught at Naval Post Grad Sch, Monterey prior to joining IBM Corp 1958. Currently respon for IBM's programs in product noise control. Active in internatl & natl standards activities; Fellow IEEE, ASA, AES, Inst of Acoustics (UK) Mbr INCE, Natl Acad of

Lang, William W (Continued)
Engg. IEEE G-AE Achieve Award 1968 ASA Silver Medal 1984 IEEE Centennial Medal 1984 Registered Professional Engineer, New York State. *Society Aff:* INCE, ASA, AES, IEEE, AAAS, IoA.

Langbein, Charles E, Jr
Home: 120 Parkside Dr SE, Winter Haven, FL 33880
Position: Sr Partner. *Employer:* Langbein & Bell, Engrs. *Education:* BME/Mech Engr/Univ of FL. *Born:* 5/12/30. Reg PE FL & PA. Native of Akron OH. USNR in Korean Conflict. Application Engr for Worthington Corp, Proj Engr for IMCC, Sr Partner of Langbein & Bell, Engrs since 1962, a cons mech-elec firm. Mbr NSPE; Fellow Mbr FL Engg Soc, Chmn Awards Committee 1979-80. Chmn Ctte on Energy 1975-76 and FES Bd of Dir 1976-79. Mbr Dir 1978-79. V. Pres 1980-81, ASHRAE; Mbr ASPE. Chmn REAC (Regl Energy Action Committee, FL Energy Office) Region VII, 1978-80, NSPE Awards Committee 1980-81. *Society Aff:* NSPE, FES, ASHRAE, ASPE.

Langdon, Paul E, Jr
Business: Greeley and Hansen, 222 So Riverside Plaza, Chicago, IL 60606
Position: Partner. *Employer:* Greeley & Hansen. *Education:* MS/Civil Engr/CA Inst of Tech; BS/Engr/CA Inst of Tech. *Born:* Nov 1931 Evanston Ill. San Engr U S Pub Health Serv 1954-57. With Greeley & Hansen since 1957: Principal Asst Engr 1957-68, Assoc 1968-71, Ptnr 1971-, Business Dev Sponsor 1981-, Business Sponsor 1977-1981, Engineering Sponsor 1975-1977 . Proj Mgr on water supply & waste-water proj's for Tampa Fla, St Petersburg Fla, Sioux Falls S D, Portsmouth Va, Westchester Cty N Y, Homestake Mining Co, Panama R P, Phila Pa, Metro San Dist of Greater Chgo, Atlanta, GA. Mbr ACEC, ASCE, AWWA, NSPE, AAEE, WPCF, Reg in 5 states. *Society Aff:* AAEE, ACEC, ASCE, AWWA, WPCF, NSPE,

Langdon, Terence G
Business: Depts. of Materials Science & Mech. Eng, University of Southern California, Los Angeles, CA 90089-1453
Position: Prof. *Employer:* Univ of S CA. *Education:* BSc/Physics/Univ of Bristol; PhD/Physical Metallurgy/Imperial College, Univ of London; DSc/Physics/Univ of Bristol. *Born:* 1/24/39. Res Metallurgist, Univ of CA, Berkeley, 1965-67; Visiting Scientist, US Steel Corp, 1967-68; Res Fellow, Cavendish Lab, Univ of Cambridge, 1968-69; Res Metallurgist, Univ of British Columbia, 1969-71; Associate Prof, Univ of S CA, 1971-76; Prof of Mtls Sci, Mech Engg & Geol. Sci, Univ of S CA, 1976-. Fellow Inst of Physics 1972, Inst of Metallurgists 1975. Amer. Ceramic Soc. 1981. Numerous publications on the mech properties of metals and ceramics. *Society Aff:* AAAS, ACS, AGU, AIME, AIP, ASM, IOM, IOP

Langenberg, Frederick C
Business: 2015 Spring Rd, Commerce Plaza, Oak Brook, IL 60521
Position: Chrmn & CEO *Employer:* Interlake, Inc. *Education:* PhD/Metallurgical Engg/Pennsylvania State Univ; BS/Metallurgical Engg/Lehigh Univ; MS/Metallurgical Engg/Lehigh Univ; Visiting Fellow/Metallurgical Engg/MIT. *Born:* July 1, 1927. BS, MS Lehigh 1950-51; PhD Penn St Univ 1955; Visiting Fellow MIT 1955-56; Exec Program Carnegie-Mellon Univ 1962. Phi Beta Kappa, Tau Beta Pi, Sigma Xi. 1975-79, Chrmn & CEO, The Interlake Corp, 1983-present. Pres & CEO, The Interlake Corp. 1982-83. Pres, The Interlake Corp, 1979-82; Pres AISI; 1970-75 Pres Jessop Steel Co; 1968-70 Pres Trent Tube Div Colt Indus; 1956-68 V P R&D Crucible Steel Corp; prior 1956 U S Steel Corp. Dir Amer Iron & Steel Inst; British Iron & Steel Inst, Internatl Iron & Steel Inst, Amer Soc for Metals (Trustee), AIME, Amer Defense Preparedness Assn. David Ford McFarland Award 1973 Penn St Chap ASM; ASM Fellow 1970; ASM Pgh Nite Lecturer 1970; ASM Andrew Carnegie Lecturer 1976. *Society Aff:* ASM, AIME, AISE, AISI.

Langford, Ivan, Jr
Business: 1450 West Belt Dr. North, Suite 108, Houston, TX 77043
Position: Pres. *Employer:* Langford Engr, Inc *Education:* BS/Civ Engr/TX A & M. *Born:* 5/11/26. Native of Bryan, TX. Served in US Marine Corp in Pacific Theater with 27th Marines, 1943-1946. Grad from TX A & M College in 1950. Employed as Field Rep for Consulting Engr firm on construction of water supply and distribution facilities, 1950-1954. Design draftsman 1954-1955. Superintendent of commercial construction for Elec Contracting firm, 1955-1957. Proj engr for Consulting Engr Firm, 1957-1968. Founded Langford Engr, Inc, Consulting Engrs, in 1968. Firm offers consulting engr services to both public entities and private dev. Areas of specialization include: surface water treatment, water supply and distribution; waste-water collection and treatment; storm water drainage and streets. *Society Aff:* ASCE, AWWA, WPCF, NSPE.

Langhaar, Henry L
Home: Box 550, Lake Rd, Corydon, IN 47112
Position: Prof of Theoretical & Applied Mech. (Emeritus) *Employer:* Univ of Illinois. *Education:* PhD/Math/Lehigh Univ; MS/ME/Lehigh Univ; BS/ME/Lehigh Univ. *Born:* Oct 14, 1909 Connecticut. Test engr, seismographer, & math instr in 1930's. Struct engr Consolidated-Vultee AirCraft Corp 1941-47. Assoc Prof & Prof Theoretical & Applied Mechanics, Univ of Illinois since 1947. Fellow ASME, Mbr of Performance Test Code Ctte on Model Testing. Author 'Dimensional Analysis & Theory of Models' John Wiley & Sons 1951, 'Energy Methods in Applied Mech' John Wiley & Sons 1962, & co-author 'Engrg Mech' McGraw-Hill 1959. Tech articles on fluid mech, dimensional analysis, aircraft structures, plate & shell theories, elasticity & viscoelasticity, buckling theory, vibrations, numerical methods, strength of cooling towers, & population theory. Awarded von Karman Medal of ASCE in 1979. *Society Aff:* ASME, AAAS.

Langhetee, Edmond J, Jr
Home: 765 Jewel St, New Orleans, LA 70124
Position: Vice Chmn of the Bd. *Employer:* Louisiana Land & Exploration Co. *Education:* BS/-/LSU. *Born:* Sept 29, 1928 New Orleans La. s. Edmond Joseph & Hilda T (Dureau) L. M. Rosalie A D'Amic Oct 9, 1949. c. Edmond Joseph III, Jeanne, Leslie, Russell. With The Louisiana Land & Exploration Co New Orleans 1949- , V P 1962-65, Sr V P 1965-67, Exec V P 1967-74, V Chmn Bd 1974- , Dir 1971- ; Chairman & Dir Land Offshore Exploration Co; President & Dir Jacintoport Corp; Pres & Chairman Dir Kaluakoi Corp. Past Pres petro clubs New Orleans, mem Houston; New Orleans Geol Soc; Houston, New Orleans Dir 1976-78, Mobile Chambers Commerce; Amer Inst Mining & Met Engrs; IPAA; Independent Petroleum Association of American; Amer Petro Inst (Dir); Natl Ocean Indus Assn (Dir); Mid Continent Oil & Gas Assn. Democrat. Roman Catholic. Clubs: New Orleans Country; Metairie (La) Country. Enjoy fishing, golf & hunting. *Society Aff:* AIME, API, IPAA, NOIA.

Langhoff, Richard R
Home: 5 Pleasant View Ave, Mattapoisett, MA 02739
Position: Senior Research Scientist, Metallurgy, Materials and Process Section *Employer:* Raytheon Company. *Education:* BS/Metall Engg/Univ of Pittsburgh. *Born:* Apr 13, 1936. Secondary education in Pittsburgh, PA public schools. U.S. Air Force 1956-58. BS Met Engrg 1964 Univ of Pittsburgh. Employment history includes Continental Screw Co., New Bedford, MA. and current employer. Chmn RI Chap ASMINT 1973-74; Secy Mbrship CHe ASMINT 1974-75; Who's Who in Engineering 1980. Commercial Instument rated pilot. Enjoys classical music, and amateur astophotography. *Society Aff:* ASMINT, ASTM, AWS, IMS.

Langlinais, Stephen J
Home: Rt 2, Box 747, Erath, LA 70533
Position: Assoc Prof. *Employer:* Univ of S LA. *Education:* MSAE, Agri Engr/LSU; BSAE/Agri Engr/USL. *Born:* 11/1/43. Native of Erath, LA; 1966-1969 Proj Engr with Soil Conservation Service; 1969 to present-Assoc Prof at Univ of S LA in Lafayette La; 1971 to 1979 Chief Design Engr & Bd of Dirs with Rayne Plane Inc, a Natl Farm Equipment Mfg; 1972 to present - Pres of S J Langlinais & Assoc Inc, Consulting Engrs & Land Surveyors; 1974 to present - VP of 3-L Land Dev Corp;

Langlinais, Stephen J (Continued)
1978, served as Pres of LA Soc of Agri Engrs; 1964 Who's Who Among Students of American Universities & Colleges; 1976 Who's Who in the South; 1969 P E (Lic Prof Engr) 1971 P L S (Prof Land Surveyor). *Society Aff:* ASAE, LES, LSPS, NSPE

Langston, Joann H
Home: 14514 Faraday Dr, Rockville, MD 20853
Position: Director, Study Pgm Mgt Office *Employer:* U.S. Army *Education:* JD/Law/Univ of Maryland School of Law; PhD (ABD)/Econ/Univ of Maryland; BA/Mathematics/College of New Rochelle *Born:* Mar 1932. BA College of New Rochelle; PhD course work completed at Univ of Maryland; J.D. University of Maryland. Native of Staten Island NY. Military Oper Res Analyst: J H U/APL; Tech Ops/CORG; ORI, GEOMET Inc; Deputy Dir Office of Prog Planning & Eval CPSC 1974-1977. Associate Exec. Director CPSC 1977-1980; Director, Study Program Mgt. Office, Headquarters Dept of the Army 1980-present. Council Mbr ORSA 1976-79; Pres WORC 1968; Bd of Dir MORS 1969-73; Steering Ctte NGC 1965-72; fellow, Washington Academy of Sciences 1984-present. *Society Aff:* ABAA, WAS, ORSA, MORS

Lanius, Ross M, Jr
Home: 4200 Ridge Rd, North Haven, CT 06473
Position: Professor. *Employer:* Univ of New Haven. *Education:* BSCE/Civil Engg/Univ of DE; MS/Civil Engg/Univ of CT; MSCIS/Computer & Information Science/Univ of New Haven. *Born:* Jan 1937. Native of Wilmington Del. Hwy Engr with Conn Hwy Dept 1958- 59. Struct Engr with New Haven area cons from 1960-70. Joined the Cvl Engrg fac at the Univ of New Haven Sept 1970. Chmn of the Cvl Engrg Dept 1975-82 and since 1985 Active in Amer Soc of Cvl Engrs, President of Connecticut Section 1973-74. Sec ASCE Structural Div Ctte on Wood since 1981. Reg Prof Engr. Community Activity - Scoutmaster BSA. Hobbies - backpacking, jogging, photography & carpentry. *Society Aff:* ASCE, ACI, ASEE, FPRS, APT.

Lanker, Karl E
Home 1290 Wheeling Rd, Lancaster, OH 43130-8701
Position: VP-Engrg *Employer:* Sims Consulting Group *Education:* MBA/Bus/OH Univ; BSIE/Ind Engg/Rutgers Univ. *Born:* Aug 1925. Native of Absecon N J. USNR 1943-46, Pacific theater operation. Corp Indus Engr & Planning Mgr for Anchor Hocking Corp. Facils Planning Mgr for Lancaster Colony. With The Sims Consulting Group, Mgmt & Engrg Cons Firm, since 1969, in phys distribution. Respon for mgmt sys & engrg implementation of matl handling designs, sys, layouts, programs, material logistics & proj mgmt. Reg P E Ohio, N J & Calif. Cert with IMMS & IMC, IIE, Facilities Design & Planning Dir. Marquis Who's Who in Midwest and Finances Industry. Association of Consulting Management Engineers (ACME). Pub article on Productivity, co-author Warehouse Modernization and Layout Planning Guide, Naval Supply Systems Commands. NAVSUP Publication 529, December 1978, Contributing Author Material Handling Handbook, J Wiley, NY, 1984. Guest Lecture Amer Mgmt Assoc and Instructor OH Univ. Enjoy golf, music & photography. Cert Mgmt Consultant (CMC) with the Inst of Mgmt Consultants (IMC). *Society Aff:* IIE, IMC, ACE, IMMS, RES.

Lankford, William T, Jr
Home: 200 Mayfair Dr, Pittsburgh, PA 15228
Position: Conslt *Employer:* Self-employed *Education:* DSc/Metallurgical Engg/Carnegie-Mellon Univ; BS/Metallurgical Engg/Carnegie-Mellon Univ. *Born:* 11/1/19. BS & DSc Carnegie-Mellon Univ. Native of Rockwood, TN. Joined US Steel Corp Nov 1945 & engaged in res in plastic deformation as related to formability, fracture toughness, fatigue & other aspects of the mech behavior of metals. Became Asst Dir, Steel Products Dev 1963; Mgr Steel Processing 1967; Assoc Dir 1975; Director-Research Planning, 1981. Mbr AAAS & AISI. The Metallurgical Soc of AIME, Iron & Steel Soc of AIME, ASM, AISE & ASME. ASTM Richard L Templin Award 1949, ASTM Charles B Dudley Medal 1959; AIME Howe Memorial Lecturer, 1972. Fellow ASME 1968. Fellow ASM 1972. Fellow TMS of AIME 1977. Natl Pres of Carnegie-Mellon Univ Alumni Assn 1974. Trustee, Carnegie-Mellon University, 1980-1985. P E - PA. *Society Aff:* ASME, TMS, ISS, AISE, ASM, AAAS.

Lanni, Michael A
Business: 3900 Lankershim Blvd, Universal City, CA 91608
Position: Chief of Elect. Plant *Employer:* MCA Inc *Education:* BSEE/Elect Engr/Newark College of Engr NJIT *Born:* 05/05/40 Born NYC; raised in Jersey City. Studied at Newark College of Engrg. (NJIT), Entered Air Force S.A.G.E. System directly from College. Ist Lt. Settled in CA 1965; reg P.E. CA. Awarded the "Archimedes" award by the CA proff. Engrg Soc 1980. Serves on Panels 15 & 16 of National Elect Code Bd and is Regional VP of American Inst of Plant Engrs; Cub Scout Leader in pack 175 and Asst Scout Master in Troop 621. Married with three sons. *Society Aff:* AIPE

Lanning, Wayland H
Business: 1005 17th St-Rm 1750T, Denver, CO 80202
Position: Asst VP-Network Services. *Employer:* Mountain Bell Telephone. *Education:* BS/EE/Univ of Denver. *Born:* Dec 7, 1925 Oak Park Ill. BSEE 1949 Univ of Denver. US Army Air Force as Navigator 1943-46. Since joining Mountain Bell 1949 have held a num of tech & admin pos incl 10 yrs as Engrg Dir in Corp Engrg Dept. Appointed to present pos Jan 1, 1979. Avocations incl golf & sailplane flying. Served 6 yrs on Univ of Colo Engrg Dev Council, Chmn 2 yrs 1975-77.

Lannom, Edward H, Jr
Business: Jere Ford Highway, Dyersburg, TN 38024
Position: President. *Employer:* Forcum-Lannom Inc. *Education:* BCE/Civil Eng/Cornell Univ *Born:* July 1924 Union City Tenn. Served with Amer Field Serv 1944-45. Prof Engr Merrit-Chapman & Scott, Palatka Fla 1946-48; Union Bag & Paper Corp Savannah Ga 1948-50; Forcum-James Co Dyersburg Tenn 1950-53; Forcum-Lannom Inc Dyersburg Tenn 1953, Pres 1972. Named Outstanding Business Man Dyer Cty 1973; Reg P E Ga, Tenn, Ky, Ark, Miss. Mbr NSPE, Natl Dir 1975-78; SE V P PEC 1974-77; Mbr TN Bd of Architect & Engrg Examiners 1981-85. ASCE; Assoc Genl Contractors; Assoc Bldrs & Contractors. Episcopalian. Dyersburg Country Club. Colonial Country Club (Memphis). W Tenn Srs Golf Assn. Engrs Club (Memphis). Named Tenn Soc of Prof Engrs "Engr of the Year" 1984. Tenn Assoc of Prof Surveyors (Mbr). Reg L S Tenn.. *Society Aff:* AGC, NSPE, ASCE, NCEE, ABC

Lanouette, Kenneth H
Home: 96 Bayberry Lane, Westport, CT 06880
Position: President. *Employer:* Industrial Pollution Control Inc. *Education:* BEng/Civil/Univ of So CA. *Born:* Sept 1925. BE (Civil) Univ of So Calif 1945. US Navy 1943-46 & 1951-53. LCDR USNR ret. Sales Engr & Mgr Dorr Oliver 1947-69 spec in san engrg & indus processing. 1966-69 Asst to the Pres of Dorr Oliver. 1970 formed Indus Pollution Control, cons engrg firm spec in designing waste water treatment plants for indus. Authored sev papers on heavy metal removal & other pollution control subjects. Gold Medal AIDIS convention Puerto Rico 1959. Backpacking, singing, tennis. 1983 formed IPC Systems, manufacturer's of activated carbon systems. *Society Aff:* AIChE, ASCE, WPCF, NSPE.

Lanphier, Basil T
Home: 1703 Golf Rd, Reading, PA 19601
Position: Group VP. *Employer:* Carpenter Tech. *Education:* BS/Met/Penn State. *Born:* 4/5/22. Specialist in Stainless Steels and High Temperature alloys. *Society Aff:* AMS.

Lantos, Peter R
Business: 1500 Market St, Philadelphia, PA 19101
Position: Director R&D. *Employer:* ARCO/Polymers Inc. *Education:* PhD/Chem Eng/Cornell; BChE/Chem Engg/Cornell. *Born:* 1924. B Chem E & PhD Cornell Univ. Chem Warfare Service 1945-46. Plastics res with Genl Elec, fibers res with duPont. Sev supervisory pos with duPont Textile Fibers Dept & New Products Div,

Lantos, Peter R (Continued)
then Tech Dir Celanese Plastics Co. V P Sun Chem (R&D, Planning, Comm Dev), & Genl Mgr of Plastics Div, Rhodia Inc. VP, R&D ARCO Polymers, Inc, Philadelphia. Lic P E NY. Adj Prof Engrg Mgmt N J Inst of Tech. Joined ARCO 1976, respon for R&D & Commercial Dev - serve on Operating Ctte. *Society Aff:* ACS, AIChE, CDA, AMA, SPE, AAAS.

Lantz, Thomas L
Home: 124 E Cardinal, Wheeling, WV 26003
Position: Plant Hydraulics & Lubrication Engr. *Employer:* Wheeling-Pgh Steel Corp. *Education:* MBA/Bus/Univ of Pittsburgh; BSME/Mech Engg/WVA Univ. *Born:* June 12, 1936. Reg P E WV. Wheeling-Pgh Steel Corp 1959- , currently Steubenville Ohio Wks. 6 yrs - Mech Engr Benwood Works; 2 yrs - Elec Foreman Yorkville Works; 20 yrs in present capacity at Steubenville Works. Respon for specifying all lubricants & hydraulic fluids, coord Plant Oil Consolidation Program, reclamation of waste oils, advising on new hydraulic equip & redesign of old equip, hydraulic troubleshooting. Past Convention Sites Selection Chmn of Amer Soc of Lubrication Engrs & Past Mbr Bd of Dir 1974-84. P Chmn ASLE Upper Ohio Valley Section. Elected a Fellow in ASLE May 1979. Authored a Chapter in Revised Handbook of Lubrication Engg Published by CRC Press in 1983. Mbr of AISE Hydraulics and Lubrication Committee. Married - 2 children. AISE Board of Directors-1982. & 1987. Part-time Instructor Belmont Technical College, St. Clairsville, Ohio. *Society Aff:* ASLE, AISE.

Lanyon, Richard F
Business: 100 E. Erie St, Chicago, IL 60611
Position: Asst Dir Res. & Dev. *Employer:* Metropolitan Sanitary District *Education:* MS/CE/Univ of IL-Urbana;BS/CE/Univ of IL-Urbana *Born:* 9/4/37 Experienced in operation and maintenance of waterway systems for flood and pollution control, planning and design of flood control systems, management of environmental engrg research and pollution control and enforcement programs. *Society Aff:* ASCE, WPCF, APWA, AWRA.

Lanz, Robert W
Home: 4517 New Franken Rd, New Franken, WI 54229
Position: Assoc Prof *Employer:* Univ of WI-Green Bay *Education:* PhD/Eng. Mech/Univ of WI; MS/Eng. Mech/Univ of WI; BS/ME/Univ of WI *Born:* 8/19/37 Instructor in Engrg Mech at Univ of WI 1963-69. Then joined faculty at the Univ of WI, Green Bay as Asst. Prof of Engrg. Currently Assoc Prof. While at the univ, involved in Applied Energy Research projects. Lectured and published papers in proceedings of National and International Conferences on these subjects. Pres of R.W. Lanz & Associates, Inc., a Consulting Mech Engrg firm specializing in Forensic Engrg and Energy Conservation. National Secretary for AEE 1979-1981. *Society Aff:* ASHRAE, NSPE, WSPE, AEE

Lanzano, Ralph E
Home: 17 Cottage Ct, Huntington Station, NY 11746
Position: PE (Professional Engineer) *Employer:* City of NY - Dept of Environ Protection *Education:* BCE/Sanitary Engg/NYU *Born:* 12/26/26. I have been one of the more important engrs involved in water pollution control with NYC since 1960. I have contributed many population & quantity of flow, appurtenance design, infiltration/inflow analysis reports. I was an honor grad from NYU with BCE degree in 1959, & have been and am a mbr of many engg societies. I am actively engaged in alumni fund raising with NYU, & had been active with ASTM as a sustaining mbr a few yrs. I am in Who's Who in the East, Dictionary of International Biography, Intl Who's Who in Engrg (1st Edition), Who's Who in the World (7th Edition). *Society Aff:* ASCE, WPCF, ASTM, USITT, APHA, NSPE, AWWA.

Laplante, Donald G
Business: Suite 401, 116 Albert St, Ottawa Ont KIP5G3 CD
Position: Exec Dir *Employer:* Canadian Coun of Prof Engr(CCPE) *Education:* BS/Honors in Bldg and Public works, Civil Engr/Ecole Polytechnique, Montreal; Grad Dipl/Public Adm/Carleton Univ, Ottawa *Born:* 06/16/27 After grad with a Bachelor's degree in civil engrg from l'Ecole Polytechnique, Mr. Laplante joined Motreal's Dominion Bridge Co Ltd as a design engr. In 1964 he joined the Fed Dept of Indus as Indus Dev Officer of the Iron and Steel Div in Materials Branch. In 1968 he became Head of that branch's Constr Section. In 1970 he was seconded for nine months to the Prices and Incomes Commission. He returned to the dept's Materials Branch and became Chief of the Iron and Steel Div. One year later, in 1971, he was promoted to Dir of the Metals and Minerals Group of the Resource Indus and Constr Branch. In 1973 Mr. Laplante became Gen Dir of the Regional Office Branch of Indus, Trade and Commerce and in 1977 was appointed Dir Gen of the Mfg Indus Branch of the Foreign Investment Review Agency. In July 1984 Mr. Laplante became Exec Dir of the Canadian Coun of Prof Engr, headquartered in Ottawa, Ontario, Canada. Mr. Laplante is respons for the operation of the Council and its services. In addition he is respons for liaison with the fed govt, with the Council's 12 constituent assocs and with other natl and intl bodies. *Society Aff:* APEO, OIQ

LaPlante, John N
Home: 2522 W Winnemac Ave, Chicago, IL 60625
Position: Deputy Commissioner *Employer:* City of Chgo, Bureau of Traffic Engineering *Education:* MS/Transportation/Northwestern Univ; BS/Civil Engr/IL Inst of Tech. *Born:* July 16, 1939. Native of Chgo Ill. Served as a Bio-Environmental Engr with the US Air Force 1962-65. Traffic Engineer with the City of Chicago Bureau of Traffic Engineering and Operations since 1965, head of Planning Section 1975, Engineer of Traffic Planning 1978, Asst. Commissioner Dept. of Public Works 1984, Deputy Commissioner and Head of Bureau of Traffic Engineering & Operations 1986. Pres Ill Sect of ITE 1975. Chmn ITE Dist IV 1975. Pres Ill Engrg Council 1975-76. Recipient ISPE Young Engineer's Award 1971 and WSE Charles Ellett Award 1973. Active in local politics & community affairs, as well as Unitarian Church activities. *Society Aff:* ITE, ASCE, NSPE, WSE, TRB, WTS.

Lapostolle, Pierre M
Business: BP 5027, 14021 Caen Cedex, France
Position: Conseiller Scientifique. *Employer:* Grand Accelerateur Natl d'Ions Lourds. *Born:* 1922. Ancien Eleve Ecole Polytechnique & Ecole Nationale Superieure des Telecommunications. Docteur es-Sciences Physiques. Centre National d'Etudes des Telecommunications electron tubes & microwaves 1945-54. CERN (European Org for Nuclear Res): proton synchrotron; synchrocyclotron; storage rings - 1954-71. Centre National d'Etudes des Telecommunications Scientific Dir 1972-77. Grand Accelerateur Natl d'Ions Lourds: Scient Advisor 1977- . Mbr Comite Consultatif de la Recherche Scientifique et Technique 1973-76. Medaille BLONDEL 1958. Fellow Mbr IEEE. *Society Aff:* SEE, IEEE.

Lapsins, Valdis
Home: 3312 Highgrove Place, Kettering, OH 45429
Position: Owner *Employer:* V L Associates Consltg Engrs *Education:* MS/Civ Engrg/State Univ of IA; BS/Civ Engrg/State Univ of IA *Born:* 3/23/26 Born in Latvia. After completion of college courses in CE and City Planning, served two years in the US army. Served as City Planning Dir for Kettering, OH 1955-1959. With John W. Judge Engrg Co 1959-1984, most recently as VP and Chief Engr. Since 1984 owner of VL Associates, Consulting Engineers. Reg PE in OH. Reg Land Surveyor in OH, KY, TN, WV, GA and FL. Mbr of Chi Epsilon and Sigma Xi. Served as P of the Dayton Soc of PE and P of the Dayton Opera Assoc. P of the Dayton Consltg Engrs Assoc. Active in civic affairs. *Society Aff:* NSPE, ASCE, ACEC.

LaQue, Francis L
Home: Claridge Dr, Verona, NJ 07044
Position: Retired. *Education:* BSC/Chem & Met Engg/Queen's Univ, Kingston Ont Canada; LLD/Chem & Met Engg/Queen's Univ, Kingston Ont Canada. *Born:* July 21, 1904. Native of Gananoque Canada. Corrosion Engr, V P & Mgr Dev & Res and Special Asst to Pres, Internatl Nickel 1927-69. Pres Natl Assn Corr Engrs; ASTM; Electrochem Soc; Amer Natl Standards Inst; Internatl Org Standards. Speller Award NACE; Coonley Medal ASA; Acheson Medal Electrochem Soc; Moore Medal, Standards Engrg Soc. Aston-Polk Medal Anss Arch Colwell Medal SAE. Enjoys golf & sailing. *Society Aff:* ASTM, NACE, SNAME, ASM, SAE, ECS.

Laramee, Richard C
Home: PO Box 121, Eden, UT 84310
Position: Scientist. *Employer:* Thiokol Corp. *Education:* MS/Ind Engg-Bus Admn/Univ of UT; BS/Civ Engg/RPI. *Born:* 12/6/30. NYC. Married Valice M Schnarr Children Elaine R, Brian D. Nine yrs Sr Stress Analyst (Aircraft, Solid Rocket Motors SRM) at Goodyear Aerospace Corp. Nineteen yrs Proj Engr, Marketing Supervisor & Scientist responsible for dev & application of new mtls & design concepts to 6000 deg vectoring, SRM nozzle systems on launch vehicles & missile products at Thiokol Corp. Author, lectr & patentee in field. PE (CE) OH. UT Chapter CHrmn, SAMPE. Chrmn Master Bldg Task Force for Church position. Elder, Deacon in Presbyterian church. Former Bd Mbr - Akron Jr Chamber of Commerce & 1960 Natl PGA Golf tournament. Enjoy skiing, golfing, music & home projs. Mbr AIAA. *Society Aff:* SAMPE, AIAA.

Lardner, Thomas J
Business: Dept of Civ Engg, Amherst, MA 01003
Position: Prof. *Employer:* Univ of MA. *Education:* PhD/Appl Mechs/Poly Inst of NY; MS/Appl Mechs/Poly Inst of NY; BAeroEngg/Aero Engg/Poly Inst of NY. *Born:* 7/19/38. Native of Brooklyn, NY. US Army, 1961-1963; Asst Prof Applied Math, MIT, 1963-1968; Assoc Prof Mech Engg 1968-1973, MIT; Prof Theoretical and Appl Mechanics, 1973-1978, Univ of IL; prof civ engg 1978- Univ of MA. Fulbright Lectr, Nepal 1965-66. Consultant, NIH Fond Fdn, engg firms. Numerous publications in appl mechanics & biomechanics. *Society Aff:* ASME, SIAM, ΣΞ, ASEE.

Larew, H Gordon
Home: 2500 Hillwood Pl, Charlottesville, VA 22901
Position: Prof, Civ Engr. *Employer:* Univ of VA. *Education:* PhD/Soil Mech/Purdue Univ; MS/Soil Mech/Purdue Univ; BS/Civ Engg/WV Univ. *Born:* 6/5/22. Independence, WV, June 5, 1922, son H G & Lula M Larew, married M J Thompson, Nov 22, 1946; children: Jane Jo, H Gordon III, Elizabeth Thompson. Apr-Nov 1946, asst engr, NY Central System, NYC; Dec 1946-Feb 1956, instr, civ engg & engg mechanics, Purdue Univ; summers 1951-56, structures des with consulting engrs & arch in WV & Ind proj engr for earth dam, Jackson, O, & res engr on experimental test rd, Ind; Feb 1956-1961, assoc prof civ engg (soil mechs & fdn engg), Univ VA; 1961-to date, Prof of Civ Engg, UVA, 1952-to date, consultant to num engg firms on soil mechanics, fdn & earth dam prob. 20 contributions to tech journals. *Society Aff:* ASCE, ASEE, AAA, TBΠ, XE.

Large, Richard L
Business: 4333 S. 48th Street, Lincoln, Nebr 68516
Position: Pres *Employer:* R L Large & Assoc, Inc. *Education:* BS/ME/Univ of NB. *Born:* 4/29/31. Formed R L Large & Assoc, Consulting Engrs, in 1971. Primarily concerned with Ind and Vehicle Accident Reconstruction, Failure Analysis and Products Liability. 1958 to 1964 was Performance Engr and Operations Supervisor of a nuclear and conventional power plant. 1964 to 1970 was with a consulting engg firm and in 1970 was pres of the firm. Reg PE in NB & CO. Patentee of an electromech patient handling unit and of a self-storing wheel chair tray. *Society Aff:* ASME, ANS, NSPE, PEN, ASM, SAE, NFPA.

Larios, Christus J
Business: 67 Maiden La, Box 3707, Kingston, NY 12401
Position: VP *Employer:* Brinnier & Larios, P.C. *Education:* PhD/Community Services/Pacific Western Univ; MA/Engg/State Univ of NY; BS/CE/Cornell Univ. *Born:* 09/29/26 Native of Kingston, NY. Worked on Philadelphia Airport from 1949 to 1956. Heads all engrg for the firm that specializes in Municipal Engrg. Has had article in Water Works Engrg in 1961 on the Use of Hudson River Water and another in the New York State Legislators Annual on the Trials and Tribulations of Federally Funded Projects. Active in the community he now serves on the Boards of Ulster Savings Bank, Kingston Hospital, Ulster County Community Coll Foundation and the YMCA Trustees where he sits as Pres. The YMCA has honored him three times, last in 1980 with the Distinguished Service Award. Wrote "Just What the Patients Ordered" and a "Professionals Responsibility to His Community and Mankind." *Society Aff:* ASCE, NSPE, WPCF, ACEC, CSE, WRA.

LaRobardier, Lamont M
Business: Resource Management & Engrg Assoc, 852 Huron Rd, Franklin Lakes, NJ 07417
Position: Principal *Employer:* Resource Management & Engrg Assoc *Education:* Prof Degree/Indust & Mgmt Engrg/Columbia Univ; MS/Mech Engrg/Columbia Univ; BS/Mech Engrg/Columbia Univ. *Born:* 2/20/26. Prof Degree in Indus & Mgmt Engrg; Native of E St Louis IL. US Navy 1944-46 & 1951-53. Design engr for Colgate-Palmolive. Production Planning & Inventory Control Mgr for Curtiss-Wright Corp. Principal, Mgmt Services, in Arthur Andersen & Co. Corp Dir worldwide inventory mgmt Otis Elevator. Eltra/Allied Corp respon for dir of energy conservation cost reduction, value analysis & productivity improvement. Cons in productivity & profit improvement; developer of advanced production planning & inventory control sys. Lic PE NJ. Visiting Lectr for grad courses in production & financial planning & control, Stevens Inst of IIE VP-Finance, Bd of Tr 1974-76. Ed Bd APICS. Fellow IIE. Bd of Tr Advisory Ctte, NJIT 1980 - Pres. CVS (certified value specialist) by Society of American Value Engineers. CPIM (Certified in Production & Inventory Management) by American Production & Inventory Control Society. 1985 Value Engineer of the Year Award by SAVE. *Society Aff:* IIE, APICS, SAVE

LaRochelle, Donald R
Business: 436 Main St PO Box 140, Lewiston, ME 04240
Position: President. *Employer:* AL & H Prof Assoc *Education:* BS/Civil Engr/Univ of ME. *Born:* 11/26/30. Asst City Engr for City of Lewiston ME 1953-55. VP & Treas Aliberti, LaRochelle & Hodson Engrg Corp 1955-75, Genl Mgr 1975-77. Pres 1977-82, Chrmn of Bd 1982-Present. Chmn of Bd ALH Const Mgmt 1975- . Mbr ASCE, NSPE/PEPP, ACEC P Pres 1972-73 Cons Engrs of ME., Chmn Mbrship Dev Ctte 1975-77 ACEC. Chmn Construction Management Ctte 1980-82 ACEC. Chrmn Budget & Finance Ctte 1982-84. VP of ACEC 1985-1987 Speaker for seminar on Constr Mgmt conducted by ACEC. Publ article in "Consulting Engr" Magazine on Constr Mgmt. Fellow of ACEC. *Society Aff:* ASCE, NSPE, ACEC.

LaRochelle, Pierre L
Business: Dept of Civil Engrg, Universite Laval, Quebec, Quebec G1K 7P4 Canada
Position: Professor of Civil Engrg. *Employer:* Universite Laval. *Education:* PhD/Soil Mech/Univ of London; MSc/Soil Mech/Univ Laval; BSc/Civil Engg/Univ Laval. *Born:* Aug 1928. Have been involved in teaching & res at Universite Laval from 1960. Was head of the Cvl Engrg Dept between 1963-67. Was co-founder of the Canadian Geotech Journal & of the 'Assn Quebecoise des Transports et des Routes'. Author of many papers, & spec in sensitive clay problems & dam construction. Have been acting as expert cons on many proj's for engrg firms & for the Quebec & Canadian gov's. Presently mbr of the Cons Bd for the Baie James dam projects & for dam const in Tunisia. Fellow Mbr of the Engrg Inst of Canada & have been recipient of the Canadian Geotech Soc Prize in 1974, of the R F Legget Award in 1977 & of the Queen Elizabeth Jubilee's Medal in 1978. Recipient of Hogentogler Award 1985 of ASTM. Presently Chairman of the International Subcommittee on Landslides of the International Society of Soil Mechanics and Foundation Engineering. *Society Aff:* EIC, CGS, ASCE, ASTM.

Larock, Bruce E
Business: Civil Engg Dept, Davis, CA 95616
Position: Prof. *Employer:* Univ of CA. *Education:* PhD/Civil Engg/Stanford Univ; MS/Civil Engg/Stanford Univ; BS/Civil Engg/Stanford Univ. *Born:* 12/24/40. Grew up in Kensington, CA and attended El Cerrito High. At UC Davis since 1966. Married Susan Gardner 1968; two daughters, Lynne and Jean. Sr Visiting Fellow 1972-3 at Univ of Wales, Swansea. Sr. U.S. Scientist award from Alexander V. Humboldt Foundation 1986-7 to study at Technical University, Aachen, W. Germany. Co-author with D G Newnan of *Engineering Fundamentals*, 2 editions. Author of over 55 technical papers. *Society Aff:* ASCE, ΣΞ.

Laronge, Thomas M
Business: 10439 NE Fourth Plain Rd, PO Box 4448, Vancouver, WA 98662
Position: Pres. *Employer:* Thomas M Laronge, Inc. *Education:* MS/Chemistry/Drexel Univ; BA/Economics/Lafayette College. *Born:* 5/3/43. Native of Cleveland, OH. Taught college chemistry in Phila 1965-1969. Asst Technical Dir, Liquid-Solids Separation, Technical Specialist & Utility Industry Mgr - Betz Labs; Regional Mgr - IWW Div of Envirotech; Prod Mgr & Mgr of Ind Water Treatment - Eimco PMD; Asst VP & VP - Betz Environmental Engrs; Pres - Betz Converse Murdoch Inc - Western Group. Formed Thomas M Laronge, Inc in 1978 as a nationwide, industrially oriented, technical consulting service co & continues to serve as its Pres. Enjoy music, theater, photography, geology, paleontology & sailing. *Society Aff:* AIChE, NACE, ACS, ASM, WPCF, AWWA, ASTM, CTI.

Larrowe, Vernon L
Home: 1219 Share Avenue, Ypsilanti, MI 48198
Position: Senior Research Engr *Employer:* Environ Res Inst of Mich (ERIM). *Education:* PhD/Elec Engg/Univ of MI; MS/Elec Engg/Univ of IL; BS/Elec Engg/Univ of KS; AS/Engg/KS City Jr College. *Born:* Feb 1921. m. Florence Glinicki 1966. Radio Operator US Army Air Corps WW II. Employed by Univ of Mich Willow Run Labs 1951-73. Hd, Analog Computer Lab 1957-68. Also taught courses in Analog & Digital Computer Tech. Chmn Midwestern Simulation Council 1958-59. V Chmn Bd of Dir of Simulation Councils Inc (SCS) 1959-60. Life Mbr SCS (now Soc for Computer Simulation) 1969. With ERIM since its formation 1973. Working with airborne remote-sensing sys and high density digital tape recording sys. Hobbies: amateur radio, private flying, photography. *Society Aff:* IEEE, AAAS, SPIE, SCS.

Larsen, Dale G
Business: Box U, Vernal, UT 84078
Position: Service Engr. *Employer:* Western Co of North America. *Education:* BS/Geol Engg/SD Sch of Mines and Tech. *Born:* 1/11/56. Native of Colome, SD. Received a BS degree in Geol Engg from SD Sch of Mines and Tech. Held several positions in Theta Tau Natl Engg Fraternity while an active mbr in college. Began working for the Western Co of North America in June 1978. Current position is Service Engr. Current position with Theat Tau as an alumni is Western Regional Dir. Diversions: Collect matchbooks and play harmonica. *Society Aff:* ΘT.

Larsen, Lester F
Home: 1205 N 42nd St, Lincoln, NE 68503
Position: Professor Emeritus. *Employer:* Univ of Nebraska - ret. *Education:* BS/B Sci in Agri Eng/Univ of NB; MS/M Sci in Agri Eng/Univ of NB. *Born:* Sept 1908. Native of Plainview Nebr. Travelled for Inteernatl Harvester Co 1933-37. Grad Asst Agri Engrg Dept Univ of Nebr 1937-39. Asst Prof S Dak State Coll, teaching & res 1939-43. Extension Agri Engr Univ of Nebr 1943-46. Engr-in-Charge Nebr Tractor Test 1946-75. Completed over 800 tractor tests & publ the results. Over 40 papers & circulars in related fields. Awarded outstanding individual of the year Mid-Central Sect ASAE. Awarded Cyrus Hall McCormick Medal 1976 ASAE. Enjoy contacts with engrs & friends worldwide after 4 overseas tours. After retirement in 1975 prepared book "FARM TRACTORS 1950-1975–. Currently preparing a Tractor Museum on East Campus, Univ of Nebr, consisting of 35 historical tractors to date. *Society Aff:* ASAE.

Larsen, William E
Business: Civil & Agricultural Engineering, Montana State University, Bozeman, MT 59717
Position: Prof Agri Engrg. *Employer:* Montana State Univ. *Education:* PhD/Agri Engg/Univ of IL, Urbana; MME/Mech Engg/Univ of DE; BS/Agri Engg/Univ of NB, Lincoln. *Born:* Aug 1928. PhD Univ of Illinois, MME Univ of Delaware, BSAE Univ of Nebraska. Res, teaching & extension-Univ of Dela 1954-59. Extension Agri Engr Univ of Ariz - dev extension prog on machinery mgmt - 2 ASAE Blue Ribbon awards for Extension methods - Chmn ASAE Machinery Mgmt Ctte 1959-68. Agri Engrg at Montana State Univ since 1968. Admin, teaching & res on hay harvest methods. Reg P E Ariz & Montana. ASAE Sr Mbr; P Chmn Tractors Ctte; Chmn PNW Region 1975. Mbr NSPE, ASEE. *Society Aff:* ASAE, NSPE, ASEE.

Larsen, William L
Home: 335 N Franklin Ave, Ames, IA 50010
Position: Prof, Dept of Matls Sci & Engrg. *Employer:* Iowa State Univ. *Education:* PhD/Metallurgical Engrg/OH State Univ; MS/Physics/OH State Univ; BME/Mech Engrg/Marquette Univ. *Born:* 1926. Lt USNR - ret. P E Iowa 4236. Res Metallurgist E I duPont, Wilmington Del 1956-58; Ames Lab - AEC & Asst Prof Iowa State U 1958-61; Metallurgist & Assoc Prof 1962-69; Assoc & Prof 1969- . Mbr & Natl Ctte-man ASM; Mbr & Cert Corrosion Specialist Natl Assn of Corrosion Engrs; Mbr ASEE, ASTM, Iowa Archeological Soc, Natl Collegiate Honors Council. Failure analyst, cons, forensic metallurgist. *Society Aff:* ASM, NACE, ASTM, ASEE, NCHC

Larson, Carl S
Business: 1308 West Green St, 207 Engineering Hall, Urbana, IL 61801
Position: Assoc Prof & Asst Dean *Employer:* Univ of IL *Education:* PhD/ME/Univ of IL; MS/ME/Univ of IL; BS/ME/Univ of IL *Born:* 09/23/34 -Berwyn, IL. Faculty mbr since 1956. Have taught all undergraduate and several graduate courses in machine design. Have advised MS and PhD graduate students since 1965. Administrative duties: Asst Dean of the Coll since 1974; responsible for admission, scholarships, and transfer progs with affiliated schs. Research-responsible for more than 15 funded projects with various governmental agencies and private industries during past ten years. Publications--over 15 published articles. Honors--Tau Beta Pi, Pi Tau Sigma, Departmental outstanding teacher 1974, 1976. Other profl activities include: conslt for numerous industrial corp, technical committee mbr for ASME and SAE, expert witness, Advisor for Tau Beta Pi and Pi Tau Sigma. *Society Aff:* ASEE.

Larson, Charles F
Business: 100 Park Ave, New York, NY 10017
Position: Exec Director. *Employer:* Industrial Research Inst. *Education:* MBA/Mgmt/Fairleigh Dickinson Univ; BS/ME/Purdue Univ. *Born:* Nov 1936. Native of Indiana. Proj engr for Combustion Engrg Inc 1958-60; Secy of Welding Res Council 1960-70, Asst Dir 1970-75; Exec Secy Pressure Vessel Res Ctte 1960-75; Admin dozens of coop res projs; edited hundreds of tech reports & books. Joined Indus Res Inst 1975; concerned with advancing the mgmt of indus res. Chmn ASME Pressure Vessels & Piping Div 1969-70; Chmn PVP Honors Ctte 1972-76; Assoc Editor of Journal of Pressure Vessel Tech 1974-75. Co- editor, Innovation and U.S. Research, A.C. S. Symposium Series 129, 1980. Reg P E NJ. *Society Aff:* ASME, AAAS, SRA, ARD, NSPE

Larson, Clarence E
Home: 6514 Bradley Blvd, Bethesda, MD 20034
Position: Consultant. *Employer:* Self. *Education:* PhD/Chemistry/Univ of CA; MS/Chemistry/Univ of MN. *Born:* 9/20/09. Native of Minn. Undergrad educ at Univ of Minn; began res work at Univ of Calif. in radiochem. Prof of Chem & Chmn of the Chem Dept Univ of the Pacific. During WW II, dir res & dev proj assoc with the chem of the electromagnetic method for the Separation Plant at Oak Ridge - later becoming Mgr of Corp Res. 1961 returned to Oak Ridge as Pres of the Nuclear Div.

Larson, Clarence E (Continued)
1969 appointed a Commissioner of the Atomic Energy Comm by Pres Nixon. Hobbies: scuba diving & underwater photography, tennis; lic amateur radio operator. Many pats in nuclear field. Fellow ANS, Amer Inst of Chemists & recently elected to the Natl Acad of Engrg. Many awards for contrib to the nuclear energy field. Mbr Tau Beta Pi, Sigma Xi, Phi Lambda Epsilon. *Society Aff:* ACS, ANS, AAAS, TBΠ, ΣΞ.

Larson, Dennis L
Business: 507 Shantz Bldg, Univ. of Arizona, Tucson, AZ 85721
Position: Assoc Prof *Employer:* Univ of AZ *Education:* PhD/Agr. Eng/Purdue Univ; MS/Agr. Eng/Univ of IL; BS/Agr. Eng/IA State Univ *Born:* 2/3/40 Raised on a North Central IA farm. Received a PhD in agricultural engrg from Purdue Univ. Served with US Army in West Germany 1964-66. Was a farm machinery design eng for John Deere 1966-68. Provided advice in ag. engr to National Univ, Medellin and ag. experiment station, Bogota, Colombia 1970-72. With Univ of AZ since 1973. Presently Assoc Prof, responsible for courses in farm machinery design, solar and biomass energy use and systems analysis. Research in agricultural energy use and alternative energy source, including solar power plant, application. *Society Aff:* ASAE, ASEE, ASES

Larson, Frank R
Home: 115 Wilson Rd, Bedford, MA 01730
Position: Ch, Matls Dev Lab. *Employer:* US Army Matls & Mech Res Ctr. *Born:* June 1924. PhD Brown Univ Matls Sci; BS Tufts Univ. Native of Medford Mass. Ltjg USNR WW II. Engr Genl Elec, Thompson Lab, Lynn Mass 1946-51, design & testing of high temp alloys. Metallurgist Watertown Arsenal Labs 1951-56: mech metallurgy, phase transformation, metal deformation & processing, phys metallurgy of steel & titanium alloys. Phys Metallurgist for Army Matls Res Agency 1956-61: high strength steels & titanium failure analysis; Supr Phys Metallurgist Metals & Ceramics Lab 1961-73; Phys Sci Administrator, Matls Dev Lab 1974- . Charles A Coffin Award (G E) for Creep & Stress Rupture 1952; Dept of Army Meritorious Civilian Service Medal 1968; Secy of Army's Res Fellow 1968. Mbr AIME, ASTM, ASM, BIM, Sigma Xi.

Larson, Gale H
Business: 168 N 1st East, Logan, UT 84321
Position: President. *Employer:* Valley Engrg Inc. *Born:* Sept 17, 1938. BSCE 1963, MSCE 1965, post grad work 1968 & 1969 Water Qual. Presently holding the pos of Principal Engr in Valley Engrg Inc, with staff of approx 45 people with 3 offices in State of Utah, founded 8 yrs ago. P Pres Northern Chap Utah Soc of Prof Engrs; State Treas Utah Soc of Prof Engrs 1976-77; State Dir Water Pollution Control Fed 1976-77; Mbr Amer Cons Engrs Council ACEC; listed in Who's Who in West. Lic P E Utah, Idaho & Oregon, with license pending in Wyoming.

Larson, George H
Home: 419 Oakdale Dr, Manhattan, KS 66502
Position: Prof (Emeritus) *Employer:* KS State Univ *Education:* PhD/Agri Engrg/MI State Univ; MS/Agri Engrg/KS State Univ; BS/Agri Engrg/KS State Univ *Born:* 1/28/15 Served in US Navy as Commissioned Engrg Officer 1943-45. Taught courses in Agri Engrg 1946-84. Hd of dept of Agri Engrg, KS State Univ 1956-70. Res in power and machinery area - utilizing LP-Gas in tractors and for weed control by flaming, operating costs of field machinery, and equip systems for pest control. Cons and Prof for USAID, sponsored projs; ICA, Bogota, Colombia (1970-72); Ahmadu Bello Univ, Nigeria (1972-74); Central Luzon State Univ, Philippines (1978-80). Short term cons to Univ of Cairo, Egypt and Paraguay, S.A. in Agri Mechanization. *Society Aff:* ASAE, NSPE

Larson, Harry T
Business: PO Box 3310 618 B318, Fullerton, CA 92634
Position: Sr Scientist. *Employer:* Hughes Aircraft, Ground Sys Group. *Education:* MS/Elec Engg/Univ of CA, Los Angeles; BS/Elec Engg/Univ of CA, Berkeley. *Born:* 1921 Berkeley Calif. Radar Officer AAF WW II. Starting in 1949, engrg, engrg & software mgmt, application, & teaching in computers, data processors, displays, peripherals, & command & control sys for use in sci, aircraft, business, indus, military, & space. In charge of Gemini/Apollo Mission Control conceptual design, & dev of Display Subsystem. Sr Scientist, Hughes Aircraft, in command and control since 1978. Fellow IEEE. Centennial Medal, IEEE. Natl Chmn IRE PG on Electronic Computers 1954-55. Bd of Gov AFIPS. Dir Social Implications of Computers activities for 14 yrs. Editor, pubs on computer & data processor design & application. *Society Aff:* IEEE, SID.

Larson, Kenneth H
Home: 3078 Sunnyside Terrace, Stoughton, WI 53589
Position: Environmental Eng *Employer:* Warzyn Engrg, Inc. *Education:* B/CE/Univ of MN *Born:* 1/23/47 Native of Caledonia, MI. Municipal eng with Minneapolis Public Works 1970- 1972. Served in Peace Corps in Cameroun, West Africa, in urban planning program from 1972-73. Environmental engr with MI Pollution Control Agency and private consultant in St. Paul, MI from 1973-77. With Warzyn Engrg, Inc. of Madison WI since 1977; as Ashland Branch Office Mgr from 1978-81 and presently as environmental eng in Madison Home Office. Lake Superior Chapter VP of WSPE, 1980-81. Professional registration in MI, WI, and MI. *Society Aff:* NSPE/WSPE, ASCE, WPCF

Larson, Marvin W
Business: 31550 Trillium Trail, Pepper Pike, OH 44124
Position: Pres. *Employer:* Larson International Services. *Education:* Adv Mgr/Bus/Harvard Bus School; BS/Chem Engg/MI State Univ. *Born:* 5/11/19. Graduated at top of Engg Class at MI State. Foreman in first general purpose synthetic rubber plant on Pearl Harbor Day held various tech & supervisory position in synthetic rubber program during WWII. Project Manager to design, build and start up first PVC plant in Japan in 1951-52. Dir of MFG Engg & Dev in international Dept of BF Goodrich Chem. In this capacity had leading position in organizing, building and operating fifteen plants in twelve countries. Area Dir for Latin America 1972-78. Dev Dilvte Ethylene Process for Vinyl Chloride Manufacture. Organized and started International Consulting firm in 1978-79. *Society Aff:* AIChE.

Larson, Maurice A
Business: Dept of Chem Engr, Iowa State Univ, Ames, IA 50011
Position: Prof & Chrmn. *Employer:* IA State Univ. *Education:* BS/ChE/IA State Univ; PhD/ChE/IA State Univ. *Born:* 7/19/27. Employed as a Chem Engr at Dow Corning Corp 1951-1954. Faculty of Chem Engg, IA State Univ, 1954-present. Currently Anson Marston Distinguished Prof & Chrmn of Chem Engg. Author of publications in Crystallization & Fertilizer Tech. Co-author of book - Theory of Particulate Processes. Chrmn IA Sec AIChE-1971, Chrmn, Div of Fertilizer & Soil Chem ACS-1975, Shell Visiting Prof, Univ College London-1971-72. NSF Faculty Fellow Stanford Univ-1965-66, Natl Acad of Sci Visitor Czechoslovakia, Poland-1974, Consultant IIT Kharagpur, India-1968. *Society Aff:* AIChE, ACS, ASEE.

Larson, Thurston E
Home: 706 LaSell Drive, Champaign, IL 61820
Position: Asst Ch & Hd of Chem Sect. *Employer:* Illinois State Water Survey. *Education:* PhD/Sanitary Chem/Univ of IL; BS/Chem Engg/Univ of IL. *Born:* March 1910. Native of Illinois. Employed by State of Ill 1932- . Lic P E Ill. Pres 1970-71 & Hon Mbr 1974 AWWA; SAME Publ Prize 1957; Medal for Outstanding Service 1966; Res Award 1972; Amer Chem Soc Pollution Control Award 1976. Author more than 80 pubs on corrosion, water chem, analytical methods, water treatment. Adv Ctte PHS 1962 Drinking Water Standards, the EPA 1973 Standards, & the NAS Ctte on the 1972 Water Qual Criteria. Mbr, Natl Acad of Engg, 1978. *Society Aff:* AWWA, ACS, NACE, WPCF.

Larson, Vernon M
Business: 505 City Pkwy W, Rm 400, Orange, CA 92668
Position: Asst VP. *Employer:* Factory Mutual Engg Assoc. *Education:* BS/Industrial Engr/Univ of CA. *Born:* 2/9/21. Native of IA. Served with Army Air Force 1942-45. Grad from UCLA in 1949 with degree BSIE. Joined Factory Mutual Engg Assoc as a fire protection engr in 1950 in CA. Have spent entire engg career with this organization, which is a leader in industrial loss prevention engg. Currently Asst VP and Dist Mgr in Los Angeles area, a position held since 1966. Past Chapter Pres of Society of Fire Protection Engrs. Elected Dir of SFPE in 1977. Nominated to grade of Fellow in Institute for Advancement of Engg in 1973. PE Fire Prof Engr., Cal.. *Society Aff:* SFPE.

Lashmet, Peter K
Home: 15 Center View Dr, Troy, NY 12180
Position: Assoc Prof *Employer:* Rensselaer Polytechnic Institute (RPI) *Education:* PhD/ChE/Univ of DE; MSE/ChE/Univ of MI; BSE/ChE/Univ of MI; BSE/Eng Math/Univ of MI. *Born:* 08/28/29 On faculty of RPI since 1965; tenured 1968; Exec Ofcr of Dept since 1977. Process engr, M. W. Kellogg Co, 1952-1953. With Air Products & Chems, 1958-1965, as mgr, cryostat engrg, and senior specialist. Served with chem corps AUS, 1953- 1955. Patentee and conslt in field. Mbr AIChE (treas, local section, 1969-1970). Mbr ACS (Div of Industrial & Engrg Chem, chrmn, 1975; chairman-elect, 1974; chrmn, prog committee 1971-1973). Representative of ACS to ANSI B78 Standards Committee, to 1980. Registered PE in PA and NY. *Society Aff:* AIChE, ACS

Laster, Richard
Home: 23 Round Hill Rd, Chappagua, NY 10514
Position: Pres and Chief Exec Officer *Employer:* DNA Plant Tech Inc. *Education:* BChE/Chem Engg/Polytech Univ. *Born:* Nov 1923. With DNA Plant Tech since 1982, Pres and CEO, Mbr Bd of Dir. With Genl Foods since 1944-82. Started in Engrg Res - moved into mfg & oper 1958, followed by mkting & then becoming Pres of Maxwell House Coffee Div & V P Genl Foods Corp 1971. Appointed Exec V P Genl Foods Corp 1973 & elected to Bd of Dir of Genl Foods Corp. 1972 Award of Food & Bioengrg Div AIChE & Disting Alumnus Award from Polytechnic Inst of N Y 1976. Elected to Chappagua Sch Bd 1971 - Pres 1973-74. Mbr AIChE, AAAS, ACS, AIC, Tau Beta Pi, Sigma Xi, Phi Lambda Upsilon. Enjoy classical music, photography, tennis, skiing & nature. Trustee Polytechnic Inst of NY 1979; Trustee Suny College of Purchase NY 1979; Chmn SUNY College of Purchase Foundation; VP United Way of Westchester 1978-81; Bd of Dirs Firestone Corp 1979-1985, Bd of Dirs Sopro Food Inc 1982-1986, Bd of Dirs Bonater Inc 1984. Bd of Dirs Lombardy Hotel 1986. Trustee American Health Foundation 1980-1983, Mbr Exec Comt-Center for Natl Policy 1983, Mbr NY Acad of Sciences-1984. *Society Aff:* AIChE, AAAS, ACS, AIC, IFT, NYA of Science.

Lataille, Jane I
Business: 85 Woodland Street, Hartford, CT 06102
Position: Chemical Acct. Consult. *Employer:* Industrial Risk Insurers *Education:* MS/EE/Rensselaer at Htfd. Grad. Ctr.; BS/Physics/Worcester Polytechnic Institute *Born:* 10/23/53 Began as field engineer for Industrial Risk Insurers. Subsequently held positions of supervisor and regional staff consultant. Currently a Chemical Account Consultant. In addition to degrees, also have background in petrochemicals, chemistry, insurance and risk mgmt. Licensed PE in Fire Protection Engrg in CT. Have served on five NFPA cttees and the Underwriters' Labs Technical Advisory Group for Intrinsically Safe Electrical Equipment. Have chaired two cttees and held every office in Connecticut Valley Ch SFPE. On Publications cttee for New England Chapter SFPE and the National Society. Published in NFPA Handbook, IFHH, and Plant Engineering. Enjoy gardening, hiking and dancing. *Society Aff:* SFPE, NCEE, NFPA

Latanision, Ronald M
Business: Dept of Mtls Sci & Engg, Cambridge, MA 02139
Position: Prof. *Employer:* MA Inst of Tech. *Education:* PhD/Met Engg/OH State Univ; BS/Met/PA State Univ. *Born:* 7/2/42. Born in Richmondale, PA. Postdoctoral NRC-NAS Fellow at the Natl Bureau of Stds in 1969. Res Scientist 1970-73, Acting-Hd Mtls Group (1974) Martin Marietta Labs. Sr Sci Award Alexander von Humboldt Fdn at Max-Planck-Institut fur Eisenforschung in Dusseldorf, 1974-75. Joined the faculty at MIT in July 1975. Director HH Uhlig Corrosion Laboratory, 1982- ; Director Materials Processing Center, 1984- . Res interests in chem stability of mtls, hydrogen embrittlement, Materials Processing. A B Campbell Young Author's Award, NACE, in 1971. Henry Krumb Lecturer, AIME, 1984; McFarland Award, Penn State, 1986; Shell Dist Prof of Materials Sci, 1983-1988; Sci Adv, US House of Reps, Ctte on Sci and Tech 1982-1983. Editor: *Advances in the Physics and Mechanics of Surface*; *Surface Effects in Crystal Plasticity*; *Atomistics of Fracture*. *Society Aff:* NACE, AIME, ASM, ECS.

Latham, Allen, Jr
Business: 400 Wood Rd, Braintree, MA 02184
Position: Founder *Employer:* Haemonetics Corp. *Education:* BS/ME/MIT. *Born:* May 1908 Norwich Conn. Process engr duPont, W Va 1930-35. Participated in dev of Polaroid Corp - Dir, Treas, Genl Mgr 1936-41. With Arthur D Little Inc, becoming Sr V P 1942-67. Founder & Pres Cryogenic Technology Inc 1967-72. Founder & Pres Haemonetics Corp 1972- . 33 pats relating to gas compression apparatus, cryogenic apparatus, blood processing. Engrg Soc of New Eng Award 1970. Author num pubs affil with prof engrg orgs. Mbr NAS Ctte on Impacts of Stratospheric Change. Hobby: Landscape Gardening. *Society Aff:* NAE, ASME, AIChE, ISA, NAS, AAAS.

Latham, Charles F
Business: 102 W. Second St, P.O. Box 661, Hopkinsville, KY 42240
Position: Branch Mgr *Employer:* Howard K. Bell Consulting Engrs, Inc *Education:* BS/CE/Univ of KY *Born:* 11/27/27 A native of Lexington, KY. Grad from high and served in US Army before entering Univ of KY and started part-time work with Howard K. Bell Consulting Engrs, Inc in 1948. Received coll degree in 1953. Has continuous employment with the firm. Has served in all capacities from surveying to admin including bookkeeping. Presnetly a Sr VP & Branch Office Mgr. Also, serving as Proj Mgr for $30 million proj of wastewater facilities. Active in civic affairs and athletics. *Society Aff:* AWWA, ACEC, SAME

Lathrop, Jay W
Business: Elec Engg Dept, Clemson, SC 29631
Position: Prof. *Employer:* Clemson Univ. *Education:* PhD/Physics/MIT; MS/Physics/MIT; BS/Physics/MIT. *Born:* 9/6/27. Born and raised in the state of ME (Orono). Electronic scientist, Natl Bureau of Stds/Harry Diamond Labs 1952-58 where I developed techniques for microminiaturization of solid state circuits, receiving Dept of the Army Meritorious Civilian Service Award for this work. Mgr of Advanced Tech, Components Group, TX Instruments 1958-68 where I was responsible for the dev of monolithic integrated circuit fabrication tech. Prof of elec engg at Clemson since 1968, active in both energy conversion and integrated circuit reliability res. Co-inventor of solar chem converter system of energy conversion. Fellow of IEEE. Consultant on solar energy and reliability. *Society Aff:* IEEE, ISES, ΣΞ

Lathrop, Kaye D
Business: Stanford Linear Accelerator Center, PO Box 4349, Stanford, CA 94305
Position: Assoc Director *Employer:* Stanford Univ *Education:* PhD/ME/CA Inst of Tech; MS/ME/CA Inst of Tech; BS/CE/US Military Acad. *Born:* Oct 8, 1932 Bryan Ohio. m. 1957; 2 children. Reactor Physics & Safety BS US Military Acad 1955; U S Atomic Energy Comm fellowship Caltech 1958-61; MS 1959; R C Baker Found Fellowship 1961; PhD (mech engrg phys) 1962. Staff mbr reactor math Los Alamos Sci Lab 1962-67; staff mbr & group leader reactor phys methods dev, Nuclear Analysis & Reactor Phys Dept, Genl Atomic Div, Genl Dynamics Corp 1967-68; Group Leader 1972, Asst Div Leader 1973, Assoc Div Leader 1975, Alternate Div Leader 1977, Div Leader 1978, Assoc Dir for Engg Sciences 1979 Los Alamos National

Lathrop, Kaye D (Continued)
Laboratory, Assoc Dir, Tech Div, Stanford Linear Accelerator Center, 1984 Fellow Amer Nuclear Soc 1972; E O Lawrence Meml Award 1976; Bd of Dirs 1973-76, 1977-81, Treasurer 1977-79 & Executive Committee 1977- 81, Amer Nuclear Soc. National Academy of Engineering. *Society Aff:* ANS, APS, NAE.

Latona, Joseph D
Business: 1868 Niagara Falls Blvd, Tonawanda, NY 14150
Position: VP/Sec *Employer:* Krehbiel Assoc Inc. *Education:* BS/CE/IN Inst of Tech; MS/ Environ.Eng/Univ of NY at Buffalo *Born:* 1/20/41 Native of Buffalo, NY. CE for Conable, Sampson, Vankuren, Huffcut and Gertis; Brown-Devlin Assoc; and Krehbiel-Guay-Rugg-Hall, from 1966-1977, specializing in all types of municipal, land development, industrial and private projects. Assumed current responsibility as VP/SEC for Krehbiel Assoc, Inc in 1977. In 1980 became the Chief Executive Officer of the company and initiated a major reorganization of the firm's staff and policies. Enjoys racquetball and running. Publications include Status of Wastewater Treatment in Erie County New York- J.D. Latona and N.E. Hopson. *Society Aff:* ASCE, AWWA, NSPE, NYSSPE, WPCF, ACEC

Latter, Robert F
Home: 179 Pittsford Way, New Providence, NJ 07974
Position: Retired *Education:* BS/EE/MI State Univ. *Born:* 4/2/21. in Lansing, MI. Lic P E NY 1949. Long Lines career began as a student engr in Chgo on Sept 30, 1946. Prior to attending the Operating Engrs Engrs Training Program at Bell Labs in 1956, held var pos in Engg, Traffic & Plant Depts of Long Lines & AT&T Genl Depts. Apptd Genl Dept's Sys Planning engr 1965 & Engg Dir - Transmission sys in Long Lines in 1978. Retired Sept 1, 1984. Sr Mbr IEEE, Mbr AIAA, ARRL & amateur radio soc's & clubs. Mbr New Providence Methodist Church, Bd of Trustees. Pres Pittsford Civic Association 1986-87. *Society Aff:* IEEE, AIAA, ARRL

Lattin, Clark P, Jr
Home: 439 Riverside Dr, Portsmouth, VA 23707
Position: Ret. *Education:* BME/-/Univ of DE. *Born:* 2/-/16. 22 yrs Foster Wheeler Corp, culminating as VP & Dir. 5 yrs Chem Const Corp. 3 years Kellogg International Corp. subsidiary of M.W. Kellogg Co Pres. 8 years Pres M.W. Kellogg, Div of Pullman Inc. 3 years Pres & CEO Pullman Inc.

Laube, Herbert L
Home: 120 Windcrest Dr, Camillus, NY 13031
Position: Author & Lecturer. *Education:* BS/ME/IA State Univ. *Born:* 1899 St Paul Minn. Engr Parker Ice Maching Co 1924-27; Carrier Corp 1927-46; V P Intl Div 1934-41; V P Engrg Div 1941-46; Pres Remington Corp 1946- 65. Sr Cons, Climate Control Div The Singer Co 1965-70. Dev incremental sys of air cond for multi-room bldgs. Dir Empire State Chamber of Commerce 1954-71. Tau Beta Pi, Pi Tau Sigma, Sigma Alpha Epsilon. Patentee field of air cond. Fellow 1972 ASHRAE 1977-present-Pres, The Comfort-Meter Co. *Society Aff:* ASHRAE.

Lauber, Thornton S
Home: 1005 Seminole Rd, Scotia, NY 12302
Position: Prof *Employer:* Rensselaer Polytechnic Inst *Education:* PhD/EE/Univ of PA; MS/EE/IL Inst of Tech; BS/EE/Cornell Univ *Born:* 1/5/24 Cornwall, Ont Canada. Citizen of USA. Married, 4 children. Served with US Army Signal Corps 1944-46. 1946-50, Commonwealth Assoc in substation design and system planning. 1951-58, GE. Co., Pittsfield, MA, Large Power Transformers in design and heat transfer development. 1958-69, G.E. Co., Philadelphia, Power Circuit Breakers as senior analytical eng. 1969-81 RPI, prof of electric power engrg. Tenured in 1970. Appointed S.B. Crary Prof of Engrg in 1971. *Society Aff:* IEEE, ASME, ΣΞ.

Laudise, Robert A
Home: 65 Lenape La, Berkeley Hgts, NJ 07922
Position: Dir Physical & Inorganic Chem *Employer:* Bell Telephone Labs *Education:* PhD/Chem/MIT; BS/Chem/Union *Born:* 9/2/30 in Amsterdam, NY. Held positions of Staff Member, Head Crystal Chemical Research Dept. Ast. Dir and Dir of Materials Research at Bell Labs. Research Interests Electronic Materials and Physical and Inorganic Chem. Presently Dir Physical & Inorganic Chem Res. *Society Aff:* ACS, AACG, ACS, IEEE, AMS.

Laughlin, John S
Home: 48 Graham Rd, Scarsdale, NY 10583
Position: Chmn Dept of Medical Physics. *Employer:* Mem Sloan-Kettering Cancer Ctr. *Education:* DSc (Hon)/-/Willamette Univ; PhD/Physics/Univ of IL; MS/Physics/Haverford College; AB/Physics/Willamette Univ. *Born:* Jan 1918. Appointed to Univ of Ill Coll of Medicine as Asst & then Assoc Prof of Radiology. Appointed Chmn of Dept of Medical Phys & Prof of Biophys Mem Sloan-Kettering Cancer Ctr 1952. Fellow Amer Phys Soc, Amer Coll of Radiology. Served as Pres of Amer Rad Soc, Health Physics Soc, & Amer Assn of Physicists in Medicine. Hon DSc Willamette Univ 1968, & Univ of Ill Coll of Engrg Disting Alumni Serv Award 1971. Res & teaching in therapeutic use of high energy electrons & x-rays, dev of computerized treatment planning, dev of quantitative nuclear medicine scanning tech, & metabolic studies with cyclotron produced radionuclides. *Society Aff:* ACR, HPS, RRS, AAPM, ASTRO, SNM.

Lauletta, Paul A
Home: 2109 Shaw Woods Dr, Rockford, IL 61107
Position: Metallurgical Engineer *Employer:* Private Practice, Consultant *Education:* BS/Met Engg/Purdue Univ. *Born:* 1/23/20. Native Chicago, IL. Met Dept Carnegie-IL Steel Corp, 1941-42. Met Lab, Univ of Chicago, 1942-46. Res on uranium & other metals used for nuclear energy. Various engrg & operational positions, Joslyn Stainless Steels Div, 1946-76. Appointed Div Pres in 1973. With Techalloy Co, Inc since 1976. Became Gen Sales Mgr-Midwest in 1978. Appointed VP and General Mgr, Techalloy IL in 1981. Responsible for operations and sales of stainless steel and nickel alloy wire produced by Techalloy IL plant of Techalloy Co, Inc. Reg PE, IL. Private consulting practice since 1986. *Society Aff:* ASM, AWS, WAI.

Laumann, Robert C
Business: 11395 Chester Rd, Cincinnati, OH 45459
Position: President. *Employer:* Technical Equipment Sales Co. *Education:* ME/Mech Engr/Univ of Cincinnati. *Born:* Nov 11, 1931. Presently Pres, Tech Equip Sales Co, Midwest Machine Tool Distributor. 1976-79; 80-82 Natl Dir, Soc Mfg Engrs SME. 1972 Tr, Dayton Engrg & Sci Ctr; Chmn Affil Soc Council of Dayton 1972; Chmn Region IV SME 1975. Dayton Chap SME 1969. Dir Amer Machine Tool Distributor Assoc 1983-1986. Mbr Dayton Engrs Club & Numerical Control Soc. Cert Mfg Engr. *Society Aff:* SME, AIMTECH, AMTDA.

Launder, Brian E
Business: Dept of Mech Engg, Davis, CA 95616
Position: Prof. *Employer:* Univ of CA. *Education:* DSc/-/Univ of London; ScD/Fluid Mechanics/MIT; SM/ME/MIT; BSc/ME/Imperial College. *Born:* 7/20/39. Born and raised in London, only son of Elizabeth A and Harry E. Launder. Numerous scientific publications on turbulence, convective heat transfer and computer-assisted learning. Books: "Mathematical models of turbulence–, Academic (1972); editor of: "Studies in Convection–, Academic, Vol 1 (1975), Vol 2 (1977); "Turbulent Shear Flows, Springer-Verlag:Vol. 1 (1979). Vol 2 (1980). Apptd Lecturer in Mech Engg at Imperial College 1964, later Reader in Fluid Mechanics (1971). Emigrated to CA and present position Sept 1976. Lived with Dagny Simonsen since 1968; daughter Katya b 1970, son Jesper b 1973. *Society Aff:* ASME, IMechE.

Launer, Milton L
Business: 725 N Park Ave, Fremont, NE 68025
Position: Asst Genl Manager. *Employer:* Fremont Dept of Utilities. *Born:* Oct 1918 Fremont Nebr. Attended Midland Coll, Fremont & Rutgers Univ, New Brunswick N J. Apprentice radio technician 1939. Substation Operator Fremont, Loup Power Dist of Columbus Nebr 1940-42. US Army 1942-45: radio instructor Signal Corps

Launer, Milton L (Continued)
school Ft Monmouth N J; powerhouse engr on 30 MW floating steam power plant, Corps of Engrs, Europe & Philippines. With City of Fremont Dept of Util since 1946: engrg functions through 1961; Asst Genl Mgr since 1962. Reg P E Nebr 1964-. Chmn Nebr Sect AWWA 1975-76. Hobbies: photography, travel.

Laurence J Sauter
Home: 3639 Conifer Lane, Ormond Beach, FL 32074
Position: Owner, Mgmt Conslt *Employer:* Self Employed *Education:* PhD/Public Adm/Walden Univ; MBPA/Bus & Pub Adm/SE Univ; BS/Indus Eng/Chase Coll *Born:* 02/27/24 Experience includes Marine Corps Veteran of WW2, Chief Indus Engr with Gen Elec Co, program mgr with Air Force at W-P AFB, OH, Sr Value Engrg with the Air Force Sys Command, Andrews AFB, MD, Chief Quality and Reliability Engr, Defense Logistics Agency, Cameron Station, Alex, VA, Dir of Value Engrg & Logistics, Naval Air Sys Command, Wash DC, Mgmt Conslt, Food Mgmt Inc, Cincinnati OH, and Adj Prof at Geo Wash Univ, Univ of Wash DC, and SE Univ *Society Aff:* ASPA, AIAA

Laurence, Robert L
Business: Dept. of Chemical Engineering, University of Massachusetts, Amherst, MA 01003
Position: Prof, Head. *Employer:* Univ of MA. *Education:* PhD/ChE/Northwestern Univ; MS/ChE/Univ of RI; BS/ChE/MA Inst Tech. *Born:* 7/13/36. Native of W Warwick, RI. Res engr with Elec Boat Div Gen Dynamics, E I duPont de Nemours, and Monsanto Co. Univ teacher since 1965 at Johns Hopkins Univ & Univ MA. Current res in polymer reactor engg, fluid mechanics, diffusion in polymers, & applied math. Active in rugby football as coach, referee, & administrator. *Society Aff:* AIChE, ACS, ASEE, SPE, AIC.

Laurent, Pierre G
Home: 38 rue Victor Hugo, Brest, 29200 France
Position: Honorary Genl Inspector - ret. *Employer:* Electricite de France. *Education:* -/-/Ecole Polytechnique - Paris, France. *Born:* Feb 1904 Brest. Ecole Polytechnique - Paris. Served with French Thomson Houston 1925-28, Genl Elec (Schenectady) 1928-29, Alsthom (Belfort) 1929-39, Cie Generale d'Electricite (Paris) 1940-46, Electricite de France - Study & Res Dept - in Paris 1946-68 and tech adviser 1968-71. Fellow IEEE, Eric Gerard & Blondel medalist, former Chmn IEC Tech Ctte on Insulation Coord, CIGRE Tech Ctte on rotating machines. Former V P Federalist Union of European Nationalities FUEN, Former Pres Union for the Protection of Environ in Brittany. *Society Aff:* IEEE, SEE.

Lauridsen, David H
Home: 155 Turrell Avenue, S Orange, NJ 07079
Position: VP Div Mgr *Employer:* Kemper Insurance Group *Education:* Juris Doctor/Law/Seton Hall Univ; BS/Agr Engr/Univ of CT. *Born:* 12/21/38. in Hartford, CT. Joined Factory Ins Assn in 1960 following grad from Univ of CT. Advanced at FIA from Inspector to Engr, Sr Engr, Engr-in-charge, and Field Mgr. Joined Kemper Ins as Div Engg Mgr, HPR Dept. Joined INA 1978, VP in charge of field operations for INA Loss Control Services. Joined GHR Energy, Corp as Corporate Risk Mgr and later added the title of Acting General Council. Joined Atlantic Companies in 1983 as Assistant Vice Pres in charge of Loss Control. Joined Kemper Group as VP-Div Mgr in 1987. Completed Law School earning a Juris Doctor degree in 1975. Mbr of Bar, State of NJ. Mbr ASSE, CPCU, and SFPE. Licensed PE in California. Enjoys handyman projs and is a breeder of Great Dane dogs. *Society Aff:* ASSE, SFPE, CPCU

Lauriente, Michael
Home: 6608 White Gate Rd, Clarksville, MD 21029
Position: Program Mgr. *Employer:* US Dept of Transportation. *Education:* Dr Eng/Metallurgy/Johns Hopkins Univ; MS/Metallurgy/MI Tech Univ; BS/Metallurgy/MI Tech Univ. *Born:* 6/26/22. born in Trail B C Canada - native of Chicago Ill. Dr Eng Johns Hopkins Univ; MS & BS Mich Tech Univ. Metallurgist with Internatl Harvestor Co prior to WW II. US Navy 1942-46; Lt Cmdr USN - ret. Cons for Ballistics Res Lab, Aberdeen Proving Ground - shaped charges 1949-55. Adv Engr Westinghouse Aerospace Div 1955-70 - magnetics, solid-state devices, satellite boom, Proj Mgr in Office of Sys Engrg of the Office of the Secy of Transp 1971-82 Respon for the Dept's R&D policy toward resolution of natl transp safety & security problems. Assumed current respn as Tech Mgr in Flight Directorate of the NASA Goddard Space Flight Center for overall management of Space Shuttle environment information for customers of Shuttle, 1982. Natl Matls Adv Bd; Solid State Sci Panel of the Natl Res Council; ASM Fellow; P E Md; Sigma Xi; Phys Met, NDT Matls Engrg; Fracture Mech; Welding; Flammability & Toxicity of Combustible Matls. *Society Aff:* ASM, ΣΞ, NMAB.

Laushey, Louis M
Business: Dept Cvl/Environ Engrg, Cincinnati, OH 45221
Position: Emeritus Dean, College of Engrg, and Emeritus Prof Dept of Cvl & Environ Engrg *Employer:* Univ of Cincinnati. *Education:* ScD/-/Carnegie Inst of Technology; MS/Civil Eng/Carnegie Inst of Tech; BS/-/PA State Coll. *Born:* May 1917. Teaching exper: Carnegie (Instr to Assoc Prof); Norwich Univ Hd of Cvl Engrg & Univ of Cincinnati (Wm Thoms Prof & Hd of Cvl Engrg) Dean (Emeritus), College of Engrg and Gaier Prof of Engrg, Univ. of Cincinnati. Served with Navy Civil Engrg Corps 1942-44. Chmn Dist 9 Council, & Pres Ohio Council ASCE. Dir, Dist 9, ASCE, V P, Zone II, ASCE. Mbr: Sigma Xi, Pres Carnegie Tech 1951; Chi Epsilon; Tau Beta Pi; Phi Eta Sigma; Sigma Tau; Lic Prof Engr PA & VT. *Society Aff:* ASCE, ASEE, SAME, AGU, ASME, IAHR.

Lauterbur, Paul C
Business: 1307 W. Park St, Urbana, IL 61801
Position: Professor *Employer:* Univ. of Illinois *Education:* Ph.D./Chemistry/Univ. of Pittsburgh; B.S./Chemistry/Case Institute of Technology *Born:* 05/06/29 Born in Sidney, OH. Employed at Mellon Inst 1951-53, 1955-63. Military service at Army Chemical Center 1953-55. Faculty of Chem Dept, State Univ of NY at Stony Brook 1963-85. Faculty of Univ of IL at Urbana-Champaign since 1985, with appointment in Dept of Medical Info Science, College of Medicine, and Dept of Chemistry. *Society Aff:* AAAS, ACS, APS, SMRM, ISMAR

Lautzenheiser, Clarence
Home: Rt 1 Box 58, Medina, TX 78055
Position: VP Emeritus, Advisor to the Pres Southwest & Research Inst. *Employer:* Retired; Consultant *Education:* BS/Metallurgy/MIT. *Born:* May 1921. 10 yrs with Dow Chem Co in fields of met, welding, corrosion, & nondestructive testing. 1962-, Southwest Res Inst in inspection, maintenance, & repair of nuclear reactor sys: design, welding, failure analysis & repair of bldgs, pipelines, & off-shore structs; qual assurance; design & fabrication of equip. Retired 1985- , Southwest Res. Inst. VP Emeritus, Private Consultant. Cert Level III in accord with SNT-TC-1A in ultrasonic, radiographic, liquid penetrant, & magnetic particle inspection. Reg P E Texas, recog as Corrosion Specialist by NACE, & author num tech papers. Active on code & prof cttes & Past Natl Pres ASNT. Fellow, ASME; Fellow, ASNT. *Society Aff:* ASNT, ASM, ASME, AWS, NACE.

Lauver, Robert M
Business: Whippany Road, Room 14A-401, Whippany, NJ 07981
Position: Director *Employer:* Bell Labs. *Education:* BS/EE/Univ of CT; MS/EE/NYU. *Born:* 8/25/34. Native of Bridgeport, CT. Served with USAF, 1956-1959. Engaged in military engg at Bell Labs since 1959-1980. Currently responsible for Military Systems Engg. Fellow-IEEE. *Society Aff:* IEEE, AAAS, NSIA, TBΠ, HKN.

Lavan, Zalman
Business: 3100 S. State St, Chicago, IL 60616
Position: Prof *Employer:* Ill. Inst of Tech *Education:* PhD/MAE/IIT; MS/MAE/IIT; BS/MAE/Polytechnic Inst of Brooklyn. *Born:* 12/23/27 Assistant Prof-1965, IIT, Assoc Prof-1968, IIT, Visiting Assoc Prof- 1971, Technion Israel, Prof-1976, IIT. Teaching and conducting research in thermal science, fluid mechanics and solar energy. Founded Solar Energy Research activities at IIT. Lead the 1975 IIT student SCORE team to win the national competition. Published over 30 Journal papers. Active in professional societies. Solar Energy Consultant to Dept of Energy. *Society Aff:* ASME, ISES.

Lavens, Edmond V
Business: One Seagate, Toledo, OH 43666
Position: Chief Eng Facilities Design *Employer:* Owens IL Inc. *Education:* BS/CE/Structural/Tri State Univ *Born:* 2/25/38 Native of Northwestern OH. Served with the USAF 1957-61. Complete Coll 1964. Field Eng with Mick and Rowland Inc 1964-68 specializing in Hwys and Bridges. Started with Owens Ill Inc in 1968 as a Facilities Design Engr. Had various positions such as field eng, project eng, purchasing mgr major construction, & project mgr. Before assuming my current responsibility as chief eng facilities design in 1976. Reg. Eng. in six states. Pres Lyon OH. Village Council since 1978. Part Time instructor Univ of Toledo. Pres Toledo ASCE 1980, VP Toledo ASCE 1978-79. Dir 1981. Enjoy skiing, hunting, fishing. *Society Aff:* ASCE, NSPE, OSPE, TSPE

Laverty, Finley B
Home: 314A W. California Blvd, Pasadena, CA 91105
Position: Ret. 1981 (Cons Engr) *Education:* SB/Civil Engg/MIT; BS/Chemistry-Math/Occidental. *Born:* July 1901. Engaged in the $1.5 billion flood control & water conservation dev of L A Cty Flood Control Dist for 32 yrs. Ch Hydraulic Engr & Asst Ch Engr. Pioneered barriers to seawater intrusion, defining boundaries of pollution, & reclamation of waste water. Author articles on flood & debris control, water conservation, prevention & removal of water contamination; co- author ASCE Hydrology Manual. Since 1962 internatl cons engrg practice in same tech areas & cont activity in prof dev of engrs, educ, & water conservation policy. Hon Mbr 1971 ASCE, Diplomate 1969 AAEE. Water Resources Award 1955 AWWA. *Society Aff:* ASCE, AAEE, AWWA, AGU, AAAS, ACEC.

Lavi, Yeshayahu
Home: 6 Bat-Yiftach st, Zahal-Tel-Aviv, Israel 61100
Position: Consultant. *Employer:* Self. *Education:* BSc/EE/Columbia Univ *Born:* June 1926. Native of Berlin Germany. Since 1933 in Israel. Israel Defense Forces in var communications & electronics commands. 1957-62 Ch Signal Officer & Electronics IDF. 1964-68 Dir Genl Ministry of PT&T Israel. 1968-70 Dir Genl Armament Dev Auth Israel Ministry of Defense. 1970-73 Dir Genl Ministry of Defense Israel. From 1977 Consultant-Telecommunications and Mgmt. 1973-76 (V P Genl Telephone Electronics Corp). Co-founder & Chmn Israel Sect IEEE. Fellow IEEE 1970. Israel Defense Prize 1968. IEEE Centenial Medal of Honor 1984. *Society Aff:* AEAI, IEEE

Laviolette, Leo E
Business: 150 Gloucester St, Ottawa, Canada K2P 0A6
Position: Dir of Public Policy. *Employer:* Canadian Automobile Assn. *Education:* BSc/Civ Engg/Queen's Univ. *Born:* 7/30/35. Born, raised & educated in Ontario, Canada. Employed as planning studies engr (1959-65) & deputy dir (1966-70) of dept of traffic engg services for the Capital City of Ottawa. Gained consulting engg experience with Toronto firm, 1965-66. Serving Canadian Automobile Assn as dir of govt & public affairs and dir of public policy, 1970-present. Served as intl dir, representing Canada, to the intl Inst of Transportation Engrs, 1977-79. Elected VP of the Intl Inst of Trans Engrs for 1981. *Society Aff:* ITE, TRB, TRF, TIRF.

Law, E Harry
Business: Dept of Mechanical Eng, Clemson, SC 29631
Position: Prof *Employer:* Clemson Univ *Education:* PhD/Applied Mech/Univ of CT; MSE/Aero Eng/Princeton Univ; BAE/Aero Eng/Rensselaer Poly Inst *Born:* 12/5/40 Prof Law has taught courses in the areas of applied mechanics, vibrations, and system dynamics, and control at Clemson Univ since 1971. His research is focused on rail vehicle dynamics. He has been active and has published in this area since 1965. As a founder and principal of the consulting firm of Acorn Assoc, Prof Law has consulted extensively on rail vehicle dynamics problems. His industrial experience includes employment with United Aircraft Research Laboratories and the Boeing Co. Prof Law has held Visiting Assoc Professorships at MIT (1978) and AZ State Univ (1977/78). *Society Aff:* ASME

Law, Harold B
Home: 145 Van Dyke Rd, Hopewell, NJ 08525
Position: Retired. *Education:* PhD/Physics/OH State Univ; MS/Physics/OH State Univ. *Born:* Sept 7, 1911. Entire career with RCA, early work with television pickup tubes, particularly the image orthicon. From 1949 on, engaged with color display tubes, particularly the shadow-mask color tube. IEEE awards: Vladimir K Zworykin Award 1955, Lamme Award 1975 'For outstanding contrib in dev color picture tubes, incl the fabrication tech which made color television practical'. Other awards: Award, Television Broadcasters Assn 1946; David Sarnoff Outstanding Team Award 1961; Frances Rice Darne Award, Soc for Info Display 1975. For recreation, partial to farming & woodworking. *Society Aff:* APS, IEEE.

Law, John
Business: Elec Engr Dept, Moscow, ID 83843
Position: Prof. *Employer:* Univ of ID. *Education:* PhD/EE/Univ of WI; MS/EE/Univ of WI; BS/EE/Case Western Reserve Univ. *Born:* 12/8/30. Prof of Elec Engr Univ of ID, Born 12/8/30 Cleveland, OH. Married Wilma L (Tailon), four children, PhD in Elec Engr Univ of WI 1962, formerly chief engr Res Div of Carrier Corp 1963 to 1974. Taught EE at MT State Univ, Univ of Wi, Bogazici Univ (Istanbul Turkey). Sr mbr IEEE. Natl Exec Councillor of Tau Beta Pi. Natl Engg Honor Soc. Specialty power elec engg. *Society Aff:* IEEE.

Lawler, Charles A
Business: 5619 Fannin Box 8098, Houston, TX 77004
Position: Vice Chairman of the Board *Employer:* Bovay Engrs Inc. *Education:* BSCE/Structural Engrg/Univ of TX. *Born:* Apr 1921. Native of Luling Texas. Design Engr with Humble Oil & Refining Co (now Exxon) 1942-44. US Naval Res, Civil Engrg Corps 1944-46. 1946-47 Design Engr with a natl engr contractor. With Bovay Engrs Inc since 1947, holding var pos incl Design Engr, Sr Engr, Proj Engr, Partner, Exec V P, President, Dir & Chairman Exec Ctte. Assumed current respon as Vice Chairman 1981. Lic P E Texas, N M, Wash. Mbr NSPE, ASCE, ACEC, SAME, Houston Engrg Soc, Rotary Internatl, Chapelwood United Methodist Church. *Society Aff:* NSPE, ASCE, ACEC, SAME.

Lawley, Alan
Business: Drexel University, Dept. of Matls. Engrg, Philadelphia, PA 19104
Position: Prof of Matls Engg. *Employer:* Drexel Univ. *Education:* PhD/Phy Met/Univ of Birmingham, UK; BSc/Phy Met/Univ of Birmingham, UK *Born:* 8/29/33. Native of Birmingham Eng. Post-doc Fellow, Univ of Penna 1958-61. Mgr Met & Solid State, Franklin Inst Res Labs 1961-66. Head Dept of Matls Engrg Drexel Univ Phila Pa 1969-79, Prof. Matls. Engrg 1969- . Mbr AIME, ASM, ASEE, EMSA, APMI, Inst. of Metals (London). Active on prof societies. Reviewer of res proposals for U S Gov. Cons to gov & indus. Num tech articles & contrib to textbooks. Visiting Prof Open Univ (UK) 1975. Visiting Eng, Cabot Corp, 1982. Active in powder met, composite matls, engrg educ. Fellow ASM; Pres, The Metallurgical Soc of AIME (1982), and AIME (1987). Editor, Int. Journal of Powder Metallurgy. *Society Aff:* AIME, APMI, ASEE, ASM International, EMSA, IM

Lawrance, Charles H
Home: 1340 Kenwood Rd, Santa Barbara, CA 93109
Position: VP *Employer:* Lawrance, Fisk, & McFarland, Inc. Consulting Engineers *Education:* MPH/Public Health/Yale Univ; BS/Publ Health Engg/MIT. *Born:* 12/25/20. Malaria-Control-in War-Areas, Arkansas, prior to WW II service in

Lawrance, Charles H (Continued)
Marine Corps. 8 yrs san engrg with Conn & Calif Depts of Health. 20 yrs with Koebig & Koebig Inc L A, as design engr, Ch San Engr & Dir of Environ Engrg; 4 yrs, Engr - Mgr Santa Barbara Cty Water Agency; 3 yrs Prin Engr J.M. Montgomery, Pasadena. Since 1983, VP, Lawrance, Fisk & McFarland, Inc. Cons Engrs, Wastewater Mgmt Water Resources and General Civil Engrg. Norman Medal ASCE 1966. Author num prof papers & articles & of epic poem on St Francis Dam disaster. Musician & writer. *Society Aff:* ASCE, AWWA, WPCF, AAEE, APWA.

Lawrence, Arthur F
Home: 3134 Walma Dr, Orchard Lake, MI 48033
Position: Area Sales Mgr. *Employer:* MTS Sys Corp, Bx 24012, Minneapolis. *Born:* Dec 1928. BSEE Northeastern Univ Boston 1952. Reg P E NM. Mbr IEEE, ASTM, SAE, SESA. SESA: Chmn Applications Ctte 1972-74, Mbr Exec Ctte 1974- , Chmn So Mich Sect 1976- . 1952-55 US Army Ordnance, Springfield Armory, Instrumentation & Facil Engr; 1955-61 Sandia Corp, Staff Mbr; 1961-69 B&F Instruments, Ch Sys Engr, Ch Engr, Mgr of Sys Sales; 1969- , MTS Sys Corp, Regional Sales Mgr (Phila Region), Area Sales Engr (Mich area).

Lawrence, Harold R
Business: PO Box 3-PSL, Las Cruces, NM 88003
Position: Director. *Employer:* Phys Sci Lab, N M State Univ. *Education:* Masters/Structural Engr/MIT; Bachelors/Civil Engr/MIT. *Born:* Sept 9, 1925 Irvington NJ. 4 1/2 yrs military service as naval aviator. 1956-59 on White House Staff, Asst to the 1st Pres Sci Advisor. 1959-60 NASA, served as U S delegate to the 1st United Nations conf on the peaceful use of outer space. 1960-68 staff, Jet Propulsion Lab Caltech. Participated in the Ranger & Surveyor proj's. Editor Amer Soc of Aero & Astronautics vol on the Tech of Lunar Exploration. Since 1968, Dir of Phys Sci Lab, N M State Univ. Enjoy skiing. *Society Aff:* AIAA, AMA, ΣΞ.

Lawrence, Harvey J
Home: PO Box 941, Los Gatos, CA 95031
Position: Staff Engr. *Employer:* Lockheed Missiles & Space Co Inc. *Education:* BS/Chem Engr/CA Inst of Tech; MBA/Acctg/Santa Clara Univ *Born:* 1/25/21. Served with RCAF & US Marines as dive bomber pilot 1941-45. Plant Chem Engr & Group Production Mgr for Proctor & Gamble 1947-59. Plant Mgr for Alberto-Culber 1959-60. Staff engr in Product Assurance 1960-76, Lockheed Missiles & Space Co, ASQC cert Qual Engr & Reliability Engr; Reg P E CHem Engg & Qual Engg. Dir ASQC 1966-74. Special interests incl amateur radio (K6KZ), travel. Retired 1976. *Society Aff:* ASQC, AIChE

Lawrence, James H
Business: Dept of Mech Engrg, P.O. Box 4289, Lubbock, TX 79409
Position: Prof *Employer:* TX Tech Univ *Education:* PhD/ME/TX A&M Univ; MS/ME/TX Tech Coll; BS/ME/TX Tech Coll *Born:* 2/9/32 in Beatrice, NB. Attended public schools in Lubbock, TX. Worked as design engr for GE Co during the summer of 1957. Served as instructor in the Mech Engrg Dept at TX Tech Univ from 1956-60; appointed as Asst Prof in 1960; promoted to Assoc Prof in 1964 and Prof in 1971; Dept Chairperson in 1972-86. Served as Gen Chrmn of the 1972 Annual Meeting of the American Soc for Engrg Education. VChrmn of the Natl Mech Engrg Dept Hds Ctte, 1981. Chrmn of the Natl Mech Engrg Dept Hds Ctte 1982; Chrmn, Mech Engr Division of ASEE 1984. *Society Aff:* ASME, ASEE

Lawrence, Joseph F
Home: 44 Revere Rd, Port Washington, NY 11050
Position: President. Retired *Employer:* The Lawrence Co. *Education:* MChE/Chem Eng/Polytechnic Inst of NY; BChE/Chem Eng/Polytechnic Inst of NY. *Born:* Mar 1918. Native of N Y. Design Engr Lummus Co to 1943. Manhattan Proj 1943-45. Proj Engr Foster Wheeler Corp 1945-49. Involved with chem plant & refinery design & const. Genl Mgr of the Tilco-Fin Tubing Div of The D E Kennedy Co 1949-52. Spec in the dev of welded steel fin tubing for high temp applications. Lawrence Co since 1952 involved with special applications of extended surface for high temp use in heat recovery & energy conservation. Mbr AIChE, ACS, & Chemists Club. Reg P E NY. Enjoy golf, photography & music. Retired 1986. *Society Aff:* AIChE, ACS, Chems Club, AAAS.

Lawrence, Leo A
Business: PO Box 1970 W/C-91, Richland, WA 99352
Position: Fellow Scientist *Employer:* Westinghouse Hanford Co *Education:* MS/Physics/Univ of Denver; BS/Pysics/Humboldt State *Born:* 06/21/38 Fellow scientist with Westinghouse Hanford Co specializing in the in- reactor performance of fuels for liquid metal reactors since 1967. Tech interests include chem behavior, performance of advanced designs, and revision of fabrication specifications. Program Mgr, Irradiation Testing of Thermionic Fuel Elements for Space Power Applications. Publ approximately 40 tech papers on oxide fuel irradiation behavior. Mbr of ANS Material, Science and Tech Div Exec Ctte, past program chrmn and Secy/Treas, Chrmn for 1986-1987. Enjoy golf and skiing. Board Mbr Washington Jr Golf Assoc., & Catholic Family & Child Services. *Society Aff:* ANS

Lawrence, Robert F
Home: 247 Cascade Rd, Pittsburgh, PA 15221 *Education:* B/Elect Engr/Pratt Inst. *Born:* May 1921. Entire career has been in the field of generation, transmission & distribution of elec power. With Westinghouse Elec Corp since 1943. Retired 1985, Mgr of Transmission & Distrib Sys Engrg, Directed group of tech consultants. Fellow IEEE since 1963. Reg P E Pa. Author over 81 pub tech papers & articles. Served on IEEE, NEMA & Internatl Tech Cttes; currently IEEE Divisional Director, Division VII, Power Engrg. Division, 1983-1985. *Society Aff:* IEEE, CIGRE.

Lawrence, William H
Home: 44937 N. 5th St East, Lancaster, CA 93535
Position: Consultant: Engrg & Mgmt (Self-employ) *Employer:* Lawrence Assoc *Education:* MBA/Bus Org & Mgt/Univ of Chicago; BSME/Mech Engr/Purdue Univ. *Born:* 7/26/27. A O Smith Corp Milwaukee WI. US Navy (Electronics WW II) & US Army (Ordnance Shop Officer, Korea). AF Rocket Propulsion Lab 1955-81. Current pos: Aerospace Engrg & Mgmt Conslt Oct 1981-. Previous Postions: 26 yrs top mgmt experience AF Rocket Propulsion Lab. including Chief, Operations office; Deputy Chief, Liquid Rocket Div; Chief Tech Support Div. Experienced at Planning Research & Development programs, Managing both in-house and contracted R&D, and organizing and motivating teams of engineers and technicians. Mbr JANNAF Propulsion Info Agency, AIAA. Served Space Shuttle Main Engine and External Tank source selection. Presently consulting on aerospace engrg mgmt of Corp Independent Res and Dev; proposal preparation and corp planning; facility design and development. *Society Aff:* AIAA.

Lawrence, Willis G
Home: 1261 Sugar Sands Blvd, 118, Riviera Beach, FL 33404
Position: Ret. (Dean Emeritus) *Education:* BS/Glass Tech/ALfred Univ; PhD/Ceramics/MIT. *Born:* Dec 17, 1916. Work Exper: 1942-47 Abex Corp, res ceramist on foundry problems; N Y State Coll of Ceramics at Alfred Univ: 1947-50 Prof of Res, 1950- 64 Chmn Dept of Ceramic Res, 1964-66 Chmn Dept of Ceramic Engrg, 1966-73 Asst Dean, Dean 1973-80. Mbr Amer Ceramic Soc, NICE, Amer Foundrymen's Soc. John Jeppson Award Amer Ceramic Soc 1972, Outstanding Alumnus Award Alfred Univ 1975. Reg P E NY State Lic 053007. Author 25 sci papers & 1 book dealing with ceramic sci. Pat 3641229 on Permeable Ceramic Mold Feb 8, 1972. Retired 1980. *Society Aff:* Am Ceram Soc, AFS, ASEE, ASTM.

Lawrenson, Peter J
Business: Switched Reluctance Drives Ltd, Springfield House, Hyde Terrace, Leeds LS2 9LN, West Yorkshire England
Position: Chief Exec. *Employer:* Switched Reluctance Drives Ltd *Education:* DSc/EE/Manchester Univ; MSc/EE/Manchester Univ; BSc/EE/Manchester Univ. *Born:* 12/03/33. Prescot UK. Res engr GEC (then AEI) 1965-61. Lecturer 1961-65,

Lawrenson, Peter J (Continued)
Reader 1965- 66, Prof 1966- (Head 1974-84). Dept of Elec & Electronic Engrg Leeds Univ. Res in elec machines & electromagnetics. Dir 1980- , Chief Executive, 1986- , Switched Reluctance Drives Ltd. Author more than 100 papers & pats & 2 bks. IEE tech paper premium 1956, 1965 & 1967 & 1981. Fellow IEEE 1974. Elected Fellowship of Engineering 1980 Fellowship of Royal Soc 1982. James Alfred Ewing Medal 1983. Council and Bds IEE 1966-VP 1987. On cttes UK Sci Res Council 1970- . *Society Aff:* RS, FoE, IEEE, IEE

Lawrie, Duncan H
Business: 222 Digital Computer Lab, 1304 W Springfield, Urbana, IL 61801
Position: Prof *Employer:* Univ of IL at Urbana-Champaign *Education:* PhD/Comp Sci/Univ of IL; MS/Comp Sci/Univ of IL; BS/EE/Purdue Univ; BA/Math-Phys/DePauw Univ *Born:* 4/26/43 Prof Lawrie has been active in the area of computer architecture since 1966, when he worked on the ILLIAC IV computer. Since then he has contributed to the design of several other machines, including the Burroughs Scientific Processor. He specializes in research and consulting in the area of design, construction, and use of very large computing systems. He is a Fellow of the IEEE. *Society Aff:* IEEE, ACM.

Lawroski, Harry
Business: 2375 Belmont Ave, Idaho Falls, ID 83401
Position: Consultant *Employer:* N. Lawroski & Assoc. *Education:* PhD/ChE/PA State Univ; MS/ChE/PA State Univ; BS/ChE/PA State Univ *Born:* 10/10/28. Dalton, PA; Res & Dev in petrol refining at PA State, 1950 to 1958; Assoc Engg at Argonne Natl Lab, 1958 to 1963; Technical Mgr for Design & Construction of Plutonium Reactor at Argonne Natl Lab, 1963 to 1968; Assoc Div Dir & Superintendent of Operations, Exper Breeder Reactor, Argonne Natl Lab, 1968 to 1973; Gen Mgr of QA & Environmental Services, Nuclear Services Corp, 1973 to 1976; Asst Gen Mgr, Nuclear Fuel Processing & Nuclear Waste Tech, Allied Chem Corp, ID, 1976 to 1979; Consultant, 1979; Pres, Am Nuclear Soc, 1980-1981; Treasurer, ANS, 1973 to 1977; Chrmn Nuclear Engg Div, AIChE; 1974; Fellow, ANS; Fellow, AIChE, Fellow, Am Inst of Chemists; Sigma Xi; Tau Beta Pi; Sigma Tau; Phi Lambda Upsilon; Rotary; Elks; Author of 35plus Scientific Articles; Two Patents. *Society Aff:* AIChE, ANS.

Lawson, Lowell W
Home: 104 Crest St, Summersville, WV 26651
Position: Treasurer. *Employer:* Bright of America Inc. *Born:* Aug 12, 1936. Native of Sophia W V. BS Chem E West Vir Univ; continued selective courses at W Vir State Coll. Process Engr, Production Supr, mbr of 2 significant start-up teams, & Sr Engr with Monsanto Co. With Bright of Amer Inc since 1968 as Plant Mgr & Asst V P of Mfg. Mbr Bd of Dir 1968- . Present pos since Aug 1976. Hold Airline Transport Pilot lic with instructor ratings.

Lawson, Robert T
Business: 55 Mitchell Blvd-3030, San Rafael, CA 94902
Position: Exec Vice Pres. *Employer:* Harding-Lawson Assocs. *Education:* BS/Civil Engg/Univ of WA. *Born:* July 1926. BSCE Univ of Washington. Native of Chehalis Wash. Navy V-12, 1944-46. Company grade infantry officer US Marine Corps 1950-51. Dames & Moore 1948-60; Partner 1957-60. Principal with Harding-Lawson Assocs since 1960. Earth dam design, foundation engrg for structures of all types. Dir ASCE 1975-78. *Society Aff:* ASCE, SAME, CEAC, SEAOL.

Lawver, James
Home: 6334 Oak Sq. E, Lakeland, FL 33813
Position: Consultant *Employer:* Self *Education:* Met E/Extractive Metallurgy/CO School of Mines; ScD/Metallurgical Eng/CO School of Mines *Born:* 5/28/19 Spent three years as a mining engineer working for the Braden Cooper Company in Chile, South America and the Bureau of Strategic Warfare in Brazil. Worked for International Minerals & Chemical Corporation for approximately 13 years, first as a research eng and then mgr of minerals research dept in FL. From 1962 to 1975 he was Prof and Dir of the Mineral Resources Center and Associate Head of the Dept of Civil and Mineral Eng and Dir of Metallurgical Graduate Studies at the Univ of MN. Prof Lawver was a Fulbright Scholar Lecturer at the Univ of Queensland in Australia in 1973. From 1973 to date has been the Technical Mgr for International Minerals & Chemical Corp. He received the Richards Award from AIME in February 1980, and in 1981 received the Ridge Chapter Combined Eng of the Year and Outstanding Achievement Award in FL. In 1985 he retired from International Minerals & Chemical Corporation and formed a consulting engineering company James Lawver Inc. *Society Aff:* AIME, TBΠ, ΣΞ

Lawyer, Donald H
Business: PO Box 626, Manor, PA 15665
Position: Pres. *Employer:* Integrated Services, Inc. *Education:* Grad Work/Structural Design/Univ of CA; BS/Structural Design/UT State; Accreditation in/Fallout Shelter Analysis Planning & Design for Protective Construction/Dept of Defense. *Born:* 3/12/24. Mr Lawyer manages and directs a bldg team which is a consortium of independent, interdisciplinary profl cos, combining to build and finance bldg projs. He uses the most advanced technological engg equip including an in-house computer. These advanced engg technologies has enable Mr Lawyer to offer design flexibility meeting the changing needs of individual bldgs, reducing the cost of construction to give better value for the bldg dollar in terms of function, environment, first cost, maintenance, and the time needed to construct a bldg. Mr Lawyer is a reg Civ Engr in the States of PA, OH, & WV. Lawyer is a registered surveyor in the state of Pennsylvania, and a member of the American Arbitration Association.. *Society Aff:* NSPE, PSPE, ASCE, PEPP.

Lay, Maxwell Gordon
Business: 500 Burwood Rd, Vermont South, Victoria Australia 3131
Position: Executive Director. *Employer:* Australian Rd Res Bd. *Education:* BCE/Civil/Melbourne Univ; MEng/Civil/Melbourne Univ; PhD/Civil/Lehigh Univ. *Born:* Oct 1936. Worked for State Elec Comm of Victoria, finally as Engr-in- Charge Major Steel Struct 1958-68. Worked for BHP Co, finally as Engrg Res Mgr 1968-75. Exec Dir Australian Rd Res Bd 1975- . Councillor Inst of Engrs Australia 1970- . Moisseiff Medalist ASCE 1968. Chmn & mbr of cttes of Standards Assn of Australia. Australian delegate to meetings of the OECD Road Res Program. 1980 Warren Medal of Inst of Engrs Australia. Author: Source Book for Australian Roads, 3rd Edition, Structural Steel Fundamentals, Handbook of Road Technology. *Society Aff:* IEAUST, ASCE, ΣX, CIT, FTS.

Layne, Jack E
Business: 1600 W 12th Ave, Denver, CO 80254
Position: Deputy Dir - Engg & Construction. *Employer:* Denver Water Dept. *Education:* BS/Civ Engg/Univ of CO. *Born:* 5/5/31. Native of CO - with Denver Water Dept 1949-1968 - Chief of Operations - with Blade & Veatch as Proj Mgr 1968-1973 with AWWA 1974-1977 Dir of Technical and Profl Activities with CH2M-Hill 1978 - Proj Mgr Island of Trinidad - currently with Denver Water Dept. Chrmn-Rocky Mtn Sec AWWA 1971. Active on several Natl AWWA committees. Fishing, golf, jogging, racquetball. *Society Aff:* ASCE, AWWA.

Layton, Donald M
Business: Dept of Aeronautics, Monterey, CA 93940
Position: Prof *Employer:* Naval Postgrad School. *Education:* BS/Eng/US Naval Acad; BS/Aero/Naval Postgrad Sch; MS/Aero/Princeton Univ; MS/Mgmt/Naval Postgrad Sch. *Born:* 9/23/22. Active duty Naval Aviator 1945-68. NPS fac 1965- . First Dir Navy Aviation Safety Progs. Former Editor, Hazard Prevention, The Journal of the Sys Safety Soc 1970-77. Assoc Fellow AIAA. Mbr AIAA tech cttes on Lighter-Than-Air. Res in aircraft controls & Helicopters. Reg P E (Safety) CA. Editor, Journal of Hydronautics 1978-81; Pres, Monterey Council Navy League 1978-80 Prof of Aeronautics, Acting Chrmn, Aeronautics NPS 1982-84. Author of

Layton, Donald M (Continued)
"Helicopter Performance" and "Airplane Performance" *Society Aff:* AIAA, SSS, ASSE, AAUP, ASEE, ΣΞ

Lazaro, Carlos A
Home: 128 Padre las Casas, Hato Rey, PR 00918
Position: Vice President. *Employer:* Carlos Lazaro Garcia. *Education:* ME/Civil Eng/Cornell Univ; BS/Civil Eng/Univ of PR. *Born:* May 1946. BS Univ of Puerto Rico. ME Cornell Univ. Main respon: struct design of all types of structures & coord with other tech engrs & architects in the preparation of plans. Supervision of const projects. Mbr ASCE, ACI & Inst of Engrs, Architects & Surveyors of P R. VP P R Chap ACI 1974-75. Actual Pres P R Chap ACI. *Society Aff:* ACI, ASCE.

Lazo, John R
Home: 70 Parmelee Dr, Hudson, OH 44236
Position: Dir-Ind Engr. *Employer:* Republic Steel Corp. *Education:* BS/Ind Engr/PA State Univ. *Born:* 5/3/33. Dir-Ind Engg for Republic Steel Corp since 1976. Previous experience includes six yrs with Booz, Allen & Hamilton as Proj Mgr of multi-disciplined consulting teams assigned to domestic & foreign clients in primary metals, OEM, airlines, candy, furniture, clothing & machine tool businesses; twelve yrs with US Steel Corp in varied ind engg & operating line mgt positions. Dir of the Steel Div of IIE for 1978-79 term & currently a mbr of the Ind & Profl Advisory Council for the Dept of Ind & Mgt Systems at PA State Univ. *Society Aff:* AISI, IIE, AISE.

Lazurenko, Lydia B
Home: 1141 S Renaud, Grosse Pte Woods, MI 48236
Position: Staff Project Engg. *Employer:* General Motors Corp. *Education:* MS/Mech Engg/Wayne State Univ; BS/Aero Engg/Wayne State Univ. *Born:* 6/6/32. Native of Ukraine, US citizen since 1955. BSAEE, MSME from Wayne State Univ. 1955-72 engaged in gas turbine design for both aerospace and sutomotive applications. Since 1973 with General Motors, working on vehicle aerodynamics. Reg PE in state of MI, serving on Boards of Registration for PE (first woman) and arch since 1978. Mbr of Soc of Women Engineers Council of Section Representatives, Past VP of Detroit SWE Section, Chmn of Affiliate Council of Engg Soc of Detroit. Active in NCEE, NSPE, and SAE. Mbr of MI Assoc of Professions and Soc of Ukrainian Engineers. *Society Aff:* SWE, NSPE, SAE, NCEE.

Leach, James L
Business: Foundry Lab, Urbana, IL 61801
Position: Prof. *Employer:* Univ of IL. *Education:* BS/Engrg/Univ NM; MS/Engrg/Univ of IL; PhD/Admin/IL State Univ *Born:* 4/9/18. Native of Lawrenceville, IL, worked for the TX Co during summers of high sch & college; design engr for Engg Res & Gage Sec, US Ordnance Dept; combustion control engr, instrument installation & calibration, Bureau of Docks and Yards for USN in southwest Pacific; instr in physics Pre-flight Sch, pilot, US Army Air Corps WWII; 8 summers with Natl Sci Fdn in India; 2 tours to Kenya with Action; consultant for Acad for Educational Dev in Iran. Prof of Mech Engrg 1 yr at Indian Inst of Tech, Kharagpur, India. Retired USAF Lt Col. *Society Aff:* AFS, ASM, SAE

Leadon, Bernard M, Jr
Business: 216 Aerospace Bldg, Gainesville, FL 32611
Position: Prof. *Employer:* Univ of FL. *Education:* PhD/Fluid Mechanics/Univ of MN; MS/Aero Engg/Univ of MN; BS/Math/College of St Thomas. *Born:* 11/29/17. Born in Farmington, MN 1917. Engrs Aide Pacific Gas & Elec Co 1941. Taught fluids two yrs Univ of MN, propulsion res at Cornell Aero Lab three yrs. Organized Rosemount Aero Lab for Univ of MN 1946-8. Taught and conducted res Univ of MN 1946-57 on compressible flow, heat and mass transfer. Sr Staff Scientist, Convair, 1957-64. Prof and consultant Univ of FL 1964-present. Music. *Society Aff:* AIAA, APS.

Leal, L Gary
Business: Dept of Chem Eng, Pasadena, CA 91125
Position: Chevron Distinguished Prof of Chemical Engg *Employer:* Calif Inst of Tech *Education:* PhD/ChE/Stanford Univ; MS/ChE/Stanford Univ; BS/ChE/Univ of WA *Born:* 3/18/43 Native of Bellingham, WA, member of Cal Tech faculty since 1970. Primary research activities in fluid mechanics, Computational hydrodynamics and polymer physics. Technical consultant. Author/co-author of more than 100 technical papers. Editorial Advisory Bd, J. Colloid & Interface Sci; consulting editor, AICHE Journal; Editorial Cttee, Am. Review of Fluid Mechanics; Associate Editor, Intenational J. of Multiphase Flow. Allan P. Colburn Award, AICHE 1978; Guggenheim Fellow 1976-77; Dreyfuss Foundation Teacher-Scholar Award 1975-79, Fellow of the American Physical Soc 1984, Member of the NAE 1987. *Society Aff:* AICHE, ASEE, Soc Of Rheology/AIP, APS/AIP, British Society of Rheology.

Leaney, David B
Business: 1690 West Broadway, Vancouver, Brit Columbia V6J 1X9. Canada
Position: Chmn of the Bd, retired *Employer:* D W Thomson Cons Ltd. *Education:* BASc/Mech Engr/Univ of British Columbia *Born:* 11/25/25. Joined the Bldg Engrg Cons Practice of D W Thomson 1949, spec in heating, ventilating & air cond & allied subjects. Practice now includes all bldg serv incorporating civil-sanitary & elec disciplines. Elected chrmn DWTC Ltd June 1984. Respon for the fiscal oper of a multi-discipline cons practice. Elected Fellow ASHRAE 1974 and Fellow of Chartered Inst of Building Services UK 1981. Reg in British Columbia, Yukon Territory, WA State and UK. Retired from active practice January 31, 1985. Appointed Adjunct Professor. School of Architecture, University of British Columbia 1985. *Society Aff:* ASHRAE, ACEC, CEBC, CIBS

Leatherdale, John W
Home: 44 Myrtle St, North Plainfield, NJ 07060
Position: Pres *Employer:* Trace Technologies, Inc. *Education:* BSChE/-/NCE. *Born:* Oct 1947. Grad work New Jersey Inst of Tech. Cons Engr actively engaged in the fields of Environ Pollution & Energy Conservation with emphasis on air pollution control. Conducting ongoing res & dev of sampling and analysis of incinerators and other hazardous waste environments. *Society Aff:* AIChE, ACS, APCA, NSPE.

Leatherman, John E
Home: 6375 Little Uvas Road, Morgan Hill, CA 95037
Position: Engr Mgr *Employer:* General Electric Co. *Education:* MS/Cybernetic Sys/San Jose State Univ; MS/Mat Sci/San Jose State College; BS/Mat Sci/San Jose State College. *Born:* 10/24/40. Crozier Engg Award, US Army Dev and Proof Services, US Army Commendation Medal, Test & Evaluation Command, Article, "Journal of Metals," Program Chrmn for Golden Gate Welding & Metals Conference; Outstanding Young Mbrs Award American Soc for Metals; Chrmn Santa Clara Valley Chapter of the American Soc for Metals; Intl ASM Chapter Advisory Ctte; Overall Conference Chrmn for the "Golden Gate Welding & Metals Conference," CA PE Licenses (Metallurgical and Nuclear) and MD PE License (ME-MET). *Society Aff:* ASM, AWS.

Leavenworth, Richard S
Home: PO Box 378, Lake Swan, Melrose, FL 32666
Position: Prof. *Employer:* Ind & Systems Engg. *Education:* PhD/Ind Engg/Stanford Univ; MS/Ind Engg/Stanford Univ; BSIE/Ind Engg/Stanford Univ. *Born:* 9/30/30. VP, Region IV, IIE (78-80) VP, Intl Operations (82-84) IIE; Editor, *The Engineering Economist*; Past Chrmn, Engg Economy Div, ASEE; formerly mbr, ctte on Economic Analysis Transportation Res Bd; Co-author with E L Grant, "Statistical Quality Control–, (McGraw-Hill) and with E L Grant & W G Ireson, "Principles of Engineering Economy," (John Wiley). Lecturer and consultant in Quality Control and Capital Budgeting procedures. Professor of Industrial & Systems Engrg, Univ of Florida, Gainesville. *Society Aff:* IIE, ASQC, ASEE

Leavitt, Sheldon J
Business: 4400 Colley Ave, Norfolk, VA 23508
Position: Principal in Practice *Employer:* Leavitt Associates *Education:* BS/Structures/Univ of IL *Born:* 10/14/22 Practice established 1953. Certified, National Council of Architectural Registration Bds. Commander, Civil Engineering Corps, USNR retired. Member: Tau Beta Pi, Sigma Xi, Phi Kappa Phi, Chi Epsilon. University Valedictorian. Chairman Design Review Committee, V Chrmn Building Code Appeals Bd, Norfolk; Past President, Tidewater Chapter, American Institute of Architects. Architect Chrmn, and General Chairman VA State Joint Co-operative Committee of Architects, Engs and General Contractors; Norfolk Downtown Plan Study Team. Architect and eng for: Temple Oheb Shalom, Baltimore (Walter Gropius, Consulting Architect); Metropolitian Baseball Park (in associatioin with Madigan-Praeger), historic restoration Chamber of Commerce Headquarters, Sewells Point Elementary School, Norfolk; Microbiological Assocs. Production Facility Walkersville, MD, Freemason Street Garage; Norfolk VA. Fellow ASCE; Diplomate, Natl Acad of Forensic Engrs; Mbr AIA. *Society Aff:* ASCE, NSPE, AIA, NAFE, ACEC, CSI, SAME, CRI

Leavy, George C
Home: 44 Radcliff Dr, Huntington, NY 11743
Position: Mgr A-10 Tech Engg. *Employer:* Fairchild Republic Co. *Education:* BAE/Aero Engg/Polytechnic Inst of Brooklyn; -/Bus Adm/Hofstra Univ. *Born:* 9/2/27. Grad studies in Bus Admin Hofstra Univ. Joined Fairchild Republic 1956. Assignments included Mgr A-10 Technical Engr. Ch of Wts for the A-10A Aircraft studies, & in the dev of wt estimation methods & computer program. Previously a East Coast Aeronautics for 7 yrs as Ch Wt Engr & at EDO Aircraft for 5 yrs participating in a var of prototype dev programs. Active mbr SAWE & author of sev papers on aircraft wt prediction & control. Currently Mgr A-10 Engg. *Society Aff:* SAWE.

Lebedow, Aaron L
Business: One Wacker Dr, Chicago, IL 60606
Position: Chairman/Sec *Employer:* Technomic Consultants *Education:* MBA/Mktg/Univ of MI; BS/IE/IIT *Born:* 8/19/35 Native of Chicago, IL. 1958-1960: Served in the USAF as a contract price negotiator for the Air Material Command. 1960-1961: Assist Mktg Mgr-Imperial Eastman Corp. 1961-1966: Mgr-Coplan Associates-the mktg consulting arm of IIT Research Inst. 1966-Present: Co-Founder of Technomic Consultants, an international, mktg consulting firm; specifically responsible for Electromechanical and International practice areas. Member of Who's Who in America; ITT's Alumni BD of Dir; Inst of Mgmt Consultants; American Mktg Assoc, Tau Beta Pi.

Lebenbaum, Matthew T
Business: Walt Whitman Rd, Melville, NY 11747
Position: VP, Adv Technology (Retired) *Employer:* Eaton Corp, AIL Div. *Education:* BA/EE/Stanford Univ; MS/EE/Stanford Univ; MS/EE/Stanford Univ. *Born:* Nov 1917. Native of San Francisco Calif. Assistantships Stanford & MIT 1938-41. Sys planning engr Amer Elec Power 1942. Res Assoc Harvard Univ Radio Res Lab 1942-45 with service on loan in England at TRE/RAF 1943 & ABL-15 1944. Eaton Corp, AIL Div., 1945-1981. Respon for advanced microwave & IR tech in receiving sys & R&D in semiconductor devices. Specialties in low noise amplification & radiometry as applied to radio astronomy. Fellow IEEE 1964; Mbr Phi Beta Kappa, Tau Beta Pi & Sigma Xi. *Society Aff:* ΦBK, TBΠΣΞ.

Lebens, John C
Home: 4 Briarcliff, St Louis, MO 63124
Position: Pres *Employer:* Lebens Assocs. *Education:* MS/Elec/WA Univ; BS/Elec/WA Univ. *Born:* July 1911. Native of St Louis Mo. Married Jeannette Lewald. One son Charles. Employed Bussmann Mfg Div McGraw Edison Co 1935-64. Ch Engr 1948. V P Engrg 1956. Contrib Editor 'Elec Mfg' 1942-48. Author over 100 tech articles. Guest Lectr Moore Grad Sch of Elec Engrg Penna Univ 1950-56. Pres Lebens Assocs (cons) 1957- . Pres Bridge Holding Corp (financial) 1974-78. Reg P E. Fellow IEEE. Mbr NSPE, MSPE, Sigma Xi, Tau Beta Pi, St Louis Engrs Club, St Louis Elec Bd of Trade. P Mbr NEMA, ASM, SAE. Hobby: tech writing. *Society Aff:* IEEE, NSPE, NSPL.

LeBlanc, Joseph U
Home: 902 Holik Dr South, College Station, TX 77840-3022
Position: Exec Dir *Employer:* LeBlanc Consulting Services *Education:* D.Eng/Proj Mgmt/TX A&M Univ; MBA/Fin & Mgmt/TX A&M Univ; MCS/Comp Sci/TX A&M Univ; BS/Math & Gen Eng/Southwestern LA Inst. *Born:* 3/17/37. Cajun from Erath, LA; Lower studies in LA; Grad studies (MCS, MBA, DENG) at TX A&M Univ with honors (Alpha Pi Mu, Upsilon Pi Epsilon, Sigma Iota Epsilon); Reg PE, Cert Data Processor, Arbitrator; Listed in Who's Who in South and Southwest, Personalities of the South, Outstanding Community Leaders, Men of Achievement, Who's Who in Tech Today; Married to Elizabeth D. Holder, 11 June 1960 (Mark, 22 Nov 1962; Kathy, 11 Jan 1965); Principal/Dir (1981-): managerial, technical, financial consulting in public/priv sectors; Prog Mgr (1979-81): R&D mgmt, marine logistics, environ impact statements, oceanographic res; Program/Project Mgr (1977-79): planning/cost eng, project/fiscal/logistical management; Executive Dir (1972-77): management/industrial eng consulting, economic/system analysis, sales/marketing, analysis;. Operations Mgr (1971-72): operations, scheduling, systems; Eng Fellow (1970-71): Production/inventory/cost systems; EDP Consultant (1968-70): Business/scientific applications; Ld Systems Analyst (1966-68): Real time geological applications; Mgr (1963-66): R&D/acquisition of C3 Systems; Programmer-Analyst (1960-63): cartographic/geodetic applications; Commissioned USAF 11 Aug 59, presently Lt Col and Engrg Mgr in USAFR. *Society Aff:* IIE, NSPE, PMI, ASEM, AACE, ICCP, ROA

LeBlanc, William J
Business: 223 Quail Run Dr Ste. F, Baton Rouge, LA 70808
Position: Owner *Employer:* LeBlanc & Assaf and Assoc *Education:* BS/Mech Engr/LA State U. *Born:* 4/6/24 Worked as eng-draftsman trainee with Ogden and Woodruff Consulting Engs. Continued training and work with similar firms specializing in mechanical building systems. Spent three years as respresentative for Worthington Corporation Air Conditioning Div. Continued consulting work and opened own firm, 1966 doing mechanical building systems. Acquired partner in 1977 and added electrical capability to firm. Participated in AIA Research Corporation national energy research project for HUD leading to various potential energy standards for buildings. Traveled to China September 1981 for People to People exchange of information with Chinese Univ of Construction. *Society Aff:* ASHRAE, AEE, ASPE, NFPA, ACEC.

LeBold, William K
Business: Engg Admin Bldg, W Lafayette, IN 47907
Position: Dir of Ed Res and Information Systems Prof Elec *Employer:* Purdue Univ. *Education:* PhD/Psychology/Purdue Univ; MS/EE/Northwestern Univ; BS/EE/Univ of MN. *Born:* 8/23/23. Conducted extensive studies of engg students, alumni, faculty, curricula, computer productivity, problem solving, and instructional tech as: Dir of Education Res and Information Sys and Prof of Eng. Purdue (1954-); Res Assoc, NAE (1971-2); Proj Coordinator, ASEE Goals of Engg Educ Study (1963-7); and Visiting Prof, UCLA (1961-2). Employed as Asst Prof & Staff Counselor, Univ of IL-Chicago (1947-53) & as Engr, Automatic Electric & Commonwealth Edison (1945- 7). Served on NRC Engrg Infrastucture Panel Minorities in Engg; NRC Task Force on Retention of Minorities, Committee on Minorities in Engg, NSF Ctte on Equal Opportunities in Sci and Tech, Examiner-Consultant, North Central Assn; as Chair of ASEE Educational Res & Methods Div, Information Sys Div, & Engg Manpower Statistics Committee and as Pres-Mbr, W Lafayette Community Sch Bd, (1966-70) Consultant: GE, Alfred Sloan Foundation, Engr Manpower Commission,

LeBold, William K (Continued)
Natl Action Council for Minorities in Engrg, National Research Council Reviewer: Natl Sci Foundation. *Society Aff:* ASEE, APA, AERA.

Lebow, Irwin L
Business: 2800 Bellevue Terr NW, Washington, DC 20007
Position: Consult *Employer:* Self *Education:* PhD/Phys/MIT; SB/Phys/MIT. *Born:* 4/27/26. in Boston, MA. Served in Navy 1944-46. On the staff of the MIT Lincoln Lab from 1951 to 1975, working in many communications & comp related areas, rising to mbr of the Lab's Steering Committee in 1970. Served as Assoc Hd of Satellite Communications Div 1970-72 & of Data Systems Div 1972-75. Chief Scientist - Assoc Dir Tech, Defense Communications Agency 1975-1981 with responsibilities for advanced technical planning and R&D for Defense Communications System & World Wide Military Command & Control System. VP, Engg, American Satellite Co. 1981-84. VP, Systems Research and Applications Corp. 1984-87. Independent consult 1987- . Coauthor of *Theory and Design of Digital Machines*, McGraw Hill (1972). Fellow of Am Physical Soc & IEEE. *Society Aff:* IEEE, APS, AFCEA

Lechner, Bernard J
Home: 98 Carson Rd, Princeton, NJ 08540
Position: Independent Consult. *Employer:* Self. *Education:* BSEE/EE/Columbia Univ *Born:* 1/25/32. Mr Lechner's early work at RCA Lab included res on video tape recorders and giga-hertz computer circuits. In 1962 he received an RCA Lab Achievement Award, and was a recipient of the 1962 David Sarnoff Team Awd in Sci for his contrib to high-speed computer circuitry. In 1962 he was appointed hd of a group engaged in res on digitally controlled visual displays. He received an RCA Lab Achievement Awd in 1967 for his work in the display field. From 1966 until late 1971 Mr Lechner was Hd of the Peripheral Equip Res Group of the Data Processing Applied Res Lab at RCA Lab. He was appointed Hd of the Community Info Sys Res Grp of the Communications Res Lab at RCA Labs 1971. He directed a grp doing pioneering work on two-way cable TV systems and Pay-TV systems. In 1974 he was apptd Hd of the Color Television Res Grp of the Sys Res Lab at RCA Labs. In 1977 he was apptd Dir of the Video Sys Res Lab at RCA Labs. In 1985 he was apptd to the position of staff VP, Advanced Video Sys Res. He was responsible for RCA's res on HPTV, EPTU, solidstate cameras and other advanced video topics. In 1987, Mr. Lechner left RCA Laboratories to establish his own consulting firm specializing in TV & advanced video systems & the management of technology transfer. *Society Aff:* SID/IEEE, SMPTE.

Leckie, Frederick A
Business: Dept. of Tam, 104 S. Wright St, Urbana, IL 61801
Position: Prof *Employer:* Univ of IL *Education:* PhD/Eng Mech/Stanford Univ; MS/CE/Stanford Univ; BSc/CE/St. Andrews *Born:* Leckie, Frederick Alexander, mech. engr., educator; b. Dundee, Scotland, Mar. 26, 1929, came to US, 1978; s. Frederick and Mary Baxter (Barclay) L; m. Alison Elizabeth Wheelwright, Mar. 30, 1957; children - Gavin Frederick, Gregor Wheelwright, Sean Charles, B.S., St. Andrews U., 1949; M.S., Stanford U., 1955, Ph.D., 1957; M.A. Cambridge U., 1958. Cons. civil engr. Mott, Kay & Anderson, London, 1949-51; research asst. Hanover (Ger.)U., 1957-59; lectr. in mech. scis. Cambridge (Eng.)U 1958-68; fellow, tutor, dir. studies Pembroke Coll., 1959-58; prof. engrging. U. Leicester, Eng., 1968-79; prof. theoretical and applied mechanics U. Ill., Urbana, 1984-, head dept., 1978-. Mem. editorial bd. various tech. jours., including engring. scis.; author: Matrix Methods in Elastomechanics, 1963, Mechanical Vibrations, 1963, Engineering Plasticity, 1968. Trustee Stoneygate Sch., Leicester, 1974-78, Bedales Sch., Hampshire, 1974-78. Fellow Am. Acad. Mechanics; mem. ASME. *Society Aff:* ASME, AAM

Lecureux, Floyd E
Business: 6000 J St, Sacramento, CA 95819
Position: Assoc Prof. *Employer:* CA State Univ. *Education:* PhD/ME/MI State Univ; MS/ME/MI State Univ; BE/ME/GM Inst. *Born:* 12/23/38. Native of Byron, MI. Taught at MI State Univ, 1963-1979. Advanced degrees are in Fluid Dynamics area of Mech Engg. Have been doing teaching and res in comp sci and especially in the CAD/CAM area and some consulting work in areas of thermodynamics and heat transfers. Enjoy most outdoor activites. *Society Aff:* ASME, SME, NCS, SIGGRAPH.

Ledbetter, James L
Business: Ga Dept of Natural Resources, Suite 1252, 205 Butler St. S.E, Atlanta, GA 30334
Position: Commissioner, Dept. of Natural Resources *Employer:* State of GA *Education:* MA/Sanitary Eng/GA Tech; B/CE/GA Tech *Born:* 11/7/34 Served as Water Quality Eng; Chief of Industrial Wastewaters; and Dir of Water Quality Surveys with the GA Water Quality Control BD from 1965 to 1971. Became Deputy Dir of GA Environmental Protection Div in 1972 and appointed as Dir January 1975. Appointed Commissioner of Ga. Dept of Natural Resources May 1984. Been an active member of ASCE, WPCF, AAEE, and the Assoc of State and Interstate Water Pollution Control Administrators (ASIWPCA) . Enjoy golf, jogging, and reading. *Society Aff:* ASCE, AAEE, WPCF, ASIWPCA.

Lederer, Jerome F
Home: 468-D Calle Cadiz, Laguna Hills, CA 92653
Position: Adjunct Prof USC. *Employer:* Univ of So Calif. *Education:* ME/Aero Engg/CA Eng in NY Univ; BSc/Aero Option/Coll of Engrg/NYU. *Born:* 1902 NYC. Aero Engr U S Air Mail Service 1926-27; Ch Engr Aero Insur Underwriters 1929-40; Dir Safety Bureau CAB 1940-42; Asst Genl Mgr & Ch Engr Aero Insur Underwriters 1943-47 with concurrent war services; Dir Flight Safety Foun 1947-67; concurrently Dir CornellGuggenheim Aviation Safety Ctr; Dir Office Manned Space Flight Safety NASA 1967-70; Dir Safety NASA 1970-72; Adj Prof USC 1974- . Mbr NAE, ASME, SAE; Fellow AIAA, Royal Aero Soc; Hon Mbr Inst Navigation. Guggenheim Medal 1961, Wright Bros Mem Award 1965, NASA Exceptional Serv Medal 1969, FAA Dist Serv Medal 1972, etc. P Pres Soc Air Safety Investigators; Pres Emer Flight Safety Foun. *Society Aff:* NAE, ASME, AIAA, RAeS, SAE, SSS, ISASI.

Lederman, Peter B
Home: 17 Pittsford Way, New Providence, NJ 07974
Position: VP *Employer:* Roy F. Weston, Inc. *Education:* PhD/ChE/Univ of MI; MSE/ChE/Univ of MI; BSE/ChE/Univ of MI. *Born:* 11/16/31. Dr Lederman has made significant contributions in res, engg, and dev, as well as mgt of, critical efforts related to energy and resource conservation, and environmental control. He is recognized for his expertise in hazardous, solid, liquid, and gaseous pollutant control. He published the first comprehensive evaluation of LNG. He was a maj contributor to the dev of toxic and hazardous mtls control and clean-up strategy. Application of computers in engg design was an active area of interest in his early career. He has published over 75 articles in technical, policy and profl areas and was responsible for the Confidentiality Guidlines of AIChE and a maj contributor to the first Guidlines for Profl Employment. He is active in community activities and serves on the Exec Bd of the Watchung Area Council, Boy Scouts of Am and chairs NAM's Hazardous Waste Task Force. He is a Fellow of the AIChE and a Diplomate of Amer Acad of Engrs. *Society Aff:* Fellow AIChE, ASME, NSPE, AAEE, AAEE, WPCF.

Lederman, Samuel
Home: 835 Pond La, Woodmere, NY 11598
Position: Prof of Mech & Aerospace Engr *Employer:* Polytechnic Inst of NY *Education:* MEE/Comm/Polytech Inst of Brooklyn; DPl Ing/Comm/Univ of Munich *Born:* 03/04/20 Educated in Germany and the US. Accepted a position at the Polytechnic Inst of Brooklyn. He has been involved in research activities ranging from aircraft structures to plasma dynamics, from subsonic and hypersonic wind tunnel instrumentation, control, data acquisition and processing to the development and application of diagnostic techniques as diverse as pressure transducers, microwaves, electrostatic probes and electron beams. Along with his duties as Prof of Mech and

Lederman, Samuel (Continued)
Aerospace Engrg he has been devoted since 1968 to the development and application of laser based diagnostic techniques such as Spontaneous Raman Scattering, CARS, LDV etc. to the field of fluid mechanics, combustion and magnethohydrodynamics. *Society Aff:* AIAA, ΣΞ.

Lee, Bansang W
Business: Coll of Eng, Piscataway, NJ 08854
Position: Assoc Prof *Employer:* Rutgers Univ *Education:* PhD/Elect/Princeton Univ; MSc/EE/Princeton Univ *Born:* 9/26/49 *Society Aff:* AVS

Lee, Bernard S
Business: 3424 S State St, Chicago, IL 60616
Position: Pres. *Employer:* Inst of Gas Tech. *Education:* DChE/CE/Poly Inst of Brooklyn; BChE/CE/Poly Inst of Brooklyn. *Born:* 12/14/34. in China. Naturalized: 1956. Reg PE in NY & IL. With Arthur D Little, Inc, 1960-65: proj leader of pilot plant studies; directed projs from lab study through semi-works to design of commercial plants. From 1965 to present at Inst of Gas Tech in various positions from Supervisor to VP, Process Res, Exec VP &, now, Pres. Work being carried on at IGT includes res in synthetic fuels from coal, shale, peat & biomass, fuel cells, solar energy, combustion, transmission & distribution of gas, & admn of Gas Engg Dept of IL Inst of Tech. Married, father of 3. Outside interests include tennis & music. *Society Aff:* AIChE, AIME, AAAS, ACS.

Lee, Chang-Sup
Business: Ship Res Station, KIMM, Daeduck Science Town, PO Box 1 Chungnam Korea
Position: Research Scientist *Employer:* Korea Inst of Machinery and Metals *Education:* PhD/Naval Arch and Marine Eng/MIT; SM/Naval Arch and Marine Eng/MIT; BS/Naval Arch and Marine Eng/Seoul Nat Univ Korea *Born:* 4/9/47 in Inchon, Korea. Assist Research Scientist for Shipbuilding and Ocean Engrg Lab, Korea Institute of Science and Tech 1973-75. Study at MIT, 1975-79. Research Scientist in Ship Research Station, Korea Inst of Machinery and Metals, since 1979, specializing in Ship Hydrodynamics, especially in Propeller Design, Propeller Performance Analysis and Cavitation. Responsible for operation of Cavitation Tunnel. Associate Member, SNAME 1977. SNAME Captain Joseph H. Linnard Prize for the best paper of the year 1979. *Society Aff:* SNAME

Lee, Charles A
Business: 3214 Tazewell Pike, P.O. Box 5477, Knoxville, TN 37918
Position: President. *Employer:* Charles A Lee Assocs Inc. *Education:* MS/Civil Engg/Lehigh Univ; BS/Civil Engg/Univ of WY. *Born:* Nov 16, 1915. Hydraulic engr David Taylor Model Basin, Navy Dept, Wash 1940-42, 1946-48; Ch Paper & Tissue Process Dev Kimberly-Clark Corp Neenah Wisc 1948-58; Cons Paper Indus 1958-59; Genl Mgr Formex Co, Div Huyck Corp, Knoxville Tenn 195963; Owner, Pres Charles A Lee Assocs Inc Knoxville 1963- . Lt Cmdr USNR 1942-46. Patentee in field. ASCE Collingwood Prize 1950; TAPPI Fellow Award Feb 1971; TAPPI Engrg Award Oct 1974. NEC Cert 2281-10; Reg P E Tenn 004132, Wisc E04338. Listed Who's Who in South & Southwest. *Society Aff:* TAPPI, ASCE.

Lee, Chin-Hwa
Business: 111 Link Hall, Syracuse, NY 13210
Position: Assoc Prof. *Employer:* Syracuse Univ. *Education:* PhD/EE & Comp Sci/Univ CA; MS/EE & Comp Sci/Univ CA; BS/Electrophys/Chiao- Tung Univ. *Born:* 8/18/47. He was born in Hsi-An, China. Received the PhD degree from the UC Santa Barbara. From 1970 to 1975 he was a res asst in the dept of Elec Engg & Comp Sci, UCSB. From Apr to July, 1975, he was employed part-time as a post grad researcher & lectr at UCSB. Since Fall, 1975 he has been a mbr of the Elec and Comp Engg Dept, Syracuse Univ as an Assoc. Prof. His current res interests is in image processing comp and communication. *Society Aff:* IEEE, ACM.

Lee, David A
Business: Building 640, Wright-Patterson AFB, OH 45433
Position: Head, Math, Dept. *Employer:* USAF Inst of Tech *Education:* PhD/Mech/IL Inst of Tech; ScM/Applied Math/Brown Univ; BS/EE/Univ of MO-Columbia *Born:* 11/7/37 in Fort Smith, AR. While in graduate school, worked part time for Antenna- Microwave section of Westinghouse's Electronics Div, and for IL Inst of Tech Research Inst. Was successively Research Mathematician, Group Leader, and Dir of the Applied Mathematics Research Laboratory, US Air Force Aerospace Research Laboratories, 1963-75. Visiting Prof, von Karman Inst for Fluid Dynamics, Brussels, 1969-70. Present position since 1975. Author of several articles in IEEE Transactions, AIAA publications, and Journal of Applied Physics. *Society Aff:* AIAA, SIAM, NYAS, ΣΞ

Lee, E Bruce
Home: 1705 Innsbruck Pkwy, Minneapolis, MN 55421
Position: Prof. *Employer:* Univ of MN. *Education:* PhD/Mech Engr/Univ of MN; MS/Mech Engr/Univ of ND; BS/Mech Engr/Univ of ND. *Born:* 2/1/32. Design of inertial navigation systems, Honeywell Inc 1955-59. Dev of control sys for Aerospace vehicles, Honeywell Inc 1960-63. Assoc Prof of Elec Engg (Univ of MN) 1963-65. Prof of Elec Engg (Univ of MN) 1966-present; Prof of Control Sci (Univ of MN) 1966-present; Prof of Computer Sci 1969-present (Univ o MN) Prof of Operation Res 1974-present (Univ of MN); Acting Hd of Dept of Computer Sci 1969-70 (Univ of MN); Hd, Dept of Elec Engg 1976-present (Univ of MN). *Society Aff:* IEEE, SIAM.

Lee, E Stanley
Business: Dept. of Industrial Engg, Darland Hall, Manhattan, KS 66506
Position: Prof. *Employer:* KS State Univ. *Education:* PhD/ChE/Princeton Univ; MS/ChE/NC State Univ; BS/ChE/China. *Born:* 9/7/30. Research engr. Phillips Petroleum Co., Bartlesville, Okla., 1960-66; asst. prof. Kans. State U., Manhattan, 1966-67, asso. prof., 1967-69, prof. indst. engring., 1969-present; prof. U. So. Calif., 1972-76; cons. govt. and industry. Dept. Def. grantee, 1967-72; Office Water Resources grantee, 1968-75; EPA grantee, 1969-71; NSF grantee, 1971-now; Dept. Agr. grantee, 1978-now. Dept. Enrgg grantee 1979-83. Mem. Am. Inst. Chem. Engrs., Soc. Indst. and Applied Math., Ops. Research Soc. Am., Sigma Xi, Tau Beta Pi, Phi Kappa Phi. Author: Quasilinearization and Invariant Imbedding, 1968; Coal Conversion Technology, 1979; Operations Research, 1981, editor Energy Sci. and Tech., 1975; asso. editor Jour. Math. Analysis and Applications, 1974, Computers and Mathematics with Applications, 1974. *Society Aff:* AIChE, SIAM, ORSA.

Lee, Edgar K M
Home: 3143 Alani Dr, Honolulu, HI 96822
Position: President. *Employer:* Engrg Design Group Inc. *Education:* MS/Structural Engg/Univ of CA; BS/Civil Engg/Univ of CA. *Born:* Nov 1934 Honolulu Hawaii. Tau Beta Pi, Chi Epsilon. Proj Engr Chin & Hensolt Inc San Francisco 1958-67. Proj Engr Alfred A Yee & Assocs Inc Honolulu 1967-69. Sr Struct Engr Shimazu, Shimabukuro & Fukuda Honolulu 1969-72. Joined Engrg Design Group Inc (formerly Kim, Kimura, Lee Assocs Inc) 1972 as V P. Assumed current office of Pres 1973. Reg Civil & Struct Engr Calif & Hawaii. *Society Aff:* ASCE, ACI, PCI, SEAOH.

Lee, G Fred
Home: 2305 Brookwood Drive, Fort Collins, CO 80525
Position: Prof *Employer:* CO State Univ *Education:* PhD/Env Eng & Env Sci/Harvard Univ; MSPH/Env Sci-Env Chem/School of Public Health; BA/Env Science/San Jose State *Born:* 7/27/33 Prof, Dept of Civil and Environmental Engrg, 1961-73. Dir Ctr for Environmental Studies, Univ of TX at Dallas, 1973-78. Prof of Civil Engrg, CO State Univ, 1978-present. Diplomate American Academy of Environmental Engs, Registered PE-TX. Published over 300 professional papers, reports, etc., on sources, significance, fate, and developing control programs for chemical contaminants in liquid and solid wastes. *Society Aff:* AAAS, ASCE, ASTM, WPCF

Lee, George C
Home: 288 Countryside Ln, Williamsville, NY 14221
Position: Dean of Engrg *Employer:* SUNY at Buffalo *Education:* PhD/Civil Engrg/Lehigh Univ; MS/Civil Engrg/Lehigh Univ; BS/Civil Engrg/Natl Taiwan Univ *Born:* 07/17/33 Civil Engng. faculty, SUNY/Buffalo since 1961; department chairman 1973-77, Dean of Engng. 1978-present. In 1977/78 Head of Engng. Mech. Section, Natl Science Foundation. Author of texts on *Structural Analysis and Design, Design of Single-Story Rigid Frames,* and *Cold Region Structural Engineering.* Research areas: structural engineering and biomechanics. Supported by NIH, NSF, ONR and steel industry. Recipient of Adams Memorial Award from American Welding Society in 1974, Superior Accomplishment Award of Natl Science Foundation in 1977, Engineering Achievement Award of Chinese Institute of Engineers, USA, in 1980, and Engineering Educator's Award of Erie-Niagara Chapter of the New York State Professional Engineers Society in 1983. Author of over 50 journal articles. *Society Aff:* AAAS, ASCE, AWS, ASEE, WRC, SSRC, ΣΞ, XE, TBΠ, USNCB

Lee, Griff C
Business: PO Box 70787, New Orleans, LA 70172
Position: Conslt Engr *Employer:* Griff C Lee Inc *Education:* MS/CE/Rice Univ; BE/CE/Tulane Univ *Born:* 8/17/26 US Navy, WWII; eng for Humble Oil (Exxon), specializing in offshore platform design. Joined McDermott in 1954 as design eng; worked as principal design engr; chief eng; group VP of engrg, VP & group executive of Research & Development/Technical Services Divisions, now is self employed as a consulting engr. Member of National Academy of Engrg and Marine Bd of National Research Council. Honorary member of ASCE Serves on Offshore Committee for API; Committee on Offshore Technology for Det norske Veritas; Amer Bureau of Shipping-Special Ctte on Offshore Structures: Offshore Operators Ctte; MIT Civ Engr Visiting Ctte; Rice Univ BD of Engrg Advisors. Tulane School of Engrg; Visiting Committee, Univ of TX Civil Engrg. Has lectured at numerous universities and seminars. *Society Aff:* ACI, ASCE, AWS, SPE of AIME

Lee, Harry R
Home: 1915 Stanford Dr, Anchorage, AK 99504
Position: Pres. *Employer:* H4M Corp *Education:* MS/Civil Engr/Univ of WA; BS/Civil Engr/Univ of WA. *Born:* 7/21/29. Prin AK Test lab 1957 to 1978, Pres 1970 to 1978. Chrmn of the Bd Denali Drilling 1970-present. Reg Engr AK 582E, WA 6314 & Partner Dowl Engrs 1973-78, Retired Dowl Engrs 1979, Founder H4M Corp 1978 for Geophysical and Geotechnical Services. Pres AK Society of Professional Engrs 1963, Chrmn Municipal Budget Commission 1970, Chrmn School Budget Commission 1978, Chrmn Geotechnical Commission 1976 (Commissions Municipality of Anchorage) Participant in numerous tech & professional conf, Contributor to tech publications. *Society Aff:* ASCE, NSPE, AAAS, ASPE, SEG.

Lee III, Thad S
Business: PO Box 218340, Houston, TX 77218
Position: Exec Dir *Employer:* Nat'l Assoc of Corrosion Eng *Education:* Master of Science/Metallurgical Eng/Univ of Florida; Bachelor of Science/Metallurgical Eng/Univ of Florida *Born:* 10/07/48 Involved in corrosion research & engrg, 1973-85, specializing in marine corrosion research and use of electrochemical test methods, with Cabot Corp. and LaQue Center for Corrosion Technology. Became involved with association mgmt 1985 and assumed current responsibilities in 1987 for management of 15,000 member professional society engaged in publishing, education and training and industry standards development. Served as chairman of ASTM Cttee G01 on Corrosion of Metals from 1984-85, chairman of ISO/TC 156/WG6 on Corrosion Evaluation Methodologies from 1981-86 and chairman of NACE Research Cttee in 1985. Member of NACE, ASTM and ECS and reg Corrosion Engr in CA. *Society Aff:* ASTM, ECS

Lee, James A
Home: 650 Main St, Altamonte Springs, FL 32701
Position: President. *Employer:* Self-employed JIM LEE Assocs. *Education:* BSAE/Univ of GA. *Born:* 12/25/27. Native of Nahunta, GA. US Army Engrs 1946-47. Field Engr J I Case Co, spec in field application, design, sales of tractors, hay tools, planting & harvesting equip 1951-61. Dist Engr Marlow Pumps Div ITT, spec in design, field application & mktg, ultra high concentrate spray equip & pumps 1961-66. FL Tractor Corp Engrg & Sales 1966-68. S E Dist Mgr Rain Bird Sprinkler Corp - field application, design, mkt res, sales irrigation equip. Conducted Irrigation Design Schs throughout S E 1968-74. Reed Irrigation Intl - irrigation design & sales, S E US, Caribbean, S Amer, W Indies 1974-75. Chmn S E Region ASAE 1974- 75. Elected Regional Dir ASAE 1976-77. Established & is Pres of JIM LEE Assocs, Cons Engrs, spec in irrigation, drainage & urban hyd & Irrigation World-Mkting Firm-all irrigation president of Jim Lee Sales Inc. Natl Dir, The Irrigation Assoc. Past Pres F/A. Irrigation Soc. *Society Aff:* ASAE, NSPE, FES, IA.

Lee, John A
Home: 8835 Lakemont, Dallas, TX 75209
Position: Consulting Engineer *Employer:* Self Employed *Education:* BS/Mech Engg/Univ of TX. *Born:* Dec 1912. Native of Coleman Texas. Design engr with Wyatt Industries, Houston & Dallas employed in pressure vessel & struct design 1936-41. With Mobil Res & Dev Corp since 1941. Retired Dec 1977, consulting work in field of res laboratory design 1978-1984. Mgr of Engg & Const at Field Res Lab, Mobil Res & Dev Corp 1966-1977. Secy of North Texas Sect ASME 1940. Chmn of North Texas Chap ASME Petro Div 1958. Chmn Liaison Ctte of Petro Div 1959. ASME Fellow 1972. Hobbies: tennis & bridge. *Society Aff:* ASME, ASHRAE.

Lee, John J
Business: 245 Park Ave, New York, NY 10017
Position: Pres. *Employer:* Barber Oil Corp. *Education:* MS/CE/Yale Univ; BS/CE/Yale Univ. *Born:* 9/26/36. Native of NY. Proj Mgr for construction & operation of chem plants overseas for W R Grace Co, 1959-64. Entered private consulting practice & formed Purvin & Lee, Inc, serving as Pres 1964-75. Firm specialized in organization & mgt of grass roots energy, transportation & mining enterprises. Purvin & Lee, Inc acquired by Barber Oil Corp (NYSE) 1975. Barber is a diversified energy natural resource co with operations in oil & gas, coal, gilsonite & shipping. Pres Trinidad Corp 1975-76. Exec VP & CEO Barber Oil Corp, 1975-78; Pres & CEO Barber Oil Corp, 1978-present; Chrmn Am Gilsonite Co, 1978-present. Yale Engg Award - 1977. Tau Beta Pi. *Society Aff:* AIChE, ACS, API, YSEA.

Lee, Ju P
Business: Dept of ME, WSU, 667 Merrick, Detroit, MI 48202
Position: Prof. *Employer:* Wayne State Univ. *Education:* PhD/Engg Mechanics/Univ of MI; SM/Mech Engg/MIT; BS/Mech Engg/Natl Wu-Han Univ, China. *Born:* 1/10/19. Graduated from Natl Wu-Han Univ, Wu Han, China with a BS degree in Mech Engg in 1941. Came to USA in 1945 as a grad asst in MIT and completed SM degree in Mech Engg in 1947. Transferred to Univ of MI and finished PhD degree in Engg Mech in 1950. Major Professional experiences are as follows: 1952-1955: Assistant Prof, IA State Univ, Ames, IA; 1955-1958: Sr Res Engr, Ford Motor Co, Dearborn, MI; 1959-1966: Assoc Prof, Wayne Univ; 1966-Now: Prof Wayne State Univ. *Society Aff:* ASEE.

Lee, Lloyd A
Business: 130 Jessie St, San Francisco, CA 94105
Position: Vice President. *Employer:* URS/John A Blume & Assocs, Engrs. *Education:* BS/Civil Engg/Univ of CA, Berkeley. *Born:* Mar 22, 1925. Participated in & supervised num engrg design & rehab proj's, incl the struct analysis & remedial design for 10 Veterans Admin hospitals & 2 Navy hospitals in sev locations throughout the U S. Principal in charge of struct restoration of the century-old Calif Capitol. ASCE. SAME. *Society Aff:* ASCE, SAME.

Lee, Low K
Home: 4479 Deerberry Ct, Concord, CA 94521
Position: Sr Staff Engr., retired *Employer:* TRW Defense & Space Sys Group.

Lee, Low K (Continued)
Education: PhD/Industrial Mgmt/CA Western; MS/EE/Univ of CA; BS/EE/Univ of CA. *Born:* Feb 1916. Native of Oakland Calif. Field Engr US Navy, installing & maintaining radar equip 1940-44. Design of radar displays at Raytheon Mfg Co 1944-47. Dev of miniaturization tech for missile electronics equip at N Amer Aviation 1947-50. Organized & managed a lab at Stanford Res Inst, to do res in the mech of electronic equip fabrication 1950-55. Organized & managed a lab at Ramo-Wooldridge Corp to dev fabrication tech for miniaturized electronic equip used in guided missiles. Organized & managed the Product Assurance Staff at TRW 195774. Assistant Director Product Assurance 1962-1974, Sr Staff Engr 1974-1978, Retired 1978, Consultant 1978. Fellow IEEE 1959. *Society Aff:* IEEE.

Lee, Quarterman
Business: 911 Main, Kansas City, MO 64105
Position: Prin. *Employer:* Lee & Rome. *Education:* BS/Mech Engg/Purdue Univ. *Born:* 9/16/44. Born in Tallahassee, FL. Worked in Mfg Engg at Ford Motor Co, structural design for McDonnell-Douglas Corp. Positions in Environmental, Maintenance, Plant and Industrial Engg for Rockwell-Intl Corp. In private practice since 1977. Frequent participant in natl and regional Plant Engg Conferences. Spoken and published on overhead crane design. *Society Aff:* NSPE, PEPP, AIPE.

Lee, Reuben
Home: 301 McMechen-Apt 1204, Baltimore, MD 21217
Position: Consulting Engr. *Employer:* Self-employed. *Education:* BSEE/Elec Engrg/West Virginia Univ. *Born:* Nov 8, 1902 Shirland, Derby, England. Westinghouse Elec Corp 1924-65: control engr 1924-28, radio transmitter design 1928-34, electronic transformer design 1934-45, adv engr 1945-57, cons engr 1957-65. Westinghouse Order of Merit 1960. IEEE Life Fellow. IEEE Magnetics Soc Electronics Transformers Tech Ctte (Chmn 1959-64). Engrg Soc of Baltimore. Author 'Electronic Transformers & Circuits'; co-author 6 other tech books, 70 tech papers. 26 U S pats. IEEE delegate to Popov Soc Congress, Moscow USSR 1975. *Society Aff:* IEEE.

Lee, Richard SL
Business: Department of Mechanical Eng, Stony Brook, NY 11794
Position: Prof of Eng *Employer:* State Univ of NY at Stony Brook *Education:* PhD/Eng & Applied Physics/Harvard Univ; MS/ME/NC State Univ; BS/ME/Nat Taiwan Univ *Born:* 7/19/29 NC State Univ: Asst and Asso Prof of Mechanical Engrg 1960-64. State Univ of NY at Stony Brook: Assoc Prof of Engrg 1964-68, Prof of Engrg, 1968-present; Chairman of Dept of Mech, 1969-75. Univ of Queensland, Australia: Visiting Prof of Mechanical Engrg, 1970, Univ of Karlsruhe, West Germany: Guest Professor of Fluid Mechanics, 1977-81. Nuclear Research Center at Karlsruhe (KFK), West Germany: Distinguished Visiting Research Scientist, 1981. Dir of a natl center of excellence in engg in Taiwan (the Inst of Appl Mech of Natl Taiwan Univ, Taipei, Taiwan), 1986-88, on leave from State Univ. of NY at Stony Brook. Recipient of the Alexander von Humboldt Award of the West German Gov't, 1980. Recipient of the Most Outstanding Teacher Award from the 1974 and 1981 graduating classes of the Coll of Eng and Applied Sciences of the State Univ of New York at Stonybrook. *Society Aff:* ASME, AIAA

Lee, Robert L
Home: 3387 Lone Pine Road, Medford, OR 97504
Position: Consultant *Employer:* Self *Education:* BS/Civil Engg/OR State Univ. *Born:* Nov 1922. Reg CE Oregon. Native Oregonian. Army Corps of Engrs 1943-46. Chmn Pacific N W Sect AWWA 1969, AWWA Fuller Award 1973, Internatl Dir AWWA 1974-77, Diplomate AAEE, Trustee AWWA Res Foun 1975-80, AWWA Dist Div Chmn 1974-75. *Society Aff:* NSPE, AAEE, AWWA.

Lee, Roger O
Business: 4th Ave SW & Winter Show Rel, Box 937, Valley City, ND 58072
Position: Secy-Treasurer. *Employer:* L W Veigel & Co P C. *Education:* BSCE/Civil/Indiana Inst of Technology. *Born:* June 1932. BSCE Indiana Inst of Tech. Grad School Univ of N Dakota. US Army Signal Corps Europe 1955-57. Engr L W Veigel & Co, P C since 1970 spec in Cty Hwys & Bridges. Taught at Valley City State College. Mbr Natl Soc of Prof Engrs, N Dak Assn of Prof Engrs, Natl Assn of Cty Engrs, N Dak Assn of Cty Engrs. Pres N Dak Assn of Cty Engrs 1975-76, presently serving on the Bd/Dir. Reg Engr & Land Surveyor N Dak; Reg P E Minn. Enjoy hunting, golf, photography & fishing. *Society Aff:* NSPE, NACE.

Lee, Sammie F
Business: 702 Starks Building, Louisville, KY 40202
Position: Pres *Employer:* Lee Engineering Co. *Education:* MS/Civil Engg/Univ of KY; BS/Civil Engg/Univ of KY. *Born:* Dec 1928. Native of Sulphur Kentucky. USAF 1951-53. Resident Engr Kentucky Dept of Hwys to 1956. Engr Portland Cement Assn to 1962. Cty Engr Jefferson Cty Ky & mbr Louisville-Jefferson Cty Planning Comm 1962-69. 1969-81 as Principal of multidisciplinary cons firm. Pres KSPE 1969-70; V Pres NSPE 1975-77; Pres, Natl Soc of Profl Engrs, 1979-80; Fellow ASCE; former Fellow, Amer Consulting Engrs Council ACEC, ITE, AAEE, APA, AICP & NAFE, Reg P E Kentucky & 6 other states. Reg L S Kentucky. Mbr, Kentucky Board of Registration for Professional Engrs & Land Surveyors, 1983 to 1987 Mbr Louisville Area Chamber of Commerce. P Pres, P Deputy Dist Governor, Louisville Lions Club. Mason, Scottish Rite, Shrine. *Society Aff:* NSPE, ASCE, ITE, APA, AAEE, NAFE.

Lee, Thomas H
Home: 44 Chestnut St, Boston, MA 02108
Position: Philip Sporn Prof of Energy Processing *Employer:* M.I.T. *Education:* BS/ME/Natl Chiao Tung Univ; MS/EE/Union College; PhD/EE & Appl Phy/Rensselaer Polytechnic Inst. *Born:* 5/11/23. Shanghai China. Professor and Business Exec (Engr/Scientist); m. Kin Ping Lee; c. William, Thomas, Richard; p. Yin Ching & N T Ho Lee (dec); Mgr Genl Elec Co Fairfield. Mbr Natl Acad of Engrg; Pres Power Engr Soc, IEEE; Fellow IEEE; author PHYSICS & ENGRG OF HIGH POWER SWITCHING DEVICES pub by MIT; Pres Nether Providence (Pa) Parent-Teacher Org 1962-64; V P Community Classes, Nether Providence Pa 1964-67; Mayors Sci Adv Bd Phila Pa; 30 U pats; Annual Achievement Award Chinese Inst of Engrg; 5 IEEE Paper Awards. Author 62 sci papers Receipient of IEEE: Power Life Award, Haruden Pratt Award Centinnial Medal. *Society Aff:* IEEE, NAE, APS.

Lee, William J
Home: P O Box 483, College Station, TX 77841
Position: Prof & Holder of Noble Chair in Petroleum Engg *Employer:* Texas A & M Univ *Education:* PhD/ChE/GA Inst of Tech; MS/ChE/GA Inst of Tech; B/ChE/GA Inst of Tech *Born:* 1/16/36 Sr Res Engr, Exxon Production Res 1962-68. Technical Advisor in charge of major fields studies, Exxon, 1971-77. Prof of Petroleum Engrg, MS State, 1968-1971; TX A&M, 1977-present. Noble chair in Petroleum Engg, Texas A&M. Sr VP Engrg and Exec Conslt, SA. Holditch & Assoc, 1979-present. Distinguished Lecturer, SPE 1978-9. Continuing Education Lecturer AAPG, 1977-present & SPE, 1970-present. Chrmn, SPE Pressure Transient Testing Ctte, 1984, SPE Continuing Education Committee, 1979-80 & SPE Formation Evaluation Committee, 1972. Author Well Testing (Textbook), 1982. SPE Dist Faculty Achievement Award, 1982. Video-tape lecturer for SPE, 1982, 1986, Tenneco Teaching Award, 1983. Halliburton Ed Fdn Award of Excellence, 1982. TX A & M Assoc of Former Students Dist Teaching Award, 1983. SPE Reservoir Engg Award, 1986. SPE Region Service Award, 1987. SPE Education & Accreditation Cttee, 1983-87, Chmn, 1986. *Society Aff:* SPE, ΣΞ

Lee, William S
Business: P O Box 33189, Charlotte, NC 28242
Position: Chrmn & CEO *Employer:* Duke Power Co. *Education:* BS/CE/Princeton Univ. *Born:* 6/23/29. Native of Charlotte NC. Navy Civil Engrg Corps. With Duke Power since 1955. Named Chairman in May 1982. Bd/Dir Duke Power, Liberty Corp & J A Jones Const Co. JP Morgan Tau Beta Pi, Phi Beta Kapp, Chi Epsilon,

Lee, William S (Continued)
ASME Geo Westinghouse Gold Medal 1972, Fellow ASME & ASCE, National Academy of Engineering, member. *Society Aff:* ASCE, ASME, ANS, NSPE.

Leeds, J Venn, Jr
Home: 10807 Atwell, Houston, TX 77096
Position: Prof. *Employer:* Rice Univ. *Education:* JD/LAW/Univ of Houston; PhD/EE/Univ of Pittsburgh; MS/EE/Univ of Pittsburgh; BA/BSEE/EE/Rice Univ *Born:* 10/26/32. Previously Mbr, Atomic Safety & Licensing Bd Panel, US Nucl Regulatory Commission; Consultant to several ins cos on causes of fires & product liability cases; Previously consultant to Exxon Res, GeoSpace Corp, TX A&M Univ; Tulane Univ; Boston Uni; PE (TX); Mbr State Bar of TX; Sr Engr at Westinghouse Electric Corp, Bettis Atomic Lab; Previously Master of Richardson College, Rice Univ. *Society Aff:* IEEE.

Leeds, Winthrop M
Home: 212 Overdale Rd, Pittsburgh, PA 15221
Position: Independent Cons Engr. *Employer:* Westinghouse Elec Corp-ret 1970. *Education:* PhD/Physics Engg/Univ of Pittsburgh; MS/Elec Engg/Univ of Pittsburgh; BS/Elec Engg/Haverford College. *Born:* Aug 1905. Long career as high voltage circuit breaker designer & developer with Westinghouse Elec Corp. 100 pats. Respon for design of first 500 kV circuit breakers put in service in U S, making use of sulfur hexafluoride gas for both insulation & arc quenching. Mgr Power Circuit Breaker Engrg 1961; Mgr New Products Engrg 1963; Cons Engr 1968- . IEEE Natl Prize Paper 1943; IEEE Res Comm Chmn 1958-60; IEEE B G Lamme Gold Medal 1971. Prof Engr Pa. Pres Pgh Chap PSPE 1972-73. Like golf, bridge, chess. *Society Aff:* IEEE, NSPE.

Leeming, Wilson
Business: 4747 Harrison Ave, Rockford, IL 61101
Position: Mgr, Matls Engg. *Employer:* Sundstrand Corp, Advanced Tech Grp. *Education:* MS/Eng Mechanics/Univ of IL; BS/Met Eng/Univ of IL *Born:* 5/30/31. Employed by Sundstrand since 1957 in project, R&D supervisory capacities in Matls Engrg. Res & publs in heat treatment, stress corrosion, fatigue & fracture. Patent in diffusion bonding. Mbr ASM since 1957, Rockford Chap Chmn 1963. Natl ASM Ctte mbrships have included Vol II, Handbook (1963), Technician (1966), & Howe-Grossman Award. Mbr ASTM since 1963 Mbr of Cttes E9 (Fatigue) & E-24 (Fracture Testing). Secy E-9 sub on Contact Fatigue 1967-71. Pres Rockford Area Engrg Study Council 1968-70. Mbr SAE since 1984; Sec of Ctte E (Carbon, Low Alloy and Specialty Steals), SAE/AMD Co representative Mbr, AIA/Materials & Structures Ctte. *Society Aff:* AMS, ASTM, SAE, AIA

Leep, Herman R
Home: 4707 Valley Station Road, Louisville, KY 40272
Position: Assoc Prof *Employer:* Univ of Louisville *Education:* PhD/Ind Engrg/Purdue Univ; MMAE/Mech & Aeros Engrg/Univ of Delaware; BME/Mech Engrg/Univ of Louisville *Born:* 10/21/40 Native of Louisville, KY. Joined Univ of Louisville as Asst Prof in 1973. Became full-time mbr of Dept of Industrial Engrg in 1979. Promoted to Assoc Prof in 1983. Teaches courses in mfg processes, advanced mfg, and engrg econ. Res interests are in areas of performance evaluation of metal cutting fluids and computer-aided mfg. Chapter Chrmn, SME 1984-85. Registered Prof Engr in KY. Mbr of Alpha Pi Mu. Enjoys traveling. *Society Aff:* IIE, SME, TBΠ, AΠM

Lefebvre, Arthur H
Business: Mech Engg, W Lafayette, IN 47907
Position: Reilly Prof of Comb Engg, Sch of Mech Engg. *Employer:* Purdue Univ. *Education:* DSc/ME/London Univ; PhD/ME/London Univ; BS/ME/London Univ. *Born:* 3/14/23. Joined the Aero-Engine Div of Rolls-Royce in 1952. Occupied Chair of Aircraft Propulsion at Cranfield Inst of Tech, Bedford, England from 1961 until appointed Hd of Sch of Mech Engg in 1971. Appointed Hd of Sch of Mech Engg at Purdue in 1976 & Reilly Prof of Combustion Engg in 1979. Holds several patents relating to combustion equip. Author of over 50 publications on both fundamental & practical aspects of combustion & heat transfer. DSc degree awarded in 1975 in recognition of published work on combustion & heat transfer in gas turbines. *Society Aff:* RAeSoc, IMechE, RSA, ASME, AIAA.

Lefebvre, Michael J
Home: 1543 Rustic Ridge Court, Green Bay, WI 54301
Position: Assoc, Dir of Corporate Systems *Employer:* Foth & VanDyke And Assoc Inc. *Education:* BS/CE/MI Tech Univ *Born:* Mr Lefebvre graduated, with honors, from MI Tech Univ, Houghton, MI in 1970. He immediately joined the firm of FOTH & VAN DYKE and Assoc, Inc as a Project Engr and has since advanced to an Assoc of the firm and presently acts as the Dir of Corporate Systems. Mr Lefebvre has been responsible for the planning, design and construction of many types of Public Works Projects. His present major responsibilities include the dev, testing and operation of the firm's info sys on an IBM 38 computer and intergraph CADD System. Mr Lefebvre has been active in ASCE and has served in varying capacities as a mbr of the Bd for the Fox Valley Chap including that of Chap Pres and presently is a President Elect for the WI Sec Bd. In addition, Mr Lefebvre is active in WSPE, NSPE, and is a mbr of the Tau Beta Pi Assoc and Chi Epsilon. Mr Lefebvre is a Registered Prof Engr in nine midwestern states abd US cert by the Natl Council of Engrg Examiners. *Society Aff:* ASCE, NSPE, WSPE, NCEE.

LeFevre, Elbert W
Business: CE/UA/BEC 4190, Fayetteville, AR 72701
Position: Prof Dept of Cvl Engrg. *Employer:* Univ of Arkansas - Fayetteville. *Education:* PhD/CE/OK State Univ; MSCE/CE/TX A&M; BSCE/CE/TX A&M. *Born:* 7/29/32. Mbr ASCE, Pres Mid-South Sect 1971. District 14 Dir 1983-86 Mbr ASEE, Pres Midwest Sect 1976-77. Mbr TRB, Sigma Xi, Chi Epsilon, Mbr NSPE, Pres ASPE 1979- 80. Mbr Westark Area Council Bd, BSA, Razorback Dist Chmn 1974, 1975. Freemason. Mbr Springdale Bd of Adjustments. Principal, Engrg Services Inc genl cons. Asst Prof Texas Tech Univ 1958-63. Assoc Prof, Prof, Prof & Hd, Interim Dean, Prof Univ of Arkansas 1966- . Married, 4 children. ASPE Nat'l Director 1980-82, 83-85; Vice Pres NSPE PEE 1983-84, Vice Pres NSPE SW Region 1985-87; Member, ABET 1984-87; Member NCEE 1984-; Member Ark State Bd of Reg for PE and LS 1984-. *Society Aff:* ASCE, NSPE, TRB, ASEE, NCEE, ABET.

Lefkowitz, Irving
Home: 3532 Meadowbrook Blvd, Cleveland Heights, OH 44118
Position: Prof of Systems Engrg and Chemical Engrg *Employer:* Case Western Reserve Univ *Education:* PhD/Control Engg/Case Inst of Tech; MS/Instrum Engg/Case Inst of Tech; BChE/Chem Engg/Cooper Union School of Engs *Born:* 7/8/21 Native of NY, NY; Control Engr and later Dir of Instrumentation Research for J.E. Seagram & Sons, 1943-1953. With Case Inst of Tech since 1953: Research Assoc in Process Automation until 1957 and faculty member 1957 to present. Currently Prof of Systems Engrg & Chem Engrg, Dir of Research Program in the Control of Industrial Systems, 1958-86. Was NATO Postdoctoral Fellow at Imperial Coll in London 1962-63 and Research Scientist at the Intl Inst for Applied Systems Analysis 1974-75. Chrmn of Tech Committee on Systems Engrg of Intl Federation of Automatic Control 1978-81. Research interests include computer control of industrial systems and hierarchical approaches to control of complex systems. Chrmn of Tech Ctte on Control of Industrial Systems, IEEE Control Systems Society. Fellow IEEE; Fellow AAAS. Recipient 1982 Control Heritage Award of the American Automatic Control Council.. *Society Aff:* IEEE, AAAS

Lefond, Stanley J
Home: 29983 Canterbury Cir, Evergreen, CO 80439
Position: Pres *Employer:* Industrials Minerals, Inc *Education:* MS/Geo/Univ of MI; BS/W Geology Opt/Univ of AK; BME/Mining Engr/Univ of AK *Born:* 11/13/17 My engrg career started in the Central Engrg Dept of Diamond Alkalie Co, Cleveland,

Lefond, Stanley J (Continued)
OH, from 1952-1966. Responsibilities consisted of supervising mineral evaluation progs including cost estimates of mining, construction and studies in mineral reserves. Later with US Borax Co, Los Angeles, CA, and Amer Metals Climax, Denver, CO, in the same capacity. I am now a conslt in the field of industrial minerals. Former mbr of the Bd of Dir of the Society of Mining Engrs, Chrmn of the Industrial Minerals Div, Distinquished Mbr of the Society of Mining Engrs, Registered PE in TX. Recip of Hal Williams Hardinge Award for excellence in Indus Minerals. *Society Aff:* SEG, AIPG.

Legget, Robert F
Home: 531 Echo Drive, Ottawa KIS IN7, Ontario Canada
Position: Retired (Consultant Self-employed) *Education:* D Eng/Hon/Univ of Liverpool; M Eng/Univ of Liverpool; B Eng/Univ of Liverpool Eng. *Born:* Sept 29, 1904 Liverpool Eng, of Scottish parents. Grad in civil engrg from Univ of Liverpool 1925; emigrated to Canada 1929 after training with consultants in Westminster & Scotland. Engaged on heavy const work in eastern Canada until 1936 when started 11 yrs of teaching at Queen's Univ & Univ of Toronto; invited in 1947 by Canada's Natl Res Council to come to Ottawa & estab a Div of Bldg Res; served as Dir until retirement 1969. Chmn Natl Bldg Code Ctte 1948-70; Pres, ASTM 1965; Pres, Geolog Soc AM 1966; author 10 bks bks & many papers; editor 3 bk. Officer of the Order of Canada. *Society Aff:* ASCE, FICE, FEIC, FGSA.

Leggett, Lloyd W, Jr
Home: 13411 Kingsride, Houston, TX 77079
Position: Engg Advisor in Synthesis Fuels. *Employer:* Exxon Co,USA *Education:* BS/Chem Engg/Rice Univ; BA/Chem Engg/Rice Univ. *Born:* July 21, 1935 Midland Texas. With Humble Oil & Refining Co 1960-70, Hd Econ & Coord Dept 1969-70. Sr Adv Corp Planning, Esso Inter-Amer, Coral Gables Fla 1970-73. Adv Corp Planning Exxon Co USA, Houston 1973-75, Coord Assoc in Refining 1975-76. Currently Engg Advisory in Synthetic Fuels of Exxon Co, USA, an Exxon affiliate. Reg P E Texas. Mbr Amer Inst Chem Engrs, Natl Dir 1969-71, Chmn Fuels & Petrochem Div 1968 Local Arrangements Chmn for 83rd Natl Mtg 1977, elected Fellow 1984, S Texas Sect Chmn 1968, Outstanding Young Chem Engr 1966 Distinguished. Service Award 1979. Soc Prof Chemists & Engrs Pres 1966. Lt USNR 1958-60. *Society Aff:* AIChE.

Legnos, John P
Business: 30 Gillett St, Hartford, CT 06105
Position: President. *Employer:* Legnos-Cramer Inc. *Education:* BS/ME/Purdue Univ. *Born:* Dec 12, 1919 Hartford Ct. Educ in NYC elem & secondary schools; P Pres Conn Soc of Prof Engrs, Hartford Cty Chap. Hon Mbr Pi Tau Sigma. *Society Aff:* NSPE, CEC, ASHRAE, ΠΤΣ.

Lehan, Frank W
Home: 1696 E Valley Rd, Santa Barbara, CA 93108
Position: Retired. *Education:* BS/EE/Caltech. *Born:* Jan 1923. BSEE 1944 Caltech. US Army Signal Corps 1942-44. Jet Propulsion Lab Caltech 1944-54 ending as Hd of Electronics Res Sect. Respon for early dev of Spread-Spectrum Radar & Communications Sys. 1954-58 Space Tech Lab of TRW as Deputy Dir of Electronics Res Div. 1958-66 Space Genl Corp as Exec V P, later Pres. Cons 1966-67. 1967-69 Asst Secy of Transp U S. 1969-77 Cons. 1960 IRE award for Space Electronics. Fellow AIEE. Mbr Natl Acad of Engrg. Reg EE Calif. *Society Aff:* IEEE, AIAA, TBΦ, ΣΞ.

Lehman, Frederick G
Business: 323 High St, Newark, NJ 07102
Position: Distinguished Prof. *Employer:* New Jersey Inst of Tech. *Education:* BS/Civil Eng/City College; SM/Civil Eng/MIT; ScD/Civil Eng/MIT. *Born:* 1918. Participated in design & const of 1st lrg wind turbine in USA in Rutland, Vt. During WW II was struct engr with Curtiss-Wright Corp. Started teaching at NCE NJIT in 1947 in Civil Engrg. 1st Dir of Computing Ctr 1961-67. Chmn Civil & Environ Engrg Dept 1967-75. Acting Dean of Engg 1975-77. Currently Disting Prof. Major engrg involvement is in planning & transp. Served as Chmn Aesthetics Ctte ASCE and Secy of Urban Planning & Dev Div. *Society Aff:* ASCE, ASEE, NSPE.

Lehman, Lawrence H
Business: 411 Theodore Fremd Ave, Rye, NY 10580
Position: President. *Employer:* Berger, Lehman Assocs, PC. *Education:* MBA/Org Behavior/Iona College; BCE/Civil Eng/N.Y. University *Born:* 4/30/29. Progressively Proj Engr, Proj Mgr, Ch Engr, Partner Vollmer Assocs NYC. Since 1968 President Berger, Lehman Assocs. Director, Louis Berger Int Educational Div Ch exec officer of cons firm spec in the planning & design of urban transp sys & facils. Reg P E NY, NJ, CT, MA, IN, IL, KY. Fellow Amer Soc C E; Mbr ACEC, SAME, TRB, NSPE, AREA, ARTBA, IABSE, WPCF, Natl Panel Amer Arbitration Assn. Listed in Who's Who in America Recipient Third Award US Steel Co Intl Bridge Design 1966; PC Inst Engrg Excell Award 1974. NY Assn Cons Engrs Engrg Excell Award 1975, 1979. *Society Aff:* NSPE, ACEC, SAME, ASCE, ARTBA, IABSE, ARSA, WPCF.

Lehmann, Gerard J
Home: 105 Ave Victor Hugo, Paris France 75116
Position: Consulting Engr (Self-employed) *Education:* Engg/-/Ecole Centrale de Paris. *Born:* Apr 6, 1909. Grad from Ecole Centrale des Arts et Manufactures Paris. 1957-75 Sci Dir, Compagnie Generale d'Electricite Paris. 1940-57 Sr Engr ITT & Sci Dir ITT Paris Lab. IEEE Life Fellow, P Chmn French Sect. Societe Francaise des Electroniciens, P Pres 1963. Societe des Ingenieurs Civils de France, P Pres 1967. Union des Assns Scientifiques et Industrielles Francaises, P Pres 1969. Cons Engr. *Society Aff:* ISF, SEE.

Lehmann, Gilbert M
Business: Valparaiso Univ, Valparaiso, IN 46383
Position: Prof of ME. *Employer:* Valparaiso Univ. *Education:* PhD/ME/Purdue Univ; MS/ME/IL Inst of Tech; BS/ME/Valparaiso Univ. *Born:* 8/4/33. Taught at Valparaiso Univ in the Mech Engr Dept from 1956 to present. Acting head of Mech Engr 1967-1968. Dean of the College of Engr 1972-1978. Acting head of Mech Engr. 1986. Area of expertice - Data Acquisition and Instrumentation. *Society Aff:* ASME, ASEE, ΣΞ.

Lehmann, Henry
Business: PO Box 4840, Court St Plant 3-51, Syracuse, NY 13221
Position: Genl Mgr. *Employer:* bGen Electric Co *Education:* BE (EE)/Elec Eng/Tulane Univ *Born:* 8/21/29. in Europe in 1929 and moved to the USA in 1941. Joined Gen Elec Co in 1948. Grad from GE Adv Engg Prog in 1951 and supervised portions of that program. Held sys engg and mgmt positions in fire control, adv guidance, ballistic missile defense, Managed the GE Integration Reliability portions of the NASA Apollo lunar landing program, managed the Business & Professional Operation of GE/Time Inc's General learning Corp, and joined GE's Heavy Military Equipment Dept in 1967, as Mgr-Engrg. Has authored papers on Reliability & Quality, Sys Approach to Education, IEEE Recommended EE curriculum, etc. Married to Elizabeth Lichtenberg and has 4 sons Leonard, Richard, Daniel & David. Currently General Mgr - Military Electronic Systems Operations. Is GE liaison to Syracuse Univ and mbr of Syracuse Univ College of Engrg Advisory Committee.

Lehmann, W Kemp
Business: 490 Balt-Annapolis Blvd, Glen Burnie, MD 21061
Position: Pres, Chmn of the Bd. *Employer:* C M Kemp Mfg Co. *Born:* Aug 7, 1929. BS Chem E Univ of Maryland 1953. Native of Baltimore Md. Army Chem Corps 1953-55. Service Tech, engr, Pres, Ch C M Kemp Mfg Co. Adsorption equip for removal of moisture & CO(2) from compressed gases esp products of combustion. Precision combustion of liquid & gaseous fuels for the purpose of generating atmo-

Lehmann, W Kemp (Continued)
spheres. U S pat 3,225,516. Relax with classical music, fishing, marine & shoreline life in Caribbean & Bermuda.

Lehnhoff, Terry F
Business: 106C Mech. Engr, Univ of MO-Rolla, Rolla, MO 65401
Position: Prof. *Employer:* Univ of MO. *Education:* PhD/Mechanics/Univ of IL; MS/ME/Univ of MO; BS/ME/Univ of MO. *Born:* 7/7/39. Native of St Louis, MO. Res engr for Caterpillar Tractor Co, Peoria, IL 1962-65 specializing in structures & fatigue. Joined the Mech & Aerospace Engg Dept of the Univ of MO-Rolla in 1968. Res areas include: Mech equip design and testing, & finite element analysis applied to solid mechanics problems. Consultant to Detroit Tool Engg in Lebanon, MO 1973-1986, Cooper Ind-Gardner Denver in Quincy IL 1978-81, Eaton Corp in Battle Creek, MI, 1980-83, and Rockwell Hanford Operations, Richland, Washington, 1981-82. Pres, Enmeco Inc., Rolla, MO, 1983-present. *Society Aff:* ASME.

Leib, Francis E
Home: Apt B 501 St. B. Village, Gibsonia, PA 15044
Position: Vice President. *Employer:* Copperweld Corp. *Education:* BS/EE/Univ of IL. *Born:* June 8, 1905. Native of Springfield Ill. Var engrg assignments Commonwealth Edison 1926-36, before joining Copperweld Corp. Active in wire applications & const practices for Rural Electrification 1936-41. From 1941-45 military liaison Signal Corps communications. 1945-60 Ch Engr & co-inventor Alumoweld wire for the lightning protection of electric power transmission lines. U.S. Mbr CIGRE-Intl Extra High Voltage Transmission Assn. 1960-70 Exec V P in charge Wire & Cable Div. Retired Copperweld 1970. IEEE Life Mbr & Fellow 'for Contrib to Overhead Line Matls & Const Practices'. Post retirement Keystone Resources V P, copper mining in Utah, recycling metals in Calif & dir Environ Engrg in Pgh. Retired Keystone 1976, now Resident at Retirement, Village at St. Barnabas, Gibsonia, PA. 15044. *Society Aff:* IEEE.

Leibovich, Sidney
Business: Sibley Sch of Mech & Aerospace Engg, Ithaca, NY 14853
Position: Prof. *Employer:* Cornell Univ. *Education:* PhD/Theoretical Mechanics/Cornell Univ; BS/Engrg/CIT. *Born:* 4/2/39. Asst & Assoc Prof of Thermal Engrg, Cornell Univ 1966-78, Prof of Mech & Aerospace Engrg, Cornell Univ, 1978-present. General Partner Flow Analysts Associates, 1979-present. MacPherson Prize, CA Inst of Tech, 1961. NATO Postdoctoral Fellow in Math 1965-66, Univ College London, Exxon Production Res Co, 1972-73. Sr Visiting Fellow, Univ of St Andrews (Scotland), 1977. Visiting Scientist, Weizmann Institute of Science (Israel). Editor: *Nonlinear Waves*. Cornell Univ Press, 1974. Fellow, Amer Physical Soc; Fellow Mbr, ASME, Assoc Editor: SIAM Journal on Applied Mathematics, 1972-75; ASME Journal of Applied Mechanics, 1976-83; Journal of Fluid Mechanics, 1982- ; ACTA Mechanica, 1986- . Contributions to the technical literature in magnetohydrodynamics, nonlinear wave propagation in fluid systems, stability theory, vortex flows, air-sea interaction, hydrodynamics of oil spills. *Society Aff:* ASME, APS

Leibson, Irving
Home: 1200 California St, Apt #24B, San Francisco, CA 94109
Position: President *Employer:* BOLD Technologies *Education:* DSc/ChE/Carnegie Inst of Tech; MS/ChE/Carnegie Inst of Tech; MS/ChE/Univ of FL; BS/ChE/Univ of FL. *Born:* Sept 1926. Chem engr to Supr Humble Oil & Refining Co Baytown Tex 1952- 61; Mgr Process Engrg 1961-63, Tech Mgr 1963-65, Dir R&D 1965-67, Genl Mgr ABS Div 1967-68; V P Dart Indus Chem Group Paramus N J 1969-74; Mgr Commercial Ventures 1974-75, Investment Dept; Mgr Process & Environ, Sci Dev 1975-78, VP & Mgr Res & Engg 1978-1982. VP Mktg & Tech 1982-1986. Bechtel Group Inc.; President BOLD Technologies 1987-present. part-time Prof Rice Univ 1957, Univ of Md 1954. Dist Commr E Harris Cty Dist Boy Scouts Amer 1958-61. AUS 1953-54. Reg P E NJ, Texas, Calif. Fellow AIChE (Dir 1967-69, V P 1973, Pres 1974); Publ award S Tex Sect 1957. AIChE Founders Award 1977. EJC (Dir 1969-76); Mbr EMC 1969-76. Co-Chmn Intersoc Comm on Guidelines to Prof Employment; V Chmn Intersoc Task Force Energy 1973; Associate World Coal Study 1978-80; Associate Coal Industry Advisory Board of International Energy Agency 1980- ; Chairman, Council on Alternate Fuels, 1985-7. Chairman, Coal and Slurry Technology Association 1986-7 Charter Member, National Coal Council 1985-present Member, Advisory Board Center for Chemical Process Safety ACS; patients, pubs in field. mbr US Natl Ctte of World Energy Conference. World Trade Club, San Francisco, Round Hill Country Club. *Society Aff:* AIChE, ACS.

Leidel, Frederick O
Business: 1527 Univ Ave, Madison, WI 53706
Position: Prof Emeritus of Gen Engg; Assoc Dean, Engg. *Employer:* Univ of WI. *Education:* BS/ME/Univ of WI. *Born:* 12/3/16. Native of Madison, WI. Design engr for Hamilton Std Propeller Div of United Aircraft Corp during WWII. On Univ of WI faculty since 1946, teaching engg graphics, specializing in descriptive geometry and graphical analysis. Assoc Dean (Academic) and freshman adviser 1966-82, responsible for freshman students and freshman academic progs. 1979 recipient of Benjamin Smith Reynolds Award for excellence in teaching. Active in Masonic organizations since 1945. Prof. Emeritus since 1982; adviser to students planning to transfer to Engrg at UW-Madison. *Society Aff:* ASEE.

Leidheiser, Henry, Jr
Business: Sinclair Lab 7, Bethlehem, PA 18015
Position: Prof. *Employer:* Lehigh Univ. *Education:* PhD/Chemistry/Univ of VA; MS/Chemistry/Univ of VA; BS/Chemistry/Univ of VA. *Born:* 4/18/20. Prof of Chemistry; Alcoa Foundation Prof; Dir, Corrosion Lab; Zettlemoyer Center for Surface Studies, Lehigh Univ, 1968- , Dir Center 1968-84; VA Inst for Scientific Res, 1949-68; Dir 1960-68. Author & editor of 7 books, 230 publications in technical literature. Recipient of awards from NACE, Humboldt Senior Scientist Award, 11th World Congress of Metal Finishing, South African Corrosion Institute, and the Electrodeposition Division of the Electrochemical Society. Fields of specialization: corrosion of metallic coatings, delamination of polymer-coated metals, paint adherence, electrical measurements of polymer-coated metals, Mossbauer spectroscopy, positron annihilation. *Society Aff:* AAAS, ACS, ECS, NACE.

Leigh, Donald C
Business: Dept of Engrg Mech, Lexington, KY 40506
Position: Prof *Employer:* Univ of KY *Education:* PhD/Math/Univ of Cambridge (England); BASc/Engrg Physics/Univ of Toronto *Born:* 2/25/29 Native of Toronto, Canada. Sr Aerophysics Engr with General Dynamics-Fort Worth, TX 1954-56. Supvr of tech computations with Curtiss-Wright Turbomotor Div 1956-57. Lecturer and assistant prof in Dept of Mechanical Engrg, Princeton Univ 1957-65. Assoc Prof 1965-68 and Prof since 1968 in Dept of Engrg Mechanics, Univ of KY to present. Chrmn 1971-80. Author of many tech articles and book *Nonlinear Continuum Mechanics*, McGraw-Hill, NY 1968. Enjoy tennis. *Society Aff:* ASME, SOR, SNP

Leigh, Robert E
Business: 1615 Downing St, Denver, CO 80218
Position: Pres *Employer:* Leigh, Scott & Cleary, Inc *Education:* MS/CE/Northwestern Univ; BS/CE/Northwestern Univ *Born:* 9/26/31 Pres of Leigh, Scott & Cleary, Inc, a traffic engrg and trans planning firm in Denver, CO, which was established in 1975. Leigh's background extends over 25 yrs in public and private engrg and planning work. He was an officer in the US Navy Civ Engr Corps with public works and construction experience in domestic and foreign assignments. He spent 6 1/2 yrs with the Denver Intercounty Regional Planning Commission as Ch Planner. He was a Prin Assoc with Wilbur Smith & Assoc in New Haven, CT for 5 yrs and Prin Assoc and Deputy VP of Alan M. Voorhees & Assoc in Denver, CO for 5 yrs. Native of Omaha, NB. *Society Aff:* ITE, AICP, ACEC

Leinbach, Ralph C
Business: 101 W Bern St, PO Box 662, Reading, PA 19609
Position: Group VP. *Employer:* Carpenter Tech Corp. *Education:* BS/Met Engg/Lehigh Univ. *Born:* 11/24/28. Mr Leinbach joined Carpenter in 1955 as a Met in the R & D Lab after previous employment as a met in the Atomic Power Div Res Ctr of Babcock & Wilcox Co. He was promoted to melting met in 1956. In 1957, while holding the same title, was transferred to Bridgeport, CT, where Carpenter Steel had acquired a facility that same yr. In 1961 he returned to Reading as plant Met. He became mgr-mill met in 1965, chief met in 1968, asst vp met in 1970 & vp-met in 1971. In 1975 he became vp-technical & div vp-techical in 1976. In Jan 1979 he became Group VP-Carpenter Steel Div. Mr Leinbach was born in Esterly, completed his publi education in Mt Penn, PA. Grad from Lehigh Univ, Bethlehem, PA in 1954 with a BS degree in Met Engg. He completed additional courses at the MA Inst of Tech & the Univ of Pittsburgh. *Society Aff:* ASM, AIME, AISI, ADPA, SME, AVS, MS, AWS.

Leipziger, Stuart
Business: 5500 Wabash Ave, Terre Haute, IN 47802
Position: Fae Chem Engrg Dept *Employer:* Rose-Hulman Inst of Tech *Education:* BS/ChE/IIT; PhD/ChE/IIT *Born:* 4/17/38 Native of Chicago, IL. Lecturer in Chem Engrg Dept at IL Inst of Tech 1964- 65. Joined Gas Engrg Dept in 1965. Have taught undergraduate and graduate courses in areas of natural gas liquefaction, energy conservation, systems analysis, engrg mathematics, fluid properties, and mass transfer. Have advised graduate students in thesis research in areas of fluid properties, mass transfer, fluidization, and second law analysis. Have served as chrmn of Gas Engrg Dept since 1977. Joined Chemical Engrg Dept at Rose-Hulman Inst of Tech in September 1984. Currently Prof of Chem Engg. *Society Aff:* ASEE, AIChE, ΣΞ.

Leissa, Arthur W
Business: Dept of Engineering Machanics, 155 W Woodruff, Columbus, OH 43210
Position: Prof *Employer:* Ohio State Univ *Education:* PhD/Engr Mech/OH St Univ; MSc/ME/OH St Univ; BME/ME/OH St Univ *Born:* 11/16/31 Asst Prof (1958), Assoc Prof (1961) and Prof (1964) of Engrg Mech at OH State Univ, and currently there. Author of two books, Vibration of Plates (1969) and Vibration of Shells (1973). Researcher in elasticity, plates, shells, vibrations, buckling and fibrous composite materials, with approximately 140 publications. Visiting Prof (1972-73), Federal Inst of Tech in Zurich, Switzerland. Visiting Prof (1985-86), U.S. Air Force Acad, Colo Springs. Engr and/or conslt with Sperry Gyroscope Co, Boeing Airplace Co, North American Aviation, Battelle Memorial Inst and Kaman Nuclear. Mbr, editorial advisory bds, Applied Mechanics Reviews, Journal of Sound and Vibration, Composite Structures and Intl Journal of Mech Sci. Fellow, ASME, American Academy of Mech (President, 1987-88). Mbr, ASEE American Alpine Club. *Society Aff:* ASME, AAM, ASEE

Leiter, Martin
Business: 18800 N W 2nd Ave, Miami, FL 33169
Position: President. *Employer:* Shiskin-Leiter & Assocs Inc. *Education:* BS/Civil Eng/Univ of IL; MS/Environ Eng/Univ of KS. *Born:* Jan 23, 1941 Brooklyn N Y. Publ thesis 1965 titled 'The Effects of Temp, Sludge Age, & Loading in Anaerobic Digestion'. Exper has been in field of public health, R&D, private utils, treatment plant oper & environ engrg cons. Reg P E Fla. Diplomate Amer Acad of Environ Engrs. Pres ShiskinLeiter & Assocs Inc, environ engrg cons firm 1972- . *Society Aff:* ASCE, AAEE, AWWA, WPCF, FICE.

Leitmann, George
Business: Dept of Mech Engg, Berkeley, CA 94720
Position: Prof. *Employer:* Univ of CA. *Education:* PhD/Engg Sci/Univ of CA, Berkeley; MA/Physics/Columbia Univ; BS/Physics/Columbia Univ *Born:* 5/24/25. From 1950-57 at US Naval Ordnance Test Station, China Lake, as Physicist ad Head of Aeroballistics Analysis Sec engaged in rocket exterior ballistics res. From 1957-present on UC faculty, full prof of engg sci since 1963 engaged in teaching and res in dynamics, dynamical systems, optimal control, differential games. Author of 12 books and 175 technical papers, one patent. Consultant to industry, govt. Winner of 1977 Pendray Aerospace Literature Award and of 1984 Mechanics and Control of Flight Award of AIAA, mbr of Natl Acad of Engrg, of Intl Academy of Astronautics and of Academy of Sci, Bologna. Fellow, AIAA. Winner of Levy Medal of Franklin Inst and Von Humboldt Senior Scientist Award. Fellow, AIAA. 1981-86, Associate Dean for Graduate Affairs; Associate Dean for Academic Affairs of College of Engineering, 1986-present. *Society Aff:* AIAA.

Leland, Samuel C
Home: 102 Coolidge Hill, Cambridge, MA 02138 *Education:* BS/Engg Sci/Harvard Univ. *Born:* 4/4/23. Native of RI. Served as commissioned officer in US Navy during WWII. Reg PE in several states. Fellow, Am Soc of Mech Engrs. Served as Chrmn-Boston Sec, Natl Nominating Committee, Chrmn-Profl Practice Committee, and Policy Bd for Profl Affairs. Also served as Pres, Soc of Harvard Engrs and Scientists. Author of publications on combustion control, air condensers, water treatment, and fuels handling. With Stone & Webster Engg corp 1946-88. Mech Engr, Asst Chief Mech Engr and mgr of Projs Admin Dept. Developed effecient nuclear and fossil fueled elec generating stations. Performed overall mgmt of major nuclear, fossil, indus and fluidized bed powr install design, const and assoc serv. Designed water treatment and waste treatment plants. Performed independent consulting for design and construction of Omnimax Theater for Boston Museum of Science and for ethanol plant for Johnson Products Inc. *Society Aff:* ASME, NSPE, MSPE, HES, FE.

LeMay, Iain
Business: Met Lab, Gen Purpose Bldg, Saskatoon, Sask, Canada S7N 0W0
Position: Prof. *Employer:* Univ of Saskatchewan. *Education:* PhD/Mtls/Univ of Glasgow; BSc/ME/Univ of Glasgow; ARCST/ME/Univ of Strathclyde. *Born:* October 30, 1936 Helensburgh, Scotland. On teaching staff, Univ of Glasgow, 1957-63. With Univ of Saskatchewan since 1963 in charge of met teaching & res. Served as Visiting Prof in Brazil & Argentina on a number of occasions since 1970. Technical Advisor, Mech & Met Engg, to NUCLEBRAS, Rio de Janeiro, Brazil, 1975- 76. Pres, Met Consulting Services Ltd, Saskatoon, from 1978. Current res activities primarily in creep at high temperature, fatigue & fracture, as well as in failure analysis. Fellow, ASM, 1977. Enjoy sailing & reading. *Society Aff:* ASM, ASME, ASTM, IMS, CIM, NACE.

LeMay, Richard P
Business: School of Mgmt, Troy, NY 12181
Position: Assoc Dean. *Employer:* RPI. *Education:* PhD/Ind Eng/IA; MS/Ind Eng/IA; BS/Ind Eng/Northeastern. *Born:* Nov 1935 Manchester N H. PhD & MS Univ of Iowa 1970, 1968; BS Northeastern Univ 1961. Prof exper with Natl Security Agency, Raytheon, Bethlehem Steel, & Proctor & Gamble. Teaching exper at Univ of Iowa 1966-70 & RPI 1970- . Currently Assoc Dean School of Mgmt RPI, Director (1976-77) Hospital & Health Services Div IIE, Samaritan Hosp Bd of Dir, Chmn TKE Bd of Dir. Mbr Alpha Pi Mu, Sigma Xi, Epsilon Delta Sigma; TKE Key Leader award. *Society Aff:* IIE, APM, ΣΞ, TKE.

Lembke, Walter D
Business: Univ. of Illinois Agr Eng Sciences Bldg, 1304 W Pennsylvania, Urbana, IL 61801
Position: Prof of Agr Engrg *Employer:* Univ of IL *Education:* PhD/Agr Eng/Purdue; MS/Agr Eng/Univ of IL; BS/Agr Eng/Univ of IL *Born:* 12/26/26 I was born in Joliet, IL and spent my early life on a farm in Will County. I served on active duty in the US Naval Reserve from 1952 to 1956. I received BS and MS degrees from the Univ of IL and the PhD from Purdue Univ. I am a reg PE in IL and IN. My major professional interests are drainage, irrigation, hydraulics & hydrology. I have served as chrmn of the Soil and Water Div of ASAE and have served as chrmn of the Central IL Section of ASAE. I am married (Donna Jean Christiansen) and we

Lembke, Walter D (Continued)
have three children. I have been a visiting lecturer in Indonesia & the United Kingdom. *Society Aff:* ASAE.

LeMehaute, Bernard
Business: 4600 Rickenbacker Causeway, Miami, FL 33149
Position: Prof Applied Marine Physics & Ocean Engrg *Employer:* Univ of Miami. *Education:* Dr Sc/Hydrodynamics/Univ of Grenoble, France, 57; Degree of Adv Techs/Hydro and aero dyn/Univ of Paris, France, 53; License as Sci/-/Univ of Toulouse, France, 51; Engr, Ecole Natl/Sys/Eletrotechnique & Hydraulique/Univ of Toulouse, France, 51 *Born:* 3/29/27. Educated in France. Taught at Ecole Polytechnique, Montreal and Queen's Univ, Kingston, Ontario. In 1961 came to the United States & in 1966 founded an Ocean Engineering Company, Tetratrech Inc. in Pasadena, California. He is now Prof of Applied Marine Physics & Ocean Engrg at the Univ of Miami. He is the author of a textbook "Hydrodynamics and Water Waves–, also published in Japanese and Russian, and of more than 100 papers in coastal engrg & water waves. In 1979, he was given the international coastal engineering award from ASCE for the design of coastal structures, theoretical developments and education. *Society Aff:* ASCE, IAHR.

Lement, Bernard S
Home: 24 Graymore Rd, Waltham, MA 02154 *Employer:* Lement & Associates *Education:* Dr of Sci/Met/MA Inst of Tech; BS/Met/MA Inst of Tech. *Born:* 2/11/17. From 1938 to 1968 doing met R & D and engg for ind & governmental labs, and univ teaching and res. Since 1968 private Mtls Engg Consultant & Dir of Lement & Assoc doing ind consulting involving mtl selection, processing, inspection & testing; and legal consulting including expert witness testimony nvolving accident reconstruction, failure analysis of mtls and safety engg. Author over 30 technical publications & participation in over 1,000 engg investigations. Reg PE (MA), Certified Safety Profl (ASSE) and ASM Fellow. Past Class A nationally rated chess player. New England ranked Sr Singles tennis player. *Society Aff:* ASM, AIME, ASTM, ASSE, AWS.

LeMessurier, William James
Business: 1033 Massachusetts Ave, Cambridge, MA 02238
Position: Chrmn, CEO *Employer:* LeMessurier Consultants Inc *Education:* SM/Structures/MIT; ABcl/Structures/Harvard Univ *Born:* 06/12/26 Native of Pontiac, MI. Founded Goldberg-LeMessurier, Structural Engrs, 1952. 1956-1968, Assoc Prof at Harvard Graduate Sch of Design and, at MIT Dept of Architecture. Since 1973, Lecturer at Harvard Graduate Sch of Design. Appointed Adjunct Professor in 1982. Founded LeMessurier Assocs Inc., 1961. Merged in 1973 to form Sippican Conslts Intl. Buy-out by employees of Sippican Consultants Operations from TSC Corporation and Formed employee-owned LeMessurier Consultants, Inc., Sept 1985. Presently, Chrmn and CEO. Structural engrg designer of major projects such as NY's Citicorp Ctr, Boston's New City Hall, Boston's Federal Reserve Bank, Inter First Plaza, One Main Center, Dallas, TX, 383 Madison Ave., NYC Columbus Center, N.Y. Canary Wharf Tower, London, England, and King Khalid Military City in Saudi Arabia. One of the originators of the "staggered truss sys" for steel framed bldgs. Co-authored "Design of Steel Structural Mbrs," part of McGraw-Hill's 2nd Edition 1979 "Structural Engrg Handbook." Author technical paper: "A Practical Method of Second Order Analysis," Parts 1 and 2, AISC's "Engrg Journal," 1976-1977. Registered PE. *Society Aff:* NAE, ASCE, ACI, AISC

LeMieux, Henry F
Business: PO Box 27456, Houston, TX 77027
Position: Chmn of the Bd & CEO. *Employer:* Raymond Internatl Inc. *Education:* BS/Civil Engr/Tulane Univ; BE/EE/Tulane Univ. *Born:* Aug 20, 1926 Greenville Miss. m. Marjorie Elizabeth Hunter; 4 children. Joined Raymond Internatl 1949. Elected Dir 1966; Pres 1968; Ch Exec Officer 1970; Chmn of the Bd 1976. Mbr Amer Soc of Civil Engrs, Amer Soc for Testing & Matls, La Engrg Soc, Soc of Amer Military Engrs, Beavers; President of The Moles 1976. Tau Beta Pi at Tulane Univ; awarded Charles Thompkins Award & Harold A Levey Award by Tulane Univ. Lic P E La, Fla & NY. Hobbies: yachting & skiing. Outstanding Alumnus of the School of Engrg, Tulane Univ, for 1977. The Moles 1980 Member Award for outstanding achievement in construction. *Society Aff:* ASCE, ASTM, SAME, LES.

Lemley, William K
Home: 425 Elmont Rd, Virginia Beach, VA 23452
Position: Senior Cvl Engr *Employer:* MMM Design Group *Education:* BS/CE/OK State Univ *Born:* 10/12/22 Native of VA, served in US Navy in World War II. Employed by MMM Design Group, Norfolk, VA. Served in office in Rome, Italy as Asst Chief Cvl Engr in charge of all cvl engrg work for a new city in Libya and visited oil camps in Libyan Desert to investigate construction problems and recommend corrective measures. Had responsibility for design of utilities, drainage, Pentagon Mall parking, traffic maintenance and construction phasing for a portion of Washington subway sys. Assigned to Houston, TX, office 1976-1979, first as Chief Sanitary Engr and later as Chief Cvl Engr on design of King Abdulaziz Military Academy, Saudi Arabia. *Society Aff:* ASCE, NSPE

Lemlich, Robert
Business: Location 171, Cincinnati, OH 45221
Position: Prof of Chem Engg. *Employer:* Univ of Cincinnati. *Education:* PhD/ChE/Univ of Cincinnati; MChE/ChE/Poly Inst of Brooklyn; BChE/ChE/NYU. *Born:* 8/22/26. Native of NYC. Served in US Navy 1944-46. Res engr with Allied Chem & Dye Corp 1948-49. On faculty of Univ of Cincinnati since 1952. Awarded Fulbright lectureships to Israel 1958-59 & Argentina 1966. Also awarded 15 res grants and have 125 publications including book "Adsorptive Bubble Separation Techniques–, Academic Press, 1972. Am Reg PE & part-time consultant. Am Fellow of AIChE and AAAS. Received annual Sigma Xi Award for Distinguished Res at the Univ of Cincinnati, 1969. Was designated Chem Engr of the Yr for the AIChE OH Valley Sec, 1979. *Society Aff:* AAAS, ACS, AIChE, ASEE.

Lemons, Jack E
Business: Dept of Biomtls SDB49, Birmingham, AL 35294
Position: Prof and Chrmn Dept of Biomaterials *Employer:* Univ of AL, at B'ham *Education:* BS/Met/Univ of FL; MS/Met Sci/Univ of FL; PhD/Met Sci/Univ of FL *Born:* 1/20/37. Jack Lemons, born in FL was educated through the PhD in Met & Mtls Sci at the Univ of FL. Ind machine shop in addition to heavy construction. Teaching experience was at the Univ of FL, Clemson Univ, and the Univ of AL at Birmingham. Currently as Prof and Chrmn of the Dept of Biomtls both teaching and res continues. Approximately 100 technical papers and seven book chapters have been published. Interests now centralize on application of synthetic mtls for biological applications. *Society Aff:* SB, ORS, IADR, ASM, AIME

Lemons, Thomas M
Business: 72 Loring Ave, Salem, MA 01970
Position: President. *Employer:* TLA-Lighting Cons Inc (Self). *Education:* BS/Elec Engg/Purdue Univ. *Born:* 9/15/34. Reg P E Mass. Founded TLA-Lighting Cons Inc 1970, after 13 yrs in Application & Dev Engrg at Sylvania Lighting Products. Fellow IES & Fellow USITT. Mbr Internatl Comm on Illumination, NSPE, MSPE, SMPTE & USITT with service in var offices or on var cttes. 7 pat awards. *Society Aff:* IES, NSPE, MSPE, SMPTE, USITT, CIE.

Lempert, Frank L
Business: One Broadway, Cambridge, MA 02142
Position: Vice President, Sales. *Employer:* Badger America Inc. *Education:* MS/Chem Engg/Carnegie Mellon Univ; BS/Chem Engg/Carnegie Mellon Univ. *Born:* May 25, 1929 NYC. BS & MS 1954 chem engrg Carnegie Mellon Univ. US Army Signal Corps 1951-53. Process engr & process mgr with the M W Kellogg Co 1954- 63. With Badger since 1963. Worked in London office as Exec Sales Engr 1968-71, & in The Hague office as Dir of Sales 1971-75. Assumed present respon as V P Sales for Badger Amer in the fall of 1975. Respon for all sales activities in the

Lempert, Frank L (Continued)
U S & for coord of Badger's sales efforts in Canada & Latin Amer, together with advertising & pub relations efforts in the US. Apptd a Dir of Badger Amer, Inc, June 1978. Enjoy rebuilding old automobiles. *Society Aff:* AIChE, ΣΞ.

Lempert, Joseph
Home: 140 Spring Grove Rd, Pittsburgh, PA 15235
Position: Consultant. *Education:* BS/Physics/MIT; MS/EE/Stevens Inst of Tech. *Born:* July 1913 N Adams Mass. Engr Westinghouse Lamp Div Bloomfield NJ 1936- 44, Sect Mgr 1944-53. At Westinghouse Electronic Tube Div Elmira NY as Sect Mgr for electronic imaging tubes & X-ray tubes 1953-56, & Mgr of Advanced Dev 1956- 58. At Westinghouse Res Labs Pgh, as res engr on electronic imaging tubes & on electron beam welders 1958-66, & adv physicist from 1966-78, working on electronic tube devices & magnetohydrodynamic energy conversion. Westinghouse Special Patent Award 1969 for Secondary Electron Conduction Camera Tube. Fellow IEEE 1968, Consultant 1978- . *Society Aff:* AIEE, AIP.

Lena, Adolph J
Business: Willowbrook Ave, Dunkirk, NY 14048
Position: Chrmn of Bd. *Employer:* AL Tech Specialty Steel Corp. *Education:* DSc/Met Engr/Carnegie Mellon; MS/Met Engr/Carnegie Mellon; BS/Met/Penn State. *Born:* 10/10/25. A native of Latrobe, PA, Dr Lena was employed by the Allegheny Ludlum Steel Corp as Res Metallurgist, Dir of Res, VP of Res, and VP & Gen Mgr of the Bar Products Div. Dr Lena has published 21 papers, is credited with the dev of several special alloys, and is an Honorary Fellow of the American Soc of Metals where he had been Natl Secy and a Trustee. Dr Lena has twice received Allegheny Ludlum's Frank Lounsberry Award for outstanding res accomplishment and Penn State's David Ford McFarland Award for distinguished contributions to metallurgy. Dr Lena became Pres and later Chrmn of the Bd of AL Tech Specialty Steel Corp when the corp was formed in 1976. *Society Aff:* ASM, AIME, AISE.

Lenard, John F
Business: 1066 Storrs Rd, Storrs, CT 06268
Position: Consulting Engg. *Employer:* Lenard Engg Inc. *Education:* MS/Sanitary Engr/Univ of CT; DiplIng/Sanitary/Hungarian Inst of Tech. *Born:* 6/13/31. In 1971 Mr Lenard established the firm of Lenard Engg, PC as a consulting engg organization. Prior to establishing the firm, Mr Lenard was Chief Engr for the Southeastern CT Water Authority & Proj Engr for the consulting firms of Tippetts-Abbett-McCarthy-Stratton & Hydrotechnic Corp. In these capacities Mr Lenard was in charge of the design & engg supervision of many engg projs encompassing a wide spectrum of the civ engg field. *Society Aff:* ASCE, WPCF, NSPE.

Lenczyk, John P
Business: R & D Center, 9921 Brecksville Rd, Brecksville, OH 44141
Position: Res Assoc/Supvr *Employer:* BF Goodrich Chem Co *Education:* PhD/Chem Engg/SUNY-AB; MS/Chem Engg/SUNY-AB; BS/Chem Engg/SUNY-AB. *Born:* 3/3/40. Worked in the Chem industry in Western NY State for 6 yrs before returning to school to obtain the terminal degree at SUNY/AB. Was awarded an NDEA Fellowship for 3 yrs support of grad education. Moved to OH in 1970 to teach at the Univ of Akron. Has worked summers or consulted for NASA and Firestone. Has published in the area of thermodynamics and heat transfer. For the last 4 yrs he has been doing research for the B.F. Goodrich Chemical Co. at the R&D Center in Brecksville, Ohio on vinyl chloride monomer. *Society Aff:* AIChE, ΣΞ.

Lender, Adam
Home: 4124 Briarwood Way, Palo Alto, CA 94306
Position: Consulting Scientist *Employer:* Lockheed Palo Alto Research Labs *Education:* PhD/Elect Engg/Stanford Univ; MS/Elect Engg/Columbia Univ; BS/Elect Engg/Columbia Univ. *Born:* 9/15/21. Joined Bell Labs 1954; did res in digital voice communications. 1961 to 1984 with GTE Lenkurt, as Head of Advanced Dev. Since 1984 Consulting Scientist at Lockheed Palo Alto Laboratory. Teaches at Santa Clara Univ. Invented the Duobinary process, which he later expanded into Correlative techniques. Received Hon Mention Award of IEEE Contech Group. Granted 30 US patents as well as many foreign pats. Fellow IEEE, active in IEEE affairs. 1970-77 Editor and 1978 to 1984 Editor-in-Chf of IEEE Transactions on Communications. Awarded 1984 Stuart Ballantine Medal by the Franklin Inst in Phila. Awarded 1983 IEEE Communications Soc Donald McLellan Meritorious Service Award. IEEE 1984 Centennial Medal. Enjoys modern history & piano playing *Society Aff:* IEEE, AIAA.

Lenel, Fritz V
Business: RPI, Troy, NY 12181
Position: Prof of Met Engrg Emeritus. *Employer:* Rensselaer Poly Inst - ret. *Education:* PhD/Chemistry/Univ of Heidelberg, Germany. *Born:* July 1907. Emigrated to U S 1933. 14 yrs in indus, 10 of these with Delco Moraine Div G M Corp, Dayton Oh, spec in dev production of struct parts by powder met. Since 1947 with RPI, Troy NY. Chmn of Matls Engrg Dept 1965-69. Dir res proj's in both applied & fundamental aspects of powder met. Indus cons. Fellow Amer Soc for Testing & Matls 1958 & Amer Soc for Metals 1971. Disting Serv to Powder Met Award, Metal Powder Indus Fed 1968. Powder Metallurgy Pioneer, Metal Powder Indus Fed 1984. *Society Aff:* ASTM, AIME, ASM, Metals Soc.

Leney, George W
Home: 5335 Tomfran Dr, Pittsburgh, PA 15236
Position: Air Pollution Engineer *Employer:* Allegheny County *Education:* MA/Geology/Univ of MI; MS/Elec Engr/Univ of MI; BS/Elec Engr/Univ of MI. *Born:* 11/13/27. Wausau Wis. m. Arax Tefankjian (deceased) 1955; c. Sara, Janet, John, Ruth. Gulf Oil Co, Shell Oil Co, Ch Geophysicist Hanna Mining Co 1956-63, Ch Geologist H K Porter Co 1966-76, V P & Dir Pacific Asbestos Corp 1972-75, Regional Geologist U S Dept of Energy 1977-81, Consulting Geologist 1976-77 & 1981-86. 7 tech publ. Robert Peele Mem Award AIME 1965. Who's Who in East 1972-. Presbyterian - Trustee Hamilton Presby Church 1975-78. Dir and Pres Armenian-American Club of Pgh 1981. Whitehall Country Club. Registered P E (PA). Air Pollution Engineer Allegheny Co. Bureau of Air Pollution Control 1986-. *Society Aff:* SEG, SExG, AIME (SME), GSA, APCA

Lenherr, Frank E
Home: 3450 Vista Ave, Cincinnati, OH 45208
Position: Mgr, CFM56 Proj Engg - Retired. *Employer:* Genl Electric Co. *Education:* Masters/Engg/Harvard Univ; BS/Phyiscs/Harvard Univ. *Born:* Aug 12, 1925 NYC. Worked at Northrop Aircraft for 3 yrs & then returned to G E in Lynn Mass & Cinn Ohio. Also transferred to the Computer Div in Phoenix & Okla City, where ran 3 plants. Then returned to the aircraft engine to take a staff job. On first job, managed to put into existence three small compressors - T58, T85 and T64. Then went one & dev with an after-burner with an aferburning. Mbr Inst of Aeronautical Sci, Newcomen Soc. P Mbr Soc of Automotive Engrs. Enjoy skiing, sailing & water skiing. *Society Aff:* NS.

Lennex, Richard B
Home: 6940 Stockport Dr, Lambertville, MI 48144
Position: Partner. *Employer:* Finkbeiner, Pettis & Strout Ltd. *Born:* May 8, 1924. BSCE Univ of Toledo. Native of Toledo Ohio. US Army Jan 1943 to Nov 1945. Field Engr Ohio State Hwy Dept 1949-52. Joined Finkbeiner, Pettis & Strout Ltd Feb 1952 as resident engr. In charge of const from 1960-73. Joined Promotional Dept 1970 to 1977. Returned fulltime const'n 1977 to date. Part-time instr in Sewer, Hwy & Water-line Const Univ of Toledo. Enjoy swimming & skiing.

Lennox, William C
Business: Faculty of Engrg, Waterloo Ont N2L 3G1 CD Canada
Position: Dean of Engrg *Employer:* Univ of Waterloo *Education:* PhD/Applied Mech/Lehigh Univ; MASc/Applied Math/Univ of Waterloo; BASc/Engrg Phys/Univ of Waterloo *Born:* 05/22/37 Visiting Prof Lehigh Univ. Visiting Prof Saudia Arabia.

Lennox, William C (Continued)
Visiting Prof Harvey Mudd Coll. Prof and Chrmn of Civil Engrg 1979-82. Prof and Dean of Engrg 1982-. Author of over 70 papers and reports. Conslt work-NASA, Gen Atomic, Lockhead. Enjoys sailing. *Society Aff:* ASEE, ΣΞ, CSPE

Lenz, Ralph C
Home: 2899 O'Neall Rd, Waynesville, OH 45068
Position: Sr Research Engr. *Employer:* Univ of Dayton Res Inst. *Education:* MS/Indus Mgt/MIT; Aero Eng/Structures/Univ of Cincinnati. *Born:* Oct 1919 Beatrice Nebr. Aero Engrg with Glenn L Martin Co 1940-44. USAAF 1945-46. Struct Engrg USAF 1947-53. R&D planning USAF 1954-74. With Univ of Dayton 1975- . Principal activities are res & cons on tech forecasting & assessment. Sloan Fellow MIT 1958-59; Chmn AIAA Tech Ctte on Mgmt 1974-75; Founding Co-Editor 'Tech Forecasting' journal 1969-71; Natl Hon Mbr Triangle Frat 1976. Hobbies: sailing & antique automobiles. *Society Aff:* AIAA, AFA.

Lenz, Richard R
Home: 967 Wood Creek Dr, Milford, OH 45150
Position: Principal & VP *Employer:* KZF, Inc. *Education:* MS/Structural Engr/Univ of Cincinnati; BSCE/Civ Engg/Univ of Cincinnati. *Born:* 9/20/43. in New Orleans, grew up in Cincinnati. Started profl career with Shaw, Lenz & Assoc, came to KZF in 1968. Experienced in surveying, hgwys, bridges, water & wastewater, bldg structures. Formerly chief structural engr, and Proj Mgr of large ind, institutional & public works projs. Presently VP and Dir of Oper. Principal in firm and mbr of the Bd of Dirs. Served local section of ASCE in every local office including Pres, District Chairman. Served as mbr of Bd of Engrs & Scientists of Cincinnati & Pres of local community assn. Enjoys tennis, skiing, woodworking & family. Mbr of Cincinnati Rotary Club. *Society Aff:* ASCE, ACEC.

Lenz, Robert E
Home: 515 Oak Valley Dr, St Louis, MO 63131
Position: Consultant-Engrg Mgmt. (Self-employed) *Education:* Bs/Chem Engg/WA Univ of St Louis. *Born:* Apr 1919. Culver Military Acad; Monsanto Co: 34 yrs res, dev, engrg - overseas assignments; directed planning, design, const grass roots petrochems complex Chocolate Bayou Tex; Dir Engrg Hydrocarbons Div; Dir Engrg Tech & Services, Corp Engrg Dept (4 engrg dirs, 275 people for advanced tech, purchasing, accting, costs, schedules, forecasts); Dir Corp Energy Planning. Consulting: project planning, mgmt, contracting, costs - CHF Engr planning $2 billion petrochems complex for undev area Saudi Arabia. Reg P E Ohio, Tex, Mo (State Dir MoSPE); one of first 3 elected to new Fellow-Grade AIChE (Chmn Natl Prof Dev Comm). *Society Aff:* AIChE.

Lenz, Robert W
Business: Polymer Sci and Engg Dept, University of Massachusetts, Amherst, MA 01003
Position: Prof *Employer:* Univ of MA *Education:* PhD/Polymer Chem/State Univ of NY; MS/Textile Chem/Inst of Textile Tech; BS/ChE/Lehigh Univ *Born:* 4/28/26 1951-53 Chicopee Mfg Co, research chem. 1955-63 Dow Chem Co, Senior Research Chem. 1963-66 Fabric Research Labs, Inc - Assistant Dir. 1966-date Univ of MA, Polymer Science and Engrg Program. 1972-73 Guest Prof, Univ of Mainz, Germany. 1975 Visiting Prof, Royal Inst of Tech, Stockholm, Sweden. 1979-80 Senior Hurrboldt Award and Guest Prof, Univ of Freiburg, Germany. 1977 Chrmn, Gordon Conference on Polymers. 1984 Chrmn Gordon Conference on Liquid Crystal Polymers. 1983 Faculty Fellowship Award, Univ of MA. *Society Aff:* AIChE, ACS

Lenzen, Kenneth H
Business: 4010 Learned Hall, Lawrence, KS 66045
Position: Assoc Dean & Professor. *Employer:* Univ of Kansas. *Education:* BS/CE/Northwestern Univ; MS/Structural Eng/Northwestern Univ; PhD/Engr Sciences/Purdue Univ. *Born:* Sept 18, 1921. Res engr Portland Cement Assn 1946-47; Res Assoc Northwestern Univ Evanston 1947-49; Instr & Res Assoc Purdue Lafayette Ind 1949- 55; Assoc Prof Mech & Aerospace Engrg Univ of Kansas Lawrence 1955, Prof 1960- , Chem of Dept 1965-67, Assoc Dean Engrg for Grad Studies 1974- ; V P Cadre Corp; cons in field. Danforth Assoc 1961- ; Fulbright grantee 1962-63. OAS grantee Peru 1974. Reg P E Mo, Kan, Ill, NJ, Ind. Fellow ASCE (J James R Croes Medal), RCRBSJ, AREA, AAUP, ASEE, Chmn of cttes. Natl Assn Prof Engrs, Sigma Xi, Sigma Tau, Tau Beta Pi; 35 prof papers; patentee in field. *Society Aff:* ASEE, ASCE, AREA, ΣΞ.

Leon, Benjamin J
Business: Dept. of EE, Univ. of Kentucky, Lexington, KY 40506
Position: Prof of Elec Engrg and Dept. Chairman. *Employer:* Univ. of Kentucky *Education:* ScD/EE/MIT; SM/EE/MIT; BS/EE/Univ of TX. *Born:* 3/20/32. m. 1954; 4 children. Native of Austin Tex. Tech Staff Lincoln Lab MIT 1954- 59; Tech Staff Hughes Res Labs 1959-62; Purdue Univ Assoc Prof 1962-65, Prof 1965-80; Univ. of Ky., Prof 1980-, Chairman, 1980-84. Lectr UCLA 1960-61; Visiting Prof Cornell 1968-69; Visiting Elec Engr Defense Comm Agency 1975-76. Visiting Prof SMU 1986-7. IEEE offices - VP- Educational Activities. Chrmn Education Activities Bd, Mbr, Exec Committee & Bd of Dirs Chairman Member Activities Council; Member US Activities Bd. - Editor Trans on Circuit Theory, Chmn Ckt Theory Group, Standards Bd Mbr, TAB Tech Plan Chmn, TAB Tech Plan Chmn, TAB Finance, Prof Activities Chmn for Wash D C. Pension Ctte Chrmn, Ed.- Impact, The newsletter of Professional Activities. Fireman for Wabash Township Volunteer Fire Dept. P.E. Ky. *Society Aff:* IEEE, AAAS, AAUP, ASEE, ACUTA, ADPA, BICSF, NSPE

Leon, Harry I
Business: 924 Bowen St, NW, Atlanta, GA 30318
Position: Pres. *Employer:* Leon & Assoc, Consulting Engrs. *Education:* PhD/ME/CO State Univ; MS/Chem-Nucl Engg/Univ of WA; BS/Aero Engg/Univ of FL. *Born:* 7/5/34. Power plant staff engr for Boeing Aircraft Jet Transport Div 1956-60. Designed nucl reactor core for first commercial HTGR & fission product removal system at Gulf Gen Atomics 1960-66. Researched energy systems at CO State Univ 1966-70 & at Los Alamos Scientific Labs 1970. On faculty, Univ of MS 1970-74 and GA Inst of Tech 1979. Since 1973, Pres of Leon & Assoc, Engrs: responsible for engg numerous ind air, water, & noise pollution abatement facilities, resource recovery installations, & energy conservation systems in southeastern US. Consultant to maj engg firms. NCEE certificate. US Patent Agent. Enjoy jogging & woodworking. *Society Aff:* NSPE, AIChE, ASME, ISES.

Leonard, Bruce G
Business: P.O. Box 33068, Raleigh, NC 27606
Position: VP *Employer:* Kimley-Horn & Assocs, Inc *Education:* BS/CE/Duke Univ *Born:* 3/29/40 Native of Catawba, NC. Hgwy designer and transportation planner with Harland Bartholomew and Assocs 1961-1970 in Memphis, TN, Releigh, NC & Durham, NC. Joined Kimley-Horn aand Assoc, Inc in 1970 as Senior Design Engr. Has directed transportation planning, parking, traffic engrg, hgwy design and environmental impact project for the firm in most Southern states. Has served as pres of Eastern NC Branch, ASCE and southern Section, ITE. Active in youth baseball and soccer programs. *Society Aff:* ASCE, NSPE, ITE, AAA

Leonard, Edward F
Business: 355 Engrg Terrace, New York, NY 10027
Position: Prof of Chem Engrg. *Employer:* Columbia Univ. *Education:* BS/ChE/MIT; MS/ChE/Univ of PA; PhD/ChE/Univ of PA *Born:* July 1932. Columbia Univ since 1958, Prof since 1967. Dir Artif Organs Res Lab since 1968. AIChE Colburn Award 1968. Prin ints: transport phenomena & application of chem engrg to problems in biology & medicine. Cons to indus (Procter & Gamble, Baxter-Travenol) & gov (NIH, FDA, ERDA). Appointments at Mt Sinai School of Medicine, St Luke's Hosp N Y. 100 pubs, editor 3 symposium volumes. *Society Aff:* AIChE, ASAIO, BMES

Leonard, Frederick U
Business: 3333 Michelson Dr, Irvine, CA 92730
Position: Mgr Projects. *Employer:* Fluor Engr & Constructors Inc. *Education:* BS/Chem Eng/Univ of UT. *Born:* June 1918 Salt Lake City. Plant engr & oper supr for oil co. With Fluor since 1953 in process engrg, & mgr of process & instrument engrg for foreign subsidiary. Pres & Genl Mgr during the org phase of a foreign subsidiary co. Presently Engrg Mgr with supervisory respon for major projects. Outside ints incl golf, skiing, & music. *Society Aff:* AIChE.

Leonard, Jackson D
Business: 560 Sylvan Ave, Englewood Cliffs, NJ 07632
Position: Pres. *Employer:* Leonard Process Co. *Education:* BS/CE/PA State. *Born:* 9/2/15. Sunbury, PA; Employed Allied Chem 1937-40; E I duPont de Nemours 1940-49; Nerck & Co 1949-1950. Established Leonard Process Co in 1950 with goal of developing chem process tech. Developed and patented new processes for catalysts for production of aliphatic amines, and licensees of this tech are now located in 16 different countries; also developed & licensed tech for oxidation of titanium tetrachloride to titanium dioxide; new tech now developed & patented for production of Formic Acid from CO & H2O will be licensed world wide. Mbr of Methodist Church; Masonic organization. Married & four children: Michael John 33, Patricia Anne 35, Jessica Taft 10, Melinda Day 7. Author of many articles dealing with chem plant maintenance, utility conservation, recycling of wastes. *Society Aff:* ACS, AIChE, NSPE, NJSPE.

Leonard, Joseph W
Business: White Hall-Willey St, Morgantown, WV 26506
Position: Dir, Coal Res Bur-Prof, COMER. *Employer:* West Virginia Univ. *Education:* BS/Mining Engg/Penn State; MS/Mineral Preparation/Penn State. *Born:* Dec 1930. Dean, College of Mineral & Energy Resources (1978-81) Dir, Mining Industrial Extension (1978-81) Dir Coal Res Bureau & Prof Mining Engrg W. V2. Univ. Singly or jointly publ 100 tech papers - edited 13 major reports, books or vols involving many phases of mineral tech & engrg. Sole inventor of processes or devices which resulted in 25 foreign & domestic pats. Mbr 4 ranking scholastic hon soc's, 9 prof insts or soc's. Author/editor *Coal Preparation* - the authoritative handbook of its field. A present or past internatl cons for over 50 individuals & firms on coals, ash & coke & an accredited prof chemist. The Howard N Eavenson Award 1969 AIME. Secy Bd of Dir, Min Engrg Handbook AIME; Bd of Dir Min Proc Handbook AIME. Secy-Treas of Central Appalachian Section- AIME. Distinguished member AIME (1980). William N. Poundstone Research Professor (1980), Comer, West Virginia University. *Society Aff:* ΤΒΠ, ΣΞ, ΦΓΥ, ΣΓΕ.

Leonard, Richard D
Home: 3909 La Cresta Dr, San Diego, CA 92107
Position: Manager of Plant Engg. *Employer:* Convair Div of Genl Dynamics Corp. *Education:* BSEE/Elec Engg/San Diego State. *Born:* Mar 1925. Army Infantry WW II. Reg P E. Oil exploration until 1951; then joined Convair as elec engr. Num mgmt pos in Plant Engrg - present title 1971. Respon for maintenance of all Convair facils in San Diego; 6 million sq ft of plant. This includes pollution control & energy mgmt as well as const of facilities. AIPE Western Region V P 1970-72, Richard Morris Medal 1971 & Hon Life Mbr 1976. *Society Aff:* AIPE, NMA.

Leonard, Ronald K
Home: Rte 1, Box 128B, Horicon, WI 53032
Position: Mgr, Engineering *Employer:* John Deere Horicon Works. *Education:* BS/AE/IS State College; MS/Engg/MI State Univ. *Born:* 7/10/34. Tech Dir ASAE 1970-72; SAE, ORVC 1974-77. ASAE Finance Ctte 1975-80. (ASAE Finance Dir 1983-85) P E 4749 Iowa. Cotton Harvesting machine design & test exper 15 yrs. Product Engrg mgmt, Consumer Products 10 yrs, Product and Production Engrg Mgr 1 year. *Society Aff:* SAE, ASAE.

Leonard, Roy J
Business: Civil Engrg Dept, Lawrence, KS 66045
Position: Prof *Employer:* Univ of KS *Education:* PhD/Geotech Engr/IA State Univ; MS/CE/Univ of CT; BS/CE/Clarkson Coll *Born:* 8/17/29 Asst Prof Civil Engrg, Univ Del, Newark, 1957-59; Assoc Prof Civil Engrg, Lehigh Univ, Bethlehem, PA 1959-63, 65-66; NSF fellow, Imperial Coll, London, 1963-65; cons E. H. Richardson & Assoc, Newark, DE 1957-65; spl projects engr, Dames and Moore, NYC, 1965; Prof Civil Engrg, Univ KS, Lawrence, 1966-; cons KS City Testing Lab, Inc, 1967-83; cons Burns & McDonnell, KS City, 1967-82. Pres, Alpha-Omega, Geotech, Inc, Kansas City, KS. Registered PE NY, PA, KS, MO, and CO. Listed in Who's Who in America (Marquis), Who's Who in the Midwest (Marquis). Who's Who in Technology and American Bar Association. Register of Expert Witnesses. *Society Aff:* ASCE, AIME, AEG, USCOLD, NSPE, ASTM, ACI, NAFE, DFI

Leonards, Gerald A
Business: Civil Engrg, W Lafayette, IN 47906
Position: Professor. *Employer:* Purdue Univ. *Education:* BSCE/-/McGill; MSCE/Soil Mechanics/Purdue; PhD/Soil Mechanics/Purdue. *Born* Apr 1921. MSCE, PhD Purdue; BSCE McGill. Native of Montreal. Lectr at McGill Univ; Soils Engr with Dept of Transp Canada 1943-46; Purdue 1946- : Prof 1958, Hd Civil Engrg 1965-68, Hd Geotech Engrg 1951-65, 1972-76. Cons geotech engr U S & abroad. ASCE Norman Medal 1965; Hwy Res Bd Best Paper Award 1965. Golf & tennis. *Society Aff:* ASCE, ISSMFE.

Leone, William C
Business: P O Box 9519, El Paso, TX 79985
Position: President & Director. *Employer:* Farah Mfg Co Inc. *Born:* May 1924. BS, MS & DSc Carnegie Inst of Tech. Theta Tau Most Promising Sr Engr 1944. US Navy Ltjg 1944-46. Fac of Carnegie Tech 194653. Indus Sys Div Mgr Hughes Aircraft 1953-59. Founder, V P & Genl Mgr Remex Electronics 1960-68. Pres & Dir of Rheem Mfg Co & of City Investing Co, incl World Color Press & Hayes Internatl 1968-76. Pres & Dir Farah Mfg Co Inc 1976- .

Leonhardt, Fritz
Business: Schottstr 11B, D7000 Stuttgart 1, Germany
Position: em Prof Dr Ing Dr Ing h c mult. *Employer:* Cons Engr & Ptnr Leonhardt und Andrae. *Education:* DrIng/Dipl Ing/Univ Stuttgart; -/-/Purdue Univ Lafayette, IN. *Born:* July 1909 Stuttgart. Bridge engr Autobahnen Auth Berlin, Cologne 1934-38; cons engr Stuttgart 1938. Prof Univ of Stuttgart 1957-74, rector 1967-69. Designer bridges crossing Rhine River, cable-stayed bridges Kniebr Dusseldorf, Nordbr Mannheim, crossing Rio Parana, Pasco-Kennewick br crossing Columbia River, etc; Stuttgart TV Tower, Frankfurt Telecommunication Tower, etc, many high-rise bldgs, tentroofs Expo 67 Montreal, Olympiade Munich 1972. Dev Taktschiebeverfahren (incremental launching method), sev prestressing sys, rubber pot bearings. Res shear & torsion of reinf & prestr concrete. Author. Mbr num natl & internatl assns & comms. Fritz-Schumacher-Prize; Gold Ehrenmunze of Osterr Ing u Arch Verein, Werner-von-Siemens-Ring, Emil-Morsch-Denkmunze, Medaille d'Or Gustave Magnel (Belgium), Hon Mbr ACI, Grashof-Denkmunze VDI, Freyssinet, Gold Medal of Inst of Struct Eng London and other Medal. Hon Mbr CEB & IABSE, Grosses Bundesverdienstkreuz. Hon. Member of the Academy of Sciences, Heidelberg, Foreign Assoc of the Natl Acad of Engrg, USA, Member Swiss Acad Tec Sc. SATW. *Society Aff:* VDI, VBI, VPI, DAI, AIV, IABSE, FIP, IASS, CEB.

Leontis, Thomas E
Home: 3590 Hythe Ct, Columbus, OH 43220
Position: Mgr, Magnesium Res Ctr. (Retired) *Employer:* Battelle-Columbus Labs. *Education:* PhD/Metallurgy/Carnegie-Mellon Univ; MS/Metallurgy/Carnegie-Mellon Univ; ME/Mech Engg/Stevens Inst of Tech. *Born:* March 1917 Plainfield N J. Affil for 27 yrs with The Dow Chem Co in magnesium res & tech admin. Joined Battelle 1971 & now respon for all contract res on magnesium. Retired in 1982 Author 29 tech papers & inventor 12 U S pats. Contrib to tech books & encyclopediae. Fellow ASM & Mbr AIME & AAAS. Tr ASM 1964-72; Secy ASM 1966-68, V P ASM

Leontis, Thomas E (Continued)
1970, Pres ASM 1971, Pres ASM Found for Educ & Res 1972. Active in local church admin. Enjoy reading classics & world travel. Mbr Pi Delta Epsilon, Tau Beta Pi, Phi Kappa Phi, and Sigmi Xi. Cited in Amer Sci, Metallurgy conslt with specialty: Magnesium, Metallurgy Technology Marketing & Economics. Archen of the Ecumenical Patriarchate of Constantinople and mbr of Order of St Andrew the Apostle. Archen is the highest recognition that the Catholic Church gives to a layman. *Society Aff:* ASM, AIME, AAAS.

Leopold, Reuven
Home: 12415 Plantation Ln, N Palm Beach, FL 33408
Position: VP. *Employer:* Pratt & Whitney Aircraft. *Education:* PhD/Ocean Engg/MIT; Mar Mech Engg/Ocean Engg/MIT; MS/Naval Arch/MIT; MBA/BA/GWU. *Born:* 5/5/38. Dr Leopold is VP, Advanced Systems, of the Govt Products Div of Pratt & Whitney Aircraft, United Tech Corp. In this position he is responsible for the strategic planning, R&D & domestic marketing for the co. The Govt Products Div is respon for the dev & design of all military jet engines (for the F-14, F-15 & F-16) as well as liquid & fuel rockets. During the preceding eight yrs Reuven Leopold was the Technical Dir of the Ship Design Div of the Naval Ship Engg Ctr. By virtue of the fact that the Naval Ship Engg Ctr is the organization which designs all US Naval Surface ships as well as all US Naval submarines, Dr Leopold was the Navy's Technical Dir for all surface ship & submarine design. *Society Aff:* SNAME, ASNE, AIAA, ADPA.

Lepore, John A
Home: 731 Timber Trail Ln, Springfield, PA 19064
Position: Prof & Chrmen. *Employer:* Univ of PA. *Education:* PhD/Civil/Univ of PA; MSME/ME/Univ of PA; BSCE/CE/Drexel Univ. *Born:* 2/19/35. in Phila, PA. Upon grad with an undergrad degree, began employment at NY Shipbuilding Corp in 1961 as Nuclear Engr, working on analyses and design of nuclear reactor systems for commercial and naval propulsion. Proj engr with GE Co, responsible for technical direction of engg and analysis group. Responsible for design and dev of component for USAF Missile System. With Univ of PA since 1968 and assuming the position of Dept Chrmn in 1978. Am reg in the State of PA in the area of structural engg. Am Danforth Associate in the Danforth Fdn, Winterstein Asst Profl Chair in 1970. *Society Aff:* ASCE, ASEE, EEI, AAUP.

Leppke, Delbert M
Home: 1039 Maple, Evanston, IL 60202
Position: Sr Tech Mgr *Employer:* Fluor Daniel-Power Sector. *Education:* BS/EE/SD State Univ *Born:* 10/16/25. Attended special nuclear sch at Oak Ridge Natl Lab 1951-52. Joined predecessor of Fluor Daniel in 1952. Successively held positions of nuclear engr, chief nuclear and became V.P. in 1970. Active in forming Chicago Sec and Power Div of the American Nuclear Soc serving as officer and mbr of exec bd for several yrs. Received ANS/Chicago Section Meritouous Service Award & SDSU Distinguished Engineer Award. Currently Senior Technical Manager for Fluor Daniel-Power Sector. *Society Aff:* ANS, NSPE, WSE.

Leps, Thomas M
Home: P.O. Box 2228, Menlo Park, CA 94026-2228
Position: President & Cons Civil Engr. *Employer:* Thomas M Leps Inc. *Education:* AB/Civil Engg/Stanford Univ; MS/Soil Mechs/MIT. *Born:* Dec 3, 1914 Keyser W Va; m. Catherine M; 1 son Timothy M. Mbr Tau Beta Pi, Phi Beta Kappa. With Southern Calif Edison Co 15 yrs on planning, design & const of hydro & steam power plants. Cons for past 26 yrs to engrg firms, contractors & gov agencies on design, const, safety & seismic analysis of electric power projects, particularly in the field of dams. *Society Aff:* ASCE, USCOLD, NAE.

Lerner, Bernard J
Business: PO Box 39, Glenshaw, PA 15116
Position: President. *Employer:* Beco Engrg Co. *Education:* PhD/Chem Engg/Syracuse Univ; MS/Chem Engg/Univ of IA; BChE/Chem Engg/Cooper Union. *Born:* Apr 28, 1921. 1969 to date: Pres Beco Engrg Co, engaged in all aspects of air pollution control tech, spec in custom-design of wet scrubbers & absorption sys. 1961-69: Pres Patent Dev Assocs Inc, engrg res lab for contract res & dev of client & in-housse tech. 1954-59: Group Leader, Chem Engrg Res, Gulf Res & Dev Co Pgh Pa. Asst Prof Chem Engrg Univ of Texas Austin. Undergrad degree Cooper Union; grad degrees State Univ of Iowa 1947 (MS) & Syracuse Univ 1949 (PhD). *Society Aff:* AIChE, ACS, APCA.

Leroy, Gerard L
Business: 112 Boulevard Houssmann, 75008 Paris, France
Position: Secretary General *Employer:* CIGRE *Education:* -/EE/Ecole Breguet Paris. *Born:* 7/15/28. Has joined EdF in 1950 for various studies on insulations mtls for large generator windings. Then in charge of a new test bed for big asynchronous motors. From 1968 in charge of the UHV Lab proj at Renardieres. From 1976 to 1980 Asst Chief of "Service Materiel Electrique–, in charge of Res Dev on Electrical Equip for production & distribution. 375 people including 80 grad engrs. Since 1981 Gen Secy of CIGRE: Intl Conference on Large High Voltage Elec Sys. *Society Aff:* IEEE, SEE.

Leslie, John H, Jr
Home: 2801 Cornelison, Wichita, KS 67203
Position: V.P. of Operations & Research *Employer:* Cerebral Palsy Res Fnd of KS-Wichita. *Education:* PhD/Indus Engg & Mgmt/OK State Univ; MS/Mech Engg/Wichita State Univ; BS/Indus Engg/Wichita State Univ. *Born:* 7/28/38. & raised in Wichita. Taught in the Coll of Engg, Wichita State Univ from 1962-77. Chmn of Indus Engg from 1964-77. Currently serving as the co-dir of an engg res activity designed to dev tech for the employment of the severely physically handicapped (and V.P. of Operations, & Research CPRFK). This effort has resulted in Wichita State Univ (and the Cerebral Palsy Res Fnd) being named as a Rehab Engg Ctr in 1976 Mbr Natl Indus for the Severely Handicapped (NISH), serve as secretary of the Exec Ctte - also a mbr of the Vocational Comm of Rehab Intl. Hobbies: hunting, fishing, & sport car restoration. Pres Wichita Chap IIEE 1971-72 chapter yr. Served a HEW-Rehab Services Admin Fellow, advising Egypt & Poland on Vocational Rehab-Nov 1978. Advisor to Egyptian Government on handicapped person vocational evaluation, Summer 1981. *Society Aff:* ΤΒΠ, ΑΠΜ, IIE, RI.

Leslie, John R
Business: RR 1, Kleinburg, Ontario L0J 1C0 Canada
Position: Retired (Mgr, Elec Res Dept) *Employer:* Ontario Hydro *Education:* BS/Engrg/Univ of Toronto. *Born:* June 9, 1917. P E; Fellow IEEE; V P Canadian Natl Ctte CIGRE. Employed 1941-45 with Natl Res Council of Canada. British Admiralty & Royal Canadian Navy in mine design & mine sweeping res. Employed at Ontario Hydro 1945-1980; carried out res prog in radio interference, fault location, infrared, ultrasonics, & power line carrier. Leonard C Wason medal Amer Concrete Inst 1950. P Chmn Toronto Sect IEEE. *Society Aff:* IEEE.

Leslie, William C
Home: RR2, Box 64AA, Palmyra, VA 22963
Position: Prof of Matls Engrg, Emeritus *Employer:* Univ of Michigan. *Education:* PhD/Met Engg/OH State Univ; MSc/Met Engg/OH State Univ; B Met E/Met Engg/OH State Univ. *Born:* 1/6/20. in Dundee Scotland. Married 1948. Metallurgist US Steel Corp Res Lab 1949- 53. Adj Prof Brooklyn Poly 1952-53. Assoc Dir Res, Thompson Prod Inc 1953-54. Metallurgist 1954-57, Sr Scientist 1957-63, Asst Dir Phys Met 1963-68. Mgr Phys Met 1968-72 US Steel Fund Res Lab. Prof Univ of MI 1973-85, Emeritus, 1985. Corps of Engrs US Army 1943-46. Battelle Visiting Prof OH State 1964-65. Disting Alumnus OH State Univ 1967. Krumb Lectr AIME 1967, Chmn 2nd ICSMA 1970, Carnegie Lectr ASM 1970, Fellow ASM 1970, Campbell Lectr ASM 1971. Chmn Inst Metals Div AIME 1971-72, Dir TMS-AIME 1971-72, Dir & VP AIME 1975-76. Sauveur Lectr, Jeffries Lectr ASM 1975. Garofalo Lectr, Northwestern Univ 1977 Visiting Prof Univ of Melbourne, Australia,

Leslie, William C (Continued)
1979 NSF-CSIR Exchange Scientist, India, 1981. Author, "Physical Metallurgy of Steels–, McGraw-Hill, 1981. Howe Lecturer, AIME, 1982. Distinguished Alumnus Lectr., Ohio State Univ. 1984. Fellow Inst. Metals, 1984; Hon. Member, ASM, 1986. Author 101 papers in phys met. *Society Aff:* AIME, ASM, AAAS, ISIJ, Inst. Metals

Lesnewich, Alexander
Business: 100 Mountain Ave, Murray Hill, NJ 07974
Position: Dir, Filler Metals Res & Dev. *Employer:* Airco Welding Products Div. *Education:* BS/Metallurgical Engg/Rensselaer Polytechnic Inst; PhD/Metallurgical Engg Rensselaer/Polytechnic Inst. *Born:* Apr 1923. Native of Ridgefield Pk N J. Army Air Corps 1942-46. With Airco since 1952, spec in met & welding res. Had been Dir of Met Res & Dir of Welding Res in Corp Labs. Presently Dir of R&D with Welding Products Div. Pres AWS 1978-79. Lincoln Gold Medal 1958, Adams Mem Lectr 1970, Meritorious Cert. Sigma Xi, Tau Beta Pi, Phi Lambda Upsilon. Enjoy sailing, cabinet making & photography. *Society Aff:* AWS, ASM.

Lessells, Gerald A
Business: Thornall St, Edison, NJ 08818
Position: Mgr Environ Servs. *Employer:* J M Huber Corp. *Education:* BS/Chem Eng Practice/MIT. *Born:* Aug 5, 1926 Wallasey England. AUS 1944-46. Process & product dev on org chem, plastics, coatings & inks with 4 co's. Current pos since 1969. Lic P E Ohio & Ill. Contrib articles to tech journals. Patentee in field. Amer Chem Soc reviewer, 'Indus & Engrg Chem'. Mbr Amer Inst Chem Engrs, Natl Dir 1973-75; Fellow 1975; Service to Soc Award 1974 for 20 yrs activities in equal oppor for minorities; Minority Affairs Coordinator 1975- . Life Mbr NAACP 1955- . Mbr MIT Educ Council 1962- . Conf's Chmn Natl Assn of Printing Ink Manufacturers 1974- . Mbr Adv Bd, NJ Inst Tech Fnd, 1980- . *Society Aff:* AIChE, ACS.

Lessen, Martin
Home: 12 Country Club Drive, Rochester, NY 14618
Position: Consulting Engr Yates Mem Prof of Engrg (EMERITUS) *Employer:* Univ of Rochester. *Education:* ScD/ME/MIT; MME/ME/NY Univ; BME/ME/CCNY. *Born:* Sept 6, 1920 NYC. Diesel Design US Navy 1940-46. Aero Res Sci NACA 1948- 49; Assoc Prof & Prof Aero E Penn State 1949-53; Prof & Chmn Appl Mech Univ of Penna 1953-60; Prof & Chmn Mech & Aerosp Sci Univ of Rochester 1960-70, Yates Mem Prof of Engrg 1967-83. On leave to USONR London 1976-79; NSF Sr Postdoc Fellow Cambridge Univ 1966-67. Fellow ASME, APS, A.A.A.S; Founding Chmn Energetics Div ASME. P E NY. Hydrodynamic stability of shearing layers, transition to turbulence, thermoelastic shock. Enjoy music, opera, ballet, stage. *Society Aff:* ASME, APS, AAAS.

Lessig, Harry J
Home: RR 4-Oakdale Dr, Springfield, IL 62707
Position: President. *Employer:* H J Lessig & Assocs. *Born:* Nov 1915 Phila Pa. Reg P E; pioneered in indus application of stat qual control sys with War & Navy Depts. Cons to indus in implementation of total qual control concept. From 1959-66 was Dir of Qual & Reliability Engrg with Genl Time Corp; also taught in Engrg curriculum at Ill Valley Jr Coll. From 1966-78 with Stewart-Warner Corp, Hobbs Div. Internationally known lecturer & author of num papers on engrg & mgmt aspects of prod qual & reliability. Special int in Boy Scouts; mbr since 1927; Pres 1973 ASQC. *Society Aff:* ASQC.

Lesso, William G
Business: Dept of Mech Engr, Austin, TX 78712
Position: Prof *Employer:* Univ of TX *Education:* PhD/Oper Res/Case Inst of Tech; MS/Econ/Case Inst of Tech; MBA/Mgt/Xavier Univ; BS/ME/Univ of Notre Dame *Born:* 3/23/31 Worked as a bearing designer, Clevite Corp 1953-1954, 1956-1958, served in USAF 1954-1956. Worked for Gen Elec as jet engine lub system designer. Member of the Operations Research Faculty, Dept of Mech Engr, Univ of TX, 1967-present. Assistant Dean, 1978-1981. E.C.H. Bantel Prof, 1979-present. Consultant in the fields of operations research applied to electric utilities, oil and gas industry. *Society Aff:* ORSA, TIMS, ASEE, AIDS

Lester, Dale Sr
285 Lake Joyce Dr, Land O Lakes, FL 34639
Position: Owner *Employer:* Self *Education:* BS/CE/Univ of KY *Born:* 10/29/22 Architect, engrg co exec; b. Mayfield, KY, s. Paul Lester and Lottie Pearl Hawes Lester; student Allegheny Coll, 1943-44, Murray State Univ, 1946-47; U.K. 1948-49; m. Carolyn Ray, Mar 16, 1948; children--Dale, Paul; Supvr constrn KY Dept Hwgys, Paducah 1949- 51; area engr F.H. McGraw & Co, Paducah, 1951-53; structural supvr Giffels & Vallet, Portsmouth, OH, 1953-55; structural designer Rust Engrs, Birmingham, AL, 1955-56; VP prodn mgr Watson & Co, Tampa, FL 1956-76, pres 1976-78; vice chrmn 1978-84; Consulting business for myself in 1984; cub scout master Boy Scouts Am, 1959; maj United Fund 1961. Served with USAAF, 1943-45, Fellow FL Engr Soc; mem Natl Soc PE, ASCE, Baptist, Clubs: Kiwanis 1957-60 (past dir); Rotary 1961-present; Palma Ceia Golf & County, 1965-83. *Society Aff:* FES, NSPE, ASCE

Lester, Horace B
Home: 1350 Eastover Dr, Jackson, MS 39211
Position: President. *Employer:* Lester Engrg Co Inc. *Education:* Grad/Military Command/US Army Command & Genl Staff Sch; Constructive/Engr Command/US Army Engr Sch; Credit Completed/Civil Engr/Interntl Corresponence Sch. *Born:* Sept 1919. Reg P E (Civil). Correspondence school educ. 50 yrs exper in design of civil engrg works with special recog in engrg econ & proj dev. 3 consecutive terms in the Miss Legis with major ctte assignments. Instrumental in dev of Miss's Pollution Abatement Prog. Formulated concepts for Barnett Reservoir & Hinds-Rankin Flood Control Works near Jackson, Miss. Initiated and guided to formal dedication the Pearl River Boatway as a counterpart to the Natchez Trace Pkwy. Acknowledged Prof. planner in an Indus Complex in Central Miss. Colonel, Army Engr Reserve. Last assignment, Office, Ch of Engrs. *Society Aff:* ASCE, ACEC, NSPE, NACE.

Lester, John W
Home: 28 McKinley Cir, Tempe, AZ 85281
Position: Sr Engg Specialist. *Employer:* Garrett Turbine Engine Co. *Education:* MS/ME/Univ of MN; BS/ME/Clarkson University *Born:* 9/1/30. Native of Binghamton, NY. Engr, GM Res Labs, 1952-1954. Navy commission, 1955. Engg Instr, US Naval Acad, 1957-1959. Holds rank of Captain in Naval Reserve. Liaison Officer for Naval Acad in AZ. Mbr US Naval Inst. sr Engg Specialist, Garrett Turbine Engine Co, specializing in gas turbine engine design & dev since 1961. Mbr SAE, 1954. Founder & first chrmn of SAE's AZ Section, 1963. Mbr SAE's Natl Bd of dirs (1977-1980). Chrmn, AZ Council of Engg & Scientific Assns, 1968. Mbr ASNE, Mbr Tau Beta Pi & Pi Tau Sigma. Lay Leader in local church. Enjoys backpacking. Reg PE, Mechanical, AZ. *Society Aff:* SAE, ASNE, NRA, USNI.

Lester, William
Business: 220 E 42 St, New York, NY 10017
Position: Sr. VP *Employer:* NPS Corp *Education:* MS/ME/Univ of Pittsburgh; BS/ME/Univ of Pittsburgh *Born:* 8/18/24 Native Erie, PA. Served Flying Officer 8th AF 1942-45. Taught Univ of Pittsburgh. Managed major projects Blaw-Knox Co for MTR, Savannah River Plant, Twelve Atlas Missle Bases, JPL Large (Mars) Antenna. 1963-78 managed all Westinghouse nuclear power plant projects including procurement, construction, initial operation. 1978-81 VP Intl Operations and mbr Boards of Directors Burns and Roe Co. Currently Sr VP and mbr Executive Ctte NPS Corp. Active in Engrg, construction, manufacturing, material distribution, automation, communications and transportation. Chairman AAES Intl Affairs Council, Chairman US Ntl Commission Union Pan American Engrg Societies (UPADI), Mbr AAES Liaison Ctte, Presidents Science and Tech (office) Advisor. *Society Aff:* AAES, ASME, AAAS, ANS, NPS

Leung, Paul
Business: 12400 E Imperial Hwy, Norwalk, CA 90650
Position: Principal Engr. *Employer:* Bechtel Power Corp. *Born:* Mar 1923. MS Rice Univ; BS Natl Sun Yat-Sen. Joined Bechtel org in 1953. Promoted to present pos 1967. Respon incl: turbine thermal cycle optimization & econ studies of fossil & nuclear power proj's. Appointed by ASME as Standing Ctte Mbr Performance Test Codes. Taught Power Engrg at Univ of Calif L A & Calif State Univ L A. ASME Prime Movers Award 1975, 1974, 1972, 1971, 1970. Fellow ASME 1973. P E Calif & Texas.

Leuschen, Thomas U
Home: 14118 Frances St, Omaha, NB 68144
Position: VP *Employer:* Leo A. Daly *Education:* MS/Transp Engr/IA State Univ; BS/CE/IA State Univ *Born:* 3/26/44 Native of Panama, IA. Traffic engr for Alan M. Voornees and Assocs, McLean, VA from 1968 to 1972. With Leo A. Daly since 1972. Appointed captain of the civil engrg team and VP in 1981. VP, SAME, Omaha Post, 1977. Pres, Engrs Club of Omaha, 1978. *Society Aff:* NSPE, ITE, SAME

Leutzinger, Rudolph L
Home: 1521 N Holder, Independence, MO 64050
Position: Prof Emeritus of Mech Engr, Ret. *Employer:* Univ of MO - Columbia. *Education:* PhD/Mechanical Engr/Univ of IA; MS/Applied Mech/Univ of MI; BS/Aero/IA State Univ; AA/Chemistry/Graceland College. *Born:* June 17, 1922 Dallas Ctr Iowa. Chmn Dept of Genl Engrg Univ of Missouri KC 1962-74; teaching Mech Engrg Univ of Missouri Columbia, Rolla; teaching Aerosp Engrg: Kansas Univ, Iowa State Univ, TX A&M. Aeronautical cons: TWA on jet engine DC-9 wake, for Boeing on flutter, & McDonnell on inlets & 'the spin jet'. Res: for NASA on inflow gas turbine & inlets, Univ of Mich & MRI on dynamics & aerodynamics of missiles & projectiles. Patents on jet engines. Reg P E TX & MO; 10 Yr Award from Appl Mech Review; Pres Mid-Amer Engrg Guidance Council 1974; Secy KC Sect AIAA. *Society Aff:* NSPE-MSPE, ASME, AIAA, ASEE, ΣΞ, ΑΓΡ, ΣΓΤ.

Le Van, James H
Home: 8135 Beechmont Ave., Apt. E-236, Cincinnati, OH 45255
Position: Engr Dir (Capt) (0-6). *Employer:* U S Public Health Service -ret. *Education:* CE/Civil Engg/Lehigh Univ; SM/San Engg/Harvard Univ. *Born:* Jan 1905. San engr, cons engrg firms. Commissioned san engr officer, Regular Corps, U S Public Health Serv 1931. Supervised san surveillance of vessels, railroads, air lines, & buses. In WW II, Ch San Engr, War Shipping Admin. Supervised const of water supply & sewerage sys, new city of Levittown Pa. Designed & placed in oper water & sewage treatment plants, Indian Service. Officer in charge of Scioto River Investigation and of Aedes Aegypti Control Unit. Detailed to US Coast Guard, Ch San Engr Officer. Awarded Coast Guard Commendation Medal. Retired 1969. At present, Marine Sanitation Cons. Prof Engr-GA, PA. Tau Beta Pi, Scabbard and Blade. Army & Navy Club (Wash). Beth Offrs Club. *Society Aff:* ASCE, APHA, SAME, RSH, ASTM&H, AAEE, WPCF, FWOA

Levens, Alexander S
Business: Dept of Mech Engg, Berkeley, CA 94720
Position: Prof of ME Emeritus. *Employer:* Univ of CA. *Education:* CE/Struct Design/Univ of MN; MSCE/Struct Design/Univ of MN; BSCE/Struct Design/Univ of MN. *Born:* 2/12/04. Native of MN. Taught at Univ of MN, College of Engg 1927-1941 (Asst/Assoc Prof). Taught at Univ of CA, College of Engg 1941-1969 (Assoc/Prof) Lectr at Univ of WI, UCLA, Univ of IL, Imperial College of Sci & Tech (London); Technion (Haifa, Israel), Tel Aviv Univ (Israel); Special courses in Nomography - Naval Ordnance Test Station, Inyokern, CA. The Boeing co, Seattle, US Naval Radiation Lab, San Francisco, Consultant to the Boeing Co, GM Corp, Author of several texts in Engg Graphics and Design; Nomography, Graphical Methods in Res. Dir, Dynamic Graphics, Berkeley; (Comp Graphics). Still active in text writing, res, and conlt. *Society Aff:* ASEE, ASEE (RWI), ASEE (EDG).

Levenson, Milton
Business: Bechtel Power Corporation, P.O.Box 3965, San Francisco, CA 94119
Position: Engineering Consultant/Special Assistant to Dir *Employer:* Bechtel Power Corporation *Education:* BChE/Chem Engg/Univ of MN. *Born:* 1923. Chem Engrg degree Univ of Minn. Nuclear engrg activities have included fuel cycle, reactor engineering and safety, & LMFBR. Member Natl Acad of Engrg, Fellow Amer Nuclear Soc, Fellow Inst of Chem Engrs. Robert E Wilson Award 1975 (AIChE); U S Advisor at the Geneva Conferences on Atomic Energy 1958, 1964, 1971. Past Pres. of American Nuclear Society. Founding Director of the Nuclear Power Division of the Electric Power Research Institute. Over 100 publications & book chapters. Three patents. Married 1950 to Mary Beth Novick. Children: James L, Barbara G, Richard A, Scott D, Janet L. *Society Aff:* ANS, AIChE, AAAS.

LeVere, Richard C
Business: Holophane Div./Manville, P.O. Box 5108, Denver, CO 80217
Position: Market Mgr, Tech Services *Employer:* Manville Corp-Holophane Div. *Education:* BEE/Elec Engg/Polytechnic Inst of Brooklyn. *Born:* June 1927. BEE Polytechnic Inst of Brooklyn (NY) 1950. Teaching exper in Armed Forces, Polytechnic Inst of Brooklyn & Pratt Inst. Author num tech papers, articles & tech workshop material. Fellow Illuminating Engrg Soc & Mbr its Energy Mgmt & Chrmn Public Affairs Cttes. P Chmn PIES Design Practice, Educ Ctte & LI Chap. Mbr sev U L Indus Advisory Conferences. PIES Vice Pres for Res & Technical Activities & Dir (1983- 86). President IES (1987-88). *Society Aff:* IES, ASHRAE.

Leverenz, Humboldt W
Home: 2240 Gulf Shore Blvd, N., Apt K4, Naples, FL 33940
Position: Staff V P - RCA Corp - ret. *Employer:* *. *Education:* AB/Chemistry/Stanford Univ. *Born:* July 1909. Grad study Univ of Muenster, Germany & Harvard Bus School. Native of Chgo Ill. Commenced R&D with RCA 1931. Devised & applied practical phosphors for TV & radar kinescopes, & fluorescent lamps. Also, ferrites for TV sets. Was RCA Dir of Phys & Chem Res 1954-57, Asst Dir of Res 1957-59, Dir of Res 1959-61, Assoc Dir of RCA Labs 1961-66, & Staff V P 1966-74. Brown Medal, Franklin Inst 1954. Mbr Natl Acad Engrs, Amer Chem Soc, Sigma Xi. Fellow Amer Phys Soc, Optical Soc Amer, IEEE, AAAS, Franklin Inst. Special int in critical matls, ecological aspects of matls, & higher educ. Enjoy classical music, tennis, travel, bridge, fishing & malacology. *Society Aff:* NAE, APS, OSA, IEEE, ACS, AAAS, FI, ΣΞ.

Leverton, Walter F
Home: 1061 Glenhaven Dr, Pacific Palisades, CA 90272
Position: Consultant Tech & Mgmt (Self-employed) *Education:* PhD/Physics/Univ Brit Columbia; MA/Physics/Univ of Saskatchewan; BA/Math and Physics/Univ of Saskatchewan. *Born:* 12/24/22. Saskatchewan Canada. Joined Aerospace 1960. Assumed respon of VP & Genl Mgr of Satellite Sys Div 1965. Group VP Dev 1978, Conslt technical and mgmt 1979. Patents in fields of crystal growing & semiconductor matls & devices. Tech articles publ in Physical Review & Journal of Applied Phys. Previously, Task Force mbr on num Defense Sci Bd panels. Mbr Amer Phys Soc; Fellow IEEE. V Chmn Exec Ctte NTC for 1977. Chrmn of CA South Coast Air Quality Bd Advisory Ctte 1981-1983. *Society Aff:* APS, IEEE.

Levi, Enrico
Home: 110-20 71 Rd Apt 620, Forest Hills, NY 11375
Position: Prof *Employer:* Polytechnic Univ *Education:* PhD/EE/Polytechnic; MEE/EE/Polytechnic; Dipl Eng/EE/Technion; BSC/EE/Technion; BSC/ME/Technion *Born:* 5/20/18 Native of Milan Italy, studied and taught at the Technion, Israel, and is Prof of Electrophysics at the Polytechnic Univ. He is also a PE and Pres and Treasurer of Enrico Levi Inc. He is author of *Electromechanical Power Conversion, Polyphase Motors: A Direct Approach To Their Design*, and a contributor to *Standard Handbook for Electrical Engrs, The Encyclopedia of Physical Science and* Technology, *Open Cycle MHD Electric Power Generation.* Wrote more than 50

Levi, Enrico (Continued)
papers mainly on plasmas, and holds 2 US Patents. 2 more are pending. In 1976-8 served as Sr Tech Advisor to US DOE, and in 1980 received the IEEE Charles J. Hirsch award and a Lady Davis Fellowship. *Society Aff:* IEEE, ΣΞ, ΤΒΠ.

Levi, Franco
Home: Corso Massimo d'Azeglio 100, Torino, Italy 10126
Position: Professor of Civil Engrg. *Employer:* Politecnico Torino. *Education:* DrIng/Prof/Politecnico Torino. *Born:* 9/20/14. Ingenieur des Arts et Manufactures Paris. Dr Engg Milan 1937. Ordinary Prof of Strength of Mtls Politecnico Torina Hon mbr ACI. Pres European Concrete ctte 1957-68; now pres h causa. Pres Federation Intl de la Precontrainte 1966-70; now Hon Pres. Medal Trasenster, Univ of Liege. Hon mbr Assn Espanola Pretensado. Prize "Torino" of the Assn of Engrs of Turin. Gold Medal, Ital Min of Educ for Educ Sci & Arts. National mbr of Acad of Sci of Turin. Prof at "Cours des Hautes Etudes de la Construction" Paris. Refraker of the editorial group CEE for Eurocode EC2. *Society Aff:* CEB, FIP, ACI, AICAP.

Levi, Ned S
Business: 190 W Glenwood Ave, Philadelphia, PA 19140
Position: Vice President. *Employer:* Richmond Oil, Soap & Chem Co Inc. *Education:* MBA/Operations Mgmt/Wharton Sch Univ of PA; Bs/Chem Engg/Carnegie Inst of Technology Carnegie-Mellon Univ. *Born:* May 1947. Native of Phila Pa. Pub Health Engr for Air Mgmt Services, the air pollution regulatory agency of the City of Phila. Since 1973 with the Richmond Oil, Soap & Chem Co Inc. Became V P 1973 with primary respon for managing mfg & production, along with other respon in sales & in res. In addition to other duties became V P of Richmond Chem Ltd, the wholly owned subsidiary of Richmond Oil, in 1975. Treas, Center City Residents Assoc, VP, Camp Council Inc, a co-ed camp for underprivileged children. Hobbies: photography, golf, tennis, & camping. Married, with two children. *Society Aff:* AIChE, AATCC.

Levin, Jordan H
Business: 207 Agri Engrg Bldg, East Lansing, MI 48824
Position: Res Leader/E Lansing Location Leader. *Employer:* U S Dept of Agri, Agri Res Service. *Born:* Nov 1920. BS Agri Engrg, BS Mech Engrg Rutgers Univ; MS Mich State Univ. Born Beverly, N J. US Army 1942-45. Worked for G E & Hercules Powder Co, joined USDA in res 1947 in Fla. From 1949-72 was Leader of Fruit & Veg Harvesting & Handling Res in U S with proj's in Hawaii, Calif, Wash, N Dak, Fla, Ga, Mich etc. Presently is Location Leader ARS in E Lansing Mich & Res Leader Fruit & Veg Mech - Mich. Fellow ASAE, 3 ASAE paper awards & USDA's Disting Serv Award (highest USDA award) - also 1976 Mich Engr of Yr Award from Mich ASAE. Life Master at bridge; enjoys tennis & sports.

Levin, Robert E
Business: 60 Boston St, Salem, MA 01970
Position: Senior Scientist. *Employer:* GTE Sylvania. *Education:* PhD/EE/Stanford Univ; Engr/EE/Stanford Univ; MS/EE/Stanford Univ; BS/EE/Stanford Univ. *Born:* Oct 11, 1931 Orange Calif. m. Karen N Andree 1958; c. Kristen & Erik. NSF Fellow 1954-56, Genl Elec Swopes Fellow 1956-58; Sigma Xi, Tau Beta Pi, Phi Beta Kappa. 1958-63 Assoc Prof of EE Calif State Univ at San Jose; 1963- , GTE Sylvania, R&D in optics, visual psychophysics, radiometry, photochemotherapy. 60 tech papers & 31 pats. Reg P E Calif & Mass. IEEE (SM), IES (Fellow), OSA, ASEE, SMPTE, USITT, Amer Soc Photobio. Hobbies: history of musical comedy, mbr Soc Amer Magicians & Intl Br Magicians. *Society Aff:* OSA, IEEE, IES, USITT, SMPTE.

Levine, David A
Business: Deer Park, NY 11729
Position: President. *Employer:* AIL-Cutler Hammer. *Education:* PhD/Mathematics/NY Univ; MS/Mathematics/NY Univ; BA/Liber Arts/Univ of Chicago. *Born:* Apr 1929. PhD & MS New York Univ; BA Univ of Chicago. Prior to Korean War, engr aide, theoretical aerodynamics NASA, Langley Field Va. Army Corps Engrs 1951-53. Res, fluid dynamics & meteorology, Courant Inst of Math Sci 1953- 59. Bioengrg res & collaboration, Brookhaven Natl Lab 1959- . Engr-analyst Republic Aviation Corp, spec in supersonic flow, shock waves & aerospace radiation hazards. Fac Engrg Coll of SUNY at Stony Brook 1965-70 & fac Queens Coll of CUNY 1970-75, res in bioengrg. Presently Engr-Systems Analyst-C3 & Computer Graphics, Software Engr. Enjoy swimming & classical music. *Society Aff:* ACM, IEEE.

Levine, David J
Home: 6454 Center Street, New Orleans, LA 70124
Position: Commercial Sales Mgr. *Employer:* Carrier Distrib Co-Div Carrier Corp. *Born:* Dec 1928. BSME Tulane Univ, MBA Loyola Univ. Tau Beta Pi, Beta Gamma Sigma. Mbr ASHRAE, ASME, NSPE, LES. Reg M E La. Org & chaired first 2 ASHRAE Tulane Symposiums on Air Cond. Served ASHRAE New Orleans Chap as Pres & received Chap's Cary B Gamble Award for Outstanding Contrib to the Chap. Past ASHRAE Mbrship Chmn - present Dir & Regional Chmn. Respon for 90-75 in Region VII. Respon for Commercial Sales & Mkting Policy for Carrier Distrib Co New Orleans. Designed Vent & A/C Sys for first ocean-going drilling vessel. Pres Congregation Gates of Prayer.

Levine, Duane G
Business: Office of Environmental Affairs for, Exxon Corp, Exxon Research & Engineering Co, Florham Park, NJ 07932
Position: Mgr Exxon's Worldwide Environmental Affairs *Employer:* Exxon Corp *Education:* MS/Chem Engg/Johns Hopkins Univ; BES/Chem Engg/Johns Hopkins Univ. *Born:* 7/5/33. Native of Baltimore, MD. Joined Exxon Res and Engg Co in 1959. Original res in combustion, electrochemistry, and fossil fuels processes and quality until 1967. Dir bd range of res activities related to fuels and lubes processes, product quality, and emission control through 1976. Served as site manager for Exxon Research s Baytown Div on Synthetic Fuels through 1978. Dir of Corp Res-Sci Labs including physical sci, chem sci, solid-state sciences, engineering sciences, solid-state sciences, engg sciences, and fuels sciences from 1979 to 1984. Currently, responsible for worldwide environ affairs in Exxon Corp. Tau Beta Pi, Sigma Xi, Phi Lambda Upsilon. *Society Aff:* AIChE, AIC, AAAS, API, ICI, IPIECA, ACS.

Levine, Jules D
Business: PO Box 225986, MS 158, Dallas, TX 75265
Position: Mgr *Employer:* Texas Instruments *Education:* PhD/Nuc Eng/MIT; BS/Mech Eng/Columbia *Born:* 6/24/37 Native of NY City. Mgr at RCA Labs, Princeton, NJ from 1963 - 79, specializing in surface phenomena, energy conversion, displays, and electron emission devices. At Texas Instruments since 1979 as Mgr, Energy Systems, specializing in photovoltaics. Became Senior Member of the Tech Staff, 1981. Became Mgr, Solar Cell Dev in 1984. Chrmn IEEE Princeton Section, 1977. Founder and Chrmn of IEEE Power Engrg Society, Princeton Section, 1973-76. IEEE Fellow, 1980. Enjoy playing string instruments and creating large paintings and sculpture. *Society Aff:* IEEE, AVS

Levine, Neil M
Business: Riverside-Lankershim Bldg, 10911 Riverside Dr, No Hollywood, CA 91602
Position: Managing Partner - VINES & ASSOCS. *Employer:* Hudson Assocs Inc. *Education:* BS/Bus/NY Univ; MBA/Bus/NY Univ; Doctor of Laws/Hon/Valley Univ of Law. *Born:* 1927. BS in Bus Admin & MBA NYU. Fulbright grant. 1950-55 chem engr with Carol Holding (Monsanto). Asst Mgr Ira Haupt & Co 1955-60, spec in econ studies for venture funding. V P New Eng Indus 1960-64 in Natural Resource Div. Sr Staff Engr 1964-66 Imperial Chem Indus. Formed (U S affil) N Levine & Co 1966-73 cons engrs for joint N A chem projects. Since 1973 V P Energy Div of Hudson Assocs engrg cons to indus. Keystone Award 1965; Lombard (Engrg) Disting Award 1967; Sloan Gold Medal 1967. Mbr AIChE, AIME; ad hoc Ctte on Energy Conservation; Indus Forum. 1976-present-Manging Partner VINE &

Levine, Neil M (Continued)
ASSOCS & partner of GOLDEN SQUARE CONSULTING LTD (UK); 1978-mbr Assoc of Mgmt Consultants. Awarded Doctor of Laws (Hon) Valley Univ of Law 1978. 1979-mbr of Bd of Dirs of the Energy Fuels Fnds (NY). 1977-Advisory mbr of Interagency Fuels Task Force Committee - ERDA. *Society Aff:* AIME, AIChE.

Levine, Robert S
Home: 19017 Threshing Place, Gaithersburg, MD 20760
Position: Chief, Fire Research Resources Div. *Employer:* Nat Bur Stds. *Education:* ScD/Chem Engg/MIT; SM/Chem Engg/MIT; BSc/Chem Engg/IA State Coll. *Born:* 6/4/21. Native of Des Moines, IA. Specialize in combustion-rel problems. 17 yrs in Rocketdyne div of Rockwell, R&D on liquid rocket engines, especially combustion stability and performance. Left as Assoc Res Dir for 5 yrs as Chief, Liquid Rocket Tech, NASA Hq. 3 yrs Remote Sensing, NASA Langley Res Center. Since 1974, Center for Fire Res, National Bureau of Standards. *Society Aff:* AIAA, ACS, NFPA, SFPE, TCI.

Levine, Saul
Home: 9910 Fernwood Rd, Bethesda, MD 20817
Position: VP & Group Exec Consltg Group *Employer:* NUS Corp. *Education:* MS/Nucl Engr/MIT; BS/Elec/MIT; BS/-/US Naval Acad *Born:* 7/1/23. Commd ensign USN, 1945; submarine service, 1946-53; proj officer US Naval Reactors in charge nucl power plant for USS Enterprise, 1955-58; chief system design sect nav br Spl Projs Office for Polaris missile system, 1958-62; ret, 1962; asst dir reactor tech reactor licensing div AEC, 1962-70, asst dir Office Environ Affairs, Wash, 1970-73, proj staff dir reactor safety study, 1972-75, dep dir div reactor safety res, 1973-75; dep dir Nucl Regulatory Comm, Office of Nucl Regulatory Res, 1975, dir, 1976-1979; Vice Pres & Gnl Mgr. NUS Corp; lectr nucl safety MIT, summers, 1967-; US AEC del to UN Conf on Environ, Stockholm, 1972; v chrmn Com on Safety of Nucl Installations, OECD, 1977-. Recipient Distin. Service award AEC, 1975. *Society Aff:* ANS, AAAS

Levine, William S
Business: Dept of EE, College Park, MD 20742
Position: Prof of EE *Employer:* Univ of MD. *Education:* PhD/Controls/MA Inst of Tech; SM/EE/MA Inst of Tech; SB/EE/MA Inst of Tech. *Born:* 11/19/41. in Brooklyn NY. Design engr for Date Tech Inc 1962-1964. Res and Teaching Asst at MIT 1964-1969. Univ of MD since 1969, Prof since 1981. Primary teaching area is control systems. Have consulted for IBM, Satellite Systems Engrg, ETS Inc and others. Current res is in control theory, computer aided control system design, neuromuscular controls and filtering. Married, with two children. *Society Aff:* IEEE.

Levingston, Ernest L
Business: PO Box 1865, Lake Charles, LA 70602
Position: Pres. *Employer:* Levingston Engrs, Inc. *Education:* BS/Mech Engr/LA State Univ. *Born:* 11/7/21. Native of Lake Charles, LA. Served in WWII in Seabees and US Navy Hydrographic Office. Engg Sec Supervisor, Cities Service Refinery, 1946-57, Group Leader, Bovay Engrs, 1957-59. Chief Engr, Augenstein Construction Co, 1959-60. Pres, Levingston Engrs, Inc 1961 to present. Reg PE, LA, MS, AR, MD, PA, NJ, TX, TN, DE, WA, DC, OK, CO. Mbr LA Bd of Commerce and Indus; Natl Examination Ctte and mbr Bd of Trustees, Inst for Certification of Engg Technicians; Advisory Bd Sowela Tech Inst; past Pres Lake Charles Chapter LA Engg Soc; past mbr Lake Charles Planning and Zoning Commission, Regl Export Expansion Council. *Society Aff:* ASME, NSPE.

Levinson, Herbert S
Home: 40 Hemlock Rd, New Haven, CT 06515
Position: Prof and Transp Conslt *Employer:* Polytechnic Univ *Education:* BS/CE/IL Inst Tech; Cert/Hwy Traffic/Yale Univ *Born:* 9/25/24 Army Air Force Weather Serv 1943-46; Junior Traffic Engr, Chicago Park Dist 1949-51; Assoc Engr, Principal Assoc, Sr VP, Wilbur Smith & Assoc 1952-81; Independent Transp Consultant since 1980. Visiting lecturer City Planning, Yale 1962-80. Prof of Civil Engrg, Univ of CT 1980-86. Professor of Transportation Polytechnic Univ of New York 1986-. Recipient ITE (New England Sec) Transportation Engr of the Year Award 1977; Tech Council Award 1981. Author or co-author *Elementary Sampling for Traffic Engrs* (1962); *Future Hgwys and Urban Growth* (1961); *Parking in the City Center* (1965); *Transportation and Parking for Tomorrow's Cities* (1966); *Bus Use of Hgwys*, 1975, *Urban Trans Perspectives and Prospects*, 1982. *Society Aff:* ASCE, ITE, APA, TRB

Levis, Alexander H
Business: M.I.T., Room 35-410/LIDS, Cambridge, MA 02139
Position: Senior Research Scientist *Employer:* Mass. Institute of Technology *Education:* Sc.D./Mech Eng/M.I.T.; M Sc/Mech Eng/M.I.T.; SM/Mech Eng/MIT; BS/Mech Eng/MIT; AB/Physics-Math/Ripon College *Born:* 10/03/40 Born in Yannina, Greece. Served on the faculty of Electrical Engrg of the Polytechnic Inst of Brooklyn (1968-1973). 1973-79 he was with Systems Control, Inc., in Palo Alto, CA as manager of the Systems Research Dept. Since 1979 with the MIT Laboratory for Information and Decision Systems doing research on applying system theory to complex decision processes. Since 1975, has held a number of positions in the IEEE Control Systems Society including President, 1987, and IEEE Director in the American Automatic Control Council. Associate editor and editor for several journals including *Automatica* (1980-85) and *IEEE Trans. on Automatic Control*. *Society Aff:* IEEE, AAAS, AIAA

Levis, Calvin E
Business: PO Box 3012, 1650 Manheim Pike, Lancaster, PA 17604
Position: Chairman. *Employer:* Huth Engrs Inc. *Education:* BS/Civil Eng/PA State; MS/Civil Eng/PA State *Born:* 1/3/26. Navy & Merchant Marine 1943-46. Designer with C A Maguire & Assoc, Boston MA 1950-53. Instr PA State 1953-56. With Huth Engrs since 1956, VP 1959; assumed respon as Pres 1967; Chmn 1984. Firm's services have been expanded from surveying & municipal eng to all phases of civil and environ engg, planning, architecture and landscape arch. Fellow ASCE, Diplomate Amer Acad of Environ Engrs, Fellow ACEC, Inventor Lagco Flume, Enjoys sailing, travel, golf and fishing. *Society Aff:* ASCE, AAEE, ACEC, AWWA, NSPE

Levis, Curt A
Business: 2015 Neil Ave, Columbus, OH 43210
Position: Prof Em *Employer:* OH State Univ *Education:* PhD/EE/OH State Univ; AM/Appl Phys/Harvard Univ; BS/EE/Case-Western Res Univ *Born:* 4/16/26 in Berlin, Germany. Immigrated US 1938, naturalized 1945. US Navy 1944-46. Engr WSRS Inc 1948-49. Asst Prof 1956-60, Assoc Prof 1961-63, Prof 1963-85, Prof Em 1985- , OH State Univ. (Dir ElectroSci Lab 1961-69). Sr Postdoc Fellow, Natl Ctr for Atmospheric Res 1976-77. Guest Worker, Inst. of Telecom. Sci, US Dept. Com., 1986-7. Married 1958 Katharine Slaven; children: Alan Patrick (1961), Linda Irene (1963), Susan Irene (1966). *Society Aff:* IEEE, HKN, ΤΒΠ, ΣΞ

Levy, Bernard S
Business: 3001 E Columbus Dr, E Chicago, IN 46312
Position: Sr Product Cons, Steel Prod Res Div. *Employer:* Inland Steel Co, Res Labs. *Education:* MBA/Mgmt/Columbia Univ; BSc/Met/MIT. *Born:* Jan 1938. Joined Inland Steel 1960 & worked in var pos in Qual Control & Res. Currently working as Sr Product Cons in the area of wrought steel products. Active res ints incl product application, machinability, metal formability & their underlying struct property relationships. Co-editor of Journal Applied Metworking of the Amer Soc for Metals since late 1977. AISI award best tech paper at a regional mtg 1964. Mech Working & Steel Processing Ctte AIME Highest Award for the best paper at the 15th Mech Working & Steel Processing Conf. *Society Aff:* ASM, SAE, SME, NAMRT.

Levy, Harry
Home: 18121 Leatherwood Way, Irvine, CA 92715
Position: Asst Ch Design Engr for Naval Arch. *Employer:* Long Beach Naval Shipyard. *Education:* BE/Mech Engg/Univ of So CA. *Born:* June 26, 1925 Cleveland Ohio. Grad Glenville H S. Army Air Corps 194346. Attended Univ of So Calif 1946-50; Bach of Engrg degree June 1950. Worked as Naval Architect at San Fran Naval Shipyard 1950-57. Attended 2 summer sessions at Univ of Calif Berkeley & studied Naval Arch. Transferred to Long Beach Naval Shipyard 1957 & proceeded to advance to present pos. Chmn L A Metro Sect Soc of Naval Architects & Marine Engrs 1975-76. Enjoy skiing & tennis. *Society Aff:* SNAME.

Levy, Matthys P
Business: 110 E 59th St, New York, NY 10022
Position: Partner. *Employer:* Weidlinger Assocs, Cons Engrs. *Education:* CE/Applied Mech/Columbia Univ; MS/Applied Mech/Columbia Univ; BSCE/Civil Engg/City College of NY. *Born:* Sept 15, 1929 Basle Switz. Oper Officer, Const Battalion, Army Corps of Engrs 1952-54. Currently Partner, Weidlinger Assocs, Cons Engrs NYC; has been with firm since 1956. Has served as Visiting Critic at Yale Univ & is presently Adj Prof of Arch Columbia Univ. Bd of Dir NYACE 1975 & of Haus-Rucker 1975-76 . Lincoln Arc Welding Award 1961. Ptnr in charge of many award winning designs & author of num tech papers & articles. Co-author book on struct analysis. Reg P E 13 states. Fellow ASCE; Mbr Arch League of NY. Mbr Inst of CE, ACI. *Society Aff:* ASCE, ICE, ACI.

Levy, Ralph
Business: KW Engg, Inc, 4565 Ruffner St, San Diego, CA 92111
Position: Vice Pres of Engg. *Employer:* KW Engg, Inc. *Education:* PhD/Appl Scis/London Univ; MA/Physics/Cambridge Univ. *Born:* Apr 1932. 1953-59 with GEC Stanmore, 1959-64 Mullard Res Labs, 1964-67 Univ of Leeds. Came to USA 1967 to join MDL. V P Res 1970-84, respon for res & new prod dev in passive microwave components. Since 1984, VP Engg at KW Engg, responsible for res and engg. Author approx 50 papers in prof journals & 12 pats; author book 'Circuit Theory', Oliver & Boyd, Vol I 1970, Vol II 1973. Elected Fellow IEEE 1973, Chmn Boston Chap MTT Soc 1973-74, Chmn MTT Soc Tech Ctte on Microwave Network Theory 1976-80 , Chrmn, Tech. Program Ctte, 1983 MTT-S Intl Microwave Symposium, Ed, IEEE Transactions on Microwave Theory and Techniques, 1986- . Mbr IEE (London). *Society Aff:* IEEE, IEE.

Levy, Salomon
Business: 3425 S Bascom Ave, Campbell, CA 95008
Position: Pres *Employer:* S Levy Incorp *Education:* PhD/Mech Engrg/UC Berkeley; MS/Mech Engrg/UC Berkeley; BS/Mech/UC Berkeley *Born:* 04/04/26 From 1953 to 1977, Dr Levy worked with Gen Elec Co in overall nucl pwr plant design, safeguards analysis, dev of small nucl pwr plants and test reactors, heat transfer, fluid flow dev in boiling water reactors and in the def of nucl pwr sys for plant proposals. Earlier he served as Mgr-Design Engrg for the Atomic Pwr Equipment Dept, Gen Mgr-Nucl Fuel Dept, and Gen Mgr, Boiling Water Reactor Operations, where he was repons for all the engrg and mfg of Gen Elec nucl pwr bus. At that time he was in charge of an organization of over 4, 000 persons. In Sept 1977, he left Gen Elec to form his own independent engr conslt firm. He presently consults for several pwr utilities, natl labs, EPRI, the Res Div of the US Nucl Regulatory Comm and several pwr equipment manufacturers. Dr Levy was a Conslt to the Staff of the Kemeney Comm, and Mbr of Indus Advisory Board to Three Mile Island-2 Recovery Operations. He is a member of at least four oversight cttes for nuclear plants & is a dir of IE Industries. He has authored more than 40 publ tech papers. *Society Aff:* NAE, ASME, ANS

Levy, Sander A
Business: 4th & Canal Sts, Richmond, VA 23219
Position: Dir, Dept of Ingot Casting, Tech & Met Serv. *Employer:* Reynolds Metals Co. *Born:* Sept 8, 1939 Scranton, Pa. s. Samuel B & Ida Grass Levy. Lehigh Univ Bethlehem Pa: 1962 BS Met Engrg & BA Appl Sci; 1963 MS Met Engrg; 1965 PhD Met Engrg. 1965-66 Bell Tele Labs Murray Hill NJ. 1966-68 US Army, Capt Ordnance Corp, Pitman Dunn Res Labs, Frankford Arsenal, Phila Pa. 1966-68 Instr Evening Div, Drexel Inst of Tech. 1968- , Reynolds Metals Co, Met Res Div. Mbr ASM, AIME. Chap Chmn Richmond Chap ASM 1972-73. Adj Fac Vir Commonwealth Univ 1969- . Voted outstanding young mbr Richmond Chap ASM 1974. *Society Aff:* ASM, AIME.

Lewin, Leonard
Business: Dept of Elec Engrg, Boulder, CO 80309
Position: Coordinator, M S Telecommun Prog. *Employer:* Univ of Colorado. *Education:* DSc/Hon/Univ of CO. *Born:* 1919. Native of Southend Eng. Employed by British Admiralty WW II. With Standard Telecommunication Labs 1946-68, Hd of Microwave Dept, later Asst Mgr Transmission Lab & subsequently Sr Principal Res Engr. With Univ of Colo since 1968, spec in Field Theory & coordinator of M S Telecommunications Prog 1974- . In 1963 awarded the IEEE Internatl Microwave Prize & the W G Baker Award. Author sev books on waveguides & mathematics. Former Chap Chmn Boulder/Denver SMTT, & Fellow IEEE. Chartered Engr & Mbr British IEE. Enjoy music, nature study, chess. *Society Aff:* IEEE, IEE, BIS.

Lewins, Jeffery
Business: Engg Department, Trumpington St, Cambridge, England CB2 1PZ
Position: University Lecturer in Nuclear Engineering *Employer:* Univ of Cambridge. *Education:* DSc/Nuclear/London. *Born:* Nov 30, 1930. Married, 3 children. Commissioned Royal Engrs with active service Korea; Gold Medallist RMA Sandhurst, Silver Medal Inst Royal Engrs. Grad studies MIT 1956-59 MSc, PhD. DSc (Eng) London 1979. Attachments to AERE Harwell, Brookhaven NL, Battelle Northwest Lab & Culhati. Visiting Prof Univ of Wash 1964-65 1985-86, Alexandria 1976. Res ints in reactor kinetics, variational & optimization methods. Books: *Importance*, Pergamon 1965; *Reactor Control*, Pergamon 1977. Editor: *Adv Nucl Sci Tech* 1965- ; Research Studies in Nuclear Technology 1987- *Society Aff:* ANS, BNES, INUCE.

Lewis, Arthur E
Business: PO Box 808 L-207, Livermore, CA 94550
Position: Fossil Energy Group Leader *Employer:* Lawrence Livermore Nat'l Lab. U of Calif *Education:* PhD/Geo/CA Inst of Tech; MS/Geo/CA Inst of Tech; BS/Math/St Lawrence Univ *Born:* 1/11/29. Jamestown NY. US Army 1951-53. 1958-60 Curtis Wright Corp Santa Barbara CA, Sr Engr; Res in fields of high temp matls & elec ceramics. 1960-62 Hoffman Electronics Santa Barbara CA, Sr Scientist; 1962-67 Fairchild Camera & Instrument Co Palo Alto CA, Tech Staff; 1967- Lawrence Livermore Natl Lab; Group leader - Fossil Energy-Proj Leader for Oil Shale Retorting R&D. *Society Aff:* AIME

Lewis, Bernard T
Home: 12728 Lincolnshire Dr, Potomac, MD 20854
Position: Head Industrial Engg Branch. *Employer:* Naval Facilities Engg Command. *Education:* PhD/Mgt/Pacific Western Univ; MA/Mathematics/Columbia Univ; BS/Genl Engg/US Military Academy. *Born:* 2/7/23. Native of Boling, TX. Served with Army Parachute Infantry 1943-1947 and Army Transportation Corps 1950-52. Industrial Engr for American Machine and Foundry, Western Elec, and Ronson specialing in industrial engg applications in mfg. Supervisory Industrial Engr with Naval Facilities Engg Command since 1958. Responsible for systems and procedures dev, manuals preparation, systems installations, and monitoring systems effectiveness of Navy's Real Property Maintenance Mgt System and Engineered Performance Standards at shore facilities. Pres, DOD Chap & AIPE 1975. Enjoy gardening and reading. *Society Aff:* AIPE.

Lewis, Brian J
Home: 10614 Hunter Station Rd, Vienna, VA 22180
Position: Consultant (Self-employed) *Education:* MS/Engrg/UCLA; Grad Cert/Public Health Engr/King's Coll, UK; BSc/CE/Univ of Durham, UK *Born:* 9/8/29 Native of England, came to USA in 1952. Worked for consultants in Pasadena, San Francisco and Seattle prior to establishing Lewis-Redford-Engrs, Inc. In 1961, Bellevue, WA.

Lewis, Brian J (Continued)
Elected WA State House, 1966, and Senate, 1968. Joined Roy F. Weston, Inc, 1973, was successively Division Mgr, VP Client Services, Group VP & Pres thru Nov 1978. Now Management Consultant affiliated with The Coxe Group, Philadelphia. Pres, CEC of WA, 1969. Pres Philadelphia Section ASCE, 1977. Former member & Chrmn ASCE Committee on Public Affairs & Legislation. Member, Past Chrmn, ASCE Committee on Public Communications. One of Three Outstanding Young Men of WA, 1963. Married, 5 children. *Society Aff:* ASCE, ACEC, ITE, SMPS, NSPE

Lewis, Clark H
Business: Aerospace & Ocean Engg Dept, Blacksburg, VA 24061
Position: Prof. *Employer:* VPI & SU. *Education:* PhD/Engg Sci/Univ of TN; MS/ME/Univ of TN; BSME/ME/Univ of TN. *Born:* 11/6/29. Native of TN. With ARO, Inc, Tullahoma, TN from 1951 to 1968. Supv Theoretical Gas Dynamics Sec, Von Karman Gas Dynamics Facility. Joined faculty VPI&SU Oct 1968 as Assoc Prof. Full prof since 1970. Consultant to US Navy, USAF and several aerospace industries. Consulting and res interest areas include reentry aerodynamics, computational fluid mechanics, chemically reacting flows and planetary entry physics. *Society Aff:* AIAA, ASME, AIP.

Lewis, David S
Business: Pierre Laclede Ctr, St Louis, MO 63105
Position: Chmn of the Bd. *Employer:* Genl Dynamics Corp. *Education:* BS/Aeronautical Engr/GA Inst of Tech. *Born:* July 1917. Hon Degree Clarkson Coll of Tech; BS Ga Tech. Prior to joining Genl Dynamics 1970, was Pres & Ch Oper Officer of McDonnell Douglas Corp. Joined Glenn L Martin Co of Baltimore. Became Ch of Aerodynamics at McDonnell Aircraft Corp 1946 & 4 yrs later entered the Design Dept. 1952, apptd Ch of Prelim Design of the Airplane Engrg Div. Promoted to Mgr of Sales 1955 & the following yr was named a Proj Mgr. Subsequently became Mgr of Proj's & V P of Proj Mgmt. In 1959 elected a Sr V P, & in 1961 Exec V P. Elected Pres 1962, a pos he continued to hold in the new McDonnell Douglas Corp, which was formed in 1967 by the merger with Douglas Aircraft Co. Dir of BankAmerica Corp, Director: the Mead Corp. and Ralston Purina Fellow AIAA, Mbr Exec Comm & Bd of Governors of the Aerospace Indus Assn, Mbr NAE, mbr var civic & social clubs. *Society Aff:* AIAA, NAE.

Lewis, David W
Home: 10 Hancock Rd, Hingham, MA 02043
Position: SR. V P - Industrial Divisions *Employer:* Chas T Main Inc. *Education:* BS/ME/Lafayette College. *Born:* Sept 1927. Raised in W Orange NJ, where attended pub schools prior to WW II. Served in the Military Police in the S Pacific prior to attending college. Joined Chas T Main Inc 1956 as a mech engr. Received current assignment of V P 1972 covering fiscal, mkting, & engrg respon for the Div. Activity in engrg soc's incl the Cons Engrg Council of New Eng; Res & Engrg Council of the Graphic Arts Indus; Graphic Arts Tech Found. At present Reg P E 6 states. Outside ints incl electronics (amateur radio), tennis & fishing. *Society Aff:* CECNE, R&E, MSPE, GATF.

Lewis, Edward R
Home: 852 Shari La, East Meadow, NY 11554
Position: Prof *Employer:* Hofstra Univ *Education:* M/ME/NYU; M/IE/NYU; B/ChE/CCNY; Cert/Naval Arch/MIT *Born:* 7/21/19 Naval Architect, NY Naval Shipyard, 1942-1945; US Navy, Electronic Techician, 1945-1946; Research Engr in Structural Mechanics and Mechanical Engrg, Naval Applied Science Lab, 1946-1960; Assoc Prof, Hofstra Univ, 1960- 1973; Prof of Engrg, Hofstra Univ, 1973-present; author of five texts and over 40 papers and research reports; as chrmn of the Metropolitan NY section of the SESA was one of the principal organizers of the First World Congress In Experimental Mechanics; reviewed articles for several technical journals; directed Inst in Earthquake Engrg for Defense Civil Preparedness Agency; consultant to several Consulting Engrg Firms in the area of Materials Engrg; member of Sigma Alpha, ASEE, SESA. Listed in American Men and Women of Science. *Society Aff:* SESA, ASEE

Lewis, Edward V
Home: 97 Plymouth Dr North, Glen Head, NY 11545
Position: Consultant. *Employer:* *. *Education:* MS/Naval Arch/Webb Inst of Naval Arch & Stevens Inst of Tech; BA/Math/NB Wesleyan Univ. *Born:* Jan 6, 1914 E Hampton NY. Married Adelyn C Sar 1939; 2 children. Naval Architect with Geo G Sharp N Y 1936-51. Hd of Ship Div, Davidson Lab, Stevens Inst of Tech 1951-61; Res Prof 1959-61. Res Prof & Dir of Res Webb Inst of Naval Arch Glen Cove N Y 1961-78 NAVSEA Res. Prof, US Naval Academy, Annapolis, MD 1979. Hon VP, Fellow Soc of Naval Architects & Marine Engrs. Linnard Prize 1955, Davidson Medal 1966, Mbr Tech & Res Steering Ctte. Technical Editor, revised edition of *Principles of Naval Arch.* Mbr Royal Inst of Naval Architects London; Phi Kappa Phi; Sigma Xi. *Society Aff:* SNAME, ASME.

Lewis, Edwin R
Business: Dept of Elec Engg & Comp Sci, Berkeley, CA 94720
Position: Prof *Employer:* Univ of CA. *Education:* PhD/EE/Stanford; Eng/EE/Stanford; MS/EE/Stanford; AB/Biol/Stanford *Born:* 7/14/34. From 1961 to 1967, served as Staff Engr for Gen Precision's Lab for Automata Res, Glendale, CA, where conducted studies in theoretical neurophysiology. Since 1967, has been a mbr of the Faculty of Elec Engg and Comp Sciences, Univ of CA, Berkeley. Since 1970, has been Dir of the Biomedical Engg Training Prog at Berkeley. Presently, is Prof of Elec Engg and Comp Sciences. Primary res interests are neurosensory biophysics and signal processing. Since 1984, has been Javits Neuroscience Investighator (NIH). Author, three books and approximately fifty journal articles. *Society Aff:* IEEE, ASA, AAAS, ARO.

Lewis, Heydon Z
Business: Box 2327, Littleton, CO 80161
Position: Pres *Employer:* Thermo-Scan Engrg Inc *Education:* PhD/EE/Univ of CO; MS/EE/Univ of IL; BS/EE/Univ of AR *Born:* 7/11/35 Came to CO in 1960 to work for Martin-Marietta in the Titan program control system design. Taught at the Univ of CO while working on doctorate. Was Chief Engr for the CO Div of Communications and developed a curriculum for Telecomunications Engrg and Management while on the engrg faculty of the Univ of Pittsburgh at Johnstown. In 1978, started firm, Thermo-Scan Engrg offering engrg services in thermal imaging and roof inspection-design. Developed and taught course on Roof Inspection as faculty member of Roofing Industry Educational Inst. *Society Aff:* ASHRAE, IEEE, NSPE, ACEC, CSI.

Lewis, Jack R
Home: 11300 Yarmouth Ave, Granada Hills, CA 91344
Position: Consultant *Employer:* Materials Tech *Education:* PhD/Met Engg/Stanford Univ; BS/Met/Stanford Univ. *Born:* 7/30/20. Born 1920 in Eureka, KS. Served in US Army in WWII. Met Engr with GE Co 1951-1961, specializing in nuclear fuel element dev. With Rockwell Corp 1961-85 as mgr, advanced materials progs with Rocketdyne Div. Self-employed consultant since 1985, specializing in mtls & processes for rocket engines, marine propulsion devices, metal matrix composites, structural ceramics, and solar power systems. Married Donnie Doan Lewis (1952); one daughter; Melinda. Enjoy wood carving. Fellow, ASM. *Society Aff:* ASM, AIME, AIAA, ΣΞ, ACerS.

Lewis, Jack W
Home: 9104 Red Branch Rd, Columbia, MD 21045
Position: President. *Employer:* ARCTEC Inc. *Education:* MS/Naval Engg & Mech Engg/MA Inst of Tech; BS/Naval Engg & Mechanical Engg/US Coast Guard Academy. *Born:* Feb 11, 1937. Naval Engrg & MSME MIT; BS US Coast Guard Acad. Served aboard Coast Guard Cutters for 4 yrs prior to grad school. After MIT, designed icebreaking ships for Coast Guard. In 1970 resigned commission as a LCDR

Lewis, Jack W (Continued)
to form ARCTEC Inc, a firm of cons engrs & architects who spec in cold regions engrg. Have managed firm since its formation. Managed proj's assoc with historic MANHATTAN voyages. ASNE Jimmie Hamilton Award 1969, SNAME Capt Joseph Linnard Prize 1970. Enjoys skiing, canoeing, sailing & equestrian sports. *Society Aff:* SNAME, ASME, ASNE.

Lewis, James P
Home: 7803 Aleta Dr, Spring, TX 77379
Position: Pres/Owner *Employer:* Project Tech Liaison Assoc, Inc. *Education:* BS/Mech Engg/CA Inst of Tech. *Born:* 4/21/33. Native of Red Cloud, NB. With Richfield Oil (1955-62) as Dist Mech Eng respon for oil and gas production equipment. With Cosmodyne (1962-69) as Asst Chief Eng respon for analytical support of projects, tech support of marketing/international licenses, and engg services for cryogenic systems. Tech dir for Distrigas (1969-72) respon for LNG import terminal design and operation. Mgr for LNG Projects for Transco Energy (1972-78). Established PTL in 1978 for LNG, LPG, and natural gas engg including project management, process engg, cryogenics, shipping, environmental engg, economics, safety, reg complicance, and personnal training. Manages Dev of co Natural Gas proj. Dir of KL Energy Servs, and Winship Trading. Reg PE in TX and British Columbia. *Society Aff:* AIChE, ASME, SNAME, SPE, NFPA, EERI.

Lewis, Oliver K
Business: ASHRAE, 1791Tullie Circle, NE, Atlanta, GA 30329
Position: (Tech. Info & Metric Coordinator) *Employer:* Am Society Heating, Refrig, & Air Cond Engineers *Education:* BS/ME/GA Inst of Tech. *Born:* Dec 16, 1920 Atlanta Ga. Tenn Eastman Corp 1943-47, U235 electromagnetic separation process design engr; Oak Ridge Natl Lab 1947-52, area engr; Albany Architects & Engrs; 1952-56, Ch Mech Engr; Oliver K Lewis & Assocs Inc Cons Engrs 1956-1980 hvac & elec design; Martin Marietta Aerospace - New Orleans, Sr. Engr 1980-81. Mbr ASTM, ASME, ASHRAE. P Mbr & Chmn Ga State Bd Regis for Prof Engrs & Land Surveyors. Mbr USMB Metric Speakers Bureau Tau Beta Pi. *Society Aff:* ASME, ASHRAE, ASTM, USMA, ANMC.

Lewis, Orval L
Home: 569 N Post Oak Lane, Houston, TX 77024
Position: Retired *Education:* BSME/-/TX Tech Univ; MBA/-/Univ of South CA. *Born:* 7/27/16. Reg P E Calif, & Texas. Primary field is mgmt of engrg & const of process indus plants. Participated in engrg dev of gas repressuring of oil fields, fluid catalytic cracking units in refineries, first commercial power plant using fluidized petro coke, test facils for space vehicles, nuclear power test facils, desalination, liquid metals fast breeder reactor prog, shale oil processing, substitute natural gas. Pres of Engrg, Sci & Tech services co. Owns & operates farm & ranch property. Active in church, political party & community affairs. V P Indus ASME 1974-78, ASME Exec Ctte 1974-76, Regional V P ASME 1966-68. Fellow ASME 1965. Engr of Month L A 1964, Natl Engr of Distinction 1970, Kansas Bankers Assn Award for Soil Use & Conservation 1973, Pres ASME 1978-79. Chrmn ASME Legal Affairs 1982-87. *Society Aff:* ASME, ADPA.

Lewis, Peter A
Home: RD-2, Box 212, Pleasant Valley Road, Titusville, NJ 08560
Position: Manager - R&D *Employer:* Public Service Electric and Gas Company *Education:* MS/Management/Newark College of Engineering; BS/EE/Lehigh University *Born:* 02/18/38 Manager, Energy Utilization Research and Development for Public Service Electric and Gas Company in Newark, NJ. Programs include assessment of new technologies, demonstration of systems and equipment for energy conversion, delivery and end-use. He has directed a program to test and evaluate advanced load levelling batteries. He has testified on energy matters before a number of Government organizations including the New Jersey Assembly and the U.S. House of Representatives Cttee on Science & Technology Subcommittee on Energy Research and Production. Mr. Lewis is active in his community and currently serves as a member of the Commission on Technology Education for NJ. *Society Aff:* IEEE, ES, NSPE

Lewis, Ronald L
Business: P.O. Box 2967, Houston, TX 77252-2967
Position: Mgr, U.S. Offshore Div; VP of Engg *Employer:* Pennzoil Co *Education:* PhD/Petro Engr/Univ of TX; MSc/Petro Engr/CO School of Mines; BSc/Petro Engr/CO School of Mines *Born:* 4/5/34 After completing high sch and receiving my Petro Engrg degree at the CO Sch of Mines, I served in the US Army Corps of Engrs stationed in Alaska. My petro engrg career began with Mobile Oil in 1956. My academic experience includes positions at the CO Sch of Mines and the Univ of SW LA. My education includes an MSc in 1963 and a PhD in 1968 both in Petro Engrg. In 1970 I started working for Pennzoil as an Advanced Petro Engr and have progressed to the present position of VP of Engrg and Mgr of the U.S. offshore Div for the oil and gas div My qualifications include being a reg petro engr in the state of LA. I have written several booklets and articles on the subjects of drilling tech, fluid flow in porous media, and well log analysis. *Society Aff:* SPE, SPWLA, API.

Lewis, W David
Business: 7008 Haley Ctr, Auburn, AL 36830
Position: Hudson Prof of History & Engrg. *Employer:* Auburn Univ. *Education:* BA/History/PA State; MA/History/PA State; PhD/History/Cornell *Born:* 6/24/31. BA, MA, Penn State; PhD Cornell. Previous pos at Hamilton Coll, Eleutherian Hills-Hagley Found, Univ of DE, SUNY at Buffalo. Mbr Soc for the Hist of tech (Prog Chmn 1973; Exec Ctte 1975-78 Advisory Council, 1979-1985). Dir Auburn Univ Proj on Tech, Human Values & the Southern Future 1973-78. Num pubs in hist of tech, esp on hist of indus res labs, hist or iron & steel industry, history of aviation. Married Patricia L. Freeman, Auburn AL 1986. Children by previous marriage Daniel, Virginia, Nancy. Episcopalian. Hobbies: music, sci fiction. *Society Aff:* SHOT

Lewis, Willard Deming
Business: Alumni Bldg 27, Bethlehem, PA 18015
Position: President. *Employer:* Lehigh Univ - President's Office. *Education:* PhD/Physics/Harvard Univ; BA/Mathematics/Oxford Univ, England; AB/Physics/Harvard College. *Born:* Jan 1915 Augusta Ga. 7 hon degrees. Bell Tele Labs 1941-62 to Exec Dir Res-Communication Sys. A founder & Managing Dir for Sys Studies, Bellcomm Inc 1962-64. Pres Lehigh Univ 1964- . Present & former outside activities incl Fellow IEEE, AAAS; Mbr Natl Acad of Engrg, Amer Phys Soc, Sigma Xi, Phi Beta Kappa, Tau Beta Pi; Mbr NRAC 1964-71, Chmn 1967-69; Mbr DSB 1967-69; Chmn Natl Acad of Sci Comm on Useful Appli of Satellites 1967-69; Chmn Natl Acad of Engrg Comm on Power Plant Siting 1970-72; Mbr Bd of Dir Bethlehem Steel Corp, Penna Power & Light Co, Fairchild Indus Fischer-Porter Co, & Zenith Radio Corp; mbr Univ Club NYC, Saucon Valley Country Club Bethlehem Pa. *Society Aff:* IEEE, AAAS, APS.

Lewis, William H, Jr
Home: 1205 W Nancy Creek Dr, Atlanta, GA 30319
Position: President *Employer:* Measurement Systems Incorporated *Education:* BS/Met Engr/VA Poly Inst. & State Univ *Born:* Sept 1934. Native of Arlington Va. Joined Lockheed Georgia Co upon grad from VPI (BS in Met E) 1956. Except for 3 yrs in USAF as Aircraft Maintenance Officer, with Lockheed until 1986, spec in nondestructive testing. Engr Tech Services 1978-83. Chairman of NDE Task Force for Lockheed Corporation. 1981-1986. Served as Dir Engrg, GETEX Div, Lockheed, 1983-86. Became Pres and CEO of Measurement Systems Inc. in 1986. Has patented device to monitor aircraft inflight for detection of struct failure. Chmn ASNT Tech Council 1977-78. Fellow ASNT 1974. Natl Dir ASNT 1976-78. Co-Founder and Mbr, Bd of Dir, Appl Technical Services, Inc. Editor, "Prevention of Structural Failures - The Role of Fracture Mechanics, Failure Analysis, and NDT–,

Lewis, William H, Jr (Continued)
published 1978, ASM. PE, CA. Member, Cttee on Compressive Fracture, National Acad. Sci. 1981-83. *Society Aff:* ASNT, ASM, NMA, AIAA

Lewis, William M, Jr
Business: 740 Fifth St, P.O. Box 1383, Portsmouth, OH 45662
Position: Pres & Chief Engr *Employer:* W. M. Lewis & Assoc, Inc *Education:* MBA/Bus Adm/OH Univ; BS/EE/OH State Univ *Born:* 7/19/27 Native of Portsmouth, OH. From 1953-1958, start-up coordinator and general power coordinator for Goodyear Atomic Corp's gaseous diffusion plant in Piketon, OH. In 1958 formed W. M. Lewis & Assocs, a consulting engrg firm specializing in electrical transmission and distribution facilities. Since 1966 firm involved with hydroelectric projects from filing of preliminary permits with Federal Energy Regulatory Commission through project management of hydroelectric power plant and associated transmission and distribution systems. Particular interests are working with young people in various activities, travel and hard work. *Society Aff:* IEEE, NSPE, PEPP, AEE.

Lewis, Willis H
Business: P.O. Box 121467, 2146 Belcourt Ave, Nashville, TN 37212
Position: President. *Employer:* Lewis & Kuhlman Engrs Inc. *Education:* MS/San Engr/Harvard Univ; BS/Civil Engr/VA Poly Inst. *Born:* Sept 1919. Employed Tenn Dept Pub Health 1941-52; Polglaze & Basenberg Engrs 1952-60. Assoc Prof C E Dept Miss State Univ 1960-62. Cons Engr Nashville Tenn 1962- . Military Service Army San Corps 1944-47 included assignment to Inst of Inter Amer Affairs field party in Venezuela S A. Life Mbr AWWA; Chmn Ky-Tenn Sect AWWA 1969-70; Fuller Award Ky-Tenn Sect AWWA 1974. Mbr AWWA Bd of Dirs 1977-80. Mbr WPCF, NSPE, CEC; Diplomate Amer Acad Environ Engrs. Reg P E Ala, Ky, Miss & Tenn. *Society Aff:* AWWA, WPCF, NSPE, AAEE, ACEC.

Lherbier, Louis W
Home: RD 2, McDonald, PA 15057
Position: Manager- R&D. *Employer:* Cyclops Corp. *Born:* March 1934. Native of Pgh Pa. BS Met Engrg Carnegie-Mellon Univ. Joined Universal-Cyclops Specialty Steel Div, Cyclops Corp 1956 as a metallurgist. Concentrated on specialty steel R&D programs involving air & vacuum melting, hot working, & alloy dev & eval. Assumed current respon as Mgr - R&D 1974. Respon for Specialty Steel Divs process & prod dev & improvement & cost reduction programs. Chmn 2nd Internatl Symposium on Superalloys, Seven Springs Pa 1972. Chmn AIME High Temp Alloy Ctte 1973. Chmn Pgh Chap ASM 1974.

Li, Che-Yu
Business: Bard Hall, Ithaca, NY 14853
Position: Prof. *Employer:* Cornell Univ. *Education:* PhD/CE/Cornell Univ; BSE/CE/Taiwan College of Engg. *Born:* 11/15/34. Taught at Cornell Univ since 1962. Spent one yr (1965) at US Steel Res Ctr (high strength steels) and two yrs at Argonne Natl Lab (1968-1970) directing res progs on nuclear mtls. Current res on mech properties of mtls, emphasizing applications at elevated temperatures & energy and microelectronics related mtls problems. *Society Aff:* ASM, AIME, APS, ASTM.

Li, Ching-Chung
Business: Dept of Elec Engg, Pittsburgh, PA 15261
Position: Prof of Elec Engg & Comp Sci. *Employer:* Univ of Pittsburgh. *Education:* PhD/EE/Northwestern Univ; MS/EE/Northwestern Univ; BS/EE/Natl Taiwan Univ. *Born:* 3/30/32. in Changshu, Kiangsu, China. Jr Engr, Westinghouse Elec Corp, Summer 1957. With Elec Engg Faculty, Univ of Pittsburgh, since 1959; currently Prof of Elec Engg & Comp Sci. Visiting Assoc Prof, Univ of CA, Berkeley, Spring 1964. Visiting Principal Scientist, Alza Corp, CA, Summer 1980. Faculty Res Participant, Pittsburgh Energy Tech Ctr, Summer 1982, 1983, 1985. Published over 100 papers in biomedical image processing/pattern recognition, industrial applications of pattern recognition, physiological systems modelling, nonlinear/adaptive control systems. IEEE Fellow, 1978. Mbr, IEEE Comp Soc Machine Intelligence & Pattern Analysis Technical Committee, 1975-. Administrative Committee, IEEE Systems, Man & Cybernetics Soc, 1977-79; Chrmn, Biocybernetics Committee, 1972-79; Chrmn, Cybernetics Committee, 1979-. AACC Bio-Medical Engg Committee, 1978-1980. Member, Biomedical Pattern Recognition Technical Committee, International Association of Pattern Recognition, 1983- ; Chairman, 1987- . Editorial Advisory Board, Journal of Cybernetics and Information Sciences, 1976-1979. Asso Editor, Pattern Recognition, 1985- . *Society Aff:* IEEE, AAAS, PRS, BES, ΣΞ.

Li, James C M
Business: Dept of Mech Engg, University of Rochester, Rochester, NY 14627
Position: Prof. *Employer:* Univ of Rochester. *Education:* PhD/Physical Chem/Univ of WA; MS/CE/Univ of WA; BS/CE/Natl Central Univ Nanking China *Born:* 4/12/25. Nanking, China. S Vie Shao & In Shey (Mai) Li. BS Natl Central Univ Nanking 1947; MS Univ of WA 1951, PhD 1953. Res Assoc Univ of CA Berkeley 1953-55. Carnegie Inst of Tech Pgh 1955-56. Phy Chemist Westinghouse 1956-57. Scientist US Steel Corp Monroeville PA 1957-59, Sr Scientist 1959-69, Mgr Mtls Res Ctr Allied Chem Corp Morristown NJ 1969-71, A.A. Hopeman Prof. Univ. of Rochester 1971- . Visiting Prof Columbia 1964-65, Adj Prof 1965-71. Cons Allied Chem US Steel, Dow Chemical, Xerox, Ames lab, Naval Research Lab. IBM, Oak Ridge SW Res. Inst., etc. Humboldt Award, Ruhr Univ Bochum W Germany, 1978-9. Am. S. Metals (Chrmn Flow & Fracture Activity 1971-76, Chrmn Seminar Comm 1976-9, Fellow 1979, Chrmn, Mtls Sci Div 1982-4, Chairman, Joint Commission on Metallurgical Transactions, 1986, Am Inst Met Engrs (Mathewson Gold Medal 1972 R F Mehl Medal & Inst Metals Lectr 1978 Fellow 1985, Am. Phys. Soc. (Fellow 1980). *Society Aff:* AIME, ASM, APS, ASME.

Li, Ruh-Ming
Business: Engg Res Ctr, Foothills Campus, Ft Collins, CO 80523
Position: Assoc Prof and Vice President & Gen. Mgr. *Employer:* CO State Univ and Simons, Li & Assoc., Inc. *Education:* PhD/Civ Engg/CO State Univ; MS/Civ Engg/CO State Univ; BS/Hydraulic Engg/Taiwan Chen Kung Univ. *Born:* 11/7/43. Civ Engr with Taiwan Power Co 1966-1969. Assumed present position as Assoc Prof of Civ Engg, CO State Univ, 1977. Recognized as leader & expert in math modeling of watershed & river systems by governmental agencies, educational inst, privte consulting firms, other natl & intl organizations. Opened Simons, Li & Associates, Inc., an Engineering Consulting Firm in January-1980. Principal interests are: hydraulics, hydrology, watershed mgt, erosion & sedimentation, math modeling, river mechanics, system engg, non-point sources pollution control, water resources dev, & stochastic processes. *Society Aff:* ASCE, AGU, ΣΞ.

Li, Sheng S
Business: 227 Benton Hall, Gainesville, FL 32611
Position: Prof. *Employer:* Univ of FL. *Education:* PhD/Elec Engr/Rice Univ; MS/Elec Engr/Rice Univ; BS/Elec Engr/Taiwan Cheng- Kung Univ. *Born:* 12/10/38. Sheng S Li received the BSEE degree in elec engg from Taiwan Cheng Kung Univ in 1962, the MSE and PhD degrees from Rice Univ in 1966 and 1968, respectively. He joined the faculty of the Elec Engg Dept of the Univ of FL in Gainesville, FL, in 1968, as an Asst rof, and is currently Prof in the same Dept. He has spent one yr (1975-1976) at the Natl Bureau of Stds as an electronic engr. He has been a consultant to the industrial and govt labs. His interests include solar cells, photodetectors, transport and defect properties in semiconductor mtls and devices. Dr Li is a Sr mbr of IEEE, mbr of Electro-chem Soc (ECS), American Physical Soc (APS), Sigma Xi and Eta Kappa Nu. *Society Aff:* IEEE, APS, ECS.

Li, Shu Tien
Home: PO Box 8286, Rapid City, SD 57709-8286
Position: Chrmn/Bd, Li Inst of Sci & Tech *Employer:* Self - Cons Engr. *Education:* PhD/CE/Cornell Univ; Eng D/Hon/China Acad. *Born:* 2/10/00. Native of Hopei, China; naturalized USA. Exec Dir No China River Comm 1928- 37; Bd Chmn & Ch Engr Great Northern Port 1929-34. Pres natl Tangshan Engg Coll 1930-32, then

Li, Shu Tien (Continued)
Natl Peiyang Univ 1932-37; organized 7 coll's & univ's 1937-49. Deputy Pres Yellow River Comm 1943-47. Reg P E, ROC & 7 U S states. P Pres, Hydraulic Engg Soc of China; P Pres, Phi Tau Phi Hon Scholastic Soc of China & USA. Prof of CE & Exec Dir, Interdisciplinary Council for Geotech. SD Tech 1962- 70, then Prof Emer. Since 1972, he has founded Li Inst of Sci & Tech and its Graduate School, the World Open Univ. with 4 Graduate Facilities and 33 Divisions in Humanities, Science, Engineering, and Administration, all leading to the Doctorates each with 2 dissertations accepted for publication by national or international learned societies.. Fellow ASCE, AAAS, F. & Hon.M. ACI. Academic Council, Natl Acad Peiping; Academician, China Acad. Recip't, ROC 1st Class Hyd Medal, Victory Decoration, Amer Concrete Inst. Disting Service Award, etc. Author, 17 books & over 800 papers & articles. *Society Aff:* FASCE, FAAAS, FACI, IABSE, AREA, IAHR, ASTM, A.M.S.

Li, Tingye
Business: AT&T Bell Laboratories, Crawford Hill Laboratory, Holmdel, NJ 07733
Position: Dept Head. *Employer:* Bell Labs. *Education:* PhD/Elect Eng/Northwestern Univ; MS/Elect Eng/Northwestern Univ; BSc/Elect Eng/Univ of Witwatersrand *Born:* 7/7/31. Joined Bell Labs in 1957. Engaged in res in fields of microwaves, antennas, lasers & optical communications. Currently Hd of Lightwave Sys Research Dept, respon for res on subsys and sys for optical fiber communications. Has publ over 65 papers & obtained 13 pats in above fields. IEEE W R G Baker prize 1975; CIE Achievement Award 1978; IEEE David Sarnoff Award 1979; CAAPA Achievement Award 1983; Fellow IEEE; Fellow OSA; Fellow AAAS; Mbr Sigma Xi, Eta Kappa Nu, Phi Tau Phi, CIE; Elected to National Academy of Engineering 1980; Northwestern University Alumni Merit Award 1981. *Society Aff:* IEEE, OSA, AAAS, CIE/USA, CAAPA

Li, Wen-Hsiung
Business: Dept of Civ Engg, Syracuse, NY 13210
Position: Prof. *Employer:* Syracuse Univ. *Education:* PhD/CE/Univ of Mancheste; BS/CE/Chiao-Tung Univ. *Born:* 11/5/18. Native of Kwang-Tung Province, China. Served as railway engr with the Chinese Govt in 1942-45. Won by nation-wide competitive examination the Sino- British Educational Fund fellowship for grad study in Civ Engg in England. Taught at Johns Hopkins Univ in 1948-59, and at Syracuse Univ since 1959. Awarded the Eddy Medal for Noteworthy Res by FSIWA (now WPCF). Author of about 40 res journal papers, and textbooks *Engineering* Analysis and *Differential Equations of Hydraulic Transients, Dispersion and Groundwater Flow* (published by Prentice-Hall), co-author of *Principles of Fluid Mechanics* (published by Addison-Wesley), and author of *Fluid Mechs in Water-Resources* Engrg (to be published by Allyn and Bacon). *Society Aff:* ASCE.

Liao, George S
Business: 12400 E Imperial Hwy, Norwalk, CA 90650
Position: Engrg Specialist *Employer:* Bechtel Power Corp. *Education:* MS/ME/OK State Univ; BS/ME/Taiwan Univ. *Born:* Mar 1931. Participated in design & const of sev steam power proj's with Taiwan Power Co. Received 1 yr training in Germany for boiler design & power plant oper. Acting Ch of Mech & Elec Sect when left for USA 1964 for post-grad study. Upon completion of MSME, joined Arthur G McKee & Co. Worked as proj engr on an oil refinery conversion proj for Imperial Oil of Canada. Since 1968 with Bechtel Power Corp, L A Power Div. As Mech Engrg Specialist, respon for engrg of large coal-fired steam power proj's. Recipient of 1972 ASME Prime Movers Award. Enjoy photography & classical music. *Society Aff:* ASME.

Liao, Paul F
Business: 331 Newman Springs Road, Red Bank, NJ 07701
Position: Division Manager *Employer:* Bell Communications Research *Education:* Ph.D./Physics/Columbia University; M.S./Physics/Columbia University; B.S./Physics/M.I.T. *Born:* 11/10/44 Born in Philadelphia, PA, November 10, 1944. He received his B.S. degree in Physics from the Mass Inst of Technology in 1966, and a PhD degree in Physics from Columbia Univ in 1973. In 1973, Dr. Liao joined Bell Labs in Holmdel, NJ. As a member of technical staff, he conducted research on nonlinear optics, laser spectroscopy, and laser materials. In 1980, he was appointed Head of the Quantum Electronics Research Dept at Bell Labs. In 1983, he became Div Mgr, responsible for Physics and Optical Science Research at Bell Communications Research, Inc. and in 1987 assumed the position of Div Mgr of the Photonic Science and Technology Research Div. Dr. Liao has authored or co-authored more than 75 papers and has been awarded 12 patents. He serves as the Pres of the Joint Council on Quantum Electronics and as the Pres of the IEEE Lasers and Electro-optics Society. He has been an associate editor of Optics Letters and topical editor for the Journal of the Optical Society B: Optical Physics. He is co-editor of the Academic Press Series entitled Quantum Electronics. He is also program co-chairman for the 1987 Conference on Lasers and Electro-optics. Dr. Liao is a fellow of the American Physical Society, a fellow of the Optical Society of America, a fellow of the IEEE, and a member of the American Vacuum Society. *Society Aff:* IEEE, APS, OSA, AVS

Liao, Thomas T
Business: Coll of Engrg and Applied Sciences, Dept of Tech & Society, Stony Brook, NY 11794
Position: Prof *Employer:* State Univ of NY & Stony Brook *Education:* EdD/Sci Ed/Columbia Univ; MS/Physics/Adelphi Univ; BA/Physics/Bklyn Coll *Born:* 5/1/39 Started educational career as a teacher of physics and general science in NY City's public schools (1961-68). Served as Assoc Dir of Engrg Concepts Curriculum Project (1968-72). Directed NSF sponsored Socio-Tech Instructional Modules and Tech, People and Environment projects (1972-80). Founder and co- editor of Journal of Educational Tech systems (1972-present). Graduate Program Dir of Masters in Tech Systems (1978-86). Currently working on projects which deal with minorities in engrg and development of microcomputer courseware. Chairperson of Dept of Tech & Soc. *Society Aff:* ASEE, NSTA.

Libby, James R
Business: 4452 Glacier Ave, San Diego, CA 92120
Position: President. *Employer:* James R Libby and Assoc. *Education:* BSc/Civil Engg/OR State College. *Born:* Apr 1927. Conducted prestressed concrete res at U S Naval Civil Engrg Res & Eval Lab 1950-51. Engr for the Freyssinet Co 1951-56. Private engrg practice since 1956. Currently Pres of James R Libby and Assocs. Author 3 prof reference books on prestressed concrete. Co-author, with N D Perkins, of reference book on prestressed concrete hwy bridges. Fellow ACI, Fellow, ASCE. Reg P E 12 states. *Society Aff:* ASCE, ACZ, SEAOSD, CEAC.

Libby, Paul A
Business: Dept of Ames B-010, University of California, San Diego, CA 92093
Position: Prof. *Employer:* Univ of CA. *Education:* PhD/Applied Mech/Poly Inst of Brooklyn. *Born:* 9/4/21. *Society Aff:* AIAA, APS.

Libertiny, George Z
Business: One Parklane Blvd. Suite 728, Dearborn, MI 48126
Position: Principal Research Engineer *Employer:* Ford Motor Co *Education:* Ph.D./Mechanical Eng/Univ. of Bristol, Bristol, England; BS./Mechanical Eng./Univ. of Strathclyde, Glasgow, Scotland *Born:* 06/14/34 Dr. Libertiny was born in Hungary, came to the US in 1963 and became a naturalized citizen in 1974. In addition to working for Ford Motor Co as a Principal Research Engineer, Dr. Libertiny is an Adjunct Prof of Mech Engrg at the Univ of Mich, Dearborn. Dr. Libertiny is a Fellow of ASME received the R. R. Teetor Award (SAE), Forest R. McFarland Award (SAE) and the Outstanding Engineer Award (Mich Soc of Prof Engrs). He is chairman of the Education Cttee at the ASME Design Div. He contributed more than 30 articles to publications of various professional societies and he is patentee in field. He testified as an expert witness in the fields of design, stress analy-

Libertiny, George Z (Continued)
sis, testing and materials. He is listed in Marquis' Who's Who in America. *Society Aff:* ASME, SAE, ASEE, SESA, I. MECH. E, ΣΞ

Liboff, Richard L
Business: Elec Engg, Phillips Hall, Ithaca, NY 14853
Position: Prof. *Employer:* Cornell Univ. *Education:* PhD/Phys/NYU; BA/Phys-Math/Brooklyn College. *Born:* 12/30/31. Came to Cornell in 1964 where presently is Prof of Appl Phys, Elec Engg & Appl Math. In summer of 1965 was Visiting Prof at Stanford Univ and in summer of 1969 was Chrmn of the first Intl Meeting on Kinetic Theory. In 1972 was awarded a Solvay Fellowship in support of a one-yr Visiting Profship at Universite Libre de Bruxelles. In same yr was elected Fellow of American Physical Soc. In 1984 he was awarded a Fulbright Scholarship in Support of a Visiting Profship at TelAviv University. Has written over eighty scientific articles and authored books: 1) "Introduction to the Theory of Kinetic Equations" (J Wiley, 1969), translated into Russian (MNP, Moscow, 1974), second printing (Krieger, 1979). 2) "Introductory Quantum Mechanics" (Holden Day, 1980). 3) Co-editor with N Rostoker of a volume on Kinetic Theory (Gordon and Breach, 1970). 4) "Waveguides, Transmission Lines and Smith Charts" Co-author with G. C. Dalman (MacMillan, 1986) Principal investigator for ONR contract till 1974 and for AFOSR contract till 1981 and presently for ARO contract. *Society Aff:* APS, ΣΞ, AAAS, IEEE.

Libove, Charles
Business: Dept of Mech & Aero Engg, Syracuse, NY 13244
Position: Prof. *Employer:* Syracuse Univ. *Education:* PhD/Mech Engg/Syracuse Univ; MApplMech/Appl Mech/Univ VA; BCE/Civ Engg/CCNY. *Born:* 11/7/23. Educator and researcher in structures, stress analysis and applied mechanics. Author of reports and articles on corrugated plates, built-up columns, sandwich construction, swept wing stresses, elastic stability, creep buckling, microelectronic packaging and complementary energy. Twice recipient of Pi Tau Sigma (Syracuse Univ Chapter) outstanding teacher award. Recipient of NASA, NSF, and RADC res grants, NSF Postdoctoral Fellowship (1967-68), and summer profship at Pratt & Whitney, E Hartford, CT (1979 and 1980). Consultant in accident and product liability cases. Previous employers: Tri-State College (1955-58), Brush Labs Co (1953-55) Natl Advisory Committee for Aeronautics (1944-53). Husband of Rosa Greenspan Libove. Father of Joel and Fred. *Society Aff:* AIAA, ASME, ASCE, AAUP.

Libsch, Joseph F
Business: 262 Whitaker Lab 5, Bethlehem, PA 18015
Position: Vice Pres - Research. (Emeritus) *Employer:* Lehigh Univ. *Education:* ScD/Metallurgy/MIT; MS/Metallurgy/MIT; BS/Metallurgy/MIT. *Born:* 5/7/18. in Rockville, CT. Mbr fac (met) Lehigh Univ 1946-83, Hd Met Dept 1960-69, Hd Matls Res Ctr 1962-69, VP Res 1969-83 Pres 1974 & Tr 1968-70, 1973-75, Amer Soc for Metals; Mbr PA Governor's Sci Adv Comm 1972-79, & Bd PA Sci & Engg Found; indus cons; Teaching Award Amer Soc for Metals 1954; R R & E C Hillman Award (Lehigh Univ) 1965; Honorary Member ASM 1985. mbr AIME, Amer Soc for Engg Educ, ASM, Sigma Xi. Contrib to num met engg journals. Trustee Fed Mat Soc 1979-82. *Society Aff:* ASM, AIME, ASEE.

Licht, Kai
Business: 1835 Dueber Ave S W, Canton, OH 44706
Position: Mgr, Prod Acceptance, Bearing Operations *Employer:* The Timken Co. *Education:* BS/Mech/Copenhagen Mechine Tech Inst. *Born:* June 1926, Copenhagen Denmark. BS Mech Engrg Copenhagen Machine Tech Inst 1952. Addl courses at Univ of Ct, Univ of Akron, Walsh Coll & through the Amer Soc of Qual Control. With The Timken Co from 1954 as Mech Engr, Asst Ch Inspector, Ch Inspector, & Ch Qual Control Engr. Taught num statistics courses through ASQC. V P ASQC 1976-77, Fellow ASQC, Cert Qual Engr, Exec Secy - Canton Joint Engrg Council 1975-80. U S citizen 1962. Hobby: sailing.

Licht, William
Business: Dept of Chem & Nuc Engg, Cincinnati, OH 45221-0171
Position: Prof Emeritus Chem Eng *Employer:* Univ of Cincinnati. *Education:* PhD/ChE/Univ of Cincinnati; MS/ChE/Univ of Cincinnati; ChE/Univ of Cincinnati. *Born:* 9/29/15. Prof at Cincinnati 1952-85, Prof Emeritus 1985- , Hd of Dept of Chem & Met Engg 1952-1967; Visiting Prof Univ of MN 1968, 1972. Consultant in chem process design, and in air pollution control. Papers & books in purification of gases, design of particulate collection systems, fluidized-bed tech etc. Lectr for AIChE, APCA, and EPA special courses. "Engineer-of-the-Year" 1973 Cincinnati. Chrmn Air Pollution Bd, City of Cincinnati 1970-72. Cohen Award for Teaching Excellence 1981. *Society Aff:* AIChE, APCA, ΣΞ.

Lichtenstein, Abba G
Business: 17-10 Fair Lawn Ave, Fair Lawn, NJ 07410
Position: Chairman of the Board & Chief Engineer *Employer:* A G Lichtenstein & Assocs Inc. *Education:* BCE/Civil-Structural/OH State Univ; Dr Engr (Honorary)/-/Ohio State Univ. 1984 *Born:* 10/29/22. Brown Scholar, Tau Beta Pi, Sigma Xi, Disting Alumnus Award OSU 1973. ASCE Life Member 1987. ASCE NJ Civil Engr of Yr 1987 Chrmn of ASCE Subcommittee on Rehabilitation of Existing Bridges; Chrmn, ASCE Specialty Conference on Bridge Inspection and Maintenance, Atlanta, May, 1984. NJSPE-Engr of Yr 1976 (Bergen Cty NJ): Engr of Yr 1981, North Jersey Branch, ASCE. Started cons firm 1963. Chmn of Bd & Chief Engr A.G. Lichtenstein & Assocs Inc., Fair Lawn, NJ; Watertown CT, Framingham MA, and Langhorne, Pa. Principal, Lichtenstein Engrg Assoc, PE NYC; Field of expertise - Bridges, Dams, Hydraulics. Taught courses on "art" of Bridge Eval. - Specialty on Rehab of Historic Bridges. Expert witness on bridge collapses (Silver Bridge over OH River) and Sunshine Bridge, Tampa, Fl. Bridge designs nominated & won sev awards. Member of the Planning Board, Tenafly, N.J. *Society Aff:* ASCE, NSPE, ACI, PCA, IBTTA, IABSE, ARTBA, ASTM, PCI, SIA, NY Academy of Sciences, Military Engineers N.Y.

Lichtenwalner, Hart K
Business: , Waterford, NY 12188
Position: Mgr Strategic Planning *Employer:* GE Silicone Products Div *Education:* PhD/CE/Lehigh Univ; MS/CE/Lehigh Univ; BS/CE/Lafayette College. *Born:* 10/1/23. Native of Easton, PA. First ind assignment (1943) as process engr at GM Res Labs, working on high performance fuels. After grad degrees, joined Silicone Products Dept of GE in 1950, as process dev engr engaged in chlorosilane process scale-up. Subsequent moves to mgr of process dev, product dev, res & dev, intermediates mfg, & various silicone product businesses. Assumed direction of European operations in 1977. Since 1980, mgr of strategic planning and ventures, based in USA. Co-author of "Silicones" in *Encyclopedia of Polymer Science and Technology*. Licensed PE in NY. Fellow of AIChE. *Society Aff:* AIChE, ACS, NYAS, ΣΞ.

Lichter, Barry D
Business: School of Engrg, Box 16 Sta B, Nashville, TN 37235
Position: Prof of Materials Sci and Mgmt of Technology *Employer:* Vanderbilt Univ *Education:* ScD/Metallurgy/MIT 1958; SM/Metallurgy/MIT 1955; SB/Metallurgy/MIT 1953. *Born:* 11/29/31 Native of Boston. Research Staff, USAF Cambridge Research Labs, 1958-60. Staff Metallurgist, ORNL, 1960-62. Post-doctoral Fellow, UC Berkeley, 1962-64. Assoc Prof, Univ WA, 1964-68. Vanderbilt faculty since 1968. Prof of Materials Science and of Mgmt of Tech, 1972-present. Significant publications in high-temp oxidation, thermochemistry, elec props and corrosion. NSF Fellow in Science and Society, 1975-76. Major role in developing courses bridging engrg and the humanities. Fellow of the National Project on Engrg Ethics and Philosophy, 1979- 80. Visiting Scientist, Natl Bureau of Standards Corrosion Group, 1983. Mbr of the Advisory Ctte of NSF Program in Ethics and Values in Science and Engg (EVS), 1985-87. *Society Aff:* TMS/AIME, ASM, NACE, ΣΞ, ECS, NYAS, ASEE.

Lichty, William H
Business: 1700 W Third Ave, Flint, MI 48502
Position: Dir Research & Graduate Studies *Employer:* GMI Eng & Mgt Institute *Education:* MA/Education/Univ of MI; BSE/Mech Engr/Univ of MI. *Born:* Nov 1921. H S Traverse City, Mich. B.S.M.E., M.S. Univ. of Mich. Design & dev engr Cadillac Car Div. Faculty Gen Motors Inst 1949, Automotive chassis and transmission design, vehicle dynamics and performance. Head, analysis and design group 1960 also taught thermodynamics and human engrg. Prof & Chmm Mech Eng Dept 1970. Assoc Dean 1977. Dir Research & Grad Studies 1982. Chmn Metric Study Comm, Amer Soc for Eng Educ 1970-75, Chmn Sector Comm for Eng Educ, Amer Natl Metric Council 1973-76, P Chmn ASME Saginaw Valley Section, mbr Soc of Research Admin. Lives on small farm, enjoys music, golf & travel. *Society Aff:* ASME, ASEE, SRA.

Lick, Wilbert J
Business: Dept of Mechanical Engg, Santa Barbara, CA 93106
Position: Prof. *Employer:* Univ of CA. *Education:* PhD/Engg/Rensselaer Poly Inst; ME/Engg/Rensselaer Poly Inst; BA/Engg/Rensselaer Poly Inst. *Born:* 6/12/33. in Cleveland, OH. Asst Prof, Engg, Harvard Univ, 1959-66. Sr Res Fellow, CA Inst of Tech, 1966-67. Prof of Engg and Earth Sciences, Case Western Reserve Univ, 1967-79. Chrmn, Earth Sciences, 1973-76. Prof, Mech Engg, Univ of CA, 1979, Chrmn, Mech Eng 1982-84. *Society Aff:* ASME, IAGLR, AGU, SIAM.

Lidman, William G
Business: KB Alloys, Inc, PO Box 14927, Reading, PA 19612-4927
Position: Director, Product Management *Employer:* KB Alloys, Inc. *Education:* BS/ME/Univ of MI. *Born:* 11/22/21. in Rochester, NY. Was Aero Res Scientist for NASA from 1943 to 1952. At Sylcor Div of Gen Tel & Electronics from 1952 to 1960, serving as Engg Sec Hd & Hd of Mfg Engg Dept. From 1960 to 1971 was Technical Dir of Gen Astrometals Corp. Since 1971 to 1986 was with KBI, Div of Cabot Corp. As Group Mgr for Mftg and Met. Res. and Dev. and Product Mgr. From 1986 to present with KB Alloys, Inc. as Dir, Product Mgmt. *Society Aff:* ASM, RESA, AIME.

Liebenow, Wilbur R
Business: 222 East Little Canada Rd, St Paul, MN 55117
Position: VP Engrg *Employer:* Short Elliott Hendrickson, Inc *Education:* B/CE/Univ of MN *Born:* 8/6/30 Native of Planview, MN. Served with Army Corps Engrs 1954-56. Soils engr for Twin City Testing & Engr Lab. Freeway design with Ellerbe 1956-61. With Short Elliott Hendrickson since 1961. Assumed current responsibility as VP Engrg in 1977. Enjoy camping, travel and woodworking. *Society Aff:* ASCE, AWWA, CSI, ASTM

Lieberknecht, Don W
12812 Marcy St, Omaha, NE 68154 *Education:* BSc/Civil Engg/Univ of NB; -/Civil Engg/Grad Courses Univ of IL. *Born:* 3/30/34. Grad work Univ of IL. Native of Omaha, NB. Served in Navy Civil Engr Corps 1955-59; retired as Cdr from Naval Reserve. US Army Corps of Engrs as Soils engr 1959-62; cons engr firm 1962-63. With NB Testing Labs Inc Omaha as soil & fnd engr and principal 1963-85. Corps of Engineers since 1985. Reg P E NB, IA, SD, WY. Active in NSPE, NB Soc of Prof Engrs; held offices since 1967, pres of Eastern chap 1974-75. Active Mbr Amer Soc of Civil Engrs and ACEC-NB. *Society Aff:* NSPE, ASCE, SAME, ACEC, APWA.

Liebman, Jon C
Business: 3219 Newmark Lab, 208 N Romine St, Urbana, IL 61801
Position: Prof of Environ Engg. *Employer:* Univ of Illinois. *Education:* PhD/San Eng/Cornell; MS/San Engg/Cornell; BS/Civil /Univ of CO. *Born:* Sept 10, 1934. US Navy commissioned officer 1956-61. MS 1963, PhD 1965 Cornell Univ in San Engrg. Asst Prof & Assoc Prof The Johns Hopkins Univ 1965-72. Western Elec Fund Award ASEE 1968-69. Prof Univ of Ill 1972- . Assoc Hd of CE, 1976-78. Hd of CE, 1978-84. Res in appli of oper res to urban & environ problems. *Society Aff:* ASCE, ASEE, ORSA, AEEP.

Liebowitz, Harold
Business: School of Engineering & Applied Science, 725 23rd St., NW, Washington, DC 20052
Position: Dean *Employer:* George Washington University *Education:* Doctorate/Aero Engr/Poly Inst Bklyn; Master/Aero Engr/Poly Inst Bklyn; Bachelor/Aero Engr/Poly Inst Bklyn *Born:* 6/25/24 Dean and Professor, School of Engrg. & Appl Sci, 1968 to present, and Dir, NASA-George Washington Univ Joint Inst of Advanced Flight Sciences and others; Asst Dean, Grad School and Exec Dir, Engrg Experiment Station, Univ Colorado (1960-61); Research Professor, Catholic Univ (1962-68); Advanced to Head, Sructural Mechanics Branch, and Engrg Advisor (1948-68), Off of Naval Res'ch; Consult to Indust, Govt & NATO; author of over 115 technical papers & books; editor & founder of two internat'l journals: Engineering Fracture Mechanics, and Computers and Structures; Fellow, AAAS; Honorary Fellow, Soc of Engrg Sci; Fellow, Amer Acad of Mechanics, ASM, AIAA; Honorary Member, Japan Soc for Strength & Fracture of Materials; served as Pres of Society of Engrg. Sciences Washington Soc of Engrs & Acad of Mechanics. Recipient of numerous honors and awards including election to US Natnl Acad of Engrg. and was its Home Secretary; Member, Cosmos Club; Pres, Woodmont Country Club. *Society Aff:* AAAS, AIAA, NAE, AAM, ICFI, AAUP, AIME, ASEE, ASM, AST, ASUP, ACM, MTS, SES, SEM, SME, WAS, WSE, ΘΤ

Liechti, Charles A
Home: 204 Sand Hill Circle, Menlo Park, CA 94025
Position: Hd, Device Physics Dept. *Employer:* Hewlett-Packard. *Education:* PhD/EE/Swiss Federal Inst of Tech; Diploma/Physics/Swiss Federal Inst of Tech. *Born:* 3/12/37. Charles A Liechti received the PhD degree in electrical engg in 1967 from the Swiss Federal Inst of Tech, Zurich, Switzerland. He is now Head of the Device Physics Dept of Hewlett-Packard Labs.He has been responsible for the dev of GaAs field-effect transistors, integrated circuits, solid-state lasers and optical couplers. Dr Liechti received several outstanding paper awards from the Institute of Electrical and Electronic Engineers, and is a fellow member of the IEEE.In 1979, he had been the National Lecturer for the Microwave Theory and Techniques Soc. *Society Aff:* IEEE.

Liedl, Gerald L
Business: CMET Bldg, West Lafayette, IN 47907
Position: Head and Prof. *Employer:* Purdue Univ. *Education:* PhD/Materials/Purdue Univ; BSMETE/Metallurgical Engg/Purdue Univ. *Born:* 3/2/33. Native of Fergus Falls, MN. On staff of Purdue Univ since 1958. Assumed current position in 1978. Research includes structure-properties relation of materials, x-ray diffraction and electron microscopy. Consultant on applications of x-ray and electron microscopy and failure analysis. Organized and directed x- ray and microstructural analysis facilities. Enjoy horticulture and sports. *Society Aff:* AIME, ASM, ASEE, MRS.

Lien, Jesse R
Business: One Stamford Forum, Stamford, CT 06904
Position: Pres, VP Engrg *Employer:* GTE Labs, GTE Service Corp *Education:* M/Physics/Boston Univ; B/Physics/Reed Coll *Born:* 11/15/17 Currently VP - Engrg and Pres of GTE Labs. Overall responsibility for coordination of research and development programs for GTE. Joined GTE Sylvania as an engrg section head in 1953 at its Electronic Defense Labs in CA, became its Dir in 1957 and VP & Gen Mgr of the Western Div in 1961. Elected Sr VP and General Mgr of the Electronic Systems Group in 1969 and to my present positions in 1977. Prior to joining GTE, I was with the Air Force Cambridge Research Center and a member of the technical staff at the Radiation Lab of MIT. *Society Aff:* AAAS, AIAA, IEEE

Lienhard, John H
Home: 3719 Durhill, Houston, TX 77025
Position: Prof of Mech Engr. *Employer:* Univ of Houston *Education:* PhD/Mech

Lienhard, John H (Continued)
Engr/Univ of CA; MS/Mech Engr/Univ of WA; BS/Mech Engr/OR State College. *Born:* 8/17/30. in St Paul MN and educated on the West Coast. Presently Prof of Mech Engg at the Univ of Houston and formerly on the faculty of Univ. of KY, WA State Univ, the Univ of CA at Berkeley, and the Univ of WA. Wrote textbooks in statistical thermodynamics and heat transfer, and about 120 technical articles. Maj contributor to the hydrodynamic theory of boiling heat transfer. Fellow of ASME (1978), winner of the Charles Russ Richards Memorial Award in 1979, and of the Heat Transfer Memorial Award in 1980. Performer of vocal music. *Society Aff:* AAAS, ASME, ASEE, SHOT, ΣΞ, ΦΚΦ.

Liggett, James A
Business: Hollister Hall, Ithaca, NY 14853
Position: Prof. *Employer:* Cornell Univ. *Education:* PhD/Civ Engg/Stanford Univ; MS/Civ Engg/Stanford Univ; BS/Civ Engr/TX Tech. *Born:* 6/29/34. After receiving a PhD from Stanford in 1959, worked for Chance Vought Aircraft in Dallas, TX, and taught at the Univ of WI, Madison, before joining Cornell in 1961. Was a res scientist at the Natl Ctr for Atmospheric Res in Boulder, CO (1967-1968). Worked in Cali., Colombia, at Universidad del Valle and Corporacion Autonoma del Valle del Cauca (1974-1975) sponsored by NSF and Fulbright. Held Erskine Fellowship to Univ of Canterbury, New Zealand (1977). Res in computational hydraulics, open channel flow, groundwater, lake circulation, numerical methods. Commercial pilot. Visiting Prof., Univ of New South Wales, 1981. Visiting Distinguished Scholar, Univ. of Adelaide, 1982. *Society Aff:* ASCE, ΣΞ, IAHR, SES.

Light, Frederick H
Home: 15 Hemlock Rd, Lansdowne, PA 19050
Position: Retired *Born:* Aug 1936, MSME Penn State Univ. Native of Lebanon, Pa. Employed by Phila Elec Co 1936. Early exper as engr in hydro-elec & steam generating power plants spec in equip testing & plant econ. Advanced to Station Economy Div, spec in oper planning, analysis of plant performance, & coord of oper with interconnected utils, & became Supt 1968. Fellow ASME; Sr Mbr IEEE. P E Penna 11619.

Lightner, Max Wm
Home: 80 Lebanon Hills Dr, Pittsburgh, PA 15228
Position: Admin V P - Res & Tech - ret. *Employer:* U S Steel Co. *Education:* BS/-/Penn State; MS/-/Carnegie Tech. *Born:* Jan 1908. Res engr Carnegie Tech 1930-33; Asst Ch Metallurgist Homestead Works Carnegie-Ill Steel Corp 1933-37, Ch Metallurgist 1937-40, Asst to Genl Supt 1940-42; V P Oper Heppenstall-Eddystone Corp 1942-44; Asst Mgr R&D Carnegie-Ill Steel Corp 1944-45, Mgr 1945-51; Mgr R&D U S Steel Corp 1951-54, Asst V P Res & Tech 1954-56, V P Appl Res 1956-58, Admin V P Res & Tech 1968-72. McFarland Award Penn State Chap ASM 1950; Regional Mtg Paper Award AISI 1957; Disting Alumnus Award Penn State 1964; Fellow ASM 1971; Fellow Met Soc AIME 1972; Eisenman Award ASM 1973. *Society Aff:* ASM, AISI, AIME, AIC.

Lightowler, Joseph, Jr
Business: 17 S 7th St-Box 2464, Fargo, ND 58102
Position: Chrmn of Bd *Employer:* Lightowler Johnson Associates *Education:* BME/Steam Power/Univ of MN. *Born:* March 1916. Minneapolis native. Employed 1937-46 Babcock & Wilcox, Barberton Ohio; 1946-53 Pfeifer & Shultz, Engrs, Minneapolis; 1954-67 Johnson & Lightowler Fargo N D; 1967- , Koehnlein Lightowler Johnson Inc, Fargo N D, Pres since organized Lightowler Johnson Associates, Bd Chrmn. P Pres CEC/ND. Mbr NSPE, ASHRAE. Activities confined mainly to steam generation, air pollution control & genl cons work. Cons to state agencies for phys plants since 1948. Cons on design of lignite & low grade fuel firing. Hobbies: charity work, golf, fishing. *Society Aff:* NSPE, ASHRAE, CEC/ND.

Ligomenides, Panos A
Business: Electrical Engrg Dept, University of Maryland, College Park, MD 20742
Position: Full Prof *Employer:* Univ of MD *Education:* PhD/EE/Stanford Univ; MSc/EE/Stanford Univ; MSc/Rad Engrg/Univ of Athens- Greece; BSc/Physics/Univ of Athens-Greece *Born:* 4/3/28 Native of Pireus, Greece. Served with Greek Navy 1952-54. Engr with Greek Telephone Co on Short and Micro-wave telephony. Research Engr with IBM, 1958-64. Consultant to IBM. Control Data Corp and other major Companies and Government Agencies, 1964-now. Prof at UCLA (1964-69), Stanford Univ (1969-70) and Univ of MD (1971-present). Fulbright Prof 1970-71. Alumni Distinguished Visiting Prof at EE Univ of MD, 1971-72. Outstanding Educator of America, 1973. Ford Found Fellow. OECD Fellow. Visiting Prof in USA, Europe and South America. Pres of Computer & Cybernetic Engrg Consultants, a co specializing in microcomputer- based and Cybernetic system R&D. Senior Member IEEE, 1970. Enjoy music and tennis. *Society Aff:* IEEE, AAAS, ΣΞ, NAFIPS

Ligon, Claude M
Home: 9560 Highwind Ct, Columbia, MD 21045
Position: Commissioner *Employer:* Maryland Public Service Commission *Education:* PhD/CE/Univ of MD; MS/CE/Univ of MD; BS/CE/Univ of IL; BS/Math/Morgan State Univ *Born:* 6/28/35 Claude M. Ligon, Lt Col (USA, Retired) has over 23 years of Civil Engrg and Transportation Planning experience with the US Army Corps of Engrs. In recent years he has been the Mgr of the Transportation and Civil Engrg Systems Div for AMAF Industries, Inc, a research, development, and engrg firm with corp Headquarters in Columbia, MD. Currently he is serving as a Commissioner on the Maryland Public Service Commission. He is registered PE in the State of MD, the District of Columbia, and the Commonwealth of VA, and has been granted Fellow status in the Inst of Transportation Engrs. Both with the Corps of Engrs and in private industry, Colonel Ligon has managed extensive civil engrg and transportation projects including elements of the entire project Life Cycle. *Society Aff:* ITE, ASCE, TRB

Likins, Peter W
Business: 205 Low, New York, NY 10027
Position: Provost of the University *Employer:* Columbia Univ. *Education:* PhD/Engg Mechanics/Stanford; SM/Civil Engg/MIT; BS/Civil Engg/Stanford. *Born:* 7/4/36. in Tracy, CA; educated at Stanford and MIT; employed as an engr, engg prof, and academic adminstrator successively at the Caltech Jet Propulsion Lab, UCLA, and Columbia; active as a technical consultant to more than a score of aerospace organizations in the USA and Europe; and author of texts and numerous res journal publications. *Society Aff:* AIAA, ASEE.

Likwartz, Don J
Business: Suite 1100, 717 N. Harwood, Dallas, TX 75201
Position: VP Operations *Employer:* Evergreen Oil Corp *Education:* MS/Petro Engr/Univ of WY; BS/Petro Engr/Univ of WY *Born:* 12/05/40 Native of Rock Springs, WY. Employed by Amoco (Standard of Indiana) for 13 years in staff and supervisory engrg positions both in the US and internationally. Specialized in offshore drilling and production operations. Joined Sedco-Hamilton Production Services in 1978 as VP-Project Development. Directed all technical and support staff in design, fabrication and installation of floating production facilities. In 1980 assumed current position as VP- Operations for Evergreen Oil Corp, an independent oil and gas exploration and production co. Mbr, SPE Reprint Committee, 1979-present. VP, Chicago SPE Chapter 1976. Enjoys music, hunting and fishing. *Society Aff:* AIME

Liley, Peter E
Home: 3608 Mulberry Dr, Lafayette, IN 47905
Position: Prof. *Employer:* Purdue Univ. *Education:* PhD/Phys/Imperial College; DIC/CE/Imperial College; BSc/Phys/Imperial College. *Born:* 4/22/27. Born in Barnstaple, N Devon, England, 1927. Served with Royal Corps of Signals, 1945-48. Studied undergrad Phys & did grad work specializing in thermodynamics in CE, both at Imperial College, London, 1948-55. CE, British Oxygen Engg, London, 1955-57. Emigrated to USA 1957. Currently is Prof of ME and Sr Researcher, Ctr for Info & Numerical Data Analysis & Synthesis, Purdue Univ. Author sec 3, Physical & Chem Data, Perry's Chem Engrs Handbook & Engg Manual, McGraw-Hill, NY, etc, Consultant in thermodynamics. *Society Aff:* InstPhys.

Lillard, David H
Business: 1500 Meadow Lake Pkwy, Kansas City, MO 64114
Position: Partner/Personnel Adm *Employer:* Black & Veatch Consulting Engrs *Education:* BS/CE/Univ of MO-Columbia *Born:* 7/31/30 Longtime resident of Prairie Village, KS. Industrial packaging and materials handling engr for Firestone Steel Products Co, 1952-54. With Black & Veatch since 1954. Partner since 1971. Structural design engr and Specification coordinator for Power Division; Office Engr and Personnel Administrator since 1968. Responsible for personnel policies, coll relations, employment, development, compensation, and employee relations. Past society offices include VP, NSPE; Chrmn, NSPE/PEPP; Chrmn, Committee on Federal Procurement of Architectural/Engrg Services; Pres, Consulting Engrs Council of MO. Enjoy golf and jogging. *Society Aff:* ASCE, NSPE, ASEE, ACEC, MSPE, PEPP.

Lilley, David G
Business: Schl of Mech and Aerospace Engrg, Stillwater, OK 74078
Position: Prof *Employer:* Oklahoma State Univ *Education:* PhD/Chem Engrg/Sheffield Univ, England; MSc/Math/Sheffield Univ, England; BSc/Math/Sheffield Univ, England *Born:* 11/01/44 Born in England in 1944, obtained PhD from Sheffield Univ in 1970, currently Prof at Oklahoma State Univ. Previous academic career includes Sheffield Polytechnic, Cranfield Inst of Tech, Univ of Arizona, and Concordia Univ. res interests in swirling flows. Pub over 60 reviewed res papers. Co- authored three textbooks. Co-editor of Abacus Press Energy and Engrg Science Series of textbooks. Registered Prof Engr in Oklahoma. Chartered Engr in Britain. *Society Aff:* ASME, AIAA, CI, ASEE.

Lilley, Eric G
Business: P O Box 482, Fort Worth, TX 76101
Position: Deputy Dir, Logistics & Sys Engrg, JVX Program *Employer:* Bell Helicopter Textron. *Education:* BS/Mech Engg/Southampton Tech College. *Born:* 2/3/34. Born in England; naturalized US citizen 1972. Grad in Mech engg Southampton Tech Coll. Reg Chartered Engr - Brit Council of Engg Insts. Assoc Fellow AIAA; Assoc Fellow Royal Aeronautical Soc. Performed duties of VP of Publications & Editor of Journal for SOC allied Wt Engg 1976-79. Presently serving as Regional, VP and Mbr of Natl Board of Dirs of Amer Helicopter Soc. Fixed-wing pilot - Royal Air Force. Engg exper spans 25 yrs incl Proj Engg, Mfg, Struct Analysis, Cost Engg & Mass Properties Engg. Assumed current assignment with Bell Helicopter Textron Aug 1973 *Society Aff:* AMA, AHS, AIAA, RAES, SAWE.

Lillibridge, John L
Home: 19168 Bob-O-Link Dr, Hialeah, FL 33015
Position: Retired *Education:* BS/Mech Engg/OK A&M; BS/Civil Engg/TX A&M. *Born:* Nov 1924. Native of Dover Okla. Served as enlisted man US Army 1943-46. Rejoined Army in Corps of Engrs 1950. Var troop commands, engrg & staff assignments in Korea, Germany, Vietnam, Hawaii. Grad US Army War College. 1970- 73 area engr for const of first anti-ballistic missile site (Safeguard) Langdon N D. Retired as Colonel 1973. Eastern Airlines Vice President 1973-1986. Wheeler medal SAME 1973. Retired 1986. *Society Aff:* ASCE, SAME.

Lilly, Edmund D
Business: 4433 Bissonnet 214D, Bellaire, TX 77401
Position: Pres. *Employer:* Lunar Engg. *Education:* BSChE/Chem Engg/Univ of AR; BS/Elec Engg/Univ of AR. *Born:* 11/8/24. Reg PE, TX. Married, three children. Native of AR, graduated from the Univ of AR after two yr service break in Navy. Prof career has included: Plant Process Engg and Economic Evaluations; Plant Operation; Teaching at Univ level as well as operator training programs; Tech writing for training courses, startup and operating manuals; Res and Dev; Major field is Plant Design and Construction having served as Project Engg, Project Manager and Manager of Engineering for several E & C companies. *Society Aff:* AICRE.

Lilly, Gary T
Home: P.O. Box 436, Daniels, WV 25832
Position: Pres *Employer:* Lively Engineering, Inc. *Education:* BS/CE/WV Inst of Tech *Born:* 9/27/41 Born and raised in Beckley, WV. With Lively Mfg & Eqpt Co, Glen White, WV, a designer and builder of coal processing facilities from 1965 until 1974. Last position held was chief engr. Formed G. T. Lilly Engrg, Inc in 1974, performed consulting and design services for coal companies and contractors in the coal industry. Now am pres of Lively Engineering, Inc., also co-owner, designer & builder of coal processing facilities. Am responsible for design, procurement of equipment and materials, and sales. Past Pres of Appalachian Chapter and Past State Dir of WVSPE. Hobbies include golf, fishing and snow skiing. *Society Aff:* NSPE, ASCE, AIME

Lim, David P
Home: 7761 E Adams, Tucson, AZ 85715
Position: Zoning & Subdivision Examiner. *Employer:* City of Tucson. *Education:* BS/CE/Univ of AZ; BS/Agri/Univ of AZ. *Born:* Jan 29, 1923 Tucson Ariz. Attended all local schools through high school; Univ of Arizona: BS Agri 1951; BSCE 1960. Taught City Planning for Civil Engrs at the Univ of Ariz. Entered the employ of the City of Tucson as a Planning Analyst in Feb 1960; became Planning Dir Apr 1, 1971; Dir of Community Dev Apr 10, 1972; Jan 20, 1975 appointed as the first Zoning & Subdivision Examiner for Tucson & presently in that position. Pres Ariz Chap Amer Soc of Civil Engrs 1973-74. V P Ariz Chap ASCE 197273. Secy Ariz Chap ASCE 1968-69. *Society Aff:* ASCE.

Lim, Henry C
Business: Biochemical Engg, University of California, Irvine, CA 92717
Position: Prof and Chmn, Biochem Engg *Employer:* Univ of CA, Irvine, CA *Education:* PhD/Chem Eng/Northwestern Univ; MSE/Chem Eng/Univ of MI; BS/Chem Eng/OK State Univ *Born:* 10/24/35. Native of Seoul, Korea. Naturalized US citizen, 1969. Three children, David, Carol and Michael. Process dev engr at Pfizer, Inc Groton, CT 1959-63 specializing in fine organic chemicals, fermentation product separation and purification, and antibiotics. Asst prof, 1966-70, Assoc prof 1970-74, and Prof 1974-87 of chem engg, Sch of Chem Engg, Purdue Univ, W Lafayette, IN. 1987- Prof & Chmn, biochem engg, Univ of Calif, Irvine, CA. Author of numerous res papers and a book "Biological Wastewater Treatment: Theory and Application," 1980. Current res interests in biological reactor engg; modeling, optimization and control of biological reactors, recombinant cell kinetics, and engineering of recombinant cell reactors. *Society Aff:* ACS, AIChE, ASM

Limb, John O
Business: 600 Mtn Ave, Murray Hill, NJ 07974
Position: Dept Hd. *Employer:* Bell Labs. *Education:* PhD/EE/Univ of W Australia; BE/EE/Univ of W Australia. *Born:* Born in Western Australia. Worked at the res lab of the Australian Post Office, 1966-1967, & joined Bell Labs, Holmdel, NJ, 1967. Investigated the efficient coding of picture signals to reduce channel capacity requirements, publishing many papers & patents in the area. Developed models of vision to describe the resolution capability of the eye. Currently hds the Telecommunications Services Res Dept at Bell Labs, Murray Hill, NJ, studying the application of telecommunications services in the office environment. Fellow IEEE, L G Abraham Prize Paper Award 1973. *Society Aff:* IEEE, ARVO, OSA, ACM.

Limpe, Anthony T
Business: Stemton Group, Inc, New York, NY 10017
Position: President & Ch Exec Officer. *Employer:* Stemton Group, Inc. *Education:* MSME/ME/Brooklyn Polytechnic; BS/ME/Purdue Univ. *Born:* 7/18/34. Spec in heat transfer. Joined Combustion Engrg as Design Engr after 18 mos of Cadet Training. Last pos with Combustion Engrg as Tech Editor. With Petro- Chem Dev Co Inc from 1962, to 1980; Became V P/Asst Genl Mgr 1969. Pres & Ch Exec Officer since

Limpe, Anthony T (Continued)
1970. Since August, 1980 became President & Chief Executive Officer of Stemton Group, a consulting management organization. Author sev tech papers on heat transfer & process furnace designs. Mbr API, AIChE, ASME, NSPE, Sales Exec Club of N Y, The President's Assn. Wife - Emily. Son - Stephen. Enjoy tennis & skiing. *Society Aff:* API, NSPE, AIChE, ASME.

Lin, Cheng S
Home: 1611 The Strand Ave, San Jose, CA 95120
Position: Sr Engr. *Employer:* Genl Electric Co. *Education:* PhD/Struct Engg/Univ of CA, Berkeley; MS/Civil Engg/Lehigh Univ; BS/Civil Engg/Natl Taiwan Univ. *Born:* Nov 1941. PhD Univ of Calif Berkeley; MS Lehigh Univ; BS Natl Taiwan Univ. Reg P E Calif. Assoc Mbr ASCE, winner Moisseiff Award 1976 for paper 'Nonlinear Analysis of RC Shells of Genl Form' - Journal of the Struct Div March 1975. Engr for Taiwan Hwy Bureau, engaged in the design of reinforced concrete bridges. With Bechtel Power Corp L A Div, involved in the dynamic analysis & design of nuclear power plants. Currently Sr Engr with Nuclear Energy Sys Div, G E, spec in seismic & dynamic analysis. *Society Aff:* ASCE.

Lin, Hung C
Business: Elec Engrg Dept, College Park, MD 20742
Position: Professor. *Employer:* Univ of Maryland. *Education:* DEE/EE/Polytech Inst of Brooklyn; MSE/EE/Univ of MI; BSEE/EE/Chiaotung Univ. *Born:* 8/8/19. Central Radio Works & Central Broadcasting Admin China 1941-46. RCA Labs 1948-56. CBS Hytron Div, Mgr of the Semiconductor Appli Lab 1956-59. Westinghouse Elec Corp, Mgr of Advanced Dev at Molecular Electronics Div 1959- 69. Presently Prof at the Univ of Md. 37 U S patents, author of the book 'Integrated Electronics' & some 70 tech articles. IEEE Fellow 'for contrib to semiconductor electronics & circuits & pioneering of integrated circuits'. Mbr Sigma Xi & Phi Tau Phi. Westinghouse Patent Award. Chinese Inst of Engrs Achieve Award, IEEE Ebers Award. *Society Aff:* IEEE, ΣΞ.

Lin, Pen-Min
Business: Schl of Elec Engrg, West Lafayette, IN 47907
Position: Prof *Employer:* Purdue Univ *Education:* PhD/EE/Purdue Univ; MS/EE/NC State Univ; BS/EE/Taiwan Univ *Born:* 10/17/28 in China. Came to US 54. Married Louise Lee 62. Naturalized 71. Three daughters. With Purdue Univ since 56. Teaching and research in circuit theory and applications of graph theory. Coauthor of the book *Computer-aided Analysis of Electronic Circuits*, published by Prentice-Hall 75, translated into Russian 80, Polish 81, and Chinese 83. Did pioneering work on symbolic network analysis. Assoc editor of IEEE Transactions on Circuits and Systems 71 to 73. Elected IEEE Fellow 81. Enjoy listening to classical music and playing Chinese violin in Peking opera. *Society Aff:* IEEE

Lin, Ping-Wha
Home: 506 S Darling St, Angola, IN 46703
Position: Prof. *Employer:* Tri-State Univ. *Education:* PhD/Env Engg/Purdue Univ; MS/Env Engg/Purdue Univ; BS/Civ Engg/Chiao-Tung Univ. *Born:* 7/11/25. Born in Canton, China. Married, wife, Sylvia Y.C. MAK, son, Karl, daughter, Karen. Appointments include: Consulting Engr, Lockwood Greene Engrs, Inc, Ammann & Whitney, John Graham & Co, etc, NY, 1951-59, 1960-61; Environmental Engr, World Health Organization, 1959-60, 1962-66; Proj Mgr 1979-1981; Prof Tri-State Univ, Angola, IN, 1966-79. 1981 Contributor to number of profl journals. Fourteen patents accredited to him. His invention "A Sulfur Dioxide Removal and Waste Products Reclamation Process" has passed the evaluation by the Natl Bureau of Stds under the Fed Nonnuclear Energy & Res & Dev Act. In 1982, he was awarded an individual grant of $130,000 by the Dept of Energy for pilot plant dev of the process. It has been proved that the process has nearly 100% 50 x removal efficiency and its by-product can be used for construction material and for wastewater treatment. Pres, NE Ind Br of ASCE, 1974; Pres, Tri-State Univ, Club, Sigma Xi. Hobby: Travel, music. *Society Aff:* ASCE, AWWA, ΣΞ, ACS

Lin, Shu
Business: Dept of Electrical Engrg, Honolulu, HI 96822
Position: Prof *Employer:* Univ of HI *Education:* PhD/EE/Rice Univ; MS/EE/Rice Univ; BS/EE/Natl Taiwan Univ *Born:* 5/20/36 in Nanking, China. From 1959 to 1961 he served on active duty as a Radar Officer in the Chinese Air Force in Taiwan. Since 1965 he has been on the Faculty of the Univ of HI, Honolulu, where he is now a Prof of Electrical Engrg. He spent the academic year 1978-1979 as a Visiting Scientist at the IBM Thomas J. Watson Research Center, Yorktown Heights, NY and the academic year 1982-1983 as a prof at TX A&M Univ, Coll Station, TX. He is the author of *An Introduction to Error-Correcting Codes* (Prentice-Hall, 1970, Englewood Cliffs, NJ), co- author of Error Control Coding: Fundamentals and Applications (Prentice-Hall, 1983 Englewood Cliffs, NJ), and has published more than one hundred technical papers in various professional journals. Dr Lin is Fellow of IEEE and a member of the Bd of Govers of the IEEE Information Theory Group. *Society Aff:* IEEE, ΣΞ

Lin, Tung H
Business: 4531 Boelter Hall, Los Angeles, CA 90024
Position: Professor. *Employer:* Univ of California. *Education:* DSc/Engg Mech/Univ of MI; SM/Aeronautical Engg/MIT; BS/Civil Engg/Chiaotung Univ China. *Born:* May 26, 1911 Chungking China. Awarded a natl fellowship to study in U S by Tsing Hwa Univ China 1933. Taught Aeronautical engrg at Univ of Detroit 1949-55. Since then, have been teaching as a Prof of Civil Engrg at the Univ of Calif L A. Fellow Amer Soc of Mech Engrs & Amer Acad of Mech. Author book 'Theory of Inelastic Structs' publ by Wiley 1968 and more than a hundred scientific and technical articles. *Society Aff:* ASME, AAM, ASCE.

Lin, Tung Yen
Business: 315 Bay St, San Francisco, CA 94133
Position: Bd Chmn, T Y Lin Internatl. *Employer:* T Y Lin Intnl. *Education:* LLD/-/Chinese Univ of Hong Kong; MS/CE/Univ of CA, Berkeley; BS/CE/Chiaotung Univ. *Born:* Nov 14, 1911 Foochow China. s. Ting Chang & Feng Yi (Kuo) L. BSCE Tangshan Coll, Chiaotung Univ 1931; MS Univ of Calif Berkeley 1933; LLD Chinese Univ Hong Kong 1972, Golden Gate Univ 1982, and others. Naturalized 1951. Ch Bridge Engr, Ch Design Engr Chinese Gov Rys 1933-46; Asst, then Assoc Prof Univ of Calif 1946- 55, Prof 1955- (Berkeley Citation Award 1976), Chmn Div Struct Engrg 1960-63, Dir Struct Lab 1960-63. Chmn of Bd T Y Lin Internatl, cons engrs 1953- , InterContinental Peace Bridge Inc 1968- . Cons to State of Calif Def Dept; also to indus. Chmn World Conf Prestressed Concrete 1957, Western Conf Prestressed Concrete Bldgs 1960. Hon. Mbr ASCE (Wellington Award, Howard Medal); Mbr NAE, (Bldg Res Council Quarter Century Citation) Academia Sinica, Internatl Fed Prestressing (Freyssinet Medal), ACI, (Hon Mbr), PCI (Medal of Hon). AIA Inst Honor; Pres's Natl Medal of Sci, 1986; ACEC Award of Merit 1987. *Society Aff:* ASCE, ACI, PCI.

Lin, Wen C
Home: 1422 Marina Circle, Davis, CA 95616
Position: Prof. *Employer:* Univ of CA. *Education:* PhD/Elec Engr/Purdue Univ; MS/Elec Engr/Purdue Univ; BS/Elec Engr/Natl Taiwan Univ. *Born:* 2/22/26. From 1950 to 1954 he was an Engr with the Instrumentation Lab, Taiwan Power Co. From 1956 to 1961 he was an Engr with the Gen Elec Co and Sr Engr with Honeywell Co, Elec Data Processing Div. From 1965-1978 he was an Asst, Assoc and Full Prof at Case Western Reserve Univ, Cleveland, OH. Since 1978 to present he is a Prof in the Dept of Elec and Computer Engg, Univ of CA at Davis. He has edited a book and authored two books as well as published over 60 technical papers in Microcomputers, digital system, pattern recognition and signal processings. *Society Aff:* IEEE.

Linaweaver, F Pierce
Business: EA Engg, Inc, Hunt Valley/Loveton Center, 15 Loveton Circle, Sparks, MD 21152
Position: Pres *Employer:* EA Engineering, Inc. *Education:* PhD/San Engrg & Water Res/Johns Hopkins Univ; BES/CE/Johns Hopkins Univ *Born:* 8/22/34 Pres, EA Engg, Inc since Feb. 1987. Partner, Rummel, Klepper & Kahl Consulting Engrs 1978-1987. Individual Consulting Engr 1974-1978. Dir Public Works, Baltimore City 1969-1974. Assoc Prof Environmental Engr, Johns Hopkins Univ, 1967-1968. White House Fellow, 1966-1967. Other: Research Assoc, Johns Hopkins Univ; Sanitary Engr, US Air Force; Civil Engr, Baltimore City; Registered PE, MD, 1963. Currently: Trustee, Johns Hopkins Univ; Dir, T. Rowe Price Mutual Funds. Former: Senior Warden, Church of Redeemer; Pres, Johns Hopkins Alumni Assoc; Trustee, Chesapeake Research Consortium; Member, Baltimore Friendship Airport Authority; Pres's Science Urban Advisory Committee, Pres's Air Quality Advisory Bd, HEW. Authored numerous technical papers on water use. *Society Aff:* ASCE, AAEE, NSPE, AWWA, WPCF, AAAS, ACEC, APWA, TBΠ, ΣΞ

Lindahl, Harry V
Business: Ashland City, TN 37015
Position: Pres. *Employer:* State Industries, Inc. *Education:* BS/Arch Engg/Univ of IL. *Born:* 8/17/26. Native of Nashville, TN. Have practiced engg and arch (primarily structural, industrial and construction) in the Southeastern US, Nevada and the Caribbean for 30 yrs. In addition to own private practice, became affiliated with State Ind in 1975, with respon for design & construction of their multi- plant expansion program. Concieved, designed and built State Ind waste water polution control facility in 1976-77. Currently designing their energy producing solid waste incineration program, for construction & implementation in 1979-80. *Society Aff:* NSPE, AIA.

Lindauer, George C
Business: Speed Scientific Sch, Louisville, KY 40208
Position: Prof. *Employer:* Univ of Louisville. *Education:* PhD/ME/Univ of Pittsburgh; ScM/Nuclear Engg/MIT; MLS/Library Sci/Long Island Univ; BS/ChE/Cooper Union. *Born:* 11/5/35. Raised in Richmond Hill, NY. Sr engr for Bettis Atomic Power Lab designing reactor fuel elements from 1957-1964. Res engr for Brookhaven Natl Lab specializing in fluidized bed heat transfer from 1964-1971. Prof of Nuclear Engg and engg librarian of the Univ of Louisville from 1971-1982. Prof of Mech Engg of Univ of Louisville from 1982 to present. Married in 1959. Three children. Leisure activities include gardening, do it yourself, classical music. *Society Aff:* ASME, ANS.

Lindemer, Terrence B
Business: P O Box X-Bldg 4501, Oak Ridge, TN 37831-6221
Position: Research Staff Member. *Employer:* Oak Ridge Natl Lab. *Education:* PhD/Met Eng/Univ of FL; BS/Met Eng/Purdue Univ. *Born:* Feb 17, 1936. m. Suzanne Teagle; 2 children. Worked in Res Depts of Inland Steel Co (blast furnace oper) & Solar Aircraft Co (diffusion bonding of refractory metals). At ORNL since 1966, presently group leader,thermodynamics, in Chem Tech Div. Investigate phase equilibria & reaction kinetics at 1000-2500 K in the Th-U-Pu-C-O-N and Si-C-O-N systems for application to engineering systems. 80 pubs, 4 pats. Mbr Tau Beta Pi & Sigma Xi. Fellow Amer Ceramic Soc. *Society Aff:* ACerSOC, TBΠ, ΣΞ, MS

Linden, Henry R
Business: Suite 830 South, 8600 W Bryn Mawr Ave, Chicago, IL 60631
Position: Executive Advisor; Frank W. Gunsaulus Distinguished Prof of Chemical Engrg. *Employer:* Gas Res Inst. Illinois Institute of Technology. *Education:* PhD/ChE/IL Inst of Tech; MChE/ChE/Poly Inst of Brooklyn; BS/ChE/GA Sch of Tech. *Born:* 2/21/22. From 1977 until retirement in 1987, Pres and mbr of Bd of Dirs of Gas Res Inst, which he helped organize. Now serves as Executive Advisor. Formerly served the Inst of Gas Tech in various mgt capacities ending as Pres and Trustee. Also, organized IGT's wholly-owned subsidiary Gas Devels Corp (now GDC, Inc) and was chief exec officer until 1978. Since 1954, he has also been a mbr of the faculty of IL Inst of Tech and currently is the Frank W. Gunsaulus Distinguished Professor of Chemical Engineering. He is a mbr of the bd of Dirs of Sonat Inc, Southern Natural Gas Co, Reynolds Metals Co, UGI Corp, Applied Energy Systems Inc, Larimer and Co, and Resources for the Future, Inc and is a mbr of the Energy Engineering Board of the National Research Council and the Energy Res Advisory Bd of the Dept of Energy. Linden began his work in the energy field with Mobil Oil Corp. Active in numerous profl and trade organizations. During this and the four preceding Admin, he has served on many gov advisory bodies. He has written and lectured extensively in the field of US and world energy problems, received numerous awards for his tech and analytical work in fossil fuel area, and has more than 200 publ and pats to his credit. *Society Aff:* NAE, AIChE, ACS, ΣΞ, IE

Lindenmeyer, Carl R
Business: Coll of Engrg, Dept of Ind Engrg, Clemson, SC 29631
Position: Prof *Employer:* Clemson Univ. *Education:* MS/Tech/Western MI Univ; BS/Ind Engg/Northwestern Univ. *Born:* 3/31/37. and raised in IL. Worked as Industrial Engr for Teletype Corp and was VP of Mfg at TASA Wire Corp. Has held professional positions in engg related higher education since 1961, including Mech Engg at Univ of NB and in Industrial Engg at Western MI Univ. Presently Prof of Indust Engrg at Clemson Univ. Sr Mbr of the Inst of Industrial Engrs and 1979-80 Pres, Greenville-Spartanburg Chapter. Researcher and Conslt in methods engrg and work measurement, ergonomics, statistical quality control, facilities planning and design, and microcomputers for indus engrg application. Pres and Prin Consultant, C R Lindenmeyer and Assocs, consultants to manufacturing and engrg management. *Society Aff:* IIE, ASEE, HFS, ATIE, MTM, ΣΞ, AΠM.

Linder, Clarence H
Home: 1334 Ruffner Rd, Schenectady, NY 12309
Position: Retired. *Employer:* VP G.E. Co *Education:* MS/EE/Univ of TX; BS/EE/Univ of TX. *Born:* 1/18/03. Assoc with Gen Elec began in 1924. Gen Mgr Major Appliance Div 1951, VP 1952, VP Engrg Services 1953, VP and Grp Exec Elect Utils Group 1960. Retired Gen Elec 1963. Tech and Prof affiliations incl: Pres AIEE 1960-61, Pres UET 1962-63, Pres IEEE 1964, Founding Mbr Natl Acad of Engrg 1964, Pres NAE 1970-73, Pres EJC 1966-68. *Society Aff:* IEEE, ASME, NSPE, NAE, ASEE, AAAS, NYAS.

Linderoth, L Sigfred, Jr
Home: 110 Burning Tree Rd. (POB 1753), Pinehurst, NC 28374
Position: Prof. Mech. Engr. *Employer:* Retired (1977) *Education:* Mechanical Engr./Honorary/Iowa State University (1951); Bachelor of Science/Aero Engr./Mass. Inst. of Technology *Born:* 12/19/07 Born in New York, NY 12/19/07. Graduated from East Orange, NJ High School, 1926. Received the degree S.B. Aero Engrg from MIT in 1930. Director of Engrg & Prod Mgr, Peerless Tube Co., Bloomfield, NJ 1930-36. Engrg Consultant, United Shoe Machinery Corp., Research Div 1936-49. Prof Mech Engrg, Iowa State Univ. 1949-58. Dir of Engrg, Continental Can Co., Research Div, 1958-62. Dir of Engrg, Battle Creek Packaging Corp., 1962-65. Prof of Mech Engrg, Duke Univ School of Engrg and Technical Dir of Duke Univ Medical Center Hyperbaric Environment Lab 1965-77. Retired in 1977. *Society Aff:* ASME, NSPE

Linders, Howard D
Business: 2121 Hudson Ave, Kalamazoo, MI 49008
Position: Chrmn of the Bd & CEO *Employer:* Walker Parking Consultants/Engineers (Formerly Carl Walker & Assoc. Inc.) *Education:* BS/CE/Univ of MI; AB/Math/Western MI Univ *Born:* 10/1/32 Kalamazoo, MI native. US Army Construction Engr in Korea, 1956-1958. Structural Engr for Louis C. Kingscott & Assocs, Inc, 1958-1965. Joined Carl Walker & Assocs, Inc in 1965. Resp have included design, project mgmt, client relations, office mgmt, Chief Operating Officer. Appointed Pres and Chrmn the Bd July 1981, Chrmn & CEO June 1986. Firm has offices in

Linders, Howard D (Continued)
Kalamazoo, MI; Elgin, IL; Minneapolis, MN; Indianapolis, IN; Denver, CO; Houston, TX; Philadelphia, PA; and Tampa, FL. Pres of Consulting Engrs. Council of MI 1982, Board of Directors Prestressed Concrete Institute, 1983; 1984. Wife, Leatha, 2 daughters, Sheryl and Sandra. *Society Aff:* NSPE, ACEC, PCI

Lindgren, Arthur R
Business: 7300 W Lawrence Ave, Chicago, IL 60656
Position: Mgr - Marketing Services. *Employer:* Magnaflux Corp. *Education:* BS/EE/Northwestern. *Born:* Dec 1923 Michigan. Joined Magnaflux Corp 1950 as field engr, spent 16 yrs in Indianapolis sales area, moved to Chicago 1968 as Midwest Dist Sales Mgr & in 1974 took Mkting Staff pos as Mgr - Mkting Services. Mbr ASNT, AFS, AMA, serving on local as well as natl pos with ASNT & AFS. Has written many papers on var nondestructive testing subjects, incl magnetic particle, ultrasonic, penetrant, & electromagnetic. Presently respon for product dev, promotion, service, order processing, co customer training dept, mkting admin. *Society Aff:* ASNT, AFS.

Lindgren, Leroy H
Business: 21 Worthen Rd, Lexington, MA 02173
Position: Vice President *Employer:* Rath & Strong Inc. *Born:* Aug 1918. BSME Illinois Inst of Tech 1941; grad studies Lewis Inst & Yale. Adj Assoc Prof Boston Univ Engrg School. 14 yrs with Rath & Strong as cons in Qual Control, Mfg & Computer Sys. The past 7 yrs as cons to special groups of Natl Acad of Sci, the Dept of Transp & the Environ Protection Agencies. More recently an estab liaison between the automotive indus & the gov agencies in the eval of alternative legis pols. 10 yrs with Joy Mfg in sev mgmt & engrg pos. Developer of the mining drills for percussion drilling using tungsten carbide. Prior yrs with G E & Sperry Products as a mfg engr. ASQC - Qual Engr status; SME - Natl Dir 1974-75; national figure skating judge since 1938.

Lindholm, John C
Business: Engg Tech Dept, Seaton Hall, Manhattan, KS 66506
Position: Prof Engg Tech. *Employer:* KS State Univ. *Education:* PhD/Mech Engg-Design/Purdue Univ; MS/Mech Engg/Univ of KS; BS/Mech Engg/KS State Univ; BS/Bus Admin/KS State Univ. *Born:* 11/3/23. Native of Cheney, KS. Tool Maker Boeing Airplane Co 1942-43. Pilot Army Air Corp World War II 1943-45. Retired Maj US Air Force. Design engr General Electric 1949-52. Sr Engr Midwest Res Inst 1952-54. Instructor Mech Engg: KS Univ 1954-57; Purdue Univ 1957-60. Assoc Prof Mech Engg KS State Univ 1960, Prof 1974, Dept Head Engg Tech, 1980. Visiting Prof Assiut Univ, Egypt, 1964-66. Year-in-Industry participant DuPont Corp 1971-72. Reg PE - MO since 1954. ASME KS City sec: Treasurer 1976, Sec 1977, VChrmn 1978, Chrmn 1979, Region VII Operating Board 1980-82, Region VII Vice-president 1982-86, ABET-TAC Evaluator 1982-86. SESA KS Sec: Sec 1978, Chrmn 1979 & 1980. Hobbies: Photography, hiking, gardening. Honor Soc Mbr: Pi Tau Sigma, Phi Kappa Phi, Tau Beta Pi, Tau Alpha Pi. *Society Aff:* ASME, ASEE, SESA.

Lindholm, Ulric S
Business: 6220 Culebra Road, P. O. Drawer 28510, San Antonio, TX 78284
Position: Technical Vice President *Employer:* Southwest Research Institute *Education:* Ph.D./Applied Mechanics/Michigan State University; M.S./Wood Technology/Michigan State University; B.S./Wood Technology/Michigan State University *Born:* 09/13/31 Born and attended public schools in Washington, D.C. Served in U.S. Navy from 1955-57. Joined Southwest Research Inst in 1960 and assumed current position as VP of the Engrg and Materials Sciences Div in 1985. Responsible for R&D program development and management in the areas of structures, mechanics, and materials. Fellow member of ASME and AAAS. Hobbies include woodworking and cabinet making. *Society Aff:* ASME, ASM, AAAS

Lindquist, Claude S
Business: Elec Engr Dept, Long Beach, CA 90806
Position: Prof. *Employer:* CA State Univ. *Education:* PhD/Elec Engr/OR State Univ; MS/Elec Engr/OR State Univ; BS/Elec Engr/Stanford Univ; BA/Humanities/Univ of Redlands. *Born:* 3/13/40. Grad in 1969 and joined Collins Radio Co as design eng. He has been with the elec engg dept at CA State Univ, Long Beach since 1971. He is design conslt to a number of cos including E&H Electronics and is currently involved in the design of based microcomputer systems. *Society Aff:* IEEE.

Lindquist-Skelley, Sharon L
Home: 8410 Orcutt Ave, Hampton, VA 23605
Position: Engr. *Employer:* Newport News Shipbldg & Drydock Co *Education:* BS/ME/Univ of Hartford; BS/Educ-Math/Old Dominion Univ. *Born:* 8/23/48. New London, CT. Taught in VA Beach City Schools. Undergrad Res Grant, Argonne Natl Lab, 1975. EIT reg in CT 1975. Analytical engr for Pratt and Whitney specializing in acoustical analysis of prototype commercial engine. Currently with Newport News Ship dealing mainly with acoustical problems along with some vibrations & flow problems. Profl activities include SWE: Hartford section Pres 1957-77; Engrs Week Ctte, CT '75 & '76; SWE MAL Rep 1979-80; SWE Audit Ctte Chair 1979-80; SWE Natl Exec Ctte 80-81; SWE Natl Secy 1981-82, ASME Student Activities Chair Eastern VA Section 80-81, ACM Vice-Chair Southeastern VA Chapter 1979-80; Pi Tau Sigma; Planning '80 high sch workshop on Engr/Math/Sci options for women. *Society Aff:* SWE, ASME, NSPE, ACM, AWIS, AWC, AWM.

Lindroos, Arthur E
Home: 20 North Briarcliff Rd, Mountain Lakes, NJ 07046
Position: Dir of Engg. *Employer:* Penick Corp - Unit of CPC Intl. *Education:* DEng/ChE/Yale; MS/ChE/Worcester Poly Inst; BS/ChE/Worcester Poly Inst. *Born:* Aug 1922. Native of Worcester Mass. USNR 1944-46. Process dev in nuclear waste recovery & air cleaning for Vitro Corp 1949-51. Process dev in specialty organics & synthetic polymers 1951-67 for Airco Chemicals; Engrg Mgr 1967-71. Respon for design, const & oper of specialty chem plant as V P 1971-76 of the Techni-Chem Co. Presently involved in process and proj engrg for fermentation, pharmaceutical, narcotic, and botanical products. *Society Aff:* AIChE, ACS, ΣΞ.

Lindsay, Philip J, Sr
Business: 518 Bulldog St., S.E, Lacey, WA 98503
Position: Manager of Construction & Director of Physical Plant *Employer:* The Founder's Group, Inc *Education:* BS/Civ Engg/LA Tech Univ. *Born:* 3/16/35. Native of ID Springs, CO. Served with USMC 1954-58. Proj engr with Southwestern Elec Power emphasizing HV & EHV design. Proj mgr for R K Hawkins Engrs emphasizing HV design & construction supervision. Partner in Consulting from 1977 to 1986 emphasizing pipelines, power & surveying. Presently Mgr. of Const. & Dir. of Physical Plant for Capitol Downs Thoroughbred Racetrack, Olympia, WA. Pres, LA Sec, ASCE 1978- 1979 Sec-treas District 14 Council 1980-83 (ASCE). Enjoy classical music, fishing, hunting, & snow skiing. *Society Aff:* ASCE, SAME, AISC, NSPE.

Lindsey, Kenneth R
Business: 1710 Seamist Dr, Houston, TX 77008
Position: Sr VP. *Employer:* Binkley & Holmes, Inc. *Education:* BS/Civ Engg/Univ of MO Rolla. *Born:* 1/24/37. Native of Sullivan, MO served with Army Corp Engrs, 1959 through 1962 with last assignment as Co Commander. 1962 through 1976 as design engr, proj engr and proj mgr for large consulting firm for projs in wastewater treatment and collection, water supply and distribution, solid waste disposal, and hydraulics. Assumed current responsibility as VP for Engg in 1976. As such, I set technical policy and direct effort of diversified engg staff providing planning, design, and construction mgt services. *Society Aff:* ASCE, ASTM, AWWA, NSPE, WPCF, SAME.

Lindsey, Larry M
Home: 3612 Ripple Creek Dr, Austin, TX 78746
Position: Owner *Employer:* Larry M. Lindsey, Consulting Engrs *Education:* BS/Mech Engg/TX Tech Univ. *Born:* 10/1/40. Native of Lubbock, TX. Early yrs experience included an Utilization Engr for Lone Star Gas Co; Design and Dev Engr for an oil field equip co; Design Engr for Gen Dynamics in the Environmental Control Group; Design Engr for a Dallas, TX mech contractor; Territory Mgr for Lennox Industries, Inc; VP and Gen Mgr of an Austin, TX mech contractor; Pres of own construction and mech contracting co; and presently is owner of Larry M. Lindsey, Consulting Engrs. Am involved in the application of technical and practical expertise to solve energy mgt/conservation opportunities in both new and existing facilities. *Society Aff:* TSPE, NSPE, PEPP, ASHRAE.

Lindt, Thomas
Business: Dept of Materials Sci & Engrg, Pittsburgh, PA 15261
Position: Prof *Employer:* Univ of Pittsburgh *Education:* PhD/ChE/Univ of Delft (Holland) *Born:* 7/8/42 Since 1972 he has been working in the area of polymers processing. First at Shell Research in Delft and Amsterdam, followed then by a visiting and later regular appointment at the Univ of Pittsburgh where, currently, he is Prof of Polymer Engrg. He has published extensively in conventional and reactive processing of polymers. In 1980 and 1982 he initiated and chaired the First and Second Intl Conference on Reactive Processing of Polymers in Pittsburgh. He has been a consultant to government & industry in the polymer field. *Society Aff:* AIChE, SPE, SOR, PPS

Lindvall, Frederick C
Home: 1224 Arden Rd, Pasadena, CA 91106
Position: Retired. *Employer:* *. *Education:* PhD/Elec Engg/CA Inst of Tech; BS/Railway Eng/Univ of IL. *Born:* May 29, 1903 Moline Ill. 1928-30 Genl Elec Co. 1930-69 Caltech, Instr to Prof Elec & Mech Engrg 1930-45; Chmn Div of Engrg & Appl Sci 1945-69; Prof Emer 1969- . V P Engrg Deere & Co Moline Ill 1969-72. Pres Lindvall, Richter & Assocs Los Angeles 1972- . Fellow IEEE, ASME. Mbr ASEE, Pres 1957-58; Mbr Sigma Xi, Pres 1968-69. Reg Engr Calif, Elec & Mech. Mbr Natl Acad of Engrg. *Society Aff:* IEEE, ASME, ASEE, ΣΞ, NAE.

Lineberry, Michael J
Business: PO Box 2528, Idaho Falls, ID 83403
Position: Assoc Div Dir *Employer:* Argonne Natl Lab *Education:* PhD/Engrg Science/CA Institute Tech; MS/Mech Engrg/CA Institute Tech; BS/Engrg/UCLA *Born:* 09/11/46 Native of Pomona CA. Joined Argonne upon completion of grad schl in 1972. During 1972-83, work was in fast reactor phys at the zero power plutonuim reactor. Present position is mgr of fuel cycle and examination programs, respon for fuel cycle work in Argonne's Integral Fast Reactor Initiative, and for programs at the hot fuel examination facility. Mbr ANS since 1974 and currently Chrmn, Scholarship Policy and Coordination Ctte. Recipient, US DOE E O Lawerence Award, 1983. *Society Aff:* ANS

Linehan, John H
Home: 1515 W Wisconsin Ave, Milwaukee, WI 53233
Position: Prof. *Employer:* Marquette Univ. *Education:* PhD/Mech Engr/Univ of WI; MS/Mech Engr/RPI; BS/Mech Engr/Marquette Univ. *Born:* 7/8/38. Native of Chicago, IL. Joined Mech and Biomed Engg Depts at Marquette Univ in 1968. Presently Prof at MU. Holds adjunct appointment as Prof of Biomedical Engg and Physiology at the Medical Coll of WI. Res interests include Heat & Mass Transfer, Fluid Mechanics and dev of instrumentation for medical uses. *Society Aff:* ASME, APS, BMES, Microcirculatory Soc.

Ling, Frederick F
Home: 30 Mellon Ave, Troy, NY 12180
Position: Chrmn of ME. *Employer:* Rensselaer Poly Inst. *Education:* DSc/ME/Carnegie-Mellon Univ; MS/ME/Carnegie-Mellon Univ; BS/ME/Bucknell Univ; BS/CE/St Johns Univ (Shanghai). *Born:* 1/2/27. Native of Tsingtao, China; came to US in 1947. Asst Prof Math, Carnegie- Mellon (1954-56). With Rensselaer Poly Inst since 1956, now William Howard Hart Prof and Chem Dept Mech Engg, Aero Engg and Mechanics. Mbr Natl Acad Engg; Fellow, AAAS; Fellow, ASME; Fellow, ASLE; Fellow, Am Acad Mechanics. Sr Fellow, NSF (1970); Outstanding Educator in Am (1974); Joseph Marie Jacquard Medal, Paris (1970); Natl Award, ASLE (1977); Wm H Wiley Distinguished Faculty Award (1979). Visiting Prof Inst of Tribology, Univ of Leeds, England (1970). Mayo D Hersey Award, ASME (1984). VP, Res, ASME (1979-81). Mbr Board of Governors, ASME (1981-83). *Society Aff:* ASEE, APS, NTAS, SES.

Lingafelter, John W
Business: 1700 Dell Ave, Campbell, CA 95008
Position: Sr VP. *Employer:* Quadrex Corp. *Education:* MS/Materials Sci/Stanford Univ; BS/Metallurgical Engg/Univ of AZ; Profl/Metallurgical Engr/Univ of AZ. *Born:* 9/3/27. Twenty-seven yrs' nuclear experience, including 22 yrs in nuclear materials, process and quality engg. Engaged in fuel fabrication, materials and process dev at Hanford Operations. Managed the dev of improved mfg processes for nuclear core components and nondestructive examination dev for remote inspection of irradiated fuel at Gen Elec. Dir overall turnkey plant engg effort for Monticello Nuclear Power Plant. Co-founder and sr VP of Quadrex Corp. Active in quality assurance, mgt reviews, fuel fabrication, NDT, and overseas operations. Licensed PE, CA. *Society Aff:* ANS, ASQC, ASM.

Link, Edwin A
Business: RFD 1, Box 196, Fort Pierce, FL 33450
Position: V Pres & Trustee. *Employer:* Harbor Branch Foundation Inc. *Born:* 7/26/04. Holds over 27 pats for inventions in the field of aviation & ocean engg - best known for his 1929 dev of the Link Flight Trainer During WWII, the Link Trainer was used to teach flying to more than half a millon airmen throughout the world. From the early 1960's, has devoted the major part of his time to oceanology res & exploration. Designed & built an oceanographic res vessel, R/V Sea Diver. Constructed an underwater habitat for shallow water, an aluminum diving chamber for deep submergence & conjuction with John Perry, the submarine Deep Diver. The 2nd generation of this type submersible designed by Mr Link is Johnson-Sea-Link, now owned & oper by Harbor Branch Fdt, Inc. Honors incl the Franklin Inst's Howard N Potts Medal, the Wakefield Gold Medal RAS, the Natl Bus Aircraft Assn Award, the Arnold Air Soc's Paul T Johns Trophy, Lindbergh Award.

Link, Fred M
Home: Robin Hill Farm, Pittstown, NJ 08867
Position: Consultant (Self-employed) *Education:* BS/Penn State Univ; EE/Penn State Univ. *Born:* Oct 1904 York Penna. Resident N J since 1930. Mbr TBPi, EKN, PiKa. Founded, managed & owner Link Radio Corp 1931-51; prior assn with N Y Tele & DeForest Radio Co; later assoc with DuMont Labs & RCA; independent consultant (communications) since 1965 with key radio communications groups. Fellow, Life Mbr IEEE; Hon Life Mbr VTS/IEEE & AdCom VTS mbr; Hon Life Mbr APCO; Pres of Radio Club of Amer; mbr of VWOA, DeForest Pioneers, AFCEA, AHSA, ESHBA & an Amateur since 1919 as 30V, 3BVA & W2ALU. Life Mbr QCWA. Former Mayor Westwood, N J & an active show horse breeder & exhibitor. Robin Hill Farm, Pittstown N J. *Society Aff:* IEEE, APCO, QCWA, AHSA, RCofA.

Linke, Simpson
Home: 383 the Pkwy, Ithaca, NY 14850
Position: Prof. Emeritus *Employer:* Cornell Univ. *Education:* MEE/EE/Cornell Univ; BSEE/EE/Univ of TN. *Born:* 8/10/17. Native of Jellico, TN. With Knoxville Utilities Bd 1941-42. Radar officer, Army Signal Corps, 1943-46 in US & Korea. Discharged as Capt. Mbr, Cornell Elec Engg Faculty since 1949. Sabbatical leaves as Sr Elec Engr with Phila Elec Co, 1956-57, & Prog Mgr for Elec Power Res with NSF (RANN) 1971-72. Supervisor, Cornell A-C Network Calculator 1953-60, Asst Dir,

Linke, Simpson (Continued)
Cornell Lab of Plasma studies 1968-75, co-principal investigator, Cornell Energy Proj 1972-73. Organized & chaired Cornell Intl Symposium on the Hydrogen Economy 1973, & seminar on superconducting Magnetic Energy Storage in 1977. Coordinator, Elec Engg graduate studies 1981-1984. Professor Emeritus July 1, 1986. *Society Aff:* IEEE, CIGRE, ΣΞ, AAUP.

Linkletter, Graeme A
Business: Box 1600, Charlottetown PEI, Canada C1A 7N3
Position: Supervisor, Engg Section *Employer:* PEI Dept of Agriculture *Education:* MSc/Agr Engr/McGill Univ; BSc/Agr Engr/McGill Univ *Born:* 3/15/39 Native of Prince Edward Island, Canada. Have been employed with Dept of Agriculture since 1960. Major part of work deals with Farm Bldg Design with specilization in Potato Storage bldgs and systems. Married with three children in Univ. *Society Aff:* ASAE, CSAE, APPEI.

Linnert, George E
Home: 14 Black Skimmer Rd, Hilton Head Island, SC 29928
Position: North American Representative. *Employer:* The Welding Inst (U K). *Born:* Dec 1916. Educated in Chicago tech schools and colleges. Joined Met Dept of Republic Steel Corp Chgo 1935. Moved to Res Labs of Rustless Iron and Steel Corp Balto 1941 (which became part of Armco Steel Corp 1947). Employed by Armco Steel as Mgr of Welding Res Middletown Ohio until early retirement 1973, when present pos as North Amer Rep for The Welding Inst was assumed. Pres of AWS 1970-71. Fellow ASM 1971. Fourteenth Adams Mem Lectr AWS. Received Samuel Wylie Miller Mem Medal 1972 AWS. Active on High Alloys Ctte WRC. Author of many tech papers and sev textbooks entitled 'Welding Metallurgy'. *Society Aff:* ASM, AWS, IIW, IOF.

Linsley, Ray K
Home: 280 Swanton Blvd, Santa Cruz, CA 95060
Position: President *Employer:* Linsley Kraeger Assoc, Ltd.. *Education:* BS/Civil Eng/Worcester Polytech; DSc/Hon/Univ of Pacific; D Eng/Hon/Worcester Tech. *Born:* 1/13/17. Native of Bristol CT. TVA 1937-40, US Weather Bureau 1940-50, Prof of Hydraulic Engg Stanford Univ 1950-75. Chrmn, Hydrocomp, Inc 1966-78. Pres. Linsley, Kraeger Assoc Ltd, 1979- Honorary Mbr ASCE, Collingwood Prize 1941; Julian Hinds Award, 1978; Fellow Amer Geophys Union, Pres Sect of Hydrology 1956-59; Mbr Amer Meteorological Soc, Natl Acad Engg 1976- . Sr author Applied Hydrology 1949; Hydrology for Engrs, 3 ed 1982; Water Resources Engg, 3 ed 1979. Commissioner Natl Water Comm 1968-73. Reg C E CT, CA, GA, IL. *Society Aff:* ASCE, NSPE, AGU, AMS, AWRA, AIH.

Linstedt, K Daniel
Home: 6647 Apache Ct, Longmont, CO 80501
Position: Senior Project Manager *Employer:* Black & Veatch *Education:* PhD/Sanitary Engg/Stanford Univ; MS/Sanitary Engg/Stanford Univ; BS/Civil engg/OR State Univ. *Born:* 11/6/40. With Univ of CO 1967-81. Currently Senior Proj. Mgr. Black & Veatch Consltg. Engrs. Served as Pres Rocky Mountain Water Pollution Control Assn 1973-74. Dir, Water Pollution Control Fed, 1975-78, Chmn Water Pollution Cont. Fed Water Re- use Ctte 1974-79. WPCF Bedell Award 1973; ASEE Dow Outstanding Campus Activity Coord Award 1975. AWWA Resources Div Award, 1977. Bd of Pres's Assocs at Bethel College & Seminary of St Paul, MN 1974-79. Diplomate, American Academy of Envir. Engrs., 1978-.; Delegate, USA National Comm, Intern. Assoc. For Water Poll. Research & Control, 1979-. Consultant, U.N. Development Programme, Madras, India, 1984, 85. *Society Aff:* WPCF, AWWA, IAWPR, AAEE.

Linstromberg, William J
Home: 1583 St. Joseph Circle, St. Joseph, MI 49085
Position: Sr Dev Engr. *Employer:* Whirlpool Corp./Retired *Education:* BS/ME/Univ of MO. *Born:* 9/3/24. Native of Beaufort, MO. Served in Army Air Corp 1943-46. Married Shirley JoAnn Smith of Hickman Mills, MO in 1949. Five children. Service & Dev Engr for Servel, Inc 1948-56. Senior Dev Engr with Whirlpool Corp since 1956; responsible for the design & dev of automatic ice makers & dispensing equip for refrigerators. Granted 39 US Patents. Received 1968 Technical Achievement Award presented by Tri-State Council for Sci & Engg. Mbr Tau Beta Pi, Pi Tau Sigma, Sigma XI & ASHRAE Fellow. Lutheran. Hobbies: woodworking and collecting old woodworking tools; mbr of EAIA & M-WTCA. Retired 1987. *Society Aff:* ASHRAE, ΣΞ.

Linvill, John G
Home: 30 Holden Ct, Portola Valley, CA 94025
Position: Prof EE., Dir. Center for Integrated System *Employer:* Stanford Univ. *Education:* ScD/EE/MIT; SM/EE/MIT; SB/EE/MIT; AB/Math/William Jewell. *Born:* Aug 1919. From 1949-51, Asst Prof EE MIT. At Bell Labs 1951-55. Joined Stanford fac 1955: Prof & Dir Solid-State Electronics Lab 1957-64, Dept Chmn EE 1964-80. Author: Transistors & Active Circuits, Models of Transistors & Diodes. Inventor of Optacon, reading aid for blind. Co-founder, Dir, Telesensory Sys. Held IEEE offices; served on adv cttes to DOD & NASA. Fellow IEEE 1960; Hon Dr Appl Sci Univ of Louvain 1966; Natl Acad Engrg 1971; Fellow Amer Acad Arts & Sci 1974; IEEE Educ Medal 1976; Medal of Achievement, AEA, 1983. *Society Aff:* IEEE, NAE, AAAS, AAAS.

Linville, Thomas M
Home: 1147 Wendell Ave, Schenectady, NY 12308
Position: Prof Engr - ret. *Employer:* Genl Electric Co - ret. *Education:* EE/Elec Engg/Univ of VA. *Born:* Mar 3, 1904 Wash D C. Univ. Virginia, E.E. 1926 Grad of Harvard Advanced Mgmt Prog, G E Advanced Engrg Prog, UCLA Modern Engrg Prog. Worked for G E 40 yrs: motor dev, Mgmt Consultation Services, successively Mgr of Exec Dev, Corp Res Oper, Corp Res Appli. Reg P E. At var times: Dir AIEE, Dir EJC, mbr Natl Res Council, mbr visiting bds at Clarkson & Norwich, Chmn Dev Council for Sci at RPI. Fellow IEEE, ASME, AAAS. Life Mbr ASEE. Pres NSPE, Natl Pres Tau Beta Pi, Pres Schenectady Indus Dev Council, Pres WMHT-TV, Chmn Schenectady City Planning Comm, Pres Schenectady Museum, Tr Sunnyview Hosp. Enjoy family, golf, & leisure. *Society Aff:* IEEE, ASME, ASEE, NSPE, AAAS, NYAS, ARRL, TBΠ

Lior, Noam
Business: 111 Towne Bldg 6, Univ of Penna, Philadelphia, PA 19104-6315
Position: Prof *Employer:* Univ of PA *Education:* PhD/ME/Univ of CA, Berkeley; MSc/ME/Technion-Israel; BSc/ME & Nucl Engrg/Technion- Israel *Born:* 3/11/40 1965-66 Instructor, Dept of Mech Engrg, Technion, Israel. 1973-current Asst, Assoc, and since 1985 Prof of Mech Engrg at the Univ of PA. 1986- Graduate Group Chairman Mech. Engrg and Appl. Mechanics. Summer employment at Stal-Delaval, Stockholm, Sweden; several yrs, as hd of engrg dept in Israel, and as res engr at the Seawater Conversion Lab of the Univ of CA, Berkeley. Chrmn of Heat Transfer Div (1975-77), Basic Engrg Section (1977-81), Profl Divs (1978-79) of the ASME Phila Section; Dir of AS/ISES (1978-80) and Chrmn of its Engrg Div (1978-1982). Maj Res Interests: Heat Transfer and Thermodynamics, Solar Energy, Energy Conversion, Water Desalination, Thermo- fluid Measurements, Combustion. Hd numerous projs. About 70 publications. *Society Aff:* ASME, ASHRAE, ISA, IDA, ASES, ISES

Liou, Ming-Lei
Business: 1600 Osgood St, N Andover, MA 01845
Position: Supervisor. *Employer:* AT&T Bell Labs *Education:* PhD/EE/Stanford Univ; MS/EE/Drexel Univ; BS/EE/Natl Taiwan Univ. *Born:* 1/6/35. Taught in the Dept of EE, Drexel Univ from 1958 to 1961. Employed by the Stanford Electronic Labs from 1961 to 1963. Joined the AT&T Bell Labs in 1963. Worked in the areas of system theory, numerical analysis, optimization, & computer-aided design of circuits & various component in analog & digital transmission systems. Currently involved in the dev of communication circuits and subsystems & in microprocessor

Liou, Ming-Lei (Continued)
applications. Recipient of two best paper awards sponsored by the IEEE Circuits & Systems Soc. Transactions Editor of the same soc from 1979 to 1981. Elected to IEEE Fellow in 1979. *Society Aff:* IEEE, ΣΞ.

Lipovski, Gerald J
Business: Dept of Elec Engrg, Austin, TX 78712
Position: Prof of Elec and Computer Engrg *Employer:* Univ of Texas. *Education:* PhD/EE/Univ of IL; MS/EE/Univ of IL; BS/EE/Univ of Notre Dame; AB/Pre-Engg/Univ of Notre Dame. *Born:* Jan 1944. Native of Coleman Alberta Can. Design engr for Genl Elec DC Motors & Generators 1966; Asst Prof EE Univ of Fla 1969-74, Assoc Prof EE & Computer Sci 1974-76; Assoc Prof EE Univ of Texas at Austin 1976-1982, Prof ECE Univ of Texas at Austin, since 1982, doing res in digital sys architecture, microcomputers, non-numeric processors, parallel computing. Chmn Tech Ctte on Computer Arch of the Computer Soc IEEE 1974-75. Chmn spec int group on Computer Arch of ACM 1976-77. Chmn Exec Ctte of the Conf on Digital Hardware Languages 1973-1979. Dir Euromicro 1976-1979. Member, Governing Bd, Computer Society IEEE, 1977-1979. Mbr Tau Beta Pi, Eta Kappa Nu, Phi Kappa Phi. Enjoy classical music, hobby electronics & microcomputers. *Society Aff:* IEEE, ACM.

Lippe, Bernard
Business: 100 Church St, New York, NY 10007
Position: Pres *Employer:* Bernard Lippe & Assocs, Inc *Education:* B/CE/City Univ of NY *Born:* 06/28/29 Since 1970 Mr. Lippe has focused his experience in heavy and bldg design and construction towards managing his firm which specializes in construction mgmt, analysis of construction claims and disputes, and other engrg services to owners, contractors, architects, and governmental agencies. He is a registered PE in several states and has lectured at numerous univ, before AACE, ASCE, AMA and RETC. His articles have appeared in several profl publications. He founded Bernard Lippe & Assocs, Inc in 1970. Previously, he was engaged in design, estimating, construction supervision and mgmt of major heavy and bldg construction projects in the United States. *Society Aff:* NSPE, NYSPE, PEPP, ASCE, AACE.

Lippe, Robert L
Business: 2133 McCarter Hgwy, Newark, NJ 07102
Position: Pres. *Employer:* Argus Coatings Corp. *Education:* MS/CE/Princeton Univ; BE/CE/Yale Univ. *Born:* 5/8/23. Primarily concerned with dev of ind chem coatings for the protection of decoration of various substrates. *Society Aff:* AIChE, ACS, YSE.

Lippmann, Seymour A
Business: 1305 Stephenson Highway, The Tech Center, Troy, MI 48084
Position: Res Assoc *Employer:* Uniroyal Tire Co. *Education:* BS/Chem Engrg/Cooper Union Inst of Tech *Born:* 11/23/19. Mr Lippmann's principle activities have been in the thermodynamics of rubbers; physical properties of filled elastomers; dynamics of maneuvering vehicles; internal tire mechanics; vibratory interactions of roads, tires 8 vehicles; sound generated by tires; relations of human annoyance to physical stimuli; control of vehicles. He has written many papers & patents. He has chaired Technical committees of SAE & Rubber Mfg Assn. The SAE has awarded him its Fellow grade, & Sigma Xi has awarded mbrship. He manages the Tire-Vehicle Systems Labs at Uniroyal and is a res assoc. *Society Aff:* SAE, IEEE, APS

Lippold, Herbert R, Jr
Business: Rockville, MD 20852
Position: Dir *Employer:* Natl Ocean Survey, NOAA *Education:* BS/CE/Univ of NH; BS//New England Coll *Born:* 4/9/26 Methuen, MA. Two years Army Air Corps. Joined Coast and Geodetic Survey (now NOS), doing research and surveying in Atlantic, Pacific, Arctic Oceans and most of US. Commanded three vessels. Established worldwide satellite triangulation geodetic network; conducted liaison with Air Force for horizontal and vertical control of intercontinental ballistic missile sites. Conducted Pacific tide and seismic seawave warning system liaison with Navy. Supervised ship construction. Dir, Pacific Marine Center. Assoc Dir for Fleet Operations, directing NOAA's research and hydrographic survey fleet. DOC Scientific and Technical Fellow; SAME Fellow and Bd Member. Dir, Natl Ocean Survey. *Society Aff:* ASNE, SAME

Lipscomb, Joseph W
Business: P O Box 1963, Harrisburg, PA 17105
Position: Director, Internatl Division. *Employer:* Gannett Fleming Corddry & Carpenter. *Born:* Jan 1928. MS San Engrg Harvard Univ; BSCE Virginia Polytechnic Inst. Joined Gannett Fleming 1959 and served as Proj Mgr & Office Mgr until assuming current responsibility as Dir of Internatl Div in 1973. Currently responsible for all internatl activities of the firm, incl all phases of engrg, prelim studies, designs, const supervision, and facil oper as well as liaison with internatl funding orgs. Diplomate of Amer Acad of Environ Engrs.

Lipuma, Charles R
Home: 50 Poplar Dr, Morris Plains, NJ 07950
Position: Manager Technology Support *Education:* BS/ChE/NJ Inst of Tech; MS/ChE/NJ Inst of Tech. *Born:* 8/16/31. Exxon Res. & Engr. 1953 to date. Positions include Mgr. Exxon Research & Development Lab, Baton Rouge, La.; Project & Planning Mgr., Synthetic Fuels Research; Genl Mgr ER&E Petroleum & Synthetic Fuels Engr Dept; Mgr Technology Support. With Exxon Corp. HQ, 1967/70 Senior Refining Advisor; 1978/82, Mgr Refining Planning. Member Hanover Twp Planning Bd; Member Advisory Committee to ChE Dept NJ Inst Tech. *Society Aff:* AIChE, API.

Lisack, John, Jr
Home: 12039 Stoneford Dr, Woodbridge, VA 22192
Position: Exec VP *Employer:* National Assoc of Personnel Consultants *Education:* MBA/Bus Adm/Univ of MA; BSCE/Civ Engrg/Univ of MA *Born:* 5/22/45 Native Allendale, NJ. Currently Exec VP of Nat Assoc of Personnel Consultants with overall operational responsibilities. Formerly Dir, Mbrship, Projs and Federal Relations-ASEE with responsibilities for the Annual Conference, Member Services and Adm of all Grants and Contracts. Also formerly: Financial Analyst-OMB, Fairfax County VA; Exec Dir, Property Adm, Real Equity Investments Inc; Mgr, Land Development, Foxvale Construction Co; Assistant Adm and Utilities Engr, Prince William County, VA; US Army Corp of Engrs-Reserves. Other past and current activities include: Juror for James F. Lincoln Arc Welding Foundation Awards ($10,000); Chmn of Bd, Greater WA Society of Assoc Execs; Who's Who Among Students in America Univ & Colls; Univ MA-Joseph Loiko Award; George Barber Scholarship; Marshall of Graduating Class-School of Engrg; Samuel Samuals Award- Univ MA. Chairman of Bd of Trustee Washington Assoc Res Foundation. *Society Aff:* ASAE, CESSE, ASEE

Lischer, Vance C
Home: RR 2 Box 48A, Cook Station, MO 65449
Position: Retired *Education:* BS/Civ Engg/Wash Univ. *Born:* 10/5/06. Initially with St Louis County Water Co (1927-44), 4 yrs in Distribution Div, 13 yrs in charge of production pioneered in preventive maintenance prog, later adopted for military installations. Twenty-six yrs, consulting practice with Horner & Shifrin as Partner, Co-Partner & Exec VP. Specialized in water supply, hydraulics, pumping & water resources, projs for Corps of Engrs, industry & municipalities. Foreign experience under Point IV Prog in Karachi, Pakistan; Conducted seminar on water distribution system design in Jamaica. Participated in Symposium on rural water supply in Thailand, both sponsored by AID. Mbr MO Clean Water Commission (1974-79). Mbr Natl Water Pollution Control Advisory Bd (1954-55). Numerous awards, chairmanships, committee assignments, publications from & for mbr profl societies. *Society Aff:* ASCE, ASME, AWWA, WPCF, NSPE.

Liss, Robert B
Home: 4408 N Rosemead, Peoria, IL 61614
Position: Staff Engineer. *Employer:* Caterpillar Tractor Co. *Education:* MS/Metallurgy/Univ of MD; BS/Metallurgy/Univ of MN. *Born:* Aug 1, 1923 Minneapolis, Minn. June 1956 MS Univ of Md, Phi Kappa Phi Hon Frat; June 1949 BS Univ of Minn. US Army 1942-46. Bureau of Standards & Naval Res Lab 1949-56 spec in X-ray diffraction & alloy dev. With Caterpillar Tractor Co since 1956. Chmn of co matls ctte & Corp Metric Matls Ctte. Respon for interplant matls activity coord. *Society Aff:* ASM, ASTM.

Liss, Sheldon
Business: Liss Engineering, Inc, 2862-A Walnut Ave, Tustin, CA 92680
Position: Prin Engr. *Employer:* Liss Engg. *Education:* BME/Engg/NY Poly Inst. *Born:* 8/4/36. Designed and supervised the engg of several hundred illumination and elec power installations in commercial, industrial, institutional and recreational projs. In 1969 founded the firm of Liss Engg and is presently principal. Licensed as an Elec Engr in the states of: CA, NV, AZ, HI, UT, NM and CO, and is a reg Fallout Shelter Analyst with the Dept of Defense. *Society Aff:* ACEC, NSPE, IES, IEEE.

List, E John
Business: 1201 E CA Blvd (138-78), Pasadena, CA 91125
Position: Prof. *Employer:* CA Inst of Tech. *Education:* PhD/Applied Mechanics/Caltech; ME/Civil Eng/Univ Auckland; BS/Math/Univ Auckland; BE/Civil Eng/Univ Auckland *Born:* 3/27/39. in New Zealand. Naturalized US citizen. Res appointment at Caltech, 1965- 66. Faculty School of Engg, Univ of Auckland, 1966-69. Asst Prof of Environmental Engg Sci at Caltech, 1969, Prof since 1978. Exec. Officer Env. Engg. Sci 1980-85. VChrmn, Caltech Faculty 1979-81. Res on turbulent fluid motions with buoyancy, modeling of hydrologic systems, analysis of pressure transients. Consultant on ocean outfall systems to Southern CA Edison, 1972-present. Extensive industrial consulting on hydraulic surge problems. Bd Chmn, Flow Science Inc, Pasadena, CA. Editor, Journ Hydraulic Engg. ASCE. Co-author of "Mixing in Inland and Coastal Waters," Academic Press, 1979. Active in ocean sports. *Society Aff:* ASCE

List, Hans
Business: Kleiststrasse 48, A-8020 Graz Austria
Position: Chrmn *Employer:* AVL-List, Ges m.b.H. of Austria *Education:* Dr of Tech Sci/-/Tech Univ Graz; Dipl-Engr/-/Tech Univ Graz. *Born:* 4/30/96. Born in Graz, Austria. Designer in diesel engine factory in Graz 1926-1945 Prof at Tungshi Univ China, Technical Univ Graz & Technische Hochschule Dresden. Founded AVL Graz in 1948. Chairman of AVL which is a private company for res & dev of internal combustion engines (sp diesel). 750 employees. 1970 enlarged by mfg of measuring instruments for engines, test stands, ballistic measuring equip, medical instruments. *Society Aff:* SAE, VDI, SIA, FVV.

List, Harold A
Home: 38 W. University Ave, Bethlehem, PA 18015
Position: Pres *Employer:* Railway Engineering Assocs *Education:* MS/Mech Engrg/Univ of PA; BS/Mech Engrg/Carnegie Inst of Tech (CMU). *Born:* 07/08/23 Shop Ofcr, Military Railway Service, WWII. Jr. Engr, Baldwin Locomotive Works; worked on last conventional steamers, experimental steam-turbine- electrics, and first Baldwin diesel-electric road locomotives. Taught ME subjects at Univ of PA. Developed public utility power plant control hardware and systems while at Leeds & Northrup. Supervised some of the earliest applications of computers to control rolling mills and furnaces at Bethlehem Steel. Have invented a method for adding axle-steering to conventional freight car trucks. Presently developing similar trucks for transit cars and locomotives. Also working on an all-electric system to replace conventional railroad air brakes. *Society Aff:* ASME, ISA, NSPE, AREA.

List, Harvey L
Home: 56 Roundtree Rt 9W, Piermont, NY 10968
Position: Prof Emeritus *Employer:* City Univ of NY. *Education:* DChE/CE/Poly Inst of NY; MS/CE/Univ of Rochester; BChE/CE/Poly Inst of NY. *Born:* 9/5/24. Resident of Piermont, NY. Prof Emeritus of Chem Engg at CUNY since 1955. Sr Process Engr, Exxon Res and Engg Co 1950-1955. Served US Naval Air Corps 1943-46. Consulting chem engr 1955 to present. Reg PE in NY, NJ, FL. Fulbright Prof, Taiwan, 1963-64. Co-author and author of books, "Material and Energy Balances" and "Petrochemical Technology" published by Prentice Hall. Principal of Rickian, Inc., Forensic Chemical Engineers, 1986 to present. President, List Associates, Inc 1980 to present. Editor *International Petrochemical Developments.* 1980 to present. *Society Aff:* AIChE, ACS, NAFE.

Litchford, George B
Home: 32 Cherry Lawn Lane, Northport L I, NY 11768
Position: Pres of Litchford Electronics. *Employer:* Private Cons to Gov & Indus. *Education:* BA/Physics/Reed College, Portland, OR. *Born:* Aug 12, 1918 Long Beach Calif. Long & disting career in the dev of many of the nation's present aviation sys. Was instrumental in dev the only microwave landing sys presently in oper the Navy C-Scan. Elected Fellow AIAA & IEEE. Has performed hundreds of hrs of flight res & is active in both electronics & avionics. Presently working on the Litchford B-CAS Collision Avoidance Sys. P E NY; 1974 recipient of the Pioneer Award AES. Married to the former Doris Cahill; daughter Jane, son George Jr. AIAA 1978 Recipient of the Wright Brothers Medal and deliverer of the Wright Brothers Lecture Los Angeles, CA, Aug, 1978. 1981 Recipient of the IEEE LAMME Medal for Sys Engrg for Outstanding Contributions in the dev of elec sys for air navigation and air traffic control. *Society Aff:* IEEE, AIAA, ION.

Lithen, Eric E
Home: 37 Boylston St, Garden City, NY 11530
Position: Pres *Employer:* Marine Technical Assocs, Inc *Education:* MS/ME/Stevens Inst of Tech; BS/Marine Engrg/US Merchant Marine Academy *Born:* 04/02/30 Native of Metropolitan NYC. Engr ofcr on-board experimental US Navy destroyer with advanced design power plant, 1953-1955. Design engr with Gibbs & Cox Inc, 1955-1964, specializing in high pressure/temperature steam propulsion plants for military vessels. Responsible for design and construction of prototype aircraft gas turbine propulsion plant for USCG Cutter HAMILTON with Propulsion Sys, Inc, 1964-1966. Formed and directed engrg firm, Marine Applications Co, Inc 1966-1974. Mgr, marine engrg with John J. McMullen Assocs, 1974-1981. Responsible for machinery plant design of several frigate type gas turbine powered military vessels. Formed marine engrg firm, Marine Technical Assocs, Inc in 1981. Registered Prof Engr in NY. *Society Aff:* SNAME

Litt, Mitchell
Business: Dept of Bioengrg, 220 S. 33rd St TB/D3, Philadelphia, PA 19104
Position: Prof *Employer:* Univ of PA *Education:* DESc/ChE/Columbia Sch of Engrg; MS/ChE/Columbia Sch of Engrg; BS/ChE/Columbia Sch of Engrg; AB/Chem/Columbia Coll *Born:* in Brooklyn, NY, s Saul and Mollie (Steinbaum); m. Zelda Sheila Levine, Sept 6, 1955; children-Ellen, Beth, Steven, Eric. Res engr Esso Res and Engrg Co 1958-61; faculty Univ of PA 1961-, Assoc Prof Chem Engrg, 1965-72, Prof, 1972-, Prof Bioenegr, 1977-; Vis Prof Environ Medicine Duke 1971-72; spl res biorheology trans processes, chemically reacting sys, med aspects of engrg. Current Int applic of physical chem & transport processes to biological processes. In particular, maj work has been on the rheological properties of mammalian secretions and their role in physiological function. Phi Beta Kappa, Sigma Xi, Tau Beta Pi, Phi Lambda Upsilon, Theta Tau, Co-editor Rheology of Biological Systems, 1973. Contbr articles to PE. Chrmn, Dept Bioengrg 1982- ; Dir, Center for Dental Bioengrg 1981- . *Society Aff:* AIChE, ACS, ASEE, BES, IEEE-EMBS

Little, Charles Gordon
Home: 6949 Roaring Fork Trail, Boulder, CO 80301
Position: Dir, Wave Propagation Lab. *Employer:* Natl Oceanic & Atmos Admin. *Education:* PhD/Radioastronomy/Univ of Manchester; BSc/Physics/Univ of Manchester. *Born:* 11/4/24. 1944-47 Indus R&D with UK radio tube manufacturers. 1954-58 Prof of Geophys & Deputy Dir Geophys Inst, Univ of Alaska. 1958-65 radio propagation res Natl Bureau of Standards, incl Dir Central Radio Propagation Lab 1962-65. 1965- , var sr res pos ESSA/NOAA. Since 1967, Dir Wave Propagation Lab NOAA, dev new methods of remotely measuring atmos & oceans. Mbr Natl Acad of Engrg; Fellow IEEE; Chmn US Natl Ctte for URSI 1979-81, Fellow AMS. *Society Aff:* AMS, IEEE, AAAS.

Little, Doyle
Business: PO Box 1947, Odessa, TX 79760
Position: Pres. *Employer:* OPI Inc. *Education:* MS/Ch Engg/Rice; BA/Ch Engg/Rice; MS-Mgmnt/-/MIT. *Born:* 2/12/36. Pres & CEO of OPI, Inc. Manufactures oilfield equipment. Vice Pres & Gen Mgr of The Wester Co of NA. Production Mgr and assistant to Pres, Texas Eastman Co. *Society Aff:* AIChE, AIME, SPE.

Little, John W
Home: 819 So Lk Starr Blvd, Lake Wales, FL 33853
Position: V.P. Chairman Exec. Comm. *Employer:* Waverly Growers Coop. *Education:* -/Mech Eng/ICS. *Born:* Apr 1909. Public schools Birmingham Ala; correspondence mech engrg. Goslin - Birmingham Inc (manufacturer of heavy mech & process equip) 1929-71: Ch Engr 1937-49 (9 pats on design of filters & evaporators), V P Mfg 1949-57, Pres 1957-71. Mbr ASME 1946- , speaker 1954 RDC, Chmn Natl Agenda Comm 1957, V P 1959-60, Fellow 1962. Hon Mbr Pi Tau Sigma 959, Ala Pi Omicron Chap. After retirement 1971 moved to Lake Wales Fla; elected to Bd of Dir Waverly Growers Coop 1972- , Pres 1975-80. Mbr Bd of Dir Fla Citrus Mutual 1976-V.P. 1979- , Taxpayers League of Polk Cty 1975-80 . *Society Aff:* ASME.

Little, Robert R
Business: Computer Science Dept, University, MS 38677
Position: Chairman of Dept. *Employer:* University of Mississippi *Education:* PhD/EE/LA Tech Univ; MS/EE/MS State Univ; BS/EE/MS State Univ. *Born:* 2/27/40. Native of Grenada, MS; Married, Four children; Field Engr (1962-63) and Systems Engr (1965-67) with IBM; Taught Computer Sci at MS State (1967-69 and 1973-77) and at LA Tech (1969-73); Programmer (1963-1965), Systems Analyst (1973-75) and Hd, Info Services (1975-76) all at MS State; Hd, Ind Engg and Computer Sci Dept (1977-80) at LA Tech; Assumed present position as Chairman, Computer Science Dept., School of Engineering, University of Mississippi, in 1980; Mbr of Upsilon Pi Epsilon (Natl Comp Sci Honor Soc) and Tau Beta Pi; Reg PE in MS. *Society Aff:* ACM

Littlefield, F Winston
Business: 11 E Forsyth St-530, Jacksonville, FL 32202
Position: Sr V P & Mgr, Southeast Region. *Employer:* Sverdrup & Parcel & Assocs Inc. *Education:* BS/Civil Engg/KS State Univ. *Born:* Nov 1917. US Army WW II - C of E Alaska/Europe. Joined Sverdrup & Parcel same yr. Engr (design, proj & mgmt levels) involved in design of hwys, expressways, bridges, commercial & indus facils, civil & pub works, environ & water mgmt. Respon for genl cons work & const mgmt services, municipal & gov authorities. Have been an officer of S&P since 1968. As Sr V P & Mgr of Southeast. Region, respon for opers of offices in Jacksonville, Gainesville Fla , Nashville Tenn, & Greensboro, NC. Reg in 6 states. Mbr IBT&T. S.A.M.E. *Society Aff:* IBT&T, S.A.M.E..

Littmann, Walter E
Business: PO Box 24647, Dallas, TX 75224
Position: VP Res & Engg. *Employer:* Security Div, Dresser Ind, Inc. *Education:* ScD/Metallurgy/MA Inst of Tech; MS/Metallurgy/MA Inst of Tech; BS/Metallurgical Engg/Univ of Cincinnati. *Born:* 11/14/26. Native of OH. Employed by The Timken Co from 1953 to 1976 serving in Metallurgical Res Mgt positions. Assumed present position in Oct 1976, responsible for res and dev of drilling tools used in drilling of earth and rock for oil, gas, and geothermal energy. Served in various technical committee capacities in the SAE, ASM, and ASME. Reg PE in the State of OH. Authored or co- authored more than 20 technical papers dealing with metallurgy, grinding, measurement and control of residual stress, the metallurgy of tapered roller bearings, and bearing damage analysis. *Society Aff:* ASM, ASME.

Littrell, John L
Home: 7349 Margerum, San Diego, CA 92120
Position: Pres *Employer:* LSW Engrs *Education:* BS/EE/Univ of AZ *Born:* 1/13/49 Consulting mechanical/Electrical Engr in the construction industry. Registered PE in (9) states, specializing in heating, ventilating, air conditioning, plumbing and piping design in industrial, commercial, institutional, and medical applications. Assumed present position as Pres and Chief Exec Officer of LSW Engrs in 1978. Previous position as VP and Chief Mechanical Engr. ASHRAE Chapter Treasurer 1980, Chapter Secy 1981. *Society Aff:* ASHRAE, NSPE, SAME, ACEC, ASPE, IES

Litzinger, Leo F
Home: 188 Bulrush Farm Rd, North Scituate, MA 02066
Position: Owner *Employer:* Litzinger & Co. Engrs *Education:* MS/ME/Univ of NM; BS/ME/Carnegie Inst of Tech. *Born:* 5/3/41. Founded Litzinger & Co. Engrs to provide consulting and design services to the energy industry. Principal Engr for several LNG & LPG projs involving design, operation, logistics, & estimating. Supervisory Engr with Chicago Bridge & Iron from 1967 to 1978. Primary design responsibility for Nineteen Liquefied Gas Facilities, including first US LNG Import Terminal for Distrigas in Boston. Heat Transfer and Proj Engr for ACF Ind from 1963 to 1967. Worked in Rover Prog to develop nucl powered rocket vehicles. Reg PE in IL, MA, & RI. Mbr AIChE, Dir CSA. Task group mbr of AGA Purging Committee. Authored several technical papers. *Society Aff:* AIChE, CSA, ACEC, AGA.

Liu, Bede
Business: Prospect St, Princeton, NJ 08540
Position: Prof of Elec Engrg. *Employer:* Princeton Univ. *Education:* DEE/Elec Eng/Polytech Inst of Brooklyn; MEE/Elec Eng/Polytech Inst of Brooklyn; BSEE/Elec Eng/Natl Taiwan Univ. *Born:* Sept 1934 China. Equip engr Western Elec Co 1954-56. Mbr tech staff, Bell Tele Labs 1959-62. Princeton Univ, Dept of Elec Engrg, since 1962. Fellow IEEE. Res int: digital & optical signal processing, communications, circuit & sys theory. *Society Aff:* IEEE.

Liu, Chung L
295 DCL, U of Ill, 1304 West Springfield Ave, Urbana, IL 61801
Position: Prof *Employer:* Univ of IL *Education:* ScD/EE/MIT; SM/EE/MIT; BSc/EE/Cheng Kung Univ *Born:* 10/25/34 Chung L. Liu was born in Canton China on Oct 25, 1934. He was Assistant and Associate Professor of Electrical Engineering at MIT, & is now a Prof of Computer Science at the Univ of Illinois. His research areas include combinational mathematics, analysis of algorithms, & CAD of integrated circuits. *Society Aff:* IEEE, ACM

Liu, Chung-Yen
Business: 405 Hilgard Ave, Los Angeles, CA 90024
Position: Prof. *Employer:* Univ of CA. *Education:* PhD/Aero/CA Inst of Tech; MS/Engg/Brown Univ; BS/ME/Taiwan College of Engr. *Born:* 10/2/33. Asst Prof - Prof, UCLA 1962-Present

Liu, David C
Business: 1550 Northwest Hwy, Park Ridge, IL 60068
Position: Partner. *Employer:* Dames & Moore. *Education:* MS/Civil Engg/OR State Coll; BS/Civil Engg/Kung Shang Coll *Born:* 7/5/20. China. Attended Univ of WA 5 quarters after MS from OR State. Reg P E CA and 6 other states. Engr with Dames & Moore since 1951, Partner since 1961. As consult, performed foundation studies for plants, bldgs, land reclamation, airports, etc. Served as firm's res engr for 3 yrs in Los Angeles after service in several ofcs, was Pacific Mgr in Tokyo for 4 yrs,

Liu, David C (Continued)
managing principal for 6 yrs in Chgo (Park Ridge) before becoming Sr Consult in 1982. *Society Aff:* ASCE, WPCF, SSA, AGU, NSPE

Liu, Donald
Home: 213 King George Rd, Warren, NJ 07060
Position: VP *Employer:* American Bureau of Shipping *Education:* PhD/ME/Univ of AZ; MS/Naval Arch & Marine Engrg/MIT; BS/Naval Arch & Marine Engrg/MIT; BS/Marine Transp/US Merchant Marine Acad *Born:* 1/17/41 Native of Santa Cruz, CA. Sailed as Third Officer aboard various cargo vessels in US Merchant Marine. Joined American Bureau of Shipping in 1966 working on development of computer programs for ship structural analysis. Assigned to Univ of AZ in 1968 for three years to conduct research and develop finite element method computer program system for analysis of marine structures. Assumed current responsibility as VP. Current work involves directing research programs in fields of ocean engrg and structures. Recipient of SNAME Joseph F. Linnard Prize 1980. Author of numerous technical papers on ship & offshore structures and response. *Society Aff:* SNAME, AWS

Liu, Hao-Wen
Business: 139 Link Hall, Syracuse University, Syracuse, NY 13244
Position: Prof Mechanical Engineering, Solid State Sci & Tech. *Employer:* Syracuse Univ. *Education:* PhD/Applied Mechanics/Univ of IL; MS/Mech Engg/Univ of IL; MS/Mgmt/Univ of IL; BS/Mech Engg/Univ of IL. *Born:* 8/20/26. Teaching & res in metal fatigue, fracture mechanics & solid mechanics. Asst Prof, Dept of Theo & Applied Mechanics, Univ of IL, Urbana, IL, 1959-61. Res fellow & sr res fellow, Dept of Aero Engg, CA Inst of Tech, Pasadena, CA, 1961- 63. Joined the faculty of Mtls Sci of Syracuse Univ, Syracuse, NY in 1963. Prof of Mtls Sci since 1968. Prof of Mech Engrg 1984. Elected & appointed as Chrmn of Solid State Sci and Tech for 1977-84. Mbr of the editorial bd of the Int J of Fatigue of Engg Mtls & Structures 1978-present, and Int. J. of Theo. and Applied Fracture Mechanics 1985-present. Awarded a sr visiting fellowship at Cambridge Univ 1976-77. *Society Aff:* ASTM, AIME, ASME, ΣΞ.

Liu, Henry
Business: Civil Engineering, Univ of Missouri-Columbia, Columbia, MO 65211
Position: Prof of Civ Engrg and Natl Gas Pipeline Co. Professor of Engr. *Employer:* Univ of MO. *Education:* PhD/Fluid Mechanics/CO State Univ; MS/Fluid Mechanics/CO State Univ; BS/Hydraulics/Natl Taiwan Univ. *Born:* 6/3/36. Native of Peking, China. US citizen. With Univ of MO-Columbia since 1965. Current position is Prof of Civ Engg and Natural Gas Pipeline Company Professor of Engineering. Specialize in fluid mechanics/hydraulics. Res contributions in flow measurement by the integrating float method, electrokinetic potential fluctuations, novel forms of Bernoulli equation, dispersion in rivers, generated pressure inside buildings, and hydraulic capsule pipeline (HCP). Inventor of electromagnetic means of pumping (U.S. Patent No. 4437799) and injection of capsules (U.S. Patent No. 4334806). Presented several papers on fluid mechanics education at ASEE natl meetings. Organized and taught eleven continuing education short courses on wind effects on bldgs. Chrmn, Aerodynamics Ctte, ASCE, 1976-79. V Chmn, Exec Comm, Aerospace Division, ASCE, 1987-88. Co-recipient of the 1980 ADCE Aerospace Division Award, recipient of Award at 4th Intl Symposium on Freight Pipeline for "immense contribution to the dev of the art, sci, and tech of capsule pipelines–. Chrmn, ASCE Specialty conference "Advancements in Aerodynamics, Fluid Mechanics and Hydrolics," Minneapolis, MN, 1986. Chrmn, NSF/WERC Symposium on High Winds and Building Codes, Kansas City, MO, 1987. Bd of Dirs, WERC, 1986-89. VP, International Freight Pipeline Society, 1985-88. Assoc Editor, Journal of Pipelines, 1985- . *Society Aff:* ASCE, IAHR, AWEA, ΣΞ, WERC.

Liu, Jane WS
Business: Dept of Computer Sci, 1304 W. Springfield Ave, Urbana, IL 61801
Position: Prof *Employer:* Univ of IL *Education:* ScD/EE/MIT; EE Deg/-/MIT; MS/EE/MIT; BS/EE/Cleveland State Univ *Born:* 12/11/35. Worked for Radio Corp of America from 1960-62 and for the Mitre Corp from 1962-64 on the design of high speed digital circuits and memory systems. After receiving ScD in 1968 from MIT worked for the U.S. Dept of Transportation in Cambridge, MA on design of communications systems in air traffic control systems. Joined Univ of IL in 1974. Research activities are in the areas of computer networks and distributed systems. *Society Aff:* IEEE-CS

Liu, Joseph T C
Business: Div of Engg, Providence, RI 02912
Position: Prof of Engg. *Employer:* Brown Univ. *Education:* PhD/Aeronautics/CA Inst of Tech; MSE/AeronEng/Univ of MI; BSE/AeronEng/Univ of MI; BSE/Math/Univ of Mi. *Born:* 11/9/34. Engg, Convair-Fort Worth, 1958-60. Res Assoc, Gas Dyn Lab, Princeton Univ, 1964-66. Assnt Prof of Engg, Brown Univ, 1966-69. Assoc Prof of Engg, Brown Univ, 1969-73. Prof of Engg, Brown Univ, 1973-present. On sabbatical leave at Imperial College of Sci and Tech, London, England, 1972-73 and 1979-80. Areas of res interests; nonlinear hydrodynamic stability, coherent structures in turbulent shear flows, sound generation by fluid flows, fluidized bed instabilities, hypersonic flow theory. *Society Aff:* APS/DFD, AMS, ASME, AIAA.

Liu, Lee
Business: Box 351, Cedar Rapids, IA 52406
Position: Chrmn of Bd and Pres *Employer:* IE Industries Inc. *Education:* BS/EE/IA State Univ. *Born:* 3/30/33. Joined IA Elec Light and Power Co in 1957, assumed chief executive officer responsibility in 1983. In 1986, he became the President and Chief Executive Officer of the parent company, IE Industries Inc. Has been actively involved in America's commercial reactor prog. Served on Steering Committee of the Utility Nuclear Waste Mgmt Group. A member of IEEE. Lecturerd at univ level on energy related matters as well as written numerous articles on both utility engr and mgt. In 1984 he was recognized by Iowa State Univ for his professional achievements in engrg. *Society Aff:* IEEE.

Liu, Philip L F
Business: Hollister Hall, Sch of Civ & Environmental Engg, Ithaca, NY 14853
Position: Prof *Employer:* Cornell Univ. *Education:* ScD/Hydrodynamics/MIT; SM/Civil Engrg/MIT; BS/Civil Engrg/Natl Taiwan Univ *Born:* 12/11/46. Born Fuchow, China in 1946. Appointed Asst Prof in Sch of Civ & Environmental Engrg at Cornell in 1974. Promoted to Assoc Prof with tenure in 1979 and Full Prof in 1983. Appointed Assoc. Director of School of Civil & Environmental Engrg in 1985. Appointed Assoc. Dean for Undergraduate Programs in Engrg College in 1986. Conduct res projs sponsored by NSF & Sea Grant in the area of Wave hydrodymics, coastal currents & ground water flows. Visiting asst prof at Univ of DE, 1976. ASCE Huber Res Prize, 1978; Justice Asst Prof at Cornell, 1978-1979. Guggenheim Fellow 1980; Visiting Assoc at Caltech 1980-81. Visiting Prof. Technical Univ. of Denmark 1987-88. *Society Aff:* ASCE, AGU

Liu, Ruey-Wen
Business: Notre Dame, IN 46556
Position: Prof *Employer:* Univ of Notre Dame *Education:* PhD/EE/Univ of IL; MS/EE/Univ of IL; BS/EE/Univ of IL *Born:* 3/18/30 in Kiangsu, China. Since 1960, he has been with the Univ of Notre Dame, and became a Prof in the Dept of Electrical Engrg in 1967. He has held visiting professorships at the Univ of CA at Berkeley (1965-66, 1977-78 and 1984-85), the Natl Taiwan Univ of the Republic of China (Spring 1969), the Universidad de Chile, Santiago de Chile (Summer 1970), and the Inst of Mathematics, Academia Sinica of the Republic of China (Summer 1976 and Summer 1978). Is a member of IEEE and Sigma Xi. He is an IEEE Fellow. *Society Aff:* IEEE, ΣΞ

Liu, Tony C
Home: 1284 Towlston Rd, Great Falls, VA 22066
Position: Structural Engineer *Employer:* U.S. Army Corps of Engineers *Education:* PhD/Structural Engg/Cornell Univ; MS/Civil Engg/SD Sch of Mines; BS/Civil Engg/Chung-Hsing Univ. *Born:* July 1943. Reg P E Calif. Miss. Recipient of two American Concrete Institute's Wason Research Medals (1974 and 1983). With U.S. Army Corps of Engineers since 1976, responsible for staff review and approval of major structural design and repair and rehab projects including dams, navigation locks, flood control structures, and hydroelectric powerhouses. Chairman of American Concrete Institute's technical committee on Rehabilitation. Elected Fellow of American Concrete Institute in 1984. Published more than 60 technical papers related to concrete repair and concrete structural design. *Society Aff:* ASCE, ACI, USCOLD

Liu, Wing Kam
Business: Dept of Mech Engr, Evanston, IL 60201
Position: Assoc Professor *Employer:* Northwestern University *Education:* PhD/CE/CA Inst of Tech; MS/CE/CA Inst of Tech; BS/Engg Sci/Univ of IL. *Born:* 5/19/52. in Hong Knog. Currently Assoc Prof in Mech and Civil Engrg at Northwestern University. Academic Experience: Assistant Professor of Northwestern University, Evanston, IL (1980-1983). Melville Medal (1979) of Am Soc of Mech Engrs. Listed in Who's Who in Technology Today, U.S.A. (1981), Listed in Outstanding Young Men of America (1981), N.S.F., NASA, ARMY, ONR. Research Grants (1981-present). 1983 Ralph R Teetor Award of Amer Soc of Auto Engrs; 1982 Listed in International Who's Who in Engineering; 1984 Listed in Who's Who in Frontier Science and Technology USA. Pi Tau Sigma Gold Medal (1985) of ASME. Authors and coauthors of 100 technical papers, Consultant to Argonne National Lab (1981-present), USA Ballistic Research Lab. *Society Aff:* ASCE, ASME, AAM, ICE.

Liuzzi, Leonard, Jr
Home: 3 Shaker E1, Loudonville, NY 12211
Position: Supervisory Metallurgist. *Employer:* Dept of Army, Watervliet Arsenal NY. *Born:* July 1931. BSME Lowell Univ; grad studies MIT, U of Denver, RPI. Native of Albany NY. USAF 1951-55. With Gov since 1961, as interdisciplinary M E & metallurgist serving as Project Leader in Mfg Methods Tech Programs in the metal forming & processing area. Chmn Failure Analysis Ctte. Currently Program Leader dir a group of engrs & other personnel engaged in estab new tech (rotary forging) in the manufacture of cannon forgings, involving the first major change in 50 yrs. Mbr ASM Mech Working & Forming Div, Forging Ctte, SAE Fatigue Ctte & Military Tri-Service Adv Group for Manufacturing Tech. Co-chmn NAMRC IV, Dept of Army Commendations 1966, 1973, 1975, 1976. *Society Aff:* ASM.

Livingood, Marvin D
Home: 2603 Landor Ave, Louisville, KY 40205
Position: Sr. Engr. Partner *Employer:* RCI Limited Engg Syst. Consultants *Education:* PhD/ChE/MI State Univ; MS/ChE/OK State Univ; BS/ChE/OK State Univ. *Born:* 8/15/18. Kansas, early schooling OK. Instr, MO Sch of Mines (undergrad & grad courses, responsible charge of lab dev). Asst prof & res asst prof, MI State. Consultant to MI industry, responsible for dev of valuable byproduct recovery processes (MI Canners' Assn award). With Du Pont 1952-1983 first as res engr. Responsible for maj process dev & process safety evaluation; co consultant in this field since 1965. Environmental engr & sr environmental engr since 1971, responsible for dev of waste treatment processes & compliance with environmental regulations. Conslt in engrg systems, chemical process safety, and energy conservation in private and governmental organizations, 1983- . Interests: career dev, classical music performer, constructive politics. *Society Aff:* AIChE, AES, NSPE, ICE.

Livingston, John A
Business: 420 Lexington Ave-1632, New York, NY 10170
Position: Consultant (Self-employed) *Education:* -/Naval Arch/Webb Inst of Naval Arch; -/Marine Engg/Webb Inst of Naval Arch. *Born:* Aug 1902 NYC. Pres Goldschmidt Corp 1938-68 - marine designer, worldwide pats (used on 8000 ships). Pres Goldschmidt Chem Corp 1938-68 - pioneers var chem processes incl original synthetic resin films for producing waterproof plywood (aircraft & boat hulls). Pres Cerium Metals Corp 1941-59 - leading manufacturer rare earth metals - worldwide pats for production ductile iron. Chmn Emer Webb Inst of Naval Arch. Mbr VP Soc of Naval Archs & Marine Engrs. *Society Aff:* SNAME, ASNE.

Lloyd, Davie L
Business: 3254 Winbrook Dr, Memphis, TN 38116
Position: Pres. *Employer:* Health Engg Res Inc. *Education:* BSME/Mech Engg/Christian Brothers College-Mem TN. *Born:* Mar 1931. Served on Bd of Governors ASHRAE Memphis Chap. Served on Bd of Dir, Cons Engrs of Tenn & Cons Engrs of Memphis. Served as Chmn, Energy Ctte ASHRAE Memphis Chap. Served on ad hoc ctte for State of Tenn regarding energy conservation matters for state bldgs. 'Featured Engr' of Memphis Chap ASHRAE for 1976 for energy conservation work. Formed TLM Assocs Inc, Cons Engrs 1964, serving as V P with respon charge of Mech Engrg. Sold financial interest in TLM Assoc Inc in 1978, and organized firm of Health Engg Resources Inc, wich he serves as Pres. Also serving as Exec VP of Drexel Toland and Associates, a specializing in the development of medical office buildings for "not-for-profit" hospital campuses. Serves as coordination of design and construction. *Society Aff:* NSPE, ASHRAE.

Lloyd, Fredric R
Home: 402 Westwood Rd, Indianapolis, IN 46240
Position: Vice Pres, Production Operations. *Employer:* Eli Lilly & Co. Retired. *Education:* PhD/Chem Eng/Purdue Univ; BS/Chem Eng/Purdue Univ. *Born:* 1923. BS 1944, Reg P E Ind. With Eli Lilly since 1949 - Production Mgmt, V P European Opers. Currently in charge of processing & pharmaceutical mfg facils outside continental U S. Mbr Sigma Xi, ACS, AIChE, Lt Governor's Sci Adv Ctte. Disting Engrg Alumnus Purdue Univ 1973 Shreve Distinguished Fellow Purdue U. 1984. Retired. Director Fluid Mechanics W. Lafayette, Ind. *Society Aff:* AIChE, ACS, NSPE, ΣΞ, AAAS.

Lloyd, John R
Business: Aerospace & Mech Engg, Notre Dame, IN 46556
Position: Prof. *Employer:* Univ of Notre Dame. *Education:* PhD/ME/Univ of MN; MSME/ME/Univ of MN; BSE/Engg/Univ of MN. *Born:* 8/1/42. Educated at Univ of MN, Dev Engr for the Procter & Gamble Co, Cincinnati, OH 1966-1967, Asst Prof - Prof Univ of Notre Dame, 1970-present. Published in numerous technical journals. Have served as Mbrship Dev Chrmn of Heat Transfer Div ASME, currently serve as chrmn K-11 Committee on Heat Transfer in Fire and Combustion Systems of Heat Transfer Div ASME. Received first Outstanding Paper Award of ASME Heat Transfer Div 1977, Received Melville Medal ASME 1978. *Society Aff:* ASME, ASEE, The Combustion Institute.

Lloyd, William
Business: 2215 W Old Shakopee Rd, Bloomington, MN 55431
Position: Utilities Supt. *Employer:* City of Bloomington. *Born:* 6/5/25. A non-graduate - 2 yrs college - with cont educ in sanitary field. Native of Bemidji MN. 33 yrs water-wastewater const & oper in communities from 3500 to 85000 pop. 1972-Trustee & officer; 1976 Chmn; 1977 to Present, Secretary- treasurer N Central Sect AWWA. Chairman State of MN Water Wastewater Operator Cert Council. Water Cond Adv Bd. Key Water Official for Mgmt of Health & Water Resources in MN Emergency Resource Mgmt Plan. Current pos - 1960 - oper of water-wastewater sys with 650 miles combined piping - water treatment - 24000 customers, 50 employees, 5.7 million dollar budget, community 70% developed. *Society Aff:* AWWA, WPCF.

Lo, Arthur W
Home: 102 Maclean Circle, Princeton, NJ 08540
Position: Prof of EE & Computer Science. *Employer:* Princeton Univ. *Education:* PhD/EE/Univ of IL; MA/Physics/Oberlin College; BS/Physics/Yenching Univ. *Born:* May 1916 Shanghai China. Tech staff & group leader RCA Labs 1951-60. Mgr Ad-

Lo, Arthur W (Continued)
vanced Tech Dev & Mgr Explor Dev IBM 1960-64. Prof of Elec Engrg & Computer Sci, Princeton Univ 1964- . Num tech papers & US pats, 2 books - field of digital electronics & computer struct. Fellow IEEE. *Society Aff:* IEEE, ΣΞ.

Lo, Donald T
Business: 1259 S Beretania St, Honolulu, HI 96814
Position: President. *Employer:* Lo Engrg Co Inc. *Education:* BS/Civil Engg/Rose-Hulman; BA/Economics/Univ of HI. *Born:* Nov 1919. Private practice since 1947, principally in Hawaii - also Thailand, Okinawa, Taiwan, Philippines, Samoa & Japan. 1945-47 with York & Sawyer on const of Tripler Genl Hosp Hawaii. 1944 with Bethlehem Steel San Fran. Progressive Arch Award 1961, 1962. Lectr Univ of Hawaii Engrg Dept 1954, 1956 Notable projects: Hawaii State Capitol, Federal Courthouse and Office Building and Honolulu International Airport. *Society Aff:* ACEC, ASCE, NSPE, ACL, PCI.

Lo, Robert N
Business: 59 Eisenhower Lane South, Lombard, IL 60148
Position: Gen Mgr. *Employer:* Xytel Corp. *Education:* PhD/Chem Engg/Univ of MO, Columbia; MS/Chem Engg/Univ of MO, Columbia; BS/Chem Engg/Cheng Kung Univ. *Born:* 2/12/44. in China. Grew up in Taiwan. Came to the US of America in 1968. Res Engg for Argonne Nation Lab, working on coal gasification and liquefaction. With Xytel Corp since 1978. Respon, as Gen Mgr, for the design and manufacturing of process res systems for petrochemical petroleum refinging and new energy researches. Enjoy gardening. *Society Aff:* AIChE.

Lo, Yuen T
Home: 704 Silver St, Urbana, IL 61801
Position: Director, Electromagnetics Laboratory *Employer:* Univ of Illinois - EE Dept. *Education:* PhD/EE/Univ of IL. *Born:* 1920. PhD & MS Univ of Ill; BS Natl Southwest Assoc Univ China. Instr, Natl Tsing Hua Univ China. Proj engr Channel Master Corp 1952-56. Prof in Elec Engrg Univ of Ill - present. Cons to Westinghouse Elec Corp, Andrew Corp, Natl Observatory, Amer Electronic Labs Inc, IBM, Emerson Elec, Raytheon Co. Harris Corp, Whirlpool Corp, JPL, TRW, Ford Aerospace & Communications Corp. Halliburton Award for Engrg Education Leadership. Fellow IEEE. IEEE John T Bolljahn Award 1965. Disting Lecturer IEEE/AP-S. IEEE AP-S Best Paper Awards 1964 and 1979. *Society Aff:* IEEE, URSI, NAE.

Lobo, Cecil T
Business: 5500 Wabash Ave, Terre Haute, IN 47803
Position: Prof of Civil Eng *Employer:* Rose-Hulman Institute of Tech *Education:* PhD/CE/Purdue Univ; MS/CE/Univ of Notre Dame; BE/CE/Gujarat Univ India *Born:* 9/22/34 b. in Magalore, India to Anthony V. and Priscilla V. Lobo. Raised in Bombay where I attended St. Stanislaus' High School and St. Xavier's College. Worked two years in construction of canals and buildings as an Assistant Field Engineer. Proceeded on to higher education at the Univ of Notre Dame where I served as a Teaching Asst and then to Purdue Univ gaining further experience as a Teaching Asst and then as a Research Asst. Joined the faculty of Civil Engrg at Rose-Hulman Inst of Tech and reached the rank of Prof in 1971. Was Prof-in- Charge of Civil Engrg from 1971 to 1982. Was married in 1972 and have two children, Colin (13 yrs.) and Trevor (9 yrs.) Received Outstanding Educator of America Award in 1975 and Awards for Outstanding Service to ASCE in 1975, 1978, 1979, 1982 and 1983. Received Inland-Ryerson Outstanding Teacher Award in 1984. *Society Aff:* ASCE, ASEE, ΣΞ, ACI.

Lobo, Walter E
Business: 497 Lost District Dr, New Canaan, CT 06840 *Employer:* Retired *Education:* BS/Chem Engg/MIT. *Born:* Mar 29, 1905 Brooklyn N Y. Ch Chem & Factory Supt Ingenio Manuelita, Colombia 1928; M W Kellogg Co 1929-57 as process engr, then Dir of Chem Engrg Div; in private consulting practice 1957 to date. Author 24 papers; holder of 13 pats in furnace & chem process fields. Mbr 8 natl prof soc's; Fellow & recipient Founders Award AIChE 1970, Dir 1954-57; EJC Dir 1956-58; United Engrg Trustee, Treas, V P & Pres 1964-66; Engrg Found Bd of Dir; Pan Amer Fed of Engrg Soc's (UPADI) V P 1964-67. *Society Aff:* AIChE, ASME, ACS, CEPP, NSPE.

LoCasale, Thomas M
Home: 29 Stringer Dr, Doylestown, PA 18901
Position: Sr VP Engr *Employer:* Aydin Monitor Systems *Education:* MS/EE/Univ of PA; B/EE/Villanova Univ *Born:* 3/24/38 Engr with Univac, 1960-65, developed nanosecond logic and thin film memories. In 1965, joined Monitor Sys ($M co now the $150M Aydin Corp) as a circuit design engr. Presently Sr VP of Aydin Monitor Sys. Dir of Engg for all Signal Recovery and Processing Equip including PSK Modems, PCM Synchronizers, Decommutators, Multiplexers, Error Correction Codecs and Bit Error Rate Testers. Have two patents, authored several papers. Secy Phila Chapter AFCEA 1975-79. Sessions Chmn for several Telemetry Conferences, Speaker for IEEE student chapters. Active in community. Enjoy ballroom dancing and chess. *Society Aff:* IEEE, AFCEA, Old Crows.

Lochner, Harry W, Jr
Home: 573 Jackson Ave, Glencoe, IL 60022
Position: Pres. *Employer:* H W Lochner, Inc. *Education:* MBA/Bus/Harvard Bus Sch; BS/Ind Admn/Yale Univ. *Born:* 7/29/36. Native of Chicago, IL. Served in USAF 1958-59. Joined H W Lochner in 1961 - Exec VP in 1970 & assumed Presidency in 1974. Responsible for gen mgt of the firm, participates in the review of all engrg assignments, supervises marketing, proj mgmt & client relations. Special interests include dev of route location criteria, socioeconomic measurements & environmental impacts of civ works projs. Also major engr proj and prog mgmt. Reg PE in fourteen states. Enjoys tennis, sailing & sking. *Society Aff:* ASCE, NSPE, ARTBA, TRB.

Lock, Gerald S H
Home: 11711-83 Ave, Edmonton, Alberta Canada
Position: Prof *Employer:* Univ of Alberta. *Education:* BSc/ME/Univ of Durham; PhD/ME/Univ of Durham. *Born:* 6/30/35. Native of England; now Canadian citizen. Served engg apprenticeship with N Eastern Marine Engg (England) before entering Univ. Emigrated to Canada (1962) as Asst Prof, now Prof. Dept of Mech Engg, Univ of Alberta: formerly Dean of Interdisciplinary Studies. Past pres, Canadian Soc for Mech Engg. Awarded Queen Elizabeth II Silver Jubilee Medal (1977), Fellow Eng. Inst. Canada, Can. Soc. Mech. Eng. *Society Aff:* CSME, IMechE, ASME, EIC, NECIES, AAAS.

Locke, Carl E
Business: 4010 Learned Hall, Univ. of Kansas, Lawrence, KS 66045
Position: Dean, School of Engrg *Employer:* Univ of KS. *Education:* PhD/Chem Engg/Univ of TX, Austin; MS/Chem Engg/Univ of TX, Austin; BS/Chem Engg/Univ of TX, Austin. *Born:* 1/11/36. After MS worked six yrs in R&D and market dev of anodic protection systems for Coninental Oil Co, Ponca City, OK. Returned to School for PhD and worked in Thermal Analytical Instrumentation simultaneously. Joined Chem Engg faculty at Univ of Oklahoma, in 1973. Joined Univ of Kansas as Dean of Engrg in January 1986. Present res interests are corrosion of steel in concrete. Teach courses in Corrosion. Inveterate jogger and handball player. *Society Aff:* AIChE, NACE, ASTM, ASEE, NSPE.

Lockie, Arthur M
Home: 3630 Mt Hickory Blvd, Hermitage, PA 16148
Position: Consulting Engr. *Employer:* Self-employed. *Education:* MEE/Power Engg/RPI; EE/Power Engg/RPI. *Born:* Sept 1909 Buffalo N Y. Instr in EE at RPI 1931-36. Design engr Westinghouse Distrib Transformer Div 1936; Mgr Long-Range Major Dev 1957; cons engr on transformer tech 1970 to retirement 1974. Now self-employed cons on power distrib. P Pres: Midwestern Chap PSPE; Sharon Sect IEEE; Mahoning Valley Tech Soc's Council. Life Fellow IEEE, Mbr Transformer

Lockie, Arthur M (Continued)
Ctte, Chmn Insulation Life Sub-ctte. Div Mgr ElecElectronics Insulation Conf. Westinghouse Special Patent Award & Product Engrg Master Design Award. Mbr Sigma Xi & Tau Beta Pi. Author num papers on transformer application. *Society Aff:* IEEE, NSPE.

Lockwood, Daniel C
Business: 1000 S Fremont, Alhambra, CA 91802
Position: Vice President and Chief Engr. *Employer:* C F Braun & Co. *Education:* BS/Engr/Stanford Univ; Engr/Mech Engr/Stanford Univ. *Born:* Mar 1918. BA 1939, Degree of Engr 1941, both from Leland Stanford Jr Univ. With C F Braun & Co since 1941. Assumed current respon as Ch Engr in 1969. Mbr Bd of Dir of Fractionation Res Inc, Heat Transfer Res Inc, & Particulate Solids Res Inc. *Society Aff:* AIChE.

Lockwood, Robert W
Home: 2136 E Vine Ave, West Covina, CA 91791
Position: Chmn of the Bd. *Employer:* Diversified Mgmt Internatl. *Education:* LLD/Engr/Northrop Univ; MBA/Bus Mgt/Univ of CA-Berkeley; BS/Bus Mgt/Univ of CA-Berkeley. *Born:* June 1924. Past pos incl: V P United Calif Bank in charge of Corp Planning, Corp Finance & Indus Engrg; V P, Asst to the Pres & VP of Academic Affairs, & Prof of Mmt, Grad School of Mgmt, Northrop Univ., Assistant to the President and Chief Operating Officer, Bradston Hurricane International, Inc.; and President, Diversified Management International. Internatl Pres of Amer Inst of Indus Engrs & mbr of Bd of Dir. Listed in Who's Who in Banking, Who's Who in Amer & was on the Bd of Examiners for Indus Engrg, Calif. *Society Aff:* IIE.

Loebel, Fred A
Home: 4530 Pine Tree Dr, Boynton Beach, FL 33436
Position: Retired *Education:* BS/ME/Univ. of Wis. *Born:* 07/16/16 Project Engr. Std. Oil Cal. 38-42. Aqua-Chem, Inc. 43-69. Worked primarily on boilers & sea water desalting plants. VP Engrg 1950-59. Pres. 1960-69. Retired 1969 when Aqua-Chem, Inc. (NYSE) bought by Coca-Cola. Holder of approx. 50 US & foreign patents on boilers & sea water desalters. Life Fellow, ASME. Distinguished Service Award, Univ. of WI. *Society Aff:* ASME

Loehr, Raymond C
Business: Civil Engrg Dept, University of Texas, 8.614 ECJ Hall, Austin, TX 78712
Position: Hussein M. Alharty Centennial Chair in Civil Engrg *Employer:* Univ of Texas - Coll of Engrg *Education:* PhD/Civil Eng/Univ of WI; MS/Civil Eng/Cast Inst of Tech; BS/Civil Eng/Case Inst of Tech. *Born:* May 1931. Previous pos: Asst Prof Case Inst of Tech; Assoc Prof, Prof Univ of Kansas; Liberty Hyde Bailey Prof of Engrg, Cornell Univ; with Cornell 1968-1985. Dir Environ Studies Prog, Cornell, 1972-80; with Univ of Texas since 1985. Mbr NAE; Mbr NAS-NRC sub-cttes since 1972; mbr Sci Adv Bd EPA; dev Agri Waste Mgmt Prog at Cornell. Author 3 bks; editor 3 books; Assoc Editor 2 journals; author over 180 tech articles. Rudolph Hering Medal ASCE 1969; Water Conservationist of the Yr, Kansas Wildlife Fed 1967; Diplomate Amer Acad Environ Engrs; cons to FAO, UNEP, EPA, indus & fed orgs. *Society Aff:* ASCE, WPCF, NAE, SETAC

Loehrke, Richard I
Home: 1901 Rangeview Dr, Ft Collins, CO 80524
Position: Assoc Prof. *Employer:* CO State Univ. *Education:* PhD/ME/IL Inst of Tech; MS/ME/Univ of CO; BS/ME/Univ of WI. *Born:* 5/11/35. in Milwaukee, WI. Worked as res engr for GE Co and for Sundstrand Corp. Taught at IL Inst of Tech and currently at CO State Univ. Res interests are in the areas of heat transfer and fluid mechanics. ASME Robert T Knapp Award, 1977. *Society Aff:* ASME, AIAA.

Loewen, Erwin G
Home: 44 Westwood Drive, E Rochester, NY 14445
Position: Vice Prec. RD&E *Employer:* Milton Roy Co. *Education:* ScD/Mech Eng/MIT; ME/Mech Eng/MIT; MS/Mech Eng/MIT; BMech Eng/Mech Eng/NY Univ. *Born:* 4/12/21. Frankfurt Germany. Bembridge School, Isle of Wight, UK; Asst Prof Mech Engrg MIT 1952-55; Tech Dir Taft-Pierce Mgr Co Woonsocket R I 1955-60; Dir Gratings & Metrology R&D, Bausch & Lomb Inc. Rochester N Y 1960-1982. Gen Mgr; Grating Operations, Bausch & Lomb Inc 1983- . Fellow ASME; Mbr Sigma Xi, AAAS, Fellow Optical Soc of Amer, Internatl Inst for Prod Engrg Res. Chmn ASME Comm B- 89, Dimensional Metrology. *Society Aff:* ASME, OSA, SPIE, CIRP, AAAS, SME

Loewenstein, Paul
Business: 2229 Main St, Concord, MA 01742
Position: V Pres & Tech Director. *Employer:* Nuclear Metals Inc. *Education:* ME/-/Stevens Inst of Technology. *Born:* Jan 1921 Germany. Present pos: V P & Tech Dir Nuclear Metals Inc. Worked at NMI & its predecessors, starting with the MIT Met Lab (Manhattan Proj - AEC) since 1946. Main areas of int: plastic working of metals, powder met, melting & casting, fabrication of nuclear fuel elements & util of depleted uranium. Author num tech papers & contrib of chaps in reference bks on beryllium, zirconium & nuclear fuel elements. Mbr ASM, AIME, APMI, SAMPE. Mbr Extrusion & Drawing Activity Ctte, Mech Working & Forming Div ASM. Hobbies: skiing, sailing, tennis. *Society Aff:* ASM, AIME, APMI, SAMPE, SME.

Loewenstein, Walter B
Business: 3412 Hillview Ave, Palo Alto, CA 94303
Position: Deputy Director, Nuclear Power Division *Employer:* Electric Power Research Inst. *Education:* PhD/Physics/OH State Univ; BS/Physics, Math/Univ of Puget Sound. *Born:* Dec 23, 1926. Prior to joining EPRI Nov 1973, Dir of the Applied Phys Div, Argonne Natl Lab (ANL), Argonne Ill. Joined ANL 1954 as an Asst Physicist, advancing successively to Assoc Physicist, Sr Physicist, Assoc Dir EBR-II Proj, Dir EBR-II Proj, Dir Appl Phys Div 1973. Author more than 20 pubs & holds 2 nuclear reactor pats. Fellow Amer Phys Soc & Amer Nuclear Soc. Served 3 yrs on the Atomic Energy Comm's Adv Ctte on Reactor Physics. *Society Aff:* ANS, APS.

Loewer, Otto J
Business: Agri Engr Dept, Lexington, KY 40546
Position: Professor *Employer:* Univ of KY. *Education:* PhD/Agri Engr/Purdue Univ; MS/Agri Engr/LA State Univ; BS/Agri Engr/LA State Univ; MS/Agri Economics (1980)/Michigan State Univ *Born:* 2/20/45. Native of Wynne, AR. Employed by the Agri Engr Dept, Univ of KY since 1973. Cooperative Extension service responsibilities included assisting farmers in layout, design and determination on the economic feasibility of grain harvesting, delivery, handling, drying and storage systems. Res efforts are in the analysis of total farm production systems. Teaching a course on the simulation of agri, biological and engr systems. *Society Aff:* ASAE.

Loewy, Robert G
Business: JEC 4010, Troy, NY 12181
Position: Inst Prof. *Employer:* Rensselaer Polytechnic Inst. *Education:* PhD/Engg Mechs/Univ of PA; MS/Aero Eng/MIT; BAE/Aero Eng/RPI. *Born:* Feb 1926. Sr Vibrations Engr Glenn L Martin Co. (1948-49) Res engr, Staff Tech Asst to the Hd, Aeromechanics Dept, & Principal Engr Cornell Aeronautical Lab. Dynamics engr, (1949-55) Ch of Dynamics & Ch Tech Engr Vertol Div, Boeing Co. (1955-62) Univ of Rochester Assoc Prof ('62) & Prof of Mech & Aerospace Sci, (1965), Dir of the Space Sci Ctr, (1966) and Dean of the Coll of Engrg & Appl Sci (1967-74). AF Ch Scientist (1965-66). RPI VP & Provost (1974-78). Cons to indus & Gov. Bd of Governors - Ctr for Naval Analyses. Chmn of the AF Sci Adv Bd, Chmn of the FAA Tech Adv Panel, Chmn of the NASA Aero Adv Committee; Bd of Dir Rochester Appl Sci Assn, (1966-74) Genl Tech Corp, (1966-68) Mohasco Ind. (1974-pres.) Vertical Flight Foundation (1986-pres.) Spec areas of interest incl: rotary wing & vertical take-off & landing aircraft, unsteady aerodynamics & struct dynamics. Exceptional Civilian Serv Awards USAF 1966, 1975 & 1985. L. Sperry Award AIAA '58. Hon Fellow 1966 AHS, Fellow AIAA '74, Fellow AAAS '81. Distinguished

Loewy, Robert G (Continued)
Public Service Medal NASA '83. Mbr 1971 NAE. *Society Aff:* ASEE, AIAA, AHS, NAE, AAAS.

Loferski, Joseph J
Business: Div of Engrg, Providence, RI 02912
Position: Professor of Engrg. *Employer:* Brown University. *Education:* ScB/Physics/Univ of Scranton; MSc/Physics/Univ of PA; PhD/Physics/Univ of PA. *Born:* Aug 1925. Mbr Tech Staff RCA Labs, Princeton N J 1954-60. Res on radiation effects on semiconductor matls & devices & on conversion of solar energy into electricity; 130 publications. Joined fac of Brown Univ in Jan 1961; Prof of Engrg since 1966 & Chmn of Div of Engrg 1968-74. Assoc. Dean Grad. School 1980-83. . Cons to Dir of European Space Tech Ctr, Noorwijk Holland 1967-68. U S Natl Acad of Sci Exchange Fellow at Inst of Nuclear Res, Warsaw Poland 1974-75. Science Counselor, US Embassay, Warsaw 1985-1987. Member Editorial Boards of the Journals "Solar Energy Materials–, and "Energy Conversion" Freeman Award, Providence Engin. Soc. 1974; W.R. Cherry Award, IEEE Photovoltaic Conf. 1981. Fellow IEEE, AAAS; Sigma Xi, Tau Beta Pi. *Society Aff:* IEEE, APS, AAAS, ΣΞ, TBII.

Lofgren, Richard C
Home: 2806 Boyer East, Seattle, WA 98102
Position: Assoc. *Employer:* R W Bech & Assoc. *Education:* BS/Civil Engg/IL INst of Tech. *Born:* 1/8/25. A licensed civil and structural engg Mr Lofgren participates in the planning, management and dev of consulting engineering services. Since 1957 he has been with R W Bech & Assoc, Engineers and Consultants. The Firm has ten offices in the US. His major tech work has been in the field of hydroelectric power, dams and water resources but also includes water and sewer systems, bridges and buildings. He served as Pres of the Seattle Section, American Soc of Civil Engineers in 1976. Born, LaCrosse, Wisconsin. Naval veteran WWII 1942-46. Wife, Frances, two children, one grandchild, Benjamin. *Society Aff:* ASCE, APWA, AWRA, WPCF, APA.

Lofredo, Antony
Home: 38 Skylark Rd, Springfield, NJ 07081
Position: Manager Production Engg. *Employer:* Airco Inc. *Education:* BChE/NY Univ. *Born:* 9/28/26. Mbr Tau Beta Pi. Entire prof career has been spent with current employer in a var of assignments. Spec in process engrg for chem & cryogenic processing plants with increasing tech managerial respon from Supervising Engr to Dir of Process Design & Dev. More recent assignments have incl VP of Engr, Genl Mgr of Tech Dev. and Genl Mgr of Cryogenic Plants Marketing. Current respon is in production engg targeted to improve efficiency and energy utilization. *Society Aff:* TBII.

Loftfield, Richard E
Home: 11602 Oriole Pl, Chardon, OH 44024
Position: President. *Employer:* Huron Chemicals of America. *Education:* BChE/-/OH State Univ; Grad work/Chem engg/Univ of MN. *Born:* June 1921. Bach of Chem E Ohio State Univ 1943. Officer US Navy 194445. Mbr AIChE, ACS, Reg Prof Engr Ohio. Holder of pats in var areas of electrochem; author of papers on electrochem iron, & uses of dimensionally stable anodes for electrochem operations. Has devoted 35 yrs to the area of cost & energy reduction for chem processing, esp pertaining to chloro alkali, sodium chlorate, sodium hypochlorite, seawater electrolysis, iron, copper & manganese dioxide, with particular emphasis to the dev, production & use of dimensionally stable electrodes. Assumed present pos of Pres of Huron Chem of Amer in 1978. Enjoys sailing, flying & golf.

Loftness, Robert L
Business: 1800 Massachusetts Ave, Washington, DC 20036
Position: Dir, Wash Office. *Employer:* Elec Power Res Inst. *Education:* DSc/CE/Swiss Fed Inst of Tech; MS/Chemistry/Univ of Wa; BS/Phys Chemistry/ Univ of Puget Sound. *Born:* 2/14/22. Res chemist, Univ of CA. Radiation Lab 1943-45; US Army, 1945-46; Res Engr, No Am Aviation, 1949-51; Sci Attache, Am Embassy, Stockholm, 1951-53; Proj Engr, No Am Aviation, 1953-60; Commercial Dir, Dynatom, Paris, France, 1960-61; Dir, Eastern Region, Atomic Intl Div, No Am Rockwell, 1961-70; Dep Dir, Atomic Energy Office; US Dept of State, 1970-73; EPRI, 1973-. Author, "Nuclear Power Plants," Van Nostrand, 1964; "Energy Handbook," Van Nostrand Reinhold, 1978, 2nd Ed, 1984. Married, 3 children. *Society Aff:* ANS, AAAS.

Loftus, William F
Business: 120 Charlotte Pl, Engelwood Cliffs, NJ 07632
Position: Pres, CEO *Employer:* William F. Loftus Assocs *Education:* B/CE/Manhattan Coll *Born:* 09/09/33 in Brooklyn, NY. After 12 years in the pile driving construction field, started own firm in 1968. Over 2000 jobs completed under his supervision varying from one day consultation to total design and field responsibility of $12,000, 000 foundation projects. A lecturer, holder of 3 patents relating to pile instrumentation. A special conslt to The Corps of Engrs on all dams/ locks on Mississippi & Red Rivers. Was the 1981-1982 and is the 1984-1985 Pres of the Deep Foundations Inst and Author of several papers and articles concerning pile foundations. Expert witness on several major foundation law suits. *Society Aff:* DFI, ASCE, ASPE, AWPI.

Logan, Horace P
Business: P O Box 1492, El Paso, TX 79978
Position: Sr VP. *Employer:* El Paso Natural Gas Co. *Born:* Apr 1920. BSME New Mexico State Univ. Native of Oklahoma. Attended Oklahoma Baptist Univ & Peabody College, Nashville Tenn, prior to WW II; NMSU also prior to WW II & for 1 1/2 yrs after. Worked for Boeing Aircraft, Seattle Wash for 1 yr after graduation; Southwestern Pub Serv Co, Roswell N M 9 mos; with El Paso Natural Gas Co for last 29 yrs. V P & Supt of Oper for 11 yrs, respon for all phases of gas production & transmission & sales. Sr VP respon for all phases of operations & engg since 5/1/77. Enjoy music, golf, fishing & all sports. Belong to ASME; Reg P E & Surveyor in New Mex.

Logue, Joseph C
Home: 52 Boardman Rd, Poughkeepsie, NY 12603
Position: IBM Fellow. *Employer:* IBM. Retired *Education:* MEE/Electronics/Cornell Univ; BEE/Electronics/Cornell Univ. *Born:* Dec 1920. BEE & MEE Cornell Univ. Prof of Elec Engrg at Cornell 1944-51. From 1949-51 on special assignment to Brookhaven Natl Lab. Joined IBM in 1951. During 1950's was respon for transistor circuit dev which lead to the IBM 7000 series computers. During 1960's had respon for memories, tapes, displays, digital communications products, & the Mod 90 computer initial dev. Was Staff Dir of the IBM Corp Tech Ctte 1967-70. In 1971 was apptd an IBM Fellow to do res & dev in areas of his choice. IEEE Fellow. Natl Acad/Engrg 1981. Enjoy flying, photography, woodworking. *Society Aff:* NAE, IEEE, AAAS

Lohman, Robert L
Home: 8213 Stone Trail Dr, Bethesda, MD 20817
Position: retired 1987. *Born:* Mar 1923, Brooklyn, NY. SB MIT. Amer Airlines, New York 1945-48, prelim design & econ. Martin Marietta, Baltimore 1948-67: aerodynamics until 1958, Tech Dir, early Apollo studies and advanced re-entry systems, Asst Tech Dir, Gemini Launch Vehicle. NASA Headquarters Washington, 1967-87: manned planetary and space station studies until 1971. Chief of Developmant, NASA/ESA Spacelab 1971-85. NASA Exceptional Service Medal 1974, Outstanding Leadership Medal 1984. Retired 1987. AIAA Assoc Fellow, Nat Capital Sect Council 1986-87. Nat Air & Space Museum docent since 1980. Married Kathleen McCloghrie of Bath, England 1945, son Robert Mark, daughter Sarah Elizabeth.

Lohmann, Carlos A J
Business: Av. Alfredo Egydio de Souza Aranha, 100, 04726 Sao Paulo - S.P - Brasil
Position: Exec VP *Employer:* CAELL - Cons. & Aplic. de Eng. Electrica Ltda. *Education:* BS/EE/Mackenzie S Paulo; BS/CE/Mackenzie S Paulo *Born:* 10/19/34 b in Rio de Janeiro. Obtained engrg degrees in Sao Paulo, Brazil. Completed tech education through assignments in USA and West Germany. Has always worked in electrical engrg, including teaching, instrumentation, operation, system design, equipment applications, data-processing and management. In 1966 founded CAEEL, a Brazilian engrg services co, and developed a successful widely used data-processing system for utility distribution networks. Was Dir of IEEE Region 9 1970/1 and elected Fellow Member in 1981. Exec VP of ELETROPAULO 1981-83 and Member of the Board in 1982. Returned to CAEEL Feb. 1986. Maintains active participation in many Sao Paulo professional and business societies. *Society Aff:* IEEE

Lohmann, Gary A
Business: P O Box 1595, Des Moines, IA 50306
Position: Manager, Reliability *Employer:* John Deere Des Moines Works. *Education:* MA/Business/Univ of Northern IA; BS/Metallurgical Engr/Univ of MO at Rolla. *Born:* 11/17/42. Commissioned Officer in US Army Corps of engrs 1965-67. Met engr in charge of heat treat oper at John Deere Waterloo Tractor Works from 1967-73 & was primarily respon for estab control sys on atmos furnaces. Mgr, matls Engg at John Deere Des Moines Works 1973-77. Dept was respon for recommending matls for product design and providing process controls for metallurgical processes. Mgd Quality mgmt Program from 1977-79 and was respon for reviewing & recommending changes in factory quality procedures. 1975-76 Chmn Des Moines Chap ASM; 1975-77 Mbr ASME Natl Long Range Planning Ctte. Manager, Quality Engrg, JDDMWKS 1979-82. Assumed current responsibility as Manager, Reliability, JDDMWKS, 1982. *Society Aff:* ASM, ASQC.

Lohmann, M R
13930 Summerstar Drive, Sun City West, AZ 85375
Position: Dean, Emeritus Div of Engg, Tech & Arch - Retired. *Employer:* Oklahoma State Univ. *Education:* BS/ME/Univ of MN; MS/IE/Univ of Pittsburgh; PhD/IE/ Univ of IA. *Born:* Sept 1914. Native of Lake Elmo Minn. Indus engr Aluminum Co of Amer 1937-41. Prof of Indus Engrg Okla State Univ 1941- 1977. Dean of Div of Engrg, Tech & Arch 1955-77 . ECPD - Pres 1970-72, V P 1968-70; ASEE - Pres 1967-68; IIE - V P for Prof Relations 1970-72, V P for Educ 1962-63. ASEE James H McGraw Outstanding Educator Award; NSPE Engr of the Yr; IIE Fellow; Okla Engr of Yr. Cons on engrg & genl mgmt problems to more than 50 co's, agencies of state & fed gov, & prof soc's. Admin of internatl assistance programs in Pakistan & Brazil. Cons on programs in Egypt & Peru. Lectr & conf leader. ECPD-Grinter Award 1977, OSPE-Engg in Mgt Award 1977, Okla State Univ - Bennet Award - 1981. *Society Aff:* IIE, ASEE, NSPE.

Lohtia, Rajinder Paul
Business: Regional Engineering College, Kurukshetra, Haryana, India
Position: Prof of Civil Engrg. *Employer:* Regional Engrg College. *Education:* PhD/Struct Engg/Univ of Saskatoon, Can; Master of Engg/Struct Engg/Univ of Roorkee, Roorkee India; BSc Engg/Civil Engg/Panjab Engg Coll Chandigarh, India *Born:* Oct 1938. PhD Univ of Saskatchewan Canada 1969; BSc Engrg (Hon) Punjab Univ India 1960; Master of Engrg (Structures) Roorkee Univ India 1963. Taught Structural Engrg 1960-72 at Roorkee Univ India as Lectr & then as Reader. Since 1972, working as Prof in Civil Engrg at Kurukshetra, Haryana, India. Author about 40 res papers in reputed journals. Amer Concrete Inst Wason Medal April 1973. Central Bd of Irrigation & Power Award April 1974, 1978 Attended Univ of Saskatchewan, Saskatoon Canada 1966-69 & taught (part-time) Struct Engrg courses there. Meritorious record throughout. Engaged in res & design since 1962. Cons in field of design of Struct & Concrete Tech. mbr Inst of Engrs, India, Icl India. Fellow and Council Member of I-E. (India). *Society Aff:* IE, ISTE, ICI

Loigman, Harold
Home: 150 Lantern La, King of Prussia, PA 19406
Position: Exec VP *Employer:* SITE Engineers Inc *Education:* MS/CE/Univ of PA; BS/CE/Univ of Pittsburgh *Born:* 1/25/30 Native of Phila Pa. Served in US Navy during Korean War and Naval Reserve Construction forces. Current rank Captain, Civil Eng Corps, USNR. Supervisory Engr, Army Corps of Eng 1955-67, worked on design and construction of major civil and military projects. 1968-72, VP, Valley Forge Labs Inc., specializing in geotech and materials consulting and testing. Since 1972, VP, SITE engrs, inc., a civil eng consulting firm and subsidiary of Day and Zimmerman Inc. Outstanding performance awards from Army Corps of Engrs 1965-66. Hobbies: Flying, sailing, tennis, photography, and music *Society Aff:* ASCE, NSPE, SAME, ISSMFE.

Lommel, James M
Home: 917 Vrooman Ave, Schenectady, NY 12309
Position: Consultant, Information Systems. *Employer:* General Electic Co., Research and Dev Center *Education:* PhD/Applied Phsyics/Harvard; MS/Metallurgy/IL Inst of Tech; BS/Metallurgy/IL Inst of Tech *Born:* 2/7/32 Native of Skokie, IL. Joined General Electric Research Lab in 1957 as staff member. Research areas include: physical metallurgy, grain growth, magnetic thin films, magnetic recording materials, magnetic materials and first order phase change materials. Consultant for computational and programming support for R&D and administrative groups, and tech info services. Active in IEEE Magnetics Soc as Past Pres and Mbr of Admin Comm. Chairman of 1981 INTERMAG Conf, and Chairman of 1974 Magnetism and Magnetic Materials Conf. Mbr Amer Inst of Physics Publishing Policy Comm. Active in community theatre and Regional arts center. *Society Aff:* IEEE, AIME

London, Alexander L
Business: School of Engrg-ME Dept, Stanford, CA 94305
Position: Emeritus Prof of Mechanical Engg. *Employer:* Stanford Univ. *Education:* MS/Mech Engg/Univ of CA; BS/Mech Engg/Univ of CA. *Born:* Aug 31, 1913. Standard Oil Co of Calif 1937, Univ of Santa Clara 1938, Stanford Univ 1939-78. US Navy Bureau of Ships 1943-46. Argonne Natl Lab, Reactor Engrg 1954. Dir of a 24-yr prog at Stanford Univ supported by the Office of Naval Res in the area of heat transfer & thermodynamics. Chmn ASME Gas Turbine Div 1967. Co-author *Compact Heat Exchangers* McGraw Hill 1964. *Laminar Flow Forced Convection in* Ducts Academic Press 1978. ASME Gas Turbine Power Award 1964 and R Tom Sayer Award. ASME Heat Transfer Div Mem Award 1962. Mbr Natl Acad of Engrs 1979. *Society Aff:* ASME, ASNE, ASEE, AAUP.

London, William B
Business: 17130 Dallas Pkwy, Suite 130, Dallas, TX 75248
Position: Senior VP *Employer:* Graham Assocs, Inc *Education:* BS/CE/TX A & M Univ *Born:* 07/21/28 A native of Hereford, TX, London graduated from TX A & M Univ in 1949, with a BSCE Degree and a commission in the US Corps of Engrs. He served two years (1951-53) in Korea. London joined Graham Assocs, Inc as a partner in 1978, and has helped manage the firm's growth from twenty to over one hundred employees, while also managing major water and wastewater projects. Prior to 1978, London was with STRAM Engrg, of the CRS Group in Houston, as Senior VP; before that was Senior VP of Forrest & Cotton, Inc of Dallas. *Society Aff:* ASCE, ASEC, NSPE (TSPE), AWWA, WPCF, TWCA

Long, Adrian E
Business: Civil Engrg Dept, Queen's Univ, Belfast BT7-1NN, Northern Ireland
Position: Professor of Civil Engrg. *Employer:* Queen's University. *Education:* PhD/Structural Engg/Queen's Univ; BSc/Civil Engg/Queen's Univ; DSc/Civil Engg/ Queens Univ. *Born:* 4/15/41. Native of N Ireland. Bridge design engr FENCO Tornoto & Asst prof Queen's Univ Kingston Canada prior to apptment as Lectr in Queen's Univ Belfast 1971. Assumed current respon as Prof of Civil Engg 1976 & Hd of Dept 1977. Active in res into behavior of reinforced and prestressed concrete

Long, Adrian E (Continued)
slabs and on "wave power as an alternative energy source." Mbr ACI-ASCE Ctte 445. Miller Prize of ICE London 1967; State of the Art of Civil Engg. Award of ASCE 1974; Raymond C Reece Award of ASCE 1976. Henry Adams (1981) and Murray Buxton (1982) Bronze Medals of the Institution of Structural Engineers, London. Patents on methods of assessing in-situ-strength of concrete'. Enjoys golf, travel and church activities. European Editor of the International Journal of Engineering Structures (1986). Vernon Harcourt Lecturer 1986/87 (I.C.E. London) *Society Aff:* ICE, ACI, CS, ISE

Long, Carl F
Business: Thayer School of Engineering, Dartmouth College, Hanover, NH 03755
Position: Prof & Dir, Cook Eng'g Design Center *Employer:* Thayer Sch of Engg. *Education:* DEng/Engg Mechanics/Yale Univ; SM/Civ Engg/MA Inst of Tech; SB/Civ Engg/MA Inst of Tech. *Born:* 8/6/28. Instr Civ Eng, Thayer Sch Engg, Dartmouth Coll, 1954-57, asst prof, 1957-64, assoc prof, 1964-70, Prof, 1970- ; assoc dean, Thayer Sch, 1972; Dean, Thayer Sch, 1972-84; Dean Emeritus and Director Cook Engg Design Ctr, Thayer Sch, 1984- ; Plant engr, West Elec Co, summer 1956 & 57; Mem, Hanover Town Planning Bd, 1963-75, Chmn, 1966-74; Trustee, Mt Washington Observatory, 1975- ; Overseer, Mary Hitchcock Memorial Hosp., 1973- ; Mem, Vis Com ABET, 1973-81; Dir Controlled Environment Corp & VP Operations, 1975-81; Pres & Dir, Q-S Oxygen Processes, 1979-87; Dir, Micro-Tool Co Inc, 1984- ; Pres & Dir, Roan of Thayer Inc, 1986- ; Dir, Roan Ventures Inc, 1987- ; Dir, Micro Weigh Systems Inc, 1987- ; Registered PE, NH. *Society Aff:* ASCE, ASEE, AAAS.

Long, Carl J
Business: 100 Fifth Ave, Pittsburgh, PA 15222
Position: Bd Chmn *Employer:* Carl J Long & Assocs. *Education:* BS/EE/Carnegie-Mellon Univ. *Born:* Dec 1908. BS Carnegie-Mellon U. After grad, worked in Photoelasticity Res Lab, Carnegie Tech. In 1938 opened office as elec & lighting cons. During WW II served in Army Air Force. Now permanent Dir of Air Force Assn. In 1964 apptd by Gov Scranton to serve on Pa Bd of Educ Adv Comm on school bldg standards. In 1965 received first annual ASE Award. In 1966 elected to bldg industry's Hall of Fame. IES - Disting Serv Award 1970, V Pres 1969-70, Board of Dir 1971-73. IES Pres 1976-77. Bd of Trustees AEF 1969-84. Elec, Industry Man-of-Year, 1982. Carnege-Mellon U. Merit Award, 1986. ACEC Fellow, 1986. *Society Aff:* ACEC, IESNA, NSPE, SAME, AES.

Long, Carleton C
Home: 1100 Linden Ave, Boulder, CO 80302
Position: Retired. *Education:* PhD/Physic Chem/Univ of CO; MA/Physic Chem/Standord Univ; BS/Chem Engg/Univ of CO. *Born:* June 1909 Boulder Colo. Joined St Joe Minerals Corp (Zinc Smelting Div) in 1935; Dir of Res 1937 to retirement 1974. Pres The Metallurgical Soc (TMS) of AIME 1960. P Mbr Exec Ctte EJC. Fellow TMS/AIME & Amer Soc for Metals. 1970 Extractive Met Lectr TMS/AIME. Hon Mbr & past V P & Dir AIME. James Douglas Gold Medal (AIME) 1976. Enjoy mountains & amateur radio (ex-W3MBF, now WOPJW). *Society Aff:* AIChE, AIME, TMS, ASM, NSPE, ACS, AAAS.

Long, Chester L
Home: 18 Barbour Dr, Newport News, VA 23606
Position: Chief Engr, Testing Div. *Employer:* Newport News Shipbldg & Dry Dock Co. *Education:* MS/Marine Engg/Univ of MI; BS/ME/Univ of WA. *Born:* 3/14/28. USN 1946-48. Grad cum laude from Univ of WA with a BS in ME in 1952. Received MS in Marine Engg from Univ of MI (1954) on a scholarship from the Soc of Naval Architects & Marine Engrs. Mbr Tau Beta Pi. Employed by Newport News Shipbldg in 1954 as draftsman; 1958, Jr Design Supervisor; 1960, Sr Design Supervisor; 1961, Assoc Engr 1963, Asst Engr; 1966, Engr; 1969, Asst Chief Design Engr; 1973 to present, Chief Engr, Testing Div. Newport News Shipbldg Trial Dir on sea trials of all maj naval (surface & submarine) & commercial ships since 1968. Mbr Am Bureau of Shipping (Engg Committee) & Chem Transportation Advisory Committee. Recipient of Pres's Award (1960) from the Soc of Naval Architects & Marine Engrs. Mbr of SNAME Natl Papers & Ship's Machinery Committees. Past Chrmn of Hampton Roads Section of SNAME. Sports: tennis & racquet ball. *Society Aff:* SNAME, ASNE.

Long, Francis M
Business: Box 3295 Univ Sta, Laramie, WY 82071
Position: Prof, Hd. *Employer:* Univ of WY. *Education:* PhD/EE/IA State Univ; MS/EE/Univ of IA; BSEE/EE/Univ of IA *Born:* 11/10/29. IA farm boy. Served with US Army Corps of Engrs, Korea, 1953-1955. Special citations: ROK Chief Engr and the Seoul Elec Co. Joined EE Faculty, Univ of WY, 1956. Currently Dept Hd. Received G D Humphrey Outstanding Faculty Award-1973; Western Elec Fund Award for excellence in undergrad teaching-1977. Served on ASEE Bd of Dirs, 1977-1979; VP 1978-1979. Co-organizer, IEEE-BME Denver Chapter. Co-organizer Rocky Mtn Bioengg Symposium. Chrmn, First, Second and Third Intl Conf on Wildlife Biotelemetry. VP Alliance for Engg in Medicine and Biology, 1975-1982, Pres 1983-84. Summers at seven industries. Received ASEE Biomed Division Outstanding Biomedical Engineering Educator Award, 1981. ASEE Electrical Engineering Division, Chairman, 1986-87. IEEE Educational Activities Board, 1987. *Society Aff:* IEEE, ASEE, ISHM

Long, Gary
Business: 101 NW 28th St, Gainesville, FL 32607 *Employer:* Univ of Florida *Education:* PhD/CE/TX A&M Univ; MS/CE/TX A&M Univ; BS/CE/Bradley Univ. *Born:* 01/09/43 Post doctoral study in Law at Notre Dame Law Sch. Began career with DeLeuw, Cather & Co, Consltg Engrs, Chicago, 1961-65, doing transportation engrg and planning studies. Principal Investigator on transportation research projects at TX Transportation Inst, 1966-71. Independent conslt, then Principal Engr, Wilbur Smith and Assocs, Consltg Engrs and Planners, Houston, 1973-75. Cvl Engrg faculty of Univ of Notre Dame, 1975-77. Now, Univ of Florida, Gainesville, focusing on research and teaching graduate courses in transportation engrg. Served as Asst Dean for Graduate Studies at the Graduate School. Expert witness to law firms in 12 states. Registered PE, Certificate in Data Processing, author of over 30 technical publications, recipient of ITE 1972 Past Pres' Award, et al. *Society Aff:* ITE, TRB, TRF, ORSA, ΣΞ

Long, J Dewey
Business: 203 S. 8, P O Box 151, Fredonia, KS 66736
Position: Consulting Agri Engr retired. *Employer:* Self *Education:* MS/Agri Engg/Univ of CA; BS/Agri Engg/IA State College. *Born:* Apr 13, 1899 Decatur Iowa. m. Izil Isabel Polson Aug 3, 1927 Topeka Kansas. Employment: Univ of Calif Davis 1922-40; Douglas Fir Plywood Assn Tacoma Wash 1940-47; U S Dept of Agri - Foreign Agri Serv, Bogota Colombia 1948, Res Serv, Beltsville Md 1949-52; Ill Inst of Tech 1953-55 - Damodar Valley Corp, Calcutta India; private cons 1956-86, Dacca East Pakistan 1965-67, Manila Philippines 1967-69; farm owner; printing plant owner Fredonia Kansas 1970-1982. President Wilson County Historical Society 1982-85, Fredonia, KS. Life Fellow ASAE, Pres 1945-46; NSPE; Kansas Engrg Soc. Reg P E Kansas & DC. *Society Aff:* ASAE, NSPE, KSE.

Long, James L
Home: 3814 Hearthstone Rd, Camp Hill, PA 17011
Position: VP & Chief Engineer-Water Supply Section. *Employer:* Gannett Fleming Water Resources Engrs, Inc *Education:* BS/Civil Eng/Lehigh Univ; MS/Mech Hydr/State Univ of IA *Born:* June 1938. BS, Civil Engrg, Lehigh Univ, MS, Mech & Hydr, State Univ of IA. Addl post-grad courses - Hydraulic Transients, Water Treatment Optimization, Instrumentation & Control. Reg P E Penn, Cert Water Treatment Plant Operator Penn. Mbr Chi Epsilon, Tau Beta Pi. Mbr AWWA, ASCE, Penn Water Works Operators Assn. Exper: Principal areas of work have included hydraulic & water supply fields. Respon for planning & design of all types of water supply facils; raw water intakes from reservoirs & rivers, pumping stations, water treatment plants, transmission systems, open canals. Spec work in water treatment Pilot Plant Research, process design, hydraulic control systems, hydraulic transients, & instrumentation & control. Respon for start-up oper of water supply & treatment projs. Chairman of AWWA-PA Sec 1984-85. Director-AWWA 1986-1989. Inst. in Water Treatment - PA. State University. *Society Aff:* AWWA, ASCE.

Long, Maurice W
Home: 1036 Somerset Dr N W, Atlanta, GA 30327
Position: Consultant. *Employer:* Self. *Education:* PhD/Physics/GA Inst of Tech; MS/Physics/GA Inst of Tech; MSEE/EE/Univ of KY; BEE/EE/GA Inst of Tech. *Born:* 4/20/25. Radar consultant. Liaison Scientist in London with U S Office of Naval Res 1966. Held var res, academic & admin pos at Georgia Tech incl Principal Res Scientist, Prof of EE, & Dir of Engrg Exper Station 1946-75. Chmn Governor's Sci Adv Council Ad Hoc Ctte on Tech Growth in Ga 1974. Fellow IEEE; Mbr U S Natl Comm F, Internatl Union of Radio Sci; Bd Mbr Ga Tech Res Inst 1968-1982. Author 'Radar Reflectivity of Land & Sea' D C Heath & Co 1975; Second Edition, ARTECH House, 1983. Author 'Statistical Properties of Data', Chap. 13, *Techniques of Radar Reflectivity Measurement*, ARTECH House, 1984, N.C. Currie, Editor. *Society Aff:* IEEE, –Old Crows–

Long, Richard L, Jr
Business: P.O. Box 3805, Las Cruces, NM 88003
Position: Assoc Prof *Employer:* NM State Univ *Education:* PhD/ChE/Rice Univ; BA/ChE/Rice Univ *Born:* 6/5/47 Res Engr: E.I. DuPont 1974-78, Asst. Prof ChE, Lamar Univ: 1979-81. Asst Prof ChE, NM State Univ: 1981. Sigma Xi Res Award 1973. Consultant - Temple - Eastex Corp. Reg PE: 39445, TX, Publications, primarily in biophysics and interfacial phenomena. Author, *A Guide to Writing and Problem Solving for Chemical Engineers*, Long et al, 1983. Born in Kansas City, MO; Military: 1 Lt EN USAR, Cmdr 377th Engrg, Cmdr 440th CL, 1977-79; Reg PE: 8107, NM, Mbr Phi Lambda Upsilon, Omega Chi Epsilon; Chrmn, Rio Grande Sec, AIChE, 1984, AIChE Stud Chap Advsr 1979-present. Amateur flutist, runner. *Society Aff:* AIChE.

Long, Richard P
Home: 31 Westgate Ln, Storrs, CT 06268
Position: Prof and Hd Civ Engg Dept. *Employer:* Univ of CT. *Education:* PhD/Civ Engg/Rennsselaer Poly Inst; MCE/Civ Engg/Rennsselaer Poly Inst; CE/Civ Engg/Univ of Cincinnati. *Born:* 11/29/34. Grew up in Allentown, PA. Served as a Lt in the Corps of Engrs, US Army, for 3 1/2 yrs. Married to the former Mary Elizabeth Doyle of Albany, NY. Two children, Marybeth and Christopher. Current res interests are in field consolidation, behavior of clays, subsurface drainage, geosynthetics, soil reinforcement. *Society Aff:* ASCE, ASEE, TRB.

Long, Robert L
Business: GPU Nuclear Corporation, One Upper Pond Road, Parsippany, NJ 07054
Position: VP-Nucl Assurance *Employer:* GPU Nucl Corp *Education:* PhD/Nucl Engrg/Purdue Univ; MSE/Nucl Engrg/Purdue Univ; BS/Elec Engrg/Bucknell Univ *Born:* 09/09/36 VP and Dir-Nucl Assurance Div. Respon include Quality Asurance, Nucl Safety Assessment, Training & Education, and Emergency Preparedness. Actively involved in advisory committees of Electric Power Research Institute and Institute for Nuclear Power Operations. Served as Dir-Training & Education - GPUN. Was a Prof and Chrmn of the Chem & Nucl Engrg Dept at Univ of NM. Has extensive reactor operations experience at Argonne Natl Lab, White Sands Missle Range, Sandia Lab, Univ of NM, Atomic Weapons Res Establishment in England, and Indian Point Nucl Station. Written numerous publ and presented lectures all over US and SE Asia. *Society Aff:* ANS, AIF

Long, Stuart A
Business: Dept of Elec Engg, Univ. of Houston, Houston, TX 77004
Position: Prof & Chrmn *Employer:* Univ of Houston. *Education:* PhD/Appl Phys/Harvard Univ; MEE/EE/Rice Univ; BA/EE/Rice Univ. *Born:* 3/6/45. Dr Long has previously worked at Gen Dynamics Ft Worth & at the Los Alamos Scientific Lab. He is the author of over 50 publications in journals, and conf proceedings. He has served as an elected Ad Com mbr of the IEEE Antennas & Propagation Soc. He is presently Prof and Chrmn of the Dept of Elec Engg at the Univ of Houston. He was organizer and gen chrmn of the 1983 Internation Symposium on Antennas and Propagation and Natl Radio Science Meeting. *Society Aff:* URSI, APS, ΦΒΚ, ΤΒΠ, IEEE.

Long, Warner D
Business: 4414 Old LaGrange Rd, PO Box 115, Buckner, KY 40010
Position: Exec VP. *Employer:* Catalyst Tech Inc. *Education:* BChE/CE/Univ of Louisville. *Born:* 2/28/39. Native of Louisville, KY. Attended Duke Univ & Univ of Louisville. Grad 1963. Joined Eastman Chem Products, Kingsport, TN, as product dev & service engr. Two patents resulted. Field service engr for Catalysts & Chems, Inc, Louisville, KY, in 1965, performing start-up & trouble shooting duties. Became regional mgr of sales to engg contractors. In 1973 co-founded Catalyst Tech Inc, service co handling catalysts in chem plants & refineries. Duties are finance, equip & systems design, environmental services, & admin of four subsidiary operations. Two patents resulted. Active in Young Life. Relax with tennis, skiing, boating, & family. *Society Aff:* AIChE.

Long, William G
Home: 1407 Brookville Lane, Lynchburg, VA 24502
Position: Manager, Ceramics Section. *Employer:* Babcock & Wilcox. *Born:* Jan 1931. MS & BS Univ of Illinois. Native of Amboy, Ill. Sr Res Engr with Atomics Internatl Div of Rockwell Internatl 1962-67. Sr Matls Engr with Cannon Elec Div of ITT 1967-70. With Babcock & Wilcox since 1970. Assumed current respon as Mgr of the Ceramics Sect in the R&D Div of Babcock & Wilcox in Oct 1975. Has served as Editor of the SAMPE Quarterly since its creation in 1969. Represented SAMPE during the formation of the Fed of Matls Soc's. Hobbies: tennis, support of local swimming & baseball programs for children.

Longe, Frederick J
Business: 3 Northway Lane, Latham, NY 12110
Position: President & Genl Mgr. *Employer:* McManus, Longe, Brockwehl Inc. *Education:* BS/CE/Union College. *Born:* Oct 1920. US Navy 1942-45. Pres & Genl Mgr McManus, Longe, Brockwehl Inc, Genl Contractors since 1947. Former Pres Norflor Const Corp of Orlando Fla since 1974. Mbr ACI, NYSPE. Lic Prof Engr N Y State. Lic Genl Contractor Fla. Active church & community affairs. *Society Aff:* ACI, ASCE, CEC.

Longenecker, Philip L
Business: 13905 McCorkle Ave, Chesapeake, WV 25315
Position: VP *Employer:* Carbon Industries *Education:* BS/Mining/PA State Univ *Born:* 3/23/47 Registered Prof Eng (Mining) in PA & WV. Certified mine foreman in WV & OH currently employed as VP Dev & Surface Mining, in charge of surface operations and property acquisition for Carbon Industries, A Unit of ITT. *Society Aff:* AIME, WVCMI, KMI

Longerbeam, Gordon T
Business: PO Box 808, L-153, Livermore, CA 94550
Position: Deputy Dept Hd, Electronics Engg Dept. *Employer:* Lawrence Livermore Lab-U of Calif. *Education:* BS/EE/Univ of UT. *Born:* 11/11/34. US Naval Officer 1957-60. With Univ of Calif, Lawrence Livermore Lab since 1960. 9 yrs as proj engr & Group Leader in Nuclear Explosives Test Prog spec in dev of instrumentation sys for nuclear measurement. Also worked in Plowshare Prog, non-nuclear energy & resource progs, magnetic fusion energy res, & laser fusion res. Since 1972 Div Leader of Nuclear Energy Sys Div respon for staff of 160 (55 engrs) in support of major DOE/LLL progs in Nuclear Explosives Res, Laser Fusin, & Laser Isotope

Longerbeam, Gordon T (Continued)
Separation. Since 1979, Deputy Dept Hd, Dept, a dept of about 1000 engrs and technicians supporting major DOE programs in natl defense, energy, environmental and biomedical research. Ex Sr Mbr IEEE, AFCEA Gold Medal Award 1957. Chmn East Bay Sub-Sect IEEE. Dabble in backpacking, tennis & skiing.

Longhouse, Alfred D
Home: 378 Elmhurst St, Morgantown, WV 26505
Position: Prof & Chem, Dept of Agri E - Retired 1976. *Employer:* West Virginia Univ. *Education:* PhD/Agri Engg/Cornell Univ; MSA/Agri Mech/Cornell Univ; BS/ Agri Educ/Cornell Univ. *Born:* Feb 1912. Native of Cassadaga N Y. Instr to Asst Prof of Rural Org WVU 1938-45. Leave of Absence 1940-41 to serve as Spec Rep to U S Office of Educ. Apptd first Chmn, Dept of Agri Engrg WVU 1945-76. Respon for undergrad & grad progs in Agri & Forest Engrg. Principal res ints are in poultry housing, ventilation & waste mgmt. Regional Dir ASAE 1969-71. Reg P E. Mbr AAAS, ASEE, WV Soc of Prof Engrs. Life Fellow ASAE. Mbr Tau Beta Pi, Sigma Xi, Gamma Sigma Delta, Phi Kappa Phi, Phi Delta Kappa. *Society Aff:* ASAE.

Longini, Richard L
Business: Carnegie Mellon Univ, Pittsburgh, PA 15213
Position: Prof of EE & Urban Affairs Emeritus. *Employer:* Carnegie-Mellon Univ. *Education:* PhD/Physics/Univ of Pittsburgh; MS/Physics/Univ of Pittsburgh; SB/ Physics/Univ of Chicago *Born:* 3/11/13. Univ of Chicago 1930-34 - admitted to grad school 1933. Indus res 1934-62. Carnegie Mellon Univ Prof of Solid State Electronics 1962-75, Prof of Elec Engrg & Urban Affairs 1975- . Fellow APS 1959, Fellow IEEE 1968. Dev first x-ray Image Intensifier 1947. Invented met tech used for first mass produced R-F transistors (Japanese & European production). Applying engrg principles to medical & urban problems. Supervises doctoral student(s). About 35 PhDs. About 50 US patents. *Society Aff:* APS, IEEE, AIEE

Longley, Thomas S
Home: Rt 1, Box 252, Aberdeen, ID 83210
Position: Asst Res Prof *Employer:* Univ of ID *Education:* MEngrSc/CE/Melbourne Univ; BS/Agri Engrg/CO State Univ *Born:* 12/5/53 He has specialized in irrigation and water mgmt related res since coming to the Univ in 1976. He grad with a BS in Agri Engrg from CO State Univ in 1974 and with a Master of Engrg Sci from the Univ of Melbourne Australia in 1977. He is currently working in the areas of deficit irrigation on crops grown in southeastern ID, low pressure irrigation for energy conservation and soil sealing for ponds and water conveyance structures. *Society Aff:* ASAE, ASCE

Longman, Richard W
Business: Dept of Mech Engrg, New York, NY 10027
Position: Prof of Mech Engg. *Employer:* Columbia Univ. *Education:* PhD/Aerospace Eng/Univ of CA, San Diego; MA/Math/Univ of CA, San Diego; MS/Aerospace Eng/ Univ of CA, San Diego; BA/Physics & Math/Univ of CA, Riverside *Born:* 9/2/43. Professor of Mechanical Engineering, Columbia University. Specialty: statellite attitude dynamics and control, and applications of control theory to spacecraft, robotics, subway operation, cam design, etc. Member Board of Directors, previously First VP, VP Technical, VP Publications, American Astronautical Soc (AAS). Managing Editor, *The Journal of the Astronautical Sciences*, 1976-1984. Fellow of the AAS, Associate Fellow AIAA. Best Paper Award, ASME Mechanisms Conference. Member AIAA Astrodynamics Committee. Temporary positions at universities: MIT, Newcastle (Australia), National Cheng Kung (Taiwan), Bonn (Germany), Darmstadt (Germany); and in industries: RAND, Bell Laboratories, NASA Langley, Goddard, European Space Operations Center, Lockheed LMSC, Martin Marietta, Xerox, MEC Systems, Designatronics. *Society Aff:* AIAA, AAS, BIS, ASME.

Longwell, John P
Business: Rm 66-554, Cambridge, MA 02139
Position: Prof. *Employer:* MA Inst of Technology. *Education:* ScD/Chem Eng/MIT; BS/Mech Eng/Univ of CA, Berkeley *Born:* 4/27/18. in Denver, CO. Employed at Exxon Res & Engg Co: 1959-77 Dir Central Basic Res Lab; 1969-1972, Sr Scientific Advisor. 1974-77, E. R. Gilliland Prof Chem Engg MIT 1977- . Pres The Combustion Inst 1966-70. Sir Alfred C Egerton Gold Medal for Contrib to Combustion Sci 1974. Mbr: Natl Acad of Engg 1976, NASA Adv Ctte on Aeronautical Propulsion, Chrmn NRC ctte on Advanced Energy Storage Sys. Amer Inst of Chem Engrs Award for Chemical Eng Practice 1979. *Society Aff:* ACS, AIChE

Looby, Lawrence F
Business: 703 Curtis St, Middletown, OH 45043
Position: Manager, Market Dev. *Employer:* Armco Steel Corp. *Education:* BS/Metallurgical Eng/MI Tech. *Born:* June 19, 1931 Canton Ohio. BS Michigan Tech 1953. New Jersey Zinc Co Palmerton Pa - Genl Foreman, Mech Oxide Furnaces. Army Ordnance, Frankfurt Arsenal 1956-57, magnesium single crystal res. With Armco Steel Corp since 1955, currently Mgr Mkt Dev, respon for transp mkts. 28 yr Mbr ASM. 11 yr Mbr SAE, P Chmn Surface Coating Ctte; P Chmn Matls Engrg Activity. Now, Mbr at Lrg Engg Activity Now, mbr of Large Engr Act Bd. Enjoy skiing, flying, sailing, fishing & lapidary. *Society Aff:* ASM, SAE, ASBE.

Loofbourow, Robert L
Business: 3601 Park Ctr Blvd, Minneapolis, MN 55416
Position: Consulting Engr. *Employer:* Multipurpose Excavation Group. *Education:* AB/Geology/Stanford; EM/Mining Engg/Stanford. *Born:* 1908 Oakland, Calif. Geology 1929 & Mining Engr 1931 Stanford Univ. Through 1944 miner, timberman, shaftman, engr, supr at mines in Ariz, Calif, Quebec, Ontario, Philippines, Nevada, Washington, Manitoba, Mexico. 1945-53 Mine Engr, Mgr Mining Div, Longyear Co, Minneapolis. 1954- , cons in mine planning, control of rock & water, underground const. 1968- , Managing Ptnr Multipurpose Excavation Group, planning prod of indus minerals & underground space for storage of fluid fuels, compressed air, pumped storage, etc. 7 pats, 25 related tech articles, chapter on Mine Water Control in Mining Engrg Handbook AIME 1972. Mbr AIME, CIM, NSPE, ASCE. Reg Philippines, Minn. *Society Aff:* AIME, CIM, ASCE

Look, Dwight C, Jr
Business: 204 Mech Engr Bldg, Rolla, MO 65401
Position: Prof. *Employer:* Univ of MO. *Education:* PhD/M & A Engg/Univ of OK; MS/Phys/Univ of NB; BA/Math/Central College. *Born:* 8/25/38. Smith Ctr, KS, Aug 25, 1938. Grad high sch, St Louis, MO, 1956. Married, 1960, Patricia Ann Wellbaum. Two sons, Dwight (III) & Douglas. Academic Appointments: Undergrad Asst Math Dept, 1958-1960; Teaching Asst, Phys Dept, 1960-1963; Evening College Instr, Math Dept, TX Christian Univ, Spring, 1967; Special Instr, A & ME Dept, Univ of OK, Spring 1969. NDEA Fellow, Univ of OK, 1967-1969; since 1969 on faculty at Univ of MO-Rolla, presently prof of ME. Ind Experience: Aerosystems Engr, Ft Worth Div of Gen Dynamics, June 1963-Sept 1967. *Society Aff:* ASME, ASEE, AIAA, SPIE, ΣΞ

Loomis, Albert G
Home: 85 Parnassus Rd, Berkeley, CA 94708
Position: Consulting Chem Engr. *Employer:* Chem Dept, Univ of CA, Guest Scientist. *Education:* PhD/Chemistry/Univ of CA; AM/Chemistry/Univ of MO; AB/ Chemistry/Univ of MO. *Born:* Feb 1893. Natl Res Council Fellow 1920-21. Phys chemist U S bureauu of Mines 1919-28 & 1948-63. Ch Chem Engr Gulf R&D Co & Sr Fellow Gulf Oil Corp Mellon Inst Industrial Research 1928-35. Asst, then Assoc Dir Shell Dev Co 1935- 45 in charge/petro prod res. Cons chem engr 1963-76. Guest Scholar, Chem Dept, Univ of Calif 1976- . Disting Serv Medal U S Dept of the Interiro *1963. AIME Anthony F Lucas Gold Mdal 1972. Mellon Institute of Industrial Research. Society Aff:* ACS, AIME, SPE.

Loomis, Herschel H, Jr
Home: 4086 Pine Meadows Way, Pebble Beach, CA 93953
Position: Prof *Employer:* Naval Postgrad Sch *Education:* PhD/EE/MIT; MS/EE/Univ of MD; B/EE/Cornell Univ *Born:* 5/31/34 Born in Wilmington, DE. Active duty USN 1957-59. Worked as Electronics Engr for Westinghouse Corp, MIT Lincoln Lab and Lawrence Livermore Lab. Joined Univ of CA, Davis Campus as Asst Prof of Elec Engrg in 1962. Served as Dept Chair 1970-75, and Prof of Elec and Computer Engrg 1974-1983. Naval Electronics Systems Command (NAVELEX) Res Prof of Elec Engrg at the Naval Postgrad Sch, Monterey, CA 1981-1983. Currently Serving as Prof of Electrical and Computer Engrg, Naval Postgraduate School. Cons to Lawrence Livermore Lab and Signal Sci, Inc. Capt, US Naval Reserve; holder of Joint Service Commendation Medal. Reg Elec Engr, CA. *Society Aff:* IEEE, ACM, USNI, ΣΞ, TBΠ.

Lopata, Stanley L
Business: 350 Hanley Indus Ct, St Louis, MO 63144
Position: Board Chrmn. *Employer:* Carboline Co. *Education:* AB/ChE/WA Univ *Born:* Aug 30, 1914. Native of St Louis Mo. Sales & Service Engr for Duriron Co for 21 yrs, spec in fluid flow & heat transfer. Founded Carboline Co while still working for Duriron. Has been Board Chairman since 1979 & has developed tech in corrosion prevention through use of specialty protective coatings. Has been active in NACE, AIChE & many charitable orgs in St Louis. Hobby: Limoges enamels & Majolica pottery. *Society Aff:* ACS, ASM, NACE, AIChE.

Loper, Carl R, Jr
Business: 1509 University Ave, Madison, WI 53706
Position: Prof, Metallurgical Engrg. *Employer:* Univ of Wisconsin - Madison. *Education:* BS/Metallurgical Engg/Univ of WI; MS/Metallurgical Engg/Univ of WI; PhD/Metallurgical Engg/Univ of WI. *Born:* July 1932. Metallurgist Pelton Casteel 1955-56. Res engr Allis Chalmers 1961. With Univ of Wisconsin since 1956. Author over 200 papers in casting & joining processes. Co-author *Principles of Metal Casting*. Foundry Educ Foun Fellow, Ford Foun Fellow, Wheelabrator Fellow. AWS Adams Mem Award 1963. AFS Best Paper Award 1966, 1967, 1975, 1976, 1985, 1986. AFS Howard F Taylor Award 1967. AFS John A Penton Gold Medal 1972. Award of Scientific Merit, Silver Medal, Portuguese Foundry Association 1978 Pres, Intl Commission on Compacted Graphite Cast Iron 1977- . Honorary Life Member, AFS, Fellow, ASM. Past Officer in ASM, AWS, AFS. Presently directing res in cast irons, nucleation & supercooling of metals & of aqueous sys, solidification of weld metals, failure analysis. Enjoys music of all types, golf, court games. *Society Aff:* AFS, ASM, AIME, AWS.

Lopez, Arthur M
Home: 9912 Dorothy Pl N E, Albuquerque, NM 87111
Position: Mgr Shroud Manuf Oper. *Employer:* General Electric. *Education:* BS/ChE/Univ of NM; MBA/ChE/NMH Univ. *Born:* Jan 11, 1939. BSChE Univ of New Mexico. ACF Industries 1960-67, Sr Plastics Engr; Motorola 1968, Sr Mech Engr; G E, Sr Engr & Mgr Methods Engrg 1968-present. Presently respon for mfg planning of plastic components for jet engines incl tooling concepts, facil requirements, methods, manpower/cost/mfg proposals. Mfg method dev includes struct laminates, compression/injection molding, potting etc using epoxies, urethanes, phenolics, polyimides, silicones, nylon etc with glass, carbon, Kevlar, asbestos & other reinforcements. P Pres N Mex Chap SAMPE, Natl Dir SAMPE. Hobbies: camping, fishing, gardening, & auto mech. *Society Aff:* ASChE, SAMPE.

Lopez, J Joseph
Business: 213 Truman NE, Albuquerque, NM 87108
Position: Principal. *Employer:* Uhl & Lopez Engrs Inc. *Education:* BSEE/EE/1937 *Born:* Feb 1913. Native of N J. Elec engr AT&T Co 1937-49; elec engr & Mgr Reynolds Elec & Engrg Co 1950-55; Mgr Nevada Test Site 1951-55; Principal Uhl & Lopez Engrs Inc since 1956 (cons elec engrg firm). Pres NMSPE 1956; S W Reg Dir NSPE/PEPP 1957. P Pres N Mex Symphony Orchestra; Albuquerque Rotary Club; Albuquerque Child Guidance Center. Bd Mbr St Joseph Hosp, UCF. Hobbies: classical music, ski, golf. *Society Aff:* NSPE, CEC, NMSPE

Lopresti, Philip V
Business: P.O. Box 900, Princeton, NJ 08540
Position: Consulting Mts *Employer:* AT&T *Education:* PhD (EE)/Elec Engrg/Purdue U.; MSEE/Elec Engrg/ U. of Notre Dame; BSEE/Elec Engrg/U. of Notre Dame *Born:* 09/27/32 Born in Johnstown, PA. Served in U.S. Army Ordnance Corps in 1955 and 1956. Taught Electrical Engrg at U. of Notre Dame, Ill Inst of Technology and Northwestern Univ. With AT&T Labs since 1970. Currently consultant in computer-aided manufacture and test of communications devices and systems. Member, IEEE Publications Bd, 1986-87 and IEEE Cttee on Engrg Accreditation Activities, 1986-. IEEE Circuits and Systems Society Darlington Prize, 1978. Married, three children. *Society Aff:* IEEE,ΣΞ

Lord, Harold W
Home 336 Corte Madera Ave, Mill Valley, CA 94941
Position: Consulting Elec Engr. (Self-employed) *Education:* BSEE/Elec Engg/CA Inst of Technology. *Born:* Aug 20, 1905 Eureka CAlif. Joined G E Co Schenectady N Y in 1926 as Test Engr. Employed as Electronics Engr & Res Assoc in serv G E Co labs until retirement in 1966. Moved to Calif & entered private practice, cons in areas of electronic circuitry & electromagnetic devices. Issued 97 U S pats. Life Fellow IEEE & Tech V P for Sci & Electronics 1962. IEEE Centenial Medal, 1984. IEEE Magnetics Soc Achievement Award, 1984. *Society Aff:* IEEE.

Lord, William
Business: Colorado State Univ, Ft. Collins, CO 80523
Position: Prof *Employer:* Elec Eng Dept *Education:* PhD/EE/Nottingham Univ; BSC/EE/Nottingham Univ *Born:* 2/18/38 Born in New Eastwood, Nottinghamshire, UK. Early education at Newton-le- Willows Grammar Sch. Ph.D. in elec engrg in 1964. Subsequent teaching experience in networks, power systems, energy conversion, electromagnetics, control systems with Univ of TN 1964-66, Clarkson Coll of Tech 1966-67, CO State Univ since 1967. Industrial consulting experience with IBM (Boulder), Inland Motor Div of Kollmorgen Corp and the Gas Res Inst. Current res on electromagnetic field/defect interactions funded by EPRI and ARO. Chartered Engr and Fellow of the Inst of Elect Eng Sr. Member of the Inst of Electrical and Electronics Engrs. *Society Aff:* ASEE, IEEE, IEE, ASNT, ΣΞ

Lordi, Francis D
Home: Star Rt Rt 5S, Rotterdam Jct, NY 12150
Position: Mgr, Matls & Processes Engg. *Employer:* Genl Elec Co. *Education:* BS/Metallurgical Eng/Columbia Univ; AB/Chemistry-Math/Columbia Univ; High School/Coll Prep/Geo Washington HS, NYC. *Born:* 1/3/27. Mr. Lordi joined General Electric on the Rotational Engineering Program in 1951 after graduation from Columbia Univ. He worked in a variety of engineering and engineering management positions with the GE Foundry Dept., the Knolls Atomic Power Laboratory, and the Large Steam Turbine and Generator Div. Materials and Processes Laboratory. He is currently Mgr. Materials and Processes Engineering in the Gas Turbine Division providing services in materials and process development, materials applications and manufacturing support. *Society Aff:* ASM.

Lorenz, John D
Business: 1700 West Third Ave, Flint, MI 48502
Position: Assistant Dean, Graduate Studies & Research *Employer:* GMI Engrg & Mngmt Inst *Education:* PhD/IE/Univ of NB; MS/ME/Univ of NB; BS/ME/Univ of NB *Born:* 7/2/42 Native of Dunbar, NB. Systems Analyst at Univ of NB, 1966-73. Appointed Assistant Prof of Industrial Engrg at GMI Engrg & Mngmt Inst, 1973. Promoted to Associate Prof, 1974. Promoted to Prof, 1978. Named Dept Head, 1984. Named Asst Dean, 1986. Recipient Distinguished Teaching Award from GMI Alumni Association. Danforth Assoc. Regional Educational Affairs Chrmn, IIE 1975-77. Chapter Pres, IIE 1979. Bd of Directors, AIIE Chap. 98, 1985-87. Member

Lorenz, John D (Continued)
of Sigma Xi, Alpha Pi Mu, Pi Tau Sigma, and Tau Beta Pi honorary societies. Consultant in areas of manufacturing and assembly systems design. Author papers in field. Enjoys running, bicycling, and automobile racing. *Society Aff:* IIE, SME, ASEE.

Lorenzen, Coby
Home: Country Club Dr, Carmel Valley, CA 93924
Position: Professor Emeritus. *Employer:* Univ of Calif Davis. *Education:* MS/ME/Univ of CA; BS/ME/Univ of CA. *Born:* Nov 1905. Native of Oakland Calif. Res engr Natl Adv Comm for Aeronautics 1929-31. Res Asst Univ of Calif 1931-34. Res Engr U S Forest & Range Exper Sta 1935-37. Res engr, Prof of Agri Engrg Univ of Calif Davis 1937-69; Chmn Dept of Agri Engrg 1963-68; principal res in Agri Mechanization. Dev 1st commercial mech harvester for canning tomatoes. Life Fellow ASAE. Mbr Davis Planning Comm 1963-67. Engr in Residence N C S U Raleigh 1975-78. 1976 Recipient of John Scott Medal (Dev of Tom Harv). 1981 Recipient of Cyrus Hall McCormick Medal, (ASAE). Reg P E Calif. Hobby: horology. Enjoy golf. *Society Aff:* ASAE.

Lorenzi, Silvio J
Business: 307 4th Ave-1400 Bank, Pittsburgh, PA 15222
Position: President. *Employer:* Lorenzi, Dodds & Gunnill Inc. *Education:* BS/Civil Engg/Univ of Pittsburgh. *Born:* 8/10/20. BS Univ of Pittsburgh. Native of Pittsburgh, PA. US Army WWII 1942-45. Design engr for Blaw Knox Co, Chem Plants Div. Started own cons engg & arch firm 1953 with Robert M Dodds, Engr & Edward F Gunnill, Architect in Pittsburgh. In 1963, opened office in Wash DC area. Fellow ASCE since 1959. Served as Pres of the PA Soc of Prof Engrs in 1978-79, Reg in num states as a P E. Also reg as a Landscape Architect in PA, as a Prof Planner in NJ & as a Surveyor in MD. Cert by the Natl Council of State Bds of Engg Examiners. *Society Aff:* NSPE, ASCE, ACEC, SAME.

Lorenzini, Robert A
Home: Box 288, Rte. 43, Hancock, MA 01237
Position: Consultant *Employer:* Self *Education:* M.Eng./B.Eng./Yale Univ. *Born:* 07/18/17 Currently a management and engineering consultant for capital goods and energy-related industries, Robert Lorenzini has formerly been Sr. VP & Dir Foster Wheeler Corp, & Group VP Capital Goods Operations, Midland-Ross Corp. H is a Fellow of ASME, Trustee of Clarkson College of Technology, & Chair of the Welding Research Council. He has completed graduate courses at Brooklyn Polytechnic, Stevens Institute and Northeastern University and holds a number of patents relating to steam generation, heat exchange and manufacturing processes. *Society Aff:* ASME

Lorey, Frederick D
Home: Box 413, Rd 3, Corning, NY 14830
Position: Dir of Melting Tech. (Ret) *Employer:* Corning Glass Works. *Education:* MS/Cer Engg/OH State Univ; BCerE/Cer Engg/OH State Univ. *Born:* Dec 4, 1924. Harvard AMP. Native of Portsmouth Ohio. Army Air Corps 1943-46. With Corning Glass Works since 1950. Assumed current respon as Dir of Melting Tech 1960, which includes all phases of hot glass manufacture for delivery to forming equip as well as pollution control & energy conservation. Also utilize org dev skills. OSU Outstanding Alumnus Award, P Chap Pres NYSSPE. Reg P E Ohio, W Vir, N Y. Enjoy hunting, sailing, Chinese astrology. Retired July 1986.

Loria, Edward A
Home: 1828 Taper Dr, Pittsburgh, PA 15241
Position: Metallurgical Consultant *Employer:* Self *Education:* MS/Met Engg/Carnegie Inst of Tech; BS/Met Engg/Carnegie Inst of Tech. *Born:* 4/29/17. BS-MS Carnegie Inst of Tech. Grad met studies MIT. Prod/ res matallurgist U S Steel 1944-46. Fellow/Sr Fellow Mellon Inst to 1950. Sr Res Engr Carborundum Co to 1953. Central Met Dept, Crucible Steel, Prod Met Engr; Stainless Steels/Superalloys to 1957; Vacuum Melt Products/ Titanium to 1959. Met Dev Mgr Climax Molybdenum Div of AMAX to 1963. Supr Res Metallurgist U S Bureau of Mines to mid-1964. Supr/ Mgr Matls & Process Res, Natl Steel Corp to mid-1975. Tech Dir Roll Manufacturers Inst to mid-1977. Div Met Stainless Univ- Cyclops Spec Steel Div-Cyclops to mid-1984 (retired). Now consultant to this firm, renamed Cytemp Spec Steel Div-Cyclops, and Niobium Products Co. Expertise in making, shaping & treating all types of iron & steel, superalloys, Ti, Mo, and Va. Over 125 tech pubs. AIME Herty Award 1967, participant in ASM Engrg Matls Achievement Award 1970, ASM Fellow 1973, ASM Bain Award 1982, ASM Andrew Carnegie Lecture 1984 *Society Aff:* TMS-AIME, ASM.

Loring, Paula L
Home: 30 Lakin St, Needham, MA 02194
Position: Consultant. *Employer:* Arthur D Little, Inc. *Education:* MBA/Bus/Boston Univ; MS/Elect Eng/Univ of IL; BS/Elect Eng/RPI. *Born:* June 1947. Native of Long Island N Y. Engr with Honeywell Info Sys from 1969-74, spec in sys arch, performance eval & analysis of high level languages. With the MITRE Corp from 1974-79. In 1976 became task leader of high order language standardization study. In 1977-79 provided software Engg support for acquistion of command, control & communications systems. With ADL since 1979 in info sys. Pres 1978-79, First VP 1977-78. Secy 1976, Treas 1975 Soc of Women Engrs. Hobbies: bridge, restaurant-ing, music. *Society Aff:* SWE, IEEE.

Loser, Rene F
Business: 230 Crossways Pk Dr, Woodbury, NY 11797
Position: Exec VP & Gen Mgr *Employer:* Sulzer Biotech Systems *Education:* MS/Mech Engg/Tech Inst St Gallen, Switzerland. *Born:* Oct 1939. Native of Mosnang Switzerland. Immigrated to the U S 1968. Received U S citizenship 1975. Employed with Chemap AG, Maennedorf Switz 1965-68 as Sales Mgr Processing Equip for Chem Indus. 1969-73 Sales Mgr with Chemapec Inc, the U S subsidiary of Chemap AG. 1974-1986 V P & Ch Exec Officer with Chemapec Inc. Since 1986, Executive VP and General Mgr with Sulzer Biotech Systems. Enjoy soccer, skiing & tennis. Completed courses at Wharton School, Univ of Penna & MIT (Cambridge Mass). *Society Aff:* AIChE, ASM, SIM, TCA, ACS

Loth, John L
Home: PO Box 4094, Morgantown, WV 26504
Position: Prof. *Employer:* WV Univ. *Education:* PhD/Mech Engg/Univ of Toronto; MASc/Mech Engg/Univ of Toronto; BASc/Mech Engg/Univ of Toronto; ING/Mech Engg/HTS Amsterdam, The Netherlands. *Born:* 9/14/33. Prof of Aerospace Engg at WV Univ. Started in 1967. Pres Dynamic Plan Inc PO Box 4094 Morgantown WV. Started in 1972. Asst Prof of Aerospace Engg at Univ of IL from 1962 to 1967. Received Doctorate Degree (Technical) from the Univ of Toronto (Canada). Holder of five US patents, author of 61 technical articles. Active as consultant in the field of Aerodynamics, Aircraft Propulsion flight testing, Combustion of gases and solar energy, Commercial rated pilot. *Society Aff:* AIAA, ASEE, AOPA, COMB-INST, ASME, ΣΓΤ.

Lott, Jerry L
Home: 1228 Benson Dr, Norman, OK 73071
Position: Consulting Engr. *Employer:* Independent Consulting Engineer. *Education:* PhD/ChE/Univ of OK; MChE/ChE/Univ of OK; BS/ChE/Univ of NM. *Born:* Feb 1935. MS & PhD Univ of Oklahoma; BS Univ of New Mexico. Chem engr for Continental Oil Co, spec in pilot plant res & dev 1957-60. With Univ Engrs Inc 1965-76 with respon for dev of a novel desalination process. Assumed respon of V P 1971 for mgmt of a 75,000 gallon per day pilot plant for desalting sea water. Assumed present pos as cons engr with Wesson & Assocs Inc 1976. Respon for process economics, process evals & heat transfer analyses. *Society Aff:* AIChE, ACS, ΣΞ.

Lotta, Joseph G
Home: 1410 W Sandison St, Wilmington, CA 90744
Position: Subsystem Mgr. *Employer:* Hughes Aircraft Co. *Education:* BS/Aero Engr/Univ of Notre Dame. *Born:* 3/21/28. BS Aero Engrg Univ of Notre Dame 1951; post-grad studies El Camino College 1963, CA State at Long Beach 1964-65, Univ of S CA 1964-65. US Naval Aviation 1946-48. Wt Control Engr Boeing Co Seattle 1951-62. With Hughes Aircraft Co El Segundo CA since 1962. Supr of Space & Communications Group Mass Properties Engrg 1962-73. Hd of Mass Properites Engrg 1973-79. Fellow Soc of Allied Wt Engrs 1954- ; Honorary Fellow, Soc of Allied Wt Engrs 1983- ; V P - Tech Dir 1976-79, Dir 1974-78; Most Outstanding Paper (Revere Cup) Award 1971; Space-Missiles-Metrology Chmn 1974-76; co-author SAWE 882-1971, 1050-1975, & 1134-1976; Author SAWE 1692-1986. Elected to Coll of Fellows of Inst for Advancement of Engrg 1975. *Society Aff:* SAWE.

Lottes, Paul A
Business: Argonne Natl Lab, 9700 S Cass Ave, Agronne, IL 60540
Position: Sr Mechanical Engr. *Employer:* Argonne National Lab. *Education:* PhD/Heat Transfer/Purdue; MS/Heat Transfer/Purdue; BS/Mech Engr/Purdue. *Born:* Aug 1926 Wilkinsburg Pa. Cons for Harrison Radiator Div of G M Corp 1959, Clearing Machine Corp 1959, Nuclear Prod Div of Amer Car & Foundry 1956- 59, Amer Machine & Foundry 1957, Genl Atomics Div of Genl Dynamics 1955. U S Delegate to 2nd U N Internatl Conf on the Peaceful Uses of the Atom 1958, & 3rd U N Conf 1964. Visiting Prof of Nuclear & Mech Engrg Univ of Ill 1961; Bd of Editorial Advisors, Nuclear Engrg & Design Journal 1966- ; Assoc Dir Reactor Engrg ANL 1968-69, Lab Dir's Office 1969-71. Mbr ASME, Fellow Mbr ANS 1968, Fellow Mbr ASME 1977. *Society Aff:* ASME, ANS.

Lottridge, Neil M
Home: 590 Westchester Rd, Saginaw, MI 48603
Position: Asst Chief Engr. *Employer:* Central Foundry Div, G M Corp. *Born:* Jan 25, 1931. MS Met E Wayne State Univ; Met E Univ of Cincinnati. Res metallurgist 1952-62 G M Res Labs, Asst Supt 1962-63 Buick (Foundry) Motor Div, Admin Engr 1963-66 & Asst Ch Engr 1966-76 Central Foundry Div. Spec in new applications for casting, conversion of forging to casting, & new casting alloys. Active in SAE, ASM, AFS. Enjoy boating, fishing, auto mech, carpentry, home decorating.

Lotz, Walter E, Jr
Home: 912 Dalebrook Dr, Alexandria, VA 22308
Position: Consultant *Employer:* Self *Education:* PhD/Phyisics/Univ of VA; MS/EE/Univ of IL; BS/Engg/USMA. *Born:* Aug 1916. Commissioned Officer Army Signal corps 1938-74. Key pos: Dir Army Res; Asst Ch of Staff, Communications-Electronics Vietnam; Asst Ch of Staff, Communications-Electronics Army; Cmdr Army Strategic Communciations Command; Cmdr Army Electronics Command; Deputy Dir Genl NATO Integrated Communication Agency. Retired July 1974 Lt Gen. 1975-77: Staff of Secy of Defense, respon for testing of all major weapons sys. Deputy Proj Manager, DCA support, TRW, Inc 1978-1980. Consultant 1980 - DATE. Director, Genisco Technology Corp, Rancho Dominguez Ca 1980 to DATE. Hon DSc Florida Inst of Tech; Disting EE Alumnus Univ of Illinois; Fellow IEEE; Fellow Radio Club of Amer. *Society Aff:* AUSA, AFCEA, RCA, IEEE.

Lou, Jack YK
Business: Ocean Engineering Program/Dept. of Civil Engr, College Station, TX 77843
Position: Professor *Employer:* Texas A&M Univ *Education:* Ph.D./Appl. Mech./Polytechnic Institute of Brooklyn; S.M./Naval Arch./Mass. Inst. of Technology; B.S./Naval Arch./CNCT. *Born:* 10/11/31 Worked for about ten years in marine related industries. Taught Ocean Engrg at Columbia Univ for six years. Joined Texas A&M Univ in 1974 and was the Head of the Ocean Engrg Prog 1983-87. A Fellow of ASME, where he served as chairman of several cttees in the Ocean Engrg Div and as Assoc Editor for the Journal of Energy Resources Technology for four years. Received Gold Certificate Award and Board of Governor's Award from ASME and an Outstanding Faculty Award given by the Student Engineers' Council of Texas A&M Univ in 1985. *Society Aff:* ASME, ASCE, SNAME

Louckes, Theodore N
Business: 1014 Townsend, Lansing, MI 48921
Position: Asst Chief Engr. *Employer:* Oldsmobile - GMC. *Education:* BS/ME/Genl Motors Inst. *Born:* 1930. Native of Lansing Mich. US Air Force 1951-54. With Oldsmobile Div, Engrg Dept since 1957. Have had engrg respon in all areas of automotive product design, dev & testing. Assumed current respon in 1974. Received SAE Natl Award for outstanding engrg contrib to automotive safety 1974 (Ralph Isbrand Medal). Active in fund raising for the Amer Cancer Soc. Received Safety Award for Engg Excellence, US Dept of Transportation, 1978. Enjoy golf & music. *Society Aff:* SAE.

Loucks, Daniel P
Business: Hollister Hall, Ithaca, NY 14853
Position: Prof *Employer:* Cornell Univ. *Education:* PhD/Environ Sys/Cornell Univ; MS/Forest Mgt/Yale Univ; BS/Forest Mgt/Penn State Univ. *Born:* 6/4/32. US Navy, Naval Aviator 1955-59 - Captain U.S. Naval Reserve. Prof. Civil and Environmental Engrg, Cornell University. Chmn Dept of Environ Engreg Cornell Univ, 1974-1980; Associate Dean, College of Engineering, 1980-1981; teaching & res in the applications of sys analysis tech to natural resource & environ qual mgmt problems. Cons to var Fed & Internatl agencies and governments. ASCE Huber Prize 1970, ASCE Julian Hinds Award, 1986. Author of numerous papers and book chapters in environmental and water resources systems analysis. *Society Aff:* ASCE, WPCF, TIMS, AWRA, AGU, AAAS, IAHR.

Louden, J Keith
Home: 257 Brook Farms Rd, Lancaster, PA 17601
Position: President. *Employer:* The Corporate Director Inc. *Education:* BBA/Ind Engrg/Ohio State Univ *Born:* 03/04/05 Duquesne Pa. Reg PE Ohio & Penna. Fellow A Effectivemer Soc Mech Engrs, Amer Mgmt Assn, Internatl Acad of Mgmt & Soc for Advancement of Mgmt. Author num articles for tech & mgmt journals. Author 7 bks: 'Wage Incentives', 'The Corporate Director", 'Making It Happen', 'The Effective Director in Action', 'Managing at the Top'. 'Think Like a President', 'How to Make it Happen', 'The Director'. Recipient Worcester Reed Warner Medal 1956 & Gilbreth Medal 1949. Has served as an officer & dir of sev corps & an officer in sev tech & mgmt soc's. Has served on Bd of Dir of 28 Corps, currently serving as a mbr of bd of dir of 4 corps. *Society Aff:* ASME, SAM, AMA

Loudon, Jack M
Home: 145 Brampton Rd, Syracuse, NY 13205
Position: Consulting Engr Stds. *Employer:* Self Employed *Education:* BS/Elec Engg/Univ of Denver. *Born:* 6/19/21. At GE Co. Worked on radar design, to 1955. Joined Stds in 1955 & assumed respon & title in 1968. Respon for gov specification interpretation, design & prod safety stds. Mbr Corp Stds Adv Bd, Stds Dev & Metric Councils, Group Div metrication, natl & internatl stds coordinator. Represented co on EIA Cttes & on var indus & gov task forces. Chmn ANMC Engg Sector on Electronics 76-83. Std Engrs Soc Fellow, Outstanding Secr Mbr, Chmn 78-85 Legislative Committee. Mbr Publication Committee 1985-. Mbr IEEE, Mbr IEEE Standards Committee SC 14-Quantities and units. 1978 EIA Engg Award of Excellence 1981-82 SES Section Chairman. Retired from GE Co June 1983. Self employed consulting engineer. Professional engineer-New York In 1983 1) Speaker at SES Natl Conference - Dayton Ohio (Sept) (Dayton, Ohio) 2) Speaker at DOD Standardization Conference (Oct) (Leesburg, VA) 3) Chrmn Task Force - "SES Ethics" - Now Published in Directory. (Have done considerable consulting in Product Safety & Specification work) *Society Aff:* SES, IEEE.

Loughren, Arthur V
Home: P O Box 404, Kailua-Kona, HI 96745
Position: Consulting Engr. *Employer:* Self. *Education:* EE/-/Columbia Univ; AB/

Loughren, Arthur V (Continued)
Science/Columbia Univ. *Born:* Sept 15, 1902. Vacuum tube & radio receiver engrg with GE Co & RCA 1925- 36. Radio receiver, television receiver & sys engrg, military electronics, color television R&D at Hazeltine Corp 1936-56; V P in charge of res 1954-56. V P in charge of Applied Res Div, Airborne Instruments Lab of Cutler-Hammer Inc 1956- 62. Pres Key Color Lab 1962-75. Mbr Natl Television Sys Comm 1939-41, V Chmn of second NTSC 1950-54. Pres IRE 1956, Dir 1953-58. Fellow IEEE, SMPTE. Sarnoff Medal SMPTE 1953, Liebmann Prize IRE 1955, Consumer Electronics Award IEEE 1974. EF Dir 1963-75. Other prof, indus & gov adv cttes; Armstrong Medal, Radio Club of America, 1981.. *Society Aff:* IEEE, SMPTE.

Louis, Joseph E
Business: 778 North First St, San Jose, CA 95112
Position: Pres *Employer:* Louis & Diederich, Inc. *Education:* BS/Civil/San Jose State Univ *Born:* 5/11/41 Native of Homs, Syria. Grad of San Jose State Univ with BS Degree in Civil Engr, 1978. Employed by various cities and consultant firms in the Bay Area. Pres and founder of Louis & Diederich, Inc., consulting engrs in San Jose, CA since 1977; specializing in land dev, transportation, and hydraulics. VP of local chap of CSPE. Chairman of the Ethical Committee (1979-81) for CSPE. Enjoy music and tennis. *Society Aff:* ASCE, NSPE, APWA, ECSJ

Love, Allan W
Home: 518 Rockford Pl, Corona del Mar, CA 92625
Position: MTS VII *Employer:* Satellite System Div Rockwell Intl *Education:* PhD/Physics/Univ Toronto; MA/Physics/Univ Toronto; BA/Math & Physics/Univ Toronto *Born:* 5/28/16 Born and educ in Toronto, CD. Served overseas for 5 yrs with Royal Canadian Air Force. Spent two years 1946-48 as Res Officer in the Radio Physics Lab of the Commonwealth Scientific And Industrial Res Organization in Sydney Australia. Came to the US in 1951 and engaged in mining geophysical exploration until 1957 with Newmont Exploration Ltd in Jerome AZ. Entered the aerospace field in 1957 with Wiley Electronics Co in Phoenix, AZ and moved to North American Aviation (now Rockwell Intl) in CA in 1963. Fellow, IEEE, & Pres of IEEE Antennas & Propagation Soc, 1984. *Society Aff:* IEEE

Love, John
Business: Engr Bldg 2024, Columbia, MO 65201
Position: Prof of MAE. *Employer:* Univ of MO-Columbia. *Education:* PhD/Flow/OK State Univ; MS ME/Heat Trans/Univ of MO-Columbia; BS ME/Mech Eng/Univ of MO; AB/Chem/Univ of MO. *Born:* 1/26/16. Educator; b Paris, TX. s John and Stella Maude (Reynolds) L; m Marjorie O Zoeller, May 18, 1942. Conductor, MO KS TX RR, Parsons, KS, 1939-49; asst prof mech and aerospace engg, Univ of MO, Columbia 1951-54; assoc prof mech engg, engg, Gen Elec Co, Norwood O, 1954-58. Engg cons Black & Veatch, cons engrs, 1973-. NSF sci faculty fellow, 1962-63. Reg PE, MO Mem ASTM (chrmn energy com 1973-), Columbia Audubon Soc (dir 1966-68), Sigma Xi, Tau Beta Pi, Pi Tau Sigma, Alpha Chi Sigma. Mem editorial bd Elsevier Sci Pub Co, 1974-. *Society Aff:* ASME, ASTM, ΣΞ, AIAA.

Love, Tom J, Jr
Business: 865 Asp, Norman, OK 73019
Position: Prof. *Employer:* Univ of OK. *Education:* PhD/Mech Engg/Purdue Univ; MS/Mech Engg/Univ of KS; BS/Mech Engg/Univ of OK. *Born:* 10/2/23. Proj Engr, Colgate, 1947-52, Sr Res Engr, Midwest Res Inst, 1952-56, Faculty Univ of OK, 1956-present. Dir, School of Aerospace, Mech and Nuclear Engg, 1963-72, Halliburton Prof of Engg, since 1972, G L Cross Res Prof of Aerospace, Mech and Nuclear Engg, since 1973. Author book "Radiative Heat transfer" and approximately 50 technical papers. Mbr Bd of Dir of Local Fed Savings and Loan Assoc, OK City since 1970, Mbr Bd of Dir, Sverdrup/ ARO Inc Tullahoma, TN 1977-1981. Recipient, AIAA 1984 Thermophysics Award. *Society Aff:* ASME, AIAA, ASEE, OSA, ASTM, OSPE

Love, William J
Business: C/O Hazen & Sawyer - Cairo, ARE, 360 Lexington Ave, New York, NY 10016
Position: VP *Employer:* Malcolm Pirnie, Inc. *Education:* MS/CE/MIT; BS/Mil Art & Engr./USMA West Point *Born:* 6/2/25 After commissioning as a 2nd Lt from West Point in 1945, served 26 yrs until retiring in 1971 as a Colonel, Corps of Engrs. Last assignment was a Dist Engr, Baltimore, Commanding and supervising over 1400 employees, Military and Civilian, including approx 100 engrs. From 1971-1978 was Gen Mgr of the Hampton Rds Sanitation Dist supervising over 400 employees of whom 23 were engrs. Since 1978, a corporate VP of Malcolm Pirnie, Inc providing consulting svcs in envir engr. Currently, while still an MPI Officer, serving as Project Mgr in Cairo, Egypt for a joint venture designing water and waste water facilities, for the 3 canal cities. Supervise 125 Americans and Egyptians of whom 40 are engrs. *Society Aff:* NSPE, AAEE, SAME

Lovejoy, Stanley W
Home: 3 Cottonwood Dr, Dix Hills, NY 11746
Position: Electric Production Results Supr. *Employer:* Long Island Lighting Co. *Education:* BS/ME/Northeastern. *Born:* Apr 1923 Boston Mass. BSME Northeastern Univ 1945. P E NY. Field engr Westinghouse in Steam Turbine Div. With Long Island Lighting Co since 1951 spec in oper, testing & efficiency analysis for sys of 4000 MW. Fellow Mbr ASME; Chmn PTC-22 Gas Turbine Power Plants; Mbr ASME Performance Test Code Standing Ctte, ANSI B-133 Gas Turbine Ctte, ISO TC-70 Gas Turbine Ctte. Received ISA's Philip T Sprague Award 1972 in recog of contrib to power plant control. Has presented num tech papers on Variable Pressure & Low Excess Oper of drum-type boilers. Married Jewel Fitchko 1947. Three children. *Society Aff:* ASME.

Lovelace, Alan M
Business: 400 Maryland Ave SW, Washington, DC 20546
Position: Deputy Administrator. *Employer:* Natl Aeronautics & Space Admin. *Education:* PhD/Organic Chemistry/Univ of FL; MS/Organic Chemistry/Univ of FL; BS/Chemistry/Univ of FL. *Born:* Sept 1929. Native of St Petersburg Fla. Dir of Air Force Matls Lab 1967- 72; Dir of Sci & Tech, Air Force Sys Command 1972-73; Principal Deputy Asst Secy of the Air Force 1973-74; Assoc Admin for Aero & Space Tech NASA 1974-76; Deputy Admin of NASA 1976-1980. AIAA Fellow; AAS Fellow; Mbr NAE; AIAA vonKarman Lectr 1974; Arthur S Flemming Award; Univ of FL Distinguished Alumnus Award 1979. Current position: Acting administrator Jan 1981 - present. *Society Aff:* AIAA, AAS, NAE.

Loveland, John R
Business: 1304 Buckley Rd-1181, Syracuse, NY 13201
Position: President *Employer:* O'Brien & Gere Engrs Inc. *Education:* MBA/Personnel/Syracuse Univ; BCE/Civil Engr/Syracuse Univ. *Born:* Sept 1937. Native of Syracuse N Y. Employed continuously by O'Brien & Gere Engrs Inc since June 1960 in successively respon pos of Designer, Proj Engr & Prin Engr. In April 1971 apptd V P of Civil Engrg Div with final engrg & client respon for a wide range of studies & design projects. In Oct 1975 appointed Group V P Civil Engrg, Genl Engrg & Bus Admin Divs. In Nov 1980 elected President, Tau Beta Pi Natl Exec Council; Chmn Town of DeWitt Environ Conservation Comm; Amer Soc of Civil Engrs ASCE; Water Pollution Control Fed WPCF; Amer Water Works Assn AWWA. *Society Aff:* ACEC, ASCE, NSPE, AWWA, WPCF.

Lovell, Donald J
Business: D J Laboratories, Stow, MA 01775
Position: Exec. Dir. (Self-employed) *Education:* MS/Physics/Univ of WI; PhD/Physics & Math/Univ of WI. *Born:* June 8, 1922. Studied at MIT, Dartmouth & Penn State. Taught at Norwich Univ, Univ of Conn, The Univ of Michigan, Univ of Wisconsin, & the Mass College of Optometry. Principal Engr at Photoswitch, Barnes Engrg & Bendix. Led the BAMIRAC program at Michigan. Now consulting

Lovell, Donald J (Continued)
for var orgs, particularly regarding the calibration of cryogenically cooled optical devices and computer determinations of optical system performance. *Society Aff:* OSA, SPIE.

Loven, Andrew W
Home: 514 Starlight Crest Dr, La Canada, CA 91011
Position: President. *Employer:* Engineering-Science Inc. *Education:* PhD/Phys Chem/Univ of NC at Chapel Hill; BS/Chem/Maryville College. *Born:* 1/31/35. Dev Mgr with Westvaco Corp, spec in wastewater water and air pollution control 1963-71. Specialty in environmental and activated carbon technology. Varied technical & management positions with Engineering- Science Inc since 1971, presently President. Respon for all activities in water & wastewater treatment, hazardous waste management, solid waste mgmt, environ impact & other environ cons activities. Reg P E num of states. Diplomate Amer Acad of Environmental Engs. Mbr Water Pollution Control Fed, Sigma Xi, Amer Inst of Chem Engrs, ASTM, Amer Chem Soc. Natl Soc of Prof Engrs, Amer Public Works Assoc, Society of American Military Engrs *Society Aff:* WPCF, ΣΞ, ACS, AIChE, NSPE, ASTM, AAEE, APWA, SAME.

Lovering, Earle W
Business: 1821 S 54th Ave, Cicero, IL 60650
Position: Corp Dev Engr. *Employer:* Chicago Extruded Metals Co. *Education:* BS/Phys Met/MIT. *Born:* 5/14/17. Arlington, MA. Met for Revere Copper & Brass 1938-1946. Did pioneer work on extrusion uranium 1943-45. Dir of Dev Engg, Scovill Mfg Co 1954-74. Gerente De Metalurgia (Met Mgr) Eluma-Isam, Sao Paulo, Brazil 1975-77. Presently in charge of process dev, Chicago Extruded Metals Co. Am specialist in copper alloy casting, extrusion, annealing, drawing, rolling. Numerous papers on these subjs. Active ASM New Haven Chapter Chrmn 1953-54. Enjoy languages -fluent French & Portuguese. Chief Met, Seymour Mfg Co 1946-1954. Registered Professional Engineer in Connecticut and Illinois.. *Society Aff:* ASM.

Lovgren, Carl A
Business: 800 Theresia St, St Marys, PA 15857
Position: VP, Marketing. *Employer:* Airco Speer Carbon-Graphite Div, AIRCO, Inc. *Education:* BS/CE/Northeastern Univ. *Born:* 6/22/19. Native of Rockport, MA. Presently involved worldwide in application of artificial graphite mtls to the steel industry. Author of intl papers on various phases of electric furnace steel production. Formerly involved in intl consulting for civ & ind dev progs. *Society Aff:* AIChE, AIME, ANS, AISE.

Low, Dana E
Business: TAMS Bldg, 655 Third Ave, New York, NY 10017
Position: Partner. *Employer:* Tippetts-Abbett-McCarthy-Stratton. *Education:* MS/Civil Engg/Thayer Sch of Engg; AB/Engg/Dartmouth College. *Born:* Dec 1932. Native of Summit N J. US Navy 1955-58, Ch Engr USS Hawkins (DDR 873). Jr Engr, Howard Needles Tammen & Bergendoff 1958-60. With Tippetts- Abbett-McCarthy-Stratton since 1960, Assoc Partner 1972, Partner since 1973. Respon for internatl proj activities in fields of transp, regional planning & resources dev. Fellow of Inst of Transportation Engrs & Amer Soc of Civil Engrs. Active in Internatl Rd Fed activities. Was Chmn for 2 years. Reg P E 9 states. *Society Aff:* ITE, ASCE.

Low, George M
Business: 400 Maryland Ave SW, Washington, DC 20546
Position: Deputy Administrator. *Employer:* NASA. *Education:* Master/Aero Engr/Rensselaer Poly Inst; Bachelor/Aero Engr/Rensselaer Poly Inst. *Born:* June 1926. MS, BS, RPI. Joined Natl Adv Comm for Aeronautics 1949. Assigned to Hdqtrs office when NASA estab 1958. Served as Ch of Manned Space Flight Prog & Deputy Assoc Admin for Manned Space Flight. Joined NASA Manned Spacecraft Ctr as Dep Dir 1964. Became Mgr of Apollo Spacecraft Prog 1967. Named Deputy Admin NASA 1969. Became Pres of Rensselaer Polytechnic Inst, 1976. Num awards & honors. Hon Dr degrees from RPI (engrg) & Fla (science) 1969. Lehigh (Engg) 1979. Space Flight Award 1968 AAS, Paul T Johns Trophy 1969 Arnold Air Soc, NASA Disting Serv Medals (2) - 1969 Arnold Air Soc, Louis W Hill Space Transp Award 1969 AIAA, Astronautics Engrg Award 1970 AIAA. Natl Space Club's Robert H Goddard Memorial Trophy 1973, Natl Civil Suc League Career Suc Award 1973, Rockefeller Public Suc Award 1974, Natl Academy of Engg Founders Award 1978. Dir, Genl Elec Co. Prof Engg NY. *Society Aff:* NAE, AIAA.

Low, Herbert M
Home: 6549 Overhill Rd, Prairie Village, KS 66208
Position: V P & Mgr - Internatl Oper - ret. *Employer:* J F Pritchard & Co. *Education:* EE/-/KS State Univ; BS/EE/KS State Univ. *Born:* 1902 Emporia Kansas. Disting Serv Award 1957. Latin Amer 5 yrs, elec const, oper & design. Phillips Petro Co 9 yrs, elec engr, power engr, refinery engr. J F Pritchard & Co 26 yrs, Ch Engr respon for design of power plants, oil refineries, chem plants, gas plants; V P Power Div; V P Internatl Oper, incl 3 yrs V P & Dir Pritchard Internatl resident in France & Algeria. Retired 1968, followed by cons on spec air pollution problems. Reg P E 6 states. Life Fellow ASME, Life Mbr IEEE. Hobbies: studies at seminary & wine making. *Society Aff:* IEEE, ASME.

Lowe, John, III
Business: 26 Grandview Blvd, Yonkers, NY 10710
Position: Consulting Civil Engineer *Employer:* (Individual) *Education:* MS/Civil Engg/MIT; BS/Civil Engg/College of City of NY. *Born:* 3/14/16. Resides in Yonkers N Y. Taught Univ of Maryland 1937-40 & MIT 1941-44. Physicist at David Taylor Model Basin 1945. with TAMS 1945-1983. Initially Hd, Soil & Rock Engrg Dept. Became Ptnr 1962. Specializes in geotechnical engineering and dam projects. Individual consult. to various clients on dam projects; both while partner of TAMS and since Jan '84. Mbr Natl Acad of Engrg. Gave Karl Terzaghi Mem Lect in 1971, Nabor Carrillo Lecturer 1978 & 2nd USCOLD Lecture 1982. Author over 30 tech papers & contrib in 3 engrg handbooks. *Society Aff:* ASCE, USCOLD, NAE, ISSMFE, ISRM, MOLES.

Lowe, Philip A
Home: 11316 Roven Dr, Potomac, MD 20854
Position: Pres *Employer:* Intech Inc *Education:* PhD/Engrg/Carnegie Mellon Univ; MS/Engrg/Univ of RI; BS/Engrg/Univ of UT *Born:* 2/25/39. Officer USN Civ Engr Corps 1962-64. Sr Engr, nucl reactor design, Bettis Atomic Power Lab, Westinghouse, 1964-1970. Mgr of Experiments, Combustion Engg Corp, 1970-1974. US Govt (Atomic Energy Commission, ERDA, Dept of Energy) 1974 to 1980. NVS Corp 1980-1986 CEO B&P Energy Inc 1982- . Pres Intech Inc. 1985- . Taught at Univ of UT, Univ of RI, RPI, Carnegie-Mellon Univ. Directed independent evaluations of the Dept of Energy's mgt of progs & operations. Presently directs corporate consulting activities in the use of advanced technologies such as combustion and conversion of coal, including pollution control and economic analyses. Directs the drilling & production of oil. Reg PE-PA. Mbr of fed Sr Exec Service & chrmn of County Recreation Advisory Bd, ANS Alternative Technologies & Sci. Committee, Advisor to Gas Research Institute, Fellow American Soc of Mech Engrs, Mbr Soc of Petroleum Engrs, Mbr American Soc of Engrs Technology Opportunities & Planning Committee, Advisor to the Center for Research on Sulfur in Coal. *Society Aff:* ASME, ANS, SPE, WCC

Lowe, Wilton J, Jr
Home: 6414 Jadecrest, Spring, TX 77389
Position: Manager Consulting Services *Employer:* Chevron USA *Education:* BS/Petrol Eng/LA State Univ *Born:* 7/27/36 Native of Port Allen, LA. Served with USAF from 1958-1961 spent the last 20 yrs working for Chevron USA in various engrg capacities until present assignment as drilling supt in 1975. Served local SPE AIME chapter as Dir for 3 yrs. Currently Tech Editor for SPE, As well as 1981 SPE Convention Drilling Session Co-Chrmn. Enjoy golf, fishing & LSU football. *Society Aff:* SPE, API.

Lowell, J David
Business: Lowell Mineral Exploration, RR 3 Box 197, Nogales, AZ 85621
Education: MS/Geology/Stanford Univ; E Geol/Geological Engr/Univ of AZ; BS Min Engr/Mining Engr/Univ of AZ *Born:* Feb 1928. Native of Arizona. Employed as engr & mine foreman (Asarco); field geologist; Proj Ch & Dist Geologist USAEC; Ch Geologist, Mine Mgr Intl Ranwick; V P Southwest Ventures; & Sr Staff Geologist & Dist Geologist Utah Const Co. Entered cons in 1961 & worked for 100 mining & oil co's & foreign govs in North, Central & S Amer, Orient, & Near East. Specialize in porphyry copper explor & have participated in discovery or dev of sev major copper orebodies in U S, Canada & Philippines & publ 25 tech papers which are mainly concerned with porphyry copper geol. Daniel Jackling Award 1970 AIME, Disting Lectr 1974 CIMM, Assoc Editor 'Economic Geology' 1975-80, Thayer Lindsley Disting Lectr SEG 1977, SEG Silver Medal 1983. *Society Aff:* AIME, MMS, SEG

Lowen, Gerard G
Home: 484 Eisenhower Ct, Wyckoff, NJ 07481
Position: Herbert Kayser Professor of Mechanical Engineering *Employer:* City College of NY. *Education:* DrIng/Mech Engg/Tech Univ Munich; MSME/Mech Engg/ Columbia Univ; BSME/Mech Engg/City College of NY; PE/NY State. *Born:* 10/25/21. Came to US from Germany in 1939. Studied Mech Engg in evening while working as a toolmaker. Taught at the City College since 1954. Main academic interest is in Theory of Machines. Author and Co-author of numerous publications in field of mechanism balancing, mechanism vibrations, safety and arming and timing devices. Consultant in machinery and fuzing problems to industry and govt agencies. Expert witness in machine safety and product liability cases. Active in intl standardization work, ASME Mechanisms Committee, ASME Policy Bd, Gen Engg Dept and ASME Council on Codes and Standards. ASME Mechanism Cttee Award 1984. Fellow ASME. ASME Machine Design Award 1987. *Society Aff:* ASME, ASEE, AAAS, SME, VDI.

Lowen, Walter
3201 Moore Ave, Binghamton, NY 13903
Position: Prof, T J Watson School *Employer:* State Univ of N Y - Binghamton. *Education:* BS/ME/NC State Univ; MS/ME/NC State Univ; Dr Sc/Nuclear/ETH *Born:* 5/17/21. Alumni Serv Award & Excell of Teaching Award Union College (Schenectady NY); Div & Dept Chmn Union College. 1967-77. 1977-1983 Prof & Dean, Sch of Advanced Tech (SAT), State Univ of NY at Binghamton. Designed initiated SAT. 1983-present - Prof of Sys Sci Dept - The Thomas J Watson School of Engrg, Applied Sci and Tech. Author of "Dichotomies of the Mind–. John Wiley & Sons, 1982. BS & MS in ME - NC State Univ 1943, '47. Dr Sc in Nuclear Engrg. Eidgendssische Technische Hochschule (ETH) Zurich, Switzerland 1962. Founding Mbr Volunteers in Intl Tech Assistance (VITA) Inc. Mbr Sigma Xi, Tau Beta Pi, ASME, ANS, ASEE, AAAS, NY Acad of Sci, Amer Assn of Univ Profs, World Future Soc, World Acad of Sc. *Society Aff:* ASME.

Lowery, Anthony J
Business: P O Box 975, St Charles, MO 63301
Position: President. *Employer:* Cadmus Corp. *Education:* MS/Chem Engg/Univ of IL; BS/Chem Engg/Carnegie Inst of Tech. *Born:* 6/22/32. Technologist Shell Oil Co 1955-62 Wood River Refinery; Sr Engr 3M Co 1962- 64 - Central Res Pilot Plant, St Paul Minn; VP Findett Corp 1964-73; Pres Cadmus Corp 1973- ; US Army, The Engrs' Sch, Ft Belvoir VA 1956. AIChE - Treas, Alton Wood River Sect; Secy St Louis Sect. *Society Aff:* TBΠ, ΣΞ, ΑΧΣ, ACS, WPCF, AIChE, NSPE, BFA.

Lowery, Gerald W
Business: 1710 Goodridge Dr, McLean, VA 22102
Position: Corp. VP & Dir Renewable Energy progams *Employer:* Science Applications Int'l Corp *Education:* PhD/Mech Engr/Auburn Univ; MS/Mech Engr/Auburn Univ; BS/Mech Engr/Auburn Univ. *Born:* 5/26/41. Native of Thomasville AL. VP of engg cons firm in Atlanta GA from 1969-73. Asso. Prof of ME at Univ of TX at Arlington 1973-75. FL Solar Energy Ctr as Dir of Res, Dev & Demonstration Div Jan 1976-Nov, 1977. Respon incl supervision & dir or prof & career-service personnel engaged in all aspects of solar energy activities. Received Outstanding Fac Award for Gulf-Southwest Sect of ASEE 1975, finalist for Outstanding Engrg Teacher Award 1975 & 1976. Joined Sci Applications Int'l Corp in Nov 1977. Overall respon within the organization for dev and coordination of all SAI activities in renewable energy. Developing near- term business strategy for moving more into the systems integrator role for alternative energy applications. Served as Proj Mgr on DOE-funded program providing sys studies and analysis of advanced solar heating and cooling systems. *Society Aff:* ASES, ASME.

Lowery, Thomas J
Home: 102 Briarwood Dr, Manchester, CT 06040
Position: Sales Mgr Cert Packaging-Professional. *Employer:* Preferred Plastics Inc. *Born:* Sept 30, 1944. 2 yrs Fairleigh Dickinson Univ; Amer Mgmt Assn courses NYC. 2 yrs in testing & design lab in Chgo (Genl Box Co). Attended Univ of New Haven (Matls Handling & Pkging). With Preferred Plastics since 1971, spec in design & testing. Current mbr of the Conn Joint Engrg Council. Pres of the Conn Chap of the Soc of Pkging & Handling Engrs. Curriculum advisor to the Univ of New Haven on pkging & matl handling. Soc of Plastic Indus 1973 Grand Eastern engine up, Eastern Regional VP form SPHE Award (Best in Show) for matl handling & 4 gold awards for Best in Category 1974-75. Enjoy golf & gardening. *Society Aff:* SPHE.

Lowi, Alvin Jr
Home: 2146 Toscanini Dr, San Pedro, CA 90732
Position: Prin *Employer:* Alvin Lowi and Assoc *Education:* Residence(PhD)/Engrg/UCLA; MSME/ME/GA Inst of Tech; BME/ME/GA Inst of Tech *Born:* 07/21/29 Educ GA Tech, UCLA and U PA. Engrg Officer in Korean conflict. Reg PE, principal in indep conslt practice since 1965 in design, test, engrg, res and mgmt in aerospace, automotive, transit, fuels, chem, water purification and forensic industries. Founder of several bus ventures based on more than 30 domestic & foreign patents on engines, refrigeration, distillation and insulation. Author, numerous tech papers. Mbr Pi Tau Sigma, Scabbard and Blade. *Society Aff:* SAE, ASME, NSPE

Lowton, Robert B
Home: 2361 Loma Park Ct, San Jose, CA 95124
Position: Mgr, Advanced Engg. *Employer:* GE. *Education:* BS/-/Yale Univ; BE/ME/ Yale Univ. *Born:* 1/17/26. Born & raised NY, NY. USN 1943-47. Kellex Corp - 1948-1964 (Became Vitro Corp in 1951) - Job engr, Constr. engr, proj engr, chem & Nuclear plant design/construction. GE - 1964-present Resident site mgr, project mgr, mgr of advanced engg - nucl power systems engg dept. *Society Aff:* ANS.

Lowy, Stan H
Home: 1016 Walton Dr, College Station, TX 77840
Position: Professor Emeritus *Employer:* Texas A&M Univ (ret) *Education:* MS/Aeronautical Eng/Univ MN; BS/Aeronautical Engrg/Purdue. *Born:* 3/10/22 Native of NYC, NY. Served with Army Air Corps 1943-46. Taught at OR State, Portland State and Univ of OK. Industrial experience with Allison Div of General Motors, Boeing Aiplane Co, Hughes Aircraft, Willamette Iron & Steel Co, Peters Co, A. Young & Son Iron Works, Stan H. Lowy & Assoc. With Texas A&M since Feb 1964, current responsibility since Jan 1980. ASEE/AIAA outstanding Aerospace Engrg Educator Award 1977, AIAA National Outstanding Faculty Advisor Award 1987. *Society Aff:* AIAA, ASEE, ΣΞ, ΣΓΤ.

Loyalka, Sudarshan K
Business: Nuclear Engg Dept, Columbia, MO 65211
Position: Prof. *Employer:* Univ of MO. *Education:* PhD/Mech Engr-Nuclear/ Stanford Univ; MS/Mech Engr-Nuclear/Stanford Univ; BE/Mech Engr/Univ of Rajasthan *Born:* 4/11/43. Received early education in Pilani, India. Was a recipient of the Bd of Secondary Education and Univ of Rajasthan gold medals for ranking first in Intermediate in Sci (1960) and Bachelor of Engg examinations (1964). Completed grad studies in Nuclear Engg at Stanford Univ in Palo Alto, CA (1967). Since then has been at the Univ of MO in Columbia, where a Prof of Nuclear Engg. During the yrs 1969-71, was a visiting scientist at the Max Planck Inst fur Stromungsforschung, Gottingen, W Germany. Published more than eighty papers in kinetic theory of gases, neutron transport, nuclear reactor physics and thermal hydraulics, mechanics of aerosols and other related areas. Received Univ of MO- Columbia Chancellor's Award for Outstanding Res in Physical and Mathematical Sci (1982), the Sigma Xi Res Award (1982) and the Byler Award 1985. Fellow, APS 1982, ANS 1985. *Society Aff:* ANS, APS, ACSΣΞ, AAAS.

Lozier, John C
Home: 21 Park Rd, Short Hills, NJ 07078
Position: Consulting Engr. (Self-employed) *Education:* AB/Physics/Columbia. *Born:* Feb 1912. At Bell Labs 1936-77. R&D in Transmission, Computing & Control Sys, incl Telephone Sys, Missiles & Satellites, Process Controls for Manufacturing Transistors, Crystals, etc, Digital Computer Design & Appli to Control Sys. Presently Control Sys Cons. Fellow & Life Mbr IEEE, former Chmn GAC & Pres Amer Control Council, P Pres Internatl Fed of Automatic Control. MacKay Visiting Prof at Univ of Calif Berkeley 1958; Chevalier of the Legion of Merit by French for Telstar. Mbr Tech Adv Ctte to Bd of Governors of Argonne Natl Labs 1962-64. Distinguished Mbr IEEE Control System Society and Computer Society. IEEE Distinguished Member 1983, Rec. Heritage Award IEEE-CSS 1987. *Society Aff:* IEEE.

Lu, Le-Wu
Business: Fritz Engrg Lab 13, Lehigh University, Bethlehem, PA 18015
Position: Prof CE *Employer:* Lehigh Univ. *Education:* PhD/Civil Engg/Lehigh Univ; MS/Civil Engg/IA State Univ; BS/Civil Engg/Natl Taiwan Univ. *Born:* June 1933. Native of Soochow, China. Served as Grad Asst Iowa Engrg Exper station, Ames Iowa 1955-57. With Lehigh Univ since 1957 & held the following pos: Res Asst, Res Assoc, Instr, Res Asst Prof, Res Assoc Prof & Dir Plastic Analysis Div, Prof & Dir Building Systems Div, Fritz Engrg Lab. ASCE L S Moisseiff Award 1967; Sr Fulbright-Hays Lectureship 1975. *Society Aff:* ASCE, ACI, ASEE, SSRC, EERI, CTBUH

Lu, Susan S
Business: 4242 W Dempster St, Skokie, IL 60076
Position: CEO *Employer:* Five Seas Trading Co. *Education:* MS/Physical Chem/ Loyola Univ; BS/Chemistry/Mundelein College. *Born:* 7/4/34. Have worked at USDA the summers of 1960, 1961. Worked as a research chemist in the Research Foundation from 7/1962-12/1963 the research finding abstruct published in Federal Proceedings. Research work done in chemical kinetics of vapor phase reactions (1967-68), the result presented at the ACS meeting (6/1969). Have assisted research work in Air Pollution control research project in 1974. 1974-87 Have done works in; Marketing R/D and intl business development; Promoted U.S. export business; Import marketing materials confined to the resource materials of importance, benefit to the U.S. industrial and economic growths and domestically needed materials; R/D in the fields of energy, basic materials and sources. Management works in --- consulting, planning, economic growth. *Society Aff:* AIChE, ΣΞ.

Lubetkin, Seymour A
Home: 11280 Aspen Glen Dr, Boynton Beach, FL 33437
Position: President *Employer:* Environmental Technology, Inc. *Education:* Master Civil Engr./Sanitary Engineering/N.Y.U.; Master Science/Elec. Engr/Newark Coll of Engr (now NJIT); Bach. Science/Mech. Engr/Newark Coll of Engr (now NJIT) *Born:* BSME & MSEE Newark College of Engr; MCE NYU. Tau Beta Pi. P E AZ, FL, IN, MD, NJ, NM, NY, MA, OH, PA & WV. Licensed WWTP Operator (NJ). Passaic Valley Sewerage Commissioners 1950-1978 Ch Engr 1954-1978. Weston Consultants 1978- 1984, V Pres 1979-1984. Environmental Technology, Inc. Pres. Mbr NJ Water Pollution Control Assn since 1951, Board of Control WPCF 1975-1978. Dr. Heukelekian Indust. Waste Award (NJWPCA 1973), Hatfield Award (WPCF 1983). Mbr Adv Panal Water Resources Res Inst (1975-1978). Arbitrator of Amer Arbitration Assn. (1985-). Mbr Palm Beach County Water & Sewer Adv Bd. (1986-). Has been on advis. committee to EPA, Interstate San. Comm. (NY-NJ), NJDEP, Pollution Engineering Mag., AMSA, WPCF etc. Author of many technical articles on pollution control. Diplomate of AAEE. *Society Aff:* WPCF, AAEE, TBΠ

Lubinski, Arthur
Business: 4469 S. Gary Ave, Tulsa, OK 74105
Position: Consultant *Employer:* Self-employed *Education:* Ingenieur Civil Mecanicien et Elecricien/University of Brussels, Belgium; Candidat Ingenieur Civil/ University of Brussels, Belgium *Born:* 03/30/10 Native of Antwerp, Belgium. Fought with the French forces of the interior in W.W.2. Immigrated to the USA in 1947. Naturalized in 1952. Special research associate, Amoco Production Co, 1950-75. First to use mathematics and applied mechanics to problems pertaining to drilling for oil and gas; managed research and technology transfer to offshore and arctic operations. Awards for cttees chairmanship or technical achievements by ASME (fellow) (1965 and 1969), Soc. of Petr. Engrs (Distinguished Lecturer 1964), International Assoc. of Drilling Contractors (1958), Offshore Technology Conference (1976). One book, 41 publications, 7 patents. *Society Aff:* NAE, ASME, SPE, API

Lubinus, Louis
Home: Rt 3, Box 24, Brookings, SD 57006
Position: Ext Agri Engr. *Employer:* SD State Univ. *Education:* BS/Agri Engr/SD State Univ. *Born:* 12/7/21. Native of Salem, SD. Grad from SDSU in 1947 with BS in agri Engrg. Served with Army Corps of Engrs from 1943-46. Employment since 1947 has been as an Extension Specialist in Agri Engr at SDSU. Primary responsibility has been teaching-training Ext Service field personal, bldg mtl suppliers, farm contractors, etc principles and methods of farm structures and livestock environment, rural domestic and livestock waste handling, and mtls handling with emphasis on complete engineered systems. Have served as ASAE Committee 402 (Beef Housing) Chrmn. Currently NC Regional dir ASAE. Awarded Distinguished Service Award, SD Ext Specialists Assn. *Society Aff:* ASAE.

Lubman, David
Home: 14301 Middletown Ln, Westminster, CA 92683
Position: Sr Systems Engr. *Employer:* Hughes Aircraft Co. *Education:* MS/EE/Univ of S CA; BS/EE/IL Inst of Tech. *Born:* 8/3/34. Staff Engr at Hughes Aircraft Co 1960-67. Sr Scientist at LTV Res Ctr, Anaheim CA 1967-68. Sr Scientist at Bolt Beranek and Newman, Van Nuys, CA 1968- 69. Independent consultant from 1969-present, specializing in sound power measurements and architectural acoustics. Sr Systems Engr at Hughes Aircraft Co. Fellow of ASA, 1972 and Chmn of its technical comm on Architectural Acoustics 1980- . ASA Medals and Awards Committee 1987- . ASA Mbrship Comm, 1975-1987. Comm on Noise, 1972-. Dir of INCE, 1976-79. Chmn, NCAC Acoustical Laboratory Committee 1977- . Visiting Scientist, National Research Council of Canada, 1976. Visiting Lecturer, CA State Univ of Los Angeles and Fullerton and Chapman College, Dept of Mathematics, and Univ of CA at Santa Barbara, Depts of Speech and Hearing; Orange County Chapter of ASA, Founding member and Vice Chairman 1985-87, Chairman, 1988. *Society Aff:* ASA, ASTM, INCE, IEEE, NCAC.

Luborsky, Fred E
Business: PO Box 8, Schenectady, NY 12301
Position: Res Staff Scientist. *Employer:* GE Co. *Education:* PhD/Physical Chemistry/ IIT; BS/Chemistry/Univ of PA. *Born:* 5/14/23. in Phila, PA & joined the GE Co in 1951. He worked in their Instrument Dept for 6 yrs & is currently with their Res & Dev Ctr in Schenectady, NY. His main fields of interest have been in the magnetic, met and electrochem aspects of hard mtls, soft mtls & thin films. He has been the key technological leader in developing GE Lodex permanent magnets, plated mag-

Luborsky, Fred E (Continued)
netic disks, plated wire memory and amorphous metals for large transformers. He is currently concerned with studies on magnetoptic alloys. He has over 180 papers & 20 patents. He was Pres of the IEEE Magnetics Soc and has served in many capacities in the Magnetics Soc & in the Conf. Magnetism & Magnetic Mtls. He is a Coolidge Fellow of The G E Res & Dev Center & is a member of the National Academy of Engineering. *Society Aff:* AAAS, IEEE, APS, ACS, NAE

Lucas, Chester L
Home: 2365 Broadmont Ct, Chesterfield, MO 63017
Position: VP *Employer:* Sverdrup Corp *Education:* BS/CE/Duke Univ *Born:* 5/28/15 Native of MA. Graduate of Phillips Exeter Academy and Duke Univ. Served in US Navy 1942-45 and 1950-53, officer in Civil Engr. Corps, USNR to rank of Commander. Experience with NC State Hgwy Commission, TVA, Panama Canal, VA Bridge Co, field engrg. Rader Engrg, Miami FL, Project Mgr and VP, Panama Office. 1953-60 Nicaragua Power & Light, Const. Mgmt Advisor. 1960-63 McGaughy, Marshall and McMillan, Pres of Subsidiary Rome, Italy, Athens, Greece 1963-75. Sverdrup & Parcel, VP Intl Operations and mktg, 1977-81. Life member ASCE. Duke Univ Distinguished Alumnus Award 1979. Enjoy golf, photography, writing. *Society Aff:* ASCE, ACEC, NSPE, SAME

Lucas, Glenn E
Business: Dept of Chem & Nucl Engg, Univ of CA, Santa Barbara, CA 93106
Position: Asst Prof. *Employer:* Univ of CA. *Education:* ScD/Nucl Engg/MIT; SM/Nucl Engg/MIT; BS/Nucl Engg/UCSB. *Born:* 3/8/51. Native of CA. Received the Barker Fellowship in natl competition as a student at MIT. Doctoral dissertation has been published in Outstanding Dissertations in Energy (Garland Publishing, 1979). Author/co-author/co-editor 3 books; 45 journal publications. Teaching responsibilities & res expertise in mech met & mtls engg of nucl systems. Active participant in the natl fusion mtls dev prog & light water reactor fuels improvement progs. *Society Aff:* ANS, ASM, ΣΞ, AAAS, ASTM.

Lucas, J Richard
Business: Virginia Tech, 213 Holden Hall, Dept. of Mining & Minerals Engrg, Blacksburg, VA 24061
Position: Hd, Dept of Mining & Minerals Engrg. *Employer:* Vir Poly Inst & State Univ. *Education:* BS/Math & Physics/Waynesburg College; BS/Mining Engg/W VA Univ; MS/Mining Eng/Univ of Pittsburgh; PhD/Mining Eng/Columbia Univ. *Born:* May 3, 1929. Employed Va Poly Inst & S U 1961- ; Ohio State 1954-61; Joy Mfg 1952-54; Crucible Steel 1948-54. Currently Hd, Mining & Minerals Engrg, VPI & SU, in teaching, res, & admin. Performed significant res in mining engrg spon by the OCR & Bur of Mines, Dept in Int; ERDA; BCR & other agencies of indus & gov. Serves as cons & performs res in min econ, mining tech, computer applications & health & safety. Active in orgs incl AIME, ASEE, AAUP, AAAS, NSPE, & others. Adv activities for fed & state govs, & educ insts. *Society Aff:* AIME, AAAS, ASEE, AAUP, NSPE.

Lucas, Jay
Business: 2225 W 7th St, Little Rock, AR 72201
Position: Owner. *Employer:* Lucas & Assoc. *Born:* 12/12/27. Attended Oklahoma State Univ. Formed Lucas & Assocs, Cons Engrs 1966, a firm spec in engrg for commercial bldgs for arch & indus clients. Reg in Ark, Okla, Texas, La, Miss, Ala, Tenn. Former Pres, V P, Secy, Illuminating Engrg Soc; former V P, Secy, Amer Cons Engrs Council. Recipient 1971 state award Illum Engrg Soc Lighting Competition Edwin F Guth Mem Award. 1971 2nd place recipient, regional award. Have served on city electrical bd since 1972. Mbr U S Chamber of Commerce, Rotary, NFPA. Helped write city elec code & state elec code, yet to be passed by state legis body.

Lucas, John W
Home: 865 Canterbury Rd, San Marion, CA 91108
Position: Asst Solar Thermal Power Sys Proj Mgr. *Employer:* Caltech Jet Propulsion Lab, Ret. *Education:* PhD/ME/UCLA; MS/ME/UCLA; BS/ME/UC Berkeley. *Born:* Mar 1923. Native of Pomona Calif. US Naval Reserve 1943-46. NSF. Post- doc Fellow at Max Planck Soc's Fritz Haber Inst, W Berlin Germany 1953-54. Var pos, incl Exec Asst to the Dir, Caltech Jet Propulsion Lab, Pasadena Calif 1954-85 Now retired. Assoc Fellow Amer Inst of Aeronautics & Astronautics (Dir 1973-76, V P Tech 1975-76); Mbr Sigma Xi. Editor, Heat Transfer & Spacecraft Thermal Control, vol 24, AIAA Progress in Aeronautics & Astronautics series 1971; Thermal Characteristics of the Moon, vol 28, 1972; Fundamentals of Spacecraft Thermal Design, vol 29, 1972. Enjoy water & snow skiing & tennis. *Society Aff:* AIAA, ΣΞ, IAE.

Lucas, Michael S P
Home: 2714 Circle Rd, Manhattan, KS 66502
Position: Prof *Employer:* KS State Univ *Education:* PhD/EE/Duke Univ; MS/EE/Duke Univ *Born:* 6/3/29 in Lincolnshire, England. Served with Royal Signals 1948-50. Worked as Plant Superintendent in Maracaibo, Venezuela with Cia Anonima Nacional Telefonos De Venezuela 1954-57. Methods Engr with General Telephone of SE 1957-59. Res Assoc, Duke Univ 1959-64. Prin Scientific Officer, Signals Res & Dev Establishment, Christchurch, England 1964-67. Lecturer in Elec Engrg at Univ of Edinburgh, Scotland 1967-68. In Elec Engrg Dept, KS State Univ 1968-present; full Prof since 1970. On leave of absence in 1974-5, as Sr Res Fellow with the dept of Scientific & Ind Res, New Zealand. In 1981 was Visiting Prof at the Natl Inst of Astronomy, Optics, and Electronics, Tonantzintla, Puebla, Mexico. Mbr of the Advisory Cttee & Editorial Review Ctte for the IEEE Instrumentation and Measurement Soc. January-June 1987, Invited Professor at the Ecole Polytechnique Federale de Lausanne, Switzerland. *Society Aff:* IEEE

Lucas, William R
Home: 6805 Criner Rd S E, Huntsville, AL 35802
Position: Aerospace Consultant. *Employer:* Retired. *Education:* PhD/Chem Metallurgy/Vanderbilt Univ; MS/Chem Metallurgy/Vanderbilt; BS/Chemistry/Memphis State Univ; LHD/Honorary/Mobile College; DSc/Honorary/Southeastern Inst Tech; DSc/Honorary/Univ of AL/HvL *Born:* Mar 1922. Native of Newbern Tenn. Served with USN 1943-46. Instr in chem Memphis State Univ 1946-48. Redstone Arsenal as matls engr 1952-55, Army Ballistic Missile Agency in Matls Engrg 1955-60. NASA-MSFC 1960- , Matls Div 1963-66; Dir Propulsion & Vehicle Engrg Lab 1966-68; Dir Prog Dev 1968-71; Dep Dir 1971-74; Dir 1974-86. NASA Except Sci Achievement Medal 1964; Two NASA Except Serv Medals 1969; NASA Disting Serv Medal 1972. AIAA Alabama Sect Oberth Award 1965; AIAA Alabama Sect Toftoy Award 1976. Fellow ASM. Fellow AIAA; Fellow, Amer Astronautical Soc, Mbr, Amer Chem Soc, Sigma Xi, Natl Acad of Engrs. Tau Beta Pi Presidential rank of Distinguished Executive, Sep., 1980; Roger W. Jones Award for Outstanding Executive Leadership, The American University, May 1981, NASA Distinguished Service Medal 1981. The Space Flight Award of American Astronautical Society, 1982. Space Award for Outstanding Contributions in field of Space, VFW, 1983. Elmer A. Sperry Award of AIAA, 1986. Third NASA Distinguished Service Medal, 1986. *Society Aff:* Fellow ASM, Fellow AIAA, Fellow AAS, Mbr ACS, Mbr NAE

Luce, Walter A
Home: 663 Banbury Rd, Dayton, OH 45459
Position: Consulting Engineer *Employer:* Previous Employer The Duriron Co Inc. *Education:* MS/Metallurgical Eng/OH State Univ; BChE/Chem Eng/OH State Univ. *Born:* Dec 7, 1920 Mt Morris N Y. Welding engr Curtiss-Wright Corp 1943. US Navy 1944-45. With The Duriron Co since 1947 as matls engr, Ch Metallurgist, Tech Dir & named V P Tech 1968. Active in ASM, NACE & ASTM. Pres Alloy Casting Inst 1967-68. Author num tech articles incl Indus & Engrg Chem's Annual Review for Stainless Steels 1951-67. Co-holder of sev alloy pats. Received Disting Alumnus Award from Engrg Coll, Ohio State 1970. Named ASM Fellow 1972. Formerly VP Research and Technology, The Durion Co. Inc, Dayton, Ohio. *Society Aff:* ASM, NACE.

Lucey, John W
Business: Aero & Mech Engrg, Notre Dame, IN 46556
Position: Assoc Prof *Employer:* U of Notre Dame *Education:* PhD/Nuclear Engrg/MIT; SM/Nuclear Engrg/MIT; BS/Chem Engrg/Notre Dame *Born:* 08/21/35 Taught at Notre Dame since 1965. Respon for Energy Sequence in Mech Engrg. Res interests in numerical methods and interaction of tech and public policy. North Central Regional Dir, Sigma Xi 1982-89. Chrmn, Education Div ANS 1983-84, Math and Computation Div ANS 1976-77; Nuclear Engrg Div ASEE 1981-82; Illinois - Indiana Sect ASEE 1979-81. Chmn, Publications Cttee ASEE, 1986-7. *Society Aff:* ASEE, ANS, AAAS, HPS, ΣΞ

Lucht, David A
Business: WPI, Worcester, MA 01609
Position: Director *Employer:* WPI-Center for Firesafety Studies *Education:* BS/FPE/Illinois Institute of Technology *Born:* 02/18/93 Has served in engineering and policy leadership positions in several capacities: Officer, director and cttee chair, SFPE; State Fire Marshal, state of Ohio, appointed by Ohio Governor John J. Gilligan; appointed by President Gerald R. Ford as first Deputy Administrator, U.S. Fire Administration, following confirmation by U.S. Senate. Served as VP of major FPE consulting firm; headed development of first graduate degree program in fire protection engineering in U.S., at Worcester Polytechnic Institute. Reg PE, OH & MA. *Society Aff:* SFPE

Luck, Leon D
Home: S.E. 900 High St, Pullman, WA 99163
Position: Prof Emeritus *Education:* Engr/Civil Engg/Stanford Univ; MS/Civil Engg/Univ of MN; BS/Civil Engg/WA State Univ. *Born:* 4/25/21. Native of Spokane, WA. Asst Mine Engr for Pend Oreille Mines and Metals Co. Before and after WWII. Served in US Navy, active duty 1943-46; Retired reserve since 1959. At WA State Univ, 1947-1983; Chrmn of Civil Engg Dept, 1972-76; Exec Secretary, Univ Senate, 1971-72, 1976-79. Chrmn, Civil Engg Div, ASEE, 1972-73. Pres, WA Soc of Prof Engrs, 1977-78. Natl Dir, ASCE, 1979-82. Natl Dir, NSPE, 1985-present. *Society Aff:* ASCE, NSPE, ASEE.

Lucky, Robert W
Business: Crawford Corner Rd, Holmdel, NJ 07733
Position: Director *Employer:* Bell Labs *Education:* PHD/EE/Purdue Univ; MS/EE/Purdue Univ; BS/EE/Purdue Univ *Born:* 1/9/36 Joined Bell Labs in 1961. 1965 made head of advanced data communications dept. 1979 made Dir, Electronic and Computer Res Lab. Author of "Principles of Data Communications: McGraw-Hill, 1968, "Computer Communication" IEEE Press, 1976. 12 patents. Fellow, IEEE. Member Natl Academy of Engrg. US Air Force Scientific Advisory Board. Pres IEEE Communications Society 1978-79. VP IEEE 1978-79 Exec VP IEEE 1981 Editor, Proceedings of the IEEE 1974-76. Consulting Editor Plenum Press. *Society Aff:* IEEE

Ludgate, Robert B
Business: 111 N Sixth St, Reading, PA 19601
Position: Prin. *Employer:* Robt Ludgate & Assoc. *Education:* -/Civ Engg/Clarkson College. *Born:* 8/14/39. Due to family circumstances I left college after 2 yrs & never returned. However I worked at engg jobs (surveyor, draftsman, inspector, technician) and studied because I wanted an engg career. Eventually I found opportunities to do hgwy & bridge design & I am very proud of several landmark projs (on I87 & US Rte 17 in NY). After 14 yrs I felt qualified & the NY Bd of Examiners found me qualified to be a PE. 3 yrs later I founded my consulting firm which has established a solid (but very local) reputation for stormwater, bridge, and dev work. *Society Aff:* NSPE, ASCE, ULI.

Ludwig, Ernest E
Home: 12495 E Millburn, Baton Rouge, LA 70815
Position: Prof., Chem Engg. *Employer:* Louisiana State U. *Education:* MS/ChE/Univ of TX; BS/ChE/Univ of TX. *Born:* 3/6/20. Native Austin, TX. Asst Engrg Mgr & Process Mgr. Dow Chem, TX 1942-60; Proj/Process Mgr Rexall Drug & Chem (Dart Ind) & Gen Works Mgr Odessa, TX Polyolefin Plant (1960-67); VP Coolymer Rubber & Chem (1967-69); Consultant, pres Ludwig Consulting Engrs, Inc for chems, petrochems & polymers plant design, engrg & mgt (1969-85). Sole proprietor consultant (1985-present). Author 3 volumes *Applied Process Design for Chemical and Petrochemical Plants*, & *Applied Project Management for the Process Industries*; Fellow AIChE, Certified Cost Engr AACE; Dist Engrg Grad 1984 from Univ of Texas: Adj Prof Chem Engrg Louisiana State Univ at Baton Rouge; Specialist industrial accident investigation and industrial safety and industrial plant design. *Society Aff:* AIChE, ASME, ACS, ACE, FEC, AACE, AASE

Ludwig, George A
Home: Nott Rd-Bx 427, Rexford, NY 12148
Position: Manager - Product Engrg. *Employer:* General Electric Co. *Education:* BS/ME/Syracuse Univ. *Born:* May 21, 1929. ME Syracuse Univ 1951. Since 1951, employed by G E Co. Has had assignments with the Aircraft Engine Div & the Corp Res Ctr. Since 1953, has been with the Gas Turbine Prod Div, holding pos as Mgr - Rotating Components & Advanced Mech Design, Mgr - Mech & Matls Dev Engrg & Mgr - Prod Engrg. Current respon incl mgmt of all the Div Product Line Products. P E NY. *Society Aff:* NYSPE.

Ludwig, John H
Home: 43 Alston Pl, Santa Barbara, CA 93108
Position: Consultant. *Employer:* Self-employed. *Education:* ScD/Industrial Hygiene/Harvard Univ; MS/Industrial Hygiene/Harvard Univ; MS/Civil Engg/Univ of CO; BS/Civil Engg/Univ of CA, Berkeley. *Born:* Mar 7, 1913 Burlington Vt. Struct & hydraulic engr 1934-49; meteorologist Army Air Corps 1943-46; Water Pollution Control & Air Pollution Control 1949- . U S Bureau of Reclamation & Corps of Engrs 1936-49; U S Public Health Service & Environ Protection Agency 1951-72. EPA Except Serv Gold Medal 1971; Superior Service Medal DHEW 1967; Gordon Fair Commendation of the Amer Acad of Environ Engrs 1973; Disting Engrg Alumnus Colorado 1976. Elected to Cosmos Club, Phi Beta Kappa & Natl Acad of Engrs. Married to Gilda M Silva; 2 sons, Howard & Robert. *Society Aff:* AAAS, NAE, AAEE.

Luenberger, David G
Business: Terman Cntr 306, Dept of Engrg-Economic Systems, Stanford, CA 94305-4025
Position: Prof & Chairman. *Employer:* Stanford Univ. *Education:* BS/EE/CA Inst of Tech; MS/EE/Stanford Univ; PhD/EE/Stanford Univ *Born:* 9/16/37. Mbr Stanford faculty 1963-. Chrmn of Engg-Economic Systems 1980-. Prof of Engg-Economic Systems 1971-. Technical asst to the dir, Office of Sci and Tech, Exec Office of the Pres, 1971-1972. Visiting Prof MA Inst of Tech 1976. Guest Prof. Tech. U. of Denmark, 1986. Director, Optimization Tech, Inc, Fellow IEEE. Author of *Optimization by Vector Space Methods* (Wiley, 1969); *Introduction to Dynamic Systems* (Wiley, 1979); and *Linear and Nonlinear Programming, Second Edition* (Addision-Wesley, 1984). *Society Aff:* AAUP, AFA, IEEE, TIMS.

Luenzmann, David I
Home: 1391 Lakeshore Dr, Cleveland, WI 53015
Position: VP *Employer:* Donohue Engrs & Architects *Education:* BS/CIE/Univ of WI *Born:* 3/20/39 Wisconsin native. Operated independent land surveying co prior to US Army service as a Topographical Surveyor. Work as an iron worker spurred interest in structural engrg. Div Engr for WI Public Service Corp; staff structural engr with several consulting firms in Minneapolis/St. Paul area; founder of consulting engrg firm in 1971. Joined present firm in 1976 and advanced to current VP posi-

Luenzmann, David I (Continued)
tion in 1981. Responsible for mgmt of Structural, Operation and Maintenance, Electrical, Mechanical, Process and Instrumentation and Control Engrg Depts and Architectural Dept. Married, 3 children. *Society Aff:* ASCE, NSPE

Luerssen, Frank W
Business: 30 West Monroe St, Chicago, IL 60603
Position: Chairman & CEO. *Employer:* Inland Steel Industries, Inc. *Education:* BS/Physics/PA State Univ; MS/Met Eng/Lehigh Univ *Born:* 8/14/27. in Reading, PA. With Inland Steel Co since 1952. Now Chrm and CEO, Inland Steel Industries, Inc.; was VP Steel Mfg 1977-78; VP Res 1968-77, Mbr AIME, ASM, AISE, NAE, SAE, WSE, AISI. Bd of Governors, Assoc. Colleges of Ind; Dir. Chicago Assn. Commerce and Industry, Trustee Northern Univ. Fellow ASM, 1975. F L Toy Award 1955 Trustee, Chairman, Dir, United Way of Chicago; Dir. Continental Illinois Corporation, Bd. of Gov, of Argonne Nat'l Labs., Serves on Listed Companies Advisory Comm. of N.Y. Stock Exchange; Dir, NWU/Evanston Research Park, Inc. *Society Aff:* NAE, AIME, ASM, SAE, AISE, AISI

Luetje, Robert E
Home: 21254 Woodfarm Dr, Northville, MI 48167
Position: General Manager - Technology. *Employer:* Kolene Corp. *Education:* BS/Met Eng/Univ of IL. *Born:* 4/20/36. BS Met Engg Univ of IL. Reg Patent Agent USPO. With Armco Steel for 25 yrs; 12 yrs in R&D & patent work. Responsible for mkting & tech liaison in Automotive Market. Active in ASM at chapter level, serving as Chmn Dayton Chap 1970; on natl level, as Chmn Chap Adv Ctte & Chap & Mbrship Council & Chmn ASM Long Ranch Planning Ctte. Chrmn ASM Council for Prof. Int. General Chrmn of Musical Mission Gp 6 yrs. Other ints are outdoor sports, golf & church activities *Society Aff:* ASM International, SAE, NACE, AISI, Eng. Soc. of Det., ASTM.

Luettich, Richard A
Business: 470 Atlantic Avenue, Boston, MA 02210
Position: Vice President *Employer:* Edwards and Kelcey, Inc *Born:* May 28, 1930. BSCE Univ of Illinois. Postgrad Univ of Miss 1967. P E Maine. Prof Land Surveyor Maine. Traffic Engr Champaign Ill 1955-57; Traffic Engr Maine St Hwy Comm 1957-63; Planning & Traffic Engr MSHC 1964-67; Deputy Ch Engr MSHC 1967-72; Deputy Commissioner MDOT 1972- . Mbr AASHTO, *NASAO, ITE, APWA, Amer Welding Soc, TRB, ARBA; office holder or ctte chmn in sev adv cttes or comms to Fed Gov; P Pres New England Sect ITE; recipient Outstanding Traffic Engr of 1974 Award; mbr St Bd of Regis for Prof Engrs; as Deputy Commissioner of Maine DOT was respon for operational phases of Transp Dept, incl hwys, ports, airports & ferry service; Acting Commissioner of Maine DOT 1979-1980; Vice President and Principal-in-charge, of Edwards and Kelcey, Inc., 50 man regional office of large multi-disciplined consultant and engineering firm, 1980 to present.*

Luh, Peter B
Business: Dept of Elect and Systems Engrg, Storrs, CT 06268
Position: Associate Professor *Employer:* Univ of CT *Education:* PHD/Applied Math/Harvard Univ; MS/Aeronautics and Astronautics/MIT; BS/EE/Natl Taiwan Univ *Born:* 12/21/50 Peter B. Luh received his B.S. degree in Electrical Engineering Natnl Taiwan Univ in 1973, M.S. in Aeron & Astron Engrg, M.I.T., 1977; and Ph.D. in Appl Math, Harvard Univ in 1980. From 1980 he has been with the Univ of Conn, and currently is an Assoc Prof in the Dept of Elect & Systems Engrg. His major research interests include Hierarchical Planning and Control of Large Scale Systems, Distributed Decisionmaking, Manufacturing Systems, and Multitarget Tracking. He has been a principal investigator and consultant to many industry and government funded projects in these areas. He served on Program Cttee and Operating Cttee of several major national intersociety conferences. He is a member of IEEE, Sigma Xi, and listed in Who's Who in the East. *Society Aff:* IEEE, ΣΞ

Luhnow, Raymond B, Jr
Home: 6510 Rainbow, Shawnee Mission, KS 66208
Position: Pres. *Employer:* Retired *Education:* MBA/Bus Adm/Univ of MO '64; BS/EE/Univ of IL '44. *Born:* 5/3/23 Native of Kansas City, MO. Served in USNR in World War II. Joined Burns & McDonnell in 1946. Served in progressive capacities ranging from design engr through principal-in-charge of projects. Initial assignments were related to design of electrical and mechanical systems for electric power stations, water and wastewater facilities, military installations, airports, and aircraft maintenance facilities. Partner in firm in 1964, established Burns & McDonnell NY office in 1966. Returned to Kansas City as VP 1971. Elected Pres Jan. 1, 1974. Registered PE and Planner. Retired May 31, 1982. *Society Aff:* NSPE, SFPE.

Lukacs, Joseph
Home: 2351 Uxbridge Dr N W, Calgary, Alberta T2N 3Z8 Canada
Position: President. *Employer:* Western Research & Dev Ltd. *Education:* MSc/Petroleum Engg/Univ of Alberta; BSc/Petroleum Engg/Univ of Alberta. *Born:* Mar 1934 Hungary. Left Hungary as a Univ student in 1956. In 1965, after several yrs of experience in the Petro indus, both in the production & gas processing fields, joined Western Research & Dev Ltd, a cons engrg firm spec in environ engrg, instrumentation & process control. Presently Pres of the firm spec ints are environ control through process tech, instrumentation, & automation. Dir 1972-75 APCA *Society Aff:* APEGGA, SPE, APCA.

Lukat, Robert N
Business: 1594 Evans Dr, SW, Atlanta, GA 30310
Position: Vice President/General Manager. *Employer:* Atlanta Saw Company *Education:* MS/Metallurgy/GA Inst of Tech; BChE/Chem engg/GA Inst of Technology. *Born:* Feb 1946. Employed since grad in 1971 by Southern Saw Service Inc as Sr Engr in charge of Qual Control, Product Res & Dev. Respon for dev of titanium carbide surface benefaction of cutting edges for meat processing. Contrib to ASM Metals Handbook vol on Failure Analysis. Serves on Qual Control Ctte of Heat Treating Div of ASM, Treas of Atlanta Chap ASM; Mbr AWS, USSA, AMI. Enjoy gardening, woodworking & hunting. In Nov 1978 became GM of S.S.S. subsidiary, Atlanta Saw Co., in charge of international marketing of cutting edges for meat processing.. *Society Aff:* AWS, ASM, SME.

Lukenda, James R
Home: 76 N Clark Ave, Somerville, NJ 08876
Position: Vice President. *Employer:* Pauls Trucking Corp. *Education:* MBA/Econ/Rutgers Univ Sch of Bus; BS/Bus/Rutgers Univ. *Born:* May 1929. Adjunct Asst Prof Mkting, Pace Coll 1969-72. With Pauls Trucking Corp since 1971 as Sr VP. Sr Assoc for Walter F Friedman & Co, spec in Matl Handling, Transp & Phys Distrib. Mgr of Engrg Services for Associated Transport. Internatl Matl Mgmt Soc. Pres IMMS - N J Chap Inc 1964-65; Region II V P IMMS 1974-76. Editor - New Jersey Chapter News; Professionally Certified Material Management; Professionally Certified Material Handling.. Tech Paper Contest Winner IMMS - N J Chap Inc 1975-76. *Society Aff:* IMMS, NCPDM, APMHC.

Lukens, John E
Business: c/o Asian Inst of Technology, G.P.O. Box 2754, Bangkok 10501 Thailand
Position: Coordinator, Natural Resources Program *Employer:* Asian Inst of Technology *Education:* PhD/Civ Engg/Cornell Univ; BS/Civ Engg/Worcester Polytech Inst. *Born:* Feb 3, 1939. PhD Cornell Univ in Civ Engrg 1969, BSCE Worcester Polytech Inst 1962. Res engr with HRB-Singer, & later with Raytheon, working with applications of remote sensing 1966-71. With the Div of Arch Studies at RI Sch of Design 1971-80, teaching civ engrg & phys planning subjects. VP Geodata Systems Intl, Nairobi, Kenya, 1981-84; Coordinator INRDM Program & Mgr Remote Sensing Lab, Asian Inst of Tech, Bangkok, Thailand, 1984- . Fulbright Fellowship in New Zealand 1979. *Society Aff:* ASP & RS, ASCE.

Lukrofka, L J
Business: 11610 Truman Rd-Bx 380, Independence, MO 64051
Position: Executive Vice Pres. *Employer:* Missouri Water Co. *Education:* BS/Civil Engr/MO School of Mines-Rolla. *Born:* June 1, 1922 near Arlington Mo. Public schools Rolla Mo. US Army 1942, discharged Capt 1946. Sanitary engr City of Springfield Mo 1950-53. Supt of Oper, Mo Water Co Independence Div, June 1953. Mgr Independence Div Apr 1972. V P & Genl Mgr Independence Div Apr 1974. Exec VP Independence Div. Oct 82 Reg P E Mo. MSPE Western Chap; NSPE; ASCE; AWWA; Mo Water & Sewerage Conf; Natl Assn of Water Co's. Kiwanis, Elder Fairmount Christian Church; Gideons Internatl. Christian Bus Mens Committee, Chi Epsilon, Jackson Cty Chamber of Commerce; Director of Commerce Bank of Independence; Trustee - Independence SAN & hospital., Independence Chamber of Commerce. *Society Aff:* ASCE, NSPE.

Luks, Kraemer D
Business: 600 S College, Tulsa, OK 74104
Position: Prof. of Chemical Engrg *Employer:* Univ of Tulsa. *Education:* PhD/ChE/Univ of MN; BSE/ChE/Princeton Univ. *Born:* 9/6/41. Native of Frenchtown, NJ. Taught 12 yrs at Univ of Notre Dame (1967-1979) and 8 yrs at Univ of Tulsa (1979-present). Tenured 1970. Promoted to rank of full prof in 1975. Distinguished Visiting Prof at Univ of Tulsa in 1977. Named Univ Professor of Chemical Engrg in 1986. Author of more than 90 publications in the area of theoretical and experimental thermodynamics. Consultant to petrol industry in areas of enhanced oil recovery and natural gas processing. *Society Aff:* AIChE, APS, SPE, ACS.

Lula, Remus A
Home: 923 Oakwood Pl, Natrona Heights, PA 15065
Position: Consultant. *Employer:* Self Employed. *Education:* MS/Met/Carnegie Inst of Tech; MS/Mining & Met/Univ of Bucharest. *Born:* 8/31/15. Native of Rumania, naturalized US citizen. Started engg career in Rumania followed by 1 yr with Shell Oil in Venezuela. Joined Allegheny Ludlum Steel Co in 1950 as a Res Met. Progressed to supervisory & mgt positions: asst dir res, mgr product res to asst to VP res-technical dir. Currently a consultant: R.A. Lula Consulting, Inc. Field of activity stainless steels, high strength steels, high temperature alloys, energy related tech. Patents, publications, co awards & ASTM Sam Tour Award. *Society Aff:* ASM, ASME, MPC.

Lum, Franklin, YS
Business: 1833 Kalakaua Ave, Suite 208, Honolulu, HI 96815
Position: President *Employer:* Ferris & Hamig Hawaii Inc *Education:* BS/ME/Univ of HI *Born:* 11/2/40 Native of Honolulu, HI. Served with US Army as a field artillery officer 1964-68. Consulting engr with Ferris & Hamig, Inc, since 1968, specializing in HVAC, plumbing and fire protection systems. Became Pres of firm in 1986. VP of HI Chpt of ASHRAE 1981-1982 & Pres in 1983-1984. VP of HI Chpt of ASPE; 1984-86 & Pres 1986-87. Projects of noteworthy are the NASA Telescope Facility at Mauna Kea, renovation of the Kahala Hilton mechanical plant, the new 1056 room Tapa Tower project at the Hilton Hawaiian Hotel and the new HI Med Service Assoc Center which won an honors award from ACEC. *Society Aff:* ASHRAE, ASPE, ACEC

Lum, Walter B
Home: 1817 Halekoa Dr, Honolulu, HI 96821
Position: Consultant *Employer:* Walter Lum Assocs Inc. *Education:* BSCE/Civil Engg/Purdue Univ; Cert/Soil Mechs/Harvard Univ. *Born:* 3/24/22. BSCE Purdue Univ 1945. Cert from Harvard Grad Sch in Soil Mech & Found Engg 1965. Structural Engg with City & Cty of Honolulu 1950-57. Cons Soil & Found Engr 1957- . P Mbr Hawaii Legislature's Sci Adv Ctte; P Mbr Hawaii Bd of Regis for Engrs, Archs & Land Surveyors; P Pres Cons Engrs Council of Hawaii; P Pres Hawaii Section of ASCE; Hawaii Engr of Yr 1975; P Pres Assn of Soil & Found Engrs. Enjoys tennis as recreation. *Society Aff:* ASCE, ACEC, NSPE, ASFE.

Lumsdaine, Edward
Business: 4901 Evergreen Rd, Dearborn, MI 48128
Position: Dean of Engrg *Employer:* Univ of MI - Dearborn *Education:* ScD/Mech Engrg/NM State Univ; MSME/Mech Engrg/NM State Univ; BSME/Mech Engrg/NM State Univ *Born:* 09/30/37 Res Engr for Boeing 1966-67. Taught at SD State Univ 1967-72, Univ of TN 1972-77. Sr Res Engr/Prof, Physical Sci Lab/NM State Univ 1977-78; Dir, NM Solar Energy Inst 1978-1981. Dir/Prof, Energy, Environ, & Resources Center/Univ of TN 1981-82. Dean/Prof at UM-D since 1982. Visiting Prof in Taipei (China), Cairo (Egypt), Doha (Qatar). Res projs in turbomachinery, aeroacoustics, solar energy, fluid mechs, heat transfer. Authored manual on industrial energy conservation for developing countries and manual on creative problem solving. Developing lifelong engrg education program and computer courseware; indus/univ cooperation. Conslt to many companies. *Society Aff:* AIAA, ASME, ASTM, ASEE, ISES

Lunardini, Robert C
Business: 114 Office Plaza, Jackson, MS 39206
Position: Owner. *Employer:* Southern Consultants, Inc. *Education:* MS/Civ Engr/Univ of MI; BS/Civ Engr/MI Tech Univ. *Born:* 2/18/33. in Holyoke, MA. Grad MI Tech 1954, served with US Army Corps of Engrs, 1954-56. First Lt in 2nd Logistical Command. Grad Univ of MI, 1957. Chief Field Engr for Testing Service Corp, Lombard, IL. In charge of all testing and sampling, 1957-1959. Soils Engr, Michael Baker, Jr, Inc, Beaver, PA, 1959-1960. Planning Engr, Michael Baker, Jr, Inc, Jackson, MS, 1960-1962. Chief Planning Engr, 1962-1964. Formed Southern Consultants, Inc in July 1964. Owner and principal, Municipal Planning & Engg, Water Works & Pollution Control, Streets & Highways, Drainage. *Society Aff:* ASCE, NSPE, SAME, WPCF, AIP, MES.

Lund, Daryl B
Business: 1605 Linden Dr, Madison, WI 53706
Position: Prof. *Employer:* Univ of WI-Madison *Education:* PhD/Food Sci-ChE/Univ of WI; MS/Food Sci/Univ of WI; BS/Math/Univ of WI. *Born:* 11/4/41. Joined Dept Food Sci, Univ WI, as Food Engr in 1968. Responsible for teaching & res in food engrg. Res interests include heat transfer in foods (eg fouling of heat exchangers & thermal process design), energy in food processing (energy analyses & solar energy) and effects of heat on food components. Author of over 100 technical papers and contributor to five books. Active in Food Engrg Divs of AIChE, ASAE, IFT. Enjoy camping, backpacking, travel, squash & golf. *Society Aff:* AIChE, ASAE, IFT.

Lund, Robert E
Business: Box A, Monaca, PA 15061
Position: Dir of Extractive Res & Dev. *Employer:* St Joe Minerals Corp. *Education:* BSChE/ChE/Univ of CO. *Born:* 6/6/20. Served as a radar maintenance officer with the US Navy 1944-46. Indus exper has all been with St Joe Minerals corp (formerly St Joseph Lead Co). As res engr & Sect Hd, dev improved smelting processes. Assumed current pos as Dir of Res 1974, with respon for dir the res functions for extractive met. Dir (VP 1974-75 & 1978) Amer Inst of Mining, Met, & Petro Engrs; Pres of the Met Soc - AIME - 1977. Fellow of the Met Soc - AIME and Amer Soc for metals. *Society Aff:* AIChE, AIME, ASM, CIM.

Lundahl, Lloyd E, Jr
Business: P O Box 3166, Tulsa, OK 74101
Position: V Pres - Operations. *Employer:* Agrico Internatl. *Born:* Sept 9, 1923. Graduated from the Univ of Minnesota with BSME in 1947. Have held var engrg, mgmt & exec pos in the petrochem indus & its related design & const indus with Shell, Bechtel, W R Grace, Great Northern Oil, Northern Petrochem Co, & Chem Const Corp. Presently V P Internatl Oper for Agrico Chem Co.

Lundberg, Lennart A
Business: 7835 SE 30th, PO Box 186, Mercer Island, WA 98040
Position: Pres. *Employer:* A H Lundberg Inc *Education:* MS/ChE/Univ of WA; BS/

Lundberg, Lennart A (Continued)
ChE/Univ of WA; BS/General/Univ of WA. *Born:* 4/18/24. Started career in Chem Engg and the Pulp and Paper Industry in 1948 upon graduation from Univ of WA. Served as field engg for start-up of first two Weyerhauser sulphate pulp mills. Joined A H Lundberg in 1951. Has worked on design and dev of chem and mech pulping systems with emphasis on chem recovery and pollution control. Became Pres, A H Lundberg Inc in 1968. Resp for company activities worldwide. Mbr AIChE, TAPPI, and Dir & Past Pres of WA Pulp and Paper Foundation. Interested in promoting growth and innovation in the Pulp and Paper Industry. A. H. Lundberg Inc incorporated 1954. Lennart became Pres of A. H. Lundberg Inc in 1968; 1977-1983 served as Pres of A.H. Lundberg Assoc, Inc. *Society Aff:* AIChE, TAPPI, WAP&P

Lundgren, Raymond
Home: 4036 Happy Valley Rd, Lafayette, CA 94549
Position: Pres *Employer:* Lundgren Engrg Conslt *Education:* MS/CE/UC Berkeley; BS/CE/UC Berkeley. *Born:* 10/1/24. BSCE 1950; MSCE 1953 Univ of Calif Berkeley. Res engr & lectr at U C Berkeley 1948-52. Principal with Woodward-Clyde Cons and Chairman of the Board 1953-83. Has a broad exper in soil mechanics, foundation cons, proj mgmt, & quality assurance on studies for power plants, dams, reservoirs, bldgs, etc. Currently consulting engineer in geotechnical practice. Amer Soc for Testing & Matls, ASFE, ASCE, ACEC. Hobbies: fishing, tennis, golf, skiing. *Society Aff:* ASCE, ACEC, ASTM, ASFE.

Lundin, Bruce T
Home: 5859 Columbia Rd, North Olmsted, OH 44070
Position: Director (Retired). *Employer:* NASA - Lewis Research Ctr. *Education:* BSME/Mech Engg/Univ of CA. *Born:* 12/28/19. With NASA & predecessor agency NACA since 1943. Dir of NASA's Lewis Res Ctr 1969-77. Fellow AIAA, AAS & Royal Aeronautical Soc. Mbr Natl Acad of Engg, Amer Soc for Public Admin, Tau Beta Pi & Sigma Xi. Cleveland Tech Soc Council Tech Award 1953; NASA Medal for Outstanding Leadership 1965; NASA Pub Serv Award 1971 & 1975; NASA Disting Serve Medal 1971 & 1977; Hon Dr of Engg Univ of Toledo 1975; Natl Space Club Astronautics Engr Award 1976.

Lundy, Harry L, Jr
Home: 5062 Durham Rd W, Columbia, MD 21044
Position: Pres *Employer:* Columbia Builders, Inc. *Education:* MBA/Business/Univ of Pittsburgh; MS/CE/Northwestern Univ; BS/CE/Lafayette Coll *Born:* 3/15/41 in Camp Hill, PA. Worked for a major intl consulting firm, specializing in Civil Engrg. With a large builder for three yrs. Started a consulting firm in 1975 and then became Owner and Pres of a builder co in Columbia, MD. Lectured at building conferences and published several civil engrg papers. PE - In PA, WV and OH. *Society Aff:* NSPE, ASCE

Lunnen, Ray J, Jr
Home: 1125 Haymaker Rd, State College, PA 16801
Position: Instructor. *Employer:* PA State Univ. *Education:* MSEE/EE/AF Inst of Tech; BS/Engg/US Naval Acad; PhD candidate/EE/PA St Univ. *Born:* 1/16/31. Native of Connellsville PA. Grad from US Naval Academy 1955. Served with the USAF to rank of Col 1955-1979. As an Electronics Engr. During service was Dir of Communications and Control Div, Rome Air Dev Ctr, and Asst Deputy for AEngg, Defense Communications Agency. Presently on a full time appointment on the faculty of the Elec Engg Dept of the PA State Univ. Completing PhD requirements. Sr mbr IEEE, Reg PE. *Society Aff:* IEEE, ΣΞ, AFCEA.

Lunsford, Jesse B
Star Route Box 435, Lottsburg, VA 22511
Position: Retired. *Education:* -/Elec Eng/Univ of AR *Born:* Aug 14, 1890 Kentucky. Equiv coll, not graduate. 40 yrs exper whole field elec, mech, electronics engrg. 7 yrs indus, incl 2 yrs Night Test Floor Foreman & 3 yrs Hd of Exper Lab, large elec manufacturer. 33 yrs Navy, first as reserve officer, later civilian. Tech supervision var naval labs, standardization of naval machinery, matls; joint Army-Navy standardization of electronic components. Last 5 yrs, Hd Engr for 'Program Planning for Long Range Res & Dev' under dir Chief BuShips, directives of CNO (Ch, Naval Operations). Special citations: Navy, Army & ASA for res, dev, standardization. Retired 1951. Life Fellow IEEE. *Society Aff:* IEEE, AIEE

Lupfer, Dale E
Business: P O Box 437, Sugar Land, TX 77487
Position: President. *Employer:* Lupfer Automation, Inc. *Education:* BSME/Mech Engg/Univ of KS. *Born:* Mar 1923. Served as Naval Aviator in WW II. Served in var capacities in Res Dept of Phillips Petro Co for 18 yrs in dev of advanced process control sys. Process automation cons from 196871. Sr V P of Control Automation Tech Co 1971- 76. Pres of Lupfer Automation, Inc Jan 1, 1978- . Reg P E TX, OK & CA. 43 pats in field of process automation. Fellow Instrument Soc of Amer. Recipient of Amer Automatic Control Council Best Paper Award. Recipient of Instrument Soc of Amer Best Paper Award. *Society Aff:* ISA, AIChE.

Lurie, Robert M
Business: Nyacol Products Inc, Ashland, MA 01721
Position: President. *Employer:* Nyacol Products Inc *Education:* ScD/Chem Engg/MIT; BS/Chem Engg/MIT. *Born:* Feb 1931 Boston Mass. Group Leader W R Grace Cambridge Mass 1955-60; res engr & Group Leader Ionics Inc Cambridge Mass 1960-62; Dir Matls Dev Avco Sys Div, Lowell Mass 1962-70; Pres Nyacol Products Inc, Ashland Mass 1970- , manufacturing colloidal metal oxides. Res Mgmt Assn Boston Mass, Bd of Governors 1972-76, Chmn 1974; Mbr Fire Retardant Chem Assn, Amer Assn of Textile Chemists & Colorists, ACS, AIChE. *Society Aff:* FRCA, AIChE, ACS, AATCC, SPE, ΣΞ, RMA.

Luss, Dan
Business: Dept of Chem Engrg, Houston, TX 77004
Position: Prof & Dept Chmn. *Employer:* University of Houston. *Education:* BSc/ChE/Technion, Israel; MSc/ChE/Technion, Israel; PhD/ChE/Univ of MN *Born:* May 1938. PhD Univ of Minnesota; BS & MS Technion Israel. Native of Tel- Aviv Israel. Served with Israeli Army Corps Engrs 1960-62. Asst Prof Univ of Minn 1966-67. Joined Univ of Houston 1967. Assumed current pos of Chmn in 1975. Spec in chem reactor design & control. Mbr Edit Bd of Catalysis Reviews, Sci & Engrg. AIChE Colburn Award 1972; Profl Progress Award 1979; Wilhelm Award 1986; IEC Div of ACS, honor scroll 1968; Elected to Natl Acad of Engrg 1984. S Texas Sect AIChE Best Paper Awards 1969, 1970, 1975. *Society Aff:* AIChE, ACS, ASEE

Lustgarten, Merrill N
Business: ECAC N Severn, Annapolis, MD 21402
Position: Scientific Advisor. *Employer:* IITRI. *Education:* MSEE/EE/Columbia Univ; BSEE/EE/NC State; AB/Chem/Brooklyn College. *Born:* 2/15/22. in NYC. Assigned to Manhattan Proj, Los Alamos, NM, during WWII. Chief of Frequency Coordination Div (US Army Signal Corps) at White Sands Missile Range 1950-1956. Employed by Rand Corp 1956-1961; Co-authored reports on electromagnetic compatibility (EMC), leading to formation of the Electromagnetic Compatibility Analysis Ctr (ECAC). Since 1962, employed at ECAC as advisor & contributor to ctr's capability dev effort. Maj areas included: cosite analysis, EMC figure of merit, degradation of performance of radar operators due to interference & RF propagation modelling. *Society Aff:* IEEE.

Lutes, Loren D
Business: Dept of Civil Engineering, PO Box 1892, Houston, TX 77251
Position: Prof and Chairman *Employer:* Rice Univ *Education:* BS/CE/Univ of NE; MS/Engr Mech/Univ of NE; PhD/App Mech; CA Inst of Tech. *Born:* 12/01/39 Involved in teaching and research concerning the application of probability theory to structural engrg problems, particularly structural dyanmics and fatigue analysis. In addition to teaching at Rice, has held visiting appointments at the Univ of Chile and the Univ of Waterloo, Ontario. Was a co- recipient in 1965 of the Wason Research Medal of the American Concrete Inst, and in 1983 of the ASCE State-of-the-Art of Civ Engrg Award. *Society Aff:* ASCE, ΣΞ.

Lutken, Donald C
Business: P O Box 1640, Jackson, MS 39205
Position: Chmn/Pres. *Employer:* Mississippi Power & Light Co. *Education:* BS/Engg/U.S. Naval Academy. *Born:* 3/26/24. BS US Naval Acad. Native of Jackson MS. Employed with MP&L in 1949 as engr after duty in Navy. Progressed through pos of Plant Supt, Supt of Prod, Ch Engr, VP - Ch Engr. Pres 1970; Pres & Ch Exec Officer 1971. Chmn and Chief Exec Officer 1984; Chmn/Pres & Chief Exec Officer 1986; Bd of Dir of Middle South Services; Southeastern Elec Exchange; Southwest Power Pool; Unifirst Fed S&L. Mbr Miss Engrg Soc, Amer Soc of Mech Engrs (Fellow 1972), IEEE, Edison Elec Inst. Enjoys hunting & fishing.

Lutz, Harry E
Business: 1500 Meadow Lake Pkwy, Kansas City, MO 64114
Position: Staff Engineer *Employer:* Black & Veatch, Cons Engrs. *Education:* BCE/Civil Engg/Rensselaer Polytechnic Inst. *Born:* Sept 1924 Kansas City Mo. Grad of New York State Maritime Acad; USMS & USNR 1942-46. With Interpace Corp 1951-70 as design engr, plant mgr & Regional Engr in design & manufacture of concrete pipe both in U S & S Amer. Joined Black & Veatch 1970 serving as project mgr of solid waste & regional wastewater studies. Reg Engr in 10 states. P Pres ASCE & SAME Kansas City. Sailing enthusiast of both full-size & model sailboats, as well as an amateur cabinetmaker. *Society Aff:* AWWA, WPCF, ASCE, SAME, APWA, NSPE.

Lutz, Raymond P
Home: 10275 Hollow Way, Dallas, TX 75229
Position: Exec Dean, Grad Studies & Res. *Employer:* Univ of Texas at Dallas. *Education:* PhD/IE/IA State Univ; MBA/Mgmt/Univ of NM; BS/ME/Univ of NM. *Born:* 2/27/35. Engr Sandia Corp; taught Univ of NM, NM State Univ, Univ of OK. Editor *The Engrg Economist* 1973-75. Eugene L Grant Award ASEE 1972. Chmn Engg Economy Div ASEE 1973-74. Chmn Engg Econ Div - Sys Engg Group IIE, P E NM, OK. Fellow AAAS; Sigma Xi. Cons & author in operations management. Fellow IIE; Bd of Trustees, VP Industry & Mgmt Divs, IIE 1978-80; Editor and Chief, Handbook of Finance Economics & Accounting (John Wiley & Sons) Editor *Industrial Mgmt* 1983- Mbr-Shipbuilding Techy Cttee, Natl Research Council. Pres, Bd, United Cerebral Palsy of Dallas County. Fred Crane Distinguished Service Award, IIE, 1987. *Society Aff:* IIE, ASEE, ORSA, TIMS, AAAS, ΣΞ

Lux, George R
Home: 1503 Greendale Dr, Blacksburg, VA 24060
Position: Assoc Prof. *Employer:* VA Tech. *Education:* MS/Engg Graphics/IIT; BS/EE/Valparaiso Univ; -/Reliability Engg/Univ of AZ. *Born:* 6/17/37. Native of Chicago, IL. Worked as Elec Engr in industry 1959-63. Engr for USAF 1963-66. Taught Mech Engg for ten yrs at Valparaiso Univ. Elec Engr Consultant for Several companies and organizations. Recipient of NSF study awards & NSF educational equipment grant. Mbr of exec committee, St Joseph Valley sec ASME (1972-76; vchrmn 75-76). Bd chrmn, Engg Supply Store Inc (1970- 76). Currently teaching Engg Fundamentals at VA Tech. Chrmn of Design Grahics Div (Southeast sec) of ASEE. (1978-80, 1987-88; VChrmn 1981-82, 1986-87) Secy of VA Sec of ASME. Mbr Tau Beta Pi. Reg PE in IL, IN, PA, VA. Listed in "Who's Who in the South and Southwest" & "Who's Who in Society–. Mbr of the Order of the Engineer. *Society Aff:* AFCEA, ASEE, ASME, SAME.

Lux, John H
Business: PO Box 8266, Rancho Santa Fe, CA 92067
Position: Chmn & Chief Exec Officer. *Employer:* AMETEK Inc. *Education:* PhD/Chem Engg/Purdue Univ. *Born:* 2/3/18. S. Carl Harrison & Mary Emma (Dunn) L. Asst Dir R&D The Neville Co 1943-46; V P, cons Atomic Basic Chem 1946-47; Dir Res Witco Chem Co 1947-50; Mgr New Prod Dev GE Co 1950-52; V P Shea Chem Co 1952-55; Pres, Dir Haveg Indus Inc Wilimington Del 1955-66, Haveg Corp, Tourlux Mgmt Corp P R; Chmn/Bd Hemisphere Prod Corp P R, Reinhold Engrg & Plastics Co Norwalk Calif, Amer Super-Temps Wires Co; Pres Ametek Inc 1966-69, Chmn/Bd, Ch Exec Officer 1969- . Mbr Amer Inst Chem Engrs, Amer Chem Soc, Metropolitan Club, Sigma Xi, Phi Lambda Upsilon. Home: P.O. Box 1889, Rancho Sante Fe, CA 92067 *Society Aff:* AIChE, ACS.

Lux, William J
Business: 1855 Eden Ln, Dubuque, IA 52001
Position: Vice President & Principal Consultant *Employer:* WJ Lux Consulting Ltd *Education:* BSME/Mech Engr/MI Tech; Professional Degree//MI Tech. *Born:* 6/29/25. Native of Saginaw, MI. Served in Army Air Corps 1942-45. Filled Design, Dev, and Engg Mgt positions from 1950 to 1971 with Caterpillar, Cummins, and Wabco. Joined Deere in 1971 with world-wide responsibility for Engg of Deere's industrial product line. Chrmn, Bd of Regents, Loras College. Interest in products liability and reliability. Enjoy philosophy and theology. Since 1985, consultant in products liability. *Society Aff:* SAE, ASM, NSC, ADPA.

Luz, Herbert M
Home: 14 Canterbury Hill, Topsfield, MA 01983
Position: VP Mfr & Engg. *Employer:* W R Grace & Co. *Education:* MBA/Bus Adm/Northeastern Univ, Boston MA; BS/Chem Engg/Northeastern Univ, Boston, MA. *Born:* 3/5/31. General Foods Corp 1954, process engg, Central Res Div. Naval OCS, served 3 yrs Pacific & Atlantic - Lt, Operations Officer, 1954-57. Atlantic Gelatin Div General Foods Corp, Res Process Engg 1957-60. Joined W R Grace & Co, Dewey & Almy Chem Div Cambridge MA 1960. Process Engg, Mgr Process Dev Dept 1962. Project Mgr 1966. Asst to Div Pres 1967. R&D Mgr 1968, VP, R&D 1969. VP Mfr Organic Chem Div 1971. Also respon for Div Engg Dept. Reg PE, MA. 1986 VP Polymer Operations. *Society Aff:* AIChE.

Lyddan, Robert H
Home: 2427 Silver Fox Lane, Reston, VA 22091
Position: Retired. *Education:* BS/Civil Engg/Univ of KY. *Born:* 12/3/10. Native of Irvington, KY. Two yrs with Interstate Commerce Commission on railroad land evaluation. Forty four yrs with US Geological Survey, specializing in topographic mapping: engaged in field surveys in eastern states, Puerto Rico and Alaska 1933-45; Chief of program planning for Topographic Div 1946-54; Atlantic Region Eng 1955; Asst Dir of USGS 1956-67; Chief of Topographic Div 1968 until retirement in 1977. Dept of Interior Distinguished Service Medal 1960. Pres of American Congress on Surveying and Mapping 1957. *Society Aff:* ASCE, ASP, ACSM.

Lykoudis, Paul S
Business: Sch of Nucl Engg, W Lafayette, IN 47907
Position: Prof & Hd. *Employer:* Purdue Univ. *Education:* PhD/ME/Purdue Univ; MS/ME/Purdue Univ; BS/Mech-EE/Natl Tech Univ of Athens. *Born:* 12/3/26. Born in Preveza, Greece Dec 3, 1926. Came to the US in 1953 & became naturalized in 1964. On Nov 26, 1953, married Maria Komis. They have one son, Michael. Dr Lykoudis joined the Purdue Univ faculty in 1956 & taught in the Sch of Aero, Astro & Engg Sci until 1973 when he became Hd of the Sch of Nucl Engg at Purdue Univ. Served as Assoc Hd of the Sch of Aero, Astro & Engg Sci from 1971-1973. He served as dir of the Aerospace Sci Lab from 1968-1971. Was Visiting Prof of Aero Engg at Cornell Univ from 1960-1961. Has contributed over 70 papers in the fields of heat transfer, fluid mechanics, magneto-fluid- mechanics, astrophysics & fluid mechanics of physiological systems. *Society Aff:* ANS, APS, AIAA, AAS.

Lyle, Samuel P
Home: 218 Spring St, Huntingdon, TN 38344
Position: Retired (Self-employed) *Education:* MS/Agri Engg/IA State Univ; BS/Agri

Lyle, Samuel P (Continued)
Engg/KS State Univ. *Born:* Feb 12, 1892. Hd Agri Engrg Depts, Univ of Georgia 1924-30, Arkansas A&M Coll 1922-24. U S Dept of Agri, Fed Extension Serv 1930-62, incl Sr Engr 1930- 40, Principal Scientist 1941-58, Dir Agri Programs 1958-62. P Pres, Life Fellow John Deere Gold Medal, Amer Soc Agri Engrs. Mbr Coord Ctte Engrs Joint Council 1962-63. Mbr Amer Soc for Engrg Educ. Fellow AAAS. P Chmn Farm Div, Natl Safety Council. Hon Mbr Natl Inst Farm Safety. Mbr., Soil Conservation Soc of America, Mbr., Council for Agricultural Science and Technology. Sigma Xi, Phi Kappa Phi, Gamma Sigma Delta, Epsilon Sigma Phi. *Society Aff:* ASAE, EJC, ASEE, AAAS, NSC, NIFS, CAST, SESA.

Lyman, W Stuart
Home: 15 N Bridge Terr, Mt Kisco, NY 10549
Position: Senior Vice President *Employer:* Copper Dev Assn Inc. *Education:* MS/Met/Univ of CA; BS/Met/Univ of Notre Dame. *Born:* 4/13/24. Mt Vernon, SD. USNR 1944-46. Hd advance planning unit, office of the chief engr, European Command 1946-47; edit & control officer 1948-49. Teaching asst, res engr, Univ of CA 1949-51. Staff met, MAB, Natl Acad of Sci 1951-54. Dept consultant, asst div chief, div chief, Battelle Meml Inst 1955-64. Mgr - technical & market services, Vice Pres and Senior Vice Pres, Copper Dev Assn Inc since 1964. Responsible for technical aspects of Assn's industry-wide progs of application engg, stds dev & info service as well as market res & statistics. Established & hds worldwide Copper Data Ctr. Active in solar energy systems engg & dev since 1970. PE. *Society Aff:* ASM, AIME, ASME, ASTM, SAE, ΣΞ

Lynch, Edward P
Business: 9700 S Cass Ave, Argonne, IL 60439
Position: Chem Engr. *Employer:* Argonne Natl Lab. *Education:* BS/CE/Carnegie Inst of Tech. *Born:* 8/13/19. Fellow of the AIChE. Reg PE in IL, KY, OH, & PA. Formerly Chief Process Engr of the midwest branch of a maj engg co. Formerly Mgr of Engg & construction of the Blockson Chem Div of the Olin Corp. Presently consultant through the ANL to the US Dept of Energy & Technical Mgr of certain DOE projs. Author of several technical papers on Engg Applications of Symbolic Logic & author of the book *Applied Symbolic Logic* Wiley-Intersci (1980). *Society Aff:* AIChE, APCA, ASL, MENSA.

Lynch, Robert D
Business: College of Engrg, Villanova, PA 19085
Position: Dean of the College of Engrg. *Employer:* Villanova Univ. *Education:* PhD/Structures/Univ of Notre Dame; MSCE/Structures/Univ of PA; BCE/Civil Engg/Villanova Univ. *Born:* Aug 27, 1930. Served as a line officer in US Navy 1953-56. Struct engr with Blauvelt Engrg Co 1956-60. Sr Struct Engr with McCormick, Taylor & Assocs Inc 1960-62. Joined the Civil Engrg Fac at Villanova Univ 1962. Dev & taught both grad & undergrad courses in the areas of struct analysis & design, struct mech, numerical methods & computer applications. Named to present pos as Dean of Coll of Engrg in Aug 1975. NSF Fac Fellowship 1966; Lindback Found Award for Disting Teaching 1975. *Society Aff:* ASCE, NSPE, ASEE, ΤΒΠ.

Lynn, Phillip W
Business: 5410 Homberg Dr, Knoxville, TN 37919
Position: President. *Employer:* PLUS engrg Inc.. *Education:* MS/San Engr/Univ of TN; BS/Civil Engg/Univ of TN. *Born:* Sept 1942 Woodbury Tenn. Mbr ASCE, NSPE, TSPE, AWWA, WPCF, LES, Knoxville Tech Soc, Chi Epsilon, Tau Beta Pi. Chmn KTS Water & Wastewater Ctte 1973; Mbr PTI Water/Wastewater Mgmt User Requirements Ctte; Dir Knoxville TSPE; Reg Engr LA, GA & TN. Engr Shell Oil Co Norco Refinery 1967-70; Ch Engr Frank Foster & Assocs Inc 1970-72; Dir of Wastewater Control, City of Knoxville 1972- 76; Pres PLUS engrg 1976- . Young Engr of Yr 1975 Knoxville TSPE. *Society Aff:* ASCE, ACEC, NSPE, AWWA.

Lynn, R Emerson
Business: 140 W 19 Ave, Columbus, OH 43210
Position: Prof. *Employer:* OH State Univ. *Education:* PhD/CE/Univ of TX; MS/CE/Univ of TX; BS/CE/Purdue Univ. *Born:* 3/16/20. Native of Elkhart, IN. Technical Service & lower mgt for Footwear & Self- Sealing Fuel Tanks - Uniroyal 1942-1946. Sr Res Engr & Mgr of Chem Engg Res-B F Goodrich 1952-1960. Responsible charge for Process Dev of first synethetic cis- polysoprene process. Mgr of Prog Planning & E P Rubber Dev Proj - B F Goodrich Chem Co - 1960 to 1967. Instituted teaching & res in Polymers and Polymer Processing - OH State Univ 1967-present. Elected Sr Mbr of SPE in 1969 & Fellow of AIChE in 1979. Enjoy boating, fishing, & classical music. Active in USPS (US Power Squadrons). *Society Aff:* AIChE, SPE, ACS.

Lynn, Scott
Business: Dept. of Chem. Engg, University of Calif, Berkeley, CA 94720-9989
Position: Prof. *Employer:* Univ of CA. *Education:* PhD/ChE/CA Inst of Tech; MS/ChE/CA Inst of Tech; BS/Appl Chem/CA Inst of Tech. *Born:* 6/18/28. Thesis res at Caltech directed by D M Mason, W H Corcoran; followed by post-doctoral yr at Delft TH under H Kramers. Twelve yrs ind R&D experience at Dow Chem, Western Div, in chlor-alkali cell dev, inorganic process res, polymer processing, etc. Joined Chem Engg Dept at UC Berkeley in 1967; res interests focus on process synthesis; proj areas include chem energy storage, removal of pollutants from stack gases, desulfurization of gas streams, clean-up of coal processing waters, etc. *Society Aff:* AIChE, ACS, ECS.

Lynn, Walter R
Business: 217 Hollister Hall, Ithaca, NY 14853
Position: Prof. *Employer:* Cornell Univ. *Education:* PhD/Systems Engg/Northwestern Univ; MS/Sanitary Engg/Univ of NC; BS/Civ Engg/Univ of Miami. *Born:* 10/1/28. BSCE, Univ Miami; MSSE, Univ NC; PhD, Northwestern Univ; PE, NY; Reg Land Surveyor, FL. Supt Sewage Treatment, Univ Miami, 1951-53; Assoc Prof Civ & Ind Engg, 1959-61, Univ Miami (1954-61); Dir Res Ralph B Carter Co, 1957-58. Cornell Univ, 1961-; Prof Civ & Environ Engg, 1964-; Dir Ctr Environ Quality mgt, 1966- 76; Dir Sch Civ & Environ Engg, 1970-78. Wash Water Supply Study, NAE/NRC, 1976- 84; Cons Who, Searo, 1978, Rockefeller Fdn, 1976-80. Chrmn Ithaca Urban Renewal Agy, 1965-68. AUS, 1946-48. Natl Acad Engrs Mex (corr), 1977; Assoc Editor ORSA, 1968-76, Jour Environ Econs & Mgt, 1975-Member, Cornell University Board of Trustees, 1980-85; Member, Cornell Research Foundation, 1978- ; Director, Program on Science, Technology & Society, 1980-; Chrmn Bd on Water Science and Tech, NAE/NRC, 1982-1985. Chrm NY State Water Res Planning Council, 1986- . *Society Aff:* AAAS, ΣΞ, ASCE, XE, ΦΚΦ

Lyon, Oscar T, Jr
Home: 712 W Encanto Blvd, Phoenix, AZ 85007
Position: Consulting Civil Engineer (Self-employed) *Education:* BS/CE/Univ of AZ. *Born:* 11/8/18. in Warren AR. Served in SW Pacific Theater, US Navy Civil Engr Corps 1943- 46. m. Patricia A Haley 1945. Employed by AZ Hwy Dept 1946. Served as Resident Engr, Area Engr, Asst Dist & Dist Engr in var dists of State of AZ. Ch Deputy State Engr since 1974. State Engr 1977-80. Retired 1980. Mbr Theta Tau, ASCE, NSPE, Order of Engr. Natl Dir ASCE, Dist 11, 1971-74. Mbr Natl Bd, Order of Engr. *Society Aff:* NSPE, ASCE, ASPE.

Lyons, Jerry L
Business: 1801 Lilly Ave, St Louis, MO 63110
Position: Manager, Engrg Research. *Employer:* Chemetron Corp, Fluid Controls Div. *Born:* Apr 2, 1939. Grad of Okla Inst of Tech. Native of St Louis Mo. Reg P E & Cert Mfg Engr in product design. Proj engr Harris Mfg Co 1964-70 St Louis Mo, designing aerospace components, valves & flight control sys. Proj engr Essex Cryogenics Indus 1970-73 - designed fluid control devices. Assumed current pos as Mgr Engrg Res at Chemetron, dev new concept in flow control, & is Pres of Yankee Ingenuity Inc, cons firm, St Louis since 1973. Author 3 books & num articles in field of fluid controls & 2 addl books to be published in near future. Chmn SME for the regis of mfg engrs in Missouri, ctte mbr of the Certification Prog with SME In-

Lyons, Jerry L (Continued)
ternatl. Lectr at num soc's & univs. Recipient of the St Louis Engrs Club 'Award of Merit' & the Prof Dev & President's Award of SME.

Lyons, John W
Business: National Engineering Laboratory, NBS, Tech, B119, Gaithersburg, MD 20899
Position: Dir, Natl Engrg Lab *Employer:* Natl Bureau Stds, Dept Commerce *Education:* PhD/Physical Chem/Washington Univ-St Louis; AM/Physical Chem/Washington Univ-St Louis; AB/Chem/Harvard *Born:* 11/05/30 As Dir of the Natl Engrg Lab and past Dir of the Inst for Applied Tech, Dr. Lyons is responsible for research in applied mathematics, bldg tech, electronics and electrical engrg, energy conservation, engrg standards development, fire, mfg engrg and chemical engrg. Dr. Lyons was Dir of the NBS Center for Fire Research until December 18, 1977, and was responsible for providing scientific and technical knowledge applicable to the prevention and control of fires. From 1955 until his appointment as Dir of the Center for Fire Research on October 1, 1973, Dr. Lyons held various positions in research and development at Monsanto Co. 1981 Presidential Rank Awd, Distinguished Exec; 1983 Acting Deputy Dir Natl Bureau of Standards. Presidential Management Improvement Award-1977; U.S. Dept. Commerce Gold Medal-1977; Edward Uhler Condon Award for his book *Fire*-1986. Author of three books: *Viscosity and Flow Measurements*; *The Chemistry and Uses of Fire Retardants*; *Fire*. *Society Aff:* AAAS, AIChE, ACS, NFPA, WAS, NAE

Lyons, Robert E
Home: 10301 Rossmore Ct, Bethesda, MD 20014
Position: Chief, Comp Sys Div. *Employer:* Dept of Defense. *Education:* PhD/EE/Univ of MD; EE/EE/MIT; MS/EE/Univ of MD; SB/EE/MIT. *Born:* 11/18/28. Native of Weymouth, MA. Served with US Army during Korean War. Have worked as elec engr in computers & communications in the Dept of Defense - Natl Security Agency & Defense Communications Agency. Also employed by IBM Federal Systems Div for 13 yrs. Currently Chief of Computer Systems Div, Defense Communications Eng Ctr. *Society Aff:* IEEE, AFCEA

Lyse, Inge M
Home: Tyholtvegen 74, 7000 Trondheim, Norway
Position: Professor Emeritus. *Employer:* Technical Univ of Norway. *Education:* Dr Tech/Civil Engg/Norway Inst of Technology; Dipl Engg/Civil Engg/Norway Inst of Technology. *Born:* Oct 22, 1898 Lysebotn, Norway. Chartered Civil Engr, Tech Univ of Norway 1923; Dr Technologie 1937. Res engr Portland Cement Assn 1927-31, Res Prof & Dir Fritz Engrg Lab Lehigh Univ 1931-38. Prof of Reinforced Concrete Tech Univ of Norway 1938-68. UNESCO Expert to India, Pakistan, Venezuela. Levy Medal of the Franklin Inst 1937, Croes Medal of the Amer Soc of Civil Engrs 1937, Knight of the Royal Norwegian Order of St Olav 1966. Hon Mbr Amer Concrete Inst 1962, Internatl Matl Assn RILEM 1971, Norwegian Acad of Tech Sci 1974, Norwegian Concrete Assoc, 1980. Foreign Associate Natl Acad of Engrg of the USA, 1981, Hon Doctor of Engr, Lehigh Univ 1981. *Society Aff:* ACI, ASCE, NATS, KNVS.

Lysmer, John
Business: 440 Davis Hall, Berkeley, CA 94720
Position: Prof of Civil Engrg. *Employer:* Univ of Calif - Berkeley. *Education:* PhD/-/Univ of MI; MSc/-/Tech Univ of Denmark. *Born:* Aug 18, 1931, Denmark. Naturalized U S citizen 1972. In cons practice 1954-61, mainly with Ove Arup & Partners London UK & Lagos Nigeria. Joined the fac at U C Berkeley in 1965. Well-known teacher, researcher, author & cons in the areas of theoretical soil mech & soil dynamics. Mbr of Exec Ctte of the Geotech Engrg Div of ASCE 1972-79. Middlebrooks Award of ASCE 1967. Huber Res Prize of ASCE 1976. *Society Aff:* ASCE, SSA, EERI.

Ma, Joseph T
Home: 15541 Toyon Dr, Los Gatos, CA 95030
Position: Prof *Employer:* Santa Clara Univ *Education:* PhD./ME/Iowa State University; M.S./IE/Purdue University; B.S./IE/Texas Tech. University; B.S./ME/National Sun-Yat-Seng U. *Born:* 08/15/25 Born Foochow, China. US citizen since 1954. Asst Prof, Iowa State Univ, 1956-59. Joined IBM 1959. Retired 1985. Highest technical position attained: Sr Technical Staff Member. Held various managerial positions in six different divisions. Received invention and outstanding innovation awards. Specialty: computer peripherals, disk drives, prints, displays, and electromechanical engrg. Current position: Adjunct Prof of Mech Engrg, Santa Clara Univ and consulting engineer. Holder of patents and publications. Reg PE. Enjoy sports and travel. *Society Aff:* ASME, IEEE.

Ma, Yi H
Business: West St, Worcester, MA 01609
Position: Prof & Dept Hd. *Employer:* Worcester Poly Inst. *Education:* ScD/ChE/MA Inst of Tech; MS/ChE/Univ of Notre Dame; BS/ChE/Natl Taiwan Univ. *Born:* 11/7/36. Came to US in 1961 for grad studies. After receiving ScD from MIT, started teaching at Worcester Poly Inst in 1967. Assumed current responsibility as the Head of the Dept of Chem Engg in 1979. Consults regularly for US Govt & Inds. Currently consultant for Cabot Corp. Enjoys skiing & water skiing. *Society Aff:* AIChE, ACS, AAUP, ΣΞ.

Maas, Robert R
Home: 17112 Via Andeta, San Lorenzo, CA 94580
Position: Proj Mgr *Employer:* Lockheed Missiles & Space Co *Education:* BSEE/Elec/USAF Inst of Tech *Born:* 08/14/23 Native of Council Grove, KS. Served 26 years with US Air Force as Command Pilot and Elec Engr, specializing in nuclear weapons and missile range instrumentation work. With Lockheed Missiles & Space Co since 1967. Currently Proj Mgr for Prod Assurance providing quality and reliability engrg support for Navy Missile Programs. ASQC certified Quality and Reliability Engr. PE, State of CA. P Pres ASQC 1983. Mbr IEEE since 1946. Named Who's Who in CA 1981-1982. Was recipient of the Presidential Award in recognition of his outstanding contributions to the Amer Soc for Qual Control in 1986. Received the Benjamin L. Lubelsky Annual Award in recognition of his outstanding contribution to the field of qual control, March 1986. Enjoy golf and fishing. *Society Aff:* ASQC, IEEE

Maas, Roy W
Home: 9813 Longwood Circle, Louisville, KY 40223
Position: VP *Employer:* Watkins & Assoc Inc *Education:* MS/Comm Development/Univ of Louisville; BS/CE/Vacparaiso Univ *Born:* 11/12/31 Started career with J. Stephan Watkins Consulting Engrs in 1955 as Field Engr on hgwy construction in IN. U.S. Navy Engrg Officer 1956-1958. Hgwy design engr with Watkins and Assocs Inc on interstate and primary hgwys in MI and WV, 1958-1962. Design Engr for IN Hgwy Commission, 1963. Assumed present position as Branch Mgr and VP in 1963. Responsible for design and construction review of municipal engrg projects for Watkins & Assocs, Inc. Registered PE in KY, IN, IL, VA and WV. Registered Land Surveyor in KY. *Society Aff:* ASCE, NSPE, ACEC, AREA, ARTBA

Maass, Robert H
Home: Dellwood Pkwy S, Madison, NJ 07940
Position: Proj Dir. *Employer:* Exxon Res & Engr Co. *Education:* BS/Aero Engr/Worcester Poly Inst. *Born:* 10/11/23. Native of Booklyn, NY. Was with the USN from 1943 to 1946 graduating from US Naval Acad (Engg Reserve), & served as engg officer aboard the USS NJ. Joined Exxon Res & Engg in 1946 & have been in various engg positions in mech and process design & mtls res & dev. For the past 16 yrs have held admin posts in proj mgt supervising contracting, estimating, cost control, inspection & as proj mgr of various maj petrol & petrochem projs. Current position is Proj Dir of engg & construction of a recently completed world class grass root olefins plant at Baytown, TX. Enjoy tennis & fishing. *Society Aff:* ASME.

Maatman, Gerald L
Home: 101 Carriage Rd, Barrington, IL 60010
Position: Chrmn of Bd. *Employer:* Natl Loss Control Service Corp. *Education:* BS/Fire Protection & Safety Engg/IIT. *Born:* 3/11/30. Elected to Tau Beta Pi & Salamander Honorary Fraternities, presented with Finnegan Award by Chicago Chap, SFPE outstanding contribution to field of fire protection. Reg PE in IL, TN, & LA. Prof background-Prof & Chrmn Dept of Fire Protection Engg, IIT 1959-66, joined Kemper Ins Group 166 VP Engg & elected Pres of newly formed cons subsidiary 1968 with advancement to present position 1974. Serve as cons to City of Chicago, Chicago Public Bldg Comm & numerous other private & public sector clients. Mbr of Bd of Dirs, Economy Fire & Casualty Co and Federal Kemper Insurance Co. *Society Aff:* SFPE, ASSE.

Mabie, Hamilton H
Business: Dept Mech Engrg, Blacksburg, VA 24061
Position: Prof of Mech Engrg. *Employer:* Virginia Polytech Inst & St Univ. *Education:* PhD/Mech Engg/Penn State Univ; MS/Mech Engg/Cornell UNiv; BS/Mech Engg/Univ of Rochester. *Born:* Oct 21, 1914 Rochester N Y; m. 1941; 3 children. From Instructor to Assoc Prof Mech Engrg Cornell 1941-60; R/D Engr Sandia Corp N Mexico 1960-64; Prof Mech Engrg Va Polytech Inst & State Univ 1964- . Life Fellow ASME; Amer Soc Lubrication Engrs; Soc for Experimental Mechanics; Amer Soc Engg Educ; Mech & Kinematics; fatigue of mech components; instrument bearings & lubrication. *Mechanisms & Dynamics of Machinery*, John Wiley & Sons, 1987. *Society Aff:* ASLE, SEM, ASEE, ASME.

Mabius, Leonard J
Business: US Army Commun Sys Agcy, Ft Monmouth, NJ 07703
Position: Tech Director. *Employer:* US Army Commun Systems Agency. *Born:* Sept 1939. MS in Telecommunications Univ of Colorado, BSEE Rutgers Univ. Native of Steelton Penna. Army (1962-67) in tactical & DCS communications; Philco-Ford 1967-68 as Proj Engr/Mgr; Electronics Engr 1969-71 with Army Communications Command doing sys engrg design; 1971-74,Army Communications- Electronics Engrg Installation Agency Ft Huachuca AZ as Tech Dir; 1974- Tech Dir Army Communications Sys Agency Fort Monmouth N J working in communications/ engrg/acquisition. Army Commendation Medal 1964. Sr Mbr IEEE; Mbr AFCEA; Assoc Mbr in Res Inst of America.

MacAdam, David L
Home: 68 Hammond Street, Rochester, NY 14615
Position: Sr Res Associate. *Employer:* Eastman Kodak Co (retired). *Education:* PhD/Physics/MIT; BS/Engg Physics/Lehigh Univ. *Born:* July 1, 1910 Philadelphia. Eastman Kodak Res Labs 1936-75. Optical Soc of Amer Fellow, Pres 1962, Editor 1964-75, Adolph Lomb Medal 1940, Frederic Ives Medal 1974. Inter-Soc Color Council Godlove Award 1963, Honorary Mbr 1975. Federation of Socs for Paint Tech Mattiello Memorial Lecture 1965; Royal Photographic Soc Hurter & Driffield Lecture & Medal 1966. Association Internationale de la Couleur, Judd Medal 1983 Colour Group of Great Britain, Newton Medal 1985. Internatl Comm on Illumination, US Expert on Colorimetry 1969-75; Life Mbr of US Natl Ctte; coauthor 'Handbook of Colorimetry' 1936, The Science of Color 1953; Editor Sources of Color Science1973; Author: *Color Measurement: Theme and Variations*1981 (2nd edition, 1985). About 100 contribs in various publs (see Journal of Optical Soc of Amer 65, 481 (1975) for list). Prof of Optics Univ of Rochester 1976- ; Editor of Springer Series in Optical Scis (Springer-Verlag, Heidelberg, Germany) 1975-. *Society Aff:* OSA, AIP, ISOC.

MacBrayne, John M
Home: 33 Mountain St, Camden, ME 04843
Position: Consultant (Self-employed) *Education:* SB/Bus & Engg Adm Inistation/MA Inst Tech. *Born:* 1909 Quincy Mass. Instructor M I T 1931-34; Lt US Navy 1944-46; Indus Engr Dennison Mfg Co; Asst Ch Indus Engr Talon Inc; Asst Budget Dir Johns- Manville; Successively Ch Indus Engr, Dir of Genl Servs, Asst to Exec V Pres Union Camp Corp; Business respons incl Dir of Const, Indus & Package Engrg, Pres of Amcreco subsidiary, Bds of Dirs Allied Container & Cartoneras Union; retired 1975. Currently Reg P E. Mbr Internatl Servs Org. IIE Regional V Pres, Dir 1956-57. TAPPI Chmn Indus Engrg Ctte 1954-58, Chmn Engrg Div 1962-64, Dir 1966- 69, Fellow 1969, Engrg Div Award 1972. *Society Aff:* TAPPI.

Macchi, Anselmo J
Business: 44 Gillett Street, Hartford, CT 06105
Position: Partner. *Employer:* Macchi Engrs. *Education:* BCE/Civil Engg/Univ of CA. *Born:* 12/18/12. 1936 USN C E C Off 1945. Reg P E 11 States. Invented Tube Slab Const. Private Practice Cons Engr 1945. 1969 Study for Permanent Traffic Connection between Sicily & Italian Penninsula over the Strait of Messina. Study report on English Channel. 1966 Internatl Competition: Award from Danish Ministry of Public Works for design of Permanent Traffic Link across the Great Belt Denmark. Proposal to stabilize Leaning Tower of Pisa. Publs: Low-cost Concrete Parking Garage ENR 4/55, Slab Costs Cut with Paper-Formed Hollows ENR 5/52. Tech Socs: NSPE, ASCE, ACI, ASTM, CEC, British Concrete Soc. Mbr.: New York Athletic Club. Civic activities: Bldg Bd of Review, City of Hartford & Town of West Hartford Conn. *Society Aff:* NSPE, ACI, ASCE, ASTM, CEC.

MacCollum, David V
Home: 1515 Hummingbird Lane, Sierra Vista, AZ 85635
Position: Safety Engr & Cons. *Employer:* Self. *Education:* BS/Bux & Tech/OR State Univ. *Born:* June 1923. BS Oregon St Univ. Reg P E AZ & CA. Pres ASSE 1975-76. Mbr, Secy Labor's Const Safety Adv Ctte 1969-73. Named 'Engr of Year' Ariz Soc Prof Engrs 1970. Received ASSE/Vets of Safety First Place Tech Award - outstanding contrib to safety engrg literature 1969. Dev design criteria for rollover protection sys for tractors & other const equip while with Corps of Engrs 1957. Served as Dir of Safety US Army STRATCOM respon for safety in design & operation in 25 foreign countries. In private practice since 1972 as cons on const & tunneling safety. *Society Aff:* ASSE, SSS, NSPE, NFPA, SME of AIME, NSC, IFS.

MacCrone Robert K,
Business: R. P. I, Troy, NY 12180
Position: Prof. *Employer:* RPI. *Education:* DPhiE/Physics/Univ of Oxford; MSc/Physics/Univ of Witwatersrand. *Born:* in South Africa. Member of APS, ACerS, MRS & NICE. Awarded Univ of Witwaterstant Post Graduate Scholarship on British Oxygen Fellow and Education Service Award for America Plastics Lustidare. Prof interests include glasses, ceramics, EPR, catalysis, superconductivity and magnetism. *Society Aff:* ACerS, APS, MRS.

MacDonald, Bryce I
Home: 80 Hulls Highway, Southport, CT 06490
Position: Corporate Environmental Issues Proj *Employer:* General Electric Company *Education:* BChE/Chem Engg/Cornell Univ. *Born:* 2/26/24. Engg & Mfg positions with Genl Elec Co 1946-62; Dir of Engg Glidden-Durkee Div SCM Corp 1962-68; present pos since 1968; corp staff espon for engg matters & for tech & economic assessment of environmental issues. Dir EJC 1973-76, Pres Elect EJC 1976. Pres EJC 1977-78. Chmn Environ Div AIChE 1974, Dir AIChE 1976- 78. Fellow, AIChE 1979. PE New York State Director, Engineering Index 1980-1984. Trustee, United Engineering Trustees 1980-1984. *Society Aff:* AIChE

MacDonald, Digby D
Business: 116 West 19th Ave, Dept. Metallurgical Engrg, Columbus, OH 43210
Position: Prof *Employer:* OH State Univ *Education:* PhD/Chem/Univ of Calgary, Canada; MSc/Chem/Univ of Auckland, NZ; BSc/Chem/Univ of Auckland, NZ *Born:* 12/7/43 Native of New Zealand. Taught at Univ of Calgary (1975-77) and Victoria Univ, Wellington, NZ (1972-75). Carried out research at Whiteshell Nuclear Research Est, Atomic Energy of Canada Ltd (1962-1972) and at SRI Intl (1977-79). Currently, Prof of Metallurgical Engrg and Dir, Fontana Corrosion Ctr, OH State Univ. Divisional Editor, Electrochemical Society for corrosion and Member of

MacDonald, Digby D (Continued)
Publications Committee, Natl Assoc Corrosion Engrs. Published over 130 papers and 1 book, *Transient Techniques in Electrochemistry*. Acts extensively as consultant to industry and government. *Society Aff:* NACE, ECS

MacDonald, Donald H
Home: RR 1 Niagara River Pkwy, Niagara-on-the-Lake Ontario, Canada L0S 1J0
Position: V President & Director. *Employer:* Acres Consulting Services Ltd *Education:* PhD/Soil Mechs/Univ of London; DIC/Soil Mechs/Imperial College, UK; MRP/Regional Planning/Cornell Univ; BASc/Civil Engg/Univ of Toronto. *Born:* 6/15/22. Employed by Toronto Transp Comm, worked 4 yrs on subway design & const. Attended Imperial Coll London on an Athlone Fellowship spec in soil mech & fdn engrg; joined H G Acres & Co 1955 & has served in var capacities incl Pres of H G Acres (1964-69) & VP of Acres Cons Servs Ltd (1971-); engaged mainly in geotech & heavy civil engrg, power, transp & indus projs in Canada & abroad. VP (North Amer) 1969-73 Intl Soc for Soil Mech & Fdn Engrg; Mbr Exec Ctte 1968-76 Canadian Natl Ctte of ICOLD; Pres Cons Engrs of Ontario 1978-79; Pres Tunnelling Assoc of Canada, 1981-82; Mbr of numerous cttes of, incl UNESCO, Sci Council of Canada. Natl Res Council of Canada, APEO, GSA. Fellow of EIC, ICE, ASCE and GSA. Awards: APEO Gold Medal, 1978, & Legget Award of Canadian Geotechnical Soc, 1978. *Society Aff:* APEO, EIC, CGS, ASCe, GSA, ICE.

MacDonald, Harold C
Business: The American Rd, Dearborn, MI 48121
Position: VP - Engrg and Research *Employer:* Ford Motor Co *Education:* BS/ME/MI State Univ *Born:* 6/20/17 Native of Virginia, MN. Employed as Deisgn Engr in automotive industry prior to World War II. Served with U.S. Navy as Engrg Officer 1942-1945. Returned to automotive industry as Designer and Design engr. Joined Ford Motor Co in 1948 as Designer, Draftsman, Project Engr and advanced to VP - Car Engrg; VP - Product Development; and assumed my current position as VP - Engrg and Research in 1975. Am responsible for Ford's staff activities worldwide in product engrg and research. National Pres SAE, 1980-1981. Enjoy participation in church activities, especially as a speaker to outside groups, and enjoy golf. *Society Aff:* SAE, ESD

Macdonald, J Ross
Home: 308 Laurel Hill Road, Chapel Hill, NC 27514
Position: W R Kenan Jr, Prof of Physics. *Employer:* Univ of North Carolina. *Education:* DSc/Physics/Oxford Univ; DPhil/Solid State Physics/Oxford; MS/Elec Engg/MIT; BS/Elec Engg/MIT; BA/Physics/Williams College. *Born:* Feb 27, 1923. USNR radar officer 1944-46; Rhodes Scholar Oxford 1948-50; Armour Res Foundation 1950-52; Argonne Natl Lab 1952-53; Mbr of Tech Staff; Dir Physics Res Lab; Dir Central Res Labs, & V Pres Corp R&D Texas Instruments 1953- 74. Mbr Natl Acad of Engrg & Natl Acad of Scis. Mbr Council of NAE 1971-74. Mbr Exec Comm Assembly of Engrg NAS 1975-78. Mbr 1971-74, Chmn 1973-74 NAS Ctte on Motor Vehicle Emissions. Mbr, Ctte on Satellite Power Sys, NAS 1979-81. Chmn NAS Numerical Data Adv Bd 1970-74. Sr paper award (1957), Achievement Award (1962) Inst of Radio Engrs. *Society Aff:* NAS, NAE, APS, ECS, AES, IEEE.

Macdonald, Malcolm J
Home: 2101 244th Av S.E, Issaquah, WA 98027
Position: Deputy Superintendent, Engrg and Utility Sys *Employer:* Seattle City Light *Education:* MSCE/Structural Dynamics/Univ of IL; BCE/Structures/Renssalear Polytechnic Inst; BS/Elec Engr/US Naval Acad. *Born:* Aug 1934. Native of Buffalo N Y. BS, U.S. Naval Acad; BCE, RPI; MSCE, Univ of IL; PE License CA, WA. Public Works Officer & ROICC Chelsea Naval Hospital Mass; Instructor in Struct Dynamics & Radiation Shielding CEC Officers School Port Hueneme Calif; ROICC Adak Alaska; R/D Officer Defense Atomic Support Agency; Planning Dept & Acquisition Dept Head, Pacific Div Naval Facilities Engrg Command; Commanding Officer US Naval Mobile Const Battalion FOUR, Pacific Best of Type & SAME Peltier Award 1975; SAME Moreell Medal 1972; Indus College of Armed Forces 1977; P Pres Aleutian Post SAME. Dir, Installations and Facilities, Assistant Secy of the Navy; Commandine, Officer, Navy Public Works Center, Norfolk, VA. Dir Energy Resource Planning, Seattle City Light. *Society Aff:* ASCE, SAME.

MacDonnell, W D
Business: 38481 Huron River Dr, Romulus, MI 48174
Position: Consultant. *Employer:* Kelsey-Hayes Co. *Education:* BS/Met/MIT *Born:* 11/3/11. in Boston, Began career with Bethlehem Steel Co as combusion engr in open hearth dept in Lakawanna NY & progressed to Asst Genl Mgr of Johnstown PA plt; moved to Great Lakes Steel 1957 as Asst to Exec VP & progressed to pres & Ch Exec Off; joined Kelsey-Hayes Co 1962; Pres, 1962-76; Chrmn 1976-77; now Consultant to co. Enjoys hunting, fishing & golf. *Society Aff:* NAE

MacDougall, Louis M
Business: 2800 E 13th St, Ames, IA 50010
Position: Mgr Mtls Engg. *Employer:* Sundstrand Hydro-Transmssion. *Education:* MS/Met Engr/Univ of WI; BS/Met Engr/MI Tech Univ. *Born:* 8/18/36. Native of Evanston, IL. Married while in college. ROTC commitment after college. Joined Sundstrand Corp Aviation Div in Rockford, IL, after active duty as proj met engr. Obtained MS while employed in Rockford. Left met group supervisor position in Rockford to join the Hydro-Transmission Div in Ames, IA, as Supervisor of Mtls Lab. Became Mgr of Engg Services, including design and test lab depts. Currently Mgr of Mtls Engg. Married, six children. Chrmn ASM Rockford Chapter 1970/71. Chrmn ANSI (Am Natl Stds Committee) Z-11 Subcommittee on Lubricants, 1980, Secretary of Subcommittes B-V, N-VII, N-X ASTM Ctte D-2 on Petroleum Prod and Lubricants. *Society Aff:* ASM, NFPA, ASTM, ASCE, SAE.

Macduff, John N
Home: 503 Wen Dover Circle, Grants Pass, OR 97526
Position: Prof Mech Engg Emeritus. *Employer:* Duke Univ. *Education:* MME/Mech Engg/NY Univ; BS/Naval Arch & Marine Engg/Webb Inst. *Born:* 6/20/12. Instructor NY Univ 1938-41; Genl Electric A B C Courses 1941-43, GE Res Lab 1942-46; Asst & Assoc Prof Brooklyn Polytech Inst 1946-50; Prof Rensselaer Polytech Inst 1950-56; 1956-78 Prof Mech Engg at Duke Univ (Chmn 1956-63). Mbr ASME Machine Design Div Exec Ctte 1961-65, Chmn 1965. Mbr Sigma Xi, Fellow ASME. Author (with John Curreri) of Vibration Control & articles pub in ASME Transactions, SAE Transactions, ASHRAE, Sound & Vibration, ISTVS Proceedings. Reg NY & NC. *Society Aff:* ASME, PE

Mace, Louis L
Business: 2227 152 Ave N.E, Redmond, WA 98052
Position: VP *Employer:* S J Groves *Education:* MBA//Geo WA Univ; BS/CE/OR State Univ *Born:* 3/5/35 After graduating from OR State in 1959, went directly to work in the construction industry. Over the past 20 years have worked on or been associated with nearly every major hydro project in the Northwest. Including such projects as Bonneville Second Powerhouse, Grand Coulee, Dworshac Dam, Lower Monumental, The Dalles & John Day. In 1974 Lou became VP of S. J. Groves & in 1981 Gen'l Mgr of their heavy constr division. *Society Aff:* ASCE, USCOLD

MacGregor, Charles W
Home: 112 Jerusalem Rd, Cohasset, MA 02025
Position: Consulting Engr. *Employer:* Self-employed. *Education:* PhD/Mech Engg/Univ of Pittsburgh; MS/Mech Engg/Univ of Pittsburgh; BS/Elec Engg/Univ of MI; BS/Engg Math/Univ of MI. *Born:* 5/25/08. Following industrial experience at Westinghouse, Pittsburgh, became instr, mech engg dept at MA Int of Tech in 1934, spending the next seventeen yrs there, becoming full prof 1942, and Hd of Materials Div and the MIT Gun Design Group. Next periods were devoted to the positions of VP in charge of Engg and Scientific Studies at the Univ of PA, Engg Consultant and Mgr of Advance Tech IBM; and Consulting Engr. Author, contributor, editor various papers and books. Levy Medal, Franklin Inst 1941; Dudley Medal, ASTM 1941;

MacGregor, Charles W (Continued)
Naval Ordnance Dev Award 1945; Fellow Amer. Acad. of Arts and Sciences and Franklin Institute. *Society Aff:* AIME, ASM, ASME, ASTM, NSPE, NYAS

MacGregor, James G
Business: Dept of Civil Engrg, Univ of Alberta, Edmonton, Alberta T6G 2G7 Canada
Position: Prof of Civil Engrg. *Employer:* Univ of Alberta. *Education:* PhD/Structures/Univ of IL at Urbana; MS/Structures/Univ of IL at Urbana; BSc/Civil Engg/Univ of Alberta. *Born:* Feb 14, 1934. Mbr of Civil Engrg Dept Univ of Alberta since 1960. Principal MKM Engineering Cons spec in behavior & strength of reinforced concrete structs. Dir Amer Concrete Inst 1973-76. Councillor Assn of Prof Engrs, Geologists & Geophysicists of Alberta 1973-76. Mbr ACI & ASCE Tech Cttes. Chmn Canadian Standards Assn Ctte on Reinforced Concrete Design. Canadian Delegate Comm Euro-Internatl du Beton, Fellow Royal Soc of Canada Academy of Sciences, ASCE State-of-Art Award 1968, ACI Wason Medal 1972, ACI Reese Medal 1972, ASCE Stateof-Art Award 1974, ASCE Reese Award 1976-79, ASCE CAN-AM Amnity Award 1979, ASCE Norman Medal 1983, Reinforced Concrete Research Council Boase Medal 1985, ACI Kelly Award 1986, Assn of Prof Engrs, Geol & Geoph of Alberta Centenial Award 1986. *Society Aff:* ACI, ASCE, CSCE, IABSE.

Machlin, Eugene S
Business: 120th St & Amsterdam Ave, New York, NY 10027
Position: Prof *Employer:* Columbia Univ *Education:* ScD/Phys Met/MIT; MS/Phys/Case Inst of Tech; BS/Mech Eng/CCNY *Born:* 12/29/20 Teacher. Researcher of knowledge in diverse fields of materials interest. Nurturer of new companies. Consultant to large corporations. Director of several corporations. Inventor of Udimet 700. *Society Aff:* AIME, ASM, MRS

Maciag, Robert J
Business: 333 Jay St, Brooklyn, NY 11201
Position: Vice President for Academic Operations *Employer:* Poly University. *Education:* MS/Metallurgy/Poly Inst of Brooklyn; BMetE/Metallurgical Engg/Poly Inst of Brooklyn. *Born:* 4/1/40. Native of Lewiston, ME. Served on Faculty of Poly Inst of Brooklyn 1961- 1971. Held various admin posts including Dir of Continuing Education. Assoc Provost for Planning. Assumed current position in 1986. Consultant to Brooklyn Union Gas Co, Con Edison of NY chrmn NY Chap ASM, 1974; Exec comm mbr since 1967; ASM-NY Chap Education Award 1971. *Society Aff:* ASME, ΣΞ, ΑΣΜ, NACUBO.

MacInnis, Cameron
Home: 242 Ford Blvd, Windsor, Ontario N8S 2E5 Canada
Position: Dean of Engrg *Employer:* Univ of Windsor. *Born:* March 1926 West Bay N S Canada. BSc Dalhousie, BE (Hons) Tech Univ of Nova Scotia, PhD Durham Univ England. Res Engr with Ontario Hydro 1948-59, latterly as head of the concrete res group respon for major concrete studies for the Niagara & St Lawrence power projs; Res Assoc Durham Univ 1959-62; with the Univ of Windsor since 1963 lecturing in Concrete Tech, Transportation Engrg & Soil Mech. Specialist cons in field of concrete tech. Author of numerous tech papers. ACI Wason Medal for Materials Res 1975. Fellow of Engrg Inst of Canada, Fellow of Amer Concrete Inst. Fellow of Cdn. Soc. for Civil Engg. *Society Aff:* EIC, APEO, ACI, ASTM.

Macintyre, John R
Home: 231 Queensbury Dr 4, Huntsville, AL 35802
Position: Ret. (Engrg Consultant) *Education:* BS/EE/Univ of MI. *Born:* June 1908. Entire prof career with General Electric Co in instrumentation dev, design, consultation & engrg mgmt. 10 US pats, 12 pub tech papers. Contrib to book 'Applied Elec Measurements'. Life Fellow IEEE (AIEE 1956); Life Fellow ISA (1983). Life Mbr NSPE. Reg Engr in Mass (to 1973) & in Alabama (to 1980.) Held various offices in local & natl prof socs (IEEE, AIEE, Mass Soc P E, Ala Soc P E, ISA, AOA, Huntsville Assn of Tech Soc). Listed in American Men of Sci. Hobbies: golf, travel & ballroom dancing. *Society Aff:* IEEE, ISA, NSPE.

Mack, Donald R
Business: W2F, Fairfield, CT 06431
Position: Prog Manager-Technical Education. *Employer:* General Electric Co. *Education:* PhD/Sys Sci/Poly Inst of Brooklyn; MS/Sys Sci:Poly Inst of Brooklyn; BSEE/Elec Engg/Univ of WA. *Born:* 3/14/25. US Army 1943-46. With Genl Elec Cp since 1948; currently Manager-Technical Education. Adjunct Prof at RPI 1969-79; mbr Phi Beta Kappa, Tau Beta Pi, Sigma Xi, IEEE, ASEE. Mbr IEEE Educational Activities Bd. 1977-1979, IEEE Educational Society Administrative Committee 1981-present P E NY State. Publs incl papers on stress analysis, sys dynamics, engg education, & 1 pat. Active Unitarian (listed in Who's Who in Religion). Leisure activies incl ensemble singing, rowing & hiking. *Society Aff:* IEEE, ASEE

Mack, Edward S
Business: PO Box 4-2265, Anchorage, AK 99509
Position: Pres. *Employer:* Tectonics, Inc. *Education:* BSCE/Civil Engg/Univ of ME. *Born:* 1/29/35. Structural Design Engr with Douglas Aircraft Co from 1957 to 1961. Prin Engr SNAP 10 with Atomics Intl from 1961 to 1963. Proj Engr and Test Engr with Rocketedyne from 1963 to 1970. Supervisor, Civil Design Engg with ITT Arctic Services from 1970 to 1972. Mgr, Construction engg with RCA Alascom from 1972 to 1976. Pres and Chrmn of Bd with Tectonics, Inc. Consulting Engrs and Land Surveyors from 1976 to present. Responsible for admin and operation of engg firm offering civil, structural, cadastral, elec and mech engg services. *Society Aff:* ASCE, NSPE, CSI, ICBO.

Mack, Harry R
Home: 45605 Harmony Ln, Belleville, MI 48111
Position: Associate *Employer:* Smith, Hinchman & Grylls *Education:* BS/Mech Engg/Yale Univ; BS/Ind Admin/Yale Univ. *Born:* 8/21/24. Native of New Haven, CT. Technical Engr for Procter & Gamble Co, Methods Engr for Bassick Co, Mgr of Applications Engg at S I Handling Systems, Sr Consultant at Stevenson, Jordan & Harrison Inc, Sr Mfg Asst for ITT Corp, Operatons Services Mgr at LTV Corp, Hood Engrs since 1972-79, elected VP in 1975. Made substantial contribution in raising co's Engg News Record ranking from 429 to 154 in five yrs. Smith, Hinchman & Grylls Associates, Inc Nov. '79 to present. Speaker & chrmn for American Mgt Assoc and others. Registered Professional Engineer in AL, CA, CT, FL, GA, IL, MI, NY, OH, Penna., WI holds NCCA certificate 4384. *Society Aff:* NNA, MSPE, ASME, IIE, MAP, ESD, APICS

Mack, James A
Business: 1370 Ontario St, Cleveland, OH 44113
Position: Pres & Gen Mgr Chems Div. *Employer:* Sherwin-Williams Co. *Education:* MBA/-/Western New England College; BSChE/CE/MI Tech Univ. *Born:* 7/13/37. Native of MI. Worked as proj engr for Monsanto, Plastics Div. Moved to Maumee Chem Co as process engr. Later, became Mgr Plant Techical Services. As part of the Sherwin-Williams Co held positions of Product tMgr, Bus Dev Mgr & Gen Mgr Marketing. Current, Pres & Gen Mgr - Chems Div, Sherwin-Williams Co. *Society Aff:* AICE, SOCMA, DCAT, DCMA.

Mack, Michael J
Home: 1433 Olympic Drive, Waterloo, IA 50701
Position: Dir. Retired. *Employer:* John Deere Product Engg Ctr. *Education:* MS/Mech Engr/Univ of IA; BS/Aeronautical Engr/Univ of Notre Dame. *Born:* 7/3/25. Native of IL. Naval Officer during WWII period - instr in Aircraft engine theory. With John Deere Co since 1946. Assignments have been in the design and dev of agricultural tractors and components. Current assignment (since 1971 - Dir of John Deere Product Engg Ctr. This ctr is assigned the worldwide engg responsibility of John Deere componets and agricultural tractors. Retired July, 1987. *Society Aff:* SAE, CRC.

Mack, Roger G
Business: 905 Penna Blvd, Feasterville, PA 19047
Position: Operations Manager *Employer:* Abar Corp. *Education:* MS/Met/RPI; BS/ME/Univ of MA. *Born:* 1/4/34. Native Northampton, MA. Past Chrmn Southern New England Section, SAE. Served on various committees for ASM & APMI. Community contributions have included schoolbd, organize & served as first Pres-Enfield Scholarship Fdn, Rotary, Church Collector, Girls Ponytail Softball. Reg PE, MA. Involved 15 yrs at Am Bosch in selection & processing of US & European mtls for diesel fuel pumps (Mtls Lab Mgr). Developed gas nitriding process for maraging steel. Maj interest, heat treating processes from standpoint of better energy utilization. Employed by ABAR to promote this interest in field of ion nitriding & carborizing. Amateur cabinet maker. *Society Aff:* ASM, SME, ΣΞ, AWS.

Mackay, William B F
Business: Nicol Hall, Queen's Univ, Kingston Ontario, K7L 3N6 Canada
Position: Prof Emeritus of Metallurgical Engrg *Employer:* Queen's Univ *Education:* PhD/Phy Met/Univ of MN; MS/Phy Met/Univ of MN; BMetE/Met Engr/Univ of MN; BSc/Elec Engg/Univ of Manitoba. *Born:* 5/21/14. Resident Kingston Ontario. Served in RCAF 1940-45 held rank of Wing Cdr; Univ of Minnesota 1946-56 Asst Prof Metallurgy; Atlas Steels 1956-66 held various positions Ch Metallurgical Engr, Mgr of R/D, etc, involved with dev & prod of tool, stainless & specialty steels, titanium & zirconium alloys; Queen's Univ Prof & Head Metallurgical Engrg Dept 1966-76, Acting Dean Faculty of Applied Sci 1976-77. Assoc Dean Faculty of Applied Sci 1978-81. Prof Metallurgical Engrg 1981-84. Part time Prof of Mechanical Engrg, Royal Military College of Canada 1980-. Cons in field of failure analysis, res interests in alloy dev & metal processing. Prof Emeritus Met Engrg 1984 continuing as Special Lecturer at Queens Univ, Dept. Metallurgical Engrg. *Society Aff:* IM, CSME, EIC, ASM, APEO.

Mackay, William K
Business: 20 Royal Terrace, Glasgow G3 7NY, Scotland UK
Position: Partner. *Employer:* Jamieson Mackay & Ptnrs, Cons Engrs. *Education:* BSc/Civil Engg/Univ of Glasgow *Born:* 6/6/30. in Peru-British. FICE (UK), FIHE (UK), FITE (USA), FASCE (USA). Founder Partner of Jamieson Mackay & Partners Cons Engrs spec in transportation studies & engg design projs throughout the UK & in the Far East, Middle East & South Amer; jointly respon for res into transportation & town form, particularly in relation to new & expanding cities. UK 'expert' delegate to OECD & other intl agencies. Awarded UK Inst of Municipal Engrs Medal for paper on Roads in Urban Area & ASCE Wellington Prize for paper on Transportation & Urban Renewal. *Society Aff:* FICE, FIHT, FITE, FASCE, MCIT, MConsE.

MacKenzie, Horace J
4629-88th St, Lubbock, TX 79424
Position: Prof Emeritus of Indus Engrg. *Employer:* Texas Tech Univ. (Retired) *Education:* MS/IE/OK State Univ; BS/IE/TX Tech Univ. *Born:* Jan 1920 San Francisco Calif. MS Oklahoma St Univ, BS Texas Tech Univ. Reg P E Texas. Sr Mbr IIE; Mbr Tau Beta Pi. Served in US Navy 1944-46; Prof of Indus Engrg at Texas Tech Univ since 1949. Prof. Emeritus 1985. Advisory Bd Alpha Pi Mu 1978-81; Natl Pres of Alpha Pi Mu 1976-78, Indus Engrg Honor Soc; Natl V Pres Alpha Pi Mu 1974-76; Regional Dir Alpha Pi Mu 1972-74. *Society Aff:* IIE, ΤΒΠ, ΑΠΜ.

MacKenzie, John D
Business: 6532 Boelter Hall UCLA, Los Angeles, CA 90024
Position: Prof of Engrg. *Employer:* Univ of Calif Los Angeles. *Education:* PhD/Physical Chem/Univ of London; BSc/Chemistry/Univ of London. *Born:* Feb 18, 1926. Naturalized 1963. Lecturer Princeton Univ 1954-56; Res Fellow Cambridge Univ England 1956-57; Res Assoc General Electric 1957-63; Prof Materials Sci RPI 1963-69; Prof of Engrg UCLA 1969- . Author of over 200 papers, Editor of 6 books, hold 13 US pats; Editor-in-Ch of J of Non-Crystalline Solids since 1963. Meyer Award Amer Ceram Soc 1964; Toledo Award Amer Ceram Soc 1974; Lebeau Medal French High Temp Soc 1969; Natl Acad of Engrg 1976. *Society Aff:* A. Ceram. Soc., APS, ACS, ECS, NAE.

Mackey, Thomas S
Home: 1210 Sunset Ln, Texas City, TX 77590
Position: Pres. *Employer:* Key Metals & Minerals Eng Corp. *Education:* PhD/Engg/Rice Univ; JD/Law/S TX Sch of Law; MS/Engg/Columbia Univ; BS/Engg/Manhattan College. *Born:* 7/14/30. Native of NJ. Pres of Consulting Engg Firm (Key Metals & Minerals Engg Corp) with worldwide clients, including firms in Indonesia, Thailand, Singapore, Bolivia & Japan. Former plant mgr of the world's largest tin smelter in TX City for 13 yrs. Holder of several patents, numerous publications. Previous Chrmn 1980 World Symposium on Lead-Zinc-Tin, ANO 1990 Chrmn sponsored by Met Soc of which a Dir. Law partner in firm Thomas S Mackey P.C. Designer, builder & operator of plants producing tin, lead, tungsten, molybdenum, copper, & ferro alloys, including ferro tungsten, molybdenum, vanadium, columbium. Who's Who in Intl Engrg. Fellow Amer Inst of Chemists, Inst of Mining and Metallurgy. Reg PE, Lisc Geologist. Pres Airtrust Intl Corp. One of ten outstanding men in Amer, US Jaycees; Paul Harris Fellow Rotary, Intl, Dir US Natl Bank, TX, NM Power Co. President Neomet Corp, Dir Siltec and reactive metals and alloys. *Society Aff:* AIME-TMS, FAIC, NSPE, ACS.

Mackie, Charles H
Business: PO Box 368, U S Hwy 9, Barnegat, NJ 08005
Position: President. *Employer:* Charles H Mackie Assoc, Inc. *Education:* Diploma/Civil Engg/ICS Scranton, PA. *Born:* Aug 8, 1928. Served U S Navy Construction Battalion 1946-47. Field & Office Engr with J George Hollerith, Surveyor, N Y City 1948-52. Engr with Edwards & Kelcey Inc Engrs & Cons, Newark New Jersey 1952-71, Assoc of firm 1964. V P 1969 & Mgr of Philadelphia Office 1968-71. Org firm Charles H Mackie Assocs Cons Engrs 1971 with engrg offices located in Barnegat, N J & Norwalk, Conn. Reg P E & Land Surveyor N J , N Y, Conn, Pa. *Society Aff:* ASCE, NSPE, AWWA, APWA, WPCF.

Mackin, Italo V
Business: 126 Craighead St, Pittsburgh, PA 15211
Position: President. *Employer:* Mackin Engrg Co. *Education:* BS/Civil Engg/Carnegie Inst of Tech. *Born:* 3/6/24. U S Army Corps Engrs (Pacific Theatre) 1943-45. Design & Proj Engr with Richardson, Gordon & Assocs Cons Engrs 1950-60. Private practice in Cons Engrg 1960. Pres of Mackin Engrg Co since 1963. Past Pres-Elect-Cons Engrs Council of PA 1980-81. Bd/Dir CEC/Pa 1971-76, 1978-1982. Enjoy music & sports; *Society Aff:* ACEC, ABCD, ACI, ASCE.

Macklin, Harley R
Business: 201 North Market, Wichita, KS 67201
Position: Mgr of Production *Employer:* KS Gas & Electric Co *Education:* BS/EE/KS State Univ; AA/Engrg/Hutchinson Jr Coll *Born:* 10/9/38 Native of Hutchinson, KS, raised on a farm until age 13. Went to work as an engr for KG&E after grad from KSU. Assumed present position of Mgr of Production in 1974. Pres of WPES, ISA, KG&E Employees Assoc. Leader of 4-H Club for 16 yrs. Enjoy automobiles, motorcycles, hunting & camping. *Society Aff:* NSPE, KES, WPES, ISA

Macklin, Harold L
Business: 275 Duncan Mill Rd, Don Mills, Ontario M3B 2Y1 Canada
Position: Director *Employer:* Marshall Macklin Monaghan Limited. *Education:* BASc/Engg/Univ of Toronto. *Born:* 5/4/20. Native of Cobourg, Ontario, Canada. Served with Royal Canadian Engrs 1942- 45. With Dept of Civ Engg, Univ of Toronto 1945 to 1984 (retired June 30, 1984), latterly part-time. Founding partner Marshall Macklin & Monaghan, 1951, consulting engrs, and land surveyors. Founding Dir Marshall Macklin Monaghan Limited 1957, Pres 1960 to 1974, Chrmn of the Bd 1975 to 1985, Now retired director. Pres Canadian Soc for Civ Engg 1973-74, and Pres Engg Inst of Canada 1984-85. Recipient of the Queen's Silver Jubilee

Macklin, Harold L (Continued)
Medal in 1977 & the Julian C Smith Medal in 1978, both by the Engg Inst of Canada. *Society Aff:* EIC, CSCE, ASEE, ASP, CIS.

Mackrell, James J, Jr
Home: 1675 Calle Lilas, Rio Piedras, PR 00927
Position: v Pres & Genl Mgr Bristol Alpha Corp. *Employer:* Bristol-Myers Co Industrial Div. *Born:* 1942. BSChE Univ of Delaware 1964. Process Engr with Monsanto Co St Louis Missouri 1964-65; US Army 1965-67, 2nd Lt to 1st Lt, Vietnam 1965-66, Edgewood Arsenal 1966-67, Awards Bronze Star, Army Commendation Medal, Vietnam Serv Award, Armed Forces Expeditionary Medal; Process to Proj Engr Bristol-Myers Co Internatl Div 1967-72 working on capital projs in Italy & Puerto Rico; Sr Proj Engr to Proj Engrg Mgr 1972-74 with J T Baker Chemical Co; V Pres & Genl Mgr Bristol Alpha Corp (Bristol-Myers Co Indus Div) 1974- , Barceloneta Puerto Rico.

MacLean, W Dan
Business: PO Box 3100, Kansas City, KS 66103
Position: V Pres Refining. *Employer:* Hudson Oil Co. *Born:* May 1933 Saskatchewan Canada. MASc Birmingham Univ U K, BASc Univ of Toronto Canada chem engrg. 1961-65 Ch Process Engr Celanese Canada, authored sev tech papers; 1965-70 Mgr supply planning & co-ordination B P Canada with special respon for petrochem dev; 1970-74 Tech Dir Golden Eagle Canada during design const & start-up of a 100,000 B/D grassroots refinery; joined Hudson Oil, a leading natl independent gasoline marketer 1974 with respon for dev a refining oper; led the proj team organizing Hudson Refining Co & negotiating the acquisition of a 19,000 B/D Oklahoma refinery.

Maclean, Walter M
Home: 24 Harbor Way, Sea Cliff, NY 11579
Position: Prof, since October 1987 *Employer:* U.S. Merchant Marine Academy *Education:* D Engrg/Naval Arch/Univ of CA, Berkeley; M Engrg/Naval Arch/Univ of CA, Berkeley; BS/Mech Engrg/Univ of CA, Berkeley. *Born:* 3/15/24. Married D. Hansen, 2 children. Employment as: Marine Engr 7 yrs in Intercoastal Transpacific and Round the World Servs; Draftsman, Mech Engr and Naval Architect M Guralnick 1955-59; Jr Engr in Ship Struct res at U.C. Berkeley 1959-1965; Prof of Engrg Webb Inst 1965-72; Mgr Req Dev Lab NMRC 1975-80, Actg Dir NMRC 1981-87; Prof USMMA 1987- . Prof. Activities: Member SNAME Panel HS-2 Impact Loading and Response 1961-75, Chmn 1976-86; Member SNAME Hull Structure Ctte 1976-86, Chmn 1986- ; Member SNAME Ship Technical Operations Ctte 1980- ; Mbr ISSC 1970 and 1973 Ctte 8 Slamming and Impact, 1988 Ctte II.2 Transient Dynamic Effects; SNAME Scholarship Ctte 1980-87; Chmn SNAME Education Ctte 1978-79; NY Metropolitan Sect SNAME Chmn 1979-80; Long Island Section ASME Chmn 1970-71; Mbr Ship Structure SubCtte, Tech Prog Ctte Chmn SSC/SNAME Extreme Loads Response Symposium 1981, Co-Chmn Maritime Innovation '84 Symposium; U.S. Rep to ICMES Stdg Ctte 1984- , Mbr Tech Ctte 7, 1983- . Prof Qualifications: Reg Prof Mech Engr Calif, Ch Engr Stm Vessels Unlimited, US C G Lic. *Society Aff:* SNAME, ASNE, ASME, SMPE NY ACAD OF SCI, ΠΤΣ, ΤΒΠ, ΣΞ.

MacLennan, Robert G
Business: 5619 Fannin St, Houston, TX 77004
Position: Senior VP *Employer:* Bovay Engrs, Inc *Education:* MS/Math/RPI; B/CE/NYU *Born:* 10/23/31 Native of NY City. In 1953, began twenty-three years of service as an officer in the U.S. Army Corps of Engrs in positions of engrg design and management, construction supervision, technical analysis, planning and education. Key assignments included: Assistant Prof at the U.S. Military Academy, Chief of Construction at field army level, Battalion Commander, Systems Analyst in the Office of the Secretary of Defense, and Commander of the Engr District in Albuquerque. Joined Bovay Engrs, Inc, a full service, diversified consulting firm with officer nationwide, in 1976 as Executive Assistant in Albuquerque. Subsequently named VP and Mgr of the home office in Houston, and in 1978 named Corporate Mgr of Operations, responsible for the six offices of the firm. In 1979 named Senior VP, in 1980 named a Dir of Bovay Intl Ltd., and in 1981 named Corporate Mgr of Marketing and Strategic Planning. *Society Aff:* ΤΒΠ, NSPE, ASCE, SAME, PSMA

MacMillan, Douglas C
Home: Colony Dr PO Box 834, East Orleans, MA 02643
Position: Retired. *Education:* BS/NA & ME/MIT. *Born:* July 1912. Native of Dedham Mass. Federal Shipbuilding & Dry Dock Co Engr 1934-41; George G Sharp Cons Naval Arch, Ch Mar Engrg 1941, Ch Engr 1946, Tech Mgr 1949; George G Sharp Pres 1951-68, Chr 1969-70; Cons Nav Arch & Mar Engr private practice 1970-72; Quincy Shipbuilding Div Genl Dynamics, Asst to Genl Mgr & Cons 1972-77. Mbr Natl Acad of Engrg. Fellow & Honorary VP SNAME, VP 1957-72. Mbr ASNE, Mbr Soc of Sigma Xi. Dir Atomic Indus Forum 1965-67. Prof Engr N Y & N J. David W Taylor Gold Medal SNAME 1969, Elmer A Sperry Medal ASME 1969. *Society Aff:* SNAME, NAE, ASNE.

MacMullin, Robert B
Home: 5137 Woodland Drive, Lewiston, NY 14092 *Education:* SB/Chem Engg/MA Inst Tech. *Born:* Sept 17, 1898 Philadelphia Penna. Bowdoin 1914-16, 1st Gas Reg't CWS, AEF WW I; 1920-45 Mathieson Alkali Works (now Olin) Mgr R/D; 1945-46 Sci Cons TIIC, FIAT, U S Dept Comm Germany; 1946-71 Sr Partner R B MacMullin Assocs Cons Engrs - 1971-85, Assoc Emeritus. Field: R/D, design, erection, oper of chem & electrochem plants, especially chlor-alkali, soda ash, electrowinning metals, organics. Scope world-wide. P E NY 21297. Fellow AIChE 1972; Amer Inst Chem 1971, Assn Cons Chem & Ch Engrs Hon 1972. Awards: Prof Achieve AIChE 1955; 28th Schoellkopf Medalist ACS 1958; 66th Perkin Medalist Soc Chem Ind 1972; Electrochem Engr & Tech Medalist Electrochem Soc 1976. Cert of Commendation, 1979, Japan Soda Industry; Ch Engr Practice Award, AIChE 1982; Founders Award, AIChE 1983. Associations Avocations: geology, trail sys Chemists Club, NY, Hon mbr 1981. Autobiog. Odyssey of a Chemical Engineer, pub. 1983 *Society Aff:* ACS, AIChE, AIC, AIME (TMS), ECS, AAAS.

Macnee, Alan B
Home: 1911 Austin Ave, Ann Arbor, MI 48104
Position: Prof of Elec Engrg. *Employer:* University of Michigan. *Education:* ScD/EE/MIT; SM/EE/MIT; SB/EE/MIT. *Born:* Sep 1920. Staff mbr MIT Radiation Lab 1943-46; Res Asst & Assoc MIT Res Lab of Electronics 1946-49; Res Assoc Chalmers Inst Tech Gothenburg, Sweden 1949-50 & Visiting Prof 1961-62. With Univ of Mich since 1951, Prof since 1959. George Eastman Fellowship in EE 1946; IRE B J Thompson Mem Prize 1951; ASEE Western Elec Teaching Award 1968-69; IEEE Fellow. AAAS Fellow (1980). Tech interests: elec circuit design, computeraided design, device modelling & circuit synthesis. Enjoy gardening, wildlife & fishing. Senior Res Assoc Goddard Space Flight Center, NASA, 1971-72; Mbr Technical Staff, Sandia Labs, 1980-81. *Society Aff:* IEEE, AAUP, AAAS, TBP, HKN, ΣΞ, ΦΚΦ.

MacNeill, Charles E
Home: 19105 Canadian Ct, Gaithersburg, MD 20879
Position: Materials & Process Engr. *Employer:* Food & Drug Administration, HHS. *Education:* BS/Chem/Northeastern Univ; Metallurgy/-/Northeastern Univ. *Born:* 4/30/26. Army Air Forces 1943-45. Taught matls engrg courses at Univ of Akron & Barberton Tech Sch. Matls & process engr spec in advance composite matl res. Dir & Pres of Inst for Composite Matls. Natl VP Soc for Advancement of Matl & Process Engrg. Amer Soc for Metals Matls Sci Div-Biomatls. ASTM Ctte F-4 on Surgical Implants. Matls Engrg Mgr Fairchild Industires. Respon for medical device matls engrg with Center for Device and Radiological Health, FDA, HHS. Registered prof eng CA. Amer Soc for Testing and matls-Man of Sci and Engg with sensitivity Award, Fairchild Industries Publication Award. Food and Drug Admin Comm Ser-

MacNeill, Charles E (Continued)
vice Award. Active Search & Rescue pilot in Civil Air Patrol. *Society Aff:* ASM, ASTM

MacNeill, John S
Business: 74 North West Street, Homer, NY 13077
Position: President. *Employer:* John S MacNeill Jr P C, Cons Engrs. *Education:* BCE/Structural Eng/Cornell Univ. *Born:* 1/24/27. Served in USN 1944-46. Field Engr School Const 1950-55; Design Engr Alleghany Homes 1955-57. Lic P E NY, ME, NH, VT, PA, MT, WV, NJ, MA. Lic L S NY, ME, NH, PA. Started cons engrg & surveying firm in 1957; Pres John S MacNeill Jr P C Inc 1976. Pres NYSSPE/CNY 1973/74, Pres CEC/CNY 1975/76, Pres CEC/NYS 1982- 83. Pres NYSSPE 1986/87, Pres Cortland Co Lic Land Surveyors Assoc 1978-79; Mbr Adv Ctte Tomkins-Cortland Community College Const Tech Prog. Treas. Coll Devel Fund Suny, Cortland. Fellow ASCE; Mbr Soc Amer Mil Engrs, ACSM, WPCF; Cornell Soc of Engrg; Pres. Cortland Cty Chamber of Commerce 1983; Pres Baden-Powell Council BAS 1980- 81; Mbr Rotary, Drum Major Fraser Highlanders Pipe Band of Roine, N. Y. Enjoy wilderness camping & fishing & hiking. *Society Aff:* NSPE, ACEC, ASCE, WPCF.

MacNichol, Edward F, Jr
Business: Dept. Physiology, Boston Univ School of Medicine, 80 E. Concord St, Boston, MA 02118
Position: Prof of Physiology. *Employer:* Boston Univ Sch of Medicine. *Education:* PhD/Biophysics/Johns Hopkins Univ; AB/Physics/Princeton Univ *Born:* 10/24/18. in Toledo, OH. MIT Radiation Lab 1941-46. Author-Vol Editor Rad Lab Series. Grad student Johnson Fnd Univ of PA 1947-49. Johns Hopkins: PhD Biophysics 1952, Prof Biophysics 1963-68; Natl Inst of health: Dir natl Inst of Neurological Diseases & Stroke 1968-72, Acting Dir Natl Eye Inst 1968-69; Marine Biological lab; Assist Dir Res Servs 1973-76, Dir Lab Sensory Physiology 1973-1985; Prof Physiology Boston Univ Sch of Medicine 1973- ; Physiology of retina of eye & invertebrate photoreceptor organs. Microspectrophotomoetry of single visual sense cells; design of instrumentation for neurophysiology & vision res. Editor IEEE Transactions BME 1963-67. Army-Navy Cert of Appreciation 1948. Fellow IEEE 1964. morlock Award IEEE 1965. Co-Editor *Sensory Processes* 1978-81, Exec Council, NRS Committee on Vision 1975-78. IEEE Centennial medal 9-16-84. *Society Aff:* ARVO, IEEE, AAAS, APS, AM. Physiol. Soc., Biophys Soc., Neurosci. Soc.

Macovski, Albert
Business: Durand Room 109, Stanford, CA 94305
Position: Professor. *Employer:* Stanford Univ. *Education:* PhD/EE/Stanford; MEE/EE/Polytech Inst of Brooklyn; BEE/EE/CCNY *Born:* May 1929. Engr at RCA Labs 1950-57; Faculty of Bklyn Poly 1957-60; Staff Scientist at SRI 1960-71; NIH Special Fellow at UCSF 1971-72; Stanford Faculty since 1972; presently Prof of Elec Engrg & Radiology. V K Zworykin Award 1973. IEEE Fellow 1975. OSA Fellow 1976. Over 100 issued pats. Current res in diagnostic imaging sys incl x-rays & ultrasonics. Enjoy music & outdoors. *Society Aff:* IEEE, AAPM, OSA, SMRM

MacPherson, Herbert G
Home: 102 Orchard Circle, Oak Ridge, TN 37830
Position: Consultant. *Employer:* Inst for Energy Analy; Oak Ridge. *Education:* PhD/Physics/Univ of CA; AB/Physics/Univ of CA. *Born:* Nov 1911. Native of Calif. With Carbon Prods Div of Union Carbide 1937- 56, Dir of Res 1950-56; with Oak Ridge Natl Lab 1956-70; Dir of Molten Salt Reactor Prog 1956-60; Dir of Reactor Div 1961-63, Asst & Assoc Lab Dir 1960-64, Deputy Dir 1964-70; Prof of Nuclear Engrg Univ of Tennessee 1970-76; Cons to Atomic Energy Comm on Emergency Core Cooling 1972-74; Acting Dir Inst for Energy Analysis 1974-75. Mbr Breeder Reactor Resource Group, NRC Ctte on Nuclear & Alternative Energy Sys 1976. ANS Fellow 1965, Dir 1969-70. Mbr, Natl Acad of Engg 1978. *Society Aff:* APS, ANS.

Macpherson, Hugh R
Home: 5669 Laurelwood, Salt Lake City, UT 84121
Position: Mgr Special Vehicles. *Employer:* Hercules Inc. *Born:* March 1926. BSChE Univ of Colorado 1950. With duPont Textile Fibers Dept in prod supr, sales dev & indus fibers res; started with Hercules Inc 1959 in solid rocket motor dev; worked as Materials Dev Supr & as a Prog Mgr with small rocket motors & gas generators on such progs as the NASA Ranger, USAF 823 & Athena, Army AMRAD & Navy Polaris progs; spent time as a Mgr in New Enterprises Dept & Advanced Composite Dev Dept. Presented papers on rocket motor progs & magnetometer metal detectors. Mbr of ACS; P Chmn & Natl Dir SAMPE.

Macpherson, Monroe D
Home: 141 Jefferson Ave, Westfield, NJ 07090
Position: Project Mgr *Employer:* J.J. McMullen Assoc *Education:* MBA/Corp Fin Mgmt/NYU; BSE/Nav Arch/Mar Engrg/Univ of MI *Born:* 9/5/27 *Society Aff:* SNAME

MacRae, Alfred Urquhart
Business: Crawford Cor Rd, Holmdel, NJ 07733
Position: Lab Dir, Satellite Transmission *Employer:* AT&T-Bell Labs *Education:* PhD/Physics/Syracuse Univ *Born:* 4/14/32. 1960-67 Basic res on fundamental properties of surfaces with emphasis on low energy electron diffraction. 1967-1973 Basic & applied res on ion implantation & its application to devices and integrated circuits. 1973-1981 Integrated circuit process dev, CMOS and high voltage; microprocessor dev. Honors - Fellow; Am Physical Soc. Fellow; IEEE. Scientific mbr; Bohmische Physical Soc; Pres elect, Electron Devices Soc IEEE. *Society Aff:* IEEE, Am Phy Soc

Madara, James R
Business: Gannett Fleming of California, Inc, 624 S. Grand Ave. Suite 1000, Los Angeles, CA 90017
Position: VP. *Employer:* Gannett Fleming Corddry and Carpenter, Inc. Inc. *Education:* BS/Civil Engg/Lafayette. *Born:* 12/8/28. After graduation joined Gannett Fleming Corddry and Carpenter Inc. Served as Prof Engr, Proj Mgr & Sr Proj Mgr in Hwy & Bridge Design. Aptd Gen Mgr 1974 coordinating Bus Dev Section. Made VP, in charge of Bus Dev, in Sept 1976. Prof Reg six states. Past President of Planning & Design Div., American Road and Transportation Builders Association mbr ASHE, ASCE, Hwy Advisory Council & Hwy Res Bd. PPres of Engrs Society of PA. Enjoys golf & bowling. *Society Aff:* ARTBA, ASCE, NSPE, ESP, SAME, ASHE.

Madden, Dann M
Business: 200 S.W. Market St, Portland, OR 97201
Position: VP/Regional Mgr *Employer:* CH2M HILL *Education:* BS/CE/OR State Univ *Born:* 9/19/32 Native of and educated in OR. US Army Service 1951-54, infantry, Korea. Survey and design experience with City of Corvallis, OR. Construction management of major pulp/paper mill expansion for ITT Rayonier, Hoquiam, WA. Port Engr at Grays Harbor, WA, 1963-67. Sr design engr in Ports/Harbors and pulp and paper projects in Portland, OR. With CH2M HILL since 1968. Several positions from project mgr to Regional Mgr in 1980. OR Section Pres of ASCE 1979-1980. *Society Aff:* ASCE, WPCF, AWWA, ACEC, APWA

Madden, Roger D
Business: 2025 M St, NW., Rm 5322, Wash, DC 20554
Position: Deputy Chief, Special Services Div *Employer:* Federal Communications Commission. *Education:* BEE/Univ of Louisville *Born:* 3/9/35. Pres of IEEE Vehicular Tech Soc 1978-1980. Sr Mbr, IEEE. Fellow, Radio Club of Am. Deputy Chief Special Services Div. Private Radio Bureau, FCC. Formerly Deputy Chief, Policy Development Div, staff engr in Office of Plans & Policy, FCC, Chief, Chicago Regional Spectrum Mgt Office, FCC. Instrumental in dev & implementation of automated spectrum mgt. Experience prior to FCC includes employment with Magnavox, GE, & Motorola. *Society Aff:* IEEE.

Maddex, Phillip J
4433 Colbath Apt 2, Sherman Oaks, CA
Position: Vice President. *Employer:* Met-L-Chek Co. *Education:* BChE/Chem Engr/OH State Univ. *Born:* Oct 1917 Mechanicsburg, Ohio. Factory Engr General Elec Co Owensboro, Ky 1941-47. Res Engr process for titanium metal extraction, Battelle Mem Inst Columbus, Ohio 1947-49. Natl Lead Co Sayerville, N J 1950-52. Plant Mgr Titanium Metals Corp Henderson, Nev 1952-59. Ch Engr Mining & Processing borax & potash, U S Borax & Chemical Corp Los Angeles, Calif 1959-65. Cons Engrg & Economics for Ocean, River & Rail Transport of raw materials & products 1965-73. V P & Gen Mgr Met-L-Chek Co Santa Monica, Calif manufac & sale of Inspection Penetrants 1973- . *Society Aff:* AIME, ASNT, ASM.

Maddin, Robert
Business: 3231 Walnut St, Phila, PA 19104
Position: Univ Prof. *Employer:* Univ of PA. *Education:* PhD/Met/Yale Univ; MA/Honor Causas/Univ of PA; BS/Met Engg/Purdue Univ. *Born:* 10/20/18. Prof Maddin spent six yrs at the Johns Hopkins Univ as Asst Prof and Assoc Prof, and came to the Univ of PA in 1955 as Prof of Metallurgy. He was Dir of the Sch of Metallurgy from 1955 to 1972, and was appointed Univ Prof in 1973. During this period he has also been Visiting Prof at the Univ of Birmingham and Oxford Univ. His res has been concerned with the effect of defects on the mech properties of solids, diffusion in solids and, in more recent yrs, the mech properties of amorphous solids as well as their transformation to the crystalline state. Since 1972, he has been concerned with the early history of Metallurgy, the manner in which metals were first extracted from their ores, the beginning of the Iron Age and the early trade in metals. *Society Aff:* AIME, ASM, AAAS.

Maddock, Thomas S
Business: 18552 Mac Arthur Blvd, Suite 200, PO Box 19608, Irvine, CA 92713
Position: Pres, Chief Executive Officer. *Employer:* Boyle Engrg Corp. *Education:* MBA/Bus Admin/Stanford Univ; MS/Civil Engg/MIT; BS/Civil Engg/VA Polytechin Inst. *Born:* Mar 23, 1928 Norfolk, Va. RADM (Rear Admiral) Civil Engrs Corps, U S Naval Reserve. Pres & Ch Exec Officer of Boyle Engrg Co ranked 76th out of largest 500 engrg-architect firms by Engrg News-Record Magazine in 1981. Prof reputation earned primarily in field of water resources dev for municipal, industrial & agri purposes in western U S & overseas. Personally respon for preparation of tech & economic feasibility reports in support of proj financing & dir of design of $200 million of major water supply facilities for delivery of 1,000,000 acre-feet of water annually in central & southern Calif. Reg P E Calif, Ariz, Nev, Fla, Va & Texas, WA, NC, WV, ID, CO, DC, NE, UT, OR. Fellow ASCE, NSPE/CSPK, Mbr Amer Water Works Assn, SAME & Cons Engrs Assn of Calif, Board of Directors Calif. Chamber of Commerce, Fellow, LA Institute for Advancement of Engineering. *Society Aff:* ASCE, NSPE, AWWA, CRAC, SAME, TCID.

Maddox, Judith H
Home: P O Box 124, Aiken, SC 29801
Position: Area Supervisor. *Employer:* duPont de Nemours-Savannah River Lab. *Born:* Oct 18, 1939. BS Chem Engrg Univ Fla 1962. Life Mbr & Sr Mbr Soc of Women Engrs (SWE). Have held following offices: 1976-77 V P, 1973-75 Treas, 1972-73 Exec Ctte & 1971-72 Council of Sect Reps.

Maddox, R N
Business: School of Chemical Engineering, 423 Engineering North, Oklahoma State University, Stillwater, OK 74078
Position: Dir, Physical Properties Lab. *Employer:* Oklahoma State University. *Education:* PhD/Chem Eng/OK State Univ; MS/Chem Eng/Univ of OK; BS/Chem Eng/Univ of AR *Born:* Sep 1925. BS Univ of Arkansas; MChE Univ of Okla; PhD Okla State Univ. On faculty at Okla State since 1954, Dept Head 1958-77. Tech. interests include natual gas processing & conditioning including sweetening, physical properties of liquids & process design particularly for computer applications. Joseph Stewart Award ACS 1971, Outstanding Engr in Okla 1972. Hobbies are golf & fishing. Andre' Wilkins award (outstanding Chemical Engineer in Oklahoma) from Tulsa Chapter of American Institute of Chemical Engineers, 1981. Hanlon Award (1985) and Citation for Service (1987) Gas Processors Association. *Society Aff:* ACS, AIChE, SPE

Madeheim, Huxley
Business: 321 E 18th St, New York, NY 10003
Position: Chief Engr. *Employer:* Huxley Madeheim Engrs. *Education:* ME/Mech Engg/Stevens Inst of Tech; MBA/Mgt/NYU. *Born:* 11/11/05. In private practice 1931. Retired as full Prof Mgt Baruch College CUNY after 40 yrs of teaching including course in Safety. Licensed PE NY 1932, CT 1939, PA 1943 CA (safety) 1976. *Society Aff:* ASME, ASSE, NSPE, SAE, SAM.

Madigan, James E
Business: 3913 Algoma Rd, Green Bay, WI 54301
Position: Ch of Bd. *Employer:* FEECO Internatl, Inc. *Education:* BSChE/ChE/Univ of Madison. *Born:* May 1923. Native of Green Bay, Wisc. Served with 106th cavalry WWII. Mbr NSPE, Amer Chem Soc, Inst Chem Engrs, Rotary Internatl, Chrmn/Bd Univ Bank, Knight Holy Seputchre & Past Pres of Serra Internatl. Bd Mbr Ferti-plant Engrg Co Pvt Limited, Bombay, India. *Society Aff:* ACS, AIChE, NSPE.

Madix, Robert J
Business: Dept of Chem Engg, Stanford, CA 94305
Position: Prof. *Employer:* Stanford Univ. *Education:* PhD/Chem Engrg/Univ of CA-Berkeley; BS/Chem Engrg/Univ of IL *Born:* 06/22/38 In 1961 Prof Madix entered grad sch in ChE at the Univ of CA, Berkeley. He was a teaching asst & a recipient of a Natl Sci Fdn Pre-doctoral Fellowship in 1962. He was awarded the PhD degree in 1964 after which he spent one yr on a Natl Sci Fdn Postdoctoral Fellowship in Gottingen, Germany, with Prof Carl Wagner at the Max Planck Int for Physical Chemistry. In 1965 he joined the Dept of Chem Engrg at Stanford Univ and began to study the role of surface structure & composition in determining the kinetics and mechanism of reactions on surfaces. He was dept chmn 1983-1987. His res is published in such journals as: Surface Sci, Journal of Catalysis, Journal of Inorganic Chemsitry, Transactions of the Farady Soc, Journal of the Phys & Chemistry of Solids, Chem Physics Letters, Journal of High Temperature Chemistry & the Intl Journal of Chemical Kinetics. He has received a Sr US Scientist Award from the Humboldt Fdn of West Germany & the Irving Langmun Distinguished Lecturship Award of the Surface & Colloid Sci Div of the American Chem Soc. Chrmn Dept Chem Engrg, Stanford Univ, 1983-present; Prof Chem, Stanford Univ, 1981-present. He serves on the editorial boards of J Phys Chem, Langmuir, J Molec Cat & is assoc editor of Catalysis Reviews. *Society Aff:* ACS, AVS.

Madsen, Niels
Home: Oxenbjergvej20, DK5700 Svendborg, Denmark
Position: Prof Emeritus. *Employer:* Univ Rhode Island. *Education:* PhD/ChE/Columbia, Univ; MS/ChE/Stevens Inst, NJ; BChE/ChE/Cooper Union, NY. *Born:* Jun 8, 1905. Chem Engr Foster-Wheeler 1944-47; Teacher Indus Chem Essex County Tech H S Newark NJ 1948-54; Res Asst Columbia U 1954-57; Assoc Prof & Prof Univ of Rhode Island & since 1975 Prof Emerit; Visiting Prof at Tech Univ of Eindhoven, The Netherlands 1962-63 & 1969-70. Paints, plays chess. Listens to classical music & enjoys ballet. *Society Aff:* AIChE (Fellow)

Maeder, Paul F
Business: Providence, RI 02912
Position: Prof. *Employer:* Brown Univ. *Education:* DiplIncEth/ME/Swiss Fed Inst of Tech; PhD/Engg/Brown Univ. *Born:* 6/29/23. Native of Basel Switzerland. Teaching and res at Brown Univ since 1947. Asst Prof 1948. Assoc Prof 1951. Prof Engg 1954. 1958-64 President College Hill Industries. 1962-68 Chrmn Div of Engg Brown Univ. 1968-71 Assoc Provost. 1971-77 VP Finance and Operations Brown Univ. 50 papers and publications. 12 patents. Specialty: fluid mechanics, transonic

Maeder, Paul F (Continued)
flow, wind tunnels, MHO geothermal energy. *Society Aff:* AAAS, AIAA, ASME, NSPE.

Maga, John A
Home: 7023 13th St, Sacramento, CA 95831
Position: Air Pollution & Environ Cons (Self-employed) *Education:* MS/San Engr/Univ of CA; BS/San Engr/Univ of CA; -/Meteorology/US Naval Post-Grad Sch. *Born:* Jan 1916. Sanitary Engr prior to WWII. Served as engr & meteorologist with U S Navy 1942-45. Sanitary Engr with Calif Dept of Health on water supply & water pollution control until 1955. Ch Calif Bureau of Air Sanitation 1955-68. Exec Off Calif Air Resources Bd 1968-73. Dept Secy for Resources, State of Calif 1973-75. Cons engr on air pollution & environ since 1975. Reg engr in Calif. Recipient of S Smith Griswold award (APCA). *Society Aff:* APCA, WPCF.

Magarian, Robert J
Home: 3 Monroe Pl, Rockaway, NJ 07866
Position: Pres & Gen Mgr. *Employer:* AM Plastics, Inc. *Education:* BS/ChE/Worcester Polytechnic Inst. *Born:* Aug 23, 1947 Worcester, Mass. Married-one daughter. Supr E I duPont 1969-71 producing polymer intermediate for Dacron manufacture. Process Engr Globe Mfg Co 1972-74 producing natural rubber thread, respon for new lubrication & cure sys. Majority owner & Pres Freespan Co 1974-present, manufacturer of synthetic rubber fibre. Have also served as cons to sev mfg cos 1974-75. Mbr AIChE, SPE, AMA, ACS. 1978-1980 Natl Sales Mgr Fluorocarbon Co, Custom Plastic Fabricator.. Hobbies include athletics, reading (science fiction) & chess. 1980- Present AM Plastics, Inc. Manufacturer of High Performance Plastic Parts *Society Aff:* AIChE, SPE, ACS, AMA.

Magee, Christopher L
Business: 20000 Rotunda Dr 2053, Dearborn, MI 48121
Position: Dir, Vehicle Concepts Res Lab *Employer:* Ford Motor Co. *Education:* PhD/Mats Sci/Carnegie Mellon; MBA/Bus/MI St Univ; MS/Met/Carnegie Tech; BS/Met/Carnegie Tech. *Born:* Jul 19 1940. Native Pittsburgh PA. Mbr of res staff of Ford Motor Co since 1966 spec in materials res, dev &application and currently vehicle technology. Has held current position since 1984 & has held var offices of ASM & AIME since 1966. Winner of 1972 Howe Medal of ASM & 1972 Alfred Noble Award of ASCE. Enjoys golf, tennis & sailing. *Society Aff:* ASM, AMS, SAE, ASME

Magee, Richard S
Business: Dept of Mech Engg, Castle Pt Sta, Hoboken, NJ 07030
Position: Prof of ME. *Employer:* Stevens Inst of Tech. *Education:* DSc/Mech Engg/Stevens Inst of Tech; ME/Mech Engg/Stevens Inst of Tech; BE/Gen Engg/Stevens Inst of Tech. *Born:* 3/26/41. Joined the Stevens faculty in 1968 as an Asst Prof in the Mech Engg Dept, appointed Prof of Mech Engg in 1976. Named Dir of the Stevens Energy Ctr in 1981. Dir of the Incineration Div for the Univ/Indus Cooperative Res Center for the Mgmt of Hazardous and Toxic Waste (1983). Teaches in the areas of combustion, heat transfer, energy conversion and fluid mechanics. Publish primarily in the area of fire and flammability, eg, flame spread, ignition, burning and extinguishment of various polymeric materials. Mbr of NAS Ad Hoc Committee on Fire Safety Aspects of Polymeric Materials. Chrmn of Energy Sub- Code Comm NJ Construction Code. Licensed PE in NJ. Current res interests include coal char gasification and incinerations of hazardous wastes. Married, wife Janet, two children, Karen 13, Brian 8. *Society Aff:* ASME, ASEE, ΤΒΠ, ΣΞ, COMBUSTION INST.

Mager, Artur PhD
Home: 1353 Woodruff Ave, Los Angeles, CA 90024
Position: Group V P Engineering, The Aerospace Corp (Retired). *Employer:* Self, Consultant. *Education:* PhD/Aeronautics & Physics/CA Inst of Tech; MS/Aeronautical Eng/Case Inst of Tech; BS/Aeronautical Eng/Univ of MI. *Born:* 9/21/19. in Poland. With the Aerospace Corp as Group VP, Engg respon for all engg efforts including admin, scientific computing & data processing. Reg Prof Engr OH. Mbr natl Acad of Engg. Fellow & Pres 1980/1981 Amer Inst Aeronautics Astronautics; Fellow AAAS; Trustee West Coast Univ. Received Disting Alumni Award Univ MI 1969. Author numerous papers in area of swirling & boundary layer flows, supersonic burning, thrust vectoring, space vehicle design & propulsion. *Society Aff:* NAE, AIAA, AAAS.

Maggard, James E
Business: 673 Bays Water Way, Lexington, KY 40503
Position: Exec VP *Employer:* Watkins & Assocs, Inc *Education:* BS/CE/Univ of KY *Born:* 12/25/33 Native of Cumberland, KY. Served as navigator in USAF 1956-59. Construction engr for B&O Railroad from 1959-1962. Design engr on municipal projects for Cincinnati consultant 1962-64. Joined Watkins and Assocs in 1964. Has served in various capacities on major highway, industrial and institutional projects in several states. Presently, Exec. VP in Charge of Client Relations. Became member of Bd of Dirs in 1975. Served in 1984-85 as President of Bluegrass Chapter of Kentucky Society of Professional Engineers. Currently serving as State Director. *Society Aff:* NSPE, SAME.

Magil, Paul A
Business: 3176 Pullman St, Suite 107, Long Beach, CA 90803
Position: Pres *Employer:* Paul Alan Magil & Assoc *Education:* BS/EE/Drexel Univ *Born:* 5/30/46 Graduate studies in modern control systems and computers at CA State Univ at Long Beach. Computer-aided design of EMC filters at Stoddart Components, Inc. Broadcast and recording console design at Quad-Eight Electronics. Theatrical, institutional, presentation and multi-unaval audio, audio-visual and video systems, alarm, access control and video surveillance systems, television distribution systems design at George Thomas Howard Assocs and currently Principal of Paul Alan Magil and Assocs since 1978. Registered EE in CA, NV, NJ. Applied for registration in LA and NY. *Society Aff:* SMPTE, AES, NARAS, NSPE, USITT, ASIS

Magill, Joseph H
Business: 841 BEH, School of Engrg, Pittsburgh, PA 15261
Position: Prof *Employer:* Univ of Pittsburgh *Education:* PhD/Chem Tech/Queens Univ; BSc/Chem Tech/Queens Univ; DIC/Chem Physics/Imperial Coll of Sci & Tech; C. Chem, FRSC/Chem/Royal Inst of Chem *Born:* 12/16/28 Prof Materials Engrg 1975 and Prof Chemical Engrg 1980 -; Univ of Pittsburgh. Adjunct Prof Chemistry, Univ of Pittsburgh, 1980 -. Assoc Prof Materials Engrg 1968-1975, Univ of Pittsburgh. Research Fellow: Polymer Science: Mellon Inst 1962-64; 1965-68. Sr Research Fellow Univ of Bristol, England 1975- 76 (Sabbatical). Sr Tech Officer British Nylon summers (ICI) Ltd 1959-1965. Chem Technologist, Chemstrand (Monsanto) Mfg, N Ireland 1957-58. Research Fellow, Natl Coal Bd Imperial Coll, London 1956-57. Assistant Lecturer, (part-time), Northampton Polytechnic, London, 1957-58. Current research interests: physical properties and structural aspects of polymers and small molecules; flammability, thermal stability and toxicity of polymer degradation products. Sr Sci Awd, Alex von Humboldt Fdn, FDR, 1984-5; Univ Harburg- (Hamburg) & Univ Mainz (Sabbatical). *Society Aff:* AIP, ACS, MS, CI, NATAS, SSP.

Magison, Ernest C
Business: 1100 Virginia Dr, Fort Washington, PA 19034
Position: Manager, Regulatory Affairs. *Employer:* Honeywell, Inc-Industrial Controls Div. *Education:* BS/EE/Tufts Univ. *Born:* Oct 1926. Reg P E Penna. Adjunct Prof Drexel Univ Eve Col; ISA Standards & Practices Award 1974; Former VP, ISA Standards & Practices Dept; active in IEC TC31, & ISA, NFPA, & Natl Res Council cttes. Many publs & talks re elec safety, OSHA, radiation pyrometry. Books 'Electrical Instruments in Hazardous Locations'; "Intrinsic Safety–; with W Calder "Electrical Safety in Hazardous Locations–. With Honeywell since 1948 in variety

Magison, Ernest C (Continued)
of prod & dev engrg supervisory assignments. Former Parliamentarian, ISA. *Society Aff:* ISA.

Magliolo, Joseph J
Home: PO Box 836, Friendswood, TX 77546
Position: Pres. *Employer:* J Magliolo & Co. *Education:* MS/CE/Univ of TX; BS/CE/Univ of TX. *Born:* 10/17/22. Resides in Friendswood, TX. Is 1979-80 Pres of Intl Soc of Plastics Engrs. Started from Univ of TX in 1949 & spent twelve yrs with Monsanto Co in sales production & technical services. In 1962 joined S W Chem & Plastics as VP for sales and production. Became Exec VP & chief operating officer until the sale of the company in 1978 to Thiokol Corp. Now engaged in consulting & sales rep work in the plastics industry. Hobby tennis. *Society Aff:* SPE, AIChE.

Maguire, John J
Home: 8234 Brookside Rd, Elkins Park, PA 19117
Position: Director *Employer:* Betz Laboratories, Inc. *Education:* BS/ChE/Univ of PA. *Born:* Dec 1912. Employed Betz Labs Inc Trevose, Pa Nov 1934. Tech Dir 1941. V P 1958. Exec V P 1966. Pres 1968. Chmn of Bd 1975. Six patents, twenty tech pubs. ACS Div of Environ Chem Bartow Award 1957. Chmn 1962, Disting Service Award 1968. NACE Corrosion Specialist 268. Editor 1st (1942) through 8th (1980) editions 'Betz Handbook of Industrial Water Conditioning'. *Society Aff:* AIChE, ACS, AWWA.

Mahan, William E
Home: 4035 90th Ave S E, Mercer Island, WA 98040
Position: President. *Employer:* Mahan Howe & Assocs. *Education:* BS/Civil Engg/Univ of WA. *Born:* Aug 1931. BS Univ Wash. Native Hoquiam, Wash. Served USAF 1950-54. Struct design engr in cons field since 1956. Mbr Amer Cons Engrs Council, ASCE & Struct Engrs Assn of Wash. Started private practice 1972. Formed present firm 1973. Past Pres Seattle Chapt Struct Engrs Assn. Chmn Soils & Structures Ctte Seattle Sec ASCE. *Society Aff:* ACEC, ASCE, SEAW, ICBO.

Maher, Laurence T
Home: 5 Cutler St, Hopedale, MA 01747
Position: Sales Mgr. *Employer:* Indus Heat Treating Inc *Born:* Oct 18, 1922. Alumnus Holy Cross Coll Worcester Mass; certificate Physical Metallurgy MIT. Married, seven children. Over 30 yrs experience in heat treating field. Active in ASM, Past Chmn Worcester Chap, Past Natl Mbrship Ctteman, former Session Chmn Wm Hunt Eisenman Heat Treat Seminar. Enjoys golf when time allows. Mbr APMI Boston Chapter.

Maher, Thomas F
Business: Box 4-4432 USL, Lafayette, LA 70504
Position: Prof & Dept Head Agri Engrg. *Employer:* Univ Southwestern Louisiana. *Education:* PhD/Agri Engg/OK State Univ; MS/Agri Engg/TX A&M Univ; BS/Agri Engg/OK State Univ. *Born:* Jun 7, 1926. U S Navy WWII. Taught Tex A & M 1950-53 & Univ of S W La since. Cons for Gulf Coast Agricultural Survey Service for 10 yrs. Past Pres La Sec ASAE. Mbr Phi Kappa Phi, Sigma Tau & Alpha Zeta Natl Hon Orgs. Reg P E La. *Society Aff:* ASAE, NEA.

Mahida, Vijaysinh U
Home: 405 Tanglewood Ct, Rochester, MI 48309
Position: Pres, CEO *Employer:* Metco Services, Inc. *Education:* DrEng/San Engrg/Tech Univ of W. Germany; DrBusAdm/Taxation/IN Northern Univ; MBA/Finance/IN Northern Univ; BA/Accounting/Walsh Coll; BE/CE/Univ of Baroda-India *Born:* 3/17/37 Naturalized American citizen. Worked in W Germany (1958-64), India (1964- 67), Canada (1967-68) and USA (1968 to date). Specialized in Proj Mgmt (planning, design, construction) and Mgmt Accounting, Sanitary/Hydraulic/Civ/Value Engrg. Assumed current responsibility in 1984. Reg PE and Land Surveyor in several states. Certified Public Accountant-MI. Land surveying mbr, MI State Bd of Reg for Land Surveyors. Enjoy intl travel. *Society Aff:* WPCF, AAEE, ASCE, IE, VDI, AICPA.

Mahoney, John R
Home: 26 Salem Rd, Fishkill, NY 12524
Position: Senior Systems Analyst. *Employer:* IBM Corp. *Education:* BS/Electronics/Univ of RI. *Born:* Dec 1919. Dev Engr Westinghouse 1942-44 Eletronic Design; loaned to Manhattan Proj 1944-45 Equip Dev. Asst Supt Plant Engrg Union Carbide, Oak Ridge Gaseous Diffusion Plant, spec in Instrumentation & Control of Process Sys 1945- 64. Dir Tech Hoke Inc 1964-67, V P 1967-69 spec Computerized Bus Controls. Sr Sys Analyst 1969-present IBM Corp Computerized Bus Control. Charter Mbr Instrument Soc of Amer, V P Dist III 1960-61; V P Standards 1963-64; Dir Standards 1965-75; Pres-elect Secy 1975-76. Pres 1977. Dir EJC 1970-75. 10 U S Patents. Married, 10 children. Elected ISA Fellow 1963. ISA Standards Award 1973. *Society Aff:* ISA, APICS.

Mahoney, Patrick F
Business: 40 Steuben St, Albany, NY 12207
Position: PE/Managing Partner (Self-emp) *Education:* BS/Civil Engg/RPI. *Born:* 7/3/40. Design work and super of municipal water and sewer projects with Benjamin L Smith & Assoc--1964-70. Assumed respon as mgr partner of Smith and Mohoney-- 1970, specializing in municipal and environmental engineering; land planning; resource recovery; and energy conservation. Published papers in Journal of American Water Works Association; Public Works Magazine; Proceeding of the 6th Mineral Waste Utilization Symposium. Major speaking engagements at The Energy Bureaus "Energy from Waste" conference (1976); NYS Energy Conference (1978); ASCEs National Conference on Environmnetal Engg (1979). A founding mbr of the NY Stagge Solid Waste Management Assoc. *Society Aff:* ASCE, NSPE, WPCF, AWWA, NSWMA.

Mahoney, William P
Business: 500 Waukegan Rd, Deerfield, IL 60015
Position: Pres *Employer:* Kitchens of Sara Lee *Education:* M/BA/IIT; BS/IE/IIT *Born:* 11/24/35 Born and raised in the Chicago area. Married with five children. Work career began as a management trainee with the Ford Motor Co (one year). Spent 16 years with Motorola, initially in computer systems progressing to financial management and finally to div controller (chief financial officer) of their consumer electronics div. Joined Sara Lee in 1974 as VP of finance and administration. In 1977, named Pres of USA operations and in 1979 named Pres of total Sara Lee Co. *Society Aff:* AFFI, IFFA, IEA, CCFR, WPTF

Maierhofer, Charles R
Home: 666 Marion St, Denver, CO 80218
Position: Cons Engr. *Employer:* Self. *Education:* CE/Engrg/Univ of TX; AS/Sci/TX Lutheran College *Born:* 12/08/11. 1935-72 US Bureau Reclamation: 1935-44 surveys reservoirs, damsites, canals; field-office engr major concrete, earth dams; physical-chem res cement, concrete, additives. 1944-48 Hd Drainage Engr 700,000 ac irrig proj--soil-water phys, chemistry; hydrology; salinity exchanges; designs. 1948-72 Ch Drainage Engr BuRec--oversight, direction, tech respon all drainage, domestic foreign-- recon, planning, estimates, design, construction, O&M-Cons Pakistan, Afghanistan, Israel, Canada, Puerto Rico, Dominican Republic. Lectr grad soil sci CO State Univ 1972-present freelance cons, drainage: Philippines; World Bank-Iraq, Syria, Jordan, Mexico, Yugoslavia, India. Expert witness litigation re drainage irrigated lands and salinity problems. Reg PE; Fellow ASCE; M USICID. Past Chrmn Exec Ctte USICID & Comms Drainage Irrigated Lands ASCE, ASAE. Disting Service US Dept Interior. ASCE Tipton Award. Life Member, Hon. Member ASCE. *Society Aff:* ASCE, USCID

Mailloux, Robert J
Home: 98 Concord Rd, Wayland, MA 01778
Position: Physicist. *Employer:* Rome Air Dev Ctr. *Education:* PhD/Applied Phys/Harvard; MS/Applied Phys/Harvard; BS/EE/Northeastern Univ. *Born:* 6/20/38.

Mailloux, Robert J (Continued)
From 1965 to 1970, he was with the NASA Electronics Res Ctr & then joined the AF Cambridge Res Labs. He is presently with the Rome Air Dev Ctr. Throughout this period, he has conducted studies of phased array radiation & periodic structures. He was awarded 7 (seven) patents, published numerous journal articles & contributed chapters to five texts on radiation and diffraction. He was given the AF Marcus O'Day Award & two IEEE Special Achievement Awards for his journal articles. He was a mbr of the IEEE/AP-S (Antenna & Propagation Soc) Administrative Committee & Chrmn of the Boston Chapter of AP-S he was elected VP of AP-S in 1982, and Pres in 1983. He was elected Fellow of the IEEE for his contributions to array antennas. He has been a Natl Lectr for AP-S and technical activities chairman of URSI Commission B. He teaches part time at Tufts Univ, and is an Adjunct Prof at the Univ of MA, Amherst. *Society Aff:* ΣΞ, ΤΒΠ, HKN.

Mainhardt, Robert M
Business: PO Box 196, Bollinger Canyon Rd, San Ramon, CA 94583
Position: Pres. *Employer:* Tracor MBA *Education:* BS/Mech Engr/Polytechnic Inst of Brooklyn, NY; MS/Mech Engr/Univ of WA; Grad Studies/Mech Engr/MA Inst of Technology. *Born:* 2/28/22. Pres, Tracor/MBA; President/Chf Exec Officer/Chrmn of the Bd; MBAssocs, 1960-80. 1956-60 VP, Aeroject Genl Nucleonics; 1952-56 Administrative Hd, Theoretical Div, Univ of CA Radiation Lab; 1950-52 Staff, Office of the Genl Mgr, US Atomic Energy Comm; 1947-52 Consultant, Oak Ridge Operations Office USAEC, Los Alamos Scientific Lab. Military Record: Civilian Field Representative, USA Army, Corp of Engrs. Co-Author: *Modern Nuclear Technology* McGraw-Hill, NY, 1960 *Refractories* McGraw-Hill, NY, 1949. Married 10/18/47 to Mary Janice Bubb. Two daughters, Kathryn Lynn Parker and Mollie Janice Mainhardt. *Society Aff:* AIEE, AIAA, AICE, ADPA, EIA, AEA, AOC.

Mains, Robert M
Home: 1419 Westwind Dr, St Louis, MO 63131
Position: Emeritus Prof of Civil Engr. *Employer:* Washington Univ. *Education:* PhD/Civil Engr/Lehigh Univ; MS (CE)/Civil Engr/Univ of IL; BS (CE)/Civil Engr/Univ of CO. *Born:* 1/18/18. in Denver, CO: Instructor of Mech, MO Sch of Mines 1940-41: Asst Dir, Fritz Engg Lab, Lehigh Univ 1941-44, PhD 1946: Project Supervisor 1944-46, Asst Group Supervisor 1951-55, Applied Physics Lab, John Hopkins Univ, gun dirs, guided missiles: Assoc Prof Civil Engg, Cornell Univ 1946-51: Consulting Engrr, Knolls Atomic Power Lab 1955-60, Sr Mech Engr, Advanced Tech Lab, 1960-65, Gen Elec Co: Dept Chrmn 1965-69, Prof of Civil Engg 1965-84, Emeritus Prof. of Civil Engr. 1984-date, WA Univ, St Louis. Tau Beta Pi, Sigma Xi, Chi Epsilon. Naval Ordnance Dev Award (special merit), Irwin Vigness Award (IES). Special interest, structural dynamics. Reg. NY, PA, MD, MO. *Society Aff:* ASCE, ASME, SESA, IES, ASEE, NYAS.

Maisel, Daniel S
Business: PO Box 271, Florham Park, NJ 07932
Position: Sr Planning Adv. *Employer:* Exxon Chem Co. *Education:* ScD/ChE/MA Inst of Tech; SM/ChE/MA Inst of Tech; BS/ChE/MA Inst of Tech. *Born:* 5/3/20. Following completion of education, I have been associated with the Exxon Chem Co in a variety of positions largely in res. Most recently I served as mgr of the Exxon Brussels Belgium Res Labs & since 1976 have been sr planning advisor responsible for long range res & planning for the corp. Outside activities - trustee Overlook Hospital Summit NJ, Bd of Ed Union NJ, Fellow AIChE, Secy-Fuels & Petrochem Div, Chrmn Govt Prog Steering Committee AIChE. I hold 27 US Patents and have numerous other technical publications. *Society Aff:* AIChE, AMA.

Makdisi, Faiz I
Business: PO Box 75, Jeddah, Saudi Arabia
Position: VP. *Employer:* MIDCO Ltd. *Education:* PhD/Civil Engg/Univ of CA, Berkeley; MS/Civil Engg/Univ of CA, Berkeley; BS/Civil Engg/American Univ of Beirut, Lebonon. *Born:* 8/24/48. A native of Lebonon and a resident of CA. In 1970 he earned a bachelor of engg degree from the American Univ of Beirut; later he earned a MS degree and PhD from the Univ of CA at Berkeley. Early in his career, Dr Makdisi participated in a res program at UC-Berkeley to study the behavior of hydraulic fill embankments during earthquakes. His work there earned him the Norman Medal Award of the ASCE in 1977. He was project engg with Woodward-Clyde in San Francisce, where he was involved in the evaluation of seismic stability of a number of earth and rock fill dams, in addition to studies of the properties of offshore soils under cyclic loading. He is currently VP of MIDCO Saudi Arabia, respon for Geotechnical Engg of a number of Transportation and Building projects. *Society Aff:* ASCE, SSA, ISSMFE, EERI, ICE, USCOLD, NSPE.

Makhoul, John I
Business: 10 Moulton St, Cambridge, MA 02238
Position: Principal Scientist *Employer:* Bolt Beranek & Newman, Inc *Education:* PhD/EE/MIT; MSc/EE/OH State Univ; BE/EE/American Univ of Beirut *Born:* 9/19/42 in Deirmimas, Lebanon, is now resident in Cambridge, MA. Since graduation in 1970, has been with Bolt Beranek & Newman Inc, Cambridge, leading research in speech communication with computers, digital signal processing, time series analysis, and pattern recognition. In 1980, was elected to the position of Principal Scientist and was appointed Mgr of the Speech Signal Processing Dept. Speech-area representative in US-USSR NSF-sponsored exchange program, 1978. 1978 Sr Award, IEEE ASSP Society; ASA Fellow, 1978; IEEE Fellow 1980, 1982 Tech Achievement Award, IEEE ASSP Soc. Who's Who in Tech Today, Amer Men and Women of Sci. *Society Aff:* IEEE, ASA, ACM, ACL, EURASIP

Mal, Kumar M
Business: Sintertech International Marketing Corp, PO 144, Ghent, NY 10715
Position: President *Employer:* Sintertech International Marketing Corp. *Education:* Dr Sci/Powder Met/Univ of Saarbrucken, W Germany; MEng/Met Sci/Univ of Saarbrucken, W Germany; BSc/Sci/Agra Univ, India *Born:* 7/9/39. in India, Met & Mtl Sci education was carried out in W Germany. Taught mtl sci courses in Germany. Carried out res on cemented carbides under the guidance of Prof Dr Dawihl. Immigrated to USA in 1965. Worked as Chief Met till 1969 at Adamas Carbide Corp. Joined Chromalloy in 1969 as Chief Met of Sintercast Div & assumed current responsibility as VP for Operations in 1978. Since 1981 has assumed current responsibility as Pres of Sintertech Intl Mktg Corp. Have written several technical papers in the field of powder met & have received several patents for the inventions in the area of heat & corrosion resistant mtls. Co-author with Prof. Dr. H.H. Hausuer, a book entitled "Handbook of Powder Metallurgy–. *Society Aff:* APMI, ASM

Malah, David
Business: Electrical Eng. Dept., Technion City, Haifa, Israel 32000
Position: Assoc. Prof. *Employer:* Technion *Education:* PhD/Electrical Eng./University of Minnesota; MSc/Electrical Eng./Technion-Israel; BSc/Electrical Eng./Technion-Israel *Born:* 3/31/43 Born in Poland, 1943, and emigrated to Israel at age 5. B.Sc. and M.Sc. degrees from Technion in 1964 and 1967, respectively. Ph.D. from the Univ. of Minnesota in 1971. Served as an Electronics Engineer at the IDF (1964-66). During 1971-72 was an Asst Prof at the Univ. of New Brunswick, Canada, and joined the Technion in 1972, currently an Assoc. Prof. During 1979-81 was on sabbatical and leave at the Acoustics Research Dept. of AT&T Bell Labs. Since 1975 is in charge of the signal processing lab of the EE Dept, Technion. Fellow of the IEEE as of Jan. 1, 1987. *Society Aff:* IEEE

Male, Alan T
Home: 24 Morris Street, Export, PA 15632
Position: Mgr Processing Research. *Employer:* Westinghouse Electric Corp. *Education:* PhD/Indust Metallurgy/Univ of Birmingham; BSc/Metallurgy/Univ of Birmingham. *Born:* Sept 1937. Reg P E Penna. Chartered Engineer, England. Fellow of Institute of Metals. Native of Birmingham England. Mbr of teaching staff Univ Birmingham 1960-67; 1968-70 was Supr of Westinghouse team operating the Experimental Metals Processing Facility of the Air Force Materials Lab; major tech em-

Male, Alan T (Continued)
phasis was on extrusion & forging of aerospace materials; assumed current respon as Mgr of Processing Res at Westinghouse Res Labs 1970; Respon for initiation & conduction of progs concerning processing of matls used in elec power generation, transmission and utilization. Authority on friction, lubrication and microstructional effects in metalworking operations. Member of ASM and SME. *Society Aff:* ASM, SME, IM.

Male, Kenneth J
Home: 2219 Garden Dr, Schenectady, NY 12309
Position: Pres. *Employer:* C T Male Assoc, P.C. *Education:* BS/Civ Engg/Union College. *Born:* 10/12/23. Native of Niskayuna, NY. During WWII, served as P.51 Fighter Pilot with 8th Air Force; as Base Utilities Officer, Nordholz Air Base, Germany, 9th Air Force. Instr in surveying - Union College 1949. Founding Partner of C T Male Assoc in 1950; Managing partner from 1955-1971; Elected Pres and Gen Mgr of PC in 1971. Licensed Land Surveyor and PE. APWA - Delegate from Upstate NY Chapter '74-'75. ACEC - Dir from NY State '75-'76; Treas 78-80. Enjoy swimming, golf, bowling and skiing. Civic: Greater Colonie Chamber of Commerce. Director 1973-1977 VP 1975, 1976, 1981, Director, Manufactures Hanover Trust Co./CORP Region, President - Niskayuna, NY Rotary Club 1981-82. *Society Aff:* ACEC, ASCE, NSPE, WPCF, ΣΞ, AWWA, APWA, AMER ARBITRATION ASSOC

Maler, George J
Business: Engrg & Appli Sci, Boulder, CO 80309
Position: Assoc Dean E & Prof EE. *Employer:* Univ of Colorado. *Education:* MS/Elec Engg/Univ of CO; BS/Elec Engg/Univ of CO. *Born:* Aug 1924. Native of Denver Colo. Active duty US Navy WWII & Korean War. USNR Lt Cdr. Engr Mtn States Telephone Denver; Mbr Elec Engrg faculty Univ of Colo since 1947; Assoc Dean of Engrg for Undergrad Studies & Spec Servs & Prof of Elec Engrg Univ of Colo since 1966. Reg P E Colo. Dir Mtn Bell Regional Communications Engrg Sch 1960-70; Dir NSF supported Engrg Concepts Curriculum Proj 6-week Summer Insts 1969-73. Rocky Mtn Section Rep. ASEE CPD Division 1980. - ASEE Guidance Cmtee 1981-84.- Member Colorado Engineering Council, Author: Introductory Circuit Analysis (with S I Pearson) 1965. (ASEE Continuing Studies Div: Natl Bd/Dirs 1973-77), Recipient Univ of Colo Disting Engrg Alumnus Award 1976. *Society Aff:* ASEE, IEEE, NSPE, AAAS.

Maletta, Anthony G
Business: 85 Metro Park, Rochester, NY 14623
Position: President *Employer:* Sear-Brown Assocs P C. *Education:* BS/Mech Engg/New Jersey Inst of Tech. *Born:* Jan 4, 1930 Paterson New Jersey. Reg P E New York, West VA, Fl, OH, CT. US Army Corps of Engrs 1953-55 Sanitary & Drainage Projs. Curtiss Wright Corp 3 yrs Materials Application Engr; 8 yrs concurrently V Pres & Dir of Materials Testing, Fact Tech Serv & Sole Proprietor Cons Practice; joined Sear-Brown Assocs Engrs & Architects V Pres 1970 became Pres in 1977 chairman in 1987. Pres of Cons Engrs Council of New York State 1968-69. Dir of Cons Engrs Council 1969-70. V Pres Amer Cons Engrs Council 1974-76. Dir of Prof Services Managers Assoc. 1985-88. 1987 recipient of the ACEC Past Presidents Award. *Society Aff:* ASCE, PSMA, NSPE, ASTM.

Maley, Wayne A
Home: 2592 Stratford Dr, St. Joseph, MI 49085
Position: Director Membership & Education. *Employer:* American Society of Agricultural Engineers *Education:* BS/Agri Engr/IA State Univ. *Born:* 3/9/27. Born and educated in IA. Power use advisor Southewestern Elec Cooperative, IL 1949-53. Field Engr, American Zinc Inst, IN, galvanized farm bldgs. From 1959-1977 various commercial assignments with US Steel, midwest and Pittsburgh; patents for fence bldg machine, nesting bulk bin; farm fencing, steel bldgs, tractor cabs and noise reduction, environmental and mining machinery. Self- employed in energy conservation equipment sales. Currently Director, Membership and Education for American Society of Agricultural Engineers. Active bd mbr, profl societies, and church. On ASTM organizing committee for soild waste. Recreation: sailing and music. *Society Aff:* ASAE, ΣΞ.

Malik, Mazhar Ali Khan
33 Central Ave, Findon Valley, Worthing Sussex, England BN14 ODS U.K
Position: Sen. Vice President *Employer:* Optimum Planning International *Education:* PhD/Ind Eng/NYU; MS/Ind Eng/NYU; BS/Mech/Punjab Univ., Pakistan *Born:* 02/05/34 Married to Nargis Sadiq, 5 children, originally from Pakistan. BS Mech Engr, 1st Div, 1st pos in Punjab Univ 57; MS Ind Engr 70, PhD: 75, NYU. Evaluation Engr NY Port Authority 69-76; Researcher in Economics 1977 and then Chrmn Industrial Engrg Arab Dev Inst Libya 78-81; mbr of the Libyan Acad of Tripoli, Libya. Consltg Engr. National Oil Corp., Libya, 82-84 Sen. Vice Pres. Opt. Plan. Int., USA/UK, 85-to date. Have taught part time and advised on Engr Educ in USA and Libya. Fellow IIE; Mbr: ASME, ASEE, IIE (Australia). Have contributed in following specializations: Ind Engrg, Educ, Const Engr & Engr Economy, Manufact & Prod, Occup Safety & Health, Operations Res., Plant & Facilities Engr, Res & Dev. Publications: 23 papers, 64 articles/communicat, 32 seminars, 11 conferences in various countries of NA, Europe, Africa, Asia and in Australia. First person to walk up 112 Stories of World Trade Center, NYC. *Society Aff:* IIE, ASME, ASEE, IIE, PEC

Malik, Om P
Business: 2500 University Drive N.W, Calgary, Alberta Canada T2N IN4
Position: Associate Dean-Academic Faculty of Engg. *Employer:* The Univ. of Calgary *Education:* Ph.D./Elec. Engg./Imperial College Univ of London, U.K.; M.E./Elec. M/E. Design/University of Roorkee, India; N.D.E.E./Elec. Engg./Delhi Polytechnic, India *Born:* 04/20/32 After graduation in 1952 worked on design, construction and operation in various electric utilities in India for a period of nine years. Worked as an Engineer with English Electric Co., England for a short period in 1965-66 before joining the Univ of Windsor, Canada in 1966. Teaching at the Univ of Calgary, Alberta, Canada since 1968 and have been the Assoc Dean, Academic, Faculty of Engrg since 1979. Chairman, Student Activities, IEEE Canadian Region, 1978, 79, 80; Chairman, IEEE Western Canada Council, 1983, 84; Confederation of British Industries Scholar 1959-60; IEEE Centennial Medal 1984; IEEE Western Canada Council Merit Award 1986; Fellow IEE 1977; Fellow IEEE 1987. *Society Aff:* IEEE, ASEE, IEE, APEO, APEGGA

Malina, Joseph F, Jr
Business: Univ of Texas ECJ 4.2, Austin, TX 78712
Position: CW Cook Prof Enivr Engrg & Chmn Dept Civil Engrg *Employer:* Univ of Texas at Austin. *Education:* PhD/Civil Engrg/Univ of WI; MS/Civil Engrg/Univ of WI; BCE/Civil Engrg/Manhattan College *Born:* 8/24/35. Proj Engr involved in design & oper of aerated lagoon Hamlin NY 1957; Res Engr Rex Chainbelt Milwaukee WI 1959; the Univ of TX at Austin: Asst Prof 1961- 64, Assoc Prof 1964-70, Prof 1970- ; Dept. Chrmn 1976- ; C.W. Cook Prof. in Env. Eng. 1980- . Recipient of Ridgeway Awd Met Sect ASCE 1957; (Awd of Excellence Halliburton Educ Fdn (1981), Gordon M Fair Awd presented by the AAEE 1983); Arthur S. Bedell Awd (1985)-WPCF; Chi Epsilon; Tau Beta Pi. (Phi Kappa Phi, Omicron Delta Kappa, Fellow, ASCE). Served as Secy 1972, V Chmn 1973 & Chmn 1974 of Environ Engrg Div TX Sect ASCE; SecyTreas 1975, V Chmn 1976 Chmn 1977 of Environ Engrg Div of ASEE. V Pres 1974, Pres 1975 Dir 1981 of TX Water Pollution Control Assn. Cons to numerous indus, engrg firms, local, state, federal & internatl agencies. Reg in TX; Diplomate of Amer Acad of Environ Engrg; Planner 1978, Secy 1979, V Chmn 1980, Chmn 1981, Prof Coordination Ctte, (V Chmn 1983, Chmn 1984, Awds Ctte) Environ Engrg Div, ASCE; Board of Control 1981-84; Chmn, 1979-83, Toxic Substances Ctte, Chmn, Pub. Comm 1986- , Water Pollution Cont Fed, Eng. Accred. Comm/ABET 1982. *Society Aff:* ASCE, WPCF, NSPE, AEEP, AAEE, AWWA, IAWPRC, ASEE.

Maling, George C Jr
Business: IBM Corp. Dept C18 Bldg 704, PO Box 390, Poughkeepsie, NY 12602
Position: Sr Physicist. *Employer:* IBM Corp. *Education:* PhD/Physics/MIT; EE/Elec Engg/MIT; SM/Elec Engg/MIT; SB/Elec Engg/MIT; AB/Physics/Bowdoin College. *Born:* Feb 24, 1931. Special interests incl digital signal processing, physical acoustics & noise control engrg. Fellow Acoustical Soc of Amer, Inst of Elec & Electron Engrs, Audio Engrg Soc, Amer Assoc for the Advancement of Sci. Dir & 1975 Pres of the Inst of Noise Control Engrg. 1976-79 Chmn Amer Natl Standards Inst Ctte S1 on Acoustics. P Chmn Acoustical Soc Ctte on Noise. 1980-1983 Member, Executive Council, Acoustical Soc of Amer. IBM Acoustics Lab 1965- . Reg P E New York State. *Society Aff:* INCE, IEEE, ASA, AES, AAAS.

Mall, Shankar
Business: AFIT/ENY, Wright-Patterson Air Force, Base, OH 45433
Position: Professor *Employer:* Air Force Institute of Technology *Education:* PhD/ME/Univ of WA; MS/ME/Banaras Hindu Univ; BS/ME/Banaras Hindu Univ, India. *Born:* 6/10/43 Have 20 years of teaching and research experience in India, Japan and States. Author of 45 technical papers. Active in research in field of fracture and fatigue, composites, fracture dynamics, wood fracture and failure of bonded joints. *Society Aff:* ASME, AAM, ASEE, TBΠ, ΣΞ, SES

Mallard, Stephen A
Business: PO Box 570, 80 Park Plaza, 14A, NJ 07101
Position: Sr VP - Planning & Res; Pres of PSE&G Res Corp *Employer:* Public Serv Electric & Gas Co. *Education:* ME/Elec Engg/Stevens Inst; MS/Elec Engg/Stevens Inst. *Born:* Sept 1924. Native of New Jersey. Served in US Navy 1944-46 (Pacific Theatre). Taught elec engrg at Stevens 1948-51. Joined PSE&G 1951; various field & genl office assignments in Elec Distrib & Sys Planning; Mgr of Sys Planning Dept 1971; Genl Mgr Planning & Res 1974; VP-Sys Planning 1977; Senior V.P. - Planning & Research and President of PSE&G Research Corp. 1980; responsible for interconnections, planning, and research for the company's electric and gas systems; sey.Fellow of the IEEE, CIGRE, Tau Beta Pi, Eta Kappa Nu. Licensed P E New Jer, Ethics Committee of Essex County Bar Association, Chairman Board of Essex County Chapter of American Red Cross. Enjoy photography, travel & history. Dir in Inst of Gas Tech 1983- . *Society Aff:* IEEE, NSPE, TBΠ, HKN.

Mallik, Arup K
Home: 39 Linnwood Rd, Morgantown, WV 26505
Position: Assoc Prof. *Employer:* WV Univ. *Education:* PhD/Ind Engg/NC State Univ; MS/Ind Engg/NC State Univ; BME/ME/Jadavpu Univ. *Born:* 11/14/43. Native of Calcutta, India. Author of a text book "Engineering Economy With Computer Applications" published by Engg Tech Inc 1979. Assoc Prof of Ind Engg at WV Univ. Teaching & res interest include production planning, scheduling, computer based economic analysis & energy modelling. *Society Aff:* IIE, NSPE, AACE, ORSA.

Malone, Frank D
Business: 5th Ave & 24th St, Bessemer, AL 35020
Position: Supr Safety, Medical & Workmns Comp. *Employer:* Pullman Standard Div of Pullman Inc. *Born:* March 1929. BA Birmingham Southern Coll. Native of Gardendale Alabama. Served with Army Signal Corp 1953-54; Radio Operator for 516th Signal Co. With US Steel 1955-63 assigned to Indus Relations & Safety; with Pullman Standard since 1963; assumed current respon as Supr of Safety, Medical & Workmens Compensation 1967. Pres of Alabama Chap ASSE 1975-76. Elected Region VI V Pres ASSE 1976-78. Enjoy sports & gardening.

Malone, William J
Business: PO Box 1480, Hot Springs, AR 71901
Position: Pres. *Employer:* Malone & Hollingsworth, Inc. *Education:* BSCE/Civ Engg/Univ of AR. *Born:* 11/8/41. Native of Plumerville, AR. Worked in Physics Div of FBI Lab 1961-63. With Affiliated Engrs Inc 1966-1975, as VP from 1973-75 in charge of Preliminary Engg. Formed Malone & Hollingsworth, Inc in Nov 1975. Currently serving as Pres of Hot Springs Regional Chapter of NSPE. *Society Aff:* ASCE, NSPE.

Maloney, Laurence J
Business: 530 Bedford Rd, Suite 250, Bedford, TX 76022
Position: Principal *Employer:* Maloney Assoc Conslt Engr's *Education:* BS/Elec Engg/Univ of TX. *Born:* 12/23/34. IES: VP, Admin/Operations, 1979-1981; VP, Southwestern Region, 1977-1979; pres, W TX Sec, 1976; Mbr, Sports Lighting Committee, 1977-. Bd of Dirs 1982-85, Fellow 1984. IEEE: Industry Applications Society; Production/Application of Light Committee; Commercial Power Systems Cttee; Elec Space Heating/Air Conditioning Cttee; On Cttee, Gray Brook, Elec Pwr Systems/Commercial Bldgs; Red Book, Elec Power Systems/Industrial Bldgs; White Book, Elec Systems/Hospitals, Chrmn, Lighting Chapter. Experience: TX Elec Service Co, 1959-1971; Mgr, Arlington eng, Distribution Stds Engr, Div Distribution Supt; Friberg, Alexander, Maloney, Gipson, Weir, Inc 1971-84; chief Elec Engr, Lighting Design Conslt, Maloney Assoc, 1984-. *Society Aff:* IESNA, IEEE, NSPE/TSPE

Malouf, Emil E
Home: 132 Dorchester Drive, Salt Lake City, UT 84103
Position: Private Conslt *Employer:* Self-employed *Education:* BS/-/Univ of Utah. *Born:* 3/23/16. in Ogden, UT;m. 1942, 4 children. Ch explosive manuf Ogden Arsenal US Army 1940-44. Res Chem Kennecott Copper /corp 1947-52; Prof Dev Engr 1952-65; Ch Hydrometallurgyl Sect 1965-72; Mgr Special Projs 1972-77. Adj Prof Univ of UT USA 1965- . USA 1944-46 1st Lt. AIME Engg Achievement Award (Earll McConnell AwARD 1972). Numerou pubs & pats on rhenium recovery processes, bio-extractive metallurgy of copper, copper leaching processes from mine waste & copper precipitation units for efficient high volume recovery of copper from sulutions. Retired from Kanecott & am now woking as a private consulant. *Society Aff:* AIME, SME.

Malozemoff, Plato
Business: 300 Park Ave-12th Floor, New York, NY 10022
Position: Chmn & Ch Exec Officer. *Employer:* Newmont Mining Corp. *Education:* MS/Metallurgical Eng/MT Sch of Mines; BS/Mining Eng/Univ of CA. *Born:* Aug 26, 1909 St Petersburg Russia. 1957 Doctor of Engrg (Honorary) Colorado School of Mines. 1932-34 Res at Montana School of Mines; 1934-40 Metallurgist Alaska Juneau & Pan Amer Engrg; 1940-43 Mgr of Mines Argentina & Costa Rica; 1944-45 Engr for war agency of US Govt; 1945- Newmont Mining Corp: 1952 V Pres, 1953 Dir, 1954 Pres & CEO, 1966 Chmn & CEO. Distinguished Mbr AIME 1972 Rand Memorial Gold Medal; Mining Club; Institution of Mining & Metallurgy London, 1974 Gold Medal; Mining & Metallurgical Soc of Amer, 1976 Gold Medal; Natl Acad of Engrg. *Society Aff:* ΦBK.

Malsbary, James S
Home: 280 Edwin Avenue, Glendale, MO 63122
Position: Cons Engr (Self-employed) *Education:* BSEE/Elec Corp/Purdue Univ. *Born:* 06/17/05 Came to Wagner Elec Corp St Louis MO 1933 where conducted dev in high voltage impulse testing & in elec-mech control. Hold a number of pats on elec control & mech inspection devices. Respon for design & const of semi-automatic motor test stands & large self-regulating power transformers. Fellow Inst Elec & electron Engrs. Now cons on problems of elec power control & surge voltage protection. Enjoys amateur radio & fishing. *Society Aff:* IEEE, ARRL.

Maltby, Lewis J
Home: 1208 Chavez St, Burbank, CA 91506
Position: Staff Asst to Sr V P. *Employer:* Menasco Mfg Co. *Born:* Jun 1921. B Aero Engrg Univ Minn. Native of Mahtomedi, Minn. Naval Air Corps Aircraft Maintenance Officer WWII. Dynamic Analyst & Struct Design Engr for Chance Vought 1946-53. Design Engr & Mgr Special Projs for Menasco Mfg Co 1953-66. Ch Design

Maltby, Lewis J (Continued)
Engr & Dir Engrg Controls Div, Hydraulic Res & Mfg Co 1966- 72. V P & Pres Western Hydraulics Inc 1972-74. Present duties in Corporate Marketing for Menasco. Mbr Triangle Fraternity & AIAA. Delivered paper on space docking mechanism at AIAA Third Manned Spaceflight Meeting.

Malthaner, William A
Home: 3001 7th Ave West, Bradenton, FL 33505
Position: Communication Sys Cons (Self-employed) *Education:* BEE/Communications/Rensselaer Polytechnic Inst. *Born:* Jul 1915 in Ohio. Mbr Tech Staff Bell Telephone Labs 1937-75; head various res & dev depts 1958-75. Cons 1975- . Respon dev commercial telephone switching & signaling sys, early digital computers, naval anti-aircraft fire- control computer & radar sys. Pioneered res in electronic switching sys, stored prog control, data processing & transmission sys. 37 U S patents. Mbr IEEE, AAAS, PSA & honor socs Tau Beta Pi & Sigma Xi. Mbr Communication Switching Ctte IEEE 1962-75. Elected Fellow IEEE 1962. *Society Aff:* IEEE, AAAS, PSA, TBΠ, ΣΞ, RESA.

Mamrak, Sandra A
Business: Dept Computer & Information Science, 2036 Neil Ave, Columbus, OH 43210
Position: Assoc Prof *Employer:* OH State Univ *Education:* PhD/Computer Sci/Univ of IL; MS/Computer Sci/Univ of IL; BS/Math/Notre Dame Coll of OH *Born:* 9/8/44 Native of Cleveland, OH. Taught at OH State Univ since receiving PhD in 1975. Worked summers at Natl Bureau of Standards (1975-79), Bell Telephone Labs (1980) and Lawrence Livermore Natl Labs (1981-84). Primary research interest is software support for distributed processing computer systems. Has also done research in areas of computer performance evaluation and status of women in academic computer science. Hobbies include world travel, racquetball and jogging. *Society Aff:* ACM

Mancini, Gerold A
Business: McKee Rd-POB 1111, Rochester, NY 14603
Position: Mgr of Quality Control. *Employer:* Monroe Forgings Inc. *Education:* PhD/Metallurgical Engr/OH State Univ; MSci/Metallurgical Engr/OH State Univ; BA/Chemistry/Univ of Rochester. *Born:* May 18, 1927. Native Rochester, N Y. Res in Corrosion & Allotropic Transformation in metals. Alloy dev in powder metallurgy of dispersion strengthened alloys for high temp applications at E I duPont deNemours Co; alloy dev of high strength alloys for glass coated pressure vessels & mgr physico-chem res including corrosion R & D for alloys & ceramics, Pfaudler Co 6 yrs. Current respon include metallurgical engrg & metallurgy, quality & reliability control for forgings mfg from low alloy, ferrous & non-ferrous including stainless steels & super alloys for high temp applications since 1964. *Society Aff:* ASM.

Mandel, Herbert M
Business: 100 Forbes Ave, Pittsburgh, PA 15222
Position: Sr VP *Employer:* Parsons Brinckerhoff Quade & Douglas, Inc. *Education:* ME/CE/Yale Univ; BS/CE/VA Polytechic Inst *Born:* 5/11/24 Native of Port Chester, NY. Served in US Army 1943-46 (Armored Force) and 1950-52 (Corps of Engr). Structural engr with Madigan Hyland 1949-50; with PBQ&D since 1950. Engaged initially in tunnel design, then in movable and long-span fixed bridge design and bridge rehabilitation. Resident engr, Chicago Northwest Corridor Transit Study, 1961; Atlanta Transit Study, 1962. Project Mgr, Newport Bridge, RI, 1963-70 Project Mgr, Rt.H-3, HI. 1970-74. Currently Project Dir for design and construction of Pittsburgh Light Rail Transit System; Regional Mgr, PBQ&D Central Region. Elected VP 1974, Sr VP 1978. Mbr of AREA Ctte 15 (Steel Structures). *Society Aff:* ASCE, AREA, IABSE, ABCD

Mandel, James A
Business: Dept of Civ Engg, Syracuse Univ, Syracuse, NY 13210
Position: Prof (Civ Engg). *Employer:* Syracuse Univ. *Education:* PhD/Civ Engg/Syracuse Univ; MS/Civ Engg/Carnegie Inst of Tech; BS/Civ Engg/Carnegie Inst of Tech. *Born:* 12/25/34. Native of Pittsburgh PA. Design Engr with Richardson, Gordon & Assoc (Pgh, PA) 1956-61. Sr Stress Engr with Goodyear Aerospace Corp (Akron, OH) 1962-64. Faculty mbr in Dept of Civ Engg, Syracuse Univ 1967-present. principal Investigator on sponsored res projs in fiber reinforced materials, Fracture Mechanics, stress waves, curved bridges, and castellated beams. Reg PE 1960-present. *Society Aff:* ASCE, AAM, ACI.

Mandel, Leonard
Home: Rochester, NY 14627
Position: Prof of Physics *Employer:* Univ of Rochester *Education:* PhD/Physics/Univ of London; BSc/Physics/Univ of London; BSc/Math Phys/Univ of London *Born:* 5/9/27 Appointments held since graduation: Technical Officer, Research Labs of Imperial Chemical Industries, 1951-54; Lecturer & Sr Lecturer, Imperial Coll, London, 1954-1964; Prof of Physics, Univ of Rochester, 1964 - present. Known principally for research contributions to the subjects of optical coherence & quantum optics. Author of over 200 papers and editor of 5 books. Together with Emil Wolf, organized series of Rochester Conferences on Coherence & Quantum Optics since 1966. First recip of Max Born prize in 1982 from OSA. Assoc Editor, Journal of the Optical Society of America 1970-76, 1982-83; Assoc Editor, Optics Letters 1977-79; Board of Directors, Optical Society of America 1984-87; Board of Editors, Physical Review, American Physical Society 1987- ; *Society Aff:* APS, OSA

Mandelbrot, Benoit B
Business: P O Box 218, Yorktown Heights, NY 10598
Position: IBM Fellow. *Employer:* IBM Corp-Res Ctr. *Born:* Nov 1924. PhD Maths Univ Paris; diploma Ecole Polytech Paris; MS Caltech. Native of Poland & Citizen of France. Taught at Harvard, MIT & Yale. With IBM since 1958, IBM Fellow since 1974. Spec is study & control of elec noise, economic fluctuations & var other irregular phenomena in nature. Author 'Fractals: form, chance & dimension' 1975-77. Fellowships include IEEE, Econometric Soc, Amer Statistical Assn & Inst Math Statistics.

Mandigo, James A
Home: 112 Martingale Rd, Lutherville, MD 21093
Position: Retired *Education:* BS/Civ Engg/KS Univ; Grad/5-yr Assoc Course/Army Command & Gen Staff College. *Born:* 10/25/11. Entered into private practice in 1975 after 40 yrs of World-wide assignments in aviation facilities res, planning, maintenance and operation on over 100 airports, terminal facilities and accesses. Served in ETO with the Corps of Engrs in WWII. Retired as Lt Col. Reg PE in MO, MD, NY. Lifetime service with the Boy Scouts of America. Returns these lifetime experiences through service to the public and the profl arena. Wife, Helen; son, Jim, Jr, and three grandsons. Currently retired. *Society Aff:* ASCE, NSPE, SAME.

Mandil, I Harry
Home: 701 Heathery Lane, Naples, FL 33963
Position: Retired; Consultant. *Employer:* MPR Assoc, Inc. *Education:* DSc/Sci (Hon) /Thiel College; Grad/Reactor Tech/Oak Ridge Sch of Reactor Tech; MS/EE/MA Inst of Tech; BS/EE/Univ of London. *Born:* 12/11/19. Retired December 1985. Principal Officer, MPR Assoc, Inc, Wash, DC 1964-85; Dir Reactor Engg Div Bur Ships, also Chief Reactor Engr Br Naval Reactors Dev Div, 1954-1964; Asst to Tech Dir Naval Reactors Br Reactor Dev Div, AEC, 1949-1954; Asst to Pres, Norcross Corp, in Charge of Field Engr 1946-49; US Navy 1942-1946; Design process controls for textile mills & chem plants, Norcross Corp, 1941-42. Dev of nuclear power for generation of electricity & for application to ship propulsion; author of numerous papers in the field. Distinguished Civilian Service Award, US Navy, 1959. PE 1799, DC. *Society Aff:* ASME, ANS, NYAS.

Maneri, Remo R
Home: 5808 Siebert St, Midland, MI 48640
Position: Chrmn of Bd & CEO *Employer:* Quantum Composites, Inc. *Education:*

Maneri, Remo R (Continued)
BS/Chem Engg/Case Inst of Tech. *Born:* 8/16/28. in Cleveland, OH. Case Mgmt Dev Prog 1956, Harvard AMP Prog 1969. US Army Signal Corps Battlefield Surveillance 1954-56. With Dow Corning Corp since 1950 in Prod Dev, Mfg, Commerical Dev, Mkt Res, Tech Mgmt; Group VP for Businesses. Pres, Dow Corning USA; Executive Vice President, Area Operations since 1981. Chrmn of the Bd & CEO Quantum Composites, Inc since 1982. Elected Alpha Chi Sigma, Sigma Xi, Tau Beta Pi; Mbr Comerica Bank - Midland Bd of Dirs, Mbr Advisory Bd of Dir Duro-Last Inc.; Mbr, Dow Corning Corp Bd of Dirs. *Society Aff:* ACS, ΑΧΣ, ΣΞ, ΤΒΠ.

Manetsch, Thomas J
Home: 2836 Chateau Way, Holt, MI 48842
Position: Prof of Systems Science. *Employer:* MI State Univ. *Education:* PhD/EE/OR State Univ; MS/EE/Univ of WA; BS/EE/WA State Univ. *Born:* 1/21/32. Born in Spokane, WA, Engg Duty Officer US Navy 1955-58. Three yrs in industry as a design and res engr. Asst prof of Engg UCLA 1964-66. Assoc Prof/Prof of Systems Sci, MI State Univ 1966-present. Author or co-author of over 50 publications dealing with Systems Science and applications to socioeconomic systems. Coordinated natl level simulation modelling for planning rural dev in Nigeria and Korea. Currently studying decision support methodology and sys approaches to food and nutrition problems in developing countries. Elder, Reformed Church in America. *Society Aff:* IEEE, ASA.

Maneval, David R
Home: 460 Lone Pine Drive, Fairbanks, AK 99709
Position: Prof *Employer:* Univ of AK *Education:* PhD/Mineral Prep/PA St Univ; MS/Chem/PA St Univ; BS/Education/PA St Univ *Born:* 12/18/28 Native of Williamsport, PA. Taught and did research in Dept of Mineral Preparation at Penn State (1953-1963), Dir of Research and Development in former PA Dept of Mines and Mineral Industries, later Deputy Secretary of PA. Dept of Environmental Resources (1963-1971), Science Advisor of Appalacian Regional Commission, Washington, DC (1971-1978), Assistant Dir, Office of Surface Mining, U.S. Dept of the Interior (1978-1981), research areas include ore dressing and mine water treatment; administered engrg programs dealing with mine reclamation, mine subsidence and surface coal mining. AIME "Environmental Conservation Distinguished Service Award" 1980. Enjoys camping, photography and Scout work. *Society Aff:* AIME, ACS, ΣΞ

Maney, C Thomas
Business: AD/DLX, Eglin AFB, FL 32542
Position: Dir Plans & Res. *Employer:* USAF, AD. *Education:* MS/Physics/Univ of KY; BS/EE/Univ of KY. *Born:* May 1924. Grad EE Purdue. Meteorologist, US Army; Creative Engrg, General Elec Co. US Army 1943-46; Design Engr Genl Elec 1955-57; Prof U of Ky 1957-61, 1963-66; USAID Indonesia 1961-63. USAF Dev Planning ASD 1966-71, ADTC 1971-78. Dir Plans & Res, Armament Lab 1978- . Respon tech direction USAF tactical weapon technology planning. Appointed NATO AGARD 1972-76. Mbr Missile Tech Ctte AIAA 1975-77. PE FL 1978-, OH 1966, Sr Mbr IEEE, Mbr UVS, ADPA, AIAA, ITEA. *Society Aff:* AIAA, IEEE, ADPA, UVS, ITEA.

Manfredi, Robert R
Business: 400 E Pratt St, Village of Cross Keys, Baltimore, MD 21210
Position: Principal *Employer:* RTKL Assoc Inc *Born:* May 1936. P E lic. Native of N Y. Has resided from 1963 to 1980 in Washington, D C area with his wife Carol, and children Kim and Rob & now resides in Baltimore MD. Served in USAF 1955-59 and attended U of Md and George Washington U. Served as Natl First V P for Amer Soc of Plumbing Engrs, Mbr Natl Acad of Sciences BRAB, Amer Natl Standards Inst. Currently Director of Mechanical & Electrical Engineering for RTKL Assoc Inc. A National Architectual & Engineering Firm with 400 employees. Enjoys tennis & skiing. *Society Aff:* ASHRAE, NSPE, SMPS, SAME, ASPE

Manfredonia, Savery S
Home: 21 Black Birch Ln, Scarsdale, NY 10583
Position: Partner. *Employer:* W A Di Giacomo Assocs. *Born:* July 1920. BSEE N Y Polytechnic Inst. Native of Scarsdale N Y. Field Engr with Reynolds-Newberry, Engrs & Constructors 1953-55; supervised const of a $30,000,000 Control Bldg for Portsmouth Atomic Energy Plant. As Ch Dept Head 1955-58 for Wyatt C Hedrick, Architects & Engrs, was respon for engrg & design of Keflavik, Iceland Airbase for USA. Ch Electrical Engr 1958-60 for Metcalf & Eddy; respon for design of Ballistic Missile Early Warning System in Thule, Greenland. With Jaros, Baum & Bolles for 11 years; respon for engrg & design of coml & inst bldgs. Also engineered & designed hardened coaxial cable transcontinental network consisting of a series of underground communication facils for providing greater capacity to handle communications with optimum reliability while simultaneously protecting the network in case of nuclear attack. For this effort, NY.

Mangan, John L
Business: Bldg 59E - Rm 117, Schenectady, NY 12345
Position: Mgr-Business Planning and Integration, Turbine Bus Group *Employer:* Genl Elec Co. *Education:* BS/Mech Engg/Carnegie Inst of Technology. *Born:* 5/24/20. Army Corps of Engrs 1942-45. Joined G E in 1945. Exper includes design, application & sales engg and mkting mgmt for steam & gas turbines & generators for utility & indus plants. Boeing Co 1964-66, program & mkting mgmt for small indus gas turbines. Rejoined G E in 1966. Currently respon for strategic & Bus plans dev for steam & gas turbine & turbine generator bus. Mbr Bd of Governors 1983-1985 and VP Power Dept 1976-79 - ASME. Chmn US Natl Ctte 1975-1981 & VP, mbr of Bd. 1977-81 - Intl Combustion Engine Council (Paris). PE in NY and MA. *Society Aff:* ASME.

Manganaro, Francis F
Home: 8514 Paul Revere Ct, Annandale, VA 22003
Position: Vice Commander. *Employer:* Nava Sea Systems Command. *Education:* PMS/Naval Arch & Mar Eng/MIT; BS/EE/USNA; -/Civil Eng/Univ of RI; -/Met & Bus/Harvard Univ. *Born:* Feb 1925 Providence R I. 2 years Civil Engrg at U of R I. BSEE U S Naval Acad. PMS M I T. AMP Harvard U. RAd, U S Navy-designated engrg duty. Ch Engr destroyers & submarines. Assigned submarine design & tech logistics off. Served as Submarine Repair Officer, Dept Head for inspection, engrg, planning, and prod in Naval Shipyards. Commanded Puget Sound Naval Shipyard engaged in overhaul, conversion, & nuclear refueling of submarines, destroyers, cruisers, and aircraft carriers 1972-76. Reg PE in Conn. Mbr SNAME, ASNE, Soc of Sigma Xi, and Tau Beta Pi. *Society Aff:* SNAME, ASNE, ΤΒΠ, ΣΞ.

Manganello, Samuel J
Home: 143 Castle Dr, Pittsburgh, PA 15235
Position: Section Supervisor. *Employer:* USS, a Division of USX. *Education:* MS/Met Engr/Univ of Pittsburgh; BS/Met Engr/Univ of Pittsburgh. *Born:* 1/2/30. Metallurgical trainee at Wheeling Steel 1951. Metallurgist in Army Ordnance 1952-53. With US Steel Res Lab since 1953 (alloy steels, forgings, ordnance prods, heavy-prod processing, railroad prods, rolls, plant equip). Currently involved with plates and bars. Helped develop submarine-hull & armor steels & scale-retardant coating for steel. 12 pats, 14 publs. PE. Mbr AIME, ADPA, ASM; Mbr, Chrmn Pittsburgh Chap 1972-73, Natl Mbrship Ctte 1973-76. AIME Roll Tech Ctte 1980-present. Chrmn, Editorial Advisory Committee, Iron & Steelmaker Magazine. Active in community service & bands. *Society Aff:* ASM, AIME, ADPA.

Manganiello, Eugene J
Home: 329 Hillcrest Dr, Leucadia, CA 92024
Position: Ret. (Deputy Dir, NASA-Lewis Res Ctr) *Education:* BSE/Elec Engg/CCNY; EE/Elec Engg/CCNY. *Born:* Jun 8 1914. Res Engr, aeronautical propulsion starting in 1936 at NACA (now NASA) Langley Res Center, transferring to newly established NACA Lewis Res Center in 1942. Progressed through section Head, Branch

Manganiello, Eugene J (Continued)
Chief, Asst Chief of Res to Deputy Dir in 1961. Ret 1973. Fellow AAAS, AIAA (Dir 1963-64). Mbr SAE (Natl Pres 1972). Honorary Mbr Pi Tau Sigma. Reg PE in Ohio. 50th Anniv CCNY Gold Medal for Outstanding Govt Service 1970. NASA Exceptional Service Medal 1969. Enjoys sailing, swimming, ice skating, and reading. *Society Aff:* SAE, AIAA, AAAS.

Mangelsdorf, Clark P
Business: 949 Benedum Hall, Pittsburgh, PA 15261
Position: Assoc Prof of Civil Engrg *Employer:* Univ of Pittsburgh *Education:* ScD/Structures/MIT; MS/Structures/MIT; BS/CE/Swarthmore Coll *Born:* 10/28/28 Primary area of professional interest is in structural mechanics and design, with most recent research activity in the behavior of steel rod mine roof support systems (trusses). Served as consultant to industry on shell and adhesive problems. Has long sought the improvement of the educational process, with particular emphasis on design. Before joining Univ of Pittsburgh, was on the faculty of Swarthmore Coll. *Society Aff:* ASCE, SME-AIME, SSRC

Manges, Harry L
Home: 1424 University Dr, Manhattan, KS 66502
Position: Prof. *Employer:* KS State Univ. *Education:* PhD/Agr Engg/OK State Univ; MS/Agr Engg/KS State Univ; BS/Agr Engg/KS State Univ. *Born:* 6/18/28. Native of Hutchinson, KS. Agricultural Engr with Soil Conservation Service, USDA, 1949-1956. Employed by KS State Univ since 1956 as researcher and teacher, reaching rank of Prof in 1977. Have authored or co-authored numerous published articles on irrigation, water resources, and animal waste mgt to minimize pollution of the environment. Maj prof for 20 MS student. Mbr of Gamma Sigma Delta, Sigma Xi and Tau Beta Pi honorary. Reg PE in KS. Served on state, regional, and ntl comm of ASAE and SCSA. Mbr of local church bd and Kiwanis Club. Enjoy athletics. *Society Aff:* ASAE, ASEE, SCSA, IA.

Maniar, Gunvant N
Business: P O Box 662, Reading, PA 19603
Position: Genl Mgr, Res & Dev labs. *Employer:* Carpenter Technology Corporation. *Born:* May 27 1932 Karachi, Pakistan (U S Citizen). BS 1953-Chemistry & Physics, Gujarat U, India; MS 1958; Metallurgy, Stevens Inst of Tech; MSE 1960- Met Engrg, U of Michigan. Carpenter Tech Corp Genl Mgr, Res & Dev Lab 1977-Mgr, Physical Metallurgy & Analytical Chemistry 1975-77; Mgr, Physical Metallurgy 1973-75; Supervisor, Applied Physics 1964-73; Supervisor, Chemistry & Physics 1962-64. Crucible Steel Co-Staff Metallurgist 1960-62; U Mich-Res Asst 1958-60. Chmn ASM LV Chap 1973-74; Chmn Handbook Comm ASM 1978-80; Mbr Handbook Ctte ASM 1974-77; Chmn ASTM Comm E4 1978- ; First Vice-Chmn ASTM, Ctte E4 1973-77. Chmn Bern Twp Gen Auth 1974-77; U Mich Res Inst Fellowship 1959. Mbr ASM, AIME, EMSA, ASTM. Published 30 papers on Physical Metallurgy of Specialty Steels in Tech Journals. Co-editor of 3 STP's published by ASTM. Awarded 2 US pats.

Maniktala, Rajindar K
Business: 290 Elwood Davis Rd, Suite 307, Liverpool, NY 13088
Position: Pres *Employer:* Maniktala Assocs., P.C. *Education:* MS/Structural Engrg/ Univ of NB; BS/CE/Univ of NB *Born:* 5/10/36 Registered PE in NB, NY, OH, PA, WA, DC and MA. He was employed by the Bridge Design Division, St of NB Department of Roads for 7 yrs and by a major environmental engrg form as head of its Structural and Arch Dept. for over eleven years. Assumed current responsibility as Pres and Chief Engr of Maniktala Assocs, Engrs, Architects and Planners in 1980. He has served as Chairman of the AWWA Ad Hoc Ctte on Post-Tensioned Tanks, and is currently serving as a member on the AWWA Standards Ctte on Prestressed Concrete Water Tanks and PTI's Ad Hoc Ctte on Water Tanks, Silos and Oil Storage Tanks. Served as Pres of Rotary Club North Syracuse, 1974-75. Club Director and Chairman, Internal Service Ctte, 1975-78, and Community Service Ctte, 1979-82. Mbr Zoning Board of Appeals. *Society Aff:* ASCE, ACI, AWWA, PCI, PTI, USCOLD

Maninger, R Carroll
Home: 146 Roan Dr, Danville, CA 94526
Position: Consultant *Employer:* Self *Education:* BS/Phys/CA Inst of Tech. *Born:* 12/24/18. Vitro Corp of Am (1948-1953) as Dir of Physical Res & Dev. Gen Precision Inc (1953-1962) as Mgr, Sunnyvale Branch & later as Dir of Res, Librascope Div. From 1962, Lawrence Livermore Lab as Hd, Electronics Engg Res Div; Deputy Hd, Electronics Engg Dept; Hd, Environmental Studies Group; Hd, Tech Applications Group. Technical & mgt activities in electronic systems for missile ordnance, chem process plants, nuclear res & dev & pollution monitoring and control. Elected Fellow of IEEE "For Contributions to the Development of Measurement Techniques Used in the Nuclear and Environmental Sciences." *Society Aff:* IEEE, AAAS, ASA.

Manjoine, Michael J
Home: 25 Lewin La, Pittsburgh, PA 15235
Position: Consultant *Employer:* Retired *Education:* MS/ME/Univ. of Pittsburgh; BS/ ME/Iowa State Univ.; BS/EE/ Iowa State Univ. *Born:* 04/29/14 50 years experience in research on properties of material up to elevated temperatures. 30 years as consultant on testing, stress analysis, design and mechanical metallurgy. 20 years as consultant on design of nuclear reactors and power generating equipment. 52 published papers or books on materials under plastic flow, fatigue, creep and multiaxial stress. *Society Aff:* ASME, ASM, PVRC, ASME, MPC.

Manly, William D
Home: Rt #1 Box 197A, Kingston, TN 37763
Position: Exec VP, retired *Employer:* Cabot Corp *Education:* MS (1949)/Met/Univ of Notre Dame; BS (1947)/Met/Univ of Notre Dame; - (1949-55)/Met/Univ of TN *Born:* 01/13/23 William D Manly was a Dir of Cabot Corp and Exec VP of its Metals Group until 1986. He began his bus career as a metallurgist at Oak Ridge Natl Labs, Nuclear Div, Union Carbide Corp, Oak Ridge, TN. Manly is a grad of the Univ of Notre Dame where he received both BS and MS degrees in met. He was the 1972-73 Pres of the Amer Soc for Metals and is a Fellow of the ASM, AIME, and Amer Nuclear Soc, as well as a recipient of a merit award from the Amer Nuclear Soc. In 1974 he was recipient of the Notre Dame Coll of Engrg Honor Award. In 1981 he was made an Honorary Mbr of the Amer Soc for Metals. He is now a consultant to Oak Ridge Natnl Labs & Dir of Boston Edison Corp. *Society Aff:* ASM, AIME, ANS, NACE

Mann, Lawrence, Jr
Business: Dept of Ind Engg, Baton Rouge, LA 70803
Position: Prof *Employer:* LA State Univ. *Education:* PhD/Ind Engg/Purdue Univ; MS/Ind Engg/Purdue Univ; BS/Mech Engg/LA State Univ. *Born:* 2/12/26. Native of Baton Rouge, LA. Served with Army Air Corps 1943-46. Mech Engr for Esso Standard Oil Co 1950-1957 specializing in process plant design, inspection and maintenance. Sr design engr for Lummus Co 1957-1959. With LA State Univ to date. Assumed Chairmanship of Ind Engg Dept in 1977. Res interest is in process plant maintenance mgt. Enjoy reading, painting and stamp collecting. *Society Aff:* IIE, ASEE, ΣΞ, LES.

Mann, Robert W
Business: 77 Mass Ave, Rm 3-144, Cambridge, MA 02139
Position: Whitaker Prof of Biomedical Engrg. *Employer:* Massachusetts Inst of Tech. *Education:* ScD/Mech Engg/MA Inst of Tech; SM/Mech Engg/MA Inst of Tech; SB/ Mech Engg/MA Inst of Tech. *Born:* Oct 6 1924 Brooklyn N Y. Asst, Assoc, Prof of Mechanical Engrg, Germeshausen Prof, Prof of Engrg, Whitaker Prof of Biomedical Engrg, all at M I T. Supervisor of Design Div in Dynamic Analysis & Control Lab; Head of Design Div in Mechanical Engrg Dept; Director of Bioengineering Programs Whitaker College, MIT; pioneer in biomedical engrg, especially limb prosthesis, sensory aids for the blind and squovial joints; elected to N A E on basis of con-

Mann, Robert W (Continued)
tribs to engrg design & biomedical engrg; Inaugural mbr Inst of Medicine; Elected to Natl Acad of Sciences, Fellow; Fellow of the Amer Acad of Arts & Sciences. Fellow of the IEEE; Fellow of the American Association for the Advancement of Science; Fellow of the ASME; Awards: Alfred P Sloan, Talbert Abrams Photogrammetric Soc of Amer; Citation for Sensory Aids from the Blind of Mass; IR-100 Innovation; Goldenson United Cerebral Palsy for Tech in Rehabilitation; Germeshausen & Whitaker Chairs at M I T; ASME Gold Medal & Lissner Award in Biomedical Engg; Engg Societies of New England Award; Sigma XI National lecturer; ALZA distinguished lecturer; MIT Bronze Beaver; J.R. Killian Faculty Achievement Award, MIT. Tau Beta Pi, Pi Tau Sigma, Sigma Xi; Reg PE. Cons to medical, engrg, & legal firms. *Society Aff:* ASME, IEEE, AAAS, NAS, NAE, IOM, ORS, RESNA, AAAS

Mann, Sheldon S
Business: 3605 Warrensville Ctr Rd, Cleveland, OH 44122
Position: Exec VP *Employer:* Dalton.Dalton.Newport *Education:* BS/Bus Admin/ Miami Univ *Born:* 6/8/27 Graduated from Miami Univ in June, 1950 with a BS degree in Business Administration. After working in the fields of finance and administration, I joined the Dalton firm in 1957 as Business Mgr and over the years became a major Principal concentrating in the areas of finance and general mgmt. I became the Gen Mgr of the firm, and upon the death of our past Pres Calvin B. Dalton, assumed Exec VP position and Chief Operating Officer. *Society Aff:* PSMA, AMA, NAA

Mannheimer, Walter A
Home: Rua Dona Mariana 53, Rio de Janeiro, Brazil 22280
Position: Head, Materials Department. *Employer:* Cepel, Electric Energy Res Center. *Education:* PhD/Mater Sci/Carnegie-Mellon; MSc/Met Eng/Carnegie-Mellon; BSc/ Chem Eng/Univ of Brazil. *Born:* 01/01/32 Rio de Janeiro, Brazil. BSCE U of Brazil; MSc, PhD (Metallurgy and Materials Science) Carnegie-Mellon. Formerly engr & tech dir, Fekima S A; general cons; cons, Science & Tech, FINEP-Ministry of Planning. Previous pos: Prof & Head, Metallurgy Dept, Federal U of Rio de Janeiro 1967-74. Mbr, Advisory Ctte, Brazilian Natl Res Council, 1980-83. Pres, SC-1:9 (Corrosion) ABNT (Brazilian Assoc of Tech Norms), 1977-1987; currently Head, Materials Dept., CEPEL Electrical Energy Res Center & Adjunct Prof, Fed U of Rio de Janeiro; Brazilian Delegate to SC-15 (Dielect) CIGRE (Intl Con on Large High Voltage Elec Sys); NACE accredited corrosion specialist; Brazilian Metals Assn CVRD Prize 1972, 1975. About 50 tech pub. Enjoys classical music & tennis. *Society Aff:* ABRACO, ABM, NACE, CIGRE, RMS, ASM.

Mannik, Jaan
Business: 3539 Glendale Ave, Toledo, OH 43614
Position: Pres *Employer:* Mannik & Smith Inc *Education:* BS/CE/OH Northern Univ; MS/Struct Engrg/OH State Univ. *Born:* July 2 1930. Design Engr with Alden E Stilson in Columbus, Ohio 1955-57, Michael Baker Jr 1957-59, Bridge Bureau, O D O T 1959-62. Stark Co Bridge Engr 1962-64. T C Biebesheimer Engrg Co Toledo Ohio 1964-71. Partner in Mannik, Schneider & Associates 1971-1982. Pres of Mannik & Smith, Inc since 1983. PE in Ohio & Mich. Mbr of ASCE, NSPE, PEPP, AREA, ACI, AISC. Outstanding Young Men of Amer 1967, Tau Beta Pi, AISC Natl Design Award 1969. Trustee, Toledo Area Chamber of Commerce. Hobbies: photography and skiing (mbr of Natl Ski Patrol, Inc). *Society Aff:* ASCE, NSPE, AREA, ACI, AISC.

Manning, Francis S
Business: 600 South College, Tulsa, OK 74104
Position: Prof of Chemical Engrg. *Employer:* University of Tulsa. *Education:* PhD/Chem Eng/Princeton Univ; MS/Chem Eng/Princeton Univ; AM/Chem Eng/ Princeton Univ; B Eng/Chem Eng/McGill Univ. *Born:* Sept 1933 Barbados W I. B Engrg (Hons) McGill U; MS, AM, PhD Princeton U. Reg PE in Ok, Penn, & Texas. Taught 9 years at Carnegie-Mellon U before joining U of Tulsa. Currently Prof of Chemical Engrg. Author of one book & 60 tech publs. Awards: Barbados Scholar 1951, R M Hunt Silver Medal AIME 1969. A. Wilkens Mem Award Tulsa Sec AIChE 1983. Chmn Allegheny Sect ASEE 1967. Chrmn Tulsa Sect AIChE 1981. Res interests: thermodynamics, chemical kinetics, indus pollution control & enhanced recovery of oil & gas. *Society Aff:* AIChE, ASEE, AIME, TBΠ.

Manning, Melvin L
Home: 405 State Ave, Brookings, SD 57006
Position: Dean & Prof Emeritus, & Consultant. *Employer:* S D State U & Keene Corp. *Education:* DrEng/Engg/SD State Univ; MS/Elect Engr/Univ of Pittsburgh; BS/Elect Engr/SD State Univ. *Born:* 11/26/00. Native of Miller, SD. Prof at Cornell Univ, IL Inst of Tech, Univ of Pittsburgh, and SD S Univ. Dev Engr. Transformers-Westinghouse Electric Corp, Sharon PA; Ch Engrg Kuhlman Electric Co; Dev Engr McGraw-Edison Co, Canonsburg PA; Dean of Engrg, SDSU; Life Fellow, IEEE; Life Mbr Phi Kappa Phi, Sigma Xi, ASEE. Mbr Sigma Tau, Eta Kappa Nu, CIGRE. Prize paper Award Conference Electrical Insulation; Mbr IEEE Transformer Ctte, plus 6 IEEE Working Groups & Cttes dealing with electrical insulation & dry-type transformers. Holder of basic pats for silicone-insulated dry-type transformers & C3F8 gas usage for such. Intl Exec Ser Corps - Three Engg assignments - overseas - Ingenienia Electrica Industrial, SA. Mexico City (June 1974 and Jan-April 1975) and Intl Electric Co Seoul Korea (Apr-Jun 1979)-Laid out complete Tech for mfg and design 01 class 220 degree C-Dry-Type tranformers 01 distribution and power ratings- explosion-Free-The first for Korea. Recently, he was requested to assist the U. S. Navy Systems Command for transformer problem on air craft carriers. Areas of specialization No. 24-25-26-91 & 47. *Society Aff:* IEEE, CIGRE, ASTM, ASEE.

Manowitz, Bernard
Business: 2 Center St, Upton, NY 11973
Position: Chmn Dept Applied Sci *Employer:* Brookhaven Natl Laboratory. *Education:* MS/ChE/Columbia Univ; BS/ChE/Newark Coll Eng. *Born:* 3/6/22. Clinton Labs, Pile Engrg. 1947- : Brookhaven Natl Lab, design & const of BNL Radiochemcial Waste Disposal Sys, engrg res on fuel reprocessing & applied radiation. Respon Environ Res Progs: Atmospheric Scis, Oceanographic Scis, Land & Freshwater Environ Scis Admin Progs in Energy Scis, Environ Scis, Energy Tech, and Energy Sys Analysis. 1955-73 Editor Intl Journal of Applied Radiation; Dir Isotope & Radiation Div ANS; Chmn AEC Publication Policy Review Bd 1973; Chmn NAE Cttes: Radiation Sources 1970-72, Food Radiation Res 1972-74; Chmn Nuclear Engrg Div AIChE 1963-64; Radiation Industry Award ANS 1971; D.Sc. (Hon.) L.I.U. 1982. *Society Aff:* ANS, AIChE, AGU.

Mansfield, Vaughn
Home: 1002 Hartsville Pk, Gallatin, TN 37066
Position: Energy Consultant. *Employer:* I C Thomasson & Assoc; Vanderbilt U. *Education:* ME/-/Vanderbilt Univ; BS/-/Vanderbilt Univ. *Born:* Oct 8 1909 Viola TN. s. Horace Clifford, Ella (Parker) M. m. Bonnie Butler Jan 19 1963; 2 sons, 3 daughters. Asst Supt Bldgs, Vanderbilt U 1933-40; Ch Engr So Coal Co 1940-50; Ch Combustion Engr Peabody Coal Co, St Louis 1950- 59; Asst to Pres 1959-68; V P Dir Res 1968-74; Energy Cons 1974- . Dir Bituminous Coal Res, Solid Fuel Admin, WWII. 1st recipient Dist Alumnus Award Vanderbilt U 1969. Mbr AIME, Fellow ASME, Inst Fuel London, 44 pats in field. *Society Aff:* ASME, AIME.

Manson Benedict
Home: Apartment B-206, 108 Moorings Park Drive, Naples, FL 33942
Position: Inst Prof Emeritus. *Employer:* Mass Inst of Technology. *Education:* PhD/Phys Chem/MIT; MS/Phys Chem/MIT; BChem/Chemistry/Cornell Univ. *Born:* Oct 1907. Dev hydrocarbon separation processes for M W Kellogg Co prior to WW II; in charge process dev for Oak Ridge U-235 gaseous diffusion plant during war. Taught Nuclear Engrg at MIT 1951-78, Dept Head 1958-71. Atomic Energy Comm in charge operations analy 1951-52; Mbr Advisory Ctte Reactor Safeguards

Manson Benedict (Continued)
1948-58; Genl Advisory Ctte 1958-68, GAC Chmn 1962-64. Mbr Natl Acad of Sci & of Engrg, Amer. Acad. Arts & Sci, Amer. Philosoph. Soc.; AIChE Walker Award 1947, Amer Nuclear Soc Pres 1962-63, Compton Award 1969, Wilson Award 1966, Founders Award 1966, Fritz Medal 1975, AEC Fermi Award 1972, Natl Medal Sci 1976. *Society Aff:* AIChE, ACS, ANS.

Mansur, John C
Business: 1648 S Boston Ave, Tulsa, OK 74119
Position: Consulting Engr. *Employer:* Mansur Danbert Williams. *Education:* BSEE/Elec/OK Univ. *Born:* 9/3/40. Reg Engr in OK, TX, CO; Reg Land Surveyor in OK. Certified Plant Engr. Regional VP - Southwestern Region - illuminating Engg Society. Winner of local lighting award - "Most Creative–. VP and part owner of Consulting Engg Firm. Formerly with Public Service Co of OK, McDonnel-Douglas Aircraft Co, Crest Engg. *Society Aff:* IES, AIPE, IEEE, OSPE, NSPE.

Mantell, Charles L
Home: 447 Ryder Rd, Manhasset, NY 11030
Position: President. *Employer:* C L Mantell Assocs. *Education:* PhD/ChE/Columbia Univ. *Born:* Dec 9 1897. Cullimore Medal, N J Inst of Tech; Fellow AIChE; PE. Books: Engrg Materials Handbook, Electrochemical Engrg, 4 English, 2 Spanish, Polish, Czech, Japanese & Asian editions; Adsorption, 2 English, Japanese translation; Batteries & Energy Systems 2 editions; Natural Resins, English & Dutch editions; Carbon & Graphite Handbook, 3 English, 1 Russian edition; Solid Wastes, English, Russian. Our Fragile Water Planet, 1 English, 1 German; Tin, 3 English, 1 German edition; Calcium; Water Soluble Gums, 2 English editions. *Society Aff:* AIChE, AIME, ACS, AES.

Manvi, Ram
Home: 588 Glen Holly Dr, Pasadena, CA 91105
Position: Prof, Engrg *Employer:* CSLA *Education:* Ph.D/ME/WA State Univ; BE/ME/Osmania Univ *Born:* 01/26/39 Naturalized US Citizen in 1977. Over 25 years of teaching, res, and consltg experience in engrg, in India, Europe, Middle East, and USA. Served as Dean of Engrg and Tech from 1981-85. Reg Professional Mech Engr in CA, and consult for major aerospace companies in Thermal Sys Engrg. Enjoys travel, classical music, and hiking. *Society Aff:* ASME, ASEE, SME

Manz, August F
Business: Linde Div, Danbury, CT 06817
Position: Mgr-Regulations Tech *Employer:* Union Carbide Corp *Education:* MS/EE/NJ Inst of Tech; BS/EE/NJ Inst of Tech *Born:* 3/7/29 Joined the Linde Div of Union Carbide Corp in 1957, as a development engr in the Welding Products Dept. He has worked with all the arc welding systems and is the inventor of the "HOT WIRE" welding processes. He holds over thirty (30) US patents and more than a hundred foreign patents on power supplies, devices and other items for welding. Is a member of the Bd of Dirs of the AWS, the first official AWS historian, and author of the *Welding Power Handbook*. Was the recipient of a 1972 Meritorious Certificate Award of the AWS. Chrmn AWS Labeling and Safe Practices Ctte; Chrmn NEMA Electric Welding-Precautionary Labeling Ctte; Chrmn NFPA Safe Practices in Welding and Cutting SIB; Kean College of NJ Advisory Ctte-Indus Studies; Mbr VICA Welding Ctte. *Society Aff:* AWS, IEEE, CGA, NEMA, NFPA, VICA

Marable, James R, Jr
Home: 3675 Tree Bark Trail, Decatur, GA 30034
Position: Mgr, Civ Div. *Employer:* Simons-Eastern Co. *Education:* MS/Hydraulic Engg/LA State Univ; BCE/Civ Engg/GA Inst of Tech. *Born:* 10/5/24. Native of Atlanta, GA. Named Engr of the Yr in Private Practice by Metropolitan Atlanta Engrs & GA Soc of PE in 1974. Presently Mgr of Civ Div for Simons-Eastern Co, formerly Chief Engr of branch office & VP of subsidiary architectural corp. Past Pres & Life Mbr GA Engg Fdn, a philantropic fdn of forty-three GA Engg Socs. Reg PE in GA, FL, SC, NC, TN, PA, TX, OK, AL, VA. Designed structures for Raburn Plaza Garages, Wash, DC & Eastern Airlines Jet Base, Atlanta, GA. Elder in the Presbyterian Church in the US. Enjoy golf and fishing. Listed in Who's Who in the South & Southwest 1977-1979. *Society Aff:* ASCE, GEF.

Marachi, N Dean
Home: 110 Bando Court, Walnut Creek, CA 94595
Position: Principal *Employer:* The Mark Group *Education:* Ph.D./CE/Univ of CA - Berkeley; M.S./Geotech. Engrg/Univ of CA - Berkeley; B.S./CE/OR St Univ *Born:* 12/27/41 Started career as project engr with Converse Davis and Assocs in 1969 and was promoted to Senior Engr and later to Assistant Chief Engr during the next 3 1/2 years. Taught courses in Foundation Engrg, Earth Dam Design, and Earthquake Engrg at the Aryamehr Univ during 1973. Founder and Chrmn of the Bd of Tehran Berkeley Consulting Engrs through 1978 where directed two PSAR and two site selections for nuclear power plants, geothermal resources evaluations for five potential zones, and numerous geotechnical, geological and seismological studies for major industrial developments. Also Dir of Energy Technology Consultants (ENERTEC). VP and Mgr of of Converse Consultants San Francisco 1982-84. Founding partner and principal of The MARK Group since 1984. Enjoys skiing, swimming, and classical music. *Society Aff:* ASCE, SSA, EERI, ASTM, CSI.

Marburger, Thomas E
Home: 3300 NE 36th St-1115, Fort Lauderdale, FL 33308
Position: V P Engrg & Construction; Retired. *Employer:* Baltimore Gas & Elec Co, Ret. *Education:* BS/EE/Johns Hopkins. *Born:* Jan 1901. Grad Baltimore Poly Inst; Grad Johns Hopkins Univ. with honors; Mbr Tau Beta Pi; Life Mbr & Fellow IEEE; V P & Dir AIEE 1961-63; Dir IEEE 1966. Licensed Prof Engr MD. V P Baltimore Gas & Electric Co 196066. Dir Materials & Equip Div, Defense Electric Power Adm, U S Govt (GS17) 1951. Lt Md Natl Guard, resigned 1926. Mason: Scottish Rite 33rd Degree, York Rite Knight Templar, Shriner. Mbr German Soc of MD, Johns Hopkins Club & Maryland Club of Fla. 2 daughters, 6 grandchildren, four great grandchildren.

Marcatili, Enrique A J
Home: 2 Markwood Ln, Rumson, NJ 07760
Position: Department Head. *Employer:* Bell Laboratories. *Born:* 1925 in Argentina. Elec & Aeronautical Engr from Cordoba U, Argentina 1947-48. Aircraft design engr & Univ Prof in Cordoba until 1953. With Bell Telephone Labs since then. Conducted res on millimeter waveguide until 1961 & on optical communications after becoming Head of Transmission & Circuit Res Dept. Author of tens of papers & holder of tens of pats. Fellow IEEE 1967; co-winner W R G Baker Prize Award 1975; Mbr NAE 1976.

Marchello, Joseph M
Home: 5000 Edgewater Dr, Norfolk, VA 23508
Position: President & Prof. *Employer:* Old Dominion University *Education:* BS/ChE/Univ of IL; PhD/ChE/Carnegie-Mellon Univ. *Born:* Oct 6 1933. Native of East Moline Ill. Taught at Oklahoma State U 1959- 61, & U of Md 1961- . Chmn Chem Engrg Dept U of Md 1967-73; Provost, Div Math & Phys Sciences & Engrg 1973-78. Univ of MO 1978-1985 Chancellor & Prof.; Old Dominion University 1985- President & Prof. *Society Aff:* AIChE, ACS, NSPE, AAAS.

Marchetti, Robert J
Business: 30 W Superior St, Duluth, MN 55802
Position: Sr V P - Engg & Operations. *Employer:* Minn Power & Light Company. *Education:* BSEE/Power & Elec/MTU. *Born:* 12/1/25. Native of Ironwood MI. IBM Customer Engr 1950-51. Joined MP&L as staff engr 1951 proceeding to Asst Statistician, Relay & Communication Engr, Sys Planning Engr, Mgr of Engrg, Ch Engr and VP Power Supply. Assumed current respon as Sr V P Power Supply & Engg in 1978. Respons include power plant & Sys Engg, const, maintenance, & operation.

Marchetti, Robert J (Continued)
Active in MAPP Power Pool, Mbr IEEE, Duluth Chamber of Commerce, Duluth Rotary. *Society Aff:* IEEE.

Marchinski, Leonard J
Home: 206 Tower Ln, Narberth, PA 19072
Position: Director, Structures Technology. *Employer:* Boeing Vertol Company. *Education:* BS/AE/Notre Dame; MS/-/Drexel Inst of Tech. *Born:* Nov 5 1925. BSAE U of Notre Dame 1947. MS Drexel Inst of Tech 1959. 25 years exp in stress analysis & engrg mgmt. Background includes both fixed & rotary wing aircraft. Has worked a fixed wing aircraft such as Convair F-106, Martin B-57, Avro CF-105, Republic F-84. All rotary wing exp is at Boeing Vertol on the CH-46, CH-47, HLH, & YUH-61A (UTTAS). Engrg mgmt exp includes V P of an engrg cons firm. Presently directs the activities of the Structures Tech org at Boeing Vertol involving stress analysis, weights, cost engrg, & structural test labs. Has been quite active in Amer Helicopter Soc, presently holding post of Director at large . *Society Aff:* AHS, AIAA, AAAA, SAWE.

Marchman, James F, III
Business: Aero & Ocean Engrg Dept, Blacksburg, VA 24061
Position: Professor. *Employer:* Virginia Polytech Inst. *Education:* PhD/ME/NC State Univ; BS/AE/NC State Univ. *Born:* 5/15/43. On VPI faculty since 1968. Dir of VPI Subsonic Wind Tunnels. Res in subsonic aerodynamics & hear transfer, aircraft wake turbulence, vortex and low RE. Aerodynamics. Mbr AIAA Student Activities Ctte; Assoc Fellow AIAA, Mbr ASEE. Coord undergrad acad pro & advising in AOE Dept. Faculty advisor of VPI AIAA; Chmn College of Engg Curriculum Cttee. Mbr AIAA Educational Activities Ctte. Also, Private Pilot. *Society Aff:* AIAA, ASEE.

Marconi, William
Business: 460 McAllister St, San Francisco, CA 94102
Position: Chief, Bureau of Engineering *Employer:* S F Dept of Public Works. *Education:* BCE/Civil Engrs/Univ of Santa Clara; Certif/Traffic Engrg/Yale Univ. *Born:* 10/6/26. Mount Vernon, NY. Army service WWII. Has worked for S F Dept of Public Works since 1948 and is presently Chief, Bureau of Engineering. Is in charge of S F's Bureau of Engineering which includes mechanical, electrical, structural, traffic, highways, construction and park engineering and surveying and mapping. Served as Secy-Treas, VP & Pres of Western District of the Inst of Traffic Engrs & as Dist Dir for the Inst of Transportation Engrs. Has also served as VP, Pres & Immediate Past Pres of the Inst of Transporation Engs. Has authored numerous res papers on traffic engg & is a reg PE in CA *Society Aff:* ITE, APWA.

Marcot, Guy C
Business: 11E Orange Grove, 1813, Tucson, AZ 85704
Position: President. *Employer:* M M & M Consulting Services. *Education:* ChE/Chem Engg/Columbia Univ; BS/Chem Engg/Columbia Univ. *Born:* Apr 1913. March 1975- , Pres M M & M Cons Services. Specialty-Magnetic Materials & Media. June 1972-March 1975 V P Commercial Dev for Orrox Corp in Opelika Ala. Magnetic Iron Oxides, Process Dev & Plant Design; Prod Dev. April 1968-June 1972-Mgr Magnetic Materials for Bell & Howell. May 1963-Jan 1968 Mgr Commercial Dev for Amer Cyanamid Co, Pigments Div. Professional Societies: SMPTE, AES, IEE, A Ceramic S, AIChE, N Y Acad of Sci. *Society Aff:* SMPTE, AES, IEEE, ACS, AIChE, ACS, NYAS.

Marcue, Donald R
Business: 550-15th St, P O 840, Denver, CO 80201
Position: Commercial & Indus Serv & Utilization Mgr. *Employer:* Public Service Company of Colo. *Education:* BS/Gen Engrg/IA State Univ. *Born:* Oct 1923. BS General Engrg Iowa State U 1948. With Public Service Co of Colo since July 1948. Wide exp in sales engrg & appl of power & gas. Grad Air Force Installations Engrg Course. Dir Colo Mining Assn. Regional VP Illuminating Engrg Soc 1973-75 & Dir 1974-75. Reg P E Colorado. Sr VP IES 1979-80. Pres IES 1980-81.

Marcus, Gail H
Business: Office of Res, US Nuclear Regulatory Commission, Washington, DC 20555
Position: Deputy Dir *Employer:* Nuclear Regulatory Comm *Education:* ScD/Nuclear Engrg/MIT; SM/Phys/MIT; SB/Phys/MIT *Born:* Currently respon for long-range planning & policy dev for the NRC's res program & for regulatory review & control of NRC rulemaking. Previously: conducted experimental res on radiation damage at MIT and US Army Elec Command; managed tech and policy analysis at Analytic Services and Congressional Res Service covering all areas of sci and tech. ANS Fellow. Served on ANS Bd of Dirs and Exec Ctte. Mbr of ANS delegation to China, 1983. Currently on MIT Corp Visiting Ctte in Nuclear Engrg, on America Management Association R&D Council, & on Lowell U. Industrial Advisory Council. Author of numerous papers on energy policy, nuclear and other energy techs, and risk assessment. *Society Aff:* ANS, AAAS

Marcus, Harris L
Business: Mech Eng Dept IMS&E Program, Austin, TX 78712
Position: Professor. *Employer:* University of Texas. *Education:* PhD/Matls & Sci/Northwestern; BS/Met/Purdue. *Born:* 7/5/31. Harry L. Kent Prof Mech Engrg/Dir of Center for Materials Sci & Engrg, U of Texas at Austin 1975-present; Grp Leader Fracture & Metal Physics Grp 1971-75, Mbr Tech Staff 1968-71, Rockwell Internatl Sci Center, Thousand Oaks, Ca; Mbr of Tech Staff Texas Instruments 1966-68. Main areas of res: fracture & fatigue of struct alloys with emphasis on environmental effects; surface phenomena & influence of 2-dimensional chem networks on physical & mech properties of matls; and electronic packaging of materials. These areas are aimed at improving processing producibility & reliability of matls. Mbr Bd/Dir TMS-AIME 1984-86, 1976-78, Prog Vice Chmn Metals Sci 1976-77, Prog Chmn 1977-78; Mbr Exec Board IMD-TMS-AIME 1974-76; Fellow ASM, Mbr ASTM, APS, ASME. PE in Texas. Secy/Treas of US/PRC Bilateral Metallurgy Symposium (1981) *Society Aff:* TMS-AIME, APS, ASM, ASTM, ASME, MRS.

Marcus, Michael J
Home: 8026 Cypress Grove Ln, Cabin John, MD 20818
Position: Div Chief *Employer:* Fed Comm Commission *Education:* SB/EE/MIT; ScD/EE/MIT *Born:* Native of Newton, MA. Served in US Air Force as Proj Officer in underground nuclear test detection res 1972-75. Mbr of Tech Staff, Inst for Defense Analyses 1975-79 dealing with comm electronic warfare. Joined Office of Sci and Tech, Fed Comm Commission in 1979 becoming Chief, Tech Analysis Div in 1981. Respon for in-house tech studies in support of telecomm policy and deregulatory initiatives. Visiting Associate Prof of Elect Engrg MIT, 1986. *Society Aff:* IEEE

Marcuse, Dietrich
Business: Crawford Hill Lab, Holmdel, NJ 07733
Position: Distinguished Member of Technical Staff. *Employer:* Bell Laboratories. *Education:* PhD/Elec Engg/Technisch Hochschule, Karlsruhe. *Born:* Feb 1929. Native of Germany. Naturalized American 1966. Worked at Siemens & Halske in Berlin 1954-57. Joined Bell Labs in 1957. Taught at U of Utah 1966-67. Spent 1 month in 1975 as Visiting Fellow at Australian Natl Univ in Canberra. Fields of interest: Quantum Electronics, Electromagnetic Theory, Optics. Author of 4 books and approx 170 tech papers. Fellow of Inst of Electrical & Electronics Engrs & the Optical Soc of Amer. Enjoy canoeing & travelling. *Society Aff:* IEEE, OSA.

Marcuvitz, Nathan
Home: 7 Ridge Drive E, Great Neck, NY 11021
Position: Professor of Appl Physics. *Employer:* Polytechnic University. *Education:* PhD/Electrophysics/Polytechnic Inst of NY. *Born:* Dec 29 1913. Radiation Lab at MIT 1942-46; Polytechnic Inst of Brooklyn 1946-66; School of Engrg & Sci, N Y U 1966-73; Polytechnic Inst of N Y 1973- . At PIB Prof of Elec Engrg 1951-65; Dir of Microwave Res Inst 1957-61; Prof Electrophysics 1961-65; V P for Res & Acting

Marcuvitz, Nathan (Continued)
Dean, Grad Center 1961-63; on leave 1963-64 & served as Asst Dir Defense Res & Engrg Res, Dept of Defense, Washington D C. Dean of Res & Dean of Grad Center 1964-65. First Inst Prof at PIB 1965-66. In 1966 appointed Distinguished Prof of Appl Phys, School of Engrg & Sci at N Y U. Returned to newly merged Polytechnic Inst of N Y as Prof of Appl Phys 1973. Elected to National Academy of Engineering, 1979; Institute Prof. 1979; Fellow IEEE; Mbr Amer Physical Soc, Eta Kappa Nu, Sigma Xi, Tau Beta Pi. *Society Aff:* NAE, IEEE, APS.

Margiloff, Irwin B
Business: 26 Windmill Rd, Armonk, NY 10504
Position: Principal (Pres) *Employer:* Margiloff & Associates *Education:* BChemE/Cornell Univ. *Born:* 11/19/31. Since 1982 chief executive & consultant with Margiloff & Associates (business and technical services); formerly with Publicker with responsibility for development (fuels, biotechnology); with Scientific Design Co in process design, proj engg, & commercial dev (petrochemicals, cement); with Allied Chem (chlorine, alkalies); & at Joliet Arsenal (explosives). Thirteen patents in chem processes & equipment, electronic memory systems, transportation systems. Mbr & Trustee (for ten yrs) of City Club of NY, & active in other civic affairs; lectures & papers on civic, technical & profl topics. Formerly Chrmn of AIChE's Profl Dev Committee; Past Pres, Cornell Soc of Engrs, formerly Chrmn, Metropolitan Engrs' Council on Air Resources. Formerly AIChE repr to and Chrman, AAES Comm on Productivity and Innovation and alt repr to Public Affairs Council. *Society Aff:* AIChE, ASME, IEEE, CMRA.

Margolin, Harold
Business: 333 Jay St, Brooklyn, NY 11201
Position: Professor of Physical Metallurgy. *Employer:* Polytechnic University *Education:* DEng/Metallurgy/Yale Univ; MEng/Metallurgy/Yale Univ; BEng/Metallurgy/Yale Univ. *Born:* Jul 12 1922. Married Elaine Marjorie Rose 1946; Children: Shelley, Deborah & Amy. 1949-56 Res Assoc, Res Scientist, N Y U, Bronx N Y. 1956-62 Assoc Prof N Y U, 1962-73 Prof N Y U, 1973- Prof Polytechnic University Fields: Titanium, Physical Met, Plastic Deformation and Fracture. Cons at various times to: General Electric, Detroit Diesel Allison Div General Motors, Timet, Reactive Metals, AIRCO AMMRC, Whittaker, Amer Cynamid, Gould Inc., Grumman Aerospace Corp. Rocketdyne Div N Amer Rockwell. Chmn N Y Chap ASM 1961-62. Chaired various natl cttes, AIME, ASM. Bd/Dir Acta Met 1968-70. Awards: 1967 N Y Chap ASM Educ Award, 1975 Fellow ASM. 121 pubs & 2 pats on Titanium Alloys 1981 Polytechnic Chap. Sigma Xi Research Award. 1983 Theodore Krengel Visiting Prof Dept of Materials Engrg Tech, Israel. *Society Aff:* AIME, ASM, ASTM, ΣΞ, AAAS.

Margolin, Stanley V
Business: 600 Maryland Ave, SW, Washington, DC 20024
Position: Conslt & Pres *Employer:* Network Consulting, Inc *Education:* BS/CE/MIT; MS/CE/MIT. *Born:* 8/29/28. Consulting for Arthur D Little, Inc and others. Had responsibility at ADL over 31 years for maj studies dealing with process industries. These studies involved tech & economic feasibility of new pocesses & tech. Responsible for environmental studies assessing impact of regulations on ind. Led maj studies of environmental impact for most Am Steel Corps. Prior employment with E I DuPont as design engr. *Society Aff:* AIChE, AIME-TMS, AIME-ISS, FMS.

Margolis, Donald L
Business: Dept of Mech. Eng, Univ of Ca, Davis, CA 95616
Position: Prof *Employer:* Univ of CA. *Education:* PhD/ME/MIT; Mech Engg/ME/MIT; MSc/ME/MIT; BS/ME/VPI. *Born:* 11/13/45. After receiving my PhD from MIT in June, 1972, I joined the Mech Engg faculty at the Univ of CA at Davis. I have developed an expertise in the area of modeling, simulation and control of dynamic systems. I teach undergrad and grad level courses and perform res in this area and have authored over 60 technical publications in 7 different natl and intl journals. *Society Aff:* ASME, SAE.

Margon, Irving B
Business: 50 UOP Plaza, Des Plaines, IL 60016
Position: Corp VP Marketing. *Employer:* Procon International Inc. *Education:* Doct Studies/-/Univ of Southern CA; MSCHE/Chem Engg/Univ of Southern CA; BSCHE/Chem Engg/ College of City of NY. *Born:* 10/22/23. Native of NY, but has divided adult life between CA and Europe. Following Doctoral Studies at USC, worked for Parsons, Holmes and Narver, and Bechtel on various refinery, missile, nuclear and petrochemical assignments as engg. Grad to sales and management assignments. Moved to London with Bechtel as Sales Mgr in 1962. Joined Conoco assignments in Ireland and Spain, joined Procon in 1971 as Sales Manager and later Managing Dir of English profet center. Returned to IL in 1976 with Procon as International VP Marketing and Sales. In Jan, 1981, made Pres of joint venture Canadian Co, Shawinigan Procon Co, Calgary, Alberta, in addition to other marketing duties. Enjoys golf, bridge, and stamp collecting. *Society Aff:* AIChE, IP, SCI, OIC.

Mariacher, Burt C
Home: 7232 W Cedar Circle, Lakewood, CO 80226
Position: Director, Process Div. *Employer:* Colo School of Mines Research Inst. *Education:* EM/CO Sch of Mines *Born:* 1918. Following service WWII, spent 7 years in various postions in the mining indus relating to minerals processing plant engrg, const & opers. Joined Colo School of Mines Res Inst 1952. In the positions of Proj Engr & Mgr of the Met Div, conducted & supervised applied res incl pilot plant opers employing minerals separation, hydrometallurgy, & pyrometallurgy processes to dev methods for recovery of metals & indus minerals. Dir, Met Div 1969. SME-AIME Bd/Dir 1974-76. Distinguished Mbr Award 1976. AIME Dir-VP 1978-79. *Society Aff:* AIME.

Mariella, Raymond P
Home: 1800 Old Meadow Rd, Unit 1005, McLean, VA 21102 *Education:* DSc/Chemistry/Carnegie Inst of Tech; MS/Organic Chem/Carnegie Inst of Technology; BS/Chemistry/Univ of PA. *Born:* Sept 1919. Post Doctorate at U of Wisc; Asst Prof Northwestern U; Prof & Chmn of Chem Dept of Loyola U of Chicago; Currently: Exec Dir, Amer Chem Soc. Dean of the Grad School of Loyola. Past exp includes: Bd/Dir Natl Chemical Exposition; Fellow AAAS, 10 years TV public service sci progs; Sci Advisor to Governor of Ill; Faculty Trustee of Lincoln Acad of Ill; Chmn of Natl Ctte on Professional Relations & Mbr of Council Policy Ctte of Amer Chem Soc; Exec Ctte of Council of Grad Schools in U S A; Secy-Treas Midwestern Assn of Grad Schools, Mbr of Bd/Dir ACS; ACS Comm on Public, Profl & Mem relations. Currently: Mbr, Bd/Dir, Council of Engg & Scientific Soc Execs. *Society Aff:* ACS, AAAS, CESSE.

Marienthal, George
Home: 10202 Parkwood Drive, Kensington, MD 20895
Position: Manager, Program Development *Employer:* Computer Sciences Corporation *Education:* MS/IE/Stanford Univ; MBA/Bus Admin/Am Univ; BS/Gen Engr/US Naval Acad. *Born:* 11/15/38. Native of Atlanta, GA. Served five yrs in USAF. Sr Res Assoc for Logistics Mgt Inst 1967-1971. Dir of Regional Liaison & Dir of Fed Activities at US Environmental Protection Agency 1971 to 1975. Deputy Asst Secy of Defense for Energy, Environment & Safety 1975-1981. VP of Survival Tech Inc, 1981-1984. Dir of Office of Water Policy, US Environmental Protection Agency 1984-85. Deputy Asst Secy of Agriculture for Administration 1985-1986. Deputy for Advanced Programs, Titan Systems, Inc 1986-1987. Currently, Manager Program Development of Computer Sciences Corporation. Responsible for corporate marketing of telecommunications and automatic data processing systems. *Society Aff:* IIE, ADPA.

Marini, Robert C
Business: One Center Plaza, Boston, MA 02108
Position: Pres. *Employer:* Camp Dresser & McKee Inc. *Education:* SM/Sanitary Eng/Harvard Univ; BS/Civil Eng/Northeastern Univ. *Born:* 9/29/31. in Quincy, MA. Officer-in-chrg of several CDM's large water, wastewater, and solid waste projs. Joined the original Camp Dresser & McKee partnership as a jr engr in 1955. Served as proj engr until 1964. From 1961-64, instructor in hydraulics at the Lincoln Inst of Northeastern Univ. Named a partner in 1967, a Sr VP of CDM in 1970, Pres of Environmental Engg Div in 1977, Exec VP in 1982 and President in 1984. Fellow Amer Soc of Civil Engrs; Diplomate Amer Acad of Environmental Engrs; Tau Beta Pi, Young Engr of the Yr 1966; "Man of the Year" Award of NEAPWA, 1982. Distinguished Eagle Scout Award, Boy Scouts of America, 1986. Member Civil Engineering Adv. Comm. Worcester Polytechnic Institute 1985-, University of MA - 1986 - Member, Corporate Board and Board of Overseers Northeastern University, Boston 1983-. Reg P E in ME, MA, NY, RI, NH, CT, MI, VT, CA, NC, WA, CO, FL, WI, TN, AZ, LA, IL, OH, & VA. *Society Aff:* WPCF, AWWA, APWA, AAEE, ASCE, IAWPRC.

Marino, Miguel A
Business: 119 Veihmeyer Hall (LAWR), Univ Calif, Davis, CA 95616
Position: Prof *Employer:* Univ of CA *Education:* Ph.D./Engrg Systems/U.C.L.A.; M. S./Hydrology/NM Inst of Mining and Tech; B. S./Petro. Engrg./NM Inst of Mining and Tech. *Born:* 11/10/40 Native of Cienfuegos, Cuba. Naturalized U.S. citizen. Asst. geohydrologist, NM State Engrs Office, 1965. Asst. hydrologist, IL. State Water Survey, 1965-69. Postgraduate Research Engr, UCLA, 1969-72. Joined Univ of CA at Davis as Asst. Prof in 1972; Assoc Prof in 1976; Prof of Civil Engrg, Agric. Engrg, and Water Science since 1980. Teaching of undergrad. and grad. courses in groundwater hydrology and water resources systems analysis. Recipient of various research grants. Recipient of the Richard R. Torrens Award of ASCE, 1986, and Best Research-Oriented Paper published in the Journal of Water Resources Planning and Management, ASCE, 1986. Water resources consultant. Editor of Journal of Water Resources Planning and Mgmt of ASCE, Assoc. Editor of Water Resources Bulletin of AWRA, and mbr of several cttes of ASCE, AGU, and AWRA. Enjoys world literature and fishing. *Society Aff:* ASCE, AWRA, AGU, ΣΞ, AAAS, IWRS

Marion Johnson PE
Business: Polychem Technology, Inc, 1209 Decker Drive, Suite 207, Baytown, TX 77520
Position: Pres. *Employer:* Polychem Technology, Inc. *Education:* BS/Che/Univ of TX; BA/Chemistry/Univ of TX. *Born:* 5/3/31. Consultant and Engr for clients in the Plastics and Petrochemicals Industry. His field includes Business Planning, Mergers/Acquisitions, Chemistry, Process Engg, Patents, Operations, and Legal. His maj focus is on polymer chemistry and polymerization process, with related applications of this tech. With Union Carbide for seventeen yrs, he served many upper level posts in operations, planning and control, and engg in the Chem and Plastics Div. In 1970, he established his consulting and engg practice. His operating co also trades in petrochemical and plastic raw mtls. *Society Aff:* AIChE, NSPE, SPE, ACS, USNI.

Mariotti, John J
Business: 1700 W Third Ave, Flint, MI 48502
Position: Professor of Industrial Engrg. *Employer:* GMI Engineering & Management Institute. *Education:* MSIE/Indust Engrg/Wayne State Univ; BSIME/Indust & Mech Engrg/Univ of MI. *Born:* Nov 10 1924. Native of Rochester N Y. Combat Engrs 1943-46. Ford Motor Corp 1951-61. Methods, Process Prod Engrg & Price Administrator. Developed 'Slave' Starter manufacturing sys and 'Motorcraft' aftermarketing prog. Lecturer Wayne State U. Grad Engrg prog 1961. With GMI since 1961-cons, teaching & developing courses & seminars. Papers published in SAE, ASME, IE, MHI, CICMHE, IMMS, Decision Sciences & Personnel Admin Journals. Advisor, Indus Engrg Journal 1971-72. Dir, Facils Planning Div IIE 1976 (advisor 1975). Natl Ctte SAE. Natl College Indus Ctte for Material Handling Educ 1964-68, 1974-77, Elected Fellow IIE 1979. Published Chapter in Industrial Eng Handbook 1982, Also production Handbook 1986 both with Wiley Brothers and Coauthor of textbook in 1984; McGraw Hill. *Society Aff:* IIE, SAE, ASEE.

Mark, Hans M
Business: , Washington, DC 20330 *Employer:* Dept of the Air Force *Born:* June 1929 Mannheim Germany. Naturalized U S Citizen 1945. AB U C Berkeley 1951; PhD MIT 1954. Taught at Boston U, MIT, U C Berkeley & U C Davis. Chmn & Prof, Dept of Nuclear Engrg, U C Berkeley 1964-69. Dir NASA Ames Research Center 1969-77; Undersecy of The Air Force 1977-79, Secy of The Air Force 1979- ; Cons Prof of Engrg Stanford U 1973- . Outstanding Engrg Teacher, U C Berkeley 1966; NASA Distinguished Serv Medal 1972 and 1977. Distinguished Public Service Medal, Dept. of Defense 1981, Fellow AIAA, Amer Physical Soc; Mbr Nat Acad of Engrg.

Mark, Melvin
Business: 93 Union St, Suite 400, Newton Centre, MA 02159
Position: Engrg Conslt *Employer:* Self *Education:* ScD/Engrg/Harvard Univ; MS/Mech Engrg/Univ of MN; BME/Mech Engrg/Univ of MN. *Born:* Nov 15, 1922. Native of St Paul Minnesota. Instructor North Dakota State Univ 1943-44; US Army Air Force 1944-45; Instructor Univ of Minnesota 1945-47; teaching & res Fellow Harvard 1947-50; Proj Mgr General Electric Co 1950-52; Dept Head Raytheon Co 1952-56; Pres Cambridge Dev & Engrg Corp 1956-59; Lecturer MIT 1955, Brandeis 1958; Dean of Faculty Lowell Tech 1959-62; Prof & Dean of Engrg Northeastern Univ 1962-79. Provost and Sr VP 1979-84; conslt 1984-; Lincoln Arc Welding Foundation Prize 1947. Author of numerous articles on thermodynamics & heat transfer & 3 books on thermodynamics. Reg P E Mass & Minnesota. Life-Fellow ASME. Life Mbr ASEE, Chmn New England Sect ASEE 1976-77. Mbr of Sigma Xi, Tau Beta Pi, Pi Tau Sigma, Phi Kappa Phi. Enjoy skiing & tennis. *Society Aff:* ASME, ASEE, ΤΒΠ, ΦΚΦ, ΠΤΣ, ΣΞ.

Mark, Robert
Business: Dept of Civil Engrg, Princeton, NJ 08544
Position: Prof of Civil Engrg/Arch *Employer:* Princeton Univ *Education:* B/CE/City Univ of NY *Born:* 7/3/30 Native of City Island, NY. Stress analysist, Combustion Engrg Nuclear Power Div 1952-55; project engr, ARDE Assocs 1955-57; research staff engr, Princeton Univ Plasma Physics Lab 1957-1962. Teaching and research in the Dept of Civil Engrg at Princeton since 1962. Joint appointment with the School of Architecture at Princeton since 1973. Author of more than 100 publications dealing mainly with optical methods of experimental stress analysis and applications ranging from deep-undersea research vessels to Gothic cathedrals. Two films: Canadian Broadcasting (CBC): *The Nature of Things*: "The Cathedral Engineers–, 1983; *NOVA*: "The Masterbuilders–, 1987. Technical editor, *Experimental Mechanics* 1969-1973. Tech consultant Exxon Res & Engg, Grumman Aerospace, Naval Civil Engg Lab, NY City Transit Authority, RCA, Res Cottrell, Washington Cathedral. Aff: SEM, SHOT, Sigma Xi, Tau Beta Pi, Soc of Architectural Historians. *Society Aff:* SEM, SHOT, ΣΞ, ΤΒΠ, SAH.

Markels, Michael, Jr
Business: 6850 Versar Ctr, Springfield, VA 22151
Position: President. *Employer:* Versar Inc. *Education:* DES/Chem Eng/Columbia Univ; MS/Chem Eng/Columbia Univ; BS/Chem Eng/Columbia Univ. *Born:* Feb 1926. Native New York City. Technologist Shell Oil Co 1948-52 performing economic & engrg studies; Res Assoc Columbia Univ Chem Engrg Dept 1952-57; joined Atlantic Res Corp 1957; Dir Advanced Tech Dept Res Div 1961; Asst Genl Mgr Res Div & Deputy Dir Corp Res Lab Susquehanna Corp 1968; Pres & Ch Exec Officer Versar Inc 1969; activities center on environmental, ecological, energy & tech cons projs for govt & private indus. Recipient Natl Capital Sr Engr-of-the-Year Award 1967. Grad Sch of Bus Adm, Harvard Univ. Smaller Co Mgmt Program completed 1979. *Society Aff:* AIChe, ACS, AIAA, NSPE, ANS.

Markle, Joseph
Business: 110 East 42nd Street, New York, NY 10017
Position: Sr Partner-French, Fink, Markle & McCallion *Employer:* Self employed. *Education:* LLB/Law/Yale Law Sch; BA/History & Economics/Yale College. *Born:* 4/29/00. Served with Army Corps of Engrs 1942-45; Legion of Merit; retired as Col. Award SAME Gold Medal (1969) and again in 1980 for Disting Serv to the Soc for bldg the NY City Post Scholarship Fund prin to over $ 800,000 with 63 scholarships annually to engg students in 45 colleges'. Prin now exceeds $1,000, 000 with scholarships awarded annually to 74 engg students in 45 colleges. SAME Charter Fellow. Dir & Pres Scholarship Fund since 1963, made honorary member in 1980. Sr Partner French, Fink, Markle & McCallion representing since 1924 engg, arch, contracting firms - typical projs: Natl Gallery (WA), Rockefeller Center, UN Lincoln Center, Bermuda Base, Oak Ridge; military establishments in Europe, Middle East, Far East, North Africa; civil works, flood control, hydro-elec power projs. *Society Aff:* ABA, NYSBA, AUSA, SAME.

Marko, Hans
Business: Arcisstr 21, 8000 Munchen 2, Germany
Position: Prof. *Employer:* Lehrstuhl fur Nachrichtentechnik *Education:* Abitur/-/Honterus-Gymnasium; DiplIng/-/Technische Hochschule Stuttgart. *Born:* 2/24/25. H Marko has been prof of Communications Techniques at the Technische Univ Munchen, FR Germany, since 1962. He is also dir of the school's Inst for Communications Tech &, since 1965, he has been a mbr of the Admin Committees of the Soc for Communications Techniques (VDI) & of the German Soc for Cybernetics. He received the diploma in communication techniques & the DrIng degree from the Technische Hochschule Stuttgart, Germany, in 1951 & 1953 respectively. In 1952 he joined the Std Elektrik Lorenz AG as a dev engr. There he founded & directed the Dept of Fundamentals for Transmission Techniques, and he has worked on projs concerning multichannel telephony, H Data Transmission & radio techniques. In 1959 he became chrmn of the Committee on Info & Systems Theory of the Soc for Communications Techniques within the Verband Deutscher Ingenieure. He has lectured at the technical Univs of Stuttgart, Karlsruhe and Munchen. *Society Aff:* NTG, DAGM, IEEE, DGK.

Marks, Albert B
Business: 3717 East 43rd St, Tulsa, OK 74135
Position: Self-employed *Education:* BS/Engrg Physics/Univ of Tulsa *Born:* 1/15/18 1971-: Industrial/Consumer Product Res, Conception, and Development; Mechanical/Electromechanical Applications Engrg. Rockwell, 1962-: Mechanical Design, Apollo/Saturn Test Facilities; Project Engr, Space and Lunar Hardware Research; Design Specialist, Ground Equipment and Major Procurement and Construction Specifications. Dresser 1953-: Production Mgr, Neutron Generators; Senior Project Supervisor, Early JPL Lunar Geology Analytical Instrumentation Studies, Oilwell Logging R&D, Industrial Applications: Nuclear, Acoustic, and Electromechanical Tecnologies; Standards Development. Amoco 1945-: Research Engr, Oil Exploration, Production, and Surveying Instrumentation. Active in Regional/National Affairs of SAE, ASM, and AIAA -- Regional Past-Chrmn of each. Registered PE, OK; Past-mbr, Bd of Dirs, SAE. *Society Aff:* AIAA, ASM, IEEE, NSPE, SAE

Marks, Allan F
Home: 16 Vineyard Dr, San Rafael, CA 94901
Position: Principal Engr. *Employer:* Bechtel Petroleum, Inc (Retired) *Education:* BSE/ChE/Univ of MI. *Born:* 7/6/21. Native of Detroit, MI. Attended Albion College & grad Univ of MI. Served with AUS 1943-1946. Grad Schools, IL Inst of Tech, Univ of Louisville & Univ of CA (Berkeley). Engr with Koppers Co & Union Carbide. With Bechtel since 1952. Retired in January 1986. Doing private consulting. Mbr - Am Chem Soc, Am Inst of Chem Engrs, CA & Natl Soc of PE & Fluid Power Soc. Sr Mbr Instrument Soc of Am. Honors: ISA - Distinguished Soc Service Award-1978 and NSPE/CSPE-Archimedes Engg Achievement Award-1979. Elected Dist 11 VP ISA - 1979-1981. Resolution 1838 - CA State Assembly, honoring Allan Marks, dated July 27, 1979. Reg Professional Engineer in Chemical and Control Systems in California. *Society Aff:* AIChE, ISA, NSPE/CSPE, FPS, ACS.

Marks, Colin H
Business: Dept of Mech Engg, College Park, MD 20742
Position: Prof. *Employer:* Univ of MD. *Education:* PhD/ME/Univ of MD; MS/ME/Carnegie Mellon Univ; BSME/ME/Carnegie Mellon Univ. *Born:* 10/8/33. in Cardiff S Wales. Moved to USA in 1946 and lived in Pittsburgh, PA until 1957. Served with Army Corps Engrs 1957-1959 working on dev of mobile, gas- turbine nuclear power plants. Taught & carried out res at Univ of MD from 1959 to present. Res studies have included bubble dynamics, oil-spill containment, hydrodynamic jelly fish & debris barriers, ground vehicle aerodynamics, fluid damping, and fire res. Taught wide variety of mech engg courses but specialize in fluid mechanics, heat transfer and thermodynamics, and ocean engg. *Society Aff:* ASME.

Marks, David H
Business: Dept of Civil Eng, MIT, Room 1-290, Cambridge, MA 02139
Position: Prof. *Employer:* MIT. *Education:* PhD/Env Eng/Johns Hopkins Univ; SM/Env Eng/Cornell Univ; BCE/Civ Eng/Cornell Univ *Born:* 2/22/39. Commissioned Officer, US Public Health Service 1964-1967. Prof of Civ Engrg, MIT. Assoc Dir of Parsons Lab, MIT 1977-1983. Presently Prof & Head of Civ Engrg, Water Resources Div and Ctr for Construction. Conslt to Camp Dresser McKee. Winner of 1977 Huber Prize for Res, ASCE. *Society Aff:* ASCE, AGU, ORSA, WPCF, AIH.

Marks, Robert H
Business: 335 E 45th St, New York, NY 10017
Position: Director of Publishing. *Employer:* Am Inst of Phys. *Education:* SB/Civ Engg/MIT. *Born:* 5/2/26. Joined AIP in 1970 & appointed a corporate officer in 1971. Responsible for AIP's publishing prog including scientific journals, advertising sales & Phys Today, the monthly news magazine for physicists. Was Mgr Public Relations for Michel-Cather Inc, an industrial adv agency. Had 12 yrs editorial experience at McGraw-Hill Inc becoming Managing Editor of Power Magazine. Before that spent 9 yrs in engg. Mfg & sales for the Permutit Co Inc becoming Dist Mgr. Author of numerous articles & technical papers on power, energy, water & waste treatment & publishing. Past Trustee, EI Inc; Mbr, Bd of Dir, Treas, Pres, NFAIS. *Society Aff:* AIChE, ASCE, ASME, NACE, AWWA, CESSE.

Markwardt, L J
Home: 12 Lathrop Street, Madison, WI 53705
Position: Cons Engineer. *Employer:* Self employed. *Education:* CE/Univ of WI; BSCE/Univ of WI. *Born:* Nov 26, 1889 Lansing Iowa. Instructor Drawing & Descriptive Geometry Univ Wisc 1914-17; Res Engr & Asst Dir Forest Prods Lab to 1960; Cons Engr 1960- . Coauthor Descriptive Geometry, author sects on timber in 4 engrg handbooks; Author many tech bulletins & papers on wood as an engrg material. P Pres ASTM; Honorary & Life Mbr ASTM, ASCE, NSPE, WSPE. Recipient Edgar Marburg Lecture Award ASTM, Walter C Voss Award for eminent achievement ASTM, Hitchcock Award for outstanding achievements in wood engrg FPRS, Superior Serv Award USDA, Honorary Citation engrg achievement Univ Wisc. *Society Aff:* ASCE, ASTM, NSPE, FPRS.

Marlowe, Donald E
Business: 1 Dupont Circle 400, Washington, DC 20036
Position: Exec Director. *Employer:* American Soc for Engrg Education. *Education:* BSE/Civil Engg/Univ of Detroit; MSE/Applied Mech/Univ of MI *Born:* March 1916. Res in experimental stress analysis. Spent WWII as Res Engr in undersea warfare, becoming Assoc Dir US Naval Ordnance Lab Whiteoak Md 1950; joined Catholic Univ of America 1955 as Prof of Mech Engrg & Dean of School of Engrg & Arch & in 1970 V Pres for Admin. Exec Dir of Amer Soc for Engrg Educ1975-1981. Cons on ordnance dev 1955-75. Pres ASME 1969-70; Pres NCEE 1966- 67; EJC Exec Comm 1962-63; NSPE Dir 1963-66. SHOT Adv Council 1975-80. Honary doctorates in Science & Engineering, (Merrimack, Villanova, Stevens, Mil-

Marlowe, Donald E (Continued)
waukee). Received numerous other honors from military sos , univs & community govt in D C. *Society Aff:* ASME, IEEE, NSPE, ASEE, AAAS.

Maroney, George E
Business: 48 Woerd Ave, Waltham, MA 02154
Position: Engrg Mgr *Employer:* W. H. Nichols Co. *Education:* PhD/Engrg/OK State Univ; M.S./Engrg/OK State Univ; B.S./Engrg/OK State Univ; A.A./Liberal Arts/Hutchinson Junior Coll *Born:* 12/21/37 Native of KS. Draftsman, Designer, Design Engr; Fluid Power Division, Cessna Aircraft, 1958-1966. Research Engr, Barnes Drill Co, 1967. Technician, Project Assoc, Project Engr, Research Engr, Assistant Adjunct Prof, Program Mgr; Fluid Power Research Center and School of Mechanical and Aerospace Engrg, OK St Univ, 1964-1979. Consultant, 1970-1980. KS Army National Guard, 1956-1978, Artilleryman and Aviator. Graduate, Command and General Staff School. Elected, U.S.A. representative to Organization for Intl Standardization. Member, Editorial Bd, Basic Fluid Power Research Journal. Regional Dir, Fluid Power Society. Dir, 1982 National Conference on Fluid Power. Author, 148 publications. *Society Aff:* ASME, SAE, NSPE, FPS, NFPA

Marple, Virgil A
Business: Mechanical Engrg Dept, 111 Church St, SE, Minneapolis, MN 55455
Position: Prof *Employer:* Univ of MN *Education:* PhD/ME/Univ of MN; MS/ME/Univ of So CA; B/ME/Univ of MN *Born:* 8/16/39 Native of Wendell, MN. Coop Engr for John Deere Dubuque Tractor Works 1960- 62. Engr for Aeronutronic, Div of Ford 1962-65 involved in several aerospace projects. Sr Engr for Fluidyne Engrg Corp 1965-67 involved in wind tunnel design. Prof at the Univ of MN, Mechanical Engrg Dept 1970-present. Currently is Dir of the Mech Engrg Co-op Intern prog at the Univ of MN. Research interests are in the areas of particle tech and aerosol physics. Holds several patents for inertial impactors of various designs, including "respirable" and "uniform deposit" impactors, one patent for a "precision aerosols divider" and one patent for a dust generator. Awards include "outstanding paper of 1974 published in the AIHA Journal." *Society Aff:* ASME, AIHA, AAAS, IPI, FPS, AAAR, GAEF

Marques, Fernando D S
Business: 504 E St Joseph St, Rapid City, SD 57701
Position: Assoc Prof *Employer:* SD Sch of Mines and Tech *Education:* PhD/Met Eng/Imperial Coll of Sci & Tech, Univ London; DiC; MS/Met Eng/Imperial Coll of Sci & Tech, Univ London; PhD; Dipl. Eng/Industrial Eng/Instituto Superior Tecnico, Univ Lisbon; BS/CE/Univ of Coimbra, Portugal *Born:* 10/15/46 Native of Portugal. Finished a three year course in Chem Eng in 1967 at the Univ of Coimbra. Holds the degrees: Dipl Eng (MS) and PhD: Univ of Lisbon and DIC (MS) and PhD: Imperial Coll of Sci and Tech, Univ of London. Taught undergrad and grad courses in Physical Chem, Physical and Mech Met and Materials Sci and Engrg in Portugal, England and USA. Has been a tenured Assoc Prof of Met Eng at SDSM&T since 1980. Has publ over 25 papers in intl proceedings and prof magazines and has given invited lectures in many countries. His current teaching and res interests are: structure/strength relationships in metallic and composite materials; phase transformations, strengthening mechanisms, deformation behavior of multiphase materials, analytical electron micros copy and failure analysis in engrg materials. He is also a mbr of the Amer Soc for Testing of Materials, Materials Res Soc, Sigma Xi, Societe Francaise de Metallurgie, Editorial and Adv Bd of Portugaliae Physica, British and Amer Soc for Metals, a reg chartered engr in England, Portugal and USA and a fellow of the Royal Microscopical Soc. He has been a mbr of the Intl Congress on the Mechanical Behavior of Materials. Enjoys flying, sailing, fishing and classical music. *Society Aff:* AIME, ASM, FRMS, CE/PE, MIM

Marquis, Eugene L
Business: Civ Engg Dept, College Station, TX 77843
Position: Assoc Prof & Assoc Res Engr. *Employer:* TX A&M Univ. *Education:* PhD/Structural Engr/TX A&M Univ; MS/CE/Univ of MO; BS/CE/Univ of MO. *Born:* 6/28/26. Native of Walker, MO. Served USNR Air Corps 1944-46. Design Engr, CONVAIR, Ft W - responsible for nose of B36 & YB60. Designs on OH Turnpike & TX flood control projs for consulting engrs. Owner of Johnson & Marquis, San Antonio, til 1970 - responsible for design of $50,000,000 structures. Chief Engr for 1040-man Composite Recruiting Training Ctrs, Lackland AFB. TX A&M since 1970. Author of 36 reports & publications & one chapter of Manual on Wastewater Operations. Faculty advisor to ASCE Chapter. ASCE Wellington Prize 1977. Enjoys hunting, sailing, fishing. *Society Aff:* ASCE, CEC.

Marr, Richard A
Business: Stanley Bldg, Muscatine, IA 52761
Position: VP *Employer:* Stanley Consultants *Education:* MS/CE/Univ of IA; BS/CE/IA St Univ *Born:* 12/29/36 in Iowa City, IA. Served two years as platoon leader and co commander with Army Corps Engrs upon graduation from IA State in 1958. Joined Stanley Consultants in 1960 as a structural design engr in power group. Carried various responsibilities in structural design and appointed chief structural engr in 1977. Appointed Dir of Planning and Design Services for Central Division of Stanley Consultants in 1978 and assistant division head in 1979. Elected VP in 1978. Appointed assistant div head of Power Div of Stanley Cnslt in April, 1982 and Div Head in Sept, 1982. Appointed head of Civil/Architecture Group in Operations Div in April, 1987. State Pres of IA Engrg Society/NSPE in 1978. Chairman of Iowa Engineering & Land Surveying Examining Board in 1986-87. Hobbies include singing, boating and golf. *Society Aff:* NSPE, ASCE, NCEE

Marr, W Allen
Home: Laurel Dr, Lincoln, MA 01773
Position: Research Assoc *Employer:* MA Inst of Tech *Education:* PhD/CE/MIT; MS/CE/MIT; BS/CE/Univ of CA - Davis *Born:* 10/10/48 Staff engr for Woodward-Clyde Consultants, Lambe and Assoc 1970-1972. Recipient of Woodward-Clyde Fellowship in Applied Earth Sciences by intl competition. Instructor in Civil Engrg at MIT 1972-1974. Research Assoc at MIT 1974 to present. Responsible for contract research to develop improved design methods for earth dams and offshore structures. Teach graduate subject on design of earth structures. Led research team which developed new method to predict displacement of foundations of offshore structures loaded by waves. Intl consultant to governments and industry on design of earth dams, major excavations and offshore structures. Have managed project teams of more than 20 professionals. Author or co-author of more than 12 professional papers in referreed proceedings, 13 MIT Research Reports and 80 consulting reports. Married to Victoria Julian Marr. Enjoy farming and stock market. *Society Aff:* ASCE, NSSMFE

Marris, Andrew W
Home: 3372 Embry Cr, Atlanta, GA 30341
Position: Regents' Prof Emeritus Retired *Employer:* GA Inst of Tech *Education:* DSc/Univ of London; PhD/ME/Univ of New Zealand; MS/Math/Univ of London; BS/Math/Univ of London; BS/Physics/Univ of London *Born:* 4/20/24 in Eversley Hampshire, England. Asst Prof, Dept of Civil Engrg, Univ of British Columbia, 1954-1960. Visiting Assoc Prof, Univ of TX, Austin, TX, 1960- 62. Prof (1962-69), Regents' Prof, (1969-87). Retired 1987, Regents' Prof Emeritus, School of Engrg Science and Mechanics, GA Inst of Tech. Main areas of research are vector field theory and differential geometry of deformation and flow. General interests: wildlife conservation, photography, and music. *Society Aff:* SNP.

Marsal, Raul J
Business: Inst de Ingenieria UNAM, Ciudad Universitaria Mexico D F, Mexico 20
Position: Res Professor. *Employer:* Univ Nacional Autonoma de Mexico. *Education:* MS/Soil Mechs/MIT; CE/Civil Eng/Univ of Buenas Aires. *Born:* Jan 1915. Studied for the doctorate in Harvard Univ during 1944; In the period 1940-42 worked in road const & design of water treatment plants in Argentina; 1945-59 part time to the design & const of dams in Mexico & part time to res on the sub-soil of Mexico

Marsal, Raul J (Continued)
City; Cons to the Comision Federal de Electricidad & Res Prof at the Univ of Mexico in the lapse 1960-72; since 1972 full time Res Prof at the Univ of Mexico. Pres Mexican Soc of Soil Mech 1969; Doctor Honoris Causa Univ of Mexico 1964; ASCE Middle Middlebrooks Award 1968; V Pres of the ISSMFE for North Amer 1973-77; ASCE F Mbr. *Society Aff:* ASCE, SMIA, CAI.

Marschall, Albert R
Business: Commissioner Public Buildings Service, 18th & F St. NW, Washington, DC 20405
Position: Rear Admiral-USN. *Employer:* General Services Administration *Born:* 05/-/21 New Orleans LA. BS USNA, BSCE & MSCE RPI. Commissioned June 1944; served as Public Works Officer US Naval Acad; Cdr 3rd Naval Const Brigade/30th Naval Const Regiment & Commanding Officer Southeastern Div Naval Facilities Engrg Command (NAVFACENGCOM); 1970 became Deputy Cdr Pacific Div NAVFACENGCOM/CDR 3rd Naval Const Brigade/Officer in charge of const Vietnam; 1971 Dir Shore Establishment Div Office of Naval Opers; 1972 became V Cdr & 1973 Cdr Naval Facils Engrg Command & Ch Cvl Engrs. 1967 SAME Goethals Medal. 1977-79 V Pres Hyman Const Co, Bethesda, MD; 1979-present Commissioner Public Buildings Service; 1981-Honorary AIA Membership. Enjoy golf & fishing.

Marschall, Ekkehard P
Business: Dept of Mech Engr, Santa Barbara, CA 93106
Position: Prof. *Employer:* Univ of CA. *Education:* DrIng/Mech Engr/Technische Hochschule Hannover; DiplIng/Mech Engr/Technische Hochschule Hannover. *Born:* 3/18/35. Held teaching and res position at the Technische Hochschule Hannover, Univ of CA at Santa Barbara, Universitat Stuttgart. Spent one yr as NATO-fellow at the Univ of MN, Minneapolis. Authored and co-authored more than 60 reviewed scientific papers. Worked as consultant for various industrial cos. Pursues actively res in the gen area of heat transfer with and without phase-change. *Society Aff:* ASME, AIChE, ΣΞ.

Marsh, Alan H
Home: DyTec Engineering, Inc, 5092 Tasman Dr, Huntington Beach, CA 92649
Position: Pres. *Employer:* DyTec Engg, Inc. *Education:* MS/Engg/Univ of CA; BS/Elec Engg/MA Inst of Tech; BA/Phys/Williams College *Born:* 3/21/32. Acoustical engr at Douglas Aircraft Co from 1956 to 1976 - primarily involved with noise control projs for DC-8, DC-9, and DC-10 airplanes and with res and dev progs. Founded DyTec Engg, Inc, in 1976 to conduct studies and provide engg consulting services in noise control and related fields. Mbr Acoust. Soc. of America's Technical Comm on Noise. Mbr and chrmn of Working Groups for American Natl Stds Inst S1 Comm on Acoustics. Vice chrmn of S1 Comm and Individual Mbr of S1 Comm on Acoustics and S12 Comm on Noise. Mbr Soc of Automotive Engrs' Comm A-21 on Aircraft Noise. Founding mbr, Officer, and Dir of Inst of Noise Control Engrs. Fellow of the ASA and Assoc Fellow of the AIAA. *Society Aff:* INCE, ASA, AIAA

Marsh, B Duane
Business: Old Ridgebury Rd, Danbury, CT 06817
Position: VP Tech *Employer:* Union Carbide Corp Specialty Polymer & Composites *Education:* PhD/ChE/Univ of WA; MS/ChE/Univ of WA; BS/ChE/Univ of WA *Born:* 6/22/38. Post-Doc at Univ of Wisconsin in area of rheology. Employed at Union Carbide since leaving Univ of Wisc (1967). Involved with Process Dev and Res in Chem and Plastics Div as a res engg, group leader and Assoc Dir. Currently, Gen Mgr R & D for the Carbon Products Div of Union Carbide. 1969 to 1981 - Chrmn of the Bd & Gen Mgr - Marsh & Basgier, Inc, PC Engrs - Surveyors - Planners, VA Beach, VA. Currently VP Tech, for Specialty Polymers & Composites Div of Union Carbide. *Society Aff:* ASCE, NSPE, VAS, VSPE

Marsh, Warner G
Business: 3331 Stoneshore Rd, VA Beach, VA 23452
Position: Conslt *Employer:* Holly Spain, Sutton & Marsh, PC *Education:* ICS/Civ Engg-Hgwy Engg/. *Born:* 3/9/24. 1946-1948 Road Design Table Hd, Asst (Location & Design) VA Dept of Hgwys, Richmond, VA. 1949 Design Engr (Hgwys) Phillip D Freeman, Consulting Engr, Norfolk, VA. 1949-1957 Engr & Mgr Frank D Tarrall, Jr & Assoc, Surveyors & Engrs, Norfolk, VA. 1957-1959 Proj Engr - Rader & Assoc St Petersburg, FL. 1959- 1962 Private Practice - Marsh & Assoc, Surveyors & Engrs, Tampa, FL. 1962-1966 Asst County Engr - Hillsborough County, Tampa, FL. 1966-1967 City Engr - Gainsville, FL. 1967-1968 Proj Engr - Heidt-Burney & Assoc Tampa, FL. 1968-1969 Engr & Mgr - Frank D Tarrall, Jr & Assoc, Surveyors & Engrs, VA Beach, VA. 1969 to 1981 Chrmn of Bd and Gen Mgr - Marsh & Basgier, Inc PC, Engrs - Surveyors - Planners, VA Beach VA, 1981 to present Vice-President - Holly, Spain, Sutton & Marsh, PC - Engrs - Surveyors - Planners, VA Beach, VA & Richmond, VA. 1982-VA State Bd of Archts, PEs, Surveyors & Cert Lscape Archs Exam Bd. 5 yr term, Sr Mbr, Engrg Section. *Society Aff:* ASCE, ACSM, NSPE, VAS, SAME, PEPP, ECHR, VSPE.

Marshall, A Frank
Business: 6100 Hillcroft, Houston, TX 77081
Position: VP-Mgr, Onshore Engrg *Employer:* McClelland Engrs, Inc *Education:* MS/CE/Univ of FL; BS/CE/Univ of FL *Born:* 11/30/34 Native of Tallahassee, FL. Research engr with FL Dept of Transportation from 1959-1962. Joined McClelland Engrs in 1962 as a geotechnical engr; advanced to engr mgr in 1968 and to sr engr mgr in 1978. In 1979, was promoted to VP and is responsible for all onshore geotechnical activities of the Houston Div. Past Pres of ASCE Houston Branch. Past VP for Technical Affairs for ASCE TX Section. Wife Heather and sons, Murray (1967) and Graham (1971). Enjoy beach sailing and fishing. *Society Aff:* ASCE

Marshall, Andrew C
Business: 2147 Wilmington Dr, Walnut Creek, CA 94596
Position: President *Employer:* Marshall Consulting, Inc. *Education:* BS/Mech Engg/Univ of CA. *Born:* 4/5/21. Lt (jg) US Navy, Aircraft Maintenance Engr, 1944-1946 United Airlines, Mtls and Process Engr, 1946-1950. Hexcel Corp: Chief Engr, 1950-1952; Regional Sls Mgr, 1952-1958, Gen Sls Mgr, 1958-1964; VP-Applications Engg, 1964-1970; VP- Mktg, 1970-1974; VP/Gen Mgr Sports Div, 1974-1978. Principal in Marshall Consulting, 1978-present. Pres-Univ of CA Engrg Alumni Assoc, 1971-1972. VP-Ski Industries America, 1978-1979. Dir-US Assoc of Ski Mfrs, 1974-1978. Recipient- "Special Award for Meritorious Service–, SAMPE, 1970. Author of numerous technical articles in field of Structural Sandwich Construction, 1956 to present. *Society Aff:* SPE, SAMPE.

Marshall, Dale E
Home: 5411 Marsh Rd, Haslett, MI 48840
Position: Agri Engr. *Employer:* US Dept of Agri. *Education:* MS/Agri Engg/MI State Univ; BS/Agri Engg/MI State Univ. *Born:* 8/13/34. From Gregory, MI grain - livestock farm. Jr Proj Engr, 1961-63, Farmhand designing forage wagons, and grinder-mixers. Proj Engr, 1963-66, Chore-Time designing controlled hog feeder that was granted US, German, British and French patents. Agri Engr, 1966-69, AERD-ARS USDA, designing citrus harvestig equip; 1969-present; ARS USDA, designing and improving fruit and vegetable harvesting and handling equip for grapes, pickling cucumbers, rhubarb and peppers. Natl Pres, Alpha Epsilon, 1976. Received 1978 Doerfer ASAE award for Engg Concept of the Yr. Selected as Specialist for 1979 Agriculture-USA exhibit in Moscow, USSR. Received 1980 award for Exceptional Creative Research in Agricultural Engineering from Pickle Packers, Int'l, Inc. & Great Lakes Vegetable Growers Conference. *Society Aff:* ASAE, ASHS.

Marshall, Francis J
Business: Sch of Aero & Astro, Grissom Hall, W Lafayette, IN 47907
Position: Prof A & AE. *Employer:* Purdue Univ. *Education:* DrEngSci/Aerodynamics/NYU; MS/Mechanics/Rensselaer Poly Inst; BME/ME/CCNY. *Born:* 1923 in The Bronx. Attended PS86 & Stuyvesant HS. US Army (ETO 75th Div, Combat Infantry Badge) 1943-46. Engr with Western Union, GE (Proj Hermes), Wright-Aero. Group Leader, Lab for Appl Sciences, Univ of Chicago 1955-1960. Visiting Engr, US NUWC, Pasadena, CA 1966-67. Fellowship at NASA Langley 1969 & 1970. Taught IL Inst of Tech, Chicago, 1957-1959 and Purdue Univ since 1960. Present interests: computational aerodynamics and computer-aided-design. *Society Aff:* AIAA, ASEE.

Marshall, John F
Business: 3013 27th Street, Metairie, LA 70002
Position: Sr Vice President. *Employer:* J J Krebs & Sons Inc. *Education:* BS/Civil Engr/Tulane Univ. *Born:* Aug 9, 1928 Cairo Illinois. Grad from Tulane Univ in Civil Engrg. Previously employed by Phillips Petroleum Co, Upham Engrg Co & Portland Cement Assn; has been assoc with J J Krebs & Sons Inc Civil Engrs & Surveyors for over 9 yrs. A veteran of WWII & Korea. Mbr of ASCE, Louisiana Engrg Soc, Natl Soc of Prof Engrs, Louisiana Land Surveyors Assn, Amer Public Work Assn. Reg P E in Louisiana, Texas & Mississippi & P Pres Louisiana Sect ASCE. *Society Aff:* ASCE, LES, NSPE, LLSA, ACSM, APWA.

Marshall, Peter W
Business: PO Box 2099, Houston, TX 77001
Position: Civil Engr Conslt *Employer:* Shell Oil Co. *Education:* MS/Civil Engg/Univ of FL; BS/Civil Engg/Univ of FL. *Born:* 12/6/38. Since 1962 in offshore struct design; currently, Hd Office Cvl Engg respon for integrating R/D effort, feasibility studies & design criteria for deep water structures. Former Chmn ASCE Ctte on Reliability of Offshore struct. Former Chmn of ASCE & AWS cttes on Tubular Struct. ASCE Alfred Noble Prize 1970. AWS Davis medal 1975. IIW Houdremont Lecturer, 1984. Mbr of API & Marine Bd Advisory Groups. *Society Aff:* ASCE, AWS, MTS.

Marshall, W Robert
Home: 610 Walnut St, Madison, WI 53705
Position: Director, Univ. Ind Research *Employer:* Univ of Wisconsin-Madison. *Education:* PhD/Chem Engr/Univ of WI; BS/Chem Engr/Ill. Inst. Tech. *Born:* May 19, 1916 Calgary Alberta Canada; U S citizen naturalized March 20, 1944. Chem Engr with E I Du Pont 1941-47; Assoc Prof 1948-53, Prof & Assoc Dean 1953-71, Dean 1971-81 Coll of Engrg Univ of Wisconsin-Madison. Dir Univ Ind Research UW-Msn 1981. Mbr Natl Acad of Engrg. Fellow AIChE. Walker Award 1953, Prof Progress Award 1959 & Founders Award 1973 all from AIChE. Dir 1956-58 & Pres 1963 AIChE. Pres Assoc Midwest Univs 1962-63. Lic P E Wisc. Recipient Gold Medal from VDG, German Engrs Soc 1974. Hon doctors degree, Ill. Inst. of Tech. 1984. Conducted extensive res in atomization & spray processes. Dev basic design principles for atomizers & spray dryers. Respon for the admin of the Coll of Engrg Univ of Wisconsin- Madison 1971-81; responsible for University-Industry Research program. *Society Aff:* AIChE, NAE, AAAS.

Marshek, Kurt M
Business: Dept. of Mech. Engrg, U of Texas at Austin, Austin, TX 78712
Position: Prof. *Employer:* Univ of TX at Austin *Education:* PhD/ME/OH State Univ; MS/ME/Univ of WI; BS/ME/Univ of WI. *Born:* 10/13/43. Reg PE in TX, & OH. He has worked in the gear, heat exchanger, & telephone equipment industries. He has been involved in the res, design & dev of machine elements: chains, sprockets, gears, bearings, universal joints, and couplings & has taught courses in stress analysis, friction & wear, & mech design. He received the Soc of Automotive Engrs Ralph R Teetor Award, the UT College of Engineering Faculty Leadership Award, and the Halliburton Education Foundation, Award of Excellence, all in recognition of outstanding achievement and professionalism in education, research, and service to students. *Society Aff:* SME, SAE, ASME, AGMA.

Marsocci, Velio A
Business: Dept. of Elec. Eng, Suny at Stony Brook, Stony Brook, NY 11794-2350
Position: Prof of Engrg & Clinical Prof. of Health Sci. *Employer:* State Univ of NY *Education:* Eng ScD/EE/NYU; M/EE/NYU; B/EE/NYU *Born:* 6/7/28 Native of Queens, NY. Served in US Navy as Electronic Technician 1946-49. Instructor in Electrical Engrg, NYU, 1954-56. Asst Prof of EE, Assoc Prof of EE, Stevens Inst of Tech, 1956-1965. Assoc Prof of Engrg, Prof of Engrg, 1965-date, Assoc Dean of Engrg, 1974-76, presently Prof of Engrg and Clinical Prof of Health Sciences, SUNY at Stony Brook. Pres, Suffolk County Chapter of NYSSPE, 1972-74. Certified by Natl Council of Engrg Examiners and registered PE in NY and in NJ. Teaching and research primarily in the fields of solid-state electronics and of bioelectronics. *Society Aff:* IEEE, NSPE, NYAS, ASEE, AAUP, AIP.

Marsten, Richard B
Business: 140th St & Convent Ave, New York, NY 10031
Position: Exec Dir Bd on Telecomm & Computer Applications *Employer:* Nat'l. Academy of Engrg.!Nat'l Research Council *Education:* PhD/-/Univ of PA; SB/SM/MIT. *Born:* 10/28/25. RCA 1956-69; radar sys 1956-61, space applications & communications 1961- 69, Ch Engr RCA AstroElectron Div 1967-69-respon for all engrg of sppacecraft, sys & AGE. Mbr Broadcast panel; Chmn Points-to-Point Communications Panel; NAS Study on Space Applications 1967-68; Dir Communications Progs Div NASA HQ 1969- 75; Acting Dir Data Mgmt Progs Div NASA HQ 1973-75. Mbr FCST panel, Automation Opportunities, Health Care 1972. Mbr Telecommunications Res panel, NAE Ctte on Telecommunications 1973. White House citation sustained outstanding performance 1972. NASA exceptional serv Medal 1974. NASA group achievement award 1974 NASA Apollo/Soyua Space Medal 1976: Int'l Women's Year citation 1975. Dean School of Engrg CUNY 1975-79, professor, engrg, 1979-81; Mgr, space program, Off. Tech Assessment, U. S. Congress 1980-81; Exec. Director, Board on Telecommunications & Computer Applications, Nat'l Acad Engr/Nat'l Research Council, 1981--. Mbr Tau Beta Pi, Sigma Xi, Eta Kappa Nu. N Y Acad Sci. Fellow IEEE, AIAA: Chmn., nat'l technical cmte, comm. systems, 1967-69; program chmn, 1st Communication Satellite Systems Conf 1966. IEEE: Comm policy bd. 1972-76, 1978-81; adv EASCON 1973-75, EASCON program Chmn 1970; Chmn on Telecomm. Policy 1978-80, consulting member 1981--; member steering cmte, U.S. Technology Policy Conference 1978--. Chmn, OTA Telecomm Advisory Panel 1979-80; member Telecomm Technology Panel 1979-80.. *Society Aff:* AAAS, IEEE, NY Acad Sci, IEEE, IEEEAES.

Martel, Eugene H
Business: 1 Broadway, Cambridge, MA 02142
Position: Senior Vice President, Marketing & Sales *Employer:* The Badger Co Inc. *Education:* BS/Chem Eng/Univ of Rhode Island. *Born:* 9/15/34. P E Mass. Entire career with Badger, initially as Process Engr, subsequently Mgr Process Engrg at Badger Ltd (London); V Pres Badger-Tomoe (Japan); Pres The China Badger Co (Taiwan); Vice Pres. ASNA Pacific. Assumed present position Jan. 1981 and now respon for all Badger Sales activities worldwide. Mbr AIChE. Pub sev tech articles & have 1 pat in processing of aromatics. Elected to Bd of Dirs the Badger Co, July 1979. *Society Aff:* AJChE.

Martens, Hinrich R
Business: 4250 Ridge Lea Rd, Buffalo, NY 14226
Position: Dir of Univ Computing Service and Prof Elec Engrg *Employer:* State Univ of New York at Buffalo. *Education:* PhD/-//MI State Univ; BS/-/Univ of Rochester; MS/-/Univ of Rochester. *Born:* 4/21/34. Native of Luebeck Germany. Asst Prof 1962-65; Assoc Prof 1965-70; Prof 1970-SUNY/Buffalo Dept of Mech Engrg & Elec Engrg. Acting Dir Univ Computing Servs 1975-76 SUNY/Buffalo. Dir Univ Computing Servs 1981- . Review Editor J- MSAC-ASME 1975-78. SUNY Chancellors Award 1974, ASEE Western Elec Award 1974. *Society Aff:* ASEE, ASME, IEEE.

Martenson, Dennis R
Home: 8140 46 1/2 Ave N, Minneapolis, MN 55428
Position: Assoc. *Employer:* Toltz, King, Duvall, Anderson & Assoc *Education:* MS/Sanitary Engr/Univ of MN; BS/Civ Engr/Univ of MN. *Born:* 10/15/42. Native of Eau Claire, WI. Has worked for a large mfg in pollution control, a regional pollu-

Martenson, Dennis R (Continued)
tion control agency & several consulting engr firms. Presently a proj mgr & assoc in the environmental engr div of Toltz, King, Duvall, Anderson & Assoc, Inc, E/A/P. Very active in profl activities. Pres, MN Sec, ASCE 1976- 77; MN's Young Civ Engr of the Yr 1975; A Principal contributing author to the joint ASCE/WPCF manual of practice on Wastewater Treatment Plant Design (1977); and a principal contributing author to the WPCF a manual of practice on Preliminary Treatment for Wastewater Facilties (1980). Dir, District 8, ASCE, 1985-88. *Society Aff:* ASCE, AWWA, NACE, WPCF, ISA

Martin, Arthur H
Home: 6541 N Cibola, Tucson, AZ 85718
Position: Sr VP *Employer:* RGA Consulting Engrs *Education:* BS/CE/IA State Univ *Born:* 2/3/28 Native of Ames, IA. 3 years Bridge Design IA State Highway Dept. 3 years active duty, 27 years inactive duty, Civil Engrg Corps, US Naval Reserve. Rank of Capt, 10 years command. 7 years Gen Supt of Heavy Construction Co building highways, bridges and structures. 17 years to present as principal of consulting engrg firm, currently Dir of Operations, 80 employees engaged in Structural, Civil, Mechanical, and Electrical Engrg. Engr of the Year-1980, AZ Society PE. Engr of the Year 1981, So Chapter ASPE. Pres AZ Consulting Engrs Assoc 1980. YMCA Bd of Dirs 15 years - Tucson. *Society Aff:* ASCE, NSPE, SAME

Martin, Bruce A
Home: 2599 Eastbourne Dr, Salt Lake City, UT 84121
Position: Sr Prin Cost Engr *Employer:* Kennecott Corp *Education:* -/Mech Engrg/Univ of Houston *Born:* 08/10/39 Native of Houston, TX. Tech Dir of the Cost Control Div within the Amer Assoc of Cost Engrs. Founding Pres of the Alberta Chapter of the Proj Mgmt Inst. Authored several papers on the subject of Cost Control for publication in the journals of AACE and PMI. Outstanding Young Men of Amer Recognition in 1973. Awarded Natl Defense and Faithful Service Decorations for Active Military Duty during the Berlin Crisis in 1962. Received Certified Cost Engr title in 1977. Organized Mining & Minerals Tech Ctte within AACE. *Society Aff:* AACE

Martin, C Samuel
Business: Sch of Civ Engg, Atlanta, GA 30332
Position: Prof. *Employer:* GA Inst of Tech. *Education:* PhD/Civ Engg/GA Tech; MS/Civ Engg/GA Tech; BS/Civ Engg/VPI. *Born:* 5/22/36. Pursued grad work at GA Tech, culminating in MS in CE in 1961 & PhD in 1964. Worked as hydraulic Engr at Newport News Shipbldg & Dry Dock Co 1959-60. Asst Prof at GA Tech 1963-66. Ford Fdn Faculty Resident as Harza Engrg 1966-67. Assoc Prof GA Tech 1967-74, Prof 1974-present. Guest Prof, Univ of Karlsruhe 1970-71. ASME John R. Freeman Scholar & Fulbright Scholar 1970-71. Tech cttes served on are Res Ctte, Hydraulics Div, ASCE & Exec Ctte, Fluids Engrg Div, Mbr- at-Large, Basic Engrg Group, ASME; Fellow, ASCE; Fellow, ASME; Alexander von Humboldt Senior US Scientist Award, Univ of Munich, 1984-1985. ASME Reg PE, GA. *Society Aff:* ASCE, ASME.

Martin, C Samuel
Business: School of Civil Engrg., Georgia Tech, Atlanta, GA 30332
Position: Professor *Employer:* Georgia Tech *Education:* PhD/Fluid Mechanics/Georgia Tech; MS/Fluid Mechanics/Georgia Tech; BS/Civil Engrg/Virginia Tech *Born:* 5/22/36 Native of Staunton, VA. Taught at GA Tech as Asst. Prof. 1963-67, Assoc. Prof. 1967-74, Prof. 1974-Present. Guest Prof., Univ Karlsruhe, W. Germany, 1970-71. ASCE Stevens Prize, 1977. Fellow ASME, ASCE. ASME John R. Freeman Scholar 1970-71. Alexander von Humboldt U.S. Senior Scientist 1984-85, Technical Univ Munich. *Society Aff:* ASCE, ASME, ANS

Martin, D Robert
Business: Apartado Postal No 5021 K, Monterrey, Nuevo Leon Mexico
Position: Manager Industrial Sales. *Employer:* Eutectic Mexicana S A. *Born:* Sept 12, 1922 New York City. Grad US Merchant Marine Acad 1944. 1st Engr Steam, 3rd Engr Diesel any Horsepower any Ocean unlimited lic. 1946-59 The Texas Co (Texaco) Asst to Supr Tech Serv Foreign Opers 1959-62; Coml & indus cons in Latin Amer; 1962- Eutectic Mexicana Mgr Indus Sales. 1967-68 Chmn ASM Northeast Mexico Chap. Directed Translation to Spanish ASM courses. Mbr Amer Welding Soc, Assn Mexicana Ingenieros Mecanicos y Electricistas. Daughters: Marlene, Dolores, Linda, Nancy, Gale. Wife:Marlene, Colombian Consul Monterrey Mexico.

Martin, Daniel W
Home: 7349 Clough Pike, Cincinnati, OH 45244
Position: Acoustical Conslt, Editor-in-Chief *Employer:* Self-Employed *Education:* PhD/Physics/Univ of IL; MS/Physics/Univ of IL; AB/Physics & Math/Georgetown College (KY); *Born:* Nov 1918. Physics Asst Illinois 1937-41; Engr RCA 1941-49; Mathematics Instructor Purdue Extension 1941-46; Baldwin Piano & Organ Co Engr, 1949-56, Res Dir 1957-83, Conslt 1984-; taught acoustics, music psychology Univ Cincinnati 1964-74. Author audio chap Standard Handbook for Electron Engrs. Pubs on musical instruments, audio, speech sys, room acoustics. Inventor 26 pats. Reg Prof Engr Ohio. Fellow Acoustical Soc (Pres. 1984-85). Fellow Audio Engrg Soc (Pres 1964). AES Award 1969. Fellow IEEE (Chmn Audio Group 1956, Editor 1953-55). Audio Achievement Award 1959. Musician. Pres United Presbyterian Men 1977. Editor-in-Chief Jour *Acous Soc of Amer* 1985-. *Society Aff:* ASA, IEEE, AES, NCAC.

Martin, Douglas H
Home: 119 Heather Dr RD 7, Allentown, PA 18103
Position: President. *Employer:* Douglas Martin & Assocs. *Born:* Aug 1929 Hackensack N J. BSChE from M I T, MChE & MBA from N Y Univ. Reg P E in N J, N Y, PA. Fellow AAAS. Mbr ACS, AIChE, NSPE, N J Acad Sci. Alumni Officer M I T. R/D Engr M W KelloggPetrochemicals & Petroleum Catalyic Processes; Dev Engr M W Kellogg; Process Engr M W Kellogg 1952-61; N J Zinc, Div Gulf & Western in Corp Planning & Marketing-left as Asst to V Pres, Pres Winfield Inc (Marketing Cons since 1970); also Pres Douglas Martin & Assocs since 1975 & Pres Allentown Minerals (1976). Holds sev pats in petroleum & petrochem field.

Martin, Edgar T
Home: 922 26th Place South, Arlington, VA 22202
Position: Telecommunications Cons. *Employer:* Self employed. *Education:* M Forensic Sc./-/George Washington Univ; LLM/-/Georgetown Univ; JD/-/George Washington Univ; MA/-/Amer Univ; MS/Elec Engg/VA Polytechnic Inst; BS/Elec Engg/VA Polytechnic Inst. *Born:* May 1918. Grad US Army Command & Genl Staff Coll. Reg P E in Va. Mbr Va Bar. Colonel SigCorps AUS (Ret). Elec Engr Radford Ordnance Works 1940- 41; served with US Army Signal Corps 1941-46; Comm Engr with Off of Military Govt for Germany 1946-49 & US High Comm in Germany 1949-52; Ch Central Frequency Staff 1952-54; Ch Engr 1954-58 & Engrg Mgr 1958-75 Broadcasting Serv US Info Agency; Telecommunications Cons 1975- . Fellow IEEE (1971). *Society Aff:* AAFS, ASIL, ABA, VTLA, IEEE.

Martin, George H
Home: 1320 Westview Ave, East Lansing, MI 48823
Position: Assoc. Prof. Emeritus *Employer:* Retired *Education:* Ph.D./Mech. Engin./Northwestern Univ.; M.S./Mech. Engin./Illinois Institute of Technology; B.S./Mech. Engin./Illinois Institute of Technology *Born:* 06/08/17 Received B.S. and M.S. degrees from Ill. Inst. Tech. 1941, 1944; Ph.D. Northwestern Univ. 1955, Design engineer Teletype Corp. Chgo. 1941-42; instr. Ill. Inst. Tech. 1943-47 and assist. prof. 1947-50; lecturer Northwestern Univ. 1950-55; assoc. prof. mech. enginr. Mich State Univ. 1955-79. Author Kinematics and Dynamics of Machines, McGraw-Hill 1982 and various papers on kinematics and machine design. Registered PE, Ill.; Life fellow ASME; member of Sigma Xi, Pi Tau Sigma. Hobbies: piano, classical music, Packard automobiles, swimming. *Society Aff:* ASME

Martin, Glen L
Business: CH2M Hill, P.O. Box 22508, Denver, CO 80222
Position: Dir, Organizational Development *Employer:* CH2M HILL, Inc *Education:* PhD/Civil Engg/Univ of AZ; MS/Civil Engg/OR State Univ; BS/Civil Engg/ND State Univ. *Born:* 1/21/32. Native of Cando, ND. Civil Engr, Red River Engrs, Fargo, ND, 1957; Instructor & Asst Prof OR State Univ 1957-62; Grad Res Assoc & Grad Teaching Assoc The Univ of AZ 1962-65; Assoc Prof (1965-70), Prof (1970-75), Hd of Dept & Dir of Res (1969-75), Civil Eng & Engg Mech MT State Univ; Dean College of Engg and Prof of Civil Engrg San Diego State Univ 1975-79. Corp Dir, Career Dev and Training (1979-83), Corp Dir, Org Dev, CH2M HILL, Corvallis, OR, 1983- . Pubs in Soil Mech & Fnd Engg, Lunar Soil Mech, Soil Erosion, Slope Stability, Civil Engr Education, Continuing Prof Dev, & Highway Info Sys. Reg P E OR, CA & MT. *Society Aff:* ASCE, NSPE, ARTBA, ASEE, TBΠ, XE, ΣΞ.

Martin, Howard W
Home: 1107 Tall Pine Dr, Friendswood, TX 77546
Position: Vice President, Process Engineering *Employer:* S&B Engrs, Inc. *Education:* MS/ChE/Univ of WI; BS/ChE/Rice Univ. *Born:* 10/27/29. Raised in Houston, TX. Held various engg, process tech & personnel dev positions for Monsanto Co, 1952-74, and Olin Corp, 1974-79. Joined S&B Engrs, Inc of Houston, TX as Mgr of Process Engg, 1979. Elected Fellow, AIChE 1979. Taught cooperative indus-Univ plant design courses: Univ of Houston, 1971-73; Yale Univ, 1977-79. Author or co-author of a number of tech publs. Chmn AIChE Equip Testing Procedures Ctte 1974-77. Mbr Tau Beta Pi, Phi Lambda Upsilon. *Society Aff:* AIChE.

Martin, Ignacio
Business: 1509 FD Roosevelt Ave, San Juan, PR 00920
Position: Partner. *Employer:* Capacete, Martin & Associates. *Education:* MSCE/Structural/Univ of IL; BSCE/Genl/Univ of Havana. *Born:* Aug 1928. Bridge Engr of Cuban Natl Dev Comm 1952-54; Partner Saenz, Cancio, Martin 1952-59; Partner Capacete, Martin since 1959. Part time lecturer Univ of Havana 1952-56, & Univ of Puerto Rico 1960-65. Mbr ACI Bd/Dir 1973-76; P Pres Soc of Structural Engrs of Puerto Rico; Pres (1984-85) of ACI; Grp Coordinator of Concrete Volume of Tall Bldg Monograph; PE in Puerto Rico, Florida, Penn, District of Columbia, MA; author of several papers on structural engrg; Tau Beta Pi; Mbr ASCE, ACI, EERI, PCI, PTI, IABSE, IASS, Soc of Structural Engrs of Puerto Rico, Colegio de Ingenieros y Agrimensores de Puerto Rico. *Society Aff:* ASCE, ACI, EERI, PCI, PTI, IABSE, IASS.

Martin, J Bruce
Home: 644 Doepke Ln, Cincinnati, OH 45231
Position: Sr Assoc *Employer:* Indumar, Inc. *Education:* PhD/Chem Engg/OH State Univ; MS/Chem Engg/OH State Univ; BS/Chem Engg/Auburn Univ. *Born:* 2/2/22. With Indumar, Inc. since 1982. The Procter & Gamble Co. 1949-82. Process dev, tech service, personnel & training, organization dev., Industrial Chem Mkt Research. Adjunct Assoc Prof, Auburn Univ 1983- . Dir DIChE 1968-70. Chrmn Mktg Div 1985. Pres Engg Soc of Cincinnati & Cincinnati Tech & Sci Societies Council 1972-73. Distinguished Engg Alumnus, OH State 1970, Dist Engr of the Year, Engrs & Scientists of Cincinnati 1982. *Society Aff:* AIChE, ACS, ASEE.

Martin, Jay R
Business: Dept of Chem Engrg, Easton, PA 18042
Position: Assoc Prof *Employer:* Lafayette College *Education:* PhD/ChE/Princeton Univ; BS/ChE/Lafayette Coll *Born:* 01/27/44 Staff and Sr Scientist at Textile Res Inst, Princeton, NJ, Nov 1971 - Aug 1976. Presently Assoc Prof at Lafayette College. 18 refereed publications and co-inventer on one patent. Worked extensively in the field of flammability and thermal behavior of polymeric materials. *Society Aff:* AIChE

Martin, John B
Business: Univ of Cape Town, Rondebosch, South Africa 7700
Position: Dean of Engrg *Employer:* Univ of Cape Town *Education:* DSc (Eng)/Civil/University of Natal; PhD/Structures/University of Cambridge; BSc (Eng)/Civil/University of Natal *Born:* 04/20/37 Native of Durban, South Africa. 1962-72 On faculty of Div of Engrg, Brown Univ; 1973- Prof of Civil Engrg, Univ of Cape Town; 1983- Dean of Faculty of Engrg; 1984- Dir of Information Tech; Also Chairman, Council for Nuclear Safety, Mbr, South African Council for Profl Engrs, Dir, Applied Mechanics Research Unit, UCT. *Society Aff:* ASME, ASCE, SAICE

Martin, John J
Home: 7818 Fulbright Ct, Bethesda, MD 20034
Position: Vice President & General Manager *Employer:* The Bendix Corporation *Education:* PhD/ME/Purdue Univ; MS/ME/Notre Dame; BS/ME/Notre Dame. *Born:* Oct, 1922 Detroit Mich. During US Navy service 1944-46 attended Naval Acad, Harvard Univ, MIT. North Amer Aviation 1951-53; Bendix Aviation Corp 1953- 59; Inst for Defense Analyses 1960-69. Lectured at Royal Aircraft Est, UK, & Australian Weapons Res Est 1963. Staff of President's Sci Adviser 1969-73. Assoc Dep to Dir of Central Intelligence for Intelligence Community 1973-74. Special Asst to DCI for Intelligence Info Systems 1974. Prin Dep Asst Secr Air Force (R&D) 1974-1976. Asst Secy Air Force (R&D) July 1976-1979. Numerous tech papers. Pat 'Ballistic Missile Defense System'. Book 'Atmospheric Reentry' (1966) translated into Russian. VP & Genl Mgr of Advanced Tech Center, 1979-present. *Society Aff:* ESD, SAE, ΠΤΣ, ΣΞ, ΤΒΠ.

Martin, Leslie D
Business: 1701 E Lake, Glenview, IL 60025
Position: Consulting Engr *Employer:* The Consulting Engrs Group, Inc *Education:* BS/CE/Univ of NB *Born:* 11/22/31 Formerly employed by the Ken R. White Co, Denver, CO, and the Portland Cement Assoc, Skokie, IL. Papers relating to precast, prestressed concrete published in Journal of the American Concrete Inst and Journal of the Prestressed Concrete Inst. Technical editor and a contributing author of the PCI Design Handbook, first, second and third editions (Prestressed Concrete Inst, Chicago). Co-author of *Connections for Precast, Prestressed Concrete Bldgs, Including Earthquake Resistance.* Principal Investigator on Research projects sponsored by Federal Highway Administration and Natl Science Foundation. Past Chrmn of the PCI Technical Activities Committee. *Society Aff:* ISPE, NSPE, ACI, PCI

Martin, Rodney O
Business: Box 1333, Syracuse, NY 13201
Position: Director of Farm Systems, R & D. *Employer:* Agway Inc. *Education:* MS/AE/Univ of ME; BS/AE/Univ of ME. *Born:* Nov 25, 1927. Native of Gorham, Maine. BS & MS in Agricultural Engrg from Univ of Maine. Served with 8th Submarine Squadron U S Navy. Served on Agricultural Engrg staffs at Univ of Maine & Penn State Univ. In 1957 appointed Dir of Dairy Cattle Housing Proj for Charles Hood Dairy Foundation of Boston. Mgr of H P Hood & Sons Dairy Farm Engrg Service 1961-69. Assumed current respon as Dir of Farm Systems R & D for Agway Inc in 1970. S&E Tech Dir of Amer Soc of Agricultural Engrs 1975-77. P Chmn NAR-ASAE. *Society Aff:* ASAE.

Martin, Thomas L, Jr
Business: Illinois Institute of Technology, Chicago, IL 60616
Position: President Emeritus. *Employer:* Illinois Inst of Technology. *Education:* PhD/EE/Stanford; MS/EE/RPI; BS/EE/RPI. *Born:* 9/26/21. Native of Memphis. Served with Army Signal Corps 1943-46. Faculty mbr at Univs of NM, AZ; Dean of Engg at AZ, FL and SMU; Pres of IL Inst of Tech 1974-1987. Author of 6 books; mbr of Natl Acad of Engg & Dir of Stewart-Warner Corp, Amsted Industries, Inland Steel, Cherry Electrical Products Corp, Commonwealth Edison Co, Kemper Mutual Funds, Sundstrand Corp & Museum of Sci & Indus. Reg PE in AZ & FL. *Society Aff:* IEEE, ASEE.

Martin, William C
Business: 620 Euclid Av, Lexington, KY 40502
Position: Owner. *Employer:* William C Martin, Architects-Engrs. *Born:* April 27, 1924. BSCE with Architectural option 1950. Major work in structural & architectural design. US Army Air Force 2nd Lt WWII, R & D Command Korean War (Cape Canaveral, Florida). Ch Architect J Stephen Watkins, Cons Engrs, 11 years. Private practice own Architect-Engr since 1961. Major practice in structural & architectural renovation, res & reports. Reg P E KY, MD. Reg Architect NCARB, KY, IN, OH, IL, TN, IA, MN, WV, VA, NC, NY, PA, SC, MD, MO, NJ, MA, MI, GA, CT. Mbr AIA, Kentucky Soc of PEs, NSPE, NFPA, ASTM, Triangle Fraternity.

Martin, William R
Home: 4007 Deanna Drive, Kokomo, IN 46901
Position: VP. *Employer:* Cabot Corp *Education:* BS/Met Engg/Univ of Cincinnati. *Born:* 9/16/34. Educated in OH, served two yrs in Army at White Sands Missile Range (1957- 1959). Metallurgical Dev Engr and later Mgr of Engg Mtls at Oak Ridge Natl Lab (1959-1976) with technical expertise in mech property res, mtls processing, powder metallurgy and directed prog to develop high temperature design data for industry. Dir of Dev Div at Y-12 plant of Union Carbide Nuclear iv (1976-1978). VP of Intl Mtls Group Operations at Mtls Res Corp from 1978-81. International Operations Mgr of WEAR Technology Div. Cabot Corp from 1981. Alpha Chi Sigma (1959), ASM Natl Award (1968) for Mtls Application, ASM Fellow (1976). *Society Aff:* AAAS, ASM, ISHM, IPMI, ASTM.

Martin, William R
Business: Dept of Nuclear Engrg, Ann Arbor, MI 48109
Position: Assoc Prof *Employer:* Univ of MI *Education:* PhD/Nuclear Eng/Univ of MI; MSE/Nuclear Eng/Univ of MI; MS/Phys/Univ of WI- Madison; BSE/Eng Phys/ Univ of MI *Born:* 06/02/45 William R Martin is currently an Assoc Prof of Nuclear Engrg at the Univ of MI. He has had experience with the Naval Reactors program as a commissioned officer (1969-1973) and with Combustion Engrg Inc (1976-1977), where he was respon for developing advanced computational methods for nuclear reactor analysis. His current specialty is the dev of algorithms for scientific computation on supercomputers. He is currently a conslt to Lawrence Livermore and Los Alamos Natl Labs, and has organized three short courses for IBM Corp in the area of advanced scientific computation. Martin co-authored (with J J Duderstadt) the textbook TRANSPORT THEORY (Wiley-Interscience, 1979) and has authored approximately 50 journal papers and conference proceedings in scientific computation. *Society Aff:* ANS, APS, SIAM, ACM

Martinec, Emil L
Business: 440 S Finley Rd, Lombard, IL 60148
Position: Prof. *Employer:* Midwest College of Engrg. *Education:* BS/Mech Engr/IL Inst of Tech; MS/Mech Engr/Univ of ID; MSA/Finance/Northwestern Univ; DSc/ Eng Mgmt/Midwest College of Engr *Born:* MS Univ of ID, MBA Northwestern Univ, BS IL Inst of Tech. Born Chicago, IL; Joined Argonne Natl Lab as Asst Engr 1955; experiment rep at MTR in ID 1956-57; Assoc Engr 1961; mbr of EBWR-100 MW Proj 1961-63; asst proj mgr on AARR 1964-69; RAS Div Dir's staff 1970-72; Asst Div Dir 1973-1979. Dir of Program Admin, Energy & Environ Tech Programs 1979-present. Joined Midwest College of Engrg in 1968; Chmn of Engrg Mgmt 1978 Dept 1969-1979; Assoc Dean 1970; Academic Dean 1971-73. VP for Academic Affairs 1982-present. Cert ASQC Qual Control Engr; P E IL. Mbr ASQC, ASEE, ASEM, ASME; Paper Review Chmn ASME Mgmt Div 1974, Mbr Exec Ctte 1976-present. Chrmn 1980; Mbr Prof Dev Bd 1975-1985, Chrmn 1980; Mbr Exec Ctte Gen Engrg Group 1980-present Mbr Editorial Bd, Engrg Mgmt Intl Elservice Publishers; Mbr, Accreditation Team, N Central Assoc of Colls and Schs. Enjoy classical music, amateur radio & gardening. *Society Aff:* ASME, ASEE, ASQC

Martinez, Pedro O
Business: 1840 N.E. 153rd St, North Miami Beach, FL 33162
Position: Pres *Employer:* Stinson and Martinez, Inc *Education:* BS/EE/Univ De La Habana *Born:* 11/27/27 in Cuba, American Citizen by naturalization. Pres of Stinson and Martinez, Inc, a FL Corp engaged in the practice of professional engrg in the fields of Mechanical and Electrical Engrg. Received Bachelor of Science in Electrical Engrg from the Univ of Havana. Registered Engr in the states of FL and NY. Past Chrmn of Miami Section IEEE, past Pres Southeastern FL Section IES, Ex-member National Bd of Dirs LDA-IES, Past Pres Cuban Engrs Association, Recipient of Edwin F. Guth Memorial Award - IES and several other lighting design awards. Voted Engr of the Year - FES Miami Section 1980. *Society Aff:* IEEE, IES, FES, NSPE, NYAS

Martinez, Roberto O
Home: 5905 Thames, Austin, TX 78723
Position: Principal *Employer:* Martinez and Wright Engrs *Education:* B.S./CE/Univ of TX A&M; M.S./CE/Univ of TX - Austin *Born:* 5/9/38 Native of Ozona, TX, served with U.S. Army (Artillery) 1960-61. Plan Review and Planning Engr with TX Hgwy Dept 1961-1964. Employed by U.S. Air Force 1964- 75, as Programming and Planning Engr (4 years) and as Pavement's Engr (7 years). Project Mgr with U.S. Dept of Commerce/EDA 1975-1980. Since Oct. 1980, Principal with Martinez and Wright Engrs, a Civil-Engrg Consultant Firm. Registered as a professional civil engr in the State of TX. Enjoys visiting with friends, helping others and traveling. *Society Aff:* TSPE, ACEC

Martino, Joseph P
Business: Research Inst, Univ of Dayton, Dayton, OH 45469
Position: Senior Research Scientist. *Employer:* Univ of Dayton Research Inst. *Education:* PhD/Mathematics/OH State; MS/Elec Engrg/Purdue; AB/Physics/Miami. *Born:* 1931, Warren Ohio. On active duty with US Air Force 1953-75, retiring in grade of Colonel. Assignments included tech positions at Avionics Lab, Office of Sci Res, & R&D Field Unit (Bangkok, Thailand), & as Dir of Engrg Standardization at Defense Electronics Supply Center. Currently developing prog in tech forecasting at Univ of Dayton. Fellow of IEEE; Assoc Fellow AIAA; Fellow AAAS. *Society Aff:* IEEE, AIAA, TIMS, AAAS

Martis, Jerome M, Sr
Home: 618 E 10th St, Berwick, PA 18603
Position: Consulting. *Employer:* Martis Assocs. *Education:* BS/ME/Bucknell Univ; Associate/EE/Bucknell Univ Jr Coll; Summer Course/Photoelacticity/MIT; Radar Systems & Radar Components//Raytheon Wayland Labs. *Born:* 11/11/20. Native of Berwick, PA, but Maj Projs for Natl Needs had me locate in various part of the US, such as - Los Alamos Scientific Labs, Los Alamos, NM; Maj Space Projs had me act as Sr Scientist and Sr Engr with a citation for Engg Achievement from NASA and Jet Propulsion Labs. Engg background to get into this work as a Sr Consultant Engr is Structural Dynamics and mathematics of Complex Control Dynamics of Space Craft and Radar Dishes. I Organized major engg groups on special projects of Natl Needs within projected concepts and estimated budgets. *Society Aff:* NSPE, PEPP, PASPE.

Martocci, Anthony P
Home: R. D. 5, Bingen Rd, Bethlehem, PA 18015
Position: Research Engr *Employer:* Bethlehem Steel Corp *Education:* MS/Metallurgy/Stevens Inst of Tech; BS/ChE/Lafayette Coll *Born:* 12/5/34 Native of Roseto, PA. Worked in Lackawanna, NY Plant of Bethlehem Steel several years before returning to graduate school for M.S. in Metallurgy. Worked for Curtiss-Wright Aircraft Co while in graduate school. Returned to steel industry at central research labs of Bethlehem Steel in 1965. Worked 12 years in Product Development, Coated Steel Products, the last seven as Supervisor of the Vacuum Coatings Group. The past 10 years have been spent on studies of energy conservation in existing steel processes, particularly at reheat furnaces. Pres of Lehigh Valley Chapter, ASM, 1975-76 also Pres of Lehigh Valley Chapter, NSPE, 1975-76. Received 4

Martocci, Anthony P (Continued)
US patents. Presented several papers and articles on Temperature Measurement and Furnance Instrumentation and Oper. *Society Aff:* ASM, AISE, NSPE, PSPE.

Marur, Hanumanthaiya
Business: 7244 N Genesee Rd, Genesee, MI 48437
Position: Township Engr. *Employer:* Charter Township of Genesee. *Education:* Post Grad Studies/Soil Mech & Hgwys/MI State Univ, USA; ME/Structural Engr/ Bangalore Univ, India; BE/Civil Engg/Mysore Univ, India. *Born:* 11/23/40. Native of Marur, Bangalore Dist, India. Jr Engr for Mysore PWD. Taught in RV College of Engg, Bangalore, India 1965-67. Project, Design and Chief Engr for Kraft Engg, Flint, MI 1969-76. Presently Township Engr and Dir of Public Works for Charter Township of Genesee since 1976. Responsible for planning, designs, implementation, financing, maintenance and operation Public Works. Subject of biographical record, Who's Who Marquis Publication Third Edition, ABI 1977-78, Dicitionary of Int'l Biography Vol XVI, Cambridge, England. The Rotary Int'l The Four Way Test Award - 1979, Resolutions of Appreciation for Exemplary Engg Service - Davison and Genesee Townships. Enjoy classical music, doing social service. Reg PE in MI and OH. *Society Aff:* NSPE, ASCE, AWWA.

Marushack, Andrew J
Business: PO Box 11368, 180 E First S, Salt Lake City, UT 84139
Position: Sr.V.P. Transmission *Employer:* Mtn Fuel Supply Co. *Education:* BS/CE/Univ of OK; BS/Gen Engg/Univ of WY. *Born:* 5/19/35. Native of Rock Springs, WY. Held various positions in Gas Supply Engg dept of Mtn Fuel, including Chief Engr, Gas Supply Operations. Maj emphasis centered on design, equip selection and start-up of compressor stations, underground storage facilities and gas treating plants. Assumed current responibilities as Sr. V.P. Transmission in November, 1980. Operational responsibilities include gas production, transmission, treating, storage and engineering. Hobbies include hunting, fishing, woodworking and music. *Society Aff:* NSPE, WES.

Marx, Christopher W
Home: 71 Cherry St, Cheshire, CT 06410
Position: President. *Employer:* Christopher Marx Assocs Inc. *Education:* ScB/Engrg/Brown Univ *Born:* Jan 1, 1930. Jr engr Brown & Blauvelt, NYC 1954-56; Proj Engr: Severud Elstad Krueger NYC 1954-56, H A Pfisterer New Haven CT 1964-66; Assoc R Besier Old Saybrook CT 1966-69; Principal of own firm New Haven CT 1969- . P E NY, CT, MA, ME, NH. Mbr ASCE, NSPE, ACEC, ACI, AISC, NAFE, AAA. CT Engrs in Priv Pract. President 1977-79. Pub in journal of AISC, 1st quarter, 1976. 5th Annual HUD Award design excellence for Sheffield Manor, New Haven. CSA/AIA Design Award Camp Laurel Program Facility, Lebanon, CT. Chmn, Cheshire High School PTO, 1976-77. Chesire Congregational Church, Sr Deacon 1975-76, Bd of Trustees 1976-77, CT Society of Prof Engrs-President 1982-83. Listed in Who's Who in Engrg since 1977, ACEC Vice Pres 1983-85. Wife: Adelaide H Marx; Children: Timothy, Cynthia, & Sarah. *Society Aff:* ACEC, NSPE, ACI, ASCE, AISC, NAFE, AAA.

Marxheimer, Rene B
Home: 81 Skyline Drive, Daly City, CA 94015
Position: Prof of Elec Engr. *Employer:* San Francisco State Univ. *Education:* EE/Elec Engr/Ecole Polytechnique Federale, Lausanne; MS/Elec Engr/Univ of CA. *Born:* 3/14/23. in Cologne and educated in Paris. Electrical and Lighting Designer for Kaiser Engrs and Bechtel Corps. Specialist in St and Bridge Lighting for CA State Dept of Public Works; design, test and inspection of first outdoor fluorescent lighting installation in USA (Richmond-San Rafael Bridge). Responsible for developing electr engr curriculum at San Francisco State Univ, teaching courses in Electric Power and Systems. First Assoc Dean of Engr and now Prof of Elec Engr. Reg Elec Engr. Active in Continuing Education in Engr. Enjoys outdoor activities and cultural pursuits. *Society Aff:* IEEE, NSPE, ASEE, ISF, ISA.

Maryssael, Gustave J Ch
Home: Virreyes 1070, Mexico 11000DF
Position: Retired, 1971. *Education:* -/Civil & Mining Engg/Univ of Brussels; -/Elec Engg/Univ of Ghent. *Born:* Oct 1903 in Verona, Italy. Belgian citizen. Mining Civil Engr, Univ of Brussels, 1926; Elec engr Ghent Univ 1927. Joined the operating dept of Sofina Brussels in 1929. Appointed Genl Mgr of The Mexican Light & Power Co, Ltd in 1937, then Pres until 1960 when Co was bought by Mexican Govt. Agent for Simon Carves, Ltd until 1966 when assumed mgmt of the Indus Promotion Dept of Banco Nacional de Mexico, SA. Retired in 1971. On board of several indus companies. Chmn of Apasco, SA. de C.V. Belgian Counselor for Foreign Trade in Mexico. *Society Aff:* AIBr, IEEE, AMIME.

Mash, Donald R
Home: 9840 SW 57th Av, Portland, OR 97219
Position: Dean, Multnomah School of Engrg. *Employer:* University of Portland. *Born:* Nov 21, 1923. BS Carnegie Tech; MS, PhD Stanford Univ. Reg P E Calif. Developed a comprehensive engrg prog at Univ of Portland emphasizing cooperation with local indus & community colleges. Indus background includes appl res & engrg respons at Standard Oil Co of Calif, US Steel Corp, & Fansteel. Working on new materials & processes, made significant contribs to nuclear, aerospace & electron industries at American-Standard, McDonnell-Douglas, Los Alamos, Oak Ridge & his own organization Ampro. Published widely in materials sci & nuclear engrg. Taught at Stanford, San Jose State, Loyola U of los Angeles, & Univ of Portland where he has completed 4 years as Dean, School of Engrg.

Maslan, Leon
Home: 1215 W 76th St, Kansas City, MO 64114
Position: Owner. *Employer:* Leon Maslan & Co. *Education:* BSCE/Civil Engr/Univ of IL. *Born:* 6/5/13. Worked for various federal agencies as an engr on dams, flood control, irrigation works, airways installation. Served four yrs in the Corps of Engrs in the European Theater during WWII. Worked for the Fed Housing Authority, the City of Kansas City, MO and local Architects as plan examiner and field engr. In 1948, opened own office for the practice of Architecture and Engg. Designs all types of bldgs, special structures, fdns, storm water systems and streets. *Society Aff:* NSPE, ASCE, ACI, CSI, AIA.

Mason, Clyde P
Home: 5601 Country Dr, Apt. 204, Nashville, TN 37211
Position: Consulting Engineer. *Employer:* Self. *Education:* BS/CE/NC State Univ; BS/HE/NC State Univ. *Born:* June 30, 1905 in Hyde County, NC. Married Leeomo Galliher June 6, 1931; 3 children. Cons civil engr, Lexington Ky since 1949; designer of sewage treatment plants at Stanford, Manchester, McKee, Mt Vernon, London, Booneville, & Barbourville Ky; water plants for McKee, London, Booneville & Mt Vernon Ky; instructor U of Ky 1942-43. Reg PE in Calif, Ky, Ohio, W Va, Va, NC. Mbr of Natl Soc Prof Engrs, Water Pollution Control Fedn, Aircraft Owners & Pilots Assn, Phi Kappa Phi. Election officer Towers Precinct 1973-75. *Society Aff:* NSPE, WPCF, ACEC.

Mason, David M
Business: Stauffer Building III, Stanford, CA 94305-5025
Position: Prof of Chem Engg & Chemistry. *Employer:* Stanford Univ. *Education:* PhD/Chem Engg/ CA Inst of Tech; MS/Chem Engg/CA Inst of Tech; BS/Chem Engg/CA Inst of Tech. *Born:* 1/7/21. Los Angeles CA; married, 1953 to Honora MacPherson, Chem Engg, Chemistry, BS CA Inst Tech, 43, MS, 47, PhD(chem engg), 49. Chem engg, Standard Oil Co, 43- 46; instr chem engg CA Inst Tech, 49-51, supvr appl phys chem group, jet propulsion lab, 52-55; assoc prof chem engg, Stanford Univ, 55-57, Prof Chmn Dept, 58-72.Nat Sci Found fellow Imp Col, Univ London, 64-65. Liaison Scientist Office of Naval Research, London 1972-73. USNR, 43-46, Lt Comdr. Distinguished Alumni Award, CIT 1966. Endowed Named Lectureship in Chem Engg at Stanford 1975, Fellow. AIChE 1978. Found-

Mason, David M (Continued)
er's Award AIChE 1984, Mbr Tau Beta Pi and Sigma Xi. *Society Aff:* AIChE, ACS, ASEE, Elect Soc.

Mason, Donald R
Business: Depts of Chem/Chem Eng, Environmental Sci/Eng, Melbourne, FL 32901
Position: Dept Hd & Prof *Employer:* FL Inst of Tech *Education:* PhD/ChE/Univ of MN - Mpls; BS/ChE/Univ of IL - Urbana; Quatrieme/French/Alliance Francaise - Paris; Dip/French/Alliance Francaise - Paris *Born:* 08/19/20 1942-44; EI du Pont, Arlington, NJ 1944-46, USNR, ETM 2/c. 1946-49 Univ MN, Mpls, grad student Chem Engr, Minor Elject Engr. 1949-56, Bell Telephone Labs, Murray Hill, NJ. (1952-53, Post-doctoral Fulbright Scholar, ENSIC, Nancy France). 1956-65, Assoc/Prof Chem Engr, Univ Michigan, Ann Arbor. (1963-64, Sabbatical, Visiting Prof of Physics, Ecale Karmale Superieure, Paris France). 1965-77, Harris Semiconductor Melbourne FL. 1977-present Prof & Dept Hd, FIT, Melbourne, FL. 30 publications, 8 patents. CSTR's, semiconductor materials and processes, industrial water treatment. *Society Aff:* AIChE, AAAS, IEEE, ACS, ECS, APS, FAS

Mason, Edward A
Business: Div of Engg, Providence, RI 02912
Position: Prof of Engg. *Employer:* Brown Univ. *Education:* PhD/Phys Chem/MIT; BS/Chem/VPI. *Born:* 9/2/26. Native of Alexandria, VA. Served in US Navy 1945-46. Res Assoc at MIT 1950- 52; Natl Res Fellow at Univ of WI 1952-53; Asst Prof at PA State Univ 1953-55; Assoc Prof 1955-60, Prof 1960-67, Dir of Inst for Molecular Physics at Univ of MD 1966-67; Prof of Engg and Prof of Chemistry at Brown Univ 1967-present; Newport Rogers Prof of Chem 1983-present; Visiting Prof at Leiden Univ, Netherlands, 1981-82. Assoc Editor of Physics of Fluids 1963-65, Journal of Chem Physics 1964-66, Case Studies in Atomic Physics 1971-76, Journal of Membrane Science, 1981-present. Scientific Achievement Award of Wash Academy of Sciences in 1962. Author of four books and over 290 res papers in scientific and engg journals. *Society Aff:* AAAS, APS, AAPT, RS, ΣΞ.

Mason, Edward A
Home: 145 Hillcrest Ave, Hinsdale, IL 60521
Position: VP, Res. *Employer:* Amoco Corp. *Education:* ScD/Chem Engg/MIT; MS/Chem Engg/MIT; BS/Chem Engg/Univ Rochester. *Born:* 8/9/24. Rochester, NY. USN, 1943-45, Asst Prof, Chem Engg at MIT, 1950-53. Dir of Res for Ionics, Inc, 1953-57. Returned to MIT in 1957 as Assoc Prof of Nuclear Engg; Prof in 1963, and Dept Hd, 1971-75. In Jan, 1975 appointed Commissioner on the Nuclear Regulatory Commission. In 1977 joined Std Oil Co (IN) as VP, Res with responsibilities for res coordination throughout the corp, for Std's Res Dept, and for Alternative Energy Dev. Councillor, NAE, 1978-84; Dir, American Nuclear Soc, 1972-75; Mbr ACRS, 1972-75. AIChE Robert E Wilson Award, 1978. Wife Barbara, six children. Enjoy sports--golf, tennis, skiing, sailing, skeet. Director: Commonwealth Edison Co; Cetus Corporation; Crerar Library, X-MR, Inc; Gene-Trak Systems. Member advisory committee to: MIT, Univ of Chicago, Univ Texas, Univ-California (Berkeley) *Society Aff:* NAE, AIChE, ANS, AAAS, ACS, AIC, AAAS.

Mason, Henry Lea
Home: 7008 Meadow Ln, Chevy Chase, MD 20815
Position: Consultant (Self-employed) *Education:* ScD/Eng Mech/Univ of MI; ME/Mech Engg/Rutgers; BS/Mech Engg/Rutgers. *Born:* 8/22/00. Taught at Rutgers, Columbia, Iowa State, U Virginia, 1925-55. Res Engr, Dir Res Taylor Instrument Co 1935-46. Genl engr, phys sci administrator, editor, cons, US Natl Bureau of Standards 1952-76. Terminology Chmn Internatl Fed Automatic Control 1963-69. Phi Beta Kappa, Sigma Xi, Tau Beta Pi. Fellow ASME, Sr ISA . *Society Aff:* ASME, ISA.

Mason, James A
Home: 159 Timber Dr, Berkeley Heights, NJ 07922
Position: retired *Education:* BS/Chem Engrg/IL Inst of Tech. *Born:* in Norfolk, VA. Granted BS in Chem Engg at IL Inst of Tech in 1950. Mbr of Tau Beta Pi & Phi Lambda Upsilon honorary socs. Joined Mixing Equip Co upon grad as Jr Engr, specialized in Fluid Mixing Tech & the application of mixers to chem & other operations. Extensive experience with fermentation and fertilizer industries. Was Director and/or Chief Tech Support for Sales Grp in NYC area until retirement in 1986. Active on AIChE Natl Admissions Ctte. Involved as singer in church choir & other vocal groups. Father of three sons. *Society Aff:* AIChE.

Mason, James A, Jr
Home: 138 Curtice Park, Webster, NY 14580
Position: Manufacturing Technical Specialist *Employer:* Xerox Corp. *Education:* MS/Ind Engg/Penn State; BS/ME/Notre Dame. *Born:* 5/21/34. Belleville, IL. After 4 yrs in the machine tool industry, held positions in product design & mfg engg in the aerospace industry. With Xerox since 1964 in mgt roles in Mfg Engg, Plant Engg & Prog Mgt. Served on several standing committees of ASME Council, including Chrmn, Investments & Constitution & By-Laws. Proj Mgr on ASME career guidance prog for minority jr high students. Charter mbr of Task Force that founded ME3 (Minority Engg Education Effort). Asst Treas, ASME 1974-1979. Also a Mbr of IIE, Tau Beta Pi, Alpha Pi Mu. Reg PE in NY. Hold patents on roll forming process. Adjunct Faculty, Manufacturing Technology, RIT (Rochester Institute of Technology). *Society Aff:* ASME, IIE, TBΠ, AΠM.

Mason, James M
Home: 488 Grove St, Framingham, MA 01701
Position: Manager Plant Engrg Dept. *Employer:* Raytheon Co-Equip Dev Lab. *Education:* Masters/Engg Mgmt/Northeastern Univ; Bachs/Electrical Engr/OH Univ. *Born:* May 1928. MS from Northeastern Univ; BS from Ohio Univ. Native of Framingham, Mass. WWII Navy. Design Engr for Dupont Co prior to joining Raytheon Co in 1958. Along with Plant & Facil engrg respons, is actively involved in the co's Energy Mgmt progs. Reg PE in Mass. State Pres of Mass Soc Prof Engr 1972- 73, currently Natl Dir to NSPE & awarded 'Fellow' Memebership Status in MSPE in 1976. Mbr of Policy Ctte of New England Engrg Journal. P Chmn of Raytheon Co Plant Engrs Council & Mbr AIPE, ESNE. Enjoys square dancing & fraternal activities. *Society Aff:* NSPE, AIPE.

Mason, Kenneth M
Home: 14 New England Dr, Rochester, NY 14618
Position: VP & Genl Mgr. (Retired) *Employer:* Eastman Kodak Company. *Education:* BA/English, Chemistry, Biology/Wash. & Jefferson College. *Born:* Sept 1917 Rochester, NY. Grad work business admin & chem at U of Rochester. Lt US Navy 1943-46, Head Processing Div, Photo Sci Lab, NAS, Anacostia. Joined Eastman Kodak 1937. Assignments included: 1937-43 Cine Processing Div, Res Lab, Film Developing & Film Planning Depts; 1946-51 Staff Engr, M P Film Dept; 1951-74 mgmt positions in Chicago, NY, & Hollywood-Motion Picture & Audiovisual Markets Div. Became Asst VP & Genl Mgr of div in 1974 & VP in Dec 1978. Retired 10/1/82. P Pres, Soc of Motion Picture & Television Engrs. Honorary Mbr SMPTE & Hon. Fellow BKSTS (British Kinematograph, Sound & TV Society). Hobbies golf, all sports. Currently chairman, Brd. Trustees, W&J College. *Society Aff:* SMPTE, AMPAS, UFA, TEA.

Mason, Maughan S
Home: 18910 Cyril Pl, Saratoga, CA 95070
Position: Programmer *Employer:* IBM. *Education:* MS/Physics/Brigham Young Univ; BS/Physics/Brigham Young Univ. *Born:* April 5, 1931. Sr Life Mbr Soc for Computer Simulation (Pres 1962-64); Secr AFIPS 1965-68; Session/group organizer/chmn tech confs (FJCC, IFIP Congress, Summer Computer Simulation Confs). Tech papers in US & Europe (Third IFAC Congress, London 1966; Modern Computational Techniques ---, Paris 1962). Pioneered computer-aided art: Journal & book coverage, plus one man show; traveling exhibit US & Europe; Inst of Contemporary Art Exhibit London 1968. 20 yrs simulation exper, including facility establishment & mgmt, tech/marketing support, etc. *Society Aff:* SCS, ΣΠΣ.

Mason, Raymond C
Business: 235 Morgan Ave, Dallas, TX 75203
Position: Chmn of the Board. *Employer:* Mason-Johnston & Assoc Inc. *Education:* BS/CE/Univ of Texas. *Born:* 8/28/19 BSCE Univ of TX. Served with Army Corps of Engrs from 1941-50 with the last assignment being Div Soils Engg, MO River Div. Formed Mason-Johnson & Assocs in 1950, specializing in Engr Geology, Soil Mechanics & Found Engr with proj varying from earth dams in France, Surinam & US, airfields in SA, to multistory bldgs, Missile Bases & Nuclear Power Plants. A Comnmissioner in the Boy Scouts, Lions Club, instructor in Grad School of Engr of Southern Methodist Univ. Award of Merit 1972 AEG, Chmn of Soils & Foundation Div, -ASCE, Secr 1972 CEC-NT 1979- 80, Mbr of ACEC, ICOLD, ISRock Mechanics. Engineer of the Year, 1986 Texas Society of Professional Engineers. *Society Aff:* ASCE, NSPE, AEG, ICOLD, TSPE.

Mason, Ted D
Business: 80 E Jackson, Chicago, IL 60604
Position: Chief Mechanical Officer *Employer:* Santa Fe RR. *Education:* MS/Ind Mgr/MIT; BS/Civ/Univ of CO. *Born:* 6/20/25. 30 yrs experience and knowledge - rr engg, maint and operation. Past positions had 3 fields of responsibility - maint of way, maint of equip, and coal dev. Coal dev had to do with opening up new field-build rail line, environmental prob, as well as designing operation - acquisition of equip etc. Present position is head of Mechnical Department (Locomotives and Cars). *Society Aff:* AREA, LMOA.

Mason, Warren Perry
Home: 50 Gilbert Pl, West Orange, NJ 07052
Position: Sr Res Assoc, School of Mines, Retired. *Employer:* Columbia University. *Education:* PhD/-/Columbia Univ; MA/Phyiscs/Columbia Univ; BS/EE/Univ of KS. *Born:* 1900. From 1921-65 was with Bell Telephone Labs where he was head of Piezoelectric & Mechanics Res. Since his retirement he has been Visiting Prof & Sr Res Assoc both at Columbia & George Washington Univ, Washington DC. Now retired from both. Recipient of many outstanding awards: Arnold Beckman Award of the Instrument Soc 1964, Distinguished Alumni Award, Univ of Kansas, 1st C B Sawyer award for quartz crystal devices, Benjamin Lamme Award of IEEE 1967, Gold Medal of the Acoustical Soc 1971, 1st Hon Mbr of the British Inst of Acoustics. Fellow & P Pres of Acoustical Soc, Fellow of IEEE, Physical Soc, Soc of Engrg Sci; Mbr of Sigma Xi, Tau Beta Pi, & Rheological Soc. *Society Aff:* IEEE.

Masri, Sami F
Home: 1705 E California Blvd, Pasadena, CA 91106
Position: Professor. *Employer:* Univ of Southern California. *Education:* BS/AeA/Univ of TX; MS/AeA/Univ of TX; MS/ME/CA Inst of Tech; PhD/ME/CA Inst of Tech. *Born:* 12/9/39. With the Civil Engrg Dept at USC since 1966; Assoc Dept Chmn 1973-1984. Conslt in the field of stuct dynamics. Active mbr in sev res & engrg socs. *Society Aff:* ASME, ASCE, AIAA, IEEE, AAAS, ΣΞ.

Massa, Frank
Business: 280 Lincoln Street, Hingham, MA 02043
Position: Bd Chrmn & Engr Consultant. *Employer:* Massa Products Corp. *Education:* MSc/EE/MIT; BSc/EE/MIT. *Born:* 4/10/06. Life Fellow IEEE; Fellow Acous Soc of Amer; Mbr Amer Inst of Physics; Exec Council A S A , Admin Council I R E; Mbr sev tech cttes in various natl Engrg Socs. Author sev textbooks & numerous articles on electroacoustics. Electroacoustic res RCA-Victor 1928-40; Dir Acous Engrg Brush Dev Co 1940-45 (respon for many basic devs & designs of sonar transducers used in antisubmarine warfare & harbor defense sys.) Mbr sev US Navy & N D R C cttes on underwater sound. Founder Massa Labs Inc 1945 (now Massa Products Corp). Awarded over 140 US pats. *Society Aff:* IEEE, ASA, AIP

Massalski, Thaddeus B
Business: , Pittsburgh, PA 15213
Position: Prof. *Employer:* Carnegie-Mellon Univ. *Education:* DSa (hon causa)/-/Tech Univ of Warsaw; DSc/Met/Univ of Birmingham; PhD/Phys Met/Univ of Birmingham; BSc/Phys Met/Univ of Birmingham. *Born:* 6/29/26 Native of Poland. Served in the Polish Corps of the British Army during WWII. Subsequently completed advanced educ in Great Britain. Came to US as Post Doctoral Fellow at the Univ of Chicago and later to Pittsburgh to hd the Metal Physics Prog at the Mellon Inst. Author of books and res articles. Editor of Progress in Mtls Sci. Editor in chief , ASM/NBS Prog on Phase Diagram evaluations. Fellow ASM (1976). Guggenheim Fellow 1956. AIME Hume-Rothery award (1980). Foreign Mbr Pol Acad (1980). Prof of Metal Physics and Met Eng at Carnegie-Mellon Univ since 1967. *Society Aff:* AIME, ASM, APS, AAAS.

Massand, Nanik P
Home: 85-18 212th St, Hollis Hills, NY 11427
Position: Pres. *Employer:* Massand Assoc, Inc & Nanik Massand, PC *Education:* MS/Civ/Roorkee Univ; BS/Struct/Poona Univ; IS/Sciences/Bombay Univ. *Born:* 11/28/32. Migrated to US in 1965. Pres of Massand Assoc Inc since 1973 and Nanik Massand, PC since 1981 engaging in Architectural/Civil Engrg, Surveying and Const mgmt serv. Supervising Tech Specialists, specifications and Engrg Documentations, Inspections, Scheduling, Cost Controls. Also working Village Engr, Village of Lawrence, Nassau County, NY; Panelist Amer Arbitration Assoc. *Society Aff:* ASCE, AAA, PEPP, NSPE.

Massart, Keith G
Home: 5520 S W Lander Place, Seattle, WA 98116
Position: Owner *Employer:* Keith G. Massart & Assoc. *Education:* BS/Mech Engg/Univ of WA. *Born:* 4/30/16. Army CAC 1941-46, awarded Bronze Star Medal. Owner Keith G Massart & Associates 1981, spec in all types of mech consulting. Previously Exec VP the Massart Co, Mechanical Contractors. P E State of Washington. Mbr of Wash Soc of P Es & N S P E. Pres Seattle Mech Contractors Assn 1959 & 1965. Pres Puget Sound Chap ASHRAE 1961. Dir ASHRAE 1976-79. Disting Serv Award ASHRAE 1975. Fellow Grade ASHRAE 1980. Dir Queen City Savings Assn 1976 to 1982. V Pres Bd of Advisors Washington School Building Sys 1976. V Pres of Bid Exchange Bd of Dirs. Formerly listed in Who's Who in the West. *Society Aff:* ASHRAE, NSPE.

Massey, Charles G
Business: 1270 First National Center, Oklahoma City, OK 73102
Position: Pres-Owner *Employer:* Ramsey Engrg Inc *Education:* B.S./Petro Engrg/TX A & M Univ *Born:* 8/3/31 Served on detached duty as job-site representative of U.S. Army Frankfurt residence engr in 1954. Employed as a petroleum engr by Amoco Production for ten year period ending in 1966. Registered PE, State of OK, PE 6579. Joined Ramsey Engrg Inc as a petroleum consultant in 1966 and acquired co-ownership and presidency in 1972. The firm is a professional engrg entity engaged in petroleum consulting, offering services to both the oil and gas industry & to individual mineral owners. *Society Aff:* SPE, AIME, PEPP, NSPE

Massey, Gail A
Business: Dept of Electrical & Computer Engrg, San Diego State Univ, San Diego, CA 92182
Position: Prof *Employer:* San Diego State Univ *Education:* PhD/EE/Stanford Univ; MS/EE/Stanford Univ; BS/EE/CA Inst of Tech *Born:* 12/2/36 Native of El Paso, TX. Engr with Raytheon Co, Santa Barbara, CA (1959-63) developing infrared and laser components and systems. Sr engrg specialist with GTE Sylvania, Mountain View, CA (1963-72) responsible for laser systems, high power lasers, and nonlinear optics. Assoc-full prof, OR Graduate Center, Beaverton, OR (1972-80) with programs in lasers, nonlinear optics. Prof of electrical engrg, San Diego State Univ (1981-) leading research in ultraviolet lasers and photoemission. Fellow, Optical Society of America and Sr member IEEE. Many journal publications in lasers, optics, electron beams. *Society Aff:* IEEE, SPIE, OSA, ASA

Massey, James L
Business: Inst. Signal & Info. Proc, ETH-Zentrum, 8092, Zurich, Switzerland
Position: Prof fur Digitaltechnik *Employer:* Swiss Federal Institute of Tech Zurich Switzerland *Education:* PhD/Elec Engr/MIT; SMEE/Elec Engr/MIT; BScEE/Elec Engr/U of Notre Dame. *Born:* Feb 11, 1934 Wauseon Ohio. BSEE maxima cum laude Univ Notre Dame 1956, SM & PhD EE M I T 1960 & 1962. 1956-59 communications officer US Marine Corps. 1962-1977 Dept EE Univ Notre Dame; 1966-67 Visiting Assoc Prof Dept EE M I T; 1971-72 Guest Prof Tech Univ Denmark. 1977-78 Visiting Prof EE MIT; 1978-80 Dept System Science, UCLA 1980- Telecommunications Institute, Swiss Federal Institute of Technology, Zurich. 1963 Best Tutorial Paper Award Natl Electron Conf. 1964 Paper Award of the IEEE Group on Info Theory. 1975 Western Elec Fund Award from ASEE. 1975 Internatl Telemetry Conf Award for contribs to the field of Coding Theory . 1987 IEEE W.R.G. Baker Award. Fellow IEEE. President, Int. Assoc. for Cryptologic Research (IACR). Transactions on Info Theory'. Mbr of Eta Kappa Nu, Sigma Xi & Tau Beta Pi. *Society Aff:* IEEE, IACR.

Massieon, Charles G
Home: 101 Decatur Rd, Marquette Hts, IL 61554
Position: Res Staff Met. *Employer:* Caterpillar Tractor Co. *Education:* Met Engr/Met/CO Sch of Mines. *Born:* 3/5/30. Born & raised in Peru, IL. Joined Caterpillar Tractor Co after grad in 1952. Have worked in the proj area of the Res Dept since. Am a mbr of the SAE subcommittee on Bushing & Bearing Alloys, & the WRC subcommittee on Hard Facing and Wear. Authored an SAE paper on Heat Treatment & Residual Stress Effects on Fatigue Strength, & appeared on an SAE panel on Hardenability Calculations. Teach a mtls sci course at IL Central College. Like fishing, hiking, photography, & reading. *Society Aff:* ASM.

Masson, D Bruce
Business: Matls Sci & Engrg, Pullman, WA 99164
Position: Professor & Chairman. *Employer:* Washington State Univ. *Education:* BS/Chem Eng/WA State Univ; MS/Chem/Univ of Chicago; PhD/Chem /Univ of Chicago. *Born:* Dec 19, 1932. Currently Prof Dept of Mechanical Materials Engrg Washington State Univ Pullman WA; Past posits: Visiting Prof Univ of Strathclyde Glasgow Scotland; Asst Prof of Mech Engrg Rice Univ Houston TX; Res Assoc Argonne Natl Lab Argonne IL. Mbr Amer Soc Metals. AIME. Natl Mbr of Latin-Amer Affairs Ctte ASM. *Society Aff:* ASM, AIME, NACE.

Mastascusa, Edward J
Business: Electrical Engrg Dept, Lewisburg, PA 17837
Position: Prof & Chrmn *Employer:* Bucknell Univ *Education:* PhD/EE/Carnegie Mellon Univ; MS/EE/Carnegie Mellon Univ; BS/EE/Carnegie Mellon Univ *Born:* 6/27/38

Masters, Robert Wayne
Home: 6423 Walters Woods Dr, Falls Church, VA 22044
Position: Pres & Chmn of Bd. *Employer:* Antenna Res Assocs Inc. *Education:* PhD/EE/Univ of PA; MS/EE/OH State Univ; BS/EE/Univ of AL. *Born:* May 25, 1914 Fort Wayne, Indiana. Married Audrey June Holstein 1941, two sons both deceased. Holder 23 U S Patents on antennas and r f transmission systems. IEEE Fellow Award 1962. Mbr Sigma Xi, Tau Beta Pi, Eta Kappa Nu. Reg Prof Engr state of Ohio. Employed RCA 1941-49, Ohio State Univ Prof 1949-58, Boeing Airplane Co Res Supr 1958-60, Melpar Inc Mgr Antenna Dept 1960-64, Westinghouse Elec Fellow Engr 1965-68, Antenna Res Assocs Pres-Chmn 1968- . Inventor Superturnstile and VHF Traveling Wave TV Broadcast antennas, Chief Cons for Empire State TV Broadcast Antenna System 1950-52. Chmn Columbus, Ohio Sect 1954. Ohio State University Distinguished Alumnus Award 1987. *Society Aff:* ΤΒΠ, HKN, ΣΞ.

Masubuchi, Koichi
Business: Rm 5-219, Dept of Ocean Engg, Cambridge, MA 02139
Position: Prof. *Employer:* MA Inst of Tech. *Education:* PhD/Naval Arch/Univ of Tokyo; MS/Naval Arch/Univ of Tokyo; BS/Naval Arch/Univ of Tokyo. *Born:* & raised in Japan. Worked at the Transportation Technical Res Inst (currently Ship Res Inst), Ministry of Transport, Tokyo, Japan from 1948 to 1958 (last position: Chief of Welding Mechanics Section, Welding Div). Worked at Battelle Meml Inst, Columbus, OH from 1958 to 1962 as a Visiting Engr. Worked at the Ship Res Inst, Tokyo from 1962 to 1963. Worked at Battelle Meml Inst from 1963 to 1968 (last position: Technical Advisor at Mtls Joining Div). Worked at MA Inst of Tech as Assoc Prof (1968-71) & Prof (1971 to present). The present title is Prof of Ocean Engg & Mtls Sci. *Society Aff:* SNAME, AWS, ASM, ASME.

Masuda, Senichi
No. 415, 3-2-1, Nishigahara, Kita-ku, Tokyo, Japan 114
Position: President *Employer:* Fukui Institute of Technology *Education:* PhD in Eng/Elect Eng/Univ Tokyo; B Eng/Elect Eng/Univ Tokyo *Born:* 4/12/26. Prof Emeritus of Univ of Tokyo & Pres Fukui Inst of Technology. Maj speciality is applied electrostatics, electrostatic precipitation and pulse corona induced plasma chemistry. Pres Inst of Electrostatics Japan. Fellow of IEEE, Corresponding mbr of VDI. Frederick G. Cottrell Award & Frank A. Chambers Award for notable contribution to electrostatic precipitation. Current work on "Boxer-Charter ", "Ceramic-Based Ozonizer–, "Cryogenic Ozone Production–, "Nanosecond Pulser" for electrostatic precipitation and DeSOx/DeNOx and mercury vapour control at Masuda Research, Inc. Award of Merit and Excellent Paper Award from IEEE, Award of Merit from ESA. *Society Aff:* IEEE, IEJ, ESA, IEEJ, VDE, APCA.

Matare, Herbert F
Home: P O Box 49 177, Los Angeles, CA 90049
Position: Pres ISSEC. *Employer:* Self. *Education:* PhD/Solid State/ENS Paris; Dr Ing/Electronics/Techn Univ, Berlin; MS/Physics/Techn Univ Aachen; BS/Physics/Techn Univ Aachen. *Born:* Sept 22, 1912 Aachen, W Germany. American citizen 1967. Directed microwave receiver lab Telefunken Germany 1939-45. Prof Electron Aachen; Tech Dir semicond lab F&S Westinghouse, Paris Fr 1945-51. First Europ transistor patents. Pres Intermetall Corp Germany; Cons U S Army Electronics Command to 1955; Dir semicond R&D G T E Sylvania NY 1955-60; Head solid state electronics McDonnell-Douglas to 1967; Scientific Advisor Rockwell Internatl to 1969; Visit Prof UCLA and Calif State Univs Fullerton; Pres ISSEC Los Angeles to present. Published over 100 papers and 2 books plus book articles on semiconduct devices, materials, solid state electronics. Over 60 patents. Fellow IEEE, NY. Mbr APS, ECS, Thin Film Div & Solid State APS, AAAS, NY Acad of Sci. Hobbies: classical music, astronomy and biology. *Society Aff:* AVS, ECS, APS, AAAS, NAS.

Mater, Catherine M
Business: 101 S.W. Western, Corvallis, OR 97333
Position: VP *Employer:* Mater Engrg, Ltd *Education:* BS/Bus Pol Sci/OR State Univ; MS/Civil Engrg/Univ of OR *Born:* 5/14/53 Principal of Mater Engrg, Ltd, Corvallis, OR (Private consulting firm). VP - Construction Engrg Management Division. Current responsibilities encompass contract and construction management for engrg projects including engrg economic analysis and project scheduling. Consultant to private enterprise and government agencies in Engrg, comprehensive planning and economic development. Consultant to private/public industry and organizations in managing public involvement required in technical projects (Facilitory Engrg). Selected Outstanding Young Woman of America - 1979. Pres of Society of Women Engrs for State of OR - 1980. Listed in "Intnatl Who's Who in Engrg (1984), the eighth ed of "The World Who's Who of Women–, and the 20th ed of "Who's Who in the West–. Enjoys private flying (working toward pilots license), hang gliding, raquetball and cross-country skiing. Has Master's degree in Civil Engrg from OR State Univ. *Society Aff:* NSPE, PEO, PEPP, SWE.

Mater, Milton H
Business: 101 SW Western Blvd, PO Box O, Corvallis, OR 97339
Position: Pres. *Employer:* Mater Engg, Ltd. *Education:* BSME/ME/CCNY. *Born:* 4/2/15. Named OR Engr of the Yr, 1964, for pioneering contribution to sawmill automation; recipient 1973 Gottschalk Award of Forest Products Res Soc for outstanding service to forest products industry. Winner of Lincoln Arc Welding Fdn Award for advanced sawmill machinery design, 1961. Author of over 100 papers on sawmill design; speaker at intl forest products confs. Col, US Army Reserve (Ret); Awarded Presidential Legion of Merit, 1971. Past Pres, Southwest OR Museum of Sci & Industry. Past Chrmn: Forest Products Res Society Pollution Abatement Committee; Lumber Mfg Committee; Energy Conservation Committee. Recipient of 20 patents in engg design. Moderator: FPRS Seminar computerization of Sawmill Functions 1984. Tech Chrmn ASME Natl Mtg Wood, Pulp & Forest Products Ctte 1984. *Society Aff:* ASME, NSPE, FPRS.

Matera, James J
Home: 8 Pinewood Drive, W Boyleton, MA 01583
Position: Dir & Chief Engr Water Division *Employer:* Black & Yeatch *Education:* Mech Eng/-/MIT; Civil Eng/-/MIT; Graduate Science/-/Harvard. *Born:* May 1912. Native of Boston, Mass. Graduate of Lowell Div MIT Mech and Civil Engrg. Career Civil Engr for Commonwealth of Mass dealing with water supply since 1932. Reg Prof Engr Mass. Pres Mass WWA 1964. Mbr and Asst Treas Boston Metro Planning Council 1964-74. Editor Journal of New Eng WWA 1964-1987. Editor Emeritus, 1987- . P Pres NEWWA, Chmn New Eng Sect AWWA 1973. Mbr and Secy Mass Bd of Certification Water Supplies 1971-74. Award of Merit NEWWA 1971, Fuller Award AWWA 1975. Diplomate Amer Acad Env Eng. Dir AWWA, Mbr Ex Comm. Reg Fallout Shelter Analylist. Technical writer, lecturer and instructor, Hobbies: history, golf & gardening. *Society Aff:* AWWa, NSPE, NEWA, SAME, AAEE.

Mates, Robert E
Business: Dept. of Mech. & Aero. Engr., 337 Jarvis Hall, Amherst, NY 14260
Position: Professor *Employer:* University at Buffalo, SUNY *Education:* PhD/Mech. Engr./Cornell University, Ithaca, NY; Masters/Mech. Engr./Cornell University, Ithaca, NY; BS/Mech. Engr./Univ. of Rochester, Rochester, NY *Born:* 05/19/35 Native of Buffalo, NY. Member of the faculty of SUNY Buffalo since 1962. Presently Prof of Mech and Aerospace Engrg and Research Assoc Prof of Medicine. Served as Ch of Mech and Aerospace Engrg for six years. Research interests are in cardiovascular biomechanics, primarily coronary blood flow. Publications include approximately 35 research papers, 5 edited Symposium Proceedings. Served as General Chairman for the 40th ACEMB in 1987 and as Program Ch for the 1988 ASME Winter Annual Meeting. Fellow of ASME. *Society Aff:* ASME, BMES, APS, ASEE, AHA

Mathe, Robert E
Home: 3918 So. Pinehurst Circle, Denver, CO 80235
Position: Pres *Employer:* PRC Engrg Consultants, Inc *Education:* MS/CE/Univ of IL; BS/Sci/US Military Acad, West Point *Born:* 4/21/20 in Oshkosh, WI. Graduated from the US Military Academy in 1943. Served in US Army Corps of Engrs until 1967. Achieved rank of Brigadier General. While in the US Army, pre-eminent positions were: Division Engr, Sacramento, CA' Commander of the 36th Engr Group, Korea; and Engrg Commissioner for District of Columbia. Pres of HOK Assocs from 1968 to 1972. Chief of Infrastructure Division, Project Analysis Dept, Inter-American Development Bank from 1972 to 1978. Pres of PRC Engrg Consultants, Inc from 1978 to present. Married since June 8, 1943 to Elinor V. (Metzen) Mathe, five children. *Society Aff:* ASCE, ICOLD, NSPE, SAME, APWA

Matheny, Robert E
Home: 689 Lake Howard Dr, NW, Apt 202, Winter Haven, FL 33880
Position: President *Employer:* Remtech Inc. *Education:* BSEE/EE/Rutgers Univ. *Born:* 2/13/16. Active career in contracting chem engg in NY area for over 30 yrs, involved in selling, designing & constructing plants & complexes for heavy chem plants such as sulfuric acid, phosphoric acid, granular fertilizers, etc. All on world wide basis. Background ranges from proj engr, proj mgr, operation mgr, resident European gen mgrcorporate VP & Mgr. International Sales & Licensing while associated with Chem Const Corp, Titlestad Corp, Dorr-Oliver Inc. Jacobs Engg Group Inc.Presently located in FL developing Consulting Business Providing Large Engineering Project Management services. Jacobs Engg Group Inc. Licensed PE in NJ, NY, CT, MI & FL. Hold Natl Engg Exam. Certificate of qualification. *Society Aff:* AIChE.

Mather, Bryant
Business: POBox 631, Vicksburg, MS 39180
Position: Chief, Structures Lab. *Employer:* U S Army Engr Waterways Exper Sta. *Education:* DSc/Hon/Clarkson College; AB/Geology/Johns Hopkins. *Born:* Dec 27, 1916 Baltimore Md. Grad work JHU 1936-38 & 1940-41, American Univ 1938-39, Curator of Minerals Field Museum Chicago 1939-41. Corps of Engrs 1941- . ASTM Award of Merit and Fellow 1959; Sanford E Thompson Award 1961; Frank E Richart Award 1972; Secy-Chmn Ctte C-9 on Concrete 1952-66; Chmn Ctte C- 1 on Cement 1968-74; Chmn E-39 on Fire Hazard Standards 1973-75; Dir 1970-78, Pres 1975-76. ACI: Pres 1964; TAC 1951-60; Dir 1958-67; Henry C Turner Medal 1973 Charles S Whitney Medal for Conc Lab 1974; Honorary Mbr 1969. Chmn Concrete Div Transp Res Bd NAS-NRC 1963-69. Roy W Crum Disting Service Award 1966. Meritorious Civ Serv Award Dept of Army 1965; Decoration for Exceptional Civ Serv 1968. Fellow Am Assn Adv Sci 1960. Mbr US Ctte on Large Dams 1959. Chmn USA Tech Adv Group for ISO Tech Ctte 71 on Concrete 1973-date. P.E. (Miss) 1969. *Society Aff:* ASTM, ACI, AIME SPE, AAAS, ESA, SAME, ΣΞ, USCOLD

Mathes, Kenneth N
Home: 2052 Baker Av, Schenectady, NY 12309-4132
Position: Cons - Elec Insulation (Self-employed) *Education:* BSEE/Elect Eng/Union College *Born:* June 1913. Native of Schenectady, NY. Private Consultant-Electr Insulation, Employed GE Co 1935-78, R&D & application of insulating materials. K N Mathes Lab at GE Lynchburg, Va. Life Fellow IEEE, P Chmn IEEE Group on Electrical Insulation, Fellow ASTM, P Chmn ASTM D-9 on Electrical Insulation, ASTM Arnold H Scott Award of Merit 1976, P Tech Advisor on Electrical Insulation to USNC of Internatl Electrotechnical Comm (IEC), Chief USA delegate to IEC on Electrical Insulation since 1962, Fellow AAAS. Reg PE in NY. Schenectady Engr of Year 1969. Hobbies include Pers Computer, photography & cabinet making. Enjoy classical music & opera. *Society Aff:* ASTM, IEEE, ACS, AAAS

Matheson, Neil, III
Home: 709 Carroll Dr, Garland, TX 75041
Position: Mgr Facilities & Industrial Engg. *Employer:* Graham Magnetics Inc. *Education:* MBA/Admin Mgmt/North TX State Univ; BSChE/Chem Engg/TX A&M Univ. *Born:* 12/20/34. BSChE from TX A&M Univ; MBA (Admin Mgmt) from North TX State Univ. Prod Engr Union Carb Corp-olefins mfg; Process Control & Dev Engr Rexall Drug & Chem co-polyolefin mfg; Plant Engr TX Insturments Inc. VP of Mfg for Cobaloy, Div of Graham Magnetic Inc 11/73-5/78; VP of Operations for Coloy Div of Graham Magnetics 5/78-12/78; Dir Corp Facilites, Graham Magnetics 12/78-10/79; Mgr acilities & IE, Graham Magnetics 9/79- . Enjoys horses, photography, dirt motocycle riding & chess. *Society Aff:* AIChE.

Mathews, Bruce E
Home: 2111 Dyan Way, Maitland, FL 32751
Position: Asst Dean, Coll of Engg. *Employer:* Univ of Central FL. *Education:* PhD/Elec Engg/Univ of FL; MSE/Elec Engg/Univ of FL; BEE/Elec Engg/Univ of FL. *Born:* June 1929. BEE, MSE, & PhD from Univ of Fla. Served in the US Air Force 1953-55. Worked for North Amer Aviation 1955-56. Res Assoc 1956-64 & Assoc Prof 1964-69 at the Univ of Fla. Took position at Univ of Central FL in

Mathews, Bruce E (Continued)
1969 & became Prof & Chmn of Elec Engrg Dept in 1971. Became Asst. Dean, College of Engineering in 1978. Mbr IEEE, ASEE, Eta Kappa Nu, Tau Beta Pi, Sigma Tau, Phi Kappa Phi, Outstanding Educators of Amer. Awarded Engr of the Year, 1972 by Orlando Sect of IEEE. Tech interests include electromagnetic field theory, Control systems, & signal processing. *Society Aff:* IEEE, ASEE, HKN, ΣΞ, ΦΚΦ, ΤΒΠ.

Mathews, Max V
Business: 2D-554 Mountain Av, Murray Hill, NJ 07974
Position: Dir, Acoustical & Behavioral Res Ctr. *Employer:* Bell Laboratories. *Education:* ScD/Elec Eng/MA Inst of Tech; MS/Elec Eng/MA Inst of Tech; BS/CA Inst of Tech. *Born:* Nov 1926. ScD & MS from MIT; BS from CIT. Native of Columbus, Neb. Joined Bell Labs in 1955; became Dir of Acoustical & Behavioral Res Center in 1962. Personal res is in sound & music synthesis with digital computers. Also Sci Advisor to Inst for Res & Coordination Acoustics/Music (IRCAM) in Paris, France. Fellow IEEE & ASA. IEEE David Sarnoff Gold Medal Award 1973. Mbr of Natl Acad of Scis 1975. Mbr of Natl Acad of Engg 1979. Enjoys music, sailing, & skiing. *Society Aff:* ASA, AES, IEEE, NAS, NAE.

Mathews, Richard A
Home: Ravine Lake Rd, Bernardsville, NJ 07924
Position: V P - Operations Planning & Engrg. *Employer:* New York Telephone Co. *Education:* BS/EE/Rutgers Univ. *Born:* April 1924 in Trenton. Served in US Naval Reserves 1944-46. Pres of Tau Beta Pi & mbr of Phi Beta Kappa. Initially joined NJ Bell Telephone Co in 1948, transferring to NY Telephone Co in 1950. Various assignments in traffic & engrg prior to going to AT&T. Returned in 1959 as Genl Mgr until 1961. Roles with Amer Co as Asst V P of Marketing & subsequently inital Dir of Bell System Data Training Prog at Cooperstown. During 1964 appointed V P -Planning & in 1967 Genl Dir BIS at BTL with current title occurring on Jan 1973. Also Reg P E in NY.

Mathews, Warren E
Home: 1010 Centinela Ave, Santa Monica, CA 90403
Position: Retired *Education:* PhD/Physics/Caltech; MS/Elec Engrg/MIT; BS/Elec Engrg/MIT; BA/Physics/OH Wesleyan Univ. *Born:* Nov 1921. Served with Army Signal Corps 1944-46. Joined Bell Telephone Labs radio res dept 1946. With Hughes Aircraft Co since 1950, in positions including Dir of Planning, Dir Infrared Labs, Mgr Missile Systems Div, Mgr Equip Engrg Divs, Asst Grp Exec Electro-optical & Data Sys Grp, Staff VP, Product Effectiveness, Staff VP, Technical Management Planning, Retired 1986. Mbr of Natl Exec Ctte, Infrared Information Symposia 1960-65; Genl Chmn, WINCON (IEEE) 1971. Fellow IEEE; Assoc Fellow AIAA; Hon Mbr Beta Gamma Sigma. Arranger & performer of choral music; enjoys tennis. *Society Aff:* IEEE, AIAA.

Mathias, Robert James
Business: 2245 de LaSalle, Montreal Quebec, Canada H1V 2K7
Position: Plant Engineer. *Employer:* Y & S Candies Div. *Born:* July 25, 1928 Plymouth, England. Attended St Boniface Coll. Served 5 years apprenticeship with Her Majesties Dockyards; also attended Dockyard Trade Coll. Immigrated to Canada 1954. Maintenance foreman with Ronalds Federated Graphics 9 years, 2 years Canron as Railway Dept Prod Supt, 9 years with Y & S Candies. Region V P Amer Inst of Plant Engrs.

Mathur, Umesh
Business: PO Box 4633, Tulsa, OK 74104
Position: Pres. *Employer:* Consul-Tech, Inc. *Education:* Post Grad Dipl/Petrol Refining & Petrochem/Indian Inst of Petrol; BTech/Chem Engg/Indian Inst of Tech. *Born:* 7/13/45. Chem engr in Tulsa, OK. Worked as Refinery Technologist for Burmah-Shell Refineries, 1967-1971. Technical Dir of Tulsa area govt planning agency for water quality mgt under EPA prog, 1974-1977. Pres of Consul-Tech, Inc since 1977. Specialize in process design & simulation of petrol & chem systems both for steady state & dynamic conditions. Evaluation of control strategies for process startup, shutdown & emergency conditions. Math & statistical methods for design & optimization. Thermodynamics of phase equilibria. Hydrological and water quality modeling. Database mgt. Serve on EPA committee (municipal construction). Reg engr, OK. Hobby: teaching math. *Society Aff:* AIChE, WPCF, ACS, APCA.

Matisoo, Juri
Business: P.O. Box 218, Yorktown Heights, NY 10598
Position: Engrg Mgr *Employer:* IBM Corp *Education:* PhD/EE/Univ of MN; MS/EE/MIT; BS/EE/MIT *Born:* 2/10/37 Dr Matisoo is mgr of engrg for exploratory cryogenic tech at the IBM Thomas J. Watson Research Center, with responsibility for device, circuit and memory development in the Josephson program, a cryogenic tech for high-performance computers. He joined the Research Center in 1964 and has concentrated on high performance device development. He received the IEEE's 1978 Jack A. Morton Award for outstanding contributions in the field of solid-state devices for "pioneering the Josephson computer tech–. He is a fellow of the IEEE and a member of the American Physical Society. *Society Aff:* IEEE, APS

Matotan, William I
Business: 230 Truman St NE, Albuquerque, NM 87108
Position: President. *Employer:* William Matotan & Assocs, Inc. *Education:* BSCE/Civil/Univ of NM; MSCE/Structural/IL Inst of Technology. *Born:* 11/11/21. Pres of William Matotan & Assocs, Inc, Con Civil Engrs, since 1957. Mbr of Cons Engr Council, Water Pollution Control Federation. Co-author of 'Albuquerque Plant Designed with Computer in Mind', Water & Wastes Engrg Jan 1976. Reg PE in New Mexico, Texas, Colorado, & Arizona. *Society Aff:* CEC, WPCF

Matschke, Donald E
Business: 2 Salt Creek Ln, Hinsdale, IL 60521
Position: President. *Employer:* D E Matschke Company. *Education:* PhD/ChE/Northwestern Univ; MS/ChE/Northwestern Univ; BS/ChE/Northwestern Univ. *Born:* March 1933. Native of Chicago area. NSF Fellow & Resident in Res at NU. Cons & subsequently Dir of Process & Environmental Res for Chicago Bridge & Iron Co, 1966-71. Joined Bauer Engrg Inc in 1971 as principal & served as Chief of Staff & Dir of R&D until formed own engrg firm in 1975. D E Matschke Co supplies process & environmental engrg services to indus, municipal, & govt clients. Currently advisor to Northeastern Ill Planning Comm & Genl Chmn Cook County Clean Streams Ctte. Enjoys sailing & tennis. *Society Aff:* AIChE, AIME, ACS, WPCF, AWWA, AAAS.

Matson, Howard O
Home: 2801 E Harrison 60, Harlingen, TX 78550
Position: Hd Engrg & Watershed Planning. *Employer:* Soil Conservation Service. *Education:* MSc/Eng/Univ of CA; BSc/Eng/Univ of NB; BA/English/Cotner College. *Born:* Oct 1902. Native of Lincoln, Nebr. Instructor & Ext Spec in Agricultural Engrg at Univ of Ky; Proj Engr, Soil Erosion Service, USDI, Lindale Tex; with Soil Conservation Service USDA, Forth Worth, Tex successively Head, Engrg Div; Head, Water Conservation Div; Head, Engrg & Watershed Planning Unit. Retired Dec 1966. VP ASAE; Chmn Soil & Water Div ASAE; Chmn SW Sect ASAE. Life Fellow ASAE. Enjoys travel, fishing, bridge & classical music. *Society Aff:* ASAE, SCSA.

Matsuda, Fujio
Business: 1110 University Avenue, Room 408, Honolulu, HI 96826
Position: Exec Dir *Employer:* Res Corp of the Univ of Hawaii *Education:* ScD/Civil Engrg/MA Inst of Tech; BS/Civil Engrg/Rose Polytechnic Inst *Born:* 10/18/24. in Honolulu, HI. Univ of HI 1942-43, 1946-47; Rose Polytechnic Inst, Indiana BS w/Hons 1949; ScD from MIT in 1952. Pres Univ of HI Sept 1974-May 1984. VP Bus Affairs Univ of HI May 1973- Aug 1974. Dir, Dept of Transportation, State of HI 1963-73. Pres Shimazu, Matsuda, Shimabukuro & Assocs, 1960-63. VP Park &

Matsuda, Fujio (Continued)
Yee Ltd, 1956-58. Univ of HI: Dir, Engrg Experiment Station, 1962-63; Chmn, Dept of Civil Engrg, 1960-63; Prof of Engrg, 1962-65 & 1974-84; Assoc Prof of Engg,1957-58, 1959- 62; Asst Prof of Engrg, 1955-57. Res Asst Prof CE Univ of IL, 1954-55. Res Engr MIT, 1952-54. Hydraulic Engr USGS, 1949. Awards: Tau Beta Pi, Sigma Xi, Rose Polytech Inst Honor Alumnus Award, 1971, HI Engr of Yr, 1972, AOCI (Airport Operators Council Internatl) Distingusihed Service Award, 1973, US Dept of Transportation Award for Exceptional Public Service, 1973; Univ of HI. Distinguished Alumnus, 1974; American Society of Civil Engineers Parcel-Severdrup Engineering Management Award, 1986; Rose Hulman Inst of Tech; Hon DEng, 1975. Soka University (Tokyo, Japan), Honorary Doctorate, 1984. Mbr Natl Acad of Engrg. *Society Aff:* ASCE, NSPE, NAE

Matsukado, William M
Home: 3325 S.W. 173rd, Aloha, OR 97006
Position: Project Manager *Employer:* CH2M-Hill *Education:* MS/Mech Engg/CA Inst of Tech; BS/Mech Engg/Univ of HI. *Born:* 11/28/42. Native of Honolulu, HI. Res Mech Engr for Naval Civil Engg Lab in Port Hueneme, CA (1964-66) developing secondary creep law for sea-ice. Continued res work at US Bureau of Mines, Albany, OR (1966-69) investigating mech properties of superconducting materials. Began process design work on Zirconium extraction at Teledyne Wah Chang Albany in 1969. Responsible for design of Chlorination Process and plant construction. In 1977, joined group of engrs in forming and establishing Western Zirconium. Responsible for overall plant design and cost analysis. In 1980 joined CH2M-Hill, consulting engineers, as a Project Manager with over lapping Mechanical design responsibilities. *Society Aff:* AIPE, ASHRAE, ASME, NSPE.

Matsumoto, Goro
Home: 6-1, 5jo-10chome, Sumikawa, Minami-ku, Sapporo, Japan 005
Position: Professor *Employer:* Hokkaido Institute of Tech. *Education:* Ph.D./Elect. Comm. Engin./Tohoku University; M.S./Elect. Comm. Engin./Tohoku University; B.S./Elect. Comm. Engin./Tohoku University *Born:* 09/22/23 Native of Miyagi Prefecture, Japan. Studied at Tohoku Univ in the area of electric contact phenomena 1950-64. Following appointment as full prof of Hokkaido Univ, his studies have been directed toward the biomedical electronic, especially to contribute the progress of biotelemetry using the thin film microelectronic technique. A charter member and a bd mbr of the International Soc on Biotelemetry from 1972. Retired Hokkaido Univ; emeritus professor. Still studying biomedical engrg at the Hokkaido Institute of Technology. *Society Aff:* IEEE, ASTM

Matsumoto, Michael P
Business: 1210 Ward Ave, Honolulu, HI 96814
Position: Principal Secy/Treas. *Employer:* SSFM Engrs, Inc. *Education:* MS/Civ Engg/Univ of IL; BS/Civ Engg/Univ of HI. *Born:* 7/3/45. Born & raised in Honolulu, HI. With SSFM Engrs, Inc since grad in 1968. Current responsibilities include serving as Vice President of Corp & proj engg for several projs at any one time. Mbr: CECH, SEAOH, ASCE (Past Pres 1978), Chi Epsilon (Natl Civ Engg Honor Soc), Governor's Environmental Council (Advisory). Active in community & church. Hobbies: tennis, martial arts. *Society Aff:* ASCE.

Matsumoto, Yuki
Business: 931 Univ Ave, Ste 305, Honolulu, HI 96826
Position: Pres *Employer:* Yuki Matsumoto & Assoc Inc *Education:* BS/EE/Univ of HI *Born:* 5/10/28 on island of HI. Wife Jennifer and three children. Electronics techician and electrical engr, United Air Lines Inc. 15 yrs. Instructor of Electronics, Electro Tech School. Pres of consulting firm since 1972. Pres, IES HI Chapter, 1977; Pres, Cech 1978; Natl dir, CECH 1979; Chairman,Legislative ctte CECH 1980; Minuteman, ACEC 1980; Mbr, Mayor's Construction Industry Advisory Council, City & County of Honolulu. Hobby: modelmaking. *Society Aff:* CECH, HSPE, IES, CSI

Mattei, Peter F
Home: 9954 Holliston Ct, St. L, MO 63124
Position: Executive Director. *Employer:* Metropolitan St Louis Sewer Dist. *Education:* BSCE/Univ of MO, Rolla; Honorary BSCE/Univ of MO, Rolla; Honorary BS/Eng Mgmt/Univ of MO, Rolla *Born:* Mar 10, 1914. Attended Soldan High School; Reg PE in Missouri. Pres Engrs of the State of Missouri 1976-77 MSPE; Mbr Tau Beta Pi March 31, 1973. Certified Charter Mbr of Acad of Civil Engrs UMR Nov 11, 1972. Chi Epsilon Hon Engrg Soc. Engr of the Year MSPE St Louis Chapt 1968. Award for Outstanding Achievement 'In recognition of Prof Leadership, Engrg Ability & Personal Integrity as evidenced by the overwhelming passage of $95,000,000 Clean Water Bond Issue' - 1962 MSCE. Mbr Amer Soc of Civil Engrs, Natl Soc of PE's, Missouri Soc of PE's, Engr's Club St Louis Chapt - President Mo Soc Prof Engrs 1978. Pres Univ of MO Rolla Alumni Assoc 1973-76. Mbr RCGA Sci & Tech Comm. Pres of Peter Mattei & Assoc Consltg Engrs.. *Society Aff:* NSPE, ASCE, SAME, APWA, WPCA, NSIE.

Mattern, Donald H
Home: 2524 Craghead Ln, Knoxville, TN 37920
Position: Retired *Education:* CE/Prof/IA State Univ; MS/Hyd Power/IA State Univ; BS/Civil/Penna State Univ. *Born:* July 1905 Johnstown, Pa. 4 years with Consumer Power Co in Operating Dept; 1 year with cons engr in Keyser, W Va. Since Jan 1934, with Tennessee Valley Authority in various design capacities. From 1954 to retirement in 1970 Chief of Proj Planning Branch with duties of conception, preliminary design & cost estimates, proj economics for hydro & steam power devs in the Valley. Natl Dir of ASCE 1958-60; VP 1961-62; Prof Recognition Award 1963; Hon Mbr ASCE 1968. *Society Aff:* ASCE.

Matthews, Allen R
Home: 107 Rolling Ct, Lexington Park, MD 20653
Position: President - Cons & Lecr. *Employer:* MATTCON, Inc *Born:* 8/23/18. 1940-42 Westinghouse Elec Corp; successively Mgr, Dir, VP Tamar Electonics Co, Raytheon Co, Boeing Corp, Admiral Corp 1959-62. Mgr advanced dev No Am Rockwell Autonetics 1962-64; Ch Sci Airborne Warning & Control System McDonnell-Douglas Co 1964-70; Ch Sci Advisor Weapons System Test Div Naval Air Test Center, Patuxent River MD 1970-75, Chief Scientist 1975-80. Pres, MATTCON Inc 1980- . Consultant, teach courses in T&E, and Exectuive Director ITEA. Founder, Pres/Treas of Intl Test and Evaluation Association (ITEA) 1980-81, Tech Prog Chmn Intl Electronic Warfare Symposium 1970, 1973. USAF 1942-46, 1947-58 Korea. Mbr IEEE, AIAA, Assn Old Crows (Natl Dir), Sigma Xi, Tau Beta Pi, Rotarian. Res: traveling wave tube dissertation, intercept receivers, radome ray tracing techniques for side lobe patterns, Radar, Test & Evaluation, Electronic Warfare, Antennas, Propagation, Mission Sys Engrg, Independent R&D. *Society Aff:* IEEE, AIAA, AOC, ITEA

Matthews, Charles S
Business: P O Box 2463, Houston, TX 77001
Position: Cons Petroleum Engr. *Employer:* Shell Oil Co. *Education:* PhD/Chemistry/Rice Univ; MS/ChE/Rice Univ; BS/ChE/Rice Univ. *Born:* Mar 1920 Houston, Tex. Joined Shell Dev Co 1944, Chief Reservoir Engr 1956; Mgr Engrg Head Office 1966; Dir Production Res 1967; Sr. Petroleum Engr Conslt 1973. Active in reservoir engrg & non-petroleum sources of energy. Twenty pubs in tech journals, seven patents. Co-author first Society Petroleum Engrs monograph 'Pressure Buildup & Flow Tests in Wells'. SPE disting lecturer 1967- 68, SPE Uren Award 1975 for disting achievement in tech of petroleum engrg, API Reserves Ctte, Amer Men & Women of Science, National Academy of Engineering, 1985. *Society Aff:* SPE.

Matthews, Edwin J
311 Adams St, Fayetteville, AR 72701
Position: Assoc Prof Ag Engg. (Emeritus since 1983) *Employer:* Owner of Matthews Engineering *Education:* MS/Agri Engg/NC State Univ; BS/Agri Engg/Univ of TN.

Matthews, Edwin J (Continued)
Born: 11/15/19 Native of Humboldt, TN. Army Air Corps, 1942-45. Instr and Asst Prof of Agri Engg, Univ of TN, Knoxville 1946-56. Assoc Prof Agri Engg, Univ of AR, Fayetteville, 1956 to date. Tenured since 1958. Chrmn TN State Sec ASAE in 1951- 52 and 1955-56. Chrmn AR State Sec 1974. Chrmn Natl ASAE, A-116 Outstanding Paper Awards Committee in 1976-77. Proj leader or Co-leader of three AR Agri experiment Station res projs involving Agri field machine dev and utilization. Agri field machine design consultant and expert witness. Enjoy catching fish. Regional Dir, ASAE 1986-88, Consult Engr. 1983. *Society Aff:* ASAE, ASPE

Matthews, James B
Business: Coll of Engrg and Applied Scis, Kalamazoo, MI 49008
Position: Dean *Employer:* Western MI Univ *Education:* PhD/Aero Engrg/Univ of AZ; MS/Mech Engrg/MIT; BS/Mech Engrg/Rose-Hulman Inst of Tech *Born:* 05/15/33 Have 29 years experience in engrg education at three insts with service as Engrg Dean at Ross-Hulman Inst, Bradley Univ and since 1983, at Western MI Univ. Primary background in Mech Engrg with emphasis in Aerospace and engrg mechs. Consltg experience includes design and construction of coal preparation plants and equip, investigations related to prod liability litigation, design of machinery, dev of tech education programs, and energy mgmt. Reg PE in MI, IN, IL, and Certified Energy Mgr. Mmbr Mich State Bd of Prof Engrs. *Society Aff:* NSPE, ASME, SAE, ASEE, NCEE, AEE

Matthews, S LaMont
Business: 2300 N.W. Walnut Blvd, P.O. Box 428, Corvallis, OR 97339
Position: VP & Dir, Indus & Energy Sys *Employer:* CH2M Hill, Inc *Education:* BS/ME/OR State Univ *Born:* 9/12/34 With CH2M Hill, Inc since 1956. Has overall corp responsibility for activities in corrosion mitigation; electric utility services; electrical, and mechanical engrg; industrial systems; instrumentation and control; and power generation. Mr Matthews has also authored a number of publications on boiler air quality control, energy conservation, pump applications, and solid waste energy recovery systems. *Society Aff:* ASME, CECO, ASCE

Mattock, Alan H
Business: More Hall FX-10, Seattle, WA 98195
Position: Prof Civil Engrg. *Employer:* Univ of Washington. *Education:* PhD/Engg/Univ of London; MSc/Engg/Univ of London; BSc/Engg/Univ of London. *Born:* Jan 1925. Fellow ACI, ASCE & Inst Civil Engrs (Britain). Reg Struct Engr IL. Since 1964 Prof of Civil Engg Univ Wash. Prior to 1964, Principal Engr Struct Res Lab, Portland Cement Assn, Skokie Ill; also other positions in design, construction & teaching in England & S America. Primary interest is behavior of concrete structures. Dir ACI 1976-79. Korn Award PCI 1975; Reese Award ASCE 1972; Lindau Award ACI 1970. Wason Medal ACI 1967. *Society Aff:* ASCE, ACI, PCI, ICE, CS.

Mattson, James B
Home: 4140 Xerxes Ave S, Minneapolis, MN 55410
Position: Principal Engr. *Employer:* Univ of MN. *Education:* Bachelor/Agri Engg/ Univ of MN. *Born:* 12/29/25. With Univ of MN's Physical Planning office since 1961. Currently Planning Coordinator for Res & Experiment Stations. Currently Chrmn of N Central Region of Am Soc of Agri Engg. *Society Aff:* ASAE.

Mattson, Roy H
Home: 9060 N Riviera Dr, Tucson, AZ 85704
Position: Dept Head, Elec Engrg Dept. *Employer:* University of Arizona-Tuscon. *Education:* PhD/EE/IA State Univ; MS/EE/Univ of MN; BS/EE/Univ of MN. *Born:* 12/26/27. Wife, June; 7 children: Kristi, Lisa, Greta, Linnea, Marla, Brent, Brian. IEEE Fellow 1971 'for contribs to education & solid state & medical electronics res'; AAAS Fellow 1972; mbr of IEEE Education Activities Bd; mbr of EE Div of ASEE; mbr of Natl Soc of Prof Engrs; Chmn IEEE Validation of Educational Achievement Prog; mbr of Sigma Xi. Reg PE in Ariz. Mbr of Amphitheaer School Bd 1971-76; Dept Head Elec Engrg at U of Ariz 1966- . Travelled to USSR for Papov Society Exchange in 1972. Assoc Prof Elec Engrg U of Minn 1961-66; Assoc Prof Elec Engrg Iowa State U 1959-61; Asst Prof Elec Engrg Iowa State U 1956-59; Tech Staff Bell Telephone Labs, New Jersey, 1952-56; Editor IEEE Transac on Educ 1970-74. *Society Aff:* IEEE, ASEE, NSPE, AAAS.

Matula, David W
Business: Dept of Computer Sci and Engg, Dallas, TX 75275
Position: Prof of Computer Sci. *Employer:* Southern Methodist Univ. *Education:* PhD/Engg Sci(Op Res)/Univ of CA, Berkeley; BS/Engg Physics/Washington Univ, St Louis. *Born:* 11/6/37. Born Nov 6, 1937 in St Louis, attending Engg School at Washington Univ receiving BS in 1959. Persued graduate work at UC Berkeley 1960-66 receiving Ph.D. in Eng. Science/Operations Res. Taught Computer Sci at Washington Univ 1966-74 and then became Prof 1974-present and Chmn 1974-79 of Computer Sci and Engg at Southern Methodist Univ in Dallas. Has been consultant for Monsanto, IBM, CDC. Visiting appointments at Stanford, Naval Postgraduate School, Univ of Texas, the Univs of Frankfurt and Karlsruhe in West Germany and the Univ of Aarhas in Denmark. Author of over 60 scientific articles. Principal investigator on 7 Natl Sci Fdn grants covering 1973-87. *Society Aff:* ACM, SIAM, ORSA, IEEE Comp. Soc.

Matula, Richard A
Business: The Institute of Paper Chemistry, Box 1039, Appleton, WI 54911
Position: Pres *Employer:* The Institute of Paper Chemistry *Education:* BS/ME/Purdue Univ; MS/ME/Purdue Univ; PhD/ME/Purdue Univ. *Born:* Aug 1939. Has taught at U C Santa Barbara, Univ of Mich, Drexel Univ & Louisiana State Univ. Has had indus & cons exper at govt labs & in the automotive, chem & aerospace industries. At Drexel, held various administrative positions including Chmn of Thermal & Fluid Sciences, Dir of Environmental Studies Inst, & Chmn of Mech Engrg & Mechanics Dept, Dean of Engrg at Louisiana State Univ. Presently President of the Institute of Paper Chemistry. Prof interests in areas of combustion kinetics & energy conversion & utilization. Mbr Editorial Bd of the Journal of Hazardous Waste, Mbr Tech Assessment Panel, Ctte on Demilitarizing Chem Mutions and Agents, Natl Res Council. Mbr and Vice-Chmn Bd of Dir of Engrg Dean's Council, Amer Soc of Engrg Ed. Mbr, Commission on Ed for the Engrg Prof, Natl Assoc of St Univ and Land Grant Colls. *Society Aff:* ASME, CI, ASEE, TAPPI

Matuszeski, Richard A
Home: 1304 San Pablo NE, Albuquerque, NM 87110
Position: Metallurgical Engineer. *Employer:* Genl Elec Co - Aircraft Engine Grp. *Education:* BS & PE/NM Tech. *Born:* Jun 1922 Pittsburgh, Pa. BS & PE degrees from New Mexico Inst of Mining & Tech; also former Mbr of Inst Bd of Regents 1955-60. Reg PE, Curriculum Advisor in Welding & Met, Tech Vocational Inst. Past Chap Chmn Amer Soc for Metals; Who's Who in Engineering 1959; Who's Who in New Mexico 1957 & 1962. Tech Soc Award 1969, & Managerial Award 1972 from GE Aircraft Engine Group. Advisory Ctte SAE Aerospace Materials Div. Interested in model railroading, radio, carpentry & mineral collecting. With GE, Albuquerque Plant Operations since 1968. On Sept 22, 1981 Amer Soc for Metals Bd of Trustees elected Richard A. Matuszeski *FELLOW* in recognition of his distinguished contributions to the field of metals and materials. *Society Aff:* AIME, ASM.

Maurer, David L
Business: 8901 N Industrial Rd, Peoria, IL 61615
Position: Exec VP. *Employer:* Randolph & Assoc, Inc. *Education:* MSCE/Structural/Bradley Univ; BSCE/Civ Engg/Bradley Univ. *Born:* 5/3/44. Native of IL. Attended pre-engg schools majoring in architecture and math and physics. Teach engg structural courses part time at Bradley Univ while performing functions with Randolph & Assoc, Inc as Exec VP, Assoc Stockholder, and Man. of Operations. Background experience includes geotechnical and transportation in addition to structures. Formerly with Hanson Engrs, Inc, Daily & Assoc, Inc, and Casler, Houser, Hutcheson. Recipient of Central Il. Engineer of Yr. Award in Peoria in 1983. Very active in profl and technical societies. Has provided expert witness testimony periodically. Registered professional engineer and structural engineer in IL.. *Society Aff:* NSPE, ASCE.

Maurer, Karl G
Home: 3106 Ninth St N, Fargo, ND 58102
Position: Chmn & Prof of Mech Engineering. *Employer:* North Dakota State University. *Education:* PhD/Engg Mech/Univ of KS; MS/Engg Mech/Univ of KS; BSME/Mech Engg/Drexel Univ. *Born:* Aug 1929 in Philadelphia, Pa. Worked as mechanic & draftsman prior to Korean War. Served with US Army Intelligence 1951-53. Engrg Trainee with Penn Meter Co. Design engr for GD/ Convair-Astronautics 1959-60. Instructor at Univ of Kansas 1960-65. Asst Prof at Rose-Hulman Inst of Tech 1966-67. Assoc Prof & Asst Dean in Coll of Engrg & Architecture, North Dakota State Univ 1967-72; Prof & Hd ME Dept N Dak State U since 1972. Mbr ASEE, Sigma Xi & ASME. Enjoy reading, travel, camping & fishing. *Society Aff:* ASEE, ASME.

Maurer, Robert D
Business: Sullivan Park, Corning, NY 14831
Position: Res Fellow *Employer:* Corning Glass Works *Education:* PhD/Physics/MIT; BS/Physics/Univ of AR *Born:* 7/20/24 Native of Arkadelphia, AR. US Army 1943-46. MIT staff 1951-52. Corning Glass Works 1952- . Mgr of applied physics 1963-78. Responsible for research on applications of glasses and glass-ceramics. Research fellow 1978-. Contributor to a range of physical research programs. Pioneer in optical communications. George W. Morey Award of American Ceramic Society. Morris Liebmann Award of IEEE. AIP prize for industrial physics. Ericsson Intl prize in telecommunications. Ind. Res. Inst. Achievement Award, Opt. Soc, Am. IEEE-LEOS John Tyndall Award. *Society Aff:* IEEE, APS, ACS

Mawardi, Osman K
Business: University Circle, Cleveland, OH 44106
Position: Prof of Engrg *Employer:* Case Western Reserve Univ. *Education:* PhD/Applied Phys/Harvard Univ; AM/Applied Phys/Harvard Univ; MSc/Elec Engrg/Cairo Univ, Egypt; BSc/Elec Engrg/Cairo Univ, Egypt *Born:* Dec 1917 Cairo, Egypt. Mbr Faculties of MIT (1951-61) & Case Western Reserve (1961-). Biennial Award Acoustical Soc of Amer 1952; Guggenheim Fellow 1955; Mbr Inst for Advanced Study, Princeton, 1970. CECON Medal of Achievement, 1979. IEEE Centenial Medal of Merit, 1984. Cons to Indus DOD & ERDA. Since 1973, Pres Collaborative Planners, a cons engrg firm for planning large scale energy systems. Specialist in acoustics (electroacoustics, holography); plasma dynamics (fusion); energy systems. Interest in oenology & art. Trustee Cieveland Inst of Art, P Pres Cleveland Soc for Contemporary Art. Fellow several prof societies. *Society Aff:* IEEE, APS, ASA, AAAS

Maxson, Marshall W
Home: 239 Dutch Lane, Pittsburgh, PA 15236
Position: Chief Metallurgist. *Employer:* United States Steel Corp. *Education:* AB/Chem/Northwestern Unii; MBA/Bus/Univ of Chicago. *Born:* Aug 1913. Elected to Sigma Pi Sigma Hon Physics Fraternity. With USS Corp since 1935 as metallurgist. Presently Ch Metallurgist, Homestead Works, with corporate respon for ordnance & forging prods. Elected ASM Fellow in 1975 for continuous dev of improved steel practices from the 'All Basic Open Hearth Furnace' through the ladle vacuum carbon deoxidation process (VCD). This pioneering work during the dev of the Basic Oxygen Process was respon for the met practices developed for the use of this equipment. *Society Aff:* ΣΠΣ, ASM, AIME.

Maxted, Wesley R
Home: 2135 Brierbrook, Germantown, TN 38138
Position: Pres *Employer:* Wes Maxted, Inc *Education:* BSME/Mgt/Purdue Univ *Born:* 4/19/26 Plant Engr - Assistant Chief Engr Palmer Bee Co, Detroit, MI 1950-1958. Chief Engr Rapistan Inc, a consulting co in Memphis, TN 1979-present. Entire career (33 years) devoted to Materials Handling Engrg. Charter Member of the Intl Materials Management Society. Holds or has applied for six patents pretaining to materials handling systems and devices. Has directed the following disciplines: consulting, project management, concepting, estimating, controls, installation and after installation service. Registered in MI and TN. *Society Aff:* NSPE, CET, IMMS, TSPE, MSPE, ACEC.

Maxton, Robert C
Home: 2125 Argonne Dr, Minneapolis, MN 55421
Position: Technical Dirs. *Employer:* MN Electric Steel. *Education:* BS/Metallurgical Engrg/Carnegie - Mellon Univ. *Born:* 4/14/30. native of Pittsburgh, PA. Officer in the Army Ordnance Corps 1953-55, Paris, France. Sales Engr with Vanadium Alloys Steel Co - tool steels. Plant Metallurgist with Colonial Div, Vasco, a Teledyne Co, Monaca, PA. With MN Electric Steel Castings Co since 1969, Div of Evans Products Co. Assumed current respons as Technical Dir in 1972. Past Chrmn of Beaver Valley Chapter ASM (1965- 66), Chrmn MN Chapter 1978-79. Vice Chrmn, Technical & Operating Conference, Steel Founders' Society of America - 1976. Chrmn Twin Cities Chapter Amer Foundrymen's Society - 1982-83, Chrmn Carbon and Low Alloy Res Ctte, Steel Founder's Society of America - 1980-84. Avid Alpine skier & sailor. Enjoys automobile restoration. *Society Aff:* ASM, AIME, ASQC, AFS, ASNT.

Maxwell, Barry R
Business: 225 Marts Hall, Lewisburg, PA 17837
Position: VP for Administration *Employer:* Bucknell Univ. *Education:* PhD/ME/Univ of NM; MS/ME/Bucknell Univ; BS/ME/Bucknell Univ. *Born:* 1/20/37. Native of Westfield, NJ. Taught for 16 yrs in The Mech Engg Dept of Bucknell Univ before being named Dean of Engg. Recipient of Western Elec & Lindback Teaching awards. Conducted res & published in the areas of Heat Pumps, Stirling Engines, & Laser Doppler Velocimetry. *Society Aff:* SAE, ASME, ASEE, TBΠ.

Maxwell, Clyde V
Home: 4150 Crane Blvd, Jackson, MS 39216
Position: Consulting Engineer. *Employer:* Self. *Education:* BS/CE/Univ of MS *Born:* Oct 1912. Native of Pickens, Miss. In private practice before WWII. Officer in charge of a Navy Const Battalion Unit with rank of Lt. In private practice since WWII. Founder & Pres to 1969 of Engrs Labs Inc. P Pres, Natl Dir Miss Soc of PE's. P Pres Cons Engrs Council of Miss. Mbr NSPE, ASCE. Reg PE in Miss & Louisiana. Semi-Retired. *Society Aff:* NSPE, ASCE.

Maxwell, James D
Business: Soil Conservation Service, b7235 So. 300 West, Midvale, UT 84047
Position: District Conservationist *Employer:* Soil Conservation Service, USDA. *Education:* BS/Civil Engg/UT State Univ; AS/Civil Engg/Southern UT State College. *Born:* July 11, 1932 Cedar City, Utah. US Army 1952-54. Raised on farm. Mbr Sigma Tau Hon Engrg Fraternity. Soil Conservation Service 29 years as Field Engr, Hydrologist, State Planning Staff Ldr, State Environmental Specialist & District Conservationist, Pres Utah Sect ASCE 1973-74 & Mbr Utah Engrg Council. Mbr Soil Conservation Soc of Amer. & Nat'l Assoc Cons. Districts. Certificate of Merit SCS 1972 & 1978-86. Mbr of Bishopric & High Council in LDS Church 11 years. Reg ldr Boy Scouts of Amer 25 years; Wood Badge Trained. Sing in church choir. Enjoy hunting, fishing, camping & hiking. Married, father of seven. *Society Aff:* NACD, SCSA

Maxwell, Marvin V
Home: 2724 Ptarmigan Dr 3, Walnut Creek, CA 94595
Position: Head Elec Utility Div (RET). *Employer:* Harza Engineering Company. *Education:* BS/Eng/Univ of MO *Born:* Feb 1901 Carthage, Mo. Eta Kappa Nu, Tau Beta Pi, Pi Mu Epsilon. Specialty: (Westinghouse) design heavy elec equip (ww generators, synch motors, synch condensors) - appli, installation, maintainence of elec

Maxwell, Marvin V (Continued)
equip; (Bureau of Reclamation) system engrg; (Harza Engrg Co) engrg, supr const, mgmt dev foreign elec utils. Engagements: (Westinghouse) Engr & Service Mgr (37); (USBR) Assoc Engr (3); (Harza) Head Elec Utils Div (8). Societies: ASME Life Fellow, Chmn Chicago Sect, Washington Awards Reps; IEEE Life Fellow; Western Soc of Engrs, VP, Trustee; Ill Engrg Council Pres. Hobbies: hunting, fishing, golf. *Society Aff:* ASME, IEEE

Maxwell, William L
Home: 106 Lake Ave, Ithaca, NY 14850
Position: Prof. *Employer:* Cornell Univ. *Education:* PhD/Operations Res/Cornell Univ; BME/Mech Engr/Cornell Univ. *Born:* 7/11/34. Author of 45 published technical papers and 5 books. Consultant to many industrial cos. Specialize in the analysis and design of scheduling systems for production, mtl handling systems, and info processing systems. Currently interested in the design of highly automated computerized mfg systems. *Society Aff:* IIE, ORSA, TIMS, SME.

Maxworthy, Tony
Business: OHE430, Los Angeles, CA 90089-1453
Position: Prof & Chrmn ME, Prof AE *Employer:* Univ of S CA. *Education:* PhD/ME/Harvard Univ; MSE/ME/Princeton Univ; BSc/ME/Univ of London. *Born:* 5/21/33. Native of London, England. Entered US in 1954 to complete higher education. Res asst in Combustion Lab Harvard Univ (1955-60) Sr Scientist at Jet Propulsion Lab 1960-67. Res in air-fuel combustion, MHD & rotating flows. Assoc Prof (1967- 1970), Prof (1970-present) M.E. and A.E. Depts., Chrmn (1979-present) ME Dept, USC, Res on rotating- stratfield flows, geophysical fluid dynamics, unsteady aerodynamics etc. Chrmn div of fluid dynamics, APS (1977-78). Editorial Bd, Physics of Fluids, Geophysical Fluid Dynamics, Dynamics of Oceans & Atmospheres, ZAMP. Numerous consultancies. Fellow, APS. Visiting Fellow ANU Canberra; NCAR Boulder; Clare Hall, Cambridge Univ; Von Humboldt Sr. Science Award, West Germany 1981-82; Sr Queen's Fellow, Australia 1984; Prof Associe, Univ de Grenoble, 1982-83. *Society Aff:* ASME, APS, AMS, AGU.

May, Adolf D
Home: 1645 Julian Dr, El Cerrito, CA 94530
Position: Professor & Research Engr. *Employer:* University of California. *Education:* PhD/Transportation/Purdue; MS/Transportation/IA State; BS/Civil Engg/So Methodist Univ. *Born:* March 1927. Taught at Clarkson Coll of Tech & Mich State Univ 1952-59. Res Engr at TRW Computers & State of Ill 1959-65. Currently teaching & res at Univ of Calif - since 1965. Prof activities include Chmnship of Group 3 Council, Transportation Res Bd; & Chmnship of Zone IV Council, Amer Soc for Engrg Educ. Enjoys world travelling, tennis & working with youth. *Society Aff:* TRB, ITE, ASEE, ORSA.

May, Curry J, Sr
Business: Hartsfield Atlanta Intl Airport, Atlanta, GA 30320
Position: Asst VP-Engg. *Employer:* Delta Air Lines, Inc. *Education:* Masters/Bus Admin/GA State Univ; BS/Engg/VA Military Inst. *Born:* 4/18/33. Born in Washington, GA, grad from Darlington School, Rome, GA, VA Military Inst, earning BS in Engg. Served in US Army Corps of Engrs; started with Delta in Dec 1956. Was a Systems Engr in Atlanta; transferred to San Diego for 2 1/2 yrs as Delta's rep at Convair during mfg and certification of CV-880. Returned to Atlanta in mid-1960's and received subsequent promotions. Named Dir - engg in Oct 1972 and Asst VP Engg Nov 1973. Leads Engg activity in dev of specifications and contracts for new and leased aircraft. Chrmn - Aerospace Coordinating Comm of SAE. Mbr AIAA - Atlanta. *Society Aff:* SAE, AIAA.

May, James W
Home: 3908 Elfin Rd, Louisville, KY 40207
Position: Mgr Tech Relations Clean Air Grp. *Employer:* Amer Air Filter Co (Retired). *Education:* BS/Mech Engr/Univ of KY; Masters/Mech Engr/Univ of KY. *Born:* Nov 1907 Anaconda, Montana. Taught 12 years in Coll of Engrg, Univ of Ky. Joined Amer Air Filter 1943 as Dir of Res. Also served as Dir of Tech Training & Mgr of Tech Relations, Clean Air Group prior to retirement in 1971. Past mbr Bd/Dir Ky Soc of PE's; Bd/Dir Engrg Joint Council; Distinguished Alumni Award Univ of Ky; Fellow & P Natl Pres Amer Soc of Heating, Refrig, & Air Conditioning Engrs. Listed in Who's Who in America & American Men of Sci. *Society Aff:* ASHRAE, ASME.

May, John F
Home: 286 Princeton Rd, Sterling Junction, MA 01565
Position: Pres *Employer:* Kaehrle Traffic Assoc Inc *Education:* BS/Civil Engg/Tufts Univ; Cert/Traffic Eng/Bureau of Highway Traffic, Yale Univ. *Born:* Sept 1929. Certificate from Bureau of Highway Traffic, Yale Univ. Native of Groton Mass. US Navy 1952-54, Ensign & Ltjg. Currently LtCdr, USN Ret. Civil Engr with Ill Div of Highways, 1954-62. District Traffic Engr with Penn Dept of Highways 1962-67. Senior Field Engr with Ill Div of Highways 1967-70. Asst Regional Traffic Engr With NY Dept of Transportation 1970-76. Proj Mgr with Wallace-Champagne Assocs 1976. Proj Mgr with Greenman-Pedersen Assocs 1976-1980. Partner & VP Champagne Assocs 1981. 1981 to 1984 Pres, Kaehrle Traffic Assocs, Inc. Pres NY Upstate Sect, Inst of Transportation Engrs, 1974. Vice Chmn, District to 1, ITE 1975. Chmn, District 1, ITE 1976. Registered Prof Engr - IL, PA, NY, MA, CT, and CA (Traffic Engr). *Society Aff:* ITE, TRB, NSPE.

May, Melville M
Home: 9 Legend Ln, Houston, TX 77024
Position: President. *Employer:* Meldon Assocs Inc. *Education:* BS/ChE/Purdue. *Born:* Aug 1927. BSChE from Purdue Univ 1948. Native of San Antonio, Texas. Manufacturers Rep since 1954 spec in solids processing machinery & air & noise pollution control equip. Became Pres of Meldon Assocs Inc, Houston, in June 1970. *Society Aff:* AIChE.

May, Thomas H
Home: 4604 South Rd, Harrisburg, PA 17109
Position: Bureau Dir. *Employer:* PA Dept of Trans. *Education:* ME/Civ Engg/Yale Univ; MGovAdm/Gov Adm/Univ of PA; BSCE/Civ Engg/Lehigh Univ. *Born:* 2/3/32. Dir, Bureau of Transportation Planning Statistics, PA Dept of Transportation. Technical engr with Exxon 1956-1961. Chief of Operations & Maintenance, PA Dept of Forests & Waters 1962-1969. PA Dept of Transportation 1969-. Serves on several committees of the Transportation Res Bd, Natl Acad of Sciences. Author of various papers. Pres of ASCE Central PA Sec 1977. Enjoys sailing, camping, woodworking, home remodeling. *Society Aff:* ASCE, ITE, APWA, TRB, AASHTO.

May, Walter G
Business: PO Box 45, Linden, NJ 07036
Position: Scientific Advisor. *Employer:* Exxon Research & Engrg Co, Inc. *Education:* ScD/ChE/MIT; MSc/Phys Chem/Univ of Saskatchewan; BSc/ChE/Univ of Saskatchewan. *Born:* 1918. Asst Prof, Univ of Saskatchewan for 3 years. Employed by Exxon Res & Engrg since 1948; one-year loan to Inst for Defense Analyses (1959); 3 yr loan to Exxon Nuclear Co Inc. Industry-based Prof at Stevens Inst of Tech, Mech Engrg, 1965-74; Industry-based Prof at RPI, Nuclear Engrg. 1976: Mbr National Academy of Engineering AIChE, ASME, Chmn ASME Cryogenics Ctte 3 years. Publs in Combustion, Coal Gasification, Fluid Solids, Solid Propellants, LNG Safety, Gas Centrifuges. 24 Patents. *Society Aff:* NAE, AIChE, ASME, Comb Inst.

Maybeck, Edward M
Business: Maybeck Engineering P.C, 75 College Avenue, Rochester, NY 14607-1033
Position: President *Employer:* Maybeck Engineering P.C. *Education:* BS/ME/Univ of Rochester. *Born:* Sept 19, 1934. BSME from U of Rochester 1958. Native of Glen Rock, NJ. US Army Reserves 1958-64, Active Duty 1958. 1958-64 worked for Donald M Barnard, PE, cons Engr, Rochester, NY. 1964-66 Proj Engr, Xerox Corp, Rochester, NY. 1966-75 Partner, Barnard & Maybeck, Cons Engrs, Rochester NY.

Maybeck, Edward M (Continued)
1975-80, Secy/Treas, Barnard & Maybeck, Cons Engrs, PC. 1980-84 Program Director, American Institute of Architects. 1985- President, Maybeck Engineering P.C. 1974-76 Natl 1st VP ASPE; 1974-78 Natl Pres ASPE 1978-80 Rochester Chap Pres ACEC; 1972-74 Rochester Chap Pres CSI; 1972-80, Bd/Dir Eastside Community Center, Rochester, NY. 1957 Charles L Newton Prize Univ of Rochester. 1973 Merit Award for Excellence in Specifications, CSI. 1976 Hon Mention for Excellence in Specifications, CSI. 1984 Award of Merit, ASPE. 1983 Technical Commendation, CSI. *Society Aff:* ASPE, ASHRAE, CSI, NSPE.

Mayer, Armand
Home: 51 rue Raynouard, Paris 75016, France
Position: Ret. (Engr General of Mines) *Education:* Grad/-/Ecole Poltechnique Scole des Mines. *Born:* 01/16/94 In Gov't Serv to rank of Engr General of Mines-ret 1954-then Cons Engr Practice in Geotech Engrg. Ended WWI as Artillery Lt with 58th US Field Artillery Brigade. Volunteer WWII-Col-Aviation Engr French Army. Honors: Commander of the Legion of Honor, Croix de Guerre, Amer Legion of Merit, Hon Chmn of both French Cttes of Soil & Rock Mechanics of the French Portland Cement Res Center, Former Vice-Chmn of ISSMFE, former Vice Chmn of the French Ctte on Large Dams, Hon Mbr ASCE etc. *Society Aff:* SFMS, SFMR, ASCe.

Mayer, Edward H
Business: P O Box 2900, Long Beach, CA 90801
Position: Chief Engr. *Employer:* THUMS Long Beach Co. *Education:* MS/Petroleum Engg/Univ of Southern CA; BS/Petroleum Engg/Stanford Univ. *Born:* Jan 30, 1928. Reg PE in Calif, Certificate P-1119. Petroleum Engr Chevron USA 1951-57; District Petroleum Engr, Exxon Corp USA 1957-61; Petroleum Engr, Occidental Petroleum Corp 1961-65; Sr Engr, then Chief (from 1969) Engr, THUMS 1965- . Mbr Soc of Petroleum Engrs of AIME. Served on following SPE cttes: Symbols & Metrication 1951- , Chrmn 1963-71; Distinguished Lecturer 1971-73, Chrmn 1973; Continuing Education 1973-76, Chrmn 1976, Meeting Policy 1980-85. Recipient of Soc of Petroleum Engrs' Certificate of Service Award in 1967. Director Society of Petroleum Engineers, 1987-. *Society Aff:* SPE, AAPG, API.

Mayer, Francis X
Business: PO Box 2226, Baton Rouge, LA 70821
Position: Engg Advisor. *Employer:* Exxon Res & Dev Labs. *Education:* BS/ChE/Univ of Tulsa. *Born:* 3/20/30. Grad from the Univ of Tulsa, with a BS degree in 1952. Joined the staff of Exxon Res & Dev Labs in Baton Rouge, LA as a chem engr in June, 1952. All of his technical career, except for a 2-yr period in the US Army Chem Corps, has been engaged in res & dev activities related to refining & petrol processes. At Exxon, worked on a number of petrol & chem process devs. Special interests are reactor engg & the study of reactor systems. At Exxon, held a number of technical & administrative positions: Engg Assoc - 1964; Section Hd - 1966; Asst Dir - 1967; Mgr - Lab Services - 1973. Presently an Engg Advisor & hds up the technical effort on hydroconversion of heavy feeds. Holder of 23 US patents. *Society Aff:* AIChE, ACS.

Mayer, James W
Business: Dept Materials Sci Engr, Cornell University, Ithaca, NY 14853-1501
Position: Prof. *Employer:* Cornell Univ *Education:* PhD/Phys/Purdue, 1960; BS/ME/Purdue, 1952 *Born:* 4/24/30. Served with Army Ordnance Corps, 1952-1954. Mbr of Technical Staff of Hughes Aircraft Co, Res Labs from 1959 to 1967. At Hughes, specialized in semiconductor nuclear particle detectors, carrier injection in semiconductors, and ion implantation. Joined CA Inst of Tech in 1967 and was Prof of Elec Engg and Master of Student Houses. Joined Cornell Univ in 1980 and is Francis N Bard Prof of Materials Sci. Current res includes, backscattering spectroscopy, ion beam modification of materials and thin film reactions. Co-author of four and co-editor of three books dealing with particle-solid interactions. Over 300 publications in scientific journals. Los Angeles County and NAUI Scuba instructor at Caltech 1970-1980. Fellow APS and IEEE. Councillor of Bohmische Physical Soc. Received MRS Van Hipple Award in 1981, Silver Medal of the University of Catania and was elected to Natl Acad of Engrg, 1984. Married (Elizabeth B) with 5 children. *Society Aff:* APS, IEEE, MRS, AVS, NAE, ECS

Mayer, John P
Home: 1018 Willowvale, Seabrook, TX 77058
Position: Chf Staff Engr. *Employer:* NASA - Johnson Space Center. *Born:* May 1922, Binghamton, NY. BS U of Mich 1944. Flight Res NASA Langley Res Center, Virginia & Flight Res Center, Edwards AFB, Calif to 1958; original mbr Proj Mercury Space Task Grp 1958, Chief of Mission Planning & real time computer formulation for US manned space flights incl Proj Apollo; Asst Dir for Data Systems 1974- at Johnson Space Center, Houston Texas; NASA Superior Achievement Award 1966; NASA Exceptional Service Medal 1968 & 1969; Amer Inst for Aeronautics/Astronautics Mechanics & Control of Flight Award 1969; U of Mich Distinguished Alumnus Award 1970; Inst of Navigation Thurlow Award 1972.

Mayer, Orland C
Home: 2830 S Cloverdale Rd, Boise, ID 83709
Position: Director of Industrial Dev (Ret). *Employer:* Idaho Power Co. *Education:* BS/Elec Engg/Univ of ID. *Born:* 1904 Mora, Minn. Taught in public schools 1925-27. Joined Idaho Power Co, Boise, Idaho as Engr Trainee in 1929. Held various positions including Power Engr, Mgr Coml & Ind Dept, Dir of Indus Dev. Retired 1969. Served on the Idaho State Board of Engrg Examiners 1950-75, Chmn 13 years, Secy 8 years. Pres Natl Council of Engrg Examiners 1973-74. Recipient of NCEE Distinguished Service Certificate 1966, & DSC Special Commendation Award 1974. VP & Dir Western Zone NCEE 1968-69 & 1975-76. VP Natl Soc of PE's 1952-56. Pres Idaho Soc of PE's 1947-49. Outstanding Engg Grad Award, Univ of ID 1966. Statesman Newspaper Award, Portrait of a Distinguished Citizen Award 1969. *Society Aff:* NCEE, NSPE, ISPE, AIDC.

Mayer, Robert
Home: 52 Whitemarsh Rd, Ardmore, PA 19003
Position: Pres. *Employer:* Instrumentation Consultants, Inc. *Education:* MS/EE/Univ of PA; BS/EE/Univ of PA *Born:* 3/28/18 Employed by Sylvania Electric Products as a development engr 1940-41, 46- 47, trained electronic technicians in the Navy during 1941-45, instructor in electrical engrg at the Univ of PA 1947-49, Research Instrumentation Section Head at Honeywell in 1949-51, 53-62, was on active duty with the Navy in 1951-52 at the Armed Services Electro Standards Agency (present rank - captain USNR- ret). Have been with Suntech (Sun Co) since 1962 as a specialized instrumentation engr, section mgr, and consultant. Presently Pres of Instrumentation Consultants, Inc. Fellow of the IEEE, Pres of the IEEE's Industrial Electronics and Control Instrumentation Society (IECI) 1974-75, received IECI "Achievement" award 1978. Registered in PA as an electrical engr and in CA as a control engr. Have three publications and twenty-eight patents. *Society Aff:* IEEE, ISA.

Mayers, Albert J, Jr
Business: 5500 Florida Blvd, Baton Rouge, LA 70806
Position: Secretary-Treasurer *Employer:* Mayers, Crosby & Firesheets, Inc. *Education:* BS/Mech Engr/LA State Univ. *Born:* Nov 19, 1924. Attended The Citadel, Charleston, SC. Served with US Army 1943-46; France & Germany WWII. Engr & Assoc with Ogden, Woodruff & Dyer, Inc 1949-56. Private practice of mech & elec cons engrg 1956- . Chap Pres ACEC & ASHRAE. Enjoys fishing, baseball & football. *Society Aff:* ACEC, ASPE, CES, ASHRAE.

Mayes, Paul E
Business: 1406 W. Green St, Urbana, IL 61801
Position: Prof *Employer:* Univ of IL *Education:* PhD/EE/Northwestern Univ; MS/EE/Northwestern Univ; BS/EE/Univ of OK *Born:* 12/21/28 Research assistant (1950-52) and research assoc (1952-54), Microwave Lab, Northwestern Univ. Assis-

Mayes, Paul E (Continued)
tant Prof (1954-58), Assoc Prof (1958-63), Prof (1963-), Elec Engr, Univ of IL, Urbana. Fellow, IEEE, 1975. Served as adviser to 8 PhD and more than 35 MSEE candidates. Research on antennas resulting in more than 20 journal articles and 10 patents. Consultant to several industrial firms. *Society Aff:* IEEE

Mayfield, F Drew
Home: 2032 W Magna Carta, Baton Rouge, LA 70815
Position: Proprietor. *Employer:* Drew Mayfield & Assoc. *Education:* PhD/ChE/Univ of TX; MS/ChE/Univ of TX; BS/ChE/Univ of TX; BS/Chem/TX A&I Univ. *Born:* 5/21/15. Res Phillips Petroleum Co, Bartlesville, OK 1940-41. Res, Dow Chem Co Midland MI 1941-43. Plant Engr Dow Chem, Freeport, TX, 1943-46. Hd of Plant Process Engr. Celanese Corp of Amer, Bishop, TX, 1946-53. Design & operation of Chem Plant; Foster Grant Co Inc, Baton Rouge, LA 1953-57. VP Foster Grant co Inc 1957-74. Cons Engr 1974- . Charles E Coates Award 1961. Fellow AIChE. Reg PE in LA & TX. Dir City natl Bank, Baton Rouge. Mbr ACS, Sigma Xi, Phi Lambda Upsilon, Tau Beta Pi. *Society Aff:* ΣΞ, ΤΒΠ, ΦΛΥ, ACS, AIChE.

Maynard, Charles W
Business: 1500 Johnson Dr, Madison, WI 53706
Position: Prof *Employer:* Univ of WI *Education:* PhD/Applied Physics/Harvard Univ; MS/Applied Physics/Harvard Univ; BS/EE/Univ of MD *Born:* 10/18/26 Birthplace, Maynard, AR. Served as Electrician in USNR during World War II. Worked at the Bettis Atomic Power Lab in Pittsburgh 1957-1961. At the Univ of WI in Madison since 1961, promoted to Prof Nuclear Engrg 1965, Fellow of American Nuclear Society and Chrmn of its Education Division 1979. AWU Fellow at the Sandia Labs 1969. Exchange Scientist Kurchatov Atomic Energy Inst, Moscow 1976. ABET Accreditation Visitor. Consultant to government and industry on the Physics and Shielding of fission and fusion reactors. *Society Aff:* ANS

Maynard, Hal B
Home: Rt 1 Box 123A, Eastsound, WA 98245
Position: Mgr Office & Staff Ser *Employer:* Orcas Power & Light Co *Education:* Assoc/IE/IM/Univ of TN *Born:* 06/01/26 Native of TN. Attended Univ of TN, Troy State Univ, Carson-Newman Coll and continuing courses related to bus. Employed by Rural Elec Cooperatives in mgmt capacity since 1955. Sr Mbrship in IIE since 1964. P Pres East TN Chapter IIE- winner Chapter Dev Award. P Pres Central AL Chapter IIE-Winner Chapter Dev Award and special Chapter award for dev and leadership. Charter mbr and past officer Soc for Advancement of Mgmt, Montgomery AL Chapter. Pioneer in automated mgmt info sys for rural elec cooperatives. 1,000 in US. Employed as Mgr of Office & Staff Services for Orcas Power & Light Co. *Society Aff:* IIE

Maynard, Murray Renouf
Home: 157 Dunvegan Rd, Toronto Ontario, Canada M5P 2N8
Position: Chairman *Employer:* DAF Indal Ltd *Education:* BASc/-/Univ of Toronto. *Born:* March 15, 1919. From 1942-46 served in Canadian & British Navies in Atlantic, Pacific & Mediterranean war zones. Decorated with MID on Normandy Beaches. After 3 years as R&D engr with Union Carbide, formed Dominion Aluminum Fabricating Ltd, a co specializing in structural aluminum & marine helicopter systems. Chmn of the Bd of Regents of Notre Dame of Canada, Saskatchewan, & founder of the Toronto Progressive Conservative Forum. Appointed a Fellow of the Engrg Inst of Canada in 1974. Married to Patricia Crawford of Toronto. Eight children. Chrmn of the Canadian Welding Bureau. *Society Aff:* FEIC.

Mayne, David Q
Business: Electrical Engrg Dept, Exhibition Rd, London, England SW7 2BT
Position: Prof *Employer:* Imperial Coll *Education:* DSc Eng/Control/London Univ; PhD/Control/London Univ; MSc/EE/Univ of Witwatersrand; BSc/EE/Univ of Witwatersrand *Born:* in South Africa. During 1950-1959 worked as a lecturer at the Univ of Witwatersrand and as a research and development engr. From 1960 lectured at Imperial Coll becoming a reader in 1967, a prof in 1971 and Head of Dept in 1984. Was a visiting research fellow at Harvard Univ in 1971 and have spent several summers at the Univ of CA, Berkeley in collaborative research. Awarded a Senior Science Research Council Fellowship in 1979. Elected a Fellow of The Royal Society for 1985. Have written over 100 papers on control theory and design and am joint author (with D H Jacobson) of *Differential Dynamic Programming*, Elsevier, NY 1970 and joint editor w R. Brockett of *Geometric Methods or System* Theory, D. Reidel, Dordrecht, 1973. *Society Aff:* IEEE, IEE

Mayo, Howard A Jr
Home: 2051 Log Cabin Road, York, PA 17404
Position: Consulting Engineer *Employer:* Self *Education:* BS/ME/Worchester Polytechnic Institute Worcester, MA *Born:* 07/30/25 Born Framingham and educated Bolton and Hudson, MA. schools. Lt. jg USNR, Norfolk, VA. employed by S. Morgan Smith Co., York, PA in 1946, bought by Allis-Chalmers Corp 1960, retired 1985. Served in numerous Engrg & Marketing Management positions. Developed the tube turbine, including world's largest at corps of engineers installations, Ozark and Webber's Falls, AK and highest capacity reversible units at H.S. Truman Plant, MO. Life Fellow ASME & Engr of the Year Award 1982. Presented numerous technical papers; received more than 25 US and foreign patents & contributed to several texts & handbooks. Pilot & scuba certificates, Eagle Scout, on troop cttee. Past Commander and Education Officer, York Power Squadron. *Society Aff:* ASME, NSPE, PSPE

Mayo, John S
Business: AT&T Bell Laboratories, 600 Mountain Avenue, Murray Hill, NJ 07974
Position: Exec Pres. *Employer:* AT&T Bell Laboratories. *Education:* PhD/EE/NC State Univ; BS/EE/NC State Univ; MS/EE/NC State Univ. *Born:* 2/26/30. Joined Bell Labs 1955 working on res on transistorized digital computers & use of digital computers in defense sys; dev of transmission sys utilizing pulse code modulation techniques. Served as Dir & Exec Dir in Ocean Sys developing electron equip. Also Exec Dir in Toll Electronic Switching. VP, Electronis Technology. Assumed current respon as Exec VP Network Sys in 1975. Fellow of IEEE & has served on a number of Inst Cttes. Mbr of the honor socieities Phi Kappa Phi & Sigma Xi & has served in various prof capacities including Chmn of the Intl Solid State Circuits Conf, Chmn of the NSIA Study of Global Communications, Mbr of the NSIA COMCAC Advisory Ctte & Mbr Bd/Dir Natl Engrg Consortium, Inc. Holds a number of pats & has authored numerous tech articles. 1977 Outstanding Engg Alumnus at NCSU, 1978 Alexander Graham Bell Medal, 1979. Mbr Natl Acad of Engg. He is a member of the Corporation of Polytechnic Univ., and a mbr of the Brd of Dir of Johnson & Johnson. He is also mbr of the Coll of Engg Advisory Brd of the Univ. of Calif., Berkeley. *Society Aff:* NSIA, NAE, IEEE, AAES.

Mazzola, Michael C
Business: 41 Berkeley St, Boston, MA 02116
Position: Pres *Employer:* Franklin Inst of Boston. *Education:* SM/Civ Engr/Harvard Univ; DE.T (Hon)/Engrg Tech/Wentworth Inst. *Born:* 3/16/25. Native of Wayland, MA. Civ Engr for Chas T Main, Inc, 1951-54. Hd, Dept of Civ Engg Tech, Franklin Inst of Boston, 1954-75, Assoc Dean of Faculty, 1957-59, Dean of Faculty, 1959-75, Dir 1975-1981. President 1981-present. VChrmn, Technical Inst Div, ASEE, 1968-69, Chrmn, 1969-71. Mbr, Advisory Committee Engg Tech Education study, ASEE-NSF, 1969-71. Mbr, subcommittee on Civ Engg Technician & Tech Educaton of ASCE Committee on Engg Education. VChrmn Technical College Council, ASEE, 1974-76, Chrmn, 1976-78. Mbr, Bd of Dirs, ASEE, 1976-78. Reg Profl Civ Engr. Mbr Ad Hoc Visiting Committee, ECPD. Recipient of ASEE, James H McGraw Award 1982. *Society Aff:* ASEE, ASCE, ACI.

McAdams, William A
Home: 220 Ingleside Rd, Fairfield, CT 06430
Position: Immediate Past Pres & Mbr Policy Com Int *Employer:* Electrotech. Comm; Cons - Int Stds & Trade *Education:* BS cum laude/Math & Physics/WA College. *Born:* 2/24/18. BS physics & math, Wash Coll & Nuclear/radiation physics courses at U of Chicago, other Inst. Expert Health Physicist with supervisory & cons positions at Hanford, other locations. With duPont 1940 to 1946, GE from 1946 to 1980 with respon nuclear/rad protection & Corp positions on natl & internatl standards & certification. Present position since 1980 with resp for presiding over commission, representing it in natl and internatl bodies. First Chrm Amer Bd Health Physics (cert. HP experts); Pres US Natl Ctte IEC 1968-74; Chmn Bd NFPA 1976-78; Dir 19 years to 1986; Pres & Chrmn ASTM 1977-78; Chmn US Natl Ctte CEE 1962-71; Dir Amer Natl Standards Inst 1967-70 & Amer Natl Metric Council 1976- 77; Chmn of officer NEMA Codes/Standards Ctte 1960-80. Alumni Citation in Sci 1960; ASA Gold Standards Medal 1965; ANSI Citation Internatl Standards 1974. Leo B. Moore Award for Standards work 1977; James McGraw Award for Electrical Men, 1980. GE Centennial Steuben Award 1978, Charles Steinmetz Award, IEEE 1983. Natl Inventors Hall of Fame Medal 1984. IEEE Fellow 1976. Fellow SES 1974. ASTM Fellow 1977. Author about 150 papers and invited speeches. *Society Aff:* IEEE, NFPA, ASTM, HPS.

McAfee, Jerry
Business: P O Box 1166, Pittsburgh, PA 15230
Position: Chairman & Chief Exec Officer. *Employer:* Gulf Oil Corporation. *Education:* ScD/Chem Engg/MA Inst of Technology; BS/Chem Engg/Univ of TX. *Born:* Nov 1916. Started career as res chem engr with Universal Oil Prod Co Chgo 1940. 1945 joined Gulf Oil Corp at Port Arthur, Texas refinery & cont career with Corp & subs in var tech & exec capacities to present. Assumed current pos as Chmn & Ch Exec Off of the Corp in Jan 1976. Mbr of the Amer Chem Soc, Natl Acad of Engrg, & Amer Inst of Chem Engrs, having served as natl Pres of the latter in 1960. *Society Aff:* ACS, NAE, AIChE.

McAfee, Naomi J
Home: 13 Seminole Av, Catonsville, MD 21228
Position: Mgr, Engrg Operations *Employer:* Westinghouse Electric Corp. *Education:* BS/Phys/Western KY State Coll *Born:* 10/27/34 Joined Westinghouse Elec Corp in 1956 upon grad from Western KY State Coll with a BS in Phys. Mgr Engineering Operations Defense & Elec Sys Ctr, Westinghouse Elec Corp, Baltimore, MD. Prior to this was Dir Corp Strategic Resources, Pittsburgh, PA. Fellow and PPres of Soc of Women Engrs & is a mbr of the advisory coun to the Sch of Engrg & Sci, Princeton Univ, Princeton, NJ; a mbr of the advisory coun to Univ of CA, Davis & a past Exec Secy of the ASQC; a mbr of the IEEE & Pres for the Reliability Soc; PPres Fed of Organizations for Prof Women. Likes duplicate bridge, reading, hiking, and tennis. *Society Aff:* IEEE, SWE, ASQC

McAleer, William K
Home: 651 Arden Ln, Pittsburgh, PA 15243
Position: Consultant. *Employer:* self *Education:* MS/IE/Univ of Pittsburgh; BS/Management Eng/Carnegie Inst of Tech *Born:* 08/14/21 Native of Pittsburgh, PA. Proj Engr, Mine Safety Appliances Co, 1942-1946. Sales Engr, McNally Pittsburgh MFG Corp; Designers and Constructors of Coal Preparation Plants 1946-1955. From 1955-1967 with HB Mayhard Co, Internatl Mgmt Consultants as staff consultant, Mgr Latin American Div and VP. Provided technical and mgmt asst to many Latin Americans and US industrial firms. 1967 to 1983 Exec VP and Pres, Peter F Loftus Corp, consulting and design engrs. 1983 to present: Consulting Mgmt Engr (own practice). PE, Chrmn ASME Mgmt Div, 1975-76. VP Andrew Carnegie Society 1975. *Society Aff:* ASME, ESWP, AMA

McAlpine, Gordon A
Business: 29100 Northwestern Hwy, Suite 400, Southfield, MI 48034
Position: Mgr, Auto Indust *Employer:* Computervision Corp *Education:* Bachelor/ME/Lawrence Inst of Tech *Born:* 11/06/21 Gordon McAlpine is a leader in the CAD/CAM field. He was respon for the dev of Precision Numerical Controlled Machine tools at Ex-Cell-O Corp and the implementation of Computer Integrated Mfg at ITT and RCA. His education includes Lawrence Inst of Tech, RCA and ITT mgmt training at Wharton Sch, PE courses at the Soc of Mfg Engrs. He is a reg PE, Certified Mfg Engr, listed in "Who's Who–, and an author of several tech papers, as well as guest lecturer at seminars and univs. He is a past Intl Pres of Computer and Automated Sys Assoc of SME, and a mbr of the Intl Bd of Dirs of SME. During 1983 he recieved the "Intl Distinguished Contributions Award" by the Inst for the Advancement of Engrg, Los Angeles, CA. Gordon was presented the 1984 Space Shuttle Tech Award by NASA and Rocketdyne Div of Rockwell Intl for his contributions to productivity improvements within the space shuttle program. *Society Aff:* SME, CASA, ESD, SAVE

McArthur, Elmer D
Home: 1604 SW 19th Dr, Boynton Beach, FL 33435
Position: Retired *Education:* MS/Phyiscs/Union College; BS/EE/Union College. *Born:* May 1903. Electronics res GE Res Lab 1926-47. Mgr High Frequency Res, GE Res Lab 1947-58; Mgr Engrg & Res, Electron Power Tube Dept, GE 1958-65; Consultant, Electron Power Tube Dept G E 1965-67. Awarded 64 patents, author 22 papers, 1 book. Civilian Advisor to Office of Dev Res & Engrg in Electron Tubes, Dept of Defense 1953-64. Awarded Charles A Coffin Award 1946; Natl Electronics Conf Award (prize paper) 1954; Certificate of Commendation, US Navy 1947. Mbr Sigma Xi 1948; Fellow IEEE 1945; Life Mbr IEEE 1971. *Society Aff:* ΣΞ, IEEE.

McAvoy, Thomas J
Business: Dept of Chem and Nuclear Engrg, College Park, MD 20742
Position: Prof. *Employer:* Univ of MD *Education:* PhD/ChE/Princeton Univ; MA/ChE/Princeton Univ; BS/ChE/Brooklyn Polytech. *Born:* 4/25/40. Native of Queens, NY. Taught at Univ of MA 1964-80 current at Univ of MD. Have published numerous articles in area of process dynamics & control. Current interests in this field involve all aspects of distillation column control. A new, recent interest involves online optimization expert systems and smart sensors. Enjoy boating & fishing. *Society Aff:* AIChE, ACS, ISA.

McBean, Robert P
Business: P.O. Box 8405, Kansas City, MO 64114
Position: Group Supvr *Employer:* Black & Veatch, Consulting Engrs *Education:* Ph.D./CE/Stanford Univ; M.A.Sc./CE/Univ of British Columbia; B.A.Sc. /CE/Univ of British Columbia *Born:* 5/6/39 Following professional practice and graduate studies, joined Civil Engrg faculty at Univ of MO-Columbia in 1968. Tenured Assoc Prof in 1971. Outstanding engrg Teacher Award in 1971. Faculty-Alumni Gold Medal Award 1972. With Black & Veatch since 1974, supervises Structural Analysis Group, specializes in seismic and transient load stress analyses, and design of complex or unique structures. Author numerous technical papers and reports. Chrmn MSPE PE in Education 1974. Chrmn Structural and Construction Committee Kansas City Section ASCE 1979. Registered PE in CA, KS, MO, NV, UT. *Society Aff:* NSPE, ASCE, ΣΞ

McBride, David L
Business: Cushwa Center for Industrial Development, Youngstown State University, Youngstown, OH 44555
Position: Dir, Ind Dev Ctr. *Employer:* Youngstown State Univ. *Education:* ScD/Metallurgy/MA Inst of Tech; BS/Metallurgy/MA Inst of Tech. *Born:* 9/19/34. 17 yrs with Youngstown Sheet & Tube Co in various managerial positions of increasing responsibility in res and operations; last position held: Dir, Production Planning and Control, responsible for allocating raw materials and planning short and long range raw material and production facility requirements to support sales forecasts and production schedules. Currently Dir, Cushwa Ctr for Industrial Dev at Youngstown State Univ, responsible for utilizing the Univ's technical and academic resources to assist local business and industry in product dev, venture analysis, productivity improvement, and gen mgt. *Society Aff:* APICS.

McBride, Guy T, Jr
Home: 2615 Oak Dr 13, Lakewood, CO 80215
Position: President Emeritus *Employer:* Colorado School of Mines. *Education:* ScD/Chem Engg/MIT; BS/Chem Engg/Univ of TX, Austin. *Born:* Dec 1919 Austin Texas. Instructor Chem Engrg MIT 1942-44; Res Assoc MIT 1946-48. Assoc Prof Chem Engrg & Dean of Students Rice Univ 1948-58. Texasgulf Inc 1958-70 first as VP Res, then VP Agricultural Chems Div. Pres & Prof Mineral Engrg Colo School of Mines 1970-1984. Pres Emeritus Colo School Mines 1984- Mbr AIChE, ACS, AIME; Bd Mbr ECPD 1961-63. Reg PE Colo. *Society Aff:* AIChE, ACS, AIME.

McBride, J A
Home: P.O. Box 1482, Canon City, CO 81212
Position: VP. *Employer:* E R Johnson Assoc, Inc. *Education:* PhD/Chemistry/Univ of IL; MSc/Chemistry/OH State Univ; BA/Chemistry/Miami Univ. *Born:* 3/29/18. Elementary & secondary schools in Dayton, OH. Phillips Petrol Co, 1944-65; Asst Supt Engg Experiment Station, 1951-52, Mgr Dev Rocket Fuels Div, 1952-57, CPP Technical Dir, Atomic Energy Div, 1959-65. Chief Design & Dev, Astrodyne, Inc, 1958-59. Dir, Div of Mtls Licensing, USAEC 1965-70. Joined Johnson Assoc as VP in 1970; current activities include gen consulting & technical services in the Nuclear Fuel Cycle & Quality Assurance Services. Chrmn, Nuclear Engg Div, AIChE, 1966. Reg PE - Quality Engg (CA). Licensed private pilot. *Society Aff:* AIChE, ANS, ASQC, ACS.

McBride, Philip R
Home: 7339 Green Clover Cove, Germantown, TN 38138
Position: Dir of Engrg *Employer:* Memphis Area Transit Authority *Education:* MS/CE/Univ of TX; BS/CE/Univ of NM; AA/Math/Taft Jr Coll *Born:* 5/7/42 Native of Grants, NM. Design Engr & Facilities engr for LTV Aerospace Corp 1966-1970. General Mgr of Greater Lafayette Public Transportation Corp, 1971-1974. Dir of Planning/engrg for Memphis Area Transit Authority 1975 to present. Also served a Project Mgr of Memphis ARZ project from design and implementation to data collection and project analysis. Project mgr for Memphis transit Capital Improvement Programs including for Construction of new 23 acre operational and maintenance facility. Pres of Mid-South Section of ASCE, 1981. Asst Area 4 Governor, District 43, Toastmasters Intl. Enjoy jogging, skiing and coaching boys Optimist basketball. *Society Aff:* ASCE, ITE

McBride, Robert R
Business: 7220 Langtry St, Houston, TX 77040
Position: Consultant *Employer:* McBride-Ratcliff & Assoc Inc *Education:* BS/CE/Univ Southwest LA *Born:* 12/17/32 Native of LA. Served as field engr in offshore construction for Magnolia Petroleum Co prior to partnership in contracting business in Lafayette, LA. Joined Palmer & Baker Engrs and served as chief soil and materials engr in construction of major port project in South America. Consultant to MD State Roads Commission and co-author of Materials Testing & Inspection Manuals. Principal and geotechnical consultant for ETCO Engrs, Houston, TX prior to management of Houston office of Woodward Clyde Consultants. Formed McBride- Ratcliff & Assocs, Inc in 1976 and served as Chrmn of the Bd to 1986. Pres of Houston Branch ASCE 1976, Sam Houston Chapter TSPE 1975, Chairman Ctte on Standards of Practice, ASCE, 1986. *Society Aff:* ASCE, NSPE, ACEC, ASFE.

McBrien, Thomas H
Home: 5 Michael Av, Scituate, MA 02066
Position: Consultant, Safety Prog Mgmt. *Employer:* Scott Wetzel Services Inc *Education:* EdM/Health/Boston Univ; BS/Biology/Central CT State College. *Born:* Feb 1930. Awarded fellowship for doctoral studies at the Center for Safety Educ, NYU. Native of Hartford, Conn. 20 years of varied experience in administration of Indus Health & Safety Progs at both the Corporate & Divisional levels with following companies: Underwood-Olivetti; SCM Corp; Westinghouse; United Brands; ITT-Grinnell; Polaroid. Taught Industrial Accident Prevention at U of Md. While Chmn of the Electronic Industries Assn's Safety & Health Ctte, appeared before US Dept of Labor in Washington DC at hearings on proposed OSHA Legislation. *Society Aff:* ASSE.

McBryde, Vernon E
Business: Rm 309, Old Engrg Bldg, Fayetteville, AR 72701
Position: Professor *Employer:* Univ of AR. *Education:* PhD/Engg/GA Inst of Tech; MSIE/Engg/Univ of AR; BSIE/Engg/Univ of AR. *Born:* 2/3/33. Native of AR. Served with USAF 1951-55. Ford Fdn Fellow at GA Tech 1961-63. Prof of Engg at Univ of AR 1958-61, 1963-67, and 1975-present. Hd, Ind Engg, MT State Univ, 1967-70. Dean and VP at AR Tech Univ 1970-73. Pres, Productivity Intl, 1973-present, Assoc Dean/Engr, Univ of Ark 1977-80. Assumed present position 1980. Ext consulting in Quality Control, Production, and Inventory Control. Fellow in APICS, Certified Quality Engr and Former Regional Dir, ASQC. Production engr with GE, Systems Engr with Clary Corp. Reg PE in AR. *Society Aff:* ASQC, NSPE, APICS, IIE, ASEE.

McCabe, Charles L
Business: 1020 W Park Av, Kokomo, IN 46901
Position: Vice President & Genl Manager. *Employer:* Cabot Corp, High Technology Materials Division. *Education:* DSc/-/Carnegie Mellon Univ; MS/-/Carnegie Mellon Univ; BS/-/Dickinson College; ScD/-/Dickinson College. *Born:* Oct 1922. Served with US Army 1945-46. Instructor in chem at Harvard U 1948-51. Prof, Dean of Graduate Studies & VP for Res at Carnegie-Mellon U 195165. 1965 was appointed Deputy Asst Secy of Commerce for Sci & Tech in the US Dept of Commerce. Pres of Koebel Diamond Tool Co Detroit 1966-70; Pres of Teledyne Firth Sterling, Pittsburgh 1970-75; VP of Cabot Corp & Genl Mgr of the Stellite Div 1976- . Mbr of the Bd of Dickinson Coll. Active in Amer Inst of Mining, Met & Petroleum Engrs & the Amer Soc for Metals.

McCabe, Joseph
Home: 601 E 20th St Unit 14A, New York, NY 10010
Position: Retired *Education:* BA/Liberal Arts/Catholic Univ; B.C.E./Civil Engineering/RPI; MS/Physics/Fordhum U; M.C.E./Environmental Eng./New York Univ. *Born:* Dec 1913. BA from Catholic U; BCE from RPI; MS from Fordham U; MCE NYU. Reg PE in NY & NJ. Fellow ASCE Diplomate, Amer Acad of Environmental Engrs. Chi Epsilon, Tau Beta Pi, Sigma Xi. Fellow ASCE. Manhattan Coll School of Engrg, New York 1942- 66: served as Prof, Head of Civil Engrg Dept, Dir of Sanitary Engrg Grad Prog. Assoc at cons engrg firm of Hazen & Sawyer, NY City, 1966-69. With ASCE as Dir of Educ Services, 1969-78. Consultant in environmental engineering, 1979 to 1987. Retired. *Society Aff:* ASCE.

McCabe, Warren L
Home: 45 Highland Farms, Black Mountain, NC 28711
Position: Retired *Education:* PhD/ChE/Univ of MI; MS/ChE/Univ of MI; BS/ChE/Univ of MI *Born:* 8/7/99 Asst & Instructor, ChE, MIT, 1923-25. Instructor, Asst Prof, Assoc Prof, 1925-1936. Univ of MI Prof, Chem Engrg, 1936-47, Carnegie Inst Tech, Administrative Dean, 1953-64, Polytechnic Inst of Bklyn. Visiting Prof of Chem Engrg, Reynolds Prof, NC State Univ 1965-72. Member, Advisory Council, Princeton Univ 1949-65. Visiting Comm Chem Engrg Carnegie Tech, 1962-67; Visiting Comm Reactor Engrg, Brookhaven Natl Lab 1964-68; Fulbright Lecturer, Natl Taiwan Univ, Taipei, Taiwan Summer 1965. Dir of Research, The Flintkote Co 1949-53, VP 1949-53, Whippany, NJ. *Society Aff:* AIChE, NAE

McCall, James L
Business: Battelle-Columbus Labs, 505 King Ave, Columbus, OH 43201
Position: Sr. Manager, Materials Research *Employer:* Battelle Memorial Institute *Education:* BMetE/Metallurgical Engr/OH State Univ; MS/Metallurgical Engr/OH State Univ. *Born:* 6/3/35. Joined Battelle-Columbus Labs 1958; currently Sr. Manager, Materials Research. Fellow and Past Bd. of ASM, Past Mbr Bd of Dirs Intl Metallographic Soc. Reg PE in OH. Mbr Ed Advisory Bd "Metallography–, ed of

McCall, James L (Continued)
Microstructural Science series, co-ed of 11 books. *Microstructural Analysis: Tools & Techniques* 1974, *Specimen Preparation Techniques: Optical & Electron Microscopy* 1975, *Interpretive Metallography*, 1977 & *Metallography in Failure Analysis*, 1979. *Society Aff:* ASM, AIME, IMS, EBSA.

McCall, Thomas F
Business: 307 E 63rd St, Kansas City, MO 64113
Position: V Pres & Mgr Process Engineering. *Employer:* The C W Nofsinger Co. *Education:* BS/ChE/KS Univ. *Born:* 12/2/34. Grad studies ChE OH State Univ, Univ of KS. Process Dev Engrg Unio Carbide Charleston WV 1957. 1st Lt Military Train Off USAF 1957-59. R & Process Dev Engr Spencer Chem Co MerriAM Ks - now Gulf Oil Corp 1959-64; C W Nofsinger Co KS City MO: Process Engr 1964-67, Mgr 1967-73, VP & Mgr 1973--Process Engrg Dept. Co representative to the Tech Advisory Committee of Fraction Research, Inc. Mbr AIChE, Tau Beta Pi, Sigma Tau, Delta Tau Delta. *Society Aff:* AIChE.

McCall, William M
Business: 5334 S. Prince St, Littleton, CO 80166
Position: Director of Highways/Engineering *Employer:* Arapahoe County . *Education:* BS/Civil Engr/CO State Univ; AA/Sci/Northeastern Jr College. *Born:* 12/21/40. PE; CO, CA, NB, WO, Engr Asst County of Los Angeles 1964-69. Consulting Eng in private practice 1969-76; County Engr of Arapahoe County, CO. 1976-present. Past engrg instructor & visiting lecturer at Metro State College, Denver, CO. Currently enrolled at Univ of CO. Grad Sch in Master's Program in Public Admin. Present Mbr of APWA, ITE, Cons Engrs Council, Natl Assoc of County Enggs, Pres, Colo. Assoc of Rd Supervisors and Engrs 1981, VP of CARSE, 1980. Interests: hunting, camping, photography, skiing, tennis. *Society Aff:* APWA, ITE, ACEC, NACE.

McCalla, William J
Business: 5301 Stevens Creek Blvd, Santa Clara, CA 94022
Position: Section Manager *Employer:* Hewlett-Packard *Education:* BS/EE/UC Berkeley; MS/EE/UC Berkeley; PhD/EE/UC Berkeley *Born:* 11/28/43 Dr. McCalla was a Member of Technical Staff at Bell Telephone Laboratories, Inc. 1972-73. He was Section Head in charge of Computer Aided Design with Signetics, Inc. 1973-76. Since January 1979, he has been with Hewlett-Packard Co. where he is a Section Manager responsible for VLSI design tools and design system technology. Dr. McCalla has consulted for Stanford Univ and has taught graduate seminars on computer-aided circuit simulation both there and at Univ of CA Berkeley. He is a member of Eta Kappa Nu, Tau Beta Pi, the IEEE Computer-Aided Network DEsign (CANDE) Cttee (of which he was Secretary/Treasurer and Chairman) and a Fellow of the IEEE. Dr. McCalla was a co-founder of the IEEE International Conference on Computer-Aided Design (ICCAD) and served as Technical Program Chairman in 1983 and General Chairman in 1984. He has served on the IEEE Circuits and Systems Society ADCOM and on the ICCD/ICCAD Joint Coordinating Cttee. *Society Aff:* IEEE

McCalla, William T
Business: 6600 75 1/2 Ave No, Brooklyn Park, MN 55428
Position: Proj Dev Engr *Employer:* Con-Force Structures, Inc. *Education:* BSCE (1956)/CE/Univ of WA; BSC (1952)/Commerce/OH Univ *Born:* 01/19/29 Native, Hicksville, OH. PE -- MN, CA, WI, OR. Fellow - ACI. Past Pres - MSPE. P Chrmn, PE in Construction - MN & Natl Dir Elect (NSPE). Pres. MFES (Minn. Fed of Engrg Soc). Mbr & P Chrmn (4 years), PCI Pile Comm. Chrmn, PCI Pole Ctte. Mbr, ACI Piling Comm (16 years). P mbr (4 years) PCI, Tech Activities Ctte. P Pres - OR, BC & IA-MN ACI Chap. Mbr Univ of MN, Annual Concrete Conf Planning Comm. Co-Chrmn & Moderator (5 years) of PEC/State Bd of Engrg. Registration, Professional Dev. Patent holder of prestressed concrete transmission poles. Bladholm Bros - 13 years. Enjoy: kayaking, camping, opera, engrg. *Society Aff:* NSPE, IEEE, TBΠ, ACI, PCI

McCallick, Hugh E
Business: College of Tech, Houston, TX 77004
Position: Dean Emeritus *Employer:* University of Houston. *Born:* 9/2/20. Postgrad Univ of TX. Dean College of Tech Univ of Houston 1964-80, faculty mbr 1947-60, 1964- , Exec VP Capitol Radio Eng Inst WA 1960-64; Exec VP ERA Inc WA 1954-; Dir Indus Polytechnic Progs in India 1963-66; V Chmn Educ Dev Center Consortium Algeria 1976- . Cons to US, Brazil, Ecuador, Japan, Costa Rica, Peru, Guatemala, Colombia, Pakistan, India, Venezuela, Algeria, Mexico. Bd Mbr Engrs Council for Prof Dev 1978-79. Mbr ASEE, Advisory Council for Tech-Vocational Educ in TX. Mbr NAE, NAS for BOSTID. Mbr James H McGraw Award Com. Mbr Houston Engg & Sci Soc. Awards: James H McGraw Award 1970; Chester H Carlson Award 1977; Soc for Mfg Engrs Interprof Cooperation Award 1976. Chrmn of Bd. - ICATER.- International Center for Applied Technology Education & Research - Prof & Dean Emeritus - 1980-.

McCallister, Philip
Home: 16819 Cranford Ln, Grosse Pointe, MI 48230
Position: Chief Engrg Div. *Employer:* US Army Engineer Dist Detroit. *Born:* Jul 1928. BSCE Univ f Mich; MPA course work Wayne St Univ. With Army Corps of Engrs 1947. With Niagara Mohawk Power Corp 1951-58 as Design Engr & Const Supr on hydroelectric power const; with N R Gibson Cons Engr 1952-58 as Test Engr, then Principal (part-time); with Corps of Engrs Detroit Dist 1958- , formerly as Hydraulic Engr & Supr Hydraulic Engr, as Ch Flood Control Section & Ch Planning Branch; assumed present position 1972 as Sr Civilian Engr in charge of: Programming for const, planning & design. Enjoys sailing. Reg P E Mich & N Y St.

McCallum, Gordon E
Home: 3533 Twin Branches Dr, Silver Spring, MD 20906
Position: Consultant *Employer:* Engineering-Science. *Education:* ScD/Sanitary Eng/Clemson Univ; BS/Civil Engg/Univ of IA; CE/Sanitary Engg/Uni of IA. *Born:* Apr 1905. Local & St Govt Mich prior to WWII. Comm Engr. Officer US Pub Health Service 26 yrs; retired as Asst Surg Genl (Rear Adm) 1965. Admin Fed Water Pollution Control Prog 195565; Dev Fed Res, Training & Financing Progs. Principal Engrg-Sci & ES Internatl 1966- . Cons ES 1985-. Cons Natl Wildlife Federation; P Pres Fed Water Quality Assn & Conference of Fed San Engrs; active 10 prof & sci orgs; Life Mbr ASCE, AWWA; Hon Mbr WPCF. Public Works Man of the Year (APWA) 1962. Featured ENR Sept 1956 & Public Works Oct 1962. Emerson Medal (WPCF) 1970. *Society Aff:* ASCE, NSPE, WPCF, AAAS, AWWA

McCann, Gilbert D
Business: 286-80, Pasadena, CA 91125
Position: Professor of Applied Science. *Employer:* California Institute of Technology. *Education:* PhD/EE/CA Inst of Tech; MS/EE/CA Inst of Tech; BS/EE/CA Inst of Tech. *Born:* Jan 1912; native of Glendale Calif. 1938-1946 participated in Manhattan Proj at Westinghouse Elec Corp, was Dir of Res on atmospheric elec & directed Westinghouse's computer prog for WWII tech. Has been Prof at Caltech for 33 yrs & dr of its Booth Computing Center 1961-71. Recipient of Eta Kappa Nu Award for Outstanding Elec Engr 1942 & Fellow IEEE. Res interests include the application of computers to engrg & sci, information sci & information processing in living nervous sys. Have cons in genl area of computer sci for various orgs. Enjoys breeding Arabian horses, photography & woodcarving. *Society Aff:* IEEE, ACM, ΣΞ.

McCardell, William M
Business: 1251 Ave of Americas, New York, NY 10020
Position: President *Employer:* Exxon Minerals Company *Education:* MS/Chem Engr/CA Inst of Techology; BS/Chem Engr/Rice Univ. *Born:* Nov 1923. USAF 1943-46 (Capt). Mbr AIChE, AIME, Amer Meteorological Soc, Chmn Gulf Coast Section AIME 1961. Reg P E Texas. Joined Humble Oil & Refining Co Prod Res Div 1949, Ch Res Engr 1959, Mgr Corp Planning Dept 1962, Marketing V Pres 1968, Exxon Corp Marketing V Pres 1970-75, Esso Eastern Inc Exec V Pres 1975-77, Exxon

McCardell, William M (Continued)
Corp VP 1977-1980, Exxon Minerals Company President 1980-Present. *Society Aff:* SME-AIME, AIChE, SPE-AIME, AMS.

McCarl, Henry N
Business: Sch of Business/UAB, Univ Station, Birmingham, AL 35294
Position: Assoc Prof. *Employer:* Univ of AL at Birmingham. *Education:* PhD/Mineral Economics/Penn State; MS/Geology/Penn State; BS/Earth Sci/MIT. *Born:* 1/24/41. Prof McCarl currently serves as Assoc Prof of economics and geology at the Univ of AL in Birmingham. He has been active in the AIME since 1961 and served as the Chrmn of the Ind Minerals Div in 1979 and Mbr of the Bd of Dirs of the Society of Mining Engrs of AIME from 1979-1982. His specialities include economic analysis of mining and mineral production, valuation of mineral property and energy economics. He has served as Chief of the Energy Economics Div of the AL Energy MGT Bd and as Sr Lecturer in Energy, Resources and Environmental Economics at the Academy of Economic Studies in Bucharest, Romania, as part of the Fulbright-Hays Prog. He is also a certified Prof Geologist (AIPG 2150) and served as Chrmn of the Birmingham Planning Commission 1979-86. In 1975, he was nominated by the AL section of AIME as their candidate for Young Engr of the Yr in Birmingham in recognition of his work with AIME and in the Birmingham Community. *Society Aff:* AIME, AIPG.

McCarley, Mack B
Business: 670 Lupton Bldg, Chattanooga, TN 37411
Position: Chief Office Service Branch. *Employer:* Tennessee Valley Authority. *Born:* Nov 1939; native of Hamilton Ala. BSCE Univ of Ala; Post Grad Studies Cvl Engrg Univ of Tenn; Esthetics in Public Utilities Univ of Wis; Public Prog Mgmt US Cvl Service Comm. Employed by TVA as Cvl Engr 1962, assumed current respon as Ch of TVA'S Office Service Branch 1974. Pres Tenn Valley Section ASCE 1972; Chmn Dist VI Council ASCE 1970; ASCE representative to ASCE-ECE Conference 1972; Chmn ASCE Public Relations Ctte 1974. Mbr Chi Epsilon Engrg Hon Frat. Outstanding Young Engr Tenn Valley Section ASCE 1970; listed in Dictionary of Internatl Biography; selected for Amer Biographical Inst's 'Personalities of the South' 1972. *Society Aff:* CEC, ASCE, TSPE, NMA.

McCarthy, Danny W
Business: School of Engg, New Orleans, LA 70118
Position: Prof. *Employer:* Tulane Univ. *Education:* DEng/ChE/Tulane Univ; MEng/ChE/Tulane Univ; BS/ChE/Tulane Univ. *Born:* 9/6/50. Currently a full-time faculty mbr of the Chem Engg Dept, and dir of M.E. Co. Fields of Res are Computer Control, Numerical Methods, and Petroleum Production. *Society Aff:* AIChE, NSPE, LES, IEEE, STE.

McCarthy, Eugene L
Home: 3607 E 47th Pl, Tulsa, OK 74135
Position: Conslt *Employer:* Yuba Heat Transfer Corp. *Education:* MSCE/CE/Univ of OK; BSCE/CE/TX College of Arts & Ind; BA/Chemistry/Univ of TX. *Born:* 5/6/23. Chicago IL. 1947 Chemist Southern Alhalai Corp Corpus Christi TX. 1949-1956 Chief Engr Black Sivalls & Bryson OK City OK. 1956-1962 VP Engg Black Sivalls & Bryson OK City OK. 1962-1965 VP Operations Black Sivalls & Bryson OK City OK. 1965-1966 VP Engg Yuba Heat Transfer Corp Tulsa OK. 1966-1982 Pres Yuba Heat Transfer Corp Tulsa OK. 1983- Conslt Yuba Heat Transfer Corp Tulsa OK. *Society Aff:* AIChE.

McCarthy, Gerald T
Home: 94 Colt Rd, Summit, NJ 07901
Position: Retired Sr Partner - Consultant. *Employer:* Tippetts-Abbett-McCarthy-Stratton. *Education:* BS/CE/Penn State Univ. *Born:* May 1909. Army Corps 1931-38 on Miss & N E flood control. Spec S Amer. Ptnr Parsons Klapp 1938-47; Ptnr-Tippetts-Abbett-McCarthy-Straton 1948-74 in entire free wrld water resources & transportation projs. NAE 1973; Hon Mbr ASCE 1971; Chmn USCOLD 1959-64; V Pres ICOLD 1964-67; Pres ICOLD 1967-70; Dir USCID 1952-58; Pres AICE 1961; Dir IRF 1960-74; Dir ARBA 1966-70, Hon Mbr 1976; Dir Far East Amer Council Comm Ind 1954-74; Diplom Amer Acad Environ Engrs; Disting Alumnus Award 1971; Tau Beta Pi; Chi Epsilon. Life mbr Amer Waterworks Assn. Enjoys music & golf. *Society Aff:* NAE, ASCE, ACEC, AAEE, USCOLD.

McCarthy, James J
Home: Box 147 RD1, Riegelsville, PA 18077
Position: Metallurgical Engineer consultant. *Employer:* Self. *Education:* BS/Met Engr/Lehigh Univ. *Born:* July 1922; native of Bethlehem Pa. Attended MIT. Served USNR WWII. P Chmn Lehigh Valley Chap ASM; P Mbr & Chmn Natl ASM Chap Advisory Ctte & Chap & Mbrship Council. Reg P E St of Penn. *Society Aff:* ASM.

McCarthy, John J
Home: 59 Edgewood Ave, Albany, NY 12203
Position: Assoc. Prof. Emeritus. *Employer:* Retired. *Education:* PhD/Met/RPI; MMetE/Met/RPI; BMetE/Met/RPI. *Born:* 6/24/23. Native of Troy, NY. Mbr of the RPI faculty for thirty-seven yrs. Retiring in 1985. Consultant to several ind cos. Darrin Counseling Award 1975. Golf & baseball enthusiast. Trivia buff. *Society Aff:* ASM, AWS.

McCarthy, Roger L
Business: 750 Welch Rd, Palo Alto, CA 94304
Position: Principal Design Eng and VP *Employer:* Failure Analysis Asso *Education:* PhD/ME/MIT; Mech E/ME/MIT; Sci M/ME/MIT; BSE/ME/Univ of MI; BA/Philosophy/Univ of MI *Born:* 11/28/48 Native of Battle Creek, MI. Engr with Proctor & Gamble Co, Inc, Cincinnati, OH, 1973-74. Prog Mgr Foster-Miller Assocs, Waltham, MA, 1976-78. With Failure Analysis Assocs since 1978, assumed current responsibility as Principal Design Engr and VP in 1980. Respon for Design Analysis Group, supervising FAA's design investigations. *Society Aff:* ASME, ASM, ASHRAE, ASTM, NSPE, SAE

McCarthy, Rollin H
Business: 19B Strawberry Hill Rd, Ithaca, NY 14850
Position: Consultant-Manufacturing Engrg. *Education:* MME/Ind Engg/Cornell Univ; ME/Ind Engg/Cornell Univ; AB/Econ & Physics/Cornell Univ. *Born:* Lt SC WWI. Taught Cornell & Univ of Nev. Manufac Engr Western Elec Co for wide variety of communications proceses & prods including military radar electronic devices & 1st transatlantic telephone cable repeaters; Dir of Plant Design & Const of W E Co's bldgs. Chmn Wood Indus & Indus Engrg Div & Fellow ASME. Army Certificate of Appreciation for 1945 cons in Germany. Dir Bldg Planning & Const for Ford Foundation N Y C Hqrs. Initiated adoption of IE as major option in Univ curricula in Ireland. Cons for IESC Indonesia; VITA Bolivia. *Society Aff:* ΦΚΦ, ASME, ASEE

McCarty, James E
Home: 5 Spyglass Hill, Oakland, CA 94618
Position: Consulting Civil Engr *Employer:* Self *Education:* BS/Civ Engg/Univ of CA. (Berkley) *Born:* 3/7/22. Mr James McCarty's profl career has been with the City of Oakland starting as an Engg Draftsman in 1947 & advancing to City Engr in 1961 & Dir of Public Works/City Engr in 1968. In his career Mr McCarty has been involved in hydraulic design, municipal facility construction & engg mgt involving a medium sized City of 350,000 people. Notable achievements of the dept over this period of time include dev of arterial sts, storm drainage and sewer systems, coordinating of three maj railroad overcrossings; initiating a maintenance mgt system, utility coordination & implementing sidewalk repair & weed abatement progs. Mr McCarty retired from public service in 1983 and is currently a consulting civil engineer specializing in municipal engrg and emergency mgmt. Mr McCarty has been active in the Am Public Works Assn serving as pres of the assn in 1978- 79 & in the Am Soc of Civ Engrs in which he served as VP 1984-86, as SF Sec pres & as chrmn

McCarty, James E (Continued)
of the Technical Council on Lifeline Earthquake Engg. He currently chairs Cttee on Govt Affairs. He has also served as an officer in the League CA Cities and in the Univ of CA, Berkeley. Engg Alumni Soc. He is currently a member of the Bd of Trustees of the Public Works Historical Soc. *Society Aff:* APWA, ASCE, EERI, PWHS

McCarty, Perry L
Business: Dept of Civ Engg, Terman Eng. Center, Stanford University, Stanford, CA 94305-4020
Position: Silas H. Palmer Prof of Civil Engrg *Employer:* Stanford Univ. *Education:* ScD/Sanitary Engg/MIT; MS/Sanitary Engg/MIT; BS/Civ Engg/Wayne State Univ. *Born:* 10/29/31. Mbr Stanford Faculty since 1962, previously on Faculty of MIT from 1959 to 1962. Mbr Natl Acad of Engg. Former VChrmn, Environmental studies Bd; Former mbr Comm on Natural Resources; Chrmn, Comm to Review Potomac Estuary Pilot Plant study; mbr, Commission on Physical Sciences, Mathematics Resources; all of the Natl Acad of Sci, Natl Res Council. Former Chrmn: AWWA Res Committee, AWWA Water Quality Div, Gordon Res Conf on Environ Sciences. Author over 200 Technical Papers, Reports. Co-author, "Chemistry for Environmental Engineering–. Awards: Harrison P Eddy, WPCF (1962 & 1977); Thomas Camp (1975), WPCF; Walter Huber, ASCE (1964); Simon W Freese, ASCE (1979). Hon Mbr, AWWA (1981), Fellow, AAAS (1980). *Society Aff:* ASCE, WPCF, AWWA, IAWPR, AAAS, AEEP, ΤΒΠ, ΣΞ, ΟΔΚ.

McCashen, Leo W
Business: 3033 N 3rd, Pheonix, AZ 85003
Position: General Manager - Network. *Employer:* Mountain Bell. *Born:* Native of Denver. Oct 1957 Univ of Colo BSEE. Reg Prof Engr Colo. US Navy 1942-45. Engr Dir Mountain Bell 1965-67; Idaho Plant Mgr 1967-70; Ariz Ch Engr 1970-73; assumed current position July 1973. Enjoys golf & fishing.

McCauley, Roy B
Home: 845 Linworth Rd E, Worthington, OH 43085
Position: Dir Center for Welding Res *Employer:* Ohio State University. *Education:* AB/Physics Chemistry/Cornell College; MS/Physical Metallurgical/IL Inst of Technology. *Born:* 2/9/19. Employed Columbia Steel 1958-59; taught IIT 1940-50; Grad Teaching Asst Chem Engrg to Asst Prof & Chmn Met Engrg Ohio State 1943-50. Active: Amer Welding Soc, V Pres & Dir 1963-66, Pres 1966-67. Pres 1960-63, Chrmn Dept Welding Engg OH State Univ 1953-79. Internatl Inst of Tech Ctte Pres Educ 1964- . Expert nondestructive testing 1962- ; expert weld defects 1968- ; Sub-comm Chmn Destructive Testing 1974- . Reg P E, Ill & Ohio. Cert Mfg Engr, Soc Mfg Engrg. Chmn Symposium Welding Educ US 1961, Yugoslovia 1965, London 1967, Sweden 1971, Australia 1976; US Acad of Sci Sr Sce to Romania 1969 & 1974. Natl Meritorious Certificate Amer Welding Soc, Adams Distinguished Teaching Award 1958, Samuel Wyle Miller Gold Medal Amer Welding Soc 1978. Mehl Lectr Amer Soc Nondestructive Testing, Silver Cert ASM, R D Thomas Internatl Achievement Award AWS, listed Who's Who in Amer; Internatl Cons Mfg Met & Quality Assurance. *Society Aff:* AWS, IIW, ASNT, ASM Intl.

McCawley, Frank X
Business: 4900 Lasalle Rd, 2401 E St. NW, Washington, DC 20241
Position: Physical Scientist *Employer:* US Bureau of Mines. *Education:* BS/Chem/Univ of Scranton. *Born:* 5/18/24. Native of Scranton, PA. Served in Air Corp 1942-45. Chem for Chicago Dev Corp. 1949-59. Specialized in electro refining of titanium scrap. With Bureau of Mines since 1959. Project Leader responsible for R&D on electrodeposition of refractory metals and platinum metal coating, and corrosion resistance of commercial alloys in high salinity geothermal brines. Sr. Coordinator for field studies performing mineral surveys in Federal public lands. Colonel in Air Force Reserve. *Society Aff:* AIME, ASM.

McClain, J H
Home: 33728 Viewcrest Dr. N.E, Albany, OR 97321
Position: President *Employer:* J H McClain, PE, Inc *Education:* BS/ChE/OR State Univ. *Born:* Oct 1915. Reg P E. Mbr AIChE, AIME, ASM, RESA, Charter Pres Chem Engrs Oregon, P Pres Oregon Sect AIME; Supervisory Chem Engr with USBM 14 yrs; Mgr Mfg, Dir Extractive Met Div, Teledyne Wah Chang Albany 25 years. Who's Who Nucleonics. Contribs to Literature: Reactor Handbook, Vol 1, Chap IV, 'Zr/Hf Separations'; Nuclear Absorber Materials for Reactor Control, Chap VI, 'The Preparation of Hafnium Oxide'; The Met of Hafnium, Chap III, 'Extraction from Ores'; Peaceful Uses of Atomic Energy, Vol VIII, 'Zirconium Metal Prod'; Progress in Nuclear Energy - Series V, 'Met of Fuels'; The Encyclopedia of the Chem Elements, Zirconium Chapter, 1968. Retired 1980 now Pres J. H. McClain, PE Inc Conslt. *Society Aff:* AIChE, AIME, ASM, ΣΞ.

McClanahan, Buford A
Business: P O Box 460, Albany, OR 97321
Position: Research Metallurgist. *Employer:* Teledyne Wah Chang Albany. *Education:* BS/Met Engr/CA State Poly Univ. *Born:* 2/5/40. US Army Artillary 2nd Lt 1962-65. BS CA State Poly Univ, San Luis Obispo, CA 1969. With Teledyne Wah Chang Albany since 1969 in R&D with primary respon in process dev engrg for reactive metals, including melting, forging, extrusion, rolling, forming & surface finishing. Mbr ASM & AWS. Extrusion & drawing activity - ASM. Co-authored papers include, "Production of Extruded Tube Hollows for Titanium-3A1-2.5V Hydraulic Tubint–, & "Eccentricity in Warm Extruded & Cold Reduced Zircaloy Tube Hollows–. Other interest incl recreation & remodeling. In Nov 1979 changed to extrusion process enggr with primary responsibility to design and order tooling for extrusion presses and cold pilger mills; and to write processing for reactive and refractory metals. *Society Aff:* ASM, AWS, ATA.

McClarran, William H
Home: 15 Evergreen St, Barrington, RI 02806
Position: Retired *Born:* June 16, 1912 Wooster, Ohio. Graduated Case Western Reserve U with BS 1935 & MS 1939. Engr with Muskingum Conservancy District, Ohio Flood Control Proj 1935-37. Field Engr - Factory Mutual Engrg Div Cleveland, Ohio 1939-43. Served in Seabees in Pacific Theater 1943-46 returning to Factory Mutuals in Boston. Joined Allendale Insurance in 1948 as engr, then Asst VP, Asst VP & Mgr Engrg Service, VP/Asst Dir Engrg, VP/Regional Operations, Engrg at Large. Reg PE Ohio, Mass & RI. Pres, Soc of Fire Protection Engrs 1975 & 1976.

McClellan, Dallas L, Jr
Business: 122 W Bay St, Savannah, GA 31401
Position: VP - Chief Electrical Engineer. *Employer:* Rosser White Hobbs Davidson, et al. *Education:* BEE/Elec Engg/GA Inst of Tech. *Born:* March 1927. Native of Savannah, Ga. USNR 1944-46. BEE Ga Inst of Tech 1951. Elec Engr - Thomas A Hutton AE 1953-67; White Hobbs & McClellan 1967-74 - Chief Elec Engr; Rosser White Hobbs Davidson McClellan Kelly Inc 1974- . P Pres Coastal Empire Sect IES & Savannah Chap GSPE. P Dir CEC Ga. Winner IES SE Region Applied Lighting Competition 1964. *Society Aff:* IES, ECEG.

McClellan, Thomas J
Business: Civil Engineering Dept, Corvallis, OR 97331
Position: Professor of Civil Engrg. *Employer:* Oregon State University. *Education:* M Engg/Civil Engg/Yale Univ; BS/Civil Engg/OR State Univ. *Born:* Aug 1920. M Engrg Yale Univ; BSCE Ore State Univ. Asst City Engr Corvallis, Ore. Ore State Univ CE faculty since 1948 (Instructor to Professor). Cons to architectural firms, cons engrs, Highway Dept, USDOT, legal firms, State Attorney Genl Office. Qualified Instructor on Radiation Shielding USOCD. Seminar lecturer, AISC. Mbr Ore Bd of Engrg Examiners 1965- ; Pres OBEE 1971-73; Pres Amer Soc CE Ore Sect 1962; Pres Ore Chap Amer Concrete Inst 1975- ; VP Western Zone, Natl Council Engrg Examiners 1976; ASCE EDEX Ctte 1975- ; Amer Arb Assoc Panel 1974- ; tech publs ACI & ASCE journals. Hobbies incl music, golf. *Society Aff:* ASCE, NSPE, ACI, NCEE.

McClelland, Bramlette
Business: 6100 Hillcroft, Houston, TX 77081
Position: President. *Employer:* McClelland Engineers, Inc. *Education:* BS/Civ Engrg/Univ of AR; MS/Civ Engrg/Purdue Univ *Born:* 12/16/20 Design Engr, City of Houston 1943-46; Partner, Greer & McClelland 1946-55; Pres McClelland Engrs Inc 1955- . Past Pres currently Dir Terra Ins Co; Past Pres, Assn of Soil & Fdn Engrs. Directs an active internatl practice with emphasis on indus & offshore fdn design. Author of over 20 tech papers on engrg geology & fdn engrg. Presented 9th Karl Terzaghi Lecture, 1972 Annual Meeting, ASCE; Distinguished Alumnus, Purdue Univ 1965 & Univ of AR 1972; ASCE State-of- the-Art of Civ Engrg Awd 1971; James Laurie Prize 1954. Mbr: Natl Acad of Engrg, Natl Res Coun, Marine Bd. Awarded Honorary Doc in Engrg, Purdue Univ, 1984. Honorary Member, ASCE. Editor, "Planning and Design of Fixed Offshore Platforms." *Society Aff:* ASCE, ASFE, ACEC, NSPE

McClelland, James E
Business: 1311 W 2nd St, Little Rock, AR 72201
Position: VP. *Employer:* McClelland Consulting Engrs, Inc. *Education:* BS/Civil Engg/Univ of AR. *Born:* 7/30/43. Reg PE in AR, LA, and TX in the areas of civil and sanitary engg, Educated at the Univ of AR at Fayetteville. Following grad from the Univ with a BS in Civil Engg in 1967 (including honors such as the Dean's List and ASCE Outstanding Sr Award), Jim entered the US Army Reserve and received further training at Ft Belvoir, VA, the Army Engr School. For the next 9 1/2 yrs Jim was employed by a large consulting firm in AR and gained design and field experience in civil and environmental projs of many types. Proj responsibilities varied from structural, hydraulic, process and instrumentation design of water and wastewater treatment plants, to water distribution and wastewater collection systems. Also included in his proj lists are airports, city parks, subdiv, drainage, roadway designs, and water supply lakes. Not only has Jim had proj respon for these types of projs, he is alert to methods of financing theses types of public facilities, through the multitude of state & fed grant and loan programs. Currently VP of McClelland consulting Engg, Mr McClelland is in charge of the Little Rock Branch Office of the person chosen twice as AR, most outstanding Young Professional Engr. *Society Aff:* ASCE, NSPE, WPCF.

McCloud, Robert J
Business: 1500 Meadow Lake Pkwy, Kansas City, MO 64114
Position: Executive Partner *Employer:* Black & Veatch Engrs - Architects *Education:* MS/Environ Health Engrg/Univ of KS; BS/CE/Univ of MO *Born:* 11/28/22 Native of the Midwest. World War II veteran. Field engr and surveyor with Phillips Petroleum Co from 1954 to 1957. Associated with Black & Veatch since 1957 as a field engr, designer, project engr, and project mgr on sanitary projects and water works. Partner since 1977. Executive Partner since 1983. Registered PE in 16 states and the District of Columbia and Licensed Land Surveyor. *Society Aff:* ASCE, AAEE, SAME, NSPE, WPCF, AWWA, KES, KCE

McClung, Robert W
Business: ORNL, P.O. Box X, Oak Ridge, TN 37831-6151
Position: Group Leader. *Employer:* Martin-Marietta Corp *Education:* BS/Chem Engrg/Univ of TN. *Born:* Oct 1928. Native of Memphis, Tenn. Radiographic Supr at Y-12 Plant, Oak Ridge 1950-52. Nondestructive Testing dev engr in US Air Force at Wright Patterson AFB 1953-55. With Oak Ridge Natl Lab since 1955. Grp Ldr of NDT Dev Grp since 1960. Pres of ASNT 1969-70; Coolidge Award 1962; Fellow of ASNT 1973; Lester Honor Lecture ASNT 1974; Chmn of ASTM Ctte E-7 on NDT 1972-76. ASNT Achiv Award 1977; Reg Prof Engr 1979, ASTM Award of Merit 1980, Fellow ASTM 1980, Fellow ASM 1981; Outstanding Engineering Alumnus, Univ of Tenn 1980. Honorary Member ASNT 1986. Enjoy tennis, reading, rockhounding. *Society Aff:* ASNT, ASTM, ASM.

McClure, Alan C
Business: 2600 S Gessner, Suite 504, Houston, TX 77063
Position: Pres. *Employer:* Alan C McClure Assoc, Inc. *Education:* MS/Naval architecture & Marine Engg/MIT; BS/Naval architecture & Marine Engg/Univ of MI. *Born:* 8/8/23. Thirty-seven yrs' experience in naval architecture and marine engg. The first twelve yrs were spent in dev of nuclear power plants for submarines and surface ships and res and dev related to submarines. This was followed by five yrs with Proj Mohole drilling platform. The next four yrs were spent in production engg with Continental Oil Co. The ensuing yrs have been spent in the consulting field, primarily in work connected with the offshore gas and oil industry. *Society Aff:* SNAME, NECIES, MTS, RINA.

McClure, Andrew F, Jr
Home: 1333 Greystone Dr, Pittsburgh, PA 15241
Position: Executive Environmental Engineer *Employer:* NUS Corp *Education:* BS/ChE/Carnegie-Mellon Univ. *Born:* 12/28/32. Native of Economy Borough, PA. Field engr with Calgon Cor, cons in the treatment of boiler & cooling water followed by proj engrg & mgmt of indus waste water projs. Joined Betz Environmental Engrs in 1972 as proj mgr. Transferred to Three Rivers Group, Betz Converse Murdoch in 1975 as a VP in charge of indus projs. Joined Dravo Corp in 1979 as Manager-Process Engrg, Lectro-Quip Dept. Joined NUS Corp in 1982 as Manager, Water/Wastewater Mgmt. Mbr AIChE, AISE, WPCF, Engrs Soc of western PA, TAPPI. *Society Aff:* AIChE, AISE, WPCF, ESWP, TAPPI.

McClure, Eldon R
Business: Lawrence Livermore National Laboratory, PO Box 808, Livermore, CA 94550
Position: Program Leader, Precision Engg Prog *Employer:* Lawrence Livermore National Laboratory *Education:* BS/Mech Engg/WA State Univ; MS/Mech Engg/OH State Univ; DEng/Mech Eng/Univ of CA, Berkeley. *Born:* 1933. Formerly Asst Prof of Mech Engrg at Oregon State U; before that, Res Engr at Battelle Memorial Inst; before that, flight test engr at Boeing Airplane Co. Initially trained in thermodynamics & heat transfer with prof exper in aircraft performance testing; graduate studies centered first on aircraft propulsion, made transition to machine design & eventually earned Doctorate in Automatic Control Systems. Specialized for many years in engrg metrology with appls in machine tool design. In recent years, has been involved in engrg mgmt for projs involving nuclear weapons dev, laser fusion, magnetic fusion, laser isotope separation, automotive res, geothermal power, high efficiency central power systems. Mbr of sevl Amer Natl Stds Inst committees and subcommittees. V Chrmn of ANSI B46 Classification & Designation of Surface Qualities committee. Natl Sci Foundation Sci Faculty Fellow. On Bd of Dir ASPE. *Society Aff:* ASME, SME, AAAS, ASPE.

McCluskey, Edward J
Business: Center For Reliable Computing-ERL460, Stanford, CA 94305-4055
Position: Professor, Elec Engrg & Computer Sci *Employer:* Stanford Univ, ERL 460 *Education:* ScD/EE/MIT 1956; BS/EE/MIT 1953; MS/EE/MIT 1953; AB/Phys, Math/Bowdoin (Summa Cum Laude) 1953 *Born:* 10/16/29 Bell Tele Labs, Whippany NJ 1955-59; Assoc Prof of Elec Engrg, Princeton Univ 1959-63, Prof of Elec Engrg, Princeton Univ 1963-66; Dir Computer Ctr, Princeton Univ 1961-66; Prof Elec Engrg & Computer Sci, Stanford Univ 1967- ; Dir, Ctr for Reliable Computing, 1976- ; Dir Digital Sys Lab, Stanford Univ 1969-78. AB in Math & Physics, Bowdoin Coll, Brunswick ME 1953; BS & MS 1953 & ScD 1956 at MIT. Fellow IEEE (pres, Computer Soc 1970-71), AAAS, AFIPS (Dir Exec Ctte), ACM. Patentee: Multivalued Integrated Injection Logic Circuitry & Method. Honors: Fellowship, Tokyo & Kyoto U-Japan, 1978; Phi Beta Kappa, Sigma Xi, Eta Kappa Nu, Tau Beta Pi. Editor & author. *Society Aff:* AAAS, ACM, IEEECS

McCollam, William, Jr
Business: Edison Electric Institute, 1111 19th St, NW, Washington, DC 20036
Position: President. *Employer:* Edison Electric Inst. *Education:* MSCE/Civ Engrg/MIT; BS/Engrg/US Mil Acad; BS/Arts & Sci/LA State Univ. *Born:* 03/15/25. BS La State U 1943; BS Engrg US Mil Acad 1946; MSCE MIT 1954. Reg PE NY. Commd 2nd Lt CE US Army 1946, advanced through grades to Lt Col 1958. Resigned 1961. Exec Asst to pres AR Pwr & Light Co 1961-64; VP 1964-67; Sr VP & Dir 1967- 70; Exec VP & Dir, New Orleans Public Service Inc 1970-71; Pres, Ch Exec Officer & Dir 1971-78. Pres, Edison Electric Institute, Wash, DC 1978- . Dir Middle South Utilities Inc, Sys Fuels Inc, Middle South Energy Inc, Middle South Services Inc, LA Pwr & Light Co 1971-78; Decorated Bronze Star. Chmn of Natl Elec Reliability Council 1975-78. Recipient of 1975 LA Engrg Soc. A B Paterson Engrg in Mgmt Award.

McCollom, Bruce F
Home: 112 Cherokee, Bismarck, ND 58501
Position: Vice President *Employer:* Bartlett & West Engineers, Inc. *Education:* EnD/Civil Engg/Univ of KS; MS/Civil Engg/Univ of KS; BS/Civil Engg/Univ of KS. *Born:* Sept 1944. PE & LS in several states. Mbr ASCE, NSPE, AWWA . ASCE Edmund Friedman Young Engrs Award for Prof Achievement 1974. Formerly with McDonnell- Douglas & State Highway Comm of Kansas. Exper in R&D, design & construction related to streets, highways, sewers, water systems, bridges & structures. Also land survey & subdivision plotting. With Bartlett & West since 1975 in charge of SD and ND operations. Since 1979 Project Manager for the $105 Million WEB Rural Water Development Project in SD. Since 1981 Project Manager for the $113 Million Southwest Pipeline Project in ND. *Society Aff:* ASCE, NSPE, AWWA.

McCollom, Kenneth A
Home: 1107 W Knapp St, Stillwater, OK 74075
Position: Dean Emeritus - Coll of Engr, Arch & Tech *Employer:* Oklahoma State University. *Education:* BS/Elec Engrg/OK State Univ; MS/Elec Engrg/Univ of IL; PhD/Elec Engrg/IA State Univ. *Born:* 6/17/22. Engr with Phillips Petroleum Co designing instrumentation and control systems. Electronics Grp Ldr, Sect Chief & Branch Mgr. of Atomic Energy Div, Idaho Test Station. Prof of Elec Engrg, Asst Dean, Assoc Dean & Dean of Engrg, Architecture & Technology at Oklahoma State U 1964-86. ASEE Chester F Carlson Award for innovation in Engrg Educ 1973. Cons to USNRC as an Administrative Judge and mbr of Atomic Safety & Licensing Bd Panel since 1972. Natl Dir NSPE; Pres OSPE. Member, Oklahoma Board of Registration for Engineers and Land Surveyors, 1986-. Outside Director on Board of Corken International, Inc., 1987- . *Society Aff:* IEEE, ANS, OSPE, NSPE, SME, NCEE

McColly, Howard F
Home: 225 Kensington Rd, East Lansing, MI 48823
Position: Professor Emeritus. *Employer:* Michigan State University. *Education:* MS/Agri Engr/IA State Univ; BS/Agi Engr/IA State Univ. *Born:* April 1902 Ames, Iowa. John Deere, Moline Ill 1926-28. Head Agri Engrg Dept, ND State Univ 1929-39. Chief Engr & State Engr ND State Water Conservation Comm 1939-41. Chief Engr Water Facilities Prog USDA 1941-46 Denver, 80 engrs in 17 Western states. Res Agricultural Engrg Ctte to China 1946-49. Ctte of 4 mbrs. Established educ & res. Prof Agricultural Engrg Mich State Univ 1949-70. Power & Machinery Area. US Patent no. 3,316,694 for rolling-compressing hay wafering machine. Published 2 text books. ASAE Life Fellow. Gold Medal (McCormick) 1968. Professional Achievement Citation Coll of Engrg, IA St Univ 1970. *Society Aff:* ASAE.

McConnaughey, William E
Home: 9621 Appaloosa Dr, Sun City, AZ 85373
Position: Cons/Hazardous Matls Safety (Self-employed) *Education:* BS/Chem Engg/Univ of NB. *Born:* 3/28/21. Grad work in mech engg at Univ of MD. Native of Lincoln, NB. 18 yrs in various Navy R&D activities with final position as Hd of Chem Div at USN Engg Experiment Station (30 prof chem engrs & chemists). Tech Specialty was atmosphere control in submarines & other closed environments. 15 yrs with US Coast Guard as Sr Technical Advisor for hazardous matls transportation. Respon for tech support of Coast Guard's regulatory missions for hazardous matls safety. Consultant to NAS Committee on maritime hazardous matls & panel on Risk Analysis. Registered Professional Eng (DC & CA). Capt in Naval Reserve (res & engg duty). Enjoys hiking & photography. *Society Aff:* AIChE, ACS, ACGIH, TBП.

McConnell, Lorne D
Home: 34 Meadowbrook Ln, Chalfont, PA 18914
Position: Engrg Consultant (Self-employed) *Born:* 1/8/26. BSEE Univ Saskatchewan 1948; Univ New Brunswick 1950-54; UCLA Sch of Bus Admin 1962-63. From 1948-61 was employed by Canadian Westinghouse. Since joining ITE in 1961, has served as Supervisor - Design Oil & Air Circuit Breakers, Mgr Engrg, Dir - Res Operations, Dir Tech Services - Res Div. Gould-Brown Boveri. Through 1979, Became independent Engr Consultant in 1980. Fellow IEEE. *Society Aff:* IEEE.

McConnell, William A
Home: 1759 Arlington Blvd, Ann Arbor, MI 48104
Position: Chief Test Engr (Retired) *Employer:* Ford Motor Co. *Education:* Dr Engrg/Engrg/Univ of NE; ME/Mech Engrg/Univ of NE; BSME/Math Physics/Univ of NE; AB/Engl, Math, Physics/Williams College *Born:* 1/3/18. McConnell is Chief Engr for Proving Ground, Lab Testing, and Engg Services in Ford's North Am operations. Previously with GM Res, Proving Grounds, and Harvard Univ's Psycho-Acoustic Labs, he held several managerial posts at Ford, including: Labs mgr; Asst Dir Product Res; Dir, Systems Res; and Dir, Emissions Certifcation, before assuming his present position. Author of numerous articles relating to automotive testing, hgwy tech, and transportation, he has held several offices in: SESA; SAE (Fellow, current-ly dir); ESD (Fellow, past dir); TRB (including exec committee); and is a mbr of Phi Beta Kappa, Sigma Xi, and Sigma Tau. Chairman, Ann Arbor Transportation Authority.. *Society Aff:* SAE, ESD

McCool, Alexander A, Jr
Home: 8727 Edgehill Rd, Huntsville, AL 35802
Position: Dir, Safety, Reliability & Quality Assurance Office *Employer:* NASA/MSFC, EPO1. *Education:* BS/ME/Univ of Southwestern; MS/ME/LA State Univ. *Born:* 12/10/23. Served in USN Reservew 1942-46. Began engg career 1951 at Waterways Experiment Station, Vicksburgh, MS. As Res Engr, developed design criteria from model & prototype tests of dams. Joined the von Braun missile team in 1954, Huntsville, AL. In 1960, team transferred to NASA by act of Congress and was involved heavily in this nation's space effort. Within the timeframe 1960-70, held key positions in propulsion with responsibility for all Saturn/Apollo rocket engines, mech subsystems, & systems design. Formerly directed large engg organization incuding structural & mech design, stress and thermal analysis, & res & dev requirements for complex aerospace systems and subsystems. Presently directing safety, reliability & quality assurance for Marshall Space Flight Center, NASA. *Society Aff:* AIAA.

McCool, Donald K
Business: Agr Engr Dept, Smith Engr Bldg, Pullman, WA 99164-6120
Position: Agr Engr *Employer:* USDA-ARS *Education:* PhD/Agric Eng/OK St Univ; MS/Agric Eng/Univ of MO; BS/Agric Eng/Univ of MO; BS/Agric/Univ of MO *Born:* 05/22/37 Born in St Joseph, MO; son of Wm Hobart and Frances Lela (Klepper) McCool. Married Donna Jean Lemon, 1972; children: James, Robert, and John. Reg profl engr in the state of MO. Employed as a research engr by USDA, ARS, Soil and Water Conservation Res Div, South Plains Br, Water Conservation Structuress Lab, Stillwater, OK, 1961-71; Asst Prof, Agri Engrg Dept, OK State Univ, Stillwater, 1966-71; Agri Engr, USDA, ARS, Western Region, Pacific Northwest Area, Palouse Conservation Field Station, Pullman, WA, 1971-present; Assoc Agri Engr, Agri Engrg Dept, Washington State Univ, Pullman, 1971-present. Current res interests are in hydrologic conditions leading to runoff and erosion-sedimentation events. Recipient of annual chapter award, Inland Empire Chapter, Soil Conserva-

McCool, Donald K (Continued)
tion Soc of America, 1979, Engr of Yr Award, Inland Empire Section, ASAE, 1980. *Society Aff:* ASAE, ASCE, SCSA.

McCormac, Jack C
Business: Civil Engr. Dept, Clemson, SC 29634-0911
Position: Alumni Prof of Civil Engr. *Employer:* Clemson Univ. *Education:* SM/Structural Engr/MIT; BS/Civil Engr/Citadel. *Born:* 7/6/27. Native of Columbia, SC. Served with Army Air Corps 1946-47. Engr for duPont Co 1951-1955. Asst Prof to Alumni Prof of Civil Engr. at Clemson 1955-present. Wrote textbooks published by Harper & Row. *Structural Analysis* (4th ed), *Structural Steel Design* (3rd ed) and *Design of Reinforced Concrete* (2nd ed). Wrote textbooks published by Prentice Hall (now Simon & Schuster) *Surveying (2nd ed)* and *Surveying Fundamentals.* Numerous translations. Two new books being published by Harper & Row. *Society Aff:* ASCE, XE, TBΠ, ΦΚΦ

McCormick, Barnes W
Business: 233 Hammond Bldg, University Park, PA 16802
Position: Boeing Prof of Aerospace Engrg. *Employer:* Pennsylvania State University. *Education:* BS/Aeronautical/Penn State Univ; MS/Aeronautical/Penn State Univ; PhD/Aeronautical/Penn State Univ. *Born:* 7/15/26. Assoc Prof of Engrg Res, PSU, 1954-55; Chief of Aerodynamics at Vertol Co, 1955-58; Head Dept of Aeronautical Engrg, Wichita State U 1958-59; with PSU since 1959, Head, Dept of Aerospace Engrg, 1969-1985, currently Boeing Professor of Aerospace Engineering. Interest areas include low-speed aerodynamics, vertical flight, propeller design, hydrodynamics, design & performance of aircraft. Past Tech Dir & Editor of the Journal, American Helicopter Soc; Assoc Fellow AIAA; Mbr ASEE; Recipient 1976 ASEE Aerospace Div - AIAA Educ Achievement Award; Former mbr, US Army Aviation Systems Command Sci Advisory Grp; Cons, past & present to many industrial firms; author of texts 'Aerodynamics of V/STOL Flight and 'Aerodynamics, Aeronautics and Flight Mechanics'; certified ground instructor & instrument rated pilot. AIAA VP for Ed, 1984-88; Member, Congressional Aeronautical Advisory Committee 1985-87. *Society Aff:* AIAA, AHS, ASEE, SAE

McCormick, Frank J
Home: 124 Pine Dr, Manhattan, KS 66502
Position: Prof Emeritus of Civil Engrg *Employer:* Kansas State University. *Education:* MS/Struct Eng/IA State Coll; BS/Arch Eng/IA State Coll *Born:* Dec 1906 Danville, Iowa. BS & MS Iowa State Univ, Ames. Additional study in Appl Math. Bridge Design, Iowa State Highway Comm 1935-39. Taught in Dept of Appl Mech at KSU 1939-75 when dept was dissolved. Life Mbr ASCE; Fellow SEM. *Society Aff:* ASCE, SEM

McCormick, John E
Business: 323 High St, Newark, NJ 07102
Position: Assoc Chrmn. *Employer:* NJ Inst of Tech. *Education:* PhD/Chem Engg/Univ of Cincinnati; BSc/Chem Engg/IA State College. *Born:* 8/18/23. in WI. Served 3 yrs US Navy 1943-46. Engr with Linde Air Products 1948- 1953. Univ of Cincinnati 1953-56. Esso Res & Engg 1957-1962. Newark College of Engg/NJ Inst of Tech 1962-date, Prof, Assoc Chrmn. Consultant to chem process industry. Received Robert W Van Houten Award for Excellence in Teaching 1972. Contributor to *Handbook of Separation Techniques.* Registered PE in New Jersey *Society Aff:* AIChE, TBΠ.

McCormick, Norman J
Business: Dept of Nuclear Engrg, BF-10, Univ of WA, Seattle, WA 98195
Position: Prof *Employer:* Univ of WA *Education:* PhD/Nuclear Engrg/Univ of MI; MS/Nuclear Engrg/Univ of IL; BS/Mech Engrg/Univ of IL *Born:* 12/09/38 Prof of Nuclear Engrg, Univ of WA, 1975-, Assoc Prof, 1970-75, Asst Prof 1966-1970. North Amer Editor, Progress in Nuclear Energy, 1980-85. Scientist, Sci Applications, Inc, Palo Alto, CA, 1974-75. Recipient of Natl Acad of Sci Exchange Fellowship, 1971, 1973, and Natl Sci Fdn Postdoctoral Fellowship, 1965. Author of *Reliability and Risk Analysis* (Academic Press, 1981). Conslt, patentee in field. Res specialties: neutron and photon transport for both direct and inverse problems, reliability and risk analysis. *Society Aff:* ANS, OSA, SRA

McCormick, Robert H
Business: 112A Fenske Lab, University Park, PA 16802
Position: Prof Emeritus of ChE *Employer:* Penna State Univ (PSU). *Education:* MS/Chem Engg/Penna State Univ; BS/Chem Engg/Penna State Univ. *Born:* 4/28/14. Native of State College, PA. Joined PSU in 1935 as a res asst. Promoted to Prof of Chem Engg in 1964. Dir or co-dir of over 60 advanced degree theses. Taught courses in Chem Engg. Early res activities were industrially oriented towards upgrading motor and aviation fuels, producing synthetic rubber and new petrochemicals. Res involved dev of many chem and phy separational processes. design and construction of equipment, and new methods of analyses. Later res included extensive testing and development of may types of fuels for diesel and auto ignition engines for quality and exhaust emissions. Co-author of many tech articles, reports, and patents. Contribution to four books. Fellow in AIChE and AIC. *Society Aff:* AIChE, ACS, ASEE, AIC, AAUP.

McCormick, William J
Business: 4300 Sigma Rd, Dallas, TX 75234
Position: Sr VP. *Employer:* Purdy-McGuire, Inc. *Education:* MS/Mech Engg/OK A&M; BS/Mech Engg/TX A&M. *Born:* 2/24/28. Assocs with consulting engg bus continuously since 1950. Prof socieities: Natl Soc of PE's; TX Soc of PE's; Amer Soc of Heating, Refrig & Air Conditioning Engrs; Soc of Fire Protection Engrs; Soc of Amer Military engrs. Military: Colonel, Corps of Engrs, USAR (Ret). *Society Aff:* NSPE, TSPE, ASHRAE, SFPE, SAME.

McCoy, Byron O
Home: 270 Grove St, Rutland, VT 05701
Position: Consulting Engr (Self-employed) *Education:* CE//Thayer School of Civil Engrg; BA/Engrg/Dartmouth Coll *Born:* 3/27/12 1937-42 William P. Creager, Consulting Engr, Buffalo, NY. Hydraulic Engr - Design studies-large dams; research, editing--Engrg for Dams--, Hydroelectric Handbook--. 1942-57 Chas. T. Main, Inc, Consulting Engrs, Boston, MA. Hydraulic Engr and Assoc in firm - Design, construction supervision, investigations, estimates-hydroelectric developments (150-500,000 Kilowatts). 1957- VT Electric Power Co, Inc, Rutland, VT. VP, Pres (Retired 1978), Dir (Continuing) Planning, supervision of construction, operation, maintenance-high-voltage (115- 345Kilovolts) transmission grid serving VT electric utilities - contracting wholesale power purchases and sales, planning generation and transmission needs. 1978- Self-employed, Consulting engrg-Hydroelectric redevelopments, power supply. 1977- VT Bd of Registration for PE. Chrmn 1981-. *Society Aff:* ASCE, NCEE, NSNA, VSE, VEA

McCoy, Donald S
Business: 7900 Rockville Rd, Indianapolis, IN 46224
Position: Staff Vice President. *Employer:* RCA Corporation. *Born:* Prior to present position as Staff Vice President, 'SelectaVision' Engrg & Mfg, Dr McCoy was Dir of Consumer Electronics Res at the David Sarnoff Res Center, Princeton, NJ, where he was respon for directing RCA's res efforts on the HoloTape video playback system & subsequently res on 'SelectaVision' Disc. After obtaining his PhD in Electrical Engrg at Yale U, he joined RCA at the David Sarnoff Res Center in 1957 where he engaged in res on a variety of subjects including magnetic video tape recording, stereophonic disc recording, psychoacoustical testing, stereophonic broadcast systems, seismic detection systems & colorimetry of Color TV systems.

McCoy, George T
Business: 10 W Orange Ave, South San Francisco, CA 94080
Position: Pres *Employer:* Guy F. Atkinson Co *Education:* AB/Engrg/Stanford Univ *Born:* 2/3/20 Native of Olympia, WA. World War II - US Navy (1941-1946) Ensign

McCoy, George T (Continued)
to Lieut Commander. 1946-1952, CA Division of Hgwys. 1952-present, Guy F. Atkinson Co. Chief Executive Officer of one of the world's foremost construction companies. (1980 volume of 850 millon dollars). *Society Aff:* PMI

McCoy, Herbert E
Home: 132 Balboa Cr, Oak Ridge, TN 37830
Position: Met. *Employer:* Union Carbide. *Education:* PhD/Met Engg/Univ of TN; MS/Met Engg/Univ of TN; BS/CE/Univ of TN. *Born:* 6/4/35. Native of Oak Hill, WV, attended VPI & grad from Univ of TN. Employee of Oak Ridge Natl Lab since 1958 with res responsibilities in a number of mtls areas resulting in 120 publications & reports. Functioned as Group Leader, Proj Mgr, & lead engr in several areas. Lectr, Univ of TN in mtls courses. Independent consultation in product liability, fabrication quality, control, welding, & electroplating. *Society Aff:* ASM, AIME.

McCoy, William R
Home: Route 1, Anderson, SC 29621
Position: Executive VP. *Employer:* Enwright Associates, Inc. *Education:* MS/Water Resources Engr/Clemson Univ; BS/CE/Clemson Univ. *Born:* 7/2/40. PE in SC & Fla; Co-founder of Enwright Assocs in 1968. Worked in SE Asia & Australia 1966-67; Pres Cons Engrs of SC 1975; Natl Dir (SC) of Amer Cons Engrs Council 1975; Chmn PE's in Private Practice (SC) 1975; enjoys outdoor activities. *Society Aff:* ACEC, NSPE, WPCF.

McCoy, Wyn E
Business: 1835 Dueber Ave, S.W, Canton, OH 44706
Position: VP-Engrg & Research (Retired) *Employer:* The Timken Co *Education:* BS/IE/OH State Univ *Born:* 8/25/17 Entire professional career-The Timken Co. 1940-42 Sales Trainee; 1942-46 US Army, Manhattan Atomic Project at Columbia Univ and Oak Ridge, TN; 1946-64 Sales Engr, Chicago District Mgr, Sales Promotion Mgr, Assistant General Mgr-Sales- Industrial Division; 1964-70 Assistant Chief Engr and Chief Engr-Industrial Application Division; 1970-73 Group Mgr-Sales-Industrial, Railroad and Steel Divisions; 1973-77 VP-Marketing; 1977-80 VP Intl Operations; 1980-present VP- Engrg and Research; 1976-1982, Bd of Dirs. Retired 1982. *Society Aff:* SAE, IRI, AISE

McCrate, Thomas A
Business: 3131 South Dixie Dr, Ste 412, Dayton, OH 45439
Position: Project Mgr *Employer:* Shaw, Weiss & De Naples *Education:* MS/CE/Univ of Dayton; BS/CE/Univ of Dayton *Born:* 8/4/48 Native of Portageville, MO. Worked as surveyor on an Interstate Hgwy Project and an industrial construction site 1966-1969. Teaching Graduate Assistant at Univ of Dayton 1971-1973. Project Engr at Betz-Converse-Murdoch 1974-1978. Project Mgr in Sanitary Engrg Dept at Shaw, Weiss & De Naples 1978 to present. Served as Contact Member to the Univ of Dayton Student Chapter of ASCE 1979-1980. Currently Secy of Dayton Section of ASCE. ASCE Mead Prize 1980. *Society Aff:* ASCE

McCready, Lauren S
Home: 169 Gallows Hill Rd, West Redding, CT 06896-1411
Position: Self-employed *Employer:* L S McCready Engrg Enterprises *Education:* MS/History of Science/Polytechnic Inst of Brooklyn; MME/Mech Eng/NY Univ; BSME/Mech Eng/NY Univ. *Born:* July 1915. Unlicensed seaman then tanker engr. Merchant Marine Cadet Corps Instructor, Washington. Head Dept of Engrg, US Merchant Marine Acad 1942- 70; helped found Acad. Prof of Marine Engrg. USCG Chief Engineers License. Senior Reactor Operator's License NS SAVANNAH. Founded Natl Maritime Res Center 1970 as Dir. Ret Govt 1974: Rear Admiral US Maritime Service. Marine engrg educ cons & marine safety specialist. Dir of Ships' Operational Safety, Inc. US Dept of Commerce Gold Medal 1968. Two terms Council, Soc of Naval Architects & Marine Engrs, Tech & Res Ctte. Hobbies: modelmaking, painting, aviation, tech history, guns, foreign languages. *Society Aff:* SNAME, IME, Newcomen Soc.

McCreary, Harold James
Home: 481 Homer Ave, Longwood, FL 32750
Position: Consulting Engineer. *Employer:* Western Electric Co, et al. *Born:* Dec 10, 1899 Shelton, Neb. Educ in Scottsbluff, Neb. BSEE U of Neb 1923. Invented & designed the magnetic cross valve & was awarded a Prof degree in EE for this thesis by the U of Neb & a prof membership in HKN. While working in Calif was awarded a membership in the Royal W Sorenson Fellowship. Since moving to Fla has retired but attends annual banquets of IEEE as a Life Mbr, 50 year Mbr & Fellow. Life Mbr 50 year Western Soc of Engrs. Inventor Cathode Ray Color TV Pat 2013162; tele instruments, apparatus & circuits; author 70 other pats. Was active in AIEE, IRE & NEC in Chgo area, serving as prog chmn at times. Secy 1951 of Natl Electronics Conf & a dir for 3 other years. Currently a cons engr in electronics, communication & magnetics for Western Elec Co, Automatic Elec Co, Ramo Wooldridge.

McCreery, C Wayne
Home: 4017 Brookshire Ct, Columbus, OH 43227
Position: President. (Retired) *Employer:* Glass Industry Consultants, Inc. *Education:* BS/Physics, Math/Ball State; MS/Mech Engr/Toledo Univ; MS/Adv Mgmt/Northwestern Univ; Ms/Adv Glass Tech/Toledo Univ. *Born:* 7/30/08. Grad Northwestern Mgmt 1961. Advanced Engrg Univ of Toledo (toward Masters); Univ of Cincinnati Architectural Coll (Depression prevented completion). Retired after 31 yrs service with owens IL. Last 15 yrs was Ch Engr & Tech Mgr of Libbey Div. Previously was Ch Process Engr during dev of TV glass pictures tubes & the entire prod sys. Prior to that was Ch Proj Engr respon for the high speed mechanization of Glass Tumbler & Stemware prod sys. Reg PE OH & IN. Fellow ASME, ACS, NSPE. Life Mbr Optimists Intl. (Master Mason) Fort Industry Masonic Lodge Toledo, OH. Mbr Brookwood Presbyterian Church, Columbus, OH. Sole or Joint Inventor of 8 patents, all assigned to Owens IL. New Patent- Issued Aug 1, 1978 "Automatic Temperature Control Sys for Press & Blow Glass Forming Machines" C W McCreery. *Society Aff:* NSPE, ASME, ACS.

McCreery, Robert H
Home: 321 E High St, Portland, IN 47371
Position: VP Metallurgical Engg and Raw Materials. *Employer:* Teledyne Portland Forge. *Education:* BS/Met E/Purdue. *Born:* Sept 11, 1925 Muncie Ind. Educ: U of Kentucky, Ball St U, & grad from Purdue U 1948. Following grad, worked as a metallurgist, Genl Foreman Heat Treating, & Prin Metallurgist Intl Harvester. Then served as Plant Metallurgist at Borg Warner Transmission Div prior to joining Teledyne Portland Forge as Ch Metallurgist in 1960. Past Natl Trustee of the ASM International, a Fellow of ASM International, a mbr of the Amer Soc of Mech Engrs, licensed PE, P Pres of the Rotary Club of Portland, & an Elder of the Presbyterian Church and City Councilman Portland. *Society Aff:* ASM International, ASME, ASTM

McCubbin, John G
Business: P O Box 8400, Kingston Ontario, Canada K7L 4Z4
Position: Research Engineer. *Employer:* Aluminum Company of Canada, Ltd. *Education:* BSc/Metallurgical Eng/Queen's Univ. *Born:* July 1935 North Bay, Ontario 1958. Worked on metallurgical uses of atmospheric gasses with Linde Co, Newark, NJ; 1960, served as troubleshooting engr then supervisor, Casting Div, Alcan Arvida, Quebec; 1965 Ch metallurgist, Alcan, Isle Maligne, Quebec respon for Metallurgical aspects or DC Casting process & prods; 1968 joined Alcan Res Centre, Kingston Ontario working projs related to DC casting of aluminum alloys; trace element effects on Al decoating/recycling of Al. Since 1963 supervised developed work on aluminum brazing. Mbr Canadian Council ASM 1970 becoming Chmn 1975. Executive mbr of several local recreational and church grps. Hobbies include square dancing, lapidary & boating. *Society Aff:* ASM, AIME, APEO, SAE, AWS

McCubbin, T King, Jr
Business: 104 Davey Lab, University Park, PA 16802
Position: Prof of Physics *Employer:* The PA State Univ *Education:* PhD/Physics/Johns Hopkins Univ; B/EE/Univ of Louisville *Born:* 6/1/25 Baltimore MD, 1925. US Navy, 1944-1946. Research Assoc, Johns Hopkins, 1951-1954. Member of research staff, MIT, 1954-1957 Physics teaching staff, PA State, 1957-present. Currently Prof of Physics. Dir-at-large, OSA, 1968-1970. Prof Etranger, Univ of Dijon (France), 1974. Current research: Molecular Spectroscopy and Laser Spectroscopy. *Society Aff:* APS, OSA

McCue, J J Gerald
Home: 20 N Hancock St, Lexington, MA 02173
Position: Retired *Education:* AB/Physics/Harvard College; PhD/Physics/Cornell Univ *Born:* 1913 South Orange, NJ. Taught at Hamilton Coll 1941-44, Smith Coll 1946- 49. Staff Mbr at MIT: Radiation Lab 1944-45, Lab of Nuclear Sci & Engrg 1949-51, Lincoln Lab 1951-79. Principal concern in recent years, defense against intercontinental missiles. Chmn IEEE Sonics & Ultrasonics Grp 1962-64; Ed, IEEE Spectrum 1968-70. Pres's Certificate of Merit 1946; Fellow, IEEE 1971. Mountaineer. *Society Aff:* APS

McCulloch, James P
Business: McLeod & Norquay, Ltd, Vancouver, BC, Canada V6A 3L2
Position: Pres. *Employer:* McLead & Norquay, Ltd. *Education:* BASc/Met Engg/Univ of BC. *Born:* 1/20/20. Vancouver, Canada. 1943-1959, Ind Res-B C War Metals Res Bd; Plant layout planning, Cockshutt Plow Co, Brantford, Ontario; Mill, Service Met-Atlas Steels Lts, Welland, Ontario; Foundry Superintendent, Vulcan Engg, Winnipeg, Manitoba; Foundry Met, Canadian Summer Iron Works, Vancouver, B C; Mgr, Pacific Air Pollution Control, Vancouver. In 1959 became part owner & Gen Mgr of McLeod and Norquay Ltd (Commercial heat treating); assumed presidency in 1979. Consultant in mtls selection, production met, failure analysis; Profl musician. *Society Aff:* APEBC, ASM.

McCullom, Cornell Jr
Business: 3 West 35th St, New York, NY 10001
Position: VP *Employer:* NACME, Inc. *Education:* MBA/General Mgmt./Long Island University; MS/Meteorology/University of Utah; BS/Engineering/Mil./United States Military Academy *Born:* 04/05/32 Completed 21 1/2 years in U.S. Army and reached the rank of Colonel with specialties in Telecommunications, R & D, Meteorology and General Mgmt. Served as member of the staff and faculty at the U.S. Military Acad 1969-72 & 1974-78 and as an adjunct faculty mbr at the Universities of Alaska and Maryland. A veteran of Vietnam and responsible for all U.S. Army telecommunications throughout Thailand from 1973-74. After leaving military service, was Senior Mgr of Operations and Mgr of Facilities Acquisition and Planning, ITT Corporation prior to assuming responsibilities as VP, NACME, Inc. Was instrumental in dev of programs at West Point to increase minority student enrollment as well as minority staff and faculty assignment. *Society Aff:* NACME, AAES, JETS

McCullough, Charles A
Home: 320 Ross Way, Sacramento, CA 95825
Position: Consulting Engr - Water Resources (Self-employed) *Education:* BS/Civil Engr/Univ of CA, Berkeley. *Born:* 1/6/19. California Dept of Water Resources 1948-82. Recent Assignment: Chief, Div of Flood Mgmt, managed California's progs for flood warning; flood plain mgmt; dev, maintenance & operation of flood control facilities; & snow surveys & water conditions forecasts. Earlier assignments: Chief of Statewide Planning Branch, 1971-75; Engr to the Calif Water Comm 1968-71; & District Engr of the San Francisco Bay District 1961-68. Reg civil engr CA. Fellow ASCE, P Pres Sacramento Sect. Mbr, Intl Comm on Large Dams. Native of CA, married & participates in community remedial educ & sport fishing. *Society Aff:* ASCE, USCOLD

McCullough, Clarence R
Home: 2002 Skyline Pl, Bartlesville, OK 74006
Position: VP of Mfg. *Employer:* Phillips Petroleum Co. *Education:* BS/Mech Engg & Petroleum/OK State Univ. *Born:* 3/1/08. Worked 41 yrs for Phillips Petroleum Co with varied engg, supervisory, operatonal and admin plant and home office assignments in numerous locations both natl and intl. Retired in 1971 as VP of Mfg Dept responsible for six refineries, ten chem plants and 5,800 employees. Proj Dir Consultant Emerson Mgt Consultants in 1976. Active community and church. Past pres Okmolgee, OK Chamber of Commerce. Past Dist Governor of TX-OK Dist, Kiwanis Intl. Active American Petroleum Inst environmental committees 15 yrs and Dir Refining Div 1970. Hobbies fishing and golfing. Married Lillian Lee 1929. One son and three grandchildren. Methodist, Kiwanian, Sigma Chi. *Society Aff:* OSPE, NSPE.

McCullough, David W
Home: 831 Clark Ave, Billings, MT 59101
Position: Sr VP *Employer:* HKM Assoc *Education:* BS/CE/MT State Univ *Born:* 9/7/22 BSCE in 1947. Employed upon graduation with consulting engrg firm Morrison- Maicile, Inc. Hclcna, MT as junior design engr & chief of survey party. Advanced through constr project inspector to project constr engr. Head of transportation engr division of M & M. Created and headed branch office of M & M in Billings, MT in 1957. Responsible for supervising design, constr and circut contacts for all types of civil engrg project. Advanced in MM to VP & minor stockholder. Resigned from MM in 1970 to enter prof corp of Hurlbut, Kersich & McCullough (HKM Assoc). Principal in charge of all phases of municipal type engrg. *Society Aff:* ASCE, NSPE

McCullough, Hugh
Business: 1818 Market St, Phila, PA 19103
Position: Pres, Engr & Constr Div. *Employer:* Day & Zimmermann, Inc. *Education:* BS/ME/Drexel Univ; Dipl/ME-Power Plant/Drexel Univ; Dipl/ME-Machine Design/Drexel Univ. *Born:* 6/27/17. Native Philadelphian. BSME-Drexel Univ. Reg PE in the states of PA, NJ, and NC TX. 1946-46 US Army - Corps of Engrs, WWII, 1st Lt, European Theater & Pacific Theater. In 1936, joined Day & Zimmermann, Inc as an engg clerk & advanced to ind engg asst. In 1946, rejoined Day & Zimmermann, Inc & advanced pogressively to positions of increasing responsibility from mech designer to chief mech engr in 1951; & seven yrs later became proj mgr responsible for coordinating all activities of complete proj dev. In 1962, construction mgr; 1966, vp-construction; 1973, sr vp-Engg & Construction Div responsible for overall mgt of constructin projs, & direction of marketing activities of the div. As of 1974, pres-Engg & Construction Div responsible for all engg design activities & construction services of the corp. Elected a mbr of the Bd of Dir of Day & Zimmermann, Inc in 1976, and appointed Corporate VP.

McCully, Robert A
Home: 33 Zerman Dr, New Monmouth, NJ 07748
Position: Director Product Management *Employer:* North Amer Philips Lighting Corp. *Education:* BS/IE/Northeastern Univ. *Born:* May 21, 1937 Lynn, Mass. Married, 5 children. Honorable Discharge US Army Reserve, 1st Lt. 1960-62 field engr, Factory Insurance Assn; 1962-66 Mgr Commercial Engrg, Champion Lampworks; 1966-74 Mgr Marketing Engrg, ITT Lamp Div; 1974-76 Fuorescent Prod Mgr, Lightolier; 1976-77 Fluorescent Prod Mgr, 1977-78 Fluor Hd Prod Mgr, 1978-79 Mktg Engg Mgr, Director, Product Marketing 1980-81, Dir, Mktg Fixture Div 1982-86 Dir Prod. Mgt NAPLC. Lecturer & Author on Lamps & Lighting; P Chmn IES Light Sources Ctte; P Chmn IES Lighting Competition Ctte; P Chmn IES Lighting Design Awards Ctte; IES Northeastern Regional Vice Pres 1969-71; IES Dir 1971-73; Treasurer, National Lighting Bureau 1980-81. Chairman IES Publications Committee, NEMA Bd. of Governors 1984-86. *Society Aff:* IES

McCune, Francis K
Home: 470 Magellan Dr, Sarasota, FL 33580
Position: Vice President (retired). *Employer:* General Electric Co. *Education:* BS/EE/Univ of CA, Berkeley. *Born:* April 10, 1906 Santa Barbara, Calif. Joined GE 1928; Asst Engr West Lynn Works 1938; Genl Engrg Staff Apparatus Dept 1946; VP Atomic Energy 1954; VP Engrg 1960; Retired 1967. Past Chmn Mgmt Ctte - mbr Operations Ctte, AIEE. P Pres Atomic Indus Forum (Hon Dir); P Pres Amer Natl Standards Assn; Howard Coonley Medal (stds) 1969; Mbr NAE, & Woods Hole Oceanographic Institution. *Society Aff:* ASME, IEEE, ANS, NAE.

McCune, William E
Business: Agri Engrg Dept, College Station, TX 77843
Position: Professor. *Employer:* Texas A&M University. *Education:* MS/Agri Engg/TX A&M Univ; BS/Agri Engg/KS State Univ. *Born:* Sept 1917, Leavenworth, Kansas. Teaching & res, Agricultural Engrg Dept, Texas A&M U 1940-45. Joined Central Power & Light Co as Agricultural Engr, organized Rural Service Dept. Western Div Sales Mgr respon for energy sales & indus engrg assistance. Prof & Dir of Texas Agri-Business Elec Council, Agricultural Engrg Dept, Texas A&M U since 1959. Agri-indus cons to elec utility companies of Texas. Respon for teaching, employee training & res in elec power applications in agriculture. Recipient of 1973 ASAE 'George W Kable Electrification Award'. Reg PE in Texas. Mbr ASAE, ASEE, NSPE & TSPE. *Society Aff:* ASAE, ASEE, NSPE, TSPE, ΓΣΔ.

McCurdy, Archie K
Business: Dept of Electrical Engrg, Worcester, MA 01609
Position: Assoc Prof *Employer:* Worcester Polytechnic Inst *Education:* Ph.D/Physics/Brown Univ; M.Sc/E.E./Worcester Polytech Inst; B.S. /Math/Eastern Nazarene Coll *Born:* 8/1/31 Mbr of the EE teaching faculty at WPI since 1955. Teaching specialties are solid state electronics and transport theory. Most recent res is thermal and acoustic propagation at low temperature in elastically anisotropic crystals. Recent publications analyze phonon focusing and thermal transport in crystals resulting from elastic anisotropy. Presentations of this res have been given at more than a dozen scientific meetings including 3 international conferences. *Society Aff:* APS, ΣΞ, HKN, NYAS, ΦΔLU

McDaniel, Willie L
Business: PO Drawer DE, Mississippi State, MS 39762
Position: Dean. *Employer:* MS State Univ. *Education:* PhD/Elec Engg/Auburn Univ; MS/Elec Engg/MS State Univ; BEE/Elec Engg/Auburn Univ. *Born:* 9/19/32. Native of Montgomery, AL. Mbr of Engg Faculty at MS State Univ for the past 28 yrs having served at every rank from instructor to Prof. Assumed duties as Dean of College of Engg in 1978. Current professional interests include energy res admin as well as "deaning–. Outside interests involve church work, local community affairs, work with alumni, and after dinner speaking. Great interest in raising, training, and hunting bird dogs. *Society Aff:* IEEE, NSPE, ASEE.

McDavid, Frederick R
Home: 9321 Convento Terr, Fairfax, VA 22031
Position: President. *Employer:* McDavid Grotheer & Co. *Education:* BME/Mech Engg/NC State Univ. *Born:* Dec 1924. Native of Sanford, NC. Served with US Army 1944-46. Process Engr US Steel Corp 1948-50. Design Engr Walter Hook Assocs, Charlotte, NC 1950- 51. Design Engr, H K Ferguson Co, Cleveland, Ohio 1951-53. Pres present co (originally a partnership) 1954- . Respon for Genl Mgmt & Engrg. Mbr ASME 1948- ; Mbr Amer Cons Engrs Council 1958- ; Chmn Fairfax County Va Bd of Plumbing Examiners & Appeals 1968-1981. Organizer & Chmn Troop 1533 BSA 1965-68. *Society Aff:* ASME, CEC.

McDavid, Frederick R, Jr
Business: PO Box 88, 122 S. Main St, Harrisonburg, VA 22801
Position: Partner. *Employer:* Systems II. *Education:* BS/Elec Engg/VPI & SU. *Born:* 10/28/50. Grew up in Northern VA. Attended VA Tech, Blacksburg, VA for 4 yrs. Worked for US Army, Harry Diamond Labs, Washington, DC for about a yr. Worked for McDavid Co Engg Consultants, Washington, DC for several yrs. Engr with Harrisonburg branch of John F Lawrence & Assoc, Consulting Engrs for 1 1/2 yrs. Established private practice in Harrisonburg in 1976 with partner, Richard I Covington. Other interests include automobiles and Satellite TV systems. Professional Registration current in 15 states. Chap Pres 1984-85, Skyline Chap, VSPE Mbr of ASHRAE, ASPE, NSPE and Shenandoah Valley Assoc of Architects. *Society Aff:* NSPE, ASHRAE, ASPE, SVAA

McDavitt, Marcellus B
Home: Santa Anita Club de Golf, Apartado Postal 31-9, Guadalajara Jalisco Mexico
Position: Independent Consultant. *Employer:* Self. *Education:* MSEE/Elec Engg/MIT; EE/Elec Engg/Univ of VA. *Born:* Aug 1903 in Texas. AT&T Co 1925-34; Bell Telephone Labs 1934-68; independent cons since 1968. Early exper was with dial telephone switching systems dev. With Bell Labs, was successively Dir of Switching Dev, Dir of Transmission Dev & VP for Customer Prods Dev. Have consulted on telecommunications devs in US, Italy & Taiwan. Hon Dir: Midlantic Natl Bank/Merchants, Neptune, NJ & Riverview Hospital, Red Bank, NJ. IEEE prize paper 1956. Enjoys golf & music. *Society Aff:* IEEE.

McDermott, Gerald N
Business: 6110 Center Hill Rd, Cincinnati, OH 45224
Position: Senior Engineer. *Employer:* Proctor & Gamble Company. *Education:* Sanitary Engg/San Engg/MIT; SM/San Engg/Harvard Univ; BCE/Civil Engg/Univ of WY. *Born:* Sept 23, 1921. Chmn Ohio Water Pollution Control Conf 1975. Chmn Indus Waste Ctte, Water Pollution Control Federation 1976. Joint Ctte ASCE, APWA, & WPCF on Financing & Charges for Wastewater Systems 1970-73. Chmn, Chem Indus Ctte, Ohio River Valley Sanitation District 1975-77. Chmn Joint Task Grp on Standard Methods for Water & Wastewater Analysis, BOD. Army Corps Engrs 1943-45. Dir U S Public Health Service, Water Pollution Control Res Prog for 20 years. Since 1966 with Proctor & Gamble Co in water pollution control & solid wastes from mfg facilities. Enjoys music, tennis, skiing, & gardening. *Society Aff:* ASCE, ACS, WPCF, ASTM.

McDermott, Joseph M
Home: 523 Monarch Ln, Addison, IL 60101
Position: Manager, IDOT Traffic Systems Center *Employer:* IDOT Traffic Systems Center *Education:* MSCE/Trans. Eng/Northwestern Univ; BS/CE/Univ of Detroit *Born:* 10/28/39 Native of Cleveland, Ohio. With Illinois Dept. of Transportation since 1963. Heads up Traffic Systems Center, which provides electronic surveillance, control and motorist information systems for improving freeway traffic flows in the Chicago metropolitan area. Registered prof engr, Illinois and Ohio. Mbr, former Chrmn, Natl Res Council, Transportation Res Bd, Cttee on Freeway Operations. Mbr (former Chrm), ASCE Urban Transportation Div Exec Cttee. Mbr (former Chrmn), ASCE Ctte on Traffic Operations. ASCE Frank M. Masters Transportation Engrg Award, 1980. ASCE Arthur M. Wellington Prize, 1981. Served on advisory panels for the National Cooperative Highway Research Program. Chrmn, Plan Commission, Village of Addison, Illinois. Enjoys golf, racquetball and photography. *Society Aff:* ASCE, ITE, TRB

McDonald, Dan T
Home: 10042 Bayou Glen, Houston, TX 77042
Position: President & General Manager. *Employer:* Consolidated Storage, Inc. *Education:* BS/ChE/TX Tech; BS/ChE/MIT. *Born:* April 1919. Magnolia Petroleum Co 1939-53, starting as engr & ending as Ch Process & Design Engr, Natural Gas Dept. C W Murchison Interests 1953-58, VP & Dir Murmanill Corp & 5 subsidiaries in natural gas prod, processing & marketing fields. Exec VP & Genl Mgr Goliad Oil & Gas Co & its 9 subsidiaries in Alberta, Canada 1958-72, & Genl Mgr & Dir of Alberta Underground Storage 1960- 72. Pres & Dir of Anlin Co of NJ & Anlin Co of Ill 1973-75. Pres & Genl Mgr Consolidated Storage, Inc May 1975- . Mbr AIChE. Pres & Dir (past) Canadian Natural Gas Processing Assn; Pres & Dir (past)

McDonald, Dan T (Continued)
Propane Gas Assn of Canada; Distinguished Engr Texas Tech Univ 1976. Reg PE Texas & Kansas. *Society Aff:* AIChE, TBΠ

McDonald, Donald
Business: Provost and Vice President, for Academic Affairs, Texas A&M University, College Station, TX 77843
Position: Provost & Vice Pres for Academic Affairs *Employer:* TX A&M Univ. *Education:* PhD/Civ Engg/Univ of IL; MS/Civ Engg/Univ of IL; BCE/Civ Engg/ Auburn Univ. *Born:* 10/16/30. Native of Montgomery, AL. Served as Lt in USAF, 1952-54. Bridge design engr for AL Hgwy Dept. With Lockheed Missiles & Space Co in Sunnyvale, CA as Sr Res Engr & Res Specialist, 1959-62, & in Huntsville, AL as Dept Mgr, 1967-72. Taught civ engg as Asst Prof & Assoc Prof at NC State Univ, 1962-67. Prof of Civ Engg at TX A&M Univ, 1973-present. hd of dept, 1979-86, Appointed Provost & VP for Academic Affairs, 1986-present. Reg PE in NC, AL, TX. Married Ann Chaney in 1960 & has 3 children. *Society Aff:* ASCE, AIAA, NSPE, ASEE.

McDonald, Gerald W G
Business: 8 Woodcroft Rd, Summit, NJ 07901
Position: President. *Employer:* The IONA Company. *Education:* BSc/Chem/Univ of Glasgow Scotland. *Born:* July 29, 1927. Married, 2 children. Educ: Hamilton Acad, Univ of Glasgow Scotland; graduated BSc (Hons) 1949. Awards: Joseph Black Medal, Muirhead Prize. Reg PE in IL and NJ-52 Trinidad Leaseholds (now Texaco) Asst Plant Superintendent; 1952-63 Universal Oil Products Co (developers & licensers of petrochem processes) - Customer Service Coordinator; 1963-68 Amerada Hess Corp (petro refiner) - VP Mfg; 1968-76 Commonwealth Oil Refining Co (petrochem manufacturers) - VP, Mfg, Engrg & Dev; 1976 The IONA Co, Petrochemical Engrs & Consultants - President. Publications: 4 US Patents relating to alkylation, catalytic cracking & hydrocracking. Several tech articles relating to petroleum & petrochemical processing. Mbr AIChE, API, Bankers Club. Hobbies: classical music, sailing, tennis. *Society Aff:* AIChE, API.

McDonald, Henry C
Business: PO Box 808, Livermore, CA 94550
Position: Assoc Dir for Engg. *Employer:* Lawerence Livermore National Lab *Education:* AB/Physics/Univ of CA-Berkeley. *Born:* 7/2/28. Grad work in elec engg and appl sci from UC-Berkeley and UC-Davis. Native of Long Beach, CA. Served with the US Navy, 1946-1948. Employed by the Univ of CA, Lawrence Livermore National Lab since 1953 as an elec engr specializing in nuclear instrumentation and control. Assumed current position as Assoc Dir in 1973 responsible for the electronic, mech and plant eng support to the Lab's scientific progs. Sr mbr of IEEE. Enjoy tennis and bicycling. *Society Aff:* IEEE.

McDonald, James R
Business: 60 Broad St, New York, NY 10004
Position: Dir, New York Operations *Employer:* RCA Global Communications, Inc. *Education:* BEE/EE/Manhattan College. *Born:* Dec 1932. Served in the US Army Signal Corps 1954-56. With RCA Global Communications 1957- . Held positions as Plant Operations Engr, Telex Engrg Leader, Mgr Switched Services Engrg & present position as Dir, Operations Engrg which is respon for Design Engrg, Central Office Engrg, Const, Fabrication Shops & Installation, & Tech Training which supports the company's internatl communications services. Mbr of IEEE. Enjoys photography tour bicycling & golf. Dir Atlantic operations responsible for the Company's operation in Washington DC, Miami, New Orleans, Puerto Rico & Dominican Republic. Also the NY Electronic & Teleprinter Shops & Field Technicians; Dir NY operations responsible for the Companys operation in New York. *Society Aff:* IEEE.

McDonald, Jerry L
Business: 690 Central Bank Bldg, Huntsville, AL 35801
Position: Consulting Structural Engineer. *Employer:* McDonald Engrg Company - Owner. *Education:* BS/Civil Engg/GA Inst of Technology. *Born:* Feb 8, 1932 Newbern, Tenn. Elementary & high school educ received in schools in Dyersburg, Tenn. Served in the US Air Force from 1951-53. Attended Georgia Inst of Tech, Atlanta from 1954-57 & received a BSCE with highest honor 1958. Has been a cons structural engr in Huntsville, Ala since 1964. 1969 McDonald Engrg Co was formed & specializes in structural design of bldgs. Reg PE in Ala, Tenn, & Georgia. Mbr of Cons Engrs Council of Ala, Amer Concrete Inst, Amer Inst of Steel Const. *Society Aff:* CEC, ACI, AISC.

McDonald, John C
Home: 586 Van Buren Pl, Los Altos, CA 94022
Position: VP, Strategic Plng. *Employer:* TRW Vidar. *Education:* PE/Engg/Stanford Univ; MS/Engg/Stanford Univ; BS/Engg/Stanford Univ. *Born:* 1/23/36. He is a Fellow in the IEEE and is active in the Computer & Communications Socs. He regularly teaches courses on digital telephony at UCLA & the Univ of MD. He has written papers for many IEEE journals & has presented the results of his work to local chapters & natl confs. He holds 14 patents & is a reg PE. *Society Aff:* IEEE, USITA.

McDonald, John F
Business: Elec Systems Engg Dept, CIE 2004, Troy, NY 12181
Position: Full Prof. *Employer:* RPI. *Education:* PhD/EE/Yale Univ 69; MEng/EE/ Yale Univ 65; BSEE/EE/MIT 63. *Born:* 1/14/42. Narberth, PA. Elec Engr, Comp Engr. Prof Exp: Mbr tech staff, Bell Tel Labs 65; from lectr to asst prof eng & comp sci, Yale Univ 69-74; Assoc Prof, Dept Elec, Computer & Systems Eng, RPI 74-86, Full Professor, RPI, 86- , Consult, Underwater Sound Lab, CT, 69-71; CTMS Inc, NY, Argonne Natl Lab, 75, Westinghouse Hanford Engrg Dev lab 76-77, GE Corp R&D Ctr, 78-81, TV Data Corp, 78, Teledyne Gurley Corp, 79-81. Sparticus Computer Corp. 81, Mbr: Sigma Xi; Inst Elec & Electronics Engrs; Assn Comput Mach; Am Inst Phys. Res: Computer design; communication; detection estimation systems res; control system studies; info theory & coding; digital test set generation; VLSI design automation; phased array design; microprocessor systems; medical instrumentation. *Society Aff:* IEEE, AIP, HKN, ΣΞ.

McDonald, Patrick H, Jr
Business: Dept of Civil Engrg, Raleigh, NC 27695-7908
Position: Professor. *Employer:* North Carolina State Univ. *Education:* PhD/Mechs/Northwestern Univ; MS/Mech Engg/Northwestern Univ; BS/Engg/NC State Univ. *Born:* Dec 25, 1924 Carthage, NC. Instructor NCSU 1947-48; Instructor, Clemson U 1948-50; Royal Cabel Fellow, Northwestern U 1950-53; NCSU Assoc Prof 1953, Prof 1957, Head Engrg Sci & Mechanics 1960-76, Harrelson Prof 1965- , Civil Engrg 1976- . Pi Tau Sigma Gold Medal ASME 1957; Res Award, SE Sect ASEE 1959; Res Award, Sigma Xi 1960. Exec Chmn, Southeastern Conf Theoretical & Applied Mechanics 1968-70. NC Soc of Engrs, AAM, ACDM; Pi Tau Sigma, Tau Beta Pi, Sigma Xi, Phi Kappa Phi. *Society Aff:* ASME, ASEE, SES, SEM

McDonald, Robert O
Business: 1500 W Dundee Rd, Arlington Heights, IL 60013
Position: Eng Mgr. *Employer:* Honeywell Inc. *Education:* BS/ME/Northeastern Univ. *Born:* 9/21/29. Married with five children. Joined Honeywell Inc in 1950 & has served in field, marketing and engg mgt positions. Chief engr in the mid 1950's in the Boston & NJ offices responsible for temperature control design for commercial, indus & institutional bldgs. Market & product mgr for Commercial Div temperature control systems from 1964 to 1972. Currently mgr of System Analysis for Commercial Div of Honeywell at Arlington Heights, IL. Chairman Region VI & mbr of the Bd of Dirs ASHRAE 1977 to 1980. Recipient of ASHRAE's Distinguished Service Award in 1976. *Society Aff:* ASHRAE.

McDonalds, Henry S
Whippany Road, Whippany, NJ 07981
Position: Assistant Director. *Employer:* Bell Laboratories. *Education:* Dr Engg/Elec Engg/Johns Hopkins Univ; MS/Elec Engg/Johns Hopkins Univ; BEE/Elec Engg/ Catholic Univ. *Born:* Oct 1927. US Navy WWII. Faculty Johns Hopkins 1953-54. Joined Bell Telephone Labs Res Area - worked on speech, vision, hearing, computer simulation, digital filters, computer graphics, digital signal processing, digital electronic switching. Currently Asst Dir Consultant for Systems Architecture Research. Cons to Dept of Defense. 8 patents. Fellow IEEE. Mbr of ACM, SIgma Xi. Resident of Summit, NJ. *Society Aff:* IEEE, ACM, ΣΞ, AAAS

McDonnell, Archie J
Business: 106 Land/Water Res Bldg, University Park, PA 16802
Position: Professor of Civil Engrg. *Employer:* The Pennsylvania State Univ. *Education:* PhD/CE/Penn State Univ; MS/CE/Penn State Univ; BS/CE/Manhattan College. *Born:* June 1936. Native of NY City. Mbr of Penn State Civil Engrg faculty since 1963. Res interests include stream pollution modeling & acid-mine drainage effects. Dir, Envir. Resources Res. Inst. PSU. Respon for the mgmt & prog planning of the Inst's Five Environ Resources Res Centers. ASCE CROES Medal 1976. *Society Aff:* ASCE, IAWPR, WPCF, AWRA.

McDonnell, John T
Home: 511 Longwoods La, Houston, TX 77024
Position: Energy Conslt *Employer:* Self. *Education:* BS/Chem Engr/Oklahoma Univ *Born:* 1/5/26. Tulsa, OK. Commissioned in US Navy; employed by Warren Petroleum Co, which was later acquired by Gulf Oil Corp. Among other positions held with Gulf and/or its subsidiaries: VP (1967-69) & Pres (1969-70), Iberian Gulf Oil Co, Madrid, Spain; Pres, Warren Petroleum Co, 1970 until transfer to Houston as present VP. Mbr of API, AIChE, MCOGA; set up Energy Conslt bus in 1983; hold directorships constlt in numerous other prof & charitable grps. Enjoy golf & travel. *Society Aff:* ACS, NGMH, AICHE, TIPRO

McDonough, George F, Jr
Home: 1902 Fairmont Rd, Huntsville, AL 35801
Position: Director, Systems Dynamics Laboratory *Employer:* Natl Aero & Space Admin. *Education:* PhD/Structural mechanics/Univ of IL; MS/Structural mechanics/ Univ of IL; BCE/Civ Engg/Marquette Univ. *Born:* 7/3/28. Chicago, IL. Served in USN 1945-47, 1951-52. Engg instr/prof/lectr Univ of IL, San Jose State Univ, Univ of AL in Huntsville. Sr Engr, E H Plesset Assoc; Sr scientist, United Aircraft Corp. With NASA Marshall Space Flight Ctr since 1963. Served as chief or deputy in structures, controls & flight dynamics, systems engg, environmental applications & data systems. Deputy Assoc Dir for Engg 1978-1981. Director, Systems Dynamics Laboratory 1981- . Responsible for pointing and control, structural dynamics, aerodynamics and wind Tunnel testing and atmospheric sciences. Chairman Huntsville Air Pollution Control Bd. Founding Chrmn, Westbury Civic Assn.

McDougall, John
Home: 1893 Pine Ridge La, Bloomfield Hills, MI 48013
Position: Retired Exec VP & Dir Intl Auto Operations *Employer:* Ford Motor Co *Education:* BSME/Mech & Ind Eng/U of D *Born:* 7/3/16 Belfast Ireland. Graduated from Ford Trade and Engrg Prep Schools 1934. Attended U of D as co-op student until 1941 also other tech schools. Entered Ford Mechanical Design Dept August 1935. Project engr on B-24 Bomber project from 1941 to 1946 various management positions in engrg until appointment as Dir Mfg Research 1961. Asst Gen Mgr General Parts Division Jan 1965 gen mgr same division 1966 to 1971. VP Manufacturing Ford of Europe June 1971 to Oct 1973 Pres Ford of Europe 1973 to 1975 Chrmn of the Bd of Ford of Europe Oct 1975 to July 1977. Elected to Corp Bd of Dirs Sept 1977. Executive VP Operations Staffs WHQ Dearborn July 1977. Oct 1978 Exec VP Ford North American Operations. March 1980 Executive VP Ford Intl Operations until Aug. 1982. Retired Aug 1982. Member Bd of Trustees Taylor Univ Upland Ind. Member & Chrmn Bd of Trustees Beaumont Hospt Royal Oak MI. Member of Trustees MI Opera Theatre, Detroit. Reg PE in MI. Conslt to several corps (intl) *Society Aff:* SAE, ESD, ASME

McEldowney, Robert, Jr
Business: 120 Highway 22, Clinton, NJ 08809
Position: Chrmn of the Bd *Employer:* Studer & McEldowney, P A. *Education:* CE/Civil Eng/Princeton Univ; BS/Civil Eng/Princeton Univ. *Born:* April 8, 1919. Native of Johnstown, Penn. In WWII served as Radio-Radar officer in Marine Corps Aviation. Formerly employed by Taylor Wharton Iron & Steel Co, a steel foundry in engrg-mgmt capacity. With Studer & McEldowney, P A since 1950, as Pres 1973-1982. Firm is in genl Civil Engrg & Land Surveying fields. Active in NJSPE ctte work since 1950. Active mbr of CEC NJ. Fellow, ACEC. Also in ACSM, Tau Beta Pi, Sigma Xi. Other interests include church & civic work, golf, photography & travel. *Society Aff:* NSPE, ACEC, ACSM, TBΠ, ΣΞ

McEleney, Patrick C
Home: 33 Fabyan St, Arlington, MA 02174
Position: Physicist. *Employer:* US Army Materials & Mech Res Cntr. *Education:* BS/Physics/Boston Coll *Born:* April 1925. Attended Boston Coll & Northeastern U Grad Schools. Physicist in Army Materials & Mechanics Res Center, Nondestructive Testing Advanced Res Branch, 1952- . Author of 31 govt reports on various subjects & presented numerous papers at tech soc meetings throughout the country. Chmn ASTM Subctte E.07.07 Electromagnetic Methods & several other sections & task grps in Ctte E.07. P Chmn ASNT, Boston Sect (1962-63). CCD Instructor, St James Church, Arlington. Adj Irving W Adams Post, 36 Amer Legion, Roslindale. Chrmn ASNT (Natl) Tech Council, Res Div, Problem Identification & Evaluation Committee and Career guidance committee. Fellow ASNT 1983 C W Briggs award recipient ASTM. *Society Aff:* ASTM, ASNT, SAE

McElhaney, James H
Business: Biomedical Engg Dept, Durham, NC 27707
Position: Prof. *Employer:* Duke Univ. *Education:* PhD/Biomechanics/W VA Univ; MS/Mech Eng/Univ of PA; BS/Mech Eng/Villanova Univ *Born:* 10/27/33. Currently Prof and Chmn of Biomedical Eng Dept. Duke Univ; Assoc Editor, J Biomech Engg; Sectional Editor, J Bioengg; Editorial Bd Mbr, J Biomechanics; Chrmn ANSI-Z89 Committee on Industrial hd Protection Author of 72 papers in Biomechanics of Trauma & Protection, Res interests - Hd & Spine Injury & Protection, Sensory Feedback in Artificial Limbs. Formerly Hd Biomechanics Dept, Hgwy Safety Res Inst, Univ of MI. *Society Aff:* ASME, SAE, AAM

McElveen, W Harry, Jr
Business: 6451 N Fedrl Hwy-Rm 1020, Fort Lauderdale, FL 33308
Position: Div Mgr - Bus Services. *Employer:* Southern Bell Telephone & Telegraph. *Education:* BME/Mech Engg/GA Tech. *Born:* Oct 1936 Sylvania Ga. Honor graduate of Ga Tech 1958, Tau Beta Pi, Phi Kappa Phi, Phi Eta Sigma, Pi Tau Sigma. With Westinghouse in steam turbine design & Callaway Mills Co as Plant Engr prior to employment with Southern Bell 1962. Selected Outstanding Young Engr of 1968 by Indian River Chap of Fla Engrg Soc. Past Chmn Broward County IEEE; Past Mbr IEEE RAB Policy & Planning Ctte, PE, mbr FES & NESPE. Div Mgr in Bldg & Equip Engrg for SE Florida Area in 1972. Div Mgr respon for current Network Planning & Trunk matters in SE FL through 1978. Presently Div Mgr. Respon for Bus Installation & Repair Services in SE FL. *Society Aff:* IEEE, HSPE, FES.

McEntyre, John G
Home: 160 Pathway Lane, W Lafayette, IN 47906
Position: Professor Emeritus (Land Surveying) *Employer:* Purdue Univ. *Education:* PhD/Land Surveying/Cornell Univ 1954; MSCE/Structural/KS State Univ 1948; BSCE/Civ Engg/KS State Univ. 1942. *Born:* 11/3/20. Prof of Land Surveying Ret. Prof. Emeritus, Purdue Univ. Consultant-Advisor Cadastral Survey of Afghanistan 1963-65. Educational Consultant, Kabul Univ, Kabul, Afghanistan, 1965-67. Author of book, Land Survey Systems (Wiley), 1978. Co-author "Land Surveying and Registration–, *Journal, Surveying and Mapping*, ASCE, 1963. Author one manual and co-author two manuals "The Perpetuation of Corners in Indiana" and "Law and

McEntyre, John G (Continued)
Surveying" and "Establishment of Unwritten Boundaries and the Land Surveyor–. Contributing Author, The Surveying Handbook (Van Nostrand Reinhold), 1987. Past Chrmn Land Surveys Div American Congress Surveying and Mapping 1978. Past Pres IN Soc Profl Land Surveyors 1974 and KS Sec ASCE 1961. Distinguished Service Award ISPLS 1975 Presidential Citation ASCM 1979, Surveying Excellence Award ACSM/LSD 1980, Earl Fennell Award (Surveying and Mapping Ed), ACSM, 1984. *Society Aff:* ACSM, ASCE, NSPE, ASEE, ASPRS, URISA.

McEwen, Robert B
Home: 3512 Wilson St, Fairfax, VA 22030
Position: Research Physical Scientist *Employer:* US Geological Survey. *Education:* PhD/Photogrammetry/Cornell Univ; MS/Civil Engg/Univ of NH; BS/Civil Engg/Univ of NH. *Born:* July 1934. Native of Mass, served as a USAF pilot, taught at the U of NH, & was a NSF Sci Faculty Fellow at Cornell. Reg PE in NH. Mbr Amer Soc for Photogrammetry and Remote Sensing, American Society of Civil Engineers, & Sigma Xi. Dir in 1975 of ASP Remote Sensing & Interpretation Div & mbr of natl Bd. Chief of Remote Sensing & Space Applications for Office of Res & Tech Standards in USGS Topographic Div from 1971-77; Team Leader of the Digital Applications Team from 1977-1980; Deputy Chief of Research from 1981-86; currently Research scientist in Office of Geographic and Cartographic Research, National Mapping Division, USGS. *Society Aff:* ASCE, ASPRS, ΣΞ

McFadden, Peter W
Engrg, U-37, Storrs, CT 06268
Position: Dir of Dev for School of Engg; Prof, Mech Engg *Employer:* University of Connecticut. *Education:* PhD/ME/Purdue; MS/ME/U Conn; BSE/ME/U Conn. *Born:* Stamford, Conn. BS 1954 & MS 1957 from Univ of Conn; PhD 1959 from Purdue Univ Asst Instructor, 1954-56, Prof & Dean of School of Engrg 1971-85, Dir of Dev, 1985- ; Univ of Conn; Instructor 1956-59, Asst Prof 1959-62, Assoc Prof 1962-65, Prof & Head School of Mech Engrg 1965-71, Purdue Univ Post doctoral res 1960-61, Swiss Federal Inst Bd mbr, Conn Prod Dev Corp. CT Acad of Sci and Engrg, Mbr PE in Indiana. Primary tech interests: heat & mass transfer, optics, cryogenics, energy including solar energy. *Society Aff:* ASEE, ASME.

McFarlane, Harold F
Business: PO Box 2528, Idaho Falls, ID 83403-2528
Position: ZPR Program Mgr *Employer:* Argonne Natl Lab *Education:* PhD/Engr Sci/CA Inst of Tech; MS/Engr Sci/CA Inst of Tech; BS/Phys/Univ of TX *Born:* 04/23/45 Born Hagerstown, MD; reared Del Rio, TX; Attended Univ of TX (Austin) 1963- 67, Phi Beta Kappa, graduated magna cum laude; attended CA Inst of Tech 1967-71; married Mary Ellen Newberry 6/22/68, one son Matthew born 1/7/70; Res Assoc, Caltech 1971; Asst Prof Nuclear Engrg, NY Univ 1971-72; Joined Argonne Natl Lab 1972 as asst nuclear engr in the Applied Physics Div, promoted to Nuclear Engr 1974, Group Leader 1977, Sect Hd 1980, Program Mgr 1983. *Society Aff:* ANS

McFarlane, Maynard D
Home: P O Box 125, Sun City, AZ 85372
Position: Research Scientist. *Employer:* Robertshaw Controls Co (ret). *Born:* 1895 London, England. Naturalized US 1941. Educated: McGill U (Canada), U of London (England). Reg PE (E-23) in Calif. War service 1914-21: Infantry & Royal Flying Corps, Cpt. Co-inventor, Bartlane picture transmission system. Established facsimile res labs NY & Calif. War service 1942-46: Res Associate, Digital MIT (radar); Chief, Calif Engrg unit, Office of Sci R&D; cons Natl Defense Res Ctte on radar navigation. Res Scientist, Robertshaw Controls Co spec in fields atomic energy & communication. Awards: Fellow AIEE, IRE, IEEE, AAAS; Army-Navy Certificate of Appreciation. *Society Aff:* IEEE.

McFate, Kenneth L
Business: Natl Food & Energy Council, 409 Vandiver W., Suite 202, Columbia, MO 65202
Position: Prof of Agri Engrg & Pres, NFEC. *Employer:* Natl Food & Energy Council, Inc. *Education:* BS/Agri Engg/IA State Univ; MS/Agri Engg/Univ of MO. *Born:* Feb 1924. Native of LeClaire, Ia. 2nd Lt, Navigator Army Air Corp WWII. Ext Agri Engr 1951-56, Iowa State; Dir, Elec Energy Utilization Project of Mo Farm Elec Council/UMC Columbia, Mo 1956-76. Prof of Agri Engg, Univ of Missouri - Columbia, MO, 1956-86. 1986-present, Prof Emeritus, AgE, UMC. Appointed Exec Mgr of Natl Food & Energy Council, Inc, Jan 1976: only natl organization focusing on providing ample amounts of electric energy for Food Chain and the selective mktg of electricity on Amer farms. In ASAE: Past Chmn, Elec Power & Processing Div; Dir of ASAE 1970-72; Chmn Standards Ctte 1970-75. ASAE Fellow 1974; Recipient of George W Kable Outstanding Service, elec applied to agri, Award 1974; Outstanding Individual of Year, Mid-Central Region, 1975. NSC Safety Award, 1975. Outstanding Service to 4-H Award, 1982; Leader, Food & Energy Del, Peoples Rep of China, 1983. Reg PE in Ia - 2875, Mo - E-14104. Deacon & Elder, First Presbyterian Church, Columbia. VP Penreico, Inc 1979-80. Chrmn, ASAE Energy Ctte, T-11. Pres, NFEC 1976-Present. Hobby antique autos. *Society Aff:* ASAE.

McFee, James H
Home: 246 Lake St, Lancaster, OH 43130
Position: Retired. *Education:* BS/ME/OH State Univ; BS/IE/OH State Univ. *Born:* Jan 30, 1913 Lancaster, Ohio. s. Alverd S & Edith (Wolfe) McFee. m. Helen L Rauch, Oct 19, 1935; children - Mary Kay (Mrs. David Adams), Molly Ann (Mrs. John Mott). With Anchor Hocking Corp, Lancaster 1932-1978. Served with AUS 1941-46; ETO. Decorated Bronze Star. Reg PE in Ohio. Ohio soc PE's, Tau Beta Pi. Kiwanian. Office: W Fair Ave, Lancaster, Oh 43130, Retired Jan 1978, Now active in community groups. *Society Aff:* NSPE.

McFee, Raymond H
Home: 5163 Belmez, Laguna Hills, CA 92653
Position: Retired *Education:* PhD/Physics/MIT; SM/Physics/MIT; SB/Physics/MIT. *Born:* Mar 1, 1916. Raised Boston area. Dir of Res, Electronics Corp of Amer 1946-56. Dir of Res, Aerojet-Genl Corp 1956-64. Mgr Lunar & Planetary Sciences, Caltech JPL 1964-67. Dir of Matl Sciences, Douglas Advanced Res Lab 1967-70. Principal Engr/Scientist Mc Donnell Douglas Astronautics Co 1970-81. Specialist in appl optics & infrared tech. Fellow OSA; former Chmn OSA Aeronautics & Space Optics Tech Grp. Assoc Fellow AIAA; Mbr APS. Hobbies: choral music & furniture making. *Society Aff:* OSA, AIAA, APS.

McFee, Richard
Home: Rd 1 (Box 172), Union Springs, NY 13160
Position: Professor of Elec Engrg. *Employer:* Syracuse University. *Education:* PhD/EE/Univ of MI; MS/Physics/Syracuse Univ; BE/EE/Yale Univ. *Born:* Jan 24, 1925. Army Signal Corps 1944-46. Res Assoc at U of Mich Medical School 1949-51. Engr Electro-Mechanical Res, Inc 1951-52. Mbr of Tech Staff Bell Telephone Labs 1952-57. Prof of Elec Engrg at Syracuse U 1957- . Res in bio- medical engrg, circuit theory, electric power, electronics, physics. *Society Aff:* IEEE, ΣΞ, AAAS.

McFeron, Dean E
Home: 3634 42nd Ave, NE, Seattle, WA 98105
Position: Professor Emeritus of Mech Engrg. *Employer:* University of Washington. *Education:* PhD/Mech Engg/Univ of IL; MS/Mech Engg/Univ of CO; BS/Mech Engg/Univ of CO. *Born:* Dec 24, 1923. Native of Portland, Oregon. US Naval Reserve, active duty 1943-46, inactive 1946-70, CDR, USNR. Instructor, U of Colo 1946-48; Instructor- Assoc Prof, U of Ill 1948-58. Argonne Natl Lab 1957-58. Univ of Wash 1958- , Prof of Mech Engrg, Assoc Dean & Dir Engrg Res 1962-65, Acting Chmn Mech Engrg 1969-70. NSF Faculty Fellowship, Stanford Univ 1967-68. Soc Mfg Engrs, Region VII, Educ Achievement Award 1970. Co-author, Outstanding Tech Applications Paper Award, ASHRAE 1974. Natl Dir, 1971-76, Pres-elect 1977, Natl pres 1978, Sigma Xi. *Society Aff:* ASME, ASEE, ΣΞ.

McGarraugh, Jay B
Business: Columbia, MO 65201
Position: Prof Civ Engr. *Employer:* Univ of MO. *Education:* PhD/Struct Engr/Purdue Univ; MSCE/CE/Purdue Univ; BSCE/CE/Purdue Univ; AS/- /Vincennes Univ. *Born:* 6/18/40. Native of Monroe City, IN. Has taught Civ Engg at the Univ of MO-Columbia since 1967 specializing in structural engg. Has been recognized by both students and faculty for numerous teaching awards. Is active in Chi-Epsilon and ASCE student societies. Conducts res in reinforced concrete area. Enjoys golf and all spectator sports. *Society Aff:* ASCE, ASEE.

McGarvey, Billie J
Home: 902 Lynn Hill Court, Vienna, VA 22180
Position: Director of Facilities. *Employer:* Natl Aeros & Space Agency. *Education:* MS/Civil Engr/Univ of CO; BS/Civil Engr/Univ of CO. *Born:* 8/11/23. Native of Fort Worth TX. USAF 1943-78. Maj Gen; Planning, programming, design, construction, oper & maintenance & financial mgmt of Air Force facilities & supporting utilities. Sr mgmt jobs at every level in civil engrg from base engr to Dir of world-wide Air Force Civil Engineering. Served in Scotland, England, Japan, Pacific & numerous bases & locations in the US. Dir of Civil Engrg for US Air Force 1974-75. Natl VP Soc of Amer Military Engrs 1975. Resp for planning, programming, design and construction of major research and test facilities and launch and recovery facilities for the Natl Space Transportation Sys, the Space Station and Aeronautics facilities, 1978 to present.

McGaughy, John B
Home: 5905 Studeley Ave, Norfolk, VA 23508
Position: Chrmn of the Bd *Employer:* MMM Design Group *Education:* BS/CE/Duke Univ *Born:* 11/5/14 Norfolk, VA. s John Bell and Frances Vivian Coleman. m. Charlotte Edna Schwartz July 20, 1940 deceased Dec 3, 1978. Married Page Cook Sept 26, 1981. One son John Bell. Various positions with US Govt 1938-44. Partner in Lublin, McGaughy & Assocs, Architects & Consulting Engrs 1945-65 Pres McGaughy, Marshall & McMillan 1965-81. Chrmn of Bd, MMM Design Group 1981-date. Offices Norfold, VA; WA, DC; Frankfurt, Germany; Dublin, Ireland. Member VA Metropolitan Area Study Commission; Consultant, Office of coal Research; VA Engr of Year Award 1970; Fellow: ASCE. Am Road Builders Assoc (Dir); Prestressed Conc Inst; Nat Society of PE (VP 1957-59); VA SPE (Past Pres). Clubs: Univ Club, WA, DC; Harbor, Norfolk Yacht, Norfolk; Cedar Point, Suffolk; Princess Anne Country Club, Virginia Beach, Va. Phi Delta Theta, Theta Tau Fraternities. *Society Aff:* ASCE, NSPE, SAME, VSPE

McGeady, Leon J
Home: 404 Ctr St, Bethlehem, PA 18018
Position: Consultant: Physical, Mechanical Metallurgy. *Employer:* Self. *Education:* PhD/Metallurgical Engg/Lehigh Univ; MS/Metallurgical Engg/Lehigh Univ; BS/Metallurgical Engg/Lehigh Univ. *Born:* 7/5/21. Native of E PA. With Lafayette College since 1949. Performed res for Pressure Vessels Res Committee and allied agencies 1955 to 1972. Was consultant to US Naval Res Lab at Wash, DC 1964 to 1973. Specialties have been physical and mech metallurgy and engg mtls with emphasis on failure analysis. Western Elec Award of ASEE for outstanding teaching, Jennings Award of AWS for contribution to welding. Dir of Engg at Lafayette College 1975-86. *Society Aff:* AIME, ASM, AWS, ASEE.

McGee, Dean A
Business: P O Box 25861, Oklahoma City, OK 73125
Position: Hon Chrmn of Bd, Dir, and Chrmn of the Exec Cttee *Employer:* Kerr-McGee Corp. *Education:* BS/Mining Engg/Univ of KS. *Born:* 3/20/04. 1927-37 Phillips Petroleum Co Ch Geologist; 1937 - Kerlyn Oil/Kerr-McGee Corp; 1963-1983 Chmn & Ch Exec Officer; 1983-pres Hon Chrmn of Bd, Dir, & Chrmn of Exec Cttee; Kerr-McGee, engaged in prod of oil & gas, refining, marketing, offshore drilling, exploration for & mining of uranium, coal & other natural resources. Honorary Mbr of AIME, AAPG (also Public Serv, Human Needs, & Sidney Powers awards), Pi Epsilon Tau (Okla Univ School of Petr Engrg). Tulsa Univ's Colo of Engrg & Physical Scis Hall of Fame Award. Colorado School of Mines Honorary Doctor of Engrg Degree. Kans Univ Distin Alumni Award-Geology & Dist Engrg Serv Award. Dir of Amer Mining Congress 1973-79. Dir of Amer Petroleum Inst 1955-83 Hon life mbr, Gold Medal for Dist Achievement; Mbr of Natl Petroleum Council 1960- 83. Life Trustee of CA Inst of Tech. 25 Yr Club of the Petroleum Industry, hon life mbr. Amer Inst of Architects, hon life mbr. All Amer Wildcatters. TX Mid- Continent Oil & Gas Association-Distinguished Service Award. Amer Inst of Professional Geologists. Natl & OK Soc of PE. Natl OK City Geo Soc, hon life mbr. Soc of Petroleum Engrs of AIME, Dist Mbr and Pub Serv Award *Society Aff:* AAPG, AIME, AIPG, NSPE, SEPM, SPE, AGI, API, NPSE, OSPE

McGee, George W
Business: 54C/030, Tucson, AZ 85744
Position: Staff Engr. *Employer:* IBM Corp. *Born:* April 1925. Native of Colby Kansas. Grad Coll of Advanced Instruments; A D from B C C Fla. 25 yrs in Metrology & Measurement Sci incl Technician & Engr in Pri Stds Lab Martin-Marietta Aerosp Denver Colorado; Metrologist & Staff Engr IBM Corp Boulder Colorado & Boca Raton Fla; duties included estab & maintaining Metrology Lab & Staff Engr in Component Tech. Sr Mbr of ISA for 12 yrs & Dir of Metrology Div of ISA 2 yrs 1975-76. Mbr of ANSI C-100 Ctte. Mbr IEEE Fundamental Elec Standards Ctte of the G-IM. ISA Prog Chrmn, ISA/82. Presently Staff Engr in Procurement Mfg Engrg, P C Card Assembly Tech. *Society Aff:* ISA.

McGee, Thomas D
Business: Mtls Sci & Engr Dept, Iowa State University, Ames, IA 50011
Position: Prof. *Employer:* IA State Univ. *Education:* BSc/ME/IA State Univ; BSc/Ceramic Engg/IA State Univ; MSc/Ind Engg/IA State Univ; PhD/Cer Engg & Met/IA State Univ. *Born:* 6/9/25. Tripoli, IA. US Naval Reserve 1944-1954, Lt 1948-1954 Res Engr, A P Green Refractories Co Mexico, MO 1954-1956 Res Engg Supv. 1956-61 Asst Prof, 1961-65 Assoc Prof 1965- Prof, IA State Univ. Teaches grad & undergrad courses in Ceramic Engg and glass tech, Res interests: Glass, High-Tech Ceramics, BioCeramics. Major college & univ committee responsibilities. Fellow, Am Ceramic Soc, Mbr Phi Kappa Phi, Sigma Xi. ABET 1979-85; AAES 1985-88; President NICE 1986-87; General Secretary, Keramos Fraternity (Cer. E. honorary). Author: Principles & Methods of Temperature Measurement, Wiley 1987. 6 Patents, 50 Technical papers. *Society Aff:* ACerS, NICE, SGT, ASEE

McGee, William D
Home: 778 Raymond Drive, Idaho Falls, ID 83401
Position: Sys Engr. *Employer:* Argonne Natl Lab. *Education:* PhD/Str-Eng Mech/NC State Univ; MSCE/Str-Hydraulics/W VA Univ; BSCE/Civil Engg/W VA Univ. *Born:* 7/1/39. With Argonne Natl Lab since Feb 1975 engaged in civil-struct design, reactor core sys engg & proj mgmt as related to Argonne's Liquid Metal Fast Breeder Reactor prog; spent 2 & one-half yrs on Civil Engrg Faculty of Washington St Univ & 6 yrs of civil-struct design prior to joining Washington State. Recipient of ASCE's T Y Lin Award 1975 & PCI's Martin P Korn Award 1974 for res on prestressed concrete. Mbr ASCE, PCI, ACI & ACI Comm. 216, Fire Resistance and Fire Protection of Structures. *Society Aff:* ASCE, PCI, ACI.

McGill, Thomas C
Business: MS 116-81, Pasadena, CA 91125
Position: Fletcher Jones Prof of Applied Physics *Employer:* CA Inst of Tech. *Education:* PhD/EE/CA Inst of Tech; MS/EE/CA Inst of Tech; BS/EE/Lamar State College of Tech; BS/Math/Lamar State College of Tech. *Born:* 3/20/42. Pt Arthur, TX, Mar 20, 1942. Married Toby E Cone, Dec 27, 1966. Children: Angela Elizabeth, born 1974 Sarah Elizabeth, born 1980. Awarded a BS (Math), 1963, & BS (EE), 1964, by Lamar State College of Tech. Obtained MS (EE), 1965 & PhD (EE), 1969 from the CA Inst of Tech. NATO Postdoctoral Fellow, Univ of Bristol, Bristol, England 1969-70; NRC Postdoctoral Fellow, Princeton Univ, Princeton, NJ

McGill, Thomas C (Continued)
1970-71. Asst Prof of Applied Phys, CA Inst of Tech 1971-74, Asoc Prof 1974-77, Prof 1977-present. Fletcher Jones Prof of Applied Physics. Alfred P Sloan Fdn Fellow 1974-78. Mbr of the Am Physical Soc Study Group on Solar Photovoltaics 1977-78, & the Advanced Res Proj Agency's Mtl Res Council 1979-present; consultant to Hughes Aircraft Co, & Xerox. Mbr of IEEE, APS, IOP, AVS, AAAS, and Sigma Xi. *Society Aff:* IEEE, APS, AVS, AAAS, IOP.

McGillem, Clare D
Business: School of Elec Engrg, Purdue Univ, W Lafayette, IN 47907
Position: Prof of Elec Engrg. *Employer:* Purdue Univ. *Education:* PhD/EE/Purdue Univ; MSE/Eng Mech/Purdue Univ; BSEE/EE/Univ of MI. *Born:* Oct 9, 1923 Clinton Mich; m. 1947. Res Engr Diamond Chain Co Inc Ind 1947-51; Proj Engr & Div Head Radar US Naval Avionics Facility 1951-56; Engrg Supr AC Spark Plug Div G M Corp 1956-60; Prog Mgr GM Defense Res Lab 1960-63; Prof EE Purdue Univ 1963- ; Assoc Dean Engrg & Dir Engrg Exp Sta 1968-72. President, Technology Associates, Inc. 1977- . Pres, Vetronics, Inc 1984-7. Active in teaching & res in communication, radar & signal processing. Co-author of 3 textbooks on signal & sys theory and 1 on modern comm engrg. Fellow IEEE. Mbr Sigma Xi, Eta Kappa Nu & Tau Beta Pi. US Navy Meritorious Civilian Serv Award 1954. Pres IEEE Geosci Electronics Society 1976. IEEE Centennial Medal 1984. *Society Aff:* IEEE

McGillivray, Ross T
Business: 8430 N 40th St, Tampa, FL 33604
Position: VP *Employer:* Armac Engrs, Inc *Education:* MS/Soil Mech/MIT; BCE/CE/GA Inst of Tech. *Born:* 4/8/42 Registered engr, FL 17920, RI 3309. MIT Research Engr 1968-1970, in charge of soil mechanics lab. Staff Engr, Lambe Assocs of Concord, MA, 1970-1972, Chief Engr, Pittsburgh Testing Lab Tampa District 1972-1974, started Armac Engrs, Inc in 1974. Design mine waste retention dams to above 100 feet. Pioneered use of geophysical instruments for sink hole analyses by FL Engrs. Univ of South FL adjunct lecturer in soil mechanics, 1978-1981. Co-chair ASCE-FLA Dam Safety Committee 1975-1976. Chair ASCE-FLA Structures-Geotechnical Group, 1974-1976. Chair Evaluation Committee of North Side Mental Health Center 1977. *Society Aff:* ASCE, FES, AIME.

McGinnis, Charles I
Business: 9666 Olive Blvd, St Louis, MO 63132
Position: Sr VP & Gen Mgr *Employer:* Fruco Engrs, Inc *Education:* M/CE/TX A&M Univ; BS/CE/TX A&M Univ *Born:* 1/31/28 Kansas City, MO. Served as a Regular Officer in the US Army Corps of Engrs from 1949 until retirement in 1979 as a Major General and Dir of Civil Works. Served as Division Engr, Southwestern in Dallas, TX from 1974 to 1977. From 1972 to 1974 was assigned as VP, Panama Canal Co and Lieutenant Governor, Canal Zone Government. From 1971 to 1972 served as Dir of Engrg and Construction, Panama Canal Co. Served as District Engr in St Paul, MN from 1969 through 1971. Chaired NSPE Policy Task Force from 1979 to 1981. *Society Aff:* ASCE, SAME, NSPE

McGinty, Michael J
Business: 1242 E. 49th St, Cleveland, OH 44114
Position: Gen. Mgr. Machinability & Testing Ctr. *Employer:* Cleveland Twist Drill Co. *Education:* BS/Metallurgical Engr/Case Inst of Tech; MS/Management/Case Western Res Univ *Born:* 8/14/32. Native of Cleveland Ohio. Reg P E Ohio. Commissioned Officer US Navy 1956- 59. Joined Cleveland Twist Drill Co (Acme-Cleve Corp) 1954; assumed present pos 1985; past respon incl high speed steel alloy dev, brazing, carbide cutting tool manufacture, design & application of cutting tools & Quality Assurance. Presently Chmn Carbide Tool Standardization Engrg Ctte of M C T I. P Natl Chmn ASM Educ Council (1974-75) & of Prof Dev Ctte of ASM (1972-74). Active in civic & political areas. President, Bd of Trustees Parma Community Gen Hospital. Mbr Reamer Tech Cttee, Milling Cutter Tech Comm. of MCTI, and of ANSI/ASC B212 Committee. *Society Aff:* ASM, MCTI, SME

McGoldrick, James P
Business: 221 East Walnut, Pasadena, CA 91101
Position: Vice President. *Employer:* James M Montgomery Cons Engrs Inc. *Born:* Feb 19, 1934. Native of California. BSCE Univ of Santa Clara 1957. Deputy Mgr of James M Montgomery Cons Engrs Inc Internatl Dept; directs business dev & engrg projs in the Middle East & Europe; major projs involve water & wastewater master plan, feasibility studies & water sys designs for undev domestic & foreign areas incl the Yemen Arab Republic & Saudi Arabia; worked on long-term overseas assignments in Guyana, South Amer & Lahore Pakistan. Prof career has been with var Calif cons engrg firms, Orange County Flood Control District, City of Santa Clara & private practice in Pasadena Calif. Mbr of Amer Waterworks Assn & Amer Soc of Prof Engrs. Enjoys golf.

McGovern, Sharon A
Home: 25400 Avenida Escalera, Valencia, CA 91355-2802
Position: Mgr, Adv Composite Mat Res and Applications *Employer:* Northrop Corp. *Education:* PhD/Chemistry/IN Univ; MS/Chemistry/E TX State Univ; BS/Chemistry/E TX State Univ. *Born:* 12/14/42. Prior to becoming involved in the aerospace industry, worked in public school teaching, univ teaching, organicsynthetic and nuerochemical res. With Vought Corp 1974-80. Served as production NDT and Process control engr. Prog mgr for IR & D progs govt contracts and had mgt responsibility for quality assurance and mtl and process dev for all advanced composite progs. With Northrop Corp since April, 1980, has management responsibility for Advanced Composite Matls and process Research and of production support groups for F-18A and F-5g aircraft. Active in Soc for Advancement of Mtls and Process Engrg on local and natl level and in the Society of Women Engrs. *Society Aff:* SAMPE, SWE

McGrath, Daniel F
Business: P.O. Box 1101, Pampa, TX 79065
Position: Q C Mgr. *Employer:* Ingersoll-Rand Oilfield Products Co. *Education:* BS/Met/Penn State U *Born:* 12/27/33. Native of Pittsburgh, PA. Met Engr for the Heppenstoll Co, specializing in the mfg of steel forging from 1956-1978. Chief Met of Heppenstall Co from 1965 to 1978. Currently with Ingersoll Rand Oilfield Products Co. as Quality Control Mgr. Chrmn of the Met Committee of the Open Dir forging inst from 1974 to 1978. Current chrmn of the test methods section of the steel forging committee of ASTM. Reg PE of PA & TX. *Society Aff:* ASM, ASTM, ASME, AIME, SAE

McGrath, Philip I, Jr
Business: 3300 First Avenue North, Birmingham, AL 35202
Position: Mgr Tech Services Dept. *Employer:* US Pipe & Foundry Co. *Education:* BS/Mech Engr/Drexel Univ. *Born:* Nov 1922. Native of Burlington N J. Served as Combat Infantryman in ETO during WWII attaining rank of Major. With U S Pipe (a Jim Walter Co) 1948- ; assumed current pos of Mgr - Tech Servs Dept 1976 with respon over co's Qual Assurance, Process Engrg & Prod Design & Dev Sects; formerly Asst to Dir of Quality Assurance, Process Control Engr & Quality Control Engr in sev of the cos 7 foundries. Reg P E. Natl Dir ASQC 1975-76. ASQC Testimonial Award 1976. Mbr AWWA, ASTM, ASQC, AFS, DIRPRA trade assoc & ANSI A21 committee. *Society Aff:* ASQC, AWWA, AFS, ASTM, ANSI.

McGrath, William L
Home: 4005 Gulfshore Bd N 200, Naples, FL 33940
Position: Self-employed *Education:* BS/Arch Engg/Univ of MN. *Born:* 1974-76 Exec Asst to Chmn Carrier Corp; 1972-74 V Pres of Carrier Internatl Corp & Pres of Caricor Ltd. Served in Carrier Air Conditioning Co as Exec V Pres, Div Mgr, Ch Engr & earlier as Dir of Advanced Engrg & as mgr of numerous R/D progs 1943-71. Honeywell Inc 1934-43. Served as V Pres of ANSI, Dir 1964-72. Dir of ASA & was active in the formation of the new Standards Inst & the drafting of its charter & By-Laws. Fellow & P Natl Pres of ASHRAE. Fellow of the Inst of Heating & Ventilating Engrs (U K). Mbr ASME. Has been V Chmn of the Building Res Adv Bd

McGrath, William L (Continued)
(BRAB) of the Natl Acad of Scis & Engrg & Dir of that agency. Chmn of the Federal Const Council, an agency of BRAB & the Natl Acads. Served on the Building Adv Ctte of US Chamber of Commerce & the Building Codes & Standards Ctte of the Natl Conf of States. Holds more than 140 US pats & numerous foreign pats. *Society Aff:* ASME, CIBS, ASHRAE.

McGrath, William R
Business: 2773 Old Yorktown Rd, Yorktown, NY 10598
Position: Partner *Employer:* McGrath, O'Rourke & Assoc *Education:* BCE/Civil Engg/RPI; Cert/Traffic/Yale Univ. *Born:* Oct 1922 Stratford Conn. Certificate in Traffic Engrg Bureau of Highway Traffic Yale Univ 1951. BCE, RPI, 1950. Dir Dept of Traffic & Parking New Haven Conn 1955-1963; Transportation Coordinator Boston Redev Authority Boston Mass 1963-68; Commissioner Dept of Traffic & Parking Boston Mass 1968-71; Pres Inst of Transportation Engrs 1973. Mbr Inst of Trans Engrs, Amer Rd and Trans Builder Assoc, Transportation Res Bd & Mbr Chi Epsilon, Tau Beta Pi, SigmaXi, Recipient 1980 Burton W. Marsh award.. *Society Aff:* ITE, ARTBA, TRB.

McGraw, Woody W
Home: Route 1 Box CF 53, Crestwood, KY 40014
Position: Engineer. *Employer:* Dept of Housing & Urban Dev. *Born:* Aug 1942. Native of Louisville Kentucky. MSCE 1965 & BSCE 1964 from Univ of Kentucky. Foundation Engr for Ky Highway Dept 1965-69; designed earth/rock dams, levees & coffer dams for US Army Corps of Engrs 1969-72; with HUD since 1972; respon for HUD review & admin of all multi-family & subdiv const in Kentucky. Served as Pres, V Pres, Secy & Treasurer of Local ASCE, chmn of various ASCE cttes. Winner of D V Terrell Award & outstanding U K CE grad award. Author of sev papers given at Natl ASCE meetings covering soil design & problems.

McGregor, Leslie S
Home: 65 Westmount Rd, Apt. 512, Waterloo, N21 5G6 Ontario Canada
Position: Ret Ch of Motive Power & Car Equip. *Born:* Sept 1911 Edinburgh Scotland. B Eng from McGill Univ 1936, concurrent with rly machinist apprenticeship. Canadian Army R C E M E Corps Overseas 1940- 45 Lt Col. Various Mech Dept tech & admin positions with Canadian Natl Rly 1937- 72; Ch of MP & CE 1961-72; Railway Loco & Car Equip cons projs Nigeria, Cuba, Bangladesh 1973-75. Chmn of Rail Transportation Div ASME 1967. Mbr of Genl Ctte AAR Mech Div 1964-72. Pres Canadian Railway Club 1970. Fellow ASME 1970.

McGroddy, James C
Business: T J Watson Res Ctr, Yorktown Hgts, NY 10598
Position: Dir Semiconductor Sci & Tech. *Employer:* IBM. *Education:* PhD/Phys/Univ of MD; BS/Phys-Electronics/St Joseph's Coll. *Born:* 4/6/37. Native of NY. Thesis res on metal physics. Joined IBM in 1965, studied hot electrons, microwave instabilities in semiconductors. Visiting Prof, Technical Univ of Denmark 1970-71. Later studied non-linear optical effects in semiconductors. Mgr of Injection Laser Dev 1973-74. Presently responsible for dept encompassing basic, appl res and tech of semiconductors and super conducting devices. Areas of responsibility include advanced silicon FET and bipolar process tech and engg, GaAs tech, Josephson Junction Tech, Lithography, packaging, solar energy as well as fundamental physic, met and mtls sci of semiconductors. Author of more than 30 papers on solid state devices and physics. Nine issued patents. *Society Aff:* IEEE, APS, SID.

McGuire, John P
Business: 1 Bridge St, Irvington, NY 10533
Position: Pres. *Employer:* Consolidated Tech. *Education:* MS/Environmental Engg/MIT; BCE/Environmental Engg/Manhattan College. *Born:* 8/3/37. Native of NYC; served as pollution control engr with Union Carbide Corp, S Charleston, WV 1960-1963; assumed position as Asst Superintendent, Public Works Dept of Westchester County 1965-1966; founded & is currently Pres of Consolidated Tech, Irvington, NY specializing in all aspects of ind & municipal pollution control from lab analyses through design & operation of waste treatment plants; Assoc of Eder Associates, Locust Valley, NY, design engr; founded & owns Ilahe Farms, Inc which provides farmland recycling of waste treatment plant sludges. *Society Aff:* NSPE, WPCF.

McGuire, Michael J
Business: PO Box 54153, 1111 Sunset Blvd, Los Angeles, CA 90054
Position: Director of Water Quality *Employer:* The Metropolitan Water Dist of So. CA *Education:* BS/Civ Engr/Univ of PA; MS/Environ Engr/Drexel Univ; PhD/Environ Engr/Drexel Univ. *Born:* 6/29/47 Sanitary Engr with Phila Water Dept 1969-1973. Res Assoc, Drexel Univ 1973- 1977. Prin Engr, Brown and Caldwell Consltg Engrs, Pasadena, 1977-1979. Assumed position as Water Quality Engr for The Metropolitan Water Dist of So CA in 1979. Promoted to Water Quality Mgr 1984. Promoted to Director of Water Quality 1986. Am respon for managing feasibility studies and implementation planning of water quality control strategies in Metropolitan's sys serving 13 million customers. Completed joint editorship of two volume proceedings of activated carbon symposium (1980) and another proceedings of a subsequent activated carbon symposium pub in 1983. Numerous tech pubs and symposia presentations. AWWA Academic Achievement Award 1978. PE registered in CA, NJ and PA. Voted Diplomate of the AAEE, 1984; mbr, Bd of Trustees, AWWA Res Fdn, 1983-86. Trustee, Board of Governors, California-Nevada Section AWWA, 1984-87. *Society Aff:* AWWA, ACS, WPCF, ASCE, ΣΞ, IAWPRC

McGurr, Aloysius W, Jr
Business: 303 Naperville Rd, Wheaton, IL 60187
Position: President & Principal Engr *Employer:* A. McGurr, Ltd *Education:* BS/Civ Engg/Purdue Univ. *Born:* 4/6/40. Native of Akron, OH. Served with US Army Corps of Engrs 1962-65. Office, planning & cost engr with H K Ferguson Co, Cleveland, 1966-68; City Engr, City of Woodstock, IL, 1968-70; Municipal engr, city of Wheaton, IL, 1970-72, city engr, Wheaton, 1972-78; self-employed consulting engr 1978 to present. Reg PE in IL. *Society Aff:* ASCE, APWA-IME, AWWA, NSPE.

McHail, Rex R
Home: 702 Stone Road, Rochester, NY 14616
Position: Photogrammetrist. *Employer:* Bausch & Lomb Inc. *Born:* BS in Forestry from Pennsylvania State Univ. Civil Federal Serv; Forester US Forest Serv; Photogrammetrist US Geological Survey; Cartographer Navy Hydrographic Office; Proj Engr Navy Photographic Interpretation Center; Principal Engr, Prog & Prod Mgr Bausch & Lomb. 2 terms Bd of Dir Amer Soc of Photogrammetry - also Exec Ctte Council Photointerpretation Award Ctte Chmn, received Ford Bartlett Award, Dir of the Remote Sensing Div & received ASP Press Citation. Assoc Editor 'Photogrammetric Engrg & Remote Sensing' journal. Mbr Optical Soc, Amer Congress Surveying & Mapping, Soc of Amer Foresters & the Canadian Inst of Surveying. Listed in 2nd ed 'Engrs of Distinction' President of the American Society of Photogrammetry 1980.

McHargue, Carl J
Business: Metals, Ceramics Div, Oak Ridge-POB X, TN 37831-6118
Position: Program Mgr. *Employer:* Martin Marietta Energy Sys Inc *Education:* D Eng/Physical Met/Univ of KY; MS/Met Engg/Univ of KY; BS/Met Engg/Univ of KY. *Born:* Jan 30, 1926 Corbin Kentucky. Served in US Army. With Oak Ridge Natl Lab 1953- , as metallurgist, grp leader, prog mgr (Matls Sci Progs) & sect hd (Matls Sci Sect); in present pos programatic respon for 60 prof studying materials for advanced energy conversion sys. Over 100 pubs in deformation, phase transformations, radiation effects & ION implantation. Fellow of Amer Soc for Metals. Fellow of The Metallurical Society of Amer Inst of Mining, Metallurgical & Petroleum Engrs, Materials Res Soc, Sigma Xi also Prof Met Engr., Univ Tenn since 1962. 1 US patent. *Society Aff:* ASM, AIME, ΣΞ, MRS.

McHenry, Keith W, Jr
Business: Amoco Oil Company, PO Box 400, Naperville, IL 60566
Position: VP, Res & Dev *Employer:* Amoco Oil Co. *Education:* PhD/ChE/Princeton Univ; BS/ChE/Univ of IL *Born:* 4/6/28. Native of Champaign, IL & Milwaukee, WI. Entire ind career with Std Oil Co (IN) & subsidiaries in R&D. Joined co in 1955 & assumed present position, VP, R&D, Amoco Oil Co in 1975. Currently responsible for R&D on petrol products and processes & synthetic fuels. Mbr Ind Advisory Comm, Coll of Engrg, Univ of IL, Chgo; Pres-Elect, Ind Res Inst; Mbr Natl Acad of Engg; Chrm, Resource Devel. Comm. Chem Eng Univ of Illinois Urbana; US Army 1946-7. *Society Aff:* AIChE, ACS, AAAS, API

McHugh, Anthony J
Business: 108 Roger Adams C-3, 1209 W. California, Urbana, IL 61801
Position: Full Prof. *Employer:* Univ of IL. *Education:* PhD/ChE/Univ of DE; MS/ChE/Univ of DE; BChE/CE/Cleveland State Univ. *Born:* 10/30/43. Native of Cleveland, OH. Undergrad co-operative education work experience in paint, steel, and chemicals industry. Mbr of Tau Beta Pi Engg Honor Soc, Mbr Bd of Reviewers, Polymer Engg and Sci. Mbr of the faculty of Chem Engg, Lehigh Univ, 1971-1979. Currently Full Prof of Chem Engg, Univ of IL at Urbana. Res activities and publications in the fields of crystalline polymer sci and colloid sci, res specialist in oriented crystallization of polymers and particle chromatography. consultant for several chem cos. Hobbies: reading, classical music, tennis and handball. *Society Aff:* AIChE, ACS, APS, Soc Rheol, ASEE.

McHugh, Edward
Business: Div of Special Progs, Rm 126, Old Main, Potsdam, NY 13676
Position: Dean. *Employer:* Clarkson College. *Education:* BME/ME/Univ of RI. *Born:* 7/11/15. Providence, RI. Married Hilda Van Camp, two children. Residence Potsdam, NY. Positions, Clarkson College of Tech, Potsdam: Presently Dean, Special Progs and Dir, Summer Sessions, Prof, Mech Engg. Dir, Res, 1940-1958, Chrmn, Mech Engg, 1948-1958, Dean, Engg, 1953-1958, Dean, Faculty, 1958-1961. Directorships: St Lawrence Gas Co, St Lawrence County Economic Dev Council and Chamber of Commerce, Reg PE, NY, Treas, Assoc Colleges, St Lawrence Valley, Pres and Dir, Clarkson Dev Corp, Pres and Dir, Golden Knight Enterprise, Inc. *Society Aff:* ASME, PE, ASEE.

McHuron, Clark E
Business: 274 Mocking Bird Circle, Santa Rosa, CA 95405
Position: CEO *Employer:* Clark E McHuron Consulting Engrg Geologist *Education:* /Geol/Brown Univ; AB/Geol/Syracuse Univ *Born:* 9/13/18 Outstanding Alumnus Award North Syracuse High School, NY. Taught at Brown Univ. Army map service early World War II. Served on US Navy Destroyer as navigator 1944-46. Chief's office to project geologist to district geologist for US Bureau of reclamation 1946-53. Consultant to heavy construction, particularly on underground openings. Worked on all five continents on total of over 500 miles of tunnels. Expert witness on ground conditions. Member of various consulting bds and bidding teams for major heavy construction projects. Consultant and CEO since 1953 on probable impact of ground conditions on construction. Enjoy travel, people and golf. *Society Aff:* AEG, GSA, USCOLD, ISRM, AAA

McIntosh, Robert E, Jr
Business: Dept of Electrical & Computer Engg, University of Massachusetts, Amherst, MA 01003
Position: Prof *Employer:* Univ of MA. *Education:* PhD/EE/Univ of IA; SM/App Physics/Harvard Univ; BS/EE/Worchester Tech. *Born:* 1/19/40. Dr McIntosh was a microwave engg at Bell Telephone Labs from 1962-65 and joined the faculty of the Univ of MA in 1967 where he is now a prof in the Electrical and Computer Engg Dept. He is a Fellow of the IEEE and recipient of an IEEE Centennial Medal. He served the Antennas and Propagation Soc as International Symposiur Chmn, International Correspondent, Editor of the AP-S Newsletter, Assoc Ed of the Transactions and Sec/Treas, VP and Pres. He was also Sec/Treas and Pres of the Geoscience and Remote Sensing Society. He is now an Assoc Editor of *Radio Science.* Mbr of the Tech Activities Bd and the Tech Transfer Ctte of the IEEE. He is also a mbr of USNC Commissions B, C and H of URSI, the APS, Sigma Xi, TBP, PKP, EKN and has published extensively in the areas of wave propagation, plasmas, communications, microwaves and remote sensing. *Society Aff:* IEEE, APS, OSA.

McIntyre, Kenneth E
Business: 245 Summer St, Boston, MA 02107
Position: Mgr of Water Res Proj *Employer:* Stone & Webster Engrg Corp *Education:* MSCE/Civil Engg/Harvard Univ; MSBA/Bus Admin/Geo Washington Univ; BS/-/US Military Acad. *Born:* March 1926. Native of Randolph VT. MSCE from Harvard Univ, MSBA from George Washington Univ, BS from USMA. Retired as Brig Gen after serving 30 yrs in US Army Corps of Engrs. Major assignments incl: Deputy District Engr Louisville KY, District Engr Huntington W VA, Deputy Dir of Civil Works Washington D C and Div Engr, Atlanta, GA. Commanded Engr const troop units in Okinawa (air fields), Korea (air fields), Thailand (highways) & Vietnam (const & combat engr support for northern region of South Vietnam). From 1976-1979 was respon for design & const of military const for Army & Air Force installations & planning, design, const & oper of Corps of Engrs Civil Works projs (incl Tennessee-Tombigbee Waterway) throughout Southeastern US. Presently proj mgr for engrs and const serv for hydroelectric and other water resources proj both in US and overseas. *Society Aff:* ASCE, SAME.

McIsaac, Paul R
Home: 107 Forest Home Dr, Ithaca, NY 14850
Position: Prof. *Employer:* Cornell Univ. *Education:* PhD/EE/Univ of MI; MSE/EE/Univ of MI; BEE/EE/Cornell Univ. *Born:* 4/20/26. Served with the US Navy, 1944-46. Studied at the Univ of Leeds, England, 1951-52 as a Rotary Fdn Fellow. Employed as Res Engr in the Microwave Tube Div of the Sperry Gyroscope Co, Great Neck, NY, 1954-59. Joined Cornell Univ in 1959; currently Prof of Elec Engg. Assoc Dean of Engg 1975-1980 Visiting Prof at Chalmers Univ of Tech, Gothenburg, Sweden, 1965-66. Res interests are in electromagnetic theory & microwave theory. . *Society Aff:* AAAS, ASEE, IEEE, ΣΞ.

McKannan, Eugene C
Home: 2512 Vista Dr, Huntsville, AL 35803
Position: Deputy Laboratory Director; materials & Processing Laboratory *Employer:* NASA. *Education:* BS/Physics/West Chester State Coll; MS/Engg/Univ Ala in Htvl. *Born:* April 16, 1928 Philadelphia, Pa. Awarded NASA Exceptional Scientific Achievement Medal for contribs to Apollo-Saturn Prog; US Army Guided Missile Center, spectroscopy, 1950-52; E I duPont Experimental Station, rheology, 1952-61; NASA Marshall Space Flight Center, Alabama, space lubrication, vacuum- radiation effects, met, space processing of materials, 1961- . Principal Investigator, Apollo 14 & Skylab space experiments; Mbr Amer Inst of Physics, Rheology Soc; Amer Soc for Metals; ANSI, Ctte B31-10; Author of 24 journal articles & co-author of 4 books related to materials. *Society Aff:* AIP, ASM, ANSI.

McKean, A Laird
Home: 1 Lincoln Ave, Ardsley, NY 10502
Position: Assoc Director - R&D. *Employer:* Conslt - Elec Cable Sys *Education:* MEE/Elect Engrg/Polytech Inst of Bklyn; BEE/Elect Engrg/Polytech Inst of Bklyn *Born:* 4/25/16. Res engr with Phelps Dodge Cable & Wire Co since 1936, throughout entire career. Major accommplishments primarily in power cable field, except during the 1950's when focusing on communication cable area. Significant contribs related to underground EHV transmission cable devs, & extruded dielectric res. Presently Elec Cable Conslt. Mbr Industry Tech Cttes; IEEE, ASTM, CIGRE; respon for developing several natl standards. Mbr Eta Kappa Nu, Fellow IEEE. Contributed numerous papers & articles. Patentee in field. Enjoy bridge & choral singing. *Society Aff:* IEEE, ASTM, CIGRE

McKee, Larry A
Business: 1450 Seabord Dr, Baton Rouge, LA 70810
Position: Pres. *Employer:* Professional Engg Consultants Corp. *Education:* BS/Civil Engg/LA State Univ. *Born:* 9/3/32. in rural St Tammany parish LA; reared and educated in E Baton Rouge Parish, LA, earning BS degree in Civil Engg from LA State Univ. Early engg experience gained as design engr for consulting firms before joining Owen & White, Inc of Baton Rouge where ultimately promoted to VP & Chief Design Engr. Acquired total ownership of PE Consultants (PEC) in May 1970 which now serves Fed, State, Municipal and private clients throughout the South & MO. Reg PE in AL; MS; LA; TX; MO; & MN. Married to former B J Thaxton; 2 children - Tracey & Kevin. *Society Aff:* ASCE, SAME, NSPE, LES, WPCF, AWWA, ACEC, CEC/L.

McKelvey, James M
Home: 9861 Copper Hill Rd, St Louis, MO 63124
Position: Dean of Engineering. *Employer:* Washington University. *Education:* PhD/Chem Engg/Washington University; MS/Chem Engg/Washington University; BS/Chem Engg/Rolla (Univ of MO) *Born:* Aug 22, 1925 St Louis, Missouri, son of James Grey & Muriel (Morgan) McKelvey. BS Univ of Missouri-Rolla (1945); MS (1947) & PhD (1950) from Washington Univ, St Louis. Res engrg for E I DuPont de Nemours & Co, Inc 1950- 54; Asst Prof of Chem Engrg at Johns Hopkins Univ 1954-57; mbr of faculty of Washington Univ, St Louis, 1957- . Assumed current respon as Dean of the School of Engrg & Appl Sci of Washington Univ, St Louis 1964. Recipient, Distinguished Educator Award, Soc. of Plastics Engrs., 1979. *Society Aff:* AIChE, ACS, SPE, ASEE, MSPE, ACS, AAUP

McKenna, John D
Business: ETS Inc, Suite C-103, 3140 Chaparral Dr, Roanoke, VA 24018
Position: Pres. *Employer:* ETS, Inc *Education:* MBA/-/Rider College; MS/ChE/NJIT; BS/ChE/Manhattan College. *Born:* 4/1/40. Native of NY City. Currently Pres of ETS, Inc., a pollution control consulting firm specializing in fabric filtration emission control and sulfur dioxide removal. Formerly served as Pres of Enviro-Sys & Res, pollution control manufacturer of fabric filter dust collectors, mechanical collectors and scrubbers. Prior experience includes Proj Dir for the Cottrell Environmental Sys Div of Res Cottrell; supervised activities of sr professionals, was government and industry consultant for comprehensive pollution control problems and directed dev activities in SO2 removal. Other positions included Proj Leader, Princeton Chemical Res; and Technical Asst to Pres, Eldib Engg & Res, Inc. Author tech papers and chapters in 3 books. *Society Aff:* APCA, AIChE.

McKenzie, Bruce A
Home: 917 S 14th St, Lafayette, IN 47905
Position: Extension Agr Engr/Prof Agr Engrg. *Employer:* Purdue University, Agr Engrg Dept. *Education:* BS/Agri Engg/Purdue; MS/Agri Engg/MI State Univ. *Born:* Jan 1927. US Navy WWII. Purdue U since 1950. Primarily respon for extension programs in grain drying, storage, handling; livestock production mechanization; & materials handling/process design. Currently 30% resident teaching, 70% Extension, plus Proj Ldr, Agri Engrg Exten. Extensive exper U S, Brazil, Argentina, Venezuela. 1976 USDA Superior Service Award; 1975 Hovde Award honoring contrib to Indiana Rural People; 1973 Natl Metal Bldg Award ASAE; 1968 Outstanding Teacher Award, Agri Engrg; 1968 Career Recognition Award as Outstanding Extension Specialist. Hobbies: billiards & wood working. Wife Irene, son Terry, daughter Michele. *Society Aff:* ASAE.

McKetta, John J, Jr
Business: Chemical Engrg Dept, Univ. of Texas, Austin, TX 78712
Position: Joe C. Walter Chair in Chem Engrg. *Employer:* Univ of Texas at Austin. *Education:* BS/ChE/Tri-State College; BSE/ChE/Univ of MI; MS/ChE/Univ of MI; PhD/ChE/Univ of MI. *Born:* Oct 17, 1915 Wyano, Pa; son of John J & Mary (Gelet) McKetta. Honorary D Eng, Tri-State Coll 1965; ScD U of Toledo 1973; Drexel U. 1974; m. Helen Elisabeth Smith Oct 17, 1943; children: Charles William, John J III, Robert Andrew, Mary Anne. P Pres AIChE; Mbr NAE. Reg PE Texas & Mich. Served as Prof & Chmn Chem Engrg Dept, Dean of Engrg, U of Texas at Austin & Exec Vice Chancellor of UT system. Presently has the Joe C. Walter Chair in Chem Engrg. Author & co-author of over 495 tech articles, 47 books & the most recent 45 volume Encyclopedia of Chem Processing & Design. Recipient of natl awards: for contribs to natl issues concerning energy, environment, & economics, 'Triple E Award', 1976; ASEE Lamme Award for 'Outstanding Engrg Educator in the USA for 1976'; & the 1975 'Service to Soc Award' of the AIChE. *Society Aff:* AIChE, ASEE, NSPE, NAE.

McKiernan, John W
Home: 1709 Cardenas, Albuquerque, NM 87110
Position: Division Supervisor. *Employer:* Sandia Laboratories; Retired *Education:* MSME/-/IA State; BSME/-/MO Univ. *Born:* Jan 12, 1923 Hannibal, Mo. Served as pilot in 15th AF, received Air Medals & DFC. Employed by E I DuPont 1947-48, Iowa State 1948-51, Sandia Labs 1951-85. Served as officer in local Sect & Region. VP of ASME 1970-74, 1978-80 & mbr of Exec Ctte 1970, 1972 & 1978. *Society Aff:* ASME, ΠΤΣ, ΤΒΠ, PROFESSIONAL ENGINEER.

McKillop, Allan A
Business: Dept of Mechanical Engrg, Davis, CA 95616
Position: Prof *Employer:* Regents of Univ of CA *Education:* PhD/ME/Univ of CA-Berkeley; ME/ME/MIT; MS/Agr Engrg/KS State Univ; BS/Agr Engrg/Univ of CA-Berkeley *Born:* 7/24/25 Native of Berkeley, CA. Taught and did research in Food Engrg, 1951-1962. Founding member Coll of Engrg Univ of CA, Davis, 1962. Dept Chrmn, 1974-1978. Teach and do research in heat transfer and fluid mechanics. *Society Aff:* ASME

McKim, Herbert P, Jr
Home: 243 North Front Street, Wilmington, NC 28401
Position: Pres *Employer:* McKim & Creed Engrs PA *Education:* M/BA/UNC-Wilmington; M/CE/NC State Univ; BS/CE/NC State Univ *Born:* 8/21/51 Born and raised in Wilmington, NC. Proud son of a community leader and prominent architect. Worked for small consulting structural engrg firms in Raleigh, NC and Greensboro, NC prior to joining with a coll classmate, Michael W. Creed, to form the business of McKim & Creed Engrs, PA in 1978. Active in professional societies, technical societies, and community civic clubs. Enjoy all sports activities, especially tennis, golf, and racketball. *Society Aff:* NSPE, ASCE, SAME, ACEC

McKim, Paul A
Business: P.O. Box 2521, Houston, TX 77252
Position: Sr VP *Employer:* Texas Eastern Corp *Education:* PhD/Chem Engr/LA State Univ; MS/Chem Engr/LA State Univ; BS/Chem Engr/LA State Univ *Born:* 2/1/23. Milford, Conn., Since 1978 Sr. Vice Pres of TX Entern Corp, 1968-1970 VP of Atlantic Richfield Co & 1974-78. Pres of Arco Polymers, Inc, a subsidiary of Atlantic Richifeld. 1949-62 with Ethyl Corp in Baton Rouge, La - last position Asst Genl Mgr of R&D. Served as Ltjg in US Navy 1944-46. Attended advanced mgmt prog, Harvard Business School, & exec prog at Aspen Inst of Humanistic Studies. *Society Aff:* AIChE, SCI, API

McKinlay, Alfred H
Home: 130 McDougall Rd, Pattersonville, NY 12137
Position: Consultant (Self-employed) *Education:* BA/Indust Mgmt/Union College. *Born:* Oct 16, 1929 Amsterdam, NY. Packaging engr with Alco Prods, Schenectady 1951-54. Sr packaging engr with GE's Central Packaging Lab 1954-68. Conslt in Materials Handling & Plant Layout Operation 1968-75. Appointed Conslt in Distribution Engrg 1975 serving all GE plants & other industries as Cons on physical distribution problems. Opened consulting practice June 1981 specializing in packaging, handling, warehousing, plant layout and facility planning. Pres SPHE 1978-79. Chrmn SPHE 1980-81. Chrmn ASTM Comm D-10. Conf Chmn for PI, AMA, SPHE. Lecturer at Univ of WI, Geo Wash Univ, Univ of Pitts. Prof engr (CA)-mfg

McKinlay, Alfred H (Continued)
eng; Cert Professional-Packaging/Handling (SPHE); Cert Mfg Eng-Mfg Mgmt (SME). *Society Aff:* SPHE, PI/INT'L, ASTM, IIE.

McKinley, Donald W R
Home: 1889 Fairmeadow Cres, Ottawa, Ont Canada K1H 7B8
Position: Vice President (retired). *Employer:* National Research Council. *Education:* PhD/Physics/Univ of Toronto; MA/Physics/Univ of Toronto; BA/Physics/Univ of Toronto. *Born:* Sept 22, 1912. Mbr Assn of PE's of Ontario. Fellow of IEEE, Engrg Inst of Canada, Royal Soc of Canada. Joined Natl Res Council of Canada as res officer in 1938. Dir of Div of Radio & Electrical Engrg of NRC 1962-68. VP of NRC 1968- 74. Author of 55 sci papers on meteors & on electronics; 1 book: 'Meteor Sci & Engrg' 309 p., McGraw-Hill 1961. *Society Aff:* EIC, IEEE, RSC, APS, AAS.

McKinley, Robert W
Home: RFD 1, Box 742, Hancock, NH 03449
Position: Management Consultant *Born:* 08/11/18 in Lowell, MA. Studies Elec Engg, Architecture, Accountingm Mktg at MIT, NY Univ, George Washington Soc & PPG Industries. Specialized in application engg: curtain walls. Lighting, Glass & glazing. Bldg Costs, Value Engg, Energy Conservation, Std, Cert, Test Methods, Bldg Codes, Solar Energy Utilization, Construction mgmt, Product Publications, Govmt Relations, Product Liability; Founder Safety Glazing & Insulating Glass Cert Council; Fellow ASTM, IES, Director ASTM, Man of the Year ASTM Sealants Director- Building Research Inst. Quarter Century Award NAS-NRC-BRAB.

McKinnell, William P, Jr
Business: P O Box 269, Littleton, CO 80160
Position: Research Director. *Employer:* Marathon Oil Company. *Education:* PhD/Met Eng/OH State U; MS/Met Eng/OH State U; BS/Met Eng/Missouri School of Mines; BS/Chem Eng/Univ of MN *Born:* Dec 1924. Native of Springfield, Mo. Metallurgist with Chrysler Corp, Detroit 1947-50. Asst Prof Va Poly Inst, Blacksburg, Va 1950-51. With Marathon Oil Co since 1956, first as Res Engr, Dept Mgr 1960-67, Mgr Commercial Dev Div 1967-73, Res Dir since Jan 1973. In charge of Petroleum R&D staff of 300. Fellow AIChE. Mbr AIME, ASM, NACE, Tau Beta Pi, Sigma Xi. Distinguished Engrg Alumnus, Ohio State U 1974. Trustee, Swedish Hospital. Dir Centennial State Bank. Chair, Engg Dev Council Univ of CO. Councilman, City of Littleton 1959-63. Married, four children. *Society Aff:* AIChE, AIME, ASM, NACE

McKinney, Chester M
Business: P O Box 8029, Austin, TX 78713
Position: Director, Appl Research Labs. (Retired) *Employer:* Univ of Texas at Austin. *Education:* PhD/Physics/Univ of TX; MA/Physics/Univ of TX; BS/Physics/East TX State. *Born:* Jan 29, 1920 Cooper, Tx. Capt in US Army Air Force (radar officer) 1942- 46. Taught Texas Tech Univ 1950-53. Applied Res Labs, Univ of Texas at Austin1946-50, 1953- . Dir 1965-1980. Published over 45 papers & 50 tech reports in underwater sound, sonar systems, microwave devices, & dielectric waveguides. Sr Life Mbr IEEE. Fellow Acoustical Soc of Amer. V.P. Ac. Soc Amer. 1984-85; Pres. Ac. Soc Amer. 1987-88; David Bushnell Award ADPA 1985. US Navy Distinguished Public Service Award 1980; Liaison Scientist, US. Office of Naval Research, London, 1983-84. Honary Fellow British Inst of Acoustics. Exec Council Mbr, Acoustical Soc of Amer 1974-77. Mine Advisory Ctte, NRC, 1959- 72. Ships Board, Naval Res Advisory Ctte 1975-78. US Underwater Sound Advisory Ctte (3 terms), Naval Studies Bd, NRC:NAS, 1979- . *Society Aff:* IEEE, ASA, BIA, AAAS, ΣΞ.

McKinney, Leon E,
Home: 841 Green Lantern, Ballwin, MO 63011
Position: President *Employer:* McKinney Associates *Education:* MS/Nuclear Engrg/US Air Force Inst of Tech; BS/Military & Civil Engr/US Military Acad; /Industrial Mobilization of the US/Industrial Coll of the Armed Forces *Born:* 9/10/30 in Kingsport, TN. Served 24 years in Army Corps of Engrs in US, Germany, Greenland and Korea. Operated two land-based nuclear power plants. Deputy District Engr in Wilmington, NC. Commanded Construction Battalion in Korea. Assistant Dir of Civil Works in WA. District Engr, St Louis, responsible for 1,000 personnel, engrs of numerous disciplines, construction and design work totalling over $265 million. Married to Linda Johnson (BA Cornell Univ, MAT, Webster U), son Leon, Jr (Purdue Univ, BS & MS Aerospace Engr) McDonnell Douglas, CA. *Society Aff:* SAME, ASCE

McKinney, Ross E
Home: 2617 Oxford Rd, Lawrence, KS 66044
Position: Prof. *Employer:* Univ of KS. *Education:* ScD/Sanitary Engr/MA Inst of Tech; SM/Sanitary Engr/MA Inst of Tech; BSCE/Civil Engg/Southern Methodist Univ; BA/Mathematics/Southern Methodist Univ. *Born:* 8/2/26. Served with USNR, 1943-46. Partner, Hagan & McKinney, Surveyors, Dallas, TX, 1948. Acting Head, Sanitary Sci Dept, Southwest Fdn for Res & Education, San Antonio, TX, 1951-53. Taught at MIT, 1953-60. Joined the faculty at Univ of KS in 1960. Chrmn of Civil Engg Dept, 1963-66. Distinguished Prof since 1966. Consultant to over 100 organizations. WPCF: Harrison P Eddy Medal, 1962; ASCE: Rudolph Hering Medal, 1964; US Presidential Citation, 1971; Nat'l Academy of Engg, 1977, Region VII EPA Environm Quality Award, 1978; Thomas R Camp Medal, WPCF, 1982, Advisory Prof, Tongji Univ, Shanghai, PRC, 1986. *Society Aff:* ASCE, WPCF, AWWA, AAAS, APWA, AAEE, ASEE, ACS, SAM, IAWPRC, AAUP.

McKittrick, David P
Business: Suite 2200, 1700 N Moore St, Arlington, VA 22209
Position: President *Employer:* The Reinforced Earth Company *Education:* M.S./C.E./Purdue Univ; B.E./CE/McGill Univ *Born:* 4/29/38 He is a Prof Engr and a Mbr of the ASCE Mr. McKittrick is a civil engrg graduate of McGill Univ in Montreal, Canada, holds a masters degree in civil engrg from Purdue Univ, and has completed advanced managerial studies at the Harvard Business School. Prior to joining The Reinforced Earth Co in 1973, he served as a senior soils engineer with Stone & Webster in Boston, Massachusetts and as VPres of Engrg with American Drilling in Providence, Rhode Island. *Society Aff:* ASCE, ISRM, ISSMFE, ICLM, IHE

McKnight, Larry E
Business: 1805 E Carnegie Ave, Santa Ana, CA 92705
Position: Consultant. *Employer:* METTEK. *Education:* BS/Met Engg/Univ of WA. *Born:* 4/6/37. Resident of S CA, partner in firm of Mettek, Mtls Engg & Tech Labs, and consultant to several aerospace, petrochem, & utility firms. Reg Met & corrosion engr, mbr of Tau Beta Engg Honor Soc, "Fellow" of the Am Soc of Metals, & an active participant in technical socs & confs. Previously, res engr, the Boeing Co, conducting studies on hydrogen detection. Res & field engr, Chicago Vitreous Corp, conducting studies of ceramic/metal products. Chief Met Engr, Testing Engrs, Inc, responsible for failure analysis, and lab services. Enjoys boating, snow skiing, golf. *Society Aff:* ASM, AWS, AISC.

McKnight, Samuel William
Home: 341 Elm Ave, Quebec, Canada H9W 5W9
Position: Manager Application & Prod Planning. *Employer:* Westinghouse Canada Limited. *Education:* BSC/Elec/Univ of New Brunswick; BSC/P Eng/Ontario. *Born:* May 19, 1919. Took educ in the Maritimes. Served one year in Canadian Army in RCEME, discharged as Lt. Spent 38 years in Lamp & Lighting Div of present co in Sales, Application & Prod Design. Mgr of Engrg for 3 years. Spent many years in dissemination of lighting knowledge instructing Sect & Chap mbrs in fundamentals and progress of IES in Canada & USA. Have written & presented many tech papers to the Electrical Assn on indoor & outdoor lighting equip. Feature designs of airport equip to meet MOT & FAA specifications as well as tunnel lighting were recent accomplishments. Natl award from engrg soc's: Fellow 1971- Illuminating Engrg Soc; TC4.4 1972- Comm Internationale Eclairage. Natl Engrg Soc's: 1963- Canadian Elec

McKnight, Samuel William (Continued)
Manufacturer's Assn; 1962- Canadian Standards Assn; 1951- Illuminating Engrg Soc; 1943- PE's of Ontario. Active in golf & curling. *Society Aff:* IES, FIES, CIE.

McKnight, Val B
Home: 6831 N Michele Ln, Peoria, IL 61614
Position: Utilities Consultant. *Employer:* Caterpillar Inc *Education:* MEE/Power/Tech Univ of Budapest. *Born:* 9/14/26. Native of Budapest, Hungary. Naturalized in 1971. Taught Inst of Tech, Budapest, 1950-54; Mgr of Engg-Ministry of Construction Industry, 1954-56; Supervising Engr-Canadian British Aluminum Co Quebec Canada 1957-66; Utilities Consultant-Caterpillar Inc, 1966-; Innovator of Yr 1955. Reg PE and Certified Plant Engr. Worldwide responsibilities for plant engg operations; telecommunications and utilities. VP of Mfg Radio Council, Member of Bd of Dir and Pres of AIPE. "Plant Engr of Yr" 1977/78. Listed Who's Who in the Midwest, served in Energy Resources and Conservation Task Force of NAM and Chamber of Commerce and Illinois Manufacturers Association (IMA) (IL) to develop policy issues affecting industry. Served on the Government and Public Affairs Committee of the Illinois Society of Prof Engrs (ISPE). Served on IEEE's Liquid Insulation Committee - EPA Work-Group to develop rules and regulations to implement the Toxic Substances Control Act (PL94-469) Member of AIPE's Editorial Board. Hobbies: photography and travel. *Society Aff:* AIPE, IEEE, AEE, NSPE, IES, EIC

McLain, Milton E
Business: 2734 S. Cobb Industrial Blvd, Smyrna, GA 30080
Position: Nuclear Engr Specialist *Employer:* Applied Physical Tech Inc *Education:* PhD/Nuclear Engg/GA Inst of Tech; MS/Chemistry/Univ of ID; BA/Chemistry/Emory Univ. *Born:* 2/20/34. Born in Marietta, GA Feb 20, 1934. Father: Milton E McLain; mother Edna Elizabeth (Strauss) McLain. Grew up in Buford, GA and grad in 1951 from Buford HS (class valedictorian). Entered Emory Univ that yr, graduating in 1955 (chemistry maj). Have worked more than 20 yrs in the nuclear tech ind. Consultant to nuclear utilities and hospitals in matters of nuclear engg and health physics. *Society Aff:* ANS, HPS, ΣΞ.

McLain, William R
Home: 2814 Kenway Rd, Nashville, TN 37215
Position: Chairman & Chief Executive Officer. *Employer:* Kusan, Inc. *Education:* -/-/GA Tech. *Born:* Dec 7, 1916 Huntsville, Ala. Georgia Tech Chem Engrg 1941. 1941-42 Tennessee Eastman Corp - Chem Engrg; 1942-44 Chem Warfare Service - Huntsville Arsenal, Chief Civilian Engr, Smoke-Incendiary Div; 1944-46 US Naval Service (Ltjg); 1946- , Kusan, Inc - Founding Pres to 1969, Chmn & Chief Exec Officer 1969- . Full Mbr: ASME, AIChE, Amer Chem Soc; Soc Plastics Indus - Past Dir; Past VP, Dir, Young Pres' Organization; P Pres, Toy Manufacturers of Amer; Past Dir, Natl Assn of Manufacturers; Newcomen Soc; Soc of Plastics Engrs; Dir, First Amer Natl Bank & Steel Heddle Mfg Co. *Society Aff:* ASMe, AIChE, ACS, SPI, SPE.

McLaren, Malcolm G
Home: R D 3, Milford, NJ 08848
Position: Prof & Chmn - Dept of Ceramics. *Employer:* Rutgers, The State Univ of N J. *Education:* PhD/Ceramic Eng/Rutgers Univ; MS/Ceramic Eng/Rutgers Univ; BS/Ceramic Eng/Rutgers Univ. *Born:* July 1928. Native of Wilkinsburg, Pa. VP & Tech Dir of Paper Maker's Importing Co, Easton Pa 1954-62. Served in US Air Force 1955-57 as 1st Lt in the Aeronautical Res Labs WPAFB. Dept of Ceramics, Rugers U 1962- . Assumed present duties of Prof & Chmn 1969. Pres of Ceramic Educ Council 1971-72. VP of Amer Ceramic Soc 1974-75. Trustee 1974- . Secy-Treas Ceramic Assn of NJ 1969- . Cons to indus; Mbr of Expert Panel on Food Additives & Contaminants, World Health Organization, Geneva. Pres, Amer Ceramic Soc 1979-80; US Dept of Energy. Chrmn of Panel on long range res for glass, cements & ceramics 1978. Editorial Bd, Ceramurgia Internatl, Italy 1976-present. Recipient "Primio Assoc Brasileira de Ceramica" Award. Rio, Brazil 1978 60 Publications. Honorary Life Mbr Brazilian Ceramic Soc. 1979 Sao Paulo, Brazil. Bleininger Award, Pitts Sec, A. Cer. S. 1979, Founder's Day Award, Phila Sec. A. Cer. S, 1982, McMahon Lecture, Alfred Univ 1982, Food and Drug Adm (FDA) Commissioner's Special Citation for Indiv, 1984, Natl Sci Foundation, Prin Investigator for Ctr for Ceramic Res, 1982-87, Mbr Bd. of Dirs, Certech and Ceramic Magnetics. Distinguished Life Member Amer Ceramic Soc 1987. *Society Aff:* A. CER S., ΤΒΠ, ΣΞ, NICE, ASTM.

McLaren, Robert W
Business: Dept of Elec Engg, Columbia, MO 65211
Position: Prof. *Employer:* Univ of MO. *Education:* PhD/Elec Engr/Purdue Univ; MS/Elec Engr/Univ of IL; BS/Elec Engr/Univ of IL. *Born:* 8/31/36. Has been associated with the Univ of MO for 21 yrs. Has been involved in various res and teaching activities. Res areas have included medical image data processing and recognition, learning control systems, and biological system modeling. Current res activities include image processing and artificial intelligence applications. Has had over 50 publications. Teaching activities include automatic control and artificial intelligence. Summer employment has included employment at NASA, Emerson Elec, Rome Air Dev Ctr, Gen Elec and McDonnell-Douglas. *Society Aff:* IEEE, ISA, ΣΞ, ΤΒΠ, MAA.

McLaren, T Arthur
Business: 1870 Harbour Rd, North Vancouver, British Columbia Canada V7H 1A1
Position: President & Managing Director. *Employer:* Allied Shipbuilders Ltd. *Education:* BASc/Mech Engg/Univ of BC. *Born:* Sept 2, 1919 Montrose, Scotland. Employed by West Coast Shipbuilders Ltd 1941-45 as Tech Asst, promoted to Shipyard Mgr 1946-48. Founded Allied Shipbuilders Ltd 1949 & President & Managing Dir of the firm & associated co's to date. Reg PE in British Columbia 1947- , obtaining Letson Memorial Prize for best thesis submitted by PE registrants. Fellow Royal Inst of Naval Architects, Inst of Marine Engrs; Mbr ASME; Mbr & Past Chmn, Pacific Northwest Sect, Soc of Naval Architects & Marine Engrs; Mbr Lloyd's Register of Shipping - Canadian Ctte, Amer Bureau of Shipping; Pres Assn of PE's of British Columbia 1976-77. *Society Aff:* APEBC, RINA, IME, SNAME, ASME.

McLaughlin, Edward
Business: College of Engineering, LSU, Baton Rouge, LA 70803
Position: Prof & Dean of Engineering *Employer:* LA State Univ. *Education:* DSc//Univ of London; PhD/Chem Physics/Univ of London; DIC/Thermodynamics/Imperial College; MSc/Physical Chemistry/Univ of London. *Born:* 10/16/28. Ballymena, Ireland. Joined faculty Chem Engg Dept, Imperial College, Univ of London in 1956. Was Reader in Chem Physics and Asst Dir of Dept until 1970. Moved to LA State Univ as Prof of Chem Engg in 1970. Chrmn of Dept, 1979-87. Dean, College of Engineering 1987-. Reg PE in Chem Engg. Chrmn of Baton Rouge Chap of AIChE 1979. Collector of old scientific instruments. Spare time farmer and forester in W Feliciana Parish. *Society Aff:* AIChE.

McLaughlin, James R
Business: P O Box 1569, Fargo, ND 58107
Position: President. *Employer:* Ulteig Engineers, Inc. *Education:* BS/CE/Univ of ND. *Born:* June 1925. With Ulteig Engrs, Inc since 1948. Field Engr for distribution & transmission power lines to 345 KV. VP Civil Engrg Dept, 1964, in charge of design for paving, water & sewer. Became Genl Mgr of home office in charge of basic operation of firm of 61 persons. In 1975 became Pres - have offices also in Minneapolis, Minnesota, & Bismarck, ND. Reg PE in 8 states. Fellow in ASCE; Mbr ND Registration Bd for Engrs 1976-81. Active in Boy Scout program; Natl VP, NCEE 1981-82. *Society Aff:* ASCE, ACEC, IEEE, NSPE.

McLaughlin, John F
Business: Dean of Engr. Office Purdue University, W Lafayette, IN 47907
Position: Assoc Dean of Engrg *Employer:* Purdue University. *Education:* PhD/Civil

McLaughlin, John F (Continued)
Engg/Purdue UNiv; MS/Civil Engg/Purdue Univ; B/Civil Eng/Syracuse Univ. *Born:* 9/21/27. Native of NY City. Army Air Corps 1945-47. Purdue U, faculty mbr 1951- . Asst Head, Civil Engrg School 1965, Head 1968-78. Dir, Amer Concrete Inst 1966- 68, 1975-77; ACI VP 1977-79; ACI Pres 1979-80; Asst Dean, Schs of Engg, Purdue Univ 1978-80. Assoc Dean, Schools of Engrg Purdue Univ 1980-. Dir Amer Soc for Testing and Materials 1984-86. Fellow ASTM 1984. Hon Mbr ACI, 1984. *Society Aff:* ASCE, ASTM, ACI, ASEE, NSPE.

McLay, Richard W
Home: 18 Athens Dr, Essex Junction, VT 05452
Position: Prof. *Employer:* Univ of VT. *Education:* PhD/Engg Mech/Univ of WI; MS/Engg Mech/Univ of WI; BS/Mech Engg/Univ of WI. *Born:* 4/21/37. Prof of Mech Engg at Univ of VT. Pres of COMtech, Inc, a firm involved in finite element analysis, mech failure studies, and composites. PhD Univ WI 1963. Five yrs with Boeing Res. Projs with AEC, EPA, NSF, Natl Safety Council, AFOSR, several industries. Co-editor two engg conf volumes. Fellow Commoner at Churchill College, Cambridge Univ, England during sabbatical yr 1976-1977. Pres ASME Green Mtn Sec. Restoring 1949 MGTC. Researching life of Scottish great- grandfather. Officer St Andrews Soc (Scottish) of VT. *Society Aff:* ASME, AIAA, ASEE, SME.

McLean, Arthur F
Home: 860 Arlington Blvd., Ann Arbor, MI 48104
Position: Mgr. Matls. Engineering *Employer:* Ford Motor Company *Education:* British Higher National/Mech. Engrg./Bristol College of Technology; British National/Mech. Engrg./Bristol College of Technology *Born:* 04/16/29 After completing military service as Engineering Officer, worked on R&D of aerospace systems in England, Canada and USA. Joined Ford in 1961, with responsibilities for engine performance/control systems/subsystems; developed computer technology for engine modelling. Headed turbine research and directed pioneering developments in ceramic turbine technology and low emissions combustion. As manager of materials engineering, heads up R&D in the areas of materials science, manufacturing processes and component/system design and development related to ceramics and other advanced materials for turbine engines, diesel engines, gasoline engines, manufacturing applications and advanced batteries. Has over forty publications in these fields. Enjoys classical music and travel. *Society Aff:* ACS, ASME, SAE, I Mech E

McLean, James D
Business: 1602 Bldg, Midland, MI 48640
Position: Research Associate *Employer:* Dow Chemical Company. *Education:* PhD/Analytical Chemistry/MI State Univ; BS/Chemistry/Univ of MI. *Born:* Nov 23, 1940. Socony-Mobil Fellow, Electrochemical Soc Fellow MSU. With Dow Chem Co since 1967 in Inorganic Materials Science and Characterization Sect. of Analytical Labs. Mbr ACS, Sigma Xi, NY Acad of Sci, Instrument Soc of Amer. Amer Assoc Advan Sci. Awards: Top Honors John C Vaaler Award, Chem Processing Magazine 1974 & 1980; Vernon A Stenger Award, Dow Chem Co 1974 & 1982; Arnold O Beckman Award, Instrument Soc of Amer 1975. IR.100 Award, Research and Development Magazine 1985. Res interests: Polarographic Trace Analyses of organics; Square wave voltammetry; design of electroanalytical methods & instrumentation. Enjoy wine-tasting, gourmet cooking, & nature. *Society Aff:* ACS, ISA, ΣΞ, NYACADSCI, AAAS, SEAC

McLean, True
Home: 2551 Windward Way, Naples, FL 33940
Position: Prof Emeritus - Retired, 1966. *Employer:* Cornell University. *Education:* PE/Electrical Engrg/State Univ, NY; EE/Radio Engg/Cornell Univ. *Born:* Jan 22, 1899 New York, NY. Mbr Sigma Xi, Eta Kappa Nu. Life Mbr, Fellow IEEE. Cons Brookhaven Natl Lab, Philco Corp, Cornell Aeronautical Lab, GE Co. *Society Aff:* IEEE, AAAS, CAP, AOPA.

McLean, William G
Home: 333 Fifth Av, Scranton, PA 18505
Position: Dir of Engrg (Emeritus). *Employer:* Lafayette College (Retired). *Education:* Dr Engg/Hon/Lafayette College; ScM/Math/Brown Univ; BSEE/Elec/Lafayette College. *Born:* Mar 1910. Native of Scranton, Pa. Taught in Scranton schools 3 years. In Mech Engrg Dept at Lafayette Coll 1937-44. Sr physicist & asst to supt of spec prod div at Kodak 1944-46. Head Engrg Sci Dept at Lafayette 1946-75. Dir of Engrg at Lafayette Coll 1962-75. Co-author 2 books in Engrg Mech. Active in ASME, ANSI & NSPE. Chmn, Hugh Moore Pky Comm, Int in codes & standards activities. VP 1970- 72, Chmn CTA 1976-78 ASME, Dir 1965-67 NSPE. Member and Past Chairman PA State Registration Bd for PE. Mbr Phi Beta Kappa, Tau Beta Pi, Pi Tau Sigma, Eta Kappa Nu, Kappa Delta Rho. *Society Aff:* ASME, ASEE, NSPE, AAUP, AAAS

McLean, William N
Business: 4100 S Kedzie Ave, Chicago, IL 60632
Position: Director of Engrg. *Employer:* Crane Co. *Born:* April 9, 1924. MS & BSME from Northwestern U. Native of Chicago, Ill. Served in Army Air Force 1943-46. Nuclear reactor engr with Atomic Energy Comm 1951-53 & Amer Electric Power 1953-55. With Crane Co since 1955. In 1968 assumed role of Dir of Engrg for indus valve design. Chmn of ANSI Ctte B16 on valves. Also Chmn of ISO TC153 on valve design & testing. Chmn of API manufacturers valve ctte. Mbr of ASME Boiler Code Nuclear Power Ctte & Chrmn Valve Design Work Group. Mbr of ANSI Nuclear Standards Mgmt Bd.

McLeish, Duncan R
Home: 2871 N. Ocean Blvd, Apt 316 Diano, Boca Raton, FL 33431
Position: Retired *Education:* BS/Mech Eng/Purdue Univ. *Born:* 5/21/17. in Fort Wayne, IN. Pi Tau Sigma, Tau Beta Pi. With Firestone Indus Prod Div of Firestone Tire & Rubber Co 1939-48, as maintenance Engr & Asst to Plant Engr. With Ali Lily & Co since 1949 as Mgr, Plant Engg & Maintenance, 1200 Kentucky Ave Plant, since 1965. Current Assignment Mgr Plant Engg & Utilities for Indianapolis Plants. Eff Dec 77 (now retired). Reg PE IN since 1949. VP ASME for Genl Engg Dept 1975-77, re-elected for 2nd term 1977-79. Currently a Mbr of Operating Bd Prof Development ASME. *Society Aff:* ASME.

McLellan, Alden, IV
Business: NJ Dept of Environ Protection CN-402, Trenton, NJ 08625
Position: Assistant Commissioner *Employer:* State of NJ *Education:* PhD/Theoretical Physics/Univ of NV-Reno; MA/Theoretical Physics/Univ of CA; BA/Physics/Univ of CA-Berkeley *Born:* 3/7/36 Highest ranking scientist in NJ State Government. Serve the state on all engrg, scientific, and technological matters. Provide direct scientific and engrg advice to NJ's Dept of Environmental Protection with an operating budget over $175,000,000 and over 3,400 employees. Spokeperson for the state on engrg and scientific matters. PE. Bd and Trustee Membership: Governor's Science Advisory Committee; PSE&G Research Corp; NJ Academy of Science; NJ Energy Research Inst; Water Resources Research Inst; NJ Dept of Environmental Protection; Drexel Univ Advisory Committee; Rutgers Univ Graduate School, Assoc Prof. Two books, 48 scientific and engrg publications. *Society Aff:* AIAA, AAAS, ANS, NSPE, APS

McLellan, Rex B
Business: Rice University, Dept. of Mech. Eng. & Materials Science, Box 1892, Houston, TX 77251
Position: Prof *Employer:* Rice Univ *Education:* PhD/Mat Sci/Leeds Univ England; BMet/Metallurgy/Sheffield Univ England *Born:* 11/21/35 Has conducted research in the area of the thermodynamics and kinetics of important solid state metallurgical systems with an emphasis on interstitial solid solutions. The thermodynamic and kinetic behavior of metal-hydrogen systems has been a thrust area. This work has resulted in about 200 publications in leading journals. In addition to basic research, McLellan has been heavily involved in industrial consulting in the areas of fatigue failures, corrosion, and products liability litigation. Hobbies are pistol shooting and

McLellan, Rex B (Continued)
squash. Born in England but naturalized US citizen since 1967. *Society Aff:* ASM, AIME

McLendon, B Derrell
Business: Agri Engg Ctr, Athens, GA 30602
Position: Assoc Prof. *Employer:* Univ of GA. *Education:* PhD/Agri Engg/Cornell Univ; MS/Agri Engg/Univ of GA; BSAE/Agri Engg/Univ of GA. *Born:* 12/26/41. A native of Cuthbert, GA. Actively involved in retail farm equip dealership prior to beginning grad work Instr and grad student in Agri Engg Dept of Univ. of Ga. from 1965-69. A Natl Sci Fdn Sci Faculty Fellow at Cornell Univ from 1969-71. Currently Assoc Prof in Agri Engg Dept at Univ of GA. Teaching in electromechical machines, electronics and control systems. Responsible for developing grad and undergrad courses and teaching lab for microprocessor systems. Research involves sensor development and application of microcomputer/microprocessor based controls for production and processing systems. *Society Aff:* ASAE, ASEE.

McLennan, Ian M
Business: 26th Floor, 140 William St, Melbourne, Victoria, Australia 3000
Position: Former Chrmn. *Employer:* Broken Hill Proprietary Co Ltd. *Education:* BEE/-/Melbourne Univ. *Born:* 11/30/09. Sir Ian McLennan spent most of his working life with The Broken Hill Proprietary Co., Ltd., Australia's largest co. He joined in 1933 as cadet engr, Gen Mgr 1950 Chrmn 1971. Retired 30th Nov, 1977. Chrmn, Australia and New Zealand Banking Group Ltd. 1977-1981. Mbr, GM Australian Advisory Council 1978-82. Chrmn, Elders IXL Ltd. 1980-85. Currently: The Foundation Pres. Australian Acad of Tech Scis; Hon Chrmn Australia Japan Business Cooperation Ctte. Hon Doctorates of Engrg from Melbourne & Newcastle Univs & Hon Doctorate Sci from Univ of Wollongong. Awarded 1978 Charles F Rand Meml Gold Medal from AIME. Award 1981 Bessemer Gold Medal from Metals Soc London. Award 1981 Dist Intl Mgr Gold Medal from Australian Inst of Mgmt. *Society Aff:* AIME, NAE.

McLeod, John H, Jr
Home: 8484 La Jolla Shores Dr, La Jolla, CA 92037
Position: Editor, Consultant. *Employer:* Self-employed *Education:* BS/EE & ME/Tulane Univ. *Born:* Feb 27, 1911. Applications Engr, Taylor Instruments Co 1940-43. R&D Engr (3 patents) Leeds & Northrup 1943-47. Naval Air Missile Test Center 1947-56 (administered dev, large-scale simulation facility). Design Specialist, General Dynamics Astronautics 1956-63. Editor 'Simulation' (tech journal Soc for Computer Simulation) 1963-74. Cons, simulation systems 1963- . Currently Publications Advisor SCS; Assoc Editor (simulation) 'Behavioral Sci'; Tech Editor 'Simulation in the Service of Soc'. Founder Simulation Council (SCS) 1952. Co-founder San Diego Symposium for Biomedical Engineering 1961. EAI Senior Scientific Simulation Award 1965; National Science Foundation grant, Ethics in Simulation, 1983; TIMS Outstanding Service Award 1986. SCS representative to AAAS. Author 'Simulation: Dynamic Modeling of Ideas & Systems with Computers' 1968; 'Computer Modeling & Simulation: Principles of Good Practice' 1982; co-author 'Large-Scale Models for Policy Evaluation' 1977. Registered Professional Engineer, California. *Society Aff:* SCS, IEEE, AAAS

McLeod, Robert J
Home: 5213 Varnum St, Bladensburg, MD 20710
Position: General Manager. *Employer:* Washington Suburban Sanitary Comm. *Education:* BSCE/Engg/Univ of MD; LLB/Law/Columbus Univ of Law; JD/Law/Catholic Univ of America. *Born:* Nov 1916. Native Grafton, W Va. Army Counter-intelligence Corps 1941-46, Lt Col serving Chief of Intelligence, Manhattan A-bomb proj. Joined engrg staff of Washington Suburban Sanitary Comm 1938, rose to Genl Mgr 1967. Thrice Chmn Washington Metro Area Regional Sanitary Bd, 2 terms Md Water Resources Comm, former Natl Dir for AWWA, headed several AWWA Natl Cttes. Top 10 public works men of the year selection by APWA; Man of the Year in Chesapeake Sect, AWWA, 1970. Mbr ASCE, WPCF & Diplomate of Amer Acad of Environmental Engrs. Retired Dec 31, 1976. *Society Aff:* ASCE, AWWA

McLucas, John L
Business: 800 Independence Av, SW, Washington, DC 20591
Position: Administrator, Fedl Aviation Admin. *Employer:* Dept of Transportation. *Born:* Aug 1920 Fayetteville, NC. PhD Penn State U; MS Tulane U; BS Davidson U. Served as officer, USN, WWII. VP & Tech Dir Haller, Raymond & Brown 1950-57; Pres of HRB-Singer 1958-62. Served as Deputy Dir of Defense Res & Engrg (Tactical Warfare Progs), DOD 1962-64. Appointed Asst Secy Genl for Scientific Affairs, NATO 1964-66. 1966-69 Pres, MITRE Corp; Under Secy USAF 1969-73; Secy USAF 1973-75; Sworn in Admin, FAA, Nov 1975. Enjoys golf & theater.

McMahill, William F
Home: 106 Lathrop Rd, Columbia, MO 65201
Position: Assist Dir Continuing Engineering Education *Employer:* University of Missouri-Columbia. *Education:* PhD/Higher & Adult Ed/Univ of MO-Columbia; MS/Engr Mgmt/WA Univ-St Louis, MO; BS/Chem Engg/Univ of MO-Columbia. *Born:* Sept 1925. Native of Mo. WWII veteran (Marine Corps). Small business owner & magr 1949-57. Chem Engr, Mallinckrodt Inc, St Louis 1961-67. 167967- , U of Mo. 1967-71 developed Public Service Prog for business & indus, metropolitan St Louis, Mo. 1971- , respon for mgmt & dev activities of statewide tech information network available to all organizations & citizens within the state, also includes linkages with natl level information networks & tech information depository systems. Mgmt respons for dev & maintenance of statewide tech transfer progs. Natl officer of AIChE (current chmn of natl ctte on public relations). Program Development of Continuing Engineering Education, Statewide.. *Society Aff:* AIChE.

McMahon, Charles J, Jr
Business: 3231 Walnut St, LRSM K1, Philadelphia, PA 19104
Position: Prof, Matl Sci & Engg. *Employer:* University of Pennsylvania. *Education:* ScD/Metallurgy/MIT; BS/Metallurgical Eng/Univ of PA *Born:* 7/10/33 Native of Phila, Pa. Served as line officer US Navy 1955-58. Instructor, physical met at MIT prior to receiving ScD. Ford Foundtion Indus Resident in Engrg Practice 1968-69. Overseas Fellow, Churchill Coll, Univ of Cambridge 1973- 74. Visiting Prof Univ of Gottingen 1983-84. Fellow, ASM, TMS, AIME & Inst of Metallurgists, London. Prof, Matls Sci & Eng, Univ of Pa. Teaching, res & consulting activities in field of physical met, especially strength & fracture of solids. Awards: ASM, Henry Marion Howe Medal 1975; AIME, Champion H Mathewson Gold Medal 1976. National Academy of Engg 1980 Albert Sauveur achievement award ASM, 1981. Alexander Von Humboldt Sr US Scientist Award 1983-84. Enjoys tennis & classical music. *Society Aff:* ASMI, TMS-AIME, AAAS

McMahon, John
Business: 2301 Bowen Rd, Elma, NY 14059
Position: Pres. *Employer:* McMahon Engg Co. *Education:* BSME/Mech Engg/Univ Buffalo. *Born:* 3/21/25. Native of Buffalo NY. US Army 1943-46. BSME 1949 supervisor at Bell Aircraft (Bell Aerospace) in Design, Fabrication & test of inertial instruments and navigators. Obtained patent for improved gas spring 1974. Founded McMahon Engrg, Co 1973. Specializing in Seismic Analysis of Machinery and Finite Element Analysis. *Society Aff:* NSPE.

McMahon, John F
Home: 13 Sayles, Alfred, NY 14802
Position: Dean Emeritus. *Employer:* Retired. *Education:* BSc/Ceramics/Alfred Univ. *Born:* Oct 1, 1900 Cohoes, NY. 1923-25 Plant Engr; 1925-36 Ceramic Engr, Canadian Dept of Mines, Ottowa Canada; 1936-65 associated with NY State Coll of Ceramics, Alfred U; Dean of Coll from 1948-65. Recipient of numerous ceramic awards & honors, & honorary Doctoral degrees from Clemson & Alfred U's. Hon Life Mbr of Amer Ceramic Soc, Canadian Ceramic Soc, & British Ceramic Soc. Reg PE in NY. *Society Aff:* ACS, CCS, BCS.

McMahon, Robert E
Business: 1377 Midway Rd, Menasha, WI 54952
Position: Chrmn *Employer:* McMahon Associates Inc. *Education:* BSCE/Civil Engrg/Univ of WI *Born:* 1924 Menasha, Wis. US Army Engr & Air Corps WWII. Joined McMahon Assoc Inc, Cons Engrs 1949. Appointed as Dir 1958; became Pres of firm 1971; Chrmn, Bd of Dirs 1983. Prof experience in Public Works Engrg, Water & Wastewater Systems. PE in Wisc, Mich, Ill. *Society Aff:* NSPE, ACEC, AWWA, WPCF

McMahon, Thomas E
Business: P.O. Box 1517, Salina, KS 67402
Position: Cons Engrs, Planners & Architects. *Employer:* Bucher, Willis & Ratliff. *Education:* MS/Env Health Engg/Univ of KS; BS/Civil Engg/Univ of KS. *Born:* 7/22/26. in Ness City, KS. Mbr of Amer Acad of Environmental Engrs. Reg PE in KS, AK, WA & MN. P Pres of KS Water Pollution Control Assoc. *Society Aff:* KWPCA, WPCF, NSPE, KES, AWWA, AAEE.

McManis, Kenneth L
Home: 4717 Purdue Dr, Metairie, LA 70003
Position: Dept Chrmn. *Employer:* Univ of New Orleans. *Education:* PhD/Civ Engr/LA State Univ; Grad Studies/Civ Engr/Tulane Univ; MS/Civ Engr/LA State Univ; BS/Civ Engr/Univ of Southwestern LA. *Born:* 10/20/41. Native of Lake Charles, LA. Production engr with Mobil Oil Co's offshore operations, 1963-64. Consultant civ engr employed by W S Nelson, Engrs & Architects, Inc and Fromherz Engrs in New Orleans, 1966-68; work involved designing offshore platforms, industrial bldgs, docks, bridge design, and drainage studies. Res Assoc/Asst, Engg Res Div, LA State Univ, Baton Rouge. Recipient of Michael Claus Meml Award for grad res. LA State Univ, Baton Rouge, 1975. Prof/Dean of Engg and Industrial Tech at Delgado College in New Orleans, 1968-78. Assoc Prof/Chrmn of Civ Engg at the Univ of New Orleans, 1978 to present. *Society Aff:* ASCE, ASEE, XE, ΣΞ, LES.

McMannon, Richard J
Business: Vultee Blvd, PO Box 210, Nashville, TN 37202
Position: Senior Metallurgical Engineer. *Employer:* AVCO Aerostructures Division. *Born:* March 1927. BS from Thomas More Coll; BS from U of Kentucky. Native of Covington, Ky. Served with US Army 1944-45. Associated with GE, Genl Dynamics & Combustion Engr. With AVCO Aerostructures Div since 1962 as Met Engr. Assumed additional duty of Div Welding Engr 1973. Past Chap Chmn ASM; past mbr Chap Advisory Ctte ASM. Enjoys golf, woodworking, & gardening.

McManus, John A
Business: 1000 Shuttle Meadow Ave, New Britain, CT 06052
Position: Dir. *Employer:* Water Dept. *Education:* BS/Civil Engg/ Worcester Polytech Inst. *Born:* 9/7/37. in Providence, RI. Served with US Army Signal Corps 1959-60. Civil Engg with CA Water Res Dept for 5 yrs and hd of Pawtucket RI Water Dept for 3 yrs. Dir of New Britain, CT, Water Dept since 1969. PPres of CWWA and PChmn of CT Section of AWWA. PE in CT, NY, CA & RI. *Society Aff:* AWWA, CWWA.

McManus, Ralph N
Home: 12907 - 66 Ave, Edmonton Alberta, Canada T6H 1Y6
Position: President. *Employer:* McManus Engineering Ltd. *Born:* Dec 1918. PhD Illinois 1952; BSc & MSc Alberta. Native Albertan. Formerly Prof of Civil Engrg, Alberta. Principal of T Lamb, McManus & Assoc, Cons Engrs to 1972. Now President of McManus Engrg Ltd. Prime engrg respon over 20 years for highway bridges in reinforced & prestressed concrete, plate girder, truss & suspension plus major pipeline suspension bridges in British Columbia & Alaska. P Pres & Life Mbr Alberta Assn of PE's. Past Pres of Canadian Council of PE's. Past Chmn Canadian Joint Ctte on Reinforced Concrete Design. Licensed in Alaska & currently proj mgr, Dickson Dam. Enjoys golf.

McManus, Terrence J
Business: 2402 West Beardsley Rd, Phoenix, AZ 85027
Position: Mgr, Corp Environmental Affairs *Employer:* Intel Corp *Education:* MS/Environmental Engineering/Cornell University; BS/Civil Engineering/Union College *Born:* 05/03/48 After Cornell in 1973, joined as proj engr with Betz Environmental Engrs (currently BCM Engineers) which was a 500 employee environmental consulting firm. By 1980, rose to position of Asst VP & Asst Dir of Industrial Services, which was a dept of 50 employees (1980-83). In 1983 joined Intel Corp, a two billion dollar semiconductor & computer manufacturer, with both domestic & international operations, as Mgr of Corp Environmental Affairs worldwide. Responsible for insuring Intel manufacturing facilities operate in compliance with governmental regulations in ten countries. Reg Civil Engr NY, CA, IN, WA, OR and PA. Chairman, Solid and Hazardous Waste Management Cttee for ASCE 1986-87, and Secretary of Environmental Engrg Div, 1987-89. *Society Aff:* ACSE

McMaster, Galen M
Home: 725 Fillmore, American Falls, ID 83211
Position: Res Prof & Supt. *Employer:* Universtiy of Idaho. *Education:* MS/AE/Univ of ID; BS/AE/Univ of ID. *Born:* June 1928. Native of Twin Falls, Idaho. MS U of Idaho 1964; BS U of Idaho 1950. Enlisted service, US Army Combat Engrs 1950-52. Agricultural Engr, US Bureau of Reclamation. County Agent (Settlers Assistance Frog), Washington State Coll. With U of Idaho at the Aberdeen Res & Extension Center since 1955. Taught irrigation & conservation engrg at U of Idaho 1962-63 as Asst Prof. Respons include res in irrigation, crop water use & application of chems through irrigation systems. Hobbies: fishing, hunting, camping. *Society Aff:* ASAE.

McMaster, Howard M
Business: 5055 Antioch Rd, Overland Park, KS 66203
Position: Regional Managing Principal & VP. *Employer:* Woodward-Clyde Consultants. *Education:* MS/Geotechnical/Univ of NB; BS/Civil Engg/Univ of NB; Cert/Geotochnical/Harvard. *Born:* July 31, 1921. Attended Harvard Graduate School & Oak Ridge Inst of Nuclear Studies, USAEC. Served as Commanding Officer of several Military Construction Units for US Army Corps of Engrs from 1941-46. From 1946-57 taught Geotechnical Engrg & Structural Engrg in graduate & undergraduate Schools of Engrg at U of Nebraska. With Woodward-Clyde Consultants since 1957 & presently serves as Regional Managing Principal encompassing those offices located in St Louis, Kansas City & Omaha. Reg PE in 49 states & District of Columbia. *Society Aff:* CGZ, MSPE, ASCE, MSPE.

McMaster, Robert C
Home: 6453 Dublin Rd, Delaware, OH 43015
Position: Consulting Engineer. *Education:* PhD/Elec Engg/CA Inst of Tech; MS/Elec Engg/CA Inst of Tech; BS/Elec Engg/Carnegie Mellon Univ. *Born:* May 1913. Regents Prof Emeritus 1977 of Welding Engrg & Elec Engrg at Ohio State U in Columbus, Oh. Cons engr in nondestructive testing & materials evaluation. Res in area of ultrasonics & sonic power; holographic, eddy current, ultrasonic, X-ray & other nondestructive tests. Past Natl Pres ASNT, Gold Medal (1977), Tutorial Citation (1973), Mehl Lecture (1950) Coolidge Award (1957); Editor of Nondestructive Test Handbook & author of more than 100 tech articles, 19 US Patents. Mbr of US NAE. Fellow, 1973, Hon Life Mbr ASNT. Adams Lec 1965 AWS. *Society Aff:* AWS, ASM, IEEE, ASTM.

McMasters, Jesse L
Home: 1317 Denson Dr, Pauls Valley, OK 73075
Position: Engrg Section Head *Employer:* Self *Education:* BS/Agric Engrg/OK St Univ; BA/Agric/Cameron St Univ; *Born:* 10/10/35 A native of Duncan, OK. Completed prep school at Empire High in 1954, received honorable discharge from USAF in 1958. Graduated from OK State Univ in 1963 with BS in Agricultural Engrg. Registered in the state of OK as a PE and land surveyor. Began career with Soil Conservation Service in 1963 at Chichasha OK. Retired from SCS-USDA with 33 years of Government Service. I have experience in watershed planning, design

McMasters, Jesse L (Continued)
and construction; irrigation and drainage; rural and urban land planning, environmental assesments, impact statements, solid waste (landfill) Design, Manage and Operation. processes. I am active in all professional societies and have served in officer positions in some. *Society Aff:* NSPE, ACSM, ASAE, SCSA, ASCE, OSLS.

McMillan, Brockway
Home: , Sedqwick, ME 04676
Position: Retired. *Education:* PhD/Mathematics/MA Inst of Tech; BS/Mathematics/MA Inst of Tech. *Born:* 1915 Minneapolis, Minn. Taught at MIT & Princeton 1936-41. Active duty, USNR 1943-46. Bell Labs 1946-61. Asst Secy, R&D, USAF 1961-63k. Undersecy USAF 1963-65. Bell Labs 1965-79, VP 1969-79, Fellow IEEE 1962, NAE 1967. Mbr IEEE, Amer Math Soc Math Assn of Amer, Inst Math Statistics, Soc Indus & Appl Math (Pres 1959-60), AAAS. Home: Sedqwick, Maine. Tech interests: communication theory, random processes. *Society Aff:* AAAS, IEEE, AMS, MAA, IMS, SIAM, NAE.

McMillan, Hugh D
Business: 16720 Park Row, 16720 Park Row P O Box 219089, TX 77218
Position: President. *Employer:* McMillan Equipment Company. *Education:* BS/Mech Engg/TX A&M. *Born:* Sept 15, 1925. Design Engr, Coastal Equip Co 194749; Sales Engr, J R Dowdell & Co 1949-55; founded McMilllan Equip Co 1955. Mbr ASHRAE since 1947, served as Pres Houston Chap 1959, Natl Bd/Dir 196770; Dir & Regional Chmn 1972- 75, Chmn Charter & By-Laws Ctte, Finance Ctte, Treas 1976-77; Distinguished Service Award 1976. VP 1977-78; Pres-elect 1978-79; Pres 1979-80. Member Exec. Committe of AAES 1980. Listed in Who's Who in the South & Southwest Who's Who in the World and Who's Who in Finance & Industry. Mbr Natl & Texas Soc of PE's. Recently designed & furnished heating & ventilating equip on the Trans-Alaska Pipeline. *Society Aff:* ASHRAE.

McMillan, Hugh H,
Business: 2739 Elston Ave, Chicago, IL 60647
Position: VP *Employer:* Paschen Contractors, Inc. *Education:* BS/EE/Valparaiso Univ *Born:* 12/23/36 Was Chief Executive Officer of the Metropolitan Sanitary District of Greater Chicago after 17 years in the Operations & Maintenance and Engrg Depts. Was deeply involved in the development of PL 92-500 and testified before legislative committees relative to the Clean Water Act. Served on numerous committees of the USEPA and IL EPA. Held positions as member of the Bd of Control of WPCF; Bd Member of AMSA and the IL Water Pollution Control Assoc and the Executive Council of the Inst of Water Resources (APWA) member of the USPEA Management Advisory Committee (MAG). Now VP of Paschen Contractors, Inc Chicago, IL and Pres of Caribbean Contractors, Inc. St. Petersburg, FL since Sept 1983. *Society Aff:* IEEE, WPCF, AMSA, IWPCF, SAME, AAEE

McMillan, John Robertson
Business: 550 S. Flower, Rm 711, Los Angeles, CA 90071
Position: Energy Producer *Education:* BS/Mech Engg/CA Inst of Technology. *Born:* Sept 1909. Barnsdall Oil Co 1929-43; VP Fullerton Oil Co 1943-54, Pres 1954; Exec VP Monterey Oil Co 1954-61; Pres Transwestern Pipeline Co 1956; Pres Monterey Gas Transmission Co 1961-63; Pres Reserve Oil & Gas Co 1963-73, Bd Chmn 1974-80. Chmn Petroleum Branch AIME 1954; Pres, AIME 1968; Pres Assocs of Calif Inst of Tech 1968-70; Dir WOGA 1963-80, Pres 1972-73;Dir API 1969-1980; Mbr Natl Petroleum Council 1972-79; Dir Amer Mutual Fund. SPE, Distinguished Service Award 1971. Fellow, Inst for Advancement Engrg 1974 Calif Inst of Tech Alumni Distinquished Service Award 1980. *Society Aff:* AAPG, AIME, SPE.

McMinn, Jack H
Home: 104 Palm Ave, San Francisco, CA 94118
Position: Vice President. *Employer:* Shearson Lehman Brothers Inc *Education:* MCE/Structural Engrg/Cornell Univ; BS/Civil Engrg/Cornell Univ. *Born:* Nov 28, 1922 Tulsa Okla. p. John Craig & Maybelle (Barber). Married Charlotte Loraine Wagenknecht Oct 18, 1958. Children, Melissa Loraine & Daniel Kimpton. Proj engr, Standard Oil Co, San Francisco, 1947-50; Asst Genl Mgr, Western Pine Supply Co, Emeryville, Ca 1952-54; Soils Engr 1955-62; regional editor, Engrg News Record 1962-68; VP Bartle Wells Assocs, municipal financing cons 1968-78. VP Shearson Lehman Brothers Inc. Investment banking 1978- present. Ordnance Corps, AUS, 1943-46; Engr Corps AUS 1950-52. Reg PE Calif, Ill, NY. Licensed genl contractor CA. Fellow ASCE, Natl Dir 1972-75. Tau Beta Pi, Chi Epsilon, Unitarian. *Society Aff:* ASCE.

McMorries, Bill R
Business: 6300 Canyon Dr, Amarillo, TX 79109
Position: Pres. *Employer:* McMorries & Assoc, Inc. *Education:* MS/Civ Engg/Univ of OK; LLB/Law/American School of Law; BS/Civ Engg/Univ of OK. *Born:* 1/9/27. Native of W TX; taught engg courses at Univ of OK during 1949-50 school yr; office engr for Hasie & Green, Engrs, specializing in municipal engg, 2 1/2 yrs; chief engr, Randall Bldg Co, 1 1/2 yrs; pres, McMorries & Assoc, Inc, Municipal Engrs, Amarillo, since 1954; VP McMorries & Glovier, Inc, Municipal Engrs, Dumas, since 1960; VP McMorries & Street, Inc., Municipal Engrs., Perrtgton. *Society Aff:* ASCE, NSPE, MENSA, INTERTEL.

McMulkin, F John
Business: P O Box 460, Hamilton Ontario, Canada L8N 3J5
Position: Vice President - Research. *Employer:* Dominion Foundries & Steel, Ltd.; Retired *Education:* BS/Metallurgy/MI Tech Univ; ME/Metallurgy/MI Tech Univ. *Born:* 1915. Metallurgist - Algoma Steel, ORF, Hamilton Bridge Fellowship, developing armour plate welding & fabrication. Dir of R&D, then VP of Res, Dofasco, starting 1946. Respon for developing first use of BOF steelmaking in Norht Amer, also the pioneering of blast furnace oil injection. Author of papers on heat treating, steelmaking & associated beneficaiation of raw materials. Awards: EIC. John Galbraith Prize; AIME Acid Converter & Basic Oxygen Steel Award, Charter Distinguished Mbr of ISS, Metallurgical Soc Howe Memorial Lecturer; CIMME Airey Memorial Award; ASM Eisenman Award, Elected Fellow of the Soc; Michigan Tech Distinguished Alumnus Award, Honorary Doctor of Engrg, Fellow Engineering Institute of Canada, Life Member EIC. *Society Aff:* EIC, AIME, ISS, CIM, ASM.

McMurray, William
Business: Corp R & D, P.O. Box 43, Schenectady, NY 12301
Position: Electrical Engr *Employer:* Gen Elec Co *Education:* MS/EE/Union Coll; BSc/EE/Battersea Polytech, England *Born:* 8/15/29 With Gen Elec Corp R & D since 1953. Has been responsible for the development of solid-state electric power control and conversion circuits such as rectifiers, inverters and dc converters using Thyristors and Transistors. Has published a number of tech papers and holds 15 US Patents in this field. Author of the book *The Theory and Design of Cycloconverters* and contributor to the book *Principles of Inverter Circuits*. Mbr of The Static Power Converter Committee of the IEEE-Industry Applications Society. Licensed PE in NY State. In 1978, he received The William E. Newell award for outstanding achievement in power electronics. In 1984, he was awarded the IEEE Lamme Medal. *Society Aff:* IEEE, IEE

McMurren, William H
Business: 1 Morrison-Knudsen Plz, Boise, ID 83729
Position: Chrmn & Chief Executive Officer. *Employer:* Morrison-Knudsen Company, Inc. *Education:* BSCE/BSCE/TX A&M Univ *Born:* 10/20/27 Ontario, OR. s Sarena Elbert & Louise (Baker) McMurren. m Carlyn Dorthy Stenberg, Oct 17, 1953; children: Catherine Lynn, John Henry. With Morrison- Knudsen Co 1955- , Exec VP 1969-72, Pres & Chief Exec Officer 1972-84, Chrmn & Chief Exec Officer, 6-1, 1984- , also Dir. Dir Albertson's, Inc, Dir Westinghouse Elec Corp, Ida; 1st Natl Bank, Boise. Served with AUS, 1945-46, Soc Amer Military Engrs, Mbr Amer Bureau of Shipping, San Francisco, CA, Amer, Moles, Beavers Pres & (Dir). Trustee, Public Works Historical Soc. Clubs: Houston Club; Hillcrest Country. Home: 4902 Roberts Rd Boise, ID 83705. *Society Aff:* SAME, PWHS

McMurtry, Burton J
Business: 3000 Sand Hill Rd. Bldg 4, Suite 210, Menlo Park, CA 94025
Position: General Partner. *Employer:* Dennis, Jamieson & McMurtry & TVI Mgmt *Education:* BA/-/Rice Univ; BS/EE/Rice Univ; MS/EE/Stanford Univ; PhD/EE/Stanford Univ. *Born:* March 1935. Native of Houston, Texas. WIth GTE Sylvania from 1957-69; microwave tube engr 195761; laser & optical detector engr & mgr 1962-66; electronic systems engrg mgr & founder of Sylvania's Electro Optics Organization 1967-69. Alfred Noble Prize from 5 engrg Soc's 1964. More than 20 publs in reviewed prof journals or books. Fellow IEEE 1969 for tech mgmt contribs in electrooptics & lasers. Venture capital investor since 1969 providing capital & assistance to new technology-based companies. *Society Aff:* IEEE, APS, OSA.

McNair, John W, Jr
Business: L B & B Bldg, Waynesboro, VA 22980
Position: Consulting Engr. *Employer:* John McNair & Assoc. *Education:* BS/Foestry/Penn Stat U; BS/Civ Engr/VA Poly Inst & St U. *Born:* 6/17/26. Army Infantry and Corps of Engrs Officer 1944-46, 1951-53; Instr, Sch of Engg, Univ of VA 1955-58; Fndr and Owner John McNair & Assoc 1958-present. Civ and Sanitary practice, Mid-Atlantic states; Pres VA Soc of PE 1966-67; mbr (1969-present) and Pres (1975-76), VA Bd of Architects, PE and Land Surveyors; Councilman (1968-72) and VMayor City of Waynesboro, VA. Elder, Presbyterian Church; Rotarian; Sailor. Chief Exec Officer, the Brucheum Group (1983-present), Multidiciplinary Conslts. *Society Aff:* ASCE, ASHRAE, NSPE, AAEE.

McNall, Preston E, Jr
Business: 507 E Michigan St, Milwaukee, WI 53202
Position: Director of Engineering. *Employer:* Johnson Controls, Inc. *Born:* June 1923. PhD, ME & MSME from Purdue U; BSME from U of Wisc. Native of Madison, WI. Taught Mech Engrg courses at Wisconsin & Purdue. Served with US Navy Seabees 1944-46. Joined Honeywell, Inc in 1951 as Res Engr. Was Ch Engr- Appl Res. Joined Kansa State U as Dept Head Mech Engrg & Assoc Dir Inst for Environmental Res in 1965. Joined Johnson Controls in 1971. Currently Dir of Engrg. Specialty is thermal comfort & air quality for human beings. Published over forty papers on tech subjects. Mbr ASME; Fellow ASHRAE; served on several natl ASHRAE cttes, & received 2 awards for excellent technical papers.

McNally, William D
Business: Lewis Res Cntr, Cleveland, OH 44135
Position: Chief, Computer Services Div *Employer:* NASA. *Education:* BS/CE/Carnegie-Mellon; MS/CE/Carnegie-Mellon; PhD/CE/Carnegie-Mellon *Born:* 7/7/39. Native of Pittsburgh. Full career in turbomachinery and computer res at NASA, Lewis Res Center. Currently Chief of Computer Services Div. Previously respon for developing a computerized sys for flow analysis & design, for 3- dimensional, transonic, viscous compressor & trubine blading. SAE Charles Manly Medal 1974. Currently responsible for managing the major computer resoruces of Lewis Res Ctr. *Society Aff:* ASME, AIAA

McNamara, Edward J
Home: 1723 Oriole St, New Orleans, LA 70122
Position: Pres *Employer:* Conslt Engr Inc *Education:* BS/CE/Tulane Univ; -/Bus Mgmt/Harvard Grad Sch of Bus. *Born:* 06/23/16 Involved in minerals development projects throughout career Proj Mgr first offshore mining project to produce brimstone in 50 ft water. Design and construction of two (7 mile, 9 mile) undersea pipelines to transport molten sulphur (285 degrees F) ashore. Development of molten sulphur transport sys inland U.S. rivers; coastwise, and to Europe and UK. Project responsiblity for development of grass roots copper, gold, silver mining project in mountains region of New Guines (12000 ft elevation) including 74 mile slurry pipeline for copper concentrate. Project responsibility for engrg and construction grass roots nickel-cobolt mine and refinery ($300mm) in Australia-150 mile railroad, thirteen major bridges, two tunnels) mine development, industrial townsite, etc. Development and construction of two(2) uranium recovery plants and initial feasibility and process development for mining low grade gold ores in Nevada (1981). Appointed to LA State Bd of Reg for P.E. & L.S. 1978. Elected Treas Natl Council Engrg Examiners 1984. Mining Consultant to Chinese National Coal Development Corp.-P.R.C., in development studies and feasibility for Shen Mu Coal Field (1984-85) and 350 km electrified railway. Tau Beta Pi. Eminent Engr Award. L.E.S. Award A.B. Patterson Medal for Engr in Mngmt, Paul Harris Fellow, NO Rotary Club. *Society Aff:* ASCE, NSPE, LES, SAME, NCEE

McNamara, Patrick H
Business: 646 E St, SE, Washington, DC 20003
Position: Pres *Employer:* McNamara & Assoc, Inc *Education:* MBA/Mgmt/Central MI Univ; BCE/ChE/Cornell Univ *Born:* 06/10/39 Established new firm 1985. Twenty years of tech and managerial accomplishments with Dow Chem; three years Air Force (Captain). Founder and chrmn, The Natural Gas Consumers Info Center; Chrmn, The Petrochem Energy Group; Natl Dir, The Amer Inst of Chem Engrs; Reg PE; Reg lobbyist, Wash. Headed combustion and extraction section on oil shale res. Proj mgr for design and construction of latex plant in Canada. Series of supvry and engrg positions in several plastics production and pilot plants. Promoted dev and acceptance of Guidelines for Professional Employment, and professional career dev programs in AIChE. *Society Aff:* AIChE

McNamara, Robert F
Business: PO Box 3330, Omaha, NE 68103
Position: Pres *Employer:* Northern Engrg Intl Co *Education:* MS/ChE/IL Inst of Tech; BChE/ChE/Polytechnic Inst of NY. *Born:* 2/21/29. Native of New York, NY. From 1952-63 was employed by Ford Bacon & Davis Inc and Arabion American Oil Co. Joined Northern Natural Gas in 1963. Has held various positions in corporate planning, design and construction. Elected VP and Chief Engr in 1975, became Pres of Notation Engrg i n1983. *Society Aff:* AIChE, AGA.

McNamee, Bernard M
Business: Drexel University, 32nd and Chestnut St, Phila, PA 19104
Position: Prof of Civil Engr *Employer:* Drexel Univ *Education:* PhD/Struct/Lehigh; MSinCE/Struct/Univ of PA; MBA/Bus/Drexel Univ; BS in CE/Civ/Drexel Univ *Born:* 09/13/30 Director of Architectural Engineering Program, Drexel Univ., 1984 to present. Prof, Head of Dept of Drexel Univ 1976-83; Drexel Univ 1955-83; Chrmn Univ Curriculum Ctte 1975 to 1984; Pres Phila Section ASCE, 1982-83; past mbr of Subctte on Fatigue, ASCE; Column Res Council; Sigma Xi; Tau Beta Pi; Phi Kappa Pi; Chi Epsilon; Lindback Teaching Award, 1971; Married Five Children; Enjoys Music, Golf and Tennis. *Society Aff:* ASCE, ASEE, ACI, SSRC, ASAE

McNamee, Michael A
Business: Box 3354 Univ Station, Laramie, WY 82071
Position: Extension Agricultural Engineer. *Employer:* Univ of WY. *Education:* MS/Soils/Univ of WY; BS/Agri Mech/Univ of WY. *Born:* May 9, 1929 Tecumseh, Neb. Raised on a dryland general farm near Pine Bluffs, Wyoming. US Army 1950-52. Employed as Extension Agricultural Engr, U of Wyoming 1956- . Affiliate Mbr ASAE 1957, Assoc Mbr, 1963. Chmn Rocky Mountain Region ASAE 1964. Dir, Regional Council ASAE 1973-75. Chmn, Advisory Ctte, Western Regional Agricultural Engrg Service 1973-75. *Society Aff:* ASAE.

McNealy, Delbert D
Home: 6523 Los Altos, El Paso, TX 79912
Position: Principal Engr. *Employer:* Intl Bwdy & Water Comm. *Education:* BSCE/Civ Engr/Univ of ID. *Born:* 9/1/25. Educated in ID. Served in 345th Inf in WWII. Eighteen yrs with CA Dept of Water Resources in various positions in water quality, hydrology & flood control activities. Two yrs in E Pakistan, as Proj Mgr for consultive services with E Pakistan Water & Power Dev Authority. Eleven yrs as Principal Engr supervising planning, design, construction & operations & maintenance for US Sec of the Intl Boundary & Water Commission. Past Secy Sacramento

McNealy, Delbert D (Continued)
Chapter CSPE and past charter VP Sacramento Engrs Club. *Society Aff:* ASCE, NSPE.

McNeill, Robert E
Business: 7310 Ritchie Hwy, 912, Glen Burnie, MD 21061
Position: President. *Employer:* McNeill & Baldwin, Inc. *Education:* BE/Elec Engg/Johns Hopkins Univ. *Born:* Dec 1921 Detroit, Mich. US Army 1942-45. CoFounder McNeill & Baldwin 1951. Pres & Ch Electrical Engr 1963- . Mbr IEEE, MSPE, Engr Soc of Baltimore, Bldg Congress of Baltimore. Fellow ACEC 1976. Dir ACEC 1971-75 incl. Reg PE 7 states & DC. Enjoys sailing, photography, music, travel. *Society Aff:* IEEE, ACEC.

McNichols, Roger J
Business: Industrial Engineering Dept, University of Toledo, 2801 W Bancroft, Toledo, OH 43606
Position: Prof. Industrial Engineering *Employer:* Univ of Toledo. *Education:* PhD/Ind Engr/OH State Univ; MSc/Ind Engr/OH State Univ; BIE/Ind Engr/OH State Univ. *Born:* 9/25/38. Joined the faculty of TX A & M Univ in 1966. Became hd of grad engg progs in Maintainability Engg conducted for the Army at Red River Army Depot. Tenured Prof at TX A & M 1974. Pres, McNichols, Street and Assoc, Inc, consulting engrs 1969-76. Prof of Ind Engg at the Univ of Toledo in 1976. Became Assoc Dean of Engg in 1977 to 1980 and was responsible for grad progs and res engg. Presently Professor of Industrial Engr., U Toledo. Profl interests in the areas of reliability, maintainability, quality control, mathematical modeling, numerical control, and applied statistics. Reg PE in TX & OH, certified diver. *Society Aff:* IIE, ASQC, MAA, ΣΞ, ASEE, ΦΚΦ, ΤΒΠ, IEEE, ΑΠΜ, ΠΜΕ.

McNiel, James S
Home: 28 Hampshire Rd, Wayland, MA 01778
Position: Senior Research Advisor *Employer:* Mobil Res and Dev Corp. *Education:* Phd/CE/Univ of TX; MS/CE/Univof TX; BS/CE/Univ of TX. *Born:* 7/4/21. 1950, Univ of TX at Austin (Tau Beta Pi, Omega Chi Epsilon, Phi Lambda Upsilon, Beta Theta Pi, Silver Spurs, tennis). US Army Air Corp, 1943-1945, aircraft pilot. Profl exp: Mobil Res & Dev Corp 1949-1974, improved oil recovery methods, production tech; Mgr, Exploration & Production Res Div, 1964-1974. Pres, Mobil Tyco Solar Energy Corp, 1974-80; dev of silicon ribbon photovoltaics. Mbr AIChE, AIME (Distinguished Lectr, SPE, 1976-77). Married 1945, 3 lovely daughters, dog, cat, & tennis rating of 5.5. *Society Aff:* AIChE, AIME.

McNown, John S
Business: Civil Engrg Dept, Lawrence, KS 66045
Position: Professor of Civil Engrg *Employer:* University of Kansas (on leave). *Education:* DSc/Fluid Mech/U de Grenoble; PhD/Hydraulic Engg/Univ of MN; MS/Hydraulic Engg/Univ of IA. *Born:* Jan 15, 1916. Taught at U of Minnesota, Iowa, Mich & Kansas; Dean, School of Engrg & Architecture, U of Kansas 1957-65. Principal Consultantships with Bureau of Public Roads, Sandia Corp, Bell Telephone Labs, NSF, Ford Foundation, World Bank, & UNESCO. Fulbright Res Scholar, France 1950-51; Croes Medal, ASCE 1955; Fellow, Amer Acad of Mechanics; Advisory or planning boards in Zambia, Nigeria, Guyana, E Africa, Singapore, Thailand, Algeria & Zaire. Reg PE in Kansas. On two-yr assignment as Hydraulic Engineer, Land Planning Unit, Min of Agriculture, PO Box 162, Mbabane, Swaziland. *Society Aff:* IAHR, ASEE, PIANC, ΣΞ, ACS.

McNutt, Willliam J
Home: 233 Eleanor Rd, Pittsfield, MA 01201
Position: Pres *Employer:* Berkshire Transformer Consultants, Inc. *Education:* MS/EE/IL Inst of Tech; BS/EE/Tufts Univ. *Born:* 8/31/27. Worked for GE since 1952 in design & dev of power transformers. Mgr - ADE 1967-1986. Pres of Berkshire Transformer Consultants, Inc 1987. Fellow of IEEE & active in the IEEE Transformers Ctte, of which he is Past-Chairman. Mbr of ANSI & NEMA Transformer Cttes. CIGRE - U.S. Rep to SC-12. Author of numerous tech papers. Reg PE Mass. Tau Beta Pi. Etta Kappa Nu. Sigma Xi. PES Prize Paper Award 1974-1984. MSPE Berkshire County Outstanding Engrg Award 1975. *Society Aff:* IEEE, CIGRE.

McPeters, James B
Business: Box 628, Charleston, WV 25322
Position: District Superintendent. *Employer:* Cabot Corporation. *Born:* Nov 16, 1931. BS New Mexico Inst of Mining & Tech 1958. Petroleum Engrg with honors. Appalachian Sect Chmn SPE of AIME 1971. Bd/Dir SPE of AIME 1974-77.

McPherson, Donald J
Business: 300 Lakeside Dr, Oakland, CA 94643
Position: Vice President & Dir of Tech. *Employer:* Kaiser Aluminum & Chem Corp. *Education:* ScD/OH State Univ; PhD/Metallurgy/OH State Univ; MSc/Metallurgy/OH State Univ; B Met E/Metallurgy/OH State Univ. *Born:* 1921. Native of Columbus, Oh. USNR, Pacific Area in WWII. Metallurgist with Carnegie-Illinois Steel Corp, Battelle Memorial Inst & Argonne Natl Lab prior to 1950; IIT Res Inst 1950-69 advancing from res metallurgist, specializing in alloy dev, to VP. With Kaiser Aluminum & Chem Corp since 1969 as VP - Tech. Over 75 tech papers & talks. Active on NSF, NMAB & other govt cttes. Many positions in ASM & AIME including ASM Trustee 1970-72, Met Soc Bd 1969-72, Campbell Memorial Lecturer 1974. Hon DSc from Ohio State, 1975. Major hobby: baseball card collecting. *Society Aff:* AIME, ASM, ACerS, Inst Met.

McPherson, John C
Business: Rts 523 & 31, Flemington, NJ 08822
Position: Chief Engineer. *Employer:* Eaton Corporation. *Born:* Attended Villanova U - BSME 1950. Majored in Mech Engrg but specialized in Fluid Power for past 25 career years. Holder of 5 patents in Fluid Power related parts, including power brake booster, transmission valving, torque converter & clutches. Cons speaker to many schools & tech soc's. Mbr of ANSI serving on the Fluid Power Systems & Components B-93 Ctte since its inception. Currently the Vice-Chmn of the B-93 Group. Mbr of Natl Fluid Power Assn, representing the 'Material Handling Group' on both sealing devices & cylinder subgroups. A Life Mbr of the Fluid Power Soc having served as Philadelphia Chap Pres (cpt 23) 1968-69 & progressing through the Natl Fluid Power Soc organization to Natl Pres 1972-73.

McRae, John L
Home: 416 Groome Dr, Vicksburg, MS 39180
Position: Pres. *Employer:* Engr Dev Co Inc. *Education:* BS/Civil Engg/Northwestern Tech Inst, Evanston, IL. *Born:* 9/16/17. Native Mississippian. Registered PE, in MS. Retired after 30 yrs in Geotechnical Engg (Soil Mechanics) with US Army Corps of Engrs; now continuing with dev, mfg and sale (intl market) of Gyratory Testing Machine (GTM) for testing soils and bituminous pavements. Spinoff from work with corps (ASTM method D 3387). Hold several patents on soil and bituminous pavement compaction and testing machines. Enjoy church (Baptist deacon), yard and garden, fishing and hunting. *Society Aff:* ASCE, NSPE, ASTM, AAPT.

McReynolds, Leo A
Business: 1413 Hillcrest Dr, Barlesville, OK 74003
Position: Ret. (Consultant) *Employer:* Self *Education:* MS/Phy Chem/Univ of SD; BS/Chem Eng/SD Mines. *Born:* April 23, 1914. Joined Phillips Petroleum Co in 1939. For more than a decade efforts have been in the field of environmental conservation. Assumed respon as Mgr for Environment, Consumer Protection & Standards Div in 1975. Was respon for coordinating & expediting Phillips' efforts in these important areas. Mbr NSPE, Reg PE in Oklahoma; mbr & held numerous offices in API, Coordinating Res Council, Natl Assn of Manufacturers, ASTM & Pres SAE in 1978. Numerous publs & patents. Retired 1979. Now part-time consltg. *Society Aff:* SAE, CRC, ASTM, API.

McRoberts, Keith L
Business: 212 Marston Hall, Ames, IA 50011
Position: Professor & Chairman. *Employer:* Iowa State U of Sci & Tech. *Education:* BS/Indust Engg/IA State Univ; MS/Indust Engg/IA State Univ; PhD/Indust Engg/IA State Univ. *Born:* July 1931 in Clinton, Iowa. Employed as Methods Engr, E I DuPont; Operations Analyst, USAF (civilian scientist); Visiting Lecturer, U of Bradford, England; Dir of Operations Analysis Res Standby Unit. Elected Fellow of IIE, 1976; Mbr of Phi Kappa Phi, Sigma Xi; ASEE; Operations Res Div Dir, IIE 1973- 74; mbr of Traffic Law Enforcement Ctte of the hHighway Res Bd; Dir & Officer of Central Iowa Chap, IIE. Mbr Hospital Bd of Trustees, Mary Greeley Hospital, Ames, IA. Mbr of Bd of Dirs IA State Meml Univ; Dir of Career Dev, IIE, 1979: Respon for academic, res, & extension prog in Indus Engrg at Iowa State U. *Society Aff:* IIE, ASEE, ΣΞ, ΦΚΦ.

McRuer, Duane T
Business: 13766 S Hawthorne Blvd, Hawthorne, CA 90250
Position: President & Technical Dir. *Employer:* Systems, Technology, Inc. *Education:* MS/Elec Engrg/CIT; BS/Elec Engrg/CIT. *Born:* 1925. Pres, Systems Tech, Inc since 1958. Formerly with Northrop Aircraft, Inc & Control Specialists, Inc. R&D in manual & automatic control of aerospace & automotive vehicles, vehicle dynamics, vehicle handling qualities, & man-machine & biological control systems. Co-author 'Aircraft Dynamics and Automatic Control', Princeton, Univ Press 1974; 'Analysis of Nonlinear Control Systems', Wiley, 1961. Awards: Franklin Inst Louis E Levy Gold Medal, 1960, AIAA Mechanics & Control of Flight Award, 1970 & HFS Alexander Williams Award, 1976. Prof Soc's: P Pres Amer Automatic Control Council 1970-72; Fellow IEEE, AIAA, AAAS, HFS & SAE. Reg PE in Calif. Reg Prof Control Systems Engr, 11. *Society Aff:* IEEE, AIAA, SAE, HFS, AAAS.

McShane, William R
Business: Polytechnic University, 333 Jay St, Brooklyn, NY 11201
Position: Prof of Trans and IE *Employer:* Polytech University, Dept of Trans and IE *Education:* PhD/Sys Engrg/Polytech Inst of Brooklyn; MS/Sys Engrg/Polytech Inst of Brooklyn; BEE/Elec Engrg/Manhattan Coll. *Born:* 7/30/43 Prof of Transportation IE, Hd of Dept of Trans and IE, Dir of Polytechnic's Transportation Training and Research Center (TTRC). Teaching and Research in faculty positions at Polytechnic since 1968. Other experience in engrg consulting firms and utilities. Areas of expertise: traffic operations, traffic capacity, transportation noise, microcomputers in trans. Former Sec of TRB Committee A3A10 (Highway Capacity and Quality of Service), Member of ITE Committees. Co-Author of paper given D. Grant Mickle Award of TRB Group 3 (January 1980). Author of other articles in transportation journals. PE in NY and CA (Traffic Engrg). Principal investigator or key participant in research projects in areas of expertise. Fellow, ITE, Senior Member, IIE. *Society Aff:* IIE, ITE, IEEE.

McSorley, Richard J
Business: 333 Ravenswood Dr, Menlo Park, CA 94025
Position: Sr Consultant *Employer:* SRI Intl *Education:* MBA/Bus/Univ of San Francisco; BS/IE/Univ of CA *Born:* 3/7/33 Operations management; development of long range manufacturing strategies; design and implementation of manufacturing facility programs; material planning and systems; material handling; production scheduling; labor standards; capital equipment selection; new product introduction into manufacturing; product cost reduction; MIS specifications and form design. Established production, materials, and manufacturing engrg groups to produce computer peripheral and rotating memory products; designed mfg process, sized factories, engrd labor standards, specified the equip and tooling, designed assembly clean-rooms, located and qualified prospective vendors, and redesigned the products to reduce costs. Managed the manufacturing, manufacturing engrg, industrial engrg and facilities for major manufacturing companies. *Society Aff:* NSPE, IIE

McTish, Matthew J
Business: McTish, Kunkel & Assoc, h6120 Clubhouse Lane, Wescosville, PA 18106
Position: Pres *Employer:* McTish, Kunkel & Assoc *Education:* BS/CE/Univ of Pittsburgh *Born:* 7/18/29 Formerly a structures engr with North American Aviation, Columbus OH (1956); Mgr of Highway Div, Swindell-Dressler Co, Pittsburgh, PA (1967); VP, Wiesenberger Assoc, Allentown, PA (1975); Pres, McTish, Kunkel & Assoc, Wescosville, PA (to present time). Specialties include municipal engrg, transportation engrg and planning. *Society Aff:* ASCE, NSPE, ASHE, APWA

McVey, Eugene S
Business: Thornton Hall, University of Virginia, Charlottesville, VA 22901
Position: Prof of EE *Employer:* Univ of VA *Education:* PhD/Elect Engr/Purdue U; MS Engr/Eng Sci/Purdue U; BSEE/Elect Engr/Univ Louisville *Born:* 12/6/27 From 1950 to 1956, worked at Naval Avionics Facility, Indianapolis, IN; Served as Group Leader at Farnsworth Electronics Co, Fort Wayne, IN, where I worked from 1956 to 1957; Was an instructor and Assistant Prof from 1957 to 1961 at Purdue Univ; Currently Alice M & Guya Wilson Prof of Electrical Engrg and Co-Chrmn of the Control and Information Systems Research Group at the Univ of VA. Published over 100 technical papers, have eighteen patents and have been advisor for twenty two PhD recipients. Member of Sigma Xi, Eta Kappa Nu, Tau Beta Pi, and the American Assoc of Univ Profs. Research, educator and engrg practice in electronics, automatic control and image processing. *Society Aff:* IEEE, ΣΞ, AAUP

McVickers, Jack C
Home: 21604 Stableview Drive, Gaithersburg, MD 20760
Position: VP, Mfg & Engg. *Employer:* McGraw-Edison Co. *Education:* MS/Mech Engr/Univ of Pittsburgh; BS/Mech Engr/IL Inst of Tech. *Born:* 11/7/32. Native of Chicago, IL. Served with Navy Seebees in Asia 1951-55. Automation engr Gen Motors. Joined Westinghouse 1958. Awarded Westinghouse Advanced Mech MS Fellowship. Mfg engrg, prog/plant mgt in defense/space electronics, gyroscapes, electro-optical systems and semiconductors. Dir, Westinghouse Mfg Dev Labs in US and Europe. Corp Dir, Prod Tech. Mgr, Mech R&D Div, Corp Res and Design Ctr. McGraw-Edison Co 1977-79 as VP, Mfg & Engg, 1979-81 Brown Boveri, Electric, VP. Joined Martin Marietta Aerospace in 1981 as Dir of Prod Improvement.

McWee, James M
Business: 711 N Alvarado St, Los Angeles, CA 90026
Position: Vice President/Treasurer *Employer:* LeRoy Crandall & Assocs. *Education:* BSc/Civil Engr/UC Berkeley; Cert of Engrg Mgmt/Engrg/UCLA. *Born:* 9/15/28. in Los Angeles, CA. Vice President and Treasurer, LeRoy Crandall & Assocs, Consulting Geotechnical Engrs, Los Angeles, a subsidiary of Law Engrg Testing Co. Fellow of ASCE; P Pres of Assoc Mbr Forum, LA Sect; Past Treas and VP of LA Sect, ASCE. Mbr ACEC and ASFE. Reg Civil Engr in CA, IL, OR & NM. *Society Aff:* ASCE, ASFE, ACEC, AEG, IAE.

McWhirter, James H
Home: 3885 Brookside Dr #505, Murrysville, PA 15668
Position: Fellow Engineer *Employer:* Westinghouse R&D *Education:* MS/Elec. Engrg./Carnegie Mellon University; BS/Elec. Engrg./Columbia University *Born:* 07/24/24 Born in Mercer, PA. Served as naval engrg officer in World War II. Transformer design and dev engr for Westinghouse Electric 1948-65. Research engineer at Westinghouse R&D Center from 1965 until present. Work has focused on analysis and optimal design in a variety of areas including transformers, Tokamak fusion reactors, electromagnetic systems, and semiconductor devices using mathematics, numerical analysis, and computers as tools. Current professional society activities include serving as a Regional Coordinator of Governmental Relations for the IEEE. *Society Aff:* IEEE

McWhirter, John R
Home: 75 Ledgebrook Dr, Norwalk, CT 06852
Position: VP and General Manager Agricultural Products Div *Employer:* Union Carbide Corp. *Born:* Dec 29, 1937. BSChE U of Ill 1959; MS & PhD Penn State U 1961 & 1962 in Chem Engrg. Employment: E I DuPont 1962-63 Res Engr; Mixing

McWhirter, John R (Continued)
Equip Co 1963-66 Mgr of R&D; Union Carbide Corp 1966- , Genl Mgr sincwe 1973, Vice President since 1976. Career activities: extensive R&D & tech contribs in areas of mass transfer, fluid dynamics & mixing, & biochemical oxidation. 10 years exper in wastewater treatment & water pollution control & holder of numerous patents in field. Inventor of UNOX oxygen activated sludge sys for secondary treatment. Awards & Honors: Chem Engrg Personal Achievement Award 1970; Chem & Engrg News Chemical Innovator 1971; Chem Engrg Kirkpatrick Award to Union Carbide for UNOX dev 1971; Western NY ACS, 1976 Schoellkopf Medal.

McWhorter, John C
Home: P O Box 222, Starkville, MS 39759
Position: Professor of Agri & Bio Engrg. *Employer:* Mississippi State University. *Education:* BS/Agri Engg/Clemson Univ; MS/Agri Engg/TX A&M Univ. *Born:* Aug 1916. Native of Spartanburg, SC. Engr with Soil Conservation Service, Texas 1941-47 & served in US Army 1942-45 (as infantry officer). With Mississippi State U since 1947 in teaching & res in Soil & Water Engrg. Reg PE in Miss. Merit Award for Teaching, 1970, Gamma Sigma Delta. Fellow, 1973, ASAE. Faculty Achievement Award for Classroom Teaching, 1976, Miss State U Alumni Assn. *Society Aff:* ASAE.

McWilliams, Bayard T
Business: P.O. Box 177, Yardley, PA 19067
Position: President *Employer:* McWilliams Engineering Inc. *Education:* MS/Aero Engg/Poly Inst of NY; MBA/Bus Admin/Univ of WA; BS/Mech Engg/PA State Univ. *Born:* Dec 1924. Ebensburg, Pa. Served with U.S. Army 1943-1946; Officer, Corps of Engineers 1944-1946. Test Engineer, Sales Engineer and Field Engineer with Wright Aeronautical Corp.; Research Engineer with United Aircraft Corp.; Industrial Engineer with Pilotless Aircraft Division of Boeing; V.P. and Chief Engineer W.P. Damon Corp.; Chief Development Division, LOH Program Mgr. and Deputy to Asst. C. of S., RDT&E Army Aviation Material Command, 1958-1963; Engineering Group Manager, Technical Consultant at Naval Air Propulsion Center 1963-1982; President, McWilliams Engineering Inc. 1971-date; Forensic Engineering Expert, National V.P. ASME, V.P. Board of Prof. Practice & Ethics ASME; Fellow, ASME; Associate Fellow AIAA; Engineer of the Year - Mercer County; First Chairman Central Jersey Engineering Council; Registered P.E. in N.J. and PA; Vice-Chairman Engineering Affairs Council of AAES. *Society Aff:* ASME, AIAA, AHS, NSPE, AAAS

McWilliams, Thomas G, Jr
Business: Widener University, School of Engineering, Chester, PA 19013
Position: Professor & Dean of Engrg *Employer:* Widener University *Education:* PhD/Chem Engrg/Univ of MD; BES/Chem Engrg/The Johns Hopkins Univ *Born:* Jan 1934. Native of St Mary's County, Md. Instructor of Met, U of Md 1958-63. Asst Prof, Old Dominion Coll 1963-66. Current position at Widener University since 1980, West Va Inst of Tech Chrmn Dept of Chem Engr 1966-80 (promoted to Prof, 1970). Assumed chairmanship of the Dept of Chem, 1971-76. Specialization in materials, thermodynamics & vapor-liquid equilibria. Chmn, Hampton Rd Chap ASM 1966; Chmn, West Va Chap ASM 1972-73. Citation of Merit, Charleston Sect, AIChE, 1973. Aptd to Federation of Matls Socs Govt Liaison Committee as AIChE representative 1979-82. Elected Mbr-at-Lrg Charleston Sect, AIChE 1979, Secretary 1980, Mbr Tau Beta Pi. Vice Chmn Mid-Atlantic Section of ASEE 1985-86. Bd of Dir, Univ City Sci Center, 1984-86. *Society Aff:* AIChE, TBΠ, ASEE

Mead, Carver A
Business: 1201 CA Blvd 256-80, Pasadena, CA 91125
Position: Gordon and Betty Moore Prof of Computer Sci *Employer:* Caltech. *Education:* PhD/EE/Caltech; MS/EE/Caltech; BS/EE/Caltech *Born:* Received the BS degree in 1956, the MS degree in 1957, and the PhD in 1960, all from the CA Inst of Technology, Pasadena. Has been a mbr of the faculty of that inst since 1957. Research has contributed to the understanding of tunneling in solids, current-flow mechanisms in thin dielectric films, metal-semiconductor barriers, band energies in semiconductors, and electronic processes in insulators. Has proposed and demonstrated the operation of a number of new solid-state electronic devices and holds several US patents. Fellow of the American Physical Society and mbr of Sigma Xi, Mbr of National Acad of Engrg, foreign member of Swedish Acad of Engrg Sci (IVA). He is the recipient of the T. D. Callinan Award (1971), the Electronic Achievement Awd (1981) and the Harold Pender Award (1984). His current res focus and teachings are in the area of VLSI design, ultra concurrent systems and phsics of computation. *Society Aff:* NAE, IVA

Mead, Lawrence M
Home: 16 Bay Crest, Huntington, NY 11743
Position: Aerospace Conslt & Retired Senior VP, Departmental Operations. *Employer:* Grumman Aerospace Corporation. *Education:* CE/Civ Engrg/Princeton Univ; BS/Civil Engrg/Princeton Univ. *Born:* May 1918. BSE 1940, CE 1941. Princeton University. Harvard Bus AMP course 1964. Joined Grumman 1941. Stress Analyst on WWII fighters F-6F, F-7F, F- 8F, F-9F. Proj Engr, XF10F, F11F-1F. Proj Mgr A-6A Intruder, proposal to Fleet intro 1957-1963. Headed design studies leading to F-14 fighter 1964-67 Proj Mgr- Gulfstream III Business Jet- 1977-80. Elected VP, F-14 Mfg Dir, 1969. VP Tech Operations 1972-75 directing tech depts. Elected Sr VP 1975, Dir of corp technical depts, Engrg, Logistic Support, Flight Test, Training, Software, Retired 1983, Aerospace Conslt 1983-. Fellow AIAA, Mbr SOLE, SAMPE, AUSA, Past 1981-1983 Director ADPA, AeroTech Council AIA, Phi Beta Kappa, Sigma Xi, Tau Beta Pi. Director N.Y. City Hall of Science. Dir & Past Chairman, Long Island Forum for Technology. Trustee Village of Huntington Bay 1974-1980. Sail, ski, & tennis. *Society Aff:* AIAA, SOLE, SAMPE

Mead, William J
Business: 563 Main St, Ridgefield, CT 06877
Position: Director of Manufacturing. *Employer:* Combe Incorporated. *Education:* MS/Mgmt/Stevens Inst; ChE/Chem Engg/OH State; ChE/Chem Engg/OH State. *Born:* Dec 29, 1927. Reg PE NY, Ohio, Indiana, Ill. Now VP of Combe Inc. With Combe Inc since 1971 in charge of mfg & engrg worldwide, in the production of drug & cosmetic items. Formerly VP Operations of the Chemway Corp, VP Mfg of the Alberto-Culver Co, & Tech Dir of Home Products Internatl Ltd. Mbr AIChE, ACS, SCC, AAAS. Editor-in-Chief 'Encyclopedia of Chem Process Equip'. Patentee in field. Fellow Amer Inst of Chemists. *Society Aff:* AIChE, ACS, AIC, SCC.

Meadows, Paul
Home: 3704 Ivydale Dr, Annandale, VA 22003
Position: Retired *Education:* BS/Petrol Engrg/TX Tech. *Born:* 11/15/22. Native of Ralls, TX. Served with Army Air Corps 1943-46. Joined Bureau of Mines, US Dept of the Interior, as Petrol Engr in 1949, serving at Bartlesville, OK, Dallas, TX, Wichita Falls, TX, & since 1966 at Wash, DC. Held increasingly responsible engrg & mgt position concerned with petrol & natural gas, energy, & mineral supply/demand problems & policy. Received the Dept of the Interior's Meritorious Service Award in 1973 & Distinguished Service Award in 1979. Retired from Bureau of Mines 08/31/80. *Society Aff:* SPE.

Meagher, George Vincent
Business: 10 Park Lawn Rd, Toronto Ont M8Y 3H8 Canada
Position: President *Employer:* Dilworth, Secord, Meagher & Assoc Ltd. *Education:* BEng/Mech/McGill Univ, Montreal; BSc/Physics/Dalhousie Univ, Nova Scotia; Eng Dip/-/Dalhousie Univ, Nova Scotia. *Born:* 4/23/19. in Halifax, Canada. Pres Dilworth, Secord, Meagher & Assocs Ltd, Cons Engrs, Toronto; Chmn, Champlain Power Products Ltd, Constructors, Toronto. Vice- Chairman Tata-Dilworth, Secord, Meagher & Assocs, Cons Engrs, Bombay. Fellow Engg Inst of Canada. Mbr Assn of PE's of Ontario. *Society Aff:* APEO, ACEC, EIC.

Meagher, Ralph E
Business: P O Box 356, South Bend, IN 46624
Position: R E Meagher (Self-employed) *Education:* PhD/Physics/Univ of IL; MS/Physics/MA Inst of Tech; BS/Physics/Univ of Chicago. *Born:* Sept 1917. Grp Ldr, Radiation Lab, MIT 1941-45. Designed first production PPI indicators for radar. U of Ill faculty 1948-58. Ch engr for ORDVAC & ILLIAC I digital computers. Res Prof of Physics & Elec Engrg & later Head, Digital Computer Lab, U of Ill. Since 1958 cons in electronics & computers. Predoctoral Fellow, Natl Res Council 1945-48. Pres's Certificate of Merit (for radar work) 1947. Editor, IRE Transactions on Computers 1954-56. Fellow in IEEE, Fellow in Amer Physical Soc. *Society Aff:* IEEE, APS, ACM.

Means, James A
Business: SAMTO/CA, Vandenberg AFB, CA 93437
Position: Tech Dir *Employer:* Space & Missile Test Org *Education:* PhD/EE/Univ of CA-Santa Barbara; MS/EE/Univ of AZ; BS/EE/Univ of AZ *Born:* 10/11/37 Is the Tech Dir at the Space and Missile Test Organization, Vandenberg Air Force Base. He has specialized in electronic engrg and computer science and has a doctorate from Univ of CA, Santa Barbara. Dr Means has been responsible for numerous design and development projects related to range instrumentation and test and evaluation since he started his federal career with the Navy at Point Mugu in 1962. He holds two US patents, has published in the IEEE Transactions, teaches computer science for Chapman Coll, has published numerous technical articles, and is a frequent speaker at tech symposia. *Society Aff:* AIAA, ITEA, IFT, ΣΞ

Mebus, Charles F
Business: 707 Transportation & Safety Bldg, Harrisburg, PA 17120
Position: Engr Mbr - Bd of Claims *Employer:* Commonwealth of Pennsylvania *Education:* BS/San Engg/PA State Univ; BS/Chem/PA State Univ. *Born:* June 15, 1928 Abington, Pa. Son of George B & Estelle C (Negus) Mebus. Graduated Penn State U 1949, BS Chem & 1951, BS Sanitary Engrg. Served on active duty with the US Army Corps of Engrs from Nov 1951 to Oct 1953, part of this period was duty with the Army of Occupation in Germany. Except for active duty professional career was spent in the employ of George B Mebus, Inc - Cons Engrg, doing public works engrg for various municipalities & municipal authorities in eastern Penn until Dec, 1978. Mbr ASCE, NSPE, & Mbr Penn House of Reps 1965-78. Chief Clerk, PA House of Representatives 1979-1980. Asst to Majority Leader - House of Reps. 1981; Dep Secy of Rev for Adm - Jan, 1982 to Mar, 1983; Dep Secy of Gen Serv for Pub Works - Mar, 1983 to June, 1983; Engr Mbr of the Bd of Claims - June, 1983 to present. *Society Aff:* ASCE, NSPE.

Mechlin, George F
Business: P.O. Box 11426, Pittsburgh, PA 15238-0426
Position: Retired. *Education:* PhD/Physics/Univ of Pittsburgh; MS/Physics; Univ of Pittsburgh; BS/Physics/Univ of Pittsburgh. *Born:* 7/23/23. Radio & radar instructor for US Naval Reserve 1944-45. Res Asst & Lecturer, 1945-48 for U of Pittsburgh. Pre-Doctoral Fellow, Atomic Energy Comm 1948-49. 1945-87, Westinghouse Elec Corp. VP of R&D. Received USN Meritorious Public Service Award in 1961. Recipient of John J. Montgomery Award. Received Westinghouse "Order of Merit" in 1961 for his engrg contribs to the dev of underwater & surface launching & handling systems for the Polaris missile. Mbr of Sigma Xi, Amer Physical Soc, Marine Tech Soc, AIAA, Amer Ordnance Assn, & Amer Oceanic Organization (Pres 1968). Mbr of Natl Acad of Engrg & Chrmn of Marine Bd of Assembly of Engrg, NRC 1976-79. Mbr, Exec Ctte, Assembly of Engrg of NAE (1981-). Chrmn, Planning Group, Assembly of Engrg (1981-), NAE. Dir of Pittsburgh Broadcasting Co (1980-). Mbr of Res Adv Ctte of US Coast Guard, 1973- 5. Mbr of Naval Res Adv Ctte, Lab Adv Bd for Naval Ships 1975-78. *Society Aff:* NAE, APS, AIAA.

Meckler, Milton
Home: 16348 Tupper St, Sepulveda, CA 91343
Position: Pres./CEO *Employer:* The Meckler Group *Education:* BSE/CHE/Worcester Polytechnic; MSE/CHE/Univ of MI *Born:* 12/29/32 Native of Red Bank, NJ WPI Peel Prize (1953). Res Asst, ERI, Univ of MI (1955). Dev Engr, Esso Res & Engg Co (1956), & Surface Combustion Corp (1957- 58). Design Engr & Assoc, Meckler Engg Co & Meckler, Hoertz Assoc (1958-63); Sr Mech Engr, DMJM (1963-65). Prin, Silver Meckler and Assoc (1965-1969); Hellman, Silver, Lober & Meckler, Inc (1969-70). Meckler Assoc, Inc (1970-74), Pres/CEO, Meckler Associates and Envirodyne Energy Servs, (1974-75), both subsidiaries of Envirodyne Industries (OTC). The Energy Group, subsidiary Welton Becket Assoc (1975-80). Organized The Meckler Energy; Group, Meckler Engrs; Group, Meckler Constructors Group, and Meckler Sys Group, Co-founder-CA; Solar Tech, Inc (1980). Mbr Tau Beta Pi and Sigma Xi. Contributor to several well known engg handbooks. Publ over 130 tech papers/articles author of "Energy Conservation in Bldgs & Industrial Plants–, McGraw-Hill (1981) and Editor "Pumps and Pump Sys–, ASPE, 6 vol, MSG Envir & Energy series and "Retrofitting for Energy Conservation" Van Nostrand Reinhold (1984). Reg PE in Mech, Chem, industrial and mfg branches CA and other states. Licensed CA General Building and Engineering Contractor, Certified CA Energy Auditor, Certified Mfg Engr (SME) and holder NCEE Certificate of qualification. Appointed an instructor in 1981 and an Adj Assoc Prof in 1982, and an Adj Prof in 1983 on the faculty of the Mech and Chem Engrg Dept, Sch of Engrg and Computer Scis at CA State Univ Northridge. *Society Aff:* ASME, ACS, SME, ASHRAE, AIChE, NSPE, ASM, ASPE, SSS, AIC, AIIE, ASTM, ICBO, AIC, NAFE, CSI, SAMC, APCA.

Medbery, H Christopher
Home: 1245 Roble Rd, Millbrae, CA 94030
Position: Consultant *Employer:* James M Montgomery Consulting Engrs. *Education:* BS/CE/Univ of CA, Berkeley. *Born:* 10/13/09. Born in 1909 & completed high sch in SD. Three yrs of surveying & mapping prior to college grad. Thirty-eight yrs with San Francisco Water Dept retiring in 1974 as Asst Gen Mgr & Asst Chief Engr. Married to Marian Massie in 1940 and have three children. Activities in AWWA include Chrmn of the CA Sec, dir 1960-63 & Pres of AWWA in 1969 - Fuller Award & Honorary Mbrship. Fellow of ASCE & Pres of San Francisco Sec in 1957. Reg Engr in CA. Publications in AWWA Journal & Trade Magazines. Currently affiliated with James M Montgomery Consulting Engrs. Avocations - golf, travel, fishing. *Society Aff:* ASCE, AWWA, AAEE.

Meditch, James S
Business: Dept of Elec Engr FT-10, Univ. of Washington, Seattle, WA 98195
Position: Prof Dept of Elec Engrg. *Employer:* Univ of WA, Seattle. *Education:* PhD/EE/Purdue; SM/EE/MIT; BS/EE/Purdue. *Born:* 09/30/34. in Indianapolis, IN. BS & PhD from Purdue Univ 1956 & 1961, respectively; SM from MIT 1957. Staff Engr, Aerospace Corp 1961-66; Assoc Prof of EE, Northwestern Univ 1965-67; Sr Res Scientist, Boeing Sci Res Labs, Seattle 1967- 70; Prof of EE, U C Irvine 1970-77; Prof of EE, Univ of WA 1977-, Chmn of EE, 1977-85. Editor, Proc IEEE 1983-1985. Fellow IEEE 1976-, Dist. Mbr IEEE Control Sys Society 1983, IEEE Centennial Medal 1984. Res interests: computer-communication networks. *Society Aff:* IEEE, AAAS, ASEE.

Medwadowski, Stefan J
Business: 111 New Montgomery, San Francisco, CA 94105
Position: Consulting Engr (Self-employed) *Education:* PhD/CE/U of CA, Berkeley; Dipl Eng/CE/U of London. *Born:* 1/17/24. Born in Poland, served in the Home Army during WWII; participated in the Warsaw Uprising in 1944. Engrg educ in Warsaw, Rome, London and Berkeley (PhD, 1956). Founded consltg engrg firm in San Francisco in 1958. In addition, has held appointment as Adj Prof at the Univ of CA at Berkeley. Resp for design of many structures, a number of which received awards for excellence in design. Designed the structure of the 10-m Keck Telescope. Publ numerous papers on analysis and design of structures. VP, IASS, and Editor-in- Chief of the "Bulletin of IASS–. Chmn ACI-ASCE Shell Ctte, 1979-83. Particularly interested in the relationship between structural action, from and esthetics. *Society Aff:* ASCE, ACI, IASS, SEAONC.

Mee, C Denis
Home: 105 Stonybrook Rd, Los Gatos, CA 95030
Position: IBM Fellow *Employer:* IBM. *Education:* DSc/Physics/Univ of Nottingham; PhD/Physics/Univ of Nottingham; BSc/Physics/Univ of London. *Born:* 12/28/27. 1951-54 worked in magnetic res lab Steel Co of Wales. Later joined MSS Recording Co, engaged in program to develop data recording tape. 1957 joined CBS Labs in Stamford, CT, directing progs in magnetic recording. 1962 joined IBM as mgr magnetics devices group, Yorktown NY. 1967-83 Mgr of Advanced Tech, GPD Laboratory, San Jose Ca. Fellow of IEEE. Received Achievement Award in 1964 for outstanding contribs to magnetic recording. Author of book *The Physics of Magnetic* Recording. Also editor/author of "Magnetic Recording" Vol I Tellnology 1987 McGraw-Hill. Eds. C.D.Mce and E.D. Daniel. Appointed IBM Fellow 1983. *Society Aff:* IEEE.

Meecham, Wm C
Home: 927 Glenhaven Dr, Pacific Palisdes, CA 90272
Position: Prof. *Employer:* UCLA. *Education:* PhD/Phys/Univ of MI; MS/Phys/Univ of MI; BS/Phys/Univ of MI. *Born:* 6/17/28. I have published over 100 papers in the fields of Acoustics & Fluid Dynamics as well as in a number of other fields of applied phys. I have done a large amount of res in the field of the effects of aircraft noise on communities and have appeared on local & natl TV, in virtually all maj US & overseas newspapers (all many times) & had news stories in Time, Newsweek & many other natl magazines. I have done correlation experiments on the tech aspects of noise production by jet aircraft. Together with associates, I have done much work in the difficult field of theories of turbulent fluid flow, deriving certain math characteristics for the first time. *Society Aff:* AIP, APS, AIAA, ASA.

Meedel, Virgil G
Business: 8404 Indian Hills Dr, Omaha, NE 68114
Position: Area Manager, Western Region *Employer:* HDR Infrastructure *Education:* MSc/Civil Engr/Univ of NE; BSc/Civil Engr/Univ of NE *Born:* 11/26/27. Various positions with Union Pacific Railroad for 10 yrs. Design & Proj Engr with Kirkham, Michael & Assocs. With Henningson, Durham & Richardson since 1965. VP since 1979. Area Manager with respon for fiscal & personnel control & for proj mgmt of civil engineering projects. Sigma Tau; Hamilton Award for Proficiency in Humanistic Studies; PE's of NB Outstanding Service Award, 1971; Pres, NB Sect ASCE, 1973; Dir, District 17, ASCE 1975-78, VP Zone III ASCE 1978-79. *Society Aff:* ASCE, AREA

Meenaghan, George F
Home: 2-B, Hammond Ave, Charleston, SC 29409
Position: VP for Acad Affairs & Dean of the College. *Employer:* Citadel *Education:* BS/Chem Engg/VPI; MS/Chem Engg/VPI; PhD/Chem Engg/VPI. *Born:* 6/7/29. Native of Holyoke, MA. Taught Chem Engg at Clemson Univ, 1955-68. Chmn & Prof Chem Engg, TX Tech Univ 1969-74. Dean for Res & Dir of Res Services, TX Tech Univ Complex, 1975-77. Associate VP Res, 1977-79. Respon for res for Univ, Medical Sch & Museum; all extramural grants contracts; contract negotiations & interdisciplinary res progs. VP for Academic Affairs & DEan of the College, The Citadel, 1978-present. Respon for all acad matters in college. As prof still teach, publish, etc. Enjoy WWII history & golfing. *Society Aff:* AIChE, ASEE, TBΠ, ΣΞ.

Meghreblian, Robert V
Business: Cabot Corp, 950 Winter St, Waltham, MA 02254
Position: VP Cabot Corp, Dir, Corp Plan & Dev *Employer:* Cabot Corporation. *Education:* PhD/Aero and Math/CA Inst of Tech; MS/Aeros/CA Inst of Techn; BAeE/-/Rens Poly Inst. *Born:* Sept 6, 1922. US Navy 1941-46; Atlantic & Pacific campaigns; discharged as Ltjg. Pioneer in dev of rockets & satellites, Jet Propulsion Lab 1947-49 & leader in space sciences with respon for first payloads flown to Moon, Venus, & Mars 1958-68; Deputy Asst Lab Dir, Tech Divisions 1968-71. Major contributor to theory of Nuclear Reactors & nuclear propulsion, Oak Ridge Natl Lab 1952-58 & Assoc Prof Caltech 1960-61. Currently VP Res & Engr, Cabot Corp, Boston 1971-79; VP Cabot Corp and Pres, DISTRIGAS Corp, 1979-1985. 1986-1987 VP and General Mgr, Cabot Crystals Bus. Unit; 1987-present VP, Dir Corp Plan & Dev; 1975-present, Curator Cabot Corp Collection. Mbr Sigma Xi, AIAA, & Fellow ANS. Guggenheim Fellow 1949-51. Author *Reactor Analysis*, McGraw-Hill; Accomplished artist. *Society Aff:* ANS, AIAA

Megna, John C
Home: 7825 SW 148 St, Miami, FL 33158
Position: Vice President. *Employer:* Amer Hospital Supply Corp, DADE Div. *Education:* BS/Chem Engg/Univ of WI; BS/Biochem Engg/Univ of WI. *Born:* Oct 1927. BSChE & Biachemical Engrg U of Wisconsin. Native of Milwaukee, Wisc. Served in US ARmy 1946-47. Extensive exper in pharmaceutical, biochemical & food industries in Res, Dev & as a mgmt cons. Joined Amer Hospital Supply Corp in 1971 as VP of Mfg of DADE Div. Respons also include DADE's Finishing Operations, Purchasing, Production Inventory/Control, Mfg Engrg activities & a separate mfg unit in Puerto Rico & Gibbstown, NJ. Granted 9 US & foreign patents. *Society Aff:* ACS, AIChE.

Megonnell, William H
Business: 1111 Nineteenth St, NW, Washington, DC 20036
Position: Dir Legislative Affairs (Environment) *Employer:* Edison Electric Inst. *Education:* MS/Eng/Harvard; MS/SE/Penn State; BS/CE/Penn State. *Born:* June 1923. Native of Dauphin, Pa. Army Air Corps 1943-46. Commissioned in US Public Health Service 1951; served in Federal air pollution prog from first law in 1955 until retirement as Asst Surgeon Genl in 1973. USPHS Commendation Medal 1964, & Meritorious Service Medal 1967. APCA S Smith Griswold Award 1973. Presently advisor to investor-owned electric utilities on environmental legislation. *Society Aff:* APCA, AAEE, CFEE.

Mehring, Clinton W
Home: 1821 Mt. Zion Dr, Golden, CO 80401
Position: Pres *Employer:* Tipton and Kalmbach, Inc *Education:* MS/CE/CO Univ; BS/CE/Case Inst of Tech *Born:* 02/14/24 Native of New Haven, IN. Served in US Army 1943-1945. Served as design engr in earth dams and canal div of US Bureau of Reclamation from 1950 to 1956. With Tipton and Kalmbach, Inc since 1956. Served as design engr on projects involving power plants, tunnels and concrete dams 1956-1959. Asst resident engr on Cumbaya Hydroelectric Project in Quito Ecuador 1959-61. Asst Chief Design Engr on Link Canals Project in Lahore, Pakistan 1961-65. VP 1966-1973. Exec VP 1973-1979. Assumed present position in 1979. Responsible for operation of company's activities in USA and abroad. *Society Aff:* ASCE, USCOLD, USICID, ACI

Mehrkam, Quentin D
Home: 321 Highland Ln, Bryn Mawr, PA 19010
Position: Pres *Employer:* Ajax Electric Co. *Education:* BA/Metallurgy/Lehigh Univ. *Born:* June 7, 1921 Allentown, Pa. Employed by Ajax Elec 1980 Co, Huntingdon Valley, Pa since 1946. Elected to Bd/Dirs in 1949. Elected VP, R&D in 1964. Served on many local & natl cttes for the ASM. Elected Chmn Phila ASM Liberty Bell Chap 1968-69. Served as Chmn of the Salt Bath Activity of the ASM Heat/Treat Council - 1974- 76. Fellow ASM 1976. ASM Eisenman Medal 1977 Delaware Valley Man of the Year-1978. Elected Sr VP Ajax Elec 1980. Elected Pres. Ajax Electric 1984 and Chrmn of Bd. Elected Chrmn of Bd, Central Panel Co 1985. *Society Aff:* ASM.

Mehta, Prakash C
Home: 412 Hickory Ln, Munster, IN 46321
Position: Proj Mgr. *Employer:* Salisbury Engg Inc. *Education:* MS/Civil Engg/IL Inst

Mehta, Prakash C (Continued)
of Tech; BS/Civil Engg/Rajasthan Univ. *Born:* 7/12/40. Reg PE in the states of IN and IL and Reg Structural Engr in the state of IL. Twenty yrs of varied professional experience include structural analysis ad design, construction mgt, preparation of reports on geotechnical study of construction sites and teaching soil mechanics and structural engg. Assumed current responsibilities as proj mgr of geotechnical group at Salisburg Engg in 1976. The responsibilities include correlating subsurface investigation data, engg analysis, preparing soils reports, structural design of fdns and mgt of non-destructive testing div. Treasurer, Calumet Chap, IN Soc of Professional Engrs. Enjoy photography and gardening. *Society Aff:* NSPE.

Mei, Kenneth K
Business: Dept of Elec Engg & Comp Sci, Berkeley, CA 94720
Position: Prof. *Employer:* Univ of CA. *Education:* PhD/EE/Univ of WI; MS/EE/Univ of WI; BS/EE/Univ of WI. *Born:* 5/19/32. Received B.S., M.S. and Ph.D in electrical engineering in 1959, 1960, 1962 respectively of the University of Wisc. Madison, Wisc. Over 40 publications in scientific Journals. Received Best Paper Award & Honorable Mention to Best Paper Award of IEEE Antennas & Propagation Society in 1967 & 1974 respectively. Was member of the Administration Committee of Antennas & Propagation Society, & Associate editor of its transactions. Interest in computer applications in antennas & MMIC (monolithic microwave integrated circuits). *Society Aff:* IEEE, URSI, EMS.

Meier, Donald R
Home: 1650 Seven Oaks Drive, Morristown, TN 37814
Position: Retired *Employer:* General Elec. Co. *Education:* BSME/Aeronautics/IA State Univ. *Born:* Nov 1910. Employed GE Co 1933-71. Advanced Engrg Course 1933-36. Designed electric traction machinery 1936-48. Mgr mech design 1945-48; lab devs 1948-50. Mgr Locomotive Engrg 1950-62, incl dev of gas turbine-electric & straight elec locomotives; locomotives for export; introduction new design concepts, domestic diesel elec locomotives. Mgr Locomotive Business Planning 1962-71. ASME Life Fellow, author 15 papers; mbr Railroad Div Ctte 1962-75. Reg PE (Pa). Transportation cons 1972- . Plan for marginal rail Lines 1973-76. Mbr school Bd 1952-69. Study of U.S. Secondary Schools Performance Statistics 1985-87. *Society Aff:* ASME

Meier, France A
Business: Dept of Indus Engrg, Arlington, TX 76019
Position: Professor of Indus Engrg. *Employer:* University of Texas at Arlington. *Education:* BS/Indus Eng/TX Tech; MS/Indus Eng/Univ of Houston; PhD/Eng/Wash Univ (St Louis, MO). *Born:* Aug 11, 1928. Educ: BSIE 1951, Texas Tech; MSIE 1959 U of Houston; PhD 1966 Washington U (St Louis). Prof activities: Prof of Indus Engrg, U of Texas at Arlington; Plant, Indus Engrg Steven Mfg Co, St Louis 1961-62; Asst Prof of Indus Engrg, Lamar U, Beaumont, Texas. Con activities at Texas Instruments, General Dynamics, & City of Dallas. Prof Soc's: Amer Inst Indus Engrs Regional VP 1975-77; Co-CHmn 1977 Natl Conf, Inst Dir of Student Chap Affairs 1972-74; Pres of Ft Worth Chap 1969-70. Pres of North Texas Chap of ORSA 1968. Mbr of ASEE, The Inst of Mgmt Sci, Regional Dir of Tau Beta Pi 1979-81, Alpha Pi Mu, Sigma Xi, Upsilon Pi Epsilon. Sponsored Res with US Dept of Transportation & US Post Office. *Society Aff:* IIE, TIMS, ACM, ASEE.

Meier, Reinhard W
Business: Mail Station 3458, Special Products Division, WILD Heerbrugg AG, Heerbrugg, Switzerland CH 9435
Position: Prog Mgr. *Employer:* Wild Heerbrugg Ltd *Education:* PhD/Physical Chem/Swiss Fed Inst of Tech Zurich; Dr. sc. nat./-/- *Born:* 01/11/34 and educated in Switzerland; naturalized US citizen since 1968. Research Work with the US Army Electronics Labs (1963/64) and Xerox Corp (1964/70), primarily in the fields of photography and optical holography. Mgmt position with Ciba-Geigy Photochemical Ltd, Fribourg, Switzerland 1970-1977 (head of quality control, mgr of photographic coating plant). Ofcr of SSC Steril Catgut Corp (medical products) in Neuhausen Switzerland 1977/80, in charge of production and engrg. 1980-1984 head of the Industrial Sys Engrg Dept of Gammaconslt Inc in Winterthur, Switzerland. Since 1984 Mgr for dev prog in Optronics at Wild Heerbrugg Ltd, Switzerland. *Society Aff:* OSA

Meier, Wilbur L, Jr
Business: University of Houston System, 4600 Gulf Freeway, Houston, TX 77023
Position: Chancellor *Employer:* Univ of Houston System *Education:* PhD/-/Univ of TX, Austin; MS/-/Univ of TX, Austin; BS/-/Univ of TX, Austin. *Born:* 1/3/39 Dr. Meier has had prof and consulting positions in a variety of government agencies and private companies. Since 1967, he has held profl and administrative positions of increasing responsibility in engrg at Texas A & M, Iowa State, Purdue, and Penn State Universities and the Univ. of Houston. He assumed his current position as Chancellor of the Univ of Houston System in 1987. He is chief executive officer of a system of four universities with a total enrollment of 44,000 students and a faculty and staff of 4200. He is a lic engr and has been active in profl society activities. He is currently Pres-Elect of IIE and mbr of the Bd of Gov, AAES. *Society Aff:* IIE, NSPE, ASEE, AAES.

Meindl, James D
Business: Vice President for Academic, Affairs and Provost, Rensselaer Polytechnic Institute, Troy, New York 12180
Position: Vice President for Academic Affairs and Provost *Employer:* Rensselaer Polytechnic Institute *Education:* PhD/Elec Engg/Carnegie-Mellon Univ; M.S/Elec Engg/Carnegie-Mellon Univ.; *Born:* 1933 Pittsburgh, Pa. (m. Frederica Ziegler 1961, children Peter b. 1970, Candace b. 1972). US Army Electronics Command 1959-67 (active duty 1959-61): Chief, Microelectronics Branch 1962-65; Dir, Integrated Electronics Div 1965-67. Lecturer Monmouth Coll 1960-67. Arthur S Fleming Comm Award 1967. At Stanford 1967-1986. Vice President for Academic Affairs and Provost at Rensselaer Polytechnic Institute 1986-present. Fellow, IEEE & AAAS. Editor IEEE Journal of Solid-State Circuits 1966-71. Chmn 1969-78 ISSCC & 1975 ACEMB. Outstanding paper awards 1970, 1975, 1976, 1977, 1978 ISSCC, IEEE Electron Devices Society J. J. Ebers Award 1980. Author 'Micropower Circuits' (Wiley 1969) & over 200 tech papers. Current res interests focus on the application of integrated circuit tech in medical electronics. *Society Aff:* IEEE, AAAS, AAMI, AAUP, AIUM, NAE, BES.

Meinhardt, John E
Business: 905 E St Paul Ave, Waukesha, WI 53186
Position: Mgr, QC & Met. *Employer:* WI Centrifugal Inc. *Education:* BS/Met Engg/Univ of WI. *Born:* 10/10/31. Native of Clintonville, WI. Specialized in the process met of high performance alloys & refractory metals since 1955. With WI Centrifugal, Inc since 1969 as mgr of all technical & quality functions. Active in local chapters of metals societies. Assisted in establishing a scholarship fund for students interested in met engg. Enjoy fishing, golf, & antique collecting. *Society Aff:* AIME, ASM, ASME, ASTM.

Meisel, H Paul
Business: 800 Hoyt St, Broomfield, CO 80020
Position: Product Line Mgr. *Employer:* FMC Corp. *Education:* Bach/Elec Engg/Univ of WI; Bach/Bus Admin/Univ of WI-Milwaukee. *Born:* 11/19/38. Mr. Meisel is with the Semiconductor Products Division, FMC Corp, Broomfield, CO. He received the BS degree in elec engg from the Univ of WI and a degree in business admin from the Univ of WI, Milwaukee. He is presently a Product Line Mgr for power conversion equip at the FMC Corp, SPD & was formerly Mgr of the Applications & Stds Engg depts for the div. Prior to joining FMC, he was with Otis Elevator Co, Gettys Mfg Co, and Allen-Bradley Co in various capacities in res, dev, and applications engg. Mr. Meisel is a Reg PE in WI & CO. He is pres of the Industry Applications Soc. He has served as Mbrship Chrmn, Treas, Secy, & Chrmn of the Milwaukee Chap of the Industry Applications Soc. He served as Area Chrmn & Council Mbr at

Meisel, H Paul (Continued)
Large for Region 4 & is active on three technical cttes. He was Chrmn of the 1973 IAS Annual Meeting. *Society Aff:* IEEE, NSPE.

Meisel, Jerome
Home: 156 Linden Rd, Birmingham, MI 48009
Position: Prof. *Employer:* Wayne State Univ. *Education:* PhD/EE/Case Inst of Tech; MS/EE/MIT; BS/EE/Case Inst of Tech. *Born:* 8/9/34. Native of Cleveland, OH. Served as an Asst Prof at Case Inst from 1961-65. From 1965-66 I was a Mbr of the Technical Staff at Bell Telephone Labs in Holmdel, NJ. Since 1966 I have been on the Elec Engg faculty at Wayne State Univ. Current res interests are in applying modern system and control concepts to bulk interconnected elec power systems. I am the author of a textbook entitled *Principles of Electromechanical Energy* Conversion. (McGraw-Hill, 1966), and am a Reg PE in the state of OH. *Society Aff:* IEEE.

Meisen, Axel
Business: Univ of Brit. Columbia, Applied Science, Canada V6T 1W5
Position: Prof, Dean, Faculty of Appl Sci. *Employer:* University of British Columbia. *Education:* PhD/Chem Eng/McGill Univ; MSc/Chem Eng/CA Inst of Technology; BSc/Chem Eng/Imperial College of Sci & Techn. *Born:* Oct 17, 1943. Brief assignments with North Thames Gas Bd (UK), Cerro Corp (Peru, SA) & Esso Res & Engrg (USA). Joined Dept of Chem Engrg, U of BC in 1969 & developed res in air pollution control, Claus process for sulphur prod, problems related to exploitation of tar sands, natural gas processing & petroleum refining Mbr of Environmental Control Grp, Imperial Oil Enterprises Ltd, 1974-75. *Society Aff:* CSChE, APEBC, ASEE.

Meiser, Kenneth D
Business: 2700 S Westmoreland, Dallas, TX 75233
Position: Executive Vice President *Employer:* Plastics Manufacturing Company. *Education:* BS/ChE/Univ of Pittsburgh *Born:* 11/22/09. Ch Chemist, Plaskon Co 1932-38. Process Dev & Mfg Spray Dried Amino Adhesives for Airplanes & gilder Structures 1939-45. Ch Process Engr Plastics Div LOF Glass 1946-51. Plt Mgr Plastics Operation, Allied Chem 1951-56. Asst Dir Engg Plastics Div. Allied Chem 1956-61. VP, R&D Plastics Mfg Co Dallas 1962-81. 1982-present Exec VP. Respon - Engg, Dev,. Amino plastic, Mfg & Molding, Polyester & Thermoplastic fabrication. Patents in amino resins & their applications. Local Chmn AIChE, PE in OH. Enjoy golf & music. *Society Aff:* AIChE, ACS, SPI

Meister, Charles L
4766 Hichory Shores Blvd, Gulf Breeze, FL 32561
Position: Pres - Consulting Engr. *Employer:* Meister & Assocs, Consulting Engrs. *Education:* BCE/Civil Engrg/Univ of FL. *Born:* Jan 1, 1936. Graduate Studies (Soils/Foundations) Georgia Tech 1967. 7 years Interstate/Turnpike Design, Const Mgmt, Precast/Prestressed Concrete Design/Mfg & Consulting with an Architechtural-Engrg-Planning Firm. Principal, Law Engrg Testing Co, Soil/Foundation Consultants/Materials Testing Engrs 1965- 69. Ardaman & Assocs, Inc, Orlando, Fla 1970-75. Elected Dir/Exec VP 1971. Became Principal in charge of Pensacola Branch Office 1974. Founded Meister & Assocs, Inc Oct 1975. ASCE Fla Sect Dir/Branch Pres 1964; NSPE/FES Chap Pres 1969-70, Dir 1971-75; ACEC/FICE Dir 1973-75; FICE Engrg Labs Forum Chmn 1973-75. Pres NW FL Chap-FL Engrg Soc (NSPE) 1978-79. Elected VP FL Engrg Soc (& Dir) 1980-81. Appointed to Governor's Ctte on Dam Safety 1980 (Florida). FES Membership Dev Ctte, NW Region 1980-81. HFES Bd of Dir 1980-81. FICE Chapter Liasion Representative 1981-1982. FICE Bd of Dir 1982. FES Exec Ctte 1982. ELF/FC&PA ACI Certification Program, and Joint Tech Ctte 1983. FICE Membership Ctte 1983-1984. FICE Hazardous Waste Ctte 1983-84. Florida Dept of Transportation Advisory Ctte, revision of Uniform Minimum Standards 1983-1984. FICE Conditions of Practice Ctte, Vice Chrmn 1984. *Society Aff:* ASCE, ACEC, NSPE, FICE, FIDIC, FES, ASTM.

Meitzler, Allen H
Home: 3055 Foxcroft Rd, Ann Arbor, MI 48104
Position: Principal Research Scientist *Employer:* Ford Motor Co *Education:* PhD/Physics/Lehigh Univ; MS/Physics/Lehigh Univ; BS/Physics/Muhlenberg Coll *Born:* 12/16/28 Born in Allentown, PA. After completing graduate study, joined Bell Telephone Labs, NJ, as Member of Tech Staff. Worked in ultrasonic, piezoelectric, ferroelectric, and electro-optic device development (1955-72). From '72 to present, Principal Research Scientist at Scientific Research Lab, Ford Motor Co contributing to automotive sensors and electronic systems development. Fellow of ASA (63). Fellow of IEEE (81). Chrmn of IEEE Group on Sonics and Ultrasonics (63). Chrmn of IEEE Subcommittee on Piezoelectric Crystals (1967-78). Married, three sons. *Society Aff:* IEEE, ASA, APS, SAE, ACS

Melcher, James R
Business: 29 Fairlawn Ln, Lexington, MA 02173
Position: Prof. *Employer:* MIT. *Education:* PhD/Elect Engr/MIT; MS/Nuc Eng/IA State Univ; BS/Elect Eng/IA State Univ *Born:* 7/5/36. James Russell Melcher, Straton Prof of Elec Engrg and Physics and Dir of Lab for Electromagnetic and Elect Sys. MIT, is a native of IA. His MS thesis won the Am Nucl Society's first Mark Mills Award. Since joining the MIT faculty in 1962, Dr Melcher's res & consulting interests have centered on continum electromechanics, the interactions of electromagnetic fields with continuous media, & especially on electrohydrodynamics. He is the author of about 80 journal publications, the MIT Press monograph *Field-Coupled Surface Waves*, co- author of *Electromechanical Dynamics* (with H H Woodson), published by John Wiley & Sons, & author of *Continuum* Electromechanics (MIT Press). He holds 9 patents & has been the principal in 3 films made for the Natl Committee on Elec Engrg Films. He received the Am Soc for Engrg Educ's. 1969 New Engl Sec Western Elec Fund Award for Excellence in instruction of engrg students, the IA State Univ. Outstanding Young Alumus Award in 1971, is a Fellow of IEEE, in 1971-72 was a Guggenheim Fellow at Churchhill College, Cambridge Univ; in 1981 received the Profl Achievement Award from IA State Univ and is a mbr of the Natl Acad of Engrg. *Society Aff:* NAE, IEEE, APS, AIEE, ASME, ACS, ASEE

Melchor, Jack L
Home: 26000 Westwind Way, Los Altos Hills, CA 94022
Position: Self Employed. *Employer:* Venture Capital. *Education:* PhD/Physics/Notre Dame; MS/Physics/Univ of NC; Bs/Physics/Univ of NC. *Born:* 7/6/25. US Navy 1943-46. Venture Capitalist. Previous employers: Sylvania, Melabs, Hewlett-Packard. Fellow, Past Sect Dir & Chmn IEEE; Notre Dame Sci Award 1969. Corp Dir, Rolm Xebec, Dir, Inc, Xitel, Heuristics. *Society Aff:* IEEE

Melden, Morley G
Business: 345 E 47 St, New York, NY 10017
Position: Exec VP *Employer:* Illuminating Engrg Soc of North America *Education:* MBA/MGT/NYU GBA; BS/EE/Case Inst Tech of CWRU *Born:* 5/29/26 Native of Cleveland, OH. Manufacturing and design engr, Picker X-ray Corp. 1948-53; Assoc Editor *Factory Management and Maintenance*, 1953-56; Co-Author *Practical* Automation and contributor to the *Industrial Engineering Handbook* 1956. US Army, HQ, military district of WA and corps of engrs school. Mgr. of Operations Improvement program for American Elec Power System. Presently Exec VP and chief operating officer of the illuminating engrg society of North America. Interests include photography, fine arts, finance, and interdisciplinary problem-solving. *Society Aff:* IES, IEE, TII

Melena, Franklin R
Business: 6907 Brookpark Rd, Cleveland, OH 44129
Position: Secretary - Treasurer. *Employer:* Wheeler & Meleena, Inc. *Education:* BS/Civil/Penn College. *Born:* 1927 Garfield Heights, Ohio. 1952-55 Design Engr for McDowell Co; 1955- Secy-Treas, Wheeler & Melena, Inc, specializing in Civil & Municipal Engrg. Presently serving as Municipal Engr for Garfield Heights, Brook-

Melena, Franklin R (Continued)
lyn Heights, Cuyahoga Heights & Seven Hills, Design Engr for Strongsville. Chmn of Standards Ctte to unify specifications for sewer & water const in NE, Ohio. Dir, Fenn Coll Engrg Alumni Assn. Dir, Euthenics, Inc. P Pres & Charter mbr of Cuyahoga County Municipal Engrs Assn. Mbr ASCE, Ohio Soc of PE's, Cleveland Cons Engrs Assn. Name appeared in Who's Who in the Midwest USA & Canada for prof recognition. *Society Aff:* CCMEA, ASCE, OSPE, CCEA.

Melese-d'Hospital, G B
Business: PO Box 81608, San Diego, CA 92138
Position: Senior Technical Advisor. *Employer:* General Atomic Co. *Education:* Baccalaureate/Maths/Univ of Toulouse; MS/Aeronautics/Univ of Paris; PhD/Aeron Eng/Johns Hopkins Univ. *Born:* 10/17/26. Hd, Nuclear Power Reactors Thermal Design, French AEC 1954-57. Asst & Assoc Prof Mech Engrg, V Chmn Nuclear Engrg Prog, Columbia Univ 1957-60. With Genl Atomic since 1960, working on advanced thermal nuclear reactors, fast breeders & fusion reactors. In charge of Intl Progs for Gas Cooled Fast Reactors 1972-79. Lectr U C San Diego 1960-62. Mbr AAAS, ASME, NSPE, Sigma Chi, Assoc Fellow AIAA. Fellow ANS, Dir 1972-75, Chmn Power Div 1975-76. On Editorial Bd Nuclear Engrg & Design, Annals of Nuclear Energy Nuclear Technology. Mbr, Bd of Dirs US Nat Committee for the World Energy Conference, 1979-82. *Society Aff:* AAAS, AIAA, ANS, ASME.

Melhorn, Wilton F
Business: Dept of Mech Engr Tech, Purdue Univ, W Lafayette, IN 47907
Position: Prof. Emeritus *Employer:* Purdue Univ. *Education:* MS/Met Engr/VPI; BS/Met Engr/Lehigh Univ. *Born:* 1/23/15. Native, York, PA. PE, OH. Prior to present assignment, was Sr Met for four yrs, for thirteen yrs was Chief Met & Dir of Quality, Tube Turns Div, Chemetron Corp, Louisville, KY, duties were admin, technical control, supervision for release of raw mtl for production, technical supervision of production operations, specifications, documentation & certification of products to end customer. Other industrial experience with US Steel, Bethlehem Steel & Brandt- Warner Mfg Co. Served in res & dev, US Army Ordnance 1942-45. Retired Lt Col. Taught at VPI & Univ of Cinncinnati. Retired, May 1980. *Society Aff:* ASM.

Melosh, Robert J
Home: 5611 Inverness Dr, Durham, NC 27712 *Employer:* Duke Univ. *Education:* PhD/CE/Univ of WA; MSCE/CE/Rensselaer Poly Inst; BCE/CE/Rensselaer Poly Inst. *Born:* 6/18/26. Struct Dynamicist, Boeing 1952-1962; Res Engr, Philco-Ford Corp, 1962-1972; Prof, VPI & SU, 1972-1974; VP, Staff Scientist MARC Analysis Res Corp 1974-1978; Prof & Dept Chrmn, Duke Univ, 1978; 70 papers, Natl Acad Sci Design Studies Panel, Assoc Editor, Computers and Struct Journal, Boeing Scholarship Award, Sigma Xi, Pi Delta Epsilon, McKinney Prize, Chi Epsilon. *Society Aff:* ASCE, AIAA.

Melott, Ronald K
Home: 11650 SW Bel Aire Ln, Beaverton, OR 97005-5908
Position: President *Employer:* Melott and Associates, Inc. *Education:* BS/Applied Science/Portland State University *Born:* 05/08/39 Native Oregonian. Served in US Army and US Air Force Reserve. Fire Fighter, Fire Officer, Fire Inspector, and Staff Officer Portland Fire Dept. 1961-72. Sr. Fire Protection Specialist and Chief Fire Prevention Specialist National Fire Protection Assn. Exec Secretary of Fire Marshals Assn of North America. Currently an independent fire protection consultant/fire protection engineer/fire investigator/expert witness. Helped originate the Fellowship of Christian Firefighters. Enjoy music and sports. Registered PE in OR, WA and CA. *Society Aff:* ASME, NSPE, SFPE

Melsa, James L
4951 Indiana Ave, Lisle, IL 60532
Position: Vice-President, R&D *Employer:* Tellabs, Inc. *Education:* PhD/EE/U of AZ; MS/EE/U of AZ; BS/EE/IA State *Born:* 7/6/38. BSEE Degree Iowa State Univ 1960; MS & PhD Univ AZ 1962 & 65. Currently VP R&D Tellabs Inc, Lisle IL. 1973-74 Prof & Chair Elec Engrg Univ Notre Dame. Previously Prof Information & Control Sciences S. Methodist Univ & Asst Prof Prof Elec Engrg Univ AZ. Research in speech encoding & digital sense processing. Published over 100 papers, authored or co-authored nine books. Mbr Tau Beta Pi, Pi Mu Epsilon, Eta Kappa Nu, Phi Kappa Phi, & Sigma Xi. Fellow IEEE. *Society Aff:* IEEE

Melton, Eston E, Jr
Business: Box 3297, E Camden, AR 71701
Position: VP. *Employer:* MB Assoc. *Education:* BCE/CE/Univ of FL. *Born:* 5/5/31. Avon Park, FL. Attended high sch at Gordon Military College, Barnesville, GA. Served in Field Artillery during Korean War. Discharged as a Capt. Process engr with Union Camp Corp, Savannah, GA, 1958-1962. Moved to Atlantic Res Corp, Alexandria, VA, 1962-1975 holding Div Dir of Propulsion and Special Products Divs. VP-Operations, Adams Labs, Fairfax, VA 1975-1977. Pres of Rabun Ridge, Inc, Rabun Gap, GA, 1977-1979. Presently Corporate VP-Mfg. MB Assoc, San Ramon, CA. *Society Aff:* AIChE.

Melvin, Howard L
Home: 375 Rockgreen Pl, Santa Rosa, CA 95405
Position: Chief Consulting Engr - retired. *Employer:* Ebascco Services, Inc. *Education:* MS/Elec Engg/MA Inst of Tech; MS/Elec Engg/Harvard Univ; EE/Elec Engg/WA State Univ; BS/Elec Engg/WA State Univ. *Born:* Dec 2, 1890. Fellow & Life Mbr AIEEE. PE NY, Ore. GE Co Test 1911-12. Instructor EE, WSU 1912-16. Asst EE Utah Power & Light 1917-20. Ch Elec Engr Wash Water Pr 1920-25. Ebasco Services, Inc, NYC 1925-59; 8 years Elec Engr; 6 years sponsor Engr EB&S Power Companies in Northwest; 2 years chief sponsor Engr; 18 years Chief Consulting Engr. Following retirement 4 years power systems cons. Principal specialties: high voltage transmission, power system planning, power pool operation in the US, Central & South Amer, Australia, Japan, Greece. Active in Edison Elec Inst, World Power Conf, CIGRE, Internatl Conf Large Elec Systems VP for US. Hobbies: golf, fishing, stereo photography. *Society Aff:* IEEE, CIGRE.

Melvin, Stewart W
Business: 203 Davidson Hall, Ames, IA 50011
Position: Professor & Extension Agricultural Engr. *Employer:* Iowa State University. *Education:* PhD/Agri Eng/IA State Univ; MS/Agri Eng/IA State Univ; BS/Agri Eng/IA State Univ. *Born:* Nov 1941. Native of Bloomfield, IA. Field Engr with Soil Conservation Service - USDA, 1963-64. Graduate Res 1964-69 in agricultural hydrology & water movement in soils. Drainage & water movement in porous media teaching & res, Colo State, 1970. Present work in area of soil, water & waste mgmt & conservation. Research in subsurface irrigation, soil compaction. Chmn Iowa Sect ASAE, 1972. Young Engr of the Year, Mid-Central Region ASAE 1976. International experience in South America & U.K. Visiting professor Silsee College, Bedfordshire, U.K., 1985-86. Professional Engineer. Enjoy outdoor sports. Married, wife Carol; daughter Catherine, son Christopher. *Society Aff:* ASAE, SCSA.

Mende, Robert G
Business: SNAME, One World Trade Ctr, Ste No 1369, New York, NY 10048
Position: Secretary And Exec Dir *Employer:* The Society of Naval Architects & Marine Engineers *Education:* BSc/Naval Arch & Marine Eng/Webb Inst of Nav Arch. *Born:* Dec 1926. BSc Webb Inst of Naval Architecture; grad NY State Maritime Acad. Engg duty officer USN 1951-53. Proj Engr Foster Wheeler Corp 1953-56. Dist Mgr - naval architect Bird-Johnson Co 1956-62. Sr naval architect J J Henry Co 1962-69. Secy (now Exec Dir) Soc of Naval Architects & Marine Engrs since 1969. Mbr Marine Engg council Underwriters Labs, SNAME, ASNE, ASME, Northeast Coast Inst of Engrs & Shipbuilders, ASAE, Council of Engg & Scientific Soc Execs, World Trade Ctr Club. Fellow, Royal Inst of Naval Architects & P Pres Webb Alumni Assn. Director, Friends of World Maritime University and China-U. S. Exchanges. *Society Aff:* SNAME, ASME, RINA, NECIES, ASNE, ASAE, CESSE

Mendel, Jerry M
Business: Univ. of Southern California, 1 Univ Park, Los Angeles, CA 90089-0781
Position: Prof and Chrmn *Employer:* Univ of S CA. *Education:* PhD/EE/Poly Inst of Brooklyn; MS/EE/Poly Inst of Brooklyn; BS/ME/Poly Inst of Brooklyn. *Born:* 5/14/38. Native of NYC, NY. Res engr with McDonnell Douglas Astro Co, 1963-1974, specializing in control & estimation theory. With USC since 1974. Chrmn of EE- Systems Dept. Heavily engaged in signal processing research, including: applying higher-order statistics to a wide range of systems and signal processing problems, seismic processing, and intelligent (i.e., AI) signal processing. More than 185 publications including four books. President of IEEE Control System Soc, 1986. Fellow of IEEE. Numerous awards, including Distinguished Member IEEE control systems society, and IEEE Centennial Award. Assoc Ed for Automatica. Enjoy bridge. *Society Aff:* IEEE, SEG, EAEG.

Mendel, Otto
Home: 68-52 Groton St, Forest Hills, NY 11375
Position: Conslt *Employer:* Self employed *Education:* BS/ME/Haifa Technion *Born:* 06/05/18 More than 35 years of experience in the field of estimating, cost control and construction mgmt, involving waste disposal plants and projs serving the petrochem, fertilizer, metallurgical and power generating indust. He has written numerous articles on estimating, cost control and related subjects, has lectured widely and has rendered testimony as an expert witness in legal proceedings, and has authored a book "Practical Piping Handbook" published by PennWell Books. He served on the bd of dirs of AACE, and is a Certified Cost Engr (CCE). He is presently in private practice as a Mgmt Conslt. *Society Aff:* AACE

Mendenhall, Irvan F
Home: 3704 Prestwick Dr, Los Angeles, CA 90027
Position: Principal *Employer:* Foregoing-Sole Proprietor *Education:* BS/Civil Engg/UC, Berkeley. *Born:* June 21, 1918 Compton, Ca. 1941, BSCE U Calif. Employed Calif. Division of Highways, 1942-45 US Navy, Civil Engr Corp, Lt. 1946, formed civil engrg firm & merged in 1950 with DM&J to become DMJM. Served DMJM as Pres. or Chrmn of Bd 1960-1983. Dir, Natl Bd Direction, ASCE 1973-77; VP 1978-79, Pres 1980-81. Chmn AAES 1982. Mbr ACEC, SAME. Reg Civil & Structural Engr, Calif. Cons to the City of Los Angeles on Hyperion sewerage imprumts., also on Rihand Dam India, and many other major projects USA & abroad. Honored as Engr of the Yr 1967; recipient of Annual Achievement Award, 1979 Los Angeles Chamber of Commerce; recipient 1981 Outstanding Engr Merit Award, Inst for the Advancement of Engrg; received Golden Beaver Engrg Award for 1981 from The Beavers. Jan 1984 retired from DMJM and became consltg Engr. *Society Aff:* ASCE, ACEC, SAME, CEAC

Mendenhall, Wesley S, Jr
Home: 209 Circleview Dr, South, Hurst, TX 76054
Position: Regional Administrator. *Employer:* Fedl Highway Admin. *Education:* BS/Civil Engg/SD Sch of Mines & Technology. *Born:* Aug 21, 1936. BSCE South Dakota School of Mines & Tech, 1958; graduate study Arizona State U & U of Mississippi. Genl highway engr direct Federal & Federal-aid highway progs of FHWA in Arizona 1961-69. Mgr demonstration projs for FHWA 1969-72. Administered direct Federal prog operations in eastern US 1972-73. Respon for natl const & maintenance policy & progs 1973-75. Principal engr respon for policy & progs all areas of highway construction, maintenance, foreign operations & agency emergency preparedness 1975-May 1979. Mbr ASCE, NSPE, APWA. Reg PE in Arizona. Respon for admin of Federal-aid highway program in the states of TX, LO, AK, OK, & NM since May 1979. Enjoy fishing & traveling. *Society Aff:* ASCE, NSPE, APWA, ASTM.

Menemenlis, Christos
Business: Univ of Patras, Patras, GREECE 26000
Position: Professor *Employer:* Univ. of Patras *Education:* Doctor/El. Eng/University of Athens, Greece *Born:* 06/11/26 1953-69 Public Power Corporation of Greece Involved in the development of the 400 kv transmission system of Greece; 1969-79 Hydro-Quebec Institute of Research (IREQ). Involved in the study and construction of the Ultra High Voltage laboratory of IREQ. Leader of Ultra High Voltage transmission research; 1979- Prof High Voltage Engrg. *Society Aff:* IEEE

Menkel, Bruce E
Business: P.O. Box 159, 235 Industrial Dr, Franklin, OH 45005
Position: Pres. *Employer:* Bruce Menkel & Assoc. *Education:* Bachelor's Degree/Mech Engg/Univ of Dayton. *Born:* 1/7/42. Master of Sci, Industrial Hygiene (to be received in Jan, 1982) Univ of Cincinnati. April, 1965 to Oct, 1967 worked as a design engr in plant engg dept Inland Mfg Div of Gen Motors Corp, Dayton, OH. Oct, 1967 to May, 1970 worked as proj engr at James R Ahart and Assoc, Inc. May, 1970 to present, Pres of Bruce Menkel and Assoc. Inc. Reg PE and Certified Industrial Hygienist. *Society Aff:* ASME, AIHA, NSPE, APCA, ASHRAE.

Mense, Allan T
Home: 601 Braxton Pl, Alexandria, VA 22301 *Employer:* Dept of Defense *Education:* Ph.D./Nuclear Engr/Plasma Physics/University of Wisconsin-Madison; M.S./Nuclear Engr/University of Arizona; B.S./Nuclear Engr/University of Arizona *Born:* 11/29/45 Dr. Mense is the Deputy chief Scientist for the strategic Defense Initiative Organization in the Dept. of Defense. He has technical oversight responsibility for the entire SDI R&D program. Prior to SDIO, Dr. Mense was a Sr Scientist in the Beam and Energy Sciences Group at McDonnell Douglas Astronautics Co. in St. Louis where he managed programs & conducted research in the areas of high power microwaves, relaturistic electron beams, & plasma physics considerations applied to systems engrg of future fusion reactor systems. Prior to this, Dr. Mense was a senior staff mbr on the Science & Technology Cttee in the U.S. House of Representatives where he followed over $1 Billion in government programs performed by the Dept. of Energy. Prior to this, Dr. Mense was a scientist in the Fusion Energy Div of the Oak Ridge National Lab in Tenn, where he worked on computer modeling of plasma behavior. *Society Aff:* AIAA, IEEE, ANS, APS

Menster, Paul C
Home: 1103 Dorsh Road, South Euclid, OH 44121
Position: Principal *Employer:* Paul C. Menster Constructing Engineer Inc. *Education:* BSME/Thermodynamics/Case Inst of Tech. *Born:* Aug 1921 Louisville, Ohio. Carrier Corp, Syracuse, NY 1948-51 as field application engr. Roth Bros, Youngstown, Ohio 1951-55 as proj engr. Smith & Oby Co, Cleveland, Ohio 1955-60 as proj engr. Private practice 1960-1986 forming own firm & as a principal. Currently a practicing consultant. Instructor in air conditioning & refrigeration courses at Fenn Coll, Cleveland, Ohio 1958-61. Mbr of prof & tech soc's related to fields of engrg practice. Officer of APEC, 1975-82. Interests include participation sports & gardening. *Society Aff:* NSPE, CSI, ASHRAE, APEC, ACEC, OACE

Mentzer, John E
Business: PO Box 37, Sunfield, MI 48890
Position: Consulting Engineer *Employer:* Wagester & Barger *Education:* MS/Agr Engg/Purdue Univ; BS/Agr Engg/Purdue Univ. *Born:* 3/22/31. Native of Mt Vernon, IN. After college 3 yrs in USAF as helicopter pilot. Following service earned a MS & remained at Purdue for 11 yrs as Ext Agr Engr. In 69 was employed by Reynolds Metals as proj engr. In 1972 assumed position of mgr, Grain Bin Div of Super Steel Products Co & in 1973 moved to position as chief engr. Dir of Finance ASAE 1977-1979. Reg PE No 9531, IN. *Society Aff:* ASAE.

Mentzer, John R
Home: 557 Clarence Ave, State College, PA 16803
Position: Prof, Dept Hd. *Employer:* PA State Univ. *Education:* PhD/Phys/OH State Univ; MS/EE/Penn State Univ; BS/EE/Penn State Univ *Born:* 6/16/16 Native of Arch Spring, PA. Design Engr Westinghouse corp 1942-1946; radar receiver design. Hd, Theory Sec, Ordnance Res Lab at PA State Univ, underwater Acoustics 1946-1948; Hd, Radar Scattering Group, OH State Univ Antenna Lab 1948- 1952; Staff

Mentzer, John R (Continued)
mbr MIT Lincoln Lab 1952-1954, Assoc Prof of EE PA State univ 1954- 1956; Prof of Engr Sci, in charge, PA State Univ 1956-1974, Prof & Hd, Dept of Engg Sci & Mech, PA State Univ, 1974-present. Hobbies-golf, piano, chess. *Society Aff:* IEEE

Menzel, Erich H
Business: D-3300 Braunschweig, W-Germany
Position: Prof *Employer:* Techn Univ *Education:* Dr Ing/Physics/Techn Univ Stuttgart; Dipl. Ing./Physics/Techn Hochschule Danzig *Born:* 8/13/18 Native of Danzig, Germany. Taught in Danzig public schools. Served in German Army 1940-44. Asst Prof Univ Tubingen and Techn Univ Darmstadt. Since 1961 Prof. Techn. Univ Braunschweig. Since 1966 editor of journal *OPTIK.* More than 100 papers on Fourier optics and surface physics of metal crystals. Enjoy Turkoman carpets. *Society Aff:* AOS, DPG, DGaO, DGE

Menzie, Donald E
Home: 1503 Melrose Dr, Norman, OK 73069
Position: Assoc Executive Dir, Energy Res Ctr & Prof, Petrol & Geological Engr. *Employer:* Univ of OK. *Education:* PhD/Petrol & Natural Gas Engg/Penn State Univ; MS/Petrol & Natural Gas Engg/Penn State Univ; BS/Petrol & Natural Gas Engg/Penn State Univ. *Born:* in DuBois, PA. Served as Marine Engr 1943-1946. Instr in Petrol and Natural Gas Engg, Penn State Univ 1946-1951. Res Asst, Penn Grade Crude Oil Assn 1946- 1951. Asst Prof, 1951-1955, Assoc Prof 1955-1964, Prof 1964-present, Petrol and Geol Engg, Univ of OK. Pres and Owner, Petrol Engg Educators, 1971-present. Engg Consultant, 17 different cos or organizations, 1951-present. Dir, Sch of Petrol and Geol Engg, Univ of OK, 1963-1972. Res Petrol Engr, Info Systems Prog, Univ of OK. Assoc Executive Dir, Energy Resources Ctr, Univ of OK, Jan 1979-present. *Society Aff:* AIME, NSPE, AAPG, ASEE, API.

Menzies, George E
Business: Stelco Tower, Hamilton Ontario, Canada L8N 3T1
Position: Divisional Foreman-Environmental Control. *Employer:* The Steel Co of Canada, Ltd. *Born:* Jan 1939. B Eng from McMaster U. Mbr APEO; native of Hamilton, Ontario. Environmental Control Engr with The Steel Co of Canada, Ltd since 1962 specializing in the evaluation & control of all aspects involved in industrial hygiene. Mbr of a recent Task Force for Ontario Govt on Occupational Hearing Loss. Past mbr of Canadian Standards Assn on nose control & City of Hamilton's Noise Pollution Comm. P Pres of both Southern Ontario Sect AIHA & Hamilton Chap of ASHRAE. Past Dir ASHRAE & Region 2 Chmn which covers most of Canada. Enjoy tennis, skiing & photography.

Mercer, Harold E
Home: 4313 Langtry, Houston, TX 77041
Position: Owner *Education:* BS/ME/Rice Univ *Born:* 12/31/25 Born in Houston TX. Graduated Hoston Public School system June 1942. Served USAAF 1944-1945 at Tenden Univ of Houston, AZ State Teachers Coll, Rice Univ. Employed City of Houston 1946-49 ES & Robt in Atkinson 1949-50 US Aberchombie Co 1950-52. JB Dannenbaum Co. 1952, 1954 Continental Oil Co 1954-63. Coulson And Assoc 1963-67. Part Owner Wikinson & Assoc Engr And Mercer Brown Engr 1967-81 Former Cubmaster, Scoutmaster, Advisor, Little league Coach, PTA Officer (Life Time Membership). *Society Aff:* TSPE, TSA, HCA, HESS, ASCE

Mercer, Kenneth K
Home: 3615 Meadow Lake Ln, Houston, TX 77027
Position: President. *Employer:* Mercer Engineers, Inc. *Born:* Feb 11, 1922 in Ohio. BSChE from Ohio State U in 1943, followed by graduate work at NYU & Columbia U. Employed by Standard Oil Dev Co in the butyl rubber prog. Followed by special assignment respons with the Allis-Chalmers Mfg Co. Since 1951 manufacturer's rep in Houston, Texas; founded Mercer Engrs in 1960 as Pres & Owner. One div of Mercer Engrs provides special processing machinery & const materials for the chem process indus. The other div supplies equip for use in the manufacture of plastic prods. Reg PE in Texas. Mbr AIChE, NACE, & Soc of Plastics Engrs. P Pres of Houston Assn of Manufacturers Agents.

Merchant, Dean C
Business: 1958 Neil Av, Rm 404, Columbus, OH 43210
Position: Professor *Employer:* The OH State Univ *Education:* PhD/Geodetic Sc./OH State Univ; MS/Geology/OH State Univ; BSCE/Civ Engrg/Univ of IL. *Born:* 3/8/28. Served in US Naval Aviation 1946-49, & US Air Force 1951-53. Field Engr for Aero Service Corp; Res Engr for Fairchild Camera & Instrument Corp; Asst Prof of Civil Engrg, Syracuse Univ; NSF Sci Faculty Fellow; presently Prof, Dept of Geodetic Sci, OH State Univ. Current primary interest in analytical photogrammetry. Dir, Amer Soc Photogrammetry 1966-69; NSF Faculty Fellowship 1966. Grand Award Winner (1975) for the Talbert Abrams Award - for authorship and recording of current & historical engrg & scientific devs in photogrammetry. Col USAF Reserve, Retired. Winner, ASCE Surveying and Mapping Award, 1981. *Society Aff:* ASPRS, ASCE, XE.

Merchant, M Eugene
Business: Metcut Research Associates Inc, 11240 Cornell Park Drive, Cincinnati, OH 45242
Position: Dir, Advanced Manufacturing Research *Employer:* Metcut Research Assoc *Education:* DSc/Physics/Univ of Cincinnati; MS/Physics/Univ of Cincinnati; BS/Mech Engg/Univ of VT. *Born:* May 1913. Native of Springfield, Mass. Cincinnati Milacron Co-op. Fellow at U of Cincinnati. With Cincinnati Milacron Inc 1936-83. Joined Metcut Research Assoc, on retirement from Milacron, in present position of Director, Advanced Manufacturing Research. Conducted, supervised & directed basic & applied res on mfg processes, equipment & systems & the future of mfg tech. Developed basic theory of the mechanics of the cutting process. Developed concept of systems approach to mfg as a basis for its overall optimization & automation. Visiting Prof of Mech Engrg, U of Salford, England. Mbr NAE. Hon Mbr of SME, Pres 1976-77. Hon Mbr of ASME, VP 1973-75. Pres Federation of Materials Soc's, 1974. Hon Mbr Internatl Inst for Production Engrg Res, Pres 1968-69. Pres ASLE 1952-53. Cincinnati Engr of the Yr 1955. ASME Richards Memorial Award 1959. ASLE Natl Award 1959. SME Res Medal 1968. I. Mech. E.(U.K.) Tribology Gold Medal 1980. George Schlesinger Prize of City of Berlin, 1980. Otto Benedikt Prize of Hungarian Academy of Sciences 1981. Hon Mbr Belgian Society Mech Engrs. Hon Fellow I Prod E (UK). Reg Prof Engrg, OH. Reg Mfg Engrg, CA. Certified Mfg Engrg. DSc, honoris causa, U of Vermont 1973, U of Salford (UK) 1980. Phi Beta Kappa, Tau Beta Pi, Sigma Xi. First recipient of M. Eugene Merchant Manufacturing Medal of ASME/SME 1986. Member Slovenian Academy of Sciences and Arts. *Society Aff:* SME, ASME, ASM, ASLE.

Merckx, Kenneth R
Business: 2101 Horn Rapids Rd, Richland, WA 99352
Position: Staff Consultant. *Employer:* Advanced Nuclear Fuels Corp *Education:* Post Doc/MIT; Phd/Eng Mech/Stanford; MS/Eng Mech/Stanford; BS/Mech Eng/Northwestern. *Born:* July 1926 Chicago, Ill. Post-doctoral at MIT & JCGS - Richland. Served in US Army 1944-46. Staff Cons on material behavior, design methods, & failure evaluation for Advanced Nuclear Fuels Corp since 1972. Previously cons for Westinghouse Hanford Co 1970-72, Battelle Northwest 1965-70, & GE 1953-65 on nuclear fuel & reactor equip design & behavior modelling. Fellow of ASME, Mbr ANS Past Mbr ASME, Fracture Prevention Res Ctte, Review Chmn Joint Internatl Creep Conf, & Advisor Nuclear Structural Engrg, PE state of Wash.. *Society Aff:* ASME, ANS

Meredith, Dale D
Business: Dept of Civil Engg, Buffalo, NY 14260
Position: Prof. *Employer:* State Univ of NY at Buffalo. *Education:* PhD/Civil Engrg/Univ of IL at Urbana; MS/Civil Engrg/Univ of IL at Urbana; BS/Civil Engrg/Univ of IL at Urbana. *Born:* 3/24/40. Native of Odin, IL. Asst Prof of Civil Engrg, Univ of IL at Urbana, 1968- 73. With SUNY at Buffalo since 1973. Acting Chmn, Dept

Meredith, Dale D (Continued)
of Civil Engrg, 1977-87. Prof of Civil Engrg since 1979. Teach courses in engrg economics, civil engrg systems analysis, water resources planning and dev, hydrology, water resources systems, and water pollution control systems. Respon for grad program in water resources and environmental engrg. Consultant to government, industry and education insts on water resources, pollution control and hydrology. *Society Aff:* AGU, ASCE, AWRA, TIMS, WPCF, IWRA, AGWSE.

Meredith, William Thomas
Home: Mere'Queen Farm, Rt. 8, Box 140, Lynchburg, VA 24504
Position: Chrmn of the Bd and Pres/COO of PBCS, Inc *Employer:* Parsons Brinckerhoff Const Serv, Inc *Education:* -/Genl Acad/William & Mary; -/Nuclear Physics/Catholic Univ; -/Civil Engg/Univ of MD. *Born:* 10/15/19. Currently, General Meredith is Chrmn of the Bd, Pres and Chief Oper Off of Parsons Brinckerhoff Const Services, Inc. Gen Meredith was proj mgr for const mgt of the Phila Commuter Rail Connection involving coordination of the activities of over 45 separate prime contractors constructing a $320-million rail connection in downtown Phila. Presently, Principal-in-charge Sunshine Skyway Bridge, Tampa FL., Other positions held with Tumpane Co, Al Khobar, Saudi Arabia as VP for engrg/const and asst general mgr; Frankland and Leinhard, consulting engrg firm in NY as principal and VP and Pres of arch/engrg subsidiary, Bosari Intl. Graduate USAF Inst of Tech; Command and Staff Coll; & Air War Coll. 32 yrs military service - rose through ranks, Corps of Engrs in Burma; progressed through all phases of engrg & const in USAF to rank of Brigadier General. Last assignment Asst for Real Property Maintenance. Office of Asst Secy of Defense (Installations & Logistics). Father of PRIME BEEF & RED HORSE; commanded RED HORSE unit of Southeast Asia. Received Newman Medal from SAME for leadership in these projects and 1986 Gold Medal for outstanding contributions. Decorations include Legion of Merit with two oak-leaf clusters; AF Commendation Medal with one oak-leaf cluster; Army Commendation Medal. Reg.PE in West VA. and PA; Rotarian, Natl.Dir and VP for Industry Liaison Matters of SAME; former Regional VP AIPE. Society Aff: SAME, AIPE. *Society Aff:* SAME, AIPE.

Mergler, Harry Winston
Business: Case Western Reserve Univ, 10900 Euclid Ave, Cleveland, OH 44106
Position: Professor of Electrical Engrg. *Employer:* Case Western Reserve U. *Education:* PhD/Control Engineering/Case Institute of Technology *Born:* June 1, 1924 Chillicothe, Ohio; s. Harry Franklin & Letitia (Walburn) Mergler. MIT 1946, Case Inst of Tech 1956; BS Physics, MS & PhD Engrg; Tau Beta Pi, Sigma Xi. Fellow IEEE. m. Irmgard E Steudel 1948; ch. Myra, Marcia, Harry F. Aeronautical Res Scientist, NACA 1948-56; Prof, Case Inst of Tech 1957- . Pres, Digital/General Corp 1968-72. Visiting Scientist USSR 1958, Visiting Prof Tech U of Norway 1961. Awarded Leonard Case Professorial Chair in 1974. Cons to the Natl Aeronautics & Space Admin & the NSF. Awarded LAMME Medal by the IEE 1978 Pres Industr Electronics & Control Instrumentation Soc (IEEE) 1977-79. Elected to the National Academy of Engineering, 1980. Recipient of the Case Institute of Technology Gold Medal for Scientific Achievement, 1980. *Society Aff:* IEEE, SNAME.

Meriam, James L
Home: 4312 Marina Dr, Santa Barbara, CA 93110
Position: Visiting Professor *Employer:* Univ. of Calif, Santa Barbara *Education:* Ph.D./Mech. Engr./Yale University; M. Eng./Mech. Engr./Yale University; B.E./Mech. Engr./Yale University *Born:* 03/25/17 Professor of Engrg mechanics, Chairman Div of Mechanics and Design, Asst Dean, Graduate Studies, Univ of CA, Berkeley 1942-63. Lt. (j.g.) U.S. Coast Guard 1944-45. Dean, School of Engrg and Professor Duke Univ 1963-72. Professor of Mech Engr, CA Polytechnic State Univ 1972-80. Retired 1980. Currently Visiting Professor of Mech Engrg, University of CA, Santa Barbara. Received "Award for Advancement of Basic & Applied Science-, Yale Engrg Assoc 1952, "Distinguished Educator Award," Mechanics Div ASEE 1978, Life Member and Fellow ASEE and ASME. Author "Mechanics" 1952, 1959, "Statics and Dynamics" 1966, 1971, "Engineering Mechanics" 1978, (and 1986 with L. G. Kraige). Hobby, boat building and sailing. *Society Aff:* ASEE, ASME

Merims, Robert
Business: 2 Park Ave, New York, NY 10016
Position: Asst Vice Pres Engrg. *Employer:* Scientific Design Co. *Education:* BS/Chem Engr/Univ of WI. *Born:* 4/26/23. Education interrupted by service in WWII (1943-45). Initially process engr at Foster Wheeler Co 1947-53. At Scientific Design Co since 1953 in variety of tech and managerial assignments. In charge of Proj Engr Group. Group respon for tech coordination of projs & administrative control of them with regard to cost & schedules. In present position 11 yrs. *Society Aff:* AIChE, ACS.

Meriwether, Ross F
Business: 1600 NE Loop 410, Suite 122, San Antonio, TX 78209
Position: Pres. *Employer:* Ross F Meriwether & Assoc, Inc. *Education:* MS/Mech Engg/LA Tech Univ; BS/Mech Engg/LA Tech Univ. *Born:* 6/19/30. Native of Shreveport, LA. US Army Artillery, 1949-53, 1st Lt Propulsion Engr, Gen Dynamics, 1956-60. Acting Asst Prof of Mech Engg, LA Tech, 1960-61. SW Res Inst, 1961-68. Establishhed present firm 1968. ASHRAE: Chrmn R & T 1976-77, Chrmn Stds 1979-80, Distinguished Service Award, 1977, Fellow, 1983. Reg PE, TX. *Society Aff:* ASHRAE.

Merkert, Clifton S
Business: 2045 W Hunting Park Ave, Philadelphia, PA 19140
Position: Manager of Engineering. *Employer:* FMC Corp - Drive Division. *Education:* BS/Met Engg/Lehigh Univ. *Born:* Aug 20, 1918 N Y City. Foundry Metallurgist Genl Motors 1940-43. Navy Officer WWII. Employed as Foundry Metallurgist Link-Belt Co 1946 which became FMC Corp; held various tech positions & is currently Mgr of Engrg of Drive Division, respon for design of geared speed reducers & variable speed drives. Reg P E Penna. Active in Amer Gear Manufacs Assoc; mbr Tech Division, Exec Ctte; Chmn Material Ctte & Spur, Helical & Herringbone Gear Ctte. Received AGMA Tech Award 1974. Elected Fellow Amer Soc for Metals 1970, Elected Bd of Directors AGMA 1981. *Society Aff:* ASM, ASTM, AFS.

Merklin, Joseph F
Business: , Manhattan, KS 66506
Position: Prof *Employer:* Kansas State Univ *Education:* Ph.D/Phys. Chem/Univ of MN; B.S./Chem/Manhattan Coll *Born:* 8/6/35 Born in New York. After receiving Ph.D., worked for Eastman Kodak Co. Have been at Kansas State Univ. since 1967. Received Coll of Engrg Award for Excellence in Undergraduate Teaching in 1974. Res interest include combustion and coal conversion kinctics, Nature and Properties of Defects in solids and kinetics of chemical reactions. Received 1981 Phi Kappa Phi Scholar Award, Kansas State Univ. *Society Aff:* RSC, APS, ACS, NYAS, AAAS

Mermel, Thaddeus W
Home: 4540 43rd St N W, Washington, DC 20016
Position: Consulting Engr, Private Practice. *Employer:* formerly Bureau of Reclamation. *Education:* BS/Elec-Civil/Univ of IL; -/CE/Univ of CO; -/CE/Geo WA Univ. *Born:* Sept 12, 1907. BS Engrg Univ of IL 1930. Civil Service career employee 43 yrs. Engrg & design of elec features at Hoover Dam, Grand Coulee Dam, Shasta Dam & many others; admin of const & equipment contracts & specification preparation; supervised research program advanced to Asst to Commissioner of the Bureau of Reclamation (Dept of Interior) as advisor on engrg matters, review of contract awards & representative of Commissioner on many tech & admin cttes (governmental, scientific & internatl). Active in prof societies; Chairman of several Engineering Foundation Conferences, Chmn of Ctte on World Register of Dams. Recipient of Dept of Interior Gold Medal for Distinguished Service. Since 1973 in

Mermel, Thaddeus W (Continued)
private consulting practice on assignments to Iran, Saudi Arabia, Pakistan, India, Nepal, Indonesia etc. Conslt Contract Adm and Procurement, Mbr Amer Arbitration Assn. Consultant for World Bank 1973- present. Author Register of Dams in US, supervision of Dams by State Authorities, Catalog of Dam Failures, and many tech articles.. *Society Aff:* ASCE, IEEE, USCOLD, WSE, ASME, ICID.

Meroney, Robert N
Business: B209 Engrg Res Center, Fort Collins, CO 80523
Position: Prof *Employer:* CO State Univ *Education:* PhD/Mech Engrg/Univ of CA, Berkeley; MS/Mech Engrg/Univ of CA, Berkeley; BS/Mech Engrg/Univ of TN, Knoxville *Born:* 10/4/37 Dr. Meroney has had over 20 yrs of experience in teaching and research on basic and applied problems of fluid mechanics. He has conducted research in atmospheric transport, fossil and nuclear power plan siting, wind power, urban air pollution environments, drying in porous media, and wind engrg. Dr Meroney is the author of more than 200 papers and reports and has been principal and co-principal investigator of projects exceeding 5.0 million dollars in value in the past 22 yrs. *Society Aff:* ASME, ASEE, AMS, APCA, AAAS, AIAA

Merrill, Roger L
Home: 4786 Dierker Rd, Columbus, OH 43220
Position: Consultant - Research Mgmt. *Employer:* Retired. *Education:* MS/EE/OH State Univ; BS/EE/OH State Univ. *Born:* Aug 1917. Design Engr Autocall Co 1939-41. US Army Signal Corps 1941-45. Employed by Curtiss Wright in Missile Res 1946-49. Joined Batelle as res engr in 1949; served as Mgr Engrg Phys Dept, Asst Dir & Assoc Dir prior to appointment in 1970 as Dir of Battelle Columbus Div with staff of 2,500 scientists, engrs & supporting specialists; assigned as Corporate Dir Natl Security, Space & Transportation Res in 1973 with respons involving Battelle labs in Columbus, Richland Wash, Frankfurt Germany & Geneva Switzerland. Serves on Bds of Franklin Univ. Retired from Battelle 10/31/79. Disting Alumnus Award by Coll of Engrg Ohio St Univ. *Society Aff:* TBΠ.

Merrill, William H, Jr
Home: 80 Huntington Ct, Williamsville, NY 14221
Position: Partner *Employer:* Bissell Merrill Assoc *Education:* B/CE/Syracuse Univ; BS/CE/RPI *Born:* 10/31/32 Born in Rochester, NY., and has resided in the Buffalo NY area his entire life. After being educated in the public school system he received degrees from R.P.I., and Syracuse Univ in CE, class of 1949. Upon graduation, he was associated with Merrill Construction Co, a family owned business involved in Real Estate Dev and Construction. Many residential and commercial developments were built. For the past 16 yrs. he has been a partner in Bissell Merrill Assoc Engrgs, formerly Senior Bissell, and Bronkie. Here he was responsible for the engrg work performed by his firm. Many waste water monitoring systems were developed and patented by him. In addition, he has published many articles concerning industrial wastes and their treatment in various tech journals. *Society Aff:* F-ASCE, NSPE, WPCF, AWWA, NAREB

Merritt, LaVere B
Business: 368 CB BYU, Provo, UT 84602
Position: Prof *Employer:* Brigham Young Univ *Education:* PhD/CE/Univ of WA; MS/CE/Univ of UT; BS/CE/Univ of UT *Born:* 3/11/36 Born and raised in Western Wyoming. Worked with US Forest Service for 2 yrs 1963-64. Has taught at Brigham Young Univ in Civil Enrg since 1964 except for 3 yrs 1967-70 when at Univ of WA as grad student. Currently Dept Chairman 1986-. Has specialized in sewer and water system design, water quality assessment, and multidisciplinary water quality studies. Pres Ut Sect ASCE 1977, Pres Ut Water Pol Cont Assoc 1978, Nat Dir WPCF 1981-84, Nat Dir ASCE 1982-85, Reg Prof Engr in Utah. *Society Aff:* ASCE, AWWA, WPCF, AEEP, ASEE, NALMS.

Merryman, John, Jr
Business: 5500 Chemical Rd, Baltimore, MD 21226
Position: Vice Pres & Gen Mgr Mfg - Industrial Chemicals *Employer:* W R Grace & Co - Davison Chem Div. *Education:* BS/Chem Engr/VA Polytechnic Inst. *Born:* Nov 10, 1922. Worked actively on Manhattan Project 1943-47; joined Davison Chem Corp Apr 1947, which was subsequently acquired by W R Grace & Co in 1954, served in various job respons & assumed current respon as V Pres Mfg for Indus Chems Oct 1973. Mbr AIChE. *Society Aff:* AIChE.

Mersereau, Russell M
Home: 5890 Pinebrook Rd NE, Atlanta, GA 30328
Position: Regents' Professor *Employer:* GA Inst of Tech *Education:* ScD/EE/MIT; SM/EE/MIT; SB/EE/MIT *Born:* 08/29/46 Res Assoc at MIT from 1973-1975. On faculty at GA Tech since 1975 with primary teaching and res interest in multi-dimensional digital signal processing. Received 1976 IEEE Browder J. Thompson Memorial Prize Award, 1976 ASEE South-Eastern Sectin Res Unit Award and 1984 IEEE Key to the Future Award (Ed Soc). Fellow IEEE. Co-author of text *"Multidimensional Digital Signal Processing."* Enjoy classical music, cooking, and gardening. *Society Aff:* IEEE

Merva, George E
Business: Dept of Agri Engg, E Lansing, MI 48824
Position: Prof. *Employer:* MI State Univ. *Education:* PhD/Agri Engg/OH State Univ; BSAE/Agri Engg/OH State Univ. *Born:* 8/20/32. Native of OH. Raised on farm in SE OH. Served in US Army as X-Ray Technician, 1953-1955. Asst Prof of agronomy, OARDC, Wooster, OH, 1960-1963. With MSU since 1967. Primary duties are teaching soil and water conservation engg in the Prof Agri Engg curric, with res in plant-soil-atmosphere relationships and agricultural drainage. Author of "Physioengineering Principles–, AVI. Contributor to "Modification of the Plant Environment–, ASAE Monograph Series. *Society Aff:* ASAE, AIH

Merz, Hans A
Home: Im Heugarten 31, Moencholtorf, Switzerland 8617
Position: Dep Head *Employer:* Basler & Partners AG *Education:* MS/Civ Engg/MIT; Dipl ETH/Civ Engg/Swiss Fed Inst of Tech. *Born:* 10/14/46. in Basel, Switzerland. Head of dept of safety planning & protective structures with current consulting firm. Initiator & maj scientific consultant to the natl seismic risk mapping proj in Switzerland. Mbr of Swiss delegation to 1976 UNESCO intergovernmental conf on earthquake mitigation. Consultant to the Swiss Nuclear Regulatory Commission on seismic risk aspects & impulsive loads. 1977 Co-winner of Moisseiff Award of ASCE for contribution to seismic risk anaylsis. Since 1975 consultant to Swiss Dept of Defense Explosives Safety Bd in matters of fabrication, transportation & storage of explosives & munitions. Conslt to Swiss Traffic Safety Bd. *Society Aff:* SIA, ASCE, SRA.

Merz, James L
Business: Dept of Elec & Computer Eng, Santa Barbara, CA 93106
Position: Professor *Employer:* Univ of CA, Santa Barbara *Education:* PhD/Applied Physics/Harvard Univ; MA/Solid State Phys/Harvard Univ; -/Solid State Phys/Univ of Gottingen; BS/Physics/Univ of Notre Dame *Born:* 04/14/36 Experience: Bell Labs, Murray Hill, NJ, Mbr of the Tech Staff in the Semiconductor Electronics Res Dept from Fall 1966 to Fall 1978; Harvard Univ, Cambridge, MA, Visiting Lecturer on Applied Physics on the Gordon McKay Endowment from December of 1971 through June of 1972; Univ of CA, Santa Barbara, CA, Dept of Elec Engrg from Fall 1978 to present. Honors: Fullbright Fellow, 1959-60; Woodrow Wilson Foundation Fellow, 1960-61; Danforth Foundation Fellow, 1969-64: Chrmn, Dept of Electrical and Comp Engrg July 1982 - Jan 1984; Assoc Dean for Res Devel, Coll of Engrg Feb 1984-April 1986; Director, Compound Semiconductor Research Center, April 1986-present. *Society Aff:* ΣΞ, IEEE, ECS, APS, MRS, SVHS

Meshii, M
Business: Dept. of Materials Science & Engineering, Northwestern Univ, Evanston, IL 60201

Meshii, M (Continued)
Position: Professor. *Employer:* Dept Materials Sci & Engrg N W Univ. *Education:* PhD/Matls Sci/Northwestern Univ; MS/Metallurgy/Osaka Univ; BS/Metallurgy/Osaka Univ. *Born:* Oct 1931; native of Hyogo Japan. Instr 1959, Asst Prof 1960, Assoc Prof 1964, Prof 1967, Chrmn 1978 at Dept of Materials Sci & Engrg Northwestern Univ. Visiting Scientist 1970-71 Natl Res Inst for Metals Tokyo. Guest Prof 1985 Osaka Univ., Fulbright Grantee 1956; Japan Soc Fellow 1958. Henry Marion Howe Medal 1968 ASM; Kosekisho Award 1972 Japan Inst of Metals. Pres 1973-74, 1983-84 Midwest Soc of Electron Microscopists Inc. ASM Fellow 1983; Chrmn, Materials Science Div, ASM; Fellow 1985 Japan Society for Promotion of Science. Enjoys skiing, tennis & classical music. *Society Aff:* ASM, TMS-AIME, APS, EMSA, MRS

Mesler, Russell B
Home: 1629 Dudley Ct, Lawrence, KS 66044
Position: Prof. *Employer:* Univ of KS. *Education:* PhD/CE/Univ of MI; MS/CE/Univ of MI; BS/CE/Univ of KS *Born:* 8/24/27. Native of Kansas City, MO. Served in the USN, 1945-46. Employed as process engr, Colgate Co 1949-51. Awarded a fellowship to attend the Oak Ridge Sch of Reactor Tech, 1951-52. Appointed teaching fellow, Univ of MI, 1952-53, & later served as res assoc & proj engr, Ford Nucl Reactor. Asst prof, Univ of MI, Joined the Univ of KS faculty in Chem & Petrol Engg in 1957 & appointed Warren S Bellows Distinguished Prof in 1970. Faculty Res Participant, DuPont, Aiken, SC summer, 1981. Spent sabbatical leave at Berkeley Nucl Labs, England, 1975-76, 1984-85. Robert T Knapp Award, ASME, 1967. Fellow AIChE, 1977. *Society Aff:* AIChE, ACS, ANS, ASEE.

Mesner, Max H
Home: 9 Wynnewood Dr, Cranbury, NJ 08512
Position: Staff Engg Scientist - Retired. *Employer:* RCA Corp, Astro-Electronics Div. *Education:* BS/EE/Univ of MO; AS/Engg/KS City Jr College. *Born:* Apr 16, 1912. Res Engr RCA Labs 1940-58; early dev of airborne radar & color TV; RCA Astro-Electronics Div 1958- , principally Mgr Space TV Sys - 35 to 65 design engrs developing spaceborne TV sys; Tiros, Ranger, Apollo; very high resolution TV for ERTS. Taught EE nights as Rutgers 13 yrs. 14 US pats, 17 pub tech papers on Space TV. Fellow IEEE, Assoc Fellow AIAA, mbr Sigma Xi, SPIE, HKN, Tau Beta Pi; on AIAA natl ctte. Held offices Lions Internatl & local church boards and district boards. Municipal Welfare Director. *Society Aff:* ΣΞ, HKN, TBΠ, IEEE, AIAA.

Messenger, George C
Home: 3111 Bel Air Dr, 7-F, Las Vegas, NV 89109
Position: Consultant. *Employer:* Self-employed. *Education:* MS/Electronics/Univ of PA; BS/Physics/Worcester Polytech; PhD/Engg/California Coast Univ *Born:* July 20, 1930 Bellows Falls Vt. Engr Philco Corp 1951-57, Prod Mgr Special Components 1957-58; Head Dept Device Electronics Hughes Aircraft Co 1958-59, Asst Mgr Dept 1959-60, Mgr Engrg Dept 1960; Mgr Transistor Div Transitron Corp 1961-63; Engrg Fellow Northrop Corp 1963-68; Engrg Cons 1968- ; V Pres & Bd of Dirs Amer Inst of Finance 1968- . Mbr APS & RESA; Fellow IEEE 1976 for 'Contributions to the Determination of Radiation Damage to Semiconductors & Advances in Semiconductor Device Technology'. Semiconductor & solid state electronic devices; radiation effects in semiconductors; system hardness design & hardness assurance; reliability of semiconductor devices & electronics systems. *Society Aff:* IEEE, RESA.

Messer, Philip H
Business: 81-711 Highway 111, Indio, CA 92201
Position: District Production Supr *Employer:* Union Oil Co of CA *Education:* MS/Petro Eng/Stanford Univ; BS/ME/Stanford Univ *Born:* 12/19/46 Four years Oil and Gas Engrg experience, including drilling, production, reservoir and secondary recovery. Twelve years geothermal reservoir engrg experience with Union Oil Co including Nine years in Reservoir Engrg & Three years Production Engg and Operations. Geothermal experience has been in hot brine resources. Published various Oil and Gas and geothermal technical papers for SPE of AIME and others. Reg petroleum engr in CA. Three years varsity football at Stanford University. Enjoy tennis, running and desert living. Married with two children. *Society Aff:* AIME, SPE

Messina, Edward J
Business: 67 W 66th St, New York, NY 10023
Position: Director Film Operatons. *Employer:* American Broadcasting Companies Inc. *Born:* Jan 14, 1933; native of N J. BS Photographic Engrg RIT NY. USN in Korea as a Photographic Engr. Photographer & mapping 1953-56, 1956-58 Pathe Color Labs as Head of Processing color materials 1958-59 Thickol Inc. Reaction Motor's Div Photographic Engr X-15 Rocket Engine. 1959-63 US Naval Test Facilities, Head of Photometric Div involved in photography of rockets, missiles, tracksleds & arresting devices. 1963-69 ABC-TV News, tech respon for world wide camera opers, 1969- , ABC-TV Dir of Firm Opers Respon for all film programming technically for the television network. *Society Aff:* SMPTE (Fellow)

Messina, Philip
Home: 38 Heather Ln, Colonia, NJ 07067
Position: Prod Mgr *Employer:* Engelhard *Education:* MChE/Chem Engg/Poly Inst of Brooklyn; BChE/Chem Engg/Cooper Union; Bus Mgt Cert/Bus/Johns Hopkins Univ. *Born:* 5/20/25. Native of NYC. Married. Six children. Military service - infantry 1944-46 as 2nd Lt & Army Chem Corps 1950-52 as 1st Lt. Res, Process Design & Proj Engr with US Ind Chem for 1946-50 & 1952-54. Employed by W R Grace & Co from 1954-74 in various capacities terminating as Commercial Dir of the European Div involved in multi-natl marketing & dev. Joined Chemico as Dir of market dev for 1974-76 & then employed by Wilputte. VP 1977-1981. Chrmn MD Sec AIChE in 1962-63. Co- inventor of four US patents. Process Mktg, CE Lummus 1981-1982. Prod Mgr, Engelhard 1983 to present. *Society Aff:* AIChE, ACS, AMA.

Metcalf, Charles M
Business: 1625 Eye St, NW, Washington, DC 20006
Position: Chief Tunnel Engineer. *Employer:* Sverdrup & Parcel & Assoc Inc. *Education:* BS/Civil Engr/Univ of WI. *Born:* 1919. USN WWII aircraft maintenance, opers & Liaison. With Sverdrup & Parcel since 1945. Experienced in engrg design & const supervision & mgmt of: major subway & highway tunnels; water crossings; major bridge, highway, indus & railroad projs; & oil loading & cargo handling facilities in US & abroad. Prof knowledgeable about wave action, beach erosion & pile corrosion controls. Was V Pres Mgr, transrortation & public works div 1970-76. Fellow Mbr ASCE & SAME. Reg in 10 states & D C. *Society Aff:* ASCE, SAME.

Metcalfe, Arthur G
Box 85376, San Diego, CA 92138-5376
Position: Director Research *Employer:* Solar Turbines Inc. *Education:* PhD/Metallurgy/Cambridge Univ England; MA/Natural Scis/Cambridge Univ England; BA/Natural Scis/Cambridge Univ England. *Born:* June 1922 London England. Now US citizen living in San Diego Calif. Res Physicist for Wickman (England) specializing in cemented carbides; Res Metallurgist for Deloro Stellite (Canada) working on cobalt-base alloys; Supr of Phys Met at Armour Res Foundation; with Solar Div of Internatl Harvester 1959- , since 1981 Solar Turbines Inc assumed position of Assoc Dir of Res 1961 with respon for all advanced res projects specializing in new processes & new material concepts for energy sys. Dir of Res since Oct 1982. Chmn San Diego Chap ASM 1962, Fellow ASM 1974. *Society Aff:* ASME, ASM, AIMME.

Metcalfe, Tom Brooks
Business: PO Box 4-4130, Lafayette, LA 70504
Position: Head Dept of Chem Engrg. *Employer:* Univ of Southwestern La. *Education:* PhD/ChE/GA Inst of Technology; MS/ChE/Univ of TX; BS/ChE/Univ of TX. *Born:* Feb 1920 on a farm in Alum Creek Texas. T U Taylor Award Univ of Texas 1941. Lt in US Naval Reserve WWII. Res Fellow Ga Tech 1952. 15 yrs indus experience includes Dow Chem & Shell Oil; Radiological Safety Officer for Shell in

Metcalfe, Tom Brooks (Continued)
Houston. Faculty positions at Univ of SW La, W Va Inst of Tech & Univ of Houston. La State Coordinator for JETS. Natl Chmn AIChE Acad Dept Heads Forum. Mbr & Public Speaker Mayor's Ctte on Property Zoning (Houston). Reg P E; Cons to Dept of Interior, Bureau of Public Roads. Dept of Commerce, Office of Saline Water. Fellow, Amer Inst of Chem Engrs. Author: Radiation Spectra of Radionuclides, Noyes Press; Chemical Engineering as a Career, USL Press. *Society Aff:* AIChE, ASEE.

Mettler, Ruben F
Business: 23555 Euclid Ave, Cleveland, OH 44117
Position: Chrmn of the Bd & Chief Exec Officer. *Employer:* TRW Inc. *Education:* PhD/Elec & Aero Eng/Caltech; MS/Elec Engg/Caltech; BS/Elec Engg/Caltech. *Born:* 1924 Shafter Calif; BS, MS & PhD CalTech in EE. 1949-54 tech staff Hughes Aircraft; 1954-55 Cons to Dept of Defense; joined TRW in 1955 as Program Dir for Thor missile, 1957 assumed overall tech supervision of Atlas, Titan, Thor & Minuteman progs, named Exec V Pres of Ramo-Wooldridge (TRW) 1958, Pres of Space Tech Labs (TRW subsidiary) 1962, Exec V Pres TRW Inc 1965, Asst Pres 1968, Pres 1969. Elected to Chrmn of the Bd & Chief Exec Officer of TRW Inc in 1977, also in 1978 became chrmn of the Exec Committee, Bd of Dirs of Goodyear Tire & Rubber Co, Merck & Co, Bank of Amer N A. *Society Aff:* NAS, NAE, IEEE, AIAA.

Metz, Barry A
Home: 3442 Wilmette Ave, Wilmette, IL 60091
Position: President *Employer:* Wright Chem Corp - Chicago. *Education:* BS/Chem Eng/IIT. *Born:* Nov 1935; native of Chicago Ill. One of first to be licensed as Reg P E in field of Corrosion. Author of 24 papers & presenter of numerous speeches in field of corrosion & water mgmt. Mbr AIChE, NACE & Alpha Chi Sigma. Certified by NACE as corrosion specialist. Avid golfer & tennis player. *Society Aff:* AIChE, NACE, ΑΧΣ.

Metz, Carl A
Home: 1550 N Lake Shore Dr, Chicago, IL 60610
Position: Prin. *Employer:* Dolio & Metz Ltd. *Education:* BS/Civil Engg/Univ of IL. *Born:* 11/30/92. From 1915-25, structural designer for Staley Mfg Co, Decatur, IL, Stone & Webster, R H Folwell, Corn Products Refining Co, Rock Island RR, Chicago. Entered private practice in Chicago, 1925 as C A Metz Engg Co, architectural engrs; 1948-1978 partner in Shaw Metz & Assoc, Architects & Engrs, Chicago. Mbr: ASCE, Western Soc of Engrs, ACI Clubs: Chicago Yacht Club, Union League, Builders, Architects, Sun City Country Club, 33rd Degree Mason. In 1978 received from the Univ of IL College of Engg the Alumni Honor Award for distinguished service in engg "For contributions advancing the knowledge of structural engg, his integration of Architects and Engrs in the planning and designing of large structures for his concern to insure safety in engg design.–. *Society Aff:* ASCE, NSPE, ASTM, AIA, WSI, ACI.

Metz, Donald C
Home: 908 Longwood Loop, Mesa, AZ 85208
Position: Dean Emeritus - Engrg Technology. *Employer:* southwest State Univ (Retired) (Minnesota) *Education:* DSc (Hon)/Tech/Capitol Inst of Tech; MSIE/Industrial Engg/Purdue; BSEE/Elec Engg/Purdue. *Born:* Dec 1908; native Midwest USA. Engr in indus prior to WWII. Army Field Artillery 1940-46; Retired Col AUS. Admin Engrg Tech Purdue 1946-51; Univ of Dayton 1951-63; Tech Coll (Principal) Ibadan Nigeria 1963-65; Western Mich Univ 1965-67; Southwest St Univ (Minn) 1967-73. Retired Dec 1973. Reg P E Ohio. Mbr Tau Beta Pi, Eta Kappa Nu, Tau Alpha Pi, Iota Lambda Sigma; NSPE, ASEE (Chmn Assn Mbr Council; Bd of Dirs 1973-75). Cons Tech Manpower USOE; BLS. Commissioner EMC. Mbr Working Group Tech Manpower Pres's Ctte Scientists & Engrs. DSc (Hon) Capitol Inst of Tech. *Society Aff:* ASEE, NSPE, TROA, ROA.

Metzger, Frederick B
Home: 1 Eagle Lane, Simsbury, CT 06070
Position: Chief-Acoustics. *Employer:* Hamilton Standard. *Education:* BSME/Mech Engg/Penn State Univ. *Born:* Aug 1933. Grad Studies Wayne St Univ, RPI & Univ of Conn. Staff Engr USAF Hqrs Air R/D Command 1956-57; Vehicle Stability Res Engr Ford Motor Co 1958-59; with Hamilton Standard Div of United Technologies Corp 1959- , currently respon for R/D progs on propellers, fans & other co products. Mbr Pi Tau Sigma ME Honorary. ASME Gas Turbine Power Award 1972; SAE Manly Medal 1973. Associate Fellow AIAA and Member SAE Aircraft Noise Measurement Cmte. Member Aerospace Industries Association Aircraft Noise Committee. Enjoys photography, music & tennis. *Society Aff:* AIAA.

Metzger, Frederick L
Business: 1701 E Broadway, Toledo, OH 43605
Position: Mgr - Distributor Packaging Dept. *Employer:* Libbey-Owens-Ford Co. *Education:* BA/Economics/Univ of MI. *Born:* July 1921 Toledo Ohio. Post Grad Study in Indus Engrg Univ of Mich. With USN 1943-46 as Lt jg in Amphibious Group. With Libbey-Owens-Ford Co since 1946 as Indus Engr & Packaging Engr; became Mgr of Distributor Packaging Dept in 1968. Charter mbr Toledo Chap IIE, Chap Pres 1960-61; charter mbr Toledo Chap SPHE, Chap Pres 1974-75; Natl Bd Mbr SPHE 1975-76. Hobbies include boating & collecting glass canes. *Society Aff:* SPHE.

Metzger, Sidney
Home: 10500 Rockville Pike, Rockville, MD 20852
Position: Pres *Employer:* Sidney Metzger Assoc Consltg Engrs *Education:* MEE/Elec Engg/Poly Inst of Brooklyn; BS/Elec Engg/NY Univ. *Born:* 2/1/17. 6. in NY City. US Army Signal Corps Labs 1939-45 Hd Radio Relay Branch. ITT Labs 1945-54 Div Hd Radio Relay Div (commercial & military). RCA Labs 1954-57; RCA Astro-Electronics Div. 1957-63 Early Satellite Projs; Dept Hd Communications Engg for Projs SCORE, TIROS, RELAY. Communications Satellite Corp 1963-82. VP & Ch Scientist. Retired. Mbr NAE; IEEE Award in Intl Commun 1976. IEEE Eascon Aerospace Electronics Award 1975. Joint Telecommun Advisory Comm 1971-79, Chrmn '75-'77. Fellow IEEE; Fellow AIAA; AIAA Aerospace Communications Award 1984. Aeronautics & Space Engrg Bd of NAE, 1981-86; Tau Beta Pi; Sigma Xi. *Society Aff:* NAE, IEEE, AIAA.

Metzler, Dwight F
Business: 900 S.W. 31st St, Topeka, KS 66611
Position: Private Consultant *Employer:* Semi-Retired *Education:* SM/San Engg/Harvard Univ; CE/San Engr/Univ of KS; BSCE/Civil Engg/Univ of KS. *Born:* March 1916. Ch Engr 1948-62 Kansas State Bd of Health; Exec Secy, Kansas Water Resources Bd 1962-66; Directed NY's environmental progs as deputy commissioner, NY State Dept of Health 1966-70, & deputy commissioner, NY State Dept of Environmental Conservation 1970-74; Secy, Kansas Dept of Health & Environment 1974-1980. Professor, Dept of Civil Engrg, U of Kansas 1948-66. Adviser Govt of India, 1960, USSR, 1962, WHO, 1964-1986. P Pres, Amer Public Health Assn; Mbr NAE; Member Assembly of Engr, Natl Research Council 1977-80. Hon Fellow Royal Soc of Health. Distinguished Service Award, U of Kansas 1970, Dist Engrg Serv Award, Univ of KS 1984, Sedgwick Medal, APHA 1981, Hon Mbr, Water Pollution Control Federation 1983 *Society Aff:* NAE, ASCE, APHA, AWWA, WPCF.

Metzner, Arthur B
Business: Dept of Chem Engrg, Newark, DE 19711
Position: Professor of Chem Engrg. *Employer:* University of Delaware. *Education:* DASc/Hon/Catholic Univ, Leuven; ScD/Chem Engg/MIT; BSc/Chem Engg/Alberta. *Born:* 4/13/27. Naturalized. Res Engr Defence Res Bd, Canada 1948; Colgate-Palmolive Co 1951-53. Instructor, MIT 1950-51; Polytechnic Inst of Brooklyn 1951-53. At DE since 1953. H Fletcher Brown Prof 1962- ; chmn of chem engg dept 1960-77; Fluid mechanics, polymer engg. Soc of Rheology (US & British); Fellow AIChE. Wilmington Sect Award ACS 1958, Colburn Award, AIChE 1958; ASEE Chem Engg Lectureship 1963; Walker Award AIChE 1970, Hon doctorate, Catholic

Metzner, Arthur B (Continued)
Univ, Leuven, Belgium. Lewis Award, AIChE 1977, Bingham Medal, Soc of Rheology 1977; Elected to Natl Acad of Engg 1979. Editor, Society of Rheology 1985-. *Society Aff:* AIChE, NAE, S of R

Meurer, Siegfried
Home: AM Riedlerberg 17, Kreuth, W Germany D-8185
Position: Ret. (Mbr of the Bd of Mgmt) *Employer:* Man Maschinenfabrik Augsburg-Nuernberg AG *Education:* Doctor Eng/ME/Tech Univ Dresden; Doctor hc/ME/Tech Univ Karlsruhe; Prof/-/Tech Univ Hachen *Born:* 05/09/08 Industrial Activity: 1938-1945 Res Engr Man-Nuernberg; 1945-1950 Engr for airplane engines Arsenal of Aeronautique Paris; 1950-1962 Chief Engr diesel engine res dept Man-Nuernberg; 1962-75 Mbr of the Bd of Mgmt Man-Angsburg responsible for the res of the co; 1975 Retired. *Society Aff:* VDJ, SAE

Mewes, Dieter
Home: Leipziger Str. 9, Kleinostheim, W Germany D-8752
Position: VP Engrg *Employer:* Degusia Corp *Education:* Apl Prof/ChE/Tech Univ; Habilitation/ChE/of Berlin; Dr Ing/ChE/of Berlin; Dipl Ing/ChE/of Berlin *Born:* 12/18/40 1972 Habilitation in Chem Engrg Dep-Tech Univ of Berlin. Major res work in heat and mass transfer and fluid flow phenomena and rheology. 1973 Joined Degussa AG in Frankfurt, Engrg Res Dept for Process Engrg. Res work in extraction and scale-up of unit operations, proj engrg for pilot plants and large production plants. 1978 Gained title of an "Apl Prof" at Tech Univ of Berlin. 1981 VP, Engrg for Degussa Corp in Mobile, AL. Responsible for proj and constr, as well as maintenance engrg. 1981 Dir of Process Engrg Res Dept of Degussa AG in Frankfort, W Germany. Publications: 28 in Chemical Engrg Magazine; 2 Books: *Mass Transfer and Chemical Relation* H. Hrauer, D. Mewer, Sauerliuder Vlg 1971 and D. Mewer, Th. Pilhofer: *Extraction*, Vlg Chemie, 1978. *Society Aff:* AIChE, DECHEMA, VDI

Meyer, Andrew U
Home: 746 Ridgewood Rd, Millburn, NJ 07041
Position: Prof. *Employer:* NJ Inst of Tech. *Education:* PhD/EE/Northwestern Univ; MS/EE/Northwestern Univ. *Born:* 4/21/27. Between 1961 & 1965 he was a Mbr of the Technical Staff at Bell Tel Labs at Whippany, NJ, where he was engaged in research on automatic control problems pertaining to satellite attitude control & missile auto-pilots. Prior to his graduated studies, between 1950 & 1957, he held engg positions with Assoc Research, Inc, Sun Elec Corp & Armour Res Fdn of IL Institute of Tech, all in Chicago, IL. In 1965, Dr Meyer joined Newark College of Engg of NJ Inst of Tech, Newark, NJ, as an Assoc Prof. Since 1968 he is Prof of Elec Engg. During the academic year 1969-1970 he was a Visiting Prof at Middle East Technical Univ, Ankara, Turkey. Dr Meyer's res interests are in automatic control systems & applications to biomedical engg. Since 1972 he is also on the adjunct res faculty of the College of Medicine & Dentistry of NJ. His publications include co-authorship (with Jay C Hsu) of the book "Modern Control Principles & Applications–, McGraw-Hill, 1968. *Society Aff:* IEEE, SIAM, AAAS, ASEE, IMACS, AAUP, ΣΞ.

Meyer, James L
Business: PO Drawer 5444, Alexandria, LA 71307
Position: President. *Employer:* Meyer, Meyer, LaCroix & Hixson, Inc. *Education:* MS/Civil Engg/LA Tech Univ; BS/Chem Engg/LA Tech Univ. *Born:* March 31, 1934. Native of Alexandria, La. Reg PE in Civil & Chem engrg. Licensed Land Surveyor. Jr Engr for Gulf Oil Corp. Served with US Air Force 1957-60. Design Engr, Louis J Daigre Assocs 1960-62. Principal Engr Daigr, Meyer & Watts, Inc 1966-68. Since 1968, owner & Pres of Meyer, Meyer, LaCroix & Hixson, Inc, Consulting Engrs & Land Surveyors, & relatd (parent) companies. Mbr Tau Beta Pi, NSPE, La Engrg Soc, AWWA & WPCF. Pres, La Engrg Soc 1976-77; Mbr, LA St Bd of Reg for Prof Engrs and Land Surveyors 1976-85. *Society Aff:* NSPE, NCEE, ACEC, TBΠ, AWWA, WPCF.

Meyer, John E
Home: 27 Flintlock Rd, Lexington, MA 02173
Position: Professor, Nuclear Engrg. *Employer:* MIT. *Education:* PhD/Mech Engg/Carnegie Inst of Technology (Pittsburgh, PA); MS/Mech Engg/Carnegie Inst of Technology (Pittsburgh, PA); BS/Mech Engg/Carnegie Inst of Technology (Pittsburgh, PA). *Born:* Dec 1931. Native of Pittsburgh, Pa. Westinghouse Bettis Atomic Power Lab, W Mifflin, Pa 1955-75. Work included design of naval nuclear reactors, computer prog dev, & advanced fuel element dev. Supr, 1960-65; Advisory Engr, 1965-74; Mgr 1974-75; Cons, 1975. Visiting Lecturer, Nuclear Engrg Dept, U C Berkeley 1968-69. Prof, Nuclear Engrg Dept, MIT since 1975. Fellow ANS 1969. Mbr Natl Prog Ctte, ANS 1972-75. Secy Mathematics & Computations M&C Div, ANS, 1976- 77. Exec Ctte M&C Div, ANS, 1977-80, Vice Chairman M & C Div, ANS 1980-81, Chairman M&C Div, ANS 1981-82. Family includes wife Gracyann, & 4 children (Susan (Mrs P A Heydon), Karl, Karen (Mrs B.E. Gleasman) & Thomas). *Society Aff:* ANS, ASME, ΣΞ.

Meyer, L Donald
Business: PO Box 1157, Oxford, MS 38655
Position: Supervisory Agri Engr. *Employer:* Agricultural Res Service (USDA) *Education:* BS/Agri Engr/Univ of MO; MS/Agri Engr/Univ of MO; PhD/Agri Engr/Purdue Univ *Born:* April 1933 Concordia, Mo. Conducted erosion-control res including dev of equip for applying simulated rainstorms, mechanics of the soil erosion process, & practices for controlling agricultural or construction-site erosion at Purdue U. 1955-73 & USDA Sedimentation Lab since 1973. Currently Res Ldr at USDA SL in Soil Erosion Processes. Past Chmn of ASAE Soil & Water Div (1972-73) & ASAE Dir of Publications (1968-69). Fellow ASAE, Mbr SCSA, SSSA, Sigma Xi, Alpha Gamma Sigma, Alpha Zeta, Gamma Sigma Delta, & Reg PE. Enjoys fishing, photography, & competitive sports. *Society Aff:* ASAE, SCSA, SSSA, ASA

Meyer, Robert E
Business: 14250 SW Allen Blvd, Beaverton, OR 97005
Position: President. *Employer:* Robert E Meyer Engrs, Inc. *Born:* April 1927. BSCE from Oregon State U 1952. Reg PE in Ore, Wash, & Colo. Founded his firm in Beaverton in 1960 & serves as Pres & Genl Mgr. Robert E Meyer Engrs, Inc has a staff of 40 including planners, 12 registered engrs, registered architects, landscape architects & surveyors, & supporing technicians. They serve municipalities & other govt agencies, commercial & industrial firms & private developers. Pres Cons Engrs Council of Ore 1970-71. VP ACEC 1974-76. Enjoys skiing, fishing & golf.

Meyer, Vernon M
Business: 200 Davidson Hall, Ames, IA 50011
Position: Prof. *Employer:* IA State Univ. *Education:* PhD/Agri Engg/Univ of MN; MS/Agri Engg/Univ of MN; BAE/Agri Engg/Univ of MN. *Born:* 12/20/24. Native of New Prague, MN. Engr Appraiser trainee for Fed and Bank following grad. Agri Engr for the MN Farm Bureau Service Co specializing in farm equip sales and service. Served on the staff of the Univ of MN from 1954-58. Since 1958 have been ext agri engr at IA State Univ. Presently subj matter leader with special emphasis on livestock facilities engg and energy conservation. Chrmn, Mid-Central Region, ASAE - 1977. Have received 14 blue ribbons in ASAE Ext Exhibits Competition. Enjoy golf, bowling and fishing. *Society Aff:* ASAE, WPCF.

Meyer, Walter
Business: 1026 Engrg, Columbia, MO 65201
Position: Professor & Chmn, Nuclear Engrg Prog. *Employer:* University of Missouri-Columbia. *Education:* PhD/Chem Engg/OR State Univ; MChE/Chem Engg/Syracuse Univ; BChE/Chem Engg/Syracuse Univ. *Born:* Jan 19, 1932. Prof & Chmn, Nuclear Engrg, U of Missouri since 1972. Academic career: 1958-64 Chem Engrg Oregon State U; 1964-72 Nuclear Engrg, Kansas State U. Cons to Argonne Natl Lab, the Boeing Co, Kerr McGee, Bendix Corp, & others. Feb 1971- Aug 1972 mbr of the Kansas Governor's Nuclear Energy Council; 1974-79, Chmn of the Public Informa-

Meyer, Walter (Continued)
tion Ctte of the Amer Nuclear Soc; 1976-77, Chmn Nuclear Engrg Div of the ASEE; 1977-78 Chrmn Nuclear Engg Div AIChE; 1978-79 Chrmn in Natl Nuclear Engg Dept Hds Organization; 1974, received the Amer Nuclear Soc's Natl Special Award for Public Information; 1962-63, 1975- 78 Touring Lecturer for ACS; NSF Scie Faculty Fellow at Oregon State U & MIT. BS & MS, chem engrg, Syracuse U, 1956 & 1957; PhD Oregon State U 1964; Elected to Board-of-Directors of the American Nuclear Society (1981-84); Co-Chairperson, No on 11 Committee (1980); Co-Director Energy Systems & Resources Program of the College of Engineering, University of Missouri-Columbia (1974 to present); Co- founder Energy and Public Policy Center of the University of Missouri.. *Society Aff:* ACS, AIChE, ANS, ASEE, AWS, ΣΞ, ΤΒΠ

Meyer, Wayne E
Business: Natl Cntr 2(PMS 400), Washington, DC 20362
Position: Project Manager, AEGIS Shipbuilding. *Employer:* US Navy, Sea Systems Cmmd. *Education:* MS/Aero-Astro/MIT; BS/Elec-Elex Eng/MIT; BS/Elec-Elex Eng/Univ of KS *Born:* April 1926. Enlisted US Navy, Brunswick, Mo; Apprentice Seaman July 1943. On continuous active duty; selected Rear Admiral Jan 1975; BS Elec/Elex Engrg; KU; MIT; Naval PG Scol; MS Aero/Astro; MIT. Held numerous posts afloat in Cruiser, Destroyers; in electronics & operations. Expert in shipboard tactical missilery; 25 years exper; last 15 years devoted to dev of shipboard combat systems engrg. Served many posts ahsore including Dir of Engrg 3 1/2 years at Naval Ship Missile System Engrg Stattion Port Hueneme, Ca; taught atomic weapon engrg for 3 years; Mgr, AEGIS Combat Sysytem Proj last 10 yrs. Authored numerous articles on Anti-Aircraft Warfare; Heavy demand as speaker on Prog Mgmt. Received 1977 Gold Medal ASNE for Engg accomplishments; 1981 Distinguished Engg Serv Awd KU; Assumed duties as Deputy Commander for Weapons and Combat Sys, Naval Sea Sys Command in Sept 1983; holds Naval Ordnance Engr Certificate No. 99; 1983 recipient of Missile Sys Awd for distinguished serv from the Amer Inst of Aero and Astronautics. *Society Aff:* ASNE, AIAA

Meyerand, Russell G
Home: 1 Churchill Ln, Kirkwook, MO 63122
Position: Mgr Electricl Engrg (Retired). *Employer:* Union Electric Co, St Louis. *Education:* BS/Elec Engg/MIT; MS/Elec Engg/MIT. *Born:* April 1902 Milwaukee, Wisc. Reg PE in Mo, Iowa, & Ill. Had respon for electrical design of Power Plants, Substations, Transmission, Protective Relaying & Communication facilities. Life Fellow IEEE, Life Mbr MSPE, Mbr NSPE, P Chmn EEI, Elec Systems & Equip Ctte, AEIC Switching & Switchgear Ctte & Light & Power Chmn ANSI-C-37 Ctte. Author of papers & articles. After retiring from U E Co became a cons engr for Moloney Elec Co. *Society Aff:* IEEE, NSPE, MSPE.

Meyerhoff, Robert W
Business: Sterling Forest, Suffern, NY 10901
Position: Principal Project Manager *Employer:* Inco Research and Development Center *Education:* PhD/PChem/IA State Univ; MS/PChem/IA State Univ; BS/Phys/Marietta College. *Born:* 11/1/35. Employed by Union Carbide from 1962-1979 & worked in area of cryogenics, last position being that of Corporate Res Fellow. Began current job Feb 1979 and am responsible for R&D projs related to diversification activities in energy related areas. Fellow of ASM, past chrmn of Helium Div of CSA & mbr of Sigma Xi. Enjoy wood working & stained glass. *Society Aff:* ASM, APS, AAAS.

Meyers, Franklin D
Business: PO Box 389, Clarkston, MI 48016
Position: Owner *Employer:* F. D. Meyers Associates *Education:* BS/Civil Engr/MI Technological Univ. *Born:* Oct 17, 1934 St Joseph Mich; s. Henry David & Marie (Weinheimer). BS Mich Tech U 1957; postgrad Wayne St U 1960-65; Ga Inst Tech 1965. m. Carole Peterson, Nov 6, 1954; children - David, Linda, Lori, Lisa, Donna. Field engr, Columbia So Chem Corp, Barbertson, Oh 1957-60; Principal facility planner Detroit Metropolitan Area Regional Planning Comm 1960-63; ch planner Macomb County Planning Comm, Mt Clemens 1963-64; Exec Dir Inter County Hwy Comm SE Mich, Center Line, 1964-85; Owner, F. D. Meyers Associates. 85- . Lecturer, public transportation (USSR), Wayne State U 1970. Who's Who in the Midwest, 1976; Mbr ASCE (Mich sect Pres 1970-71), ASPO. Director, National Board ASCE 77-79. *Society Aff:* APWA, APA, ASCE, ICMA, ASCE

Meyers, Marc A
Business: Dept of Met and Mat Engr, New Mexico Tech, Socorro, NM 87801
Position: Assoc Dir, Ctr for Explosives Tech Res & Assoc Prof of Metallurgy *Employer:* New Mexico Tech *Education:* PhD/Metallurgy/Univ of Denver; MSc/Mat Sci/Univ of Denver; BSc/Mech Engrg/Fed Univ of Minas Gerais, Brazil *Born:* 08/10/46 Previously associated with the facilities of SD Sch of Mines & Tech and the Military Inst of Engrg (IME) in Rio de Janeiro Brazil. He has publ more than 70 scientific and tech articles and is the co-editor with Dr L E Murr of the compendium *Shock waves and High-Strain-Rate Phenomena in Metals: Concepts and* Applications published by Plenum Press in 1981. He is also the co-author of textbook "Mechanical Metallurgy: Principles and Applications," publ by Prentice- Hall, 1984. Co-editor of "Frontiers in Materials Technologies–, to be publ by Elsevier in 1984. Mbr Bd of Reviewers, Metallurgical Transactions. Editor of Chapter "Mechanical Testing–, Desk Edition, Metals Handbook, 1985. Co-editor "Metallurgical Applications of Shock-Wave and High-Strain Plate Phenomona–, with L.E. Murr & K.P. Standhamrer, M. Debbery 1986. Was a visiting advisor at the Army Research Office, Research Triangle Park, NC 1985-1987. *Society Aff:* AIME, ASM, ΣΞ, ΑΣΜ

Meyers, Sheldon
Home: 3506 Dundee Dr, Chevy Chase, MD 20015
Position: Deputy Ass't Secretary *Employer:* US Dept of Energy. *Education:* MBA/Mgt & Finance/NYU; MSE/Heat Transfer/Univ of MI; BMarE/Marine Power/State Univ of NY. *Born:* 9/6/29. Since 1978 have been responsible for the Nucl Waste Mgt Prog. From 1977 to 1978 directed the licencing of all nucl facilities, except for power reactors, at the Nucl Regulatory Commission. From 1975-1977 was deputy asst Administrator for Solid Waste at EPA; 1972-1975 was responsible for pollution abatement at fed facilities & managed EPA's prog for implementing the natl environmental policy act as dir of EPA's office of fed activities. From 1969-1972 was responsible air pollution control R&D & regulation dev as dir of the control systems div & asst dir of stationary air pollution source prog. From 1958-1969 served in progressively responsible technical & admin positions with the US Atomic Energy Commission. *Society Aff:* ASME.

Meyninger, Rita
Home: 300 Winston Dr, Cliffside Park, NJ 07010
Position: Sr VP *Employer:* Enviresponse, Inc *Education:* Dr Eng Sci/Civil/NJ Inst of Tech; MSCE/Civil & Sanitary/NYC; BSCE/Civil/Newark Coll of Engrg. *Born:* First Regional Dir, NY, Federal Emergency Mgmt Agency, federal agency responsibile for preparedness and response to all natl emergencies. Appointed by Pres Carter to coordinate federal response to emergency at Love Canal. VP, Resource Planning, Hydroscience (subsidiary Dow Chem Co). Responsible for environmental planning for government and industry. Sanitary engr, water resources planning, Clinton Bogert Assocs. Started career with Edwards and Kelcey on field construction of NJ Turnpike soil stabilization procedures. Later joined Army Corps of Engrs. First woman admitted to cvl engrg prog, Newark Coll of Engrg. Alumni Achievement Honor Roll Award, NJ Inst of Tech, 1980. Who's Who of American Women, 1983-1984; The World Who's Who of Women 1984. *Society Aff:* ASCE, WPCF.

Michael, Harold L
Business: Civil Engrg Bldg, West Lafayette, IN 47907
Position: Hd, Sch of Civil Engr (since July 1, 1978) *Employer:* Purdue University. *Education:* MSCE/Trans Engrg/Purdue Univ; BSCE/Civ Engrg/Purdue Univ *Born:* 7/24/20 Native of Columbus, IN. Prof of Hgwy Engg & Dir Joint Hgwy Res Proj at

Michael, Harold L (Continued)
Purdue with 36 yrs in these activities. Maj teaching & res interests are traffic operations, transportation planning & transportation safety. Chrmn, PE's in Educ, NSPE 1965-67. Pres, Inst of Transportation Engrs 1975. Chrmn, Exec Ctte, Transportation Res Bd, 1976. Chrmn, Natl Advisory Ctte on Uniform Traffic Control Devices 1970-74. VChrmn, Am Rd and Transportation Builders Assoc 1978- 82. Engr of the Yr, IN 1972. Roy W. Crum Award for Outstanding Achievement in Transportation Engg 1978, Theodore M. Matson Meml Award for Outstanding Contributions to Traffic Engg 1979. James Laurie Prize for Outstanding Contributions to Trans Engrg 1981, G Brooks Earnst Lecture Awd 1981, George S Bartlett Awd for Significant Contributions to Trans Engrg 1982, Burton W Marsh Awd for Distinguished Serv to Inst of Trans Engrs 1984, Hon Mbr Inst of Trans Engrs 1980, Hon Life Mbr Indiana Constructors 1983, Hon Mbr Amer Soc of Civil Engrs 1985. Reg PE IN. Elected NAE 1975. Rotarian Dist Governor 1979-80. Lutheran. Hobbies are stamp & coin collecting & fishing. *Society Aff:* NSPE, ASCE, ITE, ASEE, AREA, ARTBA, APWA.

Michaelides, Efstathios E
Business: MAE, Newark, DE 19716
Position: Prof (tenured) *Employer:* Univ of DE *Education:* PhD '81/Engrg Sci/Brown Univ; MS '79/Engrg Sci/Brown Univ; BA '73/Engrg Sci & Econ/Oxford Univ, England; MA (hon) '83/-/Oxford Univ, England *Born:* 02/13/55 Native of Thessaloniki, Greece, I have been in DE since 1980. I am mbr of the Multiphase Flow Ctte, ASME and have published approx 55 articles in journals, conference proceedings and books. My specialization is in multiphase flows, energy conversion and conservation and heat transfer through flexible tubes. I have consulted with Du Pont, CDC, and TASA. Married since 1982 to Maria Laura Garcia Silva. *Society Aff:* ASME, ASHRAE, SES

Michaels, Alan S
Home: 1800 Gough St, San Francisco, CA 94109
Position: Prof Chem Engg & Medicine. *Employer:* Stanford Univ. *Education:* ScD/ChE/MIT; SM/ChE/MIT; SB/ChE/MIT. *Born:* 10/29/22. in Boston, MA; joined MIT Chem Engg Faculty (Asst Prof) 1948; Assoc Prof 1956; Prof 1960. Assoc Dir, MIT Soil Stabilization Lab, 1950-1959. Pres (Founder) of Amicon Corp, Lexington, MA, 1962-1970. Pres (Founder) of Pharmetrics, Inc, Palo Alto, CA 1970-1972. Pres, Alza Res and Sr VP Alza Corp, Palo Alto, CA 1972-1976. Prof Chem Engg and Medicine, Stanford Univ 1977 to present. Teaching, res, consulting activities in fields of surface/colloid chem, membrane tech, biomedical/biochem engg Author of ca 130 articles, 40 patents. *Society Aff:* AIChE, ACS, AAAS, ASAIO, ISAO.

Michaelson, Stanley D
Home: 1446 Circle Way, Salt Lake City, UT 84111
Position: President *Employer:* Stanley D. Michaelson & Co. *Education:* /Grad Study Metallurgy/Lehigh Univ; BS/Mining Engg/Lehigh Univ; EM/Mining Engg/Univ of Montana. *Born:* 9/4/13. Native of NY 1934-35, Institute of Res Fellow, Lehigh Univ 1935-1941, Engr in charge-Mining Lab, Met Engr, Field Engr-Allis Chalmers Mfg Co, Milwaukee and Salt Lake. US Army 1941-45 (Lt Col). 1946-54, Spec Engr, Chief Engr-Coal Mines, Chief Engr-Raw Materials, US Steel Corp, Birmingham 1954-1976 Chief Engr-Metal Mining Div & Dir Central Engg-Kennecott Copper Corp 1976-Present, Consulting Mining & Met Engr, President, Stanley D. Michaelson & Co., and Adjunct Prof Mining Engg-Univ of UT. VP and Dir-AIME; Pres, Soc of Mining Engrs of AIME, AIME Richards Award 1962, AIME Hon Mbr 1978. *Society Aff:* AIME, SME, MMSA, ΣΞ.

Michal, Eugene J
Business: 5950 McIntyre St, Golden, CO 80403
Position: Pres. *Employer:* Amax Extractive R&D, Inc. *Education:* ScD/Process Met/MIT; SM/Mineral Engg/MIT; BS/Met Engg/Univ NV. *Born:* 10/23/22. Education & early experience in metal mining ind in NV. WWII: degaussing and compass adjusting, with active duty in S Pacific. Profl career in R&D in hydrometallurgy of U, V & Ti, direct reduction of iron ores, electric furnace steelmaking, non-ferrous smelting & refining & technical mgt. Presently responsible for planning, admn & coordination of all corporate extractive res and dev. Past chrmn of CO sec of Soc of Mining Engr, & Mbr of Council of the Inst of Mining & Met, London. *Society Aff:* AIME, AMA, IMM, AIChE.

Michalak, Ronald S
Home: 922 Avon Ct, Grosse Pointe Woods, MI 48236
Position: Engg Mgr. *Employer:* GM. *Education:* MBA/Mgt/Univ of Detroit; BS/Met Engg/MI Tech. *Born:* 7/9/40. Born & raised in the Detroit Metropolitan area. After grad from MI Tech I have served industry in res, dev & production at various levels from entry level engr to chief met to mfg mgr. In my current position with Fisher Body div of GM, my engg group is responsible for all metallic components used in GM bodies. This responsibility finds me serving on several natl committees concerned with met dev & Natl Govt consulting. *Society Aff:* ASM, SAE, ADDRG.

Michalski, Ryszord S
Business: Dept of Computer Sci, 1304 W Springfield, Urbana, IL 61801
Position: Assoc Prof *Employer:* Univ of IL *Education:* PhD/Computer Sci/Tech Univ of Silesia; MS/Computer Sci/Leningrad Polytech Inst; BS/Electronics/Warsaw Tech Univ; CE/Krakow Tech Univ *Born:* 05/07/37 From 1962 to 1970 he was a res scientist, and later, a leader of the Pattern Recog Group at the Inst of Automatic Control of the Polish Acad of Scis. In 1970 he joined the Dept of Computer Sci, Univ of IL at Urbana-Champaign, and the Princ Invest of a proj on computer inference funded by the Natl Sci Fdn. Since 1976, has been a speaker of the Dist Visitor Prog of the IEEE Computer Soc. His res interests include inference systems, knowledge-based computer consltg systems. He is the author of approx 60 res and tech papers published in the USA and abroad. Is a mbr of the Am Assoc for Artifical Intelligence, Sigma Xi Soc, Classification Soc, Pattern Recognition Soc, Assoc for Computer Machinery, Polish Inst of Arts and Sciences in America.

Michel, Basil J
Home: 3 Trailwood Circle, Rochester, NY 14618
Position: President *Employer:* Omicron Consulting Assoc Div of J.S.J. Group Ltd. *Education:* BS/ChE/Univ of Rochester; MS/ChE/Univ of Rochester. *Born:* Oct, 1936. Production Engr, Procter & Gamble 1959. Started Mixing Equip Co 1960 as a Res Engr. Became Application Engr 1962. Assumed full departmental respon as Dir 1970. Appointed VP, Marketing 1975. VP Research and New Product Dev 1980. Co-author of several papers in field of mixing. Full Mbr of AIChE; held several offices in Rochester Sect including Chmn 1967-68. Full mbr soc Sigma Xi, Water Pollution Control Federation; reg PE. Since 1981 major respons presently include Consulting on Engrg design in Chem processing, waste water treatment, pulp & paper & other fluid processing applications and marketing New Product Development Consulting. *Society Aff:* AIChE, ΣΞ, WPCF.

Michel, Henry L
Business: One Penn Plaza, New York, NY 10119
Position: President & Ch Exec Officer. *Employer:* Parsons Brinckerhoff Inc *Education:* BS/Civil Engg/Columbia Univ. *Born:* June 1924. Served with Army Signal Corps 1943-46. Active in engrg design & construction for power plants with United Engrs & Constructors; for oil refineries M W Kellogg; for airfields, commercial & institutional projs with SSVK. Founder & Pres of Enconi, Spa, an internatl consulting engrg firm located in Italy with engrg projs throughout Africa & the Middle East. Partner, Dir, Pres & Ch Exec Officer of Parsons Brinckerhoff a totally integrated, multi-natl engrg/architectural/planning firm, headquartered in NY. Mbr of numerous tech socs & author of publs & books. Enjoy the theater, music, the arts, & summer recreational activities; tennis, swimming, fishing, & sailing. *Society Aff:* ACEC, AIM, ASCE, AWWA, CUEC, IRF, SAME, NYSPE.

Michel, Robert C
Business: 299 Williams Ave, Hackensack, NJ 07601
Position: President. *Employer:* The Kraissl Co, Inc. *Education:* Sm/Chem Engg/MIT; SB/Chem Engg/MIT. *Born:* July 14, 1927. Brooklyn Tech HS 1945; Reg PE in NJ. USNR as electronic technician 1945-46. Application & design engr 1951-60 for The Kraissl Co, Inc, Hackensak, NJ, manufacturers of indus pipeline strainers, duplex 3-way valves, fuel ol pumps, air compressors & vacuum pumps. New Applications Engr 1961-62 for Res Cottrell, Bound Brook, NJ, designers & builders of electrostatic precipitators, mech cyclones, scrubbers & other air pollution control equip. Returned to The Kraissl Co, Inc 1962 & served in various corporate offices including Dir, VP, Exec VP, Pres. & Chairman. Mbr AIChE, ASME, NSPE, Tau Beta Pi, Sigma Xi; P Pres Bergen County Soc of PE's; P Pres of River Edge Bd of Educ. Enjoy boating, sailing, photography. *Society Aff:* AIChE, ASME, NSPE, ASM, SAME, USPS, TBΠ.

Michels, Frank G
Business: 636 California St, San Francisco, CA 94108
Position: VP & Genl Mgr. *Employer:* Stauffer Chem Co of WY. *Education:* MSChE/Chem Engg/NY Univ; BSChE/Chem Engg/Polytechnic Inst of Brooklyn. *Born:* 11/26/31. Reg PE NY. Mbr AIChE; P Mbr AASE. E I DuPont - 2 yrs Operation Engr. Stauffer Chem - 3 yrs Res Engr; 6 yrs Process-Proj Engr; 2 yrs Principal Proj Engr; 5 yrs Mgr Proj Engg; 6 yrs Ch Engr; 1 yr to date VP & Genl Mgr WY Div *Society Aff:* AMA, AIChE.

Michelson, Ernest L
Home: 6833 N Kedzie Ave, Chicago, IL 60645
Position: Administrative Manager. *Employer:* Mid-America Interpool Network. *Education:* BSEE/EE/IL Inst of Tech. *Born:* March 1908. Native of Chicago, Ill. Served with the Commonwealth Edison Co of Chicago from 1929 to retirement in 1973, in activity related to planning of the high voltage generation $&& transmission systems. Head of Sys Planning Dept 1962-71; Asst VP 1971-73. After retirement, joined Mid-America Interpool Network (MAIN) as Administrative Mgr. MAIN is one of the 9 regional groups set up to assure reliability and adequacy of electric power in the US. Fellow IEEE. *Society Aff:* IEEE.

Michie, Donald
Business: Dept of Comp Sci, Urbana, IL 61801
Position: Prof Of Machine Intelligence, Univ of Edinburgh *Employer:* Univ of Edinburgh & Univ of Illinois *Education:* DSc/Biological; Sci/Oxford Univ; PhD/Genetics/Oxford Univ; MA/Human Anatomy and Physiology/Oxford Univ *Born:* 11/11/23 Specialised profl competence Heuristic problem-solving and machine learning. From 1965 onwards: Director of Experimental Programming Unit and Hd of Machine Intelligence, Edinburgh University. Visiting Prof in Comp Sci, Univ of IL (1976). Herbert Spencer Lectr, Univ of Oxford (1976). Consultant in biomedical computing, Sloan Kettering Inst, NY (1976). Visiting Res Prof in Comp Sci, Carnegie-Mellon Univ (1977). Visiting Res Prof in Comp Sci, Case Western Reserve Univ (1978). Samuel Wilks Memorial Lectr, Princeton (1978). Visiting Prof in Comp Sci, Stanford, Univ (1978). Currently Prof of Machine Intelligence, Univ of Edinburgh (from 1967) and Adjunct Prof of Comp Sci, Univ of Illinois (from 1979). Author of *On Machine Intelligence* (1976) and *Machine Intelligence and Related Topics* (1982), and Chief Editor of *Machine Intelligence* series, Vols 1 through 10. *Society Aff:* FBCS, FZS, FRSE, AISB, LMS.

Mickelson, Cedric G
Home: 18738 May Ave, Homewood, IL 60430
Position: Consultant. *Employer:* Self employed. *Education:* MS/Met Engg/Univ of WI; BS/Chem/Univ of WI. *Born:* 5/6/12. and married in Milwaukee WI. Employed at Am Steel Foundries in E Chicago IN from July 1936 through Feb 1973. Held positions of res met, chief met and asst mgr of res with maj responsibilities in fields of melting, heattreating and testing. On many natl committees of AIME-ASM-ASTM-SAE and Steel Founders Soc of Am. During WWII was steel casting rep on the war production bd and at the same time the casting rep on the US - Canadian - British Met Mission. Enjoy travel. *Society Aff:* ASM, AIME.

Mickey, Forrest R
Business: 3000 Youngfield St, Suite 273, Lakewood, Lakewood, CO 80215
Position: President. *Employer:* Mickey and Schneider Inc, Cons Engr *Education:* BSCE Colo. St. Univ. 1959 *Born:* Jan 1937. Graduate study at U of Colo, Denver Center. Grew up in southern Colo. Served 2-1/3 years in US Army Corps of Engrs as a Lt. in a construction battalion. Held key positions with several major consulting offices continually from April 1962 until entering own practice in May 1971. Reg PE in Colo since 1963. Also Reg PE in New Mexico, Texas, Oregon, Montana, Wyoming & Nebraska. Mbr ACI, ACEC, & Colo Structural Engrs Assn of Colo (bd mbr).

Mickle, David Grant
Home: 3251 Court Drive, Stuart, FL 33494
Position: Consultant. *Education:* MS/Transp Engg/Harvard; BS/CE/Univ of MI. *Born:* July 6, 1908. P Pres & Honorary Mbr Inst of Transportation Engrs. Hon Mbr Inst of Transportation (APWA) & Amer Road Builders Assn. Life Mbr ASCE & APWA. Deputy Fedl Highway Administrator 1961-63; Chmn US Delegation, Pan American Highway Congress 1963. Exec Dir Highway Res Bd, NAS 1964-66. VP & Pres Automotive Safety Foundation 196769. Chmn, Exec Ctte, HRB 1970. Pres Highway Users Federation 1970-75. As a public official (city, state, & federal) & officer of various prof grps, made significant contribs to development of public policies, standards, & procedures & planning coordination in transportation. Assoc Editor, Handbook of Highway Engrg, 1975. Distinguished Alumnus Award, U of Mich, 1953; Matson Award, ITE 1958; Roy Crum Award, HRB 1973; George Bartlett Award, ASHTO, 1974. Reg PE Mich, DC & Md. *Society Aff:* ASCE, ITE, APWA.

Mickle, Marlin H
Home: 1376 Simona Dr, Pittsburgh, PA 15201
Position: Prof of Elec Engr *Employer:* Univ of Pittsburgh *Education:* PhD/EE/Univ of Pittsburgh; MS/EE/Univ of Pittsburgh; BS/EE/Univ of Pittsburgh *Born:* 07/05/36 Held engrg positions with IBM and Westinghouse and was Prog Dir, Sys Theory and Applications, Natl Science Foundation. Served as Conslt to the Natl Science Foundation, Westinghouse, TASC, Contraves, III Sys, AUC, Texas Inst, Compunetics, Battelle, and others. Founding Co-Editor, Journal of Interdisciplinary Modeling and Simulation, and Assoc Editor, IEEE SMC Transactions. Currently Co-Chairman, Pittsburgh Modeling and Simulation Conference, and formerly VP (Conferences), IEEE SMC Society. Pres, Mickle Computer Technologies, Inc; VP, Power Resources, Inc. Mbr, Bd of Dirs: Power Resources; U.M. Home, Mt. Lebanon, PA; University Research and Development Associates, Inc.; Co-Author or Co- Editor of several books and author or co-author over 100 other technical publications. *Society Aff:* IEEE

Mickley, Harold S
Home: 11 Pequot Trail, Westport, CT 06880
Position: Director & Exec VP - Technical. *Employer:* Stauffer Chemical Co. *Education:* ScD/Chem Eng/MA Inst of Tech; MS/Chem Eng/CA Inst of Tech; BS/Chem Eng/CA Inst of Tech. *Born:* Oct 1918 Seneca Falls, NY. Naval Ordnance WWII. Chem Engr, Union Oil Co 1941-42 followed by long career at MIT 1942-70, advancing to Ford Prof of Engrg, Chmn - Faculty, & Dir, Center Advanced Engrg. Pres, Tyron, Inc 1955-57. Stauffer Chem Co since 1967 as Dir; 1971, VP; 1972, Exec VP - Technical 1981, Vicechairman 1981-83. Ret 1983. New York Academy of Science, Connecticut Academy of Science. Fellow AIChE. Fellow - American Association for the Advancement of Science Author. 50 patents. *Society Aff:* AIChE, AAAS, NAE, ACS, SCI, AAAS, ECPD.

Middendorf, William H
Home: 1941 Provincial Ln, Ft Mitchell, KY 41011
Position: Professor of Elec Engrg. *Employer:* University of Cincinnati. *Education:* PhD/Elec Eng/OH State; MS/Elec Eng/Univ of Cincinnati; BEE/Elec Eng/Univ of VA. *Born:* March 1921. Mbr U of Cincinnati teaching staff, 1948-. Director of UC/NEMA insulation lab. Also Dir of Engrg & Res, Wadsworth Elec Mfg Co, 1966-86, Senior Consultant, 1986-. Mbr Genl Engrg Ctte of Low-Voltage Distribution Sect of NEMA & mbr of 3 Underwriters' Labs Industry Advisory Councils. Publs are principally on design, electrical insulation & mgmt. Holds 27 US Patents. Reg PE Ohio. Fellow IEEE 1968. Distinguished Engineer Award 1978. U of Cincinnati Public Service Award 1981. Public service includes 20 years association with St Elizabeth Medical Ctr (465 beds), Covington, Ky. Pres, Bd of Trustees, 1977-80. *Society Aff:* IEEE, ASEE, TBΠ, BKN, ΣΞ

Middlebrook, R David
Business: 116-81 Caltech, Pasadena, CA 91125
Position: Prof of EE. *Employer:* CA Inst of Tech. *Education:* PhD/EE/Stanford Univ; MS/EE/Stanford Univ; MA/EE/Cambridge Univ; BA/EE/Cambridge Univ. *Born:* 5/16/29. With Caltech since 1955. Publications include numerous papers, and books on solid-state device theory and differential amplifiers. Spent a yr lecturing and consulting at some two dozen univs and cos in seven European countries, and have since made frequent professional visits to Europe. Elected Fellow of the IEEE "For contributions to electronic circuit analysis." Res interests are now in circuits and systems, and particularly in power processing electronics. Especially interested in design-oriented circuit analysis and measurement techniques, and have conducted short courses in both Europe and the US. *Society Aff:* IEEE.

Middlebrooks, Eddie Joe
Home: 1737 E 1400 North, Logan, UT 84321
Position: Dean, College of Engrg. *Employer:* Utah State University. *Education:* PhD/Environ Engg/MS State Univ; MSE/Environ Engg/Univ of FL; BCE/Civil Engg/Univ of FL. *Born:* Oct 16, 1932. Post doctoral study was done at U C Berkeley. His res activities center around water & wastewater treatment processes design, evaluation & modification; & algal growth kinetics & assessment of eutrophication. He has taught at the U of Fla, U of Arizona, Mississippi State U, & Utah State U. He was named an Outstanding Educator of America in 1973 & has received the Harrison Prescott Eddy Medal from the Water Pollution Control Federation. Reg PE in Arizona, Mississippi & Utah. Cons to both engrg & industrial firms, the nation of Iraq, & the UN Industrial Dev Organization. He is a P Pres of the Assn of Environ: mental Engrg Profs, a Fellow in the ASCE, & is active in numerous other scientific & prof soc's. He is author & co-author of over 150 publs incl 7 books. *Society Aff:* ASCE, WPCF, AAAS, AEEP.

Middleton, David
Home: 127 E 91 St, New York, NY 10128
Position: Physicist, Appl Mathematician, Educator *Employer:* Self-employed *Education:* PhD/Physics/Harvard Univ; AM/Physics/Harvard Univ; AB/Physics/Harvard College. *Born:* 4/19/20. Phi Beta Kappa 1941; Sigma Xi 1944. Physicist, Appl Mathematician, Elec Engr, Educator. Adj Prof EE and/or Appl Physics, Communications at Columbia U (1960-61); Rensselaer of Conn (Hartford Grad Ctr), 1961-70; U of Rhode Island 1963 -; adj Prof of Mathematical Sci Rice Univ, 1979- ; visiting prof at various times: Johns Hopkins U; U of Texas; U of Denver (supr of doctoral theses). NATO Advanced Study Insts Genoble 1966; Copenhagen 1980; Luneburg 1984; US Delegate to URSI (Lima) 1975. One of the founders of Statistical Communication Engrg 1942- . Cons, Contractor for US Govt agencies, indus, non-profit organizations; Exec office of Pres. Naval Res Adv Comm (NRAC) & other adv bds. Author: Intr to statistical communication theory (1960) (McGraw Hill), Reprint ED Peninsula Pub. Los Altos, CA, 1987, Russian ED (1961-62), Topics in communication theory (1965) McGraw-Hill; Reprint ED, Peninsula Pub. Los Altos, CA, 1987; Russian ED (1966); Sci Ed of VV Ol'shevskii's Statical Methods in Sonar, Acous Inst Acad Sci USSR. Over 135 papers; Natl Electronics Conference Award (1956); various prize papers IEEE, ECS (1978-80); Int'l EMC prize paper (1979). Fellow APS, (Life) IEEE, AAAS, AS, Explorers' Club (1978-), Cosmos Club (1965-) *Society Aff:* APS, IEEE, ASA, IMS, AMS, OSA.

Middleton, James R
Home: 16011 Laurelfield, Houston, TX 77059
Position: Pres *Employer:* Jr Middleton PE *Education:* BS/EE/Univ of IL; MS/EE/Univ of IL. *Born:* Oct 1922 Bloomington, Ill. Native of ElDorado, Ark. Signal Corps Officer North Africa, Europe. Built Air Force radio facilities, including theater air traffic control center from which Berlin air lift was controlled. Conceived & applied first 'line of control' instrument display concept, first fully integrated chemical plant multi-computer system. Directed Monsanto's dev of ac2, dc2, now Fisher Controls. President Process Systems, Tex-A-Mation, and JR Middleton PE. Past Pres Instrument Soc of Amer. Mbr Bd Dirs Electrical Engrg Alumni Assn, U of Ill. Reg PE Missouri, Texas, Cal. *Society Aff:* ISA.

Middleton, William W
Home: 6 Cornwall Cir, St Davids, PA 19087
Position: Consultant *Employer:* Self employed *Education:* BS/EE/Penn State Univ. *Born:* 10/24/20. Signal Corps USA WWII. LtCol (ret) USAR. Graduate Comd & Genl Staff Col. Employed Bell of Pa 1947-1983 in various engrg capacities. Dir Region 2 IEEE 1975-76; mbr US Activities & Regional Activities Bds, Chrmn IEEE Mbr Conduct Ctte 1984-87, IEEE USAB Controller 1984, Chrmn IEEE Reg Activ Bd Awards & Recog Ctte 1983-87, 1984 Recipient IEEE Haraden Pratt Award & IEEE Centennial Medal, Bd of Governors IEEE Communications Soc 1974-76; Chmn Wire Communications Ctte 1973-74; Mbr of other prof & tech organizations. Dir NSPE 1974-77; Chmn NSPE Ethical Practices Ctte 1975-77. Chmn NSPE Engg Technicians and Technologists Committee 1979-81. Chrmn NSPE Constitution & Bylaws Ctte 1984-85, Vice Chrmn AAES Ethics Ctte 1984-86. Reg PE Penn & Delaware. Pres, Pa Soc PE's 1975-76. *Society Aff:* IEEE, NSPE, SAME, NFPA, AAAS.

Miedtke, Duane R
Home: 3016 Rankin Rd NE, St Anthony Village, MN 55418
Position: President. *Employer:* DRM Consultants Inc. *Born:* Sept 14, 1931 Fairmont Minn. BCE & Grad Studies Univ of Minn. Named Young Engr of Year 1970 by NW Section ASCE; V Pres NW Section ASCE 197374. Mayor of St Anthony Village Minn. Founded DRM Cons Inc 1972; prior employment by Cons Engrg firms in Minn & Hawaii 1960-72. Pres Suburban League of Municipalities 1973-74; Bd of Dirs Assn of Metropolitan Municipalities 1974-76; serves on various local govt advisory cttes; Guest Lectr for Architectural Engrg Univ of Minn.

Miele, Joel A Sr
Business: 81-01 Furmanville Ave, Middle Village, NY 11379
Position: Prin *Employer:* Miele Assocs *Education:* BCE/Struc/Polytech Inst of Brooklyn *Born:* 05/28/34 Pres-Elect of NYSSPE; Engr of the Year for NYSSPE 1983; Fellow of the ASCE; Diplomate of NAFE; Chrmn, NYC Community Bd; Mbr of the Queens Borough Bd; Pres of the Bd of Creedmoor State Hospital; Mbr of the Bd of the Queens Borough Public Comms Corp, V.P. of Peninsula Hospital Center and Nursing Home; Trustee of the Queens Borough Public Library; Captain, Civ Engr Corps, Reserve of the US Navy. Dir & Mbr Exec Cttee, Amer Parkinson Disease Assoc. *Society Aff:* ASCE, NSPE, ASTM, SAME, NJSSPP, NAFE, NYSAP

Mielenz, Richard C
Home: Rt 1, Box 103, Brigham Rd, Gates Mills, OH 44040
Position: Pres *Employer:* Richard C. Mielenz, P.E., Inc. *Education:* PhD/Geological Sci/Univ of CA; BA/Geological Sci/Univ of CA. *Born:* 12/18/13. Burlingame, CA. Geologist, Std Oil Co of CA, 1939-41. Petrographer then Hd, Petrographic Lab, Bureau of Reclamation, Denver, CO, 1941-56, in testing & res on engg mtls, mainly concrete & concrete-making mtls. Dir of Res, Master Builders, Div of Martin Marietta Corp, 1956-64, Cleveland, OH; 1964-78, directing res & dev on products for

Mielenz, Richard C (Continued)
concrete construction. Consultant in engg geology & petrography, 1951-79, mainly concrete & aggregates. Pres, Richard C. Mielenz, P.E., Inc., 1979-present. Active in ACI committees & bds; Pres, 1977; also ASTM, formerly VChrmn, Committee C-9 on Concrete & Concrete Aggregates. Reg PE, State of OH. Numerous publications and honors. *Society Aff:* ACI, ASTM, GSA, MSA, NSPE, TRB.

Mielke, William J
Business: 419 Frederick St, Waukesha, WI 53186
Position: Senior VP, CEO *Employer:* Ruekert & Mielke, Inc *Education:* BS/CE/Univ of WI *Born:* 05/20/47 Graduated from Univ of WI in 1971 with a BS Degree in Cvl Engrg majoring in wastewater treatment, worked for WI Dept of Natural Resources as a District Field Engr, currently Senior VP and CEO of Ruekert & Mielke, Inc in charge of all wastewater treatment Facility, well and water pumping station design and construction. Past Pres of the WI Society of PE Waukesha Chapter. Past Pres of the Professional Engineers in Private Practice for Wis. In 1981 was selected as the "Young Engr of the Year" by the WI Soc of PE. Currently chairman of the NSPE PEPP Professional Selection Committee. Registered as a PE and a Land Surveyor, Certified Soils Tester, Licensed Water and Wastewater Treatment Facility Operator and Private Pilot. *Society Aff:* NSPE, PEPP, AWWA, WPCF, ACEC.

Migliore, Herman J
Business: P.O. Box 751, Portland, OR 97207
Position: Prof, Mech Engrg *Employer:* Portland St Univ *Education:* D Eng/ME/Univ of Detroit; ME/ME/Univ of Detroit; B/ME/Univ of Detroit *Born:* 07/13/46 Res mech engr for Naval Civil Engrg Lab in areas of advanced design of ocean equip, stress analysis, underwater welding experimentation, analysis of ocean cable systems, 1969-77. Doctoral intern at Chrysler Corp, 1973-75. Faculty mbr at Portland St Univ. Responsibilities include course dev and instruction in general design, computer aided design, advanced stress, finite element and microprocessor applications. Res activities in nonlinear numerical methods, Galerkin and Finite Element, advanced design, dev of intelligent sensor sys, and forensic engrg. Authored 20 papers and over 20 reports in mech engrg res and dev. *Society Aff:* ASME, ASEE, ΣΞ

Mihalasky, John
242 Starmond Ave, Clifton, NJ 07013
Position: Prof of Indus Engg. *Employer:* NJ Inst of Tech. *Education:* EdD/Bus Ed/Columbia Univ; MIE/Ind Engg/NYU; MBA/Economics/Rutgers Univ; MS/Mgt Engg/Newark College of Engg; BS/Mech Engg/Newark College of Engg. *Born:* 9/5/29. Before entering the academic world, with Westinghouse in the areas of tool, jig, and fixture design, product and process design, quality control, and methods engg. Mihalasky is a Fellow of the American Soc for Quality Control and Life fellow of the Soc for Advancement of Mgt. He also holds a PE from the State of CA. Prof Mihalasky is certified by the SME in Mfg Engg and Robotics, by ASQC in Quality Control and Reliability Engg, and by the Intl Mtls Mgt Soc as a Mtls Hdlg Professional. A 1973 recipient of the Soc for Advancement of Mgt's Phil Carroll Award, a recipient of several NSF grants, and a 1976-77 Natl Acad of Sci Exchange Scientist to Czechoslovakia and a 1986 Fulbright Award recipient, Mihalasky has lectured and consulted internationally and is the author of over 100 technical and professional papers. *Society Aff:* ASME, IIE, SME, ASQC, IMMS, ASEE, ASSE, AAAS, ASEM

Mihran, Theodore G
Home: 898 Ashtree Lane, Schenectady, NY 12309
Position: Physicist. *Employer:* General Electric Co R/D Center. *Education:* PhD/EE/Stanford Univ; MS/EE/Stanford Univ; AB/Engg/Stanford Univ. *Born:* June 1924 Detroit Mich. US Navy 1944-46. In 1950 joined G E Res Lab (now R/D Center); Visiting Prof at Cornell Univ 1963-64; Lectr at Union Coll 1953 & 1961 & Chalmers Univ (Gothenburg Sweden 1965). Author of 30 tech papers & 7 pats in following fields: microwave tube analysis, plasma wave theory, large signal klystron computer simulation, field-effect transistor modeling & microwave oven design. IEEE Fellow 1964; Assoc Ed IEEE Trans on Electron Devices 1970-73. Fortescue Fellowship 1948. Sigma Xi, Tau Beta Pi, Phi Beta Kappa. Violinist, Schenectady Symphony 1950- . Enjoy tennis, skiing & bridge. *Society Aff:* IEEE, APS, IMP.

Mikhail, Edward M
Business: Sch of Cvl Engrg, West Lafayette, IN 47907
Position: Professor of Photogrammetry. *Employer:* Purdue University. *Education:* BSCE (Hon)/Civ Engg/Cairo Univ; PhD/Photogrammetry/Cornell Univ; MS/Photogrammetry/Cornell Univ. *Born:* Sept 1935. Instr at Cairo Univ 1957-59; Res Analytical Photogrammetrist Aero Service Corp & Canadian Aero Service Ltd 1963-65; with Purdue snce 1965 in charge of grad instruction & res in photogrammetry, data adjustment, remote sensing & coherent optics in mapping, metric aspects of image processing, and appl. of A.I. and expert systems to mapping; cons to industry & govt. Author or co author of five textbooks: 'Observations & Least Squares' (1976) "Photogrammetry" 3rd ed., Univ Press, 1980; "Analysis and Adjustment of Survey Measurements–, Van Noslrand Reinhold, 1981; "Surveying: Theory and Practice" 6th ed, McGraw-Hill, "Intro to Surveying–, Mc Graw-Hill, 1985; contributing author, 4 chapters; & numerous tech papers & reports. Co-holder of pat 3820895. Active in Amer Soc for Photogr & Remote Sens. (ASPRS); ASPRS Pres Citation for work as pres of its Great Lakes Region 1970; Active in International Soc. of Photogr and Remote Sensing; twice winner of the Wild Photogrammetric Res Award 1961,62; winner of a NATO Postdoctoral Fellowship 1971; winner of 1976 Photogrammetric Award (Fairchild) Winner of the Coveted Von Humboldt Award. Enjoys classical music, walking & jogging. *Society Aff:* ASPRS, ACSM

Mikkola, Donald E
Business: Dept. of Metallurgical Eng, Mich. Tech. Univ, Houghton, MI 49931
Position: Prof *Employer:* Mich Tech Univ *Education:* PhD/Mat Sci/Northwestern Univ; MS/Mat Sci/Northwestern Univ; BS/Metallargical Eng/MI Tech Univ *Born:* 07/30/38 Native of Champion, MI. Engrg trainee with Jones and Laughlin Steel (now LTV) 1959. Joined faculty of Met Eng at MI Tech Univ 1964. Promoted to Prof 1972. Active in res on structure-propery relationships in matls utilizing x-ray diffraction and election optical techniques, particularly in studies of intermetallics and the dynamics of plastic deformation. Was a Fulbright Research Scholar at the Helsinki Univof Tech, Esp, Finland 1973-74. Currently mbr of the Bd of Dir of Accreditation Board of Engineering and Technology. *Society Aff:* TMS, ASM, ACA, ΣΞ, ΤΒΠ, ΦΚΠ

Mikochik, Stephen T
Business: 333 Jay St, Brooklyn, NY 11201
Position: Assoc Prof *Employer:* Polytech Inst of NY *Education:* MS/CE/Rutgers Univ; B/CE/Manhattan Coll *Born:* 10/10/25 Native of New York City. Taught at Manhattan College 1946-1954. Teaching at Polytech Inst since 1954. Dev the complete undergraduate and graduate course sequence in geotechnical engrg, beginning with three courses and expanding to 16 courses. Began the doctoral prog in geotechnical engrg in 1971. Granted leave of absence from 1960-1962 to accept an NSF Science Faculty Fellowship at Columbia Univ. Perform consulting services for organizations such as NY Port Authority, American Elec Power Services, Inc, Dame's & Moore, and others. Enjoy classical music and bird watching. *Society Aff:* ASCE, TRB, ASSMFE

Milam, Max
Business: Univ of Nevada, Reno, NE 89557
Position: President. *Employer:* University of Nevada - Reno. *Born:* July 13, 1930. BA Okla Baptist Univ; MA & PhD Univ of Okla. Previous experience in printing & printing mgmt, as Univ Prof, Ch Fiscal Officer of Arkansas & Ch Financial Adviser to Winthrop Rockefeller. Pres UNR as of Aug 1, 1974.

Milano, Nicholas P
Business: 2108 N Natchez Ave, Chicago, IL 60635
Position: Director of Metallurgy. *Employer:* Gear Products Division-Household International *Education:* BSME/Engrg/Marquette Univ; MS/Metallurgy/WI. *Born:* July 1922; native of Milwaukee Wis. Army Corps of ENgrs 1943-46. With Internatl Harvester Co 1946-71; 5 yrs Ch Metallurgist, 8 yrs Corp Ch Engr Materials Specification, also Met & heat treat cons in Europe; joined Ill Gear Walllace-Murray Corp 1971. Dir of Met. Mbr ASM, SAE, ASTM, AWS; numerous SAE & ASM Natl Cttes; P Chmn Milwaukee ASM, P Chmn SAE Genl Materials Council. Pub & lectrs on Met & Heat Treat. Fellow of ASM; PPres ASM. Distinguished Engr Citation Coll of Engg Univ of WI 1979. Enjoys tennis, golf & swimming. *Society Aff:* ASM, SAE, GRI.

Milano, Vito Rocco
Home: 4009 Clagett Rd, Hyattsville, MD 20782
Position: Naval Analsyst. *Employer:* Center for Naval Analysis. *Education:* BS/Eng/US Naval Acad; BS/ME/Webb Inst; MS/Naval Arch/Webb Inst; PhD/Ocean Engr/Stevens Inst. *Born:* Nov 1929 N Y City. Prof Engr N J. PhD (Ship Hydrodynamics) Stevens Inst of Tech; MS (Naval Arch) & BS (Marine Engrg) Webb Inst; BS USN Acad. Over 20 yrs Engrg Mgmt/Exec experience with specialty in naval ship design & const. Accomplished extensive res & cons on ice mechanics, ice breaker design, dev & acquisition of Strike Cruiser & AEGIS Destroyer lass ships. Mbr SNAME, Ship Prod, By Laws & Publs Cttes; Mbr ASNE, Natl Council & Chmn Papers Ctte 1974-76. SNAME Joseph H Linnard Prize recipient 1974. *Society Aff:* SNAME, ASNE.

Milbradt, Kenneth P
Home: 50 Norfolk Ave, Clarendon Hills, IL 60514
Position: Assoc Prof *Employer:* IL Inst of Tech *Education:* MS/CE/IIT; BS/ME/IIT *Born:* 05/08/24 Is Assoc Prof of Civil Engrg at the IL Inst of Tech and has been at the Inst since 1946. He has been active in ASCE by chairing a Natl Specialty Conf, being a mbr of a Structural Div Ctte, assisting and participating in an IL Section Seminar, as delegate to the Local Section Conf, mbr or chrmn of various IL Section cttes, and during the past five yrs, as Dir, Secretary and Pres Elect for the IL Section. He has published in ASCE, ACI, SESA and ASME journals. Prof Milbradt has acted as conslt to architects, engrs and various govt agencies, including the AEC. He has analyzed over 700 structural failures. He is reg prof and reg Structural engr in IL. *Society Aff:* ASCE, ACI, ΣΞ, XE, SESA, ASEE

Miles, John B
Home: 102 Westridge Dr, Columbia, MO 65201
Position: Prof. *Employer:* Univ of MO. *Education:* PhD/ME/Univ of IL; MS/ME/Univ of MO-Rolla; BS/ME/Univ of MO-Rolla. *Born:* 2/2/33. Educator; born in St Louis, Feb 2, 1933; S Aaron Jefferson & Annabelle (John) M.; Married Beverly Jean Bartlett, Feb 8, 1958; children: John David, Andrea. Asst prof applied sci S IL Univ, 1958-63; assoc prof ME Univ MO, 1963- 68, Prof, 1968-; res engr GE Co, 1965-66, NASA, U, 1971; vis prof Stanford, 1971; cons; dir NSF Summer Inst Fluid Dynamics, Univ MO, 1971. Served with AUS, 1957. A P Green Fellow, 1955; NSF faculty fellow, 1960-62; NASA-Am Soc Engg Edn sci faculty fellow, 1965, 68; Ford Fdn fellow, 1965; NRC fellow, 1970; Univ of MO Faculty/Alumni Award, 1978 res grantee NSF, NASA, USAF; Mbr Am Inst Aeros & Astro, ASME, Am Soc Engg Edn, Sigma Xi, Tau Beta Pi, Theta Tau, Pi Tau Sigma, Phi Kappa Phi. Contbr articles to tech journals. Sabbatical fellow at Solar Energy Research Institute, 1980-81. *Society Aff:* ASME, AIAA, ASEE, MSPE.

Miles, Richard B
Business: Dept of Mech & Aero Engrg, D-414 Engrg Quadrangle, Princeton, NJ 08544
Position: Prof *Employer:* Princeton Univ *Education:* PhD/EE/Stanford Univ; MS/EE/Stanford Univ; BS/EE/Stanford Univ *Born:* 07/10/43 Grew up in Orinda, CA and attended Stanford Univ as an undergraduate and graduate student. Was a Frannie and John K. Hertz fellow from 1969 until 1972 when he completed his PhD and joined the faculty of Princeton Univ in the Mech and Aero Engrg Dept. Presently conducting res in the area of nonlinear optics and lasers with a particular emphasis on laser diagnostics and molecular energy transfer. Teaching has been in the field of optics, applied physics and energy conversion. Promoted to Associate Prof in 1978 and Full Prof in 1982. Currently Chrmn of the Princeton Univ Engrg Physics Program. *Society Aff:* IEEE, AIP, OAS.

Miley, Delmar V
Home: 255 12th St, Idaho Falls, ID 83401
Position: Associate Scientist. *Employer:* Aerojet Nuclear Co. *Born:* Nov 1936 Winslow Ariz. Met Engr & MS Met Engr Colo Sch of Mines. Supr Met Lab at Sundstrad Aviation; 10 yrs at AEC's Rocky Flats Plant; provided lab support for dev of ingot sheet beryllium process; currently Assoc Sci at ERDA's Idaho Natl Engrg Lab supporting reactor safety & other energy-related progs. Publs & presentations in fields of metallography & mech testing. Mbr ASM & IMS. Served on natl MEI Ctte of ASM & active in local ASM educ progs. Judged at Internatl Metallographic Exhibit. ENjoys fishing & photography.

Miley, George H
Business: 214 NEL, 103 S Goodwin Ave, Urbana, IL 61801
Position: Professor, NE and EE. Director, Fusion Studies Lab *Employer:* Univ of IL. *Education:* PhD/Chem-Nuc Engg/Univ of MI; MS/ChemE/Univ of MI; BS/ChemE/Carnegie Mellon. *Born:* 8/6/33. Nuclear Engr at Knolls Atomic Power Lab 1959-61. Joined faculty at Univ of IL in 1961 where he is currently Professor of Nuclear and Electrical Engg & Dir of the Fusion Studies Lab. Author of numerous technical articles and two books: *Direct Conversion of Nuclear Radiation Energy & Fusion Energy Conversion*. Holds five patents. Div offices in ANS, IEEE, & ASEE; Fellow, ANS 1978, APS 1979, IEEE 1987; Guggenheim Fellow 1986; U of IL Ctr Adv. Studies 1986; Fulbright Award 1983; ASEE Western Elec Award 1976; NATO Sr Fellow 1975-76; ASEE-NE Teacher Award 1973. Enjoys tennis & jogging. *Society Aff:* ANS, IEEE, APS, ASEE, AVS.

Miley, Stan J
Home: 2812 Hillside Dr, Bryan, TX 77801
Position: Assoc Prof *Employer:* Texas A&M Univ *Education:* PhD/Aerospace Eng/MS St Univ; MSME/Mech/San Diego St Coll; BS/Aerospace Eng/San Diego St Coll *Born:* 04/02/39 Native of San Diego, CA. Dynamics engr with General Dynamics-Convair. Aeroacoustics engr with Bell Helicopter Co. Asst Prof at MS St Univ, performing res on cooling drag of general aviation aircraft. Assoc Prof at Texas A&M Univ, performing res on aircraft propeller aerodynamics. *Society Aff:* AIAA, ASEE, AHS

Militello, Sam
Business: PO Box 23646, Tampa, FL 33630
Position: Asst VP. *Employer:* Greiner Engineering Sciences, Inc. *Education:* BCE/Civil Engg/Univ of FL. *Born:* 9/24/37 Native of Tampa, FL. Served in FL Natl Guard 1961-67. With Greiner Engg Scis 1961- , specializing primarily in struc & civil engg; currently Proj Dir respon for large multi-disciplinary design projs. PPres FL Sect ASCE 1974-75 & West Coast Branch of FL Sect ASCE 1969-70; also PPres Tampa Chap FL Engg Soc (FES/NSPE). Currently state Dir for Tampa Chapter, FL Engg Soc. Reg engr in FL, GA & LA. *Society Aff:* ASCE, NSPE, FES, FTA.

Militzer, Robert W
Business: 850 Ladd Rd, Walled Lake, MI 48088
Position: Vice President *Employer:* Ex-Cell-O Corp *Education:* BSME/Mech Engg/Lawrence Inst of Technology. *Born:* Sept 1916. Advanced Business Admin Northwestern Univ. Reg P E, Certified Mfg Engr. Holds Mbrships in Soc of Mfg Engrs, Soc of Auto Engrs, NSPE & Soc of Plastics Engrs. Appointed V Pres & Genl Mgr of Micromatic Div of Ex-Cell-0 Corp 1972. During WWII Lt in USN with active service in Pacific Theater. Enjoys golf & water sports. Past President SME 1979, Past Director SME 1971-1979, Past Director MSPE Detroit Chapter, Rec'd Alumni

Militzer, Robert W (Continued)
Achievement Award, Lawrence Tech, 1973. *Society Aff:* SME, NSPE, MSPE, ASDCE, SPE, SAE.

Milks, Donald E
Home: 701 S Main St, Ada, OH 45810
Position: Prof of Cvl Engrg. *Employer:* Ohio Northern University. *Education:* PhD/Civil Engr/Univ of AZ; MS/Civil Engr/Univ of AZ; BCE/Civil Engr/Clarkson College of Tech. *Born:* June 1932. Amer Bridge Div of US Steel 1954-55; Proj Engr on Climatic Testing of recoilless rifles with US Army Ordnance Corps 1956-57; Grad Asst & Instr Univ of Ariz 1957-65; with Ohio Northern Univ 1965- . Pres Toledo Sect ASCE 1970-71; Pres Ohio Council ASCE 1974-75. Dev Individualized Instruction learning materials for struc design courses. Cons work on struc analysis & design. *Society Aff:* ASCE, AISC, ACI, ASEE.

Millar, G H
Business: MotorTech Inc, 530 Fentress Blvd, Daytona Beach, FL 32014
Position: VP, Engg. *Employer:* Deere & Co. *Education:* PhD/ME/Univ of WI; BS/ME/Univ of Detroit. *Born:* 11/28/23. Joined Deere & Co in 1963, Dir of Res. In the position, organized and directed Technical Ctr until 1969. Was appointed Asst Gen Mgr of Waterloo Tractor Works in charge of design and dev of John Deere engines. In 1972, following overseas assignment in Lausanne, Switzerland, was named VP, Engg. In 1975, was elected to Natl Academy of Engg, is Fellow of Soc of Automotive Engrs, mbr of College of Fellows of Engg Soc of Detroit and is active in numerous other professional societies. In 1977, received Honorary Doctorate of Sci Degree from Univ of Detroit. *Society Aff:* ASME, ASAE, NAE, SAE, NSPE.

Millar, Julian Z
Home: 72 Blackburn Pl, Summit, NJ 07901
Position: Consultant. *Employer:* self-employed. *Education:* BSEE/Elec Engg/Univ of IL. *Born:* July 1901. Attended grad schs of Geo Wash Univ & Columbia Univ. Western Union Telegraph Co 1923-65, Asst V Pres R/D 1953-65; Corp Dir & Cons HAzeltine Corp 1966- . Reserve Officer US Signal Corps 1923-61 reaching rank of Col. Prof Engr N Y State 1937- . Student & Assoc Mbr AIEE; Assoc Mbr 1930 & Sr Mbr 1945 IRE. Life Mbr IEEE; Chmn Prof Group-Communications 1954-55; Fellow IRE 1956. Special Award Communications Tech Group IEEE 1961; Fellow AIEE 1962; Achievement Award Communications Tech Group IEEE 1967; Disting Alumnus Award Univ of Ill 1970; Natl Security Agency Appreciation Award 1971; North Jersey Section Award IEEE 1972. Fellow Radio Club of Amer 1973. Meritorious Serv Award USAF 1960. Commendations & citations US Army. Fellow AAAS 1980. *Society Aff:* AAAS, SMPTE, AFCEA, RCOA

Millard, Charles F
Home: 803 Shaw Ct, Towson, MD 21204
Position: Partner. *Employer:* Whitman, Requardt & Assoc. *Education:* BSCE/Civ Engr/Bucknell Univ *Born:* 12/16/17. Native Pennsylvanian has been Marylander since 1950. With PA Dept of Hgwys and Corps of Engrs before WWII. Served in European & Asiatic Theaters, Army Corps of Engrs, structural designer & field engr for several yrs before joining Whitman, Requardt & Assoc in 1950. Became Assoc in 1958 & partner in 1964. Responsible for structural & geotechnical activities of WR&A. Specializes in industrial & commercial bldgs, shipyards, marine structures & tunnels. Reg Engr in DE, DC, GA, LA, MD, NJ, PA, VA, & WA. Mbr, Civitan & Engg Soc of Baltimore. Retired as of Dec 31, 1982. *Society Aff:* ASCE, NSPE, SAME

Millensifer, Tom A
Home: 2489 S Brentwood St, Lakewood, CO 80227
Position: Mgr Catalyst Purchases, Worldwide. *Employer:* Gulf Chem & Metallurgical Co, TX City, TX. *Education:* BS/ChE/Denver Univ. *Born:* May 1933 Denver Colo. 1 yr Phillips Petroleum Borger Texas, returning to Denver as Chem Engr for S W Shattuck CHem Co where held subsequent positions as Plant Engr, Production Mgr, V Pres Sales, Admin & mbr corp Bd of Dirs. Has been involved in all aspects of production, sales, distribution: Molybdenum, Uranium chems & corp admin; 1974-79 VP & Tech Dir Keystone-Lenway with respon for a new venture involving extraction, recycling of metallics from spent petroleum, petrochem catalysts & other residues from petro & chem opers. Assumed present position Feb 1979 upon acquistion of Keyston-Lenway's bus by Gulf Chem. JPL Enterprises Inc. Active in church, civic affairs; avid philatelist, fisherman, gardener. *Society Aff:* AIChE, APS, ATA.

Miller, Albert H
Business: 308 Walnut St, Newport, AR 72112
Position: Pres. *Employer:* Miller-Newell Engrs, Ltd. *Education:* MSAgE/AgE/Univ of MO; BSAgE/AgE/Univ of AR. *Born:* 11/30/32. Sales Engr, Delta Irrigation Co, Memphis, TN, 1955-57; Field Engr, Short & Brownlee Construction Co, Kansas City, MO and El Dorado, AR, 1957-60; VP, H D Kantor & Son, Clarksdale, MS, 1960-61; Pres, Miller Engrs Co, Inc, Clarksdale, MS, 1961-63; Pres, Miller-Newell Engrs, Ltd, Newport, AR, 1963 to present. Chrmn of State Sec ASAE 1971; Award for State Outstanding Agri Engr 1971; Pres of Newport Area Chamber of Commerce 1971; Pres of Newport Rotary Club 1974; Treas for American Consulting Engrs Council of AR, 1979. Pres of American Consulting Engineer Council of Ark, 1980, Natl Dir ACECA 1981-85, Natl Chrmn, Political Action Ctte ACEC 1984; Named Engr of the Year by AR Soc of PEs, 1982. *Society Aff:* NSPE, ASHRAE, ASAE, ACEC.

Miller, Carl E
Home: 36 Great Meadow Lane, Unionville, CT 06085
Position: Consultant. *Employer:* Combustion Engineering Inc. *Education:* EE/Univ of Cincinnati. *Born:* Oct 8, 1906 Lockport N Y. Educ: Univ of Cincinnati EE; Northwestern Univ Business Admin. Married Stella Johannigman; 3 children. Fellow ASME Mbr AIME; Tau Beta Pi, HKN. Reg P E in W Va. Major activities include fuel supply & power generation, fossil & nuclear; organization & mgmt; Fuels Cons Marshall Plan, Europe; Tech Adv on fuels Battelle Inst; combustion engrg; market res & long range planning; facilities & planning involved in CE's participation nuclear steam generation field; mgmt cons Argentina & Mexico Internatl Exec Service Corps. *Society Aff:* ASME, AIME.

Miller, Daniel R
Home: 701 Fulton St, Mountain Home, AR 72653
Position: Retired *Education:* BS/Elec Engg/Univ of WI. *Born:* 1913. With Genl Elec Co 1941-75; 4 yrs of G E's Advanced Engrg Prog; G E Lab 1941-47. KAPL 1947-75; various supr positions with concurrent activity as individual contributor 1941-55. Cons Engr 1955-75. Pub classical papers on flow- induced collapse of reactor fuel plates & on thermal ratcheting of pressure vessels; co-authored paper on flow-induced vibration of blade in flow channel; was KAPL's principal trouble-shooter on mech probs expecially those involving vibration, noise, mech malfunction or material failure. Held mbrships in IEEE, ASME, ANS, AWS, SESA, ASTM, ASM, NSPE; Fellow ASME. *Society Aff:* ASME, ANS.

Miller, David R
Business: c/o ATE-DMJM Joint Venture, P O Box 2303, Cincinnati, OH 45201
Position: Sr Vice President. *Employer:* Daniel, Mann, Johnson & Mendenhall. *Education:* BS/CE/UT State Univ. *Born:* 7/8/27. Sr VP, Prin & Dir Daniel, Mann, Johnson & Mendenhall Los Angeles, CA. Sr Cons for Cvl Engg. Mgr, DMJM Southwest Region. *Society Aff:* ASCE, AMA, AAEE.

Miller, Don W
Business: 206 W. 18th Ave, Columbus, OH 43210
Position: Prof & Chrmn of Nuclear Engr *Employer:* The Ohio State Univ *Education:* PhD/Nuclear Engr/OH State Univ; MS/Nuclear Engr/OH State Univ; MS/Physics/Miami Univ OH; BS/Physics/Miami Univ OH *Born:* 03/16/42 Native of Westerville, OH. Junior Engr with North Amer Aviation (1964). Teaching Asst at Miami Univ. With the Nuclear Engrg Program at the OH State Univ since 1966: Research Associate (1966-68), Univ Fellowship (1968), Teaching Assoc (1969-71), Asst Prof (1971-74, Assoc Prof (1974-80), Prof (1980-) and Chrmn of Nuclear Engrg (1977-). Dir of the Nuclear Reactor Lab (1977-). Author or co- author of 70 publications, presentations, and reports. Awarded 4 U S and 3 foreign patents. Principal or co-principal investigator on 30 R&D projects sponsored by industry and government. Chrmn, ASEE Nuclear Div (1978-79) and ANS Southwest OH Section (1979). Named Alumni Mbr of Tau Beta Pi (1980) and Honorary Mbr of Texnikoi (1980). 1979 Columbus Technical Council Person of the Year. Conslt to both industry and government. Chmn ANS Educ Div (1987-88), Chrmn- Elect Nuclear Engrg Dept Heads Org (1985/86). Coll of Engrg Res Award (1984). *Society Aff:* ANS, IEEE, ASEE

Miller, Donald R
Business: Apex Tower, High St, New Malden, Surrey KT3 4DJ England
Position: Mgr of Engrg *Employer:* Esso Eng (Europe) Ltd *Education:* BSE/ChE/Univ of MI. *Born:* 8/26/36. Native of Detroit, MI. Joined Exxon Res & Engrg Co in 1959. Worked in a variety of positions involving planning, dev, design, & startup of new process facilities. Technical & Admin Mgr of Exxon's Benicia, CA refinery in 1973-75. Asst Gen Mgr of Exxon Res & Engr co from 1975-80. Assumed current position in 1980. Responsible for Mng Exxon Res & Eng Co's European branch office, consisting of 200 people who provide a wide range of engr serv to Esso's 17 refineries in Europe. Active in Boy Scouting Prog. Hobbies: boating, fishing, & other water sports. *Society Aff:* AACE.

Miller, Dwight L
Home: 5932 N Trenton Ln, Peoria, IL 61614
Position: Asst Dir. *Employer:* N Regional Res Ctr. *Education:* BS/ChE/Univ of IL. *Born:* 10/22/13. Native of IL. Associated with fermentation & synthetic chem industries as Res Chem Engr & Chief Dev Engr before joining Northern Res Ctr in 1958. Has designed several large operating commercial chem plants for the production of such organic chem as alcohols, rubber chem, alphatic mines & agri products. Has extensive publications in these fields. Admin and responsible for maj devs on energy and products derived from various agri raw mtls. Recognized world authority on these subjs. Fellow - Am Inst of Chem Engg. *Society Aff:* AIChE, TAPPI, ACS.

Miller, Earle C
Business: 7 Terrace Dr, Worcester, MA 01609
Position: Consulting Engineer. *Employer:* Self. *Education:* BS/ME/Johns Hopkins Univ. *Born:* 7/20/13. Johnsville MD. 2 yrs in papermaking, then into power. 30 yrs with Riley Stoker Corp: engr, Mgr R&D, Dir Sales Engrg. 10 yr Chas T Main, Inc 1969-79. Spec: steam generation, fuels. Author many papers; inventor with over 40 issued pats. Hobbies: golf, horticulture, hunting, fishing. Mbr Worcester Engrg Soc. Hon statur Puu Tau Sigma. EJC Bd of Dir. ASME rep & V Chmn Coord Ctte on Energy 1975. Fellow, Pres 1976-77, Ch Comm on Org 1974-75, VP Power Dept 1975, Ch Comm on Reg Affairs 1967-68, VP Region I 1966-70, Ch Steering Comm on Energy & Environ 1974-75 ASME-Chrmn Pan American Federation of Engg Associations (UPADI) Tech Comm on Energy. *Society Aff:* ASME.

Miller, Edmond T
Home: P.O. Box 158, Pelham, AL 35124
Position: Prof & Chrmn. *Employer:* Univ of AL at Birmingham *Education:* PhD/Civ Engrg/TX A&M Univ; CE/Civ Engrg/MIT; MSCE/Civ Engrg/GA Tech; BCE/Civ Engrg/GA Tech. *Born:* 12/9/33. Soil & Fdn Engr for Law Engg Testing Co, Atlanta, GA, 1956-57. Served as Asst Engr Officer, 929 Engr Group (Construction) in 1957. Fdn Engr & Branch Mgr, Law Engg Testing Co, Tampa, FL until 1961. Taught civ engg at the Univ of AL in 1963-64 & 1967-75 with rank increasing to Prof. VP, Wm S Pollard Consultants, Inc, Memphis, TN, 1975-76. Independent consultant in traffic engg in Memphis, TN, 1976-77. Prof. and Chrmn, Dept. of Civil Engg, Speed Scientific Sch of the Univ of Louisville, 1977-80. Assumed present position at the University of Alabama at Birmingham in January, 1981. Interim Dean, Sch of Engrg, 1984. *Society Aff:* ASCE, NSPE (ASPE), ASEE.

Miller, Edward
Business: 1250 Bellflower Blvd, Long Beach, CA 90840
Position: Prof. *Employer:* CA State Univ. *Education:* DEnggSci/Mtls/NYU; MS/Met/NYU; BChE/CE/CCNY. *Born:* 7/30/32. Taught at NYU 1959-1968, CA State Univ 1968 to present. Proj Dir of res on electronic mtls, thermodynamics of metallic systems. Independent consultant 1959 to present on corrosion, failure analysis, propeties of polymers and design of polymeric parts. *Society Aff:* ASM, AIEE.

Miller, Edward
Business: 323 King Blvd, Newark, NJ 07102
Position: Acting Dean of Engineering *Employer:* NJ Inst of Tech. *Education:* BSME/ME/NJ Inst of Tech; MME/ME/Univ of DEL ME/Appl Math/Stevens Inst of Tech; MAeroE/Aero Engg/NYU; MA/Higher Ed/Columbia. *Born:* 3/10/22. in Newark, NJ. Served in US Navy during WWII. Ind experience included work at Crucible Steel Corp, Westinghouse, and Curtiss-Wright Corp. Employed by the NJ Inst of Tech since 1948. Presently holding that rank of full prof and the position of Acting Dean of Engineering. Previous positions held include that of Dir of the Experimental Stress Lab, Asst & Assoc Chrmn, coordinator of Mech Tech, and Assoc. Dean of Engineering. Numerous sectional and regional positions in the ASME including Chrmn of the North Jersey Sec. *Society Aff:* ASEE, ASME, NSPE.

Miller, George E
Home: 307 Takisa Circle, Dayton, OH 45415
Position: Chf System Engr. *Employer:* Aero Sys Div, Wright-Patterson, AFB. *Born:* April 1924. BS from Purdue U, graduate work Ohio State U. Native of Louisville, K. Served with Army Air Corps as pilot 1942-45 & with Air Force 1951-53. Propulsion engrg for Air Force specializing in engine installations & new proposal evaluations. Assigned current position as Ch Engr on Maverick Missile Sys 1967. Respon for tech leadership & for organizing & controlling the overall engrg effort required throughout the dev of a complex aeronautical weapon sys having operational capabilities & characteristics not currently available in the US Air Force inventory.

Miller, Gerald, R
Business: 163 West 1700 South, Salt Lake City, UT 84115
Position: Prof, VP *Employer:* Univ of UT, Cerametel, Inc *Education:* PhD/Mat Sc, Appl Phys/Cornell Univ; MS/Mat Sc, Appl Phys/Cornell Univ; B/Met E/Cornell Univ *Born:* 12/06/39 Has been a faculty mbr of the Univ of Utah since 1965 with res interests in solid state physics and physical ceramics. In 1971-1972 he was Visiting Science Res Council Prof at the Univ of Edinburgh. Since 1978 he has also served as vp of Ceramatel, Inc a high technology ceramics firm located in Salt Lake City. For Cermatel he directs all technical operations. *Society Aff:* APS, ACerS

Miller, Henry A
Home: 318 Denson Dr, Auburn, AL 36830
Position: State Conservation Engr. *Employer:* USDA - Soil Conservation Service. *Education:* BS/Agri Engg/Auburn Univ. *Born:* 3/10/26. Native of Madison County, AL. Two yrs US Navy, Amphibious Forces Pacific Theater of Operations. Auburn Univ June 1947 to Dec 1950. Entered on duty with USDA-SCS Jan 1951, Talladega, AL. Served as Dist and Area Engr at several locatins in AL - Huntsville, Tuskegee, Dadeville, Decatur, AL. Transfered - reassigned Dec 1963 to State Office Staff as SCS State Construction Engr. Promoted Feb 1966 to Asst State Conservation Engr - responsibilities on conservation operations and construction engr. Nov 1970 promoted State Conservation Engr. Chrmn AL Sec of ASAE 1977-78. *Society Aff:* ASAE, SCSA, OPEDA, NACD.

Miller, Herbert F
Home: 7346 E. Edgewood Ave, Mesa, AZ 85208
Position: Consultant. *Education:* MS/Agri Engg/TX A&M; BS/Agri Engg/TX A&M.

Miller, Herbert F (Continued)
Born: Aug 1919 in Texas. MS & BS Texas A&M, US Army, Ordnance officer, 7th Armored Div, European Theater & Dev & Proof Services, APG, MD, 1942-47. With Texas Agricultural Experiment Station, Cotton Mechanization Res 1948-52. Engr, US Cotton Field Station, Shafter, Ca 1952-54. Res Administration, Harvesting & Crop Processing, USDA, Beltsville, Md 1954-60. Product Planning Engrg, Deere & Co, Moline, Ill 1960-1980. Dir of Finance, ASAE, 1975-77. Enjoy bridge & golf. *Society Aff:* ASAE.

Miller, Jack C
Business: 422 Second Ave S.E, Cedar Rapids, IA 52401
Position: Principal *Employer:* Self Employed - Jack C Miller & Assocs *Education:* BS/Arch Engr/IA State Univ *Born:* 12/01/36 Born in Sheboygan Wisconsin, became native of Des Moines, IA. Received Commission upon graduation. Served two yrs in the Army Corps of Engrs. Active duty was in VA, TX, MS, Greenland and Thailand. Joined Brown Healey Bock, Architects & Engrs in 1963 in Cedar Rapids, IA. Became a partner in Brown Healey Bock in 1972. Established Jack C. Miller & Associates Consulting Struct Engrs in 1978. This office provides struct engrg service to architects and design-build contractors. Major projects to date include the Waterloo Public Library, Ecumenical Housing Proj, IA City and the Penick & Ford Co (Div of Univar) renovation in Cedar Rapids, IA, Fourth & Grand Ave Parking Ramp, Des Moines, IA, Central Fire Station, Cedar Rapids, IA. *Society Aff:* ACEC, ACI, CEC-IA

Miller, James E
Business: 1500 N Beauregard St, Alexandria, VA 22311
Position: Project Mgr *Employer:* Howard Needles, Tammen & Bergendoff *Education:* BS/CE/Univ of Missouri, Rolla. *Born:* 06/15/16 Served with Counter Intelligence Corp, Army of the US, 1945-1946. Joined Howard Needles Tammen & Bergendoff, New York, NY 1950. Present position 1964. Specialty is reinforced concrete construction - Bridges, Tunnels, Highways and Airport Runways. Fellow of ASCE and mbr of NSPE, ACI & ASTM. Fisherman, both fresh and salt water. *Society Aff:* ASCE, ASTM, ACI.

Miller, James P, Jr
Home: 6617 Church Ave, Ben Avon, PA 15202
Position: Prof & Grad Coord CE *Employer:* Univ of Pittsburgh *Education:* PhD/CE & Econ/Univ of Pittsburgh; MS/CE/Univ of Pittsburgh; BS/ChE/Univ of Pittsburgh *Born:* 7/12/21 Engr for Koppers Co 1943-44. Lt US Navy 1945-1946. Research Assoc Mellon Inst of Industrial Research 1946-1947. Prof and Grad Corrd CE, Univ of Pittsburgh 1947 to date. Engr for Kilbuck Twp and Emsworth Boro PA Pres of Ben Avon Boro Council. Pres Water Pollution Control Assoc of PA and mbr Bd of Dir. Chairman of Commonwealth of PA Bd of Certification of Water and Sewage Treatment Plant Operators. Received WPCA Ted Moses High Hat Award. Received Chrostwaite Award, Diplomate Am Academy of Envir Engrg. Mbr of Sigma XI, Chi Epsilon, Sigma Tau and Order of Omega. Biography list in Who's Who-in the East American Men of Science, Who's Who in Tech and Community Leaders of America. *Society Aff:* NSPE, WPCF, AWWA, AAEE

Miller, Jan D
Business: 216 W.C. Browning Bldg, Salt Lake City, UT 84112
Position: Prof of Metallurgy *Employer:* Univ of Utah *Education:* PhD/Met Eng/CO School of Mines; MS/Met Eng/CO School of Mines; BS/Min Proc/PA State Univ *Born:* 04/07/42 Prof J.D. Miller is well known for his teaching and numerous technical contributions in the areas of Min Proc and hydrometallurgy. Recently his attention has been devoted to the dev of solvent extraction techniques for gold recovery from alkaline cyanide solution as well as the dev of an air sparged hydrocyclone for fast flotation separations in a centrifugal field. A recognized leader in his profession, Dr. Miller is a recipient of the Marcus A. Grossman Award presented by the Metallurgical Society for significant hydrometallurgy res and is the eleventh recipient of the Van Diest Gold Medal given periodically by the CO School of Mines to alumni who have distinguished themselves in the mineral industry. Dr Miller is active in the AIME and is a P Chrmn of the Mineral Processing Div and a past mbr of the Bd of Dirs of SME/AIME. He was selected as a Henry Krumb lecturer for the AIME in 1987. *Society Aff:* SME, TMS, ACS, FPS

Miller, John A
Home: 7 E Lancaster Ave, Ardmore, PA 19003
Position: Elevator Consultant. *Employer:* Self Employed. *Education:* Diploma/Mech Engg/Wyomissing Poly Inst. *Born:* 6/13/15. Since 1935 have been continuously employed in all phases of the Elevator and Escalator Industry. Reg PE PA no 13723 MD no 3898, DC no 5007 M. Twenty yrs in Private Practice providing consulting engg services to the elevator industry, Owners, Govt Agencies, Architects and other Engrs. Also provides accident investigation and ct testimony services. Adjunct instr 12 yrs Univ of PA. Grad School of Architecture. *Society Aff:* IEEE, NAEC, NAESA, NSPE.

Miller, John C
Business: 6601 W Broad St, PO Box 27003, Richmond, VA 23261
Position: Mgr New Product Dev *Employer:* Reynolds Metals Co. *Education:* MBA/Bus Admin/VA Commonwealth Univ; BS/Met Engr/VA Polytechnic Inst & State Univ *Born:* 1/9/48. in Spartanburg, SC. Grew up in Richmond, VA. With Reynolds Metals Co since 1967. Served as Co-Op student 1967-71. Res Engr in Ingot Casting Res 1971-76, specializing in filtering & gas testing. Dev Proj Dir in Prod Dev Div in Casting & Machining Tech 1976-81 involving machining studies of hypo- and hypereutectic aluminum silicon alloys & various casting dev progs. Prod Mgr in Reclamation Div in Sales and Mktg 1981-83 specializing in tech service and prod dev for the foundry and steel indus. Current responsibility in Recycling and Reclamation Div in Tech Group involving dev of new products from recycled aluminum and dev of new business opportunities. *Society Aff:* ASM, SDCE, TMS AIME, AFS

Miller, John G
Home: 1202 E Wyomissing Blvd, Reading, PA 19611
Position: Retired *Education:* MSE/Power/Purdue Univ; BSME/Mech Engg/Purdue Univ. *Born:* Aug 1910. Native of Knox County, In. With Detroit Edison 1936-42; Served US Navy, Bureau of Ships, Design Branch 1942-46; Gilbert Assocs 1946-47; Metropolitan Edison Co 1947-71, holding positions of Supt of Prods, Asst Chief & Ch Engr, VP & Ch Engr; VP Generation Operation, GPU Service Corp 1971-75. Since retirement in 1975 served as Cons to GPUSC & other clients until 1984. Mbr EEI Prime Movers Ctte 1947-72 (Chmn 1969-71) & EEI Ctte on Advance Dev & Res Projs. Mbr Power Generation Ctte of AEIC 1971-75. Reg PE Pa, Mi, In. Fellow ASME. Mbr PSPE, NSPE. *Society Aff:* ASME, PSPE, NSPE.

Miller, John H, II
Home: 1220 Willivee Dr, Decatur, GA 30033-4119
Position: Regional Civil Engineer GS-13 *Employer:* U.S. Dept Housing & Urban Dev *Education:* MSCE/Civil Engrg/GA Inst of Tech; BCE/Civil Engrg/GA Inst of Tech *Born:* 01/20/53 Native of Tallahassee, FL. Officer in Civil Engr Corps, USNR, assigned to COMRNCF. Asst Proj Dir in res while in grad sch, pub results. Prof Engr (GA 12172 & NC 10259). Cooperative educ student with FL DOT in Estimates Div and Geodetic Survey. Design Engr for Ardaman and Assocs consltg engrs, specializing in design and constr control for earth dams. Civil Engr for FERC, supervising licensing of numerous major hydroelec devs in NC & SC and billion dollar project under constr in SC. Currently Regional Civil Engr for HUD, responsible for site engrg for all HUD sponsored housing in southeastern US. Former Secy, GA Sect, ASCE; wrote and pub Operating Manual. Chrmn, Ctte on Missions, Univ Hghts United Methodist Church. Enjoy guitar and running. *Society Aff:* NRA, USNI, SBL, ASCE, NSPE

Miller, Joseph S
Home: 2013 Wyandotte, Olathe, KS 66062
Position: Engr *Employer:* Black & Veatch *Education:* MSNE/Nuclear Engrg/KS State Univ; BSME/ME/Univ of AK; BSIE/Indust Engrg/Univ of AK *Born:* 05/26/48 Native of Sacramento, CA. Engr of NUS Corp for three years specializing reactor phys and thermalhydraulics. Served as Sr Engr for EG&G ID at ID Natl Engrg Lab, 1976-77. Specialize computer code verification and dev for nuclear power reactors. Present position is Nuclear Engr at Black & Veatch. Respon computer simulation for both nuclear and fossil power plants. Respon for fire protection of power plants. V Chrmn of MO/KS Section of ANS, 1983, 84 Chrmn of MO/KS Section of ANS. Bd of Dirs Blue Valley Optimlst, 1984. Webelos boy scouts den leader 83-85, Youth Basketball Coach - Sportsmanship award 1984-85. VP Blue Valley Optimist, 1985. Listed in 84-85 edition of "Who's Who in Frontier Sci and Tech. *Society Aff:* ANS, ASME

Miller, Lawrence C
Home: 4023 Shadow Circle NE, Olympia, WA 98506
Position: Regional Engr - Paving & Trans. *Employer:* Portland Cement Assoc. *Education:* BS/CE/WA State Univ. *Born:* Nov 22, 1926, Bellingham, Wash. Served in US Navy 1944-46. Jr Design Engr for Weyerhaeuser Co, Tacoma, Wash 19151-55. Structural Design Engr for Horace J Whitacre & Assocs, Tacoma, Wash 1956. Joined Portland Cement Assn 1957 as Field Engr in Spokane, Wash for eastern Wash & northern Idaho. Present position: Regional Engr Paving & Transportation covering five Pacific Northwest states since 1971. Mbr Tau Beta Pi, ACI, ASCE. Pres Spokane Sect ASCE in 1965. Pres Tacoma Sect ASCE 1975. Chairman of Pacific Northwest Council ASCE 1981-82. VP for District 3 Washington State Good Roads Assn. Recreational activity is boating. *Society Aff:* ASCE, ACI, APWA, ΤΒΠ.

Miller, Lawrence E
Business: 1025 Connecticut Ave, N.W., Suite 712, Washington, DC 20036
Position: Sr VP *Employer:* Leo A. Daly Co *Education:* BS/CE/IA St Univ *Born:* 01/07/18 Boyhood spent in IA. Except for World War II yrs, entire career has been with architect-engr firms with past 30 yrs in employ of Leo A. Daly Co. Has served as Dir of the Washington, DC office and presently is Sr VP in the Internatl Div of the co. Pres, Eastern Chapter, NB Society of Prof Engrs, 1954; Pres, NB Society of Prof Engrs, 1958; and Pres, NB Section, ASCE, 1960; Chrmn, Mgmt Group C of ASCE, 1981-82; ASCE Harland and Bartholomew Award in 1980. *Society Aff:* ASCE, NSTE, SAME, APA

Miller, Nathan E
Business: 841 N Broadway, Milwaukee, WI 53202
Position: Water Superintendent. *Employer:* City of Milwaukee. *Education:* EE/-/Univ of WI. *Born:* Feb 16, 1917. Reg PE in Wisc. Electromechanical design of solenoids, switchgear, motor starters, water system telemetering & automation. Work exper at Allis Chalmers Mfg Co, Stearns Elec Co, & Madison Water Utility as an engr between 1943-61. Operations Mgr to Asst Superintendent to Superintendent of Milwaukee Water Works. Started with Milwaukee Water Works in 1961 & became Superintendent in 1972. Respon for the operation of water utility serving over 900,000 consumers in Milwaukee & 10 suburbs. Active in Amer Water Works Assn & local service clubs. Chmn Wisc Sect AWWA 1973. *Society Aff:* ASCE.

Miller, Oscar O
Home: 1103 E Broad St, Westfield, NJ 07090
Position: Consultant *Education:* PhD/Met Engg/Univ of Pittsburgh; MS/Chemistry/NYU; BS/CE/Grove City College. *Born:* 3/15/06. Native of New Centerville, PA. Parents: Edward Cyrus Miller & Anna (Miller) Miller. Married Margaret Patton 1928; one son, Kristofer. Worked on tinplate and tin cans for US Steel 1928-36 & Mellon Inst (for Continental Can) 1936-39. Employed by US Steel's fundamental lab 1939-50 on physical met & military armor welding (Latter: 1941-45). Supervised steel res for Intl Nickel 1950-61, wrote and/or edited co engg publications on alloys 1961-72. Editor: Alloy Digest 1970- present. ASM: Fellow 1977; NJ Chapter's Chrmn 1962-63 & Secy-Treas 1970-present. Metal Sci Club of NY: Pres 1966-67. Hobby: language use. *Society Aff:* AIME, ASTM, ASM.

Miller, Owen W
Business: 111 Elec Engrg Bldg, Columbia, MO 65211
Position: Prof of Industrial Engg. *Employer:* Univ of Missouri-Columbia. *Education:* DSc/Industrial Engg/WA Univ; MSc/Industrial Engg/WA Univ; BSc/Industrial Engg/WA Univ. *Born:* 2/17/22. in St Louis, Faculties of WA Univ 1954-64 & Univ of MO-Columbia 1964- . Asst Chief Indust Engr, Amer Steel Foundries, E St Louis, IL 1950-54, Plant Supt, Steven Mfg Co, St Louis 1961. Editor, Res Abstracts, IIE Transactions, 1972-80. Pres Mid-MO Chapter ASQC 1978-79, Pres St Louis Chap IIE 1958-59. PE MO, 1958. Tau Beta Pi, Alpha Pi Mu, Sigma Xi. NSF Sci Faculty Fellowship, 1958. IMS Film Awards 1965-71. Co-prin dir, sevl MO Regional Medical Prog Projs (HEW) 1965-72. Assoc Scientist, Ellis fischel State Cancer Res Center, 1975-76. Publications: Productivity & Quality Enhancement, computers in engg & medicine, Industrial consultant to many cos, consultant to EPA, consultant to ASF- Luke/Williams AFB. Dir Grad Studies, KS City Program 1978- . Proj Co-Dir. Integrated Mfg Tech Proj 1983-, Proj Co-Dir Productivity & Quality Enhancement Prog 1983-4. Coordinator, Off-campus Prog, Coll of Engrg, 1982-. Int. Dir. Engr. Ext. 1986-1987, Fellow, ASQC 1981. Dir, Bus & Industry Productivity Center 1978. Areas of Specialization: Indus Engrg, Productivity, Quality Control, Reliability. *Society Aff:* ASQC, IIE, ΣΞ.

Miller, Paul L, Jr
Business: Mech Engg Dept, Manhattan, KS 66506
Position: Prof. *Employer:* KS State Univ. *Education:* PhD/ME/OK State Univ; MS/ME/KS State Univ; BS/ME/KS State Univ. *Born:* 6/27/34. Taught at KS State Univ as Instr through Prof, 1957-1975. Hd of the Dept, 1975-87. Served with US Army, 1957-58. Reg pE, KS. Served as Consultant to several cos & US govt. Contributor of articles in the field to profl journals. Dir technical res projs. Served as mbr of committees for ASHRAE & mbr of Bd of Dirs of ASEE, 1979-81. Specialist in Heat Transfer & Rm Air Diffusion. *Society Aff:* ASME, ASEE, ASHRAE, ΠΤΣ, ΤΒΠ.

Miller, Ralph W
Business: Cross Keys - Quad 130, Baltimore, MD 21210
Position: President. *Employer:* Miller, Schuerholz & Assocs, Inc. *Education:* BSME/Mech Engrg/GA Tech *Born:* Aug 1922. BSME from Georgia Tech in 1943. In WWII, served as Aircraft Maintenance Officer in China & India. 1946-56, Proj Engr with Whitman, Requardt & Assocs, Cons Engrs, Baltimore. In 1956, established Miller, Schuerholz & Assocs, cons engrg firm specializing in mech & electrical design for building projs. VP ASME 1968-70. Pres, Cons Engrs Council of Md 1975-76. *Society Aff:* ASME, ASHRAE, NSPE, ACEC

Miller, Raymond E
Business: School of Information and Computer Sci, Atlanta, GA 30332
Position: Director and Professor *Employer:* Georgia Institute of Technology *Education:* PhD/EE/Univ of IL; MS/Mathematics/Univ of IL; BS/EE/Univ of IL; BS/ME/Univ of WI, Madison. *Born:* Oct 1928. Served with US Air Force 1951-53. Director, School of Information and Computer Science, Georgia Institute of Technology. Res Staff Mbr, Mathematical Sciences Dept, IBM Watson Res Center & Asst-Dir of Dept. 1957- 81. Taught on visiting appointments at U of Ill, Cal Tech, U of Conn, NYU, U C Berkeley, & Yale. Fellow IEEE; Chmn IEEE Tech Ctte on Switching & Automata Theory 1969-72; Mbr ACM; Editor-in-Chief, JACM 1972-76; Mbr-at-Large, ACM Council 1976-82. Chmn ACM Mgmt Bd 1978-82. Author of over 50 tech papers, 4 patents, & 2-volume book 'Switching Theory' (Wiley). Prof interests in engrg, computer sci, & mathematics. Hobbies include tennis, fishing & gardening. *Society Aff:* IEEE, ACM, AAAS.

Miller, Rene Harcourt
Business: Rm 33-411, Cambridge, MA 02139
Position: Prof of Flight Transp *Employer:* MIT, Dept of Aero & Astro. *Education:*

Miller, Rene Harcourt (Continued)
MA/Engg/Cambridge Univ; BA/Engg/Cambridge Univ. *Born:* May 1916. 1937-40 G L Martin Co; 1940-45 Chief of Aero & Dev McDonnell Aircraft Corp; 1952-54 VP Engrg Kaman Aircraft Corp. Recipient of Klemin Award, AHS; Decoration for Meritorious Civilian Service, US Army; Sylvanus Albert Reed Award, AIAA. Mbr NAE. Honorary Fellow, Amer Helicopter Soc; Honorary Fellow, Amer Inst of Aeronautics & Astronautics (Pres 1977-78); Fellow, Royal Aeronautical Soc. Faculty MIT since 1944; presently H N Slater Prof of Flight Transportation Dept of Aeronautics & Astronautics. *Society Aff:* AIAA.

Miller, Richard K
Home: 498 S Main St, Madison, GA 30650
Position: Consultant *Employer:* Private Practice *Education:* BSME/Mech Engr/ Purdue Univ *Born:* 10/16/46 Consulting engr and educator in the fields of industrial noise control and robotics. Author of 18 books and over 200 articles and technical papers. Instructor for seminars sponsored by NECA, AEE, and others. Founder of Richard K. Miller & Assoc, Inc (1972) which grew to be the largest consulting firm in the field of industrial noise control in the Southeast US. Sold interest in the firm in 1980. Now consults in private practice in the areas of robotics, advanced mfg tech and strategic planning. *Society Aff:* RI/SME, AEE, AAAI, IIE

Miller, Richards T
Home: 957 Melvin Rd, Annapolis, MD 21403
Position: Naval Architect & Engr *Employer:* Self Employed *Education:* Naval Eng/ Naval Construction & Engrg./MIT; BS/Naval Architecture & Marine Eng/ Webb Inst of Naval Architecture. *Born:* 1/31/18. US Navy 1940-68, retired Capt. Tours included Hd, Preliminary Design Branch, BUSHIPS; CO & Dir Mine Defense Lab; Dir, Ship Systems Engrg & Design Dept, NAVSEC. Designed yachts, tugs, oceanographic res ships, mine sweepers, torpedo boats, destroyers, OTEC Platforms. Oceanic Div., Westinghouse 1968-78; Fellow Engr., Mgr. Ocean Engrg., Advisory Engr. Currently consulting naval architect & engr. Fellow, Hon VP (Life), Mbr Council SNAME. Mbr Sigma Xi. Mbr Tech Comm. Amer Bur. of Shipping. Capt Joseph H Linnard prize SNAME, 1964. Navy Legion of Merit, 1968. W. Selkirk Owen Award of the Webb Alumni Assoc., Webb Inst of Naval Architecture, 1983. Reg PE MD. Co-author "Sailing Yacht Design–. Sail, Paint. *Society Aff:* SNAME, ASNE.

Miller, Robert E
Business: 475 Archuleta Rd, Las Cruces, NM 88001
Position: Pres. *Employer:* Southwest Engr, Inc *Education:* BS/EE/Rose Poly Inst. *Born:* 1/12/20. Army Signal Corps 1943-46. Chief Construction Engr, NM State Hgwy Dept. Mgr, Associated Contractors of NM, 1955-67. Pres, Southwest Engg, Inc, 1971 to present. Licensed Professional Engr, NM & AZ. *Society Aff:* ASCE, ASTM, NSPE, APWA, ACIL.

Miller, Robert E
Business: 216 Talbot Lab, 104 S. Wright St, Urbana, IL 61801
Position: Prof of T & AM. *Employer:* Univ of IL. *Education:* PhD/T & AM/Univ of IL; MS/T & AM/Univ of IL; BS/Aero Engg/Univ of IL. *Born:* 10/4/32. Taught in Dept of Theoretical and Appl Mechanics at the Univ of IL at Urbana-Champaign from 1955 to present. Maj res interests are solid mechanics, finite elements, dynamics and numerical methods. Consultant to several industrial firms and govt installations. Hobbies are running, model aircraft, golf and fishing. *Society Aff:* AIAA, AAM, ASEE, ASCE.

Miller, Robert R
Home: 9240 Shirley Dr, La Mesa, CA 92041
Position: Manager, Plant Engrg & Energy Mgmt *Employer:* Rohr Industries. *Education:* Technical/Air Cond & Refrig/CA State Poly Univ. *Born:* 8/13/28. US Army Military Intelligence 1951-1953. Proj Engr & VP T H Parry & Assocs Consulting ME's Pasadena, CA 1956-1967. With Rohr Industries since 1967 and currently manager of plant engrg & energy management. Responsibilities include corp energy mgt policy and the design of all bldgs and mfg facilites. Plant engg projs coupled with an active energy mgt prog have amounted to over $7,000,000 in cost avoidance since 1974. Reg PE State of CA, Pres San Diego Chap ASHRAE 1973-74, recipient of the Energy Conservation Engr of the Yr, Pres San Diego Chap AIPE 1984. AIPE Plant Engr of the Yr 1979. Who's Who in Technology Today 1981. San Diego Engr of the yr 1983. *Society Aff:* AIPE, ASHRAE, ASPE, AEE, APEM.

Miller, Russell L
Home: 17 Taylor Woods, Kirkwood, MO 63122
Position: Consultant. *Employer:* Safety & Loss Pres Services. *Education:* BS/ChE/IA State Univ. *Born:* 5/16/13. Russell L Miller PE CSP, born and raised on a midwest farm, grad from IA State Univ in 1936 with a BS in ChE. He joined Monsanto and worked successively as process Dev & Design Engr, Production Supervisor, Plant Mgr, Corp Dir of Safety and Property Protection. He retired in 1978 to start consulting. He worked actively in AIChE and in the European Fed of Chem Eng to establish Profl Forums for exchange of Technical info important to qualified chem Engrs working in the field of Process Safety and Loss Prevention. He was elected a "Fellow" in the AIChE in 1977. *Society Aff:* AIChE.

Miller, Stewart E
Home: 67 Wigwam Rd, Locust, NJ 07760
Position: Consultant *Employer:* Self-employed *Education:* SM/Elect Eng/MA Inst Tech; SB/Elect Eng/MA Inst Tech. *Born:* 9/1/18. Joined staff at Bell Telephone Labs in 1941 & was concerned with microwave radar design from 1941-45. At the conclusion of WWII, resumed design work on coaxial-cable carrier systems until 1949, when he joined the Radio Res Dept at Holmdel. His work was concerned with circular electric waveguide communication, microwave ferrite devices, & other components for microwave radio systems. As Dir, Guided Wave Res, he headed a group which did work leading to a current millimeter-wave sys dev. Beginning in 1962, his interest & that of the Guided Wave Res grp shifted to the optical region & to exploration of the use of lasers & other lightwave devices for voice, picture, & data, transmission. Retired from Bell Labs in 1983. Became a self-employed consultant in 1984. Mbr Tau Beta Pi, Eta Kappa Nu; Assoc Mbr Sigma Xi; Fellow IEEE; Fellow OSA; Mbr NAE; Fellow AAAS. Rec. 1972 IEEE Morris N Liebmann A; Co-recip 1975 IEEE WRG Baker Award; recip 1977 Franklin Inst Stuart Ballantine Medal. *Society Aff:* NAE, IEEE, OSA, AAAS.

Miller, Wally, PE
Home: 323 W Michigan, LaGrange, IN 46761
Position: Professional Engineer. *Employer:* Self. *Education:* BS/Civil Engr/Purdue; MA/Business/Ball State *Born:* 1/11/37. BSCE Purdue Univ; MA Ball State U. 1960-63 Proj Engr, Foster-Forbes Glass Co, Marion, In. 1964-66 Indus Engr Supr, Fruehauf Corp, Fort Wayne, In. 1967-68 Master Mechanic, Fruehauf Corp, Middletown, Pa. 1969 Production Mgr, Mobile Aerial Tower Co, Fort Wayne, In. 1969- , VP, Dir & Chf Operating Officer Miller's Merry Manor, Inc, Warsaw, In. 1973- , Pres Designmasters Inc, LaGrange, In. Mbr NSPE, ASCE, ASHRAE, ASPE, IIE, NCEE. Fellow Amer Coll of Nursing Home Administrators. Mbr (as prin), ACEC. *Society Aff:* NSPE, ASCE, ASHRAE, ASPE, NCEE, ACEC

Miller, Walter J
Business: 1400 Tremont, Hillsboro, IL 62049
Position: Vice President & Corp. Secy *Employer:* Hurst-Rosche Engineers, Inc. *Education:* BSCE/Tri-State Univ. *Born:* Sept 7, 1928. Native of Racine, Wisc. Served with US Army during Korean War. Vice Pres. & Corp Secy for HurstRosche Engrs, Inc. At home office in Hillsboro, Ill. Reg PE in 3 states. Licensed land surveyor 6 states. Mbr Amer Right of Way Soc, NSPE, Ill Rec Land Surveyors Assn. In charge of all field survey, engrg const & soils work. Licensed Commerical & instrument Pilot. Enjoy horseback riding & hunting. *Society Aff:* NSPE, ARW, IRLSA.

Miller, William E
Home: 1051 Camino Velasquez, Green Valley, AZ 85614
Position: Consultant Process Control *Education:* BSEE/EE/Univ of CA. *Born:* 12/18/17. Native Californian. Joined GE in 1939. Grad GE's 3-yr Advanced Engrg Prog and Gen Mgmt Course. Received GE's Corp Charles A. Coffin Award in 1951 for "Extraordinary ingenuity & resourcefulness in devising new electric drive system for tandem cold strip steel mills–. PE NY since 1959. Elected Fellow IEEE 1962. Managed GE's Metal Systems Engrg & Drive Systems Engrg contributing to intl leadership in comp automation of metal rolling mills. Retired GE Dec 82. Presently, Secretary Am Automatic Control Council (IEEE, ASME, AIAA, AIChE, AISE, and ISA affiliated). Chrmn/VChrmn of several Intl Fed Automatic Control (IFAC) Technical Symposia/Congresses, presently VP and Chrmn of Exec Bd of IFAC. Elected Fellow, Instrument Soc of Amer (ISA) in 1984. Leader Citizen Ambassador Automatic Control Engrs to China in 1987. *Society Aff:* IEEE, NSPE, ISA, AISE.

Miller, William F
Business: 333 Ravenswood Ave, Menlo Park, CA 94025
Position: Pres *Employer:* SRI International *Education:* PhD/Physics/Purdue Univ; DSc/Honorary/Purdue Univ; MS/Physics/Purdue Univ; BS/Physics-Math/Purdue Univ *Born:* 11/19/25 William F. Miller was educated as a nuclear physicist and practiced nuclear physics and nuclear engrg at the Argonne Natl Laboratory. He then turned to computer science and computer engrg. He was a pioneer in laboratory automation and was an early worker in picture processing by computer. At Stanford Univ he engineered the data analysis and systems for the Stanford Linear Accelerator Center as well as developed computer science and engrg programs in the Computer Science Dept. He became VP and Provost and now serves as Pres of SRI Intl. He also serves on several Bds of multinational corporations. *Society Aff:* AAAS, IEEE, AAAS, AMA, SIAM, NAE.

Miller, William J, Jr
Home: 9 Manor Dr, Moore's Lake, Dover, DE 19901
Position: Exec Director *Employer:* Delaware River & Bay Authority. *Education:* BS/CE/Drexel Univ *Born:* Feb 1917. Delaware native. 1936-61 Delaware State Highway Dept, Dir of Operations & staff assignments. 1961- , Exec Dir, Delaware River & Bay Authority, a bi-state agency of Delaware & NJ. Respon for const/ operation Delaware Memorial Bridges & Cape May-Lewes Ferry. 1941-46 US Army Engrs, USA & Alaska. Mbr Bd/Dir Internatl Road Federation. P Pres Internatl Bridge, Tunnel & Turnpike Assn. 1975-76 Pres Delaware Soc PE's. 1976- , Pres Delaware Motor Club (AAA). Mbr Bd/Dir Delaware Racing Assn & Delaware Safety Council. Delaware Engr of the Year 1968. *Society Aff:* ASCE, NSPE, ITE

Miller, William R
Business: 2424 W. 23rd St, Erie, PA 16512
Position: V.P. - R&D *Employer:* American Sterilizer Co. *Education:* M.S./EE/M.I.T.; B.S./EE/M.I.T. *Born:* 12/16/28 BS and MS MIT 1952; USAF 1952-54. Started with G.E. 1949 as co-op student. Mgr, Advance Eng'g. 1965; Mgr, Speed Variator Product Eng'g. 1966; Mgr, Propulsion Eng'g. 1971. Joined AMSCO 1972 as Genl Mgr, Sys Div and Pres, Guilbert Subsidiary. Named VPres, Res & Dev 1975. Hold fifteen patents, eight published articles. Mbr: (Former Chrmn) Gannon U. Eng'g. Advisory Bd; St. Vincent Health Cntr Incorporators; Advisory Bd - Medical Sterilization, Inc; Mgmnt Cttee - Precision Scientific, Inc. Ben Franklin Advisory Bd of NW Pa. Elect. PE (Pa.), Honorary Societies Tau Beta Pi, Eta Kappa Nu, Sigma Xi. Who's Who in the East; Who's Who in Technology. Wife: Marlene Daughters: Claudette and Kathleen. Enjoy sailing, tennis, bridge. *Society Aff:* AAMI, HIMA

Millet, Richard A
Business: 3 Embarcadero Plaza, San Francisco, CA 94111
Position: Vice President/Managing Principal *Employer:* Woodward-Clyde Consultats. *Education:* MS/Civil Engg/Rensselaer Poly Inst; BS/Civil Engg/Renselaer Poly Inst. *Born:* July 18, 1940; native of Phila Pa. With Army Corps of Engrs 1964-66. Woodward-Clyde Cons from 1966- ; elected Assoc 1972 & became V Pres 1975; main respons lie in mgmt & tech supr of geotech studies for nuclear power plant projs & eaarth & rockfill dam projs. Dir N J Section ASCE 1975-77; Mbr ASCE & ASTM. Reg P E 7 states. *Society Aff:* ASCE, ASTM.

Millheim, Keith K
Home: 1816 South Carson Apt 126, Tulsa, OK 74119
Position: Research Supvr *Employer:* Amoco Production Co *Education:* MScE/Petro Engrg/Univ of OK; BScE/Petro Engrg/Marietta Coll *Born:* 02/15/41 After completing his Univ training worked for Contentinental Oil Co. in Billings, Montana as a production engr. Left Conoco and worked for 5 1/2 years in Australia as an Independent Conslt. Returning to the US in 1974 worked as head Project Engr for CER Geonuclear Corp on a joint industry massive frac project. Joining Amoco in 1975, Keith has headed the directional drilling effort for the Co. Keith has been leading an effort to develop a critical drilling facility, which includes the first engrg simulator for drilling and realtime drilling of critical wells from a central location. He is also mng a team of engrs and drilling personnel that are drilling critical wells in the US. Keith has published more than twenty technical papers on drilling, is the 1981 SPE Drilling Chrmn; Keith was the 1984 SPE Forum Chrmn on Directional Drilling and is the Co-author of a Drilling textbook being publ by the SPE. Is the 1981-82 SPE Editor for drilling, and was the 1964 Graduate SPE Paper Award Winner. *Society Aff:* SPE, IADC.

Milligan, Mancil W
Business: 414 Dougherty Hall, Univ of TN, Knoxville, TN 37916
Position: Prof of Mech & Aero Engrg *Employer:* Univ of Tennessee - Knoxville. *Education:* PhD/Engrg Sci/Univ of TN; MS/Mech Engrg/Univ of TN; BS/Mech Engrg/Univ of TN. *Born:* Nov 21, 1934; native of Shiloh, Tenn. Res Engr for Boeing Co Seattle 1956-57 & 1958-59; applied res in heat transfer & fluid mechs. With Univ of Tenn Knoxville 1959- ; dev more than 15 new course offerings. Conducted res for NASA, USAF, Union Carbide Nuclear Co & several other sponsors. Head of Mech & Aerospace Engrg Dept 1973-82. Natl V Pres Tau Beta Pi 1970-74; Natl V Pres Region IV ASME 1977-79. Professor of Mechanical & Aerospace Engineering 1967 to present. *Society Aff:* ASME, ASEE.

Milligan, Robert T
Home: 2 Vista Del Mar, Orinda, CA 94563
Position: Consultant *Employer:* Bechtel Inc *Education:* PhD/ChE/OH State Univ; BS/ChE/Univ of IL. *Born:* 12/22/19. Native of Taylorville, IL. Process Engr & Supervisor at Shell Dev Co for twenty five yrs in charge of four groups providing engg design services. As Asst Mgr Technical of Shell Chem's Polymer Div, was responsible for coordinating technical activities on polypropylene, polystyrene, epoxy resins and related products. With Bechtel Corp since 1973 as Mgr Pollution Control Tech & Mgr Chem Process Dev Currently Consultant for Bechtel Inc. Reg PE in CA. Past Chrmn N CA Section AIChE. Elected Fellow of Am Inst CE 1979. *Society Aff:* AIChE.

Milligan, William R
Home: 6425 Noble Dr, McLean, VA 22101
Position: Systems Specialist. *Employer:* General Services Administration. *Education:* MSA/Computer Systems/Geo Wash Univ; BS/Bus & Mgt Sci/Univ of MD; AA/Bus & Mgt Sci/Univ of MD. *Born:* 6/8/44. Native of Phila, PA. Have a total of 22 yrs Information Resource Management experience. Significant experience with systems relating to Retail Merchandising, Drug Inventory Mgmt, On Line Credit Authorization & Mailing Lists & Ins Applications. Developed a natl effluent (chem waste) info system for EPA. The system utilizes complex tech to handle chem (scientific) data by using commercial sortware tech. *Society Aff:* SME, IIE, ACM, WSE.

Millikan, Charles V
Home: 3142 S Gary Ave, Tulsa, OK 74105
Position: Petroleum Engr Consultant (Self-employed) *Education:* MS/Oil & Gas/ Univ of Pittsburgh; BS/Sci/OK State Univ. *Born:* 1896 Noble County Okla Territo-

Millikan, Charles V (Continued)
ry. 1922-61 Amerada Petro Corp Head of Petro Engrg; Petro Engr Cons 1961- . AIME DeGolyer Medal 1969; Hon Mbr AIME 1966; AIME-Spe Disting Lectr 1963. Certificate of Appreciation (Production) API 1954; Certificate of Appreciation (Res) API 1952. AIME Bd of Dirs 1947-52, V Pres 1951-52. Okla St Bd for Registration of Prof Engrs 1945-51. AIME Lucas Medal 1944; AIME Chmn Petro Div (SPE of AIME) 1930. Southern Hills Country Club; Presbyterian; AIME AAPG-NSPE. *Society Aff:* APE, AIME, AAPG, NSPE.

Milliken, Frank R
Home: Contentment Island Rd, Darren, CT 06820
Position: Retired. *Education:* SB/Mining Engg/MA Inst of Tech. *Born:* 1/25/14. Prior to joining Kennecott held various positions with Peru Mining, Genl Engg Co Salt Lake City; natl Lead Co (Asst Mgr Titanium Div. Joined Kennecott Copper 1952 as VP in charge of co's mining opers. In 1961 was elected Pres & Ch Exec Office. Lite Mbr of the Corp of the MIT; a mbr of the Bus Council & Disting Life Mbr ASME 1966. *Society Aff:* NAE, ΤΒΠ.

Milliken, William F, Jr
Home: 245 Brompton Rd, Williamsville, NY 14221
Position: Pres *Employer:* Milliken Research Assocs, Inc. *Education:* BS/Aero Engr & Math/MIT *Born:* 4/18/11 After grad MIT, worked 12 yrs in aircraft industry in Flt Test and Aerodynamics at Chance Vought, Vought-Sikorsky, Boeing Aircraft, Avion (Northrop). At Cornell Aero Lab/Calspan, 1946-1976. Successively Head Flt Res Dept, Vehicle Dynamics, Full-Scale Div, Transportation Res Div Initiated Res in Aircraft and Automobile Dynamic Stability and Control. Chrmn SAE Vehicle Dynamic Ctte for 5 yrs. Recipient Laura Taber Barbour Air Safety Award. Edward N. Cole Award, 1985; the IME Starley Premium Award, 1983. 1981 SAE Fellow. Professional Engr NY State. Started Milliken Res Assoc Inc in 1976. Licensed Pilot, licensed Competition Driver. Past Governor Sports Car Club. Dir Moog Inc and Trico Products Corp. *Society Aff:* SAE, SCCA

Millman, George H
Home: 504 Hillsboro Pkwy, Syracuse, NY 13214
Position: Consultant *Employer:* Self *Education:* PhD/Physics/Penn State Univ; MS/Physics/Penn State Univ; BS/Physics/Univ of MA. *Born:* June 1919 Boston Mass. Army Air Corps Electronic-Radar Officer 1942-46; military educ: Yale Univ Communications, Harvard-MIT radar. G E Co 1952- . Mbr: APS, AGU & N Y Acad of Sciences. Fellow IEEE; Mbr Comm F & G URSI & US Study Group 6 CCIR. Adjunct Prof Dept of Elec & Computer Engrg Syracuse Univ. Approximately 150 publs & tech presentations in fields of radio wave propagation, tropospheric phenomena, ionospheric radio phys, radar astronomy & radar sys analysis. Hon societies: Phi Kappa Phi, Sigma Pi Sigma, Pi Mu Epsilon & Sigma Xi. *Society Aff:* APS, AGU, IEEE, NYAS

Millman, Jacob
Home: 7 Adrienne Place, White Plains, NY 10605
Position: Charles Batchelor Prof emeritus. *Employer:* Columbia Univ (retired). *Education:* PhD/Physics/MIT; BS/Physics/MIT; -/Physics/Univ of Murich, Germany. *Born:* 1911. Taught Elec Engrg CCNY 1936-42 & 1945-52; Radiation Lab MIT 1942- 45; Columbia Univ 1952-75, Dept Chmn Elec Engrg & Computer Sci; now Charles Batchelor Prof emeritus. Educ Medal IEEE 1970; Great Teachers Award Columbia Univ 1967. Author of 'Electronics' 1941, 'Pulse & Digital Circuits' 1956, 'Vacuum Tube & Semiconductor Electronics' 1958, 'Pulse, Digital & Switching Waveforms' 1965, 'Electronic Devices & Circuits' 1967, 'Integrated Electronics' 1972, 'Electronic Fundamentals & Applications' 1976 & "Microelectronics" 1978. *Society Aff:* IEEE.

Millman, Sidney
Business: 335 E 45th St, New York, NY 10017
Position: Consultant, Rutgers Univ *Employer:* Amer Phys Soc *Education:* PhD/Physics/Columbia Univ; MA/Physics/Columbia Univ; BS/Physics/City College, NY. *Born:* Mar 15, 1908 Dawid-Gorodok Poland. nat; m; c. 1. (hon) Lehigh 1974. Asst Phys Columbia 1933-35, Tyndall Fellow 1935-36, Barnard Fellow 1936-37, Res Asst 1937-39, mbr sci staff radiation lab 1942-45; instr CCNY 1939-41; Queens Coll N Y 1941-42; res physicist Bell Labs 1945-52, Dir Phys Res 1952-65, Exec Dir Res Phys & Acad Affairs 1965-73. Secy AIP 1974-80. Fellow APS, AAAS, & IEEE. Consultant, Bell Labs 1980-84, APS 1980- Nuclear spins & magnetic moments; vacuum tubes such as magnetrons & traveling wave tubes; solid state phys. Consultant 1985- , Rutgers Univ, Center for Math, Sci and Comp Educ. *Society Aff:* APS, IEEE, AAPT, AAAS.

Mills, George S
Home: 914 Lanyard Ln, Kirkwood, MO 63122
Position: Sr Staff Engr - Crew Systems. *Employer:* McDonnell Aircraft Co (McAir). *Education:* BS/Aero Engg/Univ of AL. *Born:* 12/4/21. Naval Aviator 1942-45. Experimental Test Pilot McAir 1952-61. Grad USN Test Pilot Sch 1954. Mgr Flight Safety Engrg (McAir) 1964-67. Ch Pilot Flight Simulators (MDEC) 1967-71; Proj Sys Integration Engr (MDAC) 1971-73; Sr Staff Engr Crew Sys (McAir) F-18 Crew Station Design (Pilot Services) 1973-75; Advanced Crew Station Concepts 1975-1980;. Chmn St Louis Sect AIAA-1976-1977; Pres, McDonnell Douglas St Louis Mgmt Club (1977-78); Verification Engineer, Amraam MACS Program, 1980-1982; IFFN/Mils, 1982-1984; MSI, 1984-Present, McDonnell Aircraft Co. *Society Aff:* AIAA, SETP.

Mills, Norman T
Business: Director of Marketing, Flo-Con Systems, Inc, 1404 Newton Dr, POB 4014 Champaign, IL 61821
Position: Director of Marketing *Employer:* Flo-Con Systems, Inc. *Education:* MS/ChE/Univ of DE; BS/ChE/Purdue Univ *Born:* and raised in Hammond, IN. Aerial photographer in US Navy during Korean War, from 1950 to 1954. Research engr with Standard Oil Co (IN), Linde Div of Union Carbide, Allison Div of General Motors. Inland Steel since 1966, retired (1987) Assoc Dir of Research Flo-Con, Director of Marketing, since 1987. Pres, Iron & Steel Society of AIME 1981/82, Distinguished Member of ISS/AIME, mbr of Bd of Dir of AIME 1980-83. VP of AIME 1982/83. Pres-elect, Pres, Past-Pres of AIME 1984-87. Member Bd. of Governors & Executive Committee, AAES, 1985 & 1986. Mbr of Advisory Council to US Bur Mines Generic Mineral Tech Center for for Pyrometallurgy. VP ISS Fnd. Enjoy all sports, reading, all music. Married with three sons and three daughters. *Society Aff:* AIME, ISS, AISE, ΤΒΠ, ΣΞ, ΩΧΕ, ΦΛΥΑSMI

Mills, Robert G
Home: 150 Prospect Ave, Princeton, NJ 08540
Position: Dir, Emeritus, Interdepartmental Prog in Plasma Science & Fusion Tech *Education:* PhD/Physics/Univ of CA (Berkeley); MA/Math/Univ of MI; BSE/Elec Engg/Princeton Univ *Born:* 1/20/24. Dir, emeritus, Interdepartmental Prog in Plasma Science & Fusion Tech at Princeton Univ. Has been associated with Princeton's controlled thermonuclear res prog for three decades. USN, 1944-45. Natl Res Fellow, '52, Univ of Zurich. Author of numerous publictions on engg aspects of fusion res. Fellow of IEEE, past pres of IEEE Nuclear & Plasma Sciences Soc. *Society Aff:* IEEE, ANS, APS, AAAS.

Mills, Robert N
Business: 3135 Easton Turnpike - W2H2, Fairfield, CT 06431
Position: Retired *Employer:* GE Co. *Education:* Bachelor/EE/Univ of MN; Bachelor/Bus Admin/Univ of MN. *Born:* 5/18/24. Education through first yr of college in Normal, IL. 1943-1946 Army AF service as radar countermeasures officer. Joined GE following grad from MN in 1949. After responsibilities in mfg & engg design, moved to Corporate personnel staff in 1967. Present position in 1970 which includes responsibility for GE's college recruiting progs. Mbr of Kappa Mu Epsilon, Beta Gamma Sigma, Tau Beta Pi and ASEE (VP-Finance 1977-80, Pres-Elect 1984-

Mills, Robert N (Continued)
85. Was loaned in 1973 to the Pres of the Natl Acad of Engg to aid in establishing the natl minority engg effort. *Society Aff:* ASEE.

Mills, Robert W
Home: 621 Lakewood, Lincoln, NE 68510
Position: Consulting Engr *Employer:* Univ of Nebraska. *Education:* BSc/ME/Univ of NB; ME/Admin/Univ of NB. *Born:* Aug 1909 Nebraska. Prof Emeritus Univ of Nebr; currently Consulting Engr, Nbr Technical Assistance Center. ME & BS Univ of Nebr. Taught Univ of Nebr areas of: machine shops & production, engrg mgmt & indus safety. Various summer employment with Boeing, Chicago Export Packaging, Lipton Tea. Considerable Methods Improvement, Wk Simplification & in Supr. Primarily for hospitals, natl food processors, mobile homes, farm equipment manufacs & electronic plants. Active in prof societies. V Pres Region VII 1970-74 ASME. *Society Aff:* AAES, ASME, ASEE, LEC.

Mills, Roger L
Home: 3603 Vigilance Dr, Rancho Palos Verdes, CA 90274
Position: Senior Staff Programmer *Employer:* TRW Defense & Space Systems Group *Education:* MS/Math/Southern Methodist Univ; BS/EE/Southern Methodist Univ. *Born:* July 31, 1924. Math Instr SMU 1948-50; Northrop Corp 1951-53; Programmer BINAC, CPC & MADDIDA; N Amer Aviation 1953-61; Programmer & Programming Instr IBM-701, 704, 7090; TRW Sys 1961-present, Scientific & Sys Programmer, Mgr of Programmer Training, APM for software dev, Proj Sys Engr; Mgr Software Sys Dev, TRW EnergyProd Group; currently Software Quality Assurance, TRW DSSG. Mbr SHARE PL/I Proj 1964-70; Sat morning H S programming instr 1964-69; team teacher of computing for L A Math Teachers 1966; Chmn ACM Prof Standards & Practices Ctte 1970-73, 80- present; ACM Regional Representative 1973-76; Mbr of Prog Ctte NCC 1977; Secretary/Historian for ACM '81 committee; ACM National Lecturer 1980- present. *Society Aff:* ACM, AWC

Millsaps, Knox
Business: Dept of Engg Sciences, 231 Aero, Univ of Fl, Gainesville, FL 32611
Position: Prof *Employer:* Univ of FL. *Education:* PhD/Math & Theoretical Phys/CA Inst of Tech; BS/English Literature/Auburn Univ. *Born:* 9/10/21. Profl Appointments: OH State Univ, Auburn Univ, MA Inst of Tech, CO State Univ, Univ of FL. Governmental Positions: Chief Mathematician, Wright Air Dev Ctr; Chief Scientist, AF Missile Dev Ctr; Exec Dir, AF Office of Scientific Res; Chief Scientist, Office of Aerospace Res. *Society Aff:* AMS, SIAM, APS, AIAA, AAAS, MAA, SES.

Milroy, A Clayton
1500, 10060 Jasper Ave, Edmonton Alberta T5J 4A2 Canada
Position: Exec Dir. *Employer:* APEGGA. *Education:* BSc/Civ Engg/Univ of Alberta. *Born:* 8/26/24. A C (Clayton) Milray, PE, was born in Calgary, residing later in Redver, Saskatchewan, & Edmonton, Alberta. Served with Infantry & RCAF during WWII. Grad with BSc, Civ Engg, 1950, Univ of Alberta. Was first grad engr plan checker for the City of Calgary. Was design engr in both architectural & consulting engg offices. Commenced own bus in 1955 specializing in structural design & bldg dev. Joined the staff of the Assn of PE, Geologists & Geophysicists of Alberta (APEGGA) in 1969 as Asst Registrar & is now Exec Dir and Registrar. Served four yrs as Registrar for the Alberta Soc of Engg Technologists (ASET). Active in military reserve commanding signals & engg units. Past Pres of Military Engrs Assn of Canada (MEAC). Currently Honorary Col, 8th Field Engr Regiment. *Society Aff:* EIC, MEAC.

Milstein, Frederick
Business: Dept of Mech & Env Engr, Santa Barbara, CA 93106
Position: Prof. *Employer:* Univ of CA. *Education:* PhD/Engrg/UCLA; MS/Engrg/UCLA; BS/Engrg/UCLA *Born:* 5/14/39. Post-doctoral Res in magnetism, CNRS Lab, Grenoble, France, 1966-67; Res in Mechanics & Mtls, Rand Corp, Santa Monica, CA, 1967-69 as a physical scientist and 1969-71 as a resident consultant. Full-time Univ of CA faculty mbr since 1969 (first at UCLA, then at UCSB). Promoted to Full Prof of Mech Engg in 1978. Assoc Dean, College of Engg, 1973-75. Chairman, Dept. of Mech. & Environmental Engr., Jan 1981-Mar 1982. Fellowship awards include Guggenheim Fellowship for Theoretical Studies in Mtls Sci; Sr Fellow, to Weizmann Inst of Sci, Israel; Visiting Fellow, Clare Hall, (Univ of Cambridge, England); NATO Sr Fellow for res in theoretical strength of solids. Consulting for Civ Engg Lab, Pt Hueneme, CA. Author of approx 60 journal articles in mtls sci. Principal investigator of Natl Sci Fdn res grant on large strain elastic behavior of crystalline mtls and of EG & G grant for mechanical properties of mercuric iodide single crystals. *Society Aff:* ASME, ASM, ASEE, APS

Milstein, Laurence B
Business: Univ of Calif, San Diego, bDept Elec Engg & Comp Sci, La Jolla, CA 92093
Position: Prof *Employer:* Univ of CA, San Diego. *Education:* PhD/Elec Engg/Poly Inst of Brooklyn; MSEE/Elec Engg/Poly Inst of Brooklyn; BEE/Elec Engg/City College of NY. *Born:* 10/28/42. From 1968 to 1974 employed by Space and Communications Group of Hughes Aircraft Co. From 1974 to 1976 with Dept of Electrical and Systems Engg, Rensselaer Poly Inst. Since 1976 with Dept of Electrical Engg and Comp Sciences. Univ of CA, San Diego. Main area of res is digital communication theory with special emphasis on spread spectrum communications. Consultant to industry in radar and communications. Former Assoc Editor for Communication Theory for *IEEE Transactions on Communications*. Mbr Commission C. URSI. Mbr Eta Kappa Nu and Tau Beta Pi. Fellow of IEEE. Member, Board of Governors of Communications Society of IEEE. *Society Aff:* IEEE, URSI.

Minch, Richard H M
Home: 1144 Dopp Street, Waukesha, WI 53186
Position: Instr Fluid Power & Metallurgy. *Employer:* Waukesha County Technical Inst. *Education:* BS/Ind Ed/Univ of WI-Stout; MS/Voc Ed/Univ of WI-Stout. *Born:* July 24, 1941; native of West Bend Wis. Taught at Fort Atkinson VTAE Sch 2 1/2 yrs; employed by WCTI 1966 as Instr of Fluid Power & Metallurgy. P Bd Mbr WAVAE; P Pres WCTEA; P Educ Chmn FPS. Enjoys gardening, fishing, hunting & Nordic Skiing. *Society Aff:* FPS, ASLE.

Mindlin, Raymond D
Home: 89 Deer Hill Dr, Ridgefield, CT 06877
Position: James Kip Finch Prof Emeritus. *Employer:* Columbia University (retired). *Education:* DSc/Hon/Northwestern Univ; PhD/Pur Sci/Columbia Univ; CE/Civil engg/Columbia Univ; BS/Civil Engg/Columbia Univ; BA/Physics & Chemistry. *Born:* Sept 17, 1906 N Y City. Grad Ethical Culture Sch 1924; Northwestern Univ Hon DSc 1975. From Asst to Prof Cvl Engrg Columbia Univ 1932-67; James Kip Finch Prof of Applied Sci 1967-75, Emeritus 1975- . Mbr NAE, NAS, Amer Acad Arts & Sci, Conn Acad Sci & Engrg, ASME (Hon Mbr, Timoshenko Medal, ASME Medal), ASCE (Res Prize, von Karman Medal), Soc Experimental Stress Analysis (co-Founder, Pres, Murray & Frocht Awards), Acoustical Soc Amer (Fellow, Trent-Crede Medal), US Govt: Naval Ordnance Dev Award, Pres Medal for Merit, Sawyer Award, National Medal of Science. Columbia Univ: Varsity 'C', Illig Medal, Egleston Medal, Sch of Mines Medal, Great Teacher Award. *Society Aff:* NAS, NAE, AAAS, CASE, ASME, ASCE, ASA, APS, SESA.

Minear, Roger A
Business: Inst. for Env. Studies, Univ of Illinois at Urbana-Champaign, 408 Goodwin St, Urbana, IL 61801
Position: Professor and Director *Employer:* Univ of Illinois. *Education:* PhD/Civil Engg/Univ of WA; MSE/Sanitary Engg/Univ of WA; BS/Chem/Univ of WA. *Born:* 6/16/39. Native of Seattle, WA. Served on the faculty of OR State Univ (1966). IL Inst of Tech (1970-73), The Univ. of TN (1973-84), and the Univ. of Illinois (1985-present) as Dir, Inst. for Envir Studies. Chrmn, AWWA Student Activities Ctte (1974-78), Treas (1977-82), Chrmn (1983-84) ACS Environ Chem Div, pres of AEEP (1980), mbr, editorial advisory bd, Environ/Sci Technol (1977-82), ed adv.

Minear, Roger A (Continued)
bd. Environmental Technology Letters (1986-), ACS Books (1985-), mbr Environmental Studies Board, Nat. Res. Council (1983-86). Author of more than 70 technical papers, reports, and book chapters. Conslt in environ field to several natl corps and res organizations plus two natl labs. Enjoys travel, fishing, gardening and alpine backpacking. *Society Aff:* AAAS, ACS, ASCE, AWWA, WPCF, ASLO, AEEP, ΣΞ, IAWPRC.

Miner, Gayle F
Business: 455 CB, Provo, UT 84602
Position: Prof *Employer:* Brigham Young Univ *Education:* PhD/EE/Univ of CA-Berkeley; MS/EE/Univ of UT; BS/EE/Univ of UT *Born:* 6/13/33 Originally from Belle Fourche, SD. Served with Army Guided Missiles School, Ft. Bliss, TX, during Korean conflict. Taught guided missile ground control systems 2 yrs. Joined electrical engrg dept at Brigham Young Univ 1960, with work experience at General Dynamics/Pomona, Lawrence Livermore Laboratory, and Radio Astronomy Laboratory, Berkeley. Presently responsible for electronics option program in EE dept of BYU. Elected Sr Mbr of IEEE with 8 yrs as IEEE Student Branch Counselor. Mbr Tau Beta Pi (life) and Eta Kappa Nu. Developed voice controlled devices for handicapped. *Society Aff:* IEEE, ASEE, ΣΞ.

Miner, J Ronald
Business: Oregon State University, Corvallis, OR 97331
Position: Prof & Assoc Dir *Employer:* Office of Intl Res & Dev *Education:* PhD/ChE/KS State Univ; MSE/CE/Univ of MI; BS/ChE/Univ of KS. *Born:* 7/04/38 Born in Scottsburg Indiana. Faculty of Iowa State Univ. 1967-72. Dept of Agricultural Engrg Oregon St Univ 1972-present, hd 1976-86. Acting Assoc Dean and Dir of Academic Programs 1983-84. 1986-present Assoc Dir, Intl Res and Dev. Environmental Engr United Nations Development Program Singapore 1980-81. Ninety five Tech Publications. Frequent consultant in livestock pollution litigation. Chrmn PNW region ASAE 81-82. Frequent banquet speaker. Author of books of childrens sermons. *Society Aff:* ASAE, ASEE, WPCF.

Miner, R John
Home: 869 Fifth Ave SE, Rochester, MN 55901
Position: Dir, Elec Div. *Employer:* Rochester Public Utilities. *Education:* BS/Elec Engg/Univ of Toledo; MS/Engg Sci/Univ of Toledo. *Born:* 12/4/47. Native of Toledo, OH. Asst engr, Toledo Edison Co, 1969-72; Asst Prof of Elec Tech, Univ of Houston, 1973-76; Consultant with Bovay Engrs, Inc, Houston, 1973-76. With Rochester (MN) Public Utilities since 1976; Dir, Elec Div since 1978 responsible for engg, power production, transmission & distribution, and system operations. Officer and 1978 Young Engr of the Yr, Southeast Chap, MN Soc of PE. Listed Who's Who in the South and Southwest, 1975. Mbr IEEE/ABET Ad Hoc visitors list for accreditation of curricula in Engg Tech. Reg PE, TX, MN. *Society Aff:* NSPE, IEEE, TBΠ, HKN

Miner, Ronald E
Business: 6801 Brecksville Rd, Independence, OH 44131
Position: Senior Director *Employer:* LTV Steel Company, Research Center *Education:* PhD/Metallurgy/Case Western Reserve Univ; MS/Metallurgy/Case Western Reserve Univ; BS/Metallurghy/Case Western Reserve Univ. *Born:* 3/13/42. Joined Republic Steel Res Center June 1964; worked in various areas of Advanced Physical Met Res until 1974; respon for R/D for flat-rolled & tubular prods alloy dev at Republic. Numerous pubs & presentations. Active in ASM, AIME, AISE. Became Assistant Director of Research in January 1983 and since July 1984 to the present, Senior Director of Research. *Society Aff:* ASM, AIME, AISE, SME

Minet, Ronald G
Business: 221 E Walnut St, Pasadena, CA 91101
Position: Chrmn, CEO. *Employer:* KTI Corp. *Education:* DrEngSci/ChE/NYU; MS/ChE/Stevens Inst Tech; BChE/ChE/CCNY. *Born:* 8/13/22. USAF 1943-46, Pacific Theater, Foster Wheeler, 1946-1952, United Engrs, Chief Process Engr 1953-1960, FMC Corp, Coal Processing Facility 1960-1963, CTIP Roma, 1963-1969, Chem Projs Int 1969-1972, KTI Corp 1972 to present, Pres, Chrmn of Bd, Chief Exec Officer. *Society Aff:* AIChE, ACS, AAAS, AOCS, AIME.

Minges, Merrill L
Business: AFWAL/MLB, Wright-Patt AFB, OH 45433
Position: Chief, Non-Metallic Materials Div. *Employer:* United States Air Force. *Education:* PhD/Chemical Engr/Ohio State Univ; MS/Chemical Engr/Mass Inst Tech; BS/Chemical Engr/Mass Inst Tech; Dipl/Graduate Business School/Stanford Univ *Born:* Sept 1937; native of Denver Colo. PhD Ohio St Univ; BSc & MSc MIT. Established heat transfer/thermophysics lab while on active duty with Air Force 1960-63; in civilian capacity advanced from group leader to Branch Chief of Space & Missile Sys Support Branch, then Elastomers & Coatings Branch AFML 1964- 75, respon for high temperature materials dev & evaluation, hypersonic heat transfer & space environ res for advanced rocket, satellite & re-entry vehicle sys. Sloan Exec Fellow Stanford Univ 1971-72. Internatl Bov Bd Thermophysics Conferences. International Organizing Committee, European Thermophysics Conference Assoc Fellow AIAA. Ed Bd Internatl High T-High P Journal. Editorial Board - International Journal of Thermophysics. Editor, Compendium on Thermophysical Properties Measurement Methods Chrmn CODATA Task Group. Mbr Sigma Xi & Ohio Acad of Sciences. Tau Beta Pi. Numerous AF Scientific Achievement Awards, Outstanding Engr Award Natl Engrs Week 1976; Asst Chief and Chief System Support Div, AFML (1976-1978) and Chief Electromagnetic Mtls Div, AFML 1979-1985. Overall responsibility for AF Mtls res prog in electromagnetics, laser hardened mtls, laser device optics, electronic mtls. Chief Non-Metallic Matls Div (1986-) overall responsibility for AF Matls res. prog. in organic composites, mechanics, carbon/carbon composites, organic synthesis and non-structural materials. Member Department of the Air Force Senior Executive Service. *Society Aff:* AIAA, ΣΞ, TBΠ

Mingle, John O
Business: Fairchild Hall, Manhattan, KS 66506
Position: Executive Vice President *Employer:* Kansas St. Univ. Research Foundation *Education:* J.D./Law/Washburn Univ; Ph.D/ChE/Northwestern Univ; M. S./ChE/KS St Univ; B.S./ChE/KS St Univ *Born:* 5/6/31 Native Kansan. Korean War Veteran. Training Engr, Genl Elect Co., 1953-54. Assist., Assoc., Prof., Kansas State Univ., 1960-62-65- . Dir Inst for Computational Res in Engrg, Kansas St Univ., 1969- . Exec VP, KS State Univ Res Fdn, 1983-Present. Patent Attorney, 1984-present. Visiting Prof Univ of Southern Calif., 1967-68. Black & Veatch Distinguished Prof., Kansas State Univ 1973-78. Consultant to: Gulf General Atomic, Wilson and Company Engrs, Argonne Natl Lab, Regional VChrmn, NSPE., 1978-79. Regional Vice President, SUPA, 1985-87, Midwest Section Chairmn, ASEE, 1985-86. Res Fields: Nuclear Heat Transfer, Chem Calculations, Transport Theory, Computational Engrg, Forensic Engrg. Mbr Kansas, Patent, and Federal Bars. Forensic Engrg Consultant. Principal Biographical Listing: Who's Who in America. *Society Aff:* AIChE, NSPE, ASEE, ANS, ABA, SUPA

Minich, Marlin D
Home: 601 W North St, Ada, OH 45810
Position: Prof. *Employer:* OH Northern Univ. *Education:* PhD/Civ Engr/Case Western Reserve Univ; MS/Engg Mechanics/Case Inst of Tech; BCE/Structures/Fenn College. *Born:* 8/13/38. Native of Shelby, OH; spent 20 yrs in Cleveland, OH. Aerospace engr for NASA-LEWIS Res Ctr, Cleveland specializing in Finite Element Structural Analysis. Faculty mbr at Cleveland State Univ 12 yrs with rank of Assoc Prof. Assumed current position of Full Prof in Fall 1979; Chmn Civil Engrg Dept 1986. Do extensive legal consulting. Have appeared in court as expert witness on structural analysis failures. Publications: "Cantilevered Unsymmetric Fiber Composite Laminated Plates," –Doubly-Curved Variable-Thickness Isoparametric Hetergeneous Finite Element." Enjoy fishing, camping, canoeing & woodworking. *Society Aff:* ASEE.

Minneman, Milton J
Home: 8815 Hidden Hill Ln, Potomac, MD 20854
Position: Director, Mobility *Employer:* Office of Secretary of Defense *Education:* PhD/EE/Poly Univ; MSEE/EE/Univ of PA; BEE/EE/Cooper Union. *Born:* July 1923; native of New York N Y. Elec Engr RCA; Engrg Sect Head Martin-Marietta; Ch Electrophys Res Fairchild-Republic Co; Tech Dir Bulova Sys & Instruments; Tech Asst to Exec V Pres AIL Div Cutler-Hammer; Dir of Engrg Genl Instruments Corp; 1973-1981 Special Asst for Plans & Analysis, since 1981 Director, Mobility Office of the Under Secretary of Defense Res & Engrg, Office of the Secy of Defense Wash D C. Mbr Exec Bd Long Island Section IEEE, President Washington Society of Engineers. *Society Aff:* IEEE, ΣΞ.

Minnich, Eli B
Home: 1125 Eric Dr, Harrisburg, PA 17110
Position: Pres *Employer:* Minnich Engineering & Assoc Inc *Education:* BS/EE/PA St Univ *Born:* 11/17/26 Attended Public schools in York, PA. Grad from PA St Univ in June 1954. Registered in Pa, GA, DE, SC, NJ, RI & MD. Previous employers. Buchart Assoc, Barton Assoc, Nicholas Cowley & Assoc, Inc. Pres and Chief EE engr for Minnich Engineering & Assoc, Inc since 1968. P Pres Reg Chapters of IES, CSI and Harrisburg Chapter-PSPE. Married to Gloria Jean Eckenrode; children - Timothy and Patrick. *Society Aff:* NSPE, PSPE, IES, CSI, CEC.

Minor, Joseph E
Business: PO Box 4089, Lubbock, TX 79409
Position: P. W. Horn Prof *Employer:* Texas Tech Univ *Education:* PhD/CE/TX Tech Univ; ME/CE/TX A&M Univ; BS/CE/TX A&M Univ *Born:* 6/2/38 Native of Corpus Christi, TX. Served as commissioned officer in US Army Corps of Engrg (2 yrs). Employed 1962-69 at SW Research Inst (San Antonio) as Sr Research Engr. Moved to TX Tech Univ in 1969. Received PhD 1974. Currently P.W. Horn Prof of CE and Dir of Inst for Disaster Research and Glass Research and testing Laboratory. Awarded Fulbright Fellowship for Study in Austrailia (1978) and served as President, Tex Section ASCE. Involved in research, extension, and consulting in wind engrg with special emphasis on arch glazing systems. *Society Aff:* ASCE, NSPE, AMS.

Minor, Paul S
Business: 11800 Sunrise Valley Dr, Reston, VA 22091
Position: President. *Employer:* Centec Corporation *Education:* BS/ChE/W VA Univ; MS/ChE/Carnegie-Mellon. *Born:* 7/24/35. Union Caride Nuclear Co 1957-58. US Navy Engrg Officer 1958-61. Grad Sch 1961-62. PPG Chems Div 1962-68; Dev plant start-up, proj mgmt, planning & marketing. Parkson Corp 1969-70; Mgr of Applications Dev. 1970-71 Head of EPA Indus Tech Transfer Program. Private Cons Chem Engr 1972-74. 1975 Pres of Centec Cons Inc specializing in energy & environ engrg & advanced computer tech. Speaker at numerous tech & mgmt meetings & author of many tech papers. *Society Aff:* AIChE, WPCF, APCA.

Minozuma, Fumio
Home: 3-10-705 Shibuya 3-Chome, Shibuyaku, Tokyo 150, Japan
Position: Visiting Lecturer *Employer:* Seikei Univ. and Tokyo Science Univ. *Education:* Dr of Engg/-/Kyoto Univ; Bach of Sci/-/Kyoto Imperial Univ. *Born:* Sept 28, 1916. Res mbr & unit leader on radio waves propagation at Naval Tech Inst 1942-45. After WWII 2nd Res Mbr & Unit Leader Radio Res Lab, Ministry of Educ 1945-48. Moved to Radio Regulatory Bur, Ministry of Posts & Telecommunications 1948-64 as Dir, Mgr, Section Ch & Asst Ch Frequency Allocation, domestic & internatl admin divs for mgmt, rules & engrg standards making & engrg investigation, etc. Mgr of Space Tech Div 1965-72 & Sr Fellow Engr 1972-75 at Hitachi Ltd., Prof of Tokyo University of Agriculture and Technology 1950-55. IEEE Centennial Medal 1984, Award of the Minister of MPT 1985, Emperor's Decoration Zuihosho by Emperor of Japan 1985. Chmn CISPR,IEC Japanese Natl Comm & mbr of several Natl Govt Comms: Telecommunication, Indus & Sci. Dir, Chmn & mbr at present & past of Tech Cttes of engrg societies: Electronics, Elec, Communications, TV & Automotive, etc in Japan. Fellow IEEE, N Y USA 1964 & Life Fellow 1985. *Society Aff:* RCC, MPT.

Minter, Jerry Burnett
Business: 6 Kinsey Pl, Denville, NJ 07834
Position: President. *Employer:* Components Corporation. *Education:* BS/EE/MIT *Born:* 1913 Ft Worth Texas. Circuit dev engr Boonton Radio Corp 1935; Ferris Instruments Corp 1936-38 dev Radio Noise & Field Strength Meters; founded Measurements Corp 1939 & ser as V Pres & Ch Engrg until sale in 1953 to McGraw Edison; Pres Components Corp 1946- . Holds 14 Pats Fellow IEEE; Fellow & P Pres of both Audio Engrg Soc & Radio Club of Amer; Mbr ASM, Life Mbr SMPTE, Member SPIE, AOPA. Received Armstrong Award from Radio Club of Amer 1968. *Society Aff:* IEEE, AES, SMPTE, ASM, AOPA, SPIE

Mintzer, Olin W
Home: 7805 Hayfield Rd, Alexandria, VA 22310
Position: Prof Res CE *Employer:* Ohio St U, U.S. Army Engr Topographic Labs *Education:* BSCE/Civil Engrg/Univ of TN; MSCE/Highway Engrg/Purdue Univ *Born:* 06/06/16 Native of Spokane, Washington; in Ohio for past 30 years, except for two years in India, Prof, Punjab Engrg Coll, Chandigarh. In charge of expedition for Corps of Engrs to MacKenzie Valley, Canada, 1951. After 33 years active, Reserve, and Natl Guard duty, retired as Colonel; received Legion of Merit 1971. Past VP, Franklin County Chapter, Ohio Society of PE; Past Pres, Ohio Chapter, American Society of Photogrammetry. Terrain Analysis Specialist under Intergovernment Personnel Agreement, 1980-81, with Army Engr Topographic Labs, Ft Belvoir, VA. Since 1974, Prof, Cvl Engrg, The Ohio State Univ. Photography is only hobby. Prof Emer, 1982, The Ohio State Univ; Res CE, U.S. Army Engr Topographic Labs, Ft. Belvoir, VA 1983 to date. *Society Aff:* ASCE, NSPE, ASP, TRB, ROA

Miorin, Anton F
Home: 414 Northway Rd, Harrisburg, PA 17109
Position: Vice President *Employer:* Gannett Fleming Corddry & Carpenter, Inc. *Education:* MSE/Sanitary Engg/Univ of MI; BCE/Sanitary Engg/Manhattan College. *Born:* Apr 24, 1929. Ptnr, Senior Project Manager, Mgr Industrial Services Section, Gannett Fleming Corddry & Carpenter Inc; Administration & Management, Environmental Engineering, Energy Development, Plant & Facilities Engineering. Reg P E in Penn. Mbr WPCF, ASCE; Chmn WPCF & ASCE Joint Ctte to Revise Sewage Treatment Plant Design Manual of Practice. Pubs: "Acid Mine Drainage Abatement Plans for Selected Areas Within the Susquehanna River Basin–, paper presented at 1968 Water Pollution Control Assn of Penn conference; "Considerations in the Design of Coal Mine Drainage Abatement, Collection & Treatment Facilities–, paper presented at 1972 Penn St Univ coal mining seminar; "Financial Planning of Industrial Pollution Control Facilities–, paper presented at 1977 US EPA Natl Conference/Treatment & Disposal of Industrial Wastes. *Society Aff:* WPCF, ASCE.

Miranda, Constancio F
Home: 19330 Warrington, Detroit, MI 48221
Position: Professor of Cvl Engrg. *Employer:* University of Detroit. *Education:* PhD/Struct Engrg/OH State Univ; MSCE/Civil Engrg/Notre Dame; MA/Math/Univ of Detroit; BSCE/Civil Engrg/Univ of Bombay. *Born:* 12/4/26. m Joan Mary Menezes 1957; 3 sons 1 daughter. Prof Cvl Engrg; Cons Eng Cvl, Struc & Sys Engrg. Publs: contrib to number of prof journals, symposia, confs, etc. Fellow Instn of Engrs India; ASCE; Disting Alumnus Ohio St Univ 1973. *Society Aff:* ASCE, ASEE, ASEI (India)

Miro, Sami A
Business: 8 Inverness Drive East, Suite 224, Englewood, CO 80112
Position: Pres *Employer:* S. A. Miro, Inc. *Education:* BS/CE/Univ of CO *Born:* 01/10/47 Recipient of Natl Science Foundation grant. Practicing PE with emphasis on commerical structure. VP of INDEVCO Corp, a consortium of five multidiscipline engrg firms. Mgr of AAA consltg engrs located in Ireland. Honorary citizen of

Miro, Sami A (Continued)
Boulder, CO. Past pres of Rocky Mountain Cedars Club. Founder and pres of S. A. Miro, Inc, consltg engrs, specializing in cvl and structural engrg. *Society Aff:* ACEC, ASCE, ACI, NSPE, PEC

Misch, Herbert L
Business: The American Road, Dearborn, MI 48121
Position: Vice President Environ and Safety Engrg *Employer:* Ford Motor Company. *Education:* BS/Mech Engg/Univ of MI. *Born:* Dec 1917. Named to head Ford Motor Co's newly established Environ & Safety Engrg Staff 1972; is Ford's principal spokesman on environ & vehicle safety activities. Worked for Packard Motor Car Co Attaining the position of Ch Engr & at Cadillac Div of G M as Dir of advanced planning before joining Ford in 1957 as Asst Ch Engr. Elected a V Pres at Ford 1962 & has headed Engrg & Res Staff & Engrg & Mfg Staff prior to current appiontment. Is past mbr of Bd of Dirs SAE & Mbr ASBE, Tau Beta Pi & Phi Kappa Phi. Mbr 1973 CRC. *Society Aff:* SAE, ASBE, NAE, CRC.

Mischke, Charles R
Home: 1815 Maxwell Ave, Ames, IA 50010
Position: Prof. of Mech, Eng. *Employer:* Iowa State Univ *Education:* Ph.D./Mechanical Eng'g/University of Wisconsin; MSME/Mechanical Eng'g/Cornell University; BSME/Mechanical Engineering/ Cornell University *Born:* 03/02/27 Served as Prof & Ch, Mech Eng, Pratt Inst, 1957-64, Prof of Mech Eng, Iowa State Univ 1964-to date. Authored many technical papers and the following books: Elements of Mechanical Analysis, Addison-Wesley 1963; Introduction to Computer-Aided Design, Prentice-Hall, 1968; Mathematical Model Building, Iowa State University Press, 1980; Standard Hanbook of Machine Design, contributor and Co-Editor-in-Chief (with Joseph E. Shigley), winner of the Association of American Publisher's Award for the best book in technology and engineering published in 1986. Active on A.S.M.E. cttees. Developer of the Iowa CADET algorithm for computer-aided design, and early contributor to stochastic design methodology. Consultant. *Society Aff:* ASME, ΣΞ, AGMA, ΦΚΦ, ΠΤΣ

Miser, Hugh J
Business: A-2361, Laxenburg Austria
Position: Exec Editor. *Employer:* Intl Inst for Applied Systems Analysis. *Education:* BA/Math/Vanderbilt Univ; MS/Math/IIT; PhD/Math/OH State Univ. *Born:* 5/23/17. Taught math 1938-46 (IIT, OH State Univ, Lawrence College, Williams College); conducted operations & systems res studies as worker & mgr 1945, 1949-69 (Hq USAF, Res Triangle Inst, MIT, the Mitre Corp, Travelers Res Inst); Prof of Ind Engrg & Operations Res, Univ of MA, Amherst, MA, 1969-80 (Hd, IE/OR Dept, 1976-78); Exec Editor, Intl Inst for Applied Systems Analysis, Laxenburg, Austria, 1979-present. Author of numerous papers, books, reports. Editor, *Operations Research* (1968-74). Pres of ORSA (1962-63), Geo E Kimball Medal (1975). *Society Aff:* ORSA, TIMS, AMS, IMS, CORS, MAA, ASA, SIAM.

Mishu, Louis P
Business: 2720 Nolensville Rd, Nashville, TN 37211
Position: Pres *Employer:* Geotek Engrg Co, Inc. *Education:* PhD/Geotech Engr/ Purdue Univ; MS/CE/Purdue Univ; BS/CE/Univ of Baghdad *Born:* 4/5/29 Born in Baghdad, Iraq. Has lived in USA since 1961 and an US citizen since 1973. Am a licensed professional engr and consulting engr in geotechnical and materials engrg. Also, have been grad engrg instructor of TN St Univ and Univ of TN (Nashville Campus). Am presently pres and principal of Geotek Engrg Co, Inc a position held since the co was founded in 1974. Hobbies include rose gardening and playing chamber music (oboe). *Society Aff:* ASCE, SAME, NSPE, ASME, ASTM

Misiaszek, Edward T, Sr
Business: Engg School, Clarkson Univ, Potsdam, NY 13676
Position: Assoc Dean Engg. *Employer:* Clarkson Univ *Education:* PhD/CE/Univ of IL; MCE/CE/Clarkson Univ; BSCE/CE/Clarkson Univ. *Born:* 5/23/28. Native of Utica, NY. Taught Lehigh Univ, Univ of IL, Clarkson Univ. Served Corps of Engrs 1952-1958. Consultant area of geotechnical engg. Natl Pres Tau Beta Pi 1976-78 and as VP 1974-1976. VP Northeast Zone NCEE 1979-82. Chrmn Zone I ASEE 1979-81. Chrmn Student Dev Com ECPD 1979-81. Serving on Bd of Dir ASEE, ECPD & NCEE. Recipient ASEE, Western Electric Fund Award 1971, Recipient Tau Beta Pi Excellent Teaching Award 1972. Mbr NYS Bd for Engg & Land Surveying 1971-1981. Enjoy trout fishing & gardening. *Society Aff:* ASCE, ASEE, NCEE, ABET, NSPE, NYSSPE.

Missimer, Dale J
Business: PO Box 12777, San Rafael, CA 94913
Position: President. *Employer:* Polycold Sys Inc. *Education:* BS/EE/Rice Inst. *Born:* Aug 1925 Los Angeles. Mbr Amer Vacuum Soc; Fellow ASHRAE, Pres S Calif ASHRAE 1959-60; Pres L A IES 1962. Received ASRE-Wolverine Award 1956. ASHRAE Data Book Chap & Section Ed 1958-60; Chmn ASRE/ASHRAE Low Temp Tech Ctte 1955- 65. Several pats on cryogenic refrigeration processes; has published on same & cascade refrigeration sys. Pres Missimers Inc 1962-68, then Conrad-Missimer Div G/W 1968-72, both mfg environ test chambers; now President Polycold Systems, A Mfg of mixed refrigerant cryogenic apparatus. Spare time spent bicycling, hiking, reading, photography. *Society Aff:* AVS, AIP, ASHRAE

Mistelske, Emerson W
Business: 1220 Glenwood Ave, Minneapolis, MN 55405
Position: VP. *Employer:* Environmental Process Inc. *Education:* BS/Chem Eng/UCF MN *Born:* 2/11/24. Maj work in food engg. Intl consulting in developing milk supplies to heavily populated areas in India developed processes for reconstituting dried powder milk in Trinidad for people in Tobago & Jamaica. Developed improved systems for process cheese in USA by use of blending a variety of cheeses. Currently part owner in consltg firm specializing in plant design, equp design & construction of food plants. Pres of Melo Corp. Mbr of Melo Corp that patented method to convert whole milk to spread, a state of the art method to reduce consumer cost of a cheese replacement. Currently, designed an ice cream factory for Peoples Republic of China. This is a joint venture and will start operating July 1987. Location Xiaman, China. *Society Aff:* AICH

Mistrot, Gustave A
Business: 1200 Milam, 3200 Entex Bldg, Houston, TX 77002
Position: Sr VP *Employer:* Keplinger and Assoc, Inc. *Education:* BS/Petrol Engr/ A&M Coll of TX *Born:* 5/16/29 Head of Intl Operations of Keplinger and Assoc, Inc an energy consulting firm with numerous offices in the US and overseas. He has performed or supervised engineering and geological studies on fields and exploratory areas in the US, Western Europe, the Middle East, S Amer and Australia, involving reserve estimates, economic appraisals, field development and depletion programs, secondary and enhanced recovery, exploration potential, gas deliverability, gas storage and unitization. He has authored or co-authored several articles and publications, taught short courses and served on technical committees of the SPE and the Natl Council of Engrg Examiners. *Society Aff:* SPE, AAPG, NSPE

Misugi, Takahiko
Home: 1-2-301 Keyaki-daira Miyamae-ku, Kawasaki, Japan 213
Position: Director *Employer:* Fujitsu Lab. LTD. *Education:* Ph.D./Elec. Eng./Osaka University; B.S./Physics/Osaka University *Born:* 02/17/27 He was born in Nishinomiya, Japan in 1927. He graduated Osaka Univ. dept of physics in 1950. He joined Kobe Kogyo in 1951. For twenty years, he was the engr to develop microwave tubes. He joined Fujitsu Labs in 1968 with the merge of Kobe Kogyo in Fujitsu LTD. Since 1971 he was a mgr of r. & d. in semiconductor devices and materials, including Gunn diode, IMPATT, GaAsFET, HEMT, Semiconductor lasers and detectors. Since 1983, he is a dir of Fujitsu Labs LTD. In 1986, he was elected as a Fellow of IEEE. *Society Aff:* IEEE, JSAP, IEICE, IEEJ

Mitchell, Charles E
Business: Colorado St Univ, Fort Collins, CO 80523
Position: Prof *Employer:* Dept of Mech Engrg *Education:* PhD/Aero & Mech/ Princeton Univ; MA/Aero & Mech/Princeton Univ; BSE/Aeronautical/Princeton Univ *Born:* 4/14/41 Born in Newark, NJ, raised in Lincoln Park, NJ. Research assistant at Princeton Univ 1963-67. Summer employment at Aerojet Gen Corp in liquid and solid rocket motor development. Moved to CO to take position with CSU in 1967. Promoted to prof in 1980. Analytical research in the areas of combustion stability, combustion driven waves, ignition, in situ oil shale extraction modeling and basic combustion phenomena. Work supported by NASA, NSF, DOE, Air Force. Consultant to RDA, CGA, others. Teach graduate and undergrad courses in combustion, gas dynamics and thermodynomics. Married, 3 children. Avid skiier, runner, rock climber. *Society Aff:* CI

Mitchell, Doren
Home: Mitchell Ln, Martinsville, NJ 08836
Position: Dept Head - Retired. *Employer:* Bell Tel Labs. *Education:* BS/Princeton. *Born:* 1905. Joined AT&T after Princeton 1925. Transferred to Bell Labs 1934; early assignments included studies of transmission on long radio & wire circuits including voice operated devices. During WWII worked on special military transmission sys & problem of laying wire from airplanes; founded Somerset (N J) Mechanic's Sch which provided war time training to residents of the county. Since then has worked on radio, data transmissions, satellite communication & conference sys. Retired 1970 as Head of Conference Sys Dept at Murray Hill. Fellow IEEE & Fellow AAAS. Received Gov's Award from St of Ohio 1963. Holds 85 pats & has pub numerous papers. *Society Aff:* IEEE, AAAS.

Mitchell, J Dixon, Jr
Business: 1299 Northside Dr NW, Atlanta, GA 30318
Position: Southern Zone Engr. *Employer:* Westinghouse Elec Corp Lamp Div. *Education:* BSEE/EE/Purdue Univ; Assoc of Sci/Electrical/Kansas City Jr College. *Born:* July 1918; native of Sulphur Springs Texas. Joined Westinghouse Lamp Div 1939; helped develop several new sizes & colors of fluorescent lamps & contributed to early improvements in fluorescent lamps; currently providing engrg assistance to various govt agencies & private indus on special applications of light, ultraviolet & infrared energy. Frequent instr for lighting training courses. 1941-46 with US Army Signal Corps & Major USAF. Mbr IES 1939- ; served as Genl Secy, 2nd V Pres, 1st V Pres & Internatl Pres 1963- 64; Fellow IES; Disting Service Award 1974 IES. Reg P E. *Society Aff:* IEEE, NSPE, IES.

Mitchell, J Kent
Business: 332 Agr Engr Sci Bldg, 1304 W Pennsylvania, Urbana, IL 61801
Position: Prof *Employer:* Univ of IL. *Education:* PhD/Agri Engr/Univ of IL; MS/ Agri Engr/IA Stat Univ; BS/Agri Engr/IA State Univ. *Born:* 3/5/35. Native of Oskaloosa, IA. Served with US Army Air Defense Command 1957-1959. Agri Engr, Soil Conservation Service, USDA, for three yrs. With Univ of IL since 1964. Engaged in small watershed hydrology and soil erosion res; and teaching of hydrology, soil and water conservation engg, and surveying courses. Author or co-author of 74 technical papers, text chapter, and text. Active in Boy Scouts and United Methodist Church. *Society Aff:* ASAE, AWRA, SCSA, ASEE, IWRA.

Mitchell, James Curtis
Home: 2771 Fanwood Ave, Long Beach, CA 90815
Position: Program manager-CAD/CAM Sys. *Employer:* Information Sys Ctr-Rockwell International. *Education:* BS/Forestry/Univ of WA; BS/Biology/UCLA. *Born:* Sept 1923. Native Long Beach, Calif. Naval Air Corp WWII. Prog Mgr for CAD/CAM at Rockwell. With Rockwell since 1955. Taught Systems Analysis at U Tenn & UCLA. Loaned to subsidiary to develop & install regional information systems for Calif, Louisiana, Fla & an Alaska native corp. Owner of Resources Planning & Mgmt Consultants of Calif. Applied multidisciplinary approach to natural resources plans for Togo & other West Africa nations, Kuwait & Arab nations, & Iran. Pioneer in CAD/CAM & informations systems. Internatl Pres SAWE 1971-73. Fellow SAWE. *Society Aff:* SAWE.

Mitchell, James K
Business: University of California, Dept Civil Engg, Berkeley, CA 94720
Position: Prof of Civil Engg. *Employer:* Univ of Calif - Berkeley. *Education:* ScD/Civil Engg/MIT; SM/Civil Engg/MIT; BCE/Civil Engg/RPI. *Born:* April 1930. Soil engr for Waterways Experiment Station, 1955. Officer in Army Corps of Engrs, 1956-58. At U C Berkeley since 1958. Respon for courses & res in soil behavior & soil stabilization. Reg Civil Engr in Calif. Chmn, Soil Mechanics & Foundations Div, ASCE, 1971. Chmn, US Natl Ctte for Internatl Soc for Soil Mechanics & Foundation Engrg, 1971. ASCE Norman Medal, 1972; Middlebrooks Award, 1962, 1970, 1973; NASA Medal for Exceptional Scientific Achievement, 1973, NAE, 1976; ASCE Terzaghi Lecturer, 1984; ASCE Terzaghi Award 1985; Transp. Res. Bd Exec ctte, 1983-1987; President, San Francisco Section ASCE, 1987. *Society Aff:* ASCE, TRB, CMS, ISSMFE

Mitchell, James T
Business: 2600 1 Main Pl, Dallas, TX 75250
Position: Director. *Employer:* Purvin & Gertz, Inc. *Born:* Oct 1939. BS & MS Chem Engrg from U of Texas 1962/63. Joined Purvin & M W Kellogg Co, NY city, 1963-66. Joined Purvin & Gertz, Inc in 1966. Mgr of the London office, 1972-74. Mgr of the Dallas office, 1974-76. Currently VP & mbr of Bd/Dirs.

Mitchell, Stephen C
Business: 549 W. Randolph, Chicago, IL 60606
Position: Exec VP *Employer:* Lester B Knight & Assoc, Inc *Education:* MBA/Finance/Univ of Chicago; MS/CE/Univ of NM; BS/CE/Univ of NM. *Born:* 9/27/43 Served to 1st Lt. US Army Corps of Engrs, 1969-1972; proj mgr/dir of planning Bauer Engrg 1972-1975, Chicago; proj mgr/managing Assoc/VP/Sr VP and currently Exec VP Planning & Dev, Lester B Knight & Assoc, Inc 1975-present, Chicago. Have served as Dir/Treas/Pres IL section, ASCE. Chmn Mbr activity exec comm ASCE. Pres, Chicago Post SAME. Dir, District 8, ASCE. Zone III VP, ASCE. Mbr, Gov Comm on Sci and Tech, State of IL. Ecomomics Club of Chicago. *Society Aff:* ASCE, SAME, NSPE.

Mitchell, Thomas P
Business: Dept Mech & Env Engr, Santa Barbara, CA 93106
Position: Prof *Employer:* Univ of CA *Education:* PhD/Engr Sci/Ca Inst of Tech; MSc/Math Sci/Nat Univ of Ireland; BSc/Math Sci/Nat Univ of Ireland; BE/CE/Nat Univ of Ireland *Born:* 11/27/29 Research Fellow, CA Inst of Tech 1956-57. Faculty mbr, Dept Theoretical and Applied Mech, Cornell Univ, 1957-66. Research Scientist, Gen Motors research laboratories, Santa Barbara, CA 1963-64. NAS-NRC research scientist, NASA Goddard Space Flight Center, Greenbelt, MD 6/65-9/65; 6/ 66-9/66. Visiting Prof, Cornell Aeronautical Lab, Buffalo, NY 1962-63. Prof, Univ of CA, Santa Barbara, 1966-Present. Chrmn Dept Mech and Environmental Engrg, Univ of CA, Santa Barbara 1972-1977. *Society Aff:* AIAA, APS, ΣΞ

Mitchell, William L
Home: 3904 Silverspring Dr, Austin, TX 78759
Position: Manager of Engineering Services. *Employer:* IBM Corp. *Born:* April 1931. BS, MS & Engr D in metallurgical engrg from U of Kentucky. Reg metallurgical engr. Res Metallurgist for DuPont Co working in uranium & aluminum alloy dev. Employed by IBM Corp since 1957. Became Mgr of Materials Engrg, Lexington, Kentucky facility in 1961; Materials Tech, Austin, Texas facility 1967, & assumed current position in 1973. FounderChmn Bluegrass Chap of ASM 1963-65; received Howard F Taylor award for best tech paper from AFS 1966. Hobbies are piano, IC engine repair, & golf.

Mitra, Sanjit K
Business: Dept. of Electrical & Computer Engg, University of California, Santa Barbara, CA 93106
Position: Prof *Employer:* Univ of CA, Santa Barbara *Education:* PhD/EE/Univ of CA-Berkeley; MS/EE/Univ of CA-Berkeley; MSc/Radio Physics/Univ of Calcutta-India; BSc/Physics/Utkal Univ-India *Born:* 11/26/35 Taught at Cornell Univ 1962-65. Mbr of Technical Staff at Bell Labs 1965- 67. With Univ of CA since 1967. Transferred to Santa Barbara campus in 1977. Dept Chrmn 1979-1982. Visiting Prof - Indian Inst of Tech, New Delhi, 1972; Kobe Univ, Japan, 1973; Univ of Erlangen-Nuernberg, West Germany, 1975. Australian Natl Univ, Canberra, 1982; Tampere Univ of Tech 84. Conslt to Lawrence Livermore Natl Lab since 1975. Published over 270 technical papers, 5 books and hold 2 patents. Recipient ASEE Terman Award 1973 & AT&T-Foundation Award 1985. Elected Fellow of the IEEE 1973. Elected Fellow of AAAS 1982. Distinguished Fulbright Lecturer Award for Brazil 1984 & Yugoslavia 1986. Visiting Prof Award from Japan Soc for Promotion of Sci 1973. Honorary Doctor of Technology Degree, Tampere Univ of Technology, Finland 1987. *Society Aff:* IEEE, ASEE, AAAS, ΣΞ, HKN, EURASIP

Mittl, Robert L
Home: 3 Hearthstone Way, Convent Station, NJ 07961
Position: General Manager - Projects. *Employer:* Public Service Electric & Gas Co. *Born:* Nov 1931. Mech Engrg Degree - 1954 Stevens Inst of Tech; MS in Nuclear Sci 1958 Carnegie Inst of Tech. Graduated Oak Ridge School of Reactor Tech 1958. Public Service Elec & Gas Co 1954- . On loan to: Argonne Natl Lab 1958-60; Genl Atomic Div Genl Dynamics 1960-62. Engr & Construction Dept: Sr Engr 1965; Asst to Ch Mech Engr 1970; Ch Mech Engr 1971-72; Asst to Mgr of Engrg 1973-74; Genl Mgr of Projects 1974- . Mbr ANS, Prime Movers Ctte - Edison Electric Inst, Nuclear Power Divisional Ctte - EPRI.

Miyairi, Shota
Business: 2-2 Kandanishiki-cho, Chiyoda-ku, Tokyo, Japan, 101
Position: Prof *Employer:* Tokyo Denki Univ. *Education:* Dr/Power Elect/Tokyo Inst of Tech *Born:* 4/25/17 After being associated with Nagano Tech Coll of Japan, he joined the Tokyo Inst of Tech (TIT) in 1958 as an assoc prof and was promoted to prof in 1960. He retired from TIT in 1978 and changed his post to Tokyo Denki Univ. At present he serves as the dean of the above univ. His work has been mainly concerned with EE. For 1981-1982 he was the Pres of the Inst of EE of Japan (IEEJ). Since Jan 1980, he has been the Fellow of IEEE. *Society Aff:* IEEE, IEEJ

Mize, Joe H
Business: 1511 N Glenwood, Stillwater, OK 74074
Position: Regents Prof *Employer:* Oklahoma State U. *Education:* PhD/IE/Purdue; MSIE/IE/Purdue; BSIE/IE/TX Tech. *Born:* 6/14/34. Native of Big Spring, TX. Indus Engr, White Sands Missile Range 1958-61. Taught at Auburn Univ 1964-69; at AZ State Univ, 1969-72; with OK State Univ since 1972. Author of 5 textbooks, numerous tech articles. Editor of Prentice- Hall Intl Series in Indus and Systems Engr. Advisor to NSF, numerous other federal agencies. Reg PE; Mbr ASEE, IIE, NSPE, OSPE, SME, APICS. Executive VP, IIE, 1978-80. President, IIE 1981-82. Recipient of IIE *Book of the Year* award, 1979. Recipient of IIE *Innovative Achievement* Award, 1977. IIE Fellow Award, 1983. Recipient of Purdue Univ *Distinguished Engg Alumnus* Award, 1978. "Outstanding Engineer in Oklahoma–, 1981, OSPE. "Who's Who in America–, 41st, 42nd and 43rd Editions. *Society Aff:* ASEE, NSPE, IIE, SME, APICS

Mjosund, Arne
Home: 11 Vista Grande, Northport, AL 35476
Position: Prof. Emeritus *Employer:* Retired. *Education:* PhD/Opr Res/Johns Hopkins; Dipl NHH/Econ & Bus Admin/Norw Sch of Econ & Bus Admin *Born:* 11/13/21. Native Norwegian. Res Assoc with a North-Norwegian Res Inst, 1952-55. Developed and led a Statistics-Operations Res Group in the Royal Norwegian Navy, 1952-66. Sr Instructor and Res Assoc in Industrial Engg Dept, Johns Hopkins Univ, 1961-63. Prof, the Univ of AL, Dept of Industrial Engg, 1966-1983. Prof, Coll of Business and Industry, MS State Univ, 1984. Acting Dept Hd, 1968-69 and 1977-78. Retired 1/1/87. Res interests: Stochastic Processes, Operations Res and Comp Interface. Hobbies: Painting, Photography, Hiking. *Society Aff:* IIE, ORSA, TIMS

Moak, Charles E
Business: PO Box 12128, Jackson, MS 39236
Position: General Manager. *Employer:* Pearl River Valley Water Supply Dist. *Education:* MSSE/Water Supply Eng/Univ of NC; BSCE/Civil Engg/MS State Univ. *Born:* June 1932. MSSE from UNC; BSCE Mississippi State U. Field engr, Miss State Bd of Health. Reg PE, State of Miss. Private Practice Civil Engrg, Forest, Miss. Asst Mgr & currently, Genl Mgr, Pearl River Valley Water Supply District. P Pres, Alabama/Miss Sect, AWWA. Favorite hobbies are gardening & golf. *Society Aff:* NSPE, AWWA.

Moats, Erwin R
Home: 1141 Taft Ave, Cuyahoga Falls, OH 44223
Position: Ret. *Employer:* Goodyear Tire Co. *Education:* B.S./Mech Eng/Akron University *Born:* 11/25/19 Native of Akron, Ohio. Served with Army Air Corps 1942-46. Staff engr for Goodyear Tire specializing in design of steam and utility power equipment and piping. Field construction engr Luxembourg plant 1949. Mgr maintenance Topeka Kansas plant 1951. Returned to corporate engrg Akron 1955 as Section Head Mech Engrg, then Mgr mechanical section. Responsible for Goodyear worldwide power plant design and installation. Developed power plant operation review program which later became basis for corporate energy conservation program. Promoted to Mgr, Corp Engrg responsible for all design and construction of Goodyear plants worldwide. Elected Fellow ASME in 1984. *Society Aff:* ASME

Mochau, Alfred, Jr
Home: 8 Riverlyn Terr, Ft Smith, AR 72903
Position: VP - Engrg and Information Services *Employer:* Baldor Electric Co *Education:* MS/System Sci/Polytechnic Univ; BS/EE/Univ of MA, Amherst. *Born:* 3/22/36. A native of Providence, RI, Al had been with GE since 1953 in numerous engrg and some manufacturing assignments. A graduate of GE's four yr Apprentice Course (Drafting) in 1957, and its three yr Advanced Course in Engrg in 1967, Al was named Manager - Advance Engrg (San Jose, CA) in 1972; Mgr - Engrg, San Jose Motor Plant in 1975; Mgr - Engrg, Small AC Motor Dept, Nashville, TN in 1976; and Mgr-Special Purpose Computer Ctr in 1980. Present Appointment-Oct, '84. Al is a mbr of Tau Beta Pi, Eta Kappa Nu, Phi Kappa Phi, and a licensed PE in NY State. *Society Aff:* IEEE, ACM.

Moe, Dennis L
Business: Agri Engrg Dept, Brookings, SD 57006
Position: Professor & Head Agri Engrg Dept. *Employer:* South Dakota State U. *Education:* DS/Sci/Augustana; MS/Agri Engg/SD State Univ; BS/AG Engg/SD State Univ; ASSO /Agri/SD State Univ. *Born:* 4/1/17. Dennis Moe holds three degrees from SD State U & a doctorate from Augustana College. Reg PE in SD. He has been on the faculty at SD State U for 33 years advancing from instructor through the ranks to a Prof & Head of the Agricultural Engrg Dept in 1956. He has been a consulting engr for years to various engrs in public service including the State Engr of SD. He has also done consulting work for architects, engrs, indus educ units & others. He has been a prolific writer of over 80 publications, tech papers & referred journal articles. At present time time he holds many offices - ASAE, ASEE, ECPD & others.

Moe, Robert E
Home: 1841 Bonnie Castle Dr, Owensboro, KY 42301
Position: Consulting Engr, Electronics (PE). *Employer:* Retired *Education:* BS/EE/Univ of WI. *Born:* April 1912. GE Co various electronic design grps 1934-49. Mgr Receiving Tube Engrg 1950-67. Mbr of NTSC in 1939 on TV Synchroniz-

Moe, Robert E (Continued)
ing standards. Mbr 1962- 67 Internatl Electrothechnical Comm on vacuum tube standards. Author of Sect 5 on Electronic Tubes, Standard Handbook for Electrical Engrs, 10th ed. Now independent cons in electronic computers & peripherals. Currently a mbr of Kentucky Bd of Registration for PE's & Land Surveyors. Fellow IEEE, Regional Dir, 1956. Reg PE in NY & Kentucky. Mbr, Data Processing Mgrs Assn, NSPE. On Bd/Dir, WNIN Public Television Station. P Pres of Kentucky Assn for Mental Health. City Commissioner 1959-63. Enjoy golf, photography, quadraphonic sound, computer programming.. *Society Aff:* NSPE, IEEE, DPMA, TBΠ, HKN.

Moebius, William H
Business: 10375 Richmond Ave, Houston, TX 77042
Position: President. *Employer:* Enserch Engineers & Constructors *Education:* BS/IA/Yale. *Born:* Oct 1930. Native of New Haven, Conn. Served in US Army 1951-53. Worked as Indus Engr, Rockbestos Prods 1953-54. Field Engr for C F Braun & Co 1956-57. With Scientific Design Co 1957-84. Appointed Dir of Const in 1967. VP - Procurement & Const in 1974-79. Apptd Pres of Belmont Constructors in 1979. Mbr ASCE. Accepted presidency of Enserch Engrs & Constructors in Aug 1984. Enserch is a high-tech engrg and construction co serving the process and utility industries. *Society Aff:* ASCE.

Moeckel, Wolfgang E
Home: 29033 Lincoln Rd, Bay Village, OH 44140
Position: Consultant (Self-employed) *Education:* BS/Phys/Univ of MI. *Born:* 2/11/22. Family immig from Germany, 1927; settled in Dearborn, MI. After grad from Univ of MI in 1944, joined Cleveland Lab of Natl Adiv Committee for Aeronautics (NACA). conducted res on aircraft engines, rockets, supersonic and hypersonic aerodynamics and propulsion. Promoted successively to Chf, Adv propulsion Div when NACA became NASA. Conducted studies of space flight and non-chemical propulsion starting in 1956. My Div developed the ion propulsion sys now avail for space missions and conducted research on other adv propulsion and power concepts (MHD, thermionics, atomic hydrogen, fusion, high power lasers). Became Chf Sc of NASA Lewis Res Ctr in 1978. Retired from NASA in 1980. *Society Aff:* AAAS, AIAA, APS.

Moehrl, Michael F
Business: RACOM Corp, 201 W. State St, Marshalltown, IA 50158
Position: VP *Employer:* RACOM Corp *Education:* BS/CE/IA St Univ *Born:* 8/23/41 Native of Marshalltown, IA. Served as an officer in Army Corps Engrs 1965- 67 serving as Post Commander & Post Engr of Camp Parks, CA. Engr-in-training with MN Hwy Dept in areas of construction & location surveying, design of hgwys and planning. Served as Asst County Engr of Marshall County, IA 1967-1973. Promoted to County Engr in 1974 and served until 1978 supervising design, construction and maintenance of over 900 miles of roads & 350 bridges. With RACOM Corp since 1978, assuming current responsibility as VP of Sales in communication systems and design in 1979. Served in the following capacities with IA Section of ASCE from 1977 to present: Transportation Conference Chrmn, Dir, VP, and Pres. Enjoy raising Appaloosa horses and hunting. *Society Aff:* ASCE, NSPE

Moeller, Dade W
Home: 27 Wildwood Dr, Bedford, MA 01730
Position: Assoc Dean for Continuing Ed *Employer:* School of Public Health, Harvard U. *Education:* PhD/Nuclear Engg/NC State Univ; MS/Env Engg/GA Tech; BS/Civil engg/GA Tech. *Born:* Feb 27, 1927. US Navy, 1944-46. Commissioned Officer, US Public Health Service 1948-66. Field Engr, Los Alamos Scientific Lab, 1949-52; Environmental radiation specialist with headquarters office, Bureau of Radiological Health, 1952-54; Res Engr, Oak Ridge Natl Lab 1956-57; Chief, Radiological Health Training, Taft Sanitary Engrg Center, 1957-61; Officer in Charge, Northeastern Radiological Health Lab 1961-66. At Harvard U since 1966. Diplomate, AAEE; Mbr NCRP; Chmn, Amer Bd of Health Physics, 1967-70; Pres, Health Physics Soc, 1971-72; Chmn, Advisory Ctte on Reactor Safeguards, USNRC 1976; Fellow, Amer Nuclear Soc; Elected to Natl Acad of Engg, 1978; Fellow, Amer Public Health Assoc. *Society Aff:* APHA, HPS, ANS, AAAS, AIHA.

Moen, Walter B
Home: 140 Mtn Ave, Berkeley Hgts, NJ 07922
Position: Staff Exec *Employer:* Am Soc of Mech Engrs. *Education:* MS/ME/Columbia Univ; BME/ME/Pratt Inst. *Born:* 2/13/20. Have held responsible positions in res & dev, mft, engg admin and plant mgt. With ASME since 1975 having responsibility in several positions for Soc technical activities. Authored over 20 published technical papers and awarded 15 patents. Reg in NJ, NY, OH, PA & VA. Served with US Maritime Service in WWII. VP ASME, 1967-69. *Society Aff:* ASEE, AWS.

Moffat, Robert J
Business: Dept of Mech Engrg, Stanford, CA 94305
Position: Prof *Employer:* Stanford Univ *Education:* Phd/ME/Stanford Univ; Engr/ME/Stanford Univ; MS/ME/Stanford Univ; MS/Engr Mech/Wayne State Univ; BS/ME/Univ of MI *Born:* 11/29/27 Grosse Pointe, MI. General Motors Research Labs, 1952-1962, in heat transfer and automotive gas turbines research. MS (Engr Mech) 1961, Wayne State Univ. MS and Engrs Degrees (M.E.) 1966. Stanford; PhD (M.E.) 1967, Stanford. Assoc Prof (Mech Engr) 1967-1970. Prof and Chrmn, Thermosciences Div, 1971-85. Special fields of interest: Heat transfer and experimental methods. Active in teaching, research and consltg in these areas. Fellow of ASME, Pres, Moffat Thermosciences, Inc. Approximately 150 publications, and 8 patents. *Society Aff:* ASME, ISA, AAUP, AIAA

Moffatt, Charles A
Business: College of Engg, Morgantown, WV 26506
Position: Assoc Prof. *Employer:* WV Univ. *Education:* PhD/Mech Engr/Tulane Univ; MS/Mech Engr/Tulane Univ; BS/Mech Engr/Univ of CA. *Born:* 12/29/40. Taft CA. BSME 1963 Univ of CA Berkeley, MSME 1966, PhD 1968 Tulane Univ New Orleans. Res Engr to Sec Mgr of Biomedical Engg, Tech Inc, San Antonio, TX 1968- 70. Assoc Prof of Mech Engg and Mechanics WV Univ 1970-. Res Scientist, Accident Investigation Div, US Dept of Transportation 1974-75. Recipient Ralph R Teetor Award of SAE, 1975. Mbr Univ Senate. Publications in biomechanics, numerical methods, and reconstruction of automobile collisions. *Society Aff:* ASME.

Moffet, John A
Business: 1925 N. Lynn St, 300, Arlington, VA 22209
Position: Pres & Chrmn of the Bd *Employer:* Moffet, Ritch & Larson, P.C. *Education:* BS/EE/Swarthmore Coll *Born:* 07/22/15 Radio and Television consltg engr; born Philadelphia, PA; son of Andrew and Jane Stuart (Oetter) Moffet; BS in Engrg with honors, Swarthmore (PA) Coll, 1937; married G. A. Elizabeth Morris, December 19, 1942; three children - Gwendolyn Ida, John Andrew, II, William Morris; VP, Chief Engr of William L. Foss, Inc, Washington, 1946-52; Partner, firm Silliman, Moffet & Kowalski, Washington, 1952-77; Pres, Chrmn of the Bd, Moffet, Ritch & Larson, P.C., Arlington, VA, 1977 to present; Decorated Commendation Ribbon, USAAF; Lt Colonel USAF Reserve Retired; Registered PE, VA, Washington, DC. *Society Aff:* AFCCE, SBE, IEEE, NSPE

Moffitt, David C
Home: 5455 SW Brendon Ct, Beaverton, OR 97005
Position: Environmental Engr *Employer:* USDA-Soil Conservation Service *Education:* BS/CE/CO St Univ *Born:* 03/07/43 Native of Fowler, CO. An environmental engr with the West Technical Service Ctr, USDA-Soil Conservation Service, Portland, OR since 1976. Past assignments with SCS include field engrs position in NM 1966-1969, hydrologist in UT 1969- 1974, and environmental engr in CA 1974-1976. Present duties include coordination of complex water quality and waste mgmt concerns for 13 western states. Also responsible for preparing reference and training material for the agency use at the natl level. PE in CA. Winner of ASAE's

Moffitt, David C (Continued)
FMC Young Designer Award in 1980. Special interests include church administration and photography. *Society Aff:* ASAE

Mohamed, Farghalli A
Business: Department of Mechanical Engineering, University of California, Irvine, CA 92717
Position: Prof *Employer:* Univ of California *Education:* PhD/Mat Sci & Engr/Univ of CA-Berkeley; MS/Mat Sci & Engr/Univ of CA- Berkeley; BS/Metallurgical Engr/Cairo Univ-Egypt *Born:* 09/25/43 Conducted research on mech behavior of materials and taught at the Univ of Southern CA, Los Angeles from 1972 to 1980. Specializing in physical & mech metallugry. Also experienced in ceramics tech and industrial manufacturing. Published over 70 papers primarily in mech behavior. Contributed to understanding of deformation mech at high temperatures. *Society Aff:* ASM, ΣΞ

Mohler, Harold S
Home: 6 Springcreek Manor, Hershey, PA 17033
Position: Chairman/Ret. *Employer:* Hershey Foods Corp/Ret *Education:* B.S./Ind. Engr/Lehigh University *Born:* 03/08/19 Native of Ephrata PA. Matriculated at Lehigh Univ in Class of 1941. After freshman year worked for two years & enlisted in USAF. Returned to Lehigh in 1946 & graduated in '48. Was recruited from campus by Hershey Foods Corp. Starting as a staff Ind Engr I moved thru the ranks & in 1965 became Pres & CEO. I became a Fellow in ASME in 1985 & was awarded Outstanding Engr in Mgmt Waward by the PA Soc of Prof Engrs in 1981. *Society Aff:* ASME, AIIE, NSPE

Mohr, Milton E
Business: 5454 Beethoven St, Los Angeles, CA 90066
Position: President & Chief Exec Officer. *Employer:* Quotron Systems, Inc. *Education:* BS/Elec. Eng/Univ of Nebr; Honorary Dr. of Eng/Elec. Eng/Univ of Nebr *Born:* April 9, 1915 Milwaukee. BS U of Nebraska 1938. Communications res Bell Telephone Labs, Inc, NYC 1950-54; VP Thompson Ramo Woodridge, Inc, Canoga Park, Cal; Genl Mgr RW Div, Dir of TRW Computers Co 1954-64; VP, Mgr, Indus sys div Bunker Ramo Corp 1964-65; VP, Genl Mgr def systems div 196566, Pres, Chief Exec Officer Corp 1966-70; Pres, Chief Exec Officer Quotron Systems, Inc 1970- . Recipient hon mention as outstanding young elec engr Eta Kappa Nu. Fellow IEEE; Mbr AIAA, Pi Mu Epsilon, Sigma Tau, Eta Kappa Nu. Hon Dr of Eng, 1959, Univ of Nebr., 1948. *Society Aff:* IEEE, AIAA, ΠΜΕ, HKN, ΣΤ

Moir, Barton M
Business: PO Box 220, Davis, CA 95616
Position: Senior Industrial Engineer. *Employer:* Hunt-Wesson Foods, Inc. *Born:* June 21, 1937 San Francisco, Ca. BSIE in Operations Res from U C Berkeley. Post Graduate work at Case-Western Reserve, Loyola of Chicago, Cornell, & U C Davis. Worked for Procter & Gamble, Litton Industries, United Air Lines, Safeway Stores, & currently Sr Indus Engr with Hunt Wesson Foods, a div of Norton Simon, Inc. Was principal of Barton M Moir & Assocs, Cons Indus Engrs. Chmn of 1972 Engr's Week, Washington , DC. Sr Mbr & Dir of the Transportation & Distribution Div of the IIE. Reg PE in Illinois & Calif by examination.

Mojica, Juan F
Business: PO Box 996, Monterrey, Nuevo Leon Mexico
Position: TQC Manager *Employer:* Hylsa S. A. de C. V. *Education:* PhD/Physics/Univ of MO-Rolla; MS/Physics/Univ of MO-Rolla; BS/Metallurgy/Univ of MO-Rolla; Bs/Physics/Instituto Tecnologico De Monterrey. *Born:* 9/15/41. Involved during graduate work in suface characterization using auger electroscopy. Involved, after joining a steel co, in res, dev & implementing of new processes in the area of cold reduction. Fields of interest are: coefficient of friction, thermal effects, math modeling and metallurgical properties of cold rolled materials. Promoted to QA Manager in a company manufacturing heavy industrial equipment & pressure vessels. Currently working as Total Quality Control Manager at Hylsa Steel Works in Monterrey. *Society Aff:* ASM, AISI, IMS of AIME, ASQC, ISS of AIME.

Mok, Perry K P
Home: 13810 Vista Dr, Rockville, MD 20853
Position: Partner *Employer:* Kraas and Mok Consulting Engrs *Education:* BS/CE/Chu Hai Univ *Born:* 01/17/32 Native of Canton, China. Design engr for the Hong Kong firm of Chau & Lee, Architects & Engrs before coming to the US. Since 1956, structural engr for the Washington, DC firm of James M. Gongwer, Consltg Engr and successors. In 1970, became a partner of the present firm of Kraas and Mok, Consltg Engrs. Engaged in the practice of structural engrg specializing in institutional bldgs. Responsible fo the structural design of bldgs in the eastern, southern and north central states. Also bldgs located in Paris, Athens, Cairo and Central Amer. Enjoy classical music and photography. Graduate studies in Structural Engrg at The Catholic Univ of Amer. *Society Aff:* ASCE, NSPE, ACEC, ACI

Molinder, John I
Home: 4233 Piedmont Mesa, Claremont, CA 91711 *Employer:* Harvey Mudd Coll *Education:* PhD/EE/CA Inst of Tech; MS/EE/Air Force Inst of Tech; BS/EE/Univ of NB *Born:* 06/14/41 Born in Erie, PA. Officer, US Air Force, 1963-67. Sr engr in comms sys res group Jet Propulson Lab, Pasadena, CA, 1969-70. Part-time lecturer CA State Univ, Los Angeles, 1970-74. Faculty mbr Harvey Mudd Coll since 1970. Current position is Prof and Chrmn of Engrg. Part-time position with Jet Propulsion Lab in telecomms and data acquisition engrg office since 1974. Served as representative (detailee) from Jet Propulsion Lab to NASA Hdqtrs, Wash, DC, 1979-80. Visiting Prof of Elec Engrg, CA Inst of Tech, 1982-83. Mbr IEEE. Reg PE, CA. *Society Aff:* IEEE, ASEE

Moll, Godfrey Joseph
Home: 2502 S Delaware Ave, Springfield, MO 65804
Position: VP & Genl Mgr. *Employer:* Syntex Agribusiness, Inc. *Education:* BS/ChE/Newark College of Engg. *Born:* 4/4/27. in Irvington, NJ. Graduated fro Summit, NJ high sch in 1944; attended Newark Coll of Engg for 1 yr before serving in Army Air Force until Jan 1948, then accepted job with CIBA Pharmaceutical & then decided to attend night coll. Married Muriel Birch in 1948, had 3 children & completed requirements for a BSChE in 1956. Joined Hoffman-Taff in Springfield, MO in 1967 as Dir of Mfg; acquired by Syntex in 1969, became VP of Operations, promoted to Asst Genl Mgr in 1975. Mbr Amer Amgmt Assn. Relax through travel VP Syntex Agribusiness & Genl Manager of Nutritions & Chem Div of January 1, 1978. *Society Aff:* ΣΞ.

Moll, John L
Home: 4111 Old Trace Rd, Palo Alto, CA 94306
Position: Senior Scientist; Director IC Structures. *Employer:* Hewlett-Packard Co. *Education:* PhD/Elec/OH State Univ; BSc/Eng Physics/OH State Univ. *Born:* 12/21/21. Wauseon, Oh. RCA Labs, Lancaster, Pa 1943-45. Bell Telephone Labs, Mbr of Tech Staff, 1952-58. Stanford U, Prof of Elec Engrg 1958-69; Fairchild Cam & Instrument, Tech Dir Opto-Electronics Div 1969-74. Since 1974 with Hewlett Packard Co, Dir of Integrated Circuits Structures Res. Life Fellow IEEE; Mbr APS, Sigma Xi, Franklin Inst, NAE. Howard N Potts Medal (Franklin Inst) 1967; JJ Ebers Award, IEEE Electron Devices Group, 1971. Dist Alumnus award Coll Engrg, Ohio St U, 1970. Dr. h.c., Faculty Engrg, Katholieke U. Leuven (Belgium), 1983. Author: Physics of Semi-Conductors, 1964. Guggenheim Fellow, 1964. Co-Author: "Computer Aided Design in VLSI Development," '85. *Society Aff:* APS, ΣΞ, NAE, NAS

Moll, Richard A
Business: 432 N Lake St, Madison, WI 53706
Position: Assoc Prof. *Employer:* Univ of WI. *Education:* PhD/Met & Mtl Sci/Lehigh Univ; MS/Met & Mtl Sci/Lehigh Univ; BS/Met Engr/IIT. *Born:* 9/2/35. Dr Moll is a full Prof at the Univ of WI in the Depts of Engg & Applied Sci & Met & Minerals Engg. He is Dir of Telecommunications with responsibility for all engg & sci courses broadcast over the 200 sites available throughout the State of WI. He is the author of a number of publications in the product safety & liability, met, & welding fields; including a video cassette correspondence course "Introduction to Materials Science–. He is recipient of the Pi Tau Sigma Distinguished Teaching Award, & both the Adams Meml Award, the 7th Dist Meritorious Award and the Fred L. Plummer National Educational Lecture Award. *Society Aff:* AWS, ASTM, ASSE, ASM, ASEE.

Molstad, Melvin C
Home: Priestley House, 224 West Tulpehocken St, Philadelphia, PA 19144
Position: Emeritus Professor. *Employer:* University of Pennsylvania. *Education:* BA/Chemistry/Carleton College; BS/Chemical Engg/MA Inst of Tech; PhD/Chem Engg/Yale Univ. *Born:* Nov 1898. US Army 1918; Chem Engr USDA 1920-21, 1923-26; Yale, instructor 1926-29; DuPont Ammonia Corp, engr 1929-31; Yale, Asst Prof 1931-39; U of Penn, Assoc Prof 1939-42, Prof 1942-69, Dept Chmn 1951-61, Emeritus Prof 1969- . Fulbright lectureships, Norwegian Tech Inst 1954-55; U of Tokyo 1961-62; Helsinki Inst Tech 1969. Fellow AIChE. Honor Award, Philadelphia Sect Amer Inst Chemists; Honor Award, Indus & Engrg Chem Sect, Amer Chem Soc. Recreation - classical music & hiking. *Society Aff:* AIChE, ACS, ASEE, ΦΒΚ.

Moltrecht, Karl Hans
Home: 1394 Zollinger Rd, Columbus, OH 43221
Position: Assoc Prof *Employer:* The OH St Univ *Education:* MSE/MFG Engr/Univ of MI; B/ME/OH St Univ *Born:* 2/7/20 Karl Hans Moltrecht was born in Berlin, Germany. He came to the USA in 1925 with his family and served in the US Navy in WW II. Prior to this war he worked as a machinist, receiving his formal education afterwards. In addition to his present teaching position, he has taught at the Univ of WA (Seattle) and at the Univ of MI. He has held several engrg positions in industry and is the Assistant Editor of the past three editions of *Machinery's* Handbook. He is also the author of *Machine Shop Practice*, now in its second edition. He has lectured widely on this subject and has written papers and articles on manufacturing engr. Also, he is a Registered Professional Engr. *Society Aff:* ASME, SME, ASM

Molzen, Dayton F
Business: 2701 Miles Road S.E, PO Box 3632, Albuquerque, NM 87190
Position: Pres. *Employer:* Molzen-Corbin & Assoc. *Education:* BS/Civ Engr/KS State Univ. *Born:* 1/6/26. reared in Newton, KS. Served in Air Corps (AUS) 1942-45 & USAF 1951-53 as engg & maintenance officer. Served as proj engr KS Hgwy Comm 1950-1951, Wilson & Co Engrs, Albuquerque, NM & Salina, KS. Founded D F Molzen and Assoc 1960, Albuquerque, NM & Molzen-Corbin & Assoc 1975. Provide engg services for municipalities throughout NM & W TX, as well as federal & state agencies. Served two terms as Pres of NM/CEC. Received Engg Excellence Award 1969 for White Rock, NM Wastewater Treatment Plant. Exec Driector NM/CEC 1985-present. *Society Aff:* ASCE, NSPE, ACEC, WPCF, APWA.

Monahan, Bernard P
Home: 147B 136th St, Queens, NY 11694
Position: Vice President *Employer:* Crow Construction Co. *Education:* MS/Civil Engg/Polytechnic Inst of NY; BS/Civil Engg/CCNY. *Born:* 11/11/36. Grad work NY Univ. Mgr, Estimator, Designer & Supervisor on Major construction projs which include highways hospitals, bus terminals, airports, sewage treatment plants, railroad structures, & office bldg fnds. Responsible as engrg mgr for Const of a 65 million dollar addition to the Yonkers Sewage Plant and respon as resident engrg mgr for construction of a 25 million dollar addition to the Sprain Brook Parkway in Westchester. Reg PE in NY, NJ, MA, Washington DC & PA. Author Construction Rock Work Guide - Wiley - NY 1972; and authored chapters on Roadway Decking and Const Ramps for Handbook of Temp Structures - McGraw Hill - NY 1984. Prof activities: 1976 Pres of the Metropolitan Sect of ASCE; in past yrs served as VP, Pres, Dir & other posts in this sect; served as Secy on original ctte that organized the Joint Urban Manpower Prog, which resulted in the training of minorities for tech positions. 1976 Dir of NYC Chap of the NY State Soc of PE's. Served on Exec Comm of Const Div, ASCE 1979-1983. Currently serving as Dir for District I of ASCE. *Society Aff:* ASCE, NYSSPE.

Monahan, Raymond E
7500 Cahill Rd, Minneapolis, MN 55435
Position: Pres/Consult *Employer:* R.E. Monahan Assoc *Education:* BAEROE/44/Univ of Minn; BS/48/Univ of Minn *Born:* 1/28/23 Native of Minneapolis, MN. Served in Army AF 1944-6. Instr in Eng Graphics at Univ of MN 1946-51. Spent several yrs in fluid dynamics, heating & ventiliating, wind tunnels, tooling, & computers as a design engr. Initiated and managed an eng stds dept. Spec int now in consulting in standardization & configuration mgt. Enjoy bowling & all major sports. Outstanding Section Mbr 1969 SES, Pres 1972-74 SES, Received SES Leo B. Moore Medal in 1986 for Outstanding Contributions in the Field of Standardization. *Society Aff:* SES, ASTM.

Moncla, E E
Business: 3900 Jackson St, PO Box 1894, Monroe, LA 71210
Position: VP. *Employer:* Ford, Bacon & Davis. *Education:* BS/ME/Univ of Southwestern LA. *Born:* 5/30/26. Native of Moncla, LA. Served with the US Navy in the S Pacific in 1944 & 45. Has engineered, designed & constructed thousands of miles of pipelines throughout the US, Canada & in the Gulf of Mexico. Has been accepted in State & Fed courts as an expert in pipeline matters. Was Resident Mgr on numerous pipeline suspension bridges including the then longest pipeline suspension bridge and the first across the MS River. Assumed responsibility in 1976 as VP of the Pipeline Services Dept employing over 350 engr, designers, right of way, mtl & survey personnel working for over 50 different client cos throughout N America. Pres LA Engg Soc 1977-78. VP NSPE 1979-80 Engr in Const. Enjoy: hunting, theater, fishing, auto mechanics, farming interests. *Society Aff:* NSPE, LES, ASME, Am R/W, AAA.

Mondolfo, Lucio F
Home: RR1 Box 432, Clinton, NY 13323
Position: Adjunct Prof *Employer:* Rensselaer Polyt Inst. *Education:* Dr Engr/Mech & Ind/Polytechnic Milano; BS/Engrg/Univ, of Bologna *Born:* 8/20/10 educated in Italy. Immigrated to USA in 1939, citizen 1945. Worked 20 yrs in Non Ferrous metals industry and for 20 yrs prof and dept head of Met Engr dept at IL Inst of Tech. Currently Adjunct Prof at RPI and consultant to various metallugical companies, in USA, Europe, S America. Author of 5 books and some 50 papers, mostly on Solidification and Aluminum Alloys. *Society Aff:* AAAS, AIME, ASM, ΣΞ

Monette, Robert W
Business: 143 Main St, PO Box 174, Biloxi, MS 39533
Position: Pres. *Employer:* Gulf South Engrs, Inc. *Education:* BS/Biological engr/MS State Univ. *Born:* 3/29/49. Native of Meridian, MS. After grad from MSU served with MS Air and Water Pollution Control Commission as Regional Engr for S MS Dist. In this capacity, directed all air and water pollution control activity in the twenty-two county area. Supervised cleanup operations of oil spills and emergency disposal of hazardous mtls. Since entering consulting engg have performed 201 Facility Planning in 12 municipalities. Served as "Project Engineer" on the US EPA 208 Water Quality Mgt Study of the MS Gulf Coast. Have addressed regional and natl conf of the success of this prog. Hobby: Long Distance Runner. *Society Aff:* NSPE, MES, WPCF, MWPCA.

Money, Lloyd J
Home: 404 New Jersey Ave, SE, Washington, DC 20003
Position: Chief Engr, Information Sys *Employer:* TRW, Inc *Education:*

Money, Lloyd J (Continued)
PhD/EE/Purdue Univ; BS/EE/Purdue Univ; BS/EE/Rice Univ *Born:* Sept 1920. Native of Seminole, Oklahoma. Served with US Navy 1942-45. With Hughes Aircraft Co for 17 years serving in several positions, including Assoc Dir, Advanced Progs, specializing in air-borne fire control systems design. With US Dept of Transportation, 1971-84 respon for tech progs in energy conservation & advanced transportation sys design & analysis. Have also served as Dir, Office of Univ Res & Assoc Administrator for R&D, Urban Mass Transportation Administration. Presently respon for new and advanced designs of information processing sys including data base machines, etc. Mbr IEEE, NY Acad of Sciences & AAAS. *Society Aff:* IEEE, AAAS, NYAS.

Mongan, David G
Home: 321 Holly Hill Rd, Reisterstown, MD 21136
Position: Senior VP *Employer:* Kidde Consultants, Inc. *Education:* MS/Civ Engrg/Univ of MD; MBA/Finance/Loyolla College; BSCE/Civ Engrg/Univ of MD. *Born:* 7/30/47. D G Mongan PE is a native of MD & joined Kidde Consultants in 1973 as traffic & transportation engr. He is currently Senior VP for Mktg & responsible for all marketing functions. He is a Past-Pres of Baltimore-Wash sec of the Inst of Transportation Engrs; Past Secretary/Treasurer of the MD Section ASCE, received the Young Engr of the Yr Award in 1978 from the Baltimore Post SAME and in 1981 from the Maryland Society of Prof Engrs. He has a BSCE and a MS from Univ of Md and MBA from Loyola. *Society Aff:* ITE, ASCE, NSPE, TRB, SAME.

Monical, R Duane
Business: 5820 Massachusetts Av, Indianapolis, IN 46218
Position: President. *Employer:* Monical Associates, Inc. *Education:* MSCE/Structures/Purdue; BSCE/Construction/Purdue. *Born:* April 30, 1925. USNR 1943-46; US Army Reserve 1948-53. Jr Engr with NYCRR 1949-51; Jr Engr with Southern Railways 1951; Structural Engr with Pierce & Gruber, Structural Engrs, Indianapolis 1952-54; Co-founder of Monical & Wolverton, Cons Structural Engrs, Indianapolis, 1954; Founder of R Duane Monical & Assocs 1964, now Monical Assocs, Inc; VP of Zurwelle-Whittaker, Inc, Miami Beach, & VP & coFounder of Zurwelle-Whittaker, Monical & Assocs, Inc, Davie, Fla, both cons engrg & land surveying firms. P Chmn of Indiana Administrative Bldg Council 1973-74; Past VP of Cons Engrg Council/US 1971-73. Mbr ASCE, NSPE, CSI, ACEC, PCI, Amer Arbitration Assn, Indianapolis Scientific & Engrg Foundation, & Meridian Street Preservation Comm. Reg PE in 21 states. Reg PE & Land Surveyor in Indiana. PPres of Amer Consulting Engrs Council 1978-79, Past Chrmn of Indiana State Bd of Registration for Prof Engrs & Land Surveyors, Pres of Central IN Sec of IN Soc of Prof Engrs 1984. P.Pres of Conslt Engrs of IN, 1968-69. *Society Aff:* ASCE, NSPE, ACEC, CSI, PCI, AAA.

Monismith, Carl L
Business: Department of Civil Engineering, Rm 115 McLaughlin Hall, University of California, Berkeley, Berkeley, CA 94720
Position: The Robert Horonjett prof of Civil Engrg *Employer:* Univ of CA-Berkeley *Education:* MS/CE/Univ of CA-Berkeley; BS/CE/Univ of CA-Berkeley *Born:* 10/23/26 Since 1951 mbr of Dept of CE, Univ of CA, Berkeley; served as Dept Chrmn 1974-1979. Served as Pres of AAPT in 1968; Pres of San Francisco Section, ASCE 1979-80; Chrmn of the Pavement Design Section, TRB 1973-1979. Serves or has served as consultant to a number of organizations including: US Army Corps of Engrs, Vicksburg, MS; Chevron Research Corp; Woodward Clyde Consultants; the Asphalt Inst, and Bechtel Corp. Has published approximately 175 papers and reports in his areas of research which include pavement design and rehabilitation and asphalt paving technology. Has received the following awards: State of the Art (ASCE, 1978), K.B. Woods Award (TRB, 1972); W.J. Emmons Award (AAPT, 1961, 1965, 1985) Rupurt Myers Medal Univ of New South Wales, Aust. (1976). Mbr, Natl Academy of Engrg (1980); Registered Professional Engr, CA. *Society Aff:* ASCE, TRB, AAPT, ASTM, ASEE, AAAS

Monke, Edwin J
Business: Agr Engg Dept, Purdue Univ, W. Lafayette, IN 47907
Position: Prof. *Employer:* Purdue Univ. *Education:* PhD/Civ Engrg/Univ of Il; MS/Agr Engrg/Univ of Il; BS/Agr Engrg/Univ of Il *Born:* 6/7/25. Native of Harvel, IL. Served with US Army (1943-45) as training cadre. On agri engg staff at Univ of IL and Purdue where he is Prof in the soil and water resources area with respon in teaching and.research. Was div chmn in both SCSA (1978) and ASAE (1978-79). Fellow ASAE. Reg PE in IN. Experience with overseas projects and working with federal agencies. Enjoys classical music and gardening. *Society Aff:* ASAE, NSPE, SCSA, AGU, ASEE

Monroe, Edward W
Business: 875 Greentree Rd, Pittsburgh, PA 15220
Position: Manager, Pollution Control Div. *Employer:* Gannett Fleming, et al. *Education:* MS/Sanitary Engg/Univ of MI; BS/Civil Engg/Univ of Pittsburgh. *Born:* July 21, 1937. BSCE U of Pittsburgh; MSCE U of Mich. Native of Pittsburgh, Pa. Sanitary Engr with Penn Dept of Health, Div of Sanitary Engrg 1960-65. With Public Works Dept of a large engrg firm & joined Gannett Feming Corddry & Carpenter, Inc in 1966. Assumed current respons as Mgr of the firm's Pittsburgh Office, Pollution Control Div in 1968. Became Partner in the firm in 1973. Reg PE in Penn & a Diplomate in AAEE. *Society Aff:* ASCE, WPCE, AAEE.

Monroe, Paul S
Business: 27 Lenape Trail, Chatham, NJ 07928
Position: Consultant (Self-employed) *Education:* BS/ME/Rutgers Univ. *Born:* Dec 10, 1916 Youngstown, Ohio. Draftsman, DeLaval Steam Turbine Co 1935- 36; Designer, Merck & Co 1940, Sr Engr, 1941-44; Supervisory Engr Kellex Corp, New York 1944-45; Operations Engr Oak Ridge Natl Lab 1945; Asst Chief Engr Pfaulder Co 1945-46, Chief Engr 1946-50; Chief Proj Engr Vitro Corp of Amer 1950-53; Dir Engrg Scientific Design Co, Inc, a Halcon Internatl, Inc subsidiary 1953-63, VP 1957-63; Exec VP SD Plants, Inc 1958-63; VP Halcon Internatl, Inc 1963-81. Retired 1981.

Monsaroff, Adolph
Home: 4 Merton Crescent, Hampstead Quebec, Canada H3X 1L6
Position: Director. *Employer:* Office of Industrial Research. *Education:* BASc/Chem Engg/Univ of Toronto. *Born:* Feb 12, 1912. Plant Supt Mallinckrodt Chem Works, Toronto 1936-44; Plant Mgr Monsanto Canada Ltd, Montreal 1945-50, VP 1951, Exec VP & Dir 1959; VP & Managing Dir Domtar Chemicals Ltd 1964-67, Pres 1968-76. Pres Soc of Plastics Indus Canada Ltd 1960-62; Chmn of the Bd Chem Inst of Canada 1954-55; Pres Chem Ins of Canada 1974-75; Mbr Bd/Dir Mfg Chemists Assn 1974-77. Fellow Chem Inst of Canada; Fellow AIChE; Mbr Order of Engrs of Quebec; VP Domtar LTD 1977. Dir of Office of Industrial Res 1978-date. *Society Aff:* CIC, AIChE, AAAS.

Monsees, Melford E
Home: 8510 High Dr, Leawood, KS 66206
Position: Coordinator-Grad Engg. *Employer:* Univ of MO-Columbia. *Education:* MS/Engrg Mgmt/Univ of MO-Columbia; Grad Studies/Civil Engrg & Engrg Mgmt/MIT; ATTD/Col/CO State Univ; ATTD/Sci/KS City Univ; ATTD/Sci/Central Wesleyan Coll *Born:* 03/12/08 Over 35 yrs in engg and exec positions with US Corps of Engrs. Now, Coordinator of MS Degree progs for Civil, Electrical, Ind and Mech engrs, and Assoc of Richard Muther, an intl consulting engg firm. Recipient of awards, including Secretary of Army Fellowship for res and study at MA Inst of Tech. Meritouous Civilian Service from Washington, DC for dev of electronic computer sys for engg. Mbr, MIT Educational Council, author of papers and articles for Civil Engg, Consulting Engr, Prof Engr and Engg Educ magazines; also authored portions of a textbook and a book on engg mgt. *Society Aff:* ASCE, NSPE, ASEE, XE, AIC

Monseth, Ingwald T
Home: 320 Oak Terrace, Moberly, MO 65270
Position: Southwestern Region Engrg Manager. *Employer:* Westinghouse Electric Corp. *Education:* BSEE/Power/Univ of MN. *Born:* May 1902. Harvard U Advanced Mgmt Prog 1952. Westinghouse Award of Merit 1939; Eta Kappa Nu Award, U of Mo, Rolla 1963. Mo State Bd of Registration. IEEE Activities: Fellow 1939; Chmn St Louis Sect 1942; Natl VP & Bd/Dir 1958-60. Chmn of Southwest Natl Dist Meeting, St Louis 1952. Co-author of book 'Relay Systems', McGraw-Hill Publisher & author of many tech articles. Reg PE in Mo. *Society Aff:* IEEE.

Monson, James E
Business: Harvey Mudd College, Engr Dept, Claremont, CA 91711
Position: Prof of Engineering *Employer:* Harvey Mudd College. *Education:* PhD/EE/Stanford Univ; MS/EE/Stanford Univ; BS/EE/Stanford Univ. *Born:* 20 June 1932 Native of Oakland, CA. Mbr Technical Staff, Bell Tel Labs, Hewlett-Packard Co Natl Sci Fdn Fellow, Ford Fdn Resident in Engg Practice, Western Elec Co, Visiting Prof, Trinity College, Dublin, Fulbright Res grantee, Univ Velko Vlahovich, Titograd, Yugoslavia, Japan Soc for the Promotion of Sci Fellow, Tohoku Univ, Sendai, Japan. Reg PE, CA. Mbr, Phi Beta Kappa, Sigma Xi. Pres, governing bd, Claremont Unified Sch Dist. *Society Aff:* IEEE, ASEE, AAUP.

Montecki, Carl R
5544 Central Ave, St. Petersburg, FL 33707
Position: Pres *Employer:* Glace & Radcliffe, Inc *Education:* BS/ME/MIT *Born:* 04/05/19 Born in New Bedord, MA. Parents, Emma (Moss) and James. Married Oct 4, 1941 to Alice Isherwood, New Bedford. Daughter Alyson born 1951. Served to Major, AUS 1941-46. Staff Engr Tuscarora Oil Co, Harrisburg, PA 1946-53. Sr Engr, Glace & Glace, Harrisburg, 1953-57. Exec Vp to Pres Glace Engrg Corp, St Petersburg, FL, 1957-68. Pres Glace & Radcliffe, Inc, St Petersburg, FL, 1968 to date. Reg PE- FL, PA, CT, LA. Prof Land Surveyor, FL. Chrmn, Planning & Zoning Ctte, North Redington Beach, Fl 1972 to date. *Society Aff:* TBΠ, NSPE, WPCF, APWA

Montecki, Carl R
Business: 5000 US Highway 1 N, PO Box 607, Ormond Beach, FL 32074
Position: Dir of Marketing *Employer:* Briley, Wild & Assoc, Inc. *Education:* BS/CE/Bradley Univ *Born:* 8/31/34 Native of Chicago, IL. Was involved in land development construction projects in Central US and FL as an Engr and Development Mgr; research engineer and planner for Chicago Dept of Transportation; City Commissioner in IL; mbr Volusia County FL Planning Bd; FL Inst of Consulting Engrs Legislative Ctte; is responsible for program development, capital projects, contracts, consulting contracts for Briley, Wild & Assoc, Inc since 1974, specializing in water, wastewater, industrial waste treatment, drainage and land development. Enjoys water sports and motorcycles. *Society Aff:* NSPE, FES, APWA, ITE, SMPS, FICE

Monteith, Larry K
Business: Box 7901, Raleigh, NC 27695-7901
Position: Dean, Sch of Engg. *Employer:* NC State Univ. *Education:* BS/EE/NC State Univ; MS/EE/Duke Univ; PhD/EE/Duke Univ. *Born:* 8/17/33. Native of Bryson City, NC, Hd of NCSU Dept of Elec Engg, 1974-1978. From adjunct Asst Prof to Full Prof, 1965-present. Technical staff, Bell Telephone Labs, 1960-62, Res staff, Res Triangle Inst, 1962-64; RTI Sr Scientist, 1964-67; RTI Group Leader, 1967-68. Teaches on undergrad & grad levels in solid state electronics. Res activities include electronic properties of solids, solid state sensors & radiation effects. Author of 20 published technical articles & 30 other papers & reports. *Society Aff:* IEEE,ΠΚΠ, NSPE ASEE, ΣΞ, HKN, TBΠ.

Montemarano, Joseph A
Business: 1040 Hempstead Tnpk, Franklin Square, NY 11010
Position: Owner - Principal. *Employer:* Montemarano Associates. *Education:* BS/Civil Engrg/New York Univ; ADS/Structural Eng/State Univ NY *Born:* June 6, 1932. Native of Brooklyn. Reg PE in NY, NJ, Conn & Mass. Cert Cost Engr, Natl. Former Commissioner, Nassau County Planning Bd 1964-66; Panel Mbr, Amer Arbitration Assn; Commissioner of Appraisals, Supreme Court of NY; Civil Engr, Construction Div, Bd of Educ City of NY 1956-63; since 1963, private consulting Engineering practice. Instructor, Adelphi U and Hofstraun Basic and Advanced Cost Estimating; Lectured on Construction Costs before professional groups. Community Service Award 1967. Related Fields - Architecture, Environmental, Forensic Engineering. Diplomate - Natl Acad of Forensic Engrs, Diplomate - Amer Acad of Environ Engrs, Commissioner - Town of Hempstead Landmarks Preservation Commission. *Society Aff:* NSPE, ACEC, ASCE, AACE, AAEE, NSAE, NAFF

Montgomery, Arnold H
Business: 3300 Lexington Rd, SE, Winston-Salem, NC 27102
Position: Director of Engineering. *Employer:* Western Electric Co, Inc. *Born:* March 1928 in Denver, Colo. BSEE from U of Colo. held engrg positions with Mountain State Telephone & Telegraph & Boeing Airplane Co. Joined Western Elec in 1952 as Field Engr in Defense Activities working on radar, bombing - navigational airborne systems, Titan ICMB installations, etc. Progressed to level of Dir - Patent Licensing, & assumed present assignment as Dir of Engrg, North Carolina Works, in 1972. Mbr of IEEE, current Pres of Computer Aided Mfg - Internatl, Inc, & on the Bd/Dirs for NC State U Engrg Foundation. Principal hobby is home repair.

Montgomery, Donald J
Home: 2391 Shawnee Tr, Okemos, MI 48864
Position: Res Prof. *Employer:* MI State Univ. *Education:* PhD/Theoretical physics/Univ of Cincinnati; ChE/Organic option/Univ of Cincinnati. *Born:* 6/11/17. Instructor in Elec Engg, Univ Cin, 42-43; physicist, NACA, 43-45; asst prof physics, Princeton Univ, 45-46; sci liaison officer, ONR, London & res fellow, Univ Manchester, 47-48; chief, spec prob branch, BRL, APG, MD; head, gen'l phys sec, Textile Res Inst, Princeton, NJ, 50-53; assoc, prof, prof, res prof, physics; prof engg res; prof & chrmn, res prof, metallurgy, mech & mtls science, MI State Univ, 1953-present; Fulbright lect, physics, & Guggenheim Fellow, Univ Grenoble, 59-60; spec asst to dir, Off Grants & Res Contracts, NASA, Washington, DC, 64-65; visiting res physicist, Space Science Lab, Univ CA (Berkeley), 65-66; Fulbright sr researcher and visiting prof, macroeconomics, Univ Augsburg, & mbr, Intl Inst for Empirical Socioeconomics, Augsburg, 74-75; res and teaching in materials sci and chem physics; res and teaching in tech and society, engg and public policy. *Society Aff:* ANS, APS, AAM, ASEE, AAPSS, PSO, NYAS, AAAS

Montgomery, Douglas C
Home: 1415 Withmere Lane, Dunwoody, GA 30338
Position: Prof of Indus & Sys Engg. *Employer:* Georgia Inst of Technology. *Education:* BSIE/Indust Engg/VPI; MS/Indust Engg/VPI; PhD/Indust Engg/VPI. *Born:* June 1943. Instructor in IEOR at VPI 1965-69. Full-time and consulting exper in chem & mfg industries. With Ga Tech since 1969, Prof since 1978. Fields of interest include engrg statistics, quality control, & indus applications of operations res. Mbr IIE, ASA, ASQC, ORSA, & TIMS; Natl Dir Production Planning & Control Div IIE, 1974-75, Mbr Publications Mgmt Bd ASQC, 1978-present. Chairman Statistics Division ASQC 1981 Mbr of the Editoral Bd of *IIE Transactions, Journal of Quality Technology* and *RAIRO-Operations Research.* Mbr Alpha Pi Mu, Phi Kappa Phi, Sigma Xi. Author/Coauthor of 4 textbooks & numerous papers in various engg and statistics journals. *Society Aff:* IIE, ASA, ASQC, ORSA, TIMS.

Montgomery, G Franklin
Home: 2806 Kanawha St, NW, Washington, DC 20015
Position: Consulting Engineer. *Employer:* Self-employed. *Education:* BSEE/Elec Engg/Purdue Univ. *Born:* May 1, 1921. Educated in Washington, DC; Radio engr at Naval Res Lab 1941-44. Service in Army Signal Corps 1944-46. Electronics engr at Natl Bureau of Standards 1946-78. Chief of Instrumentation, Measurement Engrg, & Product Engrg Divs 1960-75; Sr Engrg Advisor to Dir, Center for Consumer Product Tech 1975-78. Fellow IEEE, Mbr AES, Amer Radio Relay League, Eta

Montgomery, G Franklin (Continued)
Kappa Nu, Tau Beta Pi, Sigma Pi Sigma, Sigma Xi. Reg PE in DC. 40 prof publs, 7 patents, several poems. *Society Aff:* IEEE, AES, ARRL.

Montgomery, Max C
Home: 1114 O'Callaghan Dr, Sparks, NV 89431
Position: Chief of Plant Operations *Employer:* St. of NV. Dept of Prisons. *Education:* BS/CE/Univ of WY; AA/Engrg/Sheridan Jr. Coll *Born:* 07/26/37 After graduation from the Univ of WY, I worked for the Wyoming Highway Dept until 1963 as a Highway and Traffic Engr. From 1963-83, I have been employed by the USDA - Forest Service in various positions in their Engrg organization through CA, ID, & NV. At the present, I am Chief of Plant Operation for the State of NV. Dept. of Prison in charge of construction & maintenence and coordination with other governmental agencies. I also manage a small Conslt firm which specializes in Land Mgmt planning, Water Right and Enviornmental designs. *Society Aff:* NSPE

Montgomery, Ronny D
Business: PO Box 3516, Baton Rouge, LA 70821
Position: Chemical Engineer. *Employer:* Self -- Lisevin & Assocs. *Born:* Sept 17, 1934 Flat Rock, Ill. Educated at Lincoln High School, Vincennes U & U of Tulsa (BSChE 1959); graduate work at La State U. Elected to Pi Epsilon Tau. Mbr AIChE, NSPE, CEC & La Engrg Soc. Reg PE in La. Started own engrg practice in June of 1973. Enjoy reading, theater, handball & hunting.

Moody, Arthur
Home: 600 E. Cathedral Rd., Apt A314, Phila, PA 19128
Position: Retired *Education:* M.E./Mech Eng/Princeton; A.B./Physics/Princeton *Born:* 11/08/12 Spent most of career in Product Development and Research, especially on Turbomachinery. Retired as Chief Engineer of the Trane Co., now a Div. of American Standard. Served as Chairman of the Fluids Engrg. & Process Engrg. Div. of ASME. Previously involved in compressor and gas turbine research and development, at De Laval and Elliott Cos. Awarded 7 U.S. patents. Author of technical papers. Taught engrg. subjects at U. of Del, Princeton, and U. Wis. River Falls. Honored by ASME and Wis. Soc. Prof. Engrs. *Society Aff:* ASME, NSPE, ASEE,SIGΞ

Moody, Frederick J
Business: 175 Curtner Ave, San Jose, CA 95125
Position: Engineering Consultant *Employer:* GE Co - Nuclear Energy Division. *Education:* PhD/Mech Engrg/Stanford; MSME/Mech Engrg/Stanford; BSME/Mech Engrg/Univ of CO *Born:* April 1935. 20 years with GE Nuclear Energy Div. Assignments included formulating analytical mehtods for analysis of safety systems & design of containment structure. Contribs include 2-phase flow model for predicting both reactor blowdown from hypothetical pipe rupture & resulting forces. Alfred Noble award 1967. ASME George Westinghouse Gold Medal, 1980. 14th Edition 'Who's Who in the West', Marquis, Inc. Part time instructor, GE tech courses; part time lecturer, Mech Engrg, San Jose State U. Co-Author, *The Thermanl Hydraulics of a Boiling Water Nuclear Reactor*, Amer Nuclear Soc Monograph, 1977. Enjoys teaching adult Sunday school, & creative writing. ASME Fellow, 1982. *Society Aff:* ASME

Moody, Herbert R
Home: 5 Blackwell Pl, Philadelphia, PA 19147
Position: Retired (Self-employed) *Education:* SB/Chem Engg/MIT. *Born:* Feb 12, 1919. US Army Officer from 1941-46; separated as Major. Employed by Rohm & Haas Co from 1946- . Assignments: 1946-54 Dev Engrg Supr; 1954-61 Res Supr; 1961-65 Asst Dir of Res; 1965-70 Asst Plant Mgr; 1970-77 Pres & Genl Mgr of Micromedic Systems, Inc, formed as a joint venture between Battell Dev Corp & Rohn & Haas Co. In addition to general automated analytical instrumentation for biomedical labs, we have just introduced the first totally automatic radio- immune assay sys available commercially, a revolutionary new tool for medicine. Mbr ACS, AIChE, Reg PE in Pa. Patents in chemical processing & a few publs. m. Lois Mae Tomhave; 2 children: Carl (31), Irene (29). Hobbies: golf, painting, music. Listed in Amer Men of Sci, & Who's Who in the East 1977-79 Dir Hazard Control-Rohm & Haas Co 1979-Retired-Self-employed (Consultant). *Society Aff:* ACS.

Moody, Lamon L, Jr
2845 Ray Weiland Dr, Baker, LA 70714
Position: Pres & Chmn Bd of Directors. *Employer:* Dyer & Moody, Inc, Cons Engrs. *Education:* BS/Civil Engg/Univ of Southwestern LA. *Born:* Nov 1924. Native of Bogalusa, La. US Marines South Pacific WWII. Proj Engr & Projs Mgr Texas Co in NY & W Africa. Ch Engr Kaiser Aluminum & Chem Corp. Owner Dyer & Moody, Inc, Cons Engrs specializing in petro-chemical plant & public works projs. Reg PE in La, Texas, Arkansas & Miss. Reg Land Surveyor in La. and Texas. Charles Kerr Award & other awards of La Engrg Soc. Award for Excellency Amer Congress on Surveying & Mapping. Land Surveyor of Year Award from La Land Surveyors Assn. State Chmn NSPE/PEPP. Pres LLSA. Business Leader of Year, Pres of Baker Chamber of Commerce, Exec Committee of LA Good Rds & Transportation Chrmn LesLegislative Committee. President La. Engrg. Soc. 1981-83; Received the 1981 A. B. Patterson Award for Engineer in Management and 1985 Leo Odom Award For Services To The Profession from the La. Engrg Soc. *Society Aff:* LES, NSPE, ASCE, ACSM, ACEC, CEC/L, LSPE

Moody, Willis E, Jr
Business: Ceramic Engg, Atlanta, GA 30332
Position: Prof. *Employer:* GA Inst of Tech. *Education:* PhD/Ceramic Engg/NC State Univ; JD/Law/Woodrow Wilson College of Law; MS/Ceramic Engg/NC State Univ; BS/Ceramic Engg/NC State Univ. *Born:* 3/30/24. WWII service with Army Air Corp, 1943-1946, as navigator. Ceramic Engr at Elec Anto-Lite Spark Plug Div, 1949-1950, & Lab Equip Corp, 1950-1951. Instr of Ceramic Engg & Met at NC State Univ, 1951-1956. Prof of Ceramic Engg at GA Inst of Tech, 1956-present. Res participant at Oak Ridge Natl Lab, summer, 1954 & 1955. Ceramic Educational Council Pres, 1963: Am Ceramic Soc Trustee 1965-1968; Clay Minerals Soc Councilor 1969-1971, & Chrmn Annual Meeting, 1963, 1970, 1979; Chrmn, Mtls Div, ASEE 1971; Natl Inst of Ceramic Engrs Pres, 1980; State Bar of GA, Amer Bar Assoc, AAES Governor 1979-1982. Special interest in technical ceramics & clays. *Society Aff:* NICE, ACerS, CMS, ABA.

Mooney, Malcolm T
Business: 150 Hayes St 6th Fl, San Francisco, CA 94101
Position: Resident Contruction Manager *Employer:* Geo. Hyman Construction Company *Education:* MS/Mgmt/Naval PG Sch Monterey, CA; BS/Civil Eng/Tufts Univ. *Born:* May 1930. Civil Engr Corps, US Navy since 1952. Career includes tours in Seabees, public works administration, construction prog mgmt & execution. Awarded Moreell Medal SAME 1968 for dev of Level of Effort Construction Mgmt Sys for CPAF contract in RVN. Recent duties were as Resident Officer in Charge of Construction, San Diego 1969-71, Exec Officer Construction Battalion Center Port Hueneme 1971-74, Officer in Charge of Construction Guam & Marianas 1974-76 & Commanding Officer, Navy Public Works Center, San Francisco Bay providing maintenance, repair, alterations, transportation, utilities & housing support to major military activities in Bay area $50 mill/year volume. Pres of Guam Post SAME 1975 & SFPost 1978. Reg PE in VT & CA. Presently Asst Mgr Construction Operations for DeLeuw-Greeley-Hyman JV who are Const Mgr for SFran Clean Water Program. *Society Aff:* SAME, APWA.

Moor, William C
Business: Coll Engrg & Appl Scis, Tempe, AZ 85287
Position: Associate Professor. *Employer:* Arizona State University. *Education:* BSIE/Ind Engg/WA Univ (St Louis); MS/Ind Engg/WA Univ (St Louis); PhD/Ind Engg/Northwestern Univ. *Born:* 1/17/41. Native of Collinsville, IL. Associated with AZ State Univ since completing PhD. Currently Assoc Prof with faculty of IE.

Moor, William C (Continued)
Respon for curriculum dev & student evaluation as well as teaching. Has held natl and local office in IIE. Avocations philately & bowling. *Society Aff:* IIE, ASEM.

Moore, Arthur D
Home: 718 Onondaga, Ann Arbor, MI 48104
Position: Writer, lecturer, cons, inventor (Self-employed) *Education:* MS/EE/Univ of MI; BS/EE/Carnegie-Tech *Born:* 1/7/95. Instructor to Full Prof Univ of MI 1916-63; Emeritus 1964. Natl Pres Tau Beta Pi 1924-30. Ann Arbor city Council 1940-57. Contributed to field mapping. Inventor, fluid mappers. Dev of electro- electro- & magnetospherics; new forms of electrostatic generators & effects. 1969 Donald P Eckman Award for distinguished career in education, by Instrument Soc of Amer. Founder, Electrostatic Soc of Amer, 1970. Author, many tech papers; article *Electrostatis*, Scientific Amer. Books: *Fundamentals of Elec Design*; *Electrostatics*; *Invention, Discovery & Creativity*; editor, *Electrostatics & its Applications*. Electrostatic demonstrations given for 100 colleges, universities; also, for Franklin Inst; Natl Bureau Stds; Natl Res Council, Ottawa; Clarendon Lab, Oxford; & Cavendish Lab, Cambridge. Favored pursuit: inventing, writing.

Moore, Gordon E
3065 Bowers Ave, Santa Clara, CA 95051
Position: Chrmn of Bd & Chief Executive Officer. *Employer:* Intel Corporation. *Education:* PhD/CA Inst of tEch; BS/Chemistry/Univ of CA. *Born:* Jan 1929 San Francisco, Ca. Joined tech staff of Applied Physics Lab, Johns Hopkins U in 1953 doing basic res in chemical physics. In 1957, was a co- founder of a Fairchild Semiconductor Corp. Was Mgr of the Engrg Dept through 1959, & Dir of R&D until mid-1968. Co-founded Intel Corp in July 1968 & was Exec VP until April 1975. Pres & CEO of the Corp April 1979-87; CEO/Chrmn 1987- ; Mbr NAE; Fellow IEEE since then. *Society Aff:* NAE, IEEE.

Moore, Harry F
Business: 3637 Columbia Ave, Lancaster, PA 17603
Position: Pres. *Employer:* Emtrol, Inc. *Education:* BS/EE/Tri-State Univ. *Born:* 10/29/23. Native of Williamsport, PA. Fighter Pilot WWII. Power Applications Eng, PA Power & Light Co, 1948-1957. Mgr Williamson Assoc control panel mfg 1957-1961. Emtrol, Inc 1961 to 1981, founder & Pres 1981 to present, Chrmn of Bd of Dirs. Responsible for design and mfg of ind control & instrumentation sys including many innovations in control of mtl handling sys. Formed subsidiary consltg firm, H F Moore & Assoc in 1962, Elec Engr for Mining & Mineral Processing Indus. Expanded practice in 1977 to Mech Engg & Energy Conservation, Commerce & Indus. Moore Engrg Co Inc 1981 to present, founder & Pres. Consltg Engrs on Electrical and Mech Sys for Commerce, Govt, & Indus. Enjoy fishing, boating, and golf *Society Aff:* NSPE, IEEE, AEE.

Moore, J Robert
Business: 188 Jefferson Ave, Memphis, TN 38103
Position: President *Employer:* Harland Bartholomew & Assoc Inc.. *Education:* BS/Civil Engg/Univ of IL. *Born:* July 1924. Navigator, US Air Corps, 1943-45. 1948-50, Asst Superintendent of Highways, Clinton County, Ill. 1950-56, private consulting field, design of engrg projects for municipalities. 1956-66, Proj Engr/Assoc Partner, Harland Bartholomew & Assocs. 1966-79 - Partner, 1979 - Pres & Chief Exec Officer, Harland Bartholomew & Assocs, Inc. Fellow AICE & ASCE. Mbr numerous professional assns. PE in 16 states. Land surveyor in 3 states. Pres, Memphis Branch ASCE 1965-66. Featured Engr of the Year 1973 Memphis Branch ASCE. *Society Aff:* ASCE, NSPE, ACEC, APWA, WPCF, AREA, TRB, AWWA, XE.

Moore, James W
Business: CE Dept, 340 Engrg Building, Fayetteville, AR 72701
Position: Prof *Employer:* Univ of AR *Education:* PhD/Envir Health Engr/Univ of KS; MS/CE/Univ of KS; BS/CE/Univ of KS *Born:* 10/18/38 James W. Moore is currently a Prof of CE at the Univ of AR. Since joining the academic community in 1971, he has been active in the areas of teaching, research and public service. Specializing in Environmental Engrg, he has conducted numerous research projects with both local and national scopes. He has published extensively with some 50 published reports and articles during this period. Dr. Moore has served as a consultant to 80 local and national firms during the past ten yrs. Prior to his current position, Dr. Moore was employed in the water utility and electric power production industries and as a consulting engr in private practice. *Society Aff:* ASCE, ASEE, WPCF, AWWA

Moore, John A
Home: 1227 Broadway, Galveston, TX 77550
Position: Consultant *Education:* BS/ChE/Rose Polytechnic Inst. *Born:* 1913. Tau Beta Pi, Alpha Chi Sigma. With Union Carbide since 1935, mostly in engrg. Fields of competence include heat transfer, fluid mechanics, & pressure vessel design. Fellow AIChE; Mbr ASME. Faculty member, Texas Maritime Acad of Texas A&M U. Charter Mbr & P Pres, Process Heat Exchanger Soc. Exec Ctte, AIChE heat transfer div 1962-64. AIChE equip testing procedures ctte 1955- 59. Appear in 'Extended Surface Heat Transfer'. Author & co-author of several publs in the field of heat transfer. *Society Aff:* AIChE, ASME.

Moore, John B
Dept of Systems Engrg, Research School of Physical Sciences, Australian National Univ, GPO Box 4, CANBERRA ACT 2601 Australia
Position: Professorial Fellow *Employer:* Australian National Univ. *Education:* PhD/EE/Univ of Santa Clara; ME/EE/Univ of Queensland; BE/EE/Univ of Queensland. *Born:* 4/3/41. John B. Moore was born in China in 1941. He received his bachelor & masters degrees in EE in 1963 & 1964 respectively, & his doctorate in EE from the Univ of Santa Clara, CA in 1967. He was appointed Sr Lectr at the Univ of Newcastle in 1967, promoted to Assoc Prof in 1968 & Full Prof (personal chair) in 1973. Since 1982, he has been a Professorial Fellow in the Dept of Systems Engg, Research School of Physical Sciences, Australian National Univ. He has held visiting academic appointments at the Univ of Santa Clara 1968; at the Univ of Maryland 1970; at Colorado State Univ & Imperial College 1974; at the Univ of California, Davis, 1977; the Univ of Washington, Seattle, 1981; at Cambridge Univ 1985, & at the Univ of Singapore in 1985. He was a Visiting Springer Prof in the Dept of Mechanical Engg, Univ of California, Berkeley 1987. He has spent periods in industry as a design engineer & as a consultant. Dr Moore's current research is in control & communication systems. He is co-author with Brian Anderson of two books: *Linear Optimal Control* (Prentice-Hall, 1971) & *Optimal Filtering* (Prentice-Hall, 1979). He is a Fellow of the Australian Academy of Technological Sciences, and a Fellow of the IEEE. *Society Aff:* IEEE, ATS.

Moore, John L
Home: 3891 Anderson Rd, Gibsonia, PA 15044
Position: VP and Petroleum Engr *Employer:* Mellon Bank *Education:* MS/IE/Univ of Pittsburgh; MS/Energy Resources/Univ of Pittsburgh; BS/Petro & Nat Gas Engr/Penn State U *Born:* 12/20/34 Began as petroleum engr in the Rocky Mountains for Mountain Fuel Supply Co. Worked in Gulf offshore drilling operations for Conoco. Returned to his native city of Pittsburgh to become Superintendent of Production and Storage for a Consolidated Natural Gas subsidiary, & Sys Chief Production Engr for the CNG Service Co. In 1983 joined Mellon Bank, N.A., Corporate Banking as VP & Petroleum Engr in the Energy Div. Advisory mbr to the Gas Research Inst Ctte on Unconventional Gas. Chrmn of the Devonian Shale Task Group for the Natl Petroleum Council's study on Unconventional Gas Sources. Registered professional engr, inducted into Omega Rho, Intl Engrg Honor Society in 1980, SPE Natl Dir (Region IX), 1982-85. *Society Aff:* SPE of AIME, IOGA, POGAM, OOGA

Moore, John R
Business: 700 Royal Oaks Dr, Monrovia, CA 91016
Position: V.P. Bus Dev *Employer:* Tactical & Electronic Systems Group, Northrop Corp. *Education:* UCLA Executive Program, 1956, 1957 *Born:* July 5, 1916 St

Moore, John R (Continued)
Louis, Mo. BS 1937, Washington U; DSc (Hon) 1963, West Coast U. Fellow Award from IEEE 'For his contribs to field of automatic inertial navigation systems.' Other awards: Outstanding Mech Engrg Graduate, Washington U, 1960; Meritorious Public Service Citation, MS Dept of Navy, 1961; Thurlow Award, Inst of Navigation, 1962; Distinguished Alumni Award, Washington U, 1964; Outstanding Achievement Award, IIE, 1967 Nat'l Academy of Engrg. 1978. Other experience: associated with Norht Amer Aviation, Inc, & later with the merged co of North Amer Rockwell Corp, from 1948-70. Held offices of Pres, Autonetics Div; Exec VP, North Amer Aviation; Pres, Aerospace & Systems Group, North Amer Rockwell; Corp VP, Aerospace Activities, McDonnell Douglas Corp. 1970-1978. Held offices of VP subsidiaries and President, Actron Div. Consultant to Northrop, TRW, Varian Assoc. and Teledyne, 1978- 1979. Northrop Corp., 1979-present. Asst to Sr. VP Advanced Projects, VP Business Dev, Tactical & Electronic Systems Grp. *Society Aff:* IEEE Fellow, AIAA Assoc Fellow, NAE

Moore, Kenneth L
Business: 350 Hanley Industrial Court, St Louis, MO 63144
Position: Sr. VP, Corp Dev *Employer:* Carboline Co. *Education:* MBA/Mktg/Univ of DE; BSE/Chem Engrg/Univ of MI; BSE/Met Engrg/Univ of MI. *Born:* 2/6/32. Worked as Process Development Engr, R&D, Atlantic Refining Co for 3 years. Joined Tidewater Oil Co (now Getty) as the corrosion engr, & became Refinery Tech Group Leader. Started with Sun Oil Co in 1966 as Proj Mgr in Commercial Dev, & concurrently began MBA prog. Successively became Mgr New Prod Planning, Mgr Petrochemical Dev, Mgr Marketing Planning & Economics, Midwestern Marketing Region Mgr, Sales VP, and Pres of Sun Tech (the Res, Dev and Engrg arm of Sun). Transfered to Carboline Co (a subsidiary of Sun Co.) in 1984 as Sr V.P., Corp Dev. Amateur cabinetmaker, who enjoys making antique reproductions. *Society Aff:* AIChE, TBΠ.

Moore, Noel E
Business: Rose-Hulman Institute of Technology, 5500 Wabash Ave, Terre Haute, IN 47803
Position: Prof and Chairman of Chem Engrg *Employer:* Rose-Hulman Inst of Technology *Education:* PhD/ChE/Purdue Univ; SM/ChE/MIT; BS/ChE/Purdue Univ *Born:* 12/23/34 A native of Van Wert, OH. Served in US Army (Chem Intelligence) 1958-1960. Faculty, Univ of KY 1964-1968. Faculty, Rose-Hulman Inst of Tech 1968-present. Dir, Vigo County Air Pollution Control Agency 1969-1972. Mbr, Vigo County Air Pollution Control Bd 1980-1982. Dir, Custom Chem Lab 1979-1982. Registered Professional Engr, IN Conslt (full-time) E.I. duPont 1982-1983.. *Society Aff:* AIChE, ASEE, APCA, ISA

Moore, Peggy B
Home: #1 Neumann Way N43, Cincinnati, OH 45215
Position: Prog Mgr. *Employer:* GE. Aircraft Engins *Education:* MS/Mgt/Rollins; BS/Chemistry/Univ of AL. *Born:* 2/19/37. Native of Tampa, FL. Mtls & Process Chemist with USAF in Mobile, AL 1960- 67. Consultant Mtls & Process Engr & Supervisor Mtls Lab with McDonnell Douglas Astro Co-Titusville FL 1967-74. Responsible for dev transition of new processes for Army Missile Weapons Systems. With GE Co Aircraft Engine Group - Cincinnati, OH since 1974. managing progs which transition new processes to mfg production. Natl Pres SAMPE 1976-77. *Society Aff:* SAMPE, ACS, ASM.

Moore, Ralph L
Home: 5 Colesbery Dr, New Castle, DE 19720
Position: Sys Consultant. *Employer:* Self *Education:* MS/Mech Engr/Case Western Reserve; MSME/Drexel Univ; BME/Mech Engr/Univ of DE. *Born:* May 1924. Native of Wilmington, Del. Served in US Navy in WWII. Mbr Tau Beta Pi, Phi Kappa Phi, Sigma Xi. Instrument Applications Engr with DuPont, 1950-57. Taught at Case Western Reserve U 1957-59. Sys Consultant specializing in dynamic analysis of chem processes for DuPont since 1959. ISA VP of Dist II, 1966-68; VP of Educ & Res 1975-77. ISA Journal Award for best article 1958, The Dynamic Analysis of Automatic Process Control. ISA Donald P. Eckman Education Award, 1984, ISA Golden Achievement Award, 1986. Elder in Community Fellowship Wilmington, Del. Books: *Basic Lecture Notes & Study Guide: Volume 1, Measurement Fundamentals: Volume 2 Process Analyzers & Recorders*, 3rd ed. *Neutralization of Waste Water By PH Control. Society Aff:* ISA, NSPE.

Moore, Raymond P
Home: 19191 Harvard Apt 307, Irvine, CA 92715
Position: Retired *Education:* Electrical Engr./Electrical/Rensselaer Polytechnic Institute *Born:* 01/23/95 Graduated from Rensselaer 1917 (Sigma Xi) Cadet engineer for Buffalo General Electric Co (now part of Niagara Mohawk). Rose to Chief Mech Engr and retired in 1960. Elected to grade of Fellow of the ASME as a result of inovative design work such as: unit arrangement (one boiler per turbine); powdered coal from units mills; use of bulldozer and carryalls for storing coal; automatic burner lighters (patent); condensate control (patent); fluid ash handling; one turbine driven boiler feed pump per unit. Elected Engineer of the Year 1960, Erie County Ch Prof Engrs. Became consultant for Rogers Engrg Co San Fancisco. *Society Aff:* ASME, IEEE

Moore, Robert C
Business: PO Box 469, Millburn, NJ 07041
Position: President *Employer:* Robert C Moore, Inc, Consulting Engineers *Education:* BS/Civil Engr/Lehigh Univ. *Born:* 7/27/21. Served in Army 1943-46. With Elson T Killam Assoc, Cons Environmental & Hydraulic Engr 1946-1980. Became VP in 1957, Exec VP in 1970, now Pres, Robert C. Moore, Inc, Consulting Engrs. Reg PE in 5 states. Diplomate AAEE, ASCE; Fellow ACEC, ASCE; Greely Award 1971; Pres CEC/NJ 1975; Past Chairman ASTM C-17. VP, ACEC, 1981- . Specialty: valve engg; sludge disposal & treatment. Mbr Rock Spring Club; Pres Memorial Center for Women, West Orange, NJ. *Society Aff:* WPCF, ASTM, AWWA, NSPE, ACEC, TBΠ.

Moore, Robert J, Jr
Business: 1180 Walnut Ave, Chula Vista, CA 92011
Position: Pres *Employer:* RJM Marine Consultants, Inc *Education:* Law Degree// Univ of Guadalajara; MS/IE/Bernadean Univ; MBA/Bus Mgmt/Bernadean Univ; BS/Bus Admin/LAU *Born:* 02/10/32 From Jan 51 thru Dec 73. Tacoma Boatbuilding Co, Inc - Tacoma, WA. Worked at most shipbuilding crafts while attending univ, rising to the position of PV Production; prox 1000 prod employees. 1973 thru 1976 - resident marine conslt to Mexicn Gov't. 1977 to present - founder and pres of: RJM Marine Conslts, Inc, offical marine consltg firm to Astilleros Mexicanos, operators of three shipyards. Seven in house employees, engrg joint ventures at four Mexican ports and two US ports, plus rep at D.F. (Mexico City). Hobbies: golf, tennis, skiing (snow & water), music (jazz). *Society Aff:* NSPE, SNAME, PEPP, AWS, ASTM, CCNSPE, NCIES

Moore, Robert L
Business: Commanding General, US Army Missile Command, Redstone Arsenal, AL 35898
Position: Commanding General *Employer:* US Army Missile Command *Born:* July 1930 Bluefield, Va. BS Architecture, Va Polytechnic Inst; BSCE, Mo School of Mines; MS Business Administration, George Washington U. Commissioned in US Army in 1952. Currently major General serving as Commanding General, US Army Missile Command. Previous command & staff assignments include/Chief of Staff US Army Dev & Res Command, Staff Officer to Deputy Chief of Staff for Logistics; Commander, 87th Engr Batallion, Vietnam; Military Asst to Vice Chief of Staff; Dir of Plans & Analysis, US Army Materiel Command; Buffalo Dist Engr; Exec to Dev Team, Army Materiel Acquisition Review Ctte; Commander, 2nd Infantry Div Support Command, Korea. Mbr 5 advisory Bds of the Internatl Joint Comm; Coastal Engrg Res Bd; Commissioner of Great Lakes Basin Comm & Upper

Moore, Robert L (Continued)
Mississippi River Basin Comm. Great Lakes Regional VP SAME. *Society Aff:* SAME.

Moore, Roland E
Business: San Jacinto & McKinney, PO Box 2521, Houston, TX 77001
Position: VP & Chief Engr. *Employer:* TX Eastern Transmission Corp. *Education:* BS/ME/Univ of AR. *Born:* 5/1/21. Native of Trenton, NJ. Served with the Army Air Corps 1942-1945. Joined TX Eastern Transmission Corp in 1950 & has held increasingly important engg positions, assumed current responsibility as VP & Chief Engr, in 1971. *Society Aff:* AGA, ANSI, ASME, ASTM, NSPE, SGA.

Moore, W Calvin
Home: 360 Tri-Hill Dr, York, PA 17403
Position: Mgmt & Res Cons (Self-employed) *Education:* B.S./Chem Engg/Univ State of NY *Born:* Oct 21, 1910 Oklahoma City, Okla. Sanitary Engr, Dale Mabry Field, Tallahassee, Fla 1942-44; employed by Union Carbide Corp 1944-59. Positions held at latter include Asst Chief Engr Oak Ridge Gaseous Diffusion Plant 1949-53; Asst Plant Supt (Tech) Y-12 Plant, Oak Ridge, Tenn 1954-58. Proj Mgr, Experimental Beryllium Oxide Reactor Prog, General Atomic, San Diego, Ca 1959- 62. VP & Dir of Engrg & Res, York Div, Borg-Warner Corp, York, Pa Oct 1962 until retirement on March 1, 1976. Mbr AIChE, ASHRAE, Associate Member, Int. Inst of Refrigeration. Charter member, ANS Amer Nuclear Soc, Rotary Internatl. Hobbies are music & electronics. m. Erma McKee; 3 children Annalee Mikell, Susan M Jordan, & Lawrence C Moore. Publs include articles on reactor dev, air- conditioning equip, & mgmt. *Society Aff:* AIChE, ANS, ASHRAE, IIR, NYAS, ISES

Moore, Ward F
7555 Eads Avenue, La Jolla, CA 92037
Position: President. *Employer:* Moore Business Assoc, Inc. *Education:* BS/Chem Eng/Cornell Univ; MBA/Bus Adm/Univ of Buffalo. *Born:* 3/24/22. Chronology of Career: VP Mtl Sys Div, Union Carbide Corp, NYC, 1945-1971; Corp Consultant, La Jolla, CA, 1971-72; Pres, Filtrol Corp, 1972-1981, Columbia Cement Corp, Columbus, OH, 1973-1981, Horizon Coal Corp, Zanesville, OH, 1977- 1981, Pres, MBA Inc, 1981-date. Mbr Bd Dir; Mat'l Sci Inc 1971-1974; Kanecki Berylco Inc 1973-1978; Filtrol Corp. 1972-1980; Exotic Mat'ls Inc 1971-1982; Electrometals Inc 1982-1986; Income Savings and Loan Assn 1982-1983; Atec, Inc 1982-1985; Crested, Inc. 1984-1985. Phoenix Data Inc. 1986-date. US Army, Ordnance Dept, AUS 1943-1946 Cornell Club of NY, 1958-1981; San Diego Yacht Club, 1969-date; La Jolla Country Club, 1969-date. La Jolla Beach and Tennis Club 1979-date. Desert Horizons Country Club, 1982-date. *Society Aff:* AIChE, CDA.

Moore, Wendell R
Home: 524 28th St, West Des Moines, IA 50265
Position: Senior Field Representative. *Employer:* American Plywood Assn. *Education:* BS/Agr Engr/IA State Univ. *Born:* March 15, 1925. BS from Iowa State U 1950. Native of Alburnett, Iowa. Served with the Navy in 1945-46. Agricultural Engr for Opekasit, Inc Lebanon, Ohio from 1950-60. Now employed by Amer Plywood Assn as Sr Field Rep in Iowa, Nebraska & SD. Chmn, Structures & Environment Div, ASAE 1975-76. Mbr Tau Beta Pi, Phi Kappa Phi. Reg PE in Iowa. *Society Aff:* ASAE.

Moore, William B, Jr
Home: P.O. Box 1300, Kilmarnock, VA 22482
Position: Retired *Education:* MS/ChE/Univ of Louisville; BS/ChE/Univ of Louisville *Born:* 9/18/24. Midshipman Sch, Camp Endicott, RI 1945; Highest rank Lt Civil Engrg Corps. Ch Engr & Faculty Mbr Univ of Louisville, Speed Scientific Inst of Indus Res 1947-49; Reynolds Metals Co: Tech Serv Engr 1949-50; Asst to VP Prod Dev 1950- 51; Mgr Mktg 1951-57; Mktg Dir 1957-61; Regional Genl Mgr 1961-69; Genl Mgr- Architectural & Bldg Prods Div 1969; VP 1972- ; Chmn-Tilo Co. Reg PE KY. Hon mbr, AIA; Mbr, Sigma Tau; mbr, Omicron Delta Kappa; mbr, Theta Chi Delta; mbr, Kappa Mu Epsilon; mbr, Phi Eta Sigma. *Society Aff:* Hon AIA

Moore, William E, II
Business: 1329 Quarrier St, P. O. Box 753, Charleston, WV 25323
Position: President *Employer:* Ferro Products Corp *Education:* BS/ME/MIT; BS/CE/MIT *Born:* 09/14/25 1952-53 Structural Engr, PBI Industries; 1953- Structural Engr & Pres Ferro Products Corp; 1962-63 First Pres, Charleston Branch, ASCE; 1965-79 Mbr and Secy, WV Bd of Registration for PE; 1969-71 Chairman, Secy Committee, NCEE; 1970-77 Pres Advisory Council, WV Inst of Tech; 1973-75 Chrmn, Natl Engrg Certification Committee, NCEE; 1975-81 Dir, The Madeira Sch, Greenway, VA; 1976- 1982 Visiting Committee, Coll of Engrg WV Univ; 1980- Tau Beta Pi Fellowship Bd; 1981- Committee on Specifications, American Inst of Steel Construction. *Society Aff:* ASCE, AISC, NSPE

Moore, William G
Business: 5601 E Highland Drive, Jonesboro, AR 72401
Position: Div Engrg Mgr *Employer:* FMC Corp *Education:* BS/Agri Engg/Clemson Univ. *Born:* 3/6/31. Genl Mgmt Course, AMA. Native of Olanta, SC with farm background. Sales Engr for Dresser Industries, 1953. Served with Corps of Engrs, US Army Europe 1954-55, discharged as 1st Lt. Employed with Lilliston Corp 1955-80. Promoted to Chief Engr 1957, VP Engrg, & Corporate Dir 1967, then VP for Manufacturing & Engrg 1977. Joined FMC in 1980. Responsible for worldwide engrg activities for the A G machinery div and all manufacturing & materials activities for the Jonesboro Operation. Special interest in value analysis techs, return on asset m gmt practices and genl mgmt. P Pres & Fellow, ASAE. Enjoys reading, flying, music & hunting. *Society Aff:* ASAE, ASM, NSPE

Moore, William J M
Business: National Research Council, Ottawa, Canada K1A 0R8
Position: Prin Res Officer. *Employer:* National Research Council. *Education:* MEng/Elec Engg/McGill Univ; BASc/Elec Engg/Univ of British Columbia. *Born:* May 1924 Edinburgh, UK. With Natl Res Council since 1948, working in automatic control, guided missiles, & since 1958, in Electrical Instrumentation & Measurement, mainly at the power frequencies. Concerned particularly with dev & appl of current ratio techniques to precision measurements using the current comparator. Chmn IEEE Ottawa Sect, 1966-67; Chmn 1969 Electric & Electronic Measurement & Test Instrument Conf; Pres IEEE Group on Instrumentation & Measurement, 1974; Fellow IEEE 1976. Mbr Assn of PE's of Ontario Chairman, Power Systems Instrumentation and Measurements Committee of the Power Engineering Society, 1981-82 Chairman, Morris E. Leeds Award Committee, IEEE, 1981-82. IEEE Centennial Medal 1984, Morris E. Leeds Award Recipient, 1987. *Society Aff:* IEEE.

Moore, William W
Business: 500 Sansome St, San Francisco, CA 94111
Position: Founding Partner. *Employer:* Dames & Moore. *Education:* MS/Civil Engg/ CA Inst of Tech; BS/Civil Engg/CA Inst of Tech. *Born:* Jan 1912. Employed 1935-37 by R V Labarre & F J Converse, west coast soil mechanics consultants. 1938 formed partnership with Trent R Dames for consulting engrg insoil mechanics applied to engrg practice. Served on 3 advisory cttes for BRAB; Advisory Group on Engrg Considerations & Earthquake Sciences for Calif Legislature's Joint Ctte on Seismic Safety; Engrg Criteria Review Bd for San Francisco Bay Conservation & Dev Comm. Over many year, have been involved in engrg for earthquake safety & dev of marginal lands. Enjoy photograhing travels around world & sailing on San Francisco Bay & Pacific Ocean. Dir 1966-69 ASCE, VP 1975-78 ASCE, Pres FIDIC 1970-72, EJC Bd Mbr 1971- 76. P ACEC 1964-65. Received ACEC's Special Bicentennial Amicus Award 1976 Mbr Natl Acad of Engr. Pres Applied Technology Council 1973-74. Chmn Bldg Scismic Safety Council 1979-81. *Society Aff:* ASCE, ACEC, NAE.

Moore, William W, Jr
Business: 1626 Cole Boulevard, Golden, CO 80401
Position: Managing Partner. *Employer:* Dames & Moore. *Education:* MBA Fin/Univ of CA, Berkeley; MS/Civil Engg/Stanford Univ; BS/Civil Engg/Stanford Univ. *Born:* 8/25/39. Native of CA. Structural engg for Standard Oil Co of CA, 1963-65. Soil and foundation engg for Dames & Moore, Chicago, 1967-70, London, 1970-77. Became Managing Partner, London Office in 1974. Led Dames & Moore work on numerous refinery and petrochemical projects in Europe and Middle East including a major industrial dev and 26-berth port. Assumed current position as Managing Partner of Dames & Moores Denver office (125 persons) in 1977. Was project manager of an 18 month multi discipline planning and development study for Sinai Pennisula - lived in Egypt during the project. Enjoy skiing and sailing. *Society Aff:* CEC, ASCE, NSPE, ICE.

Moorhouse, Douglas C
Business: 600 Montgomery St, San Francisco, CA 94111
Position: President and CEO *Employer:* Woodward-Clyde Consultants. *Education:* BS/Civil Engg/UC, Berkeley. *Born:* Feb 1926. Graduate study Arts & Sciences & School of Business Administration, Harvard U. CEO & Principal, Woodward-Moorhouse & Assocs 1962-73; Pres, Woodward-Clyde Consultants 1973- ; CEO & Pres Woodward-Clyde Consultants 1976- . Geotechnical engrg, 20 years exper in site investigation & foundation engrg for indus commercial projs. Reg PE in 13 states & DC. Wesley W Horner Award, ASCE 1969. Mbr ASCE, Amer Geophysical Union, ACEC, ASFE Natl Academy of Engrg. *Society Aff:* ASCE, AGU, ACEC, ASFE

Mooz, William E
Home: PO Box 1714, Santa Monica, CA 90406
Position: President. *Employer:* Met-L-Chek Co. *Education:* SB/Metallurgical Engg/MIT. *Born:* Feb 1929. Various positions in res, engrg, & administration with Titanium Metal Corp of Amer 1950-60; Economic Engr, US Borax 1960-62; Exec VP & Dir, G B Smith Chem Works 1962-63; Sr Cost Analyst, RAND Corp since 1964; Pres & Dir Met-L-Chek Co since 1964. Dir NDT Europa BV since 1964; Dir, Richard Williams Internatl 1967-70; Pres & Dir Colo Resort, Inc 1971-72. Reg PE in Calif. Mbr AIME, ASNT. Active in whitewater river running, hiking, philatelic res. *Society Aff:* AIME, ASNT.

Morabito, Bruno P
Home: 302 Saltmakers Rd, Liverpool, NY 13088
Position: Consultant *Employer:* Retired *Education:* BCE/Civil Engg/Syracuse Univ. *Born:* Feb 10, 1922 Italy. BSC (Tau Beta Pi) & additional mgmt studies, Syracuse U. Consultant to Syracuse/Onondaga County Planning Agency, Retired Group Vice Pres. Aeronca Inc. & General Manager of Environmental Systems Group. Formerly Employed Machinery & Systems Div, Carrier Corp, Syracuse, NY 35 yrs. In respon charge of design HVAC Systems for commercial buildings. Later, Mgr Application Engrg, then Mgr Centrifugal Sales, then Mgr of Marketing respon for establishing application engrg, prod mgmt & pricing policies. Represented Employer in trade assns (Carrier in ARI and TEC and AERONCA in ADC) ASHRAE Fellow, & PPres. Published many papers & received ASHRAE DSA & Wolverine Diamond Key Award 1961. Television Lay Lector for his Church & on Bd at Beaver Meadows Golf & Recreation Club. Enjoys golf, bowling, music & dancing. *Society Aff:* ASHRAE, NMA.

Moraff, Howard
Business: Coll of Vet Medicine, Ithaca, NY 14853
Position: Director, Vet Medicine Comp Facil. *Employer:* Cornell University. *Education:* PhD/Neurophysiology/Cornell Univ; MS/Elec Engg/Columbia Univ; BS/Elec Engg/Columbia Univ; AB/Elec Engg/Columbia Univ. *Born:* Feb 6, 1936 NY. Electronics R&D Officer, USAF, Wright AFB 1958-61 (Capt). Proj Engr, GE Co 1961-64; Dir, Cornell Veterinary Medical Computing Facility 1967- ; Prog Dir, Div of Computer Res, NSF 1973 & 1974. Elected Chmn, ACM Special Interest Group for Biomedical Computing 1976-79. Prof interests: biomedical computing & computer graphics. Mbr Tau Beta Pi, Eta Kappa Nu, Sigma Xi, Phi Kappa Phi, ACM, AAAS. Sr mbr, IEE. Hobbies: kinetic art (several museum exhibits), sailing, folk dancing. *Society Aff:* IEEE, AAAS, ACM.

Moran, H Dana
Business: Solar Energy Research Institute, 1617 Cole Blvd, Golden, CO 80401
Position: Mgr, Indust Aff *Employer:* Solar Energy Res Inst. *Education:* MS/Eng Mgmt/UCLA; BS/Aero Eng/Northrop Univ. *Born:* in Quincy, MA. Served US Air Force 1944-48, Northrop Aircraft (Aerodynamics, sTructures, Prelim Design) 1951-56; Aerospace Industries Assn (Eng Rep West Coast) 1956-61; Weber Aircraft (Prog Dir/Asst to VP) 1961-63. Battelle Columbus Labs (West Coast Operations Manager/Program Dir) 1963-77. With Solar Energy Res Inst (SERI) since 1977 (formerly Deputy Manager for Commercialization, currently Mgr, Industry Affairs of SERI). Asso Fellow, P Dir & Dir of Energy Activities, AIAA. Elected Outstanding Alumnus of Northrop Univ for 1969; Recipient AIAA Distinguished Service Award 1970. Recipient SERI Distinguished Performance Award 1986. Private pilot; active in genl aviation & model railroading. Tau Bet Pi, Fellow, AAAS (1981). *Society Aff:* AIAA, ASES, AAAS, LES, AOPA.

Moran, John H
Home: 9053 Roanoke Rd, Stafford, NY 14143
Position: Consulting Engineer *Education:* BS/EE/Case-Western Reserve *Born:* 09/22/23 Philadelphia, PA. US Navy 1942-1946 Navy V-12 Program. Development Engr Allis-Chalmers Co Power Transformer Dept 1947-1955. Joined Lapp Insulator Co, manufacturer of porcelain insulators 1955. Chief Elec Engr 1973, ret., 1986. Built and developed High Voltage Lab to international prominence. Founding mbr, incorporator and dir of Electrostatic Society of Amer 1974. VP 1976-1980. US mbr Scientific Committee Third Intl Symposium on High Voltage Engrg, Milan, Italy, 1979. Retired Chrmn HVTT Subcommittee IEEE. Ret. US delegate IEC TC 42 (H.V. Testing). Fellow IEEE 1980. Registered PE New York, OH. Lifelong participant in Boy Scouting. *Society Aff:* IEEE, PES, ESA.

Mordell, Donald L
Home: RFD 1, North Troy, VT 05859
Position: Consultant. *Employer:* *. *Born:* 1920. Scholar at St John's Coll, Cambridge; Mechanical Sciences, Tripos 1941. Dev Engr, Rolls Royce Ltd 1941-47; Assoc Prof Mech Engrg, McGill U, Montreal 1947-51; Prof, McGill U 1951-70; Chmn Mech Engrg Dept 1953-57; Dean, Faculty of Engrg 1957-68; President Ryerson Polytechnical Inst, Toronto 1970-74; Counsellor, Corporate of Engrs, Quebec 1965-68; Cons Canadian Aeronautics & Space Inst 1967-70; Pres Canadian Soc for Mech Engrg 1971-73; Pres Engrg Inst of Canada 1974-75; Mbr Exec Commonwealth Engr's Council 1974- ; Mbr Ctte on Educ & Training, World Federation of Engrg Organizations (WFEO) 1975- .

Moreau, James W
Home: 1639 Kamole Street, Honolulu, HI 96821
Position: Coordinator Energy Conservation *Born:* 5 Feb, 1921 Glenwood, Minn. Bachelor in Engrg from Coast Guard Acad 1942; BCE RPI 1953; Master Engrg Admin from Washington U, St Louis 1959. Federal Exec Inst 1971. Fellow in SAME; Mbr SNAME & ASNE. Various Civil Engrg assignments in Coast Guard including Dist CE in Juneau & St Louis & Public Works Officer at CG Acad. Chief of Civil Engrg CG Hq; Chief of Engrg Div Honolulu & Chief Office Engrg CG Hq Washington, DC, Senior Program Engineer with Hawiian Dredging & Constr Co. A Dillingham Corporation..

Moreno, Theodore
Business: 611 Hansen Way, Palo Alto, CA 94303
Position: Corp VP. *Employer:* Varian Assoc. *Education:* ScD/EE/MIT; AM/EE/Stanford; AB/EE/Stanford. *Born:* 9/2/20. Native of Stanford, CA. AB 1941 from Stanford with great distinction, mbr Phi Beta Kappa, Tau Beta Pi, Sigma Xi. Res

Moreno, Theodore (Continued)
Engr with Sperry Gyroscope Co 1942- 1946. Res Assoc MIT 1946-1949. Res Physicist Hughes Aircraft Co 1949-1951. At Varian Assoc 1951- as Sr Engr, Mgr R&D, Vp (1960) and Div Mgr, Group V-P, Corp VP. Originally specialized in microwave tech, author "Microwave Transmission Design Data." Numerous articles, patents, etc. *Society Aff:* IEEE.

Moretti, Peter M
Business: Dept of Mech & Aero Engrg, EN218, Stillwater, OK 74078-0545
Position: Prof. *Employer:* OK State Univ. *Education:* PhD/Mech Engg/Stanford Univ; MS/Mech Engg/CA Inst of Tech; BS/Mech Engg/CA Inst of Tech. *Born:* 4/13/35. Fulbright Scholar to Germany 1958/59. Proj Mgr at Interatom in Germany, 1964-68. Sr Engr at Westinghouse ARD, 1968-70. Prof. OK State Univ, Stillwater, since 1970. Three summers as a Faculty Fellow at NASA/Ames Res Ctr and Stanford Univ. One yr as a Prog Mgr at US Dept of Energy, Solar Tech Div, Wind Systems Branch. Res in Fluid Mechanics, Vibrations, and Heat Exchangers. Consulting in Flow-Induced Vibrations, and Wind-Energy System operation and economics. Registered PE *Society Aff:* ASME, Mensa, IDE.

Morey, Charles V
Home: 1516-301 S Lakeside Dr, Lake Worth, FL 33460
Position: Self-employed. *Employer:* Editing of AEIC Tecnical Reports. *Education:* BS/EE/Worcester Polytechnic Inst. *Born:* Sept 1903. BS from Worcester Polytechnic Inst. Native of Fall River, Mass. With Consolidated Edison Co of NY 1925-68. In respon charge of meter engrg, engrg & design of underground & overhead transmission & distribution facilities; & of all technical services including res during career. Introduced new concept of sampling techniques applied to customer meter testing. Originated new concept of determining instrument transformer accuracy, adopted as ANSI standard. Had successfully developed a smaller & lower cost gas meter & also a watt-hour meter, both now in general use. Granted patent for a frequency control system for power generation. Author of more than 20 published tech papers. Chmn of various state & natl tech cttes. Fellow of IEEE. *Society Aff:* IEEE.

Morgan, Albert R, Jr
Home: 8504 S. Winston, Tulsa, OK 74136
Position: Pres. *Employer:* Therma Technology Inc *Education:* BS/Chemistry/Univ of CA; Post Grad/ChE/Univ of CA. *Born:* 2/1/22. Native of Oakland CA. Served to Capt, Army Ordnance 1943-1946. FMC Corp 1948-1968 (Dir Res, Chem Div 1963-1968). Panacon Corp 1968-1972 - VP Operations. Jim Walter Corp 1972-1973 Asst to Pres. Thermatech Inc 1973-1977 Pres & Dir. Dyna Tech Inc 1977-1979 Pres & Dir. Therma Tech 1979- Pres & Dir. *Society Aff:* AIChE, ACS.

Morgan, Bernard S, Jr
Home: 2808 Bentley St, Huntsville, AL 35801
Position: Sr Systems Analyst *Employer:* Teledyne Brown Engrg *Education:* PhD/Instrumentation/Univ of MI; MSE/Aeronautic/Univ of MI; MSE/Instrumentation/Univ of MI; BS/-/US Naval Acad. *Born:* 6/30/27. Retired, Colonel USAF. Recent assignments Chief, Command Control, Reconnaissance Div. Hq USAF; Vice Commander, ARL; Dir Aerospace Mechs Div, FJSRL, USAF Acad. was Commander, AF Geophysics Lab - was VP, Sys Technology & Integration-BDM Corp. Interests in /sys technology & analysis of military operations. Past President, AACC; Former Bd/Dir, MORS; Fellow IEEE. Enjoy jogging & gardening & personal computing. *Society Aff:* IEEE, ASME, AIAA, ΣΞ.

Morgan, Donald E
Business: Indus Engrg Dept, Cal Poly Univ, San Luis Obispo, CA 93407
Position: Dept Head, Indus Engrg Dept. *Employer:* California Polytechnic State U. *Education:* PhD/Ind Engg/Stanford Univ; MS/Ind Engg/Stanford Univ; BS/Elec Engg/OR State Univ. *Born:* Sept 1, 1917. Worked for Westinghouse in Pittsburgh & Baltimore as Applications Engr from 1940-46. Became partner in Intermountain Surgical Supply Co in Boise, Idaho until 1960. Was Lecturer at Stanford U in Indus Engrg until 1966. Carried on engrg cons as an individual mbr of various groups until 1968. Became Dept Hd of Indus Engrg at Cal Poly until present. Took 2 years leave to become Prof & Head of Indus Engrg at U of Singapore to dev & operate a new undergraduate & graduate prog in Indus Engrg. *Society Aff:* IIE, ASEE, HFS.

Morgan, Eugene P
Home: 123 Anahola St, Honolulu, HI 96825
Position: Consulting Engr (Self-employed) *Education:* BS/Agri Engrg/Univ of IL. *Born:* Feb 1925. Native of Ill, in Hawaii since 1956. Served USNR 1945-46. Product Design Engr J I Case Co 6 years. Hawaiian Sugar Planters Design Engr 3 years. Dir Field Equip for Amfac cane sugar plantations 8 years. VP IWELD Mfg 4 years. Engrg Cons in developing foreign countries 7 yrs to present. P Chmn ASAE Hawaii Chap. Past Vice Chmn & Chmn ASAE Pacific Region. Mbr ASAE, SAE. Interest in stocks & real estate. *Society Aff:* ASAE, SAE.

Morgan, George H
Home: 7201 E. Gum, Evansville, IN 47715-4344
Position: Technical Sales *Employer:* Brake Supply *Education:* MS/ME/MO School of Mines; BS/ME/LA St Univ *Born:* 7/23/35 Developed the Ausco Fail-Safe Braking Sys popular on construction, minig, oil-field, and industrial equipment. Twelve patents in the field. A variety of SME & SAE papers published. Registered patent agent. Registered Professional Engr. *Society Aff:* SAE, ΤΒΠ, ΠΤΣ, FPS.

Morgan, Howard L
Business: W-83 Dietrich Hall CC, Philadelphia, PA 19174
Position: Prof of Decision Sciences. *Employer:* University of Pennsylvania. *Education:* PhD/Opns Res/Cornell Univ; BS/Physics/CCNY. *Born:* Nov 14, 1945. Native of New York City. Assoc Prof Decision Sciences & Mgmt, Wharton School, & Computer Sci, Moore School, U of Pennsylvania since 1972. Taught at Caltech & Cornell. Active res in database mgmt, mgmt sci, & computer languages. Implementor (with R Conway) of PL/C compiler. Mbr of ORSA, TIMS, ACM, AAAS, SMIS. Elected to ACM Council 1976-79. Editor, Mgmt Applications of Communications of the ACM. Assoc Editor, ACM Transactions on Database Systems. Chmn, ACM Curriculum Ctte on Computer Educ for mgmt.

Morgan, J Derald
Business: PO Box 3449, New Mexico State U, Las Cruces, NM 88003
Position: Dean of Engrg. *Employer:* New Mexico State Univ. *Education:* PhD/EE/AZ State Univ; MS/EE/Univ of MO-Rolla; BS/EE/LA Tech Univ. *Born:* 3/15/39. EE, TX Eastman Div Eastman Kodak Co, 1962-63; Instr Univ of MO-Rolla, 1963- 65; Instr AZ State Univ, 1965-68; Assoc Prof Univ of MO-Rolla, 1968-72; ALCOA Fdn Prof of EE, 1972-75, Assoc Dir Ctr for Intl Progs, 1970-78; Emerson Elec Prof, 1975-; Chrmn, Elec Engg Dept, Univ of MO-Rolla, 1978-1985; Dean of Engg College 1985- ; consulting to various firms. Bd of Dirs, MO Partners of the Am. Vice Chairman IEEE Power System Engg committee. Fellow of IEEE 1984 IEEE Centennial Medal. St Louis IEEE Sec 1978 Award of Honor. Author: Power Apparatus Testing Techniques, 1969; Dir, Univ of MO-Rolla/Dept of Natural Resources Energy Conf 1974-79. Electromagnetic and Electromechanical Machines Harper & Row 1985. *Society Aff:* IEEE, NSPE, ASEE.

Morgan, Jack B
Business: Research Ctr, Brackenridge, PA 15014
Position: Senior Supervising Metallurgist. *Employer:* Allegheny Ludlum Steel Corp. *Education:* BS/Physics/Univ of Pittsbugh. *Born:* Jan 12, 1922. US Army WWII. Assoc with ALCOA Res in nondestructive testing prior to 1956. With Allegheny Ludlum since 1956. Respons include instrumentation, material evaluation, nondestructive examination & mech testing. Author of numerous papers in NDT field. Mbr ASNT, ASTM, ASM; Fellow ASNT, P Chmn of Tech Council & Natl Dir of ASNT 1971-73. P Chmn of Amer Iron & Steel Inst Tech Ctte on NDT. Has been active in Tech Soc work, assisted in Handbook preparations, known for work in

Morgan, Jack B (Continued)
area of standards & techniques. Hobbies include farming & music. *Society Aff:* ASME, ASNT, ASTM, ASM.

Morgan, Leland R
Business: 810 W Main St A, Visalia, CA 93277
Position: Pres. *Employer:* Mormec Engg, Inc. *Education:* BS/Environmental Engg/CA Poly San Luis Obispo. *Born:* 11/10/48. Native of Taft, CA. Served in Air Force 1968-72. With consulting firms until 1978. Formed my own co Mormec Engg, Inc 1978. Enjoy woodworking and fishing. *Society Aff:* ASHRAE, NSPE.

Morgan, Melvin W
Home: 36 Beecher Pk, Bangor, ME 04401
Position: Soils Engineer. *Employer:* Maine Dept of Transportation. *Education:* BS/Civil Engg/Univ of ME. *Born:* Sept 26, 1932. Served in Army Signal Corps as Radio Officer 1955-57. With Maine Dept of Transportation since 1957. Assumed current respon in charge of the Soils Sect in 1967. Respon for all geotechnical engrg & engrg geology in MDOT. Formerly Mbr Transportation Res Bd Ctte on Embankments & Earth Slopes. Mbr ASCE, PPres ME Sect ASCE, Mbr Boston Soc of Civil Engrs Sect & ISSMFE. Have presented numerous papers at meetings of the Soils Engrs of the Northeastern State Highway Depts. Active Mbr of the International Supreme Council Order of DeMolay and past Exec Officer in ME. *Society Aff:* ASCE.

Morgan, Robert P
Business: Campus Box 1106, Washington Univ, St Louis, MO 63130
Position: Prof of Tech and Human Affairs *Employer:* Washington University. *Education:* PhD/Chem Engg/RPI; Nucl Engg/Nucl Engg/MIT; SM/Nucl Engg/MIT; BChE/Chem Engg/Cooper Union. *Born:* 2/26/34. Brooklyn, NY. Asst Dir MIT Oak Ridge Engrg Practice School, 1958-59. Instructor of Chem Engrg, RPI 1960-64. Asst Prof of Nuclear & Chem Engrg, Univ of MO 1965-68. Prof. With WA Univ since 1968 as Dir, Center for Dev Tech. Chrmn, Dept of Tech and Human Affairs, 1976-1983. Promoted to Full Professor, 1974. Principal author of Sci and Tech for Dev: The Role of US Universities" and "Renewable Resource Utilization for Development," Author of "Sci and Tech for Intl Dev: An Assessment of U.S. Policies and Programs–, Co-editor of "Fuels and Chemicals from Oilseeds,–, and author of some 100 articles and reports. Has served as Secy-Treas & V Chmn, Intl Div of ASEE, Mbr of NAE Panel on Role of Engrg Schools in tech assistance, Mbr of NASA Advisory Comm on Tech Transfer, Mbr of Ctte on Sci Engr & Public Policy, AAAS, Natl Sigma Xi Lecturer, Chrmn of Engrg and Pub Policy Div of ASEE, and Mbr of Res Grants Ctte, Natl Res Council. Received 1978 Chester F Carlson Award from ASEE for innovation in engg education. Married to Nancy Hutchins; children: Thomas A & Jonathan A. *Society Aff:* ASEE, AAAS, AAUP, FAS

Morgan, Samuel P
Business: 600 Mountain Ave, Murray Hill, NJ 07974
Position: Mbr of Technical Staff *Employer:* AT&T Bell Laboratories. *Education:* PhD/Physics/Caltech; MS/Physics/Caltech; BS/Cil/Caltech. *Born:* July 14, 1923 San Diego, Calif. m. Caroline Annin 1948; children: Caroline 1951, Lesley 1954, Alison 1957, Diane 1960. Mbr tech staff, Bell Labs, 1947-59; head mathematical physics dept 1959-67. Applied mathematics, theory of waveguides & microwave antennas; 35 publs, 8 patents. Dir computing sci res 1967-82; Dir computing tech 1969-70. Supervision of res in operating systems, programming languages, theory of algorithms, numerical methods, & computer- communication networks. Mbr. tech. staff 1982- . Applied quencing theory performance of computer - communication networks. Mbr AAAS, Amer Physical Soc, Soc Indus & Appl Math, Assn Computing Machinery, Sigma Xi; Fellow IEEE. *Society Aff:* AAAS, APS, ACM, ΣΞ, IEEE.

Morgan, Stanley L
Business: 270 Northfield Rd, Bedford, OH 44146
Position: Executive Vice President. *Employer:* Ben Venue Labs, Inc. *Education:* BS/Chem Engg/Case Inst of Technology. *Born:* Jan 1918. With Ben Venue Labs since 1940. Assumed current respons as Exec VP in 1964. Also Tech Dir, Reg PE in Ohio. Fields of specialty: freeze drying & pharmaceuticals under aseptic conditions Pharmacontical Engg. *Society Aff:* ACS, AIC, AIChE, IPE, AVS, NYAS, ASTM.

Morgen, Ralph A
Business: 7878 Grow Lane, Houston, TX 77040
Position: President *Employer:* Worley Engineering, Inc. *Education:* BChE/Chem Engg/Univ of FL. *Born:* Sept 1928. Began prof career with Internatl Minerals & Chem Corp as Jr Engr; later with Black, Sivalls & Bryson, Inc in USA, Europe, & Asia. Specialized in oil & gas process equip design & const of offshore production facilities. In 1970 became Regional VP ISC World Trade Corp based in Tokyo. Joined Worley Engrg, Inc, as Exec VP in Jan 1974. Dir of Worley Internatl Engrg Group, Ltd, Worley Engrg, Inc, P&W Offshore Services, Ltd, Worley Engrg, Ltd. Served as a pilot in Air Force, enjoy sports & flying. *Society Aff:* AIChE, SPE.

Morgenstern, John C
Home: Suite 1120, 1901 N. Ft. Myer Dr, Arlington, VA 22209
Position: President *Employer:* JCM Associates Ltd. *Education:* BSEE/Elec Engr/MIT *Born:* 1/21/32. Vienna, Austria. Designed & analyzed automatic control systems at MIT's Servo Mechanism Labs. Designed early solid state power supplies at ultrasonic corp. System engineered automatic checkout equip & designed systems using early atomic clocks at Natl Co. In charge of space borne automatic control system at Itek. At the Mitre Corp from 1961 to 1978 in positions of progressively greater responsibility in the concept dev, requirement determination, design & systems engg of large scale military command, control and communication systems. Dir for strategic & theatre command & control systems, office of the Asst Secy of Defense (C3I) with responsibility for system definition, R & D and acquistion of strategic C3 system, C3 system for all threaters of operation, C3 system for crisis management and for the tactical nuclear forces. Upon leaving the Department of Defense, formed JCM Associates to provide consulting services to private industry in the area of command, control, communications and intelligence. *Society Aff:* IEEE, AIAA, AFCEA, ADPA.

Morgenstern, Norbert R
Business: Dept of Civil Engrg, Univ of Alberta, Edmonton Alb Canada T6G 2G7
Position: Professor of Civil Engineering. *Employer:* University of Alberta. *Education:* PhD/Soil Mechanics/Univ of London; DIC/Soil Mechs/Imperial College of Science; BASc/Civil Engg/Univ of Toronto. *Born:* May 25, 1935. Lecturer in Civil Engrg at Imperial Coll of Sci & Tech 1960-68; Prof of Civil Engrg, U of Alberta 1968-1983; Univ Prof 1983- ; private cons in Engrg Earth Sciences 1961- ; author of over 170 publs on geotechnical engrg; mbr of editorial bds for several journals; mbr of prof cttes in Canada & USA, awards include British Geotechnical Soc Prize in 1961 & 1966; Walter L Huber Civil Engrg Res Prize in 1971; Fellow of the Royal Soc of Canada in 1975; Canadian Geotechnical Soc Prize in 1977; Legget Award of Canadian Geotechnical Soc, 1979; Rankine lecture, British Geotechnical Society, 1981; D.Eng (hc), Univ of Toronto, 1983; Univ Res Prize, Univ of Alberta, 1984; Centennial Award, Assoc Prof Engrs, Geologists and Geophysicists of Alberta; Fellow, Engineering Institute of Canada, 1985. *Society Aff:* ASCE, EIC, CIMM.

Morgenthaler, Charles S
Business: 6386 York Rd, Baltimore, MD 21212
Position: VP *Employer:* Envirodyne Engrs, Inc. *Education:* BS/CE/Lafayette Coll *Born:* 2/12/29 Native of Harrisburg, PA. Served with US Army Chem Corps, 1950-1952. Sr Bridge Designer with Gannett, Fleming, Corddry and Carpenter, 1952-1955. Assistant to the Resident Engr with D.B. Steinman, Consulting Engrs, on construction of Mackinac Straits Bridge, 1955-1957. With Envirodyne Engrs, Inc and predecessor firms since 1957. Assumed current responsibility as VP in charge of Batlimore Regional Office in 1977. On Bd of Dirs of CEC MD and Planning and Design Div of ARTBA. Registered professional engr in ten states and District of Columbia. *Society Aff:* ASCE, NSPE, APWA, IBTTA, ARTBA

Morgenthaler, Frederic R
Business: MIT Rm 13-3102A, Cambridge, MA 02139
Position: Prof. *Employer:* MIT. *Education:* PhD/Elec Engr/MIT; SM/Elec Engr/MIT; SB/Elec Engr/MIT *Born:* 3/12/33. A native of Shaker Hts, OH, Dr Morgenthaler is currently Prof of EE at MIT beginning his teaching as an Asst Prof in 1960. In 1965 he was promoted to the rank of Assoc Prof & in 1968 to the rank of Prof. He also holds an appointment to the Harvard-MIT Div of Health Sci & Tech. Appointed Cecil H. Green Prof of EE 1984-86. His res & grad teaching has centered on the field of microwave magnetics, the electrodynamics of waves & media, & medical ultrasonics. Dr Morgenthaler has served as a consultant to the US Govt as well as to private industry. He is the author of nearly 100 scientific publications & papers presented at technical confs & has been granted approximately one dozen patents. IEEE Fellow 1978. *Society Aff:* IEEE, AAAS, ΣΞ, ΤΒΠ, HKN.

Moriarty, Brian M
Business: 600-5th St NW, Washington, DC 20001 *Employer:* De Leuw Cather, Co. *Education:* MBA/Statistics/CA Univ-Fullerton; BSE/Engr Physics/Univ of MI; BSE/Engr Math/Univ of MI *Born* 9/9/36 Qualified in Assurance Sciences. Active in Sys Safety discipline since 1961 with Rockwell Industries (Missile development), Litton (Shipbuilding) and TRW (Systems Engrg). Perform Architectural/Engrg design project work at De Leuw Cather/Parsons since 1977. Active in Sys Assurance (Quality Assurance, Reliability and Maintainability, & Sys Safety) on Northeast Corridor Improvement Project (NECIP), and Wash Metro Proj. PE in Safety and Quality. Safety Consultant in Program Plans, safety analytical methods and techniques and safety training. Instructor in USC Safety Management Graduate Program (ISS) -Human Factors and System Safety. Qualified Air Controller and Engrg Duty Officer with US Naval Reserve for 25 yrs. Former Pres SSS. *Society Aff:* SSS, HFS, ASQC, ASNE, NSPE, SRE.

Moricoli, John C
Home: 7571 Alpine Way, Tujunga, CA 91042
Position: Vice President. *Employer:* W Coast Plastics Equipment Inc. *Education:* BS/Bus Admin/Kent St Univ *Born:* Nov 1920. Successively Mgr Material Control, Prod Control & Plastic Mold Sales, Gougler Machine Co Kent Ohio; Saes Mgr McLean Dev Labs Copique N Y; Tooling Mgr Zenith Plastics Gardena Calif; 1958 V Pres W Coast Plastics Equipment Inc. Mbr Soc of Plastics Engrs S Calif Sect; Soc of the Plastics Indus; AAAS; Internatl SPE 2nd V Pres & Treas; S Calif Sect Pres, V Pres & Secy, Intl Councilman. Author Engrg articles by SPE, SPI, ASTME. 1975 received Willard Lundberg Memorial Award by S Calif Sect SPE for outstanding contribs to Sect, Soc & Plastics Indus 1980 Society of Plastics Industry Distinguished Service Award. 1981 Southern California Society of Plastics Engineers President award for continued outstanding contributions to the Southern California Section and Plastics Industry in the west. 1982 Soc of Plastics Engrs Intl Presidents Cup for Outstanding and Meritorius Service. Western Section Society of the Plastic Industry Man of the Year Award for outstanding service to the industry in the West 1985. Mbr Plastic Pioneers Assoc 1981, elected to Bd of Governors 1984. 1980 Colonel AUS Retired; Active Duty, Natl Guard and Reserve Service from 1937 thru 1980. *Society Aff:* SPE, SPI

Morin, Herman L
Business: 2401 Burnet Ave, Syracuse, NY 13206
Position: Pres. *Employer:* Robson & Woese, Inc. *Education:* BS/Mech Engg/Univ of MI. *Born:* 8/16/24. Engg Officer US Navy, 1945-46; Grad Univ of MI Marine Engg 1947; Design Engg, Robson & Woese, Inc Consulting Engrs 1947-50; Ship Repair and Construction officer Boston Naval Shipyard, responsible for repair and overhaul of numerous naval vessels including cruiser SS Salem, 1951-52; US Navy Tech Officer at General Electric Co, Schenectady as part of special US Navy technical team that built submarines Nautilus and Seawolf, 1952-53; 1954 to Present Robson & Woese, Inc serving as HVAC Dept Hd, Project Dir, VP and in 1976 became Pres. Extensive engg experience in design of hospitals, laboratories, military installations, industrial complexes, central plants, major utilities, medical centers, scientific and other building types. Dev first mech guide specification in Central NY, also electrical heating guide and stds Pioneer in HVAC, Central Plants & Energy Recovery conservation in Central NY, first CSI mech guide specification in Central NY. *Society Aff:* ACEC, ASHRAE, NSPE, CSI

Morino, Luigi
Home: 9 Chauncy St. Apt. 73, Cambridge, MA 02138
Position: Prof AM Eng *Employer:* Boston University *Education:* D/Aerospace Eng/Univ of Rome; D/ME/Univ of Rome *Born* 7/21/38 Res activity in computational aerodynamics, including: Dev method known as Green's Function Method (or Morino's Method) for subsonic, transonic and supersonic steady and unsteady aerodynamics around complex aircraft configurations. Supervised dev of corresponding codes: SOUSSA (steady, oscillatory, unsteady subsonic aerodynamics: nonlinear for transonic flows) Extenseion to aerodynamics of windmill, helicopters, bldgs Grants and contracts from NASA, NSF, ERDA, DOE and ARO Currently studying rotational flows. *Society Aff:* AIAA, ASEE, AAAS

Moritz, Karsten H
Home: 7 Edgewood Rd, Summit, NJ 07901
Position: Pres *Employer:* Karsten Moritz, Inc., Consultants *Education:* MA/Classics/Univ of Chicago; BS/ChE/Univ of TX; BA/Sociology/Univ of Chicago *Born:* 5/15/29 Employed by Exxon Research and Engrg Co in 1957 after receiving Chem Engrg degree from the Univ of TX in 1957. Worked as research engr on variety of catalytic reactor developments in pilot plant stage. Promoted to Research Assoc in 1964. In the late '60's head of research section responsible for development of ER&E catalytic desulfurization processes "GO fining" and "Resid fining–. Promoted to Director of Process Development at the Exxon Research and Development Laboratories in 1972 and to Dir of Exploratory Res in 1975. Responsible for long range catalysis and reactor research in fuels processing. Named Mgr of the Products Research Div of ER&E in 1978. Div employs over 100 professional scientists and engrs, conducts research and development on all facets of petroleum fuel and lubricant products. Named Deputy Mgr, Corp Planning, 1982. Retired from Exxon in 1986. Founded Karsten Moritz, Inc., consultants in technology resources. *Society Aff:* SAE

Morkoc, Hadis
Home: 1803 Golfview Dr, Urbana, IL 61801
Position: Prof. *Employer:* U. of Illinois *Education:* Ph.D./EE/Cornell University; MS/EE/Istanbul Technical Univ.; BS/ EE/Istanbul Technical Univ. *Born:* 10/02/47 After receiving his PhD degree from Cornell, he held positions at Varian and Bell Labs before joining the Univ of Ill in 1979. His research over the years covered many aspects of heterojunction and thin film materials and devices including superlattices. Together with his colleagues at the Univ of Ill and elsewhere, he authored and co-authored some 500 book chapters, reviews and journal articles. Dr. Morkoc received the Electronics Letters Best paper Award in 1978 for his work in InGaAsP and InGaAs field effect transistors. His research on high speed devices and GaAs on Si have been covered widely in technical and popular journals. He is currently spending a sabbatical year at Caltech and the Jet Propulsion Lab. *Society Aff:* IEEE, AAAS, APS

Morkovin, Mark V
Home: 1104 N Linden Ave, Oak Park, IL 60302
Position: Prof of Mech & Aero Engrg. *Employer:* Ill Inst of Tech - Chicago Ill. *Education:* PhD/Appl Math/Univ of WI; MS/Math/Syracuse Univ; BA/Math/Univ of So CA. *Born:* July 1917. Spent roughly 1/3 of career as res engr & internal cons in fluid dynamics in aerospace indus: Bell Aircraft Corp 1943-46, Martin Co 1958-67; and the rest in univ teaching & res: Univ of Wis, Brown Univ, Univ of Mich, Johns Hopkins Univ, IIT. Served with Office of Naval Res 1946 on special assignment: setting up Applied Mechs Reviews & various res progs. Cons widely to indus

Morkovin, Mark V (Continued)
& govt. Active in ASME (Fellow), AIAA (Fellow) & APS (Fellow). Received 1st Fluid & Plasmadynamics Award of AIAA 1976; Sr Award, von Humboldt Fnd 1977- 78; 1-month Sr-Exchange Lecture Tour US-USSR Acad Sci 1979. Member, National Academy of Engineering. *Society Aff:* ASME, AIAA, APS, ΣΞ

Morley, James Q
Home: 7000 Old State Rd, Evansville, IN 47710
Position: Pres. *Employer:* Morley & Assocs, Inc. *Education:* BS/Civil Engr/Tri State College *Born:* 10/23/41. in Bruceville, IN. 1941 Const Engr Ind St Hwy Comm. Joined faculty of Vincennes Univ 1967, dev 1st curriculum & taught Cvl Engrg Tech Assumed position as Ch Civil Engr Biagi-Hannan & Assoc Inc 1970; Founded Morley & Assocs, Inc, Consulting Engrs, La Surveyors 1976, Design activities urban renewal, indus waste treatment, site dev, parks & swimming pools, drainage control & flood protection facilities, parking garages, struc design & land surveys. Pres Ind Sect ASCE 1974; Ind Soc of Prof LS; Amer Soc for Testing & Matls. Conference Chmn 1976 S Ind Earthquake Seminar Pres OH Valley Flood Control Assoc 1977, chmn Dist 9 Council ASCE 1980. VP Buffalo Trace, BSA 1987. *Society Aff:* NSPE, ASCE, ASTM ISPLS

Morley, Richard J
Home: 231 Shunpike Road, Chatham, NJ 07928-1818
Position: Reginonal VP., Regional Engineering Manager *Employer:* Allendale Mutual Ins. Co. *Education:* BS/ME/Cornell University, Ithaca NY *Born:* 09/20/25 Dick Morley, New York regional engrg mgr, joined Allendale in 1967. Over the years he has held various positions including account engr, mgr, regional engrg mgr and regional VP. Dick started his professional career in 1949 with the Industrial Risk Insurers, serving as a field engineer and enginer-in-charge. His assignments have included combustion safeguard specialist, nuclear specialist and special hazards engr. Dick holds a BS degree from the Sibley School of Mech Engrg, Cornell Univ. His professional memberships include member, SFPE (charter member NJ chapter, mbr New England Chapter, Qualifications Bd & past chairman Qualifications Bd) and the NFPA. Currently member of the Bd of Dir SFPE. *Society Aff:* SFPE, NFPA

Morlok, Edward K
Business: 220 South 33rd Street, Philadelphia, PA 19104-6315
Position: UPS Fdn Prof of Transportation, and Prof of Systems Engg *Employer:* Univ of PA *Education:* PhD/CE Trans/Northwestern Univ; CIT/Trans Econ & Mgt/Yale Univ; BE/ME/Yale Univ; MA (Hon)/-/Univ of Penna *Born:* 11/3/40 Primary interests include: transportation sys design and mgmt, with particular emphasis on service planning and innovative tech and operations. Current res on joint public/private provision of transit and on shipper/carrier planning in deregulated environment. Publications include two books and over 60 articles. Mbr of Natl Assembly of Engrg Ctte on Transportation; and its Panels on Innovation and on Hazardous Materials. Ed of McGraw-Hill Book Co Series in Transportation and Assoc Editor of *Transportation Research.* P Pres of Transportation Research Forum. Received Alexander von Humboldt Fdn US Senior Scientist Award, 1980-81. Consultant to transport agencies and companies. *Society Aff:* TRF, TRB, ASCE, AAAS, AAUP

Mormino, Paul S
Home: 95 Stanford Dr, Westwood, MA 02090
Position: Group Vice Pres Admin & Personnel, retired *Employer:* Chas T Main Inc. *Education:* BS/Structural/MIT. *Born:* Jan 15, 1914. Struc Engr for Curtiss-Wright Co 1940-45 specializing in aircraft design. With Chas T Main Inc 1945- ; assumed current responsibility as Group V Pres Admin & Personnel 1969. Reg P E in several states. Mbr of Natl Soc of Prof Engrs. Elected Dir 1978-The C.T. Main Corporation. Ret Nov 1984. *Society Aff:* NSPE.

Morral, F Rolf
Home: 2075 Arlington Ave, Columbus, OH 43221
Position: Consultant (Retired) *Education:* PhD/Metallurgy/Purdue Univ; BS/Electrochem E/MIT; Perito/Eocuelo Industrial Terresso/Spain *Born:* June 1907. R/D at Continental Steel Co; Mellon Inst; Section Ch at Amer Cyanamid & Kaiser Aluminum Res Labs; Asst & Assoc Prof Penn St Univ & Syracuse Univ; Cons at Battelle Memorial Inst on Cobalt Information Center project. Active in section affairs of ASM & AIME & tech councils in Spokane, Wash & Columbus Ohio. Columbus Tech Man of the Year 1971. Fellow AAAS, Sigma Xi, Ohio Acad of Sci, Amer Chemists Inst, Spanish Acad of Sci Madrid. *Society Aff:* ASM, AIME.

Morral, John E
Business: U-136, Rm 111, 97 N. Eagleville Rd, Storrs, CT 06268
Position: Prof *Employer:* Univ of CT *Education:* PhD/Metallurgy/MIT; MS/Metallurgical Engr/OH St Univ *Born:* 8/3/39 Taught at the U of IL from 1968-71 and at the U of CT from 1971 - present. Have written papers in the areas of precipitation Theory (spinodal decomposition; nucleation and growth, coarsening), solution Thermodynamics, multicomponent diffusion, cellular structures, subscale formation and stress corrosion cracking. Current interests are coating interdiffusion and selective oxidation. *Society Aff:* AIME, ASM, ASEE, AAUP

Morrell, T Herbert
Home: 764 E South St, Owatonna, MN 55060
Position: Mgmt Consultant (Self-employed) *Education:* BS/Mech Engg/Univ of KS. *Born:* 7/31/16. Ammunition Engr Remington Arms during WWII; Design Engr Oliver Tractors, 1944-51; Chief Engr; Oliver Tractors 1951-65; Mgr of outside prods for Oliver 1965-70; VP & Dir of Engg Owatonna Mfg Co Inc 1970-77. Has participated in ASAE, SAE & FIEI stds activities 1950-77; Mgmt Consultant 1977- ; Chmn SAE Tractor Technical Cttes 3 yrs; Mbr SAE Automotive Council 3 yrs; V Chmn FIEIEngg Policy Ctte 3 yrs. Awards received: Hon Mbr FFA, Cert of appreciation SAE & Fellow Mbr ASAE. Enjoys golf, gardening, fishing, hunting, music & church work. *Society Aff:* ASAE, SAE.

Morrin, Thomas H
Home: 654 23rd Ave, San Francisco, CA 94121
Position: President (retired). *Employer:* Morrin Associates. *Education:* BS/Elec Engg/Univ of CA, Berkeley; Grad/Communication/Annapolis, MD. *Born:* 11/24/14. Cmmr, USN, 1945-48; Off, USNR, 1938-1958. U.S. Navy, Pacific Fleet, 12 medals, including Navy Commendation & Bronze Star & 15 battle stars; 1945-47 Comdg. Officer, Office Nav Res; 1937-38, Westinghouse Mfg. Co.; 1938-41 Pacific Gas & Electric Co; 1947-48 Asst to Pres & Genl Mgr Microwave Processing Div Raytheon Mfg Co; 1948-68 Genl Mgr & V Pres Eng Res Stanford Res Inst; 1968-69 Pres Univ City Sci Inst (A consortium of 19 Univs & Colls in DE Valley); 1969-73 Pres Morrin Assocs, Wenatchee WA & San Francisco; 1973 retired & some consultanting. Fellow IEEE & Fellow AAAS; Mbr RESA; past mbr many bds of dirs. *Society Aff:* IEEE, AAAS, RESA.

Morris, Alan D
5817 Plainview Road, Bethesda, Maryland 20817
Position: Sr Partner *Employer:* Morris & Ward Consulting Engrs *Education:* DrEngr/EE/Johns Hopkins Univ; BE/EE/Johns Hopkins Univ *Born:* 1/4/31 Research Assoc, Inst of Cooperative Research, the Johns Hopkins Univ, 1951- 55; LT (jg), US Navy, Aeronautical Engrg Officer, 1955-58; Research Scientist, Operations Research Incorp, 1958-61; Associate Prof of Physics, the Amer Univ, 1961-65; Dir, Operations Research Program, Center for Technology and Administration, the Amer Univ, 1965-68; Sr Partner, Morris and Ward, Consulting Engrs, 1968 to present. Reg PE MD, DC. *Society Aff:* ACEC, IEEE, ASCE, WSE, NFPA, CEC/MW

Morris, Albert J
Home: 26520 St Frances Dr, Los Alto Hills, CA 94022
Position: President. *Employer:* Biosis *Education:* EE/Electronics/Stanford; MS/Electronics/Stanford; BS/EE/UC Berkeley. *Born:* 1919. Currently Pres, Biosis P Pres Genesys Systems Inc., Energy Systems Inc, Radiation at Stanford Inc. Papers publ

Morris, Albert J (Continued)
in med electronics, high power electronics, educ delivery sys & instructional TV with particular emphasis on continuing educ of engrs. Fellow IEEE, AAAS; Reg P E Calif. Cons to Schs of Engrg Ga Tech, IIT, Univ of Md & many others. P Chmn San Francisco Sect IEEE; San Francisco Council WEMA; & past V Pres WEMA; P Chrmn of Bd WESCON. Mbr Sigma Xi. Cdr USNR (rtd). Enthusiastic tennis player. *Society Aff:* IEEE, ΣΞ, AAAS.

Morris, Ben F
Business: 10666 N Torrey Pines Rd, La Jolla, CA 92037
Position: Dir of Engrg *Employer:* Scripps Clinic and Res Fnd *Education:* BS/CE/VA Military Inst *Born:* 1/11/48 Dir of Engrg-Scripps Clinic & Research Fdn, and internationally recognized medically related complex which includes Scientific Research Labs, Hospital and Clinics. Responsible for all plant opeation, maintenance, Bio-Medical Engrs, Design and Construction. Awarded Edmund Friedman Young Engr Award for Prof Achievement for the yr 1979 by Amer Soc of CE. P Pres AMF-ASCE Phoenix, AZ; P. Pres AMF-ASCE San Diego, CA; Past Chrmn Western Reg Younger Mbr Council ASCE; Outstanding AMF-1978 WRYMC; Past Chrmn Younger Member Publications Committee ASCE; Who's Who in the West; Eagle Scout; Asst. Scoutmaster BSA. *Society Aff:* ASCE, ASHE, NFPA, CSHE, AMA

Morris, C Robert
Business: 1019 19th St, NW, Suite 1110, Wash, DC 20036
Position: Assoc Exec Dir. *Employer:* Natl Assn of Water Cos. *Education:* BS/Civ Engg/Drexel Univ. *Born:* 6/24/27. Native of Phila, PA area. Started in engg work as staff engr for Mtn Water Supply Co. Employed by Gen Waterworks Corp as Dist Engr, Staff Engr, Chief Engr, VP - Engg & Operations, & Sr VP - Operations, 1956-1978. 1978-present, Assoc Exec Dir, Natl Assn of Water Cos, Wash, DC, involved in coordination of technical matters & legislative affairs. Officer & dir of AWWA, NAWC, & WWOA of PA. AWWA Water Utility Man of the Yr - 1974, Mgt Div Award - 1978. WWOA of PA Harry J Krum Award - 1974. Dir - Penn Federal S & L Assn. Enjoy sports, particularly golf. *Society Aff:* AWWA, NAWC, WWOA of PA.

Morris, Don F
Business: PO Box 2968, Tulsa, OK 74101
Position: Sr VP and CEO *Employer:* Dresser Engg Co. *Education:* BS/Chem. E./OK State Univ. *Born:* 6/8/32. Process engr for Phillips Petrol Co 1955-1958. Process/Proj for Dresser Engg Co 1958-1968; Chief Process/Proj Engr 1968-1975; elected VP Dresser Engrg Co in 1975. Elected Sr VP Dressser Engrg Co in 1983; Appointed CEO Dresser Engrg Co 1984. Responsible for mgt of Dresser Engrg Co *Society Aff:* AIChE, NSPE, EST, OSPE.

Morris, Fred W
Business: 137 Ash Dr. Ln, Portola Valley, CA 94025
Position: Pres & Chrmn *Employer:* Tele-Sciences Associates. *Education:* BS/Elec Engrg/CAL Inst of Tech; DSc/Engrg/Capital Inst of Tech *Born:* 2/28/22. Native of CA. Served as US Army Signal Corps Officer, WWII. Asst Prof of Elec Engg, Univ of Southern CA, 1947-50. Electronic Scientist, US Army electronics res 1949-54. Exec-Consultant/Dir including: Radiation Inc; EMTECH Inc; DS Kennedy Inc; Page Communications Engrs, Inc; Hiller Aircraft Corp; GTE Corporation. With Stanford Res Inst, MIT & Defense Analysis Inst in '50s and '60s. Deputy Asst to the Pres Assoc Dir to Telecommunications Mgt, The White House, 1964-66. Mbr, Presidential Task Force on Telecommunications Policy, 1967-69. Pres-CEO & Dir, TRT Telecommunications Corp, 1972-75. VP & Dir, COMSAT Gen Corp, 1975-78. Founder in 1969 and continues as Chrmn, Tele-Sciences Assocs. - natl & intl telecommunications engg-mgt-financial consultants. Reg PE, State of CA since 1948. Fellow, East-West Center (Honolulu). Honor, De l'Ordre Grand-Ducal de la Couronne de Chene (Grand Duchy of Luxembourg). *Society Aff:* AIAA, IEEE, AFCEA.

Morris, James G
Home: 741 Glendover Dr, Lexington, KY 40502
Position: Prof. *Employer:* Univ of KY. *Education:* PhD/Metallurgy/Purdue Univ; BSMetE/Metallurgy/Purdue Univ. *Born:* 3/20/28. Native of Marietta, O Consultant to ARCO Aluminum Co and Kaiser Aluminum & Chem Corp. Res Engr for Dow Chem Co, Kaiser Aluminum and Chem Corp & Olin Corp. Author of over 50 published papers. Extensive res experience in the physical metallurgy of aluminum alloys. Editor of books on Thermomechanical Processing of aluminum alloys and on Textures in Non Ferrous Metals and Alloys. Patents on the development and processing of aluminum alloys. *Society Aff:* I of M, AIME, ASM, ASEE, ASAS.

Morris, John W
Home: 3800 N Fairfax, Arlington, VA 22203
Position: Pres *Employer:* J.W. Morris Ltd. *Education:* MS/Civil Eng/Univ of IA; BS/CE/US Mil Acad; BS/Mgmt/Univ of Pittsburgh *Born:* Sept 1921 Princess Anne Md. Grad US Military Acad 1943 commissioned Corps of Engrs; commanded troops from platoon in WWII to 18th Engr Brigade (15, 000 men) Vietnam; Deputy Commandant US Military Acad; Dir design & const numerous air bases & army posts plus Ark River Navigation Sys; Dir Army Cvl Works prog 72-75, mbr Water Resources Council of Representatives. Reg P E. Fellow ASCE; mbr USCOLD, V Pres Permanent Inter Natl Assn of Navigation Congresses. Mbr Natl Acad of Engrg. Ch of Engrs, Lt Genl July 1976 - Sept 1980-, Pres J W Morris Ltd; Prof Univ of MD Sept 1984, Const Man of Yr 1977. Recipient Palladium Medal 1984, ACEC Award of Merit, Pres, CEO, Planning Research Corp, Engrg. 1986- . *Society Aff:* NSPE, SAME, PIAN C, ASCE, USCOLD

Morris, Kenneth B
Home: 388 Cedar Dr W, Briarcliff Manor, NY 10510
Position: Hd own consltg engrg firm *Employer:* Kenneth Morris, PE, PP, ASA *Education:* Grad School/Coll & Univ Admin/Univ of NB; -/Computers/IBM Sch for Coll Pres; -/Architecture & Plann/Inst of Design; BCE/Struct Engrg/Manhattan Coll *Born:* 02/12/22 Hd of own Consltg Engrg firm; Sr Officer, East River Savings Bank; Pres Gramercy Greenwich Corp (formed by several univs); VP Cooper Union Univ; Bus Mgr and Dir, NY Univ; Own Architectural-engrg and Planning firm, NYC & Augusta, GA (intl); Chief Engr and Asst to Pres, Kretzer Construction Corp & subsidiaries; Engr, Burns & Roe, Inc. Adj Prof-teach at Pace, NYU & Fordham Univs, also taught at IBM Corp Hdq; Licenses: PE (5 states), Professional Planner (city planning) P Pres & Chrmn (4 terms), East Side Chamber of Commerce (eastern half of Manhattan). Mbr of Bd, The Salvation Army. Speaker on TV, Radio, Natl Conventions. Articles in business and professional publications. Senior Member American Society of Appraisers. *Society Aff:* NSPE, NJSPP

Morris, Larry R
Business: Alcan Research Lab, P.O. Box 8400, Kingston, Ont K7L 4Z4
Position: Dir Res. *Employer:* Alcan R&D Ctr. *Education:* PhD/Mtl Sci/Univ of Toronto; BSc/Met Engg/Univ of Toronto. *Born:* 9/25/38. in Toronto Ontario. Worked as Plant Engr for Aluminum Co of Canada 1962-65. After return to Univ joined Alcan R&D Ctr in 1969. Research manager for NKK in Japan 1984-85. Assumed current position as Alcan R&D Ctr Dir of Res in 1985. Areas of expertise: solidfication, alloy dev, deformation. Developed new range of ultra fine grained aluminum alloys & superplastic aluminum alloy. Chrmn, Kingston Chapter ASM, 1977. Collect old instruments & antiquarian books on tech. *Society Aff:* CIMME, AIME.

Morris, M Dan
Business: 9 Northfield Ln, Westbury, NY 11590
Position: Tech Communications Cons/Editor. *Employer:* Self. *Education:* AB/Chinese History/Cornell Univ; CE/Civil Engr/Cornell Univ *Born:* Feb 1922 in New York City. Reg PE. Served Western Pacific, Army Corps of Engrs 1945-47. Fellow ASCE, Past Natl Chmn Const Div 1973-75, P Pres Metropolitan Sect 1972-73. ExCom Cornell Soc Engrs 1954. Natl Pres 1979-81. Mbr Overseas Press Club of Amer; Sr Mbr Soc for Tech Communication; Competition winner 1st places 1975 (publs) & 1976 (Booklets); Author: 'Earth Compaction' (McGraw-Hill, NY, 1959);

Morris, M Dan (Continued)
'Okinawa-A Tiger by the Tail' (Hawthorn, NY, 1970); 'Civil Engrs in the World Around Us' (ASCE 1974). Communication Cons to US Office of Personnel Mgt. & indus. Editor, John Wiley Series of Practical Construction Guides. Taught over 400 commn courses. *Society Aff:* ASCE, ASEE, STC, CSE

Morris, Robert G
Business: 1700 W 3rd Ave, Flint, MI 48502
Position: Prof & Chmn Industrial Engrg. *Employer:* General Motors Institute. *Education:* BS/Math/Duquesne Univ; MA/Mgmt/Case Western Res, Univ; MS/Indus Eng/Univ of MI. *Born:* Dec 21, 1921 Cambridge Springs, Mass. m. Jeanne B Bovaird. c. Kathleen, Mary Jo; p. John F & Sarah Morris; p-in-law. Hazel B Discher. Gannon Coll 1946; Duquesne U, BA in Ed 1948-50; Case Western Reserve U, MA 1950-54; U of Mich, MSIE 1966. General Motors Inst Prof, Chmn Indus Engrg 1970- ; Prof of Indus Engrg 1961-70, Faculty Mbr, Math Dept 1955-61; Records Mgr, Good Year Atomic Corp 1954-55; Teacher, Prin Rice Ave Sr HS Girard, Pa 1950-54. Amer Soc for Quality Control, all local offices, Chmn Saginaw VAlley Chap, Mbr Natl ASQC Systems Ctte, ASQC Soc Fellow 1973; Amer Soc for Engrg Educ Indus Engrg Div, Newsletter Ed 1973-74, Secy 1974-75, Prog Chmn 1975-76, Chmn-Elect 1976-77: Am Inst Ind Eng Dir Elect Mgmt Div 1978-79, Dir 1980-81, Chap 98 Pres 1973-74; Mbr Planning Commission 1979-84; Phi Delta Theta. Gneral Motors Inst Alumni Distinguished Teaching Award 1971; IIE Engr of Year 1976; Recipient of many awards for numerous publs & journals. *Society Aff:* ASQC, IIE, ASEE.

Morris, Robert L
Business: 111 New Montgomery, San Francisco, CA 94105
Position: Executive Vice President. *Employer:* Wildman & Morris, Incorporated. *Education:* BS/CE/Univ of CA, Berkeley. *Born:* May 1920. Native of Calif. Lt, Civil Engr Corps, US Navy 1942-45. Firm of Wildman & Morris established in 1953. Complete services are provided in structural & civil engrg, architecture & planning. Fellow & P. Natl Dir of ASCE & has served nationally on & been Chmn of Structural Div's Exec Ctte, Awards Ctte, Session Progs Ctte & Publs Ctte & mbr of Tech Activities Ctte. PPres of San Francisco Sect ASCE. P Chmn of San Francisco Bay Area Engrg Council. Mbr SEAOC, CEAC, SAME, ACI, SSA, ACEC, ASCE & EERI. *Society Aff:* SEAOC, CEAC, SAME, SSA, ACEC, ACI, ASCE, EERI

Morris, Robert L
Home: 413 Sound Beach Ave, Old Greenwich, CT 06870
Position: VP. *Employer:* ITT Continental Baking Co. *Education:* -/Bus Adm/Univ of Chicago; -/Chem E/Univ of Cincinnati; MSChE/Chem Engg/Univ of Penna; BSChE/Chem Engg/Drexel Univ. *Born:* 8/24/32. Native of Phila, PA. At Proctor & Gamble Co (1958-68) in various process dev, computer applications and engg res positions. Dir of Computing Services at Kraft Corp (1968-71). Joined Continental in 1971 as Dir of Process Dev and then Dir of Res. Served ITT (1977-79) as Tech Dir, Food & Chem Products before returning to Continental as VP Tech Affairs. Mbr Bd of Dir of Fundacion Chile. Former Dir, Computing Div of AIChE. Mbr, Scientific Advisory Committee, American Inst of Baking. Mbr, Food, Nutrition and Dental Health Advisory Ctte, Amer Dental Assoc, Rep to Indus Res Inst. Chrmn, Adm Comm of Res Inst of Food Engrg. *Society Aff:* AIChE, IFT, ARD.

Morris, William Page
Home: 2941 Chevy Chase Dr, Houston, TX 77019
Position: Retired. *Employer:* *. *Education:* EM/-/CO Sch of Mines. *Born:* 9/21/07. in Ansted, W VA. Safety & haulage engr US Potash Co, Carlsbad, NM 1932-40; mining supt Intl Minerals & Chem Corp, Carlsbad 1940-49; Resid Mgr Duval Corp, Carlsbad 1950-54 successively VP, Exec VP, Pres, Houston 1954-72, now retired. Mbr Amer Inst Mining & Met Engrs, Beta Theta Pi, Sigma Gamma Epsilon, Tau Beta Pi. Episcopalian Club: River Oaks Country. *Society Aff:* SME.

Morrison, Harry
Home: 2610 NE 51st St, Lighthouse Pt, NJ 33064
Position: Information Movement Consultant *Employer:* Self-employed *Education:* BSEE/Communications/Univ of MI. *Born:* April 2, 1924 in Detroit. BSEE from U of Mich. Mbr AIEE, Engrg Soc of Cincinnati, IEEE & NJ Com Tech Sect; Secy-Treas Com Tech Chapter, Atlanta. Joined AT&T Long Lines in 1948 & progressed to Radio Engr & Building Engr in NY in 1964; in 1968 transferred to Atlanta as Transmission Engr; became Engrg Mgr-Transmission & Engrg Mgr-Prg Implementation. Returned to NY in 1971 as Engrg Dir - Methods. Became Director of Switching Planning 1974, and Director of Systems Engineering 1980 until retirement 1986, Currently self-employed Information Movement Consultant. *Society Aff:* IEEE.

Morrison, John E, Jr
Business: PO Box 748, Grassland Soil & Water Res Lab, Temple, TX 76501
Position: Research Engineer. *Employer:* USDA-SEA-AR. *Education:* PhD/Agri Engg/Univ of KY; MS/Engg Mech/Univ of MI; BS/Agr Engg/MI State Univ. *Born:* Aug 1, 1939 Ionia, Mich. While a student, designed first tractor safety roll-bar. Employed by Massey-Ferguson, Inc 1961-66 as a res engr, receiving patents on no-tillage & precision planters. Served as res engr for Eaton Corp on truck & tractor transmissions from 1966-68. Earned MS in Engrg Mechanics from U of Mich 1968. Specialized in planting & market preparation systems for tobacco, vegetable & field crops. While located at Lexington, KY with USDA-SEA-AR, 1968- 78. Awarded 1974 Young Designer Award by the Natl ASAE. Earned PhD in Agri Engrg from Univ of KY 1978. Currently Leading Conservation-Tillage Research for Dryland Agriculture. Active Mbr ASAE, SCSA & Sigma Xi. *Society Aff:* ASAE, SCSA, ΣΞ

Morrison, John S, Jr
Business: 1911 Arch St, Philadelphia, PA 19103
Position: VP *Employer:* Pennoni Assoc Inc *Education:* BS/CE/Cornell Univ *Born:* 1/17/26 Native of Phila, PA. Served in US Navy during WW II, 1943-1946. Structural Engr for United Engrs and Constructors; Chief Engr, Phila Redevelopment Authority; VP, Engrg and Planning The Korman Corp, firm specializing in residential, commercial, industrial development; VP, Pennoni Assoc Inc, Phila, PA. Consulting Engrs since 1974. Enjoy scouting, camping and gardening. *Society Aff:* NSPE, ASCE

Morrison, M Edward
Business: 13000 Baypark Rd, Pasadena, TX 77507
Position: VP & General Manager *Employer:* AKZO Chemie America *Education:* BS/ChE/IA State; MS/ChE/CA Inst of Tech; PhD/ChE/CA Inst of Tech. *Born:* 7/18/39. Native of Sigourney, IA. Served USAF 1966-1969. Res Dept Mgr for Polymers and Synthetics fibers for Am Enka Co from 1969-1972. Technical Mgr & Plant Mgr for Polymers & Synthetic fibers of Am Enka from 1972-1975. Mgr of New Bus Dev for Akzona 1975-1977. VP & Gen Mgr for Ketsen Catalysts from 1977 to present. Hydrotreating and Fluid Cat Cracking Catalysts are developed, manufactured & sold for use with petrol, chem, coal liquids & oil shale. Reg PE in state of OH & NC. Enjoy tennis & racquetball. *Society Aff:* AIChE, ACS.

Morrison, Malcolm C
Home: PO Box 765, Silverado, CA 92676
Position: Owner *Employer:* OMMS Engrg *Education:* PhD/ChE/CaltEch; BS/ChE/Caltech *Born:* 4/12/42. in Pittsburgh, PA; raised in Lubbock, TX. Sr Scientist for Havens Intl from 1969-1970, specializing in reverse osmosis product dev. Group Leader for Calgon Corp from 1970-1972 overseeing dev of membrane products. Chief Engr (1972-1976), then VP (1976-1979) for Chem Systems Inc. VP, Production & Engg (1979-84) for Puropore. Consultant 1984-present in all aspects of engg and mfg of microfiltration products. Treasurer, Golden State Chapter of the Filtration Soc. Enjoy hunting, hiking, four wheeling and paint ball war games. *Society Aff:* AIChE, ASQC, ΣΞ, FILTRATION SOCIETY.

Morrison, W Bruce
Business: 2007 SE Ash St, Portland, OR 97214
Position: Dir *Employer:* MFIA *Education:* BS/ME/OR State Univ. *Born:* 9/30/11. Mbr Tau Beta Pi, Phi Kappa Phi, Alpha Delta Sigma. GE Co Schenectady 1935-36; Engr for James Smythe, Inc Spokane, WN 1937; & for Meier & Frank Co, Portland, Or 1938-41; Design Engr for Amry Engrs 1941-44; Cons Engr firm W Bruce Morrison & Assocs (sole proprietorship) 1945-65, incorporated 1965, name changed to Morrison, Funatake & Assocs, Inc 1969. Firm specializes in Elec & Mech design for bldgs, central heating & cooling plants, & distribution systems. Mbr Bd of Boiler Rules State of Ore; P Pres Ore Chap ASHRAE; P Natl Dir ASHRAE; Distinguished Service Award of ASHRAE. Fellow ASHRAE. Designed first heat of light installation State of Ore. Articles published in Heating, Piping & Air Conditioning & Heating & Ventilation. Licensed PE Elec & Mech State of Ore 1840. Also licensed in Wash, Calif, Alaska, Iowa & Wisc. *Society Aff:* ASHRAE, ACEC, NSPE.

Morrison, William J
Business: 2 Penn Plaza, Rm 1005, New York, NY 10001
Position: Engineer of Buildings-Design & Const. *Employer:* New York Telephone. *Born:* Sept 1924. BCE Manhattan Coll. Reg PE in 12 states. Designed elec generation plants & bridge work for 5 years. Headed Architectural & Structural design organization for Wester Elec Mfg '& Res factories. Proj Engr of larger mfg plants. Presently head of Design & const Div for NY Telephone bldgs. Active in NY Bldg Congress & NY Const Users Council.

Morrissey, Charles D
Business: 1899 L Street NW, Washington, DC 20024
Position: Sr VP. *Employer:* Frank E Basil, Inc. *Education:* MSCE/Structural/NY Univ; BSCE/Structural/Manhattan College. *Born:* April 28, 1926. Taught civil engrg & structural design at NYU & Manhattan Coll. Joined Praeger Kavanagh Waterbury in 1951, named partner in the firm in 1961, assumed respon of Exec VP of URS/Madigan-Praeger in 1971 & presidency in 1974. Joined Frank E Basil, Inc Consulting Engrs in Jan 1979. Named Sr VP and Dir of Engg in Nov 1979. Respon for design of numerous civil engrg & structural projs including marine & harbor facilities, bridges, subways, bldgs, & special structures. Served as cons to DOD on design of hardened facilities. Named Engr of the Year in Private Practice by NY Chap of NSPE. *Society Aff:* ASCEC, ASCE, ACI, SAME, ΣΞ.

Morrow, Charles T
Business: 228 Agri Engrg Bldg, University Park, PA 16802
Position: Professor *Employer:* Pennsylvania State Univ. *Education:* PhD/Eng Mechs/Penn State Univ; MSAE/Agri Eng/Penn State Univ; BSAE/Agri Eng/W VA Univ. *Born:* 6/20/41. Worked at the USDA Eastern Regional Res Lab in 1965. Major res activities include physical properties of biomaterials, dev of horticultural mechanization equip, instrumentation, microclimate modification & Computer applications. Teaching respons in stress analysis, fluid power, physical properties & instrumentation. Reg PE in PA. *Society Aff:* ASAE

Morrow, Darrell R
Business: Center for Packaging Engineering, P. O. Box 909, Piscataway, NJ 08854
Position: Prof *Employer:* Rutgers Univ *Education:* PhD/Polymer Physics/PA St Univ; MS/Mat Sci/Clarkson Col of Tech; BS/Engr Physics/Lehigh Univ *Born:* Dir of the Ctr for Packaging Engrg, Rutgers Univ. Prof of Packaging Engrg in the Coll of Engrg being active in the creation, teaching and development of many plastics courses and packaging courses a both the undergraduate and graduate levels. Educational and administrative activities include development of research programs involving a broad range of packaging applications, with an emphasis on plastics and the establishment of the Ira S. Gottscho Packaging Info Ctr. Serves on profl, technical organizations and advisory boards to other packaging education programs in the US and is a technical conslt in packaging on an international basis. *Society Aff:* AAAS, ASTM, JETS, ASEE, SPE

Morrow, JoDean
Business: 104 S. Wright St, Urbana, IL 61801
Position: Prof Theoretical & Appl Mechanics Dept. *Employer:* Univ of IL. *Education:* PhD/Theoretical & Appl Mechanics/Univ of IL; MS/Theoretical & Appl Mechanics/Univ of IL; BS/Civ Engg/Rose-Hulman Inst of Tech. *Born:* 10/16/29. Woodbine, IA. Army Corp of Engrs 1950-52. Engaged in res and teaching on mech behavior and properties of engg mtls since 1953 in Dept of Theoretical and Appl Mechanics, Univ of IL. Faculty of Engg, Kyoto Univ, Japan, 1969, and guest of Ukranian Academy of Sciences, 1978. Recipient of ASTM Dudley Medal, 1973, SAE Certificate of Appreciation, 1975, and ASTM Award of Merit, 1978, for res on the cyclic deformation and fatigue properties of mtls. Fellow of ASTM. *Society Aff:* ASTM, SAE, JSMS.

Morrow, Robert P
Home: 420 Parkside Ave, Buffalo, NY 14216
Position: Principal & Owner. *Employer:* R P Morrow & Assocs. *Education:* BS/Mech Eng/Cornell Univ. *Born:* Sept 1920 Erie, Pa. BSME Cornell U 1942. Lab technician summers of 1940 & 1941 for Socony Vacuum Oil Co; Jr Engr to Chief Proj Engr at the Buffalo Refinery of Mobil Oil Co 1942-59, except 10/42-3/46 served in US Naval Reserve as Ensign to Lt Sr Grade; 1959- , Cons Engr in private practice. Received 1973 Engrg Excellence Award from Cons Engrs Council. Former Dir & present Mbr NSPE & Cons Engrs Council. Commercial multiengine airplane pilot with instructor rating in flight & instruments & private glider pilot. *Society Aff:* NSPE, CEC.

Morse, Richard S
Home: 193 Winding River Rd, Wellesley, MA 02181
Position: *. *Employer:* Corp Dir; Consultant *Education:* SB/Physics/MIT; DSc/Hon/Clark; DEng/Hon/Brooklyn Polytech. *Born:* Aug 19, 1911. S B MIT 1933. Technische Hochschule, Munich, Germany 1933-34; DSc (Hon); D Engrg (Hon). Disting Civilian Service Medal 1961. Mbr NAE. Scientific Staff, Eastman Kodak Co; Pres Natl Res Corp -- pioneered indus applications of high vacuum tech, organized Minute Maid Corp; Chmn Army Scientific Bd; Dir of Res & Asst Secy of the Army (R&D); Senior Lecturer Sloan School MIT; Genl Advisory Ctte, ERDA; Tech Advisory Bd, US Dept Commerce; Trustee Aerospace Corp, Current activities: Dir Compugraphic Corp, Dresser Industries, Inc, Woods Hole Oceanographic Inst; Boston Museum Sci, Tracer Tech Corp, Ch. PMC/Beta Corp. *Society Aff:* ACS, NAE, ACS.

Morse, T F
Business: Division of Engineering, Providence, RI 02912
Position: Professor *Employer:* Brown University *Education:* PhD/Eng(ME)/Northwestern; MSc/ME/RPI (Hartford); BSc/Mech Eng/Univ of Hartford; MA/History/Duke; BA/English Lit/Duke *Born:* 2/28/32 Born Brooklyn, New York. Research Engr, Pratt and Whitney Aircraft, East Hartford, CT, 1956-58. Senior Research Engr, Aeronautical Research Assocs of Princeton, Inc., 1961-63. Assistant Prof of Engrg, Brown Univ, 1963-65. Assoc Prof of Engrg, Brown Univ, 1965-68. Prof of Engrg, Brown Univ, 1968-present. Current res in progress is 1) Optical fiber sensors, novel fabrication techniques for optical fibers. Director, Brown University Lab for Lightwave Technology. *Society Aff:* APS, OSA

Mortada, Mohamed
Home: 4820 Millcreek Rd, Dallas, TX 75244
Position: Pres *Employer:* Mortada International *Education:* PhD/Petr Engr/Univ of CA-Berkeley *Born:* 3/14/25 Married: Donna Davis, Oct 25, 1958; Dallas, TX. Children: Sofia Alexandra Mortada. 1946-1947; Instructor, Univ of Cairo, Egypt. 1952-1953; Petroleum Engr, Mobil Oil, Los Angeles, CA. 1954-1963; Sr Res Engr, Res Assoc, Sr Res Assoc; Mobil Oil; Dallas, TX and New York, NY. 1963-1967; Tech Adv, Ministry of Oil; Kuwait, Kuwait. 1968-present; Pres, Mortada Intl, Dallas, TX. Recipient of Alfred Nobile Prize of the Founder Engrg Socs, 1956. Recipient of the Rossiter W Raymond Awd of the AIME, 1957. Distinguished Lecturer for the Soc of Petroleum Engrs, 1971. Reg PE State of TX. Mbr SPE; Mbr AAPG. Author

Mortada, Mohamed (Continued)
of numerous papers on reservoir engrg, petroleum economic analysis, and computer applications. *Society Aff:* SPE of AIME, AAPG, RESA

Mortensen, Richard E
Business: 7731 Boelter Hall, U.C.L.A, Los Angeles, CA 90024
Position: Assoc Prof. *Employer:* UCLA. *Education:* PhD/EE/U CA, Berkeley; MS/EE/MIT; BS/EE/MIT. *Born:* 9/29/35. Native of Denver, CO. Employed by GE Co 1955-57 as part of MIT co-op course work experience. Mbr technical staff, Space Tech Labs, 1958-61. Faculty mbr at UCLA since 1965. Served as consultant to TRW, Aerojet-Gen Corp, & Comp Software Analysts, Inc on aerospace guidance & navigation problems. Author of more than 40 published papers and 1 textbook. Current research interest: Electric Power System Optimization. *Society Aff:* TBΠ, HKN, ΣΞ, IEEE, MAA

Mortimer, Richard W
Business: 32nd & Chestnut Sts, Philadelphia, PA 19104
Position: Assoc VP Acad Affairs *Employer:* Drexel University. *Education:* BS/Mech Engr/Drexel Univ; MS/Mech Engr/Drexel Univ; PhD/Apl Mechs/Drexel Univ. *Born:* Dec 1936. Native of Phila, Pa. Mbr of faculty at Drexel since 1965. Presently Assoc VP for Academic Affairs; served as Dept Chmn & Assoc Dean of Graduate School. Published or presented 50 papers & reports in field of structural impact & wave propagation involving composite & isotropic materials. Recipient of res funding from NASA & AFML. Co-recipient of 1973 ASNT Achievement Award; recipient of 1975 Dow Outstanding Young Faculty Award - Mid-Atlantic Sect ASEE; ASEE-NASA Res Fellow 1967 & 1968. Elected to Haverford Township School Bd in 1975 and 1977. Elected Chrmn of ASME's Mech Engrg Dept Head's Ctte in 1983. Member of Engineering Accreditation Commission of Accreditation Board for Engineering and Technology (ABET). *Society Aff:* ASME, ASTM, ASEE, SME

Morton, George A
Home: 1122 Skycrest Dr 6, Walnut Creek, CA 94595
Position: Consultant LBL, Uni Calif, Berkeley (part time) *Employer:* Retired *Education:* PhD/Physics/MA Inst of Tech; MS/Genl Sci/MA Inst of Tech; BS/Elec Engg/MA Inst of Tech. *Born:* Mar 24, 1903 New Hartford, NY. m. 1934; c. 4. Engrg Res Lab, GE Co 1926- 27; Res Assoc & Instructor MIT 1927-33; Res Physicist RCA Labs 1933-55, Assoc Dir Phys-Chem Lab 1955-60, Dir Conversion Devices Lab 1960-68; Retired 1968. Overseas Premium Award, British Inst Elec Engrs 1937; V K Zworykin Award IRE 1962; David Richardson Medal Amer Optical Soc 1967; Nuclear & Plasma Sci Merit Award, IEEE 1974; Fellow APS, IEEE; Mbr Sigma Xi. Books, co-author *Television* 1940 & 1954; *Electron Optics & the Electron Microscope* 1945; co-author *Instrumentation for Environmental Monitoring: Radiation* VI, editor 1983. *Society Aff:* APS, IEEE, ΣΞ.

Morton, Lysle W
Home: Assembly Point, Lake George, NY 12845
Position: Ret. (GE Co, Schdy, NY) *Education:* BS/Elec Engg/Univ of MN. *Born:* 6/1/01. Specialized in dev, design, application, marketing and standardization of mercury-arc and semiconductor electrical power conversion equips. These efficient converters have largely replaced rotating types. Starting in 1934 various managerial, standardization and consulting assignments were performed until retirement. Professional activities were highlighted as follows: IEEE Fellow (51), Life Mbr (72), first recipient of IEEE Stds Bd "Stds Award" (73); Chrmn of both American Natl Stds Inst's C-34 (52-68) and Intl Electrictechnical Commission's TC - 22 (54-70) committees on static converters. Following retirement served clients as Consultant, "Power Conversion Technologies" (66- 70). Also participate actively in natural resource preservation, receiving 2 awards. *Society Aff:* IEEE.

Moschytz, George S
Business: Sternwartstr 7 ETH Zentrum, 8092 Zurich, Switzerland CH-8092
Position: Prof. *Employer:* Swiss Fed Inst of Tech. *Education:* PhD/EE/Swiss Fed Inst of Tech; Cert/EE/Swiss Fed Inst of Tech. *Born:* 4/18/34. RCA Labs Zurich 60-62; Bell Telephone Labs, Holmdel Supervisor 62-72; since 1973 full prof of network and transmission theory and dir of the Inst of Signal & Information Processing, SFIT. Awards: outstanding paper Electronic Components Conference 1969; Fellow IEEE 1978. Books: Linear Intergrated Networks: Fundamentals (1974) and Design (1975), Van Nostrand Reinhold Co NY, Active Filter Design Handbook (1981), John Wiley, NY. MOS Switched-Capacitor Filters: Analysis and Design, IEEE Press Book, 1984, New York. *Society Aff:* IEEE, SEV.

Moses, Fred
Home: 2257 South Overlook Rd, Cleveland Heights, OH 44106
Position: Professor of Civil Engrg. *Employer:* Case Western Reserve Univ. *Education:* PhD/Civil Eng/Cornell; BCE/Civil Eng/CCNY. *Born:* 12/16/39. Faculty Case Western Reserve Univ (Case Inst) since 1963 teaching structures & mechanics courses. Visiting Faculty NTH (Norway) 1966, Technion (Israel) 1971. Imperial College (London) 1981. Res activity in structures - optimization, safety & reliability, offshore & hwy bridge structures. Reg PE Ohio. Consulting related to safety & computer applications. Winner 1971 ASCE Moisseiff Award. Editorial Bd - Journal of Structural Mech; Journal of Engrg Optimization, Journal of Structural Safety & Journal of Applied Ocean Research; Founder of Bridge Weighing Sys, Inc, a co which markets in-motion truck weighing tech. *Society Aff:* ASCE, IABSE, ASEE, AAM

Moses, Gregory A
Business: 1500 Johnson Dr, Madison, WI 53706
Position: Prof *Employer:* Univ of WI *Education:* PhD/Nuclear Engr/Univ of MI; MSE/Nuclear Engr/Univ of MI; BSE/Nuclear Engr/Univ of MI *Born:* 4/7/50 Native of Kalamazoo, MI. Assumed position of Assistant Prof of Nuclear Engrg at the Univ of WI in 1976. Assoc Prof with tenure in 1980. Prof in 1984. Research interests include inertial confinement fusion and computational physics with an emphasis on transport phenomena. Consultant to industry and US natl laboratories in the area of inertial confinement fusion. *Society Aff:* ANS, APS, ΣΞ

Moses, Joel
Business: MIT 38-401, Cambridge, MA 02139
Position: Hd of the Elec Engg & Comp Sci Dept *Employer:* Dept of Elec Engg & Comp Sci. *Education:* PhD/Math/MIT; MA/Appl Math/Columbia; BA/Math/Columbia *Born:* 11/25/41. Native of Israel, naturalized 1960. Prof of Computer Sci & Engg in Dept of Elec Engg & Computer Sci MIT since 1977; Assoc Dir, Lab for computer Sci MIT 1974-78. Former Chmn, Special Interest Group on Symbolic & Algebraic Manipulation, ACM. In charge of dev of MACSYMA sys for manipulation of mathematical formulas. Assoc Hd, EECS 1978-1981, Hd, EECS, from 9-1-1981. Co- Editor: The Computer Age; A Twenty Year View, MIT Pres 1979; Dir Analog Device, Inc. incl 1982. Elected NAE 1986. *Society Aff:* ACM, IEEE

Moses, Kenneth L
Home: 1404 W Arlington Ln, Schaumburg, IL 60193
Position: VP & Chief Engr *Employer:* Protection Mutual Ins Co. *Education:* BS/EE/Lehigh Univ. *Born:* 12/7/24. Allentown, PA s Jacob Meyer and Alice Katie (Ritter). Married Jean Elizabeth Kulp, Aug 31, 1946; children - Robert Kenneth, Barbara Jean. Served with USNR 1943-1946. Mbr Natl Fire Protection Assn, Aircraft Owners and Pilots Assn, Newcomen Soc, Theta Chi, Republican, Lutheran, Mason. Fact Mut Engr Div Chgo 1948-52, fieldman Protection Mutual Ins Co Chgo 1952-54, Detroit 1954-60, Chgo 1960-64, VP, Chief Engr Park Ridge, IL 1964. *Society Aff:* SFPE.

Moses, Warren G
Business: 929 Howard Ave, New Orleans, LA 70113
Position: Chrmn of Bd *Employer:* Warren G. Moses & Co, Inc *Education:* BE/ME/Tulane Univ *Born:* 06/27/17 Practice as a Consltg Engr since 1947. Chrmn of Bd of Warren G. Moses & Co, Inc, that practices mechanical and electrical engrg;

Moses, Warren G (Continued)
Chrmn of the Bd of Cappel, Tousley, & Moses, Inc, that practices consltg engrg in all disciplines; responsible charge of mechanical, electrical and telecommunications phases of design of construction projects value at $900 million in last ten years alone. World War II active service as B-29 Flight Engr; charter mbr of original society that became Amer Consltg Engrs Council; Commission mbr of Downtown Development District of the City of New Orleans. Past Chrmn of the Grouth Mgmt Prog of New Orleans that developed a master plan for the Central Business District. *Society Aff:* ACEC, ASHRAE

Mosher, Frederick K
Business: 10 Stelton Rd, Piscataway, NJ 08854
Position: Partner. *Employer:* Brownworth, Mosher & Doran. *Education:* -/ME/NYU; -/ME/Lafayette Univ; -/ME/Newark College of Engg. *Born:* 8/25/43. Consltg Engr; born in Middletown, NY, Aug 25, 1943; parents Fred J and Ruth M (Werlau) Maiden, student NY Univ 1970-1972, Lafayette Univ 1973-1974; married Gail J Berry, May 11, 1968; children Scott F, Kerri L, Dean K. With Mayo, Lynch & Assoc, Architects & Engrs, Hoboken, NJ, 1962-64, designer, 1964-69; mech designer, Louis Golderg & Assoc, Metuchen, NJ, 1969-74 assoc, 1975; partner Brownworth, Mosher & Doran, Piscataway, NJ, 1976- . PAST Pres, St Luke's Lutheran Church, WA, NJ, 1975-81. Served with US Army Security Agency, 1965-71. Recipient Natl Award for Engrg Excellence Amer Consltg Engrg Council, 1979 & 1981. Fellow Amer Soc Cert Engrg Technicians; mbr NJ Conslt Engrg Council (Chrmn Engrg Excellence Awards Ctte), Amer Soc Military Engrs, IEEE, ASHRAE (3rd place award for alternative or renewable energy utilization 1982), Natl Soc Prof Engrs, Construction Specification Ins, Warren County, NJ Uniform Construction Code Bd of Appeals. *Society Aff:* NSPE, ASHRAE, ACEC, ASCET, SAME, CSI

Moshier, Glen W
Home: 119 Highgrove Ln, Chesterfield, MO 63017
Position: Group Dir Corp Engrg *Employer:* Ralston Purina Co *Education:* BS/ME/Univ of OK *Born:* 8/10/40 Broad based background in production, design and construction engrg. Four yrs with the Western Electric Co in production engrg in metal working operations and in fabrication of electromechanical devices. Twelve yrs with Johnson & Johnson; last positions were Chief Engr and Plant Mgr. Assumed present position of Group Dir, Corporate Engrg, Ralston Purina Co in 1978. Current responsibilities include design, const mgmt, environmental engrg, and corporate energy conservation. Active in church activities, serving as an elder and adult teacher. *Society Aff:* ΠΤΣ, AMA.

Mosier, Frank E
Business: 200 Public Square 40-D, Cleveland, OH 44114
Position: Pres & Chief Operating Officer *Employer:* The Standard Oil Company *Education:* BS/Chem Engrg/Univ of Pittsburgh *Born:* 7/15/30 Native of Kersey, PA. BSChE Univ of Pittsburgh. Served with US Army 1953- 55. With Std Oil Co (OH) since 1953. Assumed respon as VP for Supply & Distribution in 1972; respon for Pipeline Trans added in 1976 with title of VP, Supply & Trans. Elected Sr VP of Mkting & Refining in Jan 1977 & became Sr VP, Supply & Trans in 1978. Elected Corp Dir, Dec 1980. In Feb 1982, Sr VP, Downstream Petroleum. Jan 1985, elected exec VP. April 1986, pres and chief operating officer. Serves on Bd Dir of Amer Petroleum Inst and on Bd of governors of 25 Year Club of the Petroleum Industry. Dir of Soc Corp and the Centerior Energy Corp. Trustee, Univ Pittsburgh, John Carroll Univ, Greater Cleveland Council Boy Scouts of America, Center of Science & Industry, and Cleveland Council on World Affairs. Vice chrm, Bd of trustees, Fairview General Hospital. Mbr, Greater Cleveland Roundtable. Leisure interests include golfing, fishing, hunting, American history. *Society Aff:* API, NPRA, AIChE, AMA

Mosley, Ernest T
Home: 3 Beechtree Lane, Plainsboro, NJ 08536
Position: Principal. *Employer:* Raamot Assocs, PC. *Education:* MS/Civil Engg/Univ of IL; BS/Civil Engg/Univ of TX. *Born:* May 1931. MS from U of Ill; BS from U of Texas. Field Engr on heavy const projs for Palmer & Baker, Inc for 2 years; with Raymond Internatl for 15 years as Field Engr, Job Supt, Construction Cost Estimator, Genl Supt & Chief Foundation Engr. With Raamot Assoc, PC since 1971 - respon for engrg conracts, geotechnical analysis & design, report writing & field work. Pres Syracuse Sect ASCE 1975-76; Chmn ASTM Subcttes on Deep Foundations. Enjoy skiing, bicycling. *Society Aff:* ASCE, ASTM.

Moss, Alan M
Business: DRDAR-LCU, Dover, NJ 07801
Position: Tech Dir *Employer:* ARRADCOM. *Education:* BEE/Elec/CCNY; MSc/Elect Eng/Stevens Inst of Tech *Born:* 2/12/31. Native of NYC. Attended Bronx HS of Sci. Served with Army Ord Corps 1952- 54. Remained at Picatinny Arsenal (now ARRADCOM) as a civilian engr. Designed and developed fuzing, arming & safing systems for army nucl warhead sections. In 1969 attained a supergrade position as deputy dir, nucl engg directorate. Currently in the Sr Exec Service & direct 4000 employees responsible for R&D of all non-nucl ammunition systems. *Society Aff:* IEEE, ADPA, NYAS, TBΠ

Moss, Clarence T, Jr
Business: 405 Lexington Ave, New York, NY 10174
Position: VP. *Employer:* Niagara Blower Co. *Education:* MBA/Mgt/Baruch College; MS/ME/Univ of Buffalo; BS/ME/Univ of Buffalo. *Born:* 12/20/25. in Lockport, NY. Served in USAF in 1944-45. Joined present firm on grad from college, working through various levels of engg & mgt. Author of two chapters of the volume on heat transfer equip in the Process Equip Series Active in local service organizations & high sch Bd of Ed. Enjoys horseshoes, gardening, cabinet making & stained-glass. *Society Aff:* AIChE, ASHRAE, NACE, NSPE.

Moss, Frank H, Jr
Business: 359 City Hall, San Francisco, CA 94102
Position: Deputy Director for Engineering and City Engineer *Employer:* San Francisco Dept of Public Works. *Education:* BCE/Civil Engg/Univ of Santa Clara. *Born:* June 19, 1934 Kobe, Japan of US father & Eurasian mother. Lived in Shanghai, China until was 14 years old, then in Tokyo, Japan until 17. Served 2 1/2 years in the US Army in Okinawa as Personnel Officer & Asst Troop Commander for an ordnance group. Began employment with City & County of San Francisco in 1960, engaged in structural engrg of municipal structures. Transferred to Sanitary Engrg Group in 1969, & was involved in the dev of master plan for wastewater mgmt, primarily in the hydrology area, & implemented the installation of computer based data acquisition sys. Returned to structural engrg to engage in implementing master plan const 1974-78. Recipient 1976 Wesley W Horner Award, ASCE. Now Deputy Dir for Engrg, City & County of San Francisco, Dept of Public Works.

Moss, Gerald
bTroy, NY 12181
Position: Prof Biomedical Eng. *Employer:* Rensselaer Polytechnic Inst. *Education:* PhD/Biochemistry/Union Univ; MD/-/Albany Medical Coll; MS/Physical Chemistry/NY Univ; BA/Chemistry/NY Univ. *Born:* 2/1/31. *Society Aff:* BMES, AAAS, AAMI, APS, ACS.

Moss, Ralph A
Home: 6309 NE 159th St, Vancouver, WA 98665
Position: Pres *Employer:* Self Employed *Education:* MS/IE/St Louis Univ; BS/EE/Univ of CO *Born:* 6/21/25 in Jersey City, NJ. Served in WW II 1944 to 1945. From 1947 to 1950 Motorola Field Engr in 2-Way Radio Communications. Entered Federal Service as Electronic Technician DOD. Designed Microseismic Detector/Amplifier for US Bureau of Mines. Acheved Professional Status in Government, responsible charge of engrg projects for Corps of Engrs, 1968 to 1978. Retired from Federal Service 1979 to found RAM Engrg. Currently Pres of firm. Has authored numerous tech papers & served in Engrg societies. *Society Aff:* NSPE, ACEC

Mossman, Gary L
Home: 18627 Pt. Lookout, Houston, TX 77058
Position: General Manager. *Employer:* Thiokol Corp/Spec Chem Div. *Education:* ME/Enn Eval/IA State Univ; BS/Engg Oper/IA State Univ; BS/Chem Engg/IA State Univ. *Born:* 10/17/40. IA native. Industrial Engg Faculty Mbr-IA State Univ 1964-65. Held various e ngg and marketong positions with Salsbury Lab 1966-72. Joined Southwest Specialty Chem as VP of Marketing - Pres from 1973-79. Following merger with Thiokol Corp, became General Manager of new Specialty Chem Div, with respon for dom and international operations. Experience included respon for plant design, construction, startup, pilot plant, commercial development, marketing, corp planbubgm acquisitions, and general management. AIChE Bd of Dir 1968-70. Pub articles on marketing, PERT, and engg. Enjoys all outdoor activities. July 1, 1980 became a partner in Flint Assoc, Mgt/Tech consulting firm and April 1981 Pres of Ram Mfg Inc, an affiliated co, Mfg/Mkt of industrial lasers. *Society Aff:* AIChE, ASME TCC

Mostaghel, Naser
Business: 3012 Merrill Engrg Bldg, Salt Lake City, UT 84112
Position: Prof of CE *Employer:* Univ of UT *Education:* PhD/Structures/Univ of Ca-Berkeley; MS/Structures/Univ of CA-Berkeley; BS/Gen Engr/Abadan Inst of Tech *Born:* 7/23/39 Earned PhD in Structural Engrg and Structural Mech from U C Berkeley in 1968. Taught at Pahlavi Univ in Iran for eleven yrs, spending a sabbatical year at Earthquake Engrg Res Ctr, UCB. Founded Earthquake Engrg Ctr at Pahlavi Univ. CE Dept Chmn 1974-1976, vice-dean for academic affairs, 1979-80. Consult to Atomic Energy Org. of Iran. Joined CE Dept U of Utah in 1980. Significant res includes stability of rotating columns, linear response of structures subjected to earthquake, liquefaction analysis, dev of base isolators. Current res includes dev of bounds on response of nonlinear structures subjected to earthquake loading, dev of base isolators, and formulation of new definitions of earthquake intensity. *Society Aff:* ASCE, EERI.

Mostovoy, Sheldon
Business: 10 W 33rd St, Rm 206 Perlstein Hall, Chicago, IL 60616
Position: Assoc Prof *Employer:* IL Inst of Tech *Education:* PhD/Metallurgical Engr/IL Inst of Tech; BS/Mechanical Engr/IL Inst of Tech *Born:* 10/14/33 Worked for 17 yrs in industrial research (Materials Research Lab, Glenwood, IL) on problems involving fracture mechanics and mechanical behavior of materials. Inventor and developer of "Constant-K" test methods for metals, adhesives and composites. Developed fracture tests for adhesive bonding for aircraft structures (ASTM Standard). Characterized high rate fracturing behavior of adhesives and metals. Developed complex mech working machine for lowering the transition temperature of refractory metals. Determined the embrittling effect of lead on low alloy steels. Patents on elimination of lead embrittlement and working leveller. Introduced automated test materials in both research and teaching. *Society Aff:* ASM, ASTM, SESA, AIME

Mosure, Thomas F
333 East Federal St, Youngstown, OH 44503
Position: Chmn. *Employer:* Mosure & Syrakis Co, Consltg En & Planners, Mosure & Assoc, Inc, Arch. *Education:* MS/Civil Engg/Univ of Pittsburgh; BE/Civil Engg/Youngstown State Univ. *Born:* 5/21/33. Native of Youngstown, OH. While attending college, Asst Engg for Mahoning Co, Oh. Youngstown State University instructor for total of 4 yrs teaching highway design, steel structure theory, sewage and other designs. In 1960, was Mahoning Co Chief deputy engg, and in 1961 became Chief Engg for a Canfield, Oh, Consulting Engg Firm. Since 1963, has been chmn and pres of a consulting Eng co and architectural co directing & overseeing all aspects of company operations of 107 employees and work. Reg PE in OH, PA, KY, IN, NC and WV. Reg Prof Surveyor in OH and PA, Outstanding Civil Engr, 1984, ASCE Local Chpt. *Society Aff:* NSPE, ASCE, OACE, ARBA.

Motard, Rodolphe L
Business: Dept of Chem Engg.-Box 1198, Washington Univ, St Louis, MO 63130
Position: Prof & Chairman *Employer:* WA Univ. *Education:* DSc/ChE/Carnegie Mellon Univ; MS/ChE/Carnegie Mellon Univ; BSc/ChE/Queens Univ. *Born:* 5/26/25. Native of Ottawa, Canada. In process dev with Shell Oil Co, Houston, TX, 1951-57, specializing in fluid catalytic cracking, fluidization and reactor design. Joined Dept of Chem Engg, Univ of Houston in 1957. Became Chrmn of the Dept at WA Univ in 1978. Res interests are: computer application to process engg including process design, process simulation, process synthesis, process dynamics; physical property estimation; data mgt in design; ChE sys engg. *Society Aff:* AIChE, NSPE, ACS, AAAS.

Motayed, Asok K
Business: Sheladia Associates, Inc, 15825 Shady Grove Road, Rockville, MD 20850
Position: Senior Vice Pres *Employer:* Sheladia Assocs, Inc *Education:* MS/Environ Engr/Rutgers Univ; BSC/CE/Jadavpur Univ, India *Born:* 12/30/44 Prior to immigrating to the US in 1971, worked for a consltg engrg firm in structural and water engrg. 1973-1979, worked with Dames and Moore, starting as Asst Engr and advancing to Project Engr. Joined Sheladia Assocs in 1979, 1981-84; Sr VP, Intl Div in engaged in development and mgmt of resource development and technical assistance projects in the developing countries. Currently, SR VP responsible for an Operating Division, & also responsible for firm-wide activities in the area of water resources & energy. *Society Aff:* ASCE, IWRA, AMA.

Mote, C D, Jr
Business: Dept of Mech Engr, Berkeley, CA 94720
Position: Prof & Chairman *Employer:* Univ of CA, Dept Mech Engrg *Education:* BSc/ME/CA Univ Berkeley; MSc/ME/CA Univ Berkeley; PhD/ME/CA Univ Berkeley *Born:* 2/5/37. Came to the Univ of CA, Berkeley in 1971 from Carnegie Tech where he had been since 1964. Chairman of Depart of Mechanical Engrg at Berkeley, 1987- . Res interests lie in dynamical systems, control, finite element, instrumentation, vibration and acoustics, design and biomechanics. Recognized for res contributions on the vibration control and optimal design of band and circular saws, and on snow ski design and snow skiing injuries. Mbr of San Francisco Sec of ASME Exec Comm for six yrs, Sec Chrmn 1978-79. VP, ASME, 1986-88. Chairman ASME Noise Control and Acoustics Division 1979-1982. Vice President and Secretary of the International Society for Skiing Safety 1977-1985. Distinguished Teaching Award, Univ of CA, 1971, Blackall Machine Tool Award ASME 1975, Fellow Int Academy of Wood Sci 1978. Fellow ASME 1984. *Society Aff:* AAAS, ASB, ASA, FPRS, ASTM, ISSS.

Mott, Harold
Home: 4025 Windermere Drive, Tuscaloosa, AL 35405
Position: Professor *Employer:* Univ of AL *Education:* PhD/EE/NC St Univ; MS/EE/NC St Univ; B/EE/NC St Univ *Born:* 6/16/28 Harris, NC, son of Volna Logan Mott and Lela Jane (Jackson) Mott. Married Elizabeth Irene Hunter 1959. 1 son John Harold. Engr Wright Machinery Co, Durham, NC 1953-54. Instructor NC State U 1954-60. Assoc Prof 1960-64, Professor 1964- U of AL. Summer Positions Western Elec, Oak Ridge Natl Labs, Boeing Co. Dir research contracts for NASA. Consultant Troxler Labs, Boeing Company, US Army Missle Command, JMA, Inc. *Society Aff:* IEEE

Motter, Eugene F
Home: 903 Princeton, Midland, TX 79702
Position: Engrg Mgr-SW Reg *Employer:* Cities Service Co. *Education:* BS/ME/KS St Univ; Geo/Southwest LA Univ; Exec Develop/Cornell Univ *Born:* 3/24/27 Native of Gaylord, KS. Served in US Air Force 1945 - 46. Joined Cities Service 1950, Present Position 1973; Bd Mbr NM St Bd Professional Engrs 1965 - 67; Mbr Hobbs, NM City Council 1964 -67 (Major Pro-Tem 1967), Mbr Midland, TX Zoning Commission 1978 - 80. Dir of Permian Basin SPE - various committees SPE. *Society Aff:* SPE, NSPE, API

Mottern, Robert W
Business: Kirtland AFB East, Albuquerque, NM 87185
Position: Member of Technical Staff. *Employer:* Sandia Natl Laboratories *Education:* BS/Educ & Chem/East TN State Univ. *Born:* Feb 1924 Johnson City, TN. Chemist, Manhattan Engrg Dist, Oak Ridge, TN 1944-47; Chemist, Texaco Res Labs Glenham, NY 1947-51; Staff Mbr, Sandia Labs 1951- . Currently working nuclear security systems & film image analysis. Fellow Amer Soc Nondestructive Testing; Mbr Amer Assoc for the Advancement of Sci; Mbr Shroud of Turin Res Proj. Performed X-Ray examination of Turin Shroud, busy extracting data and giving talks on own time. Enjoy gardening, bridge, & reading. *Society Aff:* ASNT.

Motzkus, A Robert
Home: 4689 Deer Creek Rd, Salt Lake City, UT 84117
Position: Engineering Superintendent. *Employer:* Mountain Bell Telephone Co. *Born:* Dec 19, 1938. Native of Salt Lake City, Utah. BSEE U of Utah 1962. Employed by Mountain Bell as an Engr in 1962; appointed Engrg Supt in 1967. Areas of specialization include Engrg Mgmt, Budget & Control of Capital Expenditures, Current & Long Range Planning. Principal in Land Dev Co & Proprietor of an Excavation Co. Reg PE in Utah. Affiliated with NSPE; Pres Utah Soc of PE's 1975-76. Utah's Young Engr of Year in 1971. Married & father of 6. Active in Mormon Church, Scout Master, Little League coach. Enjoys hunting, fishing, skiing in the outdoors.

Moulder, James E
Business: 1139 Olive St, St Louis, MO 63101
Position: Pres & Chrmn. *Employer:* Booker Assoc, Inc. *Education:* MS/Civil Engrg/Univ of MO; BS/Civil Engrg/Univ of MO *Born:* 08/29/26 Native of Camdenton, MO. Served in US Army, 1947-49. BS and MS in Civ Engg, 1953 & 1955, from Univ of MO-Columbia. Entered consulting engg practice in 1956 with Smith & Gillespie Engrs, Jacksonville, FL. Joined Booker Assoc, Inc, of St Louis, MO in 1961 as Mgr, Civ Engg, promoted to VP in 1963, Exec VP in 1968, Pres in 1973 and Chrmn of the Bd in 1977. MO Honor Award for Distinguished Service in Engg, 1977, Award of Merit, St. Louis Engineers' Club, 1981, Faculty/Alumni Award, UMC Alumni Assoc, 1979. Reg PE in MO, FL, IL, KS, MD, KY, LA, MN, OK, OR, TN, VA, WV, IA, AR, IN. Natl VP of ACEC, 1984-86, Pres of MSPE, 1984, St. Louis Chptr Prof Eng in Private Prac & Eng of Yr Awd, 1981 & 1982, MO Coordinating Bd for Higher Ed, 1981, Chrmn of UMC Coll of Engrg Advisory Coun, 1984; Past Pres, Consulting Engrs Coun of MO; 1982 Received UMC Alumni Assoc's Distinguished Ser Awd. *Society Aff:* ACEC, NSPE, NAHRO, SAME

Moulder, Leonard D
Home: 6511 Saulsbury Ct, Arvada, CO 80003
Position: Senior Industrial Engrg Supervisor. *Employer:* Coors Porcelain Company. *Education:* AB/Politcal Sc/Baldwin-Wallace. *Born:* April 1916 LaGrange, Oh. Served in Natl Guard 1936-39. Joined Fairmont Food Co, Cleveland in 1941 & worked until 1952 in various positions from process operator to supt of ice cream plant. Indus Engr, Ford Motor Co 1952-55. With Coors Porcelain Co, Golden Colo since 1956. Assumed present duties as supr of Indus Engrg in 1958. PE State of Colo; Sr Mbr IIE; Pres Rocky Mountain Chap IIE 1967-68; Dir, Mgmt Div IIE 1974-76. Active in Denver Goodwill Industries: Bd/Dirs 1971-82, Exec Ctte 1976. Active in local church. VP, Region X Director University Chapter 1980-82, Amer Inst of Industrial Engrs 1978-79 & 1979-80 Pres, Goodwill Bd of Dirs, 1977. Enjoy camping, fishing & classical music. Received Fellow Award-American Inst of Industrial Engrs - 1981. *Society Aff:* IIE, NSPE.

Moulton, Ralph W
Business: 105 Benson Hall BF-10, Seattle, WA 98195
Position: Prof Dept of Chem Engrg *Employer:* Univ of Washington. *Education:* PhD/Chem Engg/Univ of WA; MS/Chem Engg/Univ of WA; BS/Chem Engg/Univ of WA. *Born:* June 1912. Native of Seattle, Wash. Employed by Union Oil Co of Calif from 1937-41. Mbr of faculty of Dept of Chem Engrg at the U of Wash since 1941. Dean, Joint Center for Grad Study at Richland, Wash, representing both the U of Wash & Washington State U. Mbr Amer Chem Soc, ASEE, AIChE, Sigma Xi, Tau Beta Pi, & other prof societies. Mbr, Engrg & Accreditation Ctte of ECPD. Have specialized in critical flow of steam-water mixtures. *Society Aff:* AIChE, ACS, ASEE.

Mouly, Raymond J
Home: 4230 Centre Ave, Pittsburgh, PA 15213
Position: Staff Engineer. *Employer:* PPG Industries, Inc. *Education:* Ing Dip/EE/Ecole Superieure o' Electricite (Paris); Lic es Sc/Sci/Univ of Paris; BSc/Sci/Univ of Montpellier; BPh/Philos/Univ of Montpellier. *Born:* Nov 1922. Ing Dipl ESE (Paris); Lic es Sc (Sorbonne). PE Pa. Served with First French Army 1943-45. Mgr of Control Engrg for Corning Glass Works 1961-71; joined PPG Industries as Staff Engr in 1972. Interests focus on analysis & dev of glass making processes. 9 US patents. Author of numerous tech publs. Editor of 'Automatic Control in Glass' 1973. Survey Lecturer, 8th Internatl Glass Congress (London, 1968). Genl Chmn of 1st IFAC Symposium on Automatic Control in Glass (Purdue U, 1973). Dir Glass & Ceramics Group of ISA. ISA Fellow. Sr Mbr IEEE. Mbr NSPE. *Society Aff:* NSPE, IEEE, ISA.

Mourier, Georges
Home: 29 Route de Versailles, Pt Marly France 78560
Position: Scientific advisor to Director *Employer:* Thomson-CSF. Electron Tubes *Education:* Dr Sci/EE/Fac des Sciences - Orsay; Ing Dipl/Phys/Ecole de Physique et Chimie; Baccalaureat/Philophie Math/Paris. *Born:* 7/12/23. Most of my activity was devoted to the study of the interaction of free particles & electromagnetic fields, & to the applications to microwave tubes and RF heating of plasmas. I started in the res labs of CSF Co, Paris. later I spent three yrs in the US, doing res & teaching (Columbia & PIB). I returned to the same lab in France, & became interested in nucl fusion res. Since then I have also been teaching in different engg schools & at the Univ. Much of my work is devoted to the study of high power electron tubes at microwave or millimetric frequencies. I have contributed about 45 papers & confs in electron tube theory & plasma physics, & hold some 30 patents. *Society Aff:* SEER, IEEE.

Moustafa, Saad E
Home: 7156 Thornley Dr, New Orleans, LA 70126
Position: Assoc Prof. *Employer:* Univ of New Orleans. *Education:* PhD/Structural Engr/Univ of CA; MS/Structural Engr/Cornell Univ; BS/Civ Engr/Cairo Univ. *Born:* 5/25/38. Native of Egypt, came to the US in 1964, became a US citizen in 1975. Taught at Cairo Univ & the Univ of New Orleans. Sr Res Engr with Concrete Tech Corp, Tacom WA 1969-1971. Mgr of Structural Res with Concrete Tech Assoc 1972- 1975. Consulted in the design of high-rise bldgs in Cairo & New Orleans. T Y Lin Award 1971 & 1978 from the ASCE. Martin P Korn Award 1977 from the PCI. Enjoy music & fishing. *Society Aff:* ASCE, ACI, PCI, ΣΞ.

Mow, Maurice
Business: 930 CSUC, Chico, CA 95929
Position: Prof *Employer:* CA State Univ. *Education:* PhD/Applied Mech/RPI; MS/Applied Math/RPI; BCE/Civil Engr/RPI *Born:* 6/24/40. Since graduation from RPI in 1968 with the PhD in Mechanics I've been engaged as follows: 1968-1971 mbr of the Tech staff at TRW Systems, Redondo Beach, CA; 1971-1973 VP Doch Corp, Nanuet, NY; 1973-1976 Partner, Design & Planning Assoc Middletown, NY; 1976-1978 Asst Prof Univ of ME, Orono, ME; 1978- pres Prof CA State Univ, Chico. Also conslt on software dev for reinforced concrete design. *Society Aff:* ASCE, ACI, ASEE

Mow, Van C
Business: Columbia University, 630 West 168th St., BB14-1412, New York, NY 10032
Position: Prof Mech Engg & Orthopaedic Bioengrg *Employer:* Columbia Univ

Mow, Van C (Continued)
Education: PhD/Mech/RPI; BAE/Engrg/RPI *Born:* 01/10/39 Native of Chengdu, The People's Republic of China. After serv as postdoctoral mbr Courant Inst of Mathematical Sci, NYU, and Mbr Tech Staff, Bell Telephone Labs, came to RPI as Assoc Prof in Mechs in 1969. Founded the Biomechs Res Lab in 1969. Appointed Full Prof of Mech and Biomedical Engrg in 1976. Appointed John A Clark and Edward T Crossan Prof of Engrg in 1982. Dir, N.Y. Orthopaedic Hospital Res. Lab., Columbia Presbyterian Med Cntr; Prof of Mech Engrg and Bioengrg, Columbia Univ, 1986. Major res focus has been dev of models for articular cartilage and synovial fluid deformational properties, proteoglycan biorheology, and collagen-proteoglycan interactions. Has over 175 publs. Honors include Fellow ASME; Kappa Delta Award, ORS, 1980; William H Wiley, Distinguished Faculty Award, RPI, 1981; Honorary Prof, Chengdu Univ of Sci and Tech, 1981; Melville Medal, ASME, 1982; Pres, Orthopaedic Res Soc, 1982-83; Chrmn, NIH Review Ctte on Orthopaedics and Musculoskeletal Sys, 1982-84; Honorary Prof, Shanghai Univ of Sci and Tech, 1983; Chrmn, Bioengrg Div, ASME, 1984-85. Japanese Soc. for Promotion of Science Fellowship, 1986. Honorary Prof. Shanghai Jiao Tong Univ, 1987. Fogarity Senior Int. Fellowship, 1987. Alza Distinguished Lectureship, Biomedical Eng. Soc. 1987. ASME HR Lissner Award for Contributions to Bioengineering, 1987. *Society Aff:* ASME, ORS, ASB, ASCE, AAAS, AAOS

Mowrer, David S
Business: PO Box 446, 108 Flint Rd, Oak Ridge, TN 37831-0446
Position: Senior Engineer *Employer:* Professional Loss Control *Education:* BS/Fire Prot./Illinois Institute of Technology *Born:* 04/30/49 Native of Springfield, MO. Five years experience with Insurance Services Office of Missouri 1971-76 in engrg dept evaluating municipal fire protection. 1976-79 worked in Houston, TX as Senior Engr with Brown & Root as Discipline Staff Engr for fire protection design of complete naval shipyards, power plants, and petrochemical plants. Joined Professional Loss Control, Inc. in 1979 as a Senior Fire Protection Engr specializing in providing engrg consulting services to heavy industry. Current responsibilities include fire detection & suppression system design & testing, fire hazards analysis, training, and risk assessment. Pres, TN Valley Chapter SFPE 1985-87. *Society Aff:* SFPE

Moxey, Richard T
Business: PO Box 7, New Britain, PA 18901
Position: Pres. *Employer:* Moxey Assoc, Inc. *Education:* BSEE/Elect Eng/Drexel Univ *Born:* 9/20/32. Native of Phila, PA. Reg PE in States of PA, NJ, NY, VA, MD, DE. FL. Draftsman for Walker-Yeomans Assoc & L W Moxey 3rd to 1957, Proj Engr for L W Moxey 3rd to 1961, Partner 1962-1970. Secy, Moxey Assoc 1970-75, Pres 1975-date. Secy. P, Pres Bucks County Chapter PSPE, Rec for Quality & Quantity of Light Comm (IES), Who's Who in the East, former Zoning Officer, New Britain Borough (NTPD) Code Making Panel 1S. VChrmn PA/PEPP in private practice. PA/PEPP Energy Committee. Enjoy classical music, and model trains. *Society Aff:* NSPE/PSPE, IEEE, IES, NFPA, PA/PEPP, NSPE/PEPP.

Moxley, Stephen D
Home: 5810 Criner Rd, Huntsville, AL 35802
Position: Vice Pres & Genl Manager. *Employer:* Avco Corp, Electronics Div. *Education:* BS/EE/Univ of AL; SM/EE/MIT. *Born:* Dec 1, 1927. BSEE U of Ala; SMEE MIT. Native of Birmingham, Ala. Design engr for Reliance Electric, Cleveland; & geophysical res engr for Continental Oil Co, Ponca City, Okla prior to joining Avco Electronics Div in Cincinnati in 1955. Presently head this Div in Huntsville -design & manufacture of various electronic prods. Holder of 3 patents. Reg PE in Ala & Ohio. Chmn, Huntsville Sect IEEE 1971. Attended MIT Sloan School Sr Executives Prog 1972. Genl CHmn Huntsville United Way Drive in 1974. Elected Fellow IEEE 1974 for contribs to geophysical prospecting & air traffic control.

Moxon, John
Home: PO Box 275, Oley, PA 19547
Position: Retired - Chmn & President. *Employer:* Carpenter Technology Corp. *Education:* MCS/Finance/Amos Tuck; AB/Econ/Dartmouth; LLD/Hon/Albright. *Born:* May 26, 1906. AB Dartmouth Coll 1929; MCS Tuck School of Business Administration 1930; Cours de Civilization, U de Paris 1927-28; Ecole des Sciences Politiques, Paris 1927-28; LLD Albright Coll 1966. Guaranty Trust Co NY 1930-44; Carpenter Steel Co Secy, Treas, & Dir 1944; VP, Exec VP, then Pres 1959-71; Chmn 1971; Ret Nov 1971. Life Mbr ASM (hon). Main job most of business career was in managing people, including engrs. Matter of fact a high point in my career was a talk I gave Wash Soc of Engrs some time ago on subject of 'Why Engrs Don't Get More Money'. The Soc pointed up my thesis by recording me on a recorder which was wholly non-functioning. *Society Aff:* AISI, ASM.

Moy, Edward A
Business: 522 N Broad St, Woodbury, NJ 08096
Position: President. *Employer:* Edward A Moy, Cons Elec Engrs, Inc. *Education:* BS/Elec Engr/Drexel Univ. *Born:* Nov 16, 1928 Philadelphia, Pa. 1951-54 active duty with US Army Corps of Engrs; 1954-57 Elec Engr for Bellante & Clauss - Architects & Engrs. Respon for electrical design & supr of all projs from Phila office. 1957- , Owner/Pres Edward A Moy, Cons Elec Engrs specializing in electrical design for commercial, industrial const, water & waste water treatment plants, & scrap processing facilities. Reg PE in Pa, NJ, Del, WVa, Ky, DC, NY, MA, ME, IA, IN, VA, IL, Conn., & OH. *Society Aff:* IES, BOCA, NSPE, AEE.

Moyer, Harlan E
Business: 1600 SW Western Blvd, Corvallis, OR 97339
Position: Pres. *Employer:* CH2M Hill, Inc *Education:* BS/Civil Engr/Univ of NV. *Born:* 12/20/26. Native of Napa, CA. Served in Naval Air Corps 1944-46. PE with CH2M Hill since 1952. Proj Mgr on Civil and Water Pollution Control Projs 1952 to 1977. VP & Reg Mgr of Redding, CA office 1974-77. Assumed Presidency of CH2M Hill Jan 1, 1978. Enjoy skiing & outdoor activities. *Society Aff:* ASCE, NSPE.

Moyer, Horace B
Home: 2514 Greenleaf Blvd, Elkhart, IN 46514
Position: Mgr, Pharmaceutical & Food Equipment Dev. *Employer:* Miles Laboratories, Inc. *Education:* BS/Chemistry/Penn State Univ. *Born:* Aug 1917. Started with Atlas Chem Co - Acid, Bomb Loading, Powder Line Supr; Asst Chief Engr for American Home Foods, Proj Mgr of multimillion dollar food processing plants; Plant Engr, Production Mgr at Whitehall Labs. Relocated plants from eastern & southern states to Midwest. With Miles Labs since 1953, 15 years as Mgr of Food & Pharmaceutical Engrg, worldwide basis. Established engrg depts (prod res, redus, design). Reg PE Indiana. ISPE Outstanding Service Award 1962, Pres 1964-65; Natl Dir since 1965, served on numerous NSPE cttes. Order of Engr, charter mbr Natl Bd of Governors, The *Engineer of The Year* 1978 in IN. NSPE V Pres of Central Region 1981-82 - 1982-83. *Society Aff:* NSPE.

Moyer, Kenneth H
Home: 4 Green Briar Ln, Cinnaminson, NJ 08077
Position: Pres *Employer:* Magna-Tech P/M Labs *Education:* MS/Met Engr/Poly Inst NY; BS/Met Engr/Poly Inst NY. *Born:* 10/1/29. Resident of Cinnaminson, NJ. Married, four children. Hd heat treatment dept paint line & quality control US Hoffman Machinery Corp 1953-54; project engr Intl Nickel Res Lab 1954-55; jr engr quality control lab, Grumman Aircraft Engg Corp 1955-58; mgr quality control lab Sylvania-Corning Nucl Corp 1958-60; proj engr Beryllium Corp 1960-62; mgr beryllium operations Gen Astrometals Corp 1962- 66; mgr spec alloys and mgr new product development 1966-1985; Pres & Principal Investigator, Head of all Research & Development for Magna-Tech P/M Labs 1985-present; Served with USN 1948-50; Reg PE PA & NJ; Author. *Society Aff:* AAAS, ASM, AIME, IEEE, SME.

Moyer, Richard B
Business: 101 W Bern St, Reading, PA 19603
Position: NDT Specialist *Employer:* Carpenter Technology Corp. *Education:* BS/Engg-Physics/Lehigh Univ. *Born:* Nov 6, 1919 Reading, Pa. Served in various elec engrg capacities at Curtiss Wright Corp, Textile Machine Works, Gilbert Assocs & present employer. While at Carpenter established the NDT prog involving personnel training, qualification, & certification; equip procurement, maintenance, & calibration; specification review; & procedure preparation. 1984 Fellow of ASTM & Past Chmn of ASTM Ctte E-7 on Nondestructive Testing, received the 1984 ASTM Award of Merit; Fellow 1974 of ASNT; serving in various natl capacities including Chmn of Steel Producers Ctte. In 1968 was named Greater Philadelphia Engr of the Year by ASNT; Chrmn in 1977-79 of Tech Ctte on NDT of AISI. *Society Aff:* ASNT, ASTM.

Mozer, Harold M
Business: 777 108th Ave. NE, P.O. Box 91500, Bellevue, WA 98009-2050
Position: Sr Consultant, Energy and Electric Power *Employer:* CH2M Hill, Inc *Education:* BS/EE/Univ of NB *Born:* 04/27/27 Native of Omaha, Nebraska. Electrical Engr with US Dept of Interior, Bonneville Power Administration, 1948-1955. Junior, Senior and Principal Engr with H. Zinder & Assocs, Conslts, 1955-1968, specializing in generation, transmission and distribution engrg for electrical utilities. With CH2M Hill, Inc since 1968 as electrical utility and energy conslt. Corporate assignments have included Dir of Electrical Engrg and Mgr of Energy and Economics Div in Seattle Region. Consltg assignments have included testimony as expert witness before federal regulatory agencies and in cvl courts. Project Mgr of Natl Power Grid Study for Congressional Research Service. Registered PE. Hobbies include photography, skiing, scuba diving and travel. *Society Aff:* IEEE, NSPE, PES

Mraz, George J
10 Tanglewood Drive, Plaistow, NH 03865
Position: Consultant *Employer:* Self-employed *Education:* Ing./ME/Bohemian Technical University Prague; B.Sc(Eng)/ME/Imperial College London (UK) *Born:* 03/23/22 Born in Czechoslovakia. Served with Royal Air Force 1943-46. Then in Czechoslovakia HVAC engineer; development engineer, SKODA-WORKS; lecturer in Mech Engrg. Since 1967 in USA, des. eng., furnace and pressure vessel design, M.W. Kellogg Co. Since 1970, Engrg mgr, Pressure Systems Division, National Forge Co. Active in development of ASME High Pressure Vessel Code. 1987 elected Fellow ASME. At present consultant. Enjoy classical music, skiing, swimming and traveling. *Society Aff:* ASME, NSPE, ASM

Muchmore, Robert B
Home: 4311 Grove St, Sonoma, CA 95476
Position: Independent Consultant (Self-employed) *Education:* EE/Communications/Stanford; BS/Elec Engg/UC Berkeley. *Born:* July 8, 1917. Tech exper includes res & dev at Sperry Gyroscop Co; performed analysis of missile tracking systems at Hughes Aircraft Co; 1954-73 with what is now TRW Defense Space & Systems Group. Positions included: Dir, Elec Div; VP & Dir, Physical Res Div; VP & Genl Mgr, Software & Information Systems Div; & VP & Chief Scientist. Now an independent cons. Author of 'Essentials of Microwaves' & contributor to review & tutorial volumes on guided missile engrg, electronics, & experimental physics. Fellow IEEE; Mbr AAAS, Sigma Xi. *Society Aff:* AAAS, IEEE, ΣΞ, ACM

Muckenhirn, O William
Business: Univ of Toledo, EE Dept, Toledo, OH 43606
Position: Prof of Elec Engg. Emeritus *Employer:* Univ of Toledo. *Education:* PhD/EE/Univ of MN; MSc/EE/MA Inst of Tech; BSc/EE/MA Inst of Tech. *Born:* 12/10/14. MIT Coop. At GE Co Schenectady, NY. Instr to Assoc Prof Univ of MN 1938- 1961 - on leave as Assoc Scientist Univ of CA Div of War Res 1944-45. 1961 to date at Univ of Toledo as Prof - Chrmn EE Dept for 9 yrs - Acting dean of Grad Sch for 1 yr Consultant with Engg Res and Dev Dept Gen Mills Mech div Mpls MN for 10 yrs. Hold 2 patents. Several Technical publications. Mbr of Bd of Dir of Central States Univs, Inc for 10 yrs and Pres for one yr. Retired 7/1/81 but will continue teaching part-time. Enjoy body surfing, dancing, and bridge. *Society Aff:* IEEE, ASEE.

Mudd, Charles B
Home: 117 Croftley Rd, Lutherville, MD 21093
Position: VP. *Employer:* Greiner, Inc. *Education:* MSE/Structures/Johns Hopkins Univ; BE/Civil/Johns Hopkins Univ. *Born:* 12/02/22. Native of Baltimore, MD. Served with US Army Air Corps 1941-45. With Greiner Engrg since 1950. Asst VP in charge of the Structures Dept 1970-77. VP in charge of Client Dev for the Baltimore office 1977-present. Mbr ASCE (Life), AWS, ACI, CEC/MD (Pres 1978-79), AISC, ARTBA, & the Engrg Soc of Baltimore (Pres 1974-75). Reg PE in 10 states & DC. Served on Maryland Bd of Registration for Prof Engrs 1982 to 1986. *Society Aff:* AWS, ASCE, ACI, CEC/MD, AISC, ESB, ARTBA.

Mudd, Henry T
Business: 523 W Sixth St, Los Angeles, CA 90014
Position: Chrmn of the Bd *Employer:* Cyprus Mines Corp. *Education:* MS/Mining Engg/MA Inst of Tech; AB/Bus Admin/Stanford Univ. *Born:* 12/26/13. Native of Los Angeles, CA. With War Production Bd during WWII as chief, Fluorspar Sec. Served in US Navy War Contracts and Settlements Div, 1944-46. Joined Cyprus Mines in May 1946 as Asst Gen Mgr and became Chrmn and Chief Exec Officer in April 1955. Currently dir of Envirodyne Industries, Inc, Rockwell Intl Corp, Southern Pacific Co, First Interstate Bank and First Interstate Bancorp. Also Chrmn of the Bd of Trustees of Harvey Mudd College; mbr Bd of Fellows of Claremont Univ Ctr; dir Community TV of Southern CA; mbr Bd of Governors of CA Community Fdn. *Society Aff:* AIME, M&MSA.

Mueller, Alfred C
Home: Route 1 Box 53J, Church Hill, MD 21623
Position: Ret. (Consultant) *Education:* PhD/Chem Engr/Univ of MI; MSE/Chem Engr/Univ of MI; BSE/Chem Engr/Univ of MI *Born:* July 1911. Joined DuPont Co in 1936 & since 1962 have been Principal Consultant in heat transfer. Fellow of ASME, AIChE, AAAS. ASME Heat Transfer Div Memorial Award 1965; D Q Kern Heat Transfer Award 1976; Retired from Dupont Co 1976. *Society Aff:* AIChE, ASME, ACS

Mueller, Charles C
Business: 178 Terrace Dr, Chico, CA 95926
Position: Prof of Civil Engrg *Employer:* CA State Univ. *Education:* PhD/Agri Engrg/MI St Univ; MS/Agri Engrg/MI St Univ; BS/Agri Engrg/CA Poly State Univ *Born:* 7/11/37. US Navy AT3 1955-58. Res & teaching in irrigation and drainage at WA State Univ, 5 yrs. At Chico State Univ since 1973. Presently active in: soil mechanics, bldg fdn design and surveying education; student-faculty-univ relationships; and enjoying music, camping, soaring, skiing, travel. Mbr ASCE, ASEE, ASTM, Sigma Xi. *Society Aff:* ASEE, ASTM, ΣΞ

Mueller, Charles W
Business: RCA Labs, Princeton, NJ 08540
Position: Fellow, Technical Staff. *Employer:* RCA Laboratories. *Education:* ScD/Physics/MIT; MS/Elec Engg/MIT; BS/Elec Engg/Notre Dame. *Born:* Feb 1912. Elctron tube engr, Raytheon Corp 1936-38; with RCA Labs since 1942. Developed secondary emission grid & beam-deflection tubes; pioneered alloy transistor, thyristor, & silicon-on-sapphire devices. At present, a Fellow of the RCA Labs Tech Staff directing ion-implantation res & dev. Fellow of IEEE 1959; David Sarnoff Individual Award in Sci 1966; 2 David Sarnoff Awards for tezam devs; JJ Ebers Award of IEEE for 'outstanding tech contribs to electron devices,' 1972; 1975 Engrg Honor Award from U of Notre Dame. *Society Aff:* IEEE.

Mueller, Edward A
Business: 1022 Prudential Dr, PO Box 5150, Jacksonville, FL 32207
Position: Exec Dir. *Employer:* Jacksonville Transportation Authority. *Education:* BCE/-/Notre Dame; MCE/-/Catholic Univ of Am. *Born:* 5/12/23. Native of Madison, WI. Professionally an engr with Ammann & Whitney, Inc; Dir Traffic & Planning Div, FL State Rd Dept, Tallahassee; Engr Traffic & Operations with the Hgwy Res Bd, Wash; Secy of FL Dept of Transportation; 1972 to present, Exec Dir Jacksonville Transportation Authority, Jacksonville, FL. Former Pres, VP, Southeastern Assn State Hgwy Officials; Chrmn, VChrmn, Secy, Engrs in Govt; FL Engg Soc, Tallahassee Chapter Engr of Yr 1972, Jacksonville Chapter 1974, award for outstanding technical achievement 1976; Pres 1977 Inst of Transportation Engrs, distinguished Service Award FL section 1976; Past Pres FL Transit Assn. *Society Aff:* EIG, FES, ITE.

Mueller, Edward E
Business: Alfred Univ, Alfred, NY 14802
Position: Prof, Dir of Placement and Co-Op Prog *Employer:* NYS Coll of Ceramics *Education:* PhD/Ceramics/Rutgers Univ; MS/Ceramics/Rutgers Univ; BS/Ceramic Engr/MO School of Mines *Born:* 7/26/24 Native of Wood River, IL. Served with USNR in WW II. Faculty positions at Rutgers Univ, Univ of WA prior to joining Pemco Corp (Baltimore) as Asst Dir of Research. Served as Dir of Ceramic Research for The Glidden Co (Baltimore) dealing primarily in specialty ceramic coatings. Joined Alfred Univ as Dean of NYS Coll of Ceramics in 1965 and returned to full-time teaching in NYC Coll of Ceramics in 1973. Also Dir of Placement and Cooperative Education Program. P Pres of NICE 1965-66. General Secretary of Keramos 1974-85. Fellow of American Ceramic Society 1964. Registered engr in NY. *Society Aff:* ACerS, NICE, Keramos, ASEE, MAPA.

Mueller, George E
Business: 2500 Colorado Ave, Santa Monica, CA 90406
Position: Chairman & Chief Executive Officer *Employer:* System Development Corp. *Education:* PhD/Physics/OH State Univ; MSEE/-/Purdue Univ; BSEE/-/MO Sch of Mines. *Born:* July 1918. Chairman of the Board and Chief Executive Officer, System Development Corporation and Senior Vice President, Burroughs Corporation since January 1981. From April 1971 to January 1981, Chairman of the Board and President, System Development Corporation. As Assoc Administrator for Manned Space Flight at NASA was respon for & directed US manned space flight prog from the beginning of Gemini flight operations through the 2nd Apollo moon landing. Co-authored a book 'Communication Satellites'. Mbr NAE; Fellow IEEE, AAAS, AIAA, AAS & a mbr of many other prof organizations. Has received numerous awards including the Natl Medal of Sci. *Society Aff:* AIAA, AAAS, AAS, IEEE, BIS, NAE.

Mueller, Jerome J
Business: 3993 E Royalton Rd, Broadview Hts, OH 44147
Position: Marine Sales Mgr. *Employer:* OH Machinery Co. *Born:* 5/20/30. in Cincinnati, OH. Res engr with Baldwin-Lima-Hamilton 1952-1956, specializing in diesel fuel injection design. From 1956-1973, with Fairbanks Morse Diesel Div, proj engr turbocharged & dual fuel engine design. Since 1973, with OH Machinery Co currently responsible for design & marketing of all marine diesel package units. Chrmn Great Lakes & Great Rivers Sec. SNAME 1976-1977. Co- authored SNAME technical paper on marine power generation 1978. *Society Aff:* SNAME.

Mueller, Marvin E
Home: 7633 Capilia Dr, St Louis, MO 63123
Position: Mgr of Procurement St. Louis *Employer:* Anheuser-Busch Inc - St Louis. *Born:* June 29, m 1927 St Louis Mo. Attended Univ of Mo. Columbia. Ser in Army Air Force during WWII. BS from Wash Univ St Louis June 1952. Adjunct Instr in Materials Mgmt at Wash Univ 1953-70; at St Louis Community Coll 1969- . Author of numerous articles on Materials Handling/Mgmt. Mbr IIE, IMMS, APICS, Local Chap, Regional " Natl Officer in IMMS.

Mueller, Robert Kirk
Business: 25 Acorn Park, Cambridge, MA 02140
Position: Director. *Employer:* Arthur D Little, International Inc. *Education:* MS/Chemistry/Univ of MI; BS/ChE/WA Univ (St. Louis); AMP/-/Harvard Bus Sch. *Born:* 7/25/13. Robert K Mueller is Director of Arthur D Little International, Inc where his work involves corporate governance and mgt aspects of multinational inst; new ventures and diversification in various industries. Chairman CF Systems, Corp.; Director HEC Energy Corp; Prior to 1968, VP Monsanto Co, mbr of its Bd of Dirs and Exec Committee. Also served as Shawinigan Resins Corp's Pres & Chrmn of the Bd. During WWII, Plant Mgr of Longhorn Ordnance Works and following became Gen Mgr of Monsanto's Plastics Div in 1952. Former positions: Dir and Exec Committee mbr of MA Mutual Life Ins Co; and of BayBanks, Inc; Dir MassMutual Income Investors, Inc; VChancellor, Intl Acad of Mgt (CIOS). Currently Fellow, Inst of Dirs (London); Life mbrs, Am Mgt Assns; a Trustee of the Cheswick Ctr (Boston); Trustee, Colby-Sawyer College, NH; Author 12 bks on mgmt and dir matters. Chrmn, faculty for Mgmt of Technology 1970 session, Salzburg Seminars in American Studies, former mbr of that organization's Bd of Dir and Exec Committee. *Society Aff:* ACS, AIChE, SCI, AAAS, NYAS.

Mueller, Thomas J
Home: 1535 Hoover Ave, South Bend, IN 46615
Position: Prof of Aerospace & Mech Engrg. & Dir of Engrg Research & Grad Studies, Coll of Engrg *Employer:* University of Notre Dame. *Education:* PhD/Gas Dynamics/Univ of IL; MS/Mech Engr/Univ of IL; BS/Mech Engr/IL Inst of Technology. *Born:* May 1934. Asst Prof ME Univ of Ill (Urbana) 1961-63; Res Scientist 1963- 64 & Sr Res Scientist 1964-65 United Aircraft Res Lab; Assoc Prof 1965-69, Prof 1969 & Prof-in-charge of AeroSpace Lab 1975 Dir, Engrg Research & Grad Studies for Engg, 1985. Univ of Notre Dame; Visiting Prof von Karman Inst Brussels Belgium 1973-74; AGARD (NATO) Lectr in Fluid Dynamics in Europe 1974; VKIAGARD (NATO) Lectr in Computational Fluid Dynamics 1976. Consultantships: ARO Inc, NASA Langley, AGARD (NATO) Paris, US Army BR&L Lockhead - Georgia Co. Minna-James-Heineman Found Grant for study & res in Europe 1973-74. ASME Fellow 1976. AIAA Assoc Fellow 1969, AIAA Educational Achievement Award 1980, AIAA Distinguished Lecturer 1980-81, over 125 tech publications. Apptd to the Edit Bd of the Springer- Verlag Journal, *Experiments in Fluids;* 1982; named Assoc Tech Ed of the ASME JFE 1983-86, Outstanding Alumnus Award, Univ IL 1986. *Society Aff:* AIAA, ASME, ASEE, ΣΞ

Mueller, Walter H
Business: Box 1595B, Indianapolis, IN 46206
Position: Dir Tech Services. *Employer:* Indpls Power & Light. *Education:* BSME/ME/Univ of WI. *Born:* 5/1/25. Currently on Elec Vehicle Ctte. Author of many trade press articles. Certified wine judge. *Society Aff:* ASHRAE, NFPA, APEC, ISEF.

Mueller, William H
Home: Old Mine Brook Rd, RD 2, Far Hills, NJ 07931
Position: Department Manager *Employer:* Exxon Res & Engg Co. *Education:* PhD/ChE/Rice Univ; BS/ChE/Rice Univ; BA/ChE/Rice Univ. *Born:* 3/12/34. Born in El Paso, TX, currently residing in Far Hills, NJ. Joined Process Engg Div of Esso Res & Engg Co in 1960. Taught math at Fairleigh Dickinson Univ evenings during 1963-1965. Was involved in refinery planning, refinery modernization & refinery startups from 1965-1969, which involved overseas assignments in Spain, France, Japan, Amuay, the Netherlands, & Lebanon. Was transferred to Exxon Corp (NY) in 1971 & spent several yrs there in Logistics Dept as Sr Advisor in Faciities Planning & as Sr Advisor in Transportation and Terminalling. Later was Sr Technical Advisor in Sci & Tech Dept. Returned to Exxon Res & Engg Co in 1979 as Asst Gen Mgr in the Petrol Dept. Presently Manager of the Systems Engineering Department. Enjoys birding as a hobby. *Society Aff:* AIChE.

Mueller, William M
Business: Hill Hall, Golden, CO 80401
Position: VP for Acad Affairs, Emeritus *Employer:* Colorado Sch of Mines. *Education:* DSc/Met Engrg/CO Sch of Mines; MS/Met Engrg/CO Sch of Mines; MET.E./Met Engrg/CO Sch of Mines *Born:* Jan 1917. Aluminum Co of Amer, Gates Rubber Co, Dow Chem Co, Denver Res Inst & ASM. Taught at Univ of Denver & Colo Sch of Mines. Mbr Bd of Trustees ASM 1964-65; named Fellow ASM 1976. Mbr US-Indonesia Indus Workshop 1971. Currently Chmn ASTM Ctte B2, Mbr Bd of Dirs. 80-82. Ed of 9 volume series on X-Ray ANalysis & 4 volume series on Energetics in Met Phenomena; author of book & papers on metal hydrides. Invited Lecturer, Inst of Iron and Steel Tech, 1980. Leader, People to People Metallurgical and Mining Delegations to China, 1984. *Society Aff:* ASTM, MMSA, AIME, ASM

Muga, Bruce J
Home: 4110 King Charles Rd, Durham, NC 27707
Position: Prof of Civil Engg . *Employer:* Duke Univ Durham N C 27706. *Education:* BS/Civil Engg/Univ of TX; MS/Civil Engg/Univ of IL; PhD/Civil Engg/Univ of IL. *Born:* Sept 1929. Reg P E Calif. Native of Grand Saline Texas. With US Navy 1951-54. Instr at Montana St Coll Bozeman Mont 1957-58. Sr Proj Engr with US Naval Cvl Engrg Lab Port Hueneme Calif 1961-67, specializing in harbor engrg & deep ocean tech prog. Cons Military Assistance Command Viet Nam on harbor const 1965-66. Instr Intensive Short courses UCLA 1967-69. With Duke Univ 1967- . Sr Author 'Dynamic Analysis of Ocean Strucs' 1970. Cons to several large organizations on numerous major cvl & marine related const projs 1970- . Chmn Exec Comm Waterways, Port, Coastal & Ocean Engrg DivASCE. Mbr Marine Sci Council NC. Mbr Monitor Res Study Council NC. *Society Aff:* ASCE, PIANC, RESA.

Mugele, Raymond A
Home: 74 Los Altos Sq, Los Altos, CA 94022
Position: Engrg Consultant *Employer:* Self *Education:* MS/Phys/Univ of CA; BS/Phys/Univ of CA; BA/Math/Univ of CA. *Born:* 4/13/14. Native of San Francisco. Studied & taught at Univ of CA; radiation res. Served with Army Engrs & AF 1941-45; field engg, photogrammetry, meteorology. Chem engr with Shell Dev Co 1945-60; proj leader for flow & spray res, chem plant design, refinery optimization & control. Sr engr for IBM Corp 1960-75; proj leader for automation studies & applications, N & S Am & Europe. Published papers, patents. Currently, engg consultant & reg PE. Photography, languages. Originator of linguistic engg. Pres, Improvement Universal Inc. *Society Aff:* ACS, AIChE, AMS, ISA, ΣΞ, ΦΒΚ, ΠΜΕ, AMS

Muirhead, Vincent U
Home: 503 Park Hill Terr, Lawrence, KS 66044
Position: Prof, Chmn Dept of Aero Engrg. *Employer:* University of Kansas. *Education:* Aero Eng/AE/CA Inst of Tech; BSAE/AE/USN Postgrad Sch; BS/Eng/US Naval Acad. *Born:* 2/6/19. Aerospace Engrg CalTech 1949; Commissioned Ensign 1941; Comdr Fleet Aircraft Ser Squadron 795, 1951-52; Asst Bur of Aeronautics Representative Dallas 1952-53; Bur of Aeronautics Representative Ft Worth 1953-54; Comdr Helicopter Utility Squadron 1, 1955-56; Ch Staff Officer Comdr Fleet Air Philippines 1956-57; ExO Naval Air Tech Training Ctr Memphis 1958-61. Comdr retired 1961. Asst Prof Univ of Kans 1961-63; Assoc Prof 1964-75; Prof & Chmn 1976. Cons Black & Veatch, K C Mo 1964- . Author: Intro to Aerospace 1972; co- author: Thunderstorms, Tornadoes & Bldg Damage 1975. Res on aircraft, tornado vortices, shock tube & waves & Venus Probes, ground vehicle drag reduction. *Society Aff:* AIAA, ASEE, NAS, AAM.

Mujumdar, Arun S
Business: Dept of Chem Engg, 3480 Univ St, Montreal, Quebec, H3A 2A7 Canada
Position: Prof. *Employer:* McGill Univ. *Education:* PhD/Chem Engg/McGill Univ; MEng/Chem Engg/McGill Univ; BChemEng/Chem Engg/Univ of Bombay. *Born:* 1/14/45. Mech Engr, Res Div, Carrier Corp, Syracuse, NY, 1969-71. Founding Prog Chrmn of Biennial Intl Symposium Series in Drying held since 1978. Res interests in drying, vibrated fluidized beds, turbulent flow and heat transfer. Editor of several books in drying theory and practice. Editor of serial publicatons: Advances in Drying (Hemisphere) and Advances in Transport Processes (Wiley- Eastern); Editor: Handbook of Industrial Drying (Marcel Dekker). Assist Editor: Drying Tech-An Intl Journal (Marcel Dekker) *Society Aff:* ASME, AIChE, CSChE, TAPPI, ΣΞ.

Mukherjee, Amiya K
Business: Div of Materials Science & Engr, Dept of Mechanical Engr, Univ. of California, Davis, CA 95616
Position: Prof. *Employer:* Univ of CA. *Education:* DPhil/Met/Oxford Univ; MSc/Physical Met/Sheffield Univ; BSc/Pure Sci/Univ Calcutta. *Born:* 6/1/36. Prof of Mtls Sci & Engg in the Univ of CA, Davis Campus since 1966; Res Met at the Lawrence Berkeley Lab; Sr Scientist at Battelle Meml Inst & Managerial Trainee in steel plants in England prior to that; Chrmn of the Flow and Fracture Activity of ASM; Mbr, Mtls Sci Council, ASM; Mbr, Committee on Phys & Chemistry of Metals, AIME; Mbr, Editorial Bd, Met Transactions; Visiting Prof to Univs of Cambridge, Oxford, Buenos Aires, & Israel Inst of Tech; Consultant to Dept of Energy; Coordinator of ASM natl symposia on rate processes, composite mtls, acoustic emission, creep & superplasticity; Recipient of Distinguished Teacher Award, Academic Senate, Univ of CA; Recipient of Special Award for Creativity in Res, Natl Sci Foundation, Fellow Amer Soc for metals. *Society Aff:* ASM, AIME.

Mukherjee, Kalinath
Business: College of Eng, E. Lansing, MI 48824
Position: Prof and Chairperson *Employer:* Michigan State Univ *Education:* PhD/Physical Metallurgy/Univ of IL; MS/Physical Metallurgy/Univ of IL; BE/Metallurgical Engg/Calcutta Univ. *Born:* 2/19/32. in Calcutta, India, Feb 19, 1932. Naturalized US citizen. Married, 3 children. BE, 1956, Calcutta Univ, MS, 1959, PhD, 1963, Univ of IL, Urbana. Res Assoc 1963-64 Univ of IL. Asst Prof of Material Sci, SUNY, Stonybrook, 1964-67. Assoc Prof 1967-72, Prof 1972-80, Dept hd of Metallurgy 1974-80, Poly Inst of NY. Prof Michigan State Univ 1980-, Chairperson, Dept. of Met., Mech. and Mat. Science, Michigan State University 1985-, Janaki Prosad Scholar, Calcutta Univ, IIM Award 1959; Distinguished Teacher Award, Poly, 1971; Distinguished Prof Award of the Student Council, Poly 1979. Distinguished Faculty Award, Michigan State University 1986. Fellow AAAS; Fellow ASM; Mbr AIME, ASEE, APS, Sigma Xi. Honorary Mbr Alpha Sigma Mu, 1978. Commissioner, Engg Manpower Commission of the Engrs Joint Council. More than 150 tech pubs; Sr Ed of Metallurgy/Materials Ed Yearbook; Co-ed of "Lasers in Metallurgy–; Co-ed of Laser Materials Processing. *Society Aff:* AAES, AAAS, TMS-AIME, ASM, APS.

Mukundan, Rangaswamy
Business: Graduate School, Clarkson University, Potsdam, NY 13676
Position: Dean of Graduate School, Div, Dir of Research *Employer:* Clarkson University *Education:* PhD/EE/Purdue Univ; MSEE/EE/Purdue Univ; ME/EE/Rajasthan Univ; BE/EE/Rajasthan Univ. *Born:* 7/26/37. Born 26 July, 1937 in Madras, India. Taught at Purdue Univ 1960-65. Joined Clarkson 1965 and assumed current responsibility as Dean of Graduate School 1987. Interested in Control System Theory & Applications. Mbr IEEE Honorary Member Cttee. *Society Aff:* IEEE.

Mulcahy, Joseph F
Business: 681 Park Ave, Cranston, RI 02910
Position: Owner. *Employer:* Mulcahy Engineers. *Education:* BS/Civil/Univ of CT. *Born:* Mar 9, 1925 Norwich Conn. With US Navy 1943-46. Design Engr for cons firm 1952-1960; founded own firm 1961. Reg P E: R I, Conn, Mass, Maine, Va. Pres R I SPE 1972-1973; Natl Dir NSPE. Mbr R I Bldg Code Stds Comm, R I Fire Safety Code Bd of Appeal & Review. Chrmn, RI Bd of Reg PE & LS, RI Engr of Yr 1981. *Society Aff:* NSPE, NCEE, CSI, BOCA, ACI.

Mulder, Leonardus T
Business: School of Engg, Fresno, CA 93612
Position: Assoc Prof. *Employer:* CA State Univ. *Education:* PhD/EE/Univ of Notre Dame; MS/EE/Delft Tech Univ; BS/EE/Delft Tech Univ. *Born:* 2/21/51. Native of The Hague, The Netherlands. Worked after MS-degree for two yrs in area of channel error correcting codes at Eindhoven Tech Univ (1974-1976). Continued with PhD-Res in source coding, data compression algorithms at Univ of Notre Dame (1976-1979). Currently Assoc Prof at CA State Univ-Fresno. Main areas of interest: information theory, digital signal processing, microprocessor- design and applied mathematics. *Society Aff:* IEEE, ΣΞ.

Mullen, Joseph
Home: 8551 Eames St, San Diego, CA 92123
Position: Sr Weight Engineer. *Employer:* Convair Div/Genl Dynamics Corp. *Born:* Dec 1920. BA & MA from USIU. With Zenith Radio Corp & Indus Res Prods Inc inChicago Ill from 1941; there received patent on electro-mech transducer; came to San Diego Calif & Convair Div of Genl Dynamics Corp in 1956 to work on Atlas & Centaur missiles & space launch vehicles. Joined Soc of Allied Weight Engrs soon after transferring to weights engrg; held all chap offices & since 1966 internatl ctte chmnships each year; elected Fellow in 1975.

Mullen, Wesley G
Business: Dept of Civ Engg, P.O. Box 5993, Raleigh, NC 27650
Position: Prof. *Employer:* NC State Univ. *Education:* PhD/CE/Purdue Univ; MS/CE/Univ of MD; BS/CE/VA Military Inst. *Born:* 11/30/22. Native of Richmond, VA. USME WWII 1943-46. Field engr, Madigan-Hyland Consltg Engrs 1951-59 on marine, bridge and hgwy construction. Instr Univ of MD 1959-61 and Purdue Univ 1961-62. Res engr with MD State Roads Commission 1963- 65. At NC State Univ since 1965 as Coordinator of Hgwy Res Prog and Prof of CE teaching construction mtls. Res in skid resistance, portland cement concrete, hgwy pavements. Active on ASTM Stds Committees, ASTM Charles B. Dudley Medal 1966. NCSU Acad of Outstanding Teachers. ASTM Award of Merit and Fellow 1980. ASCE Award for Outstanding Service 1979. Visiting Scholar, Fredrik Wachtmeister Chair for Sci and Engrg, VA Military Inst, 1980. *Society Aff:* ASCE, ASTM, ACI, AAPT, TRB.

Mullendore, Robert A
Business: 1650 South 70th St, STE 101, Lincoln, NE 68506
Position: Pres *Employer:* R. A. Mullendore Consultants, Inc. *Education:* BS/EE/Univ of NB *Born:* 4/18/48 Native of Lincoln, NE. Served State of NE through Dept of Roads 1972-1976. High mast design engr, Valmont Indus 1976, product design and development, teaching seminars, field troubleshooting. Staff EE at Garber & Work Inc 1976- 1980, doing all electrical design (HV & LV), coordination, rate studies. Formed R A Mullendore Consultants, Inc in fall of 1980. Developed computer anaysis of roadway interchange high mast lighting. Enjoy hunting, carpentry, wiring and reading. *Society Aff:* IEEE, IES, NSPE, ACEC

Muller, Marcel W
Business: , St Louis, MO 63130
Position: Prof *Employer:* Washington Univ *Education:* PhD/Physics/Stanford Univ; AM/Physics/Columbia Univ; BS/EE/Columbia Univ *Born:* 11/1/22 Born in Vienna, Austria, immigrated 1940, naturalized 1943. Married, 3 children. Served US Army 1943-46. Research Engr to Sr Scientist, Varian Assocs 1952-66, Prof of Electrical Engrg, Washington Univ 1966-. Visiting Scientist, Max-Planck-Inst, Stuttgart 1976-77. Humboldt Prize 1976. *Society Aff:* IEEE, APS, OSA

Muller, Richard S
Business: Dept of Elec Engg and Comp Sci, Berkeley, CA 94720
Position: Prof. *Employer:* Univ of CA. *Education:* PhD/Elec Engg & Physics/CA Inst of Tech; MS/EE/CA Inst of Tech; ME/ Engg/Stevens Inst of Tech. *Born:* 5/5/33. Native of Union City, NJ. Earned MS at CA Inst of Tech as Hughes Fellow. Mbr of the Technical Staff at Hughes Aircraft Co, 1955-60, Natl Sci Fdn Fellow at Caltech. Taught at Caltech as a Teaching Assoc, at the Univ of Southern CA as an instructor, and worked on device physics at Pacific Semiconductors Inc (now TRW electronics). At the Univ of CA (Berkeley) since 1962. Maj field of res: device electronics, sensor devices. Author (with T. I. Kamins) of "Device Electronics for Integrated Circuits–, published by John Wiley and Sons Inc (1977) second edition (1986). Dept of EECS-v. chrmn 1972-1975. Consultant to several firms engaged in semiconductor electronics and integrated circuits. Member: Sensors Subcommittee of Electron Devices Society of IEEE, Editorial Board of Technical Journal - SENSORS and ACTUATORS. Chmn. IEEE Sensors Research Conf, IEEE Int. Electr. Devices Mfg., Chmn: Committee on Sensors and Actuators. *Society Aff:* IEEE, EDS.

Mulligan, James H, Jr
Home: 12121 Sky Lane, Santa Ana, CA 92705
Position: Prof of EE *Employer:* Univ of Calif - Irvine CA 92717. *Education:* PhD, Columbia Univ 1948, Ms, Stevens Inst or Tech 1945; BEE Cooper Union School of Eng 1943, EE. *Born:* Oct 1920. 1st employed in transmission dev dept Bell Tele Labs; then staff mbr Naval Res Lab. Following WWII joined Allen B DuMont Labs initially concerned with TV res; later Ch Engr TV Transmitter Div. Faculty EE Dept NYU 1949-68; Chmn of Dept 1952-68. Secy & Exec Officer NAE 1968-74; Secy 1974-1978 . Res Contributions in netwrok theory & electronic circuits. Active in ECPD as mbr of Bd of Dirs & Exec Ctte; Mbr & Chmn ECPD Engrg Educ Accreditation Comm. Mbr & Chmn of many AIEE/IRE & IEEE cttes; IEEE officer 1968-71; 1976 (Pres 1971; V Pres Tech Act 1968, 1969; V Pres 1970 V Pres Prof Act 1976). IEEE Haraden Pratt Award 1974. *Society Aff:* IEEE

Mullin, Thomas E
Mech Engrg Dept, Louisville, KY 40208
Position: Prof of Mech Engrg, Assoc Dean for Acad Affairs JB Speed Sci Sch *Employer:* University of Louisville. *Education:* PhD/Mech Engg/OK State Univ; MS/Mech Engg/Univ of Louisville; BSME/Mech Engr/Univ of KY. *Born:* Feb 29, 1928. Cost Estimating Engrg for Standard Oil of N J 1951-53. Design Engr for Girdler, Gas Process Div 1953-58. Asst Prof 1958-65, Assoc Prof 1965-70, Full Prof 1970- , Mech Engrg at Speed Scientific Sch, Univ of Louisville. Prof & Chmn of Mech Engrg Dept 1972-81, Associate Dean of the J. B. Speed Scientific School 1981-. *Society Aff:* ASME, ASEE, NSPE.

Mullin, William H
Home: 16607 Yorktown Rd, Granger, IN 46530
Position: VP of Engg. *Employer:* FGC-Fayette Genl. *Education:* BS/Mech Engg/Drexel Univ. *Born:* Dec 9, 1918 Phila Pa. Previous employment includes Gulf Oil Co, Westinghouse Mfg & Philco-Ford Corp. Carrier Air conditioning Co. Held various engrg & mfg assignments relating to major appliances, military air conditioning equipment & automotive air conditioners. Received Philco Res & ENgrg Achievement Award 1952 & 1953. Mech Engrg Instr Penn St Univ, Phila Campus 1947-48. Holder of 3 pats received 1972 Best Paper on the Year Award from Appliance Engrg magazine. Ser as chmn on numerous ASHRAE, AHAM & NEMA cttes. 1960 Pres Philadela Chapter ASHRAE. Elected ASHRAE Fellow 1972, Mbr 1945- . Enjoys civic activities, skiing & public speaking. *Society Aff:* ASHRAE.

Multer, Robert K
Business: Orrs's Island, ME 04066
Position: President. *Employer:* AIDCO Maine Corp. *Education:* B/Chem Eng/Cornell Univ. *Born:* 1926. Brooklyn Tech HS; US Army 1944-46. Borg-Warner Corp Plastics Div; Dev Engr 1950-56; polymers, ashesives, coatings, process R/D, materials testing, ATCO Ceramics Corp: V Pres Engrg 1957-58, Pres 1958-75; ceramics mfg mgmt, high- temp processing, machine design. Multi- Research Corp; Founder Pres 1966- ; materials res, engrg, testing; bldg matl sales. AIDCO Maine Corp: Founder-Pres 1974- ; design, construction mgmt, and all engg for solar & other renewable energy sys, including Strata SOL Pond nonconvective solar ponds for heating, cooling, & electricity. Land development and sales. Building construction. Mbr; AIChE, Internatl Solar Energy Soc, New England Solar Energy Assn,

Multer, Robert K (Continued)
Maine Solar Energy association, ASTM, AAAS, Soc for Plastics Engrs, ASHRAE, ME AIA. Pats: High Energy Composition, Liquid Agitator. *Society Aff:* AIChE, ASTM, AAAS, ASHRAE, AIA.

Mulvaney, Carol E
Business: Technical Info Ctr, Peoria, IL 61629
Position: Technical Librarian. *Employer:* Caterpillar Tractor Co. *Education:* BA/History/Bradley Univ. *Born:* 2/1/25. Native of Peoria, IL. Started in the Res Library of Caterpillar Tractor Co in May 1948. Became the Hd Librarian in 1964. Am responsible for acquiring, analyzing, organizing, storing & retrieving technical info for the co world- wide. Serve on the Illinois Central College advisory board. Also serve on the Peoria Art Guild Bd of Dirs. *Society Aff:* ASM.

Mumma, Albert G
Home: 66 Minnisink Rd, Short Hills, NJ 07078
Position: R Adm USN (Ret) - Mgmt Cons. *Employer:* Self. *Education:* Dipl Ing/NAPMA/Ecole Nat Sup Du Genie Maritime Paris, France; MS/Marine Eng/USN Post Grad Sch; BS/Marine Tech/US Naval Acad. *Born:* June 2, 1906 Findlay Ohio. Appointed to US Naval Acad from Iowa 1922; Midshipman to R Adm USN Ret 5/1/59; ser in USS Saratoga (CV3), Chicago, Destroyers in both deck & engrg duties; During WWII was Head of Tech Intelligence Div of Commander US Naval Forces Europe, which captured German equipment & scientists; Head Machinery Design Buships, commanded David Taylor Model Basin, Mare Island Shipyard & Ch of Bureau of Ships 1955-59; retired, became successively V Pres Engrg of Worthington, Group Exec, Exec V Pres, Pres & Chmn; retired 1971. Hon Mbr, P Pres, Fellow, SNAME; Hon Mbr, P Pres ASNE; Mbr NAE. Dr Engrg NJIT. 'Jerry' Land Gold Medal. *Society Aff:* SNAME, ASNE, NAE.

Mumma, George B
Home: 145 E. Costilla Ave, Littleton, CO 80122
Position: Director *Employer:* Martin Marietta Corp *Education:* BS/EE/Auburn Univ *Born:* 8/22/31 Native of Montgomery, AL. Served in USAF 1950-52. System Safety Engr for Martin Marietta Denver since 1962. With Martin Marietta since 1958. Assumed present job as Dir, MX System Safety in 1979. Previous assignments included Titan I, II and III Missiles, Apollo, Skylab and shuttle programs. Program mgr for development of USAF System Safety Handbook. Dir System Safety Soc, Fellow mbr of System Safety Soc, and Pres Sys Safety Soc 1983-85, Prof mbr of ASSE. Reg PE (Safety) Calif. *Society Aff:* SSS, ASSE

Mundel, August B
Home: 34 Sammis Ln, White Plains, NY 10605-4726
Position: Conslt *Employer:* August B Mundel PE *Education:* MSE/EE/U of MI; BS/EE/Cooper Union Inst of Tech *Born:* 12/21/11 Reg PE, NY and NJ. Operates consltg firm: quality sys, reliability, sampling, process control, prod liability--serving govt and ind (69-pres); Chrmn, US Tech Adv Group, Intl Org for Standardization, Ctte on Application of Statistical Methods (USTAG/ISO/TC 69); p VP and Dir, Amer Soc for Quality Control; Author of more than 50 papers on products and sampling, incl Narrow Limit Gaging and Group Test Proc; Cert Quality Engr and Cert Reliability Engr, Amer Soc for Quality Control; Founding Chrmn, Amer Nat Standards Inst Z1 Ctte on Quality Assurance; 1951-1969 VP, Sonotone Corp, elect products and batteries. *Society Aff:* ASQC, IEEE, ASTM, SSS

Mundel, Marvin E
Business: 821 Loxford Terr, Silver Spring, MD 20901
Position: Principal. *Employer:* M E Mundel & Assocs. *Education:* PhD/Indus Engr/Univ State of IA; MS/Indus Engr/Univ State of IA; BS in ME/Indus Engr/NY Univ *Born:* 4/20/16. B. Apr 1916. Served as indus engr for Tung-Sol Lamp Co. Subsequently taught at Bradley Univ & then at Purdue from 1942-52. At Purdue was Prof & Chmn of Indus Engrg. Organized & served as the first Dir of the US Army Mgmt Engrg Training Agency. Have been an indus engrg cons since 1953 with assignments in the U S, Europe, & the Far East, except for 1963-65 when served as Principle Staff Officer for Work Measurement in the U S Bureau of the Budget. Pres, IIE 1979-80. Fellow SAM & IIE. Awarded Gilbreth Medal 1953 by SAM and the Frank and Lillian Gilbreth Award from IIE in 1982. Have previously been Regional V P & VP for Mgmt & Indus Divs IIE and Exec VP for Divs, Publs & Educ. *Society Aff:* IIE.

Mundt, Barry M
Business: .345 Park Ave, New York, NY 10154
Position: Principal *Employer:* Peat Marwick, Main & Co. *Education:* MBA/Bus Admin/Univ of Santa Clara; BS/IE/Stanford Univ *Born:* 6/28/36 Native of San Francisco, CA. Formerly Statistician for Aerojet-General Corp; Reliability Engr for Lockheed Missiles and Space Co; and Management Engr for C-E-I-R Inc, specializing in statistical, reliability, and quality control consulting. With Peat, Marwick, Mitchell & Co since 1965 in the Management Consulting Dept, concentrating in strategy, organization, operations mgmt & industrial engrg consulting. Elected to partnership in 1973. Treasurer of IIE from 1977-81; Pres-Elect 1981-82, 1982-83 Pres; currently Asst. Treas. Chmn, Finance Cttee, AAES. Bd of Trustees mbr, Brandon Hall School. Registered PE in CA. *Society Aff:* IIE

Mungall, Allan G
Home: 33 Woodview Crescent, Gloucester, Ontario, Canada K1B 3B1
Position: Principal Research Off (retired) *Employer:* Natl Research Council *Education:* PhD/Physics/McGill Univ; MASc/Engr Physics/Univ of British Columbia; BASc/Engr Physics/Univ of British Columbia *Born:* 3/12/28 in Vancouver, BC, Canada. Geophysicist with CA Standard, Calgary, Alberta, 1950. With Natl Research Council Physics Div since late 1950. Employed part-time at NRC as PRO since retirement in Sept, 1983; fully retired April 1986. Principal interests in experimental physics: atomic time and frequency standards and time scales, optics, microwaves, and medical physics. Elected to Fellow of the IEEE in 1980 for contributions in the design, construction, and use of primary cesium frequency and time standards. Other interests include playing, arranging, and writing music for organ, piano, brass, string and wind ensembles and full orchestra, and also typical outdoor cottage activities. *Society Aff:* IEEE, APEO, CAPAC

Mungan, Necmettin
Business: 2400, 639-5th Avenue S.W, Calgary, Alberta Canada T2P 0M9
Position: Chief Tech Advisor *Employer:* Alberta Energy Company Ltd. *Education:* PhD/Pet Engg/Univ of TX, Austin; MS/Pet Engg/Univ of TX, Austin; BA/Mathematics/Univ of TX, Austin; BS/Pet Engg/Univ of TX, Austin. *Born:* 3/1/34. in Mardin Turkey. Worked with Sinclair Oil Corp 1961-66 as Sr Res Scientist & Hd of Displacement Processes Sect. Was with the Petro Recovery Inst as Ch Res Officer 1966-78 and was respon for the res progrs. 1978-86 he was Pres of his own Mungan Petro Consultants Ltd. Since 1987, he is the Chief Technical Advisor to Alberta Energy Company Ltd. Cedric K Ferguson Medal of the SPE of AIME 1966, AIChE Best Paper award 1966, Petro Soc of CIM disting Serv award 1969, Distinguished Lecturer award for the SPE of AIME 1969. Chmn Trasactions Reprint Ctte 1968-71; Chmn Tech Publs, Petro Soc of CIM 1973-75; Mbr Res Ctte of Interstate Oil Compact Comm. Selected Distinguished Mbr of SPE 1984. *Society Aff:* SPE, CIM.

Munger, Paul R
Business: 302 Engg Res Lab, Univ. of Missouri-Rolla, Rolla, MO 65401
Position: Dir, Inst of River Studies Prof of Civ Engr. *Employer:* Univ of MO. *Education:* PhD/Engg Sci/Univ of AR; MS/CE/Univ of MO; BS/CE/Univ of MO. *Born:* 1/14/32. Native of Hannibal, MO. Served with Army Corp of Engrs 1952-54. Civil Engg faculty of Univ of MO-Rolla since 1958, Instructor to Prof. Immediate Past Pres, Natl Council of Engr Examiner, V Pres, Zone III, ASCE; Past Chmn, MO Bd of Architects, Prof Engr and Land Surveyors and Past Pres, Chi Epsilon. Assumed present position Jan 1976. In this position plan. coordinate and administer

Munger, Paul R (Continued)
res contracts in conjunction with Proj Dir for studies undertaken by Inst of River studies. *Society Aff:* ASCE, NSPE, ΣΞ, ASEE, SAME

Munir, Zuhair A
Home: Univ of Calif, Davis, CA 95616
Position: Prof and Associate Dean *Employer:* Univ of CA. *Education:* PhD/Mat. Sci/ Univ of CA; MS/Mat. Sci/Univ of CA; BS/ChE/Univ of CA. *Born:* 7/7/34. Taught at San Jose State (CA) and FL State Univs. Consulted for IBM, GE, Rockwell-Intl, the State of CA. Energy Commission, and the Lawrence Livermore Labs. Chrmn, Northern CA Metallurgical Soc (1972-73), Chrmn, Sacramento Valley Chapter of the Am Soc for Metals (1975). *Society Aff:* AIME, ASM, AmCeramSoc, ECS.

Munroe, Gilbert G
Home: 709 E Jackson Rd, Webster Groves, MO 63119
Position: Ch Wts Engr. *Employer:* McDonnell Aircraft Co. *Education:* Cert/AE/CW Tech Inst. *Born:* Nov 18, 1920 Little Rock Ark. Wts Engr at Chance Vought Aircraft Div of United Aircraft Corp, Stratford Conn Nov 1941 to Aug 1948. At McDonnell Aircraft since Aug 1948. Ch Wts Engr since Sept 1954. Mbr of Soc of Allied Wt Engrs (SAWE) since 1943, Chmn St Louis Chap SAWE 1952-53, Internatl Pres of SAWE 1952- 53, Hon Fellow 1969. Mbr Amer Inst of Aeronautics & Astronautics. *Society Aff:* SAWE, AIAA.

Munse, William H
Business: 2129 Civ Engg Bldg, 208 N. Romine St, Urbana, IL 61801
Position: Prof Emeritus Civ Engg *Employer:* Univ of IL. *Education:* MS/Civ Engg/ Univ of IL; BS/Civ Engg/Univ of IL. *Born:* 7/10/19. Native of Chicago, IL. Engr for Champaign, IL in 1941, Engr for Am Bridge Co in 1942-43, served in Navy at Los Alamos, NM in 1944-46, Res engr at Lehigh Univ in 1946-47, Prof at the Univ of IL in 1947-81, and Prof. Emeritus since 1981. Principal contributions have been res on the basic engg behavior of metals & metal structures, the introduction of this res into the classroom, and in the translation into design specifications. Res involves the static, fatigue, and brittle behavior of riveted, bolted & welded construction. AWS, Adams Meml Mbrship, 1961; ASCE, Walter L Huber Res Prize, 1962; Japan Welding Soc, Distinguished Service Award, 1976; Hon Mbr ASCE, 1983. *Society Aff:* ASCE, ASTM, AWS, AREA, TRB.

Munson, Darrell E
Home: 23 Cedar Hill Rd NE, Albuquerque, NM 87122
Position: Managing Director *Employer:* Quaker Chemical (Holland) B.V. *Education:* MChE/NYU;BChE/CCNY *Born:* 1/18/33. in Rapid City, SD. Taught Met Engg WA State Univ 1959-1960. Joined Sandia Labs in 1961 in Mtls Res. Res included studies of dynamic response of mtls & shock wave physics. Current res involves rock mechanics & constitutive modelling for various energy progs. *Society Aff:* ASM, AIME.

Munson, James I, Jr
Home: 4840 A N.W. 3rd Street, Delray Beach, FL 33445
Position: Retired *Education:* BS/Chem Engrg/SD School of Mines & Tech, 1940 *Born:* Oct 12, 1914. Process engr Permutit Co 3 yrs, Dist Rep 14 yrs, Sales Mgr 7 yrs, V P Mkting 2 yrs, V Pres & Secy Internatl Hydronics Corp 13 1/2 yrs; respon for client relations, contract reviews & co policies. Mbr AIChE, AWWA, WPCF, AAEE. Active in scouting & Community Chest. Reg P E NJ. Interested in most sports. Enjoy golf, fishing, music. Present: Sales Rep for G H Dacy Assoc, Inc, 5840 SW 114th Terr, Miami, FL 33156 Tel (305) 665-7710 (Indus & Munic Water & Wastewater Treating Equip). *Society Aff:* AWWA, WPCF, AAEE, AIChE

Munson, John C
Business: Engrg & Science Assoc, Inc, 6110 Executive Blvd, Suite 315, Rockville, MD 20852
Position: V Pres *Employer:* Engg & Sci Assoc, Inc. *Education:* PhD/Electro Eng/ Univ of MD; MS/Electro Eng/Univ of MD; BS/Electro Eng/IA State Coll; -/Navy Scholarship/MIT *Born:* 10/09/26 Native of Clinton, IA. Naval Ordnance Lab 1949-67: developing new sonar systems and techniques, specializing the application of signal processing techniques to sonar systems. In 1967 technical dir Navy portion of Practice Nine. 1968-85 Superintendent, Acoustics Div, Naval Res Lab. Since 1985, V Pres Engg and Sci Assoc, Inc. Editor, Journal of Underwater Acoustics since 1983. Taught grad courses in signal processing and circuits. Mbr Navy advisory committees. Over 60 publications; 8 patents. Active Baptist layman: Trustee Midwestern Baptist Theological Seminary, 1971-1980; Trustee Baptist Home of DC 1976-present. *Society Aff:* IEEE, ASA, RESA

Munyan, Leon J
Business: PO Box 250, Eureka, UT 84628
Position: Gen Mgr *Employer:* Sunshine Mining Co *Education:* BS/Mining Engrg/CO Sch of Mines; MBA/Mining Eng/CO Sch of Mines *Born:* 08/24/53 Supervised operation of the CO Sch of Mines experimental mine. Employed by NJ Zinc Co as a Mining Engr and Gen Mine Foreman, then Supt for three yrs. Employed in 1980 by Sunshine Mining Co as Chief Engr. Supervised all engrg, surface and underground, for largest silver mine in the US. Promoted to Gen Mgr of the Eureka, UT, Div in 1984. In charge of all operations and serv in the Tintic Mining Dist. Currently V Chrmn of AIME, Mbr of Exec Council 1975. Reg P E in ID, VA, and CO. Cert Mgmt Acct, LSIT in ID. Enjoys hunting, boating. *Society Aff:* AIME, NSPE, BGS, BAS, AMA, NAA, AMBA

Mura, Toshio
Business: Department of Civil Eng. Northwestern Univ, Evanston, IL 60201
Position: Professor *Employer:* Northwestern Univ *Education:* PhD/Appl. Math/The University of Tokyo; Bachelor/Appl. Math/The University of Tokyo *Born:* 12/07/25 Dr. Mura's areas of research and teaching are in theoretical and applied mechanics with emphasis on micromechanics. He has developed micromechanics for the last 20 years where he has brought the theory of dislocations and inclusions into engrg. More physical and microstructural considerations are taken into account in study of continuum plasticity, fracture and fatigue mechanics. The gap between metal physics and engrg mechanics has become narrower due to him. His professional efforts have been centered in the American Society of Mechanical Engineers (ASME). He has served as Sponsor of Microstructure Mechanics in ASME and Liaison between Material Division's Cttee on Constitutive Equations and the Applied Mechanics Division. He has organized two ASME symposia on Mathematical Theory of Dislocations and on Mechanics of Fatigue and edited their proceedings. He is also on the U.S. organizing committee of the Japan-U.S. conferences on composite materials. He is Fellow of ASME and American Academy of Mechanics, a member of National Academy of Engineering. *Society Aff:* ASME

Murashige, James Y
Home: 3248 Grandview Blvd, Los Angeles, CA 90066
Position: Principal *Employer:* James Y. Murashige & Assoc *Education:* BA of Arch/ KS St Univ *Born:* 03/04/32 in Hilo, Hawaii. Graduated Hilo High, thereafter attending KSU and doing tour of Army duty in Panama Canal Zone. Employed as Architectural Designer for Gentry and Voskamp, Kansas City, MO and Larry Davidson & Assoc, Los Angeles. Worked as engr for Jasper & Forker, Kevin Kelly & Assoc, Airesearch Manufacturing Co, and W & H Construction Co. Principal of James Y. Murashige & Assoc since 1969 providing structural design service of commerical, industrial and residential construction. Interests in, philately, numismatic, fishing, golf, mountaineering, classical guitar and community service. *Society Aff:* NSPE

Murata, Tadao
Business: Dept of EECS, Univ. of Ill. at Chicago, Box 4348, Chicago, IL 60680
Position: Prof. *Employer:* Univ of IL. *Education:* PhD/EE/Univ of IL at Urbana; MS/EE/Univ of IL at Urbana; BS/EE/Tokai Univ. *Born:* 6/26/38. in Japan & came to the US in 1962. After receiving his PhD in 1966, joined Info Engrg Dept (now called EECS Dept), Univ of IL, Chicago, where he is Prof & former Dir of Grad Studies. Taught at the Univ of CA, Berkeley, 1976-77, & at Tokai Univ, Tokyo, 1968-70. Served as res consultant for Nippon Elec Co in comp aided design, 1968-70 & as guest researcher for GMD, Bonn, FRG in gen net theory, 1979. Contributed about 120 technical articles to the fields of circuit- system theory & Petri net related computation models. Fellow of IEEE. Reviewer for Math Reviews and since 1981 served as a mbr of 2 panels of the Comp Sci and Tech Bd of the Natl Res Council/Natl Acad of Sci, Washington, DC. *Society Aff:* ACM, IEEE.

Murdock, James W
Home: 684 Old School House Dr, Springfield, PA 19064
Position: Assoc Prof & Asst Dept Chmn. *Employer:* Drexel Univ - ME & Mech Dept. *Education:* BS/Mech Eng/Drexel Univ. *Born:* 12/31/14. Reg PE PA. Fellow ASME Mbr ASEE. With the Naval Ship Engg Cttr 1939-74, when retired as head of its Appl Phys Dept. In 1950 joined the Fac of Drexel Univ Evening Coll & have been teaching in the evenings since that time. After retiring from the Navy, joined Drexel's Day Fac, where am now serving as Assoc Prof & Asst Dept Chmn. Served as mbr and/or chmn of tech cttes of the ASME, ASTM, ISA, AGA, ACHEMS & ANSI. Author and/or co-author of over 60 publs incl a text on fluid mech. Served as a cons in mech engg to serv corps. *Society Aff:* ASME, ASEE, ASHRAE.

Murdock, Larry T
Business: 250 East Broadway, Salt Lake City, UT 84111
Position: Partner *Employer:* Dames & Moore *Education:* BS/CE/Univ of UT *Born:* 03/04/37 *Society Aff:* ASCE, ACEC, CECU

Muroga, Saburo
Business: Digital Computer Lab, 1304 W. Springfield Ave, Urbana, IL 61801
Position: Prof *Employer:* Univ of IL. *Education:* PhD/Elec Engg/Tokyo Univ; Gakushi/Elec Engg/Tokyo Univ. *Born:* 3/15/25. Native of Numazu-shi, Japan. Res staff, Railway Technical Lab from 1947 to 1949. Engg staff, Radio Regulatory Commission of Japan, 1950-1951. Res staff, Elec Communication Lab, Nippon Telegraph and Telephone Public Corp, 1957-1960, working on res of info theory and design of parametron computer. Res staff, IBM Res Ctr, 1960-1964, working on res of threshold logic and design automation. Prof, Depts of Comp Sci and Elec Engg, Univ of IL, since 1964, teaching logic design and VLSI design automation. Published several books and more than hundred papers. Elected as IEEE fellow. *Society Aff:* IEEE, ACM, IPSJ.

Murphy, Arthur T
Business: E.I. Dupont & Co, Electronics Dept, Experimental Station, Bldg. 336/130 Wilmington, DE 19898
Position: Departmental Fellow *Employer:* E.I. Dupont de Nemours & Co., Inc *Education:* PhD/EE/Carneige-Mellon Univ; MS/EE/Carnegie-Mellon Univ; BEE/EE/ Syracuse Univ. *Born:* 2/15/29. Instr Carnegie 1951-56; Asst Prof, Assoc Prof & Hd of EE Wichita State Univ 1956-61; Visiting Assoc Prof of Mech Engrg MIT 1961-62; Visiting Prof of Control Engrg Univ of Manchester (Eng) 1968-69; Dean of Engrg, V P & Dean, Widener Univ. 1962-75; Geo Tallman Ladd Prof & Hd of Mech Engrg Carnegie-Mellon Univ 1975-79. Wm J Brown Prof of Mech Engr, Carnegie-Mellon Univ, Prof-In-Industry, Berg Electronics Div DuPont Co 1979-80. Mgr, Computer & Automated Systems, DuPont Co, Berg Electronics Div 1980-81, Res Fellow, DuPont Co, Berg Elec Div 1981-85, Sr Res Fellow 1985-87. Departmental Fellow, Electronics Dept., Dupont Co, Wilmington, DE, 1987- . Lecturer, Penn State Univ 1983-87. Cons & editorial work in sys engrg & educ. Author over 50 publs & 'Intro to Sys Dynamics'. Western Elec Fund Award for Excellence in Instruction of Engrg, ASEE 1965. *Society Aff:* ASME, IEEE, AAAS, ASEE, ΣΞ.

Murphy, Charles H
PO Box 269, Upper Falls, MD 21156
Position: Div Chief. *Employer:* US Army. *Education:* PhD/Aero/Johns Hopkins; MS/Aero/Johns Hopkins; MA/Math/Johns Hopkins; BS/Math/Georgetown Univ. *Born:* 9/1/27. Math Instr, Univ of HI, 1949-50. Aerospace Res Engr, BRL, 1950-58; Aerodynamics Engr, Douglas Aircraft co, Summer 1958. Chief, Free Flight Aerodynamics Branch, BRL, 1958-70. Chief, Launch & Flight Div, BRL, 1970 to present. Visiting Prof, Univ of IL, Oct 1960, Visiting Prof, Univ of VA, Oct 1969. Gave numerous short courses on Missile Dynamics at UCLA, Norwich Univ, & Univ of TN. 55 scientific publications. 1976 AIAA Mechanics and Control of Flight Award. Dept of the Army Decoration for Meritorious Civilian Service. Army Res & Dev Achievement Awards, 1979 & 1986. AIAA Fellow, 1981. Guest Lecturer, East China Engg Inst, Nanjing, Summer 1985. *Society Aff:* AIAA, ADPA.

Murphy, Daniel J
Home: 2625 E. Southern Ave, No. 29, Tempe, AZ 85282
Position: Prof, Emeritus. *Employer:* Univ of AZ. *Education:* PhD/Met/Columbia Univ; MS/Engg/MIT; BS/Engg/US Military Academy. *Born:* 7/16/12. Military service 1935-55 including: Ordnance Gun Mfg, Watertown Arsenal; Chief, Pitman-Dunn Labs, Frankford Arsenal; Ordnance Technical Advisor & Chief of Planning & Engg, Ordnance Ind Operations, Far East Command. Staff Mbr & Consultant, Los Alamos Scientific Lab; Prof of Met Engg, Univ of AZ; Reg PE, State of AZ. *Society Aff:* ASM.

Murphy, Earl J, Jr
Home: 410 Pinecrest Dr, Miami Springs, FL 33166
Position: Dir of Development Services *Employer:* Post, Buckley, Schuh & Jernigan *Education:* B/CE City and Planning/Cornell Univ *Born:* 6/5/27 Served as Chief of Planning for the St of RI for an FCDA Study of Narragansett Bay 1954-57. Industrial Site Planner for EBASCO Services 1958-59. Organized and Directed Dept of City Planning and Engrg for Cleveland Hts, OH 1960-74. Joined Post, Buckley, Schuh and Jernigan as Project Mgr for Land Development in 1974. Became Miami Regional Mgr in 1979 and corporate wide Dir of Development Services in 1981 coordinating design and construction of residential, commercial, industrial, recreational and institutional projects throughout 17 east coast regions. *Society Aff:* AICP, ASCE, NSPE, NCEE

Murphy, Eugene F
Home: 511 E 20th St MB, New York, NY 10010
Position: Conslt, author *Education:* PhD/Mech Engg/IL Inst of Tech; MME/Mech Engg/Syracuse Univ; ME/Mech Engg/Cornell Univ. *Born:* 5/31/13. Syracuse NY. Engrg Dept Ingersoll-Rand Co 1936-39. Taught engrg Syracuse 1935-36, Ill Inst of Tech 1939-41, Univ of Calif 1941-45. On leave 1945-48 as staff engr Natl Res Council Comm on Prosthetic Devices. 1948-83 in Veterans Admin, concerned with nationwide res prog on rehabilitative engrg incl prosthetic & sensory aids. Fulbright Lecturer Denmark 1957; Silver Medal Paris 1961; Natl Acad of Engrg 1968; Outstanding Handicapped Fed Employee 1971; VA Meritorious Serv Award 1971, VA Distinguished Career Award 1983; IIT Alumni Assoc Prof Achievement Award 1983; Alliance for Engrg in Medicine & Biology Biomedical Engrg Leadership Award, 1983. Book chapters, num papers. Editor, Bull. Prosthetics Res 1977-83. Council, Alliance for Engrg in Medicine & Biology *Society Aff:* ASME, ASHRAE, RESNA, ISPO, ASA, OSA

Murphy, Gordon J
Business: 2145 Sheridan Rd, Evanston, IL 60201
Position: Prof of EE & Computer Sci. *Employer:* Northwestern Univ. *Education:* PhD/EE/Univ of MN; MS/EE/Univ of WI; BS/EE/Milwaukee Sch of Engg. *Born:* Feb 1927. Mbr fac of Univ of Minn 1952-57, Northwestern Univ 1957- . Chmn Dept of Elec Engrg Northwestern Univ 1960-69. Cons to G M, Minneapolis Honeywell, Motorola, Woodward Governor, & other co's. Author textbooks 'Basic Automatic Control Theory' 1957, 1966 & 'Control Engrg' 1960. Currently active in dev of electronic sys & computers Named One of Chgo's Ten Outstanding Young Men in 1961; Fellow IEEE 1966. *Society Aff:* IEEE, ΣΞ, HKN

Murphy, James L, Jr
Business: 3710 N Natchez Ct, Nashville, TN 37211
Position: President. *Employer:* James L Murphy Jr & Co. *Born:* Apr 1926 Rocky Mt N C. BCE North Carolina State Coll 1951. Civil engr TVA Nashville 1951-54; Batson & Melvin Const Co Nashville 1954-55; C J Batson Const Co Nashville 1955-56; owner James L Murphy Jr & Co Nashville 1956- ; V P Resource cons 1967-. Served with USNR 1944-46. Reg P E Ala, Ky, Tenn; Reg Landscape Arch Tenn; Reg Lan Surveyor Ala, Ky, Tenn. Mbr Amer Coll of Surveyors in Tenn - 1971 Pres, Tenn Soc of Prof Engrs, Water Pollution Control Fed, Amer Soc of Planning Officials, Amer Coll of Surveying & Mapping. V Chmn Tenn St Bd of Examiners for Land Surveyors. Received Long Rifle Award, Middle Tenn Council, Boy Scouts of Amer; 1971 St George Award, Diocese of Tenn, Boy Scouts of Amer. Roman Catholic. Home: 5231 Granny White Pike, Nashville 37220.

Murphy, James Patrick
Home: 18019 Alysson Dr, Sun City West, AZ 85375
Position: Retired *Education:* BS/Civil Engrg/Univ of ND. *Born:* 8/18/14. Jr Emgr Chicago Surface Lines 1937-40; Corps of Engrs MI & NY 1940-44; Traffic Engrg Bureau Detroit 1944; Hwy Traffic Engr, European Theatre 1945; Prin in Crawford, Murphy & Tilly Inc, Cons Engrs 1946-79; P Pres IL Soc of Prof Engrs; P Pres Springfield Chamber of Commerce; P Pres IL Chap of APWA; Chmn Springfield Auditorium Auth 1965-74; IL Award 1960, Hon mbr (1979) IL Society of Prof Engrs, Life mbr IL Roy Land Survayers Assoc. Retired as chairman of Board of Crawford Murphy & Tilly Inc on Dec. 31, 1979. Mbr of Sun City West Recreation Advisory Bd (1980-84); Engr for Town of Surprise Arizona (1981-84). Who's Who in America currently. *Society Aff:* NSPE, ASCE, ITE, ACEC, APWA.

Murphy, Loyal W, III
Home: 270 S Mendenhall Rd, Memphis, TN 38117
Position: President. *Employer:* L W Murphy Prof Engrs Inc. *Education:* MS/Industriel Systems Analysis/Memphis State Univ; BE/Civil/ Vanderbilt Univ; Grad Work/Bus & Econ/Memphis State Univ. *Born:* Nov 1938. MS in Industrial Systems Analysis at Memphis state Univ. Reg P E 14 states. Served in US Army Reserve, Artillery. With large cons firm for 9 yrs. In 1969 assumed ownership of family business - practices in civil & struct engrg, surveying & land dev counseling and as a patent agent. The Tenn Soc of Prof Engrs named him Young Engr of Yr 1969, & he served that org as Pres 1975-76. Director NSPE 1980-87. VP Natl Academy of Forensic Engineers. Married Kathryn Stockton of Nashville; 2 children. Mbr Christian Church. Ints: music, astronomy, gardening. *Society Aff:* ASCE, NSPE, WAFE.

Murphy, Michael
Home: P O Box 378, Whitestone, NY 11357
Position: V Pres. *Employer:* Ecolotrol Inc - Bethpage NY. *Education:* MChE/ChE/NY Univ; BChE/ChE/Manhattan College. *Born:* Mar 1941. Native of NYC. Res engr in membrane separation processes for 2 yrs. Taught in the Dept of Chem Engrg at Manhattan Coll for 5 yrs. Joined Ecolotrol Inc 1971, currently VP. Publ a num of articles concerning treatment of metal bearing wastewaters. Past Chmn N Y Sect of Amer Inst of Chem Engrs & P Chmn Metro Engrs Council on Environ Resources. Currently serving on the AES Bd of Dir. *Society Aff:* AIChE, WPCF, AES, ΣΞ, ACS.

Murphy, Monty C
Home: 9417 Mahler Pl, Oklahoma City, OK 73120
Position: Asst Dir-Planning & Res. *Employer:* OK Dept of Trans. *Education:* MS/Civil Engr/OK Univ; BS/Civil Engr/OK State Univ *Born:* 1/6/36. Born & educated in OK. Worked four yrs for CA Div of Hgwys in planning & design areas. Returned to OK Hgwy Dept in 1962 & progressed from Design Squad leader to Traffic Studies Engr, Proj Planning Engr, Planning Engr & Asst Dir- Planning & Res. Responsible for hgwy, railroad & public transportation planning & res along with coordination of airports & waterway. Pres - OK Sec ASCE in 1978. Mbr AASHTO Standing Committee on Planning & on various Transportation Res Bd committees. Mbr OK State Univ Bd of Visitors 1977-1979. ASCE Hgwy Div subcommittee mbr. Mbr State Capitol-Medical Ctr Plng and Zoning Comm, Mbr AASHTO to Standing Comm on Railroads and Standing Comm on Pub Transp. Served on Transp Comm of National Governor's Assn 1979-1981. Reg PE. O AZ5671. *Society Aff:* APWA, AASHTO, ASCE

Murphy, Vincent G
Business: Dept of Ag & Chem Engrg, Ft Collins, CO 80523
Position: Prof & Acting Hd of Dept *Employer:* CO State Univ *Education:* PhD/ChE/Univ of MA; MS/ChE/Univ of MA; BS/ChE/Manhattan Coll *Born:* 07/30/43 Native of Bklyn, NY. Res Engr for Southern Regional Res Center of USDA specializing in simulation and optimization of food processes 1971-73. Res Assoc at Univ of MO specializing in development of implantable oxygen electrodes 1973- 75. Asst Prof of Chem Engrg of IA State Univ 1975-77. Dept of Agri and Chem Engrg at CO State Univ since 1977. Helped establish chem engrg program at CSU and currently serves as Acting Hd of Dept. Active researcher in crystallization, extrusion processing food, and production of chemicals and fuels from renewable biomass. Recipient of AIChE Outstanding Student Chap Adv Award in 1982 and CSU Engrg Dean's Council Award in 1983. *Society Aff:* AIChE, ACS, ASEE, IFT

Murr, Lawrence E
Business: Oregon Graduate Center, 19600 N. W. Von Neumann Dr, Beaverton, OR 97006
Position: Dir: Off Acad & Res Programs; Prof of Materials Sci Engrg *Employer:* Oregon Graduate Center *Education:* PhD/Solid State Sci/Penn State Univ; MS/Engg Mechs/Penn State Univ; BSEE/Electronic Engg/Penn State Univ; BSC/Physical Sci/ Albright College. *Born:* 4/7/39. Instr & Res Asst Penn State 1962-67; Asst Prof Univ of So Calif 1967-72; Prof & Hd Met & Matls Engrg New Mex Tech 1972-80; Dir John D Sullivan Ctr for In-Situ Mining Res 1972-79; V P Res New Mex Tech Res Found 1973-79; V Chmn N Mex State Joint Ctr for Matls Sci 1976. Author 370 articles & 14 books. ASTM Photographic Awards, Electron Microscopy Soc Phys Sci Award, Outstanding Educators of Amer 1975. Pres, N Mex Tech Research Fdn 1979-81, V P for Acad Affairs and Research, Oregon Grad Ctr 1981-85. Dir: Office of Acad & Res Prog. 1985-present. Elected a Fellow The Amer Soc for Metals, 1981. *Society Aff:* IEEE, ASM, TMS-AIME, IMS, EMSA, Electro Chemistry Society, ASEE

Murray, Evelyn ME
Home: 37 Blanchard Rd, Cambridge, MA 02138
Position: Staff, Off of the Asst to the Dir *Employer:* MIT Lincoln Laboratory. *Education:* BSc/Physics & Math/Univ of Southampton *Born:* 3/15/37 1st woman 'special trainee' at Marconi Wireless & Telegraph Co England; worked in Q-switching of Ruby Lasers at Tech Opers Res, Mass: specialized in Microdensitremetry, working on film analysis Hek Corp & MIT Lincoln Lab. Presently operates the MIT Lincoln Laboratory in-house Education Program and PR Prof societies: Mass. Engrs Council Chrmn 1979-81; Soc. of Women Engrs, Natl Pres 1982-3; Mbr Natl Exec Comm 1976-1983 Boston Sec Representative 1975-76, Boston Section Chrmn 1973-75; Engr Soc of New Eng, Treas 1984- . Bd of Dir, Junior Engg Tech Soc 1986- . *Society Aff:* SWE, WES, IP, JETS

Murray, Haydn H
Business: Dept of Geology Indiana University, 1005 E 10th St, Bloomington, IN 47405
Position: Prof of Geology *Employer:* Indiana Univ. *Education:* BS/Geology/Univ of IL; MS/Geology/Univ of IL; PhD/Geology/Univ of IL *Born:* Aug 1924. Native of Toulon Ill. Army Corps Engrs 1943-46 in So Pacific. Taught in Geol Dept at Indiana Univ 1952-57. Joined Georgia Kaolin Co 1957 as Dir of Appl Res. Was Exec V P of Georgia Kaolin 1964-73. Returned to Indiana Univ as Chmn of the Geol Dept 1973. Pres of Clays Mineral Soc 1965-66. AIME Hal Williams Hardinge Award for

Murray, Haydn H (Continued)
contribs in indus minerals - elected a Disting Mbr of AIME 1976 and of the Clay Minerals Soc 1980. Pres-elect of Soc of Mining Engrs 1987. Special ints- indus minerals & clays. *Society Aff:* GSL, CMS, MSGB, AIPEA, ACS, SEG, AAPG, SEPM, SME, ACerS, AIPG, MSA.

Murray, James J
Home: 809 Hudson Ave, Durham, NC 27704
Position: Retired (Dir, Engrg Sci Div) *Education:* SM/Physics/Univ of Chicago; BS/ Physics/Loyola Univ (Chicago). *Born:* 4/28/12. m. 1944; 7 children. USA citizen. Loyola Univ Chicago, 1935 Physics; Univ of Chicago 1938 SM Physics; Univ of NC, Political Sci 1964-65. ONR 1947-50; NATC (Patuxent River (MD) 1951-54; Army Res Office 1954- . Editor-Publisher High Speed Ground Transp Journal 1967-76; Editor Urban Regional Ground Transp (book) 1973; Author 16 articles 1960-76; 10 pats. Res: wing-rotor interaction. Qattara Depression Reserach *re* Utility Hydraulic Power for Sea Water Electrolysis of Magnesium-Bromine Compounds. Adj Prof MEAE Dept NCST Univ. Awarded Dept of Army R&D Team Achievement 1966 (with S Kumar). Fellow of the Intl Society of Terrain- Vehichle Sys. *Society Aff:* CI, ISTVS, NGS.

Murray, John G
Home: 4672 Waldamere Ave, Willoughby, OH 44094
Position: Consultant. *Employer:* Self. *Education:* BS/Elect Engr/Carnegie Tech; AB/ Mathematics/Washington & Jefferson. *Born:* Dec 1922, Washington Pa. LT(j.g.) USNR WW II. Lincoln Electric Co. 1947-1984. Retired as VP, Machine Engrg. (Arc Welding Equipment, 3 phase motors.) Currently EE consultant on Product Liability ref arc welding equipment. Mbr IEEE, AWS. Enjoy water sports, travel. Married, 3 children, 3 grandchildren. *Society Aff:* IEEE, AWS.

Murray, Peter
Business: Box 158, Madison, PA 15663
Position: Chief Scientist *Employer:* Westinghouse Advanced Power Sys Divs. *Education:* PhD/Met/Sheffield Univ; BS/Chem/Sheffield Univ. *Born:* 3/13/20. in Rotherham Eng. 25 yrs of exper in nuclear field. From 1949-67, served at Atomic Energy Res Estab, Harwell England. Assignments incl Hd of Met Div, with promotion to Asst Dir. Joined Westinghouse 1967, & as Chief Scientist, Power Sys, Divs is respon for tech & dev assocs of adv energy props. Royal Inst of Chem Fellow. Inst of Ceramics Fellow. British Ceramics Soc (Mbr, PPres), Amer Cer Soc (Mbr). Amer Nuclear Soc (Mbr). Natl Acad of Engg (Mbr). *Society Aff:* RIC, BCS, ANS.

Murray, Ralbern H
Business: Consolidated Natural Gas Co, Four Gateway Center, Pittsburgh, PA 15222
Position: Vice Chmn *Employer:* Consolidated Natural Gas Company *Education:* MSIE/Indus Engg/NY Univ; BSChE/Chem Engg/NY Univ. *Born:* 10/19/29. USAF Air Res Dev Command 1951-52. Employed as Utilization Engr, Mgr Indus & Commercial Mkting, & Dir Sales with Amer Gas Assn 1950-64. With Consolidated Natural Gas Co since 1964, Dir Mktg; Assumed current respons as V P - Genl Mgr of CNG Energy Co Dv with the inception of the div in 1973. CNG Energy is respon for the dev of Consolidated's coal gasification prog & consequently is engaged in res, process eval, proj planning, coal & land acquisition, & assoc engrg studies. Mbr ASME, ACS, AIChE. Speaker: ACS, AISE, AIME, ASME. Author num articles. Assumed present position (Sr V) Aug 1, 1979. Consolidated Gas Supply Corp is part of Consolidated Natural Gas Co. Supply Corp employs approx 2600 people and is engaged in the production, transmission and distribution of natural gas in a four state area. Mr. Murray was made a Director of Consolidated Natural Gas Company, Pittsburgh, PA in 1983, named Vice Chairman, Consolidated Natural Gas Company, Pittsburgh, PA 1985- . He serves on Bds of the American Gas Assoc, Gas Research Inst and Software Valley Foundation. He serves on the Bd of Trustees of Carnegie Mellon Univ and is Chrmn of GRI 1986-87 Nominating Cttee. *Society Aff:* ASME, AIChE, ACS.

Murray, Raymond L
Home: P.O. Box 5596, Raleigh, NC 27650
Position: Professor Emeritus *Employer:* Nuclear Engrg Dept N C St Univ. *Education:* BS/Sci Educ/Univ of NE; MA/Physics/Univ of NE; PhD/Physics/Univ of TN *Born:* 2/14/20. Manhattan Proj 1942-45 electromagnetic uranium isotope separation at Berkeley & Oak Ridge. Prof at N C St Univ, phys, nuclear engrg, 1950-80. Dept Head 1960-74; Prof emeritus, 1980- . Res & teaching in nuclear reactor design analysis. Author of Texts 'Nuclear Reactor Physics' 1957, 'Introduction to Nuclear Engineering' 1954 & 1961, 'Physics: Concepts & Consequences' 1970, 'Nuclear Energy' 1975 & 1980, 'Understanding Radioactive Waste' 1982 & 1983. Cons to nuclear indus. Amer Nuclear Soc: Bd Mbr, Chmn, Educ Div; Arthur Holly Compton Award 1970. Amer Soc for Engrg Educ: Chmn, Ctte on Relationships with AEC; Chmn, Nuclear Engrg Div; Glenn Murphy Award 1976. N C Soc of Engrs. *Society Aff:* ASEE, ANS, APS

Murray, Thomas M
Business: Department of Civil Engineering, Virginia Polytechnic Institute and State Univ, Blacksburg, VA 24061
Position: Betts Professor of Steel Design *Employer:* Virginia Polytechnic Institute and State Univ. *Education:* PhD/Engr Mechanics/Univ of KS; MS/Civ Engg/Lehigh Univ; BS/Civ Engg/IA State Univ. *Born:* 5/22/40. Native of Waterloo, IA. Taught at Lehigh Univ, Univ of Omaha, Univ of KS, Univ of OK 1970-86. Distinguished visiting professor, United States Air Force Academy, 1986-87. Currently, Betts Professor of Steel Design, Department of Civil Engineering, Virginia Polytechnic Institute and State Univ. Author of a number of papers on floor vibration due to human occupancy and on structural steel connections. Active in full-scale experimental testing of structures. Member of AISC and AISI specification committees. *Society Aff:* ASCE, ASEE, AISC, SSRC, ACI, SEM.

Murray, William M
Home: P O Box 70, Georgeville P Que, Canada J0B 1T0
Position: Emeritus Prof of Mech Engrg. *Employer:* Mass Inst of Tech. *Education:* ScD/Mech Engg/MIT; SM/Mech Engg/MIT; BEng/Mech Engg/McGill Univ. *Born:* Apr 24, 1910 Montreal Canada; son of Wm A & R Alison Murray. Married 1949 to Joan R Steele; d. Sydney A Murray & s. Wm S Murray. Mech Engrg Dept MIT 1935-73. P Chmn Applied Mechs Div ASME; P Pres & 1 of 4 founders Soc for Experimental Stress Analysis; Mbr aSTM; Fellow AAAS; Fellow & Hon Pres Soc for Experimental Stress Analysis. Max M Frocht Award SESA for outstanding contribs to Engrg Educ. Now retired-formerly Reg P E Mass & Quebec. Visiting Professor of Civil Engineering, Univ Houston, Spring Terms, 1974, 75, 76 & 78. *Society Aff:* ASTM, SEM, AAAS.

Murrill, Paul W
Home: Senior Vice President, Ethyl Corporation, Baton Rouge, LA 70821
Position: Chancellor. *Employer:* Louisiana State University. *Education:* BS/Chem Engr/Univ of MI; MS/Chem Engr/LA State Univ; PhD/Chem Engr/LA State Univ. *Born:* July 1934; native of St Louis Mo. With US Navy 1956-59. Worked full-time as chem engr for PPG Indus & for Ethyl Corp; taught in mech engrg, chem engrg & computer sci depts; ser as Head of LSU's Chem Engrg Dept 1967-69; ser as Provost & Dean of Academic Affairs at LSU 1969-74; ser as Chancellor of LSU 1974-1980. Author or co-author of 8 books & 60 plus articles in natl journals. Makes about 100 talks per yr. Active in civic affairs; on several corp Bds of Dirs. Winner of numerous awards & honors. Married; 3 sons. Enjoys music & gardening. *Society Aff:* ASEE, AIChE, ISA.

Murtha, Joseph P
Business: Newmark Lab, 208 N. Romine St, Urbana, IL 61801
Position: Prof of Struc & Hydraulic Engrg; Director, Advanced Construction Technology Center *Employer:* University of Illinois. *Education:* PhD/Civil Eng/Univ of IL; MS/Civil Eng/Carnegie Mellon; BS/Civil Eng/Carnegie Mellon. *Born:* 7/18/31.

Murtha, Joseph P (Continued)
US Navy Civil Engr Corps 1955-58. Res Assoc 1958; Asst Prof 1961; Assoc Prof 1963; Prof of CE 1966; Dir Water Resources Ctr 1963-66 at Univ of IL, Urbana-Champaign; Dir Harbor & Amphib Div Naval Cvl Eng Lab 1966-67; Mgr Ocean Engg Western Offshore Div Fluor 1967-69; Univ of IL 1969- . Award Fullbright-Hays Professorship in UK 1977-78 in Ocean Eng; Special Advisor, Office of Ocean Eng, Natl Oceanic & Atmospheric Agency, Wash., DC 1978. *Society Aff:* ASCE, SSA, EERI, PIANC.

Murtha-Smith, Erling
Business: Dept of Civil Engrg, U-37, Storrs, CT 06268
Position: Assoc Prof *Employer:* Univ of CT *Education:* PhD/Engrg Sci/Univ of Durham, UK; BSc/CE/Univ of Leeds, Inc *Born:* 10/13/47 Principal Investigator for National Science Foundation awards for "Collapse of Space Trusses," 1979-82; "Experimental Verification of Space Truss Behavior" 1985-87. Conslt on Collapse of Hartford Civic Center Coliseum Space Truss 1978-84, and New York State Convention and Exposition Center 1984. Mbr American Society of Civil Engrs Ctte on Special Structures. Editorial Board of Intl Journal of Space Structures. Several publications on Space Truss behavior, Structural Stability and Structural Optimization. PE licensed in CT. *Society Aff:* ASCE, AISC.

Musa, John D
Business: Whippany Road, Whippany, NJ 07981
Position: Supervisor, Software Quality *Employer:* AT&T Bell Labs *Education:* M.S./EE/Dartmouth College; B.A./Engineering Sciences/Dartmouth *Born:* 06/11/33 Software engineer and software manager on a wide variety of software projects; work included computer graphics, computer security, systems engrg, simulation, human factors. Internationally recognized expert in software reliability. Principal author of the pioneering book *Software Reliability: Measurement, Prediction, Application* (McGraw-Hill, 1987) and over 40 papers. Fellow of IEEE, cited for "contributions to software engrg, particularly software reliability." Very active in prof societies (2nd VP, VP for Publications, VP for Technical Activities and many other offices of Computer Society of IEEE). Member of editorial boards of several major publications. Speaks fluent French. *Society Aff:* IEEE, ACM

Musa, Samuel A
Home: 5708 Covehaven Dr, Dallas, TX 75252
Position: Vice President for Res and Advanced Tech *Employer:* E-Systems, Inc *Education:* PhD/Applied Physics/Harvard Univ; BS & BA/Elec Engrg/Rutgers Univ *Born:* 06/20/40 Direct & coordinate all research & dev. activities within E-Systems. Report directly to Ch of Bd & CEO. Manage Internal Res & Development programs within E-Sys and responsible for tech base planning and computer-aided engrg acquisitions. Deputy Dir, Military Sys Tech Ofc of Under Sec of Defense for Res and Engrg, July 1979- May 1983, Dept of Defense Conslt, Exec Secy of Defense Sci Bd Summer Studies. Deputy Dir for C3 Policy and Requirements Review, Nov 1978-Jul 1979. Profl Lecturer in Engrg, George Washington Univ 1978- . Proj Leader, Inst for Defense Analysis, Jun 1971-Nov 1978. Asst Prof of Elec Engrg, Univ of PA, Sep 1966-Jun 1971. Conslt to Computer Command & Control Co, 1967-1971, RCA, 1968-70. Res Scientist at Gen Precision Inc, 1965-66. Visiting Lecturer, Stevens Inst of Tech, 1965-66. Technical Editor of IEEE transactions on Geosci & Remote Sensing, 1975-80 and Mbr of Admin Ctte, Communications Soc Conference Bd, Engrg Advisory Bd, Fellow of IEEE, over 60 publications. *Society Aff:* IEEE, AOC

Musella, Marianne A
Home: 150 Route 518, Hopewell, NJ 08525
Position: Sr Engineer *Employer:* Mobil Research & Dev Corp, Engrg Dept *Education:* BS/Chem Engg/Poly Inst of Brooklyn; MS/Comp Sci/Poly Inst. *Born:* 9/18/51. Brooklyn, NY; daughter of Frank & Gloria (Mirabito); Married Paul A Capsis Jr., also Chem Engg, July 16, 1977; BS Chem Engg, Poly Inst Brooklyn, 1973; awarded Poly, James Gordon Bennett Meml, & Regents Scholarships; process engr Rohm Haas Chem Co 1973-75, designed emulsion plants, deionized water & ion exchange systems; engr/analyst, Stauffer Chem Co 1976-80; joined Mobil 1980; work in Technical Computing, Process Simulators; mbr Omega Chi Epsilon, Am Inst Chem Engrs. Senior Member-Soc Women Engrs NY; Sec Representative 1981-82; Sec Pres 1979-81, Secy 1978-79; Bd Dir 1985-87; Convention Co-Chair-1990. *Society Aff:* SWE, AIChE.

Musiak, Ronald E
Home: 32 Fox Hill Drive, Westfield, MA 01085
Position: Assoc Prof *Employer:* Western New England Coll *Education:* PhD/EE/Univ of Mass; MS/EE/VA Polytech; BS/EE/Western New England Coll *Born:* 11/27/41 Native of MA. Spent four yrs in the Navy as an Electronics Technician- communications. Honorable discharge at grade E-5. Grad first in class (Magna Cum Laude) from WNEC. Grad top in class from VPI. PhD from Univ MA 1986. Spent 10 yrs in indus as a sr design engr in the area of television communications. Awarded patent on a novel video filter. Published internal reports on TV signal conditioning. Evening instructor for seven yrs at WNEC. Switched to full-time teaching in 1976. Introduced new courses to EE curriculum. Considered top instructor in the dept. Involved in part-time consltg for a local electronics firms doing research in the design of analog filters which satisfy Nyquist's criterion. *Society Aff:* IEEE, AAAS

Muster, Douglas
Home: 4615 O'Meara Dr, Houston, TX 77035
Position: Brown & Root Professor. *Employer:* Univ of Houston-Mech Engrg Dept. *Education:* PhD/Mechs/IL Inst of Tech; MS/Mech Engg/IL Inst of Tech; BS/Mech Engg/Marquette. *Born:* Nov 1918. Engr in Panama Canal Zone 1940-43; US Army 1943-46; Asst Prof of Mech Engrg IIT before joining G E Co, G E Lab to work in areas of mech vibration, shock & acoustics related to naval ships. With Univ of Houston Dept of Mech Engrg since 1961; Dept Chmn 1962-72 & currently Brown & Root Prof. Adjunct Prof of Law. Active Cons in design problems, engrg educ, law & tech (forensic engrg) & standards. Pres ISO/TC108, 'Mech Vibration & Shock' 1965- 1983. Fellow ASME, ASA, IMechE, IOA, AAAS. Diplomate, Natl Acad Forensic Engrs. *Society Aff:* ASME, ASA, AAAS, NAFE, I Mech E, IOA, JSME

Mutmansky, Jan M
Business: Fayette Campus, PO Box 519, Uniontown, PA 15401
Position: Mining Ctr Coordinator. *Employer:* PA State Univ. *Education:* PhD/Mining Engr/PA State Univ; MS/Mining Engr/PA State Univ; BS/Mining Engr/PA State Univ. *Born:* 4/26/41. in NY and raised in the soft coal region of Southwestern PA. Has worked in mining res at the US Bureau of Mines and in mine systems analysis at the Kennecott Copper Corp. Taught mining engg at the Univ of UT, WV Univ, and the PA State Univ. Has interests in and done res on mine systems analysis, mine ventilation, mine materials transport, and tunnelling. Presently is coordinator of mining programs at Penn State's Fayette Campus. *Society Aff:* AIME.

Muvdi, Bichara B
Home: 5935 N. Sherwood Dr, Peoria, IL 61614
Position: Past Chrmn and Prof of Civil Engg *Employer:* Bradley Univ. *Education:* PhD/Mechanics/Univ of IL; MME/Mechanics/Syracuse Univ; BME/Mech Engg/Syracuse Univ. *Born:* 8/16/27. My work experience has covered a variety of functions as follows: (1) Instructor and Res Assoc, Syracuse Univ 1952-56 (2) Sr Engr, the Martin Co, 1956-58 (3) Asst Prof, MI tech, 1958-60 (4) Assoc Prof, MI Tech, 1960-63 (5) Prof and Chrmn, CE/EM Dept, Bradley Univ, 1964-1981. Prof., CE/EM Dept, Bradley Univ., 1981-present. My primary interests are in the areas of solid mechanics and materials behavior. I have published in several journals that include AISC, IAHS, ASTM, SAE, ASM and T&AM. I am co-author of the textbook: *Engineering Mechanics of Materials*, a Macmillan publication, Second Ed 1984.

Muvdi, Bichara B (Continued)
Currently co- authoring one Book on 'STATICS' and a second on 'DYNAMICS' under contract with Holt, Rinehart & Winston. *Society Aff:* ASCE, ASEE.

Muzyka, Donald R
Home: 590 Galen Hall Rd, Reinholds, PA 17569
Position: General Manager *Employer:* Cabot Corporation *Education:* PhD/Matls/Dartmouth College; MS/Metallurgy/RPI; BS/Mech Engg/Univ of MA. *Born:* 8/23/38. Native of Northampton, MA. Metallurgist at Pratt & Whitney Aircraft 1960-63; Grad Sch 1963-66; with Carpenter 1966-82; with Cabot Cor 1982- . Major tech work has been on High Temp Alloys. Written over 30 papers & issued 7 US Pats. Assumed current position of Gen Mgr, Cabot Refractory Metals 1985. Former positions with Carpenter included VP Technical 1979, Gen Mgr of R&D Labs 1976. Mgr Alloy R&D 1973-75 & Mgr High Temp Alloy Metallurgy 1975-76. P Chmn Lehigh Valley Chap ASM; P Chrmn ASM-MSDD; Mbr ASM Long Range Planning Ctte; Trustee ASM 1981- p Chmn TMSAIME High Temp Alloys Ctte. Hobbies include antique clocks. *Society Aff:* ASM, AIME, ACS, IRI, AISE, ASQC

Myers, Alan L
Business: Dept of Chem Engg, Univ. of Pennsylvania, Phila, PA 19104
Position: Prof. *Employer:* Univ of PA. *Education:* PhD/Chem Engrg/U CA (Berkeley); BS/Chem Engrg/U Cincinnati *Born:* 9/26/32. M Irmgard Weber, Nov 30, 1957; children-Andrea, Sonia. Indust Engr Andrew Jergens Co, Cincinnati, 1949-51, 55-59; Mbr faculty Univ PA 1964-present. Prof chem & biochem engg, 1971-present. Chrmn chem engg 1977-1980. Visiting Scientist Award, Japan Society for Promotion of Science, Univ. of Tokyo, 1986. Vis Prof Technische Univ Graz, Austria, 1975-76. Served with USN, 1951-55 US - USSR Exchange Scientist, Inst Phys Chem Moscow, 1969-70. Auth: (with W D Seider) "Introduction to Chemical Engineering and Computer Calculations–, 1976. Over 80 publications in thermodynamics & adsorption. *Society Aff:* ACS, AIChE, AAAS.

Myers, Basil R
Business: Coll of Engrg & Sci, Orono, ME 04473
Position: Prof of Elec Engg. *Employer:* University of Maine - Orono. *Education:* PhD/EE/Univ of IL at Urbana-Champaign; MS/EE/Univ of IL at Urbana- Champaign; BSc/EE/Univ of Birmingham, England. *Born:* Sept 11, 1922. BSc Eng (Hons) Univ of Birmingham England 1950; MSEE Univ of Ill, Urbana-Champaign 1951; PhD ibid 1959. Prof of EE Univ of Maine at Orono, Orono Maine 04469. Fellow IEE & IEEE; C Eng (London England); Mbr: ASEE, AAUP, Men's Auxiliary of SWE, Soc for Indus & A;;lied Math, Math Assn of Amer, Sigma Xi, Tau Beta Pi, Eta Kappa Nu,Hertford & Oriel Socs of the Univ of Oxford. Author: 'North of the Border' 1963, plus numerous tech, res & educ articles. *Society Aff:* ΣΞ, HKN, TBΠ, AAUP, IEE, IEEE, ASEE, MAA, SIAM, SWE.

Myers, C Kenneth
Business: PO Box 1963, Harrisburg, PA 17105
Position: Chief Engineer *Employer:* Gannett Fleming Environ Engrs *Education:* SM/Civil Engrg/MIT; BS/Civil Engrg/Penn St Univ *Born:* 5/11/31. Hon fraternities; Chi Epsilon, Tau Beta Pi. Military Service with Cvl Engrg Corps USNR 1954-57. Reg P E in PA, N Y & IL. With Gannett Fleming Environ Engrs, Inc. 1958- , holding positions as DesignEngr, Ch of Plant Design, Deputy Ch of Design Chief Engr Asst Vp & Ptnr; respon for preparation of const plans & specifications in connection with water pollution control projs. Elected Diplomate of AAEE 1970. *Society Aff:* AAEE, NSPE, ASCE, WPCF

Myers, Dale D
Business: 251 S Lake Ave, Pasadena, CA 91101
Position: Pres. *Employer:* Jacobs Engg Group. *Education:* Hon Dr/-/Whitworth College, WA; BS/Aero Engrg/Univ . *Born:* Jan 1922. Hon Doctorate Whitworth Coll Spokane Wash. Joined Rockwell Internatl (formerly N AmerAviation Inc) 1943 as aeronautical engr engaged in R/D of long-range supersonic guided missiles; in 1964 became V Pres & Program Mgr for Apollo Command & Service Modules; in 1970 left Rockwell & joined NASA as Assoc admin for Manned Space Flight with overall respon for the manned spaceflight prog; in 1974 returned to Rockwell Internatl as Pres N Amer Aircraft Opers. 1977 under Secty, Dept of Energy, US Govt 1979, became Pres and Chief operating officer, Jacobs Engg Group. 1969 received NASA Public Service Award & NASA Certificate of Appreciation; 1971 & 1974 NASA Disting Service Medal. 1979 Rec Dept of Energy Distinguished Service Medal. Enjoys flying & fishing. *Society Aff:* NAE, AIAA, AAS.

Myers, Edward J
Home: 5856 Briar Hill Dr, Solon, OH 44139
Position: Director of Technical Divisions *Employer:* American Society for Metals. *Education:* PhD/Metal Engr/Univ of MI; MSE/Metal Engr/Univ of MI; BS/Metal Engr/Univ of AZ. *Born:* Sept 1922; native of Bakersfield Calif. Met engr for Miami Copper Co & Ladish Co 1946-50. 25 yrs in USAF R/D retiring as Col in 1974; ser as Metallurgist, AF Materials Lab, Prof of Met Engrg AF Inst of Tech, Dir of Aeromechs AF Sys Command, Dir of Aerospace Sci AF Office of Sci Res. Meritorious Service Medal 1973, Legion of Merit 1974. With ASM 1975- ASM, Director of Educations, 1975-1979 ASM, Director of Technical Divisions, 1980 to present . *Society Aff:* ASM, AIME, ASEE, AFA.

Myers, Gene W
Business: 1570 Pacheco, Suite A-7, Santa Fe, NM 87501
Position: Vice Pres *Employer:* Scanlon & Assoc *Education:* BS/CE/Univ of Denver *Born:* 08/25/51 Native of Canyon, TX. Served in Peace Corps-Honduras 1973-1974. Civil Engr for Daily & Assocs, Champaign, IL and Willis E. Umholtz & Assoc, Sante Fe, NM. With Scanlon & Assocs since 1977. Assumed current role as VP in charge of Santa Fe operations in 1979. *Society Aff:* NSPE

Myers, Harry E, Jr
Business: PO Box 1784, Mobile, AL 36633
Position: President *Employer:* Betz-Converse-Murdoch, Inc *Education:* BS/CE/Auburn Univ; BS/Bus Admin/Auburn Univ *Born:* 12/04/38 Mobile, AL native. Educated in public sch. Mbr Chi Epsilon scholastic honorary fraternity and Phi Delta Theta social fraternity. Employed by Betz- Converse-Murdoch since 1961. Expertise in municipal long range planning, finance, and operations. Presently Pres and Chief Exec officer of BCM Southern Group and mbr of the Bd of Dirs of BCM. Has served: Bd of Trustees United Fund; Dir Chamber of Commerce; Pres SAME; VP AL Consltg Engrs Council; Junior and Sr Warden of St Paul's Episcopal Church; Bd of Trestees of St Paul's Day Sch. Reg in AL, FL, GA, and MS. *Society Aff:* ASCE, SAME, ACEC, AWWA

Myers, James R
Home: 4198 Merlyn Dr, Franklin, OH 45005
Position: Consultant. *Employer:* James R Myers & Assoc. *Education:* PhD/Met Engg/OH State Univ; MS/Met Engg/Univ of WI; BS/Met Engg/Univ of Cincinnati. *Born:* 6/17/33. Native of Middletown, OH. Mtls engr for the AF Mtls Lab 1957-1960. Mtls engr for USAF Aero Systems Div 1960-1962. Prof of Met at the AF Inst of Tech 1962-1979. Author of over 60 published papers on corrosion, oxidation, & thermodynamics. Consultant to numerous governmental, educational, industrial and legal organizations. Served on active duty with the USAF 1957-1959. Presently a colonel in the USAF Reserves. Has presented over 100 papers on corrosion at natl & intl confs. *Society Aff:* AIME, ASM, NACE, JSCE, BICST.

Myers, John H
Business: 1204 E 12th St, Wilmington, DE 19898
Position: Technical Director Heisler Compounding Div *Employer:* Plastics Div-Container Corp of Amer. *Education:* MS/ChE/Stevens Inst of Tech; BS/ChE/Drexel Univ. *Born:* Dec 1925 West Chester Pa. 1947-55 R/D Engr Union Carbide Plastics; 1955- 57 Design Res Engr Monsanto Co; 1957-59 Sr Dev Engr Celanese Plastics; 1959-61 Supr Applied Dev Foster Grant Co; 1961-68 Dart Industries - Rexall Chem

Myers, John H (Continued)
Co, Mgr Dev Labs, Prod Dev Mgr ABS, Prod Mgr ABS; 1968-79 Plastics Div Container Corp of Amer, Genl Mgr R/D. 1979-1980 Genl Mgr of Heisler Compounding Div. 1980 - Technical Director - Heisler Compounding Div. 1974-75 Pres Soc of Plastics Engrs; Mbr SPE, ASTM, AMA, SPI, Soc of Rheology; Pennsylvania Soc. *Society Aff:* SPE, ASTM, SofR, AAAA, SPI, AMA, BS.

Myers, Peter B
Business: 2101 Constitution Ave, NW, Washington, DC 20418
Position: Staff Dir, Bd on Radioactive Waste Mgmt *Employer:* Natl Acad of Scis. *Education:* D.Phil/Nuclear Physics/Oxford Univ; D.HumLitt/Hon/Coll of Idaho; BS/EE/Worcester Poly Inst *Born:* 4/24/26. Native of Swarthmore, PA. D Phil Nuclear Physics Oxford Univ 1950; With USNR 1943- 46. Rhodes Scholar 1947-50. Bell Tele Labs 1950-59; Staff Scientist Motorola Semiconductor Prods 1960-62; Mgr Res & Adv Tech Martin/Bunker Ramo 1963-66; Mgr Res & Adv Tech Magnovox 1966-79; with Natl Acad of Scis 1979- , Current Respon Staff Dir, Board on Radioactive Waste Mgmt, Commission on Physical Sci, Math and Resources, Natl Res Council, Natl Acad of Scis. Fellow IEEE, AAAS; Founding Mbr TIMS; Mbr ANS, ORSA, AIP, Sigma Xi. Assoc Guest Ed Proc IEEE Special Issue Integrated Electronics Dec 1964. EIA/GPD Tech Council 1974-79. *Society Aff:* IEEE, APS, ΣΞ, ANS, AAAS, TIMS, ORSA

Myers, Phillip S
Business: 1500 Johnson Dr-ME Dept, Madison, WI 53706
Position: Prof of Mech Engrg. *Employer:* University of Wisconsin. *Education:* PhD/Mech Eng/Univ of WI; BS/Mech Eng/KS State Univ; BS/Math & Commerce/McPherson College. *Born:* May 1916. Major fields of interest: air pollution, internal combustion engines, combustion heat transfer. Taught formal classes & trained over 150 grad students at Wis 1942- . Cons to govt & indus; Prof lectrs in China, England, France, Italy, Spain, Germany, Romania, Japan & India. Has published 100plus papers & holds numerous pats. Active in prof societies & mbr numerous hon societies. Ser on numerous govt advisory cttes. Several teaching & prof awards. Enjoys camping, water skiing, photography & shopwork. Arch T Colwell Award 1966 SAE, 1969 National president SAE, Bd of Dirs 1970-71 SAE, Bd of Dirs 1971-72 ECPD. Bd of Dirs Echlin Mfg Co and Nelson Industries; Mbr Nat Acad of Eng 1973. *Society Aff:* NAE, SAE, ASME, AWS, ASEE, AAAS.

Myers, Robert D
Business: 4400 W Natl Ave, Milwaukee, WI 53246
Position: Chief Welding Engineer. *Employer:* Harnischfeger Corp. *Education:* BWE/Welding Engg/OH State; -/Mech Engg/OH Univ; -/Mech Engg/IA State. *Born:* 5/23/24. Mbr Tau Beta Pi. Ser USAAF as radar maintenance specialist 1943-46. Employed by Harnischfeger Corp since 1952 in various mfg & engrg positions. Ch Welding Engr since 1964; duties incl serving as world wide cons & field service struc engr, respon for welding liaison between mfg & engrg. Reg P E. AWS past District 12 Dir; Present AWS Natl Dir at Large; Mbr AWS D14.A; Mbr several AWS Sub-cttes. AWS Certified Welding Inspector. Enjoys golf, racketball & fishing. Cert AWS Welding Inspector. *Society Aff:* AWS, ASM.

Myles, Asa H
Home: 5870 Briar Hill Dr, Solon, OH 44139
Position: Former Ch Engr Heavy Indus Div. *Employer:* Square D Co (retired). *Education:* BS/EE/OH Univ. *Born:* Aug 1906 LaRue Ohio. Advanced work John Huntington Ins & Case Tech. Elec const 1929-30 control engr Elec Controller Mfg Co 1930-43; Asst Ch Engr 1943-56 Ch Engr Heavy Indust Div Square D Co 1956-67; Mgr Prod Planning & Promotion 1967-72 (retirement). Fellow & Life Mbr IEEE, Life Mbr Assn Iron & Steel Engrs; P Mbr NEMA, NEC Panel 12, ANSI B30 Safety Code; Mbr Cleveland Engrg Soc. Author of many tech papers, coauthor Indus Switchgear & Control Handbook. Co-author *Standard Handbook for Electrical Engineers* (tenth & eleven editions). Mbr Sons of American Revolution (SAR) Mbr Theta Chi fraternity. Mbr, Natl Rifle Assoc. Holds 12 pats. Continued limited consulting work after retirement. Enjoys golf, fishing, competitive shooting, hunting, swimming, boating & music. *Society Aff:* IEEE, AISE, CES.

Mylroie, Willa W
7501 Boston Harbor Rd, NE, Olympia, WA 98506
Position: Private Consult *Education:* MS/Regional Planning/Univ of WA; BS/Civil Engg/Univ of WA. *Born:* 5/30/17. Seattle. Reg P E in Wash. Trans Engrg and Resource Plng Conslt. Affiliate Prof of Civil Engrg, Univ of Wash. 1981-83. Res & Special Assignments Engr with Wash St Hwys 1958-81. Teaching & res in cvl engrg Univ of Wash 1948-56 & Purdue 1956-58; Military & Civilian Projs Engr Corps of Engrs 1941-45. Inst of Transportation Engrs Bd of Dir 1973-75; Pres Western District 1970-71; NAS Transportation Res Bd Council 1974-76; ASCE ad hoc Visiting Ctte ECPD 1974-78; Amer Hwy & Transportation Officials Select Ctte for Res 1974-1980; Sigma Xi; Planning Assn of Wash Bd of Dir 1970-74 with Citation for Disting Service. Inst of Trans Engrs Tech Council Chrmn Award 1982. Enjoys family, boating, gardening & music. Listed in Amer Men & Women of Sci, Community Leaders of Amer, Who's Who of Amer Women, Who's Who in the West, Who's Who in America. ASCE Edmund Friedman Professional Recognition Award 1978. King County Design Comm 1981- . Thurston Cnty Shoreline Mgmt Adv Ctte 1981-84, Thurston Regional Plng Coun Trans Adv Cttee, 1983-84; Univ Wash Coll Engrg. Visiting Ctte, 1978-86 Wash State Univ Coll Engrg Adv Cttee 1977-85; Thurston Cty Plan Comm, 1966-71. *Society Aff:* ASCE, ITE, ΣΞ.

Myrick, H Nugent
Home: 2123 Winrock Blvd, Houston, TX 77057
Position: President. *Employer:* The Process Co Inc. *Education:* BS/Bio/Lamar Univ; BS/ChE/Lamar Univ; MS/Env Sci & Engrg/Rice Univ; ScD/Env Sci & Engrg/WA Univ. *Born:* Apr 30 1935. Fellow, Instr Chem Engrg Rice Univ & Sanitary Engrg Harvard Univ; Assoc Prof Environ Engrg Univ of Houston 1964-74. Organized The Process Co Inc 1971. Prof interest in environ measures, assessment studies of facilities & activities on environment, process chem/biol & concept design of such processes & control sys in air & water pollution control & solid waste materials extraction & energy conversion. Outstanding res awards from Water Pollution Control Federation & AWWA. *Society Aff:* ASCE, AIChE, WPCF, AWWA, APCA.

Myron, Thomas L
Home: 650 Broughton Rd, Bethel Park, PA 15102
Position: Division Mgr-Chems & Polymers *Employer:* Aristech Chemical *Education:* MS/Chem Engr/Univ of Pgh; BS/Chem Engr/Univ of Pgh *Born:* 06/12/23 Served with Army Air Force 1943-46. Capt Meteorologist. Grad instructor Univ Pittsburgh Chem Engr Dept 1947-49. With US Steel Res since 1950 where various positions held include Chief of Raw Materials Div, currently Division Manager Chem & Poly. Div. Past Chrmn Ironmaking Div of Iron & Steel Soc of AIME, mbr exec bd, Iron & Steel Soc former Chrmn Program & Awards Ctte. Attended Amer Mgmt Assoc, US Steel, Indus Res Inst & Harvard courses. *Society Aff:* AIChE, AIME, SPE

Naaman, Antoine E
Business: Civil Engng. Dept, University of Michigan, Ann Arbor, MI 48109
Position: Prof *Employer:* Univ of Mich *Education:* PhD/CE/MIT; MS/CE/MIT; Dip Eng/CE/Ecole Centrale, Paris, France *Born:* 7/19/40 Born in Lebanon. Engrg Education in France. Graduate studies at MIT. Four years experience as a structural design engrg. Teaching and research since 1972. Areas of interest include *Prestressed Concrete, New Concrete Materials such as Fiber reinforced concrete and Ferrocement, Structural Optimization and Construction Mgmt. PCI M. P. Korn Award 1979. ASCE TY LIN Award 1980. PCI M.P. Korn Award 1986. Book Prestressed Concrete Analysis and Design,* 1982 McGraw Hill more than one hundred technical publications. *Society Aff:* ASCE, ACI, PCI

Nachlinger, R Ray
Home: 10910 Wickersham, Houston, TX 77042
Position: President *Employer:* Ultramarine, Inc. *Education:* PhD/Mechanics Engg/Univ of TX; MSEM/Mechanics Engg/Univ of TX; BES/Mechanics Engg/Univ of TX. *Born:* 12/4/44. Native of Austin, TX. Joined Univ of Houston faculty in 1968 and was promoted to Assoc Prof in 1972. Active in res and consulting in the areas of Engg Mech, Applied Math, and Ocean Engg. Founder and Pres of Ultramarine, Inc. *Society Aff:* SNAME, ASCE.

Nachtman, Elliot S
Business: 205 W. Randolph St, Chicago, IL 60606
Position: Principal *Employer:* Tower Oil & Technology Co. *Education:* PhD/Metallurgical Eng/IIT; MS/Metallurgical Eng/IIT; BA/Chemistry/Wooster College. *Born:* 1923. Married, 3 children. Instr IIT; Visiting Prof Univ of Ill, Circle Campus; chemist, Manhattan Proj; V P Res & Dev LaSalle Steel Co; Principal Tower Oil & Tech Co; Mbr advisory cttes; Natl Bureau of Standards; Natl Matl Adv Bd; Cons Org of Amer States. 30 U S pats; Fellow ASM; mbr num cttes ASM, ASTM, ASLE, AIMME. Many publications and talks on technical subjects and mgr. of technology. *Society Aff:* ASM, ASLE, AIMME, ASTM.

Nachtsheim, John J
Home: 9504 Boyer Pl, Silver Spring, MD 20910
Position: Asst Adm for Shipbuilding & Ship Oper. *Employer:* U S Dept of Commerce-Maritime Admin. *Education:* LLB/Law/G Washington Univ Law Sch; BS/NA&ME/Webb Inst of NAV.ARCH. *Born:* Sept 26, 1925. Formerly Ch of the Office of Ship Const, Ch of the Office of R&D - all within the U S Maritime Admin. Prior to that, Ch Naval Architect, US Navy. Grad of Webb Inst of Naval Arch & Geo Wash Univ Law School. Elected Fellow in Soc of Naval Archs & Marine Engrs; also a V P for Life of that soc. Pres Amer Soc of Naval Engrs, & Hon Life Mbr. Awarded Natl Capital Gold Key Award 1970, Engr of Yr. *Society Aff:* SNAME, ASNE.

Nack, Donald H
Business: 7462 N Figueroa St, Los Angeles, CA 90041
Position: President. *Employer:* Nack & Sunderland, Engineers. *Education:* BE/Mech Engg/Univ of Southern CA. *Born:* Jan 1923 Sheboygan Wisc. US Navy Engrg Officer light cruiser 1944-46. Field engr for Stone & Webster on frequency change. Asst Ch Engr Kistner, Wright & Wright. Co-founder Nack & Sunderland 1954: Pres & Principal Cons Mech Engr, respon for mech sys design for bldgs & energy audits for energy conservation in existing facils. P Pres Cons Engrs Assn of Calif, Mech Engrs Assn, S Calif. Fellow Amer Cons Engrs Council, Inst for the Advancement of Engrg. American Society of Heating, Refrigeration and Air-conditioning Engineer (ASHRAE). Outstanding Engr Merit Award from Inst for the Advancement of Engg 1978. Enjoy sailing & electronic music. *Society Aff:* ASME, ASHRAE, ACEC.

Nadeau, John S
Business: Metallurgy Dept, Univ of British Columbia, Vancouver BC V6T 1W5
Position: Professor. *Employer:* Univ of British Columbia. *Education:* PhD/Metallurgy/U.C. Berkeley; MS/Metallurgy/U.C. Berkeley; BS/Metallurgy/Notre Dame *Born:* 7/19/29. MS, PhD Met from Berkeley; BS Met from Notre Dame. Genl Elec Res Lab for 10 yrs. Met Dept, Univ of Brit Columbia since 1970, teaching courses in Matls Sci and Fracture Mechanics. Res in the strength and failure of composite matls. *Society Aff:* A.Cer S, SAMPE

Nadeau, Leopold M
Consultant, 25 Acadie St, Aylmer, Que Canada J9J1H7
Position: Consultant (Self-employed) *Education:* BA Sc - B. Civil Engr. *Born:* Nov 1913 Montreal Quebec. Fire Protection Engr during war yrs. 1946 Corp of Engrs of Quebec, Asst Genl Secy & Genl Secy 1949. 1954 Racey, MacCallum & Assocs Cons Engrs - 1956 Dir. 1958 organized & managed newly created Canadian Council of Prof Engrs; Genl Mgr to 1979, respon for all oper of the Council. V P WFEO & YSF. Dir or mbr var Canadian & U S engrg orgs. Enjoy golf, music, reading. *Society Aff:* NSPE, EIC, FCAE, OEQ, APEO

Nadler, Gerald
Business: Department of Industrial and, Systems Engrg, University of Southern California, Los Angeles, CA 90089-0193
Position: Prof and Chmn, I & SE Dept IBM Professor of Engineering Management *Employer:* Univ of Southern CA *Education:* PhD/IE/Purdue Univ; MS/IE/Purdue Univ; BS/ME/Purdue Univ *Born:* 3/12/24 1948-49 Instructor, Purdue Univ. 1949-64 Washington Univ, St Louis (Prof and Chmn IE, 1955-64). 1964-1983 Prof Univ of Wisconsin-Madison (Chmn IE, 1964- 67, 71-75). 1983-Present Prof and Chrmn ISE, Univ of Southern CA. 1986-present IBM Prof of Engr Mgmt. President, The Planning, Design and Improvement Methods Group (Consultants). Over 160 papers, 9 books. Member, BD of Dir, Intertherm, Inc. (St. Louis) 1969-85. VP IIE 1955-56 1987-89; VP Alpha Pi Mu 1950-52; Chairman IE Division ASEE 1958-59, and others. Five invited visiting professorships, four abroad. Fellow, IIE 1969, AAAS 1984; Gilbreth Medal, Society for Advancement of Mgmt, 1961; Distinguished Engrg Alumnus Award, Purdue Univ 1975; Book-of-the-Year, IIE 1983; Who's Who in America; National Acad of Engrg 1986. *Society Aff:* IIE, ASEE, AAAS, TIMS, AAUP, IEEE-EMS

Nadolski, Leon
Business: 50 Beale St, San Francisco, CA 94119
Position: Chief Civ/Struc Engr *Employer:* Bechtel Group, Inc *Education:* DIPL Engr/Cvl Struct/Polish Univ Coll, London UK; Bus Mgmt Cert/Univ CA Berkeley; Milt Engr/Ecole Mil D'Appl Dugenie-France & Poland *Born:* 04/02/17 Born & raised near Inowroclaw, Poland. Attended military engrs academy in Warshaw, Poland. Served in the Polish Army in World War II in Poland, France, UK, Middle East & Italy, Captain G.S. (oper) H.Q. 2nd Warsaw Arm'd Div. Completed studies in London, UK. Worked England & in 1951 immigrated to the US Kaiser Engrs and since 1954 with Bechtel. Struct Design Engr, Supervisor, Project Engr & Mgr on Multiple Domestic and Foreign Projects (Canada,France). Since 1976 Chief Cvl/Struct Engr, Hydro and Commun Facil Div. Enjoy family life (4 children), horseback riding & swimming. *Society Aff:* ASCE, SAME, PAEC

Naeyaert, Roger S
Business: 3290 W. Big Beaver Rd, Troy, MI 48084
Position: Chief Executive Off *Employer:* Ellis/Naeyaert/Genheimer Assoc Inc. *Education:* BS/ME/Univ of Detroit *Born:* 8/21/22 Canada-native Detroit, MI since 1928. Toolmaker 1940-1943. US Army Field Artillery-fire direction 1943-1945. Product, process, machine design eng 1948- 1954. Sales mgr, gen mgr and Pres of special machine design-build firm 1954-962. Registered Engr MI and 34 other states. Founded present design and consulting practice in 1962 providing industrial and manufacturing engrg planning, plant layout and design and general architectural/engrg services nationwide. Served as Dir 1978-present, and Pres 1980-1981 American Consulting Eng Council/MI and VP MI Council of Professional, Scientific and Technicial Assoc. Enjoy golf, photography, carpentry and mechanics. *Society Aff:* ASME, ASTM, EDS, NSPE, SME

Nagai, Kenzo
Home: 2-5 Tsunogoro 1-Chome, Sendai, Japan 980
Position: Retired *Education:* Dr/Eng/Tohoku Univ. *Born:* March 21, 1901. Dr of Eng 1934 Tohoku Univ, Sendai Japan. Prof Tohoku Univ 1936-64. Dir of Res Inst of Elec Communication 1956-64. Prof Emer of Tohoku Univ 1964. Fellow AIEE 1971. Hon Mbr Audio Engrg Soc 1973. Mbr of Japan Academy 1977. Main res works were concerning magnetic recording method, the theory & design of elec networks & engrg educ. *Society Aff:* IEEE, AES, IECE

Nagaprasanna, Bangalore R
Home: 18592 SW 89 Pl, Miami, FL 33157
Position: Dir *Employer:* Mt Sinai Medical Center *Education:* MPS/Health Care/Long Island Univ; MBA/Quantitative Methods/St Johns Univ; BS/Indus Engrg/Central Engrg Coll; DME/Production Engrg/Tech Educ Bd; DBA/Management/Nova Uni-

Nagaprasanna, Bangalore R (Continued)
versity *Born:* 06/25/47 Born in Bangalore, India. Worked as mgmt engr at Montefiore Hospital, NY 1970-1972. Dir Mgmt Engrg, Mary Immaculate Hospital NY, 1972-1976. Joined Eastern Airlines as Proj Engr 1977-1983. Assoc Adj Prof at FL Intl Univ and Univ of Miami. Reg Engr, State of FL, P Pres and Dir Miami Chapter - IIE. Winner of Outstanding Indust Engr Award. *Society Aff:* NSPE, FES, IIE, HMSS, AHA, ACHE

Nagarsenker, Brahmanand N
Home: 6336 Foreward Ave, Pittsburgh, PA 15217
Position: Prof *Employer:* Air Force Inst of Technology *Education:* PhD/Statistics/Purdue Univ; MS/Statistics/Purdue Univ; BS/ME/Bombay Univ *Born:* 2/2/36 Native of Pittsburgh, PA. Taught at Univ of MD, Univ of Pittsburgh and Univ of WI. Currently is Prof of Statistics at the Air Force Institute of Technology which Grants BS, MS, and PhD Degrees in Engrg for the United States Air Force Personnel. *Society Aff:* IMS, AMS

Nagel, Robert H
4406 Sunset Rd, Knoxville, TN 37914
Position: Retired *Education:* MSCE/CE/Univ of TN, 1941; BS/CE/Cornell Univ, 1939 *Born:* 04/19/18 Professional engrg experience with TN Valley Authority (1939-43); asst prof of civ engrg Univ of TN (1943-44); asst bridge engr with Southern Railway Co. (1944-46); secy-treas & editor of the Tau Beta Pi Assoc (1947-82), secy-treas emeritus (1982-); partner in engrg & mgmt consltg firm Newman, Davis & Nagel (1983-). Fellow of ASCE & life mbr. Reg PE TN. Privileged mbr of NSPE. Mbr of Tau Beta Pi, Chi Epsilon, and other Honor socs. Honorary mbr of Knoxville Tech Soc. Emeritus mmbr of the natnl Bd of Governors of the Order of the Engr. *Society Aff:* ASCE, NSPE

Nagel, Theodore J
Business: 2 Broadway, New York, NY 10004
Position: Sr Exec VP & Asst to the Chrmn. *Employer:* Amer Electric Power Serv Corp. *Education:* MS/EE/Columbia Univ; BS/EE/Columbia Univ; AB/-/Columbia College. *Born:* 12/20/13. Native of Andes NY. Taught at Columbia prior to WWII. Served with US Navy in European Theater of Oper 1942-46. With American Electric Pwr since 1939 (with time out for Navy service), advancing from Asst Engr in its Planning & Oper Div to present pos in Dec 1975. (Respon for forecasting futute elec pwr demands & necessary pwr supply facils for meeting such demands in area served by co, together with tech & econ studies & evals of alternative programs for future const. Prior Duties Now as Asst to Chrmn does duties as assigned by Chrmn. Life Fellow IEEE & Mbr Natl Acad of Engg. Serves on num indus & gov cttees. Enjoy woodworking, gardening & traveling. *Society Aff:* IEEE, CIGRE, NAE.

Naghdi, Paul M
Business: Dept of Mech Engrg, Berkeley, CA 94720
Position: Professor. *Employer:* Univ of Calif - Berkeley. *Education:* D/Eng Mechs/ Univ of MI; MS/Eng Mechs/Univ of MI; BS/Mechanical Engg/Cornell Univ. *Born:* Mar 29, 1924 Tehran Iran. Naturalized 1948. 1951-58 Univ of Michigan: 1951-53 Asst Prof, 1953-54 Assoc Prof, 1954-58 Prof of Engrg Mech. 1958- , Univ of Calif Berkeley: Prof of Engrg Sci, Chmn Div of Appl Mech 1964-69. Mbr Natl Acad of Engrg, ASME; Chmn Appl Mech Div 1971-72, Fellow 1969. Acoustical Soc of Amer, Fellow 1968. U S Natl Ctte on Theoretical & Appl Mech 1972-84. Chmn Ctte 1978-80. Soc of Engrg Sci; Bd of Dir 1963- 70. Member, Society of Rheology Univ of Mich Disting Fac Award 1956; John Simon Guggenheim Fellowship 1958; Geo Westinghouse Award of ASEE 1962; Miller Res Prof in Miller Inst for Basic Res in Sci, Univ of Calif Berkeley 1963-64 & 1971-72; Timoshenko Medal of ASME, 1980; Hon Mbr ASME, 1983. The A. C. Eringen Medal of Soc. of Engg. Sci., 1986; D.Sc. *Honoris Causa*, Natl. Univ. of Ireland, Dublin, 1987. *Society Aff:* ASME, SR, ASA, SES.

Nagumo, Jin-ichi
Home: 2-19-5 Shirokanedai Minato-Ku, Tokyo, Japan 108
Position: Emeritus Prof *Employer:* Univ of Tokyo *Education:* PhD/Applied Phys/ Univ of Tokyo; M Eng/Applied Phys/Univ of Tokyo; B Eng/Applied Phys/Univ of Tokyo *Born:* 10/14/26 in Tokyo, Japan. Faculty member of Keio Univ, Tokyo, 1953-59. Since 1959, Assoc Prof and Full Prof of Faculty of Engrg, Univ of Tokyo, working in such fields as nonlinear systems analysis, biomedical engrg, human engrg, biocybernetics, especially neural modelings. Served as Chairman of Department of Mathematicial Engrg and Instrumentation Physics. Dean of Faculty of Engrg, Univ of Tokyo, and Tokyo Chapter Chrmn of Systems, Man, & Cybernetics Soc of IEEE. Currently Emeritus prof of Univ of Tokyo, Fellow mbr of IEEE. *Society Aff:* IEEE

Nagy, Denes
Business: 65 W. Division St, Chicago, IL 60610
Position: President *Employer:* Denes Nagy Assocs Ltd *Education:* BSME/Mech/Tech Univ of Budapest; MSME/Mech/Tech Univ of Budapest; MSBA/Bus Admin/Hous Comm Inst of Pest *Born:* 10/19/29 Reg PE in IN, IL, WA, WI, MA, NY, and CA and Cert Energy Mgr, approved by AEE. Listed in Marquis Who's Who and Who's Who in Engrg, and American Consltg Engrs Council Directory. Author of articles published in tech magazines. Past chrmn of various cttes of APEC. Lectured at Univ of IL; Univ of KY; Construction Specification Inst and at First Interntl Congress on Construction Communication in Rotterdam, Holland. Former Pres of the IL Architects-Engrs Council, and Ctte Chrmn which changed the IL law on Statue of Limitation. Working in the construction industry since 1955 experienced as Design Engr, Project Engr, Chief Mech Engrs, Dir of Engr, and Pres of large architectural engrg or consltg engrg firms. Experience includes all phases of hospitals, universities, master planning, schools, power plants, refrigeration plants, commercial, high rise residential indus facilities. Energy Mngmnt. *Society Aff:* ASHRAE, NSPE, ACEC, AEE, SAVE, APEC, ASME, APCA

Nahman, Norris S
Business: PO Box 44, Boulder, CO 80306
Position: Vice President *Employer:* Picosecond Pulse Labs, Inc. *Education:* PhD/EE/Univ of KS; MS/EE/Stanford Univ; BS/Electronics & Radio Engg/CA State Poly Univ. *Born:* 11/9/25. San Francisco, CA. US Merchant Marine 43-46, US Army 52-55. Instr, Assoc & Full Prof, Dir Electronics Res Lab, Elec Engg Dept, Univ of KS 55-66. Scientific Consultant, Sec Chief-Pulse & Time Domain, Natl Bureau of Stds, Boulder, CO, & Prof Adjoint EE, Univ of CO, 66-73. Prof & Chrmn EE Dept Univ of Toledo, Toledo, OH, 73-75. Section Chief-Time Domain Metrology, Sr Scientist, 79-83, Section Chief-EM Fields Characterization, 1984-1985. Natl Bureau of Stds, Boulder, CO. VP, Picosecond Pulse Labs, Inc., Boulder, CO, 86-present. & Prof Adjoint EE Univ of CO 75-present. Fellow IEEE, PE CO, URSI: Chrm US Natl Commission A 1985-87, Chrmn Intl Intercommission Working Group on Time Domain Waveform Measurements, 1981-present. *Society Aff:* IEEE, ΣΞ.

Nahmias, Steven
Business: Dept of Indus Engrg, Pittsburgh, PA 15261
Position: Assoc Prof of Indus Engrg. *Employer:* Univ of Pittsburgh. *Education:* PhD/Oper Res/Northwestern Univ; MS/Oper Res/Northwestern Univ; BS/Indus Engg/Columbia; BA/Math/Queens College. *Born:* June 1945. BA Queens College; BS Columbia; MS, PhD Northwestern. Worked as an OR analyst for IBM 1968-70. Cons with Westinghouse, Litton, O'Donnell Assocs, Westmoreland Hosp while at Univ of Pgh - since 1972. First prize in ORSA (Oper Res Soc of Amer) student paper competition 1971. Second prize in TIMS (Inst of Mgmt Sci) student paper competition 1972. Chmn Univ of Pgh Engrg Fac Org 1976-77. Chmn ORSA Student Affairs Ctte 1976-78. Author approx 30 pubs in inventory control & oper res. Visiting Assoc Prof at Stanford Univ 1978-79 acad yr. *Society Aff:* ORSA, TIMS.

Naismith, James P
Business: P O Box 3099, Corpus Christi, TX 78404
Position: President. *Employer:* Naismith Engrs Inc. *Education:* MS/Hydr Engr/ Cornell Univ; BCE/Civil Engr/Cornell Univ. *Born:* Aug 1936. Mbr of fac Cornell Univ 1960. Asst Engr Calif Water Pollution Control 196061. With Naismith Engrs Inc since 1961, design of water transmission & treatment facils, waste water facils, drainage projs, harbor facils. Pres Texas Sect ASCE 1973. Visiting Prof Texas A&I Univ 1973 & 1979. Mbr Bd of Dir ASCE, 1982-85, chmn ASCE Technical Activities Cttee 1986-present, Trustee Calallen Ind. Sch. Dist. 1985-present. *Society Aff:* ASCE, AWWA, WPCF, NSPE, ACI

Nakagawa, Ryoichi
Business: Cntr Engg Lab, Nissan Motor Co., Ltd, 1-Natushimacho, Yokosuka Japan
Position: Sr Tech Advisor. *Employer:* Nissan Motor Co, Ltd. *Education:* Dr of Eng/- /Univ of Tokyo, Faculty of Mech Eng. *Born:* 4/27/13. Born in Tokyo. From 1936-45 in Nakajima Aircraft Co, designed aircraft engines 14 and 18 cylinders aircooler, equipment to "Zero" fighters and many army and navy s planes in Japan. From 1966-79 was Exective Managing Dir of Nissan Motor Co, respon for R&D. Honorary mbr of JSAE and SAE-A and JASME. Fellow mbr of SAE. From 1977-1984 Chmn of Japan Electronic Control Systems Co. From 1973-76 Pres of JSAE, and from 1976-81 - VP of FISITA. From April 16, 1987, Vice President of the Engineering Academy of Japan (Newly Founded). Enjoy classical music and golf. *Society Aff:* JSAE, JAME, SAG, SAEA, The Engineering Acadamy of Japan

Nakanishi, Kunio
Home: 7-23 Hisaki-cho, Isogo-ku, Yokohama Japan 235
Position: Prof. *Employer:* Yokohama Natl Univ. *Education:* PhD/EE/Tokyo Univ; BS/EE/Tokyo Univ. *Born:* 7/4/22. Native of Yokohama, Japan. Grad Tokyo Univ in 1947, was a mbr of teaching staff of the same Univ 1947-52. Since 1952, joined with Yokohama Natl Univ, served as Asst Prof, Prof, Hd of EE Dept, mbr of Univ Council. Res fields are power engg especially switching phenomena in high-power switchgear. *Society Aff:* IEEE, IEE of Japan.

Nakano, Yoshiei
Home: Ikejiri 2-Chome 33 15-1101, Setagaya-Ku, 154 Tokyo, Japan
Position: Lecturer *Employer:* Hiroshima Institute of Technology, Dept. of Elect. Eng. *Education:* Dr/Sci/Tokyo Inst of Tech; BA/Sci/Tokyo Inst of Tech *Born:* 2/21/10 Native of Hiroshima, Japan. Taught in Tokyo Inst. of Technology from 1931,4 to 1934,3. From 1934,4 to 1954,10 I had been employed by Hitachi Ltd, and involved in Hitachi Industrial Research Laboratory engaging in the short circuit testing of AC circuit breakers. From 1954,11 to 1970,3, I was Professor of Tokyo Inst of Tech. and engaged in the study of high temperature plasma for nuclear fusion. One of my important works was the development of high speed vacuum switches of trigatron type. I am now studying Soliton waves in the L type repeated ladder type circuits. *Society Aff:* IEEE, JIEE

Nall, John H, Jr
Business: 210 West 6th St, Suite 600, Ft Worth, TX 76102
Position: Principal. *Employer:* Teague Nall and Perkins, Inc. *Education:* BS/Civil Engg/Univ of TX. *Born:* 8/22/44. Native of Hope, AR. Reared in Texarkana and Tyler, TX. Served with campus Crusade for Christ 1966-69. Asst Design Engr with Public Works Dept, City of Ft Worth, 1970. Design Engr and Proj Engr, Freese and Nichols, 1971-1976. Prin in Teague Nall and Perkins, Inc since 1976, responsible for various civil engg projects ranging from pump stations and water distribution to railroads, drainage, subdivision planning and design. VP and Dir of Ft Worth Chapter of TSPE (NSPE); named Outstanding Young Engr of the Yr for 1979 by the Ft Worth Chapter. *Society Aff:* NSPE, ASCE, ULI, CSI, AREA.

Nance, Richard E
Home: 505 Monticello Lane, Blacksburg, VA 24060
Position: Computer Scientist. *Employer:* Virginia Tech *Education:* PhD/Industrial Engg/Purdue; MS/Industrial Engg/NC State; BS/Industrial Engg/NC State. *Born:* July 22, 1940. Asst & Assoc Prof at Southern Methodist Univ 1968-73. Hd, Dept of Computer Sci at Vir Tech 1973-79; SRC Senior Visiting Research Associate, Imperial College, University at London 1980, Assn for Computing Machinery - External Activities Bd Chmn 1974- ; V Chmn 1969, Chmn 1970; SIGIR. Oper Res Soc of Amer - Area Editor for Computational Structures & Techniques of "Operations Res–; Tech Prog Ctte Las Vegas 1975. Amer Inst of Indus Engrs - Edit Bd of 'IIE Transactions' 1971-76; Dept Editor for Simulation, Automation and Info Sys 1976- . Inst of Mgmt Sci - V Chmn 1973-74; Chmn 1975-76; College on Simulation & Gaming. Author num papers in sci journals. *Society Aff:* ACM, IIE, ORSA, TIMS.

Nance, Roy A
Home: 206 Trinity Dr, McMurray, PA 15317
Position: Engrg Manager. *Employer:* Bettis Atomic Power Lab-Westinghouse. *Education:* BS/ME/GA Tech; -/Math & Physics/Univ of TN. *Born:* Apr 1931. Grad of Webb School & GA. Tech. Hometown: Bell Buckle, Tenn. NDT Res, Oak Ridge Natl Lab 1956-60. Mgr NDT Engrg - Budd Co. Automation Ind 1960-69. Mgr Inspection Dev, Westinghouse - Bettis Atomic Power Lab 1969- . 4 pats on NDT equipment. Specialist on Eddy Current Testing. Num NDT papers in internatl, natl & local prof soc mtgs & journals. Author & Mbr ASNT Handbook Revision Ctte. Elected ASNT Fellow 1976. Lic Engr QC in Calif. Hobbies: auto mech, gardening, church work. Elder in Church of Christ. *Society Aff:* ASNT.

Nanda, Ravinder
Business: Polytechnic University, 333 Jay St, Brooklyn, NY 11201
Position: Assoc Prof. *Employer:* Poly Inst of NY. *Education:* PhD/Ind Engg/Univ of IL; MS/Ind Engg/Univ of IL; BSc/Mech Engg/Banaras Hindu Univ. *Born:* 9/12/36. Work experience in operations planning, organizational development and manufacturing. Formerly on faculties of Univ of Miami (1962-67), NYU (1967-73), visiting faculty IIM-Bangalore 1981-82. Joined current Inst in 1973, am also co- Dir of Oper.Mgmt Prog Author, co-author, co-editor, contributing editor of over 20 publications resulting from sonsulting and academic work appearing in tech journals, handbooks and profl monographs. Consulting experience with Amer Mgmt Assoc, Port Authority of NY & NJ, TWA, HUD, City of NY, LACSA (Costa Rica), RECOPE (Costa), Univ of Costa Rica, NITIE (India), IIM-B (India), BHEL (India), TTRC, Aerotech World Trade Corp, Case & Co, Roslyn Willet & Assoc. , Suave Shoe Corp, Stevenson Jordan & Harrison. Featured speaker at tech and profl soc corps. Elected mbr Alpha Pi Mu *Society Aff:* IIE, TIMS, VITA, ISHA.

Nannis, Walid A
Business: 972 Peachtree St, NE Suite 100, Atlanta, GA 30309
Position: Pres. *Employer:* Nannis, Terpening and Associates, Inc. *Education:* MSCE/Struct/GA Inst of Tech; BS/Engg Phys/Al-Hikma Univ. *Born:* 12/3/39. in Bagdad, Iraq. 1961 grad of Al-Hikma Univ, Baghdad, Iraq. Came to US in 1964 to attend grad sch. Received MSCE degree from GA Tech in 1965 and immediately assumed duties of proj engr with Wm E Edwards Struct Engrs, Inc. (Now Nannis, Terpening & Assocs., Inc.) Became a partner and corporate officer in 1968. Serves as Pres of the corp. Is mbr of numerous profl organizations and holds numerous state registrations. Became an American citizen in 1972. Enjoys sports and music. *Society Aff:* NSPE, ASCE, CEC/G, PCI, PTI.

Naquin, Arthur J
Home: 2515 Pine St, New Orleans, LA 70125 *Education:* BE/Elec-Mech/Tulane Univ *Born:* Mar 31, 1900 Denison Texas. Student Test Engr, Genl Elec Co (Schnectady- Erie); Equip-Transp Engr, Street Railway Dept, New Orleans Public Serv 1926-41; Co Safety Counselor 1941-68. Instr Indus Safety Engrg Tulane Univ 1968-1978. Was Pres La Safety Conf; Southern Safety Conf; N o Safety Council; Veterans of Safety (Internatl). Was Genl Chmn, Ac Prevention Comm, Edison Elec Inst; Southeastern elec Exch; Natl Safety Council's Motor Transp Conf. Tau Beta Pi, Pi Tau Sigma. Fellow Amer Soc Safety Engrs. Reg Prof Engr La; Cert Safety Professional. *Society Aff:* ASSE, TBΠ

Narain, Jagdish
Business: AIU House, 16 Kotla Marg, New Dehli 110002, India
Position: Secretary *Employer:* Assoc of Indian Univ *Education:* D. Eng./Engrg/Roorkee Univ; PhD/Civil Engrg/Purdue Univ; MS/Civil Engrg/Purdue Univ; BS/Civil Engrg/Roorkee Univ; BS/Science/Lucknow Univ *Born:* Oct 1926. PhD & MS Purdue Univ; BE Roorkee Univ; BSc Lucknow Univ. Served with State Pub Works Dept U P India 1948-51, 1952-53. Res Asst & Instr in Soil Mech at Purdue Univ 1958-61. Mbr Civil Engrg Fac of Roorkee Univ 1953-58, 1962-69; teaching & res in soil mech, foundations & structs; Sr Prof & Hd of Soil Mech 1967-69; Dean of Students 1966-69. Worked on tech dev of engrg projs incl earth & rockfill dams, hwys, subway sys & flood protection works as Senior Engineer with TAMS, New York 1969-77. Vice Chancellor, Univ of Roorkee, India, and Director, Water Resources Development Training Centre 1977-82. Fellow ASCE Indian Geotech. Soc. & Inst of Engrs (India). CBI & P (India) Medal 1966, Publ 55 res papers. Awarded ASCE Norman Medal 1965, and Doc of Engrg (Honoris Causa) Roorkee Univ 1985. Conslt on Irrigation, Water Mgmt, and Technical Educ. *Society Aff:* FASCE, FIE, FIGS

Naras, Raymond H
Home: 7511 N Overhill Ave, Chicago, IL 60648
Position: Coordinating Engr *Employer:* City of Chgo, Bureau of Engrg. *Born:* Nov 8, 1927 Chgo Ill. US Army 1946-48. 1952 - BSCE Rose-Hulman Inst of Tech. Reg Prof & Struct Engr Ill. Const & Resident Engr, Interstate Hwys, Interceptor Sewers, Arterial Streets. Design Struct Engr, Filtration Plant, Elevated Hwys, Retaining Walls. Design Civil Engr: all elements Roadways & Hwys. Exec Dir of Interdisciplinary Design Group for a major urban interstate route. Ch Res & Planning Engr: planning & programming for major public works dept. Mbr ASCE; Pres Ill Sect ASCE; Mbr APWA; Mbr Oper Res Soc of Amer.

Nardo, Sebastian V
Home: 270 Foster Avenue, Malverne, NY 11565
Position: Prof *Employer:* Polytechnic Univ *Education:* PhD/App. Mech/Polytech Inst of Brooklyn; MAe.E/Aero Eng/Polytech Inst of Brooklyn; BME/ME/Polytech Inst of Brooklyn *Born:* 12/25/17 b. and educated in Brooklyn, NY. Aerodynamics eng with Chance Vought Aircraft 1942-46. Appointed Assist Prof of Aerospace eng at The Polytechnic Inst of Brooklyn in 1950. Author or approximately 40 published papers and reports in the field of structural mechanics. In addition to teaching fluid and solid mechanics courses at the Polytechnic, have acted as departmental administrative officer and advisor to students. Enjoy loafing and music-especially participating in various choral groups. *Society Aff:* AIAA, NYAS, SESA, AAUP, ΤΒΠ, ASME

Nardone, Pio
Business: 283 Rt 17 S, Paramus, NJ 07652
Position: VP. *Employer:* Burns & Roe, Inc. *Education:* MME/ME/Poly Inst of Brooklyn; BME/ME/Cooper Union Sch of Engg. *Born:* 6/17/23. With Burns & Roe, Inc, since 1951 as Mech Engr, Proj Engr and Proj Mgr. Assumed current responsibility as VP and Dir of Proj Operations in 1970. Responsible for the direction of numerous multi-million dollar engg and construction mgt projs, including large fossil and nuclear generating stations. Reg PE in NY, NJ, WV, KY & SC. ASME rep on the Gantt Medal Bd of Award. Previously Proj Engr with Foster Wheeler Corp, Proj Leader and Applications Engr with Worthington Corp and Mech Engr with NY Navy Yard. Recipient of NY State Regents' scholarship. *Society Aff:* ASME.

Narendra, Kumpati S
507 Becton Ctr, Prospect St, New Haven, CT 06520
Position: Prof. *Employer:* Yale Univ. *Education:* PhD/Applied Phys/Harvard Univ; MA/(Honorary)/Yale Univ; SM/Applied Phys/Harvard; BE/Elec/Univ of Madras. *Born:* 4/14/33. Came to the US in 1954 after receiving the Bachelor of Engg degree in India. After completing the PhD at Harvard in 1959, was a post-doctoral fellow (1959-61) & asst prof (1961-1964) there before going to Yale Univ as assoc prof. Was made prof in 1968. Recipient of the IEEE Franklin V Taylor Award in 1972. Author of the book "Frequency Domain Criteria for Absolute Stability." Fellow IEEE in 1979, Fellow IEE (UK) 1982 and Fellow Amer Assoc for the Advancement of Science 1987. Editor of the books: "Applications of Adaptive Systems Theory" (with Prof R.V. Monopoli) 1980 Academic Press. "Adaptive & Learning Systems" Plenum Press 1986. *Society Aff:* IEEE, AMS.

Naresky, Joseph J
Home: 205 W Cedar St, Rome, NY 13440
Position: Engineering Advisor *Employer:* IIT Res. Inst, Rome, NY *Education:* MS/Engg Admin/Syracuse Univ; MS/Elec Engg/Syracuse Univ; BA/Physics/Syracuse Univ. *Born:* Dec 1923. Native of Phymouth Pa. Army Air Corps 1943-46. More than 37 yrs exper in govt electronics. With RADC from 1950 to 1979, performing & managing electronics R&D in areas such as electronic test equip, solid state devices & microelectronics, data presentation & display, energy conversion, electromagnetic compatibility, & component & equip reliability. Final RADC position was as Ch, Reliability & Compatibility Div 1969-1979. Directed an R&D program in component & equip reliability & maintainability, microelectronics & electromagnetic compatibility. Assoc Fellow AIAA; Fellow IEEE; Pres IEEE Reliability Group (1977-78). Since 1979, Engineering Advisor with ITT Research Institute (IITRI) responsible for a inside range of activity concerned with the disciplines of reliability, maintainablity, electromagenetic compatiblity, and software engineering *Society Aff:* IEEE, AIAA.

Narud, Jan A
Business: P O Box 1663, MS-428, Los Alamos, NM 87545
Position: Staff Member. *Employer:* Los Alamos Sci Lab - U of Calif. *Born:* June 1925. PhD Stanford Univ 1955; BS & MS Caltech 1951. Native of Oslo Norway. Asst Prof in Div of Appl Physics & Engrg Harvard Univ 195560. Cons, Bell Telephone Labs on pulse code communication sys 1956-60. 1960-61, IBM Res Ctr, Yorktown Hts NY; Mgr, dev of high speed memories & transistor logic circuits. 1961-67 Dir, Integrated Circuit R&D, Motorola Semiconductor Prods Div, Phoenix Ariz. 1967, became 1st recipient of 'Motorola Fellow Award for Outstanding Services'. Apptd Dir of computer Aided Design, 'Integrated Circuit Ctr', Mesa Ariz & as staff mbr to V Chmn Motorola Inc. Presently staff mbr Univ of Calif, Los Alamos Sci Lab. Has publ 55 papers & sev U S pats. Has contrib 6 chapters to the book 'Analysis & Design of Integrated Circuits'.

Nasar, Syed A
Home: 1522 Lakewood Ct, Lexington, KY 40502
Position: Prof. *Employer:* Univ of KY. *Education:* PhD/Elec Engr/Univ of CA; MS/Elec Engr/TX A&M Univ; BSc/Elec Engr/Dacca Univ; BSc/Science/Agra Univ. *Born:* 12/25/32. Has been with the Univ of KY since 1968. Specializes in elec machines. Has been the author (or coauthor) of twenty books on the subject, and is the Chief Editor of Elec Machines and Power Systems: An Intl Journal. Has been the recipient of several res grants from the Natl Sci Fdn and is a consultant to a number of industrial/res organizations. *Society Aff:* IEEE, IEE; HKN, ΣΞ

Nash, Jonathon M
Business: 9201 Corp Blvd, Rockville, MD 20850
Position: Sr Engr *Employer:* IBM Corp *Education:* PhD/Engr'g Science/Univ. of Mississippi; MS/Engr'g Science/Univ. of Mississippi; BSME/Mech Engr'g/Univ. of Mississippi *Born:* 08/10/42 A Senior Engineer with IBM Corp's Federal Systems Div. Currently Support Subsystems Segment Mgr on IBM contract team developing the Federal Aviation Administration's next generation Air Traffic Control System. Responsible for five of the nine subsystems being developed. Joined IBM in 1967. Served in Army Corps of Engineers 1968-70. Was the Materials Testing Officer, established, and was first Director of Materials Testing Lab of Army Engineer Construction Agency in Vietnam. Currently major in Army Reserve assigned to a Re-

Nash, Jonathon M (Continued)
search, Dev, & Engineering Center. ASME-fellow, past chairman Solar Energy Div, P Chairman North Alabama Section, Assoc. Editor *Applied Mechanics Reviews.* AIAA - Associate Fellow; SAME - Past Chapter Pres, Tudor Medal 1976; NSPE - Alabama Young Engineer of the Year 1976; Sigma Xi - Past Chapter Pres. *Society Aff:* ASME, AIAA, ASHRAE, SAME, ΣΞ

Nash, William A
Business: Dept of Civil Engrg, Amherst, MA 01002
Position: Prof of Civil Engrg. *Employer:* Univ of Massachusetts. *Education:* PhD/Mechanics/Univ of MI; MS/Mechanics/IL Inst of Technology; BS/Civil Engr/IL Inst of Technology. *Born:* Sept 1922. Struct Res Engr, David Taylor Model Basin 1949-54. Assoc Prof, Prof & Dept Hd at Univ of Florida 1954-67. Prof Univ of Mass 1967- . Fellow ASME. Editor Internatl Journal of Nonlinear Mech; Assoc Editor Applied Mech Reviews. Author over 90 res papers & one book. Curtis W McGraw Res Award ASEE 1961. *Society Aff:* ASME, AIAA, ASEE, IASS.

Nashman, Alvin E
Business: 6565 Arlington Blvd, Falls Church, VA 22046
Position: Pres, Sys Group. *Employer:* Computer Sci Corp. *Education:* ScD/Honorary/The George Washington Univ.; ScD/Honorary/Pacific University; MS/Electrical Engineering/New York University; BS/Electrical Engineering/City College of New York *Born:* Native of NYC. BSEE City College of New York; MSEE New York Univ. In 1952, began a 13-yr assn with the Internatl Tele & Telegraph org. Joined Computer Sciences Corp 1965. Fellow IEEE. Mbr Natl Bd of Dir of the Armed Forces Communications & Electronics Assn, as well as an officer of the Wash Chap of AFCEA. Additionally, is on the Bd of Governors of the Natl Space Club. *Society Aff:* AFCEA, NSC, IEEE

Nason, Howard K
Home: 230 S Brentwood Blvd, St Louis, MO 63105
Position: President & Consultant. *Employer:* Howard K. Nason Associates *Education:* AB/Chem/Univ of KA. *Born:* July 1913. Grad courses chem Washington Univ; grad school, Bus Admin, Advaced Mgmt Program, Harvard. Monsanto res chemist; Dir Dev, Central Res Dept; Res Dir, Organic Chem Div; Genl Mgr Res & Engrg Div; V P; Pres - Monsanto Res Corp. Chmn NASA Aerospace Safety Adv Panel; Tr C F Kettering Foundation; Mbr Natl Matls Adv Bd, ERDA Task Force on Demonstration Projects. Cited by Amer Standards Assn for contribs dev standards for Army & Navy electronic equip & for work of Assn's War Ctte on Radio; Trustee Acad Sci St Louis. *Society Aff:* ACS, AIChE, AIC, ASTM

Nasser, Essam
Business: P.O. Box 181, Dokki, Egypt
Position: Hd, Elec Dept & Dir Prof of Physics *Employer:* Middle East Consultants in Cairo *Born:* Feb 3, 1931. Dr Eng 1959 Technische Universitat, W Berlin Ger; BEE Cario Univ Egypt 1952. Worked for the Siemens Co in W Germany as a power engr 1952-54 & as a res engr 1958-61 & as a group leader for insulator res 1961-63. In 1955 & 1956, worked at the res labs of the Electricite de France near Paris. In 1961, joined the fac of the Physics Dept of the Univ of Calif at Berkeley. In 1964, taught at the Univ of Detroit & from 1964-74 taught & conducted res in gaseous ionization & high-voltage engrg at Iowa State Univ in Ames Iowa. In 1976, elected Fellow IEEE. From 1974 to 1976, Prof at Amer Univ in Cairo, from 1977 to date, Hd, Elec Dept, Middle East Consultants.

Nasser, Karim Wade
Business: Saskatoon, Saskatchewan Canada
Position: Professor. *Employer:* Univ of Saskatchewan. *Education:* PhD/Civil Engg/Univ of Saskatchewan; MS/Civil Engg/KS Univ; BS/Civil Engg/American Univ of Beirut; BA/Arts & Scis/American Univ of Beirut. *Born:* 12/9/26. Design & Const Engr Trans-Arabian Pipe Line Co 1949-56; taught at Lehigh Univ 1957-60; have been teaching at Univ of Saskatchewan since 1961; author num sci papers; ACI Wason Medal for res 1973; inventor: K-slump tester, K-5 accelerated strength tester, IN-SITU Air Meter, Nondestructive Tester, and others. Cons engr & Pres NHF Engrg Ltd, Pres, Victory Const. Ltd; Pres., Saskatoon Conva-lescent Home. Member, Science Council of Canada, Pres. Canadian Club of Saskatoon. *Society Aff:* EIC, ASCE, ACI, ASTM, CSA.

Nathan, Marshall I
Business: Research Ctr-Bx 218, Yorktown Hts, NY 10598
Position: Physicist. *Employer:* IBM. *Education:* PhD/Applied Physics/Harvard; MA/Physcis/Harvard; BS/Physics/MIT. *Born:* Jan 22, 1933 Lakewood N J. Fellow IEEE, Amer Phys Soc; SecyTreas DSSP 1973-77. IBM Corp Outstanding Contribution Award for the Injection Laser 1963. IBM Res Outstanding Contrib Award for Microwave Oscillations in Germanium 1968; IEEE David Sarnoff Award 1980. *Society Aff:* APS, IEEE.

Nation, Oslin
Business: 2532 Irving Blvd, Dallas, TX 75207
Position: Pres. *Employer:* Oslin Nation Co. *Education:* BS/EE/Southern Methodist Univ. *Born:* 4/11/14. Pres of Oslin Nation Co. Exec VP of First Co. *Society Aff:* ASHRAE, NSPE.

Nau, Robert H
Business: Elec Engrg Dept, Rolla, MO 65401
Position: Prof of Elec Engrg. *Employer:* Univ of Missouri - Rolla. *Education:* MS/EE/TX A&M; BS/EE/IA State Univ; Prof Deg of EE/EE/IA State Univ. *Born:* Apr 21, 1913 Burlington Iowa. IEEE Fellow Award 'For contribs to engrg educ'. Other awards: Breast Order of Yun Hui, Pres of China 1946; Outstanding Chap Pres, Mo Soc of Prof Engrs 1963; Life Mbr Radio Club, Univ of Mo 1963; Outstanding Pres, Dept of Mo, Reserve Officer's Assn 1964-65; Univ of Mo 1969 Outstanding Teacher Award; Sigma Xi; Phi Kappa Phi; Tau Beta Pi; Pi Mu Epsilon; Eta Kappa Nu; Sigma Tau; Kappa Mu Epsilon; Sigma Pi Sigma. Reg P E by Examination: Calif, Ill, Ohio, Mo. Author 5 textbooks and 74 tech papers. Five patents awarded. Ed Award 1982 and Award of Hon 1984 by IEEE, St. Louis sec. Hon Mbr Nat Hon of Blue Key 1983. Hon Prof of Military Sci by UMR Military Dept 1983. In 1983 UMR estab The Prof Robert H. Nau Endowed Scholarship Fund which is self perpetuating from now on. Order of the Golden Shillelagh, UMR 1984. In recognition as a distinguished alumnus for superior technical and professional achievement, the Coll of Engrg, Iowa State Univ, awarded their highest honor in June 1987, entitled Professional Achievement Citation in Engineering. Listed in American Men of Science. *Society Aff:* IEEE, ASEE, ΣΞ, ΦΚΑΡΑΦ, AAUP, ATA, NSSA.

Naudascher, Eduard
Business: Inst of Hydromechanics, D-7500 Karlsruhe, West-Germany
Position: Professor. *Employer:* Univ of Karlsruhe. *Education:* Dr Ing/Hydraulics/Univ of Karlsruhe; Dipl Ing/Civil Engg/Univ of Karlslruhe; Dr Ing/Civil Engg/Univ of Karlsruhe. *Born:* Aug 3, 1929. German nationality. Dr Ing Univ of Karlsruhe Germany. Did res at Univ of Minnesota 1959-60 & taught at Univ of Iowa 1960-68. Dir Inst of Hydromech & Prof of Engrg Hydraulics at Univ of Karlsruhe, Germany since 1968; Hd of Karlsruhe Res Ctr for Transport Processes in Flows 1970-74. W L Huber Res Prize of ASCE 1968; K E Hilgard Hydraulic Prize of ASCE 1975; ASCE Hydraulic Structures Medal 1987. About 130 contribs to prof journals & books; Editor Flow-Induced Struct Vibrations 1974 and Practical Experiences with Flow-Induced Vibrations 1980; Author "Hydraulik der Gerinne" 1987; VP Intl Assn Hydraulic Res until 1983. Enjoy classical music & photography. *Society Aff:* ASCE, ASEE, ΣΞ.

Nauman, E Bruce
Dept of Chemical Engrg, RPI, Troy, NY 12180
Position: Professor *Employer:* Rensselaer Polytechnic Institute *Education:* BS/Nuclear Engg/KS State Univ; MS/Chem Engg/Univ of TN; PhD/Chem Engg/Univ of Leeds. *Born:* 10/3/37. Native of Kansas City, MO. With Union Carbide

Nauman, E Bruce (Continued)
Corp 1963 to 1977 in Bound Brook, NJ and 270 Park Ave, NY. Positions as Res Engr, Group Leader, Styrene and Polystyrene Tech Mgr, Operations Res Mgr, Strategic Planning Mgr and Product Mgr for Engg Polymers. With Xerox Corp 1977 to 1981 in Rochester, NY as R&D Director for 120 man group in process engg, materials development and analytical services. Currently with Rensselaer Polytechnic Institute. Chairman of Chemical Engrg and Environmental Engrg, 1981-1984. Now Dir, Indus Liaison. Other academic appointments include professorships at Univ of Tennessee at Knoxville, State Univ of New York at Buffalo and Univ of Rochester. Personal research interests in polymer reaction engrg, residence time distribution theory and mixing are represented by 50 technical publications and two books. Active conslt to the process and materials indus. *Society Aff:* AIChE, AAAS, ΣΞ, ACCCE, LES, ACS.

Naumann, Albert, Jr
Home: 18531 Pt Lookout Dr, Nassau Bay, TX 77058
Position: Asst Genl Mgr. *Employer:* Lockheed Engg & Mgt Services Co, Inc. *Education:* BSME/Struct/Drexal Inst of Tech; BSEE/Power/Drexal Inst of Tech. *Born:* Dec 1925. Joined Lockheed 1958. Assigned to the space program in Houston in 1963 to implement flight instrumentation dev support to NASA. Became Dir of Genl Electronics Branch in 1966 & in 1969 was promoted to Asst Genl Mgr, Aerospace Sys Div, respon for sci & engrg programs, In 1978 was assigned as Asst Genl Mgr, Business Development, responsible for formulating the strategic business plans for corporate growth and providing direction in the implementation of the total new business acquisition process. A fellow President of the Instrument Society of Amer; Sr member of Electronics Industries Assoc, American Astronautics Society; active in Amer Federation of Infor Processing, National Society of Professional Engineers, National Contract Management Assn, and National Mgmt Assn. *Society Aff:* ISA, AAS, NSPE, NMA, NCMA.

Navarro, Guillermo R
Business: Av Union 163, Guadalajara Jalisco
Position: Sales Mgr. *Education:* PhD/CE/Tecnologico Monterrey; MBA/Marketing/Autonoma de Guadalajara. *Born:* 4/15/45. Native of Guadalajara, Jal. PhD degree June 1969. I have been working in the same co (steele bus Hylsa, SA) since I left school. The co belongs to Industrial Group ALFA the biggest private group in Mexico. Responsible for the states of the Pacific Coast. As a Sales Mgr including 8 states. *Society Aff:* ASM, IMIQ, CIMEG.

Nave, W Ralph
Home: 2308 Slayback St, Urbana, IL 61801
Position: Project Leader. *Employer:* US Dept of Agri - Agri Res Serv. *Education:* MS/Agri Eng/Univ of IL; BS/Agri Eng/Univ of TN. *Born:* June 1932. Native of Mountain City Tenn. Worked as a res & design engr for Sperry New Holland 1957-64. Conducted res on roadside maintenance equip at the Univ of Ill 1965-68; with the Agri Res Serv of the U S Dept of Agri since 1968. Proj Leader for soybean prod equip incl planting & harvesting. A major contrib has been the design of improved harvesting equip. Has served on num tech cttes for the Amer Soc of Agri Engrs, incl Chmn of the Ill-Wisc Region. Enjoy photography & gardening. *Society Aff:* ASAE, ASA.

Navon, David H
Business: Elec & Comp Engg Dept, Amherst, MA 01003
Position: Prof. *Employer:* Univ of MA. *Education:* PhD/Phys/Purdue Univ; MS/Phys/NYU; BEE/EE/CCNY. *Born:* 10/28/24. From 1954-1965 he was with the Transitron Electronic Corp, a semi-conductor device mfg, serving finally as Dir of R&D. From 1965-1968 he was Assoc Prof in the Elec Engg Dept of MIT, Cambridge, MA. Since 1968 he has been with the Univ of MA/Amherst, serving as Prof of Elec & Comp Engg. During the academic yr 1974- 1975 he was a Fulbright-Hays Lectr in Israel & Turkey. In 1977-1978 he was a Faculty Fellow with the US Gen Accounting Office, serving as a consultant on solar energy. He was a visiting lecturer at the Nanting Inst of Tech and the Fudan Univ, China, IN 1981. His current res activities include numerical modeling of semiconductor devices. *Society Aff:* IEEE.

Nawab, Ahmad B
Business: 211 N. Lois Ave, Tampa, FL 33609
Position: Pres *Employer:* Argus Engineering Co, Inc *Education:* MS/Structures/Univ of RI; BE/CE/Bombay Univ *Born:* 05/05/37 Native of Bombay. Worked as Structural Engr for Brown PE and Pavlo Engrg Co. Designed highway bridges. For PBQ&D designed aerial structures and rapid transit tunnel & ventilation shaft. As Sr Engr with Davy Powergas supervised task force involved in the design of Gas gathering and metering sys. As Chief Structural Engr with Kisinger Campo-developed specification for computer prog for inspection, analysis, rating and evaluation of Bridges - similar to BRASS. Also with KCA carried out Non-Destructive Load Testing and Evaluation of existing bridges. With Argus-designed multistory condominium & office bldgs. *Society Aff:* ASCE, NSPE

Nawy, Edward G
Home: 347 Felton Ave, Highland Pk, NJ 08904
Position: Prof of Civil Engrg. *Employer:* Rutgers Univ - St Univ of N J. *Education:* DrEng/Structures/Univ of Pisa, Italy; CE/Structures/MA Inst of Technology; DIC/Concrete/Imperial College of Sci, London Univ; Dipl Eng/Civil Eng/Univ of Baghdad. *Born:* Dec 1926. Dist Prof - Prof II Rutgers Scale. Prof of CE Rutgers, Dept of Civil Engrg, Busch Campus, New Brunswick NJ 08903. Have been engaged in res in concrete struct sys since 1955. Practiced engrg since 1948. Publ in excess of 65 sci papers. Cons to indus in NY-NJ Metro Area. Concrete Sys Cons to the FAA, Wash D C. Prof at Rutgers univ since 1959, in charge of instruction & res in concrete. Henry Kennedy Award 1972 ACI; Prestressed Concrete Bridge Design Award 1971 PCI; Chi Epsilon Hon Mbr 1970 NJIT; Award of Recog 1972 ACI-NJ; Pres NJ ACI 1977. Cons to Energy Div, Genl Acctg Office, Wash, DC 1978 Chapter Activities Award, Amer Concrete Inst Bd of Governors, Rutgers Univ; NJ State Task Force on Master Plan for HigherEducation. Chairman, Dept of Civil Eng, Rutgers. *Society Aff:* ASCE, ACI, PCI, ICE, AAUP.

Nay, Ward H
Home: 10811 East OP Ave, Scotts, MI 49088
Position: Engrg VP. Retired *Employer:* Upjohn Co. *Education:* BS/Chem Engg/Tri State Univ. *Born:* 5/10/21. Native of Battle Creek, MI. Served in US Navy WWII aboard destroyer USS Downes. Discharged as Lt (JG). Held engg positions with US Steel and Sherwin- Williams prior to joining the Upjohn Co in 1946. Promoted to Corp VP in 1975. Maj involvements include fermentation tech, pharmaceutical chem engg, facilities planning. Active in Construction Users Council (VP), Natl Assn of Corporate Real Estate Exec, Torch Club Intl, Kalamazoo County Chamber of Commerce. Interests include golf, photography and camping. Retired 1986. *Society Aff:* AIChE, PMA, AACE, ASEE.

Nayfeh, Munir H
Business: 1110 W Green St, Urbana, IL 61801
Position: Assoc Prof *Employer:* Coll of Engrg Univ of IL at Urbana-Champaign *Education:* PhD/Physics/Stanford Univ; MS/Physics/American Univ of Beirut; BS/Physics/American Univ of Beirut *Born:* 12/13/45 Born in Jordan Attended High School in Al-Bireh and Ramallah. Post Doctoral Fellow at Oakridge National Lab 1974-1976. Research physicst with Union Carbide 1977. Lecturer and research assoc at Yale Univ 1978. With the Univ of IL at Urbana-Champaign since 1979. Research includes theory and experiment on laser radiation interactions atoms and molecules, theory and experiment of atomic collisions, detection of very low concentrations of atoms and molecules and laser development. Industrial Research (IR 100) a prize 1977 for co-developing single atom detection using resonance ionization spectroscopy. *Society Aff:* APS, OSA

Naylor, Henry A, Jr
Home: 1320 Bolton St, Baltimore, MD 21217 *Education:* BS/ME/Johns Hopkins Univ. *Born:* Oct 1910. With a Baltimore cons to 1937; Baltimore Cityy 1938; Whitman, Requardt & Asscs beginning 1939 through 1980 when retired. Since 1957, Partner, in charge of Mech-Elec Dept & indus projects. Fellow ASME, IME (UK), ACEC. ASME IEEE, ASHRAE, NSPE. Reg P E 10 states; formerly on Md Regis Bd. ASME: P Chmn Baltimore Sect; Prof Practice & Mtgs Cttes; P Mbr Policy Bds Communications & Prof & Pub Affairs; P Secy Region III; Natl Nominating Ctte; liaison to NCEE; 7 certificates. CEC Md: P Pres, Natl Dir, Ctte (sev) Chmn. CEC USA: P V Pres. Former AICE Councilor. Canadian Iron Ring. Active in other orgs. *Society Aff:* ASME, ACEC, IEEE, ASHRAE, NSPE.

Neal, Donald K
Business: Fire Protection Eng. Svcs, 19847 N. 48th Ln, Glendale, AZ 85308
Position: President *Employer:* Fire Protection Engineering Services, Inc. *Education:* BS/Fire Protection Eng/ILL Inst of Tech *Born:* 1/8/40 Native of Chicago. Engr with Illinois inspection and rating bureau and western actuarial Bureau 1961-71. Instructor of Fire Insurance Rating, I.I. T, Chicago 1969-71. Regional Mgr, Insurance Services Office, Denver 1972-77. Sr Fire Protection engr with Arizona Public Service Co, Phoenix. Currently President of Fire Protection Engineering Services, Inc. Doing commercial and industrial consulting engineering. Registered PE in AZ and CA. 1980-81 Charter Pres, Arizona Chapter, Society of Fire Protection Engrs. *Society Aff:* SFPE, NFPA

Neal, Donald W
Business: 1410 SW Morrison St 804, Portland, OR 97205
Position: Owner. *Employer:* Neal Engg Assocs. *Education:* BS/Civil Engg/OR State Univ. *Born:* 1/4/37. BSCE OR State Univ 1958. With US Geological Survey 4 yrs in Portland OR & Washington DC - hydraulic engr in powersite dev. With Timber Structs Inc 10 yrs in Portland OR - Ch Engr Western Region. Founded Culbertson, Noren & Neal Engrs 1972. The firm practices struct engg, spec in struct timber design. In April 1978 the partnership of Culbertson, Noren & Neal changed to a proprietorship operating as Neal Engg Assocs under the ownership of Donald W Neal. *Society Aff:* AITC, CECO, ASCE, SEAO, IASS.

Neal, Gordon W
Home: 9192 Rhodesia Dr, Huntington Beach, CA 92646
Position: President *Employer:* Pacific Energy Consultants, Inc *Education:* BSc/Mech Engg/Univ of NB. *Born:* 9/2/22. Diversified engg practice emphasizing energy conservation and utilization, covering all phases of proj dev from detailed analysis through overall mgt. Tau Beta Pi, Sigma Tau, Pi Tau Sigma, Pi Mu Epsilon honor societies. Reg PE, eleven states. Certificate of qualification, Natl Council of Engg Examiners. Proj engr, TWA cogeneration plant that received NSPE award as one of ten outstanding engg achievements in 1971. Guest lecturer ten semesters at Purdue Univ. More than twenty published articles and conf papers. Cofounder Pacific Energy Consultants in 1978; previous twenty-five yrs with consulting engg firms serving industrial, governmental and commercial clients. Principal mechanical engineer, Thermco Associates, energy engineers. *Society Aff:* ASME, ASHRAE, NSPE.

Neas, Charles C
Business: P O Box 8361, South Charleston, WV 25303
Position: Assoc Director. *Employer:* Union Carbide Corp. *Born:* Jan 29, 1921. ScD MIT 1947; BS Univ of Illinois. With Engrg Dept Union Carbide, Chemicals & Plastics since 1947. Chem Process Engrg. Became Assoc Dir 1963. Activities in AIChE incl Chmn Natl Prog Ctte 1973 & Dir 1975-77. Served on Natl Metric Sys Study Advisory Panel 1970-71. Mbr ACS, AAAs, AIC, New York Chemists Club.

Nebeker, Eugene B
Home: 400 N Rockingham Ave, Los Angeles, CA 90049
Position: President. *Employer:* Scientific Assocs Inc. *Education:* PhD/Chem Eng/Cal Tech; MS/Chem Eng/Cal Tech; BS/Chem Eng/Stanford. *Born:* Nov 1936. PhD & MS Caltech; BS Stanford Univ. Native of Santa Monica Calif. In 1969 formed Scientific Assocs & is active in the mgmt & tech aspects of all co operations. Major projs have included the design, dev & const of a high-seas oil skimmer prototype & a percussive water jet for mining hard rock. Formerly with N Amer Rockwell; worked on problems involving fluid mech, heat transfer, & air & water pollution. Reg P E Calif & Texas. *Society Aff:* AIChE, ACS, ΣΞ, AAAS.

Nedom, H A
Home: 21 Deerwood Lane, Westport, CT 06880
Position: Conslt *Education:* BS/Petroleum Engrg/Univ of Tulsa; MS/Petroleum Engrg/Univ of Tulsa. *Born:* 8/19/25. With Amerada Petro Corp 1949-71. Ch Engr in 1961. V P in 1965. Pres of IIAPCO (Independent Indonesian Amer Petro Co) & V P & Dir of Natomas Co of San Fran 1971-74. Pres & Managing Dir of Weeks Petroleum Limited 1974-82. P Pres & Dir of Soc of Petro Engrs of AIME. Past Pres, Dir and Honorary Mbr of the Amer Inst of Mining, Metallurgical & Petroleum Engrs. Disting Alumni Univ of Tulsa 1972. Citation for Service, Amer Petro Inst. Distinguished Service Award DeGolyer Medal and Distinguished Mbr of Soc of Petroleum Engrs, Special Award, Amer Assoc of Engrng Societies; Chairman of Amer Assn of Engrng Societies *Society Aff:* AIME, AAPG, SPE.

Nee, Raymond M
Home: 203 Foss Drive, Upper Nyack, NY 10960
Position: Private Consultant. *Employer:* Retired from American Cyanamid Co. *Education:* BS/Math/Lynchburg College; -/Math & Physics/VA Poly Inst; Cert/Electrical/Lowell Inst. *Born:* Aug 1903. The Detroit Edison Co. 1929-42: Boston Edison Co - Design, oper & supervision of most phases of the Dist Steam Sys from production of steam to customer utilization. 1942-68: Amer Cyanamid Co Chemical Construction Corp subsidiary 3 yrs, then Central Div 3 yrs - as Sr Power Engr: 1948-57: Chf Engrg Lederle Labs Div. 1957-68: Engrg & Const Div - Asst for 2 yrs, then Dir. 1959- 66: Exec Asst to V P. 1966-1968- ; Private Consultant. Native of Norfolk Va. Engr of Yr 1971 NSPE; V P 1969-73 ASME, Mbr Exec Ctte of Council 1969-73 ASME. 1975 Hobbs Award for Outstanding Alumni - Lynchburg College. Life Mbr ASME, ASHRAE, NSPE & Int Dist Htg Assn. *Society Aff:* ASME, ASHRAE, NSPE, IDHA.

Neel, W Hibbett
Business: P O Box 22625, Jackson, MS 39201
Position: VP *Employer:* Hensley-Schmidt Inc. *Education:* MS/Trans/GA Inst. Tech; BE/CE/Vanderbilt Univ *Born:* 7/13/41 Has had over 15 years experience in the development design and project mgmt of public works projects with responsibilities ranging from project mgr for traffic and transportation projects to mgmt of a major branch office involving water,sewer, parks, transportation and community development projects. He is an advocate for traffic operations and safety improvements of urban streets and highways. He has lectured at numerous traffic conferences. Subjects included traffic safety mgmt, railroad crossing protection, opportunities for funding, traffic control devices. *Society Aff:* NSPE, ASCE, ITE, SAME, APWA, ARTBA

Neeley, Parley R
Business: 265 North Main, Spanish Fork, UT 84660
Position: Neeley Engrs. *Employer:* Self. *Born:* Sept 1903 Coalville Utah. BS Univ of Utah. Engr, Bureau of Reclamation 1927-63 - const of irrigation works, canals, etc - 3 dams. Conducted investigation of the billion dollar Central Utah Proj. Private practice since 1963, community facilities, water supply, sewage, streets, etc. Water right engr. Mbr ACEC, CEC Utah; Amer Soc Civil Engrs. Reg Engr Utah & Wyoming.

Neely, George Leonard
Home: 114 El Camino Real, Berkeley, CA 94705
Position: Retired-partially *Education:* BS/Marine Engr/U.S. Naval Acad *Born:* 11/25/01 SAE--Fellow" Grade 1981; Inventor, 30 US patents; Reg PE in state of

Neely, George Leonard (Continued)
CA; Capt (USNR-Ret). Exec - oil industry asst mgr - product engrg Standard Oil Co of CA. Commd ensign, USN, 1922, advanced through grades to capt. USNR, 1959; served in USS Oklahoma, Pacific Fleet, 1944-45; commdr US Naval Tech Mission to Japan, 1945-46. Pres Glengineering Co, Berkeley, CA 1966-. Invention Dev Mgmt. Decorated Bronze Star US del Tripartite Confs. Naval Fuels and Lubricants NATO, 1950-65. Episcopalian. Pioneer in lubrication res. Inventor detergent type lubricating oils chem treatment frictional surfaces, others. *Society Aff:* SAE, ASTM

Neely, H Clifford
Home: 123 Heritage Hill Rd, New Cauaan, CT 06840
Position: Sr Economist - Chem Industry. *Employer:* Merrill Lynch Economics Inc. *Education:* Chem Engr/M Engr/Univ of Louisville; BChE/Univ of Louisville. *Born:* Apr 2, 1930. Attended Speed Scientific School, Univ of Louisville. US Navy 1953-56. Ltjg Asst Engrg Officer. E I du Pont 1956-62, res & dev engr Louisville & Beaumont works. 'Chem & Engrg News' 1962-72; Pittsburgh, Frankfurt Germany, New York Bureau Hd. Roger Williams Tech & Econ Services Inc; V P & Dir 1972-76. Sr Economist - Chem Indus, Merrill Lynch Economics, 1 Liberty Plaza NYC 10006 1976- . *Society Aff:* AIChE, CMRA.

Neenan, Charles J
Business: CBS Eng'g, 555 W 57 St, New York, NY 10019
Position: Assoc Dir *Employer:* CBS TV Engg & Dev. *Education:* -/EE/Manhattan College. *Born:* 1/10/23. Associate Director Plant Facilities Planning, CBS TV Eng & Dev, design of TV broadcasting facilities, 1977-87, 1949-66. Mbr of the Technical Advisory Group, representing the engg community as advisor to DOE, HUD, and AIA-Res for the dev of Natl. Energy Performance Stds 1977-79. VP L K Comstock Co Sys Div, 1970-77. Charles Neenan Assoc, Engg Consultants, 1966-70. US Navy, radar and communication, 1942-45. Manhattan College of Engg, 1946-49, 1941-42. Accomplishments incl Fellowship in the IES, 3 pats, IES VP. Operational activities, 1975-77, CH IES Theatre, TV and Film Lighting Ctte, 1964-68, as well as the publication of numerous technical papers and articles. Citizen Ambassador to the Chinese People's Republic, 1984; lighting specialist, to participate in bilateral technical exchanges with the Chinese Assoc of Light Industry. *Society Aff:* IESNA.

Neff, R Wilson
Home: 1666 W Bullard Ave, Fresno, CA 93711
Position: Retired *Education:* MS/Ind Engr/NYU; BS/Gen Engr/IA State Univ. *Born:* 3/12/20. Served in 101st Abn Div in WWII. Entered Regular Army in 1947, retired in 1968 as Colonel after serving in variety of troop, const, staff, cmd and educ asgmts, attending Engr Sch, Cmd and Gen Staff Coll, Armed Forces Staff Coll, US Army War Coll. Some asgmts Dep Dist Engr (Tullahoma, TN) 1955-57; Ofc, Chief of Engrs, 1961-63; Dist Engr (Buffalo, NY) 1964-67; Const Div, ODCSLOG, Dept of Army, 1967-68. Exec Sec, Lorain Port . Auth (OH) 1968-71. State Dir of Pub Works (OH) 1971-73. Lecturer, CA State Univ, Fresno, 1976-1980. Past pres Council of Lake Erie Ports; Mbr Engr and Opns Comm, Intl Assoc of Great Lakes Ports; Chrmn, US Sect, Niagara River Working Comm, Intl Joint Commission. Past mbr OH Commodores, Rotary Intl. Legion of Merit w/Oak Leaf Cluster, Bronze Star Medal w/two Oak Leaf Clusters, Purple Heart. Reg PE Ohio & Vermont. Author numerous technical papers, rep.

Nehal, Syed M
Business: 2117 Grant Ave, Cuyahoga Falls, OH 44223
Position: Pres *Employer:* Syed M. Nehal & Assoc Inc *Education:* MS/Eng Gen/Univ of Akron; BSEE/Power/In Inst. of Tech; BSc/Math & Physics/Aligarh Univ *Born:* 12/8/33 Came from India in 1957. Taught air conditioning 1960-62. Worked in responsible capacities for consulting eng firms 1963-1970. Formed the present professional firm in 1970, offering services for study, analysis, design and specifications in the field of HVAC, industrial ventilation, dust collection, plumbing/piping and fire protection for all types of buildings, energy audit, analysis and technical assistance with Life Cycle Costing (certified by the State of OH as Energy Auditor and Tech Energy Analyst). President of ADSPE, local chapter of NSPE 1976-77. Chairman of Eng's Week for ASME in 1979. Active with Kiwanis, Chamber of Commerce and local consulting group. Married, four children and a dog. Enjoy nature and gardening. *Society Aff:* NSPE, ASHRAE, PEPP, NFPA, ASME

Neiderer, Earl F
Business: 615 W Highland Ave, Ebensburg, PA 15931
Position: Sr VP. *Employer:* L Robert Kimball & Assoc. *Education:* BCE/Civ Engr/Catholic Univ of America. *Born:* 5/24/28. From 1952 to 1960 worked for Michael Baker Jr, Inc and PA Dept of Hgwys. Participated in water resources survey of Jordan River Valley, Kingdon of Jordan, in 1953-1954. Joined L Robert Kimball in 1960 as Dept Hd, transportation Engr. Assumed current position of Sr VP in 1972. Responsible for coordinating all engr and architectural projs. Reg engr in PA, WV, IL, NB, TN, NC, SC, GA, PR, FL, MD, OH, NJ, DE and VA. Active in state and local engr societies, Boy Scouts, Church and civic activities. Married to former Mary Cordray. Father of 2 boys and 3 girls. *Society Aff:* ACEC, ASCE, NSPE, ASTM, ASHE, SAME.

Neidhart, John J
Business: 77 Lowe Ave, Meriden, CT 06450
Position: Lighting Consultant *Employer:* Self *Education:* BS/EE/Case Inst of Technology. *Born:* May 3, 1918. Lighting Application Engr for Westinghouse 1940-56, except 1944-45 wth Navy Dept Buships. Joined Miller Co 1956 and was Manager, Application Engineering until 1980. Illum Engrg Soc Fellow since 1954, Natl Pres 1968-69, Disting Serv Award 1976. Other IES activities incl chmnship of Exec, V Pres, Office Lighting, Handbook, RQQ, Emergency Lighting, & Indus Lighting Cttes & service on Advance Planning, Finance, Design Practice, Merchandising, & School Lighting Cttes. Co-originator of Zonal-Factor Interflectance Method of Calculating Coefficients of Utilization. Introducer of now industry standard Average Brightness concept of specifying & reporting luminaire brightness. Ohio Registered. *Society Aff:* IES.

Neifert, Harry R
Home: 1745 Spring Valley NW, Canton, OH 44708 *Employer:* Mandatorily Retired from The Timbers Co. *Education:* BS/EE/Univ of MI. *Born:* Dec 1911 Flat Rock Mich. Grad school study, Applied Mech 1936; Stanford Univ Exec Dev Program 1961. Employed by The Timken Co (1835 Dueber Ave SW, Canton Ohio 44706) in 1936 as Mech Engr, Railroad Res; Supr Railroad Res 1951- 59; assumed current pos as Ch Engr Phys Labs 1959. Continuously engaged in some phase of res and/or dev pertaining to the fatigue of metals. Pioneered in fatigue analysis of ralroad car axles which led way for qualifying anti-friction bearings for rail service. Also made significant contribs in field of contact fatigue. Has served on 9 tech cttes for var engrg soc's, chairing sev. Author 18 tech papers on fatigue. Reg P E Ohio & Mbr ASME, SAE, ASTM, SESA & ADPA. ASME Fellow Award 1975. Mandatorily retired from the Timken Co. Jan. 1, 1977. *Society Aff:* ASME.

Neil, William N
Business: SE Tower-Prudential Ctr, Boston, MA 02199
Position: Vice President. *Employer:* Chas T Main Inc. *Born:* 1923. BSEE & grad studies at Northeastern Univ. Mbr Tau Beta Pi & Eta Kappa Nu. Served with US Navy 1943-47. Assoc with Chas T Main Inc since 1949 as Proj Engr & Proj Mgr. Proj's have involved installations of non-woven fabric facils, plastics fabrication facils & pulp & paper facils. Presently respon for client relations & staff consultant activities in pulp & paper & process industries. Mbr & Treas, Bd of Tr of Plastics Inst of Amer; Mbr Newcomen Soc, PIMA, TAPPI, MSPE & NSPE; Pres, Bd of Tr, 1000 Southern Artery Sr Citizens Ctr.

Neill, Charles R
Home: 5608-108 Str, Edmonton, Alta Canada T6H 2Y9
Position: Principal Engg. *Employer:* Northwest Hydraulic Consultants Ltd. *Education:* MSc/Hydraulics/Univ of Alberta; BSc/Civil Engg/Glasgow Univ. *Born:* 6/9/26. Genl work in civil engrg design & const 1946-59, incl indus, municipal & hydro projects. Hydraulic engr, Res Council of Alberta 1961-73. Assoc Prof Mem Univ of Newfoundland 1973-75. Awards: Telford Pramium, Inst of Civil Engrs 1966; Stevens Award ASCE 1974; Merit Award, Canadian Natl Ctte for Internatl Hydrologic Decade 1973. Spragins Award, Alberta Assoc Prof Engrs 1980, Dagenais Award, Can. Soc for Civ Engrg 1981. Papers on culvert & bridge hydraulics, river processes, sedimentation, ice mechanics. Editor 'Guide to Bridge Hydraulics', Univ of Toronto Pres 1973. Editor 'Ice Effects on Bridges' Roads & Transp Assoc of Canada 1981. *Society Aff:* ICE, ASCE, CSCE, IAHR, APEGGA

Neirynck, Jacques
Business: CM-Ecublens Extension EPFL, CH-1015 Lausanne, Switzerland, CH 1007
Position: Prof *Employer:* Swiss Federal Inst of Tech *Education:* Ingenieur civil electricien/-/Univ catholique de Louvain, Belgique; Doctor en sciences appliquees/-/Univ Catholique de Louvain, Belgique *Born:* 8/17/31 Worked with the SA Foraky, Brussels, 1957; Lecturer and then Prof Univ Lovanium Kinshasa, Zaire, 1957-1963; Head of Group and Vice-Dir MBLE Res. Labs Philips, Brussels, 1963-1972; Technical Dir in charge of Department of Computer Aided Design, 1970-72; simultaneously Lecturer, then Professor Univ of Louvain, respectively 1967 and 1969, teaching Circuit Analysis. Since March 1972, Professor in charge of "Chaire des circuits et systemes" with Swiss Federal Institute of Tech, Lousanne (EPFL). 1981, Fellow Grade IEEE for contributions to circuit theory, especially in the area of filter design. 1981-1984, VP & P IEEE Switzerland Section. Author of about 100 scientific publication, among which six books. Assoc Editor of Journal of the Franklin Inst; Member of Editorial BD of Intl Journal of Circuit Theory and Applications. Chrmn Bd, Presses Polytechniques Romundes. *Society Aff:* IEEE

Nekola, Robert L
Business: 625 Alpha Dr, Highland Heights, OH 44143
Position: Vice President-Technology *Employer:* Acme-Cleveland Corp. *Education:* BSME/Mech Engg/Case Inst of Tech. *Born:* 7/22/26. Native of Cleveland, OH. Served in the Navy Air Corps. Began my career as a design engr with Natl Machinery Co in Tiffin, OH in 1949. Joined the Cleveland Twist Drill Co in 1950 as Special Apprentice. Appointed Corp Mgr Mfg Process Dev for Acme-Cleveland Dev Co in 1971 and Gen Mgr of Acme-Cleveland Dev Co in 1975. I assumed my present position as Vice President-Technology in January, 1980. *Society Aff:* ASME, SME, CES.

Nelkin, Mark S
Business: Applied Phys, Clark Hall, Ithaca, NY 14853
Position: Prof. *Employer:* Cornell Univ. *Education:* PhD/Phys/Cornell; BSc/Phys/MIT. *Born:* 5/12/31. Worked at GE & Gen Atomic from 1955-1962. Has been at Cornell since 1962. Winner of Am Nuclear Soc special award in reactor phys, 1965, Guggenheim fellowship 1968. Visiting prof, College de France, Paris, 1976. Present res interests in statistical phys & applications. Most recent work is in statistical theory of turbulent fluid flow, & in theory of noise phenomena in solid state systems. Teaching interests in phys for engrs, especially Engg Phys majors: Visiting prof: Universite Paris VI, 1981. *Society Aff:* APS.

Nelson, Burke E
Home: 53 Deer Hill Rd, W Redding, CT 06896
Position: Executive Director *Employer:* Amer Soc Mech Engr *Education:* PhD/Matls Sci/Drexel Univ; MSE/Aero-Engg/Univ of WA; BSME/Mech/MI State Univ. *Born:* 1/16/38. Res engr Boeing 1959-66 & GE 1966-70, thermodynamics. Partner Nelson-Underhill Cons Engrs, 2 yrs. 1972-78, Sr Staff Engr, Perkin-Elmer Corp. Chosen 1975 ASME Congressional Fellow. Served Senate Interior Ctte, developing first ERDA budget. 1978-81 Assoc Dir Res Perkin-Elmer, respon for all optical sys res past Chrmn, ASME Public Affairs Ctte. US pat 3,892,273. Mbr Tau Beta Pi, Sigma Xi, Pi Tau Sigma, NSPE, AIAA, Triangle Frat. Reg PE PA. 20 papers publ or presented. Presently Exec Dir, ASME, responsible for all oper of the Soc. *Society Aff:* ASME, NSPE, Tau Beta Pi, Pi Tau Sigma, Sigma Xi, Triangle Frat, AIAA

Nelson, Clarence M
Business: Spencer St, Naugatuck, CT 06770
Position: Operations Mgr, Intl *Employer:* Uniroyal Chemical *Born:* Sept 1919. BChE Univ of Minnesota. Native of Park Falls Wisc. With Uniroyal Inc since 1943. Process engrg & mgmt in SB-R Latex manufacture for 17 yrs; Co-founder & Managing Dir Sto-Chem Ltd England 1960-66; Process Engrg Mgr for Const & New Ventures 1966-72 Uniroyal Chem - engaged in starting new venture in Brasil in chem manufacture and Pres of Uniroyal Do Brasil 1973-79, Operations Mgr Chem Div Intl 79 to present, Mbr AIChE since 1945, Chmn of New Haven Sect 1958.

Nelson, David B
Business: 1515 Nicholas Rd, Dayton, OH 45418
Position: Mgr, Marketing *Employer:* Monsanto Res Corp. *Education:* MBA/-/WA Univ; BS/CE/Purdue Univ. *Born:* 2/22/32. After 28 yrs experience with Monsanto Co in varied production, technical services, pilot plant & marketing assignments, he is Mgr, R&D Marketing, for the subsidiary Monsanto Res Corp (MRC), hdquartered in Dayton, OH, & is responsible for marketing of R&D contract progs in the environmental area. He is also Mgr, Environmental Services, respon for commercial services work involving environmental & occupational health activities. These include sampling, analytical & consulting services offered to other industrial concerns. *Society Aff:* AIChE.

Nelson, E Raymond
Business: 1000 S Fremont, Alhambra, CA 91802
Position: VP-Engg. *Employer:* C F Braun & Co. *Education:* MBA/Grad Sch of Bus Adm/Harvard Univ; BS/College of Chemistry/Univ of CA. *Born:* 1/16/21. in Alhambra, CA. Served in USNR 1942-1946. Joined C F Bruan & Co August 1946 as Process Engr. Worked for Gulf Oil three years. Rejoined C F Braun & Co 1954 in Chemical Engg Dept. Mgr, Chem Engg Dept 1970-1975, then to mgr, Engg Div. VP 1978. *Society Aff:* AIChE, NSPE.

Nelson, Eric W
Home: 17 Parkway Court, Allentown, PA 18104
Position: Retired *Education:* Master/Administrative Eng'g/NYU; Bachelor/Mech. Eng'g/NYU *Born:* 01/18/13 A native of New York. Graduate of NYU with BS in ME and MAdminE. Taught at Lafayette College. Product Engineer for Western Electric Co. specializing in electron tubes. Assumed responsibility for Machine and Tool Design departments and Product Engrg depts for electron tubes, mercury relays and sealed contact manufacture. VP ASME 1973-74, member of ASME Cttee on Staff and Chairman of ASME Pension Trustees. Co-author "Theory and Problems of Engrg Mechanics" *Society Aff:* ASME, NSPE

Nelson, Ernest O
Home: 6955 Overhill Rd, Mission Hills, KS 66208
Position: VP *Employer:* Panhandle Eastern Pipe Line and Trunkline Gas Co *Education:* BS/CE/KS ST Univ *Born:* 4/9/22 Native of KS. World War II - Army Corps of Engrs. Worked for Panhandle Eastern Pipe Line 1947 to present. Presently VP of Transmission - Panhandle Eastern Pipe Line/Trunkline Gas Co. Licensed Professional Engr/St of MO. Mbr of Bd of Dirs - Trinity Lutheran Hopsital and Trinity Lutheran Manor. Engrg Advisory Council - KS St Univ. Mbr Southern Gas Assoc, American Gas Assoc, and Midwest Gas Assoc. Have held several chairmanships within these organizations. Mbr of Houston Engrg and Scientific Society.

Nelson, Frederick C
Business: 105 Anderson Hall, Tufts University, Medford, MA 02155
Position: Dean of Engineering *Employer:* Tufts University. *Education:* PhD/Applied Mech/Harvard Univ; MS/Applied Mech/Harvard Univ; BS/Mech Engr/Tufts Univ. *Born:* Aug 1932. P E Mass. Tufts: Instr 1955-57, Asst Prof 1957-64, Assoc Prof 1964-71, Prof 1971- , Chmn Dept of Mech Engrg 1969-1980. Dean of Engineering, 1980- Visiting Res Fellow Inst for Sound & Vib Res, Univ of Southampton; Visit-

Nelson, Frederick C (Continued)
ing Prof Inst Natl Sci Appliquees de Lyon. Fellow ASA Mbr ASME, Mbr AAAS. Res ints: struct dynamics, acoustics, damping of mech sys. . *Society Aff:* ASME, ASA, AAAS.

Nelson, George A
Business: 435 Telfair St, Augusta, GA 30901
Position: Partner (Self-emp) *Employer:* Zimmerman, Evans & Leopold Consult Engrs *Education:* MS/Struct Engr/KS State Univ; BSCE/Civ Engg/KS State Univ. *Born:* 3/29/31. Raised in Merriam, KS. Served as Installation Engr, USAF, 1953-55. Structural & civ engr with Black & Veatch, 1957-63; Weitz-Hettelsater, 1963-64; and Patchen, Mingledorff & Assoc, 1965-71. Partner, Williams, Nelson & Assoc, Consulting Engrs, 1975-79. Self employed, 1971-75 & 79-80. Partner, Zimmerman, Evans and Leopold, 1980-present. Chrmn, N Augusta Planning Commission, 1970-77. VP, N Augusta Chamber of Commerce, 1986-1987. Pres, SC Section, ASCE, 1978. Lutheran, Mason, Acacia Fraternity, Sigma Tau. Hobbies are fishing, music & gardening. *Society Aff:* ASCE, NSPE, ACEC, CSI.

Nelson, George G, Jr
Business: International Paper Co, P O Box 311, Natchez, MS 39120
Position: Mgr Employee Training *Employer:* International Paper Co *Education:* BS/CHE/LA St Univ *Born:* 7/7/31 Native of Benton, La. Completed Navy officer candidate school, then served in the Office of Naval Intelligence 1953-1955. With International Paper Co since 1956, progressing to Chief-Pulp Research, Erling Riis Research Laboratory, Mobile AL, then to Mgr Technical and Environmental Services, Natchez Mill. Now Mgr Employee Training. Active in TAPPI, serving as Chairman Pulp Bleaching Committee, then progessing to Chairman Pulp Manufacture Division and TAPPI Pulping Conference, 1980-1981. Member Human Resource Development Committee. Registered PE in Louisiana and Mississippi. Active in church work. *Society Aff:* TAPPI, ΤΒΠ.

Nelson, Gordon L
590 Woody Hayes Dr, Columbus, OH 43210
Position: Prof & past Chrmn - Dept of Agri Engrg *Employer:* Ohio State Univ, others. *Education:* PhD/Engrg Mech & Ag Engrg/IA State Univ; MSc/Ag Engrg/OK State Univ; BAgEngrg/Ag Engrg/Univ of MN *Born:* 12/28/19 Native of W Central MN. US Navy during WW II. Agri engr for Portland Cement Assn, fac of OK State Univ, before becoming Chrmn of the Agri Engrg Dept of OH State Univ, 1969-81. Currently Prof of Agri Engrg, the OH State Univ, Columbus, OH. 8 ASAE Outstanding Paper Awds. NSF Terminal Yr & Post-doc Fellowships. Elected to Bd of Dir of ASAE & Dir of Awds. Elected to Grade of Fellow ASAE. Mbr of Bd of Dirs of The Coun for Agri Sci & Tech (CAST). Elected Mbr of Bd of Dirs of Amer Soc of Agri Engrg (ASAE) & Dir of Edu & Res Dept, ASAE. Recipient (1 per yr) of The Metal Bldg Mfrs Assoc (MBMA) Awd for distinguished work in advancing the sci and engrg of agri buildings. Award is presented by ASAE. Served for 1 yr as Chrmn of AAES Continuing Educ Comm. Awarded Massey Ferguson Education Award, 1986 by ASAE for "Advancement of engineering knowledge and practice in agriculture." *Society Aff:* ASAE, ASEE, CAST

Nelson, Harlan F
Business: Mech & Aerospace Engr, Rolla, MO 65401
Position: Prof. *Employer:* Univ of MO. *Education:* PhD/Aerospace Engg/Purdue Univ; MS/Aerospace Engg/Purdue Univ; BS/Aerospace Engg/IA State Univ. *Born:* 8/6/38. Have been on faculty of the Mech & Aerospace Engg Dept of the Univ of MO- Rolla since receiving PhD degree in June 1968. Served as a consultant on Planetary entry heating of space vehicles for Martin Marietta Aerospace Group, Denver, CO McDonnell Douglas Astro Co, St Louis, MO., Rocket Plumes for Sverdrup Technology, Inc. Arnold Air Force Station, TN, & Loral Electro-Optical Systems, Pasadena, CA. *Society Aff:* AIAA, SAE, CI, ASME.

Nelson, Harold E
Ctr for Fire Research, Washington, DC 20234
Position: Sr Research Fire Protection Engineer *Employer:* Natl Bureau of Standards. *Education:* BS/Fire Protection & Safety Engg/IL Inst of Tech. *Born:* Feb 9, 1929. Sr. Research Engineer Ctr for Fire Res, NBS. Formerly Dir, Accident & Fire Prevention Div, GSA. Major activities related to fire-safe design of bldgs; application of sys analysis to fire protection engrg; integration & synthesis of res & tech into engrg tech & applications; the safety of persons in health care facils. GAS Meritorious Serv Award 1968, Commendable Serv Awards 1965 & 1972, Dept of Commerce Silver Medal, 1982. Pres Soc of Fire Protection Engrs. Num prof papers. PE CA, Fire Protection. *Society Aff:* SFPE.

Nelson, Harold E
Home: 6815 N Wildwood Ave, Chicago, IL 60646
Position: Cnsl Engr *Employer:* Pr Practice *Education:* MS/Civ Engg/Cornell Univ; MA/Intl Affairs/The George Washington Univ; BS/Civ Engg/Univ of IL. *Born:* 11/26/17. Native of Braidwood, IL. Corps of Engrs 1940-1966: Alaskan Hgwy, Normandy Invasion, Battle of Bulge, Korea, Pittsburgh Dist Asst Mgr. Construction, Ralph M Parsons Co in Africa, Asia and US. Asst Commissioner of Public Works, Chicago: Urban Planning, Design and Const. Exec Dir and Gen Mgr, Chicago Urban Transportaton Dist maj subway proj. Sr Assoc, Kellogg corp, private practice. ASCE Harland Bartholomew Award 1977, US and foreign decorations, Phi Kappa Phi. *Society Aff:* ASCE, SAME, NSPE, WSE.

Nelson, Kenneth W
Home: 1894 Millcreek Way, Salt Lake City, UT 84106
Position: Environmental Consultant *Employer:* Self *Education:* MS/Chemistry/Univ of UT; BEd/Physical Sci/Univ of WI; Navy Course/Ind Hygiene/Harvard. *Born:* 9/27/17. Taught Chemistry, '38-'39; Res in Chem & Toxicology, FDA, '40-'41; Navy, '42-'46; Asarco Inc, Ind Hygienist, '45-'49; Chief Hygienist, '50-'58; VP, Environmental Affairs, '73-82; Past Pres, Am Ind Hygiene Assn and American Acad of Ind Hygiene; Pres, '78-81 Ind Health Fdn; visiting lecturer, Harvard Sch of Public Health, 55-81; Reg PE, UT. *Society Aff:* AAAS, AIHA, ACS.

Nelson, Nyal E
Home: 12205 Thoroughbred Rd, Herndon, VA 22071
Position: Principle Engr *Employer:* Deleuw, Cather & Co *Education:* BS/EE/WA St Univ *Born:* 02/17/16 Native of Waubay, SD. Served with Army Corps Engrs 1942-46, ending WW-II as CO of Engr Combat Bn. In 1947, joined Bur of Recl as Design Engr. In 1948, transferred to Corps of Engrs as Design Engr. In 1960, became Asst Chief of Engrg for Corps in Alaska. In 1965, moved to Chief of Engrs office in Washington, DC, serving there until 1980. Served as Chrmn of the DOD Tri-service Cost Engrg Committee 1966-1980. Served nine years on the Natl Academy of Sciences BRAB Fed'l Const Council, advancing new techniques for use in engrg, design, and construction. Joined DeLeuw, Cather & Co in 1980, serving as Principle Engr and staff specialist in Materials Engrg and Value Engrg. *Society Aff:* ASCE, SAME, AACE, SAVE, USCOLD

Nelson, Percy L
17 Plainfield St, Waban, MA 02168
Position: VP Pulp & Paper & Industrial *Employer:* Stone & Webster Engineering Corp. *Education:* BS/Elec Power/MIT; BS/Phys & Math/Williams College. *Born:* Mar 30, 1922. Noble & Greenough School - grad 1940; Williams College grad 1943, Phi Beta Kappa: MIT 1947, Tau Beta Pi, Eta Kappa Nu, Sigma Xi. US Marine Corps: 1943 Basic Training, Parris Island SC; 1944 2nd Lt, Officer Training, Quantico Va; 1945 1st Lt Pacific Theatre; 1946 Capt, to inactive reserve; 1959- : retired Reserve. Mbr ASME, IEEE, Tech Assn of the Pulp & Paper Indus, National Soc of Prof Engrs. 1947-48 With Arthur L Nelson, Engrs, Boston Mass & Blanco Tex. 1948- : With Chas T Main Inc Boston Mass - Manager Indus Steam & Power; since 1956, Assoc Mbr of firm; since 1975-82, V P of firm. Stone & Webster Engineering Corp.- V.P. for Pulp & Paper & Industrial, 1983-present. *Society Aff:* ASME, IEEE, TAPPI, NSPE.

Nelson, Peter R
Home: 32 Ichabod Rd, Simsbury, CT 06070
Position: Critical Experiments Fac Super *Employer:* Rensselaer Polytechnic Inst *Education:* ME/Nuclear Eng/RPI; BS/Nuclear Eng/RPI; AS/Eng Sci/HVCC *Born:* 2/23/49 Native of Troy, NY. U.S. Navy veteran, served aboard USS Robert E. Lee (SSBN601). Joined RPI teaching staff in 1978. Redesigned the RPI Reactor Core for reduced critical mass. Taught Reactor and nuclear physics. Senior Nuclear Eng at Combustion, Engrg, 9/81. Work included safety system design and man- machine interfacing. Author or papers including core design and reactor operator education. *Society Aff:* ANS, ΣΞ

Nelson, Richard B
Business: Sch of Engg & App Sci, Los Angeles, CA 90024
Position: Prof. *Employer:* UCLA. *Education:* ScD/Civ Engg/Columbia Univ; MS/Civ Engg/Columbia Univ; BS/Civ Engg/Columbia Univ; BA/Math/Willamette Univ. *Born:* 9/13/40. Native of Weiser, ID. As a student at Columbia Univ earned Illip Medal 1963, held NASA Predoctoral Traineeship 1964-66, Fairchild Camera Fellowship 1967 and was Carnegie Inst of WA Res Assoc 1968. Is a mbr of the UCLA Sch of Engg and App Sci faculty; Asst Prof 1968-1974, Assoc Prof 1974-1977, Prof 1977- present. Teaching & res is centered in analytical structural mechanics with emphasis on nonlinear structural analysis & automated structural design. Mbr ASCE, AIAA, Tau Beta Pi, Sigma Xi. *Society Aff:* ASCE, AIAA.

Nelson, Richard B
Home: 27040 Dezahara Way, Los Altos Hills, CA 94022
Position: Patent Agent. *Employer:* (Independent) *Education:* PhD/Physics/MIT; BS/Physics/CA Inst of Tech. *Born:* Powell Wyo 1911. M Prof Pauline (Wright) San Francisco State Univ. Physicist: RCA Mfg Co Harrison,NJ 1938-41; Nat Res Council Canada 1941-42; GE Res Lab Schenectady 1942-50. Engr: Litton Inds San Carlos,Calif 1950-51. Varian Ass 1951-77: Mgr Klystron Engr Dept 1956-60; Mgr Tube R&D Div 1960-63. Ch Engr Tube Div 1963-74. Teaching Fe;llow MIT 1935-38. Fellow IRE 1968. 39 US Patents. Authored Chapters: Encyclopedia of Electronics, McGraw-Hill Encyclopedia of Science & Technology, IEEE Handbook on UHF & Microwaves.

Nelson, Richard H
Home: 1796 Boca Raton Ct, Punta Gorda, FL 33950
Position: Assoc. *Employer:* Howard Needles Tammen & Bergendoff. *Education:* BS/CE/Purdue Univ. *Born:* June 24,1925. BSCE Purdue Univ, 1950. Chi Epsilon, Natl Honorary Soc. 1st Lt US Army Medical Service Corps, 1951-53. Awarded Bronze Star. Design & marketing environmental engr for Henry B Steeg & Assocs, Inc since 1954 specializing in water resources & pollution control; served as Ex V Pres; merged with Howard Needles Tammen & Bergendorf 1973; responsible for national marketing of environmental services; apptd Assoc 1978; respon for office mgmt & admin. Natl Dir ACEC 1971. Dir CEI 1976. Appointed Honorary Kentucky Colonel 1974. Fellow ASCE; WPCF; AWWA; NSPE. Registered PE & LS Indiana; PE WVA; PE WI; PE FL. Elder Presbyterian Church. Enjoy family, coins, boating & travel. *Society Aff:* ASCE, WPCF, AWWA, NSPE.

Nelson, Russell C
Home: 900 Moraine Dr, Lincoln, NE 68510
Position: Assoc. Dean, Graduate Studies & Res. *Employer:* University of Nebraska-Lincoln. *Education:* DSc/Metallurgy/CO Sch of Mines; MS/Metallurgy/CO Sch of Mines; BS/Met Engr/Lehigh Univ *Born:* 11/3/25. US Marine Corps 1944-46. Teaching Asst 1949-51. Mineral Engr ORNL 1951-53. Sylvania Elec Prod Inc, Chem & Met Div 1953-61, Hd Metals Res. Since 1961 Ap & Prof, MetE Option ME Dept Univ of Nebraska, responsible for initiation, devel & admin of program; courtesy appointments Dental Coll and Medical Coll. Research interest: biomaterials, P/M, mechanical behavior. Honors: Sigma Xi, Sigma Tau, Tau Beta Pi, Pi Tau Sigma. Socs: ASM, Founder Northeast PA Chaper Chmn Career Dev Comm 1971; AIME, Mbr Refractory Metals Comm 1958-62; APMI, Founder Penn-York Sec; ASEE. Chrmn of Materials Div 1980-81 and Grad. Studies Div. 1984-85. Soc for Biomatls. Consulting Engr since 1963, Calif Regist MT-128. Assoc Dean Graduate Studies & Research. *Society Aff:* ASM, AIME, ASEE, APMI, SB

Nelson, Sherman A
Business: P O Box 60, Boise, ID 83702
Position: Chief Corporate Engineer *Employer:* Trus Joist Corporation. *Education:* BSCE/Engg/State Univ of IA; MSCE/Engg/State Univ of IA. *Born:* July 4,1934. Mbr ISPE, Pres of SW ID Chapter, NSPE, Sigma XI; Fellow ASCE; Assoc Mbr SEAOC; Past Pres So Idaho ASCE. 6 years with structural consultants Iowa & Montana, Bridge, Bldg & Radio Telescope Design. In present position since 1965, responsible for tech supervision of engrg function throughout co activities in US & Canada, lightweight custom design structural components for bldgs. Presented papers on Micro-Lam lumber & Univ of Idaho Stadium cover 1972 & 1976 Natl ASCE Struct Engrg meetings. Chmn ASCE Wood Ctte, Mbr Amer Lumber Stds Committee, Design Ctte ICBO. Reg PE in 14 states & provinces. *Society Aff:* ISPE, NSPE, ΣΞ, ASCE, SEAOC, SEAOI

Nelson, Theodore W
Business: 150 East 42nd St, New York, NY 10017
Position: Retired. *Education:* Pet Engr/-/CO Sch of Mines; AMP/-/Harvard Bus Sch. *Born:* 8/28/14. CO Sch of Mines Distinguished Achievmenet in Mineral Engrg 1964; Charles F. Rand Gold Medal for outstanding achievement in minerals mgmt. Mbr: API; AAAS; ACS; AIChE; AIME; NY Acad of Scis. Mbr Adv Council for the Grad Sch of Bus, Univ of Chicago; mbr Earth Sci Adv Bd Stanford Univ; mbr Adv Council Pace Univ, NYC. Presently resides with wife & four children in Westfield, NJ & is retired Exec V P of Mobil Oil Corp, now doing limited consulting work. *Society Aff:* ACS, AIChE, SPE, AAAS.

Nelson, Waldemar S
Business: 1200 St. Charles Ave, New Orleans, LA 70130
Position: Ch of the Bd *Employer:* Waldemar S. Nelson & Company, Incorporated *Education:* BS/Mechanical & Elect. Engineering/Tulane University *Born:* 07/08/16 New Orleans, LA. Tau Beta Pi, Pi Tau Sigma, Eta Kappa Nu. With US Corps of Engrs on military construction 1941-45. Principal in consulting engrg firm for civil works, industrial & petro chemical plants, port facilities. Reg PE in 44 states. Fellow ASCE, ASME; Mbr IEEE, NSPE, SAME, AAAS, AAPA, ACI, API, APWA, others. Honorary mbr, past pres La Engr Soc; Natl Dir NSPE 1979-81; Past Chmn of LA State Board of Reg. PE. Distinguished Service Award NCEE. Past Pres Tulane Alumni Assn. Board of Advs Tulane School Engrg; Trustee Tulane Engr Foundation; P Chrmn St. Martin's Episcopal School Bd; P Chrmn St. Andrews School Board. Enjoy photography, fishing, woodworking. *Society Aff:* NSPE, ASCE, ASME, IEEE

Nelson, William E
Business: PO Box 808, Livermore, CA 94550
Position: Emergency Response Program Director *Employer:* Univ. of CA, Lawrence Livermore National Lab *Education:* MS/Mech Engg/UC, Berkeley; BS/Mech Engg/UC, Berkeley. *Born:* Jan 1933. Served in US Navy as carrier-based fighter pilot 1952-57, retired as Capt in Naval Reserve. Joined Lawrence Radiation Laboratory 1960 as design engr. Conducted res in mech properties of high explosives until 1963 when named Proj Engr in charge ofLens Devel Group. Held various tech mgmt jobs until 1980 when apptd Emergency Response Program Director. Responsible for management of R&D and operational preparation for dealing with nuclear emergenices. Founding Chmn ASME Mt. Diablo Sec. Mbr Pi Tau Sigma, Sigma Xi. Reg PE Calif. FAA Cert Flight Instructor. Enjoy skiing, tennis, back-packing & flying. *Society Aff:* ASME, ΣΞ, ΠΤΣ, ADPA.

Nemhauser, George L
Business: School of Industrial & Systems Engg, Georgia Inst of Tech, Atlanta, GA 30332

Nemhauser, George L (Continued)
Position: Chandler Prof *Employer:* Georgia Inst of Tech *Education:* PhD/Operations Res/Northwestern Univ; MS/ChE/Northwestern Univ; BChE/ChE/CCNY *Born:* 7/27/37. Taught at Johns Hopkins Univ 1961-69, at Cornell Univ from 1970-85. Dir of the Sch of Operations Res & Ind Engg 1977-1983, at Georgia Tech 1985-present. Visiting Prof (1963-64) as Chandler Prof of Industrial and Systems Engg Univ of Leeds, England. Visiting Prof (1969-70, 1975-77, 1983-84) Ctr for Operations Res & Econometrics, Univ of Louvain, Belgium, Dir of Res (1975-77). Author of more than 75 journal articles and three books. *Introduction to Dynamic* Programming, Wiley, 1966, *Integer Programming*, Wiley, 1972, *Integer and Combinational Optimization*, Wiley, 1988 (to appear), Editor-in-chief, *Operations* Research, 1975-1978. ORSA Lanchester Prize, 1977. President ORSA 1981-1982. Editor-in-chief, Operations Research Letters 1981-. Elected to Natl Acad of Engg, 1985. Avid tennis player. *Society Aff:* ORSA, TIMS, IIE, SIAM, MPS, NAE.

Nemy, Alfred S
Home: 1401 Hunting Ridge Rd, Raleigh, NC 27609
Position: VP. *Employer:* Rockwell Intl. *Education:* PhD/Met/Carnegie Mellon Univ; MS/Met/Carnegie Mellon Univ; BS/Met/Carnegie Mellon Univ. *Born:* 12/27/30. Native of New Kensington, PA. Taught at Carnegie Tech as Asst Prof (1954- 1956). Seventeen yrs (1956-1973 with TRW, Inc, in R&D (jet engine mtls and processes), engg (nucl control rod drive mechanisms), intl ventures (Japan, Italy, England), & bus planning. Gen Mgr (1973-1975) of Walworth steel foundry in Columbus, OH. Gen Mgr (1975-1978) of Rockwell Intl nucl valve plant in Raleigh, NC. VP, Operations (1978-) of Rockwell-Draper Div producing weaving machinery (looms) & related equip in six domestic & foreign (Ireland, Mexico) plants. *Society Aff:* AIAA, AIME, ASM, ASTM.

Neou, In-Meei
Home: 3444 Murdoch Ct, Palo Alto, CA 94306
Position: Mech Engrg Conslt *Employer:* Prof Emeritis of Mech & Aerospace Engrg WV Univ *Education:* PhD/Mech Engg/Stanford Univ; MS/Mech Engg/MA Inst of Tech; BS/Mech Engg/Chekiang Univ. *Born:* 1/19/17. Native of Wuhing, Chekiang, China. Worked as Asst of Mech Engg at Chekiang Univ (1941-45), MIT (1946), and Stanford Univ (1947-50); Asst Prof at Syracuse Univ (1950-55); Assoc Prof (1955-59) and Prof (1959-66) at Univ of Bridgeport; mbr of tech staff at Bell Telephone Labs (1962-63); and Prof at WV Univ 1966- 1982. Prof Emeritus of Mech & Aerospace Engrg WV Univ 1982-. Mech Engrg Conslt 1982-. Interested in mech engg design and analysis. History of Science and Tech. Dev several electro-mech sensory aids for the blind, emission control of liquid storage tanks. *Society Aff:* ASME, ASEE, AAAS.

Neptune, David B
Business: Checkerboard Square, 2E, St Louis, MO 63164
Position: Group Project Mgr *Employer:* Ralston Purina Co. *Education:* MS/CE/Univ of MO-Columbia; BS/CE/Univ of MO-Columbia *Born:* 6/9/44 St. Louis. Became mgr Product Development for Laclede Steel Co and was involved in fabrication of Truss System for World Trade Center Twin Towers in NY. Received Young eng Award of MSPE, St. Louis Chapter ub 1974 and national ASCE young Engs Award in 1975. Joined Ralston Purina Co. In 1977 as Project Mgr responsible for all phases of Project for Grocery Products Division. Promoted to Sr Project Mgr in 1980 and to Group Project Mgr in 1987. Former Pres, St. Louis ASCE and Chairman District 16 council of ASCE, former President of Univ of MO engrg Alumni Assoc, and Dir Engrs' Club of St Louis. Currently State Director MSPE. *Society Aff:* ASCE, NSPE, MSPE.

Nerem, Robert M
Business: School of Mech Engrg, Atlanta, GA 30332-0405
Position: Parker H. Petit Professor for Engineering in Medicine *Employer:* Georgia Inst of Tech *Education:* Ph.D./Aeronautical & Astro. Engr./Ohio State University; M.S./Aeronautical & Astro. Engr./Ohio State University; B.S./Aero. Engr./ University of Oklahoma *Born:* 07/20/37 Appointed as the Parker H. Petit Distinguished Prof for Engrg in Medicine at Georgia Inst of Tech. Previous positions include Assoc Dean for Research in the Graduate School of Ohio State Univ 1975-79 and Prof & Chmn of the Dept of Mech Engrg at the Univ of Houston 1979-86. Author of more than 50 refereed journal articles with numerous other proceeding publications and meeting presentations. VP of the International Federation for Medical and Biological Engrg 1985-88, V Chmn and Chmn-Elect of the U.S. National Cttee on Biomechanics 1985-88, and Co-Pres of the 1988 World Congress on Medical Physics and Biomedical Engrg. Fellow, Council of Arteriosclerosis, American Heart Assoc; Fellow, American Physical Soc; and Fellow, American Soc of Mech Engrs. 1986 Konrad Witzig Memorial Lecturer for the Cardiovascular System Dynamics Soc. *Society Aff:* AAAS, ASME, AIAA, BMES, ASEE

Nesbitt, Ray B
Home: 69 Wahackme Road, New Canaan, CT 06840
Position: Ex VP *Employer:* Exxon International *Education:* BS/ChE/TX A&M Univ. *Born:* 12/12/33. Native of Marshall, TX. Joined Exxon Corp Humble Oil & Refining Co at Baytown TX Refinery. Advanced through various engg assignments to Tech Superintendent-Chems; transferred to Enjay Chem Co in NY to 1963; held positions of Div Mgr Plastics, Pres Enjay Fibers and Laminates Odenton MD, VP Enjay. Became VP Esso Chem Europe 1970 in Brussels, Belgium; VP Exxon Chem NYC 1973. President Exxon Chem Americas in 1978. Avid sports fan, participate in tennis. *Society Aff:* AIChE, SCI.

Nessmith, Josh T
Home: 205 Sussex Dr, Cinnaminson, NJ 08077
Position: Mgr, Systems. *Employer:* RCA-Missile & Surface Radar. *Education:* PhD/EE/Univ of PA; MSEE/EE/Univ of PA; BEE/EE/GA Sch of Tech. *Born:* 6/18/23. Served US Army 1943-1946. Joined CAA as field engr in 1947 & then assigned as radar instr at CAA Aero Ctr. Joined RCA in 1952 becoming responsible for system designs of instrumentation radars, including precision AN/FPQ-6 & TRADEX radars. Responsibilities since have included positions as TRADEX/ PRESS Prog Mgr; Mgr, Systems Engg; Deputy Prog Mgr, AEGIS Weapon System. Current assignment is Mgr, systems, responsible for system design of radars, weapons systems, & specialized communications systems. Elected Fellow, IEEE in 1977. Mbr of IEEE Radar Systems Panel & Fortesque Fellowship Committee. Reg Engr in GA & NJ. *Society Aff:* IEEE, AIAA.

Nethero, Merle F
Home: 6009 Virbet Dr, Cincinnati, OH 45230
Position: Vice President. *Employer:* The H.C. Nutting Company. *Education:* CE/Civil Engg/Univ of Cincinnati. *Born:* 4/5/29. Served in US Army Corps of Engrs 1952-54 as Instructor in Soil Mechs Courses at Engr Center, Ft Belvoir VA. With The H C Nutting Co Since 1954, served as Ch of Inspection Div, Dir of Physical Lab, Asst Ch Engr, VP since 1973. Reg PE in OH, KY W VA, IN. P Pres of Cincinnati SEc ASCE, mbr & P Chmn Cincinnati/Dayton Geotech Ctte, mbr Engg Soc Cincinnati; Charter Mbr The Deep Fnds Inst. *Society Aff:* ASCE, CEO, ESC, DFE, NSPE.

Netter, Milton A
Business: 2801 W Bancroft St, Toledo, OH 43606
Position: Prof & Chrmn. *Employer:* Univ of Toledo. *Education:* PhD/Ind Engr/Univ of MI; JD/Law/Univ of Toledo; MSE/Mech Engr/Univ of MI; BSE/Mech Engr/Univ of MI. *Born:* 4/24/23. On the Univ of Toledo faculty, 1949 to present, progressing from Instr, Engg Mechanics, to Prof & Chrmn of the Dept of Industrial Engg. PE; Mbr of Bar, State of OH. Consultant for govt on mgt systems for refuse collection, energy use, etc. Technical witness for legal-engg cases. Pres, A Netter & Sons, Inc job shop in metal processing. *Society Aff:* IIE, ТВП.

Neu, Ernest L
Home: 205 Camino De Las Colinas, Redondo Beach, CA 90277
Position: Retired *Education:* MS/ChE/Univ of Caen, France; BS/ChE/Univ of Nancy. *Born:* 1915 Frankfurt, Germany. 1938-42 Res Chemist Celotex Corp; 1942-45 US Army European Theater; 1946-66 Great Lakes Carbon Corp 1946-52 Res Chemist, 1952-60 Ch Chem ist, 1960-62 Asst Tech Dir, 1962-66 Tech Dir. 1966-1981 General Manager International Div Grefco Inc. Dir 10 overseas subsideries. Patents: fil tration & insulation. Mbr numerous professional socs. Retired 1981. 1981-present - consultant. *Society Aff:* AIChE, ACS, NYAS.

Neubauer, Walter K
Business: 1304 Buckley Rd, Syracuse, NY 13221
Position: Senior Vice President. *Employer:* O'Brien & Gere Engineers, Inc. *Education:* BCE/-/Syracue Univ. *Born:* 6/16/24. in NY City. Served with Army Air Corps 1943-46. Field Engr for Frederick Snare Corp on 1st Chesapeake Bay Bridge. With O'Brien & Gere Engrs since 1954 & a prin in the firm since 1961; assumed current respon in 1971 which includes res, water supply, resource recovery, also Dir & Corp Secretary. Reg PE in 10 states. Chmn NY Sec AWWA 1976-77; Chmn AWWA Stds Committee on Concrete Pressure Pipe 1970-78; Mbr, AWWA Stds Council from 1979; Fellow, ASCE & ACEC. *Society Aff:* AWWA, ASCE, ACEC, NSPE.

Neufeld, Ronald D
Home: 6558 Bartlett St, Pittsburgh, PA 15217
Position: Prof of Civil Engg. & of Environ Health Engrg. *Employer:* Univ of Pittsburgh. *Education:* PhD/Civ Engg/Northwestern Univ; MS/ChE/Northwestern Univ; BE/ChE/Cooper Union. *Born:* 2/10/47. Origin: Brooklyn, NY. Work experience: Chem Engr with Rohm & Haas Co, corporate engrg div, 1968-70. Asst Prof, Univ Pittsburgh, 1973-77; Assoc Prof, Univ Pittsburgh, 1977-1982; Prof 1982-present; Sr Fulbright Scholar and Visiting Prof, Hebrew Univ 1983-84. Res areas: Ind Wastes; fundamentals of biological and physical/chemical processes; coal gasification and liquefaction, wastewater & solid/hazardous waste treatment; chemical industry environmental problems; coke plant, nitrification, and heavy metal wastes; Municipal wastes, water supply, water/solid waste disposal; Consultant-Environmental Engrg & environmental process res & design. Reg PE-PA. Diplomate - Amer Acad of Environmental Engrs, Dir ASCE Pittsburgh Section 1984-1987. *Society Aff:* ASCE, AIChE, WPCF, IAWPR, AAEE, XE

Neuhoff, Charles J
Business: 1732 Lyter Drive, Johnstown, PA 15905
Position: Pres *Employer:* H.F. Lenz Co *Education:* BS/EE/Univ of Pittsburgh *Born:* 3/31/37 Native of Scranton, PA. Employed by United States Steel Corporation, in engrg and supervisory capacities, to Chief Electrical Eng Johnstown Works, 1960-1971. Employed part-time by H.F. Lenz Co., Engs, Planners, and Energy Consultants, Johnstown, PA 1961 to 1971 as Electrical Eng. Fulltime 1971 to present. Project Eng and Mgr for more than 150 major Projects. Presently a Principal as Pres. 1972 State of PA's "Young Eng of the Year" Award from the PA Society of PE. Also charter member of PA Engrg Foundation. Holds patents and copyrights on several electrical devices and systems. Registered PE in PA, Ohio, WV, VA, NV, and MD. *Society Aff:* NSPE, IEEE, ASHRAE, CEC.

Neumann, Edward S
Business: H.O. Staggers Natl Trans Ctr, Evansdale Library, WV 26506
Position: Director *Employer:* West VA Univ *Education:* PhD/CE/Urban and Regional Planning/Northwestern Univ; MS/CE/Urban and Regional Planning/ Northwestern Univ; BS/CE/Mich Tech Univ *Born:* 3/6/42 Native of Harvey, IL. Served with US Army Corps of Engs at Waterworks Experiment Station, Vicksburg, MS, 1969-1970. Joined WV Univ in 1970. Teach and conduct research in Transportation systems engrg. Active in ASCE in areas of Aesthetics and Automated People Movers. Dir of Harley O Staggers National Transportation Center. Interests include Automated People Movers, Travel Demand, Systems Analysis, Urban Aesthetics, Aerial Tramways. Numerous publications and presentations. Have conducted research for State and Federal Agencies and Private organizations. Promoted to Full Professor, 1980, Dir of Natl Transportation Center, 1983. Dir, Technology Transfer Center for local roads in W VA, BD Dir Advanced Transit Assoc; active in ITE in area of low volume roadways. *Society Aff:* ASCE, NSPE, ITE, ΣΞ, ASEE, OITAF-NACS, ATRA

Neumann, Gerhard
Business: 1000 Western Ave, West Lynn, MA 01910
Position: VP & Group Exec (Retired). *Employer:* General Electric Company. *Born:* Oct 1917 in Germany. BS Ingenieur Schule Mittweide Germany. Enlisted in Gen CHennault's American Volunteer Group 'Flying Tigers' in Nationalist China. Assembled Japanese Zero fighter for US Air Force 1943; with Tech Air Intelligence & OSS until 1945; US Citizenship by special act of Congress. Joined GE in 1948: designed & invented mech features for hi-Mach lightweight jet engines; Gen Mgr Jet Engine Dept 1955, Aircraft Gas Turbine Div 1961; V Pres 1963; Group Exec 1968. Collier Trophy 1958; Goddard Medal 1969. Mbr Acad of Engrg; Hon Fellow of AIAA; Honorary Mbr Faculty, College of Armed Forces, Knight of French Legion of Honor 1978. Daniel Guggenheim Medal 1979. Golden Door Award 1981; Dr.h.e. 1982; Elder Statesman of Aviation (NAA) 1984.

Nevill, Gale E, Jr
Business: Dept of Engg Sciences, Gainesville, FL 32611
Position: Prof. *Employer:* Univ of FL. *Education:* PhD/Engr Mechanics/Stanford Univ; MSME/ME/Rice Univ; BSME/ME/Rice Univ; BA/ME/Rice Univ. *Born:* 11/17/33. in Houston, TX. Wide range of ind experience including considerable work in oil ind. Sr Res Engr at Southwest Res Inst 1960-1964. At Univ of FL since 1964, Chrmn of Dept of Engg Sci & Mechanics 1967-1972, Dir of Ctr for Creative & Optimal Design 1972-1979. Ext contract res for ind & governmental sponsors. Currently Prof of Engg Sciences, principal activities consulting and res in design automation, robotics and emergency response planning. *Society Aff:* ASME, ASEE, AAAI, IEEE, SME/RI, WFS.

Nevin, Andrew E
Home: 1177 Race St, Suite P6, Denver, CO 80206
Position: Pres *Employer:* Granada Exploration Corp *Education:* PhD/Geol/Univ of ID; MA/Geol/Univ of CA-Berkeley; BS/Geophysics/St. Lawrence Univ *Born:* 04/04/39 Dayton, OH. Exploration geologist, 1965-70, with Phelps Dodge Corp in AZ, and Cannon-Hicks Association in Canada and Mexico. Founder and Pres of Nevin Sadlier-Brown Goodbrand Ltd, Vancouver, B.C., 1971-81. Currently Pres of mineral resource development companies Granada Exploration Corp (BC) and Nevin & Bernstein, Inc (Nevada). Honors and offices include Sigma Pi Sigma and Omicron Delta Kappa (1960-61), NSF Fellowship (1964), Dir of Society of Mining Engrs of AIME (1978-83). Registered PE (BC) and Profl Geologist (Idaho). *Society Aff:* SME-AIME, GSA, CIM, GAC

Nevitt, HJ Barrington
Home: 2 Clarendon Ave-Apt 207, Toronto, Ontario Canada M4V 1H9
Position: Consultant. *Employer:* Self. *Education:* LLD/(Honoris Casua)/Concordia Univ; MEng/(Telecomms)/McGill Univ; BASc/Elec Engrg/Toronto Univ *Born:* June 1908 St Catharines, Ontario Canada. 1934-39 mfg engr Northern Elec Co Montreal; 1939-45 systems engr Canadian Pacific & DefenseCommunicatons, Montreal. 1945-47 Exec Engr RCA International Div NY. 1947-60 consultant to LM Ericsson Telephone Co Stockholm for systems engrg, marketing & managing telecommunication projs in Europe & the Americas; resident Caracas, Venezuala & Rio de Janerio, Brazil. 1960- in Toronto as consultant to private cos & government agencies on organization of engrg, marketing, financing, & gen mgmt of international operations; also investigating psychic & social effects of innovation at Center for Culture & Technology Toronto Univ; visiting prof: Univ of Stockholm Sweden; Carlton Univ, Ottawa; Concordia Univ. Montreal; Discoveries Intl, Honda Fnd, Tokyo; Hon.

Nevitt, HJ Barrington (Continued)
President Inst for Informationsentwicklung, Vienna; reg prof conslt engr Ontario. Cmeds; Fellow IEEE, AAAS, EIC & IEE; Co-author with Marshall McLuhan, *Take Today: The Executive as Dropout* (New York: Harcourt Brace Jovanovich, 1972); author, The Communication Ecology (Toronto: Butterworths, 1982), ABC of Prophecy (Montreal: Gamma Inst Press, 1985), Keeping Ahead of Economic Panic (Montreal: Gamma Inst Press, 1986), and numerous articles in business and prof. journals. *Society Aff:* IEEE, AAAS, EIC, IEE, SGSR

Newcomb, Robert W
Business: Elec Engrg Dept, College Park, MD 20742
Position: Professor. *Employer:* Univ of Maryland Microsystems Lab *Education:* BSEE/Elec Engr/Purdue Univ; MS/Elec Engr/Stanford Univ; PhD/Elec Engr/U of CA, Berkeley *Born:* June 1933. Prof Univ of Md Assuming directorship of EE Grad Studies there 1970. Taught previously at Stanford 1960-70 & Berkeley 1957-60. Res Intern at Stanford Res Inst 1955-57. Profesor Invitado and Director "El Grupo de Trabajo en Sistemas (PARCOR)," Universidad Politecnica de Madrid (Facultad de Informatico), 1984-present, Visiting Prof as Fulbright-Hays Fellow Univ Teknologi Malaysia 1976, Prof Invite Louvain Univ 1967 & Visiting Prof as Fulbright Res Fellow Univ of New S Wales 1963-64; Guest USSR Ministry of Educ 1973; External Examiner Univ of Lagos 1970-72 . Director "Microsystems and Generalized Networks" Program. Res is conducted with students & colleagues worldwide in the operator theory of networks & microsys areas with emphasis upon biomedical applications & active MOS circuits, hearing sys, computer strucs, prof ethics and university administration, & Malaysian film. Enjoy poetic works, natl epics, classical music & kronchong. *Society Aff:* IEEE, IREE, SIAM, MAA, ΣΞ, TBΠ, HKN

Newell, Allen
Business: Carnegie-Mellon Univ, Pittsburgh, PA 15213
Position: University Professor. *Employer:* Carnegie-Mellon University. *Education:* PhD/Indus Admin/Carnegie Inst of Techn; BS/Physics/Stanford. *Born:* Mar 19, 1927. BS Stanford 1949 (phys); PhD Carnegie Inst of Tech 1957 (Indus Admin). Res Sci Rand Corp 1950-61; Univ Prof CMU 1961- . Basic res in artificial intelligence, cognitive psychology, programming sys (list processing) & computer strucs. Mbe Natl Acad of Scis; Fellow IEEE, Fellow AAAS (Psychology). Recipient Goode Award Amer Fedn & Info Processing Socs (AFIPS) 1971; Recipient A M Turning Award (jointly with H A Simon) Assn for Computing Machinery 1975. *Society Aff:* NAS, IEEE, APA, ACM.

Newell, Earl D
Home: 6474 San Diego, Riverside, CA 92506
Position: Chief Materials & Process Engrg. *Employer:* Rohr Industries Inc. *Born:* June 1947 Pennsylvania. BS Waynesburg Coll. US Naval Officer WWII. With Rohr Indus Inc 1955- ; assumed current position as Ch Materials & Process Engrg 1975, respon for nonmetallic R/D Materials & Process Group including Lab for Rohr. Currently 1sr V Pres SAMPE. Hobbies: fishing, photography, classical music.

Newell, J C
Business: Prof. of Biomedical Engineering, Rensselaer Polytech. Institute, Troy, NY 12181-3590
Position: Prof *Employer:* Rensselaer Polytechnic Inst *Education:* PhD/Physiology/Albany Med Coll; MEng/EE/RPI; B/EE/RPI *Born:* 10/13/43 He was with the Missile Systems Division of Raytheon Company before joining the Trauma Center at Albany Medical Coll as a biomedical eng in 1970. He received the PhD degree in Physiology from Albany Medical Coll in 1974. He is now Prof of Physiology and Surgery at Albany Medical Coll and Prof of Biomedical Engrg at RPI. His research interests include the regulation of the pulmonary circulation in hypoxic and normoxic lung regions. He is also an investigator in the Albany Trauma Center, where he studies gas exchange in injured patients with acute respiratory failure. *Society Aff:* APS, IEEE, NYAS, BME

Newell, John R
Home: 241 NE Spanish Trail, Boca Raton, FL 33432
Position: Ret. Pres *Education:* BS/Naval Archit & Marine Eng/MIT *Born:* 6/30/12 Philips Acad. Bethlehem Steel Shipbuilding Div (1935-8); Bath Iron Works, Bath, ME. Asst. Works Mgr 1938-44, VP 1947-50. Pres (1950-65); Dir Union Mutual Life Ins Co. (1955-1966); Trustee Bath Savings Inst Bath,ME. (1950-1963); Trustee Bates College, Lewiston ME. Dir Fed. Res Bank of Boston (1963-66); Past Pres. Soc. Nav. Arch and Marine Engrs (1961-2); (Hon) DE Stevens Inst Tech; LLD (Hon) Univ of ME; LLD (Hon) Bowdoin Coll. *Society Aff:* SNAME

Newhof, Paul W
Business: 3975 Cascade Rd SE, Grand Rapids, MI 49506
Position: President *Employer:* Newhof and Winer, Inc *Education:* MSCE/Structures/Univ of MI; BSCE/Structures/Univ of MI *Born:* 4/03/33 and raised in Grand Rapids, the son of a house building contractor. I commenced full time professional employment in 1959 with the Daverman Assoc and founded my own private consulting company, as well as the Grand Rapids Testing Service, in 1965. I took a partner, Mr. Loyed E. Winer, in 1968 and formed Newhof & Winer, Inc. Architects, Engineers & Planners. *Society Aff:* NSEE, NSPE, ACI, PCI

Newhouse, David L
Home: 402 Terrace Rd, Schenectady, NY 12306
Position: Consultant, Forgings. *Employer:* Self Employed, Associated with Engg Materials & Processes, Inc. *Education:* BS/Met Engr/Purdue Univ. *Born:* 4/18/21. Native of Rush County, IN. With Naval Res Lab, Wash, DC, 1943-45. With GE C in dev & application of mtls for large turbine-generator components since 1945. Mgr of Forgings Dev since 1954, responsible for dev of mtls & processes for forgings, large turbine-generator rotors, retaining rings & other components. Associated with Engg Materials & Processes as Consultant, Forgings since 1983. Author of 17 papers. Lic for practice of PE in NY State. Appointed Fellow, Am Soc for Metals in Nov, 1976. Activities include wilderness preservation, conservation, hiking, camping. *Society Aff:* ASM, NSPE, NYAS, ASTM, ASME.

Newhouse, Russell C
Home: 13 Dale Dr, Chatham, NJ 07928
Position: Dir Advanced Radar Lab (ret 1971). *Employer:* Bell Telephone Labs Inc. *Education:* MS/Elec Engg/OH State Univ; BEE/Elec Engg/OH State Univ. *Born:* Dec 17, 1906 Clyde Ohio. Mbr of Tech Staff Bell Tele Labs 1930-71 inclusive (retired Dec 31, 1971). Dir labs supr 100 to 200 engrs; 1955-71 engaged in dev of military radar & computer sys; was respon for the electronics engrg of DEW lineacross northern Canada & of the Nike Zeus radars of the ABM defense sys demonstrated in Kwajalein 1963. Awards: Guggenheim Fellowship in Aeronautics 1929-30; Sperry Award for FM altimeter 1938; Fellow IEEE 1956; Disting Alumnus Award Ohio St Univ 1959; Award for Dist Pub Serv as an Advisor to the FAA, 1961-65. Pioneer Award Aeronautical & Electronics Sys Soc IEEE 1967. Listed Engrs of Distinction 1973. Lic PE NJ. *Society Aff:* IEEE, AAAS, TBΠ, HKN, ΣΞ.

Newhouse, Vernon L
Business: Drexel Univ, Sch of Elec Engrg, Philadelphia, PA 19104
Position: Robert C. Disque Professor *Employer:* Drexel University *Education:* PhD/Physics/Univ of Leeds; BSc/Physics/Univ of Leeds. *Born:* Jan 1928 Mannheim W Germany. Dev 1st European magnetic core computer memory at Ferranti Ltd 1952-54; worked on computer sys & memory R/D at RCA until 1957; then joined G E Res Lab where dev numerous superconducting switching amplifying & storage devices. In 1967 was named a Fellow IEEE for contributions to computer memories, joined Purdue Univ's Sch of EE doing res successivly on solid state microwave devices, acoustic surface waves & ultrasonic blood velocity measurement & flaw detection sys. Originated Purdue Univ's Clinical Engrg prog. In 1982 joined Drexel Univ as Robert C. Disque Professor of Elec & Comp Engr. Appointed Adjunct Prof. of Radiology at Jefferson Univ, April, 1983. *Society Aff:* IEEE, AIUM.

Newkirk, John B
Business: Colorado Biomedical, Inc, 6851 Highway 73, Evergreen, CO 80439
Position: Pres *Employer:* Colorado Biomedical, Inc. *Education:* DSc/Phys Metall/Carnegie Inst Tech; MS/Phys Metall/Carnegie Inst Tech; B. Met.Engr/Metall Engrg/Renss Polytech Inst; -/Crystallography/Cambridge Univ, U.K. *Born:* Mar 24, 1920 Minneapolis Minn. Served as degaussing specialist in S Pacific during WWII; hon discharge 1946 as Lt USNR. MS & DSc Carnegie Inst of Tech 1948 & 1950; Fulbright Fellow Cambridge England 1950-51. Res Metallurgist Genl Elec Co 1951-59; Prof Cornell Univ 1959-64; Phillipson Prof Univ of Denver 1964-75; Prof Univ of Denver 1975-1984; Phillipson Prof Emeritus, Univ of Denver, 1984- . Mbr ASM, Tau Beta Pi, Sigma Xi, Phi Kapppa Phi, Alpha Sigma Mu (Natl Pres 1 yr). Pres Colorado Biomedical, Inc. Cos to indus & govt in materials, met, & bioimplant devices. Hobbies: skiing, music, beekeeping, cycling. *Society Aff:* ASM, ΣΞ, TBΠ

Newlon, Howard H, Jr
Business: Box 3817 Univ Stat, Charlottesville, VA 22903
Position: Research Dir *Employer:* Va Hwy & Transportation Res Council. *Education:* MCE/Civil Engg/Univ of VA; BCE/Civil Engg/Univ of VA. *Born:* Jan 1932; native of Brandy Station Va. Concrete Design Branch of TVA prior to service in Chem Corps US Army 1954-56. Head of Concrete Res for Va Hwy & Transportation Res Council 1956-75, & Assoc Head of Council 1968-80. Lectr in Sch of Cvl Engrg & Sch of Architecture at Univ of Va. Fellow of Amer Concrete Inst & Amer Soc of Testing & Materials. ASTM Award of Merit 1970 & ACI Davis Lectr 1973. Res specialties: concrete materials, history of concrete materials & history of road & bridge bldg. *Society Aff:* ACI, ASTM, ASCE, TRB.

Newman, Joseph H
Business: 666 5th Ave, New York, NY 10103
Position: President *Employer:* Tishman Research Corp. *Education:* MChE/Chem Engg/Polytechnic Inst of NY; BChE/Chem Engg/Polytechnic Inst of NY. *Born:* Feb 1925. Pres Tishmann Res Corp which acts as cons on bldg tech & dev of new const prods & sys; Exec VP and Director of parent co, Tishman Realty & Const Co Inc. Chairman, Natl Inst of Bldg Sci 1979-1981, Director 1976-1981. Mbr NAE; Chmn of Bldg Res Advisory Bd of Natl Acad of Scis 1971-73; has been mbr of numerous cttes advising public & private sectors on energy mgmt, fire safety, urban tech, housing & bldg & const res. Before joining Tishman in 1959 was with Flintkote Co & Curtiss-Wright. *Society Aff:* AIChE, ACS, NIBS.

Newman, Malcolm
Business: P.O. Drawer –O–, 609 Middle Neck Rd, Great Neck, NY 11023
Position: Pres *Employer:* Inter-City Testing & Consulting Corp *Education:* Engr Sc. D./ME/NYU; MSCE/Engr Mech/Columbia Univ; BSCE/Structural Analysis/City Coll of NY *Born:* 6/29/31 Native of Huntington, NY. Published over 50 papers and articles in recognized engrg and scientific journals. Former full prof of mechanical engrg, Tel-Aviv Univ, 1973-1975. Chief of structural mechanics, Republic Aviation Corp until 1965. Mgr & Dir of structural dynamics, Harry Belock Assoc, 1965-1969. Dir of applied mechanics, Analytical Mechanics Assoc, 1969-1971. Staff specialist for crash dynamics, Fairchild-Hiller Safety Car program, 1971. VP, Design Anlytics, Inc, 1976-1978. Chrmn, NY Chapter, System Safety Society. With Inter-City since 1976; firm specializes in accident reconstruction, product safety & forensic engrg. Member of MENSA. *Society Aff:* AIAA, NSPE, SAE, SSS

Newnam, Frank H
Home: 6138 Del Monte, Houston, TX 77057
Position: Chmn of Bd - Retired. *Employer:* Lockwood, Andrews & Newnam Inc - ret. *Education:* BS/CE/TX A&M Univ. *Born:* Aug 28, 1909 Temple Tex. Wife, Mrs Mary Ann McLeod Newnam; 3 children: Albert H, Elizabeth Ann, & John Frank. Res P E Texas. June 1946-Sept 1974, Chmn of Bd & co-owner of cons engrg firm of Lockwood, Andrews & Newnam Inc. 1944-46 Colonel, Corps of Engrs US Army. Ch Engr for Hdqtrs SOS, China Theater. 1941-44 Lt Colonel, Major & Capt, Corps of Engrs US Army. Ch of Engrg Div, Galveston Dist. 1931-41 Soils & Pavement Engr for the Texas Hwy Dept. ASCE Pres 1968-69; Natl Dir 1962-65. Natl Dir 1960-62 NSPE. Pres TSPE 1959. Engrs Joint Council V P 1970-73, Dir 1970-75. Mbr Bd of Tr Baylor Univ 1962-69, Baylor Coll of Medicine 1962- , Univ of Houston Found 1968- , Civil Engrg Adv Ctte, Texas A&M Univ. Mbr Task Force on Water Resources & Power, Second Hoover Comm. Rotary Club of Houston, Pres 1972. Mbr Chi Epsilon & Tau Beta Pi. Author. *Society Aff:* ASCE, ACEC, NSPE.

Newnan, Donald G
Business: PO Box 5, San Jose, CA 95103
Position: Prof, Indus & Sys Engrg. *Employer:* San Jose State Univ. *Education:* PhD/Engg/Stanford; MS/Engg/Stanford; MBA/Bus/Stanford; BS/Engg/San Jose State Univ. *Born:* Dec 1928. Herbert Hoover Fellow. 1975-76 Dir, Engrg Econ Div, IIE. 1981-82 Vice Chairman, Engrg Econ Div, ASEE. Design engr in indus & US Army officer. Presently Prof of Indus & Sys Engrg at San Jose State Univ. Author 6 books, incl 'Engrg Econ Analysis' & 'Civil Engrg License Review' 8th edition. Reg civil & indus engr in Calif; cons on engrg econ analysis. *Society Aff:* IIE, ASEE.

Newton, Alwin B
Home: 136 Shelbourne Dr, York, PA 17403
Position: HVAC & Solar Cons (Self-employed) *Education:* MS/Mech Engr/MIT; BS/Mech Engr/Syracuse Univ. *Born:* 8/8/07. 55 yrs in air-cond, refrig, heating, design & dev. Began dev of solar heating & cooling in 1937. V P & Bd Mbr Coleman Co; V P Acme Indus, York Div, Borg-Warner before retirement. Over 220 US pats. Fellow and Life Mbr ASHRAE, Life Mbr ASME, Founding Mbr ISES. Since retirement, Visiting Prof Purdue Univ; cons to NSF, DOE, NBS & to num state & local govs on solar energy dev. Cons to manufacturers of HVAC & Solar equip. Managed ASHRAE cons contracts for ERDA and DOE. Past Pres of Comm E of Intl Inst of Refrig. Worldwide speaker in these var fields. Author ASHRAE Solar Collector Performance Manual, 1981. Reg Engr OH. *Society Aff:* ASME, ASHRAE, ISES, IIR.

Newton, Carroll T
120 Circuit Rd, Nokomis, FL 33555
Position: Consulting Engr (Self-employed) *Education:* MS/Civil Engrg/MA Inst of Tech; BS/Architectural Engrg/MA Inst of Tech. *Born:* May 1911. Adv Mgmt Prog Harvard Business School. Comm Officer Regular Army Corps of Engrs 1933-63; served as Dir Waterways Exper Sta, Ch R&D Div OCE, Dist Engr Los Angeles. With Swindell-Dressler Co Pgh as Genl Mgr & V P of 100 mbr org performing CE cons & design 196372. Self-employed spec in port studies & transp sys domestically & internationally 1972- . Pres Design Div ARBA 1970; active as official in local sections of ASCE & SAME num locations; prof regis CE 8 states; Freeman Hydraulic Fellowship Award from BSCE 1949-50. *Society Aff:* ASCE, SAME, AAPA

Newton, Edwin H, Jr
Business: Drawer A & B, Big Stone Gap, VA 24219
Position: Mgr of Mining Services. *Employer:* Westmoreland Coal Co - Va Opns. *Education:* MS/Ind Engg/Univ of Pittsburgh; BS/Ind Engg/GA Inst of Tech. *Born:* 8/24/43. Native of Savannah, GA. Held positions in Ind Engg with Alcoa, Ingalls Shipbldg, & Gurmman Am Aviation. With Westmoreland Coal Co since 1975. Assumed current responsibility of Mgr of Mining Services in 1979. Responsibilities include Ind Engg, Mining Engg, Health & Safety, & Maintenance. Mining Div Dir, IIE. Profl Engr, PA & VA. *Society Aff:* IIE, NSPE.

Newton, Jeffrey M
Business: 41 Res Dr, Hampton, VA 23667
Position: VP. *Employer:* Engg Inc. *Education:* MSME/Engg/Old Dominion Univ; BSME/ME/Clemson Univ. *Born:* 1/20/42. Joined Newport News Shipbldg in 1965 as a Res Engr in the Res Div; promoted in 1970 to Group Leader in the Engrg Tech Dept of the Machinery Design Div. Joined the firm of Engrg Inc as a Partner & Dir in 1974. Assumed resp as VP of Engrg in 1977, and promoted to VP of Operations in 1983, responsible for turnkey design & fabrication of customed engrg equip and systems for clients in govt & industry. Officer & Dir in Penisula Chapter VSPE. Dir

Newton, Jeffrey M (Continued)
of Penisula Chapter of VSPE 1980-81 and Officer in Hampton Roads Chapt of NCMA. *Society Aff:* NCMA, NSPE, ACEC, ASME, NPMA.

Newton, Robert E
Business: Mech Engrg Dept, Monterey, CA 93940
Position: Prof of Mech Engrg. *Employer:* Naval Postgraduate School. *Education:* PhD/Engg Mechs/Univ of MI; MS/Engg Mechs/WA Univ; BS/Mech Engg/WA Univ. *Born:* 10/16/17. Taught appl mech at Washington Univ beginning 1938. Engr in stress analysis, vibration & flutter at Curtiss-Wright Corp 1941-45. Returned to Washington Univ until assuming present pos as Prof of Mech Engrg at Naval Postgrad School 1951. Was Dept Chmn 1953-67. Visiting Prof at Univ of Wales Swansea 1968-69. Spec ints are applications of finite element method to struct & fluid dynamics problems. ASME Jr Award 1940. Fellow ASME. Visiting Prof at Ecole Nationale Superieure de Mecanique, Nantes 1981, 82. *Society Aff:* ASME, SESA, ASEE, ΣΞ.

Neyer, Jerome C
Business: 38955 Hills Tech Drive, Farmington Hills, MI 48018
Position: Pres *Employer:* Neyer, Tiseo & Hindo, Ltd *Education:* MSCE/Geotech/Univ of WA; B/CE/Univ of Detroit *Born:* 7/15/38 in Cincinnati, OH. Worked as geotechnical eng for consulting firms in seattle, WA and Detroit MI. Became partner in Halpert, Neyer & Associates in 1970 and president of Neyer, Tiseo & Hindo, Ltd. in 1978. Directed engrg activities for variety of subsurface construction projects. Taught soil mechanics at Univ of Detroit from 1973 through 1978. Pres, Southeastern Branch ASCE (1973); President, Consulting Eng Council of MI (1981); Member, St of MI Mineral Well Advisory Bd (1969-1987). *Society Aff:* ACEC, ASCE, ASTM, NSPE, ASFE

Niarchos, Demetrius G
Home: 870 U.N. Plaza, Apt 11-d, New York, N.Y 10017
Position: Retired *Education:* MS/Civil Engr/Columbia Univ; BS/Civil Engr/Robert College *Born:* in Constantinople Turkey. 1936-49, employed as civil engr on projs in Greece for Ulen & Co, Water Supply Co of Athens, Atkinson-Drake-Park and Economic Cooperation Admin. Work interrupted by 8 yr service in Greek & British Army. In 1951 joined Ford, Bacon and Davis Inc as Sr Civil Engr for projs in aluminum, iron ore, steel, Oil and chem indus. Transferred to Australiain 1959. In 1971 elected VP & Genl Mgr Australian Div. Directed engrg, design & proj mgmt of 10 new railway lines with aggregate length of 1600 miles. Studies of proposed rail lines in Australia & Algeria, aggregate length 5000 miles. Also Consultant to Brisbane Metropolitan Transit Authority 1974-84 for projs estimated at $300 million. Fellow ASCE & formerly of Inst Engrs Austr; Mbr NSPE, NYSSE; PE NY & Queensland; formerly Dir Amer Ch of Comm Australia; Managing Dir of Ford Bacon & Davis (QID) Pty Ltd. Retired Oct 1986. *Society Aff:* ASCE, NSPE

Nibley, J William
Business: 964 Kennecott Bldg, Salt Lake City, UT 84111
Position: Pres. *Employer:* Nibley & Co. *Education:* BS/ME/Univ of UT. *Born:* 3/8/21. Employed as Consulting & Application Engr by Westinghouse Elec Corp from 1943 to 1948 & DeLaval Steam Turbine Co from 1948 to 1954. Organized Nibley and Co, a UT Corp, in 1954. This firm represents mfg of process & power equip and serves the intermtn area. Elected to Eta Beta Pi & Phi Kappa Phi while attending Univ of UT. *Society Aff:* AIChE.

Nicholas George
Business: The Institute of Optics, University of Rochester, Rochester, New York 14627
Position: Prof of Optics *Employer:* The Inst of Optics *Education:* PhD/EE & Physics/CA Inst of Tech; MS/EE/Univ of MD; BS/Engrg Physics/Univ of CA-Berkeley *Born:* 10/29/37 Formerly Dir of the Institute of Optics; Prof of Electrical Engrg and Applied Physics at the CA Inst of Tech; Senior Staff Physicist at Hughes Aircraft Co; Chief, Physics Section, Emerson Research Labs; and Physicst to Section Chief at the National Bureau of Standards. In modern optics he is credited with several major advances in the areas of holographic gratings, holographic stereograms, the ring-wedge photodetector array, automatic pattern recognition, laser sensors of pollutants, and speckle sensors for remote height measurements. Prof George is the holder of 7 patents, and has over 60 published papers. *Society Aff:* OSA, SPIE, APS, AAUP, ΣΞ, ΦΒΚ, ΤΒΚ, ΦΚΦ

Nicholls, Richard W
Home: 7 Middlebury Lane, Cranford, NJ 07016
Position: President. *Employer:* Koch Oil International Co. *Education:* BS/Chem Engg/Lehigh Univ; AA/Chem Engg/Keystone Jr College. *Born:* April 1930. BSChE Lehigh Univ. Native of Blakely Pa. R&D with Merck & Co 1951-53 spec in distillation. Spent next 17 yrs with Chevron Oil Co, incl 4 yrs refinery engrg & 13 yrs internatl crude oil & product trading. Joined Koch Indus 1970 as V P of Koch Oil Internatl Co, engaged in foreign oil trading. Pres Koch Oil Internatl since Sept 1975. Married with 3 children. Interested in golf & fishing. *Society Aff:* AIChE.

Nicholls, Robert L
Business: Civil Engrg Dept, Univ of Delaware, Newark, DE 19716
Position: Professor. *Employer:* Univ of Delaware. *Education:* BS/Civil Engg/Univ of CO; MS/Civil Engg/IA State Univ; PhD/Civil Engg/IA State Univ. *Born:* 6/11/29. Proj Engr, Army Corps Engrs, Hdqtrs Far East 1953-55. Soils & Matls Engr, Gannett Fleming Corddry & Carpenter 1957-59. Univ of Delaware Civil Engrg Dept 1959- . Geotech engrg cons 1961- . Pres Del Sect ASCE 1974-75. Author 'Composite Const Matls Handbook' 1976; co-author 'Mathematical Foundations for Civil Engrg Sys' 1972 - texts. Pats & res publs: geotech engrg, const matls, engrg sys optimization. *Society Aff:* ASCE.

Nichols, Clark
Home: PO Box 352, Searsport, ME 04974
Position: Consultant (Self-employed) *Employer:* Retired *Education:* MS/EE/MIT; BS/EE/MIT. *Born:* 8/9/14. Joined Leeds & Northup Co in Phila, PA in 1935 after graduating from MIT. Until 1955 had field & factory engg assignments on Control Systems for Elec Power Generation & System Dispatching. Introduced control concepts covered by US patent & described in 1953 AIEE Power Div prize paper. After 1955 held mgt positions including Mgr of Systems Engg & Gen Mgr of Control Systems Div. Retired from L & N in 1976. IEEE activities included mbrship in Power Systems Engg & Power Generation Committees, Chrmn of System Controls Subcommittee 1956- 60. Presently an IEEE Life Fellow & Reg PE. *Society Aff:* IEEE, PES, CSS.

Nichols, Donald E
Business: 2120 8th Ave So, Nashville, TN 37204
Position: President and Ch of Bd *Employer:* I C Thomasson Assocs Inc. *Education:* BS/Mech Engrg/TN Tech Univ *Born:* Aug 1923. Native of Nashville Tenn. With I C Thomasson Assocs Inc since 1949. Served as Jr Engr, Sr Assoc, Secy-Treas, Exec V P, President & presently Chairman of the Board. Reg P E Tenn. Pres Nashville Chap ASHRAE 1961-62; Regional Chmn ASHRAE, Region VII, 1969-72; Dir ASHRAE 1969-72; Dir at Large ASHRAE 1979-80; mbr TN Tech Univ Engrg Adv Council, Chmn 1971-74; mbr TN Tech Univ Dev Council, Treasurer, 1987/88, 1978- ; mbr Engrg Jt Council Tech Assessment Panel 1971-74; mbr Bd of Commissioners of the Metro Dev & Housing Agency, V Chmn 1972; Chrmn 1977; Fellow ASHRAE 1974; ASHRAE Disting Serv Award 1975; Mbr of Distinction Tau Beta Pi; Mason, Shriner; Engr of Distinction, TN Tech Univ 1981. *Society Aff:* NSPE, TSPE, ACEC, ASHRAE

Nichols, Donald L
Business: RFD 2 Box 3695, Oxford, MI 04270
Position: Consultant (Self-employed) *Education:* BS/EE/Worcester Poly Inst. *Born:* 12/8/25. Native of Auburn ME. Conducted R&D in sonar & Elf Electromagnetics at

Nichols, Donald L (Continued)
Naval Underwater Sound Lab 1946-1954, was Chief Engr for Anti-submarine & Undersea Warfare at Gen Dynamics/Electronics throught 1963. Was Mgr of Anti-submarine Warfare Progs at Raytheon Co until 1967. Joined Naval Underwater Systems Ctr Staff in 1967, & became Assoc Technical Dir for Engrg. Retired from Govt in June 1979. Currently self employed consultant, PE, CT. *Society Aff:* ASA, ΣΞ, NYAS

Nichols, Edward
Home: Whale Rock Rd, Narragansett, RI 02882
Position: Professor. *Employer:* Univ of Rhode Island. *Education:* PhD/Engg/Purdue Univ; MS/Engg/Syracuse Univ; BS/Mech Engg/Syracuse Univ. *Born:* 11/22/19. US Navy WWII. P E Rhode Island. Prof of Engr Univ of RI since Feb 1959. Engr duPont at Savannah River Atomic Energy (H-Bomb Proj); Engr Dow Chem at Midland MI. 12 years as ASME Rep on the Engrg Soc Library Bd & P Chm ESL Bd; 3 yrs as VP of Region I of Amer Inst of Indus Engrs. Teaching & cons specialty in statistical qual control & reliability engrg. Interested in military history of WWII. Faculty Counselor to Univ of RI Student Chapter of the Soc of Women Engrs. Currently on the ASME Council on Codes & Standards; Mbr ASME Bd on Accreditation; Mbr ASME National Nominating Cttee. *Society Aff:* ASME, NYAS, SWE, USNI.

Nichols, Edward J
Home: 2002 Cool Spring Dr, Alexandria, VA 22308
Position: Pres *Employer:* Edward J Nichols & Assoc, Inc *Education:* BS/ChE/Univ of MA *Born:* 5/7/29 Mr. Nichols is the recipient of the 1983 Department of Defense Award for Outstanding Achievement in Value Engineering, recognizing the high percentages of savings and return on investment ratios resulting from Value Engineering Studies he conducted on DOD projects. He has 25 years experience in Value Engineering, Life-Cycle Costing and energy conservation. He has conducted over 600 Value Engineering Studies on wastewater treatment plants; commercial, government, industrial buildings; bridges, tunnels, subway systems; power stations, electrical distribution plants; and hardware equipment. He has trained over 4,000 design professionals in VE, LCC and energy conservation. He enjoys fishing, snorkeling and photography. *Society Aff:* SAVE, ACEC, WPCF

Nichols, Edwin Scott
Home: 3836 Cranbrook Dr, Indianapolis, IN 46240
Position: Sect Chief, Engg Prod Matls Lab. *Employer:* Allison Gas Turbine Div - GM. *Education:* BS/Mech Engr/Northwestern Univ. *Born:* 3/25/24. US Navy Engg Officer WWII. With Haynes Stellite Co as sales engr for 9 yrs. With Genl Motors since 1956 concerned with eval and use of high strength, corrosion-resistant alloys, titanium alloys & high temp matls & coatings. Now supervises Matls Engg Testing Lab respon for quality evaluations & lab support mfg for aircraft & indus gas turbine & Diesel engine matls. Also responsible for development of marine turbine blade alloys. Awards First Prize ASM Matls Awards competition - Ferrous Metals 1963; Chmn Indianapolis Chap ASM 1968; Mbr AIME High Temp Alloy Ctte 1969-73; Mbr ASM Adv Tech Awareness Council 1968-72, Chmn 1973-74; Mbr ASM Natl Nominating Ctte 1975; Mbr AIA Matls & Structures Ctte 1974-79, Chmn 1979-80; Mbr ASTM-ASME Gas Turbine Panel 1963-present. *Society Aff:* ASM, AIME

Nichols, Herbert E
Home: 776 Woodfield Dr, Cincinnati, OH 45231
Position: Mgr, Proj Engg. *Employer:* GE Co. *Education:* MSME/Mech/Univ of Cincinnati; BSME/Mech/Univ of Akron. *Born:* 2/13/31. H Nichols has had 25 yrs experience in Jet Engine design engg & mgt with the GE Co. He is currently Mgr of Proj Engg for the CF6-50 High-Bypass Engine.

Nichols, James R
Business: 811 Lamar St, Ft Worth, TX 76102
Position: President. *Employer:* Freese & Nichols. *Education:* BS/Civil Engrg/TX A&M Univ; M/Civil Engrg/TX A&M Univ. *Born:* June 29, 1923. Principal directing design of water supplies, treatment plants, pipelines, sewerage sys. wastewater treatment plants, airports & with engrg ventures incl studies & reports. Mbr num prof soc's, tech soc's & active in many civic affairs. *Society Aff:* ACES, NSPE, ASCE, WPCR.

Nichols, Kenneth D
Business: 16715 Thurston Rd, Dickerson, MD 20842
Position: Consulting Engr. *Employer:* Self-employed. *Education:* PhD/Hydraulics/State Univ of IA; MCE/Hydraulics/Cornell Univ; CE/Hydraulics/Cornell Univ; BS/US Military Acad. *Born:* 11/13/07. Served Corps of Engrs, US Army 1929-53. Highest rank Major Genl. Dist Engr, Manhattan Dist (Atomic Bom Proj) WW II. Prof. Mechanics, U.S.M.A. Chief, Armed Forces Special Weapons Proj; Dep Dir Guided Missiles OSD; Chief, R&D US Army; Genl Mgr USAEC; cons engr in atomic power field since 1955. Author: The Road to Trinity - Morrow 1987. Presently Mbr Natl Acad of Engrg, Fellow Amer Nuclear Soc, Hon Mbr ASME, Mbr PIANC, Former Dir Atomic Indus Forum, Detroit Edison Co, Fruehauf Corp. *Society Aff:* NAE, ANS, ASME, PIANC.

Nichols, Lee L
Home: 402 VMI Parade, Lexington, VA 24450
Position: Dir of Engg. *Employer:* VMI. *Education:* PhD/Elec Engg/VPI; MS/Elec Engg/OH State Univ; BS/Elec Engg/VMI. *Born:* 6/5/23. Native Richmond, VA, Served with US Army 1942-46, USAR 1942-1978, Colonel, Armor. Have taught at VMI 1947-to date, and current Head, Electrical Engg Dept and Dir of Engg. Profl Engr, VA, and practicing consulting engr in microwaves, traffic radar, and electrical safety in hospitals. *Society Aff:* IEEE, ASEE, HKN, ΣΞ, NSPE, NEEDHA.

Nichols, Richard S
Business: 2300 NW Walnut Blvd, P.O. Box 428, Corvallis, OR 97339
Position: Principal Electrical Engineer *Employer:* CH2M-Hill *Education:* B.S./Elect. Engrg./Oregon State University *Born:* 02/23/28 Native Corvallis, OR. General Electric Co. Test Program, motor design, industrial power systems engineering 1949-57. US Army Signal Corps 1950-52. With CH2M Hill Consulting Engrs since 1957. Present responsibility, Principal Electrical Engr responsible for firmwide engrg excellence. Special expertise in ac adjustable speed drive systems and National Electrical Code interpretation. Reg PE, OR, WA, ID, CA, CO, TX, VA. President IEEE Industry Applications Soc 1984-85. IEEE Bd of Dir 1987-88. IEEE Foundation Bd of Dir 1987-89. Enjoy gardening, travel to historical sites, and community service work. *Society Aff:* IEEE, NSPE,

Nichols, Robert L
Business: 811 Lamar St, Fort Worth, TX 76102
Position: VP. *Employer:* Freese & Nichols. *Education:* MS/Civil Engg/TX A&M; BS/Civil Engg/TX A&M Univ. *Born:* June 1926. Native of Fort Worth Tex. Has served as design & principal engr in charge of preparation of plans & specifications & inspection of num water treatment facils, wastewater treatment facils, water distrib sys, sewerage collection sys & other civil works. Major int in area of environ engrg. P Pres of Texas Soc of Prof Engrs. P Natl Dir Natl Soc of Prof Engrs. P Chmn NSPE/PEPP. P Chmn Municipal Panel/Texas Water Conservation Serv. P VP of Texas Water Conservation Dist. PP of Natl Soc of Prof Engrs. *Society Aff:* ASCE, NSPE, AWWA, WPCF, APWA, AAEE

Nicholson, Morris E
Business: 320 Aero Engg, 110 Union St SE, Minneapolis, MN 55455
Position: Prof. *Employer:* Univ of MN. *Education:* ScD/Phys Met/MIT; SB/Met/MIT. *Born:* 2/15/16. Indianapolis, IN, raised Cleveland, OH. Served in US Army 1941-46. Section Hd Engg Res Std Oil (IN) 1947-50. Res prof Univ of Chicago Inst for the Study of Metals 1950-56, Univ of MN Dept Hd Met 1956-62, Prof 1956-present. dir, Continuing Education in Engg & Sci 1973-present. Dir Univ Ind TV

Nicholson, Morris E (Continued)
for Education 1971-present. Principal res interests: transformations in iron & steel, alloy theory, corrosion in metals. Community Services - Boy Scouts of Am 25 yrs, Silver Beaver Award 1965, Roseville Area Schools Bd mbr 18 yrs, MN Sch Bds Assn Dir 1971-1976 916 Area Voc-Tech Inst Bd mbr 1976- Awards: Dist. Serv. Awd. ASEE, CPD Div. 1981 MSBA, Outstanding School Bd Mbr 1981. *Society Aff:* AIME, ASM, NACE, ASEE.

Nicholson, Richard H
Home: 711 Albion St, San Diego, CA 92106
Position: Div Director Qual Assurance. *Employer:* Genl Dynamics Electronics Div. *Education:* MS/Industrial Tech/Natl Univ; BS/Industrial Tech/TX State Tech Inst. *Born:* Feb 23, 1922 San Diego Calif. Mbr Natl Panel Consumer Arbitration Panel Better Bus Bureau 1973- ; Asst Prog Chmn San Diego Biomed Symposium 1974; Mbr Adv Bd San Diego City Coll. Recipient Award of Merit Toastmasters Internatl 1955. Fellow Amer Soc Qual Control (PDir-at-Large); Mbr European Org for Qual Control. Amer Defense Preparedness Assn (Chmn 1981-83 Product Assurance Tech Div). USN Natl Security Indus Assn (Qual), Aerospace Indus Assn. Army Liaison Panel. Reg P E Calif. *Society Aff:* AIAA, ASQC, ADPA, ASNT.

Nickel, Donald L
Business: 777 Penn Center Blvd, Pittsburgh, PA 15235
Position: Manager, Process Control Projects *Employer:* Westinghouse Electric Corp *Education:* BS/Math, Physics/Univ of WI *Born:* 6/19/33 Served in US Army 1954-1956. Employed at Westinghouse since 1957. Aided in Development of Generation Capacity Expansion Models and Wrote First Digital Computer Programs Available in Utility Industry. Became Project Mgr for Distribution Laboratory 1967-1971; Appointed Distribtion Systems Engrg Mgr 1972- 1976, Responsible for Distribution System Planning, and Product Development and Application. Past Chairman of IEEE Distribution SubCommittee; currently is Vice Chairman of IEEE Transmission & Distribution Cttee. Served as Technical Program Chairman, Executive Vice Chairman and Executive Chairman of IEEE/PES Transmission and Distribution Conference & Exposition in 1981, 1984, and 1986, respectively. Author of 35 technical papers and holds one patent. *Society Aff:* IEEE

Nickels, Frank J
Home: 2341 Washington Blvd, Venice, CA 90291
Position: President. *Employer:* Frank J Nickels Assocs. *Education:* MBA/Bus Admin/UCLA; MS/Naval Architecture/Stevens Inst; BS/Marine Engr/NY St Maritime College *Born:* 5/15/36. Engrg Officer aboard merchant & US Navy vessels 1959-62. Naval architect with Bethlehem Steel, Sun Shipbuilding, & Litton Indus, spec in prelim ship design. Designed the American Lancer class containerships, the world's largest & fastest genl cargo ships at the time. Founded Frank J Nickels Assocs, Naval Archs, Marine Engrs, & Marine Transp Economists 1974. Dev the short, wide, shallow hull form for tankers in restricted draft service. Lectr in Naval Arch at CSULB. Chmn Los Angeles Sect SNAME 1973-74. *Society Aff:* SNAME.

Nickolaus, Nicholas
Business: 30 Sea Cliff Ave, Glen Cove, NY 11542
Position: VP of Marketing. *Employer:* Pall Corp. *Education:* MSChE/ChE/Columbia; BSChE/ChE/CCNY; ME/ME/Rutgers *Born:* MS Chem Engg Columbia Univ 1950. US Army WWII; worked on Atomic Bomb proj in Oak Ridge TN. Res Scientist for Allied Chem 1947-49, Sales engr Envirotech - solids, liquids, separation equip 1950-58. Gen Mgr Envirotech Italy 1958-63. VP Mkting Pall corp 1963-; mfg of ultrafine cartridge filters. Chrmn Filter Cartridge Ctte AIChE 1974. Sr VP PallComp 1980. *Society Aff:* AIChE, PDA

Nickum, George C
Business: 911 Western Ave, WA 98104
Position: Chrmn. *Employer:* Nickum & Spaulding Assoc Inc. *Education:* -/English/Dartmouth College. *Born:* 3/28/10. Apprentice to Father, W C Nickum 1930-1934, partner W C Nickum & Sons 1936- 1941, managing partner 1941-1957, pres W C Nickum & Son Co 1957-1971, pres Nickum & Spaulding Assoc, Inc 1971-1977, chmrn 1977-1979. All firms engaged in practice of naval architecture & marine engr. Licensed PE, WA, 1947, mbr Marine Bd, NAE, 1965-1973. US delegate - IMCO technical committees 1964-1977, Coast Guard Distinguished Public Service Award 1977, Dartmouth College Distinguished Alumnus Award, 1963, Coast Guard Public Service Commendation, 1979. *Society Aff:* SNAME, RINA.

Nicoladis, Frank
Business: 2700 Lake Villa Dr, Metairie, LA 70002
Position: Pres. *Employer:* N-Y Assocs, Inc. *Education:* BS/Civil Engg/MS State Univ. *Born:* 10/23/35. Born in October, 1935 in Gulfport, MS. Graduated from Mississippi State Univ in 1957. Received BS in Civil engrg. Artillery Officer in USAR from 1957- 1965. Employed as Proj Engr with Fromherz Engrs and Albert Switzer & Assoc, 1957-1965. Asst Chief Engr for Lake Forest, Inc, 1965. Proj Engr for Burk & Assoc, Inc. 1965-1969. Specialized in municipal and sanitary engg. Established N-Y Assoc, Inc, Consltg Engr, Architects and Planners, and assumed present position in 1969. Responsible for directing all of firm's efforts in engg, architecture and city planning. Reg E in MS, LA, AR & TX. *Society Aff:* ASCE, WPCF, AWWA, NSPE, APWA, ASCE.

Nicolaides, Emmanuel N
Business: 2900 S W 28th Lane, Miami, FL 33133
Position: Managing Director. *Employer:* Hufsey-Nicolaides Assocs Inc. *Born:* Jan 1927. BSEE Univ of Miami. Served in the US Navy 1944-46. Served as traffic engr for the City of Miami Beach & then started as a Jr Elec Engr & worked through to Ch Elec Engr with var cons firms until became a co-founder of Hufsey-Nicolaides Assocs Inc in Apr 1965. Reg P E Fla, Ga, Tex, N Y, N J & with the Natl Council of Engr Examiners. Has served as V P & Pres of the Cons Engrs Council of Fla, on the Bd of Dir of the Fla Inst of Cons Engrs & as Natl Dir of the Amer Cons Engrs Council. Presently serving as Natl V P of the Amer Cons Engrs Council. Received an award by the Const Specifications Inst for Excell in Specifications Category H - Coliseums & Stadiums June 1973.

Nicoletti, Joseph P
Business: 130 Jessie St, San Francisco, CA 94105
Position: President *Employer:* URS/John A Blume & Assocs, Engrs. *Education:* BS/CE/Univ of CA, Berkeley. *Born:* July 21, 1921. US Navy, Pacific Area 1942-46. With URS/Blume since 1947; Pres since 1984. Principal in charge of design & cons wth respon for planning, design, & contract admin. Proj's have included coastal & waterfront structs for indus & mil facils, airport structs, & commercial, Public & inst facils. Also in charge of static & dynamic analyses for many structs. Cons Engrs Assn of Calif. Struct Engrs Assn of Calif, VP P Dir & Mbr of State Seismology Ctte. Struct Engrs Assn of No Calif, P Pres, P Dir & P Chmn of the Seismology Ctte. Amer Soc of Civil Engrs, Fellow. Soc of Amer Military Engrs. Appl Tech Council, Dir & P Pres, Earthquake Engrg Res Inst, Fellow. Amer Concrete Inst, Fellow. *Society Aff:* ASCE, SAME, EERI, ACI.

Niebel, Benjamin W
Home: 334 Puddintown Rd, State College, PA 16801
Position: Conslt Indus Engr *Employer:* Self Employed *Education:* IE/IND Engg/PA State Univ; MS/IND Engg/PA State Univ; BS/IND Engg/PA State Univ *Born:* 5/17/18 I.E., MSIE & BS. Penn State. Ch Indus Engr Lord Mfg until 1947. With Penn State 1947 to 1979. Retired 25 Prof Emeritus of Indus Engrg. Author of ten textbooks & many tech articles in area of process engrg, methods, standards, & wage payment. Holder of 5 patents. Cons to many co's in U S, Mexico, & Orient in area of productivity improvement. V P of IIE. Mbr of hon engrg soc's incl Tau Beta Pi, Sigma Tau, Alpha Pi Mu, Sigma Xi. Frank & Lillian Gilbreth and Phil Carroll Indus Engrg Awards. *Society Aff:* ASEE, SME, IIE

Niedenfuhr, Francis W
Home: 3737 Fessenden St N W, Washington, DC 20016
Position: Cons Engr. *Employer:* C3I Div of the MITRE Corp. *Education:* PhD/Eng Mechs/OH State Univ; MSc/Eng Mechs/Univ of MI; BSc/Eng Mechs/Univ of MI. *Born:* 1/23/26. Vibration & flutter analyst North Amer Aviation, Downey CA 1951-52. Prof of Engrg Mech OH State Univ 1952-66. Adj Prof Howard Univ 1967-74. Tech staff Inst for Defense Analysis 1964-66, dealing with space vehicles & optical sys. Prog Mgr, Asst to Dir, Advanced Res Proj's Agency DOD 1966-76. Lecturer in Mech Engg Univ of MD, 1977- . Named Disting Alumnus OH State Univ 1970. Tau Beta Pi, Phi Kappa Phi, Sigma Xi, Pi Mu Epsilon. Various consultancies & service on tech cttes of prof soc's. 50 pubs in structs, stress & vibration analysis, defense R&D, operations analysis of Army combat systems, meteor burst communications systems. *Society Aff:* Assoc Fellow, AIAA, IEEE.

Niederhauser, Warren D
Home: 1087 George Rd, Meadow Brook, PA 19046
Position: President *Employer:* Amer Chem Soc *Education:* PhD/Organic Univ; Univ of WI; AB/Chemistry/Oberlin College. *Born:* Jan 1918 Akron Ohio. Off Sci Res & Dev 1941-43. With Rohm & Haas Co 1943, res mgmt in surfactants, plasticizers, plastics, ion exchange resins, petro chems, fibers, & rocket propellants. Dir of Pioneering Res 1973-1983. Dir Amer Chem Soc 1976- . Dept of Defense: Alvarez Ltd War Comm 1961-63; Chem & Biol Warfare Adv Comm 1961-63. *Society Aff:* ACS, AIC.

Niedringhaus, Philip W
Home: 365 Lakeview Ave E, Brightwaters, NY 11718
Position: Consultant *Employer:* Grumman Aerospace Corp. Retired *Education:* BS/Chemistry/Ursinus College. *Born:* Oct 1924. Met at Brooklyn Polytech Inst. X-ray diffraction/electron microscopy at Rutgers. Native of Lester Pa. US Navy 1944-45. Chemist with Baker Chem Co. Chem & Mfg Res Engr with Republic aviation, spec in metal finishing of aerospace hardward. With Grumman Aerospace since 1964. (retired) Respon incl mfg & met engrg for chem processes & failure analysis of aircraft components. Taught met at SUNY at Farmingdale. ASM LI Chap Chmn 1974-75. ASM Natl Ctte Mbr 197578. Enjoys boating, fishing & fraternal activities. *Society Aff:* ASM.

Niehaus, John C
Business: 6 E Fourth St, Cincinnati, OH 45202
Position: Dir of Engg. *Employer:* SW OH Regional Transit Auth. *Education:* MA/Trans & Planning/Univ of Cincinnati; BS/Civ Engg/Univ of Cincinnati. *Born:* 12/30/35. Reg engr in OH & NY. Responsible for planning, design, construction & renovation of transit facilities & for traffic engg matters relating to transit. Previously Exec Dir of Engg Soc of Cincinnati; City Traffic Engr of Hamilton, OH; Planning & Transportation Dir of Fairfield, OH & Div Traffic Engr for ODOT. Pres of Cincinnati Soc ASCE. Past Pres of Engrg Soc of Cincinnati, community civic club, parish council, OH Sec ITE & Cincinnati Post SAME. Chrmn of natl ITE Committee 2-17; former liaison between ITE & APTA. Chairs COG committees; teaches in UC Evening College. Author of technical & public information articles. *Society Aff:* ITE, ASCE, SAME.

Nielsen, John H
Business: 1339 Washington Ave, Racine, WI 53403
Position: Sr Partner. *Employer:* Nielsen-Madsen Cons Engrs. *Education:* BS(CE)/Struct/U of WI *Born:* 2/8/24. Racine, WI. Served as navigator in USAAF from 1942-45. BS(CE) from Univ WI 1949. Principal of consulting firm 1954 to present. Served as chrmn of WI Transit Authority 1963-66. Mbr of WI Exam Bd for Arch, PE, Designers and Land Surveyors 1970-74, chrmn of land surveys sec 1974. Remonumented USPLS in Racine County WI & conducted control surveys for Racine Co. Integrated survey & mapping prog (1967-76) for which county received award from Natl Assn of Counties. Developed coastal zone cadastre for Racine Co. *Society Aff:* ASCE, NSPE, ACSM

Nielsen, Michael J
Business: Department of Design, Arizona State University, Tempe, AZ 85287
Position: Prof. *Employer:* AZ State Univ. *Education:* Masters/Product Design/Stanford Univ; Bach Product Design/Product Design/NC State Univ. *Born:* 2/19/37. in Salt Lake City, UT. Lived & educated in Northeast, Southeast, N Central, & Western regions of US. Field technician (chief-of-party) with US Forest Service, 1957-60. Served as Combat Construction Engr, US Army during Berlin Crisis, 1960-62. Field Technician (chief-of-party) with US Bureau of Reclamation, 1962-64. Product Design Analyst at Stanford Res Inst, 1967-69. Sloan Fdn Visiting Prof, Univ of Detroit, 1971-72. Prof of Design at AZ State Univ since 1969. Received Ralph R Teetor Award Outstanding Educator Award, Soc of Automotive Engrs, 1979. Elected Chair of AZ State Faculty Assembly, 1987-88. Design Consultant, 1969-Present. Enjoy travel and outdoor recreation. *Society Aff:* SAE, IDSA, AEMS.

Nielsen, N Norby
Business: 2540 Dole, Honolulu, HI 96822
Position: Prof of Civil Engrg. *Employer:* Univ of Hawaii. *Education:* PhD/Civil Eng/CA Inst of Tech; MS/Civil Eng/Tech Univ of Denmark. *Born:* Mar 1928. 1954-56 Lt Danish Navy. 1956-60 Asst Prof Univ of Southern Calif. 1960-64 Ford Found Fellow & Inst Scholar Caltech. 1964-70 Asst to Assoc Prof of Civil Engrg Univ of Illinois. 1967-68 UNESCO Expert in Earthquake Engrg, Internatl Inst for Seismology & Earthquake Engrg, Tokyo Japan. 1970- , Prof of Civil Engrg Univ of Hawaii. 1972- , Chmn of Civil Engrg Univ of Hawaii. Has publ widely in natl & internatl journals. Main res ints: structural dynamics & earthquake resistant design. Moisseiff Res Award 1972 ASCE. *Society Aff:* ASCE, ASEE, ACI, EERI.

Nietz, Malcolm L
Business: 1 Appletree Sq, Bloomington, MN 55420
Position: Sr V Pres, Dir of Engrg. *Employer:* Ellerbe Inc. *Born:* Sept 1931. BSEE Univ of Minnesota 1955. Native of Rochester Minn. Served as Communications Officer in Air Force 1956-57. Elec engr for Ellerbe 1955-56, 1958-65. Intervening time employed at Hoffman Elec Co, St Paul & Foster Elec Co, Rochester Minn. Returned to Ellerbe in 1966 holds pos of Sr V P, Dir of Engrg, & Mbr of Ellerbe Bd of Dir. Ellerbe is a firm of approx 700 people, with annual fee earnings of approx $20 million. Respon for all engrg activities of the firm. Regis in all states except Hawaii & Alaska. Cert by Natl Council of Engrg Examiners, Mbr Illum Engrg Soc & Cons Engrg Council Minn. Hobbies: skiing (mbr Natl Ski Patrol), camping, fishing.

Nightingale, Richard E
Business: Battelle Blvd-Box 999, Richland, WA 99352
Position: Program Manager *Employer:* Battelle NW, Battelle Mem Inst. *Education:* PhD/Physical Chemistry/WA State Univ; BA/Chemistry/Whitman College. *Born:* June 1926. Native of Walla Walla, Wash. Army Air Corps 1944-45. Post-doc Fellow Univ of Minn. Res in spectroscopy, struct, kinetics, chem process dev, & nuclear matls. Employed by Genl Elec Co as Sr Engr, later mgr of tech depts employing 40 to 60 engrs & scientists, conducting R&D on matls, chem engrg, & nuclear engrg. U S delegate to Geneva Conf's & mbr sev U S teams to Europe. Extensive pubs incl 1 book. Local & natl offices in Amer Nuclear Soc; Cert of Merit, Amer Nuclear Soc 1965; Fellow Amer Nuclear Soc 1968; Adv & Exec Cttes of Amer Carbon Soc. *Society Aff:* ANS, ACS.

Nilson, Arthur H
Business: 422 Hollister Hall, Ithaca, NY 14853
Position: Professor of Structural Engrg. *Employer:* Cornell Univ. *Education:* BS/Civil Engrg/Stanford Univ; MS/Structure Engr/Cornell Univ; PhD/Structure Engr/Univ of CA, Berkeley *Born:* 8/27/26. Civil & struct cons engr CA 1948-51, CT 1951-54. Asst Prof 1956-61, Assoc Prof 1961-69, Prof of Struct Engrg Cornell Univ 1969-present. Chrmn Dept of Struct Engrg 1978-1983. Visiting Res Prof Univ of Manchester England summer 1967. Lectr Ecole Polytechnique Federale de Lausanne Switz spring term 1971. Res Prof Politecnico di Milano Italy spring term 1975. Visiting Res Prof

Nilson, Arthur H (Continued)
Salford Univ England summers 1975-77. Author "Design of Concrete Structs" 1986; author "Prestressed Concrete Structs" 1987; num tech papers Consultant for special problems structural analysis and design, particularly reinforced and prestressed concrete. Amer. Conc. Institute Wason Medal in 1974, 1986, and 1987 for materials research. *Society Aff:* ASCE, ACI, PCI, IABSE

Nimmer, Fred W
Home: 1264 Sunset View Drive, Akron, Ohio 44313
Position: Consulting Engr. *Employer:* Self. *Education:* BS/EE/Univ of WI. *Born:* 9/14/02. in Norfolk, NB. Consumers Power & Commonwealth Assocs 1924-27; OH Edison Co 1927-67: Dir Engr, Genl Distrib Engr, Genl Transmission Engr - in charge of engrg & const stds. Chmn of var com Transmission & Distrib EEI. Cons engr 1967- . Mbr ASME; VP Region V. Life Mbr IEEE. Bd Mbr Multiple Sclerosis, Akron Metro Housing & Greater Akron Area Council on Alcoholism. VP Summit Metro Housing & Summit Metro Holding Corp. Pres Springfield Commercial Heat Treating Inc. Hon Elder Trinity Luth Ch. Asst Conductor UW Band. Mbr of Sinfonia honorary musical fraternity. Former mbr of bd of OH Chapter, NAEI, Goodwill Industries, YMCA Chrmn of Public Utilities Committee on revision of OH PUCO 72. Registered Professional Engr VP - Region V, ASME 1976-78. *Society Aff:* ASME, IEEE, NSPE.

Nims, John B
Business: 20000 Second Ave, Detroit, MI 48226
Position: Manager - Strategic Planning. *Employer:* Detroit Edison Co. *Education:* AB/Physics/Boston Univ; MA/Physics/Boston Univ *Born:* Dec 1924. BA & MA Physics Boston Univ. Raised in Concord Mass. Served in the 104th Div, US Army 1943-46. Reactor Physicist for Genl Elec Co at KAPL 1951- 54. With Atomic Power Dev Assocs Inc on Enrico Fermi fast reactor proj from 1954-72, as Hd of Reactor Analysis Div from 1966. With The Detroit Edison Co & Dir of Nuclear Engrg Div until 1974, Corp Strategic Planner until 1976, and Mgr Strategic Planning 1976, Asst V.P. Strategic Planning 1984. Enjoy boating, fishing, future research. *Society Aff:* ANS, APS, NASCP

Ning, Tak H
3085 Weston Lane, Yorktown Heights, NY 10598
Position: Research Staff Member *Employer:* IBM *Education:* Ph.D./Physics/University of Illinois; M.S./Physics/University of Illinois; B.A./ Physics/Reed College *Born:* 11/14/43 B.A. in physics, Reed College, Oregon, 1967; Ph.D. in physics, Univ of Ill 1971. Researched in areas of impurities in silicon, inversion layer transport, hot-electron effect in MOSFET, MOSFET technology, and silicon bipolar technology. Assumed current responsibilities as Manager of Silicon Devices and Technology in IBM Research Div in 1983. IEEE Fellow 1987.

Niordson, Frithiof I
Business: Bldg 404, Lyngby Denmark 2800
Position: Prof. *Employer:* Technical Univ of Denmark. *Education:* PhD/Appl Math/ Brown Univ; MS/Tech Phys/RIT (Sweden); BS/Tech Phys/RIT (Sweden) *Born:* 8/1/22. Born in Johannesburg, educated in Sweden and USA. Acting prof of Strength of Mtls, RIT, Sweden 1951. Own consulting engg bureu 52-58. Supervized stress and vibration analysis for STAL-LAVAL (jet-engines, power turbines). Full prof of mech engg, Technical Univ of Denmark 58. Dean of mech engg 75. Pres of the Itl Union of Theoretical and Appl Mechanics 76-80. Mbr of the Council, German Soc for Appl Math and Mechanics. Mbr of the Structures and Mtls Panel (Advisory Group on Aero Res & Dev). Secy of the Danish Ctr of Appl Math and Mechanics. Knight of Dannebrog. Centennial Yr Honorary mbr of ASME. Sports: sailing, horse- back riding, skiing. *Society Aff:* ASME

Nippes, Ernest F
Business: RPI, Troy, NY 12181
Position: Prof of Met Engg. *Employer:* Rensselaer Polytechnic Inst. *Education:* PhD/Met Engg/Renss Poly Inst; MS/Met Engg/Renn Poly Inst; BS/Chem Engg/ Renss Poly Inst. *Born:* Feb 1918. Native of East Islip N Y. Mbr of RPI fac since 1938. Chmn Dept of Matls Engrg 1961-65. Dir Office of Res & Spons Programs 1965-74. Pres Rensselaer Res Corp 1965-69. Natl Pres Amer Welding Soc 1968-69. Adams Hon Lect AWS 1958. Samuel Wylie Miller Mem Award AWS 1959. Amer Soc for Metals Teaching Award 1956. A.E. white Dist Teacher Award, Amer Soc for Metals 1983. Fellow, Amer Soc for Metals 1982. Lic Prof Engr NY. Cons to indus in met & welding. *Society Aff:* AIME, ASTM, ASM, AWS.

Nisbet, John S
Business: University Park, PA 16802
Position: IRL Prof of EE. *Employer:* PA State Univ. *Education:* BSc/EE/London Univ; MS/EE/PA State Univ; PhD/EE/PA State Univ. *Born:* 12/10/27. Worked on the design of specialized test equipment for everything from radars to gas water heaters with Nash & Thompson, Ltd 1944-51. Microwave Engr with Decca Radar, Ltd 1952-53. Broadcast Sys Engr, Canadian Westinghouse, Ltd 1953-55. Res Assoc, Dept of Engg Res, PA State. Worked on electro mechanical istrumentation 1955-57. Ionosphere Res Lab, PA State Univ. Worked on various rocket, satellite and incoherent scatter measurements of aeronomical parameters. Conducted theoretical research on the factors affecting the constitution and energy balance of the neutral and ionized inputs of planetary atmosphere. 1957- date. Apptd Dir 1971. *Society Aff:* AGU, IEEE, URSI.

Nishino, Osamu
Home: 1-24-2 Nishi-Shinjuku, Shinjuku-ku, Tokyo Japan 160
Position: Professor. *Employer:* Kogakuin Univ. *Education:* DR of Engrg/-/Univ of Tokyo; BE/-/Univ of Tokyo. *Born:* 9/11/12. Electrotech Lab, Ministry of Indus Trade & Indus 1936-53; Prof Dept of Appl Phys Univ of Tokyo 1953-58, Prof Dept of Nuclear Engrg Univ of Tokyo 1958-73; Prof Dept of Electronic Engrg Kogakuin Univ 1973-85. Fellow IEEE 1973. *Society Aff:* JIEE, IEIC, JSAP, JPS, JAES, SICE.

Nishizawa, Jun-ichi
Business: Research Inst of Elec Communication, Tohoku Univ, Sendai Japan 980
Position: Professor. *Employer:* Res Inst of Elec Commun-Tohoku Univ. *Education:* Dr/Engrg/Tohoku Univ; BS/Elec Engrg/- *Born:* Sept 1926. Res Asst 1953; Asst Prof 1954; Prof 1962 and Dir 1983 by March, 1986 Res Inst of Elec Communication, Tohoku Univ. Engaged in work at the Solid-State Electronics Field; also Dir of Semiconductor Res Inst. Won Emperor Invention Award 1966 by invention of pin diode & pnip transistor; Dir Award of Jap Sci Agency 1974 by invention of semiconductor Injection Laser in 1957; Annual Award of Medals of the Japan Acad 1974 by work about perfect crystal, graded index optical waveguide & static induction transistor; Person of Cultural Merits in Japan 1983; IEEE Jack A Morton Award, 1983 by invention and devel of the class of SIT and for advances in optoelec devices. President of IEICE of this year. Mbr APS & ECS. Fellow of former Phys Soc & the Inst of Physics & of IEEE. *Society Aff:* APS, AIP, ECS

Nishkian, Byron L
Business: 1 Holland Ct, San Francisco, CA 94103
Position: President. *Employer:* Martin, Cagley & Nishkian, Inc. *Born:* Jan 1916. BS Univ of Calif. Native of San Francisco. Civil Engr Corps USN 1942-45. 1946-47 Partner L H Nishkian, Cons Engrs (in practice 1919 -47). 1947-70 Sole owner L H Nishkian & B L Nishkian, Cons Engrs. 1970-79 Nishkian, Hammill & Assocs Inc, Cons Engrs. Martin, Cagley & Nishkian, Inc 1979-present Fellow ASCE. Fellow Amer Inst of Cons Engrs. Avocations: music, skiing, golf, travel.

Nissan, Alfred H
Business: College of Environmental Sci, Forestry, Syracuse, NY 10583
Position: Consultant & Adjunct Professe *Employer:* Westvaco & SUNY respectively *Education:* DSc/Chem Eng/Univ of Birmingham, England; PhD/Chem Eng/Univ of Birmingham, England; BSc/Chem Eng/Univ of Birmingham, England. *Born:* Feb 1914. Lectr at Univ of Birmingham 1940-47. Acted as cons on flame warfare to

Nissan, Alfred H (Continued)
Dept of Sci & Indus Res (Britain) 1942-57. Res Dir Bowater Paper Corp & Mbr of Bd of Bowater R&D Ltd (England) 1947-53. First occupant of Chair of Textile Engrg at Univ of Leeds England 1953-57. Immigrated to USA 1957 to become Res Prof of Chem Engrg RPI & Tech Adv to W Vir Pulp & Paper Corp N Y (later Westvaco). In 1962 joined Westvaco full time as Corp Res Dir & became V P in 1967. Hobby: oil painting. Dir 1968-71 TAPPI. Schwarz Mem Award of Textile Div 1967 ASME. Dir 1973-76 IRI. Hon Visiting Prof Univ of Uppsala Sweden 1974. R&D Div Award 1976 TAPPI. Alexander Mitscherlich Award of Zellcheming (Germany), 1980. *Society Aff:* AIChE, ACS, TAPPI, AAAS, (Fellow)

Nix, William D
Business: Dept of Matl Sci & Engr, Stanford, CA 94305
Position: Prof & Assoc Chmn. *Employer:* Stanford Univ. *Education:* PhD/Met Sci/ Stanford Univ; MS/Met Engg/Stanford Univ; BS/Met Engg/San Jose State College. *Born:* 10/28/36. On Stanford Fac since 1962, becoming Assoc Chmn of Matls Sci & Engg 1975. Served 1 yr as Asst to Dir of Tech, Stellite Div, Union Carbide Corp, & was Dir of Stanford's Ctr for Matls Res between 1968 & 1970. Assoc Chrmn of Matls Sci & Eng Dept, 1975-present. Author & co-author of 169 papers in the field of high temp creep of crystalline solids. 1970 Bradley Stoughton Teaching Award ASM, 1979 Mathewson Gold Medal of AIME, Elected Fellow of ASM in 1978. In 1987 he was elected to the Natl Acad of Engg. He is co-author of "The Principles of Engineering Materials" pub. in 1973 by Prentice-Hall. *Society Aff:* ASM, AIME, ASEE, MRS, ΣΞ.

Nixon, Alan C
Business: 2140 Shattuck - Rm 511, Berkeley, CA 94704
Position: Consultant. *Employer:* Self. *Education:* PhD/Chem/UC Berkeley; MSc/ Chem/Univ of Saskatchewan; BSc/Chem/Univ of Saskatchawan. *Born:* Oct 1908. Instr Berkeley (Chem) 1935-37. Petro chem & Res Supr Shell Dev Co 1937-70. Principal Investigator Navy, Air Force & NASA contracts re fuels & combustion. P Pres Amer Chem Soc; P Chmn Council of Sci Soc Presidents. Mbr AIAA, ACS, AAAS, Combustion Inst, AIC, Sierra Club. Enjoys organizational activities, writing, sports, & the outdoors. Director Teck Research Inc; MOLI Energy Ltd; N. Ca Conf UCC; (and Pres) Calsec Consultants, Inc. Author and publisher: NickNacks. *Society Aff:* AIAA, AAAS, ACS, CSA, CI, AGE, AIC

Nixon, David D
Home: 4733 W Conrad Rd, Ludington, MI 49431
Position: Principal *Employer:* Earth Systems *Education:* BS/Geol/Univ of CA *Born:* 8/22/38 1964-70 Engrg Geologist, CA Aqueduct Project. 1970-72 Project Geologist, Ludinton Pumped Storage Hydroelectric Project, world's largest. 1972-74 Foundation Engr, Keban Hydroelectric Project, built on cavernous limestone, world's most extensive foundation treatment. 1974-78 Chief of Division of Geotechnical Specialists, LG-2, world's largest underground powerhouse. 1978-79 Resident Engr, Caniapiscau Reservoir consisting of about 60 dams and dikes. 1979-80 Principal Construction Engr for repairs of Tarbela, world's largest dam. 1980-81 Consultant, Principal and founder of the firm, Earth Systems, specializing in engrg and environmental geology. Recipient of Thomas Fitch Rowland Award of ASCE 1980. *Society Aff:* ASCE, USCOLD, AEG

Nobe, Ken
Business: Depart of Chem & Nuclear Engrg, Los Angeles, CA 90024
Position: Prof & Chairman *Employer:* UCLA *Education:* PhD/Eng/UCLA; BS/ChE/ Univ of CA-Berkeley *Born:* 8/26/25 *Society Aff:* ACS, ECS, NACE, AIChE, ΣΞ, ISE

Noble, Charles C
Business: Prudential Center, Prudential Center, MA 02199
Position: Pres and Chief Exec Off *Employer:* C.T. Main Corp *Education:* MSCE/Engg/MA Inst of Tech; MAIA/Intl Affairs/George Wash Univ; BSCE/Engg/ US Military Academy. *Born:* 5/18/16. Syracuse NY. Commissioned 2nd Lt, Corps of Engrs US Army 1940. Advanced to Maj Gen. Served as Ex O Oak Ridge Manhattan Dist; Exec Secy MLC of AEC; Planner, Supreme Headquarters Allied Powers Europe, Deputy Dist Engr NY; Dist Engr Louisville Ky; Dir Atlas D&E & Minuteman ICBM Facilities Const; Ch Engr US Army in Europe, Korea & S Viet Nam; Director Construction, Office Secretary of Defense; Engineering Agent, Atlantic-Pacific Interoceanic Canal Study; Dir Civil Works OCE; CG USA Engrg Cmd Viet Nam; Pres Miss River Comm & Div Engr Lower Miss Valley; Chmn Red River Compact Comm. President, US Army Constal Engineering Research Board; Currently Dir & Vice Chrmn-Chas. T. Main, Inc, and Chas. T. Main Intl, Inc. Reg P E DC, La, Texas, NY, Nebraska, Nevada, Va, Miss & Mass. Fellow: ASCE, ACEC, SAME; Mbr NSPE, PIANC, CIGRE; US Ctte on Large Dams Mbr Exec. Comm; Wheeler Medal SAME; Chmn Amer Sect PIANC 1968. Elected to National Academy of Engineering 1981; Mbr The Beavers; The Moles; Newcomen Soc of US. *Society Aff:* NAE, ASCE, NSPE, PIANC, SAME, USCOLD, ACEC, NSS OF ΣΞ

Noble, Marion D
Business: P O Box 30, Dallas, TX 75221 *Employer:* Sunmark Exploration Co. *Education:* BS/Petrol Eng/TX A & M Univ *Born:* 1/19/26 Joined Sun Co. in 1950 and named as assistant mgr for Venezuelan Sun in Caracas in 1965. Mgr of foreign production operations. Named director of international products, then VP for exploration and production for Sun International in 1971, and elected president of that company in 1974. In 1975 he was named Pres of Sunmark Exploration Co. Member and past director of the Society of Petroleum Engs branch of AIME. Is a member of the American Petroleum Inst, the Mid-Continent Oil & Gas Assoc and the Dallas Petroleum Club. He is also past director and member of the finance committee of the World Affairs Council of Phila, PA and dir of the World Affairs Council of Dallas, TX. A mbr of the Banking/Finance Adv Ctte of Northlake Comm Coll, Irving, TX, and serves on the Bd of Dirs of Republic Bank of Irving, TX. Noble is married to the former Geraldine O. Norris of Baytown, TX and they have six children. *Society Aff:* AIME (SPE)

Nobles, Elon John
Home: 5827 Vinehill Rd, Minnetonka, MN 55343
Position: Vice Pres - Director. *Employer:* Econo-Therm Energy Sys Corp. *Education:* MS/Power & Economics/Stevens Inst of Tech; ME/Mech Engg/Stevens Inst of Tech. *Born:* 9/21/16. 30 yrs engg design & contracting, largely in high pressure - high temp synthesis field. Originated a num of designs now std items in indus, along with a num of patents. Active Mbr ASME, AIChE, API; Prof Engr. A Founder of Deltak Corp, a producer of process boiler & heat recovery equip. Recently a founder of econo-Therm Energy Sys Corp, a substantial equipment fabircating & design firm in the field of heat exchanges, fired heaters, fin-tubes, incinerators, & process equipment. *Society Aff:* ASME, AIChE, API.

Nofsinger, William M
Home: 6645 Brookside Rd, Kansas City, MO 64113
Position: President & Genl Mgr. *Employer:* The C W Nofsinger Company. *Education:* BS/Chem Engrg/Univ of KS. *Born:* Sept 1932. Native of Summit NJ. Served with US Air Force 1955-58. With The C W Nofsinger Co: 1959-66 Process Engr & Proj Engr, cons engrs specializing in refining, chemical & petrochemical fields; 1966-71 Mgr of Proj Engrg; 1972-73 Asst Gen Mgr; 1973 V Pres. 1978 Pres. Chmn Kansas City Sect AIChE 1977. Pres Rotary Club of Kansas City 1985-86. Enjoy skiing, music & spectator sports. *Society Aff:* AIChE, NSPE, ACS.

Nolan, James W
252 Howard Avenue, Des Plaines, IL 60018
Position: President. *Employer:* James W Nolan Company. *Education:* BS/ME/Purdue. *Born:* March 1927. 1950-61 application engr specializing in fire protection pumping equip for Peerless Pump. Since 1962 Pres of James W Nolan Co specialists in water supply systems. Mbr Natl Fire Protection Assn, Western Soc of Engrs, & Amer Soc of Plumbing Engrs (ASPE), IL Fire Prevention Assn, Chrmn NFPA Standpipe Systems CTTE 14, NFPA Centrifugal Fire Pump Ctte 20, & Gen

Nolan, James W (Continued)
Services Admin Airlie House Conference on Fire Safety in High Rise Bldgs; Blue Ribbon Ctte Mbr of McCormick Place Fire Investigation Ctte; presently Mbr of Mayor's City of Chicago Ctte on Bldg Code Ammendments; 1973-75 Pres Soc of FELLOW, Soc of Fire Protection Engrs (SFPE), Fire Protection Engrs; Mbr Bd/Dir EJC. *Society Aff:* SFPE, ASPE, WSE

Nolan, Ralph P, Jr
Business: Harbor Dr-POB 80278, San Diego, CA 92138
Position: Special Projects Engineer. *Employer:* National Steel & Shipbuilding Co. *Education:* Cert/Naval Architect-Marine Engr/MA Inst Tech; BS/Mech Engg/Tulane Univ. *Born:* March 1 1920. MIT Short Course Naval Architecture/Marine Engrg. Naval Officer, active duty 6 years in Engrg Billets at Sup Ship offices during WWII & Korea, including Sr Asst Sup Ship. Mgr Marine Dept of A M Lockett Co representing Babcock-Wilcox, Worthington & Griscom-Russell Cos on Gulf coast. Since 1957 at Natl Steel & Shipbuilding Co: served as Ch Marine Engr, Asst to Mgr of Engrg, Ch Materials Engr, Ch Estimator, Contracts Administrator, & Special Projs Engr. 1976 SNAME Council & Chmn San Diego. *Society Aff:* SNAME.

Nolan, Robert E, Jr
Home: 222 Fairview Rd, Clarks Green, PA 18411
Position: Pres. *Employer:* Rendon Cor. *Education:* BS/Mining Engr/Lehigh Univ. *Born:* Dec 28, 1929. Native of Scranton, PA. BSEM from Lehigh Univ. With Bellante, Clauss, Miller & Nolan since 1951. SAME: 1957 became Partner, since 1967 Partner-In-Charge of all engrg for firm, including offices in other states & Virgin Islands. Pres of Rendon Cor, Contractors and Engrs, since 1976. Resigned 1976. Reg PE in 3 states . P Pres, V Pres & Dir of Lehigh Valley Sect ASCE, presently serving on Natl Ctte. P Pres Greater Scranton Chamber of Commerce (3 terms), presently on Exec Ctte; Past Chrmn of Bd of Marywood Coll; Dir of Catholic Social Services; Dir of Scranton Natl Bank. Hobbies: fishing, tennis, skiing, travel. *Society Aff:* ASCE, NSPE.

Nolan, Robert W
Home: Box 69, Cokesbury Village, Hockessin, DE 19707
Position: Consulting Engineer. *Employer:* self employed. *Education:* BS/CE/Rensselaer Poly. *Born:* Feb 1906. 1928-51 with Newport News Shipbuilding & Dry Dock Co on design & dev of steam turbines & other machinery. 1951-71 with E I du Pont de Nemours & Co, retired as Sr Cons in Mech Engrg. Since 1971 Cons engr: stress analysis, vibration, pressure vessels, structures, and past tensioned concrete. Mbr Sigma Xi 1933. Mbr winning team, US Navy contest for sub-chaser design 1939. Soc of Naval Arch & M E Linnard Prize for best paper 1946. Fellow ASME 1968. Author 'Vessels for Storage & Processing Fluids,' Perry's Chem Engrg Handbook ed V, 1973. *Society Aff:* ASME, NSPE, ΣΞ, PCI, PTI

Nolen, James S
Home: 838 Ivy Wall Dr, Houston, TX 77079
Position: President *Employer:* J S Nolen & Assocs, Inc *Education:* PhD/ChE/Tulane Univ; BS/ChE/LA State Univ *Born:* 10/1/43. Native of Baton Rouge, LA. Grad magna cum laude from LSU in 1965. Sr Res Engr at Esso Production Res Co 1968-71, specializing in petrol reservoir simulation. With Intercomp 1971-79, serving as Gen Mgr of European Operations and VP of Res & Dev. Since 1979, Pres of J S Nolen & Assocs, a company specializing in numerical simulation of petroleum reservoirs. *Society Aff:* AIChE, SPE of AIME.

Nolte, Byron H
Business: 590 Woody Hayes Dr, Columbus, OH 43210
Position: Exten Agri Engg - Prof. *Employer:* Coop Exten Serv-Ohio State U. *Education:* PhD/Agri Engg/OH State Univ; MS/Agri Engg/Univ of MO, Columbia; BS/Agri Engg/Univ of MO, Columbia. *Born:* Dec 8 1934, Higginsville MO. Reg PE, MO. 1956-59 Officer & Pilot USAF. Field Engr with Soil Conservation Service Bethany MO, & Area Engr Springfield MO. Assumed duties as Extension Agricultural Engr (Professor) with The Ohio State Univ in 1966, responsible for statewide educational program in soil & water resource engrg: watershed dev, drainage, erosion control, flood prevention & drainage law. Received NSF Science Faculty Fellowship 1970, 12 ASAE blue ribbons for educational materials, 1st ASAE Young Extension Man Award 1973. P Chmn Ohio Section ASAE; Mbr Honor Soc of Agri, Honor Society of Agricultural Engrg and Natl Honorary Extension Fraternity. *Society Aff:* ASAE, SCSA.

Nolting, Henry F
Home: 221 Greyhound Pass, Carmel, IN 46032
Position: Pres. *Employer:* Nolting Assoc, Inc. *Education:* MS/ChE/Purdue Univ; BS/ChE/Purdue Univ. *Born:* 8/15/16. Native of Indianapolis, IN. Explosives Engr with Hercules 1940. With Std Oil Co of IN 1941-1977. Served as Mgr of Operations, Refinery Coordination, Mgr Overseas Operations, Construction & Maintenance Mgt. Assumed present position in 1977. Provide services to HPI. Dir AIChE 1958-1960. Published papers on unit operations, gen mgt, productivity improvement. Also chaired symposia on above plus alternate energy sources. Conducted seminars on maintenance & construction planning. Active in civic affairs. Enjoy golf & spectator sports. *Society Aff:* AIChE, AMA.

Nonken, Gordon C
Home: 81 Noblehurst Ave, Pittsfield, MA 01201
Position: Engineering Consultant. *Employer:* General Electric Tech Services Co. *Education:* BS/EE/KS State Univ. *Born:* July 1908. 1930 joined General Electric as test engr; 1930-41 proj engr high voltage insulation dev G E Lab MA, 1941-52 industrial engr responsible for dev of process heating for transformer plant; 1952-56 supervisor of mfg process dev sect; 1956-70 cons engr responsible for coordination of dev projs for distrib transformer engrg & mfg; 1970-73 Sr Dev Engr in transformer bus & tech planning oper; since retirement in 1973, cons for G E Mexico. G E Coffin Award 1948; IEEE Fellow 1955. Awarded 8 patents. *Society Aff:* IEEE, ΦΚΦ.

Noodleman, Samuel
Home: 3302 E Kleindale Rd, Tucson, AZ 85716
Position: Adjunct Prof. *Employer:* Univ of AZ. *Education:* SB/EE/MIT. *Born:* 1/20/14. Native of Saco ME. During WWII was at Radiation Lab MIT. Received Certificate of Merit, US War Dept for work at Radiation Lab. Past Chrmn Dayton Sec IEEE, & DC Sub-Committee RMC-PES. Life Mbr, Fellow IEEE. PE State of OH. Hold 37 US patents & a number of foreign patents. Served as VP of Std Dayton Corp Dayton OH, Welco Ind Cincinnati OH, & Inland Motor Div Kollmorgen Corp Radford VA. Retired as Sr VP Inland Div Jan 1979. Engg career devoted to innovation & dev of rotating electric machines. *Society Aff:* IEEE-PES, IEEE-IAS, IEEE-EnM.

Noonan, Mark E
Home: 2345 15th Ave, San Francisco, CA 94116
Position: Associate Sanitary Engineer. *Employer:* City & County of San Francisco. *Born:* Nov 1941. Native of San Francisco. BS From Univ of the Pacific; MS from Stanford Univ. Proj engr for Bechtel Corp specializing in refinery wastewater treatment. Since 1970 employed by City of San Francisco, primarily working on dev of City's wastewater management master plan. Wesley W Homer Award 1976. Enjoy theater & tennis.

Noor, Ahmed K
Business: GWU-NASA Langley Research Center, MS-269, Hampton, VA 23665
Position: Professor *Employer:* GW Univ *Education:* Ph.D./Structural Mechanics/Univ. of Illinois at Urbana-Champaign; M.S./Structural Mechanics/Univ. of Illinois at Urbana-Champaign; B.S. (Honors)/Structural Engineering/Cairo University, Cairo, Egypt *Born:* 08/11/38 Ahmed K. Noor is Prof of Engrg and Applied Science. He taught at Stanford Univ, Cairo Univ, Univ of Baghdad (Iraq) and the Univ of New South Wales (Australia) before joining George Washington Univ. He has

Noor, Ahmed K (Continued)
edited eight books and authored numerous papers in the fields of structural mechanics, plates and shells, large space structures, and computational mechanics. He is on the editorial board of several technical journals and served on a number of cttees of the National Academy of Engineering. *Society Aff:* ASME, AIAA, ASCE, AAM, IACM

Noordergraaf, Abraham
Business: 101 Hayden Hall, Univ. of Pennsylvania, Philadelphia, PA 19104-6392
Position: Professor *Employer:* Univ of PA *Education:* PhD/Biophysics/Univ of Utrecht; MS/Exp physics/Univ of Utrecht; MA/honor/Univ of PA; BS/Math/Univ of Utrecht *Born:* 8/7/29 Born in Utrecht, The Netherlands. Since 1957, with Univ of PA. Appointments have included: Assoc Dir, Moore School of Electrical Engrg; Chrm, Dept of Bioengrg. Presently Professor of Biomedical Engrg, of Veterinary Medicine, and of Dutch Culture. Fellow, IEEE; Charter Mbr, Biophysical Soc; Pres, Cardiovascular System Dynamics Soc, 1976-80; Founding Mbr, Biomedical Engrg Soc. Charter Mbr, Soc of Mathematical Biology; Fellow, Amer Coll of Cardiology, Royal Society of Medicine. Author: *Circulatory System Dynamics*, 1978. Contributing Author: *Biological Engrg*, 1969. Editor (with G N Jager and N Westerhof) *Circulatory Analog Computers*, 1963; (with E Kresch) *The Venous System: Characteristics and Function*, 1969; (with J Baan and J Raines) *Cardiovascular System Dynamnics*, 1978; Editor, *Biophysics and Bioengrg Series. Society Aff:* IEEE, ASEE, APS, ACC

Norberg, Hans A
Business: PO Box 726, Tulsa, OK 74101
Position: Chairman. *Employer:* Nelson Elec Unit of Genl Signal Corp. *Born:* Aug 19 1901, Astoria OR. BSEE from Univ of Minnesota 1927. 1927-31 worked General Elec Co Schenectedy NY & Philadelphia PA, switchgear design. 1931-37 Phillips Petroleum Co, elec engrg. A founder of Nelson Electric 1937- , Pres & then Chmn. Engr of Year, OK 1969. Fellow & Life Mbr IEEE. Active in civic affairs; enjoy sailing & classical music. *Society Aff:* IEEE, NACE, NEMA.

Nordby, Gene M
Business: 1100 14th St, Denver, CO 80202
Position: Chancellor *Employer:* University of Colorado at Denver *Education:* PhD/Civil Engg/Univ of MN; MSCE/Civil Engg/Univ of MN; BSCE/Civil Engg/OR State Univ. *Born:* 5/7/26. Taught Civil Engg at Univ of MN 1948-49; Univ of CO 1950-55; Purdue Univ 1956; Geo Wash Univ 1957-58; Univ of AZ 1958-62. Program Dir for Engg Sci NSF 1956-58. With Univ of Colo at Denver since 1980 as Chancellor; also Prof Civil Engr; GA Inst of Tech, 1977-80 as VP for Business and Finance; also Prof Civil Engg; Univ of OK 1962-77, first as Engg Dean & in 1969 as VP; also Prof Civil, Mech Engg. Author *Intro to Struct Mech* & over 50 papers on reinforced concrete, fatigue, epoxy resins, res admin. Edited 6 civil engg tests for MacMillan. Active in ASCE Educ & Res Cttes & accreditation activities for Dir, ABET & North Central Assn. *Society Aff:* ASCE, NSPE, ASEE.

Nordgren, Ronald P
Home: 14935 Broadgreen Dr, Houston, TX 77079
Position: Research Associate *Employer:* Shell Development Co. *Education:* Ph.D./Mech. Engr./Univ. of California; M.S.E./Engr. Mech./Univ. of Michigan; B.S. E./Engr. Mech./Univ. of Michigan *Born:* 4/3/36 Born and raised in Michigan. Research in applied mechanics at Inst. of Engr. Res., Univ. of Calif. at Berkeley, 1959-62. Joined Shell Dev. Co. in 1963 and am currently a Research Associate. My research has involved the application of mechanics and mathematics to problems in offshore engr., petroleum production, and arctic engr. Special assignment to Shell Intern., The Netherlands, 1970-71. Lecturer in Mech. Engr., Rice Univ., 1965-68 and Univ. of Houston, 1980. Assoc. Ed., J. Appl. Mech., ASME, 1972-76, 1981-85. Fellow, ASME. *Society Aff:* ASME, SIAM, ΣΞ

Norman G Einspruch, Dean
Business: University of Miami, College of Engineering, P. O. Box 248294, Coral Gables, FL 33124
Position: Dean & Prof of Elec and Comp Engrg *Employer:* Univ of Miami. *Education:* PhD/Appl Mathematics/Brown Univ; MS/Physics/Univ of CO; BA/ Phsyics/Rice Univ. *Born:* 6/27/32. Dean of the Coll of Engrg and Prof of Elec and Comp Engrg. Appointed as Dean in 1977 after 17 1/2 yrs with TX Instruments Inc of Dallas, TX; positions were Mbr, Technical Staff--Central Res Labs (CRL), 1959-62; Mgr, Electron Transport Physics Branch--Physics Res Lab, CRL, 1962-68; Dir, Advanced Tech Lab- -CRL, 1968-69; Dir of Tech--Chem Mtls Div, 1969-72; Dir--CRL, 1972-75; Elected Asst VP in 1975, Corporate Dev Mgr, 1975-76; Planning and Tech Mgr--Consumer Products Bus, 1976-77. Served as Chrmn of the Natl Res Council Panel on Thin- Film Microstructure Sci and Tech (1978-79). Author and co-author of more than 50 publs in recognized technical journals. Ed of *VLSI Electronics: Microstructure Science*, Academic Press - Vols 1 through 16. Married to Edith Melnick Einspruch (1953); they have three sons. Member Natl Res Council Panel on Impact of DOD Very High Speed Integrated Circuits (VHSIC) Program (1980-1981). Member Natl Res Council Panel of Education and Utilization of the Engineer (1981). Mbr Bd of Dirs, Ogden Corporation. Advisory Bd of Venture Mgmt Assoc. *Society Aff:* AAAS, IEEE, IIE, ASEE, APS, ASA, SID.

Norman, John D
Business: 1090 Speers Rd, Oakville, Ontario L6L 2X4 Canada
Position: President. *Employer:* Pollutech Ltd. *Education:* PhD/Chem Engrg/Rice Univ; MSc/Env Engrg/Univ WI; Bsc/Engrg/Univ Manitoba. *Born:* 5/23/35. Professionally involved in pollution control and environ field in Ontario and US since 1956. Degrees from Univ of Maintoba, Univ of Wisconsin, and Rice University. Reg P E Ontario 1957. Founded Pollutech Limited an environ consltg firm in 1969 and has been Pres ever since. Pollutech is primarily a laboratory based processed consulting firm on environmental matters involving water pollution, air pollution, occupational health, and product recovery strategies for industries and major municipalities. A large amount of the activity involves tuning up of effluent treatment plants and proposing schemes for product recovery. Previously Assoc Professor of Chemical Engrg at McMaster Univ for 7 years, Instructor in Environ Engrg at Rice Univ for 2 years, worked in a engineering firm for 5 years and in construction for 2 years. Has served on 9 advisory Bds, has authored or co-authored 10 science publication papers; has presented 26 papers on environ topics at tech conferences. Presented with Arthur Sydney Bidell award by the Water Pollution Control Federation in 1983. Diplomate of Amer Academy of Environ Engrs. *Society Aff:* WPCF, IAWPR, ΣΞ.

Norman, Joseph H, Jr
Business: 3408 Hermitage Rd, Richmond, VA 23227
Position: Partner. *Employer:* Harris, Norman & Giles. *Education:* BS/Civil Eng/Univ of VA. *Born:* 9/22/34. Native of Richmond, VA. Served as Ch Struct Designer for Baskerville & Son, Arch & Engrs until 1966. Presently Principal in the firm of Harris, Norman & Giles, Cons Engrs. Mbr ACEC, NSPE, ASCE & P Pres of CEC/ VA. *Society Aff:* ASCE, NSPE, ACEC.

Norman, Neil A
Home: PO Box 3965, San Francisco, CA 94119
Position: Proj Mgr *Employer:* Bechtel Natl Inc *Education:* MBA/Mgmt/Univ of CT; BS in ME/Power Sys/Univ of CA - Berkeley *Born:* 02/12/31 In 34 years of state of the art power sys dev, his respon have been: Proj Mgr: High Level Nuclear Waste Repository Design for Commercial spent fuel in Tuff Rock & salt; Light Water Reactor Nuclear Steam Supply Sys Design and Construction. Asst Proj Mgr: NERVA Nuclear Rocket gas reactor, design, construction & test. Design, Dev, and Test Supervision; Polaris A-3 rocket; Apollo SPS Engine; First US Axial Flow Compressor Jet Engine. Active in community service, ANS, and Natl and CA Socs of PE (NSPE-CSPE) receiving 20 state and national awards for service to the Engrg Profession. Continuing public speaker on Engrg Professionalism, Strengthening US in

Norman, Neil A (Continued)
Implementation of Engrg Tech, and Peaceful Uses of Nuclear Tech. Bechtel Fellow; VP - NSPE 1984-85. Chrmn ANS Fuel Cycle and Waste Mgmt Div 1985-86. Natl Dir - CSPE 1985-88. Pres - CSPE 1980-81. *Society Aff:* NSPE, ANS, CSPE, BAEC, AAA

Norquist, Warren E
Home: 89 Bradford Rd, Weston, MA 02193
Position: Worldwide Director, Purchasing and Materials Management *Employer:* Polaroid Corp - Cambridge. *Education:* MBA/Bus/Harvard Univ; BS/Engg/Univ of MI. *Born:* 9/21/31. Pres of Univ of Michigan Engrg Class. Harvard Bus Sch, Baker Scholar. Westinghouse Elec. Polaroid 1960-. Worldwide Dir of Purchasing and Materials Management; previous pos incl VP of Quality and Reliability, mfg mgr & tech mgr of film assembly. Formerly Bd of Dir ASQC and ANSI. Fellow ASQC; P Pres Engrg Soc of New England. Chairman Certification Board N.A.P.M. Reg P E. Writer and lecturer and author of *Creative Countertrade*, A Guide to Doing Business Worldwide, Published by Ballinger, 1987. Creator of Zero Base Pricing Purchasing Program. Taught Mfg Policy at Bostn Univ. Patent on design of Polaroid Prof Film. Married to Carol Lutz, 4 children. Hobbies: skiing & boating. *Society Aff:* NAPM, ASME, ASQC, NSPE.

Norris, Dan P
Business: P.O. Box 11680, Eugene, OR 97440
Position: Exec. VP. (retired) *Employer:* Brown and Caldwell *Education:* BS/CE/OR State Univ. *Born:* Aug 1924. Native of Oregon. Served as Naval Aviator 1946-47. Worked for Arabian Amer Oil Co & Creole Petro Corp as design engr 1951-56. Joined Brown and Caldwell as proj engr 1956. Dir studies, design, & const of a wide var of environ engrg projs. Tech articles publ in Journal AWWA; Journal WPCF; Journal Sanitary Engrg Div ASCE. 1970 Westley W Horner award ASCE. In 1972 assumed duties as V P & Engrg Mgr of Brown & Caldwell's office in Eugene, Oregon. 1983 Gascoigne Medal, WPCF. Retired as officer of Brown and Caldwell in 1986. Serve as special consultant on major projects. *Society Aff:* ASCE, AWWA, WPCF, NSPE.

Norris, James C
Business: P O Box 3, Houston, TX 77001
Position: Sr V P - Petro & Chem Engrg Div. *Employer:* Brown & Root Inc. *Education:* BS/Chem Engr/LA State. *Born:* May 1924. BS 1948 Chem Engrg Louisiana State Univ. Native of Galveston Tex. Process engr for McCarthy Chem Co, Winnie Tex, mfg organic chems. With Brown & Root Inc since 1951 as process, proj & pre-contract engr. Assumed respon as V P of Petro & Chmn Engrg in 1972. Promoted to Sr VP in 1977. Reg P E in Texas & La. Mbr Amer Inst of Chem Engrs. P Pres of Gas Processors Suppliers Assn & Genl Chmn of its 1972 issue of Engrg Data Book. Enjoys boating & fishing. *Society Aff:* AIChE, HESS.

Norris, Rollin Hosmer
Home: 35 Front St, Schenectady, NY 12305
Position: Consulting Engr (Self-employed) *Employer:* Self *Education:* MSME/Mech Engr/Harvard Univ; BS/Math/Harvard Univ. *Born:* Aug 11, 1906. Engr G E Co Schenectady NY 1932-71 (retirement). Since then, engaged part-time in cons engrg, mostly for G E Co. Grad of G E Co Advanced Engrg Prog 1935. Respon at corp level for heat transfer res & dev work & cons engrg, & problems related thereto, from 1937-71. This work included a guided missile proj; wrk on the Manhattan Dist Proj; work on reliability problems. ASME Pi Tau Sigma Medal 1941; ASME Heat Transfer Div Mem Award 1966; G E Co 'Charles Coffin Award' 1950. Life Fellow ASME. *Society Aff:* ASME.

Norris, Roy H
Business: 1845 Fairmount, Wichita, KS 67208
Position: Chairman Electrical Engrg *Employer:* Wichita State Univ *Education:* PhD/EE/OK State Univ; MS/EE/Univ of Wichita; BS/EE/Univ of Wichita; AA/EE/Parsons Jr Coll *Born:* 4/13/30 Scammon, KS. 1951-53 Supervisor of Production Planning at the Kansas Ordnance Plant, Parsons, KS. US Army 1953-55. Wichita St Univ 1959-present time. Presently am Prof and Chairman of the Electrical Engrg Dept. Also serve as Co- Director of the National Rehabilitation Engrg Center at Wichita, Kansas. Have consulted for the Boeing Co and the Cardwell Co, Wichita, KS. Specializing in Avionics. Received National Science Faculty Fellowship, 1966, The Regents Award for Excellence in Teaching, 1966, Labette Comm Coll Cardinal Citation, 1980, and the Coll of Engrg Outstanding Educator Award, 1981. *Society Aff:* IEEE, AOC, RESNA

North, Edward D
Home: Rt 3, Brooksview Rd, Lenoir City, TN 37771
Position: Retired *Education:* PhD/ChE/Univ MI. *Born:* 10/30/18. Responsible for process systems & equip of the Hot Experimental Facility for reprocessing fast breeder reactor fuels. Formerly Mgr of Engg & Construction, Nuclear Fuel Services, Inc; responsible for the expansion of the West Valley reprocessing plant. Developed the press without binder process for the mfg of nuclear fuel pellets; the std process used in the industry. Built and qualified the plant that furnishes virtually all of the fuel for nuclear powered naval vessels. Chrmn, Nuclear Engg Div, AIChE, 1977. Chrmn, Inst of Nuclear Mtls Mgr, 1964. Fellow, AIChE. *Society Aff:* AIChE, ACS.

North, Forrest H
Home: 2449 Wild Valley Dr, Jackson, MS 39211
Position: President. Retired. *Employer:* North, Beasley & Swayze P A. *Education:* BS/Mech & Elec/LA Tech Univ. *Born:* Mar 1920. US Navy, Lt Cdr - retired. Cons engr in field of commercial & inst bldgs since 1947. Pres of North, Beasley & Swayze since 1970. (Now retired.) Field of specialization is hospital & medical sys. Currently mbr of St Bd of Engrg Regis. P Pres Miss Engrg Soc, P Natl Dir NSPE. PPres MS Consulting Engr Council, P Natl Dir Elect CEC. Engr of Yr Award 1972. Mbr ACEC, ASHRAE (Fellow). Other ints incl golf, antique furniture reproductions. *Society Aff:* ASHRAE, NSPE, CEC, NCEE.

North, Harper Q
Home: 12440 Surrey Circle Dr, Fort Washington, MD 20022
Position: Assoc Dir of Res for Technical Services. *Employer:* Naval Res Lab. *Education:* PhD/Physics/Univ of CA, Los Angeles; MA/Physics/Univ of CA, Los Angeles; BS/Sci/Cal Tech. *Born:* Jan 1917. 1940-49 Res Assoc G E Res Lab. 1949-54 Dir Semiconductor Div Hughes Aircraft. 1954-62 Founder & Pres Pacific Semiconductors Inc. 1962-69 V P R&D TRW Inc. 1969-75 Mgr (& cons) Electro-Optical Dept, Northrop Corp. 1975-79, Assoc Dir of Res for Electronics Naval Res Lab. 1979-Assoc Dir of Res for Technical Services Naval Res Lab. 2-yr Chmn of Electronic Indus Assn (EIA), receiving the EIA Medal of Honor 1966. Sev yrs Chmn of Dept of Def (DDR&E) Adv Group on Electron Devices. Fellow IEEE & Amer Phys Soc. *Society Aff:* IEEE, APS.

North, James C
Business: 110 Westwood Place, P.O. Box 1848, Brentwood, TN 37027
Position: Pres *Employer:* Resource Consultants, Inc. *Education:* MS/San. Engr/Vanderbilt Univ; BE/CE/Vanderbilt Univ *Born:* 12/30/39 Since 1969, President, Resource Consultants, Inc. Brentwood, TN. multi- disciplined environmental engrg firm specializing in industrial pollution control and waste-water recovery/recycling. With complete project capabilities in sampling analysis, treatability studies, pilot plant studies, design and contract mgmt, enviromental impact statements and assesments, industrial hygiene and safety services. Registered PE in seven states. Involved for over 10 years in innovative recovery projects nationwide. Has sponsored computer software research for OSHA Hazard Communication Mgmt. Currently working on research and development of waste minimization for surface finishing industry. Born in Aurora, Ill. Married four boys, one girl, active in civic work. *Society Aff:* ACEC, ASCE, WPCF, AIChE, AES, AAEE

North, John R
Home: PO Box 70188, Sunnyvale, CA 94086
Position: Pres & Exec Engr - ret. *Employer:* Commonwealth Assocs Inc. *Education:* BS/EE/CA Inst Technology. *Born:* Feb 20, 1900 Cambridge Mass. Tau Beta Pi. Whse E&M Co 1923-24. Commonwealth Assocs Inc (cons & design engr) & predecessor co Jackson Mich 1924- 63. Genl Tech Engr to 1938; Asst EE to 1945; Ch EE to 1949; V P, Dir & Ch EE to 1956; Dir & Exec Engr to 1958; Pres, Dir & Exec Engr to 1963. V P, Dir Commonwealth Services Inc NY 1952-63. Reg P E Ill, Ind, Mich, NY, Ohio, Penn. Dir Natl Bank Jackson Mich 1960-63. IEEE AM 21, M29, F41; Dir 1945-49; V P 1949- 52; Ch Admin, Tech & Stds Cttes. ASA Stds Council 1938-60, Achieve Award 1953. EEI Bell Sys D&R Comm 1930-63; Co Chmn Plan Coord Comm 1950-63. Mbr A Mgmt Assn NSPE, MSPE, MES, SAE, Engrg Club S F. *Society Aff:* IEEE, NSPE, TBΠ.

Northouse, Richard A
Business: 7110 W Fonddulac Ave, Milwaukee, WI 53218
Position: Pres *Employer:* COMPCO *Education:* PhD/EE & CS/Purdue; MS/CS/Purdue; MS/EE/WI; BS/EE/WI. *Born:* 4/2/38 Has been active in Engrg for over 25 yrs. He has authored over 40 papers in Engrg Journals. During the 1970's, Dr. Northouse was active as a Univ Prof. Since he has also become involved in industry as the Pres of COMPCO, a retail store in Milwaukee, and as pres of Northouse Industries a computer software manufacturer and developer. He presently also does consltg to the industrial & agricultural communities. *Society Aff:* IEEE, ASM, PRS.

Northup, Larry L
Business: 112 Marston, Ames, IA 50011
Position: Prof of Freshman Engrg. *Employer:* Iowa State Univ. *Education:* PhD/Me-Aero E/IA State Univ; MS/Aero E/IA State Univ; BS/Aero E/IA State Univ. *Born:* Aug 1940. Native of Audubon Iowa. Reg Mech Engr Iowa. Sr Mbr Soc of Mfg Engrs. Taught in Aerospace Engrg Dept at Iowa State Univ 1963-73. Was a Mfg Engrg Proj Dir for Bourns Inc Ames Iowa 1973-74. Returned to Iowa State Univ as Assoc Prof of Freshman Engrg 1974. Have served as as Asst in the Engrg Dean's Office 1975-78. Dow Outstanding Young Fac Award, North Midwest Sect ASEE 1976 Promoted to Prof 1980. Elected Director-Engr Design Graphics Division, ASEE 1981. Co-Author " Engineering Fundamentals & Problem Solving" McGraw Hill 1979, "Engrg Graphics Fundamentals" McGraw Hill 1985. Served as Chrmn-Freshman Prog Constituent Ctte, ASSE 1983-84. *Society Aff:* ASEE, SME, NACADA.

Norton, David J
Business: Texas A&M Univ, College Station, TX 77843
Position: Prof *Employer:* Texas Engrg Experiment STA *Education:* PhD/ME/Purdue Univ; MS/ME/TX A & M Univ; BS/ME/TX A & M Univ *Born:* 10/23/40 Manhattan, KS. Travelled in Europe 1947-1954 with parents. Distinguished military graduate BS, MS, Texas A&M. Commission 2nd Lt. 1963. Married Concetta Galceron 1964. Active Duty Jet Propulsion Lab 1968-1970. Leaving as Capt. US ordinance Corps. Joined Texas A & M Univ 1970 as Asst. Prof, Assoc Prof 1974, tenure 1975, Prof 1980. Currently asst director for res and full prof of Aerospace Authored 20 papers and 30 reports. *Society Aff:* ASME, AIAA, ASEE, SNAME

Norton, John A
Home: 659 Grove St, Norwell, MA 02061
Position: Hydrodynamicist *Employer:* Bird-Johnson Co *Education:* MTh/Theol/Gordon Conwell Seminary; BS/Naval Arch & Marine Engr/MIT *Born:* 2/20/39 Professional career has included work with Central Technical Dept, Bethlehem Steel Co. From 1965 to 1977 with J. E. Bowker Assoc, Inc, consulting naval architects. Since that time with Bird-Johnson Co. As Hydrodynamicist/Senior Staff engr, am responsible for design of ship propeller and maneuvering devices, and for application of computer technology in the Engrg Dept. Chrmn, SNAME, New England Section, 1980/81. Reg PE, MA. Mbr SNAME panel H-8 Propeller Unsteads Hydrodynamics. *Society Aff:* SNAME, ASNE.

Norton, John O
Home: 7106 Jones Valley Dr, SE, Huntsville, AL 35802 *Employer:* Self *Education:* BS/Civ Engg/MS State Univ. *Born:* 12/12/24. Guntown, MS. Served in Pacific Theatre in Army during WWII. Attended OH Univ & received BS in CE from MS State Univ. Mbr of Tau Beta Pi. With Memphis Dist Corps of Engrs, 1949-1953. Served in Engr Avn BN in Alaska during Korean War. Joined Army Ballistic Missile Agency, Redstone Arsenal, AL, in 1956 performing construction mgt for missile systems, through 1973. 1974-1980, served as Chief, Master Planning Construction & Environmental Office, Army Missile Command, supporting missile systems res & dev & fielding of missile systems. Served in various offices of ASCE, including Pres of AL Sec, 1978-79. Now retired from Government service. *Society Aff:* ASCE, SAME.

Norton, Kenneth A
Home: 323 Bellevue Dr, Boulder, CO 80302
Position: Cons on radio & political sci (Self-employed) *Education:* BS/Physics/Univ of Chicago. *Born:* Feb 1907. Columbia Univ. Western Elec Co; Natl Bureau of Standards; Fed Communications Comm; Operational Res Group in Pentagon; Operational Res Sect in England; cons in political sci; demonstrated the irrelevance of the principle of majority rule & of Arrow's impossibility theorem. Estab in the Internatl Radio Consultative Ctte the principle that optimum use of the radio spectrum is achieved only when interference from other users rather than from noise represents the ineluctable limit to satisfactory reception. Fellow 4 orgs; recipient 4 awards incl Stuart Ballantine Medal, Franklin Inst. *Society Aff:* IEEE, APS, IMS, ASA.

Norton, Robert G
Business: Distrigas, 2 Oliver St, Boston, MA 02109
Position: V Pres, Engineering & Construction *Employer:* Distrigas Corp. *Education:* BS/Chem Eng/MIT. *Born:* July 24, 1925 Idaho Falls Idaho. Infantry Service 1943-46. Joined Cabot Corp 1951 as R&D engr. 1958 plant mgmt. 1961 internatl prod mgmt. 1968 Genl Mgr Cabot Engrg iv. 1971-1975 and 1979-present V P Engrg & Const for Distrigas Corp (Cabot LNG subsidiary). 1976 V P Mkting & Supply for Distrigas. P E Texas, La, Mass. 1980 Chairman AGA LNG Committee. 1983 U.S. representative on International Gas Union (IGU) LNG Committee. *Society Aff:* AIChE, AGA.

Norwood, Sydney L
Business: P O Box C-190, Birmingham, AL 35283
Position: Regional Mgr *Employer:* Republic Steel Corp. *Education:* BS/Chem/Bimingham So Coll *Born:* 9/18/28 Native of B'ham Ala. Served with US Navy in Far East 1950-1954. Chief Metallurgist Rudisill Fdry Co., 1957-1958; Div Metallurgist & Chief Chemist, Chief Steel Production Metallurgist, Fairfield Works, US Steel 1958-1969. Chief Metallurgist Texas Works, US Steel 1969-1980; Mgr Special Bar Products, Judson Steel Corp. 1980-1981; Regional Mgr-Southern, Tech Service-Republic Steel Corp. 1981- 1981-1982 Director at Large, Iron & Steel Society, AIME. 1974-1975 Chairman TX Section-AISE. 1976 Director-AISE. 1971-1972 Chairman SW Section AIME. 1968-1969 Chairman SE (AL) Section AIME. 1964-65 Chairman Birmingham Chapter ASM. 1972-73 Director Houston Chapter ASM. *Society Aff:* AIME, AISE, ASM

Noton, Bryan R
Business: 505 King Ave, Columbus, OH 43201-2693
Position: Research Leader *Employer:* Battelle Memorial Institute *Education:* MS/Aircraft Design/Cranfield Inst of Technology, England. *Born:* 5/14/28. Student-apprentice Bristol Aeroplane Co & Bristol Tech College. MS (Distinction in Design) Cranfield Inst of Tech. Res Engr; later Tech Asst to Dir. Aeronautical Res Inst of Stockholm Sweden 1951-63. Dir of New Products, Matls & Structs Group, Whittaker Corp Los Angeles. Visiting Prof, Aerospace Design, Stanford Univ, & Prof of Mech & Civil Engrg, Washington Univ St Louis. Deputy Dir, Materials and Structures, AIAA. Presented with AIAA Structures, Dynamics and Materials (SDM)

Noton, Bryan R (Continued)
Award (1986). AIAA Representative to U.S. National Committee for Theoretical and Applied Mechanics. Elected: Fellow (1983), SAMPE, Fellow (1986) RAeS, Fellow (1987), I.Mech.E. Specialties incl: technoeconomic forecasts of emerging matl applications, mfg cost/structural performance trade-studies, innovative design of metallic and composite lightweight structures to minimize life-cycle costs. *Society Aff:* AIAA, SAMPE, RAeS, IMechE

Notowich, Alvin A
Business: Plastico-POB 12183, Memphis, TN 38182-0183
Position: Vice President. *Employer:* Notowel Corp dba PLASTICO. *Education:* BE/Chem Eng/Vanderbilt Univ *Born:* Dec 1942. 1964-65 Tech Service Engr for a co in plastics cons, fabrication & distrib. 1965-70 PPG Indus Inc, Coatings & Resins Div, as Prod Engr & Oper Planning Services Supr. 1970 estab PLASTICO, an indus plastics distrib, fabricating & engrg co - V P & principal in charge of sales, mkting, & tech service. 1971 Memphis Jt Engrs Soc Achieve Award for 'Engr of Yr' from AIChE. 1971 Bd of Dir Memphis Sect SPE. 1972-73 Pres Memphis Sect AIChE. 1974 Mbrship Award for Mbrship Efforts in Memphis Sect AIChE. *Society Aff:* AIChE, ACS, SPE

Nouvion, Fernand F
Home: Saint-Amand 27, Paris, Seine 75015 France
Position: Retired *Born:* Nov 1905 France. Engr Ecole Superieure Electricite 1927. With French Natl Railroads 1931-70, in charge of elec diesel gas turbine rolling stock - original design for high speed motive power - world railroad speed record (205 mph) 1955. Pioneer in AC traction indus frequency. Prof Ecole Superieure Electricite. Tech Adv of India for Railroad Electrification. Now, Tech Mgr Traction-Export. Commandeur French Legion d'Honneur 1975; Prix Coignet 1948; Grand Prix 1968 Ingenieurs Civils France; Premium Paris 1881 IEE United Kingdom; Medal City of Paris; Giffard Premium 1965 French Acad of Sci; IEE Fellow. Enjoy tennis & bicycling. Life Fellow Medal and Diploma Pan American Railway Congress 1981. William Alexander Agnew Reward from The Institution of Mechanical Engineers United Kingdom 1987. *Society Aff:* IEEE, SISF, SEE, RCF.

Novacek, Charles,
Home: 22295 Maplewood, Southfield, MI 48034 *Employer:* Retired *Education:* BS/Mech & Civil/State Indus College of Enggs. *Born:* May 1928 Czechoslovakia. Grad from State Indus Coll of Engrg, Brno Czech. Reg P E Mich. Mbr ESD, ACI, NSPE, Dir of CIB, recipient of 1973 ACI Const Practice Award. MSPE Distinguished Service Award. Proj Mgr 1960-74 with Darin & Armstrong Inc. Const Mgr 1974-85, with Barton Malow Co. Enjoy classical music, painting, golf, gardening & stone carving. *Society Aff:* NSPE/MSPE, ASCE, ACI, CIB, ASQC.

Novak, Al V
Home: 55 Wakerobin Place, Hendersonville, NC 31522
Position: Tech. Director *Employer:* Crawford Filling Co. *Born:* Sept 1918. B of Sci in Chem Engrg Univ of Nebraska. Mbr ISA, AIChE, AAAS & ASME, NACE & Fellow in ISA, AAAS, & Founding Fellow at Bowling Green Univ. 30 yrs with E I duPont in var automation & mgmt pos incl plant mgmt. 6 yrs Fluor Engrs & Constructors on Arabian Gas Gathering Sys. 3yrs with Crawford Fitting Co as Technical Dir.

Novak, James M
Home: 160 Charlesworth, Dearborn Hts, MI 48127
Position: Prin Staff Engg *Employer:* Ford Motor Co. *Education:* PhD/Aero Engg/Univ of MI; MSE/Aero Engg/Univ of MI; BSE/Aero Engg/Univ of IL. *Born:* 9/1/44. in Chicago area; lived in Tucson, AZ, 1956-64. Res Engg (Ballistic Missile) Penetration Aids Dept of Conductron Corp, McDonnell-Douglas Aircraft, 1969-71. Joined Ford Motor Companys Res Staff 1973. Career has involved analysis and compter simulation in combustion, fluid dynamics, and heat trasfer. Currently interested in prediction of performance and exhaust emissions characteristics of internal combustion engines. Edmund J James Scholar, grad with Highest Honors, Univ of IL. Mbr Tau Beta Pi and Sigma Gamma Tau Engg Honorary Societies. SAE Arch T Colwell Award 1979. Married, three children. Private Pilot. Active in community youth and recreational sports activities. *Society Aff:* SAE, CI, ΣΞ.

Novak, John T
Business: Dept of Civil Engrg, Blacksburg, VA 24061
Position: Prof *Employer:* Va. Polytech Inst and State Univ *Education:* PhD/Sanitory Eng/Univ of WA; MS/CE/Univ of MO; BS/CE/Univ of MO *Born:* 1/24/42 Prof of Civil Engrg at U of MO-Columbia for past 12 yrs. Recently joined facility of VA Tech in Civil Engrg. Published about 40 papers dealing with water and wastewater treatment and sludge disposal. Past chairman of Student Acitivities Committee, AWWA. Past Pres- MO. Water Poll'n Control As'sn. Received the following awards: Hering Medal-ASCE 1979, Collingwood Prize-ASCE 1976, Best Paper Water Qual. Div. AWWA 1979, Nalco Award AEEp (Best Thesis-Advisor)-1979. Bedell Award-WPCF 1979. *Society Aff:* ASCE, WPCF, AWWA, ASLO, AEEP

November, William E
Business: 1 Lincoln Plaza, New York, NY 10023
Position: Pres. *Employer:* Self. *Education:* BS/IE/Syracuse Univ *Born:* 2/3/24. Specialize in analysis of construction & renovation projs as to best method of construction, economics, dev. Financing methods. Published in Civ Engr Journal; Consultant to Institutions and Private Owners and developers. Public Service with the Shield Inst for Retarded Children - new facilities & stds of design; Past mbr ASQC Bldg Mtl & Construction Technical Committee; Deans Staff Syracuse Univ; Tau Beta Pi; Served with AF 1943-46. Licensed PE, Licensed Profl Planner, Real Estate Broker. *Society Aff:* ASQC, NSPE

Novick, David A
Business: 1500 Walnut St, Philadelphia, PA 19102
Position: VP and Area Mgr *Employer:* Parsons, Brinckerhoff, Quade & Douglas, Inc *Education:* MS/Civil Engr/Columbia Univ; BS/Civil Engr/Columbia Univ *Born:* BS & MS Columbia Univ. Native of NYC. US Army 1944-46. Was foundation & soils engr for Tippetts-Abbett-McCarthy-Stratton NYC; Exec V P & Proj Mgr for Goodkind & O'Dea Inc Hamden Ct & Chgo Ill. Founder and Presdent, Westenhoff & Novick 1960-76. Adjunct Prof, Univ of IL, Chicago Circ 1977; Exec VP LB Knight & Assoc, 1978-1981. Mbr & Secy Ill Struct Engr Examining Ctte; Tr North Suburban Mass Transit Dist; Lectr Hwy Engrg IIT; Tr Western Soc of Engrs. Prof exper incls a wide range of projs incl: transp sys, bridges, airports, hwys, railroads, transit sys, tunnels, waterfront facils & dams. Assigments included projs in such diverse locations as Chicago, NY, Philadelphia, Atlanta, ME, Puerto Rico, Alaska, Turkey, El Salvador. Space int in structural, geotech engrg and transportation planning. *Society Aff:* ASCE, TRB, ASTM, NYSE/PSPE, CEC

Novo, A, Jr
Business: 207 Santillane Ave, Coval Gables, FL 33134
Position: President. *Employer:* Prof Assoc Cons Engrs Inc (PACE). *Education:* BS/EE/Univ of Miami, FL. *Born:* 6/13/39. Cons engg field in S FL past 20 yrs. Pres PACE Inc P Pres SE FL Sect IES. P VP Greater Miami Chap CSI. Mbr NSPE, FICE, FES, IES, ACEC, ASHRAE, AIHE, MENSA, LBA, P Pres to Alumni Bd of Sch of Engg & Arch. Univ of Miami, Natl Dir IES P Chmn Natl Lighting Design Awards Ctte IES. Instr for IES Advanced Lighting Design Course, Arch Exam Bd Course, Dale Carnegie Public Speaking & Human Relations course. Reg PE AL, GA, FL, LA, TX, MD, NJ, DC, NCEE; Mbr Bd of Dirs- Univ of Miami Gen Alumni Assoc; Engr of the Year 1983 and 1984 by Latin Builders Assoc. *Society Aff:* IES, NSPE, FICE, ACEE, MENSA.

Novotny, Donald W
Business: ECE Dept. 1415 Johnson Drive, Madison, WI 53706
Position: Professor *Employer:* University of Wisconsin *Education:* PhD/Elec. Engr./University of Wisconsin; MS/Elec. Engr./Illinois Institute of Technology; BS/Elec.

Novotny, Donald W (Continued)
Engr./Illinois Institute of Technology *Born:* 12/15/34 Faculty mbr, Univ of Wis - Madison since 1961. Currently Professor, co-director of Wisconsin Electric Machines and Power Electronics Consortium and Associate Dir of Univ-Industry Research Program. Served as ECE Dept Chairman, 1976-80. Active as a consultant to many organizations and a Visiting Prof at the Eindhoven Technical Univ, Netherlands, the Catholic Univ of Leuven, Belgium, and a Fulbright Lecturer at the Univ of Ghent, Belgium. Author of textbook and over sixty technical papers. Recipient of two teaching awards and five technical paper awards. Research interests include electric machine theory and power electronics. *Society Aff:* IEEE, AFS, ΣΞ

Nowick, Arthur S
Business: 1144 Mudd Bldg, New York, NY 10027
Position: Professor of Metallurgy. *Employer:* Columbia Univ. *Education:* PhD/Physics/Columbia; BA/Physics/Brooklyn College; MA/Physics/Columbia Univ *Born:* 8/29/23. Held academic pos at Univ of Chicago, Yale Univ & at present (since 1966) at Columbia Univ. Held indus pos at IBM Yorktown Res Ctr 1957-66. Publ over 125 papers & 3 books in fields of crystal lattice defects, anelasticity, metastable alloys, & elec properties of ceramics. Cons to indus & gov. Fellow Amer Phys Soc; Fellow AIME; Mbr Amer Cer Soc. Awards from Amer Soc for Metals & Univ of MO. *Society Aff:* APS, AIME, ACerS

Nowicki, George L
Home: PO Box 36, Worcester, PA 19490
Position: VP Domestic Operations. *Employer:* Quaker Chem Corp. *Education:* MChE/-/NYU; BChE/-/CCNY. *Born:* 12/4/26. Since 1984, and currently VP Operations for Quaker Chemical Corp. In 1981, took the position of Mng Dir of Quaker Chem (Holland) BV, and Corp VP of Quaker Chem Corp. Mbr of Bd of Dir of Sebby-Battersby Co. Began in 1979 as Pres of Selby Battersby, a subsidiary of Quaker Chem Corp, while continuing as VP of the parent co. At Quaker, held the position of VP Mfg for 9 yrs. Prior to this I headed the chemical mfg operations of Burroughs Welcome Co in Tuckahoe, NY for 18 yrs. Lic PE in NY & PA. Mbr & Pres of Central Dist No 7 Sch Bd in Hartsdale, NY from 1960-69 & Pres of the Westchester County Sch Bds Assn from 1964-66. *Society Aff:* AIChE.

Nowlin, I Edward
Home: 4851 S 68th E Ave, Tulsa, OK 74145
Position: Mgr, Prod Engrg & Mfg Engrg. *Employer:* Dover Corp/Norris Div. *Born:* b Dec 1922. BSME Texas A&M. Army Air Corp, 9th Air Force 1942-45. After college, joined Temco Aircraft as tool engr 1951-52. Began working with oil field equip manufacturers in Houston 1954-69 - Cameron Iron Works Tool Engr WKM as Ch Tool Engr & Mission Mfg as Ch Tool Engr. Joined Dover Corp/Norris Div 1969 as Mgr Mfg Services & became Prod Engr 1974. SME Dir from Region V 1976. P Chmn Region V 1972 & Tulsa Chap 90 1975. Enjoy camping, golf & bridge.

Nowlin, R L
Home: PO 2648, Minden, NV 89423
Position: Consultant *Education:* BS/Chem Engg/Univ of TX. *Born:* Sept 30, 1921 Hillsboro Texas. Tau Beta Pi, Phi Lambda Upsilon, Omega Chi Epsilon. Mbr AIChE, NSPE, Amer Soc of Enologists. 1942-43 Chem engr Dow Chem, Freeport Tex. 1943-46 US Naval Reserve Radar Maintenance Officer. 1946 Chem engr Carbide & Carbon Chems, Texas City. 1947-48 Chem engr Schenley Distillers Wine Div, Fresno Calif. 1948-52 Ch Engr Oscar Krenz Inc, Berkeley Calif; Genl Mgr 1952-55. 1956-61 Ch Engr E & J Gallo Winery, Modesto Calif. 1961-68 V Pres L & A Engrg & Equip Inc; 1968-1984, Pres (Turlock Calif). Principal int: food & beverage processing & process equip design. Reg P E Calif in Chem E & Mech E; also regis in Ga. 1985 to present, Consultant. *Society Aff:* AIChE, ASE, IFT.

Nowowiejski, Alfred J
Home: 2402 W 5075 S Roy, Roy, UT 84067
Position: Deputy Base Civil Engr. *Employer:* USAF. *Born:* Oct 7, 1924. BS Univ of Colo 1947. Native of Denver Colo. USMCR 1942-45. Field engr USBR 1947-58; USAF, Genl Engr 1958-61, var assignments incl ballistic missile const; 1961- , Deputy Civil Engr, Hill AFB. Held pos of Pres, V P, Treas in Amer Inst of Plant Engrs, Chap 109. 1970- , Reg P E Cert 18695 Mass. Enjoy hunting, fishing & camping.

Noyce, Robert N
Business: 3065 Bowers Ave, Santa Clara, CA 95051
Position: Vice Chairman *Employer:* Intel Corp. *Born:* Dec 1927 Burlington Iowa. BA Grinnell Coll 1949; PhD MIT Physics 1953. Philco Corp 1953-56; Shockley Transistor Corp 1956-57. Cofounder of Fairchild Semiconductor 1957. V P Fairchild Camera until 1968. Co-founder Intel Corp 1968 - Pres 1968-75, Chmn 1975-1979, Vice Chairman since 1979. Work in transistor & integrated circuit design & processes. Co-inventor in integrated circuit. Fellow IEEE. NAM Modern Pioneers in Creative Indus Award 1965. Stuart Ballentine Medal 1966. Mbr Natl Acad of Engrg. Tr Grinnell Coll 1962- . Visiting Ctte MIT, Stanford. Awards: IEEE Medal of Honor; Cledo Brunetti Award; Harry Goode Award by AFIPS;-1978, 1979-I.E.E. Faraday Medal of Honor; National Medal of Science 1979; Harold Pender Award by University of Pennsylvania-1980; Natl Medal of Technology-1987; Membership: Member American Academy of Arts & Sciences.

Noyes, Jack K
Home: 2008 Ravinia Circle, Arlington, TX 76012
Position: Project Weight Engineer *Employer:* Bell Helicopter Textron *Education:* BBA/Accounting/TX Tech. *Born:* Mar 1925 Kansas City Mo. Grew up in Midland Tex. US Naval Acad Congressional Apptment 1945. Studied engrg at Texas A&M & Texas Tech with BBA in 1949 - Class Prs. Taught & coached for 3 yrs. Bell Helicopter since 1952 as Wt Engr. Joined Soc of Allied Wt Engrs 1952. Have worked on local & internatl levels. Previously candidate for Pres & Exec V P. Dir for North Texas for 4 years. Editor of chapter publ - Lone Star Wt Engr for 2 yrs. Best new SAWE publ first yr & best publ the second. SAWE Operations Manual Chairman for 3 years. Enjoy bowling, water sports & flying. *Society Aff:* SAWE.

Noyes, Jonathan A
Home: PO Box 4008, Bryan, TX 77801
Position: Ret. (Consulting Engr) *Education:* BS/-/MIT. *Born:* Dec 25, 1889 Waltham Mass. VP of the Amer Soc of Mech Engrs & Mbr of Council, N Y. Fellow Amer Soc of Mech Engrs. Mbr Amer Inst of Mining & Met Engrs. Legion of Honor. Pres Duluth Engrs Club, Duluth Minn. Pres Minn Unitarian Conf, St Paul. Pres Southwest Unitarian Conf, Dallas. Amer Unitarian Assn Dir, Boston. Wife Caroline Howard Clark Noyes, Middlebury 1909. 3 sons: Baldwin Dow, Jonathan Howard, Theodore Alvan. 3 daughters: Lillian Frances, Cornelia Bonner, Priscilla Ruth Crosson. *Society Aff:* ASME, AIME.

Noyes, Robert E
Business: Mill Rd at Grand Ave, Park Ridge, NJ 07656
Position: President. *Employer:* Noyes Data Corp. *Education:* BSChE/Chem Engr/Northwestern Univ. *Born:* June 22, 1925. Pres & Publisher of Noyes Data Corp & Noyes Publications, publishing 40-50 tech & engrg books per yr; this firm founded in 1959. Author tech books on Chemicals, food, energy, etc. Publisher - Noyes Press (classics, archaeoloogy). Previous affiliations - Natl Lead Co, U S Indus Chem Co, Curtiss- Wright Corp. Hobbies: archaeology, naval history, art, sailing. *Society Aff:* AIChE, ACS, CAA, AIA, ANWA, ACS, SAMPE, MRS, AIAA

Noyes, Ronald T,
Business: Associate Professor, Agricultural Engineering Dept, Oklahoma State University, Stillwater, OK 74078-0497
Position: Extension Agricultural Engineer *Employer:* OSU *Education:* MSAE/Agri Engg/OK State Univ; BSAE/Agri Engg/OK State Univ. *Born:* Jan 1937 Leedey, Okla. 1st Lt, CE, AUS 1961-63, Cap, USAR, 1966. Ext Agri Engr Purdue Univ 1964-68; Ch Engr & VP., Engrg., Beard Indus Frankfort Ind 1968-1985. Ext. Agr.

Noyes, Ronald T, (Continued)
Engr. Okla State Univ. 1985- . Reg P E Ind, OK, IL, IA; Mbr Amer Soc of Agri Engrs (Chmn, Ind Sect ASAE, 1973); Aircraft Owners & Pilots Assn (comm., instr. & M.E. ratings); Alpha Zeta, Blue Key, Sigma Tau, Farmhouse (social), Kiwanis, Toastmasters. Presbyterian (Chmn Deacons 1976, Elder, 1979). ASAE Young Designers Award 1975; Who's Who in the Midwest 1975. Nat. Society of Prof. Engrgs. Patentee with U.S. Grain Dryer Patents issued in 1972, 1978, 1980, 1981, 1982, 1983. Enjoys flying, fishing, golf, hiking, canoeing, camping, music, bridge. Mbr Frankfort Sch Bd 1976-1982 (Pres 1978-79). Dev 26 continuous flow dryer models, 1979-1981. Member, Board of Directors, Carnegie International Corp. (1979-83), Carnegie Capital Corp, Financial Planning Services Corp, Carnegie Life Insurance Co, Carnavest, Inc, 1981-1983; Chairman, Carnegie Board of Advisors (CBA) and CBA Exec Cabinet, 1979-81. *Society Aff:* AOPA, ASAE, APRES, ASEE, NAAA.

Nuccitelli, Saul A
Business: 122 Park Central Sq, Springfield, MO 65806
Position: Pres. *Employer:* Saul A Nuccitelli, Inc. *Education:* CE/Civ Engg/MA Inst of Tech; MCE/Civ Engg/NYU; BCE/Civ Engg/NYU. *Born:* 4/25/28. Yonkers, NY. First Lt, Corps of Engrs 1951-53. Asst Prof & Res Engr, Univ of Denver, 1955-58, Cryogenic testing on metals for Titan Missile and satellites. Asst-Assoc Prof, the Cooper Union, 1960-62. Staff of MIT, 1958-60, Thesis on Dynamic Loading of Reinforced Concrete Columns. Pres, SA Nuccitelli, Inc, Consulting Engrs and Architects, 1962-present. Projs include $8 million viaduct reconstruction, $3 million dormitories. MO Consulting Engr of the Yr Award 1973. Certificate of Appreciation, 1979, by American Consulting Engrs Council, on Reconstruction of Viaducts. Pres, Ozark Chapter MSPE, 1977-78; Chrmn of the Bd-City Util, Springfield, MO; Advisory Council-MO Public Drinking Water; Rec of Cert of Appreciation, MO Municipal League, 1981. *Society Aff:* NSPE, ASCE, ΣΞ, ASTM.

Nudelman, Sol
Business: AZ Health Sciences Ctr, Tucson, AZ 85724
Position: Professor. *Employer:* Univ of Arizona. *Education:* PhD/Physics/Univ of MD; MS/Physics/IN Univ; BS/Physics/Union College of Schenectady & NY. *Born:* Aug 1922. Native of NYC. US Navy 1943-46. Instr Union Coll 1948, Knox Coll 1949-51. Res physicist Naval Ordnance Lab 1951-56, Univ of Mich 1956-61. Mgr Res for Solid State, IITRI, 1961-64. Prof Univ of R I 1964 -73, Univ of Ariz 1973- .Sigma Xi, Phi Kappa Phi, Sigma Pi Sigma. Governor SPIE, Sr Mbr IEEE, Fellow AAAS, Mbr Amer Phys Soc Res, pubs & pats: luminescence, semiconductors, infrared sensors, displays, imaging devices & their applications to diagnostic medicine. *Society Aff:* SPIE, IEEE, APS, AAAS.

Nummela, Walter
Business: 1100 Superior Avenue, 19th Floor, Cleveland, OH 44114-2589
Position: Mgr. Technical Services *Employer:* Cliffs Mining Company *Education:* ScD/Metallurgy/MA Inst of Tech; MS/Metallurgical Engg/Univ of MN; BMetEng/Metallurgical Engg/Univ of MN; -/Engg/Hibbing Jr College. *Born:* 12/6/29. in Chisholm, MN; attended the local schools and neighboring Hibbing Jr College. Metallurgical Engr for the Cleveland-Cliffs Iron Co developing processes for low grade iron ores. Returned to sch for an ScD; iron and steelmaking res for Youngstown Steel. Rejoined Cleveland-Cliffs with assignments in operations and res, currently Mgr of Technical Services. In 1977, Chrmn of the Mineral Processing Div of the Soc of Mining Engrs (SME) of AIME; past Dir of SME. Ad-Hoc Accreditation Committee for ECPD. Enjoy being with my family and outdoor activities. *Society Aff:* AIME, AISI.

Nunnally, Stephens W
Home: 474 St. Lucia Court, Satellite Beach, FL 32937
Position: Consulting Engineer *Employer:* S. W. Nunnally *Education:* PhD/CE/Northwestern Univ; MS/CE/Northwestern Univ; BS/Military Eng/US Military Academy *Born:* 11/30/27 Army Corps of Engrs (2nd Lt to Lt Col) 1949-70, Legion of Merit. Taught at Univ of FL 1971-75. Prof of CE N Carolina State Univ 1975-84; Professor Emeritus 1984. Author of CONSTRUCTION METHODS AND MANAGEMENT (1987, 1980) and MANAGING CONSTRUCTION EQUIPMENT (1977). Consultant to industry and research organizations. PE and Certified Gen Contractor. Chairman, Construction Eng Committee ASEE 1980-81. Construction Research Council ASCE 1976-. Energy Advisory Council ARTBA 1980-86. Chairman Cttee on Construction Equipment and Techniques ASCE 1985-. *Society Aff:* ASCE, ASEE, SAME

Nunziato, Jace W
Home: 6300 Montgomery, Apt 338, Albuquerque, NM 87109
Position: Dept Mgr *Employer:* Sandia Labs. *Education:* PhD/Appl Mechs/WVA Univ; MS/Appl Mechs/WVA Univ; BS/Aero Eng/RPI. *Born:* May 1942 Brooklyn N Y. BAE (1964) from RPI, MS (1966) and PhD (1969) from WVU. While at WVU, held NASA Traineeship & elected to Sigma Xi. From 1969 to 1980, mbr of the tech staff of Sandia Labs, spec in res on wave propagation in chemically reacting solids & viscoelastic solids. From 1980 to 1983, supervisor of Fluid Mechanics and Heat Transfer Div I. Since 1983, Mgr of Fluid and Thermal Sciences Dept. Author or coauthor of more than 70 tech papers & book articles, incl article in the Handbuch der Physik. Pi Tau Sigma Gold Medal 1974. Mbr ASME, Soc of Rheology, Amer Phys Soc, Soc for Natural Philosophy. *Society Aff:* ASME, APS, SRheo, SNAtl Phil.

Nurmikko, Arto V
Business: Division of Engineering, Providence, RI 02912
Position: Prof *Employer:* Brown Univ *Education:* PhD/EE/Univ of CA-Berkeley; MS/EE/Univ of CA-Berkeley; BS/EE/Univ of CA- Berkeley *Born:* 01/29/45 in Finland. Joined Brown Univ in 1975 after a stay on research staff at MIT. Research interests include study of ultrafast electronic processes and highly excited states in semiconductors by optical techniques. Has developed techniques for switching of infrared radiation at picosecond speeds. *Society Aff:* IEEE, APS

Nurse, Edward A
Business: 839 Front St, Helena, MT 59601
Position: President & Owner. *Employer:* Foundation & Matls Cons Inc. *Education:* BS/CE/MT State Univ; MS/CE/Univ of CA, Berkeley. *Born:* 6/18/28. BSCE MT State Univ; MSCE Univ of CA, Berkeley. US Navy 1946-48. Struct Engr Morrison-Maierle, 4 yrs. F & M Cons since 1959. PPres MT Sect ASCE, MT Soc of Engrs, Optimist Club, Lewis & Clark Cty Econ Dev Ctte, Helena Chamber of Commerce. Currently Pres, Cons Engrs Council MT. Past Natl Dir NSPE. Candidate for MT House of Reps. Respon for pavement & slope design 120 miles on interstate hwys; pavement investigations & designs for 77 airport projs in the Northwest. 2 res papers on pavements. 20 papers & presentations on geotech engg & pavements. *Society Aff:* NSPE, ASCE, ACEC, NACE.

Nutt, Wells E
Home: 117 Oakwood Dr, Franklin, VA 23851
Position: Mill Mgr - Fine Paper Div *Employer:* Union Camp Corp. *Education:* MS/ChE/Univ of MS; BS/ChE/Univ of MS. *Born:* 6/25/39. Home town - Crossett Ark. High sch honor grad. Phi Eta Sigma, Tau Beta Pi, Omicron Delta Kappa, Outstanding Jr & Sr chem Engr, Founder & Editor of *Ole Miss Engr.* Indus exper: 1963-81, in world's largest paper mill at Union Camp Corp, Savannah GA in pos such as Process Engr, Group Leader, Supt Process Engr Overall Asst Supt - Power, Overall Asst Supt-Pulp and Technical Director. 1981- in Union Camp Corp Franklin, Va Mill as Gen. Oper. Supt and Mill Mgr. Reg P E MS/ Mbr Amer Inst of Chem Engrs, Tech Assn of Pulp & Paper Indus (Past Chmn of Southeastern Sect, Past Chrmn of Alkaline Pulping Committee, Chrmn Pulp Manf. Div). *Society Aff:* A.I.Ch.E.,TAPPI

Nuttall, Herbert E
Business: Dept of Chem & Nucl Engg, Albuquerque, NM 87131
Position: Prof *Employer:* Univ of NM. *Education:* PhD/ChE/Univ of AZ; MS/ChE/Univ of AZ; BS/ChE/Univ of UT. *Born:* 4/10/44. Born & raised in Salt Lake City, UT. Worked part time for Kennecott Copper Corp & E I Dupont during my yrs as a grad student. After receiving my PhD, I worked for Garrett Res & Dev Co in LaVerne, CA. Areas of responsibility were dev of a flash pyrolysis process for coal conversion & math modeling of energy processes. In 1972-73, served as a visiting Asst Prof of CE at the Univ of TX at Austin while on leave from Garrett Res & Dev Co. Began teaching career at the Univ of NM in 1974. Was promoted to an Assoc Prof in 1977. Res interests include dev of in situ energy recovery processes & modeling of transport processes in porous media. Full Prof 1982. Full Prof 1982. *Society Aff:* AIChE, ACS, AGU, RMFS.

Nutter, Dale E
P.O. Box 700480, 639 W. 41st Street, Tulsa, OK 74170
Position: Consultant. *Employer:* Nutte Engg Div of Patterson-Kelly Co. HARSCO Corp *Education:* BSME/Mech Engg/OK State Univ. *Born:* July 20, 1935. Engr Douglas Aircraft Missiles & Space Sys 1958-61. Employed by Nutter Engrg Co to date. Assignments included Sales Engr, Sales Mgr, Dev Engr for hardward & computer design programs. Consultant since 1978. Patentee in field with 40 issued worldwide. Hardware designs received John C Vaaler awards in 1966 & 1972. Tech Rep for co with Fractionation Res Inc since 1963. FRI Dir since 1968. Elected by FRI engrg mbrship to Tech Committee since 1971. Elected Vice Chairman FRI Tech Committees 1984. Mbr AIChE. Reg P E Okla 1962, Tex 1969. Hobby: radio controlled model airplanes with num natl records & awards in competition. Author of papers: "Ammonia Stripping Efficiency Studies–, 69th Natl Meeting AIChE, 1971. "Weeping & Entr Studies for Sieve and V-Grid Trays in an Air-Oil Sys–, 3rd Intl Symposium of Inst of Chem Engg, 1979. "Nutter RingsTM: A Random Packing Developed for Consistent Performance–, AIChE Annual Meeting, 1986. *Society Aff:* NSPE/OSPE, AIChE, FRI, AMA.

Nutter, Roy S, Jr
Business: Dept of Elec Eng, Morgantown, WV 26506-6101
Position: Professor *Employer:* West Virginia Univ *Education:* PhD/EE/WVA Univ; MSEE/EE/WVA Univ; BSEE/EE/WVA Univ. *Born:* 04/28/44 Native of Kingwood, WV. Project Engr for NCR in Dayton OH and Wichita, KS, 1972-74, responsible for design of minicomputers and microprocessor-based peripheral controllers. Joined WVU in 1974. Assisted in development of world's first computerized underground coalmine environmental monitoring system. Developed free world's first remote control mine ventilation equipment. Developed methodology for hazard and reliability analysis of computer based monitor and control systems and sys for testing distributed based computer sys, Currently developing an expert system for underground mines, energy management computer for residential homes, developing computer based training, and developing neural networks for artificial intelligence. NASA Pre-doctural Fellowship 1966-69; Senior Mbr, IEEE; Dir, Pittsburgh Section, IEEE; mbr NSPE, Eta Kappa Nu, Tau Beta Pi, Sigma Xi; Reg PE. *Society Aff:* IEEE, NSPE, IAS, CS, ACM

Nye, John C
Business: Agri Engg Dept. Louisiana State Univ, Baton Rouge, LA 70803-4505
Position: Prof and Dept Hd *Employer:* LA State Univ *Education:* BS/AgE/KS State Univ; MS/AgE/Purdue Univ; PhD/AgE/Purdue Univ *Born:* 3/24/45. Native of Anthony, KS. Researcher & Ext engr at Purdue Univ, 1971-73 Chief Agri Permit Team, Region V, US Environ Protection Agency, 1973-74. Educator & Researcher at Purdue Univ, 1974-84, Dept Hd of Agri Engrg at LA State Univ. Res has involved the recovery of resources from livestock waste & the renovation of Pesticide contaminated waste waters. Chrmn of the Struct and Environ Div of ASAE 1985-86. Chrmn, IN Sec ASAE, 1976-77. Secy, Exec Ctte, Intl Symposium on Livestock Waste, 1975; Chrmn of Exhibits, Intl Symposium on Livestock Waste, 1980. Chrm T-9 Env. Coordinating Ctte, 1983. Chrmn SE-412, Agri Waste Ctte, 1980. Vice Chairman, Baton Rouge Chapter, LES, NSPE, 1987. Am Soc of Agri Engrs, Reg PE, State of IN & LA. *Society Aff:* ASAE, ASEE, NSPE

Nypan, Lester J
Business: 18111 Nordhoff St, Northridge, CA 91330
Position: Prof. *Employer:* CA State Univ. *Education:* PhD/Mech Engr/Univ of MN; MSME/Mech Engr/Univ of MN; BS/Mech Engr/Univ of MN. *Born:* 10/30/29. Minneapolis, MN. Served with USAF 1952-54. Employed as Instructor at University of MN, Sr Engr at General Mills Inc Mech Div, and Lockheed CA Co. With CA State Univ, Northridge since 1962. Res effort employs photo-optical methods to study rolling element bearings. *Society Aff:* ASME, ASLE, NSPE, CSPE, ASEE.

Oatney, R Dale
Home: 71 Mt Zion Rd, Lancaster, OH 43130
Position: Owner *Employer:* Engineering/Management Counsultant *Education:* OSU BS Agr. Engr.- 1948 *Born:* Feb 17, 1926. BS Agri Engrg June 1948. Native of Fairfield Cty Ohio. Dist Engr for SCS, respon for all design & const of conservation practices & flood control projs. Exec Dir of Ohio Soc Prof Engrs; Natl Dir NSPE. Consultant & Management, respon for const bidding, engrg supervision & const scheduling. Secy-Treas of Const Indus Legis Council of Ohio. Hon Mbr Alpha Epsilon, Future Farmers of Amer; Meritorious Service Award OSPE 1970; Boss of Yr Amer Bus & Prof Womens Assn 1973. Hobbies: golf, hunting & fishing. *Society Aff:* NSPE,ASAE

Oberman, Leonard S
Home: 112-20 72nd Dr, Forest Hills, NY 11375
Position: Conslt to the Adm *Employer:* United Nations Development Programme *Education:* BS/CE/Univ of IL; Cert/Welding Theory/Milwaukee School of Engrg *Born:* 3/17/18 Registered in NY and CA. Co-editor *Civil Engrg Section*, Engrg Manual (McGraw-Hill). Consulting Editor for various technical periodicals. Author and coauthor numerous published technical papers and articles. Conslt to the Adm, UNDP, 1984-present. Sr Technical Adviser, UNDP, 1978-1983. Assoc, TAMS Engrs, 1950-1978. Project Engr, Robins Engrs, 1949-1950. Research Engr, Barber-Greene, 1947-1949. Research Field Engr, Harnischfeger Corp, 1946-1947. Chief, Equipment Unit; Office, Chief of Engrs, US Army, 1942-1946. Research Engr, Ill Dept Pub Wks, 1940-1942. Pres, Met Sect ASCE, 1977-1978. Pres, New York Federation Engrg Chapter, 1978. Chrmn, Natl Ctte on LNG Facilities, 1975-1977. Chrm Natl Ctte on Port-Related Dist Systems, 1974-1975. Mbr, US Coast Guard Technical Advisory Commission (Hazardous Materials), 1975-1978. National Dir, ASCE, 1980-1983. *Society Aff:* ASCE, SAME, PIANC, MU SAN, XE

Obert, Edward F
Home: 7843 Ox Trail, Verona, WI 53593
Position: Professor, Mechanical Engineering. *Employer:* University of Wisconsin, Madison. *Education:* MS/Mech Eng/Univ of MI; ME/Mech Eng/Northwester Univ; BS/Mech Eng/Northwestern Univ. *Born:* Jan 1910 Detroit. 1929-30 engr mfg Western Elec Co Chigago; 1934-37 Office of Naval Inspection Chicago; 1937-58 professor of mech engr Northwestern Univ; 1958- , professor at Wisconsin, dept chmn 1963-67. Owner Profess Engr Cons; cons Natl Acad Scis, USAF Acad WY & Aeromed Lab AK. Recipient George Westinghouse Award ASEE 1953; G Edwin Burks Award ASEE 1971; US Army Certificate of Appreciation ABM 1970; Benjamin Reynolds Award Univ of Wisconsin 1973. Fellow ASME; Mbr SAE, ASEE, Sigma Xi, Pi Tau Sigma, Tau Beta Pi, Mark Twain Soc. Author of 3 books in thermodynamics, 1 in combustion engines; editor of Mech Engrg Series, Intext. *Society Aff:* ASME, SAE, ASEE.

Obey, James H
Business: P O Box 1041, Buffalo, NY 14240
Position: President. *Employer:* Blaw-Knox Food & Chemical Equip Co. *Education:* BChE/Chem/Lawrence Tech. *Born:* Aug 29, 1916. 1943-45 Univ of Pittsburgh Graduate Sch. 1939-43 res sci with Ford Motor Co; 1943-56 res sci with Mellon Institute; 1956-62 market dev specialist with Consolidation Coal Research Dept; 1962-64 Pres Danville Products Inc; 1964-72 sales responsibilities for Blax-Knox

Obey, James H (Continued)
chemical plants; 1972-73 sales responsi bilities for Jacobs Engineering. Pres of Wirz Y Machuca, Mexico City. Reg PE PA. Chmn Pittsburgh Sect AIChE 1962; Mbr AIChE.. *Society Aff:* AIChE

Oblad, Alex G
Business: 302 Browning Bldg University of Utah, Salt Lake City, UT 84112
Position: Distinguished Professor Metallurgical and Fuels Eng. *Employer:* University of Utah. *Education:* BA/Chem/Univ of UT; MA/Phys Chem/Univ of UT; PhD/Phys Chem/Purdue Univ. *Born:* Nov 1909. Native of Salt Lake City. 1937-57 directed res in petroleum refining, catalysis, chemicals from petroleum Standard Oil Co (IND), Magnolia Petroleum Co Dallas, Houdry Process Corp Marcus Hook PA 1947-1955; V Pres R&D & Mbr Bd/Dir Houdry Process Corp 1955-1957; 1957-69 V Pres R&D with M W Kellogg; 1970 joined Univ of Utah as professor of Metallurgy & Fuels Engrg; 1972-75 acting Dean of Coll of Mines 1976, Distinguished Professor of Metallurgical & Fuels Engrg & Prof of Chemistry. DSc Purdue Univ 1959 honoris causa; Distinguished Alumni Award Univ of Utah 1962; E V Murphee Award Ind Chem ACS 1969; Chem Pioneer Award Amer Inst Chem 1972; elected Mbr NAE 1975; active in ACS. Co-founder Int Nat Congress on Catalysis. DSc University of Utah 1980 honoris causa U. active in coal liquifaction, coal gasification, tar mining and processing, chemicals from petroleum, petroleum refining processes, fundamental and applied catalysis. *Society Aff:* ACS, AIChE, AAAS, ΣΞ, NAE, AIC.

O'Brien, Edward E
Home: Box 10, Moriches Rd, St James, NY 11780
Position: Prof *Employer:* State Univ of NY *Education:* PhD/Fluid Mech/Johns Hopkins Univ; MS/ME/Purdue Univ; BE/Engrg/Univ of Queensland, Australia *Born:* 5/16/33 in Toowoomba, Australia. Post Doctoral research in Turbulent Transport at Hopkins, then faculty position at Stony Brook. Full Prof since 1968. Research in chemically reactive turbulence supported mainly by National Science Foundation. Half year visiting scientist with CNRS in Ecully France (1978). *Society Aff:* APS

O'Brien, Eugene
Home: 52 Glen Rd, Woodcliff Lake, NJ 07675
Position: Partner *Employer:* Tippetts-Abbett-McCarthy-Stratton *Education:* BS/ME/Cornell Univ *Born:* 9/3/27 Born and raised in Granville, NY. Served in US Army 1946-47. Mechanical engr with Gibbs & Hill 1950-56. Attended Intl School of Nuclear Science & Engrg 1955. With Development & Resources Corp 1956-70. Participated in engrg for dams, hydroplants, irrigation projects in Iran, Columbia, other overseas locations. VP for Engrg 1964-70, directing engrg activities for water resource based consulting firms. Joined TAMS in 1970, Head of Water Resources Division 1970-72, Partner 1973 - present time. Provide overall direction to water resources and environmental activities of firm. Married to Jean B. O'Brien, three children, one grandchild. *Society Aff:* ASCE, ASME, USCOLD

O'Brien, Michael
Business: Agri Engr Dept U of CA, Davis, CA 95616
Position: Professor. *Employer:* University of California. *Education:* PhD/Agri Engr/IA State Univ; MS/Agri Engr/IA State Univ; BS/Agri Engr/-. *Born:* Oct 1918. Native of Iowa, 1942-45 Navy pilot; 1948-50 instructor Iowa State; 1950- , Instructor--Professor Univ of CA Agri Engrg Dept, res proj leader on fruit & vegetable harvesting, handling, transportation, sampling, grading, packaging & storage. Holder of 13 U S Patents on associated equip. Mbr of univ, coll & dept cttes. Mbr Bd of Dir ASAE 1972-74, 1980-82 and 1984-86, Chrmn food Engr Div ASAE 1978-79, Chmn Pacific Region ASAE 1975-76; Chmn annual meeting ASAE Davis CA June 22-25 1975. Fellow ASAE. Author of 4 books & over 160 tech papers. A bowler & golfer. *Society Aff:* ASAE, ASEE, ΦΔΚ, ΕΞ.

O'Brien, Morrough P
Home: P.O. Box 265, Cuernavaca, Mexico 62000 *Employer:* Univ of Calif Field Station *Education:* BS/Civil Eng/MIT *Born:* 9/21/02. Dean Emeritus & Prof of Engrg Emeritus, Coll of Engrg Berkeley, Univ of CA. Mbr NAE, Pres Amer Shore & Beach Preservation Assn. Cons engr, Genl Elec Co DSc(Northwestern; D Eng Purdue; LLD CA (Berkeley) Hon Mbr ASCE, Hon MBR ASME. *Society Aff:* ASCE, ASME, ASEE, ASBPA

O'Brien, Richard S
Business: 51 West 52 St, New York, NY 10019
Position: Retired *Education:* BS/Elec Engg/UC-Berkeley. *Born:* 1917. Continued study at Stanford through 1941. Following wartime res at Harvard's Radio Res Lab, he joined CBS; has been responsible for CBS's major television broadcast systems design projs. Received SMPTE 1955 Journal Award for paper 'CBS Color Television Staging & Lighting Practices.' Sr Mbr IEEE; Fellow of AES & SMPTE; MBR of Eta Kappa Nu & Sigma Xi; served SMPTE as Dir, as 1st Television Affairs V Pres & as Secy. Retired, 1982. Received SMPTE 1984 David Sarnoff Gold Medal Award. *Society Aff:* AES, IEEE, SMPTE, HKN, ΣΞ.

Ochab, Thomas F
Home: 25 Mountainview Rd, Verona, NJ 07044
Position: Mgr-Engg & Construction Services *Employer:* Coopers & Lybrand *Education:* ME/ME/Stevens Inst of Tech; BE/CE/Stevens Inst of Tech *Born:* 4/4/47 Native of northern NJ. Manager of construction cost control consulting practice performing work on projects throughout the US. Over 20 years diversified experience in various field and home office positions related to the construction of large office buildings, power plants, wastewater facilities, airports, hospitals, university facilities and rapid transit systems. Specialist in keeping construction projects within budget, on schedule and at desired quality. Previous experience as Project Engineer on earthen dam/reservoir project and various power plants. Has authored several articles and taught seminars on project management and construction control. Licensed PE in NJ, NY and PA. Avid golfer. *Society Aff:* ASCE, ANS, USCOLD, NSPE, CMAA.

O'Connell, John P
Business: 7601 W 47th St, McCook, IL 60625
Position: President *Employer:* Motor Oils Refining Co *Education:* Bachelor/Chem Engg/Polytechnic Inst of NY. *Born:* 2/19/38. Native of State of NY. Served with Army Corps of Engrs from 1960-62 as 1st Lt. Held numerous engg and management positions in the following companies: Getty Oil Co, Hess Oil & Chem Corp, Conoco and SRS, Inc. Present Position held; Pres, Motor Oils Refining Company. Motor Oils Refining Company is a lubricant sales and re-refining corp. In addition to the above, presently a mbr of the Admissions Committee of the AIChE. *Society Aff:* ASLE, AIChE, ASTM, IOCA, APR.

O'Connell, Joseph M
Business: Kingsland Ave-Bldg 46, Nutley, NJ 07110
Position: Senior Supervisor. *Employer:* Hoffman La Roche Inc. *Education:* BS/Mech Engg/NJ Inst of Technology; MS/Mgmt Engg/NJ Inst of Technology. *Born:* June 12 1936. Assoc Equip engr Western Elec Co responsible for design & coordination of telephone installations; proj control & estimating engr, Foster Wheeler Corp I C F Braun Corp involved in chemical plant construction; supervisor of Cost Estimating & Control Grp, Hoffman La Roche Inc, responsible for capital cost estimates. Pres NJ sect AACE 1971-72; Dir Natl AACE 1973- Treasurer AACE 1978-80. *Society Aff:* AACE.

O'Connor, James J
Home: 631 Birch Ave, RiverVale, NJ 07675
Position: Editor-Emeritus *Employer:* McGraw-Hill Publishing Co. *Education:* BEE/EE/NYU. *Born:* 8/3/17. After serving 10 yrs on construction & test engg in power plants & substations, joined editorial staff of McGraw-Hill's POWER Magazine. Lic PE, NY & NJ. Author of numerous engg articles, reports & papers. Editor & author of "Standard Handbook of Lubrication Engineering." Appeared on tv & radio on many occasions representing engg profession on matters dealing with

O'Connor, James J (Continued)
power tech. Listed in *Who's Who in America*. ASME Engg Leadership Award. IEEE Chrmn of Ind & Commercial Power Systems. IEEE Chrmn Public Relations. ASME Boiler Code Chrmn (Section VII). Consultant to Natl Coal Policy Proj. Jesse H Neal Award for "US Energy Mgt Guidelines" editorials. *Society Aff:* IEEE, ASME, CES, ASLE, NSPE.

O'Connor, John T
Business: 1047 Engineering Bldg, Dept. Civil Engineering, Columbia, MO 65211
Position: Chairman of Civil Engineering. *Employer:* University of Missouri - Columbia. *Education:* BSCE/Civ Engrg/Cooper Union; MSCE/Civ Engrg/NJ Inst Tech; Eng D/Sanitary Engrg/Johns Hopkins. *Born:* 2/11/33. 1961-75 Professor Civil Engrg Univ of IL Urbana; 1975- , Chmn Civil Engrg Univ of MO Columbia. 1969-76 Newsletter Editor, Secy ASCE Env Engr Div; Editor Env Engr Unit Operations Lab Manual AEEP. Teaching & res on water & wastewater treatment processes: metals removal, disinfection, ozonation, ion exchange, activated carbon, chemical precipitation, control of biological growths and trihalomethane formations in distribution systems, removal of radioactive pollutants, disposal of water plant residues, leachate from fly ash, effect of sanitary and hazardous waste landfills on groundwater quality, virus in water supplies, wter conservation, environmental effects of hydropower generation, trichloroethylene and pentachlorophenol adsorption on soil. *Society Aff:* AWWA, WPCF, ASCE, ACS, ASLO.

O'Connor, William W
Home: 11933 NW 28th St, Coral Springs, FL 33065
Position: Mgr Adv Dev. *Employer:* Motorola. *Education:* MBA/Marketing/Loyola Univ; BSEE/EE/Univ of IL; BA/Phys/St Marys College. *Born:* 2/10/37. in Chicago, IL. Grad with honors from Univ of IL. Mbrship Chrmn of Etta Kappa Nu & Resident House VP while at Univ of IL. Started in the Mobile Radio Engg Dept in 1961 at Motorola & became Mgr of Engg in 1971. Currently am Mgr of Advanced Dev which involved managing Engg & IC Development group to define and develop the next generation portables two-way radio products & their options. Hold 5 US patents. Served on Advisory Council to the Bd of Dirs of Motorola. Chosen for "Motorola Executive Institute–. Married, 6 children & enjoy outdoor sports. *Society Aff:* HKN, ΣΤ.

O'Dea, Thomas J
Home: 11 Robin Hill Rd, Mt Kisco, NY 10549
Position: President *Employer:* Thomas J O'Dea & Assoc., P.C. *Education:* BS/ME/NY Poly Inst. *Born:* 5/16/34. & raised in NYC. Mr. O'Dea is President and managing principal of Thomas J. O'Dea & Associates, P.C., which he established in 1974. He directs all admin & engrg functions, as well as monitoring technical developments and progress of on-going projects. Was responsible for design and specifications for the first major competitively-bid, commercial, computerized building automation system. The $7.5 contract proved to be the forerunner of current state-of-the-art building automation systems. Before starting his own firm, Mr. O'Dea was an Associate Partner and Associate Engineer at two separate prominent engrg firms, where he was responsible for design and supervision of HVAC and Plumbing, from project inception through construction. Licensed Professional Engineer in NY, NJ, NH, MA, RI, CT & ME. *Society Aff:* ASHRAE, NSPE, ACEC, NFPA.

Odeh, Aziz S
Business: P.O. Box 819047, Dallas, TX 75234
Position: Sr Scientist *Employer:* Mobil R & D Corp *Education:* PhD/Engrg/UCLA; MS/Engrg/UCLA; BS/Petro Engrg/Univ of CA-Berkeley *Born:* 12/10/25 Is a Sr Scientist at Mobil Research and Development Corp's Dallas Research Div in Dallas, TX. Except for a two year assignment as a Staff Reservoir Engr with Mobil Oil de Venezuela, he has been at the Dallas Research Div since joining Mobil in 1953. His work has involved all aspects of Reservoir Engrg and he is the author or coauthor of over 40 publications on the subject. In 1972, he served as Chrmn of the Editorial Committee and in 1975, as Chrmn of the Monograph Committee of the Society of Petroleum Engrs. He is a permanent lecturer for the Society on Reservoir Engrg. He has presented courses on Flow Test Analysis and Water Influx Theory worldwide. He was the recepient of the Society of Petroleum Engrs John Franklin Carll award in 1984, and was inducted to the National Academy of Engineering in 1987. *Society Aff:* SPE, NAE, RESA

Odle, Herbert A
Business: Manville Corp, 214 Oakwood Ave, Newark, OH 43055
Position: Research Mgr. *Employer:* Holophane, Div of Johns-Manville. *Education:* MBA/Finance/Xavier, OH; MSEE/Electrical/Univ of IL; BS/Electrical/Univ of IL. *Born:* 6/18/25. BS & MSEE, Illumination Option, Univ of IL. MBA, Xavier OH. Taught at Univ of IL on graduate fellowship; served with Army Communications during WWII; started with Holophane as design engr, currently respon for managing proj design teams, coord patent work & serving as staff consultant on optical systems design and Mgr of Photometer Testing. Reg PE OH, IN, MA also PPres local sec OH Soc of Profl Engr; Past Chmn IES Roadway Ltd Comm, IES Fellow. Mbr IES. Committee on Research for Quality & Quantity of Light. IES Ctte Group Mgr. Author of several tech papers & holder of 13 US pats. *Society Aff:* IES, AES, NSPE.

Odom, Leo M
Home: 2032 Ramsey Dr, Baton Rouge, LA 70808
Position: Chairman *Employer:* Pyburn & Odom Inc. *Born:* Aug 1905. Native of Glenmora LA. BS from LSU 1927. 1927-29 LA Hwy Comm 1929-41 civil engr with US Army Corps of Engrs. 1941-48 Chief Engr Dept of Public Works, State of LA. 1948-70 partner Pyburn & Odom Consulting Engrs; 1970- 81 Pres and Chrmn of the Bd of Dirs Pyburn & Odom, Inc; 1981-Present Chrmn of the Bd of Dirs Pyburn & Odom, Inc, 1968-75 Pres and Chrmn of the Bd Odom Offshore Surveys, Inc; 1975-Present Chrmn of the Bd Odom Offshore Surveys, Inc. Recipient Dist Serv Cert NCEE 1963, Spec Commendation Cert NCEE 1975, Dir. Southern Zone Natl Council State Bds. Mbr ASCE, LA Engrg Soc (Pres 1968-69), NSPE, LA State Bd of Registration for Profl Engrs & Land Surveyors 1944-76, Intl Assn for Hydraulic Res. 1980 Selected by L.S.U. College of Engrg as a mbr of the Engrg Hall of Distinction. Contributed articles to Profl Journals & ASCE transactions.

O'Donnell, Neil B
Business: 105 S Meridian, Indianapolis, IN 46225
Position: VP. *Employer:* Amax Coal Co. *Education:* MBA/Mgmt/Boston Univ; BSChE/ChE/Tufts College. *Born:* 12/20/24. Joined Amax Coal Co as VP of Midwest Surface Mining Operations in 1977. Prior experience included Genl Mgr and Operating respon with Revere Copper and Brass, Esso Chemical Inter-America, Kaiser Aluminum and Chemical, and Monsanto in the manufactures of organic and inorganic chemicals, alumina-aluminum, agricultural chemicals, and bauxite mining. Also have been respon for new plant design, construction, and startup as well as all administrative functions. mgmt positions included internatl assignments in Jamaica, South America, and India for a total of seven yrs. Native of the Boston, MA area and served with the US Navy from 1943-46. Enjoy participation in community activities and civic organizations. *Society Aff:* AIChE, AIME.

O'Donnell, William J
Business: 241 Curry Hollow Rd, Pittsburgh, PA 15236
Position: President. *Employer:* ODonnell & Assoc Inc. *Education:* PhD/Engg/Univ of Pittsburgh. *Born:* June 1935. With Westinghouse Res Lab & Bettis Atomic Power Lab 1957-68. Performed struct design & analyses of nuc reactor & power plant components. Adv Engr respon for reactor struct design basis. Pub 50 papers on engrg analysis methods & struct materials sci. Active ASME Boiler & Pressure Vessel Code Cttes & Pressure Vessel Res Comm. Cons Engr since 1968. Pi Tau Sigma Gold Medal 1967 ASME, Cert of Merit 1972 DIB. Received 1974 ASME Pressure Vessel & Piping Div Natl Award. Reg P E PA. Elected Fellow - ASME 1976. *Society Aff:* SESA, ASME, ASAS, ΣΞ, ASM.

Oertel, Goetz K
Business: 1625 Mass. Ave. N.W. #701, Washington, DC 20036
Position: Pres *Employer:* Aura, Inc. *Education:* Ph.D./Physics/U. of Maryland, CP; Vordiplom/Physics/U. of Kiel, Germany *Born:* 08/24/34 Aerospace Scientist, Engineer, then Chief of Solar Physics, NASA Langley Research Center and Headquarters, Washington. Assignments to President's Science Advisor, OMB, and National Science Foundation. US DOE positions including Dir of Defense Nuclear Waste, Deputy Mgr of Albuquerque Operations, and Acting Mgr. of Savannah River Operations. Also Deputy Assistant Secretary for J. Fitz, Health, Q.A.; - Washington. AS CEO of AURA, in charge of Space Telescope Science Inst & National Optical Astronomy Observatories (AZ, NM & Chile). Service on Visiting Cttees for Engrg at Penn State to 1985 and Engrg Div, Los Alamos. *Society Aff:* APS, AAS, AAAS,ΣΞ

Offner, Franklin F
Home: 1890 Telegraph Rd, Deerfield, IL 60015
Position: Professor. *Employer:* Northwestern Univ. *Education:* PhD/Biophysics/Univ of Chicago; MS/Chem/CA Inst of Tech; BChem/Chem/Cornell Univ. *Born:* April 8, 1911 Chicago. Fellow IEEE. Laureate in Tech Lincoln Acad of Illinois. Approximately 60 pats in elec control, biomedical elecs, amlification, hydraulics. Principal res interest membrane biophysics. Pres Offner Electronics Inc 1939-63; Prof Northwestern Univ 1963- . *Society Aff:* IEEE, APS, AEEGS, BS.

Offner, Walter W
Home: 5585-D Kuamoo Road, Kapaa, HI 96746
Position: President. *Employer:* X-Ray Engg Internal Kauai. *Education:* -/Elec Engg/Augsburg Germany. *Born:* Feb 1906 Munich Germany. Grad Tech Univ Augsburg Germany, postgrad work Univ of Chicago, Univ of Washington. Founder X-Ray Engrg Internat'l & branches in Australia, Hawaii, India, Mexico. Pioneer in radiographic inspection; numerous contracts in Europe, Middle East, Japan, Australia, India, Turkey, Mexico, South Amer. Dev radiographic weld quality control 1947 for the first large diameter & longest pipeline from the Persian Gulf to the Mediterranean (Aramco). Fellow ASME & Soc of Nondestructive Testing. Mbr ASM, AWS, Ship Res Council, NAS 1969, ASTM, S F Port Engrs. US Delegate to "International Institute of Welding" (IIW) also member of "American Security Council–. Author numerous articles in tech pubs incl Encyclopedias Americana & Britannica. *Society Aff:* ASM, ASME, AWS, ASNT, ASTM, NSPE.

Ogata, Katsuhiko
Business: Dept of Mech Engg, Minneapolis, MN 55455
Position: Prof. *Employer:* Univ of MN. *Education:* PhD/Engg Sci/Univ of CA; MS/ME/Univ of IL; BS/ME/Univ of Tokyo. *Born:* 1/6/25. in Tokyo, Japan. Came to US in 1952. Res Asst, Scientific Res Inst, Tokyo; Fuel Engr, Nippon Steel Tube Co, Tokyo; Prof of Elec Engg Yokohama Natl Univ. Dept of Mech Engg, Univ of MN: Asst Prof, 1956-58; Assoc Prof, 1958-61; Prof, 1961-present. Received Outstanding Adviser Award, Inst of Tech, Univ of Minn, 1981. Authored four books: State Space Analysis of Control Systems, 1967; Modern Control Engg, 1970; Dynamic Programming, 1973; System Dynamics, 1978. *Society Aff:* ASME, ΣΞ, ΠΤΣ

Ogburn, Hugh B
Home: 4340 Pahoa Ave 16-A, Honolulu, HI 96816
Position: VP. Retired *Employer:* Pacific Resources, Inc. *Education:* PhD/Chem Eng/Princeton Univ; MS/Chem Eng/Princeton Univ; BS/Chem Eng/Princeton Univ *Born:* 7/13/23. Native of Lexington, VA. Served with USNR, 1942-1946. Active in res, dev and engg mgt function in the petrol & petrochem industries with the Atlantic Refining Co, M W Kellog Co & Union Carbide Corp, 1950-1970. Adjunct Prof of Chem Engg, Drexel Univ, 1950-1961. Founded firms in consulting, intl petrol trading & shipping. Assumed current position & responsibilities in 1978. VP Pacific Resources, Inc 1978-1983. Pres, Pacific Oasis 1983; Chrmn, Pacific Oasis 1984; Presently, Conslt, Pres, H B Ogburn & Assoc. Retired 1987. *Society Aff:* ACS, AIChE, RESA, ΣΞ

Ogilvie, James L
Home: 3312 S Oneida Way, Denver, CO 80224
Position: Consultant. *Employer:* Private. *Education:* BS/Civ & Irrigation Engr/CO State Univ. *Born:* 01/04/11. Native Coloradoan. Thirty five yrs in Water Resource dev, Dept of Interior, Bureau of Reclamation Proj Mgr, Asst Regional Dir. Received Depts Distinguished Service Award. 5 yrs WWII USA Corps of Engrs-Lt Col Aviation Engr Battalion. Mgr Denver Water Dept & Secy Denver Bd of Water Commissioners. Responsible for planning, designing, constructing & operating vast municipal water system supplying over one million customers in the Denver Metropolitan Area. Reg PE in CO. Consltg Engr, Water Resources-Planning, Management, Conservation, Environment. *Society Aff:* ASCE, AWWA, AAEE.

Ogilvie, Robert E
Home: 2 Locke Ln, Lexington, MA 02173
Position: Prof. *Employer:* MA Inst of Tech. *Education:* ScD/Met/MIT; Eng/Met/MIT; SM/Met/MIT; SB/Met/Univ of Wash. *Born:* 9/25/23. Served with the USN 1943-45. Mbr of the faculty at MA Onst of Tech, 1955-, Prof of Mtls Sci & Engg 1966-; Dir of Res at Advanced Metals Res, Burlington, MA 1958-73; Visitor Res Lab Boson Museum of Fine Arts 1968-. Pres of the Microbeam Analysis Soc 1970. Pres of the New England Soc of Electron microscopy 1977. Enjoy sailing & skiing. *Society Aff:* MAS, NESEM, MS.

Oglesby, David B
Home: Engg Mechanics Dept, Rolla, MO 65401
Position: Assoc Prof *Employer:* Univ of MO. *Education:* DSc/App Mech/Univ of VA; MAM/App Mech/Univ of VA; BS/CE/VA Military Inst. *Born:* 3/3/41. Native of Charlottesville, VA. Res Engr for the Res Labs for the Engg Sciences, Univ of VA, 1968-69. Res in area of fiber reinforced composites. Assoc prof of Engg Mechanics, Univ of MO-Rolla, 1969 to present. Res & publications in area of finite element stress analysis & in-situ coal gasification. Recipient of Outstanding Teacher Award 1971, 1978, 1979, 1981, 1982, 1983, 1984, 1985 & 1986. *Society Aff:* ASCE, ASEE.

Oglesby, Sabert
Home: 1348 Panorama Dr, Birmingham, AL 35216
Position: Pres *Employer:* Southern Research Inst *Education:* MS/EE/Purdue Univ; BS/EE/Auburn Univ *Born:* 5/14/21 Native of Birmingham, AL. Served in US Army Signal Corps 1943-1946. Research Engr at Southern Research Inst 1946-1948. Instructor in Electrical Engrg at Purdue Univ 1948-1951. Assumed current responsibility as Pres of Southern Research Inst Jan 1, 1981. Responsible for co policy and direction in varied fields of research. Coauthor of two books and numerous technical articles in fields of air conditioning and air pollution control. *Society Aff:* APCA

O'Grady, Joseph G
Business: 1916 Race Street, Philadelphia, PA 19103
Position: President *Employer:* ASTM *Education:* BEE/Elec Engrg/NY Univ. *Born:* 3/21/27. Responsible for the activities of 140 tech cttes of ASTM who develop Voluntary Consensus Standards, Special Tech Pub, and other tech material which is given worldwide distribution. Formerly Exec VP, ASTM, 84-85, VP PSE&G Res Corp, 1969-83 and had the overall responsibility for the activities of the Corp Res & Test Lab in Maplewood, NJ. This multi-discipline lab represented all major engrg disciplines and, in addition, performed applied res in such areas as fuel cell tech, solar demo, and energy conservation. ASTM Dir 1973-81, Bd Chrmn 1978-79; Dir AALA 1979-81; Chrmn EEI Codes and Stds Policy Ctte 1979-81; Past Sec Chrmn N Jersey Sec IEEE 1968-69, Chrmn Metropolitan Sec Advisory Ctte 1973-74, Region I Award 1974; Mbr Bd of Dir, Natl Inst of Bldg Sci, Reg PE NJ; Commercial Instrument Pilot. *Society Aff:* NSPE, SES, CESSE, Newcomen Society, NIBS

Ohanian, M Jack
Business: College of Engineering, University of Florida, Gainesville, FL 32611
Position: Assoc Dean for Res *Employer:* Coll of Engrg *Education:* PhD/Nuclear Eng & Sci/RPI; MS/EE/RPI; BS/EE/Robert College (Turkey) *Born:* 08/07/33 Native of

Ohanian, M Jack (Continued)
Istanbul, Turkey; naturalized US citizen, 1967. With Univ of Fl since late 1963. Assumed current respon as Assoc Dean for Engrg Res, Univ of FL in 1979; has overall programmatic respon for the Coll's res program. Prior to that Chrmn of the Nuclear Engrg Scis Dept for ten years. Currently completing second three-year term on the Bd of Dirs of ANS; mbr, Engrg Accreditation Commission of ABET. P Pres, natl nuclear sci and engrg honor soc, Alpha Nu Sigma. Fellow ANS; Exceptional Service Award, ANS, 1980; FL Blue Key Distinguished Service Award, 1984; Fellow AAAS. *Society Aff:* ANS, ASEE, AAAS, TBBI, HKN, ΣΞ, ΦΚΦ, ΑΝΣ

Ohara, George T
Business: Wastewater Treatment Division, 445 Ferry Street, San Pedro, CA 90731
Position: Chief Engr. *Employer:* City of Los Angeles Bureau of Sanitation. *Education:* MS/Engg/CA State Univ; Certificate/Bus-Mgt/UCLA; BS/Engg/CA State Univ L.B. *Born:* 4/12/33. Raised & educated in S CA. Started career with the City of Los Angeles in 1961, as an Civ Engg Asst, performing municipal design. Promoted to Sanitary Engr in 1969, responsible for sewage treatment engg, des, rsch & support functions for operations & maintenance. Promoted to Chief Engr in 1977 & assumed responsibility for the City of Los Angeles Wastewater Treatment Div. Have coauthored several Sanitary Engg papers & co-recipient of the ASCE Rudolph Hering Medal. Have also instructed wastewater operations & engg at the community college & grad sch level. Active photographer & fisherman. *Society Aff:* ASCE, WPCF, CWPCA.

O'Hara, James C
Business: 145 Technology Park/Atlanta, Norcross, GA 30092
Position: Consltg Engr *Employer:* EBASCO Services Inc *Education:* BME/ME/Catholic Univ *Born:* 01/24/25 Radiological Engrg Conslt at TMI immediately after 1979 accident and during recovery operations. Conslt on health physics and radiological aspects of operation, maintenance and refueling of naval reactors while with Gen Elec at the Knolls Atomic Power Lab, Schenectady, NY for 15 years. Currently with EBASCO, conslt on radiological engrg and applied health phys training to USA and foreign nuclear power plants. Chrmn, ASME Hudson Mohawk Section (NY) 1964-65. Chrmn, ANS Atlanta Section, 1981-83. Amer Natl Standards Inst Subctte N45-8 "Nuclear Power Plant Air and Gas Cleaning Systems–. *Society Aff:* ANS, HPS

Ohlson, John E
Business: 1195 Bordeaux Drive, Sunnyvale, CA 94086
Position: Director *Employer:* Stanford Telecommunications, Inc. *Education:* PhD/Elec Engr/Stanford; MS/Elec Engr/Stanford; SB/Elec engr/MA Inst of Tech. *Born:* 5/29/40. in Seattle, WA. Prior to 1967 employed summers and part-time by Boeing Co, Renton, WA, Stanford Res Inst, Menlo Park, CA, San Jose State College, San Jose, CA. Asst Prof of Elec Engg dept, Univ of So CA, Los Angeles, 1967-71. Assoc Prof of Elec Engg at Naval Postgrad School, Monterey, CA 1971-78. Prof, 1978-81 and was Dir of Satellite Communications Lab. Consultant to Jet Propulsion Lab, Pasadena, CA, 1968-81. Joined Stanford Telecommunications in 1981 as Director of Telecommunications Programs Operations, responsible for programs in Satellite communications, spread spectrum, TDMA and signal monitoring. Mbr Sigma Xi, Eta Kappa Nu. *Society Aff:* IEEE.

Ohsol, Ernest O
Home: 711 Hyannis Port North, Crosby, TX 77532
Position: Consultant *Employer:* Ohsol Technical Associates *Education:* ScD/ChE/MA Inst of Tech; BS/ChE/CCNY; -/Ind Chem/Swiss Fed Poly Inst. *Born:* 5/28/16. Born in Wash, DC. Early education in NYC public schools. Spent yr at Fed Poly Inst, Zurich, Switzerland on undergrad scholarship. Grad studies including Practice Sch at MIT. Taught at Stevens Inst. Dev Engr with Exxon 10 yrs. Mgr of Process Dev at GE. Dir of R&D at Pittsburgh Coke & Chem. VP Haveg Industries and Chemico (NY). Also with Am Cyanamid & Jacobs Engg. Mgr European Fluid Processing for Selas in Munich, Germany. Project Coordinator for Dexter Corp. Self-employed consultant. Fellow of AIChE, Dir, 1972. Licensed PE, NJ. 27 patents. Married, 4 children. Hobbies: languages, skiing. *Society Aff:* AIChE, TAPPI, ACS.

Ohtake, Tadashi
Business: Fundamental Res Labs, 1618 IDA, Nakahara-Ku Kawasaki Japan
Position: Exec Advisor *Employer:* Nippon Steel Corp. *Education:* Dr Eng/-/Tohoku Univ *Born:* 2/13/15. Joined Japan Iron & Steel Co, Ltd in 1939, predecessor of the Nippon Steel Corp. Assumed the post of Exec Advisor of Nippon Steel Corp in 1981. VP the Japan Inst of Metal in 1979. VP, Engg Res Assn of Nucl Steelmaking in 1979. Received Meritorious Honor Award of the Japan Inst of Metals in 1956, Saburo Watanabe Award of the Iron & Steel Inst of Japan in 1972 & Nishiyama Memorial Prize of the Iron & Steel Inst of Japan in 1981. Fellow ASM in 1978; Medal of Honour with Purple-Ribbon in 1983; Murakami Memorial Award of Murakami Memorial Assoc in 1984. Enjoy playing golf & fishing. *Society Aff:* ISIJ, JIM, ASM, AIME

Oishi, Satoshi
Business: 70 South Orange Ave, Livingston, NJ 07039
Position: Partner/Pres & Chief Executive Officer *Employer:* Edwards and Kelcey *Education:* BS/CE/Univ of CT *Born:* 1/19/27 With Edwards and Kelcey since 1949. Pres & Chief Executive Officer. Responsible for design and engrg management of major transportation, recreation, bridge and tunnel projects involving multidisciplinary design teams. Honor Member, Chi Epsilon; Tau Beta Pi; Sigma Xi; Previously, responsible for rail transit, bridges & structures, architectural, planning, soils & foundations & military facilities design divisions. Director, District 1, ASCE. ASCE Director, Engg Foundation. Past Pres, MET Section ASCE. Born Japan; Married 7/2/60, Jeanette C.; Children: Michelle Y. *Society Aff:* ASCE, AIA, ACEC, SAME, ACI, APTA, IABSE.

Oizumi, Juro
Business: Chiba Inst of Techn, 2-17-1 Tsudanuma, Narashino, Chiba, Japan 275
Position: Professor *Employer:* Chiba Inst of Technology. *Education:* DOC/Eng (Electrical)/Tohoku Univ; BACH/Eng (Electrical)/Tohoku Univ. *Born:* March 1913. Native of Sendai Japan. Communication Engr for Anritsu Campany Ltd 1935-51; Prof Computer Scis Tohoku Univ 1951-76; Dir of Computer Center & of Res Center for Applied Info Scis at the Univ; Prof of Univ of Elec- Communications since 1976-78. Emeritus of Tohoku Univ 1976. Mbr of ALOHA Computer Network. Prof, of Chiba Inst of Tech, since 1978-83. Mbr of Japan Sci Council since 1972-81. Fellow of IEEE 1975. Life Fellow 1983. Honorary member of IPSJ. Former V Pres of IEICEJ & of IPSJ. Mainichi Communication Prize 1943. Official Commendation by Minister of ITIJ 1974. Ishikawa Prize 1975. Merits Prize by Acous. Soc. of J. 1984. Bestowal of the Order of the Rising Sun. Gold Rays with Neck Ribbon, 1986. *Society Aff:* IEICEJ, IPSJ, ASJ, IEEE, ACM, ASA.

Ojalvo, Irving U
Business: U. of Bridgeport, Tech. Bldg, Bridgeport, CT 06601
Position: Prof & Chmn. of M.E. *Employer:* Univ. of Bridgeport *Education:* Sc.D/M.E./N.Y.U.; M.S./M.E./M.I.T.; B.S.E./M.E./C.C.N.Y.; *Born:* 01/16/36 1960-61 Fulbright Scholarship after receiving Sc.D from N.Y.U. 1961-83 Engrg Supervisor at Republic Aviation, Grumman Aerospace and Perkin-Elmer. Predicted loss of tiles on first space shuttle launch. Developed state-of-the-art algorithm for solution of very large eigenvalue problems using Lauc 705 vectors. 1958-present: Published over 60 technical papers. 1983-present: Ballard Chair of M.E. at Univ. of Bridgeport. & chairman of M.E. at Univ. of Bridgeport. 1978-present: Consulting Engr to high tech companies and expert witness and accident reconstructionist for attorneys and insurance companies. 1985-Elected a Fellow of the ASME. *Society Aff:* ASME, AIAA, AAM, AAUP

Ojalvo, Morris
Home: 2258 Wickliffe Rd, Columbus, OH 43221
Position: Retired *Education:* PhD/CE/Leigh Univ; JD/Law/OH St Univ; M/CE/RPI;

Ojalvo, Morris (Continued)
B/CE/RPI *Born:* 3/4/24 Born NYC. Served in the Civil Engrs Corp of the USN during WW II. Taught at the City Univ of NY, Rensselaer Polytech Inst, Princeton and Lehigh Univs before employment at The OH St Univ, 1960-82. Conducted research for the Amer Iron and Steel Inst and the Natl Sci Fnd on Columbus and beams. Author of a theory of bending and buckling for thin-walled rods of open profile section and holder of a patent for a device for the improvement of torsional properties of beams. Professor Emeritus, The Ohio State Univ. 1982-; Visiting Prof, The Univ Texas at Austin, 1982-83. *Society Aff:* ASCE, SSRC

Ojalvo, Morris S
Business: Nat'l. Sci. Found, Washington, DC 20550
Position: Program Dir Div of Chem, Biochem & Thermal Engg. *Employer:* Natl Science Foundation. *Education:* PhD/ME/Purdue Univ; BME/ME/Cooper Union; MME/ME/Univ of DE. *Born:* July 6, 1923. Military serv with US Navy 1944-46. Conducted res & taught for over 20 yrs at the Univs of Penn State, Illinois, Delaware, Maryland, Purdue, George Washington & at the Natl Polytech Inst (Mexico City); 22 yrs exp to date at the Natl Sci Foundation as Prog Dir of Engrg Energetics, Special Engrg, Heat Transfer, Indus Tech & Particulate & Multiphase Processes Progs. Mbr ASME (Fellow), ASEE, SAE, Tau Beta Pi, Pi Tau Sigma. P E Maryland. Prof interests: Thermodynamics, Heat Transfer, Fluid Mechs, Energy Conversion, Noise Control, Fine Particle Processing, Res Admin. *Society Aff:* ASME, SAE, ASEE.

Okada, Minoru
Home: 15-17 Sakuragaoka-2, Setagaya-Ku, Tokyo, Japan
Position: Adviser *Employer:* Japan Civil Aviation Promotion Fdn *Education:* Doctor/Elec Engg/Univ of Tokyo; Bachelor/Elec Engg/Univ of Tokyo. *Born:* Oct 1907. Sr Res Mbr Elec-Tech Lab Ministry of Communications (1931- 1952); Dir of Elec Communications Labs (1952-56); Exec Dir (1957) Nippon Telegraph & Telephone Public Corp; Prof Univ of Tokyo (1957-68); Pres & Prof Kogakuin Univ (1968-74). Chmn Aircraft Accident Investig Comm. Ministry of Transport (1974-1980). Honorary Mbr IECE Japan (1973). Fellow IEEE (1974). Spec in the field of Elec Navigation invented MF Omni-Range just the same principle as of VOR (1934) & succeeded in its fight test (1937). Medal of Honor with Purple Ribbon Tenno (1972). Distin Servs Awards IECE Japan (1972). *Society Aff:* IECE, IEEE.

Okamoto, Hideo
Business: 3-27 Rokkakubashi, Yokohama Japan 221
Position: Prof. *Employer:* Kanagawa Univ. *Education:* PhD/EE/Osaka Univ; BS/Phys/Osaka Univ. *Born:* 9/17/18. From 1952 to 1974, Dr Okamoto had been with Central Res Inst of Elec Power Industry (CRIEPI), where he had been engaged in researches on SF6 gas insulated cables, partial discharges (corona) & cryogenic power transmission. In 1974, he changed his occupation to Prof, Dept of EE, Kanagawa Univ. Dr Okamoto has been elected a Fellow, IEEE in Jan 1, 1979 & is now Dir of the Bd in Tokyo Sec of Region 10, IEEE. In 1971, he was awarded with the Minister's Prize of Sci & Tech Agency in Japanese Govt. *Society Aff:* IEEE, IEEJ, CESJ.

Okamura, Sogo
Home: 4-12-15 Numabukuro Nakano-ku, Tokyo Japan 165
Position: Director General *Employer:* Japan Society for the Promotion of Science *Education:* Dr of Engrg/Elec Engrg/Univ of Tokyo *Born:* 03/18/18 Native of Mie Japan. Assoc Prof (1947-51); Prof of Elecs (1951-78) Univ of Tokyo. Dean of Faculty of Engrg (1973-75), Prof Emeritus Univ of Toyko (1978-), Exec Dir, JSPS (1979-), Prof Tokyo Denki Univ (1978-), Dean of Grad Sch, Tokyo Denki Univ (1979-). Am also Councilor of Sci Coun, Radio Tech Coun (1970-78), Elecs Coun Atomic Energy Comm etc. Chmn of Comm A of URSI (1978-81); V Chmn of CSTP in OECD (1976-79); V Pres of IEEJ (1975-77). Fellow of IEEE. Pres of IECEJ (1971-72). Distin Serv Awd by IECEJ 1974; Governor of Tokyo Metropolis Awd 1975 Broadcast Culture Prize, by NHK, 1979. Enjoy golf, gardening & classical music. *Society Aff:* IEEJ, IECEJ, IEEE, ITEJ, JSAP

O'Keefe, Thomas J
Business: Univ of MO-Rolla, Rolla, MO 65401
Position: Prof of Metallurgical Engrg. *Employer:* Univ of Missouri-Rolla. *Education:* BS/Met Engg/MO Sch of Mines; PhD/Met Engg/Univ of MO, Rolla. *Born:* Oct 1935. Process control Metallurgist for Dow Metal Prods 1958-61; with the Univ of Missouri-Rolla since 1961; served in rank of Instructor to Prof; presently engaged in teaching, res & cons in the area of non-ferrous metallurgy with emphasis in electrometallurgy, extraction & refining. AIME Exec Ctte 1976, AIME Bd of Dirs 1978-81, UMR Outstanding Teaching Award 1979. Certificate of Commendation Phi Kappa Theta 1970. UMR Alumni Merit Award 1971 Outstanding Educators of Amer 1975. *Society Aff:* AIME, AES, ΣΞ, ΤΒΠ, ES.

Okochi, Masaharu J A
Home: 5-51-12 Denenchofu Ohta-ku, Tokyo, Japan 145
Position: Prof *Education:* PhD/EE/Tokyo Inst of Tech; BS/EE/Tokyo Inst of Tech *Born:* 3/30/16 In 1948 started research and teaching carrier as Asst. Prof at Tokyo Inst of Technology, Tokyo. In 1959 promoted to Prof at Kanagawa Univ, Yokohama as Dean of Faculty of Engrg. In 1972 appointed as prof at Senshu Univ, Tokyo. Mainly concerned with the research of Microwave Technology and later, turning to computer, information and system sciences. Contributed to the development of design techniques for low-noise diode guns for microwave tubes, methods for frequency band compression and computeraized design systems. Elected as Fellow of IEEE for contributions to information and systems science in 1980. Enjoys classical music and ham radio. *Society Aff:* IEEE

O'Kon, James A
Business: 34 Peachtree Street NW, Suite 2900, Atlanta, GA 30303
Position: President *Employer:* O'Kon and Company, Inc *Education:* MCE/Environ Planning/NYU; CE/GA Inst of Tech *Born:* 8/8/37 Pres of O'Kon and Co, Inc firm of engrs and designers specializing in special structures. Mr. O'Kon is reg as a PE in fifteen states, including NY, IL, MA, GA, SC, NC, FL, AL, TN, KY, LA, TX, MS. Author of scientific and tech articles and contributor to scientific volumes relating to creative design and high-rise buildings and megastructures; lecturer and visiting critic at several universities; mbr of Ctte for Energy Conservation for AISC. He has designed many award winning and unique structures, including US Fed pavilion at Expo '82, Disneyworld projects, Aerial Tramway in NY City. Mr O'Kon had been included in the volume *Personalities of the South*, recently made a honorary citizen of Knoxville, TN; and is exec engr for the design of the 1982 World's Fair in Knoxville. Mr. O'Kon is a mbr of Central Atlanta Progress and the Atlanta Chamber of Commerce. *Society Aff:* CEC, SIPI, UUC, IAHS, NAI

Okrent, David
Business: 5532 Boelter Hall UCLA, Los Angeles, CA 90024
Position: Prof of Engrg & Applied Sci. *Employer:* Univ of California Los Angeles. *Education:* PhD/Physics/Harvard Univ; MA/Physics/Harvard Univ; ME/ME/Stevens Inst of Tech. *Born:* April 1922. Worked for NACA 1943-46; worked for Argonne Natl Lab 1951-71 pioneering in fast reactor physics & safety; Visiting Prof Univ of Arizona 1970- 71; Prof UCLA since 1971. Fellow Amer Physical Soc & Amer Nuclear Soc (ANS). P Chmn Math & Computations Div & Nuclear Reactor Safety Div ANS. A U A Distin Appointment Award 1970. Mbr US Nuclear Regulatory Commision Adv Ctte on Reactor Safeguards. Mbr Natl Acad of Engrg. Guggenheim Fellow, 1961-62 and 1977-78 ANS Tommy Thompson Award, 1980 Isaac Taylor Chair, Technion 1977-78; US Nuclear Regulatory Commission Distinguished Service Award, 1985. *Society Aff:* ANS, NAE.

Okress, Ernest
Business: 2400 chestnut street, Philadelphia, PA 19103
Position: Cons Engr *Employer:* Self-Employed *Education:* ScD/Electro-Physics/Math/SCT/England(with U of M/NYU/PIB); MSc/Electro-

Okress, Ernest (Continued)
Physics/Univ of MI; BEE/Elec Engg/Univ of Detroit. *Born:* 3/9/10. Native of MI. Taught US Office of Edn; Mgr Adv Dev and Westinghouse Microwave Center Westinghouse Elec Corp; Sr Engr Sperry Gyroscope Co; Mgr Plasma Physics Research Division, Amar Std Inc; Sr Physicist Brookhaven Natl Lab; Sr Physicist Schlumberger, Ltd - EMR Div; Sr Staff Physicist Genl Electric Co; Cons Engr (Self-Employed) Genl Elec Co. Univ City Sci Center & The Franklin Res Ctr. Westinghouse Outstanding Invention Awards. Cert Commendation US Office Sci R/D. NASA Certificate of Recognition. Natl Council Engg Examiners Cert Reg Engr. Fellow IEEE & APS. Editor-author: Books, papers. Patents in field. Contributions to the dev of: Electron and ion tubes and devices, particularly crossed-field electron tubes (e.g., super-power magntrons) and electron space-charge controlled tubes (e.g., high power triodes and reasonant tetrodes) and Microwave Power Engr. Electromagnetic container less levitation of solid and molten materials (e.g., tungsten, beryllium, alumina). Non-equilibrium (closed-cycle) magnetohydrodynamic power generator. Subnanosecond high voltage modulator. Electromagnetic spherical phased array thermonuclear fusion reactor. Rotating plasma energy storage. Screened plasma intense, low-emittance electron beam gun. *Society Aff:* IEEE, APS, ΣΞ, NCEE.

Oktay, Sevgin
Home: 097 Fox Run, Poughkeepsie, NY 12603
Position: Senior Engr., Mgr. *Employer:* IBM *Education:* ME/Mech. Eng.,/Columbia University; MS/Mech. Eng.,/Columbia University; BS/Eng., Sc./Antioch College *Born:* 04/17/35 Joined IBM in 1963 at the Thomas J. Watson Research Center, Yorktown Heights, NY as a Research Staff Mbr. Held various Senior Eng., managerial positions, and is a recognized expert in electronic packaging heat transfer. Recipient of IBM Outstanding Innovation Award and IBM Fifth Level Invention Achievement Award. Author or co-author of more than thirty invention publications, ten patents, more than twenty publications and two books. Actively involved as an officer of the Heat Transfer Div of the ASME, and is a Fellow of the ASME. Enjoys flying, tennis and playing musical instruments. Speaks other languages and fluent Turkish. *Society Aff:* ASME, CHMT

Okun, Daniel A
Home: Linden Road Route 7, Chapel Hill, NC 27514
Position: Kenan Prof of Env Engg Emeritus *Employer:* Univ of North Carolina at Chapel Hill. *Education:* ScD/Sanitary Engg/Harvard; MSCE/Civil Engg/CA Inst of Tech; BSCE/Civil Engg/Cooper Union. *Born:* June 1917 N Y C. US Public Health Serv 1940-42; Army 1942-46. ScD Harvard Univ. With Malcolm Pirnie Engrs 1948-52; Faculty Univ of North Carolina at Chapel Hill 1952; Head Dept of Environ Scis & Engrg 1955-73; elcted Chmn UNC-CH Faculty 1970-73. UNC Thomas Jefferson Award. Amer Acad Environ Engrs Gordon M Fair Award. Amer Soc of Civil Engrs Freese Award. New York Acad of Scis Billard Award. Water Pollution Control Federation Eddy & Fair Medals. Inst of Water Engineers (Britain) Friendship Medal, Amer Waterworks Assn Fuller Award. Catedratico Honorario Universidad Nacional de Ingenieria Peru. Mbr Natl Acad of Engrg. Mbr Just. of Medicine. Visiting Prof: Duke, Delft, Univ Coll London, Tianjn Univ. PRC. Author & internatl cons on water quality mgmt. *Society Aff:* NAE, AAEE, ASCE, IWPC, IOFM, AAAS, AAUP, AWWA.

Okwit, Seymour
Business: 180 Marcus Blvd, Hauppauge, NY 11787
Position: Pres *Employer:* LNR Communications Inc. *Born:* Aug 1929. BS from Brooklyn Coll, MS in applied Math & MS in theoretical physics from Adelphi Univ. Founder Chrmn of the Bd, CEO and Pres of LNR Communications Hauppauge N Y. Has pub 35 tech papers & holds sev pats in the field of microwave & millimeter active & passive components & sys. Fellow IEEE. P Editor of IEEE Transactions on Microwave Theory & Tech. Mbr & Chmn of the GMTT Admin Ctte. 1974 MTT Natl Lecturer etc.

Olander, Donald R
Business: Dept of Nuclear Engrg, University of California, Berkeley, CA 94720
Position: Prof of Nuclear Engrg *Employer:* Univ of CA *Education:* ScD/ChE/MIT; BS/ChE/Columbia Univ; AB/Chem/Columbia Univ *Born:* 11/6/31 Taught at UC Berkeley since 1958. Also Faculty Sr Scientist of the Lawrence Berkeley Lab. Research interests include: high temperature nuclear materials; gas-solid chem kinetics; nuclear waste management. *Society Aff:* ANS

Old, Bruce Scott
Business: PO Box 706, Concord, MA 01742
Position: Pres. *Employer:* Bruce S Old Assocs, Inc. *Education:* ScD/Metallurgy/MA Inst Technology; BS/Chem Engg/Univ of NC. *Born:* Oct 1913. Bethlehem Steel Res Dept 1938-41; Comdr USNR 1941-46; Author Bill establishing Office of Naval Res 1946; Arthur D Little Inc 1946-79 Sr V Pres Sr Metallurgist USAEC 1947-49; Press Sci Adv Ctte 1953-56; Pres Nuclear Metals Inc 1954-57; Natl Acad of Engrg 1968; Foreign Secy 1969-76. N Y Acad Sci. Fellow MIT, Amer Soc Metals, Amer Inst Chem, AAAS Patents & articles in metallurgy. Articles & books on res admin & tennis. *Society Aff:* ASM, AIME, ANS.

Oldfield, William
Home: 4561 Camino Molinero, Santa Barbara, CA 93110
Position: President. *Employer:* Materials Res & Computer Simulation. *Education:* PhD/Material Sci/Stanford Univ; MSc/Metallurgy/Manchester Univ; BSc/Chemistry/Manchester Univ. *Born:* 12/10/31. Currently specializing in applications of computer modeling. Applications include the modeling of a "liquids from coal" process, predicting coal decomposition and optimum plant conditions; and crack initiation and growth in steel and weldments, including the study of radiation damage effects. Previously involved in modeling crystal growth processes (especially dendritic growth) including the effect of zero-gravity; Ostwald Ripening" of bubble and crystal arrays in solids and liquids; hydrogen attack of coal liquifaction vessels; the study of casting solidification; and modeling of drop and particle dynamics. Early interests included cast iron solidification; the design and development of Fast Reactor fuel elements; and thermal fatigue failure of molds. 1966 Awarded ASM Grossman Award *Society Aff:* ASM, ASTM, ASCG, ΣΞ.

Oldham, William G
Business: EECS Dept, Berkeley, CA 94720
Position: Prof *Employer:* Univ of CA *Education:* PhD/EE/Carnegie Mellon Univ; MS/EE/Carnegie Mellon Univ; BS/EE/Carnegie Mellon Univ *Born:* 5/5/38 William G. Oldham studied at Carnegie Mellon Univ, and received the PhD degree in Electrical Engrg in 1963. He has worked at Siemens (1963-64) and at Intel (1974-76) on semiconductor materials research and integrated circuit development. He joined the faculty of the Univ of CA, Berkeley in 1964 where he is now a Prof in the Dept of Electrical Engrg and Computer Sciences and Director of the Electronics Research Laboratory. He is a member of the Editorial Bd of Solid-State Electronics, a Fellow of the IEEE, and a member of the National Academy of Engineering. *Society Aff:* IEEE, ECS, NAE

Olds, Joneil R
Business: Amoco Bldg, 17th & Broadway, Denver, CO 80202
Position: Div. Mech. Engr *Employer:* Amoco Production Co *Education:* BS/ME/OK St Univ *Born:* 10/11/28 in Sterling, CO. Joined Amoco, formerly Stanolind Oil & Gas, in General Office February, 1951, in gas plant construction group. Transferred to Division office in Casper, WY, in 1954 and assigned as Plant Engr at the Salt Creek Gasoline Plant; in 1956 supervising engrs and plant chemist. In 1960 assigned as Division Plant Engrg coordinator and in 1965 became Division Project Engr for design and installation of four drilling and production platforms in the ice infested waters of Cook Island, AK. In 1973 promoted to Division Engrg Supervisor. Currently Western Division Mechanical Engr, supervising Division Mechanical Engrg, Denver Region. *Society Aff:* AIME, MPE

Oldshue, James Y
Business: 135 Mt Read Blvd, Rochester, NY 14611
Position: V President Mixing Tech. *Employer:* Mixing Equipment Co Inc. *Education:* PhD/ChE/IL Inst of Tech; MS/ChE/IL Inst of Tech; BS/ChE/IL Inst of Tech. *Born:* 4/18/25 Specialist in fluid mixing in chemical, metallurgical, water-treating and related industrial process industries. Member, National Academy of Engineering: Fellow of the American Institute of Chemical Engineers; received Founders' Award of AIChE in 1981. President, AIChE 1979; active in various Board and Committee assignments of the American Association of Engineering Societies. Joined Mixing Equipment Company in 1950 as Head, Development Engineering. Successive positions include Director of Research, Technical Director and Vice President Mixing Technology. Published over 100 papers, articles, chapters and patents on general theory and application of fluid mixers. Conducted many continuing education courses for AIChE and other organizations. Lectures include participation in all world centers of Africa, Asia, Europe, North and South America. Original contribution and studies on role of fluid mixing in industrial processes and the fluid mechanics background, plus practical techniques of non-geometric scale-up parameters. Received Ken A. Rae award- AAES-1987. *Society Aff:* AIChE, ACS, NSPE, AAES, NAE, NYSPE.

Olen, Robert B
Business: 4223 Monticello Blvd, South Euclid, OH 44121
Position: Chief Engr. *Employer:* Delaval Turbine Inc Spec Prods Div. *Born:* Nov 1923 Cleveland Ohio. BSME Ohio State Univ; Grad study CIT & Akron Univ. Reg P E Ohio 26195, Wisconsin E12703. Served in US Marine Corp 1942-46. Held Engrg & Mgmt positions with various cos primarily involved with fluid power before joining Delaval Turbine Inc 1972 as Ch Engr; am respon for all engrg from concept thru prod of fluid handling equip for indus, aerospaces & defense applications. ASME, FPS (P Pres Cleveland Chap). NFPA Tech Bd & Chmn Hydraulic Valve sect USATAG to ISO/TC 131 are areas of current involvement. Recreational activities incl reading, fishing & spectator sports.

Olesen, Peter F
Business: Peter F. Olesen & Assoc, 500 W. Central Rd. Ste 201, Mount Prospect, IL 60056
Position: President *Employer:* Peter F. Olesen And Associates, Inc. Consulting Engineers *Education:* BS/Civil Engg/IL Inst of Tech. *Born:* April 23, 1933. Resident of Mount Prospect Ill. Reg P E in Illinois, Az, Fl, Ia & Wi. Served with US Army (NIKE) 1955-57, Founder & President, Peter F. Olesen & Assoc., Inc., Mt. Prospect, Ill, 1984. Prev. Exec VP Ciorba Gp, Inc., 1979-84, VP & Director Transportation Engnrg 1973-79; Ch Highway Eng. Knoerle, Bender, Stone & Assoc. Fellow: ASCE & ITE; mbr: ACEC, APWA, CECI, IAAPA, ISPE, NSPE & Rdwy Ltg Forum. Extensive exp. toll road sys., freeways & traffic engrg. theme parks, site dev. 11 yrs mbr & p Pres. Dist. 57 Bd of Educ. Mt. Prospect Ill. *Society Aff:* ACEC, ASCE, APWA, NSPE, ITE, IAAPA, Rdwy Ltg Forum.

Oleson, Calvin C
Home: 11026 Meade Drive, Sun City, AZ 85351
Position: Cons Engr. *Employer:* Wiss Janney Elstner & Assoc. *Education:* MS/Civil Engr/IA State Univ; BS/Civil Engr/SD State Univ. *Born:* Oct 1901. State Highway Const before 1936; Prof & Acting Head Civil Engrg SDSU 1936-46; District Concrete Engr Pa Turnpike 1939-40; AUS Engrs Pacific Theater 1943-45; Prin Res Engr Acting Head Field Res Portland Cement Assoc 1946-66; Cons Engr to present. Fellow ASCE. Life Mbr Fellow ACI, ASTM, H R B. Reg P E So Dakota, Illinois. Delmar E Bloem Award ACI 1974. Planning Comm Village Trustee Northbrook Ill. Numerous tech pubs. Private Flying, Photography, Golf, Pistol, Swimming. Specialist in Evaluation of Concrete in Service. *Society Aff:* ASCE, ACI, ASTM.

Olhausen, Dale D
Business: 2300 W Eisenhower Blvd, Loveland, CO 80537
Position: President/owner. *Employer:* Owner - Landmark Engineering Ltd. *Education:* BS/Civil Engr/SD State Univ; -/Graduate Work/Univ Southern CA *Born:* 1/24/35. in Hartley, IA. BSCE SD State Univ 1957; grad work at the Univ of So A (sanitary, hydraulics, struct) 1959. 1957-62 Engr for Los Angeles Cty Engr, LA CA. 1962-67 Proj Engr for Parker & Underwood Inc, cons Engrs in Greeley CO. 1967-69 Owner & engr - Rocky Moutain Engrg Service in Greeley CO. 1967- , Mgr & engr, Central Weld Cty Water Dist, Greeley CO. 1969- , Parnter & engr Hogan & Olhausen Inc, Cons Engrs, Central Weld Cty Water Dist, Greeley CO. 1969- , Owner & Engrg Landmark Engrg LTD, Cons engrs, Loveland-Greeley-CO (Past Manager, Horseshoe Lake Sanitation District & Management Sanitation District. Loveland CO). *Society Aff:* ACEC, ASCE, AWWA, WPCF, HBA, USCID, ACIL, CWC

Olin, John G
Business: The Sierra Bldg, Pilot Rd (P.O. Box 909), Carmel Valley, CA 93924
Position: Pres *Employer:* Sierra Instruments, Inc *Education:* PhD/ME/Stanford Univ; MS/ME/Stanford Univ; BS/ME/IL Inst of Tech. *Born:* 8/27/39 John G. Olin is the Pres, founder, and owner of Sierra Instruments, Inc, of Carmel Valley, CA. Sierra Instruments is a leading manufacturer of air and gas flow meters and controllers. Previously, Dr Olin was the Deputy Dir of the MN Pollution Control Agency. He received his PhD degree in Mechanical Engrg from Stanford Univ in 1966. Dr Olin has been in the air pollution and gas flow instrumentation fields for 20 years and has published over 25 papers and holds several patents in the fields. He is active in the Air Pollution Control Assoc and served as its Pres in 1983/84. In 1981, he received the IL Inst of Tech Alumni Profession Achievement Award and in 1986 the Honor I Alumni Athletic Achievement Award. *Society Aff:* ASME, APCA, ISA, SEMI, ΣΞ.

Oliner, Arthur A
Business: Polytechnic Univ, 333 Jay St, Brooklyn, NY 11201
Position: Prof *Employer:* Polytechnic Univ *Education:* PhD/Physics/Cornell Univ; BA/Physics/Brooklyn College. *Born:* Mar 1921. Res Assoc Polytech Inst of Brooklyn 1946-53; Prof of Electrophysics 1957- ; Dept Hd 1966-71; Hd, Dept of EE & Electrophysics 1971-74; Dir Microwave Res Inst 1967-82. About 150 pubs, 3 books. Fellow IEEE, AAAS, British Institution of Elec Engrs (IEE); Inst Premium (highest award of British IEE) 1964; Guggenheim Fellowship 1965-66; IEEE Microwave Prize 1967; First Natl Lectr IEEE MTT-S 1967; Outstanding Educators of Amer 1973. Sigma Xi Citation for Distinguished Res 1974; IEEE MTT-S Hon Life Mbr 1977; IEEE Microwave Career Award 1982; IEEE Centennial Medal 1984. Mbr Bd of Dir, Merrimac Indus; cons to indus. Natl Chmn IEEE Group on MTT 1959-60; Chmn URSI Comm 1 1961-64, Comm D 1981-87; Chmn NAS Adv Panel to NBS 1960-64. *Society Aff:* IEEE, OSA, AAAS, URSI.

Oliphant, Edgar Jr
Business: Foley Engineering, 7501 Front St, Kansas City, MO 64120
Position: President *Employer:* Foley Engineering *Education:* BS/Chem Engg/Univ of MO at Rolla. *Born:* 3/15/30. Employed by Standard Oil of Indiana for ten yrs in oil refinery tech service and operations. Served two yrs with the Army Chem Corps. PChmn of the Kansas City Section of AIChe. Joined J F Pritchard and Co in 1962 and served in various engg and sales positions. Prof engg in the States of MO & TX. Pres of The Pritchard Corp 1977-80. Joined Foley Engineering as President in July 1980. Awarded Honorary Degree as Chemical Engg by the Univ of MO. *Society Aff:* AIChE.

Olitt, Arnold
Home: P O Box 4395, Incline Village, NV 89450
Position: Consultant & Dir Emeritus. *Employer:* Woodward-Clyde Cons. *Education:* Grad Study/Civil Engg/Univ of CA, Berkeley; BSCE/Civil Engg/Univ of CA, Berkeley. *Born:* Oct 13, 1913, Portland Oregon. Civil Engrg fac 1943-50. Co-founder Woodward-Clyde & Assocs (now called Woodward-Clyde Cons), cons engrs, geologists & environ scientists. Past Mbr of Bd of Dir of parent co & the subsidiary firms

Olitt, Arnold (Continued)
of Woodward Envicon Material Research & Development & Woodward-Moorhouse. Also served as Chmn of the Bd of the latter firm. In addition, respon included the dev, organization & implimentation of the firm's bus dev prog. Now retired. Consults & lectures on mkting of prof services. *Society Aff:* ASCE, ACEC, SEAONC, SMPS.

Oliver, Ben F
Business: Dept Mtls Sci & Engrg, Knoxville, TN 37916
Position: Prof of Met Engrg. *Employer:* Univ of Tennessee. *Education:* BS/MET/PA State Univ; MS/MET/PA State Univ; PhD/MET/PA State Univ *Born:* Oct 19, 1927. Areas of interest: solidification, crystal growth, chem process met, ultra-high purity matls, composites, matls applications, thermodynamics. *Society Aff:* ASM, AIME, ASEE

Oliver, Bernard M
Business: 1501 Page Mill Rd, Palo Alto, CA 94304
Position: Consultant to the President *Employer:* Hewlett-Packard Co. *Education:* PhD/Elec Engg/Caltech; MS/EE/Caltech; BA/EE/Stanford. *Born:* May 1916. As mbr of tech staff of Bell Tele Labs from 1940-52, worked on high qual TV image generation for estab TV transmission standards (during WW II), automatic tracking radar error signal detection, application of info theory & PCM to TV transmission. Joined Hewlett-Packard in 1952 as Dir of Engrg. Named Vice President for Research and Development of Hewlett-Packard Company in 1957. Pres 1965, VP 1963-64, Dir 1967 IEEE. Dir IRE 1959-61; Mbr NAE. *Society Aff:* IEEE, AAS, NAE, NAS.

Oliver, Billy B
Business: Bedminster, NJ 07921
Position: VP - Planning and Design. *Employer:* AT&T - Long Lines. *Education:* BEE/-/NC State. *Born:* Feb 14, 1925 in Selma NC. BSEE North Carolina State 1954. Engaged as a Craftsman in 1942; left to obtain degree. Returned to Long Lines as a Student Engr & progressed in var engrg & sales assignments. Became N Y's Military Serv Engr March 1961. Transferred to Wash D C as Gov Serv Engr same yr. Appointed Acting Genl Mgr 1963 & in July 1964 named Proj Dir AUTOVON (Automatic Voice Network, part of the defense communications sys). Moved to Atlanta Apr 1967 as Area Ch Engr & accepted present assignment June 1, 1972. Mbr IEEE. Also Mbr of Eta Kappa Nu, Tau Beta Pi, Phi Kappa Phi & Blue Key.

Oliver, Paul E
Business: 543 Third Street, Lake Oswego, OR 97034
Position: VP *Employer:* Van Gulik/Oliver, Inc *Education:* BS/ME/Univ of CO *Born:* 1/10/34 VP and Chief Engr of a firm specializing in Mechl, Elec, Acoustical and Forensic Engrg Servs. Has responsibility for mgmnt and tech dir of most major design projects and studies. His expertise includes wood fired boiler plants, power plant design, energy conservation, pump operation and vibration. Some projects include design of a power plant and a facility energy plan for the Naval Base, Adak, Alaska, design of boiler installations, heat recovery on a solid waste incinerator, steam turbine installation and correction of pump operating problems. Wrote the Boiler Analysis Program for PCs, and has published papers on wood products plant energy conservation, wood fired power generation, and a municipal solid waste incinerator plant. PE Registration: OR, WA, AK, ID, CA, NV. *Society Aff:* ASME, ASHRAE, AEE, CECO, ACEC

Oliver, R C
Home: 2520 Red Bridge Terrace, Kansas City, MO 64131
Position: Exec Partner. *Employer:* Black & Veatch Engrs. *Born:* 3/9/12. BS Kansas Univ. Native of Kansas City MO. With Geophys Services Inc prior to WWII. USN 1943-46. Field engr fo Black & Veatch in connection with power projs primarily generating station const. With Black & Veatch in admin & mgmt pos since 1937. Current position as Consultant to BEV Hobbies: flying & misc sports. *Society Aff:* NSPE, IEEE, ASME, MSPE, ASCE.

Olivieri, Joseph B
Home: 37294 Ilene Dr, Mt Clemens, MI 48043
Position: Prof *Employer:* Lawrence Inst of Tech. *Education:* MS/Occupational Health/Wayne State Univ; BME/Mech Engr/Univ of Detroit. *Born:* 3/11/25. Detroit, MI 1925. Married, three children. Reg PE MI. Fellow grade ASHRAE and ESD. Awarded Engg Soc of Detroit Gold Award as yr's outstanding engr in 1976. Chief mech engr Gruen Assoc 1952 to 1956. Pres Olivieri Assoc Consulting Engrs 1956 to 1970. Pres OEM Assoc Architects/Engrs 1970 to 1976. Dir ASHRAE 1971-1975 and treas 1976. dir ASHRAE 1980 to 1983. Engg Editor Air Conditioning, Heating, Refrig News 1960 to present. Author "How to Design Heating and Cooling Comfort Systems–. Professor Lawrence Institute of Technology School of Architecture and school of engrg 1976-present. *Society Aff:* ASHRAE, ESD.

Olling, Gustav
Home: 2901 N Bigelow St, Peoria, IL 61604
Position: Prof & Dir, Mfg Res Inst *Employer:* Bradley Univ. *Education:* PhD/Engg Tech/Purdue Univ; MS/Numerical Control/Northern IL Univ; MA/Higher Education/Bradley Univ; BS/Numerical Control/Northern IL Univ. *Born:* 7/4/37. Native of W Germany. Completed apprenticeship training as tool and die maker. Worked in Chicago area industries, Buda Co., Ingalls & Shepard Co. (1955- 1965), as Mfg Engr specializing in automated mfg systems. With Bradley Univ since 1967. Developed mfg program with emphasis in Computer Aided Design/Computer Aided Mfg education. Currently Prof Dept of Mfg and Dir of the Mfg Res Inst. Serves as consultant to a large number of natl and intl organizations in the area of Computer Aided Mfg. Recipient of many awards including 1978 Engr of the Yr Award, IL Society of Prof Engrs. Published and presented over hundred papers in Computer Aided Mfg in the U.S., Taiwan, South Korea, Jan, Western & Eastern Europe. *Society Aff:* SME, ASEE, IFIP 5.3, AIMTECH.

O'Loughlin, John R
Business: 1201 E 38th St, P.O. Box 647, Indianapolis, IN 46223
Position: Prof. *Employer:* IN Univ. *Education:* PhD/ME/Purdue Univ; MS/ME/Univ of Pittsburgh; MBA/Finance/IN Univ; BE/ME/Youngstown State Univ. *Born:* 8/12/33. Native of Warren, OH. Teaching experience at Youngstown State Univ & Tulane Univ before accepting current position. Ind experience with eight firms via consulting & summer employment. Publications in the areas of combustion & heat transfer. *Society Aff:* ASME.

Olowu, Olayeni
Home: 12 Alhaji Tokan St, Surulere, Lagos, Nigeria
Position: Managing Dir *Employer:* Addis Engineering Ltd *Education:* BSc (Hons) London/Mech Eng/Woolwich Polytechnic. *Born:* 12/10/29 Educated in Nigeria and London, England. Worked as Mech Engr in the state ministry of works and later became head of the mechanical and electrical div. Appointed Dir of Relief Operations at end of our Civil War to plan logistics of distribution of relief materials throughout war affected areas. Established a private company in 1970 to design, develop and manufacture food processing machines which removed drudgery from traditional methods of processing. Company now designs and produces machinery for rural development, in the fields of water supply and rural transportation conferred with honorary membership of the ASME in 1980. *Society Aff:* NSE, ASME

Olsen, Harold L
Business: Donohue & Assoc, 6325 Odana Rd, Madison, WI 53719
Position: Chief Mech Eng *Employer:* Donohue & Assoc. *Education:* MSME/HVAC/Purdue Univ; BSME/HVAC/Mch Des/Cornell *Born:* Dec 1925. Chief Mech Engr Donohue & Assocs Cons Engrs, Madison Wisc. US Navy 194346. Designer aircraft support heating-cooling devices, Pacific Airline Equip Co 1948. Application Engr plant layout crushing, cement, mining Allis Chalmers 1949-51. Designer Heating, air-cond sys Harry Wilson Assocs 1952. Designer railway & bus

Olsen, Harold L (Continued)
air-cond - refrig equip, Waukesha Motors 1952. Designer heating, plumbing, ventilating, regrigeration - air-cond sys, Wisc State Div Arch 1953-55. Pres, Ch Mech Engr Olsen & Assocs 1955-79. Ch Mech eng Donohue and Assoc 1979-prsent. Chmn T-5 & T-6 CSI Ctte; V Chmn CEC-CSI Liaison Ctte; P Pres Wisc CEC; P Pres Madison CSI; Hon Mention Mech Specification CSI; Fellow CSI; P Pres Madison Chap ASHRAE. Reg PE WI, IA, IL, MN,MI, Natl. *Society Aff:* CSI, WSPE, ASHRAE, CEC, AIPE.

Olsen, Joseph C
Business: 8 S Montgomery-POB 1809, San Jose, CA 95109
Position: President. *Employer:* Engrg Consulting Services. *Education:* DScEngg/Engg/CA Western Univ; JD/Law/Santa Clara Univ; MBA/Bus/Santa Clara & Stanford; BSME/Engg/Univ of WA. *Born:* Sept 28, 1927 NYC; s. Henry & Rosetta O. Tool & diemaker's apprenticeship 1948-53; m. Dina Elizabeth Schotz Oct 26, 1954; c. Joseph Carl II, Paul Maurice. Pres multi-disciplined cons engrg firm San Jose Calif 1963- . Pres Calif Investors' Council 1971-73. USNR 1952-60. Reg P E Calif. Mbr ANSI, ASME, ASTM, NSPI, NSPE, Cal Soc Prof Engrs, Cons Engr Assn Calif, ICBO, IAPMO Engrs Club San Jose. Patentee in field. Contrib articles to prof journals. *Society Aff:* ASME, AMSI, ASTM, NSPI, NSPE, CSPE, ICBO, IAPMO.

Olsen, Richard A
Business: 62-91, B538, 1111 Lockhead Way, Sunnyvale, CA 94086
Position: Human Factors Staff Engineer Specialist *Employer:* Lockheed Space Systems Division *Education:* PhD/Exp. Psychology/PA State Univ; MS/Exp. Psychology/PA State Univ; BS/Phys/Union College. *Born:* 11/1/33. in MN, raised also in Upstate, NY. Served 3 yrs in electronics and communications on US Navy destroyer and as instr. Design display systems for Hughes Aircraft, Ground Systems Group (1960-64) before returning to grad sch. After the MS degree (1966), I served as Dir of Human Factors Res Div of the PA Transportation Inst, and (1974), Asst Prof of Human Factors of Engg, (Ind & Mgt Systems Engg Dept), teaching since 1969. Summers spent at HRB-Singer, Inc, as Res Psychologist; VTI in Sweden, as Visiting Scientist; Fed Hgwy Admin, office of Res, as Res Psychologist; and NASA/ASEE Fellow, Langley AFB, VA. At LMSC since July 1980. Consultant and Expert witness in Human Factors and Accidents. Publications and Book Chapters. *Society Aff:* HFS, APA & DIV 22, IEEE, AAAM, IEA, CSPA, LICENSED PSYCHOLOGIST, ΣΞ.

Olsen, Sydney A
Home: 2329 Pleasant Dr, Cadar Falls, IA 50613
Position: Manager, Engine Engrg. *Employer:* John Deere - Prod Engrg Ctr. *Education:* MS/Mech Engr/MI State Univ; BS/Mech Engr/SD Sch Mines & Tech. *Born:* 3/21/25 in Aberdeen SD. US Navy 1943-46. Started with Deere as Dev Engr 1951; became Ch Engr, Engines 1970 & Mgr of Engine Engrg 1973; respon for design & dev of all Deere & Co engines worldwide. Reg PE IA. SAE Mbr 1952; active in Farm, Const & Indus Machinery & Powerplant Activities, Chmn FCIM Activity 1969; Chmn Engrg Activity Bd 1973; Mbr SAE Bd of Dir 1974-76. Mbr Bd of Dir Engine Manufacturers Assn; VP Engine Manufacturers Assn 1975-76; pres EMA 1977-78. *Society Aff:* SAE.

Olson, Charles Elmer, Jr
Business: School of Nat Resources, Ann Arbor, MI 48109-1115
Position: Professor *Employer:* Univ of Michigan. *Education:* PhD/Forestry/Univ of MI; MF/Forestry/Univ of MI; BSF/Forestry/Univ of MI. *Born:* Feb 18, 1931. Served as image interpreter/photogrammetrist with US Navy 1953-56. Univ of Ill fac 1956-63, Chmn Ctte on Aerial Photography 1960-63. Univ of Michigan 1963- : Prof 1973; Dean School of Natural Resources 1974-75; Chmn univ-wide Remote Sensing Prog 1972-79. Mbr NAS/NRC Ctte on Remote Sensing Programs in Earth Resource Surveys (CORSPERS) & Chmn of its biology panel 1973- 78. U S Reporter to Comm VII (Data Handling & Interpretation) of the Internatl Soc for Photogrammetry 1968-76. Amer Soc of Photogrammetry Photo Interpretation Award 1966. Member State of Michigan Inventory Advisory Committee 1981- (appt. ends in 1987). *Society Aff:* SAF, ASP, SPIE, AAAS.

Olson, David L
Business: Off Research & Dev, Colorado School of Mines, Golden, CO 80401
Position: Dean of Research and VP of R&D *Employer:* CO Sch of Mines. *Education:* Postdoctoral studies/Met Engg/OH State Univ; PhD/Mtls Sci/Cornell Univ; BS/Physical Met/WA State Univ. *Born:* 3/17/42. Native of Spokane, WA. Tech Staff of Semiconductor R&D Lab of TX Instruments (1969-70). The OH State Univ Post Doctoral Fellow (1970-72). Dean of Research and V Pres of R&D (1986-present) Prof of Met Engrg at CO Sch of Mines (1972-present) teaching & res in welding, corrosion, reactive metals, physical met. Sabbatical studies at Norwegian Inst of Tech (1979). Bradley Stoughton (ASM) Award (1976), Comfort Adams (AWS) Teaching Award (1978), Adams Lecturer (AWS) 1984, ASM Fellow (1985), McKay-Helm (AWS) Award 1985, McKay-Helm (AWS) Award 1986, Warren F. Savage (AWS) Award 1987, Sigma Xi, Alpha Sigma Mu, Tau Beta Pi, NY Acad of Sci, University and Weldability Ctte of Welding Res Council, ASM Joining Division Council (Past-Chairman), NAS Marine Structures Ctte. Conslt in welding, corrosion, mtls selection, PE. *Society Aff:* APS, AWS, ASM, AIME, ACerS, NACE

Olson, Donald R
Business: 255 Applied Research, University Park, PA 16802
Position: Emer Prof of Mech Engr *Employer:* Retired *Education:* BS/Mech Engr/OR State; M Eng/Mech Engr/Yale; D Eng/Mech Engr/Yale. *Born:* Dec 1917 Sargent Nebr. Served on Yale fac 1951-62 as Asst & Assoc Prof. Served as Prof of Mech Engrg & Hd, Underwater Powerplants, Appl Res Lab, Penn State 1962-71. Prof & Hd Mech Engrg Dept, Penn State 1971-83. Bd of Dir of SAE 1970-72. Res efforts in combustion, heat transfer & power sys. *Society Aff:* ASME, SAE, ΣΞ.

Olson, Emanuel A
Home: 925 S 52nd St, Lincoln, NE 68510
Position: Prof Emeritus, Univ. of Nebraska *Employer:* Retired *Education:* BSc/Eng/Univ of NE. *Born:* Feb 1916. Native of Upland Nebr. Extension Agri Engr Univ of Nebr since 1939, except for 3 1/2 yrs with US Army. Respon for dev educ progs & supporting matls in household waste disposal sys, domestic water sys, crop drying & storage sys, farm machinery repair & maintenance, concrete tilt-up const with insulated sandwich panels, livestock housing sys & livestock waste mgmt sys. Mbr ASAE: served as Chmn of Farm Structs Div, Mid-Central Sect & nominating Ctte for 1975- 76, elected Fellow 1975. Received 12 Blue Ribbon awards from ASAE in Educ Aids competition. Awarded Outstanding Engrg Achievement Award 1974 by Prof Engrs of Nebr, (State Chap of NSPE), elected Life Fellow ASAE 1979, Achievement Award by Nebr Sand, Gravel and Ready-mix Concrete Assoc 1978, Mbr Nebr Environment Control Council, 1978 Cert of Achievement from livestock environ Sci Committee 1978, Retired Colonel US Army. Retired Prof Emeritus, Univ of NB 1978. 1979 ASAE Gun Logson Countryside Engineering Award 1981 "Excellence in Concrete" Award, Oregon Concrete & Aggregate Producers Association. *Society Aff:* ASAE, NSPE.

Olson, Ferron A
Business: 412 Browning Bldg, Salt Lake City, UT 84112
Position: Prof of Metallurgy. *Employer:* Univ of UT. *Education:* PhD/Fuels Engg/Univ of UT; BS/Chemistry/Univ of UT. *Born:* 7/2/21. Native of UT. Served in Army Signal Corps 1943-46. Following grad served as Res Chemist, Shell Dev Co 1956-61. Since 1961 with metallurgical Engg Dept, Univ of UT. Full Prof in 1968 and dept chrmn 1966-74. Fulbright lecturer 1974-75 and distinguised prof 1980 in Yugoslavia. Mbr Bd of Dir of ABET 1975-1982. Chrmn UT Sec of AIME 1978-79. Lecturer at Universities in Chile, Yugoslavia, Rumania and W Germany. Consultant for firms including Metallurgical Ctr in Chile. Married to former Donna Lee Jefferies and father of Kandace, Randall, Paul, Jeffery and Richard. Active mbr LDS Church. Enjoys outdoor activities. Dir of UT Mining and Mineral Resources

Olson, Ferron A (Continued)
Res Inst, 1980 to present. *Society Aff:* AIME, ASM, ΣΞ, FULBRIGHT ALUMNI ASSN.

Olson, Glenn M
Business: John Deere Road, Moline, IL 61265
Position: Mgr, Cultivation, Cotton & Planting Equip Planning *Employer:* Deere & Co *Education:* BS/Agric Engr/IA State Univ *Born:* 02/25/39 Native of IA. Employed by John Deere since 1959. Career has specialized in tillage equipment design, development and manufacture. Have been Mgr, test; Mgr, Engrg Services; Mgr, Reliability and Division Engr in Charge of Disk Design prior to present assignment. Active in ASAE Quad City Section and FIEI Tillage and Crop Production Equip Council. ASAE Section Chrmn 1980-81, Secy 1975-76 and chrmn of several ASAE and FIEI committees. Registered PE in IL. Hobbies include gardening, camping and reading. *Society Aff:* ASAE

Olson, Harry F
Home: 71 Palmer Sq West, Princeton, NJ 08540
Position: Staff Vice Pres - ret. *Employer:* RCA Labs, RCA Corp. *Education:* BE/Elec Engg/Univ of IA; MS/Physics/Univ of IA; PhD/Physics/Univ of IA; EE/Profl/Univ of IA; DSc/Hon/IA Wesleyan. *Born:* Dec 1901. Hon DSc Iowa Wesleyan. Res engr RCA Corp 1928. Staff V P Acoustic Res, RCA Labs 1965. Pioneering res in directional microphones, loudspeakers & sound reproduction. Author 135 sci articles & 10 books. More than 100 U S pats. Mbr Tau Beta Pi, Sigma Xi & Natl Acad of Sci. Fellow SMPTE, IEEE, APS, ASSE; Pres ASA 1952, Pres AES 1959. AES Potts Medal 1952, SMPTE Warner Medal 1959, Phila Scott Medal 1956; ASSE Ericsson Medal 1963; IEEE Kelly Medal 1967, IEEE Lamme Medal 1970; ASA Silver Medal 1974 ASA Gold Medal 1981. *Society Aff:* NAS, ASA, AES, AIEE, SMPTE, APS, ASSE.

Olson, John W
Business: 54 Far Pond Rd, Southampton, NY 11968
Position: John W Olson, PE. *Employer:* Consulting Engr. *Education:* MCE/Civ/Polytech Brooklyn; BSE/Civ/Polytech Brooklyn. *Born:* May 1926. BCE 1952, MCE 1958 Polytechnic Inst of Brooklyn. USAAF 1944- 46. Assoc with sev N Y engrg firms as design engr 1952-61. Principal with Olson & Cruse Bldg Corp 1961-65 doind const in N Y. 1965- , cons engr John W Olson P E, struct & arch engrg. Reg NY & Fla. Enjoy sailing. *Society Aff:* ASCE, NYYSE, ACI

Olson, L Howard
Home: 6455 Scott Valley Rd NW, Atlanta, GA 30328
Position: Assoc Prof *Employer:* GA Inst of Tech *Education:* PhD/Textile Physics/Univ of Manchester Inst Sci & Tech (England); MS/Textile Engrg/GA Inst of Tech; BSE/ME/Princeton Univ *Born:* 12/29/41 Native of GA. Reg P.E. GA 8311. Secy/Treasurer of Textile Quality Control Assn and Textile Operating Executives of GA. Computer Coordinator for Textile Engrg. Professional interests include high performance woven and knit textiles, and process control. Currently developing flexible components for NASA prototype space suit, and high strength lines for rapid payout of inertial loads. *Society Aff:* NSPE, AATT, FS, TQCA

Olson, Lee R
Business: Nyala Farms Rd, Westport, CT 06880
Position: Exec VP *Employer:* Stauffer Chem Co. *Education:* MS/Chem Eng'g/Northwestern U; BS/Chem Eng'g/Northwestern U *Born:* 1/13/29. Evanston IL. With Stauffer Chem Co since 1969. Assmed current respon in 1981. Previously Mgr Chem Engrg Res in Stauffer's Res Dept; Dir of Mfg & Group Bus Mgr in Plastics Div, VP Corp Planning & Dev, & Group VP. Previously with Internatl Minerals & Chem Co, Skokie IL 1964-69. Amer Cyanamid Co NYC & Wayne NJ 1955-64. Dow Chem Co Midland MI 1952. US Army Chem Corps Edgewood Md 1953-54. Mbr AIChE (Treas of Chgo Sect 1968); Tau Beta Pi, Sigma Xi. *Society Aff:* AIChE

Olson, Melvin A
Home: Gap Head Rd, Rockport, MA 01966
Position: Retired *Education:* BS/Chemistry/Harvard. *Born:* May 1915. Native of Rockport Mass. Served with Navy Amphibious Forces WW II. Awarded 5 battle stars, Legion of Merit with Combat V, Bronze Star with Combat V by US Navy & Mention in Dispatches by British Admiralty. R&D Lever Bros Co 1938-41 & 1945-47. Sales engr of process equip 1947-78. Founder in 1961 & Retired Pres of M A Olson Co Inc, Indus Sales Engrs. Reg P E Mass. Mbr AIChE, TAPPI & WPCF. Retired CDR, USNR. Hobbies: travel & sailing. *Society Aff:* AIChE, TAPPI, WPCF.

Olson, Robert R
Home: 17 Black Swan, Irvine, CA 92714
Position: Consultant *Employer:* Self *Education:* Petroleum Engr/Petroleum Production Engg/CO School of Mines; Program for Mgt Dev/Business/Harvard Business School. *Born:* 10/17/25. Native of Denver, CO. Served with US Navy 1944-46. 1949-53 Resident Engr for Panhandle Eastern Pipeline Co, involved with design and construction of compressor stations. CO Interstate Gas Co 1953-80. 1963-75 - Chief Engr, 1975-79 Asst VP Engrg. responsible for design and construction of new facilities. 1979-80 Asst VP, Operations. 1980-86 Manager, Projects, Fluor Engineers Inc. assigned to various Alaskan Projects. 1986- Independent Consultant. Mbr of ASME Gas Piping Tech Comm, Patentee in field, Lutheran, licensed amateur radio operator. Petroleum Engr. CO Sch of Mines 1949. Program for Mgmt Dev, Harvard Bus Sch 1966.

Olson, Robert W
Home: 5322 Falls Rd, Dallas, TX 75220
Position: V P Res & Engrg - ret. *Employer:* Texas Instruments Inc. *Education:* BS/EE/Univ of MN. *Born:* Feb 17, 1915. Fellow IEEE. Reg P E Texas 10339. Retired 1970 as V P Texas Instruments (mostly res & engrg). Pres Assn for Grad Educ & Res of No Texas (TAGER) 1966-69 (on leave from Texas Instruments). Pres Houston Tech Labs 1953-57. Sr Engr U S Navy, Bureau of Aeronautics 1940-45. Mbr Texas St Bd of Regis for Prof Engrs 1966-71, Chmn 1969-70. Outstanding Engr Dallas & Preston Trails Chap TSPE 1972. Dir Indus Res Inst (corp mbrship) 1969-70. Pres 17th Internatl Sci Fair Dallas 1966. *Society Aff:* IEEE.

Olson, Roy E
Business: Dept of Civil Engrg, Univ of Texas, Austin, TX 78712
Position: Professor. *Employer:* U of Texas. *Education:* PhD/Civil Engg/Univ of IL; MS/Civil Engg/Univ of MN; BS/Civil Engg/Univ of MN. *Born:* Sept 1931. Civil Engrg Prof at Univ of Illinois 1960-70 & Univ of Texas 1970- . Chmn Geotech Engrg Div ASCE 1973-74 & U S Natl Soc of Soil Mech & Found Engrs 1973-74. (Management Group E in ASCE 1981-84). First Minn Surveyors & Engrs Soc Fellowship 1953-54, first ASTM Grad Fellowship 1960, ASCE Walter L Huber Res Prize 1972, ASTM Hogentogler Award 1973 and 1987, ASCE Croes Medal, 1985; ASCE Norman Medal 1975. *Society Aff:* ASCE, ASTM, TRB, USNS-SMFE.

Olson, Valerie F
Home: 5828 Serrania Ave, Woodland Hills, CA 91367
Position: Mgr Tech Staff *Employer:* Hughes Aircraft Co *Education:* MS/Physics/Rutgers Univ; BS/Physics/City Coll of NY *Born:* 2/14/40 educated in NY. Early work in nuclear engrg and rocket spectroscopy. Optics activities span almost 20 years with specialties in missiles, trackers, and space sensors. Particular interest include metal mirrors, infrared materials and low cost fabrication techniques. Management responsibilities have ranged from small to large line organizations and small projects to large, multimillion dollar projects. Simultaneous Directorship of OSA and Governor of SPIE 1979- 1981. Amateur archaeologist and ardent white water enthusiast. *Society Aff:* OSA

Olson, Walter T
Home: 18960 Coffinberry Blvd, Fairview, OH 44126
Position: Conslt *Employer:* Self *Education:* PhD/Chem Eng/Case Inst Tech; MS/Chem/Case Inst of Tech; AB/Chem/DePauw Univ. *Born:* 7/4/17. m. Ruth E Barker;

Olson, Walter T (Continued)
son David Paul. Instr Case Inst of Tech 194142; res scientist NASA Lewis Res Ctr Cleveland Ohio 1942-45, Ch Combustion Branch 1945- 50, Ch Chem & Energy Conversion Res Div 1950-63, Dir Tech Utilization & Public Affairs 1963-81. Mbr num gov adv cttes, incl Dept of Defense, Natl Acad of Sci, NACA/NASA, Air Force, Ohio Bd of Regents. V Chmn United Appeal of Greater Cleveland 1965-73. Trustee, Blue Cross NE OH, 1972-84. Disting Serv Award Cleveland-Akron Sect AIAA. Career Serv Award Fed Exec Bd. Mbr Sigma Xi; Fellow AIAA; Fellow AAAS; Mbr ACS; Hon Dir, Combustion Inst; Visiting Ctte, Fenn Coll of Engrg, Cleveland State Univ; Cosmos Club (Wash D C); City Club (Cleveland) *Society Aff:* ACS, AIAA, AAAS, CI.

Olsson, John E
Business: 611 NBC Center, Lincoln, NB 68508
Position: Chrmn & CEO *Employer:* Olsson Assocs *Education:* BS/ME/Univ of NB; Marine Engrg/Steam & Diesel/NY State Maritime Coll *Born:* 5/14/26 Richmond Hill, Queens, NY. Graduated NY State Maritime Coll, 1946. USNR 1946-1947. Graduated Univ of NB, BSME, 1951. Fulton & Cramer consulting firm 1951-1956. Currently CEO, Olsson Assocs, Consulting Engrs-Architects, Lincoln, NB. Honorary member Tau Beta Pi, Pi Tau Sigma and Chi Epsilon Engrg honoraries. Registered Engr NB, SD, KS and CO. Registered Land Surveyor, NB. Past Chrmn, NB Section AWWA; Past Pres NB Consulting Engrs Assoc. Univ of NB Engrg Advisory Bd Treas. Past Pres Lincoln Chamber Commerce; Past Chrmn United Fund. Past Pres Comm Chest, Bd of Dirs National Bank of Commerce; Trustee, Cooper Foundation, Bankers Life NB. *Society Aff:* APWA, ASCE, ASME, IEEE, NSPE, APPA, AWWA

Olt, Richard G
Home: 312 Aberdeen Ave, Dayton, OH 45419
Position: Consultant. *Employer:* Self-Employed. *Education:* ME/Mech Engg/Univ of Cincinnati. *Born:* Aug 1909; m. 1938; c. 2. P E Ohio E9508. Fellow Mbr ASME 1971. Native Dayton Ohio. Over 40 yrs in indus res, dev, & production. Affiliations incl Delco Products 1929-36 & Moraine Products 1938-44, Divs of GM; Infilco 1936-37; Johnson & Johnson 1937; Optron Lab 1944-47; Monsanto Chem Co & subsidiaries 1947-61; Monsanto Res Corp 1961-75 - retired. Significant accomplishments incl dev of first commercial porous metal filters; quartz fiber untramicrobalances for microassay of radioisotopes; an absolute geometry alpha counter; an analysis sys for continuous monitoring of combustion products; unit micron diameter quartz capillaries. IR 100 Award 1969. *Society Aff:* ASME, OSA.

Olt, Theodore F
Home: 113 Euclid St, Middletown, OH 45042
Position: Retired. *Employer:* Armco Steel Corp. *Born:* May 18, 1905 Dayton Ohio. Chem Engrg degree from Univ of Cincinnati Ohio 1928. Joined Armco Steel Corp, Middletown Ohio, in Met Dept, Oper Div 1928. Transferred to Armco Res Ctr 1934 as Sr Res Engr. In 1940 was made an Assoc Dir & in 1947 was made Dir of Res. Was elected V P Res & Tech 1958. Elected V P Strategic Planning 1967. Retired March 31, 1969 as Asst to the Pres. Recipient of the Benjamin F Fairless Award, Amer Inst of Mining, Met & Petro Engrs Feb 1974.

Olver, Elwood F
Home: 402 Burkwood Ct, Urbana, IL 61801
Position: Prof. and Asst. Dean, Coll Engg. *Employer:* Univ of IL. *Education:* PhD/-/IA State Univ; MS/Agri Engr/PA State Univ; BS/Agri Engr/PA State Univ. *Born:* 4/10/22. Reared in PA. Was navigator-bombardier in WWII. Rural Engr with PA Power and Light Co, 1946-48. Teacher & researcher at PA State Univ, 1948-50 & 1952- 60. Was Educational Dir, IA Rural Elec Cooperative Assoc, 1950-52. Teacher and researcher at Univ of IL, 1960 to 1984. Group leader for Univ of IL res team in India, 1967-69 and short res trips to Africa & other developing countries. Exec Secy of the IL Electrification Council, 1970 to 1980, and was presented its merit award. Fellow in the Am Soc of Agri Engrs (ASAE). Hd of the Elec Power & Processing Div (EPP) of Agri Engg Dept, 1970 to 1982. Chrmn of EPP Div of ASAE, 1976-79. Author of book "Engineering Principles in Agriculture–. Consultant to the US Dept of Agri in Agri Energy Res, 1977-78 and 1980-84. Licensed engr in IL & PA. Currently Asst Dean, Coll Engg, Univ IL. *Society Aff:* ASAE, ΓΣΔ, ΦΚΦ, ΣΞ

O'Malley, Dennis M
Home: 14 Meadowlark Dr, Latham, NY 12110
Position: Signal System Engr *Employer:* City of Albany *Education:* MS/Traffic Engrg/Central MO St Univ; BS/Sci/Suny/Albany *Born:* 10/13/48 Native of Glens Falls, NY. Dir of Traffic Engrg Dept from 1974-1977. Worked with Albany County 1977-1979. Currently working on City's Computerized Signal System. Currently Pres of ITE Upstate NY Section, Editor of Technical Notes, Member of Policy Committee, District Dir Urban Traffic Engrs Council. Elected to "Who's Who in American Government" and Community Leaders and Noteworthy Americans–. Married to the former Evelyn A LaGrange. Two children, Sean and Colleen. *Society Aff:* ITE

O'Malley, Robert F
Home: 609 8th St, Marietta, OH 45750
Position: Traffic Field Engr. *Employer:* Ohio Dept of Transportation. *Education:* BS/Civil Engg/IN Inst Tech. *Born:* Jul 1923. BSCE 1949 Indiana Inst of Tech & special courses at Northwestern & Ohio State. Served with US Marines during WW II. Traffic Field Engr for ODOT; Served as liaison & represents Engr of Traffic in field districts in traffic engrg & admin matters. Served as Master Councilor of DeMolay, Chap Pres, Tr & Hwy Practice Sect Chmn of OSPE, Chmn of Marietta Traffic Comm & Pres f Ohio Sect Inst of Transportation Engrs. Recipient of Amer Assoc State Hwy & Trans Officials. 25 year Award of Merit. *Society Aff:* ITE.

O'Meara, Robert G
Business: 414 E 12th-City Hall, Kansas City, MO 64106
Position: Ch, Div of Distribution. *Employer:* Kansas City, Mo Water Dept. *Born:* Feb 25, 1939. BSCE S Dakota State Univ 1962. Reg P E SD & Mo. Native of Sioux City Iowa. Proj Engr Sioux Falls S D 1962-68. With Kansas City Mo Water Dept since 1966. Assumed current respon as Ch, Distrib Div in 1970. V P ASCE, S D Eastern Branch 1966; V P Southeastern Chap S D Engr Soc 1966; PEG Dir MSPE 1974-75; State Dir MSPE 1975-76; Chmn Mo Sect AWWA 1975-76; Young Engr of Yr 1974 Western Chap MSPE. Active in youth sports programs & scouting. Enjoy hunting & fishing.

O'Melia, Charles R
Business: Dept of Geography and Envir Engrg, The John Hopkins University, Baltimore, MD 21218
Position: Prof of Environ. Engg *Employer:* The John Hopkins Univ. *Education:* BCE/Civil Engr/Manhattan College; MSE/Sanitary Engr/Univ of MI; PhD/Sanitary Engr/Univ of MI. *Born:* Nov 1, 1934 NYC. Asst San Engr with Hazen & Sawyer, Engrs 1956-57. Asst Prof of San Engrg Ga Inst of Tech 1961-64. Post-doc Fellow Harvard Univ 1964-66. Assoc Prof 1966-70 & Prof 1970-80 , Univ of N Carolina at Chapel Hill Prof, 1980-The John Hopkins Univ.. Visiting Prof Caltech 1973-74. Publ award AWWA 1965-85; ASCE award for application of res to practice 1972; Disting Fac Award 1972 & Environ Sci Award 1975 AEEP. Assoc Editor Environ Sci & Tech. Mbr ASCE, ACS, AWWA, WPCF, ASLO, AEEP, Chi Epsilon, Sigma Xi. Reg P E Ga. Bd of Dirs, AEEP, 1977-79; Pres, AEEP, 1979. ASCE Simon Freese Award, 1985. *Society Aff:* ASCE, ACS, AEEP, AWWA, WPCF, ASLO, AAEE.

Omi, Hanzo
Home: 1-23-12 Kugahara, Otaku, Tokyo, Japan 146-00
Position: Counceler (Past Pres) *Employer:* Fujitsu Labs Ltd. *Education:* D of Engrg/Comp/Tokyo Inst of Tech; B/EE/Tokyo Inst of Tech *Born:* 4/5/01 Feb. 1923, he joined to the South Manchuria Railroad Co. April 1936, he moved to the Fujitsu Limited, appointed as a Dir of the Bd of the Fujitsu in 1945, a Managing Dir in 1953, and Executive Managing Dir in 1960, The Pres of the Fujitsu Labs Ltd. in 1968, a Technical Counselor of the Fujitsu in 1971, and a Counselor of the Fujitsu Labs in 1975. He is awarded The Third Class of Order Zuihosho Decoration (Nov. 1971) and The Medal of Blue Ribbon (1964) from the Emperor of Japan for his contribution to the Industry of Electronic Computers. Also, he received the Founders Medal from the IEEE. *Society Aff:* IEEE

Omori, Thomas T
Home: 1601 Parway Dr, Glendale, CA 91206
Position: President. *Employer:* TTO Internatl Inc. *Born:* Oct 1917. BS, MS & DSc Carnegie Inst of Tech 1940, 1941, 1947. Native of Prescott Ariz of Japanese parents. Solid propellant rocket res at Allegany Ballistics Lab, Cumberland Md during WW II. Rocket res & dev work for JPL 1947- 50. Served with Aerojet Genl Corp until 1966; 3 yrs in Paris as Mbr of Aerojet- NATO Propulsion Prog & 2 yrs in Tokyo as Far Eastern Mgr for Aerojet. V P Kreha Corp of Amer (petrochems) 1970-75. Current pos as Pres TTO Internatl Inc since 1975, tech mkting.

Omura, Jimmy K
Home: Apt 214, 22330 Homestead Rd, Cupertino, CA 95014
Position: Chmn *Employer:* CYLINK Corp. *Education:* PhD/EE/Stanford Univ; MS/EE/MIT; BS/EE/MIT. *Born:* 9/8/40. Born in San Jose, CA in 1940. Served as chrmn of the San Francisco Sec of the Info Theory Group of IEEE & as an AdCom mbr of the intl Info Theory Group for six yrs. Res engr for Stanford Res Inst from 1966 to 1969. Prof of Engg at UCLA from1969 to 1987. Consultant to ind & govt specializing in communication theory. Coauthor (with A J Viterbi) of "Principle of Digital Communication and Coding" and (with Simon, Scholtz, and Levitt) "Spread Spectrum Communication-Volumes I, II, and III–. Published over 100 papers & served on Natl Sci Fdn committees. Fellow of the IEEE. *Society Aff:* IEEE.

O'Neal, Orville R
Business: 50 Lawrence Rd, Springfield, NJ 07081
Position: Mgr, Purchased Products Engrg. *Employer:* Western Elec Co Inc. *Education:* BS/EE/TX Tech. *Born:* Aug 1930 Morton Texas. BSEE Texas Tech. With Western Elec since 1952. Designed transformers & Power supplies for BTL for 2 yrs, production engr 5 yrs, Mfg Engrg Supr 1959-65, all in North Carolina Works. Production Supr, Installation Supr 1 yr each. Navy Sonar Proj Engrg Mgr 1967-71, providing tech, financial, man power mgmt of major proj. Mgr Plant Design & Const 1971-73 NYC. Since 1973, Mgr Purchased Products Engrg. Respon for specifications, product & supplier evals for communications products purchased for the Bell Sys. Sr Mbr IEEE. Reg P E NC. *Society Aff:* IEEE.

O'Neil, David A
Business: P.O. Box 338, Essex, CT 06426
Position: Pres. *Employer:* Seaworthy Systems, Inc *Education:* BS/Marine Eng/US Merchant Marine Acad. *Born:* 2/7/39. Former merchant mariner with four steamship cos, Naval officer, and designer for BUSHIPS. Additional education in naval architecture, finance, and marine transportation. Instrumental as mgr for early naval and commercial applications of marine gas turbines for Pratt & Whitney Aircraft. Founded eng firm of Seaworthy Systems, Inc in 1973 which has become a leader in propulsion for commercial ships. Widely published and holder of patents in fields of advanced marine propulsion and fuels tech. On Bd of Dirs of five firms, trustee of CT River Fdn, VP of the Soc of Naval Architects and Marine Engrs, and an ardent yachtsman. *Society Aff:* SNAME, ASME, ASNE, IPEN, Newcomen, SMC.

O'Neil, Robert S
Business: 1201 Connecticut Ave NW, Washington, DC 20036
Position: Sr VP *Employer:* De Leuw, Cather & Co *Education:* MCE 1970/CE/The Catholic Univ of Amer; BSCE 1957/CE/The Univ of Notre Dame *Born:* 06/06/35 Sr VP and Dir, De Leuw, Cather & Co; Mgr, Eastern Region, 1978-present. Native of Chicago, IL; with De Leuw, Cather & Co since 1960. Currently manages the company's Eastern Region. Formerly VP and Proj Mgr, Wash Metro Proj, 1976- 79; Chief Engr, Wash Metro Proj, 1969-1976; Chief Civ Engr, Wash Office 1966-69. Design Engr, Proj Engr, Proj Mgr, Chicago Office, 1960-66. Lt USAF 1957-1960. Extensive experience in planning, designing and construction mgmt of transportation projs. Reg PE 18 states. Mbr, the Moles *Society Aff:* ASCE, ARTBA, APTA, AREA, NSPE

O'Neil, Thomas J
Business: Cyprus Minerals Co, 7200 S. Alton Way, Englewood, CO 80112
Position: Project Manager *Employer:* Cyprus Minerals Co. *Education:* PhD/Mining Eng/Univ of AZ; MS/Mining Eng/Penn State Univ/BS/Mining Eng/Lehigh Univ *Born:* March 1940. Native of Fredonia N Y. Mining engrg exper with Utah Div, Kennecott Copper 1962-64; dev engrg & mkt res with Rock Drill Div, Ingersoll Rand Co 1966-68. With Univ of Ariz 1968-81: Assoc Prof & Hd, Dept of Mining & Geol Engrg 1972-78, Prof & Hd, 1978-81; Amoco Minerals Co 1981-85; Cyprus Minerals Co 1985- . Inaugurated new Mineral Economics prog in 1976. Mbr SME-AIME, MMSA; won Peele Award for outstanding authorship. Chmn APCOM Council 1973-76 (Internatl Assn for Computer Applications in the Minerals Indus), V. P. and Distinguished Member of SME of AIME, cons to gov agencies & private indus on mine valuation, financial analysis, royalty & tax policy. *Society Aff:* SME, MMSA

O'Neill, David W
Home: Box #217, Northville, MI 48167
Position: Manager-Testing Applications *Employer:* GE Co. *Education:* BS/Engg Phys/MI Tech Univ; BS/Bus Admin/Lawrence Inst of Tech. *Born:* 4/21/37. 1959-62; engrg officer, 1st Lt USAF. GE Co, 1962-4, Dev & Evaluation Engr; transferred to Ind Testing Systems Group. 1964-P associated with Dynamometers and automotive testing systems. Returned to MI 1966 as Ind Application Engr; presently Manager-Testing Applications. Reg P E, MI 17837. MI Soc of PE, Pres 1979-80. MI Assn of the Professions, Dir 1979-81. Natl Soc of P E, Dir 1982-86, MI Profl Ed Fnd, Pres 1982-86, MSPE-PAC Trustee & Chrmn 1981-86. Daughter, Kristin L; three adopted sons, Craig A D, Mark R & Brian D. Son of Stanley P & Catherine M. now of Aiken, S.C. Native of Dearborn, MI. *Society Aff:* NSPE, SAE, MSPE.

O'Neill, Eugene F
Business: Crawfords Corner Rd, Holmdel, NJ 07733
Position: Executive Director. *Employer:* Bell Labs. *Born:* 7/2/18. MSEE Columbia Univ 1941. DSc (Hon) Bates, St John's Univ, Politecnico di Milano. Bell Labs 1941- . Principal work in long haul transmission dev; coaxial sys, submarine cable, microwave radio, satellites & millimeter waveguide. Dir Telstar satellite project 1961-65. Exec Dir Toll Transmission 1966-75. Transmission Terminals & Maintenance Div, 75-78. Network Proj Planning 1978- , Responsible for coordination of Network project devs. IEEE Fellow & Field Award in Intl Communication. Mbr Natl Acad of Engrg 1976. Hobbies: reading, walking, travel.

O'Neill, John J
Home: Castlewood Lane, P.O. Box 429, Pinehurst, NC 28374
Position: Pres. *Employer:* Jonco, Inc. *Education:* BS/Chem Engg/Univ of MO, Rolla; Chem Engg/Chem Engg/Univ of MO, Rolla. *Born:* 9/13/19. Native of NYC. Joined Olin Corp 1940. 25 Patents. Successively, Asst GM Explosives, GM Military propellants, Asst to Corp VP R&D, Dir Corp Planning, Asst to the Pres, VP Bus Planning, VP Commercial Development-Chemicals, Corp VP GM Plastics Div, Corp VP Bus Dev. Joined Kleer-Vu Industries NYC 1972 - Ex VP, Coo, mbr of Bd of Directors. Joined Vertac Inc Memphis, TN 1976 Pres, Chief Exec Officer, Mbr of Bd of Directors. In 1979 became independent Business Consultant. 1980 Asst Chrmn of Bd, COO Vertac Chem Corp, Memphis, TN, 1981 Acting COO, Acting C. Financial Officer Braegen Corp, Cupertino CA, 1982-86. Bus Conslt. Formed Jonco, Inc 1986, Pres. Reg Engg - IL. Bd of Trustees, St Mary-of-the-Woods College, Terre Haute IN. AIChe, AIC. *Society Aff:* AIChE, AIC.

O'Neill, Michael W
Business: Department of Civil Engineering, University of Houston-University Park, Houston, TX 77004
Position: Prof. *Employer:* Univ of Houston. *Education:* PhD/Civ Engg/Univ of TX; MSCE/Civ Engg/Univ of TX; BSCE/Civ Engg/Univ of TX. *Born:* 2/17/40. in San Antonio, TX. Served in US Army Medical Service Corps, 1965-67. Instr at TX Western College, 1966-67. Res Assoc, Ctr for Hgwy Res, Univ of TX, 1970- 71. Mgr of Geotechnical Engg, Southwestern Labs, Inc, of Houston from 1971-74. Joined Univ of Houston in 1974 as Asst Prof & tenured in 1978. Presently Prof. Acting Chrmn of Civil Engrg, 1984-85. Winner of ASCE 1984 State-of-the-Art in Civil Engrg Award & 1986 Huber Research prize. *Society Aff:* NSPE, ASCE, ISSMFE, ASTM

O'Neill, Patrick H
Home: 42 Dunning Rd, New Canaan, CT 06840
Position: Mining Consultant *Employer:* Self-Employed *Education:* BS/Mining/Univ of AK; BME/Mining/Univ of AK; EM/Mining/Univ of AK; Dr of Sce/Honorary/ Univ of AK. *Born:* 8/11/15. Cordova, AK; son Harry and Florence (Leahy) ONeill BME 1941, E M 1953, Doctor of Science (Hon), 1976, Univ of AK; M 2nd Sandra Dorris 1967; children - Kevin, Erin, Patrick, Timothy, Frederick. US Smelting Refining & Mining Co, Fairbanks, Alaska Engg to Dredge Supt 1939-41; 46-53; Army Air Force 1941-46; South American Gold & Platinum Co, Colombia & New York, Chief Engg to Exec VP 1953-63; merged 1963 into International Mining Corp Exec VP 1963-70, Pres 1970- 77; Pato Consolidated Gold Dredging, Ltd 1961-76 Chmn, Frontino Gold Mines Ltd 1958-76, Chmn; South American Placers Pres 1960-76; Rosario Resources Corp. Sr. V.P. 1977-1980; Rosario Resources Corp. Executive V.P. 1980-1982; Mining Consultant 1982-Present; Dir ZEMEX Corp. NY, Bruneau Mining Corp. Canada, Ireland-U.S. Council for Commerce & Industry, American Geographical Society; Dir and Chmn. Joslin Diabetes Center, Boston. *Society Aff:* AIME, CIM, MMSA, IMM, AGS.

O'Neill, Russell R
Home: 15430 Longbow Dr, Sherman Oaks, CA 91403
Position: Emeritus Prof & Dean of Engrg. *Employer:* Univ of Calif, Los Angeles. *Education:* PhD/Engg/Univ of CA, Los Angeles; MS/Mech Engg/Univ of CA, Berkeley; BS/Engg/Univ of CA, Berkeley. *Born:* 6/6/16. With UCLA since 1946. Dean 1974-83. Prior to 1946 was an engr for Dowell Inc, design & dev engr for Dow Chem Co & design engr for AiResearch Mfg Co. Chmn Maritime Transportation Res Bd, NAS-NAE-NRC 1977-81. Chmn Continuing Engrg Studies Div ASEE 1969-70; Mbr ASEE Bd of Dir 1971-72. Mbr Tau Beta Pi, Sigma Xi, ASEE, ASEM, Natl Acad of Engrg *Society Aff:* ASEE, ASEM.

Ongerth, Henry J
Home: 905 Contra Costa Ave, Berkeley, CA 94707
Position: Consultant *Employer:* Self *Education:* BS/Civ Engg (Sanitary)/Univ of Calif; MA/Public Health (Sanitary Engg)/Univ of Mich. *Born:* June 1913. Native of San Francisco Calif. Sanitary Engineer, California Dept. Public Health 1936-1978; except 1942-1943 U.S. Engineer Dept. and 1947-48 with W.J. O'Connell & Assoc. Chief, Bu. Sanitary Engineering 1968-1978. Retired 1978. Hon. Member & Diven Award, AWWA. Elected Member National Academy of Engineering, 1976. Chairman, Environmental Engineering Intersociety Board, 1972-1973. Chairman, Conference of State Sanitary Engineers, 1973-1974. Chairman, Public Water Supply Panel, Environmental Studies Board, National Academy of Sciences, 1971-1972. Member, Excutive Committee, EPA Science Advisory Board, 1974. Member, National Drinking Water Advisory Council (EPA), 1975-76. Trustee AWWA Research Foundation, 1979-1985. Member, Board of Directors, California Public Health Foundation, 1979-present. Member, Hearing Board, (S.F.) Bay Air Quality Management District, 1983-Present. *Society Aff:* NAE, AAEE, ASCE, AWWA.

Ono, Kanji
Business: 6531 Boelter Hall, University of California, Los Angeles, CA 90024
Position: Prof of Engrg & Applied Sci. *Employer:* Univ of Calif - L A. *Education:* BE/Met Engg/Tokyo Inst Tech; PhD/Mat Sci/Northwestern Univ. *Born:* Jan 2, 1938 Tokyo Japan. First son of Saburo & Yoshiko Ono. Married to Fumie Asano 1962; 3 children. Joined Univ of Calif L A as Asst Prof 1965, becoming Assoc Prof 1970 & Prof 1976. Recipient of 1968 H M Howe Medal of ASM, and of 1981 Achievement Award of ASNT. Mbr ASM, ASNT, AEWG Active res in strengthening mechanisms & fracture of material & acoustic emission, which resulted in many papers & two patents. *Society Aff:* ASM, ASNT, ASTM

Onoe, Morio
Business: Ricoh Co. Ltd, Research and Development Center, 4686, Nippa-cho, Kohoku-ku, Yokohama 223, Japan
Position: Prof. *Employer:* Univ of Tokyo. *Education:* PhD/EE/Univ of Tokyo *Born:* 3/28/26. Native of Tokyo. Interested in piezoelectric devices & filters, acoustic emission, non-destructive testing, medical electronics & digital image processing. Now an exec mgg dir of Ricoh Co. Ltd. Prof Emeritus of Univ of Tokyo retired in 1986. Was a prof & the director of Inst of Industrial Science. From 1956 to 1958, a Fulbright Exchange Scholar at Columbi Univ, NY. In 1961 and 1966, with Bell Tel Labs. Received C B Sawyer Award at Frequency Control Symposium in 1975; Inada Award in 1955, Okabe Award in 1959 & Achievement Award in 1976 from IECE & Niwa-Takayanagi Paper Award in 1977. IEEE Fellow. Pres of Japanese Soc of Nondestructive Inspection for 1976. Enjoy books & music. *Society Aff:* IEEE, ASA.

Openshaw, Keith L
Business: P O Box 60455, Los Angeles, CA 90060
Position: Sr VP. *Employer:* Union Chems Div & Union Oil Co of CA. *Education:* BE/Chem Engg/Univ of So CA. *Born:* 7/26/25. BA Chem Engg USC 1950; post-grad courses. Air Force 1944-45. Process engr with Mobil Oil 1950-51. Joined Union Oil 1951 as Refining Engr & Operation Supr. Tranferred to Res Dept & then to a chem subsidiary, Collier Carbon & Chem Corp in 1956 as a proj engr. Became P R&D 1964. Mbr Bds of Dirs of Poco Graphite Inc & PureGro Co. Full Mbr AIChE 1972. Becmae Sr VP of the Carbon Group of Union Chems Div in October, 1977. (Collier became a div of Union Oil Co in October, 1977). *Society Aff:* AIChE.

Opie, William R
Home: 119 Crawfords Corner Rd, Holmdel, NJ 07008 *Employer:* Retired *Education:* ScD/Met/MA Inst of Tech; BS/Met Eng/MT Mines *Born:* Apr 1920. ScD MIT; BS & ME Montana School of Mines; Harvard Bus School AMP. ScD Mont Tech (Hon). Native of Butte Montana. Foundry Metallurgist Wright Aero Corp 1942-45. USN 1945-46. Res metallurgist ASARCO 1949-50. Res Supr Natl Lead Co 1950-60. AMAX Inc 1960-85. Retired Consultant since 85. Respon for R&D in base metals. Num papers & pats in titanium & copper extractive met & copper alloy dev. Fellow ASM; Fellow TMS-AIME. Mbr Natl AC of Eng. *Society Aff:* AIME, ASM

Opila, Robert L
Business: 222 N Dearborn, Chicago, IL 60601
Position: President. *Employer:* Globe Engrg Co. *Education:* BS/ChE/IL Inst of Tech; Masters/Bus Adm/Univ of Chicago *Born:* June 1927. Native of Chgo Ill. With Globe Engrg since May 1973. Prior exper incl chem & food process dev & design, facils design & const in the U S & overseas for CPC Internatl & Griffith Labs. Performed related work as Proj Mgr & Div Dir for Lummus Co & Commonwealth Assocs Inc respectively. Pres IL Engrg Council; Chmn Chgo Sect & Chmn Food & Bio Engrg Div, AIChE. Dir Food & Phamaceutical Indus Div, ISA. Enjoys camping, bridge, drama & opera. *Society Aff:* AIChE, ACS, AACE, AOCS, ISA

Oppelt, Winfried
Home: Meissnerweg 67, D61 Darmstadt, 14 West-Germany
Position: Prof Dr Ing. (Emeritus) *Employer:* Technische Hochschule (Univ). *Education:* Dr Ing/-/Darmstadt Univ; Dipl-Ing/Phyiscs/Darmstadt Univ. *Born:* June 1912. Taught in Darmstadt public schools, high school & univ. Res engr DVL Berlin Adlershof 1934-37. Dev of flight controllers with Anschutz & Co Kiel 1937-42; with Siemens LGW Berlin 1942-45. Lab engr in process control with Hartmann & Braun Frankfurt 1949-56. Prof control engrg Darmstadt Univ 1956-1977. Dr Ing E h (TH Munchen) 1965. Fellow IEEE 1966. Grashof Denkmunze VDI 1971. Wilhelm - Exner - Medaille Wien 1979. VDE-Ehrenring 1980. Gairn-EEC-medal London 1982. Book: Kleines Handbuch technischer Regelvorgange 1953-72. Ueber das Menschenbild des Ingenieurs 1984. *Society Aff:* VDI, VDE

Oppenheim, Alan V
Business: Dept of EE & CS, Rm 36-625, 50 Vassar St, Cambridge, MA 02139
Position: Prof. *Employer:* MIT. *Education:* ScD/EE/MIT, 1964; SM/EE/MIT, 1961; SB/EE/MIT, 1961 *Born:* 11/11/37. In 1964, joind MIT faculty, Elec Engrg & Computer Sci. Currently full prof. Was Guggenheim Fellow at the Univ of Grenoble, France; held the Cecil H Green distinguished chair in Elec Engrg & Computer Sci, MIT; served as Assoc Hd, Data Sys Div MIT, Lincoln Lab. Fellow of the IEEE. Elected to Natl Acad of Engg, Co-Author of several texts on signal processing including *Digital Signal Processing* (Prentice-Hall) and *Signals and Systems* (Prentice-Hall). Frequent contributor to the research literature. Res interests in digital signal processing and its applications to speech, image & seismic data processing, in sys theory and knowledge-based signal processing. *Society Aff:* IEEE, NAE.

Oppenheim, Antoni K
Business: Dept of Engrg, Berkeley, CA 94720
Position: Prof of Mech Engrg. *Employer:* Univ of Calif Berkeley. *Education:* Dipl Ing/Mech Engg/Warsaw Inst of Tech (in Oxile London); PhD/Mech Engg/Univ of London; DSc/Mech Engg/Univer of London *Born:* 8/11/15. Studied Aeronautical Engrg at Warsaw Inst of Tech & received the degree Dipl Ing in 1943; PhD Univ of London & DIC Imperial Coll 1945. D.Sc University of London, 1976 Dr. Honoris Coura University of Poitiers, 1981. Served as Lectr & Asst Prof on facs of City & Guilds Coll London 1945-48 & Stanford Univ 1948- 50. Since 1950, at Univ of Calif Berkeley. Visiting Prof Sorbonne (Sr Postgrad Fellowship NSF) 1960-61 academic yr. Miller Prof at Berkeley 1961-62. Professeur Associe Univ of Piotiers 1973 and 1980. NASA Res Adv Ctte on Fluid Mech 1963-68; Dep Editor 'Combustion & Flame' (journal of the Internatl Combustion Inst) 1972 & 1973; Assoc Editor of ASTRONAUTICA ACTA (archive journal of the Internatl Acad of Astronautics) 1973-74; Editor-in-Chief ASTRONAUTICA ACTA 1974-79. Currently Co-Chmn Intl College. Dynamics of Explosions & Reactive Systems. *Society Aff:* ACS, APS, ASME, AIAA, SAE

Oppenlander, Joseph C
Business: Dept. of Civil Eng, University of Vermont, Burlington, VT 05405
Position: Prof. *Employer:* Univ of VT. *Education:* PhD/Civ Engg/Univ of IL; MSCE/Civ Engg/Purdue Univ; BSCE/Civ Engg/Case Inst of Tech. *Born:* 4/6/31. Native of Bucyrus, OH. Served with Army Ordnance Corps at Redstone Arsenal as Test Engr, 1953-55. Taught in Civ Engg with specialities in Transportation and Traffic Engg and Safety and Community Planning and Design at Purdue Univ (1955-57; 1962-69), Univ of IL (1958-62), and Univ of VT (1969-present; dept chrmn, 1969-79). VP of TRANS/OP, INC, Systems Engrs and Consultants. Author of approximately 50 publications, including "Manual of Traffic Engineering Studies–, published by Inst of Transportation Engrs. Active in technical activities of various profl societies. Past Pres of IN Sec, Inst of Transportation Engrs and Green Mtn Transportation Club. 1985 New England ITE Transportation Engineer. *Society Aff:* ASCE, ITE, ARTBA, TRB, Sigma Xi

Orava, Raimo Norman
Business: SD School of Mines, 501 E. St. Joseph St, Rapid City, SD 57701
Position: Dean of Grad Div/Dir of Res. *Employer:* S Dakota School of Mines & Tech. *Education:* BASc/Engg Physics/Univ of Toronto; MASc/Met Engg/Univ of British Columbia; PhD/Physical Met/Imperial College, Univ of London; DIC/ Physical Met/Imperial College, Univ of London. *Born:* Sept 1935. Native of Toronto Canada. Taught at Univ of Waterloo for 2 yrs before grad study in England. After grad in 1963, conducted res in mech met at Battelle - Columbus, Franklin Inst, & Denver Res Inst. The main res emphasis during the last 10 yrs has been in the field of high-velocity impact phenomena incl explosive forming & shock hardening. Since 1974, Prof of Met at SDSM&T, becoming Acting Dean of Engrg in 1975 & also Acting V P for Acad Affairs in 1976. Appted Dean of Grad Div & Dir of Res in 1977 and also Director, Mining and Mineral Resources and Research Institute in 1980. Assigned by SD Bd Regents in 1979 to manage faculty union Agreement & serve as Chief Negotiator. Interested in bridge, skiing, & travel. *Society Aff:* AIME, NCURA, AAAS, CGS, LES, SRA, SUPA

Orchard, Henry J
Business: 6730 Boelter Hall, Los Angeles, CA 90024
Position: Professor. *Employer:* Univ of California. *Education:* MSc/Mathematics/Univ of London; BSc/Mathematics/Univ of London. *Born:* May 1922 England. Taught at British Post Office Engrg School, Cambridge 1942-47. With Res Dept, British Post Office, London 1947-61, spec in network design. Fellow IEE (London) 1961. Joined GTE Lenkurt, San Carlos Calif 1961 as network cons; became Hd of Networks & Math Group 1963. Fellow IEEE 1968. Outstanding Paper Award (jointly with G C Temes) from IEEE Circuit Theory Group 1969. Joined UCLA as Prof of Engrg 1970. 8 pats. Author over 40 papers in engrg sci. *Society Aff:* IEEE, IEE.

Ordung, Philip F
Business: Dept of EE & CS, Santa Barbara, CA 93106
Position: Prof of EE & Computer Science. *Employer:* Univ of Calif - Santa Barbara. *Education:* D Eng/Elec Eng/Yale U; MSc/Elec Engr/Yale U; BSc/Elec Eng/SD State *Born:* Aug 12, 1919. Grad Asst to Prof at Yale 1940-62; First Chmn Dept of EE at Univ of Calif Santa Barbara 1962-68 & Prof 1962- .Fellow IRE 1959 for 'contrib to network theory & to Elec Engrg Educ'; Fellow AAAS 1959; Mbr New York Acad of Sci 1975; Royal W Sorenson Fellow IEEE 1962. Mbr Bd of Regents CA Lutheran College 1977- . 1985 IEEE Centenial Award. *Society Aff:* IEEE, ASEE, AAAS

Orehoski, Michael A
Home: 200 Harden Ave, Duquesne, PA 15110
Position: Consultant-Steelmaking & Metal Casting *Employer:* Self *Education:* Grad/Met Studies/Carnegie-Mellon; BS/Met Eng/Carnegie-Mellon. *Born:* July 23, 1916. Since 1935, worked in var plants & res labs for U S Steel Corp. Spec in heat treatment, hardenability, steelmaking, metal casting, continuous casting, vacuum degassing, & desulfurization. Has written num papers; inventor and/or co-inventor of num patents. Fellow ASM; Achieve Award from Amer Vacuum Soc. *Society Aff:* AIME, ASM.

Orehotsky, John L
Business: River St, Wilkes-Barre, PA 18766
Position: Prof of Engg. *Employer:* Wilkes College. *Education:* PhD/Solid State Sci/ Syracuse Univ; ME/Metallurgical Engg/Brooklyn Poly Inst; BS/Metallurgy/MA Inst of Tech. *Born:* 7/24/34. in Hartford, CT. Past affiliations include Thomas Watson, IBM Res Labs, Yorktown Heights, NY, and Sylvania Elec Products Res Labs, Bayside, NY, where res efforts involved anelasticity amorphous magnetic matls, x-ray diffraction, physical metallurgy of tungsten. Currently with Wilkes College where teaching respons include Thermodynamics, matls Sci, Physical Electronics and Solid State Devices. Current res efforts include magnetic matls, gas-metal reactions, solar energy devices, electrochemisy. Consulting activities with the Brookhaven natl Labs and the RCA Corporation. *Society Aff:* AIChE.

O'Reilly, Roger P
Business: 5520 Los Santos Way, Jacksonville, FL 32211
Position: VP. *Employer:* Ebasco Services, Inc. *Education:* BSCE/Structures/Univ of Notre Dame. *Born:* 2/5/23. NY, NY. For Ebasco Services Inc 1948-71 as draftsman,

O'Reilly, Roger P (Continued)
designer, resident engr, construction superintendent, chief estimating engr, mgr of engg, and utility consultant. 1971-1984 for Burns and Roe, Inc as VP responsible for Jacksonville office operations. 1984 for Ebasco Services as VP, Utility Conslt. Experience in the design, construction, construction mgt and engg mgt of hydro and steam elec generating facilities having a combined capacity in excess of 30, 000MW. Fellow ASCE, mbr Newcomen. Mbr Jacksonville Chamber of Commerce Committee of 100. *Society Aff:* ASCE, ASME, SAME, USCOLD

Oresick, Andrew
Home: 441 N Main St, Herculaneum, MO 63048
Position: Supr Safety, Sec & Environ. *Employer:* PPG Industries Inc. *Born:* Sept 1921. MA Indiana Univ of Penna; BA Penn State. Native of Ford City Penna. Served with Army in S Pacific 1944-45. With PPG Industries Inc since 1946. Now Supr Safety, Security & Environ Control. Certified Safety Professional. Respon for safety, fire, security, Medical & environ programs. Prof Mbr St Louis Chap ASSE, serving on its Exec Bd since 1973. Mbr Self- Insurers' Exec Ctte, Assoc Indus of Mo, serving as Chmn 1975-76. Contrib to 'Prof Safety'. Winner of 1973 Best Paper Award, spon by ASSE & Veterans of Safety, for article 'Safety Tech in Handling Flat Glass'.

Oriani, Richard A
Business: University of Minnesota, Minneapolis, MN 55455
Position: Professor and Director Corrosion Research Center *Employer:* University of Minnesota *Education:* PhD/Physical Chem/Princton Univ; MA/Phsyical Chem/ Princeton Univ; MS/Chemistry/Stevens Inst of Technology; BChE/Chem Engg/Coll of the City of NY. *Born:* July 1920. Res Assoc G E Res Lab Schenectady 1948-59. U S Steel Res Lab 1959-1980: Asst Dir Phys Chem, Mgr Phys Chem, Sr Res Cons. Professor and Director, Corrosion Research Center, University of Minnesota, 1980 -. Thermodynamics alloy theory, corrosion, irreversible thermodynamics, hydrogen embrittlement. Fellow Amer Soc for Metals, Amer Inst Chemists, Natl Acad Sci adv panels to Met Div & to Inst for Matls Res of Natl Bur of Stds; Adv Panel Matls Res Lab Univ of Ill; Educ Comm on Prof Dev AIME. Enjoy music, hiking. *Society Aff:* ASM, AIME, ACS, APS, ECS, NACE.

Orlob, Gerald T
Business: 211 Walker Hall, Dept of CE, Davis, CA 95616
Position: Prof *Employer:* Univ of CA *Education:* PhD/Hydraulic Engrg/Stanford Univ; MS/Environmental Engrg/Univ of WA; BS/CE/Univ of WA *Born:* 7/4/24 Native of Seattle, WA. Served with US Army Corps of Engrs, 1943-1946. Survey Supvr, WA Pollution Control Commission, 1949-52; Prof of CE, Univ of Ca, Berkeley, 1952-1965; Founder-Pres of Water Resources Engrs, Inc 1959-1974; Founder-Pres, Resource Management Assoc, Inc, 1974-1979; Prof of CE, Univ of CA at Davis, 1968 to date; Registered CE, CA and WA; ASCE Hilgard Hydraulic Prize, 1963; ASCE Herring Medal, 1963; AWWA Water Resources Award, 1956; AGU Hydrology Award, 1956; WPCF Eddy Medal, 1970; Fulbright-Hayes Fellow, 1973. *Society Aff:* ASCE, AGU, AAEE, WPCF, IAHR

Ormsbee, Allen I
Business: 105 Transportation Building, 104 South Mathews Avenue, Urbana, IL 61801
Position: Prof. *Employer:* Univ of IL. *Education:* PhD/Aeronautics/Caltech; MS/ Math/Univ of IL; BS/Aero Engr/Univ of IL. *Born:* 8/20/26. in Reno, NV. Mbr of Univ of IL faculty since 1946, Prof since 1957. Howard Hughes Fellow, CA Inst of Tech, 1952-1954. Visiting Prof, UCLA, 1964. Mbr AIAA Gen Aviation Sys Tech Ctte, 1974-1978. Assoc Editor, Journal of Aircraft, 1976- 1980. Res and conslt in airfoil and propeller aerodynamics, in aerial spray patterns and in missile aerodynamics. Assoc Fellow, AIAA. Mbr, NRC Panel on Causes and Prevention of Grain Elevator Explosions 1980-81. Mbr, AIAA Ctte on Academic Affairs. Member, Engrg Accreditation Commission 1985- . *Society Aff:* AIAA, ASEE, AAUP.

Ornes, Edward D
Business: 3800 Regent St, Madison, WI 53705
Position: Principal *Employer:* Strang Partners Inc *Education:* BS/ME/Univ of WI-Madison *Born:* 6/6/43 Registered PE, 1971. Principal of Strang Partners Inc (arch/ engrg firm), 1975. Past Pres of Madison Chapter of ASHRAE, Member of City of Madison's Mayors Engrg Council. Served on Madison Area Joint Engrg Council. *Society Aff:* ASHRAE, NSPE, AEE, PEPP

Ornitz, Martin N
Business: P O Box 226, Midland, PA 15059
Position: Corp VP. *Employer:* Colt Industries Inc. *Education:* MS/Metallurgy/Carnegie Inst Tech; BME/Engg/Cornell Univ; Grad Courses/- /Univ of Pittsburgh; Grad Courses/-/John Hopkins Univ. *Born:* Sept 1921. US Navy 1943-46. Served as V P of Roblin Indus & Group Pres of its Indus Prod Div 196870. V P & Genl Mgr of Natl Alloy Div of Blaw-Knox Co Pgh 1956-68. Assumed current respon as Pres of Crucible Stainless Steel Div, Colt Indus Inc 1970. ASM Fellow 1972. Recipient of the Trinks Award 1964. Hold of many pats. Reg P E Pa. Pres Concordia Club; former Bd Mbr of Montefiore Hosp. Respon for: Crucible Alloy Div; Crucible Stainless Steel Div). Enjoy golf, swimming, sailing & photography. *Society Aff:* AISE, ASM, AIME, ASTM, FIA.

O'Rouark, Terence
Home: 1710 Warm Springs Ave, Boise, ID 83712
Position: Owner. *Employer:* O'Rouark Engg, Inc. *Education:* BS/EE/Univ of ID. *Born:* Apr 1922. Native of Honolulu Hawaii. Army Corps of Engrs. 1943-46 with Manahattan Dist Oak Ridge, again in 1950-51 as Post Engr of Hospital in Hawaii. With Hawaiian Elec Co 1946-50 & 1951-55. Private practice in Boise, Idaho 1955- . Hobbies: sailing & amateur radio. *Society Aff:* NSPE, ACEC.

O'Rourke, James T
Business: Camp, Dresser & McKee Inc, One Ctr Plaza, Boston, MA 02108
Position: Pres. *Employer:* Ind Div Camp Dresser & McKee Inc. *Education:* PhD/Civ Engg/Stanford Univ; MS/Sanitary Engg/WA Univ; BS/Civ Engg/Univ of RI. *Born:* 3/11/34. Pres, Ind Div, Camp Dresser & McKee, Inc, Boston, MA, directing all CDM facilities and environmental projs for ind clients. Specialist in ind waste treatment, hazardous waste management, and high tech facilities design, providing consulting services to industries, 1975-78, VP, Bethel, Duncan & O'Rourke. 1972-75, SVP, & Chief, Ind Waste Dept, Metcalf & Eddy, Inc, Boston; '71-72, VP & Dir of Res; '70-71, Asst VP; '69-70, Sr Assoc & Dir of Res. 1967-69, Sr Assoc, Ryckman, Edgerley, Tomlinson & Assoc, St Louis, MO; '62-67, Assoc. 1964-67, Res Fellow, Dept of Civ Engg, Stanford Univ. 1962-64, Asst Prof, Environmental & Sanitary Engg, WA Univ, St Louis; '61, Instr. *Society Aff:* AIChE, ASCE, AIPE, IES.

O'Rourke, Robert B
Business: 4405 Talmadge Rd, Toledo, OH 43623
Position: Mbr & Treas - Bd of Mgrs. *Employer:* Finkbeiner, Pettis & Strout Ltd. *Education:* -/Civil/Univ of Toledo. *Born:* May 1920. Attended Univ of Toledo. Native of Toledo Ohio. Served in AUS - Coast Artillery 1942-45. Served in USAR 1945-69. Transferred to Retired Res with rank of Lt Col. Employed by Finkbeiner, Pettis & Strout Ltd 1947 as draftsman - survey crew mbr. Admitted as Jr Partner 1958; Full Partner 1965. Serve as Chmn Bd of Mgrs on rotating 3 yr basis. 7 yrs as Tr of Operator Training Ctte of Ohio. Reg PE FL, Ohio, Mich, Ky, W Va, NC. VA Fellow ASCE; Mbr AWWA, WPCF, NSPE, OACE & Military Engrs. M W Tatlock Mem Citation from Ohio Sect AWWA 1964. La Due Citation from OH Sect AWWA 1980. Dir OH Assoc of Consulting Engrs, NSPE/PEPP Central Region Vice-Chrmn. *Society Aff:* NSPE, AWWA, WPCF, ASCE, SAME.

O'Rourke, Thomas D
Business: 265 Hollister Hall, Ithaca, NY 14853
Position: Professor *Employer:* Cornell Univ *Education:* PhD/CE/Univ of IL-Urbana; MS/CE/Univ of IL-Urbana; BS/CE/Cornell Univ *Born:* 7/31/48 Native of Pittsburgh, PA. Worked on field instrumentation for large underground construction projects from 1970-75 on the Washington, DC Metro. Represented the US Dept of Transportation on research with the British Transport and Rd Research Lab during 1976-77. Teaching and research at Cornell Univ since 1978. Research and consulting projects include underground construction, foundation engrg, field instrumentation, design and protection of buried pipelines, and earthquake engg. Pres Ithaca Section ASCE 1981, ASTM Hogentogler Award 1976, ASCE Collingwood Prize 1983. *Society Aff:* ASCE, EERI, ASME, AAAS

Orr, Charles K
Business: P.O. Box 4180, Houston, TX 77210
Position: VP *Employer:* Sant Fe Drilling Co *Education:* BS/Petro Engrg/TX A & M Univ *Born:* 10/23/34 Currently employed by Santa Fe Drilling Co, a division of Santa Fe Intl Corp. Been employed since 1970 and has served in operations, marketing, contracts and sr management roles. Prior to working at Santa Fe, worked in Gulf Coast area for Penrod Drilling Co and Moviable Offshore Co. After working for TX Petro Research Corp at TX A&M, worked briefly for Standard of TX before joining H.L. Hunt Industries (Penrod) in 1958. Professional Eng - TX & LA. *Society Aff:* SPE

Orr, Leighton E
Box 291 - RD 4, Tarentum, PA 15084
Position: Tech Cons - Glass & Ceramics. *Employer:* Self-employed. *Education:* BS/Mech Engr/Univ of Pittsburgh. *Born:* Feb 11, 1907. Native of Pgh Pa. Genl Matls Testing, Pgh Testing Lab 1930-36. Testing & performance of glass & glass products, leading to Hd of Phys Testing PPG Indus Res Lab 1936-72. Involved in a var of testing work, incl optics, heat flow, strength, safety, impact, degree of temper, defects, fracture analysis, wind & water load design of windows. Fellow ASME 1967, Life Fellow 1979. Publ articles on glass strength & fracture analysis. Glass cons work 1972- 87 incl design, special testing, & investigations of causes of glass failures, for glass companies, bldg owners, architects, insurance co's & lawyers. Enjoy traveling. *Society Aff:* ASME.

Orr, Paul H
Business: 76 S Main St, Akron, OH 44308
Position: Manager, General Plant Projects Dept. *Employer:* Ohio Edison Co. *Education:* BCE/Structural Engg/Fenn College. *Born:* 1/15/38. BCE Fenn College 1961. Proj engr for one of the 7 Amer Soc of Engrs - Outstanding Civil Engg Achievement of 1973, OH Edison's 345 kV Hanna Substation. 1976 Pres of the Akron Sect of ASCE. Has done extensive engg work in both distrib & extra high voltage substations, office bldg design concept dev specifically open office plan dev, deeply involved in the dev of sys by which to manage large constr projs (nuclear) and is now Project Manager of projects involving generating plant engineering and construction.. *Society Aff:* ASCE, AACE, PMI.

Orr, Robert M
Home: P.O. Box 176, Monahans, TX 79756
Position: Owner-Petroleum Engr (Self-employed) *Education:* BS/Petrol Engrg/Univ of TX-Austin *Born:* 10/4/26 Native Texan served as a captain in U.S. Army. Employed by Gulf Oil Corp and then became pres of Buckles and Hostetler Co. Presently own and operate my own oil and gas production co and petroleum engrg consulting firm. Designed secondary oil recovery waterflood projects used in four states. Served a Pres of Permian Basin Petroleum Association and served on the energy Targets Sub- Committee of Coordinating Committee on Energy AAES. Was elected Engr of the Year- 1969 and Citizen of the Year-Monahans, TX-1981. Also served as pres of local school bd and city councilman for city of Monahans. *Society Aff:* AIME, NSPE, IPAA

Orr, Virgil
Business: P O Box 5276 Tech Sta, El Dorado, LA 71270
Position: V P for Academic Affairs. *Employer:* Louisiana Tech Univ. *Education:* BS/Chem Engg/LA Polytechnic Inst; MS/Chem Engg/LSU; PhD/Chem Engg/LSU. *Born:* Feb 1923. BS La Tech; MS, PhD LSU. Native of Glenmora La. Shift Analyst with Cities Service; Corps of Engrs, Army Map Service 1944-46; Res Engr with United Gas Corp - Phase Equilibria. Assoc Prof of Chem Engrg 152-55 wth La Tech, Prof & Dir of Engrg Res 1955-63, Dean of the Coll 1963-70. V P for Academic Affris 1970- . Asst Dir La Coord Council for Higher Educ 1970 (on leave). Exec Dir State Bd of Educ 1973 (on leave). Enjoy hobbies of fishing, camellia culture, macrame, rockhounding, & handwriting analysis. *Society Aff:* AIChE, ASEE.

Orr, William H
Business: AT&T, PO Box 1008, Indianapolis, IN 46206
Position: Hd, Advanced Products Dept *Employer:* AT&T *Education:* PhD/Eng Physics/Cornell Univ; MS/Physics/Catholic Univ; BEP/Eng Physics/Cornell Univ. *Born:* 11/3/30. Buffalo N Y. US Navy - engrg officer USS Capricornus (AKA-57) 1953-55, Instr Phys & Elec Engrg US Naval Acad 1955-57. Bell Labs 1962-83. AT&T consumer Products Laboratory since 1984. Dev thin film hybrid integrated circuit tech for telephones, authored articles, contrib to books & granted pats in field. Design and development AT&T consumer products. Mbr Sigma Xi, Phi Kappa Phi, IEEE, CHMTS, CES, CS, Inc. Corp. for Sci. & Tech., Dean's adv. com. Purdue U. Sch. Sci. Indpls. *Society Aff:* IEEE, CHMTS, CES, CS.

Orrell, George H
Home: 14208 London Ln, Rockville, MD 20853
Position: Technical Dir/Deputy *Employer:* US Army Corps of Engrs. *Education:* BS/Civ Engr/Citadel; Fellow, Natl Inst/Public Affairs/Univ of Chicago; Sr Exec Educ Prog/-/Federal Exec Inst. *Born:* 3/7/33. Native of Hazleton, PA. Served as Army Engr Officer 1954-56. Worked 4 yrs as design engr with Corps of Engrs (CE) at Wilmington, NC. Joined CE Engr Studies Ctr (ESC) at Wash, DC in 1960. Received Army R&D Achievement Award in 1964. Received Ford Fdn Career Education Award in 1965. Became ESC Technical Dir in 1967. Became supergrade exec in 1974. Have received 13 Army Commendations. In July 1979, received the Decoration of Meritorious Civilian Service-second highest Army award for performance-and became a charter mbr of the US Sr Exec Service. Enjoy church work, family, sports, & reading. *Society Aff:* SAME.

Orrok, George A
Home: 5 Cleveland Street, Cambridge, MA 02138
Position: V P Engrg & Const. *Employer:* Retired. *Born:* 1905. BS & MS in Mech Engrg Harvard 1927, 1928. After early yrs in oper & testing, spec in the design & const of Boston Edison's power stations; eventually being V P of Engrg & Const. Power Test Code Ctte of ASME & Prime Movers & Power Generation Cttes of EEI & AEIC. Life Mbr ASME, NDHA & Assoc Mbr ASCE. Have enjoyed work as a tech mech engr for 41 yrs & am enjoying retirement on old family farm.

Orth, H Richard
Home: 21 Champlain Rd, Marlton, NJ 08053
Position: Prin. *Employer:* Orth-Rodgers-Thompson & Assoc, Inc. *Education:* MCE/CE/Villanova Univ; Cert of Hgwy Traf/Traf Engr/Yale Univ; BCE/CE/ Villanova Univ. *Born:* 1/26/42. Have served as traffic engg/transportation planning consultant for past 18 yrs. Formerly with Simpson & Curtin, transportation engg firm, including five yrs (1972-77) as VP. Established own consulting firm in 1977 to serve both public and private clients First recipient of the John J Gallen Meml Award from Villanova Univ (1977). Past pres of the Mid-Atlantic Sec and Dist 2 (comprising the Mid-Atlantic and Wash-Baltimore Sections) of ITE; 3 year term a member of the International Bd of Dir. of ITE. Also served as lecturer/instr in transportation engg at the Evening College of Drexel Univ and at Temple Univ. *Society Aff:* ASCE, ITE.

Orth, William A
Home: QTRS 6550, USAF Academy, CO 80840
Position: Dean of Faculty *Employer:* US Air Force *Education:* PhD/Applied Math/ Brown Univ; MS/ME/Purdue Univ; BS/Engrg/USMA West Point *Born:* 9/28/31 in Coatesville, PA. Commissioned in Air Force in 1954. Pilot in Fighter- Interceptor squadrons in MA, England, and Germany. Assistant Prof in Mechanics, USAF Academy, during 1961-63. Instructor pilot and Chief of Wing Safety in Air Training Command. Assistant Prof in Mathematics at Academy during 1968-70. Attack Squadron Commander in Vietnam. Served in Comptroller area at Hq Strategic Air Command (SAC) prior to becoming Dir of Engrg and Construction. Assistant DES/ Civil Engrg, Hq SAC, during 1973-1974. Prof and Head, Dept of Physics, USAF Academy, 1974-78. Dean of the Faculty, 1978-present. *Society Aff:* AIAA, ASEE, AAPT, MAA

Orthmeyer, Frank B
Business: Box 1518, Grand Forks, ND 58201
Position: City Engr & Dir of Public Works. *Employer:* City of Grand Forks. *Education:* BSCE/Civil Engr/NDSU *Born:* BSCE ND St Univ Fargo 1949. Reg PE North & South Dakota. Proj Engr II-City of Bismark 1949-66. Dir of Public Works-City of Mitchell SD 1966-70. Div of Public Works-City of Grand Folks ND 1970- . Mbr Natl Water Pollution Control Fed-Bd of Control. PPres ND Amer Pub Works Assn; PPres ND Water & Wastewater Conf; PPres ND Soc of Prof Engrs-Chap III, Sec and Mbr of the ND State Bd of Registration for Prof Engr & Land Surveyors (1977-1984). *Society Aff:* AWWA, WPCF, NSPE, APWA

Ortiz, Carlos A
Business: P.O. Box 10412, Caparra Station, San Juan, PR 00922
Position: Pres *Employer:* Corporacion Geotec *Education:* MSCE/Geotechn/Univ of PR-Mayaguez; BS/CE/Univ of PR-Mayaguez *Born:* 11/2/45 Native of Santurce, PR. Served with US Corps of Engrs upon graduation (1970). Worked with US Bureau of Mines as Materials Engr and with Efrahim Murati & Assocs as a Geotechnical Engr prior to going into private practice in 1974. Acquired a partner and founded Corporacion Geotec in 1978. Past Pres of ASCE's PR Chapter. Past Pres of the PR Society of Geotechnical Engrs (1978-1980). Member of ASCE's Engrg Geology Committee. Enjoys classical music and coaching Boy's Baseball Teams. *Society Aff:* ASCE, ASTM, NSPE, AWS.

Osborn, Elburt F
330 E Irvin Ave, State College, PA 16801
Position: Emeritus VP Res *Employer:* Penn State Univ *Education:* PhD/Geology/CA Inst of Tech; MS/Geology/Northwestern Univ; AB/Geology/DePaul Univ. *Born:* Aug 13, 1911 Winnebago Cty Ill. Hon D Sci Alfred Univ, DePauw Univ, Northwestern Univ, Ohio State Univ. Petrologist Geophys Lab Carnegie Inst Washington 1938-42; phys chemist Div 1, Natl Def Res Comm 1942-45; res chemist Eastman Kodak Co Rochester NY 1945-46; Prof geochem Penn State Univ 1946-70, Chmn Div Earth Sci Coll Mineral Indus 1946-53, Dean 1953-59, V P Res 1959-70; Dir U S Bureau of Mines 1970-73; Disting Prof Carnegie Inst Wash 1973-77; NSF Sr Postdoc Fellow Cambridge Univ 1958. Amer Ceramic Soc: Pres 1964, Fellow, Hon Life Mbr, Jeppson Award 1973, Bleininger Award 1976. AIME: Hardinge Award 1974, Disting Mbr SME - class 1975. Mineral Soc Amer: Pres 1960-61, Fellow, Roebling Medal 1972. Soc Econ Geologists: Pres 1972-73. Geochem Soc: Pres 1967-68. Canadian Ceramic Soc. Geological Soc Amer, Amer Chem Soc, Amer Geoph Union. Mbr Natl Acad Engrg. *Society Aff:* GSA, MSA, SEG, GS, AGU, ACS, ACerS, AAAS, AIME, IAVCEI, CCS

Osborn, John R
2405 Sauk Pl, Lafayette, IN 47905
Position: Prof. *Employer:* Purdue Univ. *Education:* PhD/ME/Purdue Univ; MSME/ ME/Purdue Univ; BSME/ME/Purdue Univ. *Born:* 8/11/24. Native of Kansas City, MO. Served with the Navy Air Corps 1943-1945. Engr with the Thiokol Chem Corp specializing in solid rocket motor design. Branch Chief of the Combustion and Propulsion Branch of the Ballistic Res Labs, Aberdeen Proving Ground. Currently, Prof of Aero and Astro at Purdue Univ teaching propulsion and combustion courses. Conduct res on the combustion of composite solid propellants. Enjoy photography and sports car racing. *Society Aff:* AIAA, SAE.

Osborn, Richard K
Business: Cooley Bldg, Dept Nuc E, Ann Arbor, MI 48105
Position: Professor. *Employer:* U of Michigan. *Education:* PhD/Physics/Case Inst of Tech; MS/Physics/MI State College; AB/Physics/MI State College. *Born:* Mar 12, 1919. With Oak Ridge Natl Lab, Oak Ridge Tenn 1951-57. Univ of Michigan Dept of Nuclear Engrg 1957- . Also with KMS Fusion, part time 1971-1979 . *Society Aff:* APS, ANS.

Osborne, Edward A
Business: Dept of Civil Engrg, US Air Force Academy, CO 80840
Position: Prof and Head *Employer:* US Air Force *Education:* PhD/Engrg Mech/Univ of Denver; MSE/Engrg Mech/Univ of MI; BS//US Military Acad, West Point *Born:* 2/28/39 Raised in Hicksville, NY. Commissioned in US Air Force upon graduation from West Point. Served a fighter pilot and flight examiner from 1960 to 1969. Member of US Air Force Academy faculty since 1969. Awarded tenure in 1978. Served as chrmn of the engrg mechanics division and deputy head. Currently Prof and Head of the Dept of Civil Engrg. Active in the aerospace and civil engrg divisions of ASEE. Assoc Editor of ASEE's Civil Engrg Education and member of CE education policy committee. Member, Kiwanis Club of the Rampart Range. *Society Aff:* ASCE, ASEE, ASME

Osborne, Merrill J
Business: 200 East Camperdown Way, P.O. Box 1028, Greenville, SC 29602
Position: Dir of Engrg. *Employer:* Bowater Inc *Education:* BS/EE/GA Inst of Tech. *Born:* Dec 1921. Tau Beta Pi. After work at Oak Ridge, entered the paper indus in 1946 in Crossett Ark. Joined Bowater in 1952; was promoted to V P & Dir of Engrg in Jan 1970 for Bowater Southern Paper Corp; was promoted to Dir of Engrg in Sept 1970, respon for all Bowater oper in N Amer. First TAPPI Engrg Div Award 1966. Pulp & Paper Indus Award 1970 IEEE. Pres 1973-75 TAPPI; V P & Chmn 1971-73 TAPPI; Bd of Dir 1971-75 TAPPI. Mbr of ASME, ASM, TAPPI, IEEE-Sr Mbr. *Society Aff:* TAPPI, IEEE, ASME, ASM, TBΠ, HKN

Osborne, William J
Business: 1401 N Westshore Blvd, PO Box 22317, Tampa, FL 33622
Position: Pres *Employer:* Gulf Design Div/Badger Am Inc. *Education:* BS/Nucl Engg/NC State Univ. *Born:* 5/31/42. Native of Greensboro, NC. Test Engr for Duquesne Light Co. Involved in reactor evaluation 1964-66. Process Engr with Wellman-Lord, involved in design of phosphate fertilizer facilities. During nine yrs with Wellman-Lord/Davy Powergas served as Process Engr, Sr Process Engr, Process Advisor, & Sales Engr. Moved to Gulf Design Div, Badger Am, Inc 1975 as Mgr of Bus Dev responsible for world wide sales. 1978 assumed responsibility as VP responsible for all sales & marketing. 1980 assumed responsiblity as pres responsible for div operations. Mbr local & natl AIChE and AIME. Enjoy golf, tennis, & fishing. *Society Aff:* AIChE, AIME

Osepchuk, John M
Home: 248 Deacon Haynes Rd, Concord, MA 01742
Position: Cons Scientist. *Employer:* Raytheon Co. *Education:* PhD/ESAP/Harvard Univ; Am/ESAP/Harvard Univ; AB/Engg Sci & App Phys/Harvard College. *Born:* 2/11/27. Native of Peabody, MA. With Raytheon since 1950 except for 1962-1964 when chief microwave engr at Sage Labs, Natick, MA. Directed projs in microwave tubes, image tubes & physical electronics. Consults on microwave ovens (Amana) and radiation hazards. Consulting Scientist, Raytheon Res Div. Fellow IEEE & Int Microwave Power Inst. Natl Lectr, IEEE-MTT-S, 1977-1978. Active in microwave safety, stds dev & public education. Mbr of IEEE Committee on Man & Radiation,

Osepchuk, John M (Continued)
Chrmn, Sci & Tech Cttes; Mbr of Bd of Dirs. Electromagnetic Energy Policy Alliance. *Society Aff:* IEEE, IMPI, BEMS, ΣΞ, ΦBK.

Osgood, Richard M
Business: 500 West 120th Street. Microelectronics Sci Lab, New York, NY 10027-6699
Position: Professor *Employer:* Columbia Univ *Education:* Ph.D./Physics/M.I.T.; M. S./Physics/Ohio State University; B.S./Engineering/U.S. Military Academy *Born:* 12/28/43 R.M. Osgood, Jr. is a Professor of Electrical Engrg and of Applied Physics at Columbia Univ in the City of New York. He also serves as Dir of the Microelectronics Sciences Labs and the Co-Director of the Columbia Radiation Lab. He has previously worked on the scientific staff of the M.I.T. Lincoln Lab, the U.S.A.F. Avionics Lab, and the U.S.A.F. Materials Lab. He has performed research in many areas of electrical engrg, physical chemistry, and optical physics; his most extensive research has been in the development of new infrared and ultraviolet lasers, the application of laser-induced chemistry to materials preparation, and the study of molecular kinetics and spectroscopy. He is a member of the ACS, the OSA, the MRS, a fellow of the IEEE, Co-Editor of Applied Physics, and an Assoc Editor of the IEEE Journal of Quantum Electronics. He has served as a consultant to numerous industrial and governmental organizations, including the DARPA Materials Research Council and the Los Alamos Chemistry Advisory Bd. *Society Aff:* IEEE, MRS, OSA, ACS

Osman, Richard R
Business: 707 Lake Cook Road, Deerfield, IL 60015
Position: VP *Employer:* Schirmer Engineering Corp. *Education:* B.S./Fire Protection Engineering/Illinois Institute of Technology *Born:* 12/04/36 After employment with the Illinois Inspection and Rating Bureau and the Home Insurance Co, joined Schirmer Engrg in 1964. Elected VP, Engrg in 1976. Responsible for four regional engineering offices providing fire protection design and consultation services. Mbr, bd of dir of Soc of Fire Protection Engrs. 1987 recipient of Joseph B. Finnegan Award in Fire Protection Engrg. Mbr of Life Safety Code and Portable Extinguisher Cttees of the National Fire Protection Assn. Past Chairman of the Chicago Assoc of Commerce and Industry Fire Prevention Cttee. Active in local community service. *Society Aff:* SFPE

Ostberg, Orvil S
Business: 320 Park Ave, New York, NY 10022
Position: Dir, Engrg Mgmt. *Employer:* IT&T Corp. *Education:* MBA/Bus/Univ of Chicago; BS/Elec Engg/Northwestern Univ. *Born:* Feb 1921. Native of Iron Mountain Mich. With ITT in var engrg capacities since 1947. Currently Corp Dir, Engrg Mgmt, respon for engrg practices throughout the ITT Sys & the budgetary & admin dir of the corp's res & dev programs. 1st Lt US Army Signal Corps WW II; currently Lt Col Signal Corps Res Ret. Sr Mbr IEEE; Reg P E Ill. Mbr AFCEA; P Chmn IEEE Comm Tech Group N Y Sect. Active in community affairs: Trustee Fanwood/ Scotch Plains YMCA *Society Aff:* IEEE, AFCEA.

Ostergaard, Paul B
Business: 115 Bloomfield Ave, Caldwell, NJ 07006
Position: President. *Employer:* Ostergaard Assocs. *Education:* SB/Meteorology/MIT; SM/Unspecified/MIT. *Born:* Nov 1924. Native of Erie Pa. US Army Air Corps 1943-46. SB&SM, MIT 49 & 50. Actuarial student Phoenix Mutual Life 1950-53. Sr Engr (acoustical specialist) Carrier Corp 1954-59; Sr Engr to V P of Goodfriend-Ostergaard Assocs 1959-71 cons in acoustics. Paes-Ostergaard Assoc 1971- . Fellow Acoustical Soc of Amer 1967; serves on cttes for ASA, & ASME, & Natl Panel Amer Arbitration Assn. Active in church & Boy Scouts. Pres-Natl Council of Acoustical Cons 1979-80. Pres - Consulting Engrs Council of New Jersey 1986-87. Fellow - Inst. of Acoustics. Diplomat - Amer. Acad of Environ Engrs. *Society Aff:* ASA, IOA, ASME, NSPE, AIHA, ASHRAE, AAEE.

Osterling, Allen W
Home: 173 Ramblewood Rd, Moorestown, NJ 08057
Position: V Pres - Woodbury Ofc *Employer:* EMJ/McFarland-Johnson Engrs Inc. *Education:* MBA/Business-Mgt/IN Univ; BSCE/Civil/Purdue Univ; BSNS & T/ Civil/Purdue Univ. *Born:* June 13, 1926. Chi Epsilon, Tau Beta Pi, Beta Gamma Sigma. Reg P E Ill 1954. Started career with Milwaukee Railroad. Worked for Portland Cement Assn 21 yrs incl pos as Statewide Paving Engr & Mgr Ill Dist. Asst V Pres at Westenhoff & Novick in charge of bus dev 5 yrs. Prof activities incl: Ill Soc of Prof Engrs, Chap Pres, State V P, legis ctte chmn, Dir to NSPE; NJ soc of Prof Engr State Pres Dir to NSPE; Ill Engrg council Pres; Amer Rd & Trans. Bldrs Assn Bd of Dir;ASCE Fellow, Branch-Pres, APWA DE Valley Chapter Delegate, VP, SAME; AREA. Married, 8 children. *Society Aff:* AREA, ASCE, NSPE, APWA, ARTBA, SAME, ITE.

Ostermann, Jerry L
Business: P.O. Box 188, Sylvan Grove, KS 67481
Position: Pres *Employer:* Ostee Corp *Education:* BS/Ag Engrg/KS State Univ *Born:* 5/17/50 Worked for Cessna Fluid Power Division and Henry Industries before founding Ostee Corp in 1976. Hold three US Patents. Kansas Section of ASAE Chrmn and 1986 "Young Engineer of the Year–. Chrmn of Bd of Western Kansas Manufacturers Association. *Society Aff:* ASAE

Ostrach, Simon
Glennan Bldg, Cleveland, OH 44106
Position: Professor. *Employer:* Case Western Reserve Unic. *Education:* PhD/Appl Math/Brown Univ; ScM/Appl Math/Brown Univ; ME/Mech Engg/Univ of RI; BS/ Mech Engg/Univ of RI. *Born:* Dec 26, 1923. Res scientist at NACA (later NASA) for 16 yrs. Formed Div of Fluid, Thermal & Aerospace Sci at Case Western Res Univ & was its head for 10 yrs. Since 1970 Wilbert J Austin Disting Prof of Engrg. P Chmn ASME Heat Transfer Div. Best paper award at 6th Natl Heat Transfer Conf 1963. Pi Ta Sigma Richards Mem Award 1964; ASME Heat Transfer Mem Award 1975. Mbr, Natl Acad of Engg. Sigma Xi National Lecturer 1978-79, ASME Freeman Scholar 1982. D. Sc. (Honoris Causa) Technican-Israel) Inst. of Tech. 1986. *Society Aff:* ASME, AIAA, AAM, ΣΞ.

Ostrander, Lee E
Business: Ctr for Biomedical Engg, Troy, NY 12181
Position: Assoc Prof. *Employer:* RPI. *Education:* PhD/EE/Univ of Rochester; MS/ EE/Univ of Rochester; AB/Phys/Hamilton College. *Born:* 2/18/39. Lee E Ostrander is Assoc Prof of Biomedical Engg at the Ctr for Biomedical Engg, RPI, Troy, NY. He is coordinator for the Clinical Engg prog at RPI, and is currently engaged in res on computer-based analysis of patient data, and clinical mass spectrometry for transport analysis in tissue. He received a PhD in electrical engg from the Univ of Rochester in 1961 and has since been involved in teaching & res, except for a yr spent in industry on product testing. He is a mbr of the IEEE Engg in Medicine & Biology Soc, & also a mbr of ASEE & BMES. *Society Aff:* IEEE, BMES, ASEE.

Ostroff, Norman
Home: 87 Fishing Trail, Stamford, CT 06903
Position: Mgr Process Engrg *Employer:* Peabody Process Systems, Inc. *Education:* PhD/Chem Eng/Polytechnic Inst of Brooklyn; MS/Chem Eng/Polytechnic Inst of Brooklyn; BChE/Chem Eng/Polytechnic Inst of Brooklyn *Born:* 2/20/37 BS, MS, PhD Chem Engrg Polytechnic Inst of Brooklyn. Native of Brooklyn NY. Res engr Amer Cyanamid Co Norwalk Ct 1960-73; Sr Engr Engelhard Minerals & Chems Newark N J 1973-74; Principal Engr UOP 1974-78. Math Instr Bridgeport Engrg Inst 1963-70; Chmn Math Dept 1970-78; Mgr Process Engrg, Peabody Process Systems Inc, Nowalk Ct. Prof exper incls catalyst dev, ultra high temp tech, chem mfg, energy prod, pollution control, electrophotography, teaching. Pubs deal with electrophotography, high temp work, pollution control and corrosion control. Married, 4 children; outside ints incl music, photography, ice hockey & tennis. *Society Aff:* ACS, NACE, ARCA

Ostrofsky, Benjamin
Home: 14611 Carolcrest Dr, Houston, TX 77079
Position: Prof. *Employer:* Univ of Houston. *Education:* PhD/Engg Design/UCLA; MEngg/Engg Mgt/UCLA; BSME/Mech Engg/Drexel Univ. *Born:* 7/26/25. Dr Ostrofsky's experience includes analysis, design, customer liason, proj engg, ind consulting, test planning, & operations. He authored a design methodology text & was Principal Investigator for the AF Office of Scientific Res in applying this developmental process to aerospace systems. He was a mbr of the AF Scientific Advisory Bd ad hoc Committee for ADP long range requirements & holds mbrship in five honor socs. In 1978 he was awarded the Armitage Medal by the Soc of Logistic Engrs for maj contributions affecting natl practice. He is a reg PE in TX & CA. *Society Aff:* IIE, AIAA, SOLE, ORSA, ASEE.

Ostrofsky, Bernard
Business: Amoco Res Ctr-Bx 400, Naperville, IL 60540
Position: Research Assoc. *Employer:* Standard Oil Co (Ind). *Education:* BS/Physics/City College of NY. *Born:* Jan 1922. BS City College of New York 1945; Grad Sch Polytech Inst of Brooklyn 1945-49. Native of NYC. Manhattan Proj Columbia Univ 194245, Natl Lead Co 1945-53; Commonwealth Engrg Co of Ohio 1953-54 ; with Standard Oil Co (Ind) since 1954. Presently Supr Matls Res & Services Dept. Fellow Amer Soc for Nondestructive Testing; Tech Editor 'Matls Eval'; Reg P.E.; John C Vaaler Award 1970. Active in Amateur Radio (W9IQ Extra Class) & enjoy music, classical & jazz. *Society Aff:* ASNT, ASTM, ACA, EMPSA, ACS.

Ostwald, Phillip F
Business: Box 427, Dept of Mechanical Engrg, Boulder, CO 80309
Position: Prof *Employer:* Univ of CO *Education:* PhD/IE/OK State Univ; MS/IE/OH State Univ; BS/ME/Univ of NB *Born:* 10/21/31 Is affiliated with the Univ of CO, Boulder, CO, USA, in the Dept of Mechanical Engrg. He teaches courses such as Cost Estimating, Operations Research, Engrg Statistics and Manufacturing Processes. Presented over eighty seminars, lectures and short courses. Is Dir for the Lab of Manufacturing Productivity. He is cited by American Men and Women of Science, Who's Who in the West and others. He has consulted with large and small industries. Interests include cost estimating and design-economic feasibility and his concentration on cost estimating is defining that professional field. He has over sixty publications. *Society Aff:* SME, IIE, ASEE

Oswald, William J
Home: 1081 St Francis Dr, Concord, CA 94518
Position: Prof of San Engrg & Pub Health. *Employer:* Univ of Calif - Berkeley. *Education:* PhD/San Eng/UC Berkeley, CA; MS.San Eng/UC Berkeley, CA; BS/Civil Eng/UC Berkeley, CA. *Born:* July 6, 1919. Res engr U C Berkeley since 1950. Prof since 1957. Lic C E Calif. Co-owner Swanson Oswald Associates Internatl Inc San Francisco Calif. Diplomat American Academy of Environmental Engineers 1980- . Mbr Water Pollution Control Fed; Harrison Prescott Eddy Medal 1954. Mbr Amer Soc Civil Engrs; James J R Cross Medal 1958; Rudolph Herring Medal 1958; Arthur M Wellington Prize 1966. Mbr American Waterworks Assoc 1954- ; Mbr Amer Pub Health Assoc 1980- ; Mbr Inter American Assoc of Environmental Engineers 1960- ; Mbr Amer Soc Prof Engrs; Dir Diablo Chap Calif Soc Prof Engrs 1969- . Spec cons for Agency for Internatl Dev, Internatl Bank for Reconst & Dev on waste mgmt for dev countries; sub-specialties bioengrg & applied algology. Enjoys fishing, swimming, astronomy, travel. *Society Aff:* NSPE, ASCE, WPCF, AWWA.

Ota, Hajime
Home: 5708 64th Ave, Riverdale, MD 20737
Position: Res Agri Engr. *Employer:* US Dept of Agriculture (Retired). *Education:* MS/Agr Engrg/Univ of MN; BS/Agr Engrg/MI State Univ. *Born:* Aug 1916. Native of San Gabriel Calif. US Army 1942-45. Grain storage res 1950-51 USDA Redwood Falls Minn. Since 1951, USDA, Res Agri Engr, dev heat & moisture losses from laying hens & broilers for use in designing insulation, heating & cooling of commercial production houses, & later obtaining similar data for growing pigs under 50-lb wt; also rabbits. Served as a mbr of var ASHRAE & ASAE tech cttes involving struct & environ for poultry & swine prod. Chmn Wash D C-Md Sect ASAE 1973; Chmn Prince Georges Area Sci Fair Assn; Chairman of the Joint Board on Science and Engineering Education for the Greater Washington, DC area. Mbr Wash Acad of Sci, Wash Soc of Engrs, AAAS, Tau Beta Pi and other Prof Orgs. *Society Aff:* ASHRAE, PSA, ASAS, WPSA, ASAE.

Othmer, Donald F
Business: Polytechnic Univ, 333 Jay Street, Brooklyn, NY 11201
Position: Distinguished Prof. *Employer:* Polytechnic Univ. *Education:* PhD/Chem Engg/Univ of MI; MSc/Chem Engg/Univ of MI; BSc/Chem Engg/Univ of NE. *Born:* 5/11/04. Hon Doctor Engrg from Polytechnic Univ, Univ of NB, NJ Inst of Tech, Univ of Concepcion (Chile). Dev engr Eastman Kodak 1927-31; Prof Polytech University 1932, Hd Dept Chem E 1937-61, Secy Grad Fac 1948-58, Disting Prof 1961- . In charge of program to build chemical industry for all Burma 1951-1953. Lic Prof Engr NY, NJ, OH, PA. Fellow AIChE, ASME, AAAS, Am Inst Chemists, NY Acad of Sci. Panel Mbr Natl Matls Adv Bd of NRC. Dir Engrs Joint Council 1956-1959. Co-ed Kirk- Othmer Encyc Chem Tech 83 vol in 3 Eng, 1 Span Edition. Over 150 pats & 350 res & engrg journal articles, also textbooks, Mbr, Editoral Adv Cttee Chem Eng Handbook 3rd Ed.; Tyler Award AIChE; Barber Coleman Award 1958 ASEE; Disting Consultant Award 1975, Assoc Consult, Chem & Chem Engr; Honor Scroll 1970 & Pioneer Award 1977 Amer Inst Chem; Hon Life fellow 1976 NY Acad of Sci; Hon Life Fellow-Institution of Chem Engr, (London) 1979; Elected to Hall of Fame, IL Inst of Tech, 1981. Professional Achievement Award 1978; Perkin Gold Medal 1978 Soc Chem Industry; Murphee (Exxon) Award 1978 ACS; Hon Life Mbr 1976 ACS; New York City Mayor's Award of Honor for Science & Technology 1987. Consult to government agencies, and corporations, lecturer, and licensor of patents in US and 33 foreign countries. *Society Aff:* ACS, ASME, AIChE, AIC, AAAS, IChE (London). DECHEMA, NYAS

Otis, Irvin
Business: 14250 Plymouth, Detroit, MI 48232
Position: Indus Engr, Mgr Indus Engr. *Employer:* Amer Motors copr. *Education:* MBA/Mgmt/Univ of Detroit; BS/Ind Mgmt/Univ of SD. *Born:* Apr 1926. MBA Univ of Detroit; BS Univ of S Dakota. Native of Brooklyn N Y. Employed AMC since 1969. Prior managerial indus engrg pos with Chrysler, Ford Corps 1951-68. Active Mbr IIE, serving as Dir Indus & Labor Relations Divs; Engrg Soc of Detroit; Soc for Advancement of Mgmt; Indus Mgmt Soc; Inst of Practitioners in Work Study, Org & Methods (London). Founder Indus Engrg Inst of Res & Tech. Publ papers var indus engrg subjects. Author textbooks 'Cost Controls' & 'First Line Mgmt - the Foreman's Respon for Mfg Costs'. Prof Central Mich Univ. Awarded Fellow by IIE & Phil Carroll Operations Mgmt Award by SAM *Society Aff:* IIE, SAM, IPWS, ESD.

Ott, Dudley E
Home: 1849 Sonoma Ave, Berkeley, CA 94707
Position: Retired. *Born:* Apr 1911. BS Univ of Calif 1933. Power engr Fibreboard Prod Inc 1933-36; Assoc W Harry Archer & Assoc 1936-40; Ins Insp Hartford Stm Blr Insp & Ins Co 1940-43; staff engr Golden State Co Ltd 1943-48; Sr Engr Arabian Amer Oil co 1948-50; Proj Engr 1950-61, Proj Dir 1961-62, Supv Engr 1963-65 Proj Mgr 1966-73 Power & Indus Div Bechtel Corp; Prin Engr, Staff Asst to Div Mgr Engrg San Fran Power Div, Bechtel Power Corp 1973-76. Mbr study team Pakistan AEC on feasibility of nuclear power in E Pakistan 1962-63; Mbr study team UN/Internatl atomic energy agency on feasibility of nuclear power in S Korea 1963. Fellow ASME (Mbr Exec Comm 1957-66, Chmn San Fran Sect 1964-65; Chmn Sects Comm Region IX 1962-64; Mbr Exec Comm 1964-69, Chmn 1968-69 Nuclear Engrg Div; Mbr-at-Large 1967-70, V P 1971-75 Policy Bd Power Dept; California Energy Coordinator for Region IX 1978-); Mbr ANS.

Ott, Walter R
Home: 60 Maple Ave, Wellsville, NY 14895
Position: Dean *Employer:* NYS College of Ceramics at Alfred Univ *Education:* PhD/Ceramic Engg/Rutgers Univ; MS/Ceramic Engg/Univ of IL; BS/Ceramic Engg/VPI. *Born:* 1/20/43. Dean-NYS Coll of Ceramics at Alfred Univ, Previously Assoc Prof and Dir of the Grad Prog in Ceramic Engg at Rutgers Univ. Worked as a Process Engr for Corhart Refractories and Staff Res Engr for Champion Spark Plug Co 1973 SAE Ralph Teetor Award, 1974 Profl Acheivement in Ceramic Engg Award - Natl Inst of Ceramic Engg, Keramos, Sigma Xi, Outstanding Educators of America 1974. PE - PA, Pres - Ceramic Educational Council 1976. Secy Ceramic Assn of NY. Fellow - The American Ceramic Soc. Bd of Trustees-American Ceramic Soc 1980-83. Res in high performance ceramic materials. *Society Aff:* ACS, ASEE, SGT, CANY.

Otte, Karl H
Home: 1005 S Knight Ave, Park Ridge, IL 60068
Position: Mech & Indus Engrg Cons. *Employer:* Self. *Education:* SM/Mech Engg/MIT; ME/Mech Engg/Armour Inst Tech; BS/Mech Engg/Armour Inst Tech. *Born:* 1904 Chgo Ill. Mech engr in foundry dev, Hawthorne Plant, Western Elec Co; Sr Indus Engr supr Dev of Improved Methods, E J Brach & Sons; Bldg Process Engr supr process engrg. Dept Milwaukee Ordance Plant of U S Rubber Co; mech engr supr engrg R&D of equip, Purity Bakeries Serv Corp, & successor co Amer Bakeries Co Chgo 1943-63; Asst Prof Mech Engrg Univ of Ill Chgo; mech & indus engrg cons. Reg P E Ill. Fellow Amer Soc Mech Engrs (V P). Mbr Amer Mgmt Assn, ISPE, NSPE, Western Soc of Engrs Chgo. Holder of Amer, Canadian & British patents. *Society Aff:* ASME, ISPE, NSPE, ASEE, AFS, AMA.

Otterman, Bernard
Home: 2188 Seneca Dr S, Merrick, NY 11566
Position: Pres 1974 to present *Employer:* Misharon Energy Services Inc *Education:* BME/Mech Engg/CCNY; MS/Heat Transfer/UCLA; PhD/Fluids/SUNY at Stonybrook. *Born:* 1937 Lodz Poland. Naturalized citizen. m. Sandra Meyer; 3 children. BME CCNY; MS UCLA; PhD SUNY at Stony Brook. Chmn Dept of Engrg & Computer Sci Hofstra Univ & Coord Poly/Hofstra Coop Engrg Prog 1974-78. Cons (LNG) & mbr of tech staff Cobot Corp Boston 1973. Asst Prof Northeastern Univ 1968-73. Aerospace exper with Grumman 1963-64 & Rocketdyne 1959-62. Mbr ASME, ASEE, SES. Elected 1975 N Y Acad of Sci. *Society Aff:* ASME, ASEE, SES.

Otth, Edward J
Home: 3902 Winterset Dr, Annandale, VA 22003
Position: Deputy Chief of Naval Matl (Acquisition). *Employer:* Chief of Naval Matl. *Education:* NASval Engr/Naval Arch & Marine Eng/MIT; BS/Engg/US Naval Acad. *Born:* Mar 1925. MS MIT 1955; BS US Naval acad 1949. Native of Jacksonville Ill. Rear Admiral US Navy. As commissioned officer USN since 1949 has been assigned wide var of jobs as naval engr in ships, naval shipyards, Bureau of Ships, Bureau of Naval Weapons, fleet maintenance staff & major proj offices. Specialist in ship design & acquisition. Prof Mgr Guided Missile Frigate Ship Acquisition Proj Mar 1971-Oct 1976. Specine Asst for Shipbldg, Naval Sea Sys Command Nov 76-Aug 78; Prin Deputy Commander Naval Sea Sep 78-Mar 79. Sub specialty weapon sys engrg. Sev military decorations incl Legion of Merit. Mbr Soc Sigma Xi, Phi Eta Sigma, Soc Naval Arch Marine Engrs & Amer Soc Naval Engrs. Enjoy photography & woodworking. *Society Aff:* SNAME, ASNE.

Otto, Carl W
Home: 250 Austin Ave, Atherton, CA 94025
Position: Conslt *Employer:* Self *Education:* MCE/Structural/Rensselaer Polytechnic Inst; BCE/Structural/Rensselaer Polytechnic Inst; BS/Naval Science/US Naval Academy *Born:* Apr 18, 1922. Reg Civil Engr Wash, Calif, Hawaii & Idaho. More than 30 yrs exper in exec role, civil engrg, proj mgmt & prof cons practice. Respon have incl client liaison, bus dev, execution of studies, prod of designs, supervision of const, & mgmt of offices in sev locations, incl Latin Amer. Respon for many large projs related to mech/elec installations, ports, harbors, coastal engrg, water supply, cogeneration plants & bridges. Amer Soc of Civil Engrs. Soc of Amer Military Engrs. Natl Soc of Prof Engrs. Cons Engrs Assn of Calif. Engrs Club of San Fran. San Fran Bay Area Engrg Council. *Society Aff:* ASCE, SAME, NSPE

Otto, George
Business: The Maytag Co, 403 W 4th St N, Newton, IA 50208
Position: Supr - Process Engrg. *Employer:* The Maytag Co. *Education:* MS/Met Engr/Univ of WI; BS/Met Engr/Univ of WI. *Born:* July 1924. With The Maytag Co since 1954, originally as Plant Metallurgist until present pos. Areas of spec: powder met, aluminum melting & die casting & failure analysis. Mbrship and/or offices: P Chmn Ctte B-9 on Metal Powders & Metal Powder Prods of the Amer Soc for Testing and Mtls PPres of In- Plant Powder Met Assn of Metal Powder Indus Fed; Amer Soc for Metals - held appointments on sev natl cttes - contrib to handbooks on powder met & failure analysis, Reg PE. Author sev tech articles on powder met. Ints: skiing, hiking, theater. *Society Aff:* APMI, ASM, ASTM.

Otto J Nussbaum,
Consulting Engineer, 144 Penns Woods Drive, Newtown, PA 18940-1115
Position: Consultant *Employer:* Self *Education:* BS/ME/Leningrad USSR. *Born:* 1913 Budapest Hungary. Attended Technische Hochschule, Vienna Austria 1931-32; grad ME Mech-Tech Inst, Leningrad USSR 1937. Speaks English, German, Russian. Reads French & Polish. Came to U S 1937, naturalized 1942. Shift engr Cold Storage Plant 2, Moscow 1935-37; Dev Engr Kramer Trenton Co, Trenton N J 1939-50, Ch Engr 1950-64; Ch Dev Engr ITT Nesbitt, Phila Pa 1964-66; Dir of Engrg & Res Larkin Coils, Atlanta Ga 1966-68; Dir of Engrg & Res Halstead & Mitchell Div, Scottsboro 1968-1983. U N tech advisor Polish Gov, Warsaw 1960. Reg P E NY, NJ, PA, GA, FL & AL. Fellow ASHRAE; Mbr Natl Soc Prof Engrs, INternatl Inst Refrigeration, Refrigeration Service Engrs Soc. Patentee in heat transfer, air cond, refrig. Author tech articles. Contrib ASHRAE handbook. Consltg Engr, Newtown, PA since 1984. *Society Aff:* ASHRAE, NSPE, RSES, IIR

Otto, Thomas Herbert
Home: 41 Blanche Avenue, Demarest, NJ 07627
Position: President *Employer:* Thomas H Otto & Assocs., Inc *Education:* BCE/Civil Engg/Rensselaer Polytechnic Inst. *Born:* 11/11/38. in NY City. BCE from RPI, 1960. Post-grad work in civil & indus engg Columbia 1962-64. Soils & fnd engr Tippetts-Abbett-Mc Carthy-Stratton 1960-66; Asst Dept Hd, Soils & Fnd, Frederic R Harris Inc 1966-68; Dept Hd Storch Engrs 1968-69; Dept Hd & Proj Engr Van Houten Assocs Inc 1969-71. Founded Thor Engrs Newark NJ & NY City 1971: Founded & Owner of Thomas H Otto & Assocs 1977- . Served with Corps of Engrs USARADCON 1961-63. Mbr ASCE; Mbr Natl & NJ Soc of Prof Engrs, Chap Pres; Mbr. Who's Who in the East; Mbr Apha Chi Rho (social); Charter Mbr Eastern Sailing Club; Reg P E NY, NJ, CT, VT, MA. Reg Professional Planner NJ. Member Demarest, N.J. Planning Board. *Society Aff:* ACSE, NSPE, NJSPE

Ouimet, Alphonse
Home: 227 Lakeview Ave, Pointe Claire Quebec, Canada H9S 4C8
Position: Consultant *Education:* BEng/-/McGill; BA/-/Univ of Montreal. *Born:* Montreal 1908. Canadian communications pioneer, executive & consultant. Successively Res, Oper & Ch Engr Genl Mgr & Pres Canadian Broadcasting Corp, retiring 1968 after 33 yrs service. Chmn Telesat (333 River Rd, Ottawa) 1969 to 1980. Chmn of Communications Res Adv Bd of Dept of Communicatio ns. 1975-80. Chmn CBC Corp Olympics Ctte. 1973-76 Hon doctorates from Univs of Montreal, McGill, Acadia, Saskatchewan, Ottawa, Sherbrooke, Royal Military Coll & Laval. The Ross, Julian C Smith & Sir John Kennedy Medals of EIC; IEEE Fellowship & McNaughton Medal; SMPTE & ACFAS Special Awards; Canadian Prof Engrs Gold Award 1975; 1977 Intl Emmy Award; 1978 Great Montreal Award; Companion Order of Canada. 1987 Fellow Canadian Acad of Engg. *Society Aff:* FIEE, LMEIC.

Outwater, John Ogden
Business: Mech Engrg Dept, Burlington, VT 05401
Position: Professor. *Employer:* Univ of Vermont. *Education:* PhD/Cambridge Univ; ScD/ME & MET/MIT; MA/-/Cambridge; BA/Mech Sci/Cambridge. *Born:* Jan 2, 1923. E I duPont 1950-52; Universal Moulded Prods 1952-54; ILO, MIT 1954-56; Prof Mech Engrg Dept Univ of Vermont 1957- . Chmn 1957-63. Led four archaeology expeds for Wenner-Gren Foundation. Conducted res on composite matls, glass, ski accidents etc. Organized Vt Instrument Co Inc 1962. Fellow Amer Soc of Mech Engrs. P E 856 Vt. Cons U S Naval Ordnance Lab 1960-71. Cons Monsanto Res Corp 1964-79. Officer REME WWII Vermont Engineer-of-the-Year 1970. Outstanding Educator of America 1970. *Society Aff:* ASME, ASTM.

Over, R Stanton
Business: The EADS Group, P.O. Box 1887, Altoona, PA 16603
Position: Pres. *Employer:* EADS. *Education:* MS/Civ Engg/Carnegie Mellon Univ; BS/Civ Engg/Carnegie Mellon Univ. *Born:* 2/15/35. As a PE I am pres of EADS, Gen Engg Assoc, Inc in Altoona and Chmn of the Bd of the Neilan Engrs, Inc in Somerset, PA serving clients in PA & adjacent states. Managing Partner of Rackoff-EADS headquartered in Chicago, IL providing engg services in the metro area. I am a reg PE in 14 states & am certified by the Natl Council of Engg Examiners. Member - ACEC Transportation Committee 710 1985 to present. Also, mbr of ACEC PR Ctte *501 1973-76, served as chmn 1980-85, mbr of 1979-80 Cteof 100, Minuteman - ACEC/GAAP 1977 to present, National Director to ACEC for CEC/PA 1981-83. Served as pres of the Consulting Engrs Council of PA having held several positions of leadership in local, state & natl engg socs in the past. I have taught engg courses at Carnegie Mellon Univ & the PA State Univ & published res in the ACI Journal & was chrmn of the publication committee responsible for ARTBA's EIS manual. Society Aff:* ACEC,ACSM, ASCE, ASTM, NSPE, ASHE.

Overbeck, Edward M
Business: 8585 Commerce Park Dr, Houston, TX 77036
Position: Division President. *Employer:* Stubbs Overbeck & Assocs Inc. *Born:* Sept 1921. BS Texas A&M Univ. Native of Dallas Texas. Entered military service, Chem Corps, from college in 1942. Discharged in 1945 with rank of Major. 1945-51 design engr for Diamond Shamrock & Olin Chems. Proj engr for Daroid Div, Natl Lead 1951-54. Formed a cons & design engrg co 1954. Until 1972, acted as proj mgr on air & water pollution projs. Accepted active mgmt of co in 1972. Enjoy golf, fishing & hunting.

Overman, Allen R
Business: Agri Engg Dept, Gainesville, FL 32611
Position: Prof. *Employer:* Univ of FL. *Education:* PhD/Agri Engg/NC Univ; MS/Agri Engg/NC Univ; BS/Agri Engg/NC Univ. *Born:* 3/11/37. Res interests include land treatment of wastes, chem transport in soil, virus transport & survival in soil, modeling of rate processes in soil & ground water. *Society Aff:* ASAE, ASCE, AIChE, WPCF.

Owczarski, William A
Business: 400 Main St, E Hartford, CT 06108
Position: Manager, Technical PLanner *Employer:* Pratt & Whitney Aircraft. *Education:* BS/EE/Univ of MA; MS/Met E/Rensselaer Polytechnic Inst; PhD/Met E/Rensselaer Polytechnic Inst. *Born:* June 1934. Employed by Sprague Elec Co 1955-57; Genl Elec Co 1958-61; PWA (Div of United Tech Corp) 1962- . Present activity at PWA involves group technical planning in materials and manufacturing technology. Author more than 40 pubs & holder of 10 pats. Dir AWS 1969-75, thrice winner of AWS William Spraragen Award (1967, 1970, 1973); recipient AWS Comfort A Adams Lect Award 1973; recipient United Tech Corp George Mead Medal for Engrg Achievement 1974; Fellow, ASM 1978. *Society Aff:* ASM, AWS, TMS-AIME, ΣΞ.

Owen, Edwin L
Business: PO Box 20148, San Antonio, TX 78220
Position: Sr Process Metallurgist. *Employer:* Turbine Support Div. *Education:* PhD/Metallurgy/OH State; BS/Metallurgy/CO Sch of Mines. *Born:* Aug 1942. PhD Ohio State; BS Colorado School of Mines. Native of Chgo Ill. Metallurgist for Kaiser Steel Corp 1965, 1966. Asst Prof of Met Penn State Univ 1970-74. Kennecott Copper Corp (Ledgemont Lab) 1974- . At Kennecott, is Sr Process Metallurgist for the Ocean Mining Prog respon for processing of ocean nodules & Long Range Planning Ctte. Also cons in corrosion engrg to the In-Situ Mining Prog. As a mbr of the Amer Soc for Metals, has served on the Natl Public Service Ctte & Career Guidance Ctte. Also has been Chmn of the Penn State Chap of ASM and Bd of Dirs of the Alamo Chap. Dir of Quality Assurance for Chromalloy, turbine support Div which works on Turbine Engines for the Air Force and Commerical Airlines. *Society Aff:* ASM, NACE, TMS, AES.

Owen, Eugene H
Business: 9490 Airline Hgwy, PO Box 66396, Baton Rouge, LA 70896
Position: Pres *Employer:* Owen and White, Inc *Education:* BS/ME/Vanderbilt Univ *Born:* 10/1/29 Pres and Chrmn of Bd of Owen and White, Inc, since 1960. Practicing consulting engr in fields of sanitary engrg, oil and gas pipeline design, engrg economics, and water and gas utilities. Member of Bd of Dirs of First Bankshares of LA, Inc; LA National Bank; Baton Rouge Waterworks Co; LA Water Co; French Settlement Water Co; and Gulf South Research Inst. Chrmn of Bd of Trustees, Presbytery of South LA. Member of Session First Presbyterian Church; Bd of Dirs, Catholic Presbyterian Apartments; Bd of Trustees First Presbyterian Church; Bd of Trustees of LA Arts and Science Foundation. *Society Aff:* NSPE, ACEC, APWA

Owen, Robert R
Home: 32743 Upper Bear Crk Rd, Evergreen, CO 80439
Position: President. *Employer:* Eversman Mfg Co. *Education:* BS/Mech Engg/Univ of CA, Berkeley; BS/Agri Engrg/Univ of CA, Berkeley. *Born:* Aug 1921. Native of Yuma Ariz. Army Corps of Engrs 1942-46. Brigadier Genl USAR-Ret. Agri Engr for Del Monte Corp & duPont Co. Hd, engrg Dept Pineapple Res Inst Honolulu. Prod Planning Mgr, Asst Ch Engr & Genl Mgr Equip Oper, Ford Tractor 1956-68. Pres Great Western Sugar Co 1968-71. Pres Great Western Producers Co-op 1971-76. Fellow Amer Soc of Agri Engrs. Univ of Calif Centennial Citation 1968. Colo State Univ Award of Merit for Disting Service to Agri 1971. P E Calif. Mbr USDA Joint Council for Food & Agricultural Science 1983-1986. *Society Aff:* ASAE.

Owen, Walter S
Business: Matls Sci & Engrg, Cambridge, MA 02139
Position: Dept Hd, Prof of Matls Sci & Engrg. *Employer:* MIT. *Education:* DEng/Met/Univ of Liverpool; PhD/Met/Univ of Liverpool; MEng/Met/Univ of Liverpool; BEng/Met/Univ of Liverpool. *Born:* Mar 13, 1920. Alsop High School; Univ of Liverpool. Metallurgist, D Napier & Sons & English Elec Co 1940-46; Asst Lectr & Lectr in Met 1946-54, Commonwealth Fellow, Met Dept MIT 1951-52 - res staff 1954-57; Henry Bell Wortley Prof of Met Univ of Liverpool 1957-66. Thomas R Briggs Prof of Engrg & Dir of Matls Sci & Engrg Cornell Univ 1966-70; Dean of Tech Inst Northwestern Univ 1970-71; V P for Sci & Res Northwestern Univ 1971-73. Head Dept of Matls Sci & Engrg, MIT 1973-82; Prof of Phys Met 1982 to present. *Society Aff:* NAE, TMS/AIME, ASM, BISI.

Owen, Warren H
Business: Box 33189, Charlotte, NC 28242
Position: Exec. Vice President *Employer:* Duke Power Co. *Education:* Bachelor/Mech Engr'g/Clemson University *Born:* 01/08/27 A native of Rock Hill S.C., Warren Owen attended Clemson Univ, graduating with high honors in mech engrg. After joining Duke Power in 1948, he served in various positions, becoming a Sr. VP and being elected to the Bd of Dir in 1978. In 1984, he was elected exec VP, engrg, construction and production group. He is the recipient of several outstanding awards by the Instrument Soc of America, Charlotte Engrs Club, N.C. Soc of Eng, and Atomic Industrial Forum. He was elected a Fellow in ASME and was

Owen, Warren H (Continued)
recently awarded the John Landis Medal. He is married to the former Virginia Boulware and they have three daughters. *Society Aff:* ANS, ASME

Owens, C Dale
Home: P O Box 105, Gosport, IN 47433
Position: Mbr of Tech Staff - ret. *Employer:* Bell Telephone Labs. *Education:* MA/Physics/Columbia Univ; AB/Physics/IN Univ. *Born:* May 1906 Wadesville Ind. AB Physics, Indiana Univ., MA Physics, Columbia Univ. Employed 1928-71 Bell Labs as specialist on magnetic matls & cores for loading coils, broadband carrier, submarine cable, memories & telephone sets. Designed inductors for high powered radar during WW II. Author, speaker, active ctte mbr IEEE, ASTM, IEC on magnetic matls & measurements. Fellow IEEE. Mbr Phi Beta Kappa, Lions Internatl, USNC of Internatl Electrotech Comm. US rep & chief delegate to IEC ctte on ferrites & magnetic components, 1959-1982. Elder, active churchman United Presbyterian Church USA. Hobbies: Photography, travel, golf, gardening. *Society Aff:* IEEE, ASTM.

Owens, James B
Home: 746 Barat Court, Lake Forest, IL 60045
Position: Vice President *Employer:* Packer Engineering Associates, Inc. *Education:* BS/Elec Engg/Rice Univ. *Born:* 7/23/20. Native of Houston TX. Worked for Westinghouse Elec Corp on Navy & Marine elec equip during WWII. Joined I-T-E Imperial Corp in 1954. Moved through pos of VP - Engg & VP - Mktg to pos of Pres of the I-T-E Pwr Equip Group. Was Pres of IEEE Pwr Engg Soc 1973. Principal hobby is amateur radio. Became VP-Electrical Sys Group of Gould, Inc 1976. Elected Pres of Gould-Brown Boveri, a joint venture co of Gould Inc & Brown Boveri of Switzerland, 1979. Pres of IEEE 1983. Joined Packer Engrg, Naperville, IL in Jan 1984 as Dir of Elect Engrg. Currently a VP at Packer Engrg. *Society Aff:* IEEE.

Owens, Thomas C
Home: 1809 N Third St, Grand Forks, ND 58201
Position: Prof & Chmn of Chem Engrg. *Employer:* Univ of North Dakota. *Education:* PhD/Chem Engr/IA State Univ; MS/Chem Engr/IA State Univ; BSChE/Chem Engr/Univ of ND. *Born:* May 8, 1941. Sr Res Engr wth Exxon Production Res Co (Houston Tex) prior to joining Univ of N Dak 1968 as Asst Prof. Promoted to Assoc Prof 1970 & to Prof 1976. Assumed dept chmnship in 1974 after one yr leave-ofabsence with Exxon Prod Res Co (Houston Tex) working on tertiary oil recovery. Mbr AIChE, ASEE, ACS, AAAS, N D Acad of Sci, Sigma Xi. Reg P E ND. Recipient of Dow Outstanding Young Fac Award ASEE 1972. *Society Aff:* AIChE, ASEE, ACS, AAAS.

Owens, Willard G
Business: 4405 W. 29th Ave, Denver, CO 80212
Position: Pres *Employer:* Willard Owens Assocs, Inc *Education:* MS/Geol/MO School of Mines; BS/Geol/MO School of Mines *Born:* 9/29/31 Initially a water well contractor, Owens became a consulting geologist and engr. Registered as a PE in six states and as a professional geologist in two states, Owens established Willard Owens Assocs, Inc in 1969. Owens was chrmn of Denver Section, AEG, and secy of AIPG. He served on the national executive bd of both organizations. Owens has been active in case law and legislation on water in CO and WY. *Society Aff:* ASCE, AEG, NWWA, AIPG

Oxford, Carl J, Jr
Home: 288 Wimberly Dr, Rochester, MI 48064
Position: Conslt Engr *Employer:* Self *Education:* ME/Mech Engr/Univ of MI; BSE/Engr Mechanics/Univ of MI. *Born:* Jan 1918 Detroit Mich. Reg P E. R & D Proj Engr for US Navy 1941-45. With Natl Twist Drill & Tool Co since 1945; elected V P 1967, Retired 1983. Now Consltg (Mfg Engrg & Metallurgy). Conducted first comprehensive investigation of fundamental mechanics of twist drill operation; papers 1953 & 55. First recipient ASME Blackall Machine Tool & Gage Award 1955, for paper covering study of screw thread assembly strength. Active in ASME and SME. Fellow ASME and SME. Director SME 1969-71. Organizing Chairman of SME Material Removal Division. Frequent author and lecturer on Metal Cutting and related subjects. Rotarian. Hobbies: Amateur Radio, Boating, Live Multi-Channel Sound Recording of Musical Programs. Received SME Progress Award, 1978. *Society Aff:* ASME, SME, ASM.

Ozisik, Necati M
Business: Mechanical & Aerospace Eng. Dept. Box 7910, Raleigh, NC 27695-7910
Position: Prof *Employer:* North Carolina State Univ. *Education:* PhD/ME/Univ. of London; MS/ME/Univ. of London *Born:* 06/17/23 Served as sr research scientist at the ASHRAE research Lab and the Oak Ridge National Lab before joining North Carolina State Univ in 1963. Author or coauthor of eight textbooks and over 200 research papers in the area of heat transfer. Recipient of several awards, including F. Bernard Hall prize of Inst. Mech. Engrs. London; G. Westinghouse Award of ASEE; Turkish Science Award of Turkish research foundation; O. Max Gardner Award, and Reynolds awards of NCSU system and Heat Transfer Memorial Award of ASME. *Society Aff:* ASME, ASEE

Paavola, Ivar R
Home: 2100 Duckwalk Ct, St Charles, MD 20601
Position: Struct Engr *Employer:* Off, Chf Engrs, US Army *Education:* BS/Civil For/OR State Univ *Born:* 12/9/33 With US Army Corps of Engrs since 1963, serving through assignments in construction, planning, structural engrg on design of multi-purpose dams, Chief of Structures for Portland District (38 engrs and architects) (73-79). Assumed current position with responsibility for civil structural design for the West coast, AK and HI, also the Great Lakes Region. Develops structural design criteria and standards for nationwide civil projects for navigation, floodcontrol, hydropower, etc. PE in OR. Commander US Naval Reserve. *Society Aff:* USCOLD, NSPE, SAME, TROA, NRA

Pace, Danny L
Home: #12 Lake Blvd, Vicksburg, MS 39180
Position: Mgr Nuclear Design *Employer:* Systems Energy Resources, Inc. *Education:* BS/Nuclear Engrg/MS State Univ *Born:* 12/07/55 Native of Magee MS, Employed by Systems Energy Resources inc since 1977. Served as startup test supvr for four years, shift tech advisor/reactor engr for three years, then Engrg Supvr in charge of the Independent safety Review group for one year. Currently Manager-Nuclear Design engineering at the Grand Gulf Nuclear Station. P Pres of the Local Kiwanis Club. Asst scoutmaster, Young Engr of the year 1981 MES western div. Two term Secy-Treas one term chrmn of the MS section of ANS. District Comm Activities Chair, BSA, and Admin Bd member of United Meth Church. Mr Pace is a NRC Licensed Sr. Reactor Operator and a Licensed Prof Engr. *Society Aff:* NSPE, ANS

Packard, David
Business: 1501 Page Mill Rd, Palo Alto, CA 94304
Position: Chairman of the Board. *Employer:* Hewlett-Packard Co. *Education:* BA/-/Stanford Univ; EE/-/Stanford Univ. *Born:* 1912 Pueblo, Co. Phi Beta Kappa. Engr GE vacuum tube dept, Schenectady 1936-38. With William R Hewlett founded Hewlett-Packard Co 1939. HP is major manufae of electronic, biomedical, analytical & computing instrutation. Resigned from HP to accept Presidential appt as U S Deputy Secy of Defense Jan 1969. Resigned this post Dec 1971 to return to CA where elected Chmn of Bd Hewlett- Packard. Over the years has been active in many prof, edu, civic & bus organizations. Fellow IEEE, lifetime mbr ISA, mbr NAE. Recipient six hon degrees & num engrg, bus & govt awards. *Society Aff:* IEEE, NAE, ISA

Packer, Kenneth F
Business: P O Box 353, Naperville, IL 60566
Position: Chairman of the Board and CEO *Employer:* Packer Engrg Assoc, Inc. *Education:* PhD/Industrial Engg/Purdue Univ; MSE/Metallurgical/Univ of MI; BSE/

Packer, Kenneth F (Continued)
Chem/Univ of MI; BSE/Metallurgical/Univ of MI. *Born:* Aug 1924. Assoc Chem Grand Rapids Jr Col. USMC WWII 1st Lt. Assoc Metallurgist American Brake Shoe Co 1952. Univ Mich Instructor Dept Production Engrg to 1957. Staff Engr Danly Machine Specialties Chicago to 1962. Company & staff spec broadly in areas encompassed by materials related science, engrg & tech, conducting res, dev & providing cons tech assistance to var industries. Extensive testing facilities located near Chicago. Registered Professional Engineer. *Society Aff:* NSPE, ASTM, ASM, ASME, ASAE.

Packer, Lewis C
Home: 2125 S E Lakeview Dr, Apt. 16, Sebring, FL 33870
Position: Engrg Consultant (Self-employed) *Employer:* Westinghouse Elec. Corp. Now Retired. (1958). *Education:* BEE/EE/OH State Univ. *Born:* Native Wellsville, OH. Westinghouse Elec Corp. 1917, -assisted in design of 1st Electric Motor Driven Torpedo; Engr Course-Frac. HP, des engr. D-C motor-generator-Universal Motor Sec, 1919. Dev first complete line Universal (AC-DC) non-Compensated motors at the approx start of the era of elec appliances, portable tools, office equip etc. In 1921 Resident Engr new Fract HP plant in Springfield, MA. 1922 Polyphase-Universal motor sect, Dev-des and mfg procedures for first uncompensated Universal motor with exceptionly good commutation and brush life many times greater than Industry. Dev-des single field compensated Universal motors, in 1928. Moved into single and polyphase and Universal motors sec; 1927, Mgr Universal Motor Section; 1929 dev, patented first synthetic shock proof resin (plastic) motor housing for appliances - portable tools. Dev, patented constant speed governor for Mot-Gen Air-planes, etc; 1933 Sec Mgr, then mgr Motored Appliances (new) Engr Dept.- Fans, vac cleaners, appliances, etc; 1937 Mgr Motor Engrg combined Mot Appl eng and Domestic and refrig & air conditioning compressors (up to 100 HP) Dept; World War II dev-des the motors for all Tank and most of the Battleship Gyros for the Gun Stabilizers - Servo and Gryo motors for Radar Control, Dev new lines 2-Pole Single phase. Consultant 1959 to date. Life Fellow IEEE, Reg PE MA. 20 patents; recipient, highest Westinghouse awd; IEEE: "A Century of Honors" (1884-1984). Hobbies: golf & travel. *Society Aff:* IEEE.

Pade, Earl R
Business: Box 355, Pittsburgh, PA 15230
Position: Sr Engr. *Employer:* Westinghouse Elec Corp. *Education:* BBA/Industry/Univ of Pittsburgh. *Born:* Nov 18, 1930. Native Pittsburgh, Pa. Res Engr Alcoa respon area NDT control & personnel qualification. Principle Res Engr Combustion Engrg, supr long term NDT res. Presently employed Westinghouse Nuclear Technology Div with diversified assignments relating to NDT of nuclear steam supply sys. Chmn ASNT Sonics Ctte, Chmn ASNT Methods Div, Secy & V Chmn ASNT Personnel Qualification Div 1965-75, Secy ASNT Tech Council 1976. Chmn ASNT Tech Council, 1978, Natl Dir ASNT, 1978-79. Elected ASNT Fellow 1973. Author 18 training texts & Company personnel qualification manuals. *Society Aff:* ASNT.

Padgham, Henry F
Home: 806 West Glen Oaks La, Mequon, WI 53092
Position: Program Dir and VP *Employer:* CH2M Hill *Education:* BS/Civ Engrg/OR State Univ; BS/Agri/Univ Calif, Davis. *Born:* 4/12/32 Extensive experience specializing in development and operations of cost and schedule control systems of large multiple construction contract public works projects. Directed all project control services provided by the General Engrg Consultant for Metropolitan Atlanta rapid transit construction program. Have been with CH2M HILL since 1980 as Program Dir and in responsible charge of a metropolitan water pollution abatement construction program in excess of one and a half billion dollars. *Society Aff:* ASCE, NSPE, PMI.

Paduana, Joseph A
Home: 551 Pico Way, Sacramento, CA 95819
Position: Prof *Employer:* CA State Univ *Education:* PhD/CE/Univ of CA-Berkely; MS/CE/Polytech Inst of Bklyn; BS/CE/Cooper Union *Born:* 09/09/17 in Utica, NY. Served with Army Air Corps, 1942-45, Captain; awarded Distinguished Flying Cross. Structural engrg, 4 yrs, then proj engr with Moran, Proctor, Mueser & Rutledge, 6 yrs. Sr proj engr with Raymond Intl before entering teaching at CA State Univ, Sacramento, Prof of Civil Engrg. Served as Dept Chrmn 1968-69. CA reg Civil Engr; NY licensed PE. Tau Beta Pi; Sigma Xi; Chi Epsilon. Publications with Highway Res Bd. Enjoy music, reading, and exercising. *Society Aff:* ASCE, ASEE

Padulo, Louis
Business: 110 Cummington St, Boston, MA 02215
Position: Dean of Engrg *Employer:* Boston Univ *Education:* PhD/EE/GA Inst of Tech; MS/EE/Stanford Univ; BS/EE/Fairleigh Dickinson Univ *Born:* 12/14/36 Born in AL, grew up in NJ. Worked at RCA, the Mitre Corp, IBM-Watson Labs, Airesearch Corp, and GA Tech Engrg Experiment Station. Taught Physics at Fairleigh Dickinson, Electronics at Stanford and San Jose State and Mathematics at GA State, Morehouse Coll, Stanford, Harvard and Columbia. Authored two books, active in increasing minority and female participation in engrg and sci. Founded Dual-Degree Program of Atlanta Univ and GA Tech. Was Chrmn of Math Dept at Morehouse before joining faculty at Stanford. At Boston Univ since 1975 as Dean of Engrg and Prof of Mathematics and Engrg. ASEE Western Elec Fund Award, 1973, Natl Consortium for Black Prof Devel Award 1977; GE/NACME Reginald H. Jones Award 1983, ASEE Vincent Bendix Award 1984. *Society Aff:* AAAS, IEEE, MAA, ASEE

Pae, Kook D
Business: Dept of Mechanics & Materials Science, College of Engg., Rutgers Univ, Piscataway, NJ 08855-0909
Position: Distinguished Prof *Employer:* Rutgers Univ. *Education:* PhD/Mechanics/Penn State Univ; MS/Mech Eng/Univ of MO; BS/Physics/MO Valley Coll *Born:* 9/20/32. Born in Seoul, Korea. Became naturalized in 1972. Received the PhD from Penn State and assumed an Asst Profship at Rutgers Univ in 1962. Is currently a Distinguished Professor in Dept of Mechanics and Materials Science. Res interests in physical and mechanical properties, surface modification, piesoelectricity, and high pressure phenomena of polymers. Published numerous articles on the subjects. *Society Aff:* APS, SPE, ASEE, AAUP, ΣΞ, AIRAPT.

Paganelli, Thomas I
Business: Court St 3-1, Syracuse, NY 13201
Position: V P & Gen Mgr Elec Sys Div. *Employer:* General Electric Co. *Education:* BSEE/Electr/Univ of CA. *Born:* 10/13/18. Native Cle Elum, WA. Air Force 1942-46. Staff ambr Servo-Mech Lab at MIT 1946-50. With GE's Heavy Military Elec Sys Dept in var supr capacities spec in complex military elec sys. Made Genl Mgr of HMES Dept 1962 with 6000 employees, 2000 engrs & scientists. Assumed current position as Gen Mgr of Elec Sys Div & elected GE VP 1974. Fellow of IEEE 1969. Serve on var bds including EIA, Lincoln First Bank, LeMoyne Co. Golf & gardening are hobbies. *Society Aff:* IEEE, ADPA, AFCEA, ASNE, AUSA, RESA, EIA.

Page, Chester H
Home: 1707 Merrifields Drive, Silver Spring, MD 20906
Position: Retired. *Education:* PhD/Physics/Yale Univ; ScM/Physics/Brown Univ; AB/Physics & Math/Brown Univ. *Born:* Nov 1912. Native Providence, R I. Taught Lafayette Col 1937-41; with NBS 1941-1976. Formerly Chief, Electricity Div, Then Inst for Basic Standards Coordinator for Internatl Standardization Activities & NBS SI Units Coordinator. Chmn or secy of Cttes in ANSI, IEC, OIML. Active on cons cttes of Internatl Ctte for Weights & Measures (CIPM). Harry Diamond Award 1974. Charles Protens Steinmetz Award 1986. *Society Aff:* APS, IEEE.

Page, James W
Business: 3290 W. Big Beaver Rd, Troy, MI 48084
Position: VP *Employer:* Ellis/Naeyaert/Genheimer Assoc, Inc *Education:* BS/CE/MI State Univ *Born:* 05/05/44 Grad of MI State Univ is a reg PE in the State of MI. Began his profl career as a junior engr at Giffels & Rossetti (architects/engrs) in Detroit and is presently VP of Ellis/Naeyaert/Genheimer Assoc and a mbr of the company's bd of dirs. Page was Pres (1981-1982) of the Detroit Post of the Soc of American Military Engrs and past Junior VP of the Post. He has served on committees for the American Consltg Engrs Council & the Consltg Engrs Council of MI. He was the 1980-1981 Chrmn of the American Consltg Engrs Council's Engrg Excellence Awards Program Committee and also served on ACEC's Public Relations Committee. Page is a native of Grand Rapids, MI. He also served as a Dir and VP of the Consulting Engrs Council of Michigan. He is currently Chmn of the Mbshp Dev Ctte for CEC/M. *Society Aff:* SAME, ASCE, ACEC

Page, Raymond J
Home: 1151 Normandy Terr, Flint, MI 48504
Position: Consultant *Employer:* Self employed *Education:* MSIE/Ind Engrg/Purdue Univ; BSME/Mech Engrg/Purdue Univ. *Born:* Apr 1925. Additional grad study Cornell Univ. Native Chicago, Ill. US Navy 1943-46. Indus Engr Sylvania Elec Corp spec in sys analysis. Taught SUNY at Buffalo & Cornell. Joined General Motors Inst 1956. 1960 became respon for Fifth-Year Program as Dir of Fifth-Year Plans. Assumed respon as Dir of Technical Educational Services 1964, providing tech progs & services to GM Units. 1980 Appointed Dir, Tech Ed Services, GM Education and Training, Personnel and Dev Staff, Gen Motors Corp 1982 retired from GM and presently active as a conslt in tech training. Policy Board Education ASME 1970-74; Chmn Continuing Education Ctte ASME 1970-73; V P ASME 1974-76; Chrmn CES Div ASEE 1977-78 . Enjoy woodworking, sports & model building. *Society Aff:* ASME.

Page, Robert H
Home: 1905 Comal Circle, College Station, TX 77840
Position: Forsyth Chair Prof *Employer:* TX A&M Univ. *Education:* PhD/Mech Engg/Univ of IL; MS/Mech Engg/Univ of IL; BSME/Mech Engg/OH Univ. *Born:* Nov 1927. U S Signal Corp WWII. Fulltime Industrial experience Esso Res & Engrg Co. Faculty Univ Ill, Stevens Inst Tech, Rutgers Univ & TX A&M Univ. Internatl recognition for res on flow separation analysis & its application. Assoc Fellow, AIAA. Western Elctric Teaching Award ASEE 1968. Rutgers Univ Distin Teaching Award 1969. Fellow ASME 1971. Distin Service in Engrg Award Univ Ill 1973. ASME Region II Life Quality Engrg Award 1974. ASEE Mech Engg Div, Ralph Coates Roe Award 1979. Fellow AAS 1983. Hon Prof of RUHR Univ-Bochum, West Germany 1984. 1984 Distinguished service citation of ASEE. Ohio Univ Medal of Merit 1984. James Harry Potter Gold Medal of ASME 1984. Fellow AAAS 1985. Fellow ASEE 1985. ASME Dedicated Service Award 1986. *Society Aff:* AAS, AIAA, ASME, AAAS, APS, ASEE, NSPE

Page, Robert W
Business: 3 Greenway Plaza, Houston, TX 77046
Position: Pres CEO. *Employer:* Kellogg Rust Inc *Education:* BS/Civ/TX A&M. *Born:* 01/22/27. Native Dallas, TX, USN 1944-46. Taught Am Univ of Beirut, Field Engr Aramco, Proj Mgr Bechtel, in Charge of Dev Rockefeller Bros. Pres GEO A Fuller, Pres Rust Engg. Chrmn and CEO, Kellogg Rust Inc, Houston TX. Chrmn and CEO, Kellogg Rust Inc, Houston TX. Enjoy music, fishing, golf & tennis.

Pagliasotti, Robert D
Business: 10289 West Centennial Rd, Littleton, CO 80127
Position: Regional Building Services Sales Mgr. *Employer:* Johnson Controls Inc. *Education:* High Sch Grad/-/West Denver High Sch. *Born:* Apr 1934. High School Grad. Native Denver, Col. U S Army 1956-58. Johnson Control 1952- . ASHRAE 1965- . Mbr since 1969. Editor, Treas, Secy, V P, Pres Elec & Pres Rocky Mountain Chap. Dir & Regional Chmn Region IX 1972-75. Pres Miss Valley Chap 1976-77. Pres Award of Excellence Rocky Mountain Chap 1970-71. Regional Award of Merit Region IX 1974-75. Pres Award of Excellence - MI Valley Chapt 76-77. Distinguished Service Award June 1980. Enjoy golf, bowling & traveling. *Society Aff:* ASHRAE

Pai, David H
Home: 12 Peach Tree Hill Rd, Livingston, NJ 07039
Position: Sr Vice Pres *Employer:* Foster Wheeler Devel Corp *Education:* ScD/NYU; MS/CE/Lehigh Univ; BS/CE/VA Military Inst *Born:* 01/07/36 Joined Foster Wheeler in 1960 and held various positions of increasing responsibility. In January 1981, named VP of FW Energy Applications, Inc, a wholly owned subsidiary of Foster Wheeler Corp. Involved in overall technical mgmt of advanced engrg projs involving design and analysis of nuclear components and piping, thermal/hydraulic modelling and sys simulation studies, heat recovery components. May 1984 appointed Sr VP of Foster Wheeler Devel Corp, responsible for the R&D Div, Core Res Group and Contact Oper. ASME activities include VP, (Materials and Structures); past mbr and chrmn of the Exec Committee of the Pressure Vessels and Piping Div (PVPD); past mbr of the Design and Analysis Committee of the PVPD; and past mbr of the Policy Bd, Indus. Mbr of Argonne Universities Association Review Committee for the Experimental Breeder Reactor II Proj. Mbr, Engrg Tech Advisory Committee to the Bd of Trustees-NJ Inst of Tech. Author of over 30 technical papers; editor of two books; and holder of several domestic patents. 1987 Recipient of the ASME Pressure Vessel & Piping (PVP) Medal. Technical Advisor to the China National Standards Committee of Pressure Vessels. *Society Aff:* ASME, ASCE

Pai, Shih I
Business: Ipst, University of Md, College Park, MD 20742
Position: Res Prof Emeritus *Employer:* Univ of MD. *Education:* PhD/Aeronautics & Math/CA of Tech; Dr Tech/LC/Tech Univ of Vienna; MS/Aero Engr/MIT; BS/Elec Engr/Natl Central Univ China *Born:* 9/30/13. Prof in Aero Engg, Natl Central Univ 1940-47, Visiting Prof, Cornel Univ, Grad Sch of Aero Engg 1947-49, Res Prof, Inst for Fluid Dynamics and Appl Math, Univ of MD 1949-1976, Res Prof, Aero Engg Dept and Inst of Phys Sci and Tech, Univ of MD 1976 to 1983. Prof Emeritus 1983- . Visiting prof in the following univs (i) Tokyo Univ, Japan, Technical Univ of Vienna, Austria, Technical Univ of Denmark. University of Karlsruhe, Germany and Univ. of Paris VI, France. Consultants in the following co (i) Cornell Lab, Martin Co, GE Co, North American Aviation Co and Boeing Co. Fellow of Academia Sinica since 1962, Corresponding Mbr of Intl Acad of Astro. *Society Aff:* AIAA, APS

Paik, Young J
Business: 19826 South Alameda St, Compton, CA 90220
Position: President *Employer:* Paco Engrg Corp *Education:* BS/CE/IN Inst of Tech; BS/Physics/Univ of OR; BS/Physics & Math/Yunsea Univ-Korea *Born:* 03/19/30 Young J. Paik came to America in 1956 from Korea. Since then he has earned degrees in mathematics, physics and engrg. Is a Reg PE in five states and hold patents on a number of innovations in the steel industry. Married to Sue K. Paik and has three children. Paik has pioneered the welded I beam and latest patent is for the highly innovative corrugated I beam. *Society Aff:* NSPE

Pailthorp, Robert E
Business: PO Box 22508, Denver, CO 80222
Position: VP *Employer:* CH2M Hill *Education:* MSCE/Civil Eng/OR State Univ; BSCE/Environmental Engr/OR State Univ *Born:* 10/8/31 Joined CH2M Hill in 1957. Presently is VP; Regional Mgr of the Rocky Mountain Regional Office where he supervises the firm's activities in eight western and mid-western states; (District Mgr of a 20 state district and a mbr of CH2M Hill's Board of Dirs). Served as Discipline Dir for Industrial Processes until 1978. In this position he was responsible for all industrial projects involving effluent treatment, waste product recovery, and related industrial processes. His experiences have been in food processing, pulp and

Pailthorp, Robert E (Continued)
paper, wood products, seafoods, chemicals, metals, oil preservations and others. As a result of over 20 years in the industrial field, he achieved nationwide recognition as an expert in the biological concepts and treatment techniques relating to separate and combined municipal-industrial waste treatment. Active many professional groups, and has given numerous presentations and papers for publication. *Society Aff:* NSPE, WPCF, AWWA, ASCE, AAEE

Paily, Paily (Poothrikka) P
Business: Sch of Civ Engg, Atlanta, GA 30332
Position: Assoc Prof. *Employer:* GA Inst of Tech. *Education:* PhD/Hydraulics/Univ of IA; MTech/Hydropower/Vikram Univ; BSc/ME/Kerala Univ. *Born:* 5/18/42. Born in Kerala, India. asst Prof, Univ of Kerala 1965-70. Res Engr, IA Inst of Hydraulic Res 1974-75. Sec Hd, Physical Sciences Sec, Nalco Chem Co, 1975-78. Dir of Hydrology & Simulation Modeling Dept, Hazleton Environmental Sciences Corp 1978-79. With GA Tech since fall 1979. Scientific achievements include dev of thermal regime model for river systems, thermal plume model for river discharges, & oil spill transport model for N Sea. US Citizen 1978. Vikram Univ Gold Medal 1969. ASCE Karl Emil Hilgard Hydraulics Prize 1977. Enjoy feature writing. *Society Aff:* ASCE, ASME, ASEE, IAHR, IWRA.

Paine, Myron D
Business: Cessna, Aircraft Co, Pawnee Division, Dept 80, P.O. Box 1521, Wichita, KS 67201
Position: Engineer Group Leader *Employer:* Cessna Aircraft Company *Education:* PhD/Agri Engg/OK State Univ; MS/Agri Engg/Univ of IL; BS/Agri Engg/SD State Univ. *Born:* 6/27/34. Native Hamlin County, SD. Army Corps of Engrs 1957, Asst Professor SDSU 1958-63. Team leader International Voluntary Service, Laos, 1963-65. Fuel supervisor, Air America, Laos, 1965-66. Extension Specialist Animal Systems, OK Sta U, 1966-70. National Science Foundation Faculty Fellow, 1970-71. Regional Extension Specialist Feedlot Waste Management, 1971-74 ASS/ASSOC, Prof OK. STA. U. 1974-80, Cessna Aircraft Company since 1980, Elect Group Leader on Model 208, Caravan, 1982-84, P E OK. Awarded Outstanding ASAE Young Extension Engr 1974, Wonder of Engineering Achievement Award OSPE 1976. *Society Aff:* SAE

Paine, Thomas O
Business: 1800 Century Park East, Los Angeles, CA 90067
Position: President & Chief Operating Officer. *Employer:* Northrop Corporation. *Education:* PhD/Phys Metallurgy/Stanford Univ; MS/Phys Metallurgy/Stanford Univ; AB/Engg/Brown Univ. *Born:* 11/9/21. Served in the Navy. Associated with General Electric as Research Associate, Manager of Meter & Instrument Lab, Manager of TEMPO, VP Power Generation Group, Senior VP Technology and Planning. Administrator of NASA 1968-70. Presently President and Chief Operating Officer, Northrop Corporation. Hold several patents and have written several technical papers. Member NAE,Fellow AAS,etc. NASA Distinguished Service Medal, Washington Medal, Faraday Medal, John Fritz Medal, order of Al Merito Della Republica Italiana'. *Society Aff:* NAE, AAS, IEEE, AIME.

Painter, Jack T
Business: Box 6155 Tech Station, Ruston, LA 71272
Position: Prof of Civ Engr. *Employer:* LA Tech Univ. *Education:* MSCE/CE/WV Univ; BSCE/CE/WV Univ. *Born:* 7/23/30. Born and educated in WV. Taught at WV Univ before and after two yrs active duty as officer in US Navy. Currently a Commander in US Naval Reserve. Mbr of faculty of LA Tech Univ since 1956. Visiting Prof at various univs for special summer progs. Mbr of faculty of LA Tech-Rome summer prog for 15 yrs. Recipient of several teaching awards including the first James M. Robbins Award of Chi Epsilon and the first Alumni Fdn Profship Chair. Current areas of specialization are structural analysis, structure in architecture, and history of civil engrg. Hobby is the circus. Past Natl Pres of Circus Fans of America. *Society Aff:* ASCE, ACSM, SHOT, ASEE.

Painter, James A
Home: 8516 Paul Revere Court, Annandale, VA 22003
Position: Tech Adv, C4 Systems *Employer:* US Marine Corps *Education:* PhD/Comp Sci/Stanford Univ; MS/Math/Univ of Pittsburgh; BS/Math/Univ of Pittsburgh. *Born:* May 1929. Served US Army 1946-47, 1951-53. Project Mathematician for IBM 1955-59 providing computer support to engrg lab. Mgr, Special Programming for Philco Corp 1959-60. Sr Programmer 1960-73 for IBM. Tech Dir, WWMCCS, ADP, Def Commo Agy 1973-80. Responsible for tech support of Worldwide Military Command and Control System (WWMCCS) ADP, including research and development program, Tech Adv, Command Control, Communications, and Computer (C4) Sys, US Marine Corps, 1980 to present. *Society Aff:* IEEE, ACM, MAA.

Painter, John H
Business: Dept. of Electrical Engrg, Texas A & M University, College Station, TX 77843
Position: Prof *Employer:* Texas A&M Univ *Education:* PhD/EE/So Methodist; MS/EE/Univ of IL-Urbana; BS/EE/Univ of IL-Urbana *Born:* 03/27/34 Dr Painter is presently a tenured Prof of Elec Engrg, at TX A&M Univ. He is also Pres of Altair Corp, a TX Corp engaged in intl consltg in elec comm, navigation, and knowledge-based signal processing. He was a USAF Navigator, NASA Apollo Engr and researcher, and indus Sr Engr from 1955 to 1974. He has been in education since 1971, with many tech pubs. He holds patents in digital signal processing sys. Recently developed software and implementation techniques for artificial intelligence in signal processors. He lectures internatly in satellite radio nav, Kalman filtering, and computationally-based signal processing. *Society Aff:* IEEE, ΣΞ, ΦΚΦ, ΤΒΠ, HKN, ΠΜΕ

Painter, Robert A
Home: 10916 Pennway Dr, Richmond, VA 23235
Position: Dir of Utilities. Retired. *Employer:* County of Chesterfield. *Education:* BS/Civ/VPI & SU. *Born:* 4/30/18. in Pulaski, VA. Served in US Army & US Air Corps from 1941-1945. Asst City Engr, Fredericksburg, VA; Operation & Maintenance Engr, Fairfax County, VA; Asst Engr Dir, Alexandria, VA Sanitation Authority; County Engr & Dir of Utilities, Chesterfield County, VA, 1956 until retirement in 1982. Chrmn, VA Sec AWWA 1966. Enjoy hunting, fishing & gardening. *Society Aff:* NSPE, ASCE, AWWA, WPCF.

Pair, Claude H
Home: 215 Hillview Drive, Boise, ID 83712
Position: Agricultural Engineer. *Employer:* US Dept Agriculture (Ret). *Education:* MS/Elec Engg/WA State Univ; BS/Elec Engg/WA State Univ; BA/Education/WA State Univ. *Born:* July 1911. Additional year study Princeton U and MIT. Reg Prof Engr ID, OR, WA. Employed by USDA 1934-75 as Agri Engr in ASCS, SCS, ARS working on soil and water conservation. Pub 62 tech papers on results of sprinkler irrigation research. Editor-in-Chief Third, Fourth, and Fifth editions 'Sprinkler Irrigation' textbook used as worldwide reference on subject. Fellow grade mbr ASAE, Life Mbr Sprinkler Irrigation Assn, mbr Sigma Xi. Man of Year 1962, Special Citation 1970, and Award of Recognition 1974 by Sprinkler Irrigation Assn. Worldwide irrigation cons. Elected to "Idaho Water Users Association Hall of Fame" - 1975. Presented "Agricultural Engineer of Year for 1979" award by Pacific Northwest Region, ASAE. Presented "John Deere Medal" for Distinguished Achievement in the Application of Science and Art to the Soil. ASAE Summer Meeting 1981. *Society Aff:* ASAE, ΣΞ, IA.

Palazotto, Anthony N
Home: 6358 Siena St, Dayton, OH 45459
Position: Prof *Employer:* Air Force Inst of Technology *Education:* PhD/Solid Mech/NYU; MCE/Structures/Polytech Inst of Brooklyn; B/CE/NYU *Born:* 12/15/35 After receiving the PhD in 1968, pursued a research and educational career. Prior to 1968, experience was in consulting engrg related to the field of Structures. Assumed the position of Assoc Prof of Aerospace Engrg at the Air Force Inst of Technology, School of Engrg in August 1975. Obtained the rank of Full Prof in January 1978. Fellow in ASCE 1979. Assoc Fellow AIAA 1982; member of Tau Beta Pi, Sigma Xi, and Chi Epsilon Honor Societies. Res fields include: Viscoplastisity, Fracture Mechanics, Composite Materials, Shell Analysis and Finite Elements. Over 80 pubs in tech journals. Received the SESA Hefenyi Award 1982. *Society Aff:* AIAA, ASCE, AAM, ASTM.

Palermo, Peter M
Home: 3116 Waterside Lane, Alexandria, VA 22309
Position: Asst Deputy Commander/Tech Dir *Employer:* Naval Sea Systems Command *Education:* BS/CE/Manhattan Coll *Born:* 04/18/29 Employed at David Taylor Model Basin (now NSRDC) from 1953-1967, engaged in structural res and ship trials. Joined Naval Ship Engrg Center (now NAVSEA) in 1967. Presently is Asst Deputy Comm/Tech Dir Ship Design and Engrg Directorate, as such is responsible for design and life cycle technical and logistics support for all ships and submarines and their equipments. Guest lecturer on design and fabrication at George Washington Univ, Catholic Univ, MIT and foreign countries. Reg PE, Washington, DC. Recipient of US Navy Superior Civilian Service Silver Medal 1962 and 1985; US Navy Distinguished Civilian Service Gold Medal 1974; Presidential Meritorious Rank Award 1980 and 1985; SNAME David Taylor Gold Medal 1980; ASNE Gold Medal 1985; DC Council of Engrs Outstanding Naval Architect Award 1979; Association of Scientists and Engrs Silver Medal 1978; and Navy Dept Sr Exec Award 1975. *Society Aff:* SNAME, ASNE, ASTM, ASE

Palladino, Nunzio J
Business: Nuclear Regulatory Commission, 1717 H Street N.W, Washington, DC 20555
Position: Chairman *Employer:* United States Nuclear Regulatory Commission *Education:* BS/Mech Engg/Lehigh Univ; MS/Mech Engg/Lehigh Univ; Grad/Nuc Engg/Univ of TN; Cert/Bus & Mgmt Program/Univ of Pittsburgh; DEng/Hon/Lehigh Univ. *Born:* Nov 1916. Native of Allentown Pa. Engr Westinghouse Elec 1939-42,, 1945- 46; Nuclear Reactor Designer Oak Ridge Natl Lab 1946-48; Argone Natl Lab 1948- 50; Sect Mgr Westinghouse Bettis Atomic Power Lab 1950-52, Subdiv Mgr 1952-59; Prof & Hd, Nuclear Engrg Dept Penn State 1959-66, Dean Coll of Engrg 1966-1981. Past Mbr, Past Mbr Governor's Energy Council, P Mbr Governor's Sci Adv Ctte, Mbr ASEE, NSPE, NAE; Fellow ASME, ANS. Gotshall Scholar 1938-39. Capt US Army 1942- 45, Past Mbr PA Commission on Three Mile Island. *Society Aff:* ANS, ASME, ASEE, NSPE, NAE

Palm, Gordon F
Home: 602 Schoolhouse Rd, Lakeland, FL 33803
Position: President. *Employer:* Gordon F Palm & Assocs Inc. *Education:* MS/ChE/Univ of Houston; BS/ChE/ Univ of IL *Born:* May 31, 1925. Native of Iowa Park Tex. BS ChE Univ IL with honors; MS ChE Univ Houston. Asst Prof Chem Engrg Univ of Houston 1953-55. Indus exper 1951-64: phosphate fertilizer field - process dev, process design, plant design, proj mgmt, cost estimation, cost control, plant start-up, tech sales. 1964- , independent cons chem engr for phosphate fertilizer indus - process design, plant design, proj mgmt, plant oper, economics, new process eval, air & stream pollution control, solid waste disposal, financial evals, & cert of new facils performance for financiers. Fellow AIChE; Reg P E Fla. P Chmn Central Fla Sect AIChE. *Society Aff:* AIChE

Palm, John W
Home: 7703 S Hudson, Tulsa, OK 74136
Position: President *Employer:* Petro-Sulf Consulting, Inc. *Education:* MS/CE/Univ of MI; BS/CE/OK State Univ. *Born:* 10/10/21. Blackwell, OK native. Employers: 1943-52, Cities Service Res & Dev Co, advancing from Res Engr to Asst Chief Chemist, supervisor of Tallant, OK lab; 1952- 1981, Amoco Production Co, Sr Res Engr to Special Res Assoc; 1981-present, Petro-Sulf Consulting, Inc, President. Specialties: production of petrochem by hydrocarbon oxidation & hydrocarbon synthesis; insitu partial combustion of heavy oils; recovery of sulfur from sour gas, including asst to licensees in designing 230 plants. Author 15 publications, 27 patents. Served as vchrmn & chrmn of OK Sec AIChE & Tulsa Sec AIChE. *Society Aff:* AIChE, SPE.

Palmatier, Everett P
Home: 1011 Seagate Dr, Delray Beach, FL 33444
Position: Consulting Engr (Retired) *Education:* ME/Mech Engg/Stevens Inst of Tech; MS/Mech Engg/Harvard Univ. *Born:* July 1911. Carrier Corp Syracuse: Mgr Transportation Equip Dept; Dir Res, Aircraft Environ Sys. WW II Curtiss Wright Propeller Div - B36 propeller anti-icing, early rocket sys for Robert H Goddard & US Navy. 25 U S pats. Fellow ASHRAE; Disting Serv Award 1964, F. Paul Anderson Medal 1987; Reg P E Fla, N J; Tau Beta Pi. Int: enclosed cities, solar energized air cond; swimming, music. *Society Aff:* ASHRAE.

Palmer, Bruce R
Business: 500 E. St Joseph St, Rapid City, SD 57701
Position: Prof *Employer:* South Dakota Sch of Mines and Tech *Education:* PhD/Metallurgy/Univ of UT; BS/Metallurgy/CO Sch of Mines *Born:* 06/20/46 Currently Prof of Metallurgical Engrg, SD Sch of Mines and Tech, Rapid City, SD. He previously held the positions of Asst Prof 1972-1976 and Assoc Prof 1976-1981 at this institution. He is recipient of the Arthur F. Taggart Award (Mineral Processing Div-Soc of Mining Engrs), van Diest Gold Medal-1981 (CO Sch of Mines), and the Bradley Stoughton Award-1981 (American Soc for Metals). Prof Palmer is also a mbr of Tau Beta Pi Honorary Fraternity. He has published widely in the area of extractive metallurgy and is co-editor of the book, *Proceedings of the Western Regional Conference on Gold, Silver, Uranium and Coal.* Concerning his profl activities, Prof Palmer has served as Chrmn, Black Hills Section, AIME, and as Chrmn of the Accreditation Committee, Educ Bd, and the Educational Publications Committee of the Soc of Mining Engrs-AIME. *Society Aff:* AIME, ASM

Palmer, C Harvey
Business: 3400 N. Charles St, Baltimore, MD 21218
Position: Prof *Employer:* Johns Hopkins Univ *Education:* PhD/Physics/Johns Hopkins Univ; AM/Physics/Harvard Univ; SB/Physics/Harvard Coll *Born:* 12/08/19 Worked at MIT Radiation Lab on radar 1942-1945. Taught electronics, optics, and general physics at Bucknell Univ. Taught electronics, microwaves, and, now, optics at John Hopkins. Have done basic research on diffraction grating anomalies at optical and microwave frequencies, far infrared water vapor spectra, optical detection of ultrasonic waves by interferometry, and optical detection of acoustic emission transients. Developed many lab experiments, some included in a book: *Optics: Experiments and Demonstrations.* Hold Joint appointment in Electrical Engrg and Materials Sci and Engrg. Married, two children. My wife and I hike extensively, run, and climb mountains including Kilimanjaro. *Society Aff:* OSA, APS, IEEE, AAAS, ΣΞ

Palmer, James D
Business: 968 Albany-Shaker Rd, Latham, NY 12110
Position: VP & Genl Mgr - Res & Dev Div. *Employer:* Mechanical Technology Inc. *Education:* PhD/Engg Sci/Univ of OK; MS/Elec Engg/Univ of CA at Berkeley; BS/Elec Engg/Univ of CA at Berkeley. *Born:* 3/8/30. Chief Engr Motor Vehicle & Illumination Lab, Univ of CA at Berkeley, 1955- 57; prof Univ of OK, Norman 1957-63, assoc prof 1963-65, prof 1965-66, asst to dir. Res Inst, 1960-63, cons, 1966-69, dir Sch Elec Engg, 1963-66, dir Sys Res Center, 1964-66; dean sci and engg, Union Coll, Schenectady, 1966-71; pres Met State Coll, Denver, 1971-78; Admin, Res & Spec Prog Admin & Dept of Transp, Washington, DC 1978-79; VP and Genl Mgn Res Dev Div Mech Tech Inc, 1979- ; adj prof Univ of CO; cons Sys Mgmt Corp, Bd Dirs; Bd Dirs, 1968- , Exec VP VITA, 1969-71, chmn exec com 1971; trustee Auraria High edn exec com Hudson-Mohawk Assn Colls and Univ, trustee, chmn bd 1969-71; adv com US Coast Guard Acad mem 1972- , Chrmn 1978- ; mem CO Gov's Sci & Tech Adv Council; pres Denver Cath Community Services Bd.; mem

Palmer, James D (Continued)
of Bd Dirs Archdioceasan Catholic Charities & Community Services. *Society Aff:* VITA, NCCE, IEEE, HKN, WOGCS.

Palmer, Melville L
4731 Goose Lane SW, Alexandria, OH 43001
Position: Prof & Extension Agri Engr. *Employer:* Ohio State Univ. *Education:* MSc/Agri Engrg/OH St Univ; BSc/Agri Engrg/Univ of Guelph *Born:* 8/30/24. Canadian Air Force 1943-46. Asst Mgr of Farm Mechinery Dept, United Coops on Ontario 1952-53. Res Asst OARDC 1953-55. Ford Found consultant in agri engrg at Punjab Agri Univ, India April-May 1967. Extension Agri Engr in water mgmt at OSU 1955- , conducting a wide var of educ progs. Chmn OH Sect ASAE 1975. Pres OH Chap SCSA 1973 & Outstanding Mbr Award 1976. June 1979-Received Hancor Soil & Water Engg Award from ASAE. 1982-Elected Fellow in ASAE and Fellow in SCSA. Dec 1986-received Gunlogson Countryside Engineering Award from ASAE. *Society Aff:* ASAE, SCSA, AAAS.

Palmer, Nigel I
Business: 1114 Ave. of Americas, New York, NY 10036
Position: VP Technology/Bus. Dev. *Employer:* W R Grace & Co. *Education:* MA/ChE/Cambridge Univ (U.K.); BS/Chemistry/Southampton Univ (U.K.). *Born:* 6/18/33. Born & educated in England. Joined Shell Oil in Canada in 1957. Technical Dir Leesona Corp NY. 1960 to 1970, (electrochem energy conversion, fuel cells & zinc air batteries) (Technical Dir). With W R Grace & Co since 1970; Technical Dir in battery separator business, 1977 VP of Technical Services for Polyfibron Div, currently VP Technology & Business Dept, Grace Specialty Chemicals Co, New York. Mbr Exec Committee-Energy Conversion Div, AIChE 1976-1979. Enjoy sailing, skiing, classical music. *Society Aff:* AIChE, ACS, ECS, CDA, SCI.

Palmer, Roy G
Business: 585 Iowa Ave, Riverside, CA 92507
Position: Consultant (Self-employed) *Education:* BS/CE/Univ of Santa Clara *Born:* 05/10/33 Worked for City of Los Angeles, Bureau of Engrg and Dept of Airports 1956- 61; Served as Lt in Army Corps of Engrs. Chief Engr on West Coast for Sanfrod Truss 1961-65; conslt in private practice 1965 to present specializing in structural engrg for all types of structures including residential, commerical, industrial, institutional. *Society Aff:* NSPE, ASCE

Palomba, Joseph, Jr
Business: Colo. Dept of Health, 4210 E. 11th Ave, Denver, CO 80220
Position: Tech Secy. *Employer:* Colorado Air Quality Control Comm. *Education:* MPH/Indus Health/Univ of MI; BS/Sanitary Sci/Univ of Denver. *Born:* 5/18/33. Mbr APCA's Bd of Dir 1973-76; Conf Chmn of 67th APCA Annual Mtg; (1974) Mbr APCA's TE-3 Econ Ctte & A-1 Mbrship Ctte. Was first chmn & founder of the Rocky Mountain States Sect of APCA. Past Dir for the Colo Pub Health Assn & was a Charter Mbr of the State & Territorial Air Pollution Prog Admins. Pres of APCA 1977-78, PPres APCA 1978-79. Honorary membership APCA (1986). *Society Aff:* APCA

Pan, Coda H T
Home: 17 Alpine Dr, Latham, NY 12110
Position: Professor of Mechanical Engineering *Employer:* Columbia University *Education:* PhD/Aero Engg & Astronautics/RPI; MS/Aero Engg/RPI; BS/Mech Engg/IIT. *Born:* Feb 1929. Employed by Genl Elec Co 1950-61. Joined Mechanical Technology Inc & served as Dir of Res until 1973. Spec in gas bearings & high speed rotor dynamics. Adj Prof at RPI 1961- . Guest Prof Royal Tech Univ of Denmark 1971. Special Post-Doc Fellow Natl Heart & Lung Inst 1971-72. Co-founder & Tech Dir Shaker Res Corp. IR- 1100 Award 1967. 3 pats. Over 40 pubs. Fellow ASME & ASLE; Mbr AAM & APS. Enjoy classical music. Contributing author, Tribology, edited by A. Szeri. *Society Aff:* ASME, AAM, ASLE, APS.

Pandullo, Francis
Home: 2100 Wesley Ave, Ocean City, NJ 08226
Position: Consult Engr *Employer:* Sole Pract *Education:* MS/Management/Manhattan Coll; BCE/EE/Manhattan Coll *Born:* 4/2/34 Prof Pandullo is a Profess Engr lic to practice in NJ, NY, CT, DE, PA, VT, NC, MD, FL and CA. Following his retirement as Chrmn of the Bd of Dir of the Engrg Co he founded in 1961, Prof Pandullo continues his professional practice in Proj Mngmt, Forensic Engrg, Lecturing, Teaching and Writing. Prof Pandullo is the recip of the Daniel W. Mead Award for his paper, "Professional Ethics and the Municipal Engineer in Private Practice" which was pub in the Journal of the ASCE Engrs. *Society Aff:* NSPE, ACEC, ASCE.

Panek, Louis A
Home: 8 Hillside Dr, Denver, CO 80215
Position: James Westwater Prof of Mining & Geological Engrg *Employer:* MI Tech Univ *Education:* PhD/Engrg/Columbia Univ; MS/Mining Engrg/Sch of Mines-Columbia Univ; BS/Mining Engrg/MI Tech Univ; BS/Geol/MI Tech Univ *Born:* 12/03/19 Attended Cleveland OH public sch. Served with Army Corps of Engrs, Panama and Burma. Specialist in planning, executing, and analyzing programs of monitoring and testing to determine in-place properties and behavior of large rock mass structures (mines and tunnels). Devised the only quantitative method for evaluating reinforcement of horizontally bedded mine roof by bolting. Conducted geotechnical research applied to high-extraction mining by undercut- cave and by longwall methods, and to the resulting surface subsidence. Developed first scheme for determination of true rock-mass modulus in a borehole 1963. Author of about 60 technical publications. AIME Peele Award 1957. Chrmn Intersociety Committee for Rock Mechanics 1967. *Society Aff:* ISRM, SME, ASA, ΣΞ

Pankove, Jacques I
Business: Washington Rd, Princeton, NJ 08540
Position: Fellow Tech Staff. *Employer:* RCA Labs. *Education:* PhD/Physics/Univ of Paris; MS/EE/Univ of CA, Berkeley; BS/EE/Univ of CA, Berkeley. *Born:* 1922. In 1948 joined RCA Labs, where contrib to understanding, tech & evolution, of var semiconductor devices (large-area photocells, transistors, tunnel diodes, injection lasers & LEDs). In 1956 received David Sarnoff Scholarship to study at Univ of Paris. In 1962 made educ film: 'Energy GaP & Recombination Radiation'. 1968 Visiting MacKay Lectr at Univ of Calif Berkeley. (Credited wth many pats & pubs, incl textbook 'Optical Processes in Semiconductors' Prentice Hall & Dover; 3 RCA Achieve Awards, 1975 IEEE Ebers Award. Mbr AAAS, Sigma Xi; Fellow APS & IEEE. Former Assoc Ed Journal of Quantum Electronics (1968-77); Mbr Edit Bd Solid State Electronics. Tr Princeton Art Assn. *Society Aff:* IEEE, APS, AAAS.

Pankow, Charles J
Business: 2476 North Lake Ave, Altadena, CA 91001
Position: Pres/CEO & Treasurer/Ch of Bd *Employer:* Charles Pankow, Inc *Education:* BS/CE/Purdue Univ *Born:* 10/6/23 Engr S B Barnes, Structural Engrs - 3 yrs; Peter Kiewit Sons Co, Arcadia, CA - approximately 12 yrs; formed Charles Pankow, Inc 1963. Patents on Process for Manufacturing Hollow Concrete Piles utilizing slipform techniques, assigned to Kiewit, and Slipforming Vertical Air Conditioning Ducts. Authored various publications, in particular chapters on slipforming in Jospeh J Waddell's Concrete Construction Handbook. Past Pres and Bd of Dirs and Fellow-ACI. Received ACI's Roger H Corbetta Award 1974 for innovative methods in use of concrete, Purdue Univ's 1970 distinguished Alumnus Award and in 1983 was presented Purdue Univ Honorary Doctorate in Civil Engrg. *Society Aff:* ACI

Panlilio, Filadelfo
Home: 49 Aspinwall Rd, Loudonville, NY 12211
Position: Prof *Employer:* Union College *Education:* PhD/Eng Mech/Univ of MI; MS/Eng Mech/Univ of MI; BS/ME/Univ of Philippines *Born:* 02/05/18 Naturalized citizen from the Philippines. Former Chrmn Mech Dept, Univ of the Philippines.

Panlilio, Filadelfo (Continued)
Wrote award-winning paper on Limit Design. Authored book on *Strength*. Conslt to Gen Elec, ALCO Products, Watervliet Arsenal. Current interests: Finite rotatious, non-symuetrical plastic bending, 3-D isotropic fiber composites. Prof in Mech Engrg. With Union Coll in Schenectady, NY since 1955. *Society Aff:* ASM, ASME, SESA

Panoff, Robert
Business: 1140 Connecticut Ave, NW, Wash, DC 20036
Position: Principal Officer. *Employer:* MPR Assoc, Inc. *Education:* DSc/Sci (Hon)/ Allegheny College; BS/EE/Union College. *Born:* 8/16/21. Principal Officer of MPR Assoc, Inc, of Wash, DC, an engg firm engaged in high tech aspects of the energy field, with special emphasis on both fossil and nuclear power generating stations. During his career, he has also been intimately involved in the dev of nuclear powered submarines & ship propulsion systems. He has served in an advisory capacity to the Navy Dept in connection with combatant ship design & dev problems. PE 6708, DC. *Society Aff:* AIF, AAAS, SNAME, TBΠ.

Pansini, Anthony J
Home: 1916 Trinity Dr, Waco, TX 76710
Position: President. *Employer:* A J Pansini, Engrg & Mgmt Cons. *Education:* EE/Elec Eng/Cooper Union; BS/Elec Eng/Cooper Union *Born:* 1909 NYC. Engrg & managerial accomplishments for Con Edison 1926-41, & Long Island Lighting Co 1941-71; cons 1971- . Contrib to dev of high-voltage transmission & distrib sys, & of means for maximizing mgmt effectiveness. Originator of plan for damming of Long Island Sound. Taught in cont educ progs at C W Post College. Author of patent, sev tech & non-tech books & many papers. Lic P E NY & Texas. Life Fellow IEEE. History & political sci buff; several historical and political science works and papers. Play organ, travel for relaxation. *Society Aff:* IEEE, ASTM, AHA, APSA

Pantazis, John D, PE
Home: 1309 Conway Rd, Lake Forest, IL 60045
Position: President *Employer:* John D. Pantazis & Assoc LTD, Engrs *Education:* BS/EE/Chicago Tech Coll; BSAE/Arch/Chicago Tech Coll *Born:* 12/18/36 b. in Volos Greece, grew up in Athens where successfully operated own electrical constructing company. Served the Greek Army and NATO forces in Salonica Greece, honorably discharged as staff sergeant 1959-60. Educated in US, studied Electrical and Architectural Engrg. Worked as an Electrical Engr, supervisor and dept head for various consltg firms in Chicago, Il. In 1979 established own firm practicing consltg engrg in the fields of Electrical, HVAC, Plumbing, and Fire Protection, along with Energy conservation for commercial, institutional, and industrial bldgs. Married to Helena J. Mefsout have a son Dennis, a daughter Argy and live in Lake Forest, Il. *Society Aff:* ASHRAE, NSPE, ASPE.

Pantell, Richard H
Business: McCullough Bldg, Rm 308, Stanford, CA 94305
Position: Prof. *Employer:* Stanford Univ. *Education:* PhD/EE/Stanford Univ; MS/ EE/MIT; BS/EE/MIT. *Born:* 12/25/27. Worked for GE Co 1947-49. Taught at the Poly Inst of Brooklyn 1950-51. Prof at Stanford Univ 1954-present. Visiting Asst Prof at Univ of IL 1956-57. Engr at ITT 1962-1963. Visiting Prof at Univ of New South Wales 1969-70, & at Univ College, London 1976-77. Authored the books "Fundamentals of Quantum Electronics–, Wiley (1969) and "Techniques of Environmental Systems Analysis–, Wiley (1976). Presently working in the field of quantum electronics with lasers and electron beams, & has published 105 papers in archival journals. Fellow of the IEEE. *Society Aff:* AAAS, IEEE, APS, ΣΞ.

Pao, Yoh-Han
Business: Center for Intelligent Systems Research, Case Western Reserve University, 2721 Scarborough Rd, Cleveland Heights, OH 44106
Position: George S. Dively Dist Prof of Engrg *Employer:* Case Western Reserve Univ. *Education:* PhD/Appl Phys/Penn State Univ; BS/Gen Engg/Henry Lester Inst (Shanghai China). *Born:* Dir. Ctr for Automation and Intelligence Sys Res, Case Western Reserv Univ. ; Dir of Div of Electrical, Computer and Systems Engg, Natl Sci Foundation, 1978-80, on leave from Univ, Hd of Dept of EE and Applied Physics. Case Western Reserve Univ, 1969-78. Mbr of Tech Staff, Bell Telephone Lab, Murray Hill, NJ 1962-67. Post Doctoral studies at Univ of Chicago and Edinburgh Univ. Pres of Bonschul International Inc. Active in communications, pattern recongnition res and machine intelligence. Consultant to major industrial corp and federal agencies. Fellow IEEE and Optical Soc of America. Active in international education projects. Married to Helen Chung-Ying Koo. Three children: John D, Victor E, and Robin A. *Society Aff:* IEEE, AAAI.

Paolino, Michael A
Business: Dir of Engg, Lafayette College, Easton, PA 18042
Position: Dir (Dean) of Engg *Employer:* Lafayette College. *Education:* PhD/Mech Engg/Univ of AZ; MS/Mech Engg/Univ of AZ; BS/Math/Siena College. *Born:* 3/8/39. Born in Albany, NY. Served in the U.S. Army as a commissioned officer for 26 years. Retired with the rank of Colonel. Received Ph.D., Mech. Engrg., from the Univ. of AZ, 1972. Engrg. exp includes R&D at the Army Missile Command dev hypervelocity rocket tech. Registered professional engineer, VA. From April 1973 to July 1986 member of the Faculty, Dept. of Mech., U.S. Military Academy. Appointed Prof of Mech Engg and Coordinator of the Mech Engg Program. In July 1986, appointed Prof of Mech Engg and Dir (Dean) of Engg at Lafayette College. Recipient of the SAE Ralph R. Teetor Award, 1978. Mbr Phi Kappa Phi. *Society Aff:* ASME, ASEE.

Paoluccio, Joseph P
Business: 7175 Construction Ct, San Diego, CA 92121
Position: Principal. *Employer:* Paoluccio Willis Nau Assocs - Estab 1961. *Born:* 2/24/33. Principal in firm Paoluccio Willis Nau Assocs with offices in San Diego, Modesto & Sacramento. Reg CA P E: Civil, Mech, Architecture Control Sys & Fire Protection Architecture; Reg AZ P E Mech. Mbr ASCE, ACEC, ASHRAE. CSI, NSPE, CSPE, SAME. Chmn Mech & Elec Cons Engrs of San Diego; Legis Ctte CEAC 1974-76; Dir CEAC, Archs & Engrs Selection Ctte, City of San Diego 1976. Instructor, Univ of CA, San Diego; Outstanding Air Con design Award 1965; Design of Energy Conservation Residence, Natl Publs 1975. Hobbies: photography, fly fishing, skiing & music. *Society Aff:* ASCE, ACEC, CEAC, ASHRAE, CSI, NSPE, CSPE, SAME, AIA, AEE.

Paone, James
Home: 12155 Breckenridge Lane, Woodbridge, VA 22192
Position: Asst to Asst Dir *Employer:* US Dept of the Interior *Education:* MS/MBA/Geo Wash Univ; Grad/Mgmt/Indus Coll of the Armed Forces; Grad/ Mgmt/Fed Exec Inst; BS/Mining Eng/PA State Univ *Born:* 08/15/25 Son of Domenick and Grace (Antonazzo) Paone; m. Joan Westover; Children: Mary Grace, Antoinette, Patricia. Mining engr, Cambria Div, Bethlehem Mines Corp PA 1951-55; mining research engr, program mgr, US Bureau of Mines, MN and Wash, DC 1955-1972; Chief, div Ferrous Metals 1972-1974; Chief div of Environment 1974-1979; Dir, div Mineral Land Assessment 1979-1982; Asst to Asst Dir 1982-1986; private consultant 1986-; Mining cons Defense Dept 1958-68, NASA, 1961-68, Bur of Public Roads 1962-65. Served with USAF 1943-46. Recipient Dept Interior SES Meritorious Award 1976, performance awards 1963, 65, 66, 68, 1971, 1974. Dept Interior superior perf 1980. Mbr Am Legion (Comdr 1947-48), St Vincent de Paul (VP 1973-Pres-1987-) Soc Mng Engrs, Am Inst Mining, Metall and Petroleum Engrs. Over 40 publications. Office: US Bureau of Mines, 2401 E. St NW, Wash, DC 20241. *Society Aff:* AIME, SME

Papadopoulos, Spyridon G
Business: 4301 Connecticut Ave, NW, Washington, DC 20008
Position: Vice Pres *Employer:* ALPHATEC, p. c. *Education:* MEA/Eng Adm/George Wash Univ; BS/ME/Northeastern Univ; Asso MET/ME/So Tech Inst *Born:* 07/14/45 Athens, Greece, came to US, 1963; naturalized, 1969. Mech Engr, Gan-

Papadopoulos, Spyridon G (Continued)
teaume & McMullen, Inc, Boston, 1966-71; Proj Engr, Gen Engrg Assocs, Wash, DC, 1971- 74; Sr Engr, Syska & Hennessy, Inc, Wash, DC 1974-76; Chief Mech Engr, A. Epstein & Son, Inc, Chicago, 1976-77, VP, ALPHATEC, p. c., Wash, DC, 1977 to present; course dir and lecturer, environmental engrg, Prince George's Coll, Largo, MD, 1973-76. PE, DC, NY, MD, VA, IL, and NC. Order of American Hellenic Educational and Progressive Association; Pi Tau Sigma, Greek Orthodox. Researcher Psychrometrics. ASHRAE Tech Committee TC1.1, Dir of Metropolitan Wash Consltg Engrs Council 1980-82. *Society Aff:* ASME, ASHRAE

Papian, William N
Home: 1424 E-W Shadyside Rd, Shadyside, MD 20764
Position: Vice President. *Employer:* Claxton Walker Assocs. *Education:* MS/EE/MIT; BS/EE/MIT. *Born:* July 1916. Res Asst MIT Digital Computer Lab to Group Leader Lincoln Lab 1948-63. Prof of EE & Asst V Chancellor for Res, Washington Univ St Louis 1964- 71. Visiting Scientist HSMHA, HEW 1972. Semi-retired to new field with Claxton Walker Assocs, Potomac Md, in residential const cons & energy conservation. Fellow IEEE for the dev of the magnetic core memory. *Society Aff:* NSPE, IEEE, AAAS, ASHI.

Papirno, Ralph
Home: 121 Damon Rd, Needham, MA 02194
Position: Res Mech Engr. *Employer:* US Army Matls & Mechs Res Ctr. *Education:* MS/Engg Mechs/NY Univ; MA/Science Education/NY Univ; BA/Math/NY Univ. *Born:* Feb 1920. Air Force meteorologist 1942-46. Taught physics in NYC schools 1948-53. Engrg scientist NYU 1953-62. Sr Scientist Allied Res Assocs Inc 1962- 68. In present pos since 1968. Also cons on dental engrg, NYU Coll of Dentistry 1962-65; Lectr on Prosthetic Dentistry, Harvard School of Denta Medicine 1975- . Major prof ints are res in mech of solids & applications of engrg in dentistry. Henry Marion Howe Medal ASM 1958; Sigma Xi. Reg P E Mass. Married Cecile Worby 1945; two children. Like classical music, water sports, reading, cross-word puzzles, sailing home-built dinghy.

Papo, Maurice
Home: 14 Blvd Prince de Galles, Nice, France 06000
Position: Directeur Scientifique *Employer:* IBM France *Education:* DR/Physics/Polytechnique; Engineer/Electronics/Telecommunications *Born:* 02/06/28 Joined IBM in 1954, Mgr of Electronic Dev, 1959 Asst to IBM World Trade Dir of Engrg, 1960 Asst Dir of Dev IBM France, 1962 IBM France Dir of Dev, 1963 IBM Europe Dir of Laboratories, 1969 IBM World Trade Corp Dir of Standards, 1970 Dir of Intermediate Systems Product Line, 1971 IBM France Dir of R & D, 1975 IBM France Chief Scientist, 1984 IBM France Dir of Commercial, Industry and Scientific Relations, Patents Contracts and Licensing. 1986 Regional Council (Alpes Maritimes) Scientific Advisor to the President (on sabbatical from IBM) *Society Aff:* IEEE, SEE, ISF

Paprocki, Stan J
Business: 666 N Hague Ave, Columbus, OH 43204
Position: Pres. *Employer:* Mtl Concepts Inc. *Education:* MS/Met Engr/Univ of IL; BS/Met Engr/Univ of IL. *Born:* 7/10/22. Served with AF 1943-1945. Received MS degree in Met Engg from the Univ of IL in 1947. Associated with Battelle Columbus Labs from 1947 to 1974 in positions of Div Chief, Assoc Dept Mgr, Dept Mgr & Assoc Dir. Assumed responsibility for current position of Pres of Mtl Concepts, Inc, Columbus, OH in 1974. *Society Aff:* ASM, AIAA, SAMPE

Parady, William H
Home: Box 520, 2351 College Station Rd, Athens, GA 30605
Position: Retired *Education:* EdD/Voc Educ/Univ of GA; MA/Ag Engrg § Ag Educ/ Univ of FL; BSAE/Ag Engrg/Univ of GA *Born:* May 8, 1919. WW II 1940-45. Area Supr, Fla Dept of Educ 1948-52. Fla Ford Tractor Co & Tractor Div of Ford Motor Co 1952-66, assuming pos of Dist Supr, Serv Mgr, Bus Mgmt Mgr & Retail Dealer. Joined Amer Assn for Agri Engrg & Vocational Agri (now Amer Assn for Vocational Instructional Matls - working cooperatively wth all 50 states) as editor, then coordinator. Became Exec Dir 1973. Author & co-author of sev books & articles. P Chmn both Fla Sect & Ga Sect ASAE. Mbr ASAE, ASEE, AVA, Gamma Sigma Delta, Phi Kappa Phi & Rotary Internatl. Received Natl Safety Award, Natl Safety Council. Retired June 30, 1986. *Society Aff:* ASAE, ASEE, AVA, AVERA

Paratore, William G
Business: 500 Sansome St, San Francisco, CA 94111
Position: Partner. *Employer:* Dames & Moore. *Education:* BS/Civ Engg/Univ of CA. *Born:* 11/3/34. Grew-up in San Francisco, CA and spent 15 years working out of Chicago Office of Dames & Moore. During 22 yrs with Dames & Moore have supervised multi- disciplined studies for various types of industrial facilities. Directed surface and subsurface explorations; coordinated gathering of geological, seismological and hydrological data; supervised engg studies to determine site suitability; supervised the preparation of environmental reports. Supervised studies in the US, Caribbean, Central America, Canada and Middle East. Served as Dames & Moore's rep and mbr of the Policy Bd for the HOKplus4 Consortium which designed Univ of Riyad, Saudi Arabia. Presently Dir of Natl Client Prog and US Coordinator of Middle East activities. Reg PE in six US States. *Society Aff:* ASCE, NSPE, ISPE, ASCE.

Parcher, James V
Business: School of Civil Engrg, Stillwater, OK 74078
Position: Prof emeritus *Employer:* Oklahoma State Univ. *Education:* PhD/Soil Mech/Univ of AR; AM/Soil Mech/Harvard Univ; MS/Structures/OK State Univ; BS/Civil Engrg/OK State Univ. *Born:* July 1920 Drumright Okla. Engr Remington Arms Co 1941-42. US Army Marianas Islands 1942-46 & Greenland 1950-52; Col USAR, EN (Ret). CIVEN Fac, Okla State Univ 1947-85; Hd of School, 1969-83. Cons in Soil Mech & Foundations. Co-author 2 texts on soil properties & soil mech. Pres Oklahoma Sect ASCE 1972. V P PEE Div, NSPE 1981-82, S W Region 1975-77. Sec PEE Div, NSPE, 1978-80. OSPE Engr of Yr 1976. Hobbies: music (French horn player) & bowling. *Society Aff:* ASCE, NSPE.

Parden, Robert J
Business: Univ of Santa Clara, Santa Clara, CA 95050
Position: Professor, Engineering Management 1955- *Employer:* Univ of Santa Clara. *Education:* PhD/IE/Univ of IA; MS/IE/Univ of IA; BS/ME/Univ of IA. *Born:* Apr 1922. m. Elizabeth Taylor 1955; c. Patte, James, John, Nancy. To 1st Lt QMC, AUS 1943-46. Assoc Prof Ill Inst of Tech 1953-54; Dean Engrg, Univ of Santa Clara 1955-82, Prof Engr. Mgt. 1955- . Principal, Saratoga Consulting Group. ASME (Chmn Santa Clara Valley Sect 1958); ASEE (Chmn Pacific S W Sect 1960); IIE (Educ Chmn 1958-63, Dir ASEE-ECPD affairs 1963-68); ECPD, Dir 1964-65, 1966-69; NSPE; AIIE, ASEM, IEE, Sigma Xi; Tau Beta Pi; Reg P E Iowa & Calif. *Society Aff:* AIIE, ASEM, IEE

Pardo, Jaime
Home: 1552 Sandringham Dr, Columbus, OH 43220
Position: Secretary-Treasurer *Employer:* Columbus Engrg Consultants, Ltd. *Education:* BS/CE/Purdue Univ *Born:* 4/19/31 Native of Bogota, Colombia. Studies at: Colegio Mayor del Rosario, Bogota, Colombia, Honorary Scholarship, 1949. Universidad de los Andes, Bogota, Colombia. Purdue Univ, 1955, Chi Epsilon, National Honorary Civil Engrg Fraternity. Bridge Engr and Supervisor for Alden Stilson and Assocs, 1955-1963. Supervisor in the Bridge Dept for Rader and Assocs, 1963-1968. With Columbus Engrg Consultants, Ltd. since 1968. Partner and Head of the Bridge Dept. Vice Chrmn 1974-1980. Secretary-Treasurer 1980-. Responsible for the design and/or supervision of nearly 300 highway structures. Four of these structures have received national awards. *Society Aff:* NSPE, ASCE, SCI, AREA, PEPP

Pare, Ronald C
Business: Cogswell College North, P.O. Box 33547, Seattle, WA 98133
Position: VP of Cogswell Coll & Dean of Cogswell Coll North *Employer:* Cogswell College. *Education:* MS/ME/CA State Univ-Los Angeles; BS/ME/WA State Univ. *Born:* Aug 1940; m. Ann; c. Julie 8, Wayne 6; p. Eugene & Marjorie. Pullman Washington H S 1959; Univ of Washington 1959-61; Washington State Univ 1961-65 BSME; Calif State Univ Los Angeles 1969-71 MSME. VP Dean of Cogswell College North: 1973-80. Asst Prof Calif State Polytech Univ 1968-73. Capt & pilot US Army 1965-68. ASEE 1968- , Design Graphics Div Secy 1973-74, Pacific Southwest Sect V Chmn 19 74-80 ; Amer Soc of Mech Engrs 1961- . Author 5 papers on self-paced instruction 1971-74. Dow Outstanding Young Engrg Educator Award 1974. Charter Mbr Montclair Jaycees 1971-73. Pi Kappa Alpha frat; Major & Pilot, US Army Reserves 1975- ASEE Engineering Technology Division Secretary 1979-1982. *Society Aff:* ASEE, ASME.

Parente, Emil J
Home: 26931 Highwood Circle, Lagund Hills, CA 92653
Position: Senior Vice Pres. - Sales *Employer:* Fluor Engineers and Constructors, Inc. *Education:* MChE/Chem Engg/NY Univ; BChE/Chem Engg/The Cooper Union. *Born:* 8/2/30. Native of NYC. Rejoined Fluor in 1978. *Society Aff:* AICHE, API.

Parfitt, Harold R
Home: 9535 Hilldale Drive, Dallas, TX 75231 *Employer:* Retired *Born:* Aug 1921. MS MIT; BS US Military Acad. Native of Coaldale Pa. Lifetime career with Army Corps Engrs, advancing to current rank of Maj Genl. Extensive exper in charge of design & const of major military & civil const projs as Dist Engr, Jacksonville Fla 1962-65 & Div Engr Southwestern Dallas Texas 1969-73. Governor, Canal Zone & Pres Panama Canal Co 1975-79 Retired Oct 1, 1979.

Parikh, Niranjan M
Home: 1230 Park Ave. West, #215, Highland Park, IL 60035
Position: Conslt *Employer:* Self *Education:* Sc.D./Ceramics & Metallurgy/Mass. Inst. of Technology, Cambridge Mass.; B.S. and M.S./Glass Technology/Alfred Univ, Alfred, N.Y.; B.Sc./Chemistry & Physics/Univ. of Bombay, India. *Born:* Jan 14, 1929 Godhra, India. U S citizen. m. 1954; c. 3. Res engr, sci lab Ford Motor Co Michigan 1954-57; Sr Res Officer Govt of India, Bombay, Atomic Energy Res Estab 1958-59; res metallurgist IIT Res Inst 1969-61, Sr Metallurgist 1961-63, Metal Sci Adv 1963-64, Asst Dir Metal Sci 196465; Dir Metals Res 1965- 75, Amer Can Co Barrington Ill, Managing Director, R&D 1975-1984. Managing Partner, Advanced Aluminum Products, Hammond, Ind. 1984-86. Conslt 1984- . Fellow ASM, Amer Ceramic Soc, Fellow British Inst Metals, Amer Inst Chemists; Matl Adv Bd 1965-67; NSF Adv Bd 1974-76; IR-100 Award 1970, 1971. Elected 1 of 10 outstanding young men by Chgo Jaycees 1964. Powder met - fracture; corrosion & stress corrosion; composites; metal working; tribology; bio matls; matls for coal gasification & liquification; matls for nuclear powder plants; pkging matls; metal can technology; resource mgmt & recovery; ceramics; Organic coatings; plastics packing; continuous Strip casting; Aluminum Mini-Mills. *Society Aff:* ASM, AIME, Inst of Met

Paris, Demetrius T
Business: School of Electrical Engineering, Georgia Institute of Technology, Atlanta, GA 30332
Position: Prof & Dir, School of Elec Engrg. *Employer:* Georgia Inst of Tech. *Education:* PhD/EE/GA Tech; MS/EE/GA Tech; BS/EE/MS State Univ. *Born:* Sept 1928. Native of Greece. Design engr for Westinghouse Elec Corp 1952-58; Sr Electronics Engr for Lockheed-Georgia Co 1958-59. Mbr Ga Tech fac since 1959. Assumed current respon as Dir of Georgia Tech's School of EE 1969. Founder & first Pres, Southeastern Ctr for Elec Engrg Educ Inc. Author 'Basic Electromagnetic Theory' McGraw-Hill 1969. Editor IEEE Transactions on Educ 1976- 79. President, IEEE Education Society, 1986. Enjoy music, swimming. Recipient 1980 IEEE Education Society Achievement Award and one of IEEE's 1984 Centennial Medals. *Society Aff:* IEEE, ASEE.

Paris, Luigi
Home: Piazza di Spagna, 81, Rome, Italy 00187
Position: Prof of Power System Analysis & Power Engg Consultant *Employer:* Univ of Pisa *Education:* D/EE/Univ of Pisa *Born:* 11/30/27 Native of Pisa, Italy, where he got the Doctor's degree in Industrial Engrg. With Edison Co of Italy since 1951 specializing in design and construction of transmission lines. Since 1963 with Enel, Italian Electricity Board, where was Mgr of Research and Development Dept up to 1982 when he left Enel and started his activity as Power Engineering Consultant; Prof of Power System Analysis at Univ of Pisa. Since 1975 Chmn of Steering Ctte of 1100 kV Project, and Intl research program on UHV among Italy, Argentina and Brasil, Past Pres of Tech Ctte of CIGRE. Fellow of IEEE. Past Chmn of IEC Tech Ctte on Recommandations for overhead lines. *Society Aff:* IEEE, CIGRE, AEI.

Parisi, Paul A
Business: 335 E 45th St, New York, NY 10017
Position: Mgr of Special Projects *Employer:* Amer Inst of Physics *Education:* BS/CE/Cooper Union. *Born:* 4/22/28. Born in Brooklyn, NY. Civil Engg, Mgr. s. Paul & Vincenza (Pumo) Parisi, m. Joan M Galbo May 1, 1955; c. Kenneth J, Steven P, Janet M. BSCE Cooper Union 1950; Omega Delta Phi, Chi Epsilon. 1950 Jr Civil Engr Dept of Marine & Aviation NY City; 1950-52 const surveyor US Army Korea; 1952 civil engr Ammann & Whitney NY City; 1954 Adj Instr of Civil Engg The Cooper Union NY City; 1953-72 Asst Editor, Assoc Editor, Editor & Mgr of Tech Pubs ASCE; 1972-79, Dir of Publ Services ASCE; 1979-83, Asst Dir, Engg Index, NY; 1983-, Mgr of Special Projects, Amer Inst of Physics, NY. Fellow ASCE; Sr Mbr Soc for Tech Comm; Mbr Amer Soc for Information Science; Reg P E NY. *Society Aff:* ASCE, STC, ASIS.

Park, Doo H
Business: 521 Fifth Ave, New York, NY 10017
Position: President. *Employer:* Duha Cons Inc & Yuhwa Assoc, Inc. *Education:* BS/Chem Engg/MIT. *Born:* Sept 1928. BS MIT. Native of Seoul Korea. Tech Adv to Korean Gov 1959- 62. Formed Mon-Soon Engrg Co Ltd in Seoul 1963; later name was changed to Internatl Engrg Co Ltd. Assumed current respon as Pres for performing engrg cons services for chem process industries. Formed affiliated co Yuhwa Assocs Inc in 1972 & Duha Cons Inc in 1976 in NY. Enjoy classical music & opera. *Society Aff:* AIChE.

Park, Gerald L
Business: Dept of Elec Engg, E Lansing, MI 48824
Position: Prof of Elec Engg & Systems Sci. *Employer:* MI State Univ. *Education:* PhD/Elec Engg/Univ of MN; MS/Engg Sci/Stanford Univ; BME/Mech Engg/Univ of MN. *Born:* 2/7/33. Born and raised in Minneapolis, MN. Served in air technical intelligence of USAF. On faculty of MI State Univ since 1964. Since 1966 has been principal investigator for a sequence of govt and elec utility sponsored res progs in the analysis, control and instrumentation of electric power systems. In 1974, his efforts shifted to the evaluation and planning of wind elec generating systems and since 1974 on power quality. Has been consultant to DOE ORNL NASA-LEWIS, SERI and EPRI as well as an an expert witness in elec accident cases. Is a reg PE in MI and is an FCC-licensed technician. Repairs marine electronic equip as a hobby. Twice elected to the E Lansing Sch Bd and its pres in 1979-1980. *Society Aff:* IEEE, CIGRE.

Park, Joon B
Business: Dept of Biomedical Engrg, Iowa City, IA 52242
Position: Prof *Employer:* Univ of IA *Education:* PhD/Materials/Univ of UT; MS/Materials/MIT; BS Mfg Engr/Boston Univ *Born:* 06/20/84 Upon graduation Dr Park has been involved in research on Biomaterials at various universities. He's been Univ of Washington 1972-73, Univ of Il, Urbana 1973-75, Clemson Univ, SC 1975-81 and Tulane Univ, New Orleans, LA 1981-83. Now at Univ of IA as a full prof of Biomedical Engrg. He published extensively on Property-Structure Relation-

Park, Joon B (Continued)
ship of natural materials and man-made materials esp polymers, over 35 papers. He also published two books *Biomaterials: An Introdution* Plenum, 1979 and *Biomaterials Sci and Engrg* Plenum, 1984. He's active in teaching and research on biomaterials specifically the interface phenomenon between body tissue and man-made materials. *Society Aff:* SFB, ORS

Park, Robert H
Business: Fast Load Control Inc, Freeport, IL 61032
Position: President. *Employer:* Fast Load Control Inc. *Born:* Mar 15, 1902. BSEE MIT 1923; Royal Tech Inst Stockholm 1923-24 mathematical physics. Genl Elec Co Schenectady 1924-29 Central Station, Cons & Turbine Depts; Stone & Webster Boston 1930-31 elec engr; Calco Chem Div Amer Cyanamid Co Bound Brook N J 1931-40 phys res & chem engrg; Naval Ordnance Lab 1940-43 aircraft laid under-water mines, 1943-45 Bureau of Ordnance torpedoes; Emhart Mfg Co Hartford Ct 1945-53 Dir R D E; 1953-68 formed R H Park Co Inc, dev automatic machinery; 1968 formed Fast Load Control Inc, a patent holding co, worked with TVA. Navy Dept Disting Civilian Service Award & IEEE Lamme Award.

Park, William H
Business: 207 ME Bldg, Univ Park, PA 16802
Position: Prof of ME & Head, Graduate Programs. *Employer:* PA State Univ. *Education:* PhD/ME/Cornell Univ; MS/ME/PA State Univ; BS/ME/PA State Univ. *Born:* 3/2/29. Native of Carlisle, PA. Design Engr for Sanders & Thomas - high pressure hydraulics. With PA State Univ since 1953 as Instr, Asst & Assoc Prof. Promoted to Prof 1979 & am teaching machine dynamics, vibrations, control systems, & machine elements. Head Grad Prog since 1980. Res areas are road profiling, vehicle response to road roughness, passenger comfort, mobile micro-processor applications & computer aided engrg. One yrs leave as Principal Res Engr, Ford Motor Co - hybrid computation & vehicle dynamics. Various summer & consulting positions. Enjoy gardening, woodworking & sailing. *Society Aff:* ASME, SAE, ASEE, SCI.

Parker, Blaine F
Agri Engr Dept 0075, Lexington, KY 40546
Position: Professor. *Employer:* Univ of Kentucky. *Education:* PhD/Agri Engg/MI State Univ; MS/Agri Engg/VPI; BS/Agri Engg/VPI. *Born:* June 12, 1924 Gaston Cty N C. Teaching exper: 2 yrs at VPI, 3 yrs at N C State Univ & 29 yrs at the Univ of Kentucky serving as Prof & Chmn of the Dept of Agri Engrg for 16 yrs. Fellow Amer Soc of Agri Engrs. Served as Chmn of the Southeast Region, Chmn of the Structs & Environ Tech Div, Chmn of the Res Ctte & Educ & Res Dept of ASAE. Mbr of the Amer Soc for Egrg Educ & the Internatl Solar Energy Soc. Mbr Alpha Zeta, Phi Kappa Phi, Omicron Delta Kappa, Tau Beta Phi, & Sigma Pi Sigma. Listed in Who's Who in Amer. Currently engaged in solar energy and thermochemical conversion of biomass to heat energy res for the purpose of dev an agri energy sys, Two solar collector patents. *Society Aff:* ASAE, ISES, ASES, ASHRAE.

Parker, Don
4500 Park Allegra, Calabasas, CA 91302
Position: Mgr, Radar Lab *Employer:* Hughes Aircraft Co. *Education:* DSc/EE/MIT; MS/Appl Phys/Harvard Univ; BS/EE/Brigham Young Univ. *Born:* 1/14/33. Dr Parker has been Mgr Radar Lab, Advanced Missile Sys Div since 1981. He has over 25 yrs experience in res & dev of microwave components & subsys. He joined Hughes in 1976 as Mgr Microwave Dept within the Radar Lab. Previously he was Dir of the Electromagnetic Techniques Lab at Stanford Res Inst. At MIT Lincoln Lab (1964-1969) he designed high-power solid-state sources for comm satellite applications. He is author or co-author of over twenty papers on solid-state power generation & microwave components. He has been a mbr of the MTT-S Admin Ctte since 1972 and Pres in 1979. He was editor of IEEE Transactions on Microwave Theory & Techniques 1975 through 1977. He has been Awards Chmn since 1983. *Society Aff:* MTTS.

Parker, Jack R
Home: 106 The Mews, Haddonfield, NJ 08033
Position: President *Employer:* Brown & Root, (Delaware Valley) Inc. *Born:* Apr 25, 1919. Attended Polytech Inst Brooklyn Chem Engrg 1943. Founded Royalpar Indus Inc, publicly owned engrg co, Chmn of Bd to 1975. Founder, Pres, Dir Vernitron Corp 1958. Founder, V P, Dir Refinadora Costarricense de Petroleo, S A petro refinery Repub of Costa Rica. Cons Stone & Webster Engrg Corp 1956-75 in oper of power indus ctrs at N Y & Cherry Hill N J. Estab Kellex Power Services, Pullman Kellogg Co 1975. Founder, Philadelphia Foreigh-Trade Zone. Co- recipient in Fairleigh Dickinson Univ 1975 Humanitarian Award. Pres., Brown & Root (Delaware Valley) Inc. 1981-85. Pres., J. Royal Parker Associates, Inc., 1975 to present. *Society Aff:* AICHE.

Parker, Jack S
Business: 3135 Easton Turnpike, Fairfield, CT 06431
Position: V Chmn & Exec Officer.(retired) *Employer:* Genl Electric Co. *Education:* BS/Mech Engg/Stanford Univ. *Born:* 7/6/1918 Palo Alto Calif. BSME Stanford Univ 1939. 1939-50 shipbldg & chem indus. 1950-1980 , G E: 1952 aircraft nuclear propulsion & small aircraft engine dept mgmt; 1955 V P - Genl Mgr Aircraft Gas Turbine Div; 1957 mbr Exec Office, V P Relations Services; 1961 Group V P Aerospace & Defense Group; 1968 V Chmn of Bd & Exec Officer. Reg P E Calif. Mbr Natl Acad of Engrg, Fellow Amer Soc of Mech Engrs; Assoc Fellow Royal Aeronautical Soc; Fellow AIAA. Tr RPI; Adv Council Stanford Univ Grad Sch of Bus. Ints: ranching, hunting, conservation, golf. *Society Aff:* ASME, NAE, AIAA.

Parker, Jack T
Home: 3590 Norwich Dr, Tucker, GA 30084
Position: Vice President *Employer:* Soil & Material Engrs Inc. *Education:* BS/Civil Engr/Auburn Univ *Born:* 9/29/40 Born in Montgomery Alabama and graduated from Auburn University in 1962. Did graduate study toward MSCE at GA. Tech and MBP at GA. State University. Worked at Law-Engineering Testing Company from 1963 thru 1977 and was Branch Manager and assistant Vice President at the time I left to become one of 4 officers and principals of Soil & Material Engineers Inc, a firm of 325 engineers with 12 branch locations in Southeast United States. *Society Aff:* ASCE, NSPE, ACEC

Parker, James W
Business: 1133 Ave of Americas, NYC, NY 10036
Position: Regional Specialist. *Employer:* Eastman Kodak Co. *Education:* BA/Chemistry/MI State Univ. *Born:* May 26, 1945. BA Chem Michigan State Univ 1968. Joined Eastman Kodak Co 1968 & wrked as sales engr in Atlanta Ga & Houston Tex. Present capacity incl cons in audiovisual sys design. Served as Chmn of Houston Tex Sect, Soc of Motion Picture & Television Engrs. *Society Aff:* SMPTE.

Parker, Jerald D
Home: 6 Pecan Dr, Stillwater, OK 74075
Position: Professor *Employer:* Okla State Univ *Education:* PhD/M.E./Purdue; MS/M.E./Okla State; BS/M.E./Okla State. *Born:* 02/24/30 Professor Parker has been active as a writer and consultant in the field of thermal and energy engrg. He has served in several offices at the regional and national level in professional societies. He has frequently been invited to speak and to conduct short courses on HVAC, solar, heat pumps and related areas. He is a Fellow in ASME. *Society Aff:* ANS, ASME, ASHRAE, NSPE-OSPE

Parker, Karr
Business: 75 W Mohawk St, Buffalo, NY 14202
Position: Chmn of the Bd. *Employer:* Buffalo Electric Co Inc. *Education:* DSc/-/Carthage Coll; MS/-/Univ of IL; BS/-/Carthage Coll. *Born:* July 3, 1892 Sebree Ky. Student Carthage Coll; BS Univ of Illinois 1913, MS 1914, DSc 1929.

Parker, Karr (Continued)
Radio engr Marconi Wireless Telegraph Co Amer, Cleveland & Milw 1913-15; elec engr McCarthy Bros & Ford, elec engrs 1930- ; Chmn Buffalo Elec Co; Mbr Bd Dir Buffalo Savings Bank. Exec Comm Buffalo War Council. Mbr Council, Chmn Bldgs & Grounds Comm Univ of Buffalo. Mbr Buffalo C of C (P Pres), Elec League Niagara Frontier (P Pres), Assoc Elec Contractors of Buffalo (P Pres), Engrg Soc Buffalo. *Society Aff:* ASME, IEEE, NSPE.

Parker, Norman A
840 William St, River Forest, IL 60305
Position: Chancellor Emeritus *Employer:* University of Illinois at Chicago Circle *Education:* ScD/Mech Engg/Univ of CO; ME/Mech Engg/Univ of CO; MS/Mech Eng/Univ of CO; BS/Mech Eng/Univ of CO. *Born:* April 16, 1906. Native of Colorado. Held teaching & administrative positions in mech engrg at Univ of CO; at Univ of IL in Champaign-Urbana; developed new campus in Chicago & served as 1st Chancellor; served as V Pres & Proj Dir for Academy for Educational Dev Inc in Tehran, Iran, developing a master plan for the Imperial Government of Iran for training personnel for maintenance & oper of the railroads, roads, ports & shipping of the country. Served as cons on educational programs for three new univs in Algeria. Pres Assn of Urban Universities 1970; Fellow ASME; served 10 yrs on Chicago City Plan Commission; served as V Chrmn of Chicago Home Rule Commission; served as consultant on education to government of Afghanistan 1977-78. *Society Aff:* ΣΤ, ΤΒΠ, ΣΞ, ASME, ΠΤΣ, ASEE.

Parker, Norman F
Business: 611 Hansen Way, Palo Alto, CA 94303
Position: President & Chief Executive Officer. *Employer:* Varian Associates, Inc. *Education:* DSc/Engg/Carnegie-Mellon Univ; MS/Elec Engg/Carnegie-Mellon Univ; BS/Elec Engg/Carnegie-Mellon Univ. *Born:* May 14, 1923 in Fremont Nebraska. 1948-67 with North American Aviation (now Rockwell Internatl) in various positions including Pres of Autonetics Div & V Pres of North American Aviation; 1967-68 Exec V Pres of Bendix Corp; 1968 joined Varian as Pres & a Dir, in 1972 appointed Chief Exec Officer as well. Life Trustee of Carnegie-Mellon Univ, mbr Bus Adv Ctte Graduate Sch of Industrial Admin of Carnegie-Mellon; elected to Natl Acad of Engrg in 1976. *Society Aff:* IEEE, NAE, AIAA.

Parker, Norman W
Home: 1302 North Scott St, Wheaton, IL 60187
Position: VP Tech *Employer:* Motorola Inc. *Born:* Nov 3, 1922. Case-Western Reserve Univ. Fellow IEEE, winner of IEEE Consumer Electronics Award 1971, 2 1st place paper awards IEEE; NQRC Natl Chmn; NAMSRC Natl V Chmn Panel 1; NIAC Chmn Field Testing Ctte; Land Mobile-TV Sharing Ctte, Chmn Lab Testing Ctte; NSRC, Chmn several cttes & subcttes; NTSC Mbr Panels 14 & 16; CCIR, Mbr U S Preparatory Ctte, U S Delegations SG 10, SG 11 & contributor of position papers; SMPTE Mbr Ctte on Colorimetry; EIA Mbr Broadcast Systems Ctte. Numerous patents, papers & book contributions.

Parker, Paul E
Business: 1308 W Green St #207, Urbana, IL 61801
Position: Assitant Dean *Employer:* Univ of IL *Education:* MS/ME/SUNY-Buffalo; BS/ME/N.C.A. & T. State Univ *Born:* 10/23/35 Native of Jenkins Bridge, VA. Taught in the U.S. Army Engrs School during service years. Stress analyst on rocket engine development for Bell Aerosystems 1961-67. Taught mechanical engrg (mechanics and vibration), developed an evening program in engrg, developed the cooperative education program in engrg, served as M.E. Dept Chrmn, and served as Assistant to the Dean of Engrg at N.C.A. & T. State Univ during the period 1967-73. Assumed current position as Assistant Dean of Engrg in 1973. Responsible for developing the Principal's Scholars Program, while administering undergraduate programs within the Coll. *Society Aff:* ASEE, ASME, NAMEPA, JETS

Parker, Robert B
Home: , Clayville, RI 02815
Position: Owner. *Employer:* Parker Engineering Company. *Born:* 1905. 3 years Northeastern Univ. Reg PE in RI. ASME Fellow, active 21 years in region, 2 terms on Natl Nominating Ctte. 10 years designing & mgmt; 3 years Ch Engr defense plant; 33 years private practice cons for many machines & devices developed to reduce labor & increase quality, some resulting in patentable products; 10 years designing machinery for grinding ultra hard precision parts. The opportunity to encourage young engrs who have been associated with me during the years has been a source of lasting satisfaction. *Society Aff:* SME.

Parker, Robin M
Business: PO Box 939, Alvin, TX 77511
Position: Pres. *Employer:* Industrial Consultants & Contractors, Inc. *Education:* BS/Chem Engg/Auburn Univ. *Born:* 6/1/38. Native of Chattanooga, TN. Graduated Auburn Univ Auburn, AL. Project Engg for Great Lakes Chem and MI Chem El Dorado, AR 1968-74; VP-Control Systems Engg, Shreveport, LA, 1974-76; Pres-Industrial Consultants and Contractors, Incd- Alvin, TX 1977-; Dir, J A Reece Co, Inc-Clarksville, TN 1978-; Consultant to Corning Glass Process Systems on Bromine and Brominated Chemicals; Reg PE, TX. *Society Aff:* AIChE, NSPE, TSPE.

Parker, Sydney R
Business: Parker Associates, P.O. Box AQ, Carmel, CA 93921
Position: Consultant *Employer:* Univ. of CA Lawrence Livermore Lab *Education:* ScD/Elec Eng/Stevens Inst of Tech; MS/Elect Eng/Stevens Inst of Tech; BEE/Elec Eng/CCNY *Born:* 4/18/23. From 1952 to 1956 he was with the RCA Advanced Dev Group. From 1956 to 1965 he was a Prof of EE at Univ CCNY, & in 1965 became Prof at the Univ of Houston. In 1966 he joined the Faculty of the Naval Postgrad Sch & in 1970 became Dept Chrmn, leaving in 1975 to become Dean of Engrg at Rutgers. He returned to the Naval Postgrad Sch as a Prof from 1976-86. Since 1984 he has also been Visiting Prof of EE at Stanford Univ. Dr Parker has published over 100 papers and textbooks. He has served as a conslt at Rockwell Int & is currently a consultant U C Lawrence Livermore Lab. He has been an Editor for Pegammon Press & the Macmillan Co. He is editor of the journal Circuits Systems & Signal Processing, Birkhauser/Springer-Verlag. He was Pres 1980-81 of Eta Kappa Nu. He was awarded the NPGS Sigma Xi Res Award in 1977. He is a Fellow of the IEEE (1975) and was awarded the IEEE Harry Diamond Mem Award in 1984. He co-founded the Asilomar Conf on Circuits & Systems. Groups/Societies: Pres Eta Kappa Nu (Hon EE Soc) 1980-81. Circuits & Systems: Pres 1974, VP 1975, Admin Ctte 1971-72, *Society Aff:* IEEE, ΤΒΠ, ΣΧΙ, HKN, AAAS

Parker, Walter B
Home: 1490 Skylark Drive, Spring Hill, FL 33526
Position: President. *Employer:* Uniform Boiler & Pressure Vessel Law Society, Inc. *Education:* BS/Mech Engg/Univ of TN. *Born:* June 1909, native of Monteagle TN. 1933-35 X-ray technician Combustion Engrg Chattanooga TN; 1935-73 with Hartford Steam Boiler Inspection & Insurance Co as inspector, adjuster, 1940-73 Asst Ch Engr specializing in safety engrg work with State Boiler Inspection Depts in co-operation with ASME & Natl Bd of Boiler & Pressure Vessel Inspectors; 1973-1981 , Pres Uniform Boiler & Pressure Vessel Laws Society encouraging uniformity in boiler & pressure vessel laws, rules & regulations in the U S & Canada. Mbr TAPPI, Fellow ASME. Enjoy tennis & travel - Retired. *Society Aff:* ASME, TAPPI.

Parker, William H
Business: 13135 Lee Jackson Memorial Hwy, Suite 200, Fairfax, VA 22033
Position: Sr Vice pres *Employer:* Camp Dresser & McKee *Education:* MEM/Eng Mgmt/Univ of Detroit; MBA/Bus/Univ of Detroit MS/CE/Northeastern Univ; BS/CE/Univ of ME *Born:* 05/04/37 Has managed numerous major environmental engrg projs, including: Water supply, treatment, distribution, and pollution-control projs; mgmt services programs; hazardous waste programs; and solid waste mgmt/resource recovery projs for both the public and private sector. Responsibilities on these assignments ranged from planning through design, operations and maintenance, and

Parker, William H (Continued)
services during construction. Has written more than a dozen profl papers published in leading indus journals. Was Massachusetts Soc of PE "Young Engr of the Yr" for 1971. Currently manages a regional staff of more than 100 environmental engrs, planners, and mgmt conslts. Reg PE in 13 states. Mbr of Chi Epsilon, Tau Beta Pi, Alpha Kappa Psi, Sigma Psi, & Beta Gamma Sigma. Diplomate of the American Academy of Environmental Engineers. *Society Aff:* WPCF, ASCE, ACEC, NSPE, AWWA

Parkin, Blaine R
530 Stony La, State College, PA 16801
Position: Professor of Aerospace Engineering. *Employer:* The Pennsylvania State University. *Education:* BS/Mech Eng/Caltech; MS/Aeronautics/Caltech; PhD/ Aeronautics/Caltech *Born:* 7/25/22. 1943-45 USAAF, decorated: Air Medal & Cluster; 1952-56 Caltech Research Fellow in Applied Mechanics & Hydrodynamics Lab; 1956-62 RAND Corp engr, weapons effects & ground shock; 1962-64 AirResearch Mfg Co, isotope separation with gas centrifuges; 1964-72 Genl Dynamics Convair, middle mgmt in engrg; aerospace tech, systems analysis, advanced aircraft programs; 1972-82, PA State Univ as Prof of Aerospace Engrg & Dir Garfield Thomas Water Tunnel, Applied Res Lab 1972- . Prof of Aerospace Engineering & Chief Scientist Applied Research Lab., Penn State. AIAA Assoc Fellow, Chmn Central PA Sect 1976-77; ASME Fellow, Chmn Polyphase Flow Ct 1974-76, member Fluids Engineering Division Executive Committee 1986-1990; SNAME Mbr; Chrm Cavitation Cttee Amer Towing Tank Conference 1976-77. Deligate mbr, Intl Towing Tank Conference 1978. ITTC Cavitation cttee: Mbr 1978-87, Sec 1981-87. Registered Professional. *Society Aff:* ASME, AIAA, ΣΞ, ΓΣΤ, SNAME.

Parklinson, John R
Home: 1472 185th Ave NE, Bellevue, WA 98008
Position: President *Employer:* Parkinson Assocs, Inc *Education:* PhD/Pulp-Paper/Lawrence College; MS/Pulp-Paper/Lawrence College; MS/ChE/OH State Univ; BChE/OH State Univ. *Born:* 8/4/26. Pulp mill engr & consultant specializing in chem systems & heat & by- product recovery. Reg PE in OH & WA. *Society Aff:* AIChE, TAPPI.

Parks, Lyman L
Business: 650 Park Ave, East Orange, N.J 07017
Position: Dir, Bldg. Engg. *Employer:* N J Bell *Education:* BS/Mech. Eng./Rutgers University; Certificate/Fire Prot. Technology/Newark College of Engineering *Born:* 08/15/34 A New Jersey native from Westfield, graduated Rutgers Univ as mech engr and joined NJ Bell in '55 as a bldg mech engr. Held various positions in engrg obtaining P.E. license in 1960. Joined AT&T as Engrg Specialist responsible for Bell System fire protection standards and practices. Rejoined NJ Bell to manage special fire protection project in '77. Managed major energy conservation effort until made Director of all building mech & elec engrg. Mbr NFPA Technical Cttees on Air Conditioning and Smoke Mgmt. Pres NJ Ch SFPE, Covenant Christian School. Golf, bowling, automotives leisure activities. *Society Aff:* SFPE, NFPA

Parks, Robert J
Business: 4800 Oak Grove Dr, Pasadena, CA 91103
Position: Asst Lab Dir for Flight Projects. *Employer:* Jet Propulsion Laboratory. *Education:* BS/EE/CA Inst of Technology. *Born:* April 1922 in Los Angeles CA. BS Caltech. 1947-51 engr Jet Propulsion Lab; 1951-56 Ch Guidance & Control Sect; 1956-58 Guidance & Con't Div; 1957-60 Dir Sgt Prog; 1960-62 Planetary Prog; 1962-65 Asst Lab Dir Lunar & Planetary Proj; 1965-67 Surveyor Proj Mgr; 1967- , Asst Lab Dir for Flight Proj. Cons Tech Adv Panel Aeronautics Office Secy of Defence 195758; Louis W Hill Space Trans Award 1963; NASA Public Service Award 1963; Stuart Ballantine Medal, Franklin Inst of Pa 1967; Ca Air Force Assoc Man of Year 1967. Fellow AIAA, IEEE; elected to Natl Acad of Eng 1973. Married, 3 children.

Parks, Thomas W
Business: EE School, Cornell University, Ithaca, NY 14853
Position: Prof *Employer:* Cornell Univ *Education:* PhD/EE/Cornell Univ; MS/EE/ Cornell Univ; B/EE/Cornell Univ *Born:* 03/16/39 Joined Rice Univ dept of Electrical Engrg as an asst prof in 1967. Presently a Prof in the School of Electrical Engrg at Cornell University conducting research in digital signal processing and teaching courses in linear systems and communications. Received the IEEE Acoustics, Speech, and Signal Processing Soc Technical Achievement Award for 1981. Elected Fellow of the IEEE in 1982. Co-author of two books on digital signal processing. *Society Aff:* IEEE, SEG

Parks, Vincent J
Business: Civil Engrg Dept, Washington, DC 20064
Position: Professor. *Employer:* The Catholic University of America. *Education:* PhD/Solid Mechs/Catholic Univ of Amer; MCE/Civ Engrg/Cathlic Univ of Amer; BSME/Mech Engrg/IL Inst of Tech; AS/Engrg/Lewis College of S & Tech. *Born:* May 5, 1928 in Chicago IL. 1953-55 design engr with Andrew Corp Orland Park IL; 1955-61 res engr with Armour Res Foundation; now Professor of Civil Engrg at Catholic Univ. Author of 130 papers & 2 books: 'Moire Analysis of Strain' with A J Durelli, Prentice Hall 1970 & 'Advances in Experimental Mechanics' editor, Catholic Univ 1975. Hetenyi Award 1974 and Frocht Award 1981 from the Soc for Experimental Mechanics. *Society Aff:* ASME, SEM, ΣΞ, AAM.

Parks, William W
Home: 1724 E. Ridgewood Lane, Glenview, IL 60025
Position: Retired President. *Employer:* Vapor Corp *Education:* M/BA/Univ of Chicago; BS/CE/IL Inst of Tech *Born:* 12/11/21 Oak Park, IL. s. Paul Brownlee and Margery Eulalie (Rohan) P; m. Mary Patricia Gagan; one daughter Julia, Structural Designer Swift & Co, Chgo 1946- 47; devel engr Vapor Corp, Chgo, 1947-50; chief research engr, 1950-55, chief engr Vap-Air div 1956-62; chief engr Vapor Corp 1962-66; VP, gen mgr Transp Systems Div 1966-75; Sr VP 1975-76; 1976 Gen Mgr, Vapor Div, and Sr VP to 1982. 83-86 Pres; now retired; Bd Dirs Ronald Knox Montessori Sch, 1964-66; trustee Il Inst Tech 1973- ; Served as LTJG CE, USNR, 1943-46; Mbr Am Soc ME; Am Pub Transit Assn; (dir, 1972-), Instrument Soc Am, Intl Star Class Yacht Racing Assn. (mem governing comm 1965-; Pres 1974-78; Il Inst Tech Alumni Assn (Pres 1972-73); Episcopalian, Club: Chicago Yacht Club (Dir 1960-72); Patentee in field. Now consulting - Parks & Associates, 1724 E. Ridgewood Ln, Glenview, IL, 60025. Telephone (312) 724-0749. *Society Aff:* ASME, ISA

Parmley, Robert O
Business: 115 W 2nd St S, Ladysmith, WI 54848
Position: Pres *Employer:* Morgan & Parmley, Ltd. *Education:* MSCE/Civil Engineering/Columbia Pacific University; Diploma/ME Technology/American School; BSME/Mechanical Engineering/Columbia Pacific University *Born:* 4/18/37 Pres of Morgan & Parmley, Ltd., Consulting Engrs, which serves both municipal and industrial clients in general, civil, mechanical, and industrial projects. Design experience includes a wide variety of machines, structures, and systems - from dams and bridges to pollution control equipment, and plumbing systems. Registered PE in WI, CA and Canada. Certified Manufacturing Engr under S.M.E. National Certification Program. Certified Wastewater Treatment Plant Operator, State of Wisconsin. Editor-in-Chief of Standard Handbook of Fastening & Joining (McGraw-Hill 1977), Field Engr's Manual (McGraw-Hill 1981) and Mechanical Components Handbook (McGraw-Hill 1985). Published 42 technical articles in such periodicals as Product Engrg, American Machinist, Machine Design, Assembly Engrg and Plan & Print. *Society Aff:* ASME, WSPE, NSPE, SME, AIDD, CSI

Parr, James Gordon
Business: Ontario Science Centre, 770 Don Mills Rd, Don Mills Ontario, Canada M2T 2C1
Position: Director General *Employer:* Ontario Science Centre *Education:* PhD/Met Eng/Liverpool Univ; BSc/Metallurgy/Leeds Univ. *Born:* May 1927. Native of U K. Canadian citizen. Taught at univs of Liverpool, British Columbia, Alberta, Windsor 1948-72. Pres Univ of Windsor Indus Res Inst from inception to 1972. Deputy Minister of Colleges & Univs, Ontario, 1973-79. Chairman/CEO TV Ontario 1979-85. FRSC, FEIC, FASM, FRSA, P Eng (Ont). Centennial Medal 1967. Jubilee Medal 1977. Citizenship Award, Assoc. of Prof. Engrs of Ontario, 1983; Hon LLD Univ of Windsor 1984. Author *Man, Metals & Modern Magic*; co-author wth A Hanson, *The Engr's Guide to Steel*, and *Intro to Stainless Steel*; author, *Any Other Business*; *Is There Anybody There?*; Technical and other papers, and broadcasts. *Society Aff:* EIC, ASM, RSA, SHOT, RSC.

Parrish, Arthur E
Business: PO Box 5805, Columbia, SC 29250
Position: Vice Pres *Employer:* Lott Parrish and Assocs *Education:* BS/CE/NC State Univ. *Born:* 02/07/52 Native of Raleigh, NC. Civil Engr for Federal Aviation Admin (Atlanta) 1975-1976. Sr Proj Engr, Airport Div, Wilbur Smith and Assocs, 1976 through 1981. Responsible for major airport planning and design projs in US and in South Amer. Formed present consltg engrg firm and served as VP and Chief Engr since early 1981. Currently responsible for technical aspects of all engineering projs undertaken by the firm. Grad of Federal Aviation Admin Acad in Airport Planning and Airport Engrg 1977. Reg PE - NC, SC, VA, GA, FL, CO, TX, AL, TN, MS. *Society Aff:* ASCE, NSPE, ITE, SAME.

Parrott, George D
Business: PO Box 22738, 620 Euclid Ave, Lexington, KY 40522
Position: Exec VP *Employer:* Parrott, Ely and Hurt, Consltg Engrs, Inc *Education:* BS/CE/Univ of KY *Born:* 01/22/30 Native of Springfield, KY. Served in US Navy Civil Engrg Corps 1953-1956. Associated with a, consulting firm, Lexington, KY, 1956-1967. A founding principal of Parrott, Ely and Hurt, Conslt Engrs, Inc in 1968. Directs the firm's Wastewater Collection and Treatment Div. Work includes supervising the design of sewage treatment plants, pumping stations and sewer extensions for both the Kentucky and Florida (Fort Myers and Marco Island) offices. Past Pres of the Consltg Engrs Council of KY. He is also a licensed commercial pilot, instrument rated. *Society Aff:* NSPE, WPCF, AWWA, AAOPA, CEC

Parsons, Alonzo R
Business: Honeywell Plaza, Minneapolis, MN 55408
Position: Mgr, Corp Standardization Services. *Employer:* Honeywell Inc. *Education:* AB/Math/Univ of PA. *Born:* 3/9/23. Grad work NY Univ 1950-51. US Navy 1942-46, Ltjg. Joined Honeywell Inc, Indus Div, Phila 1947. Served successively as Field Sales Engr, Regional Sales Mgr - Test Instruments Div, New York. Currently Mgr Corp Standardization Services in MN. Pres Instrument Soc of Amer ISA 1966-67; Dir ISA 1966-68; elected Fellow 1971, Dir Stds ISA 1972-75. Currently on US Natl Ctte, Intl Electrotech Comm IEC as US Tech Adv & Chrmn, IEC TC 75, *Classification of Environ Conditions'*; Mbr Amer Natl Stds Inst - Exec Stds Council & Co Mbr Council. *Society Aff:* ISA, SES.

Parsons, Donald S
Home: RD 5 Box 295A, Towanda, PA 18848
Position: Retired. *Employer:* GTE Products Corp *Education:* MS/ChE/Bucknell Univ; BS/ChE/Bucknell Univ. *Born:* 5/31/30. Lived in Cranford, NJ prior to 1947. With GTE Sylvania-Towanda, PA since 1951, except for serving with US Army Ordnance Corp 1956-57. Retired in August 1986. Held positions of Product, Sr & Dev Engrs, and since 1973 until retirement, Hd, Physical Testing Lab. Adjunct Professor of Chemical Engineering at Bucknell University 1981-86. Published 5 papers, presented 12 papers; hold 4 patents and 2 GTE Sylvania Proprietary Process Dev Awards. Mbr of Bd of Dirs Bucknell Univ Engg Alumni Assn 1973-83. Elected fellow of AIChE May 1979. Mbr of Admissions Committee. Held offices of Secy, VChrmn & Chrmn in both Central PA Sec of AIChE & PA-York Sec of APMI. Listed in Who's Who in the East. Received AICE 75th Anniversary Award for Service to the Central Pennsylvania Section. Received AIChE Prof Development Certificates in 1979, 1982 & 1985 (3yr certificates). *Society Aff:* AIChE, AAAS.

Parsons, James D
Home: 437 Vanderbilt Rd, Biltmore Forest, Asheville, NC 28801
Position: Consultant *Employer:* Self Employed *Education:* MS/Civ Engg-Foundations/Harvard Grad Sch of Engg; BS/Sci/Harvard College. *Born:* 3/31/09. Res Asst in Soil Mechanics at Harvard Grad Sch of Engg 1935-37. Fdn Engr for 1939 NY Worlds Fair Corp 1937-39. Field Engr in Venezuela, SA for Parsons, Klapp, Brinckerhalt, & Douglas (Consul Engrs) 1939-40. Principal with Mueser, Rutledge, Wentworth & Johnston, Intl Fdn Consultants, 1940-75. Recipient of ASCE's Rowland & Wellington Awards in 1962 & the T A Middlebrooks Engg Award in 1978. Recipient of Outstnading Engr of the Yr Award from NC Sec of ASCE in 1979. Recipient of ASCE Honorary Mbrshp in 1983. Recipient of ASCE Honorary Mbrshp in 1983. Currently in private practice as consulting Engr in Asheville, NC. *Society Aff:* ASCE, Moles.

Parsons, John T
Home: 205 Wellington, Traverse City, MI 49684
Position: Pres & Treas. *Employer:* John T Parsons Co: *Born:* 10/11/13. Recipient: Soc of Mfg Engrs 'Engrg Citation presented to John T Parsons, industrialist & inventor, whose brillant conceptualization of numerical control marked the beginning of the second indus revolution & the advent of an age in which the control of machines & indus processes would pass from imprecise craft to exact sci'. First Jules Marie Jacquard Award by Assoc for Integrated Mfg Tech for 'outstanding tech contrib'. First to design & put into production (1945) adhesive-bonded metal-to-metal primary aircraft structs. Originator of OCS - an operation control sys integrating specifications, budgets & controls of design, matls, production & quality. *Society Aff:* SME, AIMTech

Parsonson, Peter S
Business: Sch of Civ Engg, Georgia Institute of Technology, Atlanta, GA 30332
Position: Prof *Employer:* GA Inst of Tech. *Education:* PhD/Civ Engg/NC State Univ; MS/Civ Engg/MIT; BS/Civ Engg/MIT. *Born:* 10/18/34. Native of MA. Early career as soils & mtls engr in Labrador, Venezuela, Argentina. Asst Prof at Univ of SC, 1965-69. With GA Tech since 1970. Teaching, res & consulting in traffic engg, hgwy design, microcomputer application. Developed Tech's Traffic Signal Lab & Traffic Evaluation Lab. Short courses & educational films in signalization. Res in evaluation of effectiveness of control devices. Consultant in accident reconstruction, patent infringement, before-and-after evaluations. Reg in GA (Civ), CA (Traffic) and FL. Fellow, ITE. Pres, Southern Sec ITE 1977. SSITE Hensley Award, 1975, Hoose Award, 1984. *Society Aff:* TRB, ITE

Partee, Frank P
Business: Ford Motor Company, 15201 Century Dr., STE. 608, Dearborn, MI 48120
Position: Principal Staff Engr. *Employer:* Ford Motor Co. *Education:* Master Public Health/Air Pollution-Ind Health/Univ of MI; BSChE/-/Univ of Cincinnati. *Born:* 4/23/37. Presently is Principal Staff Engr, Office of Stationary Source Environmentl Control, Ford Motor Co. This Office is responsible for compliance by the Co's mfg facilities with governmental requirements. Previously served as Dir, Div of Pollution Control, KY State Health Dept, & earlier was a Commissioned Officer, US Public Health Service, & served in a number of assignments involving interstate air pollution studies around the country, & administering prog grants. Appointed by Gov. James Blanchard to Michigan Air Pollution Control Commission in 1983; Reappointed 1986. Reg PE & a Diplomate in the Am Acad of Environmental Engrs. Pres, APCA, 1980. Native of Hamilton, OH. Married with three children. *Society Aff:* AAEE, APCA, NSPE.

Parthasarathi, Manavasi N
Home: 47 Whittier St, Hartsdale, NY 10530
Position: Mgr Dev. *Employer:* Intl Lead Zinc Res Org. *Education:* PhD/Met Engg/Univ of IL; MS/Met Engg/Univ of IL; BS/Met Engg/Benares Hindu Univ; BS/Phys/Madras Univ. *Born:* 1/13/24. Son of M K Narasimhachari. BSC (Madras) BS (Banaras), MS (Univ of IL), PhD (Univ IL). Scientific Officer, Council of Sci Ind Res Asst Hd of Res Dept, Bird & Co, India. Gen Mgr Zinc & Lead Dev Assns. Mgr Dev Intl Lead Zinc Res Org. Dir of Hindustanzinc 1968-75. Post Grad Prof of Met Bengal Eng College, Howrah 1963- 64. Mbr Faculty Engg & Tech Inst of Tech. Banaras Hindu Univ. Recipient of Gold Medal of Ind Inst of Metals 1952. Recipient of John Taylor Gold Medal of Inst of Metals (India) 1972. Distinguished Alumnus Award Banaras Hindu Univ 1973. Fellow Inst of Met, London. Fellow of Am Soc of Diecasting Engrs. Fellow Indian Stds Inst. Fellow Am Soc for Metals. Mbr Sigma Xi (1957). Listed in Marquis "Who is Who in the World–, 1978-79. Author of numerous papers. Secy Ind Inst of Metals, 1963-76. VP of Ind, Inst of Metals, 1977. *Society Aff:* ASM, SDCE, Inst of Met.

Parthum, Charles A
Business: 1 Center Plaza, Boston, MA 02108
Position: Sr Vice President. *Employer:* Camp Dresser & McKee Inc. *Education:* BSCE/Civil/Northeastern Univ. *Born:* Sept 1929 Lawrence Mass. More than 30 yrs career exper with CDM. Have specialized in environ engrg. Respon for dir the preparation of reports, designs, & services during const of projs undertaken for municipal clients & gov agencies, as well as indus & private concerns. Became an Assoc in 1964, Partner in 1967, Sr V P & Dir in 1970. Diplomate Amer Acad of Environ Engrs. Fellow Amer Soc of Civil Engrs. Fellow ACEC. *Society Aff:* ASCE, NSPE, ACEC, WPCF, AWWA.

Partlow, Harry A
Business: Manager, Industrial Hygiene, Safety, & Compliance Review, Amoco Corporation, P.O. Box 87703, Chicago, IL 60680-0703
Position: Mgr, Industrial Hygiene, Safety & Compliance Review *Employer:* Amoco Corp. *Education:* BS/Chem Engg/Auburn Univ. *Born:* Sept 1933. Native Burnsville Miss. U S Civil Serv prior to Korean War. Army Signal Corps 1953-55. Field engr for Factory Insurance Assn, spec in chem industry. Monsanto Co. 1964-1980. Dir SAF Amoco Corp 1980-1984. Assumed current respon with Amoco in 1984. Also serve as primary contact with Natl Safety orgs & gov agencies, in the mgmt of corp safety practices. Past Pres Amer Soc of Safety Engrs, & Past VP & Mbr Bd of Dir Mo Safety Council. Originated efforts that resulted in passage of Mo Schools Mandatory Eye Protection Law. Two time winner Veterans of Safety & ASSE Tech Papers Award. Received Natl Safety Council's Disting Service to Safety Award in 1979. Received Agri Chem Billy Creel Award for excellence in 1983. Named Fellow ASSE 1986. *Society Aff:* ASSE, VOS, NFPA, NSC, API.

Parzen, Benjamin
Business: 3634 Seventh Avenue, San Diego, CA 92103
Position: Consulting Engr. *Employer:* Self-employed. *Education:* BS/EE/CCNY. *Born:* Apr 1913. Postgrad studies at BPI. Reg P E NY. Fellow IEEE with citation 'for contrib to the theory & practice of wide-range synthesizers & to the precision measurement of signals of high spectral purity'. Author num tech pubs. 14 pats. From 1938-44, successively Jr, Asst, Assoc Elec Engr at US Navy Dept. 1944-52 successively, Asst Proj Engr, Proj Engr, & Sr Proj Engr at Fed Telecommunications Lab of ITT. 1952-61 successively, Dept Hd, Dir of Engrg & V P for Engrg at Olympic Radio & TV Div of Lear Siegler Corp. In 1961 founded Parzen Res Inc & was its pres until 1968. 1968- , cons engr spec in communications, instrumentation, & measurement. Author of "Design of Crystal and other Harmonic Oscillators–, Wiley, 1983. *Society Aff:* IEEE.

Pask, Joseph A
Business: Dept/Matls Sic & Mineral Engg, Berkeley, CA 94720
Position: Prof Emeritus of Ceramic Engrg. *Employer:* Univ of Calif at Berkeley. *Education:* PhD/Ceramic Engg/Univ of IL, Urbana; MS/Ceramic Engg/Univ of Washington, Seattle; BS/Ceramic Engg/Univ of IL, Urbana. *Born:* 2/14/13. Native of Chicago, IL. Taught at Univ of IL & Univ of WA. Res engr in Lamp Div, Westinghouse Elec Corp before coming to the Univ of CA in 1948 to start a program of education and research in ceramic engineering and science; Chmn of Dept of Matls Sc & Engg 1957-61, Assoc Dean for Grad Student Affairs in College of Engg 1969-1980; faculty senior scientist in Matls & Molecular Res Div Lawrence Berkley Lab; John Jeppson Award 1967, Distinguished Life Mbrship 1976, Ross Coffin Purdy Award 1979, Outstanding Educator in Ceramic Engg for 1981 in the Amer Ceramic Soc. Elected to natl Acad of Engg 1975. Berkeley Citation 1980. Alumni Honor Award for Distinguished Service in Engg, Univ of IL 1982, Gold Medal for Res and Dev, French Soc for Promotion of R&D 1979, Intl Inst for Science of Sintering, Yugoslavia 1981, Materials Advisory Board, Div of Engg and Industrial Research of Nat Acad of Sciences-NRC 1964-68. Fellow in the American Ceramic Society, Mineralogical Soc. of America, Am. Assoc. for the Advancement of Science. *Society Aff:* ACS, ASEE, NICE.

Paskusz, Gerhard F
Business: EE Dept, Bldg D, 4800 Calhoun, Houston, TX 77004
Position: Dir Min Engr Prog. *Employer:* Univ of Houston. *Education:* PhD/Engg/UCLA; BS/Engg/UCLA. *Born:* 1/21/22. in Austria. Served in US Army Signal Corps 1943-46. Assoc Dean of Engg 1968-76, Dir of Minority Prog 1974-present. Author of 5 technical books, 42 papers. Served as Dir of SCORE Inc 1972-6. Mbr, Exec Comm of IEEE Educ Soc 1976- 9. Mbr, Bd of Dirs, TSPE - PE in Educ, 1975-present. Editor, Technical Careers Newsletter, 1978-present. Active in educational methods res & minority engg education. *Society Aff:* IEEE, ASEE, NSPE, ACM, TSPE.

Pasqua, Pietro F
Business: , Knoxville, TN 37916
Position: Hd, Nuclear Engr Dept. *Employer:* Univ of TN. *Education:* PhD/ME/Northwestern Univ; MS/ME/Northwestern Univ; BS/ME/Univ of CO. *Born:* 5/30/22. in Englewood, CO. Taught at the Univ of CO & Northwestern Univ prior to employment with the Univ of TN in 1952. Organized the nuclear engg dept in 1957. Consultant to Oak Ridge Natl Lab, Sci Application Inc. Brooks Award, Outstanding Teacher College of Engg 1977; Univ Macebearer 1979. *Society Aff:* ANS, ASEE, ΣΞ.

Pass, Isaac
Business: PO Box 101, Florham Park, NJ 07932
Position: Engrg Advisor *Employer:* Exxon Research and Engrg Co *Education:* MS/ChE/Newark Coll of Engrg; B/ChE/CCNY *Born:* 12/07/32 Native of Brooklyn, NY Aeronautical Res Scientist with NACA 1954-56. with Exxon Res and Engrg Co since 1956. Approximately 10 yrs in Process Development and Process Engrg/Design for Petroleum Refining and Petrochemicals. Since 1968, have specialized in contracts engrg and proj mgmt. Promoted to Engrg Advisor in 1981. Am responsible for developing and maintaining contracting procedures and documents applicable to a broad spectrum of Exxon Corp's capital projs in various locations worldwide. Mbr of State of NJ Science Advisory Committee and Exec Bd of AIChE Engrg and Construction Contracting Committee. *Society Aff:* AIChE

Passero, Gary W
Business: 145 Lake Ave, Rochester, NY 14608
Position: Sr Partner. *Employer:* Passero-Assocs and Paramount Engrg Group *Education:* BS/Civil/IA State Univ. *Born:* Mar 1942. Grad courses Cornell & Buffalo. Reg P E NY, FL, IL & OH. Native of Rochester, N Y. Early career as engr Sear-Brown, Cons Engrs, handling design for residential & apartment projs; proj engr W C Larsen P E, Cons Engrs, handling design of municipal engrg projs. Since 1972, Sr Partner of Passero Assocs, Cons Engrs, Architects, Surveyors. Active Mbr of Rochester Sect ASCE & P Pres 1974; Cons Engrs Council N Y; Genesee Valley Land

Passero, Gary W (Continued)
Surveyors Assn; NYS Soc of Prof Engrs; Water Pollution Control Fed; Amer Concrete Inst. Holy Spirit Mens Club; Kiwanis; Edgerton Area Neighborhood Assn. Married, 3 children. Rochester Chapt CEC Pres 1983. *Society Aff:* CEC, ASCE, WPCF, RES, AWWA, GVLSA.

Pate-Cornell, M Elisabeth
Business: Dept of Industrial Engineering, Stanford, CA 94305
Position: Associate Prof. *Employer:* Stanford Univ. *Education:* PhD/Eng-Econ. Syst/Stanford U.; MS/Oper. Research/Stanford U.; Eng. Degree/Applied Math/Comp. Sci./Glenoble U. (France); BS/Mathematics/Marseilles (France) *Born:* 08/17/48 Born in Dakar (Senegal). Studied in France (Mathematics/Applied Math, Computer Science) and at Stanford Univ. in Operations Research and Engrg Economic Systems. Doctoral work on seismic risk analysis and mitigation. Taught at MIT 1978-81 in the Dept of Civil Engrg. Author of many scholarly articles on probabilistic risk analysis, and of a theory of warning systems. Consultant to the USEPA and the Electric Power Research Inst. *Society Aff:* IIE, SRA, ORSA/TIMS

Patel, C K N
Business: AT&T Bell Laboratories (1C-224), 600 Mountain Avenue, Murray Hill, N.J 07974
Position: Exec Dir, Res, Physics Div *Employer:* Bell Labs. *Education:* BE/Telecommunications/Poona Univ (India); MS/EE/Stanford; PhD/EE/Stanford *Born:* 7/2/38. At Bell Labs since 1961. Mbr Tech Staff 1961- ; Hd Infrared & Electronics Res Labs 1967-70; Dir Electronics Res Lab 1970-76; Dir Phys Res Lab 1976-1981; Exec Dir, Res, Physics Div 1981- ; Elected to the Bd of Trustees, Aerospace Corp. Los Angelos, 1979; res on hi power lasers, tunable lasers, pollution detection, spectroscopy & stratospheric measurements. Adolph Lamb Medal Optical Soc of Amer 1966; Ballantine Medal Franklin Inst 1968; Coblentz Medal ACS 1974; Honor Award Assn of Indians in Amer 1974; Lamme Medal IEEE 1976; Zworykin Award Natl Acad of Engrg 1976. TX Instrument founders Award, 1978. Townes Medal, Opt Soc Am 1982; Schawlow Medal, Laser Inst & Amer, 1984. Mbr Natl Acad of Sci; Mbr Natl Acad of Engrg Natl Fellow APS, IEEE, Amer Acad of Arts & Sci, OSA. *Society Aff:* APS, IEEE, OSA.

Patel, Purushottam M
Home: 12 Christopher Circle, New Hartford, NY 13413
Position: Sr Engr *Employer:* Stetson-Harza *Education:* BS/EE/Univ of NB *Born:* 04/05/34 Native of India. US citizen. Worked for a utility for nine yrs in India. Worked over seven yrs for a consltg engrg firm of Rist-Frost Assocs in upstate NY. Presently at the A/E firm of Stetson-Harza in Utica, NY over ten yrs. Field of experience: power, lighting & special systems for industrial, commerical, institutional and government projs. Past Pres of Mohawk-Hudson section of IES. Co-inventor of an Electrical Demand Control System. *Society Aff:* NSPE, IES

Patel, Virendra C
Business: Inst of Hyd Res, Iowa City, IA 52242
Position: Prof. *Employer:* Univ of IA. *Education:* PhD/Aeronautics/Cambridge Univ; BSc/Aeronautics/Imperial College, London Univ. *Born:* 11/9/38. in Kenya, Indian parents. Educ in England. Visiting Prof, Indian Inst of Tech, Kharagpur, 1966-67. Sr Asst in res, Univ of Cambridge, 1967-69. Conslt, Lockheed-GA, 1969-70. Recipient of Sr U.S. Scientist Awd, Alexander von Humboldt Foundation, Fed Rep of Germany; Visiting Professor, Uni Karlsruhe, F.R. G. (1980-81); Visiting Professor, Univ of Nantes, France, 1984. On Univ of IA faculty since 1971. Presently: Prof, Mech Engg; Mbr: (IA) Governor's Sci Advisory Council 1978-1983; Mbr: Resistance Comm of Intl Towing Tank Conf; Assoc Editor, AIAA Journal, 1987- ; Mbr: ASME, AIAA, ASEE, SNAME; Res Engr: Inst of Hydraulic Res. *Society Aff:* AIAA, ASME, SNAME, ASEE

Paterson, Walter H
Home: 163 Lyndhurst Ave, Toronto, Ontario, Canada M5R 3A1
Position: Consultant. *Employer:* Self Employed. *Education:* BSc/-/Queens Univ, Kingston, D.Sc (Hon.) Queens Univ. Kingston *Born:* 1909 Ontario Canada. Employed by Municipal & Hwy Engrg 1934-39; oil field const Colombia S A 1939-42. Employed by Toronto Transit Comm 1942 as Asst Engr to initiate planning of Yonge St Subway; 1949-59 Ch Engr of the Comm; 1959- 61 Ch Engr - Subway Const TTC; 1961-73 Genl Mgr - Subway Const TTC. Retired 1973. Now cons to the Urban Transp Dev Corp of Ontario, & others through private practice. Mbr FASCE, FEIC. P E Ontario. 1972 Gold Medal of Prof Engrs of Ontario, the Soc's highest tribute. 1970 Sons of Martha Medal APEO. 1970 Citation by Gov Mandel of the State of Md. *Society Aff:* EIC, ASCE, ACEC.

Patriarca, Peter
Home: 9604 Briarwood Dr, Knoxville, TN 37919
Position: Mgr, Breeder Reactor Prog. *Employer:* Union Carbide Corp-Nuclear Div. *Education:* MS/Met Engg/RPI; BS/Met Engg/RPI. *Born:* 1/7/21. Native, Utica, NY. Served with US Marine Corps, WWII. Joined UCC, Oak Ridge, TN, July 1950. Career scope: mtls res & dev for Aircraft Nuclear Propulsion, Experimental Gas-Cooled Reactor, High-Flux Isotope Reactor, Molten- Salt Reactor Experiment projs including welding & brazing, mech properties, corrosion & tribology, ceramics, nondestructive testing. Approx 50 open- literature publications, 6 patents. Assumed current responsibilities as mgr of ORNL Breeder Reactor Prog in March 1977; includes fast reactor safety, phys, high-temperature structural design, mtls, fuels, & measurements & controls. Fellow, ASM; Honorary Lifetime Mbr, AWS; Comfort A Adams Lectr, AWS Award. *Society Aff:* ASM, AWS, ANS.

Patrick, Robert J
Home: 1539 View Woods Dr, St Louis, MO 63122
Position: VP - Engg. *Employer:* Apex Oil Co. *Education:* MBA/Bus/WA Univ; BSE/Naval Arch/Univ of MI; BS/Marine Engg/US Merchant Marine Acad. *Born:* 5/1/27. MBA WA Univ. 1972; BSE Naval Arch & Marine Engg. Univ of MI 1951; BS US Merchant Marine Acad 1949. PE IL, LA, MD, MO, NC, PA, TX, WVA. VP Engg Apex Oil Co since 1976; Pres Apex marine Service 1974-76. For 10 yrs prior, was VP Engg with St Louis Ship. Earlier exper with Beth Steel Shipbldg Div, Quincy MA & MD Shipbldg in Baltimore. Chmn Great Lakes & Great Rivers Sect SNAME: Mbr ABS Western Rivers Tech Ctte; Outstanding Prof Achieve Award 1969 from US Merchant Marine Acad & Cert of Appreciation in recog of notable services from US Coast Guard 1976. *Society Aff:* SNAME.

Patsfall, Ralph E
Business: Interstate 75, Evendale, OH 45215
Position: Mgr, Mfg Tech Operation. *Employer:* G E CO - AEBG *Education:* JD/Law/Univ of Marquetto; BS/Met Engg/Univ of WI; MS/Met Engg/Univ of WI. *Born:* Sept 1922. Native of Milwaukee Wisc. USN - Engrg Officer 1944-46. Prod engr for Boeringer Engrg Products, design & manufacture of const equip replacement items. With G E Aircraft Engine Group 1952- . Managed sev engrg & mfg labs/dev shops & assumed pos as Mgr of Mfg Tech 1975. Respon for advanced value processes engrg, mfg engrg, application dev & transition to production of new matls & processes & dir major tech progs such as CAM & powder metal. Natl Pres SAMPE 1964. Chmn Div M&P Bd 3 yrs. Mbr Adv Bd of Policy - Cincinnati Council on World Affairs. Chairman Executive Committee, NF6 Committee AIA. New position, Chief Manufacturing Engr for GE Aircraft Engine Group and Mgr of the Prod Tech Dept, undertaken June, 1982. Respon for design reviews for all major new installations including two major plants and eight satellite plants - 29,000 employees. Also provides consultation in all advanced manufacturing engrg disciplines including CAP/CAM. *Society Aff:* SME, SAMPE, SAE.

Patten, Charles A
Business: 4800 Grand Ave, Neville Island, Pittsburgh, PA 15225
Position: Group VP. *Employer:* Dravo Corp. *Education:* BSME/ME/Lehigh Univ. *Born:* 5/12/20. Native of Allentown, PA. Built warships at Dravo Corp's Neville Island, PA shipyard during WWII. In charge of plant construction & new process

Patten, Charles A (Continued)
dev for Joy Mfg Co until 1955, then mfg mgr & works mgr of Joy's Mining Machinery Plants untl 1963. VP-Mfg for White Motor & Colt Ind until 1969. Presently group VP-mfg for Dravo Corp. Enjoy music, sports & railroading. *Society Aff:* ASME, ABS, ADPA, WRC, AMA.

Patterson, David H
PO Box 660268, Dallas, TX 75266-0268
Position: Asst Mgr Res & Com Svcs *Employer:* Texas Power & Light Co. *Education:* BS/Bus/SFASU. *Born:* 9/9/36. in Hillsboro, TX. BS Stephen F Austin. USAF Troop Carrier Wing; assignments-coordinator for Civil Engr Sect. Employed by TX Power & Light Co 1959. Worked for time in lighting 1963-1979 & has been in present pos since 1980. Active in tech ctte work in IES. Served on IES Indus Lighting Ctte. Areas & Energy Budgeting Procedure Tech Cttes. Served on ASHRAE Std 90-75 Panel. P Chmn Lighting Ctte. Respon for 5 tech reports n lighting. VP Southwestern Region IES 1971-73. VP Region Activities 1974-76. Sr VP (Pres-elect) IES 1976-77. Pres IES 1977-78. Married: 4 children. Hobbies: Sports Cars. Mbr of US Natl Ctte of CIE. *Society Aff:* IES, CIE.

Patterson, J M
Business: 1700 Montreal Cir, Tucker, GA 30084
Position: Principal (Self-employed) *Education:* BS/Mech/Auburn Univ *Born:* 07/24/37 Wife, Pat and four children: Meri 12, James 10, Laura 6, and Bill 4. *Society Aff:* GSPE

Patterson, LeRoy B
1617 Dubuque Ct, Clinton, IA 52732
Position: Sen Vice Pres. *Employer:* Hawkeye Chem Co. Retired *Education:* BS/Chem Engrg/KS State Univ. *Born:* Jan 19 1922 Marysville Kansas. s. William David & Hilda Vera (Stauf). m. Jean Nickerson Oct 19, 1944. Daughters Patricia (Mrs George N Schmid Sr) & Peggy (Mrs Bill A Munson). Pos held: Commercial Solvents Corp Indiana, Louisiana; Atlas Chem Indus Delaware, Mo; Hawkeye Chem Co Clinton Iowa - Sen V P. Director; Mbr Congregational Church - Deacon, Mason, Knight Templar; Member Scottish Rite, Shrine, Amer Inst of Chem Engrs - Sect Dir, United Fund Trustee, Boy Scout Council Vice President, Salvation Army Adv Bd Pres, Clinton Pollution Control Comm Chmn. US Army to 1st Lt; served in Europe; combat infantry badge. *Society Aff:* AIChE.

Patterson, Sam H
Home: 2515 Fowlers Lane, Reston, VA 22091
Position: Consulting Geologist *Employer:* Retired *Education:* PhD/Geol/Univ of IL; MS/Geol/Univ of IA; BA/Geol/Coe Coll *Born:* 08/14/18 Born in Marion, IA. Served in US Army, 1941-1946, Southwestern and Western Pacific theaters. Geologist-US Geological Survey 1947 to present. Worked on bentonite, refractory clay, bauxite, fuller's earth, and kaolin. In current responsibility serves as the Geological Survey and Interior Dept's authority on the geology of bauxite and clays. Recipient of Interior Dept Meritorious Service Award, 1976, AIME Hal William Hardinge Award, 1981. *Society Aff:* AIME, SEG, CMS

Patterson, Walter B, Jr
Business: 1801 Market St, Philadelphia, PA 19103
Position: Mgr, Business Dev. *Employer:* Sun Petroleum Products Co. *Education:* MS/Mech Engg/Univ of PA: BChE/Chem Engg/Rensselaer Polytechnic Inst. *Born:* July 1924. Native of Phila. US Navy 1943-46. With Sun Co Inc 1946- : process dev; design engr spec in pilot plat design & instrumentation; Tech Asst to Dir of R&D, respon for program planning & budgeting & econ analyses of R&D projs; Mgr Contract Serv & Purchasing Res; Mgr Mgmt Sys & Sys Planning; present pos 1973- , respon for identifying & developing new business opportunities in support of Div's diversification & growth objectives. Fellow AIChE; Chmn Natl Prog Ctte 1974; Chmn Delaware Valley Sect 1971. *Society Aff:* AIChE.

Patti, Francis J
Home: 3737 Maxwell Dr, Wantagh, NY 11793
Position: Chief Nuclear Engr. *Employer:* Burns & Roe, Inc. *Education:* Prof/Nucl Engr/Columbia Univ; MS/Nucl Engr/Columbia Univ; MS/Civ Engr/MIT; BS/Civ Engr/Drexel Univ. *Born:* 12/23/28. Early technical positions included TVA (as a co-operative student), Gen Ind & M W Kellogg. Served as instr at US Army Engr Sch. With Burns & Roe, Inc since 1955. Initially worked as structural designer. Transferred to Nuclear Engg Dept in 1957. Named Chief Nuclear Engr in 1973. Also directed Piping Engg, Mtls Engg & Chem Engg specialty groups from 1979 to 1983. Current duties include direction of both nuclear and environmental engineering activities. Main areas of technical expertise are in the design of nuclear power plants & nuclear res facilities. *Society Aff:* ANS, ASCE, NSPE.

Patton, Alton D
Business: Electrical Engineering Dept, Texas A&M University, College Station, TX 77843
Position: Professor & Director of Electric Power Institute *Employer:* Texas A&M Univ *Education:* PhD/EE/TX A&M Univ; MS/EE/Univ of Pittsburgh; BS/EE/Univ of TX-Austin *Born:* 02/01/35 Born in Corpus Christi, TX. Elec Utility Systems Engr with Westinghouse Elec Corp for 8 yrs concerned with development and application of methods for power sys analysis. Prof of EE at TX A&M Univ for 18 yrs concerned with development and teaching of courses on elec power sys and research on methods for power sys analysis. Pres of consltg firm of Associated Power Analysts, Inc since 1973. Contributor to quantitative methods for power system reliability analysis and evaluation and reliability standards. Name Fellow of IEEE in 1980 for reliability work. Chrmn of technical committees in IEEE. *Society Aff:* IEEE, NSPE

Patton, James L
Business: 1500 Meadow Lake Parkway, Kansas City, MO 64114
Position: Partner *Employer:* Black & Veatch *Education:* MS/Environ Health Eng/Univ of KS; BS/CE/Univ of KS; B of A/Pre-Eng/Kansas City, KS Jr Coll *Born:* 8/23/41 Environ Engr with Black & Veatch since 1964. Advanced from design engr to partnership. Work assignments consist of design of all phases of water and wastewater projects and project mgmt of numerous water and wastewater projects throughout the southwest and midwest. Reg PE in 9 States. Received the following project awards: Best Work of Architecture 1974-78; Certificate of Engrg Merit 1977; Engrg Excellence Award-Honor Award 1979; Engrg Achievement Award 1979; and Best AWWA Publication Award 1980. Natl V Chmn - Standards Ctte on Carbon Dioxide. Member AWWA Std. Council, Member AWWA/AGC Liaison Committee, Kansas AWWA Section Trustee. Diplomate Amer Acad of Environ Engrs, Fellow of ASCE. *Society Aff:* AWWA, ASCE, KES, NSPE, WPCF, CRWUA, CECMO, AAEE

Patton, John T
Home: 2955 McDowell, Las Cruce, NM 88005
Position: Prof & Hd, Dept of ChE. *Employer:* NM State Univ. *Education:* PhD/ChE/OK State Univ; MS/ChE/OK State Univ; BS/ChE/OK State Univ. *Born:* 5/9/31. A native of Oklahoma City; served in the Army Ordnance Corps 1954-1956. Work experience includes 3 yrs in petrochemicals with TX Eastman & 10 yrs with Exxon in oil production, new recovery methods, refinery construction, corporate planning & res advisor for synthetic fuels. In 1968 became ChE Prof at MI Tech and in 1977 Prof & Hd, Dept of Chem Engg, NM State Univ. A consultant in the areas of synfuels, biosynthesis, enhanced oil recovery & Pres of CBI, Inc, consulting firm specializing in above areas. Also the author of 25 US patents and numerous scientific publications. *Society Aff:* AIChE, SPE OF AIME, ACS, TBΠ, ΣΞ.

Patton, Robert L
Business: 7955 Redfield Rd, Scottsdale, AZ 85260 *Employer:* K-TRON Intnl *Education:* Attended Texas A&M Univ.; Ohio State Univ - BSME; Oklahoma State Univ., 1948 *Born:* May 1924. Native of Okla City, Okla. US Infantry in US & Europe 1943-45. Honeywell Inc: Commercial Div, Sales Engr 1949-52; Commerical Div Dallas, Branch Mgr 1952-58; Commerical Div Denver, Regional Mgr 1958-63; Commercial Div Mpls, Mgr of Mkting 1963-68; Indus Div Phila, Mgr Contract Serv 1968-69; Indus Div Phila, Mgr of Mkting 1969-70; Process Control Div Ft Wash, V P Mkting 1970-77. Formerly mbr of the Speaker's Bureau of ASHRAE. Pi Tau Sigma 1947. Vice Pres Intl Indus. Prod. Grp. 1977-79; Vice Pres Corp Field Mktg 1979-present.

Patton, Willard T
Home: 152 Ramblewood Rd, Moorestown, NJ 08057
Position: Mgr, Antenna, Microwave and Transmitter Systems *Employer:* RCA. *Education:* PhD/EE/Univ of IL; MS/EE/Univ of TN; BS/EE/Univ of TN *Born:* 4/20/30. in Schenectady, NY. Engaged in Antenna Res since 1952. Served with USN 1953-56 as Ship Superintendent DER Conversions Boston Naval Shipyard. With RCA since 1962. Lead that co's res in phased array antennas. Received David Sarnoff Award for Outstanding Technical Achievement 1975, Elected Fellow of IEEE 1979 for Contributions to the Dev of Phased-Array Antenna Tech. Unit Mgr Advanced Tech 1962, Mgr Equip Design 1970, Mgr Advanced Tech 1976. Mgr Antenna, Microwave, and Transmitter Systems 1983. Mbr Sigma Xi, Tau Beta Pi, Eta Kappa Nu, Phi Kappa Phi. *Society Aff:* IEEE

Patwardhan, Prabhakar K
Home: 328 Linking Rd Khar, Bombay, India 400 052
Position: Head. *Employer:* Bhabha Atomic Res Ctr. *Education:* PhD/Physics/Banaras Hindu Univ; MSc/Physics/Banaras Hindu Univ; BSc/Physics, Chemistry & Math/Agra Univ. *Born:* 12/2/27. PhD, Pioneered nuclear electronics at Tata Inst of Fundamental Res. Hd Computer Section & Hd Solid State Electronics Bhabha Atomic Res Ctr, Trombay Bombay, Govt of India. 31 yrs prof standing & specialization in nuclear electronics, data handling sys & instrumentation. Respon for a large team of scientists & engrs for R&D, prototype production & tech transfer. Mbr Internatl Panel, Ch Coord Officer Internatl Conf on Nuclear Electronics under auspices of IAEA at Bombay. Fellow IETE and IE, India. Fellow IEEE, IERE, IEE, Inst of Physics UK. Founder-Chmn Jt Chap NPS/IECI/IEEE India Council, Mbr Exec Comm IEEE India Council. Fellow & Mbr of the Senate Univ of Bombay 1969-75. Over 80 research publications presented at Natl and intl conf and published in reputed Journals. Dir, Maharashtra Electronics Corp (public sector undertaking). Involved in sci & tech pol planning of State Govt of Maharashtra. Interested in traveling chess, bridge & Indian classical music. Biographied in 8 Intl Who's Who & 4 Natl Who's Who. Won three Natl Awards & One Intl Award. *Society Aff:* IE, IETE.

Paul, Burton
Business: Dept of ME & Appl Mech, Philadelphia, PA 19104
Position: Prof of Mechanical Engineering *Employer:* U of Pennsylvania. *Education:* PhD/Applied Mechs/Polytech Inst of Brooklyn; MS/Engg Mechs/Stanford Univ; BSE/Mech Engg/Princeton Univ. *Born:* 1931. Fellow ASME. Mbr Phi Beta Kappa, Tau Beta Pi, Sigma Xi. Prof of ME, Univ of Penna 1969- ; Dept Chmn 1973-78. ASA Whitney Prof of Dynamical Engrg 1982- . Ch Solid Mech Res, Ingersoll-Rand 1963-69. Supr Engrg Mech Bell Tele Labs 1961-63. Asst Prof Brown Univ 1958-60. Res engr Bulova R&D Labs 1954-56. Author num papers on stress analysis, machine design, brittle fracture, etc & book on kinematics & dynamics machinery. Cons to indus. Chmn Transp Comm, Appl Mech Div ASME 1973-76. Assoc Ed J Appl Mech 1971-74, Mbr Ed Bd Computer Methods in Appl Mech & Engrg 1971- . *Society Aff:* ASME, ASEE, ASSE.

Paul, Charles K
Business: S & T/FENR, 509c SA-18, U.S. Agency for Intl. Dev, Washington, D.C 20523
Position: Mgr, Remote Sensing Program. *Employer:* Agency for Internatl Dev. *Education:* PhD/Civil Engr/Cornell Univ; MS/Civil Engr/Cornell Univ; BS/Civil Engr/Univ of NM. *Born:* Aug 11, 1937 Panama Canal Zone. US Coast & Geodetic Survey, oceanographic observations, ch of astronomic survey party 1961-63. US Naval Civil Engrg Lab, deep ocean marine salvage 1964-65. MS 1967, PhD 1970 Civil Engrg Cornell Univ. Asst Prof Cornell Univ, computer simulation of engrg sys 1971-72. Jet Propulsion Lab, Mgr of Land Use Mgmt Info Sys 1973-75. NASA, Mgr of Natural Resource/Land Use Prog 1976, AID, Mgr of Remote Sensing Program 1977-79. *Society Aff:* ASCE, ASPRS, AAAS.

Paul, Donald R
Business: Chemical Engineering Dept, The University of Texas, Austin, TX 78712
Position: Melvin H. Gertz Regents Chair in Chemical Engrg. *Employer:* Univ of TX. *Education:* BS/ChE/NC State College; MS/ChE/Univ of WI; PhD/ChE/Univ of WI. *Born:* 3/20/39. Res Chem Engr for E I duPont Co, summers 1960 & 61, and Chemstrand Res Ctr, Durham, NC 1965-67. Chem Engg educator (22 yrs to date) at Univ of WI (1963-65) and the Univ of TX at Austin (1967-present). Active consultant to ind cos. Published over 200 publications dealing with various aspects of polymer sci and engg, co-edited five books, is on the Editorial Bds of three maj polymer journals, and has held numerous offices in profl socs. Recipient of numerous soc & educational awards. Editor of Industrial & Engineering Chemistry Research published by American Chemical Society. *Society Aff:* ACS, AIChE, SPE, NAMS.

Paul, Francis A
Business: 381 Middlebury Rd, Middlebury, CT 06762
Position: Owner. *Employer:* Paul Assocs —. *Education:* MCE/Structural/Rensselaer; BSCE/Civil/Univ of CT. *Born:* July 1940. MCE RPI 1965. BSCE Univ of Conn 1962. Hon discharge 1st Lt US Army Security Agency. Previous employment with Esso Res & Engrg Co & with Storch Engrs. Self-employed as engrg-surveying cons (Paul Assocs, Engrs-Surveyors- Planners) 1970- . Tau Beta Pi & chi Epsilon. Officer Ct Assn of Land Surveyors. ASCE Daniel Mead Prize 1968. *Society Aff:* ASCE, ACSM.

Paul, Frank W
Business: Mechanical Engg Dept, Clemson, SC 29634-0921
Position: McQueen Quattlebaum Prof of Mech Engrg. *Employer:* Clemson Univ. *Education:* PhD/Mech/Lehigh Univ; MS/Mech/PA State Univ; BS/Mech/PA State Univ. *Born:* 8/28/38. Native of Jersey Shore, PA. Industrial positions have been held with Hamilton Standard, United Technologies and Bell Telephone Lab. Academic teaching and research positions have been held with Lehigh and Carnegie-Mellon Univs. Currently with Clemson Univ since 1977. Dir, Ctr for Advanced Manufacturing, College of Engrg, Clemson Univ. Mbr of Tau Beta Pi, Pi Tau Sigma and Sigma Xi. Research interests in control systems, automation, robotics and computer aided mfg. Member of ASME, RI OF SME, ASEE. PE, Paul Consultants *Society Aff:* ASME, ASEE, RI of SME.

Paul, James C
Business: 352 N Main St, Plymouth, MI 48170
Position: VP *Employer:* Airflow Sciences Corp *Education:* MSE/Aero Engr/Univ of MI; BSE/Aero Engr/Univ of MI *Born:* 6/8/48 Native of Manistee, MI. Teaching Fellow, Univ of MI 1971-1973. Dir of Systems Development, Intl Husky, Inc. 1973-1975. VP and Co-Founder of Airflow Sciences Corp, 1974-present. VP attached Flow Corp, 1980-present. Specialist in numerical flow simulation tech for seventeen mbr specialized consulting firm. Hobbies include Photography, Travel, Woodworking. *Society Aff:* NSPE

Paul, James R
Home: Kniebrecheweg 5, CH8810 Horgen, Switzerland
Position: Technical Mgr *Employer:* Dow Chemical Europe *Education:* BS/Petrlm Eng/Univ of MO-Rolla *Born:* 09/18/21 Born in Louisiana, US Navy, LT WWII, Field Engr, Regional Engr, Technical Sales, Regional Mgr for Dowell 1946-1963 specializing in Engrg Stimulation and Cementing Services to the Oil and Gas indus in most major producing areas of US; Technical Specialist Mgr, Enhanced Oil Recovery and Petroleum Drilling and Production Chemicals & Services for Dow

Paul, James R (Continued)
Chem Co in US, North Sea Europe, Africa, Mid East, USSR, 1963 to date; Life mbr Dir, Section Chrmn other offices in SPE- AIME, married, four daughters. *Society Aff:* SPE-AIME

Paulette, Robert G
Business: Stanley Building, Muscatine, IA 52761
Position: Project Mgr *Employer:* Stanley Consultants, Inc *Education:* BS/CE/Univ of KS *Born:* 08/30/20 Born in Halstead, KS. Entry position Sr Draftsman, Paulette and Wilson, Consltg Engrs. Engrg Officer to rank of Captain, Army Air Corps 1942-46. Project and Field Engr, INFILCO, Manufacturer Water/Waste Treatment Equipment 1945-55. Project Engrg, Chief Engr, Mgr, and Asst VP, Municipal Service Co: Mgr, Industrial Water and Waste Div, J. F. Pritchard and Co 1955-66. Project Mgr and VP 1972-77, Stanley Consits 1966 to date. Reg in nine states including OH by eminence provisions. Mbr of Bd, Johnson County Water District, KS, (35,000 customers) 1962-64. Chrmn, Muscatine Airport Commission 1969-71. Author of numerous technical papers. Hobbies are golf and woodwork. *Society Aff:* ASCE, NSPE, SAME, AAEE, CEC/Iowa

Paullin, Robert L
Home: 2917 Ellenwood Dr, Fairfax, VA 22031
Position: Ch, R&D Resources Mgmt Div. *Employer:* US Dept of Transp. *Education:* DPA/Trans Mgt/Univ of So CA. *Born:* Oct 1928. MS Univ of Calif Berkeley; BSME SDSM&T. From Mitchell S D. USAF pilot. Flight Safety Engr with Douglas Aircraft Co. CAB investigator. FAA engr, DOT Office of the Secy R&D Administrator. Dir, Office of Operations & Enforcement. P E Dist of Columbia. Assoc Prof Transp Mgmt Geo Wash Univ. DOT Secretarial Award 1974. Enjy music, sports. Mbr Sigma Tau & Triangle. *Society Aff:* SAE.

Paulling, J Randolph, Jr
Home: 119 Crestview Dr, Orinda, CA 94563
Position: Prof. *Employer:* Univ of CA. *Education:* DrEngr/Naval Arch/Univ of CA; Nav Arch/Naval Arch/MIT; MS/Nav Arch/MIT; BS/Nav Arch & Mar Engr/MIT. *Born:* 1/8/30. Mbr of faculty of Naval Arch Dept, Univ of CA, 1954 to present, (Dept Chrmn, 1967-74). Teaching & conducting res in ship structures, ship motions in waves, ship safety & steering, offshore platform design & computer applications in naval architecture. Maj consulting work has been concerned with design of ships (container ships, fishing vessels) & offshore structures, especially semisubmersible platforms & tension leg platforms, Ocean Thermial Energy PlantsOne yr, 1962-63, spent with Det Norske Veritas, Oslo Norway, developing finite-element structural analysis software. Chosen by ASCE & Japan Soc for promotion of Sci as "Eminent Ocean Engineer" for tour of Japanese Ocean Engg activities, 1978. *Society Aff:* SNAME, XIΣΞ, RINA, STG, JSNA.

Paulson, Bernard A
Business: P O Box 2256, Wichita, KS 67201
Position: President *Employer:* Koch Refining Co *Education:* BS/Chem Engr/MI State U; -/Chem E *Born:* July 12, 1928 Lakeview Mich. BSChE Mich State Univ 1949. Process engr. 1949-57 Mid-West Refineries Alma Mich. 1957-66 Refinery Mgr Kerr-McGee Corp, Wynnewood & Cleveland Okla. 1966-71 V P Coastal States Petrochem, Corpus Christi Tex. 1971-74 V P Koch Refining Co St Paul Min. 1974-, V P Koch Indus Inc Wichita Kansas. 1st Lt USAF 1955-57. Reg P E texas. Mbr AIChE, Genl Ctte Refining API, Dir NPRA. Pres Koch Refining Co 1981. *Society Aff:* AICE, API, NPRA

Paulson, Boyd C, Jr
Business: Dept of Civil Engineering, Stanford, CA 94035
Position: Professor. *Employer:* Stanford Univ *Education:* PhD/CE/Stanford Univ; MS/CE/Stanford Univ; BS/CE/Stanford Univ *Born:* 03/01/46 Prof and Assoc Chrmn of Civil Engrg in Stanford Univ Grad Prog in Construction Engrg and Mgmt since 1974. On Civil Engrg faculty at Univ of IL, 1972-1973. Visiting Prof, Univ of Tokyo, 1978, Tech Univ of Munich, 1983. Indus experience since 1962 on several heavy and industrial construction projs in the US, Australia, and Japan. Author of one book and over 60 papers. Research and teaching interests primarily in automated const field data acquisition, process control and in comp applications in const. Elected Secy, Proj Mgmt Inst, 1974- 1978. Mbr, ASCE Const Exec Ctte, Past Chrmn, ASCE Committee on Profl Construction Mgmt, Chrmn, ASCE Task Ctte on sm computer, Vice Chrmn, ASCE Const Res Council. Winner of ASCE William L. Huber Research Prize, 1980, W German Alexander von Humboldt Fdn Res Fellowship, 1983, and ASCE Const Mgmt Award, 1984, and selected for Dist Scholar Exchange Prog with the PR of China by US Natl Acad of Sci China Ctte, 1984. Mbr of Tau Beta Pi and Sigma Xi. *Society Aff:* ASCE, ASEE, IEEE-CS, ACM.

Paulson, Donald L
Business: 815 Forward Dr, Madison, WI 53711
Position: Secretary *Employer:* Arnold & O'Sheridan, Inc *Education:* BS/CE/Univ of WI-Madison. *Born:* 12/21/35 Civil Engr-Land Surveyor-Warzyn Engrg, 1958-1964. Civil Engr-Land Surveyor- Arnold & O'Sheridan, Inc, 1964-present. Corporate Secy and Dir since 1967. Designer-Tenney Park Pedestrian Bridge-1972 Natl Award for Outstanding Steel Bridge. Guest Lecturer-Univ of WI Madison 1978 to present. Mbr City of Madison Bldg Code Study Committee 1972 to present, Chrmn 1976 to present. Charter Mbr- Madison Area Surveyors Council-Pres 1970 and 1971. Mbr-WI Soc of Land Surveyors- 1974 to present-State Pres 1979, Director 1974 to 1981. Married 1957, three children, Richard, Sandra anc Cynthia. Mbr WI Registration Bd of Architects, Engrs, Designers & Land Surveyors 1983 to present, Chrmn, Land Surveyors Section, 1986 to Present. *Society Aff:* ASCE, ACSM, NCEE.

Paulson, James M
Home: P.O. Box 292, Greenbush, MI 48738
Position: Prof of Civil Engrg (Emeritus) *Employer:* Wayne State Univ. Retired *Education:* PhD/Civil Engg/Univ of MI; MSCE/Civil Engg/IL Inst of Tech; BSCE/Civil Engg/Citadel. *Born:* Jan 1, 1923 Wausau Wisc. Mbr ASCE, ASEE, Mich Soc of Prof Engrs, Amer Concrete Inst, Tau Beta Pi, Chi Epsilon, Sigma Xi. ASCE Task Ctte on Folded Plat Structs 1959-65. Detroit Chap MSPE Bd of Dir 1962-65, 1976-79. Employment: Wayne State Univ: Instr to Prof of CE 1949-67, Chmn of CE Dept 1967-72, Assoc Dean Coll of Engrg 1973-1983. Prof of CE 1983-1985. Retired 1985 Prof. Emeritus 1985- . *Society Aff:* ASCE, NSPE, ASEE, ACI.

Paulus, J Donald
Home: 216 Fallsbrook Rd, Timonium, MD 21093
Position: Partner. *Employer:* Whitman, Requardt & Assocs. *Education:* BE/Mech/Johns Hopkins Univ. *Born:* 10/30/25. Married, 1 daughter. Apdication engr Koppers Co 1949-52. Whitman, Requardt & Assocs Baltimore 1952- . Assumed current respon as partner 1979. Respon for daily mgmt of indus, commercial projs containing arch, civil, struct, elec & mech incl design of HVAC sys, matl handing & process piping sys. Reg P E MD, DE, VA, PA, NJ, MI, DC. OH., Ill., CO. NCEE Cert of Qualification. Fellow Mbr ASME, Mbr ASHRAE, NSPE & AIChE. VP ASME 1972-74, Treas 1975-76. Hobby: dog obedience training & exhibiting; traveling; duckpin bowling. *Society Aff:* ASME, ASHRAE, NSPE, AIChE.

Pauly, Bruce H
Home: 143 Kenton Rd, Chagrin Falls, OH 44022
Position: V P - Engrg & Res.(Ret.) *Employer:* Eaton Corp. *Education:* BS/Mech Engg/VPI; MS/Engg Admin/Case Inst Tech. *Born:* Nov 11, 1920 Washington D C. Engrg Officer Air Corps 1941-45, Lt Col 1944-45 Dir/Maint 2nd Air Div; Tech Branch Ch Wright Field 1945. Westinghouse Elec AGT Div 1945-52, Sect Engr 1951-52. Borg-Warner Corp 1952-55 (Aircraft Sales Mgr, Pesco Prod). Weatherhead Co 1955-69, V P Res & Engrg 1965-69. Eaton Corp 1969-82, V P Engrg & Res 1974-82. Pi Tau Sigma, Pi Delta Epsilon, Legion of Merit, Bronze Star, Croix de Guerre. Mbr AIAA, SAE, Dir 1977 NCFP, CTSC (Secy 1972-73), Cleve Engrg Soc. (Board of Governors 1979-81); SAE: Chairman "Motor Vehicle Council" 1981-82; AIAA: Co-chairman: National Aerospace Propulsion Conference-1982; "Grace

Pauly, Bruce H (Continued)
Commission" (PPSS) 1982; Executive Committee: VPI College of Engineering 1980-present. PE PA. *Society Aff:* AIAA, SAE

Pavelchek, Walter R
Home: 1050 S New St, West Chester, PA 19382 *Education:* MSChE/ChE/Univ of IL; BS/ChE/Purdue Univ. *Born:* 1/8/25. Technical Service (fluid catalytic cracking) Sinclair Refining Co 1947-50. Taught Chem Engg: Tufts Univ 1951-57. R&D cellophane, plastic films, & strapping products for Am Viscose Corp - then FMC Corp 1957-74. Continuing with FMC Corp as Proj Mgr for Ind Packaging (Phila 1974-75), Engg & Technical Superintendent at Downingtown (1975-79) (films & strappings) & Sr Proj Engr in Corporate Chem Engg (Phila 1979-86). Co-author of "Cellophone" section in Sci & Tech of Packaging Films (published 1971). Served AIChE in Phila-Wilmington Local Sec offices, including VChrmn 1968-69; Also natlly as Student Chapters Committee Chrmn (two yrs) & Dir (1973-75). *Society Aff:* AIChE, TBΠ, ΩXE.

Pavelic, Vjekoslav
Business: College of Engrg and Applied Science, Milwaukee, WI 53201
Position: Prof *Employer:* Univ of WI-Milwaukee *Education:* PhD/ME/Univ of WI-Madison; MS/Univ of WI-Madison; BS/State Univ Zagreb, Yugoslavia *Born:* 06/20/29 Nine yrs of various industrial experience, primarily in Research and Development and since 1968 a faculty mbr at Univ of WI. Consultant to various Corps, very active in ASME, had many articles published in prof periodicals, and reg PE in WI. Recipient of ASEE Western Elec Award for Excellence in Teaching, Past Pres of Milwaukee Council of Engrg and Scientific Societies and most recently consltg editor for Engrg Design for Encyclopedia of Sci and Tech. *Society Aff:* ASME, AWS, ASEE, AAUP, ΣΞ

Pavia, Edgar H
Home: 7443 Onyx St, New Orleans, LA 70129
Position: President & Chf. Engr *Employer:* Pavia-Byrne Engrg Corp. *Education:* BS/Mech Engg/LA Polytechnic Inst. *Born:* Apr 9, 1925. Native of Elizabeth NJ. US Marine Corps 1943-47, 1951-52. Founded Pavia-Byrne Engrg Corp with late William H Byrne, P E 1963. Pres & Ch Engr to date. Firm specializes in environmental pollution control. Chmn Water Qual Ctte ASME 1971-73. V P ASME 1973-75. Chmn Process Indus Div 1975-77. Policy Bd, Industry Dept 1977-80, Environ Affairs Ctte, ASME 1979-, Chmn 80-84; Centennial Award 1980. Dir Water Pollution Control Fed 1976-79. Arthur Sidney Bedell Award, 1978; Service Award, 1980; ANSI. Mbr U S TAG to ISO-TC 147 - Water Qual 1973-82. Chrmn, US, Tag-TC-147, 1978-82. Cons to UN on Water Pollution Control Problems since 1975. Author sev tech papers & holder of U S & foreign pats on water pollution control. Reg PE La, Ala, Miss, NJ, TX. *Society Aff:* ASME, NSPE, WPCF, AAEE.

Pavia, Richard A
Home: 7145 N Ionia Ave, Chicago, IL 60646
Position: Pres (CEO) *Employer:* Speer Financial, Inc *Education:* MS/Sanitary Engr/IIT; BS/Civ Engr/IIT; Adv Mgt/Bus. Adm./Univ of Chicago. *Born:* 7/18/30. Native of Chicago, IL. Served in US Army 1952-1954, held Reserve Commission with US Public Heath Service 1957-1977. Employed in private industry 1954-1958, engg consulting & construction engg. Served with local & state govt agencies from 1958-1979 including Chicago Dept of Planning, IL Bldg Authority, Metropolitan Sanitary Dist of Chicago. Employed by Dept Water & Sewers - Chicago 1966-1979, most recently as commissioner of water & sewers 6/73-9/79. Pres, IL Sec ASCE - 1978. Chrmn, Water Utility Council AWWA. Diplomate - AAEE. Enjoy music, fishing, boating & photography. *Society Aff:* ASCE, AWWA, WPCF, APWA, GFOA, AAEE.

Pavlidis, Theodosios
Business: Bell Laboratories, Murray Hill, NJ 07974
Position: MTS *Employer:* Bell Labs *Education:* PhD/EE/Univ of CA; MS/EE/Univ of CA; Dipl/Elec & ME/Natl Tech Univ of Athens. *Born:* 9/8/34. For the last 15 yrs my work has been in the gen area of image processing with maj emphasis on pattern recognition. Among my publications, there are the books: "Structural Pattern Recognition" (Springer, 1977) and "Algorithms for Graphics and Image Processing" (Comp Sci. Press, 1981). Was the gen chrmn of the Fifth Intl conf on Pattern Recognition (Miami, 1980). *Society Aff:* IEEE, ACM.

Pavlis, Frank E
Home: 3544 Congress St, Allentown, PA 18104
Position: Retired *Education:* MS/Chem Engg/Univ of MI. *Born:* Oct 29, 1916 Mich. Attended the Advanced Mgmt Prog at Harvard Univ 1957. Mbr Tau Beta Pi & Phi Lambda Upsilon. Employed by Air Products & Chemicals Inc & predecessor for 40 yrs since 1940. Held pos of Ch Engr, Tech Dir, Tech Sales Dir, Treas, V P Engrg, V P Finance and V P Internatl, currently retired. A dir of Air Products from 1952 to 1980. Emeritus Mbr ASC & AIChE. *Society Aff:* ACS, AIChE.

Pawlowski, Harry M
Business: Water Purification Div, 1000 E Ohio St, Chicago, IL 60611
Position: Engr of Water Purification. *Employer:* City of Chicago. *Education:* MS/San Engg/IL Inst of Tech; BS/Civil Engg/IL Inst of Tech. *Born:* 10/30/29. Hydraulic Engr with the Water Resources Div of the US Geological Survey, 1952. Civil Engr with the Dept of Subways and Superhighways and the Dept of Public Works, City of Chicago, 1953-60. Chief Sanitary Engineer and Engr of Water Purification with the Dept of Water and Sewers, City of Chicago, 1961-87. Presently in charge of the Chicago Water Purification Div which includes the two largest water purification plants in the world, supplying drinking water to over 4 1/2 million residents of the Chicago Metropolitan Area. *Society Aff:* ASCE, AAEE, APWA, AWWA

Paxton, Harold W
Business: MEMS Dept, 4309 Wean Hall, Carnegie Mellon, Pittsburgh, PA 15213
Position: USS Professor *Employer:* Carnegie Mellon *Education:* PhD/-/Univ of Birmingham; MSc/-/Univ of Manchester; BSc/-/Univ of Manchester. *Born:* 2/6/27. Became Asst Prof of Met Engg at Carnegie Inst of Tech, now Carnegie-Mellon Univ, in 1953 & became Hd of the Dept of Met & Mtls Sci & Dir of the Metals Res Lab 1966. Visting Prof in Met & matls Sci at Imperial College, London, in 1962- 63 and MIT in 1970. Served 2 yrs as first Dir, Div of matls Res, NSF 1971-73. Consultant to industry 1953-74. V.P Research US Steel 1974-86. Authored many papers, primarily in field of physical metallurgy, and co-authored a book, *Alloying Elements in Steel*, with late Dr E C Bain. Received a Bradley Stoughton Award 1960. Fellow Amer Soc for Metals, PPres & Fellow of The Metallurgical Soc, former Mbr of Bd of Dir & VPres and 1982 Pres, of the Amer Inst of Mining, Met & Petro Engrs; P Chrmn of the Genl Research Committee of Amer Iron & Steel Ist. 1978 Edward de-Mille Campbell Lecturer. 1982 BCRA Carbonization Science Lecturer, Australasian Institute of Mining & Metallurgy Lecturer. 1983 awarded ASM Gold Metal for Advancement of Research. Member National Academy of Engineering & Fellow AAAS. 1985 Yukawa Lecturer & Honorary Member, Iron & Steel Institute of Japan, 1987 Harold Moore Lecturer, Institute of Metals, London. *Society Aff:* ASM, AIME, NAE, AAAS, SME.

Paxton, Hugh C
Home: 1229-41, Los Alamos, NM 87544
Position: Consultant-retired. *Employer:* Los Alamos Sci Lab - U of Caif. *Education:* PhD/Physics/Univ ov CA, Berkeley; AB/Phyiscs/UCLA. *Born:* Apr 1909 Los Angeles Calif. From 1937-48, worked in nuclear physics at the College of France & Columbia Univ, with gaseous diffusion tech at SAM Lab & Oak Ridge, & on precision casting dev at Sharples Res Lab. Headed Critical Experiments Group at Los Alamos Sci Lab 1948-75; retired 1976. Fellow Amer Nuclear Soc & Amer Physical Soc; Mbr NRC Atomic Safety & Lic Bd Panel 1963-85. Nuclear safety award at ANS meeting 1972. *Society Aff:* ANS, APS.

Payne, Charles N
Home: 9999 Smitherman Dr, Apt 100, Shreveport, LA 71115
Position: Self-Employed *Education:* ScM/Naval Arch & Mar Engg/MIT; BS/-/US Naval Academy. *Born:* 2/25/19. Ensign USN, 1941, advanced through grades to Rear Admiral, 1969. Controller Bureau of Ships Systems Command, Washington 1964-68; Comdr Charleston Naval Shipyard, 1968-71, Supr Shipbldg Pascagouls, MS 1971-74. Retired 1974. Pres, Webb Inst of Naval Arch, Glen Cove, NY 1974-80, Retired 1980. Marine Consultant 1980- .Faculty, Coll William and Mary, 1949-50, U VA 1952-54, US Naval Acad 1954-57. Explorer Chm Coastal Carolina Council Boy Scouts Am 1969-70, Pine Burr Council, 1971-73. Recipient Silver Beaver Award BSA. Decorated Legion of Merit with bronze star. Author: Naval Turbine Propulsion Plants, 1957. Contbr articles to prof journs. *Society Aff:* SNAME, ASNA, ASEE.

Payne, Matthew A
Home: 40 Lloyd Rd, Montclair, NJ 07042
Position: Mgr Facility Planning *Employer:* Revlon Health Care Group *Education:* MS/IE/Case Inst of Tech; BS/ME/Case Inst of Tech *Born:* 09/29/19 Formerly: P Natl Pres IIE, Dir of Mgmt Div, Chrmn Chapter Organization and Extension Ctte. Formerly VP, Genl Mgr Lubrizol of Canada Ltd. P Pres Greater Niagara Chamber of Commerce, Greater Niagara Community Fund Pres and Campaign Chrmn,Mbr Bd of Governors of Ontario Bd of Trade. Author of Fatique Allowance in Industrial Time Study. Present: Fellow and Life Mbr IIE, Professional Engr. *Society Aff:* IIE

Payne, William H
Business: 9856 Sunland Bl, Sunland, CA 91040
Position: Operations Mgr, MLC Mtls *Employer:* E I Du Pont de Nemours & Co, Inc *Education:* MS/Cer Engr/Univ of IL; BS/Cer Engr/Univ of IL. *Born:* 4/25/41. Peoria. Ceramic engr from IL. Sigma Xi & Keramos. Prior to 1972, Mgr of Matls R&D for Interspace Corp. Pres of Natl Inst of Ceramic Engrs, VP of Amer Ceramic Soc & has been active mbr of ISHM, ASTM, SAMPE. Presented sev papers. Chaired the So Calif Sect ACS 1969, Pacific Coast Regional Mtg - ACS 1968, & Res Ctte - Tile Council of Amer 1971. Pres Alpine Village Homeowners Assn 1974. Enjoys sailing with wife & daughter. Operations Mgr, MLC Mtls, E I Du Pont de Nemours & Co, Inc (supplying dielectrics, electrodes and termination mtls to the multilayer ceramic capacitor industry). Employed by Solid State Dielectrics, Inc., Sun Valley, CA 1972 to 1986. SSD was acquired by E. I. DuPont de Nemours Co., Inc. in 1982. Positions: VP - 1972-77, President - 1978-82, Operations Manager, MCL Materials - 1982-86. Currently self-employed. 1986-87 - President, United States Advanced Ceramics Association. 1987-88 - Treasurer, American Ceramic Society. Boards of Directors: Novacap, Inc. 1986- ; Ceramic Devices Inc. 1986- ,(1987 - Chairman). *Society Aff:* NICE, ACerS, USACA

Peaceman, Donald W
Business: PO Box 2189, Houston, TX 77001
Position: Sr Res Advisor. *Employer:* Exxon Production Res Co. *Education:* ScD/Chem Engg/MA Inst of Tech; BChE/Chem Engg/College of the City of NY. *Born:* 6/1/26. in Miami, FL; grew up in NYC. Upon grad from MIT in 1951, joined the Production Res Div of Humble Oil Co. Has been assoc with Humble and Exxon up to the present, working primarily in the areas of numerical mathematics and application of computers to petroleum reservoir simulation. Author or co-author of eighteen articles, and one book, on numerical reservoir simulation. Received, jointly with Jim Douglas, Jr and H H Rachford, Jr, the Robert Earll McConnell award from AIME in 1979. *Society Aff:* SPE of AIME, ACM, SIAM, AIChE.

Peach, Lester C
Business: 3300 South Federal St, Chicago, IL 60616
Position: Prof Emeritus *Employer:* IL Inst of Tech Retired; Part-time Teacher *Education:* PhD/EE/IL Inst of Tech; MS/Physics/Univ of Chicago; BS/Math/Univ of Chicago *Born:* 01/03/22 1950-53 Process Engr with Manufacturing Res, Intl Harvester. 1954-57 Res Engr with IIT Res Inst. 1958-61 Res Engr, Dept of Electrical Engrg, IIT. 1961-62 Asst Prof; 1962-67 Assoc Prof; 1967-72 Prof and Chrmn; 1972-present Prof, Dept of Electrical Engrg, IIT. 1967-77 Assoc Dir, Lab for Atmospheric Probing. Past mbr of the Bd of Dirs of the Natl Electronics Consortium. Primary areas of interest include communications, electromagnetic field theory, and processing of meteorological and biological signals. Enjoys classical music. *Society Aff:* IEEE, HKN, TBΠ, ΣΞ, APS.

Peach, Robert W
Home: 541 N Brainard Ave, LaGrange Pk, IL 60525
Position: Principal *Employer:* Robert Peach and Assoc Inc. *Education:* MBA/Indus Mgt/Univ of Chicago; SB/Bus & Eng/MIT; -/Indus Mgt/IL Inst of Tech. *Born:* Oct 1924. Native of Chgo. Began career with Sears in 1948. Organized Qual Control Dept 1954. Assignment at Sears consists of reviewing & improving in-plant Quality progs of Sears suppliers, through a worldwide staff. Taught Q C in Evening Div of Ill Inst of Tech for 5 yrs. Served as V P Publs of the ASQC. ASQC Fellow, Cert Qual Engr, Chmn Stds Council. Chrmn, Awards Bd, Contrib to 'Juran', Q C Handbook. SubCtte Chmn ANSI Z-1 Ctte on Qual Assurance. Mbr Follow- up Services Council, Underwriters Labs. Conduct training seminars in Quality Engineering for ASQC. Mbr, Bd of Dirs, Amer Assoc of Laboratory Accreditation, Delegate to Intl Standards Organization Technical Ctte 176 on Quality Assurance. *Society Aff:* ASQC, ASTM.

Peake, Harold J
Home: 2213 Sherwood Hall Lane, Alexandria, VA 22306
Position: Consultant (Self-employed) *Employer:* George Washington University *Education:* BS/EE/Va. Poly. Inst.; MS/EE/V. of Md.; MEA/Engrg Admin/GWU *Born:* Dec 1920. Master of Engrg Admin from Geo Wash Univ. With Naval Res Lab 1942-58, R&D in naval radar, radio, & electronics. With NASA 1958-80 in spacecraft sys dev, tech applications, res prog mgmt, & public sector tech transfer. Assoc Profl Lecturer George Washington Univ. Adj Prof of Mgmt Amer Univ. Fellow IEEE. Eta Kappa Nu. Sigma Xi. *Society Aff:* IEEE, HKN, ΣΞ

Pearce, Frank G
Business: 200 E Randolph Dr, Chicago, IL 60601
Position: Retired *Education:* ScD/Chem Engg/MA Inst Tech; BS/Chem Engg/Rose-Hulmon Inst. *Born:* Sept 17, 1918. Native of Terre Haute Ind. Army Chem Corps 1942-45, in R&D for Defense Equip with AMF R&D 1945-47. Joined Exploration & Production subsid of Standard Oil Co Ind in 1947 at Tulsa in R&D Dept & became Dir, Design & Econ. Joined Chem subsid (Amoco) in 1958, first as Dir Proj Engrg & then as Dir R&D (Design & Economics). Became Corp Coord of R&D & part of long-range planning in 1962. Became Mgr (Corp) of Info Services in 1964 & Genl Mgr Info Services in 1971 (to date). Respon for computer hardware, software, telecommunications, sys analysis, design, programming & oper res. Became Consultant, Finance in 1978 and Retired in 1979. *Society Aff:* AIChE, ACS.

Pearce, Norvin D
Business: Box 1906, Grand Island, NE 68802
Position: Dir of Engrg Senior Product Engr *Employer:* York Foundry & Engine Works Blount Inc *Education:* BSc/Agri Engg/Univ of NE of Lincoln. *Born:* May 1931. Native of Arnold Nebr. US Army 1952-54 Korean conflict. Design engr for Caldwell Mfg Co, Kearney Nebr. Spec in design of aeration & crop drying equip 1958-64. Ch Engr for Hymark Indus, Henderson Nebr. Manufacture of steel grain bins, steel bldgs, matl handling & crop drying equip 1964-74. Dir of Engrg for York Foundry & Engine Works, York Nebr 1974-84. Senior Product Engr for Blount Agri Products, Grand Island since 1984. Material Handling, Grain Bias, Crop Dryers for Agri & Ind. handling equip pertaining to agri crops, fertilizer & food processing. Also respon for improving prod methods & machines. Pres Nebr Sect ASAE 1974. Hobbies: golf, fishing, reading. *Society Aff:* ASAE, NSPE.

Pearce, Philip L
Home: 78 Cricket Lane, Stamford, CT 06903
Position: Program Mgr, Engg Services. *Employer:* IBM - Americans/Far East Corp. *Education:* BS/EE/Univ of KY. *Born:* 1923 Kentucky. US Air Force WW II. After grad practices elec engrg with a util co & in the design & const field prior to joining IBM in 1955. Career with IBM has incl both mgmt & staff assignments in plant design & const, cons services in facils opers & in facils engrg. Elected Pres of Amer Inst of Plant Engrs May 1976. Held previous offices in AIPE of Exec V P, V P Policy & Mbrship, V P Tech Services, & Chap Pres. Cert Plant Engr, AIPE Fellow. Currently Vice Chairman of AAES International Affairs Council, first Elected for calendar year 1980 - relected for 1981. *Society Aff:* AIPE.

Pearlman, Bertrand B
Business: Engg Ctr, Dobbs Ferry, NY 10522
Position: Mgr of Design Engg. *Employer:* Stauffer Chem Co. *Education:* BEE/Power/Poly Inst of NY. *Born:* 11/30/36. Resident of Westchester County, NY. Twenty-five yrs experience in the Engg Field. Elec Engr for Singmaster & Breyer, specializing in Chem & Nucl plant design. Involved with primary distribution & Power Plant Control Systems at Am Elec Power. With Stauffer Chem Co since 1965. Assumed present responsibility as Mgr of Design Engg in 1969. Responsible also for Stauffer's Engg Stds & Engg Metric Policies. Pres of IEEE Profl Communication Soc in 1979, 1980 & reelected for 1981. Chrmn IEEE Pub Relations Committee 1981. Re-elected for 1982. Mbr IEEE TAB & PR Committees. Mbr of Chem Mfg Assn Engg Advisory Committee. PE. Enjoy classical music & electronics kit bldg. *Society Aff:* IEEE, AIChE, STC, NSPE.

Pearlstein, Joel P
Business: 800 Rose Lane, Union Beach, NJ 07735
Position: V P & Director of Fragrance Operations, N.A. *Employer:* Intl Flavors & Fragrances. *Born:* Sept 1936. MChE Newark College of Engrg; BChE Polytechnic Inst of Brooklyn. 1st Lt Corps of Engrs USAR. With IFF since 1968 as Genl Mgr Aroma Chem Plant since 1973.

Pearsall, George W
Home: 2941 Welcome Dr, Durham, NC 27705
Position: Prof. *Employer:* Duke Univ. *Education:* ScD/Metallurgy/MA Inst of Tech; BMetE/Metallurgy/Rensselaer Poly Inst. *Born:* 7/13/33. Dow Chem Co, Magnesium Lab, Res Engr, 1955-57. MA Inst of Tech, Asst Prof, 1960-64 (Dept of Metallurgy). Duke Univ, School of Engg, Assoc Prof 1964-66 and Prof 1966-present (Dept of Mech Engg and Materials Sci); also Acting Dean 1969- 71 and Dean 1971-74, 1982-83. Duke University, Trinity College of Arts and Sciences, Prof 1982-present (Institute of Policy Sciences and Public Affairs). Res in the areas of mech deformation and fracture in metals, plastics, and biological materials; dev of new materials and novel deformation processing techniques. Author of numerous professional publications. Consultant in the areas of failure analysis, products safety, materials processing and specification, and mech engg. Professional Engr, NC. Married; enjoy tennis and sailing. *Society Aff:* AAAS, ASM, ASME, ASTM, ΣΞ, SPE, SRA.

Pearson, Allan E
Business: Div of Engg, Providence, RI 02912
Position: Prof of Engg. *Employer:* Brown Univ. *Education:* PhD/Engg/Columbia Univ; MS/Mech Engr/Univ of MN; BS/Mech Engr/Univ of MN. *Born:* 6/18/36. Joined the faculty of Brown Univ, Providence, RI in 1963. Currently Prof of Engg at Brown Univ and a mbr of the Lefschetz Ctr for Dynamical Systems. Res interests lie in the area of control systems. A mbr of Tau Beta Pi, Pi Tau Sigma, Sigma Xi, the American Society of Elec and Electronics Engrs, and the Fulbright Alumni Assn. *Society Aff:* ASME, IEEE, AAAS.

Pearson, Frank H
Business: 1301 S 46th St, Richmond, CA 94804
Position: Asst Dir. *Employer:* San Engrg Res Lab - U C Berkeley. *Education:* PhD/Civil Engg/PA State Univ; MS/Water Resources/Univ Newcastle on Tyne, UK; BE/Civil Engg/Canterbury Univ, New Zealand. *Born:* Oct 1939. PhD Penn State Univ; MS Univ of Newcastle on Tyne England; BE Canterbury Univ New Zealand. Asst Dir & Staff Engr SERL, respon for mgmt of lab; conducts wastewater treatment res. Previously Proj Engr, Gannett Fleming Corddry & Carpenter Inc 1974-76, on treatment plant performance eval, process design, water qual mgmt planning. Dev mine drainage neutralization processes Penn State Univ 1971-74. With New Zealand Ministry of Works 1961-71; 8 yrs wastewater treatment design, 2 yrs hwy & bridge const. ASCE Croes Medal 1976; Chadwick Medal, Univ of Newcastle on Tyne 1966-67. Enjoys backpacking. *Society Aff:* ASCE, WPCF.

Pearson, Gerald L
Business: Electronics Lab, Stanford, CA 94305
Position: Prof of Elec Engrg. *Employer:* Stanford Univ. *Education:* ScD/Science/Willamette Univ; MA/Physics/Stanford Univ; AB/Physics/Willamette Univ. *Born:* Mar 1905. Native of Salem Oregon. Res physicist at Bell Tele Labs 1929- 58; Hd, Solid State Electronics Dept 1958-60. Prof of Elec Engrg at Stanford Univ 1960-70; Emer Prof 1970- . Dir of Stanford Solid-State Electronics Lab 1968-80; emer Dir 1980- . 35 U S pats & 140 tech pubs in area of solid state electronics Mbr Natl Acad of Engrg & Natl Acad of Sci. John Scott Award from City of Phila 1956; John Price Wetherill Medal from Franklin Inst 1963; Marian Smoluchowski Medal from Polish Phys Soc 1976; Solid State Science and Technology Medal Award from The Electric chemical society 1981. Fellow Institute of Electrical and Electronics Engineers and American Institute of Physics: Member Sigma Xi; Life member Franklin Institute and Telephone Pioneers of America. *Society Aff:* IEEE, APS, NAE.

Pearson, Hugh S
Business: South Cobb Dr, Marietta, GA 30067
Position: Sr Engr *Employer:* Lockheed Georgia Co. Dept 72-53 *Education:* Masters in Met/-/GA Inst of Tech; BME/-/GA Inst of Tech. *Born:* 6/8/31. Twenty yrs experience in Mech Testing & Mtls Res at Lockheed-GA Co, specializing in Fracture Analysis & Fracture Mechanics Testing. Joined Pratt and Whitney at FL Res & Dev Ctr as Specialist in Fracture in 1973. In 1974, became VP of Applied Technical Services, Inc. Rejoined Lockheed GA Co in 1980. Presently responsible for res and testing in mtls and structures especially fracture and fracture mechanics. Reg PE in GA & CA. Active in ASM in Atlanta Chapter & Natl Programming Committees & active in ASTM's Committee E-24 in Fracture. *Society Aff:* ASM, ASTM.

Pearson, J Boyd, Jr
Home: 5626 Wigton, Houston, TX 77096
Position: Prof. *Employer:* Rice Univ. *Education:* PhD/EE/Purdue Univ; MS/EE/Univ of AR; BS/EE/Univ of AR. *Born:* 6/3/30. Native of Pine Bluff, AR. Served in US Army 1952-55. Asst Prof of EE at Purdue Univ 1962-65. Assoc Prof 1965-70, Prof 1970-, Chrmn of Dept of EE 1974- 79, JS Abercrombie Prof 1979-, all at Rice Univ. Fellow IEEE, 1979. *Society Aff:* IEEE, AAAS.

Pearson, James V
Box 3147, Siloam Springs, AR 72761
Position: Prof of Engrg. *Employer:* John Brown Univ. *Education:* PhD/Elec Engr/Univ of AR; MS/Math/Univ of AR; BEE/Elec Engr/Univ of MN *Born:* 5/28/33. Native of Minneapolis, Minn. Worked in prod engrg for 3 yrs with IBM. 3 yrs with 3M Co in prod dev - info retrieval sys & memory component dev. 16 yrs teaching elec & sys engrg at John Brown Univ. Presently Chmn Div of Engrg. Cons in sys analysis & synthesis - special interest in control sys; and comm sys; educ mgmt sys, solar thermal power plant control energy recovery from solid waste. *Society Aff:* IEEE, ASEE

Pearson, John
Business: P.O. Box 1390, Ridgecrest, CA 93555
Position: Senior Research Scientist *Employer:* Naval Weapons Ctr. *Education:*

Pearson, John (Continued)
MS/Applied Mechs/Northwestern Univ; BS/Mech Engg/Northwestern Univ. *Born:* Apr 1923. US Army Corps of Engrs 1943-46. Pioneered the field of explosive metal-working. Extensive pubs & pats. Co-author (with J S Rinehart) 'Behavior of Metals Under Impulsive Loads', 'Explosive Working of Metals'. Civilian scientist with US Navy Dept since 1951 spec in explosive ordnance, terminal ballistics, & shock met. Cons to indus & gov agencies. Assumed current respon as Senior Research Scientist, NWC, in 1983. Mbr Tau Beta Pi, Pi Tau Sigma, Sigma Xi, Triangle, ASM, Fellow ASME, ASME, AIME, APS, NYAS, NSPE, CSPE. Reg P E Calif. L T E Thompson Medal 1965; Secy of Navy Cert of Recognition 1975; William B McLean Medal 1979; Navy Department Award of Merit, 1979. Secy of Navy Cert of Commendation, 1981, Superior Civilian Service Medal, US Navy, 1984, Haskell G. Wilson Award, 1985, Charter Member, Senior Executive Service of The United States of America, 1979. *Society Aff:* ASM, ASME, APS, NYAS, AIME, NSPE, CSPE

Pearson, John W
Business: 900 Bush Ave, 3M Center, St. Paul, MN 55144
Position: Vice President, Development *Employer:* 3M Co. *Education:* BChE/ChE/Univ of MN. *Born:* 6/2/18. St Paul, MN of John A & Evelyn G Pearson; BChE 1939 Univ of MN; Married Elizabeth Ann Fitch, two children Jon Oakley & Ann Elizabeth; 1939-44 3M Co Chem Engr; 1944-46 US Army, Staff mbr, Manhattan Proj, Los Alamos, NM; 1946-55 Dir Dev Engg 3M Co; 1955-62 Mgr New Products Div 3M; 1962-68Mgr Ind Finishing Dept 3M; 1968-72 Group Engg Mgr Photo Worldwide 3M; 1972-1980 Dir & Exec Dir Div Engg 3M; 1980 - present Vice President, Development, 3M; Current Mbr MN State Bd of Reg. *Society Aff:* AIChE, ACS, ESC.

Pearson, Joseph T
Business: Sch of Mech Engg, West Lafayette, IN 47907
Position: Assoc Prof. *Employer:* Purdue Univ. *Education:* BME/ME/NCSU; MS/ME/NCSU; PhD/ME/SUNY at Stony Brook *Born:* 12/7/33. Assoc Engr with Douglas Aircraft Co (1956-57) and Mgr of a Mechanical Contracting Firm (1957-59) before returning to graduate sch at NCSU in 1959. Instructor in engg at SUNY from 1961-67. Went to Purdue University in 1967 where I am engaged in teaching and research in the area of heat transfer and thermal systems design. Licensed PE in IN and TX. Fulbright-Hays fellowship for res (1973- 74). *Society Aff:* ASME, ASHRAE, NSPE.

Pearson, William B
Business: Univ of Waterloo, Univ Ave, Waterloo, Ontario Can N2L 3G1
Position: Prof of Physics and of Chemistry. *Employer:* Univ of Waterloo. *Education:* DSc/Metals/Oxford Univ; MA/Chem/Oxford Univ. *Born:* July 1, 1921. WW II - Pilot RAF, DFC. 1952-69 Low-Temp & Solid State Phys Group, Natl Res Council, Ottawa. 1969-77 Dean of Sci, 1977- Prof of Phys and of Chemistry, Univ of Waterloo, Ontario Canada. FRSC 1960, William HumeRothery Award of TMS of AIME 1975, Canadian Metal Physics Medal, 1979. *Society Aff:* RSC.

Pearson, William H
Home: 59 Montvale Dr, Scarborough, Ontario Canada M1M 3E5
Position: Ontario Proj Mgr. *Employer:* Atlas Alloys. *Education:* Mech/Metals/Byerson Inst of Technology. *Born:* Nov 1927. Grad of Ryerson Inst of Tech, Canada. Served with Toronto Scottish 1943-45. Mgr of Coulter Copper & Brass Co, spec in metal forming & processing. Mgr Supreme Aluminum Indus, specialist in metal finishing. With Atlas Alloys since 1965. Assumed current respon as Ontario Proj Mgr in 1972. Internatl Dir SME 1975-79. Exec Council ASM 1976-79. Pres AOTS, Dr Bell Award 1971. Registered Prof Engg (CA), SME Intl Award of Merit 1980. *Society Aff:* SME, ASM, CWB.

Peart, Robert M
Business: Agri Engrg Dept, West Lafayette, IN 47906
Position: Professor. *Employer:* Purdue Univ. *Education:* BS/Agri Engg/IA State Univ; MS/Agri Engg/Univ of IL; PhD/Agri Engg/Purdue Univ. *Born:* Nov 11, 1925. Lived on farm near Kewanee Ill. 2 yrs Air Force WW II. 5 yrs Eastern Iowa Light & Power Co-op, 5 yrs Agri Engrg Dept, Univ of Ill, Asst Prof. Since 1961 at Purdue. Design of agri matls handling sys - 2 pats. Design & analysis of grain drying sys - 1 pat. Leader in computer simulation of grain drying sys. Simulation of crop ecosystems - insects, alfalfa, corn, & application to pest mgmt. Active in simulation of energy input/output relationships in agri. Bd of Dirs, Third Century Venture Corp, Indianapolis (1978-83). ASAE Bd of Dir 1975-77. ASAE Paper Awards 1967, 1969, 1984. ASAE Fellow 1979. Enjoy sailing & skiing. *Society Aff:* ASAE, ASEE, SCS.

Pease, Robert B
Home: 326 Dewy St, Pittsburgh, PA 15218
Position: Exec Dir. *Employer:* Allegheny Conf on Community Dev. *Education:* BSc/Civil Engg/Carnegie-Mellon Univ. *Born:* May 1925 Atkinson Nebr. Drake Univ. Navigator 15th Air Force, European Theater, WW II. 1949-53 engr, Carnegie Inst of Tech. 1953-68 Engr, Ch Engr, Asst Exec Dir, Exec Dir Urban Redev Auth of Pgh. 1968- , Exec Dir Allegheny Conf on Community Dev. Mbr Pa State Transp Comm. Fellow ASCE. Tr Alaska Methodist Univ. Tr Allegheny Genl Hosp, Pgh. ASCE 1966 Civic Gov Award. Man of Yr Award - Good Gov, Pgh Jaycees 1964. Dor of Social Sci (hon degree) Duquesne Univ Pgh. Hon Mbr Amer Inst of Architects. *Society Aff:* ASCE, AIA, NAHRO.

Peattie, Edward G
Business: P. O. Box BK, Mississippi State, MS 39762
Position: Prof of Petroleum Engrg *Employer:* MS State Univ *Education:* MS/ChE/Univ of MI; MA/Chem & Eng/Temple Univ; BS/Chem/OH Univ *Born:* 12/12/13 Amsterdam, NY. Schooling in Cleveland OH. Army Service in India and Iran (World War II) Chem Engrg experience includes TNT plant (Trojan) and, Research at Rohm & Haas. Petroleum experience includes Aramco, Saudi Arabia (1954-60); Sohio (Project Mgr) (1960-65); Commonwealth Oil, (Puerto Rico), (Technical Group Leader), Ebasco (Principal Corrosion Engr), (1969-73). In 1973 became Assoc Prof, Chem Engrg at the Univ of Petroleum & Minerals, Dhahran, Saudi Arabia. My present position is Assoc Prof of Petroleum Engrg, MS State Univ. Textbooks in preparation are Petroleum Operations Engrg and Petroleum corrosion Control. Prof ChE in the state of OH. *Society Aff:* SPE, AIME, NACE, NSPE, ACS, AAAS, ASM

Peavy, Howard S
Business: Dept of Civil Engrg, Montana State University, Bozeman, MT 59717
Position: Prof of Civil Engr; Dir, Water Resources Center. *Employer:* Montana State Univ. *Education:* PhD/Environ Engr/OK Univ; MSCE/Civil Engr/Duke Univ; BSCE/Civil Engr/LSU. *Born:* 9/11/42. Reg P E Montana. Work Exp.: Ch Engr, Okla Dept of Pollution 1972-74; Dept. of Civil Engr, Montana State Univ Asst Prof 1974-77 (Assoc Prof 1977-83, Prof 1983-). Distriktshogskole, Stavanger, Norway, 1980; Director Montana University System Water Resource Center 1982- ; Coordinator of Environmental Engr. Program 1986- . Dow Award for outstanding young faculty, Pacific Northwest Sect of ASEE 1976. U S Jaycees Outstanding Young Men of Amer 1979. Principal Author, Environmental Engineering, McGraw-Hill, 1985. *Society Aff:* ASCE, AWWA, ASEE, WPCF.

Peck, D Stewart
Business: 3646 Highland St, Allentown, PA 18104
Position: Ret. (Consultant) *Education:* MS/EE/Univ of MI; BS/EE/Univ of MI. *Born:* 1918 Grand Rapids Mich. With G E 1940-47 in design of thyratrons, ignitrons & phototubes. Joined Bell Tele Labs 1947 on design of gas tubes. Spec in semiconductor device reliability from 1955, respon for selection of all components for the Telstar satellites; initiated the dev of practical tech for accelerated-stress testing & screening semiconductor devices in the var conditions of ionizing radiation, power or temp with bias, & humidity with temp & bias. (Title - Dept Hd, Reliability, Tech Info & Semi-conductor Devices). Specialist in the analysis of data from accelerated-stress tests. Fellow IEEE 1966; Chmn Electron Tubes Sub-ctte AIEE 1955-

Peck, D Stewart (Continued)
58. Dir Reliability Phys Symposium 1970-78. Retired BTL Oct, 1980. *Society Aff:* IEEE.

Peck, Eugene L
Home: 8513 Electric Ave, Vienna, VA 22180
Position: President *Employer:* Hydex Corp *Education:* PhD/Civ Engg/UT State Univ; MS/Meteorology/Univ of UT; BS/Meteorology/Univ of UT. *Born:* 6/2/22. Born Kansas City, KS June 2, 1922. Married 1946, 4 children. Education Univ UT BS 47, MS 51; UT State Univ. PhD (Civ Engg), 67. Profl exp Weather Officer Army AF 1942-46. Hydrologist Western Region, US Weather Bureau 1947-1967. Hydrologic Res Lab, Natl Weather Service, NOAA 1967-80, Dir 1974-80. Hydex Corporation 1980- . Honors & awards Silver Medals, US Dept Comm 1959 & 1975. Mbr Am Meteorological Soc, Am Geophysical Union, AM Water Res Assoc. Res: hydrometeorology; precipitation, snow, evaporation. Mailing address: Hydex Corporation, Suite 200, 105 Rowell Court, Falls Church, VA 22046. *Society Aff:* AMS, AGU, AWRA, WSC.

Peck, Kenneth M
Business: 625 Delaware Ave, Buffalo, NY 14202
Position: VP. *Employer:* URS Corp. *Education:* BCE/Civil Engr/Univ of the City of NY. *Born:* 3/13/42. After graduation from CCNY in 1968, joined the firm of Nebolsine, Toth, McPhee Assoc. Published papers on physical/chemical treatment of wastewater and industrial waste pretreatment. In 1972 the firm changed its name to McPhee, Smith, Rosenstein Engineers. In 1977 URS Corp purchased the firm. Presently Senior VP for URS' Buffalo, Puerto Rico and Columbia operations. *Society Aff:* WPCF, NSPE, ASCE, ACEC.

Peck, Ralph B
Home: 1101 Warm Sands Dr SE, Albuquerque, NM 87123
Position: Civil Engr: Geotechnics. *Education:* CE/Structures/Rensselaer Poly Inst; DCE/Structures/Rensselaer Poly Inst; - /Soil Mechs/Harvard Univ. *Born:* June 23, 1912. Prof of Foundation Engrg Emer, Univ of Ill Urbana. Author (with K Terzaghi) 'Soil Mech in Engrg Practice', (with W E Hanson & T H Thornburn) 'Foundation Engrg'. Cons on foundations, dams, tunnels, landslides. Mbr Natl Acad of Engrg; Hon Mbr ASCE, Dir 1962-65; Fellow Geol Soc Amer; Mbr Internatl Soc Soil Mech & Foundation Engrg, Pres 1969-73. Chmn U S Natl Comm on Tunneling Tech 1976-77. ASCE Norman Medal 1944, Wellington Prize 1965, Karl Terzaghi 1969, Presidents Award 1986; NSPE Award 1972; Moles Non-mbr Award 1973; Outstanding Civilian Serv Medal US Army 1973; Natl Medal of Sci 1974; Washington Award 1976; John Fritz Medal 1987; D Eng RPI 1974; D.Sc. Laval 1987. *Society Aff:* ASCE, NSPE, AREA, GSA, USCOLD.

Peck, Ralph E
Business: 3100 Michigan Ave No. 807, Chicago, IL 60616
Position: Prof Emeritus. *Employer:* Illinois Inst of Tech *Education:* PhD/Physical Chemistry/Univ of MN; BChE/Chem Engg/Univ of MN. *Born:* 12/8/10. Instructor in Chem Engg - Drexel Inst of Technology, 1936-39 Instructor to Prof - IL Inst of Technology, 1939-77 Chrmn, Dept of Chemical Engg - IL Inst of Technology - 1954-67 Visiting Prof: Panjsb Univ, India, 1959-60; Technion, Israel, 1962-63; Sao Paulo Univ, Brazil, Summer 1972; Chonnam natl Univ - Kwanju, Korea - 1979-80. Lecturer: Sao Paulo Univ, Brazil; Rio Technical Inst; Univ of Algeria, Boumerades. Received Westinghouse Award for Chemical Engg, Excellence In Teaching Award, IIT; Fellow of AIchE; Prof Emeritus Over 45 Publications in Natl Journals. Hobbies: fishing and gardening. *Society Aff:* AIChE, ACS, ASEE, ΣΞ, ΤΒΠ, ΦΛΥ.

Peck, William B
Business: P O Box 16858, Philadelphia, PA 19142
Position: JVX Engrg *Employer:* Boeing Vertol Company. *Education:* BAE/Aero Engr/RPI *Born:* 6/3/28. Employed by Boeing Vertol for past 35 yrs; previously was Dir of V/STOL Tech, and Deputy UTTAS Prog Mgr. Currently, Dir of JVX Engrg. Mbr Amer Helicopter Soc and AIAA. Author of Paper 508E, *Design & Test Experience with High-Speed Mechanical Drive System for Helicopter & VTOL Aircraft*, presented at SAE meeting April 1962; *A Survey of Helicopter Current Practices Relative to Fatigue*, presented at AGARD meeting in 1971. Also co-author of, *The Value of Various Technology Advances for Several V/STOL Configurations*, Paper No 39, presented at Fourth European Rotorcraft & Powered Lift Aircraft Forum, Sept 13-15, 1978, Stresa, Italy. published in AHS Journal, July 1979 and co-author of JVX Design Update, presented at 40th Anuual Forum AHS, Arlington, VA May 16-18, 1984.. *Society Aff:* AHS, AIAA

Peden, Irene C
Business: 407 EEB FT-10, Seattle, WA 98195
Position: Professor of Electrical Engineering. *Employer:* University of Washington. *Education:* PhD/Elec Engg/Stanford Univ; MS/Elec Engg/Stanford Univ; BS/Elec Engg/Univ of CO. *Born:* 9/25/25. Native of Topeka KS & Kansas City MO. DE. Pwr & Light Co 1947-49; Stanford Res Inst antenna res 1949-52; 1954-57 ABET Bd of Dirs; Hansen Microwave Lab Stanford 1958-61; Univ of WA faculty 1962-present; Assoc Dean Engg 1973-76. Res & publ in area of radio sci in polar regions. Soc Womn Engrs Achievement Award 1973; Distinguished Engr Alum Awd Univ of CO 1974; US Army Outstanding Civilian Service Medal; Bd of Dir BDM International; Fellow of IEEE 1974. IEEE V Pres Education Activities 1976 & 77; Chair Army Sci Bd; Bd of Visitors, U. CA-Davis Coll of Engrg; Eng. Dev. Comm. VCO Chmn - Engrg Accreditation Commission of the Accreditation Board of Engineering and Technology (ABET) Intl Union Radio Sci (URSI) Comm B (fields & waves); Member of Large US National Committee ORSI; AAAS Nominating Comittee 1983 & 84; Tau Beta Pi; Sigma Xi; NY Acad Sci; AGU, AAAS. *Society Aff:* IEEE, AGU, ASEE, AAAS, SWE

Pedersen, George C
Business: P O Box 570846, Perrine, FL 33257-0846
Position: President. *Employer:* Kimre Inc. *Education:* Chem Eng/Chem Eng/MIT; MS/Chem Eng/MIT; BS/Chem Eng/MIT. *Born:* Dec 1940, Miami FL. 1967-74 with Albany Internatl Corp progressing from Sr Dev Engr to Mgr of R&D to Cons. 1973 formed Kimre Inc to produce new products for mist eliminators & tower packing based on 1 of my inventions; now active in use & continued dev of these products for the chem processing, air pollution & related indus. Author & inventor with patents for new process to make hydrazine. Reg PE; Mbr AIChE, ACS, NSPE, APCA, Filtration Soc. *Society Aff:* NSPE, AIChE, ACS, APCS, ASSCT.

Pedersen, William W
Home: 3812 Applewood Rd, Midland, MI 48640
Position: VP Auditing. *Employer:* Dow Corning Corporation. *Education:* MAsters/Chem Engg/Univ of LA; Bachelors/Chem Engg/Univ of LA. *Born:* June 1916 in Brooklyn NY. BSChE & MSChE from Univ of Louisville; Advanced Mgmt Prog Harvard Bus Sch 1965. Employed by Dow Chemical in 1938 for product dev on ethyl cellulose; joined Dow Corning June 1945 & moved to Cleveland OH in 1946 to open regional sales office; joined Corporate Planning Ctte in Midland 1959; became Mgr of Engrg Products Div in 1962 & V Pres of Marketing & Distrib 1967, VP Auditing 1977. Enjoy sailing & tennis.

Pederson, Donald O
Business: EECS Dept Cory Hall, Berkeley, CA 94720
Position: Professor of Electrical Engineering. *Employer:* University of California, Berkeley. *Education:* PhD/EE/Stanford; MS/EE/Stanford; BS/EE/ND State. *Born:* 1925. 1951-53 Res Assoc at Stanford Univ; 1953-55 mbr tech staff of Bell Labs; since 1955 at Univ of CA Berkeley, Dir of Electronics Res Lab 1960-64, Chmn of EECS Dept. 1983-85. Now Prof of elect engrg. Awarded Guggenheim Fellow & IEEE Fellow in 1964; IEEE Educ Medal in 1969; elected to Natl Acad of Engrg 1974 and to Natl Acad of Science, 1982; Awarded honorary Doctorate of Applied Science, Catholic Univ. Leuven, Belgium, 1979. *Society Aff:* NAE, IEEE, ΣΞ, NAS.

Peebles, Peyton Z, Jr
Business: Elec Engg Dept, Gainesville, FL 32611
Position: Prof and Assoc Chrmn of EE Dept. *Employer:* Univ of FL. *Education:* PhD/EE/Univ of PA; MS/EE/Drexel Inst; BS/EE/Evansville Coll. *Born:* 9/10/34. Design & systems Engr for RCA 1958-1969 except for 1964-1966 when at Univ of PA as David Sarnoff Fellow from RCA. At Univ of TN 1969-1977 except 1976-1977 when at Univ of HI. Honolulu, as visiting prof. Author or co-author of 51 tech articles. Author of three books: *Communication System Principles* (1976, Addison-Wesley), *Probability, Random Variables, and Random Signal Principles*, 2 ed. (1980, McGraw-Hill), Digital Communication Systems (1987, Prentice-Hall). Consultant to various ind concerns & govt agencies. Fellow of Institute of Electrical & Electronics Engineers. *Society Aff:* HKN, TBΠ, ΣΞ, ΣΠΣ, ΦBX.

Pefley, Richard K
Business: Mech Engr Dept, Santa Clara, CA 95053
Position: Professor Mechanical Engineering Dept. *Employer:* University of Santa Clara. *Education:* Engr/Mech Engr/Stanford; MS/Mech Engr/Stanford; BA/Mech Engr/Stanford. *Born:* 1921 in Sacramento CA. Sacramento City Coll & 3 degrees in mech engrg from Stanford Univ. Industrial exper includes flight engrg for the USAF, product engrg in packaged air conditioning systems & res engrg relative to liquid metal systems for nuclear powered submarines. Since original appointment in 1951, Dept Chmn-Professor of Mech Engrg at Univ of Santa Clara. During past 5 years served as principle investigator in following areas of sponsored res: methanol as as alternative fuel for automobiles & gas turbines, human calorimetry & artificial lung design. Fellow ASME; licensed mech engr in CA. Coauthored text 'Thermo-fluid Mechanics' & pub numerous articles relative to above res topics. Elected VP ASME 1979, Apptd Permanently to Chair in Mech Engr at Santa Clara Univ Chair is Funded by Endowment. Conslt to United Nations on Alcohol Fuels Pres of Alcohol Energy Sys, Inc. *Society Aff:* ASME, SAE, ASHRAE

Pegram, Anthony R
Business: 11 Embarcadero West, Suite 202, Oakland, CA 94605
Position: Assoc *Employer:* Jordan/Avent & Associates *Education:* M/Gen Mgmt/Golden Gate Univ; B/CE/In Inst of Tech *Born:* 08/30/47 Native of Washington, DC where he attended McKinley Tech HS graduating in 1965. He began his profl career as an engr with the Bechtel Corp, San Francisco, CA. He later became Asst Prog Mgr with US Genl Services Administration and subsequently joined GE's Nuclear Energy Div as a Task Mgr for the containment design prog. Mr Pegram joined a consltg engrg firm, Jordan/Avent & Assocs, Inc in 1978 and has since become an assoc with the firm resp for operations throughout the state of CA. *Society Aff:* ASCE, NTA, NCCBPE, SEANC, EBSES

Pehlke, Robert D
Business: Dept of Matls Science & Engr, 2158A Dow Bldg-N. Campus, 2300 Hayward, University of Michigan, Ann Arbor, MI 48109-2136
Position: Professor *Employer:* The University of Michigan. *Education:* ScD/Metallurgy/MIT; SM/Metallurgy/MIT; BSE/Met Eng/Univ of MI. *Born:* 2/11/33. Ferndale MI. 1956-57 student Tech Inst Aachen Germany; 1960-63 Asst Prof Dept Matls & Met Eng Univ of Mich, 1963-68 Assoc Prof, since 1968 Prof, 1973-84 Chrmn of Dept. Cons to Metall Indus. Reg PE MI. NSF Fellow 1955-56; Fulbright Fellow 1956-57; Fellow ASM; Tech Dev Bd 1983-84; Distinguished Life Mbr Iron & Steel Soc AIME, Chrmn Process Tech Div 1976-77; Bd of Dirs 1976-79; Mbr Iron & Steel Soc, Japan; Fellow, Met Soc AIME; Mbr Soc Iron & Steel, Germany; Mbr Iron & Steel Soc, London; Mbr AFS, ASEE, NY Acad Sci; Sigma Xi; Tau Beta Pi; Alpha Sigma Mu. Intl Prs 1977-78. Author 'Unit Processes of Extractive Metallurgy' 1973; contrib numerous articles to sci journals; Chrmn ed bd AIME Monograph Series on Basic Oxygen Steelmaking. Best paper award AFS Malleable Iron Div 1971; AFS Steel Div 1970, 1979, AFS Brass & Bronze Div 1983; Gold Medal Sci Extractive Met Award, Met Soc AIME 1976; AIME Howe Mem. Lecturer 1980. *Society Aff:* AIME, ASM, AFS, ASEE.

Pehrson, David L
P.O. Box 5508, L-151, Livermore, CA 94550
Position: Deputy Assoc Dir for Elec Engg. *Employer:* Lawrence Livermore National Laboratory. *Education:* MSEE/Elec Engg/NY Univ; BSEE/Elec Engg/Univ of CA, Berkley. *Born:* Sept 1940. Mbr of tech staff with Bell Telephone Labs 1962-65 emphasizing electronic telephone switching system architecture studies; MSEE 1964 under graduate studies prog; with Lawrence Livermore Natl Lab since 1965: 8 years as proj engr & grp leader for LLL Octopus Computer Network Grp, Div Leader Computer Systems Div 1973-76 respon for all Computer Center system software planning & dev, Div Leader Electronics Dept for Fusion Energy Systems 1976-83 respon for support of Major DOE magnetic fusion energy prog & related tech & physics res. Div Leader Electronics Dept for Laser Engg Div 1983-87 respon for support of major DOE Confinement Fusion and Laser Isotope Separation Programs, Deputy Assoc Dir for Electronics Engg 1987 to present resp for Electronics Engg work in support of the lab's programs. Mbr IEEE, Phi Beta Kappa, Eta Kappa Nu, Tau Beta Pi, Sigma Xi. *Society Aff:* IEEE, Phi Beta Kapa, Eta Kap NU, Tau Beta Pi, Sigma Xi

Pehrson, Elmer W
Home: 3900 Watson Pl NW 8E-A, Washington, DC 20016
Position: retired. *Employer:* *. *Education:* AB/Geology & Mining/Stanford Univ; EM/Mining & Metallurgy/Stanford Univ. *Born:* 1896 in CA. 1918-25 US Army & Engr western mines; 1926-63 engr-mineral economist U S Government: Internal Revenue Service; 1928-63 Bureau of Mines Ch Economics Branch & Dir of Minerals Yearbook; 1939-40 Minerals Adv Ctte Army-Navy Munitions Bd; 1940-49 Interior Dept's representative on Bd's Strategic Materials Ctte; 1942 Advisor U S Delegations Pan American Mining Congress, Santiago; 1946 Internatl Tin Conference, London; 1948-65 Professor Mineral Economics at Columbia Univ. AIME Bd Mbr & Chmn Mineral Economics Ctte 1949, Mineral Economics Award 1970; Councillor M & MSA 1954-57; Marburg Memorial Lecturer ASTM 1958. Special interests: minerals in world affairs, U S mineral policy. *Society Aff:* SME-AIME.

Pei Chi Chou
Business: Mechanical Engineering Dept, Drexel University, 32nd & Chestnut Sts, Philadelphia, PA 19104
Position: Prof *Employer:* Drexel Univ *Education:* ScD/Aeronautical Engr/NY Univ; MS/Engrg Scs/Harvard Univ; BS/Aeronautical Engr/Nat Central Univ, China *Born:* 12/1/24 Billings Prof of Mechanical Engrg, Drexel Univ since 1953. Pres of Dyna East Corp, Wynnewood, PA. Holds membership in AIAA, ASME, ADPA, ASEE, AAM; is listed in Who's Who in the East, also in American Education, American Men of Sci and Dictionary of Int'l Biography. Publications appear in AIAA, J. Appld Physics, J. Applied Mechanics, J. Composite Materials, J. Spacecraft and Rockets, etc. Authored one book and edited one. Specializations include impact and wave propagation, mechanics of composite materials and ballistics. Listed also in Who's Who in America. *Society Aff:* AIAA, ADPA, ASME, AAM, ASTM, ASEE

Peikari, Behrouz
Business: Elec Engg Dept, Dallas, TX 75275
Position: Professor *Employer:* Southern Methodist Univ. *Education:* PhD/Elec Engg/Univ of CA; MS/Elec Engg/Univ of IL; BS/EE-ME/Univ of Tehran. *Born:* 3/16/39. Has Taught Engg at Univ of Tehran, Iran, Univ of CA at Berkeley and Southern Methodist Univ in Dalla. Has also served as technical consultant to Xerox Corp and TX Instruments. He is the author of the text book "Fundamentals of Networks Analysis and Synthesis" Prentice Hall 1974 and a coauthor of "Circuit Analysis" Matrix Publishing 1977. *Society Aff:* IEEE.

Peixotto, Ernest D
Business: , Washington, DC 20310
Position: Comptroller of the Army, HQ Dept of the Army. *Employer:* Department of the Army. *Education:* MS/Civil Engg/MA Inst of Tech, Natl War College, Command & Genl Staff College; BS/Engr/US Military Acad, West Point. *Born:* 7/24/29. in Ft Leavenworth KS. Command & Genl Staff Coll 1965; Natl War Coll 1970. Lt Gen, US Army. Important asignments: flood control projs Vicksburg Engr Dist, MS; nuclear reactor projs US Atomic Energy Comm, Wash DC; airfield const Gulf Engr Dist, Iran; 2 tours in Vietnam, Military Acad & Combat Batttalion; R&D Mgmt, Dir Waterways Experiment Station MS; R&D staff assignments, Ch of Engrs & Ch of R&D, Wash DC; Asst Commandant US Army Engr School Ft Belvoir; Dir, Materiel Plans & Programs, Dept of Army; Dir of the Army Budget, Dept of Army; Deputy Chief of Staff for Resource Mgmt, HQ US Army Europe; Comptroller of the Army. *Society Aff:* ASCE, SAME, ASMC.

Pekau, Oscar A
Business: Dept Civil Eng, 1455 de Maisonneuve Blvd W, Montreal Quebec, Canada H3G 1M8
Position: Associate Professor of Engineering. *Employer:* Concordia University. *Education:* PhD/Civil Eng/Univ of Waterloo; MSc/Structures/Univ of London; DIC/Structures/Imperial College; BASc/Structural/Toronto. *Born:* Feb 1941. Athlone Fellow, Bd of Trade, U K 1964-65. Design engr with Morrison, Hershfield, Millman & Huggins Ltd, Toronto until 1967; cons in structural dynamics, earthquake engrg & structural design; with Concordia Univ since 1971. FRSA, London; ASEE Outstanding Young Faculty Award 1974; ASEE; M ASCE; MEIC; CSCE; Engr, Quebec; P Engr, Ontario. *Society Aff:* ASCE, EIC, ASEE, CSCE.

Pelan, Byron J
10 Orchard Avenue, North Plainfield, NJ 07060
Position: Professor Emeritus *Employer:* Rutgers Univ. *Education:* MS/Mech Engg/Lehigh Univ; BS/Mech Engg/Rutgers Univ. *Born:* 6/5/21. US citizen by birth. Indus experience: mech design and proj engg of production equip. Teaching experience: mech design, stress analysis, and vibration courses through grad level. Soc activities: Chrmn, Middle Atlantic Sec, ASEE, 1970-71. Mbr, Exec Bd, Council of All Sections E of MS, ASEE, 1970- 71. Secy and mbr, Exec Bd, Council of All Sections in NE, ASEE, 1971-72. Active in society committees up to natl level. Author of number of papers on innovative methods for instruction of engg students. Recipient from ASEE of the Western Elec Fund Award for Excellence in Teaching, 1977. *Society Aff:* ASEE, ASME, NSPE, AAUP, ΦBK, TBΠ, ΠΤΣ.

Pelczarski, Eugene A
Business: 600 Seco Rd, Monroeville, PA 15146
Position: Chairman of Board. *Employer:* CSE Corporation. *Born:* Jan 8 1930. BS, MS & PhD in Chem E Univ of Pittsburgh. Served with USAF in Korea 1950-54; 1966 V Pres Engrg at Black Sivalls & Bryson in Kansas City; founded & became Pres of Applied Tech Corp, Pittsburgh PA in 1969; both companies are subsidiaries of Internatl Systems & Controls; Chmn of CSE Corp 1974 to present, engaged in mfg mining equip, fans, instruments & the distribution of health & safety prods pertinent to mining.

Pell, Jerry
Home: 10310 Tailcoat Way, Columbia, MD 21044-3809
Position: Sr Environmental Scientist *Employer:* Office of Clean Coal Technology *Education:* Ph.D./Meteorology/McGill University, Montreal, Canada; M.Sc/Meteorology/McGill Univesity, Montreal, Canada; B.Sc/Physics/Math/McGill University, Montreal, Canada *Born:* 02/08/42 Sr Environmental Scientist, Clean Coal Technology Prog, U.S. Dept of Energy, Washington, 1983-present. Corp Rep for UOP Inc. (Signal Cos.), 1982-83. Various environmental and coal-related capacities, U.S. Dept. of Energy, 1975-82. State of MD, Power Plant Sitting Program, 1972-75. Asst Prof, Meteorology, Rutgers Univ, 1969-71. Certified Consulting Meteorologist of the Amer. Meteor. Soc. Past Chmn, ASME Research Cttee on Cooling Towers 1973-74. Member, Air Pollution Control Assoc (APCA), and Past Chmn, APCA Meteorology Cttee 1979-81. Past Chmn, APCA Government Affairs Seminar Cttee 1981-82. APCA delegate to AAES Public Affairs Council, 1979-82. Chmn, AAES Coordinating Cttee on Environment, 1980-82. Member, AAES Coordinating Cttee on Energy, 1980-present. *Society Aff:* ASME, APCA, AMS, AAPT

Pellett, C Roger
Business: Pipe Fabrication Div, 1115 Gilman St, Marietta, OH 45750
Position: VP, Commercial Services *Employer:* Dravo Corp. *Education:* MBA/Finance/OH Univ; BSE/Naval Arch & Marine Engrg/Univ of MI. *Born:* 6/4/43. Reg PE in the State of OH, grad of the Bettis Reactor Engrg Sch. Work experience: 4 yrs, Nucl Power Proj Engr for the Naval Reactors Branch of the US Atomic Energy Commission. 2 yrs, Quality Supervisor, Dravo Corp, Pipe Fabrication Div. 7 yrs, Chief Estimator, Power Fabrication, 3 years, Chief Engr Dravo Corp, Pipe Fabrication Div, 6 years, Manager of Estimating Dravo Corp, Pipe Fabrication Div.

Pellissier, George E
Home: 907 Inman Rd, Schenectady, NY 12309
Position: Manager, Metallurgy, Adv. Tech., Dept *Employer:* Mechanical Technology, INC. *Education:* Matl Sci/Metallurgy/Carnegie-Mellon Univ; MChem/Chemistry/Cornell Univ; BChem/Chemistry/Cornell Univ. *Born:* March 1915, Springfield MA. Graduate studies/res (INCO Fellow) Carnegie Inst Tech 1938-41 & Stevens Inst Tech 1941-42. 1942-45 Grp Leader Met, SAM Labs Columbia Univ, Manhattan Proj; 1945-68 dev engr/res assoc/Div Ch R&D Lab/ Mgr Adv Applied Res with US Steel Corp; 1968-72 Lab Mgr with E F Fullam Inc; 1972- 77, Mgr Mech & Matls Dept with RRC Internatl Inc. Mgr, Metallurgy with Mech Technology Inc 1977- . Fellow ASM & AIC. *Society Aff:* ASM, ASTM, ECS, AIC, EMSA.

Pelloux, Regis M
Business: Room 8-237 77 Mass Ave, Cambridge, MA 02139
Position: Professor. *Employer:* Massachusetts Institute Technology. *Education:* Engg Dip/Mech Eng/ECP-Paris; MS/Metallurgy/MIT; ScD/Metallurgy/MIT. *Born:* Dec 1931. 1958-60 French Army military ser; 1960-68 sr scientist, The Boeing Co Scientific Res Labs; 1966-67 Visiting Assoc Professor, Dept of Mech Engrg MIT, Professor 1968-present Dept of Materials Sci & Engrg. Technical consultant: failure analysis, fatigue & fracture toughness, creep-fatigue interaction & fractography. Mbr ASM, AIME, ASTM. Interested in skiing & sailing. *Society Aff:* ASM, AIME, ASTM.

Peltier, Eugene J
Business: 8 Ladue Forest, St Louis, MO 63124
Position: Consultant. *Employer:* Self Employed *Education:* BS/Civil Engr/KS State Univ. *Born:* 3/28/10. Hon Doc of Laws KS Univ-1961; 1934-40 Res Engr to KS Hwy Comm; 1936 Lt (JG) Civil Engr Corps USNR, 1940 active duty, 1946 regular Navy: Public Works Officer & other activities; 1945 Officer in Charge of Naval Const Regt, Okinawa; 1957 Ch Bureau of Yards & Docks with rank of Rear Adm; 1962 retired, awarded Legion of Merit. 1962-75 with Sverdrup & Parcel & Assocs Inc, engrs/architects/planners of St Louis: V Pres, 1964 Sr VP, 1966 Exec V Pres, 1967 Pres, 1973 Ch Exec Officer, 1975 retired; presently Cons with firm. P Pres & Dir Amer Rd & Transportation Builders Assoc; P Pres Soc American Mil Engrs. 1962 CEC Award of Merit; 1960 APWA Top "Ten Public Works Men" of the Year Award; 1974 MSPE (St Louis) Eng of the Yr Award & 1973 Amer Inst of Steel Const Spec. 1979 National Academy of Engineering; 1980 Honorary Member Amer Soc Civil. Engineers. *Society Aff:* ASCE, SAME, NSPE, CEC.

Pence, Ira W
Business: 685 West Rio Rd, Charlottesville, VA 22906
Position: Mgr-Engg. *Employer:* Genl Elec Co. *Education:* PhD/EE/Univ of MI; MS/EE/Univ of MI; BS/EE/Univ of MI. *Born:* 6/18/39. MTS of the Univ of MI's Inst ot Sci and Technology 1960-70. Participated in design and installation of major astronomical observatory on Mt Haleakala, HI. MTS of Genl Electric's Corp Res & Dev Ctr 1970-79. Respon for solid-state microwave pwr source res, electron beam memory and lithography dev, and later for coordinating the res activity on behalf of

Pence, Ira W (Continued)
Industrial Products sector of GE. Assumed current respon for all engg within the Industrial Control Dept of GE in 1979. Respon for design, dev, and reliability of numerical and plastic molding controls and other industrial electronic products including Robots and Factory Automation. Chrmn Schenectady Chapter IEEE 1977, Area Chrmn 1978-80, Region I Committee Mbr 1977-80. *Society Aff:* IEEE.

Pence, Robert F
Home: Route 8, Box 388, Fort Worth, TX 76108
Position: Assistant Project Engineer. Retired *Employer:* General Dynamics Fort Worth Division. *Education:* BS/Mech Engr/Univ of TX, Austin. *Born:* Aug 1920, Zanesville OH. BSMechE from Univ of TX, Austin. With General Dynamics from 1943 to retirement in 1985 in structural design, structural analysis, mass properties & Proj coordination. Projects include: B-36, YB-60, B-58, F-111, Space Shuttle & F-16. Mbr Soc of Allied Weight Engrs since 1952: paper 'Designing Weight Out of the B-36' 1952 SAWE Natl Conference, Natl Dir from TX Chapter 1955-56, Natl Pres 1956-57, elected SAWE Honorary Fellow 1958. Chmn Fort Worth Regional Sci Fair 1963. Flight instructor & Operator of airstrip at home, where 32 aircraft are based. *Society Aff:* ΤΒΠ, ΠΤΣ, SAWE, AOPA, ABS.

Pendleton, Joseph S, Jr
Home: Box 122 RD2, Fleetwood, PA 19522
Position: Ret. (Metallurgical Consultant) *Education:* AB/History/Princeton Univ. *Born:* Sept 11 1917, Reading PA. 33 years with Carpenter Tech in tool & alloy met. P Chmn Tool Steel Tech Ctte AISI, P Chmn Lehigh Valley Chapter ASM, p mbr ASM Natl Chapter Ctte. *Society Aff:* ASM, ASTME, AIME.

Pendleton, Roger L
Home: 8909 Cromwell Dr, Springfield, VA 22151
Position: Consultant *Employer:* Self *Education:* BS/Engrg/Univ of ME; Post Grad/Engr Admin/George Washington Univ; Post Grad/Sys Engr/VA Poly Inst & State Univ; -/Military Sci Study/Army War College; -/Military Sci Study/Air War College *Born:* 3/2/23. Native of Carmel, ME. Reg PE (VA and 2 Addn States). With Dept of Air Force in various sr civ engrg positions, 1953-55; 1956-67; Proj Engr - Midwest Engr & Construction Off, Kansas City, MO - 1955-56. With DCA 1967-85. Served as prin tech advisor, consultant to Agency in areas of civ, electrical, structural, facility and systems engrg for command, control and communications facilities (R&D). 1985 to Present: consulting Engineer to architect-engineer firms and Defense Contractors; and published author to engrg and military journals. Veteran of WWII and Korea. Col USAR Ret. Last "active reserve" assignment with US Army Corps of Engrs as "Deputy Asst Chief of Engrs" (Received "Meritorious Service Medal with First Oak Leaf Cluster" in 1979); Pres - VA Soc of PE, 1969-70 (Received "Distinguished Service Award" in 1972); Natl Dir - NSPE, 1970-72; Treas-VA Assoc of Professions (VAP), 1973-74 (Received "Outstanding Professional of the Year Award" in 1971. Co-Chmn VA Link-Order of the Engineer (OOE), 1980-81; Pres - Springfield Kiwanis Club, 1971-72 & 1979-80, (rece ived "Distinguished Club Pres Award" in 1980); Lt. Governor 20th Div, Capital District, Kiwanis Intl, 1981-82; Listed in "Who's Who in the South & Southwest-; Member of Masons, Shrine and Royal Order of Jesters. Enjoy sports music. "Distinguished Lieutenant Governor Award" in 1982 from Kiwanis Intl. *Society Aff:* NSPE, SAME, ASCE, WSE, ROA, TROA, VAP, AFCEA

Pendleton, Wesley W
Home: 1542 Clinton St, Muskegon, MI 49442
Position: Retired (Consultant) *Education:* MS/EE/MIT; BS/EE/URI. *Born:* 4/2/14. Native of Cranston, RI. 1936-38 student engr GE; 1938-41 lab asst MIT; 1941-48 employed in insulation dev Westinghouse; 1948-50 res on cables, Gen Cable; 1950-80 Magnet wire dev, Anaconda. Fellow IEEE 1935-81. Life Mbrship 1980. Joint author of book *Materials for Electrical Insulating & Dielectric Functions*, Hayden 1974. AD. Comm. IEEE Insulation G-32. Enjoy bowling, golf, cactus raising. Retired 5/1/80. *Society Aff:* IEEE, NSI, TORCH, Platform.

Penn, William B
Home: 4616 Sunnydale Blvd, Erie, PA 16509
Position: Consultant *Employer:* Self *Education:* MS/Chem Engg/MIT; BS/Chem Engg/MIT. *Born:* 5/2/17. Consultant on Electrical insulation systems after 42 yrs with G.E.: developed sys for destroyer escort generators, aircraft generators, ac & dc motors, aerospace machines, Trident Submarine MG sets, etc. Sr Mbr ACS; Life Fellow IEEE, active in IEEE Insulation Stds, P CHmn Insulation Subctte, Rotating Machinery Ctte; Past President Elec Insulation Soc; Technical paper reviewer for Elec/Electronic Insulation Conference. Reg PE. Hobby - family photography. *Society Aff:* IEEE, ACS.

Pennacchio, Vito F
Home: 33 Blake Rd, Weymouth, MA 02188
Position: President. *Employer:* Coffin & Richardson, Inc. *Education:* BS/Civil Eng/Northeastern Univ. *Born:* May 1928. 1950-52 served with Army Corps of Engrs; 1952-53 Asst Plant Engr for Carpenter & Patterson Inc; 1953-54 civil engr with Clarkeson Engrg Corp Inc; 1954-56 Asst Town Engr, Town of Norwood MA; since 1956 with Coffin & Richardson, Currently Pres. Mbr ASCE, NEWWA, AWWA, NAWC, ASA; Mbr AWWA, NEWWA & NAWC Cttes on Water Rate Design. Reg PE. Qualified as expert on utility valuations and on water rates before courts and regulatory agencies. Authored & Presented several papers on water rates, valuation of property, and related subjects. *Society Aff:* ASCE, NEWWA, AWWA, NAWC, ASA.

Penner, Stanford S
Mail Code B-010, AMES, LaJolla, CA 92093
Position: Dir Energy Ctr. *Employer:* Univ of CA/San Diego. *Education:* PhD/Phys Chemistry/Univ of WI; MS/Phys Chemistry/Univ of WI; BS/Phys Chemistry/Union College. *Born:* 7/5/21. Allegany Ballistics Lab (1944-45); Esso Res (1946); JPL (1947-50); CalTech (1950-64); Dir. Res and Engg Div, Inst for Defense Analyses (1962-64). At UCSD, Prof and First Chrmn, Dept of Aero & Mech Engg Sciences (1964-68); Dir, Inst for Pure & Appl Physical Sciences (1968-72); VChancellor-Academic Affairs (1968-69); Dir, Energy Ctr, since 1972. Author or co-author of 265 articles and of 12 books and monographs, editor of 7 books and Founder-Editor of JQSRT (since 1960), of Energy (since 1976), and of the *Journal of Missile Defense Research* (1964-68). Mbr, Engg Sciences Sec, Intl Academy of Astronautics; Natl Academy of Engg; Appl Sci Section, Amer Acad of Arts and Sci. AIAA G Edward Pendray, Thermophysics and Energy Systems Awards; Numa Manson Medal of Intl Colloquia on Gasdynamics; Hon doctorate, Rheinisch-Westfalische Technische Hochschule, Aachen, for dist research, teaching and public service. Res interests: gasdynamics of reactive sys, combustion sci, radiative heat tranfer, reentry phenomena, energy sys. *Society Aff:* AIAA, APS, OSA, AAAS, COMBUSTION INSTITUTE, ΣΞ, NYAS

Pennoni, Celestino R
Business: 1911 Arch St, Phila, PA 19103
Position: Pres. *Employer:* Pennoni Assoc Inc. *Education:* MS/CE/Drexel Univ; BS/CE/Drexel Univ *Born:* 12/31/37. in Plains, PA. Worked for several yrs for governmental & private organizations; started firm in 1966 with one person. Firm now employes over 80 personnel providing engg, architectural & related services to Townships, Boroughs, Authorities & clients in the commercial & ind sector. Taught & appeared as guest lectr in area colleges. Serve as Authority Bd Member, American Arbitration Association & Mayor's Sci & Tech Advisory Council; active in PSPE, ASCE, CEC & various Profl & Technical Socs. Young Engineer of Year 1971-PSPE: Honor Man of Year 1973 - Drexel University: Legion of Honor - Chapel of Four Chaplains. Prof Dev Awd, NSPE, 1984; Prof Dev Awd, NJ Soc Prof Engrs, 1983. *Society Aff:* ASCE, NSPE, ACEC

Pense, Alan W
Business: Room 310, Packard Lab #19, Lehigh Univ, Bethlehem, PA 18015
Position: Prof & Assoc Dean, Assoc. Director NSF ERC Center *Employer:* Lehigh Univ. *Education:* PhD/Metallurgy/Lehigh Univ; MS/Metallurgy/Lehigh Univ; BMetE/Metall/Cornell Univ. *Born:* 2/3/34. Native of NY State. Entered teaching Lehigh Univ 1960. Currently Assoc Dean, Coll of Engrg & Applied Sci. Assoc. Dir, NSF ERC Advanced Technology large structural systems. Conducted res on welding, low alloy steels, Fracture mechanics & failure analysis. Teach physical metal of alloys, welding. Failure analysis. Chrmn AWS Handbook Committee 1974-1977. AWS Sparagan award 1963, Addams Mbrship 1966, Jennings award 1970, Addams Lectureship 1980, Hobart Medal 1982. Honorary membership, 1986. Published 75 technical papers, Coauthored 1 book. Lehigh Univ. Teaching Awards 1965, 1976, 1985. Collects antiquities. *Society Aff:* ASM, ASTM, AWS, ASEE, IIW, PVRC

Pentecost, Joseph L
Business: School of Materials Engrg, Ga. Tech, Atlanta, GA 30332
Position: Director, School of Ceramic Engrg. *Employer:* Georgia Institute of Technology. *Education:* PhD/Cer Engg/Univ of IL; MS/Cer Engg/Univ of IL; B. Cer E/Cer Engg/GA Inst of Tech. *Born:* April 1930, native of Winder GA. 1951-53 served with USMCR; 1956-59 with Melpar Inc, 1962-68 supervised materials R&D, later became Assoc Dir of Res; 1959-60 Ch Res Engr Aeronca Mfg Corp; 1968-72 Mgr W R Grace & Co, developed a process for ceramic honeycomb fabrication, process for oxide raw materials preparation; 1972 became Dir Sch of Ceramic Engrg, GA Inst of Tech. Pres of Natl Inst of Ceramic Engrs 1975-76, Trustee AM. Cer. Soc 1980-1983. PACE Award, NICE 1967. Patents pub, articles in field. Enjoy hunting, amateur radio, jewelry making. VP ACS 1983-84; Treas ACS 1984-85; Pres-Elect 1985. *Society Aff:* ACS, NICE, ASM.

Penzias, Arno A
Business: 600 Mountain Ave, Room 6A-409, Murray Hill, NJ 07974
Position: VP, Research *Employer:* AT&T Bell Labs *Education:* Ph.D./Physics/Columbia U.; MA/Physics/Columbia U.; BS/Physics/City College of NY *Born:* 04/26/33 VP of Research at AT&T Bell Labs, Arno Penzias joined Bell Labs in 1961. Participated in Echo and Telstar communications satellite experiments, discovered evidence supporting "big bang" theory, for which he shared '78 Nobel Prize for Physics. He has held adjunct positions at Princeton, Harvard and SUNY Stony Brook. He is Vice Chmn, Cttee of Concerned Scientists. Recipient of many awards and prizes and eleven honorary degrees, he has written over 80 published articles and holds several patents. *Society Aff:* AAAS, AAS, APS, NAS

Penzien, Joseph
Room 731 Davis Hall, Berkeley, CA 94720
Position: Professor of Structural Engineering. *Employer:* University of California, Berkeley. *Education:* ScD/Structural Engg/MIT; BS/Civil Engg/Univ of WA. *Born:* Nov 27, 1924. 1950-52 staff mbr Sandia Corp, Albuquerque; 1952-53 Sr Structures Engr Consolidated Vultee Aircraft Corp, Fort Worth; 1953-present faculty mbr Univ of CA, Berkeley, Dir Earthquake Engrg Res Ctr 1968-73. Reg CE in CA & WA. Mbr ASCE, Structural Engrs Assoc of CA, Earthquake Engrg Res Inst, SSA & ACI. Received 1959 NSF Post Doctoral Fellowship. Mbr, Natl Acad of Engg, 1965 Walter Huber Research Prize ASCE, 1969 NATO Sr Science Fellowship, 1973 NSF Sr Sci Fellowship, 1983 ASCE Nathan M. Newmark Medal, 1986 ASCE Alfred M. Freudenthal Medal. *Society Aff:* ASCE, ACI, SEAOC, EERI, SSA.

Peot, Hans G
Home: .6425 S Scarff Rd, New Carlisle, OH 45344
Position: Chief Avionics Division. *Employer:* Wright-Patterson AFB. *Born:* June 4 1931. BS from Wash State; MS from Air Force Inst of Tech. 195458 Officer in USAF; since 1954 employed at Wright-Patterson AFB: major tech responsibility in dev of Air Force's Airborne Warning & Control System (AWACS), involved with dev of B-1 strategic bomber from its inception, as Ch Avionics Engr on that program, brought the offensive avionics complement from conception to successful flight test in B-1 aircraft at minimum cost to government; currently Ch Avionics Div, Directorate of Systems Engrg. 1965 Exceptional Civilian Service Award; 1975 AFSC Certificate of Merit; 1959, 1969 & 1975 Outstanding Performance Awards.

Pepe, John J
Business: 6301 Richmond Avd, Houston, TX 77057
Position: Owner. *Employer:* John J Pepe, Consulting Engineers. *Education:* MSCE/Civil Eng/Univ of MN; BCE/Civil Eng/Manhattan College. *Born:* 1927. BS Manhattan Coll, MS Univ of Minn. 1950-53 1Lt US Army; 1954-60 Design Engr, Assoc Engr & Proj Engr for civil engrg projs with Frank Metkyo, Millwee & Assocs & J B Dannenbaum Cons Engrs, respectively. 1960-present, Owner of John J Pepe Cons Engrs; firm primarily concerned with all phases of civil engrg for municipal & indus projs such as, residential industrial devs, water supply, sewage, land use drainage, waste treatment, mass transit, highways and bridges. Mbr Regional Export Expansion Council, Texas Council on Marine Related Affairs, TSPE, AWWA, WPCF & P Dir of CEC Texas. *Society Aff:* ACEC, NSPE.

Peppas, Nikolaos A
Business: Sch of Chem Engg, W Lafayette, IN 47907
Position: Prof *Employer:* Purdue Univ. *Education:* ScD/Chem Engg/MA Inst of Tech; DiplEng/Chem Engg/Natl Technical Univ of Athens. *Born:* 8/25/48. Native of Athens, Greece, Tenured Full Prof since 1980. Res and teaching in the areas of diffusion in polymers, structure and morphology, polymer surfaces and biomedical mtls. Ed of the Journal Biomaterials. Mbr of the editorial bd of the Journal of Applied Polymer Sci, also Editorial Bds J Controlled Release, Biomedical Polymers, Pres of the Controlled Release Society. V Chmn Materials Division AIChE, Author of seven books, 270 res publications and 50 abstracts. Chrmn of polymer res symposia since 1976. Recipient of Materials Award of AIChE 1984, Zyma Foundation (Switzerland) International Award for Advancement of Medical & Biological Sciences, 1981, ASEE Western-Electric Fund Award 1980, Shreve Teaching Award, 1978, 1980, 1982, 1985. Purdue's Potter and Shreve Teaching Awards, 1978, 1985, Best Councelor AIChE, 1982. Extracurricular activities include opera and Greek and Roman archaeology. *Society Aff:* AIChE, ACS, APS, SPE, ASAIO, ISAO, ASEE, SR, SB, AAAS, NYAS, SES, CRS

Peppin, Richard J
Home: 5012 Macon Rd, Rockville, MD 20852
Position: Pres. *Employer:* Scantek, Inc. *Education:* MS/ME/Rensselaer Poly Inst; MS/Theor & Appl Mechanics/WV Univ; BE/ME/CCNY. *Born:* 2/18/43. Initial engrg work in heat transfer and lubrication. Since 1966 has been involved in appl mechanics (finite elements and finite elasticity), acoustics, vibrations, energy conservation, and safety and health with industrial, academic, governmental and consulting experience. Currently VP of the Inst of Noise Control Engrg, & subcommittee chairperson and task group leader in ASTM, among other activities. Formerly, Chrmn of the Tech and Soc Div of ASME. Editor of *Community Noise, Measurements for Industrial Noise Control*, and author of numerous papers and book reviews. Reg PE in NY, NJ, MD & Ontario Canada. US Designated Expert to ISO on impedance tubes. *Society Aff:* INCE, ASME, ASHRAE, ASA, IES.

Peralta, Manuel
Home: 1316 New Hampshire Ave., NW, Apt. 608, Washington, DC 20036
Position: Special Assistant for Institutions *Employer:* National Aeronautics and Space Administration *Education:* MSCE/Civil Engg/NY Univ; BSCE/Civil Engg/NY Univ. *Born:* 2/14/34. Mr. M. Peralta...has over 30 years experience in domestic and international business, engineering, and project management. His current position is with NASA as Special Assistant to the Administrator. Prior to his current position, he was a Senior Executive with Exxon responsible for managing the implementation of world-wide capital projects valued at over $15 billion. In addition, he was responsible for a multi-national company located n the U.K. providing a full range of engineering services to Exxon's European affiliates. His career also involved corpo-

Peralta, Manuel (Continued)
rate level responsiblity for Exxon's Coal and Synthetic Fuels business...and, a special assignment as Head of a Grace Commission team that reviewed the Defense Department's Weapons Acquisition Program. He is Vice-Chairman of the American National Standards Institute and Board Member of AIChE's Engineering and Construction Contracting Division...and, a former Member of the Board of the Construction Industry Institute. He holds an MCE (1960) and a BCE (1956), both from New York University. He also attended Cornell University's Advance Executive Program in 1972. *Society Aff:* ASCE, AMA.

Perepezko, John H
Business: Dept of Met Engg, 1509 Univ Ave, Madison, WI 53706
Position: Prof *Employer:* Univ of WI. *Education:* PhD/Met Mat Sci/Carnegie Mellon Univ; MSC/Met/Poly Inst of NY; BSC/Met/Poly Inst of NY. *Born:* 8/26/45. Res Scientist US Steel Corp 1968-70. With Univ of WI since 1975. Res & teaching interests in physical met with specialization in diffusional nucleation & growth reactions in solids & in liquid metals. Patents & consultant in solidification processing. Active in several technical committes of ASM, TMS/AIME. ASM Bradley Stoughton Best Young Prof Award 1978, Creativity Award (NSF) 1982, Editorial Bd-Int J. Rapid Solidification, ASM Fellow 1987. *Society Aff:* ASM, TMS/AIME, Electrochem Soc, MRS.

Peretti, Ettore A
18295 Brightlingsea, South Bend, IN 46637
Position: Asst Dir. *Employer:* Natl Consortium for Grad Deg for Min in engg. *Education:* BS/Metal Eng/MT College of Sci & Tech; MS/Metal Eng/MT COllege of Sci & Tech; ScD/Metal Eng/Univ of Stuttgart; Met E(HC)/metal Eng/Univ of MT. *Born:* April 5 1913, Butte MT. Metallurgical Engr (Honoris Causa) from Univ of Mont. 1936-40 taught courses in met engrg at Mont Coll of Mineral Sci, 1940-46 taught at Columbia Univ; 1946-78, at Univ of Notre Dame, Chmn of Dept of Met Engrg 1951-69, Asst Dean of Coll of Engrg 1970-78. 1978 - Assistant Director, Natl. Consortium for Grad. Deg. for Minorities. Mbr AIME, ASM & ASEE. Elected to Theta Tau, Tau Beta Pi, Alpha Sigma Mu (P trustee & Internatl Pres) & Sigma Xi. Author num tech pubs on met. Elected Fellow of ASM. Reg PE in Ind. *Society Aff:* AIME, ASM, ASEE.

Perez, Jean-Yves
Business: 4582 S. Ulster St. Pkwy, Suite 1000, Denver, CO 80237
Position: Principal & Exec VP *Employer:* Woodward - Clyde Consultants *Education:* MS/CE/Univ of IL, Urbana; Ineinieur/CE/Ecole Centrale des Arts et Manufactures, Paris *Born:* 3/3/45 Has over twenty yrs of experience in soil and rock mechanics and foundation engrg. He has been responsible for geotechnical investigations, quality control, design recommendations, and project supvr for high-rise buildings, power plants, tunnels and underground openings, deep supported excavations, locks, dams, cutoffs, and site developments. For the last six years has directed investigations, feasibility studies and designs for hazardous waste sites involving soil and ground water contamination. Currrently manages the Central Operating Group of Woodward-Clyde Consultants. *Society Aff:* ASCE, SAME, ISSMFE, USCOLD, AEG

Pergola, Nicola F
Home: 87-89th St, Brooklyn, NY 11209
Position: Marine Engrg Cons *Employer:* Self *Education:* MSME/Heat-Power/Columbia Univ; BSMarine Engg/Power/US Merchant Marine Acad. *Born:* 1/11/27. Resident of NYC. USCG lic engr. Formerly with Energy Transportation Corp, operator of fleet of 125,000 cubic meter lng carriers. Responsibilities ranged from engg plan approval, supervision of construction, dev of manning scales. Negotiated original union contracts. Developed planned maintenance and inventory control prog. Managed guarantee claims program and shipboard personnel labor relations. Experience also in machinery, plant and system design, procurement and ship operations. Holder of several US patents. Man of the Yr for outstanding achievement in the field of Marine Engg awarded by peers. Past chrmn of NY Metropolitan Sec of the Soc of Naval Architects & Marine Engrs. Enjoy skiing and tennis. *Society Aff:* SNAME, ASME, ASNE, SOMPENY, IMar.E.

Perini, Jose
Home: 5093 Skyline Dr, Syracuse, NY 13215
Position: Prof. *Employer:* Syracuse Univ. *Education:* PhD/EE/Syracuse Univ; BS/EE/ME/Escola Politecnica De Sao Paulo, Brazil *Born:* 3/1/28. Native of Sao Paulo, SP, Brazil. Mgr Radio Maintenance Dept, Real Transportes Aereos, Brazil, 1952-55; Asst Prof EE 1955-58, Assoc Prof EE 1961- 62, Escola Politecnica de Sao Paulo, Brazil; Consultant: GE Co 1959-61 and 1962- 69, Syracuse, NY in TV Transmitting Antennas, GE Co, S Paulo, Brazil 1962-62 in TV Receiver's Design, SETASA, S Paulo, Brazil, 1976 in TV Network Design; Pres Data Functions Corp 1963-69, Syrcacuse, NY, Asst Prof of EE 1962-66, Assoc Prof EE 1966-71, Prof EE 1971 to present at Syracuse Univ. Consultant USAF, NSWC, NUSC, Syracuse Res Co. *Society Aff:* IEEE, ASEE

Perkins, Courtland D
Business: 2101 Constitution Av NW, Washington, DC 20008
Position: President. *Employer:* National Academy of Engineering. *Education:* MS/Aero Eng/MIT; BS/Mech Engg/Swarthmore. *Born:* Dec 1912, Philadelphia PA. D. Eng (Hon) RPI, Swarthmore, Lehigh. During WWII headed Stability & Control unit of Aircraft Lab US Army Air Corps, Wright Field; joined Princeton Univ in 1945, appointed Professor in 1947, Chmn of Aeronautical Engrg Dept in 1951, Chmn of Dept of Aerospace & Mech Sciences in 1963, named Assoc Dean of Sch of Engrg & Applied Sci in 1964; appointed Ch Scientist of the USAF 1956-57 & Asst Secy of USAF for R&D 1960-61; has served on USAF Scientific Advisory Bd since 1946 and as Chrmn of Bd from 1969-72 and 1977- 78; 1963-67 was Chmn of Adv Grp for Aeronautical R&D, NATO; Mbr of Defense Sci Bd 1969-73. Pres AIAA 1964; Fellow of Royal Aeronautical Soc; French Legions of Honor, Hon Fellow AIAA; Mbr NAE. Elected Pres of Natl Acad of Engrg in 1975. *Society Aff:* NAE, AIAA, RAS, AHS, AAAS.

Perkins, Henry C
Business: Aerospace-Mech Engr Dept, Tucson, AZ 85721
Position: Prof. *Employer:* Univ AZ. *Education:* PhD/Mech Engr/Stanford Univ; MS/Engr Sci/Stanford Univ; BS/Phys/Stanford Univ. *Born:* 11/23/35. Miami, FL, married, four children. Yuba Fellow 60-61, Atomic Energy Commission Fellow 61-62. Res Assoc, Stanford Univ 62-63; Acting Asst Prof, Stanford Univ, 63-64. Assoc Prof, Univ AZ 64-67, Prof 67-. Visiting Prof Tech Univ Denmark, 71-72, US Military Acad, 78-79, Stanford Univ, 81-82. Author "Air Pollution–, 1974, co-author (W C Reynolds) "Engr Thermo–, 1970, 1977, co-author "Handbook Heat Transfer–, 1973. *Society Aff:* ASME, ΣΞ, ΤΒΠ

Perkins, Kendall
Home: 7742 Wise Av, St Louis, MO 63117
Position: Corp V Pres Engrg & Res (Retd.). *Employer:* McDonnell Douglas Corp. *Education:* D Eng/Hon/Tri State College; BSEE/Elec Engg/WA Univ. *Born:* 2/23/08. 1967-73 & 1975-78 McDonnell Douglas Corp; Corp VP Engg & Res 1941-67 McDonnell Aircraft Corp Proj Engr, Asst Ch Engr, Mgr of Engg & Engg VP respon for aircraft, spacecraft & missile engg; formerly Hd of Aircraft Scheduling Unit, War Prod Bd, Res Engr Amer Airlines; and Project Engr Curtiss Robertson Airplane Mfg Co. Mbr natl Acad of Engg. Fellow Amer Inst of Aeronautics & Astronautics. Trustee Emeritus WA Univ. *Society Aff:* NAE, AIAA.

Perkins, Richard W
Business: Dept Mech & Aero Engg, Syracuse, NY 13244
Position: Prof and Chairman, Dept of Mech and Aerosp Engrg *Employer:* Syracuse Univ. *Education:* PhD/Wood Prod Eng/SUNY Syracuse; MS/Wood Prod Eng/SUNY Syracuse; BA/Economics/Dartmouth *Born:* 4/28/32. Native of Poughkeepsie, NY. Served in US Navy 1954-1956. Taught Wood Products Engg at SUNY College of Environmental Sci and Forestry 1959-1963. Assumed current position in 1964.

Perkins, Richard W (Continued)
Tau Beta Pi Eminent Engr award 1979. *Society Aff:* ASME, ACAD MECH, TAPPI, SES.

Perkins, Roger A
Business: O/52-31, B/204, 3251 Hanover St, Palo Alto, CA 94304
Position: Sr Mbr-Res Laborator. *Employer:* Lockheed Palo Alto Res Lab. *Education:* MS/Met Engr/Purdue Univ; BS/Met Engr/Purdue Univ. *Born:* 5/28/26. Roger A Perkins is a Sr Mbr of the Lockheed Palo Alto Res Lab. He received his BS & MS Degrees in Metallurgical Engg from Purdue Univ. From 1951-1959, he was a Research Met at the Union Carbide Metals Co, Niagara Falls, NY. Before joining Lockheed in 1960, he was hd of the Refractory Metals Application Sec, Aerojet-Gen Corp, Sacremento, OK. At Lockheed, Mr Perkins is a group leader for res on high temperature corrosion & projective coatings. He is recognized intlly for his work in the field of high temperature gas-metal reactions. He is active in profl socs on a local & natl level & is a Fellow of Am Soc of Metals. *Society Aff:* ASM, AIME.

Perkins, Thomas K
Home: 6816 Stichter, Dallas, TX 75230
Position: Distinguished Research Advisor *Employer:* Arco Oil & Gas Co. *Education:* PhD/Chem Engr/Univ of TX; MS/Chem Engr/TX A&M Univ; BS/Chem Engr/TX A&M Univ. *Born:* 1/31/32. Employed by Dow Chem Co in 1952. He became instr in 1955 and asst prof in 1956 in the Chem Engg Dept of the Univ of TX. Joined Arco in 1957 and has supervised res in hydraulic fracturing, drilling, reservoir mathematics, pipeline design and operation, well completions in permafrost, arctic operations, and well, reservoir, and earth mechanics. He has received more than 30 patents and has published 38 technical papers. He served as Distinguished Lecturer in 1977-78 for SPE of AIME, and was recipient of the Lester C Uren Award in 1978. He received the A.R. Co. Outstanding Technical Achievement Award in 1980. Mbr Natl Acad of Engrg 1984. *Society Aff:* SPE, ΣΞ.

Perkins, William R
Business: Dept of Elec & Comp Engrg, 1406 W Green St, Urbana, IL 61801
Position: Professor. *Employer:* Univ of Illinois. *Education:* PhD/EE/Stanford; MS/EE/Stanford; AB/Engg & Applied Physics/Harvard. *Born:* 9/1/34. Council Bluffs IA. Instructor Stanford Univ (Elec Engrg) 1959-60; Univ of IL: Asst Prof 1961-65, Assoc Prof 1965-69, Prof of Elec Engrg & Res Prof Coord Sci Lab 1969- . Assoc, Center for Advanced Study Univ of IL 1971-72. Fellow IEEE. Major tech interests: control sys, feedback theory, large-scale systems, parameter sensitivity. Have served on many IEEE Cttes incl IEEE Control Sys Soc AdCom, TAB Pubs Ctte, Control Sys Soc Awards Ctte (Chmn 1975-78). Associate Editor at Large, IEEE Trans. on Automatic Control, 1986-87. Pres, IEEE Control, Sys Soc, 1985. IEEE Dir, American Automatic Control Council, 1984. *Society Aff:* IEEE.

Perlman, Theodore
P O Box 808, Livermore, CA 94550
Position: Consultant to LLNL. Retired *Employer:* Univ of Calif Lawr Livermore Natl. Lab. (LLNL) *Education:* BS/Mech Engg/LA State Univ. *Born:* 9/30/23. Dev Engrg with Manhattan Proj at Los Alamos & Pacific Theater of opers 1944-46; Engr Cyclotron Lab Univ of Rochester 1946-49; Nuclear Weapons Engr Sandia Corp 1950-56; joined Lawrence Livermore Lab 1956; Polaris Warhead Proj Engr, Sect Head, Div Leader of Weapons Engrg Div conducting engrg dev of nuclear warheads; appointed Div Leader of Res Engrg Div in 1973 managing engrg support of R/D activities in Chem, Materials Sci, Physics, Biomedical & Environ Scis, Energy & Hazards Control; Retired in December 1985; consultant to LLNL.

Perlmutter, Daniel D
Business: Univ of Pennsylvania, Philadelphia, PA 19104
Position: Prof Dept Chem/Biochem Engg. *Employer:* Univ of Pennsylvania. *Education:* PhD/ChE/Yale; BS/ChE/NY Univ. *Born:* May 24, 1931. R/D with Exxon & Shell Dev Co 1955-58, 1961; Asst Prof Chem Engrg Univ of Illinois 1958-64; Res Fellow Harvard Univ 1964-65; Univ of PA 1964- : Prof 1967- & Chmn 1972-77 and 1985-86. Honors have incl ASEE Chem Engg Lectureship Award for 1979, Fulbright Lecturer, Fulbright Sr Res Fellow, Guggenheim Fellowship, Election to N Y Acad of Sci, Ford Foundation Faculty Dev Grant. Major res interests are chem reactor design, automatic control, stability, optimization, environ & energy problems, gas-solid reactions. Author of textbooks: Introduction to Chem Process Control & Stability of Chem Reactors. Married, 3 children. *Society Aff:* AIChE, AAAS.

Perlmutter, Isaac
Home: 150 Trailwoods Drive, Dayton, OH 45415
Position: Cons Aerosp Propuls & Struct Matls. *Employer:* US Dept of Defense & priv indus. *Education:* SB/Metallurgy/MA Inst of Tech. *Born:* Nov 1912 Metallurgist South Chicago Wks, US Steel 1936-41; Materials Lab US Air Force Dayton OH 1941-73; Successively Proj Engr, Ch of High Temp Materials Sect 1945-58, Ch of Metals Branch 1958-73. R/D on ultra-high strength steel, alloys of titanium, aluminum, beryllium, heat & corrosion resistant materials, bearing materials. Since 1973 cons on aerospace propulsion & struct alloys. Sev papers in US & European literature on physical metallurgy. Fellow of ASM. Reg P E Ohio. *Society Aff:* ASME, AIME, ASM.

Perna, Angelo J
Business: 323 High Street, Newark, NJ 07102
Position: Prof Chem/Environ Engrg. *Employer:* New Jersey Inst of Tech. *Education:* PhD/Chem Eng/Univ of CT; MS/Chem Eng/Clemson Univ; BS/Chem Eng/Clemson Univ. *Born:* 9/15/31. Native of Brooklyn, NY. Served US Army 1952-53. Prod-Dev Engr Union Carbide Nuclear Co Oak Ridge TN 1957-60; on faculties of VP I & Univ of CT; since 1967 mbr of faculty at NJIT currently Prof of Chem & Environ Engg. Also served as Resch Coor, Found for Adv of Grad Student & Resch, Faculty Council Chmn, Dir ASEE/DELOS, 1976; Natl VP Omega Chi Epsilon 1974; Vice Chmn AIChE/SCC 1978; Vice Chmn ASEE/DELOS 1978, and Dir ASEE/CHED 1977. Mbr ACS, ASEE & AIChE. Currently Chmn AIChE/SCC, Natl Pres Omega Chi Epsilon & Chmn ASEE/DELOS. *Society Aff:* AIChE, ASEE, ACS, ΣΞ.

Pernichele, Albert D
Business: 566 West 900 South, Salt Lake City, UT 84101
Position: Pres *Employer:* Mtls Recycling Assoc *Education:* BS/Geology/Univ of IL; MS/Geology/Univ of ND *Born:* Jan 1936. US Army Transportation Corps 1955-56; Geologist with State Geol Survey of Illinois, No Dakota & Utah; Analytical Chem W Q Branch USGS 2 yrs; Engrg Geologist & Process Engr Bingham Canyon Mine Kennecott Copper Corp 1966-73; joined Dames & Moore Cons Engrs in Washington D C office 1973; dir major geotech & engrg studies for phosphate indus, environ, surface mining & reclamation studies for Govt. Also major study on processing of manganese nodules for Govt. Senior Partner of Dames & Moore and Dir of Mining for the firm. Presently Pres, Mtls Recycling Assoc, Salt Lake City, UT. Held teaching position at Univ of N Dakota. Res Fellowship Amer Chem Soc 2 yrs. Received Pele Award AIME 1972 *Society Aff:* SME

Pernoud, Rene B
Home: 4151 Wynona, Houston, TX 77087
Position: Civ Engr. *Employer:* City of Houston. *Education:* BS/ME/Univ of Houston. *Born:* 8/15/21. Almost 33 yrs of continuous experience with consultants. Ind & govt agencies in principally civil engrg & mech fields of contractual services. *Society Aff:* NSPE, HESS, TSPE, PEPP, PEG.

Perona, Joseph J
Business: Dept of Chem Engg, Knoxville, TN 37996-2200
Position: Prof. & Dept. Head *Employer:* Univ of TN. *Education:* PhD/ChE/Northwestern Univ; BS/ChE/Rose-Hulman Inst of Tech. *Born:* 5/28/30. Began profl career at Oak Ridge Natl Lab in 1956, engaged in res & dev of processes for the nucl fuel cycle. Joined faculty of Univ of TN in 1968. Became Dept Head in 1984. Sabbatical yr 1974-1975 with Energy Res & Dev Admin, Wash, DC, in the

Perona, Joseph J (Continued)
Div of Geothermal Energy. Active res interests in separations & heat and mass transfer. Reg consultant for Oak Ridge Natl Lab. *Society Aff:* AIChE, ASEE.

Perper, Lloyd J
Home: 3725 Ironwood Hill Dr, Tucson, AZ 85745
Position: Prof Engr. *Employer:* self. *Education:* MSc/Physics/OH State Univ; SB/EE/MIT. *Born:* April 23, 1921 New York N Y. C B S 1941-42; USAF Comm & Nav Lab 1942-54 military & civil serv; self-employed P E 1954- . R/D in instrument landing, doppler navigation, air traffic control, pats & applications in landing, simulation sys, automatic dir finding, magnetic support & fluorescent lighting ballasts. Fellow IEEE. Chmn Tucson Sect IEEE 1972-73. Pres the Inst of Navigation 1975-76. Pres Arizona-Sonora Desert Museum 1974-76. *Society Aff:* IEEE, ION, AIAA.

Perrigin, John G
Home: 4111 Ossa Wintha Place, Birmingham, AL 35243
Position: Sr Design Engr. *Employer:* The Rust Engrg Co Inc. *Education:* BSCE/Civil/Univ of AL. *Born:* Nov 1931. Native of Carrollton Alabama. BSCE from Univ of Alabama 1958. Served with US Air Force 1951-55. Struct Design Engr Palmer & Baker Engrs Inc 1958-61; Stress analyst & Facilities Design Engr Brown Engrg Co 1961-68; Sr Design Engr with Rust Engrg Co from 1968 to 1980. With Weyerhaeuser Co since 1980, presently Engr Mgr for Columbus Pulp and Paper Complex. Pres Huntsville Branch ASCE 1965. Dir Alabama Sect ASCE 1968-69. V Pres Alabama Sect ASCE 1974. Pres Alabama Sect ASCE 1975. Chmn Engrg Council of Birmingham 1970. Pres Rust Engrg Co Credit Union 1975. Engr of the year in Birmingham 1974. Hobbies incl hunting & fishing. *Society Aff:* ASCE.

Perrin, Shepard F, Jr
7465 Boyce Drive, Baton Rouge, LA 70809 *Employer:* Self Employed *Education:* BS/Chem Engr/Tulane U *Born:* June 3, 1922 New Orleans LA. BSChE Tulane Univ 1942. Served in USN in Pacific Theatre May 1942 to Dec 1945. Tech & Oper Supr Esso (Exxon) Baton Rouge Refinery, Internatl Sales Esso New York & Coral Gables; assigned to Esso Asia for coordination of spec refinery start up in Singapore 1972; returned to private business 1974; then from 1975 to 1981 Exec Dir Superport Authority; from 1981 to 1983 Bus Dev Dir of PYBURN & ODOM and OFFSHORE SURVEY'S self employment- indus dev. 1975-76 Pres Tulane Univ Natl Alumni Assn. Mbr Tulane Engrg Coll Bd of Advs since 1974 (Pres Engr Coll Bd of Adv 1976-78). Former Mbr, Tulane Univ Bd of Visitors. Former Mbr Tulane President's Mbr Council of Trustees, Gulf South Res Inst. Commercial and industrial real estate marketing and development. *Society Aff:* AIChE.

Perrine, Richard L
Business: Engineering I, Room 2066, University of California, Los Angeles, CA 90024
Position: Prof of Engg & Appl Sci. *Employer:* Univ of CA. *Education:* PhD/Chemistry/Stanford Univ; MS/Chemistry/Stanford Univ; AB/Chemistry/San Jose State College. *Born:* 5/15/24. Native of Mtn View, CA. Served with US Army 1943-46. Res chemist in petrol production, CA Res Corporation, 1953-59. Assoc Prof & Prof, UCLA, 1959 to present; Div Hd, Chem, Nucl & Thermal Engg, 1961-63; VChrmn, Instruction, 1963- 65; Chrmn, Environmental Sci & Engg, 1971-82. Teaching & res in phys of flow through porous media & in environmental sci & engg. Public service and consulting applications in energy & water resources, & air & water quality. Outstanding engr merit award, Inst for the Advancement of Engg, 1975, ACT-50 award in the field of chemistry, West Coast region NAACP, 1984. Enjoy trout fishing, all-yr alpine mtn recreation. *Society Aff:* AAAS, ACS, AIChE, AIME, NAEP, APCA, AEEP, AWRA, CIME, NAEE, SETAC, NYAS, WRSA, IAGLR.

Perron, Gilles
Home: 10765 Jeanne-Mance, Montreal, Canada H3L 3C5
Position: VP *Employer:* Societe d'Ingenierie Cartier Ltee *Education:* BASc/EE/Laval Univ; BA/EE/Coll La Pocatiere *Born:* 9/27/21 Native of Quebec, Canada. Spent nearly twenty years with Hydro-Quebec in operation. System Operating Engr 1954-59. Dir Sales and Contracts 1959-65. General Secretary, Corp of PE of Quebec 1965-69. VP, T. Pringle & Sons, bldg and industrial consultants, 1969-71. General Mgr, Societe Lahaye, Ouellet, town planners and architects. VP, Cartier engrg, 1972 to 1982. Gen Mgr of Consultants Dutch Inc since 1984. Assumed responsibility for projects. President of IEEE Montreal Section 1962, Eastern Canada Council IEEE 1963. Pres of Order of Engrs of Quebec 1976-78, Pres of Canadian Council of PE 1981 Mbr of the Prof Bd of Quebec since 1984. Married to Martha Gilbert, seven children. Enjoys arts and sports. *Society Aff:* CCPE, IEEE, EIC, OEQ.

Perry, Anthony J
Home: Route 3 Box 519, Moneta, VA 24121
Position: Consulting Engr. *Employer:* Self-employed. *Education:* BS/Civ Engg/MA Inst of Tech; AB/Liberal Arts/Boston College. *Born:* 9/7/05. in Boston. Thirty-five yrs with Bureau of Reclamation, US Dept of Interior, the last twenty of which were in the Power Div as Asst Chief and Chief of the Resources and Dev Branch. Special foreign assignments with State Dept consulting on hydroelectric generation and transmission in Iran, Iraq, Lebanon, Italy, Cambodia and Brazil. Retired from Bureau in 1965, engaged in private consulting engg practice specializing in power generation and transmission. With engg firms and govt agencies, including State Dept, UN, Atomic Energy Commission, Govts of Bolivia, S Korea and Argentina. *Society Aff:* ASCE, NSPE, USCOLD.

Perry, Charles R
Business: 3800 E. 42nd, #621, Odessa, TX 79762
Position: Chrmn of the Bd & CEO *Employer:* Perry Mgmt, Inc. *Education:* BS/ChE/Univ of OK. *Born:* 9/8/19. Shamrock, TX. Married, 3 children. College honors: AIChE annual chapt scholarship Award 1949, Who's Who Among Students in Am Colleges & Univs 1950-51. Recipient of 1986 Distinguished Service Citation by Univ of Okla. Former Chrmn of MBank Odessa, Tx. Present Dir. Texas Utilities Co, Colorado River Municipal Water Dist, TX Natl Research Lab Commissioner; Pres of Council for Secure America. Pres of Permian Basin Petr Assoc. Authored 12 papers (gas ind related). Reg PE Tx. Pats: 4 US, 1 Canadian. *Society Aff:* TSPE, AIChE.

Perry, Charles W
19724 Greenside Terrace, Gaithersburg, MD 20879
Position: Sr Advisor *Employer:* Office of Waste Programs Enforcement *Education:* DrEng/ChE/Johns Hopkins; MS/ChE/MIT; BS/ChE/Northeastern. *Born:* 10/1/10. Native of Massachusetts. Dr Engr Process Plants design to 1941; Polymer dev Ch Office of Rubber Dir during WWII; Petrochem planning & plant const for Phillips Petroleum Co 1945-52; Engrg Dir for divs of Olin Mathieson Chem Corp 1952-59; V Pres Refining for Kendall Refining Div of Witco Corp 1960-71; Sr V Pres Kendall-Amalie Div of Witco 1971-73; Cons for FEA on oil stimulation for 1974; Branch Ch for oil stimulation for Fossil Energy Div of ERDA/DOE 1975-. Sr Engrg Specialist, Heavy Oil and Tar Sands, US Synthetic Fuels Corp Washington, DC 1982-1986. Reg PE NY & Md. 17 pubs. Sr Advisor, Environmental Protection Agency, Office of Waste Programs Enforcement 1986--. *Society Aff:* TBΠ, AIChE, SPE

Perry, Edward B
Home: 11 Jil Marie Circle, Vicksburg, MS 39180
Position: Res Civil Engr. *Employer:* US Army Engr Waterways Experiment Station. *Education:* PhD/Civil Engr/TX A&M Univ; MS/Civil Engr/MS State Univ; BS/Civil Engr/Univ of MS. *Born:* 9/29/39. Oxford MS. Res Civil Engr US Army Engr Waterways Exper Sta 1963-. Adj Prof Civil Engr Miss State Univ. Condr res on soil dynamics, dispersive clays, reinforced earth, streambank erosion. Contbr reports, articles to tech jours, spl tech publs, encys. Mbr: ASCE, ASTM, ASEE, TRB, USCOLD, ISSMFE, DFI, Chi Epsilon, Who's Who S and SW. Spl Act Award, WES, 1976; Director's Res and Dev Achievement Award, 1979; Army Res and Dev Achievement

Perry, Edward B (Continued)
Award, 1979. m Jimmie F. McKee, 3 children. *Society Aff:* ASCE, ASTM, ASEE, TRB.

Perry, Francis J
Business: 338 State Office Bldg, Providence, RI 02903
Position: Chief of Data Operations *Employer:* RI Dept of Trans. *Education:* MS/Structures/Univ of RI; BS/Civil Eng/Univ of RI *Born:* 10/2/38 Native of Coventry, RI. Bridge Designer for RI Public Work Dept after college graduation (1960). Organized and headed Engineering data processing unit of RIDPW from 1963 to 1968. Promoted to Chief of Highway Design in 1968 where I was responsible for all Transportation Facilities Design until 1979 when I assumed my current position, responsible for all data processing of the reorganized RI Dept of Transportation (RIDOT), including Motor Vehicles and Accident & Statistics, Past President of RI section ASCE, Past Chairman of New England Council ASCE, Former Member AASHTO Design Committee and Chairman of Steering Committee (Northeast), Young Engineer of the Year (RI)-1967 Past area Director of HEEP. *Society Aff:* ASCE, NSPE, AASHTO, HEEP

Perry, Jesse T
Home: 1723 Kingsbury Drive, Nashville, TN 37215
Position: Supervisory Civil Engr. *Employer:* Nashville Distr US Army Corps Engrs. *Education:* BE/Civil/Vanderbilt Univ. *Born:* 8/3/35. Al Johnson Const Co Cheatham Dam TN 1957. LT US Army Security Agency 1958- 59 Commendation Ribbon; Design Engr TVA Knoxville TN 1960-63 Melton Hill Lock & Wheeler Powerplant; 1963-85 Nashville District US Army COE Supervisory Engr, Contract Mgmt, Planning, Design for multipurpose civil projects, TN-TOM Waterway, MS; Big South Fork Natl Area, TN and KY; Currently Chief, Hydrology and Hydraulics Branch, Cumberland River Basin, KY & TN. Reg KY & TV; Fellow ASCE, Tennessee Section, TSPE, NSPE, SAME. Instructor for Corps-wide AE Contracting, Huntsville Division. *Society Aff:* ASCE, NSPE, TSPE, SAME.

Perry, Lawrence E, Jr
Business: PO Box 12365, 30 W Church Ave, Roanoke, VA 24025
Position: Pres *Employer:* Lawrence Perry & Assoc, Inc *Education:* BS/ME/VA Polytechnic Inst & St Univ *Born:* 7/20/32 in Cumberland County, NC; attended public schools in Roanoke, VA. Attended VPI & SU, serving as class officer and VP of the cadet student body. Twenty-one yrs experience as designer of heating and air conditioning systems for commercial, institutional, industrial buildings. Mbr of Rotary Club and other community activities and serve as Deacon of church. Pres of ASHRAE Chapter, VP of CSI Chapter, and pres of VA Solar Energy Assoc. Recipient of national and regional ASHRAE awards for energy efficient design. Founded consulting engrg firm in 1975. Speaker and author of articles on energy conservation. Married 1957, three children. *Society Aff:* ACEC, ASHRAE, CSI, NSPE

Perry, Richard P
Home: 5289 Angus Drive, Vancouver B C Canada V6M 3M9
Position: President. *Employer:* Perry Engrg Ltd. *Education:* BE/Engg/Wairarapa College. *Born:* 12/27/23. Grad Wairarapa College New Zealand 1941 Mech Engrg. Served with the Royal New Zealand Air Force 1942-45. Design Engr with Anderson s Ltd in New Zealand spec in Machine Design; emigrated to Canada 1952; D W Thomson Ltd 1952-64 Design Engr building Mech Sys; Phillips Barratt & Partner 1964-73 Partner-in-chrg bldg mech Sys; Perry Engg Ltd 1973-76 Pres. Reg P E Province of BC. Letson Memorial Award 1964. Dir & Regional Chmn X ASHRAE 1975-78; Treasurer 1979-80, Vice President 1981-1982, President 1983-1984, ASHRAE, Fellow; IMechE (Great Britain), Fellow; CIBSE (Great Britain), Fellow; Respon for design of numerous large mech sys in Indus & Commercial bldgs in Canada. Enjoy golf and fishing. *Society Aff:* ASHRAE, CSME, EIC, CIBS, IMechE.

Perry, Robert R
801 Wythe St, Alexandria, VA 22314
Position: Asst Exec Director. *Employer:* Water Pollution Control Federation. *Education:* BS/Sanitary Engg/Penn State Univ; MS/Civil Engg/Purdue Univ. *Born:* Sept 16, 1926. Native of Muncy Pennsylvania. Reg Engr in Pennsylvania, Ohio & D C. Indus Waste Engr for GATX 1951-54; City Engr Meadville PA 1954-62; City Engr Hamilton Ohio 1962-64; employed in various sanitary engrg positions with the District of Columbia 1964-75; final 2 yrs as Deputy Dir Dept of Environ Servs; this dept is respon for water distrib, sewage collection & treatment & solid waste collection & disposal for D C; Asst Exec Director Water Pollution Control Federation 1975- . Mbr WPCF, ASCE, APWA. Commissioner on Interstate Comm on the Potomac River Basin 1974-75. *Society Aff:* WPCF, ASCE.

Perry, Robert W
Business: Dept of Mech Engg, Louisville, KY 40292
Position: Dist Prof of Mech Engrg *Employer:* Univ of Louisville. *Education:* PhD/Heat-Power Engg/Cornell Univ; MME/Heat-Power Engg/Cornell Univ; BME/Mech Engg/Cornell Univ. *Born:* 4/2/21. Mgr hypervelocity br Gas Dynamics Facility, Arnold Engg Dev Ctr, Tullahoma, TN, 1953-59; chief hyperaerodynamic res Republic Aviation Corp, Farmingdale, NY 1959-65; prof aero engg Poly Inst Brlyn, 1965-67; sr staff cons Liquid Metal Engg ctr, Canoga Park, CA, 1968-70; distinguished prof of mech engg Univ Louisville, 1971-. Mbr NASA res adv com for fluid mechanics, 1959-60; cons adv group for aero res and dev NATO, Brussels, 1966. Licensed, PE, NY, KY, TN. *Society Aff:* ASME, ΣΞ.

Perry, Ronald F
Business: Northeastern Univ, 360 Huntington Ave, Boston, MA 02115
Position: Assoc. Prof of Indus Engrg & Info Systems *Employer:* Northeastern Univ. *Education:* PhD/Industrial Eng & Opns Engr/Univ of MI; MS/Industrial Eng/Northeastern Univ; BS/Industrial Eng/Northeastern Univ. *Born:* June 1939 Somerville Mass. Honors: Tau Beta Pi, National Vice President Alpha Pi Mu. Prof Socs: IIE, TIMS, ORSA, DSI, ACM & SCS. Opers Res Analyst with Dept of the Army 1968-70; USPHS Doctoral Trainee with Bureau of Hospitals Univ of Michigan 1970-74; doctoral thesis res: Simulation based Planning Model for a Radiology Dept; currently Assoc Prof of Indus Engrg & Info. Systems at Northeastern Univ; teaching areas: Simulation, Mgmt Info Sys, Probability & Statistics. Res interests: Simulation, Insitutional Research, M.I.S. Design and Analysis. *Society Aff:* IIE, ORSA, DSI, TIMS, ACM, SCS

Perry, Russell L
Home: 2139 MacBeth Place, Riverside, CA 92507
Position: Retired *Education:* Mech Engr/-/Univ of WI; BS/ME/Univ of WI; BS/Agri/Univ of WI. *Born:* Aug 24, 1904 E Orange N J. Tau Beta Pi, Pi Tau Sigma, Phi Kappa Phi, Alpha Zeta, Sigma Xi. Prof Emeritus Univ Calif. Instructor Agri Engrg OR State Univ 1926-28; Instructor to Prof & Agri Engr in Expt Sta Univ of Calif, Davis, Los Angeles, then Riverside 1928-72; Emeritus 1972. Sci cons Tech Indus Intelligence Ctte England, Germany 1945. Visiting Prof Gadjah Mada Univ Jogjakarta Indonesia 1961-64 Ch of Party 1964. V Chmn Agri Engrg Univ of Calif, Davis, Los Angeles 1956-64, Davis & Riverside 1964-70. VC Soil Sci & Agri Engrg Riverside 1970-72. Co-author Agri Process Engrg. Author numerous tech papers on heat transfer & agri processing. Life Fellow ASAE. Life Mbr ASME. Mbr ASHRAE. Mbr Cal & Natl Soc Prof Engrs. Current cons interests: heat transfer, refrigeration, dehydration & processing farm prods, heating & cooling greenhouse, solar energy & energy conservation. *Society Aff:* ASME, ASAE, ASHRAE.

Perry, Vincent D
PO Box 615, West Hampton Beach, New York, NY 11978
Position: Retired. *Education:* MS/Mining Geology/Columbia Univ; BS/Mining Engg/Univ of CA-Berkeley. *Born:* Nov 1901. Native San Francisco Calif. Mining Engr Calif Mother Lode 1922-23; Mining Geologist Anaconda 1924-28; Ch Geologist Cananea Mexico 1928-37; Exploration Geologist Anaconda 1938-47 & Ch Geologist 1948-69; V Pres Anaconda 1956-69; Dir 1966-70. AIME recipient Jackling Award 1961. Engrg Achievement Award 1970. Honorary Geological Engrg Montana School

Perry, Vincent D (Continued)
of Mines 1965. Married Margaret Moore Butte 1926, 1 daughter Mary Patricia (Mrs Walt Thomas Zielinski), 4 grandchildren. Enjoy occasional Mine consultation, & travel. *Society Aff:* AIME, SEG, GSA, ΣΞ, TBΠ.

Perry, William J
Business: 495 Java Dr, Sunnyvale, CA 94086
Position: President. *Employer:* ESL Inc. *Born:* Oct 1927. PhD PA State Univ, MS, BS Stanford Univ, Recd BS & MS in math from Stanford Univ 1949 & 1950, respectively & PhD from PA State Univ 1957. Math Instructor at the Univ of ID 1950-51 & at PA State Univ 1951-54; joined Sylvania Elec Prods Inc in Mountain View CA 1954 & was Dir of Sylvanias Elec Defense Labs 1961-64; is the Pres of ESL Inc which was co-founded 1964. Is engaged as a sci cons to the Dept of Defense & Mbr of sev govt adv comm. Spec in electromagnetic sys analysis & partial differential equations. Was awarded the US Armys Outstanding Civilian Service Medal 1960 for work in the design of advanced reconnaissance sys. Was elected to the Natl Acad of Engrg 1970. Was elected V Pres of WEMA & Chmn of the WEMA San Francisco Council 1972.

Pershing, Roscoe L
Business: John Deere Rd, Moline, IL 61265
Position: Mgr, Engg Sys. *Employer:* Deere & Co. *Education:* PhD/Agri Engr/Univ of IL (Urbana); MS/Agri Eng/Univ of Il; BS/Agri Eng/Purdue Univ; AS/Pre-Eng/Vincennes Univ. *Born:* Oct 1940 Washington Indiana. Teaching Fellow & Instructor at Univ of Illinois; joined Deere & Co Tech Center as Sr Res Sci 1966; Group leader in vehicle dynamics & simulation. P E 1972 Ill. Author & Patentee in field. Natl ASAE Tech Paper Award 1970. Natl ASAE Young Res Award 1973. Exec Asst Off of the Chmn 1973-74 conducted worldwide parts distrib study Div Engr Advanced Products John Deere Ottumwa Wks (IA); now Mgr Eng Sys Dept. 1979-81 ASAE Natl Dir-Profl Dev, SAE, Gov Bd MVS, Sigma Xi Chapt Pres 1968. Part-time Instructor Black Hawk College Moline (1968-71). Agri Engr Adv Ctte Univ of Ill 1974- . Golf, tennis, singing. Methodist Lay Leader. Dads Club Baseball coach. 1980"ABET Bd. Directors 1981"JETS (Ill.) Bd. Directors. *Society Aff:* ASAE, SAE, Sigma Xi

Persons, Benjamin S
Business: 455 East Paces Ferry Rd. N.E, Suite 200, Atlanta, GA 30305
Position: Partner *Employer:* Dames & Moore. *Education:* BCE/-/GA Inst Tech; Dipl/Mil Engr/Engr Sch. *Born:* 3/3/23. in Macon, GA. Issue of Benjamin Stephen & Mary Scandrett; married Frances Neisler 1946. Issue Donna Maria, Benjamin Stephen Jr & Robert Scandrett. Graduate GA Military Acad; BCE GA Inst Tech Post-grad US Army Engg Sch - Commissioned Service; US Army Capt C.E. in European Theatre of OPNS 1942- 46. Presently US PHS Capt.-Duty in Indian Health Service; Partner Dames & Moore, cons in the appl & environ sci, CA 1950-54, IL 1954, GA & SE 1954-76, Regional Mgr for Far East, Pacific & Australia 1976-78, Regional Mgr SE USA 1979-1982. Reg PE GA, states of SE USA El Salvador, Malaysia and Reg P Geol GA, Inst of Engrs Singapore, Inst of Engrs Malaysia, Inst of Engrs Australia. Fellow ASCE, Mbr S African Inst of CE, Assn of Salvadorean Engrs & Archs, US Silver Star & US Bronze Star. Elder United Presbyterian Church & Uniting Ch in Australia, Author "Laterite, Genesis, Location, USE–. *Society Aff:* ASCE, SAME

Persson, Sverker
Home: 1135 Westerly Pkwy, State College, PA 16801
Position: Prof. *Employer:* PA State Univ. *Education:* PhD/Ag E/MI State; MS/ME/Chalmers Inst Tech *Born:* 8/3/21. Presently prof of agri engg at PA State Univ with primary res areas mechanics of cutting fibrous mtls, agricultural power & methane production from manure. Grad & undergrad teaching on power for agri systems, soil-machine systems & advanced agricultural machine design. Earlier as assoc prof at MI State Univ studies related to tractors & soil dynamics. For 16 yrs assoc prof at Royal Agricultural College, Uppsala, Sweden with several res projs. Before that, design proj leader with Bolinder-Munktell, Eskilstuna, Sweden. *Society Aff:* ASAE, AAAS, ISTVS.

Perumpral, John V
Business: Agri Engg Dept, Blacksburg, VA 24061
Position: Prof & Acting Dept. Head *Employer:* VPI & SU. *Education:* PhD/Agri Engg/Purdue Univ; MS/Agri Engg/Purdue Univ; BS/Agri Engg/Allahabad Univ. *Born:* 1/14/39. Born & raised in India. After undergrad education served the Govt of India for a yr. Joined Purdue Univ in 1964 for grad work. Completed the PhD prog in Agri Engg in 1969 with emphasize in the area of Soil-machine systems. Joined VA Tech in 1970 as Asst Prof & was promoted to Full Professor status in 1982. Acting an department head since Nov. 1986. Current res and teaching activities are in the gen areas of Agri & Forestry mechanization. *Society Aff:* ASAE.

Peskin, Richard L
Business: Dept Mechanical & Aerospace Eng, Box 909, Piscataway, NJ 08854
Position: Prof *Employer:* Rutgers Univ *Education:* PhD/ME/Princeton; MA/ME/Princeton; MSE/ME/Princeton; BSE/ME/MIT *Born:* 05/31/34 Native of Cambridge, MA. Res Staff of Princeton Univ 1959-61. Prof at Rutgers Univ since 1961. Appointed Distinguished Prof 1974. Conslt to Natl Sci Foundation; Conslt to several industrial firms. ASME Pi Tau Sigma Gold Medal winner 1964. Visiting Scientist at Natl Center for Atmos Res 1969 and Princeton Univ 1975. Currently Dir of Parallel Computer Lab at Rutgers. *Society Aff:* APS, IMACS, AAAS, IEEE-Computer Soc, SCS

Pesses, Marvin
Home: 6430 Via Rosa, Boca Raton, FL 33433
Position: CEO *Employer:* Pentad Group, Inc. *Education:* DSc/Metallurgy/London Inst of Applied Res; MS/Ceramic Eng/Univ of IL; BS/Chem Engg/Purdue Univ. *Born:* 7/18/23. Native of Rock Island, IL. 1947-48 Deere & Co Moline IL, metallurgical trouble-shooter; 1949-50 W J O'Connell & Assocs, San Fran CA, cons engr; 1950-64 Alloy Metal Prods Inc, Davenport IA, VP; 1964-71 Mercer Alloys Corp, Greenville PA, Pres; 1971-72 Ogden Metal Corp, Cleveland, OH, cons; 1972-84, Founder & Chmn Bd of The Pesses Co. Chmn of Bd of Met Corp of Amer, Pulaski PA & Genl Ecologica sys Inc. 84-Present CEO Pentad Group, Inc. Mbr ASTM (on sev cttes), AIME, ASM (former Chap Chmn), AFS, AICE, ASTM, ACS, AIChE, APM, NAIFA. 18 pats in field of solid waste recycling, synthetic ore production, nickel alloys production & recovery, with more pats pending. *Society Aff:* AIME, ASM, ASTM, AFS.

Pestorius, Thomas D
Home: 1827 Panarama Ct, McLean, VA 22101
Position: VP *Employer:* Dynametrics Electronics *Education:* M.S./Mech. Eng./US Naval Postgraduate School; B.S./Mech. Eng./US Naval Academy *Born:* 05/06/46 Native of Rochester NY. Served in Nuclear Navy 1968-79. Project Engineer Consultant for Nuclear Utility Companies 1979-81. ASME Congressional Fellow '81, ASME White House Fellow '82. Sr Policy Analyst in White House Science Office 1983-84. EBASCO Services Inc Wash D.C. Representative 1984-86. VP/owner Dynametrics 1986-present. Mbr Bd on government relations ASME, ASME trustee for Federation of Materials Societies, ASME & AAES Representative to Congressional Advisory Committee for Technology Policy. Captain in US Naval Reserve. *Society Aff:* ASME, IEEE

Pesuit, David R
Business: 70 Main St, Northampton, MA 01060
Position: Pres. *Employer:* D R Pesuit & Assoc. *Education:* BA/Appl Sci/Lehigh Univ; BS/CE/Lehigh Univ; PhD/Appl Sci/Yale. *Born:* 12/7/42. The business of D R Pesuit & Assoc is divided between energy conservation and accident investigation. In energy, using computerized heat/mass balances for power plants we pioneered seven yrs ago to assess and optimize power plant perf. This work has led to devel of microcomputer-based software for very accurate fuel comparisons -- programs which put bargaining power in the hands of purchasers at a time when oil and gas

Pesuit, David R (Continued)
prices are competitive. Also pioneered microcomputer-based monitoring and ctrl sys for small to medium-sized industrial plants where larger computerized ctrl sys cannot be cost-justified. Interest in arch and historic rehabilitation. Projs now incl a 4,000 sq ft commercial building in the center of Northampton. *Society Aff:* ΦBK, TBΠ, AIChE

Peters, Alexander R
Business: Mech. Engrg Dept, Univ. of Nebraska-Lincoln, Lincoln, NE 68588-0525
Position: Prof *Employer:* Univ of Nebr (Coll of Engrg, Tech). *Education:* BS/Mech Engr/Univ of NE; MS/Mech Engr/Univ of NE; PhD/Mech Engr/OK State Univ. *Born:* Nov 1936. Native of Kearney, Nebr. Served as an officer - pilot in USMC 1959-62. Engr, Aerodynamics, wth S&ID, North Amer Aviation, working on S-11 stage of Saturn V vehicle 1963-64. With Univ of Nebr since 1966. Teaching/res ints incl fluid mech, aerodynamics, heat transfer, & combustion. Dept Chmn 1975-1985. SAE Ralph R Teetor Award 1969; ASEE Ford Found Residency Fellow 1970 at Ford Motor Co; Pi Tau Sigma V P 1975. Pi Tau Sigma President 1980-83; Accreditation Bd for Engrg and Tech (ABET), Bd of Dirs, 1983-86; SAE (Soc of Automotive Engrgs) Engrg Ed Bd, 1984-86. Mgr, Design and Dev Engg Dept., Brunswick-Defense, 1985-86. Enjoy flying, golf, woodworking & homecraftsman projs. *Society Aff:* ASME, ASEE, SAE, AIAA, SAMPE.

Peters, David A
Business: Campus Box 1185, St Louis, MO 63130
Position: Assoc Prof. *Employer:* Washington Univ *Education:* PhD/Aero-Astro/Stanford; MS/Appl Mechanics/Wash. Univ; BS/Appl Mechanics/Wash. Univ. *Born:* 1/31/47. Native of Fairview Heights, IL. Married, three children. Engr with McDonnel-Douglas Astro Co, 1969-1970, working in structural dynamics. Res Scientist, Army Lab at Ames Res Ctr, 1970-1975, working on helicopter dynamics. With Washington Univ since 1975, teaching and engaged in res in rail-car dynamics and rotor dynamics. NASA Scientific contribution Award, 1975. NASA Tech Utilization Award, 1976. ASME Pi Tau Sigma Gold Medal, 1978. Enjoy Choral Singing and Bible Study. *Society Aff:* AIAA, ASME, AHS, ASEE.

Peters, Donald C
Business: 1425 Beaver Ave, Pittsburgh, PA 15233
Position: Chmn Exec Comm. *Employer:* Mellon-Stuart Co. *Education:* BS/CE/Marquette. *Born:* Mar 1915 Milwaukee. Prior to joining Mellon-Stuart Co 1951 as Pres, was engr, timekeeper & estimator with Siesel Const Co, Milwaukee & Ch Engr, V P & Dir of Crump Inc Pgh. Fellow Amer Soc of Civil Engrs; Reg P E Wisc & Pa; P Pres Pa Soc of Prof Engrs; Pres Pa State Regis Bd of Prof Engrs; Natl Dir 1954-57 Natl Soc of Prof Engrs - Mbr Bd of Ethical Review - Mbr NSPE/AGC Jt Coop Ctte. Enjoys fishing, gardening, hockey. *Society Aff:* NSPE, PSPE.

Peters, Ernest
Business: Dept. of Metals and Materials Engineering, 309-6350 Stores Rd, Vancover, B.C, Canada V6T 1W5
Position: Prof Extractive Met. *Employer:* Univ of British Columbia. *Education:* PhD/Metallurgy/Univ of BC; MASc/Metallurgical Engg/Univ of BC; BASc/Metallurgical Engg/Univ of BC. *Born:* Jan 27, 1926 Steinbach Manitoba. Employed by Geneva Steel Co Utah 1949- 50, Cominco Ltd Trail B C 1951-53, Union Carbide Corp N Y 1956-58, Univ of B C 1958- . EMD Author's Award (AIME) 1957. CIC Author's Award (Chem Engg Div) 1962. Alcan Award (CIMM) 1982. AIME-EMD Lecturer 1976. Benelux Metallurgie Plenary Lecturer 1977. Canada Council Killam Fellow, 1983-85. James Douglas Gold Medal (AIME) 1986. AIME - TMS Fellow (1986). Visiting Prof Univ of Calif Berkeley 1971 and 1984; Monash Univ Australia 1976. Mbr Bd of Examiners, B C Prof Engrs, & Canadian Accreditation Bd 1975-77. Cons to Cominco Ltd since 1958. Married 1949; 2 daughters: Charlotte 1956, Elizabeth 1959. *Society Aff:* AIME, CIMM

Peters, George A
Business: Consultant, 1460 Fourth Street, Santa Monica, CA 90401
Position: Consulting Engr (Self-employed) *Education:* JD/Law/UWLA; MA/Engineering Psychology/Temple Univ; BA/Engr & psychology/Univ of MA. *Born:* 11/24/24. Contrib Editor, Qual Mgmt & Engg 1968-77. Editorial Bd, Journal of Safety Res 1972-5. Editor, Hazard Prevention 1965-67 and 1977-79. CSP, CRE, PE (CA Safety, Qual). System Safety Society, Past President (1977-1979); American Society of Safety Engineers, Past Vice President (1974-1976); American Society for Quality Control, Product Liability Prevention Chairman (1977-1978); Author (books) "Automotive Engineering and Litigation–, (2 volumes), "Source book on Asbestos Discases" (3 volumes), and "Product Liability and Safety–, Editorial Boards at various engineering, safety, and health publications. Also: Licensed Psychologist (Calif) and Lawyer (Calif and Federal). *Society Aff:* SAE, ASQC, SSS, ASSE, HFS, IOSH, LABA, AAAS, AIC.

Peters, Henry F
Business: P O Box 183, South Jamesport, NY 11970
Position: Principal. *Employer:* H F Peters Assocs. *Education:* BSChE/CHE/Cooper Union. *Born:* Jan 1918. Plant const for Caye & duPont; process engr for M W Kellogg Co. 1948-72, Sr V P Engrg & Const for Sci Design Co, SD Plants Ltd London. P E Lic 18 states. Mbr NSPE, AIChE, ACS. Co-author Encyclopedia of Chemistry, The Chemical Plant. Enjoy big game sport fishing. *Society Aff:* AIChE, ACS, NSPE.

Peters, Jacques M
Business: Inst voor Werktuigkunde KUL, Celestijnenlaan 300B Heverlee, Belgium 3030
Position: Full Professor. *Employer:* Catholic Univ Leuven. *Born:* Dec 2, 1923 Liege. Mech Engr Univ of Leuven. Bach of Philosophy. Ordained priest in 1948. Prof at Cath Univ of Leuven. Hd of Dept of Mech Engrg 1965-82. Dr honoris causa of Technische Hochschule Aachen and of Aston University (Birmingham UK). Mbr Royal Belgian Acad Foreign Associate of The Nat Acad of Engineering (Washington DC). SME Res Medal 1975. Hon Mbr ASME, Visiting Prof at Univ of Calif Berkeley, spring 1969. Chmn of Internatl Inst for Prod Engrg Res 1973-74. Pres Belgian Soc of Mech Engrs 1975-80; Chmn ISO-TC29/WG 20 1974- . Field of res: machine tools - metal cutting - grinding - dimensional metrology. Sr. Schlesinger Pres-Berlin 1986. R.D. Springer Professor UC Berkeley (Cal) 1987.

Peters, Leo C
Home: 3424 Oakland St, Ames, IA 50010
Position: Assoc Prof & Mech Engrg Cons. *Employer:* ME Dept, Iowa St Univ; Self. *Born:* Sept 1931. BS Kansas State Univ; MS & PhD Iowa State Univ. Reg P E, Field-Mech Engrg Iowa, Ill & Nebr. Design & dev engr, John Deere Tractor Works, Waterloo Iowa 1953-54, 1956-61. USAF Aviation Engrs officer 195456. Mech Engrg staff Iowa State Univ 1961- . Main area of teaching & int is design, failure analysis, safety, & products liability considerations in design. Also self- employed as mech engrg cons 1967- . Mbr SAE, ASME, ASAE, ASEE & Sigma Xi. Founding Mbr ASME Ctte 'Design, Engrg & the Law'; Mbr SAE Natl Bd of Dir.

Peters, Leon, Jr
Home: 1410 Lincoln Rd, Columbus, OH 43212
Position: Prof *Employer:* OH State Univ *Education:* PhD/EE/OH State Univ; MS/EE/OH State Univ; BS/EE/OH State Univ *Born:* 05/28/23 Born in Columbus, OH, served as radar repairman in WWII. This triggered a long time interest in radar, electromagnetics and antennas. He has conducted and supervised research at the Electro Sci Lab since 1950. He has published numerous journal articles, oral papers and is co-inventor of several patents. As a Prof he has taught a variety of grad and undergrad courses. He has acted as advisor for many Masters and PhD students. He is a Fellow of the IEEE and a Mbr of Commission B, URSI. *Society Aff:* IEEE, ΣΞ, URSI

Peters, Leonard K
Business: Dept. of Chemical Engineering, University of Kentucky, Lexington, KY 40506
Position: Prof/Assoc. Dean, Grad Sch *Employer:* Univ of KY *Education:* PhD/ChE/Univ of Pittsburgh; MS/ChE/Univ of Pittsburgh; BS/ChE/Univ of Pittsburgh *Born:* 03/16/40 Native of McKeesport, PA. Research engr at Alcoa Res Labs, 1962-1968. Taught in chem engrg at Cleveland State Univ, 1971-1974. Joined Chem Engrg Dept at Univ of KY in 1974 and Dept Chrmn from 1980-84. Currently, Assoc Dean for Res, Grad Sch. Principal research interests are atmospheric transport and chemistry and physico-chemical behavior of aerosol systems. Co-investigator on MAPS tropospheric carbon monoxide monitoring experiment on two Space Shuttle flights. Involved in Natl Acid Precipitation Assessment Prog to analyze regional- scale transport of acid rain precursors and acid deposition. *Society Aff:* AIChE, ASEE, APCA, AAAR, ΣΞ.

Peters, LeRoy L
Business: 2014 P St, Sacramento, CA 95814
Position: Pres. *Employer:* Peters Engg. *Education:* BS/ME/Fresno State College; AA/ME/Modesto Jr College. *Born:* 7/16/32. Mr Peters is married, has two children, & resides in Somerset, CA. Grad of Fresno State College, 1959, BS, Mech Engg. Reg as Profl ME in CA, NV, & OR. Active mbrship in ASHRAE, NSPE, CSPE, CSI, ASPE. Past Pres of ASHRAE & CSI. Pres of Peters Engg, Consulting Mech & Elec Engg Firm, established in Sacramento in 1969. Twenty yrs experience includes work with a Stockton Consulting Engg Firm, a Sacramento Consulting Engg Firm, Faculty mbr of Humphrey's College, Univ of CA, Lawrence Radiation Lab, & a Sacramento Consulting Elec Engg Firm. *Society Aff:* ASHRAE, NSPE, CSPE, CSI, ASPE.

Peters, Max S
Business: Chem Engg, Box 424, Boulder, CO 80309
Position: Prof of Chem Engg - Coll of Engg & Appli Sci. *Employer:* Univ of Colorado. *Education:* PhD/Chem Engg/PA State Univ; MS/Chem Engg/PA State Univ; BS/Chem Engg/PA State Univ. *Born:* 8/23/20. Mercules Powder Co 1942-44. US Army 1944-46. Geo I Treyz Chem Co 1947-49; Univ of IL 1951-62, Prof of Chem Engg & Hd of Dept; Dean College of Engg 1962- 78, Prof of Chem Engg, 1978- Univ of CO. Adv Ctte Engg Div NSF; Pres's Natl Medal of Sci Ctte; Chmn CO Environ Comm; Natl Acad of Engg; Amer Inst of Chem Engrs (Pres 1968, Founders Award 1974, Lewis Award 1979); Amer Soc of Engg Educ (Geo Westinghouse Award 1959, Lamme Award 1973); Penn State Disting Alumnus Award 1974. Reg P E PA & CO. Num chem engg books & res articles. *Society Aff:* AIChE, ACS, ASEE, AACE.

Peters, Robert C
Business: PO Box 419173, 4600 E. 63rd St, Kansas City, MO 64141-0173
Position: Mgr T&D Dept *Employer:* Burns & McDonnell Engrg Co *Education:* MS/CE/Structural/MT State Univ; BS/CE/MT State Univ *Born:* 10/10/39 Manages the transmission and distribution dept which is responsible for the design and proj mgmt of substation, transmission, and distribution line projs. Is a Sr Mbr of IEEE and is chairman of the IEEE Working Group on Wood Poles and coordinator for the preparation of a Design Guide for Wood Transmission Structures. He is also a mbr of the ASCE Subcommittee on Foundation Design for High Voltage Transmission Structures which involves preparing a transmission line foundation design guide. He is a reg PE in seven states. He is also a mbr of NSPE and CIGRE. *Society Aff:* IEEE, ASCE, NSPE, CIGRE

Peters, Robert R
Business: P O Box 7095, Norfolk, VA 23509
Position: Vice Pres. *Employer:* Layne Western Co. *Born:* Nov 15, 1928 Mathews Cty Va. Educ at public schools in Norfolk Va, Fork Union Military Acad - Norfolk Div, William & Mary; advanced work in ground water hydrology at MIT. Reg P E Va. Pres-Elect AWWA 1976; P Pres Natl Water Well Assn 1974; Chmn NWWA Disaster Preparedness Ctte 1973; Natl Dir AWWA 1973; P Chmn Va Sect AWWA; P Pres Va Water Well Drillers Assn; Mbr East Carolina Engrs Club, Amer Soc of Military Engrs, & Natl Soc of Prof Engrs. Member, Virginia State Water Study Commission, 1977-81. *Society Aff:* AWWA, NWWA, ASME, NSPE.

Peters, Stanley T
Home: 925 Sladky Ave, Mountain View, CA 94040
Position: Principal Engineer *Employer:* Westinghouse/Marine Div *Education:* BS/Chem/Univ of Santa Clara. *Born:* May 16, 1934. BS Chem Univ of Santa Clara 1957. Chemist W P Fuller 1957- 60; Assoc Scientist Lockheed Missiles & Space Co 1960-62; Sr Res Engr Rocketdyne 1962-65; Group Leader Matl & Process Engrg, Dalmo Victor Div Bell Aerospace 1965-76; Group Leader, Staff Engr Chemical Sys Div/United Technologies 1976- 1981. Senior Design Engineer, Principal Engineer Westinghouse Electric/Marine Division 1981-Present. Adj Instr Stanford Univ Assn for Cont Educ 1974-76, private cons matls & processes engrg 1966-76; Editor & Publisher SAMPE Journal 1973-79. Chmn Natl Editorial & Publicity Ctte, Soc for the Advancement of Matls & Process Engrg SAMPE; Mbr Amer Soc for Metals, Amer Soc for Testing & Matls, Reg Prof Engr CA, Mfg Engg. *Society Aff:* ASTM, SAMPE, ASM.

Peterson, Carl W
Business: Exper Aero Div 1334, Albuquerque, NM 87115
Position: Supr, Exper Aerodynamics Div. *Employer:* Sandia Labs. *Born:* Apr 1942. BS, MA, PhD Princeton Univ. Joined Sandia Labs' Fluid Dynamics Res Dept 1969. Assumed present pos in Exper Aerodynamics Div 1971. Respon for exper progs in aerodynamics of re-entry vehicles, rockets, aircraft stores & fluid mech res in boundary layers, highpowered laser flow fields, atmos wind modeling, & energy syys. Sect Chmn AIAA 1975. Enjoy skiing, tennis, hiking.

Peterson, David K
Home: 3431 W AB, Plain Well, MI 49080
Position: Assoc Prof. *Employer:* Western MI Univ. *Education:* PhD/Chem Engrg/Univ Denver; MS/Chem Engrg/Univ Denver; BS/Chem Engrg/MI State Univ. *Born:* June 1940. PhD 1969, MS 1963 Univ of Denver; BSChE 1962 Michigan State Univ. Reg P E Ind 1975- . Asst Prof & Res Fellow The Inst of Paper Chem Appleton Wisc 1966-69, chem kinetics, transport phenomena, pulping processes. Assoc Prof of Chem Engrg Ind Inst of Tech, Fort Wayne Ind 1969-77; named Hd of Chem Engrg 1971; named Hd of Chem Engrg & Nuclear Engrg 1975. Prof 1977. Areas of int: transport phenomena, reaction kinetics, chem engrg & resource conservation. Mbr AIChE, ASEE, TAPPI. Assoc Prof of Paper Sci & Engg (Environmental) Western MI Univ 1977-; areas of int: Environmental control processes, solid waste mgt, and energy conservation. *Society Aff:* AIChE, ASEE, TAPPI.

Peterson, Dean F
Home: 765 E 8th North, Logan, UT 84321
Position: Prof and Dean Emeritus *Employer:* UT State Univ *Education:* DCE/Civ Eng/Rensselaer Poly Inst; D Sc (Hon)/-/Mahatma Phule Univ, India; D Sc (Hon)/-/UT State Univ; MCE/Civ Engr/Rensselaer Polytechnic Inst; BS/Civ Eng/UT State Univ *Born:* June 1913. Cons Engr./Prof Emeritus, UT State U., 1982- ; Agr and Irrigation Adviser, USAID, New Delhi, 1979-82; Ch Soil and Water, 1977-78 Dir, Agri, AID 1978-79 V P Res 1973-76; grg. Dean Engrg USU 1957-73. Office of Sci & Tech, White House 1965-66, Dir Water for Peace 1968-69. Hd, Civil Engrg, Colo State Univ 1949-57. V P ASCE 1972-74; Royce Tipton Award 1968 Julian Hinds Award 1980; Hon Mbr 1976. Icko Iben Award, Am Water Res Assoc, 1979. Mbr or chmn var panels, Natl Acad of Sci. Cons to var Fed agencies, municipalities etc. Mbr U S study teams sent to Afghanistan, Iran Israel, Brazil, People's Repub of China & USSR. Mbr Natl Acad of Engg, fellow Am Acad of Arts & Sciences, Am Assoc Adv Sci; Am Geophysical Union. Int'l Water Res. Assn. *Society Aff:* ASCE, AGU, AAAS, ASEE, ICID

Peterson, Donald L
Business: Route 2 Box 45, Kearneysville, WV 25430
Position: Agric Engr. *Employer:* USDA - ARS *Education:* PhD/Agri Engg/MI State Univ; MSAE/Agri Engg/Univ of HI; BSAE/Agri Engg/Univ of GA. *Born:* 2/24/47. Raised on a family farm on MD's Eastern Shore. Res Agri Engr with USDA, ARS since 1971. Area of specialization is mech harvesting of horticultural crops. Develop mech harvesters for apples, peaches, pecans, chestnuts and blackberries. In 1977 was recipient of ASAE's Engg Concept of the Yr award for dev work on continuously moving shake-catch harvester for tree crops. Is lead scientist of mechanization proj at USDA's Appalachian Fruit Res Station in Kearneysville, WV. Hobbies are travel and golf. *Society Aff:* ASAE, ASPE.

Peterson, Donn N
Business: Suite 202, 7601 Kentucky Ave No, Brooklyn Park, MN 55428
Position: Pres *Employer:* Peterson Engineering, Inc. *Education:* MS/ME/Univ of MN; BS/ME/Univ of ND *Born:* 1/1/42 Born & raised in ND. Founder of Peterson Engineering, Inc., a consulting firm specializing in forensic practice. Fellow of AAFS and Diplomate of NAFE. Held offices in ASME, MSPE & PEPP. Prior academic experience at Univ of MN 1970-1974 & industrial experience at GE Aircraft Engine Group 1963-1970. Completed GE Advanced Engrg Courses in 1966 & Ph.D. candidate requirements in 1974. ASME MN Section "Young Engineer of the Year" of 1976. *Society Aff:* ASME, NSPE, SAE, AAFS, NAFE, PEPP.

Peterson, Enoch W F
Home: 9940 Cove View Dr. E, Jacksonville, FL 32217
Position: Retired. *Education:* BS/Civil Eng/Univ of TX; BS/Arch Eng/Univ of TX. *Born:* Jan 1923 El Paso Tex. Struct designer for Burkhardt Steel Co, Tipton & Assocs, Phillips-Carter Osborn Inc, all of Denver. Pres Peterson-Norris-Carrillo Inc 1957-59. Owner E W F Peterson, Cons Engrg 1959-1980 . Fellow ASCE, P Pres Struct Engrg Assn of Colo, & Metro Chap PEC. Reg P E Wyoming, Florida. Cert Cons Engr Colo. Served on Jefferson Cty Bldg Code Ctte, Lakewood Contractors Lic Bd, & Lakewood Chamber of Commerce Bd. Retired 1980. *Society Aff:* ASCE.

Peterson, Gary J
Business: 815 Fourteenth St S.W, PO Box 301, Loveland, CO 80537
Position: Fab. Oper. Engrg. Mgr. *Employer:* Hewlett-Packard Co *Education:* MS/ME/Univ of AZ; BS/ME/Univ of AZ *Born:* 10/7/39 Native of Norway, IL. Served with U.S.A.F. 1957-61 and A.N.G. 1962-66. Tool & Die Apprenticeship 1964-67. Univ of AZ Grad Res Assoc 1968-70. Joined Hewlett- Packard in 1970. Held positions of R&D Engr, Proj Leader, Produc Engrg Mgr, Manufact Engrg Mgr, Section Mgr. and assumed present responsibility as FAB Operations Engrg Mgr in 1985. Served on both the NAIT and ABET Accreditation Teams. Author of tech publications on fatigue analyses probabilistic analysis, mfg systems, and hybrid micro-electronics mfg. Is both a Cert Mfg Engr and a Reg PE. SME Region VII Citation for Prof Ach 1976. CO State Univ. Dist Service Award 1983. CSU Faculty Affiliate 1982-85. Intl. Dir SME 1981-90, Sec/Treas SME 1986-87, Vice Pres. SME 1987-90. *Society Aff:* SME, ASME

Peterson, Gerald R
Business: Park & Univ, Tucson, AZ 85721
Position: Prof. *Employer:* Univ of AZ. *Education:* PhD/EE/Univ of AZ; MS/EE/Univ of AZ; BS/EE/Univ of CA. *Born:* 8/2/30. Native of Oakland, CA. Field Engr & Test Engr with GE. Served in US Army in radar field testing. Joined faculty at Univ of AZ in 1957, promoted to Prof in 1969. Consultant to Burr-Brown Res Co, 1964-66. Author of *Basic Analog Comoputation, Introduction to Switching Theory and Logical Design* 3rd Ed, (with F J Hill), *Digital Systems* 2nd Ed, (with F J Hill). Editor, *IEEE Transactions on Education*, 1972-1976. Gen Chrmn, 1976 Conf on Frontiers in Educ. Active in curricular affairs at dept, college & univ level. *Society Aff:* IEEE, ASEE.

Peterson, Grady F
Business: 3300 Lexington Rd, SE, Winston-Salem, NC 27102
Position: Dept Hd, Advance Planning & Wage Practices. *Employer:* Western Elec Co Inc. *Education:* MSIE/Indu Engg/NCA & T State; BIE/Indus Engg/NC State. *Born:* 6/7/29. Grad work NC State & VPI & SU; Reg PE. Native of clinton, NC. With Western Elec Co Inc since 1952, spec in work measurement, wage practices dev, productivity optimization & production monitoring sys dev. Became Dept Hd Indus Engr - Burlington Plant in 1975 and Dept Hd, Engg, Pers Rel - NC Works in 1978. Assumed current respon as Dept Hd. Advance Planning & Wags practices-NC Works in 1979. Disting Serv Award (Zero Defects Prof Dev), Dep to Def 1971. Co-founder Alamance Co Info & Referral Serv 1972. TV apperarances on "The Energy Problem" 1975. VP IIE 1975-77. Also active in area & regional health planning councils. Enjoy travel & sports. *Society Aff:* IIE.

Peterson, Harold A
Business: 121 W Montana Jack, Green Valley, AZ 85614
Position: Emer Prof Elec & Computer Engrg. *Employer:* Univ of Wisconsin-Madison (Retired). *Education:* BS/EE/Univ of IA; MS/EE/Univ of IA *Born:* 12/28/08. 1908 Essex IA. G E Co 1934-46. Became Prof of EE at Univ of WI 1946. Chmn of Dept 1947-67. In June 1967 named to newly endowed Chair as Prof of Electric power Engrg. Held this Chair until 1974, when named Edward Bennett Prof of Elec & Computer Engrg. Emer Prof June 1976. Author about 70 tech papers & 1 book "Transients in Power Sys" (Wiley 1951, repub by Dover 1966). Designed & built the first Transient Network Analyzer (TNA). Awards incl Benjamin Smith Reynolds Award for Outstanding Teaching of Young Engrs 1957. Spec Citation of Merit from WI Utils Assn 1965. Fellow IEEE 1947, Life Mbr 1973, IEEE Gold Medal Education Award 1978, Elected to Natl Acad of Engg 1978. IEEE Centennial Medal Award 1984. *Society Aff:* IEEE, ASME, NSPE, ASEE

Peterson, Harold O
Home: 1425 Hillcrest Dr, Melbourne, FL 32935
Position: Retired. *Education:* BSci/EE/Univ of NB. *Born:* 11/3/99. Blair Nebr. Dr Sci (Hon) Univ of Nebr 1953. Fellow IEEE. Modern Pioneers. Sigma Xi. Sigma Tau. *Society Aff:* IEEE

Peterson, Norman L
Business: Matls Sci Div, Argonne, IL 60439
Position: Group Leader Basic Ceramics Group *Employer:* Argonne Natl Lab. *Education:* PhD/Mat Sci/MIT; MS/Metallurgy/MIT; BS/Metallurgy/MIT. *Born:* 1/16/35. Joined Argonne Natl Lab in 1961. Assoc Dir for Matls Sci Div from 1968-77. Current position as Group Leader Basic Ceramics Group. Sr Scientist since 1970. NSF Fellow 1964-65 Harwell England; US Sr scientist Award (Humboldt Foundation) 1973 Stuttgart Germany. Guest Prof 1977 and 1981 KFA, Julich, Germany. 59 contributed papers at Meetings of professional societies; 104 invited talks at professional societies univs & res centers; 82 publications including nine extensive review articles (book chapters). Fellow ASM & APS. Mbr AIME & A Cer S *Society Aff:* ASM, APS, AIME, ACerS.

Peterson, Richard E
Business: 2404 Maile Way, Honolulu, HI 96822
Position: Professor. *Employer:* Univ of Hawaii. *Education:* PhD/Economics/Univ of CA, Berkeley; AB/Economics/Stanford. *Born:* Feb 1931. PhD Univ of Calif at Berkeley 1972. Currently Prof Univ of Hawaii, Honolulu. Sr Fellow East-West Ctr, Tech & Dev Inst, Honolulu 1973-74. Dept Chmn, Dept of Bus Econ, Coll of Bus Admin, Univ of Hawaii 1973-76. Res int & pubs in Bayesian decision theory, economics of info, engrg/ econ aspects of aquaculture, nonelec applications of geothermal energy, beneficial uses of thermal effluents. Eugene L Grant Award from Amer Soc of Engrg Educ 1976. *Society Aff:* WMS, AMS.

Peterson, Robert S
Home: 1109 Independence Dr, West Chester, PA 19380
Position: Sr Vice Pres. *Employer:* Fischer & Porter Co. *Education:* BSME/Mech Engg/Univ of Cincinnati; ME/Mech Engg/Univ of Cincinnati. *Born:* July 26, 1928 Highland Pk Ill. Sr Mbr Instrument Soc of Amer; President's Adv Bd ISA. Pi Tau

Peterson, Robert S (Continued)
Sigma, Mech Engrg Honorary. Sr V P Fischer & Porter Co, Horsham Pa. Chairman Board of Directors, Sci Apparatus Mfg Assn (SAMA). Mbr Bd of Dir Lab-Crest Glass, Warminster Pa. Mbr Bd of Dir Warminster Fiberglass Co Southampton Pa. *Society Aff:* ISA, SAMA.

Peterson, Thorwald R
Business: 1139 Olive St, St. Louis, MO 63101
Position: Sr VP/Proj Mgmt *Employer:* Booker Associates, Inc *Education:* MS/CE/Univ of IL; MS/Public Admin/Geo Wash Univ; BS/Military Sci/US Military Acad, West Point *Born:* 03/17/31 US Army, Corps of Engrs; retired 1979 Colonel. District Engr, Corps of Engrs, St Louis, MO, 1973-76; Dir of Engr Combat Developments, US Army, 1976-78; Exec Dir of the Engr Staff, Office of the Chief of Engrs, Washington, DC, 1978- 79. Booker Assocs, Inc, Engrs/Architects/Planners, St Louis, MO, 1979-Present. Dir of Proj Mgmt, Sr VP, Bd of Dirs. National Dir, SAME, 1977-80. *Society Aff:* ASCE, SAME, ΦΚΦ

Peterson, Victor L
Home: 484 Aspen Way, Los Altos, CA 94022
Position: Dir of Aerophysics *Employer:* NASA. *Education:* MS/Aero & Astro Sci/Stanford Univ; MS/Mgt/MIT; BS/Aero Engg/OR State Univ. *Born:* 6/11/34. With NASA-Ames Res Ctr since 1956. Res Scientist, 1956-68; Asst Chief, Hypersonic Aerodynamics Branch, 1968-71; Chief, Aerodynamics Branch, 1971-74; chief, Thermo-& Gas-Dynamics Div, 1974-84, Dir of Astronautics, 1984-85, Dir Aerophysics 1985-. Currently responsible for progs in computational & experimental fluid dynamics, entry tech, computational chemistry, aerodynamics, artificial intelligence, and Numerical Aerodynamic Simulation. Author of over 45 technical publications in fields of flight- and fluid-mechanics and supercomputing. Alfred P Sloan Fellow, 1972-73, chrmn, San Francisco Sec of AIAA, 1969. Recipient of Apollo Achievement Award, NASA Medal for outstanding Leadership and elected Fellow, AIAA, 1986. Enjoy photography, auto racing & woodworking. *Society Aff:* AIAA.

Peterson, Warren S
Business: Cabor Research Labs, P.O. Box 1462, Reading, PA 19610
Position: Mgr. Process Development, R & D *Employer:* Cabot Corp. *Education:* PhD/ChE/Polytechnic Inst of NY; MS/ChE/VA Polytechnic Inst; AB/ChE/Clark Univ. *Born:* Nov 1917. Tech admin genl, specific opers exper in melting, casting & electrolytic reduction of alumiunum. Aloca Res Labs, Asst Ch Process Met Div 1942-47; Daiser Aluminum, DMR Dept Hd & Sec Supr 1947-57; Olin Aluminum, Assoc Dir MRL 1957-71, Tech Asst to Pres 1971-72; Martin Marietta Aluminum, Mgr Aluminum Reduction Tech, Cabot Corp, Mgr Process Development R&D development new improved processes and products in field of master alloys and copper alloys. Bd of Dir TMS of AIME; former Counselor & Sect Chrmn ACS. *Society Aff:* AIME, AFS, ASM.

Peterson, William E
Business: 2108 S. W. 152nd, Seattle, WA 98166
Position: Owner *Employer:* Self *Education:* BS/CE/Univ of WA *Born:* 11/12/30 Native of Timberlake, SD until 1935; Yakima, WA until 1951 then Seattle, WA to present. Private practice since 1960 offering structural engrg and services as Tech Mgr for American Standards Testing Bureau Inc, Seattle. Discoverer of the Fountains of Youth and the Arc of the Covenant. Both discoveries came as a result of attempting to design a concrete & steel nuclear cannon and then realizing that what I was working toward was a stone pyramid. Water poured over the hot stones would form clouds. With the help of the sun, they create ozone to protect us from the suns rays. The arc is the shaft of light-energy from a pyramid. *Society Aff:* ACEC, WSEA

Petit, Richard G
Business: PO Box 1209, Seattle, WA 98111
Position: Asst Chief Engr. *Employer:* Pt of Seattle. *Education:* BS/Civ Engg/Univ of IL. *Born:* 12/15/31. Native of Rockford, IL. Served with Army Corps of Engrs in Korea 1953. Design Engr with consulting firm in Crystal Lake & Urbana, IL specializing in interstate hgwy geometrics & design. Assoc Cecil C Arnold & Assoc 1961-1969 specializing in interstate hgwy & bridge design in Seattle. With the Pt of Seattle since 1969. Assumed current position as Asst Chief Engr 1974, with responsibility for direction of all design & construction of waterfront facilities & for airport facilities at Sea-Tac Intl Airport. Pres, Seattle Sec ASCE 1977. Play trumpet in Bellevue, WA Community Band. *Society Aff:* ASCE.

Petkovic-Luton, Ruzica A
Business: P.O. Box 45, Linden, NJ 07036
Position: Group Head *Employer:* Exxon Res. and Engr Co *Education:* Ph.D/Metallurgical Engr/McGill Univ-Montreal; M.Eng/Metallurgical Eng/McGill Univ-Montreal; B. Eng/Metallurgical Eng/Univ of Belgrade *Born:* 4/17/40 in Yugoslavia. Emigrated to Canada 1969. Graduate Research on the mechanisms of static and dynamic recrystallization during hot working. Research Assoc, McGill Univ 1975-77. Joined the Corp Science Research Lab of Exxon Corp in 1977 as a Research Engr, specializing in creep and stress rupture of materials in high temperature corrosive environments. Assumed current responsibility as Head of the Physical Metallurgy Group in 1979. Author of some 30 technical papers and recipient of the award for the best paper published in Canadian Metallurgical Quarterly in 1975. Presently Secretary of the Metal Science Club of NY, member of the Ferrous Metallurgy Committee and the Shaping and Forming Committee of AIME. *Society Aff:* TMS-AIME, ASM, NACE

Petree, Frank L
Home: 315 Princeton Rd, Plainsboro, NJ 08536
Position: Nuclear Engr. *Employer:* MVA Services, Ltd. *Education:* MS/EE/MIT; BS/EE/MIT. *Born:* Dec 6, 1927. MS & BS MIT 1951. Live in Plainsboro N J. USNR WW II 1945- 46. Nuclear Engr with Phillips Petro Co 1952-61, spec in radiation instrument dev. Engrg Mgr TMC & Nuclear Data 1962-65. Pres of Data Acquisition Corp & Numerical Analysis Corp 1966-69. Ch Engr Simpson Elec Co 1970-71. Nuclear Engineering Consultant, Ebasco Services 1976-1983. Senior Consulting Engineer, MVA Services Ltd., 1984-1987. Spec in control sys for nuclear reactors, computer modelling of electrical sys for fusion reactors. IR-100 Award 1966. 5 pats. Enjoy Backpacking, tourskiing, music. *Society Aff:* ANS, NSPE.

Petrides, Fedon N
Business: 191 W Fifth St, Waterloo, IA 50701
Position: President. *Employer:* Brice, Petrides & Assocs Inc. *Education:* BS/Civil Engg/TUI; MS/Civil Engg/TUI; BS/Geo/TUI; PhD/Chem Engg/U of I. *Born:* Feb 1930. PhD (not completed) Univ of Iowa; Work as proj engr in var European countries until 1958. Joined Robert L Brice & Assocs 1959 & became V P 1964. Name of firm changed to Brice, Petrides & Assocs Inc Engrs in 1967; became Pres 1968. P Pres of Church Bd. P Bd Mbr of Chamber of Commerce & Environ Coalition. Mbr of Bd of Dir of Peoples Bank & Tr Co, Community Dev Council, Bus Dev Council, Mayor's Action Ctte for Hwys, & Mayor's Action Ctte for Downtown Dev. P. of Bd Schoitz Meml Hospital; Bd Mbr Cedar Arts Forum, Bd Mbr Waterloo Industrial Dev Corp; Mbr Environment & Energy Council. Chrmn. Aviation Task Force, Chrmn. Convention Related Business Development, Bd. Mbr. Cedarloo Hospital Council, Bd. Mbr. Family Practice Residency Program, Chrmn. Hospital Cooperative Committee, Mbr. Transportation Task Force. Enjoy sports, music, painting. City Council Reorganization Committee. *Society Aff:* ASCE, NSPE, IES.

Petrou, Nicholas V
Home: Petrou Associates, Ltd, 302 Hallsborough Dr, Pittsburgh, PA 15238
Position: President *Employer:* Petrou Associates, Ltd *Education:* BS/EE/Northeastern Univ; MS/EE/Harvard. *Born:* 8/2/17. Native of Springfield, MA. Army Signal Corps 1942-45. Have been with Westinghouse Elec Corp since 1940. Present respon as VP Human Resources-1977. Active Engg Mgr for Air Arm Div 1958-61. Fellow IEEE 1971.Now President, Petrou Associates, Ltd Management Consultant. VP Human Resources - 1977-1981 at Westinghouse. Pres Defense

Petrou, Nicholas V (Continued)
Cntr Westinghouse 1967-1977. Exec VP Defense- Westinghouse 1973-1977. Pres Petrou Assoc LTD 1981-. *Society Aff:* IEEE, AIAA, SME.

Petrovic, Louis J
Business: 80 Bacon St, Waltham, MA 02154
Position: Pres. *Employer:* Resource Engg Inc. *Education:* PhD/ChE/Northwestern Univ; MS/ChE/Northwestern Univ; BS/ChE/Case Inst of Tech; MBA/Bus Policy/Boston College. *Born:* 10/6/40. Native of Cleveland, OH. Initial work for Cabot Corp in high temperature production of T1O2 paint pigment. Prog mgr of coal R&D and, also, process engg and economic evaluation at Kennecott Copper Corp for eight yrs. Elected to current position as pres of Resource Engg Inc in 1977. Currently responsible for work in coal & mineral consulting, exploitation, and engg. *Society Aff:* AACE, ACS, AIChE, AIME, IAEE, ΣΞ, ΤΒΦ

Petryschuk, Walter F
Business: c/o Polysar Ltd, Sarnia, Ontario N7T 7M2 Canada
Position: Mgr, Rubber Mfg Operations. *Employer:* Polysar Ltd. *Education:* BASc/Chem Engg/Univ of Toronto; M Eng/Chem Engg/McMaster Univ; PhD/Chem Engg/McMaster Univ. *Born:* 9/26/36. in Ontario. Hon Grad BASc Chem Engg Univ of Toronto, M Engg & PhD Chem Engg McMaster Univ. Employed in process design, computer applications, advanced chem engrg, opers res, heavy water process dev, simulation work, managerial roles with Polysar Ltd 1959- . Hd up staff of about 600 personnel (50 professionals) oper 3 planns utilizing emulsion, suspension, & solution polymerization techs to make over 80 grades of synthetic rubber. Fellow Chem Inst of Canada. Mbr AIChE. Reg P E Ontario. Mbr Chamber of Commerce; Mbr Sarnia Planting Bd; Chrmn, 1979 Cdn ChE Conf. *Society Aff:* AIChE, CSChE, CIC.

Pettengill, Kenneth H
Business: 1501 W Elizabeth Ave, Linden, NJ 07036
Position: V Pres - Operations. *Employer:* Malmstrom Chem, Emery Indus Inc. *Born:* Jan 1926 Milton Mass. Submarine service US Navy WW II. BS MIT 1949. Chem Engr A D Little, Plant Engr ICI, Emery Indus since 1960. Mgr Process Res; Dir Chem Engrg Dept; V P Operations 1974- . Respon for manufacture, engrg, capital facils & planning. Chmn Ohio Valley Sect AIChE 1967-68, Chmn Southwest Ohio Engrg Ctr Fund Drive. U S & foreign pats. Hobbies: sports car rallying, golf.

Pettigrew, Allan
Home: 3647 Beaudesert Rd, Park Ridge, Queensland, Aust 4125
Position: Pres *Employer:* Pettigrew Engrg Pty Ltd *Education:* ME/Queensland Inst of Tech *Born:* 9/8/32 Born in Bundabery, Queensland, Australia. Educated at Toowoomba Grammar School and Queensland Inst of Tech. Cadet Engr with "Southern Cross" Engrg - Designers and Manufacturers of pumping and hydraulic equip. Specialised in Water Treatment and commenced own co as Pettigrew Engrg in 1968. In 1969 commissioned by Australian Gov to act as Expert Advisor in Water Treatment to South East Asian countries. 1976 expanded to full engrg service of consulting, design and installation of "turn-key" projects specialising in water reclamation and by- product reclamation for industry and public utilities. Has published eleven tech papers on water re-use and solids reclamation. *Society Aff:* MIE (Aust), MAWWA, FAIM, WPCF, MIWPC

Pettigrew, James L
Business: Flight Systems Engineering, Aeronautica Sys Div/ENF, OH 45433
Position: Deputy Dir Flight System Engineering *Employer:* HQ Aeronautical Systems Division/USAF *Education:* MS/ME/Clemson; BS/ME/Clemson. *Born:* Oct 1934. Native of Starr S C. Instr Mech Engrg Clemson Univ 1956-58. Active Duty USAF 1958 - USAF Pilot Rating 1959 - Pilot-in-Command of USAF B-47, B-52 & RF4C combat aircraft & num support aircraft. DOT-FAA airline transport pilot & flight instr. Assoc Prof of Aerospace Studies (AFROTC) MIT 1965-68. Reg P E SC 3347 in 1966. Ohio 43408 in 1978. With Aircraft Engrg Div, Strategic Air Command, 1971-1977 spec in aircraft propulsion. Respon for reliability & maintainability engrg for over 10,000 engines incl TF33, J57, & TF30 engines. 1977-1981 Aeronautical Systems Division, Propulsion System Program, Office- Engineering Directorate as. Deputy Director, Director, and Chief Flight Systems Engineer. Provided engineering support for over $2 billion/year propulsion system acquisition. Since March 1981 Deputy Director of ASD/ENF - Providing Flight system engineering support to Aeronautical System Divisions Acquisition Programs. Mbr ASME, SAE, AIAA, & ASQC. Dev & implemented the first operational Air Force engine condition monitoring prog for the SAC KC-135 aircraft fleet. Presented paper on prog at 11th and 13th Internatl SOLE Symposium. *Society Aff:* AIAA, MCSI, SOLE, ASQC, SAE.

Pettigrew, Richard R
Business: PO Drawer 807, Clovis, NM 88101
Position: Pres. *Employer:* Pettigrew & Assoc. *Education:* BE/Civ Engg/Vanderbilt Univ. *Born:* 11/24/27. Native of Nashville, TN. Served US Army 1946-48. Survey Party Chief, Atlantic Refining Co, Dallas, TX, 1951-53; Construction Superintendent, Inter- State Construction Co, Nashville, TN, 1953-57; Asst Hgwy Engr, CA Div of Hgwys, Bishop, CA, 1957-59; Proj Engr, NM State Hgwy Dept, 1959-61; County Engr, Lea County, NM, 1961-65; Pres, Pettigrew & Assoc, Consulting Engrs, Clovis, NM, 1965-pres. Fellow, ASCE;Natl Dir, NMSPE, 1976-78, & 1980-82; Pres NMSPE, 1979- 80; SW Region VC NSPE/PEPP 1980-82. *Society Aff:* NSPE, ASCE, APWA, ASTM.

Pettit, Frederick, S
Business: Mat. Sci. & Eng. Dept, University of Pittsburgh, 848 Belt, Pitt, PA 15261
Position: Prof. & Chrmn *Employer:* Univ of Pittsburgh *Education:* PhD/Met Eng/Yale Univ; ME/Met Eng/Yale Univ; BE/Met Eng/Yale Univ *Born:* 03/10/30 Performed postdoctoral research at Max Planck Inst for Physical Chemistry 1962-1963. Employed at Pratt and Whitney Aircraft 1963-1979 and performed research on high temperature corrosion of materials and use of coatings for protection. Appointed sr staff scientist in 1977. Joined Univ of Pittsburgh in 1979 as Prof and Chrmn of Metallurgical and Materials Engrg Dept. Current research is involved with thermodynamics, high temperature corrosion, and use of metallic coatings for protection. Approximately seventy technical publications concerned with oxidation, hot corrosion, metallic coatings and thermodynamics of high temperature processes. *Society Aff:* AIME, ASM, ECS

Pettit, Joseph M
Business: 225 North Ave, Atlanta, GA 30332
Position: President. *Employer:* Georgia Inst of Tech. *Education:* PhD/Elec Engg/Stanford Univ; Engineer/Elec Engg/Stanford Univ; BS/Elec Engg/Univ of CA, Berkeley. *Born:* July 15, 1916. Instr Univ of Calif Berkeley 1940-42; Electronic Countermeasures, Harvard, England, China 1942-45, Pres Cert of Merit; Airborne Instruments Lab N Y 1945-47; Stanford Univ, Assoc Prof 1947-54, Prof 1954-72, Dean School of Engrg 1958-72; Pres Ga Inst of Tech 1972- . Author or co-author 'Electronic Measurements' 1952; 'Electronic Switching, Timing, & Pulse Circuits' 1959-70; 'Electronic Amplifier Circuits' 1967. Dir Varian Assocs, Scientific- Atlanta; Mbr Natl Acad of Engrg (on Council 1973-79); Fellow & former Dir IEEE; Hon Mbr & P Pres 1972-73; ASEE; Charter Mbr Soc for Hist of Tech (on Adv Council & previously on Exec Council); formerly cons to Org of Amer States & Ford Foundation (Mex & Central Amer) Mbr, Natl Sci Bd, 1977-1982. *Society Aff:* IEEE, ASEE, SHOT, AAAS, NAE.

Pettit, Ray H
Home: 2124 Brentwood Ave, Simi Valley, CA 93063
Position: Prof. *Employer:* CA State Univ. *Education:* PhD/Elec Engg/Univ of FL; MS/Elec Engg/GA Inst of Tech; BS/Elec Engg/GA Inst of Tech. *Born:* 5/12/33. Native of Canton, GA. Served with Army Ordnance Corps 1955-57. Engr for Westinghouse Elec Corp 1954-58. Design Engr for Martin-Orlando Co 1960-63. Res Scientist for Lockheed-GA Co 1963-66. Prof of Elec Engg at GA Tech 1966-78. Prof of

Pettit, Ray H (Continued)
Elec and Computer Engg Consultant to govt and industry in communications ECM/ECCM. Enjoy collecting B-Western sound movies. *Society Aff:* IEEE.

Petzow, Gunter E
Business: Heisenbergstr 5, D-7000 Stuttgart 80, FR Germany, D-7000
Position: Acting VDir. *Employer:* Max-Planck-Inst fur Metallforschung. *Education:* Dipl Ing/-/Univ of Stuttgart; Dr rer nat/-/Univ of Stuttgart; Dr. h. c./-/Technical Univ of Tokyo, Japan; Dr.-Ing. E.h./-/Hanyang Univ of Seoul, Korea *Born:* 7/8/26. Prof Dr Petzow, who is an ASM Fellow a Hon Mbr of the Korean Inst for Metals and a Dr h.c. of the Tokyo Inst. of Tech, has received several awards, among them the Kuczynski Diplom 78 and the Schlumberger Award 78. He is author of 250 res papers, five books and holds several patents. He has been guest prof at the Tokyo Inst of Tech, Chalmers Univ Goteborg, Univ of Surrey, Univ Sao Carlos and at the Metals Research Inst in Shenyang, where he teaches introductory courses in equilibrium phase diagrams and advanced courses in powder metallurgy, metallography, high temp and reactor materials. He is Editor of the Zeitschrift fuer Metallkunde and Practical Metallography and Mbr of the Editorial Bd of the following journals: Metallography, International Metals Review, Powder Metallurgy International, and Ceramurgia International. *Society Aff:* DGM, DKG, GDCh, ASTM, ASM, IMS, AIME, APMI, MS, KIFM, ACerS, VWF.

Peurifoy, Robert L
Home: 301 E Brookside Dr, Bryan, TX 77801
Position: Retired. *Education:* MS/Civ Engg/Univ of TX; BSCE/Civ Engg/Univ of TX. *Born:* 12/1/02. Past director TX Sec of ASCE. Past Chrmn Civ Engg Div of Am Soc for Engg Educ (Natl). Mbr of Tau Beta Pi & Chi Epsilon Fraternities. Recipient of ASCE Construction Mgt Award, 1979. Reg PE, TX; Author of three books related to construction & construction engg education; Author of numerous articles related to construction. Have conducted seminars on constructin in several foreign countries. Mbr of Publications Committee of the Construction Div of ASCE. Recipient of NSPE Construction Engineering Educator Award, 1983. ASCE Peurifoy Construction Research Award, 1986. *Society Aff:* ASCE, ASEE, XE.

Pfeffer, John T
Business: 3230 CEB, 208 N. Romine St, Urbana, IL 61801
Position: Prof. *Employer:* Univ of IL. *Education:* PhD/Sanitary Engr/Univ of FL; MS/Sanitary Engr/Univ of Cincinnati; BS/CE/Univ of Cincinnati. *Born:* 10/2/35. Native of OH. After completing grad study, taught in the CE Dept at the Univ of KS, Lawrence from 1962 to 1967. Joined the CE Dept at the Univ of IL in 1967, reaching rank of Prof in 1969. Worked for the IL Inst for Environmental Quality during 1970-71. Active res prog in biological waste treatment and recovery of energy from organic residue. *Society Aff:* ASCE, ASEE, ASM, AAAS, WPCF.

Pfeffer, Robert
Home: 590 Albin St, Teaneck, NJ 07666
Position: Herbert Kayser Prof & Chmn, Dept Chem E. *Employer:* City Coll of City Univ of NY. *Education:* D Eng Sc/Chem Engg/NY Univ; MChE/Chem Engg/NY Univ; BChE/Chem Engg/NY Univ. *Born:* Nov 1935. Fac mbr City Coll since 1957, Dept Chmn since 1973. Cons to indus & gov. Res ints in transport phenomena with appli to biomed problems, fluid-particle sys, & air pollution control. Has been principal or co-principal investigator of 21 res grants & has publ over 70 tech papers. Visiting Prof at Imperial Coll 1969-70. Visiting Prof & Fulbright Scolar at Technion- Israel Inst of Tech 1976-77. Mbr AIChE, ASEE, Tau Beta Pi, Sigma Xi, Phi Lambda Upsilon. *Society Aff:* ТВП, AIChE, ASEE, ΦΛΥ, ΣΞ, NYAS

Pfeiffenberger, Andrew R
Home: 760 S Steele St, Denver, CO 80209-4840
Position: Principal. *Employer:* A R Pfeiffenberger & Assocs. *Education:* BS/Bus & Engg Admin/MA Inst of Tech. *Born:* Jan 26, 1928 Alton Ill. BS in Bus & Engrg Admin MIT 1949; Admin Officer & Ammunition Proof Officer, Army Ordnance Dept 1949-50; Asst Production Supt, Asst to Pres & Production Control Coordinator Bowman Biscuit Co 1950-62; Mgr Compensation & Org, United Biscuit Co of Amer 1962-63; Partner Hartzell- Pfeiffenberger & Assocs 1964-69, Pres 1969-75; Principal A R Pfeiffenberger & Assocs 1975- . Cert Cons Engr Colo; Reg P E Colo, Licensed Real Estate Broker Colo; cert Natl Council of Engrg Examiners; Fellow ASCE; Sr Mbr IIE (Dir Colo 1968-69); Mbr NMA (Natl Dir 1960-62), ACEC (Dir Colo 1972-74); Dir Silver State Savings & Loan Assn 1975-82; Exec Bd Denver Area Council BSA (V P 1974-79); recipient Silver Beaver Award BSA 1976; Who's Who in Finance & Industry; Who's Who in the West; Rotary Club of Denver (Dir 1971-73). *Society Aff:* ASCE, ACEC.

Pfender, Emil
Business: Dept of Mech Engg, 111 Church St, SE, Minneapolis, MN 55455
Position: Prof of ME. *Employer:* Univ of MN. *Education:* DrIng/Elec Engg/Univ of Stuttgart; Diploma/Phys/Univ of Stuttgart. *Born:* 5/25/25. Native of Germany. Studied physics at the univs of Tuebingen and Stuttgart, Germany. Spent sabbitical yr (1961/62) with the Plasma Physics Branch, Aerospace Res Labs, Dayton OH. Joined the Univ of MN in 1964 as Assoc Prof; full prof in 1967. Served on Advisory Panel for NSF and on several committees of the NRC. Editor of journal on plasma chemistry and plasma processing. Received Adams Meml Mbrship Award (AWS) and US Sr Scientist Award (Fed Republic of Germany). Published approximately 150 papers in arc technology, plasma heat transfer and plasma processing. Main res interests: plasma chemistry and plasma processing, arc tech, plasma heat transfer. *Society Aff:* APS, ASME, IEEE.

Pfister, Henry L
Home: 19406 Beckworth, Torrance, CA 90503
Position: Asst Prof. *Employer:* Univ of S CA. *Education:* PhD/Engg/Univ of S CA; MS/Industrial Engg/San Jose State College; BS/Mathematics/Lamor Univ. *Born:* 6/19/43. Dr Pfister has specialized in the optimization of aerospace systems using digital computers. He has been responsible for the development of on board computers for satellites, of space/ground communication networks, and of mission optimization and scheduling algorithms. *Society Aff:* ORSA, TIMS.

Pflug, Charles E
Home: 33720 Lakeshore Dr, Burlington, WI 53105
Position: Cons Engr (Self-employed) *Education:* BS/EE/Univ of NB. *Born:* Oct 1900 Nebr. G E Test Engr 1926-28; Wisc Pr & Lt Asst Div Engr, Local Mgr 1928-37; Wisc Gas & Elec 1937-39 Sub Engr; Nash Kelvinator (Amer Motors) 1939-65, Elec Engr, Mgr. Plant Engr plants, mfg processes, equip, design - built war & automobile plants in U S & Canada, dev mfg processes used generally in automobile plants. Natl cttes; Fellow IEEE; Mbr AWS; Wisc Soc Prof Engrs (Dir); Wisc Governor's Task Force - Cost Reduction. Reg P E Wisc, Mich, Calif, Neb. Retired 1965. Hobbies: travel, shop, golf, organizations. *Society Aff:* IEEE, WPE, NPE.

Pforzheimer, Harry Jr
Home: 2700 –G– Rd 1-C, Grand Junction, CO 81501
Position: President & CEO *Employer:* Harry Pforzheimer, Jr. & Associates *Education:* BS/Chem Eng/Purdue Univ; -/Law/George Washington Univ; -/Business/Case Western Reserve *Born:* 11/18/15. in Manila Philippines. Distinguished Student-Purdue Univ. Joined The Standard Oil Company, Cleveland, OH. Served Amer Inst of Chem Engrs as Chmn Cleveland Sect 1955; Genl Chmn Intl Mtg in Cleveland 1961. Progressed through Standard's Mfg, assignment to Petro Admin for War 1942-45; Finance; Long Range Planning; Natural Resources; Elected V.P. Oil Shale and Tar Sand, President White River Shale Oil Co. and Prog Dir, Paraho Oil Shale Demonstration. Retired from Standard in 1980 and accepted position as President, CEO and Chairman Paraho Development Corp. Retired Paraho 1982 and organized Harry Pforzheimer Jr. and Associates, Consultants. Serves as Dir of CO Sch of Mines Res Inst, St Mary's Hospital Advisory Bd, Intrawest Bank of Grand Junction and Wayne Aspinall Foundation. *Society Aff:* AIChE, COMA, AMC.

Pfrang, Edward O
Business: Amer Soc of Civil Engrs, 345 East 47th Street, New York, NY 10017
Position: Exec Dir *Employer:* Amer Soc of Civil Engrs *Education:* PhD/Struct Engg/Univ of IL; ME/Struct Engrg/Yale Univ; BS/CE/Univ of CT. *Born:* 8/9/29. Dr. Pfrang joined the staff of the ASCE in 1983 as Exec Dir. As Dir he is responsible for managing the Society's staff. Staff activities include technical publishing, conventions and conferences on both technical and professional affairs, educational activities, public communications programs and a number of other activities designed to improve technology, advance the technical capability of ASCE mbrs and improve the professional and economic status of civil engrs. Formerly Chief, Structures Div, Ctr for Bldg Techn of the Natl Bureau of Standards, Dr. Pfrang directed a broad range of res programs in structural engg, earthquake hazards mitigation, construction safety and geotechnical engg. *Society Aff:* EERI, ASTM, ASCE, ACI, AAAS, PCI, ANSI, NFPA.

Pfundstein, Keith L
Business: John Deere Rd, Moline, IL 61265
Position: Manager, Product Safety. *Employer:* Deere & Co. *Education:* BS/Mech Engg/Univ of IL. *Born:* 9/20/17. Native of Erie Ill. Pres Triangle Frat 1940. Ethyl Corp 1940-65, Tech Sales & Res Engr Denver, St Louis & Detroit 1940-44; Tech Serv Div 1944-65; Mgr Div's Agri Engrg Dept 1953-65. Deere & Co Prod Planning Engr Moline 1965; Mgr Prod Safety Dept 1966- . Guest lectr & seminar instr on prod safety & liability at univs, tech soc's & trade assns. Author tech papers in SAE, ASAE, ASME. Chmn Power & Machinery Div ASAE 1974, Natl Safety Council Bd of Dir current. Enjoy fishing, golf & woodworking. *Society Aff:* ASAE, SAE, NSC.

Phadke, Arun G
Business: 2 Broadway, New York, NY 10004
Position: Consltg Engr *Employer:* AEP Service Corp *Education:* PhD/EE/Univ of WI; MS/EE/IIT, IL; B Tech/(Hon) EE/IIT, India; BSc/Math/Phys/Agra Univ, India *Born:* 08/27/38 Born in Gawlior, India. Grad educ in Power Sys Engrg. Was Sys Engr with Allis-Chalmers from 1964-1967. Taught at the Univ of WI 1967-1969. Participated in the design and development of the AC/DC Power Sys Simulator at the Univ of WI. With AEP Service Corp since 1969. Responsible for research and development of microcomputer applications in Power Sys Instructor at short-term course at several universities. Was a Visiting Prof at VPI & SU, Blacksburg, VA during 1978-1979. Fellow of IEEE. Convenor of CIGRE Working Group 34.02 during 1980- 1983. Enjoy sketching and painting. *Society Aff:* IEEE

Pheanis, David C
Home: 5217 S. Monaco Drive, Tempe, AZ 85283
Position: Assoc Prof *Employer:* Arizona State Univ *Education:* PhD/EE/AZ State Univ; MS/EE/AZ State Univ; BS/Math/Case Inst of Tech *Born:* 05/31/47 Born and raised in Greenfield, OH. After graduation from Case Inst of Tech entered the USAF as a programming officer for the SAGE sys of On-Line Real-Time Aerospace Defense. Left the Air Force as a Captain to pursue a PhD degree. After completing a PhD degree, joined the engrg faculty at Arizona State Univ as an Asst Prof specializing in digital sys design. Presently as Assoc Prof working with both the hardware and software of microprocessor-based systems. Also the coordinator of the Digital Computer Lab, which has a VAX-ll/780 and a Honeywell 68/80. Conslt for several companies who use microprocessors and minicomputers. Presently doing conslt work for Sperry Flight Systems. *Society Aff:* IEEE

Phelan, Merrill D
Home: 5945 Oakland Park Dr, Burke, VA 22015
Position: VP Systems *Employer:* C. E. Smith Co *Education:* MSIE/IE/OSU; BIE/IE/OSU. *Born:* 11/03/47 Mr Phelan is a native of Columbus, OH. Upon grad from the Oh State Univ, he began a career in Systems and Data Processing. Major areas of accomplishment include Design Engrg Drawing Control; Configuration Control; Shop Floor Process Control; and, most recently, financial controls in the real estate indus. He has been involved in a new plant start-up situation, as well as assuming responsibility for the creation of a Data Processing Dept within an existing company. *Society Aff:* AICPA, VSCPA

Phelan, Richard M
Home: 4 Cornell Walk, Ithaca, NY 14850
Position: Prof. *Employer:* Cornell Univ. *Education:* MME/Machine Design/Cornell Univ; BS/ME/Univ of MO. *Born:* 9/20/21. Born & raised in Moberly, MO. Served in USNR 1943-46. Worked for the US Navy at Pt Hueneme, CA, before going to Cornell Univ in 1947 as an instr with the opportunity for doing grad work. Asst prof 1950, assoc prof 1956, & prof 1962. Author of "Fundamentals of Mechanical Design–, 3rd ed, McGraw-Hill, 1970; "Dynamics of Machinery–, McGraw-Hill, 1967; "Automatic Control Systems–, Cornell Univ Press, 1977. *Society Aff:* ASME, ASEE, SEM, AAAS.

Phelps, Boyd W
Business: 1000 Washington St, Michigan City, IN 46360
Position: President. *Employer:* Boyd E Phelps Inc. *Education:* BSCE/Civil/Purdue. *Born:* 10/3/28. Reg PE. Pres Boyd E Phelps Inc, Cons Engs with offices in Michigan City & Indianapolis IN. Past Chairman-State of Ind Admin Bldg Council. Past President - Consulting Engineers of Indiana, Inc.. *Society Aff:* ACEC, WPCF.

Phelps, Dudley F
Home: 2 Fenimore Rd, Port Washington, NY 11050
Position: Cons (Self-employed) *Education:* ME/Mech Eng/Cornell Univ *Born:* Oct 1904. Joined The J G White Engrg Corp 1940; Pres & Ch Exec Officer 1957-69; cons 1969-74; Dir 1956-74; V P & Engrg Mgr 1956; Ch Mech Engr 1950-56; V P & Dir subsid Whitengeco Venezolana S A 1957-62. Earlier prof exper wth E L Phillips & Co & Long Island Lighting Co. Also taught at Pratt Inst. Mbr Process Selection Bd, Office of Saline Water 1959-62. Dir Ninth Fed Savings & Loan Assn 1964-82. Dir Nineco Corp, Nanuet, NY 1977-82. Fellow ASME. Mbr Newcomen Soc. Reg P E NY, NJ. V P & Mbr Council ASME 1964-67; Mbr Metro Sect Exec Comm 1952-55, Chmn 1955. Clubs: Cornell NYC. *Society Aff:* ASME

Phelps, Edwin R
Home: 12000 Heatherdane, St Louis, MO 63131
Position: President *Employer:* Coadser, Inc. *Education:* BSCE/Hgwy/KS Univ. *Born:* 1/12/15. Began service in the coal industry in Southern IL in 1937. Served as Lt Com in the Navy Air Corp 4 1/2 yrs during WWII. Served 15 yrs with P&M Coal Mining Co as Chief Engr, VP-Operations and Pres. Retired from Peabody Coal Co in 1979 after 19 yrs' service as Chief Engr, VP-Engg, Sr VP-Operations, Pres, and VChrmn. Mbr of Sigma Tau Honor Engg Soc and awarded the AIME Howard N Eavenson Award in 1972 and the Erskine Ramsay Award in 1979; also received the Distinguished Service Citation from KS Univ in 1976. Chrmn Bd. Nat'l Coal Assoc. 1974-75. *Society Aff:* AIME, NSPE.

Phelps, George C
Business: 4453 S 67th St, Omaha, NE 68117
Position: Mgr., Chemical & Industrial Services *Employer:* Nebraska Testing Corporation *Education:* BS/Chem/Wichita Univ. *Born:* Aug 21, 1935. Attended Westminster Coll, Fulton Mo & Univ of Wichita. Wichita Kansas; Officer US Army Air Defense Missile Ctr 1958-60. Post grad courses in engrg at Univ of Nebr at Omaha. Passed Nebr Prof Engrs exam for Civil Engr 1969. Lectured at sev symposia in fields of expertise. Taught concrete & asphalt matls courses at Univ of Nebr at Omaha, Coll of Engrg 1971-73. Chemist at Nebr Testing Labs 1961; Lab Dir 1964, Mgr 1966, V P & Secy 1967- . Firm merged with Twin City Testing Corporation in Dec of 1986. Became manager of Chem. & Indust. Services for the new firm. Also Tech Dir & Qual Assurance Mgr at present. Pres: ACS, Instrument Soc of Amer, Const Specifications Inst. Mbr ACEC, WPCF, APCA, ACIL. Listed in Who's Who in Ecology 1973; Amer Bar Assn Directory of Tech Experts 1974-76; Fellow, Amer Inst of Chem; Who's Who in Tech, 1981-1984; Recognized Prof Conslt in Forensics. *Society Aff:* ACS, ISA, CSI, APCA, WPCF, AIC, ACEC, ACIL.

Phelps, Pharo A
Home: 269, Birchwood Dr, Moraga, CA 94556
Position: Principal *Employer:* Phelps Associates *Education:* PhD/Physics/US Naval Academy Postgrad Sch; BS/Civil Engg/Rensselaer Polytech Inst; BS/-/US Naval Academy. *Born:* Aug 1928. Navy Civil Engr Corps managing design, const & facils R&D in U S, London, Spain, Pacific & S E Asia. Completed naval serv as Deputy Commander Naval Facils Engrg Command. With Bechtel from 1974 to 1985; Applications Engrg Group Mgr until 1976; 1976-78 Mgr Engrg Dept of Bechtel's Res & Engrg Oper. 1979-80 Mgr of operations in Middle East. 1980-82 Manager of Engineering and Materials of Bechtel Research and Engineering Operation. Vice Pres. & Mgr of Defense Projects 1983-85 for Bechtel Nat'l. Since 1985 Principal, Phelps Associates, consultant for engineering management and technical issues. President of Hazco, a startup company in toxic waste disposal. Spec ints incl underground engrg. Dir SAME 1971-73, Mbr Underground Tech Res Council of ASCE/AIME, APS, Sigma Xi, Tau Beta Pi, Chi Epsilon, & Amer Underground Space Assn. Enjoy swimming, hiking & recreational flying & soaring. *Society Aff:* ASCE, SAME

Pherigo, George L
Business: 200 Century Ave, Hutchinson, MN 55350
Position: Dept Hd, Nondestructive Testing. *Employer:* Hutchinson Area Tech Inst. *Born:* Mar 31, 1938 Minn. BS St Cloud State Coll; MS Univ of Wisconsin. Var indus exper in welding & welding inspection prior to dev a welding tech rog in 1967 at the Hutchinson Voc-Tech School. Dev in 1969 a 2 yr NDT -6o:iqu 12 iy training program at Hutchinson. Mbr ASNT, AWS, ASM. Present V Chmn of Minn Sect ASNT, Chmn of Educator's Div Natl ASNT, & Mbr ASNT Standing Ctte 'Personnel Training & Certification'. Selected as the 'Outstanding Vocational Instr in Minn' 1971 & received the Natl ASNT Tutorial Award 1974.

Phillips, Albert J
Home: 620 Bowsprit Lane, Sarasota, FL 33577
Position: VP & Dir of Res - ret. *Employer:* Asarco Inc, 120 Broadway, NYC. *Education:* PhD/Metallurgy/Yale Univ; MS/Metallurgy/Yale Univ; BS/Metallurgy/Yale Univ. *Born:* Feb 4, 1902. Native of New Haven Conn. Metallurgist Scovill Mfg Co 1923- 31. ASARCO 1931-67. Cons since 1967. AIME James Douglas Medal & Hon Mbr. Fellow TMS. ASM Gold Medal & Hon Mbr. *Society Aff:* AIME, ASM, ASTM, AAAS.

Phillips, Allan L
Business: Inst de Ingenieria Agricola, Casilla 537 Chillan Chile
Position: Proj Mgr. *Employer:* FAO. *Education:* PhD/Engg/Univ of CA; MS/Appl Mechanics/MI State Univ; BS/Agri Engg/MI State Univ. *Born:* 12/31/36. Native of Deckerville, MI. Career devoted to applications of agri engg in developing areas of the world, with teaching & res experience at the Univ of Puerto Rico-Mayaguez, & UNESCO expert at the Univ of Nairobi & Central Luzon State Univ, Fulbright-Hays Sr Lectr at the Escuela Politecnica Natl, Quito, Ecudor, & Res Assoc at the East-West Ctr, Honolulu. Currently FAO Proj Mgr at the Agri Engg Inst of the Univ of Concepcion, Chile. Chrmn ASAE Hwaii Sec 1978, Reg PE in HI & PR. *Society Aff:* ASAE.

Phillips, Aris
Business: Becton Ctr, 15 Prospect St, New Haven, CT 06520
Position: Prof. *Employer:* Yale Univ. *Education:* Dr Ing/Appl Mechanics/Berlin Technical Univ; Dir Engg/Civ Engg/Athens Technical Univ. *Born:* 11/30/15. Robert Higgin Prof of Mech Engg, Yale Univ. DrIng Technical Univ Berlin, Germany, DipIng Technical Univ Athens, Greece. Native of Smyrna, Asia Minor. In USA since 1947. Faculty mbr at Caltech 1947, Stanford 1948-54, and Yale since 1954 (Prof since 1960, dir of Grad Studies in Engg, 1963-1970, dir of Undergrad studies in Mech Engrg since 1979). Fellow ASME, Fellow American Acad of Mechanics, Fellow American Association Advancement of Science, Mbr Soc of Engg Sciences. Mbr American Society Engineering Education. Editor "Acta Mechanica–. Author of three books and in excess of 100 technical publications. (wife: Bessie Barbikas Phillips, children: John Aristotle, Dean Aris). *Society Aff:* ASME, AAM, AAAS, ASEE.

Phillips, Benjamin A
Business: 721 Pleasant St, St Joseph, MI 49085
Position: Pres. *Employer:* Phillips Engg Co. *Education:* PhD/Phys Chem/Univ of TX; MA/Phys Chem/Univ of TX; BS/CE/Univ of TX; BA/Chem/TX College Mines & Met. *Born:* 8/22/12. Native of El Paso, TX. At Servel, Inc, 1939-1956, res & engg, primarily in absorption & compression refrig as Res Supervisor & Maj Product Engr. At White Sands Missile Range 1956-1958, Chief, Propulsion Branch, & Chief, Rocket, Vehicle & Warheads Labs. At Whirlpool Corp, 1958-1970, corporte res & engg in absorption refrig & absorption & compression air conditioning & heat pumps as Dir of Res & Engg of Environmental Systems. 1970, Phillips Engg Co., contract engg on new products in the energy field: absorption systems, solar air conditioning, heat actuated heat pumps. *Society Aff:* AIChE, ASHRAE, ACS, ISES.

Phillips, Chandler A
Business: Dept of Biomedical Engrg, Dayton, OH 45435
Position: Prof of Biomedical Engrg *Employer:* Wright State Univ. *Education:* MD/Medicine/Univ of So CA; Prof Engr/Elec Engg/State of CA/State of OH; AB/Biophysics/Stanford Univ. *Born:* 12/21/42. Medical Officer (Capt), US Air Force, 1970-72, served in Southeast Asia. Res Physician (Bioengg), Univ of Dayton, 1972-74. Asst Prof of Engg and Physiology, Wright State Univ, Dayton, OH 1975-79. Assoc Prof, 1979-1984. Prof, 1984-Present. Dir, Biomedical Engg Prog, 1975-1984; Deputy Dir, Natl Ctr for Rehab Engrg, 1983-1986; Emergency Medicine Consultant, 1972-Present. Grantee: American Heart Assoc, Natl Inst of Health. Contractor: US Army, Natl Aeronautics & Space Admin. Res: Skeletal and Cardiac Muscle Biomechanics and Energetics. Rehabilitation Engrg Consultant: US Air Force, US Veterans Admin. Awards: Air Force Commendation Medal, Physicians Recognition Award. Harry Rowe Mimnoe Award, Natl Aerospace & Electronic Sys Soc (IEEE). Outstanding Engrg Achievement Award, Natl Soc of Prof Engrs. Honorary Fellow, American Academy of Neurological & Orthopedic Surgeons. Other Listings: Who's Who in the Midwest, American Men and Women of Sci, Intl Biographical Directory, and Men of Achievement. *Society Aff:* AsMA, APS, ASB, IEEE.

Phillips, Don T
Home: Texas A&M Univ, Zachery Eng. Center, Dept. of Ind. Eng, College Station, TX 77840
Position: Prof of Indus Engrg. *Employer:* Texas A&M Univ. *Education:* PhD/IE/Univ of AR; MS/IE/Univ of AR; BS/IE/Lamar Univ. *Born:* Feb 6, 1942. PhD Indus Engrg Univ of Arkansas; MS Univ of Arkansas; BS Lamar Univ. Native of Texas; born in Bessemer Ala. Previously taught at Univ of Arkansas, Univ of Texas, & Purdue Univ. Currently Prof of Industrial Engrg at Texas A&M Univ. Widely published in tech journals; author of over 100 papers and 5 college textbooks, including, 'Appl Goodness of Fit Testing' (IIE Monograph), co-author 'Appl Geometric Prog' & 'Opers Res: Principles & Practice'. Foundations of Optimization and Fundamentals of Network Analysis. Active cons to indus, mbr of Edit Bd IIE Transactions, Active Researcher and Conslt in Manufacturing Sys, TEES Research Fellow, E D Brockett Prof of Engr, Who's Who in Amer, Midwest, South & Southwest and Texas.Enjoys fishing & golf. *Society Aff:* IIE, MHI, ORSA, TIMS.

Phillips, James W
Business: Theor & Appl Mechanics, 216 Talbot Lab, 104 S Wright St, Urbana, IL 61801
Position: Prof & Assoc Head *Employer:* Univ of IL. *Education:* PhD/Engg/Brown Univ; ScM/Engg/Brown Univ; BME/Mech Engg/Catholic Univ of America. *Born:* 3/8/43. Native of Wash, DC. Grad work (1964-1969) in stress-wave propagation under H Kolsky. Faculty mbr, Univ of IL, 1969-present. Contract work for Argonne Natl Labs, 1974-1978. Consultant to Rock Island Arsenal, 1975. Editor, *Mechanics*, 1975-1978. Sabbatical leave, Univ of MD, 1978-1979. Technical interests: stress waves in solids, waterhammer waves, stress analysis of wire rope, experimental mechanics, photomechanics, computer graphics. Other interests: music, choir. Married; 4 sons. *Society Aff:* AAM, SESA, TBΠ.

Phillips, John R
PO Box 388, Yazoo City, MS 39194
Position: Sr. VP-Operations (Retired) *Employer:* MS Chemical Corp. *Education:* BS/ChE/Case Inst of Technology. *Born:* 11/21/22. in Cleveland, OH. Employed by Shell Chemical Corp, Pittsburg, CA, Dec, 1943 as Control Chemiast; Res Chemist, and Jr Technoligst until Nov, 1946. Joined Girdler Corp, Gas Processes Div, Louisville, KY, as Process Engr; Operating Engr; finally Chief Operations Engr. Joined MS Chemical Corp, Yazoo City, MS, Nov, 1952 as Asst Operations Mgr; Operations Mgr; promoted to VP-Operations Feb, 1972. Formerly in charge of manufacture, maintenance, security, and related activities at facilities at Yazoo City and Pascagoula, MS, and Carlsbad, NM. Vice Chrmn, Mgmt committee, Triad Chemical, Donaldsonville, LA. Enjoys golf, viewing contact sports, lawn culture. Retired 1/1/87. *Society Aff:* AIChE, ACS, AXΣ.

Phillips, John R
Home: 911 Maryhurst Dr, Claremont, CA 91711
Position: Prof of Engrg. *Employer:* Harvey Mudd College. *Education:* D Eng/Chem Engg/Yale Univ; M Eng/Chem Engg/Yale Univ; BS/Chem Engg/Univ of CA, Berkeley. *Born:* Jan 1934 Albany Calif. Reg P E Calif. Teacher, cons, res engr. Worked for Stanford Res Inst, & for Chevron Res Co. Was Staff Officer, US Army CBR Combat Devs Agency. Visiting Prof Univ of Edinburgh Scotland, Cambridge Univ England, ESIEE Paris. Naval Postgraduate School Monterey. Started as Asst Prof of Engrg at Harvey Mudd Coll 1966; Assoc Prof 1968; Prof 1974. C F Braun Fellow in Engrg 1968. Dir of Engg Clinic, 1977. Cons to var indus & educ orgs; estab own cons firm 1973. Fields of int: desalination, energy conversion, research mgmt. *Society Aff:* AIChE.

Phillips, Joseph J
Business: Patent and Trademark Dept, Haynes International, Inc, Kokomo, IN 46902-9013
Position: Manager, Patent & Trademark Department *Employer:* Legal Div, Haynes International, Inc. Kokomo, IN. *Education:* BA/Metallurgy/OH State Univ. *Born:* Apr 1917 Youngstown Ohio. Law School Indiana Univ 1963-64. Met res Battelle Mem Inst Columbus Ohio, AEC Manhattan Proj; gov & indus res 1942-53. Union Carbide Corp Kokomo Ind, Tech Editor, Patent Engrg Liaison 1953-70. Cabot Corp, Patent Agent 1970-present. Purdue Univ Seminar Speaker 1963-1981. Chap Chmn, Natl Prof Dev Ctteman & Speaker's Directory ASM. Effective Serv Award U S Office of Sci Res & Dev 1947. U S Patent Office, Patent Agent Registration 26, 976. Mbr Natl Soc of Prof Engrs. Enjoy classical music, swimming & tennis American Association Advancement & Science, member American Association of Registered Patent Attorney and Agents. *Society Aff:* ASM, AAAS, AARPAA.

Phillips, Monte L
Business: Civ Engg Dept, Grand Forks, ND 58202
Position: Assoc Prof. *Employer:* Univ of ND. *Education:* PhD/Civ Engg/Univ of IL; MS/Civ Engg/Univ of ND; BS/Civ Engg/Univ of ND. *Born:* 12/12/37. Born & raised in ND. Engrg educator for 25 yrs - Univ of ND 1961-62; OH Northern Univ 1962-63; Univ of IL 1963-70; Univ of ND 1970-present. Profl and Consultant experience - ND State Hgwy Dept 1956-61; Woerfel Corp 1970-73 Design and mgt of field construction - ABM System; Engrs - Architects PC 1973 - 80 Structural and Geotechnical Consulting; Nodak Contracting Corp 1981- ; Proj Mgr ALCM Support Facility-ND; Engineers Architects, P.C. - Forensic Engineering, Design, & Construction Management 1982-present; Reg PE in ND. Pres - ND Soc of PE 1979-80. Pres - ND Sec - Am Soc of Civ Engrs 1979-80 & 1980-81. Chrmn - Univ of ND Faculty Senate 1978-79; National Dir NSPE 1983-present; NSF Sci Faculty Fellowship. Sigma Xi. Sigma Tau. Tau Beta Pi. *Society Aff:* NSPE, ASCE, NAFE.

Phillips, Orley O
Business: 1176 S Jackson, Denver, CO 80210
Position: Sr. V Pres *Employer:* DMJM - Phillips, Reister, Haley Inc *Education:* CE/-/Univ of CO. *Born:* Nov 24, 1903. Hon Mbr Chi Epsilon; 1970 recipient Disting Engrg Alumnus Award. Private practice, Edmund Friedman Prof Recog Award of ASCE & many other honors. Active in soc affairs, holding many natl & local offices. Reg in sev states. Long career incl engrg projs throughout the U S, South Amer, Mexico, Spain & Canada. Vast engrg exper has been achieved by personal exper & mgmt of engrg of dams, pipelines, tunnels, irrigation, water works, sewerage sys, bridges & struct engrg on all types of bldgs. *Society Aff:* ASCE, ACEC, NSPE, AWWA.

Phillips, Paul J
Business: Dept of Materials Science & Engrg, Knoxville, TN 37996-2200
Position: Prof *Employer:* Univ of TN *Education:* PhD/Phys Chem/Univ of Liverpool; BSc/Chem/Univ of Liverpool. *Born:* 11/26/42 Main interests are the Crystallization of Polymers and its influence on mech and electrical properties. In particular, a major area of research is the effect of elevated pressures (up to 5kbar) on crystallization kinetics and morphology of flexible, semi-flexible and rigid polymer molecules. A second area involves the study of crystallization in crosslinked polyethylene and its influence on electrical breakdown in cable insulation. A third area involves the interaction of crystalline polymers with fused ring aromatic molecules and, in particular, compound crystals. A native of England he taught in the Univ of London 1970-75, SUNY at Buffalo 1975-77 and Univ of UT 1977-84 before joining UT. Pastimes include the history of polymer Sci, english history, philately, postal history and the preservation of printed materials. *Society Aff:* ACS, APS, SPE, ΣΞ.

Phillips, R Curtis Jr
Home: 2803 Pounds, Tyler, TX 75701
Position: Pres-Owner *Employer:* MAP Production Co., Inc. *Education:* MS/Petro Eng/Univ of OK; BS/Petro Eng/Univ of OK *Born:* 4/20/40 Employed by Union Oil of Calif. (UNOCAL); as a Petroleum Engr upon graduation, with a masters degree in Petroleum Engrg from University of Oklahoma. Worked as field engr in drilling and production operations, Reservoir engr and secondary recovery engr became supervisor of Planning and Valuation. Three years District Engr in Casper, WY. East TX area Production Mgr for six years. Elected to SPE Bd of Dirs in September, 1979. Served six years on the API Oil Reserves Sub-Committee (West Texas) and three years on the AGA Rocky Mt. Gas Reserves Sub-Committee. Vice-Pres of Maritime Natl Bank. Est Oil & Gas Loan Dept. Operate as independent consltg petroleum engr specializing in economic analysis and appraisal of petroleum properties. Est the Petroleum Collection in Tyler Public Library. Received SPE Regional Service Award. Reg PE in 2 states. *Society Aff:* AIME, SPE, AAPG, AAPL, API

Phillips, Richard E
Business: 200 Agri Engg Bldg, Columbia, MO 65211
Position: Prof. *Employer:* Univ of MO. *Education:* PhD/Agri Engg/MI State Univ; MS/Agri Engg/Cornell Univ; BS/Agri Engg/Cornell Univ. *Born:* 12/7/36. Native of Oswego, NY. Three yrs in test and dev work for Intl Harvester Co. Holder of two patents. Nine yrs with the Univ of CT. Recipient of 14 awards for educational aids. Presently responsible for the agri engg ext prog for the state of MO. Board of Directors of ASAE. Sr mbr, ASAE. Instrument rated private pilot. *Society Aff:* ASAE, ΣΞ.

Phillips, Robert V
Home: 844 Monte Verde Dr, Arcadia, CA 91006
Position: Cons Engr, Adj Prof. *Employer:* Self, Univ of Calif L A. *Education:* BSCE/Civil Eng/UC Berkeley. *Born:* 1917 Los Angeles. Reg C E Calif. Los Angeles Dept of Water & Power 1945- 75; Ch Engr of Water Works 1967-72; Genl Mgr & Ch Engr of DW&P 1972-75. Currently cons engr & Adj Prof of Engrg UCLA. Fellow ASCE. Hon Mbr AWWA. Geo Warren Fuller Award AWWA 1976. AWWA Publs Award 1973. Mbr Tau Beta Pi; Engrg Adv Council U of Calif; Bd of Dir APPA 1972-75; Bd of Dir Elec Power Res Inst 1972- 75. Engr of Yr Award, So

Phillips, Robert V (Continued)
Calif Inst for the Advancement of Engrg 1973. *Society Aff:* ASCE, AWWA, USCOLD.

Phillips, Samuel C
Business: 9841 Airport Blvd, Los Angeles, CA 90045
Position: VP - TRW Inc and Genl Mgr, TRW Energy Products Grp. *Employer:* TRW Inc. *Education:* LLD/Hon/Univ of WY; MS/Elec Engg/Univ of MI; BS/Elec Engg/Univ of WY. *Born:* Feb 1921. Attended pub schools in Cheyenne Wy. Commissioned 2nd Lt Infantry, upon completion of ROTC & grad from U of Wyoming 1942. Transferred to Army Air Corps & earned pilot wings. During WW II, served wth Eighth Air Force in Europe. Since 1950, assoc with major res, dev & prod projs, incl atomic experiments & aircraft, space, & missile programs. Dir Minuteman ICBM Prog 1959- 63. From 1964-69, assigned to NASA as Dir of the Apollo Prog. Commander of Space & Missile Sys Org of the AF from late 1969 until appt by the Secy of Defense as Dir, Natl Security Agency in Aug 1972. Was promoted to Genl, US Air Force & assigned as Commander, Air Force sys Command in Aug 1973. Retired from US Air Force in Sept 1975. Joined TRW Inc in late 1975 and has served in his present position since early in 1976. *Society Aff:* NAE, IEEE, AIAA, AAS, API, SAMPE.

Phillips, Thomas L
Business: 141 Spring St, Lexington, MA 02173
Position: Chmn & Ch Exec Officer. *Employer:* Raytheon Co. *Education:* BS/EE/VA Polytechnic Inst; 1947 MS/EE/VA Polytechnic Inst, 1948. *Born:* 5/2/24. Elec Design Engr & Managerial in Engg Dept 1948-57; Raytheon Co, Lexington, MA. Mgr Sparrow III Missile Sys 1957-60; VP & Genl Mgr, Missile & Space Div 1960; Exec VP 1961-64; Dir 1962- ; Pres 1964-75; Chief Operating Officer 1964- 68; Chief Exec Officer 1968- ; Chrmn of the Bd 1975- . *Society Aff:* NAE, Business Council.

Phillips, Weller A
Business: 300 Lakeside Dr, Oakland, CA 94643
Position: Mgr, Technology Sales, Raw Matls. *Employer:* Kaiser Aluminum. *Education:* MS/Chem Engg/GA Sch of Technology; BS/Chem Engg/GA Sch of Technology. *Born:* Jan 1921. BS, MS Chem Engrg Georgia Tech. Res Engr Goodyear Tire during WW II. With Kaiser Aluminum since 1948. Have held pos respon for process design of alumina plants. Tech Dir, Chem Div. In current pos as Mgr, Tech Contracts, respon for sales & execution of tech & tech assistance contracts in alumina & aluminum fields. Present position: Mgr, Tech Sales, Raw Mtls. *Society Aff:* AIME, AICE.

Phillips, Wendell E, Jr
Home: 7 Huntington Rd, Edison, NJ 08820
Position: Pres *Employer:* Phillips Consulting Engr's Inc. *Education:* MBA/Mgmt/NY Univ; SB/Elec Engg/MA Inst of Tech. *Born:* 9/27/20. Engg, Engg Mgmt, and Program Mgmt positions with Fedl Tele and Radio Corp, air assocs, Inc Mack Truck, Sylvania and Raytheon 1942-71. Self-Employed Profl Engineer in Private Practice 1971-78. Joined Gibbs & Hill Inc in 1978, Retired, 1985. Formed Phillips Conltg Engrs, Inc. 1985. US Representative for Constuctora for William W. Phillips, S.A. of Quito, Equador & W. Phillips, S.A. of Panama. *Society Aff:* IEEE, NSPE, TRB, ΣΞ, TBΠ, ASME, AREA, HKN.

Phillips, William J, II
Business: 2777 Cleveland Ave, Suite 115, Santa Rosa, CA 95401
Position: Pres Civil/San Engr *Employer:* Summit Engrg., Inc., Consulting Cvl Engrs. *Education:* BSCE/CE/Univ of Santa Caara; MS/CE & Environ/Stanford. *Born:* 5/28/46. Reg CE CA. for Orange Cty CA Flood Control - Technician, proj engr 1963-68 (part time); US Navy Civil Engr Corps - constr. admin 1969-72, Charleston S C.; John Carollo Engrs - Proj reports 1973-74; Keith & Assoc. - design, winery wastewater reclamation 1974-76. James M Montgomery - Sr Engr, industrial, rural wastewater treatment 1976-78. Deeths Consulting/Summit Engrg - Sr Engr/Pres, winery & geothermal site design, wastewater mgmt. *Society Aff:* ASCE, AWWA, WPCF.

Phillips, William R
Home: 1468 Lakehills Dr, Folsom, CA 95630
Position: Pres. *Employer:* Investigative Engr Labs, Inc. *Education:* MS/Safety/Univ of S CA; BE/Met/Youngstown State. *Born:* 6/28/26. W R Phillips - born New Castle, PA June 28, 1926. BE, Youngstown State Univ; MS, Univ of S CA. Reg PE; Met Engr & Mech Engr. Former instr at PA State Univ; presently resident faculty Univ of S CA; mbr ASM Committee for Hgwy & Off- Hgwy Vehicle Activity for Mtl Selection & Design Dev. Former supervisor of Aerospace Mtls Res & Dev Lab; Patent holder on safety devs. Presently Pres of Investigative Engg Labs performing failure analysis, design review vis-a-vis system safety & mishap analysis. Mbr of ASM, NSPE, CSPE, PEPP, ASTM, SAE. *Society Aff:* ASM, NSPE, CSPE, PEPP, ASTM, SAE

Phillips, Winfred M
Business: School of Mech Engrg, Purdue University, West Lafayette, IN 47907
Position: Prof & Hd, Sch Mech Engrg *Employer:* Purdue University *Education:* DSc/Aero Engg/Univ of VA; MS/Aero Engg/Univ of VA; BSME/ME/VPI. *Born:* 10/7/40. NSF Trainee 1965-67, Res Scientist 1967-68, Univ of VA; Dept of Aerospace Engg, the PA State Univ since 1968 - Prof of Aerospace Engg 1978; Visiting Prof, Institut de Pathologie Celulaire, Univ of Paris 1976-77; Acting Chrmn, Bioengg Prog, Penn State, 1978-79; Assoc Dean Res, Penn State College of Engg, 1979-80; Head, School of Mechanical Engineering, Purdue Univ., 1980-; Dow Award, Outstanding Engg Educator; Natl Inst of Health Career Res Award; Eminent Engineer, Tau Beta Pi; Fellow AAAS, Assoc Fellow, AIAA, Fellow, ASME, Member BME, ASAIO, APS, Fellow ASEE, Member NYAS, ISB, AAMI, Sigma Xi; Research and Publications in Gas Dynamics, Fluid Mechanics; Hemodynamics; Artificial Organs. *Society Aff:* AAAS, AIAA, ASME, ASAIO, ASEE, NYAS, BMES, ISB, APS.

Phipps, Thomas T
Business: 2375 Dorr St, Toledo, OH 43691
Position: Director of Sales & Mkting. *Employer:* Midland Ross Corp. *Born:* July 15, 1925. BS Univ of Michigan Mech Engrg 1949. Employed by Surface Combustion Div of Midland Ross Corp since 1950. Present pos since 1971. Mbr ASM & AISE. Mbr Heat Treat Comm ASM.

Phister, Montgomery, Jr
Home: 605 E. Garcia St, Santa Fe, NM 87501
Position: Consultant. *Employer:* Self. *Education:* PhD/Computers/Cambridge Univ; MS/EE/Stanford Univ; BS/EE/Stanford Univ. *Born:* Feb 1926. Dir of Engrg for sales of process control computers at Thompson-RamoWooldridge 1955-60. Ch Engr for design & oper of stock market quotation service at Scantlin Electronics 1960-66. V P for dev of digital computer processors, memories, & peripherals at Xerox Data Sys 1966-72. Author 'Logical Design of Digital Computers' Wiley 1958 & 'Data Processing Tech & Econ' Digital Press, 1979. Taught computer design at UCLA 1954-64, computer econ at Harvard & Univ of Sydney 1974-75. Fellow IEEE. *Society Aff:* ACM, IEEE.

Pian, Theodore HH
Home: 14 Brattle Circle, Cambridge, MA 02138
Position: Prof *Employer:* MIT *Education:* ScD./Aeronautical Engng./MIT; M.S./Aeronautical Engng./MIT; B./Engng./Tsing Hua University, China *Born:* 01/18/19 Working in aircraft industry in China prior to arrival in America. Stress analyst, Curtiss Aircraft Division, 1944-45. Research and teaching at Mass Inst of Technology since 1946. Becoming Prof of Aeronautics and Astronautics in 1966. Visiting Prof Tokyo Univ 1974, Technical Univ Berlin 1975. Specialist in structural mechanics. Lecturing in more than 120 universities and research establishments worldwide. Received 1975 AIAA Structures, Structural Dynamics and Materials Award for fundamental contributions to general field of structural dynamics and innovative development of assumed stress hybrid model for finite element method structural analysis. Elected Honorary Member ASME in 1985. *Society Aff:* AIAA, ASME, ASEE

Piasecki, Frank N
Business: Island Ave, International Airport, Phila, PA 19153
Position: Pres *Employer:* Piasecki Aircraft Corp *Education:* PhD/Science/Alliance Coll; BS/Aero Engr/Guggemheim School of Aero of NY Univ; PhD/Aero Sciences/PA Military Coll; PhD/Aero Engrg/NY Univ *Born:* 10/24/19 Aero/mechanical engr, pilot, pioneer in development of transport helicopters, vertical lift aircraft. Founded and headed research group, P-V Engrg Forum. Flew their first successful helicopter to fly in America. Constructed and flew world's first successful tandem rotor helicopter "Flying Banana–. 1946 Piasecki Helicopter Corp designed, produced transport helicopters including H-16, world's largest transport helicopter. 1955 Piasecki Aircraft Corp was formed and series of unique experimental VTOL aircraft were developed and flown. Piasecki has designed Multiple Helicopter Heavy Lift Systems, is currently building hydrid dynamic-static lift vehicle called "HELI-STAT" to be used in timber harvesting operations. *Society Aff:* AIAA, SAE, ASPE, SETP, SAME

Picard, Dennis J
Business: 430 Boston Post Rd, Wayland, MA 01778
Position: VP, Dep Gen Mgr *Employer:* Raytheon Co, Equipment Div *Education:* BBA/EE Mgt/Northeastern Univ *Born:* 08/25/32 Began career with Raytheon Co in 1955 as a design engr, progressing through increasingly responsible positions, mainly with the Equipment Div. Specialization has been in the technical and business mgmt aspects of phased array tech applications for anti-ballistic missile, intelligence and early warning radar systems. Designated a company VP in 1977. Currently Deputy Gen Mgr of Equipment Div, assisting the Gen Mgr in the technical and operations mgmt of a 6400 employee div, including 3000 professionals. Directly responsible for Div operations, including six business directorates, several engrg development labs and a large mfg facility. Selected as an IEEE Fellow in 1981. *Society Aff:* AFCEA, AIAA, AUSA, IEEE

Piccin, Marshall J
Business: 120 Fox-Shannon Pl, St Clarisville, OH 43950
Position: Dept Chrmn. *Employer:* Belmont Technical College. *Education:* MEd/Ed/OH Univ; BS/EE/OH Univ *Born:* 7/1/29. in Lansing, OH, July 1, 1929. Educated Bridgeport, OH School Sys, Grad OH Univ, Athens. OH BSEE Degree 1952. Tau Beta Pi Hon Fraternity. Served as Engg Duty Officer US Navy. Work & Professional Experience: Firestone Tire & Rubber Co-Elect Engr 2 1/2 yrs; Wheeling Electric Co. Power Engr-5 yrs; Mobay Chemical Co Engr & superintendent - 8 yrs. Belmont Tech College 16 yrs Engg Dept Chrmn. Reg PE-OH & W VA Pres/Owner of Marshall J. Piccin & Assocs - Consulting Engg Services. *Society Aff:* NSPE, SME

Picha, Kenneth G
Business: Professor Dept of Mech Engr, Univ of Mass, Amherst, MA 01003
Position: Professor of Mech. Engr. *Employer:* Univ of Mass/Amherst *Education:* BME/ME/Ga Tech; MSME/ME/Ga Tech; PhD/ME/Univ of MN. *Born:* July 24, 1925 Chgo. m. Vivien O Crawford May 1, 1948; c. Kenneth George, Kevin Crawford, Katrina Alison. Res Scientist NACA 1948-49; Instr, then Assoc Prof Sch of Mech Engrg, Ga Inst of Tech 1949-58; Prog Dir Engrg Sci Prog NSF 1958-60; Director Sch of Mech Engrg, Ga Inst of Tech 1960-66; Dean Sch of Engrg Univ of Mass Amherst 1966-76. Dir, Office of Univ Progs ERDA 1976-1977. Director, Office to Coordinate Energy Research and Education, Univ of Mass. 1977-1980; Professor, Mechanical Engr. Univ of Mass 1980- . Cons in field 1959- . USNR 1943-46. Fellow ASME; Fellow AAAS, Fellow ASEE, Ga Inst of Tech Natl Alumni Assn, Sigma Xi, Tau Beta Pi, Phi Kappa Phi, Pi Tau Sigma. Chmn ECPD Ed & Accreditation Comm 1973-74; V Chmn of Bd Council on Post Secondary Accreditation 1976; Dir Council on Specialized Accrediting Agencies 1974-76; Alumni Assn Award Univ of Mass 1976. Exchange Prof and Conslt to Chancellor Univ of Puerto Rico, Mayaguez, PR 1982-83. Fulbright Scholar, Univ College Galway, Ireland 1986-87. *Society Aff:* ASEE, ASME.

Picha, Robert T
Business: PO Box 12677, 3756 N Dunlop St, Arden Hills, St Paul, MN 55112
Position: Consultant, Pres. *Employer:* R T Picha Co. *Education:* BS/Met Engg/Univ of MN. *Born:* 9/17/24. Native of Minneapolis, MN. Served overseas as Naval officer in WWII. Past mbr of ROA, NDT, NSPE. Met for Inland Steel Co, investment casting foundry met, Mtls engr & Met for Honeywell for 5 yrs, VP of commercial metals testing co, Dir & founder & first Pres of Flame Industries (commercial heat treating co), Owner & pres of commercil lab-R T Picha Co since 1965. Specializing in "Failure Analyses and Prevention." Expert witness in court testimony & ins product liability cases. Contributor-Author of articles in Vol 10-ASM Handbook. Chrmn, MN Chapter ASM 1973-74, exec committee 9 yrs, Treas, ASM, MN chpt 1977-79. Reg PE. Presently consultant to several mfg firms, state, and governmental agencies. *Society Aff:* ASM.

Pichon, William J
Business: 5775A Glenridge Dr NE, Atlanta, GA 30328
Position: Pres & Genl Mgr. *Employer:* Columbia Engrg & Services Inc. *Education:* BS/Civil Engg/Univ of IL. *Born:* Jan 1931. BSCE Univ of Illinois 1953. Native of Danville Ill. Field engr for Geo A Fuller Co 1953-55. Const Supt for Lueck Const Co & C G Schmidt Const Co 1955-60. Proj engr for Armour & Co (U S Steel Agrichem) 1961-65. Proj Mgr for Amer Cryogenics Inc (Div of Esso) 1965-67. Dir of Engrg & Maintenance for ICI Amer (Atlas Chem 1967-73 (2 locations). Pres of Columbia Engrg & Services Inc (Subsid of Abrams Indus Inc) Atlanta Ga 1973- . Mbr AIChE, ASCE, AOA, Sigma Phi Delta, Masonic Lodge. *Society Aff:* AIChE, ASCE.

Pickering, C J
Business: 700 Fidelity Union Life Bldg, Dallas, TX 75201
Position: Executive Director *Employer:* North Texas Regional Clearing House Assn. *Education:* MS/IE & Mgmt/OK State Univ; BS/IE & Mgmt/OK State Univ. *Born:* July 1940. Native of Oklahoma. Employed in 1964 by Texas Instruments & served as staff engr & product line controller in Dallas Texas & London England. Employed by the Fed Reserve Bank of Dallas in 1970. Served 1 yr at the Bd of Governors of the Fed Reserve Sys in Wash D C. then as V P with respon for Check Processing, ACH Processing, Transportation & Dist Planning. Employed in 1980 as Executive Director for the North Texas Regional Clearing House Assn, his current responsibilities are to manage the trade association and to develop new products for member financial institutions. Reg P E Texas. Treas & V P Dallas Chap IIE; Who's Who in Texas 1974. *Society Aff:* IIE.

Pickering, Charles W
Business: P O Box 1963, Harrisburg, PA 17105
Position: VP - Hd of Dams *Employer:* Gannett Fleming Corddry et al. *Education:* MS/Mechanics & Hydraulics/State Univ of IA, Iowa City; BS/Civil Engg/State Univ of IA, Iowa City. *Born:* Nov 1924. Reg P E Iowa, Pa. Served as Lt (jg) with US Navy 1942-46. Hydraulic Engr US Army, Omaha Dist 1948-51. With Gannett Fleming Corddry & Carpenter Inc since 1952, in pos of design, proj engr & const resident engr .Pos entails mgmt & supervision of the investigation, design, preparation of contract plans & specifications, contract admin & const inspection of water-related projs, incl dams, flood control facils, water intake & control facils, pumping stations, & the inspection, renovation, & modification of hydraulic structs. *Society Aff:* NSPE, USCOLD, ACEC.

Pickering, George E
Business: Acorn Park, Cambridge, MA 02140
Position: Sr Staff Consultant. *Employer:* Arthur D Little, Inc. *Education:* MBA/-/Harvard Univ; BA/CE/Tufts Univ. *Born:* 10/1/16. Past-Pres, VP, Treas & Chrmn, Blow Molding Div of Soc of Plastics Engrs; Mbr, Bd of Dirs - Plastics Education Fdn; Honorary mbr, Plastics Inst of Australia & winner of its John W

Pickering, George E (Continued)
Derham Intl Award (1975) for outstanding contribution to the plastics industry. Past Chrmn, Coordinating Committee on Energy of the Assn for Cooperation in Engg; Mbr MA Petrochemicals/Plastics Industry Legislative Committee. Consultant on plastics & packaging products and processing - Arthur D Little, Inc. *Society Aff:* SPE, PEF.

Pickering, William H
Business: 1401 S Oak Knoll Ave, Pasadena, CA 91109
Position: President, Pickering Research Corp *Employer:* Self *Education:* PhD/Physics/Caltech; MS/Elec Engg/Caltech; BS/Elec Engg/Caltech. *Born:* New Zealand. Appointed to fac 1936; Prof of Elec Engrg 1947; Prof Emeritus 1980. Dir of Jet Propulsion Lab 1954-76. Mbr NAE, NAS. Hon Fellow AIAA; Pres AIAA 1963. Fellow IEEE. Pres IAF 1965-66. Spirit of St Louis Medal ASME 1965; Proctor Prize RESA 1965; Hill Award AIAA 1968; Edison Medal IEEE 1972; Disting Civilian Award US Army 1959; Disting Serv Medal NASA 1965; Natl Medal of Sci 1976, National Engineering Award AAES 1984. *Society Aff:* AIAA, IEEE, BIS, NZIE.

Pickett, Eugene L
Home: 26 East Pond Road, Narrangansett, RI 02882
Position: Consultant *Employer:* Self *Education:* MSE/Sanitary/Univ of MI; BS/Civil/IA State. *Born:* 1/16/26. Native of Oneida Iowa. US Navy Civil Engry Corps 1945-76 in grades Ensign to Captain. Commanded Southern Div, Naval Facils Engrg Command, Charleston S C. Betz Converse. Murdoch Inc 1976-1981; Senior V Pres. Wakefield Data, Inc August 1983 to June 1985. Consultant 1987 in Project Management. P Regional V P, South Atlantic, Soc of Amer Military Engrs 1974-76. Fellow SAME. Fellow ASCE. P E Rhode Island, PE PA. *Society Aff:* ASCE, SAME.

Pickett, Leroy K
Home: 327 - 8 Street, Downers Grove, IL 60515
Position: Senior Proj Engr. *Employer:* J.I. Case, & a Tenneco Co *Education:* PhD/Agri Engr/Purdue Univ; MS/Agri Engr/Univ of IL; BS/Agri Engr/KS State Univ; -/Gen Ed-History/Westmar College. *Born:* 5/8/37. Native of Randolph, KS. Test & Dev Engr, Allis-Chalmers 1962-65. Grad Asst and Natl Sci Fdn Trainee, Purdue Univ 1965-68. Asst Prof, MI State Univ 1968-73. Intl Harvester - Grain Harvesting Equip Engg - 1973-84, J I Case Grain Harvesting Equip Engg - 1984-86, Advanced Engg - beginning 1986. ASAE Paper Award 1970. ASAE Quad City Sec Chrmn 1979-80. ASAE Agri Engg Editorial Bd Chrmn 1979-82. ASAE Power and Machinery Dir 1983-85, ASAE A-204 Engrg & Tech Accreditation Cttee Chrmn 1986-87. ASAE Tech VP beg June 1986. HOPE Fair Housing ctr Dir 1984 to 1986. Mbr of First United Methodist Church, Downers Grove, IL. Enjoy photography, theatre, metal & wood working, & hiking. *Society Aff:* ASAE, SAE.

Pickett, Rayford M
Business: Pickett, Ray & Silver, Inc, 333 Mid Rivers Mall Drive, St. Peters, MO 63376
Position: Pres. *Employer:* Pickett, Ray & Silver, Inc. *Education:* BS/Civ Engg/St Louis Univ. *Born:* 2/22/34. Native of Hollow Rock, TN. Served with Army Guided Missile Training Ctr, FT Bliss, TX. Supervisor with McDonnell Douglas Corp Gemini Space Prog - 1964-1969. City Engr, St. Peters, MO - 1974-1977. Pres, Pickett, Ray & Silver, Inc since 1973. Pres, Soil Consultants, Inc since Jan 1978. Dir, St Louis Div of Natl Assn of Home builders. Founding Pres, St Peters Kiwanis Club. Founding Dir, St Peters, MO Committee for Civic Progress. Dir St. Charles Cty, MO Salvation Army. Dir, Mark Twain Bank of St. Peters, MO. Hobbies: Golf. *Society Aff:* MSPE, NSPE, MARLS.

Pickholtz, Raymond L
Business: 3613 Glenbrook Road, Fairfax, VA 22031
Position: Prof & Chrmn Dept of Elec Engg & Comp Sci. *Employer:* George Wash Univ. *Education:* PhD/EE/Poly Inst of NY; MS/EE/City Univ of NY; BS/EE/City Univ of NY. *Born:* 4/12/32. Was a res engr at RCA Labs and at ITT Labs for a period of ten yrs working on problems ranging from color tv to secure communications and guidance before returning to academia. Was an Assoc Prof at the Poly Inst of Brooklyn until 1972 when he joined GWU. Taught part-time at NYU and in the physics Dept of Brooklyn College. Was a visiting prof at the Univ of Quebec in 1977. Has been an active consultant in communication to industry and govt for many yrs and has lectured extensively in this country, Canada, Europe, and S America. Pres of Telecomm Assoc, a res and conslt firm in Fairfax, VA. Was elected as Fellow of IEEE in 1980 for contributions to the design of Digital Comm Sys and to Engrg Education. Fellow of AAAs and the TC Wash, Acad. of Sciences in 1986. Visiting scholar at the Univ of CA in 1983. Holds six US patents and has published over 60 tech papers. In 1984 he was a recipient of the IEEE Centennial Medal. Member, Cosmos Club. *Society Aff:* IEEE, MAA, AAUP, AAAS.

Piech, Kenneth R
Home: 51 Hetzel Rd, Williamsville, NY 14221
Position: Pres *Employer:* Aspen Analytics, Inc. *Education:* PhD/Physics/Cornell; BS/Physics/Canisius. *Born:* 1941 Buffalo N Y. BS Canisius College 1962. PhD Cornell Univ 1967. Grad specialization in low energy atomic & molecular phys. Prof res & pubs in optics, photographic analyses, & remote sensing. Recipient 1973 Autometric Award, Amer Soc of Photogrammetry (ASP), for outstanding tech publ on photointerpretation. Chmn Engrg Applications Ctte ASP 1973-74. Deputy Chmn Hydrospheric Sci Ctte ASP 1976-77. Editor, Skylab Earth Resources Exper Pkg Summary Volume. *Society Aff:* ASP.

Pieczonka, Ted J, Jr
Business: Pieczonka Engg, PO Box 206, Orchard Park, NY 14127
Position: Owner *Employer:* Pieczonka Engg *Education:* MSCE/Sanitary/WA St Univ. *Born:* 10/9/45 Long-experienced, with a technical/professional career in water pollution control extending back to 1963. Started as laborer and then lab technician at sewage treatment plants, while securing education in biology and engrg. With these joint disciplines, emerged as qualified operator of biological processes, serving as sanitary engr with the US Army. Continued unique role as operator/engr by providing troubleshooting service at plants, while heading facilities planning and preliminary design in smaller consulting firms. Strives to link early project planning to ultimate operability. Pieczonka Engg now provides comprehensive, broad-based environmental services as well. *Society Aff:* ASCE, NSPE, WPCF, AWWA.

Piekarski, Julian A
Home: 5570 N Bay Ridge Ave, Milwaukee, WI 53217
Position: President. *Employer:* PAICE Assocs, Inc *Education:* BSME/Mechanical Engg/Marquette Univ; MSME/Industrial Engg/Univ Wisconsin; MBA/Management Science/Marquette Univ. *Born:* BSME & MBA Marquette Univ, MSME Univ of WI; Mbr Pi Tau Sigma & Tau Beta Phi. Amer Assn of Cost engrs: Natl dir 1973-75, Natl Tech VP 1976-77, Adm VP 1977-78, Pres 1978-79, currently Ppres. Received Award of Recog from AACE in June 1977 at the Annual Convention in Boston MA; Elected a Fellow of AACE in June 1982. Received the Award of Merit June 1984 at the Annual Convention in Montreal, Canada; of Grad Program of Engg Mgmt; Reg P E WI; Cert Cost Engineer; designed & patented arc welding apparatus, US Pat 2,903, 567. *Society Aff:* AACE.

Pierce, Allan D
Business: Sch of Mechanical Engineering, Atlanta, GA 30332
Position: Regents Prof *Employer:* Georgia Inst Tech *Education:* PhD/Physics/MIT; BS/Phys/NM St Univ. *Born:* 12/18/36 Rand Corp, 1961-1963; Avco Corp, 1963-1966; Dept of Mech Engrg, MIT, 1966- 1973; joined Georgia Tech as Prof of Mech Engrg in 1973, appointed Regents Prof in 1976. Fellow of Acoustical Soc since 1973 and author of many articles on sound propagation, diffraction, fluid mechanics, and vibration. Author of textbook *Acoustics: An Introduction to its Physical Principles and Applications* (McGraw-Hill 1981). *Society Aff:* ASME, ASA, IEEE

Pierce, Francis C
Home: 156 Barney St, Rumford, RI 02916
Position: Consultant *Employer:* Maguire, Group *Education:* MS/Geotech Eng/Harvard Univ; BS/CE/Univ of RI; CE/ASTP/Univ of CT *Born:* 5/19/24 Native of RI. Served U.S. Army World War II 1943-1946. Instructor of civil Engrg at Univ of RI, Univ of CT. Employed by Corps. of Engrs Loring Air Force Base, ME. Joined Maguire Assocs as geotechnical engr, 1950. Advanced as project engr, dept. head to Dir Civil Division. Elected co Assoc 1966, VP, 1969, Senior VP, 1972. Executive VP, Dir of Operations 1979, currently Mbr Bd of Dir and consultant to EAP Maguire Group. Past Pres. RISPE, ASCE, RISPA, author of technical papers. Engr of the Year Award RI, 1974. Recip of Geotechnical Engrg Award ASCE by BSCE April 1979. Meritorious Public Service Award from U.S.C.G., April, 1987. Enjoys fishing and photography. *Society Aff:* ASCE, NSPE, SNAME, ASEE, SAME, ISSMFE, ICOLD, PIAWC, DFI, AAPA, USNI.

Pierce, G Alvin
Business: Aerospace Engineering, Atlanta, GA 30332
Position: Prof *Employer:* GA Inst of Tech *Education:* PhD/Aerospace Engrg/OH State Univ; BSc/Aerospace Engrg/MIT *Born:* 12/22/31 Native of Philadelphia, PA Mech design engr for A. D. Cardwell Electronics, CT, 1953-54. Served with USAF 1954-56. Research specialist for North American Aviation Corp, OH, 1956-63, specializing in aeroelastic analyses. Taught Aerospace Engrg courses and performed research in gas dynamics at OH State Univ 1964-66. Joined faculty of GA Inst of Tech in 1966. Have taught numerous courses in aerodynamics, structures and aeroelasticity. Principal research has been in the areas of unsteady aerodynamics, structural dynamics and aeroelasticity as associated with helicopter systems. Principal advisor to twenty PhD and fifty MSc graduates. *Society Aff:* ASEE, AIAA, AHS, ΣΞ

Pierce, Harry W
Home: The Evergreens, 309 Bridgeboro Rd, Moorestown, NJ 08057
Position: Retired *Education:* M/Naval Const/MIT; B Eng/US Naval Acad *Born:* 09/30/01 *Society Aff:* SNAME, AWS

Pierce, John Alvin
Home: 51 Jason St, Arlington, MA 02174
Position: Sr Res Fellow/Appl Phys. *Employer:* Harvard Univ - retired. *Education:* BA/Physics/Univ of ME. *Born:* Dec 11, 1907 Spokane, Wash. Fellow IEEE, AAAS, Amer Acad of Arts & Sci. At Harvard 1934-74, except for WW II yrs at the Radiation Lab MIT, becoming hd of Div 11. Guided the dev of Loran & the res for & dev of the modern navigation sys Omega. Thurlow Award, Inst of Navigation 1947; Presidential Cert of Merit 1948; Morris Liebmann Prize IRE 1953; Pioneer Award PGANE, IRE 1961; Public Serv Commendation USCG 1975; Conrad Award USN 1975. *Society Aff:* IEEE, ION, AES, Am Acad.

Pierce, John R
Business: CCRMA-Music, Stanford, CA 94305
Position: Visiting Prof of Music, Emeritus *Employer:* Stanford *Education:* PhD/Physics/Caltech; MS/Physics/Caltech; BS/Physics/Caltech *Born:* Mar 27, 1910 Des Moines Iowa. Bell Labs 1936-71. Prof of Engrg Caltech 1971-80. Chief Technologist, Jet Propulsion Lab, 1980-82. Chief work in traveling-wave tubes, microwaves & communication. Echo I embodied his ideas; Telstar was based on his work. Acoustical Soc of Amer, Amer Phys Soc, IEEE, Natl Acad of Sci, Natl Acad of Engrg, Amer Acad of Arts & Sci, Amer Philosophical Soc, Royal Acad of Sci (Sweden). Honors: Natl Medal of Sci; Edison Medal, Inst of Elec Engrs; Valdemar Poulsen Medal; H T Cedergren Medal; Marconi Award; Medal of Honor IEEE. Many hon degrees Recent awards: Founders Award, Natl Acad of Engg, 1977; Marconi Intl Fellowship, 1979; IEEE Comm Soc Career Award, 1984, Japan Prize, 1985. *Society Aff:* IEEE, APS, ASA, IECEJ

Pierce, Louis F
Business: 244 Country Club Rd, Eugene, OR 97401
Position: Chief Engr & Pres *Employer:* OBEC Conslt Engrs *Education:* BS/CE/Univ of WA *Born:* 10/14/29 Commissioned in the US Navy Civil Engr Corps and served with MCB 5 in the Philippines 1952-54 during construction of Cubi Point NAF. Spent five yrs with the OR State Highway Bridge Div designing bridges and related structures, including a major highway tunnel. Employed by various structural and bridge conslts in early yrs. With OBEC Conslg Engrs since 1966 as principal partner, Chief Engr, and President; major emphasis being the design of bridges. Past Pres of Willamette Branch ASCE. Mbr of ASCE Natl Ctte on Aesthetics in Design. Active in Toastmasters Intl. Native of Pacific Northwest. Enjoy fishing, gardening and outdoor activities. *Society Aff:* ASCE, CECO, SEAO.

Pierce, Ralph
Business: 26111 Evergree Rd, Southfield, MI 48076
Position: Managing Partner. *Employer:* Harley Ellington Pierce Yee Assoc Arch, Engrs Plan. *Education:* BSEE/EE/Northwestern Univ. *Born:* 4/14/26. Test Engr, Am Elec Heating Co, Detroit, 1946-47; Sr Assoc Engr, Detroit Edison Co, 1947-52; Chief Utility Engr, Geo Wagschal Assoc, Detroit, 1952-58; Sr Partner, Pierce Yee Assoc, Detroit, 1958-73. Since 1973, Managing Partner, Harley Ellington Pierce Yee Assoc, full service A/E firm involved in master planning & design of ind, commercial, educational, municipal, medical, cemetery facilities, & utility systems; energy mgt; & engg consultation services. Reg PE in many states, Nationally Certified & PE, Ontario, Canada. Honors received include appointment US Dept of Commerce Trade Mission to Yugoslavia; after active service from 1944-46, retired from Reserves with Rank of Commander, US Navy. *Society Aff:* NSPE, IES.

Pierce, Roger J
Home: 900 Staub Ct N E, Cedar Rapids, IA 52402
Position: President. *Employer:* Triad Tech Co. *Education:* BSEE/EE/IA State Univ. *Born:* 1911 Des Moines Ia. Grad study Ohio State Univ. Started at Collins Radio Co 1934 - transmitter design, FM res. Harvard Univ 1942, radar CM res. Motorola Inc 1943 Asst Dir Engrg; military radar dev. Hawaiian Tele Co 1946 Mgr Radio Comm Div; pioneered interisland radio tele network. Collins Radio Co - Dir Space Comm Sys - designed Space Communications Sys for Mercury, Gemini, & Apollo progs. Hydrospace Sys Corp - Pres & Founder. Triad Tech Co. - President. Inventor of advanced stable ocea platform. Other pats on electronics devices. Mbr Sigma Xi. Fellow IEEE. *Society Aff:* IEEE.

Pierrard, John M
Home: R D 4, West Chester, PA 19380
Position: Engrg Fellow. *Employer:* E I DuPont de Nemours & Co. *Education:* PhD/Air Resources/Univ of WA; MS/Meteorology/TX A&M; BS/Chem/IL Inst of Tech. *Born:* Mar 1928 Chgo Ill. m. Helen A Huitt; 5 children. USAF Weather Officer 1953-57. Applied res & proj dir in many facets of atmos chem & phys at Univ of Chgo 1952-53, Texas A&M Res Found 1957-59, IIT Res Inst 1959-62, Natl Ctr for Atmos Res 1962-66. Joined DuPont 1969. Work in recent yrs has centered on air pollution & impacts of pollution control strategies. Pubs in sci & engrg journals. SAE Horning Mem Award 1972.

Pierre, Percy A
Home: 8412 Carlynn Dr, Bethesda, Washington, MD 20817
Position: Engrg Mgmt Consultant (Self-employed) *Education:* PhD/Elec Engg/Johns Hopkins Univ; MSEE/Elec Engg/Notre Dame; BSEE/Elec Engg/Notre Dame. *Born:* 1/3/39. Post doctoral Fellowship Univ of MI 1968. Asst Prof Southern Univ 1963; 1968-71 Staff Mbr RAND Corp; 1969-70 White House Fellow (leave of absence from RAND Corp) 1971-77 Dean of Engrg Howard Univ; 1973-75 served as half-time program officer with A P Sloan Fnd; 1977-81 Asst Secy of the Army for Res, Dev & Reg. 1981-present Engrg Mgmt Consultant. Author of tech articles dealing with statistical communication theory, probability theory & detection of signals in radar clutter. Honors: Senator Proximise's Award of Merit 1979, Outstanding Notre Dame Alumni, Sigma Xi Scientific Res Soc of Amer, Hon D of Engrg Univ of Notre Dame, Tau Beta Pi, White House Fellow. Former consultant to Alfred P

Pierre, Percy A (Continued)
Sloan Fnd, Ctr for Nabbal Analysis, Educ Dev Ctr, RAND Corp, Energy Res & Dev Admin, BdM Corp. Former Mbr of Bd of Trustees Univ of Notre Dame, Natl Fund for Minority Engrg Students. Sr mbr of IEEE. *Society Aff:* IEEE, ΣΞ.

Pierskalla, William P
Business: 3620 Locust Walk CC, Philadelphia, PA 19104
Position: Deputy Dean of The Wharton School *Employer:* Univ of PA. *Education:* PhD/Operations Research/Standord U; MS/Statistics/Stanford; MS/Mathematics/U of Pittsburgh; MBA/Business/Harvard; AB/Economics/Harvard *Born:* 10/22/34. St Cloud MN. 1965-68 Case Western Reserve Univ; 1968-70 Assoc Prof Southern Methodist Univ, 1969-70 acting Dir Computer Sci/Oper Res Ctr; 1970-74 Assoc Prof Northwestern Univ, Prof 1974-78, 1975-76 acting Dir Health Services Res Ctr. Assoc Editor of *Operations Research, Management of Science & Opsearch*. 1974-78. Secty 1977-80 and Pres 1982-83 of Operations Res Soc of Am, Ronald A. Rosenfeld Prof of Health Care of Decision Sciences, Wharton, Professor of Systems, Engineering, Univ of PA, 1978-83. Dir of Natl Health Care Mgt Ctr, Univ of PA, 1978-. Exec Dir of Leonard Davis Inst of Health Economics, Univ of PA, 1978-83. Deputy Dean for Academic Affairs The Wharton School Univ of PA, 1983-present. Editor of *Operations Research* 1978-82, Treasurer and Bd of Dirs of Assoc of Health Services Research 1981-86. *Society Aff:* ORSA, TIMS

Pierucci, Mauro
Home: 3106 Las Palmas, Escondido, CA 92025
Position: Prof *Employer:* San Diego State Univ *Education:* PhD/Astronautics/Polytech Inst of NY; MS/Astronautics/Polytech Inst of NY; BS/Aerospace/Polytech Inst of NY *Born:* 01/05/42 Native of Italy. From 1968 to 1979 was principal engr at GDEB Div specializing in acoustic radiation and flow noise. Joined SDSU Coll of Engrg in 1979. Author of several papers in ASA Journal, AIAA Journal, Journal of Sound & Vibr and others. Has been active in affairs of AIAA (Treas of CT Section) and ASA (mbr of Engrg Acoustic Committee since 1978. Chrmn of Natl Committee since 1981 and mbr of Natl Technical Council since 1981. *Society Aff:* AIAA, ASA, AAM

Piest, Robert F
Home: Route 5 Box 84, Columbia, MO 65201
Position: Hydraulic Engineer. *Employer:* Agricultural Research Service, USDA. *Education:* BS/Civil Engg/Univ of NB. *Born:* Nov 1 1921. After graduation, employed by US Geological Survey until 1958, conducted water resource investigations in KS, NE & MS; employed by US Agri Res Service since 1958, conducting res to quantify yields of sediment & sediment-associated chemicals & relating those yields to erosion sources & to controlling hydrologic factors. Author of num papers on this subject & coauthor of recently pub manual of practice 'Sedimentation Engineering' of ASCE, and "Field Manuel for Res in Agri Hydrology–, of US Dept of Agri, Sci & Educ Admin. *Society Aff:* AE, ΣT.

Pietsch, Joseph A
Home: 4319 Bobbitt Dr, Dallas, TX 75229
Position: VP - Engg & Res. *Employer:* Northrup, Inc. *Education:* BEE/Electronics/Univ of MN; BBA/Ind Admin/Univ of MN. *Born:* Oct 1926. 1944-46 served with US Navy. BEE & BBA from Univ of Minn 1950. 1950-78 employed by General Electric Co: 1950-53 GE Test Prog & Advanced Engrg Prog, 1953-55 Engrg Education Prog Staff, 1955-63 Dev Engr room air conditioners, 1963-66 Mgr Value Control & Advance Engrg air conditioners; 1966- 68 Mgr Prod Planning central air conditioners; 1968-78, Mgr Engrg Central Air Conditioning Bus Dept. 1978-present employed by Northrup, Inc Hutchins, TX as VP-Engr and Res Northrup Manuf HVAC and solar equipment Reg PE in TX. Fellow Mbr ASHRAE Mbr ASHRAE Stds Comm. Chmn ASHRAE TC 7 6 Unitary Air Conditioners & Heat Pumps 1974-78; Chmn ARI Unitary Heating & Cooling Sect 1975-76. 15 patents in air conditioning & appliance field. *Society Aff:* ASHRAE.

Pigage, Leo C
Home: 206 Elmwood Rd, Champaign, IL 61821
Position: Prof of Ind Engg. Emeritus. *Employer:* Univ of IL. *Education:* MME/Ind Engg/Cornell Univ; ME/ME/Cornell Univ. *Born:* 11/22/13. Native of Rochester, NY. Taught at Cornell Univ; Duke Univ; Purdue Univ and the Univ of IL Emeritus in Champaign-IL, IL. Was in charge of areas of work measurement, job evaluation and facilities planning. For over thirty-five yrs have acted as a consultant and/or arbitrator in the above fields for both mgt & labor. Interested in stamp, coin & antique collecting. *Society Aff:* ASME, IIE, ASEE, LIRA.

Pigford, Robert L
Business: 300 Wilson Rd, Newark, DE 19711
Position: Professor of Chemical Engineering. *Employer:* University of Delaware, Newark. *Education:* PhD/Chem Eng/Univ of IL; MS/Chem Eng/Univ of IL; BS/Chem Eng/MI State Univ. *Born:* 4/16/17. 1941-47 Res Engr, E I duPont de Nemours & Co Experimental Station, Wilmington DE; 1947-66 Prof of ChEm Engrg Univ of DE; 1966-75 Prof of Chem Engrg Univ of CA Berkeley; 1975- , Univ Prof of Chem Engrg Univ of DE. AIChe Wm H Walker Award, W K Lewis Award, Professional Progress Award, Founders' Award & Institute Lecturer; Francis Alison Award, Univ of DE, 1979; Mbr NAE & NAS; Visiting Profships at Univ of WI & Cambridge Univ. Prin cons work with E I duPont de Nemours & Co, M W Kellogg Co. Prin publications: sections on mass transfer, distillation & solvent extraction in Perry "Chem Engrs Handbook," –Absorption & Extraction" with T K Sherwood; "Mass Transfer" with T K Sherwood & C R Wilke; "Application of Differential Equations to Chem Engrg Problems" with W R Marshall Jr; author of technical publications in several research journals. Editor of Fundamentals Quarterly, Industrial & Engineering Chem, publ by American Chemical Society, Wash, DC, 1962-1987. *Society Aff:* AIChE, ACS, ASEE, NAE, NAS.

Pigford, Thomas H
Business: Dept of Nuclear Engrg, Berkeley, CA 94720
Position: Prof, Nuclear Engg. *Employer:* University of California. *Education:* BS/ChE/GA Inst of Tech; SM/ChE/MIT; ScD/ChE/MIT. *Born:* April 21 1922. BSChE 1943 Georgia Inst of Tech; SM 1948 & ScD 1952 in Chem Engrg, Mass Inst of Tech. 1950-57 Asst & Assoc Professor of Chem & Nuclear Engrg, MIT; 1952 Sr Dev Engr Oak Ridge Natl Lab; 1957-59 Dir of Engrg, Dir of Nuclear Power Projs & Asst Dir of Res, General Atomic; 1959 to present Professor of Nuclear Engrg, Univ of Cal Berkeley, 1959-64 & 1974-79 Department Chmn. Mbr ANS, elected to Bd/Dir 1963, elected Fellow 1971, Dir of Nuclear Fuel Cycle Div 1975, received Arthur H Compton Award 1971; Charter Mbr's Award 1979; Mbr of Natl Atomic Safety & Licensing Bd Panel, US Nuclear Regulatory Comm 1963- ; elected to NAE 1976. Specialty in design of nuclear power reactors, nuclear fuel cycles, nuclear safety & environmental analysis. Coauthor of textbook 'Nuclear Chemical Engineering.' Ctte on radioactive waste mgt, Natl Acad of Sci, 1978-; pres commissionate investigate tee accident at Three Mile Island, 1979. *Society Aff:* ANS.

Pignataro, Louis J
Business: 333 Jay St, Brooklyn, NY 11201
Position: Prof of Trans Engrg *Employer:* Poly Inst of NY. *Education:* Dr Tech Sc/Transportation/Tech Univ of Graz; MS/Appl Mechanics/Columbia Univ; BCE/Civ Engg/Poly Inst of Brooklyn. *Born:* 11/30/23. Teaching and res at the Poly from 1951 to present. Responsible for the dev of grad academic and res progs in transportation which are widely recognized nationally and internationally. Contributed numerous papers to the field of transportation planning and engg which have been published in various periodicals and journals. Author of the book, *Traffic Engineering: Theory and Practice*, published by Prentice-Hall, 1973. VP, 1975, Pres-Elect, 1976, Metropolitan Sec ASCE. Mbr of several profl societies' committees and governmental advisory panels. Selected 1974 Engr of the Yr by NY State Soc of PE. Reg PE in CA, FL, NY. *Society Aff:* ASCE, NSPE, TRB, ITE, TRF.

Pih, Hui
Home: 7108 Downing Dr, Knoxville, TN 37909
Position: Professor of Engineering Science and Mechanics. *Employer:* University of Tennessee, Knoxville. *Education:* PhD/Engr Mech/IL Inst of Tech; MS/ME/Stanford Univ; BS/ME/Natl Inst of Tech (China) *Born:* 2/14/22. 1945-48 Engr & Superintendent 53rd Arsenal, China; 1953-56 Proj Engr, Mech Analysis Intl Harvester; 1956-59 Staff Engr Mech Anal IBM; 1959-65 Assoc Prof of TAM, Marquette Univ; 1965- , Prof of Engr Sci Univ of Tenn. SESA, M Heteny Award May 1975. Mbr Tau Beta Pi, Sigma Xi, ASME, SME, Amer Acad of Mechanics, ASEE, AAAS, Soc of Manuf. Engr & AAUP. Registered Professional engineer in the State of Tennessee. *Society Aff:* ASME, SEM, ASEE, AAAS, AAM, SME, AAUP

Pikarsky, Milton
Home: 300 N State, Apt 5321, Chicago, IL 60610
Position: Dir, Transportation Res. *Employer:* IIT Research Inst. *Education:* BCE/-/City College of NY; MSCE/-/IL Inst of Technology. *Born:* March 28 1924. 1964-73 Commissioner of Public Works, City of Chicago; 1973-75 Chairman, Chicago Transit Authority. Mbr NAE; Chmn Transportation Res Bd 1975; Chmn BART Impact Ctte NAE 1976; coChmn (for Mayor Richard Daley) US Conference of Mayors, Transportation Ctte 1972-74. 1 of APWA 'Top Ten Public Works Men-of-theYear' 1969; 'Civil Engineer of 1970,' Ill Sect of ASCE; Townsend Harris Medal for Outstanding Achievement in a Chosen Profession, City Coll NY 1969; ASCE Civil Government Award 1973.

Pike, Ralph W
Business: 6053 Hibiscus Dr, Baton Rouge, LA 70808
Position: Assoc V Chancellor, Res *Employer:* Louisiana State Univ. *Education:* PhD/Chem Engrg/GA Inst of Tech; B Ch E/Chem Engrg/GA Inst of Tech *Born:* Nov 10 1935, Tampa FL. Reg PE in LA & TX. 1962-64 Res Engr Exxon Res & Eng, Baytown TX; 1964- , Professor of Chem Engrg & Systems Sci, LA St Univ, 1975- , Assoc Vice Chancellor for Res. Sponsored res by NSF, NASA, NOAA, NEH & private industry on transport phenomena, optimization energy & systems ecology. Textbooks & over 60 publications. Mbr prof & honor societies; cons to government & indus. *Society Aff:* AIChE, ACS, ASEE, AAAS

Pilcher, J Mason
Home: 125 Lookout Point, Osprey, FL 33559
Position: Retired *Employer:* Battelle-Columbus Laboratories. *Education:* BS/Chem Engg/VA Polytechnic; MS/Chem Engg/VA Polytechnic; PhD/Fuel Sci/PA State. *Born:* 1/22/13. Employed by Battelle Columbus Labs for 30 yrs working primarily in the fields of fuel utilization, air pollution control & environmental studies; 1960- 70 Ch of Environmental Mechs Div. Pub over 30 papers & book chapters on topics relating to coal preparation, combustion of solid fuels, kinetics of steam- carbon reactions, air pollution control, particle mechanics, atomization of liquid fuels & contamination control. Elected Fellow ASME 1973; Mbr Sigma Xi, Phi Kappa Phi, Tau Beta Pi & Omicron Delta Kappa. Reg PE in OH. *Society Aff:* ASME.

Pilkington, Theo C
Business: Dept Biomedical Engrg, Durham, NC 27706
Position: Professor *Employer:* Duke University. *Education:* PhD/EE/Duke Univ; MS/EE/Duke Univ; BS/EE/NCSU *Born:* 6/23/35. 1960 appointed to full-time faculty in EE at Duke Univ; 1963 Asst Prof, 1966 Assoc Prof; 1967 Assoc Prof of BME & EE, & Chmn of newly established Div of BME; 1969 Prof of BME & EE, & Chmn of Dept. of BME; 1971-79, 1979-Prof of BME and EE. Fellow by-Courtesy, Dept EE, The John Hopkins Univ Spring 1974. Res interests & sponsored res in areas of bioelectricphenomena, potential theory & inverse problems. Mbr IEEE AdCOM, Chmn of IEEE Bioelectric Phenomena Ctte; ECPD accreditation site visitor; Editor 'CRC Critical Reviews in Bioengineering 1976- 79, Editor of IEEE Transactions on Biomedical Engineering 1979-1984, Mbr. Editorial Board Prog. of IEEE, 1985-. Enjoys beagling. *Society Aff:* IEEE, BME, ASEE.

Pillsbury, Arthur F
Home: 35333 Lenard Rd, PO Box 768, Springville, CA 93265
Position: Retired. *Education:* Engrs/Civ Engg/Stanford Univ; AB/Engg/Stanford Univ. *Born:* 10/11/04. Married Mary Alice Reasoner, 24 June 1933. 5 children. Employed by Univ of CA as a Jr Irrigation Engr in 1932 to conduct field studies. Transferred to UCLA in 1939, & conducted field & lab res in areas of irrigation & drainage. Academic appointment added in 1940 in irrigation & soils, & in Engg in 1948. Consulting for Intl Boundary & Water Commission. the World Bank, the FAO, & numerous more local water agencies. Dir, systemwide UC Water Resources Ctr, & Prof, Sch of Engg, UCLA, until retirement in 1972. Mbr, State Environmental Quality Study Council, 1969-72. Recd Royce J. Tipton award of ASCE in 1977. *Society Aff:* AAAS, ASAE, ASCE, AGU, ΣΞ, TBΠ.

Pincus, George
Business: Dept of Civ Engg, Houston, TX 77004
Position: Prof *Employer:* Univ of Houston. *Education:* PhD/Structural Engg/Cornell Univ; MBA/Mgt/Univ of Houston; MSCE/Structural Engg/GA Inst of Tech; BSCE/Civ Engg/GA Inst of Tech. *Born:* 7/5/35. in Havana, Cuba. Taught at the Univ of KY 1963-1966. Was Chief of Party and Visiting Prof with the Univ of Houston Brazil prog in higher education. Became Prof with the Univ of Houston in 1969 & was Chrmn of the Civ Engg Dept in 1975- 1980. Winner D.V. Terrel Award, 1965, ASCE District 9. Winner of the Kittinger Teaching Excellence award in 1975. Published numerous articles & 3 books. Consultant to industry. Married with 3 children. *Society Aff:* ASCE, ACI, AAUP, ASEE.

Pincus, Robert E
Business: 917 W Juneau Ave, Milwaukee, WI 53201
Position: President. *Employer:* P-L Biochemicals & Premier Malt Prod. *Education:* BS/ChE/IIT, Chicago; -/ChE/Univ of MI. *Born:* Nov 1918, Chicago IL. 1941-43 Chem Engr for US Ordinance Dept; 1943-46 served in US Navy Bureau of Ships, grades to Sr Lt; 1946-58 Process Engr & Supt Pabst Brewing Co, Peoria IL; 195867 Founder & President VegeFat Inc, St Louis IL. 1962 rejoined Pabst at Milwaukee WI home office as Genl Mgr, Animal Feed Div, 1964- , elected Pres of Pabst non-beer subsidiaries P-L Biochemical Inc & Premier Malt Products Inc. Enjoy riding & jumping horses, tennis, skiing. Active in Community Chest, Multiple Sclerosis & Jewish organizations & fund raising. Full Mbr AIChE; P Chmn several years Brewery By-Products Ctte US Brewers Assoc. *Society Aff:* AIChE, ACS, IFT, AOCS.

Pinder, George F
Business: Dept of Civil Engineering, Princeton, NJ 08540
Position: Prof & Chrmn *Employer:* Princeton Univ *Education:* PhD/Geol/Univ of IL; BSc/Geol/Univ of Western Ontario *Born:* Born in Windsor Ontario, Canada to Stella & Percy S. Pinder. He married Phyllis M. Charlton. They have two children - Justin and Wendy. After graduation, he worked as a Res Hydrologist for the US Geological Survey. In 1972, he was appointed Assoc Prof of Civil Engrg at Princeton Univ. He was promoted to full prof in 1977 and became Chrmn of the Dept in 1980. *Society Aff:* ASCE, SPE, ISCME, ΣΞ, AGU

Pinkerton, John E
Home: 2002 13 1/2, Beaver Falls, PA 15010
Position: Prof of EE. *Employer:* Geneva College. *Education:* PhD/Physics/Univ of SC; MS/Physics/Univ of WI; BS/Physics/Geneva College; grad work in EE/EE/Worchester Poly Inst. *Born:* 3/2/39. Born Cortland, NY 1939. Prof Experience: Teacher Pine-Richland School 60- 61, Audubon High School 61-65 - Physics and Electronics. Asst Prof of Physics 1967-1974 Geneva College; teaching nuclear physics and electronics. Assoc Prof of Physics 1974-1979 Geneva College; teaching nuclear physics and electronics. Prof of EE 1979- Dir of microprocessing lab and EE prog. Concurrent positions 1968 Engr Westinghouse Elec Corp. 1969-71 Teaching Asst Univ of SC. 1972 NDEA fellow at Univ of SC. 1974- Pres SPL systems (custom electronic design of signal communication systems). 1975- Faculty Appointment, Res Physicist US Bureau of Mines (Theoretical Support Group) specializing

Pinkerton, John E (Continued)
in the design of computer controlled data acquisition systems using mini and microcomputers. Project Dir Ben Franklin Partnership Grants (2) for Dev of a Computer Aided Engg Program at Geneva College. Project dir Dev of a Stator workcell. Teacher of the Year Award 1986 at Geneva College. *Society Aff:* APS, AOS, IEEE, AAPT.

Pinkham, Clarkson W
Business: 2236 Beverly Blvd, Los Angeles, CA 90057
Position: President. *Employer:* S B Barnes Associates. *Education:* BS/CE/Univ of CA. *Born:* Nov 1919, Los Angeles CA. 1941-46 on active duty with US Navy; 1947 joined S B Barnes & Assocs, Cons Engrs, Los Angeles. P Pres Structural Engrs Assn of Southern CA 1971, & Structural Engrs Assn of CA 1975; Dir ACI 1975-78; Dir TMS 1986-, Mbr ASCE, IABSE, CEAC, EERI, ASTM, AWS, SSRC, SSA, ICBO, SSLC, IAE, ICBO, AISC, SEASC, TMS. *Society Aff:* ASCE, ACEC, EERI, ASTM, AWS, SSRC, SSA, TMS.

Pinnel, M Robert
Business: 6200 E Broad St, Columbus, OH 43213
Position: Supvr Printed Wiring Bd Eval Reliability Grp *Employer:* Bell Telephone Laboratories. *Education:* PhD/Matls Engg/Drexel Univ; MS/Met Engg/Drexel Univ; BS/Elec Engg/Drexel Univ. *Born:* Philadelphia Pennsylvania. 1970 joined Bell Labs, engaged in res on physical & mechanical characterization of copper alloy spring materials, interdiffusion in electrical connector materials systems & characterization of semi-hard magnetic alloys; 1975 assumed respon as supervisor of Mechanics & Materials Group of Fundamental Studies Dept. 1980 assumed current respon as supervisor for printed wiring evaluation and reliability studies. Mbr ASM, AIME, Tau Beta Pi, Phi Kappa Phi, Alpha Sigma Mu & Eta Kappa Nu. Recipient Jacque-Lucas Award 1974 from ASM & IMS. *Society Aff:* ASM, AIME, IMS.

Pinnell, Ray A, Jr
Home: 118 Rio Bravo, San Antonio, TX 78232
Position: Pres. *Employer:* Feigenspan & Pinnell. *Education:* BS/Architectural Engg/Univ of TX. *Born:* 7/5/25. USN, WWII. Former Chief engr of Lift Slab, Inc. Former Pres of Lift Slab Res & Dev Corp. Former Pres of Prestressing Industries, Inc. Present Treas of Argus Engg Corp (A Consortium of specialized Engg Cos). Present Pres of Feigenspan & Pinnell, Consulting Engrs. Principal Engg Designer of numerous notable bldg structures in USA and abroad. Enjoy sporting activities, hunting and fishing. *Society Aff:* NSPE, EERI, PEPP, ACI, CSI.

Pinnell, Steven S
Business: Pinnell Engineering, Inc, 5441 S. W. Macadam Ave, Suite 208, Portland, OR 97201
Position: Pres *Employer:* Pinnell Engr, Inc. *Education:* MS/Construction Mgmt/Stanford Univ; BS/CE/Univ of AZ *Born:* 03/27/42 Steve Pinnell has worked as a superintendent, estimator and proj engr for contractors and as a resident engr, design engr and proj mgr for consltg engrs. He has worked in the San Francisco Bay area, Alaska, and Chile. In 1975 he formed Pinnell Engrg in Portland, OR, to specialize in proj mgmt services to owners, contractors and engrs. In addition to considerable practical experience, he has taught Critical Path Scheduling and proj mgmt at over 50 seminars and univ courses and has published a major article on CPM techniques in the July, 1980 issue of Civl Engrg - ASCE magazine. *Society Aff:* ASCE, AGC, PMI.

Pino, Isaac J
Home: 7900 Viscount Blvd 349, El Paso, TX 79925
Position: Assoc *Employer:* Bohannan-Huston, Inc *Education:* BS/CE/NM State Univ *Born:* 07/16/51 Began engrg career as City Engr for City of Roswell, NM in 1975 at age 23, serving till 1978. Entered civil consltg field in 1978 with Bohannan-Huston, Inc of Albuquerque, NM. Started in general civil design, named an Assoc in 1979 in charge of Business Development. Transferred to El Paso Office in 1981 to serve as Chief Administrative Officer of branch (El Paso). Reg in NM & TX. Mbr NSPE; Southeastern Chapter Secy 1978. Legislative Committee Chrmn Albuquerque Chapt 1980. Mbr APWA: Secy NM Chapt 1977, VP 1978, State Pres 1979. *Society Aff:* NSPE, APWA

Piore, Emanuel R
Home: 115 Central Park West, New York, NY 10023
Position: retired V Pres, Ch Scientist & Dir. *Employer:* IBM. *Education:* PhD/Physics/Univ of WI; BA/Physics/Univ of WI. *Born:* July 19 1908. Mbr NAS, Treasurer & Council Mbr 1965- ; Mbr Res Ctte American Philosophical Soc; IEEE formerly BD/Dir 1955; Amer Acad of Arts & Sciences; Bd Mbr NSF 1964-72, V Chmn 1970-72; Bd Mbr NY St Sci & Tech Foundation 1968- ; Woods Hole Oceanography Inst; Ex Ctte Sloan Kettering Cancer Res Inst. Distinguished Civilian Science Medal, Navy 1955; Gold Mdeal Industrial Res Inst 1966; Kaplan Prize for Technology, Hebrew Univ 1975. Lt Com USNR 1943-46; President's Science Adv Ctte 1959-63; Office of Naval Res 1945-55; Deputy Ch & Ch Scientist 1952-55; Mbr of the Bd of Stark Duyser Lab 1971-1983. *Society Aff:* ΣΞ, ΣΠΣ

Piper, Harvey S
Home: 238 Outer Drive, State College, PA 16801
Position: Sr Research Assoc. *Employer:* Penn St Univ - APL. *Born:* Sept 1934. BS Penn State, MEE New York Univ, PhD Penn State. Mbr of Tech Staff, Bell Tele Labs 1960-63. With the Appl Res Lab, Penn State Univ since 1963. Conducting res in signal processing & underwater sound propagation. Currently serving as proj leader on a prog studying long range propagation of explosive sound in shallow water.

Piper, James E
Home: 618 E Walnut St, Titusville, PA 16354
Position: Mgr Quality Assurance *Employer:* Cytemp Speciality Steel Div Cyclops Corp. *Education:* BS/Met Engr/Grove City College. *Born:* 8/7/31. Native of Butler, PA. With Cytemp Specialty Steel Co since 1960. Activities chronologically: Complaint Metallurgist, Tool Steel and High Speed Steel Product Metallurgist, Supervisor Lab Testing, Supervisor of Inspection, Supt Quality Control, Quality Assurance Coordinator, Mgr. Qual. Engr, Director Quality Engr. Current responsibility as Mgr of Qual Assurance for four plants mfg various forms of specialty steels. Chrmn, Northwestern PA Chapter of ASM 1971/1972. Enjoy hunting and fishing. *Society Aff:* ASM, ASTM, SAE-AMS, ASQC.

Piper, Lloyd L, II
Home: 1226 Indian Trail, Hinsdale, IL 60521
Position: SVP - On-Site Technologies *Employer:* Chemical Waste Management, Inc. *Education:* MS/IE/Univ of Houston; BS/EE/TX A & M Univ *Born:* 04/28/44 Wareham, Massachusetts, son of Col. Lloyd Llewellyn and Mary Elizabeth (Brown). Married Jane Melonie Scruggs, April 30, 1965; 1 son, Michael Wayne. With Houston Lighting and Power Co in Engrg Dept 1965-1974. With Dow Chem Engrg and Construction Services, Houston, 1974-1978 as Project Engrg Mgr. With The Ortloff Corp, Gulf Coast Div, Houston, 1978-79 as Project Mgr, 79-80 as Mgr of Engrg, 80-83 as VP. During 1983-86, Pres & CEO of PLANTECH Engrs & Constructors, Inc., Houston a sub of Dillingham Corp. In 1986 named S.V.P. On-Site Technologies for Chemical Waste Management, Inc of Oak Brook, Illinois. Phi Kappa Phi. Registered PE. Pres, San Jacinto Chapter (Houston) of NSPE. Chrmn, NSPE PE in Industry Div, NSPE VP 1976- 1977. Natl Young Engr of the Year Award, NSPE 1976. Chrmn, NSPE Political Action Committee, 1980-83. *Society Aff:* NSPE, IEEE, PMI, EMS.

Piper, R Davidson
Business: 2940 Edwards St, Butte, MT 59701
Position: President. *Employer:* Piper & Assocs Inc. *Education:* BS/Mining Engr/MT College of Mineral Sci & Tech. *Born:* Oct 1913. Native of Butte, Montana. With The Anaconda Co in the mining indus from 1937-74. Assumed respon as Dir of

Piper, R Davidson (Continued)
Oper Engrg & Planning & Mgr of Analysis & Planning the overall engrg & design of the mining opers in Montana. As Pres of Piper & Assocs Inc, Mining Indus Cons, respon for the firm's overall cons assignments. Natl Dir NSPE 1973-75. V P Western Region NSPE 1975-77. Bd Dir of Mineral Res Ctr of the Montana Tech Fnd 1978-79. *Society Aff:* NSPE, AIME, WPCF, MMA, NWMA.

Pipes, R Byron
Business: College of Engineering, University of Delaware, Newark, DE 19716
Position: Dean, Coll of Engg *Employer:* Univ of DE *Education:* BS/-/LA Polytech Inst; MS/-/LA Polytech Inst; MSE/-/Princeton Univ; PhD/- /Univ of TX *Born:* 08/14/41 Currently Dean College of Engrg at the Univ of Delaware in Newark, Delaware where from 1978 he served as Dir of the Center for Composite Matls. The Center was founded in 1974 and is dedicated to the dev of a comprehensive sci and tech in the important and growing field of composite matls. Dr Pipes served a six-year membership on the Natl Matls Adv Bd of the Natl Acad of Scis and Acad of Engrg. He is the author of over fifty tech papers, editor of two symposium proceedings, edit. book series, composite materials and coauthor of two books on experimental mechanics for composite matls. He received the Gustus Larson Award from ASME in 1983 and was elected to the Nat. Acad of Engrg in 1987. *Society Aff:* ASME, SPE, SESA, ASTM, SME

Pipes, Wesley O
Business: Department of Civil Engineering, Drexel University, Philadelphia, PA 19104
Position: Prof of Civ Engrg *Employer:* Drexel Univ-Dept of Civ Engrg *Education:* BS/Biol/North TX State; MS/Biol/North TX State; PhD/Sanitary Engg/Northwestern. *Born:* Jan 28, 1932 Dallas Tex. Northwestern Univ 1959-74: Asst Prof of Civil Engrg 1959-62, Assoc Prof of Civil Engrg 1962-67, Prof of Civil Engrg 1967-74. Drexell Univ 1975-84: Betz Prof of Ecol & Prof of Biol Sci 1975-83, Prof of Environ Engrg & Sci. 1975-87, Acting Dir of the Environ Studies Inst 1982-83, Prof and Head of Civ Engrg 1983-87. Cons: U.S. Environmental Protection Agency, Commonwealth Edison Co, Smith Kline Chemical Co., Pennsylvania Power & Light, others. Assn of Environ Engrg Profs, Bd of Dir 1971-76, V P 1974, Pres 1975, Bd of Dir Award 1973; Water Pollution Cont Fed; AWWA; others. Selected as one of Chicagoland's Ten Outstanding Young Men of 1963. Author more than 100 tech articles on biol wastewater treatment, environ impacts of wastewater discharges, water supply, and bacterial indicators of pollution. *Society Aff:* ASCE, AEEP, AWWA, WPCE, ASM

Pirnie, Malcolm, Jr
Business: 2 Corporate Park Dr, P. O. Box 751, White Plains, NY 10602
Position: Chrmn *Employer:* Malcolm Pirnie, Inc *Education:* SM/Sanitary Engr/Harvard Univ; AB/Engr Sci/Harvard Coll *Born:* 03/15/17 Chrmn of the Bd of the environmental engrg/planning firm of Malcolm Pirnie, Inc, culminating more than 40 years of distinguished practice in the US and abroad. Directs long-range planning and client/government/profl organization relationships. During World War II served as Commissioned Ofcr, US Public Health Service, responsible for water supply and wastewater treatment/disposal during Alaska Highway construction. A licensed PE in 15 states, has held major offices in natl profl societies (ASCE, ACEC, AICE), is a published authority on consltg engrg practice, Diplomate AAEE. Active in community service. Born in Mt. Vernon, NY. An avid golfer. *Society Aff:* ASCE, NSPE, WPCF, AWWA

Pisetzner, Emanuel
Business: 79 Madison Ave, New York, NY 10016
Position: Partner. *Employer:* Weiskopf & Pickworth - Cons Engrs. *Education:* BCE/Civil Engg/CCNY. *Born:* July 1926. Joined Weiskopf & Pickworth 1953; made a full partner 1960. A PE in 16 states. In charge of struct design on many projs such as new Natl Gallery of Art. Bldg - Wash D C, Credit Lyonais Office - Hotel Tower - Lyon France, John F. Kennedy Memorial Library - Boston Mass; Raffles City, Singapore; Headquarters General Foods Corporation, Rye N.Y. Hdquarters-Union Carbide Corp- Danbury CT, Christian Sci Church Ctr Boston Mass. Mbr num engrg groups. ACEC Past Presidents Award 1984. Disting Serv to Professions Award N Y Assn of Cons Engrs 1976. PCI Excell in Design of Structs Award 1971 & 1973. CRSI Design Award 1974. 1st Award-Lincoln Arc Wldg Fdn-Design of Welded Structures 1978-for Space Frame Natl Gallery of Art. Mbr Chi Epsilon; V P 1972 & 1973, Dir 1967-69 ACEC; Pres 1965-66 NYACE; Treas 1963-64 NYACE; Arbitrator with AAS; contrib articles to Civil Engrg Mag & other pubs. *Society Aff:* ACEC, ASCE, NYACE, NSPE.

Piske, Richard A
Home: 5516 River Point Cove, Knoxville, TN 37919
Position: Mger-Corporate Quality Assurance. *Employer:* Allen & Hoshall, Inc *Education:* BSME/ME/Tulane Univ. *Born:* May 1924. Native of New Orleans La. Employed by Carrier Corp 1944-65. From 1955-62 Regional Engrg Mgr Southeastern Region. 1962-65 Progs Mgr Military Equip Dept, Syracuse NY. Prime projs Minuteman Wings III, IV, V & Saturn V Environ Control Sys. 1965-1981 : I C Thomasson & Assocs Inc, Cons Engrs. Allen & Hoshall Consulting Eng 1981 to present Mgr Corporate Quality Assurance. Mbr ASHRAE, NSPE, ACEC, CET, TSPE, Kiwanis P Pres Cons Engrs of Tenn, P Natl Dir Amer Cons Engrs Council. Collect & restore antiques. *Society Aff:* ACEC, NSPE, ASHRAE, CET, TSPE.

Pitler, Richard K
Home: 5718 King of Arms Dr, Gibsonia, PA 15044
Position: Senior V P - Tech Dir. (Retired) *Employer:* Allegheny Ludlum Corp. *Education:* PhD/Metallurgy/RPI; MS/Metallurgy/RPI; BS/Metallurgy/MIT. *Born:* 3/13/28. Native of Merrick N Y. 2nd Lt Ord Watertown Arsenal 1949-50. Res Metallurgist Allegheny Ludlum, working on Ti, valve steels, superalloys to 1965. Tech Dir, Special Metals Corp 1965-66. Dir - Prod Dev, Allegheny Ludlum, then Dir Tech Services & V P - Tech Dir. Respon for R&D, Mkt & Prod Dev, & licensing & know-how sales. Formerly Chmn Pgh Chap AIME-IMD, Exec Ctte Pgh Chap ASM, Mbr AISI Genl Met Ctte. Currently mbr Board of Trustees ASM, Formerly Chrmn Genl Res Ctte-AISI, Enjoy tennis, golf, swimming, bridge & painting. *Society Aff:* ASM, AIME.

Pitrolo, Augustine A
Business: US Dept of Energy, PO Box 880, Morgantown, WV 26505
Position: Director *Employer:* Morgantown Energy Tech Center *Education:* BS/ME/WV Univ *Born:* 02/14/31 Began career with Joy Mfg Co 1952 in design and development of underground mining equipment. Joined Gen Elec Co 1956. Power plant engr in design and development of aircraft gas turbines. Managed space nuclear RTG progs. Designed and developed commercial gas turbines. Since 1975, Dir, Morgantown, WV Energy Tech Center, US Dept of Energy (formerly Bureau of Mines and ERDA). Responsible for R&D in Coal Gasification, Fluidized-Bed Combustion, Unconventional Gas Recovery, Components, Combined Cycles, and Gas Stream Cleanup. Dual role as Acting Depty Prog Dir for Fossil Energy, Washington DC, for eight months. Received NASA Public Service Award 1972.

Pitsker, Peter B
Business: 1800 E Garry Ave, Suite 114, Santa Ana, CA 92705
Position: Pres *Employer:* North & Donahoe *Education:* BS/ChE/Stanford Univ *Born:* 03/14/33 Three yrs in the Refining Dept of Mobil Oil led to the joining of the Foxboro Co in 1960. Was promoted to VP, Marketing and Sales, in 1970. Interest in computer control prompted a move to a new venture, Modular Computer Systems, in 1972 as Sales Mgr. Growth was from $5 million to $26 million through 1975. After returning to CA as Dir of Marketing for Gen Automation, joined North & Donahoe, Mgmt Consltg Engrs, as VP and Dir. North & Donahoe specializes in market planning, strategic planning and mgmt consltg in the high-tech industries. Current position at North & Donahue, Pres. Involved in start-up and funding of

Pitsker, Peter B (Continued)
Triconex Corp, a devel and mfg of fault-tolerant indus ctrl sys, where he is a Bd Mbr. *Society Aff:* AIChE, ISA, RI/SME

Pitt, Charles H
Business: 416 Browning Bldg, Salt Lake City, UT 84112
Position: Prof of Met. *Employer:* Univ of UT. *Education:* PhD/Met/Univ of UT; BS/Met Engg/Univ of WI. *Born:* 8/9/29. Dr Pitt grad from the Univ of WI in 1951 in Met Engg and was employed by GE Co in the Atomic Energy industry at Hanford Works from 1951-53. He served in the US Army from 1954-56 and attended grad sch at the Univ of UT from 1956-59 receiving a PhD in Met. He was a NSF post doctoral fellow at the Univ of Cambridge 1960-61. He has been on the staff of the Univ of UT since that time and achieved the rank of Prof of Met in 1971. He has taught courses in physical met, corrosion and metal failure analysis. Res interests include effect of small amounts of impurities on the properties of metals, corrosion and electrometallurgy. Dr Pitt has served as a consultant in the area of heat treatment, corrosion and metal failure analysis. *Society Aff:* AIME, ASM, NACE, ΣΞ.

Pitt, Lawrence L
Business: PO Box 15851, Baton Rouge, LA 70895
Position: VP. *Employer:* AWC, Inc. *Education:* BSChE/-/Rose-Hulman Inst of Tech; MBA/Finance/WA Univ. *Born:* Mar 1940. BSChE Rose-Hulman Inst of Tech, MBA Washington Univ 1971. Native of Louisville Ky. Tech & mgmt pos wth Firestone Tire & Rubber Co & Mallinckrodt Inc. Served as an officer in US Army Chem Corps 1962-64. With MC/B Mfg Chemists since 1971, as Dir of Commercial Dev & V P of Oper. MC/B is a leading manufacturer of lab & fine chems AWC, Inc is to Sales Engg firm selling process controls & process equipment in the Gulf Coast & West Coast. *Society Aff:* AIChE, ACS, CDA.

Pitt, Paul A
Home: 5730 Lancaster Dr, San Diego, CA 92120
Position: V P, Engrg & Res. *Employer:* Solar Turbines Internatl. *Education:* BSME/Mech Eng/Univ CA *Born:* May 1917. Attended Calif State Polytech Coll & Univ of Calif Berkeley. Spec in mech & aeronautical engrg. Joined Solar Co 1942. Served 5 yrs as Proj Engr & named Asst Ch Engr/Dev 1947. Became Ch Engr 1950. Dir R&D of aircraft jet engine components & gas turbines. Named V P, Engrg & Res, Solar Div of Internatl Harvester 1960. Respon for dev of indus gas turbines, driven equip & pkgs 300 to 10,000 HP. 12 pats. Fellow ASME; recipient ASME Gas Turbine Power Div Award 1964. Conslt since 1980. *Society Aff:* ASME, SAE

Pitt, William A J, Jr
Home: 10131 Kendale Blvd, Miami, FL 33176
Position: Sr Hydrologist-Civil Engr. *Employer:* Post, Buckley, Schuk & Jernigan, Inc. *Education:* BS/Civil Engg/Univ of Miami; -/Mining Engg/CO Sch of Mines. *Born:* 11/29/42. P E FL.,P.H. With USGS 1967-78. Proj leader spec in discharge measurements, sedimentation, water qual, & flood delineation & control. Prin investigator on studies of the effect of septic tank effluent & landfill leachates on ground- water qual, eval of deep well injection sys for disposal of municipal & indus wastes. Also involved in the dev of water mgmt alternatives for increasign water supplies, prin by fresh water storage in deep aquifers. With PBS&J since 1978. Proj Mgr for ground and surface water supply dev, deep well injection, aquifer evaluation, wellfield design, and evaluation of environ impacts. Also in charge of flood investigations, evaluations of mining effects on the environ, drainage design for land dev, waste-water disposal, design of perc ponds & spray irrigation projs, and their effect on ground and surface water. *Society Aff:* ASCE, NSPE, FES, MGS, NGS, NWWA, AIH.

Pitts, Robert G
Home: 724 S College St, Auburn, AL 36830
Position: Prof & Hd Emeritus Dept of Aerosp Engg. *Employer:* Auburn Univ. *Education:* MS/ME/Caltech; BS/AE/Auburn Univ. *Born:* Aug 19, 1910 Perry Cty Ala. Mbr Tau Beta Pi, Phi Kappa Phi, Blue Key, Scabbard & Blade, Pi Tau Sigma, Sigma Gamma Tau; Disting Serv Award, Jr Chamber of Commerce 1943; served as pres local chap Phi Kappa Phi; Pres Univ Fac Council; served in all acad ranks, Auburn Univ Instr to Prof & Dept Hd; Secy- Treas Aerospace Dept Chmn's Assn 1976 & Chrmn of Assoc in 1977. Respon for entire aviation prog of Auburn Univ & all oper of univ airport. Fac Mbr Auburn Univ July 1, 1935-until Retired 6/30/79. *Society Aff:* ASEE, ADCA, AIAA.

Pitzer, Kenneth S
Business: Dept of Chem, Univ. of Calif, Berkeley, CA 94720
Position: Professor Emeritus. *Employer:* Univ of Calif Berkeley. *Education:* BS/Chem/CA Inst of Tech; PhD/Chem/Univ of CA, Berkeley; D Sc (hon)/-/Wesleyan University, Middletown, CT; LL D (hon)/-/University of California, Berkeley; LL D (hon)/-/Mills College, Oakland, CA. *Born:* 1/6/14. Instr to Assoc Prof of Chem Univ of Calif 1937-45; Prof 1945-61; Dean Coll of Chem 1951-60. Pres, Prof of Chem Rice Univ 1961-68, Stanford 1968-70. Prof of Chem Univ of Calif 1971- . Tech Dir Md Res Lab for OSRD 1943-44; Dir Res U S AEC 1949-51, Mbr Genl Adv Comm 1958-65, Chmn 1960-62, Mbr Adv Bd U S Naval Ordnance Test Sta 1956-59, Chmn 1958-59, Mbr Commn Chem Thermodynamics, Intl Union Pure & Appl Chem 1953-61. Dir Owens-Ill Inc; Mbr Pres' Sci Adv Comm 1965-68. Tr Pitzer Coll; Guggenheim Fellow 1951; one of the 10 Outstanding Young Men, U S Jr C of C 1950; Clayton Prize, Inst Mech Engrs London 1958; Priestley Medal Amer Chem Soc 1969; Natl Medal for Sci 1975. Fellow Amer Nuclear Soc, Amer Inst Chemists (Gold Medal 1976). Berkeley Citation (U of CA 1984); Robert A Welch Award in Chem, 1984; Honorary Fellow in the Indian Academy of Sciences, 1986; Mack Award, Ohio State University, Columbus, OH, 1986; Pitzer Lecture, Department of Chemistry, University of California, Berkeley, CA, 1987. *Society Aff:* ACS, AIC, NAS, ANS, APS, APS, FS, AAAC.

Piziali, Robert L
Business: Mech Engg Dept, Stanford, CA 94305
Position: Assoc Prof. *Employer:* Stanford Univ. *Education:* PhD/Mech Engg/Univ of CA; ME/Mech Engg/Univ of CA; BS/Mech Engg/Univ of CA. *Born:* 9/3/42. Has taught at Stanford Univ since 1970 in the Dept of Mech Engg, Design Div, Specializing in design and structural mechanics applied to bioengg. Res work includes injury mechanics of the human musculo-skeletal system, with emphasis on the knee and spine. Design projs have involved cardiology, anesthesia and pediatrics. He is currently co-editor for biomechanics of the Annals of Biomedical Engg. He is the co-director of the Stanford Inst for Engg Design of Medicine, and former chrmn of the Sch of Engg Advising Committee on Engg in Medicine and Biology at Stanford Univ. *Society Aff:* ASME, ΣΞ, AAAS, ASTM, ORS.

Plane, Robert A
Business: Potsdam, NY 13676
Position: President. *Employer:* Clarkson University. *Education:* PhD/Chemistry/Univ of Chicago; AB/Chemistry/Univ of Evansville; SM/Chemistry/Univ of Chicago. *Born:* Sept 1927. Res Scientist at Oak Ridge Natl Lab 1951-52; fac mbr Cornell Univ 1952-74: Prof of Chem 1962-74, Chem Dept Chmn 1967-70, Acting Provost 1969- 70, Provost 1970-73; Pres Clarkson Univ Potsdam N Y 1974- ; Visiting Scientist Univ of Calif at Berkeley 1969. NIH Special Fellow, Nobel Medical Inst, Stockholm Sweden 1960, Oxford Univ 1961. Author of textbooks & res papers. Cons to chem indus. Pres Assn of Independent Engrg Colleges. Chmn Comm on Independent Colleges & Univs. *Society Aff:* ACS, AAAS, ASE.

Plank, Charles A
Business: Dept of Chem Engr, Univ. of Louisville, Louisville, KY 40292
Position: Prof of Chem Engrg. *Employer:* Univ of Louisville. *Education:* PhD/Chem Engrg/NC State; MS/Chem Engrg/NC State; BSChE/Chem Engrg/NC State. *Born:* Oct 1928 Charlotte N C. Taught at N C State. Special projs engr for Olin Corp Brandenburg Ky. With Univ of Louisville since 1957. Dir of Interdisciplinary Progs in Engrg 1971-73. Chrmn Dept of Chem Engr 1973-79. Mbr Bd of Tr of Univ of Louisville & also Univ of Louisville Found 1973-75. Named Distinguished University Teacher 1985. Private cons. Enjoy sports & music. Produce radio progs on jazz. *Society Aff:* AIChE.

Plank, R David
Business: Jewell Sta-P O Box 551, Springfield, MO 65801
Position: Mgr - Design Engg and Planning Dept-City Utilities 1976 *Employer:* City Utilities. *Education:* Hon Prof/Degree of Civil Engg/UMR Rolla, MO; MBA/Bus Adm/Drury College; BA/Physics/Drury College; BSCE/CE/UMR Rolla, MO. *Born:* 1/1/36. Prof Degree of Civil Engr UMR 1975 (Hon). Dir NSPE, Pres MOSPE 1974-75. Chmn MO Sect AWWA 1976. With City Utils since 1959; assumed current pos 1976. Previously engr, Asst Mgr Water Dept, Mgr Water Department. Respon for design engg & system planning functions for electric, gas, water & transit operations of municipally owned utility serving 150,000 people. Hobbies: autos & fishing. *Society Aff:* NSPE, AWWA, APPA.

Plants, Helen L
Business: College of Engg and Tech, Indiana Purdue Univ. at Fort Wayne, Fort Wayne, IN 46805
Position: Chrmn Civil and Architectural Engr. Tech; Prof. Civil Engr. Tech *Employer:* Purdue Univ. *Education:* MSCE/CE/W VA Univ; BSCE/CE/Univ of MO. *Born:* Mar 9, 1925 Desloge Mo. Maiden name: Lester. Practiced as struct engr 1945-47. Reg P E W Vir 1953- . Served on mechs fac of W Vir Univ 1947-85. Directed doctoral prog in engrg educ & Office for Effective Instr. ASEE V P 1975-76, Dir 1974-76. Western Elec Fund Award for Excell in Instr of Engrg Students 1975. Univ of Mo Hon Award for Disting Serv in Engrg 1975. Amoco Found Outstanding Teacher Award 1971. Author three textbooks in mechanics numerous chapters and papers on teaching engrg. Currently dept chair CAET Purdue Univ at Fort Wayne. Served as UNESCO Consultant in Manila and as Conslt to Kingston Polytechnic and Brighton Polytechnic in England. Mbr of Academic Advisory Bd of US Naval Academy 1976-80. Fellow of ASEE. *Society Aff:* ASEE, ASCE, SWE, ASME.

Planz, Edward J
Home: 1119 Brookhill Rd, Tuscaloosa, AL 35404
Position: Prof *Employer:* Univ of Alabama *Education:* MS/Met Eng/RPI; BS/Chem/Univ of CT; Grad/Metallurgy/Yale Univ; Grad/X-ray Diffraction/Bklyn Polytech Inst *Born:* 05/09/20 Native of West Haven, CT. Employed as Metallurgical Engr in indus for 14 yrs. Employed as Assoc Prof of Metallurgical Engrg from 1960 to present. Active conslt for indus for past 15 yrs in physical metallurgy, welding, x-ray diffraction, failure analyses and an expertise in product liability. Dir of Junior Coll Relations and Undergrad Counseling for Engrg. PE (Reg) for the State of AL. *Society Aff:* ASM, AIC, AIME, AAS, AAUP

Plass, Harold J
Business: Dept of Mech Engrg, Coral Gables, FL 33124
Position: Univ of Miami - P O Box 248294. *Employer:* Univ of Miami - PO Box 248294. *Born:* Dec 1, 1922. PhD Stanford Univ; BS & MS Univ of Wisc. 2 yrs with RCA Victor as Television engr. Taught Appl Mech at Univ of Texas Austin for 17 yrs. Served 5 yrs as Dept Chmn Engrg Mech. Taught 11 yrs at Univ of Miami, Mech Engrg. Served 4 yrs as Dept Chmn. Recent res int: energy tech, environ, resources, econ sys studies. Fellow ASME. Mbr ASEE, Sigma Xi, ISES.

Plate, Erich J
Business: Institut F. Hydrologic U. Wawirtsch, Kaiserstr. 12, D-75 Karlsruhe 1, W. Germany
Position: Professor *Employer:* Universitat Karlsruhe. *Education:* DrIng/Civil Engr/Univ Stuttgart, FRG; MS/Hydr Engr/CO State Univ; Dipl Ing/Civil Engr/Univ Stuttgart, FRG *Born:* July 1929. Fulbright Student CSU 1954-57, Asst (Hydraulic Engr) Stuttgart until 1959; teaching & res Colo State Univ 1959-69 (Prof 1968), Argonne Natl Lab 1969-70. Present pos since 1970. Teaching, res & cons in water resources, waves, & air & water Pollution, as Dir of Institut fur Hydrologie und Wasserwirtschaft, Univ of Karlsruhe. Dean of Civil Egrg Karlsruhe 1973-75. V P IAHR 1975- , Croes Gold Medal ASCE (with J Nath) 1971, JAHR-Pres 1984-. *Society Aff:* ASCE, AMS, AGU

Platt, George
Home: 5941 Midiron Circle, Huntington Beach, CA 92649
Position: Staff Engr, Control Sys. *Employer:* Bechtel Power Corp. *Education:* BS/ChE/CCNY. *Born:* 1920. BChE City Coll of New York. Lic P E Mech & Control Sys. Control Sys staff engr, Bechtel Power Corp, power plant design. Formerly in oil refinery & chem plant design, oper. Instru Soc of Amer Stds & Practices Award 1968; Dir Stds & Practices Dept; Chmn Stds Ctte S5, Graphic Symbols or Instrumentation. V Chmn - Mech, Amer Natl Stds Inst Ctte Y32, Graphic Symbols. Internatl Org for Standardization Tech Ctte 10, Subctte 3, Graphical Symbols for Instrumentation. Mbr So Calif Meter Soc. Author num publ articles & papers. *Society Aff:* ISA, SCMA.

Platzer, Max F
Business: Dept of Aeronautics, Monterey, CA 93940
Position: Prof and Chrmn *Employer:* Naval Postgrad Sch *Education:* Dr Tech Sci/Aerospace/Tech Univ of Vienna, Austria; Dipl Eng/ME/Tech Univ of Vienna, Austria *Born:* 06/26/33 1957-60 Asst Prof, Technical Univ of Vienna, Austria. 1960-66 Supvy Aerospace Engr, NASA Marshall Space Flight Center, Huntsville, AL. 1966-70 Scientist and Section Hd, Lockheed-GA. Res Lab, Marsotta, GA. 1970-present Naval Postgrad Sch. *Society Aff:* AIAA, ASME, ΣΞ

Plazek, Donald J
Business: MME Dept, 848 Benedum Hall, Pittsburgh, PA 15261
Position: Prof *Employer:* Univ of Pittsburgh *Education:* PhD/Phys Chem/Univ of WI; BS/Chem/Univ of WI *Born:* 1/12/31 Native of Milwaukee, WI. Married Patricia L. Filkins. Seven children. After 2 yrs of post-doctoral study with professor John D. Ferry at the Univ of WI, spent 9 yrs as a fellow in independent res at the Mellow Inst in Pittsburgh, PA 1958-67, where he developed a frictionless magnetic bearing creep apparatus. In 1967 he became an Assoc Prof in materials Engrg at the Univ of Pittsburgh and was promoted to prof in 1975. His principal res activities are in correlating polymer molecular structure with mechanical behavior. *Society Aff:* ACS, APS, Soc Rheol

Plesset, Milton S
Home: 860 Orlando Rd, San Marino, CA 91108
Position: Prof of Engg Sci, Emeritus *Employer:* California Institute of Technology *Education:* PhD/Physics/Yale University; MS/Physics/Univ of Pittsburgh; BS/Physics/Univ of Pittsburgh. *Born:* Feb 1908. PhD Yale Univ; BS & MS Univ of Pittsburgh. Instr Univ of Rochester. Hd, Analytic Group Douglas Aircraft Co. Assoc Prof, then Prof of Engrg Sci, Caltech. Chmn Cavitation Ctte ASME. Mbr, Chmn, Exec Ctte of Fluids Engrg Div ASME. Mbr, Chmn Exec Ctte, Fluid Dynamics Div, Amer Phys Soc. Mbr Adv Ctte for Nuclear Safeguards (U S Nuclear Regulatory Comm). Fluids Engrg Div ASME Knapp Award 1968. Fellow ASME. Fellow Amer Phys Soc. Mbr Amer Nuclear Soc. Mbr, Natl Acad of Engrs. *Society Aff:* APS, ASME, ANS.

Pletcher, Richard H
Home: 411 Oliver Ave, Ames, IA 50010
Position: Professor *Employer:* Iowa State University *Education:* PhD/Mech. Engr./Cornell University; MS/Mech. Engr./Cornell University; BS/Mech Engr/Purdue University *Born:* 05/21/35 Native of Elkhart, Indiana. Served in U.S. Navy 1957-60, Instructor of Mechl Engrg at Cornell Univ 1962-65, and Senior Research Engr at United Aircraft Research Labs 1965-70. Faculty member at Iowa State Univ since 1967. Professor, Mechl Engrg since 1976. Currently Assoc Dir of Computational Fluid Dynamics Center. Active in ASME serving as Ch of the Heat Transfer Div

Pletcher, Richard H (Continued)
Cttee on Aircraft & Astronautical Heat Transfer & as a Tech Ed of the Journal of Heat Transfer. Mbr of editorial advisory bd for Numerical Heat Transfer. Consultant to industrial firms and a reg PE in Iowa. *Society Aff:* ASME, AIAA, ASEE, ΣΞ

Pletta, Dan H
Home: 1414 Highland Circle, Blacksburg, VA 24060
Position: Univ Disting Prof Emer-Cons Engr. *Employer:* Va Poly Inst & St U-Self. *Education:* CE/Civil Engg/Univ of IL; MS/Civil Engg/Univ of WI; BS/Civil Engg/Univ of IL. *Born:* 12/31/03. Taught civil engrg & mech at Univ of Wisc, Univ of So Dak, U S Military Acad & VPI & ST Univ. Dept Hd 1948-70. Publ 2 mechanics textbooks, 1 book on professionalism, 1 handbook chapter, articles on ethical, prof & tech subjs. Dir 1969-72, Chmn Engrg Mech Div, Pres VaSect, Committee on Prof Conduct 1973-76 - Chmn 1976 ASCE. Natl Dir 1957-65, Pres Educ Found, Pres Va (SPE), Prof Engrs in Educ (PEE) Exec Bd 1974-76, Chmn 1977 NSPE. Various Committees of ACI, ASEE, SESA, etc. VSPE Engr of Yr 1972; VSPE Distinguished Service 1971; ASEE Western Elec Excell in Teaching 1968; Engrg News Record - Citation 1972; Univ of Ill Coll of Engrg Alumni Honor Award for Distinguished Serv in Engg 1976; Hon Mbr ASCE 1979; Paxton Award, Intl Assoc of Torch Clubs 1984; NSPE Distinguished Service Award 1987. *Society Aff:* ASCE, NSPE, ASEE, ACI, ASEM.

Plewes, W Gordon
Home: 40 DeLong Dr, Ottawa, Canada K1J 8H4
Position: Prof Engr *Employer:* Self. *Education:* MSc/Civil Engg/Queen's Univ; BSc/Civil Engg/Univ of Manitoba. *Born:* 6/14/19. RCAF 1941-45. Sev yrs bridge, bldgs engr CNRailways, lectr Royal Military College. Joined Natl Res Council, Div of Bldg Res 1954, retired 1975. Active in structures, codes & stds, educ, Canadian & intl tech cttes. Doupe Gold Medal 1949; ACI Delmar Bloem Award 1974; FACI 1975; Hon Mbr ACI 1976. Hobbies: mineral collecting,, music, philosophy, gardening. Fellow CSCE 1983. *Society Aff:* IEC, ACI, CSCE.

Pline, James L
Business: 3311 W State St-Bx 7129, Boise, ID 83707
Position: Roadway Design Supervisor. *Employer:* Idaho Transportation Dept. *Education:* BS/CE/U of I; Cert/Traffic Engr/Yale Univ; Masters/Pub Adm/Boise State Univ *Born:* 11/18/31. Cert in Traffic Engrg Yale Univ 1959, Automotive Safety Found Fellowship. Boise State Univ Master, Publ Admin 1978. US Air Force Europe 1954-56. Ret Lt Col USAF Res Oct 1975. Reg P E & LS ID. Mbr ISPE & NSPE. Pres ISPE 1979. Nat'l Dir, NSPE (83 & 84) Cert for Outstanding Serv ISPE 1974 & NSPE Natl Mbrship Award 1974. Mbr of ID & Natl Guard Assn, Air Force Assn & ANG Assn of Civil Engrs. Pres Air Force Chap ID Air Natl Guard Assn 1972 & 1973. Signing Subctte of Natl Adv Ctte of Uniform Traffic Control Devices, ID Traffic Safety Comm, Inst of Transp Engrs, Intl Dir (75-78), Int'l Bd of Dir (1985-87), Tech Paper Award 1978, Intl Policy Comm and ITE Tech Council Chrmn; Active in Boy Scouts as merit badge counselor. City of Boise, Urban Design Committee. Appointed Affiliate Prof CE Dept Univ of ID (82), James L. Pline Distinguished Mbr Award (84) by Intl ITE Sect. Order of Engineer, Idaho Link 1977. Idaho Outstanding Engineer 1985. American Association of State Highway and Transportation Officials (AASHTO) Traffic Engineering Subcommittee (1969-1979) Roadway Design Subcommittee (1983-present). National Advisory Committee on Uniform Traffic Control Devices (1973-present). Transportation Research Board, Group 3 Council (1979-81) NCHRP Advisory panels. International Commission on Illumination; U.S. National Committee. *Society Aff:* NSPE, ISPE, ITE

Pliskin, William A
Home: 31 Greenvale Farms Rd, Poughkeepsie, NY 12603
Position: Sr Staff Mbr *Employer:* IBM Corp *Education:* PhD/Physics/OH State Univ; MS/Physics/OH Univ; BS Ed/Physics, Math/Kent State Univ *Born:* 08/09/20 Native of Akron, OH. Officer in USAF 1943-46. Res physicist at Texaco Res Ctr 1949-59, working on catalysis and infrared spectra of chemisorbed molecules. With IBM since 1959, advisory physicist 1960-63, sr engr 1963-79, mgr of various depts dealing with thin films and semiconductor processing, 1964-present, assumed present position of Sr Tech Staff Mbr in 1979. Specialization: Characterization of thin insulating films, Semi Conductor Processing, Thin film measurements, Infrared Spectroscopy. Honors and Awds: IBM Corp Invention Awd 1966; The Annual Electronics Div Awd of Electrochemical Soc 1973; IBM Corp. Recognition Awd 1979; IEEE Fellow 1981; IBM Div Awd 1981. Hon socs; Sigma Xi, Sigma Pi Sigma, Pi Mu Epsilon. *Society Aff:* IEEE, ECS, APS, ACS, AAAS

Plonsey, Robert
Business: Department of Biomedical Eng, Duke University, Durham, NC 27706
Position: Prof *Employer:* Dept of Biomedical Engr. *Education:* PhD/EE/Univ of CA; MSEE/EE/NYU; BSEE/EE/Cooper Union. *Born:* 7/17/24. My profl interests have been in engg education & res. Upon completion of my PhD, I emphasized electromagnetic theory, antennas, & microwaves. This led to an interest in the application of electromagnetic potential theory to problems in medicine & biology specifically electrophysiology & electrocardiography. I've contributed to the dev of a quantitative approach utilizing engg methods. This approach was also the basis of a course on bioelectric phenomena, which I dev and a text book. I've also particapted as both a mbr and leader in biomedical engg profl socs with the goal of increasing their profl activity & service. *Society Aff:* AAAS, IEEE, BMES, ASEE, APS, NAE.

Plotkin, Allen
Business: Deptartment of Aerospace Engrg & Engrg Mechanics, San Diego State Univ, San Diego, CA 92182
Position: Prof. & Chairman *Employer:* San Diego State Univ *Education:* PhD/Applied Mech/Stanford; MS/Applied Mech/Columbia; BS/Applied Mech/Columbia. *Born:* 5/4/42. NYC, 1942. Bronx HS of Sci, grad 1959. BS, MS, Columbia Univ Sch of Engg and Appl Sci, 1963, '64. PhD, Stanford Univ, 1968. Joined Dept of Aerospace Engg, Univ of MD, College Park, as Asst Prof, 1968. Promoted to Full Prof, 1977. Joined Aerospace Engrg & Engrg Mechanics Department of San Diego State U. as Professor and Chairman in 1985. NASA-American Soc of Engg Education Summer Faculty Fellow at Ames Res Ctr, 1969, '70. Assoc Fellow, AIAA. Assoc Editor, AIAA Journal, 1986-8. 1976 AIAA-Natl Capital Section Young Engr/Scientist Award, Visiting Assoc in Engg Sci, CA Inst of Tech, 1975-6. Maj res interest: Fluid mechanics - aerodynamics and hydrodynamics. Navy - American Soc of Engg Education Summer Faculty Fellow at Naval Surface Weapons Center, 1980. 1981 Engrg Sci Award of Wash Acad of Sci. Aerospace Department Chairmen's Association, Secretary-Treasurer 1987, Chairman, 1988. Contributor to World Book Encyclopedia articles on Aerodynamics, Propeller and Streamling. *Society Aff:* AIAA, ASME, SNAME, AHS, ASEE, AAM.

Plourde, Arthur J Jr
Home: 2412 Langley Ave, St Joseph, MI 49085
Position: Quality Consultant & Owner *Employer:* Quality Consultants of Michigan *Born:* 03/17/23 Attended MI State Univ 1941-1945. Active in the field of Quality Control since 1952. Employment commenced with Aerojet Gen Corp Sacramento, CA inspecting polaris and minuteman missles. Became a mbr of ASQC 1973. Sr mbr 1982 and regional Dir 1984. Obtained PE Registration from State of CA in 1977; Certificate No/ 1421, Certified Quality Engr (ASQC) in 1979 Certificate No. E7454. Own and operate "Quality Consultants of Michigan" Teaching Indust complexes Statistical Quality Control Methods & Mgmt. *Society Aff:* ASQC

Plude, George H
Home: 7199 Ward Rd, Parma, OH 44134 *Employer:* Private Consulting *Education:* BS/Naval Arch/Univ of MI. *Born:* 4/12/32. Native of NYC. Res of Cleveland, OH. Received degree in Naval Architecture from Univ of MI & served subsequently as a Naval Aviator in the US Naval Reserve. Returned to Naval Arch with Marine Consultants in 1958 and progressed through numerous design projects as Proj Engr, including design & construction of the first 1000 foot Great Lakes, the M V Stewart J Cort, to VP. Past Chrmn Great Lakes & Great Rivers Sec of SNAME. Recipient of Sperry Award, 1985, for contribution to the Advancement of the Art of Transportation, awarded by Committee of Representatives of ASME, IEEE, SAE, SNAME, AIAA. *Society Aff:* SNAME, RINA

Plumlee, Carl H
Home: 39 Lilac Lane, Carmel Valley, CA 93924
Position: Principal. *Employer:* Public Works Consultant. *Education:* DEng/Civil Engg/MT State; BS/Elec Engg/MT State; Grad/Strategy & Logistics/US Naval War College; Postgrad/Econ & Educ/MT State. *Born:* 6/14/09. Civil Engr with MT Power Co & Army Corps of Engrs 1933-39. Officer Navy Civil Engr Corps 1939-63, in pos of top direction & decision making for large naval tech & logistic orgs as Commanding Officer, Officer in Charge of Const, Opers Officer, Pub works Officer, retiring voluntarily 1963 to pursue cons engg career. Author many tech papers & reports in field of port planning & port mgmt. Fellow ASCE & ACEC & SAME. Mbr NSPE, IEEE, PIANC & other engg soc's. ASCE J James R Croes Medal 1967. Cons Eng 1964-87, ports development and operation. *Society Aff:* ACEC, ASCE, SAME, IEEE.

Plummer, James W
Business: 1111 Lockheed Way, Sunnyvale, CA 94086
Position: Exec VP *Employer:* Lockheed Missiles & Space Co, Inc *Education:* MS/EE/Univ of MD; BS/EE/Univ of CA-Berkeley *Born:* 01/29/20 Joined LMSC 1955. VP & AGM Space Sys Div 1965; VP & AGM R&D Div 1968; VP & GM SSD 1969. Under Secy Air Force 1973-76. LMSC Ex VP 1976. NAE, Chrmn NRC's Assembly of Engrg Space Application Bd 1980-; mbr NAE Peer Committee Aero/Astro 1979-; AIA Energy Task Force 1979-; Bd Mbr UC Berkeley Coll of Engrg 1981-; Defense Sys Mgmt Coll Bd of Visitors 1979-. USAF Exceptional Civilian Service Award 1976. DOD Distinguished Public Service Award with Palm 1976. Selected Eminent Engr 1981 by CA Alpha Chapt of Tau Beta Pi. Distinguished Engrg Achievements Award by Soc Mfg Engrs Region VII 1981. *Society Aff:* NAE, AIAA, AAS

Plummer, Ralph W
Business: Professor and Chairman of Industrial Engineering Dept, West Virginia University, PO Box 6101, Morgantown, WV 26505
Position: Prof and Chair *Employer:* West Virginia Univ *Education:* PhD/Eng/WV Univ; MS/IE/WV Univ; BS/IE/WV Univ *Born:* 03/30/42 Dr. Plummer has over 20 yrs experience in indus, res and teaching. His teaching has been centered in the areas of Human Factors, Safety and Industrial Hygiene. He is Coordinator of the Occupational Health and Safety Engrg Prog at WV Univ. Dr. Plummer's publications exceed forty and he has presented papers at various natl conferences. The topics have been concentrated in the areas of Human Factors, Safety and Industrial Hygiene. He has served as a conslt to indus, government and expert witnessing. His industrial experience was with IBM and with Corning Glass Works. Reg P E and mbr of the Natl Soc of P E's. *Society Aff:* NSPE, IIE, TBΠ, HFS, ΣΞ

Plumtree, William G
Home: 1414 Virginia Ave, Glendale, CA 91202
Position: Assoc Dean, Engg. *Employer:* CA State Univ. *Education:* MS/Appl Mechanics/Poly Inst of Brooklyn; MCE/Civ Engrg/NYU; BS/Civ Engg/Wayne State Univ. *Born:* 3/5/17. London Canada. Married: Elizabeth Simpson 3/29/47. Children: Wayne, Gary. 1939, surveyor, Claude Postiff; 1940-47 (supvr 1942-47) hydrographic engr, USN Hydrographic office; 1947-1950, instr, NYU; 1950-55, asst prof engr mech, NYU; 1955-58, asst prof, CA State LA. 1960-61, Acting Chrmn, Div of Engg; 1965-67, Asst Div Chrmn; 1962-70, Chrmn, Civ Engg Dept; 1970, Acting Assoc Dean, Engg 1979 Assoc Dean, Sch of Engg. Retired June, 1982. Co-author "Engineering Mechanics," 1954; "Shear Strength of Aluminum Honeycomb–, ASTM, Spec pub 201, 1956. *Society Aff:* ASCE, ASEE, SEAOC.

Plunkett, Joseph C
Business: Sch of Engg, Cedar at Shaw Ave, Fresno, CA 93740
Position: Prof of Electrical Engg *Employer:* CA State Univ. *Education:* PhD/Elec Engg/TX A&M Univ; MSEE/Elec Engg/GA Inst of Tech; BSEE/Elec Engg/Univ of TN; BS/Math & Phys/Middle TN State Univ. *Born:* 12/3/33. Prof Elec Engrg, CA State Univ, Fresno, CA. Harold and Lorraine Plunkett, Centerville, TN. Staff Engr, Martin Co, '66-'69. Dev Engr, Raytheon Co (Phased array radar) '69-'71. Res Engr, IIT Res Inst (RF Commun) '71-'72. Res Engr, TX A&M Univ, '74-'77. Chrmn, Elec Engrg Dept 1980-84, CSUF; Assoc Prof/Prof Elec Engrg, CSUF, 1977 to present. Published book contributioon and numerous articles in solid state device res. Present res interests are heterojunction bipolar transistors, ion implantation, optical and microwave commun. Mbr IEEE, Eta Kappa Nu, Sigma Xi, Reg Elec Engr, MA. Enjoy fishing, football, mtns. *Society Aff:* IEEE, ΣΞ, HKN, NSPE, CSPE, NYAS

Plunkett, Robert
Business: 107 AERO, Minneapolis, MN 55455
Position: Professor. *Employer:* Univ of Minnesota. *Education:* ScD/ME/MIT; BS/CE/MIT; D de l'U/-/Nantes. *Born:* Mar 1919. D de l'U(hc) Univ of Nantes. Native of NYC. Res Asst Elec Engrg MIT 1939-41; Army Corps Engrs 1942-46; Asst Prof ME MIT 1946-48, Rice Univ 1948-51; Cons Engr Acoustics & Appl Mech GE Co Schenectady 1951-60; Prof Aero Engrg & Mech Univ of Minn since 1960. V P 1970-74, Chmn Comm on Tech Affairs 1974-76 Board of Governors 1981-82 ASME. Mbr NAE; Fellow ASME, Acoust Soc of Amer AAAS; Assoc Fellow AIAA; Mbr SESA, ASEE. Extensive cons on sound & vibratin control, influence of matl properties on dynamic response, Composite materials, hydroacoustics. Enjoy sailing & crosscountry skiing. *Society Aff:* NAE, ASME, ASA, AAAS, AIAA, SESA, ASEE.

Pluntze, James C
Business: Mail Stop LD-11, Olympia, WA 98504
Position: Hd, Water Supply & Waste Sect. *Employer:* Dept of Social & Health Serv. *Education:* BS/CE/Univ of WA; MS/CE/Univ of WA. *Born:* Jul 27, 1931 Akron Ohio. US Army 1955-56. Dist Engr, Wash State Dept of Health 1959-66; assumed current pos, in charge of State of Wash drinking water prog 1966. Mbr, Natl Drinking Water Advisory Council 1984-86. Mbr AWWA, ASCE, WPCF, AAEE,. Married 4 children. Active in United Methodist Church, enjoy camping, music, photography, hiking & mountaineering. *Society Aff:* AWWA, ASCE, WPCF, AAEE.

Poage, Scott T
Home: 609 Edgewood, Waco, TX 76708
Position: President *Employer:* Poage Land and Cattle *Education:* PhD/Indust Engrg/OK State Univ; MS/Indust Engrg/TX A&M Univ; BS/Indust Engrg/TX Tech Univ *Born:* 12/05/31 Prod Control Engr Phillips Petroleum; Prod Control Officer USAF (Capt USAF Res Ret); Instr I E, Texas A&M; Asst Prof Okla State; Prof & Hd, IE Dept, Texas Univ at Arlington; Prof & Chmn Univ of Houston 1967-76; Natl V P IIE; IIE Fellow Award; author 'Quantative Mgmt Methods' & tech papers. Tau Beta Pi, Kappa Sigma, Ridgewood Country Club, Petro Club of Ft Worth, Episcopalian. Reg P E. *Society Aff:* IIE, ΚΣ, ASEE, ΑΠΜ; TBΠ.

Podolsky, Leon
Business: 77 Wendell Ave, Pittsfield, MA 01201
Position: Elec & Electronics Cons Engr. *Employer:* Self. *Education:* DSc/Engg/Drexel Univ; EE/Elec Engg/Drexel Univ. *Born:* Nov 3, 1910 Phila Pa. Educ Philadelphia Pub Schools; Temple Univ Med School, Res Fellow 1935-37; Drexel Univ DSc 1965; Alexander Hamilton Inst (Bus Admin) 1940. Pres U S Natl Ctte, Internatl Electrotech Comm; Internatl V P IEC. Chief U S Delegate IEC meetings 26 countries 1952-80. Asst to Pres, Cons, Sprague Elec Co 1938-81. Cons Depts Commerce & Defense, U S Govt. Cons Natl Bureau of Stds, Mbr Adv Panel 1965-71. Mbr Internatl Stds Ctte, Engr Exec Ctte Electronic Indus Assn. Fellow IEEE; Fellow Stds Engrs Soc; Mbr AAAS, Mbr Bd of Dir, ANSI, 1977-79; VChrmn, Intl Stds Council ANSI Univ Club of N Y. & Washington, D.C. Reg P E Mass & N Y. Num publs,

Podolsky, Leon (Continued)
134 inventions. Num hons & medals from govt, prof soc's, & trade assns. *Society Aff:* IEEE, SES, NSPE, MSPE, AAAS.

Poe, Herbert V
Home: 119 Ft Rutledge Rd, Clemson, SC 29631
Position: Assoc Prof *Employer:* Clemson Univ *Education:* MS/EE/A & M Coll of TX; B/EE/NC State Univ-Raleigh *Born:* 09/02/24 Instructor-NC State Coll 1944-45. Transformer Design Engr-Westinghouse 1945-47. Asst Prof Clemson Univ 1947-52. Assoc Prof Clemson Univ 1952 to date. Reg in SC. Conslt to many Elec Utilities and A & E firms summers and other. Edited and graded Natl Exams for NCEE. Elec Conslt to many Law Firms and insurance companies. Held following AIEE (pre IEEE) jobs: State Section Chrmn, District Chrmn of Student activities, Natl Student Branches Committee, Branch Counselor. Prepare and present Annual Continuing Educ Seminars "Design of Industrial Power Sys" and "Motor Applications and Variable Speed Drives." Conduct annual review courses for preparation for PE Exam. *Society Aff:* IEEE

Poehlein, Gary W
Business: Georgia Institute of Technology, Centennial Research Building, 400 Tenth St. NW, Atlanta, GA 30332-0370
Position: Dir/Sch of ChE. *Employer:* GA Inst of Tech. *Education:* BS/ChE/Purdue Univ; MS/ChE/Purdue Univ; PhD/ChE/Purdue Univ. *Born:* 10/17/36. Gary W Poehlein completed his education through high sch in Tell City, IN. His parents are Eva (Dickman) Poehlein & the late Oscar Poehlein. Upon grad from Purdue Univ in 1958, he worked as a design engr at P&G until returning to Purdue in 1961. After finishing grad work he accepted a faculty position at Lehigh Univ in the Fall of 1965. At Lehigh he moved through faculty ranks, achieveing the position of Prof of Chem Engg in 1976. He was also Co-Dir of the Emulsion Polymers Inst at Lehigh. Dr Poehlein Assumed the position of Dir of the Sch of Chem Engg at GA Tech in July, 1978. *Society Aff:* AIChE, ASEE, ACS, SR.

Poettmann, Fred H
Home: 47 Eagle Dr, Littleton, CO 80123
Position: Prof Petroleum Engrg *Employer:* CO Sch of Mines *Education:* PhD/ChE/Univ of MI; MS/ChE/Univ of MI; BS/ChE/Case Inst of Tech. *Born:* 12/20/19. Worked for Lubrizol Corp as a Res Chemist after graduating from Case. After grad from MI, worked for Phillips Petroleum Co as Mgr of Production Res, 1945. Taught grad ext courses in natural gas engg, thermodynamics, heat transfer and fluid dynamics for OK State Univ. Assoc Res Dir for Production for Marathon Oil Co, 1955-1971. Mgr of the Commercial Dev Dept. 1971-1983 Prof Petroleum Eng 1983-present, CO Sch of Mines. Authored or co-authored 45 technical publications, eight books & forty-six US & foreign patents. Recipient of the Lester C Uren Award (1966); the John Franklin Carll Award (1971) of the Soc of Petrol Engrs; and the Univ of MI Sesquicentennial Award of the College of Engg as Outstanding Alumnus (1967); Mbr Natl Acad of Eng (1978); Fellow AIChE (1974); Distinguished Mbr SPE (1983); Designated "EOR Pioneer" Soc of Petrol Engrs (1984); Honorary Member SPE (1985); Honorary Mbr AIME (1984); Halliburton Teaching Award (1986). *Society Aff:* SPE, AIChE, ACS.

Poggemeyer, Lester H
Business: 121 East Wooster St, Bowling Green, OH 43402
Position: Pres & Owner. *Employer:* Lester H Poggemeyer, PE, Inc. *Education:* BSCE/Civil Engg/Univ Toledo; Dip/Coll Prep/Troy-Luckey HS. *Born:* 12/24/40. Reg includes OH, MI, ID, KY, TN. Presently is Pres of OH Assoc of Consulting Engg and help found the Toledo Reg Assoc of Consulting Enggs. Manages a firm specializing in environmental and civil engg, arch, and surveying projs, and has dev numerous public improvements for various agencies involving local, stare, and federal funding. Active in numerous civic and prof activities, and pres is trustee of the Chamber of Comm, VP of the County Council on alcoholism, PPres of the school band boosters organ, serves as dir on a non-profit retirement and nursing home facility for the elderly. *Society Aff:* ACEC; ASCE; AWWA; ISPS; MES; NSPE; OACE; OSPE; PEPP; PLSO; TACE; TSPE; WPCF.

Polack, Joseph A
Business: Chem Engrg Dept, Baton Rouge, LA 70803
Position: Prof & Dir, Audubon Sugar Inst. *Employer:* Louisiana State Univ. *Education:* ScD/Chem Eng/MIT; MS/ChE Practice/MIT; BE/Chem Engr/Tulane *Born:* Sept 1920 New Orleans La. Employed at Chem Warfare Lab MIT during WW II, & served a yr as instr in Chem Engrg at MIT. With Exxon Corp for 22 yrs in various tech & R&D admin posts assoc with process R&D. Dir Esso (now Exxon) Res Labs 1966-70. Served as Hd of Chem Engrg at LSU 1970-76. Currently engaged in sugar process res. Dir, Audubon Sugar Inst, LSU. Addl teaching & res interests in mgmt (esp for engrs), org dynamics, & human relations in industry. *Society Aff:* AIChE (Fellow)

Polak, Elijah
Business: , Berkeley, CA 94720
Position: Prof. *Employer:* Univ of CA. *Education:* PhD/EE/Univ of CA; MS/EE/Univ of CA; BEE/EE/Univ of Melbourne. *Born:* 8/11/31. Native of Poland. After obtaining BEE worked as instrument engr in Australia. Came to US in 1957 & has been on staff of Univ of CA, Berkeley since 1958. Author of Computational Methods in Optimization (1971); co-author of Theory of Math Programming & Optimal Control (1970), System Theory (1970) & Notes for a First Course on Linear Systems (1969). Guggenheim fellow 1968, UK Sci Res Council Fellow 1972, 1976, 1979. Fellow IEEE. *Society Aff:* IEEE, SIMA, MP.

Polaner, Jerome L
Home: 30 Tiffany Dr, Livingston, NJ 07039
Position: Prof Emeritus, Mech Engg. *Employer:* NJ Inst of Tech. *Education:* MS/ME/Stevens Inst of Tech; BS/ME/Newark College of Engg. *Born:* 9/13/15. Served on faculty of Newark College of Engg/NJ Inst of Tech 1938-1983. Presently, Prof Emer & Conslt to Dept of Mech Engg conslt & instal of Labs and Res support facilities. Served with US Navy WWII. Visiting Lectr, Ext Div, Rutgers Univ 1951-1966. Received NSF Fellowship summer 1964, 1966 Univ of AZ. NASA Res Fellowship summer 1969, 1970. Jet Propulsion Lab/CA Inst of Tech. Consulting Engr specializing in Ind Safety, Product design, patent litigation, system testing and evaluation. Hobbies: travel, photography & skiing. *Society Aff:* ASME, AAAS, NYAS, NSPE, NJSPE, ASEE, AAUP.

Poli, Corrado
Business: Mech Engg Dept, Amherst, MA 01003
Position: Prof *Employer:* Univ of MA. *Education:* PhD/Engrg Mech/OH State; MAE/Aero Engrg/Rensselaer Poly Inst; BAE/Aero Engrg/Rensselaer Poly Inst. *Born:* 8/9/35. Troy, NY. Mbr of the AF Res & Dev Command, Wright-Patterson AFB, OH 1958- 1965. Taught AF Inst of Tech 1965-1967. With the Univ of MA since 1967, Hd 1978- 1981, assuming present position 1981. Originally an aerospace engg, became interested in mfg engg around 1970 and devotes most of his res efforts to the area of design for economical mfg. *Society Aff:* SME, ASME, SPE.

Polivka, Milos
Business: 533 Davis Hall, Berkeley, CA 94720
Position: Prof of Civil Engrg. *Employer:* Univ of Calif Berkeley. *Education:* MS/Civil Eng/Univ of CA, Berkeley; BS/Civil Eng/Univ of CA, Berkeley. *Born:* Dec 1917. Native of Czechoslovakia. Fac Mbr Univ of Calif Berkeley since 1948. Recent res in areas of expansive cement concrete, mass conrete, concrete for nuclear reactors, shrinkage & durability of concrete, & effect of elevated and aryogenic temperatures on properties of concrete. Fellow ACI & ASCE. Mbr ASTM, U S Ctte of the Internatl Comm on Large Dams, Sigma Xi, Tau Beta Pi, & Chi Epsilon. ACI Arthur R Anderson Medal 1975. Reg C E & Reg Qual Engr Calif. *Society Aff:* ASCE, ACI, ASTM, USCOLD.

Polk, Charles
Home: 21 Spring Hill Rd, Kingston, RI 02881
Position: Prof of Elec Engg. *Employer:* Univ of RI. *Education:* PhD/EE/Univ of PA; MS/Phys/Univ of PA; BS/EE/WA Univ; Diploma/French literature/Univ of Paris. *Born:* 1/15/20. Native of Vienna, Austria. Engr at RCA Victor and Mbr of Tech Staff, RCA Labs, Princeton (1948-1959). Taught at Univ of PA, Drexel Univ and Univ of RI (At URI since 1959 chairman E. E. deptm. 59-79). Visiting Prof, Stanford Univ (1968/69), Univ of Wisconsin/Madison (1983/84). Vchrmn (1963/64) and chrmn (1964/65) Providence, RI Sec, IEEE. Mbr, Bd of Dirs of NEREM (Boston, 1969-72), Radio Communications Committee, IEEE (1964-), Cttee on Man and Radiation, IEEE (1986-) Commissions on Fields/Waves, Radio Interference and Ionospheric Radio of URSI (1976-). Hd, Elec Sciences and Analysis Sec and then Acting Dir, Engg Div of the Natl Sci Fdn (1975-1977); recipient of NSF Superior Accomplishment Award, 1977. Publications on Antennas, Radio Propagation, E M Noise Biological effects of e.m. fields in IRE Proc, IEEE Trans, Radio Sci, Jl of Atm and Terr Phys, CRC Press Handbooks.. *Society Aff:* AAAS, BEMS, IEEE, AGU, URSI, NYAS, ΣΞ, TBΠ, ASEE

Polkinghorn, Frank A
Home: 3426 P Bahia Blanca W, Laguna Hills, CA 92653
Position: Mbr of Tech Staff - ret. *Employer:* Bell Tele Labs. *Education:* BS/EE/Univ of CA, Berkeley. *Born:* 7/23/97. Born in Holbrook, MA. Mbr Eta Kappa Nu, Tau Beta Pi, Sigma Xi, Phi Beta Kappa. Employed Radio Lab Mare Island Navy Yd 1922-24; Ch Engr A-P Radio Labs, SF 1924-25; Pacific Tel & Tel S F 1925-27; Bell Tel Labs NY & NJ 1927-62; cons 1962-67. Supervised design of radio telephone equip for transoceanic communication. After 1940 designed military equip & conducted sys analysis incl Mercury Satellite tests. Lic Engr NY. Loaned 1948-50 to SCAP supervising all civilian comm R&D & mfg in Japan. Dir IRE 1957-58; Sect Awards 1961, 1969, 1975. Fellow IEEE, AAAS. Hon Mbr Inst of Electronics and Communications Engrs of Japan. Author 25 tech papers. Hobbies: Genealogy, lapidary. *Society Aff:* IEEE, AAAS.

Poll, Harry F
Business: 12354 Lakeland Rd, Santa Fe Springs, CA 90670
Position: SR. Vice Pres - MFG. *Employer:* Powerine Oil Co. *Education:* BSChE/ChE/Purdue Univ. *Born:* 1/24/20. in Chicago, IL. Postgrad Lehigh Univ 1941, Notre Dame 1943, Caltech 1955, UCLA 1957. Operations Supr Trojan Powder Co Allentown PA & Sandusky OH 1941-42; Process Engg Supr Union Oil CO of CA 1946-63; VP & Genl Mglr Mfg Powerine Oil Co 1963- . US Navy WWII 1943-46, Capt USNR; NAtl Petro Ref Assn (Bd of Dir & V P); Amer Petro Inst (Genl Refg Comm); Pacific Energy Assoc (Bd of Dir); AIChE; num pats & pubs in petro refining tech; listed in Leaders in Amer Sci & Who's Who in Finance & Indsu & Who' Who in the World. *Society Aff:* NPRA, API, PEA, AIChE, AAAS.

Pollack, Herbert W
Home: 25 Suzanne Rd, Lexington, MA 02173
Position: Chrmn & Pres *Employer:* Parlex Corp *Education:* M/EE/NYU; B/EE/City Coll of NY *Born:* 03/27/27 Founded Parlex Corp in 1970 after being with Sanders Assocs from 1965-1970 as a Group Div Mgr. From 1955-1965 with Polarad Electronics Corp; Joined in an Engrg capacity and became VP and a Dir of the co and Pres of the Electronic Instrument Div. Prior to 1955 was in various engrg and research positions with CBS-Columbia, NY Univ, and A.B. Dumont Co. Has also been on the Graduate Sch Staff of the Polytechnic Inst of Brooklyn. Fellow, IEEE for contributions to flexible printed circuits and their application; Pres, IPC, 1984; Tau Beta Pi and Eta Kappa Nu. *Society Aff:* IEEE, IPC, TBΠ, HKN

Pollack, Louis
Business: 15321 Delphinium La, Rockville, MD 20853
Position: Owner *Employer:* Louis Pollack Assoc. *Education:* BSEE/Elect Eng/College of the City of New York *Born:* Nov 4, 1920 NYC. Principal Engrg Aide US Army Signal Corps 1941-43. With ITT Fed Labs 1943-67; last pos Dir Transmission Sys Oper with Communications Satellite Corp 1967-1984. Last position, Vice Pres World System Div formerly Exec Dir - , COMSAT Labs, respon for tech mgmt of res, dev & proj work. Presently, owner Louis Pollack Assoc, consultants for satellite system design. Fellow IEEE, P Chmn & Mbr Space Communications Ctte; Technical Adv Council, Telecomm Policy Comm. Awards Board, Comm Soc; Assoc Fellow AIAA; member Study Group IV (Space Sys) ITU 1962-1980. Registered Prof. Eng. Member Sigma Xi and NSPE. Issued 2 patents. *Society Aff:* IEEE, AIAA, NSPE, ΣΞ.

Pollack, Solomon R
Business: 220 South 33rd St, 119 Towne Bldg, Philadelphia, PA 19104
Position: Prof Bioengrg *Employer:* Sch of Engrg & Applied Sci, Univ of PA *Education:* PhD/Physics/Univ of PA; MS/Physics/Univ of PA; AB/Physics/Univ of PA *Born:* 5/7/34 After 4 yrs with Univac Division of Sperry Rand, Dr. Pollack returned to the faculty in the Dept of Materials Sci and Engrg as an Asst Prof. IIn 1977 Dr Pollack was promoted to Full Prof and became chrmn of the Dept of Bioengrg. From 1981 to 1987 he served as Assoc Dean for Graduate Educ and Res. Dr Pollack also serves as Pres of te CARA Corp., Pres of the Society for Biomaterials 1981-82, Pres of BRAGS 1984-85, and has served on the Council of a number of Professional Societies. Dr Pollack's area of res is in Bioelectric Effects in Bone. *Society Aff:* SOCIETY FOR BIOMATERIALS, ORS, BRAGS, ΣΞ

Pollard, Ernest I
Home: Hilton Rd, RD8, Greensburg, PA 15601
Position: Cons Engr. *Employer:* Self. *Education:* MSc/EE/Univ of Pittsburgh; BSc/EE/Univ of NB. *Born:* Nov 11, 1906 Nehawka Nebr. Reg P E PA 1967. Fellow IEEE 1952. Cert of Commendation US Navy 'for outstanding serv to the US Navy during WW II' 1947. Designer of large synchronous machines for Westinghouse 1929-45 & for Elliott Co 1945-56. Mgr Engg, Ridgway Div of Elliott Co 1946-49. Ch Elec Engr Elliott Co 1962 until retirement 1971. Author AIEE & ASME Transaction papers. Sev U S pats. Chmn AIEE Power Div 1957-59. Presently self-employed cons in the felds of motor & generator engrg, & torsional vibration analysis & measurement. Fellow A.S.M.E. 1981. *Society Aff:* IEEE, ASME.

Pollard, William S, Jr
Business: 1100 Fourteenth St, Denver, CO 80202
Position: Prof of Cvl & Urban Engrg *Employer:* Univ of Colorado-Denver *Education:* MSCE/Trans/Purdue Univ; BS/CE/Purdue Univ *Born:* 01/01/25 Native of Oak Grove, LA. Served in US Marine Corps Engrs 1942-1946. Taught Cvl Engrg at Purdue Univ 1948-1950 and at Univ of IL 1950-1955. With Harland Bartholomew and Assocs 1955-1971 as Chief Engr, Assoc Partner, Partner, then Pres. Pres Wm. S. Pollard Conslts, Inc 1971-1981. In 1970 received ASCE Cvl Engrg State of the Art Award. In 1969 received Purdue Univ Distinguished Engrg Alumnus Award. Adjunct prof of Urban and Regional Planning, Memphis State Univ 1974-1981. Currently Prof of Cvl Engrg, Dir of Center for Urban Trans Studies, Univ of CO at Denver and conslt in transportation engrg and planning *Society Aff:* ASCE, NSPE, SAME, ITE.

Pollock, Stephen M
Business: 10E Building 1205 Beal, Ann Arbor, MI 48109
Position: Professor *Employer:* Univ. of Michigan *Education:* Ph.D./Physics/Op. Res/M.I.T.; S.M./Physics/M.I.T. B. Eng. Phys/Cornell University *Born:* 02/15/36 Consultant at Arthur D. Little 1964-65; faculty, Naval Postgraduate School (search theory, detection, Military O.R.); since 1969 faculty, Dept of Indust and Operations Engrg, Univ. of Mich (Chairman since 1981). Teaching and research in decision analysis, applied stochastic processes, maintenance and reliability modeling. Consultant to industry, government. Past Pres, ORSA. Member, advisory bds, panels, at NSF, NRC, NIJ. *Society Aff:* ORSA, TIMS, IIE

Pollock, Warren I
Business: Louviers Bldg, Engg Dept, Louviers Bldg, Wilmington, DE 19898
Position: Mtls Engg. *Employer:* E I du Pont de Nemours & Co. *Education:* PhD/Mtls Engg/Carnegie-Mellon Univ; MS/Mtls Engg/Carnegie-Mellon Univ; BS/Chemistry/Syracuse Univ. *Born:* 5/31/30. Joined E I du Pont de Nemours & Co, Inc, Wilmington, DE, in 1955. Coordinates Du Pont Co's Mtls Engg Progs, conducted by the Engg Dept, on identification & dev of new tech for mtls selection, fabrication, operation, maintenance & repair of process equip & piping. Studies include corrosion resistant metals, fiber-reinforced plastics, linings, thermal insulation, refractories, protective coatings & welding. PastChrmn, Technical Advisory Council, The Mtls Tech Inst of the Chem Process Industries, Inc, Columbus, OH (a not-for-profit, co mbrship organization founded in 1977). Mbr of ASTM A-1, A01. 09, A01. 10, B-2 & B02.07. Active in ASM & NACE. *Society Aff:* ASM, ASTM, NACE.

Pollock, Wilfred A
Home: 2731 SW Drive, Apt 242, Abilene, TX 79605
Position: Consultant. *Employer:* Tippett & Gee Inc - Cons Engrs. *Education:* BSME/Mech/Univ of WI. *Born:* Sept 1909. Native Whitewater Wisc. Employed by Wisc Elec Power 1935-74 in areas of admin, power plant design, oper & testing. Retired as Asst VP. Currently designing fuel & ash sys for Texas lignite and coal fired plants. Presented papers on fuels, welding & pollution ontrol to AWS, ASME, APCA & APC. Mbr Oper & Econ Cttes Enrico Fermi Breeder Proj. Ch Milw Cty Air Pollution Control Bd 1948-75, mbr State of Wisc Air Pollution Adv Council 1968-74. Mbr ASME PTC Standing Ctte. Fellow ASME; Mbr APCA, ISA & ESM. Patent fluegas scrubber. Hobbies: photography, gardening & travel. *Society Aff:* ASME, APCA.

Polomsky, John V
Business: Dept of Met, Mech & Mtls Sci, E Lansing, MI 48824
Position: Asst Prof. *Employer:* MI State Univ. *Education:* PhD/Ind Education/MI State Univ; MA/Educational Admin/MI State Univ; BS/Ind Education/MI State Univ. *Born:* 8/12/30. A diversified background of education, military service (Korea), inventing, entrepreneurship, industry, profl sports, public office, consulting and natl soc office, all contribute to motivated orderly plans of action utilized to achieve goals and objectives developed in five yr increments throughout my life. Currently, VP for Profl Dev of SAVE, teaching the nations first accredited undergrad Value Engg class, consulting for several cos and the NAPM are prime functions. Writing a text, coaching and developing a curriculum in mfg are secondary time consumers. Married with five children (identical twin boys) running a resort and coaching, round out my time schedule. *Society Aff:* SAVE, ASEE.

Polve, James H
Business: 242K CB, Provo, UT 84602
Position: Prof, Mech Engrg *Employer:* Brigham Young Univ *Education:* PhD/Aerospace Engr/Univ of AZ; MS/Aeronautica Engr/Princeton Univ; ME/ME/Univ of UT; BS/ME/Univ of UT *Born:* 02/07/21 Native of UT. USAF and US Army 1942-1969. Retired Colonel USAF. Command Pilot, Test Pilot. Taught at USAF Experimental Flight Test Pilot Sch 1951-1955. Taught at USAF Acad 1955-1960 (Mathematics, Mech, Aeronautics). Dir Flight Test Engrg Air Force Flight Test Center, Edwards Air Force Base 1963-1966. Chief, Seattle Office of Supersonic Transport Development for Federal Aviation Admin 1966-1969. Present position since 1969 as Prof, Mech Engrg, Brigham Young Univ. *Society Aff:* AIAA, ASME, ASEE

Pomerantz, Herbert
Business: 35 W 35th St, New York, NY 10001
Position: President. *Employer:* Herbert Pomerantz, PE, PC. *Born:* Jan 1949. BME City Coll of New York. Married Sept 1950; 3 children. Sr Engr Voorhees, Walker, Smith & Haines NYC 1951-60; Partner Brown & Pomerantz NYC. 1965 Pres of Levittown Democratic Club. Deputy Commissioner of Pub Works Nassau Cty N Y 1965-67. Pres Herbert Pomerantz, PE, PC since 1967. Served with AUS 1946. Reg P E NY 1954- ; Reg Pa, Conn, RI, Tenn, NJ, Ohio, Fla. Mbr Soc of Prof Engrs, ASHRAE, N Y Bldg Congress, Amer Soc of Military Engrs. Home: Nassau Cty N Y.

Pomerene, James H
Business: IBM Corp, Thomas J. Watson Res Center, PO Box 704, Yorktown Heights, NY 10598
Position: IBM Fellow. *Employer:* IBM Corp. *Education:* BS/Elec Engg/Northwestern Univ. *Born:* June 1920 Yonkers N Y. BSEE Northwestern Univ 1942. Radar design engr Hazeltine Corp 1942-46. Mbr Electronic Computer Proj, Inst for Advanced Study 1946-51, Ch Engr 1951-56. Sr Engr IBM Corp 1956-67; Sr Staff Mbr IBM Armonk 1967-76; IBM Fellow 1976. Cons DOD. IEEE Press Editorial Bd. Fellow IEEE. Involved with Large scale computers, parallel processing, high availability techs. *Society Aff:* IEEE, ΣΞ, AAAS.

Pomeroy, Richard D
Home: 280 Malcolm Dr, Pasadena, CA 91105
Position: Consulting Engr *Employer:* James M Montgomery, Inc, Consulting Engrs *Born:* Dec 22, 1904. BS Caltech 1926, PhD 1931. Employed by Los Angeles Cty Sanitation Dists 1932-40. Founder and Pres of Pomeroy, Johnston & Bailey, until 1973. In 1978 he joined the firm of James M Montgomery, Consulting Engrs. Num publ papers, esp on the subjects of control of sulfide in sewerage sys, design & oper of sludge digestion tanks, design of ocean outfalls, hydraulics of sewers, corrosion of metals by water & lab methods. Harrison Prescott Eddy Award of WPCF 1946, Rudolf Herring Medal of ASCE 1972, Hon Life Mbr, WPCF, 1981. *Society Aff:* ASCE, ACS, WPCF, AWWA, IAWPR.

Pomraning, Gerald C
Business: Engg & Applied Science, 6266 Boelter Hall, Los Angeles, CA 90024
Position: Prof. *Employer:* UCLA. *Education:* PhD/Nucl Eng/MIT; Cert/Chem Eng/Delft Tech Inst; BS/Chem Eng/U of WI *Born:* Feb 1936. Ph.D. Nuclear Engrg. MIT. Engrg res/mgmt pos with G E, Genl Atomic, Sci Applications Inc. With Sci Applications since 1969; served as V P Oper & Admin. Mbr AAAS (Fellow), AMA, ANS (Fellow; Mark Mills Award), APS, (Fellow) MAA, N Y Acad Sci, Soc Engrg Sci, SIAM; 150 res papers in 22 prof journals; author book 'Radiation Hydrodynamics'; areas of expertise incl: reactor phys & engrg, transport theory, applied math, radiative transfer. Reg Prof Nuclear Engr; Prof Engrg & Appli Sci at UCLA since Fall 1976, Vice-Chairman; Mechanical, Aerospace, and Nuclear Engrg. Dept. *Society Aff:* AAAS, ANS, APS, MAA, SIAM, AMA.

Ponder, Thomas C
Business: POBox 2608, Houston, TX 77001
Position: Petrochem Editor. *Employer:* 'Hydrocarbon Processing'. *Education:* BS/ChE/U of AR *Born:* Dec 1921 Jackson County AK. USAAF 1944-46. Process Engr Pittsbgh Plate Glass Co 1948-51; Proj Engr Foster Wheeler Corp 195154; Proj Engr The Lummus Co 1954-55; Petrochemicals Editor 'Hydrocarbon Processing' a tech Journal 1955- . Mbr AIChE, ACS, AACE (Natl Secy, Natl Pres 1965). Award of Merit 1971. Guest of var foreign govmts (Brazil, Kuwait, Iraq, Abu Dhabi, Sweden, Romania, Hungary, East-Germany) to var petrochem seminars. Have led US Dept Commerce East-West Trade Sales Seminars to var Eastern Block Nations. Mbr Cert Bd/AACE. Fellow-Amer Assoc of Cost Engrs; Cert Cost Engr; Reg Prof Engr, State of TX. *Society Aff:* AIchE, ACS, AACE

Ponter, Anthony B
Business: Cleveland State University, 1983 East 24th St, MC 219, Cleveland, OH 44115
Position: Dean of Engrg *Employer:* Cleveland State Univ *Education:* D.Sc./CE/Birmingham Univ (UK); PhD/CE/Manchester Univ (UK); MSc (Tech)/CE/Univ of Manchester Inst of Sci & Tech (UK); BSc/CE/Birmingham Univ (UK) *Born:* 09/27/33 Previously Prof and Hd, Dept of Chemistry and Chem Engrg MI Tech Univ; Prof and Dir, Chem Engrg Inst Swiss Fed Inst of Tech Lausanne; Prof

Ponter, Anthony B (Continued)
Univ of New Brunswick Canada; Univ and Coll Lecturer Manchester Univ; Tech Officer and Plant Mgr Imperial Chem Industries. Trained as a chemist, chem engr and mech engr. Fellow: Inst of Chem Engrs (UK) Fellow: Chem Inst of Canada. Fellow, American Inst of Chem Engrs, Member, Canadian Soc of Chem Engrs, Chartered Engr (Counc of Engrg Insts); Pub 120 refereed papers in the areas of diffusion, surface phenomena, distillation, nucleate boiling and acoustics. Conslt to a number of industrial and government agencies. Outstanding Enginrg Educator of the Year (1985) award by Ohio Soc of Prof Engrs. *Society Aff:* AIChE, IChemE, CIC

Pool, Monte J
Business: Dept Mat Sci, Cincinnati, OH 45221
Position: Prof Met Engrg. *Employer:* U of Cincinnati. *Education:* PhD/Met/OH State Univ; MS/Met/OH State Univ; MetE/MetE/Univ of Cincinnati. *Born:* Nov 1934 Toledo OH. U of Denver & Denver Res Inst as Asst & Assoc Prof & Res Metallurgist. Since 1968 with U of Cincinnati as Assoc Prof & Prof Met Engrg. Prof-inCharge of Dept 1973-75. Chmn Pyromet Ctte AIME 1970-71; Chmn ASM Met Transactions Ctte 1975; Visit US Sci Max-Planck Inst Stuttgart-Germany Sept 1976-77. Wandmacher Excellence in Teaching Award, 1984 and 1987. Amer Soc for Metals Eisenmamm Award (Cinn Chap) 1984. *Society Aff:* ASM, AIME.

Poole, H Gordon
Home: 1812 W 12th Ave, Albany, OR 97321
Position: Retired. *Employer:* *. *Education:* -/Geology/Univ of WA; -/Metallurgy/MA Inst Tech; MS/Metallurgy/Univ of ID; BS/Min Engg/Case Inst of Tech. *Born:* 11/28/09. Grad Study U of Washington, MIT (Met) Internatl Minning & Met prior to 1941. Met US B of Mines 1941-46; Res Dir Albany Met Res Ctr 1966-71. Assoc Prof U of Wash, Case Inst 1947-54; Prof, Hd Dept Met Engrg Colorado Sch of Mines 1954-62. Tech Dir, V P Oregon Met Corp 1962-66. Var Cons priv indus & govmt agencies. Reg Prof Engr Min & Met Engrg in Utah & Wash. Fellow ASM; Legion of Honor 50 yr member AIME; Life Mbr ADPA. Enjoy Tucson AZ & Old Mexico. *Society Aff:* ASM, AIME, ADPA.

Poole, Wiley D
Home: 703 Coventry Dr, Baton Rouge, LA 70808
Position: Professor Engineering & Consultant. *Employer:* Louisana State University. *Education:* MS/Engg/LA State Univ; BS/Agri Engg/LA State Univ. *Born:* 10/5/13. Army courses MIT 1943-44. Asst Prof LSU prior to WWII: 1942-45 Mech Engr Army Air Corps; 1945-51 Assoc Prof Engrg LSU 1951- , Prof Engg LSU. Res & teaching in machinery design, dev & struc. Reg PE: mech, agri, cons & expert witness on machinery & safety for prod liability cases. Awarded 3 patents on machinery. Awarded LES Tech Accomplishment Medal 1956. Mbr ASAE, Chmn SW region 1974; Mbr LA Engrg Soc; Honor Soc Tau Beta Phi, Gamma Sigma Delta, Sigma Xi. Author &/or co-author of 47 pubs in field of agri engg. Jan 1977 Retired Prof Emeritus Engg LA State Univ. Registered Professional Eng, LA, Mechs & Agri. Pres of A&M Consultants Inc, Baton Rouge, LA. *Society Aff:* ASAE, ASME, ASSE, NSPE, LES.

Poole, William S
Business: 3811 Rawlins, Dallas, TX 75219
Position: President & Chief Engineer. *Employer:* Environmental Systems Inc. Cedar Creek Consultants/NC. *Education:* BS/Elec Engg/Univ of MN. *Born:* 11/2/23. Reg PE in TX & LA. Reg in 10 states (also with NCEE-Natl Council Engg Examiners), TX, LA, CO, MO, CA, HI, NV, NM, WA, MS. 1946-54 VP Poole Instruments Inc. Dallas Texas; 1954-69 North Amer Rockwell Corp, in charge of 35 test engineers & Propulsion sys Dev Facility; 1969-73 Ch Engr, Zetterlund Boynton Co & Assocs; 1973-75 V Pres & Ch Elec Engg Flower & Assoc; 1975-Now Environmental Sys Inc cons engrs; In addition 1976 formed William Sherwood Poole PE. Now Cedar Creek Consultants Inc. Sustaining Mbr & Pres N TX Sect IES; 1973-76 Bd/Dir Greater Dalles Electrical League; Mbr AIPE. Married 43 yrs, grown children, 5 grandchildren. *Society Aff:* IES, AIPE.

Poor, Vincent H
Business: 1101 West Springfield Avenue, Urbana, IL 61801
Position: Prof *Employer:* Univ of Ill *Education:* Ph.D./EECS/Princeton University; M.A./EE/Princeton University; M.S./EE/Auburn University; B.E.E./E.E./Auburn University *Born:* 10/02/51 Born in Columbus, GA, 1951. Educated at Auburn Univ and Princeton Univ (Ph.D., EECS '77). With the Univ of Illinois since 1977. Professor, Electrical & Computer Engrg, and Research Prof, Coordinated Science Lab, since 1984. Visiting appointments: London Univ, 1985; Univ of Newcastle, Australia, 1987. Principal research interests: statistical signal processing, multiuser digital communications. Author numerous publications including graduate text, An Introduction to Signal Detection and Estimation. Consultant to industries, government, and universities. Confounder and principal, SIGCOM, Inc. IEEE Fellow; Associate Editor, IEEE Trans. Automatic Control (1981-1982), IEEE Trans. Information Theory (1982-1985); Member, IEEE Information Theory Group BOG (1984-); Program Chair, 1986 IEEE Conf. Decision and Control; General Chair, 1989 Am. Control Conf. *Society Aff:* IEEE, URSI

Pope, David P
Business: Univ. of Pennsylvania, Philadelphia, PA 19104-6391
Position: Prof and Assoc Dean, Sch of Engrg and Applied Sci *Employer:* Univ of PA *Education:* PhD/Mat Sci/CA Inst of Tech; MS/Mat Sci/CA Inst of Tech; BSc/Eng Sci/Univ of WI *Born:* 7/31/39 Native of Waukesha, WI. After graduate school I was a Post-doctoral Fellow at CA Inst of Tech until 1968. I have been on the faculty of the Univ of PA since 1968, and Assoc Dean since July 1984. *Society Aff:* AIME, ASM, AAUP, ASEE, AAAS

Pope, James H
Business: 77 Beale St, Rm 2892, San Francisco, CA 94106
Position: Sr Civ Engr. *Employer:* Pacific Gas & Elec Co. *Education:* MS/Construction Mgr/Stanford Univ; BS/Civ Engg/OR State Univ. *Born:* 6/3/45. Native of CA. Currently, Sr Civ Engr in the Gas System Design Dept of Pacific Gas & Elec Co in San Francisco. Has had varying levels of responsibility ith PG&E & been involved in the construction of Diablo Canyon Nuclear Power Plant & the design of Geothermal Power Plants at the Geysers. Served as Pres of the San Luis Obispo Branch, LA Sec & Pres of SF Sec's Assoc Mbr Forum. In 1977, named "Outstanding Young Civil Engineer" in Pacific SW Council & in 1978, received the Edmund Friedman Young Engr Award for Profl Achievement. *Society Aff:* ASCE, ASNDT, PCGA, PCEA.

Pope, Malcolm H
Business: Ortho/Rehab, Burlington, VT 05405
Position: Prof. *Employer:* Univ of VT. *Education:* PhD/Mech Engg/Univ of VT; MS/Mech Engg/Univ of Bridgeport; HND/Mech Engg/Southall College. *Born:* 2/11/41. Jointly tenured prof of Mech Engg and Orthopaedic Surgery. Reg (chartered) engr in Great Britain. Res interests are in the biomechanics of the musculoskeletal system in gen and in the lumbar spine and in ski injuries in particular. Was recipient of the ski injury res prize given by the USSA in 1974 and 1977. Author of more than 100 res papers and is the founder and exec mbr of the New England (now Northeast) Bioengg Conference. *Society Aff:* ASME, IME, ASTM, ASB, ORS, ILSS, ISSS, TBΠ, ΣΞ, VITA.

Pope, Michael
Business: 26 Broadway, New York, NY 10004
Position: Chief Executive Officer. *Employer:* Pope, Evans & Robbins Inc. *Education:* BEE/EE/CCNY. *Born:* 5/16/24. Reg PE several states. 1944-46 US Maritime Service, licensed marine engr; with present & predecessor firm since 1946, Ch Exec Officer since 1969. Former Mbr GSA Design Standards Ctte & NY St Energy Conservation Stte. Pub on combustion, energy conservation & power plant oper. Pioneered dev of fluidized combustion (coal) tech. Distinguished Engineer Awards, ACEC Award of Honor 1979 Engineering News Record 1980 and 1973, CCNY 1981

Pope, Michael (Continued)
(David B. Steiman Medal), 1979 and 1974. Mbr ASME & IEEE; Fellow ACEC. *Society Aff:* ACEC, ASME, IEEE, NSPE.

Pope, Robert M
Business: 10 High St, Boston, MA 02110
Position: Pres *Employer:* Weston & Sampson Engrs Inc *Education:* MS/Sanit Eng/Harvard Univ; BS/CE/Univ of NH *Born:* 5/13/20 US Army Corps of Engrs. CBI World War 2 35 yrs of engrg and project mgmt on public works projects throughout US Middle & Far East; Consultant to Mgmt on Water and Waste to major US Chemical Industry. Editor of National Engrg publications. Principal of Weston & Sampson for 17 yrs. Registered Prof Engr in 9 states. Formerly Nat'l Dir of Water Pollution Control Federation and Pres of New England Water Pollution Control Assoc. *Society Aff:* ASCE, ACEC, WPCF, AWWA

Popek, Gerald J
Business: 3532 Boelter Hall, Computer Sci Dept, Los Angeles, CA 90024
Position: Assoc Prof. *Employer:* UCLA, Palyn Assoc. *Education:* PhD/App Math/Harvard; MS/App Math/Harvard; BS/Nuclear Engg/NY Univ. *Born:* 9/22/46. Native of Rutherford, NJ. Joined UCLA faculty 1973. Prin Investigator of ARPA res project concerned with distributed systems, security in computer systems, and with networks 1976-present. Managed dept computing faculty 1977- present. Lectured exten in Europe and South America on computer security, data management, operating systems and networks. Major cons clients include International Telephone and Telegraph, Honeywell, Olivetti. Member of Palyn Assoc Author of over 40 prof pub. *Society Aff:* ACM.

Popelar, Carl H
Business: 155 W. Woodruff Ave, Columbus, OH 43210
Position: Prof *Employer:* OH State Univ *Education:* PhD/Eng Mech/Univ of MI; MS/Applied Mech/MI State Univ; BS/CE/MI State Univ *Born:* 02/05/35 Native of Dundee, MI. Engr with the Martin Co 1961-62. Joined the faculty of Engrg Mech, The OH State Univ, in 1966 after receiving the PhD degree in 1965. Became Prof of Engrg Mech in 1972. Conducted research in fracture mech and structural stability and authored some forty technical publications. Served on the editorial board of Journal of Engrg Mech Div and as a reviewer for numerous technical journals. *Society Aff:* ASME, AAM

Popham, Richard R
Business: 260 Madison Ave, New York, NY 10016
Position: President. *Employer:* Laramore, Douglass and Popham. *Education:* MME/Power Engg/Brooklyn Poly; BSME/-/Purdue. *Born:* Feb 21 1917, Chicago, son of Audrey Joseph & Lida Eugenia (Colborn) Popham. 1947 married Dorothy A Hennig. 1940-49 mech engr (electric power) with Allied Chemical, United Engrs & Constructors, Ebasco Services, M W Kellogg, H K Ferguson, Lockwood Greene, & Allied Process engrs; 1949- , principal of R R Popham Engr & Pres of Laramore, Douglass & Popham, Cons Engrs. Reg PE in 13 states. Mbr ASME, ASTM & US Ctte on Large Dams. *Society Aff:* ASME, ASTM.

Popov, Egor P
Business: Davis Hall, Berkeley, CA 94720
Position: Prof Emeritus of Civil Engineering. *Employer:* University of California, Berkeley. *Education:* PhD/CE-Applied Mech/Stanford; MS/CE/MIT; BS/CE/Univ of CA, Berkeley. *Born:* Feb 1913, Kiev Russia. 1936-45 professional engrg exper, licensed in Calif as CE, ME & SE; since 1946 on Univ of Calif staff, 1957-60 served as Chmn of Div of Structural Engrg & Structural Mechanics and Dir of Structural Engrg Lab. Author or coauthor of over 180 tech papers & 2 books on mechanics. Fellow ASCE, Chmn Engrg Mechanics Div 1960-61; Fellow AAM; Chmn ASEE Mechanics Div 1961-62. Miller Research Professor 1967-68; SEM Hetenyi Award 1967; AISC Higgins Lectureship Award 1971; ASCE Howard Award 1976; Croes Medal 1979, 1982, Newmark Medal 1981; Honorary Member 1986; ASEE Western Electric Fund Award, 1977, Distinguished Educator Award, 1979; Mbr NAE. *Society Aff:* ASCE, ASME, AIAA, ASEE, SEM, SEAoNC.

Popov, Theodore R
Business: 3990 Westerly Pl, Newport Beach, CA 92660
Position: Pres. *Employer:* Popov Engineers, Inc *Education:* MS/Mech Engg/Univ of Sofia (Bulgaria). *Born:* 10/3/38. in Bulgaria and came to the US in 1968, a naturalized citizen. He grad from the Univ of Sofia in Bulgaria in 1963 with a degree in Mech Engg; married Sylvia Szakacs on May 9, 1975 and had a son, Emil, born in Jun of 1976. He had worked with Smirnensky Co in Bulgaria, Glav Proj Co in Bulgaria and DMJM in Los Angeles, Kocher & Nishimura, Inc in Los Angeles. William H Stockly & Assoc in Pasadena and presently principal of Popov Engrs, Inc in Newport Beach CA. He is a Reg P mech E in the states of AZ, CA NV, NM, TX and UT and WA. Social interests range from the Newport Beach Tennis Club to Scuba Diving. He enjoys tennis, the theatre and classical music. *Society Aff:* ASHRAE, NSPE, SAME, CSPE, CEAC, AIPE.

Popovich, Milosh
Home: 1390 NW 14th, Corvallis, OR 97330
Position: Vice Pres for Adm (Emeritus). *Employer:* Oregon State University. *Education:* MS/Mech Engr/OR State Univ; BS/Chem Engr/OR State Univ. *Born:* 9/3/17. Pittsburg, CA. 1941-45 served with army Ordnance Corps, decorated legion of Merit Medal, Bronze Star Medal, Croix de Guerre with Gold Star, Croix de Guerre with Silver Star; 1945, 1947-59 taught mech engrg at OR State Univ, 1947-54 Dept Hd, 1950-54, Asst Dean of Engg 1954-59, Dean of Admin respon for finance & admin of univ including physical & financial planning & operations, computing & genl services 1959-77, vice pres 1977-79. Chmn OR Sect ASME 1956, Fellow 1970. Dir Cascade Federal Savings & Loan Assoc 1961-81. Retired 1979. *Society Aff:* ASME, ΣΞ.

Popovics, Sandor
Business: Dept of Civil Engineering, Philadelphia, PA 19104
Position: Samuel S. Baxter Prof of Civil Engg *Employer:* Drexel Univ *Education:* PhD/Highway Eng/Purdue Univ; MS/Mat/Hungarian Acad of Sci; BS/CE/Polytech Univ, Budapest *Born:* 12/24/21 raised in Hungary, he started as Metropolitan Engr in Budapest, 1945. Res Engr and Mgr in the Inst of Bldg Scis, Budapest, until 1956. Tenured Prof, Auburn Univ, Auburn, AL, until 1968. Head Prof of Civil Engrg and Res Coordinator, No. AZ Univ, Flagstaff, AZ, until 1976. Presently tenured and chaired Prof, Drexel Univ, Philadelphia, PA. Produced four books and about 150 major publications. Fellow of ACI and ASCE. Past Chrmn of TRB Committee A2-E3. Past Vice-Chrmn of ASTM Rocky Mountain District. Reg PE in three states. Rockefeller Fellow in 1957. Listed in: Who's Who in Engrg 1964; American Men of Sci 1967 et cet. Conslt for UNIDO in China in 1980 and 1983. Exchange scientist to Bulgaria (1972 & 1980), India (1984) and Soviet Union (1986). *Society Aff:* ACI, ASTM, TRB, ASCE, RILEM.

Poppino, Allen G
Business: PO Box 20400, Oklahoma City, OK 73156
Position: Vice Chrmn *Employer:* The Benham Group *Education:* BS/Civ Engg/OK State Univ. *Born:* 10/30/25. Grad in 1950 from OK State Univ with a BS degree in Civ Engg. Licensed PE in 27 states, a Reg Land Surveyor in OK and has a Certificate of Qualification from the Natl Council of Engg Examiners. Joined Benham Engrg Co in 1950 as Chief Structural Engr; now holds office of Corporate Vice Chrmn on the Bd of Dirs. Active in various civic & engg societies. Past Pres of the OK State Univ Chi Epsilon Honorary Civ Engg Fraternity and the OK Sec of American Soc of Civ Engrs. Member of the Natl Engr's Joint Contract Documents Ctte and 15 other professional org. *Society Aff:* NSPE, ACEC, ASCE, SAME, EJCDC, ARTBA, HRB, IABSE, IASSS, ISSMFE, NYMAUDEP

Porcelli, Richard V
Home: 287 Crestwood Ave, Tuckahoe, NY 10707
Position: Consultant *Employer:* Richard Fleming Assoc, Inc *Education:* DES/ChE/Columbia Univ; MS/ChE/Columbia Univ; BS/ChE/Columbia Univ. *Born:* 5/17/45. 1970 joined Halcon Internatl Inc; initial assignment was economic evaluation of novel chemical tech, 1974 transferred to Projects Res Dept, 1975 named Proj Mgr, respon for developmental res on novel chemical processes. 1977 Named Dir of Exploratory Process Development. 1981 Dir of Economic Eval. 1985 Dir of Res. Joined Richard Fleming Assoc, Inc, in 1987.

Porcello, Leonard J
Home: 3925 N Pantano Rd, Tucson, AZ 85715
Position: Corporate VP and Manager, Defense Systems Group *Employer:* Science Applications International Corporation (SAIC) *Education:* PhD/Elec Engg/Univ of MI; MSE/Elec Engg/Univ of MI; MS/Physics/Univ of MI; BA/Physics/Cornell Univ. *Born:* March 1934, New York NY. 1955-72 Mbr res staff Univ of Mich Inst of Sci & Tech, Dir Radar & Optics Lab 1968-72, 1969-72 Assoc Prof Electrical Engrg Univ of Mich, 1972 Prof in radar & related areas; 1973-76 V Pres, Environmental Res Inst of Mich (ERIM), also Dir of ERIM's Radar & Optic Div, respon for R&D programs in imaging radar, coherent optics & related areas. Trustee, Erim, 1975; Science Applications International Corp (SAIC), Asst VP, 1976-79; VP, SAIC, 1979-1985; Corporate VP, 1985-present; Manager, Defense Systems Group, SAIC, 1986-present. Fellow. IEEE, 1976. *Society Aff:* IEEE, OSA, AAAS, ΣΞ, HKN.

Porras, Octavio R
Business: Benjamin Franklin 47, Mexico 18, DF Mexico
Position: Scholar Services Director *Employer:* Universidad La Salle *Education:* Master/Systems Engg/Instituto Politecnico Nacional; BS/Ind Engr/ITR De Chihuahua; Analyst/Systems/IBM Inst. *Born:* 4/20/28 Octavio Porras Ruiz, PE is dean & Prof, sch of Engg Univ of LaSalle, in Mexico City. He serves as consultant, - Recursos Humanos, SC, also is a prof of Systems Production - Grad Sch, IPN. Since 1962 is an External Lectr in IACE in all fields of Ind Engg & Maintenance Mgt. His 25 yrs industry experience includes appliances mfg, - Chem processes, glass bottles & copper industries. Ing Porras holds a BSIE & MSE in Mexico Tech Inst. He is a Sr mbr of IIE & Ex VP Region XIII (Mexico). *Society Aff:* IIE, APICS, CIME, ASMAC.

Porter, Alan L
Business: , Atlanta, GA 30332
Position: Assoc Prof *Employer:* Georgia Inst of Tech *Education:* PhD/Eng Psych/UCLA; BS/ChE/CA Tech *Born:* 06/22/45 Research assoc/asst prof Program in Social Mgmt of Tech, Univ of Washington 1972-74. With the Sch of Industrial and Systems Engrg, Georgia Tech 1975- . Profl interests in tech assessment, methods to evaluate program effectiveness, and engrg and public policy. Co-author of *A Guidebook for Tech Assessment and Impact Analysis*, and *Sci, Tech, and Natl Policy*. Co-founder and secy, Intl Association for Impact Assessment 1981-84. *Society Aff:* IAIA, AAAS, ASEE, TIMS, ΣΞ

Porter, Harold F
Home: Rt 1 Box 55, Hockessin, DE 19707
Position: Principal Division Consultant. *Employer:* E I du Pont de Nemours & Co. *Education:* BS/Met Engg/Univ of UT. *Born:* Nov 1911, Morgan Utah. 1934-46 Engr to Plant Mgr Wyanadotte Chemical Corp; 1946-49 Prod Supt Amer Potash & Chemical Corp; 1949-50 Cons for Wyandotte; since 1950 with E I du Pont de Nemours & Co: 1950-61 engr to Cons Grp Mgr; 1961- 63 Engrg Res Div; 1963- , Principal Cons to Principal div cons, Engrg Serv Div. Major fields: crystallization, evaporation, drying, solids separations, environment & energy. Fellow AIChE, Tau Beta Pi. Reg PE in Del. Patents, publications, Sect Ed Perrys Chemical Engrg Handbook. *Society Aff:* AIChE, TBΠ.

Porter, James H
Home: PO Box 1131 Daggett Ave, Vineyard Haven, MA 02568
Position: Pres. *Employer:* Energy & Environmental Engg. *Education:* ScD/CE/MIT; BCE/CE/RPI. *Born:* 11/11/33. Dr Porter is Pres of Energy & Environmental Engg Inc involved in the dev and commercialization of new Energy Conversion & Environmental control processes. He is co-founder, mbr of the exec bd & current pres of the Natl Organization for the Profl Advancement of Black chemist & chem engrs. He has had 25 yrs of profl experience including VP of Energy Resources Co Inc, Assoc Prof of CE at MIT & Mgr of Comp Applications & Design at Abcor Inc. He is a mbr of the EPA Sci Advisory Bd. *Society Aff:* AIChE, NOBCChE, NYAS.

Porter, Lew F
Business: US Steel Research Lab, Monroeville, PA 15146
Position: Sr Research Consultant, Physical Met. *Employer:* U S Steel Corporation. *Education:* BS/Chem Eng/Univ of WI; MS/Metallurgical Engg/Univ of WI; PhD/Metallurgical Engg/Univ of WI. *Born:* July 1918. 1940-44 Ch Chemist Ind Harbor Works, Amer Steel Foundries; 1944-46 US Naval Reserve; 1946-49 Res Metallurgist, Chain Belt Co; 1955- , US Steel, Sect Supervisor Heavy Products Div to 1976, presently Sr Res Cons, Supervisor of Basic Res in Physical Met. Papers of gray cast iron, thermomechanical treatments of steel, radiation effects, dev of submarine-hull steels & alloy steel dev. Reg PE in Wisc. Fellow of ASM 1974, recipient of Henry Marion Howe Medal of ASM 1953 & 1966. *Society Aff:* ASM, AIME, ASTM, MS.

Porter, Marcellus C
Home: 3449 Byron Ct, Pleasonton, CA 94566
Position: Vice President Research & Dev. *Employer:* Nuclepore Corporation. *Born:* May 14 1938, Louisville KY. BS, MS & ScD in Chem E from MIT. Sr Res Engr & Grp Leader for AVCO's Scientific Systems Div; 1969 joined AMICON Industrial Separations div, respon for applications res, dev, production & marketing of large scale membrane ultrafiltration equip; presently with Nuclepore. Authored over 30 papers on theory & application of ultrafiltration to industrial & Lab separations problems. Frequent lecturer on membrane separations technology, course dir for annual course on Industrial Membrane Technology sponsored by the Center for Professional Advancement.

Porter, Nancy J
Business: Sorrento Electronics, 11045 Sorrento Valley Ct, San Diego, CA 92121
Position: Mgr, Quality Reliability Engrg *Employer:* Sorrento Electronics *Education:* MSc/EE/RPI; BA/Mathematics/Central CT State Univ *Born:* 02/03/54 Born and educated in CT. Engrg career began with Combustion Engrg, Inc in 1976. Assignments included thermal-hydraulic core analysis for operating nuclear power plants, dev of first application of optical videodiscs utilized in nuclear industry, training of nuclear plant operators, and proj mgmt for a midwest utility. Currently Mgr quality & reliability engrg for Sorrento Electronics, supplier of nuclear radiation monitoring systems, analog and digital. Adj Asst Prof, Elec Engrg, RPI- Hartford (1981-1985); taught electrical and nuclear graduate-level engrg courses. Mbr, ANS; Tech Program Chair, 1988 Annual Meeting. *Society Aff:* ANS, TBΠ

Porter, R Clay,
Home: 2027 Medford Rd, Apt 183, Ann Arbor, MI 48104
Position: Emeritus Prof, Mech Engrg. *Employer:* Univ of Michigan *Education:* ME/Mech Engg/Univ of KY; MS/Mech Eng/Univ of MI; BS/Mech Eng/Univ of KY. *Born:* 10/11/02. 1925-27 Test Engr, GE Co; 1927-37 Res Engr & Asst Prof, Univ of Ky; 1937-73 on faculty of Univ of Mich, full Prof since 1949, Emeritus Prof since 1973. Res on fluid metering & wind tunnel studies on behavior of stack gas plumes. Cons & researcher for various elec utilities & indus on smoke stack design, air pollution problems & fluid metering. *Society Aff:* ASME, TBΠ, ΠΤΣ, ΣΞ.

Porter, Robert P
Business: Old Quarry Rd, Ridgefield, CT 06877
Position: Dir Electromagnetic and Acoustical Physics *Employer:* Schlumberger Tech Corp *Education:* PhD/EE/NE Univ; EE/Elec Engrg/MIT; SMEE/Elec Engrg/MIT;

Porter, Robert P (Continued)
SBEE/Elec Engrg/MIT *Born:* 09/21/42 Native of Brockton, MA. MIT class of 1963. Completed PhD in 1970. Thesis research in electromagnetics, liquid metal MHD, and holography. Worked on radar, signal processing, and inverse scattering at Mitre Corp 1966-1971. Assoc Scientist, Woods Hole Oceanographic Inst until 1977. Research on long range acoustic propagation and acoustic navigation in the ocean. With Schlumberger since 1977, responsible for the mgmt of the company's research in acoustic and electromagnetic propagation in the oil well environment. Fellow, ASA. VP Nippon Schlumberger, 1982-1984. Assoc Editor of Acoustic Speech and Signal Processing Society of the IEEE, 1981-1982. Author of about 50 journal publications; holder of 2 patents. *Society Aff:* OSA, ASA, IEEE, SEG

Porter, William A
Business: Dept of Elec & Computer Engg, Baton Rouge, LA 70803
Position: Professor *Employer:* LA State Univ. *Education:* BS/EE/MI Tech Univ; MS/EE/Univ of MI; PhD/EE/Univ of MI. *Born:* 12/11/34. Prof of Elec Engg, Univ of MI 1961-1977; Prof and Chrmn, Dept of Elec and Computer Engg, LSU 1977-82. Author of 125 journal articles, on system theory, automatic control, computer, architectures and signal processing. Assoc. Editor Circuits, Systems and Signal Processing, Guest Editor for IEEE Spectrum, Journal of the Franklin Institute, Governing Board Midwest Circuit & System Conf, Steering Committee Intl Symposium on Mathematical Theory of Networks and Systems. *Society Aff:* IEEE.

Porter, William R
Business: SUNY Maritime College, Fort Schuyler, Bronx, NY 10465
Position: Vice President Academic Affairs. *Employer:* State University NY Maritime College. *Education:* PhD/Engg Sci/Univ of Ca, Berkeley; Engr/Naval Construction/MIT; MS/Elec Engg/MIT; MA/Economics/Univ of MD; BS/Elec Engg/MIT; BS/Elec Engg/US Naval Acad. *Born:* Nov 1922. 1940-73 US Navy, ret Capt, designated for Engrg Duty, ship design & const; 1960-63 Assoc Professor MIT, 1970-74 Professor of Nav Arch; 1973-74 Dir Educ Act C S Draper Lab; 1974- , V Pres Academic Affairs Maritime College. Soc of Nav Archs of USA, Germany, Japan. *Society Aff:* SNAME, STG, ZK, ASNE.

Porterfield, Jay G
Home: R R 1, Box 137, Stillwater, OK 74074
Position: Professor Agricultural Engrg. *Employer:* Oklahoma State University. *Education:* BS/Agri Engr/IA State Univ; MS/Agri Engr/IA State Univ. *Born:* July 1921, Holton Ks. 1947-49 Instruction ISU; 1949 Ser & Prod Training Iowa Ford Tractor Co; 1950-51 Asst Prof ISU; 1952- , Professor Ok St Univ, 1958 Visiting Professor Univ of Melbourne Australia. Cons to state, federal govts (4) & multinational corps (2). Reg PE in Ok. Mbr Sigma Xi; Fellow ASAE. Pub over 100 tech papers, 45 popular articles, 2 patents, 1 lab manual & 1 IPI text. Mbr of graduate faculty 1974-78 Hd of Agri Engrg Dept Ok St Univ; 1978-79 Agri Engr USDA/SEA/CR Washington DC. Retired 1982. *Society Aff:* ASAE, ΣΞ.

Pos, Jacob
Home: 10 Mayfield Ave, Guelph Ontario, Canada N1G2L8
Position: Associate Professor. *Employer:* School of Engineering Univ of Guelph. *Education:* MSc/Civil & Agr'l Eng/Michigan State; MSA/Agr'l Mechanics/University of Toronto; BSA/Agr'l Mechanics/Ontario Agricultural College. *Born:* April, 1921, The Netherlands. BSA Ontario Agricultural College; MSA Univsity of Toronto; MSc Michigan State University. 1940-45 served in Royal Canadian Air Force. Charter Member, past President & Director CSAE; recipient CSSBI-CSAE Award, 2nd Maple Leaf-CSAE-Award, and Fellow of CSAE. Past Chairman NAR-ASAE, and Past Regional Director ASAE. Professional areas of interest include structures, animal behaviour, pollution control, waste management, resource recycling, and energy trom waste (composting & methane production) Retired Professor, University of Guelph, now President - Jack Pos & Associates Limited, Agricultural Engineering consultants. Past District Lt. Canadian Power Squadrons and active in Freemasonry. *Society Aff:* CSAE, ASAE, PCAO, APEO, OIA.

Posey, Chesley J
Business: U37, Storrs, CT 06268
Position: Emer Prof of Civ Engr (but still teaching). *Employer:* Univ of CT. *Education:* MS/Civ Engr/Univ of IL. *Born:* 6/12/06. Born Mankato, MN, 6/12/06, son of C Justin Posey, Prof of Geography, and Maude Johnston Posey, public school teacher and supervisor. Education: KS Univ BS 1926, CE 1933, Univ of IL MS 1927. Structural engg, 1927-29. Instr to dept hd, Univ of IA 1929-62. IA State Bd of Engg Examiners, 1957-62. Prof of Civ Engg, Univ of CT, 1962-present. Dir, Rocky Mtn Hydraulic Lab, 1945-present. *Society Aff:* ASCE, AGU, IAHR, AWRA.

Posey, Owen S
Home: 1212 Belle Chene Dr, Mobile, AL 36609
Position: President. *Employer:* Owen S Posey & Associates, Inc. *Education:* BS/Ee/Auburn Univ. *Born:* 10/20/24. in Haleyville, AL. Army Air Force 1943-46. Haleyville High School 1946; BSEE Auburn Univ 1949. Engg Exper: utility distribution, heating A/C contracting, aerospace facilities, city engrs, consulting engg; Pres of Owen's Posey & Assocs Inc, elec, mech & civil structural cons engg firm. Reg PE in AL, MS, LA, TX & MD. P Pres Mobile Chapter ASHRAE; P Pres Mobile Chapter IES; Mbr NSPE & Pres ASPE, Dir ACEC.

Posner, Edward C
Business: 116-81 Caltech, Pasadena, CA 91125
Position: Chf Telecom & Data Technologist, JPL, & Visiting Prof, Elect Engg. *Employer:* CALTECH *Education:* PhD/Math/Univ of Chicago; MS/Math/Univ of Chicago; BA/Physics/Univ of Chicago *Born:* 08/10/33 Native of NYC. At Bell Labs, NY, 1956-57, Univ of WI, 1957-60, Harvey Mudd Coll, 1960-61. Employed by Caltech 1961-present, Jet Propulsion Lab (Mbr of Technical Staff, 1961-62; Res Group Supvr, 1962-67; Deputy Section Mgr, 1967-73; R&D Mgr, Telecomm Div, 1969-73; Mbr, Advanced Projs & Data Processing, Telecomm and Data Acquisition Off, 1973-79; Mgr, Telecomm and Data Acquisition Planning, 1979-81; Ch Telecom & Data Technologist, 1981); also in Caltech Elec Engrg Dept (Visiting Prof. in Elec Engrg). Contributor to deep space and civil communications tech, development, mgmt, and planning and neural network devel. Co-author of major text. Editor of translation of Japanese traffic theory book; 4 patents in communication and signal and data processing; NASA maj patent award. Also involved in communications switching and traffic. Hobbies include archaeology, linquistics, and bicycling. *Society Aff:* IEEE, AAAS, ΣΞ, SIAM, AMS, MAA, AIAA, SEG, PS

Possiel, Norman C
Home: 15 Hickory Place, Murray Hill, NJ 07974
Position: Exec VP *Employer:* Edwards & Kelcey, Inc. *Education:* BS/Civ Engg/College of Engg, NJ Inst Tech. *Born:* 6/27/27. Native of NJ. 1949 to date: Associated with Edwards & Kelcey. Became an Associate in 1958. Now Exec V.P. Currently responsible for administration of operations for Edwards & Kelcey, Inc, it subsidiaries and affiliated companies. Served as officer of Consulting Engrs Council of NJ for 7 yrs, including 2 consecutive terms as Pres. Fellow of ACEC. Reg PE in NY & CT. *Society Aff:* NSPE, ACEC, ASCE.

Post, Roy G
Home: Box 17690, Tucson, AZ 85731
Position: Prof. *Employer:* Univ of AZ. *Education:* PhD/Chemistry/Univ of TX; BS/Chem Engrg/Univ of TX. *Born:* 6/24/23. Began nuclear work in Jan 1944 in the West stands of the Univ of Chicago. Worked til 1949, returned to grad sch and then to Richland Wash, the Hanford Plant. This work and teaching and res at the Univ of AZ has been in the fields of nuclear fuel processing and nuclear waste mgt. Editor, Nuclear Technology. Former chrmn AZ Sec Am Inst of Chem Engrs. Chrmn AZ Section of American Nuclear Society. *Society Aff:* ANS, ΣΞ, AAAS, NSPE.

Potter, Charles J
Business: 655 Church St, Indiana, PA 15701
Position: Chairman of the Board. *Employer:* Rochester & Pittsburgh Coal Company. *Education:* PhD/Chem Eng/W VA Univ; MS/Chem Eng/Univ of MO; BS/Chem Eng/Univ of MO; D Sci/Hon/Univ of MO; LL D/Hon/Indiana University of Pennsylvania. *Born:* 7/16/08. Greenfield MO; Chmn & Dir of Rochester & Pittsburgh Coal Co & subsidiaries; former Chmn & Dir of Savings & Trust Co of PA; former Dir of Amer Mining Congress; Dir of Bituminous Coal Res; formerly Deputy Solid Fuels Admins during WWII. Medal for Merit awarded by President Truman; Honorary Commander, Most Excellent Order of the British Empire; Erskine Ramsay Medalist & Howard N Eavenson Award, AIME; member of PA Governor's Energy Council; Governor of PA State System of Higher Education; trustee of Indiana University of PA. *Society Aff:* SME.

Potter, David S
Business: 3044 W Grand Blvd, Detroit, MI 48202
Position: VP in charge of Power Products & Defense Operations Group *Employer:* General Motors Corporation. *Education:* PhD/Physics/Univ of WA; BS/Physics/Yale Univ. *Born:* 1/16/25. Children Diana (Mrs Paul Bankston), Janice (Mrs Robert Meadows), Tom, Bill. 1943-46 served with USNR; 1946-49 Jr Physicist Applied Physics Lab, Univ WA, 1950-53 project, 1953-55 Sr Physicist, 1955-60 Asst Dir 1960-66 Hd Sea Opers Dept Defense Res Labs, Genl Motors Corp, 1966-69 Dir ab, 1969-73 Ch Engr AC Electronics Div; 1973 Dir R&D Detroit Diesel Allison Div, 1973-74 Asst Secy Navy, 1974-76 Undersecy Navy, 1976-78, VP, Environ Act Staff, Nov 1978- VP in chrg of Pub Affairs Group. 5/1/83 VP in charge of Power Products and Defense Operations Group. Genl Motors Corp. Mbr Natl Acad Engrg, Marine Tech Soc, Physics Soc, Acoustical Soc, (NOA now Defunct). Basic res in cosmic rays, magnetics, underwater acoustics *Society Aff:* NAE, AAAS, AAS, AIAA, APS, SNAME.

Potter, J Leith
Home: 804 Westwood Dr, Tullahoma, TN 37388
Position: Senior Staff Scientist *Employer:* Sverdrup Technology, Inc. *Education:* PhD/ME/Vanderbilt Univ; MS/Engg Mgt/Vanderbilt Univ; MS/Engg/Univ of AL; BS/Aero Engg/Univ of AL. *Born:* 2/5/23. Engg scientist and mgr with extensive experience in res, test facility dev, engg undergrad and grad teaching. Directed dev of hypervelocity, low-density wind tunnels, aeroballistics ranges, space simulation chambers, & other experiemental equip. Directed theoretical studies related to these facilities. Reg PE. Prof of mech and aerospace engg (part time), Univ of TN. Currently consultant to NATO-AGARD, ABET Ad Hoc Visiting Cttee, AIAA Pub Comm, mbr of USAF-AEDC Boundary Layer Transition Study Group, & NASA Technical Steering Group for the Numerical Aerodynamic Simulator. Active in profl societies & contributor to engg res literature. Married Dorothy Williams, 1957, three children. Fellow, AIAA. *Society Aff:* AIAA, ΣΞ, ΣΓΤ, ΤΒΦ, ASEE, ASEM

Potter, Merle C
Business: Mech Engr Dept, E Lansing, MI 48824
Position: Prof. *Employer:* MI State Univ. *Education:* PhD/Engr Mech/Univ of MI; MS/Aero Engr/Univ of MI; MS/Engr Mech/MI Tech Univ; BS/Mech Engr/MI Tech Univ. *Born:* 10/13/36. Native of Grand Rapids, MI. Taught at MI Technological Uiv 1957-61 and at the Univ of MI 1961-65. Have been at MI State Univ since 1965 in Mech Engg specializing in Fluid Mechanics. Received a Teacher-Scholar Award in 1968. Published three books: in Fluid Mechanics, in Applied Mathematics & a Fundamental of Engrg Review book. Have authored many articles on theoretical stability analysis of incompressible flows and experimental channel flow. Serve as faculty advisor for ASME. Reg PE. Listed as an Energy Auditor for large buildings. *Society Aff:* ASME, ASEE.

Potter, Richard C
Home: 209 No Hillcrest Blvd, Inglewood, CA 90301
Position: Prof-Emeritus *Employer:* California State Univ, Long Beach. *Education:* PhD/Engg/Purdue Univ; MSE/Engg/Purdue Univ; BSPSE/Public Service Engg/Purdue Univ. *Born:* May 19 1919. 1940-41 Asst Res Engr, Crane Co Chicago; service in WWII as Ordnance Corps Off, discharge rank of Capt, Ret Army Res Col 1974; 1946-48 Inst & Res Fellow in Mech Engr, Purdue Univ; 1952-59 Asst Dean & Prof of Mech Engrg, Kan St Univ; 1959-60 Res Staff Mbr Genl Atomic Div of Genl Dynamics; 1960 Mgr Tech Staff Dev RamoWooldridge Div of TRW; 1960-63 Mgr Prof Placement & Dev Space Tech Labs Inc; 1963-65 Dir of Inst of Indus Res Univ of Louisville; 1966-67 Pres Northrop Inst of Tech; 1967-69 Dir of Educ Services Calif St Univ - Long Beach; 1969-1982 Dean Sch of Engrg Calif St Univ - Long Beach. 1982-83 Mech Engr, Lawrence Livermore Nat'l Lab. 1983 Prof-emeritus, CA State Univ, Long Beach. Recipient of Silver Anniv Award Sports Illustrated Magazine 1964; Reg P E Kan. Mbr Amer Soc Engrg Educ, ASME, Western Coll Assn (mbr exec ctte), Sigma Xi, Phi Gamma Delta, Phi Kappa Phi, Pi Tau Sigma, Tau Beta Pi, Republican. *Society Aff:* ASME, ASEE, TMSA.

Potvin, Alfred R
Business: Division, Bld 28, Lilly Corporate Center, Indianapolis, IN 46285
Position: Director, Medical Instrument Systems Research Division. *Employer:* Eli Lilly & Co. *Education:* PhD/Bioengrg/Univ of MI; MS/Psychology/Univ of MI; MS/Elec Engrg/Stanford Univ; BS/Elec Engrg/Worcester Polytechnic Inst. *Born:* 2/5/42. Native of Worcester, MA. Began teaching Electrical Engrg at the Univ of TX at Arlington in 1966 and organized joint grad degree program in biomedical engrg with the Univ of TX Health Sci Center at Dallas, 1971-72. Prof Engrg (TX) license, 1975. Prof and Chmn of Biomedical Engrg and Prof of Elect Engrg in 1976-84. Assumed current position as Dir of Medical Investment Sys Research Div of Eli Lilly and Co in 1984. Chmn of ASEE Biomedical Engrg Div, 1979-80. Chmn, BMES Education and Public Affairs Ctte, 1976-80. Chmn, IEEE Education Ctte of EMBS, 1979-81. Elected Region 5 rep for IEEE-EMBS, 1980-82. Program Chrmn of IEEE-EMBS Frontiers of Enggrg in Health Care Conference, 1982. Elected VP for Tech Activities in IEEE-EMBS, 1982. Elected Pres of IEEE-EMBS;1983 and 1984. Program Chrmn of IEEE-EMBS Symposium on Biosensors; 1984. Mbr of Sigma Xi, Eta Kappa Nu. Editorial Board Member of IEEE Spectrum, 1987-90. VP Nat Affairs, of AEMB, 1987-89. Advisory Bd Mbr for BME at NSF, 1983-88. *Society Aff:* IEEE-EMBS, BMES, AAMI

Pound, Guy M
Home: 3003 Country Club Ct, Palo Alto, CA 94304
Position: Prof. *Employer:* Stanford Univ. *Education:* PhD/Physical Chemistry/Columbia Univ; MS/CE/MIT; BA/Chemistry/Reed College. *Born:* 4/2/20. Asst Prof, Assoc Prof & Alcoa Prof of Physical Met & Sr Staff Mbr, Metals Res Lab, Carnegie Inst of Tech, 1949-63. Dir of Metals Res Lab, Carnegie Inst of Tech, 1963-66. Guest Prof, Guggenheim Fellow & Fulbright Sr Scholar-Univ of Sheffield, 1959-60. Guest Prof-Univ of Berlin 1964-65. Guest Prof-Univ of Vienna 1972-73. Prof-Stanford univ, 1966 to present. Co-owner & mbr of Bd of Dirs of Failure Analysis Assoc, Inc, a Palo Alto engg consultation firm, 1967 till present. Reg Corrosion Engr, State of CA; Albert Eastman White Awardof Am Soc for Metals for distingished teaching of Met, 1978; Fellow of the Am Soc for Metals, 1979. *Society Aff:* AIME, ASM, ACS.

Pounsett, Frank H R
Home: 23 Parkhurst Blvd, Toronto, Ont Canada M4G 2C7
Position: Retired. *Employer:* *. *Education:* BASc/-/Univ of Toronto. *Born:* Sept 12, 1904 London England. IEEE 1926, Fellow 1947, Life Mbr 1969. Chmn Toronto Sect IRE 1945-46, Dir Canadian Region 1949-50. Assn of Prof Engrs of Ontario 1939. Res Enterprises Ltd Toronto, Ch Engr Radar Div 1940-45. Stromberg Carlson Co Ltd Toronto, Ch Engr & Mgr of Mfg 1946-52; Philips Electronics Indus Ltd, V P & Genl Mgr - Consumer Prods Div 1952-67. Centennial Coll of Arts & Tech Scarborough Ont, Dean Engrg Tech Div 1967-71. Canadian Radio Tech Planning

Pounsett, Frank H R (Continued)
Bd 1945-67, Pres 1958-62. Internatl Electrotech Comm, Chmn CNC/IECTC12 (Radio Communications) 1963-70. *Society Aff:* IEEE.

Povey, Edmund H
Business: 85 Walnut St, Watertown, MA 02172
Position: Consultant. *Employer:* Doble Engrg Co. *Education:* MS/Math & Physics/ Northeastern Univ; BSEE/Power/Northeastern Univ. *Born:* Aug 5, 1908. Native of Boston Mass. Employed by Doble Engrg Co as Field Engr, Ch Engr, V P & Cons. Fellow IEEE, Mbr ASTM DO9. Mbr IEEE Rotating Machinery Comm & Power Sys Instrumentation & Measurements Comm; also ANSI C-68. Granted sev pats on circuits for measuring loss characteristics of elec insulation. Active in IEEE & ASTM stds work. *Society Aff:* IEEE, ASTM.

Powals, Richard J
Home: 4384 Reilly, Troy, MI 48098
Position: Dir, Bus Dev. *Employer:* Clayton Environmental Consultants, Inc. *Education:* BS/Chem Engr/Univ of MI. *Born:* 4/6/47. Over a decade of experience in the multi-disciplinary areas of air pollution control, water pollution control and indus hygiene. Specialities include process-emission correlations, coke ovens, fugitive particulate emissions, non-criteria pollutants, and vegetation "kills." Author of over a dozen publications, including manuals and textbooks. Formerly with Owens-Corning Fiberglas Corp and Midwest Environmental Mgt Inc. Am currently responsible for all Business Dev and Strategic Planning Activities for both existing and new Clientele. Also Chapter Secretary, MI Air Pollution Control Assoc. *Society Aff:* AIHA, APCA, ASTM, ESD, NSPE.

Powe, Ralph E
Business: Vice President for Research, P.O. Box 6343, MS State, MS 39762
Position: VP for Res *Employer:* MS State Univ. *Education:* BS/Mech Engrg/MS State Univ; MS/Mech Engrg/MS State Univ; PhD/Mech Engrg/MT State Univ. *Born:* 7/27/44. Native of Tylertown, MS. Participated in Cooperative Education Prog with NASA at Wallops Island, VA. Faculty mbr at MT State Univ from 1970 to 1974, and joined faculty of Mech Engg Dept at MS State Univ in 1974. On Jan 1, 1979 assumed duties as Assoc Dean of Engg, Dir of the Engg and Industrial Res Station, and Dir of the Energy Res Ctr. Associate Vice President for Research, July 1, 1980-Dec. 1985, and as Acting VP, Grad Studies & Research 1985-86. Appointed VP for Research, 1986-present. Chrmn, MS Sec of ASME 1978-79; mbr, Natl ASME Comm for theory and fundamental res in heat transfer since 1973 (chmn 1983-86), Exec Ctte of Advanced Energy Sys Div (1984-88), VP, Region XI (1984-86); ASME Senior VP & Chair of Council on Member Affairs 1986; have organized and chaired a number of technical sessions at natl meetings; SAE Teetor Award 1971; over 50 publications in tech journals. *Society Aff:* ASME, ASEE, NSPE, NCURA, SRA.

Powell, Alan
Business: Mechanical Engineering Dept, University of Houston, Houston, TX 77004
Position: Chair & Professor. *Employer:* University of Houston. *Education:* DLC/Aero Eng/Loughborough College; DLC 1st Class Honors/Aero Eng/ Loughborough College; BSc Eng (1st Class Hons)/Aero Eng/London Univ; PhD/ Engrg/Southampton U *Born:* 2/17/28. Tech asst Percival Chair and Professor, Dept-Mechanical Engineering, W. Houston 1986-. Aircraft 1949-51; Lectr Southampton Univ 1951-56; Res Fellow Caltech 1956-57; Cons Douglas Aircraft Inc 1956-65; Assoc Prof Engrg Univ of Calif LA 1957-62 Prof Engrg 1962-1965, Director, Aero-sonics Lab 1957-62; Academic Senate 1964-1965; Assoc Tech Dir Acoustics & Vibration Lab, David Taylor Model Basin 1965-66 Tech Dir 1966-67 Tech Dir David W Taylor Naval Ship R&D Ctr 1967-85. Mbr Comm on Hearing, Bioacoustics & Biomech, Natl Acad Sci NRS 1961-86, Exec Council 1963-1965, Chmn 1965-1966, Advisor 1986-; Fellow Royal Aero Soc (Baden-Powell Prize, Wilbur Wright Prize); Mbr US-Japan Coop Agreement on Natural Resources-Marine Facilities 1979-; International Towing Tank Conference, Advisory Council, 1981-86, Advisor 1986-. Fellow Acoustical Soc Amer (Biennial Award, Assoc Editor journal 1962- 67, Exec Council 1967-1970, Chmn Ed Comm 1964-66, Chmn Medals and Awards Comm 1979- V P Elect 1981-1982, VP 1982-1983; Fellow The Physical Society, London; Fellow Inst of Physics London; Fellow Inst of Acoustics (UK); Assoc Fellow AIAA (Aeroacoustics Award 1980); Fellow Inst Mech Engrs; Senior Mbr IEEE, Speech and Signal Processing Soc (Admin. Ctte 1969- 71, Awards Ctte 1971-73, Constitution and By Laws Ctte, Chmn 1973-75). Inst Noise Control Engrg (initial mbr, dir 1974-1977, VP 1981-83). Mbr Amer Soc Engrg Educ (Res Council), Am Soc Mech Engrs, Soc Naval Archs & Marine Engrs, Am Soc Naval Engrs Inst, Tau Beta Pi (Hon life Mbr), Sigma Xi. Chartered Engr (Mechanical Engr, Aeronautical Engr, UK). Navy Meritorious Civilian Service Award 1970; Cited Meritorious Exec by Pres Reagan, 1982 Capt Robert Dexter Conrad Award for Scientific Achievement by Sec of the Navy 1984, Amer Mgmt Assoc (Res Council, Distinguished Service Award 1986), Hon Dr Tech U Loughborough, UK. *Society Aff:* ASA, AIAA, RAeS, IMech E, INCE, ASNE, SNAME, IOA, ASME.

Powell, Allen L
Home: P.O. Box 843, Luthrin, TX 75901
Position: retired *Education:* BS/Petroleum/Univ of TX. *Born:* July 12, 1915 Nacogdoches Tex. Attended Draughton's Business College, Schreiner Inst; Field assignments have incl shipboard duty on the USC&GS Ships COWIE, LYDONIA, the wire drag vessels, PARKER, BOWEN, STIRNI, PIONEER, BOWIE, EXPLORER. 2 1/2 yrs with photogrammetric field parties; sev geodetic party assignments. 1963, assigned to New Ships Staff. This office merged subsequently with the Facils Div & then became the Ship Const Group. In 1968 became Acting Dir Atlantic Marine Ctr. 1969, promoted to temporary grade of Rear Admiral & designated Dir AMC. 1971, returned to NOS Hdqtrs & appointed Assoc Dir of the newly created Office of Fleet Opers. 1972, appointed Dir Natl Ocean Survey NOAA. Retired 1 Aug. 1979.. *Society Aff:* SAME, USNI, ASNI, MIT.

Powell, Gordon W
Business: 490 Watts Hall, Dept of Metall Engr, Columbus, OH 43210
Position: Prof *Employer:* OH State Univ *Education:* ScD/Met/MIT; MS/Met/MIT; BS/Met/MIT. *Born:* 03/10/28 Born in Providence, RI and attended high sch in Newport, RI. following grad from MIT; I worked two yrs for Nuclear Mtls, Inc and then taught at the Univ of WI for one yr. In 1958, I was invited to teach at the OH State Univ. I have been at OSU since that time. I am married and have four sons. My pastimes include handball, weightlifting and fishing. My major publications have been in the areas of the mechanical properties of metals and alloys, diffusion, powder metallurgy and failure analysis. *Society Aff:* ASM, IMS, APMI, NACE.

Powell, H Russell
Business: 180 Allen Rd NE-217, Altanta, GA 30328
Position: V Pres, Secy. *Employer:* Hartrampf/Powell Inc. *Education:* BME/Mech/GA Tech. *Born:* Nov 1937. Native of Augusta Ga. US Army Corps of Engrs. Facils engrg, Lockheed-Ga Co 1960-63 mech & process engrg. Mech Dept Hd for Zimmerman, Evans, & Leopold, Cons Engrg 1963-66, respon for mech & process engrg. 1966-68 Assoc with Wm J Hartrampf & Assocs, Cons Engrs. 1969- , principal in firm Hartrampf/ Powell Inc. Jointly rspon for all engrg produced in fields of mech & civil engrg. Active in Catholic Church & Civic activities. Enjoy classical music & family camping. *Society Aff:* NSPE, ASCE.

Powell, Herschel L
Home: 4685 Blanding Dr, Memphis, TN 38118
Position: VP *Employer:* Archeon, Inc. *Education:* -/ME/TX Tech Univ. *Born:* 5/29/26. Served in S Pacific aboard USS Lackawanna (USNR) WWII. 1947-1953 Asst to Chief Engr in design of A C Systems for Hughes Tool Co, Houston, TX. 1953-1956 Chief Engr for Cook & Nichol, Inc, distributors for GE Heating & Air Conditioning. 1956-1969 Chief Engr fo Yager & Nenon, Inc, designing & installing Mech & Plumbing Systems world wide. Reg PE in 22 states. 1970-87, Herschel L.

Powell, Herschel L (Continued)
Powell & Associates, Partner. 1987 assumed position of VP of Archeon, Inc., Architects & Engineers. *Society Aff:* NSPE, TSPE, ASPE, ASHRAE.

Powell, Michael J
Home: Rt 11, Columbia, MO 65202
Position: Vice Pres *Employer:* MFA Inc *Education:* BS/Ag Eng/MO Univ *Born:* 02/15/44 Raised on a grain and livestock farm near Warsaw MO. Worked as a Project Engr. Responsible for Design and Construction of feed, grain, and fertilizer facilities for FS Services at Bloomington, IL 1966-1971. Accepted a similar position with MFA Inc at Columbia MO in June 1971. Promoted to mgr of Engr in Nov of 1971 and to VP of Engr, Quality Assurance in Nov of 1979 with resonsibility for all Engr, quality control, weight & measures, and safety functions and compliance with all regulation relating to same. Currently Chrmn of MO Sec ASAE, mbr of Design Committee of Mid Central Sec ASAE, and mbr Univ of MO Ag Engr Advisory Council. *Society Aff:* ASAE, NSPE

Powell, T Charles
Home: 815 Canterbury Ln, Saginaw, MI 48603
Position: Mgr., Mfg. Engrg. *Employer:* Saginaw Steering Gear Div, GE. *Born:* Mar 1938 Detroit Mich. MS MIT; BSME G M Inst. Have participated in design of std, adjustable, energy absorbing & air bag steering columns; prop shafts; suspension ball joints; hydraulic pumps; conventional & rack & pinion steering sys of both power & manual type. Awarded Alfred P Sloan Fellowship at MIT for 1973-74. Co-recipient of Ralph H Isbrand Award for contrib to automotive safety by SAE 1974. Active in var levels of church work & esp enjoy working with youth. Hobbies incl golf & domestic & foreign travel.

Powell, Wm Llewellyn
Home: 7425 Villanova, Dallas, TX 75225
Position: Partner *Employer:* Powell & Powell Engrs *Education:* BSCE/Sanitary/Univ of TX-Austin *Born:* 04/27/11 Dallas, TX. In 1932, joined father in consltg engrg practice established 1925. Five years US Navy CEC duty World War II, OinCC Navy and Marine aviation facilities. Major projects City of Dallas 150 mgd sewage treatment, Dallas Co. Park Cities MUD 25 mgd water supply and treatment; Dallas-Ft Worth Turnipke including award winning Hampton Rd Arch bridge. Current practice: hydrology, water resources, urban drainage and flood control, water and waste water treatment, bridges and engineering structures. Outstanding Cvl Engr Award, Dallas Branch ASCE 1978; active local and state profl and technical societies, including presidency terms. Active layman Episcopal Church. Reg Prof Engr in TX. 50 yr certificate. Married Mary Lee Weston 1934. 5 children, 10 grandchildren. *Society Aff:* NSPE, PEPP, ACEC, ASCE, AWWA, ACI, AREA, SAME, ASTM, WPCF, AWRA

Power, Richard J
Home: 11512 Four Penny Ln, Fairfax Station, VA 22039
Position: Productivity Principal. *Employer:* Dept of Defense. *Education:* MS/Ind Engr/OH State Univ; BIE/Ind Engr/OH State Univ. *Born:* 5/30/26. Started profl career in 1951 as an Ind Engr in Cleveland, OH area. Worked with Harris Corp, a consulting firm, and two aerospace mfg. Joined the Dept of Defense (DoD) in 1958 at Dayton, OH, as Supervisory Engr responsible for Depot Maintenance Systems Design & Deputy Dir of Ind Operations. In 1971, joined OSD staff as Technical Asst to Dir of Ind & Mgt Engg. Subsequently, became Dir of Office, & currently DoD Productivity Principal & Dir of the Defense Productivity Prog Office, responsible for productivity measurement and improvement. MTM Assn Dir 1973-74, 1977-80. VP 1975-76. Recipient of President's Award 1978. IIE Dir, Govt Div 1978-79. *Society Aff:* IIE, MTM.

Powers, John H
Home: Box 346 R D 3, Pleasant Valley, NY 12569
Position: Mgr *Employer:* IBM *Education:* MS/Ind Admin/Union Coll; BE/EE/ Stevens Inst of Tech *Born:* 03/25/43 IBM experience: 1964-73 Engrg & Mgmt assignments in component development & engrg. 1973-76 Product Mgr-technical & business responsibility for a wide variety of electronic components including hybrid & discrete devices. 1976-81 Corp Mfg Staff Mgr of Technical Services with staff responsibility for all technical functions and inter-divisional activities across entire scope of IBM's mfg operations, including Mfg Engrg, Quality Assurance, Plant Engrg & Chem Mgmt. Currently Mgr of Semiconductor Equipment & Systems Engrg at IBM's E Fishkill, NY facility. Professional Activities: Twice elected Pres of Components, Hybrids & Mfg Tech Soc of IEEE. Numerous technical papers, publications, and presentations on electronic component & mfg technologies. Past Gen Chrmn of Electronic Components Conference. Recipient of IEEE/CHMT Outstanding Contribution Award and IEEE Centennial Award. Author of textbook, "Computer-Automated Manufacturing–, published McGraw-Hill, 1987. *Society Aff:* IEEE, ISHM

Powers, Kerns H
Business: David Sarnoff Res Ctr, Princeton, NJ 08540
Position: Staff VP, Communications Res. *Employer:* RCA Corp - RCA Labs. *Education:* ScD/EE/MIT; MS/EE/Univ of TX; BS/EE/Univ of TX. *Born:* Apr 15, 1925 Waco Texas. US Navy 1942-46. With RCA Labs since 1951 working on color TV, radar, & communication sys. 1958-66 served as Tech Dir of a proj that dev the Navy's extremely low-frequency sys known as SANGUINE. Assumed pos of Dir, Communications Res Lab 1966, and present position of Staff VP, 1977, respon for res progs in satellite communications, microwave technology, video recording & transmission, data transmission, & communication tech. Life Mbr Tau Beta Pi & Eta Kappa Nu; Fellow IEEE. Navy's Forty-One for Freedom Award 1967. *Society Aff:* IEEE, SMPTE

Poynor, Russell R
Home: 122 Country Club Pl, Geneva, IL 60134
Position: Retired (Part-time Conslt) *Employer:* Packer Engrg *Education:* MS/Civ Engg/Purdue Univ; BS/Civ Engg/Univ of WI; BS/Agri/Univ of WI. *Born:* 9/9/13. Raised on farm near Waunakee, WI. Taught engg at UT State Univ & did res and teaching at Purdue Univ (agri engg) before joining Intl Harvester in 1945. Respons at Intl Harvester included-engr on tillage machinery. Gen Supv Farm Practice Res, Farm Equip Product Planning and Mgr Farm Equip Product Safety. Retired 1977. Life Fellow of ASAE. Reg PE Recipient of ASAE Cyrus Hall McCormick Medal. ASAE Pres 1967-68; Hon Mbr, Farm Conf, Natl Safety Council. *Society Aff:* ASAE, SCSA, NIFS.

Prabhakar, Jagdish C
Business: Sch of Engg., Cal State Univ, Northridge, CA 91330
Position: Prof. *Employer:* CA State Univ. *Education:* PhD/Elec Engr/SMU Dallas; MS/Elec Engr/IIT Chicago; MS(Hons)/Physics/Panjab Univ; BS(Hons)/Physics/ Panjab Univ *Born:* 9/14/25. After working with the broadcasting Div of Govt of India, I came to US in 1962 for higher education. I have published in IEEE Proceedings & Transactions on Communication oriented subjects. The latest publication was on External Degree Programs in IEEE transactions on Education in 1982. *Society Aff:* IEEE, ASEE.

Prados, John W
Home: 7021 Stagecoach Trail, Knoxville, TN 37909
Position: V P for Academic Affairs and Research. *Employer:* Univ of Tennessee. *Education:* PhD/Chem Engg/Univ of TN; MS/Chem Engg/Univ of TN; BSChE/ Chem Engg/Univ of MS. *Born:* 10/12/29. Munitions Officer USAF 1951-53. With Univ of Tenn since 1956 serving as Prof of Chem Engrg, Assoc Dean of Engrg, Dean of Admissions & Records, Acting Chancellor of Knoxville & Martin Campuses and Acting Dir, Energy Conversion Div at Univ of Tenn Space Inst. Since 1973 V P for Acad Affairs respon for educ prog planning & eval for 4-campus statewide univ sys. Named VP for Academic Affairs and Res in 1981 with additional responsibility for system-wide res coordination. Engrg resident & part-time cons, Oak

Prados, John W (Continued)
Ridge Natl Lab 1957-73; Conslt, Enrichment Tech Programs, Union Carbide Corp and Martin-Marietta Energy Systems, 1979-1986. Natl Dir AIChE 1975-77; National Director Sigma Xi 1976-82, President-Elect 1982-83, Pres 1983-84. Outstanding Teacher & Disting Engrg Alumnus awards from Univ of TN. Engineer of Distinction Award, Univ of Mississippi, 1985. Fellow of AIChE. Knoxville-Oak Ridge Chem Engg of the Yr 1977. Engg Accreditation Commission, ABET 1978-1986, Vice Chrmn 1981-84, Chrmn 1984-85. Executive Council Member, Tau Beta Pi Assoc, 1986-Present. *Society Aff:* AIChE, ACS, ASEE, AIC, ASEM.

Prager, Martin
Business: 125 E 87th St, New York, NY 10128
Position: Materials Cons (Self-employed) *Education:* PhD/Matls Sci/UCLA; Master/Metallurgical Eng/Cornell; Bach/Chemical Eng/Cornell. *Born:* May 23, 1939. Thesis on environ embrittlement of hightemp alloys. Employed as matls engr by Rocketdyne 1962-68. Spec in heat treating, joining, failure analysis. Mgr of Application Engrg for Copper Dev Assn until 1974. Assoc Dir of The Metal Properties Council, Inc from 1978. Headed studies in fatigue, welding, coating, elec, & marine areas. Currently cons in failure analysis, metallurgy, joining & corrosion to marine, aerospace & elec indus. Authored Welding Res Council Bulletins on nickel alloys, other articles on embrittlement, fatigue, elevated temperature properties, welding, & clad matls. Served on AWS handbook, MPC, IEEE Substation, & NMAB Substitution cttes. PhD UCLA Awards AWS, IEEE. *Society Aff:* AIME, NACE, ASTM, ASM, IEEE.

Prasuhn, Alan L
Business: Civil Engineering Dept - Box 2219, Brookings, SD 57007-0495
Position: Prof of Civil Engr. *Employer:* South Dakota State Univ *Education:* PhD/Fluid Mechanics/University of Connecticut; MS/Hydraulic Engr./University of Iowa; BCE/Civil Engr./Ohio State University *Born:* 02/19/38 Taught at the Univ of CT from 1963-65 and CA State Univ from 1968-78. Serve as a consultant to the Army Corps of Engineers and others, primarily in river engrg and sediment problems. Long-term involvement in the ASCE History and Heritage Program. Serve as National Director for ASCE (1986-88). *Society Aff:* ASCE, ASEE, NSPE, IAHR

Prats, Michael
Business: Apartado S.A, Caracas 101, Venezuela
Position: Head Special Studies Reservoir Engineering *Employer:* Maraven S.A. *Education:* MS/Physics/Univ of TX; BS/Physics/Univ of TX. *Born:* Dec 1925 Tampa Fla. Raised in Cuba; native language Spanish. Returned to U S 1939; in Houston since 1940. Army AF Technician B-29 Remote Control Sys 1944-46. BS & MS Phys Univ of Texas 1949 & 1951. With Shell Dev Co since 1950; R&D in petro & gas reservoir engrg, supplemental recovery processes, in-situ processes for tar sands, oil shale & coal, heat & mass transfer in earth formations. Publs in J Petro Tech, Soc Petro Engrg Journal, Journal Geo Res, Oil & Gas Journal. Mbr SPE-AIME, AAAS. Dir-at-Large SPE 1976-79; SPE Cedrik K Ferguson award 1962; Lester C Uren award 1974 current position since July 12, 1981. *Society Aff:* SPE, AAAS.

Pratt, David F
Business: 5813 Main St, Williamsville, NY 14221
Position: Partner *Employer:* Gordon & Broderick Assoc *Education:* BS/CE/State Univ of NY-Buffalo *Born:* 09/23/44 Native of Holland, NY. Served in US Army Corps of Engrs, Okinawa 1967-1969 specializing in contract administration of military and civil works projects. Worked for McFarland Johnson Engrs 1966, and 1969-1974 in highway design and construction, also resident Engr for Sewage Treatment Plant Construction. Joined McIntosh & McIntosh PC Land Surveyors to form a consltg firm. Currently VP. In 1977 became a partner in Gordon and Broderick Assocs which was purchased by the principals. McIntosh specialized in site development, subdivision and sanitary engrg. Gordon & Broderick concentrates on civil works. Am Chief Engr and operating partner for all engrg projects both firms. *Society Aff:* NSPE, ASCE, SAME

Pratt, David T
Business: Mail Stop FU-10, Dept Mech Engr, Seattle, WA 98195
Position: Prof *Employer:* Univ of WA. *Education:* PhD/ME/Univ of CA; MSc/ME/Univ of CA; BSc/ME/Univ of WA. *Born:* 9/14/34. Native of ID Falls, ID. Married Marilyn Jean Thackston, 1956. Three children, Douglas (b 1959), Elizabeth (b 1960) & Brian (b 1962). Asst Prof, US Naval Acad, 1957-64. Asst, Assoc, Prof of Mech Engg & asst dean for instruction at WA State Univ, 1968-76. Prof of Mech Engg and Adjunct Prof of Chem Engg at Univ of UT, 1976-78. Prof and Chmn of Mech Eng and Appl Mech, Univ of MI, 1978-81. Present position since 1981. Fulbright-Hays Res Schol, 1974-75. Res Dir-Supercomputing-Aerojet Propulsion Research Institute, 1986-87. Hobbies include hiking, skiing, music. *Society Aff:* ASME, AIAA, SCS, CI, SIAM.

Pratt, George L
Home: 2519 Willow Rd, Fargo, ND 58102
Position: Chrmn. *Employer:* ND State Univ. *Education:* PhD/Agri Engr/OK State Univ; MS/Agri Engr/KS State Univ; BS/Agri/NDSU. *Born:* 1/31/26. Married Patricia Jones, 1955, Children: Thomas & Nancy. Instr, Agri Engg Dept, NDSU, 1951-52. Clay Equip Corp, Cedar Falls IA, 1952-53. Asst Prof, Assoc Prof, Prof, Agri Engg Dept, NDSU, 1953-63 & 1965-present. Department Chairman 1976-present. Grad Res Asst, OK State Univ, 1963-65. US Marine Corps 1944-46. USDA Inspection Team to USSR, 1975. ASAE Metal Bldgs Award, 1977. *Society Aff:* ASAE, ΣΞ, ΦΚΦ.

Pratt, Lawrence J
Business: S Ferry Rd, Narragansett, RI 02882
Position: Asst Prof *Employer:* Univ of Rhode Island *Education:* PhD/Physical Oceanography/MA Inst of Tech; MS/CE/Univ of WI; BS/CE/Univ of WI *Born:* Sept 1952 Rochester Minn. Native of Rhinelander Wisc. ASCE J Waldo Smith Hydraulic Fellowship 1976 for res on electronic analogs of progressive & standing water waves. Presently doing res on nonlinear problems in Geophysical Fluid Mechanics. *Society Aff:* AMS, AAAS

Prausnitz, John M
Business: Gilman Hall, Berkeley, CA 94720
Position: Prof of Chem Engrg. *Employer:* Univ of California. *Education:* PhD/Chem Eng/Princeton Univ; MS/Chem Eng/Univ of Rochester; BChE/Chem Eng/Cornell Univ *Born:* Jan 7, 1928. Mbr of fac Chem Engrg Univ of Calif BerBerkeley 1955- . Author or co-author 360 articles & 6 books & monographs on chem engrg thermodynamics with spec application to separation problems. Consultant to several corporations in the chemical, cryogenic and polymer industries. Colburn Award 1962 & Walker Award 1967 AIchE. Natl Acad of Sci 1973. Chem Engrg Award of ASEE 1976. Guggenheim Fellow 1962 & 1973. Murphree Award 1979 ACS. Natl Acad of Engg 1979. Dr Ing (h.c.) Univ of L'Aquila (Italy). *Society Aff:* ACS, AIChE, AAAS

Precious, Robert W
Home: 18684 W. Spring Lake Rd, Spring Lake, MI 49456
Position: Engineering Consultant *Employer:* National Dynamics Corp. *Education:* BS/Mech. Engr./University of Michigan, Ann Arbor, MI. *Born:* 04/18/24 Served as an Engrg Officer in US Navy during WWII. Worked in Field Service and Engrg for Babcock & Wilcox Co. until 1951. Was Principal Design Engr for steam and power projects at Union Carbide Chemicals Co. for 17 years, and General Mgr, Boiler Div of Henry Vogt Machine Co. for 15 years. Served as Pres of American Boiler Manufacturers Assoc, 1976-78. In 1983, became Pres of Johnston Boiler Co. and later Pres of Energy Recovery Inc. Currently retained as Engrg Consultant for the Parent, National Dynamics Corp. Life-Fellow in ASME. *Society Aff:* ASME

Preckshot, George W
Business: 1029 Engrg Bldg, Columbia, MO 65211
Position: Prof Emeritus, Chem. Eng. *Employer:* Univ of Missouri-Columbia. *Education:* BS/Chem Engg/Univ of IL; MS/Chem Engg/Univ of MI; PhD/Chem Engg/Univ of MI. *Born:* Nov 1918. Native of Collinsville Ill. Asst metallurgist Amer Zinc Co; oil field chemist Standard Oil Co Calif; Asst & Assoc Prof Univ of Minn 1948-64; since 1964, Prof & Chmn, Chem Engrg, Prof Emeritus Sept 1985 Univ of Mo - Columbia. 1956 recipient Fatty Acid Producers & Amer Oil Chem Soc Res Award & 1957-58 Guggenheim Res Scholar to Univ of Edinburgh. Sev terms as officer, Twin Cities Sect Amer Inst Chem Engrs, officer Mo Chap Sigma Xi. Reg P E Mo & Minn. Cons to a num of indus firms. Enjoys fishing, hunting & wine making, cabinet making. *Society Aff:* AIChE, ACS, AAAS, AAUP.

Prehn, W Lawrence, Jr
Business: 7616 LBJ Freeway-715, Dallas, TX 75251
Position: Dir - Southwest Operations. *Employer:* Economics Res Assocs. *Education:* MSc/Chem Engg/Cornell Univ; BSc/Chem Engg/Rice Univ. *Born:* Apr 1920. BSc ChE Rice Univ 1943; MSc Engrg Cornell 1946. Pilot plant engr ESSO Lab Baton Rouge 1946-49; evaluation engr Atlantic Labs Dallas 1949-52; ordnance engr US Navy 1943-46 & 1952-54; res engr Stanford Res Inst 1954-62; commercial & indus dev engr Del E Webb Corp Houston 1962-63; Dir Appl Econ & Social & Mgmt Sci, Southwest Res Inst 1964-75; engrg economics cons, Business & Indus Planning Assocs (own business) 1975-76; Dir Southwest Operations, Economics Res Assocs Dallas 1976- . Involved in tech eval of mkts, feasibility, & related studies in applied economics. *Society Aff:* AIChE.

Preiss, Kenneth
Business: Ben-Gurion Univ of the Negev, Beer Sheva, Israel 84105
Position: Sir Leon Bagrit Prof. of CADCAM *Employer:* Ben-Gurion Univ of the Negev *Education:* Ph.D./Nuclear Energy/U. of London, UK; DIC/Nuclear Power/Imperial College of Science & Technology, London; BS/Civ. Eng./U. of Witwatersrand, Johannesborg *Born:* 08/11/37 30 years experience in industry and university, mostly as a univ professor. Experience in civil engrg, nuclear engrg and in computer aided mech des and manufacturing. Has been working on artificial intelligence applications in engrg since 1977. Member of national and international bodies; editorial bds of various professional journals. Over 100 technical publications. Honorary member of ASME since 1980. *Society Aff:* ASME, ACM, IEEE, SME

Prelas, Mark A
Business: 0039 Engrg, Columbia, MO 65211
Position: Assoc Prof *Employer:* Univ of MO *Education:* PhD/Nuclear Engrg/Univ of IL at Urbana-Cham; MS/Nuclear Engrg/Univ of IL at Urbana-Cham; BS/Engrg Sci/CO State Univ *Born:* 07/02/53 Born in Pueblo, CO and received BS from Co State Univ, MS and PhD from Univ of IL (UI). While at UI, discovered five "nuclear-pumped" lasers, and was a UI fellow. Joined the Univ of MO-Columbia in 1979. Developed courses and founded a res program in thermonuclear fusion, this program is supported by major grants from industry. Developed high-power laser concepts, and co-founded res company-- Nuclear-Pumped Laser Corp. Recognized in 1983 as a Presidential Young Investigator. On cooperative res assignment with the RFC-XX-M Proj at the Japan Inst for Plasma Phys in 1984, dev plasma diagnostics. *Society Aff:* ANS, IEEE, APS, ASEE

Prendiville, John F
Home: 27 Elm St, West Acton, MA 01720
Position: VP - Network. *Employer:* New England Telephone Co. *Education:* BS/Elec Engg/Northeastern Univ; MS/Elec Engg/Northeastern Univ; SM/Ind Mgmt/MA Inst of Tec Northeastern *Born:* 5/26/28. Native of Waltham MA. With New Englant Telephone since 1948. Army Signal Corps 1950-54, First Lt R A. Assumed current respon as VP Network 1978. Respon for Network Planning, Switch Service Operator Services, Distributions & Engr. Dir & Const Users Council Dir Natl Council Northeastern Univ. Tau Beta Pi, IEEE. Enjoy jogging, refereeing ice hockey, swimming & sailing. *Society Aff:* TBΠ, IEEE.

Prendiville, Paul W
Business: 1 Center Plaza, Boston, MA 02108
Position: Sr V Pres. *Employer:* Camp Dresser & McKee Inc. *Education:* MS/Sanitary/Northwestern Univ; BS/CE/Northwestern Univ. *Born:* Mar 1933 Boston Mass. MS San Engrg Northeastern Univ 1962; BSCE Northeastern Univ 1955. Joined the Camp Dresser & McKee partnership 1959 as a proj engr & was made as assoc in 1966, with respon as proj mgr on sev major assignments. In 1970, appointed a partner & at the same time a sr vice pres of Camp Dresser & McKee Inc, now the parent firm. Respon for such projs as the design of the 500 mgd water filtration plant for Los Angeles. Diplomate, Amer Acad of Environ Engrs, BSCE San Sect Award 1964. *Society Aff:* AWWA, ASCE, AAEE, WPCE.

Prentiss, Louis W, Jr
Business: Commanding Genl, Fort Leonard Wood, MO 65473
Position: Commanding General *Employer:* U.S. Army *Education:* MSE/Civil/Princeton; BS/Mil Eng/USMA. *Born:* Dec 25, 1927. Reg P E DC. Military assignments have incl troop command & staff, staff & faculty, Army Engr School; Dir, Peru Inter-Amer Geological Survey; Baltimore Dist Engr; Div Engr US Army Engr Div Europe. Deputy Chief of Staff, Eng US Army Europe, Div Engr, OH River Div. As Dist Engr (Balto), dir Hurricane Agnes recovery operations 1972 in the Susquehanna & Potomac River valleys. Assumed current respon as Commanding Genl, Fort Leonard Wood, MO. Regional V P Europe, Soc of Amer Military Engrs; Mbr Amer Soc Civil Engrs, Natl Soc Prof Engrs. Military awards incl Legion of Merit, Bronze Star, Meritorious Serv Medal, Air Medal & Army Commandation Medal. Enjoys skiing, golf, tennis & bowling. *Society Aff:* NSPE, ASCE, SAME.

Preparata, Franco P
Business: Coordinated Science Laboratory, Urbana, IL 61801
Position: Prof, EE & CS. *Employer:* Univ of IL. *Education:* DrIng/EE/Univ of Rome. *Born:* 12/29/35. in Reggio Emilia, Italy. He received the Dr Eng degree in 1959 and the Libera docenza from the Italian Univ Sys in 1969. After yrs of ind experience with Univac & Raytheon, in 1965 he joined the Univ of IL, Urbana. Since 1970 he has been a Prof of Elec Engg and a Res Prof in the Coordinated Sci Lab, Univ of IL at Urbana. He spent the academic yrs 1972-73 and 1973-74 at the Univ of Pisa, Italy, as a Chair Prof of Comp Sci. His res interests lie in the areas of computational complexity, combinatorial computing, and theory of VLSI computations. *Society Aff:* IEEE, ACM, EATCS.

Presecan, Nicholas L
Business: 75 North Fair Oaks Avenue, P. O. Box 7107, Pasadena, CA 91109-7207
Position: Sr VP. *Employer:* Engg-Sci, Inc. *Education:* MS/Sanitary Engg/Univ of CA; BS/Civ Engg/Purdue Univ. *Born:* 9/4/40. San Bernardino County Flood Control 1963, US Marine Corps from 1963 to 1966 as Aviation Electronics Officer. Joined Engg-Sci, Inc (ES) in 1968 as a proj engr conducting water resources and water reclamation studies and designs of advanced wastewater treatment facilities. Performed as principal engr and proj mgr on numerous water and sewarage works projs. Appointed VP for the midwestern region in 1972 and became The Chief Engr in 1973. VP of Int'l Div 1981-1983. Group VP and Manager Western Group 1984-1986. Sr. VP and mgr of companywide Engineering Direction 1986. Licensed Professional Engineer in 30 states. *Society Aff:* ASCE, NSPE, WPCF, AWWA, AAEE, SAVE.

Presley, Gordon C
Home: 8581 East Dry Creek Place, Englewood, CO 80112
Position: Mineral Proj Devel *Employer:* Conslt *Education:* BS/Geol/VA Polytech Inst & State Univ - Sch of Engrg & Architecture *Born:* 12/16/36 Began career as a Mining Engr for US Gypsum Co (Gypsum & Anhydrite) and Intl Minerals & Chem Corp (Bentonite). Was plant engr, asst plant mgr, plant mgr (Koalin) and sr project engr (Talc) for Cyprus Industrial Minerals Co. Formed Horton Intl, Inc, a privately

Presley, Gordon C (Continued)
owned Kaolin Co. Directed the construction of a Fuller's Earth Mine and processing plant for Anschutz Mineral Corp, and was responsible for engrg, operations and marketing. Was 1974 participant in leadership Georgia program. Dir of the Soc of Mining Engrs (1976-1978 & 1980- 1983). Chrmn of the Industrial Mineral Div of SME-AIME (1980-1981); Mbrshp Chrmn, SME-AIME 1982-85, Treas of SME-AIME, 1982-85; Mbrshp Chrmn CO Sec of AIME 1981-85; Has recruited over 1236 of SME's approx 24,000 mbrs. Since April 1984 has conslt in mineral proj devel in indus minerals, specializing in acquisitions, evaluations, proj recommendations, and plant construction. *Society Aff:* SME, AIME

Presnell, David G, Jr
Business: 200 West Broadway, Suite 701, Louisville, KY 40202
Position: President *Employer:* Presnell Assocs, Inc *Education:* BS/CE/Univ of KY *Born:* 03/02/37 Native of Ft Sill, Ok. VP and Regional Mgr for Midwest office of Vollmer Assocs, Inc until 1974. Pres and Chrmn of the Bd of Presnell Assocs, Inc since its inception. Gen Mgr for Louisville and Jefferson County Metropolitan Sewer District's $500 million sewer expansion program. Past Pres of Louisville Chapter of KY Soc of PE and Scientific Council. Outstanding Engr in Private Practice, 1974, KSPE. Activities: Bridge, tennis, boating. *Society Aff:* ACEC, KSPE, ACSM, AASHTO, APWA, AWWA, SAME

Press, Leo C
Business: 7470 N Figueroa St, Los Angeles, CA 90041
Position: Principal. *Employer:* Sampson, Randall & Press. *Education:* BS/Elect Engr/Univ of CA, LA. *Born:* July 1922. Reg elec engr in Calif, Ariz, Nev. Native of Los Angeles Calif. Army AF 1941-45. Elec engr Kistner, Wright & Wright 1948-63. Principal in Sampson, Randall & Press 1963- . Pres Assn of Cons Elec Engrs 1976-77. Chmn Mech Elec Cons Engrs Council 1975-76. Bd of Dir Cons Engrs Assn of Calif 1968-70. Adv Bd to Office of the State Architect (Calif) 1968- . Mbr IEEE & IES. *Society Aff:* IEEE, IES, ACEC.

Pressman, Ada I
Home: 1301 S Atlantic Blvd, Monterey Park, CA 91754
Position: Engg Mgr. *Employer:* Bechtel Power Corp. *Education:* BSME/-/OH State Univ; MBA/-/Golden Gate Univ. *Born:* 3/3/27. Native of Troy OH. Joined Bechtel 1955, spec in instrumentation & controls on fossils & nuclear power projs. Ch Control sys engr 1974-79 for the Los Angeles Power Div of Bechtel Power Corp. Assumed the position of Engg Mgr 1979. Dist VP of ISA 1973-75, VP Publs ISA 1976-78, Exec Ctte SWE 1975-77. VP 1977-79, Pres 1979-80. Engg Merit Award Los Angeles IAE 1975. Disting Alumni Award OSU 1974. Achievement Award YWCA 1976. SWE Achievement Award 1976. Engr of the Yr, Long Beach Engr Council 1979. YWCA Twin Award, 1981. Reg Mech Engr CA & AZ. Reg Control Sys Engr CA. *Society Aff:* SWE, ISA, ANS.

Pressman, Norman J
Home: 626 Whitby Drive, Wilmington, DE 19803
Position: Research Scientist *Employer:* E.I. DuPont de Nemours & Co. *Education:* PhD/Medical Engg/Univ of PA; MS/Systems Engg/Univ of PA; BS/EE/Columbia Univ. *Born:* 9/30/48. NY, NY, m 1971; BS, Columbia Univ, 1970; MS, Univ PA, 1972, PhD, 1976. Profl Exp: Biomed Engr, Lawrence Livermore Lab, Univ CA, 1973-1976; Hd, Quantitative Cytopathology Labs, the Johns Hopkins Med Insts, Asst Prof Pathology, Med Sch, 1976-1986; Consultant, NIH & FDA; Chrmn, Intl Coun on Automated & Quantitative Cytol, Intl Acad Cytol, 1977-; Mbr, IEEE, Biomed Engg Soc, AAMI, OSA, ACM, ISA, IAC, ASC, NY Acad Sciences, Sigma Xi; Res: Quantitative & automated cytopathology; medical imaging processing & analysis; flow-system analysis & sorting of biological cells; clinical lab comuterized info systems. Research Scientist, E.I. du Pont de Nemours & Co., Exp. Stat., Central R & D, 1986-. *Society Aff:* IEEE, AAAS, BES, ACM, AAMI, ASC, IAC, OSA, SAC

Prestele, Joseph A
Home: 1 Parkwood Dr, Atherton, CA 94025
Position: Exec Asst to the V P. *Employer:* Consolidated Edison Co of NY Inc. *Education:* ME/ME/Stevens Inst of Technology; MS/IE/Stevens Inst of Technology. *Born:* Mar 1928. BS & MS Stevens Inst of Tech. Native of NYC. US Army 195052, spec in radiation safety. Joined Consolidated Edison 1949. Consulted for Knolls Atomic Power Lab 1956-58. Served as Asst Plant Mgr in startup of nuclear power plant Indian Pt 1. Last served as Mgr, Nuclear Power Generation Dept. Presently cons in nuclear matters for the Elec Power Res Inct, Palo Alto Calif. Mbr ASME & Fellow ANS. Lic P E NY. Mbr ANS Bd of Dir & Chmn ANS Stds Steering Ctte. Represented the U S at U N meeting in Bangkok Thailand 1973, speaking on U S Light Water Reactor Prog. *Society Aff:* ANS, ASME.

Preston, Floyd W
Home: 832 Sunset Dr, Lawrence, KS 66044
Position: Prof *Employer:* Univ of Kansas. *Education:* PhD/Pet & Natl Gas/PA State; MS/Chem Engg/Univ of MI; BS/Chem Engg/Univ of MI; BS/Chemistry/Cal Tech. *Born:* Feb 11, 1923. PhD Petro & Nat Gas Penn State 1957; MS ChE Michigan 1948, BS ChE Michigan 1947; BS Chem Caltech 1944. Res engr Calif Res Corp - Std Oil Calif 1948-50, Asst Prof Petro Engrg Univ of Kansas 1955-57 - Assoc Prof 1958-64 - Prof Chem & Petro Engrg 1965- , Chmn Chem & Petro Engrg Dept Univ of Kansas 1974-79. Advisor Ministry of Mines & Hydrocarbons Venezuela 1959 & 1960. Cons in oil & natural gas to gas companies & federal agencies. *Society Aff:* AIChE, SPE, ACS, ACM, ASEE.

Preston, Frank S
Business: Engr Analysis & Design, Charlotte, NC 28223
Position: Consultant & Distinguished Lecturer *Employer:* Univ of NC at Charlotte. *Education:* MS/Elec Engr/MA Inst of Tech; BS/Elec Engr/Univ of WA. *Born:* 7/30/18. Designed equip and was the first to produce SF6. The process was adopted to the production of UF6 for atomic energy prog. Dev military counter measure equip in WWII for NDRC. Worked 33 yrs for Norden Div of United Tech in various positions including Chief Engr. Hold 12 patents and related publications. Supervised dev of ECAP and GPSS/N. Retired in 1977 and now teach engg analysis at UNCC. Chrmn Westchester Sub-sec IEEE 1957. Chrmn AOA. Comm on Effectiveness Analysis (1968). Professional Engr State of CT (No 5079). Now consulting on systems effectiveness and simulation. Stanford/ASEE Fellowships 1980, 1981 including work on giga flop computer. *Society Aff:* IEEE.

Preston, Walter B
Home: 103 Norcrest, San Marcos, TX 78666
Position: Retired-(Consulting) *Education:* B.S./Mech. Eng/Univ of Texas - Austin *Born:* 04/13/04 Graduate Univ. of TX (Austin) Tau Beta Pi, Pi Tau Sigma, Graduate Westinghouse Graduate Training Prog (Pgh). Sales Westinghouse, Instructor, Mech. Engg. Dept. Univ. of TX, Austin. Forty-one years with Texas Gulf Sulphur Co. (Now Els Aquaitaive Inc) Retired from Texas Gulf. Varied experience Management, design, construction, operation. Ch, South Texas Section (Houston) ASME. ASME activity in other offices. Elevated to Fellow grade in ASME. *Society Aff:* ASME, TSPE (NSPE)

Prewitt, Charles E
Home: 1105 Linwood Ave, Metairie, LA 70003
Position: President *Employer:* Denson Engrs, Inc. *Education:* MME/Mech Engg/Tulane Univ; BSME/Mech Engg/Univ of KY. *Born:* 10/23/47. With Texaco, Inc 1970-1974 as Gas Engr designing and operating natural gas processing and production facilities in South LA. Joined Denson Engrs, Inc in 1974. Became President in 1984. Active in design of facilities for various commercial, industrial, and govt clients. Also active in forensic engineering, specializing in investigation of accidents, fires and equipment failures. Accepted by state and federal courts as expert. Registered PE in Lousiana. Since 1974. Active in progs and activities of various engg soc and Parkway Presbyterian Church. Served as part- time professor mechancial engineering at Univ. of New Orleans, 1979-80. *Society Aff:* ASME, NSPE, LES, AEE, SAME, NAFE.

Prewitt, Judith M S
Home: 7304 Nevis Rd, Bethesda, MD 20817
Position: Research Mathematician *Employer:* Natl Institutes of Health *Education:* PhD/Computer Sci & Applied Math/Uppsala Univ; MA/Math/Univ of PA; BA/Math/Swarthmore Coll *Born:* 10/16/35 IEEE: Fellow, 1981. IEEE Computer Soc: Chrmn, Computational Medicine Technical Committee, 1981- ; Machine Intelligence and Pattern Analysis Exec Committee, 1979- . IEEE: Representative to AAAS, 1981- . IEEE Systems, Man and Cybernetics Soc: Admin Committee, 1981- . SIAM: Natl Visiting Lecturer, 1976- ; Representative to AAAS, 1979- . BMES: Bd of Dirs, 1978-1981; Alliance for Engrg in Medicine and Biology Project Advisory Committee, 1979- . ACM SIGBIO Exec Committee, 1979- . AWM: Exec Committee, 1980- . NCGA Program Committee, 1981- . Editorial Bds: IEEE Transactions on Pattern Analysis and Machine Intelligence, Computer Graphics and Image Processing, Medical Decision making, Analytical and Quantitative Cytology, Computer Graphics News. Research interests: *Society Aff:* IEEE, SIAM, ACM, BMES, AWM, AAAS, NCGA

Price, Bobby E
Business: PO Box 10348, Ruston, LA 71272
Position: Prof of Civ Engg Dir. Undergrad Stud *Employer:* LA Tech Univ. *Education:* BS/Civil Engr/Arlington State Coll; MS/Civil Engr/OK State Univ; PhD/Civil Engr/Univ of TX-Austin *Born:* 11/21/37. Native of Henderson, TX. Worked for the City of Dallas, TX, and City of Austin, TX. Assumed present position in 1967. Named Dir of Engg Grad Studies in 1978 and Dir of Undergrad Studies in 1983. Also, Dir of Water Resources Ctr since 1968. Named for outstanding teaching and service to the univ by various student groups, the Coll of Engg alumni. Served on Bd of Dirs as an officer in the LA Sec of the ASCE, the Monroe Chapter of LA Engr Soc and the Southwest Section of the AWWA including Pres of each. Presently serving as Pres-Elect of the LA Engrg Soc. *Society Aff:* NSPE, ASCE, AWWA

Price, Donald D
Home: 2009 183rd NE, Redmond, WA 98052
Position: Pres. *Employer:* Weld Met Tech *Education:* BS/AE/Lockheed Inst of Adv Studies. *Born:* 8/4/18. Native of San Fran, CA. Served apprenticeship in shipyards. Supt of Burning/Welding, Todd Corp prior to WWII. Airplane pilot WWII, Command pilot- Korea, Command pilot-Vietnam War. Prof pilot between wars. Owned, oper weld shops. Welding Res Spec-Lockheed Missiles. Chief Welding Engr, Lockheed Shipbuilding Corp. Pres, Weld Met Tech-Consultants. Pres, Stress-Relieving Tech, Inc. Author, Lecturer, Teacher, Holder of AWS QC-1. *Society Aff:* AWS, ASM.

Price, Donald R
Business: Grinter Hall, Gainesville, FL 32611
Position: VP for Res; Pres, Univ Florida Research Foundation, Inc. *Employer:* University of Florida *Education:* PhD/Agri Engr/Purdue Univ; MS/Agri Engr/Cornell Univ; BS/Agri Engr/Purdue Univ. *Born:* 7/20/39. Price joined the Cornell faculty in 1962 as an Asst Prof and Prog Leader of the Farm Electrification Council. He was promoted to Assoc Prof in 1969 and to full Prof in 1977. He served as prog mgr in the US Dept of Energy in 1977-78 while on leave from Cornell. He was assigned to Pres Carter's Reorganization Proj in 1978. He served as chrmn of a US panel to draft a natl energy policy for the food system. He is a mbr of the Bd of Dirs for the Food and Energy Council and the Natl Food Advisory Bd. November 1980, Assistant Dean for Research and Extension, Institute of Food and Agricultural Sciences, University of Florida. Sept 1982, Assoc Dean for Res Central Adm. Jan 1983, Acting Dean for Grad Studies and Res; and July 1984, VP for Res and President, University of Florida Research Foundation, Inc. *Society Aff:* ASAE, NCURA, SURA, SRA

Price, Eugene B
Home: 28 Valley Forge Dr, E Brunswick, NJ 08816
Position: Project Manager *Employer:* H-R International, Inc. *Education:* BS/ChE/Bucknell Univ; -/Grad work in Chemical Engr Yale Univ *Born:* 1/20/31. Native NYC area. Left Yale Grad Sch for Naval Service as Engg Officer aboard destroyer & minesweeper 1953-56. Fermantation recovery production supervisor with Pfizer 1957. Proj engr with Chemico 1958-64 including proj mgt of large grass-roots fertilizer plant. Special equip design engr with M.W. Kellogg, 1964-66 Proj engg group leader with Ciba-Geigy 1966-72 including mgr of multi-purpose pilot plant proj. Pres of steel fabricating shop. Chief mech & chief proj engr with Nichols Engg & Res 1974-77. Plant mgr for co recovering nonferrous metals from waste residues. Plant projs supervisor SGS Control Systems. Proj Mgr, the Hayward Robinson Co, Inc. Now H-R International Jan 1980-present. including resident mgr of group of 30-40 engr and support personnel. Acting as part of Exxon's Proj Dev Dept at the Bayonne/Bayway Refineries. Proj Mgr Fuel Gas Unit at Mobil's Joliet Refinery. Proj Mgr Tank Farm and FCCU Fractionation Revamp Proj Sun's Marcus Hook Refinery, Proj. Mgr. various projects mostly at CIBA-Geigy's Toms River Plant, Resident Engineer at office bldg and road projects. *Society Aff:* AIChE, AMA.

Price, Kenneth S
Business: 211 N Race St, Urbana, IL 61801
Position: Vice Pres - Operations. *Employer:* Clark, Dietz & Assocs - Engrs Inc. *Born:* Aug 4, 1942. PhD Purdue Univ; BS & MS Purdue Univ. Native of Kirklin Ind. Worked for Indiana State Hwy Dept 1964. Instr at Purdue Univ, School of Civil Engrg 1967-68. Served in Medical Serv Corps, US Army, Preventive Medicine Div, dir & oper environ health activities 1968-70. Environ engr Union Carbide Corp, R&D 1970-73. V Pres Oper, Clark, Dietz & Assocs Engrs Inc, respon for all corp opers of 6 branch offices & 10 internal divs. Mbr ASCE, AWWA, WPCF. ASCE Rudolph Hering Medal 1974. Enjoy golf & family activities.

Price, Ralph H
Home: 4100 Jackson Ave-Apt 312, Austin, TX 78731
Position: *. *Employer:* Retired. *Education:* BS/Chem Eng/MA Inst of Tech; AB/Chemistry/Baker Univ. *Born:* Feb 1895 Kansas. Teaching Fellowship Harvard Univ 1918; Instr MIT Sch of Chem Engrg Practice 1921-24. 2nd Lt San Corps WW I 1918-19. R&D Std Oil Co Ind & subsidiaries, 29 yrs, engaged in petro refining, petrochem dev & dept admin. Charter Mbr So Texas Sect Amer Inst of Chem Engrs. Chmn of section & Natl Dir AIChE, each 1 yr. Also served on sev natl cttes of AIChE. Fellow AIChE 1973. Mbr ACS. *Society Aff:* AIChE, ACS.

Price, Robert I
Home: 1 Oaklawn Road, Short Hills, NJ 07078
Position: Maritime consultant *Employer:* Self *Education:* NavE/Naval Arch/MIT; BS/Marine Engg/US Coast Guard Acad; BBA/Accounting/CCNY. *Born:* 9/22/21. Career US Coast Guard Officer, Commissioned 1945, Promoted VAdmiral 1978. Sea service as Deck, Navigating, Engr and Commanding Officer. Involved natlly and intlly ship, port and environmental safety progs and stds. US principal tech negotiator to United Nations Maritime Agency IMD. Duty as Capt, Port of Phila; Chief, Coast Guard Hdquarters Marine Environment and Systems; Commander, Southern CA Dist. Commander Atlantic Area responsible for all Coast Guard operations east of continental divide. Retired govt service 1981. Twice decorated Distinguished Service Medal. (1982 recipient SNAME Land Medal). Managed NY office of major naval architect and marine engrg firm. (1981-86). Now Self-Employed Maritime Consultant. *Society Aff:* RINA, SNAME, ASNE, ΣΞ, NAVY LEAGUE, PROPELLER CLUB

Pricer, W David
Home: 58 Van Patten Pkwy, Burlington, VT 05401
Position: Sr Engr. *Employer:* IBM. *Education:* MS/EE/MIT; BS/EE/MIT; BA/Phys/Middlebury College. *Born:* 7/22/35. Since 1959 participating in the dev of magnetic

Pricer, W David (Continued)
& semiconductor circuits for various computer systems applications at IBM. Awarded over thirty US Patents. Past Chrmn of the Intl Circuits Conf. Presently mbr Solid State Circuits & Tech Committee, mbr Sigma Xi, Fellow IEEE, Chrmn of the IEEE Solid State Circuits Council. *Society Aff:* IEEE.

Prickette, Gerald S
Business: 14227 Fern Dr, Suite 204, Houston, TX 77079
Position: Pres *Employer:* Gerald S. Prickette Consulting Engrs, Inc *Education:* BS/Arch Engr/Univ of TX-Austin *Born:* 03/31/45 Graduated from UT-Austin in January 1968. Worked for Chicago Bridge & Iron Co as Field Engr for one year. Left for military service. Went to work as Design Engr in Houston for consltg firm and stayed until starting my own company in 1978. Obtained Real Estate Broker's Liscense in 1980. In connection with Engrg Co. I started a Construction Co in March 1980. For recreation I enjoy tennis and racquetball. *Society Aff:* ACEC, CEC

Priddy, I Richard
Home: 326 South Rte 3, Millersville, MD 21108
Position: President *Employer:* Priddy Design Assoc. *Education:* BSCE/Civil Eng/VA Polytechnic Inst *Born:* 4/7/41. PE 4 states. Navy Civil Engr Corps 1965-69. Proj engr for land dev projs with prominent cons firms in MD suburbs of WA DC 1964-75. Established Priddy Design Assocs 1976-present. Founder & Pres Anne Arundel Council of Cons Engrs & Surveyors 1975-77; Mbr ASCE, NSPE, AWWA, NAHB. Free time spent boating & fishing. *Society Aff:* ASCE, NSPE, AWWA, NAHB.

Pridgeon, Hal L, Jr
Business: 735 E Main St, PO Box 3028, Spartanburg, SC 29304
Position: Principal *Employer:* Thomas, Campbell, Pridgeon, Inc *Education:* BS (Honors)/Arch-Engr/Clemson Univ; Grad Studies/Civ Engr/Univ of MO (Rolla) *Born:* 08/09/24 Native of Spartanburg, SC. Grad of Athens, GA, High Sch. Attended Univ of GA, Grad Clemson Univ BS (Honors) Arch-Engrg, Amer Inst of Architects Medalist, Tau Beta Pi, Grad Studies Univ MO (Rolla). Corps of Engrs, Europe, 1943-1946. Employed USDA-SCS 1942-1965. Lockwood-Green Engrs 1965-1967. Present association 1967 to date. Certificate of Merit-USDA. Prof Engr Reg: SC, NC, VA, DC, GA, FL, AL, TN, MS, AR. RLS in NC. Fellow ASCE, Mbr NSPE, Diplomate AAEE, Lions Club Charter Mbr. Methodist. Married. One son. *Society Aff:* ASCE, NSPE, AAEE, ACSPE, TBΠ

Pridgeon, John W
Business: Middle Settlement Rd, New Hartford, NY 13413
Position: Pres *Employer:* Special Metals Corp. *Education:* BS/Met Eng/MI State Univ; MS/Met Eng/Univ of TN. *Born:* 11/7/35. Advanced Mgmt Prog Harvard Bus Sch. Fellow ASM. Mbr Sigma Xi. Pres Mohawk Valley Chap ASM. V Chmn ASM Tech Divs Bd (Chmn starting Nov 1976). P Mbr ASM Long Range Planning Ctte. P Mbr Exec Ctte, Vacuum Met Div; Amer Vacuum Soc - presently on Adv Bd. Mbr ASM Spec Ctte on Natl Mbrship. Mbr ASM Power Ctte; Pres, Ajax Forging and Casting Co - 2/77-7/79; 8/79-7/81, V P-Tech, Special Metals; 7/81-82, Exec V P and Genl Mgr, Special Metals. 1982 - Pres Special Metals. Bd of Trustees, ASM - 1978 through 1981, VP & Trustee, ASM 1984-1985. *Society Aff:* ASM, AVS, AIME, AMA, AFS.

Priemer, Roland
Business: PO Box 4348, Chicago, IL 60680
Position: Assoc Prof *Employer:* Univ of IL *Education:* PhD/EE/IL Inst of Tech; MS/EE/IL Inst of Tech; BS/EE/IL Inst of Tech *Born:* 10/28/43 Author of numerous papers in the areas of control theory, network analysis, digital signal processing, and microcomputer applications in control systems and bioengrg. Has supervised many grad students at both the MS and PhD level. Conslt to various companies and institutions such as Argonne Natl lab and The IL Eye and Ear Infirmary. *Society Aff:* IEEE

Prien, John D, Jr
Home: 3861 Foxford Dr, NE, Atlanta, GA 30340
Position: Associate *Employer:* Dames & Moore *Education:* Juris Doctorate/Law/Univ of NB; BS/Civ Engg/Univ of NB. *Born:* 4/2/30. Native of Ord, NB. Served as officer in US Army Corps of Engrs, 1953-1955. Chief Civ Engr on Atlas Missile Prog, Lincoln AFB, NB, 1960-1962. Attorney/Contract Administrator, US Space Prog, Cape Canaveral, FL, 1962-1967. Exec Dir, GA Soc of PE, 1967-1975. State President, GA Soc. of Association Executives, 1972-73. VP for Business Dev, HDR, Inc, 1975-79. Dir. of Mktg., Bernard Johnson Inc., 1979-81; VP, ATEC Assoc, Inc, 1981-87. Associate, Dames & Moore, 1987- . GA "Engineer of the Year-, 1975. "Engineer of the Year, Metro-Atlanta," 1986. Founder and Pres, GA Engg Fdn, 1979. Chrmn, Natl Exec Committee on Guidance, 1979. Trustee, NSPE Education-al Fdn, 1975-. Natl Dir, GA Soc of PE, 1979-. Pres, Atlanta Post, SAME, 1982. Pres, NSPE Educ Fnd, 1984-86; Chrmn, NSPE St Soc Adm Council, 1973-75; Reg PE: GA, FL and NE; Mbr, NE and Amer Bar Assoc; Fellow, ASCE; Fellow, SAME; Mbr, "Order of the Engr-; Trustee, GSPE Political Action Ctte. Enjoy flying, wine-making, wood-working, photography, public speaking, etc. *Society Aff:* ASCE, SAME, NSPE, ACEC, ABA, SMPS.

Priester, Gayle B
Business: 800 Stoneleigh Rd, Baltimore, MD 21212
Position: Consulting Engr. *Employer:* Gayle B Priester, PE. *Education:* MS/Sanitary Engg/Harvard Univ; ME/ME/Case Inst of Tech; BME/ME/Univ of MN. *Born:* 7/1/12. Air cond sales engr Carrier Corp 1935-41 NYC, Chgo, Mpls. Case Inst of Tech 1941-46, teaching mech engg & consulting; resigned as Assoc Prof. Baltimore Gas & Elec Co, Air Cond Engr 1946-73, Ch Civ Engr 1973-1977. Consult Engr 1977-. Author many articles, handbook chapters, co-author textbook Refrig & Air Cond. Fellow ASHRAE, Past Treas, Dir, Region Chrmn, Pres 3 chapters, disting service award. Dir Acred Bd Eng'g & Tech. Past Dir Engg Soc of Baltimore. Presbyterian elder, deacon, trustee. P Pres PTA, P Officer Community Assn, Boy Scouts, United Appeal. Country Club of MD, Sigma Nu, Tau Beta Pi. Extensive traveller. PE. *Society Aff:* ASHRAE.

Priestman, John
2555 Penrhyn St, Victoria BC V8MIG2 Canada
Position: Retired. *Born:* 8/23/12. Served articled pupilage as a civ engr in England. Awarded the dignity of Mbr of the Order of the British Empire & Mentioned in Desphatches for service in Burma during 1939/46 war. Gold Medal by the Inst of Municipal Engrs of Great Britain in 1964. Emigrated to Canada in 1957, working in the consulting field, becoming VP & Gen Mgr of Ker Priestman & Assoc Ltd. Served as Natl Councillor, EIC 1973/76; Pres of CSCE 1976/77; Pres. of E.I.C. 1981. *Society Aff:* FEIC, FCSCE, PEng.

Primrose, Russell A
Business: 300 Coll Park, Dayton, OH 45469
Position: Dean *Employer:* Univ of Dayton *Education:* PhD/CE/VPI; MS/CE/VPI; BS (with honors)/CE/VPI *Born:* 05/02/32 Native of Waterloo, IA. Worked for the FBI in Wash, DC, prior to coll. As a co-op student, worked for ERDL, Ft Belvoir, VA; graduated with honors from VA Polytech Inst as Res Coordinator for the UMR (MSM). Hd of the Chem Engrg dept (1975-78) in Victoria, TX, before assuming the position as Dean of Engrg and Prof of Chem Engrg (1978-85) at the Univ of Dayton. Presently serviong as VP and Dean of the Coll, LeTourneau Coll, Longview, TX (1985). Served on the Engrs Fdn of OH, the Engrg Fdn in Dayton, and Pres of the Engrs and Sci Fdn (Dayton, OH). Program Chrmn of the Engrs and Sci Week (Dayton, OH), and active in many aspects of the assemblies of God. *Society Aff:* AAAS, AIChE

Prince, Elbert M
Business: 32 Edgewater Rd, Falmouth Fsde, ME 04105
Position: President. *Employer:* EMP Sales Inc. *Education:* BS/CE/Univ of ME. *Born:* Nov 1920. BS Univ of Maine. Native of Maine. Officer US Army Corps of Engrs; served in So Pacific Theater 1942-45. Town & city mgr for 6 yrs. Salesman Concrete Pipe Plant. Mgr, then V P of Sales for steel fabricating & concrete products mfg co until May 1974. Started sales co that date, serving as manufacturers' rep for sev natl co's who produce products for paper mills & genl hwy const. *Society Aff:* ASCE.

Prince, John L
Business: Dept of Elec & Comp Engr, Univ of Arizona, Tucson, AZ 85721
Position: Professor *Employer:* University of Arizonia *Education:* PhD/EE/NC State Univ; MEE/EE/NC State Univ; BSEE/EE/Southern Methodist Univ. *Born:* 11/13/41. Native of Austin, TX Res Engr, Res Triangle Inst, 1968-70. Mbr of the Tech Staff, TX Instruments, Inc, 1970-75. Prof, Dept of Elec and Computer Eng, Clemson University, 1975-80. Intermedics, Inc., Director of Reliability Assurance, 1980-83. Joined the Univ of Arizona as Prof of Elec and Computer Engrg. Research programs in electronic packaging and interconnections, and electronic reliability. Consultant to US Navy, Dept of Defense, and several industrial firms. Hobbies stamp collecting and classic automobiles. *Society Aff:* IEEE, ISHM.

Prince, M David
Home: 3132 Frontenac Ct, Atlanta, GA 30319
Position: Senior Staff Specialist *Employer:* Lockheed - Georgia Co. *Education:* MS/EE/GA Inst of Technology; BS/EE/GA Inst of Technology. *Born:* 3/27/26. Past pos incl: Assoc Dir of Res Sys Sci, Res Lab Dir & Dir of Res Progs-Lockheed-Ga Co 1964-71. Author book 'Interactive Graphics for Computer-Aided Design' 1971 (English) Addison-Wesley; 1973 (Japanese), 1977 (Russian) & 1982 (Chinese); var papers on computer graphics, automation, U.S. research activities, Presented invited papers in London, Paris, Berlin, Stuttgart, & Beijing. Fellow IEEE. Mbr Sigma Xi & Eta Kappa Nu. Four patents. Past activities incl: Disting Visitor, IEEE Computer Soc, Mbr IAS Instrumentation Panel & AOA Steering Ctte on Computer-Aided Design, cons to the Ga Tech Sch of Arch, 1982-83, Chrmn, Electronics Automation Program of Computer-Aided Manufacturing, Intl, Inc. Presently Engrg Project Mgr, and also Prof (part-time), GA Tech Sch of Info & Computer Science, 1981 Engr/Scientist of the Year, Lockheed-GA Co. *Society Aff:* IEEE.

Prindle, William R
Business: Natl Acad of Sci, 2101 Constitution Ave, Washington, DC 20418
Position: Exec Dir. *Employer:* Natl Mtls Advisory Bd. *Education:* ScD/Ceramics/MIT; MS/Phys Met/Univ of CA; BS/Phys Met/Univ of CA. *Born:* 12/19/26. in San Francisco, CA. Exec dir of Natl Mtls Advisory Bd, Natl Res Council, Natl Acad of Sci since 1976. VP, res & dir, Am Optical Corp 1971-1976. VP, res, Ferro Corp 1966-1971. Dir, mtls res, Am Optical Co, 1962-1965. Various R&D positions (up to gen mgr of R&D) Hazel-Atlas Glass Div, Continental Can Co, 1954-1962. Held various elective offices in glass div, Am Ceramic Soc, including chrmn & trustee. Fellow of Am Ceramic Soc, treas, 1971, VP 1972, Pres-elect 1979. *Society Aff:* AAAS, ACerS, AIME, ASM, ASTM, NICE.

Pringle, Arthur E
Home: Jacks Point, Oxford, MD 21654
Position: Retired. *Employer:* *. *Education:* BS/EE/Penn State Univ. *Born:* Oct 1899. BS Penn State Univ. Reg P E Penna. First employment Engrg Dept Bell Tele Co of Penna 1921-25; then to Pringle electrical Mfg Co as engr & in 1959 as Pres until retirement 1965. Active in NEMA 1927-65 working primarily on Codes & Stds; also mbr of their Bd of Governors 1955-65, & Treas 1961-62. James H McGraw Award 1962 Manufacturers Medal. Chmn Phila Sect AIEE 1955, Mbr IEEE Stds Ctte 1955-65, Fellow IEEE 1962, Fellow SES 1961. ASA Stds Council 1946-63, ASA Bd of Dir 1956-65. *Society Aff:* IEEE, SES.

Pringle, Oran A
Business: Dept of Mech & Aero Engg, Columbia, MO 65211
Position: Prof. *Employer:* Univ of MO. *Education:* PhD/Mech Engg/Univ of WI; MS/Mech Engg/Univ of WI; BS/Mech Engg/Univ of KS. *Born:* 9/14/23. Native of Lawrence, KS. Served with US Army 1943-45. Engg experience with Black & Veatch, consulting engrs, and McDonnell Douglas Corp. Engg faculty mbr Univ of MO since 1948. Reg PE. Ford Fdn grantee. Former Chrmn, Fastening and Joining Comm, Design Engg Div, ASME. Published numerous technical papers. Formerly on bd of dir of United Cerebral Palsy (MO). *Society Aff:* ASME.

Pringle, Weston S
Home: 324 Willamette, Placentia, CA 92670
Position: Owner. *Employer:* Weston Pringle & Assocs. *Education:* BS/Civil Engr/CA State Univ-Fresno. *Born:* 4/25/37. in Santa Cruz, CA. Fellow Inst of Transp Engrs; Mbr ASCE. Owner Weston Pringle & Assocs, Fullerton, CA since 1976. VP & Mgr Orange Cty office Crommelin-Pringle & Assocs Inc 1973-76. Previously VP Lampman & Assocs; Principal Engr Wilbur Smith & Assocs; & staff of cities of Downey & West Covina CA. Have been respon for transp planning & traffic operational studies as well as design projs. Mbr ITE ctte receiving Tech Council Award 1975. Pres Western District ITE, 1978-79. Pres So CA Sect ITE 1973-74. International Director ITE 1987-89. Sr Editor *Tech Notes* ITE. Reg CE & Traffic Engr, CA. Fellow, Inst for the Advancement of Engr. *Society Aff:* ITE, ASCE.

Prisuta, Samuel
Home: 1121 Atlantic Ave, Monaca, PA 15061
Position: Plant Mgr. *Employer:* Shasta Inc. *Education:* BS/Met Engr/Univ of Pittsburgh. *Born:* Jan 9, 1924. Native of Aliquippa Pa. US Army AF, pilot 1943-45. Formerly assoc with Jones & Laughlin Steel Corp, Vulcan Crucible (H K Porter), Teledyne Vasco (Colonial Steel Div) & now with NF&M Corp as V P of Mfg. Respons encompass all phases of Titanium processing at intermediate Y finish product stages, purchasing & selling of new & old equip, installation and/or removal of same, co policies & contract negotiations. P Chmn of the Beaver Valley Chap of ASM & presently V P of the Monaca Municipal Auth Respons encompass all phases of conditioning & machining of carbon & alloy materials including the processing of Ti as described above. *Society Aff:* BSM.

Pritchard, David C
Business: , 3333 Rice St, Miami, FL 33133
Position: Chief Engineer *Employer:* James S. Krogen & Co. *Education:* BS/Naval Arch/Univ of MI; BS/ME/Univ of British Columbia. *Born:* 7/10/42. Native of Trail BC Canada. Educated at Univ of British Columbia & Univ of MI. 2 yrs with Chris Craft designing pleasure yachts. 12 yrs with Chevron Shipping Eng Dept. involved in oil tanker & tug construction & maj conversion work. Now with James S. Krogen & Co., Consulting Naval Architect. Mbr of SNAME since 1966. Elected secy treas of Northern CA Sec 76/77, VChrmn 77/78, & Chrmn 78/79. Owns boat & enjoys sailing. *Society Aff:* SNAME.

Pritchard, David F
Business: 5085 Reed Rd, Columbus, OH 43220
Position: Ch Engr, Hydroelectric *Employer:* Burgess & Niple Ltd *Education:* MS/CE/OH State Univ; BS/CE/OH State Univ *Born:* 6/29/43 Civ Engr specializing in hydraulics, hydrology, design of hydraulic facilities and water resources planning. Served in the US Peach Corps in Iran (1967-70) as provincial govt staff engr - civ & architectural. Current work in hydroelectric power dev. Currently representative to ASCE Dist 9 and OH Council and Past Pres of Central OH Section. Mbr of ASCE Natl Ctte on Alternative Energy Systems. Chrmn of Section Ctte on Continuing Education. Chrmn, OH Council, Legislative activities Ctte. *Society Aff:* ASCE, NSPE

Pritchard, Wilbur L
Business: 7315 Wisconsin Ave, Bethesda, MD 20814
Position: President. *Employer:* SSE Telcom, Inc. *Education:* BEE/Elec Engg/CCNY; Grad/Elec Engg/MIT. *Born:* May 1923 N Y. 1943-46 Philco Corp; 1946-62 Raytheon Co, first in Equip Div, then Mgr Wayland Labs & Dir of Engrg Europe; 1962-67 Aerospace Corp, Group Dir of Communications Satellite Sys (produced DSCS-

Pritchard, Wilbur L (Continued)
1); 1967-73 V P COMSAT Corp & first Dir COMSAT Labs; U S Delegate to the Tech Subctte of INTELSAT; 1973-74 Pres Fairchild Space & Electronics Co; 1974-87, Pres Satellite Sys Engrg Inc. 1981-. 1987-Pros. SSE Telecom, Inc. Fellow IEEE & AIAA. Air Force Sys Command Award for Outstanding Achievement 1967. AIAA Aerospace Communications Award 1972. AAS Lloyd V Berkner Space Utilization Award; CCNY & CUNY medals. Mbr HKN; International Academy of Astronautics; Cosmos Club; Professional Engineer (Massachusetts, Maryland). *Society Aff:* AIAA, IEEE, AAS, BIS, IAS

Pritchett, Harold D
Business: Apperson Hall-Rm 111, Corvallis, OR 97331
Position: Const Educ & Res Coordinator. *Employer:* Oregon State Univ. *Education:* DE/Econ Planning/Stanford Univ; MS/Hydr Engrg/OR St Univ; BS/Civil Engrg/OR St Univ *Born:* Aug 17, 1930. BS & MS Civil Engrg Oregon State Univ; Degree of Engr Stanford Univ. Army Corps of Engrs 1951-53. Taught at OSU since 1957; Const Educ Prog Coordinator at OSU since 1967; Exec Dir OSU Const Educ & Res Found since 1975. PresOregon Sect ASCE 1972-73; Pres OSU-CERF 1971-74; Ford Found Fellowships in 1960 & 1963; OSU School of Engrg Award for Teaching Excell 1966; OSU Student Chap Award for Excell in Teaching Const 1971; named one of the 'Outstanding Educators of Amer for 1972'. Chmn ASEE Natl Const Engrg Ctte 1976- 77. Mbr ASCE, ASEE, AIC, ACCE, NSPE, APWA, ARBA, AMA, Toastmasters & Rotary Internatl. Current Pres of Oregon Columbia Section of ASEM (Amer Soc of Engr Mgt) 1984. *Society Aff:* ASEM, ASCE, AGC, ACCE

Pritchett, Thomas R
Home: 1430 Laurenita Way, Alamo, CA 94507
Position: VP, R&D, Aluminum Div. *Employer:* Kaiser Aluminum & Chem Corp. *Education:* PhD/Chem & ChE/Univ of TX-Austin; MA/Chem & ChE/Univ of TX-Austin; BS/Chem & ChE/Univ TX-Austin. *Born:* Sept 1925. Native of Colorado City Texas. Served with Army Signal Corps 1944-46. Res Scientist Defense Res Labs 1947-51; Res Chemist Monsanto Chem Co 1951-52. With Kaiser Aluminum & Chem Corp since 1952; presently VP & Dir of Res & Dev. Asst Univ of Texas 1949-51. Member Amer Chem Soc, Natl Assn Corrosion Engrs; Fellow Amer Soc for Metals, Amer Inst of Mining & Metallurgic Engrs; Prof Engr California; Assoc Amer Medical Assn; Dir Metal Properties Council. Res and pubs in corrosion of nonferrous alloys, metallurgy of aluminum, and casting res. *Society Aff:* ASM, AIME, AMA.

Pritsker, A Alan B
Business: PO Box 2413, West Lafayette, IN 47906
Position: Chairmen of the Board *Employer:* Pritsker & Assocs, Inc *Education:* PhD/IE/The Ohio State Univ; MS/IE/Columbia Univ; MS/EE/Columbia Univ *Born:* 02/05/33 Was a prof at AZ State 1962-69, VPISU 1969-70, and Purdue Univ 1970-81, and pres of Pritsker & Assocs from 1973 to 1986. He is Chrmn of Bd of Pritsker & Assoc & Factoral, and Pres of IN venture. He is the developer or codeveloper of GASP II, GASP IV, GASP-PL/I, GERT, Q-GERT, SLAM and SLAM II. He has published over 100 papers and 7 books. He was elected to Natnl Acad of Engg 1985. He is a fellow of IIE, the holder of IIE's Distinguished Res Award 1966, IIE's H. B. Maynard Innovation Achievement Award 1978, IIE's Operations Res Div Award 1978, and AZ State's Univ Faculty Achievement Award 1967. He is listed in Who's Who in Amer. *Society Aff:* AAAS, IIE, ORSA, TIMS, PMI, ACM/IEEE, NAE

Pritzker, Paul E
Business: 35 Hanna St, Quincy, MA 02169
Position: President. *Employer:* Geo Slack & Pritzker Assocs. *Education:* -/Elec Engg/Wentworth Inst; -/Eng Math/Northeastern Univ; -/Elec Engg/US Army-ASTP. *Born:* Feb 1925; native of Boston Mass. Served with Army Air Corps-Signal Corps WW II 1943-46. Graduated Wentworth Inst 1948; attended Northeastern Univ. Reg P E in Mass, Calif, and Ct. Chmn NSPE. Private practice 1975-76. VP NSPE 1975-76; Pres Mass SPE 1973-74; elected Fellow Mass SPE 1976 and awarded 'Engr of the Year' 1976. V Chmn Prof Advisory Ctte Coll of Elec Engrg Southeastern Mass Univ; member Coll Advisory Trustee Wentworth Inst. Licensed Pilot member of Aircraft Owwners & Pilots Assn, Natl Pilots Assn, Boston Navigators Club and US Power Squadron N Club. Former Pres Western Middlesex Chap MSPE & Trustee Temple Israel Natick; VP Architects & Engineers Lodge B'nai Brith & Amer Techn-ion Soc. Married to Janice Silverman 1944; 3 sons, 2 grandchildren. *Society Aff:* NSPE, IEEE, AACE, ACEC.

Prizer, Charles J
Business: Independence Mall W, Phila, PA 19105
Position: Group VP - Corporate Operations *Employer:* Rohm & Haas Co. *Education:* MS/CE/Drexel Univ; BS/CE/Univ of IL. *Born:* 4/24/24. in Lake Bluff, IL. Manhattan Proj, Oak Ridge 1944 & 1945. Joined Rohm and Haas in 1951 as process engr. Various assignments in res, marketing & mfg. Elected VP in 1972. Currently Group VP, Corp Operations. Enjoy travel, tennis & golf. *Society Aff:* AIChE, ACS, SCI..

Probasco, Johnny L
Business: 555 South 3rd East, Salt Lake City, UT 84111
Position: Sr Vice Pres *Employer:* Bush & Gudgell Inc *Education:* BS/CE/CO State Univ *Born:* 11/21/27 in Montrose, Co to John Reason and Elpha Irene (Woodman). Married Dorretta Leishman, Nov 10, 1967; Children-Jackie Lynette, Tamyra Sue; Stepchildren- Melanie, Deniece, Tamara, and Clay Anderson. 1951-1952-Resident Engr for C-B Engrg Co, Denver, CO. 1952-1953-Materials Engr for Mountain View Elec Assoc, Limon, CO. Responsible for design of elec distribution lines to serve local communities. 1953-1964 Chief Design Engr for Bush & Gudgell Engrs Salt Lake City, UT 1964-1977. Consultg Engr and Partner of Bush & Gudgell Inc 1977-1980-VP and Parner Bush & Gudgell Inc. 1980-Present Sr VP of Bush & Gudgell Inc. Experienced in design of wastewater treatment facilities, water treatment plants, storm drainage, highway design, and design of earthfill dams. Reg PE in UT, CO, NV, ID, and WY. *Society Aff:* NSPE, ASCE, WPCF, AWWA, APWA

Prober, Daniel E
Business: Becton Ctr, P O Box 2157, Yale Station, CT 06520
Position: Prof *Employer:* Yale Univ. *Education:* PhD/Phys/Harvard; MA/Phys/Harvard; BA/Phys/Brandeis. *Born:* 10/16/48. Native of Schnectady, NY. Asst prof, Engg and Appl Sci, Yale Univ. 1975-1980; Currently Prof. of Applied Physics. Also, visiting Professor 1979 and 1985, Physics Dept Tel Aviv University. Fulbright Fellow 1985. Current res in solid-state devices, microstructure sci and engg and superconducting devices. *Society Aff:* IEEE, APS, AVS, MRS.

Probstein, Ronald F
Home: 5 Seaver St, Brookline, MA 02146
Position: Prof of Mech Engrg. *Employer:* MIT. *Education:* PhD/Aeronautical Engrg/Princeton Univ; AM/Aeronautical Engrg/Princeton Univ; MSE/Aeronautical Engrg/Princeton Univ; BME/Mech Engrg/NY Univ. *Born:* 3/11/28. While night student at NYU, served as instructor. In 1952, became Asst Prof of Aeronautical Engrg at Princeton. Went to Brown in 1954 as Asst Prof of Engrg and Appl Math, becoming full Prof in 1959. At Brown wrote several texts on hypersonic flow and directed res in this area. Went to MIT in 1962 as Prof of Mech Engrg and around 1966 converted activities to desalination & water purification. In 1974 formed consulting engrg firm Water Purification Assocs (later Water General Corp) in Cambridge, Mass to deal with water purification problems, particularly those related to synthetic fuel production and power generation. Chrmn of Bd of Water General Corp which was acquired by Foster- Miller Inc, Waltham, Mass in 1983. In 1982 published together with R. E. Hicks first text on synthetic fuels (McGraw-Hill). Since 1983 research and teaching centered in physicochemical hydrodynamics. Text on subject to appear in 1988 (Butterworths). Freeman Award 1971 ASME, Councilor Amer Acad Arts & Sci 1975-79, Commissioner, Commission on Engrg and Tech Sys Natl Res Council,

Probstein, Ronald F (Continued)
1980-83, Mbr Natl Acad of Engrg since 1977, Amer Acad of Arts & Sci since 1961. *Society Aff:* ASME, AIChE, AIAA, AAAS, APS.

Proctor, Charles L
Business: Dept of Mech Engg, King Saud Univ; PO Box 800, Riyadh 11421, Saudi Arabia
Position: Prof of Mech Engg *Employer:* King Saud Univ *Education:* PhD/IE/OK St Univ; MS/ME/Purdue Univ; BS/ME/OK St Univ *Born:* 02/28/23 Native of Collinsville, OK. Served in the US Navy 1942-46. Held engrg and faculty positions as follows: Beech Aircraft Corp 1951-53; Purdue Univ 1953-55, McDonnell Aircraft Corp 1955-56; Univ of Toronto 1956-63; Univ of OK 1963-66; Univ of FL 1963-72; Univ of Windsor 1972-76; W. MI Univ 1975-80; Univ of Alaska 1980-85. Joined King Saud Univ. in 1984 as Prof of Mech Engg. Chrmn MHED/ASME 1982-3. ASME Centennial Award in recognition of service to the Society, 1980. Tau Beta Pi Distinguished Service Award 1968. *Society Aff:* ASME, IIE, ORSA, IEEE, AIAA.

Proctor, Stanley I
Business: 800 N Lindbergh Blvd, St Louis, MO 63167
Position: Dir, Eng Tech. *Employer:* Monsanto Co. *Education:* DSc/ChE/WA Univ; MS/ChE/WA Univ; BSChE/ChE/WA Univ. *Born:* 12/23/36. Belleville, IL. Married Carol Kroeger, Akron, OH, 1960. Joined Monsanto in 1959. Specialized in chem reaction engg & dev of comp methods for chem process engg. Assumed current responsibility in 1986. In present position directs a group of specialists in chemical engineering, environmental engineering, mechanical and materials engineering, process control and engineering computations. Reg PE - MO. Sigma Xi, Tau Beta Pi, Alpha Chi Sigma. AIChE Pres-1987. *Society Aff:* AIChE, NSPE, ACS, ASEE.

Proebster, Walter E
Business: Postfach 800 880, 7000 Stuttgart 80, Fed Rep of Germany
Position: Dir *Employer:* IBM Deutschland GmbH *Education:* DrIng/EE/Tech Univ of Munich; Dipl.-Ing/EE/Tech Univ of Munich; Abitur/- /Oberschule Munich *Born:* 4/2/28 Dir of R&D Coordination, IBM europe since 1973 after he had served for 9 yrs as the Dir of the IBM Dev Lab in Boeblingen, Germany. He holds the position of hon prof at Karlsruhe Univ since 1972 and lectures there on the subject of input/output technologies. In 1977 he has been elected as a fellow of the IEEE in recognition of his contributions and technical leadership to the dev of computer components and sys. For the period 1979-81 he is the elected chrmn of the IEEE German Section and Dir IEEE Region 8 for the period 1981-82. *Society Aff:* IEEE, VDE, NTG

Proffitt, Charles Y
Business: 125 W Barnes-Box 1265, Wilson, NC 27893
Position: Secy-Treas. *Employer:* Fenner & Proffitt Inc/Cons Engrs. *Education:* BEE/Elec Engr/NC State Univ. *Born:* 6/19/27. m. Lucille Higgins Nov 1947. Illuminating Engrg Soc - mbr "Offic Lighting" Ctt; Edwin F Guth award 1970. Natl Soc of Prof Engrs - Natl Dir 1973-81, 1987-93. Prof Engrs of NC PEPP Chmn 1968-69, Treas 1970-71, VP (Pres-Elect) 1976. Pres 1977- 78. B O Vannort Engrs 1947-48. USDA Rural Electrification Admin, WA DC, Bangor ME, Raleigh NC 1948-54. L E Wooten & Co 1954-59. Fenner & Proffitt Inc 1959- present (Principal). P E Regis NC, W Vir., TN, GA, KY. *Society Aff:* NSPE, IES, IEEE.

Proffitt, Richard V
Business: Crosstrees Hill Rd, Essex, CT 06426
Position: Pres. *Employer:* The Proffitt Group. *Education:* BME/ME/NYU. *Born:* 6/24/26. Established The Proffitt Group an assn of engg, marketing & ind design consultants with extensive food service equip, housewares & maj appliance experience able to carry out complete product progs from idea generation to pilot run 1978. Previously Dir of Engg for AMF Food Service Div 1968-77, & with GE in various engg & mfg capacities 1953-67. Reg PE-IL, a grad of GE 3 yr Mfg Training Prog 1956, GE's Creative Approach 1960, veteran WWII, holder of several patents, Previously member of UL and NSF industry advisory committees. Skier, sailor, pilot-Married 2 children. *Society Aff:* NSPE.

Profio, A Edward
Business: Dept of Chem & Nuclear Engg, Univ of California, Santa Barbara, CA 93106
Position: Prof. *Employer:* Univ of CA. *Education:* PhD/Nuclear Engg/MIT; SB/Physics/MIT. *Born:* 4/18/31. in New Castle, PA. Scientist at Westinghouse Bettis Atomic Power Lab 1953- 55. Officer in Army, 1955-57. Res Assoc then Asst Prof, MIT, 1957-64. Performed neutron physics res at Gen Atomic, 1964-69. Joined faculty as Assoc Prof then full Prof of Nuclear Engg at UCSB in 1969. Teaching and Res interests include radiation shielding, reactor physics, and biomedical engg. Dir of educational reactor facility. Author of two books and numerous papers. Received award for Outstanding Technical Achievement from Radiation Protection and Shielding Div of ANS in 1977. Interests include photography and raising Welsh ponies. *Society Aff:* ANS, AAAS, ASP, ASLMS.

Prohl, Melvin A
Home: 51 Towne Rd, Boxford, MA 01921
Position: Retired *Employer:* Medium Steam Turbine Dept - GE; Retired *Education:* MS/Mech Engg/MIT; BS/Mech Engg/MIT. *Born:* Oct 1915. Native of Melrose Mass. Employed by G E Co upon graduation, spec in the design of steam turbines for central station, indus & marine applications, with particular emphasis in mech vibration, stress analysis, elasticity, mech of matls & struct design. Included in publ papers is original work on calculation methods for critical speeds & vibration response of turbine buckets. Formerly, Mgr of Turbine & Heat Recovery Steam Generator Engrg at Lynn Mass. Retired from GE, 1981. Awarded the grade of Fellow ASME 1967. Mbr SNAME. Hobbies: hiking & skiing. *Society Aff:* ASME, SNAME.

Projector, Theodore H, PE
Business: 407 Gills Neck Rd, Lewes, DE 19958
Position: Consulting Engr. *Employer:* Self. *Education:* BS/Physics/CCNY; BS/Physics & Math/NBS Grad Sch. *Born:* Oct 1914 Stamford Ct. grad studies NBS Grad Sch 1939-47. Physicist Natl Bureau of Stds 1938-58, photometry & colorimetry. Cons engr since 1958 - photometry, colorimetry, illuminating engrg spec in visual & lighting problems in transp. Fellow IES, Mbr OSA & AAAS. Num papers & reports: photometry, colorimetry, aircraft lighting, lighting & safety in transp, incl papers on flashing lights in Illum Engrg, reports on visual collision avoidance in the air, rear lights on automobiles. IES Aviation Lighting Comm; SAE Aircraft Lighting Comm. *Society Aff:* IES, OSA, AAAS.

Promersberger, William J
Home: 55-18th Ave North, Fargo, ND 58102
Position: Prof Emeritus. *Employer:* North Dakota State University. *Education:* MS/Agr Engr/KS State Univ; BS/Agr Engr/U of MN *Born:* 5/28/12. Born near Littlefork, MN. Taught at NW Sch of Agri, Crookston MN; on faculty at ND State Univ since 1938, prof and Dept Chmn, Agri Engg 1941-77; currently Prof Emeritus; Visiting Prof at Univ Coll, Dublin Ireland 1965-66. Honors: Alpha Epsilon; Phi Kappa Phi, Life Fellow Amer Soc of Agri Engg; Hon State Farmer in ND, FFA; NDSU Faculty Lectureship 1972; Who's Who in Engg 1959; Who's Who in Midwest 1976; Outstanding Educ of Amer 1973. Mossey-Ferguson Gold Medal (ASAE)-1979. Sr author of *Modern Farm Power* (Prentice-Hall) 1962, 2nd Ed 1971, 3rd Ed 1979. MS/Agr Engr/KS State Univ; BS/Agr Engr/U of MN *Society Aff:* ASAE, ASEE

Promisel, Nathan E
Home: 12519 Davan Dr, Silver Spring, MD 20904
Position: International Cons Matls & Policy. *Employer:* Self. *Education:* MS/Electrochemistry/MA Inst of Tech; BS/Electrochemistry/MIT; Post Grad/Doctoral Metallurgy/Yale Univ; Dr Eng/Honorary/MI Tech Univ. *Born:* 1908 Mass. BS MIT 1929, MS MIT 1930, Doctoral work Yale 1933. 1930-40 Res & Asst Tech Dir Internatl Silver Co; 1940-41 cons chemistry & metallurgy; 1941-66 Navy Dept

Promisel, Nathan E (Continued)
with final simultaneous titles Dir Materials Div & Ch Materials Engr (Naval Weapons) & Navy Materials Admins; 1966-74 Exec Dir Materials Adv Bd, Natl Acad of Sciences; 1974- , international cons. Mbr adv cttes universities, Navy, Congressional Office Technology Assessment, National Academies, & industrial labs; Mbr Natl Matls Advisory Bd 1976-79; Chmn NATO res grp; US representative OECD (Materials). Chrmn US side, matls sci exchange program with USSR. Member National Academy of Engineering. 7 honorary professional lectures; Capital Area Annual Engineering Award; 4 society awards; 9 Navy Dept awards; Pres & Fellow ASM; Pres Federation of Materials Socs and Decennial Award; Hon Doctor of Engg, MI Technoligcal Univ; Hon Mbr ASTM, ASM, AIME, SAE (Aerospace Div); Fellow SAMPE; Hon Permanent Pres Intl matls Congress; Fellow British Inst of Metals. Internatl lecturer & author. *Society Aff:* ASM, AIME, SAE, BIM, SAMPE.

Pronske, Kurt N
Business: 11484 Washington Plaza W, Suite 401, Reston, VA 22090
Position: Consultant *Employer:* Self-employed *Education:* BS/CE/IA State Univ *Born:* 11/24/27 Over twenty yrs experience with Skidmore, Owings and Merrill providing site development design support to commercial, educational and institutional projects, domestic and abroad. Established my own conslt practice in 1976, providing site design services to major architectural firms and developers along the eastern seaboard. Active in church and community affairs. *Society Aff:* NSPE

Pruitt, Ralph V
Home: 135 Cedarmill Dr, St Charles, MO 63301
Position: Unit Chief - Technology. *Employer:* McDonnell Aircraft Company. *Education:* MS/Engr Mgmt/Univ of Rolla, St Louis; BSME/Aero Engg/OK State. *Born:* July 1936. 20 yrs of experience in the design, dev, operational test, evaluation & support of aircraft & spacecraft systems; trained & experienced in aircraft accident investigation, respon for McDonnell Aircrafts' F-15 System Safety prog during design & dev; principal contributor to the design & dev of the 1st practical Energy Mgmt Display system. Presently engaged in operational analysis of McAir's advanced design programs. Co-recipient of 1974 SAE Wright Brothers Award. Active in camping, scale modeling, youth & church programs.

Prussing, John E
Business: 104 S. Mathews St, Urbana, IL 61801
Position: Prof *Employer:* Univ of IL. *Education:* ScD/Instrumentation/MIT; MS/Aero & Astro/MIT; BS/Aero & Astro/MIT *Born:* 08/19/40 Born in Oak Park, IL. Staff of MIT Instrumentation Lab and Experimental Astronomy Lab 1963-67. Asst Res Engr and Lecturer Univ of CA at San Diego 1967- 69. Faculty of Univ of IL Dept of Aero and Astro Engg since 1969. Asst Dean of Engg 1976-77. AIAA Astrodynamics Technical Committee 1977-80, 1981-84. Res interests are orbital mechanics and control theory. *Society Aff:* AIAA, AAS, ASEE

Pry, Robert H
Business: 10 Gould Ctr, Rolling Meadows, IL 60008
Position: Exec VP - Res and Dev *Employer:* Gould Inc. *Education:* PhD/Phys/Rice Univ; BS/Phys/TX A&I College; MS/Phys/Rice Univ. *Born:* 12/28/23. in Dormont, PA. Attended TX A&M, TX A&I, OK Univ, Univ of Manchester, England & Rice Univ. US Army Signal Corps, 1943-1946. Res Scientist & Mgr with GE Co from 1951-1976. R&D Mgr, Mtls Sci & Engg & R&D Mgr, Electronic Sci & Engg at GE Res & Dev Ctr, Schenectady, NY. VP/R&D, Combustion Engg Co, 1976-1977. Exec VP/R&D, Gould Inc, 1977 to present. Elected officer with staff responsibility for product dev activities, in Gould Operating Divisions and, responsibility for Corp Res and Dev Labs, and New Business development. *Society Aff:* IEEE, AIP, AAAS, AIME, ASM

Pryor, C Nicholas
Home: Atlantic Ave, Newport, RI 02840
Position: Technical Dir. *Employer:* Naval Underwater Systems. *Education:* PhD/EE/Univ of MD; MS/EE/MIT; BS/EE/MIT. *Born:* 2/26/38. Native of Wash, DC area. Served with Naval Surface Weapons Ctr (Naval Ordnance Lab) in White Oak, MD from 1957 to 1975, specializing in acoustic signal processing & digital tech. Technical Dir of the Naval Underwater Systems Ctr in Newport, RI since 1975. Also taught grad level electrical engg for MIT, Univ of MD, & Catholic Univ. Principal hobby is flying. *Society Aff:* IEEE.

Pryor, William T
Home: 1136 N Powhatan St, Arlington, VA 22205
Position: Engrg & Aerial Surveys Consultant *Employer:* Retired *Education:* BS/Civil Engg/Univ of UT *Born:* 5/19/06 in Cedar City, Utah. Active Mbr LDS Church. Surveying & mapping work began during high sch; BSCE Univ of Utah 1930; Phi Kappa Phi; graduate courses in ground & aerial surveying. 1930-75 employed by Fed Hwy Admin, US DOT, & predecessors: 1930-42 Hwy location, design & const Dist 12; 1942-44 design & admin engr Alaska hwy; March 1945 assigned aerial surveying at Wash DC headquarters; 1952-74 Ch Aerial Surveys Branch. 1975-79 Engg and Survey Consultant. Terms as elected Bd Mbr of ASP & ACSM; V Pres ACSM. Drafterd aerial surveys specs for hwys. Organizer & Sec of Photogrammetry & Aerial Surveys Comm, TRB. Served on num cttes of ASP, ACSM, ASCE & ISP. SAR Mbr; Scouter BSA since 1946. Mbr ACSM, ASCE, ASP & SAME; supporting mbr TRB. 1950 Citation & Silver Medal from Secy of Commerce for outstanding contributions to hwy engrg; many other awards from professional socs & employer. Author num papers, award *Society Aff:* ASCE, ASP, ACSM, SAME, NGS, TRB

Prywes, Noah S
Business: 200 S 33rd & Walnut St, Philadelphia, PA 19174
Position: Professor of Computer Science. *Employer:* University of Pennsylvania. *Education:* PhD/Applied Physics/Harvard Univ; MSc/Elec Engg/Carnegie Mellon Univ; BSc/Elec Engg/Techicion. *Born:* Nov 28 1925. 1955-59 Mgr of Univac Larc Computing Unit, continued as cons to Univac until 1965; since 1959 at Univ of Penn, active in a variety of fields, most recently in information storage & retrieval & automatic programming, presently Professor in Dept of Computer & Information Sciences, Moore Sch of Electrical Engrg at the univ. Fellow IEEE; Mbr Sigma Xi. Written approximately 60 articles in various journals. *Society Aff:* IEEE

Przemieniecki, Janusz S
Home: 2535 Valdina Dr, Xenia, OH 45385
Position: Dean, School of Engineering. *Employer:* Air Force Institute of Technology. *Education:* BS/Mech Engg/Univ of London; Diploma/Aeronautics/Univ of London; PhD/Aeronautical/Univ of London. *Born:* Jan 30 1927, Lipno Poland. Diploma 1953 from Imperial Coll of Sci & Tech. Reg PE in Ohio. 1954-61 Hd Structural R&D Sec, Bristol Aircraft Ltd England; 1961-64 Assoc Professor of Mechanics, Sch of Engrg at Air Force Inst of Tech, Wright-Patterson AFB Ohio, 1964-65 Professor of Mechanics, 1966-69 Asst Dean & Assoc Dean for Res, 1969-present Dean. Recipient of USAF Superior Performance Award 1965; Recipient of Air Force Decoration for Exceptional Civilian Service 1978; Recipient of Presidential Rank of Meritorious Exec 1981; recipient of Presidential Rank of Distinguished Exec 1982; Fellow of Royal Aeronautical Soc; Fellow AIAA; Mbr ASEE, Ohio Acad of Sci, Natl Soc for Professional Engrs & Tau Beta Pi. *Society Aff:* AIAA, RAeS, ASEE, NSPE, OSPE.

Przirembel, Christian E G
Business: Clemson University, College of Engg, Clemson, SC 29634-0921
Position: Prof & Head, Dept of Mech Engg. *Employer:* Clemson University *Education:* PhD/Mech Engg/Rutgers Univ; MS/Mech Engg/Rutgers Univ; BS/Mech Eng/Rutgers Univ. *Born:* 3/30/42. Immigrated to U.S. from Germany in 1951. Naturalized citizen in 1958. Involved in teaching and research at Rutgers Univ 1967-81. Developed several unique wind tunnel facilities for separated flow studies. Assoc Prog Mgr of Rutgers/AFOSR THEMIS Prog-Separated Flows (1968-71). Assoc Dean for Acad Affairs, Coll of Engg, Rutgers Univ. (1977-81). Since 1981, Head, Dept of Mech Engg, Clemson Univ. Current research interests: subsonic

Przirembel, Christian E G (Continued)
wakes, resonance tubes, slender bodies at high angles of attack. Fellow, ASME, Fellow, ASEE, Assoc Fellow, AIAA. Pi Tau Sigma Gold Award (1973). ABET/ECPD (ASME) Visitors List (1977-83). ASEE, VP, Prof. Interest Council (1985-86). ASEE, Bd of Dir (1984 -86). Bd of the Engrg Soc Library (1979-81). ASME, Bd, Engrg Educ (1983 -), Bd, Public Info (1983 -). Married Donna L Faust, 2 children. Enjoy theology, music and sports. *Society Aff:* ASME, AIAA, ASEE, AAAS.

Przygoda, Zdzislaw
1262 Don Mills Rd, Don Mills, Ont Canada M3B 2Y9
Position: Consulting Engineer & President. *Employer:* Z Przygoda & Associates, Ltd. *Education:* DSc/Civil/Tech Univ, Munich; MSc/Civil/Tech Univ, DAN21G. *Born:* Feb 1913, Warsaw Poland. Cons practice in Warsaw prior to WWII; 1939-44 served with Polish Regular & Underground Forces, arrested by Gestapo & confined to concentration camps in Germany; 1945-48 Cons to 3rd US Army in Germany; 1948-50 tech expert with French War Reparations Branch; 1950-52 cons practice in Israel, prize winning design for the redevelopment of downtown area of Haifa; 1954- , cons practice in Toronto, Canada, design of feed plants & grain elevators & structural cons to architects for school & institutional buildings. Pres, Ont Cons Struc Engrs Assoc 1969-70; Pres Polish Engrs Assn in Canada Inc 1972-73; Chmn Professional Interviewing Ctte of APEO 1973-74; The Citizenship Award of APEO (Gold Medal) 1972; Canada's Centennial Medal 1967; Fellow Engrg Inst of Canada 1975; Fellow Inst of St Engrs of Great Britain 1976. Sons of Martha Medal of APEO-1977. Mbr of the Appeal Bd of APEO 1978-1980, adviser to Toronto Historical Board 1980-date. *Society Aff:* APEO, FEIC, ACEC, FISE, MEAC

Puckett, Allen E
Business: PO Box 1042, El Segundo, CA 90245
Position: Chairman of the Board and Chief Executive Officer *Employer:* Hughes Aircraft Co. *Education:* PhD/-/CA Inst of Tech; MS/-/Harvard Univ; BS/-/Harvard Univ. *Born:* 7/25/19 Res Asst in Aerodynamics Cal Tech; Lectr & Ch, Wind Tunnel Sect JPL; with Hughes Aircraft Co 1949- . Exec VP & Asst Gen Mgr since 1966. Exec VP & Gen Mgr 1976- . Mbr Natl Acad of Sci; Natl Acad of Engrg; Fellow AIAA - Pres 1972. 1980- Calif Manufacturer of the Year Award; IAS Lawrence Sperry Award; Lloyd V Berkner Award (AAS); Phi Beta Kappa, Sigma Xi. Mbr num ad hoc cttes for NASA, Pres's Sci Avd Ctte. Mbr Bd of Dir of Amer Mutual Fund, Lone Star Indus, Teleprompter Corp. 1980-Calif Manufacturer of the Yr Award; Mbr Bd of Dirs, Investment Co of Amer, Mbr of Bd of Governors of AIA. Awards: EIA Medal of Honor, IEEE Frederik Philips Award, Polytechnic Inst of NY Award for Creative Tech, Kitty Hawk Memorial Award from Los Angeles Area Chamber of Commerce, Britain's Institution of Production Engrs 1983 Intl Award.

Puckett, Hal K
Home: 1113 Sangre de Cristo, Santa Fe, NM 87501 *Employer:* Consultant *Education:* BS/ME/TX A&M. *Born:* 3/2/24. US Navy active duty 1944-46. Lockwood, Andrews & Newnam - Archs/Engrs/Planners 1946. Exec VP 1974-79. Pres Tex Cons Engrs Council 1973-74. Natl Dir ACEC 1974-76; Pres Houston Engg & Sci Soc 1978-79; Fellow Amer Cons Engrs Council; Trustee ACEC Life/Health Trust 1977-79; Mbr ASME, NSPE, TSPE, Houston Club, BraeBurn County Club, Univ Club, Rotary Club of Houston; Reg TX, OK, IN. *Society Aff:* ASME, ACEC.

Puckett, Hoyle B
Business: 376 Agri Engrg Sci Bldg, 1304 W Pennsylvania Ave, Urbana, IL 61801
Position: Res Leader/Technical Advisor. *Employer:* US Dept of Agri ARS *Education:* MSAE/Agri Engr/MI State Univ; BSAE/Agri Engr/Univ of GA. *Born:* 10/15/25. Native of Jesup, GA, served in Army AF during WWII, NC and IL Natl Guard 1950-69. Res agri engr with US Dept of Agri, Agri Res Serv 1949-to present. Proj leader 1950, Investigation Leader 1958, Res Leader 1972, Technical Advisor 1975. Serves as res leader and natl technical advisor in the field of automatic elec equip for livestock production. Chrmn EPP Div ASAE 1968-69, Technical Dir ASAE 1976-78, Fellow ASAE, Sr Mbr IEEE. *Society Aff:* ASAE, IEEE, ΣΞ, ΓΣΔ, ΦΚΦ.

Puckett, James C
Business: P.O. Box 10383, Birmingham, AL 35202
Position: Manager Treatment Plant Division *Employer:* Brasfield & Gorrie, Inc. *Education:* BE/ChE/Vanderbilt University *Born:* 4/14/44 Prof Engr-Alabama Reg No. 13491. Publs: 'Chem Model of Muscular Contraction' Biomed Engrg, Sept 1965; 'Precoat Filtration of Waste Activated Sludge' WPCF Journal, Oct 1976. Res engr E I duPont de Memours & Co 1966-70 - new fibers dev. Chem Engr/Sales, Envirotech Corp 1970-74 CPI & pollution control equip sales. Pres Puckett Const Co 1974-77 design & const indus proces & pollution control facils. Mgr, Treatment Plant Div, Brasfield & Gorrie 1977 cons of municipal water and sewage treatment plants. Design & const industrial wastewater treatment facilities. Mbr AGC, AIChE, AWWA, NSPE, WPCF. Married; 2 children. Native of Birmingham Ala. Hobbies: sailing & swimming *Society Aff:* AGC, AIChE, AWWA, NSPE, WPCF

Puckett, Russell E
Home: 3604 Midwest Dr, Bryan, TX 77802
Position: Assoc Prof *Employer:* Texas A & M Univ *Education:* PhD/EE/Pacific Western Univ; MS/EE/Univ of KY; BS/EE/Univ of KY *Born:* 03/28/29 in KY. Served in Army Air Corps and US Air Force 1946-49 and 1951-54. Engr with US Dept of Defense, TX Instruments Inc, and Central Associated Engrs, Versailles KY. Research administrator and prof of electrical engrg at Univ of KY 1957-66 and 1968-76. Assumed present position in 1977. KY Soc Prof Engrs, award for outstanding achievement in education 1974. Dir, secty, treas, and pres-elect KY Soc Prof Engrs, 1971-77. Treas, chrmn IEEE Lexington KY section 1962-70. Mbr Eta Kappa Nu, Tau Beta Pi, Tau Alpha Pi. Author of textbooks in field, patentee in field. Licensed PE in KY and TX. Interest areas: instrumentation, electronics in medicine. *Society Aff:* ASEE, IEEE, NSPE.

Pugh, Emerson W
Business: Watson Res Ctr-Bx 218, Yorktown Hts, NY 10598
Position: Mgr of Exploratory Magnetics Res. *Employer:* IBM Corp. *Education:* PhD/Physics/Carnegie Mellon; BSc/Physics/Carnegie Mellon. *Born:* May 1929. Co-author *Principles of Elec & Magnetism* Addison Wesley 1960 (2nd ed 1970). IBM res 1957-62. Managed dev of magnetic film memory array used in the 360/95 computer 1962-64. Dir of Res Tech Planning & Corp Cons 1965-75. Mgr exploratory magnetic memory res 1976- . Ed-in-Chief 'Transactions on Magnetics' 1968-70, Pres of Magnetics Soc of IEEE 1972-74, Genl Chmn of 'Intermag' 1968 & 'MMM' Conf 1976, Exec Dir of Motor Vehicle Emissions study of Natl Acad of Sci 1974. Fellow APS, IEEE & AAAS. *Society Aff:* IEEE, APS, AAAS.

Pugh, James W
Business: 167 Willis Ave, Mincola, NY 11501
Position: Consulting Engineer. *Employer:* James Pugh, PE, PC. *Education:* PhD/Bioeng/MIT; SB/Met & Matls Sci/MIT. *Born:* 1/4/46. Raised in & attend pub schs in Richmond, VA. Valedictorian of class of 306 John Marshall HS 1964. SB Met & Matls Sci MIT 1968. For Dow Chem Co Prize Best sr thesis Dept of Met & Matls Sci, Grad Class of 1968 MIT. PhD Biomed Engrg MIT 1972. Sigma Xi, Phi Lambda Upsilon. Theta Chi (social). Dir Biomech Lab, Hosp for Joint Diseases & Med Ctr, NYC 1972-79; Dir Div of Bioeng, Hosp for Joint Diseases Orthopaedic Inst, 1979-84; Asst Prof of Orthopaedics (Biomech) Mt Sinai Sch of Medicine, City Univ of NY 1973-81; Assoc Prof of Orthopaedics; teaches, consults & directs res in bioengrg, orthopaedic biomech, biomatls, and engineering. Mbr Amer Soc for Metals, Soc for Biomatls, Orthopaedic Res Soc. Now, Consulting engineer; Director, Biomedical Engineering, Metallurgy, and Material-Science, Inter-City Testing & Consulting Corp.; Adjunct Professor of Materials Science and Engineering, SUNY/Stony Brook; Visiting Professor of Bioengineering, The Cooper Union School of Engineering; Affiliate Professor of Bioengineering, Univ of Washington, Seattle. *Society Aff:* ASTM, ASME, NSPE, NYSSPE.

Puleo, Peter A
Home: 2 Clerbrook Ln, St Louis, MO 63124
Position: Pres & Owner. *Employer:* Industrial Process Equip. *Education:* BS/CE/WA Univ. *Born:* 5/1/23. Native of St Louis, MO. Received principal education in St Louis. Studied Accounting & Bus Mgt at St Louis Univ. Served in US Army for three yrs. Received BS Degree from WA Univ, St Louis, in 1949. Was Sales Engr with Fisher and Porter Co, involved with fluid flow dynamics. Founded Ind Process Equip Co in 1952, a distributor & mfgs' rep, specializing in pumping equip systems and valve instrumentation. Am a reg PE, State of MO. Active in AIChE for thirty yrs at Local & Natl level. Was Natl Mbrship Chrmn from 1969 to 1972. *Society Aff:* AIChE, ISA.

Pullen, Keats A
Home: 2807 Jerusalem Rd, Kingsville, MD 21087
Position: Electronic Res Engr. *Employer:* Aberdeen Proving Ground-USA. *Education:* Dr Eng/EE/Johns Hopkins; BS/Physics/Caltech. *Born:* Nov 1916. Native of Onawa Iowa. Engineered war materiel during WW II. Taught at Pratt Inst of Brooklyn 1945-46. Electronic res engr at (Aberdeen Proving Ground), working in guided missile & satellite electronic tracking sys & related theoretical studies 1946- . Was Chmn of Balto Sect IEEE 1966-67. Fellow AIEE & IEEE 1960. Recipient Marconi Memorial Award, Veteran Wireless Operators Association 1982. Publ 8 books on aspects of elec ciruit design; contrib to 2 handbooks; many papers, articles, reports. Awarded 5 patents, 1 more pending. US Army Matl Sys Analysis Activity Aberdeen proving Ground. Other ints incl photography & classical music. *Society Aff:* IEEE, ΣΞ, ADPA, AFCEA, AUSA.

Pulley, Frank L
Home: 6815 Reite Ave, Des Moines, IA 50311
Position: President *Employer:* Frank Pulley Assocs *Education:* BS/ME/Univ of MO *Born:* 01/04/22 Born in Cameron, MO. Attended William Jewell Coll, Liberty, MO and received engrg degree from Univ of MO, Columbia. First position with Carrier Corp, Syracuse, NY in Development Dept. Entered private practice in 1950 as engr for architectural firm. In 1953 formed present practice specializing in mech and electrical engrg for bldgs. Presbyterian, Mason, Rutarian. Interested in beef and grain farming. Enjoy classical music, antique automobiles and machinery, machine work and travel. *Society Aff:* ASHRAE, NSPE, IES

Pulling, Ronald W
Business: 1101 15th St, NW, Suite 700, Washington, DC 20005
Position: Consultant *Employer:* Tippetts-Abbett-McCarthy-Stratton *Education:* Fellow/Public Affairs/Princeton Univ; BS/CE/Univ of CA *Born:* 10/30/19 Prog dir for major transportation projects undertaken by Washington Office of TAMS, in Lisbon, Portugal; Bangkok, Thailand; Amman, Jordan; Los Angeles, CA and elsewhere throughout the world. (1976-date) Pres, R. W. Pulling Assocs. (1974-81) Senior VP, William L. Pereira Assocs. (1973-1975) Deputy Assoc Administrator for Plans, Federal Aviation Admin. (1969-1972)Chief Airport Sys Planning Div and other managerial and technical engrg positions, Federal Aviation Admin, Washington, DC; Oklahoma City, Los Angeles, Honolulu (1941- 1969). Meritorious Service Award, FAA, 1970. James Laurie Prize, ASCE, 1980. Distinguished Service Award, TRB, 1980. PE: HI, CA, FL, VA, and MD. *Society Aff:* ASCE, AIAA, NSPE, TRB, NAE.

Puls, Louis G
Home: 6665 Garrison St, Arvada, CO 80004
Position: Consultant *Education:* AB/CE/Stanford Univ *Born:* 01/30/97 in Lowell, NY, graduated Clinton NY High Sch; Stanford Univ, CA 1920. Inspector on construction Wilson Dam, Al 1923-1927. Flood Studies Tennessee River, Corps of Engrs Report 185 to Congress 1927-1929. Field Engr employed by Hugh L. Cooper on construction of Dnieprostroy USSR Dam, 1929-1932. Design of concrete dams, Bureau of Reclamation 1933-1962, supervised design of Glen Canyon and Flaming Gorge Dams. Retired as Chief Designing Engr of Bureau 1962. Consltg Bds in US, India, Brazil, Peru, Thailand and Mexico. Philadelphia Elec Co on Muddy Run Pumped Storage Project; CA Dept Water Resources on Tehachpi Crossing; Commission Federal De Electricidad, Mexico, hydroelectric power projects, 1962 to date. Commander US Navy Seabee Battalion South Pacific WW II. *Society Aff:* ASCE

Pulvari, Charles F
Business: 2014 Taylor St NE, Washington, DC 20018
Position: Pres Electrocristal Co Inc. *Employer:* Catholic Univ - Prof Emer EE. *Education:* Dip Ing/Dr of EE/Royal Hungarian Univ; PhD/EE/-. *Born:* July 19, 1907 Kalsbad. Depl Engrg Royal Hungarian Univ of Tech Sci, Budapest 1929. cp/var posts, Hungarian Indus 1929-43; Founder Pulvari Electrophys Lab Budapest 1943-49; Lectr, Postgrad, Univ Tech Sci, Budapest 1943- 49; Res Prof, Prof Dr of Elec Engrg Cath Univ of Amer 1953- prof of Emeritus. Fellow IEEE; Life Mbr & Fellow N Y Acad of Sci; Felow of The Ceramic Soc of Amer; Sigma Xi, Tau Beta Pi. Originator 1st Lt Valve TV Display Tube 1933; holder 20 pats; hon/author num papers in prof journals; IR 100 Award for Most Significant New Tech Prods of 1963; DIB; DAR Amer Medal. Studied violin at the conservatory in Budapest. Enjoys planing tennis, chess. IEEE Citation for pioneering work on ferroelectrics, organizer of the IEEE Cttee on Ferroelectrics. *Society Aff:* IEEE, NYAS, ACS, ΣΞ, TBII.

Pumphrey, Fred H
Home: 706 Cary Dr, Auburn, AL 36830
Position: Dean Emer, Sch of Engrg. *Employer:* Auburn Univ - ret. *Education:* DSc/Hon/OH State; BEE/Elec power/OH State; AB/Math & Sci/OH State. *Born:* July 1898 Dayton Ohio. Worked for Dayton P & L Co, Kansas G & E Co & Staten Island Edison Corp. Hd Prof EE Dept Rutgers Univ 1928-45. Hd Prof of EE Univ of Fla 1946-58. Dean of Engrg Auburn Univ 1958-69. Life Fellow IEEE. 59 yrs of married life; 3 children; 8 grandchildren. As Dean Emer, am active in the college & in civic projs. Was commissioned as 2nd Lt US Army, WWI and Lt Col, Signal Corp - and Army Serv Forces in WW II. *Society Aff:* IEEE, ASEE.

Purcell, Fenton Peter
Business: 60 Hamilton St, Paterson, NJ 07505
Position: Partner. *Employer:* Lee T Purcell Assocs. *Education:* BCE/Civil/Rensselaer Poly Inst. *Born:* Nov 23, 1942 Paterson N J. With Lee T Purcell Assocs, Cons Engrs - engr 1965-66, partner 1969- . V P Fenton Corp, Paterson 1970- . Served to Capt AUS 1966-69. Reg P E NY, NJ, Pa, Mass. Mbr AWWA, Water Pollution Control Fed, N J Cons Engrs Council. American Academy of Environmental Engineers. Home: 4 Highview Terr, Upper Saddle River N J 07458. *Society Aff:* AWWA, WPCF, NJCEC, AAEE.

Purcell, James P
Home: 200 Cold Springs Rd, Rocky Hill, CT 06067
Position: President. *Employer:* James P Purcell Assocs Inc. *Education:* BCE/Civil Engg/Cornell Univ. *Born:* Sept 23, 1922. USNR 1942-46. Served in Night Fighter Squadron 90 on the USS Enterprise, Dec 1944-June 1945. Hwy design engr for Army Corps of Engrs 1949-51. Pos with cons engr firms 1951-59. Sole owner James P Purcell Assocs 1959 & currently Pres of James P Purcell Assocs Inc, Purcell & Taylor P C, J P P Internatl Inc providing services of engrs, architects & planners throughout the U S & the Mideast. P Pres Ct Engrs in Private Practice 1968 & 1969. Fellow ASCE. Mbr Transp Ctte Greater Hartford Chamber of Commerce, Hwy Res Bd, Engrg Div of ARBA. Married to former Rita Cantwell; 6 children. *Society Aff:* ASCE, ACEC, NSPE.

Purcell, Leo Thomas, Jr
Business: 60 Hamilton St, Paterson, NJ 07505
Position: Partner. *Employer:* Lee T Purcell Assocs. *Education:* BSCE/Sanitary Engr/Manhattan College. *Born:* July 28, 1935 Pompton Lakes N J. With Lee T Purcell, Cons Engrs, Paterson N J 1958; engr 1962-64. Engr Interstate Sanitation Comm NYC 1958-59. Partner Lee T Purcell Assocs, Cons Engrs, Paterson 1965- . Pres of

Purcell, Leo Thomas, Jr (Continued)
Fentor Corp. Served to Lt, Med Serv Corps AUS 1959-62; US Army Commendation Medal. Reg P E NJ, NY, Pa, Fla & Conn. Reg Prof Planner NJ. Dipl Amer Acad of Environmental Engrs. Mbr ASCE, Water Pollution Control Fed N J, Water Pollution Control Assn N J. Mbr Cons Engrs Council, AWWA, Beta Theta Pi. Home: 7 Seminole Way, Chatham N J 07928. *Society Aff:* ASCE, NSPE, AAEE, FWPCA, ACEC

Purcupile, John C
Business: 865 Asp Ave, 212 Felgar Hall, AMNE Dept, Norman, OK 73019
Position: Prof *Employer:* Univ of OK *Education:* PhD/ME/Carnegie-Mellon univ; MS/EE/Carnegie-Mellon Univ; BS/Engrg/UCLA *Born:* 10/3/30 Prof, Sch of Aero, Mech and Nuclear Engrg, Univ of OK, 1980-present; Assoc Prof, Carnegie-Mellon, Pittsburgh, PA 1970-80; Sr Res Engr, Carnegie-Mellon, 1968-70; Sr Design Engr, Westinghouse Corp, 1963-68; Ch Engr, Time-O-Matic, 1961-63; Borg-Warner Corp, Design Engr (1953), Resident Test Engr (1953-55), Design Engr (1955-57), Supervisor Test Lab (1957-59), Proj Engr (1959-61). Con to the following firms: Occidental Petro, Westinghouse, Natl Sci Fdn, Corning Glass Corp, Xerox Corp, Jet Propulsion Lab, DOE, Gen Motors. *Society Aff:* ASME

Purdom, Paul W
Home: 245 Gulph Hills Rd, Radnor, PA 19087
Position: Prof *Employer:* Drexel Univ *Education:* Phd/Pol Sci/Univ of PA; MGA/Govt Adm/Univ of PA; MSE/San Engr/Univ of MI; BS/CE/CA Inst of Tech *Born:* 07/23/17 Since 1963, Drexel Univ, Dir Environmental Studies Inst 1953-63, Dire, Environmental Health Div, Philadelphia (Crumbine Award 1961). Prior work, hydrology, sanitary and structural engrg. Pres APHA 1971. Chrmn Bd ASEIB (now AAEE) 1973-75. Co-Author, Environmental Science, 1980, merrill. Editor, Environmental Health, 2nd Ed 1980, Academic Press. Co-Author, Health Planning Related to Environmental Factors, 1980, NTIS. Listed Who's Who in Amer and Who's Who in World. Registered Engr GA and PA. Since 1973, conslt mbr, Atomic Safety and Licensing Bd Panel, US Nuclear Regulatory Commission. *Society Aff:* ASCE, AAEE, APHA, APCA, WPCF, AIHA, AEEP AAAS, RSH, ΣΞ

Purdy, Paul J
Business: P O Box 120, Grandview, MO 64030
Position: Mgr, Tech Engrg. *Employer:* Pitman Mfg Div, A B Chance Co. *Born:* Oct 1945. BSME UMR. Native of Harris Mo. Prod design engr J I Case Co, Burlington/Bettendorf Iowa 1967-73; spec in design & dev of hydraulic sys, & applied pressure compensated valves to mobile const equip. 1973- , Mgr Tech Engrg Pitman Mfg Co; respons incl role as hydraulic cons to Ch Engrs of (3) product lines, design & dev of 'in house' manufactured hydraulic components, selection, testing & eval of purchased hydraulic components, & approving tech specifications for purchased matls & components.

Purdy, Richard B
Business: 90 E. Ridge Rd, Ridgefield, CT 06877
Position: Dir of Engrg *Employer:* Boehringer Ingelheim, Ltd *Education:* BS/Arch Eng/Univ of MI *Born:* 09/21/22 US Navy Aviator, WWII. 11 yrs Gen Contracting, Commerical and Instl Projs, White Plains, NY. 3 yrs Consltg Arch, National Council of YMCA, New York, NY. 14 yrs Dir of Facilities Engrg CIBA-GEIGY Corp, Ardsley, NY. 4 yrs Dir of Engrg Boehringer Ingelheim, Ltd, Ridgefield, CT. Experience: Directing corporate long range planning and masterplan development; Administering consltS, architects, engrs, contractors; facilities and maintenance operation mgmt; pharmaceutical lab & plant engrg. *Society Aff:* AIPE, ISPE

Purdy, Verl O
Business: 100 Cherry Hill Rd, Parsippany, NJ 07054
Position: Gen Mgr. *Employer:* BASF Wyandotte. *Education:* MBA/-/Univ of NC; BS/ChE/WV Univ. *Born:* 10/25/42. Native of Charleston, WV. Worked in various engg & production assignments for Goodrich - Gulf & FMC from 1964 to 1969. Joined BASF Group in 1969 as production mgr of their Charlotte, NC facility & have since served as Plant Mgr, Works Mgr. Mfg Mgr in 1978 was named a Divisional Gen Mgr. *Society Aff:* AIChE.

Purl, O Thomas
Home: 466 La Mesa Ct, Menlo Park, CA 94025
Position: VP Shareowner Relations and Planning Coordination *Employer:* Watkins-Johnson Co-Palo Alto. *Education:* PhD/EE/Univ of IL; MS/EE/Univ of IL; BS/EE/Univ of IL. *Born:* June 1924 E St Louis Ill. 1943-46 US Army Air Corps; weather officer. 1948-49 Collins Radio Co, Cedar Rapids Iowa. 1951-55 Univ of Ill, Res Asst. 1955-58 Hughes Res Labs, Culver City Calif, finally as Hd, High Power TWT Sect. 1958- , Watkins-Johnson Co, Palo Alto Calif; presently V P - shareowner Relations & Planning Coordination, respon for shareowner relations & corporate planning coordination. Mbr IEEE, Eta Kappa Nu, Sigma Xi, Phi Kappa Phi. Formerly Santa Clara Valley Subsect Chmn IEEE & Chmn Career Guidance Group. Fellow IEEE Jan 1974, for contrib to high power twts & for leadership of microwave electron device engrg. Has contrib sev tech publs & holds pats in twt tech.

Purnell, William B
Business: P.O. Box 7488, Houston, TX 77008
Position: Pres of Corp. *Employer:* Purnell Inc. *Education:* MS/Chem Eng/Rice Univ; MS/Civil Eng/Univ of Houston; MS/Mech Eng/Univ of Houston. *Born:* May 31, 1920. Native of Houston Texas. US Navy WW II, Engrg Officer. Design engr Shell Oil Co 1945-48; Prof of Engrg Univ of Houston 1949-58; Ch Engr Keystone Internatl 1958-60; Purnell & Assocs, tech mkting org to petro & petrochem indus 1960-64; V P Carmody Corp, Div of Singer, in charge of worldwide mkting 1964-67; Pres Purnell Inc, engrg & tech mkting to petrochem indus in plastics processing. Reg P E Tex. Mbr AIChE. Enjoy mineralogy & anthropology. *Society Aff:* AIChE, SPI.

Pursell, Carroll W
Business: Dept of History, Santa Barbara, CA 93106
Position: Prof of Hist of Tech. *Employer:* Univ of Calif, Santa Barbara. *Education:* BA/History/Univ of CA; MA/History/Univ of DE; PhD/History/Univ of CA. *Born:* 9/4/32. Eleutherian Mills-Hagley Fellow Univ of Delaware 1956-58. Taught at Case Inst of Tech 1963-65, & was Andrew W Mellon Disting Prof of the Humanities at Lehigh Univ 1974-76. Currently Prof of Hist of Tech at Univ of Calif Santa Barbara. Founding Mbr & 1975-1983 Secy of Soc for the Hist of Tech. Specialist in social impact of Amer tech & hist of engrg policy. Frequent contributor to 'Tech & Culture' & other prof journals. Author, editor, or contributor to over 12 books in the field. Fellow, Amer Assoc for the Advancement of Sci; former Mbr, NASA History Advisory Board. *Society Aff:* SHOT, AHA, HSS, ASA.

Purser, Nolan A
Business: 10611 Grant Rd, Houston, TX 77070
Position: Owner. *Employer:* Development Consultants. *Education:* BS/CE/TX Tech Univ. *Born:* 7/8/35. Early prof Texas Hwy Dept. Ch const engr for Coastal Embankment Proj & Chandpur Irrigation Proj with Leedshill-DeLeuw & Gov of E Pakistan (now Bangladesh) 1966-71. Founded Dev Cons 1971. Dev Cons provides engrg design, construction, operation and mgmt services to municipalities and land development projects in the Houston area. *Society Aff:* ACEC, ASCE, ACSM, ASES, AWWA.

Pursley, Michael B
Business: 1101 W. Springfield Ave, Urbana, IL 61801
Position: Prof *Employer:* Univ of IL. *Education:* PhD/EE/Univ Southern CA; MS/EE/Purdue Univ; BS/EE/Purdue Univ *Born:* 8/10/45. Has several yrs ind experience in communications systems with the Northrop Corp and the Hughes Aircraft Co. Was an Acting Asst Prof in the System Sci Dept of the Univ of CA, Los Angeles. Presently Prof with the Dept of Elec & Computer Engg and the Coordinated Sci Lab at the Univ of IL. res interests: info theory, spread- spectrum communications,

Pursley, Michael B (Continued)
applications of error-control coding. Mbr Phi Eta Sigma, Tau Beta Pi. Prog Chrmn for 1979 IEEE Intl Symposium on Info Theory, Grignano, Italy. President IEEE Info Theory Group, 1983; Fellow IEEE, 1982; IEEE Centennial Medal, 1984. *Society Aff:* IEEE, IMS.

Purswell, Jerry L
Business: 202 W Boyd-Suite 116, Norman, OK 73019
Position: Prof of Indus Engrg. *Employer:* Univ of Oklahoma. *Education:* PhD/IE/Tx Tech; BS/ME/Lamar State; MS/Engr/Univ of AL; BS/IE/Lamar State *Born:* Dec 1935. Native of Cleveled Tex. Employed by Procter & Gamble as indus engrg June 1961-Sept 1962. Joined Collins Radio Co in Dallas Sept 1962. Served as staff indus engr, group leader & mgr of staff Indus Engrg Dept. Resigned Jan 1965 to pursue PhD at Texas Tech. Completed PhD 1967. Fac appointments at Univ of Arkansas & Univ of Okla. Dev ergonomics prog at O U. Pres of IIE-OKC Chapter 1987. Res in hwy safety for sev yrs; serve as expert witness in this area in court. Standard Oil Award for teaching excell 1972. Enjoy tennis & woodworking. Res & consulting in areas of product & industrial safety & ergonomics. Expert witness in these areas. Served as Dir of Safety Stds Progs for OSHA-Dept of Labor from 7/78-1/81. Dir staff of 50 engr & specialists in dev of occupational safety stds for OSHA. Returned to Univ. of Okla. on 1/81 as Prof of Ind. Engr. Nominated and selected after international competition to become Director, occupational Safety and Health Branch of Intl. Labor organization in Geneva, Switzerland. Univ. of Okla. granted two year leave of absence for this appointment starting 8/1/81. Baldwin award for teaching excellence, 1984. Regents award for professional service, 1984, in recognition of intl and natl contributions to occupational safety and health. IIE award for research excellence, 1985. *Society Aff:* IIE, ERS, HFS, AIHA, NSPE.

Purvin, Robert L
Home: 770 Taylors Ln, Mamaroneck, NY 10543
Position: Principal. *Employer:* Robert L Purvin & Assoc. *Education:* LLD/Hon/Manhattan Coll; DSc/ChE/MIT; BS/ChE/Univ of TX; B/Chem/Univ of TX. *Born:* 6/5/17. Worked as refinery technician at Humble Oil & Refining Co 1941-45 in gen petrol refining and butyl rubber mfg. Founded and managed Purvin & Gertz, consulting engrs, from 1946-57. Organized TX Butadiene & Chem Corp, 1953; organized Goliad Corp of TX (Gas processing), 1954; and Goliad of Canada, 1955. Served as Exec VP, Foster Grant Co, Inc 1957-62. Organized and managed Purvin & Lee, consulting engrs, 1962-75. Organized Gulf Central Pipeline (anhydrous ammonia), 1968; organized Fertilizantes Fosfatados Mexicanos (phosphate fertilizer) 1965-67; and organized Distrigas Corp (LNG trading) in 1969. Served as Pres and Chief Exec Officer, Barber Oil Corp, 1957-78. Presently, a consultant. *Society Aff:* AIChE, ACS, ΦBK, ΣΞ, TBΠ, NSPE.

Puscheck, Herbert C
Home: 8106 W Boulevard Dr, Alexandria, VA 22308
Position: Asst Deputy Chief of Staff *Employer:* Hq US Army Materiel Command *Education:* PhD/Ind Engr (Ops Res)/Purdue Univ; MS/EE/Purdue Univ; BS/Gen Engrg/US Military Acad *Born:* 07/14/36 Dr Puscheck is currently Asst Deputy Chief of Staff (Program Budget) for the US Army Materiel Command. He has a staff of 150 most of whom are program and budget analysts or operations res analysts. Prior to this assignment he served as Dir of Program Analysis and Evaluation, AMC. Dr Puscheck served for twenty years as a military engr with US Army. He is a reg PE and holds degrees in gen engrg, elec engrg and Indust Engrg. He has pub several articles and has taught at the US Military Acad, US Naval Post Grad Sch, Golden Gate Univ and the Indust Coll of the Armed Forces. *Society Aff:* SAME, ORSA, WORMSC, MORS

Putman, Laurel E
Home: 35 Mountain Ave, Maplewood, NJ 07040
Position: President. *Employer:* Azote, Inc. *Education:* BS/Chem Eng/Clarkson College; MS/Chem Eng/Columbia Univ. *Born:* 1920. US Army Engrs 1942-46. G E Co 1947-48. Engr R&D Labs 1948-51. HQ Genl Staff US Army 1951-52. Superior Air Products Co (2001 Jernee Mill Rd, Sayreville N J) 1952-1980. Mbr AIChE, ASME. Pres Azote, Inc. 1980 to Present. *Society Aff:* AIChE, ASME, ACS, NSPE, ISA.

Putnam, Abbott A
Home: 471 Village Dr, Columbus, OH 43214
Position: Senior Res Engr, Semi-retired *Employer:* Battelle, Columbus Labs. *Education:* BME/Mech Eng/Cornell Univ. *Born:* Nov 24, 1920. Design & production for DRAVO Corp; instr in mech engrg at Cornell; fatigue engr at Natl Adv Ctte Aeronautics; Battelle 1946. Res in gas dynamic of combustion sys (incl flame stability, combustion-generated noise, fuel interchangeability, free burning fires), flow-induced vibrations, vapor & particulate transport in atmos & closed sys, dimensional analyses & phys modeling. Lic engr Ohio & N Y. Chartered engr Great Britain. Fellow ASME & Inst of Energy. Assoc Fellow AIAA. Mbr Combustion Inst, Sigma Xi. Author 'Combustion Driven-Oscillation in Indus'. *Society Aff:* ASME, AIAA, ΣΞ, IE.

Putnam, Allan Ray
Home: P.O. Box 1130, Orleans, MA 02653
Position: Secretary-General, World Materials Congress and Retired Senior Managing Director *Employer:* ASM International (Name Change Eff. Oct 86) *Education:* BS/Econ/Wharton Sch of Finance, Univ of PA *Born:* 07/16/20 Born in Melrose, MA. Reside in Orleans, Mass. with wife, Marion. Exec Staff of Electroplaters 1946-49; Asst Exec Secy SME 1949-59; Managing Dir of ASM International 1959-1983; Senior Managing Director 1984-85; Secretary-General World Materials Congress. Favorite recreation is sailboating on Cape Cod. *Society Aff:* ASAE, NSTA, ISIJ, IOM, NAEM, AISI, CESSE

Pyatt, Edwin E
Business: Dept of Environmental Engg Sciences, Gainesville, FL 32611
Position: Chrmn. *Employer:* Univ of FL. *Education:* DrEng/Envir Engg/Johns Hopkins Univ; MS/Environ Engg/Univ of CA; BS/Civil Engg/Caltech. *Born:* 5/13/29. USAF Medical Service Corps, 1953-55; Gilman Fellow, Johns Hopkins, 1956-58; Instr, Johns Hopkins, 1959; Asst Prof, Northwestern Univ, 1960-62; Sr Res Engr, Travelers Res Ctr, 1962-1965; Prof, Univ of FL, 1965-present; Chrmn of the dept, 1970-present. Consultant to numerous organizations. Reg PE in FL, CA and IL. Mbr of Tau Beta Pi, Sigma Tau, Chi Epsilon, Sigma Xi, Phi Kappa Phi, ODK and FL Blue Key. Mbr of Kappa Sigma social fraternity. *Society Aff:* AAAS, ASCE, NSPE, WPCF, AWRA.

Pyke, Robert
Business: 2855 Telegraph Ave, Suite 415, Berkeley, CA 94705
Position: Individual Conslt *Education:* PhD/Soil Dynamics/Univ of CA-Berkeley; BE/CE/Univ of Sydney-Australia *Born:* 08/04/42 Born in Melbourne, Australia; attended Univ of Sydney, then worked on design and construction of earth and rock-fill dams. Studied with H. Bolton Seed at Univ of CA, Berkeley and following several yrs with Dames & Moore started practice as an individual conslt in the geotechnical aspects of earthquake, offshore and embankment dam engrg. Concurrently serves as principal of Telegraph Avenue Geotechnical Assocs, a geotechnical res and development group, and TAGA Engrg Software Servs. *Society Aff:* ASCE, ASTM, SSA, EERI

Pyke, Thomas N, Jr
Home: 4887 N. 35th Road, Arlington, VA 22207
Position: Asst. Adm for Environmental Satellite, Data, and Information Services. *Employer:* Natl Oceanic and Atomospheric Administration. *Education:* MS/Elec Eng/Univ of PA; BS/Elec Engg/Carnegie-Mellon Univ. *Born:* 7/16/42. Joined NBS 1960 as student trainee. Electronic Engr 1964-69; Ch, Computer Sys Sect 1969-73; Ch, Computer Networking Sect 1973-75, Ch, Computer Sys Engg Div 1975-79, Dir, Ctr for Computer Sys Engrg 1979-81. Dir, Ctr for Programming Sci & Tech, 1981-86. Assumed current respon for the nation's civil remote sensing satellites and envi-

Pyke, Thomas N, Jr (Continued)
ronmental information dissemination in 1986. Bd of Dir, Amer Fed of Info Processing Socs 1974-76. Governing Bd, IEEE Computer Soc 1971-73, 1975-77. Dept of Commerce Silver Medal Award 1973. WA Acad of Sci Engg Sci Award 1974. William A Jump Found Meritorious Award for Exemplary Achievement in Public Admin 1975 & 1976. *Society Aff:* IEEE, ACM, ΣΞ, HKN, AAAS.

Pyle, Don T
Business: 7456 W 5th Ave, Denver, CO 80226
Position: VP & Treas. *Employer:* KKBNA. *Education:* BS/Civil Engg/CO State Univ. *Born:* Nov 1934 Chgo. Started as struct design engr with KKBNA, Inc, Cons Engrs, Denver Colo 1957. Became principal & part-owner 1965. Mbr of Computer Dept. Struct engr of record on multi-million dollar bldg projs. Spec in struct engrg & computer applications. PE CO, CT, NY. P Pres Struct Engrg Assn of CO. P Chmn Amer Concrete Inst Ctte 118 Use of Computers, P Pres, Rocky Mtn Chapter, ACI. 6th Award 1970 Natl Award Prog for Progress in Engrg Design of Arc Welded Structs. Avocations: making musical instruments & performing early classical music, VP, CEPA. *Society Aff:* ACEC, ASCE, ACI, NSPE, CSI, CEPA

Pyle, William L
Business: PO Box 345, Puunene, Maui, HI 96784
Position: Mgr, Cultivation Dept. *Employer:* Hawaiian Commercial & Sugar Co *Education:* BS/Agri Eng/Univ of CA at Davis *Born:* 02/18/42 Native of Vacaville, CA. Grad from UCD in 1965. Performed Water and Land use investigations in No CA for CA Dept Water Resources. Started at HC & S on Maui in 1967. Converted 35,000 acres of surface irrigated sugar cane to drip irrigation by 1986. Took over cultivation dept. 1983. Chrmn for ASAE, Hawaii Section in 1980. Reg in Civil and Agricultural Engrg. Serve on Bd of Governors for Hawaii Sugar Technologists. Active marathon runner, hiker, hunter, surfer, windsurfer, motorcyclist, community volunteer. *Society Aff:* ASAE, NSPE

Pyper, Gordon R
Business: Precision Park, North Springfield, VT 05150
Position: Assoc & Managing Dir, Res Special Projs *Employer:* Dufresne-Henry, Inc *Education:* PhD/Civil (Sanitary)/Univ of MI; MSE (SE)/Sanitary/Univ of MI; ScB/Civil/Brown Univ *Born:* 09/01/24 Cedar Falls, IA; grew up in RI and MA. Served in US Air Force, active and reserve, 1943-1972. Sanitary engr, IL Dept of Health 1949-53. Instructor to Full Tenured Prof, Norwich Univ, Northfield, VT 1953-73, Dept Hd, Civil Engrg 1964- 73; taught civil and sanitary engrg. Commissioner of Water Resources, State of VT 1973-77. Dufresne-Henry, Inc, consltg engrs; Sr Environmental Engr 1977-78; Hd, Water Resources Dept, 1978-79; Managing Dir, Res and Special Projs 1979- present. Res on innovative waste treatment and small water treatment systems. Pres ASCE (VT section) 1961; Pres, VSPE, 1981; Commissioner, NE Water Poll Control Comm 1973-77. *Society Aff:* ASCE, ASEE, NSPE, NEWWA, WPCF, AAEE

Quader, Ather A
Business: GM Res Lab, GM Tech Ctr, Warren, MI 48090
Position: Senior Staff Res Engr, BE Osmania Univ *Employer:* General Motors Corp. *Education:* PhD/Mech Engrg/Univ of WI; BE/Osmania Univ. *Born:* 10/10/41. Assoc Sr Res Engr GM Res Labs 1968. Sr Research Engr 1973. Staff Res Engr 1981. Present pos since 1983. Specialization: lean and stratified charge combustion in engines, exhaust pollution control methods, ignition & flame propagation, spectroscopic studies of combustion. Several papers presented at Natl and Intl meetings. Awards: The Lang Silver Medal 1956, Aligarh Univ, India; SAE Horning Mem Award for best paper on mutual adaptation of engines and fuels 1974. SAE Arch T Colwell merit award for outstanding contribution to the SAE literature 1976 and 1982, SAE Award for Excellence in Oral Presentation 1976 and 1978. *Society Aff:* SAE, ΣΞ.

Quance, Robert J
Business: Campbell Centre II, PO Box 2880, Dallas, TX 75221
Position: Consultant *Employer:* Sun Oil Co *Education:* MSC/Petro Eng/Univ of OK; BSC/Petro Eng/Univ of OK *Born:* 12/26/30 Native of Calgary, Alberta, Canada. Field engr for Mobil Oil Corp 1952-53, specializing in drilling. With Pan American Petro Corp 1955-1965 as Dist Reservoir Engr and Res Scientist. Worked for Sun Oil in Reservoir simulation 1966-70. Assumed current oil recovery scientist position in 1970 in area of reserves, improved recovery & economics. Considered an expert in water flood and tertiary oil recovery processes. *Society Aff:* SPE

Quandel, Charles H
Business: 401 N Centre St, Pottsville, PA 17901
Position: Pres. *Employer:* Charles H Quandel Assoc, Inc. *Education:* MS/CE/Lehigh Univ; BS/Engg/US Naval Acad. *Born:* 5/21/47. Reg PE. Pres of Chas H Quandel Assoc, Inc and VP of Quandel-Ewing. He grad in 1969 from the US Naval Academy with a BS degree in Engg and went on to earn a MS degree in CE from Lehigh Univ. Mr Quandel served 4 yrs with Schuylkill Products, Inc, as a struct engr for pre-stress and pre-cast concrete design and also as a mfg's consultant for sanitary and storm drainage facilities. Mr Quandel has also served as Pres of the Pottsville Rotary Club and is a mbr of the Pottsville Rotary Fdn, Inc. Mr Quandel is married and the father of three children. *Society Aff:* NSPE, PSPE, ACI, CSI, PCI.

Quaney, Robert A
Home: 1530 Finecroft Dr, Claremont, CA 91711
Position: Prof *Employer:* CA St Polytech Univ *Education:* MS/IE/Stanford Univ; BS/IE/Stanford Univ *Born:* 7/3/28 Native of Kansas. US Navy Officer and Aviator; awarded Air Medal and Korean Service Medal. Graduated from Stanford "With Great Distinction-. Sr Manufacturing Engr and Production Planning Supervisor with electronics and aerospace co's. Asst Prof, Assoc Prof and Prof of Industrial Engrg at CA State Polytech Univ; Acting Head of Industrial Engrg Dept. Distinguished teaching awards form Univ, Engrg School and Industrial Engrg students. Consultant to industry and gov't. Registered Prof Engrg. Officer and Dir of IIE and ASEE. Awarded NSF Science Faculty Fellowship. Tau Beta Pi and Alpha Bi Mu scholarship societies. Listed in *Who's Who in the West.* *Society Aff:* IIE, ASEE

Quate, Calvin F
Business: Appl Physics & Elec Engg, Stanford, CA 94305
Position: Professor. *Employer:* Stanford University. *Education:* PhD/EE/Stanford Univ; BS/EE/Univ of UT. *Born:* Dec 1923. Native of Baker Nv. Joined tech res staff at Bell Labs, Murray Hill in 1949, doing res on electron dynamics in microwave frequency region; joined Sandia Corp, NM in 1959, becoming VP & Dir of Res in 1960; Stanford U appointed him Prof of Applied Physics & Elec Engrg in 1961; Chmn of Applied Physics Dept 1969-72 & 1978-present; Assoc Dean for Natural Sci in the Sch of Humanities & Sci 1972. Res interests are in genl field of wave propagation in solids; interactions between electrons, electromagnetic and acoustic waves and acoustic imaging. Author of over 90 sci publs. Enjoys skiing & kayaking. Mbr: Tau Beta Pi, Sigma Xi, APS, NAS, acoustical Soc of Amer & Fellow of IEEE. Chmn of IEEE Awards Bd Prize Papers Comm 1971-72; Mbr of Natl Acad of Engrg. *Society Aff:* APS, NAS, ASA, NAE, IEEE.

Queneau, Bernard R
2434 Berkshire Dr, Pittsburgh, PA 15241
Position: Consultant. *Employer:* BS/Metallurgy/Columbia Univ; Met E/Metallurgy/Columbia Univ; PhD/Metallurgy/Univ of MN *Born:* 7/14/12. US Steel Res 1936-38; Asst Prof Columbia Univ 1938-41; US Navy 1941-46; Comdr USNR; US Steel 1946-77 Ch Dev MEt South Works 1946-51; Ch Metallurgist Duquesne Works 1951-57; Ch Metallurgist TCI Div 1957-63; Genl Mgr Process met 1964-77. Fellow ASM; Distinguished Mbr AIME. Author of many papers on steel melting, solidification & heat treatment. Presented the Educational Lectures at the Natl Metal Congress in 1955 on Embrittlement of Metals. A contributor to AIME book on Elec Furnace Steelmaking. Technical Editor-Iron & Steelmaker. *Society Aff:* ISS, ASM

Queneau, Paul B
Business: Hazen Research, Inc, 4601 Indiana St, Golden, CO 80403
Position: Project Mgr *Employer:* Hazen Res, Inc *Education:* PhD/Met Eng/Met/Univ of Minn; B/Met Eng/Met/Cornell Univ *Born:* 03/17/41 Kennecott Copper Corp (1967-1972) Res Engr, dev and piloting of hydrometallurgical processes; Amax Extractive Res & Dev, Inc (1972-1982) Sect Hd nickel-cobalt-copper, R&D/Sect Hd molybdenum, tungsten, and specialty metals; P B Queneau Co, Inc (1982-1983) Consltg Engr, extractive met of non-ferrous metals; Hazen Res, Inc (1983-present) Proj Mgr. providing tech services and proj mgmt relating to recovery, concentration, and refining of non- ferrous metals. Reg PE, CO; 40 tech papers and US patents; Chrmn of AIME-ASM Denver Chapt; Soc Aff: AIME, CIMM. *Society Aff:* AIME, CIMM

Queneau, Paul E
Business: Thayer School of Engineering, Dartmouth College, Hanover, NH 03755
Position: Professor of Engineering, Emeritus *Employer:* Dartmouth College, Thayer Sch of Engrg *Education:* DSc/Extract Met/Delft Univ of Tech, Holland; EM/Mineral Engrg/Columbia Univ; BSc/Min Engrg/Columbia Univ; BA/Sci/Columbia Univ. *Born:* 1911. BA, BSc, EM (Columbia), DSc (Delft). Evans Fellow Cambridge U, England 1933. WWII 2Lt Engr, advanced to Col Engr, graduate Engr Sch, Command & Gen Staff Coll; served five campaigns ETO & later in High Arctic. INCO 1934-69, Dir Res, VP, Tech Asst to Pres, Asst to Chmn; Delft Univ, Holland, 1970; Prof of Engineering, Dartmouth 1971- ; Visiting Prof U of Minn 1974. Engrg council Columbia 1965-70; Visiting Com MIT 1967-70; Bd Mbr Engrg Found 1966-76, Chmn 1973-75. Egleston Medal Columbia 1965; Fellow TMS-AIME (Dir 1964, 1968-71, Pres 1969); Mbr AIME (Douglas Medal 1968, VP 1970, Dir 1968-71); Fellow Inst Mining & Met (overseas Mbr Council 1970-80, The Gold Medal, 1980); National Academy of Engineering, 32 US Patents. *Society Aff:* AIME, CIMM, IMM, AusIMM, NSPE.

Querio, Charles W
Home: 13710 Camara Ln, Houston, TX 77079
Position: Pres. *Employer:* PB-KBB Inc. *Education:* BS/Civil Engg/Univ of MI. *Born:* 3/11/24. Franklin, KA. Worked with Dow Chem Co for 20 yrs where I had various respon connected with subsurface technologies. Became VP of TX Brine Corp in Jan 1970. Dir resp for operation and maintenance of solution mining fac. I have also served as Dir of Engg for Morton Salt Co. Assumed current resp as Pres, PB-KBB Inc on Jan 1, 1979. PB-KBB is an engineering firm devoted to subsurface systems & tech. Reg Engr in MI, TX, LA, NY, KS and OH. *Society Aff:* ASCE, AIChE, AIME, SPE.

Quicksall, William E, Jr
Business: 554 W High Av PO Bx 646, New Philadelphia, OH 44663
Position: Chairman *Employer:* W E Quicksall & Assocs, Inc. *Education:* BCE/Civil Engr/OH State Univ. *Born:* 6/25/27. Served US Marine Corp 1945-46; with Ohio Dept of Highways as Inspector & Project Engr 1952-57; Deputy Dir of Div 11 1957-59; Co-Founder of W E Quicksall & Assocs 1959; Pres of W E Quicksall & Assocs, Inc 1973-79, firm employs 70 with 20 reg engrs & surveyors. Served as Pres of Council, City of New Philadelphia 1968-71; Mbr Bd of Trustees of Union Hospital 1967-82, Pres 1973-75. Past Pres. of Bd/Dir of OH Association of Consulting Engrs. Licensed to practice engrg in Ohio, Penn, W Va, Ky, & FL. *Society Aff:* NSPE, ACEC.

Quimpo, Rafael G
Home: 2138 Ramsey Rd, Monroeville, PA 15146
Position: Prof *Employer:* Univ of Pittsburgh *Education:* PHD/CE/CO State Univ; M Eng/Hydraulics/SEATO Grad School of Engrg; BS/CE/FEATI Univ, Phillipines *Born:* 3/23/39 Native of Philippines. Worked in consulting firms in the Philippines and Thailand. Has been with the Univ of Pittsburgh since 1966. Spent sabbatical yr at Federal Univ of Rio de Janeiro, Brazil. Was US delegate to US-Japan Bilateral Semiar in Hydrology in 1971; Congress of the Int. Union of Geodesy and Geophysics at Moscow in 1971, and at Grenoble in 1975, was Visiting Scientist in the Philippines in 1976. Has conducted and directed research on the math modeling of hydrologic systems and water resources engrg. Enjoys playing the piano, chess and tennis. *Society Aff:* ASCE, AGU, ASEE

Quinn, Alfred O
Home: D'Aloquin-Quaker Mt, Wilmington, NY 12997
Position: Photogrammetric Consultant *Employer:* Self Employed *Education:* BSCE/Civil Engr/Syracuse Univ; LLB/Law/Chattanooga College of Law. *Born:* April 2, 1915. Licensed as a Prof Engr & Land Surveyor in many states; licensed to practice law in Tenn. Worked with TVA as Civil Engr and photogrammetrist 1936-43; served as a US Naval Officer during WWII 194346; received Navy Commendation. Assoc Prof of Photogrammetry in Dept of CE at Syracuse U 1946-50; Ch Engr & Dir Photogrammetry Engrg Div of Aero Serv Corp 1950-64; Pres Quinn & Assocs 1964-82. 1982 - Self employed as a consultant and beekeeper (Apiarist) . Cons on wide variety of surveying & mapping projs; lectr various univs & prof socs. Fellow ASCE, Pres, VP & Dir of Philadelphia Sect; Honorary Mbr Amer Soc of Photogrammetry, Pres, VP, Dir, & Chmn Prof Activities & Prof Conduct Cttes; Sigma Xi; Mbr, Amer Congress on Surveying and Mapping, Dir. Received "1984 Surveying and Mapping Award" from ASCE. Mmb, Bd Dir, Placid Mem Hospital; Mbr, Bd Dir Whiteface Mountain Ch of Comm. *Society Aff:* ASPRS, ASCE, ACSM, ΣΞ

Quinn, Brian
Business: 50 Washington Rd, Princeton, NJ 08540
Position: President. *Employer:* ARAP. *Education:* Doctorate/Physics/Sorbonne; MS/Mgmt/MIT; BS & MS/Aero Engr/Notre Dame; MBA/Finance/Miami (OH). *Born:* April l, 1938 Flushing, New York. Doctorate from the Sorbonne; MS from MIT; BS and MS Univ of Notre Dame; MBA Miami (Ohio) Univ. Sloan Fellow at MIT 1975. Aerospace Res Labs, Wright-Patterson AFB, Ohio: l964-69 res engr, reentry aerodynamics and dynamic stability; 1969-75 group leader, structure of turbulence and compact thrust augmentors. 1972 USAF Outstanding Scientific Achievement Award. Cons to Navy XFV-12A VTOL fighter program. More than 30 publs, 3 patents. Appointed Dir of Aerospace Sciences AFOSR in 1976 with respon for all Air Force basic res in mechanics and energy conversion. Became Pres & COD of Aeronautical Res Assocs of Princeton in 1979.. *Society Aff:* AIAA, ΣΞ.

Quinn, J A
Business: Dept. of Che, Univ of Pennsylvania, 220 s. 33rd. Street - 311a Towne Bldg, Philadelphia, PA 19104-6393
Position: Professor and Chairman, Dept of ChE *Employer:* University of Pennsylvania. *Education:* BS/ChE/Univ of IL; PhD/ChE/Princeton Univ. *Born:* Sept 1932. Joined the Chem Engrg staff at the Univ of Ill in 1958 as Asst Prof. Spent 1965-66 at Imperial Coll, London as Visiting Prof; promoted to Prof of Chem Engrg at Ill in 1966 and moved to current pos at Univ of Penna in 1971. Res interests in interfacial phenomena, membrane transport and bioengineering. Received AIChE Allan P Colburn Award in 1966 and AIChE Alpha Chi Sigma Award in 1978. Elected to Natl Acad of Engg in 1978. Apptd to Atlantic Richfield Foundation Prof (The Robert D Bent Chair) in Chemical Engg, Univ of PA, 1978. *Society Aff:* AIChE, ACS, AAAS.

Quinn, James D
Home: Box 200, Warwick, MD 21912
Position: Consultant. *Employer:* Self. *Born:* Nov 1907. Consultant. E I DuPont 1933-73; positions of const Proj Mgr and Works Engr several locations USA and Brazil; Cons plant maintenance engrg; named Principal Cons in 1963; assisting plants USA, South Amer, Mexico, Canada, Europe. Author several papers on maintenance engrg; two chapters 'Maintenance Engrg Handbook'; holds patents data collecting equipment. Mbr American Inst Plant Engr; Charter Mbr Wilmington Chapter. In ASME: Founder and Chmn Plant Engrg and Maintenance Div; Chmn Genl

Quinn, James D (Continued)
Engrg Dept; V Pres 1969-74; awarded grade of Fellow 1966. Reg Prof Engr Delaware. *Society Aff:* ASME, AIPE.

Quinn, Joseph J, Jr
Business: Quinn Equip Sales Co, PO Box 6577, Bridgewater, NJ 08807
Position: Pres. *Employer:* Quinn Equip Co. *Education:* BME/ME/Villanova Univ. *Born:* 6/18/18. VP & Sales Mgr of Bowen Engg - 22 yrs. Frequent lectr (9 times) Ctr for Profl Advancement on ind drying tech. Instr - Sch of Engg. Villanovia Univ 1947 to 1950. Lt USNR Degaussing Service - 1943 to 1946. *Society Aff:* AIChE.

Quinn, Will M
Home: Sterling Point Island, Portsmouth, VA 23703
Position: Manager - Engineering. *Employer:* GE Co, Television Business Dept. *Born:* Sept 1924. Navy Air Corps 1942-45. BEE Auburn Univ 1950. GE Co 1950-62 as Engr, Proj Engr & Engrg Mgr in television, audio and computer fields. 1962-68 Div V Pres Engrg for Norden Div of United Aircraft Corp; 1968-70 V Pres in charge of High Technology Dept of Walter B Delafield Co, an institutional brokerage co; 1970 to present Mgr-Engrg for Television Business Dept of the Genl Elec Co. Senior Mbr IEEE.

Quist, William E
Home: 18215 NE 27th St, Redmond, WA 98052
Position: Sr Research Specialist. *Employer:* The Boeing Commercial Airplane Co. *Education:* BS/Met Eng/Univ of WA; MS/Met Eng/Univ of WA; PhD/Met Eng/Univ of WA. *Born:* 5/13/35. Seattle, Washington. Metallurgist with PACCAR Inc before joining Boeing in 1959. Has specialized in alloy dev, fracture, fatigue, composites, and metals usage in aircraft design. Taught at U of W 1972-73. Mbr: Alpha Sigma Mu, Seattle Prof Engrg Assn, Fellow and Trustee of ASM International; natl cttes, Mbrship, Materials Characterization; Chrmn, Engrg Assocs Achievement Award Ctte and Young Teachers of Metallurgy, Chrmn Puget Sound Chapt of ASM 1973-74. Co-VChrmn The Pacific Northwest Metals and Minerals Conf 1974; Co-chrmn in 1977; Chrmn in 1983. Editorial Ctte, Metal Progress West. Reg PE seven patents for alloy and process development, Achievement and Recognition awards from the Univ of WA, AIME, AIAA and The Boeing Co. Sports enthusiast and active in various Christian endeavors. *Society Aff:* ASM, AΣM.

Quittenton, Richard C
Business: Box 1182, Windsor Ontario, Canada
Position: President. *Employer:* St Clair College. *Born:* July 1921, Toronto, Ontario. PhD, MASc, BASc Univ Toronto. Tech Supr Alcan, Arvida 1951; Res Dir John Labett Ltd 1955; Pres St Clair Coll 1967. FEIC, FCIC; Pres Prof Engrs Ont 1975; Dir, Canadian Council Prof Engrs 1976.

Qureshi, Abdul Haq
Home: 31216 Narragansett, Bay Village, OH 44140
Position: Prof *Employer:* Cleveland State Univ *Education:* PHD/EE/Tech Univ Aachen Germany; BS/EE/Peshawar Univ *Born:* 10/28/32 Born in Pakistan, migrated to Canada after Ph.D. in Germany. Taught at Univ of Waterloo, Calgary and Windsor. Dept Head at the Univ of Windsor for 4 yrs. Worked as res scientist at Nuclear Res Center, Julich, Visiting Prof at Ruhr Univ Bochum 1972 Consultant in High Voltage tech to various organizations. *Society Aff:* IEEE

Qureshi, Shahid U
Business: 20 Cabot Blvd, Mansfield, MA 02048
Position: VP, Research & Adv. Dev. *Employer:* Codex Corp *Education:* Ph.D./Elect. Eng./University of Toronto, Canada; M.Sc./Elect. Eng./University of Alberta, Canada; B.Sc./Elect. Eng./Univ. of Eng. & Technol., Lahore, Pakistan *Born:* 09/22/45 He received the President of Pakistan's Award for outstanding student at the graduate stage in 1968 and held the Canadian Commonwealth Scholarship, 1968-72. In 1973, he joined Codex Corp as Mbr of Technical Staff in the research dept. He was elected an Associate Mbr of the Motorola Science Advisory Bd in 1978 and was named a Motorola Dan Noble Fellow in 1984 in recognition of leadership in modem research and development. He was recently elected an IEEE Fellow "for contributions to the architecture & commercial dev of high-speed voice-band modems–, and awarded the IEEE Fink Award for his paper on "Adaptive Equalization–. Since 1980, Dr. Qureshi has represented Codex in CCITT Study Group XVII. He is currently VP of Research and Advanced Dev at Codex, responsible for technology dev in the areas of networking, software, data transmission, ISDN voice coding and digital signal processing. *Society Aff:* IEEE

Raab, Harry F, Jr
8202 Ector Ct, Annandale, VA 22003
Position: Chief Physicist. *Employer:* Navy Nuclear Propulsion Directorate. *Education:* SB/EE/MIT; SM/EE/MIT *Born:* May 9, 1926. Native of Johnstown Pa. US Navy Electronics Technician 1944-46; SB & SM EE MIT 1951. m. Phebe Duerr June 1951; children Constance, Harry III, Cynthia. Bettis Atomic Power Lab, Westinghouse - Phys Mgr (Surface Ship Reactors, Lt Water Breeder) 1951-72; Nuclear Power Directorate, Naval Sea Sys Command 1972 - Pres, respon for physics of US Naval Reactors & Lt Water Breeder. Episcopal Layreader 1956- . Fellow ANS; Mbr Tau Beta Pi, Eta Kappa Nu, Sigma Xi, Stewardship Chrmn 79- pres, Senior Warden 1983 Church of the Good Shepherd, Burke, VA. Stewardship Comm. Virginia Diocese Episcopal Church. *Society Aff:* ANS, AAAS, TBΠ, HKN, ΣΞ.

Raba, Carl F, Jr
Business: 10526 Gulfdale, San Antonio, TX 78216
Position: Chrmn and CEO *Employer:* Raba-Kistner Consultants, Inc. *Education:* PhD/Civ Engr/TX A&M Univ; ME/Civ Engr/TX A&M Univ; BS/Civ Engr/TX A&M Univ. *Born:* 12/24/37. Previous experience with TX Transportation Inst, TX A&M Univ Civ Engr Dept, US Corps of Engrs, McClelland Engrs and San Antonio College. Started consulting geotechnical engr firm in 1968 which subsequently expanded into mtls engr, geology, chemistry, environmental, concrete component tech and scientific investigations. Has specialized in geotech, materials and forensic engrg. Author or coauthor of over 30 tech pubs. Mbr of Chi Epsilon, Sigma Xi. Listed in Amer Men & Women in Science and Men of Achievement in TX. Selected "Young Engineer of the Year" in San Antonio, 1971, in TX in 1972, and "An Outstanding Young Engineer" by NSPE in 1972 and "Engineer of the Year" in San Antonio in 1977. Married to former Bunny Jean Dever and father of 3 daughters and 2 sons. *Society Aff:* NSPE, ASCE, ACIL, ASFE, ΣΞ, XE.

Raben, Irwin A
Business: 130 Sandringham South, Moraga, CA 94556
Position: Pres. *Employer:* IAR Technology Inc. *Education:* MS/ChE/LA State Univ; BS/ChE/Tulane Univ. *Born:* 10/26/22. Native New Orleans, LA. Officer US Army Air Corps 1942-46. Sr Process Engr cities Service 1947-55. Asst Ch Process Engr Hedrick Engg Corp 1955-58. Mgr Chem Engg Res, Southwest Res Inst 1958-64. Spec in process dev & saline water conversion; Mgr Air Qual Coantrol, Bechtel Corp 1964-73, spec in mgmt & dev of Air Qual Control Sys. VP Combustion Equip Assocs Inc 1973-79, respon for Mkting and Dev of Air Pollution Control Sys. Pres, IAR Technology, Inc 1979- Mgmt & Engg Consulting-Air Pollution Control Technology, Economic & long range process studies, Publ num tech papers. US Pat, Cert of Merit ACS 1963, Enjoy jazz music, traveling, gourmet cooking. *Society Aff:* AIChE, ACS, APCA, ΣΞ.

Rabii, Sohrab
Business: Moore School of Elect Engrg, Philadelphia, PA 19104
Position: Prof of Elec Engrg *Employer:* Univ of Penn *Education:* PhD/EE/MIT; MS/EE/MIT; BS/EE/Univ of Southern CA. *Born:* 12/30/37 Born in Ahwaz, Iran. Came to US in 1958. Res fellow at MIT, 1966-67. Sr Res Physicist at Monsanto Co in St Louis, 1967-69. Joined Moore Sch of Elec Engrg, Univ of PA in 1969. Presently Prof of the Elec Engrg Dept at PA. Spent sabbatical yr, 1975-76, as Fellow of the Max-Planck Soc in Stuttgart, Germany. Area of specialty, Solid-State Theory: Elec-

Rabii, Sohrab (Continued)
tronic properties of molecules and solids. Two daughters. *Society Aff:* APS, IEEE, ΣΞ.

Rabiner, Lawrence R
AT&T Bell Laboratories, bRm 2D-533, Murray Hill, NJ 07974
Position: Head, Speech Research Dept. *Employer:* Bell Labs. *Education:* PhD/Elec Eng/MIT 1967; MS/Elec Eng/MIT 1964; BS/Elec Eng/MIT 1964 *Born:* Sept 1943. Bell Labs June 1962- ; work in areas of speech communication & digital signal processing. Fellow Acoustical Soc of Amer, IEEE, Mbr Natl Acad of Engrg, Past Pres IEEE Group on Acoustics, Speech, & Signal Processing, 1974- 75. Paper award IEEE/ASSP Group; Biennial award of ASA, Eta Kappa Nu; Hon Mention for Outstanding Young Elec Engr. IEEE Piori Award, IEEE ASSP Society Award, IEEE Centenial Award, Author Theory & Application of Digital Signal Processing, Digital Processing of Speech signals, Multirate Digital Signal Proc. *Society Aff:* IEEE, ASA, NAE

Rabinow, Jacob
6920 Selkirk Dr, Bethesda, MD 20817
Position: Ch Res Engr, OERI. *Employer:* Natl Bureau of Stds-Gaithersburg Md *Education:* MS/EE/CCNY; BS/EE/CCNY. *Born:* 1/8/10. Kharkov Russia. 1938-54, Natl Bureau of Stds, Mech Engr to Ch, Electro-Mech Ordinance Div. 1954-64 Pres, Rabinow Engrg Co. 1964-72, V P Control Data Corp. 1972- , Natl Bureau of Stds. Duties incl eval of energy-related inventions. 221 US pats. Pres's Cert of Merit; War Dept Cert of Appreciation; Edward Longstreth Medal from Franklin Inst; Jefferson Medal Award, NJ Patent Law Assn; Industrial Research & Development Magazine Scientist of the Year, 1980; Mbr Natl Acad of Engrg & Fellow IEEE; IEEE's Harry Diamond Award 1977; Doctor of Humane Letters, Towson State Univ 1983. *Society Aff:* IEEE, NAE, AES, AAAS.

Rabinowicz, Ernest
Business: M.I.T., Room 35-010, Cambridge, MA 02139
Position: Professor of Mech. Eng. *Employer:* M.I.T. *Education:* BA/Physics/Cambridge University; Ph.D./Physical Chem./Cambridge University *Born:* 04/22/26 Professional career has been in the Mech Engrg Dept at M.I.T. Research interests have been the tribological phenomena of wear, friction, electrical contacts, and lubrication. Has written a couple of books (Introduction to Experimentation and Friction and Wear of Materials) and videotaped three series of lectures in these areas. Walter D. Hodson Award of the STLE in 1957, Ragnar Holm Award of the IEEE Conference on Electrical Contacts in 1983, Mayo D. Hersey Award of the ASME in 1985. *Society Aff:* ASME, STLE

Rabins, Michael J
Business: ME. Dept, 100 EPB, Texas A&M University, College Station, TX 77843-3123
Position: Prof. & Hd. *Employer:* Texas A&M Univ *Education:* PhD/ME/Univ of WI; MS/ME/Carnegie Inst of Tech; BS/ME/MA Inst of Tech. *Born:* 2/24/32. in NYC. Taught 10 yrs at NYU and five at Poly Inst of NY. Served as Univ Res Dir at US Dept of Transportation for two yrs before going to Wayne State Univ as ME Chrmn for 10 yrs. Head of M.E. Dept at Texas A&M since 1987. Consultant to over 15 cos. Author of two textbooks and over 25 monographs, reports and papers. Active in mech control systems. Mbr of Tau Beta Pi and Pi Tau Sigma. Recipient of: Lindback Award for Distinguished Teaching, NSF Faculty Fellowship, Visiting Profship at the Univ of Tokyo and the Poly Inst of Grenoble, Elected Fellow of ASME, and the US Secy of Transportation Silver Medal. *Society Aff:* ASME, ASEE, SAE.

Rabson, Thomas A
Business: 6100 Main St, Houston, TX 77005
Position: Prof *Employer:* Rice Univ *Education:* PhD/Physics/Rice Univ; MA/Physics/Rice Univ; BS/EE/Rice Univ; BA/EE/Rice Univ *Born:* 07/31/32 raised in Houston, TX. Taught at Rice Univ since 1959. Currently Prof of Electrical Engrg. Founder and Pres of Technical Enterprises Corp from 1962-1968. Natl Sci Foundation Sci Faculty Fellow 1965-1966. *Society Aff:* APS, IEEE, SPIE, OSA

Race, Hubert H
Home: Rydal Park 251, Rydal, PA 19046
Position: Retired. *Education:* PhD/Dielectrics/Cornell Univ; EE/Power/Cornell Univ. *Born:* Feb 12, 1899 Buffalo N Y. Cornell Univ Instr & Asst Prof 1921-29. Employed by Genl Elec Co 1929-64. R&D Lab, Atomic Power Lab Ordnance Dept & Mgr Dev Inst. Former Chmn sev natl engrg cttes. 6 yrs represented U S Chamber of Commerce on U S Natl Comm for UNESCO. Fellow IEEE. Compiled 3 books on mgr dev for G E Co. *Society Aff:* CSE, IEEE, AARP.

Rachford, Henry H, Jr
Home: 6150 Chevy Chase, Houston, TX 77057
Position: Prof of Math Sciences. *Employer:* Rice Univ. *Education:* ScD/CE/MIT; MA/PChem/Rice; BS/CE/Rice. *Born:* 6/14/25. High sch in Houston. 1949-1964 Humble (now Exxon) Production Res Div; numerical math, emphasis on developing computer models of petrol reservoir processes. 1964-present Rice Univ faculty, math & math sciences; computational methods for solving partial differential equations, emphasis on engg problems. In 1969 helped found a consulting/software firm now called DuPont-Rachford Engg Math (DREM) Co (Pres). Primary DREM work on batch & real time computer models of transients in pipeline networks. *Society Aff:* AIME, AIChE, SIAM, AMS.

Rack, Henry J
Business: Exxon Enterprises, P.O. Dwr H, Greer, SC 29651
Position: Manager, Metallurgy Depart. *Employer:* Exxon Ent., Mat'l Div. *Education:* ScD/Metallurgy/MIT; SM/Metallurgy/MIT; SB/Metallurgy/MIT. *Born:* Nov 1942. Native of NYC. Dow Chem Award 1964. AIME Anual Paper Award 1965. VASCO Fellow 1964-68. Scientist LockheedGa Co 1968-72, principal investigator ferrous & non-ferrous struct met. Mbr Tech Staff Sandia Labs 1972- 81 ; respon for struct matls R&D. Manager, Met. Dep. Exxon Ent.; metal matrix composites Adj Prof New Mex Inst Mining & Tech 1975-81. Chmn ASM Albuquerque Chap 197576. Mbr ASM Adv Tech Awareness Council 1975- . Young Mbrs Comm 1968-70. Howe/Grossman Award Comm 1976. Mech Behav Ctte 1972- .·AIME Titanium Commission- Chrmn 1979 to present. Enjoy camping and Shrine activities. *Society Aff:* ASM, AIME.

Rackley, Clifford W
PO Box 2511, Houston, TX 77001
Position: Chrmn. (Retired) *Employer:* Tenneco Oil Co. *Education:* BChE/ChE/GA Inst of Tech. *Born:* 9/19/23. Chrmn, (Retired) of Tenneco Oil Co, maj subsidiary of Tenneco Inc located in Houston, TX. Born in Tifton, GA, grad from the GA Inst of Tech in 1949 with a Bachelor of Chem Engg degree. Supervisory process engr Mobil Oil, Beaumont 1949-56; sr process engr Tenneco Oil New Orleans 1956-60; sr vp marketing Tenneco Oil Houston, 1960-72, Exec VP, refining & Mktg, 1972-74, Pres, 1974, Chrmn, 1978. Served to 1st Lt, USAAF, 1943-45. Decorated DCF, Air Medal. Mbr Am Inst Chem Engrs, Am Petrol Inst, PE, LA, Alumni Assn GA Inst Tech. Married Julia Susan Old, Feb 6, 1945, 3 children. *Society Aff:* AIChE.

Radcliffe, Charles R
Business: Dept of Mech Engrg, Berkeley, CA 94720
Position: Prof of Mech Engrg. *Employer:* Univ of Calif. *Born:* Feb 14, 1922. BS, MS & ME Univ of Calif Berkeley. Ltjg - Engrg, USS New Mexico WW II. Has taught Machine Design Kinematics, Dynamics of Machinery at Univ of Calif Berkeley since 1948. Res in design of Prosthetic devices for amputees, biomech of human walking & rehab engrg. Fulbright Res Scholar, Bioengrg Unit, Univ of Strathclyde Glasgow, Scotland 1966-67. Machine Design Award ASME 1976.

Radcliffe, Charles W
Business: Dept of Mech Engg, Berkeley, CA 94720
Position: Prof. *Employer:* Univ of CA. *Education:* ME/Mech Engg/Univ of CA; MS/Mech Engg/Univ of CA; BS/Mech Engg/Univ of CA. *Born:* 2/14/22. Served with US Navy in Pacific 1944-46. Since 1948 teaching machine design, advanced kinematics, mechanisms, optimization and computer-aided design at UC- Berkeley. ASME machine design award-1976 for res and dev of improved components and fitting methods for articial limbs. Current interests include application of numerical methods to problems in orthopaedic and rehabilitation engg. *Society Aff:* ASME, ISPO.

Radcliffe, Frederick A
Business: Indus Park Rd, Centerbrook, CT 06409
Position: Sr Partner Chief Engr *Employer:* Radcliffe Engrg Co *Education:* BS/Civ/MA Inst of Tech. *Born:* 4/8/27. Native of Naugatuck, CT; US Navy 1944-46 MIT 1951; Engr MA Dept of Transportation 1951; Structural Engr Bigelow Sanford Co 1952, 53; Field Engr, Proj Engr, L G Defelice Constr Co, 1954-58; Private Practice 1958-present. Collaborated with Prof Chas Miller in dev of COGO at MIT in 1961. Manage a gen civ and municipal engg practice principally in Middlesex County, CT. Reg CT, MA & RI; Biographical listing *Who's Who in the East*; Active in golfing and skiing. *Society Aff:* ASCE, NSPE, ACSM.

Radcliffe, S Victor
Home: 2101 Connecticut Ave, NW, Washington, DC 20008
Position: Vice Pres *Employer:* National Forge Co *Education:* PhD/Physical Metallurgy/Univ of Liverpool; B Eng/Metallurgy/Univ of Liverpool *Born:* 07/28/27 Applied research in British steel industry. Academic career at MA Inst Tech, and Case Inst Tech. Prof of Metallurgy and Materials Sci, and Dept Chrmn at Case. On leave 1974/75 in Washington, DC, as Sr Policy Analyst for the Sci Adviser to the Pres. 1976/79, Sr Fellow at Resources for the Future. Since Fall 1979, VP and Chrmn Asst for Corporate Development at the Natl Forge Co. Conslt to industry, and the United Nations. Study dir for Natl Acad of Scis review of materials sci and engrg, *Materials and Man's Needs*. More than 80 publications. Dir of Industrial Materials Tech, Inc. Chartered Engr (United Kingdom). Fellow, Institution of Metallurgists. Clubs: Cosmos Club, Natl Economists Club. *Society Aff:* AAAS, ACS, AIME, APS, ASMIM

Radell, Nicholas J
Home: 328 Linden St, Winnetka, IL 60093
Position: VP & Dir. *Employer:* Cresap, McCormick & Paget Inc. *Education:* BSME/Mech Eng/Univ of MI; MBA/Bus Adm/Univ of MI. *Born:* Sept 2, 1930. Reg P E Mich; Cert Pub Accountant Mich; Cert Mfg Engr, Soc of Mfg Engrs; Cert Data Processor, Data Processing Mgmt Assn. P Pres Soc of Mfg Engrs 1973-74; Mbr Bd of Dir Engrs Joint Council; VP, Engrs Joint Council; Chrmn-Elect, Assoc for Corp in Engg. Hon & Life Fellow Inst of Prod Engrs; Mbr SME, IIE, ASME, Chrmn of AAES 1984. *Society Aff:* SME, IIE, ASME.

Rademacher, John M
Home: 183 Timber Ridge, Barrington, IL 60010
Position: VP. *Employer:* Velsicol Chem Corp. *Education:* MS/Sanitary Engr/Northwestern Univ; BS/Civ Engr/Purdue Univ. *Born:* 9/21/24. Native of Hammond, IN. WWII Veteran--ETO with 13th Armored Div. Field engr for IN State Bd of Health, 1949-51. Sales & service engr - Wallace & Tiernen Co - IN Dist to 1956. Joined Federal service as Public Health Service officer at water pollution control prog's inception in 1956. Variety of assignments including Regional Administrator in Region VII, KS City, for successor agency (EPA). Joined Velsicol Chem Corp 1979 as line VP for Environmental, Health and Regulatory Affairs. Pres, Fed Water Qualtiy Assn (WPCF) 1978-79. Prof-Engineer, Dipolmate- American Academy of Environmental Engineers. *Society Aff:* NSPE, WPCF, AAEE.

Radlinski, William A
Home: 2712 Calkins Rd, Herndon, VA 22071
Position: Exec Dir, ACSM Assoc Dir, US Geol Survey (retired) *Employer:* Amer Cong on Surveying and Mapping *Education:* BA/Math/Hofstra Univ. *Born:* Aug 1921. Native of Salamanca N Y. 2 yrs grad work Georgetown Univ. Army Corps Engrs WW II; Lt Colonel Retired. Reg P E Wash D C. Army Map Serv 1946. With USGS 1949-79, spec in photogrammetric engrg res & res mgmt; with respon for dir of 13,000 employees engaged in earth sci investigations involving var engrg specialties; also respon for safety & environ protection in offshore petro opers. Retired in March 1979. Claude H Birdseye Award, Amer Soc of Photogrammetry 1969; Surveying & Mapping Award, Amer Soc of Civil Engrs 1972; Mt Radlinski, Antarctica 1962; Hon Mbr RICS England 1976. Chmn Cartography Div, Amer Congress on Surveying & Mapping 1964; Chmn Exec Ctte Surveying & Mapping Div, ASCE 1970; Pres ASP 1968; Pres Internatl Fed of Surveyors (FIG) 1973-75. *Society Aff:* ACSM, ASCE, ASP, NSPE.

Rady, Joseph J
Business: 400 Continental Life Bldg, Ft Worth, TX 76102
Position: Chrmn of Bd. *Employer:* Rady & Assoc. *Education:* CE/Hydraulics/Cornell Univ. *Born:* 10/4/99. in NY. On grad went to TX, worked for Devlin Engrs four yrs, then opened own office as Rady and Assoc, continues in practice to date. Firm has designed more than 1000 water, sewer and other public works projs in TX including Dallas- Ft Worth Turnpike and D-FW Airport. During WWII served in joint venture on design of numerous large Defense Projs. Active in Civic Affairs and Techni-cal- Profl Societies. Maintains active interest in engg education. Honorary Mbr ASCE and received ASCE History Award in 1977. Continues as Rady Assoc Bd Chrmn. *Society Aff:* ASCE, NSPE.

Radziminski, James B
Home: 300 Finsbury Rd, Columbia, SC 29210
Position: Prof *Employer:* Univ of South Carolina *Education:* PhD/CE/Univ of IL; MS/CE/Univ of IL; B/CE/Cooper Union *Born:* 04/28/37 Native of Mineola, NY. Held appointment as Asst Prof of Civil Engrg at Univ of IL 1965-72. On faculty of Coll of Engrg, Univ of South Carolina since 1972, as Associate Prof (1972) and Prof (1981). Chrmn of Civil Engrg Dept from 1979 to present. Engaged in research on earthquake resistant design of steel frame bldg structures, and on fatigue of structural mbrs and connections. Pres of Midlands Branch, South Carolina Section, ASCE, 1977-78. *Society Aff:* ASCE, EERI

Raemer, Harold R
120 Noanett Rd, Needham, MA 02194
Position: Prof Elec Engg Dept. *Employer:* Northeastern Univ. *Education:* PhD/Physics/Northwestern Univ; MS/Mathematics/Northwestern Univ; BS/Physics/Northwestern Univ. *Born:* Apr 26, 1924. 1950-52 Northwestern Univ, Teaching Asst Physics. 1952-55 Bendix Res Labs Detroit Mich, Physicist 1952-54, Sr Physicist 1954-55. 1955-60 Cook Res Labs Morton Grove Ill, Sr Engr 1955-57, Staff Engr 1957-60. 1960 Ill Inst of Tech Chgo, Asst Prof of EE (leave-of-absence, Cook Res Lab). 1960-70 Appl Res Labs, Sylvania Electronic Sys - Engrg Specialist 1960-62, Sr Engrg Specialist 1962-63, Consultant 1963-70. 1963- , Northeastern Univ Boston Mass, Assoc Prof of EE 1963-66, Prof of EE 1966- , Chmn Dept of EE 1967-77. 1962 Harvard Univ, Visiting Lectr Applied Phys. 1972-73, Harvard Univ, Hon Res Assoc Applied Phys. 1969- , US Naval Res Lab Wash D C, Consultant. Author text on communication theory & sev journal & conf articles on electromagnetic wave propagation, plasma & communication theo. *Society Aff:* IEEE, ASEE, ΣΞ, AAAS.

Raffel, David N
Business: PO Box 12608, El Paso, TX 79912
Position: Owner & Ch Engr *Employer:* Protrans Consultants *Education:* BSE/ME/Univ of MI *Born:* 07/23/29 Jr. Engr hydro turbine mfg before military service during Korean War. Engr Harza Engrg Co 17 yrs first designing mechanical features hydroelectric plants. Later responsible planning hydroelectric plants, expansion electric utility systems, financial and economic analyses, and all aspects engrg hydroelectric machinery. Became Harza's VP and Pres Argentine subsidiary. Entered private practice founding PROTRANS Conslts 1970. Planned 25 hydroelectric plants and provided full gamut engrg services for turbines and valves of 60 plants.

Raffel, David N (Continued)
Chrmn ASME Hydro Power Committee. Married Sandra de Lagasse 1960, div 1984. Have 5 children. Play piano (classical music). Strong interest astronomy, outdoor camping, hiking. *Society Aff:* ASME, NFPA, USCOLD

Raftopoulos, Demetrios D
2801 W Bancroft St, Toledo, OH 43606
Position: Prof of Mech Engr. *Employer:* Univ of Toledo. *Education:* BSCE/CE/Widener Univ; MCE/Struct Engrg/Univ of DE; PhD/Mechancs/PA State Univ *Born:* 5/30/26. in Argostolion, Greece. Worked as Field Supervisor in Construction Co, 1946-49. Served with Greek Army, 1949-51. Received a BSCE from Widener Univ in 1959. Worked as Hgwy Engr with DE Hgwy Dept, Dover, DE, 1959-61. Instr of Engr at Widener Univ, 1961-64. Received MCE from the Univ of DE in 1963 and PhD in Mechanics from Penn State Univ in 1966. Assoc Prof of Mech Engg in 1967 and promoted to Prof of Mech Engr in 1973 at the Univ of Toledo, OH. Visiting Prof at the Chair of Theoretical and Appl Mechanics, Natl Technical Univ of Athens, Athens, Greece in 1973-74. Adjunct Professor, Division of Orthopedic Surgery, Medical College of Ohio, at Toledo Sept 1979 to date. Visiting Prof, Dept of Mech Engg & Applied Mechanics Univ of Michigan Ann Arbor, 1985-1986. NSF Summer Fellow 1962, 1963, 1967 and 1968. Ford Fdn Fellow 1964-66. Grantee US Army, AEC, US Navy. Mbr of Honor Socs Tau Betta Pi, Phi Kappa Phi, Pi Tau Sigma, Sigma Xi, etc. Contributed articles in technical journals in the fields of structure mechanics, waves in solids, Earthquake Eng sound waves biomechanics and in fracture mechanics. *Society Aff:* ASME, AAM, SES.

Ragan, Ralph R
Giles Road, Lincoln, MA 01773
Position: Planning Staff Hd. *Employer:* Charles Stark Draper Lab Inc. *Education:* MS/WO/MIT; BE/EE/IA State. *Born:* Nov 1923. Native of Ridgeway Mo. USNR 194246. Staff, USN Post Grad Sch 1946-48. Gun fire control sys engrg, Dir of Polaris Guidance Sys at MIT Instrumentation Lab 1948-62. Mgr Raytheon Sudbury Oper 1962-66. Dep Dir MIT Instrumentation Lab in charge of NASA programs incl Apollo Guidance & Control 1966-73. At present, respon for planning & acquisition of new business. Ed-in- Chief AIAA 'Journal of Spacecraft & Rockets' 1971-74. AIAA Vice President Publications 1974-78. Secretary of Navy Commendation. NASA Public Service Award. Registered Professional Engineer in Massachussetts. Fellow of AIAA. Sailing, skiing, gardening and jogging. *Society Aff:* AIAA, ION.

Ragan, Robert M
Business: Dept of Civ Engr, College Park, MD 20740
Position: Chrmn. *Employer:* Univ of MD. *Education:* PhD/Water Resources/Cornell Univ; MS/Sanitary Engg/MIT; BC/Civ Engr/VMI. *Born:* 12/19/32. in San Antonio, TX. Served in US Army as officer in Corps of Engrs. Prior to entering Univ teaching, practiced as engr in both private & public sectors. Joined the faculty of the Univ f VT in 1959. Moved to Univ of MD in 1967. Prime area of interest is hydrology, especially the application of remote sensing tech to water resources planning & mgt. *Society Aff:* ASCE.

Raghavan, Rajagopal
Business: 600 So. College, Tulsa, OK 74104
Position: Prof *Employer:* Univ of Tulsa *Education:* PhD/Petro Eng/Stanford Univ; Dip/Petro Prod Eng/Univ of Birmingham-England; BSc/EE/Birla Inst of Tech-India *Born:* 07/26/43 McMan Prof of Petroleum Engrg at the Univ of Tulsa, where he has taught since 1975. Previously, he was with Amoco Production Co and has also taught at Stanford Univ. He has authored or coauthored approximately 70 papers. Has served as a consultant. He has served on Editorial Review Committee of the Soc of Petroleum Engrs (SPE) in various capacities, since 1978. He has also worked on many committees concerned with planning of technical meetings within the SPE. He is one of the recipients of the 1981 Distinguished Achievement Awards for Petroleum Engrg Faculty awarded by the Soc of Petroleum Engrs. *Society Aff:* SPE of AIME, NYAS, IP, AGU.

Ragland, John R
Business: PO Box 3008, 2101-40th St, Lubbock, TX 79452
Position: VP & Chief Engr. *Employer:* Hicks & Ragland Engr Co, Inc. *Education:* BS/EE/TX Tech Univ. *Born:* 4/5/17. Native of Electra, TX, W T Utilities Co, 1939-1948, as design engr for power lines except served with US Navy, 1941-46 as Electronic Warfare Design Engr. Retired as Capt, USNR 1969. Joined consulting firm as Chief Elec Engr for Power Systems Design - 1949. Started own practice in 1954-date as Chief Engr for all work of firm; for Power Systems and Telephone Systems. Reg PE in 28 states, plus NCEE. Have practiced in 16 states, as PE. Developed growth of consulting service from annual gross fee of $200,000 to present gross of $2,500,000. Enjoy golf & fishing. *Society Aff:* IEEE, NSPE.

Ragold, Richard E
142 Fairfield Road, Fairfield, NJ 07006
Position: President *Employer:* Edwin M Ragold Assocs, P.A. *Education:* BME/ME/Cornell Univ *Born:* 10/9/31. US Army, 1st Lt, Ordnance Corps, 1954-1956. Dev Engr, Walter Kidde Co, 1956-1957. H F Butler Co 1957-1958, Welding Engr. 1958 to present, Edwin M Ragold Assocs-Structural Engr, Proj Mgr, Partner in 1963, owner in 1977. Merged firm in 1985 with H2M/Holzmacher, McLendon and Murrell, P.C., Consulting Civil and Environmental engrg firm. Now pres of H2M/Ragold, Inc. Ragold Assoc, Founded in 1947, is a consulting engg firm speializing in structural design of bldgs, and industrial structures. Work encompases power plant and treatment plant structures, scrap processing and gen industrial facility design. Pres, Essex County Soc of Prof Engrs 1973. V.P., Consulting Engrs Coun Of N.J. 1976, Fellow-ASCE, 1983, Fellow-ACEC 1986. *Society Aff:* NSPE, ASCE, ACI, ACEC, AIPE.

Ragone, David V
Home: 37125 Fairmount Blvd, Chagrin Falls, OH 44022
Position: President *Employer:* Case Western Reserve Univ. *Education:* ScD/Metallurgical Engg/MIT; SM/Metallurgical Engg/MIT; SB/Metallurgical Engg/MIT. *Born:* 1930. Asst Prof, Assoc Prof, Prof Dept of Chem & Met Engrg, Univ of Mich 1953-62. Chmn Met, Asst Lab Dir, Genl Atomic 1962-67. Alcoa Prof of Met, then Assoc Dean of Sch of Urban & Public Affairs Carnegie-Mellon Univ. 1967-70 Dean of Thayer School at Dartmouth Coll 1970-72. Dean, Coll of Engrg Univ of Mich. 1972-1980; currently Reg P E Calif. Ints: alternative automotive power sources, energy conversion & tech-soc interactions. *Society Aff:* ASM, AIME, ACS, AAAS.

Ragsdell, Kenneth M
Business: School of Mechanical Engineering, W. Lafayette, IN 47906
Position: Assoc Prof *Employer:* Purdue Univ *Education:* PhD/ME/Univ of TX; MS/ME/Univ of MO Rolla; BS/ME/Univ of MO Rolla *Born:* in Jacksonville, IL. Raised in MO & points west of MS. Completed High Sch in St. Louis, MO. Graduate work & teaching position at OK State Univ. Joined IBM in Austin, TX in July 1968. Completed doctoral studies at Univ of TX in July 1972. Joined Purdue ME Design Faculty in fall 1972. Teaching experience at Univ of MO at Rolla, OK State Univ, Univ of TX, Purdue & Univ of AZ. Married Janet Feb 14, 1962; Father of Keith, Thomas & Matthew. Research in CAD, Optimization & Dynamics. *Society Aff:* ASME, ASEE, NSPE

Raheja, Dev G
Home: 12904 Bentley Grove Pl, Laurel, MD 20708
Position: Pres *Employer:* Tech Mgmt, Inc *Education:* MS/Indust Eng/TX Tech Univ; BS/Mech Eng/Karnatak Univ (India) *Born:* 10/22/38 Pres of Tech Mgmt, Inc, a consltg firm in safety, reliability maintainability engrg. He is a pioneer in accelerated testing methods and has made unique applications of safety and reliability analysis. Worked as sr sys assurance conslt with Booz-Allen & Hamilton, Inc, as Chief Engr for Reliability and Safety at RTE Corp, as design engr for Standard Screw Co, and Recipient of Austin Bonis Award and Cecil C Craig Award from ASQC and several

Raheja, Dev G (Continued)
other awards. Has pub over 20 tech papers and lectures internatly. Adjunct faculty for grad school at Univ of MD *Society Aff:* IEEE, SSS, ASQC, SRE

Raimondi, Louis A
Business: 110 Stage Rd, Monroe, NY 10950
Position: President *Employer:* Raimondi Associates, P.C. *Education:* MSCE/Sanitary/NJ Inst Tech; BSCE/Civil Engr/NJ Inst Tech *Born:* Oct 1932 Paterson N J. Various experience incl design engr, admin asst Boswell Engrg; hydraulics engr Bergen Cty Engrg Dept, V P & Assoc Canger Engrg Assn. Principal Raimondi Assocs Monroe N Y 1968- . Treas Ramapo Valley Testing Lab 1971- . Municipal Engr in sev N J & N Y towns. Reg P E & Land Surveyer in 6 states. Fellow ASCE socs prof engrs. Chi Epsilon. *Society Aff:* NSPE, ACEC, NSPS, ACSM

Rainer, Rex K
Business: State Capitol, Montgomery, AL 36130
Position: Finance Director *Employer:* State of Alabama *Education:* PhD/Civil Engg/OK State Univ; MS/Civil Engg/Auburn Univ; BS/Civil Engg/Auburn Univ. *Born:* July 1924. Attended pub schools in Ala & Fla. Engr L&N Railroad 1945-46. Engr Polglaze & Basenberg, Cons Engr 1946-51. Pres Rainer Co 1951-62. Engrg Educator Auburn Univ 1962- . Prof & Dept Hd, C E 1968-79. Alabama Highway Director 1979-80. Executive Vice-President, Auburn University, 1980. Special Assistant to Governor 1980-81. Chmn Const Engrg Ctte Amer Soc for Engrg Educ 1975. Chmn Const Res Council Amer Soc of Civil Engrs 1975. Pres Elect Ala Sect Amer Soc of Civil Engrs. Cons assignments with insur co's, cons engrg firms & const firms. *Society Aff:* ASCE, ASEE, APWA, AASTO, SASTO.

Rainey, Frank B
Business: 2211 W. Meadowview Rd, Suite 114, Greensboro, NC 27407
Position: Vice Pres *Employer:* Sverdrup & Parcel and Assocs, Inc *Education:* BS/CE/Bradley Univ *Born:* 07/28/27 Native of Petersburg, IL. Served with US Navy (Seabees) 1945-46 and US Air Force 1951-52. With Il Div of Highways, 1953-55. Field assignments on major bridges for Sverdrup & Parcel 1955-69. Design and Project Mgr of Civil Projects. Appointed VP and Mgr of Greensboro office in 1980. Enjoy fishing, classical music and travel. *Society Aff:* ASCE, AME

Rajagopalan, Raj
Business: Dept of Chem & Environmental Engg, Troy, NY 12180-3590
Position: Assoc Prof. *Employer:* RPI. *Education:* PhD/ChE/Syracuse Univ; MS/ChE/Syracuse Univ; BTech/ChE/Indian Inst of Tech. *Born:* 12/25/48. Res interests in transport phenomena, statistical mechanics, Colloidal Interactins, Separation Processes, Stochastic Analyses, & Image Processing. Consultant to CAChE Subcommitte, Natl Acad of Engg, 1972; Mbr Natl Sci Fdn Review Panels; Natl Prog Committee Chrmn & Coordinator, AIChE, 1978; Mbr, Natl Tech Area Ctte on Fluid-Particle Interactions, FPS; Programs Chrmn, Chem Physics of Colloidal Phenomena, ACS; Program Chrmn, Colloidal and Interfacial Phenomena, FPS; Tau Beta Pi Outstanding Teacher Award, RPI, 1977; Early Career Award, RPI, 1983. *Society Aff:* AIChE, ASEE, FPS, ACS, APS, SIAM.

Rajchman, Jan A
Home: 268 Edgerstoune Rd, Princeton, NJ 08540
Position: Retired - Now Consultant. *Employer:* RCA Labs-David Sarnoff Res Ctr (Former). *Education:* PhD/Electronics/Swiss Fed Inst of Tech Zurich; EE/Electronics/Swiss Fed Inst of Tech Zurich. *Born:* 8/10/11. Coll de Geneve, Maturite 1930; Diploma EE, 1935, Dr Tech Sci, 1938, Swiss Federal Inst of Tech. Franklin Inst Levy Medal 1947; IEEE Liebmann Mem Prize 1960, Edison Medal 1974. Univ of PA Harvard Pender Award 1977, RCA Lab Achievement Awards 48,50,51,54 NASA Cert of Recog 1975. Soc of Info Display Francois Rice Darne Mem Award 1981. Inventor (116 pats) & scientist. Contribs: electron photomultiplier, computer memories incl first read-only memories, first digital random access, electrostatic, core & other magnetic types etc. Computer logic devices; contrib to particle accelerators (betatron). As res exec respon for broad res in computers, incl all aspects of hardware & software, optical devices & ultrasonic imaging for medical diagnosis. Fellow Amer Phys Soc 1936, Franklin Inst 1972, IEEE 1956 (active on magnetics, computer & award cttes); Mbr NAE 1966, Natl Inventors Council 1974, Adv Govt Res Comm (NSF, DOD) 1969, NY Acad of Sci 1972, Assn Computingplusplus *Society Aff:* IEEE, APS, SID, NAE, SPIE, NYS.

Raju, Pal P
Home: 26 Harvard Dr, Sudbury, MA 01776
Position: Quality & Reliability Manager *Employer:* Digital Equipment Corp. *Education:* PH.D/Engg. Mech/University of Delaware; M.scc/Engineering/University of MADRAS (INDIA); B.E./Civil Engg./University of Madras (INDIA) *Born:* 06/15/37 Dr. Raju is a Quality and Reliability Mgr with Digital Equip Corp, Marlboro, Mass, where he provides total cross-functional quality and reliability integration in the development of new products. He joined this firm in 1987. He was previously employed by Westinghouse Nuclear Energy Systems (1968-76) and Teledyne Engrg Services (1976-84). Dr. Raju received his B.E. (1960) in Civil Engrg and M.Sc. in Engrg from Madras Univ and a Ph.D. (1968) in Engrg Mechanics from Delaware Univ. He is a Fellow of ASME and a member of the Society of Sigma Xi. He has published extensively on the design, analysis, and failure evaluation of structures, pressure boundary systems and components. He has also directed and taught a number of seminars on the design and fabrication of pressure vessels. Dr. Raju is a Reg PE in MA & OH. *Society Aff:* ASME, WRC, PVRC,ΣΞ

Rakow, Allen L
Box 3805, Las Cruces, NM 88003
Position: Assoc Prof of Chemical Engrg *Employer:* New Mexico State Univ *Education:* DSc/Chem Engg/WA Univ; ME/Chem Engg/Stevens Inst of Tech; BChE/Chem Engg/Rensselaer Poly Inst. *Born:* 11/16/43. Native of NYC. Res Engr with Exxon Res and Engg Co from 1966-1968. Peace Corps Volunteer and Instructor at Arya Mehr Univ. in Iran from 1968-1970. Dev Engr and Cons in Norway (Dyno Industries) from 1970-1972. Shell Fellow, NIH. Res Asst and Assoc at Washington Univ from 1972-1975. Appointed Asst Prof of Chem Engg at the Cooper Union in Sept 1975. Promoted to Assoc Prof in Sept 1979. Res Conslt at Columbia Univ Med School 1975-1978. Res has included bioengg, polymer processing, tech assessment and solar energy. Was Co-Dir of NSF sponsored proj at Cooper Union entitled "Profl Comptences Dev in the Undergrad Engrg Curriculum." Appointed Assoc Prof of Chem Engrg at NM State Univ in Aug 1981. Vice Chrmn, Engrg Section, NY Acad of Sciences, 1981. Conducting res on bioseparation, food engineering and alternative technologies. *Society Aff:* AIChE, ASEE.

Raksit, Sagar K
Home: 777 S Euclid Ave, Pasadena, CA 91106
Position: Proj Engr. *Employer:* L A County Sanitation Dists. *Education:* PhD/Sanitary Engg/Jadavpur Univ; MS/Sanitary Engg/Jadavpur Univ; BS/Civ Engg/Jadavpur Univ. *Born:* 4/15/40. Native of India; naturalized US Citizen. Taught at Univ of Southern CA & Jadavpur Univ (Calcutta, India). Was with the City of Los Angeles as Supervisor or Res in wastewater treatment & operation from '72 to '76. From '76 to '79 was with Regional Wastewater Mgt Prog, Los Angeles/Orange County Metropolitan Area (LA/OMA Proj). Currently employed by the Los Angeles County Sanitation Dists. Responsible for res progs in solid waste mgt including energy recovery projs. Have several publications on waste mgt. Enjoy golf & wilderness backpacking. ASCE Rufolf Herring medal 1977. *Society Aff:* ASCE, WPCF.

Ralph, James A
Home: 322 Tampa Ave, Indialantic, FL 32903
Position: Mgr, KSC Ground Operations Analysis. *Employer:* IBM, Fed Sys Div. *Education:* BS/Physics/Fordham Univ. *Born:* Apr 1929. BS Fordham Univ. Native of NYC. US Army Guided Missile School 1951-53. Test Engr Curtiss-Wright Corp. Joined Martin Marietta Corp 1961 as Test Conductor on Titan Prog; was later Dep

Ralph, James A (Continued)
Proj Engr on Gemini Prog & Ch - Advanced Progs. With IBM, Fed Sys Div, Kennedy Space Ctr, since 1966. Tech Dir Proj OCALA, Mgr KSC Ground Operations Analysis at Kennedy Space Ctr. Currently spec in computer-assisted resource mgmt sys. Assoc Fellow AIAA. Southeastern Regional Dir AIAA 1973-75. Genl Chmn U S Space Congress 1973. Instr, Space Tech, FL Inst of Tech. Prin Investigator for Space Shuttle Payload Program Sponsored by Canaveral Council of Tech Soc. *Society Aff:* AIAA.

Ralston, David C
Business: Rm 6130 S Agriculture Bldg, PO Box 2890, Washington, DC 20013
Position: Chief Design Engr, Engrg *Employer:* Soil Conservation Service USDA. *Education:* MS/CE/Univ of IL; BS/AE/Univ of IL. *Born:* May 1930. Native of Roscoe Ill. US Army 1952-54. Employed by the SCS- USDA since 1950: student training 1950, 1951, 1952; Area Engr at Urbana Ill 1954-58; Design Engr at Milwaukee Wisc 1958-64; Asst State Engr at St Paul Minn 1964-66; State Conservation Engr at Morgantown W Vir 1966-71; Hd, Design Sect at Portland Ore 1971-74; Ch, Soil Engr. Wash, DC 1975-76; presently Ch, Design Eng Reg P E Ill & W Vir. Mbr ASAE, ASCE; grad training at Harvard Univ in soil mech 1963. *Society Aff:* ASCE, ASAE.

Ramakumar, Ramachandra G
2623 N Husband St, Stillwater, OK 74075
Position: Prof of Elec and Comp Engrg *Employer:* OK State Univ. *Education:* PhD/EE/Cornell Univ; MTech/EE/Indian Inst of Tech; BE/EE/Univ of Madras. *Born:* 10/17/36. in Coimbatore, India. Grad with first rank from Madras Univ. TCM scholar at Cornell. Affiliated with Coimbatore Inst of Tech from 1957 to 1967. With OK State Univ since 1967. Interested in conventional and unconventional energy conversion, energy storage, power engrg, renewable energy sources dev and application in developing countries. Published extensively on thee topics. Contributed chapters to several books and handbooks. Awarded four US patents. Conslt to JPL, UNU, UNEP, Schlumberger, and Kuwait Univ. Mbr, UNEP/ESCAP expert panel on energy storage in devel countries. IEEE - sr mbr, power gen. committee and energy devel subcommittee, ASES, ISES, Reg PE, OK. *Society Aff:* IEEE, ISES, ASES, ASEE.

Ramalingam, Panchatcharam
Business: 3801 W Temple Ave, Pomona, CA 91768
Position: Prof *Employer:* CA State Poly Univ. *Education:* PhD/Ind Engg/OR State Univ; MS/Ind Engg/OR State Univ; BE/Mech Engg/Univ of Madras. *Born:* 6/15/41. Born in Tamil Nadu, India. Taught at Univ of Madras. Was regional engr at Skoda (India) mfg engr at Winkelstraeter (W Germany) and production and test engr at Dominion Engg (Canada). Am VP, RSM Co, a mgt consulting firm specializing in productivity improvement, production/inventory control and computers in mgt. Authored three books and many technical papers in systems analysis, operations res, mgt info systems and industrial engg. IIE - United Air Lines res award, 1969. Reg profl industrial engr (CA). *Society Aff:* IIE, APICS, ORSA, TIMS.

Ramamoorthy, Chittoor V
Home: 1117 Sierra Vista Way, Lafayette, CA 94549
Position: Prof. *Employer:* Univ of CA, Berkeley *Education:* PhD/Comp Theory/Harvard Univ; MS/Mech Engr/Univ of CA. *Born:* 5/5/26. Sr Staff Scientist, Honeywell Inc Waltham MA (1956-1971). Prof, Elec Engg & Comp Sciences Univ of TX, Austin, 1967-1972. Prof, Elec Engg & Comp Sciences Univ of CA, Berkeley, CA 1972-present. Fellow, IEEE (1978). Received 2 awards, from IEEE-Comp Soc. VP, IEEE Comp Soc 1979 & 1981-83. IEEE Centennial Medal 1984. Editor-in-Chief, IEEE Transactions on Software Engrg. (1984-Present). *Society Aff:* IEEE.

Ramberg, Edward G
900 Woods Rd, Southampton, PA 18966
Position: Retired. *Education:* PhD/Physics/Univ of Munich; AB/Physics/Cornell Univ. *Born:* June 1907 Florence Italy. Res Asst physics Cornell 1932-35. With RCA Corp 1935 to retirement in 1972 as Fellow of Tech Staff, joining Electronics Res Lab Camden N J in 1935 & transferring to David Sarnoff Res Ctr Princeton N J in 1942. Civilian Pub Serv WW II 1943-45. Principal contribs in phys electronics, electron optics, electron microscopy, thermoelectricity, holography, TV. Fellow IEEE 1955; Fellow Amer Phys Soc 1957; Fulbright Lectr 1960; David Sarnoff Outstanding Team & Individual Award 1961 & 1972; IEEE David Sarnoff Award 1972. Since 1974: Ctte on the UN, Friends Yearly Meeting, Phila. *Society Aff:* APS, IEEE.

Ramberg, Steven E
Business: 4555 Overlook Ave, Washington, DC 20375
Position: Research Engr *Employer:* Naval Research Laboratory *Education:* PhD/ME/Catholic Univ; MS/ME/Univ of Lowell; BS/ME/Univ of Lowell *Born:* 01/04/48 A native of Waltham, MA. Joined the staff of the Naval Research Lab in 1972, after a short stay at DOT Transportation Systems Research Center. Responsible for a variety of R & D efforts at NRL including: flow-induced vibrations, vortex wakes of bluff bodies, marine cable mechanics, wave forces on ocean structures, stratified wake flows and breaking wave mechanics. Published extensively in these areas and was the recipient of the ASCE Moissieff Award for 1979. *Society Aff:* ASME, ΣΞ, ASCE.

Ramberg, Walter G C
Home: 1711 Belfast Rd, Sparks, MD 21152
Position: Retired. *Born:* Feb 16, 1904 Florence, Italy. AB Cornell, Dr tech science Tech Hochschule, Munich, Germany. Student engineer and lab asst Westinghouse Electric 1926-28; Natl Bur of Standards 1931-59: Chf Engrg Mech Sect 1937-46, Chf Mech Div 1947-59; Scientific Attache American Embassy Rome 1959-69. 1942 Award of Washington Academy of Science for Outstanding Achievement in Engineering Sciences; 1957 Templin Award ASTM; Fellow and Hon Mbr ASME (Chrmn Applied Mech Div 1958); SESA (Pres 1948). Publications and patents relating to the mechanics in general and to the static and dynamic strength of aircraft structures and materials in particular. *Society Aff:* ASME, SESA, APS, ASTM.

Rambo, William R
Home: 856 Esplanada Way, Stanford, CA 94305
Position: Consulting Engr. *Employer:* Self. *Education:* Engr/EE/Stanford Univ; AB/Eng/Stanford Univ. *Born:* Sept 1916. Radio broadcast engrg 1939-42. Res Assoc Radio Res Lab Harvard Univ 1942-45. Res Engr Airborne Instruments Lab Mineola N Y 1945-50. Stanford Univ: Res Assoc 1951, Prof of EE 1957, Dir of Stanford Electronics Labs 1960, Assoc Dean for Engrg Res 1961. Prof Emer 1972. Advisor/cons to gov/indus 1950- . Sr Sci Advisor SRI 1974- . Currently Dir Argosystems, Inc, Chmn Engrg Coll Res Council 1970-72. V Pres Amer Soc for Engrg Educ 1970-72. Fellow IEEE, AAAS. Mbr Sigma Xi, Tau Beta Pi, ASEE. *Society Aff:* IEEE, AAAS, ASEE, ΣΞ, TBΠ.

Ramee, Paul W
Business: 4111 Montgomery St, Savannah, GA 31405
Position: Pres *Employer:* Doughty, Powers & Ramee, Inc *Education:* MS/Math/RPI; MS/CE/Univ of IL; BS/Engr/US Military Acad *Born:* 06/05/20 Phillipine Islands 1920. Graduated from US Military Academy and commissioned US Army Corps of Engrs 1941. Retired as Colonel, 1968. Served in various engrg positions from platoon to brigade commander, and as US Army District Engr, Savannah, GA. This included service in combat areas in World War II, Korea, and Vietnam. Since 1968, engaged in the practice of cvl engrg, both as principal in an engrg co and as a sole proprietor, in Savannah, GA. Cvl engrg emphasis has been on water-oriented projects, with special attention to environmental effects. Licensed in GA, SC, and NY. *Society Aff:* ASCE, SAME, NSPE, ΦKΠ

Ramey, Henry J, Jr
Petro Engrg Dept, Stanford Univ, Stanford, CA 94305
Position: Prof, Petro Engrg. *Employer:* Stanford Univ. *Education:* PhD/ChE/Purdue Univ; BS/ChE/Purdue Univ. *Born:* Nov 30, 1925. Grad from Huntington High School W Va 1943; attended Marshall Univ W Va. Navigator Army AF 1943-46. With Mobil Oil Co 1952-63; Staff Res Engr Los Angeles Div. Prof of Petro Engrg Texas A&M 1963-66. Prof of Petro Engrg Stanford Univ 1966- . Chmn of Dept 1976-86. Author of over 200 papers, 1 monograph, sev pats, chapts in books, etc. Res ints: enhanced oil recovery, transient flow in porous media, geothermal energy, natural gas production. Mbr SPE of AIME, AIChE, AGS, AGU, CIM, AAA. SPE of AIME awards: Ferguson Medal 1959, Uren Award 1973, Carll Award 1975. Lucas Medal 1983, Dist Mbr 1984. Purdue Univ Disting Engr Award 1975. Keleen & Carlton Beal Prof of Pet Eng 1980; Mbr Natl Acad of Engrg 1980. Cons to many major oil co's, util co's, & domestic & internatl fed agencies. *Society Aff:* SPE of AIME, AIChE.

Ramirez, W Fred, Jr
Dept. of Chem Engr, Univ. of Colorado, Boulder, CO 80309-0424
Position: Prof, Dept of ChE *Employer:* Univ of Colorado. *Education:* PhD/ChE/Tulane Univ; MS/ChE/Tulane Univ; BS/ChE/Tulane Univ *Born:* Feb 1941. PhD, MS, BS Tulane Univ. Native of New Orleans La. With Univ of Colorado since 1965. Chmn of Dept of Chem Engrg 1971-1979. Fulbright resfellowship to France 1976. Levey Award for outstanding achieve within 10 yrs of grad by a Tulane Univ grad 1974. ASEE Dow Outstanding Young Fac award 1974. ASEE Western Electric Award for excellence in Engineering 1980 and Croft Research Professorship Fall 1980. Faculty Fellowship 1985, University of Colorado College of Engineering Faculty Research Award 1986. Major res areas are optimal control and identification of chemical, biochemical and energy recovery processes. Publ over 50 tech res papers, 2 books, has 1 US pat *Society Aff:* AIChE, SPE, ASEE

Ramkrishna, Doraiswami
Business: Sch of Chem Engg, W Lafayette, IN 47907
Position: Prof. *Employer:* Purdue Univ. *Education:* PhD/ChE/Univ of MN; BChemEngg/CE/Univ of Bombay. *Born:* 10/29/38. Native of India. Taught at the Indian Inst of Tech, Kanpur between 1967 and 1974. Served as Visiting Prof of Chem Engg at the Univ of WI (1974-75) and at the Univ of MN (1975-76). Have been Prof of Chem Engg at Purdue Univ sce 1976. Res areas include dispersed phase systems, stochastic modeling and applications, bioengg and problems of gen appl math interests. A conslt to Pillsbury Company, Minneapolis.. *Society Aff:* AIChE, ACS.

Ramsay, H J, Jr
1580 Lincoln St, Suite 680, Denver, CO 80203
Position: Pres *Employer:* HJ Ramsay & Assoc, Inc *Education:* MS/Petroleum Engrg/Univ Tulsa; BS/Petroleum Engrg/Univ Pittsburgh. *Born:* 08/15/30 Owner of H. J. Ramsay & Assocs, Inc, a petroleum engrg consltg firm specializing in reserve estimates and property valuations. Previously Mgr of Engrg for Forest Oil Corp, VP for IIAPCO (Natomas Co), and Mgr of Technical Services for Amerada Hess Corp. Past Chrmn of Denver Section of SPE and past chrmn of SPE PE Registration Committee. *Society Aff:* SPE, SPEE, SPWLA.

Ramsay, John F
212 N. Townsend Street, Apt. 104, Syracuse, NY 13203
Position: Retired. *Education:* MA/Mathematics and Physics/Univ of Glasgow, Scotland. *Born:* May 12, 1908. R&D engr with GEC & AEI 1927-36. Res engr with Marconi's W/T Co Ltd 1936-56. Serv with British Admiralty 1941-46. Sr engr Canadian Marconi 1956-59. Cons Airborne Instruments Lab 1959-73. US citizenship May 14, 1965. Spec ints: Radio & circuit engrg 1927-41; wartime radar antenna res 1941- 46; antennas, microwaves & semi-optics 1946-73. Professional papers - 28; book chapters - 2; monographs - 31. Patents: USA - 12, UK - 29. Fellow IEEE; C Eng FIEE. IEE Ambrose Fleming Premium, IEE Premium. *Society Aff:* IEEE, IEE.

Ramsdell, Earl W
Home: RR 1 Box 650A, Windham, ME 04062
Position: Chief Engr. *Employer:* S D Warren Co-Div Scott Paper. *Education:* BS/Chem Engg/Univ of ME. *Born:* 10/3/32. native of Standish ME. With SD Warren Co since 1950 as technician, machine 2nd hand, 1st hand, forman, process engr, asst tech dir, technical dir, Utilities Manager. Assumed current respon as Chief Engr 1987. USAF 1951-55. Chmn TAPPI Engg Div, & Struct & Performance Cttes, Bd of Dirs. Mbr API Instrument Res Ctte. Maine Professional Development Council. Author text on indus statistical applications & several technical articles published in various trade magazines. Recipient of Tappi Engineering Divisions 1986 Leadership and Service Award. Mbr of Tau Beta Pi, Univ of ME Pulp & Paper Fnd. Avid outdoorsman. *Society Aff:* TAPPI, TBΠ.

Ramsey, Jerry D
Home: 6903 Lynnhaven Dr, Lubbock, TX 79413
Position: Professor. *Employer:* Texas Tech Univ. *Education:* PhD/IE/TX Tech Univ; MS/IE/TX A&M Univ; BS/IE/TX A&M Univ. *Born:* Nov 1933. Have served as Pres, Great Plains & N Mex Chapters IIE; Chmn Ergonomics Comm AIHA; Chmn Stds Adv Ctte on Heat Stress OSHA; NM State Comm, Engrs in Indus NSPE; Exec Comm Public Employees, Natl Safety Council; Vice Chairman, Bd of Dir Texas Safety Assn. Prof Indus Engrg & Biomed Engrg & Computer Medicine, Texas Tech Univ. Cons to OSHA, NIOSH, CPSC & num private firms. Contrib author to 6 books & author of over 60 papers in prof journals. Chmn Lubbock Cty Ctte for Employment of Handicapped & Bd of Dir Lubbock Goodwill Indus, Assoc VP for Acad Affairs, TX Tech Univ. Chairman, Bd of Directors, Southwest Lighthouse for the Blind. *Society Aff:* AIHA, ASSE, HFS, ASEE, IIE.

Ramsey, Melvin A
Business: Rd Boxs 190, Shohola, PA 18458
Position: Consulting. *Employer:* Self-employed. *Education:* ME/Mech Engg/Stevens Inst of Technology. *Born:* Dec 1905 Paterson N J. Worked on design & installation of heating, air cond & refrig sys in NY, Tenn, Buffalo & Buenos Aires (last 5 yrs as Pres of Arnott & Co). With Worthington Corp 20 yrs in design, application, customer problems, etc, refrigeration, air cond & Diesel pumping sys in Israel. Over 100 tech articles & 1 book publ. Taught evening classes 10 yrs at Newark Coll Engrg & grad air cond course since 1959 at Stevens. Fellow ASHRAE with Disting Serv Award 1966. Enjoy music & clock repairing. *Society Aff:* ASHRAE.

Ramsey, Paul W, Sr
Home: 3125 N Menomonee R Pkwy, Wauwatosa, WI 53222
Position: Mgr, Welding and Manufacturing R & D. *Employer:* A O Smith Corp. *Education:* MS/Metallurgy/Univ of WI; BS/Metallurgy/Carnegie-Mellon. *Born:* Feb 17, 1919. Native of Wilkinsburg Pa. Res metallurgist for N J Zinc Co (of Penna) 1941-51. US Navy 1944-46. With A O SMith Corp since 1951. Assumed current pos as Mgr of Welding R&D 1965, involving respon for welding equip & processes for mfg automotive frames, water heaters, motors, & other indus products for plants in 6 states & abroad. Pres AWS 1976; AWS Natl Meritorious Cert Award 1971; Univ of Wisc Disting Serv Citation 1974; AWS Samuel Wylie Miller Memorial Award 1980 Mbr ASM, AIME, SAE; SME P E Wisc. Enjoy canoeing, sailing, photography. *Society Aff:* AWS, ASM, AIME, SAE, SME.

Ramsey, Robert R
Business: Leo A Daly Co, 3333 Wilshire Blvd, Los Angeles, CA 90010
Position: Vice Pres. *Employer:* Leo A Daly Co. *Education:* BS/Civil Engg/Marquette Univ; MS/Struct Engg/Stanford Univ. *Born:* Apr 6, 1932. Army Corp of Engrs 1953-56. Engr for John Blume & Assocs San Fran; Engr for Bechtel Co San Fran; with Leo A Daly Co since 1961 & have been design engr, team captain, dir of const & currently V P for Special Projects. *Society Aff:* NSPE, AIA.

Randall, Clifford W
Business: Dept. of Civil Engineering, Virginia Tech, Blacksburg, VA 24061
Position: Lunsford Prof of Civil Engrg *Employer:* Vir Polytech Inst & State Univ. *Education:* PhD/Envir Health Engr/Univ of TX, Austin; MSCE/Sanitary Engr/Univ

Randall, Clifford W (Continued)
of KY; BSCE/Civil Engr/Univ of KY *Born:* 5/1/36. Native of Somerset Ky. US Coast & Geodetic Survey 1959-62. Asst Prof of CE Univ of TX Arlington 1965-68. With VPI SU since Feb 1968; Assoc Prof 1969, Prof 1972, Lunsford Prof 1981. V P Va Water Pollution Con Assn 1975-76; Pres 1976-77. Chmn Occoquan Watershed Monitoring Subcomm of State Water Control Bd 1972-84. Mbr ASCE 1966- . Chmn Water Resources Man Comm EED, ASCE 1976-77, Mbr AEEP, AWWA, IAWPR, Chi Epsilon, Phi Kappa Phi, Sigma Xi. Mbr Bd of Control, WPCF 1981-84. Over 150 publ tech articles. Indus waste treatment cons approx 50 corps. Tech Paper Awd ASCE TX Sect 1967. Serv Awd, EED ASCE 1979 & 1980. Acad Excellence Awd, AWWA 1980, Serv Awd, AEEP 1981. Mbr, Bd of Dirs, AEEP 1978-80, Secy-Treas 1979-80. Mbr USA Natl Ctte for Representation to IAWPRC. Mbr James River Water Quality Monitoring Ctte. Arthur Sydney Bedell Awd, WPCF, 1983; Chairman, USA National Committee for representation to the International Association on Water Pollution Research and Control (IAWPRC), Member, Governing Board IAWPRC, Phillip F. Morgan Cert. of Merit, WPCF, 1982. Member, Scientific and Technical Advisory Committee, Chesapeake Bay Project, 1986 Conservationist of the Year, Chesapeake Bay Foundation, Member, Control Board, Environmental Engineering Research Council, ASCE. *Society Aff:* WPCF, ASCE, AEEP, IAWPRC, AWWA

Randall, Frank A, Jr
Business: 28 Timber Lane, Northbrook, IL 60062
Position: Consulting Struct Engr. *Employer:* Consulting Structural Engr *Education:* BS/Civil Engg/Univ of Illinois, Urbana; -/Structures/Yale Univ. *Born:* 2/21/18. Resides in Northbrook Ill. Mbr Alpha Delta Phi & Presbyterian Church; Fellow Amer Concrete Inst, Fellow Amer Soc of Civil Engrs. Wason Medal for Most Meritorious Paper (ACI) 1975; ACI Construction Practice Award, 1978. Co-author Concrete Masonry Handbook (PCA) 1976. Previous struct engrg practice with the Texaco Co & M W Kellogg co, both of NYC; Frank A Randall & Sons, Struct Engrs, Chgo; Archs, Engrs Spanish bases, Madrid Spain; Corbetta Const Co of Ill & Portland Cement Assn Chgo & Skokie; all in field of bldgs & structures. *Society Aff:* ASCE, ACI, OFP, SAR.

Randich, Gene M
Business: 165 W. Wacker Dr, Chicago, IL 60601
Position: Sr Vice Pres *Employer:* De Leuw, Cather & Co *Education:* BS/CE/Purdue Univ *Born:* 09/20/29 Reared and educated in the Chicago Metropolitan area. Field engr with Chicago dept of subways and superhighways. Served in Army Corps of Engrs, 1951- 54, as officer in a topographic mapping unit. Consltg engr for De Leuw, Cather & Co in 1954, specializing in transportation;-highways, public transit and railroads. Directed major transportation programs. Sr Vice Pres and Dir of the firm in 1974. Currently responsible for business and technical operations of the firm's midwest region. Dir of Western Soc of Engrs and Builders' Club. Licensed PE in 13 states. Advocations include Alpine and Nordic skiing, golfing, woodworking and numismatics. *Society Aff:* ASCE, WSE, ITE

Randolph, Patrick A
Business: 3333 Michelson Dr, Irvine, CA 92730
Position: Chairman *Employer:* Fluor Power Services, Inc. *Education:* MS/Mech Chem Eng/Stevns Inst of Tech; ME/Mech Eng/Stevens Inslt of Tech. *Born:* Mar 1916. Native of Jersey City N J. US Army Ordnance Corps 1941-46; retired as Colonel 1967. Mech engr for Babcock & Wilcox, heat transfer engr with M W Kellogg Co. With Fluor Corp since 1951 as Proj Mgr to V P - Design & Project Engrg. Assumed current pos upon acquisition of Pioneer Serv & Engrg Co by Fluor Corp 1974. Fellow ASME, P Natl Chmn ASME Petro Div. Mbr Amer Nuclear Soc, Western Soc of Engrs, chgo Assn of Commerce & Indus, Northwest Elec Light & Power Assn, Pacific Coast Elec Assn, Wisc Utils Assn, Reserve Officers Assn. *Society Aff:* ASME, ROA, ANS.

Randolph, Robert E
Home: 1857 Meadow Downs Way, Salt Lake City, UT 84121
Position: Mgr, Composite Structures. *Employer:* Hercules Inc. *Born:* May 1933. BSME Rose Polytechnic Inst 1956. Native of Robinson Ill. Engr with duPont to 1958. With Hercules since 1958. Var design & supervisory assignments on solid rocket motors. Began adviced composites work as structures group supr 1965. Current respon is Mgr, Advanced Composite Structs. Sev papers on composite structs. Best mbr paper award 1973 SAMPE Tech Conf. AIAA. Enjoy fishing, hunting.

Raney, Donald C
Business: Box 2908, University, AL 35486
Position: Prof. *Employer:* Univ of AL. *Education:* PhD/Engg Mech/VPI & State Univ; MS/Mech Engg/Auburn Univ; BS/Mech Engg/Univ of KY. *Born:* 5/6/33. Educated in Middlesboro KY. Public School System. Served as Aircraft Maintenance Officer with USAF (1955-57). Taught at Auburn Univ and VPI while working on Advanced Engg Degrees. Industrial and Governmental experience with Lockheed Aircraft Corp, Pratt and Whitney Aircraft, Southern Res Inst, AL Power Co and the US Army Corps of Engrs Waterways Experiment Station. Served as consultant for several cos and governmental agencies. Began as Asst Prof at the Univ of AL in 1964 with present rank of full Prof. Primary teaching and res interests in fluid mechanics and hydraulics especially in numerical techniques for simulation of fluid systems. *Society Aff:* ΣΞ, ΠΤΣ, ASCE, ASPE.

Rangaswamy, Thangamuthu
10404 Scarlet Oak Ct, Louisville, KY 40222
Position: Pres *Employer:* Rangaswamy & Assoc *Education:* PhD/Struct Engg/Univ of KY; MS/Struct Engg/Univ of CA; BE/CE/Univ of Madras. *Born:* 6/12/42. in India. Naturalized citizen of US. Was Jr Engr in Parambikulam Dam Proj. Proj Engr for Watkins and Assoc in charge of design of bridges and bldgs. Sr res Assoc incharge of analysing and testing of Postensioned space frame floor system developed at Univ of KY during 1968-1970. Wrote Dissertation titled "Finite Element Analysis of Orthotrophic Frame-Plate Interaction–. Principal structural Engr for an engg co in Louisville between 1970-1974. Organized Rangaswamy, Hatfield & Assoc April 1974. Married to Kohilam Murugaiya. Have one son Chandhiran. Enjoys travel & photography. At present, the Pres of Rangaswamy & Assoc, Inc. A specialized structural engrg firm capable of providing complete inhouse engrg serv for highrise, inst bldgs, bridges and waste water treatment plants. *Society Aff:* ASCE, ACI, AISC, AWS.

Ranieri, Albert A
Home: 301 Allendale Way, Camp Hill, PA 17011
Position: Ch, Div of Mine Reclamation, Bureau of Design. *Employer:* Dept of Environ Resources-PA. *Education:* BS/Civil Engg/Drexel Inst of Tech. *Born:* May 1937. Native of DuBois Pa. m. Shirley B Graham. Children Catherine A & Robert A. Hydraulics Engr designing flood control projs for Dept of Forests & Waters, Commonwealth of Pa 1960-66. Reviewed & coordinated design of flood control, dam & park dev projs by cons firms for Dept of F & W 1966-69 & Dept of Environ Resources 1969-72. Since 1972, serving as Ch, Div of Mine Reclamation, Bureau of Design, DER respon for design of mine pollution abatement projs. Pres Central Pa Sect ASE 1975. Tau Beta Pi. Chi Epsilon. Active in church affairs. *Society Aff:* ASCE.

Rankin, Andrew W
Home: 16 Virginia Rail La Hilton Head Plantation, Hilton Head Island, SC 29928
Position: Retired *Education:* MSE/ME/Purdue U.; BSEE/EE/Purdue U. *Born:* 10/12/09 Born Phila. PA. Oct 12, 1909. BSEE and MSE from Purdue Univ in 1933. Graduate study as exchange student at Technische Hochschule, Aachen, Germany 1934. Joined General Electric Co (Large Steam Turbine Dept.) in 1935. Received GE Coffin Award in 1953, Transferred to GE plant in Fitchburg, Mass. in 1962. Became Mgr. Engrg in Fitchburg Plant 1971. Retired to Hilton Head Island in 1972. Reg. Prof. Eng. NY 1947, Mass 1963 S.C. 1972 *Society Aff:* ASME

Rankin, John P
Home: 859 Seamaster Dr, Houston, TX 77062
Position: Supervisor *Employer:* The Boeing Company *Education:* BS/EE/MS State Univ *Born:* 07/20/43 Currently supervisor of systems engrg analysis group in Boeing Technical Services organization. Recognized developer of sneak circuit analysis, sneak software analysis, and common mode failure analysis technologies. Supervised application of these technologies to over 60 projects in all types of industry. Prior experience in Seattle (commercial airplanes - Boeing) and Panama City, FL (Pulp & Paper mill - Intl Paper Co). VP and Fellow of System Safety Soc. Presented 1981 Safety Engr of the Yr Award. Founded Houston Chapters of System Safety Soc and American Nuclear Soc; past pres of both. Award for Excellence in Oral Presentation, SAE, 1980. *Society Aff:* SSS, ANS

Rao, Krishna T R
Business: Dorr-Oliver House, Link Road, Chakala Andheri East, Bombay, 400 099 India
Position: Chairman & Managing Director *Employer:* Hindustan Dorr-Oliver Ltd *Education:* MS/Mech/State Univ of IA; BE/CE/Hydraulics/Univ of Mysore, India; BSc/Chem/Univ of Mysore, India *Born:* 12/24/29 Married with two daughters. B.Sc, M.Sc., Post grad degree in Engrg. Career: Demonstrator in Chemistry, Univ of Mysore 1949-50. Civil Engr with several firms in India 1953-56. Research Asst/Research Assoc, Univ of IA, 1956- 58. Joined Hindustan Dorr-Oliver in 1958 as Sales Engr, and held several posts such as Mgr Env Engg, Project Mgr, Deputy Gen Mgr, Gen Mgr, Managing Dir in 1977-78, and Chrmn in May, 1983. *Society Aff:* IWWA, BMA, AIMO, IACC

Rao, Peter B
Business: 1210 Kenton Rd, Springfield, OH 45501
Position: Sr Product Engr *Employer:* Koehring BOMAG, AMCA Internatl *Education:* MS/Mech/OK State Univ; BS/Mech Osmanoa. *Born:* 04/08/44 Holder of four patents (US) and two others pending. Grad with MS in 1967. 1965-66 Worked in refrigeration compressor manufacturer (USHA) as a starting engr. 1967-71 Eng specialist/proj engr working on hydraulic sys design, gear pump motor design. 1971-73 VP Engr of FLUIDICS, Inc - R. D. Co (now dissolved) dedicated to producing tech such as 95% water-5% mixture waterbased fluid for coal mining indus. 1973-up Working as a senior product engr designed recycling machine and soil stabilizing machine. *Society Aff:* FPS, SAE.

Rapp, Robert A
Business: 116 W 19th Ave, Columbus, OH 43210
Position: Professor. *Employer:* Ohio State University. *Education:* BS/Metall Eng/Purdue Univ; MS/Metall Eng/Carnegie-Mellon Univ; PhD/Metall Eng/Carnegie-Mellon Univ *Born:* Feb 1934. Native of Lafayette, Indiana. Fulbright Scholar of MPI Phys Chem (Gottingen, Germany) 1959-60. Met res in USAF 1960-63, Wright-Patterson AFB. Prof in Met Engrg, Ohio State Univ 1963- . ASM Stoughton Teachers Award 1967. Guggenheim Fellowship, Ecole N S D'Electrochimie Grenoble, France 1972-73. ASM Howe Gold Medal 1973. Fellow ASM 1980. Fellow, Met Soc AIME 1982. OH State Univ Res Award 1982. Campbell Lecturer ASM 1983, OH State Univ. Eng. College Research Award 1984. Fulbright Scholar at ENS Chimie (Toulouse, France) 1985-86. NACE Whitney Award 1986. Teach courses, conduct res, and cons in areas of thermodynamics, kinetics, electrochemistry, chemical and high- temperature met, corrosion and oxidation, point defects. About 135 pubs, 6 patents. Ohio Prof Engr. Fluent in German and French. Mbr: ECS, AIME, ASM, NACE. Enjoy gardening, sports and reading. *Society Aff:* ECS, AIME, ASM, NACE

Rappaport, George
6244 Clearwood Road, Bethesda, MD 20817
Position: Corporate Consultant (Self-employed) *Education:* PhD/EE/OH State; MSc/EE/OH State; BEE/EE/City College of NY. *Born:* Dec 1919. Native of New York, NY. Ch Scientist USAF Lab Wright Field, Ohio. V Pres Emerson Radio and Phonograph Corp; Pres Warnecke Electron Tubes Inc; V Pres Scope Inc; V Pres Dewey Electronics Corp. Natl Dir Bd of Inst of Electronic and Electrical Engrs; Pres Natl Conf Aerospace Electronics. Fellow: IEEE, Honorary Doctorate Parsons Coll; Prof of Engrg George Wash Univ; Corp Cons to ind firms, philatelist, writer. *Society Aff:* IEEE, AOC.

Rappaport, Stephen S
Business: State University of NY, Stony Brook, NY 11794
Position: Professor *Employer:* Dept. Elec. Engg. *Education:* PhD/Elec. Engg./New York University; MSEE/Elec. Engg./Univ. So. Calif.; BEE/Elec. Engg./Cooper Union *Born:* 09/26/38 Mbr of Technical Staff at Hughes Aircraft and at Bell Labs before joining SUNY Stony Brook in 1968. Extensive university service. Fellow of IEEE. Member of Tau Beta Pi, Eta Kappa Nu, and Sigma Xi. Active in IEEE Communications Society. IEEE service includes: Bd of Gov, IEEE ComSoc; Ch, Technical cttee on Data Comm Systems; Assoc. Editor for IEEE Trans. on Comm.; ComSoc Conf. Board; Treasurer, L.I. Section, IEEE; Chairman, L.I. ComSoc Chapter; L.I. Section Awards Cttee; Associate Editor, Communications Magazine. Numerous technical publications on comm systems, multiple access, mobile radio, comm traffic, and spread spectrum. Substantial research funding from U.S. National Science Foundation and the U.S. Office of Naval Research. Also listed in "American Men & Women of Science," –Who's Who In the East," –Who's Who In Technology Today." *Society Aff:* IEEE, TBΠ, HKAPPN, ΣΞ

Rase, Howard F
Home: 3700 River Road, Austin, TX 78703
Position: W A Cunningham Prof of ChE *Employer:* Univ of TX. *Education:* PhD/ChE/Univ of WI; MS/ChE/Univ of WI; BS/ChE/Univ of TX. *Born:* 10/18/21. Dow Chem Co, 1942-44; Process Engr, Eastern States Pet, 1944; Process & Proj Engr, Foster Wheeler Corp, 1944-49. Joined faculty at the Univ of TX in 1952; served as Chrmn, 1963-68. Visiting Prof, Tech Univ of Denmark, 1957. Consultant on catalysis & reactor design. Author of "Project Engineering of Process Plants" (with M H Barrow)," Piping Design for Process Plants, "Philosophy and Logic of Chemical Engineering," –Chemical Reactor Design for Process Plants," & articles on catalysis & process & proj engg. Recipient of Gen Dynamics Teaching Award, Distinguished Advisor Award & Distinguished Graduate Prof Award. *Society Aff:* AIChE, ACS.

Rasmussen, Norman C
Business: 77 Mass Ave, 24-205, Cambridge, MA 02139
Position: Prof Nucl Engrg *Employer:* Mass Inst of Tech. *Education:* PhD/Physics/MA Inst of Tech; AB/Physics/Gettysburg Coll *Born:* 11/12/27. in Harrisburg, PA. US Navy 1945-46. Head, Dept of Nucl Eng, MIT, 1975-81; Prof Nucl Eng, MIT 1965-present; Assoc Prof Nucl Eng, MIT, 1961-65; Asst Prof Nucl Eng, MIT, 1958-61; Instructor Physics, MIT, 1956-58. Early res in radiation detection, activation analysis, and gamma-ray spectroscopy. Current res nuclear reactor safety, environ impact of nuclear power, reliability analysis & risk assessment. Dir US Nuclear Regulatory Comm 'Reactor Safety Study' WASH-1400. Dir, MIT Spec Summer Course, "Nuclear Reactor Safety–, 1969-present; Mbr, Defense Sci Bd, 1975-78, Bd of Dirs, Amer Nuclear Soc, 1976-80; Bd of Trustees, Northeast Utilities, 1977-present; Sr Consultant, Defense Sci Bd, 1979-present; Mbr, Natl Council on Radiation Protection and Measurements (NCRPM), 1980-85; Bd of Dirs, Atomic Industrial Forum (AIF) 1980-86. Hon ScD Gettysburg Coll. Doctor Honoris Causa, Catholic Univ of Leuven, Belgium, 1980. Fellow ANS; Distin Achievement Award, Health Physics Soc July 1976; Special Award ANS 1976; Distinguished Service Award, NRC, 1976; Elected to Natl Acad of Engrg, 1977 & Natl Acad of Sci, 1979. Theos J. Thompson Award ANS 1979; Mbr Natl Sci Board 1982-. McAfee Professor of Engineering, 1983-; Enrico Fermi Award from Dept. of Energy (U.S.), 1985. *Society Aff:* ANS, INMM, SRA.

Rassinier, Edgar A
Home: 3526 Dumbarton, Houston, TX 77025
Position: Conslt Engr *Employer:* Self *Education:* Geological Engr/-/Univ of MO, Rolla; BSEM/Petrol Geology/Univ of MO-Rolla; BS/Mining Engrg/Missouri School of Mines. *Born:* 3/16/19. in Louisville, KY. Geol Engr BSEM Univ MO-Rolla. Employed by Louisville Gas & Elec Co 1940-41; Phillips Petroleum Co 1941-52; Trunkline Gas Co as Ch Petrol Engr 1953-63; Dir Res Planning 1964-82. Active in civic & prof societies in Houston, TX. Served on num ctte of gas industry trade assocs. Natl Chmn 1972-77 Potential Gas Ctte. Dir 1968-72 SPE of AIME, VP 1974-75, Dir 1972-75 AIME. Industrial advisor on natl Gas Surveys of Fedral power Commission & Gas Res Inst. Co-author (Gas) Considine's Energy Technology Handbook McGraw-Hill 1977. Reg prof engr in TX. Life Member Tau Beta Pi. *Society Aff:* SPE-AIME; TBΠ

Rastoin, Jean M
Business: CEN Saclay - DEDR/DIR, Gif Sur Yvette Cedex 91191 France
Position: Dir, Hd of Nuclear R & D Div *Employer:* Commissariat Energie Atomique *Education:* ENSGM/-/Ecole Nationale Superieure du Genie Maritime; EP/-/Ecole Polytechnique *Born:* 01/16/32 *Society Aff:* ANS, IAEE, AIF, IAHE, IA SMIRT

Ratcliffe, Alfonso F
Business: 18111 Nordhoff St, Northridge, CA 91330
Position: Dean *Employer:* CA State Univ. *Education:* PhD/Engg/UCLA; MS/Engg/UCLA; AB/Phys/UCLA. *Born:* 10/21/28. Native of St Louis, MO. Served in US Army Transportation corps 1946-1948. Performance and Environmental Test Engr for Rototest Labs and Ogden Tech Labs specializing in electronic and electromechanical testing for ten yrs. Dir of Testing, Quality control and Dir of Process Instrumentation for 6 yrs. Professional Engineer in control system engineering. State of CA. Sr staff engr in Mattel Audio Lab and Mgr of Technical studies for Audio Magnetics Corp for 6 yrs. 1 yr as Assoc Dean School of Engr and Comp Sci, CA State Univ Northridge. Sr Staff Engr for UCLA Linguistics Lab for 1 yr, Prof of Elec Engg, CA State Univ at Northridge for 5 yrs, 6 years to date as Dean of the Sch. *Society Aff:* ASEE, IEEE, ΣΞ, TBΠ.

Ratermann, Mark J
Business: 2966 West Clarendon Ave, Phoenix, AZ 85017
Position: District Vice Pres *Employer:* International Engineering Co, Inc *Born:* 03/21/30 Born in Breese, IL, became Asst Superintendent of Highways, Design and Resident Engr for the Clinton and Champaign County Highway Depts. Moved to AZ in 1957. Licensed as Civil Engr in five states: AZ, and ID, 1965; NM, NV, and UT in 1978. Joined Earle V. Miller (EVME) in 1962 was appointed Chief Design Engr in 1965. 1970 Intl Engrg Co, Inc. (IECO) assigned Principal Engr and Chief Engr of the EVME Div of IECO. October, 1978 assumed role of District VP and Mgr of the Southwest District Office for IECO. Enjoys: Mountain cabin and Travel. *Society Aff:* APWA, ASCE, NSPE, PMI, SAME

Rath, Gerald A
Business: Vice President, Engineering, Advanced Energy Systems Co, Inc, 126 N. Water Street, Rockford, IL 61104
Position: Vice President, Engineering *Employer:* Advanced Energy Systems Co., Inc. *Education:* MS/Engg/Purdue Univ; BS/Elec Engr/IA State Univ. *Born:* 1/22/33. Native of Sioux City, IA. Communications Officer USAF 1955-57. Res engr for Delco-Remy Div of Gen Motors, 1957-66 in Advanced Engg Dept. Asst Prof of Elec Engg Tech at Purdue Univ 1966-76. Joined Wichita State Univ in Feb 1976 as Director of Engineering Technology within the College of Engg. Joined Rockford Coll in Jan 1985 as Chrmn, Engrg Dept. Joined Advanced Energy Systems Co., Inc, an engineering consulting firm in May 1987. Chrmn, Professional Engrs in Education, KS Sec, 1978-79, 1981-82; Chrmn, Engineering Technology Division of ASEE, 1981-83; Chairman, Wichita Section IEEE, 1981-82; Awards Chrmn, IEEE Region 5, 1982-84. *Society Aff:* ASEE, IEEE, NSPE, SME, AIPE.

Rathbone, Donald E
Business: College of Engineering, KSU, 146 Durland Hall, Manhattan, KS 66506
Position: Dean of Engrg. *Employer:* KS State Univ. *Education:* PhD/EE/Univ of Pittsburgh; MS/EE/Northwestern Univ; BS/EE/Purdue Univ. *Born:* 1/22/29. Born & raised in Havre, MT. He has taught at the Univ of Pittsburgh, the Univ of ID & Northwestern Univ. He was hd of Elec Engrg at the Univ of ID before becoming Dean of Engrg at KS State Univ. He worked for the Westinghouse Electric Corp and has served as a consultant to numerous ind firms & govt agencies. In 1976 he received the Distinguished Service Award from the Univ of Pittsburgh. He is a mbr of the City of Manhattan Economic Dev Ctte, 1975 -; is state legislative chrmn of the Kansas Engrg Soc, 1981-; is a mbr of the Commission on Engrg of the Nat'l Assoc of State and Land Grant Coll (Vice-Chair, 1981-85) and has been a mbr of the engrg dean's council, ASEE, 1980-83. He is a mbr of NSPE, was Vice-Pres of the North Central Region, 1984-86, and is Chairman-elect of the Professional Engineers in Education, 1987-88. He is also a member of ASEE, IEEE, Tau Beta Pi, Sigma Xi, Eta Kappa Nu, Phi Kappa Phi, Alpha Nu Sigma, and Golden Key. He rec'd his BS from Purdue Un., MS from Northwestern Un. and PhD from the Un. of Pittsburgh. *Society Aff:* IEEE, ASEE, TBΠ, ΣΞ, ΦΚΦ, NSPE, ΑΝΣ, GOLDEN KEY, HKN.

Rathe, Alex W
Home: 11-B Heritage Hills, Somers, NY 10589
Position: retired *Education:* D Eng/Charlottenburg Inst of Tech, Germany; MS/IE/Charlottenburg Inst of Tech, Germany; BS/EE/Charlottenburg Inst of Tech, Germany *Born:* 01/21/12 Prof of Mgmt, Pace Univ, 1977-82. Prof Emeritus, NYU 1949-1977. Mgmt Counsel, 1946-82. Chief Industrial Engr and Asst to Exec VP, Hudson American Corp, 1944-1946. Chief Production Engr, General Instrument Corp, 1943-44. Project Engr, Jackson & Moreland, 1941-43. Industrial Engr, Display Lighting Corp, 1937-1941. Visiting Prof, Columbia Univ and Univ of Buenos Aires. Distinguished Lecturer, Univ of Southern CA. Past Pres, Fellow, American Inst of Industrial Engrs. Fellow, American Soc of Mech Engrs. Dir, Engrs Joint Council, 1956-1960. Gantt Medal Citation. Honorary mbr of mgmt and engrg societies in Latin Amer and Asia. *Society Aff:* IIE, ASME

Rathke, Arlan E
Home: 2214 Nolen Dr, Flint, MI 48504
Position: Assoc Prof. *Employer:* GMI Engrg & Mgmt Inst *Education:* MS/Met Engg/Lehigh; BS/Met Engg/Montana Tech. *Born:* 9/29/36. Native of Western Montana. Five yrs as Applied R&D Metallurgist in iron & steelmaking for Inland Steel, E Chicago. 3 yrs as res met in steelmaking for Babcock-Wilcox. Assoc with GMI since 1968 teaching undergrad courses in materials engineering metallurgy and manufacturing processes. Also respon for operation of Gleeble Lab facility for elevated temperature properties of metallic materials. *Society Aff:* ASM, AIME.

Ratliffe, Donald E
Business: 6527 New Peachtree Rd, Atlanta, GA 30340
Position: Vice President *Employer:* Talbert, Cox & Assoc, Inc *Education:* BArch Engrg/Arch Eng/OK State Univ *Born:* 08/01/26 Has over thirty-six yrs of diversified public and private sector experience in the following areas; airport fueling systems, automation/controls including related petroleum facilities, chemical storage and handling petroleum, including the storage, loading/off-loading, spill prevention and control thereof. In others areas, his expertise covers the design and planning of office space; commercial and industrial development and in relation to this, his experience covers land acquisition, architectural coordination with engrg projects, bldg codes, standards and regulations as regulated by DOD and OSHA and the Environmental Protection Agency. In this same line, Mr. Ratliffe has a broad knowledge of cost estimating (bldgs, petroleum facilities, and light industrial); appraisals, particularly petroleum facilities and warehouses also petroleum related. Hobbies: Golf and reading. *Society Aff:* NSPE

Rattan, Kuldip S
Dept of Electrical Systems Engr, 354A Fawcett Hall, Dayton, OH 45435
Position: Assoc Prof. *Employer:* Wright State Univ. *Education:* PhD/Elec Engg/Univ of KY; MS/Elec Engg/Univ of KY; BScEngg/Elec Engg/Panjab Engg College. *Born:* 4/25/48. Native of India. After finishing PhD degree, was awarded a Post-Doctoral Fellowship (Jan 1976-Dec 1976, Found Fund for Res in Psychiatry) to do res in Neurophysiology at the Univ of KY. Remained there as a Res Associate specializing in intracellular recording from nerve cells, and determining changes in membrane properties as a result of habituation and aging. Assumed position as Asst Prof of Engg at Wright State Univ in Jan 1979 and currently holds the position of Assoc Prof. Involved with computer engg and elec. Engg programs along with teaching a grad course in digital control system and Robotics. I enjoy photography and music. *Society Aff:* IEEE.

Rau, George J, Jr
Business: 5319 Shreve Avenue, St Louis, MO 63115
Position: General Mgr *Employer:* Combustion Engrg, Inc. *Born:* Oct 1923. Native of St Louis, Mo. Served with Armed Forces 1942-45. Associated with Combustion Engrg Inc since 1941. Has worked at various jobs within the org and at present is Genl Mgr of St Louis Plant of Combustion Engrg. Association with the welding ind began over 30 yrs ago. Has been active with the St Louis Sect of AWS and was Chmn during 1967-68. Served on the Society's Natl Nominating Ctte and Vice Chmn of 1966 Fall Tech Meeting. Mbr of Arrangement Ctte for the 1976 Exposition & Tech Meetings held in St Louis, Missouri & is currently Dir of Dist 14 of Amer Welding Soc. Has been mbr of Joint Engrg Council of St Louis and is mbr of the adv ctte to study vocational educ in St Louis public schools. Has been certified by SME as mfg engr in the field of mfg mgmt.

Rau, Jim L
Business: 800 Heath Street, Lafayette, IN 47904
Position: Ch Engr Hydrostatic Steering Systems. *Employer:* TRW Inc, Ross Gear Div. *Education:* MSME/ME/Purdue; BSME/ME/Purdue. *Born:* 2/9/47. BS & MSME from Purdue Univ 1971. Native of Indianapolis IN. Co-op student for Intl Harvester Co 1965-68 Indianapolis Engine Works. Assumed Design Engr respon at TRW Ross Gear Div 1969 in charge of military application of steerin sys; 1970 transferred to Advanced Design to become Principal Engr of innovative special application steering roducts & sys 1974; 1976 Ch Enbr of Hydrostatic Steering Products. Currently Div Ch Engr Adv Design. SAE papers 720803, 740435, 750806. 1974 SAE Springer Award. SAE Ind Sect Governing Bd 1972-79. Mbr SAE Natl FCIM Ctte. *Society Aff:* SAE, ASAE.

Rauch, William F, Jr
Business: 560 N Park Avenue, Helena, MT 59601
Position: Staff Mgrr. *Employer:* Mountain Bell. *Born:* Oct 1938. BS from Montana St Univ 1960. Native of Elmhurst Ill. Served in the Army as Capt in the Transportation Corp 1960-1962. Started with Mountain Bell in the Bldg Engrg Dept advancing to Supervising Engr(Construction and Prog), moved into Financial Analysis and then served three years as Engrg Supt (Planning & Prog). Presently responsible for Mountain Bell's Bldg, Motor Vehicle and Supply operation in Montana. Became reg P E 1968. Pres elect of Montana Soc of Engrs 1975-76. Received Private Pilot's license 1976. Hobbies: skiing and hunting.

Raudebaugh, Robert J
417 Stelle Ave, Plainfield, NJ 07060
Position: Consultant *Education:* PhD/Metallurgical Engg/Purdue Univ; BS/Chem Engg/Carnegie Tech. *Born:* 4/8/10. Native of Dayton Ohio. Head of metallurgy progs at Univ of Rochester 1944- 48 and Georgia Inst of Technolgoy 1948-52; At res lab of Internatl Nickel Co as asst to dir to res 1967-74. (Exec Dir, USNC, World Energy Conference 1975-80) Member and P Pres of metal Sci Club of NY; American Society for Metals; Engrs Joint Council. Member AIME; AAAS; Sigma Xi. Received distin alumnus award from Purdue Univ. Hobbies: philately and photography. *Society Aff:* ASM INTERNATIONAL, AIME, AAAS, ΣΞ

Rausch, Donald O
Business: 134 Union Blvd, Ste 640, Lakewood, CO 80228
Position: President & CEO *Employer:* Western Nuclear Inc. *Born:* 2/2/26. DSc and EM from CO Sch of Mines. Distinguished Achievement Medal, CO Sch of Mines. Joined Western Nucl, Inc in 1979; President & CEO. Formerly with NL Industries as VP of Mineral Resources 1974-79; responsible for minerals exploration, developing and operating new mines, and operating existing mines on a worldwide basis. Before joining NL was Gen Mgr of Tintic Div of Kennecott Copper Corp. Prior to joining Kennecott in 1959 worked as a mining consultant and taught mining engg at Colorado Sch of Mines. Proj Engr with the Denver Res Ctr, serving as proj mgr. *Society Aff:* AIME, MMSA.

Rautenstraus, Roland C
Business: 914 Broadway Univ Ctr, Boulder, CO 80309
Position: Professor & Pres Emeritus. *Employer:* Univ of Colorado. *Education:* MS/Engg/Univ of CO; BS/Engg/Univ of CO; Doctor of Laws/-/Univ of NM. *Born:* 2/27/24. in Gothenburgh, NB to Christian and Emma (Stein) Rautenstraus; student CO State College 1941-43, BS in Civil Engg from Univ of CO 1946; MS 1949. Married Willie Dean Alter June 30, 1946; one son Curt Dean. With McNeil Engrs in San Diego 1941-42; mbr faculty of Univ of CO 1946; prof of civil engg 1957- ; chmn dept 1959-64; assoc dean of faculties 1964-68; VP for educational & student relations 1968-70; VP for univ relations 1970-73, exec VP 1973-74, acting pres 1974, pres 1974-1980, Prof. 1980- . Served to lt (jg) USNR 1942-46. Recipient Gold Medal Award Amer Lincoln Arc Welding Soc 1950; Robert L Stearns medal of distin service 1965; alumni recognition award 1970; Norlin Recognition Award for Distin Achievemtn 1974; Distin Engg Alumnus Award 1975; Univ of North CO Alumni Assoc Columbia Award 1975; Hon degree of Doctor of Laws, Univ of NM 1976; Rocky Mtn. Center on Environ Award, 1967-77; Alfred J Ryan Award by Profes Engin of CO, 1977; Natl Assoc of Accountants citation, 1978; Consulting Eng Council of CO Award, 1979; Univ Medal, 1979. *Society Aff:* XE, TBΠ.

Ravenis, Joseph VJ, II
Home: 6041 Ridgemoor Dr, San Diego, CA 92120
Position: Staff Scientist *Employer:* Cubic Corp *Born:* May 1932. PhD from The Johns Hopkins Univ; BSEE from The Johns Hopkins Univ; Intl studies from Syracuse Univ & Kyoto Univ, Japan. BA from Western Md Coll; Adjunct Prof at San Diego State Univ. and Rochester Inst Tech. Served with US Army 1954-56. Engr for Martin Co. Fellow Engr for Westinghouse Elec Corp Space & Defense Div. At Genl Dynamics Corp Electron Div 1965-78 specializing in design dev, test and operation of military electron sys, and prog mgmt responsibility including GPS, Satellite Navigation electronic mail handling, fingerprint, mail tracing, ECG and metal detection sys. Cubic Corp since 1978 Staff Scientist and Program Mgr on secure communications, signal processing, microcomputer, interactive video disc and CAI technologies. Founder and Pres of San Diego Section ORSA; Genl Chmn 1973 Natl ORSA Meeting, Chmn Geo Sects Ctte. Three patents and numerous publs and soc lectures. Part-time faculty mbr at San Diego State Univ since 1974. *Society Aff:* IEEE, ASA, AAAI, AAAS, ΣΞ.

Raville, Milton E
Business: Sch of ESM, Atlanta, GA 30332
Position: Dir of ESM. *Employer:* Georgia Tech. *Education:* PhD/Eng Mechs/Univ of WI; MS/Appl Mechs/KS State Univ; BS/Civil Engr/Norwich Univ. *Born:* July 1921. US Army 1943-46; horse cavalry; corp of engrs and transportation co; highest rank - Capt. Faculty of Kansas St Univ 1946-62 Prof and head, Dept of Applied Mechanics 1956-62. Prof and Dir, Sch of Engrg Sci and Mechanics, Georgia Tech, 1962-present. Cons to several co's over the years 1947- present. Mbr and office holder in several prof and honorary socs, particularly ASEE and NSPE. Named Engr of the

Raville, Milton E (Continued)
Year in Educ St of Ga 1975. Hobbies: golfing, reading and music. *Society Aff:* ASEE, NSPE, SES.

Ravindran, Arunachalam
Business: Sch of Ind Engg, W Lafayette, IN 47907
Position: Prof. *Employer:* Purdue Univ. *Education:* PhD/Ind Engg/Univ of CA; MS/Ind Engg/Univ of CA; BS/EE/Birla Inst of Tech. *Born:* 1/1/44. in India. Came to USA in 1965. Joined Purdue Univ in 1969. Currently Prof of Ind Engg. Co-Author of the text "Operations Research: Principles & Practice–, Wiley, 1976 Received the Am Inst of Ind Engrs Book of the Yr Award in 1977 for Outstanding Literary Contribution in Ind Engg. Has received numerous Best Teachers Awards in Ind Engg at Purdue Univ. Pres of Central IN Chapter of the Am Inst of Ind Engrs, 1979-80. Res interests include math programming, multi- criteria optimization & health planning. *Society Aff:* IIE, ORSA, TIMS, AIDS.

Rawlins, Charles B
Business: PO Box 150, Massena, NY 13662
Position: Tech Conslt *Employer:* Aluminum Co of America *Education:* MS/IE/Clarkson Univ; BE/ME/Johns Hopkins Univ *Born:* 07/04/28 in Annapolis, MD. Rawlins joined Alcoa Laboratories in 1949 and has specialized in overhead conductor dynamics. He is presently a Technical Consultant with the Laboratories. Current activities include research on wind-induced motions of overhead lines, and consltg for other organizations doing similar research. He is an author of a number of technical papers and holds nine patents. *Society Aff:* ASME, IEEE

Ray, Alden E
Business: 300 College Park Ave, Dayton, OH 45469
Position: Prof of Matls Engrg. *Employer:* Univ of Dayton. *Education:* PhD/Metallurgy/IA State Univ; BA/Chemistry/So IL Univ. *Born:* 2/14/31. in Centralia, IL. Prof of Matls Engrg in Sch of Engrg & Supr of Metals & Ceramics div in Res Inst, Univ of Dayton. At U D since 1961. Initiated matls Engrg Grad Prog 1971. Intl reputation for res in rare earth aloys for permanent magnets (29 pubs, 6 pats). Named Outstanding Engr in Dayton area 1971; Outstanding Prof in Sch of Engrg 1972; Prof of yr Univ of Dayton 1973; Outstanding Eng/Sci in Dayton area, 1978. Fellow of ASM, 1980. *Society Aff:* ASM, AIME, ΣΞ, AWS.

Ray, Asa M, Jr
Home: 5710 Bon Aire Dr, Monroe, LA 71201
Position: President. *Employer:* Ray Engrg Inc. *Born:* Oct 20, 1922 Louisiana. BSCE Louisiana Tech 1953 after serving in So Pacific WW II with Field Artillery, US Army. Var pos & exper at Los Alamos, Phillips Petro, with Saturn Missile Prog, as Louisiana Hwy Major Proj Engr, & with Olin Paper Indus before private practice. Started Ray Engrg, Cons Engrs 1966. Exper design & const surveillance. Mbr Bd, CEC-L; Bd LES; NSPE & Natl Mgmt Ctte PEPP Sect; Lions Club; Baptist Church; P Chmn La PEPP Sect LES. Ints: fishing, hunting, photography, music.

Ray, Gordon T
Business: 441 Ninth Ave, New York, NY 10001
Position: VP-Planning *Employer:* NYNEX Enterprises *Education:* PhD (Hon)/Engrg/Midwest Sch of Engrg; BEE/Elec Engrg/RPI *Born:* 01/31/28 Born in NY City. PhD (hon) Midwest Coll Engrg. BEE from RPI. Served in AK with US Army and Air Force. Joined NY Telephone Co 1954; subsequent assignments at BTL and AT&T, including CCITT representation. 1983 transferred to NYNEX Enterprises. Currently planning future tech strategies, product evaluation, electronic data processing sys and corp adm. Sr mbr, NSPE, IEEE. Mbr, AAAS, NY Acad Sci, Amer Soc Qual Control, Natl Assoc Purchasing Mgmt. *Society Aff:* NSPE, IEEE, AAAS, ASQC, NAPM

Rayl, Leo S, Jr
Business: PO Box 1225, Battle Creek, MI 49016
Position: Arbitrator *Employer:* Arbitration Limited *Education:* EdD/Educ Admin/Pacific States Univ; MSIM/Indl Mgt/Purdue Univ (Krannert); MPA/Public Admin/Western MI Univ; BSME/Mech Engg/Purdue Univ. *Born:* 2/22/23. Army 42-46; Maj-CE (USAR-Ret). Between 1948 and 1965 was an Industrial Engr with ALCOA, Proj Engr (facilities) with Johnson Wax, Asst to the VP with Steel Industries, and a Superintendent in mfg and Mgr of Industrial Engg with General Foods. Formerly a Sr of IIE. Affiliated with Western MI Univ in 1965; Profr Emeritus Ind Engrg (1987). Currently a Labor-Mgt Arbitrator (mbr of SPIDR, IRRA and NAC). Private Pilot. Listed in American Men & Women of Sci. KY Colonel, Who's Who in Midwest, Men of Achievement (British), Who's Who in Technology Today. *Society Aff:* ASEE, SPIDR, NAC, IRRA, IPA

Raymond, Arthur E
Home: 73 Oakmont Dr, Los Angeles, CA 90049
Position: Retired. *Education:* PhD/-/Brooklyn Poly; MS/Aero/MIT; BS/-/Harvard. *Born:* Mar 1899. Douglas Aircraft 1925-60: Ch Engr & V P Engrg 1934-60 - retired 1960. Tr Aerospace Corp 1960-71; Res Ana Corp 1965-71. Cons The Rand Corp 1960- ; Cons to Adm, NASA 1963-69. Mbr NACA 1946-56, Natl Acad of Sci, Natl Acad of Engrg. Hon Fellow AIAA, Pres IAS 1946. Guggenheim Medal 1957. Sylvanus Albert Reed Award 1965 AIAA. *Society Aff:* AIAA, NAS, NAE.

Raymond, Louis
Business: PO Box 7925, Newport Beach, CA 92658-7925
Position: Principal Consultant; Pres *Employer:* L. Raymond & Associates, Inc *Education:* PhD/Mtl Sci/Univ of CA; MS/ME/Carnegie Mellon Univ; BS/ME/Carnegie Mellon Univ. *Born:* 11/18/34. Independent Consultant (L Raymond & Assoc Inc.) servicing Engg firms in Southern CA. Owner of LRA Laboratories, Inc., a failure analysis laboratory that supports consulting activities. Awarded numerous government sponsored Research Programs. Elected to two Natl Acad of Sciences, Natl Mtls Adv Bd Ctts on Fracture toughness testing/requirements for weapon systems. Editor of ASTM-STP 962 on Hydrogen Embrittlement: Prevention and Control. Co-Editor of ASTM-STP 543 on Hydrogen Embrittlement Test Methods; served as consultant to Dept of Army, Navy & UNESCO; actively involved in techniques of life-prediction analysis and utilization of advanced mfg tech concept for the USN, US Coast Guard, Federal RR, & Federal Hgwy Administration. PE in CA (MET) FIAE (Fellow of Institute for Advancement of Engineering). *Society Aff:* FIAE, ASTM, ASM, AAAS.

Raynaud, Walter L
Home: 8911 Jackwood, Houston, TX 77036
Position: President. *Employer:* Chenault & Brady Engrs Inc. *Born:* May 22, 1935. Raised Ft Worth Tex. BSME Texas A&M 158; MME Univ of Houston 1973. Prod design engr Genl Dynamics after grad. USAF as Civil Engrg Proj Officer, Wright Field Ohio 1959-62. 5 yrs in sales with Trane & Carrier & 1 yr with mech contracting firm. Joined Lockwood, Andrews & Newnam Cons Engrs 1968 - proj engr. Moved to Chenault & Brady Engrs in 1970. Elected to Bd of Dir 173 & co pres 1975. Commercial/instrument pilot. Played tympani 2 seasons with Houston Civic Symphony. Deacon in Presbyterian Church. Reg P E Tex, La, Ohio, Colo, Mo, Ind, N Mex, Penn.

Rayner, John H
Business: Kennedy/Jenks/Chilton, 543 Byron St, Palo Alto, CA 94301
Position: VP, Dir of Marketing *Employer:* Kennedy/Jenks Engrs *Education:* MBA/Finance/Univ of Hawaii; BS/Civil Engg/San Jose State Univ. *Born:* 11/23/43. With Kennedy/Jenks Engrs since 1968. Worked on the planning & design of var domestic & indus wastewater sys. Involved in the initial studies & subsequent design & const of num ship wastewater sys at West Coast & Hawaii Naval bases. In 1973, assumed the pos of regional mgr of The Firm's Honolulu office. During 1978, he returned to the San Francisco home office to be in charge fo corp-wide marketing for Kennedy/Jenks Engrs. Former chmn of the Solar Energy Stds Ctte and Bldg code of

Rayner, John H (Continued)
Appeals Bd for the Cty of Sunnyvale. *Society Aff:* ASCE, WPCF, AWWA, SAME, SMPS.

Rea, John E Jr
Home: 2201 Alderham, Oklahoma City, OK 73170
Position: Chmn of the Bd *Employer:* RAM Engrg, Inc *Education:* BS/CE/Univ of NM *Born:* Mar 1917 Los Angeles Calif. Hydraulics Engr U S Bureau of Reclamation, Okla City 1941-42 & 1946-47. Stress & Liaison Engr Douglas Aircraft 1942-44. Served with 109th USNCB as LTjg on Guam, CEC, USNR 1944-46. Engr with Phillips & Stong Engrg Co 1947-48. Partner Engrg in Rea Engrg 1948-62. Pres of Rea Engrg & Assoc 1962-75, Chmn & CEO to July 1982. July, 1982 to present Chmn & CEO of Multi-Mode Environ Sys, Inc now changed to RAM Engrg, Inc with title of Chmn of Bd. Dir Southwestern Bank & Tr Co, Okla City. Exec Adv Bd Last Frontier Council, Boy Scouts of Amer. Mbr of So Okla City Rotary Club - Pres 1963. Mbr Consulting Engr Council of Okla, Pres 1966. Spec int in waste treatment & water reusage. Inventor of Multi-Mode Waste Treatment System; Multi-Mode sequential batch reactor designed for Choctaw, Okla, declared innovative by E.P.A., Cincinnati, and Region VI in August, 1981, completed in Aug 1982. Delegate in People to People delegation on Pollution Control to Peoples Republic of China 1986. Inventor of Aerated Batch Clarifier (ABC) (pat. appl. for) to remove Algae from lagoon water. *Society Aff:* NSPE, OSPE, CECUS, CECO.

Read, Jay R
Business: 1001 Office Park Rd, W. Des Moines, IA 50265
Position: Partner *Employer:* Brown Engrg Co *Education:* BS/ME/IA State Univ *Born:* 09/30/36 Brown Engrg Co, 1959-63, Design Engr for power plants and industrial utility projs. AMF Western Tool Inc, 1963-66, proj engr in design of Lawn and Garden Machinery. Brown Engrg 1966 to present. Became partner in 1975. Became managing partner 1981. Mbr of IA State Univ Engrg Advisory Council, Mbr IA Coal Utilization Advisory Ctte. Bd Mbr, IA Consltg Engrs Council. Past Pres, Des Moines Civilan Club, Past Conslt to proj business. Engr of the Yr, Central IA Chapt, IA Engrg Soc, 1981. *Society Aff:* ASME, ASHRAE, NSPE, ASTM, ACEC

Read, Robert H
Business: Box 905, Portland, IN 47371
Position: President *Employer:* Teledyne Portland Forge. *Education:* PhD/Met/Penn State Univ; Masters/Phys/Penn State Univ; AB/Math/IL College. *Born:* 2/15/28. in Morgan County, IL in 1928. Married Martha Smith in 1950, three children. Employed as Res Scientist at Armour Res Fdn 1956 to 1962. Became VP, Atlas Steels Div Rio Algom Mines. Canada & Corp Planner for Rio Algom in period 1962-1973. VP Sales & Met Teledyne Vasco (Steel), 1973. Exec VP & Gen Mgr, Teledyne Portland Forge, 1976 President 1980. *Society Aff:* ASM, AIME, APMI.

Read, William E
Business: OCE, ATTN: DAEN-ZC, Washington, DC 20310
Position: Asst Chief of Engg, US Army. *Employer:* U S Govt. *Education:* BS/Mil Engr/US Military Acad; MS/Civil Eng/Univ of IL; MA/Bus/Webster College. *Born:* May 1927 N C. MSCE Univ of Illinois; MA Bus, Webster Coll; BS U S Military Acad. Airfield const, Okinawa & Korea; engr co cdr Europe. Instr, later Asst Prof Engrg Mech USMA; Asst Army Attache Israel; Engr Battalion Cdr 9th Inf Div, Ft Riley & Vietnam; served on Dept of Army & Jt Staffs Wash; Cdr 4th Inf Div Support Command & later of Task Force Ivy, Vietnam; Tulsa Dist Engr; Dir Procurement & Prod Directorate, & later Dep Commanding General, Army Aviation Sys Command; Div Engr, USA Engr Div, Mo R, and Mbr Mo River Basin Comm;. Reg P E; Assumed current respon Sept 1978. *Society Aff:* SAME.

Reading, Thomas J
Home: 1520 S 93rd Ave, Omaha, NE 68124
Position: Civil Engr - Matls. *Employer:* Private Consultant *Education:* MS/Civil Engg/MIT; BSCE/Civil Engg/OH Univ. *Born:* 7/5/15. Spent 4 yrs in concrete res, 2 yrs in WWII concrete ship prog, and 32 yrs in Corps in lab, construction, and as Chief Materials Engr. in MRD Office- serving as cons, tech specialist & staff advisor. Retired in 1977 and is now a consultant. Fellow ACI & ASCE, Mbr ASTM & USCOLD. Previous mbr of Bd of Dir & chmn or mbr of sev tech cttes in ACI. Author sev papers. Frequent speaker. Recd ACI Disting Serv Award 1969 & Kennedy Award 1976 and Hon. Member in 1986. *Society Aff:* ASCE, ACI, ASTM, USCOLD.

Reams, James D
Business: AFAPL/PO, bWright-Patt AFB, OH 45433
Position: Ch, Aerospace Power Div. *Employer:* USAF, Aero-Propulsion Lab. *Born:* 1933. MS Univ of Michigan; BS Univ of Ky. Native of Richmond, Ky. USAF 1952-56. Entire prof career with USAF as Facils Engr; Flight Vehicle Power Res & Dev Engr; Advanced Long Range Planner; Adviced Mission Studies & Analysis; Ch of Energy Conversion Branch & now Ch of Aerospace Power Div. SAE Arch T Colwell Merit Award 1968. Mbr Pi Tau Sigma, Tau Beta Pi, AIAA. Enjoy gardening, golf, tennis, classical music.

Reaser, Wilbur W
Home: P O Box 1554, Sequim, WA 98382
Position: Consultant. *Employer:* Self. *Education:* BS/Mech Engg/Univ of WA. *Born:* June 1913. Native of Auburn Ks. Taught at UCLA & Caltech. Chmn subctte for icing problems, Natl Adv Ctte for Aeronautics 1953-55, Chmn Air Cond Ctte Soc of Automotive Engrs 1943-60. Ch Air cond Engr Douglas Aircraft Co 1942-61. Dir of Mech Sys, sys safety & product reliability, Apollo & related sys McDonnell Douglas Aerospace Co 1962-69. V P Sales & Engrg Natl Utils Co 1969-71. Cons Hosp Safety & Automated Radio Sys 1971- . ASME Spirit of St Louis Jr Award 1941; ASME Fellow 1962; ASME Life Fellow Mbr 1971. Enjoy classical music, fishing, golf. Volunteer Executive, International Executive Service Corps, N.Y., N.Y. Served with Fabricantes De Equipo Para Refrigeracion, San Nicolas De Lus Garza, Nuevo Leon, Mexico from 3/16/81 to 6/16/81. MBR Volunteers in Technical Assistance (VITA) Advisory Ctte - Energy Initiatives for Djibouti, Arlington, VA. Teacher: adult Bible class, Trinity United Methodist Church of Sequim. Conducting Weekly Bible study with inmates of Challam County Jail, Port Angeles, WA. *Society Aff:* ASME

Reaveley, Ronald J
Business: 2126 S 10th E, Salt Lake City, UT 84106
Position: Cons Engr - Partner. *Employer:* Reaveley Engrs & Assocs. *Born:* July 1941. MECE Univ of UT 1971; BSCE Univ of UT 1964. Currently lic to practice as a PE in UT, CA, NE, ID, & CO. Immediate P Pres Amer Cons Engrs Coun of UT. Have served as VP ACEC/UT & currently serving as Mbr Bd of Dir ACEC UT. Mbr Bldg Codes & Stds Comm ACEC/U S. P V Pres ASCE Student Chap at Univ of UT. Have spec in field of struct design of bldgs & special structs with emphasis on earthquake engrg & value engrg.

Rebane, Henn PE
Business: 360 Central Ave - 12th floor, St Petersburg, FL 33701
Position: V Pres. *Employer:* Wedding & Assocs, Architects, Inc. *Education:* BS/EE/GA Tech. *Born:* 11/14/33. Native of Estonia. BSEE GA Inst of Tech. Reserve officer in Pub Health Service. Worked at Natl Insts of Health, FL Power & Lt Co & NASA in power engrg. VP of Wedding & Associates, Architects, Inc. Pres of FL Inst of Cons Engrs 1971 & 1975. Pres FL Engrg Soc 1983/84. Enjoy stamp collecting & yachting. Chmn St. Petersburg Planning Commission & Historical Preservation Comm. 1987/88. *Society Aff:* NSPE, ASHRAE, IEEE, IES.

Rechard, Paul A
Business: PO Box 4128, Laramie, WY 82071
Position: Pres *Employer:* Western Water Conslts, Inc *Education:* CE/Civil/Univ of WY; MS/Civil/Univ of WY; BS/Civil/Univ of WY. *Born:* June 1927. P E & Land Surveyor WY; P E UT, MT, CA, Co. Registered Professional Hydrologist. Pres of

Rechard, Paul A (Continued)
Western Water Conslt, Inc since 1980. Formerly Dir of Water Resources Res Inst & Prof of CE (Hydrology) Univ of Wyoming; Asst Dir Natural Resources Res Inst, Univ of Wyoming; Principal Hydraulic Engr, Upper Colo River Comm; Dir of Water Resources, Wyoming Natural Resource Bd; Wyoming Interstate Streams Commissioner; Hydraulic Engr U S Bureau of Reclamation. Fellow ASCE, (Pres WY Sect 1968), Mbr. NSPE, AGU, Natl Water Well Assoc, AWRA (Pres Wyo Sect since 1970), AWWA, ICID, Wyoming Engrg Soc (Pres). Current res incl snow hydrology, water resources planning, impacts of surface mining, waste mgmt, watershed mgmt, water law. *Society Aff:* ASCE, AGU, AWRA, NSPE, AWWA, NWWA, ICID, AIH

Rechtin, Eberhardt
Business: The Aerospace Corporation, bPO Box 92957, Los Angeles, CA 90009
Position: President & CEO *Employer:* Aerospace Corp. *Education:* PhD/EE/CA Inst of Tech; BS/EE/CA Inst of Tech. *Born:* 1/16/26. Jet Propulsion Lab 1949-67 to pos of Dir of the NASA JPL Deep Space Network & Asst Dir of JPL. DOD Advanced Res Projs Agency Dir 1967-70. DOD Prin Dept Dir of Defense Res & Engrg 1970-71. Asst Secy Def (Telecommunications) 1971-73. Ch Engr Hewlett-Packard 1973-77. Pres, Aerospace Corp 1977- . Fellow IEEE, AIAA. Mbr Natl Acad of Engrg. Tau Beta Pi. Sigma Xi. Achieve awards from NASA, DOD, USN, USAF, AIAA, IEEE, AFCEA, UCLA and Caltech. Fields: R&D, telecommunications, sys, space sys. *Society Aff:* AIAA, IEEE, AFCEA.

Recker, Wilfred W
Business: Dept of Civil Engg, Sch of Engg, Irvine, CA 92717
Position: Prof of Civil Engrg *Employer:* Univ of CA. *Education:* PhD/Civil Engg/ Carnegie-Mellon Univ; MS/Civil Eng/Carnegie-Mellon Univ; BS/Civil Eng/ Carnegie-Mellon Univ. *Born:* 8/3/42. Joined fac of Dept of CE, State Univ of NY at Buffalo, as Asst Prof 1967. Assoc Prof 1971, Prof 1978. Sr Res Engr, Transp & Urban Analysis Dept, Res Labs, GMC from 1974-76. Joined fac of Univ of CA, Irvine, as Prof in 1979. Dir of Inst of Transportation Studies, U C Irvine. Active in research in area of transp, behavioral res, transp. operations. Dow Outstanding Young Fac Award/ASEE 1976 *Society Aff:* ASCE, ITE.

Reckert, Robert D
Business: 315 First Ave, Rock Rapids, IA 51246
Position: Pres *Employer:* DeWild Grant Reckert & Assocs Co. *Education:* BS/CE/IA State Univ. *Born:* 11/11/25. BSCE Iowa St Univ. With present firm since 1954; CEO of arch engrg, planning firm in 12 state cons practice. Reg P E & land surveyor. Author sev papers on water supply wells, waste treatment, regis of engrs & land surveyors, & prof dev. Dev uniform natl exam in land surveying. Pres., ACSM 1974-75; Chmn, Comm on Registration of Engrs, ASCE, 1978-79; ECPD, Dir 1976-79; Chmn NCEE-POL Committee 1979-80; Chmn Iowa Bd of Engrg Examiners 1973-75; NCEE 1967-75. NCEE Disting Serv Cert 1974; John S Dodds Award 1977; Disting Serv Award, Iowa Engr Society, 1981-82; ACSM Pres's Citations 1972, 1973. Hobbies: golf, family genealogy, fishing, gardening, Scottish deerhounds. *Society Aff:* ACSM, ASCE, NSPE, ACEC.

Rector, Alwin H
Home: 8448 Meadow Lane, Leawood, KS 66206
Position: Retired (Sr VP) *Employer:* Burns & McDonnell *Education:* BS/EE/KS State Univ *Born:* 06/23/15 Student Engr-Gen Elec Co 1937-40. Distribution Engr-Indianapolis Power & Light 1940-43. Served with Army Signal Corps 1943-46. Served in Gen Mac Arthur's Headquarters Tokyo, Japan 1946-47. With Burns & McDonnell Engrg Co 1947-1980. VP & Mgr Power Div 1971-76. Sr VP 1976-80. Reg PE in MO, KS, AZ, AR, IN, KY, MI, MN, MT, NB, NJ, NY, ND, OK, and WY. Retired as Sr. Vice President, Burns & McDonnell, July 1980. Enjoys Amateur Radio. *Society Aff:* IEEE, ASME, NSPE, MSPE

Redd, John P
Business: 450 S 900 E, Salt Lake City, UT 84102
Position: President. *Employer:* J P Redd & Assocs. *Education:* BS/Chemical Engg/ Univ of UT. *Born:* Dec 25, 1930 Mt Pleasant Utah. m. Erma Helquist July 16, 1949; children - Michael, David, Linda, Lowry, Audree, Kristin. Pres cons engrg firm of J P Redd & Assocs 1976; Partner cons engrg firm Redd & Redd 1962-76. Mbr Utah State Bd of Educ 1976; Mbr & Maj Leader, Utah State House of Reps 1967-75; Mbr NSPE & Utah Soc of Prof Engrs 1969- , Pres 1970-71; Mbr ASHRAE 1969- , Pres 1969-72. Utah's Engr of Yr, Utah Engr's Council 1968. Service to Mankind Award, So Davis Sertoma 1971. Mbr White House Conf on Children & Youth, AIChE. Reg P E 11 states. *Society Aff:* AIChE, ASHRAE, NSPE.

Redfern, Donald B
Business: 45 Green Belt Dr, Don Mills, Ont, M3C 3K3 Canada
Position: Chrmn. *Employer:* The Proctor & Redfern Group. *Education:* BASc/Engg/Univ of Toronto. *Born:* Nov 1927. Joined Proctor & Redfern 1948. Became Partner 1954, V P 1969, Pres & Genl Mgr 1974, Chmn 1978. Designed & supervised const of var of sewage treatment plants. Respon for engrg main content of area gov, amalgamatin & annexation studies. In a sr & exec capacity, participated in many major engrg studies. Mbr Assn of Prof Engrs of Ontario. Pres Part of 1965, 1966. Designated Cons Engr. Pres Canadian Council of Prof Engrs 1968, 1969. Fellow Engrg Inst of Canada. Enjoy farming, reading, travelling, boating. *Society Aff:* EIC, APEO, CCPE, CEO

Redhead, Paul A
Business: Div of Physics, Natl Research Council, Ottawa Ont, Canada K1A 0R6
Position: Secretary, Committee on Science & TGechnology Policy. *Employer:* National Res Council of Canada. *Education:* PhD/Physics/Cambridge Univ; MA/ Physics/Cambridge; BA/Physics/Cambridge. *Born:* May 1924 Brighton England. Sci officer with British Admiralty 1944-47. With Natl Res Council of Canada since 1947 engaged in res on electron physics, surface sci, vacuum science, & mass spectrometry; Dir Genl of Planning 1972-73, Dir Div of Phys 1973-1986, Group Dir Phys/Chem Sci Labs 1974-82. Chairman, Committee of Laboratory Director 1980-86; Secretary, Committee on Science & Technology Policy 1986- . Served as Pres Amer Vacuum Soc; Editor Journal of Vacuum Sci & Tech; presently Asst Editor-in-Chief Canadian Journals of Res. Medard W Welch Award of the Amer Vacuum Soc 1975; Fellow IEEE, APS, Royal Soc of Canada. *Society Aff:* CAP, AVS, RSC, APS, IEEE.

Rediker, Robert H
Business: 244 Wood St, Lexington, MA 02173
Position: Senior Staff, Lincoln Labs; Adj. Prof *Employer:* MIT. *Education:* PhD/Physics/MIT; SB/Elec Engg/MIT. *Born:* June 1924. Native of Brooklyn N Y. Served with Army Signal Corps AUS 1943-46. With MIT since 1950 with exception of academic yr 1952-53 at Indiana Univ. Prof of EE 1966-1976, Adj. Prof. of EE since 1976. Hd of Optics Div at Lincoln Lab 1972-1980. Senior Staff since 1980. Has served on Natl Acad of Sci cttes: Eval Panel for the Electromagnetics Div NBS 1975-78, ad hoc ctte on Matls & Processes for Electron Devices 1970-72, num IEEE cttes. Fellow IEEE & Amer Phys Soc. IEEE David Sarnoff Award 1969. Enjoys bridge & playing 'at' tennis. *Society Aff:* APS, OSA, IEEE, AIP.

Redline, John G
Business: Three Springs Drive, Weirton, WV 26062
Position: President *Employer:* Weirton Steel Div, Natl Steel Corp *Education:* BS/Eng/Lehigh Univ *Born:* 05/16/21 A native of DE. Grad from the public schs at Nazareth, PA, and received a bachelor of sci degree in engrg from Lehigh Univ. Prior to military service, he attended the Univ of PA, and during World Warr II held the rank of captain in the US Air Force, serving in the Mediterranean area. He joined Natl Steel Corp in August 1959 as asst superintendent of the cold reduction and finishing dept of Midwest Steel Div, Portage, Indiana. In June 1966 he was transferred to the Weirton Steel Div and was named pres in June 1967. On June 1, 1972 he was transferred to Great Lakes Steel Div, Detroit, as pres and held that position until 1977 when he returned to Weirton Steel Div as pres. *Society Aff:* AISE, AISI, AIME

Redmond, Robert F
Home: 3112 Brandon Rd, Columbus, OH 43221
Position: Assoc Dean, Col Eng *Employer:* OH State Univ *Education:* PhD/Physics/Oh State Univ; MS/Math/Univ TN; BS/ChE/Purdue Univ *Born:* 07/15/27 Native of Indianapolis, IN. Served with US Navy 1945-46. R & D Engr, Oakridge Natl Lab, 1950-53. Research Scientist and Group Leader, Battelle Columbus Labs, 1953-70. Prof and Chrmn, Nuclear Engrg, OH State Univ, 1970-77. Dir, Engrg Experiment Station, and Assoc Dean, College of Engrg, OH State Univ 1977- . Mbr of Bd of Trustees, Argonne Universities Association, 1972-80, and VP 1976-77. Mbr of OH Power Siting Commission, 1978-81. Chrmn, S.W. OH Section, American Nuclear Soc, 1972. Reg PE, OH. *Society Aff:* ANS, ASEE, AAAS.

Redpath, Richard J
Business: Checkerboard Sq, St Louis, MO 63188
Position: VP. *Employer:* Ralston Purina. *Education:* MME/Engg Mgt/Univ of City NY; BSME/Mech Engr/Fairleigh Dickenson Univ. *Born:* 10/27/39. VP Engg Ralston Purina Corp. Previously Natl dir Engg Johnson & Johnson. Licensed PE MO & NJ. responsible for Corporate Engg & Resource Conservation. Including Construction, Process & Machinery Design, Energy Conservation & Environmental Affairs. *Society Aff:* Ind. Adv. to NSPE

Reece, C Jeff, Jr
Business: PO Box 540, Waynesville, NC 28786
Position: Consulting Engr *Employer:* Reece, Noland & McElrath Engrs *Education:* BS/EE/Clemson Univ; AS/Pre-Eng/Oak Ridge Univ Inst *Born:* 03/20/36 Reg Engr NC, FL, GA, and NCEE Certificate. Sr Mbr IEEE, Mbr ASHRAE, ISA, IES, ACEC, and NSPE, Sr Engr, Champion Papers, Canton, NC 1957-1972, assumed position of Pres, Reece, Noland & McElrath Engrs 1973. Dir Haywood Savings & Loan Association, Waynesville, Brevard Coll and Lake Junaluska Assembly. Past Chrmn PE of Western NC. Lives with wife, Judith, and two children. *Society Aff:* NSPE/PEPP, ACEC, NCSE, ASHRAE, NFPA

Reece, John D
Home: 4430 E 107th Terr, Kansas City, MO 64137
Position: Operations Engineer *Employer:* Black & Veatch *Education:* BSCE/Civil Engg/Univ of MO, Columbia MO. *Born:* Oct 1, 1939. Field Engr Mo Water Pollution Bd, 3 yrs; Ch, Sewage Treatment Div, Pollution Control Dept, City of Kansas City Mo, 14 yrs. Operations Engineer, Black & Veatch, Kansas City, 1 year, Reg P E Mo. Secy-Treas of the Mo Water Pollution Control Assn 12 yrs. Arthur Sidney Bedell Award WPCF 1971. Award of Merit MWPCA 1970. Hobbies: camping, fishing, hunting, golf. *Society Aff:* WPCF

Reed, Charles H
Home: 73 School House Lane, East Brunswick, NJ 08816
Position: Retired *Education:* MS/Agri Engg/Cornell Univ; BS/Botany/Univ of ME. *Born:* Oct 1911. Native of Maine. Prof of Agri Engrg Rutgers Univ 1946-75, teaching farm structs, & during the last 10 yrs spec in R&D of equip for incorporating biodegradable wastes into the soil, & planning land treatment sys. Retired Jan 1976 & presently do some cons in planning land treatment sys for incorporating animal manure & sewage sludge into the soil. Life Fellow Amer Soc of Agri Engrs. Permanent Mbr of US Coast Guard Aux. Prof Engr. *Society Aff:* ASAE, USCGA.

Reed, Frederick J
Home: 3828 Birchwood Rd, Falls Church, VA 22041
Position: Tech Consultant. *Employer:* Self-Employed. *Education:* ME/Mech Engg/ Stevens Inst of Tech; MS/Mech Engg/Univ of Pittsburgh. *Born:* Apr 8, 1905; native of NYC. Mech Engr Westinghouse Sharon Works - Transformers 1926-30. Instr ME Vanderbilt Univ 1930-35. Instr, Asst Prof, Assoc Prof, Prof ME Duke Univ 1935-57. P E North Carolina 1951. Ch Engr, Dir of Engrg, Air-Cond & Refrig Inst (ARI) 1957-76. Retired 1976. Tech cons to ARI and others 1976 to Date. Life Mbr ASHRAE; Life, 50-Yr ASME. Fellow ASHRAE. Major work dev, Admin of ARI stds through ARI engrg cttes, dev & admin of ARI cert progs covering 11 classes of equip & components. Active participant in Stds for ANSI, incl Natl Plumbing Code, & in NFPA on Natl Elec Code. Mbr & Chmn Stds Ctte ASHRAE. Consultant: Qualification Solar Collector Testing Labs (ARI/NBS); Cert Program for Ind. Labs Testing for Safety (AALA/OSHA). *Society Aff:* ASME, ASHRAE, ASME, ASHRAE.

Reed, Geo Lindmiller
Business: 188 Jefferson Ave, Memphis, TN 38103
Position: Assoc & Dir. Transp. Planning Division *Employer:* Harland Bartholomew and Assocs, Inc. *Education:* MS/Transp Engr/NC State Univ; BS/Civil Engr/NC State Univ. *Born:* Mar 1937 Charlotte N C. Specialist in urban & regional transp planning, traffic engrg, transit planning. Proj Mgr Harland Bartholomew & Assocs, Greenville S C; Memphis Tenn; & South Bend Ind 1965-71 - also 1974 to present. Proj Mgr 1971-73, V P 1974, W S Pollard Cons Inc Memphis Tenn. Lectr Univ of Notre Dame 1968. Cons Fed Railroad Admin 1976. Contrib of articles to prof publs. Mbr ASCE, Inst of Transp Engrs. D W Mead award ASCE 1970. Reg P E Ind, Mich, NC, Tenn & Fla. *Society Aff:* ASCE, ITE.

Reed, Joseph R
Home: 1394 Penfield Rd, State College, PA 16801
Position: Full Prof of CE *Employer:* Penn State Univ. *Education:* PhD/Civil Engg/ Cornell Univ; MS/Civil Engg/Penn State Univ; BS/Civil Engg/Penn State Univ. *Born:* 8/15/30. Born in Pittsburgh, PA; married to the former Mary Leggett of Leonard, TX; two children: Stephanie (age 22), David (age 17); presbyterian faith; fraternal memberships; Phi Sigma Kappa, BPOE Lodge 1600, F & AM Lodge 747; employee and consultant with eleven firms since 1953; Engr Liaison Officer with USAF, 1956-59; Res Asst at Cornell Univ, 1964-66; author of more than 40 professional publications and presentations; Reg PE in TX, license 17571; chmn, State College Storm Water Authority, 1977-78; faculty Senator, 1977-81; awarded Sci-Faculty Fellowship by NSF in 1966-67; Outstanding Teacher Award by Penn St Coll of Engrg in 1982; listed in Whos Who in American Education (1963-64), Dictionary of International Biography (1972-73), Community Leader and Noteworthy Americans (1975-76), Men of Achievement (1976-77), Who's who in Technology (1986), American Men and Women of Science (1986). *Society Aff:* ASCE, ASEE, IAHR, NSPE, XE, ΣΞ, TBΠ.

Reed, Richard S
Business: 715 White Bridge Rd, Millington, NJ 07946
Position: President. *Employer:* Reed Process Co. *Education:* MS/Chem Eng/Purdue Univ; BS/Chem Eng/Purdue Univ. *Born:* Sept 1924. Engrg assignments in Europe & Indonesia with Esso/Mobil subsidiary 1952-57. Dir of Proj Eval for Cities Service Co, respon for new venture analysis. Lic P E NJ, Penna, Vir, In 1970 founded Reed Process Co which provides process engrg services for the chem indus. Also pres of Catalox Corp, a R&D co in the field of energy production processes. Patentee of processes for production of chems, plastics, & fertilizers. Author of articles in prof journals. *Society Aff:* AIChE

Reed, Robert D
Business: PO Box 7388, Tulsa, OK 74105
Position: V Pres Engrg. *Employer:* John Zink Co. *Education:* ScD/Hon/Embry-Riddle. *Born:* 5/2/05. Bd of Dir John Zink Co; Trustee John Zink Fnd; Adj Prof Univ of Tulsa; engg cons; seminar lectr; thesis cttes, grad degrees, Univ of Tulsa. Tau Beta Pi (Eminent Engr); ScD; Hall of Fame Univ of Tulsa; 361 US & foreign pats; Mbr AIChE; Reg P E; Inventor of Yr-1977 (OK Bar Assn); Hon Lectr for AIChE by Election. Author: hanbook *Furnance Operations* contributor to 'Encyclopedia of Chem Engg & Design. Papers for: Chem Engg Progress; Oil & Gas Journal; Hydrocarbon Processing; Natl Petro Refiners Assn; Naturll Gas processors

Reed, Robert D (Continued)
Assn; Amer Gas Assn; US Pub Hellth Serv; Amer Indus Hygiene Assn. *Society Aff:* AIChE, EST.

Reed, Samuel Kyle
Business: 8955 Wesley Place, Knoxville, TN 37922
Position: Prof. *Employer:* Univ of TN. *Education:* PhD/Production/Univ of Edinburgh; -/Ind Engr/Univ of TN; BS/Ind Mgt/Univ of TN; -/Basic Engr/NC State. *Born:* 4/17/22. Received 1979 IIE Fellow Award. IIE officer at local, regional & natl levels (1950-present). Regional VP 1954; mbr Natl Bd of Dirs (1954-1957). Founding mbr Assn for Bus Simulation & Experiential Learning. Ford Fdn Fellow (1962); Alcoa Fdn Fellow (1964); Foundation for Economic Ed (1963). Employed in various industrial engg capacities with US Steel, Robertshaw Controls Co, Union carbide Corp. Joined Univ of TN faculty 1958. Author many articles in profl & other journals. Co-author *Production Scheduling Simulation.* TN Natl Alumni Assn Outstanding Teacher Award 1979. President and Board Chairman, Knoxville Opera Company, 1982-1987. *Society Aff:* IIE, ABSEL, NAPM.

Reed, Stan C
Home: 2421 Sierra Lane, Plano, TX 75075
Position: V Pres. *Employer:* Bridgefarmer, Pinkerton, etal. *Born:* Mar 24, 1946. Native of McKinney Tex. BSCE w/hon Univ of Texas at Arlington May 1969. Reg P E Tex Apr 1972. Mbr Amer Concrete Inst, Chi Epsilon, Tau Beta Pi. Cert of award from James F Lincoln Arc-Welding Soc for design of Taxiway Bridges at Dallas/Ft Worth Airport. Employed by URS/ Forrest & Cotton Inc Apr 1967 until May 1976. Began as struct engr & assumed the pos of Struct Sect Mgr Apr 1974. Resigned pos with URS/Forrest & Cotton in May 1976 & formed a partnership with three other engrs in May 1976 - Bridgefarmer, Pinkerton, Reed, Brown & Assocs.

Reed, Willard H
Home: 12126 Cornish Ave, Lynwood, CA 90262
Position: Assoc Dean of Engrg. *Employer:* Calif State Univ - Long Beach. *Education:* MS/Civil Engg/Univ of So CA; BS/Civil Engg/CA State Univ, Long Beach. *Born:* 1917. Native of Globe Ariz. Course work PhD completed USC. Genl const prior to WW II. USA/USAF 1941-45, Builder WW II 1959. Asst Prof, Assoc Prof, Prof 1962 to date CSULB, Chmn Dept CE 1965-74. Assoc Dean of Engrg 1974- . Mbr ASCE, NSPE, CSPE, ASEE, L A Regional Forum, Solid Waste Mgmt, Chmn Human Res Dev Committee ASCE. Cons Engrg Const, Solid Waste Mgmt, Environ Sys, Engrg Educ. Active in area of urban & environ sys as related to water/waste water, hazardous & solid waste, Dir Women/Minorities in Engr Prog, Dir MESA/LB. *Society Aff:* ASCE, ASEE, NSPE, CSPE.

Reed, William H
Home: 771 Longwood Rd, Lexington, KY 40503
Position: Senior Materials Engineer *Employer:* IBM Corp, Information Products Div *Education:* MS/Met Engg/Univ of KY; BS/Met Engg/Univ of TN. *Born:* 2/21/47. Graduated Magna Cum Laude from the Univ of TN after Co-Opping with the Aluminum Co of America in Alcoa, TN. Joined the IBM Corporations Office Products Div in Lexington, KY as a metallurgical engg respon for material selection and failure analysis for typewriters, electrostatic copiers, impact printers, ink jet printers, and magnetic & electronic materials. Attended grad school at the Univ of KY and has served there as an Adjunct Prof in Metallurgical Engg, Worked on divisional Quality Staff in White Plains, NY Reg PE and practicing consultant. Duties at IBM included management responsibility for Metals Engineering, Polymers Engineering, Chemical Analysis, and Materials Technology during development of the Selectric 2000 line of typewriters and printers. Recent focus on resistive ribbon printers, surface mounted circuit board technology, and workstation product development. *Society Aff:* ASM, ΤΒΠ, ΣΞ.

Reeder, Harry C
Home: 2475 NW 144th, Beaverton, OR 97006
Position: Pres. *Employer:* R & W Engg. *Education:* ME/ME/OR State Univ; BSME/ME/OR State Univ. *Born:* 8/23/33. Harry C Reeder, PE began his career with the Boeing Co. His accomplishments at Boeing included dev of an air bearing that permitted wind tunnel testing of simulated jet engine without impairing the data gathering system & predictions of community response to jet airplanes. He joined a maj consulting firm where he was responsible for mech design of water & waste water treatment plants. He was proj mgr for the water supply & discharge treatment systems for a "zero discharge" 200 MW power plant & a 104,000 square foot manufacturing facility. He is currently Pres of R&W Engg, a mech-elec consulting firm located in Portland, OR. *Society Aff:* ASA, ASME, WPCF, ASHRAE.

Reed-Hill, Robert E
Business: Dept Matls Sci & Engrg, Gainesville, FL 32611
Position: Prof & Asst Chmn for Met. *Employer:* Univ of Florida. *Education:* D Eng/Metallurgy/Yale Univ; MS/Physics/Univ of MI; BS/Eng Physics/Univ of MI. *Born:* 11/19/13. Mbr of Permanent Commissioned Instructional Staff US Coast Guard Acad 1939- 60. At Univ of Fla 1960- . Author of 'Phys Met Principles', D Van Nostrand; co- editor of 2 seminar volume & 93 sci publs. Mbr Sigma Xi & Tau Beta Pi, Phi Kappa Phi, ASM, AIME, AAAS, Met Soc (London). *Society Aff:* AIME, ASM, ASTM, Met Soc, AAAS.

Reem, Herbert F
Home: 6239 Lakeview Drive, Falls Church, VA 22041
Position: Pres *Employer:* H. F. Reem & Associates *Education:* PhD/Bus Adm/Univ of Sussex; MS/IE/Columbia Univ; MS/CE/IIT; BS/CE/IIT *Born:* 12/05/21 30 years experience, US and abroad 24 years with US Govt, former Dir Office Special Project, Dir Div of Assessment and Evaluation DOE, Asst to Dir Logistics Office Secy Def other key govt assignments. Presently Energy & Financial Conslt Natl VP Assoc of Energy Engrs 1979 & 1980, Pres Natl Capital Chapter Am Inst Industrial Engrs, 1964 & 1965. Reg PE DC & DE, Chartered Engr (UK) Wallace Prize (i. struct E.) 1947, Energy Prof Development Award 1980 (AEE) Mbr Natl Energy Policy Council (AEE) Liaison with US Govt & Dir Intl & Govt Affairs (AEE). Accomplished linguist (8 languages). Various publications in English, French & German in energy field. Hon citizen Huntsville, AL, Hon LLD. Married, one daughter. *Society Aff:* ASCE, IIE, AEE, NSPE, ISE

Reents, August Curtis
Home: 4411 Dunbar Pl, Rockford, IL 61111
Position: VP. *Employer:* TECHNI-CHEM INC. *Education:* BS/Chemistry/Bradley Univ; MS/Chem Engg/IA State Univ. *Born:* Sept 15, 1921 Oak Hill Illinois. Educ: Brimfield Pub Schools, Married; wife June; children David, Doris, Donald. Res Dept Ill Water Treatment Co 1947- 52. Dir of Res 1952-68; V P Res 1968-72. Formed TECHNI-CHEM INC 1972, with partner Harold Keller. Nature of business: special applications, particularly sugar purification, in field of ion exchange. Patentee in field. P Pres Water & Waste Div Amer Chem Soc. USNR, LTjg 1944-46. *Society Aff:* ACS, AIchemE, AWA, ASIT, SIT.

Reese, Francis E
Business: Beck Building, 7811 Carondelet, St Louis, MO 63105
Position: Conslt *Employer:* Fernandez, Throdahlo, Reese, Inc. *Education:* BS/Chem Engg/Purdue Univ; Adv Mgmt Prog/Harvard Business School *Born:* 11/3/19. With Monsanto Co St Louis 1941-1984: Dir 1973-1984; with Plastics Div as Res Engr 1941-48; Ch Dev Engr 1948-53; Asst Dir of Engrg 1953-56; Dir of Engrg 1956-59; Asst Genl Mgr 1959-61; Asst Genl Mgr Hydrocarbons Div 1961-66; Genl Mgr Internatl Div 1966-68 Corp VP 1968-74; Genl Mgr Hydrocarb & Polym Div 1968-73; Managing Dir Internatl Div 1973-74; Group V P - Facils & Planning, 1975-77. Grp VP, Mgr Dir Monsanto Chem Intermediates Co 1977-79; Sr VP - Facilities & Matl, Pats on thermosetting polymer compositions & spray drying sys for polymers. Former Chmn of Prof Dev Comm of AIChE. Former Chrmn Industry Advisory Group - NSPE; Member Engrg Foundation Advisory Council- Univ of Texas at Austin. Reg P E. Mbr AIChE, (Fellow) Natl Soc Prof Engrs, ACS, AAAS (Fellow)

Reese, Francis E (Continued)
Soc Chem Indus, Tau Beta Pi, Phi Lambda Upsilon. Director, SCS/Compute. President, Fernandez, Throdahlo & Reese, Inc. 1986-. *Society Aff:* AAAS, AIChE, SCI, ACS, NSPE.

Reese, Lymon C
Home: 11512 Tin Cup Dr, 109, Austin, TX 78750
Position: Prof & Assoc Dean. *Employer:* Coll of Engrg, Univ of Texas-Austin. *Education:* PhD/Civil Engg/Univ of CA, Berkeley; MSCE/Civil Engg/Univ of TX, Austin; BSCE/Civil Engg/Univ of TX, Austin. *Born:* Apr 1917. Holder, Nasser I Al Rashid Chair, 1981-; T U Taylor Prof CE; Assoc Dean of Engrg 1972-79, Univ of Texas Austin; Chmn Dept of CE 1965-72. Formerly Asst Prof of CE at Mississippi State. BSCE & MSCE Univ of Texas Austin; PhD Univ of Calif Berkeley. Has had sev yrs of indus exper; cons to a num of co's & gov agencies; spent 2 summers teaching in India under the sponsorship of U S AID; author many articles on deep fnds including drilled shaft foundations; presented many invited lectures in major cities in U S & abroad; Fellow ASCE, received Middlebrooks Award Terzaghi Lecturer and Terzaghi Award; Honorary Member, ASCE, 1984; Mbr Natl Soc Prof Engrs; Reg P E Texas; and Louisiana Mbr U S Natl Acad of Engrg. *Society Aff:* ASCE, ASEE, NAE.

Reethof, Gerhard
720 Windsor Ct, State Coll, PA 16801
Position: Professor & Director. *Employer:* Penn State Univ-Univ Park. *Education:* SB/ME/MIT; SM/ME/MIT; ScD/ME/MIT. *Born:* July 1922. US Army 1943-46. Asst Prof & Res Engr Dynamic Analysis & Control Lab MIT 1950-55. Ch of Res, Vickers Inc 1955-58. Flight Propulsion Group G E Co Cincinnati Ohio 1958-67. Managerial pos incl controls analysis, reliability engrg, J93 engrg & acoustic engrg. Prof of Mech Engrg, Penn State Univ 1967- . Teaching & res in Mech Design, Reliability Engrg, Noise Control in Machinery, & Applications of High Industry Acoustics to Energy Related Areas. Initiated & dev Noise Control Lab at Penn State - Director. Cons to indus & gov in reliability engrg & noise control. *Society Aff:* ASME, AIAA, ASEE, ISA, ASA, INCE.

Reeves, Adam A
Home: 3146 Lakeside Dr 202, Grand Junction, CO 81501
Position: VP/Treas. *Employer:* Paraho Dev Corp. *Education:* BSCE/CE/Univ of OK; BS/Chemistry/Univ of Denver. *Born:* 12/2/16. Schooled in Denver 1922-41. US Infantry & Chem Warfare Service 1942-46. Chem engr at government's oil shale proj at Rifle, CO 1947-55 Refinery pProcess engr with Texaco 1955-66 at El Paso, TX. Process engr for vertical lime kiln design & engg, Denver, 1966-71. At merger into Paraho Dev Corp, elected their VP & Treas. Directed dev & operation of continuous vertical oil shale retorts. Produced 109,000 barrels shale oil 1974-78. Directing process engg of full-size oil shale retort. Supervised pilot plant retorting of two foreign oil shales. Inventor & co-inventor 9 vertical kiln patents for calcination & retorting. Reg Engr (TX). *Society Aff:* AIChE, ACS.

Reeves, Edward D
Home: 10 Euclid Ave. Apt 602, Summit, NJ 07901
Position: Chmn of the Bd. *Employer:* Reeves Enterprises Inc. *Education:* BA/Chemistry/Williams College. *Born:* July 26, 1909 NYC. Employed by Exxon Corp affiliates as chem engr 1930- 44 & exec 1944-64. Exec V P Exxon Res & Engrg Co 1944-58, 1962-64. Currently Bd Chmn Reeves Enterprises Inc & P Chmn Union Coll Cranford N J. P Pres Indus Res Inst; P Chmn - Dirs of Indus Res & Soc of Chem Indus. Author Mgmt of Indus Res, Reinhold 1967. Major current activities: cons on mgmt & use of tech, & working on dev of optimum community coll sys. *Society Aff:* ACS, AIChE, SAE, API.

Reeves, James W
Business: USL Box 42251, Lafayette, LA 70504
Position: Dean of Engrg *Employer:* Univ of Southwestern LA *Education:* PhD/CE/Univ of AZ; MS/CE/LA State Univ; BS/CE/LA State Univ *Born:* 10/31/31 Native of Covington, LA. Chance Vought Aircraft Co prior to the Korean Conflict. LTJG in Navy Civil Engr Corp 1953-56. Served as Co Commander and Projects Officer, in charge of several construction projects in the Philippine Islands and as Maintenance Officer at Corpus Christi Naval Air Station. Structural Designer for the LA Dept of Public Works 1956-57. With a consltg group 1957-59, working primarily in subdivision development and design of street and highways. Since 1959, with the Univ of Southwestern LA. Head of the Dept of Civil Engrg 1969-72. Dean, Coll of Engrg since 1972. *Society Aff:* ASCE, NSPE, ASEE, NCEE.

Reeves, Robert G
Home: 4040 Lakeside, Odessa, TX 79762
Position: Consultant, Engrg. & Geol. *Employer:* Self Employed *Education:* PhD/Geology/Stanford; MS/Geol-Geophysics Option/Stanford; BS/Mining Engg/NV (Reno). *Born:* May 30, 1920. Professor of Geology, 1978-85. Chrmn, Earth Sci, The Univ of TX of the Permian Basin, 1978-83. Dean, Coll of Sci & Engg 1979-84; Director, Division of Engineering, 1984-5; Staff Scientist, US Geol Survey-EROS Data Ctr 1973-78. Prof of Geol Colo School of Mines 1969-73; Univ Rio Grande Sul Brazil 1960-62. Geologist U S Geol Survey 1950-69 (econ geol, mineral resources, remote sensing). Editor-in-Chief 'Manual of Remote Sensing'; recipient Autometric Award 'for the outstanding tech publ on photographic interpretation in 1975'; Dir Remote Sensing & Interpretation Div 1972 Amer Soc of Photogrammetry. Colonel, Retired, Army Engr Corps Reserve; served as Infantry & Signal Corps officer (to Major) 1942-46. Registered Professional Engineer, Texas. Consulting engr & geologist, 1985- *Society Aff:* AIME, ASP, AAPG, SEG, GSA, NSPE/TSPE, SIPES, SBG

Reeves, William G, Jr
Home: 1021 Hillsboro Mile, Hillsboro Beach, FL 33062
Position: Vice President. *Employer:* Miami Frozen Beverages. *Born:* Feb 1926, Cartersville GA. BSAE Univ of Ga. 15 years with J I Case Co as trainee, Dist Mgr, Sales Promotion Mgr, Asst Branch Mgr & working with William Criswell developed & named the Draft-O-Matic hitch presently marketed by J I Case Co; '7 years with Perfect Circle Corp & Dana Corp as Zone Mgr; 1970-73 Genl Mgr Crane Carrier Co; 1973-74 Genl Mgr Atlanta Truck & Parts Co; 1975 assumed V Presidency of Miami Frozen Beverages Inc as Ch Operating Officer, respon for creation of all policy & programs to insure a successful oper. Pres Natl Council of Student Branches ASAE, 1948; Chmn 1970, V Chmn, Secy-Treasurer Ga Sect ASAE; Chmn 1972-73 Southeastern Region ASAE, also Secy-Treasurer. Appreciation Award, The Agri Alumni Assoc Univ of Ga; Dist Dir (5 counties) Agri Alumni Assoc 1957- 59 & 1972-74; Dir Atlanta Metro Agri-Business Council 1972-74. Silver Beaver A.

Reggia, Frank
Home: 5227 N. Garden Lane, Roanoke, VA 24019
Position: Elec Engr *Employer:* Dept Army (Retiree) *Education:* MS/EE/Bucknell Univ; BS/EE/Bucknell Univ; Grad Cert/Radar Eng/Naval Research Lab (USN) *Born:* 10/30/21 He is a Fellow of IEEE 1968. Received his BSEE (cum laude) and MSEE from Bucknell Univ, where he was elected a member of Natl Engrg Hon Soc (Tau Beta Pi). He was also elected a Fellow of the Washington Acad of Sci in 1971 and a Fellow of the American Association for Advancement of Sci in 1972. Author of more than 45 published technical reports in the microwave field, and has been issued 22 US and Canadian patents. Awards received by him include Superior Accomplishment Awards from the Natl Bureau of Standards and Harry Diamond Labs (HDL), Dept of Army Fellowship to Bucknell Univ (1970), Army Res & Dev Award (1976), HDL Inventor of the Yr Award (1977), and Federal Retiree of the Yr Award (1978) for "Outstanding Contributions to Public Service–. *Society Aff:* IEEE, ΤΒΠ, S-MTT

Reh, Carl W
Business: 222 S Riverside Plaza, Chicago, IL 60606
Position: Partner. *Employer:* Greeley & Hansen. *Education:* BS/CE/IIT. *Born:* Nov 4, 1917 Chicago Ill. 1940 Dist Sanitary Engr with Ill Dept of Public Health; Asst

Reh, Carl W (Continued)
Hd Sanitation Sect, Bureau of Yards & Docks, Navy Dept; since 1945 with Greeley & Hansen, Engrs, Partner since 1957. Clients include cities of Chicago, Highland Park, Oak Park, Westchester, Hinsdale Ill; New York; Alexandria & Richmond Va; Ashland Wisc; Fairfax County Water Authority, Alexandria Va; Sanitation Authority, Westchester County NY; States of NY & Mich; Corps of Engrs & sev industries - all projects related to water supply & pollution control. Fellow ASCE & ACEC; Diplomate AAEE; Mbr NSPE, WPCF, AWWA. Reg PE in Ill & 6 other states. *Society Aff:* ASCE, AAEE, ACEC, NSPE.

Reich, Bernard
Business: PO Box 5, Farmingdale, NJ 07727
Position: Unit Mgr *Employer:* SEMCOR *Education:* BS/Physics/CCNY; -/Elec Engg/Rutgers. *Born:* 1/7/26. BS Physics, CCNY in NY City. Grad Rutgers Univ. From 1948-1981 Physicist Supv Physical Scientist and Electronics Eng USA Electronics Command, with exception of 1961-62 Pres of Molecular Electronics Inc; currently Unit Mgr at SEMCOR. Fellow IEEE 1973; Fellow & Chartered Engr IEE (UK); Chmn NATO Grp of Experts on electronic Parts 1973-1979; Chmn NATO Special Working Group on Semiconductors & Microelectronics; 1970-1979. Chmn Joint Commanders, JTCG-RAM 1973-79. Decoration for Meritorious Civilian Service 1981, listed in: Who's Who in the East; Leaders in Electronics, Who's Who in Technology Today; International Who's Who in Engineering. *Society Aff:* IEEE, IEE

Reich, Herbert H
Business: 6 E. Mall Plaza, Carnegie, PA 15106
Position: President *Employer:* H. H. Reich Consulting Engrs, Inc *Education:* BS/Plumbing, Heating & Ventilating/Carnegie Mellon Univ *Born:* 07/19/20 Pres of Consltg Engrg Co since 1955. Reg in OH, PA & WV. Past Chrmn of Intl Activities Committee of ASHRAE. ASHRAE Ambassador-at-Large to Israel in 1977. Guest lecture at meeting of Israel Soc of Heating, Refrigeration and Air Conditioning Engrs in 1977. Currently on Bd of Dirs of PEBA (State Chapter of NEBB). Service award in 1971 by American Arbitration Association. Certified and registered as a Fallout Shelter Analyst. Certified by NEBB to perform and manage Testing and Balancing and to supervise Environmental Testing and Balancing. *Society Aff:* ASHRAE, NSPE, ACEC, NEBB

Reich, Ismar M
Home: 2136 Holland Way, Merrick, NY 11556
Position: Vice President. *Employer:* Chock Full o'Nuts Corp. *Education:* MChE/Chem Engg/Polytechnic Inst of Brooklyn; BChE/Chem Engg/CCNY. *Born:* Aug 1924. Tau Beta Pi; MChE 1955 Polytechnic Inst of Brooklyn. 1945-60 Standard Brands Inc, process dev; 1960-69 Coffee Instants Inc, V Pres Mfg; 1969- , Chock Full o'Nuts Corp, Technical Director. Food engrg, R&D, tech mgmt, extraction, dehydration, agglomeration & instrumentation; 8 patents in coffee & tea processing. AIChE, currently Secy of Long Island Sect; ACS; ACI; Inst of Food Technologists; NY Acad of Sci; AAAS & Amer Mgmt Assoc. *Society Aff:* AIChE, ACS, IFT, AIC, NYAS, AMA, AAAS.

Reichard, Edward H
Business: 959 N Seward St, Hollywood, CA 90038
Position: V Pres Tech Planning *Employer:* Consolidated Film Industries. *Born:* Feb 21 1912. ME from Stevens Inst of Tech, 1933, Hoboken, NJ. 1933-36 Design Engr Consolidated Film Industries, Div Republic Corp, NJ; 1936-67 Ch Engr (CFI) Hollywood, Calif; 1967-79; V Pres Engrg 1979- V Pres Tech Plan; Motion Picture Film & Television Processing Lab. Reg PE Cal M-3144. Tau Beta Pi, Natl Honorary Engrg; Fellow Award SMPTE 1952, Bd of Governors 1960-69, and 1980, V Pres Motion Picture Affairs 1970-79; Sections VP 1981; Mbr AMPAS, NATAS; Assoc Mbr ASC. Recipient of 11 scientific & tech awards for outstanding achievement from Acad of Motion Picture Arts & Sciences at their annual Oscar presentations. Received SMPTE Premier Award 1976 Progress Gold Medal. Married Jan 1936 to Alice J Quigley; 3 daughters, 7 grandsons, 1 grandaughter. Hobby: golf (a hacker).

Reichl, Eric H
Home: POB 786, Greenwich, CT 06836
Position: Consultant *Education:* MS/Chem Eng/Technische Hochschule-Vienna. *Born:* 12/3/13. Married, 2 daughters. 1938 Babcock & Wilcox Co; 1938-44 Winkler-Koch Engrg, res plant design const opers; 1944-46 Stanolind Oil & Gas (now Standard Oil) of Ind, process res on synthetic liquid fuels included tour of duty with US Navy as civilian tech to evaluate German synthetic oil Ind 1945; 1946-48 CA Res Corp (Standard Oil of CA), process res petrochemicals; 1948-54 Consolidation Coal Co Res Mgr, 1954-62 Dir Res, 1962-74, VP Res. 74-78 Pres Conoco Coal Dev't Co. Retired 1/1/79. ACS AICHE; Mbr Natl Acad of Engg; Mbr Energy Research Advisory Bd/DOE; Fossil Energy Advisory Comm/DOE. *Society Aff:* AIChE, ACS, NAE.

Reichle, Alfred D
Business: PO Box 2226, Baton Rouge, LA 70821
Position: Engg Advisor. *Employer:* Exxon Res & Dev Labs. *Education:* PhD/ChE/Univ of WI; MS/ChE/Rice Univ; BS/ChE/Rice Univ. *Born:* 12/19/20. Pt Arthur, TX, Dec 19, 20; m 43; c 2 - ChE. Educ: Rice Univ, BS 42, MS 43; Univ of WI, PhD 48. Military: USN 44-46. Prof Exp: Chemist, Shell Dev Co 43-44, 48-57; Staff Asst, Phillips Pet Co 57-59; Engr, Exxon Res & Dev Labs 59-60, Sr Res Engr 60-62, Engg Assoc 63-67, Sr Engg Assoc 67-71, Engg Advisor 71-present. Mbr Am Inst of Chem Engrs. *Society Aff:* AIChE.

Reid, Robert C
Business: 66-540 MIT 77 Mass Av, Cambridge, MA 02139
Position: Professor Chemical Engineering. *Employer:* Mass Inst of Tech. *Education:* ScD/Chem Engg/MIT; MS/Chem Engg/Purdue; BS/Chem Engg/Purdue; BS/Mech Eng/US Merchant Marine Acad. *Born:* June 1924, Denver Co. 1942-46 USNR; 1954-56 Dir of MIT Engrg Practice Sch, 1956-62 Dir of all MIT Practice Schs, currently Emeritus Professor of Chem Engrg. Dir AIChE 1969-71; Institute Lecturer 1967; Ed AIChE Journal 1970-76; Warren K Lewis Award 1976; ASEE Chem Eng Lectureship 1977. Olaf A. Hougan Professor in Chemical Engineering, Univ. of Wisconsin, Madison, 1980-81; Nat. Acad. of Eng, 1980. Enjoy mountain climbing & botany. *Society Aff:* AIChE.

Reid, Robert L
Business: 1515 W. Wisconsin Avenue, Milwaukee, WI 53233
Position: Dean of Engrg *Employer:* Marquette Univ *Education:* Ph.D./Mech. Engr./Southern Methodist University; M.S. /Mech. Engr./Southern Methodist University; B.S.E./Chem. Engr./University of Michigan *Born:* 05/20/42 Native of Flint, MI. Was research engineer for ARCO from 1964-65 and Union Carbide, 1966-68. Professor at Cleveland State Univ for 2 years and Univ of TN, Knoxville, for 11 years. Chairman of the Mech & Indus Engrg Dept at Univ of Texas at El Paso from 1982-87. Founded TN Energy Conservation in Housing Program in 1975. Founded and directed UTEP Manufacturing Engrg Consortium and El Paso Solar Pond Project, both in 1983. Chairman of ASME's Solar Energy Div, 1983-84. Awarded Fellow, ASME, 1985; Engr of the Year in El Paso, 1986. *Society Aff:* ASME, ASHRAE, NSPE, ASEE, ISES

Reid, William T
Home: 2470 Dorset Rd, Columbus, OH 43221
Position: Ret. (Consultant) *Education:* BSChE/-/Univ of WA. *Born:* Feb 14 1907, Racine Wisc. 1929-46 US Bureau of Mines, Pittsburgh - coal combustion, coal-ash slags & corrosion in boiler furnaces; 1946 Scientific Cons USGS, Japan; 1946-72 (retired) Battelle Memorial Inst - combustion, graphic arts, fuel cells & energyconversion systems. Prime Movers Award, ASME/EEI 1965; Percy Nicholls Award, ASME/AIME 1968; Melchett Medal, Inst of Fuel (London) 1969. V Pres Res, ASME 1971-72. Honorary Member, ASME, 1983. George Westinghouse Gold Medal, ASME, 1982. *Society Aff:* ASME, AAS, Inst Engg.

Reider, James E
Business: 100 E Broad St 902, Columbus, OH 43215
Position: President. *Employer:* International Trade Group of Oh Inc. *Born:* Jan 18 1928, Marion Oh. BS 1951 & MSc 1951 The Ohio St Univ. Married Jeannie Johnson 1952, 2 sons Brent Carleton & Carson Robert. 1953-74 Industrial Nucleonics, Grp V Pres. Pres Upper Arlington Civic Assn 1973; Bd of Trustees Silver Bay Assn, Glenn Fall NY. USNR 1941-46; USMCR 195153. Recipient Distinguished Alumnus Award, Engrg Coll OSU 1972. Rotarian. Num patents.

Reiersgaard, William L
Home: 6509 SE 29th, Portland, OR 97202
Position: Pres *Employer:* Bagaard Automation Inc *Education:* BSME/Automotive/OR State Univ; MBA/MFGR/Univ of Portland *Born:* 4/10/38. He joined Hyster Co of Portland, OR as a Design Engg, leaving to enter the US Air Force where he served as a Vehicle Maint Officer. He completed his military commitment while stationed at Goose Bay, Labrador, Canada, leaving with the rank of Capt, and resumed his employment with Hyster Co where he progressed to Manufacturing Project Engg. He then served as Manufacturing Manager for Hytorq Corp before joining Cascade Corp, manufacturers of hydraulically actuated attachments for lift trucks. He joined Freightliner Corp as Vehicle Engg Manager and was promoted to Dir of Manufacturing Engg in 1974. He was named Chief Engg in 1977. Became Pres of Bagaard Automation Inc in 1982. He was a mbr of Pi Tau Sigma and Sigma Tau, Engineering Honoraries as well as having been elected to the National Honor Soc. He is a holder of a US Patent on an automotive differential and is a Reg PE, CME in the field of robotics and has served on the Natl Bd of Dirs for SAE. *Society Aff:* SAE, SME, PACEE

Reiffman, Norman L
Business: 600 Hempstead Turnpike, W Hempstead, NY 11552
Position: Partner. *Employer:* Reiffman & Blum. *Education:* BCE/Civ Engg/Cooper Union. *Born:* 7/10/32. He organized his own consulting firm in 1961 and in 1966 was joined by George Blum, PE, to form the firm of Reiffman and Blum, specializing in the structural design of commercial, indus and institutional bldgs. He is also highly experienced in the field of Structural Steel Detailing and has been an officer in the Natl Inst of Steel Detailers. Mr Reiffman holds PE licenses in the states of NY, NJ, PA, FL, IL, MI, LA, OH and MA. *Society Aff:* ASCE, NSPE, AWS, CEC.

Reimer, Paul H, Jr
968 Postal Road, Allentown, PA 18001
Position: Prin. *Employer:* Reimer & Fischer Engrg, Inc *Education:* MS/Structural/Lehigh Univ; BS/Civ/Lehigh Univ. *Born:* 1/3/38. Native of Allentown, PA. Field and office Engr (Special structures) with Chicago Bridge & Iron Co, Oakbrook, IL; Structural Engr with Fuller Co, Catasauqua, PA; Willis & Paul Corp, Netcong, NJ; Lehigh Structural Steel Co, Allentown, PA; Chief Structural Engr with G Edwin Pidcock Co, Allentown, PA before opening private practice (1974) under name of Reimer & Fischer Engrg, Inc, Bethlehem, PA. Mbr of Chi Epsilon Honorary CE Fraternity, Past Dir and Pres of Lehigh Valley Sec American Soc of Civ Engr, member-Engineers Week Joint Planning Council, member township Building Code Board of Appeals. Past PTA & Swim Club Pres, Active Boy Scouts of America and active churchman. Enjoy music, reading, sailing, and scuba diving. Lincoln Welding Contest Honorable Mention. Reg PE PA, NY, NJ, MD, AL, OH, WA, MO, VA, GA, KY, CT, DE & MA. Member of Board of Dir. of Lehigh Valley Industrial Park and Episcopal Housing of Lehigh Valley, Inc. *Society Aff:* ASCE, ACI, NSPE, AASDO.

Rein, Harold E
Home: 345 Victory Blvd, New Rochelle, NY 10804
Position: Pres & CEO *Employer:* Systems Planning Corp *Education:* BS/Civ Engg/Purdue Univ. *Born:* 4/11/26. Native of Wilkes-Barre, PA. Served with US Army Air Force 1973-1945. Began work for Frederic R Harris, Inc in '50: Field Engr '50-'52; Design Engr '52-'54; Resident Engr '54-'56; Proj Mgr '56-'58; Asst VP '58-'59; Partner Frederic R Harris Assoc '60-'62; VP Dir Marketing '62-'65; Sr VP Western Hemisphere Operating Group '66-'69; Exec VP '69-'72; Pres & Chief Exec Officer and Dir '72- '80. Totally responsible for policy decisions in finance, marketing, planning and mgt direction. Pres of Cahn, Inc May 1980 to Aug 1981. Apptd Pres and CEO at Systems Planning Corp in Sept 1981. *Society Aff:* ASCE, NSPE, SAME, IRF, IBBTA.

Rein, Robert J
Business: 2531 Jefferson Davis Highway, Arlington, VA 22202
Position: VP *Employer:* Columbia Research Corp *Education:* BS/IE/General Motors Inst; BS/ME/US Naval Postgrad Sch *Born:* 10/15/35 Native of Pennington, NJ. Employed as an industrial engr Ternstedt Div, GMC, Trenton, NJ. Engrg Duty Ofcr in US Navy, 1959-1979. Active in aircraft carrier design and maintenance. Ship Design Mgr for several light carrier designs. With Columbia Research Corp since 1979 as Dir for Fleet Support Sys. Directs support for several major ship acquisition progs. Mbr of CRC's exec and strategic planning committees. Asst Secy-Treas, ASNE. Chaired ASNE's first West Coast Technical Symposium, 1976. Active in church and community affairs. Former trustee of Burke Centre Conservancy; past Dir of Robert Pierre Johnson Housing Corp, Commissioner, Fairfax County Redevelopment and Housing Authority. Mbr, Bd of Dirs, Yorkville Corp. *Society Aff:* ASNE, USNI, SOLE

Reinhardt, Gustav
Home: 13972 Lake Lure Ct, Miami Lakes, FL 33014
Position: Chief Met. *Employer:* Eastern Airlines. *Education:* MSc/Met/Niagara Univ; BSc/Phys/Kent State Univ. *Born:* 8/21/27. Responsible for all met analyses at Eastern Airlines and consultant to aerospace & general industry in areas of failure analyses, mtls selection, heat- treating, welding, coating, corrosion & nondestructive testing. Formerly (1967- 70) Scientist & Sr Quality Engr, Lockheed GA Co, as principal investigator in graphite reinforced composites & light weight armor mtls; (1961-67) Res engr at NASA - powder met methods for composite mtls; (1952-61) Union Carbide Metals Co in alloy dev & welding res. Hold several US patents; published numerous technical papers. Served as Chrmn of ASM Miami Chapter and Chrmn AWS South Fla Section. Founder/Editor ASM FL State Yrbook. *Society Aff:* ASM, AWS, EMSA.

Reinhart, William A
Home: 12504 S E 14th, Bellevue, WA 98005
Position: Technology Chief. *Employer:* Boeing Commercial Airplane Company. *Education:* MS/Mech Engr/OR State Univ; BS/Mech Engg/OR State Univ. *Born:* Nov 29 1925, Dallas Tex. 1949-52 taught Theoretical & Applied Mechanics at Iowa St Univ; 1952-59 engr in jet transport dev programs at The Boeing Co, 1959-62 Hd of commercial airplane propulsion res, 1962-66 Hd of supersonic transport propulsion/mechanical staff; 1966-72 Ch Engr propulsion/noise staff, 1972-77 Dir 707/727/737 Div Technology Staff. 1977-81-Technology Chief, 707/727/73 Div Product Dev. 1981-85 Technology Chief, 737-300 Program. 1985-Technology Chief, Operations & Laboratories, Boeing Commercial Airplane Co. Engineering Dept. Recip Charles Manly Memorial Award, SAE 1965; Assoc Fellow AIAA. Patentee in field. Reg PE in Wash. Married, wife Dorothy & 3 children. Enjoys travel & outdoor activities. *Society Aff:* AIAA.

Reinig, L Philip
Business: PO Box 410, 1650 Trinity Dr, Los Alamos, NM 87544
Position: President *Employer:* Los Alamos Technical Assocs *Education:* BS/EE/Gonzaga Univ *Born:* 01/06/25 Worked for the Gen Elec Co becoming engrg mgr of nuclear power plant systems. In 1965, he went to Los Alamos Scientific Lab as head of their Engrg Dept, and in 1976 founded Los Alamos Technical Assocs, Inc, an engrg consltg firm specializing in technical services in the fields of energy, environment and natl security. Reinig and his wife have six children and four grandchildren. He enjoys skiing, hunting and fishing. *Society Aff:* ANS, IEEE, NSPE

Reintjes, Harold
Business: 200 Broadacres Dr, Bloomfield, NJ 07003
Position: Pres. *Employer:* Petrocarb, Inc. *Education:* BS/ChE/Rose Hulman Inst, Terre Haute, IN. *Born:* 3/28/13. Native of Terre Haute, IN ... Resident of NJ since 1951. 1935-36 Employed by Commercial Solvents Corp. 1936-42 Employed by Corn Products Refining Co. 1942-46 Served in US Armed Forces. 1946-51 Great Lakes Carbon Corp (Assoc Dir, Dev-Res & Dev Div). 1951-present President and Chrmn of Petrocarb, Inc, and Petrocarb International, Inc, privately owned specialty engr, equipment and matls sales co. Profl Engr in NY, IL & WY. Inventor and author of many pats in process industry. Many outside interests, including golf and platform tennis. *Society Aff:* ACS, ISS, AIChE.

Reiser, Castle O
Home: 4224 E Mitchell Dr, Phoenix, AZ 85018
Position: Prof Emeritus of ChE *Employer:* Arizona State Univ. *Education:* PhD/Chem Engg/Univ of WI; PE/Refining/CO School of Mines; BS/Chemistry/CO State Univ. *Born:* Dec 21, 1912 Berthoud Colo. Petro Engr Colo Sch of Mines 1938. Prof of Chem Engg AZ State Univ 1958-80 , Prof & Chmn of Chem Engrg Ariz State Univ 1958-77, Seattle Univ 1956-58, Univ of Idaho 1947-53. Assoc Prof Okla State Univ 1946-47, Univ of Colo 1945-46. Res Engr Std Oil Dev Co 1938-41; Fellow & P Chmn Ariz Sect AIChE; P Chmn Ariz Sect ACS; P Chmn Idaho Sigma Xi Soc; Amer Soc for Engrg Educ; Tau Beta Pi; Evaporation Suppression Field Tests of Water Planning for Israel 196768. Amer & foreign pats. *Society Aff:* AIChE, ACS, ASEE.

Reisinger, Robert R
Home: 2515 Midpine Dr, York, PA 17404
Position: Ch Engr. *Employer:* Acco Indust, Inc. *Education:* Assoc/Mech Design/PA State Univ. *Born:* 4/21/35. Assoc degree PA State Univ. Reg PE PA & NJ. With Acco (Hoist & Crane Div) since 1960 spec in dev & design of elec hoists & cranes. Held pos of Design Engr, Proj Engr, Customer Relations Mgr, chief Engr & assumed current respon as mgr of Engg 1979. Author articles & papers & speaker on subjects of hoist selection methods & hoist inspection & safety. Mbr ASM, ASME, NSPE. Served on ASM Metals Engg Inst Ctte & Educ & Dev Council. Instructed PE refresher courses & in PA State evening prog. Current delegate on ANSI B30 ctte & ASME ctte on Hoists Overhead. *Society Aff:* ASM, ASME, NSPE.

Reissner, Eric
Business: B-010 UCSD, La Jolla, CA 92093
Position: Prof of Appl Mech & Math Emeritus. *Employer:* Univ of Calif San Diego. *Education:* PhD/Math/MIT; Dr Ing/Civil Eng/Techn Univ Berlin; Dipl Ing/Appl Math/Techn Univ Berlin. *Born:* Jan 1913. Native of Aachen, Germany; came to U S 1936, naturalized 1945. Assoc with MIT Math Dept 1939-69. With Univ of Calif San Diego since 1970. Known for contrib to the literature on plates & shells, variational methods, & unsteady aerodynamics. Mbr AMS, NAE, Intern Acad Astronautics; Fellow Amer Acad of Arts & Sci, Amer Acad of Mech, AIAA, ASME. Guggenheim Fellow 1962; von Karman Medal ASCE 1964; Timoshenko Medal ASME 1973, Structures Award AIAA 1984. *Society Aff:* ASME, AIAA, AMS, NAE.

Reistle, Carl E, Jr
Business: 1100 Milam Bldg 4601, Houston, TX 77002
Position: Investments & Business Cons. *Employer:* Self. *Education:* BS/Chem Engg/Univ of OK. *Born:* June 26, 1901 Denver. Petro chemist U S Bur of Mines 1922-29, petro engr 1929-33; Chmn E Texas Engrg Assn 1933-36; Engr in charge Humble Oil & Ref Co 1936-40, Ch Petro Engr 1940-45, Genl Supt Production 1945-46, Mgr Production Dept 1946-51, Dir 1948-51, Dir-in-Charge Production Dept 1951-55, V P 1955-57, Exec V P 1957-61, Pres 1961-63, Chmn/Bd Ch Exec Officer 1963-66, retired 1966. Cons & bus investments 1966- ; Dir Eltra Corp, Dir, Eltra Copr 1966-79; Reed Tool 1972 (merged w/Baker 1975) 1977; Olinkraft 1967-78, Baker Internatl; Chmn/Bd Olinkraft Inc; Dir Univ of Okla Res Inst, Norman. Anthony F Lucas Golf Medal, Amer Inst Mining, Met & Petro Engrs 1958. Mbr Amer Petro Inst, Amer Inst Mining, Met & Petro Engrs (Pres 1956), Petro Club, Sigma Xi, Tau Beta Pi, Sigma Tau, Alpha Chi Sigma. Clubs: Ramada & River Oaks Country Club. Num articles to prof journals. *Society Aff:* AIME.

Reitinger, Robert L
Home: 7419 New Second St, Philadelphia, PA 19126
Position: Associate. *Employer:* Rohm & Haas Co. *Education:* EE/Elect Engg/Univ of PA; BS/Elect Engg/Univ of PA. *Born:* Oct 1913. Reg P E Penna 1944. Rohm & Haas Co since 1935, design, const, maintenance of domestic & foreign chem plants. Pres Penna Soc of Prof Engrs 1966-67, Pres Phila Chap 1958-59. Pres Natl Soc of Prof Engrs 1973-74. Mbr IEEE & Engrs' Club of Phila. Dir EJC 1976-80, Dir ECPD 1975-79, Secy of Council 1976. Awards incl Citation from Penna House of Reps for prof service 1974; Dedicated Serv Award PSPE 1974; Engrg News Record Citation for serv to the const indus 1974. *Society Aff:* NSPE, IEEE.

Reitz, Henry M
Business: 1040N. Lindergh, St Louis, MO 63132
Position: President. *Employer:* Reitz & Jens Inc. *Education:* MS/Soil Mech/Harvard Univ; BS/Civil Engrg/WA Univ. *Born:* 3/29/22. Native of St Louis, MO. Army Ground Forces, 1942-45; Horner & Shifrin Consulting Engrs, Jr Partner 1948. Grad Sch 1948-49; faculty WA Univ, Asst Prof in Civil Engg 1949; prof & Dept Chrmn 1954-58. Consulting Engr in Soil Mechs & Fnds and Drainage. incorporated Reitz & Jens, Inc in 1969. Special personal interests; Geohydrology fnds, expansive soil behavior, lateral support, aspects of marginal terrain dev, structural properties of hydraulically placed fills, solid & hazardous wastes disposal, riverine alluvial plains and flooding phenomena - probabilistic and/or deterministic natural phenomena, ground water, evaluation of economic effects of rarer natural pheonomena, contractor advisory in heavy construction, differing site conditions, Forensic Engineering. *Society Aff:* ASCE, AGU, ACEC, ASTM, BOCA, Urban Land Institute

Reklaitis, Gintaras V
Business: Purdue University, Sch of Chem Engg, W Lafayette, IN 47907
Position: Prof & Dean of Engg, Research *Employer:* Purdue Univ. *Education:* PhD/CE/Stanford Univ; MS/CE/Stanford Univ; BS/CE/IIT. *Born:* 10/20/42. Born in Posen, Poland. Naturalized US citizen; married, two children; PhD 1969; NSF postdoctoral Fellow, Inst fur Operations Res & EDV Zurich, Switzerland; at Purdue Univ since 1970; Fulbright-Hayes lectr USSR, Spring 1980. AIChE Computing in Chemical Engrg Award 1984. Co-editor of *Computer Applications to Chemical Engineering Process Design & Simulation and Proceedings*; *International Symposium on System Engineering & Analysis*; author of "Intro to Material and Energy Balances" Wiley, 1983; co-author of "Engrg Optimization" Wiley-Interscience 1983. Co-editor-in-chief, *Computers & Chemical Engg (a Pergamon Journal) 1986- . Publications in nonlinear optimization, simulation, and process modelling. Lectr in flowsheet computations, engg optimization, process design & scheduling. Res supported by Reilly Fdn, NSF, ONR, ERDA, DOE, and private industry. Active res interests in design, scheduling, and simulation of batch/semi-continuous processes, and computer aided design and optimization. Society Aff:* AIChE, ACS, ORSA, MPS, AAAS, ASEE.

Remer, Bertram R
Home: 10602 Pinedale Dr, Silver Spring, MD 20901
Position: Sr Program Director *Employer:* ORI, Inc. *Education:* BSEE/Communications/Duke Univ. *Born:* July 16, 1924. BSEE Duke Univ. Pi Mu Epsilon. Lived in Wash D C area since 1930. Entered Navy 1943; discharged 1947 as Ensign. 3 1/2 yrs Patent Examiner in telegraphy. 5 yrs Naval Ordnance Lab, White Oak Md, in proximity fuze design & test. In 1956 began work at Bureau of Ordnance as Sidewinder Missile Fuze Proj Engr. 1959 became Sect Hd for dev of all Navy air launched weapon fuzes. In 1966 became Weapon Sys Proj Engr for Std Arm Missile. Took same job in 1970 on Harpoon Missile. Silver Medal from Assn of Scientists & Engrs of the Navy Dept for at least 7 yrs of 'Outstanding Contribs to

Remer, Bertram R (Continued)
Naval Engrg'. Retired from Fed. Gov't in 1979. Work for ORI, Inc. as Senior Program Director.

Remer, Donald S
Business: 12th St and Dartmouth, Claremont, CA 91711
Position: Oliver C. Field Prof of Engrg *Employer:* Harvey Mudd College. *Education:* PhD/ChE/Caltech; MS/ChE/Caltech; BSE/ChE/Univ of MI. *Born:* 2/16/43. Chem Engr with interest in the 4 E's - economics, energy, environment, and education. The Oliver C Field Prof of Engg. Previously with Exxon as task force leader, div coordinator, process and proj engr and economic analyst. Earlier position with Marathon Oil, Scott Paper, Allied Chem, Std Oil of CA, and Jet Propulsion Lab of Caltech. Founding partner of the Claremont Conslt Group - an engg and business mgt firm. Reg PE. Case Study editor for the journal "The Engineering Economist–. Frequent contributor to profl and tech journals and meetings. Married; three children. Likes jogging and aerobics. On the Editorial bd of the Journal "Engrg Costs and Production Ecomonics–. On the Advisory Council of the Nat'l Energy Foundation. *Society Aff:* AIChE, ASEE.

Remick, Forrest J
Business: 114 Kern Graduate Building, The Pennsylvania State Univ, University Park, PA 16802
Position: Assoc V P for Research Professor of Nuclear Engrg. *Employer:* Penn State Univ. *Education:* PhD/Mech Engrg/PA State Univ; MS/Mech Engrg/PA State Univ; BS/Mech Engrg/PA State Univ; Diploma/Nuc Eng/Oak Ridge Sch of Reactor Technology. *Born:* 3/16/31. Grad of the Oak Ridge Sch of Reactor Tech. Presently Assoc. VP for Research and Professor of Nuclear Engrg. Formerly engr with Bell Tele Res Labs, Dir of Nuclear Reactor Facil, Dir of Curtiss Wright Nuclear Res Lab & Ch of the Training Sect of the Intl Atomic Energy Agency. Director of Inst for Sci & Eng, Acting Dir of the Ctr for Air Environ studies. Formerly, Dir of Intercollege Research Programs. Fellow ANS. Formerly Mbr Nuclear Regulatory Comm's Atomic Safety ' Licensing Bd Panel, Formerly Mbr of Governor's Energy Council (Penn). Formerly Dir, Office of Policy Eval, US Nuclear Reg Comm (81-82). Presently Mbr, Adv Ctte on Reactor Safeguards, USNRC; Natl Nuclear Accrediting Bd, Inst of Nuclear Power Operations; Scientific Adv Ctte, EG&G Idaho, Inc.; Reactor Safety Advisory Committee, E.I. Dupont de Nemolls. *Society Aff:* ANS, ASME, ASEE, NCURA, NCAR.

Remley, Frederick M
Business: 400 Fourth St, Ann Arbor, MI 48103
Position: Dir Media Resources Cntr *Employer:* Univ of MI. *Education:* BS/Physics/Univ of MI. *Born:* 5/20/29. in Washington, PA. Employed in radio and tv broadcast eng by Univ of MI since 1952. Technical Dir since 1958. Have served as consultant to many universities and to US govt an various areas of audio and video tech. Have participated in a variety of natl and intl tv stds groups including ANSI C 98, several SMPTE engg committees, US delegation to CCIR meetings 1965-79. Chairman of IEC TC 60 since 1981, mbr of EBU Working Group G-2. Past VP for TV affairs of SMPTE. Active photographic hobbiest. Member, PBS Engineering Comm. *Society Aff:* SMPTE, IEEE, AES.

Remley, Marlin E
Business: Rockwell International, 6633 Canoga Ave., LA06, Canoga Park, CA 91303
Position: Tech Dept Dir. *Employer:* Energy Sys Group, Rockwell Intl. *Education:* PhD/Physics/Univ of IL; MS/Physics/Univ of IL; AB/Physics & Math/SO East Missouri State Univ. *Born:* Apr 1921. Army Signal Corps, Radar Officer, Supply Officer & Co Cdr 1942-46. With Atomics Internatl Div of Rockwell Internatl 1952- , spec in design, dev, const, oper, & safety analysis & eval of nuclear reactors for res, central station power & space power applications, & in radiation safety & special nuclear matl control. Tech Dept Dir 1958- ; current pos 1962- . Fellow ANS; Chmn Reactor Opers Div ANS 1975; Alumni Merit Award, Southeast Mo State Univ 1973. Mbr Sigma Xi, Phi Kappa Phi, Pi Mu Epsilon: Reg Prof Nuclear Engr, State of California. *Society Aff:* ANS, AIF, NMA.

Remus, Gerald J
Home: 8565 Mendota, Detroit, MI 48204
Position: Prof Engr. *Employer:* Detroit Metro Water Dept; others. *Born:* Apr 15, 1908. MSEE Univ of Mich, BSME DIT, Bus Admin Columbia. Reg P E Mich. Awards: One of Top Ten Pub Works Men 1962; Prof Engr of Yr 1965; Hon Mbr AWWA; Sesquicentennial Award Univ of Mich; Alumni Award DIT; Governor's Award 1973. Dir & Dist Chmn AWWA; Charter Mbr Detroit Engrg Soc. 44 yrs with City of Detroit, working in all phases of san engrg. Was Genl Mgr & Ch Engr of Detroit Metro Water & Sewage sys. Dev one water resource mgmt complex servicing over 4 million people wth all related indus, approx 45% of Michigan's pop. After retirement, in continuing to work for the Water Dept, Peerless Cement Co, Univ of Mich, as Tr of DIT, & on spec assignments such as improved methods of reducing sewage plant sludge, teaching, & advising on engrg problems.

Remy, Edwin D
Home: 48 Stono Dr, Greenville, SC 29609
Position: Plant Mgr, Spartanburg Plant. *Employer:* Phillips Fibers Corp. *Education:* BS/ChE/IA State Univ. *Born:* Mar 15, 1921. BS Iowa State Univ. Native of Knoxville Iowa. WW II Infantry & Chem Corps 1942-46. Opers Supr Ethyl Corp; Tech Mgr Monsanto Co; Dir Q C, Tech Dir, Plant Mgr, & present Mgr Budget & Evaluations for Phillips Fibers Corp in 1979, spec in mfg of synthetic fibers. P Sect Chmn ASQC, P Pres Spartanburg Dev Assn, V Chmn Salvation Army Bd, Bd Mbr United Way, Spartanburg Boys Home, & Indus Mgmt Assn. Hobbies: water skiing, square dance calling & teaching, photography. *Society Aff:* ASQC, AIChE, LSPE.

Renard, Kenneth G
Business: 442 E. 7th St, Tucson, AZ 85705
Position: Research Leader *Employer:* USDA Agric Res Service *Education:* PhD/CE/Univ of AZ; MS/CE/Univ of WI; BS/CE/Univ of WI *Born:* 05/05/34 Native of Sturgeon Bay, WI. Began working with USDA in a research capacity in 1957. Transferred to AZ in 1959 when he assumed responsibility as resident engr of the Walnut Gulch Experimental Watershed. Present job as Dir of USDA's Southwest Watershed Research Center in 1970. Contributes articles on hydrology/ sedimentation to USDA periodicals and soc journals. Has held numerous offices in Profl societies including his current one as Pres of the AZ Section of ASCE. *Society Aff:* ASCE, ASAE, SCSA, AGU

Rennicks, Robert S
Business: Dart & Kraft Inc, West 115 Century Rd, Paramus, NJ 07652
Position: Dir of Engr & Const. *Employer:* Dart & Kraft Inc. Chemical Plastics Group *Education:* MBA/Bus Admin/Pace Univ; BSCE/Civ Engr/VA Military Inst. *Born:* 1/13/34. Since BS in College, spent 2 yrs as officer in USAFR as Project Engr Developing Specialized Bldgs and Equip for use by USAF. Spent 8 yrs as plant engr for Amoco Oil Co, Yorktown, VA. Refinery and 5 yrs as proj engr for Amoco Chem Central Engg Dept; remaining 10 yrs served as mgr of engr services, mgr of projs and dir of engg and construction div, Dart Industries, (now Dart & Kraft) Chem-Plastics Group. Reg PE in IL and NJ. Enjoy tennis, swimming, outdoor sports and music. *Society Aff:* AIChE, EJC, HBR.

Replogle, John A
Business: 4331 E Broadway, Phoenix, AZ 85040
Position: Research Leader. *Employer:* US Water Conservation Lab, Agri Research Ser, U.S. Dept. of Agr. *Education:* PhD/Civ Engrg/Univ of Il; MS/Agr Engrg/Univ of Il; BS/Agr Engrg/Univ of Il. *Born:* 1/13/34. Native of Charleston, IL. Teaching and research, Univ of IL, 1958-63. Research Hydraulic Engg with the US Water Conservation Lab since 1964, specializing in open-channel flow metering and irrigation hydraulics. Assumed position as Research Leader for Irrigation and Hydraulics in 1974. COL in USAR, Corps of Engrs. Reg PE. Chmn, Publications Committee, Irrigation and Drainage Div, ASCE, 1973-78, and Mbr of Exec Ctte, 1984-present.

Replogle, John A (Continued)
Active in other committee activities of both ASAE and ASCE. Received ASCE Croes Medal 1977. *Society Aff:* ASCE, ASAE, AAAS.

Repscha, Albert H
Home: 27 Pine Arbor Lane Apt 206, Vero Beach, FL 32962
Position: Prof Mech Engr. *Employer:* Retired *Education:* M.S./Mech Engr/Iowa State University; M.E./Mech Engr/University of Maine; B.S./Mech Engr/University of Maine *Born:* 02/28/04 Professor of Mech Engrg at Drexel Univ, Swarthmore Coll, Los Angeles State Coll, Univ of Southwestern LA & Assuit Univ in Egypt. Research Engineer Lockheed Aircraft, Senior Design Engineer General Dynamics, Asst. Project Engineer Pratt & Whitney. Engineer U.S. Naval Civil Engrg Labs. Dir of Research Schramm Inc. Consultant, H.L. Yoh Co on air compressor dev for U.S. Air Force. *Society Aff:* ASME, ASEE

Resen, Frederick Larry
Business: P.O. Box 2678, Wilton, CT 06897
Position: President *Employer:* Larry Resen Associates *Education:* BS/CE/Univ of CO. *Born:* 7/17/23. Served USN 1941-45 (Chief Firecontrolman); Initially with Dow Corning, 1950, Oil & Gas Journal as Gulf Coast & Refining Ed; Joined AIChE in 1959 as Editor, Chemical Engineering Progress; became Editor/Publisher in 1968, Left AIChE in May, 1984, to establish industrial communications firm, Larry Resen Associates. Assumed Staff Dir's position in 1979. Served as Editorial Dir of International Chemical Engineering. Elected an AIChE Fellow in 1975. At CO was Student Body Pres & Editor, the Colorado Engineer. Wife Margaret, Daughter Emily, Son John. *Society Aff:* AIChE, ΤΒΠ.

Reshotko, Eli
Home: 2481 Wellington Rd, Cleveland Heights, OH 44118
Position: Prof of Engg and Dean, Case Ins Tech *Employer:* Case Western Reserve Univ. *Education:* PhD/Aeronautics/CA Inst of Tech; MME/Mech Engrg/Cornell Univ; BME/Mech Engg/Cooper Union. *Born:* 11/18/30. From 1951-57, was with NACA-Lewis in aerodynamic heating and compressible boundary layer studies; 1956-57 Head of the Heat Transfer Sec; 1960-64 with NASA-Lewis Res Ctr in plasma physics, MHD and elec propusion; from 1961-64 was Chief of the Plasma Physics Branch. Joined faculty of Case Inst of Tech in 1964. Is Prof of Engg since 1966, from 1970-79 served as Chrmn of the Dept of Mech and Aerospace Engg, from 1986 as Dean of Case Institute of Technology. Is consultant to industrial and govt organizations and has served on comms of professional socs and as an ECPD accreditation examiner. Is presently Chrmn of the US Boundary Layer Transition Study Group, a member of the NASA Advisory Group on Hypersonic Technology, and a U.S. Member of the Fluid Dynamics Panel of AGARD (NATO). Received the AIAA Fluid and Plasma Dynamics Award in 1980 and was elected to the Natl Acad of Engrg in 1984. *Society Aff:* ASME, AIAA, APS, AAM, AAAS.

Resnick, Joel B
Home: 10604 Trotters Trail, Potomac, MD 20854
Position: Asst VP. *Employer:* Science Applications, Inc *Education:* MSEE/Electronics/MIT; BSEE/Electronics/CCNY. *Born:* 1/24/35. Engr with MIT Lincoln Lab developing sensors & systems for strategic defense systems from 1957-1970. With US ACDA from 1970-1972 working on SALT. From 1972-1978 directed analyses in OSD (Systems Analysis) on strategic & theatre forces (offense & defense) & SALT. In 1978 joined the Intelligence Community Staff as Dir of Systems Analysis Div. In 1979 promoted to Dir, Prog Assessment Office, (GS-18), responsible for directing analysis of resource allocation across the entire intelligence community. In 1980 joined Science Applications Inc. Currently Asst V Pres and Mgr of Policy Analysis Division.

Resnick, William
Business: Dept of Chem Engg, Haifa Israel
Position: Wolfson Prof of Chem Engrg. *Employer:* Israel Inst of Tech. *Education:* PhD/Chem Engg/Univ of MI; MS/Chem Engg/Univ of MI; BSc/Chem Engg/Purdue Univ. *Born:* Sept 1922. Asst Prof Ill Inst of Tech 1947-50. Sr Proj Chem Engr Std Oil Ind 1956-60. At Technion, Israel Inst of Tech 1951-56 & since 1960: Chmn Dept of Chem Engrg 1960-73. Technion V P 1961-65, Genl Mgr Ctr for Indus Res, 1974-76. Founding mbr & first pres of Israel Inst of Chem Engrs; Fellow Inst of Chem Engrs & of Amer Inst of Chem Engrs. Sabbatical leaves at Univ of Minn 1965-66, Univ of Cambridge 1973, 1977, Technical Inst of Denmark, 1977, N Carolina St Univ 1981 UC Berkeley, UCLA, 1986-87. Israel Acad. Royal Society Visiting Research Professorship, 1977. Dir, Div of External Studies, Israel Inst of Tech 1984-86. *Society Aff:* AIChE, ACS, ASEE, IChE.

Restemeyer, William E
Business: Dept of Elect & Comp Engrg, University of Cincinnati, Cincinnati, OH 45221-0030
Position: Prof & Admin Assoc. *Employer:* Univ of Cincinnati. *Education:* DSc/Hon/Cap Inst Tech; MA/Applied Math/Univ of Cincinnati; EE/Elec Engg/Univ of Cincinnait. *Born:* 4/28/16. in Cincinati, OH. Postgrad study UCLA, MI State Univ, Univ of Cincinnati; DSc (Capitol Inst of Tech) '76. Listed in Who's Who in America, Amer Men & Women of Science, etc. Reg PE OH. Former naval officer at Naval Res Lab. Visiting Scientist for OH Acad of Sci. Cons to GE, AVCO-Crosley, Milacron & other local indus, and to gov agencies incl NASA, AEC, NATO, NSF, USOE, NY St Educ Dept. Served as officer, liaison rep & comm mbr in ASEE, ECPD, ABET, EJC, AAAS, MAA, IEEE. Served as editorial cons, referee, reviewer & contributor to num engrg journals & publishers. Former Chmn of Univ Fac & Univ Senator; recd Disting Serv Awards & currently holds joint apptment as Prof in Dept of Math Sci at Univ of Cincinnati. Officer & mbr Sigma Xi, Tau Beta Pi, Omicron Delta Kappa, Eta Kappa Nu, Alpha Sigma Lambda. *Society Aff:* ASEE.

Restivo, Frank A
Business: 3000 Lakeview Ave, St Joseph, MI 49085
Position: Mgr of Metallographic Sales. *Employer:* Leco Corp. *Born:* Native of Chgo, Ill. 1929 Ill Inst of Tech - Met. USN 1942-45 in Pacific Theatre. Sales Engr wth Buehler Ltd, spec in Metallographic Labs, for 17 yrs. Internatl Sales Mgr with Buehler Ltd for 8 yrs, respon for world mkting. Secy & Internatl Sales Mgr with Instra-Met Inc for 3 yrs. Assumed current respon 1973. Natl mbrship ctte ASM 1965-66-67. P Chmn of Chgo Chap Amer Soc for Metals 1975- 76. Enjoy music & boating.

Reswick, James B
Business: Rehab. R&D Evaluation Unit (153), VA Prosthetics R&D Center, 103 South Gay Street, Baltimore, MD 21202
Position: Dir *Employer:* Veterans Admin *Education:* ScD/ME/MIT; SM/ME/MIT; SB/ME/MIT. *Born:* 4/16/22. in Ellwood City, PA. ScD, MIT 1952, D Eng (hon) Rose Poly Inst 1968. m. Irmtraud Hoelzerkopf Dec 27, 1973. Asst Prof, Assoc Prof, Hd Machine Design & Graphis Div MIT 1948-59; Leonard Case Prof Engg, Dir, Engg Design Ctr, Case Western Res Univ 1959-70; Dir, Rancho Los Amigos Rehab Engrg Ctr, Downey, CA; Prof Elec & Biomed Engg & Community Medicine Univ of So CA; Assoc Dir for Tech, Natl Inst of Handicapped Res, US Dept of Educ; Dir, VA RR&D Evaluation Unit, Veterans Admin; engg cons on automatic control, product dev, automation & biomed engg. Mbr comm prosthetics es & dev Natl Acad Sci 1962- , chmn design & dev comm; Mbr Bd Rev Army R&D Off 1965- ; Mbr Appl Physiology & Biomed Engg Study Sect NIH 1972- . Served to LTjg USNR 1943-46; PTO. Product Engg Master Designer Award 1969; Isabelle & Leonard H Goldenson Award United Cerebral Palsy Assn 1973. Dr of Eng (hon) Rose-Hillman Inst of Tech. Fellow IEEE; Mbr NAE, IOM, AAOS, RESNA (Found Pres), Biomed Engg Soc (Sr Mbr) *Society Aff:* IEEE, RESNA, AAOS, NAE, IOM.

Retallick, William B
Home: 1432 Johnny's Way, West Chester, PA 19382
Position: Conslt Chem Engr *Employer:* Self *Education:* PhD/Chem Engg/Univ of IL; Ms/Chem Engg/Univ of IL; BSChE/Chem Engg/Univ of MI. *Born:* 1/16/25 Mbr Sigma Xi, Phi Lambda Upsilon, ACS, AIChE. Lic P E IL and PA. Phillips Petro Co

Retallick, William B (Continued)
1948-50. Consolidation Coal R&D 1953-64, economics & process res on making synthetic liquid fuels from coal. Wrote chapter entitle 'Hydroprocesses' in Encyclopedia of Chem Tech. Air Prods & Chem 1964-74, res on catalysts & catalytic processes. Joined Oxy-Catalyst (now a subsid of METPRO Inc) 1974 as VP for R&D. Since Jan 1980, Consulting Engr, Catalysts and Catalytic Processes. Enjoys rifle shooting, jogging, raquetball. *Society Aff:* ACS, AIChE, AAAS, ΣΞ, ΦΛΥ.

Reti, Adrian R
Business: Ashby Rd, Bedford, MA 01730
Position: VP. *Employer:* Millipore Corp. *Education:* ScD/CE/MIT; MS/CE/MIT; BS/CE/MIT. *Born:* 2/28/38. Native of Argentina. Consulting engr with Dynatech Corp, specializing in applied chemistry & solid-gas reaction kinetics. With Millipore Corp since 1971, working on membrane dev and applications Corporate R&D activity. *Society Aff:* AIChE, ACS, ASTM.

Reti, G Andrew
Business: 445 S Figueroa St No. 3500, Los Angeles, CA 90071
Position: Partner *Employer:* Dames & Moore. *Education:* BS/CE/MIT; MS/CE/MIT; -/Engg Mgmt/UCLA. *Born:* 3/7/30. 1954-60 worked for Creole Petro Corp in Venezuela. With Dames & Moore since 1960. Spec in fnds, dams & earth structs. Expertise in appli of computers to engrg problems, info retrieval & mgmt. Engrg Mgmt degree from UCLA 1972. Currently Partner in charge of Budgeting and Planning at Headquarters office. Reg PE CA, District of Columbia, MD, & VA. Mbr ASCE, AMA, NSPE, CSPE. Greensfelder Award 1965. *Society Aff:* ASCE, AMA, NSPE, CSPE.

Reucroft, Philip J
Business: Dept of Met Engg and Materials Sci, Lexington, KY 40506
Position: Prof. *Employer:* Univ of KY. *Education:* PhD/Chemistry/Univ of London (Imperial Coll); BSc/Chemistry/Univ of London (Imperial Coll). *Born:* 3/29/35. Natl Res Council Postdoctoral Res Fellow (1959-61). Franklin Inst Res Labs employee (1961-1969). Lab Mgr (1966-69). Joined Univ of KY as Assoc Prof of Materials Sci in 1969. Became Prof of Materials Sci in 1976. Ashland Oil Fdn Prof (1970-74). Inst for Mining and Minerals Res Fellow (1978-79). Research interests include surface properties of materials, surface and interface characterization, electronic properties of polymers. *Society Aff:* ASM, ACS, APS, Amer. Carbon. Soc., Royal Society of Chemistry.

Reusswig, Frederick W
Business: Stanley Bldg, Muscatine, IA 52761
Position: Sr V Pres. *Employer:* Stanley Cons Inc. *Education:* BS/Civil Engg/MA Inst of Technology. *Born:* 2/8/24. Native of Western, NY. Married, four children, USAAF 1943-46. Joined Stanley Cons 1963; respon for corp dev and services in mgmt and technical capabilities, market support, contracts, data processing & info services; VP 1967; Sr VP 1975; Dir 1972-77 and 1981-. Reg PE in 7 states. Exper incl civil engg, hydroelec power, proj mgmt. Fellow ACEC & ASCE (Pres Buffalo Sect 1962); Mblr NSPE, ASME (Power Test Code Ctte, Hydraulic Prime Movers 1970-75). Mbr Engg Adv Bd Univ of IA 1972-75. MIT Educ Counselor 1971-78. Mbr Mgmt Center Advisory Bd, Univ of IA 1977. *Society Aff:* ACEC, ASCE, ASME, NSPE.

Reutter, John G
Business: 9th & Cooper Sts, Camden, NJ 08101
Position: President. *Employer:* John G Reutter Assocs. *Born:* July 1910. Pres, Ch Exec John G Reutter Assocs, founded 1951 - now among top 300 in U S. CE degree Drexel. Fellow ASCE; Pres 1968-69 Cons Engrs Council of U S; Pres (2 terms) N J Cons Engrs Council; Commissioner 9 yrs N J Prof Engrs & Land Surveyors Lic Bd; Fellow Amer Inst of Cons Engrs & Natl Soc of Prof Engrs; prior to founding own firm, Ch Engr for Hungerford & Terry, leading waste treatment equip manufacturer. Lic P E, land surveyor & reg planner in 6 states & Puerto Rico. *Society Aff:* ASCE, ACEC, NSPE, AWWA, WPCF

Revay, Andrew W Jr
Business: Florida Institute of Technology, 150 W Univ Blvd, Melbourne, FL 32901-6988
Position: VP for Academic Affairs *Employer:* FL Inst of Tech *Education:* PhD/Elec Engr/Univ of Pittsburgh; MSEE/Elec Engr/Univ of Pittsburgh; BSEE/Elec Engr/Univ of Pittsburgh *Born:* 10/08/33 Native of Lower Burrell, PA. Served in the US AF 1956-59 as Navigator- Bombardier and armament-electronics officer. Taught in elec engr dept of Univ of Pittsburgh 1959-1967. FIT Melbourne, FL 1967-present. Assoc Prof-Prof of elec engrg. EE Dept Head 1971-80; ME Dept Head 1972-78, Assoc Dean for Res 1976-80. Dean of the Coll of Science and Engrg 1980-1986. VP for Academic Affairs 1986-present. Active conslt in the field of electromagnetic compatibility electromagnetic pulse, and lightning protection. IEEE Centenial Medal 1984. Married (Suzanne Roche), one son (Kenneth), one daughter (Andrea). *Society Aff:* IEEE, ASEE, NSPE/FES, ΣΞ, ΤΒΠ

ReVelle, Charles S
Business: Charles & 34th Sts, Baltimore, MD 21218
Position: Prof *Employer:* Whiting Sch of Engrg, Johns Hopkins Univ *Education:* PhD/CE/Cornell Univ; BChE/ChE/Cornell Univ. *Born:* 03/26/38 Is Prof, The Johns Hopkins Univ and Coordinator of the Operations Research Group, an interdepartmental association of Johns Hopkins faculty in disciplines relating to OR. He has written more than 80 papers on location analysis, water resources, and other topics; these have appeared in Water Resources Res, Mgmt Sci, and other journals. With his wife Penelope he is author of two textbooks: *Sourcebook on the Environment* (Houghton Mifflin, 1974) and *Environment, Issues and Choices for Society* First Edition, 1981; Second Edition, 1984, Willard Grant Press). *Society Aff:* AGU, AAAS, TIMS, ORSA, RSA.

Revesz, Zsolt
Business: PO Box 1126, Baden CH-501 Switzerland
Position: Pres *Employer:* Assoc. Cons. Engrs. (Revesz & Ass.) Cons. Serv. *Education:* PhD/Mech Engrg/Century Univ, CA; MS/Mech Engrg/Tech Univ of Budapest, Hungary; BA/Languages/Budapest, Hungary *Born:* 03/18/50 Born Hungarian, living in Switzerland from 1975. Lectured for Tech Univ, Hungary before 1975. Involved in R&D and Engrg since 1970 -- for TU Hungary before 1975, mainly for Brown, Boveri & Cie and for Electrowatt Engrg since 1975. Active in private postsecondary education from 1980. Founder and Secretary of ANS Swiss Sect. Over 30 publ (in German, English, Hungarian, Spanish, Russian) covering comp. fl heat transfer, piping analysis and computer graphics. Proceedings editor for a nuclear safety conference, author of a monograph about piping stress analysis. Presides ACE (R&A) Consult Serv mainly for numer. analysis & software development. *Society Aff:* ANS, ASME, ISCME, SGK, SVA

Rew, William E
Home: P. O. Box 35, Islamorada, FL 33036
Position: Construction Mgr *Employer:* Post, Buckley, Schuh & Jernigan, Inc *Education:* M Engr/CE/Yale Univ; BE/CE/Yale Univ *Born:* 11/24/23 Born and raised in Corning, NY. Active duty in US Army, World War II 1942- 1946. Worked as project mgr for Aramco in Saudi Arabia 1955-1961. Employed as a design and construction consltt for past twenty yrs. Currently an assoc with Post, Buckley, Schuh & Jernigan, Inc, a large south-eastern cvl engrg consltg firm. Active in engrg societies. Past pres of SAME/Canaveral Post 1974; past pres of ASCE/Cape Canaveral Branch 1975; past pres of NSPE/Forida Engrg Society/Indian River Chapter 1976. Awarded ASCE/Cape Canaveral Branch Engr of the Year 1974. *Society Aff:* NSPE, FES, ASCE, SAME

Rey, William K
Business: The University of Alabama, P.O. Box 664, Tuscaloosa, AL 35486
Position: Asst Dean. *Employer:* Univ of Al *Education:* MSCE/Civ Engg/Univ of AL;

Rey, William K (Continued)
BSAE/Aero Engg/Univ of AL. *Born:* 8/11/25. Native of NYC began teaching mathematics at the Univ of AL in 1946. Taught engg mechanics 1947 to 1952 and aerospace engg since 1952. Prof of Aerospace Engg since 1958 and, since 1976, Asst Dean for Undergrad Progs at The Univ of AL. Published res in the areas of metal fatigue and aircraft structural analysis. Mbr of: Tau Beta Pi; Sigma Gamma Tau; Pi Mu Epsilon; Omicron Delta Kappa Golden Key; Theta Tau PE Fraternity (natl officer 1957-70, Grant Regent 1963-66); NSPE Chap. Pres. 1983, State Pres. 1987; Kiwanis Club (Distinguished Pres 1979 Distinguished Lt. Gov 1986); Univ Club (Pres 1976). Active in Baptist church (deacon). *Society Aff:* AIAA, ASEE, NSPE.

Reyes-Guerra, David R
Business: 345 E 47 St, NY, NY 10017
Position: Exec Dir. *Employer:* Accred Bd for Engrg & Tech *Education:* Doc/Engrg/Lawrence Inst of Tech; MA/Engrg/Yale Univ; BA/Civ Engrg/The Citadel; Doc/Engrg/The Citadel. *Born:* 10/4/30. Born in London, England. Served as proj engr for United Fruit Co; instructor of gen engrg at Univ of IL-Urbana; exec dir of JETS; exec dir of ABET; pres of JETS; mbr Chi Epsilon, Tau Beta Pi, Citadel Hon Soc; reg PE NJ; mbr adv bds Fulbright Scholarship; OH Bd of Regents Eminent Scholars; Air Force Inst of Tech; Natl Sci Fdn SEP Review Group; mbr Engrg Manpower Comm; Comm on Educational Credit & Credentials/ACE; Comm on Recognition/COPA; US Natl Comm on UPADI, Comm on Education/UPADI, UPADI Fund; awarded Distinguished Service Awd of ASEE and Outstanding Service Awd from NY State Soc of Prof Engrs. Conslt in engrg education, intl engrg education, and civ engrg. *Society Aff:* ASCE, ASEE, NSPE

Reyhner, Theodore O
Home: Chico, CA 95929
Position: Chrmn, Div of Engg. *Employer:* CA State Univ. *Education:* PhD/Admin/NYU; MS/Civ Engg/Stanford Univ; MA/Sci Education/Columbia Univ; BS/Civ Engg/NJ Inst of Tech. *Born:* 4/19/15. Chrmn, 1978-, Hd, Civ Engg, 1967-74, Prof, Civ Engg, 1956-, CA State Univ, Chico; Assoc Prof, Civ Engg, 1953-56, MI Tech Univ; Assoc Prof, Civ Engg or Architecture, 1946-53, Univs of Denver, OR, ND; Hgwy Engr, Summers 1952-57, US Bureau of Public Rds; Res Engr, Timber Mechanics, 1944-46, US Forest Products Lab; Instr, Civ Engg, 1943-44, Lehigh Univ; Instr, Physics, 1941-43, Cooper Union Sch of Engg and Newark Coll of Engg; Mbr, Std Bldg Code Committee of American Concrete Inst, 1955-78, and Chrmn of its first Seismic Subcommittee, 1965-71. *Society Aff:* ASCE, ACI, ASEE.

Reymann, Charles B
Home: 4125 Crescent Rd, Birmingham, AL 35222
Position: Chief Materials Lab *Employer:* Hayes Internatl Corp. *Education:* BS/Indus Mgmt/Sanford Univ. *Born:* 7/1/24. Studied Mech Engrg at Johns Hopkins Univ; BS Sanford Univ in Indus Mgmt. Courses taken in met & indus pollution contrl. Have taught welding, report writing, & matls engrg. With Hayes Internatl Corp since 1951, working in met - lab dir, facil engrg, matls & process engrg, pollution control. Cons for all divs & all depts for matls, fabrication procedures, pollution control, & special facil planning. P Chmn local ASM & AWS chapters. On ASM natl ctte for 7 yrs. On bd of regional sci fair for 7 yrs. Enjoy reading, music & family activities. Mbr SAE, G-9 Cttee. *Society Aff:* ASM, SAE, ASNT, ASTM.

Reynholds, Walter H
Home: Daisy Ln Merry Hill, Poughkeepsie, NY 12603
Position: Site Telecommunications Mgr *Employer:* IBM Corp. *Education:* BS/Ind Mech Engg/Univ of CT. *Born:* 9/14/30. Native of W Hartford, CT. Served in USAF 1951-52. Joined IBM Corp in 1956 a Endicott, NY. Transferred to East Fishkill, NY in 1965. Held broad range of positions in Ind Engg & Planning. Assumed current position as Site Telecommunications Mgr in Feb 1986. Sr mbr IIE. IIE Dir of Honors & Awards since 1976. Active in church & community activities. Hobbies include saltwater fishing, antique automobiles, & furniture refinishing. *Society Aff:* IIE.

Reynolds, Bruce R
Business: , 2 Forbes Rd, Lexington, MA 02173
Position: Sr Prin Optical Design Dev Engr. *Employer:* Honeywell Electro-Optics Operations *Education:* BS/Math/Carnegie Inst of Tech & Boston Univ. *Born:* Apr 1942. Worked as a res asst in the Spectrometer Lab for Jarrell-Ash Co. Joined Baird-Atomic Inc 1961 & worked first as a mbr of the math dept & later as a mbr of the optica design group. Joined Honeywell Radiation Ctr 1969, spec in computer analysis & optical lens design on such programs as the Skylab S-192 Multispectral Scanner, the Nimbus F Radiometer, Spacelab ATNBS Spectrometer, & sev forward looking infrared sys. First recipient of SPIE Kingslake Medal & Prize 1975. Enjoy tennis, golf, gardening, & wife Susan. *Society Aff:* OSA, SPIE.

Reynolds, Don P
Home: 5 Manor Dr, Yonkers, NY 10710
Position: Civil Engr Conslt *Employer:* Self Employed *Education:* MSCE/Civil Engg/Univ of MI; BSE/Civil Engg/Univ of MI. *Born:* Dec 9, 1915 Reading Mich. m. Martha D Lok. Asst Engr Mich Hwy Dept; Asst Dir Pub Serv, City of Toledo; Engr Asst Lake Erie Water Proj, Toledo; Asst Engr Sun Oil Co; Instr Ill Inst of Tech; Civil Engr Mills, Rhines, Bellman, Nordhoff; Assoc Ed Civil Engrg Magazine; Asst Secy; Asst Exec Dir and Dir Policy, Planning, Res, Stds ASCE 1942-1982. Reg P E NY, Ohio. Daniel W Mead Ethics Prize; Council of Engrg & Sci Soc Exec Commendation for Most Innovative Prog; Mbr ASCE, ASEE, APWA, ASAE, NSPE, NYSSPE, WCPE, CESSE; Adv Councils KAB, NCIC, CIAC of AAA; Rep to ANSI, TRB, APWA, CCE, ICT. Home: 5 Manor Dr, Yonkers NY 10710. *Society Aff:* ASCE, ASEE, APWA, ASAE, NSPE.

Reynolds, Donald K
Business: Dept of Elec Engrg, Seattle, WA 98195
Position: Emeritus Prof of Elec Engrg. *Employer:* Univ of Washington. *Education:* PhD/Engg Sciences and Applied Physics/Harvard Univ; MA/Electrical Engg/Stanford Univ; BA/Electrical Engg/Stanford Univ. *Born:* Dec 9, 1919 Portland Oregon. Served as Res Assoc, Radio Res Lab Harvard 1945-48; Sr Res Engr Stanford Res Inst 1948-53. Have been teaching Elec Engrg since 1953, respectively at ITA, State of Sao Paulo Brazil, at Seattle Univ (serving as Dept Chmn) & at Univ of Washington (since 1959), Prof. Emeritus since 1983. Tech ints incl appl electromagnetics & electronic sys design. Served as Sci Attache U S Embassy in Brazil 1972-74. Former Regional Dir of Inst of Radio Engrs; Fellow IEEE. Ints incl music & ocean sailing. *Society Aff:* IEEE.

Reynolds, Edward E
Home: 31705 Forest Lane, Warren, MI 48093
Position: Senior Staff Analysis Engr. *Employer:* G M Corp. *Education:* PhD/Met Engg/Univ of MI; MS/Met Engg/Univ of MI; BSE/Met Engg/Univ of MI. *Born:* Mar 22, 1921 Utica Ohio. BSE, MS, PhD (1950) in Met Engrg Univ of Mich. Did res (1941-50) at Univ of Mich on metals at elevated temp. Did res in dev, manufacture & appli of steels & alloys at Allegheny Ludlum Steel Corp 1950-65 as Assoc Dir of Res, & at Latrobe Steel Co 1966-70 as Dir of Res. Was Mgr of Oper making specialty steels & alloys at Stellite Div of Cabot Corp 1970-72. Doing engg on automotive performance & matls as Senior Staff Analysis Engr for Current Product Engrg of G M Corp 1972- . Reg P E Calif. Mbrship in & ctte serv for ASM, AIME, ASME, ASTM & SAE; Fellow ASM; service as tech advisor for NASA, Matls Adv Bd (Natl Acad of Sci), US Army Matls & Mechs Res Council & Metals Properties Council. Publ & patented extensively. *Society Aff:* ASM, AIME, ASTM, ASME, SAE.

Reynolds, Gardner M
Business: 445 S Figueroa St-3500, Los Angeles, CA 90071
Position: Sr Partner *Employer:* Dames & Moore. *Education:* BSCE/Civil Engg/Cornell Univ. *Born:* Ithaca N Y. US Army Air Corps 1942-46. 1st Lt at Discharge; Prisoner of War 1942-45. 1948- , with cons engrg firm Dames & Moore. San Fran office 1948- 49; New York office 1949-68; L A office 1968- . Became Partner 1954. Present pos Sr Partner. Provided cons engrg on major projs (oil refineries, aluminum reduction plants, cement plants, waterfront facils etc) in eastern part of U S & overseas. Overseas assignments incl India, the Azores, Saudi Arabia, Viet Nam, Jamaica. Dir of ASCE 1966-68. Married 1945 to Kathleen Kane; 4 children - Marcia, Stephen, Timothy, Richard. *Society Aff:* ASCE, ASFE, DFI, AAA, ACEC.

Reynolds, J David
Home: 1831 Bellefonte Dr, Lexington, KY 40503
Position: Pres. *Employer:* Reynolds Engg, Inc. *Education:* MS/Civ Engg/Univ of KY; BS/Civ Engg/Univ of KY. *Born:* 2/13/44. Native of Lexington, KY. Prior to June 1968, field engg aide with KY Hgwy Dept, Construction and Res Div and Univ of KY Res Fdn. June 1968 to May 1975, associate, Harp Engrs responsible for field surveys and commercial, industrial, and residential subdivisions. May 1975 to Dec 1976, principal, John Horn & Assoc, responsible for design and surveying operations for multidisciplined engg and construction firm. Dec 1976 to present, pres, Reynolds Engg, Inc offering civ engg, surveying, and construction mgt services in both public and private markets. Past secy-treas, vp, and pres of KY Sec, ACSM. Received "Kentucky Surveyor of the Year" award for 1979. *Society Aff:* NSPE, ASCE, ACSM, ASP.

Reynolds, Joseph P
Business: Chem Engg Dept, Manhattan College, Riverdale, NY 10471
Position: Prof. *Employer:* Manhattan College. *Education:* PhD/Chem Engrg/Rensselaer Poly Inst; BA/Chemistry/Catholic Univ of America. *Born:* 5/19/35. Native of NYC. Taught in the Chem Engg Dept at Manhattan College since 1964. Holds the rank of Prof and, from 1976-1983 the position of Dept Chrmn. Current research interests are in hazardous waste incineration and air pollution control, having authored, published and presented numerous papers in these areas. Has been involved in NSF- sponsored research projects on air pollution control equipment; served as a consultant to Argonne Natl Lab on a a DOE-study involving fine particulate control technology, to Consolidated Edison on a condenser biofouling project and to private industry on a computer-aided heat exchanger design project; and is presently working on an EPA-sponsored project to develop computer software for hazardous waste incineration calculations. He has recently co-authored a text/reference book entitled: *Introduction to Hazardous Waste Incineration* (Wiley Interscience, 1987). *Society Aff:* APCA, AIChE.

Reynolds, Roger S
Business: Dept of Nuclear Engg Univ, Mississippi, MS 39762
Position: Prof. *Employer:* MS State Univ. *Education:* PhD/Nuclear Engg/KS State Univ; MS/Nuclear Engg/KS State Univ; BS/Engg Science/Univ of NV-Reno. *Born:* 10/28/43. Native of Las Vegas, NV, attended public schools in Reno NV. Active in fallout shelter design and research during graduate school years. Currently specializing in nuclear reactor design/analysis and nuclear fuel cycle research. Reg PE in MS and have held many offices in the MS engr society; including chap pres, state bd of dirs, PEE chrmn. Active in ANS, including many comm assignments as well as 1980-81 chrmn of the education div of ANS. ANS representive to Natl Council of engg Examiners and ABET bd of dirs. Member of Engineering Accreditation Division of ABET from 1985-1990. *Society Aff:* ANS, NSPE, ΣΞ, TBΠ.

Reynolds, Samuel D, Jr
Home: 1003 Neely St, Oviedo, FL 32765
Position: Fellow Engineer *Employer:* Westinghouse Electric Corp *Education:* BS/Met. Engrg./Lehigh Univ. *Born:* 12/19/31. Instr at Temple Univ and Drexel Univ Evening Colleges 1958 to 1968. Fellow, ASM. PE in State of PA. PE in Corrosion Engg State of CA. Reg NACE Corrosion Specialist. Written many papers on welding and corrosion aspects of heat exchangers. Employed in all phases of mtls engg since 1953 in nuclear and fossil power plant heat exchanger design and fabrication. Active on AWS and ASME tech ctte. Experience in fabrication and corrosion problems in many ferrous and non- ferrous alloy systems and nuclear power plant inspection. *Society Aff:* ASM, AWS, NACE.

Reynolds, William C
Business: Dept of Mech Engg, Stanford, CA 94305
Position: Prof *Employer:* Stanford Univ. *Education:* PhD/ME/Stanford; MS/ME/Stanford; BS/ME/Stanford. *Born:* 3/16/33. Earned BS, MS, PhD in ME at Stanford, 1954, 1955, 1958. At Stanford as faculty mbr since 1957. Chrmn, ME, 1972-1982. Founded Stanford's Inst for Energy Studies in 1973. Active in res on turbulent flows, experimental and theoretical. Elected Fellow of ASME, Mbr of NAE, in 1978. Fellow Am Phys Soc 1982. *Society Aff:* NAE, ASME, APS, AIAA, ΣΞ.

Reznik, Alan A
Business: Chemical & Petroleum Engineering Dept, 1236 Benedum Engineering Hall, Pittsburgh, PA 15261
Position: Dir: Petroleum Engrg Prog *Employer:* Univ of Pittsburgh *Education:* POST-DOC/Chem & ChE/Univ of Pittsburgh; PhD/CE/Univ of Pittsburgh; MS/Petrol Engr/Univ of Pittsburgh; BS/Petrol Engr/Univ of Pittsburgh *Born:* 09/25/39 Native of Pittsburgh, PA. Research scientist for Conoco, specializing in multi-phase flow in petroleum reservoir rocks. Research assoc with Calgon Corp working on polymer/water-flooding of petroleum reservoirs. Chem engrg supvr for USBM (ERDA), responsible for reactor design for conversion of coal to oil. Assumed assoc prof faculty position in 1975. Tenured in 1977. Head: petroleum engrg prog, 1981. Specialized courses include: petroleum reservoir engrg, thermodynamics, fluid dynamics, tensor analysis. Recipient of 11 Federal Res Grants, including: in- situ recovery of methane from coal and the modeling of porous media transport properties. Authored numerous papers on same. Publ a major revision of oil recovery from waterflooding, 1984. *Society Aff:* SPE/AIME.

Rhee, Dail
Home: 1729 Nalulu Pl, Honolulu, HI 96821
Position: Construction Manager. *Employer:* KG (Hawaii) Construction. *Education:* BSCE/Civil Engineer/MIT. *Born:* July 21, 1936. BSCE MIT 1960. Proj Engr R M Towin Corp up to 1968. Ch Civil Engr R M Pasons Co - Honolulu office - 1970. Dir of Engrg Donald Wolbrink & Assocs 1972. V P Gray, Rhee & Assocs of Honolulu Deputy Director, Dept. of Transportation, City of Honolulu - 1985 Construction Manager, Kumagai Gumi Construction, Honolulu Office 1980-present. Reg P E Hawaii. Mbr ASCE. MIT Educ Counselor.

Rhee, Moon-Jhong
Home: 14312 Sturtevant Rd, Silver Spring, MD 20904
Position: Assoc Prof. *Employer:* Univ of MD. *Education:* PhD/Appl Physics/Catholic Univ of Amer; MS/Physics/Seoul Natl Univ; BS/Physics/Seaul Natl Univ *Born:* 2/19/35. in Korea. Served with Korean Army. Instr, Dept of Nuclear Engg, Seoul Natl Univ. Came to US 1966. Res Fellow, Dep of Space Sci & Applied Phys, 1966-1970. Asst Prof, Assoc Prof, 1970-1983, Prof, 1983-present, Elec Engg Dept, Univ of MD, specializing Nonneutral Plasma Phys, Collective Ion Acceleration & Pulsed Power Systems. Spent Sabbatical at Seoul Natl Univ as a visiting prof, 1977-1978. Have been involved in Univ of MD Electron Ring Accelerator proj & responsible for the lab. Enjoy skiing, fishing, & tennis. *Society Aff:* APS, KPS, IEEE

Rhett, John T
Home: 3175 N 21st St, Arlington, VA 22201
Position: Federal Inspector. *Education:* MS/Civil Engrg/Univ of CA, Berkeley; MS/Interntl Rel/George Washington Univ; BS/Mil Engrg/US Military Academy. *Born:* 2/20/25. Grad West Point 1945. Master of CE, Univ of Calif. 27 yrs spent in US Army Engrs. Some of the more challenging engrg work was with Army's Nuclear Power Program and serving as Chief, US Army Construction Agency, Vietnam and as Dist Engrg, Louisville, KY. Upon retirement, supervised the Wastewater Treatment Construction Program for Envir Protection Agency, building it into largest

Rhett, John T (Continued)
Federal public works program. In 1979, appointed by Pres to head a separate Agency of Government to supervise, from Federal viewpoint, the construction of Alaska Natural Gas Pipeline. Ret 1986; Currently private consultant. *Society Aff:* ASCE, SAME, AAEE, NSPE

Rhoades, Richard G
Home: 3604 Lookout Dr, Huntsville, AL 35801
Position: Assoc. Director, Army Missile Research Development & Engg Center. *Employer:* US Army Missile Command. *Education:* PhD/Chem Engg/RPI; MS/Mngt/MIT; BChE/Chem Engg/RPI. *Born:* 8/15/38. Born and raised in MA; married with 3 children. Employed by Eastman Kodak and GR Kinney while in school; served as Army Officer 1963 - 1965 at US Army Missle Command and as propulsion res engg and project mgr form 1965 - 1973. From 1973 to 1980, Dir for Propulsion with tech and mngt respon for dev of missile propulsion. 1980 to present, Associate Director for Technology, US Army Missile Command, responsible for all technology base and new system activities in missiles and high energy lasers. Also, taught grad and undergrad courses at Univ of AL. Author of numerous tech reports/journal articles in propulsion/mngt; holder of patents. NSF Fellow at RPI and won Brooks Thesis prize at MIT. *Society Aff:* AIChE.

Rhoades, Warren A
Business: P.O. Box 35361, Houston, TX 77035
Position: VP & Genl Mgr *Employer:* Marchem Products Co of Teton, Inc. *Education:* BSME/Engg/LA State Univ. *Born:* Aug 1924. Native of Shreveport La. WW II aboard an Army Marine Repair Ship, Pacific Theater. Engr with Cooper Bessemer Corp 1948-65 in design & oper of large diesel & gas engines. Mgr of Engrg for the Engine-Compressor Div, De Laval Turbine Inc 1965 to 1980. Respon for design, sys engrg, R&D, & product engrg of diesel & gas engine & compressors. Mbr ASME, SNAME, SAE; Assoc Mbr Exec Ctte, DGP Div, ASME; ASME Diesel & Gas Power Award 1972. Reg P E Texas & Louisiana. Responsible for Engrg, Manufacture, Sales of Film Lubricated Bearings and Non Contact Seals. *Society Aff:* ASME, SAE, SNAME.

Rhode, Gerald K
Business: 300 Erie Blvd W, Syracuse, NY 13202
Position: Exec Consultant *Employer:* Niagara Mohawk Power Corp. *Education:* BS/ME/RPI; Grad/-/ORSORT. *Born:* 8/24/29. in Elmira, NY. Grad Oak Ridge Sch of Reactor Tech. US Army 1946-49. employed: Babcock & Wilcox in Atomic Energy Div 1954. Particiapted in the Indian Pt I & N S Savannah projs. Joined Niagara Mohawk Power Corp 1960. Elected VP Engg 1975. Respon for sys planning, corp & div engg; nuclear & fossil generation. Assumed present pos of VP - Sys Proj Mgmt 1977. Respon for maj construction - Nuclear, fossil and hydro generation, transmission and bldgs. Mbr ASN, AIF, CIGRE, EEI, AMA & Prof Engg Soc; also Tau Beta Pi & Pi Tau Sigma. Enjoy boating, hunting, fishing & horseback riding. *Society Aff:* NSPE.

Rhodes, Allen F
Home: 73 East Broad Oaks, Houston, TX 77056
Position: President & CEO. *Employer:* Gripper Inc. *Education:* ML/Mgmt/Univ of Houston; BS/Mech Engg/Villanova Univ; -/Mech Engg/Rice Univ. *Born:* Oct 1924. BS Villanova; MA Univ of Houston. Engr with Hughes Tool 1947- 52; Dir of Engrg to Pres McEvoy Valve Houston 1952-63; with Rockwell Mfg Pgh 1963-71, V P Res & Engrg 1966-71; V P Corp Planning & Dev ACF Indus 1971-74. Pres & Ch Exec Officer, McEvoy Oilfield Equip Co Hou 1974-79, Ex VP & COO Goldrus Marine Drilling 1979-82 Pres Warren Oilfield Services 1981-82 - both of these companies were division of Warren-King Companies; Pres & CEO Anglo Energy, Ltd. 1983-86; Pres. & CEO Gripper Inc. 1987- . Reg P E Texas. Tau Beta Pi. Fellow Inst Mech Engrs UK; Fellow & P Pres ASME; Mbr SPE-AIME Dir Amer Natl Stds Inst. Cert of Award, ANSI 1978; Robt. Henry Thurston Award, ASME 1978; Howard Conley Medal, ANSI 1980; Dist. Engr. Alumni, Villanova 1982; Elected Mbr of National Academy of Engineering 1985. Served to LTjg USNR 1943-46. Author & patentee in field. Enjoys sailing. Married; 2 sons. *Society Aff:* ASME, IMechE, ANSI, AIME, NSPE, AMA.

Rhodes, Donald R
Business: ECE Dept, Bx 7911, Raleigh, NC 27695
Position: Univ Prof. *Employer:* N Carolina State Univ. *Education:* PhD/EE/OH State Univ; MSc/EE/OH State Univ; BEE/EE/OH State Univ. *Born:* Dec 31, 1923. Ohio State Univ Res Found 1945-54; Cornell Aeronautical Lab 1954-57; Radiation Inc 1957-66. With N C State Univ as University Prof since 1966. Editor IEEE Transactions on Military Electronics 1957-65. Mbr Admin Ctte of IEEE Group on Military Electronics 1959-65. Mbr Admin Ctte of IEEE Antennas & Propagation Soc 1965-73. V P 1968, Pres 1969 IEEE Antennas & Propagation Soc. IEEE John T Bolljahn Award 1964; Benjamin G Lamme Medal Ohio State Univ 1975; NC State Univ Outstanding Teacher 1980, 1983, 1986. Fellow IEEE & AAAS. Publ 'Intro to Monopulse' with McGraw-Hill Book Co 1959 republished by Artec House in 1980, & 'Synthesis of Planar Antenna Sources' with Oxford Univ Press 1974. Participates in var choral & instrumental music groups. Research on medieval music. *Society Aff:* IEEE, AAAS.

Rhodes, Eugene I
Business: 369 Lexington Ave, New York, NY 10017
Position: Principal, Officer *Employer:* Rhodes & Basso Engrs P. C. *Education:* M/CE/Polytech Inst of Bklyn; B/CE/Polytech Inst of Bklyn *Born:* 05/21/24 Native of NYC. Served with the Army Corps of Engrs 1942-43. Structural Design Engr with Edwards & Hjorth on high rise bldgs 1945-1951. With Laramore Douglass & Popham as design engr, project engr, asst construction superintendant, and chief civil engr on power plants and industrial facilities projects. One of the founding partners of Rhodes & Basso Engrs, P. C. since 1964, developed service to government and industry specializing in power generation, facilities and forensic engrg. Active in facilities design for the pharmaceutical, steel, rubber, paper, chemical, air cleaning, and power industries. *Society Aff:* AAAS, ACEC, ACI, ASTM, NYSPE, ASCE

Rhodes, Gilbert L
Home: 4864 Haley Dr, Castro Valley, CA 94546
Position: Cons in Safety Engrg. *Employer:* Gilbert L Rhodes & Assocs. *Born:* Apr 6, 1912 Mammouth Ariz. Attended UCLA & USC. Lectr UCLA Sch of Engrg, U C Sch of Pub Health & of Medicine, Amer Mgmt Assn, Natl Safety Council Training Inst. Reg P E Calif. Fellow Amer Soc of Safety Engrs, Cert Safety Prof. Owner Gilbert L Rhodes & Assocs, Safety Cons 1970- . Supr Safety Engrg & Environ Health, Sandia Corp 1957-70. Dir Home Safety Proj, Calif State Health Dept 1953- 57; Supr Safety Engrg Indus Indemnity Co Oakland 1950-53; Safety Engr State of Calif 1942-50. Cons in Accident Prevention to St Bd of Health Calif 1957-63; P Pres Veterans of Safety Internatl; Western Region V P Amer Soc of Safety Engrs; Proctor Bd of Cert Profs. Mbr Permanent Adv Ctte Calif OSHA Stds. Inventor - pats 2,881,934 & 2,990,075. Expert Examiner State of Calif. Bd of Regis for Prof Engrs. U S Rep to Safety Engrg Curriculum Dev, Internatl Labor Office, Geneva S. *Society Aff:* ASSE, CSP, VOS.

Rhodes, John M
Home: 29 Pine, Medfield, MA 02042
Position: VP & Chief Operating Officer *Employer:* Factory Mutual Research Corp *Education:* B/ChE/Univ of DE *Born:* 07/05/18 Grew up in PA and DE. Relocated to Boston area in 1940 as Engr for Factory Mutual System. Served during WWII as Artillery Unit Commander in Africa and Europe. Rejoined Factory Mutual in 1946 serving progressively as Engr-in-Charge of Test Station, Asst Dir of Laboratories, District Mgr, Asst Chief Engr, Dir of Engrg and Laboratories. Currently Chief Operating Officer, Factory Mutual Research Corp responsible for engrg policy development, approval of loss central equipment, materials and services. Also VP of all

Rhodes, John M (Continued)
three Factory Mutual engrg organizations: Engrg Association, Corp, and Research Corp. Fellow, SFPE. Play tennis. *Society Aff:* NFPA, ASTM, SFPE

Rhodes, Robert R
Home: 709 Victoria Dr, Hilton Head, SC 29928
Position: VP. *Employer:* Barrett, Haentjens & Co. *Education:* MBA/Mgt/NYU; BSChE/Univ of TN. *Born:* 8/17/21. Also Director of Advertising. Served in USN 1943-46 & 1951-1953, now retired Capt, USNR. With Worthington Corp 1943-1974 as Mgr Marketing-Pumps. VP Marketing & Managing Dir-Worthington Australia. Responsibility in all cases included chem process pumps. VP, Chas S. Lewis & pres. Lewmet Alloys 1974-1980. Was Chrmn-AIChE Pump Sub-Committee of Equip Testing Procedures Committee 1963- 1966. Enjoy golfing & fishing. *Society Aff:* AIChE.

Rhodes, William T
Business: Sch of Electrical Engineering, Atlanta, GA 30332
Position: Prof *Employer:* Georgia Inst of Tech *Education:* PhD/EE/Stanford Univ; MS/EE/Stanford Univ; BS/Physics/Stanford Univ *Born:* 04/14/43 Palo Alto, CA. Grad research asst with Stanford Electronics Lab, summers with Hewlett Packard and Stanford Research Inst. At Georgia Tech since 1971, research and teaching in optics and signal processing. Conslt for Naval Research Lab. Fellow of OSA and SPIE. Chrmn IEEE Computer Soc Optical Processing Technical Interest Group, 1980-1981. Coauthor of *Introduction to Lasers and Their Applications* (Addison-Wesley, 1977). Dir, Optical Soc of Amer (1984- 1986); Gov of SPIE, The Intl Soc for Optical Engrg. Topical Ed, *Journal of the Optical Society of America A* (1983-1985); Assoc Ed, *Optics Letters* (1980- 1983). *Society Aff:* AAAS, IEEE, OSA, SPIE

Rhoten, Ronald P
Business: Dept of Elec Engrg, Stillwater, OK 74074
Position: Prof. *Employer:* Oklahoma State Univ. *Education:* PhD/Elec Engg/Univ of TX at Austin; MS/Elec Engg/Univ TX at Austin; BS/Elec Engg/Univ TX at Arlington. *Born:* Mar 3, 1943. Elec engrg fac mbr of Oklahoma State Univ 1969- ; specialty areas incl product safety, energy use models, optimal control. Consultant in over 500 electrical accident and fire investigations. Res grants from NSF, NASA, USAF Avionics Lab, EXXON Educ Found. Recipient Outstanding Young Engr Fac Award, Dow Chem Co/ASEE 1974; Young Engr of Yr, Okla Soc Prof Engrs 1975. Mbr IEEE, ASEE, OSPE, NSPE, IAAI, NFPA, AAAS. Reg P E Okla. *Society Aff:* IEEE, ASEE, NSPE, AAAS, IAAI, NFPA

Rhutasel, Larry J
Business: 1 Sunset Dr, Freeburg, IL 62243
Position: President *Employer:* Rhutasel and Assoc, Inc *Education:* BS/CE/Univ of IA; MS/Env Engrg/Univ of Il *Born:* 08/18/40 A native of IA, employed by IL Dept of Public Health from 1963-1969. Started engrg firm in 1969 which provides Engrg and Architectural services. Reg PE in IL, IA, MN, and MO. *Society Aff:* ACEC, NSPE, ASCE, WPCF, AWWA, AAEE

Rib, Harold T
Home: 10129 Glenmere Rd, Fairfax, VA 22030
Position: Ch, Environ Design & Surveys Branch *Employer:* Fed Hwy Admin. *Born:* Jan 1931 NY. PhD Purdue Univ; MS Cornell Univ; BCE CCNY. Signal Corps AUS 1954-56. Soils engr US Army Corps of Engrs 1953-54, 1956; Ch, Photointerpretation Sect, Keystone Mapping Co 1957-58. Joined Fed Hwy Admin as Hwy Res Engr 1958, respon for major res progs in remote sensing & exploration techs through 1974; moved to Aerial Surveys 1974, assuming pos as Ch in 1975. Visiting Prof Univ of Md since 1974. Mbr Exec Ctte & Bd of Dir and Dir of Remote Sensing & Interpretation Div, Amer Soc of Photogrammetry 1972-73; Pres Citation ASP 1971, 72, 73, 74, 76. Mbr Photogrammetry & Aerial Surveys Comm, Transp Res Bd. Mbr Remote Sensing Comm, ASCE; Mbr Commission VII, Intl Soc of Photogrammetry. Married, 3 children, enjoy classical music & sailing.

Ricardi, Leon J
Home: 12 Priscilla Circle, Wellesley Hills, MA 02181
Position: Group Leader. *Employer:* MIT Lincoln Lab. *Education:* PhD/EE/Northeastern Univ; MS/EE/Northeastern Univ; BS/EE/Northeastern Univ. *Born:* Mar 21, 1924, Brockton Mass, to Philip J and Eva I (DuBois) Ricardi. Married Lena M (Giorgio) Ricardi Jan 19, 1947; 3 children; Eva M, John P, & Richard C. Andrew Alford Cons Engrs, Boston Mass 1950-52; Proj Engr Gabriel Labs, Needham Mass 1952-54; Proj Engr MIT, Lincoln Lab, Lexington Mass, Group Leader 1954; Part time teacher Northeastern Univ 1968. Served as 1Lt US Army Air Force 1943-45. Air medal & 5 Oak Leaf Clusters; Fellow IEEE; Mbr IEEE Soc on Antennaes & Propagation; IEEE Soc on Microwave Theory & Techniques; Mbr A DCOM; Disting Lecturer 1973-75; Editor of IEEE *Transactions on Ant & Prop* 1971-74. Present position - Design & Dev, Communication antennas and Components. Office - Lincoln Lab, Lexington Mass 02173. *Society Aff:* IEEE.

Ricca, Vincent T
Home: 3117 Mountview Rd, Columbus, OH 43221
Position: Prof. *Employer:* OH State Univ. *Education:* PhD/Hydraulic Engg/Purdue Univ; MSCE/Hydrology/Purdue Univ; BS/Civ Engg/CUNY; Assoc Degree/Arch/Long Island Agri & Tech Inst. *Born:* 12/19/35. Native of NYC. Served in USNR, 1952-58, in aircraft hydraulics. Worked as Construction superintendent in NY area, hydraulics engr in NY Dept of Public Works & CA Dept of Water Resources. With the OH State Univ specializing in hydrologic & hydraulic engg since 1966. Past Pres of the Centra OH Sec of ASCE. Reg engr in OH. Consulting & res efforts have been in watershed simulation, flood profiles, flood plain mgt, acid mine drainage, erosion control & sediment pond design. Enjoy woodworking & player piano restoration. *Society Aff:* ASCE, ASEE, AWRA.

Riccardi, Louis G
Home: 8024 Ridge Blvd, Brooklyn, NY 11209
Position: Pres Plastics Division. *Employer:* Evans Prods Co. *Education:* PhD/Chem Engg/Polytechnic Inst of Brooklyn; Masters/Chem Engg/Polytechnic Inst of Brooklyn; Bachelors/Chem Engg/College of City of NY. *Born:* 10/7/25. Native of NYC. Taught at PIB adjunct prof; Sr Chem engr of organic prods for Colgate-Palmolive Co 1954-57; dir of Processing Jello Div of Genl Foods 1957-60; Genl Mgr & Coml Dev Mgr of Styrenic Profit Ctr & New Ventures for Rexene Div of Dart Indus 1960-70; Pres Plastic Div of Evans Products Co more than 10 pats & author of sev articles on distillation and esterification. Mbr ACS; AIChE; SPE; AIC; Phi Lambda Upsilon; Sigma Xi; RESA; Chem Who's Who. *Society Aff:* AIChE.

Rice, Charles E
Business: PO Box 551, Stillwater, OK 74076
Position: Res Hydraulic Engg. *Employer:* USDA-ARS *Education:* PhD/Agr Engg/Univ of MN; MS/Agr Engg/OK State Univ; BS/Agr Engg/OK State Univ. *Born:* 2/13/32. Native of OK. Res Agri Engr, USDA-SWCRD-ARS, 1961-66. Asst and Assoc Prof of Agri Engg, OK State Univ, Stillwater, OK, 1966-78, teaching and res in the water resources area. Res Hydraulic Engr, USDA-ARS, 1978-present, conduct of res on hydraulic structures for the control, conveyance, and measurement of flood waters occurring on agricultural watersheds. Adjunct Prof, Agr Engg and mbr, Grad Fac, OK State Univ. *Society Aff:* ASAE, ΣΞ, ΦΚΦ.

Rice, Edwin E
Home: 2307 Lochlevin Dr, Memphis, TN 38119
Position: Consultant (Self-emp) *Education:* BS/CHEM Eng MI Tech Univ *Born:* 8/4/18. Detroit, MI. Tau Beta Pi, Phi Lambda Upsilon. Special Projs Engr Atlas Powder Co, Wilmington, DE, 1943-46. Plant Mgr Werner G Smith Co, Wyandotte, MI, 1947-62. Mgr Technical Services ADM Co/Ashland Chem Co, Mapleton, IL, 1962-69. With Kraft Inc 1969-1980. VP Engg since 1977. Responsibility for divisional engg since 1973. Led several maj fatty chem plant expansions plus design & erection of near-pharmaceutical chem operation. Divisional Energy responsibility.

Rice, Edwin E (Continued)
Mbr Corp Energy Conservation Committee. Process & Energy Consultant to other firms. Hobby machine & woodworking shop. Lapidary. Gunsmithing, target shooting. Married, three children. *Society Aff:* AOCS, AIChE, ACS.

Rice, James K
17415 Batchellors Forest Rd, Olney, MD 20832
Position: Pres. *Employer:* James K. Rice Chartered *Education:* M.S./Chm. Eng/Carnegie Institute of Technology; B.S./Chem. Eng/Carnegie Institute of Technology *Born:* 03/12/23 Pres. Cyrus Wm. Rice & Company, 1959-73; Sr. VP. NUS Corp., 1973-76; Pres. James K. Rice Chartered 1976-; Past Director ASTM; Past Chair ASTM Cttee D-19, Water; Present Chair. ASTM Environmental Coordinating Cttee.; Past Mem. ASME Policy Bd Research; Past Chair. ASME Res. Cttee. Water in Thermal Power Systems; present mbr. Exec. Cttee.; recipient ASTM Award of Merit, Award of Merit International Water Conference; present mem. Advisory Bd International Water Conference; consultant to Utility Water Act Group, Electric Power Research Inst, investor-owned Electric Utilities, and industrial firms with large steam-power facilities. *Society Aff:* ASME, ASTM, AIC, ACS, AAAS

Rice, James R
Business: Harvard University, Division of Applied Sciences, Cambridge, MA 02138
Position: Gordon McKay Prof of Engrg Sci and Geophysics *Employer:* Harvard University *Education:* PhD/Mechs/Lehigh Univ; ScM/Mechs/Lehigh Univ; ScB/Mechs/Lehigh Univ. *Born:* 12/3/40. in Frederick, MD. Res & teaching mechs of solids at Brown Univ from 1964 to 1981, at Harvard Univ since 1981. Res contribs to crack growth & fracture processes in solids; metal plasticity; earthquake source mechs. Approx 100 pub papers. Mbr ASME; ASCE; AGU; AAAS; Assoc Editor ASME J Appl Mech; Ed Bd Engr Fracture Mech 1968-71; Quart Appl Math; Int J Engr Sci; Tech Conf Org, ASME & ASTM; NSF Sr Postdoc. Fellow & Churchill Col Overseas Fellow, Univ of Cambridge 1971-72. ASME Hess Award 1969; Pi Tau Sigma Gold Medal 1971; ASTM Dudley Medal 1969. Amer Acad Arts & Sci, 1978. National Academy of Engineering, 1980. National Academy of Sciences, 1981. *Society Aff:* ASME, ASCE, AGU, AAAS.

Rice, John F
Business: PO Box 27307, Raleigh, NC 27611
Position: Asst State Conservation Engr. *Employer:* USDA - Soil Conservation Service. *Education:* BS/Agri Engg/Univ of MO. *Born:* 12/10/32. Native of Trenton, MO. Served in USN 1952-56 with service in the Korean theater of operations. With Soil Conservation Service since 1962. Area Engr for 13 NE counties in MO 1963-71. Responsible for conservation engg Problems. Assumed current responsibility as Asst State Conservation Engr with state wide responsibility for conservation operations in NC 1972. Chrmn MO State Sec ASAE 1971, NC State Sec ASAE 1979. *Society Aff:* ASAE, SCSA.

Rice, Leonard
Business: 2785 N Speer Blvd, Denver, CO 80211
Position: Pres. *Employer:* Leonard Rice Cons Water Engrs Inc. *Education:* MS/CE/Univ of IA; BS/CE/Auburn Univ. *Born:* 1932. Naval Ser 1954-56. Assoc with cons firms in Fla, Iowa & Colorado. Established Leonard Rice Cons Water Engrs 1970 to provide servs in water resource planning, water rights, floodplain mgmt, environ analysis & water related urban problems. Firm has received natl & st awards for engrg excellence. Fellow ASCE; mbr NSPE; ACEC; AAA. Pres Cons Engrs Council of Colorado 1975-76; Chmn ASCE Water Resource Sys Ctte 1975-76. Teach Water Resources Denver Univ & Univ Colo/Denver. Reg PE Colo, Wyo, NM, Nev & Fla. NCEE Cert & Cert Cons Engr, Colo. *Society Aff:* ACEC, ASCE, NSPE, ICID.

Rice, Philip A
Business: Dept of Chem Engg. & Mat'ls Sci, 320 Hinds Hall, Syracuse, NY 13210
Position: Prof. & Chair *Employer:* Syracuse Univ. *Education:* PhD/CE/Univ of MI; MSE/CE/Univ of MI; BSE/CE & Engrg Math/Univ of MI *Born:* 8/3/36. Native of Coldwater, MI. NATO Post doctoral Fellow, Inst for Physical Chemistry, Univ of Goethingen, 1962-1963. Chem Engr, Analytic Ser, Inc Bailey's Crossroads, VA 1963-1965. With Dept of Chem Engrg & Mtls Sci, Syracuse Univ since 1965. Prof since 1976. Res Assoc Prof, Dept of Obstetrics & Gynecolgy, Upstate Medical Ctr, State Univ of NY, since 1971. Program Dir, Separation Processess, Natl Sci Fdn, 1984-1985. Res interests in transport processes in the human placenta, heat storage using phase change mtls, biochem reaction and separation processes. Co-chrmn, Environmental Engrg 1977-85. Chair, Chem. Engrg & Mat'ls Sci, 1985-present. Enjoy running & other sports, music. Rofator: Seporahm & Purification Process Program, Natl Foundation, 1984-1985. *Society Aff:* AIChE, ACS, AAUP, AAAS

Rice, Raymond J
Business: 85 Willis Ave, Mineola, NY 11501
Position: Managing Partner. *Employer:* Rice Partnership. *Education:* BS/Engg/Columbia Univ; MS/Urban Planning/Columbia Univ; BS/Liberal Arts/Columbia Univ. *Born:* 1/28/27. Founded the Rice Partnership-Engrs, Architects & Planners-1955. Pres of Urbanetics Corp-Transportation, Recreation & Community Planners. Received 1978 Honors Award for Civ Engg from NY Assoc of Consulting Engrs. Has written extensively on engg, planning and govt. Served on many governmental and industry advisory Bds. Represented 5th Senate Dist at NYS Constitutional Convention. Served as VP NY Assoc of Consulting Engrs. Chaired committee for two ACEC Natl Conventions. Has served on the Bd of Dir of Molloy Coll and serves on the Exec Ctte of the Bd of Overseers of the NY Inst of Tech. He serves on the Bd of Dir of public television station WLIW (channel 21). *Society Aff:* ASCP, AICP, NSPE.

Rice, Richard L M
Business: Louviers Bldg, Wilmington, DE 19898
Position: Stds Asst Mgr. *Employer:* Du Pont Co. *Education:* BS Admin ME/Indust Engg/Cornell Univ. *Born:* 12/31/20. Native of Phila, PA. After graduation, worked for Chance Vought Aircraft and served in US Army for two years. Joined Du Pont as a mgt engg consultant in 1950. Served as a process engr in design and from 1953-1970 as a Sr Stds Engr. In 1970, was appointed Mgr of Mgt Serv in the Design Div and returned to Stds as Asst Mgr in 1974. Currently Engrg Dept Safety & Health Coord. Past Pres of Stds Engrg Society and hold the grade of Fellow. Am a reg PE in the state of Delaware. *Society Aff:* SES, NSPE.

Rice, Stephen L
Business: Mech Engg & Aerospace Sci Dept, Orlando, FL 32816
Position: Prof & Chmn *Employer:* Univ Central FL *Education:* PhD/Mech Engrg/Univ CA Berkeley; M Engg/Mech Engrg/Univ CA Berkeley; BS/Mech Engrg/Univ CA Berkeley. *Born:* 11/23/41. Dr Rice is a native of Alameda, CA, educated at the Univ of CA, Berkeley. He has several years experience in design, res and dev, including work at the Lawrence Berkeley Lab and with the Univ of Connecticut. Presently with the faculty of The Univ of Central Florida, Dr Rice's res interests are in tribology (wear), engineering design, and computer based education). His honors include SAE's Teetor Award, the ASEE/DOW Outstanding Young Faculty Award, and a Fulbright Sr Res Fellowship. *Society Aff:* ASME, ASEE, SME, NSPE, ASTM.

Rice, Stephen O
Home: 8110 Paseo Grande, La Jolla, CA 92037
Position: Retired. *Employer:* Bell Telephone Labs. *Education:* BSc/EE/OR State College. *Born:* 11/29/07. in Shedds Ore. Graduate work in Physics at Cal Tech and Columbia Univ with Bell Labs 1930-72. Worked on problems in modulation theory, radio wave propagation, noise & channel interference. Hon D Sci from Ore St Univ 1961. IEEE Mervin Kelly Award 1965. NTC 1974 Award. Alexander Graham Bell Medal, IEEE 1983. *Society Aff:* IEEE, NAE.

Rice, William J
Business: Chemical Engr Dept, Villanova, PA 19085
Position: Prof *Employer:* Villanova Univ *Education:* PhD/ChE/Princeton Univ; MS/ChE/WPI; BS/ChE/WPI *Born:* 08/06/27 Born in Whallonsburgh, NY and reared in Clinton, MA. Instructor, Chemical Engrg at Catholic Univ, 1948-1953. Asst Prof, Assoc Prof, and Prof, Chemical Engrg, Villanova Univ, 1957-present. Mbr Sigma Xi and Tau Beta Pi. Reg PE, PA. *Society Aff:* AIChE, ASEE, AAUP

Rich, Donald G
Home: 121 Cashin Dr, Fayetteville, NY 13066
Position: Res Mgr. *Employer:* Carrier Corp. *Education:* MS/Engg/Harvard Univ; BS/Engg/Brown Univ. *Born:* 11/24/29. Native of RI. Started employment with Carrier Corp, Syracuse, NY in 1952 as res engr. Worked on systems analysis & heat exchanger designs for specialized applications. From 1959 to 1978 headed res groups in heat transfer, thermodynamics, & math. Published several papers, one of which was chosen by ASHRAE as best res paper of 1973. Hold 5 US patents. Awarded ASHRAE Distinguished Service Award in 1972 & awarded Fellow grade in 1978. Promoted to present position as Mgr in 1978. Is presently responsible for progs in mtls engg, stress analysis, chemistry, standards, and information resources. Elected to ASHRAE Bd of Dir in 1983. Enjoy tennis, fishing, bridge, & chess. *Society Aff:* ASHRAE, ASME.

Rich, Donald L
Business: 801 S Glenstone, Springfield, MO 65802
Position: Partner. *Employer:* Hood-Rich. *Education:* BS/Mech Engg/Purdue Univ. *Born:* 10/30/34. Partner in firm of Hood-Rich, Arch & Conslt Engrs, Springfield, MO. Partnership performs architectural/engg design for commercial, motels, schools and banks including all engg in-house work. Also engg for municipalities including water, sanitary and subdiv. Firm consists of 55 total employees with 15 engrs. Active in civic and profl organizations. Past pres Springfield Area Chamber of Commerce. Rotary Club and other local and State Bds. *Society Aff:* ASME, NSPE.

Rich, Elliot
Home: 1640E 1140N, Logan, UT 84321
Position: Prof. *Employer:* UT State Univ. *Education:* PhD/CE/Univ of CO; MS/CE/Univ of UT; BS/CE/UT State Univ. *Born:* 5/27/19. Brigham City, UT. Grad of Northern Vocational Sch, Toronto, Canada. Field Engr, Cahill Engrs, Ltd. Electronics Technician Officer WWII 1943-1946. US Bureau of Reclamation 1946-47. Instr Weber College, Ogden, UT 1947-57. Assoc Prof - Prof, UT State Univ 1957 to present. Hd of Civ Engg Dept 1968-1975. Assoc Dean of Engg 1975 to present. Consulting work for AEC, Insurance Cos, Architects, Mfg firms. Pres UT Sec ASCE 1977-78. Mbr UT Geological & Mineral Survey Bd 1977 to present. Pres USU faculty assn 1972. *Society Aff:* ASCE, ASEE.

Rich, Melvin J
Home: 447A Hammer Stone Ln, Stratford, CT 06497
Position: Exec-VP-Provost *Employer:* Bridge Fort Engg Inst *Education:* MS/Appl Mechanics/NYU; MSAE/Aero Engg/NYU; BSAE/Aero Engg/NYU. *Born:* 2/15/22. Lt US Army Air Force WWII. Aeronautical Engr with Curtiss Wright Airplane Co. Bendix Aviation Missile Div, and as Chief of Structures with Skiorsky Aircraft Div of United Technologies Corp. Published Forty papers on structures and materials. Reg PE (structures). Associated with Bridgeport Egg Inst since 1959 as Instructor of Mech Engg, VChrmn of Mech Engg Dept, Chrmn of Mathematics Dept, Dean of Faculty. Appointed as full prof of engrg in 1981 and currently Exec-VP-Provst. *Society Aff:* AIAA, AHS, ASEE, ASTM, AFA.

Rich, Stuart E
Business: 3300 NW 21st St, Miami, FL 33142
Position: VP. *Employer:* Rich Electron Inc. *Education:* PhD/Chem Eng/Tulane Univ; MS/Chem Eng/Tulane Univ; BE/Chem Eng/Vanderbilt Univ. *Born:* Oct 1944. Academic res involved dev of new mathematical technique for designing control sys for nonlinear multivariate processes with constrained inputs. Additional studies related to modeling of pulsatile fluid flow in elastic vessels. Later employed by Union Camp as Process Sys Group Leader with responsibility for developing computer-aided control for pulp and paper mill opers. VP of Rich Elec since 1973. In addition to overall financial & opers planning, duties involve process control sys & dev of energy conversion sys. Hobbies: piano & swimming. *Society Aff:* AIChE.

Richard, Charles W Jr
Home: 7128 Montague Rd, Dayton, OH 45424
Position: Assoc Prof *Employer:* AF Inst of Tech *Education:* MS/EE/Univ of Il; BS/EE/MIT *Born:* 07/25/30 Served with USAF 1952-1963. Has taught engrg mathematics at the Air Force Inst of Tech since 1957. Assoc Prof since 1963. Mbr of Tau Beta Pi and Eta Kappa Nu. Also retired colonel in USAF Reserve. *Society Aff:* ACM

Richard, Oscar E
Home: 1308 Springfield Dr, Belleville, IL 62221
Position: Pres *Employer:* Constrn Engrg Meteorology, Inc *Education:* BS/Math/Tufts Univ *Born:* 05/07/18 Staff Weather Officer, WWII 1942-1946. Taught Boston, MA, Math & Gen Sci, 1947. Editor: Tufts Jumbo Yrbook (1946). Extended Forecast Sec, US Weather Bureau 1951-1955, prepared 5-Day & monthly forecasts. USAF Environmental Technical Applications Ctr 1955-1982, (Chief, Engrg Meteorology Sect). Author of Tri-Service Manual (AFM 88-29, TM 5-785, NAVFAC P-89) *Engineering Weather Data* (Revised 1961, 1963, 1967, 1978). Author of several technical papers. ASHRAE Committees: T C 4.2 Mbr 1959-1966, Chrmn 1966-1975; Mbr Prog Committee 1975- 1977; Mbr Handbook Committee 1977-1980; Mbr ASHRAE Jour Ctte 1982-present; Mbr of several other technical committees. John Henry Newman Honor Key, New England Province, 1948. Lt Col USAF (Ret). Elevated to grade of "Fellow–, ASHRAE, Jan 1980. Received "Distinguished Service Award, ASHRAE, Jun 1981. *Society Aff:* ASHRAE, AMS

Richard, Terry G
Business: 155 W. Woodruff Ave, Columbus, OH 43210
Position: Assoc Prof *Employer:* Dept of Engr Mech, OH State Univ *Education:* PhD/Eng Mech/Univ of WI-Madison; MS/Eng Mech/Univ of WI-Madison; BS/Eng Mech/Univ of WI-Madison *Born:* 02/25/45 A native of Tomah WI. Was an NSF Grad Fellow in 1972. From 1973 to 1975 was a Postdoctoral Fellow at the Univ of WI doing research in fusion reactor designs and superconducting electrical energy storage. Part-time employment as a research scientist with Battelle-Columbus Labs, Kimberly-Clark Corp, US Naval Weapons Ctr, Columbia Gas Systems Corp. Research completed in the areas of optical methods of structural analysis, fracture characterization of polymeric materials, time-dependant mechanical properties of coals and shales, cryogenic materials response and fatigue. *Society Aff:* SESA, ASME, ASTM

Richards, Earl F
Business: Dept of EE, Rolla, MO 65401
Position: Prof of EE. *Employer:* Univ of MO. *Education:* PhD/EE/Univ of MO; MSEE/EE/MO Sch of Mines & Met; BSEE/EE/Wayne State Univ. *Born:* 3/11/23. Earl F Richards has twenty-eight yrs experience in Univ Teaching & ten yrs ind experience in the field of EE. His area of interest is in Modern Automatic Control Theory with applications to power systems design, system modeling, analysis & stability. He has also been active in the application of digital comp systems to control systems. He also serves as a consultant to the legal profession. He has over seventy technical publications in the area of control theory & is a mbr of many honorary socs. *Society Aff:* IEEE, HKN, ΣΞ, TAUBΠ

Richards, Earl L
Business: 1729 Penn Ave, Monaca, PA 15061
Position: VP & Tech Dir. *Employer:* NF & M Corp. *Education:* BS/Chem/TX Western; MS/Met Eng/Carnegie Tech. *Born:* March 1926. MS from Carnegie Tech; BS from Texas Western; high sch in Houston Texas. USAF in WW II; fuel element

Richards, Earl L (Continued)
dev. Westinghouse Bettis, Supr Zirconium Technologist to 1959. Joined Reactive Metals in Niles Ohio as Titanium Process Dev Engr & later Asst Ch Metallurgist. Moved to Titanium West Inc in Reno Nev as VP Engr 1968. Developed modern titanium melting methods, is responsible for tech sers & prod quality. VP Genl Mgt respon for operations. With NF & M in Monaca Pa since 1974; responsible for titanium quality & engrg ser. Mbr ASM; past chmn Warren Ohio chap. Hobbies: work, family, golf & community affairs through Kiwanis Club. *Society Aff:* ASM, ASTM.

Richards, Elmer A
Business: 708 Third Ave, New York, NY 10017
Position: Partner *Employer:* Mueser, Rutledge, Johnston & DeSimone *Education:* BSCE/CE/Lehigh Univ; MS/CE/Northwestern Univ. *Born:* 02/15/30 Raised in Somerville, NJ. Previously employed with US Bureau of Reclamation and Howard Needles Tammen & Bergendoff. With Mueser, Rutledge since 1952 - Partner of firm since 1973. Specialist in soil mechs and foundations throughout entire career. Active in projs throughout US including, bridges, dams, tunnels, bldgs, industrial installations, bulkheads, piers etc. Conslt to Corps, contractors, state and federal government, municipalities, legal firms and architects. Recognized as Specialist in Special Foundation Problems involving difficult soil and rock conditions. Mbr MOLES. *Society Aff:* ASCE, ASTM, ACBC, MOLES.

Richards, John C
Home: 4409 Chalfont Pl, Bethesda, MD 20016
Position: VP - Govt Affairs *Employer:* M.W. Kellogg Co. *Education:* MS/Engg Econ/Stevens Inst of Tech; BS/Chem Engg/Univ of UT. *Born:* June 1921. US Navy sers in European & Pacific theaters 1943-46. MW Kellogg Co R/D 1946-52; Process design & dev 1952-56, specializing in fluid solids processes; elected VP Kellogg in Iran 1967; Coml VP of Kellogg in London in 1970 - responsible for European-Middle East Sales. Currently VP - Govt Affairs - for M.W. Kellogg Co. in Wash DC. Hobbies: classical music & golf. *Society Aff:* AIChE.

Richards, Kenneth J
Business: Kennecott Minerals Company, 1515 Mineral Square, Salt Lake City, UT 84147
Position: V Pres-Process Tech *Employer:* Kennecott Corp *Born:* Nov 29, 1932. Native of L.A. Calif. BS & PhD from Univ of Utah. Process Engr for Union Oil & CF Braun. Served with USAF 1957-67 in USIA & Aerospace Res Labs. Kennecott Copper Corp 1967-present. Made Res Dir in 1974 with responsibility for R/D in electrometallurgy, mineral processing, pyrometallurgy, hydrometallurgy, proc control & analytical techniques. Mbr SME, The Metallurgical Soc, AIME, ASM, AIChE; served on sev natl AIME cttes, authored num tech papers & patents. Pres of TMS for 1981. Made V Pres in Oct 1979 and is responsible for the process tech programs in the fields of mineral processing, pyrometallurgy, hydrometallurgy, process control, process evaluation, and materials technology.

Richards, Richard T
Business: 800 Kinderkamack Rd, Oradell, NJ 07649
Position: Mgr Hydraulic Engrg *Employer:* Burns and Roe, Inc *Education:* MS/Struct Eng/Columbia Univ; BE/CE/Yale Univ *Born:* 10/14/22 Native of South Orange, NJ. Officer Navy Civil Engr Corps 1943-46. Civil and hydraulic engr with Ebasco Services Inc 1948-59. Supervising Civil Engr and Mgr Hydraulic Engrg Burns and Roe, Inc 1959-1984 (present). Author of over 40 technical papers on hydraulic engrg related to power plants and pipelines. ASCE Technical Committee mbrship. Active in community affairs, sailing and photography. *Society Aff:* ASCE

Richards, Robert B
Business: 165 W Wacker Dr, Chicago, IL 60601
Position: Senior Executive Consultant. *Employer:* De Leuw, Cather & Co. *Born:* April 1, 1914 in Milwalkee, Wisc. Graduated Armour Inst of Technology (now Ill Inst of Technology) BSCE 1939. With De Leuw, Cather Org since 1936, serving as Pres & Dir since 1966 & as Chmn & Pres since 1971. Retired 4/1/79. P Pres of AICE (now ACEC); P Dir ARBA, IRF & IRT; Tr WSE; mbr ACEC, ARBA, ASCE, AWWA, ITE, Inst of Engrs, Australia, IBTTA, IRF, NSPE, PEPP, WSE & IRT. Holds prof engr licenses in state of Illinois. Hobbies: golf & philately.

Richards, Robert W
Business: 2600 Centruy Pkwy, NE, Atlanta, GA 30345
Position: Sr VP. *Employer:* Stanley Cons Inc. *Education:* MBA/Mktg/Univ of Chicago; BSME/Purdue. *Born:* 4/4/28. Hd of Southeastern Div & Sr VP of Stanley Cons Inc; Student Engr, boiler erector & sales engr for Babcock & Wilcox Co 1952-64. With Stanley Cons since 1964. Formerly mgr of Chicago regional office. Mbr ASME; AISE; TAPPI; NSPE; AMA. *Society Aff:* ASME, AISE, TAPPI, NSPE, AMA.

Richardson, Brian P
Business: 100 Colony Square, Ste 2301, Atlanta, GA 30361
Position: Partner *Employer:* Kinetics Consulting Group *Education:* BS/ME/Univ of MA *Born:* 4/1/42 Partner/Principal of Kinetics Consulting Group in engrg for environmental/energy/industrial processes since 1979. Registered in three states. Previously Process and Utilities Dept Head for Stanley Consultants; Industrial Sales Engr for Northeast Utilities; Product Service Engr for General Electric Large Steam Turbine; Peace Corps Volunteer assigned to Public Utilities Authority, Liberia, 1964-1966. Authored over 50 reports and papers on engrg studies of: steam generation, solid and residue fuels, fuel conversions, air pollution control, powerhouse refurbishing, energy supply planning, cogeneration, and environmental impact. Design mgr for facility designs involving boiler fuel conversions, air pollution control, and condenser cooling water systems. *Society Aff:* ASME, NSPE

Richardson, C D
Business: PO Box 1500, Huntsville, AL 35807
Position: Dir Systems Tech Proj Office. *Employer:* BMD Systems Command. *Education:* BS/Aero Engr/Univ of MI. *Born:* 4/30/34. Responsible for overall direction of the Ballistic Missile Defense Systems Tech Proj res & dev effort, which includes technical & fiscal controls of an annual budget of over $100 million, involving several maj contractors & govt agencies. Also act as principal civilian engg consultant & advisor to the Commanding Gen of the Ballistic Missile Defense Systems Command. *Society Aff:* AIAA, ADPA, AUSA.

Richardson, Elmo A Jr
4875 Riverside Drive, P.O. Box 13147, Macon, GA 31208
Position: Pres. *Employer:* Tribble & Richardson, Inc. *Education:* BCE/Civ Engg/GA Inst of Tech. *Born:* 11/11/36. A native of Macon, GA. Grad of GA Tech-served in US Army Combat Engrs from 1960 to 1962. Served as proj engr and associate with Tribble & Assoc from 1962 to 1970. In 1970 became exec vp of Tribble & Richardson, Inc, Consulting Engrs. In 1978, assumed position of Pres of firm. Served as State Dir of the GA Soc of PE, the Consulting Engrs Council of GA and the GA Water & Pollution Control Assn. Served as Pres of Macon Chapter, GA Soc of PE and the Macon Area GA Tech Alumni Assn. Past Pres - Conslt Engrs Council of GA; Natl Dir - Amer Conslt Engrs Council; Natl Dir - Water Pollution Control Fed; Diplomate - Amer Acad of Environ Engrs; Mbr - Macon Rotary Club. Enjoys fishing, hunting and golf. 1987 President-elect of Greater Macon Chamber of Commerce. *Society Aff:* WPCF, AWWA, ACEC, NSPE, ACSM, AAEE

Richardson, Everett V
Engrg Res Ctr, Colorado State Uni, Ft Collins, CO 80523
Position: Prof of Civil Engrg. *Employer:* Colorado State Univ. *Education:* BS/Civil Eng/CO State Univ; MS/Civil Eng/CO State Univ; PhD/Civil Eng/CO State Univ. *Born:* Jan 5, 1924 Scottsbluff Nebr. Hydraulic & Res Engr wth U S Geol Servey 1949-68. Prof of C E & Administrator, Engrg Res Ctr, Colo State Univ 1968-83. Prof of CE and Dir Hydr Lab 1983- . Major respons & ints are river mech, sedimentation, hydraulics, irrigation, fluid measurement & rural dev. Dir Consor. for

Richardson, Everett V (Continued)
Int. Dev's Egypt Water use and Management Proj. 1977-1985, Dir Egypt Irrigation Impr. Proj. 1985- . P Chmn of Sedimentation Ctte, Hydraulics Div, Waterways Ctte & Res Ctte, Waterways, Harbors & Coastal Engrg Div of ASCE. J C Stevens Award 1961 ASCE; U S Gov Fellow MIT 1962-63. Fellow ASCE; Mbr US Cong. of Irr. & Drainage, Sigma Xi, AAAS, Sigma Tau, Xi Epsilon, & Gama Sigma Delta. *Society Aff:* ASCE, USCID, AAAS.

Richardson, Gregory N
Home: 1111 Braemar Ct, Cary, NC 27511
Position: VP. *Employer:* Civ Engg & Appl Res. *Education:* PhD/Engg/Univ CA; MS/Engg/Univ CA; BS/Engg/CA State College. *Born:* 4/18/47. CA educated & licensed. Proj engr for McClelland Engrs in 1976, specializing in offshore instrumentation & pipeline route surveys. Asst prof at NC State Univ with undergrad & grad teaching responsibilities & Dir Soil Mechanics Lab. Formed Civ Engg & Appl Res in 1979 to asst consulting engrs in instrumentation and analysis. Also provide res facilities to cos developing products for the Civ Engg market place. Extensive res experience in soil reinforcing & earth retention structures. ASCE: J James R Croes Medal 1978. *Society Aff:* ASCE, TRB, ASTM, SSA.

Richardson, John H
Business: 12700 Hillcrest Rd, Suite 125, Dallas, TX 75230
Position: Regional VP *Employer:* Henningson, Durham & Richardson, Inc *Education:* MS/Eng/Univ of TX-Austin; BS/CE/Univ of TX-Austin *Born:* 06/17/52 Responsible for all engrg and architectural proj for HDR in the South. Engrg experience includes mgmt and design of projects in the fields of traffic and transportation, land use and flood plain reclamation, water and wastewater treatment and solid waste mgmt, major highway and structural projects. With HDR since 1970. Enjoy family, travel and committment to Christian church outreach. *Society Aff:* ASCE, NSPE, SAME, CEC-T

Richardson, John L
Home: 715 Wimbleton, Birmingham, MI 48008
Position: manager Of Mktg. *Employer:* Oxy Metal Indus Corp/UDYLITE. *Education:* PhD/Chem Engg/Stanford Univ; BS/CHem Engg/Stanford Univ. *Born:* Jan 29, 1935. BS, PhD Chem Engrg Stanford Univ. Reg P E Calif. Mbr AIChE, Tau Beta Pi, Sigma Xi, Phi Lambda Upsilon. 13 yrs with Ford Motor Co (Aeronutronic Div, Newport Beach Calif) in appl res, dev, engrg, & mgmt pos. Spec areas of emphasis then incl desalination, membrane sys (reverse osmosis & ultrafiltration), & transport in chemically reacting sys. Co-dev sev membrane module prods now patented. From 1974-78, respon for businesses within Oxy which are principally concerned with the dev, design, const, & installation of sys for recovery & effluent treatment. Since 1978, respon for mktg for UDYLITE, serving the metal & plastics finishing industries with plating chems, equipment, & services.. *Society Aff:* AIChE, ESD.

Richardson, John M
Business: Energy Engineering Board, JH-424, Natl Acad of Sci, 2101 Constitution Ave, Washington, DC 20418
Position: Associate Director, Energy Engineering Board. *Employer:* Natl Acad of Sciences. *Education:* PhD/Physics/Harvard Univ; MA/Electronics/Harvard Univ; BA/Physics/Univ of CO. *Born:* Sept 5, 1921. Mid-career educ at Kennedy Sch of Gov, Harvard Univ 1966- 67. US Naval Res (radar) 1943-46. Natl Acad of Scis, 1979- ; Associate Director, Energy Engrg Bd. Natl Telecommunications & Info Administration, 1978- 79; Chief Scientist. Office of Telecommunications, 1969-78; Deputy Dir then Dir ; respon for overall agency oper in telecommunications res & engrg & spectrum mgmt & pol analysis. Natl Acad of Engrg 1968-69; Exec Secy of Ctte on Telecommunications. Natl Bureau of Stds 1952-68; res & instrumentation in radio frequency measurement stds & atomic stds of time & frequency. Fellow IEEE. U S Dept of Commerce Gold Medal for Except Serv 1964. Chairman, IEEE Committee on Communications and Information Policy, 1987- . Married; 4 children. *Society Aff:* IEEE, APS, AAAS, URSI.

Richardson, Joseph G
Home: 12434 Woodthorpe, Houston, TX 77024
Position: Partner. *Employer:* Richardson, Sangree & Sneider *Education:* SM/Chem Engg/MIT; BS/Chem Engg/TX A&M Univ. *Born:* 10/28/23. Joseph G Richardson joined the Humble Oil Co (now Exxon Co, USA) in the Production Res Div in 1948. He conducted and supervised res studies on various oil recovery projects and on operating plans for maj oil fields throughout the world. He has published numerous technical papers. He has served the Soc of Petroleum Engrs of AIME on numerous committees and was Sr Technical Editor of the *Journal of Petroleum Technology* from 1976 through 1978. He received the SPE's Uren Award in 1977 and the DeGolyer Medal in 1978. He served on the SPE Bd from 1979-81 and is a distinguished mbr of SPE. He retired from Exxon in 1986 and is a consultant. *Society Aff:* SPE.

Richardson, Lee S
Home: 210 N. Garfield, Moscow, ID 83843
Position: Prof Metallurgical Engrg, Director, Institute for Materials & Advanced Processes *Employer:* Univ of Idaho. *Education:* ScD/Met/MIT; SM/Met/MIT; SB/Met/MIT. *Born:* 3/17/29. Jr Met at ORNL 1951-2, Met at Westinghouse Elec 1955-6, Ind Staff Mbr LASL 1956-1958, Supv Met at Westinghouse 1958-1963. Joined Foote Mineral Co in 1963 as Dept Mgr, Met & Ceramics. Became Dir of Res, Engg, & Dev in 1969. Joined EG&G ID in 1977 as Mgr, Mtls Sci. Became Mgr., Hot Cell Oper. in 1981, and Tech Ldr, Matls Proc, in 1983. Joined the Univ of Idaho in 1986 as Prof. of Metal Engrg & Director of Institute for Materials & Advanced Processes. Now specializing in plasma & matls sci and engrg. *Society Aff:* ASM, TMS/AIME, ΣΞ.

Richardson, Neal A
Home: 30823 Cartier Dr, Rancho Palos Verdes, CA 90274
Position: Dir, Advanced Systems Engg; Engr Consultation *Employer:* Self *Education:* PhD/Engg/UCLA; MS/Engg/UCLA; BS/Engg/UCLA. *Born:* 3/14/26. Res & teaching staff, UCLA, 1949. Infantry officer, US Army, WWII & Korea. Asst dir of engg res inst, UCLA, 1957. Ntl Sci Fdn, Sci Faculty Fellow, 1960. TRW sr technical staff, 1962. Mgr, energy conversion, TRW Systems & Energy, 1966. Mgr, coal conversion progs, TRW Systems & Energy, 1975. Dir, advanced systems engg, TRW Automotive Worldwide, 1978. Mbr of the Technical Bd, Soc of Automotive Engrs, Inc, 1977-79. Retired from TRW 1986. Doing Engr Consultation. *Society Aff:* SAE, ΣΞ.

Richardson, Peter D
Business: Box D, Brown University, Providence, RI 02912
Position: Prof, Chmn Faculty Executive Committee; Chair, University Faculty. *Employer:* Brown Univ. *Education:* DSc/Physiology/Univ London; DSc (Eng)/Mech Engrg/Univ London; PhD/Mech Engrg/Univ London; MA/ad emdem/Brown Univ; BSc (Eng)/Mech/Univ London *Born:* Aug 1935 W Wickham England. Visiting Lectr, Asst Prof, Assoc Prof Brown Univ 1958-68, Full Prof 1968- . Prof of Engrg and Physiology 1984- . Exec Ctte, Div of Engrg 1968-70. Chmn of Ctr for Biomed Engrg (Exec Ctte) 1970- . Res in thermophys properties, heat transfer, artificial internal organs & biorheology. Approx 150 papers publ. Cons to U S Gov & indus. Sr Visiting Fellow Univ of London 1967-68; Prof d'Echange Univ of Paris 1968; Visting Prof Orta Dogu T U Ankara 1969; A von Humboldt Award 1976. DSc (Eng) Univ of London 1974. Elected Fellow of The Royal Society of London 1986. Laureate in Medicine, Ernst Jung Foundation 1987. *Society Aff:* ASME, ASEE, ASAIO, ESAO, BMES.

Richardson, William H
Business: 20 N. Wacker Dr, Chicago, IL 60606
Position: Partner *Employer:* Alvord, Burdick & Howson *Education:* BS/Civil Engr./University of Illinois *Born:* 06/10/29 Born and raised in Chicago. Attended Lane Technical High School; Wright Jr. College 1 1/2 years and Univ of Illinois-3 1/2

Richardson, William H (Continued)
years. Graduated Feb. 1952, B.S. in Civil Engrg. Immediately joined Alvord, Burdick & Howson, consulting engrs, specializing in water supply and water pollution control. Served as a Civil Engr in Design for one year and then was in responsible charge as a resident engr on construction of approximately $15 million of water facilities for the following six years. Returned to Chicago Office and served as a project engr on design, valuation, feasibility studies on water and wastewater projects, including Louisville, Racine & Kenosha, WI, Evanston, IL; Monrovia, Liberia and many others. Made a partner of the firm 1965, to present. As a partner have been in responsible charge of several hundred million dollars of water & wastewater projects from the report phase to placing in operation. Testified as an expert witness before the Public Service Comm in many states on water rates, valuations and depreciation. President, WSE, 1982-83. Octave Chanute Medal, 1981. AWWA President 1984-85, Fuller Award-1984; Distinguished Civil Engr Alumnus Award, 1984, Univ of Ill., CE Alumni Assn. VP IWSA, 1986-87. Enjoy golf, sports and travel. *Society Aff:* AWWA, ASCE, WSE, IWSA, AAEE, NSPE, WPCF

Richart, Frank E, Jr
Home: 2210 Hill St, Ann Arbor, MI 48104
Position: W.J. Emmons Distinguished Prof. Emeritus of Civil Engrg. *Employer:* Univ of Michigan. *Education:* PhD/Engg/Univ of IL; MS/Civil Engg/Univ of IL; BS/Mech Engrg/Univ of IL. *Born:* Dec 6, 1918 Urbana Ill. Res Asst & Assoc, Dept of Civil Engrg Univ of Illinois 1946-48; Asst Prof Mech Engrg Harvard Univ 1948-52;Assoc Prof 1952-54, Prof 1954-62 Univ of Fla; Prof 1962-77, Chmn 1962-69 Dept Civil Engrg Univ of Mich; W.J. Emmons Dist Prof of CE, 1977-1986, Emeritus, 1986- , Cons Moran, Proctor, Mueser, & Rutledge summers (1953-55, 1957), Office Engrs US Army, NASA. Cons to indus & gov agencies; specialty - soil dynamics & vibrations of foundations. Hon M ASCE; Mbr NAE. *Society Aff:* NAE, ASCE, ISSMFE.

Richey, Clarence B
2217 Delaware Dr, W Lafayette, IN 47906
Position: Prof Emer, Agri Engrg. *Employer:* Purdue Univ. *Education:* BSAE/Agri Engg/IA State Univ; BSME/Mech Engg/Purdue Univ. *Born:* Dec 1910. Taught farm power & machinery at Purdue & Ohio State prior to WW II. In farm machinery engrg for 27 yrs incl 5 yrs as Mgr, Res & Advanced Engrg, Ford Tractor Div & 5 yrs as Ch Engr, Fowler Div, Massey-Ferguson. Returned to teaching & res in farm machinery at Purdue 1970. Dir ASAE 1961-63. Named Fellow ASAE 1969. Editor-in-Chief 'Agri Engrs' Handbook' 1961 & author of Crop Prod Equip Sect. Inventor or co-inventor in some 80 US pats in farm equip field. 1977 recipient of Cyrus Hall McCormick Gold Metal for "Engg Achievement in Agri-. *Society Aff:* ASAE.

Richman, Marc H
Home: 291 Cole Ave, Providence, RI 02906
Position: Prof of Engrg. *Employer:* Brown Univ. *Education:* ScD/Metallurgy/MIT; MA/-/Brown Univ; BS/Metallurgy/MIT. *Born:* Oct 1936 Boston Mass. Major, Ordnance Corps USAR. Instr in Met MIT 1957- 63. Brown Univ, Div of Engrg, Asst Prof 1963-67 Assoc Prof 1967-70, Prof 1970- . Hd, Matls Sci & Engrg 1967-70, 1975-78. Pres Educ Aids of Newton Inc 1968-71. Visiting Prof univ of RI. affiliate Staff. Mbr, Dept of Medicine Miriam Hosp 1974- . Cons engr spec in prods liability, 1957- . Albert Sauveur Mem Award ASM 1968-69. Fellow Amer Inst of Chemists. Outstanding Young Fac Award ASEE 1969. Reg P E Mass & RI. Cert Ceramic Eng (NICE) affiliate Staff, Dept of orthepaedics & Fractures, RI Hospital 1978- Dir, Brown Univ Matl Res Lab. Central Electron Microscope Facility 1978-1986. President, Marc H. Richman, Inc. 1981-pres., Editor, Soviet Physics: Crystallography, 1970- , Editorial Board Metallography, 1970- , Editorial Board J. Founsie Engineering, 1985- . *Society Aff:* ASM, AIME, AWS, ACS, EMSA, ASEE, AIC, NSPE, SAE, ACA, ASTM, IAAI, ATLA.

Richman, Peter
Home: 22 Barberry Rd, Lexington, MA 02173
Position: President *Employer:* KeyTek Instrument Corp. *Education:* MS/Mathematics/NY Univ; BS/EE/MIT. *Born:* 11/7/27. Co-founded Rotek Instrument Corp Lexington MA 1960; VP until 1964 when it was acquired by Weston Instruments, Newark, NJ. Pioneered designs of full line of precision ac measuring instrumentation. VP Weston Instruments, 1964-67. Engg/mkting/mgmt cons, 1967-75 - prod definition & design, key mgmt asst, plus a var of prod devs & licenses. Founded KeyTek Instrument Corp 1975, Wilmington, MA, Pres, respon for line of instruments & sys for surge instrumentation. Prior to 1960, Ch Engr of Epsco Instrument Div; Asst Ch Engr Reeves Instrument Corp. Over 30 papers, 25 pats. Fellow IEEE, Sr Mbr ISA, past mbr Eval Panel for Elec Div of NBS. Mbr Tau Beta Pi, Sigma Xi. *Society Aff:* IEEE, ISA, ТВП, ΣΞ.

Richmond, Frank M
Home: 1127 Folkstone Dr, Pittsburgh, PA 15243
Position: V P - Research & Development *Employer:* Universal-Cyclops Spec Steel Div. *Education:* PhD/Met Engg/Univ of Pittsburgh; MS/Met Engg/Univ of Pittsburgh; BS/Met Engg/Univ of Pittsburgh. *Born:* July 7, 1922. Native of Washington Pa. Pilot USAF 1943-45. Began career with Ohio Works of U S Steel, then joined Westinghouse Res 1947. Joined Universal-Cyclops (Cyclops Corp) 1953, assuming pos (V P - Tech Dir) 1966 & present pos (V P-R&D) 1981. Respons incl R&D, Tech Info Ctr and licensing. Publ papers & presented talks on vacuum melting, stainless steels, & high temp aloys. Num tech ctte mbrships & chmnships. ASM Fellow. Ints: platform tennis, skiing, tennis, gardening, golf. *Society Aff:* ASM, AIME, AISI, ISI.

Richter, Alan F
Business: 11 Robinson St, Pottstown, PA 19464
Position: Pres. *Employer:* STV Engrs Inc *Education:* MS/Civ Engrg/Drexel Univ, 1964; BS/Civ Engrg/Drexel Univ, 1957. *Born:* 9/26/32. Native of Phila PA. Served in the US Army Corps of Engrs on Active Duty in 1958 and in the reserves until 1962. Highest rank was Capt. Have been with Sanders & Thomas since 1962 where I began as a Proj Engr. I am presently pres and chief operating officer of STV Engrs Inc, a co of 1200 persons. S TV is the parent co of Sanders & Thomas Inc, Seeyle Stevenson Value & Knecht, and Lyon Assoc Inc. *Society Aff:* NSPE, ASCE, ASHE, XE.

Richter, Donald W
Home: 4407 Riverview Dr, Middletown, OH 45042
Position: VP *Employer:* Corp Finance Assoc *Education:* PhD/Agr Structures/Cornell Univ; BS Agr Engrg/Cornell Univ. *Born:* 10/22/26. Specialist with Armco Steel Corp in agri apli of specialty steels. Transferred to Metal Prod Div & supervised Sales Engrg program for agri bldgs. In 1966 became Ch Sales Serv Engr for bldg sys; Dist Mgr 1967; Mgr Order & Serv Engrg 1969. 1976, became Mgr, Planning for Metal Products Div. 1980 became Gen Mgr, Corp Development at Corp. Hdqtrs. 1982 became Gen Mgr, Planning & Dev, Fabricated Products & Services Group. 1983 became VP, Corp Finance Assoc engaged in consulting on mergers and acquisitions, business financing and general mgmt. Received Metal Bldgs manufacturers Assn Award from ASAE 1968. Mbr Amer Soc of Agri Engrs, serving in var pos in Structs & Environ Divs, incl Div Chmn, Mbr. Planning Forum serving as Dir SW OH Chapter *Society Aff:* ASAE, PF.

Richter, John A
Home: 3414 Monteen Dr, Orlando, FL 32806
Position: Cons Engr. *Employer:* Various Clientele. *Education:* BA/Met Engr/Univ of WA Seattle, WA. *Born:* July 1914 Centralia Wash. Grad schools - SMU, TCU. Boeing 1938-52, heavy & precision mech indus 1952-58, aerospace 1958-70, V P QC Lab 1970-72, Aerospace 1973-74, cons engr 1970 to date in prod liability - personal injury - process beneficiation struct integrity - safety. Reg P E since 1947 Fla, Ga, La, MIss, S C, N C, Tenn, Tex, Wash. Sr Mbr AIME, ASM, AWS, ASTM. ASM Pres Award, AWS Dist Meritorious Award; past Natl Dir AWS. Chmn Adv Comm

Richter, John A (Continued)
for Dept of Matl Sci & Engrg Univ of Fla - Gainesville. Received Paul Harris Fellow Award - Oct 30, 1979 (ROTARY); various patents. *Society Aff:* AIME-TMS, ASM, AWS.

Ricigliano, Anthony R
Home: 4020 Quakerbridge Rd, Trenton, NJ 08619
Position: Deputy Administrator *Employer:* NJ Dept of Env Protection *Education:* MS/Environ Sci/Rutgers Univ; BS/CE/NJ Inst of Tech *Born:* 08/02/29 raised and educated in NJ. Served in the US Army Corps of Engrs during the Korean conflict (1951-1953). Employed by the State of NJ in its water pollution control program since 1959. Assumed current position of Asst Dir, Div of Water Resources in 1975 and have been responsible for managing the Federal Construction Grants Program for the planning, design and construction of wastewater facilities. Retired from State of NJ and presently in private conslt practice in the field of Environ Engrg. Served a two yr term (1978-1980) as Pres of the NJ Water Pollution Control Association; Water Pollution Control Federation Dir (1981-1983); Water Pollution Control Federation Exec Committee (1981). Licensed PE in the States of NJ and FL Sept 1, 1984. *Society Aff:* WPCF, AAEE

Rick, William B
Business: 5620 Friars Rd, San Diego, CA 92110
Position: President *Employer:* Rick Engrg Co *Education:* BS/CE/UCLA *Born:* 09/02/27 Born in San Diego. After service in USAF as sanitary engr, went into consulting practice with father, Glenn A. Rick, in 1955. Have been pres of surveying, planning, engrg & environmental consulting firm since 1970. Hon mbr & former trustee of ULI; former pres (1971) Calif council of civil engrs & land surveyors. *Society Aff:* ASCE, ULI, AICP

Rickard, Corwin L
Business: P O Box 81608, San Diego, CA 92138
Position: Executive VP. *Employer:* Genl Atomic Co. *Education:* PhD/Engrg/Cornell Univ; MS/Mech Eng/Univ of Rochester; BS/Mech Eng/Univ of Rochester. *Born:* 9/26/26. in Medina, OH. 1949-53 taught engrg courses at Univ of Rochester & at Cornell Univ. Following this, spent 3 yrs at Brookhaven Natl Lab, Nuclear Engrg Dept. Joined Genl Atomic 1956 & was a mbr of team that initiated dev of the High Temp Gas-Cooled Reactor. Became VP of Genl Atomic 1967. In 1971 became respon for Advanced Energy Sys, incl dev of the Gas-Cooled Fast Breeder Reactor, the Gas Turbine HTGR Sys, the HTGR for high temp process heat applications, for the Genl Atomic programs on controlled thermonuclear fusion. Apptd Exec VP in July 1977, with current respon for Gas-Cooled Reactor Programs well as Genl Atomic's Fusion Program. *Society Aff:* ANS, ASME; USNC-WEC.

Rickelton, David
Home: 3413 Highview Rd, Charlotte, NC 28210
Position: Consultine Engineer. *Employer:* Self. *Born:* Nov 1916. Educ Brooklyn Polytech Inst & Pratt Inst. Employed by Aeronca Inc Environ Controll Group 1940- . Successively draftsman engr, engrg cons. V P since 1964. Sales & mgmt functions. 1977-78 Genl Mgr. Auth of design & application of high velocity air cond sys. Many publ articles. Lectured extensively in U S & foreign countries. Served with the AUS 1942-46 from Pvt to Capt; retired Major USAR. Pres 1974-75 ASHRAE, Bd of Dir 1965-75. Fellow 1975. *Society Aff:* NSPE, AEE.

Ricketts, William Dale
Business: P O Box 8126, Wichita, KS 67208
Position: President. *Employer:* World-Wide Const Services Inc. *Born:* 2/14/29. Mech. Engr. 33 yrs in const of chem, petro chem-refining process industry. Assumed current respon as Pres on May 1969 of World-Wide Fabrications Inc. and Process Plant Services, Inc., which provided design, engrg, procurement & const services to the above process industries. Prior to 1969 was Pres of Litwin Const & V.P. & Dir of the Litwin Corp in Wichita, Kansas. President & C.E.O. of United States Gasohol Corporation of Wichita, Kansas; plant located in Lockeford, California. *Society Aff:* ISA, ASME, AIChE, SAME, API, AWS.

Ricklin, Saul
Business: 386 Metacom Ave, Bristol, RI 02809
Position: Chrmn. *Employer:* Dixon Indus Corp. *Education:* BS/Chem Eng/Columbia Univ; ChE/Chem Eng/ Columbia Univ. *Born:* 9/5/19. Ch Process Engr M&T Chems 1940-46. Army Air Corps 1946-47. Asst Prof Brown Univ 1947-54. Owner Ricklin Res Assocs (cons) 1954-59. VP R&D Dixon Corp 1954- 70. Pres Dixon Indus Corp (engrg plastics processing) 1970-78. Chrmn 1978- ; Exec VP & Dir NTN-RULON INDUS CO LTD (Japan) 1967- . Dir Dixon Italia sPa 1970- . Dir Entwistle Co 1975- Dir Electron Fusion Devises Inc 1979- . Chmn Fluorocarbons Div of Soc Plast Indus & Dir & Exec Soc Plastics Indus 1975- . Dir - Founder Repertory Theatre Prov RI, Dir ACS Industries 1980 -. *Society Aff:* ACS, AIChE, SPE, ASCE, ASTM.

Rickover, Hyman G
Business: Adm, Naval Sea Sys Cmd (08), Washington, DC 20362
Position: Dir, Div of Naval Reactors. *Employer:* US Navy. *Education:* MS/Elec Engg/US Naval Academy Post-Grad Sch & Columbia Univ; BS/-/US Naval Academy. *Born:* Jan 27, 1900. Grad US Naval Acad 1922; advanced through grades to Admiral 1973. Engaged in dev of all nuclear-powered ships for US Navy 1947- . Respon for dev of USS Nautilus, world's first nuclear-powered submarine, & for the first nuclear-powered elec util central station at Shippingport, Pa. Egleston medal from Columbia Engrg Alumni Assn 1955; ASME Geo Westinghouse Gold Medal 1955; Pupin medal 1958; Gold Medal from U S Congress 1959; Enrico Fermi award from U S Atomic Energy Comm 1965; Presidential Medal of Freedom 1980. Author 'Educ & Freedom' 1959; 'Swiss Schools & Ours: Why Theirs are Better' 1962; 'Amer Educ - a Natl Failure' 1963; 'Eminent Americans - Namesakes of the Polaris Submarine Fleet' 1972; 'How the Battleship Maine was Destroyed' 1976. Presently Dir Div of Naval Reactors, U S Dept of Energy, Wash D C & Dep Cdr for Nuclear Propulsion, Naval Sea Sys Command, Wash DC. *Society Aff:* NAE.

Riddell, Matthew D R
Business: 222 South Riverside Plaza, Chicago, IL 60606
Position: Partner. *Employer:* Greeley and Hansen. *Education:* MS/Sanitary Engg/ Harvard Grad Sch of Engg; SB/Engg Sci/Harvard College. *Born:* 3/13/18. Employed as Asst in Sanitary Engg by Harvard Grad Sch of Engg, Cambridge, MA 1940-41. Served as commissioned officer in US Navy 1941-46. Employed by Greeley and Hansen as asst engr 1946-48, prin asst engr 1948-57, partner 1957-85, consultant 1986- date. Respon for investigations and reports, designs and services during construction of water, wastewater and solid wastes facilities. pres, APWA Inst of Water Resources 1979-80. Pres AAEE 1980-81. *Society Aff:* AAEE, APWA, ASCE, AWWA, WPCF.

Riddick, Edgar K, Jr
Business: 1600 First National Building, Little Rock, AR 72201
Position: President *Employer:* Riddick Engrg Corp *Education:* BS/ME/Univ of AR *Born:* 07/16/23 Walnut Ridge, Lawrence Country, AR. Graduated Univ of AR 1948; Phillips Petroleum Co 1948-1955; Sr Partner Blass/Riddick/Chilcote A/E 1955-1978; Pres of Riddick Engrg Corp from 1978 to present. Prof Activities: State: Pres of ASHRAE; Chrmn PEPP; Charter Mbr AR Engrg Foundation; Chrmn ACERS; Engrg Advisory Council of Univ of AR at Little Rock; received ASME Centenial Award. National: Chrmn Coordinating Ctte on Energy of ACE (now AAES) Chrmn, ASHRAE Energy Ctte, Mbr ASHRAE Energy Council; Chrmn NSPE/PEPP Energy Conservation Mgmt Ctte; Mbr, Fed A/E Council Engrg Building Standards. Patentee in Field. Exec VP, Nathaniel Curtuis, Riddick, Hieple, Architecture & Planning. *Society Aff:* ASHRAE, ASME, NSPE

Rider, Bobby G
Business: PO Box 3986, Odessa, TX 79760
Position: Senior VP *Employer:* El Paso Products Co. *Education:* BS/Pet Ref Engg/Univ of Tulsa. *Born:* 1/18/25. Native of OK. Received degree 1950 after serving in US Marine Corps 1942- 1945. Held position of Refinery Superintendent when co was acquired by El Paso Natural Gas Co in 1956. Held managerial & officer level positions with present co in petrochem mfg & in comp services until elected VP of Admn in 1974. Assumed present responsibility as Senior VP in 1980. *Society Aff:* AIChE.

Ridge, John D
Home: 1402 NW 18th St, Gainesville, FL 32605
Position: Visiting Prof of Geology. *Employer:* Univ of Florida. *Education:* PhD/Econ Geology/Univ of Chicago; SM/Geology/Univ of Chicago; SB/Geology/Univ of Chicago *Born:* July 1909. Native of Cincinnati OH. With Corps of Engrs US Army WW II; Lt to Lt Col. now Col Retd. Prior to WW II, with U S Steel, Natl Pk Serv, Cerro De Pasco Corp. With N J Zinc Co 1946-47, Penn State Univ 1947-75 - Dept Hd 1951- 75; Univ of FL, Adj Prof 1975-80, Acting Chmn 1980-81, Visiting Prof 1980-83; Council of Economics Award 1972 AIME. Editor Graton-Sales Volumes 1964-68; Krumb Lectr 1971-72. Council Soc of Econ Geologists 1974-77. Author Annotated Bibliography of Mineral Deposits of the World, vols 1, 2, & 3 1972, 1976, 1983. First V P Intl Assn on the Genesis of Ore Deposits 1972-76; P Pres 1980-84; Pres 1976-80 Council of Economics AIME 1962, Mbr Coord Comm Soc Mineral Dep Geols 1980-1984. *Society Aff:* SEG, MSA, AIME

Ridgeway, Joseph A, Jr
Business: 140 Marconi Blvd Rm 400, Columbus, OH 43215
Position: City Engineer. *Employer:* City of Columbus/Div of Engineering and Construction. *Education:* BCE/Transportation/OH State Univ; Cert/Traffic Enggf/Traffic Inst, Northwestern Univ. *Born:* Sept 1939. Reg P E Ohio, Florida. Employed as city engineer since December 1984; with Div of Traffic Engrg City of Columbus Ohio between 1961 and 1964, holding pos of Signal Engr, Opers Engr and Asst Ch Traffic Engr. Has held var pos in Ohio Sect ITE; Uniformite Ctte Chmn, Dir, SecyTreas, V P, Pres (1975). Dist III - ITE Secy 1974-75. Mbr Amer Public Works Anns, Catholic Men's Lucheon Club (P Pres), Enjoy tennis, antique & sports cars. *Society Aff:* ITE, APWA.

Ridgway, David W
Home: 1735 Highland Pl, Berkeley, CA 94709
Position: Exec Dir, Chem Educ Matl Study. *Employer:* Univ of Calif *Education:* MBA/Bus/Harvard; AB/Economics/UCLA. *Born:* 12/12/04. Los Angeles Calif. m. Rochelle Devine 1955. Motion Picture Production, RKO Studios Hollywood 1930-41; Motion Picture Specialist WPB Washington 1942-43; Lt Cdr USN 1943-46; Prod Mgr, Producer, Encyclopedia Britann Films, Wilmette Ill 1946-60; Dir Film Activities & Exec Dir Chem Educ Matl Study, Univ of Calif Berkeley 1960- . Amer Sci Film Assn; Mgr San Fran Sect SMPTE 1971-72; author articles in 'Journal of Chem Educ', 'Sci Activities' 'Canadian Chem Educ'. 11 awards for prod of CHEM Study educ films in chem & Gold Camera Award at the U S Indus Film Festival. *Society Aff:* SMPTE, ISFA.

Ridley, Donald E
Business: 110 Norfolk St, Walpole, MA 02081
Position: Sr VP Regional Operations, Dir *Employer:* Bird-Johnson Co. *Education:* MSME/Mech Eng/Northeastern Univ; BSME/Mech Eng/Syracuse Univ; Dip/Mar Eng/SUNY Mar Col. *Born:* 7/19/27. Grad of NY State Maritime Coll, MS ME Northeastern Univ; BS ME Syracuse Univ; Dip Mar Eng SUNY Mar Col; attended MIT Sloan Sr Execs Prog. Sect Supr at Gibbs & Cox prior to joining present co as Ch Engr 1956. Assumed present pos 1982. Authored sev tech papers for SNAME, ASME, ASNE. P Chmn of the New England Sect SNAME, Honorary VP & Mbr of Executive Com. and Council, Chrmn of Natl Sections Ctte SNAME, Reg. PE, MA; Mbr, Panel on Propellers, Amer Bureau of Shipping. *Society Aff:* SNAME, ASME, ASNE.

Riebman, Leon
Home: 1380 Barrowdale Rd, Rydal, PA 19046
Position: Chairman of Board *Employer:* Amer Electronic Labs Inc. *Education:* PhD/EE/Univ of PA; MS/EE/Univ of PA; BS/EE/Univ of PA. *Born:* 4/22/20. 1944 assigned as Naval Officer to Naval Res Lab in Radar Dev. Res Assoc & Instr at Univ of Penna 1949-51. One of the founders & Pres of Amer Electronic Labs Inc 1950. AEL is a technological, broadly-based electronics co in a var of fields such as electronic warfare, communications, & CATV. IEEE activities incl Receiver Ctte & AFC Subctte & Phila Chap Awards Ctte. IEEE Fellow 1966; Phila Sect IEEE Spec Award 1971. Mbr Army Science Bd. *Society Aff:* IEEE.

Riedesel, Dale L
Business: 708 Shoshone St. East, Twin Falls, ID 83301
Position: Owner *Employer:* Dale L. Riedesel P.E. of L.S. *Education:* BS/CE/Univ of ID *Born:* 08/14/37 Past Pres ISPE; Outstanding Young Engr Award ISPE, 1969; Past State Natl Dir NSPE; Past Chapter Pres and State Dir, ISPE. Following graduation worked for C. W. Glasby & Assocs as EIT. Formed Dale L. Riedesel and Assocs in 1967 and later became Riedesel and Straubhar, Consltg Engrs merging with J-U-B Engrs, Inc in 1971 serving as VP and Transportation Projects Mgr for J-U-B Engrs, Inc six offices also serving as Area Office Mgr, Administer of various offices of J-U-B Engrs, Inc. Currently Owner of his own Conslt Engrg firm. Married (Roberta L. Riedesel), Children, Jeffrey J. and Karen L. *Society Aff:* NSPE, ISPE, IALS, ASCE.

Riegner, Earl I
Home: 926 Stoneybrook Dr, Springfield, PA 19064
Position: Asst Mgr/Engrg Labs. *Employer:* Boeing Co Verton Div. *Education:* BS/ME/Drexel Univ. *Born:* Sept 1930. BS from Drexel Univ. 25 yrs exp in test mgmt, testing & exp perimental stress analysis. Joined Engrg Labs of Boeing Co Vertol Div 1959; in present position responsible for structural, mech & environ testing. Previous exper includes participation during coml dev of photoelastic coating method of experimental stress analysis. Holds patents in photoelastic & deformation analysis; has co-authored papers on photoelastic coatings, fatigue, environ testing & model analysis. Pres of SESA 1976-77. *Society Aff:* SESA, AHS.

Riehl, Arthur M
Business: Speed Sci Sch of Engrg, Dept of Engrg Math & Comp Sci, Louisville, KY 40292
Position: Dept Chrmn and Professor *Employer:* Univ of Louisville *Education:* PhD/Water Quality Control Engg/Univ of Louisville; MA/Math/Univ of Louisville; BS/Math/IN Univ. *Born:* 5/3/34. Instruction and Res in the Speed Scientific Sch of Engg in the Univ of Louisville for twenty-five yrs and am presently Professor and Chrmn of the Engrg Math of Comp Sci Dept in the engg college since 1979. Was a Sr Computing Engr with N American Aviation in 1963 and a consultant for US Navy in computer-aided gear design from 1966 to 1975. Have developed hydrological and water quality computer models for all streams in KY and the OH River on funded res grants totalling over three-hundred thousand dollars, since 1973. *Society Aff:* ASEE, ACM, MAA, IEEE.

Rieman, Francis C
Business: 1101 San Antonio Rd202, Mountain View, CA 94043
Position: Proj Mgr. *Employer:* CSC. *Education:* MS/Computer Sci/UCLA; BS/Control Sys/UCLA. *Born:* Jan 1929 in Iona Minnesota. BS; MS from UCLA in Computer Sci Applications. Served as USAF pilot 1948-53. Currently Proj Mgr for CSC on-site at NASA Ames Res Ctr's flight simulation facilities. Entire prof career involves real-time simulation applications of computers. 1 yr VP & 2 yrs Pres of SCS; V Chmn & Chmn of 4-st SCS Western Region; 5 yrs SCS elected rep to AFIPS Bd/Dir & Exec Ctte; Sr Life Mbr of SCS; mbr AFIPS; mbr Tau Beta Pi; Sigma Xi. *Society Aff:* SCS, IEEE.

Riesenfeld, Richard F
Home: 1337 Harrison Ave, Salt Lake City, UT 84105
Position: Chrmn & Prof *Employer:* Univ of Utah *Education:* PhD/Comp & Info Sci/Syracuse Univ; MA/Math/Syracuse Univ; AB/Math/Princeton Univ. *Born:* 11/26/44 He is Prof and Chrmn of Computer Sci, and Hd of the Computer Aided Geometric Design Group (CAGD) in Computer Sci at the Univ of UT, where he joined the faculty in 1972. Prof Riesenfeld has published and consltd in the area of computer graphics and CAD/CAM. His major res focus for the last yr has been the design and implementation of a CAGD sys called Alpha-1 which advances the states-of-the-art in both high quality computer graphics and free-form surface models. *Society Aff:* ACM, SIAM, NCGA, MAA

Rietman, Noel D
Business: 410 17th St, Suite 600, Denver, CO 80202
Position: VP Northern Div *Employer:* Diamond Shamrock *Education:* MS/Math/West TX State Univ; BS/Petro Eng/TX Tech Univ; BA/Geo/TX Tech Univ *Born:* 01/06/34 Native of Amarillo, TX. With Diamond Shamrock since 1957 as production engr, mgr of petroleum engrg, mgr of production, mgr of drilling and production. Assumed current responsibility of VP and Gen Mgr, Northern Div in 1981. Responsible for all exploration and production activities in the northern div. Bd of Dirs, SPE 1980-81. Enjoy tennis, mountain climbing, running, soccer and classical music. *Society Aff:* SPE, AAPG

Rigas, Harriett B
Business: Dept of Elec Engr and Syst. Sc, East Lansing, MI 48824
Position: Chair, Elec Engr and Syst. Sc *Employer:* Michigan State Univ *Education:* PhD/EE/Univ of KS; MS/EE/Univ of KS; BSc/EE/Queens Univ Canada *Born:* 04/30/34 Worked as an engr at the Mayo Clinic, Minneapolis Honeywell and Lockheed Missile and Space Co. From 1966 she was on the faculty at WA State Univ, serving as chair from 1979-84. From 1984-87 she was Prof and Chair of Elec & Comp Engrg at the Naval Postgrad School. She is currently Prof & Chair EESS at Michigan State U. Dr. Rigas has focused on computer architectures, particularly those for high speed computation. She developed a sys for automatically patching hybrid computers and is currently working on a multiple processor architecture. In 1975-76 Dr Rigas served as Prog Dir for the Sys Theory and Application Prog at the Natl Sci Fdn. *Society Aff:* IEEE, SWE, TBΠ, ΣΞ.

Rigassio, James L
Home: 23 Colony Dr, Summit, NJ 07901
Position: Prof & Chmn Emeritus. *Employer:* NJ Inst of Technology. *Education:* BSEE/Engg/Newark College of Engg; M Eng/Engg/Yale Univ. *Born:* Aug 13, 1923. Lic PE NJ & CA. Mbr ASME; ASEE. Dev Engr for Johnson & Johnson 1949-52; Ch Engr Ethicon Div of Johnson & Johnson 1952-58; Prof at NJ Inst of Technology 1958-present. Chmn Dept of Indus & Mgmt Engrg; Labor Panel of NJ St Bd of Mediation; Cons to Sch Bds. Res in Queuing Theory & Work Methods. Listed in 'Amer Men & Women in Sci.' US Army 1943-46. Retired. *Society Aff:* ASME, ASEE.

Rigdon, Michael A
Business: Institute for Defense Analyses, 1801 N. Beauregard St, Alexandria, VA 22311
Position: Professional Staff Member. *Employer:* Institute for Defense Analyses. *Education:* BS/Mat Sci & Met Eng/Purdue Univ; MS/Mat Sci & Met Eng/Purdue Univ; PhD/Mat Sci & Met Eng/Purdue Univ. *Born:* August 2, 1942. Became charter mbr & was elected 1st pres of Alpha Sigma Mu during jr yr at Purdue; elected to full mbrship in Soc of Sigma Xi as grad student. Spent yr following completion of PhD as a Visiting Scientist at Aerospace Res Labs, Wright Patterson Air Force Base; joined Babcock & Wilcox Co 1972, joined the Institute for Defense Analysis as a Professional Staff Member in 1986. Helped found Central Va Chap of ASM & was elected 1st chmn; has served on ASM's Natl Mbrship Ctte; is an editor for Journal of Materials for Energy Systems. *Society Aff:* ASM, ACS, TMS.

Riggs, Henry E
Business: Dept of Industrial Engg & Eng Mgmt, Stanford, CA 94305
Position: Vice President for Development *Employer:* Stanford Univ. *Education:* MBA/Bus Admin/Harvard Univ; BS/Ind Engg/Stanford Univ. *Born:* 2/25/35. Served as Industrial Economist at Stanford Res Inst from 1960-1963. Joined Icore Industries in 1963 as Treasurer-Controller; became Pres of the co in 1967. From 1970-74, held the post of VP-Finance for Measurex Corp. With Stanford Univ since 1974. Assumed chrmnship of the Dept of Industrial Engg & Engg Mgt in 1978. Teach financial control and mgt courses in the Engg School and small business mgt and new ventures in the Grad School of Business. Received awards for outstanding teaching in 1979 and 1980. Other activities include consulting for tech-based cos, serving on Bds of Dir, and lecturing in exec education prog. *Society Aff:* IIE, ASEE.

Riggs, Lorrin A
Business: Hunter Lab of Psych, Providence, RI 02912
Position: Prof. *Employer:* Brown Univ. *Education:* PhD/Psychology/Clark Univ; MA/Psychology/Clark Univ; BA/Psychology/Dartmouth College. *Born:* Harput, Turkey, June 11, 1912 (parents Am. citizens); Riggs, Lorrin Andrews, psychologist, educator; A.B., Dartmouth, 1933; M.A., Clark U., 1934, Ph.D., 1936; NRC fellow biol. scis. U. Pa., 1936-37; Instr. U. Vt. 1937-38, 39-41; with Brown U., 1938-39; 41-, successively research asso., asst. prof., asso. prof., prof., 1951-, L. Herbert Ballou Found. prof. psychology, 1960-68; Edger J. Marston Univ. prof. psychology, 1968-1977; Univ. prof. Emeritus 1977- . Recipient Howard Crosby Warren medal Soc. Exptl. Psychologists, 1957; Friedenwald award Assn. Research Ophthalmology, 1966; Prentice medal Am. Acad. Optometry, 1973; Kenneth Craik Award, St. John's College, Cambridge U., 1979; Guggenheim Fellow, 1971-72. Mem. Am. (div. pres. 1962-63 Distinguished Sci. Contbn. award 1974), Eastern (pres. 1975-76) psychol. assns., AAAS (v.p., chmn. sect. I, 1964), Internat. Brain Research Orgn., Optical Soc. Am. (Tillyer medal 1969 Frederic Ives Medal, 1982), Nat. Acad. Scis., Am. Physiol. Soc., Intern. Brain Research Organization, Assn. for Research in Vision and Ophthalmology (pres. 1977), Soc. Exptl. Psychologists, Am. Acad. Arts and Scis., Sigma Xi (chpt. pres. 1962-64). Society for Neuroscience. Clubs University (Providence). Author sci. articles on vision and physiol. psychology. *Society Aff:* NAS, AAAS, AAS, APS, APA, ARVO, OSA, IBRO.

Riggs, Louis W
Business: 149 New Montgomery St, San Francisco, CA 94105
Position: Pres. *Employer:* Tudor Engrg Co. *Education:* BSCE/Civil Engg/Univ of CA/Berkeley. *Born:* 6/29/22. Employed since 1951 with Tudor Engg Co; Pres since 1963. Mbr of Bd of Cottrol of Parsons Brinckerhoff-Tudor-Bechtel Joint Ventures respon for engg San Francisco BART & Caracas Metro. Mbr of Bd of Control of Parsons Brinckerhoff- Tudor Joint Venture respon for engg Metropolitan Atlanta Rapid Transit Authority. Noteworthy projs include Salazar Bridge in Lisbon; Columbia River Bridges; port & harbor facilities; US military installations. PPres CEAC; Fellow ACEC, SAME & ASCE; winner of ASCE AP Greensfelder Construction Prize 1967; winner of Golden Beaver Award for Engg 1970. Mbr APWA; Pres SAME. *Society Aff:* CEAC, ACEC, SAME, ASCE, BRAB, APWA.

Rigney, David A
Business: 116 W 19th Ave/OSU, Columbus, OH 43210
Position: Prof. *Employer:* Ohio St Univ. *Education:* PhD/Met & Matls/Cornell Univ; SM/Appl Physics/Harvard Univ; AB/Chem & Physics/Harvard Univ. *Born:* Aug 1938 in Waterbury Conn. Post-doctoral work at Univ of Ill/Urbana. Mbr of faculty of Dept Metallurgical Engrg at Ohio St Univ since 1967; Prof since 1975. Res has been on solidification, liquid metals, magnetic resonance, magnetic properties & electrotransport; current res friction and wear. Served on Bd of Review & Exec Ctte Joint Commission of Met Trans; Publs Council of ASM; Alloy Phases Ctte; Physical Met Ctte; Hume-Rothery Award Ctte of AIME; deputy editor of *Scripta Metallurgica*; OSU Council on Academic Affairs; Governor's Council on Natural

Rigney, David A (Continued)
Areas (Ohio); Columbus Symphony Chorus; Sierra Club. Organized 1980 ASM Materials Sci Seminar bk editor: *Fundamentals of Friction & Wear of Materials.* Co-Editor: *Sourcebook on Wear Control Tech* (ASM). Delegate: US/China Bilateral Conf on Metallurgy, Beijing, Nov 1981. Chr, US/China Workshop on Wear, Shenyang, Aug-Sept 1983. *Society Aff:* ASM, AIME, ΣΞ.

Rigo, Joseph T
Business: 1385 York Ave, New York, NY 10021
Position: Pres. *Employer:* SYSDOC Inc. *Education:* BA/Journalism/Univ of ME. *Born:* May 28, 1933 in Portland Maine. Wrote weather reports for the AP 1959- 64; wrote programming manuals for IBM 1964-70; organized & managed tech writing staff for Bankers Trust Co in NY 1970-74. Established own co, specializing in computer sys tech writing & documentation 1974. Founded SIGDOC 1975. Contributer to 'Computerworld' & Auerbach Mgmt Series 'Datamation.' Hobby: whale tracking in the N Atlantic. *Society Aff:* ACM.

Rigsbee, H Ken, Jr
Business: Phillips 66 Natural Gas Company, PO Box 1967, Houston, TX 77251-1967
Position: Gas Development Director, Gulf Coast Region *Employer:* Phillips 66 Natural Gas Co. *Education:* BS/Architectural Engg/Univ of TX. *Born:* 6/23/43. A native of Austin, TX. Employed by Phillips Petroleum Co since 1966 in several engg and mgmt. assignments in OK, NC, AK and TX. Now responsible for major natural gas gathering and transmission projects of Phillips Petroleum Co and subsidiaries. A Reg PE in TX and OK. Past Pres. of National Society of Architectural Engineers. Past Secy of the Bd of the Engrs Council for Profl Dev (ABET), Former Trustee of OK Engrg Fdn. and Past pres of the OK Soc of PE. Former Dir of JETS, Inc. and past chm. Natl Exec. Com for Guidance. Currently President NSPE Education Foundation. Name Outstanding Young Oklahoman in 1977 by the OK Jaycees and OK Young Engr of the Yr in 1978 by OSPE. *Society Aff:* NSPE, ASEE, NSAE.

Rikard, Donald A
Business: 2030 Abbott Rd, Midland, MI 48640
Position: VP Manufacturing & Engineering, Mgr. Bd of Dir. *Employer:* Dow Chem Co. *Education:* BS/ChE/TX A&M Univ *Born:* 1/18/28. Native of Hearne, TX. Grad from TX A&M Univ in 1950. Joined Dow Chem Co, TX Div, in mfg. Worked in substantially all of the sixty or so production plants there before becoming gen mgr of the Oyster Creek Div in 1973. Became gen mgr of the TX Div in 1977. In 1978 was named Exec VP of Dow Chem USA with responsibility for Engg, all operations, Oil & Gas exploration, & several business depts. Became Vice President for Manufacturing and Engineering for the corporation in 1982 with responsibility for global operations, engineering, health, safety, & environmental matters. Named to the board of directors in 1982. *Society Aff:* AIChE, ACS, SCI

Riley, William F
Business: Dept of Engg Sci & Mechs, Iowa State Univ, Ames, IA 50011
Position: Prof of Engrg Mechanics. *Employer:* Iowa St Univ. *Education:* MS/Mechs/IL Inst of Tech; BS/Mech Engg/Carnegie Inst of Tech. *Born:* March 1925. 1951-54 worked as mech engr for Mesta Machine Co; 1954-66 was associated with IIT Res Inst as a res engr, mgr of the experimental stress analysis sect & sci advisor to the mechanics res div. Served as a cons for the USAID summer inst prog in India 1966 & 1970. Author or co-author of 40 tech papers & 5 textbooks. Mbr of Exec Ctte of Mechanics Div of ASEE 1974-78; mbr of Exec Ctte of SESA 1970-72; Chmn of papers ctte of SESA 1967-70. Received M M Frocht Award of the Soc for Experimental Stress Analysis 1977. Elected a Fellow of the Soc for Experimental Stress Analysis 1977. Named Anson Marston Disting Prof in Engg at IA State Univ 1978. Elected an Hon mbr of the Soc for Experimental Stress Analysis in 1984. *Society Aff:* ASME, SEM.

Rim, Kwan
Business: Dept of Biomedical Engrg, Iowa City, IA 52242
Position: Prof & Chrmn. *Employer:* Univ of IA. *Education:* PhD/Applied Mech/ Northwestern Univ; MS/Mech Engg/Northwestern Univ; BS/Mech Engg/Tri-State College; -/Mech Engg/Seoul Natl Univ. *Born:* 11/7/34. Asst Prof 1960-64, Assoc Prof 1964-68, Prof 1968-present, Chrmn 1971-74, Mechanics & Hydraulics Dept of the Univ of IA. Also Assoc Dean of Engg 1974-79 and Chrmn of the Materials Engg Div 1978-84. Currently Chrmn of the Biomedical Engg Dept. Responsible for the dev of the Biomedical Engg Prog at the Univ of IA. President of the Korea Advanced Inst of Sci and Tech, Seoul, Korea, 1982-84. SAE Teetor Award 1965; NSF SEED Professorship 1976-77. Co-chrmn, Midwestern Mech Conf, May, 1983. Rotarian; Elder of IA City First Presbyterian Chruch. *Society Aff:* ASME, ASEE.

Rimmer, Jack
Business: 1114 Ave of Americas, New York, NY 10036
Position: Group Exec/Corp Exec VP. *Employer:* WR Grace & Co. *Born:* March 1921 in NYC. BS from Brooklyn Coll. Served with US Navy 1943-45. Asst Head of Process Dev Lab & chemist in charge of new prod production at Winthrop Labs; Ch Chemist & Dir of Opers at BT Babbitt Inc. With WR Grace & Co since 1961 in various mgmt positions covering the co's internatl & domestic chemical opers, including Pres of the Pacific Div; assumed present position in 1976 with responsibility for serving world-wide chemical opers. Mbr AIChE; ACS; Japan Soc; Amer-Australian Soc. Faculty mbr of Georgetown Univ's Inst for Internatl Trade & Law; formerly faculty mbr of Hudson Valley Tech Inst's extension prog. Trustee Brooklyn College Foundation. Hobbies: hiking; tennis; golf; classical music; the theater.

Rimpel, Auguste E
Business: 35 Acorn Park, Cambridge, MA 02140
Position: VP. *Employer:* Arthur D Little Intl Inc. *Education:* PhD/CE/Carnegie Inst of Tech; CE/CE/MIT; MS/CE/MIT; BA/Chemistry/Inter-Am Univ of PR. *Born:* 8/25/39. Native of St Thomas, US Virgin Islands. Worked in plastics res at Am Cyanamid between MS & PhD degrees. Conducted PhD res work on adsorption kinetics. Joined Chem & Met Engg Sec of Arthur D Little in 1965. Directed ind dev planning effort in chem & met sectors for govts of a number of developing countries in S Am, Africa & Asia. Leave of absence from Arthur D Little from 1975 to 1978 as Commissioner of Commerce for US Virgin Islands. Assumed current responsibility in charge of ADL's activities in Tropical Africa & the Caribbean in 1978. *Society Aff:* AIChE, ACS, AIC.

Rimrott, Friedrich P J
Business: 227 Mech Engg Bldg, Univ of Toronto, Toronto Ontario Canada M5S 1A4
Position: Prof. *Employer:* Univ of Toronto. *Education:* DrIng/ME/T U Darmstadt; PhD/Engg Mech/PA State Univ; MASc/ME/Univ of Toronto; DiplIng/ME/Univ Karlsruhe. *Born:* 8/4/27. Native of Halle, Germany. NRC Postdoctorate Fellow 1959, NRC Sr Res Fellow 1969, Av Humboldt Sr Fellow 1962. Asst Prof at PA State Univ, then Univ of Toronto. Became Prof of Mech Engg, Univ of Toronto, in 1967. Visiting Prof T H Wien 1969/70, T U Hannover 1970, R U Bochum 1971, U de la Habana 1972, B U Wuppertal 1978, T.U. Wien 1986. Pres CMA 1971-72, Pres CSME 1974/75, Pres 15th Intl Congress of Theoretical and Applied Mechanics, 1980. Fellow CSME, Fellow EIC, Fellow IMechE. *Society Aff:* CSME, ASME, CASI, VDI, EIC, IMech E, APEO.

Rinaldi, Michael D
Business: 1000 Western Ave, BLDG 2103N, Lynn, MA 01910
Position: Mgr-Thomson Lab. *Employer:* GE Co-Aircraft Engine Group. *Education:* PhD/Met & Mat Sc/MIT; BS/Met & Mat/MIT. *Born:* 9/9/44. Native of Boston, MA. In 1971 initial work at GE, AEG was in the manufacturing quality control area responsible for heat treating, brazing, welding, chemical cleaning and plasma spray facilities. Since 1972 have been assoc with Thomson Lab in positions of increasing responsibility. The organization is responsible for the evaluation and selection of materials and processes for use in aircraft engines, introduction of new, innovative ma-

Rinaldi, Michael D (Continued)
terials and processes, and component failure analysis. Was appointed Lab Mgr in July, 1979. ASM Henry Marion Howe Medal, 1973. Enjoy racket sports and music. *Society Aff:* AIME, ASM.

Rinaldo, Peter M
Home: 543 Scarborough Rd, Scarborough, NY 10510
Position: Retired *Employer:* W R Grace & Co. *Education:* SM/Chem Eng/MIT; SB/ Chem Eng/MIT; BS/Math/Bowdoin. *Born:* June 21, 1922 in Evanston Ill. Served to Ltjg USNR 1944-46. Chem Engr with Dewey & Almy Chem Co in Cambridge Mass from 1947 until aquired by WR Grace & Co 1954. Presently retired. Treasurer, Boston Sect of AIChE 1948-52; Director NY Soct of AIChE 1981- mbr AIChE; ACS; Phi Beta Kappa; Tau Beta Pi; Alpha Tau Omega; Alpha Chi Sigma; Chemists' Club (NYC); Briarcliff Manor Volunteer Fire Dept. *Society Aff:* ACS, AIChE, SPI.

Rinard, Sydney L
Business: 4800 East 63rd, PO Box 419173, Kansas City, MO 64141-0173
Position: Mechanical Consultant *Employer:* Burns & McDonnell Engr Co *Education:* BS/ME/KS State Coll *Born:* 02/04/36 Native of Salina, KS. With Burns & McDonnell since 1958. Performed Mech Design for Central Station Steam Power Plants and Air Conditioning Systems. Started the Power Div Controls Dept in 1968. Has been responsible for the design and installation of all power plant control and instrument systems plus preparation of plant descriptive manuals. Currently performs consultant duties for industrial and power design of mechanical and instrumentation systems. Enjoy woodworking, auto repair, reading, and writing. *Society Aff:* ISA, ASME, ASHRAE, NSPE

Ring, Chester A, 3rd
Business: Elizabethtown Water Co, Elizabeth, NJ 07202
Position: Exec VP. *Employer:* Elizabethtown Water Co. *Born:* July 23, 1927 in Bangor Maine. BS in Civil Engrg from Univ of Maine 1950. Joined Amer Water Works Ass'n 1957; became a NJ tr 1967; V Chmn 1968; Chmn 1969; Pres of Assn'n until June 1977. Licensed & reg PE in NJ. Mbr of ASCE; licensed Water Supply & Treatment Plant Operator in NJ; mbr of the St Licensing Bd for Water Works & Waste Water Opers in NJ; NJ Dir of Natl Ass'n of Water Co's, P Pres of NJ Chap; Dir of Water Resources Ass'n of Delaware R Basin Comm; Past mbr of Natl Drinking Water Adv Council, appointed by US Dept of Environ Protection (15-mbr council).

Ring, Sandiford
Business: PO Box 14666, Houston, TX 77021
Position: Pres. *Employer:* Kemlon Products. *Education:* BS/CE/Univ of TX. *Born:* 9/4/21. Born & raised of Anglo Saxon parentage in TX. Completed public sch educ in Houston. Received BS Chem Engg Univ of TX, austin. Grad US Naval Engg College Columbia Univ, NY. Served as Chief Engr Destroyers WWII. Employed Union Carbide 1946 to 1950. Married, 3 sons. Served with Keystone Engg & affiliate Kemlon Products since 1950. Now pres of Kemlon Products & Chrmn of the Bd of Keystone Engg Co. Patentee of processes & designs relating to high pressure devices. Designer & inovator of machinery & processes for mfg plastic & ceramic products. Is a reg PE. Enjoys offshore sailing. *Society Aff:* AIChE.

Ringham, Rodger F
Home: 7002 Woodland Dr, Dallas, TX 75225
Position: Retired VP *Education:* BAeroE/Aero Engg/Univ of MN. *Born:* 4/25/20. Grad BAeroE from Univ of MN, 1942; & joined Vought-Sikovsky Aircraft as a Jr Aerodynamics & Flight Test Engr. Left Vought Aero position of VP-Engg and Product Support to join Intl Harvester as Corporate VP-Engg, 1969. Profl and Trade Assn activities include many speeches & papers. Offices include: SAE Pres, Dir & Fellow; AIAA-Dir & Assoc. Fellow; SAME-VP, Dir & Fellow Engg Mfgrs Assoc- Dir & Treas; Motor Vehicle Mfgrs Assoc-Engg Advisory Committee. Listed in Who's Who in Am. *Society Aff:* SAE, AIAA, ASAE, SAME.

Ringlee, Robert J
Home: 315 Juniper Dr, Schenectady, NY 12306
Position: Principal Consultant *Employer:* Power Technologies Inc. *Education:* PhD/Mechanics/Rensselaer Polytechnic; MS/EE/Univ of WA; BS/EE/Univ of WA *Born:* April 23, 1926. US Navy Reserve 1944-59; active ser 1944-46. Professional Engineer. IEEE Fellow; Past Chrmn of Power Sys Engrg Ctte; 3 IEEE prize paper awards; AAAS Fellow; CIGRE mbr and Expert Advisor Study Committee Nos 38 & 39. Awarded 3 US patents. Mbr of the Schalmont Central Sch District Bd of Education 1967-70; VP 1969, Pres 1970. NY State registration. Wife Helen & 3 children. *Society Aff:* IEEE, AAAS, CIGRE, NYAS

Rinker, Robert G
Business: Chemical & Nuclear Engr. Dept, Univ. of Calif, Santa Barbara, CA 93106
Position: Prof of Chem & Nuclear Engrg. *Employer:* Univ of Calif. *Education:* PhD/Chem Engg/Caltech; MS/Chem Engg/Caltech; BS/Chem Engg/Rose-Hulman Inst of Tech. *Born:* in Vincennes, IN. Full Prof of ChE. At Univ of CA/Santa Barbara since the dept was 1st formed in 1965; chmn 1973-78. BS from Rose-Hulman Inst of Technology 1951; MS 1955; PhD 1959 from Caltech; all in ChE. Served 2 yrs in Korean War. Mbr Caltech faculty 1959-65. Major res interests are in kinetics, catalysis & reactor design. Mbr of AIChE; Tau Beta Pi; Sigma Xi; Amer Assoc for Advancement of Sci. *Society Aff:* AIChE, ТВП, ΣΞ, AAAS.

Rinne, John E
Home: 16 Yale Circle, Kensington, CA 94708
Position: VP-Conslt *Employer:* (Earl & Wright Cons Engrs) Retired *Education:* MS/CE/Univ of CA Berkeley; BS/CE/Univ of CA Berkeley. *Born:* Sept 24, 1909. 1931-37 worked for cons engrs HD Dewell, AW Earl & WL Huber; 1937-69 worked for Standard Oil Co of Calif, heading civil engrg & architectural div; 1969-80 VP of Earl & Wright in San Francisco and London. Fellow ASCE; P Pres 1973, Dir 1958-60, VP 1968-69; HM SEAONC, P Pres 1951; SEAOC P Pres 1953; Fellow EERI, P Pres 1966-67; mbr SAME; P Pres IAEE; past mbr of Univ of Calif Coll of Engrg Adv Council. Author of sev papers in fields of earthquake engrg; fatigue of steel offshore structures; storage tanks; pipelines. *Society Aff:* ASCE, EERI, SEAONC.

Ripa, Louis C
Business: PO Box 158, 200 Madison Ave, Convent Station, NJ 07961
Position: Exec VP; Pres *Employer:* Par International Corp *Education:* BSCE/San Engg/NJ INst of Tech; -/Planning/Add courses Rutgers & Seton Hall Univ. *Born:* BSCE/San Engg/NJ Native of Orange, NJ. Served with US Navy 1945-47. Field engr, proj engr - turnpike & interstate projects NJ, VA, for Howard Needles, Tommen & Bergendoff 1952-59, Formed own engg, planning, architectural firm 1959. Since 1976 Exec VP Enviroplan and Pres Par Corp. Resp for administration of planning, studies and design in the fields of transportation and urban dev. Spec interest "Mass Living" concepts and creation of integrated, balanced multimodal transportation systems. Author, lecturer & pilot. Enjoy water skiing and flying. VP 1971-72, Pres 1972-73, NJ Section ASCE. Chapter honor mbr Chi Epsilon 1974. *Society Aff:* NSPE, NJSPE.

Ripling, Edward J
Business: 1 Science Rd, Glenwood, IL 60425
Position: President & Director of Research. *Employer:* Materials Research Laboratory Inc. *Education:* PhD/Metallurgy/Case Inst of Tech; MS/Met Engg/Case Inst of Tech; BS/Met Engg/PA State Univ. *Born:* Feb 25, 1921 in Lewistown Pa. BS in Metallurgy from Penn State 1942; MS from Case Inst of Technology 1948; PhD from Case Inst of Technology 1952. Metallurgist for Westinghouse Electric Corp in E Pittsburgh Pa 1942-43; Copperweld Steel Co in Glassport Pa 1943-44; Asst Prof & Res Proj Dir at Case Inst of Technology in Cleveland 1946-55; Lab Dir for Continental Can Co in Chicago 1955-60; Pres & Dir of Res for Materials Res Lab in Glenwood Ill 1960- present; Cons to Pittman-Dunn Lab at Frankford Arsenal 1957-60. Served with USNR 1944-46. Mbr Sigma Xi; Phi Lambda Upsilon; elected

Ripling, Edward J (Continued)
Fellow of ASM in Oct 1970 for work on fracture mechanics of steel and aluminum alloys; mbr ASM; ASTM; AIME; Metallurgical & Petroleum Engrs & SESA. Has contributed articles to prof journals & authored (with NH Polakowski) 'Strength & Struc of Engrg Materials.' Thirty-sixth recipient of David Ford McFarland Award for Achievement in Metallurgy (April 28, 1984). *Society Aff:* ASM, ASTM, AIME, SESA.

Rippel, Harry C
Home: 1434 Sharon Park Dr, Sharon Hill, PA 19079
Position: Private Consultant *Employer:* Self *Education:* MS/Mech engg/Drexel Univ; BS/Mech Engg/Drexel Univ. *Born:* Feb 1926. Native of Philadelphia Pa. Prof exp. 1952-87 has been at FRC providing specialized R&D & engrg services on sliding-surface bearings for indus & governmental clients that design, build or use machinery. Proj Engr on more than 100 successful bearing design applications in machine tools, turbomachinery, pumps, motors, centrifuges, antennas, telescopes, rolling mills, crawlertractors, ship propulsion shafting, ore processing equipment, textile machinery, etc. Cons to some 75 indus concerns & governmental agencies. Currently a private consultant in the field of Tribology. Author of more than 50 tech publs & 5 books on sliding-surface bearings that have become standard references for designers & users of indus equipment. Fellow ASME 1977; Fellow ASLE 1979. P E in st of Pa 1961. Mbr Pi Tau Sigma 1952; Sigma Xi 1957. *Society Aff:* ASME, ASLE, ΠΤΣ, ΣΞ.

Risbud, Subhash H
Business: 204 Ceramics Bldg, Urbana, IL 61801
Position: Assoc Prof *Employer:* Univ of IL. *Education:* PhD/Mtls Sci/Univ of CA; MS/Mtls Sci/Univ of CA; BS/Met/IIT. *Born:* 8/3/47. Native of India, Came to US in 1969 for Grad Study. Completed PhD in 1976 in Ceramics and Glass Sci. Res experience at Stanford, ind experience at WESGO"GTE Sylvania in Ceramics. Have been faculty mbr at Univ of NB, Lehigh and currently at IL in Engg. Received the Ross Coffin Purdy Award of the American Ceramic Soc for Outstanding Contribution to Ceramic Technical literature in 1977; Xerox Award for Outstanding Fac Res in 1983; Excellent Teachers List Univ of IL, Urbana 198-84. *Society Aff:* ACS, ASEE, AAAS.

Rischall, Herman
Home: 11838 Bel Terrace, Los Angeles, CA 90049
Position: Head/Lab Sers Sect/Materials & Procs. *Employer:* TRW Defense & Space Sys Grp. *Born:* 1926. BS Met E Univ of Minnesota 1948. Asst Plant Metallurgist 194852 Amer Hoist & Derrick, developed new techniques in 'nodular iron' production; 1953-55 Navy Dept, developed theory & publs on stainless steel Hot Cracking; 1956-59 Prod Dev Mgr Utica Metals, developed Udimet 700 Series Alloys; 1960-present Aerospace Indus, Project Mgr Material & Processes on major communication satellite progs. Holds Calif Prof Engr Certificate MT 562. Mbr of AFS; ASM, 1962-64 Orange County Chap Chmn ASM; Prog Chmn WESTEC 1965; Los Angeles ASM Exec Ctte 1963-64. Published num papers on 'Laser Welding' and holds sev patents on high temperature alloy sys. Married, 4 children. Hobby: tennis.

Risoli, Joseph F
Business: 32 Field Point Rd, Greenwich, CT 06830
Position: Pres. *Education:* BSCE/Engg/Manhattan College; MSEnvEngg/Env Engr/ Poly Inst NY. *Born:* 2/26/47. Attended Manhattan College, NYU & Polytech Inst in NY. Worked as PE for SE Minor & Co from 1970 to 1977. Started Joseph F Risoli, Co in 1977. In two and a half yrs we have gone from three projs to well over 60. *Society Aff:* NSPE.

Rist, Harold E
Business: 21 Bay St, Glens Falls, NY 12801
Position: Chrmn of Bd *Employer:* Rist-Frost, Assocs, PC *Education:* MCE/Structures/RPI; BCE/Structures/RPI *Born:* 8/6/19 Education: Newcomb Central School; Grad. '37; Renss. Poly. Inst., BCE '50; MCE '52. Married former Ruth Mahony; Res. Diamond Point, NY. Employment Hist.: Seelye Stevenson Value & Knecht, N.Y. City, '52-'58; Founded Harold E. Rist, Associates, Cons. Engrs. & Surveyors, '58-'60; CEO, Rist Bright & Frost, '60-'63; CEO, Rist-Frost Associates, Engineers, Architects, Planners, Surveyors, '63-'84; Chairman, Rist-Frost, '84-Pres.; CEO, Twenty-One Bay Partnership; Pres., CEO, Mechanical-Electrical Systems, Inc.; CEO, Smith Flats Partnership; Pres., CEO, Glens Falls Communications Corp.; Pres., CEC/NYS, '60-'64; Dir., CEC/USA, '68-70; V.Pres., CEC/USA, '70-'72; Commissioner, Hudson River Valley Commission, '70-'78; Dir., National Professional Services Council, '72-'84; Dir. & V.Pres., Adirondack Regional Chambers of Commerce; Dir. & V. Pres., Adirondack North Country Association; Director, Capital Region Technology Development Council. *Society Aff:* NSPE, ACEC.

Ritchie, Malcolm L
Business: School of Engineering, Dayton, OH 45435
Position: Emeritus Prof *Employer:* Wright State Univ *Education:* PhD/Engr Psychol/ Univ of IL; MA/Social Inst/Univ CA; AB/Psychology/Univ CA. *Born:* 5/3/20. Private pilot 1941. USAAF flight instructor and Night Fighter Pilot 1942- 46. Res staff of the Univ of IL Aviation Psychology Lab 1951-57. Pres of Ritchie Inc 1957-present. Prof of Engg Wright State Univ 1969-1982. Since 1951 has conducted research in mind-machine relations and consulted on the design of machines for operation by humans. Developed hybrid engr-phyc undergrad degree program in Sch of Engrg at Wright St Univ. Is now program dir of a hybrid grad prog in which students earn Masters Degree in Engrg and Doctorate in Psyc. *Society Aff:* AIAA, IEEE, HFS, APA.

Ritter, Gerald L
Business: Advanced Nuclear Fuels Corp, 2101 Horn Rapids Road, Richland, WA 99352
Position: Mgr, Fuel Engrg & Technical Services *Employer:* Advanced Nuclear Fuels Corp *Education:* MSCE/CE/Univ of CA; BS/CE/Univ of WA; BA/Chemistry/Pacific Lutheran Univ. *Born:* 8/13/39. Native of Kellogg/Wallace, ID. Began career in the Nucl Ind at the Hanford Plant near Richland, WA. Worked from 1964 to 1971 in various process engg & mgt assignments in the radioactive waste mgt & nucl fuel reprocessing areas for the GE, Isochem, & Atlantic Richfield Hanford Cos. Joined Exxon Nucl Co, Inc in 1971 & worked for 8 yrs on a commercial nucl fuel reprocessing plant proj. Was VP, Technical Dept, with Exxon Nucl ID Co, Inc (ENICO) from 1979 to 1981. ENICO was a prime contractor for the operation of the Dept of Energy's ID Chem Processing Plant. Became Mgr, Process Engrg and Dev Dept in July 1981 and in March 1985 became Mgr, Fuel Engrg & Technical Services Dept. where he is responsible for nuclear fuel design and R&D activities for Advanced Nuclear Fuels Corporation (formerly Exxon Nuclear Company, Inc.). *Society Aff:* AIChE, ANS, ΣΞ, ΤΒΠ.

Ritter, Guy F
Business: 423 Pine Ave Bldg, Albany, GA 31702-0803
Position: President *Employer:* Lindsey & Ritter, Inc *Education:* MSBE/Bldg Eng/ MIT; B Arch/Arch Eng/GA Tech; BS/-/GA Tech *Born:* 02/09/33 B. in Detroit, MI. Raised in Nashville, TN. R & D Draftsman with Temco, Inc 1950-54. Design Engr with Morris Boehmig & Tindel, Inc 1956-58. District structural engr for Portland Cement Assoc 1958-62 GA-SC. Associate with J. L. Lindsey Co, Structural Engrs, Albany GA 1962-64. Principal of Lindsey, Tucker & Ritter, Inc 1964-74 and sucessor firm Lindsey & Ritter, Inc 1974 to date. Pres 1979. Reg PE GA, SC, Nc, TN, FL and AL. Certified Value Engrg. Expert in Structural Engrg and Foundations. *Society Aff:* PCI, ASCE, ACI, IASS, CSI, ACEC, SBCC, ASAE

Ritter, John E, Jr
Business: Mech Engg Dept, Amherst, MA 01003
Position: Prof. *Employer:* Univ of MA. *Education:* PhD/Materials Sci/Cornell Univ; MS/Ceramics/MIT; BS/Chem Eng/MIT *Born:* 7/17/39. Have taught at the Univ of MA since 1965. Promoted to Prof in 1976. In 1971-72 was ASEE-Ford Fdn Resident at the Owens-IL Dev Ctr as a Technical Product Mgr for high temperature seal mtls. In 1983 I was a Sr scientist at the Atomic Research Establishment, Harwell, England, and carried out a res proj strength variability of alumina. My res expertise has been concentrated in the area of strength and fatigue of glass and ceramic mtls. Am a consultant to several industries in the area of strength and fatigue of brittle mtls. Have given numerous invited talks and have published over 80 papers. *Society Aff:* ACS, ASEE

Rittner, Edmund S
Business: 8800 Fallen Oak Dr, Bethesda, MD 20817
Position: Conslt Scientist *Employer:* COMSAT Labs *Education:* PhD/Chemistry/MIT; BS/Chemistry/MIT. *Born:* May 1919. Married with one daughter. Sect Ch, Dir of Physics Dept & Dir of Exploratory Res Philips Labs Briarcliff NY 1946-69; conducting & directing res on cathodes, infrared & x-ray detectors, transistors, thermionic converters, vidicons, dielectric triodes, solar cells, photoconductors, & thermoelectric heat pumps & generators. Executive Director for Physical Scie at COMSAT Labs since 1969 managing R&D on materials and devices including tunnel diodes, IMPATTs, FETs, AMPs, solar cells, batteries, ion engines, radiation damage, thermal coatings, & materials characterization & processing; initiated & supervised R&D leading to COMSAT solar cell breakthroughs. Fellow IEEE & APS. Enjoys classical music with musician wife & tennis. *Society Aff:* IEEE, APS.

Ritzmann, Robert W
Home: 3504 Kent St, Kensington, MD 20895
Position: Principal Conslt *Employer:* Ritzmann & Associates *Education:* Post Grad/ Nucl Eng/Oak Ridge Sch of Reactor Tech; Post Grad/Engrg Adm/Geo. Wash Univ; MS/ChE/Carnegie-Mellon Univ; BS/ChE/PA State Univ; BS/Commercial Chemistry/PA State Univ. *Born:* 7/5/24. Chem engr with Std Oil Co (IN) 1947-51, & Koppers Co 1951-57. Nucl engr specializing in civilian applications of nucl power with Atomic Energy Comm 1957-75. Served as Sr Evaluation Engr, Scientific Rep to Atomic Energy of Canada, Asst to Commissioner, Chief of the Energy Policy Branch, & Dir of the Office of Industry Relations. Dir of Industry Relations with the Energy Res & Dev Admin 1975-77. Dir of Federal Relations with the Dept of Energy 1977-79. Deputy Dir of Wash Office of the Electric Power Res Inst 1979-82. Conslt engr 1982- , specializing in energy, bus opportunities in energy indus, and bio-med applications of chem and engrg. Chrmn of Natl Capital Sec of AIChE 1980-81. Fellow AIChE. Reg PE in PA & MD. *Society Aff:* AIChE.

Rivas-Mijares, Gustavo
Home: Res Jarama, A.P. 7-A, Calle C, URB. S. Rosa Lima, Caracas, Venezuela
Position: Full Prof *Employer:* Universidad Central Venezuela *Education:* MSE/Wat & Wast water treat/Univ of MI; Doctor En Ingenieria//Univ Central Venezuela *Born:* 11/7/22 Design Engr, Inst Nacional Obras Sanitarias, Venezuela (1945-54). Ch Engr Natl Prog Rural Water Supplies, Venezuela (1954-57). Cons engrs, Pres firm same name, Caracas, Venez. Full prof, Sanitary Engrg Sch, Univ Central Venezuela (1958-84). Hd Dept Sanitary Engrg. Univ Central Venezuela (UCV) (1962-65). Dean Grad Sch, UCV, Venezuela. Foreign Assoc Mbr, Natl Acad of Engrg of the US of America (1981). Pres, Natl Acad of Sciences of Venezuela (1981-85). Have published three text books. Coauthor three other books. Have assisted to more than 100 conferences, symposia, congresses, etc in intl and natl levels. Several scientific prizes for technical & res papers. Pres Venez Sections of ASCE of WPCP. Mbr governing bd "Intl Assoc on Water Pollution Res." Pres (Intl) AIDIS (Assoc Interamericana Ingenieria Sanitaria). Venezuelan National Science Prize (1986). *Society Aff:* ASCE, WPCF, CIV, SVSP, AIDIS, IAWPRC

Rivera, Alfredo J
Business: 601 Bergen Mall, Paramus, NJ 07652
Position: Assoc Engr. *Employer:* N H Bettigole Co. *Education:* MS/Structural Design/MIT; BSCE/Civ Engg/Mapua Inst of Tech. *Born:* 3/28/29. Mr Rivera has specialized in such engg areas as structural design, the design & supervision of construction of long-span bridges (fixed & movable), and work on special structures including power substations, transmissions facilities, & railroad equip. Maj assignments have included the design of the Flushing River Bridge in Queens; structural design of dormitories for the State Univ of NY at Oneonta; rehabilitation of the 60th St Heliport, the Midtown Hgwy Viaduct & the Pelham Bay Bridge, all located in NYC. Prior to joining the Bettigole organization, Mr Rivera worked as a Structural Designer for Gibbs & Hill & Hardesty & Hanover, consulting engg firms. *Society Aff:* ASCE.

Rivera-Abrams, Carlos
Home: 1799 San Alejandro St, San Ignacio Dev, Rio Piedras, PR 00927
Position: Gen Supt Design. *Employer:* PR Elec Power Auth. *Education:* BS/EE/TX A&M College; MS/EE/Union College. *Born:* 11/6/28. As a dev engr in the GE Co & the RCA Service Co, Mr Rivera was engaged in the design of radar antennas & microwave components. In 1958 he joined the elec utility serving the island of PR, where he held positions in engg such as Construction Superintendent & Asst Hd of Planning & Res. In 1965 he developed the isokeraunic map of the island. After being the Mgr of Data Processing for five yrs he was named Hd of the Purchasing Div. Mr Rivera was IEEE Regional Dir for Latin Am during 1978-79. *Society Aff:* IEEE, CIAPR.

Rivers, Lee W
Business: Box 1219-R, Columbia Rd, Morristown, NJ 07960
Position: Dir, Corp Planning *Employer:* Allied Corp. *Education:* MChE/Chem Engg/ Univ of DE; BChE/Chem/NY Univ. *Born:* 4/28/29. NYC. Res Chem Engg Indust Chem Div, Allied Chem Corp, 50-58; sr res chem engg, 58-59; tech supvr, 59-60; mgr Baton Rouge Dev Lab, LA, 60-61; asst dir planning res, 52-53; dir, 63-65; mgr commercial dev, 65-66; gen mgr commercial dev, 66-70; dir mkt res and plan, 70-72; VP dev, 72-75; VP comm dev, 75-77; gen mgr chloring products, 77-78; VP tech, 78-79; VP R&D, 79-84; Dir Corp Plng 1984- . Am Chem Soc, Am Inst of Chemists, Comm Dev Assoc, North Am Soc for Corp Planning, Fluorocarbons; pilot plant design and operations; new product planning and commercial dev, gen management, corp planning. *Society Aff:* AIChE, ACS, CDA, LES, NASCP.

Rivers, William H
Business: 800 North 12th Blvd, St Louis, MO 63101
Position: Executive V P/ Ch Operating Officer. *Employer:* Sverdrup & Parcel and Assocs Inc. *Born:* 1921. BSCE NC State; aero cert MIT. Joined Sverdrup & Parcel 1950; have worked on engrg projs involving design & dev of aeronautical & aerospace testing facils (e.g. Saturn V, wind tunnels, space shuttle components). Other R&D work in transportation, atomic energy & oceanography, many of them one-of-a-kind; also chem food processing & indus facils. As V P/Ch Engr 1969 & now as Exec V P/Ch Operating Officer have top-level responsibilities for streamlining mgmt methods, controls & communications. P E Missouri. Mbr ACEC; Amer Def Prep Assn.

Rizzo, Edward G
Home: 34 Birchwood Terrace, Wayne, NJ 07470
Position: Quality Control Director. *Employer:* Alloy Stainless Products Co. *Born:* Nov 9, 1935; US citizen by birth. Married; resides with wife & 2 children in own home. Graduate of Jersey City Sch of Medical Radiology; attended special courses Div of Newark Coll of Engrg; presently enrolled at Rutgers Univ, Bus Admin. Alloy Stainless Prods Co Totowa NJ - QC Dir 1971present; Curtiss Wright Corp Fairfield-Woodridge NJ - Radiographic & QC Technician 1961-71. Fellow Award in non-destructive testing for over 15 yrs of working with education progs, including organizing, implementing & teaching of a fully accredited curriculum in NDE at Keane Coll Union NJ. P Chmn Soc for non- destructive testing, Met NY sect, 20 yrs; past mbr Paterson Division of the Amer Red Cross; mbr of Wayne Township First Aid Squad 19 yrs; charter mbr Passaic Valley Elks Lodge 2111.

Rizzo, Paul C
Business: 13 Thorncrest Dr, Pittsburgh, PA 15235
Position: Pres. *Employer:* Paul C Rizzo Assoc, Inc *Education:* PhD/Civil Engr/Carnegie Mellon Univ; MS/Civil Engr/Carnegie Mellon Univ; BS/Civil Engr/Carnegie Mellon Univ. *Born:* 5/13/41. Native of Pittsburgh, PA. Taught Civ Engrg courses at undergrad & grad level at Carnegie-Mellon Univ and Univ of Pittsburgh, following position as field Engr with Ft Pitt Bridge Works. Joined D'Appolonia in 1962 as Engr and assumed positions as Pres in 1977, Managing Partner and VP of Enginereal Const Intl (ECI), and Pres D'Appolonia Petroleum Inc. In March of 1984, left D'Appolonia and started a new firm, Paul C Rizzo Assocs, Inc. He now serves as Pres and CEO of this new firm which provides conslt serv in geotech engrg and earth and environ sci on hazardous waste, nuclear and underground proj in the US and overseas. *Society Aff:* ASCE, ANS, ACEC, NSPE, TBΠ, PSPE.

Rizzo, William J Jr
Business: 235 West Central Street, Natick, MA 01760
Position: President *Employer:* Rizzo Associates, Inc. *Education:* MS/Civil Engineering/Georgia Institute of Technology; MCP/City Planning/Georgia Institute of Technology; BA/Applied Science/Lehigh University; BS/Civil Engineering/Leigh University *Born:* 09/18/44 Native of Boston, MA. Planner, Engr and Project Mgr with Metcalf & Eddy, Inc. for twelve years, involved in planning, engrg and environmental projects nationwide. Planning Dir for the City of Leominster, responsible for planning and engrg within the City. Asst Secretary of Transportation for the State of MA 1982-83, responsible for major highway projects in the State. Started Rizzo Assoc, an engrg and environmental consulting firm, in 1983, which currently has an employee enrollment of almost 100. *Society Aff:* ASCE

Rizzone, Michael L
Business: 2001 N Lamar, PO Box 478, Dallas, TX 75221
Position: Retired *Education:* BSME/Mech Engg/IN Inst of Tech; -/Academic Arts/St Bonaventure College; - /Exec Dev Prog/Univ of GA; -/Ext Courses/Univ of Pittsburgh-PA State College. *Born:* 1/30/23. Native of Oil City, PA. Employed by OILWELL Div, US Steel Corp since 1949. Present position, Dir Product Engg & Dev & Officer of OILWELL Div. PE Licenses in State of TX & State of PA. Have written numerous technical papers; well known world-wide in slurry pumping tech. "Engineer of the Year Award–, ASME 1965. Elected "Fellow Member" of ASME; past Natl Chrmn of ASME-Petrol Div. Instr of IADC Sch of Drilling Pratices in Lafayette, LA. Eight patents issued in my name & two patents pending. One of two men who founded Offshore Tech Conf; was Natl Chrmn in 1971. *Society Aff:* ASME, API, IADC, AIME.

Rizzoni, Eitel M
Business: 2715 M Street, N.W, Washington, DC 20007
Position: President. *Employer:* Teleconsult Inc. *Education:* PhD/Indus Engrg/Univ of Palermo; Diploms/Elec Comm/Polytechnical of Turin; MS/Elec Comm/MIT. *Born:* 1/16/25. in Italy. PhD (cum laude) Indus Engrg Univ of Palermo; Postgraduate diploma (cum laude) Electrical Communications Polytechnic of Turin; MS Electrical Communications MIT in Cambridge. 1952-63 RCA Internatl, in charge of RCA recording studio in Rome Italy, later worked in planning design, construction & testing of multichannel radio-relay sys in Clark NJ; 1963-70 Page Communications Engrs Inc Wash DC, Assoc Dir of Engrg in charge of telecommunications progs throughout the world; since 1970 Pres of Teleconsult Inc, consulting frim specializing in tech, economic & org planning of nationwide telecommunications sys. Published in IEEE; IRE; Alta Frequenza; Consulting Engineer; Telephony, National Development. Sr Mbr IEEE. PE, District of Columbia. *Society Aff:* IEEE.

Roa, William J, Jr
Home: 1230 S Geyer Rd, Kirkwood, MO 63122
Position: Consultant *Employer:* Semi-Retired *Education:* MS/Elec Engrg/WA Univ; BS/Elec Engrg/WA Univ. *Born:* July 1913 St Louis Mo. Plant electrical engr for Alcoa 1936-47 at Mobile Ala & E St Louis in charge of electrical engrg; with Sverdrup & Parcel cons engrs 1947-77. Drafted revisions to A.R.E.A. design standard covering movable bridge electrical sys; cons to industries for engrg reports on large electrical equipment failures. Jr college guest lecturer. Chmn AIEE 1959 St Louis sect; Fellow IEEE 1965 'for contribs to the dev, design & construction of complex test facils.' Hobbies: float fishing & economic research with microcomputer. *Society Aff:* NSPE, IEEE, AAAS.

Roach, Kenneth E
Business: 380 Main St, Suite 350, Dunedin, FL 33528
Position: Regional Manager *Employer:* Black & Veatch Engineers Architects *Education:* BS/Phys/Lynchburg College. *Born:* 6/7/29. Native of Lynchburg, VA. Physicist for the Babcock & Wilcox Co Atomic Energy Div 1955-1963. Taught eighteen semesters of undergrad physics & math for the Univ of VA 1957-1963. Staff Physicist for Gen Nucl Engg, Dunedin, FL 1963- 1964. A founder & VP Southern Nucl Engg 1964-1973. Exec Scientist NUS Corp 1973- 1977. A founder, Pres, & Chrmn of the Bd of Southern Sci Applications, Inc, a co wholely owned by Black & Veatch 1977-82. Responsible for the overall operation of the firm. 1982- Regional Manager Black & Veatch, responsible for the Southern Science Office. Reg Nucl Engr, CA. Mbr of the Engg Advisory Council, Univ of FL. Enjoy sailing; classical music. *Society Aff:* ANS.

Roake, William E
Home: 2336 Harris Ave, Richland, WA 99352
Position: Tech and Mgmt Conslt *Employer:* Self-employed *Education:* PhD/Phys Chem/Northwestern Univ; MS/Phys Chem/OR State Univ; BS/Arts and Sci/OR State Univ. *Born:* 9/30/19. 1942-45 Natl Defense Res Ctte; 1948-65 GE Co, Hanford Atomic Prods Oper; 1965-70 Battelle NW; 1970-1984 (October) Westinghouse-Hanford Co, mgr of LMFBR nuclear fuel control element coolant structural material dev, design, testing base technology & bus forecasting: fusion, solar & other tech dev; non- destructive testing dev; operation of plutonium fuel fabrication facilities; dev and operation of low level waste tech, spent fuel handling tech; October 1984-present Tech and Mgmt Conslt. Fellow, ANS 1975; Chmn, Richland Section ANS 1981- 82. *Society Aff:* ANS, ACS, AIAA, ΣΞ.

Robb, Walter L
Home: 1358 Ruffner Rd, Schenectady, NY 12309
Position: Sr VP, CR&D. *Employer:* G E. *Education:* BS/Chem Eng/PA State; MS/Chem Eng/Univ of IL; PhD/Chem Eng/Univ of IL *Born:* 4/25/28. Native of New Bloomfield PA. 1951-present employed by G E Co at Knolls Atomic Power Lab, R&D center, Silicone Prods Dept, Medical Dev Operation, Presently in charge of Corporate Research & Development & Medical Sys Group. Member Natl Acad of Engrg; AIChE. Winner of 5 IR-100 awards. *Society Aff:* NAE, AIChE

Robbins, Jackie W D
Business: Dept of Civil Engrg, Ruston, LA 71272
Position: Prof of Civil Engrg. *Employer:* Louisiana Tech Univ. *Education:* PhD/Agri Engr/NC State Univ; MS/Agri Engr/Clemson; BS/Agri Engr/Clemson *Born:* Feb 1940 at Spartanburg SC; reared on a genl-type farm. Served with Army Corps of Engrs 1961-62. Const engr with Bureau of Reclamation 1961; Asst Prof with LSU 1963-65; Res Assoc with NCSU 1969-70; Assoc Prof with Univ of Mo 1970-71. Prof & Hd of Agricultural Engg with Louisiana Tech Univ 1971-87. Prof activities include both education & res. Patents and Consulting in irrigation. Publications and Consulting in animal waste mgmt. Married Betty Wright of Frankfort Ky 1963, & have 2 sons (Jay 1971, Robin 1974). Active in church & civic work. *Society Aff:* ASAE, NSPE, IA, ΣΞ, ASEE

Robbins, Joseph E
Business: 26 Broadway, New York, NY 10004
Position: President *Employer:* Pope Evans and Robbins, Inc *Education:* B/ME/City Coll of NY *Born:* 07/12/22 Joined the Marine Service and attained the rank of Lt CDR/First Engr. Worked for Burn & Roe, Power Plant and Military Facilities Consltg Engrs for 13 yrs, and was in charge of all work relating to Aeronautical and Defense Projects when P l&ft in 1960 to join Pope and Evans as a partner. Accepted presidential appointment in 1961 as Deputy for Engrg to Asst Secy Air Forces (R&D). For the past 20 yrs have been partner and since 1973 Pres of Pope Evans and Robbins, Inc (PER). *Society Aff:* TBΠ, NSPE

Robbins, Louis
c/o GPI, 325 West Main St, Babylon, NY 11702
Position: Exec VP. *Employer:* Greeman-Pedersen Inc *Education:* BS/Civ Engg/Northeastern Univ. *Born:* 8/12/33. Left Long Island to serve with Army COE - construction supervision/quality control on overseas Air Force bases. Started engg career with Lockwood, Kessler & Barlett - two yrs extensive experience design, construction inspection- interstate/arterial hgwys. Three yrs comprehensive responsibility with Charles Sells Inc - maj hgwy projs included Northway and Cross Westchester Arterial. In 1966, joined with Beecher Greenman, a long-time profl associate, to for GPI. Firm has grown in size and stature - listed in ENR Top 500 staff increased to 180 - seven regional offices established. Reg PE - NY, NJ, MA and FL. *Society Aff:* ASCE.

Robbins, Paul H
Business: 2029 K St NW, Washington, DC 20006
Position: Exec Dir - Retired. *Employer:* Natl Society of Professional Engrs. *Education:* BS/CE/Syracuse Univ; SM/CE/MIT; Dr/Engrg/Rose Polytechic Inst; Dr/Engrg/Norwich Univ *Born:* Feb 1914. Taught at Cooper Union, NYU, & Univ of Maine; in 1941 became cons on engr training Exec Office of Mayor NYC; serving during WWII with Transportation Corps US Army at NY Port of Embarkation & Pentagon. Mbr of & advisor to several task forces & cttes of US Govt. Reg P E; Exec Dir NSPE 1946- 78. Pres CESSE 1972-73. Tau Beta Pi. Dir of Fellowships 1947-79, Pres 1982-86. Dir 1967-82 & Secy 1972-80 JETS. Secy Section M (Engrg) AAAS 1974-79. Author Building for Prof Growth 1984; NSPE Awd 1984. *Society Aff:* ASEE, NSPE, AAAS

Robe, Thurlow R
Home: 28 Canterbury Dr, Athens, OH 45701
Position: Dean *Employer:* OH Univ *Education:* PhD/Applied Mechanics/Stanford Univ; MS/ME/Ohio Univ; BS/CE/Ohio Univ. *Born:* 1/25/34. Born in Petersburg, OH. Worked as Mech Engr for GE Co in Niles, OH (1954); Cleveland, OH (1955); Erie, PA (1955-56); and Evendale, OH (1959-60). Served in USAF as a fighter pilot (1956-59). Taught engg at OH Univ (1960-63). Taught at Univ of KY in Engg Mechanics Dept (1966-1980). Dean of Coll of Engr & Tech, OH Univ. (1980-present). Served as Special Asst to Pres of Univ of KY while Am Council on Education Admin Fellow (1970-71). Served as Assoc Dean for Academic Affairs in College of Engg (1976-79). Received award from ASEE for Outstanding Contribution in Res, 1967. Engg Consultant to a number of ind organizations, ins firms, and attorneys on mech and structural problems. Has published a number of articles in journals and proceedings in area of dynamics and stability. *Society Aff:* ASME, ASEE, NSPE, TBΠ, ΣΞ.

Robeck, Gordon G
Home: 614-Q Avenida Sevilla, Laguna Hills, CA 92653
Position: Retired; consultant *Employer:* Self *Education:* SM/Sanitary Engrg/MIT; BS/Civil Engrg/Univ of WI, Madison; Honorary Doctor of Science/University of Cincinnati *Born:* Feb 1923 Denver. Environmental Engineer, Married 1951 3 sons. Commissioned Officer (San Engr) USPHS 1944-74 with duty stations throughout US doing public health work at all levels of govt. Experience mainly in stream & indus waste surveys, pilot & field scale res on water treatment & for last 20 yrs tech leader & mgr of natl drinking water res USPHS (now USEPA) designed to serve as a basis for Natl Drinking Water Standards. Advisor on water probs to NAS/NAE, Dept of Interior & several prof assns. 7 publication awards, res award, outstanding service award 1979, Chmn of Water Quality Div & Hon Life Mbr all from AWWA; Res Prize, Honorary Member (1985) ASCE. Gold Metal from USEPA-1978, received Distinguished Service Citation, Univ. of Wis-Madison (College of Engrg)-1986, Mbr, Natl Acad of Engrg (1980). Member of Water Science & Technology Board 86-89. Hobbies: music & tennis. Retired from USEPA Feb '85. Now part time water consultant and advisor. *Society Aff:* ASCE, AWWA, WPCF, IWSA

Roberds, Richard M
Business: Sch of Engrg Tech and Engrg, Martin, TN 38238
Position: Dean *Employer:* The Univ of TN at Martin *Education:* PhD/Nuclear Eng/Air Force Inst of Tech; MA/Physics/Univ of KS; AB/Physics/Univ of KS *Born:* 06/22/34 Retired Air Force Colonel. During AF career, served as research physicist and was the first program mgr of the AF particle-bean weapon program. Served on AF Studies Bd summer study (1977) on pulsed power, an arm of the Natl Acad of Scis. In 1977 became Chief, Reconnaissance and Weapon Delivery Div within the AF Avionics Lab. Served as an operational AF pilot accruing over 3800 hours in fighter and trainer aircraft. Received multiple decorations. Upon retirement, took position as first Dept Head of Engrg Tech within Clemson Univ Coll of Engrg. *Society Aff:* ANS, APS, ASEE

Roberson, David A
Business: Pan Am World Services, Inc, Facility Services, TN 37389
Position: Mgr Utilities Branch. *Employer:* Pan Am World Services, Inc. *Education:* BS/EE/Univ of TN-Knoxville. *Born:* Apr 1921; native of Etowah Tenn. Reg P E: N Y, Tenn, Fla. Elec Engr with Tenn Eastman & Monsanto Chem Co at Oak Ridge 194347; Design Engr for Ebasco Services Inc N Y 1947-55; with ARO Inc 1955- ; assumed current respon as Mgr Utilities Branch 1963 for ARO which was contract operator for the Air Force Arnold Engrg & Dev Ctr. Pres TSPE 1967-68; IEEE Region 3 Exec Ctte 1971-73; Natl Dir TSPE 1974-80. Hon societies and awards: Tau Beta Pi; TSPE "Engineer of the Yr" 1972; TSPE Presidential Citation 1970; LIEEE Cert of Appreciation 1960; Who's Who in the South & Southwest 1971-72. Chrmn of NSPE Reg and Qualification for Practice Committee, 1978-80; NSPE VP, SE Region, 1980-82; Chrmn, Tullahoma Utilities Bd. Enjoys playing golf & watching football. *Society Aff:* NSPE, TSPE, IEEE, AFA.

Roberson, James E
Business: 216 S Pleasantburg Dr, Greenville, SC 29606
Position: VP *Employer:* J. E. Sirrine Co *Education:* B/ME/GA Tech *Born:* 9/2/29 Native of Greenville, SC. Served with US Air Force 1951-53. Service engr with Babcock & Wilcox 1953-56. Sales engr with B & W 1956-1963. Power staff engr with Rust Engrg 1963-1966. Joined J.E. Sirrine Co as Dept Power Engr. Currently VP of Business Development for Sirrine. Past Chrmn of TAPPI's Air Control Committee. Currently Chrmn of TAPPI's Engrg Economics and Management Committee. Member of ACEC Energy Committee. *Society Aff:* ASME, ACEC, NSPE, TAPPI

Roberts, A Sidney, Jr
Home: 5437 Glenhaven Crescent, Norfolk, VA 23508
Position: Prof. *Employer:* Old Dominion Univ. *Education:* PhD/Nuclear Engr/NC State Univ; MS/Mech Engr/Univ of Pittsburgh; BS/Nuclear Engr/NC State Univ. *Born:* 9/16/35. After first engg degree in Nuclear engrg (1957) accepted employment with Westinghouse Elec Corp, Bettis Atomic Power Lab, becoming assoc engr. Returned to full-time grad study (with fellowships) in 1960. Except for leaves, chose to be engg educator at Old Dominion Univ, 1965-present, Dept Chrmn, Mech Engrg 1971-72, Assoc Dean for grad studies, 1972-75. Was exchange engr in Romania, April, 1968; visiting Res Engr Swedish Atomic Energy Co, 1968-69; short term employment/consulting with NASA Langley Res Ctr and Norfolk Naval shipyard; Reg PE in VA. Published research papers in gaseous plasmas, electron beams, advanced power cycles, solar energy conversion concepts, thermodynamics and heat transfer; part-time practice, principal in Univ Engrg, Ltd, Norfolk, VA. *Society Aff:* ASME, ASEE, AAAS, ΣΞ, ASHRAE

Roberts, Alan J
Business: Burlington Rd, Bedford, MA 01730
Position: Vice Pres *Employer:* MITRE Corp *Education:* MS/EE/MIT; BS/EE/MIT *Born:* 09/08/26 Native of Brooklyn, NY. Assumed present position as VP of Strategic Sys in 1979. 1975-1979 Technical Dir, Aerospace Surveillance and Defense Div. 1970-1975 Assoc Technical Dir, Tactical Sys Div, responsible for deployable systems. 1968- 1970 Assoc Technical Dir, Info Sys Div. 1960-1968 Project Leader on large defense projects including coastal radars for SLBM detection, NORAD combat operation center, and SADE air defense system. 1955-1960 at MIT's Lincoln Lab, headed installation of first operational SAGE computer program capability at McGuire AFB. 1951-1955 Research Asst at MIT's Digital Computer Lab; worked on the Whirlwind I computer, responsible for operation and maintenance of electrostatic storage, assisted in installation of first magnetic core memory. *Society Aff:* IEEE, AFCEA

Roberts, Bruce E
Business: 631 E. Crawford, Salina, KS 67401
Position: Exec Partner (Retired) *Employer:* Wilson & Co *Education:* BS/CE/KS St Univ *Born:* 05/22/17 Born in Chanute, KS. Married Marcelle Braden. Two sons - Stephen Kent, Craig Arnold. Joined Wilson & Co in 1946. Became co-owner and managing partner of Wilson & Co on 1 January 1959. In 1963 organized and became co-owner and pres of Wilson-Murrow Companies, a foreign subsidiary of Wilson & Co, engaged as consultants to the Ministry of Communications, Riyadh, Saudi Arabia. Became Exec Partner of Wilson & Co in 1968. Dir of Planters State Bank, Salina; co-owner of B & R Ranch, Ltd.; P-Chrmn of KS State Bd of Engrg Examiners; Recipient of Distinguished Service Award, KS State Univ, School of Engrg in 1979 and Outstanding Service Award, KS Engrg Society in 1975. Reg Prof Engr and Land Surveyor in KS, mbr of the American Society of Prof Engrs, American Prof Engrs in Private Practice. Is also a mbr of the Masons and Elks. *Society Aff:* NSPE, ACEC, ASCE

Roberts, Craig A
Business: 631 E. Crawford, Salina, KS 67401
Position: Partner *Employer:* Wilson & Co, Engrs & Architects *Education:* BS/CE/KS State Univ *Born:* 06/05/47 Native of Salina, KS. Joined Wilson & Co, Engrs & Architects in 1969 as a design engr and completed engrg design projects for roads, drainage, highways and water and sewage facilities. Became partner in 1975 in charge of Marketing. Assumed current responsibilities in 1977 as Partner-in-Charge of Support Services Div including mktg, finance, personnel and adm. Named Outstanding EIT in 1973 and Outstanding Young Engr in 1981 by the KS Engrg Soc. Currently Mbr of the KS State Bd of Technical Professions which regulates licensing of PE, LS, RA and RLA in state. Holds PE licenses in 14 states. *Society Aff:* NSPE, WPCF, PSMA, SMPS

Roberts, David H
Home: 147 Grumman Hill Rd, Wilton, CT 06897
Position: Pres *Employer:* HayDay Glassworks *Education:* MChE/Chem Eng/NY Univ; BS/Chem Eng/City Coll, NY. *Born:* 3/31/23. Brief experience in catalytic cracking with Shell Oil, Houston, sandwiched around several yrs in US Navy Amphibious Forces, followed by 30 yrs with rapidly-growing Stauffer Chem Co covering process dev, process design & subsequently, total engrg/const mgmt of multi-divisional heavy chem, plastic, agricultural organic, food div & overseas plants plus res, engrg & hqrs facilities as VP-Engrg, respons include tech dir of extensive licensing activities in vinyl chloride, PVC, chlorinated hydrocarbons. Subsequently, Dir of Mgmt Prog, Westchester Grad Center, Polytechnic Inst of New York's Mgmt Div, and Assoc Prof of Mgmt. Enjoys family, good music, bad tennis and glassblowing. *Society Aff:* AIChE, ACS.

Roberts, Earl C
Home: 24914 S E 422nd, Enumclaw, WA 98022
Position: Principal Engineer. *Employer:* The Boeing Co. *Education:* ScD/Metallurgy/MIT; SM/Metallurgy/MIT; BS/Met Engg/MT Sch of Mines. *Born:* Nov 1921; native of Butte Mont. Employed by Alcoa Spokane Wash 1943-44. In US Navy 1944-46. Res Asst Armour Res Foundation 1946-47; Res Asst MIT 1947-52; Asst Prof Mont Sch of Mines 1952-54; Assoc Prof & Prof Univ ofWash 1954-63; Ch Metallurgist Materials & Processes lab Boeing Aerospace Co 1963- . Chmn Advisory Ctte on Met Educ ASM 1961; Chmn Educ & Dev Council ASM 1964-65; Fellow ASM 1976; Fellow Selection Committee ASM 1977-79; Nominating Committee ASM 1979. 1982-Bd of Trustess, ASM. Deceased July 2, 1984. *Society Aff:* ASM, AIME.

Roberts, George A
Business: 1901 Ave of the Stars, Los Angeles, CA 90067
Position: President. *Employer:* Teledyne Inc. *Education:* D Sc/Metallurgy/Carnegie-Mellon Univ. *Born:* 2/18/19. Was previously Pres & Chmn of Vasco Metals Corp which merged into Teledyne in 1966. He is a trustee of Carnegie-Mellon Univ from which he received DSc 1942. Honor: (Member) Natl Acad of Engrg. Fellow Met Soc & ASM; Gold Medal Award, Amer. Society of Metals; Mbr SME. Co-authored book on Tool Steels. Was Internatl Pres ASM, Pres of ASM Foundation for Educ & Res; Trustee of Council for Profit Sharing Industries, Trustee of Trade Relations Council & Chmn of Met- Ceramics Foundation. *Society Aff:* NAE, ASM, TMS-AIME, SME, AISI.

Roberts, Irving
Home: 3 Westwick Rd, Richmond, VA 23233
Position: Consultant (Self-employed) *Education:* PhD/Chemistry/Columbia Univ; MA/Chemistry/Columbia Univ; BS/Chemistry; CCNY. *Born:* 1/9/15. 1937-9 Res Asst to Harold C Urey, Columbia Univ-Iotopes. 1939-43 Designer of organic chem plants, Weiss & Downs, Inc. 1943-45 Group Leader, Manhattan Proj-Uranium Enrichment. 1945-50 Elliott Co, Jeannette, PA - Oxygen & Cryogenics Plants. 1951-56 Consultant to steel & aluminum industry. 1956-78 Reynolds Metals Co, retired as VP in 1978. 1978-present Consultant to energy industry. *Society Aff:* AIChE, ASME, ACS, AIMME, ICSOBA.

Roberts, J Kent
Business: Dept of Cvl Engrg, Rolla, MO 65401
Position: Professor Emeritus of Civil Engineering. *Employer:* University of Missouri - Rolla. *Education:* BSCE/Civil Engg/Univ of OK; MSCE/Civil Engg/Univ of MO-Rolla. *Born:* Jan 15, 1922. US Army in WWII. With Univ of Mo, Rolla 1947- ; serving as Instr, Asst Prof, Assoc Prof & Prof of Cvl Engrg; Asst Dean of Sch of Engrg 1970-80; Professor Emeritus. Prof summer employment with Mo Hwy Dept, Mo Div of Health, US Geol Survey, US Public Health Service & Natl Park Service. Fellow ASCE; Mbr Mo Soc of Prof Engrs, Pres 1965-66; Mbr Amer Soc for Engrg Educ; Mbr NSPE, V Pres 1973-75; former mbr Mo Bd for Architects, Prof Engrs & Land Surveyors. Mbr Rolla Rotary Club; Chi Epsilon, Tau Beta Pi, Sigma Xi. *Society Aff:* ASCE, NSPE, ASEE.

Roberts, J Moran
Home: 1586 Peachtree Btle Ave, Atlanta, GA 30327
Position: Chrmn (Retired). *Employer:* Robert & Co. *Education:* -/Civil Engg/GA Tech. *Born:* 12/25/07. in GA. Educ GA Tech. 1930-33 GA Hwy Dept; 1934-36 Field Engr Fedl Land Bank of Columbia, land surveys, descriptions, etc; 1936-40 City Eng Dublin GA; 1940- 41 Utility Cons Engr Ft Stewart GA; 1941 joined Robert & Co Assocs, Asst mgr Municipal Engg Dept; 1957 promoted to VP & Mgr of Dept; 1973 named Pres of Co; 1976 Chmn & Pres. Pres 1950 GA Section ASCE; Chmn 1963 SE Section AWWA; 1968-70 Natl Dir AWWA; 1970 Fuller Awardee AWWA. Reg PE: GA, SC & TN. Retired as Chief Executive Officer of Robert & Co Assoc June 1978. Retained as mbr of Bd of Dirs. *Society Aff:* ASCE, NSPE, AWWA.

Roberts, John M
Business: Dept of Mech Eng & Materials Science, Box 1892, Houston, TX 77251
Position: Prof of Materials Science *Employer:* Rice Univ *Education:* PhD/Physical Metallurgy/Univ of PA; MASc/Physical Metallurgy/Univ of Toronto; BASc/Metallurgical Eng/Univ of Toronto *Born:* 02/16/31 Naturalized US citizen 1980. Metallurgist with Aluminium Lab Ltd 1954-1956. Union Carbide and Carbon Research Fellow, Univ of PA 1956-1959. Asst Prof of Mech Eng, Rice Univ 1959-1963. John Simon Guggenheim Fellow, Univ of Paris, France 1964-1965. Associate Prof of Mech Eng, Rice Univ 1963-1965. Associate Prof of Materials Sci, Rice Univ 1965-1970. Prof of Materials Sci, Rice Univ 1970-present. Guest Prof at the Max Planck Inst for Metallforschung, Stuttgart, Germany 1972-1973. Visiting Prof and CNPq fellow at the Federal Univ of Sao Carlos, Brasil 1979-1980. SCEEE/SFRP/USAF Fellow summer of 1981. Fifty technical publications . Mbr exec committee of the Houston Chapter of ASM 1981 and 1985. Guest Prof. at the Univ of Tsukuba, Tsukuba Science City, Japan 1986-87. *Society Aff:* ASM, APS, ΣΞ, ΤΒΠ, TMS-AIME.

Roberts, Philip J W
Business: Civil Engineering, Georgia Inst of Tech, Atlanta, GA 30332
Position: Asst Prof *Employer:* GA Tech *Education:* PhD/Envr Eng/CA Inst of Tech; MS/Envr Eng/CA Inst of Tech; SM/ME/MIT; BSc/ME/Imperial Coll-London *Born:* 11/02/46 Native of England, obtained undergraduate degree there before coming to US for Grad Study. While at MIT, did research in compressible flow and gas kinetics. At CALTECH, specialized in Ocean Outfall Design and Dispersion for Thermal and Sweage Outfalls. Ocean Engr for CH2M Hill following PhD, responsible for Ocean Outfall Design, Oceanographic Data Analysis, and Transport Predictions. GA Tech since 1978, teaching in Fluid Mechanics and Hydraulics. Current research interests include Environmental Hydraulics, particulary Coastal Dispersion Modeling, Lake and Reservoir Hydrodynamics, and River Modeling. Mbr two ASCE task forces. ASCE Collingwood Prize 1980. Enjoy backpacking and photography. *Society Aff:* ASCE, IAHR, ΣΞ

Roberts, Philip M
Business: 707 Harrison Ave, Rockford, IL 61101
Position: Chief Metallurgist. *Employer:* Rockford Products Corp. *Born:* Apr 1933; raised in Niagara Falls N Y & Homewood Ill. BS MetE MTU; Alpha Sigma Mu. Inland Steel R/D 1958, mill dev & phys met; Tremec S A Queretaro Mexico (Clark Eq) 1964; Ch Metallurgist 1st automotive transmission plant of consequence in Mexico; Rockford Prods Metallurgist 1967, Ch Metallurgist 1973. Chmn Rockford ASM Chap 1975, mbr ASM Mech Working & Forming Div Council. Enjoys camping & canoeing.

Roberts, Robert J E
Business: P.O. Box 726, Wilton, CT 06897
Position: President. *Employer:* Fred T Roberts & Co. *Education:* AB/Mathematics/Syracuse University; MS/Engr. Economics/Oklahoma University *Born:* Dec 1920 Phila Pa. AB Syracuse Univ; MS Okla Univ. After attending USC started career as Machine Designer with Advance Tennis Ball Co 194041; Proj Engr Danbury Rubber Co 1941-42; with Fred T Roberts & Co since 1942 in various positions at different company facilities. Firm is an engrg/design/res co specializing in rubber & plastics industries. Ch Engr 1945-52; V Pres Engrg 1952-62; Exec V Pres 1962-70; Pres 1970-to date. Holder of numerous pats in field of elastomeric production methods, special machinery, hose, athletic equipment, etc. Fellow AAAS & ASME; Sr Mbr AIIE, SPE; SAME, ISI, Inst Rubber Indus (England), Inst. of Rubber & Plastics etc. Reg P E in Conn & N.Y.; Mbr Conn SPE/NSPE; Engrg Advisory Comm Univ of Bridgeport, St Ethics Ctte CSPE; Hobbies: sailing, skiing, fishing. Clubs: Chemists (N Y), Green Mountain, Mt Mansfield Cl. Trustee, United Engineering Trust, Selection Committee Inventors Hall of Fame, Alt. Fritz Medal Board, Board, Engineering Foundation etc. *Society Aff:* ASME, SAE, SME, AIAA

Roberts, Robert R
Business: College of Engg, Columbia, SC 29208
Position: Assoc Prof. *Employer:* Univ of SC. *Education:* PhD/Civ Engr/WV Univ; MSCE/Civ Engr/GA Tech; BCE/Civ Engr/GA Tech. *Born:* 1/17/33. Native of GA. Worked with GA Hgwy Dept 1956-1960 in roadway design. With Wilbur Smith & Assoc in Columbia, SC 1962-65 specializing in urban transportation modeling. Joined faculty of college of engrg, Univ of SC in 1965. Pres of Southern Sec Inst of Transportation Engrs 1976-77. Recipient of NSF College Teacher Res Participation Grant in 1967 & UMTA Fellowship in 1970. Performs consulting work in automobile accident reconstruction. Author of numerous technical publications. *Society Aff:* ITE, TRB, SAE, NSPE, NAFE, AAFS

Roberts, Robert W
Home: 2352 Chatham Rd, Akron, OH 44313
Position: Robert Iredell Prof. *Employer:* University of Akron. *Education:* PhD/BS/IA MS/ChE/IA/ChE/WA Univ St Louis. *Born:* 8/24/23. in Riverside, IL. Mbr AIChE, SPE, ASEE, Alcoa-Bayer Plt 1948-51; W R Grace- Cryovac 1951-59; Allehany Ballistics-Propellat Process 1962-64; Dir R/D Cadillac Plastics 1964-66; Univ of Akron 1966- , Robert Iredell Prof 1970-. *Society Aff:* AIChE, SPE, ASEE.

Roberts, Sanford B
Business: 405 Hilgard Ave, Los Angeles, CA 90024
Position: Assoc Prof. *Employer:* Univ of CA. *Education:* PhD/Engg/Univ of CA; MSCE/Structures/Univ of S CA; BCE/Civ Engg/CCNY. *Born:* 2/20/34. A native of NYC. A reg PE in CA. Taught structural engg at USC & UCLA. In recent yrs res interests have been in the structural behavior of the human skeletal system & the mech properties of biological tissues. Served as consultant to aerospace & civ engg firms. Currently, Pres of the Assn of Scientific Advisers, a consulting firm specializing in scientific consulting to the legal community. *Society Aff:* ASCE, ΤΒΠ, ΣΞ.

Roberts, Thomas D
Business: Dept of Elec Engg, Duckering Building, 306 Tanona Drive, Fairbanks, AK 99701
Position: Prof. *Employer:* Univ of AK. *Education:* PhD/Applied Phys/OR State Univ; BS/Phys/Univ of AL. *Born:* 6/21/35. Served as Army electronic technician 1953-56. Engr with Chrysler Corp & US Bureau of Mines in electronic design & instrumentation. Conducted basic plasma res with Natl Bureau of Stds 1965-66. At Univ of AK since 1966. Current Position: Prof of electrical engg, specializing in telecommunications and research director for enginering. Reg PE (AK). Avocations include Skiing, classical music & amateur radio (KL7TR). *Society Aff:* IEEE, ASEE, APS.

Robertson, Channing R
Business: Dept of CE, Stanford, CA 94305
Position: Chrmn & Prof. *Employer:* Stanford Univ. *Education:* PhD/CE/Stanford Univ; MS/CE/Stanford Univ; BS/CE/Univ of CA. *Born:* 11/26/43. Native of CA. Res Scientist with Marathon Oil Co, Littleton, CO, 1969-70. At Stanford since 1970. Prof & Chrmn of Dept of Chem Engg since Sept 1978. Former Mbr of editorial bds of *Biophysical Journal* & *Journal of Bioengineering*. Author of papers on transport processes in the mammalian kidney, fluid mechanics, heat transfer, enzyme engg, protein adsorption, whole cell immobilization & blood platelet adhesion & aggregation phenomena. Enjoys backpacking & skiing. *Society Aff:* AIChE.

Robertson, Elgin B
Home: 1330 Club Cir, Dallas, TX 75208
Position: Chrmn of Bd. *Employer:* Retired. *Education:* EE/EE/Univ of TX; DrEngg/-/Southern Methodist Univ; Distinguished Engg Grad/-/Univ of TX. *Born:* 6/4/93. Meridan TX June 4, 1893. Educ in Meridan public sch-Meridian College (Prep) & Univ of TX with degree of EE in 1915. Westinghouse Grad Sch-Lamme Design Sch- Design Engg Power Transformer Sec. Chief EE Railway & Ind Engg Co - Dist Mgr Middle West Chicago. Dallas TX & established Elgin B Robertson Inc-Plastics

Robertson, Elgin B (Continued)
Mfg Co-Pres-Chrmn Bd Retired 6-4-78. P Pres AIEE - Natl Dir NSPE. Honorary mbr- Fellow-Dir Emeritus IEEE. *Society Aff:* IEEE, NSPE.

Robertson, James B, Jr
Home: PO Box 369, St Michaels, MD 21663
Position: retired. *Education:* BSc/Naval Arch and Marine Engg/Univ of MI. *Born:* Jan 1909 Chicago. Naval Architect (ship safety) BMIN & US Coast Guard 1938-67; Cons since 1967. Work has included ship strucs, fire control, life saving, bulk cargo handling, loadlines, subdivision & damage stability. Mbr US delegations to Internatl Safety of Life at Sea Conferences 1948, 1960, 1974 & to Internatl Conference on Loadlines 1966. US representative on internatl subcttes on subdivision & stability & on bulk cargoes 1962-73. Meritorious Civilian Service Awards 1962 & 1974. SNAME Joseph H Linnard Award 1962, David W Taylor Gold Medal 1975. 1985 Halert C. Shepheard Award for Achievement in Merchant Marine Safety. Life Fellow SNAME. *Society Aff:* SNAME.

Robertson, James M
Business: Box 1097, Silverthrone, CO 80498
Position: Prof. *Employer:* Univ of IL. *Education:* PhD/Hydraulics & Fluid Mechs/ Univ of IA; MS/Hydraulics/Univ of IA; BSCE/Civ Engg/Univ of IL. *Born:* 4/18/16. IL native. Wartime res at Navy's Model Basin & Douglas Aircraft. Taught mechanics, fluids & hydraulic engg 1942-49 & was Res Prof & Water Tunnel Dir 1949-54 at PA State Univ. Prof Theoretical & Appl Mechanics at IL (fluid mechanics) since 1954. Visiting lectr/prof at KS State, TAPPI short course, TN Space Inst, CO State; fluid dynamics consultant to corps as Consolidated Papers, Caterpillar Tractor (1963-72), Westvaco; sr scientis Interscience Res Inst 1966- 73. Author grad hydrodynamics text & many papers. Enjoy ballet, classical music, ballroom dancing, skiing, sailing, gardening. *Society Aff:* ASCE, ASME, AIAA, AAAS.

Robertson, Lawrence M
Home: 320 Ash St, Denver, CO 80220
Position: Consulting Engr. *Employer:* Self-employed. *Education:* Dr of Engg/Elec Engg/Univ of CO; EE/Elec Engg/Univ of CO; MS/Elec Engg/Univ of CO; BS/Elec Engg/Univ of CO; Juris Doctor/Law/Univ of Denver; LL B/Law/Westminster Sc. *Born:* 1/20/00. Distinguished Engr Alumnus Award by Univ of CO, 1968; Habirshaw Award by IEEE 1963; Gold Medal Award by CO Engr Council; Service Award by Fed Power Comm, 1964; Prize Paper Award by IEEE, 1969; Service Award by Denver Regional Council of Govts, 1971 and by Natl Council of Engg Examiners, 1971. Was with Public Service Co of CO in Denver from 1922 to 1968; Chief Elec Engr, Mgr of Engr, VP of Engr, Sponsor of res on lightning at high altitudes in CO and EHV transmission at high altitudes. Am a Life Fellow of IEEE, VP 1945-47, Natl Dir 1956-60. Reg PE in CO, WY, WA, CA, IL, NY. On CO Bd of Reg of PE 1954-69, Chrmn 1960-69, NCEE dir 1967; ECPD 1965-69. American Legion WWI. Author of 84 published technical articles & chap of a book on environment. Pres of CO Engg Council & CO Soc of Engrs. Chrmn of Res Proj Committee of Edison Elec 1965-68. 100th Anniv Medal by IEEE 1984. *Society Aff:* IEEE, NSPE, AAAS, ΣΞ.

Robertson, Leslie E
Business: 211 East 45th Street, New York, NY 10017
Position: Senior Partner. *Employer:* Skilling, Helle et al. *Education:* BS/Eng/UC Berkeley. *Born:* 2/12/28. in CA. Presently Managing Partner Skilling, Nelle, Christiansen, Robertson. Respon for some of the most important bldg structures of our time, including the World Trade Center, NY & the US Steel Bldg, Pittsburgh. Pioneer in wind engg & currently 1st VP, Wind Engg Res Council. Fellow ASCE, recipient of its Raymond C Reese Research Prize 1975, for dev work in determination of human response to wind-induced bldg oscillation. Elected Mbr NAE 1975.

Robertson, Richard B
15354 Terrell Rd, Baton Rouge, LA 70816
Position: Pres. *Employer:* Richard B. Robertson, Consulting Engineer, Inc. *Education:* BS/ME/Purdue Univ. *Born:* May 8 1920, Ind. 1942-48 engr for Curtis Wright & United Air Lines; 1944-46 Lt US Army FA; Bovay Engrs Inc 1948-1985: Partner 1952-1962: Exec V Pres & Dir 1962-1985: also Treasurer 1962-72, in charge of power generation projects since 1952 totalling over 3000 MW. President, Richard B. Robertson, Consulting Engineer, Inc 1986 to Date. V Pres ASME 1969-71, Pres 1974-75; Fellow ACEC; Honorary Member Pi Tau Sigma & Tau Beta Pi; Distinguished Engineering Alumnus, Purdue University 1975; Reg PE in Tex, La, Ark, Wash. Mbr NSPE, ACEC, ASME. *Society Aff:* ASME, ACEC.

Robertson, Roy C
Home: 2117 Indian Hills Dr, Knoxville, TN 37919
Position: Engineering Staff Member Retired *Employer:* Oak Ridge Natl Lab, Energy Division. *Education:* MS/Mech Eng/Univ of TN; BS/Mech Eng/Univ of TN. *Born:* 8/5/16. 1944-46 Officer USNR; Assoc Prof ME at Univ of TN 9 yrs, also engrg cons; joined Oak Ridge Natl Lab (Union Carbide Corp) in 1955 in design fluid-fuel (including molten-salt) nuclear reactor tests studies of large-scale reactors, num reports, articles & guest lecturer. Analyzed environmental impacts of thermal discharges from nuclear power stations for Environ Statements of Nuclear Regulatory Comm; expert witness at hearings; engaged in geothermal energy studies and dev for Dept. of Energy. P Natl V P ASME, Life Fellow, East TN 75th annual ASME Medal, ASME Centennial Medal, Reg PE TN; P Mbr St Bd Boiler Rules. Retired. *Society Aff:* ASME.

Robins, Daniel F
Home: 4014 GreenBrook Ct, Toledo, OH 43614
Position: Principal *Employer:* Robins Engineering *Education:* BS/ME/MI State Univ *Born:* 12/19/32 Native of Ithaca, MI; resident of Toledo, OH since 1962. Served as pilot with Air Force, 1955-58; Air Force Reserve 1959-81; Ret'd as Lt. Col. With Owens IL as project engr 1962-76, design and construction of mfg facilities and systems. Asst Chief Mech Engr, Hoad Engrs 1976-1979. Supervision of 50 man dept in design of facilities and systems for electrical power plants, mfg plants. Instrumental in introducing computer analysis of piping on in-house basis. Principal of Robins Engrg since 1979. Reg PE in IN, MI, OH. *Society Aff:* ASHRAE, ASME, NSPE, ACEC

Robins, Norman A
Business: 3210 Watling Street, E Chicago, IN 46312
Position: VP, Technological Assessment & Strategic Planning *Employer:* Inland Steel Co. *Education:* PhD/Math/IL Inst of Tech; MS/Chem Engrg/MIT; BS/Chem Engrg/ MIT *Born:* 11/19/34 1956 to present Inland Steel Co, 1956 Metallurgist-Res Dept, 1960 Res Metallurgist-Res Dept, 1962 Asst Mgr-Processing Sys & Controls Div, 1967 Assoc Mgr-Processing Sys & Controls Div, 1972 Dir-Process Res-Res Dept, 1977 VP-Res. 1984 VP-Technological Assessment, 1986 VP-Technological Assessment & Strategic Planning. *Society Aff:* AISI, AIME, AIChE, MAA.

Robinson, Arthur R
Business: 2129 Newmark C.E. Lab, 208 N. Romine St, Urbana, IL 61801
Position: Prof of Civ Engg. *Employer:* Univ of Illinois, Urbana-Champaign. *Education:* PhD/Civ Engg/Univ of Ill (Urbana); MS/Civ Engg/Univ of Ill (Urbana); BCE/Civ Engg/The Cooper Union. *Born:* Oct 28 1929, Brooklyn NY. BSCE The Cooper Union; MS & PhD 1956 Univ of Ill. 1955-60 Res Assoc & Asst Professor Depts of Mechanics & Materials & of Aero Engrg & Engrg Mechanics, Univ of Minn; 1960-63 Assoc Professor Civil Engrg, Univ of Ill at Urbana, 1963- , Professor of Civil Engrg at Urbana-Champaign. Walter L Huber Civil Engineering Research Prize of ASCE 1969, Moisseiff Award of ASCE with HH West 1970. Current research includes wave propagation in solids with application to earthquake engg and fracture mechanics, and numerical methods in structural mechanics. *Society Aff:* ASCE, ASME, SSA, AAAS. #TTL‡Prof

Robinson, Arthur S
Business: 11790 Indian Ridge Road, Reston, VA 22091
Position: President *Employer:* System/Technology Planning Corporation *Education:* DEngSc/EE/Columbia; MS/EE/NY Univ; BS/EE/Columbia. *Born:* Sept 1925. 1943-46 served with US Army; Electronics Engr, Hazeltine Corp; Computer Grp Leader Columbia Univ Electronics Res Labs; Dir Adv Electronics Lab, Bendix Corp Eclipse-Pioneer Div; Dir of Res, Kollsman Instrument Corp; Mgr Adv Tech Programs, RCA Missile & Surface Radar Div: President, System/Technology Planning Corporation: planning and implementation of real time sensing, computing and control systems. Fellow IEEE, cited for 'leadership in digital & analog computing & control systems, solid-state radar, coherent electro-optical systems & medical electronic instrumentation.' 42 domestic & foreign patents. *Society Aff:* IEEE.

Robinson, August R
Home: 8473 Imperial Dr, Laurel, MD 20708
Position: Consulting Engr *Employer:* Self *Education:* MS/Irrigation Eng/CO State Univ; BS/CE/Univ of IA *Born:* 04/24/21 Native of Young County, TX. Served in Army Combat Engrs 1943-46. Conducted irrigation and drainage research, USDA and CO Exp Station 1949-63. Dir of USDA Snake River Cons Research Center 1963-69. Dir of USDA Sedimentation Lab 1969-74. USDA Natl Prog Staff Scientist, Erosion and Sedimentation, 1974-79. Consltg Engrg for erosion, sedimentation, irrigation, drainage, hydraulic structures, Conservation Systems 1979-present. Reg PE. ASAE Hancor Soil and Water Engrg Award 1980. *Society Aff:* ASCE, ASAE, NSPE, SCSA, ΣΞ, XE

Robinson, Carlton C
Business: 1776 Mass Ave NW, Washington, DC 20036
Position: Executive Vice President. *Employer:* Highway Users Federation. *Education:* BS/Civil Engg/OR State Univ. *Born:* March 25 1926, Long Beach Calif. Graduate of Bureau of Hwy Traffic. Asst Traffic Engr of Portland, Ore before joining staff of Automotive Safety Foundation in 1955; appointed Dir of foundation's Traffic Engrg Div 1961, named Dir of Program Planning 1970 when foundation merged into the Hwy Users Federation, appointed Exec V Pres Oct, 1974. Reg PE in Wash DC & Ore. Active Mbr of Inst of Transportation Engrs, served as Pres 1971. Mbr various cttes of Transportation Res Bd & Natl Ctte on Uniform Traffic Control Devices. Received 1985 Theodore Matson Award for distinguished contribution to transpotation engineering. *Society Aff:* ASCE, ITE.

Robinson, Charles S
Home: 5265 McIntyre, Golden, CO 80403
Position: Vice Pres/Gen. Manager *Employer:* Mineral Systems, Inc. *Education:* PhD/Geo/Univ of CO; BS/Chem/MI Coll of Mines and Tech *Born:* 06/23/20 Served in the US Navy during World War II after graduation from coll. On completion of grad work, *LB*after World War II,*RB* joined the US Geological Survey. Spent 10 yrs in Mineral Deposits and 7 yrs in Engrg Geology. Established my own consltg firm in 1965. Merged with Converse Ward Davis Dixon in 1980. Started Mineral Systems, Inc., 1981. A Consulting Engineering and Mining Geology firm. Author of more than 40 scientific or engrg articles or books on Geology and Engrg. *Society Aff:* AIME, GSA, SEG, AEG, IAEG

Robinson, Denis M
Business: S Bedford St, Burlington, MA 01803
Position: Sr Consultant *Employer:* High Voltage Engineering Corporation. *Education:* Univ of London, MIT *Born:* 1907, London England. Came to MIT in 1929 with PhD in electrical engrg to study underground transmission of electric power. 1931 returned to London, employed in high voltage labs of Callenders Cables Ltd & subsequently in television dev; 1939-end WWII radar dev for Royal Air Force & British representitive at Radiation Lab MIT, OBE & US Medal of Freedom; Professor of Electrical Engrg at Univ of Birmingham, England; 1946 returned to US & with Professors John Trump & Van de Graaff of MIT founded High Voltage Engrg Corp which makes very high voltage generators, Ch Exec Officer 1946-70, Chmn of Bd 1970- . Chmn of Bd of Trustees, Marine Biological Lab Woods Hole 1971- . 1970-73 Secy of Amer Acad of Arts & Sciences. *Society Aff:* NAE

Robinson, Donald A, Jr
Business: 1714 Cesery Blvd, Ste 14, Jacksonville, FL 32211
Position: Executive VP *Employer:* Liebtag, Robinson & Wingfield, Inc *Education:* B/EE/Univ of FL; B/ME/Univ of FL *Born:* 4/10/41 Native of Jacksonville, FL. Registered consulting engr in both electrical and mechanical engrg. Principal with Liebtag, Robinson and Wingfield, Inc and VP in charge of electrical engrg. Specialist in the field of energy management; responsible for lighting power budgets, daylighting, and building system computer simulations. National Society Participation - AEE Building Energy Conservation Committee, IES Lighting and Thermal Environment Committee, and IES Emergency Lighting Committee. State Society Participation - FES Energy Committee. City Participation - City of Jacksonville Building Code Adjustment Bd. *Society Aff:* AEE, ASHRAE, IES, NSPE

Robinson, Ernest L
Home: 1757 Wendell Ave, Schenectady, NY 12308
Position: retired. *Education:* ScD/-/St Lawrence Univ; MS/CE/Harvard Grad Sch Applied Sci; BA/St Lawrence Univ. *Born:* April 1890, Canton NY. 1917-18 1st Lt 302 Engrs US Army, France; 1918-19 Cpt Adjutant, 2 Engr Training Regt; 1919-55 Turbine Engrg Dept of GE Co, struct & thermodynamic improvement of steam turbines & early jet engines, elimination of destructive vibration, perfection of steam tables, reduction of fuel consumption by improved power cycles using regenerative extraction & reheating, higher pressures, temperatures & new metals to make higher temperatures possible. Author of tech & some nontech articles. ASME Melville Medal 1944; Hon M 1958. Many hobbies: books, travel, outdoors, stamps, coins, pictures & civic activities. *Society Aff:* ASCE, ASME, SAME.

Robinson, George H
Home: 4816 Freer, Rochester, MI 48063
Position: Department Head. *Employer:* Research Labs, General Motors Corp. *Education:* BChE/Chem Engg/Univ of Detroit; BSMet/Met Engg/Carnegie Tech. *Born:* June 1924, Detroit. 1943-46 Army service Pacific Theater, Squad Leader Heavy Mortar Bn; 1946 joined GM Res Labs as co-op student, res specialities in ferrous physical metallurgy, wear, fatigue, 1962 Asst Hd, Metallurgy Dept, 1971- 73 Hd Emissions Res Dept, 1973 assumed current respon as Hd Metallurgy Dept. Fellow ASM. Reg PE Mich. Wife Mildred, 7 children. *Society Aff:* ASM, AIME, SAE, AFS.

Robinson, Harold F
Home: 1111 S. Lakemont Avenue, Winter Park, FL 32792
Position: retired. *Born:* Feb 1896, Detroit Mich. BS Naval Architecture & Marine Engrg 1918 & Doctor of Engrg (Honorary) 1953 from Univ of Mich; Tau Beta Pi 1955. 2 years of shipbuilding on the Great Lakes; 44 years with Bethlehem Steel (shipbuilding), Bethlehem Pa & Quincy Mass, 1948-58 Ch Naval Architect, Tech Mgr until retirement in 1964. Several natl & internatl tech committees, served as expert witness following retirement. SNAME Permanent Mbr, Honorary V Pres & Mbr of Council. Several tech papers, co-awarded Joseph H Linnard Prize for 1. Administrative & tech cttes. Rotarian since 1940. Dir Braintree Cooperative Bank (Mass) 15 years. Enjoy photography, music, playing piano & organ. Univ Club, Winter Park Fl.

Robinson, John H
Business: 1500 Meadow L Pkwy (P.O. Box 8405), Kansas City, MO 64114
Position: Mgrg Partner *Employer:* Black & Veatch, Engs-Architects. *Education:* BS/Civil Engg/Univ of KS. *Born:* 2/14/27. Env engrg with Black & Veatch continuously since 1949, assignments in all phases of water & sewage projects, advanced from resident engr supervision to executive partnership in firm. Reg PE in 16 states. Pres, Amer Water Works Assoc. Past Pres Cons Engrs Council of MO;

Robinson, John H (Continued)
Fellow Amer Cons Engrs Council & ASCE; Chrmn Standards Council AWWA, Bd of Directors, Executive Comm AWWA; Mbr KS Engrg Soc; Mbr WPCF, MO Soc of Prof Engrg & others; Diplomate Amer Acad of Environ Engrs; and Past-Chairman, KS Sect AWWA. *Society Aff:* ACEC, AWWA, WPCF, AAEE, ASCE.

Robinson, John K
Business: PO Box 6345, Mobile, AL 36660
Position: President *Employer:* E. T. P., IUC *Education:* BS/CE/Univ of AL. *Born:* 5/10/41. Boston MA - Married with 2 children. owner of a textile fabricating plant dealing in engineered textile items. These include liners, tarps, curtains, & various other items such as football & baseball field covers. Hobbies are sailing, & RR history. *Society Aff:* AIChE.

Robinson, Kenneth E
Home: 352 Silver Lake, Mears, MI 49436
Position: Consultant. *Employer:* self employed. *Born:* 1913, Mason Mi. 1940 Ch Engr McLouth Air-Conditioning; 1943 set up & Ch Ventilation Sect, Occupational Health Div of Mich Health Dept; 1953 joined General Motors as Corporate Hd, Ventilation Sect of Industrial Hygiene Dept, worked throughout US, Canada & oversees, retired 1974; now a cons on ventilation problems relating to inplant environ conditions. Helped start & taught univ short courses on heating & ventilation continuously since 1936. Served on Alumni Adv Bd for Coll of Engrg at Mich State Univ. Served onctte writing codes, educational books & making of films & video tapes involving industrial heating & ventilation to control the industrial environ; many papers & lectures on the subjects in the US & overseas. Mbr Tau Beta Pi 1969 as Eminent Engr. Certified Safety Prof; Mbr Amer Acad of Industrial Hygiene; AIHA; Honorary Mbr ACGIH; Fellow/Life member, ASHRAE. *Society Aff:* ASHRAE, AIHA, ACGIH.

Robinson, Lee
Business: PO Box 8371, Pocatello, ID 83209
Position: Assoc Prof. *Employer:* ID State Univ. *Education:* PhD/Civ Engr/MT State Univ; MS/Irrig & Drng Engr/UT State Univ; BS/Agric Engr/UT State Univ. *Born:* 2/27/41. Reared on southern UT farm. Managed USU Drainage Res Farm - 1963. Joined ID State Univ, Sch of Engg Faculty in 1967 as an Instr; rank was raised to Asst Prof in 1968 and to Assoc Prof in 1977. Has taught and researched in Hydrology, Fluid Mechanics, and Geotechnical Engg during tenure at ISU. Has developed innovative teaching methods. A reg PE since 1972; has been consultant in agricultural, water resources, and geotechnical engg. Special expertise in properties of collapsible soils. *Society Aff:* ASCE.

Robinson, M John
Business: PO Box 8405, Kansas City, MO 64114
Position: Partner *Employer:* Black & Veatch *Education:* PhD/Nucl Engrg/Univ of MI; MS/Nucl Engrg/Univ of MI; BS/Mech Engrg/Univ of MI. *Born:* 06/19/38 From 1960 to 1965 Dr. Robinson worked as a research engr on various NASA programs at the Univ of MI. In 1966 he joined the faculty at Kansas State Univ, where he was also a licensed reactor operator for the TRIGA MARK II Reactor. In 1970 he was appointed by the Intl Atomic Energy Agency as Technical Assistance Expert and spent one yr in Sao Paulo, Brazil. He was also conslt in the area of shielding and a radiological effects conslt on the Monticello Nuclear Power Plant. Dr. Robinson is currently a partner in the Power Div of Black & Veatch Consltg Engrs and is a reg PE in six states. *Society Aff:* ANS.

Robinson, Thomas B
Business: 1500 Meadow Lake Pkwy, PO Box 8405, Kansas City, MO 64114
Position: Retired-Conslt *Employer:* Black & Veatch. *Education:* BSCE/-/KS Univ; MSCE/-/Columbia Univ *Born:* 2/28/17. KS City. BSCE, KS Univ, 1939; MSCE, Columbia Univ, 1940. Served to Lt, USN Civ Engr Corps, 1943-46. With Black & Veatch since 1940, except service; Mgg Partner 1973-1982; Chrmn of Bd, Black & Veatch Intl 1973-1982. Reg PE 9 states. Natl Acad of Engg; Pres, Fellow, AM Consulting Engrs Council; Fellow, ASCE; Dipl, AM Acad of Environmental Engrs; Hon Mgr, Chrmn, several cttes, AWWA; Sigma Xi, Tau Beta Pi, Sigma Tau. MO Hon Award for Distinguished Service in Engg, Univ of MO-Columbia; Citation for Distinguished Service, Univ of KS. *Society Aff:* ASCE, NSPE, AWWA, WPCF

Robison, Ralph E
Home: 29709 West Oakland Blvd, Bay Village, OH 44140
Position: Assoc. *Education:* MS/Struct Engg/Univ of IL; BS/Struct Engg/Univ of IL. *Born:* 8/23/25. Born 8/23/25. BSCE 1945; MS 1947 - Univ of IL. US Navy Civil Engrg Corps 1945-46. With HNTB since 1947. Served as Engr-in-Charge of the firm's Cleveland, OH office from 1972-78. Named Assoc in 1978 and Associate in Charge of the Cleveland office in 1985. Identified with the design of numerous major and minor highway and railroad bridges. ASCE Cleveland Section - Civil Eng of the year 1981; Cleveland Engrg Societies - Eng of Year 1982. Society Aff. ASCE, AREA, ASTM, AWS, IBTTA, NSPE *Society Aff:* ASCE, AREA, ASTM, IBTTA, NSPE, AWS.

Roby, Dennis E
Business: 1900 East Eldorado St, PO Box 1425, Decatur, IL 62525
Position: President *Employer:* Dennis E. Roby & Assoc, Inc *Education:* MS/CE/Purdue Univ; BS/CE/Purdue Univ *Born:* 05/28/37 Native of DeGraff, OH. Employed by Chicago Bridge & Iron Co as a supervisor in charge of structural design group responsible for foundation designs, special structures and computer applications 1961-1965. A Soil Engrg Conslt responsible for engrg analysis and recommendation reports pertaining to foundations and other special structures 1965-1967. Since 1967, owner of Dennis E. Roby & Assocs, Inc, a consltg engrg firm which provides structural, mech, agricultural, and civil engrg services for various clients in the US and in foreign countries, including industries, municipalities, large developers, commercial establishments and similar organizations. State Chrmn ISPE/PEPP 1976-77 & 1977- 78. *Society Aff:* AAA, NSPE, ASCE, AIPE, ACI, PEPP

Roche, Edward C Jr
Business: New Jersey Institute of Technology, Newark, NJ 07102
Position: Prof. *Employer:* NJ Inst of Tech. *Education:* ScD/Chem Engg/Stevens Inst of Tech; MS/Phys/Harvard Univ; ME/Gen Engg/Stevens Inst of Tech. *Born:* 5/9/33. Native of Boston, MA. Employed by Esso Res and Engg/Esso Math and Systems Inc as a Sr Analyst 1957-1967, specializing in process design and simulation. Joined Newark College of Engg in 1967 as a mbr of the Chem Engg faculty. Currently a Prof of Chem Engg at NJ Inst of Tech (formerly Newark College of Engg) specializing in mass transfer, process and plant design, and process/plant simulation. *Society Aff:* AIChE, ASEE, SCS.

Rockwell, Glen E
Home: 7181 Cedarwood Circle, Boulder, CO 80301
Position: President *Employer:* Engineering Hydraulics, Inc *Education:* MS/Hydraulics/State Univ of IA; BS/CE/Hydraulics/State Univ of IA *Born:* 07/23/34 Born in Iowa City, IA. Taught in Civil Engrg Dept of Univ of TX. Joined Engrg Conslts, Inc of Denver, CO, an intl water resources development engrg firm in 1961, became Chief Mech and Hydraulic Engr in 1966 and Pres in 1973. Became Pres of Engrg Hydraulics, Inc, in Longmont, CO, in 1979, a firm providing services in hydraulic modeling, research and consltg throughout the US. *Society Aff:* ASCE, IAHR, IWRA

Rockwell, Theodore
Home: 3403 Woolsey Dr, Chevy Chase, MD 20815
Position: Principal Officer. *Employer:* MPR Assoc, Inc. *Education:* ScD/Sci/Tri State Univ (H.C.); MS/ChemEE/Princeton Univ; BS/ChemEE/ Princeton Univ *Born:* 6/26/22. Co-founder, MPR Assoc, Inc, Wash, DC, 1964. Nuclear engr, naval reactors br AEC, also nuclear propulsion divs. Navy Bur of Ships, 1949-55; Technical Dir, 1955-64; plusHd, shield engg group Oak Ridge Nat Lab, 1945-49; Process Im-

Rockwell, Theodore (Continued)
provement engr Manhattan proj, Oak Ridge, 1944-45. Res Assoc, Johns Hopkins Univ Ctr Fgn Policy Res, 1965-66. Chrmn Atomic Ind Forum Reactor Safety Task Force, 1966-72; adv group artificial heart prog NIH, 1966; cons to Joint Congl Com on Atomic Energy, 1967; adv council dept chem engg, Princeton Univ, 1966-72. Dist Civilian Service Medal, US Navy, 1960. Dist Service Medal, USAEC, 1960. Several patents, books, tech papers. PE 807, DC. "First Lifetime Contribution Award, henceforth known as the Rockwell Award–, Amer. Nuclear Soc. (ANS) (1986). *Society Aff:* ASTM, AAAS, PA, ASPR, ANS.

Rockwell, Willard F, Jr
Business: 600 Grant St, Pittsburgh, PA 15219
Position: Director (Ret Chrmn, 1979) *Employer:* Rockwell International Corporation. *Education:* BS/Industrial Eng/PA State Univ; Techn Degree/Industrial Engg/PA State Univ. *Born:* March 3 1914, Boston Mass. Reg PE in Pa & Calif. WWII Cpt (ordnance) US Army; 1947-64 Pres Rockwell Mfg Co, 1964-71 V Chmn & CEO, 1971-73 Chmn; 1963-67 Pres & CEO Rockwell Standard Corp; 1967-79, Chmn of the Bd Rockwell Internatl Corp (name changed in 1973 from N Amer Rockwell). Fellow ASME, Honorary Mbr SME. IIE award for Outstanding Achievement in Management 1971; 1974 Henry Laurence Gantt Medal Award from ASME & Amer Mgmt Assn; honorary doctor of engrg degrees Tufts Univ 1965, & Washington & Jefferson Coll 1966; dr of Laws by Hambuth Coll, Jackson, TN, in 1967; dr of engg by Carnegie-Mellon Univ, Prof in 1978. *Society Aff:* ASME, SME, IIE.

Rockwood, William R
Business: 1735 S Alamo St, P.O. Box 9496, San Antonio, TX 78204
Position: President. *Employer:* Alamo Welding Supply Co. *Education:* BS/Chem Engg/TX A&M. *Born:* July 1920, Wharton Tex. BS Tex A&M 1940. 1 year chemistry lab instructor, Tex A&M; WWII service Army Engrs, separated Cpt & company Cdr, constr battalion duty in Central Pacific; 23 years Union Carbide gas production, Texas City & Brownsville, mgmt ethylene & oxygen plants; 1969 became owner & Pres Union Carbide distributor Alamo Welding Supply Co, San Antonio Tex, respon for sales & distribution of welding supplies, industrial & medical gases. Pres, Union Carbide distributors' Southwestern US. Trustee, Univ of the South, Sewanee Tenn. P Chmn, Red Cross Chapter & United Fund Bd, Brownsville. 1 publication. *Society Aff:* AIChE.

Roddis, Louis H, Jr
Business: PO Box 1513, Charleston, SC 29402
Position: Owner & Consulting Engr. *Employer:* Self. *Education:* MS/Naval Arch/MIT; BS/Eng/USNA. *Born:* Sept 1918. Officer in US Navy 1939-55, participated in design of USS Nautilus. Deputy Dir AEC's Reactor Dev 1955-58; Pres Pa Elec Co 1958-69 when co was awarded indus's Edison Award; Pres & V Chmn Consolidated Edison Co of N Y Inc 1969-74. Pres & CEO John J McMullen Assoc Inc 1975-76, Consulting Engr 1976- . Author of numerous tech papers including section on Nuclear (Atomic) Power in Mark's Standard Handbook for Mech Engrs. Received AEC's Outstanding Service Award 1957. Mbr NAE; Reg P E in NY, PA, NJ, DC, SC; P Dir & P Pres of both the Atomic Indus Forum & Amer Nuclear Soc; Fellow ASME, ANS & Royal Inst of Naval Archs; Mbr IEEE, ASHRAE, SNE, SNAME, AEE, HFS & NSPE. Dir of Hammermill Paper Co, Detroit Edison Co, Research-Cottrell, Inc. and Gould Inc; Cons to serveral cos and govt agencies. *Society Aff:* IEEE, SNAME, SNE, ASHRAE, AEE, NSPE, HFS, RINA, ASME, ANS.

Roden, Martin S
Home: 5659 Halifax Rd, Arcadia, CA 91006
Position: Prof Engr and Chairperson, Elect and Comp Engrg *Employer:* CA State Univ. *Education:* PhD/EE/Kensington Univ; MSEE/Elec Engr/Poly Inst of Brooklyn; BSEE/Elec Engg/Poly Inst of Brooklyn. *Born:* 8/14/42 Part time Lecturer, Poly Inst of Brooklyn and Mbr of Technical Staff, Bell Telephone Labs from 1963 to 1968. CA State Univ, Los Angeles, 1968 to present. Periods as Assoc VP for Academic Affairs, and Dean of the Sch of Engg. Presently, prof and Chair of the Electrical and Computer Engineering Department. Author of five texts: Introduction to Communication Theory, Pergamon Press, 1972, Analog and Digital Communication Systems, Prentice Hall, 1979 (second Edition, 1985), Digital and Data Communication Systems, Prentice Hall, 1982. Electronic Circuit Design (with Savant and Carpenter), Benjamin/Cummings 1987 and Design of Digital Communications Systems, Prentice Hall, 1988. Mbr of IEEE, ASEE, Tau Beta Pi, Eta Kappa Nu, Phi Kappa Phi. Fellow of the Inst for the Advancement of Engg. *Society Aff:* IEEE, ASEE.

Rodenberger, Charles A
Business: Dept of Aerospace Engrg, Coll Station, TX 77843
Position: Prof *Employer:* TX A & M Univ *Education:* PhD/Aero Engr/Univ of TX-Austin; MS/ME/So Methodist Univ; BS/Gen Engr/OK St Univ *Born:* 9/11/26 Registered PE in TX. After graduation in 1948, worked as engr in petroleum, electrical, civil, aerospace structural design, corrosion, biomedical and teaching. Served in USAF in WWII and Korean War. Entered teaching in 1960 at TX A&M Univ. Taught aerospace structures, aerospace design, systems engrg, preliminary systems design, ethics, and professional practice, engrg economics and engrg entrepreneurship. Research in hypervelocity impact, cleanup of oil spills, orthopedic devices and conducted a workshop for Gas Research Inst on basic research problems. National pres of JETS, Inc 1978-80. VP - NSPE - 1980- 81. *Society Aff:* AIAA, ASEE, ASME, NSPE, TSPE

Rodes, Elmer O, Jr
Business: PO Box 4038, Roanoke, VA 24015
Position: Partner. *Employer:* Sowers Rodes & Whitescarver. *Born:* June 17, 1923 Harrisonburg Va. BS Min Engrg VPI 1947. Ptnr in Sowers Rodes & Whitescarver Cons Engrs 1959- ; respons include admin, cvl, plumbing & fire protection. Treas & mbr of Bd of Trustees Amer Cons Engrs Council 1973-75. Pres Cons Engrs Council of Va 1967-68, Natl Dir 1968-69, recipient of CEC/VA P Pres Award 1970. Reg PE in Va, N C, Tenn, W Va. Mbr ODK, Pi Kappa Alpha, Tau Beta Pi, Rotary Club. Hobbies: boating & fishing.

Rodes, Harold P
Business: 1700 W 3rd Ave, Flint, MI 48502
Position: President Emeritus. *Employer:* General Motors Institute. *Born:* May 7, 1919 Moorestown N J. PhD Yale Univ; AB Dartmouth. Teacher & Athletic Coach Bradford H S, Vt 1941-42. Lt US Marine Corps 1942-43. Aircraft Drafting Instr San Diego Calif 1943-44; Lectr in ME & Asst Supr of War Training Univ of Calif, Berkeley 1944-45; Teaching Asst Yale Univ 1945-48; Asst Prof of Engrg & Asst Dir of Relations with Schools UCLA 1948-51; Pres Ohio Coll of Applied Sci, Cincinnati 1951-54; Pres Bradley Univ, Peoria Ill 1954-60; Pres G M Inst Flint Mich 1960-76. Ser on Bds of Amer Soc for Engrg Educ, Internatl Inst, C S Mott Foundation, United Way & Urban League. Mbr Tau Beta Pi, Phi Beta Kappa, Sigma Nu. Enjoys golf, tennis & sailing.

Rodger, Walton A
Business: 3206 Monroe ST, Rockville, MD 20852
Position: Pres. *Employer:* Nuclear Safety Assoc, Inc. *Education:* PhD/Chem Eng/IL Inst Tech; MSE/Chem Eng/Univ of MI; BSE/Chem Eng/Univ of MI; BSE/Met Eng/Univ of MI. *Born:* 3/18/18. *Society Aff:* AIChE, ANS.

Rodgers, Alston
Home: 1638 La Tierra Ln, Lake San Marcos, CA 92069
Position: Retired. *Education:* ME/Mech Engg/Stevens Inst of Tech. *Born:* October 1903 N Y City. Tau Beta Pi. Genl Elec Lamp Div lighting engr in N J, N Y, Ohio, Calif 1926-42. Lt Cdr Special Devices Div, Navy Office of Res & Inventions 1942-45. Sr Engr, Mgr G E Lighting Inst Nela Park Cleveland Ohio 1945-68. Dev many devices to demonstrate lamp & lighting principles. Widely known speaker on lighting subjects. Prof Engr Ohio. With J S Hamel Engrg Burbank Calif; planned Walt Disney World lighting 1969-71. IES Fellow Emeritus, RVP, Dir, Chmn of natl cttes.

Rodgers, Alston (Continued)
Trustee Edison Birthplace Assn. Hobbies: books, records & humorous verse. *Society Aff:* IES.

Rodgers, Colin
Business: 4400 Ruffin Rd, San Diego, CA 92123
Position: Chief, Conceptual Design *Employer:* Sundstrand Turbomach *Education:* HND/Aero & Mech/Hull College of Aeronautics; ONC/Mech/Leeds College of Technology *Born:* 09/24/29 Mr. Rodgers received his ONC.ME from Leeds College of Technology and the HND (ME. Aero) from Hull College of Aeronautics. For the last 30 years he has worked in various capacities for Solar Turbines, currently as Chief of Aerothermo and Conceptual Design in the Turbomach Div of Sundstrand. He is recognized for his years in the aerodynamic dev of radial compressors & turbines for small gas turbine engines and industrial compressor sets. He has been responsible for preliminary engine design work, supervision of gas turbine engine performance analyses, and the management of major engine, compressor, & turbine R&D programs. An outstanding author of papers on turbomachinery technology, Mr. Rodgers has been the invited lecturer at various turbomachinery courses at major universities in Europe and the US. Currently, he is a member of the AIAA & a Fellow of ASME. *Society Aff:* ASME, AIAA

Rodgers, Joseph F
Home: 5177 Eileen Dr, San Jose, CA 95129
Position: President *Employer:* Electrical Specialty Co *Education:* MS/Ind Eng/UC Berkeley; BS/Mech Eng/UC Berkeley. *Born:* May 1932; native of Napa Calif. With 1st Armored Div 1955-57. Indus Engr & Mfg Mgr with Colgate Palmolive Co; taught engrg & mgmt at Univ of Calif Extension Div; with Raychem Corp since 1967 as Mgr of Indus Engrg, Mfg, Marketing V.P. Industry & Mgt. Divs of IIE 1980. & Bus Planning. Regional X Pres IIE 1976. Active in Scouting. V P Ind & Mgt Divs of IIE 1980. *Society Aff:* IIE

Rodgers, Oliver E
Home: 507 S Providence Rd, Wallingford, PA 19086 *Education:* MS/Applied Mechs/Harvard Univ; SB/Engg Sci/Harvard. *Born:* 12/12/15. grew up in Anaconda MT. 15 yrs Westinghous Elec developing steam & aviation gas turbines; last position Asst Engrg Mgr Aviation Gas Turbine Div. 4 yrs with Studebaker Packard as Ch Engr 4 yrs with Curtiss Wright as Ch Engr Utica Div working on marine & industrial gas turbines, marine & industrial engines, missile & space components. 1960-1981. Scott Paper developing processes and prod equip. Last position Dir Tech Dev. Currently Adj Prof Univ of DE. Fellow ASME; Assoc Fellow AIAA; Treasurer AFSC. Reg P E MI & PA. *Society Aff:* ASME, AIAA, TAPPI

Rodin, Ervin Y
Business: Dept of Systems Sci and Mathematics, Box 1040, St Louis, MO 63130
Position: Prof. *Employer:* WA Univ. *Education:* PhD/Mathematics/Univ of TX, Austin; BS/Mathematics/Univ of TX, Austin. *Born:* 1/17/32. and received primary and secondary education in Budapest, Hungary. Moved to the US in 1956 and became a citizen subsequently. Married, three children. Professional activities include founding and being Editor-in-Chief of three intl journals and of a book series: *Computers and Mathematics with Applications, Mathematicl Modelling, and Applied Mathematics Letters*; and *Modern Applied Mathematics and Computer Sci.* Res interests include diffusion theory, wave propagation problems, applied mathematical education, game theory and artificial intelligence. *Society Aff:* SES, SIAM, AMS, MAA, IAMM.

Rodkiewicz, Czeslaw M
Business: Dept. of Mech Engrg, Univ. of Alberta, Edmonton, Alberta, Canada
Position: Professor Emeritus active in research *Education:* Ph.D./Hypersonics/Case Inst. of Technology; M.Sc./Gas Dynamics/University of Illinois, Urbana; Dip. Ing./Gas Dynamics/Polish Univ. College, London, Gr. Brit. *Born:* Born in Poland. Research Engr English Electric Co., Rugby, 1952-54; Dowty Equipment Ltd., 1954-55; Ryerson Inst of Tech, Toronto, Ontario, Canada, 1955-58; Faculty of the Univ of Alberta, Edmonton, 1958--. Summer 1959 with National Research Council, Offawa. Summer 1960 with Atomic Energy Canada Ltd. in Chalk River. Polish Army, 1938-39; Polish Home Army 1941-44; Polish Army of Allied Forces 1946-48; Decorated. Supervised programs leading to M.Sc. and Ph.D. degrees; served as external examiner for Ph.D. degrees. Guest lectures and addresses at various organizations and universities. Author of numerous research papers on fluid mechanics, heat transfer, hypersonic flight, lubrication, blood flow. *Society Aff:* ASME, CSME, NYAS, PTMTS, CEESSA ΣΞ, CMBES, AASVA

Rodrigue, George P
Home: 1090 Kingston Dr, Atlanta, GA 30342
Position: Regents' Professor. *Employer:* Georgia Institute of Technology. *Education:* BS/Physics/LA State Univ; MS/Physics/LA State Univ; PhD/Applied Physics/Harvard Univ. *Born:* June 1931. Worked at Clearwater Fla with Sperry Microwave Eelctronics Div of Sperry-Rand Corp 1958-68 in R/D on ferrimagnetic materials preparation & application to microwave devices, also on parametric amplifiers & microwave acoustic devices. At Ga Tech 1968- , continued ferrite res & pursued near-field antenna measurements. Genl Chmn 1974 Internatl Microwave Symposium. Pres IEEE Microwave Theory & Techniques Soc 1976; IEEE Exec Cttee/Bd Dir/V.P. Publications 1982 & 1983; IEEE Tech Activities Bd 1976, 1979 & 1980; IEEE Pubs Bd 1977, 78, & 1982-86, IEEE US Activities Bd 1986-87. Fellow IEEE. Consultant for government and industry. Enjoys, sailing. *Society Aff:* IEEE.

Rodriguez, Ferdinand
Business: Olin Hall of Chem Engg, Ithaca, NY 14853
Position: Prof. *Employer:* Cornell Univ. *Education:* PhD/Chem Engg/Cornell Univ; MS/Chem Engg/Case Western Reserve Univ; BS/Chem Engg/Case Western Reserve Univ. *Born:* in Cleveland, OH. Dev Engr, Ferro Chem Corp, Bedford, OH, 1950-54. US Army, 1954-56. Mbr of Chem Engg Faculty, Cornell Univ since 1958. Ind experience also with Union Carbide Corp, ICI (Great Britain), and Eastman Kodak Co. Author of textbook Principles of Polymer Systems (Harper and Row) and over 100 articles in profl journals. Publications also include several gospel songs. *Society Aff:* AIChE, ACS, SHPE

Rodriguez, Manuel
Business: Continental Can International, PO Box 10004, Stamford, CT 06904
Position: Director President *Employer:* Shellmar Embalagem Moderna S.A. *Education:* BChE/ChE/The Cooper Union; MBA/Internatl Finance/Harvard Bus Sch. *Born:* Dec 1932; native of Spain, raised & educ in N Y City. 1954-64 Prod/Tech supervision in heavy chem & fertilizer plants of Allied Chem, Hercules & Armour Ag at various US locations; also R/D assignment in uranium processing & corporate staff work in environ affairs; 1964-70 fertilizer plant mgmt in Central Amer, Caribbean & Indonesia & tech/business dev work in Western Europe; returned to US 1971; joined Crown Zellerbach 1972, V Pres CZ Internatl 1974-79, respon for Latin Amer interests. Joined Continental Group 1979; Gen Mgr Beverage Can Company in Venezuela 1980-84; currently Pres of Continental's Flexible Packaging Subsidiary in Brazil. Enjoys literature, music, outdoors & travel. *Society Aff:* ACS, AIChE.

Rodriquez, Harold V
Business: College Of Engineering, EGCB 108, University of South Alabama, Mobile, AL 36688
Position: Dean of College of Engineering. *Employer:* University of South Alabama. *Education:* PhD/Chem Engr/LA State; MS/Chem Engr/LA State; BS/Chem Engr/LA State. *Born:* Aug 30, 1932; native of New Orleans. US Army 1955-57 Instr of Radiological Defense, Chem Corps Sch. Instr Chem Engrg LSU; Sr Engr Mobil R/D Corp 1961-69 oil recovery by in-situ combustion; recovery of oil from oil shale; joined faculty of Univ of S Ala in Sept 1969, Assoc Prof of Chem Engrg; appointed Dir of Div of Engrg Sept 1971, Prof Sept 1973, Dean of Coll of Engrg June 1976. Reg P E in Ala. Mbr AIChE, Amer Soc for Engrg Educ, Sigma Xi, Phi Lambda Upsilon, Soc. of Amer. Military Engrs.; Natl Soc Prof Engrs; Ala Soc Prof Engrs; Phi Kappa Phi. *Society Aff:* AIChE, NSPE, SAME, ASEE.

Rodriquez, Pepe
Home: 2556 N 124th St, Wauwatosa, WI 53226
Position: Prof; Owner/Dir *Employer:* Mil School of Engrg; Pepe Rodriguez Engrg Co. *Education:* MS/Sanitary Engg/Marquette; BS/Civil/Marquette. *Born:* 10/30/46. Chicago, IL. Co-op with Milwaukee County Expressway Commission. With MSOE since 1970. Concurrently consultant with E D Wesley Co, Milwaukee. Bd mbr WI Licensing Bd of Architects & Engrs since 1973. Chrmn of same since 1976 (First woman in country). "Outstanding Teacher of the Yr-, MSOE, 1976. "Outstanding Young Women of America" Award 1976 & 1980. "American Bicentennial Commission" Award 1976. Responsible for WI state statute changes on water conservation & plumbing fixtures. Maj publications: "Water Waste Due to Flushing Devices on Urinals" & "Detergent Effects on Activated Sludge Systems" & "The Dilemma of Nuclear Power" & "An Engineer's Case for Automatic Fire Springklers" 1980- Appointed-Milwaukee County Planning Commissioner. Consultant to Milwaukee in redevelopment of minority areas in city. Comm Member; Mil. County Light Rail Transit Commission. 1984-present Comm Member: Mil. County Sewage Commission (Exec Adv Comm) 1984-present. *Society Aff:* ASPE, WPCF, ASHRAE.

Roe, Kenneth A
Business: 700 Kinderkamack Rd, Oradell, NJ 07649
Position: Chairman of Bd & President. *Employer:* Burns and Roe Enterprises Inc. *Education:* MS/Mech Eng/Univ of PA; BS/ChE/MIT; BA/ChE/Columbia Univ. *Born:* 1/31/16. MS Univ of Pa; BS MIT; BA Columbia Univ. Hon doc in ME-Stevens Inst of Technology. Employed by Burns and Roe Inc, engrg & cons co 1938- , as Mech Engr, Chem Engr, Exec Admin Engr, Exec V Pres & Bd Pres; 1971 elected Chmn of Bd, supervised co's work on many of country's major nuclear & fossil power plants, advanced energy technology projects, Clinch River Breeder, important aerospace & defense projects: Proj Mercury & LEM test Facility. Ser in US Navy WW II, Lt Cdr. Lic PE 27 states. Hon Mbr, P V Pres Reg II & P Pres ASME; Hon Mbr ASEE; Fellow AAAS, IE(Australia), IMech E(England); Mbr NAE, ANS, AIAA, IEEE, NSPE, ACS (Sr Grade) AIChE, ASCE, NCEE, & others Pres EJC; P First Chrmn & Acting Pres AAES; Bd Mgr ANEC & USNC/WEC. Mem Natl Council on Synthetic Fuels. Trustee: Manhattan Coll and Stevens Inst. of Technology. Awards: Carl Kayan Award (1977), Pupin Medal (1983). John Fritz Medal (1984) MIT Corporate Leadership Award (1987) & others. *Society Aff:* ASEE, AAAS, NAE, ANS, AIAA, IEEE, NSPE, ACS, AIChE, ASCE, NCEE, IME, ASME, SPE

Roe, Lowell E
Home: 3119 Glenn St, Toledo, OH 43613
Position: Retired *Education:* B/ME/OH State Univ *Born:* 05/14/25 Native of Toledo, OH. Attended Bowling Green State Univ and Harvard Univ while in the Navy 1943-46. Served on active duty 1951-53 and am LCDR in retired reserve. With Toledo Edison since 1948. Assigned to Enrico Fermi Fast Breeder Reactor project, Monroe, MI, 1956-61. Named VP, Facilities Development 1973 with responsibility for design, construction and initial operation of the Davis-Besse Nuclear Power Station. Was VP, Energy Supply from 1979 until retirement in 1984. *Society Aff:* ASME

Roe, William P
1041 Minisink Way, Westfield, NJ 07090
Position: VP & Dir of Res. *Employer:* ASARCO Inc. *Education:* PhD/Metallurgy & Inorganic Chemistry/Vanderbilt Univ; MS//Vanderbilt Univ; BA//Vanderbilt Univ. *Born:* 8/25/23. Chemist, Union Carbide '48-49. Res Metallurgist, Titanium Div, Natl Lead '51-56. Hd, Physical Metallurgy Sec, Southern Res Inst '56-57. Central res Labs, ASARCO Inc '57, Mgr '63, Dir of Res '69, VP & Dir of Res '74-. *Society Aff:* AIME, ASM, AAAS, RESA, ILZRO, STDA, ZI.

Roeder, Charles W
Business: 233 B More Hall FX-10, Seattle, WA 98195
Position: Prof *Employer:* Univ of WA. *Education:* PhD/-/U of CA, Berkeley; MS/CE/U of IL, Urbana; BS/CE/U of CO, Boulder. *Born:* 10/12/42. Native of Hershey, PA. Employed in the construction ind of central PA, 1960-1966. Served with the US Army 46th Engr Battalion in Ft Polk, LA & S Vietnam. Structural Engr with J Ray McDermott, Inc 1971-74. Asst Prof 1977-81, Assoc Prof 1981-85 Prof 1985-present at the Univ of WA Reg PE CO & WA. ASCE J James R Croes Medal 1979. ASCE Raymond C Reese Res Prize 1984. *Society Aff:* ASCE, AISC, EERI.

Roehrig, Clarence S, Jr
Home: 1103 Hennepin Dr, Louisville, KY 40214
Position: Specifications Engineer. *Employer:* Louisville/Jefferson Sewer Dist. *Born:* May 27, 1976; native of Louisville Ky. BCE Univ of Louisville. Served in US Navy 91st Seabee Battalion 1943-46; Inspector, Estimator, Design Engineer & Specifications Engr for MSD 1952- . Reg P E & L S Sanitary Engr state of Ky. Ky Section ASCE: Secy-Treas 1973-74, V Pres 1974-75, Pres 1975-76. V Pres now Dir Iroquois Civic Club; VChrmn 80-81. Pres Faith Lutheran Church. Louisville Branch ASCE-VP 1978-79, Pres 1979-80; district 9 Council ASCE - Secy-Treas 1979-80 Chmn 81-82; Iroquois Civic Club - Pres 1978 and 1979; Faith Lutheran Church - Pres 1978 & 1979. Enjoys camping, fishing, gardening & professional civic involvement. *Society Aff:* ASCE.

Roehrs, Robert J
Home: 4762 Barroyal Dr, St Louis, MO 63128
Position: Senior Technical Specialist, NDT *Employer:* McDonnell Douglas Corp. *Education:* BS/Prod Mgmt/Washington Univ. *Born:* 12/14/30 Mallinckrodt Chemical Works, Uranium Dir, uranium chemical and metallurgical processing R/D Dept. 1956-1961; Mgr of Quality Assurance, Nooter Corp, responsible for Q.A. functions, inspection, and non-destructive testing, 1961-1968; ASME Nuclear Survey Consultant 1968-1969; Assumed overall responsibility 1969 for all McDonnell Aircraft engineering requirements related to field of NDT/NDI. ASNT, SNT-TC-1A Level III in all methods; Reg P.E., active in development of National Codes, Standards, and Specifications for over 20 years. McDonnell Douglas Corp Presidential Citation 1972. ASNT Gold Medal Recipient, 1979. Fellow ASNT; Chairman and member of committees, Board of Dir ASNT 1973-1976; Tech Counc ASNT 1966-1978; ASTM Fellow, 1982; ASTM E7 Charles W Briggs Award, 1980; ASTM Award of Merit, 1982; Exec Counc ASTM 1966-1979; Coordinator ASNT Leak Test Handbook; ASME Section V, Subcommittee on NDE; Presentation and publication of numerous papers. Internationally recognized for work in Leak Testing. *Society Aff:* ASME, ASNT, ASTM.

Roemer, Louis E
Business: 302 E Buchtel Ave, Akron, OH 44325
Position: Prof of EE. *Employer:* Univ of Akron. *Education:* PhD/EE/Univ of DE; MSEE/EE/Univ of DE; BS/Phys/Univ of DE. *Born:* 7/5/34. Grad from the Univ of DE (BS Phys, 1955, MSEE 1963, PhD, 1967). He is currently Prof of Elec Engrg and Adjunct Prof of Bioengineering at the Univ of Akron and Res Prof of Physiology at the Northeastern OH Univs College of Medicine. Reg engr in the State of OH. *Society Aff:* IEEE, ΣΞ.

Roesner, Larry A
Business: 555 Winderley PL Su. 200, Maitland, FL 32751
Position: V P, Tech Dir of Water Resources *Employer:* Camp Dresser & McKee Inc *Education:* Ph.D./Civil Engr (Sanitary)/Univ of Washington; M.S./Civil Engr (Hydrology)/Colorado State Univ; B.S./Civil Engr/ Valparaiso Univ *Born:* Mar 14, 1941. BS Valparaiso Univ; MS Colo St Univ; PhD Univ of Wash. Reg P.E. CA, MI, VA, MD, FL. With Water Resources Engineers (WRE) 1968-1977. Camp Dresser & McKee 1977-, engaged in dev and application of hydrologic, hydraulic & water quality models 1964-, including stream & embayment water quality models with applications throughout the U.S. Lecturer on water quality modeling and urban runoff

Roesner, Larry A (Continued)
quantity and quality. Principal developer CPM's urban stormwater runoff models & nonpoint source pollution models. Applications throughout the U.S. Chairman Urban Water Resources Research Council, ASCE; member USGS task cttees on urban runoff detention facilities and urban gaging networks, Recipient of 1975 Water L. Huber Cvl Engrg Prize for res in urban hydrology. *Society Aff:* ASCE, AGU, WPCF, AIH, FES, AAEE

Roess, Roger P
Business: 333 Jay St, Brooklyn, NY 11201
Position: Prof of Transportation Engg; Dean of Engg. *Employer:* Polytechnic Univ. *Education:* PhD/Trans Planning/Polytech Inst of NY; MS/Trans Planning/Polytech Inst of NY; BS/CE/Polytech Inst of NY *Born:* 12/12/47 Born and raised in Flushing, NY. One of founding faculty mbrs of the Dept of Transportation Planning and Engrg of Polytechnic Inst, the first full academic dept in this specialty in the US. Conducted several major studies in the area of highway capacity analysis, including the final preparation of the 1985 Highway Capacity Manual. Other research interests include Public Transportation Operations and Economics. Promoted to Assoc Prof in 1976; tenured in 1979; promoted to Prof in 1983; appointed Dean of Engg in 1985. Winner of TRB's D. Grant Mickle Award in 1980. *Society Aff:* ITE, TRB, ΣΞ, XE.

Roethel, David
Business: 7315 Wisconsin Ave #525E, Bethesda, MD 20814
Position: Exec Dir *Employer:* Amer Inst of Chemists *Education:* Certificate/Nuclear Reactor Technology/Oak Ridge School of Reactor Technology; M.S./Radiochemistry/Marquette University; B.S./Chemistry/Marquette University *Born:* 02/17/26 Native of Milwaukee, WI. Served in US Air Force (U.S. Army Air Corps) in WW II. Chemical engr for nuclear submarine prog, 1952-57. Mgr, Prof Relations, ACS, 1957-72. Exec Dir, Amer Assoc of Clinical Chemists, 1968-70; National Registry in Clinical Chemistry, 1967-72; American Orthotic & Prosthetic Assoc, American Bd for Certification in Orthotics & Prosthetics, American Academy of Orthotists & Prosthetists, 1973-76; Amer Inst of Chemists, National Certification Comm in Chemistry and Chemical Engrg, 1977-; AIC Student Research & Recognition Foundation, 1982- . Editor: The Almanac of AOPA, 1973-76; The Chemist, 1977- . *Society Aff:* AIC, ACS, AAAS, GWSAE

Roettger, Jerome N
Business: 30 W Superior St, Duluth, MN 55802
Position: Genl Manager-Central Div. *Employer:* Minnesota Power & Light Co. *Education:* PRE/Engr/Gustavus Adolphus; BS/Elec Eng/Univ of MN. *Born:* 10/28/30. US Army 1951-52, held positions of Div Engr, Distrib Engr, Mgr of Engr, Chief Engineer with MN Power & Light Co. Mbr IEEE; P Pres Duluth Kiwanis Club; Mbr Duluth Chamber of Commerce. *Society Aff:* IEEE.

Roettger, Martin
Home: 11629 Edgewood Rd, Chardon, OH 44024
Position: Planning & Technical Mgr. *Employer:* Diamond Shamrock. *Education:* MSME/Mgmt/Newark College of Engg/BSME/Mechanical Engg/Newark College of Engg. *Born:* Oct 1931. MSME & ME Newark Coll of Engrg (now known as NJIT). Substitute Instr for NCE evening grad sch for brief period. Plant Engr Nopco Chem Co early 1960's, Amer Urethane (Stauffer Chem) mid 1960's; with Diamond Shamrock last 10 yrs - currently respon for Opers Analysis, Capital Budgeting Sys & planning reporting to Mgr of Bus Planning; Ch Proj Engr UOP Chem later 1960's. Coauthor 'Plant Engineers Manual & Guide', Prentice Hall. Pres Chap 61 AIPE 1970-71; Chmn Internatl Mgmt Dev Ctte AIPE 1970-76. Ch Plant Engr of the Year 1973. Prog Chmn for 7 major conferences AIPE which keynoted such as J Gellerman, F Herzberg, Sen Muskie, G Lippert, G Nierenberg, etc. Mainly in mgmt, negotiations or motivational areas. Bd of Trustees SAM 1972-74. Formed & became Pres of Diamond Shamrock Mgmt Club (E Coast) 250 mbrs affiliated with NMA. Green belt Karate currently. *Society Aff:* AIPE, SAM, ASTD.

Rogers, Benjamin T
Home: P.O. Box 2 (Old Taos Rd, rural), Embudo, NM 87531
Position: PE (Self-employed) *Education:* BS/ME/Univ of Wisconsin/Madison *Born:* 10/4/20 Military: Sgt to 1st Lt, Manhattan Engr District. Engrg Apprenticeship: Black & Veatch '46-49. Professional Registration: NM, CO, AZ, TX. Univ of CA '49-76, LANL and parallel private practice. R&D high speed photography to '56. Assoc Dir, Applied Res, R&AD Div AVCO (on leave). Principal Mechanical Engr, PHERMEX (major electron accelerator) to '74. Founding mbr, Solar Group to '76. VP Barkmann & Rogers, Consulting Engrs, '64-70. 1st National Award, Building Systems Design, 1972 (Energy Conservation). National Passive Solar Design Award, HUD/DOE, 1978. Visting Prof, Coll of Architecture, ASU, Tempe; Grad Arch Solar Option. Academic '80-81, 84. Consultant to LANL, '76-'86. People to People solar energy delegate to People's Republic of China 1986-Professional exchange delegate, Tiller Int. Foundation, to Tibet 1987. *Society Aff:* NSPE, ASME, ASM, ASHRAE, AS-ISES

Rogers, Bruce G
Home: 4303 Kenneth, Beaumont, TX 77705
Position: Prof *Employer:* Lamar Univ *Education:* PhD/CE/Univ of IL; MS/CE/Univ of IL; BS/CE/Univ of Houston *Born:* 02/20/25 in Houston, TX. Worked as surveyor, draftsman, and instrument operator on geophysical crew, 1946-52, and as engrg asst and engr for TX Highway Dept, 1952- 56. Served as teaching asst and instructor, Univ of IL, 1957-61. Teaching at Lamar Univ, Beaumont, TX, from 1961 to present. Occasional consltg work during this period. Reg PE in TX, AR. Mbr, Phi Kappa Phi, Tau Beta Pi, Sigma Xi, Blue Key. *Society Aff:* ASCE, ASME, NSPE, ASEE

Rogers, Cranston, R
Business: Turner, Collie & Braden, Inc, P.O. Box 13089, Houston, TX 77019
Position: Manager *Employer:* Turner Collie & Braden, Inc. *Education:* MS/CE/MIT; BS/CE/The Citadel. *Born:* 2/10/25. Native of Orlando, FL. Served in Infantry in Europe WWII; discharged rank of 1st Syt. Joined Charles A Maguire & Assocs 1950 as Bridge Engr; primary experience in bridge & hwy desing, highway location studies & tranportation planning; Dept Hd 1959, Assoc 1965, VP 1969; and Mgr of Transportation Group Waltham Office C E Maguire Inc. Joined Black & Veatch in 1978 as Regional manager of Boston office. Joined Turner, Collie & Braden, Inc. as manager of Transportation Projects in 1981. Special interests are prof soc activities. S.M. in GE fromMIT 1951; BS CE from The Citadel 1949. *Society Aff:* SAMe, NSPE, ITE, ARTBA, ASCE.

Rogers, Darrell O
Home: 5137 Aintree Rd, Rochester, MI 48064
Position: Supplemental Restraints Project Mgr *Employer:* CPC-Headquarters *Education:* MS ME/Mech Engrg/Univ of MI; BME/Mech Engrg/Gen Mtrs Inst *Born:* 4/20/45. With General Motors since 1963 beginning with Ternstedt Div, Columbus plant. Transferred to Fisher Body Central Engineering in 1970 - now at CPC Engineering. Experience is primarily in body structure and crashworthiness. Current responsibility is as Systems Manager for inflatable restraints on a new GM platform.

Rogers, Franklyn C
Home: 6414 13th Ave Dr W, Bradenton, FL 33529
Position: Consultant (Self-employed) *Education:* MS/CE/Columbia Univ; BS/CE/Purdue Univ *Born:* 4/24/10 NYC. Ambursen Dam Co NYC 1935-41: Design Engr Rutgers Univ 1941-52: Prof CE. Harza Engrg 1952-72: VP and Dir. Consultant to Harza and others since 1972. Chairman Bd of Experts Hemren, Bekhme, Haditha, Fatha Projects Ministry of Irrigation, Govt. of Irag, 1977 to present. Permanent Consulltant, Corpus Project, Comision Mixta Argentino-Paraguaya, 1980 to present. Ricking Medal ASCE 1976. Chairman Question J2 ICOLD 1982. *Society Aff:* ASCE, NSPE, ACEC, ΣΞ, TBΠ

Rogers, Gifford E
Home: c/o Monroe, 515 North Grace, Grand Island, NE 68803
Position: Project Mgr *Employer:* Engg Consultants, Inc. *Education:* MS/CE/Purdue Univ; BS/CE/Univ of NB. *Born:* 5/22/20. Native of Grand Island, NB. Officer, CE, US Army 1943-46, PTO. Field Engr, Soils Engr, Proj Engr, Chf Engr, United Fruit Co, Panama, Honduras, Guatemala, Costa Rica, Dominican Republic, 1948-60. Proj Mgr, Intl Dev Services, Guatemala and Dominican Republic, 1960-64. Special Consultant to Governments of Dominican Republic (1964-65) and Nicaragua (1967-69). Proj Mgr, Dev & Resources Corp, Nicaragua, 1965-67. Water Resources Specialist, Robert R Nathan Assocs, Ghana and El Salvador. 1969-83, Proj Mgr, PRC Engg Consultants, Inc, Bangladesh, Turkey, Indonesia, Sri Lanka, VP 1978. 1983-1987, Training Adv, Public Adm Svc, Bangladesh. Dipl of Honor, Cartographic Inst, Guatemala, 1962. Enjoy golf, swimming, reading. Elected Explorers Club (Fellow), 1980. PE, NE, Cert by Natl Council of Engrg Examiners. *Society Aff:* ASCE, NSPE, ASAE, AAEE, AWF, IA.

Rogers, H Daniel, Jr
Home: 7 Summit Ave, Latham, NY 12110
Position: Associate *Employer:* Ryan-Biggs Associates, P.C. *Education:* MS/CE/RPI; BS/Physic Eng/Washington & Lee Univ *Born:* 03/27/47 Native of Meriden, NH. Served in US Army Corps of Engrs (Reserve), 1969- 1977. Employed by Ryan-Biggs Associates, P.C. since 1982 as Associate in charge of bridge and highway projects. Worked previously for New York State Dept. of Transportation from 1970-1982. Pres, Mohawk-Hudson Section, ASCE, 1981-82. Chairman, NY State Council ASCE, 1985-86. Adjunct Asst Prof of Civil Engrg, Union Coll. Thomas Archibald Bedford Prize, RPI, 1971. PE-NY State. Ruling Elder-Presbyterian Church. I enjoy all kinds of music and singing with several choirs and choruses. *Society Aff:* ASCE, NSPE

Rogers, James C
Business: Dept of Elec Engrg, Houghton, MI 49931
Position: Prof EE *Employer:* MI Tech Univ *Education:* PhD/EE/Univ of WA; MS/EE/Univ of WA; BS/EE/Univ of WA *Born:* 08/21/41 Native Alaskan, born in Anchorage. 1965-67 Res Engr with Univ of WA at Byrd Station, Antarctica. Studying VLF propagation and ice elec properties. 1972-74 Asst Prof, Univ of AK, Fairbanks, Elec Engrg and Geophysical Inst. Teaching and radar studies of ice movement. 1974-78 Univ of AK, Anchorage Engrg Sch, teaching and res on detection of permafrost beneath Arctic Ocean, 1978-80 Dir of Engrg Sch-developed grad and undergrad engrg programs. IEEE AK Section Chrmn 1980, Nominated Engr of Yr by AK Soc of Prof Engrs. 1980-83, Assoc Prof, MI Technological Univ, teaching, res in signal processing and acoustics, Mbr Soc of Exploration Geophysicists Natl Res Ctte, 1983-present, prof of Elec Engrg. *Society Aff:* IEE, NSPE, SEG, ASEE

Rogers, Kenneth A
Home: 4421Romlon St No. 4, Beltsville, MD 20705
Position: Senior Eng'r *Employer:* U.S. Synthetic Fuels Corp *Education:* PhD/Chem Engg/GA Inst of Tech; MSChE/Chem Engg/GA Inst of Tech; BSChE/Chem Engg/Northwestern Univ. *Born:* 4/9/27. Served as Chief Process Engr for Liquid Carbonic Corp and Chief Engr for W F M Schultz Co. Technical Dir for Applied Engg Co. Engg Consultant for Mfg of Industrial Gases, Process Design, Control Systems. First Engr appointed to two- yr term as Engr in residence in 1977 with the Engg Socs Comm on Energy, Assoc. Prof CAL State-Long Beach, Executive Manager, ESCOE. *Society Aff:* SPE, ASME, ACS, AIChE

Rogers, Kenneth C
Home: Castle Point Station, Hoboken, NJ 07030
Position: Pres. *Employer:* Stevens Inst of Tech. *Education:* PhD/Physics/Columbia Univ; MA/Physics/Columbia Univ; BS/Physics/St Lawrence Univ. *Born:* 3/21/29. Res assoc lab nuclear studies, Cornell Univ, 1955-57; Asst Prof physics Stevens Inst Tech, 1957-60; Assoc Prof, 1960-64; Prof 1964-; Head Physics Dept, 1968-72; Acting Provost, 1971; Pres, 1971-; Hon M Engg, Stevens Inst Tech, 1964; Pres, Assn Ind Engg Colleges, 1976-78; Author papers on Plasma Physics, consultant to defense industries; Dir Public Service Enterprise Group Inc.; Dir First Jersey National Corp; Dir New Jersey Chamber of Commerce; Trustee, Independent College Fund of NJ. Mbr Sigma Xi; Newcomen Soc; Royal Soc of Arts; Cosmos Club, Washington, DC; D.H.L. (hon) St. Lawrence Univ, 1983; Fellow AAAS; Mbr: Gov's Comm on Sci and Tech, Chrmn-Sub Task Force on Telecomm, 1983-1985: ASEE; IEEE; Mbr: New Jersey Ctte on Sci & Tech 1985- . NY Acad of Sci; NJ Res & Dev Council; Reg Plan Assoc, NJ Ctte; Amer Phy Soc; AAHE. *Society Aff:* IEEE, ASEE, APS, AAAS, AAHE, AAPT

Rogers, Lewis H
Home: 2607 NW 22nd Ave, Gainesville, FL 32605
Position: Cons. *Employer:* Environmental Sci & Engg. *Education:* BSChE/Chem Engg/Univ of FL; MS/Chem/Univ of FL; PhD/Chem/Cornell Univ. *Born:* 10/1/10. From Instructor to Prof, Univ of FL. 1932-1948; Res Supervisor, Union Carbide Nuclear, Oak Ridge, TN, 1948-1952; Leader, Anal Div, Kraft Res Lab, Oakdale, LI, NY, 1952-1954; Chief Chemist, Air Pollution Fdn, Los Angeles, CA 1954-1958; from Dir of Lab to Dir of Corp Res & Dev, Automation Industries, Los Angeles, CA, 1958-1971; Exec VP, Air Pollution Control Assoc, 1971-1978; Environmental Consultant, 1978 to present. *Society Aff:* APCA, AAAS, ACS

Rogers, Stanley
Business: P.O. Box 2025, Laguna Hills, CA 92654
Position: Publisher emeritus. *Education:* AB/Physics/Occidental. *Born:* 1908 Chicago. EE at USC & MIT 1936-43. Reg Elec Engr Calif. Dev Engr Genl Communication Prod Co Los Angeles 1941. Dev Engr Lee de Forest's personal lab (radar altimeter) 1942-43. Various divs Genl Dynamics in San Diego 1943-70: Res Lab Analyst (1st person to slve engrg probs on accounting machines in pre- computer days at Genl Dyn 1944); Res Engr, Sr Res Engr, Res Group Engr 1945-60 (proj engr design & const large analog computers for San Diego & Pomona Divs, superv of Analog Computer Lab, proj engr contract data sys); Sr Data Sys Group Engr, Mgrgr Data Sys & Computers Dept Info Tech Div 1960-62; Engrg Staff Specialist, Sys Engrg Spec 1962-70 (proj engr Manned Space Sys Simulator, design studies on 0-200 MHz computer-contrld "crossbar" switches. EAI Scientific Award 1971. Publisher Simulation 1953-80. Former Bd Chmn & Secy Soc for Computer Simulation, formerly active on tech comms & editorial bds AIEE IRE IEEE AFIPS. Now also tech editor. 10 patents. *Society Aff:* SCS.

Rogers, Vern C
Business: 515 East 4500 South, Suite G-200, Salt Lake City, UT 84107
Position: President *Employer:* Rogers and Assocs Engrg Corp *Education:* PhD/Nuclear Eng/MIT; MS/Mech-Nuclr Eng/Univ of UT; BS/Physics/Univ of UT *Born:* 08/28/42 Currently Pres and founder of Rogers and Assocs Engrg Corp with 35 personnel. Engaged in contractual radioactive and hazardous waste handling, disposal, and risk assessment. Formerly VP and Mgr of Nuclear and Advanced Programs at Ford Bacon and Davis UT, Inc, from 1977 to 1980. Mgr of the Nuclear and Applied Sci Dept of the IRT Corp from 1972 to 1977. Published over 200 scientific papers. *Society Aff:* ANS, APS, ACS, HPS, SSSA, AIChE

Rogers, Willard L
Home: 5744 E Holmes, Tucson, AZ 85711
Position: Prof of Mech Engg Emeritus. *Employer:* Univ of AZ. *Education:* PhD/ME/Stanford Univ; MS/ME/Northwestern Univ; BS/ME/IA State Univ. *Born:* 4/14/17. Native of Ottumwa & Sioux City, IA. Taught Mech Engg at IA State College (1942-46): Northwestern Univ (1946-55, 1958-59): Univ of AZ (1959-81). Past vchrmn, dir, awards comm mbr, other offices, ERM Div ASEE (1965-77). Danforth teacher study grant, 1955. Regional dir ASEE Effective Teaching Inst (1967-72). Mbr ASHRAE technical committees on sound & vibration control & hydronics (1953- 55, 1959-67). Consulting activities noise & vibration control, energy systems. Author fifteen natl papers, coauthor two engg textbooks. Current maj interests solar

Rogers, Willard L (Continued)
energy & thermodynamics of energy systems. Recreation camping, hiking, travel, boating, photography, fix it. *Society Aff:* ASME, ASEE.

Rogoway, Lawrence P
Business: 4465 Willshire Blvd, Los Angeles, CA 90010
Position: Pres. *Employer:* Rogoway/Borkovetz Assocs. *Education:* BS/Civil Engr/OR State Univ. *Born:* 12/9/32. Commissioned Naval Officer 1954-56. Asst Civil Engr with Lockheed Aircraft Corp 1956-57. Civil Engr Asst with City of Los Angeles 1957-59. With predecessor firm to Rogoway/Borkovetz Assoc since 1959. Assumed responsibility as Pres and Chief Exec Officer in 1977. Native of Portland, OR, resident of Los Angeles since 1956. Responsible for developing and maintaining firms reputation for sensitivity to aesthetic as well as functional design of civil engg facilities in support of projects. Currently licensed in seven states. Enjoy travelling, music and photography. *Society Aff:* ASCE, NSPE, AWWA, CEAC

Roha, Donald M
Home: 708 Winhall Way, Silver Spring, MD 20904
Position: Pres *Employer:* D Roha and Associates *Education:* BS/Physics/Allegheny College; MA/Pub Adm/Amer Univ. *Born:* 8/28/29. Native of Meadville PA. Engr with Air Force & Martin Co specializing in dynamics test & analysis; with Vitro Labs Div of Automation Indus 1959-86. Developed test methodology for major defense sys; Hd of Engrg & Analysis Dept 1965-70; Managed co sys engrg efforts in fields of health, environ & energy 1971-74; Dept Hd Mgmt Sys 1974-79. Bd of Dirs Volunteers in Tech Assistance; Montgomery Cty Md Library Bd. Program Manager Energy Sys 1979-82; Mgr, Engrg Mgmt 1982-83; VP Vitro Corp 1983-86; Pres D Roha and assoc 86- . *Society Aff:* NAC.

Rohde, Steve M
Business: G M Technical Center, Warren, MI 48090
Position: Sr Staff Research Scientist. *Employer:* Genl Motors Research Laboratory. *Education:* BS/Elec Eng/NJ Inst of Tech; MS/Math/Lehigh Univ; PhD/Math/Lehigh Univ. *Born:* 5/18/46. Genl Motors Res Labs Warren MI 1970-current pos: supervisor of vehicle systems engrg group. Chmn ASME Tribology Exec Ctte. Chmn ASME Computer Simulation Ctte. Chmn ASME Power Transmission Systems Ctte, Assoc Editor ASME Trans. Journal of Tribology. Assoc Ed ASME Computers in Mech Engrg. Mbr ASME Ctte on Compt-Applied Mech. Mbr ASME, SAE, MAA. Honors & awards: ASME Henry Hess Award 1973, ASME Burt L Newkirk Award, 1977; ASTM, Harry Kummer Meml Award 1976. Soc of Actuaries Prize 1965. Res interest: Vehicle Systems, optimization, hydrodynamic lubrication, finite element methods, engine friction, hybrid vehicles, simulation, control. Enjoys antique car and radio restoration, boating & fishing. *Society Aff:* ASME, MAA, SAG.

Rohlich, Gerard A
Business: 8.6 E Cockrell Hall, Austin, TX 78712
Position: Prof of Environ Engrg *Employer:* University of Texas/Austin. *Education:* PhD/Civil Engr/Univ of WI-MSN; MS/Civil Engr/Univ of WI-MSN; BS/Civil Engr/Univ of WI-MSN; BS/Civil Engr/Cooper Union. *Born:* July 1910. Teaching exp at Carnegie Tech, Penn St Univ, & Univ of Wisc/Madison 1946-72. 1972-present at Univ of Texas/Austin. Visiting professorships at Univ of Calif/Berkeley, Univ of Helsinki Finland, Univ of Wash as Walker-Ames visiting prof. Hon citation from Cooper Union; WPCF Harrison Prescott Eddy Medal for res; AWWA Fuller Award; ASCE Karl Emil Hilgard Hydraulic Prize; Benjamin Smith Reynolds Award for excellence in teaching from Univ of Wisc; Hon Mbr WPCF, Gordon Maskew Fair Medal WPCF, Distinguished Service Citation, Univ of Wisc; mbr NAE 1970. Special fields: water quality mgmt; indus waste treatment; lake & stream pollution. *Society Aff:* AAAS, AEEP, AWWA, WPCF, NAE, AAEE

Rohr, Peter H
Business: PO Box 101, Florham Park, NJ 07932
Position: VP. *Employer:* Exxon Res & Engg. *Education:* MS/ME/Stevens Inst of Tech; BS/ME/Stevens Inst of Tech. *Born:* 7/22/24. Joined Exxon Res & Engg Co upon grad from Stevens Inst of Tech in 1945. Early career involved a number of technical fields & mgt of various engg activities within Exxon Engg; named Gen Mgr of Tech Dept-1966. Joined Exxon Corp, Logistics Dept, as Mgr of Supply Planning Div in 1969. Transferred to Exxon Co USA in 1974 as Mgr of the Fuel Products Planning & Technical Div, Refining Dept; Mgr of Fuel Products Operations Div in 1976. In 1978, returned to Exxon Corp as Mgr of Operations Coordination in Logistics Dept. Named VP of Engg, ER&E, in July 1979. Active in API & ASME. Enjoys skiing, sailing, and woodcrafts. *Society Aff:* API, ASME.

Rohrer, Ronald A
Business: P.O. Box 8106, Charlottesville, VA 22901
Position: Staff Scientist *Employer:* GE Co *Education:* PhD/EE/Univ of CA - Berkeley;l MS/EE/Univ of CA - Berkeley; BS/EE/MIT *Born:* 08/19/39 Staff Scientist at GE's Industrial Electronics Dev Lab and an Adjunct Prof ofElec Engrg at Southern Methodist Univ. Prior to his present position in industry, he was affiliated with a number of academic institutions where he performed res in computer-aided integrated circuit design and applying personal computers to enhance instructional productivity in undergraduate courses. Dr Rohrer is a Fellow of the IEEE "For Theoretical Contributions and Practical Software for Computer-Aided Circuit Design–. *Society Aff:* IEEE, ACM, ASEE

Rohrmann, Charles A
Home: 4707 W 7th Avenue, Kennewich, WA 99336
Position: Resident Consultant. *Employer:* Battelle Mem Inst, Pacific NW Labs. *Education:* BS/Chem Engg/OR State Univ; PhD/Chem Engg/OH State. *Born:* Dec 17, 1911 Pendelton Ore. Res chem engr EI du Pont de Nemours & Co Ohio 1939-43; plant process supr Pa 1943-48; GE Co WA 1948-64; Battelle NW 1965- present; cons chem & metallurgical div PAC NW LABS BATTELLE MEM INST 1968-70, STAFF ENGR CHEM TECH DEPT 1970-76, Res Cons, 1976-1987, Retired. Civilian with Off Sci R/D, Natl Defense Res Ctte 1944. Sigma Xi; ACS; Fellow AIChE. Chem engrg & indus chemistry of heavy inorganic and organic chemicals; coml sulfur-nitrogen chemicals; chem engrg applications in the atomic energy indus; large scale production & application of radiosotopes; sulfur oxides air pollution control; Biomass for Energy and Chem *Society Aff:* ACS, AOS, AIChE.

Rohsenow, Warren M
Business: Dept of Mech Engrg, Cambridge, MA 02139
Position: Professor of Mech Engrg. *Employer:* MIT. *Education:* DENG/Mech Engrg/Yale; MENG/Mech Engrg/Yale; BSME/Mech Engrg/Northwestern. *Born:* Feb 1921. Native of Chicago, Ill. Dir of Heat Transfer Lab & Dir of Grad Studies in Mech Engrg Dep at MA Inst Tech. Served in Navy 1944-46 at Engrg Exp Sta Annapolis; currently Lt Cdr USNR (ret). Extensive cons in variety of indus orgs in heat transfer & its application. Co-founder of Dynatech Corp 1957. Max Jakob Award 1971 ASME; Heat Trans Div Mem Award ASME. Fellow Natl Acad Engrg. Hobbies: golf; jazz piano playing. *Society Aff:* NAE, ASME, AAAS.

Roley, Daniel G
Home: RR 3, Box V-78, Chillicothe, IL 61523
Position: Senior Engr *Employer:* Caterpillar Tractor Co *Education:* PhD/Engrg/Univ of CA-Davis; MS/Agr Engrg/Univ of IL; BS/Agr Engrg/Univ of IL *Born:* 11/13/48 Native of Shelbyville, IL. Received PhD in Engrg, Human Factors major, from Univ of CA, Davis in 1975. With Caterpillar Tractor Co since 1975, 4 years in research before transfering to the vehicle test division. In Nancy France for 1 year at the Vibration Lab of Inst National de Recherche et de Securite as recipient of NSF US-France exchange of scientists post-doctoral Fellowship. Member SAE Human Factors Technical Committee chrmn of its Visibility subcommittee. Chrmn of the ASAE Central IL section, 1980. MOBDES Captain in Army Reserves at MERADCOM, Ft. Belvoir in construction equipment lab. Enjoy sports and travelling. *Society Aff:* SAE, ASAE, ISBM

Rolf H Sabersky
Business: Calif. Inst. of Tech, Pasadena, CA 91125
Position: Prof. *Employer:* CA Inst of Tech. *Education:* PhD/ME/CA Inst of Tech; MS/ME/CA Inst of Tech; BS/ME/CA Inst of Tech *Born:* 10/20/20. Born in Berlin, Germany. Came to the US in 1938. Dev Engr at Aeroject Gen Co 1943-46, regular consultant from 1949-70. Involved in dev of rocket engines and several maj propulsion systems for missiles & space vehicles. Joined faculty of the CA Inst of Tech in 1949; appointed to present position of Prof of Mech Engg in 1961. Received Heat Transfer Meml Award of ASME in 1977. Mbr, Honorary Editorial Bd, Int J of Hea & Mass Transfer. Author of two books & numerous res papers. Specialized in Heat Transfer. *Society Aff:* ASME.

Rolf, Richard L
Home: 226 Laura Drive, New Kensington, PA 15068
Position: Senior Technical Specialist *Employer:* Aluminum Co of Amer (Alcoa Tech Ctr). *Education:* MS/Civil Engr/Univ of IL; BS/Civil Engr/Marquette Univ. *Born:* Nov 1935. Native of Milwaukee Wisc. Postgrad work at Univ of Pittsburgh. 1960-present concerned with the dev of res progs related to the behavior of aluminum & ceramic strucs & components, & with design problems involving aluminum & materials for Alcoa. Presently leading Alcoa's efforts in area of material performance mechanics. ASCE Croes Medal 1966. Mbr ASCE; Amer Academy of Mechanics; American Society for Testing & Materials; Sigma Xi. Reg P E in Pa. Married. Hobbies: gardening; hunting. *Society Aff:* ASCE, AAM, ASTM, ΣΞ.

Rolfe, Stanley T
Home: 821 Sunset Drive, Lawrence, KS 66044
Position: Forney Prof Engrg & Chmn Civ Engrg. *Employer:* Univ of Kansas. *Education:* PhD/Civil Eng/Univ of IL; MS/Civil Eng/Univ of IL; BS/Civil Eng/Chicago IL *Born:* 1934. Sect Supr, Division Ch US Steel Res Lab 1962-69; Ross H Forney Prof of Engrg Univ of Kansas 1969-present, Chmn Civil Engrg Dept 1975-present. ASTM Sam Tour Award 1971; AWS Adams Engrg Educator Award 1974; Sch of Engrg Henry E Gould Excellence in Teaching Award 1972 & 1975. AISC 1980 T.R. Higgens National Lectureship Award winner. Pub 50 tech papers on fracture, fatigue, material behavior. Reg P E. Author 'Fracture & Fatigue Control in Strucs - Applications of Fracture Mechanics' Prentice-Hall 1976 revised 1987. Co-Author-Strength of Materials McGraw Hill 1981. Mbr ASCE; ASTM; ASME; SESA; ASEE; NSPE. Elected to mbrshp in Natl Acad of Engrg, 1982. Univ. of Illinois College of Engineering Alumni Honor Award for Distinguished Service in Engineering, 1987. *Society Aff:* ASCE, ASME, ASEE, ASTM

Roll, Frederic
Business: Dept of Systems, School of Engrg & Appld. Sci, University of Pennsylvania, Philadelphia, PA 19104
Position: Prof Emeritus of Civil Engrg *Employer:* Univ of Pa/Sch of Engrg & Appld Sci. *Education:* PhD/Civil Engrg/Columbia Univ; MSE/Civil Engg/Columbia Univ; MA/Civil Engg/Univ of PA; BCE/Civil Engg/College of the City of NY. *Born:* Sept 1921 NYC. Taught at Columbia; joined Univ of Pa as Assoc Prof 1957. NSF Sci Faculty Fellowship 1963; ACI Fellow 1973. ASCE Awards: St-of-the-Art 1974; Outstanding Ser 1975; Raymond Reese 1976. Chmn ACI Comm on Creep 1961-66; Pres Delaware Valley Chap ACI 1976; Phila Sect of ASCE Chmn of Struct Grp 1974- 76, Bd/Dir 1974-1984. Author of num tech papers. Speaker at natl & internatl confs & seminars. Specializes in concrete res. Hobbies: enthusiastic traveler; actor in little theatre in USA & Eng. Pres of the Philadelphia Section of the American Soc of Civil Engrs 1981-82; Engrg Conslt re: structural failures. Structural Engr of the Year 1987, Philadelphia Section of ASCE. Reviewer of technical papers for ACI and ASCE. *Society Aff:* ASCE, ACI, PCI

Rollag, Dwayne A
Home: 320 Lincoln Lane S, Brookings, SD 57006
Position: Prof & Hd Dept of Civil Engg. *Employer:* South Dakota St Univ. *Education:* PhD/Civil Engg/Purdue Univ; MSCE/Civil/SD State Univ; BCE/Civil Engg/Univ of MN. *Born:* June 1930. Native of Minnesota. Practicing cons sanitary engr for 6 yrs prior to 1965; 1965-present teaching sanitary engrg, conducting res & part-time cons with South Dakota St Univ. Pub sev papers in field of sanitary engrg. Reg P E in Minnesota, Iowa & SD. P Pres E Branch SD Sect ASCE; Past St Pres of SD Engrg Soc; currently National Director, & SD Water Pollution Control Assn; P Chmn SD Bd of Certification for Water & Wastewater Operators. Past State Pres, Amer Water Works Assn, SD Pres, Amer Soc of Civil Engrs, SD Sec, Holder of One Patent. Also mbr of Amer Soc of Engg Educators ASEE; Also mbr of Hon Tau Beta Pi; Also mbr of Hon Sigma Xi, Chi Epsilon. Hobbies: skiing; golf. *Society Aff:* ASCE, AWWA, WPCF, ASEE, NSPE.

Roller, Warren L
Business: Agri Engrg Dept OARDC, Wooster, OH 44691
Position: Prof & –Coordinator-Program Dev Office of the Dir, OARDC– *Employer:* OH Agri R/D Center & Ohio State Univ *Education:* PhD/Agri Engg/Purdue Univ; MS/Agri Engg/Purdue Univ; BS/Agri Engg/Purdue Univ. *Born:* May 1929. Native of Cass Cty Ind. Naval Officer 1951-53. Engrg Res with Ohio Agri Expt Station 1955-66; Sr Visiting Scientist with US Army Res Inst of Env Medicine 1966-67; Engrg Res Mgmt with OARDC 1969-81. Agr Engrg Dept Chrmn, OH State Univ, 1981-87. Chmn ASAE Ohio Sect 1969-70; ASAE Dir 1970-72, Fellow since 1979. Contributing editor of Chap 8 ASHRAE Handbook of Fundamentals 1972. ASAE Paper Award 1964 & 1978, honorable mention 1972 and 1984; ASAE Metal Bldg Manufae Assn Natl Award 1975. Inventor of an automatic feeding sys for swine based upon a paste-consistency feed. Inventor of a right-angle viewing device for bicycle racing competition, patented 1987. Authorship on more than 20 tech papers. *Society Aff:* ASAE, ΣΞ, ΓΣΔ, AE.

Rollins, Albert W
Business: 604 Ave H E, Arlington, TX 76011
Position: Pres *Employer:* Schrickel, Rollins & Assoc, Inc. *Education:* MS/Civ Engg/TX A&M Univ; BS/Civ Engg/TX A&M Univ. *Born:* 7/31/30. Native of Dallas, TX. Officer in US Army Corps of Engrs during Korean War, 1951-53. Served as Dir of Public Works, 1956-63, and City Mgr, 1963-67, of Arlington, TX. Partner in Schrickel, Rollins & Assoc, Inc, since 1967 specializing in gen civ engg design and municipal engg. Served as mbr of TX Turnpike Authority from 1970-72, and Chrmn of TX Mass Transportation Comm from 1972 to 1974. VP TX Soc of PE, 1978-80. Chrmn, Urban Planning and Dev Div, ASCE, 1979-80. Pres, TX Public Works Assn, 1962. *Society Aff:* ASCE, NSPE, APWA, WPCF, AWWA, SAME.

Rollins, John P
Home: 44 Bay St, Potsdam, NY 13676
Position: Prof. *Employer:* Clarkson College of Tech. *Education:* MME/Mech Engg/RPI; BS/Mech Engg/Penn State Univ. *Born:* 11/21/17. Birthplace, Clearfield, PA. Taught Mech Engg at: Wyomissing Poly Inst, 1939-40; Rensselaer Poly Inst, 1940-48; and Clarkson Univ, 1948-1981. Asst Prof 1948, Assoc Prof 1953, Prof since 1957. Co-author ASME paper No 67WADE14, *Primary Natural Frequency of Curved Cantilever Beams.* Co-author of paper *Parametric Performance Analysis of the Darrieus Aeroturbine Under Various Loading Conditions,* Journal of Energy, ASAA, May 1979. Principal editor *Compressed Air and Gas Handbook,* 3rd, 4th and 5th editions. Designed first cantilevered, silo-supported, agriculture wind turbine, constructed 1978 at Clarkson Univ. NSF Fellowships 1964-67. Consultant for a number of firms. Enjoy folk dancing and travel. NSF Fellowships 1964-67. *Society Aff:* ASME, ASEE, ΠΤΣ, ΠΜΕ, ΤΒΠ.

Rollins, Ronald R
Business: COMER, Mining Engrg, Morgantown, WV 26506-6070
Position: Prof *Employer:* West Virginia Univ *Education:* PhD/Metallurgy/Explosives/Univ of UT; BS/Fuels Eng/Univ of UT *Born:* 10/02/30 m. Monique Thaon, Sept 6, 1957; Parents: S. Warren Leroy and Marcella (Miller) R.; Children: Michael, Aline, Bruce, Scott, and Daniel. Metallurgist/Ceramist Valle-

Rollins, Ronald R (Continued)
citos Atomic Lab, Gen Elec Co, Pleasanton, CA, 1962- 64; Assoc prof Rock Mechanics and Explosives Res Center Univ of MO, Rolla, 1964- 79; Prof, mining engrg, 1979- ; Chrmn Mineral Processing Engrg Dept, 1981-83; Research assoc Argonne Natl Lab, 1965, Sandia Corp, 1966, 69, Picatinny Arsenal, 1968, Naval Weapons Ctr, 1983; Cons in field. Served with USNR, 1948-52, US Army, 1953-54. Sandia Corp Fellow, 1959-62. Enjoy tennis, golf, racquetball, and classical music. *Society Aff:* AIME, ASEE, AAAS, ΣΞ, SEE

Roloff, Raymond C
Home: 511 Goodhill Road, Kentfield, CA 94904
Position: Exec V P - Marketing. *Employer:* Bepex Corporation. *Born:* 1928. Native of Minneapolis Minn. ChE Univ of Minn 1951; Advanced Mgmt Harvard Grad Sch of Bus 1976. Specialized in chem & food process machinery for career. 1951-present Strong-Scott Mfg Co & parent co, Berwind Corp; assumed responsibility of V P - Marketing 1963 (Strong-Scott), Pres & CEO 1968 (Strong- Scott, subsidiary of Berwind), Exec V P - Marketing 1976 of the corporate grp of Bepex Corp (subsidiary of Berwind Corp). Mbr AIChE; Process Equipment Manufae Assn; Univ of Minn Alumni Club.

Rolston, J Albert
Business: PO Box 62, Granville, OH 43023
Position: Principal. *Employer:* J A Rolston & Assoc. *Education:* MS/ChE/NC State Univ; BChE/ChE/NC State Univ. *Born:* 7/27/19. Dalton, GA. B-29 navigator WWII. USAF Reserve R&D Officer Off of Aerospace Res, retired 1979 Lt Col. Instr, Lectr & Res Assoc NC State Univ & Univ of VA 1947-1961. Res Engr & Sr Res Engr Spaulding Fibre & Owens-Corning Fiberglas Corp 1962-1976. Formed consulting firm in 1976 specializing in fiberglass composites design, fabrication methods, and stress analysis. Reg PE VA & OH, NACE Corrosion Specialist. Fellow AIChE, Chrmn Mtls Div, mbr Phi Eta Sigma, Sigma Xi, NACE, ASTM, SPI. Author and publisher Index to SPI Preprints. Papers & patents on fiberglass mtls. *Society Aff:* AIChE, NACE, ASTM, SPI.

Romaguera, Mariano A
Business: Peral 16N, Condominio Torre Peral, PO Box 1340, Mayaguez, PR 00709-1340
Position: Pres. *Employer:* Mariano A. Romaguera & Assoc. *Education:* MS/Mech Engg/Univ of PR; BS/Mech Engg/MIT. *Born:* 5/4/28. Native of Mayaguez, PR. Worked as Plant Engr in Sugar Mills in S America and PR 1950-52. US Army-Korean Conflict 1952-54. Chief Engr in Sugar Mill in Colombia, SA 1954-64. VP Chg Engg R & V Dev Corp-Land dev 1964-1969. Consulting Engr to PR Sugar Corp in charge of improvements and Environmental Controls 1969- present. Pres of Mariano A Romaguera & Assoc 1972 to present - Engrs Appraisers - Consultants. Bd of Governors Coll of Engrs PR. Pres ASME SW PR Chap 1976-78; 1983-1984. Prof UPR Mayaguez Campus. Continuing Education Prog for Engrs 1977- 81; also Biomass Steam Generation research. *Society Aff:* ASME, NSPE, ASA, ISA, IREI.

Roman, Basil P
Home: 32451 Seven Seas Dr, Laguna Niguel, CA 92677
Position: Prof. *Employer:* CA State Univ. *Education:* PhD/Engg/UCLA; MS/Aeronautics/CA Inst of Tech; Dipl Ing/Mech Engr/Univ of Bucharest. *Born:* 9/4/28. Native of Roumania. Taught for two yrs Physics, Los Angeles Community College and since 1959, Prof of Mech Engg at CA State Univ, Long Beach. 1961-62 served as UN-UNESCO expert in Morocco. 1963-64 taught as Fulbright Lecturer at the Natl Univ of Engg, Lima, Peru. 1966-67 Advisor at the Univ of Costa Rica. 1969-71, Researcher at the von Karman Inst for Fluid-Dynamics, Belgium. *Society Aff:* ASEE, ARA.

Romani, Fred
Business: 5814 Heffernan, Houston, TX 77087
Position: Mgr of Engrg *Employer:* Natl Soil Services, Inc *Education:* PhD/Geotech Eng/Purdue Univ; MS/CE/Geotech Eng/Purdue Univ; Civil Eng/Natl Univ of Peru *Born:* 11/11/37 Born in Peru, South Amer. Grad with honors as Civil Engr at its Natl Univ. Twenty-yrs of experience at Technical and Mgmt levels. Experience includes Project Engrg and Mgmt for bldgs, earth dams, highways and airports, industrial plants, waterfront structures, offshore structures and earthquake engrg. At this time Mgr of Engrg and Asst Gen Mgr for Natl Soil Services, Inc, Houston, TX. *Society Aff:* ASCE, NSPE

Romano, Amardo J
Home: 10301 Stanley Circle, Bloomington, MN 55437
Position: Senior V.P. *Employer:* Edwards & Kelcey, Inc. *Education:* MS/Civil/Northeastern Univ; BS/Civil/Clarkson College of Tech. *Born:* 6/11/24 Reg P E in 7 state. Navy PT Boats. Mbr ASCE, Past Pres Minn Sect; Construction Specifications Inst, P Pres Minneapolis-St Paul Chap; Cons Engrs Council/Minnesota, P Secy & Treas; Engrs Club of Minneapolis; Past Pres. Soc. Am. Military Engineers-St Paul Post; SAME; other engrg & civic assns. V P Edwards & Kelcey Inc; mgmt Minneapolis Regional Office. Proj mgmt & admin level, directing & man major tunnels, hwys & bridge design projs, transportation & environ studies. Authored & presented papers at Univs, sev pub. Hobbies: sports, traveling, gardening, painting. *Society Aff:* MGR, ECM, MSES, CEC/MN, APW, ASCE, CSI, SAME, SETAC, SMPS

Romano, Frank A
Home: 2440 Wheatfield, Florissant, MI 63033
Position: President. *Employer:* Frank A Romano & Assoc Inc. *Education:* BS/ME/Univ of IL. *Born:* May 1923. BSME 1950 Univ of Ill. Native of Wood River Ill. Served with Army Corps Engrs 1943-46. Design Engr for Harry F Wilson Cons Engr 195058; Secy for Delaney-Shelton Cons Engrs 1958-62; Assoc for Wilson & Aach Cons Engrs 1962- 65; Partner Hamm & Romano Cons Engrs 1965-69; Pres Frank A Romano & Assoc Inc 1969. Field of HVAC, Plumbing & Fire Protection . Responsible for entire operation of Frank A Romano & Assocs Inc 1969. Reg PE in MO, IL, IA, CO, MN, OH, WI, LA, PA, NY, IN, AL, MS, NC, SC, GA & VA. Mbr: Bd of Plumber & Drainlayer Examiners St Louis County, Mo. Enjoy swing & jazz music, bowling & golf. *Society Aff:* ACEC, CECMO, ASHRAE.

Romanos, Michael C
Business: 1003 W Nevada St, Urbana, IL 61801
Position: Assoc Prof. *Employer:* Univ of IL. *Education:* PhD/Regional Sci/Cornell Univ; MS/Regional Planning/FL State Univ; DiplArchEngr/Archit Engg/Ntl Technical Univ of Athens. *Born:* 6/12/41. From 1964 until 1970, he worked as an arch and urban planner in Greece, where he became involved in the planning of underdeveloped regions. Came to the US in 1970 as a Fulbright scholar. Since 1970, he has worked as a transportation and urban planning consultant for private firms and govt agencies. Between 1973 and 1975 he taught at Cornell Univ. he joined the Univ of IL in 1975. In 1978, he founded the consulting firm of Energia. Since 1980, he is the Director of the University of Illinois Bureau of Planning Research, and an Associate Professor in the Departments of Urban and Regional Planning and Civil Engr. He has made extensive contributions in the area of community energy conservation. Books published: *Residential Spatial Structure* (1976); *Western European Cities in Crisis* (1979). *Society Aff:* TRB, AEA, RSA, ΣΞ, ΦΚΦ.

Romig, Phillip R
Home: 2032 Goldenvue, Golden, CO 80401
Position: Prof *Employer:* CO Sch of Mines. *Education:* BS/EE/U of Notre Dame; MS/Geophysics/CO Sch of Mines; PhD/Geophysics/CO Sch of Mines *Born:* 7/24/38. U.S. Navy 1960-63, Aerospace electronics 1963-64, Geophysical Exploration Research and Teaching 1964-present. Head, Dept. of Geophysics, Colo. Sch of Mines, 1983-present. *Society Aff:* IEEE, SEG, SSA, ΣΞ, AGU, EERI.

Romine, Thomas B, Jr
Business: 300 Greenleaf, Fort Worth, TX 76107
Position: Pres. *Employer:* Romine, Romine & Burgess, Inc. *Education:* BS/ME/Univ of TX. *Born:* 11/16/25. Billings, MT. Fort Worth resident since 1930. USNAS, 1943-45. Engg for contractor, 1948-50, and major A/E 1950-56. Founded present consulting firm 1956. Reg PE TX, OK, LA, GA. Fellow grade mbr ASHRAE, ACEC, APEC. Pres: APEC (International), 1974; TX Assoc Consulting Engg, 1967; FW ASHRAE, 1958. ASHRAE National Committeeman: Standard 90-75, 1974-81; Standard 100.3 (Chmn) 1976-84; TC4.1 (Chmn.) 1986-Pres. Bd Chmn: City Mech, 1976-80; City Plumbing, 1971-75; NTCOG Plumbing, 1971-75. VP FW Symphony, 1978-81. Pres Starfish Class (Sailing), 1972-75, National Champion, 1976. Computer program author, HVAC and energy analysis. Delta Tau Delta, Pi Tau Sigma, Married, three sons. *Society Aff:* ASHEAE, ACEC, APEC, NSPE, ΠTSIGMA.

Roney, Robert K
Home: 1105 Georgina Ave, Santa Monica, CA 90402
Position: Senior Vice President. *Employer:* Hughes Aircraft Co. *Education:* PhD/EE/Caltech; MS/EE/Caltech; BS/EE/Univ of MO. *Born:* Aug 1922 Newton Iowa. Lt(jg) USNR 1944-46. With Jet Propulsion Lab 1948- 50; with Hughes Aircraft Co 1950- ; Systems Analysis, Guidance & Control, System Dynamics, Mgr Space Systems Div 1968; Asst Group Exec Space & Communications Group 1970. Vice Pres 1973. Senior VP 1986. Mbr Sigma Xi, AIAA. Fellow IEEE. Pres Santa Monica Symphony. *Society Aff:* IEEE, AIAA, ΣΞ.

Rony, Peter R
Business: Dept. of Chem. Eng, Virginia Tech, Blacksburg, VA 24061
Position: Prof. *Employer:* VA Tech *Education:* PhD/ChE/U of CA, Berkeley; BS/ChE/Caltech *Born:* 6/29/39. Received secondary and undergrad education in Los Angeles, CA. Worked as res engr in the corporate res labs at Monsanto (1965-70) & Exxon Res & Engg Co (1971). Have been with VA Tech since 1971. Currently Editor-in-Chief IEEE MICRO. Write texts and magazine columns, teach, and offer short courses in the areas of digital electronics, microcomputers, and process control. Texts have been translated into Italian, German, French , Japanese, & Spanish. Research work is in traditional chemical engg fields: catalysis and chemical separations. *Society Aff:* IEEE, ACS, AIChE, ASEE, AAAS

Rooney, Thomas S
Business: PO Box 341939/Dixie Hwy, Coral Gables, FL 33134
Position: President. *Employer:* Connell Metcalf & Eddy. *Born:* Oct 1933. BSCE Cornell Univ. Pres Connell Metcalf & Eddy Architecture Engrg Planning, in charge of all aspects of Coral Gables based firm since Aug 1975. Previously Senior Vice Pres of Lester B Knight & Assocs Inc, in charge of the firm's architectural engrg operations in Chicago with assignments in indus, commercial & municipal areas from 1970-75. Various responsibilities including Mgr of Corporate Const for Olin Corp, responsible for all aspects of proj const. Div Engr for EI duPont. Superintendent with American Dredging Co. Contraction Officer for US Army Corps of Engrs.

Roorda, John
Home: 12226 Boheme, Houston, TX 77024
Position: VP (Ret'd) *Employer:* Shell Oil Co. *Education:* PhD/ChE/Purdue Univ; BS/ChE/Purdue Univ. *Born:* 1/16/23. Evanston, IL. Distinguished Engg Alumnus Award 1976 Purdue Univ. 1943-46 served in USNR as Ltjg; 1949 joined Shell Oil Co. After variety of engg assignments in Shell Chem plants became involved in economics & design of maj projs. Later nominated 1st Res Dir with Shell Chem's Woodbury Plastics Technical Ctr. Became Mgr of Shell Chem Economics & later Gen Mgr of combined Oil Products/Chem Economics. In 1974 assumed position as VP of Planning & Economics. In 1977 joined Shell Dev as VP. Returned to Corp Planning 1980. Retired in 1983. Jormed John Roorda, Inc; now quiescent. Coordinator for Exec Service Corps of Houston.

Root, L Eugene
Home: 1340 Hillview Dr, Menlo Park, CA 94025
Position: Pres-Lockheed Missiles & Space Co, Inc. *Employer:* Lockheed Corp. *Education:* MSAE/Aero/CIT; MSME/Mech/CIT; AB/Engg & Math/UOP. *Born:* July 4 1910, Lewiston Id. Alumni Distinguished Service Award 1966. Aircraft design (Aaerodynamics & flight testing) Douglas AirCraft Co Santa Monica & El Segundo; Proj Rand Spl assign & Rand Corp Santa Monica 1934-39, 1939-46, 1946-48, 1948-53; 1953-56 Lockheed Aircraft Corp, Corp Dir Devel Planning; 1956-59 V Pres Genl Mgr Lockheed Missiles & Space Div, 1959-61 Grp V Pres Missiles, Space, Propulsion & Electronics, 1961-69 Pres, Lockheed Missiles & Space Co, Inc. Pres, Lockheed Missiles & Space; 1951-52 on leave from Rand as Spl Asst fo Dep Ch of Staff for Dev, Headquarters USAF. Mbr Defense Sci Bd, Office of Dir Def for Res Engrg 1954-62, 1964-66; Mbr Scientific Adv Bd USAF, Ch of Staff 1948-56. Hon Fellow AIAA, Fellow AAS, RAS; Mbr NAE, Delta Upsilon, Beta Gamma Sigma. Reg PE in Cal. *Society Aff:* AIAA, AAS, RAS, NAE.

Roper, Val J
Home: 1290 Forest Hills Blvd, Cleveland Heights, OH 44118
Position: retired Mgr Product Planning. *Employer:* Lamp Division General Electric Co. *Education:* BS/EE/Purdue Univ. *Born:* June 1903. 1925-67 Automotive Lighting Expert, GE Co, 1955-67 Mgr Product Planning, Miniature Lamp Dept of GE. P Chmn IES Motor Vehicle Lighting Comm; SAE Lighting Comm; TC-47 US Natl Comm/ CIE; Life Mbr CIE; Fellow IES; US Rep Ctte of Experts, Brussels Working Party ISO/CIE. Author of 7 patents & num IES & SAE papers. Independent consultant 1967- . *Society Aff:* SAE, IES.

Rorschach, R I
Business: Suite 1108-320 S. Boston, Tulsa, OK 74103
Position: Pres. *Employer:* Process Tech Corp. *Education:* MS/Chem Engg/MIT; MS/Petr Ref/Univ of Tulsa; BS/Chem Engg/MIT. *Born:* 8/20/22. in Tulsa, OK. Active duty with Navy 1943-46. 1947-68, various positions in engg design, R & D, process plant operations, teaching. Co-founder in 1968 of Data Systems Corp, which became Process Tech Corp in 1974. Main fields of activity: hydrocarbon processing, pollution control, engg application of computers. Chrmn, Tulsa Sec AIChE 1962. Gen Chrmn, AIChE OK Meeting, 1979. Chrmn, Tulsa City-County Environmental Advisory Council, 1967-1978. *Society Aff:* AIChE, ACS, SPE, NSPE.

Rosard, Daniel D
Home: 2000 Valley Forge Circle, King of Prussia, PA 19406
Position: Consultant *Employer:* Self *Education:* MS/ME/Univ. of Penna; BS/ME/Drexel University *Born:* 10/17/27 Fellow of ASME, Reg PE. Worked at the Westinghouse Steam Turbine Division for over 20 years, where he held various technical and management positions with responsibilities for design, development, analysis, testing and resolution of field problems. 1979-84, pres of Power Dynamics, where he developed turbine rotor stress monitors, a boiler stress analyzer, and a power plant performance on-line monitoring system. Also conducted several power plant performance investigations, cycling studies and EPRI studies of bypass systems. Now an Engineering Consultant offering technical services and software systems to Electric Utilities. *Society Aff:* ASME

Rose, Albert
Home: 292 Stockton Rd, Princeton, NJ 08540
Position: Consultant. *Employer:* Self. *Education:* PhD/Physics/Cornell Univ; AB/Physics/Cornell Univ. *Born:* 3/30/10. 1935-75 RCA labs, originated TV Camera tubes Orthicon & Image Orthicon, initiated proj leading to modern Vidicon; other major areas of work: solid state, photoconductivity, vision, noise & solar energy. 1955-58 Dir of Res RCA Ltd, Zurich Switz; 1961 Visiting Prof, Princeton Univ; 1967 Mary Upson disting Visiting Prof, Cornell Univ; 1975 Fairchild Disting Scholar, CA Inst Tech; 1976- 79 Visiting Prof Stanford and Hebrew Univ. Boston Univ, Univ of DE, Polytechnic Inst (Mexico City). Bks *Photoconductivity* 1961 & *Vision* 1975. Fellow IEEE, APS; Mbr Soc Suisse de Physique, NAE. Awards: IEEE Morris Liebman 1945; Tel Broadcasters Assoc 1945; SMPTE 1946; SMPTE Sarnof Gold Medal 1958; Soc for Inf Display 1976; IEEE Edison Medal 1979. *Society Aff:* APS, IEEE, SPSE, NAE

Rose, Arthur
Business: 140 N. Barnard St, State College, PA 16801
Position: Self Employed Owner. *Employer:* EAGR Dev Group. *Education:* PhD/Physical Chem/Univ of Cincinnati; MS/Chem/Univ of Cincinnati; BA/Chem/Univ of Cincinnati. *Born:* 7/26/03. 1927-40 univ teaching Lehigh, Univ Hawaii & Penn State Univ, mostly Genl Chem for non-chemists and non-chemical engrg students. WWII Tech Dir Explosives Plants, Quartermaster R&D; after 1946 teaching, cons, writing and res on chem engrg. Editor Condensed Chem Dictory, other books and publications. Res on batch distillation theory, computers in engrg calculations, vapor liquid equilibrium, vacuum distillation, purification of fatty acids & esters, among others. Organizer & Pres of Applied Sci Labs Inc 1951-79; consultant to date on chromatography, consulting to chem ad bus mgmnt, small businesses. Mbr AIC, AAAS, NYAS, AIChE, AOCS, AACE, ACS (Council Policy Comm and many offices). Republican County Chmn 1966-70. Relaxation; travel, dancing, white water streams and primitive country. *Society Aff:* AIC, AAAS, NYAS, AIChE, AOCS, AACE, ACS.

Rose, David N
Business: 180 E First S, PO Box 11368, Salt Lake City, UT 84139
Position: VP, Admin. *Employer:* Mtn Fuel Supply Co. *Education:* BSEE/EE/Univ of UT; MBA/Bus Admin/Univ of UT. *Born:* 9/8/44. As Mtn Fuel's youngest VP, Nick Rose's areas of responsibility include Security, Personnel, Training & Career Dev, Data Processing, Purchasing, Warehousing, Electronic Communication, Transportation, Office & Bldg Services and Property & Rights of Way. Nick came to Mtn Fuel in 1969 with a degree in Engg & a Master's degree in Bus Admin, both from the Univ of UT. Prior to his current assignment, he was in the Co's Rate Dept for seven yrs, the last two yrs of which he served as Mgr of Rates for Mtn Fuel Supply Co. Nick is a licensed PE in the states of UT & WY. *Society Aff:* IIE.

Rose, Herbert A
Home: 405 Detwiller Lane, Bellevue, WA 98004
Position: Consulting Engineer. *Employer:* self employed. *Born:* BSEE May 1924 & BSME May 1925 from Kan State Coll. 1925-65 Design & District Engr for Westinghouse Electric Corp, design DC & AC power conversion apparatus, District Engr power equip for pulp, paper & wood products industries; 1965- , cons engr in Seattle, Wash area specializing in power drives & controls for wood products industries/plants. Fellow ISEE 1962, ASME 1970. Author Sawmill Sections for 8th & 9th editions of Standard Handbook for Electrical Engineers. 45 patents relating to power equipment. Reg PE in Cal.

Rose, Joseph
Home: 4203 Sandhurst Court, Annandale, VA 22003
Position: Deputy Manager. *Employer:* Department of Defense. *Education:* BIS/Bus Comm/George Mason. *Born:* 11/6/22. Philadelphia, PA. Career has been primarily concerned with planning, mgmt & engrg of government communications systems: served as communications officer in WWII, recalled to active duty with SAC during Korean conflict; 1948-55 Radio Engr FCC; 1955-60 Ch Radio Engr USASCC; 1960-73 Electronics Engr DCA; July 1973- Jan 1981 Deputy Mgr, Natl Communications System Dept of Defense. Reg PE; Sr Mbr IEEE; Fellow Radio Club of Amer; Mbr AFCEA. Currently employed as a Advisory Engr with the Satellite Business Systems. *Society Aff:* IEEE, RCA, AFCEA.

Rose, Joseph L
Business: 32nd & Chestnut Sts, Philadelphia, PA 19104
Position: Professor Mechanical Engrg. *Employer:* Drexel University. *Education:* PhD/Applied Mechanics/Drexel Univ; MS/Applied Mechanics/Drexel Inst of Technology; BSME/Mech Engr/Drexel Inst of Technology. *Born:* 7/5/42. Res activities include work on ultrasonics, adhesive bond inspection, signal processing & pattern recognition. Over 20 journal type res papers in structural dynamics, biomedical engrg, photoelasticity & over 60 publs on nondestructive testing. 2 textbooks also. Also active as cons for Ultrasonics Intl Inc, Krautkramer-Branson Inc, & Thomas Jefferson Univ Hospital in Philadelphia. *Society Aff:* ASME, SESA, ASA, ASNT, ASM, AIUM, IEEE.

Rose, Kenneth E
Business: Dept of Mech Engrg, Lawrence, KS 66045
Position: Prof of Mechl Engrg Emeritus; Consultant. *Employer:* University of Kansas. *Education:* MS/Metallurgy/Cornell; E Met/Metallurgy/CO Sch of Mines. *Born:* Oct 20 1915. Formerly engr trainee, Caterpillar Tractor Co; res engr, Battelle Memorial Inst; taught at Cornell, Okla; Kan since 1947 serving as Dept Chmn & Assoc Dean of Engrg; cons engr in materials & corrosion. Reg PE in Kan, MO. AIME; Sigma Xi; ASM Fellow, Bradley Stoughton Award, natl cttes on Metals Handbook, met education & internatl relations. Extensive exper in higher education in Latin America through Fulbright, Ford Foundation, UNESCO, NSF & OAS programs. Interested in pre-Columbian metal work, applications of metallurgy in corrosion & in mfg processes. Retired, 1984. Consultant. *Society Aff:* ΣΞ, ASM, AIME

Rose, Robert M
Business: Rm 4-132 MIT, Cambridge, MA 02139
Position: Professor of Materials Sci & Engrg. *Employer:* Massachusetts Institute of Tech. *Education:* ScD/Physic Met/MIT; SB/Physic Met/MIT. *Born:* April 15 1937. 1961-63 Asst Professor of Met & Ford Foundation Post-Doctoral Fellow MIT, 1966 promoted to Assoc Professor, 1972 Full Professor; 1973 Professor, Harvard Med. Sch.-M.I.T. Div. Health Sci. Tech.; Bradley Stoughton Award, ASM 1966; Kappa Delta Award, Amer Acad of Orthopaedic Surgeons 1973. *Society Aff:* ASM, APS, AAAS, TMS-AIME, ORS, NYAS

Roseman, Donald P
Home: 14501 Woodcrest Dr, Rockville, MD 20853
Position: Chief Naval Architect. *Employer:* Tracor Hydronautics, Inc. *Education:* MS in Engg/Naval Arch & Marine Engg/Univ of MI; BE/ME/Johns Hopkins Univ. *Born:* 7/17/27. Native of Baltimore, MD. Served in US Merchant Marine Cadet Corps, 1945-46. Initial naval arch & proj mgt experience at MD Shipbldg & Drydock Co, & M Rosenblatt & Son. Supervised preliminary design activities for all ship types at Bethlehem Steel Corp at Quincy & Sparrows Point yards, 1955-1966. At Hydronautcs, Inc since 1966 as principal naval arch and chief naval arch. VP 1979-83. At Tracor Hydronautics, Inc. since 1983 as Chief Naval Architect. Responsible for merchant and naval ship design studies & prog mgt. Served on various NAS/NAE advisory committees regarding ship structures. SNAME Linnard Prize 1976. Chairman of SNAME ship Design Committee since 1984. *Society Aff:* SNAME, ASNE.

Rosen, Harold A
Business: PO Box 92919 Airport St, Los Angeles, CA 90009
Position: Vice President Engineering. *Employer:* Hughes Aircraft Co. *Born:* Mar 1926, New Orleans La. BE Tulane, PhD Cal Tech. 1944-46 served in US Navy as staff engr for Raytheon, contributed to the dev of Lark, Sparrow & Hawk missiles &

Rosen, Harold A (Continued)
CW radar; at Hughes conceived spin stabilized synchronous communication satellite & led dev efforts leading to Syncom, ATS, Early Bird, Intelsat IV & Anik satellite, presently manage commercial systems div, space & communications grp. Astronautics Engr, NSC 1964; Communications Award, AIAA 1968; Mervin J Kelly Award, IEEE 1971. Enjoy skiing & tennis.

Rosenbaum, Fred J
Business: 12180 Prichard Farm Rd, Maryland Heights, MO 63043
Position: Chief Scientist *Employer:* Central Microwave Co *Education:* PhD/EE/Univ of IL; MS/EE/Univ of IL; BS/EE/Univ of IL. *Born:* 2/15/37. Born & raised in Chicago, IL. Res Asst in Ultramicrowave Group of Univ of IL (1959-1963). Res Scientist at McDonnell Aircraft Corp, St Louis (1963-1965). Joined EE Dept of WA Univ (1965). Prof. Dir of Microwave Lab. 1965-1983. Joined CMC 1983. Res interests include microwave & millimeter wave semiconductor devices, microwave ferrites, & comp aided design. Fellow, IEEE (1979). Editor, IEEE Transactions on Microwave Theory & Techniques (MTT), (1972-1974). Mbr MTTS Administrative committee (1971-present). VP, President (1981) MTTS(1980). Married, two daughters. *Society Aff:* IEEE, MTTE, EDS.

Rosenbaum, Joe B
Home: 875 Donner Way, No. 301, Salt Lake City, UT 84108
Position: Consulting Metallurgist Retired *Employer:* US Bureau of Mines. *Education:* MetEngr/-/CO Sch of Mines. *Born:* 3/1/12. Engr, Public Works Admin & Walker Mining CO before joining us Bureau of Mines 1941; 1942-46 Army Corps of Engrs, now LtC AUS retired; at Bureau labs in charge of R&D on process dev for extraction of manganese, chromium, uranium, vanadium, tungsten, beryllium, nickel & rhenium, 1962-68 Ch Metallurgist & Dir Met Res Wash DC, 1969-74 Res Dir Salt Lake City Met Res Center; retired, Cons Metallurgists US Bureau of Mines 1975-76, Consultant 1977- . Author 55 publ. Presidential Citation 1964; Disting Achievement Medal CO Sch of Mines 1968; Disting Service Award US Dept of Interior 1968; Mbr NAE 1973; Henry Krumb Lectr AIME 1974. Adj Prof, Metallurgy, Univ of UT 1968- . Chrmn Intl Atomic Energy Agency Symposium on Uran Ore Processing, Wash, DC 1975. Mbr Panel on concepts of Uran Resources and Pro, Nat Res Council, 1978; Mbr Committee on Surface Mining and Reclamation, Nat Res-Counil, 1978-79; Mbr Bd of Mineral & Energy Resources Nt Res Council 1979- , Mbr Committee on Mineral Technology Dev. Options, Nat Res-Council, 1979-1980: Mbr Panel on Titanium Availability, Nat Res-Council, 1980-81. Retired. *Society Aff:* AIME, ASM, NAE.

Rosenberg, Leon T
Business: 2562 N. 83rd St, Wauwatosa, WI 53213
Position: Consultant Turbogenerator Design & Operation. *Employer:* Allis Chalmers Corp. *Education:* MS/Elec Engg/IL Inst of Technology; BS/Elec Engg/Univ of MI. *Born:* Nov 1904, South Bend Ind. Since 1926 with Allis-Chalmers, design of synchronous motors & generators, speciality turbogenerators, 1952 Engr-in-Charge A-C design, 1956 Ch Generator Design Engr, 1960 Sr Cons generator design, 1970--cons AllisChalmers & Allis-Chalmers Power sSystems. Mbr Tau Beta Pi, Eta Kappa Nu & Sigma Xi. Reg PE Wis. Fellow IEEE 1958. Author of 42 papers mainly on design, oper & care of turbogenerators. Recipient of 1st IEEE Nikola Tesla Award for outstanding contributions to electrical power generation and utilization Jan 1976, as well as Nikola Tesla Award offered by the Yugoslavian Nikola Tesla Soc at same time. Since 9/1/77 Self-employed as a turbogenerator Consultant. *Society Aff:* IEEE, ESM.

Rosenberg, Marvin A
Home: 1501 Baylor Ave, Rockville, MD 20850
Position: Dir, Damage Control, Hull Outfit & Furnishings Div. *Employer:* Naval Sea sys Command, Dept of the Navy. *Education:* BS/Marine Engg/US Merchant Marine Acad; BMnE/Marine Engg/State Univ of NY, Maritime College. *Born:* 10/4/23. BS USMMA; BMnE NY Maritime Coll. WWII Engr Officer US Navy; 1945-52 Ch Engr in merchant ships; teacher related tech subjs, NY City; engg editor Convair- Astronautics Atlas ICBM prog, San Diego; Operations Eng in submarine nuclear propulsion prototype plants with Genl Elec Co's Knolls Atomic Power Lab; 9 yrs Sr Proj engr with M Rosenblatt & Son Inc, NY; since 1969 with Naval Sea Sys Command. Since 1979 Dir of Damage Control Hull Outfit and Furinshings Div. Cdr USNR, retired. USCG licensed Ch Engr Steam Vessels. Mbr SNAME *ASNE, SPE, & STC. Secy Subcommittee F25.03 (Outfitting) on ASTM Committee of Shipbldg. Society Aff:* SNAME, ASNE, STC, SPE.

Rosenberg, Paul
Business: PO Box 729, Larchmont, NY 10538
Position: President. *Employer:* Paul Rosenberg Associates. *Education:* PhD/Physics/Columbia Univ; MA/Physics/Columbia Univ; BA/Math/Columbia Coll *Born:* March 1910, New York NY. Physicist, engr & multidisciplinary scientist. 1934-41 Lecturer, Columbia Univ; 1941-45 Radiation Lab, MIT; since 1945 Pres Paul Rosenberg Assocs, cons to indus & government. Mbr NAE since 1970; Pres Inst of Navigation 1950-51; V Pres AAAS 1966-69, Council AAAS 1962-72; Talbert Abrams Grand Award, Amer Soc of Photogrammetry 1955; Natl Comm on Clear Air Turbulence 1955- 56; Chmn of many natl prof meetings, symposia & advisory committees: RTCA-RTCM- ION Electronics Conference 1950, NAS/NRC Navigation & Traffic Control Panel Space Applications Study 1968, Chmn NRC/NAS Cartography Panel CORSPERS 1972-77. Board of Directors, Center for Environment & Man 1976-85. Fellow IEEE, AAAS, Amer Inst of Chemists, 75 publ, 13 US patents. Assoc Fellow, AIAA. Sigma Xi. *Society Aff:* IEEE, AAAS, ION, ASPRS, AAPT, APS, ACS, AIC, SIST, NAE, OSA.

Rosenberg, Richard
Home: 5230 College Garden Ct, San Diego, CA 92115
Position: Consultant. *Employer:* Self. *Education:* BSME/ME/Univ of TN. *Born:* 11/22/26. CA resident since 1960. Served in USN, 1943-1946. Started nuclear career at Oak Ridge Natl Lab in 1951 as a design engr. Five yrs at Westinghouse (Bettis Pl) in the Naval Reactors Prog. & Gen Atomic Co 26 years, retired, March '86. Presently, in private consultation. VP ASME for Region IX, 1978-1980; ASME Bd of Govs, 1982-84, Pres ASME, 1987-88. Enjoy classical music, photography, woodworking and deep sea fishing. *Society Aff:* ASME.

Rosenberg, Robert B
Business: 8600 West Bryn Mawr Avenue, Chicago, IL 60631
Position: Sr VP, Mbr & Indus Relations *Employer:* Gas Research Institute *Education:* BSChE with distinction, MS & PhD Gas Tech/Chemical Engineering/Illinois Inst. of Technology (IIT) *Born:* 03/19/37 Chicago Ill. Vice president, research and development, since the establishment of GRI in 1977 with overall responsibility for the technical aspects of GRI's research and planning activities. Served GRI in the administrative role of Exec VP before being elected to his present position of Sr VP, Mbr & Indus Relations. Until July 1, 1977, was vice president, engineering research for the Institute of Gas Technology, with responsibility for sponsored research in energy utilization, development of advanced energy systems, and management sciences. He joined IGT's research staff in 1962 and was named vice president in 1973. From 1965 to 1969, was also an adjunct assistant professor at the Illinois Institute of Technology. He is a registered professional engineer in Illinois and has authored numerous papers, articles and speeches. *Society Aff:* AICHE, AGA, APCA, IGE

Rosenberg, Ronald C
Business: Dept of Mech Engg, Michigan State U, E Lansing, MI 48824
Position: Prof. *Employer:* MI State Univ. *Education:* PhD/ME/MIT; MS/ME/MIT; BS/ME/MIT. *Born:* 12/15/37. Prof of Mech Engrg at MI State Univ, specializing in computer-aided modeling and design of dynamic systems. Co-author of several texts including Introduction to Physical System Dynamics. Author of the ENPORT Com-

Rosenberg, Ronald C (Continued)
puter Program. Pres of ROSENCODE Assoc, Inc, an engrg software co. *Society Aff:* IEEE, ASME, ASEE.

Rosenblatt, Lester
Home: 8 E 83rd St, New York, NY 10028
Position: Chrmn of Bd & CEO *Employer:* M Rosenblatt & Son, Inc. *Education:* BSE/Naval Arch & Marine Engg/Univ of MI; BS/Engrg/CCNY. *Born:* 4/13/20. Reg PE NY & MA. 1942 Naval Arch with John W Wells Inc; 1944-45 active duty US Navy Design Div, Pearl Harbor Navy Yard; 1946 in charge of Preliminary Design, John H Wells; 1947 co-founder M Rosenblatt & Son Inc, Naval Architects & Marine Engrs, Pres 1959-1984, Chmn, CEO & N.A. Chmn NY Metropolitan Sect SNAME 1961-62, Natl Chn SNAME Mbrship Ctte 1974-78, SNAME Pres 1979 & 1980. SNAME Fellow, and Honorary life mbr. Mbr Amer Bureau of Shipping & Int'l gen'l ctte Bureau Veritas; Trustee, Webb Institute of Naval Architecture; SNAME; ASNE. Univ of MI Sesquicentennial Award 1967. SNAME Vice Admiral Jerry Land Medal 1984. *Society Aff:* SNAME, ASNE.

Rosenblith, Walter A
Business: 77 Mass Ave Rm 4-258, Cambridge, MA 02139
Position: Institute Professor . *Employer:* Massachusetts Institute of Tech. *Education:* Ing Rad/Radioelectricien/Ecole Superieure d 'Electriclte' Paris; Ing Rad/ Radiotelegraphiste/Univ of Bordeaux, France. *Born:* Sept 21 1913, Vienna Austria. 1951-57 Assoc Professor of Communications Biophysics, Dept of Elec Engrg MIT, 1957-- Professor, 1951-69 Res Lab of Electronics, Center for Communications Sciences, 1967-69 Chmn of Faculty, 1969- 71 Assoc Provost, 1970-71 Acting Director MIT-Harvard Joint Center for Urban Studies, 1971-80 Provost, 1975-- Inst Professor. Fellow Amer Acad of Arts & Sci (Mbr Exec Bd 1970-77), World Acad of Art & Sci, Acoustical Soc of Amer, AAAS, IEEE & Soc of Experimental Psychologists; Mbr NAS, NAS Inst of Medicine (Charter Mbr, Council 1970-76 , Exec Ctte 1973-76), NAE, Biophysical Soc (Charter Mbr, Council & Exec Bd 1957-61, Council 1969-72), , Sigma Xi (Pres MIT Chapter 1966- 67), Eta Kappa Nu, Amer Otological Soc. *Society Aff:* NAS, NAE, IOM.

Rosenblueth, Emilio
Business: Instituto de Ingenieria, Ciudad Universitaria DF, 04510 Mexico
Position: Professor. *Employer:* Universidad Nacional Autonoma Mexico. *Education:* PhD/Engrg/Univ of IL; MS/Civil Engrg/Univ of IL; CE/-/Natl Acad Univ of Mexico *Born:* April 8 1926, Mexico. 1959-66 Dir Inst de Ing UNAM, 1966-70 Coordinator of Scientific Res, 1964-65 Pres Academia de la Investigacion Cientifica, Mbr Bd of Governors UNAM 1971-81 . Trustee Univ Aut Metropolitana, Mexico 1974-77; Pres Internatl Assoc for Earthquake Engrg 1973-77; Pres DIRAC, Cons Engrs Mexico 1970-77; Viceminister for Planning, Ministry of Educ Mexico 1977-82; Hon Mbr ACI, ASCE, Internal Assoc for Earth Engrg; Foreign Assoc Mbr NAE, NAS, AAAS; Foreign Honorary Mbr Amer Acad of Arts & Sci. Moisseiff, Huber & Freudenthal Awards, ASCE; National, AIC & Elizondo awards; Award for Services in Engineering, Univ of Ill; Prince of Asturias (Spain) Award in Science and Technology; Professor Honoris Causa, Univ Nac de Ing, Peru; Honorary Doctor, University of Waterloo and National Univesity of Mexico. *Society Aff:* ASCE, ACI, NAE, NAS, AAAS.

Rosenfeld, Azriel
Home: 847 Loxford Terrace, Silver Spring, MD 20901
Position: Research Professor. *Employer:* University of Maryland. *Education:* PhD/Mathematics/Columbia Univ; DHL/Rabbinic Literature/Yeshiva Univ; Ordination/Ritual Law/Yeshiva Univ; MS/Math Education/Yeshiva Univ; MHL/Hebrew Literature/Yeshiva Univ; MA/Mathematics/Columbia Univ; BA/Physics/Yeshiva Univ. *Born:* Feb 1931. BA physics, ordination (Rabbi), MHL Hebrew lit, MS math education & DHL rabbinic lit from Yeshiva Univ; MA & PhD math from Columbia Univ. Dr. Tech. (Hon.) from Linkoping University, Sweden (1980). Fellow IEEE. Piore Award, 1985. 1954-64 worked in electronics indus (most recently Res Mgr Electronics Div, Budd Co) prior to joining computer sci faculty of Univ of Md. Pub 20 books, nearly 400 papers, primarily on computer analysis of pictorial information; edit journal 'Computer Graphics & Image Processing.' Pres cons firm ImTech Inc; P Pres International Assn for Pattern Recognition; P Pres Assoc of Orthodox Jewish Scientists; Director, Machine Vision Association, Society of Manufacturing Engineers. Married to Eve (Hertzberg), 3 children (Elie, David, Tova). *Society Aff:* IEEE, ACM, MAA, PRS, AAAI, MVA/SME, ΣΞ.

Rosenfield, Alan R
Business: 505 King Ave, Columbus, OH 43201
Position: Res Leader *Employer:* Battelle-Columbus. *Education:* ScD/Met/MIT; SM/ Met/MIT; SB/Met/MIT. *Born:* 9/7/31. Postdoctoral fellow, Liverpool Univ (1961-2). Resident consultant, Open Univ, UK (1974). Author of 200 scientific papers & editor of three books. With Batelle-Columbus (Phys. Metal. Sec) since 1962. Maj activities have been res & engg investigations of strength & fracture behavior of mtls. Have studied metals, glass, rocks, ceramics, & plastics. Engg accomplishments include dev of rational basis for thickness specification of eyeglass lenses and successful analysis of failure pressures for pipelines of tough steel. Born & raised in Chelsea, MA. Married with two children. *Society Aff:* AIME, ASM Intl, ASTM, JISI, OAS.

Rosenkoetter, Gerald E
Business: 801 N 11th St, St. Louis, MO 63101
Position: Exec VP *Employer:* Sverdrup Corp. *Education:* MSCE/Structures/Sever Inst of Tech WA Univ of St Louis; BSCE/Structures/WA Univ of St Louis. *Born:* 3/16/27. Taught engrg WA Univ 1956-60. Joined Sverdrup & Parcel 1951, experienced engr in complex aerodynamic test & rocket launch facilities, elctrical & mech, power generation & transmission, foods & beverages, mining & metals, all manner of industrial plants; as VP & Mgr, Industrial Div, respon for dir of 250 engrs engaged in R&D, energy systems, industrial projects, food & beverage facilities design; as corporate VP, industrial, respon for all industrial work accomplished by any of the Sverdrup Corp companies as Pres of Constr Mgmt Co fully respon for constr mgmt and constr contracting as Exec VP of Sverdrup Corp a member of the Office of the Presidency respons for all executive direction of the Corporation. Reg PE in 16 states. Mbr ASCE, Air Force Assn. Forest Hills CC. *Society Aff:* ASCE.

Rosenqvist, Terkel N
Business: Dept of Met, 7034 Trondheim-NTH Norway
Position: Prof. *Employer:* Norw Inst Tech. *Education:* DrTech/-/Norw Inst Tech; CandReal/Chemistry/Univ of Oslo. *Born:* 10/2/21. in Oslo, Norway. Res position with Inst for the Study of Metals, Univ of Chicago, IL, 1947-51. Assoc Prof NTH, Trondheim 1952-54. Prof same place 1955-. Visiting Prof MIT 1959-60, Univ of WI 1964 and 1983, OH State Univ 1976. UNESCO expert with Middle East Tech Univ Ankara 1966-68. Main res on the thermodynamics of extrative met. Author of textbook: "Principles of Extractive Metallurgy," McGraw-Hill Book Co 1973 and 1983. Enjoy wine & fish. *Society Aff:* NMS, DKNVS, TMS-AIME, IMM, NKS, DNVA.

Rosenshine, Matthew
Business: 207 Hammond Bldg, Univ Park, PA 16802
Position: Prof. *Employer:* PA State Univ. *Education:* PhD/Operations Res/State Univ of NY; MS/Math/Univ of IL; MA/Math Education/Columbia Univ; AB/Sci/ Columbia Univ. *Born:* 5/25/32. and raised in NYC. Worked as an Aerodynamist for Bell Aircraft Corp and taught part-time at the Univ of Buffalo. Between 1958 & 1968, mathematician at Cornell Aero Lab of Cornell Univ, reaching position of principal mathematician and completing PhD degree requirements during sabbatical leave awarded by the Lab. Taught part-time as a lectr in the Ind Engg Dept at the State Univ of NY at Buffalo. Since 1968, full-time faculty mbr in the Ind & Mgt Systems Engg Dept at the PA State Univ. Current rank - Prof. Also consultant to various ind corps & govt agencies. *Society Aff:* IIE, ORSA, TIMS.

Rosenthal, Howard
Business: David Sarnoff Res Ctr, Princeton, NJ 08540
Position: Staff Vice President Engineering. *Employer:* RCA Corporation. *Education:* MChE/Chem Engr/NY Univ; BChE/Chem Engr/CCNY. *Born:* Sept 15 1924, Brooklyn NY. 1944-46 served in USNR as ETM 2/C; 1947 employed by Sunray Electric Co, Warren Pa; 1948 Graduate Asst NYU; since 1949 employed by RCA, initially as Mbr Tech Staff RCA Labs working on application of phosphors in color Kinescopes & progressed to Mgr Tech Admin, 1968 joined Corporate Res & Engrg organization, 1972 appointed Staff V Pres, Engrg. *Society Aff:* ACS, IEEE.

Rosenthal, Philip C
Home: 960 Parkwood Dr, Dunedin, FL 33528
Position: Professor (retired). *Employer:* University of Wisconsin. *Education:* MS/Met Engrg/Univ of WI; BS/Met Engrg/Univ of WI *Born:* March 1912, Allis Wisc. 1935-37 Res Engr, 1941-44 Res Engr, Asst Supervisor at Battelle Memorial Inst; 1937-41 Instructor, 1945-50 Assoc Professor, 1950-64 Professor of Met Engrg, 1955-64 Dept Chmn at Univ Wisc Madison; 1964-69 Dean of Engrg Univ Wisc Milwaukee, 1970-71 Visiting Professor Univ Wisc-Madison; 1971-72 Visiting Professor Pahlavi Univ, Shiraz Iran. Tau Beta Pi, Sigma Xi, Phi Kappa Phi, Alpha Sigma Mu (P Natl Pres); Fellow ASM; P Chmn Milwaukee Chapter ASM, AFS Wisc Chapter, WSPE Milwaukee Chapter. Mbr Governor's Recycling Task Force. Co-author of book on metal casting; contributor to other tech books & papers. Corecipient of ASM Henry Marion Howe best paper award 1953. *Society Aff:* AIME, ASM, AFS, ASEE.

Rosi, Fred D
Business: Sch of Engg & Appl Sci, Thorton Hall, Charlottesville, VA 22901
Position: Res Prof Dir, Energy Policy Studies Center Ctr. *Employer:* Univ of VA. *Education:* PhD/Met/Yale Univ; ME/Met/Yale Univ; MS/Phys/Yale Univ; BE/Met/ Yale Univ. *Born:* 1/13/21. Native of Meriden, CT; Adjunct Prof NYU, 1950-57; Harvard Business Sch Certificate; Gen Dir, R&D, Reynolds Metals Co, 1975-78; Dir, R&D, Am Can Co, 1973-75; Technical Staff & VP, Mtls & Device Res, RCA Labs, 1954-72; Mbr, Natl Mtl Adv Bd, 1971-75; Mbr, Res Advisory Council, NASA, 1960-65; Consultant to CIA, AASA, ITT, & Gould, Inc; Bd of Editors, Ind Res inst, 1968-71; Lt, US Navy, 1944-46; Gould Scientific Advisory Bd; David Sarnoff Gold Medal; Fellow of Am Phys Soc & Am Inst Metallurgical Engrs; Sigma Xi, Author of 38 technical articles & 12 patents in Mtl Sci & Electronic Devices; Chrmn, Bd of trustees, Trenton State College, 1967-73; VChrmn, NJ Council of Higher Education, 1967-69. *Society Aff:* AIME, APS, AAAS.

Rosoff, Peter S
Business: 3225 Gallows Rd, Fairfax, VA 22037
Position: Manager. *Employer:* Mobil Oil Corp. *Education:* MBA/Ind Mgt/CUNY; BME/-/CCNY. *Born:* 5/19/27. 1952-53 served with US Army; Supervisor of Pressure Vessels Design & Pipe Stress Analysis for Chemico; 1969 joined Mobil as Assoc Engr in Process Equip Grp, 1974 Mgr Capital Equipment Purchasing Div, directing & controlling procurement of capital equipment for the corp's various Dev. Mgr ASME, Charter Mbr of Computer Technology Ctte, Pressure Vessel & Piping Div. Reg PE in NJ & Penn. Resident of Oakton, VA. Assumed current position 1976 as Mgr Proj Procurement and matls control. In addition to Directing and Controlling Procurement of Capital Equip, is respon for procurement on all Major new corp projs and disposal of surplus Equip and matls. *Society Aff:* ASME

Ross, Arthur L
Home: 122 Maple Ave, Bala Cynwyd, PA 19004
Position: Staff Engr *Employer:* General Electric Co. *Education:* Eng. Sc. D./ Aeronautics/New York University; M. AE. E./Aeronautics/New York University; B. AE. E./Aeronautics/New York University *Born:* 03/09/24 Employed by General Electric Co since 1954 in a number of responsible staff and managerial positions in structural mechanics. Often sought by other divisions of the company for failure evaluations, design reviews, etc. and for general support of safety and design tasks and for solutions of critical structural problems via an active aerospace technology transfer activity. Numerous GE awards and honors, numerous publications and patents. Past chairman ASME Philadelphia section and ASME Design Engrg Div. Chairman and/or member of large number of other ASME technical and administrative cttees and member of AIAA Structures Cttee. Fellow, ASME. *Society Aff:* ASME, AIAA, AAM, ΣΞ

Ross, Bernard
Business: c/o Failure Analysis Assoc, 2225 East Bayshore Rd, Palo Alto, CA 94306
Position: Chrmn of the Bd *Employer:* Failure Analysis Assocs *Education:* PhD/Aero Engr/Stanford Univ; Cert/Sup Aero/Paris, France; Dipl/Vibrations/Univ of Edinburgh; MSc/Aero Engr/Stanford Univ; B/ME/Cornell Univ *Born:* 11/10/34 Montreal, Quebec, Canada; son of Miller and Evelyn (Kauffman) Rosenberg; came to US 1951, naturalized 1957; married Shelley Spencer Jan 28, 1968; children: Jonathan, Jennifer. Stress engr Canadair, Montreal, 1956; controls engr Marquardt Corp, Van Nuys, CA, 1957-58; stress engr. Douglas Aircraft Corp, Santa Monica, CA 1959; vibrations engr, ONERA, Paris 1960; research assoc Stanford 1962-65; research engr. Stanford Research Inst, Menlo Park, CA 1965-70; Pres Failure Analysis Assocs, Palo Alto, CA 1978-82; vis lectr Univ Santa Clara, CA 1970-78; Author papers in field. Home: Atherton, CA. *Society Aff:* AIAA, ASME, NSPE, SAE, CSPE, SESA, ASSE

Ross, David S
Business: Agri Engg Dept, University of Maryland, College Park, MD 20742-5705
Position: Assoc Prof. *Employer:* Univ of MD. *Education:* PhD/Agri Engrg/PA State Univ; MS/Agri Engrg/PA State Univ; BS/Agri Engrg/PA State Univ. *Born:* 4/16/47. Native of Brockway, PA. Asst Prof, 1973-78, Agri Engg, concurrently Ext Agri Engr, 1973-present. Chrmn, ASAE Environment of Plant Structures Committee, 1977-78; previously secy and vchrmn. Chrmn, ASAE Nursery Mechanization Committee, 1981-82; formerly secy and vchrmn. Chrmn, ASAE Extension Committee, 1981-82; vchrmn, 1980-81. Chrmn, Wash, DC - MD Sect, ASAE, 1981-82; formerly Public Relations Chrmn; Secy-Treas; vchrmn. Mbr, USDA Yrbook of Agri. Committees, 1977 & 1978; coauthored three chapters. Coauthor of four regional ext bulletins on hobby greenhouses, greenhouse energy conservation and solar heating, and trickle irrigation. ASAE Blue Ribbons - Publications, 1977, 1979 & 1980. ASAE Blue Ribbons - Ext Educational Methods, 1978 & 1979; 1983 Young Engr of the Year, N Atlantic Reg, ASAE; 1983 Exc in Extension Award, Coll of Agri Alumni Chap, Univ of MD, Natl Pres of Alpha Epsilon, 1983 (Agri Engrg Hon Soc), President Elect, 1987- ; Treas of MD Ext Specialists' Assoc, 1983-87. Treasurer of ASAE North Atlantic Region, 1987- . Have Ext educational prog in horticultural crops areas on engrg topics. *Society Aff:* ASAE, CAST.

Ross, Donald E
Business: 345 Park Ave, New York, NY 10154
Position: Partner *Employer:* Jaros, Baum & Bolles *Education:* MBA/Bus Admin/ NYU; BS/ME/Columbia Univ Sch of Eng; BA/Lib Arts/Columbia Coll *Born:* 05/02/30 Served as LTJG, US Navy, 1953-1955; with Carrier Corp, 1955-1970; with Dyna-Data, 1970-1971; with Jaros, Baum & Bolles, Consltg Engrs, 1971 to present, Partner 1977 to present. Reg PE in National Council of Engineering Examiners, NY and nine other states. Past Pres, and current member of B of D, NY Association of Consltg Engrgs, Secretary Columbia Univ Engrg Council, Chrmn of ASHRAE Handbook, Chrmn of Professional Liability Ctte of ACEC, Chairman of Building Research Board Cttee on Controls for HVAC Systems, Chrmn of Builg Services Section of National Council on Tall Buildings and Urban Habitat, Member American Society of Mechanical Engineers, Fellow Amer Soc of Heating, Refrig and Air Cond Engrs. Over 20 yrs of profl experience in the design of mech systems for office bldgs, laboratories, hospitals, computer centers and hotels. *Society Aff:* ASHRAE, TBΠ, NSPE, ASME

Ross, Donald K
Business: 7912 Bonhomme Ave, St Louis, MO 63105
Position: President. *Employer:* Ross & Baruzzini, Inc. *Education:* ScD/Ind Eng/WA Univ; MS/EE/MIT; BEE/EE/Univ of MN. *Born:* April 1925, St Louis Mo. WWII & Korean conflict Deck Officer in US Navy. Mbr Tau Beta Pi, Eta Kappa Nu & Sigma Xi. Reg PE. Pres Ross & Baruzzini Inc Cons Engrs. P Pres Mo Assoc of Cons Engrs & P Pres of Assoc of Prof Material Handling Cons. Mbr IEEE & IES; Fellow ACEC. *Society Aff:* IEEE, ACEC, IES, APMHC.

Ross, Edward N
Home: 10225 89th Ave SW, Tacoma, WA 98498
Position: Environ Engg Mgr. *Employer:* Weyerhaeuser. *Education:* BE/Mech Engg/ Yale Univ. *Born:* July 1924. BE Yale Univ 1948. 1943-46 USNR. Dev pollution control (in plant measures) for Lee Paper Co, pub num papers on this subject; 1967-70 Dir of ENngrg for Simpson Lee Paper Co; 1971-74 cons, directed contract to develop EPA anti-pollution guidelines; currently with Weyerhaeuser Co. Received Technical Association of the Pulp & Paper Industry Pollution Control Award 1970, Mbr TAPPI Bd/Dir 1970-73. Married 1959 to JoEllen Reuter, 4 children. Enjoy skiing, sailing. Kalamozoo Mich County Planning Study 1969. Pres Tacoma Gymnastics Boosters 1976, Lakewood Community Council 1979.. *Society Aff:* TAPPI, WPCF, APCA.

Ross, Gerald F
Business: Sixty The Great Road, Bedford, MA 01730
Position: President *Employer:* Anro Engineering Consultants, Inc. *Education:* PhD/EE/Polytech Inst of NY; MS/EE/Polytech Inst of NY; BEE/EE/City College of NY. *Born:* Dec 14 1930. Native of Brooklyn NY. 1952-53 Res Asst Microwave Receivers, Univ of Mich; 1954-58 Sr Staff Engr W L Maxson Corp; 1958-65 Res Sect Hd Phased Array Radar, Sperry Gyroscope Co NY; 1965-1981, Mgr Sensor Systems, Sperry Corporate Res Center Sudbury Mass. Reg PE NY & Mass. USAF 1953, Res 1953, Capt.; Fellow IEEE; K C Black Award, Best Paper NEREM. More than 100 pub papers & patents. Enjoy golf & fresh-water fishing. *Society Aff:* IEEE, ROA, NSPE, ADPA.

Ross, Gilbert I
P.O. Box 526, Rye, NY 10580-0526
Position: Principal. *Employer:* Ross & Co, Engineers & Consultants. *Education:* BS/ME/Yale Univ. *Born:* March 24 1902, New York NY. 1923-25 with Bridgeport Brass Co, Conn; 1925-28 English & Mersick Co, New Haven; 1928-32 MacDonald Bros Inc, Boston; 1932-38 V Pres, Ch Engr & Dir R A Lasley Inc, NYC; 1938-40 partner Ross, Wells & Co engrs NYC; 1940-44 served to Col AUS, col Res retired, decorated Legion of Merit; 1944-46 Exec V Pres, Genl Mgr Unexcelled Mfg Co NYC; 1946 Partner /Ross & Co, Engrs & Cons NYC. Reg PE NYS, MO. Fellow ASME; Mbr AICE (Council 1960-66, Pres 1962), TROA, Fellow ACEC. Author articles, contributor handbooks, served on various prof cttes. *Society Aff:* ASME, TROA.

Ross, Hugh C
Business: Ross Engineering Corp, 540 Westchester Drive, Campbell, CA 95008
Position: President & Chief Engineer. *Employer:* Ross Engineering Corp. *Education:* BS/EE/Stanford Univ. *Born:* 12/31/23. Instr elec San Benito Cty HS & Jr Coll Hollister Cal 1950-51. Ch Engr Vacuum Power Switches Jennings Radio Mfg Corp San Jose CA 1951-62; Ch Engr Vacuum Power Switches ITT Jennings 1962-64; owner Ross Engrg High Voltage Consultants 1964-present, Pres & Genl Mgr Ross Engrg Corp Campbell CA 1964- present. Reg P E CA. Fellow IEEE, Chmn Santa Clara Valley subsect 1960-61; mbr Amer Vacuum Soc; ASM; Pacific Coast Elec Assn. Many patents on high voltage devices. Major devel in high power vacuum relays, vacuum switches, vacuum circuit breakers, high voltage voltmeters, digital & analog. Wide band voltage dividers, energy generation devices. Many articles to IEEE, Transactions & other tech journals. *Society Aff:* IEEE, AVS, ASM, PCEA.

Ross, Ian M
Business: 600 Mountain Avenue, Murray Hill, NJ 07974
Position: Pres. *Employer:* Bell Laboratories. *Education:* PhD/EE/Cambridge Univ; MA/EE/Cambridge Univ; BA/EE/Gonville & Csius College Cambridge Univ. *Born:* Aug 15, 1927 Southport Eng. Joined Bell Labs 1952, engaged in dev of wide variety of semiconductor devices; appointed Man Dir 1964 then elected Pres Bellcomm Inc 1968, Bell Sys subsidiary which provided sys planning support for Apollo manned space flight prog; returned to Bell Labs 1971 as Exec Dir of network planning div; became V P responsible for sys planning of Bell System network & planning & dev of customers serv & equip. Assumed pos as Exec V P 1976 respon for dev of customer serv, switching & transmission sys, & electonics & tech. Elected Pres of Bell Labs in Apr 1979. *Society Aff:* IEEE.

Ross, Irving
Home: 22908 Newport, Southfield, MI 48075
Position: President. *Employer:* Thermatek Inc. *Education:* BSChE/ChE/Wayne State. *Born:* 1918. Graduate work in math & economic analysis during mid 60's. Early in career did pilot plant dev work for Ford Motor Co & Sherman Labs. Pres of Thermatek Inc for past 20 yrs. Thermatek Inc builds & rebuilds indus heat furnaces. Hobbies: classical music; gardening; working in own lab. *Society Aff:* AIChE, ESD.

Ross, Jack H
Business: ASD/ENEU, WPAFB, OH 45433
Position: Chief of Clothing Division. *Employer:* USAF Air Force Systems Division. *Born:* April 1924. Native St Louis Mo. BS Textile Engrg Texas Technological Coll. Entered USAF 1950 after 2 yrs as dev engr with Bordan Mills Kingsport Tenn. Responsible for dev of textile materials for all USAF parachute sys from personnel to aircraft deceleration. Initiated USAF prog on nonflammable textile materials which resulted in increased protection to aircrew personnel. Assumed managerial responsibility for USAF prog to develop fibrous reinforcements for plastic & metallic matrix composites. Currently responsible for USAF distinctive & functional uniform prog. Has written more than 50 tech papers & reports & presented more than 40 papers on fibers, fabrics & test procedures. Currently USAF rep to NMAB Ctte on 'Flammability of Polymeric Materials.'.

Ross, James L
Business: P. O. Box 8405, Kansas City, MO 64114
Position: Partner *Employer:* Black & Veatch *Education:* BS/ME/Univ of KS *Born:* 2/10/26 Mr Ross is a native Kansan who joined Black & Veatch in 1951. His engrg experience includes construction supervision, mech design, engrg studies and project mgmt. He has managed unique projects resulting in new areas of practice for the firm. His success in managing the firm's professional services for industry clients resulted in founding of the firm's Industrial Div. He became a partner in 1975 and is a registered prof engrg in 22 states. *Society Aff:* ASME, ASHRAE, NSPE, ACEC, WPCF.

Ross, John H
Home: 9 Cobble Hills, Islington, Ontario, Canada M9A 3H6
Position: Retired. *Education:* BSc/Mech Eng/Queen's Univ. *Born:* 6/11/08. With Natl Carbon Co Ltd, Toronto; Works Engr with Eveready SA, Buenos Aires; 1938 returned to Toronto gaining varied exper with John Inglis Co Ltd, ON Hydroelectric Power Comm & Defense Indus Ltd Montreal; after 4 yrs as Works Engr with Small Arms Ltd Long Branch ON was with G Lorne Wiggs, Cons Engr Montreal; 1945 entered private practice in Toronto, 1957 Inc, active in engg design of institutional, commercial & industrial bldgs specializing in hospitals & sch. Returned to private practice 1976. Spec int in prof organizations. Distinguished Service Award 1968 ASHRAE, Fellow 1972; Presidential Citation 1979; Councillor 1967-69 AICE, ASHRAE - Presidential Representative for Canada, 1976 to date. Retired from active practice, April 1980. *Society Aff:* ASHRAE, IEEE.

Ross, Julius
Home: 2510 Virginia AvNW 710N, Washington, DC 20037
Position: Retired (Consultant) *Education:* BEE/Elect Engrg/Polytechnic Inst of Brooklyn. *Born:* 7/30/11. 1935 Radio Engr, Bartlett Arctic Expedition; 1936-38 Radio Engr, broadcasting sta; 1939-1944 US Radio Inspector & Monitoring Officer-in-Charge, engaged in inspection, enforcement & intelligence work of FCC; 1944-45 Radio broadcast facilities design & const, Office of War Information; 1945-47 Proj Engr, broadcast equip design, Federal Telephone & Radio (IT&T); 1947-75 Ch Engrg Div, Voice of America (USIA), was Exec Asst to Dir for Engrg & Planning, was largely respon for planning, design & const of world-wide high-powered broadcasting facilities of Voice of Amer having a total cost of about 160,000, 000 dollars. Reg PE 4642E Dist of Col. Enjoy tennis & piano. From 1978-79 was Tech Adv to Deputy Dir for Mgmt of ICA (Int Comm'ns Agency). Ret 2/24/79 from ICA & now private consultant. *Society Aff:* AES, WSE.

Ross, Keith E
Home: 29 Oak Drive, North Manchester, IN 46962
Position: Retired *Education:* ET/Electrical/GE Apprentic School; BS/Math & Physics/Manchester College *Born:* April 23 1909 Fellow, Certified Quality Engr & Certified Reliability Engr in ASQC. Recipient of Ed J Oakley Award by Midwest Conference Bd ASQC 1972; Alumni Award from Manchester Coll 1974; Achievement & Service Award from NE Ind Sect ASQC 1974. P Mbr 8 years Bd/Dir ASQC, presently Past Mbr Edwards & Oakley Awards Cttes, also Mbr Quality Progress Review Bd ASQC. Pioneer in quality control & retired after 37 years in quality control activities in mech & electrical prods, US weapons & meat processing. *Society Aff:* ASQC, FWEC

Ross, Monte
Home: 19 Beaver Dr, St. Louis, MO 63141
Position: Director, Laser Communications *Employer:* McDonnell Douglas Astronautics Co *Education:* MSEE/Electronics/Northwestern; BSEE/Elec Engg/Univ of IL. *Born:* 5/26/32. Chicago Ill. Dev Engr for Chance Vought; St Electrical Engr Motorola; Proj Engr, Assoc Dir of R&D & Dir of R&D, Hallicrafters; Mgr Laser Tech, McDonnell Douglas Astronautics East, Director of Laser Communications. Guest lecturer various univ; cons NSF; Fellow IEEE for leadership & contribution to dev of high rate optical communications system; McDonnell Douglas Fellow Award 1985; Technical Contribution Award 1976 St Louis Sect AIAA for contribution to dev of laser communication tech. Author *Laser Receivers* 1966, J Wiley & Sons; Tech Editor *Laser Application* series pub by Acad Press (Vol I 1971, VOl II 1974 Vol III 1977 Vol IV 1980). Num patents in field including lasers & modulation techniques. *Society Aff:* IEEE, AIAA.

Ross, Richard D
Business: 1000 Conshohocken Rd, P.O. box 578, Norristown, PA 19404 *Employer:* Chemical Waste Management, Inc. *Education:* BS/CE/Drexel Univ. *Born:* 1/28/28. Formerly Pres of Thermal Res and Engg Corp, now Trane Thermal Co, mfg of incineratin systems for destruction of liquid & gaseous wastes. Served as VP of Marketing for Rollins Environmental, Inc, which contracts for disposal of hazardous chem wastes. Formerly VP Read-Ferry Co a Private consultant to industry in areas of pollution control equip & hazardous waste mgt, including carcinogenic mtls. Author of more than 25 papers presented to various technical societies. Author/ editor of three technical books. Currently Eastern area Mgr for Chemical Waste Management, Inc., a major contract disposal organization for hazardous waste materials. *Society Aff:* AIChE.

Ross, Stephen M
Business: 300 Orange Ave, W Haven, CT 06516
Position: Assoc Prof. *Employer:* Univ of New Haven. *Education:* PhD/-/Johns Hopkins Univ; BEME/-/NYU. *Born:* 6/20/45. *Society Aff:* ASME, AmPhysSoc, New York Academy of Sciences

Ross, Stuart T
Business: Meyer Indus - POB ll4, Red Wing, MN 55066
Position: Vice President/Technical Director. *Employer:* Meyer Industries, div of IT&T Corp. *Education:* PhD/Metallurgy/Purdue; MS/Met Engrg/Purdue; BS/Met Engrg/Purdue. *Born:* July 1923, Oshkosh Wisc. BS, MS & PhD Metallurgy from Purdue. 1942-46 served with USAAF, Lt Col USAFR-ret; 10 years early automotive tech background; 1965-69 Dir Engr & Dev, UOP-Wolverine Tube; 1969-74 V Pres Advanced Products, Crucible Speciality Metals Div; 1974-76 V Pres Engrg & Res, Internatl Mill Service; currently respon for all technology at ITT Meyer Industries including design engrg, materials, processes & facilities engrg, quality control & engrg/drafting services. Fellow ASM; Distinguisned Engineering Alumni Award, Purdue 1972. 5 patents, others pending; 25 tech & mgmt publ. Failure Analyses of Metallic Structures/Weldments 20 yrs. *Society Aff:* AMS, AIME, SSPC.

Rosselli, Charles A
Home: 69 Mt Vernon St, Somerville, MA 02145
Position: Prof *Employer:* Wentworth Inst of Technology. *Education:* Doctoral Cand/ Mech Eng/Tufts Univ; MS/Eng/Northeastern Univ; BS/Eng/Northeastern Univ; Assoc/Civil Engg/Wentworth Inst. *Born:* Sept 13 1944, Somerville Mass. BS & MS from Northeastern Univ; Assoc Degree from Wentworth Inst. Field Engr & Lab Engr with Golder-Gass Assoc Cons Engrs, specializing in sub-surface soil investigations throughout New England; presently Prof Dept Civil Engrg Technology at Wentworth Inst of Technology, respon for dev of course material & instruction in civil const, civil engrg technology, civil & hwy programs, respon for entire environ prog including budgeting & grant proposal work. ASEE-Dow Chemical Outstanding Young Faculty Award 1974. Grant recipient from HEW Dept. for Wentworth Environmental Education Program 1978. *Society Aff:* ASCE, ASEE, SAME, NEWPCA, NEWWA, AAAS.

Rossi, A Scott
Business: 109 Suburban Ave, State College, PA 16801
Position: Pres. *Employer:* Self. *Education:* BS/Civ Engg/Fairleigh Dickinson Univ. *Born:* 12/9/28. Actively engaged Municipal Engg & Technical Services encompassing rds, sts, water, sewer, surveying, land dev & environmental control for approximately twenty five yrs. Veteran, USMC Korea conflict Former Pres, Universal Engg Servics, present Owner/Pres, A Scott Rossi & Assoc, Civ Engrs, Surveyors & Technicians. Engaged all aspects of Municipal Engg services & environmental engg from study phase through & including dev phase. Present projs includes storm water control, water supplies & distribution, waste water mgt, land & control surveys, land dev engg & soil erosion & sedimentation control with USDA Soil Conservation Service. *Society Aff:* NSPE, PSPE, NJSPE, PEPP, ASCE, ACSM.

Rossi, Boniface E
Business: 81 Georgetown Rd, Weston, CT 06883
Position: Pres *Employer:* BER Assocs. *Education:* ME/Mech Engg/Stevens Inst of Tech; Cert in Weld Engrg/Rensselaer Poly Inst. *Born:* March 1913. Attended sch in Hoboken NJ. 1936-39 Industrial Engr LO Koven & Bro Inc, Jersey City NJ; 1939-42 Founder & Dir Welding Div, Delehanty Inst NYC; 1943-46 Standards Materials & Processes Engr Columbia Aircraft Corp, Valley Stream NY; 1947 Engr-in-Charge Welding Lab Genl Electric, Bloomfield NJ; 1947-54 Exec Secy Pressure Vessel Res Ctte, Welding Res Council NYC; 1955-60 Editor *Welding Journal* & Secy Tech Papers Ctte, Amer Welding Soc NYC; 1960-78 Managing Dir & Editor SESA, Westport Conn. 1979-81 Editorial Cons, SESA. 1979 Pres, BER Associates, Weston, CT. 1970 Editorial Cons, Publications Assocs, Wilton Conn; 1949-51 Lecturer welding engrg Stevens Inst of Tech, Industries Training Sch; 1941- , Cons Engr; author & editor. Samuel Wylie Miller Memorial Medal, AWS 1960; Fellow, SESA. Licensed PE in NJ. *Society Aff:* SESA, ASME, ASEE, ASM, SES, AISA, ΠΔΕ

Rossie, John P
Business: 660 Bannock, Denver, CO 80204
Position: Director of Special Projects *Employer:* R W Beck & Assocs, Cons Engrs.

Rossie, John P (Continued)
Education: BS/ME/Univ of NE. *Born:* Oct 24 1917. 1942-46 active duty USNR, Asst Engrg Officer USS NJ; 1947- 53 Omaha Public Power Dist, power plant design; 1953-57 Partner Nance & Rossie cons engrs, in charge of mechanical design; 1957- , R W Beck & Assocs, respon for design of utility projs including steam power plants, substations & transmission lines, & for economic feasibility studies of generating plants & associated transmission systems. In charge of Special Studies of Co-generation and Heat Rejection Systems for Power Plants. *Society Aff:* ASME, NSPE, ASTM.

Rossington, David R
Business: College of Ceramics, Alfred University, Alfred, NY 14802
Position: Prof & Dean, Sch of Engrg *Employer:* NY State Coll of Ceramics *Education:* PhD/Physical Chem/Univ of Bristol-England; BSc/Chem/Univ of Bristol-England *Born:* in London, England. After receiving PhD was awarded a Fulbright Travel Schlrship for Post-Doctoral work in US (1956-58). 1958-60 Technical Officer with Imperial Chemical Industries Ltd, England. On faculty of NY State Coll of Ceramics at Alfred Univ since 1960. 1972-73 American Council of Educ Academic Fellow. 1976 received Suny Chancellor's Award for Excellence in Teaching. Pres: Ceramic Educ Council, 1981-82. Research interests include Surface Studies, Nuclear Waste Materials. 34 research publications. Other interests include travel and photography. *Society Aff:* NICE, ACS, ASEE, ACerS

Rossoff, Arthur L
Home: 14 Stafford Dr, Huntington Station, NY 11746
Position: retired *Education:* MEE/-/Polytechnic Inst of Brooklyn; BEE/-/City College Of NY. *Born:* April 1922. 1950-68 with Genl Instrument Corp, most recent positions of Dir of Advanced Dev Lab & V Pres Marketing; 1968-72 with Grumman Aerospace Corp on sr staff & space programs, particularly Apollo/LM; 1973-87 respon for corporate relations for Polytechnic Univ. Fellow & Mbr Bd/Dir IEEE; Mbr AIAA, NSPE, PRSA & NSEE. Reg PE by eminence NY. P Chmn Space Shuttle Task Force & Solar Energy Res Consortium of Long Island; Exec VP L I Forum For Technology; Recipient IEEE Centennial Medal, 1984. *Society Aff:* IEEE, AIAA, NSPE, ASEE

Rosson, Harold F
Business: Dept of Chem & Petrol Engg, 4006 Learned Hall, Lawrence, KS 66045
Position: Chrmn. *Employer:* Univ of KS. *Education:* PhD/CE/Rice Univ; BS/CE/ Rice Univ. *Born:* 4/4/29. Served with USAF 1951-54. With Univ of KS since 1957. Chrmn of Dept of Chem and Petrol Engg 1964-70; 1979-present. Ford Fdn Resident in Engg Practice, Upjohn Co, 1967-68. *Society Aff:* AIChE, ACS.

Rostoker, William
Home: 2052 West 108 Place, Chicago, IL 60643
Position: Professor of Met & Bioengrg. *Employer:* University of Illinois. *Education:* PhD/Metallurgy/Lehigh Univ; MASc/Metallurgy/Univ of Toronto; BASc/ Metallurgy/Univ of Toronto. *Born:* 6/21/24. 1948-50 Lecturer, Birmingham (England); 1950-51 Asst Prof IIT; 1951-65 Asst Dir metals & Sr Sci Advisor IITRI; since 1965 Prof Univ of IL, 1968-69 Acting Dean Graduate College. 1967-71 Mbr Bd/Dir Automatic Screw Machine Prod; 1966-70 Mbr Natl Mat Adv Bd; 1971-77, Cons & Mbr Army Sci Adv Panel; Fellow ASM, Kappa Delta Award, Amer Acad Orthoped Surg 1970; Visiting Prof Orthoped Surg, Rush Medical College 1972- . Author, Co-author 4 books, 130 res publ in structure & transformations, alloy dev, mfg, fracture analysis, biomedical engg, & archeometallurgy. *Society Aff:* ASM.

Roth, Bernard
Business: Mech Engrg Dept, Stanford, CA 94305
Position: Professor. *Employer:* Stanford University. *Education:* PhD/Mech Eng/ Columbia Univ; MS/Mech Eng/Columbia Univ; BS/Mech Eng/City College of NY. *Born:* 5/28/33. Taught at CCNY & Hunter Coll before joining Stanford Univ Faculty, 1971 attained present rank of Professor of Mechanical Engrg. Pub widely in area of kinematics, especially as applied to design of mechanical linkages and robotic devices. Developed new courses in robotics, kinematics, computer aided design & interpersonal relations. Lectured in many countries, Visiting Professor at Technological Univ of Delft 1968-69, Univ of the Negev 1973 and Indian Inst of Science, India 1984. The Joseph F. Engelberger Award 1986, Fellow ASME 1985. ASME Melville Medal 1967, Mechanisms Award 1982, Machine Design Award 1984; Chmn Mechanisms Ctte & Conference 1972; Chmn 2nd Internatl Symposium on Robots & Manipulators 1976; Pres IFToMM 1980-83; Assoc Editor Journal of Machine & Mechanism Theory; Chmn Design Engrg Division ASME, 1981-82. Dir Integrated Automation, and Center for Economic Conversion. *Society Aff:* ASME, IEEE

Roth, J Paul
Business: T J Watson Res Ctr, Yorktown Heights, NY 10598
Position: Research Staff Member. *Employer:* IBM. *Education:* PhD/Math/Univ of MI; MS/Math/Univ of MI; BME/ME/Univ of Detroit. *Born:* 12/16/22. 1953-55 Pierce Instructor of Mathematics, Univ of CA, Berkeley; 1954-55 Cons Shell Dev Co; 1955 Cons GE; Mbr Electronic Computer Proj, Institute for Advanced Study, Princeton NJ; 1957; Res Mbr IBM Res Div; Visiting Prof Princetion Univ Mathematics Adjunct Prof EE Dept, Columbia Univ 1973-74, Adjunct Prof EE Dept City Univ of NY, Fellow, IEEE. Best known for methods in diagnosis of computer failures, automatic logic design. Native of Detroit MI. *Society Aff:* IEEE, AAAS.

Roth, J Reece
Business: 409 Ferris Hall, University of Tennessee, Knoxville, TN 37996-2100
Position: Prof of Electrical Engineering *Employer:* Univ of TN. *Education:* PhD/Engg Phys/Cornell Univ; SB/Phys/MIT. *Born:* 9/19/37. Worked at the NASA Lewis Res Ctr from 1963 to 1978, and is Prof of Elec Engrg at Univ of TN, Knoxville. While at NASA, pioneered in the application of superconducting magnet facilities to high temperature plasma res, and initiated res on the elec field bumpy torus concept, an approach to creating a plasma of fusion interest by applying strong radial elec fields to a toroidal plasma. Elected Fellow of IEEE January, 1981. Senior Member, AIAA, 1986. Published Textbook, "Intoduction to Fusion Energy–, 1986. Research interests include fusion energy, electric field dominated plasmas, and interactions of radio frequency radiation with plasmas. *Society Aff:* AAAS, IEEE-NPSS, ANS, APS-PPD, AIAA, AIA.

Roth, John A
Business: Box 1574, Sta B, Nashville, TN 37235
Position: Professor, Chemical & Environmental Engrg *Employer:* Vanderbilt University. *Education:* PhD/Chem Engg/Univ of Louisville; MChE/Chem Engg/Univ of Louisville; BChE/Chem Engg/Univ of Louisville *Born:* May 1934, Louisville Ky. 1960-62 Res Officer (1st Lt) USAF; since 1962 with Vanderbilt Univ, Nashville Tn: 1962 Asst Professor, 1968-70 Asst Dean, 1970-72 Assoc Dean, since 1969 Professor; 1972-75 Chmn Div Chemical, Fluid & Thermal Sciences, 1974-1980; Dir Center for Environ Quality Mgmt. 1980-present, Professor of Chemical & Envir. Engr; 1986-present, V P, Chemical & Environmental Services. Participant duPont Engr Dept Year-in-Indus prog 1967-68. Cons & Res in industrial waste treatment & hazardous materials. Bd/Dir House Boating Corp of AMer, Gallatin Tenn 1971-75; Bd/Dir Dede Wallace Community Mental Health Center 1975-77. Commissioner, Ky Bureau of Environmental Protection, 1977-1978. *Society Aff:* AIChE, WPCF, ΣΞ, NSPE, ACS, TBΠ, ASEE, ES

Roth, Robert L
Business: Rt 1 Box 587, Yuma, AZ 85364
Position: Research Associate. *Employer:* University of Arizona. *Born:* April 24 1943, Bridger Mont. BSAE Mont State Univ; MSAE Univ of Ariz, agricultural background. Prod Analyst for Ford Motor Co, Tractor Div; 1971 joined Univ of Ariz as Res Assoc, currently involved in irrigation res at Yuma Branch Experiment Station.

Roth, Wilfred
Home: 185 S Cove Rd, Burlington, VT 05401
Position: Res Prof in Orthopedics *Employer:* Univ of Vermont. *Education:* PhD/Physics/MIT; BS/EE/Columbia Univ. *Born:* 6/24/22. 1943-46 MIT Radiation Lab; microwave radar & training devisces; 1947-48 Reiber Res Lab, geophysical instrumentation; 1948-49 Harvey Radio Labs, low frequency loran; 1949-50 Raytheon, automatic machine tools & ultrasonics; 1950- 55 Rich-Roth Labs, ultrasonics; 1952-55 Ultra-Viscon Corp, viscometry; 1955-66 Roth Lab Phys Res, instrumentation; 1964- , Prof Elec Engg Univ of VT, 1964-75 Dept Chmn. Fellow Acoustical Soc of Amer 1959, IEEE 1968. 25 Us patents, 115 foreign patents in instrumentation & ultrasonics. Extensive industrial cons in instrumentation & biomedical engg, ultrasonics. Hobbies: flying, sailing, skiing, tennis & squash. Emeritus Prof Elec Engg 1986. Res Prof in Orthopedics 1987. *Society Aff:* IEEE, ASA, AAAS.

Rothbart, Harold A
Business: Dept of Indust Mgmt, Dominguez Hills, CA 90747
Position: Chair of Industrial Mgmt *Employer:* California State College. *Born:* Dec 1917, Newark NJ. BS NJ Inst of Tech; MS Univ of Penn; Dr Engrg Technische Hochschule Munchen, Germany. 1946-61 Professor Mechanical Engrg & Indus Mgmt CCNY; 1961-75 Fairleigh Univ, 1961-72 also Dean Coll of Sci of Engrg; 1939-43 Marine Engr, Phila Nayv Yard. Lectured in US, Europe & Asia on high-speed machinery, mgmt & creativity. Books: 'Cams,' 'Bent Mirrors,' 'Mechanical Designs & Systems Handbook' & 'Cybernetic Creativity.' Presently Professor of Industrial Mgmt, Cal State Coll at Dominguez Hills. Enjoy tennis & chess.

Rothberg, Henry M
Business: 1 Laticrete Park North, Bethany, CT 06525
Position: President. *Employer:* Laticrete International Inc. *Education:* BS/ChE/Univ of SC. *Born:* June 1922. BSChE Univ of S Carolina. 1945-55 Design & Engrg Cons wood technology; 1956-66 Cons to Uniroyal Inc, USA & overseas divs, dev & application of polymers for concrete; Founder & Pres Laticrete Internatl Inc, mfg & marketing polymer concrete materials & technology, work involves world wide travel & dev of new materials & const techniques. Author of papers on latex & polymer concrete; US & Canadian patents on polymer concrete; recognized as 1 of world's leading experts in polymer modified concrete technology. *Society Aff:* ASChE, ACS

Rothermel, U Amel
Home: 4507 Tenella Rd, New Bern, NC 28560
Position: Professor (Retired). *Employer:* City University of New York. *Education:* MS/Indust Engr/Columbia Univ; BS/Elect Engr/Univ of WI. *Born:* June 16, 1903 Emmetsburg Iowa. Licensed P E Wis & N Y (inactive). Power Sales Engr 1928-31; Indus instrument & control sales engrg Minneapolis-Honeywell & Republic Flow Meters Co 1932-56; Prof CUNY 1956-73 - Chmn Mech Engrg Tech Dept 1959-68. Mbr ASME: P Secy Region II, P V Pres, elected Fellow 1970. Retired 1973. Active in Presbyterian Church & Civitan Club of New Bern, New Bern Preservation Fdn. Enjoys travel, golf & fishing. *Society Aff:* ASME.

Rothschild, Paul H
Business: PO Box 1035, Toledo, OH 43666
Position: VP, Worldwide Bus Dev, Closure & Plastic Prods. *Employer:* Owens-Illinois Inc. *Education:* MSE/Plastics/Princeton Univ; BS/Chem Eng/MIT. *Born:* Feb 1937; native of Morristown NJ. 2nd Lt USA Chem Corps 1959. Joined Owens-Ill Inc 1961 as Rheologist, Corporate Res; 1966-73 Section Ch, Advanced Materials Engrg, Corporate Res; 1973-76 V Pres Corporate Staff, Dir Organic Technology & Tech Dir Plastic Beverage Container Prog; assumed position of V Pres & Tech Dir Plastic Prods Div 1976. Mbr AIChE, Soc of Plastic Engrs & Soc of Rheology. Married Rona Shavin 1962; sons - Stephen (18), James (16) & Robert (14). *Society Aff:* SPE.

Rothstein, Jerome
Business: Dept of Computer & Info Science, 2036 Neil Ave. Mall, Columbus, OH 43221
Position: Prof *Employer:* Ohio State Univ *Education:* AM/Physics/Columbia Univ; BS/Physics/CCNY *Born:* 12/14/18 1938-42 AM Physics and all courses for doctorate. 1942-1957 US Army Research & Dev Lab, Belmar, NJ. 1957-1961 EG&G, Inc, Boston MA. 1961-1962 Maser Optics, Boston MA. 1962-1967 LFE, Inc, Boston MA. 1967- OSU, Columbus, OH. About 200 publications, patents, book reviews. Sr Life Mbr IEEE and Distinguished Visitor IEEE Computer Soc 1981-82. Best Paper Awards, 1976 and 1977 Intl Conf on Parallel Processing. Author of book *Communication, Organization, and Science* 1958. *Society Aff:* AAAS, IEEE, APS, BS

Rotman, Walter
Home: 17 Gerald Rd, Brighton, MA 02135
Position: Electronic Scientist. *Employer:* Mass. Institute of Technology *Education:* MS/EE/MIT; BS/EE/MIT. *Born:* Aug 1922; native of Boston Mass. Served with Army Corps 1942-45. Res scientist & BranchCh at Air Force Cambridge Res Labs (now RADC/DET), Hanscom AFB 1948-1980. Technical Staff Member at M.I.T. Lincoln Laboratory, Lexington, Ma 1980-. Specialties in electromagnetics, microwave antennas & reentry technology. Fellow IEEE 1969; Mbr URSI & Sigma Xi. *Society Aff:* IEEE, ΣΞ, URSI.

Rotolo, Elio R
Home: 4369 LaBarca Dr, Tarzana, CA 91356
Position: Sr VP *Employer:* Financial Corp of Amer *Education:* BS/Indus Eng/Lehigh Univ. *Born:* 1/2/24 Native of New York City. B.S. Indust Engrg Lehigh Univ; Principal Arthur Young & Company 1960-1970., Director manufacturing engineering ITT Corp 1970-1975., Vice Pres, Industrial Engineering Security Pacific National Bank 1975-82; Pres Rotolo & Whitney, Inc 1982 to Present., Sr VP, Financial Corp of America 1984 to present. Retired, US Army Reserve 1st Lt., infantry. Fellow American Inst of Industrial Engineers (National President 1967-68), Fellow, Inst for the Advancement of Engrg, Director Engineers Joint Council 1968-1970, Mbr, NY State Soc of Prof Engrs, Chrmn, LA County Productivity Commission; Chairman, Productivity Council of the Southwest, Director Office Automation Council, Registered Professional Industrial Engineer, California. Republican County Committeeman, Union, NJ (1955-60). Listed in "Who's Who in the World–, "Who's Who in the East–, "The West, & Finance and Industry–. *Society Aff:* IIE

Rotty, Ralph M
Business: Department of Mechanical Engineering University of New Orleans/ Lakefront New Orleans, LA 70148
Position: Professor *Employer:* Univ. of New Orleans *Education:* PhD/M.E./Michigan State University; MS/M.E./California Institute of Technology; MS/Meteorology/ California Institute of Technology; BS/E.E./University of Iowa *Born:* 08/01/23 Born St. Louis, MO; Attended Missouri Univ, 1940-43; Cadet/First Lieutenant (Weather Officer) USAAF, 1943-46. Instructor/Assoc Prof (ME) Michigan State Univ, 1949-58. Prof and Head, Mech. Engr. Dept. Tulane Univ, 1958-66. Prof (Engrg)/Dean, School of Engrg, Old Dominion Univ, 1966-72. Sr Postdoctoral Assoc, Air Resources Lab, NOAA, 1972-74. Sr Scientist, Inst for Energy Analysis, Oak Ridge Associated Univ, 1974-86. Analyzed national and world energy problems, especially related to atmosphere and potential global climate change. Developed and maintained time series for carbon dioxide emissions from fossil fuel combustion. Prof. of M.E, Univ of New Orleans, 1986-present. *Society Aff:* ASME, AAAS, AMS

Roudabush, Byron S
Business: 65 K St NE, Washington, DC 20002
Position: President & Owner. *Employer:* Byron Motion Pictures Inc. *Born:* Aug 1907; raised in Pa, moved to Wash D C 1930. Lehigh Univ, Amer Acad of Dramatic Arts. Managed chain of 5 radio stores; 1938 founded Byron Motion Pictures (Pres & Owner) - headed production & later added Motion Picture Lab; owned & managed 2 motion picture labs in N Y City. Founded Assn of Cinema Labs & served 2 yrs as Pres & 4 yrs as Secy. Wrote ACL Handbook of Recommended Stan-

Roudabush, Byron S (Continued)
dards & Procedures. Designed High Light exposure meter 1937; the fader for a contact printer 1942 & ultra-sonic film cleaner about 1950. Served 20 yrs as officer of SMPTE: Conference V Pres, Treas, Executive V Pres, Pres & P Pres. Hon Mbr Edison Pioneers. SMPTE Life Fellow. *Society Aff:* SMPTE, ACVL

Rouse, George C
Home: 2801 Pierce St, Denver, CO 80214
Position: Retired *Education:* PhD/Struct Eng/Univ of CO; MS/CE/Univ of CO; BS/CE/Univ of CO *Born:* 04/20/09 Native of Mt Vernon, NY. Supervising Structural Engr with the US Bureau of Reclamation from 1934 to 1972 specializing in structural engrg research of dams and appurtenant works, including: structural behavior measurements for static and dynamic loadings (incl subsurface nuclear blasts), structural model studies, analyses for dynamic E. Q. loadings, and rock mechanics. Served during WWII as Aircraft Service Div Officer aboard an aircraft carrier in Asiatic-Pacific area. Awarded Bronze Star with Combat "V–. Retired from USBR in 1972. *Society Aff:* ASCE(F), ΣΞ, EERI

Rouse, Hunter
Home: 10814 Mimosa Dr, Sun City, AZ 85373
Position: Carver Prof & Engrg Dean Emeritus. *Employer:* The University of Iowa, Iowa City *Education:* Dr es Sci Phys/Fluid Mech/Univ of Paris; DrIng/Hydraul/Karlsruhe Tech Univ; SM/Civ Engrg/MIT; SB/Civ Engrg/MIT *Born:* Mar 1906 Toledo Ohio. Doktor-Ingenieur Karlsruhe Tech Univ; Docteur es Sciences Univ of Paris; Honorary Dr. -lng. Univ. of Karlsruhe; SB & SM MIT. Asst in Hydraulics MIT 1931-33; Instr Columbia 1933-36; Asst Prof CalTech; Assoc Hydraulic Engr SCS 1936-39; Prof Iowa 1939- ; Dir Iowa Inst of Hydraulic Res 1944-66; Dean of Engrg Iowa 1966-72; Carver Prof & Dean Emeritus 1972- Visiting Prof. summers Colorado State Univesity 1976--- . MIT traveling Fellow, Germany 1929-31; Fulbright Res Scholar, France 1952-53; NSF Sr PostDoc Fellow, Goettingen, Rome, Cambridge & Paris 1958-59; Australian-Amer Lectr 1973. PRC Ministry of Water Conservancy Lecturer 1981 Missions: Smith-Mundt to Egypt, NSF to India, NAS to Romania, UNESCO to Thailand, SEED to S Amer. Led or participated in USSR, Japan, & PRC exchanges. Norman & Karman Medals ASCE; Westinghouse & Bendix Awards ASEE. Married 1932; 3 children; 6 grandchildren. Mbr NAE 1966; Hon Mbr ASME 1968; Hon Mbr ASCE 1973; Hon Mbr IAHR 1985. *Society Aff:* ASCE, ASME, IAHR, NAE, AAAS

Rouse, John W, Jr
Home: 1605 Northridge Dr, Arlington, TX 76012
Position: Dean of Engr. *Employer:* Univ of Texas at Arlington *Education:* PhD/EE/Univ of KS; MS/EE/Univ of KS; BS/EE/Purdue Univ *Born:* 12/7/37. Dev engr for Bendix Corp, Kansas City, MO 1959-64. Res Assoc and Instr, Univ of KS 1964-68 while earning PhD in elec engg. Joined TX A&M Univ 1968 eventually become dir of Remote Sensing Ctr and Prof of Elec Engg. Pres of IEEE Geoscience Electronics Group 1975. Served as resident consultant to NASA, Wash, DC 1975-77. Logan Prof of Engg, and Chrmn of Electrical Engrg Univ of MO- Columbia in 1978-81. Appointed Dean of Engr, Univ of TX at Arlington in 1981. US Delegate to Intl Congress of URSI in 1972, 1975 and 1978 held in Europe and S America. Mbr of Natl Academy of Sci Study Panels and Naval Studies Bd 1978-86. *Society Aff:* IEEE, URSI, ASEE

Rouse, William B
5550A Peachtree Pkwy, No. 500, Technology Park/Summit, Norcross, GA 30092
Position: Prof *Employer:* GA Inst of Tech *Education:* PhD/Systems Eng/MIT; MS/Systems Eng/MIT; BS/ME/Univ of RI *Born:* 01/20/47 Prof of Mech and Industrial Engr at Univ of IL until 1981. Prof of Industrial and Systems Engrg at GA Inst of Tech since 1981. Pres of Search Tech Inc, a research and consltg firms, since 1980. Author of *Systems Engrg Models of Human-Machine Interaction* co-author of *Mgmt of Library Networks: Policy Analysis, Implementation, and Control*, coeditor of *Human Detection and Diagnosis of System Failures*, editor of *Advances in Man-Machine Systems Research*. (A research annual). Coeditor of "System Design: Behavioral Perspective on Designers, Tools, and Organizations." Recipient of 1979 O. Hugo Schuck Award from American Automatic Control Council, 1984 IEEE Centennial Medal and Certificate. 1986 Norbart Wiener Award from Systems, and Cybernetics Soc. Past Pres of IEEE Systems, Man, and Cybernetics Soc. *Society Aff:* IEEE, HFS, IIE, AMA

Roush, Maurice D
Home: 7212 Zircon Dr S.W, Tacoma, WA 98498
Position: VP *Employer:* Daniel, Mann, Johnson & Mendenhall *Education:* BS/Gen Engrg/US Military Acad; MS/CE/Univ of IL *Born:* 2/2/28 Mr Roush, (Brig Gen Ret) served in responsible engrg positions of increasing magnitude and scope in the Army Corp of Engrs, with eleven years in the Corps Civil works and military design and construction field. He also commanded military organizations up to Brigade level, was the United Nations Command Engr in Korea and served as the last american Lt Governor of Okinawa. His last military tour was as Division Engr of the Pacific Ocean Division, responsible for Armed Forces design & construction in the Far East and Pacific Islands and for civil works throughout the Pacific. Until recently, he was assoc prof of engrg at St. Martin's College in Lacey, WA and currently is a VP for DMJM in the Far East. His specialty is engrg mgmt. *Society Aff:* SAME, USCOLD

Roush, Robert W
Business: The Benham Group, 1200 NW 63rd St, PO Box 20400, Oklahoma City, OK 73156
Position: Sr Corp VP *Employer:* Benham-Blair & Affiliates, Inc. *Education:* BSEE/Communications/OK State Univ. *Born:* 10/30/30. Native of Tulsa, OK. Served as Radar Observer - All Weather Fighter Interceptor, USAF, 1st Lt) 1951-1955. Field Engr for Schlumberger Well Surveying Corp 1957-1965. Aerosystems Engr for Gen Dynamics, Inc 1966-1969. Consulting Engr for Persons and Assoc, Inc 1969-1971. With The Benham Group since 1971. Assumed current position as Sr Corp VP and Prin Elec Engr in 1979. Adjunct Assoc Prof at OK State Univ, School of Architecture 1977-1979. Reg PE in OK, FL, CO, CA, LA and TX. Pres Central OK Chapter Illuminating Engg Society, 1973-1974. State Chairman, Professional Engineers in Private Practice, OSPE, 1980-82. *Society Aff:* IEEE - PES/IAS, NSPE/OSPE, IES.

Roussel, Herbert J, Jr
Home: 1901 Cleary Ave, Metairie, LA 70001
Position: Pres. *Employer:* Roussel Engr, Inc. *Education:* BS/Civil/Tulane; MS/Civil/Tulane; DE/Civil/Tulane *Born:* 7/13/31. Native of New Orleans, LA. Taught grad level courses in Comp Analysis of Structures and Offshore Structure Design. Founded own consulting firm in 1968. Reg E in TX, LA, MS, AL & NJ. Mbr of Tau Beta Pi, Sigma Xi & Alpha Sigma Lambda honor societies. Listed in Marquis' Who's Who in South and Southwest, Who's Who in LA, Men of Achievement. *Society Aff:* ACI, ASCE, AWS, NSPE

Row, Thomas H
Business: P.O. Box X, Oak Ridge, TN 37831-6198
Position: Director *Employer:* Oak Ridge National Laboratory *Education:* Masters Degree/Nuclear Engineering/Virginia Polytechnic Institute & State Univ.; Bachelors Degree/Physics/Roanoke College *Born:* 02/09/35 Mr. Row has been employed at Oak Ridge National Lab (ORNL) since July 1959. Initially assigned to the Reactor Analysis Section of the Reactor Div, he later transferred to the Nuclear Safety Section. He was a member of the project staff at the Nuclear Safety Pilot Plant. He served as National Coordinator of the Pressurized Water Reactor Containment Spray Systems and Absorption Technology Program, directed by the Atomic Energy Comm. He also served as Head of the Environmental Impact Section of the Energy Div. Mr. Row is Director of the Nuclear and Chemical Waste Program at ORNL since April 1981. *Society Aff:* ANS

Rowan, Charles M, Jr
4720 NE 26th Ave, Ft. Lauderdale, FL 33308
Position: Senior Vice President *Employer:* Centec Corp. *Education:* BS/CHE/OH Univ *Born:* 8/21/43. Current position with Centec Corp includes overall technical & administrative mgt for marketing & ind energy & environmental consulting efforts for private industry, Dept of Energy & Environmental Protection Agenc. Areas of responsibility include computer process control; energy audits for processing plants; performance & economic evaluations of new technologies applied in the areas of wastewater pollution control (pH, heavy metals removal, sludge concentration & acid recovery) & advanced energy conservation systems for drying, distillation, incineration, evaporation, heat recovery & VOC control. Experienced in tech transfer, tech market dev & commercialization. Frequent speaker at ind seminars. Enjoys music, golf, fishing & tennis. *Society Aff:* AIChE, AEE, AOCS.

Rowe, David T
Business: 1449 E Pierson Rd, Flushing, MI 48433
Position: President. *Employer:* Rowe Engineering Inc. *Education:* BS/CE/MI Tech Univ. *Born:* Aug 4, 1929; native of Flint Mich. Army Corps of Engrs 1951-53 at Engrg R/D Lab, Ft Belvoir Va; V Pres in charge of engrg, Gould Engr, Flint 1953-62; formed Rowe Engrg Inc 1962 - now serves as CEO. P E & Reg L S in Mich & Wis; Prof Community Planner in Mich. Pres CEC/Mich 1969; Secy of Mich Soc of Reg L S 1967-68; Chmn Educ Ctte ACEC 1973-74, Elected Fellow ACEC Sept 30, 1977. *Society Aff:* ACEC, NSPE, ACSN.

Rowe, Harrison E
Business: Castle Point Sta, Hoboken, NJ 07030
Position: Anson Wood Bunch and Professor of Electrical Engrg *Employer:* Stevens Inst of Tech *Education:* ScD/EE/MIT; MS/EE/MIT; BS/EE/MIT *Born:* 1/29/27. US Navy 1945-46. With Radio Res Lab of Bell Labs 1952-1984. Stevens Inst of Tech 1984-. Publs include numerous papers & 1 book on a variety of subjects including parametric amplifiers, noise & communication theory, modulation theory, propagation in random media & related problems in waveguide, radio & optical communication sys. Mbr Sigma Xi, Tau Beta Pi, Et Kappa Nu, URSI; Fellow IEEE. A joint holder of the Microwave Prize and of the David Sarnoff Award. *Society Aff:* IEEE, URSI

Rowe, Jack F
Business: 30 W Superior St, Duluth, MN 55805
Position: Chrmn & Chief Exec Officer *Employer:* Minn Power *Education:* BEE/Elect Eng/Univ of MN. *Born:* 5/10/27. Attended sch in Littlefork & Duluth MN. Served in US Navy in WW II. After graduation joined MN Power & Light Co as an Apprentice Engr. Through yrs served as Engr, Asst Ch Engr, Past Ch Engr, Asst to Pres, V P, Exec V P & Co Dir 1969, Ch Oper Officer 1973 & Mbr of Exec Ctte & Pres 1974. Chief Exec Officer 1978 Chrmn & Pres 1979. Past Pres, Elec Inf. Council; Pres and Mbr of the Exec Comm, No. Central Elec Assn. P Chrmn, MARCA. Served on the Exec Bd of the Natl Elec Reliability Council and P Chrmn, Upper Mississippi Valley Pwr Pool Plng comm. Mbr, AIF, sr mbr of IEEE. Mbr Engrs Club of No Minnesota and Presented with Outstanding Leadership Award in Energy Conversion Sciences by the NY Sec of ASME at annual mtg in 1979. Dir of First Bank- Duluth. Past Pres of Northeastern Minnesota Dev Assn. P Pres, Duluth Area Chamber of Commerce and Duluth Rotary. Mbr of many other Boards. Mbr of Lakeside Masonic Lodge, Scottish Rite and Aad Temple Royal Order of the Jesters of the Shrine and Kitchi Gammi Club of Duluth, Pres and Dir Kitchi Gammi Club. Awards: Outstanding Achievement Award, University of Minnesota Alumni Association, 1986; Chief Executive Officer of the Year 1986, Bronze Award, from FINANCIAL WORLD MAGAZINE; Establishment at University of Minnesota, Duluth, of Jack F. Rowe Chair of Engineering, 1986.

Rowe, Joseph E
Business: P.O. Box 430, Melbourne, FL 32919
Position: Vice Pres & Genl Mgr *Employer:* Harris Corporation *Education:* PhD/EE/Univ of Mich; MSE/EE/Univ of Mich; BSE/EE/Univ of Mich; BSE/Math/Univ of Mich *Born:* 6/4/1927 Born in Detroit, Mich. Faculty member at the University of Michigan 1955- 1974; Chairman of Electrical Computing Engineering Dept. 1968-1974. Dean of Engineering at Case Institute of Technology 1974-76; Provost (Science & Engineering) Case Western Reserve University 1976-78. Vice President Technology with Harris Corp. 1978-81; Vice President and General Manager Harris Controls Division 1981- . McGraw Research Award of ASEE, 1964; Fellow IEEE 1965; Distinguished Faculty Achievement Award, University of Michigan, 1970; National Academy of Engineering, 1977. Technical expert in integrated circuits, microwaves and communications. Currently managing the development of computer- based control systems for electric utilities, pipelines and railroads. *Society Aff:* IEEE, NAE, APS, AAAS

Rowe, Robert R
Business: 1844 - 600 Grant St, Pittsburgh, PA 15230
Position: Sr. Industry Rep - Light Const Indus. *Employer:* U S Steel Corp. *Education:* BS/Agri Engg/Univ of IL. *Born:* May 2127 Freeport Ill. 11 yrs St Louis; 21 yrs Pittsburgh Pa. BS AgEng Univ of Ill. Reg P E in Missouri. 1960-81 US Steel Corp Pgh, Marketing & Market Dev in Agri & Light Commercial & Indus Structs. 1949-60 Doane Agri Serv Inc St Louis, Mgr Engrg Div & Bldg Marketing & Cons Servs. *Society Aff:* ASAE.

Rowells, Lynn G
Home: 1029 Community Dr, LaGrange Park, IL 60525
Position: Group Manager. *Employer:* Sears Roebuck & Co. *Education:* BS/Met Eng/Univ of IL. *Born:* Feb 1924; native of Chicago. US Navy 194346. Met Engr for Imperial Type Metal & Stewart Warner Corp; joined Sears 1953; assumed urrent assignment as Group Mgr 1972, respon for testing lawn, garden & automotive prods in sears Labs. Previously mbr of Engrg Prog Ctte Engg Matls Achievement Award Selection Committee & HandbookCtte ASM Metals Park Ohio. Ser on offeror ctte on Lawn Mower & Elec Hedge Trimmer Standards CPSC. *Society Aff:* ASM.

Rowen, Alan L
Home: 251 East 82nd St, New York, NY 10028
Position: Prof of Marine Eng *Employer:* Webb Inst of Naval Architecture *Education:* MS/Mech Ocean Eng/Stevens Inst of Tech; BE/Marine Eng/SUNY Maritime *Born:* 03/22/45 Independent conslt; part-time employee of George G. Sharp, Inc at Webb Inst since 1977. Previously employed by Anglo Nordic Bulkships as mgr of construction; by Zapata Naess Shipping Co as marine engr and naval architect. Taught at SUNY Maritime 1968-71. Asst engr aboard US Flag Ships 1965-68. USNR 1965-71. USCG license as lst Asst Engr. NY PE license. *Society Aff:* SNAME, IME, RINA, ASME

Rowland, Chester A, Jr
Home: 1035 W Fairfield Ct, Glendale, WI 53217
Position: Conslt Comminution Sys *Education:* BS/Mech Engr/NM State Univ *Born:* 07/03/24 Retired Conslt Comminution Sys. Until retirement prof career was with Allis-Chalmers Corp for 37 yrs. Engrg and mgmt positions related to crushing and grinding equip. Positions covered application engrg, product sales, process dev and in-plant start-ups and studies. Currently on Adv Ctte Comminution Res Ctr Univ of UT. Conslt to the Mining Indus on grinding of ores. Author of tech articles, papers and lectures pub and presented in the US and internationally. Distinguished mbr SME of AIME. Past Chrmn WI Sect AIME, past VP Central Region SME of AIME, past mbr past Chrmn, and current mbr various AIME and SME tech and operating cttes. Also a mbr of The Canadian Inst of Mining & Metallurgy. Reg PE State of WI. *Society Aff:* SME of AIME, CIM

Rowland, Samuel W, Jr
Home: 637 Kalamath Dr, Del Mar, CA 92014
Position: Manager - Safety. *Employer:* General Atomic Co. *Born:* 1921. Grad of San Jose St Univ 1948. Proj Safety Engr Genl Dynamics Convair 1951-56; Ground

Rowland, Samuel W, Jr (Continued)
Safety Dir Convair Flight Test Centers, Edwards AFB Calif 1956-58; Proj Safety Engr Apollo Capsule Flight Test, White Sands N M 1960-61; Safety Engr John Jay Hopkins Lab for Pure & Applied Sci, San Diego Calif 1964-68; Mgr Safety, Genl Atomic Co 1969- . Elected Safety Engr of the Year San Diego EJC 1966; St Representative ASSE on San Diego EJC. Certified Safety Professional BCSP; Reg P E in Calif. Expert Examiner Calif Bd of Registraton for P E's, Safety Discipline. Prof Mbr, Regional V Pres 1975-76 & 1976- , Mbr Bd of Dirs 1975-76 & 1976- ; Liaison Representative 2nd Internatl Conference on Sys Safety 1975: ASAE.

Rowland, Walter F
Business: Sch of Engg, Fresno, CA 93740
Position: Prof. *Employer:* CA State Univ. *Education:* PhD/Civ Engr/Stanford Univ; MS/Sanitary Engr/Univ of IL; BS/Civ Engr/Univ of IL. *Born:* 11/9/31. Decatur, IL. Military: US Army (Missiles). Taught at: Univ of IL, Stanford Univ, Univ of CA-Berkeley, CO St Univ, CA St Univ-Fresno since 1967. Dir of Atmospheric Water Resources Res (CSUF), 1971-73. Sabbatical leave at Princeton Univ, 1975. Past Pres, Fresno Branch, ASCE. Reg PE (CO). Primary interests: education, water resources (dev, economics, policy), hydrology, applied hydraulics. *Society Aff:* ASCE.

Rowlands, E Carlton
Home: 2409 Hirschman Lane, Hartland, WI 53029
Position: Prof Engr *Employer:* Graef, Anhalt, Schloemer & Assoc *Education:* BSCE/CE/Univ of WI-Madison *Born:* 12/03/42 Univ. of Wis. (Madison, WI.), Field & Design Engr with Milwaukee Road Railroad Co 1966-1969, Construction and Design Engr for City of Madison, WI 1969-1973, Project Mgr for Yaggy Assocs, Inc of Rochester, MN 1973-1978. (Also NSPE Chapt Sec, VP, Pres and State Dir for SE MN Chapt.) Proj Engr Graef, Anhalt, Schloemer & Assoc, Inc Milw, WI 1978-1983; Proj. Mgr. Threshold Design Inc. Pewaukee WI. 1984-1985. Proj. Mgr. Kapur & Assoc. Inc. Milwaukee WI., 1985-1986; Proj. Mgr. Threshold Design. 1986-present. *Society Aff:* NSPE, WSPE

Rowlands, Richard O
Business: PO Box 30, St College, PA 16801
Position: Emeritus Professor of Engg Res. *Employer:* Penn State Univ *Education:* BS/Math/Univ of Wales; MS/EE/Univ of Wales. *Born:* Apr 1914 Anglesey Wales. Head of filter design group GEC Coventry England 1937-48; Sr Lectr BBC Engrg Training Dept 1948-57; Assoc Prof of EE Penn St Univ 1958-59; Prof of Engrg Res at Applied Res Lab Penn St Univ 1959-79; Chmn of Engrg Acoustics Program Penn St Univ 1967-70; Liaison Scientist ONR London England 1970-71. 20 publs, chapters in 2 books; 8 US & numerous UK pats. Chmn of Seminars on Underwater Acoustics & Signal Processing held annually at Penn St Univ. Emeritus Prof of Engg Res 1979. *Society Aff:* IEEE, ASA, ΣΞ.

Rowlands, Robert E
Home: 5401 Russett Rd, Madison, WI 53711
Position: Prof and Dir, Matls and Structural Testing Laboratory - Conslt *Employer:* University of Wisconsin. *Education:* PhD/Mechs/Univ of IL-Urban; MS/Mechs/Univ of IL-Urban; BASc/Mechs Engg/Univ of British Columbia-Vancouver, Canada. *Born:* 7/7/36. 1936 Trail B C Canada. Married; 2 sons. Engr, MacMillan, Bloedel & Powell River Co, Canada 1959-60; Sr Res Engr, IIT Res Inst Chicago 1967-74. Book Review Ed, Experimental Mechs; mbr SESA papers Review ctte & SESA ed bd; co-recipient SESA Hetenyi Award 1970 & 1976. Has lectured domestically & abroad; 90 tech publs; reviews for numerous prof & tech socs & journals. Indus Cons. Enjoys out- of-doors. *Society Aff:* SESA, ASTM, ASME.

Rowley, Louis N, Jr
Home: 36 Orchard Farm Rd, Port Washington, NY 11050
Position: Consultant (Self-employed) *Education:* Dr Eng/-/Polytechnic Inst of NY; MBA/Economics/NY Univ; ME/-/Polytechnic Inst of Brooklyn. *Born:* Sept 1909. Various engrg posts Con Edison Co of N Y 1931-37; joined ed staff McGraw-Hill's Power 1937, became Ch Ed 1954, Ed & Publisher 1957. Received Crain Award for disting editorial career 1974. Retired 1974 as Ed Chmn Power & Elec World. Held many offices in ASME, culminating in election as Pres 1967 & as Hon Mbr 1972; Pres United Engrg Trustees 1974-75; Chmn Emeritus Polytechnic Corp (Bd of Tr) . *Society Aff:* ASME, NSPE, AAAS.

Roy, Francis C
Business: PO Box 1663, Lake Charles, LA 70602
Position: VP. *Employer:* Ind Equip & Engg Inc. *Education:* MS/EE/Univ of TX; BS/EE/LA State Univ. *Born:* 11/28/26. Native of Iota, LA. Served with USAAF 1944-46. Allis-Chalmers, Milwaukee, 1949-1952, specializing in elec controls. Related Sci teacher at Jeff Davis Vo- Tech Sch, Jennings, LA 1952-1955. Assoc Prof of EE at LA Tech Univ, 1955-1965. Received Tau Beta Pi award for Outstanding Engg Teacher, 1965. Natl Sci Fdn comp specialist, 1961-62, VP of Ind Equip & Engg, Inc in Lake Charles, LA since 1965, specializing in ind switchgear & control. Reg PE in LA. Tau Beta Pi & Eta Kappa Nu. State Pres, LA Engg Soc 1979-80, Member - Louisiana State Board of Registration for Professional Engineers and Land Surveyors. (term expires 1989). Vice-Chairman of Examinations Preparation for National Council of Engrg Examiners (NCEE). *Society Aff:* NSPE, IEEE.

Roy, Guy W
Business: 596 Indian Home Rd, Danville, CA 94526
Position: President *Employer:* GWR Assocs *Born:* MS Chem Engrg U C Berkeley; MBA Golden Gate Univ; BSc Univ of Alberta. Pres of GWR Assocs, which provides consltg services to the process industries, primarily in the areas of process engrg, venture dev and analysis, and new tech dev and evaluation. Experience encompasses genl mgmt, prof mgmt, process & proj engrg, & R/D. Past company affiliations incl engrg/const & operating companies in chem & food industries. Mbr AIChE. Reg P E in CA, Past Chrmn NoRCAL Section, AIChE.

Roy, Martin H
Home: 12 Westmont Cir, Little Rock, AR 72209
Position: Engr. Supervisor *Employer:* AR Dept of Pollution Control & Ecology. *Education:* BS/Civ Engg/Univ of Detroit. *Born:* 2/11/39. Past Pres of AR Sec of ASCE. Presently Wastewater Treatment Innovative & Alternative Tech Coordinator for State of AR. Five yrs with private consulting engr firm, three yrs as field engr & two yrs as design engr. Vietnam veteran. LTC in the US Army Reserve. Married, two daughters. Hobby is home improvements. *Society Aff:* ASCE, WPCF.

Roy, Robert H
Home: 7826 Chelsea St, Baltimore, MD 21204
Position: Prof Emer of Ind Engr/Dean Emer Engr. *Employer:* The Johns Hopkins University. *Education:* BE/Mech Engg/Johns Hopkins. *Born:* Nov 1906 Baltimore. US Olympic Lacrosse Team 1928. 1928-47 Waverly Press Inc VP for Engrg & Dir; appointed to Johns Hopkins faculty 1947, Dean of Engrg Sci 1953; retired 1973. Has served as cons & labor arbitrator. Books: 'Mgmt of Printing Production' 1953; 'The Administrative Process' 1958; 'The Cultures of Mgmt' 1977; 'Operations Technology: Systems and Evolution' 1986; co-author 'Horizons for a Profession' 1967; co-editor 'Opers Res & Systems Engrg' 1960. Pres 1949-50 Engrg Soc of Baltimore; Pres. 1949-50, Founders Award 1980; Hon mbr ASEE, Pres 1966-67; Fellow IIE (Tr, V P, Gilbreth Award 1967); Bk of the Yr Award 1977; ASME; AAAS. Mbr Visitors & Governors (Trs), Washington Coll (Chmn 1976-1980). Governing Bd Patuxent Inst; Comm on Metropolitan Government; Chmn Panel on Nuclear Policy in st of Md; Dir Chesapeake Res Consortium Inc 1972-73; Mbr Md. Gov. comm on Three Mile Island 1980-87; Chmn Hazardous Waste Facilities Siting Bd. Md. 1980-84. Tau Beta Pi; Phi Beta Kappa; Omicron Delta Kappa; Alpha Tau Omega. Reg P E. DSc (hon) Washington College 1982. *Society Aff:* IIE, ASEE, ASME, AAAS.

Roy, Rustum
Business: 202 Materials Res Lab, University Park, PA 16802
Position: Dir, Matls Res Lab; Chmn Sc Tech & Soc Prog. *Employer:* The Pennsylvania State University. *Education:* PhD/Ceramics/Penn State; MSc/Chemistry/Patna Univ (India); BSC (Hons)/Chemistry/Patna Univ (India). *Born:* 7/3/24. Native of India. Affiliated with Penn State since 1950; Prof of the Solid St, Evan Pugh Professor 1981-; founder & 1st Chmn Solid St Tech Degree Prog 1960-67; founder & 1st Dir Matls Res Lab 1962-1986. Science, Tech. & Society Program, 1969-present, Chmn, 1977-present. Director 1985- . Res area: matls synthesis, preparation & characterization. Author of some 350 scientific papers & chapters in bks. Mbr NAE; Natl Matls Adv Bd; sev NRC cttes. Governor's Sci Adv Ctte Penn; Chmn Penn Matls Adv Panel; Natl Sci Foundation; Engg Adv Ctte. Founder & joint editor-in-ch *Material Res Bulletin*; Founder and Editor: Journal of Education Modules in Materials Sci and engineering. Founder and Editor-in Chf: Bulletin of Science Technology and Society. Bd of Governor's Dag Hammerskjold coll. Mineralogical Soc of Amer Award 1957 MSA, Hibbert Lecturer, London, England 1979. *Society Aff:* Am Cer Soc/ACS, APS, AAAS.

Royce, Barrie S H
Business: D416 Duffield Hall (Eq), Princeton, NJ 08544
Position: Prof *Employer:* Princeton Univ *Education:* PhD/Physics/London Univ; BSc/Physics/London Univ. *Born:* 01/10/33 Prof Royce is active in research into the properties of materials that are important in microelectronic devices, energy conversion, catalysis and lightstrong structures. He is interested in the role of atomic defects and solid state reactions on material properties. Prof Royce serves on the editorial advisory boards of Crystal Lattice Defects. He is currently the Master of Mathey College of Princeton Univ and a Past Pres of the Princeton Chapt of Sigma Xi. *Society Aff:* APS.

Royden, Thomas S
Business: 3055 W. Indian School Rd, Phoenix, AZ 85017
Position: VP *Employer:* Royden Engrg Co *Education:* BS/CE/Univ of AZ *Born:* 1/30/49 Native of Phoenix, AZ, attended Univ of AZ, graduating in 1953 with BS in CE. Worked in summers during high school and coll, in various construction crafts. After graduation, did falsework design, Superintendent, Project Engr, VP and currently Pres of Royden Construction Co. Was granted Registration in AZ in 1958. Formed Royden Engrg Co in 1968 and undertook first Statewide Bridge Safety inspections for AZ Hgwy Dept, involving the establishment of criteria, procedures, and reporting formats, and remain active in that field. Also doing design of civil projects with emphasis on Bridges and special expertise in construction costs. Also active in construction management and construction administration. *Society Aff:* SAME, DFI.

Roys, Henry E
Home: 327 S PAseo Chico, Green Valley, AZ 85614
Position: Chief Engineer/RCA Records (retired). *Employer:* RCA Corp. *Born:* Jan 1902. BSEE Univ of Colorado 1925; Tau Beta Pi, Eta Kappa Nu. Radio Dept G E Co 1925-30; RCA 1930-67 engrg dev, primarily disk & magnetic tape recording, Ch Engr RCA Records prior to retirement. Active as mbr & chmn in many cttes pertaining to audio recording standards: NAB; IRE; EIA; RIAA & Amer Standards. US Natl Mbr IEC on Sound Recording. Fellow IEEE; AES; (Pres 1963); ASA. Awards: IRE-PGA Achievement 1956; AES Emile Berliner 1957; Univ of Colo Distin Engrg Alumnus 1967; AES Gold Medal 1973. *Society Aff:* IEEE, AES, ASA

Rozovsky, Eliezer
Home: 2237 Pennsylvania St, Allentown, PA 18104
Position: Metallurgist. *Employer:* Gilbert Associates Inc. *Born:* April 1928. PhD Lehigh Univ; BS, MS Technion Israel Inst of Technology. Maintenance & Corporate Testing Engr Israel Electric Corp 1953-66; Dir Materials Testing Div D & K Inc Boston 1970; Metallurgical & Materials Engr Gilbert Assocs Inc Reading Pa 1971-present. Responsible for material quality & selection for the fossil & nuclear power plants & other energy conversion sys. Developed interaction on the natl level between Materials Researchers & Developers & designing orgs of coal conversion sys. Mbr Natl Power Ctte of ASM; Co-Chmn Natl Conf of Materials for Coal Conversion Sys Design. Num of articles in ASM Quarterly & other periodicals.

Rubin, Alan J
Business: Water Resources Center, Columbus, 1791 Neil Ave, OH 43210
Position: Prof of Civil Eng *Employer:* Ohio State Univ *Education:* PhD/Environ Chem/Univ of North Carolina; MSSE/Sanitary Eng/Univ of North Carolina; BSCE/Civil Eng/Univ of Miami *Born:* 03/20/34 Just prior to grad sch I worked for the FAA as a civil engr. In 1965 ,I joined the Civil Engrg faculty at the Univ of Cincinnati. After three yrs I moved to the OH State Univ where I am currently Prof of Civil Engrg. My work on physical-chemical treatment processes, especially foam separations, flotation, disinfection, and chemical coagulation, is internationally known. I am the editor of four books on water or environmental chemistry and the author of more than 40 scientific and technical papers. In 1984 I spent 6 months in Haifa, Israel as a Visiting Prof of Civil Engrg at the Technion - Israel Inst of Tech. My current research is on the precipitation of aluminum fluorides, aquifer restoration with surfurtants, desig of a new type of fluoridator, and inactivation of Giardia cysts. *Society Aff:* WPCF, ACS, AWWA, IAWPCR.

Rubin, Arthur I
Home: 917 Stuart Rd, Princeton, NJ 08540
Position: Account Executive *Employer:* Prudential Bache Securities *Education:* MS/Physics/Stevens Inst of Technology; Grad Study/Physics/Columbia Univ; BS/Physics/CCNY. *Born:* 12/3/27. in NY City. 2 children. Physicist Picatinny Arsenal Ord Corps 1950-55; appln engr Electronic Assocs Inc NJ 1955-56, supr 1957-59; Dir comput ctr 1959- 62; ch automatic comput Martin Co 1962-67; mgr hybrid comput sci dept Orlando Div Martin Marietta Corp 1967-69; Electronics Assocs Inc 1969, Dir comput ctr 1969-70, mgr Anal Engg Dept 1971-77. Autodynamics, Inc 1977, Mgr Simulation Engg 1977-78, dir Planning, 1978-80, Dir Proj Control, 1981-83. Prudential-Bache Securities 1983-, Acct Exec. Mbr ed Bd & Chrmn libr ctte Simulaton Coun 1964- ; mbr surv ctte Amer Fed Info Processing Socs 1966-69; secy 1968-69, Genl Chmn 1976 Summer Computer Simulation Conf; USAAC 1946-47; fmr Sr mbr Inst Elec & Electronics Engg; Sr Mbr, Soc for Computer Simulation; fmr assoc fel Amer Inst Aeronaut & Astronaut; Computers, Hardware & Software; Simulation. *Society Aff:* SCS

Rubin, Charles H
Home: 52 Strasser Ave, Westwood, MA 02090
Position: Pres. *Employer:* Mgmt Techniques (Consultants in Quality) *Education:* BSEE/Elec Engr/Drexel Inst of Tech. *Born:* Jan 3, 1924 Wilkes-Barre Pa. Resident-Inspector-in-Charge USAF Procurement Electron NE Penn 1943-54; Quality Supervision G E Co Johnson City NY 1954-58; Dir Quality & Reliability Tracerlab-Keleket Waltham Mass 1958-62; Quality Mgr Sylvania Electronics Needham Mass 1962-64; Mgr Product Assurance Sanders Assoc Bedford Mass 1964-69; Mgr Quality Assurance Polaroid Corp Norwood Mass 1969-82, Pres, Mgmt Techniques (Consultants in Quality) 1982- ; Fellow Amer Soc for Quality Control; Certified Quality Engr; Certified Reliability Engr. Co-author of text 'Fundamentals of Statistical Quality Control.' Pres Automatic Measurements Corp 1964-84; Chmn Boston Sect ASQC 1975-76, Rec of R Shaw Goldthwaite Award from ASQC-1978. *Society Aff:* ASQC.

Rubin, Izhak
Business: 6731 Boelter Hall, Univ of CA, Los Angeles, CA 90024
Position: Prof *Employer:* Univ of CA. *Education:* PhD/EE/Princeton Univ; MS/EE/Technion, Israel; BS/EE/Technion, Israel. *Born:* 5/22/42. Born in Haifa, Israel. From 1964 to 1967, served as a communications engr in the Israel Signal Corps. During 1967-68, was employed by the Israel Aircraft Industries, working in the areas of electronics & control engg. From Sept 1968 to June 1970, was an RCA Fellow & Res Asst in the Dept of Elec Engg at Princeton Univ. Since July 1970, with the faculty of the Elec Engrg Dept, Sch of Engg & Appl Sci, Univ of CA, Los

Rubin, Izhak (Continued)
Angeles. Serves as consultant to a number of industries and as pres of IRI Corp, a consulting firm in the areas of Communications Networks and Systems. Current interests are in the area of data communication networks, telecomm systems, C3 systems, comp & satellite networks, queueing, communications & network flows, metropolitan and local area networks, traffic engrg. *Society Aff:* IEEE, AIAA.

Rubinoff, Morris
Home: Green Hill Apt WA605, 1001 City Ave, Philadelphia, PA 19151
Position: Professor Emeritus *Employer:* University of Penn; Brandt-PRA Inc. *Education:* PhD/Physics/Univ of Toronto; MA/Physics/Univ of Toronto; BA/Math & Physics/Univ of Toronto. *Born:* August 20, 1917 Toronto Can to Israel & Emma (Nathanson). In US since 1946. M Dorothy Weinberg Oct 26, 1941; children Elayne, David, Robert. Prof Emeritus Electrical Engrg Univ of Pa 1964-85, faculty mbr 1950- . Chmn & Pres Pa Res Assos Inc 1960-77; Pres Brandt-PRA Inc 1976-78; Dir Radonics Inc 1968-74, Amer Technion Soc 1976- . Former res asst Univ of Toronto 1941-46, res f instr Harvard Univ 1946-48, res engr Inst for Advanced Study 1948-50, ch engr for computers Philco Corp 1957-59. V Chmn sci & electron div Amer Inst of EEs 1962- 65; P Chmn Natl Joint Computer Comm; P Chmn Phila sect Assn of Engrs & Architects in Isr. Mbr: Inst of Radio Engrs, Fellow 1962; Amer Assoc Univ Profs; ASEE; Farband; Sigma Xi; Past V P, Chmn, tech comm, Phila chap Amer Technion Soc. Recipient: Inst of Radio Engrs-Profsl Grp on Electronic Computers. *Society Aff:* IEEE, ACM, AAUP.

Rubinstein, Moshe F
Business: 405 Hilgard Ave, School of Engg & Applied Science, Los Angeles, CA 90024
Position: Prof. *Employer:* UCLA. *Education:* PhD/Engr/UCLA; MS/Engr/UCLA; BSc/Engr/UCLA. *Born:* 8/13/30. Designer, Los Angeles 1954-61; Prof of Engg Univ of CA at Los Angeles 1961; Chrmn Engg Systems Dept, 1970-1975; Prog Dir-Modern Engg for Executives 1965- 1970; Consultant - Pacific Power & Light, Northrop Corp, US Army, NASA, TX Instruments Co, Hughes, US ARmy Sci Adv Com, Kaiser Aluminum, IBM. Recipient Distinguished Teaching Award - Academic Senate, UCLA 1964; Sussmann Chair for Distinguished Visitor, Technion Israel Inst of Tech, 1967-68; Fulbright-Hays fellow, Yugoslavia & England, 1975-76. Author: (with W C Hurty) Dynamics of Structures, 1964 (Yugoslavian translation 1973); Matrix Comp Analysis of Structures, 1966 (Japanese 1974); Structural Systems, Statics Dynamics & Stability, 1970 (Japanese 1979); Patterns of Problem Solving, 1975; Concepts in Problem Solving (with K Pfeiffer), 1980. Tools for Thinking and Problem Solving 1986. Res in use of computers in structural systems, analysis and synthesis; problem solving and decision theory. *Society Aff:* ASCE, ASEE, SSA, TBΠ.

Rubio, Abdon
Home: 20250 Regan Lane, Saratoga, CA 95070
Position: Director/Engineering & Operations Department *Employer:* Electric Power Research Institute *Education:* BEE/Elect/Pratt Inst. *Born:* Feb 1927. Native of Brooklyn. US Navy 1945-46. With G E 1951-1979 Graduated Advanced Engrg Prog; then managed grp in design & dev of buckets & rotors for large steam turbines. Experienced in design & manufacture of steam & gas turbines, air circuit breakers, power transformers, capacitors, electric motors & nuclear reactors. Became Genl Mgr of nuclear control & instrumentation dept in 1975. Became Dir-Engg & Operations Dept of Elec Power Res Inst 1979. Fellow ASME 1972; ASME Service Award 1970; Metals Engrg Div Chmn 1969; ASME Metal Fracture Ctte Secy 1964; AIEE Leadership Award 1951. Reg P E NY, Mass, Calif. Hobbies: racquetball & reading. *Society Aff:* ASME.

Ruby, Kenneth W
Business: 1308 5th Street South, Fargo, ND 58102
Position: Supt Water & Sewage Utilities. *Employer:* City of Fargo. *Education:* BSChE/Chem Engr/Univ of IA. *Born:* Jan 30, 1924 Oskaloosa Iowa. Served in Navy 1943-46. Asst Supt Oskaloosa Water Dept 10 yrs. Supt Water & Sewage Utilities Fargo, ND 1959- . Chmn ND Water Pollution Control Federation 1961; Dir WPCF 1962-65. Recipient of WPCF Arthur Sidney Bedell Award 1967; WPCF William Hatfield Award 1966. Chmn N Central Sect Amer Water Works Assn 1968-69. Recipient George Warren Fuller Award 1971. Presently Dir AWWA. *Society Aff:* AWWA, WPCF

Ruby, Lawrence
Home: 663 Carrera Lane, Lake Oswego, OR 97034
Position: Prof. *Employer:* Reed College *Education:* PhD/Physics/Univ of CA; MA/Physics/Univ of CA; BA/Physics/Univ of CA. *Born:* 7/25/25. Lawrence Berkeley Lab, Staff Scientist, 1950 to 1987. Univ of CA Berkeley, Dept of Nuclear Engg, Lecturer 1959-62, Assoc Prof 1962-67, Prof 1967 to 1987. Reed College, Portland, OR, Prof. of Nuclear Science, 1987 to present. DOE "Q" and DOD "Secret" clearance. Married, 3 children. CA Reg PE. *Society Aff:* ANS, APS, AAPT.

Ruckenstein, Eli
Business: Dept of Chemical Engineering, Amherst, NY 14260
Position: Prof *Employer:* State Univ of New York *Education:* Dr Eng/ChE/Polytech Inst, Bucharest; BS/ChE/Polytech Inst, Bucharest *Born:* 08/13/25 Born in Romania. Prof at Polytechnic Inst Burcharest 1949-1969, at the Univ of DE 1970-1973 and at State Univ of NY at Buffalo 1973- . Recipient of the awards: Natl Awards of the Romanian Dept of Educ for Res in Turbulence 1958 and in Distillation 1964; George Spacu award of the Romanian Acad of Scis for Res in Surface Phenomena 1963; Alpha Chi Sigma Award of the American Inst of Chemical Engrs 1977. Recent Achievements: Electronic Mechanism of Catalyst Promotion and Poisoning, Catalytic Combustion, Kinetics of Selectivity in Heterogeneous Catalysis, Origin of Low Interfacial Tension in Multiphase Systems Containing Microemulsions, The New Separation Methods: Potential Barrier Chromatography, and Extraction by Micellar Solutions. *Society Aff:* ACS, AIChE, CS

Rudavsky, Alexander B
Business: 3334 Victor Ct, Santa Clara, CA 95050
Position: President *Employer:* Hydro Res Science (HRS, Inc) *Education:* ScD (Dr-Ing)/Hydraulics/Franzius Inst Tech Univ-Germany; MS/CE/Univ of MN; BS/CE/Univ of MN *Born:* 01/17/25 Born in Poland. Hydraulic engr for Justin and Courtney 1953-1959. Res Fellow, St Anthony Falls Hydraulic Lab, Univ of MN, 1953-1956. ASCE Freeman Traveling Scholar, 1958: study in hydraulic labs in Switzerland, Portugal, France, and England. Prof of hydraulic engrg at CA State Univ, San Jose, since 1960; full prof since 1975. Visiting Assoc Prof, Stanford Univ, 1970. Founded HRS in 1963. HRS does consulting and applied res in hydraulic engrg, with emphasis on physical and numerical modeling in fluid mechanics. Chrmn, Pipeline Div, ASCE, 1970. Tennis player. *Society Aff:* ASCE, ASME, USCOLD

Rudd, Dale F
Business: Chem Engrg Dept, Madison, WI 53706
Position: Professor. *Employer:* University of Wisconsin. *Education:* BS/Chem Eng/Univ of MN; PhD/Chem Eng/Univ of MN. *Born:* March 2, 1935. Co-author 'Strategy of Process Engrg'; 'Process Synthesis'; 'Stragtegy of Pollution Control'Petrochemical Technology Assessment. ' Recipient of Colburn Award AIChE 1971; Guggenheim Fellowship 1971; Lectureship Award ASEE 1972; Davis-Swinden Mem Lectureship 1974; PC Reilly Lectureship 1975; Merck, Sharp, Dohme Lectureship 1973. Elected Natl Acad of Engg 1978. Currently engaged in res on dev of the chems indus. *Society Aff:* NAE, AIChE.

Rudd, Wallace C
Business: 229 Valley Rd, New Canaan, CT 06840
Position: Consulting Engineer. *Employer:* Thermatool Corp. *Education:* EE/Elec Engg/RPI. *Born:* Dec 1910. Native Yonkers NY. Dev Engr for Con Edison of NY 8 yrs. Ch Engr 9 yrs Induction Heating Corp, spec in high frequency heating of metals. Founded Thermatool Corp & was V P R/D 20 yrs. Invented some 90 processes for high frequency resistance welding of pipe & structural shapes. Presently cons in this field involved in inventing new metal fabrication processes. Some 100 patents. Honorary Member AWS 1986 - Fellow IEEE 1986- "Inventor of the Year–, N.Y. Patent Law Assoc. 1986. Hobby: building live-steam, coalburning locomotives. Bd Mbr 1972 EJC. *Society Aff:* IEEE, AWS.

Ruddick, Bernard N
Business: PO Box 208, Wichita, KS 67201
Position: VP Engg. *Employer:* KS Gas and Elec Co. *Education:* BS/Elec Engg/KS State Univ. *Born:* 2/19/24. Attended Emporia State Teachers College 1941 to 1943; Carlton College 1943 to 1944; and KS State Univ 1946 to 1949. Served in Army Air Corps 1943 to 1946. Employed by KS Power and Light Co 1949 to 1954; and KS Gas and Elec Co 1954 to present. Made VP - T&D in 1973; VP - Engg 1978. Authored articles in Elec World, Transmission and Distribution and Wood Preserving magazines. Native of KS. *Society Aff:* NSPE, IEEE.

Rudee, M Lea
Business: UCSD, Dean of Engrg (B-010), La Jolla, CA 92093
Position: Dean Of Engineering *Employer:* Univ of Cal/San Diego. *Education:* BS/Metallurgy/Stanford; MS/Matl Sci/Stanford; PhD/Matl Sci/Stanford. *Born:* October 1935. Served as officer in US Navy for 3 yrs. Faculty of Rice Univ 1964-74, finishing as Prof of Materials Sci & Master of Wiess Coll. At UCSD since 1974. Provost of Warren Coll (1974-1982) & Prof of Materials Science. Dean of Engrg 1982-present. Guggenheim Fellow in Cavendish Lab & Visiting Scholar of Corpus Christi Coll Univ of Cambridge Eng 1971-72. Visiting Scientist, IBM T.J. Watson Research Center, Yorktown Heights, NY, 1987. Co-founder & P Pres Texas Soc for Electron Microscopy. Reg P E in Texas. *Society Aff:* APS, EMSA, IEEE.

Rudinger, George
Home: 47 Presidents Walk, Buffalo, NY 14221
Position: Adjunct Professor of Mech Engrg. *Employer:* State Univ of NY at Buffalo. *Education:* Ing/Engg Physics/Vienna Tech Univ. *Born:* May 1911 Vienna Austria. Res Vienna Genl Hospital 1935-38; Physicist Sydney Hospital Australia 1939-46; Principal Physicist Cornell Aeronautical Lab Inc Buffalo N Y 1946-70; Principal Scientist Textron Bell Aerospace Buffalo N Y 1971-76 (ret); Adjunct Prof of Mech Engrg SUNY/Buffalo 1976- . Res in radiology, nonsteady duct flow, blood flow, gas-particle flow. Num publs in the fields of activity. Fellow of ASME; APS; Inst of Physics London; Fellow AIAA Fellow AAAS. Mbr Amer Heart Assoc Basic Science Council; Mbr Sigma Xi. 1974 Pioneer Award Niagara Frontier Sect of AIAA. Centennial Medallion Fluids Engineering division ASME 1980; Fluids Engrg Award, ASME 1983. Dedicated Service Award ASME 1986. Exec Ctte of Fluids Engrg Div ASME 1971-76, Chmn July 1974-June 1975. Bd on Professional Dev ASME 1977-85. ASME Natl Nominating Cttee 1983-84. Board on Issues Management ASME 1985- . Reg P E NY. *Society Aff:* ASME, APS, AIAA, AAAS.

Rudkin, Donald A
Home: 217 Oak Lane, Cranford, NJ 07016
Position: Consultant (Self-employed) *Education:* MBA/Mgmt/Rutgers; BS/Ind Mgt/Rutgers. *Born:* 2/3/21. Held positions with Mobil, Merck, Colgate, & Johnson & Johnson as Mgr Human Resources Consulting Serv, Mgr-Indus Engrg, Mgr, Mgr-Org Planning, Proj Engr, Design Engr & Production Supr. Taught bus mgmt for Rutgers & Union Coll. Co-author of text 'Principles of Supervision' with Fred D Veal Jr. Exec of Assn of Intl Mgmt Cons; Mbr Bd Human Resources Planning Society; mbr Bd/Governors of North Jersey Chap; IIE. Fellow IIE 1976. *Society Aff:* IIE, AIMC.

Rudnicki, John W
Business: Dept. of Civil Engineering, Northwestern University, Evanston, IL 60201
Position: Assoc Professor *Employer:* Northwestern Univ *Education:* PhD/Engg/Brown Univ; ScM/Engg/Brown Univ; ScB/Engg/Brown Univ. *Born:* 8/12/51. Native of Huntington, WV. Res Fellow in Geophysics at CA Inst of Tech 1977- 1978. Assist Professor in Theoretical and Appl Mechanics Dept of Univ of IL 1978-1981. Assumed present position in the Dept. of Civil Engn. at Northwestern Univ. in 1981. Worked in mechanics of geological mtls and particularly the mechanics of earthquake precursory processes. In 1977 received Award for Outstanding Res in Rock Mechanics from US Natl Committee on Rock Mechanics. *Society Aff:* AGU, SSA, AAAS, SES, AAM, ΣΞ.

Ruelle, Gilbert L
Business: Dir Dem/E, Alsthom, Belfort France 90006
Position: Dir. *Employer:* Alsthom-Atlantique. *Education:* Engg/-/Ecole Nationale Superieure d'Electricite; Docteur Ingenieur/-/Universitede Liege ENSEM. *Born:* 11/2/26. From 1952-66 worked in the field of hydrogenerators design. From 1966-69 was head of turbogenerators dept. From 1966 Chief Engg for Generators and Large AC Motors. From 1969 Technical Dir for Generators and Large AC Motors. Since 1974 Dir of Elect Dept (technic and sales). Chmn of CIGRE Study Committee 11 since 1978. IEEE Fellow. Awards: First Section of SEE in 1953/Mascart Medal, Ampere Medal. Blondel medal in 1965. Montefiore Prize 1975. Doctor honoris causa of Liege Universite 1980. Dir of Engrg of Alsthom-jevmont Co in 1983 Dir of Generator Dept Alsthom-Atlantique and Dir of Advanced Electrotechnology of the same in 1984. *Society Aff:* SEE, ISF, CIGRE, IEEE.

Ruf, Jacob F
Home: 13700 Pflumm Road, Olathe, KS 66062
Position: President. *Employer:* Ruf Corp. *Education:* MS/Chem Engg/KS Univ; BS/Chem Engg/KS Univ; AS/Engg/KS Metro Jr College. *Born:* Dec 30, 1936. Res Engr Panhandle Eastern Pipe Line 1959-61; Supr Scien Sys Great Lakes Pipe Line 1961-63; Mrg Sys Dev BS & B 1963-66; Dir Data Sys Metroplan (Marc) 1966-69; Pres Mid-Continent Comput Inc 1967-69; Exec V P Inform Sys Dev Inc 1969-76. VP NLT Computer Services, Inc 1976-77. Mbr AICHE (Comm mbr Natl AICHE Meet KC 1976); Assn for Computing Machinery (Chmn KC Chap 1967). Math Judge Natl Sci Fair & KC Sci Fairs. Mbr Accrediting Study Team NATTS; Kansas & Missouri Governor's Adv Councils & other Adv Bds. Served USNR 1954-59. Contributed articles on Chem Engrg, Planning & Computer Information Sys to tech journals; Pres Intl IMPRS users groups, Hi Tech Adv Cttee Jo Co, Kansas; Chrmn Bd Interlink Data Services Inc; Author and Developer of IMPRS 4th Generation Computer Language, Author of Correlation for Helium Thermo. Properties which were instrumental in the design and development of helium extraction Plant (1960). AIChE Professional Development Recognition Certificate 1984-87. Vice Chair. JO. CO. KS. Transportation Council 1985-present. Instructor Avila College 1985-present, Adjunct Prof Engr. K.U. (KS. Univ.) 1986-present. *Society Aff:* AIChE, ACM, DECUS, IMPRSUG.

Ruf, John A
Business: P.O. Box 419173, Kansas City, MO 64141
Position: Dept Mgr *Employer:* Burns & McDonnell *Education:* PhD/Env Eng/Univ of FL; MS/Env Health Eng/Univ of KS; BS/CE/Univ of KS *Born:* 01/16/39 Native of Kansas City. Employed as Sanitary Engr with Kansas State Dept of Health from 1962-66, specializing in Water Supply and Wastewater. Served s a commissioned officer, US Public Health Service 1966-72. Specializing in Solid Waste Engrg. Served as project officer PHS/EPA Gainesville, FL Compost Plant 1966-71. Was Chief, Solid Waste Branch, US EPA Region 2 office, 1972-74. With Burns & McDonnell Consltg Engrg since 1974. Assumed current responsibility as Mgr, Solid Waste Dept in 1976. Published some ten technical articles. *Society Aff:* AAEE, WPCF, ASCE, NSPE, APWA.

Ruff, Alonzo W
Home: 101 S Yale Street, York, PA 17403
Position: Pres. *Employer:* Current Expertise Asst Inc. *Education:* -/Mech/MIT. *Born:* March 24, 1902. York Corp (2 yrs) Student Course. Asst to Mgr Harding Mfg Co; Head Equipment Dev York Corp; Engr & Ch/Bd St Onge, Ruff & Assocs; Expertise

Ruff, Alonzo W (Continued)
Asst Inc; design cons in USA, Canada & Europe. Mbr: ASME; ASHRAE; IFT; OHI; IIAR; PSPE; IIR; IARW. 65 reg patents in fields of automatic machinery; controls; air conditioning; heat pumps; dry ice; combustion; fog generators; food freezing; etc. Author; teacher; lecturer. ASME Life Mbr, Fellow & York Engr of the Yr; ASRAE Life Mbr, activities in Guide & Data Book, and cttes; listed in AN Marquis 'Who's Who in the East' & 'Industry & Commerce.' Reg Mech Engr Penn; Md; Nova Scotia. Baptist; sch dir; York Country Club; York Univ Club; YMCA; Boy Scouts; Community Improvement; etc. Hobbies: sports; travel; fishing. Mbr York Chamber of Commerce; School Dir - Spring Garden Township. *Society Aff:* ASME, ASRE, IARW, PSPE, SCORE

Rugh, Wilson J
Business: Charles & 34th St, Baltimore, MD 21218
Position: Prof. of ECE *Employer:* Johns Hopkins Univ. *Education:* PhD/Elec Engr/Northwestern Univ; MS/Elec Engr/Northwestern Univ; BS/Elec Engr/PA State Univ. *Born:* Received BS in electrical engineering from The Pennsylvania State University in 1965, and the MS and PhD degrees in electrical engineering from Northwestern University in 1967, and 1969 respectively. Joined the faculty of The Johns Hopkins University in 1969. Areas of research interests: linear and nonlinear systems and control theory. Author of two books and numerous research articles in these areas. Mbr of the editorial boards of the *IEEE Transactions on Automatic* Control and *Mathematics of Control, Signals, and Systems. Society Aff:* IEEE, SIAM, MAA.

Ruhl, Robert C
Business: 20600 Chagrin Blvd, Cleveland, OH 44122
Position: Vice President/Engineering. *Employer:* Chase Brass & Copper Co. *Education:* PhD/Metallurgy Engg/MIT; BSE/Metallurgy Engg/Univ of MI; BSE/Chem Engg/Univ of MI. *Born:* Sept 1941. BSEE Chem E & Met E Univ of Michigan 1963; PhD Met E MIT 1967. Employed by Chase Brass & Copper(subsidiary of Kennecott Copper) 1967- ; Casting Lab 1967-69, Sect Mgr 1968-69, Central Engrg Dept 1969- , Dir of Engrg 1970-76, V P Engrg 1976- . Mbr: ASM; AIME; ASM Handbook Comm 1971-76. Major prof emphasis: Productivity; automation; computers; continuous casting; primary metalworking processes. *Society Aff:* ASM, AIME.

Ruhlid, Robert R
Business: 719 Cheese Spring Rd, New Canaan, CT 06840
Position: Pres *Employer:* Facilities Management Consultants, Inc. *Education:* MBA/Mgt/Univ of New Haven, CT; BSEE/Elec. Eng./Univ. of Notre Dame, IND. *Born:* 10/24/38 Recognized as an effective leader and innovator in the fields of facilities & maintenance mgmt. Began career in field engrg and production mgmt for Shell Chemical Co; developed maintenance management pratice for the K.W. Tunlell Co. in Philadelphia. For 15 years was Div Mgr and VP for Syska & Hennessy, consulting eng firm headquartered in New York City. Dev facilities mgmt practice to include computerized facilities mgt prog. Is Pres of Facilities Management Consultants, Inc. A small business enterprise devoted to the executive devel of facilities mgrs and the implementation of mgmt systems. *Society Aff:* NSPE, AIPE, AAES

Ruhlin, William R
Business: 3 Cascade Plaza, Akron, OH 44308
Position: Pres. *Employer:* John G. Ruhlin Const Co. *Education:* BCE/CE/Univ of Akron. *Born:* 1/5/25. Akron native; entered family business 1948 after grad from college. Married 1951, he and wife Mary have two adult daughters. Succeeded his father as company pres 1960. Past pres of OH Contractors Assn & Akron Chapter of Associated Contractors of Am. Mbr Consulting Constructors' Council of Am. Mbr Advisory Bd Liberty Mutual in Cleveland. Bd mbr First Fed Savings of Akron. Mbr Advisory Bd, College of Engg & of Dev Steering Ctte, Univ of Akron. Ctte work in Jr Achievement, United Way, Area Progress Bd, Community Trusts, and other community groups. *Society Aff:* NSPE.

Ruina, Jack P
Business: E-38-620, 77 Mass Ave, Cambridge, MA 02139
Position: Professor of Electrical Engrg. *Employer:* Massachusetts Inst of Technology. *Education:* BE/EE/CCNY; MEE/EE/Polytechnic Inst of NY; DEE/EE/Polytechnic Inst of NY. *Born:* August 1923 Poland; reared NYC. Career: academic; mgmt; government. Prof Interests: Communications; Defense Technology; Defense & Arms Control Policy; Technology & Public Policy generally. Served many government cttes & adv grps in Dept of Defense; Arms Control & Disarmament Agency; Dept of Transportation; HEW; NSF; Natl Security Council. Faculty Brown Univ 1950-54; Univ of Ill 1954-63; MIT 1963- . While on leave from faculty served DOD 1959-63 in sev positions including Dir of Advanced Res Projs Agency 1961-63; Pres Inst for Defense Analyses 1964-66. V P for Spec Labs MIT 1966-70. Mbr Genl Advisory Committee a Adv & Dis (1969-73). Flemming Award as one of 10 outstanding young men in federal serv 1962. Sr Consultant, Office of Sci & Techn, Pol. Tr Mitre Corp; Bd/Dir Flow-Genl Corp. Fellow IEEE; Amer Academy of Arts & Sciences; AAAS. Mbr: Internatl Scientific Radio Union; Council on Foreign Affairs; Internatl Inst of Strateg Studies; American Academy of Arts & Sciences. *Society Aff:* AAAS, IEEE, URSI, IISS, AAS, CFR

Ruisch, Robert G, Jr
Business: 11401 Lamar, Overland Park, KS 66211
Position: Partner *Employer:* Black & Veatch Consulting Engr *Education:* BS/ME/Univ of IA *Born:* 5/25/39 Native of Waterloo, IA. Employed by Black & Veatch Consulting Engrs since 1964. Worked in the Mechanical Engrg Dept, the Field Engrg Dept, and Project Management. Became a Partner in Black & Veatch in 1979. Currently the Project Mgr for design and construction management of coal fired power generating station. Enjoys golf, private flying, racquetball, and skiing. *Society Aff:* ASME, NSPE, TBΠ.

Ruiz, Aldelmo
Home: 3816 Lake Blvd, Annandale, VA 22003
Position: Consultant - P/T *Employer:* Self-Employed-International development *Education:* MS/San Engrg/VPI & St Univ; BS/CE/VPI & State Univ; Dip/Mgmt Res for Nat Security/Industrial Coll of Armed Forces *Born:* 6/12/23 Born 6/12/23 in Yauco, PR. Served US Army 1942-45. Teaching-research fellow, VPI. Design engr, WA Suburban Sanitary Commission. Project engr, ERDL, Corps of Engrs. Chief, various engrg depts-assist mgr-tech coordinator Far East Division A&E firm. Owned consulting engrg firm. 1962-81 Agency for Intl Development. Chief engr, Taiz; dir, water supply & sanitation dept, YAR; development officer, Sana; chief engrg, Kabul; inter-regional engrg coordinator, AID. AID Mission Dir, YAR, El Salvador & Panama. Responsible US Tech Assistance programs. Chaired and participant in numerous international conferences and seminars; speaker at universities, institutes, professional clubs and television/radio appearances. Contributed articles to professional journals/publications. Author/co-author of numerous technical papers and reports. Board of Directors, Sana International School, Yemen Arab Republic. Spec int in furthering advancement of developing countries. Dist Honor Award 1966 AID. Honorable Certificate by Presidential Decree, YAR 1977. (YAR: Yemen Arab Republic). Condecoracion de la Orden De Vasco Nunez De Balboa En El Grado De Gran Cruz, Republica De Panama; Outstanding Career Award - Agency For Intnatl Development - 1981. (1987) Reg PE - DE, MD, DC, MA, PR. *Society Aff:* ASCE, NSPE, WSE, CIAPR, AWWA, USCID, DACOR, AFSA

Rumbaugh, Max E, Jr
Home: 10941 Timber Lane, Carmel, IN 46032
Position: Dir, Res Engg. *Employer:* Wallace Murray Corporation. *Education:* BS/Eng/US Military Acad; MS/Eng/Purdue Univ; MS/Bus/Purdue Univ. *Born:* Dec 11, 1937. Pres Soc of Res Admins 1973-74 & instrumental in org of its 1st internatl sect; Chrmn (1978-79) Ind Sect SAE. Corporate Officer of Midwest Applied Sci Corp, prod dev & engrg cons firm 1965-72. Artillery & Ordnance Officer US Army

Rumbaugh, Max E, Jr (Continued)
1960-63. 1972 joined Schwitzer/Wallace Murray Corp 1977 promoted to current position; fmed new Advanced Technology Grp which introduced sev new engrg techniques in design of turbochargers, fans, & fan clutches. Author of sev articles in prof & other journals. Hobbies: skiing; sailing; amateur winemaking. *Society Aff:* SAE, SRA.

Rummel, Charles G
Business: 549 West Randolph St, Chicago, IL 60606
Position: Vice Chrmn/Dir *Employer:* Lester B Knight & Assocs, Inc *Education:* BS/Architect-Eng/Univ of IL *Born:* 2/29/12 Plym Traveling Fellowship for European study. *Society Aff:* AAES, SAME, ARA, WSE

Rummel, Robert W
Business: PO Box 7330, Mesa, AZ 85206
Position: Pres. *Employer:* Robert W Rummel Assoc, Inc. *Education:* Aero Engg/-/Curtiss Wright Technical Inst. *Born:* 8/4/15. After grad worked in engg capacities at Hughes Aircraft Co, Lockheed Aircraft Co, Aero Engg Co, & Natl Aircraft Co. Was Chief Engr Rearwin Aircraft and Engines 1937-1943. Designed "Cloudster" & "Skyranger" aircraft series. With Trans World Airlines, Inc 1943-1978. Responsible for fleet planning, new aircraft progs, & forward tech planning. VP Engg 1955, VP Planning & Res 1959, VP Technical Dev 1968-1978. Formed Robert W Rummel Assoc, Inc 1978. Engages in airline, airplane & airport planning, engg and economic consulting services. Active in NAE, AIAA and SAE affairs. *Society Aff:* SAE, NAE, AIAA.

Rummel, Ward D
Home: 8776 W Mountain View La, Littleton, CO 80125
Position: Staff Engineer. *Employer:* Martin Marietta Aerospace/Denver Div. *Born:* BSChE 1959 Univ of Colorado; Teaching Asst while doing grad work in Biophysical Chemistry. Joined Martin Marietta 1962; primary tech work in quantitative nondestructive materials evaluation. Has presented & pub num papers in nondestructive testing. P Chmn; Education Chmn; Outstanding Mbr of Colorado Sect; Fellow ASNT. Prof Achievement Award Martin Marietta Aerospace 1973. Recent work in establishing NDE Reliability & in analysis of NDE Reliability data.

Rumsey, James W
Home: 557 Isla Pl, Davis, CA 95616
Position: Agri Engr Consultant (Self-employed) *Employer:* Self. *Education:* MS/Agri Engr/Univ of AZ; BS/Agri Engr/Univ of CA. *Born:* 12/5/43. Born & raised on a family farm in Woodland, CA. Served as a Civ Engr officer in USAF from 1967-72. Decorated three times for work on operations, maintenance & construction of USAF facilitieis around the world. Worked as Mgr of Field Res for world's largest co-operative fruit & vegetable food processor from 1972-77. Mbr of Natl Canner's Assn Agricultural Res Advisory Committee. Chrmn Pacific Region of ASAE. Reg PE in Calif. *Society Aff:* ASAE.

Rumsey, Victor H
Business: Univ of Calif/San Diego, La Jolla, CA 92093
Position: Professor of Applied Physics. *Employer:* University of California/San Diego. *Education:* DSc/Physics/Cambridge Univ; D Eng/Elec Eng/Univ Tohoku; MA BA/Math/Cambridge Univ *Born:* Nov 1919. Radar scientist 1941-45 TRE England & NRL USA; Atomic scientist 1945-48 Chalk River Canada; 1948-54 Assoc Prof Ohio St Univ; 1954-57 Prof Univ of Ill; 1957present Prof Univ of Calif at Berkeley & San Diego. 1960 Fellow IRE (now IEEE); 1962 Morris Liebman Mem Prize (IRE); 1965 Guggenheim Fellowship; 1971 Outstanding Educators of Amer Award. 1980 National Academy of Engineering. 1982 George Sinclair Award OH State Univ. Has made contrib in fields of antennas, especially frequency-independent antennas; atomic piles; electromagnetic waves; radio astronomy; optical & radio propagation through the turbulent atmosphere & solar wind. *Society Aff:* IEEE, URSI, NAE

Runyan, Edward E
Business: PO2988/2067 Commerce Dr, Midland, TX 79702
Position: Chairman of Board & President. *Employer:* Bell Petroleum Services Inc. *Education:* BS/Petr Eng/Univ of Tulsa; MS/Petr Eng/Univ of Tulsa *Born:* 1933. Sohio Petroleum Co 1955-62 in various engrg positions including District Engr for Permian Basin District Midland Texas; joined Permian Enterprises as V P, specialized in corrosion control problems & plant design; became V P of marketing & Pres of Agricultural Chem Div of Elcor Chem Corp, specialized in Marketing & Mgmt; became co-owner of cons engrg firm of Sipes, Williamson & Runyan, specialized in property evaluation & reservoir engrg; founded Bell Petroleum Services Inc, spec in oil field servs & environ studies. Mbr 1954 AIME; Dir SPE 1970-74; Pres SPE 1975; Dir AIME 1974-76. Permian Basin Engr of the Yr 1976. Pres Elect AIEM 1983. Pres AIME 1984. *Society Aff:* SPE, AIME, NSPE, SPWLA

Ruoff, Arthur L
Business: Dept. Materials Sci & Engg, Bard Hall, Ithaca, NY 14853
Position: Prof, Dept Dir. *Employer:* Cornell Univ. *Education:* PhD/Chemistry/UT Univ; BS/Chemistry/Purdue Univ. *Born:* 9/17/30. Specialist in high pressure science and engg including work with liquid and gas systems and with ultrapressure solid systems above 2 Mbars. Recent res work involves studies of the strength of diamond, studies of insulator to metal transitions in the megabar pressure range and x-ray crystallography above 2 Mbars. *Society Aff:* APS, ACS, MRS.

Rupp, Arthur F
Home: 113 Outer Drive, Oak Ridge, TN 37830
Position: Consultant (Self-employed) *Education:* BSChE/Chem Engg/Purdue Univ. *Born:* 1911 Indiana. Genl chem engrg with sev co's; 1943 transferred by DuPont to Univ of Chicago Met Lab Design & dev engrg on 1st operating reactor, separation of gram amounts of plutonium, Hanford design & start-up; transferred back to high explosives (chem) res to Clinton Labs (ORNL), started isotope prog 1946. Head of opers (reactor, engrg); ORNL exec; Dir Isotopes Dev Center. Ret. Mbr ACS; AIChE; Fellow ANS; recipient ANS Radiation Indus Award 1973. *Society Aff:* ACS, AIChE, ANS.

Rusack, John D
Business: 285 Main St, Catskill, NY 12414
Position: President *Employer:* John D. Rusack, P. C. *Education:* BS/ME/Duke Univ *Born:* 04/27/27 Native of Catskill, NY. Formerly Application Engr for Hyatt Bearings Div, GMC; Plant Engr, Staff Engr, Mfg Mgr for Ferroxcube Corp; and Plant and Chief Engr for Lorbrook Corp. Pres and Chief Engr of gen consltg firm engaged in mech, electrical, environmental, civil, and forensic engrg. Reg PE in NY, NJ, FL and IL. Specialist in plant and facility engrg. *Society Aff:* ASME, NSPE, SME, WPCF, AEE

Rusch, Willard VT
Business: Dept of Electr Engrg, Los Angeles, CA 90007
Position: Professor. *Employer:* University of Southern California. *Education:* PhD/EE/Caltech; MS/EE/Caltech; BSE/EE/Princeton. *Born:* 7/12/33. Prof or EE at USC 1972- ; visiting appointments at Tech Univ of Aachen (Germany), Naval Res Lab, Bell Labs, Tech Univ of Denmark, Max Planck Institute for Radio Astronomy; cons to Jet Propulsion Lab, TRW Sys Grp, Hughes Aircraft, others. 90 tech articles & papers on applied electromagnetics with particular specialization to reflector antennas. NSF Fellow; Fullbright scholar; Phi Beta Kappa; Eta Kappa Nu, Natl Dir; Sigma Xi; IEEE Fellow. Mbr Ad Com IEEE Antennas & Propagation Soc, Sr Sci Prize, Von Humboldt Foundation. *Society Aff:* IEEE.

Rusche, Peter A E
Home: 1021 Westfield Dr, Jackson, MI 49203
Position: Exec Engr. *Employer:* Consumers Power Co. *Education:* BS/EE/Tri State Univ; -/EE/Univ of MI; -/EE/MI State Univ. *Born:* 12/31/33. Peter A E Rusche PE, Exec Engr - Power Resources & System Planning Dept - responsible for dynamic

Rusche, Peter A E (Continued)
performance of Consumer Power Co's bulk power facilities & related res. Lectr & published extensively, including IEEE Transactions on Power Apparatus & Systems re: turbine generator torsional distress due to network disturbences. Is reg PE; mbr CIGRE & ASME, Sr Mbr IEEE; active on technical cttes, including the "ASME/IEEE Joint Working Group on Power Plant/Electric System Interaction." Served in various elected IEEE offices, Past Dir of Region 4. *Society Aff:* IEEE, ASME, NSPE, CIGRE.

Rush, James R
Business: Schmit, Smith & Rush, Consulting Engineers, Box 1802, Minot, ND 58702
Position: Chairman of Bd. *Employer:* Schmit, Smith & Rush Inc/Cons Engrs. *Education:* MS/Mech Eng/Univ of MN; BS/Mech Eng/Univ of MN. *Born:* May 29, 1930; born & raised in Minot ND. Served US Army Corp of Engr 1953-54 Fairbanks Alaska. Mech Engr Bloom & Oftedal Minneapolis Cons Engrs 1955- 59; formed present cons firm Fall 1959. Pres ND CEC 1967-68; mbr Prof Liability Insurance Ctte Natl CEC 1965- , Ctte Chmn 1972-74; Natl V P Amer Cons Engrs Council 1974, Sr Natl V P 1975; Dir Design Prof Financial Corp; Tr EADS Pension Plan (Natl ACEC Pension Plan). Coml pilot with instrument rating, active pilot. Hobbies: golf; sailing. Member Minot City Planning Commission - 1974 to present, Chairman 1977 - present. *Society Aff:* ASHRAE, ACEC.

Rush, Sidney
Home: 22 Tara Drive, East Hills, NY 11576
Position: Manager/Equipment Management. *Employer:* Sperry Corp/Systems Mgmt Div. *Education:* MS/Elect Engr/Univ of PA; BS/Elect Engr/Columbia Univ. *Born:* Native of NY. Served with Field Artillery during WWII. Proj Engr Franklin Inst 1948-50; EDO Corp 1950-54; Sperry 1954- . Assumed current position of Mgr/Equipment Mgmt 1972; respon for Combat Sys Equipment. Mbr Tau Beta Pi; Sr Mbr IEEE. Hobbies: tennis; bridge. *Society Aff:* IEEE.

Rushing, Frank C
Home: 6436 Belleview Drive, Columbia, MD 21046
Position: Engineering Consultant. *Employer:* Self. *Education:* BS/Mech Engg/Univ of TX; MS/Mech Engg/Univ of Pittsburgh. *Born:* July 1906. Native of Texas. BS Univ of Texas; MS Univ of Pittsburgh; Univ of Michigan; Univ of Charlottenberg (Germany). Westinghouse Res 1931-44: mechanics, dynamics, dynamic balancing, high speed machinery including uranium centrifuge, original high shock technology & machine dev for Navy; Engrg Mgr Westinghouse Motor Div 1944-64: electric motor dev & production, 1 to 2000 hp; Adv & Cons Engr Westinghouse Aerospace 1964-present: mech design of radar & electro optical equipment for aircraft & spacecraft. Mbr Tau Beta Pi; Sigma Xi; Westinghouse Order of Merit; US Navy Certificate of Commendation (for Shock); OSRD Certificate of Appreciation (Uranium Centrifuge). 30 patents. Fellow ASME; Fellow IEEE: P E. *Society Aff:* ASME, IEEE.

Rushlow, Maurice R
Business: 920 Broad St, St Joseph, MI 49085
Position: Pres. *Employer:* Wightman & Assoc, Inc. *Education:* BS/Civ Engrg/MI Stat Univ. *Born:* 4/25/33. Took additional schooling at Purdue and IL Inst of Tech. Was employed in Petroleum and construction industry from 1955 to 1971. Currently pres of Wightman & Assocs, Inc, St Joseph, MI and secretary of Wightman Pierce & Assoc, Inc, Cassopolis, MI R. W. Petrie & Assocs Benson Harbor, MI. Married, five sons & one daughter. *Society Aff:* ASCE, NSPE, MSRLS.

Rushton, J Henry
Home: 300 Valley St, 305, Lafayette, IN 47905
Position: Chemical Engineering Consultant. *Employer:* Self. *Education:* PhD/Phy Chem/Univ of PA; BS/ChE/Univ of PA. *Born:* 11/25/05 Born in PA. Prof Emeritus, Chem Engg, Purdue Univ; Dept Hd Univ of VA & IL. Inst of Tech; Prof Univ of MI. During WWII Hd of Oxygen Production Dev Prog for the armed forces, OSRD. Indus consultant for many yrs with Std Oil Co (IN), Mixing Equip Co, Commercial Solvents Co, AMAX Corp, & with other cos for shorter periods. Amer Inst of Chem Engrs, Pres 1957, Treas 1958; Awards: Walker, Van Antwerpen, Founders, Lecturer. Author of numerous publications on mixing, process design, and oxygen production. Interested in history, travel & geography. *Society Aff:* AIChE, ACS.

Russell, Allen S
Home: 164 Club Course Drive, Hilton Head Island, SC 29928
Position: Adj Prof, retired *Employer:* Univ of Pgh *Education:* PhD/Phys Chem/Penn State Univ; MS/Phys Chem/Penn State Univ; BS/Chem/Penn State Univ *Born:* 05/27/15 Born Bedford, PA. Joined Alcoa at New Kensington, PA 1940 in Alcoa Res Lab. Served in various tech mgmt positions. Elected VP-Alcoa Lab 1974, VP-Science & Tech, Alcoa 1978, VP and Chief Scientist 1981. 10 patents, 40 publ. Participant in res leading to Alcoa Smelting Process, major breakthrough in aluminum production. Fellow AIME, ASM, Penn State. Scientist of Year 1979, ind R&D magazine. Bayer Metal, 7th Int Congress Light Metals, Vienna. Gold Medal-ASM. Chem. Pioneer Award, Amer Inst Chem. James Douglas Gold Medal, AIME, Retired. *Society Aff:* NAE, AIME-TMS, ASM, ACS

Russell, David O
Business: 27 Birdsege Ave, Caribou, ME 04736
Position: Tech. Mgr. *Employer:* A.E. Staley Mfg., Co. *Education:* 5th Yr Certif/Mill Mgt/Univ of ME; BS/CE/Univ of ME. *Born:* 2/3/40. Native of ME. Previous work experience in pulp & paper, mfg of sugar, & mfg of fertilizer. Currently technical mgr. potato starch plant & responsible for all phases of operation in addition to the design of new process & related environmental & energy conservation measures. Active in rotary, sch system, church, two local treatment dist bds, & potato commission technical review committee. *Society Aff:* AIChE, TAPPA, IFT, PEng.

Russell, Frank M
Home: 14432 Linda Vista Dr, Whittier, CA 90602
Position: Principal Estimator. *Employer:* The Ralph M Parsons Co. *Education:* CE/Civil Engg/Univ of CO; MS/Civil Engg/Univ of CO; BS/Arch Engg/Univ of CO. *Born:* Feb 1908. Mbr Sigma Xi Hon Res Fraternity. Reg Mech Engr, Calif. Engr US Bureau Reclamation prior to WWII; civilian engr with Sea Bees 1942-45; Hd of Estimating Dept of Aetron, a div of Aerojet-Genl Corp 1961-69; Principal Estimator, The Ralph M Parsons Co. Overseas staff assignments include Australia, Hawaii & Greece. Charter mbr Amer Assn of Cost Engrs; Natl Publs Ctte Chmn 1962- 64; Certificate of Recognition Award 1974; Fellow 1978; Emeritus; founding mbr & Pres Southern Calif & Hawaii sects. Fellow Inst Adv Engr. Founding Mbr Nat'l Soc. Arch. Engr. Hobbies include music & photography. *Society Aff:* AACE, IAE, NSAE

Russell, Fred G
Home: 6143 Queensloch, Houston, TX 77096
Position: Vice President-Engineering. *Employer:* Delta Engineering Corp. *Education:* BS/ChE/Rice Univ; BA/ChE/Rice Univ. *Born:* Aug 1934. Native of Fort Worth, Tx. From 1957-64 worked for Amer Natural Resources Co performing petroleum & reservoir engrg functions; worked for Hudson Engrg Corp from 1964-73 as process engr, project engr & project mgr. Area of specialization at this time became design & const of oil & natural gas handling facilities. Joined Delta Engrg Corp in 1973 as VP of Engrg. Continue to work in design & const of oil & gas handling facilities as well as sour gas treating & sulfur recovery units. *Society Aff:* AIChE, SPE-AIME.

Russell, Frederick A
Home: 101 Elkwood Ave, New Providence, NJ 07974
Position: Distinguished Prof. in E.E., retired *Employer:* New Jersey Inst of Technology. *Education:* Eng Scd/EE/Columbia Univ; MS/Math & Physics/Stevens Inst Tech; EE/EE/Newark College of Engg; BSEE/EE/Newark College Engg. *Born:* April 1915. Dr Eng Sci from Columbia Univ 1953; Engr Arcturus Radio Tube Co 1935-37; taught EE at NJ Inst of Tech (Newark Coll of Engrg) 1937-44; Sr Proj Engr Columbia Univ, Div War Res 1944-45; returned to NJIT 1945- . Disting Prof in Elec Engrg 1967; Chmn of EE Dept 1956-76; Asst to Pres for Planning & Institutional Res 1976-79. Dir, Planning & Inst Res, 1979-1980. Fields: control systems, analog & digital computers, circuits & communications. Mbr Eta Kappa Nu, Tau Beta Pi, Sigma Xi, ASEE, & Fellow IEEE. Retired 1980. *Society Aff:* IEEE, ASEE, ΣΞ, ΤΒΠ, HKN.

Russell, James E
Business: Dept of Petroleum Engr, College Station, TX 77843
Position: Prof of Petroleum Engg & Geophysics. *Employer:* TX A&M Univ. *Education:* PhD/Theoretical & Appl Mechs/Northwestern Univ; MS/Civil Engr/SD Sch of Mines & Tech; BS/Civil Engr/SD Sch of Mines & Tech. *Born:* May 20 1940, Rapid City SD. 1966-67 Sr Res Engr Southwest Res Inst, San Antonio TX; 1967-71 Asst & Assoc Prof of Civil Engrg, SD Sch Mines & Tech, 1971- 76 Assoc & Full Prof of Mining Engrg. Exec V Pres & Staff Cons RE/SPEC Inc, Rapid City SD 1972-76. Proj Mgr for Rock Mechs, Off of Waste Isolation, Union Carbide Corp, 1977-78; Prof of Petroleum Engrg and geophysics, TX A&R Univ 1978- present. Chmn Black Hills Sect AIME 1975; Pres SD Sect ASCE 1976. *Society Aff:* SME-AIME, AGU, ASCE, AAM, ASTM

Russell, John E
Business: Box 1228, Baltimore, MD 21203
Position: VP. *Employer:* Maryland Casualty Co. *Education:* BS/Industrial/Univ of TN. *Born:* Oct 1920, Louisville Tenn. 31 years in safety engrg & engrg mgmt positions; head of Engrg Dept with 200 members. P Chmn Admin Engrs Ctte, Amer Insurance Assoc; presently on Bd/Dir & President of ASSE 1980-81, Chmn Prof Dev Conference 1975, Past Pres local ASSE Chap; Mbr Bd/Dir Md Safety Council; 1 of 3 persons selected to advise the Examination Ctte of Bd of Certified Safety Profs regarding examination content. Elder in Presbyterian Church. *Society Aff:* ASSE, AIHA, SSS.

Russell, John V
Business: 5456 Wayman, Riverside, CA 92504
Position: Consultant (Self-employed) *Education:* MS/Met Engg/Univ of KY; BS/Met Engg/Univ of KY. *Born:* 1/19/18. Lab Dir, Republic Steel, Chicago 1945-1958. Superintendent, Asst Superintendent Chicago 1958-1963. Chief Met, Kaiser Steel Corp, Fontana, CA 1963-1972. Self employed consultant in metals processing. Selection & failure analysis-works with mfg of welded pipe lines & offshore structures. Past Chrmn, Chicago Sec, AIME. Exec Committee, Chicago Sec, ASM. *Society Aff:* AIME, ASM, AWS.

Russell, Joseph L
Business: 2 Park Ave, New York, NY 10016
Position: Senior Vice President. *Employer:* Halcon International, Inc. *Born:* Oct 1931. BS Univ Kan; ScD MIT. 1952 joined Dow Chemical as Chem Engrg Proj Leader; 1958 joined Scientific Design Co Inc, subsidiary of Halcon Interntl Inc, as Dir Engrg Res, 1960 promoted to Asst V Pres Dir of Res, 1963 joined staff of Halcon Internatl Inc as V Pres, 1970 appointed Sr V Pres. Mbr ACS, AIC, API, AIChE & Electrochemical Soc.

Russell, Kenneth C
Rm 13-5066, MIT 77 Mass AVe, Cambridge, MA 02139
Position: Prof of Metallurgy & Prof of Nuclear Engg. *Employer:* Massachusetts Institute of Tech. *Education:* MetE/Metall Eng/CO Sch of Mines; PhD/Metall Eng/Carnegie Inst of Tech. *Born:* 2/4/36. Assoc Engr Westinghouse, specializing in thermoelectric matls; 1964- , Asst & Assoc and Prof of Metallurgy & Prof of Nuclear Engg MIT res & teaching in physical & nuclear metallurgy, 4 yrs in charge of MIT matls grad prog (140 students). 22 yrs cons in forensic metallurgy cases. *Society Aff:* AIME, APS, ASM, ANS, ASTM

Russell, Lynn D
Business: Mech Engr Dept, MS State, MS 39762
Position: Prof of Mech Engg. *Employer:* MS State Univ. *Education:* PhD/Mech Engg/Rice Univ; MS/Mech Engg/MS State Univ; BS/Mech Engg/MS State Univ. *Born:* 11/1/37. in Pontotoc, MS. Worked in space activities for 6 yrs during 1960's with NASA, TRW Sys & Lockheed; From 1969-79 served as Dean/Dir of Engg at Univ TN, Chattanooga in charge of developing new academic progs in engg. Num publ in heat transfer & Fluid Mech Reg PE AL, MS & TN. Chattanooga Engrs Club (PPres), TN Soc Prof Engrs (Pres 1978-79) & 7 others. From 1979-present, Prof of Mech Engg, MS State Univ, Res & Teaching in Energy Area. *Society Aff:* ASME, NSPE, ΣΞ, AAAS, MES, CEC, ΠΤΣ, ΤΒΠ, ΦΚΦ, ΟΔΚ

Russell, Paul L
Home: 6050 W Yale Ave, Denver, CO 80227
Position: Consulting Mining Engineer *Employer:* Self Employed *Education:* BS/Mining/Univ of AZ; MS/Mining Engrg/Univ of AZ. *Born:* Jan 1917 Ariz. Married Florence M Horricks, children Lillian & Roger. 1942-45 Mining Engr US Government; 1945-48 Mining Engr private indus, Mont, Nicaragua CA, Colo; Mining Engr Bureau of Mines, Oil Shale Dev, Mining Res, design & const of underground defense facilities as cons & in direct charge; nuclear res in weapons & peaceful uses; Res Dir USBM Denver, 140 employees- 60 prof, mining & related res. Mbr SME-AIME, Colo State Oil Shale Ctte; Federal Council on Sci & Tech Energy R&D. Reg PE Ariz & Colo. Honorary Degree Univ Ariz; Meritous & Distinguished Service Award, Dept of Interior; Who's Who in the West & Who's Who in Atoms. Author of *The History of Western Oil Shale*-January 1980 release. Technical Editor, Mining Engrg Magazine, Society of Mining Engrs. *Society Aff:* SME, CMA, DMC, SME.

Russell, Ralston, Jr
Business: 2041 College Rd, Columbus, OH 43210
Position: Prof Emeritus & Cons Ceramic Eng *Employer:* Ohio State University. *Education:* PhD/Ceramic Engg/OH State Univ; CerE/Ceramic Engg/OH State Univ; MSc/Ceramic Engg/OH State Univ; BCerE/Ceramic Engg/OH State Univ. *Born:* Nov 4 1910, Pomeroy Oh. Employed by AC Spark Plug Div, Genl Motors Corp; Genl Ceramics & Steatite Corp; Westinghouse Electric Corp, Dir Ceramic Res & Sect Mgr; since 1946 Professor Ceramic Engrg Ohio State Univ & prof Ceramic Cons. US Sci Cons in Germany 1945. P Pres Amer Ceramic Soc, Natl Inst Ceramic Engrs, Ceramic Education Council & Keramos; Fellow Inst of Ceramics Ltd; Mbr Natl Soc Prof Engrs, Canadian Ceramic Soc, AAAS, ASEE, ASTM, Tau Beta Pi, Sigma Xi, etc. Reg PE Ohio, Penn, DC. US Army Award of Merit 1951; Amer Ceramic Soc Distinguished Life Mbr 1982 and Fellow 1941; Amer Ceramic Soc Cramer Award 1974 and Bleininger Award 1987, NICE Greaves-Walker Award 1970, Keramos Role of Honor Award 1975, ASTM Honorary Member 1980. Author of 100 tech & prof publ, patentee. Interests include athletics, travel & wildlife. *Society Aff:* ASM, ACS, ASTM, ACerS, NICE, ASEE, NSPE, INSTITUTE OF CERAMICS (ENGLAND), CCS, ΤΒΠ, ΣΞ, KERAMOS, AAAS, CLAY MINERALS SOCIETY

Russell, Robert A
Home: 5000 W 68 Terrace, Shawnee Mission, KS 66208
Position: Control Department Sr Tech Coordinator, Power Div. *Employer:* Black & Veatch. *Born:* 4/14/15. Kansas City Mo. BSME 1937 Univ of KS. Since 1938 employed by Black & Veatch, since 1956 Genl Partner, specializing in design of steam-electric generating stations, since 1960 to 1977 Hd Control Engrg Dept in Power Div. Retired from active profl work as of May 1, 1980. Now Life Fellow of ASME, Life Mbr ISA. Mbr IEEE. Reg PE MO & KS. *Society Aff:* ASME, IEEE, ISA

Russell, Samuel O
Home: 4433 Belmont Ave, Vancouver, British Columbia V6R 1C3
Position: Professor *Employer:* University of British Columbia. *Education:* BS/Civil Engg/Queen's Univ N Ireland; MS/Hydraulics/Queen's Univ N Ireland; PhD/Water Sys/Queen's Univ N Ireland. *Born:* May 1932. BSc, MSc & PhD Queen's Univ, Bel-

Russell, Samuel O (Continued)
fast N Ireland. 1954-57 worked on const in England; 1957-68 worked on hydro-electric design in Canada, respon for design of hydraulic strucs at Mica Dam on Columbia River; now Professor in Civil Engrg at Univ British Columbia , Vancouver - teaching & water resources with emphasis on applications of systems engrg concepts. ASCE Hilgard Prize 1968. *Society Aff:* ASCE, CSCE.

Russell, T W Fraser
bNewark, DE 19716
Position: Chrmn, Chem. Eng. Dir, Inst of Energy Conversion *Employer:* Univ of DE. *Education:* PhD/ChE/Univ of DE; MSc/ChE/Univ of Alberta; BSc/ChE/Univ of Alberta *Born:* 8/5/34. Res Engr, Res Council of Alberta, 1956-1958. Design Engr, Union Carbide Canada, 1958-1961. Asst, Assoc, Full Prof, Chem Engg Univ of DE 1964 to present. Assoc, Acting Dean of College of Engg 1974-1979. Presently chair, Dept of Chemical Engineering & Dir, Inst of Energy Conversion Univ of DE. Consultant to El DuPont, various other industrial firms. Author of three books and over 60 technical publications. Reg PE - DE. *Society Aff:* AIChE, ASEE, ACS

Russo, A Sam
Home: 1130 Mine Hill Rd, Fairfield, CT 06430
Position: VP. *Employer:* Stauffer Chem Co. *Education:* BS/ChE/TX A&M *Born:* 4/15/31. Native of Houston TX. Joined Stauffer in 1954 as Jr Engr at Houston Mfg Plant. Held various positions in sales & marketing in the ind chem div & in food ingredients div. and of Gen Mgr of Food Ingredients. Assumed present position of VP and Gen Mgr of Sulfuric Products Div in Oct 1981. Enjoy Golf. *Society Aff:* ΤΒΠ

Russo, Joseph R
Business: Plaza Centre, 20 W. Ridgewood Ave, Ridgewood, NJ 07450
Position: Sales Manager. *Employer:* Chicago Mill & Lumber Co. *Education:* BS/Commercial Forestry/College of Forestry at Syracuse Univ. *Born:* Sept 1924, Hillburn NY. Since 1949 with Chicago Mill & Lumber Co, designers & manufacturers of heavy duty shipping & storage containers, since 1970 Sales Mgr & Container Design Dir. Prof Mbr SPHE since 1959, Eastern Regional V Pres 1976-77; Pres Internatl Packaging Council 1977 in conjunction with PMMI Exposition; Prof Mbr NIPHLE, Wash DC since 1970; Lecturer SPHE packaging courses since 1973. Boy Scout Commmitteeman. Lector St Paul's Church. Golfer, Sports Dir Ramsey Golf & Country Club, Pres, 2 yrs Ramsey Golf & Country Club; Cub Scout Committeeman & Publicity Chrmn; Natl VP, SPHE 6 yrs. *Society Aff:* SPHE, NIPHLE.

Rust, Thomas D
Business: 10523 Main St, Fairfax, VA 22030
Position: Sr Vice Pres *Employer:* Patton, Harris, Rust & Assocs *Education:* MS/Public Works/Geo Wash Univ; BS/CE/VA Polytech Inst *Born:* 07/21/41 Native of Front Royal, VA. Began engrg career with Fairfax County, VA. Chief Design Engr for Sanitary Engrg Projects. Joined Patton, Harris, Rust & Assocs in 1969 and works in Civil-Environmental field. Active in politics serving on Town of Herndon's Planning Commission & Town Council. Currently serving third term as Mayor. Mbr and currently Rector of the Longwood Coll bd of Visitors. Chosen "Young Engr of the Yr" in 1969 and "Man of the Yr" in Herndon in 1979. Listed in "Outstanding Young Men of America" and "Who's Who in the South. *Society Aff:* NSPE, WPCA, CEC

Ruston, Henry
Business: 333 Jay St, Brooklyn, NY 11201
Position: Professor *Employer:* Polytechnic Univ *Education:* PhD/EE/Univ of MI; MS/EE/Columbia Univ; BSE/EE/Univ of MI; BSE/Math/Univ of MI. *Born:* 07/23/29 Born in Lodz, Poland. Served with US Army 1952-54. Intermediate test engr, Curtis-Wright Corp Woodridge, NJ. 1955, electrical engr, Reeves Instrument Co, Mineola, NY. 1955-56, assoc research engr, Univ of Michigan Research Inst 1956- 60, asst prof electrical engrg, Univ of PA 1960-64, prof, electrical engrg and computer sci Polytechnic Inst of NY 1964-; engrg conslt 1960-; Mbr IEEE (Sr), Sigma Xi, Eta Kappa Nu, Tau Beta Pi. Co-author: *Electric networks: Functions, Filters, Analysis,* McGraw-Hill 1966, author: *Programming with PL/I,* McGraw-Hill 1978. Contributor of articles to profl journals. Hobbies: bridge and tennis. *Society Aff:* IEEE.

Rutherfoord, J Penn
Home: 116 Dogwood Drive, P.O. Box 664, Marion, VA 24354
Position: Pres. *Employer:* Rutherfoord Assoc. *Education:* BS/Elec Engg/Univ of VA; BS/Gen Engg/Univ of VA. *Born:* 3/26/14. Native of Roanoke, VA. Instr in Math and English during grad work at Univ of VA. Engg Training Prog at Gen Elec Co. Moved up in GE through line mgt. Spent final 8 yrs of 21 yr service on President's staff and Dir of Corporate Advanced Mgmt School. Retired on pension in 1958. Consulted smaller cos until present. Held positions of Dir and Exec VP at Intl Resistance Co and at Penn Controls, Inc and as VP of Burgess Battery Co. Have been active in IES, ITE, IEEE, and ASHRAE. (For more see Who's Who in America, 1969) and (Who's Who in Finance and Industry 22nd Ed. 1981-82). President VA Soc of PE 1979-80. Licensed PE Virginia. *Society Aff:* NSPE, AEE.

Ruthroff, Clyde L
Home: 1 Brook La, Holmdel, NJ 07733
Position: Consultant *Employer:* Retired. *Education:* MA/Math/Univ of NB; BSEE/EE/Univ of NB. *Born:* Feb 4 1921, Sioux City Iowa. US Navy; 1946-77, Bell Telephone Labs. Fellow IEEE. Speciality is res in microwave communications including propagation. *Society Aff:* IEEE.

Ruths, David W
Home: 2900 High Ridge Rd, Matthews, NC 28105
Position: Vice Pres - Process Division Manager. *Employer:* Luwa Corporation *Education:* BS/ChE/PA State Univ; MChE/ChE/Clarkson Coll of Tech; MBA/Business/FL Atlantic Univ *Born:* 2/12/39. Hometown, PA. Process Dev assigments with Firestone Plastics & Avisun Corp preceded Dev Mgr position with Parkson Indus Equip Co. Mfg Tech Supt for Monsanto Co, Springfield MA for vinyl acetate based & acrylic polymers; 1968 joined Parkson Corp as Dev Mgr, from 1973 VP & Tech Dir, respon tech, customer services, dev of specialty mass & heat transfer equip & sys, & new products selection. From 1977-1980. VP - Corp Dev, Parkson Corp respon for a profitable growth prog through addition of new prod and business via licenses, acqusitions, etc. From 1980 on - VP-Process Div Gen Mgr, respon for overall profitablity, performance, and growth of process equip/systems centered div of Luwa Corp and for broadening product line and accelerating growth through new products acquisitions. Enjoy camping, sailing, skiing, active Christian service work. Tau Beta Pi & Sigma Tau. Texaco Grad Fellow at Clarkson. Bd mbr of South Amer Crusades. Bd mbr of Charlotte Christian School. *Society Aff:* AIChE, ACS

Rutledge, Philip C
Home: 103 Brynwood Dr, Easton, PA 18042
Position: Partner - Retired. *Employer:* Mueser Rutledge Wentworth & Johnston. *Education:* SB/-/Harvard College; MS/Civil Engg/MIT; ScD/Civil Engg/Harvard Univ. *Born:* Feb 1906. Work in const & engrg between ScD 1939 Harvard Univ & honorary Dr Engrg 1957 Purdue. 1933-52 Instructor, Professor & Dept Chmn at Harvard, Purdue & Northwestern; since 1952 Partner Mueser, Rutledge, Wentworth & Johnston & predecessor, cons engr specializing in foundations for large projs including Vehicle Assembly Bldg for NASA Apollo proj. Chmn Bd of Cons on Earth Dams for CA Dept of Water Resources; Cons to AEC. Tech Assoc for Atlantic-Pacific Interoceanic Canal Studies. ASCE Terzaghi Lecture 1968; Mbr NAE. *Society Aff:* ASCE, NAE, ACEC.

Rutledge, Wyman C
Business: 8th & Hickory Sts, Chillicothe, OH 45601
Position: Fellow Res Dept, Cons. *Employer:* The Mead Corp Cent Res Labs. *Education:* PhD/Physics/Univ of MI; MS/Physics/Univ of MI; AB/Physics Math &

Rutledge, Wyman C (Continued)
Chemistry/Hiram College. *Born:* 12/15/24. 1944-46 US Army; 1946-47 Underwater instrumentation, Woods Hole Oceanographic Inst; 1947-50 upper atmospher res, nuclear spectroscopy, Engrg res Inst Univ MI; 1950-52 short-lived nuclear decay sch, Argonne Natl Lab; 1952-56 ultra-high vacua, electron emission, Phillips Lab; since 1956 at the Mead Corp, physicist, Mgr of instrumentation & automation, Fellow of the Research Dept, Consultant in Sys Engrg, specializing in online sensors, process control & computer sys, air & water quality inst, electrophotography, ink-jet printers & physical phenomena. Phoenix Predoctoral Award 1950-52; Columbus OH Tech Man of Year 1972. ISA Dist VP 1973-75, Fellow 1975, VP Tech Dept 1976-78, Environ Tech Comm & Intersociety Air Sampling Comm 1972-76, 1976-79; Chm, Pres Soc Advisory Comm, 1980-81 . Tappi Engrg, VP Process Con 1972-73, Fellow 1976; Amer Paper. Inst, Inst Res Comm 1973-79; Chrmn Acad & Symposium Comm, Pulp & Paper Fdn, Miami Univ; Prog Chrmn, Houston 1980 ISA and Columbus 1982 ISA. *Society Aff:* ISA, APS, TAPPI, OSA, SCI, IES, ICC, MDC.

Ryan, Harold J
Home: 69 Chestnut Ave, East Setauket, NY 11733
Position: Consulting Engineer (Self-employed) *Education:* MBA/Bus Admin/NY Univ; ScB/Electrical Engg/MIT. *Born:* March 1905, Boston Mass. Reg PE NY & NJ. Designed 1st hotel guest room air conditioning system in NY for Waldorf Astoria; Pres & Ch Engr during WWII of mech contracting co designing & installing engine test cells; Cons Engr, specialty air conditioning existing bldgs. Pres NY Chap ASHRAE 1944-45; Representative to EJC 1945; Chmn ASHRAE Tecgh Cttes: Large Bldg Air Conditioning 1961-73, Definitions & Nomenclature 1961-68. Author *Air Conditioning,* 'Heat Recovery' & 'Abbreviations & Symbols' chapters, ASHRAE Handbooks. ASHRAE Distinguished Service Award 1970, Fellow 1971. Mbr ASHRAE Handbook Ctte 1973-84, Mbr ASHRAE Terminology Ctte 1982-85. *Society Aff:* ASHRAE.

Ryan, John T, Jr
Business: P.O. Box 426, Pittsburgh, PA 15230
Position: Chairman of the Board. *Employer:* Mine Safety Appliances Co. *Education:* MBA/Bus Admin/Harvard Univ; BS/Mining Engr/Penn State Univ; DSc (Hon.)/-/Duquesne Univ; LLD (Hon.)/-/Univ of Notre Dame. *Born:* 3/1/12. 1936 joined Mine Safety Appliances Co as Sales Engr, Asst Genl Mgr, Genl Mgr, 1948 elected Exec V Pres, 1953 Pres, 1963 Chmn of Bd, Trustee Emeritus Univ of Notre Dame. Distinguished Alumnus Award, Penn State Univ; Pittsburgh Man of the Yr; Honorary Fellow Amer Coll of Hospital Administrators. Erskine Ramsey Medal AIME; Distinguished Mbr Award SME or AIME. *Society Aff:* ACS, ASME, ASSE, NSPE, MIIA, NMRA.

Ryan, Michael P
Business: 1133 Ave of Americas, New York, NY 10036
Position: Vice President - Personnel. *Employer:* Inmont Corporation. *Education:* MChE/Chem Engg/Rensselaer Polytechnic Inst; BChE/Chem Engg/Rensselaer Polytechnic Inst. *Born:* Nov 1921. BChE 1942 & MChE 1947 Rensselaer Polytechnic Inst, Troy NY. 11 Years with Allied Chemical as Dir of Advertising respon for promotion & publicity for 8 div; currently a V Pres & officer of Inmont Corp, a major manufacturer of printing inks & automotive paints, respon have included advertising, personnel & managing a business grp for the corp. Various articles have appeared in 'Harvard Business Review,' 'New York Times,' sev business journals; specialized work with Dr George Gallup Creating ideas & questions for use in his nationwide poll appearing in newspapers.

Ryan, Robert S
Business: 50 West Broad St, Columbus, OH 43215
Position: President *Employer:* Robert S. Ryan & Assocs *Education:* BS/IE/OH State Univ *Born:* 07/25/22 Pres, Robert S. Ryan & Assocs; 1975-1981 Dir of OH Dept of Energy; 1955- 1975 Sr VP of Columbia Gas System; 1952-1955 Research Assoc Battelle Memorial Inst; 1947-1952 Plant Engr, Maintenance, Intl Harvester, Published over 35 articles in technical journals. Received 1970 Distinguished Alumnus Award Coll of Engrg, OH State Univ. Mbr Amer Soc Prof Engrs; Amer Soc Prof Conslts; OH Oil & Gas Assoc; OH Coal & Energy Assoc; International Platform Association (IPA). Personal: Born 1922; married Esther L. Moore, 3 children. Residence; 6566 Plesenton Dr So, Worthington, OH 43085. *Society Aff:* ASPE, ASPC, IPA.

Ryan, Roderick T
Home: 4501 Hayvenhurst, Encino, CA 91346
Position: Consultant, Motion Picture Technology *Employer:* Self *Education:* PhD/Communications/Univ of So CA; MA/Cinema/Univ of So CA; BA/Cinema/ Univ of So CA. *Born:* 9/18/24. in Philadelphia, PA. Military service: US Navy, combat motion picture cameraman, motion picture labs at San Diego & Wash DC, Able & Baker Atomic Bomb Tests at Bikini Atoll, photographic squadrons. Prof exper, Eastman Kodak Co: Quality control Engr in Color Print & Processing Div, Phptographic Engr in Color Technology Div, motion picture engr & coordinator of engg service District Sales Mgr Regional Engrg Dir in Motion Picture & Audiovisual markets Div, reson for sales & engg activity within the Western Region of the US. Active in Boy Scouts of Amer as cub master, scout master, explorer post advisor, order of the arrow, chap advisor; collector of edged weapons. SMPTE: Governor 1969, 1983-85, VP Photo Sci 1970-77, Fellow 1964, Special Commendation 1976, 1973 & 1960. Herbet T Kalmus Gold Medal 1978, VP Motion Pictures Affairs 1980-81 Treasurer 1986, AMPAS - Scientific and Engrg Award 1982, BKSTS Fellow 1983. ASC Honorary Membership 1986. *Society Aff:* SPIE, BKSTS, ASC, AMPAS, SMPTE, ATAS.

Rybicki, Edmund F
Home: 10524 S. 68 E. PL, Tulsa, OK 74133
Position: Prof & Chrmn, Mech Engg. *Employer:* Univ of Tulsa. *Education:* BS/Civil Enggg/Case Inst of Tech; MS/Engg Mechanics/Case Inst of Tech; PhD/Eng Mechanics/Case Western Reserve. *Born:* 10/7/41. Native of Auburn, OH. PhD in Engrg from Case Western Reserve Univ. Res Scientist and Projs Mgr at Battelle Columbus Labs from 1968-1979, specializing in finite element stress analysis for composite materials, biomechanics and residual stresses due to welding. Prof and Chrmn of the Mech Engg Dept at Univ of Tulsa in Tulsa, OK. Over 65 publications and one patent. Mbr of ASME, NACE, ASEE and Reg PE in OK. Member of Tau Beta, Sigma Xi, and Phi Gamma Kappa Honary Fraternities. *Society Aff:* ASME, NACE, ASEE, NSPE.

Ryder, John D
Home: 1839 SE 12th Ave, Ocala, FL 32670
Position: Dean of Engineering, retired *Employer:* Michigan State Univ. *Education:* PhD/EE/IA State Univ; DEng/Engrg/Tri-State Univ; MS/EE/OH State Univ; BEE/EE/OH State Univ *Born:* 05/08/07 Grew up in Columbus, OH. Two yrs in Gen Elec Co electronics test, then ten yrs in charge electrical research with Bailey Controls, Wickliffe, OH; temperature control and telemetering-24 patents. Taught electrical engrg at IA State Univ, Ames, for 8 yrs, as asst prof to prof and Asst Dir, Engrg Experiment Station. Head of EE Dept, Univ of Il, 1949-54, then Dean of Engrg, MI State Univ to 1972. Pres, IRE 1955. IEEE Haraden Pratt Award, 1979; chrmn Centennial Task Force, IEEE 1978-84. Two yrs Brazil-AID High Education Team; Mbr Radio Club of Amer, Ed, IRE, 1958, 59; Ed IEEE 1963, 64; Exec VP IEEE 1974; Prof, Univ of FL, '80 Centennial Hall of Fame- IEEE- Educator 1984. *Society Aff:* IEEE, AAAS, ASEE, Rotary Club, Ocala

Ryder, Robert A
Home: 20 Hanken Dr, Kentfield, CA 94904
Position: Chief Engr. *Employer:* R Ryder Consulting Engineers *Education:* MS/Sanitary Engg/MIT; BS/Civ Engg/Purdue. *Born:* 5/28/29. Native of Lafayette, IN. Served with Navy Civ Engr Corps in Korean War. Proj Mgr with Kennedy Engrs of San Francisco from 1956 to 1972 & headed investigation & design of maj

Ryder, Robert A (Continued)
water & wastewater treatment facilities in Western US & S Am. Dir of Pacific Environmental Lab 1964-1979. Started Collins & Ryder. Consulting Engrs of Tacoma, Reno & San Francisco 1979. Author of numerous technical papers & recipient of AAWA's Best Paper of 1978 on "State of the Art in Water Treatment Deisgn, Instrumentation and Analyses–. *Society Aff:* AAEE, ASCE, AWWA, NACE.

Ryerson, Joseph L
Home: RD 2 Box 4, Holland Patent, NY 13354
Position: Pres *Employer:* Joseph L Ryerson *Education:* PhD/Elect Engg/Syracuse Univ; MEE/Elect Engg/Syracuse Univ; BEE/Elect Engg/Clarkson College. *Born:* Oct 1918, Goshen NY. 1941-46 Deputy Lab Dir, Ward Leonard Electric Co; 1946-49 Taught electrical engrg at Mohawk Coll, 1949-51 Evansville Univ; 1951-64 Rome Air Dev Center Engr; 1964-67 Tech Dir SHAPE Tech Center, 1967-69 Scientific Advisor Headquarters US Air Force; 1969-76 Tech Dir Communications Div, Rome Air Dev Center. Chmn Dept of Defense Metrication Panel 1973-75. Developed aircraft approach landing equip, 1st internatl satellite communications. Mbr Sigma Pi Sigma, Sigma Chi; Fellow IEEE, AAAS; NATO Star 1967; Air Force Outstanding Awards 1963, 1972. Who's Who in World, Who's Who in Government. Hobbies: Violin, Photography, Choral Singing. P E NY and IA. *Society Aff:* ΣΞ, ΣΠΣ, AAAS, IEEE.

Rynn, Nathan
Business: Physics Dept, Univ of CA, Irvine, CA 92717
Position: Prof *Employer:* Univ of CA, Irvine *Education:* PhD/EE/Stanford; MS/EE/Univ of IL; B/EE/CCNY *Born:* 12/02/23 NYC. USNR 1944-46. Research Engr RCA Sarnoff Research Center 1947-52, Princeton, NJ. Worked on computers and color TV. Research asst at Stanford 1952- 56. Mbr of the Tech Staff at Ramo Wooldridge, 1957-1958. With Plasma Physics Lab, Princeton Univ 1958-65, worked on controlled fusion and developed the Q-machine. With U. C. Irvine Physics Dept 1966 to present. Main interests are basic plasma physics and controlled fusion. Enjoy wooodworking, classical music and photography. *Society Aff:* IEEE, APS, AAAS, ΣΞ

Saad, Theodore S
Business: 11 Huron Drive, Natick, MA 01760
Position: President. *Employer:* Sage Laboratories, Inc. *Education:* BS/EE/MIT. *Born:* Sept 1920 West Roxbury, Ma. Microwave Engr since 1941. Worked at MIT Radiation Lab 1942-45; Submarine Signal Co 1945-49; Sylvania 1953-55; Microwave Dev Labs 1949-53; Co-Founder, Pres and Chmn of Sage Labs Inc 1955- ; Co-Founder, Vice Pres of Horizon House Inc 1958-1979 . Fellow of IEEE, Honorary Life Mbr of IEEE - MTT Soc, Natl Lecturer of MTT-S 1972. Chrmn of IEEE Pub Inf Com 1983-1985. Chrmn IEEE Electro Bd 1980, Chmn of S Middlesex Area Chamber of Commerce 1977-78; MTT Soc Distinguished Service Award 1983, IEEE Centennial Medal 1984. *Society Aff:* IEEE.

Saarinen, Arthur W, Jr
Home: 555 Winderley Place, Maitland, FL 32751
Position: Sr VP - Regional Mgr, Chief Technical Officer *Employer:* Camp Dresser & McKee Inc. *Education:* B/CE/Univ of FL *Born:* 12/9/27 Mr. Saarinen has practiced in the engineering profession for over 37 years, including 35 years with the same firm and its predecessors. He is a Sr. VP and mbr of the Bd of Dirs of Camp Dresser & McKee Inc. Specifically, he is the Chief Technical Officer of the firm and is responsible for technical development and quality management. He is also Regional Manager of the South Region of the U.S. encompassing 14 offices and some 400 employees in 13 states and the District of Columbia. He has served as Pres of the FL Inst of Consulting Engrs and the FL Pollution Control Assoc, and is a Dir and Mbr of the Bd of Ctrl of the Water Poll Ctrl Fed. He is a Diplomate of the American Acad of Environ Engrs and a Reg PE in many states. *Society Aff:* ASCE, FPCA, AAEE, AWWA, ACEC, NSPE, FES, FICE, PMI

Saba, Shoichi
Business: 1-1, Shibaura 1-Chome, Minato-Ku, Tokyo 105, Japan
Position: Adviser to the Bd *Employer:* Toshiba Corp. *Education:* B.Eng./Elec Engg/Tokyo Univ. *Born:* 2/28/19. in Tokyo, Japan. Grad Elec Engg Dept of Tokyo Univ 1941. Joined Toshiba immediately after then. Assigned to research engg in High Voltage Engg Lab of the Co. 1951 transfered to Electric Power Engg Dept of the same company. Worked as a rep and Liaison engg to General Electric Co Schenectady, NY 1956-57. Chmn of The Japan Inst of Indus Engg, 1982–. Adviser to the
e Bd of Toshiba Corp, 1987–. *Society Aff:* IEEE, IEEJ, CIGRE, JIIE.

Sabnis, Gajanan M
Business: Suite 110, 6801 Kenilworth Ave, Riverdale, MD 20737
Position: VP, Prof *Employer:* Sheladia Assoc Inc., Howard Univ. *Education:* PhD/Struct Engrg/Cornell Univ; M of Tech/Struct Engrg/Indian Inst of Tech; BE/CE/Univ of Bombay *Born:* Native of Bombay, India; naturalized US Citizen (1977). Res Engr, American Cement Corp (1968-69). Engrg Group Supervisor, Bechtel Power Corp (1970-74). Presently at Howard Univ, Wash, DC (since 1974) and also, Principal at Sheladia Assoc Inc., Riverdale, MD (since 1980). Responsible as VP (Res & Dev) for the Dept; coordinate res with education and ind. Pres, American Concrete Inst. (ACI) chapter in Wash, DC (1974); Honorary Distinguished Mbr of ACI Chapter of Maharashtra, Bombay (1980). Author of two books and over 50 technical publications. James Berkeley Gold Medal Award (1961), ACI Chapter Activities Award (1976) as First Recipient. *Society Aff:* ACI, IE, ASCE, ASEE, PCI

Sabri, Zeinab A
Home: 2144 Ashmore Ct, Ames, IA 50010
Position: Assoc Prof. *Employer:* IA State Univ. *Education:* PhD/Nucl Engg/Univ of WI; MSC/Nucl Engg/Univ of WI; BSC/EE/Univ of Alexandria. *Born:* 3/18/45. Dr Sabri is a nucl engr with strong background in licensing & environmental aspects of nucl power stations & nucl power plant design, operation & safety. From 1972 to date, Dr Sabri has consulted for Los Alamos Scientific Lab & at EPRI. She has participated in the preparation of an EPRI document on the Worldwide Status & Options of Unconventional Energy Sources. The document was prepared for the World Energy Conf (1978) under a contract with EPRI. She is currently managing an interdisciplinary team of 26 engrs, statisticians, & psychologist to analyze the impact of human factors on the operation, relaibility, economics & safety of commercial nucl power plants. Dr Sabri has chaired, organized & participated in several workshops & confs. She has served as prog chrmn for the IA/NB Sec of the Am Nucl Soc. Dr Sabri has authored or co- authored over 60 technical publications including 15 tech papers on reactor safety and reliability. *Society Aff:* ANS.

Sachs, Herbert K
Business: 667 Merrick, Rm 302.1, Detroit, MI 48202
Position: Prof *Employer:* Wayne State Univ *Education:* DrIng/Mech/Techn Univ Braunschweig; MS/Engrg Mech/Wayne State Univ; Dipl/ME/Cantonales Technikum Winterthur, Switzerland *Born:* 3/4/19 Fourteen yrs of industrial practice in the vehicle ind. Since 1958 on the faculty of Wayne State Univ. 1968, conveying of the 1st Intl Conference on Vehicle Mech at Detroit. Elected first pres of IAVSD (1977-81). Founder of the journal Vehicle System Dynamics and its co-editor in chief since 1971. Conslt and res engr for the US Dept of Trans, the Automobile Ind and expert witness in several vehicle accident litigation proceedings in fed courts. (25 publications, 2 patents); Reg PE (1958), MI. *Society Aff:* IAVSD, ASME, IMS

Sachs, John P
Business: 299 Park Ave, New York, NY 10171
Position: Pres *Employer:* Great Lakes Carbon Corp. *Education:* PhD/ChE/IL Inst of Tech; MS/ChE/IL Inst of Tech; BS/ChE/IL Inst of Tech *Born:* 3/10/26 in Dusseldorf, Germany. Raised in Chicago, IL. Studied chem engrg at IL Inst of Tech. Interrupted by service in US Army 1944-46. Received PhD from IIT 1952. Process engr for the Visking Corp, Chicago, followed by serving in various mgmt capacities and

Sachs, John P (Continued)
R&D engrg operations for Union Carbide's Plastics Div which acquired Visking in the mid-50's. Joined Great Lakes Carbon Corp as VP- Operations 1966. Elected Pres and Ch Exec Officer 1978. Also presently serving as Chrmn of the Bd of Gen Refractories Co. Fellow of the American Inst of Chem Engrs and presently serving as VP-Pres Elect of that organization. Dir-Chem Mfgs Assoc 1978-82 and a Trustee of Fairfield Univ. *Society Aff:* AIChE, ΣΞ, SCI.

Sachs, Samuel
Home: 860 Burr Avenue, Winnetka, IL 60093
Position: Pres *Employer:* Sam Sachs, PC *Education:* MS/Mech Eng/U of Ill; BS/Mech Eng/U of Ill. *Born:* 8/14/17. BSME & MSME U I. Prof Engr: Ill, Ca, DC, Wis, Mass, Kan. Distin Alumni Dept of M & I Engr U I 1972. Fellow: ASHRAE 1972; NSFPE NSPE; ISPE; City of Chicago Ventilation Code Ctte; Bd of Editors 'Heating, Piping and Air Conditioning'; Editorial Dir 'Building Systems Design'. Mbr: Standard Club; Art Inst; Brain Res Foundation; Automated Procedures for Engrg Cons Inc - Founding Mbr V Pres, Mbr of the Bd of Trustees. U I Mech Engr Res Dept 1938-42; Natl Adv Ctte for Aeronautics Langley Field, Va 1944-45; 2nd Lt AUS 1944. Joined Skidmore, Owings & Merrill 1947; Ch of Mech & Elec Div 1953; Assoc Partner 1955 Sr Consultant 1979, PEPP Sam Sachs PC 1982-Present. Projects: US Air Force Acad Colorado Springs, Col; U I Chicago Circle Campus Chicago, Ill; John Hancock Ctr Cicago, Ill; Solar Telescope Kitt Peak, Ariz; Chicago Civic Ctr Chicago, Ill, Gateway Ctr I, II, III & IV. Chicago, Ill; ARAB Intl Bank World Trade Ctr, Cairo, Egypt; Hyatt Intl Hotels: Cairo, Egypt; Lima, Peru, Riyhad, Saudi Arabia; Headquarters Groupo Alfa, Monterrey, Mex; One Magnificant Mile Bldg, Chicago, ILL; Menninger Foundation West Campus, Topeka, Kansas; Orchestra Hall, Chicago; Woodfield Center IV. Schaumburg, Ill; East-West Corp Ctr Powers Grove, Ill; Hay's Wharf Dev, London, UK. *Society Aff:* ASHRAE, NSPE, NSFPE.

Sack, Edgar A
Business: 600 W John Street, Hicksville, NY 11802
Position: Sr VP & Genl Mgr. *Employer:* Genl Instrument Corp. *Education:* PhD/EE/Carnegie Mellon; MS/EE/Carnegie Mellon; BS/EE/Carnegie Mellon. *Born:* 1/31/30. Westinghouse Elec Corp 1954-69, Res Labs Television Engrg, Mgr Electronics Dept, Mgr Solid State Devices Dept; Molecular Electronics Div Mgr of Engrg, Mgr of Operations, Genl Mgr. Genl Instrument Corp 1969 to date VP and Genl Mgr, Integrated Circuits Div, Sr VP and Genl Mgr, Microelectronics Div, VP & Genl Mgr Microelectronics Group. Pubs 30, patents 15. Fellow of IEEE, Eta Kappa Nu, Outstanding Young Elec Engr for 1969. Fellow Polytechnic Inst of NY, Merit Award, Carnegie Mellon Alumni Assn; Dir, Regional Industrial Technical Educ Council, Long Island; Mbr, Action Ctte of Long Island; Advisory Cttes, SUNY Stony Brook, Carnegie Mellon Univ and Polytechnic Inst of NY. Hobbies: travel, classical music, boating. Vice Commodore, Huntington Yacht Club. *Society Aff:* HKN, IEEE, TBΠ, ΣΞ, ΦKΦ

Sacken, Donald K
Home: PO Box 372, Ace, TX 77326
Position: Engineer *Employer:* Tx Dept. of Health *Education:* BS/Chem Engg/Clarkson. *Born:* July 1920. Tau Beta Pi and Omega Chi Epsilon. Native of Leonia, N J. Served with US Navy 1942-45. Worked in Res & Dev with American Cyanamid and Jefferson Chem 1946-56, 5 patents issued. Production (grass-roots plants) GAF and Houston Chem 1957-69. Opers Mgr Oxirane Corp 1970-75. V Pres Oxirane Corp, Genl Mgr Oxirane Chem Co (Channelview) 1976. Respon for grass-roots new petrochemical complex. 1978-80 VP Mfg Admin, Oxirane Corp. Engr TX Dept of Health, implementation hazardous Waste Control, 1981- . *Society Aff:* AIChE.

Sackinger, William M
Home: PO Box 80591, College, AK 99708
Position: Assoc Prof Geophysics and Elec Engrg *Employer:* Univ of AK Geophysical Inst. *Education:* PhD/Elec Eng/Cornell Univ; MS/Elect Eng/Cornell Univ; BS/Physics/Univ of Notre Dame *Born:* 1/10/39. Native of Bolivar, NY. Proj Engr on microwave research at Zenith Radio (1961-62), at Westinghouse Elec (1962-64). Dev image intensifiers containing channel electron multipliers at Westinghouse, 1964-66. Studied statistics of channel multipliers at Corning Glass Works (1966-70), and analyzed new display and reprographic concepts there. From 1970, conducted res and taught at Univ of AK, Fairbanks, in electronic instrumentation, high-frequency devices, and electronics. Extensive res on ice movement, ice stresses, spray ice accretion, ice islands. EE, Dept Head (1972-75), coordinator offshore res Geophysical Institute, (1972-76, 1979-83). Pres POAC Intl Comm, 1983-85, Chairman, Fossil Energy Res. Council (1986-present) Conslt on Arctic sea ice engrg problems. *Society Aff:* IEEE, IOP, IGS, EPS, AGU.

Sacks, Newton N
Business: 400 - 19th Street, John Deere Road, IL 61265
Position: Dir, Safety & Environment. *Employer:* Deere & Company. *Education:* MS/Ind Mgmt/State Univ of IA; BS/Elec Engg/State Univ of IA; BA/English/State Univ of IA. *Born:* March 1918 Sioux City, Iowa. With Deere & Co since 1941 in wide range of engrg activities including foundry engrg, indus furnace design, dust control. Mgr Production Engrg Dept, Mgr Materials Engrg Dept. Assumed current respon as Dir Safety & Environ in 1970. Respon for corp environ control, occupational safety, indus hygiene, and product safety. Fellow ASM 1969; Chmn & Dir AFS-CMI 1972; Dir Environ Div AFS 1975. Reg Prof Engr Ill Farm and Industrial Equipment Institute (FIEI) Chm- Safety Policy Advisory Comm Lt (ELT) USNR 1943-46. *Society Aff:* ASM, AFS, CMI, ASME.

Sadowski, Edward P
Home: 57 Walker Drive, Ringwood, NJ 07456
Position: Conslt *Employer:* Self Employed *Education:* BS/Metallurgy/Univ of Notre Dame. *Born:* July 16, 1922 Bayonne, NJ. Served with USAAF 1943-45. Staff Met/Matls Engr with M W Kellogg Co 1950-53; Sr Met Crucible Steel Central Res Lab 1953-55; Sr Met The Intl Nickel Co Inc Bayonne and Paul D Merica Res Labs 1955-82, is now known as INCO Res & Dev Center Conslt 1982-. Ten US patents and 20 papers on high temperature alloys, Pt metals, stainless and maraging steels, cast low alloy steels, welding and welding products underwater welding, coal gasification and liquefaction - IR 100 Award - 1963. Mbr: ASM, AWS NY Acad of Sciences. Listed: Who's Who in the East, Amer Men & Women of Sci, Intl Biography. *Society Aff:* ASM, AWS

Safar, Zeinab S
Business: Mech Engrg Dept, Boca Raton, FL 33431
Position: Assoc Prof *Employer:* FL Atlantic Univ *Education:* PhD/ME/Univ of Pittsburgh; MS/ME/Univ of Pittsburgh; BS/ME/Cairo Univ *Born:* 10/18/48 Dr. Safar was born in Egypt. Between 1968 & 1973 she was teaching fellow at the Univ of Pittsburgh. She taught at Cairo Univ, American Univ at Cairo and FL Atlantic Univ. In 1978 she was a visiting scholar at Univ of CA, Berkeley. Dr. Safar is the author of over 30 papers in the areas of lubrication and Rheology. *Society Aff:* ASME

Safdari, Yahya B
Home: 120 E High Point Rd, Peoria, IL 61614
Position: Prof *Employer:* Bradley Univ *Education:* DSc/Heat Trans/NM State; MSME/Heat Trans/Univ of WA; BSEE/Power/Aligarh Muslim Univ; BSME/Thermal Sci/Aligarh Muslim Univ, India *Born:* 7/25/30 Prof of Mech Engrg since 1969; Pres, Sun Systems, Inc., 1975-79; designed five solar homes, supervised solar installations, and tested their performance. Patented Solar Air Collector SAFTM, specialist in solar grain drying. Founding father of Solar Energy Industries Assoc. Mbr, IL Solar Adv Bd. Pres, SAF Energy Consultants, Inc. Developed Mech Engrg Curriculae at UMM AL-QURA (Makkah) Univ, Makkah, Saudi Arabia. Author of several research papers. Received a grant from Energy & Natural Resources Depart-

Safdari, Yahya B (Continued)
ment, State of IL to design and install photovoltaic (sun to electric) lighting for remote area application in the Wildlife Prairie Park. *Society Aff:* ASME, ISES, ASES, ΣΞ

Saffman, Philip G
Business: CA Inst of Tech, Pasadena, CA 91125
Position: Prof. *Employer:* Caltech. *Education:* PhD/Appl Mathematics/Cambridge Univ; BA/Mathematics/Cambridge Univ. *Born:* 3/19/31. Native of Leeds, England. Taught at Cambridge Univ and King's College London Univ. Appointed to Caltech Faculty 1964. Specialty: Theoretical Fluid Mechanics. Served as editor of Journal of Fluid Mechanics. *Society Aff:* AAAS.

Safhay, Meyer
Business: 79 Madison Avenue, New York, NY 10016
Position: Chief Engineer and Associate. *Employer:* Charles A Manganaro, Cons Engrs. *Education:* MCE/Sanitary/NY Univ; BCE/Civil/CCNY. *Born:* Sept 1925. Prof Engr in NJ. Served US Navy 1944-46. Design engr with cons firms spec in sanitary and hydro-elec design 1948-56. With Charles A Manganaro, cons engrs & its predecessor firms since 1956. Assumed current pos as Assoc & Ch Engr of this 90 plus man firm in 1963. Respon for personnel, reports, specifications and design of water, solid wastes and sewerage projs including a 450-million dollar highly sophisticated water pollution control facility serving 29 communities and several thousand industries. Various articles pub in trade journals. *Society Aff:* WPCF, ACEC, PEPP.

Sagar, Bokkapatnam Tirumala Ananda
Business: Bldg 40 DTC West, Denver Tech Center, PO Box 3006, Englewood, CO 80155
Position: Asst Chief Mechanical Engrg *Employer:* PRC Engrg Consultants, Inc *Education:* PhD/CE/CO St Univ; MS/CE/Univ of CO; BE/CE/Andhra Univ-India *Born:* 3/6/29 Served as Dir/Deputy Dir of the Central Water Power Commission, Government of India, and was largely responsible for major hydraulic structures and gates and hoists for most of the major dams and hydropower projects for two decades in India. Was also teaching in Water Resources Development Training Centre, Univ of Roorkee, India, for more than ten years as visiting expert. Serving PRC Engrg Consultants, Inc since December 1969. Responsible for hydraulic and hydro- mechanical equipment for major hydropower and irrigation projects around the world. Author of several papers in intl journals and conferences. Invited to serve as member of High Dam Consultancy Bd by U.P. State Government in India, and TOKTEN program of the United Nations. Enjoys philosophy and music and is a member of Toastmasters Intl. *Society Aff:* ASCE, USCOLD, IWRA, ΣΞ.

Sage, Andrew P
Home: 8011 Woodland Hills Ln, Fairfax Station, VA 22039
Position: First Amer Bank Professor of Info Tech & Dean *Employer:* George Mason Univ Fairfax VA *Education:* BSEE/EE/Citadel; SMEE/EE/MIT; PhD/EE/Purdue Univ. *Born:* 1933 Charleston, SC. Assoc Prof at Univ of Arizona and Prof of Electrical Engrg & Nuclear Engrg Sciences at the Univ of Fla. With Southern Methodist Univ for 7 yrs as Dir of Information and Control Sciences and Hd of Elec Engrg Dept. With Univ of VA for 10 yrs as Lawrence R Quarles Prof, Assoc Dean of Engrg and Applied Sci Chrmn, Dept Chem Engg & Chrmn Dept of Engg Sci & Systems Engr; 1984-85 Assoc VP, Acad Aff and since 1985 Dean of the School of Information Technology & Engineering, George Mason Univ. Author of 5 textbooks and numerous journal papers in various areas of Systems Engineering including Planning and Decision Support Systems, optimization and estimation theory, software systems engrg. Recipient of Frederick Emmonds Terman Award from Amer Soc for Engrg; Barry Carlton Award and Norbert Winer Award from IEEE. Editor since 1972 of IEEE Transactions On Systems, Man and Cybernetics; CoEditor, since 1979, of Large Scale Systems; Editor of International Federation of Automatic Control Journal Automatica since 1981. Elected Fellow in IEEE & AAAS. Received Centenial Medal from IEEE and (honorary) Doctor of Engineering Degree from Univ of Waterloo, Canada. *Society Aff:* IEEE, AAAS, TIMS, ASEE.

Saha, Subrata
Business: Dept. of Orthopaedics, PO Box 33932, Shreveport, LA 71130
Position: Prof. and Coor. of Bioeng. *Employer:* LSU Medical Center *Education:* Ph.D./Applied Mech./Stanford University; M.S./Eng. Mech./Tennessee Technological University; B.E./Civil Eng./Calcutta University *Born:* 11/02/42 Was Assistant Prof of Engrg & Applied Science at Yale Univ from 1977-79, before joining LSU Medical Center in 1979. Became a Fellow of ASME in 1987. Received RCDA award from NIH, Fulbright Award, US-India exchange of scientist award from NSF. Author of over 200 publications in national and international journals. Editorial bd mbr of Journal of Biomedical Materials Research, Journal of Bioelectricity, Bioetelemetry and Patient Monitoring. Organized the 4th New England Bioengineering Conference in 1976 in New Haven Connecticut; 1st and 5th Southern Biomedical Engrg Conference in 1982 and 1986 respectively in Shreveport, LA. *Society Aff:* IEEE, ASME, SEM, BME, ASCE

Sahney, Vinod K
Home: 4727 Burnley Dr, Bloomfield Hills, MI 48013
Position: Prof. *Employer:* Wayne State Univ. *Education:* PhD/Ind Engg/Univ of WI; MSME/ME/Purdue Univ; BSME/ME/Ranchi Univ. *Born:* 11/16/42. Prof of Ind Engg & Operations Res College of Engg, Wayne State Univ, Detroit. VP Corp Planning, and marketing Henry Ford Health Care Corp 1981-present, Pres, Fairlane Health Services Corp 1984-present. Also taught at Harvard Univ & Univ of WI-Madison. Consultant to numerous organizations including many hospitals, Natl Center for Health Services Res, Veterans administration, Kuwait Govt Ministry of Health, US Army Tank Command. Mbr, AAAS, APICS, Sigma Xi, Tau Beta Pi. VP IIE- Detroit Chapter 1972-1974, Div Dir IIE - C & IS 1977-1979, Div Dir IIE - Health Services 1984-1985, IIE - Publications Policy Board - 1987, IIE - Res. Adv. Bd. IIE - HSD Conf Chrmn 1976-1977, ORSA - HAS Council 1979-1982. ORSA - HAS Council Chrmn 1983-1985. Outstanding Young Men of Am 1978. Author of over fifty Publications, Fellow - IIE and HMSS. *Society Aff:* IIE, ORSA, ΣΞ, APICS, ASEE, HMSS

Saibel, Edward A
Business: 2022 Bivins St, Durham, NC 27707
Position: Retired *Education:* PhD/Math/MIT; SB/Math/MIT. *Born:* 12/25/03. Instr math and mechanics Univ of MN 1927-30, asst prof to prof mechanics Carnegie Inst of Tech 1930-57. Prof & chrmn mechanics RPI 1957-1967, Inst prof Carnegie-Mellon Univ 1967-1972, chief solid mechanics branch US Army Res Office 1972-86, adjunct prof NC State Univ 1974-, adjunct prof Duke Univ 1977-, reg PE PA, Fellow ASME, Mayo Hersey Award ASME 1978, life mbr ASLE 1972. Retired. *Society Aff:* ASME, ASLE, AMS, MAA.

Sain, Michael K
Business: Dept of Electrical and Computer Engineering, Notre Dame, South Bend, IN 46556
Position: Frank M. Freimann Prof. of Electrical Engg *Employer:* Univ of Notre Dame. *Education:* PhD/EE/Univ of IL; MS/EE/St Louis Univ; BS/EE/St Louis Univ. *Born:* 3/22/37. Dr Michael K Sain is a Fellow of the Inst of Elec & Electronics Engrs (IEEE) for contributions to the theory of multivariable control systems. He is a Past Editor of the "Transactions on Automatic Control–, a publication of the IEEE Control Systems Soc, & Past Chrmn of the Soc's Technical Committee on Linear Systems. Dr Sain is author of the book "Introduction to Algebraic System Theory" (Academic Press, 1981), technical editor of the volume "Alternatives for Linear Multivariable Control" (Natl Engg Consortium, 1978), & is associated with more than 150 technical articles - one of which received an award from the Bendix Energy Controls Div. *Society Aff:* IEEE, SIAM, ASEE.

St Clair, John M
Business: 76 Parmelee Drive, Hudson, OH 44236
Position: Exec VP *Employer:* Walter Metals Corp *Education:* BS/Metallurgy/Lehigh Univ. *Born:* April 1937. Native of Pittsburgh Pa. BS Lehigh Univ. 1959 joined Universal-Cyclops Specialty Steel Div of Cyclops Corp as Sales Trainee, 1960 Sales Rep Chicago office, 1968 Resident Sales Rep Providence RI, 1972 Asst Dist Sales Mgr Cleveland Sales office, 1974 Dist Sales Mgr Cleveland office. Active in ASM & local civic activities. Joined Walter Metals in 1981; Became Exec VP in 1986. *Society Aff:* ASM.

St Pierre, George R
Business: 116 West 19th Ave, Columbus, OH 43201
Position: Professor and Chairman Department of Metallurgical Engineering *Employer:* The Ohio State University. *Education:* ScD/Metallurgy/MA Inst of Tech; SB/Metallurgy/MA Inst of Tech. *Born:* June 2 1930. 1954-56 Res Metallurgist, Inland Steel Co; 1956-57 US Air Force; 1957- , Faculty Ohio State Univ. Cons & author. Visiting Professor, Univ of Newcastle NSW 1975; Fellow USEPA 1976; ASM Bradley Stoughton Award 1961; Fellow Met Soc AIME 1976; Fellow ASM 1982; NSF Engrg Adv Comm 1984. ASM gold Medal 1987; AIME Mineral Industry Educator Award 1987; Dir Mineral Industry Inst, 1985- ; Adv. Bd NSF Net Shape Manufacturing Ctr; Edison Materials Technology Ctr; Dir The Metallurgical Soc of AIME, 1988- . *Society Aff:* TMS-AIME, ISS-AIME, ΣΞ, ASM, ASEE, AME-AIME.

St Pierre, Philippe D
Business: Box 568, Worthington, OH 43085
Position: Manager Engineering. *Employer:* General Electric Company. *Education:* PhD/Ceramics/Imperial College; Dip ARSM/Mining/Met/Royal Sch of Mines; BS/Metallurgy/Royal Sch of Mines. *Born:* 1925. Nuffield Travelling Fellowship US & Canada 1947. 1948-55 Ceramist/ Metallurgist Canadian Bureau of Mines; 1955-60 Ceramist/Metallurgist Genl Electric Co Res Lab, since 1960 GE Mgr, since 1967 Mgr Engrg Specialty Materials. Special tech interests: powder processing, metallurgy of tungsten, high temperature ceramics & high pressure technology. Fellow ACS; GE Inventors Medal. *Society Aff:* Am Ceram Soc.

Saito, Shigebumi
Home: 15-6, Kamiuma 5-chome, Setagaya-ku, Tokyo 154, Japan
Position: High Commissioner *Employer:* Space Activities Commission of Japan *Education:* Doctor of Eng/Elec Engg/Univ of Tokyo; Bach of Eng/Elec Engg/Faculty of Eng. *Born:* Sep 17 1919 Tokyo Japan. Assoc Prof U of Tokyo 1947-57. 1957- , Prof Inst/Indus Sci & Inst/Space & Aero Sci U of Tokyo, res on low noise microwave electron tubes, parametric amplifiers & laser engrg. 1955-- Mbr Japan Sci Satellite Proj. Exec Dir Natl Space Dev Agency of Japan 1969-74. Since 1974 Commissioner Space Activities Comm of Japan. IEEE Fellow 1974; IEEE Spectrum Editorial Bd Mbr. Commendation Award Minister of Post & Telecommunication 1971. Inst of ECE of Japan the Merit Award 1976. Prime Minister's Award of World Communication Year 1983. Imperial Purple Ribbon Medal 1984. Since 1980 Prof Emeritus of Univ of Tokyo. *Society Aff:* IEEJ, IECEJ.

Sakamoto, Toshifusa
Business: 2-2 Kanda-Nishikicho, Chiyoda-ku Tokyo, Japan 101
Position: Pres. *Employer:* Tokyo Denki Univ. *Education:* Dr/Elec Engg/Univ of Tokyo. *Born:* July 1906 Nara, Japan. Lecturer, Assoc Prof, Prof, Dean of Engg of Univ of Tokyo 1929-67. Prof since 1967, Pres since 1974 of Tokyo Denki Univ. Fellow of IEEE 1972, was Pres of IFMBE 1965-67; was Pres & is Honorary Mbr of Inst of Electronics & Comm Engrg, Inst of Elec Engrg; Japan Soc of Medical Elec & Biol Engrg, Inst of Television Engrg. The Medal of Honor with Purple Ribbon from the Emperor of Japan 1963; Medal of Honor from Inst of Elec Engrg 1964 and from Inst of Elec and Comm Engrg 1958. *Society Aff:* IEEE, IEEJ, IECEJ, JSMEBE, ITVJ.

Salama, C Andre T
Business: Department of Electrical Engineering, Univ of Toronto, Toronto, Canada M5S 1A4
Position: Professor *Employer:* University of Toronto *Education:* Ph.D/Elec. Eng./Univ of British Columbia; M.A.Sc./Elec. Eng./Univ of British Columbia; B.A.Sc. (Honors)/Elec. Eng./Univ of British Columbia *Born:* 27/09/38 Employed, Bell Northern Research, Ottawa, Scientific Staff, 1966-67. Since 1967, on staff, Department of Electrical Engrg, Univ Toronto; currently Prof and Holder James M. Ham Chair in Microelectronics. Appointed, Bd of Dir, Canadian Microelectronics Corporation (1984-86); founding Ch of Bd. Research interests include design & fabrication of semiconductor devices & integrated circuits. Published extensively in technical journals. Holder of ten patents. Served as consultant to semiconductor industry in Canada and U.S. Member, editorial board, Solid State Electronics. Associate Editor, IEEE Transactions on Circuits and Systems. Fellow, IEEE. *Society Aff:* IEEE, ECS

Salama, Kamel
Business: Mech Engrg Dept, Houston, TX 77004
Position: Prof. *Employer:* Univ of Houston. *Education:* PhD/Phys/Cairo Univ; MSc/Phys/Cairo Univ; BSc/Phys & Math/Cairo Univ. *Born:* 4/1/32. Born in Egypt. Lectr phys at Cairo Univ until 1965. Two yrs on leave at Upsala Univ, Sweden. Immigrated to the US 1966 to work as research consultant at the Ford Scientific Lab in Dearborn, MI. Joined Rice Univ 1968 as sr research scientist in Dept of Mech Eng & Mtls Sci. 1973, joined the faculty of Mech Eng Dept, Univ of Houston as assoc prof. 1977, Prof & Dir of Mtls Engg prog. Over 100 technical publications in the areas of Mtls Sci & Nondestructive Testing. Mbr of TMS-AIME, ASM, ASNT & APS. *Society Aff:* TMS-AIME, ASM, APS.

Salane, Harold J
Business: Civ Engg Dept, Columbia, MO 65211
Position: Prof. *Employer:* Univ of MO. *Education:* PhD/Civ Engr/Univ of TX-Austin; MS/Civ Engr/Rice Inst; BS/Civ Engr/Univ of TX-Austin. *Born:* 1/7/22 in Schenectady, NY. Served as pilot instr with USAF 1944-46. Employed by Gulf Oil Corp as a construction engr 1949-51, and by Dow Chem Co as a civ engr 1951-58. Held interim teaching appointment at Rice Inst 1960-63. Joined Univ of MO-Columbia 1965, and presently a prof of civ engg. Awarded an NSF Sci Faculty Fellowship to attend Univ of TX-Austin 1963-65. Did postdoctorl study in Manchester, England, with a Leverhulme Fellowship 1966. Employed as a consulting engr in offshore structures with Brown and Root in Houston 1978-81. Recreational interests are golf and fishing. *Society Aff:* ASCE, SEM, ASEE.

Salati, Octavio M
Business: 308 Moore Sch D-2, Philadelphia, PA 19104
Position: Prof Elec Engr & Dir, Tv Systems. *Employer:* Sch Engrg & Applied Sci U Penn. *Education:* PhD/Elec Eng/Univ of PA; MS/Elec Eng/Univ of PA; BS/Elec Eng/Univ of PA; Grad Studies/-/Brooklyn Poly. *Born:* 12/12/14. Native of Philadelphia, PA area. Student Engr RCA Corp 1937-39; div engr CG Conn Ltd 1939-42; Senior Engr Hazeltin Corp 1942-48; Moore Sch of Elec Engrg, U of Penn 1948- (Res Assoc 1948-50, Asst Prof 1950-63, Assoc Prof 1963-75, Prof 1975-84) Emeritus Prof 1984. Dir of Continuing Engrg Studies 1970-76; Dir of TV Sys 1970- . Fellow of IEEE Past Chmn Philadelphia Sect (Philadelphia Sect Award 1972); Faculty advisor of student winning First Prize Paper award (GEMC-1970); Natl Dir Etakappa Nu 1963-65, Chmn Philadelphia Chap; Sigma Xi, AAAS, ASEE; Holder of 5 US pats (one on BNC connector) & Canadian patent on Microwave Absorber. Teaching of electronics, signal theory, musical acoustics & electronics & current res on education tech. Listed in: Amer Men & Women in Sci, Who's Who in East. Consultant on Microwave Communications sys, Cable TV Sys. Visiting Professor, Phalavi University, Shiraz, Iran 1969, Visiting Professor, Northwest Telecommunications Institute Xi'an, Peoples Republic of China 1981 lecturer on "Engineering of Microwave Communications Systems' for IEEE. Lecturer on Modern Microwave Measurements. Lectures in India, Scotland, Belgium. Partner of: STELCOM -

Salati, Octavio M (Continued)
Comm Consltg Eng Firm. Presently consultant on Electrical & Magnetic properties of composite materials and metal laminates. *Society Aff:* IEEE, HKN, ΣΞ, AYH, AMC, PTC.

Saleh, Fikri S
Business: 505 Oberlin Rd, Rm 218, Raleigh, NC 27605
Position: President. *Employer:* Fikri Saleh & Assocs, Cons Engrs. *Education:* MS/Civil Engg/NC State Univ; BS/Structural Engg/Cairo Univ, Egypt. *Born:* April 1928 Cairo, Egypt. MS NC State U; BS & Diploma of higher studies in structural engrg from Cairo U, Egypt. US citizen. Served as structural engr in the Proj Directorate, Ministry of Military Production, Egypt 1955. Established a cons engrg firm in Cairo, Egypt 1955; established another branch office in Raleigh, NC 1975. Mbr: ASCE, ACI, NSPE & Chi Epsilon. *Society Aff:* ASCE, ACI, NSPE, XE.

Salem, Eli
Business: 2720 US Hgwy 22, Union, NJ 07083
Position: VP-Technical Dept. *Employer:* Graver Water Div *Education:* MBA/Finance/NYU; MChE/CE/PIB; BChE/CE/Bklyn College & CCNY. *Born:* 1/21/36. Joined Graver Water Div in 1957 as CE in Technical Dept; held various positions including that of Technical Mgr. Appointed VP - Technical Dept, Feb, 1978. Spent entire career in process design of water treatment systems, specializing in ion exchange, coagulation and filtration, plus other process technologies. A number of papers were authored and patents issued which have been commercially utilized in the industry. Enjoys tennis. *Society Aff:* ASTM, AIChE, AWWA, WPCF, ASME.

Saletta, Gerald F
Home: 10 W 32nd St, Chicago, IL 60616
Position: Assoc Dean of Engrg *Employer:* IL Inst of Tech *Education:* PhD/EE/IL Inst of Tech; MS/EE/Notre Dame Univ; BS/EE/Notre Dame Univ *Born:* 5/30/36 Native of Chicago, IL. Taught at Notre Dame Univ prior to joining staff at IL Inst of Tech in 1962. Formerly Acting Chrmn of Elec Engrg Dept; original dir of the Univ's ITFS (Instructional Television Fixed Service) TV Sys. Assumed Asst Dean of Engrg duties in 1975. Profl interests include integrated circuit fabrication, electronic and digital circuit applications. Consultant to engrg corps. *Society Aff:* ASEE, IEEE

Salgo, Michael N
Home: 137-32 76th Ave, Flushing, NY 11367
Position: Dir of Engrg *Employer:* A-J Contracting Co Inc *Education:* MS/Civil Engg/VA Poly Inst; BS/Eng/Northwestern Univ. *Born:* 1/7/14. Reg PE NY & 6 other states. WWII Navy duties included planning & design for Charleston Naval Shipyard (new concepts for drydock & shop structures) & special facilities (floating structures, artificial harbors, etc). As Dir CBS Facilities Engg (1957--74) had major involvement to produce Broadcast Facilities during major expansion period; technical breakthoughs made projs economically feasible. ASCE: Treasurer National Society P Pres MET Sect & Natl VP; Past Chmn Res Council on Performance of Structures, active on Evaluation Existing Transportation Structurers & their overall natl priorities; Past Chmn Tech Council on Res; Chairman Man Group A; as Manager Transportation & System, Div for Lk Comstock & Company was involved in Transportation Engg-Evaluation at Structures & Railroad Engg etc. As Sr Assoc for Ammann & Whitney was Dir of Broadcast Engrg. As Director of Engrg for A-J Contracting in charge of construction management for conversion of former Navy Receiving Station to NY City Jail. *Society Aff:* ASCE, NSPE, SAME, USNI, TROA

Saline, Lindon E
Home: Rt 2-Dresbach, La Cresent, MN 55947
Position: Retired *Education:* PhD/Elec Engg/Univ of WI; MS/Elec Engg/Univ of WI; BEE/Elec Engg/Marquette. *Born:* March 1924 Minneapolis, Minn. US Navy 1942-46; Cutler-Hammer 1946-47; General Electric 1948-1984. Tech & Mgmt positions in power systems, computers, aerospace & defense bus & corporate staff. Reg PE in NY, Ariz., Fla., MD. Fellow ASEE, IEEE, AAAS. Numerous ctte roles; author; lecturer; a leader in natl minority engrg effort. Retired 1984. *Society Aff:* IEEE, ASME, NSPE, ASEE.

Salinsky, John L
Home: 8988 Florence St, Brewerton, NY 13029
Position: Chief Mech Engr. *Employer:* Sargent, Webster, Crenshaw & Folley. *Education:* BSME/ME/RPI. *Born:* 6/24/24. Served in Army 43-46. Worked for GE, Carrier Corp, before joining present firm. Dept Hd with firm since 1954 specializing in HVAC, PLBG, & Energy Conservation Design. *Society Aff:* ASME, ASHRAE, NSPE, ASPE.

Salis, Andrew E
Home: 4105 Shady Valley Court, Arlington, TX 76013
Position: Dean of Engineering. *Employer:* U of Texas at Arlington. *Education:* PhD/EE/TX A&M Univ; MS/EE/Auburn Univ; BS/EE/Auburn Univ; Prof (ME)/EE/Auburn Univ *Born:* Oct 10, 1915. Instructor, Assoc Prof Texas A&M U 1940-51; US Navy WWII; Senior Engr, Proj Engr, Group Engr, Senior Group Engr for Radiation System, Genl Dynamics; Prof, Head of Dept, Dean of Engrg U of Texas, Arlington 1951- . Established first hurricane detection, spherics & radar network for Gulf Coast. Chmn of IEEE Regional Convention SWIEEECO (1963). GE Prof. Chmn of TX Prof Engr in Education Committee 1979-80. Dir of Construction Research Center UTARL. (1972-1981). VP Region of TX Soc Prof Engr's 1982-1984 Dean Emeritus UTARL. *Society Aff:* ASEE, NSPE

Salisbury, Marvin H
Business: 1501 E Main St, Griffith, IN 46319
Position: Pres. *Employer:* Salisbury Engg Inc. *Education:* BS/Civ/Univ of IL. *Born:* 10/8/23. Reg PE in the states of IN, IL, MI, WI, and OH. Before forming the firm, I served as an asst res engr with Std Oil div of the American Oil Co, then as Chief Soils Engr for DeLeuw, Cather & Co. My assignments have included numerous structural and hgwy subsurface explorations, Geotechnical engg, mtls investigation, concrete mix design, quality control and studies of defective concrete, construction inspection and mgt. I hold mbrships in the Hgwy Res Bd, Natl Soc of PE, In Soc of Civ Engrs, American Soc for Testing and Mtls, American Concrete Inst, American Society on Nondestructive Testing and American Welding Soc. *Society Aff:* ASCE, ASTM, NSPE.

Salkovitz, Edward I
Business: 800 N Quincy St, Arlington, VA 22217
Position: Dir, Research Programs *Employer:* Office of Naval Res. *Education:* DSc/Physics/Carnegie Inst of Tech; MS/Physics/Carnegie Inst of Tech; BS/Physics/Carnegie Inst of Tech. *Born:* 9/3/17. in Braddock, PA. Beginning in 1942, worked at the Naval Res Lab in metal physics Served as Assoc Superintendent of the Metallurgy Div (1957-60). Dir of Mtl Sciences Div of Adv Res Proj Agency (1964-65). Chrmn, Met and Mtl Engg Dept and Prof of Physics, Univ of Pittsburgh (1965-70, 73); designated the Sch of Engg Prof ('73). Chief Scientist, London Branch, Office of Naval Res (1970-72). Dir, Mtl Sciences Div, ONR (1974-81). Responsible for Navy's basic res on mtls. Is Adjunct Prof of Engg at Pitt; has taught metallurgy, physics and sci policy, has 75 publications, editor Naval Research Reviews plus - 2 books, awarded Navy Meritorious Service Award. Collect books, enjoys art and music. Dir, Research Programs (1980-). Responsible for Navy's Basic and Applied Research-All Disciplines, (Plans, Budgets, Administers) *Society Aff:* AAAS, AIME, ASM, MRS.

Sallet, Dirse W
Business: Dept of Mech Engg, College Park, MD 20742
Position: Prof. *Employer:* Univ of MD. *Education:* Dr Eng/Engrg Sci/Stuttgart Univ (Germany); MS/Mech Engrg/Univ of KS; BS/Mech Engrg/George Washington U, Wash DC *Born:* 8/10/36. in Wash DC. Res Assoc, Univ of Stuttgart 1963-1966, subsequently with the US Naval Ordnance Lab. Asst Prof, Mech Engg Dept, Univ of MD 1967, Assoc Pof 1970, Full Prof 1976. Past teaching duties include numerous diverse undergrad subjects plus grad courses in hydrodynamics, compressible flow, viscous flow, thermodynamics and applied math; supervised many MS and several PhD theses. 1973-1974 on leave from MD to do res at the Aerodynamische Versuchsanstalt Gottingen and the Max-Planck-Institut fur Stromungsforschung in Gottingen, Germany. 1980-81 on leave from Md. to research at the Center for Nuclear Research Kernforschungszentrum karlsruhe, Germany. Current res and consulting deal with two-phase flow, thermodynamics, heat transfer, thermal-hydraulic modeling, valve performance, flow induced vibrations and vortex flow. Reg PE in MD. *Society Aff:* ASME, APS

Salmon, Vincent
Home: 765 Hobart Street, Menlo Park, CA 94025
Position: Acoustical Consultant (Self-employed) *Education:* PhD/Theoretical Physics/MA Inst Tech; MA/Physics/Temple Univ; BA/Physics/Temple Univ. *Born:* 1/21/12. In chg R&D, Jensen Mfg Co, Chicago 1939-49; in chg sonics prog Stanford Resch Inst 1949-76; with Indus Noise Services 1971-76. Fellow & P Pres Acoustical Soc of Amer; member & P Pres Inst of Noise Control Engg; mem & P Pres Natl Council of Acoustical Cons: Fellow IEEE (resigned) & P Chrmn Prof Groups on Audio & on Sonics and Ultrasonics; Charter Fellow & P Western VP Audio Engg Soc; mem Natl Acad of Scis Comm on Bioacoustics and Biomechanics, & Panel on Submarine Noise Mgmt. Recipient, Acous Soc Amer Biennial Award & Silver Medal in Engrg Acoustics; Reg PE in CA, IL: Consulting Prof, Stanford Univ, 1977- . *Society Aff:* ASA, INCE, NCAC, AES

Saltarelli, Eugene A
Business: 4 Research Pl, Rockville, MD 20850
Position: Sr VP. *Employer:* NUS Corp. *Education:* MS/Mech Engr/Northwestern Univ; BME/Mech Engr/Univ of Detroit. *Born:* 2/22/23. Native of Buffalo, NY. Attended Buffalo State Teachers College prior to WWII. Served as Army Air Force Pilot 1942-1945. Rocket Motor Design Engr for Bell Aerospace, 1950-1956. With Bettis Atomic Power Lab, Westinghouse Elec Corp, 1956-1967. Served as Sr Design Engr and various mgt positions responsible for design of Navy Nuclear Propulsion Plants. Currently, Sr VP, Engg & Operating Services Group, NUS Corp (1967-present). Responsible for nuclear and fossil engg plant design, training of staff personnel, & service & maintenance functions in support of the electric utility field. Enjoy music and all sports. *Society Aff:* ANS, ASME, NSPE, AIF.

Salter, Winfield O
Business: 66 Luckie St. NW. Suite 400, Atlanta, GA 30303
Position: Senior Vice President. *Employer:* Parsons, Brinckerhoff et al, Inc. *Education:* MSCE/Soil Engg/Columbia Univ; BSCE/Trans Eng/MIT. *Born:* 3/12/28. in Medford, MA. Joined Parsons Brinckerhoff in 1952; became partner and sr officer in 1968. Served as Deputy Ch Hwy Engr to 1959; Proj Mgr of planning & design of San Francisco Bay Area Rapid Transit Sys; Proj Dir on planning & design of Atlanta Rapid Transit Sys; Prin-in-charge of design of Pittsburgh Light Rail Transit Sys reconstruction; and Proj Dir of Engg of Portland (OR) Light Rail Transit Sys. Tech Dir of firm's public trans work. Mbr US Natl Ctte on Tunneling Tech. *Society Aff:* ASEE, NSPE, SAME, IBTTA, ACEC

Salton, Gerard
Business: Computer Science, Cornell University Upson Hall, Ithaca, NY 14853
Position: Professor of Computer Sci. *Employer:* Cornell University. *Education:* PhD/Appl Math/Harvard *Born:* March 8, 1927. Inf Retrieval, Linguistic Analysis, Library Applications. Mbr: ACM, IEEE, ASIS, ACL, AAAS. Automatic Inf Org & Retrieval 1968; The Smart System-Experiments in Automatic Document Processing 1971; Dynamic Inf & Library Processing 1975 (Publs). Introduction to Modern Info Retrieval, 1983. Asst Prof of Applied Mathematics Harvard U 1958-65; Assoc Prof (1965-67), Prof (1967-), Chmn (1971-77), Dept of Computer Sci Cornell U. Fellow AAAS; Guggenheim Fellow, 1963; Mbr ACM Council 1971- ; Phi Beta Kappa. *Society Aff:* ACM, ACL, ASIS, AAAS, IEEE/CS

Saltzberg, Burton R
Home: 91 Southview Terrace, Middlitown, NJ 07733
Position: Member of Technical Staff. *Employer:* Bell Laboratories, Inc. *Education:* ScD/EE/NY Univ; MS/EE/Univ of WI; BEE/EE/NY Univ. *Born:* June 1933. BEE, ScD New York U; MS U of Wisc. Served with Army Signal Corp 1955-57. With Bell Labs since 1957. Engaged in devel & analy of data communications systems. Fellow IEEE 1976. *Society Aff:* IEEE.

Salvadori, Mario G
Home: 2 Beekman Place, New York, NY 10022
Position: Principal, Weidlinger Assocs. *Employer:* Weidlinger Associates. *Education:* Dr CE/Structures/Univ of Rome (Italy); Dr Math/Applied Mat/Univ of Rom (Italy). *Born:* 3/19/07. in Rome, Italy. Dr Civil Engg & Dr Mathematics Univ of Rome, Italy. Taught Columbia Univ 1940-date. Prof Civil Engg, Columbia Univ (Emeritus). Prof Archit, Columbia Univ (Emeritus). Principal Weidlinger Assocs, Cons Engrs 1953-date. Author of 14 books on engg, architecture & applied mathematics, translated into 13 languages. *Society Aff:* ASCE, ASME, NAE, AIA.

Salvendy, Gavriel
Business: Sch of Industrial Engg, W Lafayette, IN 47907
Position: NEC Prof. *Employer:* Purdue Univ. *Education:* PhD/Human Factors/Engr Production; Univ of Brimingham, England; MSc/Human Factors/Engr Production; Univ of Birmingham, England; Grad Dipl/Ergonomics & Work Design/Engr Production; Univ of Brimingham, England. *Born:* 9/30/38. Gavriel Salvendy is NEC Named prof of Industrial Engg at Purdue Univ and Chrmn of the Grad Schs Interdisciplinary Program in Human Factors. Gavriel's main areas of research include Human-Computer Inter & Ind. Prod. He has authored over 160 refereed publications including Editor of the Handbook of Industrial Eng. & Handbook of Human Factors. Gavriel is the recipient of the Phil Carroll Award for Outstanding Contributions in Work Measurement and Methods Engg of IIE and twice the recipient of the Fulbright-Hays Distinguished Prof Award. Prof Salvendy has lectured widely in over 26 countries on how to effectively measure and improve productivity and quality of working life. Gavriel is the Chairman of the International Commission on Human Aspects in Computing headquarterd in Geneva, Switzerland. *Society Aff:* IIE, APA, ERS, HFS, WSM, ACM.

Salvin, Robert
Home: 704 3rd St, Dunellen, NJ 08812
Position: Ch Facilities Engr *Employer:* Personal Products Co. *Education:* BS/ME/Rutgers Univ *Born:* 7/6/26 Native of Dunellen, NJ. Served in WWII with USN 1944-46. With Johnson & Johnson Co since 1952. Transferred to Personal Products Div in 1967. Assumed current responsibilities of Ch Facilities Engr in 1979. Also responsible for PPC Energy Conservation Program. VP & mbr of Exec Ctte NSPE 1980-81. NSPE PE in Ind (PEI) Chrmn 1980-81. NSPE/PEI Secy 1978-79, NSPE/PEI NE VChrmn 1975-77. NSPE Chrmn Pension Improvement Ctte 1976-78. NJSPE Chrmn PEI 1972-74. Chapter Pres AIPE 1970-71. NSPE Chrmn Profl Employment Ctte 1982-84, NSPE N.E. VP 1984-86, Pres NICET 1984-85. Recipient of Plant Engr Award. Elected Eminent Engr Tau Beta Pi 1980. Reg PE NJ, Certified Plant Engr. NSPE Chm Site Selection 1985-86. *Society Aff:* NSPE, ASME, AIPE, NICET.

Salzer, John M
Home: 909 Berkeley Street, Santa Monica, CA 90403
Position: President. *Employer:* Salzer Technology Enterprises, Inc. *Education:* ScD/Elec Engg/MIT; MS/Elec Engg/Cast Inst of Tech; BS/Elec Engg/Case Inst of Tech. *Born:* Sept 1917 Vienna, Austria. Served with Army Ordnance & Signal Corps 1941-45; Res Assoc MIT Digital Computer Lab, one of the first large computer projs 1948-51; at Hughes Aircraft designed system with airborne digital computer

Salzer, John M (Continued)
1951-54; Asst Dir of Magnavox Res Lab working on digital communication 1954-59; Dir of Intellectronics Lab of TRW involved in inf sciences, computers & displays 1959-63; Vice Pres for Tech & Planning Singer Librascope 1963-68; Pres of Salzer Tech Enterprises, Inc., Cons in Tech Assessment, computers, electronic packaging & integrated circuits 1968- . Taught courses at Case, MIT & UCLA; pioneered in sampled-data systems & digital controls. Active in IEEE Computer Soc, Awards Cttes, P Pres of Controls Group, & Fellow. Mbr: ISHM, SEMI, Eta Kappa Nu, Tau Beta Pi, Sigma Xi. Listed in 3 Who's Who. Hobbies include tennis. *Society Aff:* IEEE, ISHM, SEMI.

Samaras, Demetrios G
Home: 855 Hillside Dr, Palm Harbor, FL 33563
Position: Program Manager (retired). *Employer:* Air Force Office of Sci Res. *Education:* DSc/Eng Scis/Athens Univ of Engg; MS/EE & ME/Athens Univ of Engg. *Born:* 1911 Smyrna. Postgraduate: London, Berlin 1937-39; 3 First Prizes, 3 Fellowships. Aeroengineer GrAF 1934-35; Ch Autocontrol 1936-37; Founder Gasdynamics Lab, NCR-Canada 1946-48. WPAFB: Ch Propulsion Branch 1949-53; Ch Energy Release 1954-60; Prof-OSU, Graduate Center 1952-57. AFOSR: Res-Admins (Thermonuclear Space Power-Propulsion) 1961-70; Prog-Mgr Plasmas 1971-79 . Res Axialturbojets (supersonic compressors, combustors) 1931-45; invited by British Government establish Jet-Propulsion 1942; Res Nuclear Plasmas, (fission, fusion) 1950-60. Fellow: RAeS, F-BIS, AF-AIAA, SM-IEEE, ANS, AAIF, AMA. Publs: Nuclear Propulsion 1958,; Theory, Applications Ionflow Dynamics 1962. Publs- 150. Military Service: RCAF 1941-45, 3 medals. *Society Aff:* IEEE, ANS, AIAA.

Sambach, Warren A
Business: 173 Hillside Avenue, Williston Park, NY 11596
Position: Consulting Engr. *Employer:* Warren A Sambach Cons Engrs. & Planners *Education:* BME/Structural/Pratt Isst. *Born:* Jan 8, 1920. Var Positions C H Smith Constructing Co 1939-41; USAAF Tech Rep 1942-45; Reg PE in NY, NJ, Conn, Penna, Vt. Mbr: Amer Arbitation Assoc, Natl Soc Prof Engrs, Engrs Alliance, Amer Cons Engrs Council/NYS. Pres ACEC, Long Island Chap 2 years; Treasurer ACEC/NYS, Alt Dir, Dir-ACEC/NYS to ACEC/US; Recipient 2 awards for Church Design; Trustee Lutheran Church; Mbe Bd of Education Nassau Christian Sch. Married, 3 sons; golf & painting. *Society Aff:* ACEC/NYS, NYSSPE & NATL, AAA.

Samborn, Alfred H
Business: 1001 Madison Ave, Toledo, OH 43624
Position: Chrmn of the Bd. *Employer:* Samborn, Steketee, Otis & Evans, Inc. *Education:* BSCE/Structures/Univ of Toledo. *Born:* 4/30/17. Ward Products Corp, Cleveland, OH -- Jr Engr 1939-40. Builders Structural Steel Co, Cleveland, OH -- Profl Engr 1940. Giffels & Vallet, Inc, Detroit, MI - - Profl Engr 1940-48. USNR - Apprentics Seaman to Sr Lt, WWII, Seabees 1942-46. Samborn, Steketee, Otis and Evans, Inc, Toledo, OH -- Profl Engr and Pres 1948- 77. Currently Founder & Emeritus, Chrmn of the Bd of Samborn, Steketee, Otis and Evans, Inc. 1984-present Prof of Civil Engrg, Coll of Engrg Univ of Toledo. Received a "KY Colonel" Citation in 1978, commissioned by KY Governor Julian M Carroll. Received OH Association of Consulting Engrs Distinguished Consultant Award, 1979. Honorary PhD Adrian Coll, Business 1982 & NSPE Award 1982. *Society Aff:* ASEE, ASCE, NSPE, AAEE.

Sammet, Jean E
Business: IBM Corp, 6600 Rockledge Drive, Bethesda, MD 20817
Position: Programming Language Tech Mgr *Employer:* IBM Federal Sys Div. *Education:* MA/Math/Univ of IL *Born:* Have managed groups & have also done independent research in programming. Technical specialties are programming languages, software engineering & symbolic computation. Have published many papers in these subjects & also a 785 page book 'PROGRAMMING LANGUAGES: History & Fundamentals'. Have taught courses at var colls & Univs. Was Pres of Assoc for Computing Mach 1974-76 & also served in many other capacities. In Oct 1975 was one of first 2 people elected honorary mbr in Upsilon Pi Upsilon (natl computer honorary soc). Elected mbr of Natl Acad of Eng in 1977. Recd honorary ScD from Mt Holyoke College in 1978. Editor-in- Chief of *Computing Reviews*, 1979-1987. *Society Aff:* ACM, MAA

Sample, Frank C
Business: 2419 N Seventh St, West Monroe, LA 71291
Position: Owner. *Education:* BS/Civil Engg/LA Tech Univ. *Born:* 12/9/37. and raised in southern rural KS. Started engg career with Santa Fe Railroad in 1958. Received degree in 1965. Worked as Safety Engr in Dallas, TX. Moved to Wichita, KS to assume design engr responsibilities with Professional Engg Consultants. Became Proj Engr on Esler Field Airport in 1968 in Alexandria, LA. Started private practice in 1969 in West Monroe, LA. Owner of Frank Sample and Assoc since 1973. Enjoy work, hunting, fishing, farming, dancing and raising six children (ages 15 thru 20). *Society Aff:* NSPE, CSI, SAME, LES.

Samuels, Alvin
Business: 1040 1st NBC Bldg, 210 Baronne St, New Orleans, LA 70112
Position: VP *Employer:* Ironite Products Co *Education:* BS/Gen Engrg/Univ of IL *Born:* 4/6/34 Raised in Paragould, AR. Cadet Corps Commander Western Military Acad, 1952. Drilling Engr, Magnolia Petroleum Co, 1956-59. Drilling Consultant for E. A. Polumbus & Assoc through 1963. Staff Drilling Engr, Shell Oil Co through 1972. VP and co-founder of Ironite Products Co, responsible for res & dev of mtls and methods of reacting hydrogen sulfide. Enjoy reading nonfiction and vigorous physical activity. *Society Aff:* SPE, API

Samuels, Leonard E
Home: 10 Alexandra Court, Glen Iris, Australia 3032
Position: Superintendent Metallurgy Div. *Employer:* Materials Res Labs, Austrln Dept Def. *Education:* DSc/Metallurgy/Melbourne Univ; MSc/Metallurgy/Melbourne Univ; BMetE/Metallurgy/Melbourne Univ. *Born:* Feb 1922 Brisbane, Australia. With Materials Res Labs since 1943; Res Scientist in NSW Branch 1943-62; assumed current position, responsible for the met prog of the Labs, in 1962. Res specialties are in mechanisms of abrasive machining & effects on surfaces produced & in mechanical methods of metallographic poliching. Pres Australian Inst of Metals 1971; Dir Internatl Met Soc 1975-79. Fellow Institution Metallurgists, ASM Inst Engg Aust. Battelle Visiting Prof Ohio St U 1976. Melb U Syme Res Prize 1958; AES Leather Award 1960; AIM Silver Medal 1971, IMS Sorby Award 1980. *Society Aff:* ASM, IMS, IEA, AIM.

Samuels, Reuben
Home: 195 Harwood Place, Paramus, NJ 07652
Position: Chairman of the Board, Thomas Crimmins Contracting Co & Assoc Contracting Co *Employer:* Thomas Crimmins Contracting Co. *Education:* BSCE/Civil Engg/Dartmouth; MSCE/Soil Mech Eng/Harvard. *Born:* 1926 Suffern, NY. Soil & Foundation Engr, Mueser, Rutledge, Wentworth & Johnston, NYC. Attended MIT from 1943-44; BSCE Thayer Sch Engrg, Dartmouth; MSCE Harvard 1948; Doctoral Candidate, teaching asst to Profs Terzaghi & Casagrande Harvard 1951-54. Fellow ASCE. Mbr: ASTM, ACI, Boston Soc of Civil Engrs, Moles, Sigma Xi & Harvard Club of NY. Natl Ctte on Deep Foundations, ASCE; P Pres of the Moles, P Pres Harvard Engg Soc; P Pres Dartmouth Soc of Eng; Arbitrator, Amer Arbitration Assoc; Overseer of Thayer Sch of Engrg Dartmouth Coll. Mbr of Geologic Technical Site Invest Sub Ctte; Parent Ctte; US Natl Ctte on Tunnelling Tech, Natl Acad of Sci, Chairman. *Society Aff:* ASCE, ASTM, ACI, ΣΞ

Samulon, Henry A
Home: 575 Muskingum, Pacific Palisades, CA 90272
Position: Consultant *Employer:* Self employed *Education:* Diplom Ing/Electronics/Swiss Federal Inst of Technology. *Born:* Dec 1915. Diplom Ingenieur (equivivalent MS) Swiss Federal Inst of Tech, Zurich 1939; Graduate studies; Instructor. Electronics Lab of General Electric Co Syracuse, NY, Tech Mgmt 1947-55. Active in color TV standards efforts for NTSC; TRW Los Angeles Tech Mgmt, later Genl Mgr of Electronics Systems Division, contributor to early space electronics 1955-74, TRW Corp Vice Pres 1964-74, Xerox Vice Pres Electronics Div (containing RD&E & Mfg) 1974-81. Retired from Xerox June 1, 1981 and is now a consultant. Publs, 2 patents. Wincom chmn 1972. Fellow IEEE. *Society Aff:* IEEE.

Sanborn, David M
Home: 4100 River Cliff Chase, Marietta, GA 30067
Position: VP-Engrg & Mfg *Employer:* Amer Combustion, Inc. *Education:* PhD/Mech Engg/Univ of MI; MSE/Mech Engg/Univ of MI; BSE/Mech Engg/Univ of MI. *Born:* Dec 6, 1942 Detroit, Mich. V.P.-Engrg & Mfg of Amer Combustion, Inc (1985-). VP of E-Tech, Inc (1978-85), an energy conservation prods co. Asst Prof (1969-75) & Assoc Prof (1975-77) Sch of Mech Engrg Georgia Inst of Tech. Interests include combustion, energy conservation, bearings, seals, rotor dynamics, design. Approximately 25 papers on lubrication, bearings & seals have been published. Jointly awarded the 1975 Melville Medal for the best ASME paper & received the 1975 Sigma Xi Res Award for Junior Engrg Faculty. Married in 1966. Children born in 1970, 1972 & 1974. *Society Aff:* ASME, ΣΞ, ASHRAE.

Sances, Anthony
Business: Dept of Neurosurgery, 8700 W. Wisconsin, Milwaukee, WI 53226
Position: Prof & Chrmn BioMed Engrg *Employer:* Medical Coll of WI *Education:* PhD/BioMed Engrg/Northwestern; MS/Physics/DePaul Univ; BS/EE/Am Inst Tech *Born:* 7/13/32 Prof & Chrmn of BioMedical Engrg at Marquette Univ and Dept. of Neurosurgery at the Med Coll of WI for 20 yrs. Published over 250 papers and 5 books. Twenty-five yrs of experience in biomedical, clinical, biomathematical, and neurophysiological research and consultation. Design of protective structures, mechanical, electronic and electromechanical systems. Computerized systems for evaluating spinal cord and brain injured patients. Biomechanics of trauma. Analysis of electrical trauma. Neurophysicologic investigations. Electrostimulation. Crash analysis. *Society Aff:* NES, IEEE, BMS, AAMI, BES, ASB, ISA.

Sandall, Orville C
Business: Dept of Chem & Nuclear Engg, Santa Barbara, CA 93106
Position: Prof *Employer:* Univ of CA. *Education:* PhD/CE/Univ of CA; MSc/CE/Univ of Alberta; BSc/CE/Univ of Alberta. *Born:* 7/4/39. Native of Alberta, Canada; naturalized 1970. Has taught in the Chem and Nuclear Engg Dept at the Univ of CA, Santa Barbara since 1966. Res interests are in the areas of turbulent heat and mass transfer, gas absorption and multicomponent distillation. *Society Aff:* AIChE.

Sandberg, Harold R
Business: 233 N Michigan Ave, Chicago, IL 60601
Position: President. *Employer:* Alfred Benesch & Company. *Education:* MS/Structures/Univ of IL, Urbana; BS/Civil Engg/Univ of IL, Urbana. *Born:* 12/28/19. Field Engr Dravo Corp 1942-44. CEC officer USNR 1944-46; joined Alfred Benesch & Co in 1947 as structural engr, assoc & proj engr 1956; V Pres 1964, Exec Vice Pres 1967, Pres 1971-83, Chairman of the Board 1983. Responsible for design of many large buildings & bridges. Very interested in high rise buildings. Past Pres Structural Engrs Assoc of IL, the Univ of IL Civil Engrg Alumni Assoc; Consulting Engrs Council of IL; Chicago Ctte on High Rise Bldgs. Currently Chairman ACI-ASCE Comm 343, Concrete Bridge Design, and AREA Comm 8, Concrete Structures, Member Dean's Advisory Board, U of IL. *Society Aff:* ASCE, ACI, NSPE, AREA, ACEC, SAVE

Sandberg, Irwin W
40 North IH-35, Austin, TX 78701
Position: Professor *Employer:* Univ of Texas at Austin *Education:* DEE/Elec Comms/Poly Inst of Brooklyn; MEE/Elec Comms/Poly Inst of Brooklyn; BEE/Elec Comms/Poly Inst of Brooklyn. *Born:* Jan 23, 1934 New York, NY. Tech aid Bell Telephone Labs Inc Murray Hill, NJ summer 1954, Mbr tech staff 1958-67, head systems theory res dept 1967-72, Mbr Mathematical Sciences Res Center 1972-1985, Professor and Holder of the Cockrell Family Regents Chair in Engineering *1 at the University of Texas at Austin 1985- . Westinghouse Fellow, 1956, Bell Telephone Laboratories Fellow, 1957, 1958, US delegate URSI Munich 1966, recipient best paper award Asilomar Conf 1970, US Natl Inst Rep 1972 NATO advanced study inst on network & signal theory (Bournemouth, England), Distinguished Invited Speaker at Asilomar Conf 1973 & 1974. Mbr: IEEE (admin ctte group on circuit theory 1969-70, Vice Chmn group on circuit theory 1971-72, Fellow 1974), AAAS, Eta Kappa Nu, Sigma Xi, Tau Beta Pi, Distinguished Staff Award, Bell Laboratories, 1983, IEEE Centennial Medal, National Academy of Engineering, Recipient First Technical Achievement Award of the IEEE Circuits and Systems Society (1985).* *Society Aff:* IEEE, AAAS.

Sandell, Dewey J, Jr
Home: Reis Cir, Fayetteville, NY 13066
Position: VP. *Employer:* Carrier Corp. *Education:* PhD/Physical Chem/MIT; BA/Chemistry/Univ of MT. *Born:* 3/17/22. Carrier Corp, 1953 to present. VP, Asst to Pres - Tech. Pres Machinery & Systems Div. VP - Dir, Res Div. Res Dir; Div Dev Dept, Central Engg Staff. Ind R&D experiences: Cambridge Corp (Cryogenics). H L Johnston Co; E I DuPont. Academic - MIT: OH State Univ. Physical chemistry - Cryogenics - Energy Thermo. *Society Aff:* ACS, AIChE, ASHRAE, AAAS.

Sandell, Donald H
Business: 150 S Wacker Dr, Chicago, IL 60606
Position: Sr Assoc *Employer:* Harza Engrg Co *Education:* BS/EE/Univ of MN *Born:* 9/18/26 Native of Minneapolis, MN. Sr Staff Engr 1952-65 for Aluminum Co of American specializing in elec conductor applications of aluminum in transmission and distribution lines and substations. Joined Commonwealth Assoc Consulting Engrs as Proj Mgr for Transmission Lines and Substations. Later assumed the position of VP and Mbr of the Bd of Dirs. Responsible for construction mgmt of company projects 1965-72. Present position - Sr Assoc, Harza Engrg Co, Area Operations Mgr, Heavy Construction Projects. Patents on Extra High Voltage Conductors and Substation bus. Enjoy music, swimming & skiing. *Society Aff:* IEEE, ASTM, CIGRE, WSE.

Sander, Duane E
Business: Box 2220, Brookings, SD 57007
Position: Professor *Employer:* SD St Univ *Education:* PhD/EE/IA St Univ; MS/EE/IA St Univ; BS/EE/SD School of Mines *Born:* Native of SD. US Army from 1964-1966. Employed as Prof of EE at SD State Univsince 1967. Specializing in bioengrg and analog circuit design. On Bd of Dirs and cofounder of Daktronics, Inc, since 1968. Pres of Medical Engrg servs Affil, Inc, since 1980. *Society Aff:* IEEE, NSPE, ISHM, AAMI

Sander, Louis F
Business: Dept of Mech Engg, Villanova, PA 19085
Position: Chrmn. *Employer:* Villanova Univ. *Education:* PhD/Met/PA State Univ; MS/Mtls/Marquette Univ; BS/ME/Marquette Univ. *Born:* 8/1/33. Native of Rockville Ctr, NY. Engg officer, US Navy 1955-58. Engg Inspector for Mobil Oil 1958-59. Instr, Marquette Univ 1959-61. With Villanova Univ since 1961. Leave of absence and Ford Fdn Fellowship 1964-66. Chrmn of ME Dept since 1971. Elected to Pi Tau Sigma, Tau Beta Pi, and Sigma Xi. Fellow, American Inst of Chemists. Serves as active consultant for numerous corps and recently engaged in res related to ground support systems for US Naval Air Engg Ctr, Lakehurst, NJ. *Society Aff:* ASME, ASM, AIME, ASEE, ASTM.

Sanders, Charles F
Business: Energy Systems Assoc, 15991 Red Hill Ave, Tustin, CA 92680-7388
Position: Exec V Pres *Employer:* Energy Systems Assoc *Education:* PhD/Chem Engg/Univ of Southern CA; MChE/Chem Engg/Univ of Louisville; BChE/Chem Engg/Univ of Louisville. *Born:* 12/22/31. Engr with Esso Res & Engg Co 1955-62, Envi-

Sanders, Charles F (Continued)
ron Health officer with US Army Environ Lab 1956-58; Prof and Dean of Engg at CA St U, Northridge; Chmn, CSUC Council of Engg Deans 1974-75; Chmn CA Engg Liaison Ctte 1975-76; Pres San Fernando Valley Engrs Council 1975-76. Pres & CEO Rusco Industries 1980-81. V Pres, Energy Systems Associates 1982-present. Dir & V Chrmn, MacBay Energy Corp, 1985-87. Dir, Datametrics Corp, 1982-86. Mbr, Bd of Dir, Rusco Industries. Mbr, Bd of Dir, San Fernando Valley Child Guidance Clinic. Mbr: AIChE, ASEE, Combustion Inst, NSPE, CSPE. Cons on heat & mass transfer, kinetis & combustion. m. Marie A Galuppo 1956; 3 children Karen, Craig, Keith. *Society Aff:* AIChE, ASEE, NSPE.

Sanders, John C
Home: 15305 Forest Park Ave, Strongsville, OH 44136
Position: Asst Ch Wind Tunnel & Flight Div. *Employer:* Natl Aeronautics & Space Admin. *Education:* BS/Mech Eng/VA Poly Inst; MS/Mech Eng/VA Poly Inst. *Born:* Oct 29, 1914 Roanoke, Va. Worked as Indus Engr for Aluminium Co Amer Alcoa, Tenn till Nov 1939. Joined Natl Advisory Ctte for Aeronautics & have remained with that org & successors until retirement June 1974. Became Branch Ch in charge of 30 engrs working on aircraft engine controls; Organized the NACA subctte on engine controls; became Asst Ch of Advnanced Systems Div in charge of 100 engrs in 1960 & retained this position to retirement. Headed ASME subctte on propulsions in Aviation & Space Div for sev years; read papers before prof socs in US & England. Hold several patents on controls; presently a cons on controls. *Society Aff:* ASME, AIAA.

Sanders, John H
Business: PO Box 511, Kingsport, TN 37662
Position: VP. *Employer:* Eastman Kodak Co, Eastman Chems Div. *Education:* BS/CE/Auburn Univ. *Born:* 4/16/21. Native of Birmingham, AL. Served in WWII as field artillery pilot with 17th, 101st, & 82nd airborne divs in Europe. Discharged as Capt. Joined TN Eastman Co (mfg of chems, fibers, & plastics) in 1946. Moved into chem sales with the co in 1949 & became Pres of the marketing co, Eastman Chem Products, Inc, in 1976. Assumed current responsibility as VP, Eastman Kodak Co, & Asst Gen Mgr, Eastman Chems Div, in 1979. Distinguished Alumnus in Chem Engrg at Auburn Univ, 1977. Enjoys amateur radio & barbershop quartet singing. Married former Mary Helen Wilson. Dir AIChE 1980-82, VP AIChE 1983, Pres, AIChE 1984. Mbr Bd of Govs AAES. Mbr R&D Foundation, Res Adv Council, & Alumni Advisory Engrg Council of Auburn Univ. *Society Aff:* AIChE, SCI, AAES.

Sanders, Karl L
Home: 3407 S Walker Ave, San Pedro, CA 90731
Position: Sr Tech Specialist *Employer:* Northrop Corp *Education:* BS/Aero Engr/ Polytech Inst Bremen, Germany *Born:* 02/13/22 Currently respon for the mass properties methodology dev. Primary background in conceptual and preliminary design of fighter and attack aircraft. From 1959-73 at Ryan Aeronautical Co, San Diego. Respon for configuration design and analysis of XV-5A liftfan res aircraft and related advanced studies. From 1973-1976 Head of Advanced Aircraft Design Branch of the Naval Weapons Ctr engaged in analysis and evaluation of advanced navy aircraft proj. During 1942- 1945 participated in design of ME 163 and ME 262 first combat jet fighters. *Society Aff:* AIAA, SAWE

Sanders, Lon L
Home: 6677 June Ln, Westhills, CA 91307
Position: Sr Mbr Tech Staff *Employer:* ITT Gilfillan *Education:* BS/EE/Newark Coll of Engrg; -/EE/Columbia Univ *Born:* 6/13/27 Performed Electronics Engrg for Hoffman Radio Corp., Los Angeles, CA (1948) and A B Du Mont Labs, Clifton, NJ (1950). Since 1953, Design and Sys Engrg on aircraft landing sys, Ground Control Approach radar and Air Surveillance radar for ITT Gilfillan, Los Angeles, CA. Pioneering dev of first Scanning Beam Microwave Landing Sys (MLS) (1957). Active in natl cttes for standardization of MLS (1968-76). Inventor on four patents and author of 11 technical papers. Engr of the Year - San Fernando Valley, 1977. Fellow of Inst for Advancement of Engrg, 1979. Fellow of IEEE, 1980. *Society Aff:* IEEE, ION

Sanders, Wallace W, Jr
Business: 104 Marston Hall, Ames, IA 50011
Position: Prof of Civil Engrg & Assoc Dir, Engrg Res Inst. *Employer:* IA State Univ. *Education:* PhD/Structural Engr/Univ of IL; MEngr/Civ Engg/Univ of Louisville; MS/Structural Engr/Univ of IL; BCE/Civ Engg/Univ of Louisville. *Born:* 6/24/33. Louisville, KY. Mbr of faculty at Univ of IL from 1959 to 1964. With dept of Civ Engg of IA State Univ since 1964 & Prof since 1970. Active in res on behavior of metal components & structures & design of bridges. Contributor to numerous technical & profl publications. Mbr of committees of ASCE, AISC, ECCS, IIW, AREA & AWS. Recipient of 1971 AWS Adams Meml Mbrship Award & 1978 ASCE Raymond C Reese Res Prize. *Society Aff:* ASCE, AWS, ASEE, AREA, IABSE.

Sanders, William T
Business: , New York, NY 10027
Position: Assoc Prof. *Employer:* Columbia Univ. *Education:* EngrScD/Mech Engg/ Columbia; Master/Nuc Engg/NY Univ; BSc/Chem Engg/Purdue. *Born:* 6/13/33. in Owensboro, KY. Attended Oak Ridge Reactor School in 1954-55, then worked in nuclear reactor engg for Combustion Engg and for American Machine nnd Foundry Co. Joined Columbia Faculty in 1962: teaching in many areas of Mech Engg, res in defects in solids, in mechanics of solids, in numerical methods, in energy storage, in solar energy. Hobbies: reading, classical music, canoeing, hiking, amateur radio. *Society Aff:* ASME, ΤΒΠ

Sanderson, Robert L, Jr
Home: 6511 O/Donnell Lane, Billings, MT 59106
Position: Pres. *Employer:* S/S/G Engineering, Inc. *Education:* MS/Civil Engg/MT State Univ; BS/Civil Engg/MT State Univ; BS/Mech Engg/ Mt State Univ. *Born:* 1/20/42. Born and raised in Billings, MT. Earned college expenses working as land surveyor. Construction engg for Naval Facilities Engg Command 1966-67 at Seattle and Puget Sound Naval Shipyard. Purchased interest in Engineering, Inc in 1969, became Pres and majority stockholder in 1976. Firm employs 50 persons in Billings, Bozeman, and Sheridan, WY engaged in Civil Engg and land surveying. Pres of MT Chapter ACEC 1978-79; Pres of MT Soc of Engg (NSPE)-1979-80. Outstanding Young Engineer Award 1972. Director Billings Area Chamber of Commerce 1984 to present. Raise horses, hunt, fish ski, travel. *Society Aff:* ASCE, ACEC, NSPE, APWA

Sandgren, Find
Business: , W Lafayette, IN 47907
Position: Prof ME. *Employer:* Purdue Univ. *Education:* Cand Poly/EE/Poly Inst of Denmark; BS/ME/Copenhagen Tech; MBA/Bus/Univ of Chicago. *Born:* 6/15/22. *Society Aff:* IEEE, ASEE.

Sandhu, Ranbir S
Business: Dept of Civ Engrg, 2070 Neil Ave, Columbus, OH 43210
Position: Prof *Employer:* OH State Univ *Education:* PhD/Structural Mech/Univ of CA, Berkeley; M Eng/Civil Eng/Univ of Sheffield, UK; BSc (Hons)/Civil Eng/East Punjab Univ, India; BA (Hons Math)/Math/Univ of Punjab, Lahore, Pakistan *Born:* 1/19/28 Worked as Asst Engr/Sr Design Engr/Deputy Dir Designs, Irrigation Dept, Punjab India from 1950-63. Participated in the design of Bhakra Dam, 740 ft high concrete dam and appurtenant structures. Developed methods of stress analysis for gravity dams and concrete tunnel linings. Assoc Prof at Punjab Engrg Coll, 1963-65. Assoc Prof at The OH State Univ, 1969-73. Prof, 1973 to date. About 80 res papers. Special interest in variational methods, finite element methods, structural mechanics, mechanics of composites, flow through porous media. Visited NGI Oslo in 1977 under a Sr Scientist Fellowship, was Guest Prof at Univ of Stuttgart in

Sandhu, Ranbir S (Continued)
1980, was conslt in India under UN Dev prog in 1981. *Society Aff:* ASCE, AAM, SES

Sandler, Stanley I
Business: Univ of Del, Dept of Chem Engrg, Newark, DE 19716
Position: H.B. duPont Prof of Chem Engrg *Employer:* Univ of DE *Education:* PhD/Chem Eng/Univ of MN; BChE/Chem Eng/City Coll of NY *Born:* 06/10/40 Prof Sandler is the HB du Pont Prof of Chem Engrg at the Univ of DE. He has been the recipient of the AIChE Prof Progress Award, Humboldt Foundation Res Fellowship at the Tech Univ of Berlin (W), and a Dreyfus Foundation Faculty-Scholar Grant. His res areas include applied thermodynamics, phase equilibria, and statistical mech. He is author or editor of seven books and 105 res papers, and a conslt for chem and oil co. *Society Aff:* AIChE, ASEE, ACS

Sandor, George N
Home: 1717 NW 23rd AV Apt 3C, Gainesville, FL 32605
Position: Res Prof & Dir Mech Engg Des Lab *Employer:* U of Fla, Dept of Mech Engr. *Education:* Dr of Engg Sci/Mech Engg/Columbia Univ; Diplom/Engg/ Technological Univ of Budapest. *Born:* Feb 24, 1912 Budapest, Hungary. Naturalized USA 1949. m. Magda B 1964. Design Engr, Vice Pres & Ch Engr, Bd/Dir. In academia since 1961: Assoc Prof, Engrg & Applied Sci Yale U; Alcoa Foundation Prof of Mech Design, Div Chmn, Center Dir, Rensselaer Polytechinic Inst; Res Prof U of FL. Life Fellow: ASME, Machine Design Award 1975, Design Engg Division Mechanisms Committee Award, 1980, Life Fellow; Pi Tau Sigma, Tau Beta Pi; Sigma Xi; New York Academy of Scis; NSPE; ASEE. 140plus papers. Reg PE. FAA Airline Transport Pilot's License. Doctor Honoris Causa Technological Univ of Budapest, 1986, ASEE's Ralph Co-Ats Roe Award, 1985. *Society Aff:* ASME, ASEE, NSPE, AAM, NYAS

Sandquist, Gary M
Business: Nuclear Engrg, Mech Engrg Dept, Salt Lake City, UT 84112
Position: Prof & Dir *Employer:* Univ of UT *Education:* Post Doctoral/Nuclear Engrg/MIT; PhD/Nuclear & Mech Engrg/Univ of UT; MS/Engrg Sci/Univ of CA; BS/ME/Univ of UT *Born:* 4/19/36 Prof of Mech & Indus Engr Dept, Dir Nuclear Engrg Program and Nuclear Engrg Lab at Univ of UT. Reactor Supervisor for training (AGN-201) and res (TRIGA) nuclear reactors. Commander in US Naval Reserve. Res Prof of Surgery, Univ of UT Medical College. Reg PE in UT, MN & CA. American Bd. Certified Health Physcist Serve as short mission expert in Nuclear Sci for IAEA. Principal Scientist for Rogers & Assocs Engrg Corp. Mbr of Educ Div Exec Ctte for ANS. Consultant for 10 companies, fed govt labs, and state agencies. Published over 225 journal articles and technical reports. Author of 2 books. *Society Aff:* ANS, ASME, ΤΒΠ, ΠΤΣ, ΣΞ

Sandretto, Peter C
Home: 190 Highland Ave, Montclair, NJ 07042
Position: Retired. *Education:* BS/EE/Purdue; EE/-/Purdue Univ. *Born:* April 1907. Command & Staff US Army; Mbr Tech Staff Bell Labs 1932; Dir Comm Lab United Air Lines 1932-42; US Air Force, Active & Reserve duty 1942-67, Rank to BGen; Vice Pres & Tech Dir Internatl Telecommunications Labs to 1960; Deputy Dir Defense Group ITT to 1963; Pres ITT Kellog 1963; Dir Engrg Mgmt ITT 1972. Author of 3 books: Principles of Aeronautical Radio Engrg, 1942; Electronic Avigation Engrg, 1958; The Economic Mgmt of Res & Engrg, 1968. Miscellaneous awards & honors: Eta Kappa Nu, Honorary Electrical Engrg Soc; Bronze Star; Legion of Merit; R & D Award, NJ Wing AFA; Volare Award, Airline Avionics Inst; Fellow IEEE; Charter Engr, British Institution of Electrical Engrs; Assoc Fellow, American Inst of Aeronautics & Astronautics; Honary Vice Pres, Inst of Navigation; Honorary Ceremonial Speaker, Ausschuss Fur Funkortung. *Society Aff:* IEEE, ION, IEE, AZAA, RION.

Sandstrom, Donald J
1224 Vallecita Dr, Santa Fe, NM 87501
Position: Dep Div Leader *Employer:* Los Almos Natl Lab *Education:* MS/Engr Sci of Mtls/Univ of NM; BS/Met Engr/Univ of IL. *Born:* 7/26/37. Born & raised in Chicago, IL. Upon receiving my bachelors degree in 1958 I was employed as a metallurgical engr for ACF Industries in Albuquerque, NM. In 1961 I joined the staff at Los Alamos Natl Lab, attended grad school & obtained my MS. I have been active in the technical soc and was named a Fellow of ASM in 1985. I chaired a Ctte for the Natl Acad of Sci on Kinetic energy penetrators and have remained active in armor/ anti-armor dev. I am the Deputy Div Leader for the Materials Sci and Tech Div at Los Alamos. My special areas of interests and expertise are materials sci and engg of the actinides and refractory metals. I am an avid skier and participant in outdoor sports. *Society Aff:* ASM, AIME.

Sandvig, Robert L
Business: Sch of Mines & Tech, Rapid City, SD 57701
Position: Professor & Head Dept of Chem Engrg. *Employer:* South Dakota Sch of Mines & Tech. *Education:* PhD/Chem Engg/Univ of CO; MS/Chem Engg/Univ of Cincinnati; BS/Chem Engg/SD Sch of Mines & Technology. *Born:* Sept 1923, married, 3 children. US Navy 1944-46. Chem Engr Darling & Co 1948-49; taught at South Dakota Sch of Mines & Tech in Chemistry 1946, 1949-53, in Chem Engrg 1953- . Chem Engrg Dept Head 1973- . Mbr: AIChE, ACS, ASEE, Sigma Xi. P Pres South Dakota Academy of Sci. Indus exp with Dow Chemical Co (Midland & Rocky Flats), Hercules, Rockwell Internatl (Rocky Flats). Mbr: Presbyterian Church, Masonic Lodge, Elks Club, Tau Beta Pi. *Society Aff:* AIChE, ACS, ASEE.

Sandza, Joseph G
Business: GPO Box 2242, San Juan, PR 00936
Position: Chmn. *Employer:* Caribtec Laboratories Inc. *Education:* PhD/Chemistry/Fordham Univ; MS/Chemistry/Fordham Univ; BS/Chemistry/ Polytechnic Inst Brooklyn. *Born:* 2/4/17. Native of NY, NY. Post Doctoral Fellowship at Northwestern Univ 1942-44 (ORSD); Dept Hd Lederle Labs (American Cyanamide Co) 1944-48; Tech Dev Hoffman- La Roches 1948, Stauffer Chem Co, Eastern Res Center, Asst Dir 1948-63. Cons Economic Dev Admin, Commonwealth of Puerto Rico 1963-69. Pres of Caribtec Labs Inc San Juan, Puerto Rico 1969-84, Chmn 1984- . Engaged in chem & environ cons, process & product dev. Chmn Div of Sci & Technology, World U San Juan, Puerto Rico 1968- 82. GPO Box 2242 San Juan, Puerto Rico 00936. *Society Aff:* ACS, AIChE, AIC, ΣΞ.

Sanford, Keith C
Business: 420 Broadway, Sacramento, CA 95818
Position: President. *Employer:* Sanford-Alessi Associates, Inc. *Education:* BS/AC & Refr/CA Poly Tech. *Born:* Oct 28, 1921. Reg PE in Calif, Hawaii, Nevada & Oregon. Served in Air Force in WWII 1942-45. Mbr: ASHRAE, CEAC, ACEC. With St of Calif as Design Engr 1948; joined present firm in 1949, Vice Pres 1962, Pres 1966. This was the first mech engrg firm in this area, established in 1947, specializing in mech systems for commerical industry buildings. *Society Aff:* ASHRAE, CEAC, ACEC.

Sanford, Robert J
Business: Code 5831 Naval Res Lab, Washington, DC 20375
Position: Head Struc Rlbty Sect Marine Tech Div. *Employer:* Naval Research Laboratory. *Education:* PhD/Solid Mechs/Catholic Univ of Amer; MSE/Solid Mechs/ George Washington Univ; BME/Mech Engg/George Washington Univ. *Born:* 1939. Specialties include optical stress analysis, fracture mech & Non- destructive evaluation Employed at Naval Res Lab 1960- , Head of the Structural Reliability Sect, Marine Technology Div 1974- . Respon for tech direction & mgmt of basic & applied progs in experimental stress analysis & non-destructive testing. NRL Res Publ Award 1972 & 1979; SESA, M Hetenyi Award 1974. SESA, Fellow 1980. Concurrent position: Visiting Prof, Univ of MD: SESA, American Acad of Mechanics, ASME. *Society Aff:* ASME, SESA, AAM, ΣΞ.

Sanghera, Gurbaksh S
Business: Prudential Ctr, Boston, MA 02199
Position: Civil Engr *Employer:* Chas T. Main, Inc *Education:* BS/CE/Punjab Univ-India; BS/Math/Punjab Univ-India *Born:* 2/20/23 Born in Punjab, India. Educated in Lahore, now in Pakistan. Got training in Dam Engrg in USBR, Denver, CO. Worked on Fdn Grouting of Hirakud Dam and Design of Bhakra Dam in India 1951-56. With Hindustan Steel and Bokaro Steel on design and construction of steel plants in India. Deputy Ch Engr, Bokaro Steel 1965-69. Worked with Justin & Courtney, Inc. (now O'Brien & Gere) Philadelphia on design of Water Resources Projects, 1969-79. With Chas T. Main, Inc. since 1980 on design of Hydro Power Projects. *Society Aff:* ASCE, USCOLD

Sangrey, Dwight A
Business: Dept of Civil Engg, Porter Hall, Pittsburgh, PA 15213
Position: Prof & Hd of Civil Engg. *Employer:* Carnegie-Mellon Univ. *Education:* PhD/Geotechnical Eng/Cornell; MS/Civil Engrg/MA; Civil Eng/Lafayette *Born:* 1940 Lancaster, Penn. Reg PE. Field & proj engrg with H L Griswold Cons Engrs & Shell Oil Co prior to faculty position at Queen's U, Kingston; Cornell 1970- 79 Carnegie-Mellon 1979- . Recipient of Tau Beta Pi & Chi Epsilon excellence in teaching awards. Mbr: ASCE, ASTM, ASEE, NRC-TRB, Internatl Soc Soil Mech & Foundation Engrg, Canadian Geo Soc, Internatl Soc Rock Mech Author Consultant. Specialization in prof education, soil dynamics, marine geotechnical engrg, engrg geology, construction robotics, regional economic dev. *Society Aff:* ASCE, ASEE, ASTM, NRC-TAB

Sangster, William M
Business: College of Engineering, Atlanta, GA 30332
Position: Dean, College of Engineering. *Employer:* Georgia Institute of Technology. *Education:* PhD/CE/Univ of IA; MS/CE/Univ of IA; BS/CE/Univ of IA. *Born:* Dec 9, 1925 Austin, Minn. Served in US Navy 1944-46. Served successively as Asst Prof, Assoc Prof, Prof, Assoc Dir Engrg Experiment Station, Assoc Dean of Engrg U of Mo, Columbia 1948-67. Res Engr Boeing Co, Seattle summer 1951, 59, 60. Dir Sch of Civil Engrg, Georgia Tech 1967-74; Dean of Engrg 1974- . Res in engrg aspects of fluid mechanics. Bd/Dir: ASCE 1965-67, EJC 1976-80, ABET 1975-76, 1978-84, UET 1975-83, EF 1983-, EI 1980-85. Pres ASCE 1974-75. Vice Chairman AAES-EMC 1986- . *Society Aff:* ASCE, NSPE, ASEE

Sanks, Robert L
Home: 411 W. Dickerson, Bozeman, MT 59715
Position: Prof Emeritus; Consultant *Employer:* Retired *Education:* PhD/Sanit Engg/Univ of CA; MS/Struct Engg/IA State Coll; BS/Civ Engg/Univ of CA; AA/Science/Fullerton Jr Coll. *Born:* 2/19/16. Served with several consultants, US Army Engrs, & as res engr at Univ of CA from 1940 to 1946. Taught at Univ of UT 1946-1958, & Gonzaga Univ 1958-61. NSF Fellow 1961-63 & Assoc Specialist III at Univ of CA 1963-66. As Prof (Dept Civ Engg & Engg Mechanics, MT State Univ), developed & was in charge of grad environmental engg prog & lab. Sr engr (half or third-time) for Christian, Spring, Sielbach & Assoc, Billings, MT 1977-1982. Pres, Intermtn Section ASCE 1957. Fuller Award AWWA 1976. Chairman, Montana Section Amer. Water Works Assoc. , 1981. Author (or Editor) of three books & many papers. Hobbies: outdoor sports, photography & music. Retired from active teaching 1981. Consulting engineering since 1944. *Society Aff:* AWWA, ASCE, WPCF, ΣΞ, AAEE, AEEP

Sannuti, Peddapullaiah
Business: P.O. Box 909, Busch Campus, Piscataway, NJ 08854
Position: Prof of Elec Engrg *Employer:* Rutgers Univ *Education:* PhD/EE/Univ of IL; MTech/Control Sys/Indian Inst of Tech-India; BE/EE/Govt Coll of Engrg-India *Born:* 4/2/41 Presently Prof of Elec Engrg at Rutgers Univ. Accumulated 19 yrs of Profl experience. Also involved in the field of mathematics. Principle expertise involves, "Singular Perturbation Methods in Control." *Society Aff:* IEEE

Santilli, Paul T
Business: 505 King Ave, Columbus, OH 43201
Position: VP, Gen Counsel. *Employer:* Battelle Meml Inst. *Education:* JD/Law/Capital Univ; MS/CE/OH State Univ; BS/CE/OH State Univ. *Born:* 2/10/29. Columbus, OH. Served with USAF 1951-1953, Proj Engr, Guided Missiles and Warheads Sec, Special Weapons Br. With Battelle Meml Inst since 1953: Chem Engr; Contracts Administrator; Chief Counsel, Battelle Northwest Labs. Assumed current position as Gen Counsel, 1968, & VP in 1972. Am responsible to the Pres for all legal functions of this res & dev corp. Serve on Bd of Dirs of Scientific Advances, Inc, Battelle Commons Co, & Mbr Council on Patents, U.S. Chamber of Commerce Mbr. Alumni Adv Bd, Ohio State Univ. Mbr Tau Beta Pi; Am, OH, & Columbus Bar Assn; Columbus Chamber of Commerce; Kiwanis; OH State Alumni Assn. Enjoy golf, music, travel. *Society Aff:* TBΠ.

Santini, William A, Jr
Business: 833 Butler Rd, Kittanning, PA 16201
Position: Pres, CEO, Chrmn of the Board *Employer:* Phoenix Mtls Corp. *Education:* MCE/CE/NYU; BSCE/CE/Notre Dame. *Born:* 2/1/31. Born 1931, married. 1948-52 BS degree in chem engr from the Univ of Notre Dame. 1953-54 US Army, 1954-56 MCLE at NYU while on Teaching Fellowship, 1956-58 Westinghouse Elec Corp semi-conductor div as dev engr, 1958-63 TX Instruments Inc, advancing to mgr of engr of the Silicon Mtls Dept. 1963-64 consulting engr. 1964-68 founder & pres of Pittsburgh Mtls & Chem Corp. 1968 to present founder Chairman, CEO & pres of Phoenix Mtls Corp. 1977 to present Chairman & CEO of Silicon Electro-Physics Inc. Prof-Engr Licensed, Texas since 1959.. *Society Aff:* AIChE, ECS.

Santoli, Pat A
Home: 1201 Union Ave, Natrona Heights, PA 15065
Position: Senior Metallurgist. *Employer:* Allegheny Ludlum Industries Inc. *Born:* July 1928. BS Case Inst of Technology (now Case-Western Reserve U); Master in Automotive Engrg from Chrysler Inst of Engrg. Native of Cleveland, Ohio. Proj Officer at Wright Air Dev Center with activities in High Temperature Alloys. Reg PE in Ohio. Proj Engr for Allegheny Ludlum with responsibilities for Extrusion & Drawing Dev of Stainless & Refractory Metals. P Chmn of Extrusion & Drawing Activity for ASM Tech Div on Mech working & Forming.

Santry, Israel W, Jr
Home: 7131 Twin Tree Lane, Dallas, TX 75214
Position: Pres Chief Engineer *Employer:* I W. Santry, JR. P.E. *Education:* MSCE/Civil Engrg/Southern Methodist Univ; BS/Civil Engrg/Univ of CA- Berkeley *Born:* 5/24/17 Employed by California Division of Highways, California Water Service Co., and City of Oakland between 1940 and 1942. Taught Civil Engineering at Southern Methodist University 1942-1969, serving as chairman of department 1964-1969. Consulting engineer 1947-1969. President, I. W. Santry, Inc., 1969-1980 and Vice President Boyle Engineering Corporation 1977-1980. President, I. W. Santry, P.E. 1980-present. Executive Secretary Texas Section, ASCE, 1947-1953, President 1960-61. Editor, Sanitary Engineering Division Journal, ASCE. Editor of Transit of Chi Epsilon Fraternity 1960-1972, Vice President 1974-1976 and President 1976-1978. Board of Directors, Boys Clubs of Greater Dallas, 1978-present. *Society Aff:* ASCE, ACEC, AWWA, WPCF, XE, TBΠ, ΦΚΦ

Sanzenbacher, William P
Home: 1185 Alta Mesa Dr, Moraga, CA 94556
Position: Consultant (Self-employed) *Education:* BSE-CE/Civ-Univ of MI *Born:* 6/11/10. Currently Self-Employed Consultant. Formerly: - Managing partner Sanzenbacher-Miller & Scott Engrs & Archts Toledo, OH; VChrmn & Civ Engr Mbr OH Rail Trans Authority, Mbr Toledo Area Transit Authority; Pres OH Soc PE, Dir Natl Soc PE, Pres Caltec, Inc Comp Service; Dir United Savings & Loan Assoc, Mbr Toledo Public Library (Pres); Pres Toledo Bd of Educ; published in OH engr, Consulting engr, Torch Intl. Currently Dir Bostleman Intl, Montserrat West Indies.

Sanzenbacher, William P (Continued)
Enjoy profl soc activities. Reg engr & land surveyor OH, AZ, Reg Civ Engr CA. *Society Aff:* ASCE, NSPE, ICE.

Sapega, August E
Business: Hartford, CT 06106
Position: Prof. *Employer:* Trinity College. *Education:* PhD/Elec Engr/Worcester Poly Inst; MS/Ind Engr/Columbia; BS/Elec Engr/Columbia. *Born:* 12/10/25. Trinity College, Instructor/Prof 1951-present. Engg Dept Chrmn 1971-1982. Res interests in selenium diodes, and instruction in computing. Responsible for academic computing facilities since 1965. Served as mbr of CT Bd of Reg PE and Land Surveyors, 1965-1979. *Society Aff:* ASME, IEEE, SWE, ASEE.

Saperstein, Zalman P
Business: 1500 Dekaven Ave, Racine, WI 53401
Position: VP, Res Dev *Employer:* Modine Mfg Co. *Education:* MS/Engg/UCLA; BS/Engg/UCLA. *Born:* 3/28/31. Chicago, IL. Mar 28, '31; married; 3 children; BS Engr 1954, MS Engr 1962 (Both UCLA); Lectr Mtls Engr UCLA 1957-61; Instr Welding Met IL Inst of Tech; Sr Engr '56-'60, C F Braun & Co (mfg dev); Group Engr 60-64, Douglas Aircraft Missiles Space Systems (mtls & processes); Mgr Welding & Fabrication, IIT Res Inst '64-'69; Chief Engr, '69-'78, the Am Welding & Mfg Co (gas turbine engine component mfg); Since '78, Modine Mfg Co (mfg of heat exchangers); Elected fellow ASM, 1977. *Society Aff:* ASM, AWS, SAE, SME, ΣΞ.

Sarapu, Felix R
Home: 278 Kalalau St, Honolulu, HI 96825
Position: Consulting Structural Engr. *Employer:* Self-employed. *Education:* BSC/Struc Engg/Univ of Technology Stuttgart, W Germany. *Born:* 7/16/20. in Estonia. BSC Stuttgart Univ of Technology, W Germany. With prominent Sydney, Australia Structural Consulting firm, WHB&C, 1950 - 60, as Proj Eng and Sen Assoc. Engr the world's then highest lift slab bldg in 1958. Taught 3 yrs at Sydney Univ of Technology. With Richard R Bradshaw Inc in Los Angeles 1962. Cons in Australia 1963. Rejoined R R Bradshaw org as Sr VP in Honolulu Office 1964. Respon for numerous high-rise hotels, public bldgs, aptments, etc. Biggest proj 1900-room Sheraton Waikiki Hotel. Designed hotels in South Pacific, Tahiti, Fiji, Amer Samoa, etc. Elected Fellow of the Inst of Engrs, Australia 1971. Hobbies are photography & horticulture. *Society Aff:* ACI, SEAOH.

Sarasua, Jose I
Home: 824 Ariege Dr, St. Louis, MO 63141
Position: Sr. Engr. *Employer:* Belcan Corp *Education:* MBA/Bus Adm/S Illinois Univ Ed; BSBC/Constru/Univ of FL *Born:* 10/18/46 Native of Havana, Cuba. Field Cost Engr with United Engr & Constructors Inc (1969-72). Constr analyst with FL Pwr & Light Co in Miami Headquarters (1972- 74). With Monsanto Co 1974-87. Served 5 yrs as supvr of cost, estimating and planning & scheduling for Monsanto Co subsidiary in Sao Paulo, Brazil, SA. Assigned to St Louis headquarters in 1979 as an engrg specialist. Presently working at Belcan's Engrg Div in Cincinnati. Certified Cost Engr (CCE) since 1977. Has held various offices at local AACE level. Past Region IV Dir of AACE & mbr of Bd of Dir. Enjoys swimming and all sports. *Society Aff:* AACE

Sarazin, Armand
Home: INSA 156 Ave de Rangueil, Toulouse Haute-Garonne, France 31077
Position: Director General of INSA. *Employer:* French Ministr of Ed, U of Toulouse. *Education:* Doctor of Sciences/-/French Universite; Ingenieur/Ecole Natl Superieure d'electricite et de Mecanique/Nancy- *Born:* 1924. Electrical Engr (ENSEM, Nancy). PhD Physics Lab of the Ecole Normale Superieure, Paris. Physicist at CERN Geneva 1953-56. Prof at the U of Algiers T Dir-Assoc of the Inst for Nuclear Studies of Algiers 195661. Prof at the U of Lyon & Dir of the Inst for Nuclear Physics 1961-76. Appointed dir of the INSA in Toulouse April 1976. Interests: expermental nuclear physics, electronics, systems, value analysis, cooperative education. Was awarded the 'Medaille Blondel' in Paris 1969. Fellow IEEE. Mbr of the American, British, Canadian, French, Italian, & Japanese Socs of Physics, & of the French Soc of Electrical & Radiolectronic Engrs. *Society Aff:* IEEE, SEE, SFP, SAVE.

Sarchet, Bernard R
Business: 301 Harris Hall, Engrg Mgmt Dept, Rolla, MO 65401
Position: Prof & Chrmn *Employer:* Univ of MO *Education:* MS/ChE/Univ of DE; BS/ChE/OH State Univ; Advanced Management Prog/Harvard Business School *Born:* 6/13/17 Native of Byesville, OH. Served with Koppers Co, Inc., 1941-67 as design engr, general foreman, engrg supervisor, plant mgr of two plants, asst sales mgr, mgr of dev, mgr of panel dept, and dir corp commercial dev. Joined Univ of MO-Rolla 1967 to found Dept of Engrg Mgmt as prof and chmn until 1981. 1981-date Robert B. Koplar Prof of Engrg Mgt and Dir of Teleconference Activities. Spent 1971-73 as chief of party in Saigon, Vietnam developing an engrg school. Served 1975-79 as Exec Dir of External Affairs. Grad 2400 BS & MS students in Engrg Mgmt. Founding Pres of ASEM. Winner of four Freedom Foundation Awards. Advanced mgmt program at Harvard Business School. *Society Aff:* ASEM, ASEE, AIChE, IIE

Sard, Eugene W
Home: 49 Harriet Lane, Huntington, NY 11743
Position: Consultant, Receiver Systems and Technology Dept. *Employer:* AIL, Div Eaton Corp. *Education:* MSEE/Electronics/MIT; BSEE/Electronics/MIT. *Born:* Dec 21, 1923 Brooklyn NY. Radar officer with the USNR 1944-46. With present Co (formerly, Airborne Instruments Lab) 1948- . Elected Fellow of the Inst of Electrical & Electronics Engrs in 1973 'for contributions to the field of low-noise microwave, millimeter wave, & infrared receivers'. Mbr Sigma Xi. *Society Aff:* IEEE.

Sargent, Donald J
Business: PO Box 2200, Hwy 52 NW, W Lafayette, IN 47906
Position: Vice President-Technical Director. *Employer:* E/M Lubricants Inc. *Education:* BS/Chemistry/Wayne State Univ; Bus Dipl/-/Alexander Hamilton Inst. *Born:* April 1936. BS Wayne Univ; Graduate Studies in Surface & Colloid Chem. Res Staff for BASF-Wyondotte, specializing in detergent res. Res Assoc & Mgr of Sales & Dev for Climax Molybdenum Co, specializing in solid lubrication. With E/M Lubricants Inc as Tech Dir 1974- , assumed current responsibility as Vice Pres in 1975. I am responsible for Res & Dev, Quality Control & Glass Indus Group. Chmn of ASLE, Solid Lubricant Ctte; Chmn of NLGI, Education Ctte; Chmn ASTM Subcties. Enjoy gardening, wood carving, tropical fish & reading. *Society Aff:* NLGI, ASTM, ASLE, SAE, SAMPE.

Sargent, George W
Home: 193 Old Black Rock Tpk, Fairfield, CT 06430
Position: Chrmn - Physics Dept Retired *Employer:* Bridgeport Engineer Institute. *Education:* MS/ChE/WPI *Born:* Feb 1923. BSChE 1944, MSChE 1949. US Navy 1944-46. Cm Physics 13 years; Admin Physical Scis & Cm Physics 7 years both at BEI (evenig sch only). Days: Rubber Chemist 5 years US Steel, 15 years G E Co, Adv Mfg Eng 16 years GE Retired, July 1986; BSA troop ctte Cm 7 years; neighborhood assoc Pres 2 years; hobbies- backpacking, canoeing, sailing. *Society Aff:* AAPT

Sargent, Gordon A
Home: 2881 Graig Ct, Lexington, KY 40503
Position: Prof. *Employer:* Univ of KY. *Education:* PhD/Physical Metallurgy/Imperial Coll, London Univ UK; BSc (Engg)/Metallurgy/Imperial Coll, London Univ UK. *Born:* 4/8/38. Educator, b Winterton, Lincolnshire, England; married Amy Skinner Sept 17, 1966; children - Andrew, Mark, Maria, Anne, Adrian, Elizabeth, Susan; Res Fellow, Mellon Inst 1962-1967, asst Prof Univ KY 1967-72, assoc Prof 1972-76, full prof 1976-Present Chairman Department of Metallurgical Engineering and Materials Science, University of Ky 1981- , also Associate Director Materials Engineering Division, Institute for Mining and Minerals Research Laborato-

Sargent, Gordon A (Continued)
ry. Pres of Alpha Sigma Mu, Metallurgy and Materials Hon Soc 1978-79. Chrmn Bluegrass Chap ASM 1970. *Society Aff:* ASM, AIME.

Sargent, Lowrie B, Jr
Business: Alcoa Laboratories, Alcoa Center, PA 15069
Position: Senior Scientific Associate. *Employer:* Aluminum Co of America. *Education:* PhD/Physs Chem/PA State Univ; MS/Chemistry/Lehigh Univ; BS/Chemistry/Washington & Jefferson College. *Born:* Sept 1918. Native of Fredericktown, Penn. With res labs of Alcoa 1942- . Mgr of Lubricants Div for 18 years, Senior Scientific Assoc 1975- . Pres ASLE 1962; Assoc Editor ASLE Publs 1960-74; Fellow ASLE 1970. Chmn Ctte D-2 of ASTM 1966-75; Bd/Dir ASTM 1974-77; Fellow ASTM 1974; ASTM D-2 Award 1974. Chmn Ctte Z-11 of ANSI 1966-75. Bd of Trustees W&J Coll 1961-66. Mbr Sigma Xi, Phi Lambda Upsilon & RESA. Publs in tribology. *Society Aff:* ASLE, ASTM.

Sargent, Robert G
Business: 431 Link Hall, Syracuse University, Syracuse, NY 13210
Position: Prof. *Employer:* Syracuse Univ. *Education:* PhD/Ind Engr/Univ of MI; MS/Ind Admin/Univ of MI; BSE/Elec/Univ of MI; Assoc of Sci/Pre-Engr/Port Huron Junior College *Born:* 6/14/37. in Pt Huron, MI. Attended public schools in St Clair, MI. Educated at Univ of MI for BSE(EE), MS, & PhD. Was electronics engr for two & one-half yrs at Hughes Aircraft Co before doing grad studies. Was Gen Chrmn of 1977 winter simulation conf. Was, Department Editor, Simulation Modeling and Statistical Computing, Communications of the ACM (CACM) was, Chairman, TIMS College Simulation and Gaming, 1978-1980. Chrmn, Board of Directors, of Winter Simulation Conference, 1979-1981. Dir, Soc for Comp Simulating, 1984- . National Lecturer for ACM. Currently working in simulation, applied operations res, and modelling computer systems for performance evaluation. Assistant Associate, and Full Prof (Chrmn of Ind Engr & OpRes Dept 1982-83)., Syr. Univ. 1966- . *Society Aff:* ACM, IIE, ORSA, TIMS, SCS

Saridis, George N
Business: , Troy, NY 12180-3590
Position: Prof *Employer:* Rensselaer Poly Inst *Education:* PhD/Optimal Control/Purdue Univ; M/EE/Purdue Univ; DipEng Mech stl/ME&EE/Natl Tech Univ of Greece *Born:* 11/17/31 Born in Athens, Greece. Taught at NTU Athens as an instructor 1955-63. Taught at Purdue Univ as Asst Prof 1965-70, Assoc Prof 1970-75, Prof 1975-81. Prof and Director Robotics and Automation Lab ECSE RPI 1981-present. Author of book *Self Organizing Control of Stochastic Systems.* Author of over 200 journal articles and conference papers. Fuzzy Automatic, editor of *Advances in Automation & Robotics.* Founding President of IEEE Council of Robotics & Automation. IEEE Centennial Medal Award. *Society Aff:* IEEE, ASME, ASEE, SPIE, RI/SME, MVA/SME, NYAS

Sarkaria, Gurmukh S
Business: 220 Montgomery Street, San Francisco, CA 94104
Position: Vice President. *Employer:* Internatl Engineering Co Inc. *Born:* Nov 1925. MS Harvard; MCE Brooklyn Poly Inst; BSCE Punjab Univ India. Native of India. 4 years with the Bureau of Reclamation; 4 years with Bhakra Dam design group in India; with IECO 1956- , Vice Pres & Ch Engr 1969-75. Major projs for which served as Proj Mgr or Proj Engr include Yuba River Dev, Ross High Dam, Princeville Proj (Kauai, Hawaii) & Pacoima Dam studies. As Ch Engr had overall responsibilities for engrg of of all IECO projs. Am Genl Coordinator for IECO & Electroconsult of Italy for the engrg of the Itaipuu Hydroelectric Proj between Brazil & Paraguay, and resident in Rio de Janeiro 1975- . Have pub sev articles on dams & hydroprojs. Enjoy hunting, gardening & golf.

Sarofim, Adel F
Business: 77 Mass Ave Rm 66-466, Cambridge, MA 02139
Position: Professor of Chemical Engineering. *Employer:* Massachusetts Inst of Technology. *Education:* BA/Chemistry/Oxford Univ; SM/Chem Eng/MIT; ScD/Chem Engg/MIT. *Born:* 10/21/34. Faculty MIT 1961- ; prof of Chemical Engg 1972- . Visitng Prof Univ of Sheffield 1971; Chevron Visiting Energy Prof Caltech 1979. Panel on Hazardous Trace Substances, Office of Sci & Technology 1971-72. Ctte on Principles of Protocols for Evaluating Chems in the Environ, Natl 1972-73. Ctte on health & Ecological Effects of Increased Coal Utilization, HEW 1977. Energy Engg Bd, Nat. Res. Council, 1983- . Co-author of book, Radiative Transfer, 1967. Recipient Kuwait Prize, Petrochem. Eng., 1983; Edgerton Medal, Combustion Inst, 1984; Hottel Lecturer, Combustion Inst, 1986; Lacey Lecturer, Caltech, 1987. Author of about 100 tech papers on radiative transfer, kinetics of pollutant formation, coal combustion, freeze desalination. *Society Aff:* AIChE, ACS, SES.

Sarriera, Rafael E
Business: Box 11095, Santurce, PR 00910
Position: Principal. *Employer:* R E Sarriera Assoc. *Education:* BSCE/CE/Univ of PR. *Born:* 6/12/26. Civ Engr, reg PR. Parents Enrique and Carmen (Rodriguez). Married to Mercedes Riancho, 1954; 1 son, Gerardo R. BSCE, Univ of PR, 1946. Experience: Field Engr, Long Construction Co, 1947-48; Structural Engr, PR Water Resources Authority, 1948-52; Design Engr, PR Ind Dev Co, 1952-56; Chief Engr, Guillermety & Ortiz, 1956-59; Private Practice, Principal, RE Sarriera Assoc, Santurce, PR, 1959-date. Military Service: Served as 2nd Lt, AUS, 1946-47; Natl Guard, 1st Lt, 1951-53; Army Reserve, Capt, 1953-58. Mbrship: Fellow, ASCE, (Past Pres, PR Section); Diplomate, AAEE; Mbr, NSPE, WPCF, AWWA, IAWPR, AIDIS, CIAPR. *Society Aff:* AAEE, ASCE, NSPE, AWWA, WPCF, ACI, IAWPRA AIDIS, CIAPR

Sartell, Jack A
Business: 10701 Lyndale Ave So, Bloomington, MN 55420
Position: Research Manager *Employer:* Honeywell Inc. *Education:* BMet E/Metallurgical Engg/Univ of Minnesota; MS/Metallurgical Engg/Univ of Minnesota; PhD/Physical Metallurgy/Univ of Wisconsin. *Born:* June 18, 1924. BMetE 1949, MS 1951 Univ of Minnesota; PhD 1957 Univ of Wisconsin. Res Engr at Alcoa Res Labs 1952. Employed at Honeywell Res Center 1957-87. Presently Consultant, Materials and Research. Mbr: Sigma Xi, AIME, American Soc for Metals. Fellow of American Soc for Metals. International Society for Hybrid Microelectronics. Res areas: oxidation kinetics, properties of beryllium, magnetic materials, plate wire memory, magnetic bubble memories, electronic packaging matls, reliability physics. *Society Aff:* ISHM, ASM INTL, AIME, ΣΞ.

Sarto, Jorma O
Home: 7560 Honeysuckle, Orchard Lake, MI 48033
Position: Research Engineer, Power Train Systems *Employer:* Chrysler Corporation. *Education:* MS/Automotive Engr/Chrysler Inst; BS/ME/MI State Univ. *Born:* Jan 26, 1917. Private Pilot License 1941; Chrysler Mgmt Training 1955- 60; Detroit Indus Mission 1964; Toastmasters Internatl 1966-72. Fellow ASME; Mbr SAE; Reg PE Mich; Inventor, 48 US Patents Granted; Co-author of 2 SAE Tech Papers. Joined Chrysler Corp in 1942. Product dev has included Advance Fuel Metering Systems, Vehicle Low Temperature Starting, Emission Controls, Fuel Systems for Aircraft Engines, Gas Turbines & Domestic Oil Furnaces & Turret Controls for Military Tanks. Served as Pres of MSU Engrg Alumni Assn. Married, 3 daughters. Pres Finlandia Male Chorus of Detroit. Enjoy singing, dancing, flying, sailing, travel & carpentry. *Society Aff:* ASME, SAE.

Sasman, Robert T
Business: PO Box 409 29W002 Main St, Warrenville, IL 60555
Position: Hydrologist. *Employer:* Illinois State Water Survey Division. *Education:* BS/Soil Conservation/Univ of WI, Madison. *Born:* July 1923. Col US Army 1943-46, Col US Army Reserve Retired 1975; Meritorious Service Medal 1975; Honor Graduate US Army Command & Genl Staff Coll 1966. With Ill Water Survey 1951- , in charge of Northern Regional Office 1957- . Responsible for water resources inventory & res in northern Ill. Author of 27 publs on water resources in northern Ill.

Sasman, Robert T (Continued)
Dir II Sect AWWA 1975-78, Sect Chmn 1974-75, AWWA Water Works Man of the Year in 1976, AWWA Diamond Pin Club Mbr 1976, Editor Sect Publ 1965-68. Hobby- photography. *Society Aff:* AWWA, NWWA.

Satija, Kanwar S
Business: 30 S. 17th St, Philadelphia, PA 19101
Position: Mgr *Employer:* United Engrs & Constr *Education:* PhD/Mech & Hyd/Univ of IA; MS/Hydr/Punjab Engrg Coll; BS/Mech-Civ/Punjab Engrg Coll; Dipl/German-French/Punjab Univ-India *Born:* 7/1/42 Native of India. Taught at Punjab Engrg Coll Chandigarh, India 1962-67. Wrote a book on Machine Drawing 1964. Grad work at the Univ of IA 1968-70 along with part-time consulting/design. Worked at Chicago 1970-73 with Pioneer Engrg and Sargent & Lundy working on hydraulic, water-resources and site dev aspects of Power Plants, including hydroelectric, fossil and nuclear. Continued this work with United Engrs, first as Mgr Specialty Engrg and currently Proj Engrg Mgr. Also work as Adjunct Prof at Drexel Univ teaching hydropower courses. Chrmn Student Relations ASME. Mbr Power Div ASME. Enjoy swimming, bridge and badminton. (Participant: Plant Engrg Handbook). *Society Aff:* ASME

Sato, Risaburo
Business: Dean, Faculty of Engineering, Tohoku Gakuin University, Chuo, Tagajo, Japan 985
Position: Dean of Faculty of Engrg *Employer:* Tohoku Gakuin Univ *Education:* PhD/Elec Comm Engg/Tohoku Univ; BE/Elec Comm Engg/Tohoku Univ. *Born:* 9/23/21. Born in Furukawa City, Miyagi Prefecture, Japan. From 1949 to 1961, Asst Prof & from 1961 to 1984, Prof of Tohoku Univ. Was the invited Prof of SRI in 1969. Now, the Dean of Faculty of Engrg of Tohoku Gakuin Univ & Hon Prof of Tohoku Univ. Relating to IECE of Japan, the vp (1974-1976), the chrmn of Tech Group of Circuit & Systems Theory (1971-1975) & the Chrmn of Tech Group on EMC (1977-1981) & the chrmn of IEEE EMC-S Tokyo chapter (1981 -). Relating to Govt, a mbr of the Sci Council of Japan & a mbr of the Radio Technical Council of the Ministry of Post & Telecommunications. A fellow mbr of IEEE. *Society Aff:* SCJ.

Satterfield, Charles N
Business: 66-572, Cambridge, MA 02139
Position: Prof *Employer:* MA Inst of Tech *Education:* ScD/ChE/MIT; BS/Chem/Harvard Univ *Born:* 9/5/21 A mbr of the MIT Chem Engrg Faculty since 1946, and co-currently visiting lecturer at Harvard, 1948-57. Chrmn or mbr of numerous adv bds to govt agencies and profl societies. Author or co-author of over 120 technical papers and five books including most recently, *Mass Transfer in Heterogeneous* Catalysis, 1970 and *Heterogeneous Catalysis in Practice,* 1980 (translated into Russian, 1984). Consultant to major companies in the petroleum, chem and pharmaceutical industries. 1980 recipient of the AIChE Wilhelm Award for "distinguished and continuing contributions to chem reaction engrg." *Society Aff:* AIChE, ACS, AAAS.

Satterlee, George L
Home: 11035 State Line, Kansas City, MO 64114
Position: Director of Public Works *Employer:* City of Kansas City, Missouri *Education:* BS/Civil Engr/Univ of MO. *Born:* Dec 1930. Graduate work at Purdue Univ & Univ of Mo. Lt, Army Corps of Engrs 1952-55. Mo St Highway Dept 1955-1986; Kansas City Dist Engr 1970-86. Employed as Director of Public Works, Kansas City, MO. 1986- . Responsible for directing all Public Works activities in Kansas City, Mo. Pres Kansas City Sect ASCE 1971; St Dir MSPE 1972-75; Reg PE in Mo; Tau Beta Pi; Chi Epsilon; Episcopal Church. Leisure time spent enjoying family activities with wife & 2 children. *Society Aff:* NSPE, ASCE, APWA, ARTBA

Sauer, Enno T
Home: 2317 Clarkwood Road, Louisville, KY 40207
Position: President - Retired 7/1/80 *Employer:* Rohm & Haas Kentucky Inc. *Education:* MS/Chem Engg/MIT; BS/Chemistry/Univ of Richmond. *Born:* June 6, 1915. With Rohm & Haas Co 1937-80. Supervisory positions in Semi-Works Philadelphia plant, Toronto plant & Corporate Production Control 1937-60; Asst Plant Mgr Knoxville plant 1960-62; Plant Mgr Louisville plant 1962-80; Pres Rohm & Haas Kentucky Inc 1972-80. Commissioner Kentucky Environ Quality Comm 1973-86. Bd/Dir Louisville Chamber of Commerce, Kentucky Chamber of commerce, Associated Industries of Kentucky. Bd/Dir Concordia Publishing House of Lutheran Church. Mbr of KY Governor's Commission on Hazardous Waste 1979-81. Bd/Dir Action Now Secy-Treas Louisville Rotary Club, 1982-5. *Society Aff:* AIChE.

Sauer, Gilbert F
Home: 4675 Hathaway Dr, Franklin, OH 45005
Position: Mgr-Marketing. *Employer:* Armco Inc. *Education:* MS/Ag Engg & CE/Univ of IL; BS/Ag Engg/Univ of IL. *Born:* 6/30/27. in Decatur, IL, served in US Navy in WWII. On staff of Univ of IL after grad. Joined Armco in 1953 as a Product Design Engr. Was Chief Engr and then Mgr, Design Engg - Bldg Systems, before assuming present responsibility. Am now responsible for Armco's corporate agricultural and construction marketing progs. Dir of Publications ASAE, Bd of Governors FEMA, mbr of AISI, FIEI. *Society Aff:* ASAE, SAE.

Sauer, Harry J, Jr
Home: College Hills, Rt 4, Rolla, MO 65401
Position: Prof. *Employer:* Univ of MO. *Education:* PhD/ME/KS State Univ; MS/ME/MO Sch of Mines & Met; BS/ME/MO Sch of Mines and Met. *Born:* 1/27/35. Native of St Joseph, MO. Teaching positions at KS State Univ & the Univ of MO-Rolla since 1957, currently Prof of Mech & Aerospace Engg and Dean of Grad Studies. Consultant to ind & govt. Author or co-author of over one hundred technical publications & six books. Chrmn, St Louis Section SAE 1977-78; recipient of SAE Ralph R Teetor Education Award in 1968; recipient of St Louis Chapter ASHRAE Hermann F Spoehrer Meml Award in 1979; ASHRAE's Distinguished Service Award in 1981; ASHRAE's E. K. Campbell Award of Merit in 1983; ASHRAE Award for Best Res Technical Paper of 1974; ASHRAE 1986 Best Paper Award; elected Fellow in ASHRAE in 1979; Chrmn, Engg Div, MO Acad of Sci, 1975-77; and mbr of ASME, ASEE, NSPE, NCURA, SRA, Pi Tau Sigma, Tau Beta Pi, Sigma Xi, Governor's Commission on Energy Conservation. *Society Aff:* ASHRAE, ASME, SAE, ASEE, NSPE, NCURA, SRA.

Saul, William E
Business: College of Engineering, University of Idaho, Moscow, ID 83843
Position: Dean of Engrg and Prof of Civil Engrg *Employer:* Univ of ID *Education:* PhD/Civ Engg/Northwestern Univ.; MS/Civ Engg/MI Tech Univ; BS/Civ Engg/MI Tech Univ. *Born:* 5/15/34. 7-55 - 9-59 Mech Engr, Div, Production Dept, Shell Oil Co, New Orleans, LA. 9-59 - 6-60 Grad Student & Teaching Asst, MI Tech Univ. 8-60 - 9-62 Instr of Engg Mechanics, MI Tech Univ, 9-62 - 9-64 Grad Student, Northwestern Univ. 9-64 - 8-67 Asst Prof of Civ Engg, Univ of WI, Madison, WI. 9-67 - 8-72 Assoc Prof of Civ Engg, Univ of WI, Madison, WI. 1970-71 - Alexander von Humboldt Foundation Fellow, Germany 1970-71 - Fulbright grant, Germany 9-70 - 9-71 Visiting Prof, Inst for Statik & Dynamik der Luft-& Raumfahrtkonstruktionen der Universitat Stuttgart. 9-72 - 1-84 Prof of Civ & Environmental Engg, the Univ of WI, Madison, WI. 5-76 - 7-80 Chrmn, Dept of Civ & Environmental Engr, Univ of WI, Madison, WI. 1984-date, Dean of Engrg and Prof of Civil Engrg, the Univ of Idaho, Moscow, Idaho. *Society Aff:* ASCE, ACI, ASEE, IABSE, NSPE, ISPE.

Saunders, Byron W
Business: 304 Upson Hall, Ithaca, NY 14853
Position: Ex Dean of the Univ Faculty (Retired). *Employer:* Cornell University. *Education:* MS/Eng Economics/Stevens Inst of Technology; BS/Elec Engg/Univ of RI. *Born:* June 1914. Native of Providence RI. Engr with Narragansett Electric Co & Douglas H Paton Co (Marine Communications) prior to WWII; Mfg Dev with

Saunders, Byron W (Continued)
RCA during WWII. Cornell Univ 1947-79 Asst Prof, Assoc Prof & Prof of Indus Engrg & Operations Res 1956- . Acting Chmn & Chmn, then Dir of Sch of Indus Engrg/Operations Res 1962-75, Dean of Faculty 1974-78. Mbr ASME, ASEE, elected Fellow IIE 1975; P Chmn CIC-MHE; M H Div of ASME; IE Div ASEE. Trustee of SEA Educ Assoc. Bd of Dirs/Wilderness Corp. *Society Aff:* IIE, ASME, ASEE, AAUP.

Saunders, F Michael
Business: Environ. Engrg, Georgia Tech, Atlanta, GA 30332-0512
Position: Assoc Prof *Employer:* GA Inst of Tech *Education:* PhD/Envir Engrg/Univ of IL; MS/CE/VA Poly Inst; BS/CE/VA Poly Inst *Born:* 4/3/44 1968-74 Univ of IL; 1974 Wiley and Wilson Conslt Engrs; 1974-present Sch of Civil Engrg (Envir Engrg), GA Inst of Tech; Teaching, res and conslt in indus and municipal water and wastewater treatment with special interests in sludge reclamation, treatment and disposal, biological wastewater treatment and in situ groundwater reclamation. Author of over 50 tech presentations, articles and reports. Reg PE, GA; Diplomate, Amer Acad of Environ Engrs; USA Regional Editor and Intl Ed Bd, *Environmental Technology Letters*; Organization Committee, 1992 Biennial Conf. of IAWPRC, President, Assoc of Environ Engrg Prof, 1985; Chrmn, Water Poll Ctrl Fed Res Ctte, 1983-86; Chair, Amer Soc of Civil Engrs Water Poll Mgmt Ctte of Environ Engrg Div, 1986- 1987; Mbr, Water Poll Ctrl Fed Publ Ctte, 1987-present; Chrmn, ASCE Natl Conference on Environ Engrg 1981. *Society Aff:* ASCE, IAWPRC, ACS, WPCF, AEEP, AES, AWWA, AAEE

Saunders, Robert M
Business: Sch of Engg, Irvine, CA 92717
Position: Prof of EE. *Employer:* Univ of CA. *Education:* DEngr/EE/Tokyo Inst of Tech; MS/EE/Univ of MN; BEE/EE/Univ of MN. *Born:* 9/12/15. Electric Machinery Co, Minneapolis: 1938-42; Univ of MN: EE faculty 1938- 44; USNR (active duty) 1944-46; Univ of CA Faculty: Berkeley, 1946-65; EE Dept Chrmn 1959-63, Irvine, 1965-, Dean 1965-73. MIT Visiting Assoc Prof 1954-55; Univ of Manchester (England): Simon Fellow, 1960; ECPD: Engrg Educ & Accreditation Committee 1965-71, Chrmn 1969-70, Bd of Dirs 1971-75; IEEE: Fellow '63, Bd of Dirs, 1973-79, Pres 1977; AAES: Organizing Ctte 1977-80, Exec Ctte 1982-84; Chrmn Bd of Governors 1983; AAES: Chairman Awards Committee 1984-85; Chairman Nominating Committee 1984-85. National Research Council, Committee on Education and Utilization of the Engineer, 1983-84; Elected Fellow, American Assoc. Advancement of Science 1985; Consultant: Honeywell, GM, GE, Rohr, Aerospace Corp, Hughes Aircraft; NSF: Engrg Advisory Committee, 1968-71; US Army Transportation Sch Bd of Visitors, 1970-73; USN: Secy's Advisory Bd on Education & Training, 1972-79; *Society Aff:* IEEE, ASEE, AAES.

Saunders, Walter D
Business: Martin & Saunders Inc. Structural Engrs, 1700 Adams Ave Ste 210, Costa Mesa, CA 92626
Position: Pres *Employer:* Martin & Saunders, Inc. *Education:* BS/Civil Eng/Univ of So CA. *Born:* Aug 1923. US Navy 3rd Amphibious Force 1942-46. Reg CE & SE in Calif, SE in Guam. Part-time lecturer USC Sch of Architecture 1950-55. Manhattan Beach Planning Comm 1955-58; Design & Proj Engr for Brandow & Johnston, Str Engrs 1950-57. John A Martin & Assoc 1957-74: Assoc Branch Office Mgr Orange County 1963-70, Production Mgr 1970-72, Dir Spec Projs & Computers 1972-74. Managing partner of Martin & Saunders, Structural Engineers 1974- , specializing in Seismic, Dynamic & Special Analyses, Structural Design Reviews, Investigations, & reports. PPres Bd/Dir Applied Technology Council, Bd/Dir Structural Engrs Association of CA, Bd/Dir Structural Engrs Assoc. of Southern CA. *Society Aff:* ASCE, ACI, ASTM, CRSI, AITC, AISC.

Sauter, Harry D
Home: RR 4, Princeton, IN 47670
Position: Dir Quality Assurance. *Employer:* Hansen Mfg Co Div Inc Corp *Education:* BSEE/Power/Purdue Univ *Born:* 9/6/24. Native of Grandview, Ind. Served with US Navy 1944-46. Customer Engr IBM Corp 1950-51; Prod Engr, Engrg Servs Mgr 1951-59 AVCO Corp. Senior Applications Engr 1956-63; Senior Reliability Engr 1963-77; Dir Quality Assur 1977-1983, Potter & Brumfield Div AMF Inc 1963-77, Mgr QA Hansen Mfg Co 1984-present. Fellow Mbr, Certified Quality Engr, Region 9 Dir-at-Large, Amer Soc for Quality Control; P Chmn local sect of ASQC. Dir Midwest Conference Bd ASQC. Taught Quality Control Courses for Vincennes Univ, local ASQC sect & in-plant. Served on C83.1 Ctte Amer Natl Standards Inst, 4 yrs. Taught Sunday Sch & Bible in local Baptist Church for 27 yrs. Hobbies: woodcraft, gardening, fishing. *Society Aff:* ASQC

Savage, Charles F
Home: 3609 Palmer Court, Clovis, NM 88101
Position: RETIRED *Education:* B.S./E.E./Oregon State University *Born:* 02/24/06 Div Engr, General Electric Co. 1928-71. Exec Dir Council on Engrg Laws 1971-77. Staff Consultant, EJC 1977-80. IEEE Centennial Medal; GE Phillipe Award; Steinmetz Medal; Inventor Award. Licensed P.E. in NY, MA, OR. Fellow AAAS, Chrmn Sect M.; AIAA Assoc. Fellow, Charter Mbr. ANS, Mbr SAE, Mbr SWE, Bd of Dir, IEEE, Instrument Society of America, AIAA. Chairman, several Sections and Cttees. Patents. Radio Amateur W2WN IEEE Fellow Citation: "For contributions in the field of electrical measurement, esp aircraft engine, guidance & control. *Society Aff:* ADPA, IEEE, ASME, CESSE, NSPE, ASEE, HKNU, TBΠ

Savage, John E
Business: Dept of Computer Sci, Box 1910, Providence, RI 02912
Position: Prof of Computer Science. *Employer:* Brown University. *Education:* PhD/Elec Engg/MIT; ScB/Elec Engg/MIT; ScM/Elec Engg/MIT. *Born:* 9/19/39. John Edmund Savage, computer science educator, researcher; b. Lynn Sept 19, m. Patricia Joan Landers, Jan. 29, 1966; Children - Elizabeth, Kevin, Christopher, Timothy. Sc.B., MIT, 1962, Sc.M., 1962, PhD., 1965. Mem. tech. staff Bell Telephone Labs., Holmdel, N.J., 1965-67; prof. computer sci. Brown U., Providence, 1967-, chmn. computer sci. dept., 1985-; vis. prof. U. Paris, 1980-81; cons. in field. Author: The Complexity of Computing, 1976; (with others) The Mystical Machine, 1986. Patentee data scambler, 1970, means and methods for generating permutations of a square, 1976. Fulbright-Hays grantee, 1973: NSF fellow, 1961, Guggenheim fellow, 1973. Mem. Assn. Computing Machinery, 1987. The Complexity of Computing (reissued). IEE, Sigma Xi, Tau Beta Pi. Avocations: reading; skiing. Home: 65 Humboldt Ave Providence RI 02906 Office: Brown U Dept Computer Sci Box 1910 Providence RI 02912 (401) 863-1833. *Society Aff:* IEEE, ACM.

Savage, Marvin
Business: PO Box 177, Huntingdon Valley, PA 19006
Position: Pres. *Employer:* Savage Assoc. *Education:* BSChE/CE/Drexel Univ. *Born:* 2/22/35. Native of Phila Area. Involved in design, application & sale of bulk solids drying blending & processing equip since 1962. Author of several papers on the subj. Also involved in consulting in this field. Consultant for Franklin Inst Res labs. Pres of Savage Assoc, a Sales agency specializing in bulk solids processing equip. *Society Aff:* AIChE.

Savage, Rudolph P
Home: 8300 Kingsgate Rd, Potomac, MD 20854
Position: Pres *Employer:* Shorelines, Inc. *Education:* BS/Engr/NC State Univ. *Born:* 4/4/27. Graduate work at Texas A&M; completed Third Internatl Course in Hydraulic Engrg at Delft, Netherlands in 1960. Worked as a res hydraulic engr for the US Army Corps of Engrs' Coastal Engrg Res Center from 1951 to 1983. As Chief, Res Div from 1972 to 1983 doing & supervising res on beach erosion, storm surge generation & coastal ecology. Now doing conslt on coastal problems. 18 reports & publs; Natl Capitol Award for prof achievement in 1962; ASCE Huber Res Prize in 1966; DA R&D Award 1969. Served on ASCE WWH&CE Publications & Coastal Engrg Cttes and Hyd Div Research and Exec Cttes. *Society Aff:* ASBPA, ASCE

Savant, Clement J, Jr
Home: P.O. Box 967, Culver City, CA 90232
Position: Full Professor *Employer:* Calif State Univ *Education:* PhD/EE/CA Inst of Tech; MS/EE/Calif. Inst. of Tech; BS/EE/Calif. Inst of Tech *Born:* 08/09/26 Native of Butte, MT. In 1944-45 served with the US Navy in WWII. In 1952 worked as an Assoc Prof of Elec Engrg at the Univ of Southern CA. Worked as res engr for Jet Propulsion Lab, N American Rockwell, Northrop Corp, Hughes Tool Co. At the Whittaker Corp became Director of Engrg of Tech Products Div, (with 350 engrs) then gen mgr, and finally VP and Group Exec for Dynasciences, a Whittaker Corp Subsidiary. Was Pres and Chrmn of the Bd of MCA Tech, a subsidiary of MCA, Universal City. In 1972 purchased 50% of Neo Flasher Elec and was named Chrmn of the Bd and Exec VP later sold the company. In 1976 returned to CA State Univ Long Beach as Assoc Prof of Elec Engrg. Authored 8 textbooks, 1 of which will be published in 1989. Won the School of Engrg Outstanding Prof Award three consecutive yrs and in 1981 was awarded the CSULB Univ-wide award for Outstanding Prof. In 1983 went to Cal State Univ Los Angeles as Prof of EE. *Society Aff:* IEEE, TBΠ, ETTAKN, ΣΞ, ASEE

Savic, Michael I
Home: 4 Sawmill Dr, Wilbraham, MA 01095
Position: Prof Elec Engg & Dept Chrmn. *Employer:* Western New England College. *Education:* EngScD/Elec Engr/Univ of Belgrade; DiplIng/Elec Engr/Univ of Belgrade. *Born:* 8/4/29. Born in Belgrade, Yugoslavia, on Aug 4, 1929. Received the Dipl Ing and the Eng Sc D degrees in elec engg from the Univ of Belgrade, in 1955 and 1965 respectively. From 1956 to 1966 was with the Elec Engg Dept at the Univ of Belgrade, Yugoslavia, and in 1967 with Yale Univ. Since Sept 1967 has been at Western New England College, Springfield, MA and is now Prof and Chrmn of the Dept of Elec Engg. Maj res activities have been in the areas of ultrasound, industrial and biomedical electronics. *Society Aff:* IEEE, ACC.

Saville, Thorndike, Jr
Home: 5601 Albia Rd, Bethesda, MD 20816
Position: Consultant *Education:* AB/Engg Sci & Appl Physics/Harvard; MS/Civil Engg/Univ of CA, Berkeley. *Born:* 1925 Baltimore, Md. Res Asst UCB 1947-49; hydraulic engr, Beach Erosion Bd & Coastal Engrg Res Center 1949-81; Chief, Res Div 1964-71; Tech Dir 1971-81. Conslt res & applications in coastal engrg: shore processes, winds, waves, tides, surges, currents, as applied to navigation, flood & storm protection, beach erosion control, effects on coastal ecology. Served with USAAF 1943-46. Reg PE in DC. US Commissioner to PIANC 1970-78. (Hon. Mem. PIANC Commission 1985) Mbr Natl Acad of Engr (1977-); Dir, Am Shore & Beach Preservation Assoc (1976-). ASCE Moffatt-Nichol Award 1979. Fellow ASCE, AAAS, Washington Acad of Sci; Mbr AGU, IAHR, PIANC. Author of over 50 papers in scientific & tech literature. Active in ASCE & PIANC. *Society Aff:* ASCE, IAHR, AAAS, NAE, PIANC, MTS, AGU, ASBPA, WAS.

Savitsky, Daniel
Business: Davidson Laboratory, Stevens Institute of Tech, Hoboken, NJ 07030
Position: Dir. *Employer:* Stevens Inst of Tech. *Education:* PhD/Oceanography/NYU; MS/Fluid Dynamics/SIT; BCE/Civ Engg/CCNY. *Born:* 9/26/21. Currently Dir of Davidson Lab, Stevens Inst of Tech and Prof in the Ocean Engg Dept. Specialties are in hydrodynamic design and research relating to marine vehicles, especially high speed craft. Serves on various govn't committees to establish research and dev programs. Is chrmn of the High Speed Marine Vehicle Ctte of the Int'l Towing Tank Conference; Mbr of Technical and Res Steering Ctte of SNAME. *Society Aff:* SNAME, ASNE, ΣΞ.

Savitt, Sidney Allan
1050 George St, New Brunswick, NJ 08901
Position: President. *Employer:* S A Savitt Assocs. *Education:* DChE/Chem Eng/PINY; MChE/Chem Eng/PINY; BS/Chem/CCNY. *Born:* April 1920. Sigma Xi, Phi Lambda Upsilon. Reg PE NJ. Cons Chem Engr. Specialty: extractive metallurgy of processed strategic & exotic alloys; separation & recovery of precious metals; design & process layout in dev of commercial plants; environmental & pollution control; economic feasibility studies. Reviewer for Engrg & Scientific publs; career guidance in Engrg - AIChE, NYAS. Mbr AIChE, ASME, ASEE; Fellow Amer Inst Chemists, NY Acad of Sciences (Bd of Governors 1971-73). Amer Inst of City of NY - Exec Dir - Science Fairs; PINY Alumni Bd/Dirs - President-1980. State of N.J. Governor's Science Advisory Committee 1981-83; Chair, VChrmn, Engrg Sect, NY Acad of Sci 1983-7; Tau Beta Pi; Eminent Engr Award 1983. Amer Inst of Sci & Tech Pres 1984. Polytechnic Univ. Distinguished Alumnus -1985. *Society Aff:* AIChE, NYAS, ASME, NSPE.

Savolainen, Ann W
Home: 4853 Cordell Ave, Apt 1513, Bethesda, MD 20014
Position: Chief, Policy & Pub Mgmt Br. *Employer:* US Nuclear Regulatory Comm. *Education:* AB/Physics/Miami Univ. *Born:* 12/24/15. Career in nuclear field began in 1948, USAEC; 1949-51 Knolls Atomic Pwr Lab; 1951-71 ORNL compiling reports on dev of homogeneous, pebble-bed, molten- salt, & gas-cooled reactor concepts & evaluation of all reactor concepts; 1961- 71, Secy ANS Stds Ctte; 1973-76 elected Mbr ANS Bd/Dirs; 1971-75 Dir Nuclear Prog Amer Natl Stds Inst; 1974, elected Chmn, Intl Organization for Standarization Tech Ctte 85 on Nuclear Energy; 1974-75 ISO liaison with Intl Atomic Energy Agency, Vienna; 1975-77, Cons USNRC. 1977- , Chief, Policy & Publications Mgmt Branch, Div Technical Info, USNRC. 1981 elected Fellow ANS. *Society Aff:* ANS, STC, AAAS.

Sawabini, C T
Business: Box 8373, San Marino, CA 91108
Position: Consulting Petroleum Engr *Employer:* Sawabini Services, Inc. *Education:* BS/Petroleum Engr/Univ of OK; MS/Petroleum Engr/USC; PhD/Engg (Pet Engg)/USC. *Born:* 10/1/35. Prior to establishing Sawabini Services, Inc., Dr. Sawabini had been retained as a consultant by independent oil producing companies and professional engrg consulting firms. He worked with the petroleum industry for 21 yrs. He first roughnecked for Signal and Santa Fe Drilling companies. Later he worked in reservoir engrg and secondary recovery for Standard Oil of Calif and THUMS Long Beach. Dr. Sawabini spent 7 yrs working and consulting in Saudi Arabia, Libya, Algeria and Iran. He served for one yr as Dean, College of Engrg Northrop University. He is a registered professional petroleum engr in the states of California, Texas, Oklahoma and Louisiana. *Society Aff:* SPE

Sawyer, James E
Business: PO Box 23646, Tampa, FL 33622
Position: Exec Vice President. *Employer:* Greiner Engrg Sciences, Inc. *Education:* BS/Arch/GA Tech. *Born:* Sept 28, 1931. BS Georgia Tech 1952; graduate study 1952-63. Served Army Corps of Engrs, US & Korea 1953-54. Joined Greiner in 1955 as structural engr; served as VP & Chief Engr of Tampa office 1971-75; assumed present duties as Exec VP of Greiner Engrg Sciences, Inc in 1975. Served as Dir ASCE 1970-72; Fellow ASCE, FES; Davis Medal, 1964; Fla Sec ASCE Engr of the Year 1971; Fla West Coast Engr FES of the Year 1974. *Society Aff:* ASCE, NSPE, ACEC, ARTBA.

Sawyer, R Tom
Business: Box 188, Ho-Ho-Kus, NJ 07423
Position: Editorial Chairman. *Employer:* Turbomachinery Publications. *Education:* MS/-/OH State Univ; BEE/-/OH State Univ. *Born:* June 1901. Service & Sales Engr GE Co 1923-30. Delivered 1st diesel locomotive sold in USA 1925, NY City pollution ordinance to eliminate steam locomotives, started use of diesel locomotives. With Amer Locomotive Co 1930-56. Engineered the world's 1st gas turbine locomotive with mechanical drive, delivered to USArmy 1954. Publications: 1945, The Modern Gas Turbine; 1946, Applied Atomic Power; 1960, Internatl Gas Turbine Magazine; 1963- , Gas Turbine Catalog. Organized in 1947 & was 1st Cmn of Gas Turbine Div of ASME; Treas 1956- ; Hon Mbr ASME 1969. Also Mbr of IEEE,

Sawyer, R Tom (Continued)
SAE, AIAA, PE, I Mech E, ANS, SXI, FEF & Hon Mbr Gas Turbine Soc Japan 1979. *Society Aff:* IEEE, SAE, AIAA, EEF, GTSJ.

Sawyer, Robert F
Business: Dept of Mech Engg, Berkeley, CA 94720
Position: Prof. *Employer:* Univ of CA. *Education:* PhD/Aerospace Sci/Princeton; MA/Aero Engr/Princeton; MS/Mech Engr/Stanford; BS/Mech Engr/Stanford. *Born:* 5/19/35. Prof of Mech Engg, Chair, Energy and Resources Grp, Sr Faculty Scientist, Lawrence Berkeley Lab, Univ of CA, Berkeley. Teaching & res in combustion, air pollution, propulsion, fire safety. Dir & Secy, the Combustion Inst. Reg PE. Dir KVB, Inc 1974-8. Mbr, CA Air Resources Bd, 1975-6. Rocket propulsion res. AF Rocket Propulsion Lab, 1958-61. *Society Aff:* SAE, ASEE, AIAA, ASME, CI.

Saxe, Harry C
Home: 1026 Burning Sprgs Cir, Louisville, KY 40223
Position: Dean, Speed Scientific School. *Employer:* Univ of Louisville. *Education:* ScD/Civil Engg/MA Inst of Tech; MSE/Civil Engg/Univ of FL; BCE/Civil Engg/City College of NY. *Born:* Mar 1920. ScD MIT; MSE U of Fla; BCE CCNY. Native of NY. Served with Army & Air Force Engrs 1942-48. Disch rank of Capt-Reserve. Instructor U of Fla 1949-50; Res Asst MIT 1950-52; Assoc Prof 1952-56 & 1957-59 at Ga Inst of Tech, Polytechnic Inst of Brooklyn & U of Cincinnati. Structural Engr Praeger- Kavanagh, NY 1956-57. Prof & Dept Chmn, U of Notre Dame 1959-69; Acting Dean of Engrg at Notre Dame 1960-61 & 1966-67; Visiting Prof & NSF Fellow at Imperial Coll, London 1965-66; Dean, Speed Scientific School 1969- , & Pres Inst of Indus Res U of Louisville 1969-74. Mbr Ky State Bd of Registration 1972- . *Society Aff:* ASCE, ASEE, NCEE, ASTM, NSPE, IABSE, SAME.

Saxena, Kanwar B
Business: 1014 Reynolds Rd, Charlotte, MI 48813
Position: President *Employer:* Global Plastics Corp. *Education:* BS/Chemistry/Agra Univ India; BS/Chem Engr/MI State Univ. *Born:* July 18, 1932. Native of Ajmer, Rajasthan, India. 1956. 1955-57 served as Chemist, Barley Earhart Co; 1957-60 served as Process Engr & Proj Engr for Govt of India, Fertilizer Corp; 1960-61 served as Mgr of Elec Chemical Plant of Birla Bros, India; 1961-68 served as Mfg Supr of phillps Petroleum Co's Petro chemical plant in India then decided to immigrate to USA. Had extensive mgmt raining at Phillips Plants in USA & Administrative staff coll, Hyderabad, India. An active mbr of Soc of Plastics & Rotary Internatl. *Society Aff:* SPE.

Saxena, Narendra K
Business: Dept of Civil Engr, University of Hawaii, Honolulu, HI 96822
Position: Professor *Employer:* Univ of Hawaii. *Born:* 10/15/36. Dr-tech from Tech Univ, Graz, Austria; Dipl-Ing from Tech Univ, Hanover, W Germany; BSc from Univ of Agra, India. Project - In charge "Satellite Triangulation of India" 1969; taught geodetic sci courses at OH State Univ 1969- 70; Adjunct Asst Prof, OH State Univ 1973-74; Asst Prof Photogrammetric & Geodetic Engg, Dept of civil Engg, Univ of IL, Urbana, 1974-78. Chmn Marine Geodesy Ctte, Marine Tech Soc, 1974-78; Mbr ASCE, AGU, MTS, ACSM. Univ of Hawaii, Dept of Civil Engg, since 1978. Fellow, Marine Technology Society, 1984. Adjunct Research Professor of Oceanography, Naval Postgraduate School, Monterey, since 1984. CHOC Research Chair in Mapping, Charting & Geodesy (MC &G), Naval Postgraduate School, Monterey, 1985 and 1987. *Society Aff:* ASCE, AGU, MTS, Tsunami Society

Saxer, Richard K
Business: Vice Commander, Aero Sys Div, Wright-Patterson AFB, OH 45433
Position: Major General *Employer:* USAF *Education:* PhD/Met Engrg/OH State Univ; MA/Aeromech/AF Inst of Tech; BS/US Naval Acad *Born:* 8/28/28 Maj Gen Saxer has served as an Assoc Prof of Engrg Mech at the AF Acad and the AF Inst of Tech. From 1970-74 he was Dir/Commander of the AF Mtls Lab. In 1974, he was Deputy for Reentry Sys at the Space and Missiles Organization. In 1977 he became Deputy of Aeronautical Equip, and in 1980 became Deputy for Tactical Sys at the Aeronatucial Sys Div, Wright-Patterson AFB, Ohio. In 1981 he became Vice Commander of that organization. Gen Saxer is married to the former Marilyn Doris Mersereau of his hometown, Toledo, OH.

Saylor, Wilbur A
Business: 3364 E 14th St, Los Angeles, CA 90023
Position: President. *Employer:* Kennard & Drake Labs. *Education:* BS/Chem Engr/Penn State Univ; Grad Work/Met/Carnegie Tech, UCLA. *Born:* Aug 1907. BSChE Penn State U; graduate work Carnegie-Mellon & UCLA. 57 years as Metallurgical Engr & Production Supr for US Steel. Developed large diameter double submerged arc welded line pipe 1946. Reg Prof Mech Engr, Chem Engr & Met Engr in Calif. Fellow ASM. For 30 years associated with sophisticated fabrication problems, atomic energy work, wind tunnels, pressure vessels & pipe. *Society Aff:* ASTM, AWS, ASM.

Sayre, Clifford L, Jr
Home: 1415 Ladd St, Silver Spring, MD 20902
Position: Professor of Mech. Engr. *Employer:* Univ of MD. *Education:* PhD/ME/Univ of MD; MS/ME/Stevens Inst of Tech; BSME/ME/Duke Univ. *Born:* 6/14/27. Naval officer, USS Hobson DMS-26, 1947-49, David Taylor Model Basin 1952- 53; Proj engr, Experimental Towing Tank, SIT, 1950-51; Asst Prof to Prof, Univ of MD, 1955-present; Visiting Prof, Univ of the Philippines, Manila, 1967-68; Assoc Dean of Engg, Univ of MD, 1976-79; Western Electric Award for Teaching, ASEE Middle Atlantic Section, 1979. Reg P E, MD. *Society Aff:* ASME, SNAME, ASNE, ASEE, NCGA

Sayre, Philip R
Business: 750 E Main St, Branford, CT 06405
Position: Executive Vice President *Employer:* George Schmitt & Co *Education:* MS/Mgmt/MIT; BS/Chem Engrg/MIT *Born:* 11/28/32. BSChE, MS, Mgmt, AP Sloan Fellow all at MIT. Native of Mass, but lived in Calif prior to entering MIT. From 1953-54 sponsored res at MIT-Fluidization of Solids; 1954-55 US Steel Corp; 1955-58 US Air Force - Jet Fighter Pilot & engrg officer; Genl Tire & Rubber Co as dev engr, polymers & polymerization processes; various engrg positions to tech mgr of Chem Div in 1965. In 1967 founded Emeloid-Guilford, in CT, a sub of Addressograph Multigraph. In 1972 named Sloan Fellow at MIT & res associate MIT Energy Lab. Sprague-Textron in 1973, Gen Mgr, Exec VP 1976, Pres 1977. 1982 Pres, CEO Balzers, Div, Oerlikon Buhrle, USA, 1984 Exec VP Geo Schmitt & Co. Lt Col, Command Pilot, USAFR Retired. – AICHE, ACS, AGA, AACE, Res Off Assoc. *Society Aff:* AIChE, ACS, AACE, AGA.

Sayre, Robert D
Business: PO Box 9457, Richmond, VA 23228
Position: President. *Employer:* Sayre & Assocs, PC. *Education:* MCE/Civil Engg/Univ of VA; BSCE/Civil Engg/SD Sch of Mines & Technology. *Born:* Oct 1928. Area Engr, E I DuPont de Nemours, Savannah River Proj, SC 1952-53; Chief, engrg lab, Wasington District, Corps of Engrs, Washington, DC 1953-56; Materials Engr, Parsons, Brickerhoff, Quade & Douglas in Richmond, Va 1956-58; Chief engrg & Corporate Secy, Froehling & Robertson, Inc, independent testing lab, Richmond, Va 1958-68. Founder & Pres of Sayre & Assocs, PC, Geotechnical Engrs, Richmond, Va 1968- . Pres of Va Soc of PE's 1972-73; Pres of CEC/ Va 1980-81. *Society Aff:* ASCE, NSPE, ASFE, ACEC.

Scala, Eraldus
Business: 177 Port Watson St, Cortland, NY 13045
Position: Pres *Employer:* Cortland Cable Co *Education:* DEng/Metallurgy/Yale Univ; MS/Metallurgy/Columbia Univ; BS/Chemistry/CCNY. *Born:* 6/22/22. Res metallurgist, Chase Brass & Copper Co, 48-52, training dir, 52-53, hd phys metall sect, 53-55; mgr, mat dept, res & adv develop div, Avco Corp, 55-61; prof metall & mat sci, Cornell Univ, 61-68; dir, US Army Mat & Mech Res Ctr, MA, 68-70; Prof

Scala, Eraldus (Continued)
Mat Sci & Engg, Cornell Univ, 70-74 Guggenheim fel, Delft Tech Univ, 67-68; consult, Aerospace Corp; Man Labs; AF Ballistic Missile Re-entry Systs, Battelle Mem Inst; adv bd mat, NASA, mem res adv comt mat; mat adv bd, Natl Acad Sci USA, 43-46. Am Inst Aeronaut & Astronaut; Fellow - Am Soc Metals; Am Inst Mining, Metall & Petrol Engg Physical metallurgy; high-temperature mtls and composites; liquid metals. High strength fibers and composites, Cables and Ropes, Marine Tech soc, Cable Engg and Mfg, Marine Cables. 1974 to present, Consulting Engr, Mtls Sci & Eng, Pres, Cortland Cable Co. *Society Aff:* AIME, ASM, MTS.

Scanlan, Robert H
Business: Dept. C.E. Latrobe Hall, Johns Hopkins University, Baltimore, MD 21218
Position: Professor Civil Engineering. *Employer:* The Johns Hopkins University. *Education:* SB/Math/Univ of Chicago; SM/Math/Univ of Chicago; PhD/Math/MIT; Dr es Sci/Mechanics/Sorbonne-Univ of Paris. *Born:* Aug 15, 1914 Chicago Ill. Held a number of posts in indus, government, & Univs- notably as Prof of Engrg at RPI (Troy, NY), Case-WRU (Cleveland, Ohio) & through 1982, at Princeton where he was Dir of the Structs & Mechanics Prog. 1984-present he has been Prof of Civil Engrg. at the Johns Hopkins University, Baltimore, MD. Recipient of the ASCE Natl State-of-the-Art in Civil Engrg Award (1968) & the AISC T R Higgins Natl Lectureship Award (1976). ASCE Wellington Prize (1986) and Newmark (1986). Has been cons to agencies of the US, Canadian, Japanese, German, & French governments & to private firms. Areas of specialty include structural dynamics, vibrations, fluid-struct interaction, & the behavior of structs notably long-span bridges- under wind, earthquake, & wave loadings. Author of some 120 tech papers in refereed journals. He was elected to the U.S. National Academy of Engineering in 1987. *Society Aff:* ASME, ASCE

Scanlan, Sean O
Business: University College, Upper Merrion St, Dublin 2 Ireland
Position: Professor of Electronic Engineering. *Employer:* University College Dublin. *Education:* DSc/EE/Natl Univ of Ireland; PhD/EE/Univ of Leeds; ME/EE/Univ College of Dublin; BE/EE/Univ College of Dublin. *Born:* Sept 1937 Dublin Ireland. Fellow IEEE, IEE, Inst of Mathematics & its Applications. Inst of Engineers of Ireland. Mbr Royal Irish Acad. Employed at Mullard Res Labs Surrey England & Univ of Leeds (Prof of Electronic Engrg 1968- 73) before becoming Professor of Electronic Engrg in Univ Coll Dublin. *Society Aff:* IEEE, IEE, IMA, IEI, RIA.

Scardino, A J, Jr
Business: 105 Royal Oak Blvd, Pass Christian, MS 39571
Position: President. *Employer:* Sigma Assocs LTD. *Education:* BS/Health & Safety/USL - Lafayette; PhD/Safety & Ergonomics/Kensington; MS/Safety Engineering/Kensington. *Born:* 6/19/34. Certified hazard control mgr, Master level; Certified fire & explosion investigator. Reg PE in CA, SF 479. Prof Mbr ASSE; Chrm Board of Governors Safety and Health Hall of Fame International, Mbr Bd of Dir NFPA Fire Science and Technology Educators Section, Mbr AAFS, Mbr ASTM D-21; Blue Key Intl Hon Fraternity 32D498. Mbr: ANSI A-10 ctte for construction & demolition, exec ctte construction sec Natl Safety Council. Chmn ANSI A10-15 subctte 'dredging', Qualified expert in both federal & state courts. Pres Sigma Assoc Ltd cons, fire & arson investigators, risk mgmt, loss control, accident reconst, case analysis, Human Factors Engg. *Society Aff:* ASSE, VOS, SSS, NFPA, NAFI, NSC, IAAI, AAFS

Scarola, John A
Business: 300 South St. Paul, Dallas, TX 75201
Position: Sr VP *Employer:* ENSERCH Corp *Education:* BE/CE/Yale Univ; BS/Naval Sci/Yale Univ *Born:* 12/24/24 Joined Ebasco Services Inc in 1947 as a sr draftsman. Elected VP, Projects, in 1972; Exec VP, Operations, 1976; Pres & Ch Exec Officer, 1978; and to my current position of Sr VP of the Engrg and Construction Div of ENSERCH Corp, the parent company of Ebasco Services Inc, in 1980. I am a reg PE in 31 states. *Society Aff:* ASME, ASCE

Scavuzzo, Rudolph J
Business: Dept of Mech Engg, Akron, OH 44325
Position: Prof *Employer:* Dept of Mech Engr. *Education:* PhD/ME/Univ of Pittsburgh; MSME/ME/Univ of Pittsburgh; BSME/ME/Lehigh Univ *Born:* 1/21/34. Born and raised in North Plainfield, NJ. 1955-64 - Worked with Westinghouse Elec Corp, Bettes Atomic Power Labs. Experiments in stress analysis of reactor components, naval shock analysis of mech systems and seismic analysis of nuclear power plants. 1964-70 - Univ of Toledo, Assoc Prof of Mech Engg. res interests in seismic analysis, naval shock analysis with interest in machine design and stress analysis. 1970-73 - Rensselaer Poly Inst, Hartford Grad Ctr, Assoc Prof of Mech Engg. 1973-84 - Univ of Akron, Prof and Hd, Dept of Mech Engg, 1984- present Univ of Akron, Prof of Mech Engr. *Society Aff:* ASEE, ASME, AAM, PVRC, AIAA.

Schacht, Paul K
Business: 1914 Albert Street, Racine, WI 53404
Position: Manager, R&D *Employer:* Racine Hyd Div Dana *Education:* BA/Chemistry/Carthage College *Born:* May 31, 1938. Post graduate work Univ of Wis. Chemist- Racine Hydraulics 1961-70; Lab Supervisor- Rexnord Corp R&D 1970-1981. Mgr R&D 1982-Present-Dana Corp. Racine Div. Natl Pres FPS 1974; Chmn Natl Fluid Power Assns Fluids Standards Ctte. Chaired writing of ANSI document B93.5-1979 'Practice for the Use of Fire Resistant Fluids for Fluid Power Systems'. US delegate Internatl Standards Org & mbr of subctte 6. Chrmn NFPA Tech BD. Authored papers & articles covering various facets of lubrication. *Society Aff:* FPS, NFPA, ASLE

Schad, Theodore M
Home: 4138 26th Road North, Arlington, VA 22207
Position: Conslt on Water Resources *Employer:* Self-Employed *Education:* BS/Civil Engg/Johns Hopkings Univ. *Born:* 8/25/18. Born in Baltimore, MD. Employed Baltimore & Seattle Districts, US Army Corps of Engrs; Denver, Pendleton & Salem, OR & Wash DC offices, Bureau of Reclamation; Budget examiner on water resources progs, US BUreau of Budget during the Eisenhower Admin; Senior specialist, Engg & Public Works & Deputy Dir, Congressional Res Service, Library of Congress; Staff Dir US Senate Select Ctte on Natl Water Resources 1959-61; Exec Dir Natl Water Comm 1968-73; Executive Secretary, Environmental Studies Bd, NAS/NAE 1973-77; Deputy Executive Director, Commission on Natural Resources. NAS/NAE, 1977-83. Exec. Director, National Ground Water Policy Forum, The Conservation Foundation, 1984-86. Hobbies: sailing, operatic & classical music, woodworking. Fellow ASCE, Honorary Mbr AWWA 1970, Trustee, National Speleological Foundation, Member, National Academy of Public Administration. Private practice since 1986. *Society Aff:* ASCE, AWWA, AGU, AAEE, AIH, USCOLD, USCID, AWRA

Schadler, Harvey W
Home: 110 Woodhaven Drive, Scotia, NY 12302
Position: Manager Metallurgy Lab. *Employer:* GE Research & Development Ctr. *Education:* PhD/Metallurgical Engr/Purdue Univ; BS/Metallurgical Engr/Cornell Univ. *Born:* Jan 4, 1931 Cincinnati Ohio. Presently Mgr of the Met Lab at GE's Res & Dev Center where about 50 staff & assoc staff scientists & about 20 technicians & hourly workers are engaged in identification & synthesis of new metals & alloys, evaluation of their properties & dev of processes required to exploit them. Served for 2 years as a mbr of the Natl Sci Foundation's Advisory Panel on the Materials Res Labs. Mbr: AIME, ASM. Recipient of the Alfred E Geisler Award of the Eastern NY Chap of the ASM. Fellow ASM. *Society Aff:* ASM, AIME, AAAS.

Schaefer, Adolph O
Home: 1351 Butler Pike, Blue Bell, PA 19422
Position: Emeritus Executive Director. *Employer:* The Materials Properties Council Inc. *Education:* BS/Chem Engg/Univ of PA. *Born:* March 1901. Res Metallurgist Midvale Steel & Ordnance Co 1922; Engr of Tests The Midvale Co 1932, Exec Engr

Schaefer, Adolph O (Continued)
1942, Vice Pres Engrg & Mfg 1951; Pres Pencoyd Steel & Forge Co 1956; Vice Pres Met, Struthers Wells Corp 1960; Exec Dir The Metals Properties Council Inc 1966 to date. Pres (1955), Fellow & Honorary Mbr ASM; Fellow & Honorary Mbr ASTM; Fellow ASME. Mbr: AIME, AWS, AISI, Chmn Indus Advisory Comm, Dept of Metals Sci Univ of P. P Pres Engrg Alumni Soc. Emeritus Mbr Bd Genl Alumni Soc (P Vice Pres). The Franklin Inst - Sci and Arts Ctte. ASME - Pressure Vessels and Piping Awd. ASME - J. Hall Taylor Medal. Univ of PA - Gallery of Distinguished Engg Alumni, Robert Yarnall Awd, Alumni Awd of Merit. Franklin Inst Sci & Arts Cttee; ASME Pressure Vessels & Piping Award and J Hall Medal; Univ of Penna Gallery of Distinguished Engrg Alumni, Robt Yarnall Award, Alumni Award of Merit. *Society Aff:* ASTM, ASME, ASM, AWS, AISH-AIME, NACE, ADPA, TME

Schaefer, David L
Business: 841 Chestnut St, Phila, PA 19107
Position: Pres. *Employer:* Ballinger-Meserole Co. *Education:* MBA/Ind Met/CUNY; BChE/CE/CUNY. *Born:* 12/29/38. Born & raised in NYC. Spent 1st 10 yrs of my career in I E positions relating to mtls handling/mtls mgt with Revlon, FMC, GAF, & CBS Records. Have been an IE consultant since 1973. Joined Ballinger-Meserole as VP in 1977. Promoted to Pres in 1978, am responsible for all phases of co operations including marketing & engg. Active in profl socs; currently VP - IMMS. Taught at Rutgers Univ & Galssboro State College. Frequent lectr on Mtls handling/warehousing. Proflly certified in Mtls Handling & Mtls Mgt (IMMS). Enjoy tennis & skiing. *Society Aff:* IIE, IMMS, WERC, NCPDM, APMHC.

Schaefer, Edward J
Home: 2279E 250N, Bluffton, IN 46714
Position: Chairman of the Board. *Employer:* Franklin Electric Co Inc. *Born:* 7/10/01. in Baltimore, MD. BSEE 1923 Johns Hopkins Univ. Tau Beta Pi 1922. Design Engr GE Schenectady NY & Fort Wayne IN. Co-founder of Elec Motors & Specialities Co 1943 & Franklin Elec Co Inc 1944. Hon Dr Engg, IN Inst of Technology. Pres & Chmn of Bd 1947-67, Chmn of Bd 1967- . Recipient Coffin Award, GE Co 1937. Fellow IEEE. Dir of Old-First Natl Bank in Bluffton IN; Mbr Bd of Trustees, Ind Inst of Technology 1974. Recipient of about 100 patents on elec & mech devices. Fellow, Bd of Trustees, Johns Hopkins Univ.

Schaefer, Jacob W
Home: 115 Century Lane, Watchung, NJ 07060
Position: Executive Director. *Employer:* Bell Laboratories. *Education:* DrS/Hon/OH State Univ; BS/ME/OH State Univ. *Born:* 6/27/19. –Distinguished Alumnus" 1966. Hon ScD 1976 all from OH State Univ. Joined Bell Labs in 1941. While serving in Army Ordnance, invented "Command Guidance" from which evolved NIKE family of missile sys. Upon return to Bell Labs, help mgmt positions in military dev, including Dir of Field Test Station on Kwajalein Island. Awarded Army's Outstanding Civilian Service Medal. Now responible for Military System Division. P Pres of Bd of Education, current Chmn of Planning Bd; Pres, Bancroft Sch & Community, Inc. Mbr of the Natl Acad of Engrg. A Fellow of IEEE. *Society Aff:* IEEE, ΣΞ, TBΠ, NAE.

Schafer, Albert C
Business: 515 North Russell St, Urbana, OH 43078
Position: VP & Genl Mgr. *Employer:* Grimes Div, Midland Ross Corp. *Education:* BE/-/Ohio State *Born:* June 1935 & raised in Urbana Ohio. BIE 1958 Ohio State Univ & also employed by Grimes Mfg Co as Proj Engr. Promoted to Sales Dept in 1961. Mgr of contracts Admin Feb 1967. Dir of Admin June 1971. Assumed responsibility as Vice Pres in March 1975. Co Tequired by Midland Ross Corp, Cleveland, OH in Sept 77 now operating as Div. Assumed respon as Div Genl Mgr Jan 78. Promoted to VP & Genl Mgr June 78. Promoted to GRP VP July 1983, responsible for Grimes Aero Gp. consisting of Grimes Divn, Grimes E.L. and Grimes Galley Prod Div. Founding mbr OSU ISE Dept Alumni Advisory Ctte, Chmn 1975-76. Enjoy golf, tennis & sailing. *Society Aff:* ADMA.

Schafer, Richard K
Home: 7 Hollyvale Drive, Rochester, NY 14618
Position: Television Technical Liaison. *Employer:* Eastman Kodak Co. *Born:* Sept 12, 1936. BA Physics Univ of Penn. Native of Philadelphia, Penn. Currently assigned to Photographic Technology Div as co interface rep on tech matters relating to video technology. Previously, tech specialist for pro motion picture production methods & materials. Fellow SMPTE. Mbr SPSE.

Schafer, Ronald W
Business: Sch of EE, Atlanta, GA 30332
Position: Regents' Prof. *Employer:* GA Inst of Tech. *Education:* PhD/EE/MIT; MScEE/EE/Univ of NB; BScEE/EE/Univ of NB; -/Math & Phys/Doane College. *Born:* 2/17/38. Native of Tecumseh, NB. From 1968 to 1974, was a mbr of the Acoustics Res Dept at Bell Labs, Murray Hill, NJ, specializing in res on speech analysis and synthesis, digital signal processing techniques, & digital waveform coding. Since 1974, John O McCarty/Audichron Prof of EE at GA Inst of Tech. Co-author of the textbooks, *Digital Signal Processing & Digital Processing of Speech Signals.* Formerly Assoc Editor of the IEEE Transactions on Acoustics, Speech and Signal Processing & pres of that soc during 1978-79. Fellow of IEEE & the Acoustical Soc of Am. Co-recipient of IEEE, E.R. Piore Award, 1980. *Society Aff:* IEEE, ASA, ΣΞ, ΦΚΦ.

Schaffer, Stanley G
Home: 435 Sixth Ave, Pittsburgh, PA 15219
Position: Retired *Employer:* Duquesne Light Co. *Education:* BSME/Mech Engrg/PA State Univ. *Born:* 6/8/19. Past Pres & Dir Duquesne Light Co; P Pres Pen Electric Assn. Dir: First Fed Savings & Loan Assn of Pittsburgh Mbr; Natl & Penn Soc of Prof Engrs; Dir Engrs' Soc of Western Penn; APCA; Pres ITI; Mbr PA Manufacturers Assoc. Fellow ASME. Trustee, The Penn State Univ; (Member) Penn State Alumni Council; 1972 recipient, Distinguished Alumnus Award of Penn State; At PSU: Tau Beta Pi, Sigma Tau, Pi Tau Sigma. BSME 1941 *Society Aff:* NSPE, PSPE, ASME, APCA, ITI, ESWP.

Schaffhauser, Robert J
Business: 40 W 40th St, New York, NY 10018
Position: VP-Tech *Employer:* American Standard Inc *Education:* PhD/Polymer Physics/Princeton Univ; MA/Phy-Chem/Princeton Univ; BS/Chem/Fordham Univ *Born:* 5/10/38 1964-79 Allied Corp. After 2 yrs military service (US Army, Capt) joined Allied in R&D. Rose through positions of increasing responsibility in res, comml dev, marketing, mfg and planning to Gen Mgr of Engrg Plastics. Joined American Standard in June 1979 as VP-Tech. Also Pres Cole-Resdevel Corp, Bd Mbr SPE, Adv Bd of "Inform–. *Society Aff:* SPE, SME, ACS

Schaffner, Charles E
Business: 11 West 42nd St, New York, NY 10036
Position: Consultant *Employer:* Syska & Hennessy Inc. *Education:* MCE/Civil Engg/Poly Inst of Brooklyn; BCE/Civil Engg/Cooper Union; BSSE/Civil Engg/Univ of IL *Born:* 7/21/19. Jr engr Moran Proctor, Freeman & Mueser, NYC, 1941; instr Cooper Union, 1941-44; mbr faculty Brooklyn Poly Inst, 1946-70, prof engrg, vp admin, 1962-70; vp Syska & Hennessy, Inc, 1970-73, sr vp 1973-76, exec vp, 1976-83, Vice Ch-1984-86, Consultant 1987-; also dir Chrmn nat adv bd. NYC Mayor's Energy Policy Adv Group, mbr 1978- ; Natl Inst of Bldg Scis - Exec Comm of Consultative Council - Charter Mbr, 1978- ; Dir 1982- ; AM Assoc Engr Socs - organizing Comm, 1978-80; comm school dist Locust Valley, NY, 1956-59, pres 1958-59; commr edn pres Central Dist S Oyster Bay, NY 1959-63. AAES - Bd of Dirs 1979-81, chrmn, Educ Affairs Coun 1979-80. Trustee, Cooper Union 1975-78. Served with AUS, 1944-46. Named Outstanding Alumnus, Cooper Union, 1956 Distinguished Alumnus Award; Poly Alumni Assn, Alumnus of yr 1972. Mbr Operation Democracy, ASCE (Civil Engr of Yr 1969). ASTM, AM Arbratinn Assn, ASEE. VP

Schaffner, Charles E (Continued)
& Mbr of Bd of Dir and Exec Comm 1965-67; 1974-77. Pres 1979-80. Engr Const Award 1967, 1979, NY Bldg Congress - Pres 1979-83, Chrmn, Council of Pres. 1983-87, Chairman, Council of Business and Labor 1987- ; Natl Res Council Bldg Res Adv Bd 1972-79, Chrmn 1976-77; Natl Acad of Engr Comm on Educ & Util of Engrs - 1982-84. Tau Beta Pi, Chi Epsilon, Omega Delta Phi. *Society Aff:* ASCE, ASTM, ASEE, ACI, NSPE.

Schairer, George S
Business: PO Box 3707 M S 10-47, Seattle, WA 98124
Position: Vice President - Research Consultant. *Employer:* The Boeing Co. *Education:* MS/Aero Engg/MIT; BS/-/Swarthmore Univ. *Born:* May 19, 1913 Wilkinsburg Penn. With Boeing since 1939, Vice Pres Res since 1959. Contributed to designs of B-24, B-17, B-29, B-50, B-47, B-52, Stratoliner, Stratocruiser, 707, KC-135, 727, 737, 747, 767, SST, YC-14, & helicopters made by Vertol. Special interest in aerodynamics. Retired July 1978; now Consultant to The Boeing Co. Former Mbr of NACA, USAF, SAB, DIA, SAC, & URIA; various cttes. Honors: AIAA, Wright Brothers Lecture, Sylvanus Albert Reed Award, Daniel Guggenheim Medal, Honorary Fellow; ASME, Spirit of St Louis Medal; RAeS, Wilbur & Orville Wright Memorial Lecture. Hobbies: swimming, sailing. *Society Aff:* AIAA, NAE, NAS, AHS, SNAE.

Schalliol, Willis L
Business: School of Electrical Engrg, W Lafayette, IN 47907
Position: Educator, Retired *Employer:* Purdue Univ. *Education:* PhD/Met Engg/Stanford Univ; MS/Met Engg/Stanford Univ; BS/Met Engg/Purdue Univ. *Born:* 12/20/19. Elkhart, IN. Army Liaison Pilot 1943-46. Mtls Engr Westinghouse Sunnyvale, CA 1948-49. Nuclear Mtls Engr GE Richland, WA 1950-53. Dir Engg NIBCO Elkhart, IN. 1953-59. Dir Res Bendix S Bend, IN 1959-63. Dir R&D CTS Res W Lafayette, IN 1964-69. Mbr Advisory Bd Ceramic Engg Dept Univ of IL 1966-69. Exec Officer Dept Chemistry Purdue Univ W Lafayette, IN 1969-74. Assoc Coordinator Cooperative Engg Education Purdue Univ 1975-80. Assoc Prof Sch of Mtls Engg Purdue 1976-80. Dir Critical Needs Prog Cooperative Engg Education Purdue 1976-80. Mgr Industrial Relations, School of Electrical Engineering, Purdue 1980-85, Acting Asst. Prof. Mechanical Engg. Technology Purdue 1985-86. Retired 1986.

Schaper, Laurence T
Business: P.O. Box 8405, Kansas City, MO 64114
Position: Partner *Employer:* Black & Veatch *Education:* MS/Civil Eng/Stanford Univ; MS/Env Health Eng/Univ of KS; BS/Agri Eng/KS State Univ *Born:* 10/6/36 Employed by Black & Veatch Constg Engrs since 1962. Partner resp for water, wastewater, solid waste and hazardous waste projects. Has participated as speaker at univ and technical soc sponsored seminars and programs. Current profl and technical soc resp include ASCE (fellow) - past pres, Kansas City Section. AWWA - Steel Pipe Ctte and Past Chrmn of the MO Section. Elected to two terms on the Prairie Village City Council (1978-82). Reg PE in thirteen states. Diplomat - AAEE. *Society Aff:* ASCE, NSPE, AWWA, WPCF, APWA, AAEE

Schapiro, Jerome B
Business: 158 Central Ave Box 6, Rochelle Park, NJ 07662
Position: President. *Employer:* Dixo Company Inc. *Education:* BChE/ChE/Syracuse Univ. *Born:* 2/7/30. Proj Engr, Propellants Branch US Naval Air Rocket Test Station 1951-52; Lt USAF US assigned as safety officer; Chem & Radiological Lab Army Chem Center MD 1952-53; Dixo Co Inc 1954-. Past Chrmn ASTM Ctte D-12 (Detergents). Received ASTM Award of Merit 1970. USA Spokesman ISO/TC-38/SC-1/WG-8 (Drycleaning). Mbr US Tech Adv Group ISO/TC-38/SC-11 (Care Labeling). Past Chrmn & Ch of Delegation. ISO/COPOLCO (Consumer Policy Questions). Past Chrmn Consumer Council ANSI, current member ANSI Consumer Interest Council's Exec Ctte. Past Vice Chairman Bd/Dir ANSI. Mbr: ACS, AIChE, AOCS, AATCC, SES & Past Mbr AATCC Exec Ctte on Res current Chrmn AATCC Ctte RR-43 Drycleaning. Past Chrmn ANSI Intl Consumer Policy Adv Comm, Current V Chrmn AATCC Intl Test Methods CTTE, Past Leader, US Delegation ISO/TC-38/SC-2 Laundry, Drycleaning and Finishing Tests (for Textiles); Past Mbr ASTM Standing Ctte on Intl Standards; Past Mbr ANSI Intl Standards Council. *Society Aff:* ASTM, ACS, AIChE.

Scharf, Louis L
Home: 1129 W Oak, Ft Collins, CO 80521
Position: Prof of Elec Engr. *Employer:* Univ of Colorado, Boulder. *Education:* PhD/EE/Univ of WA; MS/EE/Univ of WA; BS/EE/Univ of WA. *Born:* 8/5/41. Alumnus of Univ of WA (Seattle). Prof of Elec Engg at Univ. of Colorado and consultant to several industries. Main technical interests are digital signal processing, communications & estimation theory. *Society Aff:* IEEE.

Scharp, Charles B
Home: 1406 Newport Place, Lutherville, MD 21093 *Employer:* Retired, Consulting Engineer *Education:* MS/ME/Johns Hopkins Univ; BE/ME/Johns Hopkins Univ *Born:* 6/30/25. Reg PE MD. Native of Baltimore. Served with US Navy 1943-45. Staff Engr for Baltimore Gas & Electric Co specializing in performance evaluation of power plant equipment, with CO 1948-1986. From 1974 until retirement in 1986 was in responsible charge of Performance Sect of Electric Production Dept with 20 prof level engrs. Also responsible for air quality compliance and chemical treatment programs. Chmn of PTC Ctte 6 on Steam Turbines from 1967 to 1986. Chmn: Board on Performance Test Codes, ASME VP - Performance Test Codes 1983-1987. TAG of IEC/TC 5 on Steam Turbines. Fellow ASME 1971. Senior Mbr ISA. *Society Aff:* ASME, ISA

Scharres, John W
Business: 2 N Riverside Plaza, Chicago, IL 60606
Position: President. *Employer:* Scharres & Assocs Inc Cons Engrs. *Born:* 1912. Attended Lewis Inst & Ill Inst of Technology. Practicing engr since 1940. Reg PE Ill 1946. Ch Engr, E R Gritschke & Assocs 1950, Pres 1961; President Scharres & Assocs Cons Engrs 1961- . Mbr: NSPE, IEEE, ASHRAE, Power Engrg Soc, Const Specifications Inst, Panel of Arbitrators (American Arbitration Assn), Cons Engrs (Mech Speciality contractors Liaison Ctte, Chicago Area). Paper: 'An Engrs View on Specification Writing', presented Const Specification Inst Regional Conf 1967. Article: 'Throttling is Good Hot Water Pressure Control' pub in 'Heating, Piping & Air Conditioning' 1960. *Society Aff:* IEEE, ASHRAE, CSI, ACEC, PES, NSPE

Schaub, James H
Business: Dept of Civil Engrg University of Florida, Gainesville, FL 32611
Position: Distinguished Service Professor *Employer:* Univ of Florida *Education:* Ph.D./Civil Engineering/Purdue University; M.S./Civil Engineering/Harvard University; BSCE/Civil Engineering/Virginia Polytechnic Institute *Born:* 01/27/25 Native of Moundsville, WV. Military Service WW II and Korea. Prof Practice 1948-55 DC Dept of Highways, Oregon Highway Dept, Palmer & Baker, Inc. Civil engrg faculty VPI 1955-58; WV Univ 1960-69 Chmn & Assoc Dean; Univ of FL 1969-87, Chmn & Distinguished Service Prof, 1984-. Conquest Chair in Humanities, VA Military Inst, 1986. Consulting for public agencies, industry, universities. ASCE - Pres, WV Ch, various education division assignments, W. H. Wisely Award, 1986; Director-at-Large APWA 1983-; Eminent Career Award, College of Engrg, Univ of Florida 1987; NSF Faculty Fellowship 1975-76; author/co-author books in engrg and humanities, professionalism and ethics, technical papers. Married Malinda K. Bailey 1948. *Society Aff:* ASCE, NSPE, APWA

Schaufelberger, Don E
Business: PO Box 499, Columbus, NE 68601
Position: Pres and CEO *Employer:* Nebraska Public Power District. *Education:* BS/Elec Engrg/Univ of NE *Born:* Oct 1925, Lincoln, Nebraska. BSEE Univ of Nebraska. Navy 1944-46, Electronics Technician. 1949 Field Engr, Consumers Public Power District (CPPD); 1952 Sys Planning Engr; 1958 Ch Engr; 1964 Operations Dir; 1968 Asst Genl Mgr. 1971 Asst Genl Mgr for Nebraska Public Power Dist,

Schaufelberger, Don E (Continued)
CPPD successor; 1972 Deputy Genl Mgr. Assume Genl Mgr's respons in absence; have genl supervision in areas of tech respon incl power supply, transmission, distribution, operations, maintenance, rates-contracts, system planning. Pres & CEO 1983. Reg PE in Nebraska. Chmn Nebraska Sect IEEE 1965; 1st Chmn MidContinent Area Power Pool and Mid-Continent Reliability Council 1972-74. V Chrmn Mid Continent Area Power Pool 1985-86. Bd of Trustees N. Amer Elect Reliability Council 1972-74, 1983-84. Enjoy golf, hunting. *Society Aff:* NSPE, IEEE, IES, AWPA

Schaumburg, Frank D
Business: Apperson Hall, Corvallis, OR 97331
Position: Prof & Head, Dept of Civil Engrg. *Employer:* Oregon State University. *Born:* Jan 15, 1938. PhD & MSCE from Purdue Univ; BSCE from Arizona State Univ. Native of Watseka, Ill. Officer in US Army Corp of Engrs 1961-62. Res Engr with British Columbia Res Council 1966; Asst Prof at Oregon State Univ 1967-70; Assoc Prof 1970-76; Prof 1976; Civil Engrg Dept Head 1972- . Special res interests: environmental trade-offs human dimension of engrg. Leisure interests: tennis, fishing, miniature building.

Schawlow, Arthur L
Business: Dept of Physics, Stanford, CA 94305
Position: Professor of Physics. *Employer:* Stanford University. *Education:* PhD/Physics/Univ of Toronto; MA/Physics/Univ of Toronto; BA/Physics/Univ of Toronto. *Born:* May 5, 1921, Mt Vernon, NY. Postdoctoral Fellow, Columbia Univ 1960; Res Physicist, Bell Labs 1951-61; Prof of Physics, Stanford Univ 1961- (Dept Chmn 1966-70, 1973-74, J G Jackson-C J Wood Prof, 1977). Coinventor of the laser & author of 'Microwave Spectroscopy' (with C H Townes). Member US Natl Acad of Sciences; Fellow Optical Soc of Amer (Pres 1975, Ives Medal 1976), APS (Mbr Council 1966-70, President 1981), AAAS, American Philosophical Society, IEEE (Liebmann Prize 1964), Inst of Physics (GB, Thomas Young Medal 1963), Ballantine Medal, Franklin Inst 1962; Calif Scientist of the Year, 1973; Marconi Fellow, 1977; Nobel Prize, 1981; Arthur L. Schawlow Medal, Laser Inst of America, 1982; Golden Plate Award, Amer Acad of Achievement, 1983. Honorary doctorates: Ghent, 1968; Ll.D, Toronto, 1970; DSc Bradford, 1970. DSc AL, 1984. D.Sc Trinity College, Dublin, 1986. Jazz history. *Society Aff:* IEEE, APS, OSA, AAAS, AAAS, AAUP, APS

Schechter, Robert S
Business: Petro Engrg Dept, Austin, TX 78712
Position: Professor, Petroleum Engrg. *Employer:* Univ of Texas at Austin. *Education:* PhD/ChE/Univ of MN; BS/ChE/TX A&M Univ. *Born:* Feb 26, 1929. Native of Houston, Tx. Lt in Chem Corp 1951-53; Asst Prof Chem Engrg, Univ of Texas at Austin 1956-60, Assoc Prof 1960-63, Prof 1963- . Administrative Dir Center Statistical Mechs & Thermo 1968-72; Chmn Chem Engrg Dept 1970-73; Chmn Petroleum Engrg Dept 1975-78; E J Cockrell Jr Prof Chem & Petroleum Engrg 1975-1979; E.J. Cockrell Jr. Chair in Engineering 1979-1984; Getty Oil Chair in Engineering 1984- . Recipient of Outstanding Teacher Award, Univ of Texas, 1969; Outstanding Paper Award,1973. Donald Katz Lectureship Award 1979 Named to Order of Palmes Academique 1980. Claude Hocott Research Award 1984, General Dynamics Teaching Award 1986. Reg PE Texas. Mbr AIChE, Amer Chem Soc, Soc of Petroleum Engrs, Sigma Xi, Tau Beta Pi, Pi Epsilon Tau, NAE. Author, contributor numerous articles to prof journals. Res specialties include surface rheology, oil recovery processes, & hydrodynamic stability. Married, Mary E Rosenberg 1953; children: Richard 31, Alan 30, & Geoffrey 24. *Society Aff:* AIChE, SPE-AIME, ACS, NAE.

Scheer, John E
Home: 105 W Georgianna Dr, Richboro, PA 18954
Position: Sr Product Engr. *Employer:* SPS Technologies, Inc. *Education:* BS/Met Engg/Drexel Univ. *Born:* 4/6/47. Native of Phila, PA. Grad from Drexel in 1969, followed by grad studies in Mtls Sci. Mtls dev engr with GE, RESD, and Fansteel Cos, specializing in failure analyses and high temperature process dev. With SPS since 1973. Spent five (5) yrs as a R&D metallurgist responsible for fastener alloy evaluations and fastener failure analyses. Assumed current responsibility as Sr Product Engr, Aero & Ind Products Div in 1979. Active in ASM as Vice Chairman of Phila Chapter. Hobbies includes hunting & sports. *Society Aff:* ASM.

Scheibal, Charles A
Business: 116 St. Louis St, Edwardsville, IL 62025
Position: Pres *Employer:* Flagg & Assoc, Inc. *Education:* MBA/Bus Adm/So IL Univ; BS/CE/WA Univ *Born:* 8/28/36 Native of Glen Carbon, IL. Began engrg career as resident engr & proj mgr for IL Div of Highways, specialized in urban freeway projects. Ch Construction engr for consulting engrg firm in NY & Ch Engr for major Land Dev Co. Assumed current presidency of Flagg & Assoc, Inc., a civil, sanitary & structural consulting engrg firm in 1973, directing all activities of firm since. Pres- Madison Co Chapter ISPE, 1967; SIUE Liaison - ASCE (St. Louis Chap) 1977-80. ACEC natl business mgmt chrmn 1979-80. *Society Aff:* ACEC, CECI, ASCE, NSPE, ISPE, IAHE

Scheid, Vernon E
Home: 33 Rancho Manor Dr, Reno, NV 89509
Position: Prof *Employer:* Univ of NV *Education:* PhD/Econ Geol/Johns Hopkins Univ; MS/Geol Engrg/Univ of ID; AB/Geo/Johns Hopkins Univ *Born:* 9/5/06 Born in Baltimore, MD. Grad Baltimore Poly Inst (Advanced Course). Taught geology and mineral resources at Johns Hopkins Univ and Univ of Idaho (Chrmn, Dept Geology-Geography). During WWII worked for US Geological Survey on strategic mineral investigations. 1951-52: Dean, Mackay Sch of Mines; Dir, Nevada Bureau of Mines and Geology; Chrmn and Dir, Nevada Oil and Gas Conservation Commission. Presently Prof Mineral Economics. Served UN as mining engrg consultant. Private consultant: geology and mineral deposits. Actively worked with US Congressmen for natl mineral policy. AIME Robert Earll McConnell (Engrg Achievement) Award. Distinguished Mbr Soc Mining Engrs. *Society Aff:* AIME, SEG, GSA, AGU, AGID

Schell, Allan C
Home: 21 Wedgemere Ave, Winchester, MA 01890
Position: Chief Electromagnetic Scis Div. *Employer:* RADC Hanscom AFB. *Education:* ScD/EE/MIT; MS/EE/MIT; BS/EE/MIT. *Born:* April 1934 New Bedford Mass. Fulbright student at Tech Univ of Delft Holland. Lieutenant USAF 1958-60; Conducted res on atennas & signal processing at Air Force Cambridge Res Labs 195675. Visiting Assoc Prof at MIT 1975; Guenter Loeser Memorial Award 1965; IEEE G-AP Best Paper Award 1966. Editor IEEE Transactions on Antennas & Propagation 1968-71; Chmn IEEE Boston Sect 1976; Editor IEEE Press 1976-1979 Director Electro, 1978-Pres Director, IEEE 1981- 1982. URSI Reg PE in Mass. *Society Aff:* IEE, URSI.

Scheller, William A
Business: Dept of Chem Engg, Lincoln, NE 68588-0126
Position: Prof of Chem Engg. *Employer:* University of Nebraska. *Education:* PhD/Chem Engg/Northwestern Univ; BS/Chem Engg/Northwestern Univ. *Born:* June 1929 Milwaukee Wis. Served US Marine Corps Reserves 1948-52; design engr, group leader, supervisor of engrg res Calif Res Corp (now Chevron Res Co) 1955-63; Became Assoc Prof Univ of Neb 1963, Prof 1969-, Prof & Chmn of Dept 1971-78. Fellow of Graduate Faculty, Mbr of Coll of Engrg Exec Ctte, Mbr of Graduate Council. Reg PE, Ne, Ca and Fla. Guest Prof Univ of Erlangen-Nurnberg West Germany 1970. Tech Advisor to Neb Gasohol Ctte 1971- ; Director Fermentation Foundation 1983-; Received 'Outstanding American Educator' award 1973; MASUA Lecturer & Fellow 1974; Technical Leadership Award 1979; Service Recognition Award 1984; Res interest & publs - thermodynamics, energy conversion,

Scheller, William A (Continued)
heterogenous catalysis, process economics, biochemical engrg. *Society Aff:* ACS, AIChE, ASEE, NSPE.

Schellhase, Marion W
Business: 1830 NASA Rd 1, Houston, TX 77058
Position: Proj Mgr. *Employer:* Lockheed Engrg & Mgmt Service Co. *Education:* MA/Math/Univ of TX; BS/Phys/Univ of TX. *Born:* 1/23/30. Native of Comfort, TX. Attended Trinity Univ, San Antonio, TX, and the Univ of TX, Austin, and received BS in Physics and MA in Math. Worked in Military Physics Res Lab at Univ of TX from 1955-1960 testing aircraft fire control systems. Moved to Boeing Co, WA. and New Orleans, LA in Transport Aircraft and Saturn V Progs in simulation facilities from 1960-1966. With Lockheed Electronics Co at Johnson Space Ctr since 1966 with engg and mgt responsibilities in Spacecraft Simulation Facilities on Apollo, and Shuttle progs. Currently Proj Mgr in Simulation Facility supporting Shuttle Avionics Systems Engg and integration testing of flight system. *Society Aff:* AIAA.

Schenck, A Carl
Business: PO Box 7097, Norfolk, VA 23509
Position: Owner. *Employer:* A Carl Schenck & Assoc. *Education:* BS/Civ Engg/Univ of AL. *Born:* 7/31/10. Native of Phila, PA. Undergrad Asst, Univ of AL; Instr, Univ of VA Extension School 1940-42. Field, Office and Resident Engr, Stone and Webster Engg Corp 1934, 1936-42. Inspector, Corp of Engrs, 1935. Chief Engr and VP, Carpenter Construction Co, Inc, 1942-1963. Principal, A Carl Schenck and Assoc, 1963- present, Construction Mgt and Engg Consultants to Industry, Attorneys, and Insurance Cos. Past Pres-Tidewater Chapter, Va Soc of Prof Engrs. Past Pres, Engrs Club of Hampton Rds. Past Pres, VA Branch, Associated Gen Contractors of America. Mbr Tau Beta Pi; Theta Tau; Chi Beta Pi. Licensed PE. *Society Aff:* NSPE.

Schenker, Leo
Business: Room 3J 608, Crawford S Corner Rd, Holmdel, NJ 07733
Position: Exec Dir *Employer:* Bell Labs. *Education:* PhD/Civil Engg/ Univ Of MI; MS/Civil Engg/ Univ of Toronto; BSc/Civil Engg/ Univ of London, England. *Born:* 1/3/22. Served in Royal Air Force (Great Britain) 1941-45. Res Engg with Hydro-Electric Power Commission of Ontario. Joined Bell Labs in 1954. Dev circuitry for TOUCH-TONE (R) dialing system. PICTUREPHONE, (R) TRIMLINE (R) and repertory dialers. Responsible for dev of Automatic Repair Svc Bureau. Exec Dir, Network Sys Planning Div, June 1983. Seven Patents. Fellow IEEE 1979. *Society Aff:* IEEE.

Schepman, Berne A
Business: 3000 Sand Hill Road, Menlo Park, CA 94025
Position: President & Chief Operating Off. *Employer:* Envirotech Corporation. *Education:* BS/Chem Engr/Northwestern Univ. *Born:* Dec 1926 Bushton Kansas. BSChE 1950 Northwestern Univ. Tau Beta Pi. ESSO Standard Oil Co Linden NJ 1950-53; dev engr, Dir of Testing Lab, Asst Pres, Exec Vice, & Pres Eimco Corp 1953-69. Became Pres of Envirotech Corp when founded in 1969 through the acquisition of Eimco & other companies. Envirotech designs, mfgs, & markets a broad line of continuous processing equipment, air quality control equipment, instruments & controls, & underground mining machinery throughout the world. Mbr AIChE, AIME & WPCF. *Society Aff:* AIChE, AIME, WPCF.

Scher, Robert W
Home: 550 Chagrin Blvd, Chagrin Falls, OH 44022
Position: VP, Manufacturing *Employer:* Lubrizol Corp. *Education:* MS/CE/Univ of MI; BS/CE/Purdue Univ. *Born:* 10/27/31. Native of Cleveland, OH. Married, four children. Joined Lubrizol in 1954 as a process engr in the pilot plant. Named dir of process dev (1964), dir of the lubricants testing lab (1972), dir of res & dev (1973), & gen mgr, mfg (1975). V.P. Mfg (1980) Was chrmn, Cleveland Sec, AIChE, & have served on several natl committees. Interested in travel, natural sciences, history. Enjoy sailing, golf. *Society Aff:* AIChE.

Scherer, Harold N, Jr
Business: American Electric Power, 1 Riverside Plaza, P.O. Box 16631, Columbus, OH 43216--663
Position: Sr VP-EE And Deputy Chief Engr. *Employer:* Am Elec Power Serv Corp. *Education:* MBA/Finance/Rutgers Univ; BE/EE/Yale Univ. *Born:* 4/5/29. Joined the Am Elec Power Serv Corp in 1963 after twelve yrs with PSE&G Co. Various positions in engrg & EE mgt with AEP. And deputy chief engineer responsible for all elec engrg matters including distribution, transmission, communications & generation. Assists in management of engrg for matters in all disciplines. Author of numerous technical papers. Fellow of IEEE, IEEE William Habirshaw Award, Former Pres NJ. Jr Chamber of Commerce, Mbr City Council Plainfield, NJ & Watchung Hills Sch Bd. Young Man of the Yr 1963, Plainfield, NJ. mbr Bd of Dirs: AEP Service Corporation, Ohio Power Co, Vice Chair, American National Standards Inst, Tau Beta Pi, Beta Gamma Sigma. *Society Aff:* IEEE, CIGRE.

Scherer, Richard D
Business: 2400 N. Woodlawn, Suite 200, Wichita, KS 67226
Position: Asst. District Mgr. *Employer:* BE&C Engineers (Boeing subsidiary) *Education:* BSEE/Elect Engrg/Oklahoma St Univ *Born:* 1/22/26 Assistant District Manager, BE&C Engineers, Inc. (Subsidiary of Boeing). BS OK State Univ. Native of McCune Kansas. Served US Army Air Corps 1944-46; attend OK State Univ 1946-52. Design Engr for Boeing Co. I progressed through mgmt ranks to Plant Facilities Mgr 1953-71; became an Assoc of R S Delamater Cons Engrg firm in 1971 & formed Delamater, Freund & Scherer PA. Rejoined Boeing Company's subsidiary BE&C Engineers 1977 to present. Natl Dir Natl Soc of Prof Engrs 1972, PPres Kansas Engr Soc. Mbr ASME, IEEE. Kansas State Appeals Bd SS Mbr. Kansas State Univ Adv Bd Sch of Engrg. Received Meritorious Achievement Award Federal Government Design 1953. Hobbies: stamp collection & fishing. *Society Aff:* IEEE, ASME, NSPE, KES, WPES

Scherich, Erwin T
Home: 3915 Balsam St, Wheat Ridge, CO 80033
Position: Consltg Civil Engr *Employer:* Self employed *Education:* MS/CE/Univ of CO; BS/CE/Univ of NB *Born:* 12/6/18 Native of NB. Served US Army 1941-1945. Professional career: Consltg Civ Engr - Hydraulic Structures 1984-present, civil engr, 1948-until retirement 1984, US Bureau of Reclamation. Chief, Div of Tech Review 1978-until retirement 1984; Chief, Dams Branch 1975-1978; Head, Spillways & Outlets Section 1974-1975. Provide consultative services for design and technical review of spillways, outlet works, hydropower waterways, diversion works, and other structures appurtenant to concrete and embankment dams. Also, provide safety of dams inspections and evaluations. Recipient, Dept of Interior Meritorious Service Award 1975 and Distinguished Service Award 1984. Professional Engrs of Colorado Engineer of the Year Award 1982 and Alfred W. Ryan Award 1984. BS, NB Univ, 1948; MS, CO Univ, 1951; Registered PE-CO. State Pres, PE of CO (NSPE) 1977- 1978. Presently, Dir, NSPE. Fellow, ASCE. Member, USCOLD (US Committee on Large Dams). *Society Aff:* NSPE, ASCE, USCOLD

Schermerhorn, R Stephen
Business: P.O. Box 6672, Denver, CO 80206
Position: Pres & Chrmn *Employer:* Stratigems, Ltd. *Education:* MSME/Product Design/Stanford Univ; BSME/Heat Transfer/Stanford Univ *Born:* 3/29/41 Studies at WI in Human Cybernetic Systems concurrent with heading the Human Systems Lab. Faculty fellow to NASA for Skylab B. Headed environmental divisions for Burns and McDonnell and R. W. Beck. Founder and CEO of Impact, Environmental Consultants. Founded Stratigems, a forensic engineering and government relations firm. Currently representing ASME on the ANSI Acoustical Standards Management Bd and AAES PAC. Serving as ASME V.P. Government Relations. National Energy Advisory Committee of ACEC and Energy Chrmn of the White House Conference on Small Business. Numerous articles and papers, symposia, and

Schermerhorn, R Stephen (Continued)
patentee in the field. Registered PE in 16 states and territories, Certified Consulting Engr, and Fellow, ASME. Leisure activities - Ski Patrol Mountaineering and Avalanche instructor, crime lab detective, MENSA, musician, fiction writer. *Society Aff:* ASME, ANSI, ACEC, NSPE, SMPS, IAI

Scherr, Harvey H
Home: 1406 Feather Ave, Thousand Oaks, CA 91360
Position: Technical Dir. *Employer:* USN Ship Weapons Engg Station. *Education:* BME/Machine Design/NYU; BEE/Control Systems/NYU. *Born:* 8/6/32. Born in Pittsburgh, PA & raised & educated in NYC. I worked for the Ford Instrument Co Div of Sperry Rand Corp for 9 yrs after grad from NYU. With the Ford Inst Co, I worked my way from a Jr Engr to a Principal Engr. My assignments varied from mech servo designs to systems engg. I joined the Govt in 1964 as a Div Hd. I am now the technical dir in charge of a workforce of 2700 Civilian employees. Over 500 are engrs. The station is involved in support of ship installed surface weapons.

Scherr, Richard C
Home: 9695 Ash Ct, Blue Ash, OH 45242
Position: Section Head *Employer:* The Procter & Gamble Co *Education:* BS/ChE/Univ of MI *Born:* 9/18/46 Native of Detroit, MI. Grad Univ of MI 1968 and worked for Procter & Gamble until 1971. 1971-73 worked as Sales Engr. Returned to Procter & Gamble Engrg Div in 1973. Assumed position of Mgr, Air Pollution Control in 1975. Assumed present position, Mgr, Environmental Control, Soap, Food and Intl Div in 1979, which includes responsibility for Co Environmental Policy and supervision of a staff of 10-20 profls. Elected to the Bd of Dirs of APCA in 1981. Elected to the bd of the East Central Section of APCA in 1980. Will be Chrmn of East Central Section of APCA 1982-83. *Society Aff:* APCA

Schetky, Laurence McD
Home: 11 Westport Rd #47, Wilton, CT 06897
Position: Chief Scientist - Dir *Employer:* Memory Metals, Inc. *Education:* PhD/Met/RPI; MMetE/Met/RPI; BChE/Ch.E./RPI. *Born:* 7/15/22. US Navy Pacific Theater 1944-1946. Taught and directed res at RPI 1948- 1953. Lectr & dir Mtls Lab, Instrumentation Lab MIT 1953-1959. VP, Dir Res, Alloyd Electronics, Cambridge, MA 1959-1963. Technical Dir, Met, Intl Copper Res Assoc, New York, NY 1963-1983. Chief Scientist - Dir Memory Metals, Inc. Stamford CT, 1983-present. Awards: Alcoa Fellowship 1948-50. Amer Foundrymen's Soc Exchange Lectr 1968. Fellow ASM Internatnl. Fellow British Institute of Metals. Editor Beryllium Tech Vol I, II; Editor, Physical Chemistry of Copper: 12 vols; Editor, Copper in Iron & Steel; Ext publications on joining tech, refractory metals, copper met, vapor phase deposition, shape memory alloys. Extra curricular, sailboat crusing. *Society Aff:* ASM, AIME, Met. Soc. (Brit).

Schetz, Joseph A
Business: Aero and Ocean Engg Dept, Blacksburg, VA 24061
Position: Prof & Dept Hd. *Employer:* VPI & SU. *Education:* PhD/ME/Princeton Univ; MA/ME/Princeton Univ; MSE/ME/Princeton Univ; BS/Naval Arch & Marine Eng/Webb Inst of Naval Arch. *Born:* 10/19/36. Born in Orange, NJ Oct 19, 1936. Grad first in class Dwight Sch, 1954. BS Webb Inst of Naval Arch, 1958. MSE, 1960, MA, 1961, PhD, 1962, Princeton Univ. Sr Scientist, Gen Appl Sci Labs, 1961-1964. Assoc Prof of Aero Engg, Univ of MD, 1964-1969. Prof and Dept hd, Aero and Ocean Engg Dept, VPI & SU, 1969-present. Guest Prof, Inst for Theoretical Aerodynamics, Aachen, Germany, 1971. Consultant, Appl Physics Lab, Johns Hopkins Univ, 1964-present. Active in local Republican politics, Boy Scouts and the Catholic church. *Society Aff:* AIAA, ASME, SNAME.

Scheuing, Richard A
Home: 37 Quaker Path, Cold Spring Habor, NY 11724
Position: Dir of Res. *Employer:* Grumman Aerospace Corp. *Education:* PhD/Aero & Astro/NYU; MS/Aero Engg/MIT; BS/Aero Engg/MIT. *Born:* 8/19/27. Native of Lynbrook, NY. Undergrad schooling under Grumman scholarship. Joied Grumman Aerospace Corp in 1948. Involved in theoretical & experimental aerodynamic res. Hd of Fluid Mechanics Res Sec from 1956 to 1970. Principal Investigator for sev AF, Navy, & Natl Sci Fdn res contracts during that time. In 1961, assumed the duties also of Deputy Dir of Res. Feb, 1977 became Dir of Res. Responsible for the entire spectrum of res activities necessary to achieve Grumman Corp objectives. Reside with wife Doris & 3 children in Cold Spring Harbor, NY. Active in sailing (racing & cruising) & skiing. *Society Aff:* AAAS, AIAA, ANS, APS.

Schick, William
Home: 5610 Post Road, Bronx, NY 10471
Position: Asst Dean & Professor of Elec Engrg. *Employer:* Fairleigh Dickinson University. *Education:* EdD/Science Ed/Columbia Univ; MSEE/EE/NYU; BEE/EE/CCNY. *Born:* Jan 1922. BEE City Coll of New York; MSEE New York Univ; EdD Teachers Coll Columbia Univ. Native of New York City. Design Engr United Transformer Co, Sperry Gyroscope Co, Todd Products Co, New Jersey Electronics specializing in transformers & electric filters 1944-57. With Fairleigh Dickinson Univ 1957- . Rank of Prof of Electrical Engrg & Asst Dean. Directing the development of computer graphics and computer aided design laboratories. Senior author, 'FORTRAN for Engrg 1972. Hobby: classical music. *Society Aff:* IEEE, AAVP.

Schiff, Anshel J
Business: Dept. Civil Engrg, Stanford, CA 94305
Position: Consulting Prof. *Employer:* Stanford Univ. *Education:* PhD/Engg Sci/Purdue Univ; MS/Engg Sci/Purdue Univ; BS/ME/Purdue Univ. *Born:* 9/24/36. Joined Sch of Aero & Engg Sciences, Purdue Univ, as Asst Prof in 1967. In 1972 joined Sch of Mech Engg, Purdue Univ and was promoted to Assoc & Full Prof in 1973 & 1978, respectively. In 1987 Joined Dept of Civil Engrg, Stanford Univ. as Consulting Prof. Founder and principal of Precision Measurements Ind, Inc, which is involved in ind education, engg consulting and mfg of instrumentation. Interests include vibrations, structural dynamics, system identification, instrumentation, measurement and data analysis and earthquake engg. *Society Aff:* ASME, ASCE, ASEE, AAAS, EERI.

Schiff, Daniel
Business: Assurance Technology Corp, 84 South St, Carlisle, MA 01741
Position: Chief Scientist *Employer:* Assurance Technology Corp *Education:* PhD/Phys/Univ of IL; MS/Phys/Univ of IL; BS/Engg Phys/OH State Univ. *Born:* 9/16/25. Sr Scientist, Westinghouse Atomic Power Div, 1953-1956. Mgr, Phys Dept, Raytheon Wayland Labs, 1956-1959. Dir R&D, High Temperature Mtls, Inc, Boston, 1959-1964. Pres, Res Consultants, 1964-84. Chief Scientist, ATC, 1984 to present. Work in satellite design and analysis, radiation effects, mechanical/thermal analysis. *Society Aff:* NSPE, ASSE, APS, AAAS, IEEE, ASME.

Schiffer, Francis H
Business: 80 Park Plaza, Newark, NJ 07101
Position: Manager - Cost and Scheduling *Employer:* Public Service Electric & Gas Co. *Education:* BSME/Mech Engg/Univ of WI. *Born:* April 1914. Native of Madison Wis. Served with Army Corps Engrs 1941-46, Army Reserve 1946-63. PSE&G Co 1946- , assumed current responsibility as Manager Cost and Scheduling 1979. Responsible for cost control, cost info & related activities for PSE&G Co Engrg & Const Dept. Reg PE in NJ. Dir of American Assn of Cost Engrs 1969-71, Sec of Amer Assoc Cost Enrs 1971-79. AACE Certification Board 1979-82. CCE. Enjoy amatuer radio, wood working, & lapidary work. *Society Aff:* SAME.

Schilling, Donald L
Business: 140th St & Convent Ave, New York, NY 10031
Position: H. G. Kayser Chair Professor of Electrical Engrg. *Employer:* City College of New York. *Education:* PhD/EE/Polytechnic Inst of Brooklyn; MS/EE/Columbia; BEE/EE/CCNY. *Born:* June 11, 1935. Currently Prof of Electrical Engrg at the City Univ of NY, City Coll. Has co-authored 4 textbooks, 'Electronic Circuits: Discrete & Integrated' 1968, 'Principles of Communications Systems' 1971, & 'Digital & Analog Systems Circuits & Devices' 1973, & 'Digital Integrated Circuits' 1976. Is Dir of Publs & Editor of the IEEE Transactions on Communications; serve on the Bd of Governors of the Communications Soc, & am a Fellow of the Inst of Electrical & Electronic Engrs. Current interest- delta modulation & phase locked systems. More than 50 publs in the Communications area & a cons to many firms. *Society Aff:* IEEE, HKN, AAUP, ΣΞ.

Schimpeler, Charles C
Business: 1429 S Third Street, Louisville, KY 40208
Position: Principal. *Employer:* Schimpeler-Corradino Associates. *Born:* BSCE 1960, MSCE 1962 Univ of Kentucky Lexington Ky; PhD (Urban Planning & Engrg, Operations Res) 1967 Purdue Univ Lafayette Ind. Professional Exper: Tech Dir & Dir of Planning Process Louisville Metropolitan Comprehensive Transportation & Dev Prog Louisville Ky 1964-69; Cons Louisville & Jefferson County Air Bd, responsible for major indus & related commercial dev planning, & economic & financial planning, transportation systems planning & comprehensive urban, land use , & environ planning 1969-73; Adjunct Prof Inst of Community Dev Univ of Louisville, Louisville Ky 1969- ; Principal Schimpeler-Corradino Assocs Louisville Ky, firm consists mainly of divs specializing in transportation, urban & regional planning, systems planning, civil engrg, & environ mgmt & engrg 1964- .

Schinzinger, Roland
Business: Univ. CA, Dept. of Electrical Engrg, Irvine, CA 92717
Position: Prof of EE *Employer:* Univ of CA. *Education:* PhD/EE/Univ of CA; MS/EE/Univ of CA; BS/EE/Univ of CA. *Born:* 11/22/26. Roland Schinzinger is an engineer with experience in industry, consulting, teaching, and public service. His primary interests are in energy systems and operations research applied to utility systems in general. He has worked in design and development, testing, and computer simulation. His research has been in optimization methods, power systems, water system emergencies, and engineering ethics. His consulting activities range from electromechanical devices and energy conservation to the mechanization of production planning. His teaching experience includes interdisciplinary courses at all levels. He has authored or coauthored over 50 technical papers in addition to numerous reports. He is coauthor of Ethics in Engrg (McGraw-Hill, 1983). *Society Aff:* AAAS, IEEE, ORSA, ASEE, TIMS

Schirmer, Chester W
Business: 3701 W Lake Ave, Glenview, IL 60025
Position: Pres. *Employer:* Schirmer Engg Corp. *Education:* BS/Fire Prot Engr/IL Inst of Tech. *Born:* 3/22/28. Joined present co as engr in 1952, became VP in 1957, and Pres in 1964. Reg PE in IL and eleven other states. Chrmn Automatic Sprinkler and Exposure Comm; Mbr, Rack Storage of Materials and System Concepts Comm of Natl Fire Protection Assoc. Serves on the Bd of Dir, Natl Fire Protection Assoc. Mbr of Steering Comm and Chrmn, Test Planning Comm of the Rack Storage Fire Protection Comm, Mbr of Intl Conference of Bldg Officials and Southern Bldg Code Congress, and professional mbr of the Bldg Officials and Code Administrators Intl. *Society Aff:* SFPE, ASSE, NSPE, ISPE.

Schirmer, Robert M
Home: 1436 Hickory, Bartlesville, OK 74003
Position: Research Engineer. *Employer:* Phillips Petroleum Co. *Education:* BS/ChE/SD Sch of Mines & Technology. *Born:* July 15, 1920 Aberdeen SD. Married- 1 child. Res Engr for Wright Aeronautical Corp during WWII; served US Navy 1945-46; with Phillips Petroleum Co 1946- . Group Leader on over 20 Navy & Air Force contracts concerning effect of fuel composition on combustion characteristics of aircaaft turbine fuels. Granted over 40 US Patents on burner design & fuel composition & have many publs in this field. Reg PE in Ok. SAE Manly Memorial Medal 1970; Chmn SAE Midcontinent Sect 1974. Enjoy gardening. *Society Aff:* AIChE, SAE, CI, AIAA, NSPE.

Schlabach, Tom D
Business: 600 Mtn Ave, Murray Hill, NJ 07981
Position: Dept Hd. *Employer:* Bell Labs. *Education:* PhD/Physical Chemistry/MI State Univ; BS/Chemistry/Baldwin-Wallace College. *Born:* 7/4/24. Native of Cleveland, OH. Joined Bell Labs in 1952 engaging first appl res in corrosion and electroplating and then in composites and printed wiring mtls and processes. Authored a book on the latter subj, "Printed and Integrated Circuitry–, in 1963. In 1965, became hd of the Met Engg Dept concerned with the dev and application of metals and alloys and with engg phys. Since 1971, has been additionally involved with mtls conservation and has contributed to various natl advisory studies on this subj. *Society Aff:* TMS-AIME, ASM, ACS, ASTM, AAAS.

Schlatter, Rene
Business: Guterl Special Steel Corp, 695 Ohio St, Lockport, NY 14094
Position: Manager of Melting and Forging *Employer:* Guterl Special Steel Corp. *Education:* MS/Met Engg/Univ of Pittsburgh; BS/CE/Inst of Tech. *Born:* 12/7/30. & educated in Switzerland with advanced degree from Univ of Pittsburgh. Worked in Foundry Tech for 8 yrs. 1962-1965 Res Met with Special Metals Corp, New Hartford, NY in field of vacuum melting of superalloys. 1965-1979 with Latrobe Steel Co, Latrobe, PA as sr process res met, mgr of process res & since 1973 mgr of res & dev, specializing in mfg of specialty steels & alloys for advanced tech applications. Established modern mfg tech for special steels and special purpose alloys and organization of Metallurgical Quality Control System, inspection & Res & Dev at Eletrometal Acos Finos SA in Brazil during 1979-1981. *Society Aff:* ASM, AIME, ASTM.

Schlegel, Walter F
Business: 17 North Ave, Norwalk, CT 06851
Position: VP. *Employer:* CPI Plants, Inc. *Education:* BE/Chem.E/Yale. *Born:* 5/12/33. Native of Shelton, CT. Married Janet Koenen Schlegel in 1955. Have seven children. Worked for BF Goodrich Chem Co in production, plant engg, dev, & central engg. Worked 19 yrs at Crawford & Russell, ultimately as Mgr of Process Engg. Joined CPI Plants, Inc as Partner, Dir, Officer, & VP. Mbr of Norwalk Symphony on Bd of Governors, VP Orchestral Production, First Trombone. Mbr church vestry & jr warden. Enjoy basketball, music & motorcycling. *Society Aff:* AIChE.

Schleif, Ferber R
Business: 800 So Fillmore St, Denver, CO 80209
Position: Electric Power Consultant. *Employer:* Self-employed *Education:* BS/Elec Engg/WA State Univ. *Born:* March 1913 Oroville, Washington. Reside in Colo. Employed by Bureau of Reclamation 1936-74 in const, power opers, planning & res. Appointed Ch Electric Power Branch Div of Res 1962. Important work has been power system control & stabilization. Engaged in cons work 1974- . Fellow IEEE 1972; Distinguished Service Award by Dept of Interior 1970. Mbr Tau Beta Pi, Sigma Tau, & Sigma Xi. Enjoys woodworking, music, & electronics. *Society Aff:* IEEE.

Schlesinger, Stewart I
Business: P O Box 92957, Mail Stop M5-011, Los Angeles, CA 90009
Position: General Mgr Network Control Systems Div. *Employer:* The Aerospace Corporation. *Education:* PhD/Math/IL Inst of Tech; MS/Math/IL Inst of Tech; BS/Math/IL Inst of Tech. *Born:* April 1929 Chicago. Worked with early computers at Los Alamos Scientific Lab developing one of first nationally used software systems (DUAL for IBM 701) 1951-56. Joined Aeronutronic Div of Ford Motor Co 1956, started computing group & served as Mgr Mathematics & Computation. With Aerospace Corp since 1963, first as Dir of Mathematics & Computation Center, then as Genl Mgr Info Processing Div then as Genl Mgr Mission Info Systems Div, then as Genl Mgr Satellite Control Div, and since 1986 as Genl Mgr Network Control Systems Div. Served as Genl Chmn & Prog Chmn for 1973 & 1975 Summer Computer Simulation Conferences. Chmn Tech Ctte on Model Credibility, VP (1978-79), Pres

Schlesinger, Stewart I (Continued)
(1979-82) and currently Vice President-Conferences, Soc for Computer Simulation. *Society Aff:* SCS

Schlieger, James H
Home: 3806 Milbourne Ave, Flint, MI 48504
Position: Parts Rep. *Employer:* Fraza Equip Inc. *Education:* Sales Course/-/Dale Carnegie; Mgt Course/-/Dale. *Born:* 3/29/29. Grad of Flint Northern High Sch, Grad of Dale Carnegie Mgt Course Grad of Dale Carnegie Sales Course. Previously was store mgr of J & R Auto Stores for 25 yrs. Currently Pres of Flint-Saginaw Valley Chapter of Mtl Mgt Soc. *Society Aff:* IMMS.

Schlimm, Gerard H
Business: 3400 N Charles Street, Baltimore, MD 21218
Position: Dir, Part-Time Undergraduate & Special Programs *Employer:* Johns Hopkins University, Whiting School of Engineering *Education:* BS/Civil/Univ of MD; MS/Engr Mgmt/NJ Inst of Tech; PhD/Structures/Univ of MD. *Born:* 5/26/29. Served with the US Army Signal Corps 1951-1953. Mechanical Equipment Engr, Esso Research & Engrg Co, 1957-1960. Taught at Univ of Maryland and US Naval Academy before moving to Hopkins in 1966. Past Pres of Maryland Section, ASCE and Engrg Soc of Baltimore; Mbr Governor's Science Advisory Council 1975- present; Chmn Civic Design Ctte of Baltimore 1974-78; Chrmn, Governor's Hart-Miller Island Commission 1981-current; Executive Councillor, Tau Beta Pi Assoc 1982-1986. Registered PE. *Society Aff:* ASCE, ASEE, NSPE, ESB, ΤΒΠ

Schlink, Frederick J
Home: RD 4, Box 209, Washington, NJ 07882
Position: Technical Director and Editor, Retired - 1983 *Employer:* Consumers' Research Inc. *Education:* ME/-/Univ of IL; BS/ME/Univ of IL. *Born:* 10/26/91. Native of Peoria, Ill. Assoc. Physicist & Tech Asst to Dir Natl Bureau of Stds, 1913-19; Physicist in charge Instruments Control Dept. Firestone Tire & Rubber Co, 1919-20; Mech Engr-Physicist Eng Dept Laboratory, Western Electric Co (now Bell Labs), 1920-22; Asst Secy Am Stds Assn, 1922-31; Member of Board of directors, Am Nat Stds Inst, 1977-81; Tech Dir and Editor (Science & technology), Consumers' Res 1929-. Contributor of articles in encyclopedias & economic, scientific, tech & other journals; author & co-author of several books on consumer technical and economic problems. Prof socs: ASME, Life Fellow; IEEE, Life member; SAE, 35 year member; Fellow, Am Physical Soc; Life Fellow, Franklin Inst; Life Fellow, Sigma XI Scientific Res Society ofNorth Am; Fellow, AAAS. Awards: Distinguished Grad. in Mech Engr, 1969; Alumni Honor medal award for distinquished service in eng, 1971; Edward Longstreth Medal, Franklin Inst. *Society Aff:* ASME, IEEE, ANSI, APS, AAAS, ΣΧΙ

Schlintz, Harold H
Home: 4321 N Moroa, Fresno, CA 93704
Position: President. *Employer:* Harold Schlintz & Assocs Inc. *Education:* BS/Engg/Univ of WI. *Born:* March 1918. Asst Prof Fresno State Coll 1955-62. Mbr: ASCE, SAE, Cons Engrs Assoc Calif. Reg as Civil Engr, Mech Engr & Safety Engr in Calif. Former Dir of the Cons Engrs Assoc of Calif. Chmn Expert Witness Ctte of CEAC 1970-74. Co-authored 'The Cons Engr as an Expert Witness' (CEAC). Formed the cons engrg firm Harold Schlintz & Assocs in 1954. Architect reg. in CA 1986. *Society Aff:* ASCE, CEAC, SAE

Schloemann, Ernst F
Business: 131 Spring St, Lexington, MA 02173
Position: Consulting Scientist. *Employer:* Raytheon Res Div. *Education:* PhD/Theoretical Physics/Univ of Goettingen (Germany); MS/Theoretical Physics/ Univ of Goettingen (Germany) *Born:* 12/13/26. Ernst Schloemann is a Consulting Scientist at Raytheon's Res Div where he has worked since 1954. From 1961-62, he was a visiting assoc prof at Stanford Univ, & in 1966, he was a visiting prof at the Univ of Hamburg (Germany). He has been granted twelve US patents & several foreign patents & has published over 120 scientific papers on subjects such as ferrite microwave mtls & devices, magnetostatic & magnetoelastic waves, magnetic printing, bubble memories, & the recovery of nonferrous metals from waste. He is a fellow of the Inst of Elec & Electronics Engrs & a fellow of the Am Physical Soc. *Society Aff:* IEEE, MAG (IEEE), MTT (IEEE), APS, ΣΞ

Schloemer, Robert E
Business: 345 N. 95th St, Milwaukee, WI 53226
Position: Principal. *Employer:* Graef-Anhalt Schloemer Inc. *Education:* MS/Structural/Univ of WI; BS/Civil/Univ of WI. *Born:* Jan 7, 1923 West Bend Wis. Reg PE in WI, FL, AZ, KS, KY, GA, Reg Structural Engr in Ill, Reg Civil Engr in Calif. US Air Corp WWII. Instructor Univ of Wis, Ext 1951-53; City of Milwaukee Bridge Div 1953-55; in Cons Engrg 1955-61. Present firm of Graef-Anhalt-Schloemer & Assocs Inc formed in 1961. *Society Aff:* NSPE, ASCE, ACEC.

Schlueter, Robert A
Business: Dept of Elec Engg & Systems Sci, E Lansing, MI 48824
Position: Prof. *Employer:* MI State Univ. *Education:* PhD/Systems Sci/Poly Inst of Brooklyn; MSEE/EE/Rensselaer Poly Inst; BSEE/EE/Rensselaer Poly Inst. *Born:* 5/11/42. Professor at Mich State Univ, Dept. Elec Engrg & Systems Sci. Research interests in the modelling and control of power systems. Current research includes characterization of the region of stability and instability for various types of power system stability problems including voltage collapse, steady state stability, and transient stability. Dr Schlueter is working on methods of identifying all single and multiple contingencies that cause these stability problems to occur. Adaptive and nonlinear controls that can avoid and operate during the onset of these stability problems are being investigated. *Society Aff:* IEEE, HKN, ΣΞ.

Schmatz, Duane J
Home: 1335 Charlesworth, Dearborn Hgts, MI 48127
Position: Prin Staff Engr. *Employer:* Ford Motor Co. *Education:* MS/Met Engr/Univ of WI; BS/Met Engr/Univ of WI. *Born:* 12/23/32. With Ford Motor Co since 1956 active in various areas of met res including high strength steels, laser processing of mtls, and brazing of aluminum. ASM Henry Marion Howe Award 1961 & 1964. Named Outstanding Young Engr by the Engg Soc of Detroit in 1962. *Society Aff:* ASM.

Schmerling, Erwin R
Home: 9917 La Duke Drive, Kensington, MD 20895
Position: Deputy Dir/Space Sci *Employer:* Natl Aeronautics & Space Admin-Goddard Space Flight Ctr *Education:* PhD/Radio Physics/Univ of Cambridge; MA/ Radio Physics/Univ of Cambridge; BA/Physics/Univ of Cambridge. *Born:* July 1929. Came to US in 1955; naturalized in 1962. Taught in the Electrical Engrg Dept of Penn State Univ & worked in Ionospheric Physics. With NASA Headquarters 1964-1982, Chief of Space Plasma Physics responsible for research in the ionospheres & magnetospheres of the earth & planets & interplanetary space. Active in prof socs, Chmn of US Comm III of the Internatl Union of Radio Sci 1969-72, Advisory group for Aerospace Research & Development & a Fellow of IEEE. Advanced Management Program, Harvard Business School 1968, Senior Executive Program, Federal Executive Inst 1975. 1983-Visiting Scholar, Stanford Univ, CA; Jan 1984- , Asst Dir/Space Sci, NASA Goddard Space Flight Ctr, currently on assignment to assist with the dev of Space Data and Computing Systems. COSPAR (ctte on space res) Chrmn, subcttee C1 "On Earth's Atmosphere, Ionisphere" 1982. *Society Aff:* IEEE, AGU

Schmertmann, John H
Home: 2926 NW 14th Place, Gainesville, FL 32605
Position: Prin. *Employer:* Schmertmann & Crapps, Inc-Cons Geotech Engrs *Education:* PhD/Civil Engg/Northwestern Univ; MS/Civil Engg/Northwestern Univ; BS/Civil Engg/MIT. *Born:* Dec 1928. Postdoctorate work at Norwegian Geotech Inst Oslo 1962-63. Soils Engr US Mueser Rutledge Wentworth & Johnston, NYC 1951-

Schmertmann, John H (Continued)
54. Res Engr US Army Corps Engrs SIPRE, Wilmette Il 1954-56. Prof Univ of Fla Dept of Civil Engrg 1956-79. Adjunct Prof, 1979-present, Schmertmann & Crapps, 1979-present. On leave as Tech Dir Soil Mech & Foundation Engrg Inc Palo Alto CA 1964-65; sabbatical Visiting Scientist Div of Bldg Res NRC Ottawa Canada 1971-72. Res contrib: settlement prediction methods, soil shear strength behavior, in situ testing methods, structures on shrink-swell clay. ASCE Collingwood Prize 1956, Norman Medal 1971, State-of-the-Art Award, 1977 Middlebrooks Award, 1981. Married, 4 children. Mbr Natl Acad of Engrg (elected 1984), Reg PE FL & AZ. *Society Aff:* ASCE, ASTM, NAE.

Schmid, David M
Home: 621 B Hamilton St, Norristown, PA 19401
Position: Chrmn, Pres. *Employer:* Tech Alloy Co, Inc. *Education:* BS/-/Carnegie Mellon Univ; Dr of Engg/-/Carnegie Mellon Univ. *Born:* 1/16/15. *Society Aff:* ASM Fellow, AWS, WA

Schmid, Walter E
Business: 8150 Leesburg Pike Suite 700, Vienna, VA 22180
Position: Pres *Employer:* NKF Engg Assoc, Inc. *Education:* BSEE/Electrical Eng./ JUVENTUS Inst of Tech, Zurich Switzerland. *Born:* 6/27/32. Native of Switzerland. Proj and Application Engr with Brown Boveri, Switzerland & Canada (1955-1962). Proj Engr & Hd, Elec Engg Dept at Sun Shipbldg & Drydock Co, Chester, PA (1962-1968). Dept Mgr & Dir of Elec Engg with Litton Ship Systems, Culver City, CA (1968-1972). Mgr & VP in charge of Wash Office, & Dir of Geo G Sharp, Inc of NYC (1972-1978). Assumed current position of Pres & part-owner of NKF Engg Assoc, Inc, of Vienna, VA and Arlington, VA, in 1978. Chrmn, Chesapeake Section of SNAME (1978/79). *Society Aff:* SNAME, ASNE, IEEE, NSPE.

Schmidlein, Joseph A
Business: Berdan Ave, Wayne, NJ 07470
Position: Pres. *Employer:* Arizona Chem. Co *Education:* MS/CE/Columbia Univ; BS/CE/Columbia Univ. *Born:* 2/1/18. Worked one yr for TX Co then joined Am Cyanamid Co in 1941. Have been employed with them or subsidiaries ever since. First 16 yrs in res & rect in varied assignments in production, sales, & mgt. Since 1972 Pres of Arizona Chem Co, owned 50% by Am Cyanamid & 50% by Intl Paper Co. *Society Aff:* AIChE, ACS, ΣΞ, AAAS.

Schmidt, Albert J
Home: 312 Brompton Rd S, Garden City, NY 11530
Position: Asst VP Engg. *Employer:* Mutual of New York. *Education:* BSCE/-/Dartmouth College. *Born:* Oct 1921. BSCE Thayer Sch of Engrg Dartmouth Coll. Reg PE in NY, NJ, & Fla. Mbr NSPE. Resident of Garden City NY. Proj Engr with Eipel Engrg Cons New York NY for 15 years. In private practice as partner of Crow, Lewis & Wick Architects & Engrs for 4 years. Have been in charge of Engg for Mutual of New York for the past 12 yrs. Responsible for multi-million dollars worth of const. Enjoy classical music & golf. *Society Aff:* NSPE, NYSPE, NJPE, FSPE.

Schmidt, Alfred D
Business: PO Box 529100, Miami, FL 33152
Position: VP. *Employer:* FL Power & Light Co. *Education:* BS/ME/Univ of FL. *Born:* 5/5/18. Reg Engr - State of FL. Served as Lt in USN, rejoined FPL in Production Dept 1945. 1951 became Plant Superintendent. 1954-72 held various positions including Superintendent of Operating Power Plants & Regional Superintendent of Power Plants. 1972 named Dir of Power Resources. Assumed current responsibility of VP Power Resource in 1974. Ind group mbrships include Chrmn, Steam Generator Owners Group; AEIC Power Generation Ctte & AIF Ctte on Power Plant Design, Construction and Operation. *Society Aff:* ASME, NSPE, FES.

Schmidt, Alfred O
Home: 634 Prospect Ave, State College, PA 16801
Position: Machine Tool Consultant *Education:* DSc/ME/U of Mich; MSc/ME/U of Mich; ME/ME/JNG. Schule JLMENAU *Born:* May 1906. ScD, MSE Univ of Mich; ME Ingenieurschule Jlmenau. Engr Carl Zeiss Optical Works Jena Germany 1929-38. Ch Res Engr Kearney & Trecker Corp Milwaukee Wis 1943-61. Held Professorships at Colorado State Univ, Marquette Univ, Penn State Univ, Univ of Roorkee (India), Univ of Rhode Island, & Univ of Wis-Milwaukee. Korea Advanced Institute of Science Seoul, 1979; Busan National University Busan, Korea, 1980. Machine Tool Expert United Nations Indus Dev Org 1967 to 1982 in Israel, Pakistan, Kenya, Argentina, Brazil, & Sri Lanka; Taiwan, RoC. Fellow ASME; Fellow SME, ASEE. Gold Medal ASTE 1959. Mbr Sigma Xi, Pi Tau Sigma, Tau Beta Pi. Enjoy swimming & skiing. *Society Aff:* ASME, SME, ASEE

Schmidt, Charles M
Business: 7333 Fair Oaks Rd, Cleveland, OH 44146
Position: Pres *Employer:* Schmidt Assocs, Inc *Education:* BS/ME/St Univ of IA *Born:* 4/29/37 Native of Charles City, IA. Plant Engr, Chem Division, Pittsburg Plate Glass, Barberton, OH, 1960-1963; Assistant chief engr, Karl R. Rohre Assocs, Consulting Engrs, Akron, OH, 1963-1965; Chief Mechanical Engr, Noble W. Herzberg Assocs, Consulting Engrs, Cleveland, OH, 1965-1967; Pres, Schmidt Assocs, Inc, Consulting Engrs, Cleveland, OH, 1968-present; Firm specializes in industrial coal-fired power plants and environmental controls. Special Consultant, OH Dept of Public Works, 1973-1980; Judge, East OH Energy Conservation Award, 1974. Registered PE OH, IA, KY, IL, IN, MI and PA. Two (2) patents in field; numerous magazine publications in field. *Society Aff:* ASME, ASHRAE, NSPE

Schmidt, Frank W
Home: 130 Kennedy Street, State College, PA 16801
Position: Professor of Mechanical Engineering. *Employer:* The Pennsylvania State University. *Education:* PhD/Mech Eng/Univ of WI; MS/Mech Engg/Penn State; BS/ Mech Engg/Univ of WI. *Born:* 3/16/29. Maintenance Officer USAF; Instructor & Asst Prof Univ of Wis-Madison 1956- 62. Joined Penn State Univ 1962. Natl Sci Foundation Fellow & Academic Visitor Imperial Coll London 1965-66. Sabbatical leave Imperial Coll London 1973. Sabbatical leave 1980 U.K. AERE, Harwell Eng and Imperial College London. Secy 1972-75, Div Chrmn 1978, Mbr Exec Ctte 1975-80 Heat Transfer Div ASME. Papers Award ASAE 1973. Mbr Editorial Bd Heat Exchanger Design Handbook (Hemisphere Pub Co). *Introduction To Thermal* Sciences F. W. Schmidt, R.E. Henderson, C.H. Wolgemuth, J. Wiley (1984) Editor-in-Chief - International Journal Heat and Fluid Flow; Butterworth Pub Co. Fellow ASME (1984). Review of Tech Papers & Proposals for: Int J Heat & Mass Transfer; J Heat Transfer & Fluids Engrg, Transactions of ASME; Applied Mechanics Reviews; & NSF. Mbr, Organizing Ctte of Turbulent Shear Flow Symposuims. Book: *Thermal Energy Storage and Regeneration*, F. W. Schmidt, A. J. Willmott (Hemisphere Pub Co, 1981). *Society Aff:* ASME, ΣΞ.

Schmidt, John P
Home: 11 Honey Lake Dr RD 2, Princeton, NJ 08540
Position: Senior Vice President *Employer:* Arco Chemical Company *Education:* ScD/Chem Eng/MIT; BChE/Chem Eng/Renssalear. *Born:* April 1933. Native of Holyoke Mass. Field Engr for DuPont 1955-58. With Halcon Internatl 1963-73, eventually as Vice Pres - Res & Dev. Participated in dev & commercialization of new processes for manufacture of propylene oxide, cyclohexanol, & aniline. Transferred to Oxirane 1973. Executive Vice Pres, responsible for marketing, opers, & commercial dev.Joined Arco Chemical in 1980 as Senior Vice Pres, Technology and Development, joined ARCO Chemical in 1980 as, Sr VP, Tech and Development, Holds 10 US Patents.

Schmidt, Robert
Home: 2437 Windemere, Birmingham, MI 48008
Position: Prof *Employer:* Univ of Detroit. *Education:* PhD/Structural Engr/Univ of IL; MS/Civil Engr/Univ of CO; BS/Civil Engr/Univ of CO. *Born:* 5/18/27. Son of

Schmidt, Robert (Continued)
Aquilina Konotop and Alfred Schmidt. Married to Irene Hubertine Bongartz. Father of Ingbert Robert. Draftsman, 1943-44. Student at Technische Hochschule Karlsruhe, 1947-49. Asst Prof of Theoretical and Appl Mechanics, Univ of IL, Urbana, 1956-59. Assoc Prof, Univ of AZ, Tucson, 1959-63. Prof of Engg Mechanics at Univ of Detroit since 1963; Chrmn of Civil Engg Dept 1978-80. Editor of Ind Mathematics, Journal of Ind Mathematics Soc, since 1969. Pres of Ind Math Soc, 1966-67, 1981-84. Reviewer for Appl Mechanics Reviews since 1959. Prin investigator in four NSF res proj. Recipient of the first Gold Award of the Indust Mathematics Soc, 1986. Author of 108 technical papers in the field of Appl Mechanics. Walking, bicycling, and biosophy are among my many interests. *Society Aff:* ASCE, ASME, IMS, AAM, ΣΞ, AAUP

Schmidt-Tiedemann, K J
Business: Vogt-Koelln-Strasse 30, D 2000 Hamburg 54, Germany (West)
Position: Geschaeftsfuehrer. *Employer:* Philips GmbH' Hamburg. *Education:* Dr/-/Universitat Hamburg. *Born:* July 1929 Dresden Germany. Studied mathematics, physics, philosophy. Dr rer nat from Hamburg Univ 1957. Joined Philips Res Labs Hamburg Germany 1958. Res in gas discharges & solid state physics. Visiting Prof Tech Univ of Copenhagen Denmark 1963-64. General Manager of A M lab 1968- . Geschaeftsfuehrer (mbr of the Exec Bd) of Philips GmbH 1974- . Prof (Physics) Univ Hamburg 1981- . Mbr of German Physical Soc, European Physical Soc, German Soc of Electrical Engrs, Fellow IEEE. Enjoys tennis, skiing, music, painting. *Society Aff:* DPG, EPS, VDE, IEEE.

Schmit, Lucien A
Home: 712 El Medio Ave, Pacific Palisades, CA 90272
Position: Prof. *Employer:* UCLA. *Education:* MS/Civil Engg/MIT; BS/Civil Engg/MIT. *Born:* 5/5/28. Struct Engg, Grumman Aircraft 1951-53; Research Engg, Aero & Structures Research Lab, MIT, 1953-58; Faculty Mbr, Case Inst of Tech, 1958-70; Prof of Engg and Applied Sci, UCLA, since 1970-. Research interests optimum design of structures. finite elements analysis, fiber composite structures. Mbr, Sci Adv Board, USAF 1976-84. Recipient of: Walter E Huber, Civil Engg Research Prize, ASCE, 1970; Design Lecture Award AIAA, 1977; Structures, Structural Dynamics, and Materials Award, AIAA, 1979. *Society Aff:* AIAA, ASCE, ΣΞ, ASME.

Schmitt, Frederick C
Home: 47 Odessa Dr, East Amherst, NY 14051
Position: VP *Employer:* Hartman & Schmitt Consulting Engrs, P.C. *Education:* D of Philosphy/Applied Mech/Drexel Univ; MS/Structural/Drexel Univ; BS/CE/Drexel Univ *Born:* 3/25/42 Worked on three research projects (1965-71) involving reinforced concrete and steel; research results used in A.I.S.C. Code. Taught Civil Engrg at Philadelphia Univ (1971-73, 1977-79). Project Engr at large Philadelphia A/E Firm (1973-79); responsible for design of ten different structures (railroad shipping and airmail facilities, crane runways, manufacturing plants, research labs, highrise hospitals and office buildings, sports complexes; involved steel, concrete and prestressed construction. VP of Consulting Engrg Firm (1979- present) specializing in design of cofferdams, underpinning, shoring and bracing; and inspection, evaluation, protection, and restoration of seriously damaged buildings, towers, bridges and marine structures. *Society Aff:* ASCE, NSPE, AISC, ACI, ISSMFE

Schmitt, George F, Jr
Home: 1500 Wardmier Dr, Dayton, OH 45459
Position: Advanced Development Mgr. *Employer:* USAF Mtls Lab. *Education:* MBA/Bus Admin/OH State Univ; MChE/CE/Univ of Louisville; BChE/CE/Univ of Louisville. *Born:* 11/3/39. Native of Louisville, KY. Advanced Development Mgr for Laser Hardened Materials; formerly principal engr for USAF in dev of mtls resistant to rain, sand, and ice erosion, focal point for protective coatings; Participant in numerous intl and natl erosion and radome confs, 50 publications and reports; SAMPE-national president. natl secy, long range planning chrmn, mbrship chrmn, ways and means chrmn; ASTM-chrmn, committee G-2 on erosion and wear, 1976 to 1980, recording secy, 1972-76; affiliate societies council of Dayton (44 technical societies)-chrmn, 1978-79, vchrmn, 1976-78; midwest chapter of SAMPE, all chairs and natl dir, 4 yrs; married, 2 sons; hobbies-music (vocal and instrumental), gardening, wine making, travel. *Society Aff:* SAMPE, AIAA, ACS, SPIE, ASTM.

Schmitt, Gilbert E
Home: 2804 Greenlee Dr, Austin, TX 78703
Position: Consultant (Self-employed) *Education:* BS/EE/Univ of TX. *Born:* Aug 1906. Native of Seguin Texas. Employed 1928-39 with CP&L Co. Resigned to assume total respon of engrg & Oper of large hydro & steam elec sys in Texas as Ch Engr & Asst Genl Mgr. Retired 1973 to devote time in energy problems as cons. V P AIEE; Fellow IEEE; Eta Kappa Nu; mbr Exec Bd NERC & FPC Regional Adv Council. Awards, Texas Interconected Sys & Texas Senate. Listed in Who's Who in Southwest, Men of Achievement & Internatl Register of Profiles. Enjoy photography & sports. *Society Aff:* IEEE.

Schmitt, Hans J
Business: RWTH - Aachen, Melatener Str 25, 5100 Aachen, W Germany
Position: Dir, Prof. *Employer:* Institut fur Hochfrequent Technik *Education:* Dr rer nat/Appl Physics/Univ of Gottingen; Dipl-Phys/Appl Physics/Univ of Gottingen. *Born:* 8/3/30. Born in Dortmund, FRG, res on electromagnetic techniques, plasmas, microwaves and magnetics; Asst Prof Harvard Univ 1957-63, Engg and Applied Physics, Sperry Rand Res Center til 1965; since 1965 Philips Res Lab, Hamburg: Dept Head Applied Physics: with groups in Microwaves, Optics and Solid State Physics. Guggenheim Fellow 1962; Since 1980 Prof and Dir Inst of High Frequency Technique; Fellow IEEE 1978; Councillor, German Section IEEE; European Microwave Management 1973-77. *Society Aff:* IEEE, VDE, DPG, URSI.

Schmitt, Harrison H
Business: U.S. Senate-NM, Washington, DC 20510
Position: Cons Geologist (Self-emp) *Education:* PhD/Geology/Harvard; BS/Geology/Caltech. *Born:* July 3, 1935 Santa Rita NM. Grad from Western H S, Silver City NM; Bach of Sci in Sci from Caltech 1957. After receiving a Fulbright Fellowship, studied at the Univ of Oslo in Norway 1957-58; Doctorate in Geology Harvard Univ 1964. Became NASA astronaut 1965 & participated in the flight of Apollo 17 as lunar module pilot. Beginning in 1974, assumed new duties as the NASA Asst Administrator for Energy Progs - resigned 1975. Elected U.S. Senator from NM 1976. Amer Inst of Mining, Met, Petro Engrs Ceremonial Medallion & Cert of Hon Mbrship 1973; Hon Fellow Geol Soc of Amer 1973; Hon Life Mbr New Mexico Geol Soc 1973; Hon Mbr Norwegian Geographical Soc 1973. *Society Aff:* GSA, AGU, AAPG, AIME.

Schmitt, Murray H
Home: 43 Maplewood Pl, Kitchener, Ontario N2H 4L4 Canada
Position: Designated Cons Engr. *Employer:* Murray H Schmitt *Born:* BASc 1952 Univ of Toronto. Native of Kitchener Ont. Engrg Dept City of Kitchener 1950-52; Warren Bitulithic Ltd 1952-55; Waterloo Regional Mgr & Assoc of The Proctor & Redfern Group 1955-75; Murray H Schmitt Inc, 1976 to present; Philips Engrg & Planning Ltd - 1977-79; Corp of the City of Cambridge - 1980; Conestoga-Rovers & Assocs - 1980 to present. APEO 1953; MEIC 1961; Camp 15 Warden 1962 to present. Elder St Andrew's Pres Kby Church 1963 to present, EIC Natl Council 1963-73, EIC V P Ontario 1970-73, Bd of Managers St Andrew's Presby Church 1964-74; Pres Kitchener Horticulture 1970; MCSCE 1970; Designated Cons Engr 1974; Fellow Engrg Inst of Canada 1974. MCGS 1981. AWWA 1982. Enjoy horticulture, H O railroading & woodworking.

Schmitt, Neil M
Business: E-204, Fayetteville, AR 72701
Position: Dean of Engrg *Employer:* Univ of AR. *Education:* PhD//Southern Methodist Univ; MS/EE/Univ of AR; BS/EE/Univ of AR. *Born:* 10/25/40. Native of Pekin, IL. Served with Army Air Defense Artillery Unit 1964-1966. Systems Engr for IBM 1966-1967. Design Engr (Radar) for TX Instruments 1967- 1969. Elec Engg Faculty, Univ of AR - Fayetteville 1970-1979. Assoc Dean of Engg, Univ of AR - Fayetteville 1980-1982. Dept Hd of Elec Engrg, Univ of AR- Fayetteville 1982-1983, Dean of Engrg and Dir of the Engrg Experiment Station, 1983-present. Active res in Cardiovascular Systems & Emergency Health Care. Consultant to govt and private agencies in health care. *Society Aff:* IEEE, NSPE, ASEE.

Schmitt, Roland W
Business: GE Corporate R&D, P.O. Box 8, Schenectady, NY 12301
Position: Sr VP Corp R&D *Education:* PhD/Physics/Rice Univ; MA/Physics/Univ of TX; BS/Phys-Math/Univ of TX *Born:* 7/24/23 Dr Schmitt, a native of Seguin, TX, joined the GE R&D Center in 1951. In 1957, he assumed his first managerial position. After a series of assignments of increasing responsibility, he was appointed VP of corp R&D in 1978, and Sr VP of corp R&D in 1982. He directs the R&D Center, which has some 2100 scientists, engrs, and supporting personnel. He also is chrman of the Natl Science Bd. He is a mbr of the Natl Acad of Engrg and serves on its council and is VP and mbr of the Bd of Dir of the Indus Res Inst. *Society Aff:* IEEE, APS, NAE, NSB, IRI.

Schmitz, Eugene G
Home: 3061 Vesuvius Lane, San Jose, CA 95132
Position: Pres & Chief Engr *Employer:* Schmitz Engrg Assocs *Education:* BSME/Mech Engrg/AZ State Univ *Born:* 9/17/29 As a professional consultant to major corps and government agencies, have designed/developed, modernized or improved the facilities of over fifty companies throughout the U.S. Founded and operated a high-rel assembly plant for Ford Aerospace in Tijuana, Mexico. As program administrator with Memorex, put major products into volume production. Developed Division Manufacturing Standards and was lead project engr of Methods/Manufacturing Engrg for SRS Division of Philco-Ford. Senior Project Engineer & Prog. Mgr with major consulting firm. Stetter Associates. Project Engineer w/FMC Corp., Ordnance Div. Engr'g in Weapons & Armored vehicles. Registered PE, CA and Certified Mfg Engr, Plant Layout & Material Handling Systems. Hobbies: gardening and computer programming. As a professional writer, am author/coauthor of over 100 proposals for various corps. *Society Aff:* NSPE, PEPP, IIE, SME, MVA, NRA, ADPA.

Schmitz, Roger A
Business: 202 Administration Bldg, Notre Dame, South Bend, IN 46556
Position: Keating-Crawford Prof, VP & Assoc Provost *Employer:* Univ of Notre Dame. *Education:* PhD/Chem Eng/Univ of MN; BS/Chem Eng/Univ of IL. *Born:* Oct 1934. Native of Carlyle, Ill. US Army 1953-55. Joined Univ of Ill Chem Engrg fac 1962. Joined Univ of Notre Dame fac 1979. Appointed VP & Assoc. Provost, 1987. Elected to Natl Academy of Engrg 1984. Received Wilhelm Award of AIChE for Res in 1981, Westinghouse award ASEE 1977; Univ of IL campus award for teaching excell 1975; Colburn award AIChE for res 1970; Guggenheim Fellowship 1968-69. Res ints in dynamics and control of chemically reacting systems. *Society Aff:* AIChE, ACS, ASEE.

Schnabel, James J
Business: 4909 Cordell Ave, Bethesda, MD 20814
Position: Principal *Employer:* Schnabel Engrg Associates *Education:* BCE/Geotechnical/Rensselaer Polytechnic Inst *Born:* 02/18/30 Principal and Chief Exec Officer of Schnabel Engrg Associates, P.C., Consulting Geotechnical Engrs since founding in 1964. Practicing consulting geotechnical engr since 1957 for numerous projs located within the Middle Atlantic States. Reg Prof Civil Engr in VA, MD, District of Columbia, NJ, DE and W VA. P Pres of Assoc of Soil and Foundation Engrs for USA. Former Dir of Consulting Engrs Council for Metropolitan Washington and National Capital Section of ASCE. Former Chrmnand mbr of National Capital Section ASCE Geotechnical Engrg Ctte. Chairman Ctte for Soc and Rock Properties, Trans Res Board. *Society Aff:* ASCE, TRB, ACEC, ASFE

Schnebel, Verne H
Home: 4181 Cornell Rd, Cincinnati, OH 45241
Position: Associate. *Employer:* Camargo Associates, Ltd. *Education:* BS/Chem Eng/Univ of Pittsburgh. *Born:* 5/26/12. Merck & Co Inc: Supr Pilot Plant, Dev Engr, Design Engr, Proj Leader, Proj Mgr; Lehn & Fink Prods Inc: Mgr Central Engrg; Processes Res Inc: Proj Engr, Proj Mgr, VP 1967-77. P E OH, Camargo Associates, Ltd, Associate, Treasurer 1977-. *Society Aff:* AIChE.

Schneider, Darren B
Home: 119 W Queen Ann Rd, Greenville, SC 29615
Position: Retired *Education:* BS/EE/KS State Univ. *Born:* May 1922 St Francis Ks. Joined G E 1944 & remained with them. Completed 3 yr G E Advaced Engrg Prog 1948. Joined Indus Controls & dev Numerical Controls for Mach Tools. Mgr Engrg, Numerical Controls & Equip 1959-72. Mgr, Production Engg, Greenville Plant at retirement in 1985. Chmn EIA Stds Ctte on Numerical Control 1964-69. Chmn ANSI Info Processing Sys Std Bd 1967-68. Ch U S delegate ISO Tech Ctte 97, NC Stds, 1964, 1967, 1968. IEEE Fellow 1971. Who's Who in Commerce & Indus 1962. Enjoy gardening, golf, music. *Society Aff:* IEEE.

Schneider, Edward P, Jr
Home: 6 Bopp Lane, St Louis, MO 63131
Position: President. *Employer:* Lark Engrg Corp. *Education:* Juris Doctor/Law/St Louis Univ; BS/Chem Engg/MO Sch of Mines & Metal. *Born:* Jan 1918. Chem Engr Monsanto Co, Sauget Ill 1942-43; Prod Supv 194548; Foreign Sales 1948-49; Tech Supv Chem Warfare Serv 1943-45. Mgr Brake Fluid Div of Wagner Elec Corp Berkley Mo 1951-64. Pres Lark Engrg Corp St Louis Mo 1964- . Dir Fed Enterprises Inc Nixa Mo 1968-1980. Alpha Chi Sigma Natl Pres 1968-70, Bd of Tr Educ Found 1964-76; Pres Air Pollution & Control Assn St Louis Mo 1976-77; Chmn Indus Dev City Berkeley Mo 1958-62; Secy Planning & Zoning City Town & Country Mo 1961-64; Chmn Trustees, Bopp Lane Subdiv 1959-64; St Louis C of C (Chmn Air & Water Pollution Comm 1967); Reg P E; AIChE, Phi Alpha Delta, Alpha Phi Omega, Mason (32 degree). Recipent 'Plaque of Appreciation for Civil Serv' City of Berkeley & Town & Country. *Society Aff:* ΑΧΣ, ΦΑΔ, APCA, AIChE, NACE

Schneider, Frederick W
Business: 80 Park Plaza, PO Box 570, Newark, NJ 07101
Position: Exec VP-Operations *Employer:* Pub Service Elec & Gas Co. *Education:* MS/Mech Engg/NJ Inst of Engg; BS/mech Engg/NJ Inst of Engg *Born:* Dec 20, 1923. Employed by PSE&G in 1949 holding a var of line supervisory pos in the Elec Generation Dept through 1964. Loaned in 1962 to Edisonvolta Util Co, Italy, to assist in initial startup of LaSpezia Generating Station. In 1965 moved to Elec Engrg Dept, progressing to Mgr of the dept 1971. This dept respon for the design of all sub, switching & generating stations for the co. In 1974 advanced to VP-Production overseeing all of the co's elec & gas production facils. In 1981 was given title "Asst to Sr VP - Energy Supply and Engrg–. In 1982 was elected Sr VP - Corporate Planning, and Aug 1984 was elected to present pos. Ctte serv incl EEI Prime Movers Comm, Penna Elec Assn, US/USSR tech exchange ctte for the design & oper of thermal power plants, & currently AEIC Power Generation Comm. (Prof Engr & Mbr ASME). *Society Aff:* ASME.

Schneider, Gilbert G
Business: 1611 N San Fernando Bl, Burbank, CA 91504
Position: Exec V P. *Employer:* Enviro Energy Corp. *Education:* BA/Math/Wofford College; BChE/Chem Engr/Rensselaer Polytech Inst. *Born:* Oct 1923. Native of Woodstock NY. 22 yrs with Western Precipitation, L A Calif, involved with application, dev, sales, service of scrubbers, mech collectors, bag filters, & electrostatic precipitators. Estab & managed W-P Div, Joy (SA), Johannesburg S Africa. Managing Dir Joy Trading Corp Ltd, London Eng. Pats on paper mill high velocity agglomerator scrubber. Presently respon for initiation of tech assistance progs, design, mkting, troubleshooting of electrostatic precipitator apparatus. Pres Indus Mkting

Schneider, Gilbert G (Continued)
Sys Inc, L A Calif. Bd/Dir Cyclotech Inc L A Calif. Guest lectr on air pollution control apparatus in Grad Sch of Chem Engrg, USC & West Coast Univ. *Society Aff:* AIChE, AIME, APCA.

Schneider, Irving
Home: 224-05 Hillside Ave, Queens Village, NY 11427
Position: Pres/Consulting Engr. *Employer:* Irving Schneider, PE, PC. *Education:* BS/Engg-Sci/CCNY. *Born:* 10/24/16. Formerly Assoc (in charge, machinery), Hardesty & Hanover, consulting engrs; chief engr, Washex Machinery Corp; sr proj engr, Silent Hoist & Crane Co; proj engr, Loewy-Hydropress Div, BLH Corp Responsible for mach'y design: some of largest movable bridges ever built, including Delair RR Lift Bridge (heaviest in US) & Burrard Inlet Lift Bridge — Can Natl Rways (largest in Canada). Contbr, machinery reqts: AREA & AASHTO Std Specifications for Movable Bridges. Designer special machy: steel mills/hvy bridge expansion rollers. Served 80th Inf Div, ETO, WWII. Awarded Combat Inf Badge, Certif of Merit. *Society Aff:* ASME, NSPE, ADPA.

Schneider, Morris H
Business: W181 Nebraska Hall, Lincoln, NE 68588-0501
Position: Assoc Dean Engg & Tech *Employer:* Univ of Nebraska. *Education:* PhD/Ind Engg/OK State; MS/Ind Engg/KS State; BSME/Mech Engg/Univ of NB; BS/Education/Univ of NB. *Born:* 11/26/23. Native of Sutton Nebr. Taught at Kansas State, TX Tech, Univ of Nebr. Teaching, res & cons in Engrg Mgmt & Mfg. Active in ASEE & IIE at Regional & Natl level. Chmn Indus Engrg at Nebr 1970-1986. Assoc Dean 1986 to date. Outstanding Indus Engr State of Nebr 1973-74, Outstanding Educator of Amer 1974-75. Reg P E. *Society Aff:* IIE, ASEE, NSPE.

Schneider, Robert W
Business: 3539 Glendale Ave, Toledo, OH 43614
Position: President. *Employer:* Mannik & Schneider, Inc. *Education:* BS/Civil Engr/Univ of MO at Rolla. *Born:* Nov 1934. Engrg Aid with Ill Hwy Dept 1955-56 summers. Design Engr with HNTB in Kansas City 1957-60. Sales Engr with Armco Steel Corp in Madison Wisc, Cincinnati & Toledo Ohio 1960-71. Partner in MSI since 1971. Pres since 1975. Prof Engr in 5 states plus Natl Council. Active in num tech & civic soc's. OSPE V P 1972-74. NSPE Natl Dir 1973-74. Sharonville, Ohio Jaycee of Yr 1964. Ohio Young Engr of Yr 1968. Outstanding Young Men of Amer 1970. Hobbies incl fishing, golf, auto racing. *Society Aff:* NSPE, ASCE, WPCF, AWWA.

Schneider, Sol
Home: 100 Arrowwood Ct, Red Bank, NJ 07701
Position: Consultant (Self-employed) *Education:* MS/Physics/NY Univ; BA/Physics/Brooklyn College. *Born:* 2/24/24. US Army 1943-46. Consultant, SRI Int. 1983-present, Adjunct Professor, Southeastern Center for Electrical Engineering Education 1980-86; US Army Electronics Tech & Devices Lab 1948- 1980; spec in gaseous electronics & pulsed power tech. Secy of Army's Special Act Award 1963; Cited in Proj Hindsight 1966; Mbr Exec Ctte of Gaseous Electronic Conf 1961-65 (Secy 1964); Mbr Exec Ctte Electron Phys Div APS 1965; Fellow IEEE 1973; Mbr Steering Ctte, DOD Pulsed Power Workshop 1976. US Army R&D Achievement Award 1963-78; US Army Sci Conference Madallion 1978; Chmn, Exec Ctte, IEEE Pulse Power Modulator Symposia 1957-1981; Chairman Emeritus, IEEE Pulse Power Modulator Symposia 1982-present, Mbr, Tech Ctte, IEEE Intl Pulsed Power Confs 1976-1980; Assoc Mbr, DOD Advisory roup on Electron Devices 1974-80, Mbr US Navy Pulsed Power Technical Advisory Panel 1978-80, Mbr US Air Force Panel on Pulsed Power - Education 1978-80; Mbr, SDI Pulse Power Technical Advisory Committee 1985-present. Enjoys camping, hiking, classical music, duplicate bridge, Num publs & pats *Society Aff:* IEEE, APS.

Schneider, William C
Home: 11801 Clintwood Pl, Silver Spring, MD 20902
Position: Vice President Development *Employer:* Computer Sciences Corp. *Education:* DEng/Aero Eng/Catholic Univ of Amer; MS/Aero Eng/Univ of VA; BS/Aero Eng/MA Inst of Tech. *Born:* 12/24/23. Native of NYC. Served in the USN 1942-46. Joined the NACA as an Aeronautical Res Scientist in 1949. Subsequently joined the USN Bu Aer where I worked on guided missile dev projs and later on the Navy's Space Program. in 1961 joined ITT as Dir of Space Dev Labs. In 1963 joined NASA. Served as Deputy Program Dir Gemini; Gemini Mission Dir; Apollo Mission Dir (Missions 2 thru 8); Dir Skylab Program; Deputy Assoc Admin for Manned Space Flight; Dir of Proj Mgmt (GSFC); Assoc Admin for Space Tracking and Data Acq. Assumed current position 1980; Respon for Systems Development at CSC. received NASA's Exceptional Service medal, Distinguished Service Medal (2), Outstanding Leadership Medals, the Collier Trophy for 1973, the Natl Space Club's Astronautical Engr Award, the AAS Space Flight Award VP & Fellow, AAS Assoc Fellow AIAA, Mbr EMCS, AFCEA, AUS. *Society Aff:* AIAA, AAS, AFCEA, EMCS.

Schneider, William J
Home: Box 474, McLean, VA 22101
Position: Consulting Hydrologist *Employer:* Self-employed *Education:* BCE/Ohio State University, 1950 *Born:* 9/20/21 Native of New York. Served with Air Corps & Engrs 1942-46. With U S Geological Survey 1947-1980 in var of assignments. Adviser to U N on dev of water resources of Mekong River 1967-72. Team leader of Interagency Task Force on Dev of Phosphate Resources in Southeastern Idaho, 1975-77.Task Force Leader for Presidential Study of dev of petroleum resources of National Petroleum Reserve in Alaska as mandated by PL94-258. Consultant to UNDP, OAS, NPS, and private industry 1980 to present. ASP AUtometrics Award 1967. AWRA Boggess Award 1973.

Schneidewind, Norman F
Business: 2822 Raccoon Trail, Pebble Beach, CA 93953
Position: Prof of Computer Sci; Tech Dir of Computing *Employer:* Naval Postgrad Sch - Monterey. *Education:* Phd/Operations Res/Univ of Southern CA; MSOR (Engr)/Operations Res/Univ of Southern CA; MBA/Comp Sci/Univ of Southern CA; MSCS(Engr)/Comp Sci/ San Jose State Univ; BSEE/Elect Engg/Univ of Calif. (Berkeley). *Born:* 4/12/28. BSEE Univ of CA, MSCE (Engrg) San Jose State Univ, MSOR (Engrg), MBA, DBA Univ of So CA. Eta Kappa Nu, Tau Beta Pi, Sigma Xi, Alpha Pi Mu, Beta Gamma Sigma. Cert in Data Processing. Recipient of US Commissioner of Customs commendation for ADP Panel work. Dept of Adm Sci, Naval Postgrad Sch. Mbr IEEE & ACM. Chrmn of IEEE Working group for quality metrics standard. Teach courses in computer networks, software engg, and microcomputers. Res areas are software engg & computer networks. Indus exper incls work as tech mgr & eng for SDC, PRC, CUC, UNIVAC & Hughes Aircraft. *Society Aff:* IEEE, ACM, ΣΞ.

Schnizer, Arthur W
Home: E Walpole, MA 02032
Position: VP. *Employer:* Bird & Son, Inc. *Education:* PhD/Chem/Northwestern Univ; BS/Chem/Baylor Univ. *Born:* 1/16/23. Native of Des Plaines, IL. Dev Engr PPG Industries 1943-6. US Army Chem Corps 1946-7. Res Chemist, Group Leader, Section Hd, Dir Chem Res, Dir Engg Res, Mgr Technical Ctr, Celanese Chem Co 1950-1969. Dir Chem & Polymer Res, Celanese Res Co 1969-1971. Chief Scientist & Technical Dir, Day & Zimmermann Inc 1971-6. VP R&D Bird & Son, Inc 1976-. Fields of interest: Petrochems, polymers, bldg mtls, process economics, gen & technical mgt. *Society Aff:* AIChE, AAAS, ACS, SCI.

Schnobrich, William C
Business: 3146 NCEL, 208 N. Romine, Urbana, IL 61801
Position: Prof *Employer:* Univ of IL *Education:* PhD/CE/Univ of IL; MS/CE/Univ of IL; BS/CE/Univ of IL *Born:* 11/26/30 in St. Paul, MN. Served as a Nuclear Weapons Effects specialist while in the Civ Engr Corps of USN 1955-58. Was a researcher with Space Tech Labs, Los Angeles, 1962. Joined the permanent faculty of Civ Engrg Dept, Univ of IL, 1962. Consultant on number of reinforced concrete

Schnobrich, William C (Continued)
plate and shell design and failure evaluation problems. Active on technical committees of many natl and intl profl societies. Prominent researcher in applying Finite Element Methods to predict behavior of reinforced concrete. Recipient of Von Humboldt US Sr Scientist Award 1977. Fellow ACI, 1982. *Society Aff:* ASCE, ASME, ACI, IABSE

Schoech, William J
Business: 176-A Gellersen Engg Ctr, Valparaiso, IN 46383
Position: Professor of Mech Engg., Director Manufacturing Labs *Employer:* Valparaiso Univ. *Education:* ASME/IE/Purdue Univ; MSIE/IE/PA State; BSEE/EE/Valparaiso Univ. *Born:* 3/14/44. Native of FL. Taught Elec Engg at Valparaiso Univ 1970-1974 before joining the Mech Engg Dept as Assoc Prof and Dir of the Mfg Processes Labs. Chrmn of the Mech Dept in 1976-1981. Presently holds rank of Assoc Prof of Mech Engg and teaches courses and labs in the areas of Mfg. Processes, Production Systems, Numerical Control, Automation/Robotics, and Interactive Computer graphics. *Society Aff:* ASM, SME, TBΠ, AΠM.

Schoenfeld, Robert L
Business: 1230 York Ave, New York, NY 10021
Position: Hd, Electronics Lab. *Employer:* Rockefeller Univ. *Education:* BA/Psychology/NYU; BS/Electronics/Columbia U; MEE/Electronics/Poly Inst Bklyn; DEE/Electronics/Poly Inst Bklyn. *Born:* 4/1/20. Schoenfeld lives in NY. He is married with 5 children. His entire career has been in bio-engg, starting as a Res Assoc in Neurology in 1947 at Columbia Medical Sch. He was chrmn of the IEEE EMB group in 1965, served as editor of its Transactions & a mbr of its administrative committee. He has taught part and full time at Poly Inst of NY 1947-83. For the last 23 yrs he has been hd of the Lab of Electronics at Rockefeller Univ. This group pioneered in use of computers for the acquisition of "real- time" bioelectric data. Currently Schoenfeld is developing microprocessor based lab systems for control and data acquisition in biological experiments. *Society Aff:* IEEE, ACM.

Schoenfeld, Theodore M
Business: 86C Empress Plaza, Cranbury, NJ 08512
Position: Vice Pres. *Employer:* Ramco Mfg Co - Roselle Pk NJ. *Education:* Grad Cert/Indus Eng/Stevens Inst of Tech; Grad Cert/Pub Admin/NY Univ; BS/Biology/College of City of NY. *Born:* 7/10/07. Reg P E (Calif). Newspaperman 1930-32. Asst Sys Dir City of NY to 1942. US Army 1943. US Dept of State 1945-46. Ch Indus Engr with MGM Intl to 1949. Mgmt cons 1950-74. Currently VP of Ramco Mfg Co. Respon for engrg, mkting & res. Sr Mbr & Natl Div Dir IIE 1975. Fellow, Amer Inst of Chemists. Mbr Ethical Culture Soc. Author many tech papers & articles. Lectures frequently on the relationship between humanism & indus pres, Brooklyn Soc for Ethical Culture, 1978-1983, VP 1984-86; 1986-present, President, Princeton Ethical and Humanist Fellowship; Recipient of Natl John C. Vaaler Chem Processing Award with honors in 1978 & with top honors in 1980 for outstanding contributions to safety in chem processing indus; Co-Dev of Ramco Spra-Gard Safety Shield and Ramco Econo- Gard widely used by Chem Proc Ind Tech paper on multinationals judged by Inter Rev Comm as one of five outstanding art on subj and pub in World Symposium, 1977. *Society Aff:* IIE, SPE, AIC.

Schoenman, Richard L
Business: P O Box 3707, Seattle, WA 98124
Position: Ch Engr - Flight Controls Tech. *Employer:* The Boeing Co. *Education:* MS/EE/Univ of WA; BS/EE/Univ of WA. *Born:* Sept 1927. Native of Seattle Wash. Worked with Bonneville Power Admin, was Asst Prof of EE Union College Schenectady NY, and has been with The Boeing Co since 1955. Currently respon for tech staff support for all commercial aircraft progs in the field of aerodynamic stability & control & flight control sys. Mbr SAE Aerospace Control & Guidance Sys Ctte, AIAA, Reg P E, holds a private pilot's license. *Society Aff:* AIAA.

Scholer, Charles F
Business: Sch of Civ Engg, W Lafayette, IN 47907
Position: Assoc Prof. *Employer:* Purdue Univ. *Education:* PhD/Civ Engr/Purdue Univ; MSCE/Civ Engr/Purdue Univ; BSCE/Civ Engr/KS State Univ *Born:* 5/31/34. Native of Manhattan, KS. Worked for consulting engg firm, Burgwin & Martin, Topeka, KS prior to service as a civ engr in the USAF 1958-60. Briefly served as an asst county engr, Riley Co KS, before resuming studies for the PhD. Has been a prof of civ engg since 1965 with res in the areas of portland cement concrete, past chairmanships include ACI 201 - Durability 221 Aggregates, ASTM C-27 Precast Concrete Products and Concrete Materials Research Council. Dir Am Concrete Inst 1984-87, Dir Hwy Extension Res Proj Indiana Counties and Cities. Consultant on problems of concrete and concrete construction, low volume roads. Co-chairman of Third 1985 & the Fourth Intnl Conference of Concrete Pavement Design & Rehabilitation, 1989. *Society Aff:* ASCE, ACI, ASTM, APWA, TRB.

Scholtes, Robert M
Business: P O Drawer CE, Mississippi State, MS 39762
Position: Head. *Employer:* Civil Engrg, Miss State Univ. *Education:* PhD/Civil Engg/GA Inst of Tech; MS/Civil Engg/MS State Univ; BS/Civil Engg/MS State Univ; AA/-/Pekkinston Jr College. *Born:* May 28, 1928. Native of Biloxi Miss. Taught at Miss State Univ 10 yrs prior to his grad work at Ga Inst of Tech. Returned to MSU in 1963 to resume respons as Prof of CE & later appointed Hd. Served in prof orgs as Dir & V P (1974-), Southeastern Region NSPE; Dir, V P, Pres Elect & Pres MES; Dir of Mid- South & Miss Sects of ASCE & Chmn of Southeast Region Civil Engrg Div of ASEE. Enjoy woodwork & golf. *Society Aff:* ASCE, NSPE, ASEE.

Scholtz, Robert A
Business: Elec Engrg Dept, Los Angeles, CA 90089-0272
Position: Prof, Dir of Comm Sci Inst *Employer:* USC *Education:* PhD/EE/Stanford Univ; MS/EE/USC; BS/EE/Cincinnati Univ *Born:* 1/26/36 Hughes Aircraft Co: MS Fellow, 1958-60; Doctoral Fellow 1960-63; Sr Staff Engr (part-time) 1963-78 specializing in missile radar signal processing. Univ of Southern CA: 1963-present; Prof of Elec Engrg; specializing in signal design, coding, and processing for communications and radar. Consultant to LinCom Corp and Axiomatix, Inc. IEEE: 1980 Fellow; on Bd of Governors of Communication Soc 1981-83, and of the Info Theory Group, 1981-; Finance Chrmn for 1977 Natl Telecomm Conference (NTC '77); Prog Chrmn for the 1981 Internatl Symposium on Info Theory (ISIT '81). Rcvd: 1982 Dist Alumnus Award, U of Cincinnati, 1983 Leonard G Abraham Prize Paper Award, IEEE Comm Soc, 1984 Donald J Fink Prize Award, IEEE. *Society Aff:* IEEE, ION

Scholz, Paul D
Business: College of Engr, University of Iowa, Iowa City, IA 52242
Position: Assoc Dean. *Employer:* Univ of IA. *Education:* PhD/ME/Northwestern Univ; MS/ME/Northwestern Univ; BS/ME/Univ of WA. *Born:* 11/11/36. Worked as a mech engr at the US Naval Ordnance Test Station, Pasadena Annex, 1960-62 (now the Naval Oceans Systems Center, San Diego). B.S.M.E. Degree 1960 (Univ of Washington); M.S. (1965) and Ph.D. (1967) from Northwestern Univ. Have been on the facuty of the Univ of IA since 1967. Assumed current position as Assoc Dean of the College of Engg July 1, 1979. Am a mbr of ASME, ASEE, AAAS, Tau Beta Pi, Pi Tau Sigma (honorary mbr), Sigma Xi, and Omicron Delta Kappa. Res areas of interest include aerosol dynamics and plasma radiation. *Society Aff:* AAAS, ASME, ASEE.

Schoneman, John A
Business: 2000 Westwood Dr, Wausau, WI 54401
Position: Chrmn & Ch Exec Officer *Employer:* Wausau Insurance Co. *Education:* MBA/Finance/Univ of Chicago; AMP/Adv Mgmt Prog/Harvard Grad Sch; BS/Safety Engrg/IIT *Born:* 7/9/28 Fire protection engr with MI & IL Inspection Bureau 1947-55; US Army Ordnance Corp 1951-52; Atlantic Mutual Insurance Co, NY 1955-76 - various positions including Exec VP and Trustee. Joined Wausau Insur-

Schoneman, John A (Continued)
ance Companies as Exec VP in 1976, elected Pres and CEO 1977 and Chrmn and CEO 1981. Married 1950, Mary F. Schoneman; four children, Judith, James, Robert, Carolyn. *Society Aff:* SFPE, AFI, AMRC, AIPLU, IA

Schonfeld, Fred W
Home: 4 Acoma Lane, Los Alamos, NM 87544
Position: Mgr, Plutonium Phys Met (CMB-5). *Employer:* Los Alamos Sci Lab. *Education:* MS/Metallurgy/Univ of MN; BS/Phys Met/WA State. *Born:* Nov 1918 Spokane Wash. Trainee Dow Chemical; Instr, Asst Prof Wash State; joined Los Alamos Sci Lab in Feb 1947 - Asst Scientist, Staff Mbr, Alternate Group Leader, Group Leader, Mgr, Plutonium Phys Met, Test Engr, Supervisor R&D Foundry. Have spec in behavior of Pu alloys, with emphasis on phase relationships & phys properties. Fellow ASM, ACI. P Chmn Nuclear Met Ctte AIME. Hobbies: gardening & carpentry. *Society Aff:* ASM

Schoofs, Richard J
Business: 1675 School Street, P. O. Box 67, Moraga, CA 94556
Position: President & CEO *Employer:* Schoofs Inc. *Education:* Master Engrg/Chem Engrg/Yale Univ; BS/Chem Engrg/Univ of Wis (Madison) *Born:* Pres of Schoofs Inc, a firm spec in Desiccants, Adsorbents & Catalysts. Genl Mrg Schoofs & Assocs inventors, patent developers & conslts in adsorption & catalyst tech. Other diverse business ints. BS Chem Engg Univ of WI; Master of Engg Yale Univ; Fulbright Scholar at Technishche Hogeschool Delft, The Netherlands. Mbr of the staff & fac US Army Chem Corps School 1954-55; Engr Shell Dev 1958-63; Sales Mgr Union Carbide Corp 1963-68. Former Partner in CATCO, a catalyst regeneration co. Mbr Tau Beta Pi, AIChE, Univ of WI Foundation. *Society Aff:* AiChE

Schoppe, Conrad J
Home: 8111 Rampart, Houston, TX 77081
Position: Res Div Supr. *Employer:* Tennessee Gas Pipeline Co. *Education:* BS/Genl Engg/Univ of Houston. *Born:* 1921. Native of Childress, Tex. Ch Engr for Houston Pipe & Steel. US Army 1944-46. Engr with Tennessee Gas Pipeline Co 1946-56, when assumed current respons as Hd of Met Div. Respon for met res, NDT, failure analysis, & arctic pipeline specifications. Have made significant contribs to pipeline indus through extensive res in pipeline welding & stress corrosion cracking in pipelines. Chmn Texas Chap ASM 1965-66. Served on local & natl tech cttes in ASM, AWS, AGA, ASNT, & API 1104. Enjoy hunting & ranching. *Society Aff:* ASM, AWS, ASNT, API, API-AGA, PFWP.

Schott, F W
Business: Rm 7732 Boelter Hall, Los Angeles, CA 90024
Position: Prof. *Employer:* Univ of CA. *Education:* PhD/EE/Stanford Univ; Engr/EE/Stanford Univ; AB/Phys/San Diego State College. *Born:* 10/2/19. Native of Phoenix, AZ; Engr Asst, San Diego Gas & Elec Co 1943-44; USN (Radar) 1944-46; Asst Prof to Prof Univ of CA at Los Angeles 1948-present; Electronic Scientist USN Electronics Lab 1949-50; Res Physicist Hughes Aircraft Co 1956; Swiss Fed Inst of Tech (Zurich) 1957-58; Consultant Elec Machinery 1975-present. *Society Aff:* IEEE, ASEE.

Schottman, Robert W
Home: 270 Trailwood Dr, Watkinsville, GA 30677
Position: Asst Prof *Employer:* Univ of GA *Education:* PhD/Soils & Water Engrg/Cornell Univ; MS/Agri Engrg/Cornell Univ; BS/Agri Engrg/Univ of IL *Born:* 6/5/44 Grew up on cattle and grain farm near Montrose, IL. Join res and teaching staff at ND State in 1972. Moved to Univ of MO in 1975 to teach and serve as state extension irrigation specialist. Took present res and teaching position at Univ of GA in Sept, 1981. Primary res interest is in improvement in application and distribution of state's irrigation water resources. Chrmn elect Mid Central Region ASAE 1981. Enjoy long distance running and the study of American history. *Society Aff:* ASAE

Schoustra, Jack J
Business: Ertec, Inc, Long Beach, CA 90807
Position: President. *Employer:* Ertec, Inc. *Education:* MSc/Civil Engg/Delft Techn Univ; BSc/Civil Engg/Delft Techn Univ. *Born:* Jan 8, 1931 Netherlands. Held prof engrg & mgmt pos wth Racey, MacCallum & Assocs Ltd Toronto 1956-62 and Converse, Davis & Assocs Pasadena Calif 1962- 70. Founded Fugro Inc Cons Engrs & Geologists in 1970 & has been Pres & Chmn since. Firm employs 400, incl 120 engrs & geologists. Co-author (with R F Scott) of 'Soil: Mechanics & Engrg' McGraw Hill 1968. Former Chmn Geotech Group, L A Sect ASCE. Mbr CEC, ASCE, SEAOC, Royal Dutch Engrg Inst, Intl Soc of Soil & Snow Mechanics, Old Ranch Tennis Club, Alamitos Bay Yacht Club. *Society Aff:* ASCE, CEC.

Schrader, Ernest K
Business: City-County Airport, Walla Walla, WA 99362
Position: Civ Engr *Employer:* US Army Corps of Engrs *Education:* MS/CE/Clarkson Coll of Tech; BS/CE/Clarkson Coll of Tech *Born:* 9/16/47 Served as Capt with US Army. Previous employment as Asst Superintendant for Turner Construction, CT; & Asst to Res Engr for Libby Dam & Hgwy Construction with the Corps of Engrs, MT; currently Concrete and Construction Mtls Specialist for the Walla Walla Dist, Corps of Engrs. Provide private consulting for special designs and concrete construction problems. Pres of Schrader-Johnson LTD. Received 1976 Army R&D Commendation, 1980 Tudor Medal from SAME for outstanding engr under 36, 1980 Div Engr Commander's medal, 1983 Greensfelder Const Prize (ASCE), and 1984 Wason Medal (ACI). Selected as 1983 Engr of the Year for US Army Corps of Engrs and Nom for Engr Const Man of the Year. Speaker at various Intl Congresses and Symposiums. Authored approximately 40 articles in technical and intl publications. Mbr of six ACI technical cttes. *Society Aff:* ASCE, SAME, ACI, USCOLD

Schrader, George F
Business: P O Box 25000 Univ. of Central Florida, Orlando, FL 32816
Position: Assoc. Vice Pres. for Research. *Employer:* Univ of Central FL. *Education:* BS/Mech Engg/Univ of IL; MS/Mech & Ind Engg/Univ of IL; PhD/Indust Engg/Univ of IL. *Born:* July 21, 1920. 1986-pres, Assoc. Vice Pres for Research 1977-1986, Assoc Dean Coll of Engg, Univ Central Fl. 1969-77, Chmn Indus Engrg & Mgmt Sys, Fla Tech Univ. 1966-69 Dir, State Tech Service, Univ of Nebr. 1962-66 Hd, Dept of Indus Engrg, Kansas State Univ. 1961-62 Prof of Indus Engrg Okla State Univ. 1953-61 Asst Prof of Mech Engrg Univ of Ill. 1973-76 Secy, Treas, V Chmn, Chmn Council of Indus Engrg Acad Dept Hds. Amer Inst of Indus Engrs Exec VP Chap Opers 1979-81, 1976-78. VP Region IV 1964-65 VP Region IX. Mbr NSPE, FES, ASEE, ASQC. *Society Aff:* IIE, ASEE, NSPE.

Schrader, Gustav E
Home: 31649 Trillium Trail, Pepper Pike, OH 44124
Position: Vice President, Manufacturing *Employer:* TRW Inc. *Education:* BSME/Design/Rennsselaer Polytechnic Institute; MSME (Incomplete)/Mfg.Eng. Mgmt./Rennsselaer Polytechnic Institute *Born:* 01/06/23 Native of Bristol, CT. Joined TRW in 1967. Assumed current responsibilities in 1980. Responsible for design and development of electronically controlled steering products. Co-written three books and many technical articles on manufacturing and engrg topics. Filed for or was issued eight patents. Received Engineer of the Year Award in 1981 from the Cleveland Engineering Society. Member of ASME, SME, CES, NSPE, ADPA, AIA, and NSIA. Active in many civic organizations including the American Red Cross, Cleveland Health Education Museum, Western Reserve Historical Society, Society for the Blind, Museum of Natural History, Shaker Lakes Historical Society, Cleveland Council on World Affairs, and Cleveland's Mayor's Operation Volunteer Effort. *Society Aff:* ΣΞ, TBΠ, NSPE, ASME, SME

Schrader, Henry C
Home: 10234 Democracy Ln, Potomac, MD 20854
Position: VP *Employer:* URS Corp. *Education:* MS/CE/Structural Dynamics/Univ of IL; BS/CE/Univ of IL; Dipl/Intl Econ/Natl Defense Univ *Born:* 1/5/18 Native of

Schrader, Henry C (Continued)
Chicago, IL. Commissioned Officer, Corps of Engrs, US Army, served in all ranks from Second Lt to Maj Gen, March 1942 through May 1973. Commander, 18th Engr Brigade in Vietnam May 1970-June 1971. Commander, Computer System Command, Ft. Belvoir, VA July 1971-May 1973. Joined the architectural engrg firm of Dalton-Dalton-Newport as Principal in June 1973. Engr in charge of design of redev of Natl Naval Med Ctr Bethesda Maryland 1974-75. Presently Vice Pres of mktg for URS Corporation, Virginia Beach, office. Bd of Dir, ARTBA; Bd of Dir, High Speed Rail Assn. Hold Distinguished Service Medal w/Oak Leaf Cluster, Legion of Merit w/3 Oak Leaf Clusters, etc. Hold PE licenses in ILL, MD, VA, & Wash, DC. *Society Aff:* NSPE, ASCE, SAME, HSRA, ARTBA, TRB.

Schrady, David A
Business: Monterey, CA 93940
Position: Dean of Info & Policy Sciences. *Employer:* Naval Postgraduate School. *Education:* PhD/Ops Res/Case Inst of Tech; MS/Ops Res/Case Inst of Tech; BS/Mgt Sci/Case Inst of Tech. *Born:* Nov 1939. Asst Prof of Opers Res, Naval Postgrad School 1965. Assoc Opers Res Prog Dir, Office of Naval Res 1970-71. Assoc Prof Naval Postgrad School 1971. Chmn Dept of Opers Res & Admin Scis, Naval Postgrad School 1974. Dean of Info & Pol Scis and Dean of Acad Planning, Naval Postgrad School 1976. Res specialization, publs, in mathematical inventory control. Secy-Treas Western Opers Res Soc 1975-76. Mbr Bd of Dir of Military Opers Res Soc 1973-77. Treas Opers Res Soc of Amer 1976-79 Pres, Military Operations Res Society 1978-79. . *Society Aff:* ORSA, TIMS, MORS.

Schreck, Seymour
Home: 3 Deerfield Lane, Huntington Station, NY 11746
Position: Prod Mgr. *Employer:* Berkey Tech Co. *Education:* BChE/Chem Eng/CCNY. *Born:* 9/29/27. B Chem E CCNY 1949. USNR 1945-46. Civilian employee USAF 1949-55, Photo Processing Sect of Aerial Recon lab; Prog Ch at Fairchild Space & Defense Sys until 1959, working on specialized & in-flight processing tech & equip. More than 15 yrs in chem, matls & equip dev of saturated web film processing. 7 yrs, VP Townley Chem Co & PoroKem Inc. Served as Natl Dir of SPIE & NY Chap Pres. Prod Mgr for Berkey Tech Co. Play tennis & bridge.

Schregel, Peter F
Business: 1945 Sheridan Dr, Buffalo, NY 14223
Position: VP. *Employer:* McFarland-Johnson Engg, Inc. *Education:* BS/Civ Engr/MI State Univ; AAS/Building Construction/SUNY, Alfred. *Born:* 8/13/39. Worked two yrs for Merritt-Chapman and Scott Construction Co, Inc in surveying and field engg on the PANSY Niagara Hydro Power Project. Returned to MI State Univ and began working for McFarland-Johnson Engrs, Inc in 1963 as a surveyor, inspector, and Jr Engr in the Buffalo and Binghamton offices. Worked in Charleston, WV as a Proj Engr and Sr Proj Engr primarily designing Interstate Highway Projs. Returned to my native Western NY in 1975 as VP and Office Mgr. Presently responsible for various civil consulting projects including Highways, Bridges, Flood Control, Buildings and Airport projs. *Society Aff:* ASCE, NSPE, CEC, ABCD.

Schreiber, William F
Business: E15-387 MIT, Cambridge, MA 02139
Position: Prof of EE. *Employer:* MIT. *Education:* PhD/Applied Phys/Harvard; MS/EE/Columbia; BS/EE/Columbia. *Born:* 9/18/25. in NYC & attended the NYC public schools. Jr Engr at Sylvania 47-49, Res Scientist, Technicolor Corp 53-59. MIT 59 to present. Fellow IEEE. Since 1948, my maj profl interest has been in image processing & transmission systems, particularly those designed for practical applications. I have designed several significant systems now in commercial use, including the Assoc Pres Laserphoto facsimile & the Autokon electronic process camera. Cofounder of ECRM, mfg of optical character recognition machines & laser scanners, Industrial Consultant. *Society Aff:* IEEE, SPIE, TAGA, SMPTE.

Schreier, Stefan
Business: Mech Engrg Dept, Storrs, CT 06268
Position: Assoc Prof *Employer:* Univ of CT *Education:* PhD/Aero Engrg/Univ of MD; MSE/Aero Engrg/Princeton Univ; BA/Math/PA State Univ *Born:* 7/21/31 Vienna, Austria. Educated public schools Austria, Switzerland, NY and PA. Taught at Univ of MD, George Washington Univ and Univ of CT. Consulted for ind and govt. Maj fields of interest include gas dynamics and boundary layer theory. Author of *Compressible Flow* (Wiley-Intersci, 1981). Ind experience includes work for Martin-Marietta, Aerojet Gen, GE, NASA, Grumman and Avco. Presently engaged in experimental work in low speed aerodynamics.

Schreiner, Warren C
Home: 511 Bayou Knoll, Houston, TX 77079
Position: Dir of Development. *Employer:* Pullman Kellogg. *Education:* MS/ChE/NYU; BShE/CCNY. *Born:* Aug 1921. MSChE NYU; BChE CCNY. Native of Houston Tex. US Army Air Force 1942-45, 1st Lt; decorated DFC, Air Medals. Has been with Pullman Kellogg since 1945 & currently Dir of Dev, respon for new process & mech engrg dev in fields of energy conversion, refining, petrochem & chem processes. Holds num pats & authored many articles. Chmn of Engrg Found 1963-66. Dir of Engrg Found 1958-70. Served on ed bds of I&EC Process Design & Fundamentals Quarterlies. Mbr ACS, Sigma Xi, AAAS; Fellow AIChE, AIC. *Society Aff:* AIChE, ACS, AIC, AAAS, ΣΞ.

Schrider, Leo A
Business: 7555 Freedom Ave, North Canton, OH 44720
Position: VP Engrg *Employer:* Belden & Blake Corp *Education:* BS/Petro Engrg/Univ of Pittsburgh; MS/Indus Engrg/WV Univ. *Born:* 2/21/39 and raised in Pittsburgh, PA. Field Engr for Shell Oil Co. in LA 1961-64. Spent 17 years with Federal Government mostly in Oil, Gas and Coal Research. Served in variety of positions including Mgr of In Situ Coal Gasification Project in WY, and Deputy Dir of the Morgantown Energy Research Center. Joined Belden & Blake Corp in 1981 as responsible individual for engrg development of oil and gas properties throughout US. Past Region IX Dir for SPE of AIME and served as SPE's Career Guidance Committee Chrmn and Northern WV Section Chrmn. Adjunct professor at WV Univ. Over 35 professional publications. Enjoys skiing and running. Mbr of Bd of Dir of Belden and Blake Energy Co, a publicly traded limited partnership on Amer Stock Exchange. Mbr of Bd of Dir of Belden and Blake International. *Society Aff:* SPE, OOGA.

Schriever, B A
Business: 2000-15th St, Ste. 707, Arlington, VA 22201
Position: Mgmt Consultant. *Employer:* Schriever & McKee Inc. *Education:* MA/Eng/Stanford Univ; BS/Eng/TX A&M. *Born:* Sept 1910. General, USAF Ret. Own firm Schriever & McKee Inc. Chmn of Bd. Aeronautical Engrg degree Texas A&M; MS Stanford. Served in AF 33 yrs - ret full Gen 1966. In 1954 commanded AF Western Dev Div at Inglewood Calif. Directed dev of ICBM & AF's initial space progs. Assumed command of ARDC 1959, AFSC 1961 until retirement 1966. Respon for R&D testing, engrg & procurement actions to insure USAF having best equip then & for the future. Now has own mgmt cons firm & serves as cons to indus & govt in tech & mgmt areas. Serves on many corp bds. Mbr NAE. *Society Aff:* NAE, AIAA, AFA.

Schrock, Virgil E
Business: University of California, Dept of Nuclear Engra, Berkeley, CA 94720
Position: Professor *Employer:* Univ of CA *Education:* ME/ME/Univ of CA, Berkeley; MS/ME/Univ of WI; BS/ME/Univ of WI *Born:* 01/22/26 Native of San Diego, CA. Taught mech engrg at Univ of WI (1946-48), Univ of CA, Berkeley (1948-1958) and nuclear engrg Univ of CA, Berkeley (1958-present). Res specialization is heat transfer, fluid mech and thermal aspects of nuclear reactors. Served as Asst Dean of Engrg (1968-1974). Chrmn of ASME Heat Transfer Div (1976-1977), Editor *Journal of Heat Transfer* (1978-1983). Chrmn ANS Thermal Hydraulics Div (1982-83). Consultant to government and industry. Enjoy tennis, skiing, mountain-

Schrock, Virgil E (Continued)
eering and sailing. Fellow ASME (1977). Fellow ANS (1981), Recip. 1983 ASEE Glenn Murphy Award, Japan Soc Prom Sci Res Fellow 1984; ANS Thermal Hydraulics Div. Best Paper Award 1985; ASME Heat Transfer Memorial Award 1985. *Society Aff:* ANS, ASME, ASEE, ΤΒΠ, ΣΞ, ΠΤΣ.

Schroder, Klaus
Business: 409 Link Hall, Syracuse, NY 13210
Position: Prof of Mtls Sci. *Employer:* Syracuse Univ. *Education:* PhD/Metal Physics/Univ Gottingen. *Born:* 11/1/28. Dissertation on Plastic Properties of Metals in 1954 Gottingen, W Germany. 1955-58 res officer, Tribophysics; work on yielding in Cu-alloys. 1958-60 Univ of IL at Urbana; res on specific heats. 1961-62 res assoc, Univ of Gottingen. From 1962 in College of Engg at Syracuse Univ. Now Prof of Mtls Sci. Res on thin films, elec and magnetic mtls and magnetic memories. 1966-68 guest scientist at Natl Magnet Lab. Wrote three patents, about sixty publications and two books: "Electronic, Magnetic and Thermal Properties of Solid Materials–, & "Electrical Resistivity of Binary Metallic Alloys–. *Society Aff:* ASM, APS, AAUP, ΣΞ

Schroeder, Alfred C
Home: Apt. I-114 Pennswood Village, Newtown, PA 18940
Position: Mbr of the Tech Staff. *Employer:* SRI *Education:* MS/EE/MIT; BS/EE/MIT. *Born:* Feb 1915. Univ of Berlin; RCA Victor Color TV Res Staff 1937-42; MTS RCA Res Lab Princeton NJ 1942- . 70 pats, more than half for devs in color TV; included is the pat on the shadow-mask color tube used in all commercial color TV sets worldwide. Participated in dev of the NTSC color sys; contrib to many of its fundamental concepts & decisions that went into its adoption in the U S. Publ 18 tech papers, incl one on my original theory of the mechanism of color vision. Fellow IEEE; Mbr Optical Soc of Amer, SMPTE, AAAS, Franklin Inst. Honored with 6 RCA Labs Achievement Awards for outstanding work in res; the David Sarnoff Gold Medal of the SMPTE, & the Vladimir K Zworykin Award of the IEEE. *Society Aff:* IEEE, OSA, SMPTE, AAAS.

Schroeder, Manfred R
Business: Buergerstrasse 42-44, Univ Goettingen, D-3400 Goettingen, Germany FDR
Position: Dir, Drittes Physikalisches Inst. *Employer:* Universitaet Goettingen. *Education:* PhD/Physics/Univ of Goettingen; Dipl/Physics/Univ of Goettingen. *Born:* July 1926. 'Diplom-Physiker' & 'Dr rer nat' (PhD) Goettingen Univ. Native of Germany; came to U S in 1954 to join Bell Tele Labs. Hd, Acoustics Res Dept 1958; Dir Acoustics, Speech & Mechs Res 1963. Prof of Physics & Dir Drittes Physikalisches Institut, Goettingen Univ 1969; cons to Bell Labs on Auditory Analysis & Speech Synthesis Res. Specialties: speech, hearing, architectural acoustics. 44 U S pats. Served on Natl Stereophonic Radio Ctte. Fellow IEEE, Acoustical Soc & Audio Engrg Soc (Gold Medal 1972). Mbr Goettingen Acad of Sci 1973. Foreign mbr of Max Planck Soc 1974. Computer Art Competition First Prize 1969. Mbr, Natl Acad of Engg, Wash, DC (1979). Baker Prize Award, Inst of elec and Electronics Engrs (1976). IEEE Acoustics, Speech and Signal Society (Senior Award 1979). Books: "Number Theory in Science and Communication" (Springer, Berlin 1984); "Speech and Speaker Recognition" (Karger, Basel 1985); "Chaotic Waves, Random Rays and Concert Halls" (Springer, Berlin 1985). *Society Aff:* ASA, IEEE, AES.

Schroeder, W L
Business: College of Engg, Oregon State University, Corvallis, OR 97331
Position: Assoc Dean of Engr. *Employer:* OR State Univ. *Education:* PhD/Civ Engg/Univ of CO; MS/Civ Engg/WA State Univ; BS/Civ Engg/WA State Univ. *Born:* 1/3/39. Native of Kelso, WA. Served in US Corps of Engrs, 1963-65 after grad from WA State Univ. Completed grad studies at Univ of CO & began teaching in 1967 in the Civ Engg Dept at OR State Univ. Currently Prof & Assoc Dean of Engg for Res. Past Pres of OR Sec, ASCE & OR Sec Geotechnical Engg Technical Group. Currently Secretary, ASCE Geotechnical Division Exec Ctte. Author of *Soils in* Construction published by John Wiley. *Society Aff:* ASCE, ASTM, ISSMFE, AGC.

Schubert, George H
Business: 282 Richmond St, Providence, RI 20903
Position: President/Owner. *Employer:* Schubert Heat Treating Co. *Born:* BS Chem The Citadel 1941, BS Met Engg Lehigh Univ 1955. Native of Providence RI. Earned rank of Major - Ordnance Corps 1951-55. VP Schubert Heart Treating Co 1955-72. Pres/Owner 1972- . Tr of ASM 1973-76. Also Pres & owner Sargeant & Wilbur Heat Treat Corp, 1977- .

Schubert, William L
Home: 7 Eighth St, Downers Grove, IL 60515
Position: Evaluation Dev Mgr *Employer:* Case-I.H., J1 Case Co *Education:* BS/Agr/Eng/Univ of IL *Born:* 10/04/38 Grew up on farm in IL near St Louis, MO. Early interest in machinery lead to Agri Engrg degree. Some work completed toward MS degree in Agri Engrg. Completed Univ of IL extension Mgmt Dev course. Have progressed thru various mgmt positions in design, test and devel for Agri equip. Currently Mgr Evaluation Dev for Case-IH. Enjoys outdoor activities and "do-it-yourself" projects. *Society Aff:* ASAE, SAE

Schudt, Joseph A
Business: 3920 West 216th St, Matteson, IL 60443
Position: Owner *Employer:* Joseph A Schudt & Assocs *Education:* BS/CE/Notre Dame Univ; MA/Env Systems Mgt/Governors State Univ *Born:* 4/16/38 Native of Chicago, IL. With Joseph Schudt & Assocs since 1957, starting as a field man, performed subdivision calculations and survey boundaries. Later, as a project engr in charge of design of land development projects. As a partner and Owner in 1972, assumed responsibility for operation of the Consulting Engr and Survey Firm. Was a Staff Member at Prairie State Coll and Community Prof at Governors State Univ. Past Pres of IL Society of PE and National Dir of National Society of PE. Registered Engr in IL, IN, MI, MO, WI, FL and KY. Registered Land Surveyor in IL. *Society Aff:* AWWA, ASTM, WPCF, ISPE, NSPE, ASCE

Schuermann, Allen C, Jr
Business: 322 Engineering N, Stillwater, OK 74078
Position: Prof & Head of Indus Engrg & Mgmt *Employer:* OK State Univ *Education:* PhD/IE/Univ of AR; MS/Math/Wichita State Univ; BA/Math/Univ of KS *Born:* 9/28/43 in Denver, CO. Operations research analyst, Boeing Co, Whichita KS, 1965- 1969; senior operations research analyst, Boeing Computer Services, Wichita KS, 1971; Assistant prof of industrial engrg, Wichita State Univ, 1971-1976; assoc prof of industrial engrg, Wichita State Univ, 1976-1979; assoc prof and chrmn of industrial engrg, Wichita State Univ, 1979-1984; prof and head of indus engrg and mgmt, OK State Univ, 1984-present. Research interest in educational software for microcomputers. Secy IE div ASEE. Consultant to aerospace, light manufacturing and electric utility industries. Member editorial bd of Computers and Industrial Engrg and IIE book reviewer. Enjoy sailing and racquetball. *Society Aff:* IIE, ASEE, APM, PKP, PME

Schuler, Gerald A
Business: Mohawk Bldg, Portland, OR 97204
Position: Regional Safety Engr. *Employer:* Federal Hwy Admin. *Education:* BS/ME/OR State Univ. *Born:* 2/4/24. US Navy during WW II. Cert Safety Prof. Prof Mbr Amer Soc of Safety Engrs. Regional V P 1969-75. Interim V Chmn Mgmt Div ASSE 1975-77. Mbr Pub Employee Sect Exec Ctte, Natl Safety Council 1971- . Chmn Portland Fed Safety Council 1957-58, 1976-77. Assoc of Fed Safety & Health Pros Reg VP 78- . Mbr Veterans of Safety. Regional Safety Engr Fed Hwy Admin 1955- . Respon for managing Safety Prog for engrg, const & inspection of direct fed & fed aid hwys in northwest states & Alaska. Creative Communication Award 1967. CA Reg PE Safety. *Society Aff:* ASSE, VOS, ASFE.

Schuler, Theodore A
Home: 5907 Adelia Dr, Knoxville, TN 37920
Position: Asst Dir, Dept of Engrg *Employer:* City of Knoxville *Education:* ME/CE/Univ of Louisville; BS/CE/Univ of Louisville *Born:* 07/01/34 Native of Louisville, KY. Married Jane Bandy, July 29, 1979, with 4 children: Marc, Elizabeth, Eric, Ellen. Served with Civil Engrs Corps, USNR 1957-60. Bridge and highway design with constr supervision, 1960-65, with Brighton Engrg Co. With Hensley Schmidt, Inc 1965-81. For ten yrs, supervised highway, drainage, airport, industrial park, water & sewer, and urban renewal projs in Chattanooga, TN office. In 1975, Mr Schuler was named a firm principal with responsibilities as asst vp managing the Knoxville, TN office. Work included design and constr supervision of highways, drainage, dams, and water & sewer facilities. Since 1981 has been in charge of planning and tech serv for the Dept of Engrg, City of Knoxville. Interested in sailing and metaphysical subjects. Listed in Who's Who in South and Southwest. *Society Aff:* ASCE, APA

Schulke, Herbert A, Jr
Home: 138 Borden Rd, Middletown, NJ 07748
Position: VP & Dir of Telecommunications. *Employer:* Chase Manhattan Bank. *Education:* PhD/EE/Univ of ILL; MS/Comm Engr/Univ of ILL; BS/Mil Engr/USMA West Point. *Born:* Nov 12, 1923. BS U S Military Acad West Point; MS & PhD EE Univ of Illinois. 1970 Dep Dir for Opers in the Defense Communications Agency. 1973 Dir of Communications-Electronics (J-6) Joint Chiefs of Staff. Recipient Disting Serv Medal, Legion of Merit w/OLC, Bronze Star Medal, Army Commendation Medal w/20LC, Fellow IEEE. Mbr Sigma Xi, Armed Forces Communications & Electronics Assn. Retired US Army Dec 1974 - Major General. Retired Chase Manhattan Bank Fed. 1984-Vice President. President, Aero-TELE-COMP, Inc. *Society Aff:* IEEE, AFCEA, AUSA

Schulman, James H
Home: 5628 Massachusetts Ave, Bethesda, MD 20816
Position: Consultant *Employer:* Self *Education:* PhD/Chem/MIT; BS/Chem/MIT *Born:* 11/15/15 Res Assoc MIT (1941-44) and Sr Engr Sylvania Elec (1944-46). Established res group on luminescence at Naval Research Lab (1946) where successively was Head Dielectrics Branch, Superintendent Optical Physics Div, and Assoc Dir Res for Materials and Gen Sci (1946-74). Chief Scientist London Branch Office of Office of Naval Res (1974-78) and Acting Technical Dir Office of Naval Res Washington (1978-79). Over 100 papers and 16 patents mainly on luminescence, radiation effects in solids, and dosimetry; co-author of book. Major honors: Navy Award for Distinguished Achievement in Science and Navy Distinguished Civilian Service Award. Current interests: science policy and mgmt. Hobbies: music, books, art. *Society Aff:* APS, OSA, AAAS, RESA, WAS, PSW

Schulte, Dennis D
Business: Dept of Agri Engrg, Univ of Manitoba, Winnipeg, Manitoba R3T 2N2 Canada
Position: Asst Professor. *Employer:* Univ of Manitoba. *Born:* Oct 1945 Hartington Nebr. BSc Agri Engrg Univ of Nebr; MSc & PhD Cornell Univ. Mbr Alpha Epsilon & Sigma Tau Engrg Hons. O J Ferguson Award - outstanding sr engr at Univ of Nebr 1968. Held Environ Protection Agency Fellowship at Cornell Univ. Commissioned officer US Army Res. Presently Asst Prof & Res Coord of Agri Waste Mgmt Univ of Manitoba; mbr of Expert Panel on Farm Animal Wastes - Assoc Ctte on Sci Criteria for Environ Qual - Natl Res Council. Chmn N Central Region Amer Soc of Agri Engrs. Mbr Canadian Soc of Agri Engrg, Water Pollution Control Fed, Amer Soc of Engrg Educ.

Schultz, Albert B
Business: Dept. Mech Eng. & Appl. Mech., 3112 G. G. Brown Lab, Ann Arbor, MI 48109-2125
Position: Vennema Professor *Employer:* Univ of Mich *Education:* Ph.D./Mech. Eng./Yale University; M. Eng./Mech. Eng./Yale University; B.S./Mech. Eng./University of Rochester *Born:* 10/10/33 Currently Vennema Professor of Mech Engrg and Applied Mechanics, Univ of Mich. Faculty Mbr, Univ of Delaware 1962-65, Univ of Ill, Chicago, 1965-83. Published extensively on musculoskeletal biomechanics. Served as Assoc Editor or Editorial Bd Mbr for Clinical Biomechanics, Journal of Biomechanics, Journal of Biomechanical Engrg, Journal of Orthopaedic Research, Spine. Formerly Pres of the International Soc for the Study of the Lumbar Spine and of the Amer Soc of Biomechanics, and Chairman of the Bioengineering Div of ASME and of the US National Cttee on Biomechanics. Phi Beta Kappa, NIH Javits Neuroscience Investigator. *Society Aff:* ASME, AAAS, ASB, ORS, ISSLS

Schultz, Andrew S, Jr
Home: 66 Village Walk, Ponte Vedra Beach, FL 32082
Position: Spencer T Olin Professor of Engrg, Emeritus *Employer:* Cornell University. *Education:* PhD/Ind Eng/Cornell Univ; BS-ME/Admin Eng/Cornell Univ *Born:* Aug 1913. Employed by N J Bell Telephone Co before graduate sch. US Army 1941-46; Cornell Faculty 1941- . Dean College of Engrg 1963-72; Spencer Olin Prof of Engg 1972 Emeritus, 1980. Acting Dean 1978; VP Res Logistics Mgmt Inst 1962-63; full-time Cons to Western Electric Co Engrg Advisory Ctte 1960; Operations Res Office Johns Hopkins Univ 1953; Chmn Accident Prevention Study Sect NIH 1962-67; Dir S I Handling Systems Inc; Chicago Pneumatic Tool Co; 1972-86 Logistics Mgmt Inst (Chmn Bd of Trustees 1972-84) Lexington Growth Fund; Lexington Res Fund 1970-84 Zurn Indus. 1970-1986. *Society Aff:* IIE, ASEE

Schultz, Jay W
Business: Sterling Forest, Suffern, NY 10901
Position: Manager of Research. *Employer:* Inco Research & Dev Center. *Education:* PhD/Met Eng/Univ of MI; MS/Met Eng/Univ of MI; BS/Met Eng/MI Tech Univ. *Born:* 2/9/37. in Detroit. Joined Intl Nickel's P D Merica Res Lab in 1965 as Res Metallurgist; apptd Mgr Matls Sys Sect 1970, Mgr Corrosion Sect 1972; Chem Res Mgr 1973; Mgr of Research 1978; apptd to current position of Dir in 1980 with respon for mgmt of center. Publs & patents pertain to high temperature corrosion & heat resistant alloys. P Mbr EPRI Corrosion Adv Ctte & ASM Matls Sys & Design Div, Power Act Ctte. Mbr ASM, NACE, Sigma Xi-RESA. Enjoy reading, hunting, fishing, skiing & golf. *Society Aff:* ASM, NACE.

Schultz, Roger L
Home: 15 Lafayette Ct, Greenwich, CT 06830
Position: VP. *Employer:* Robert Heller Assoc Inc. *Education:* MBA/Finance/Cornell Univ; BChE/CE/NYU. *Born:* 1/31/34. Native of NY. Worked in spectrum of admin capacities in chem process & other technically oriented industries from 1961-1972. Ran offshore consulting firm in Bermuda from 1972-1974. Assumed present post as VP, Robert Heller Assoc, Inc (1974) specializing in Ind Relations consulting in the chm process & related industries. Interests include music & yacht racing. *Society Aff:* AIChE.

Schultz, Sol E
Home: 4241 E Mercer Way, Mercer Island, WA 98040
Position: Consulting Engineer. *Born:* 12/23/00. Graduate of Dresel Evening Sch 1923; Ch Electrical Engr Port of NY Authority 1937-39; Cons Engr TVA, FPC, SEC & Power Authority State of NY 1939- 40; Ch Engr Bonneville Power Admin 1940-54; Partner & Mgr H Zinder & Assocs 1954-68; Senior Cons Engr C H 2 M/Hill 1968- . *Society Aff:* IEEE.

Schulz, Helmut W
Home: 611 Harrison Ave, Harrison, NY 10528
Position: Pres. *Employer:* Dynecology Inc. *Education:* PhD/ChE/Columbia Univ; CE/ChE/Columbia Sch of Engg; BS/Chemistry/Columbia College. *Born:* 7/10/12. Berlin, Germany. Grad from Columbia as chem engr, joined Union Carbide, 1934. Directed synthetic organic res for 35 yrs, retiring as Dir of Tech, Europe & managing dir of Carbide's European res subsidary. Directed rocket propulsion tech for Defense Dept, '64-'67, & helped plan Natl Inst of Educ, '71. Columbia Univ since

Schulz, Helmut W (Continued)
1972: Sr Res Scientist; Adj Prof; Dir, Urban Tech Ctr; Dir, Fossil Energy Lab. NSF consultant; Fellow, AIChE. Founder, pres Dynecology, Inc; founder, chrmn Brandenburg Energy Corp. Inventions: gas centrifuge for uranium enrichment; reinforced grain rocket motors; Simplex coal/biomass gasification; Hydrex regenerative railway propulsion; ethylene hydration process; 56 patents. Honors: $100,000 Atomic Energy Prize for gas centrifuge cascade. *Society Aff:* AIChE, ACS, NYAS.

Schulz, Richard B
Home: 2030 Cologne Drive, Carrollton, TX 75007
Position: EMC Consultant *Employer:* Self *Education:* MSEE/Elec Engrg/Univ of PA; BSEE/Elec Engrg/Univ of PA *Born:* May 1920 Philadelphia Pa. Res Assoc at U of Penn 1942-45; Partner Schulz & Weisbecker Co 1946-47; Owner of ElectroSearch 1947-55; Prog Dev Coordinator Armour Res Foundation 1955-61; Ch Electro-Interference Sect United Control Corp 1961-62; Ch of Electrocompatibility Boeing 1962-70; Staff Engr Southwest Res Inst 1970-74; Scientific Advisor IIT Res Inst (ECAC) 1974-83; Mgr EMC/SID Design and Text Xerox IPD 1983-87; EMC Consultant 1987- . Reg PE; hold 3 patents. Author of 75 pub papers & chapters of 2 books. Editor of IEEE Transactions on EMC. Ch of IEEE EMC Group & Internatl EMC Symposium 1968. IEEE Fellow. L. G. Comming award. IEEE Centennial Award. Recipient of 15 other awards. Bridge Life Master. *Society Aff:* IEEE

Schumacher, Earl W
Business: 8675 NW 53rd St Ste 201, Miami, FL 33166
Position: Assistant Operations Engineer. *Employer:* Metropolitan Dade County. *Education:* BS/Agri Eng/TX A&M. *Born:* June 26, 1929 Scranton Pa. Served 2 years in Army Corps of Engrs 1955- 57. Active reservist LTC EN. P Pres Fla Sect ITE 1971. 22 years 8 months with Dade County. Supervision of operational traffic progs for entire County 8 in present position 1968- . Interested in genealogical work. *Society Aff:* ITE.

Schurman, Glenn A
Home: 840 Powell St #302, San Francisco, CA 94108
Position: Corp. V.P., Production & Resources. *Employer:* Chevron Corp. *Education:* PhD/ME/CA Inst of Tech; MSc/ME/CA Inst of Tech *Born:* 9/6/22 Dr. Schurman joined Standard Oil Co of CA in 1950 and worked initially in offshore drilling, tech and petroleum production problems. He was Mgr of Production Res at the La Habra, CA lab before transferring to Chevron Oil Co in 1963. Since then, he has held various mgmt positions in LA, TX & CO before moving to London where he has been in charge of Chevron's North Sea operations. He transferred to Chevron Corporation's Home Office in San Francisco in 1981 and currently is Corporate Vice President, Production & Resources. Dr. Shurman holds a BS degree from WA State College and a PhD from CA Inst of Tech, both in Mech Engrg. *Society Aff:* ASME, SPE, NAE

Schuster, Robert L
Home: 1941 Golden Vue Dr, Golden, CO 80401
Position: Civ Engr/Geologist *Employer:* US Geological Survey *Education:* PhD/CE/Purdue Univ; MS/CE/Purdue Univ; MS/Geol/OH State Univ; Dipl/Soil Mech/Univ of London; BS/Geol/WA State Coll *Born:* 8/29/27 Native of Chehalis, WA. Geologist in arctic res with Snow, Ice, and Permafrost Res Establishment, 1952-55. Instructor, Sch of Civ Engrg, Purdue Univ, 1956-60. Assoc Prof and Prof, Dept of Civ Engrg, Univ of CO, 1960-67. Natl Sci Fdn Sci Faculty Fellow, Univ of London, 1964-65. Prof and Chrmn, Dept of Civ Engrg, Univ of ID, 1967-74. Chief, Engrg Geology Branch, US Geological Survey, 1974-79; currently civ engr/geologist, US Geological Survey Chrmn, Exec Cttes: Geotechnical Engrg Div, American Soc of Civ Engrs, 1981, US Natl Soc for Soil Mechanics and Fdn Engrg, 1981, Engrg Geology Div, Geological Soc of Amer, 1984- 85. *Society Aff:* ASCE, AEG, IAEG, GSL, ISS, MFE, USNS/SMFE.

Schutz, Harald
Home: 610 Regester Ave, Baltimore, MD 21212
Position: Retired, part time Instructor. *Employer:* Johns Hopkins U, Evening Coll. *Education:* Dr. of tech. Science - Electrical *Born:* June 24 1907 Vienna Austria. Degree Electrical Engr 1929, Doctor of Technical Sci 1934 Vienna Inst of Technology. Fellow IEEE; Mbr Engrg Soc of Baltimore, Scientific Council, Md Acad of Scis. P Chmn IEEE tech group on Aerospace & Navigational Electronics. Reg PE Md.

Schutzer, Daniel M
Business: Naval Intelligence, Washington, DC 20350
Position: Technical Director. *Employer:* Department of the Navy Naval Intelligence. *Education:* PhD/Elec Engg/Syracuse Univ; MSEE/Elec Engg/Syracuse Unv; BSEE/Elec Engg/CCNY. *Born:* April 25, 1940. Over 21 years of diversified exper in government, indus, & univ. Emphasis has been in the fields of data processing, radar, sonar, communications, & computer sci. Have previously worked for IBM, Bell Labs, Sperry Rand, & The Defense Communications Agency. *Society Aff:* IEEE, ACM, AMA, ΣΞ, ΤΒΠ, ΗΚΝ, ΠΜΕ

Schwab, Glenn O
Home: 2637 Summit View Rd, Powell, OH 43065
Position: Professional Agricultural Engineer. *Employer:* Ohio State Univ. *Education:* PhD/Agri Eng/IA State Univ; MS/Agri Eng/IA State Univ; BS/Agri Eng/KS State Univ. *Born:* Dec 1919. Native of Gridley Kansas. Served in US Army 1942-46 with rank of Capt. Teaching & Res in soil & water in Agricultural Engrg Iowa State Univ 1946-56, Ohio State Univ 1956-85. Co-author of 3 texts in soil & water engrg. Major interest in agricultural drainage. ASAE Drainage Engrg Award 1968. ASAE John Deere Medal 1987. Cons work in India, Pakistan, Yugoslavia & New Zealand & for Government & indus in the US. Enjoy woodworking, gardening, rock collections, & photography. *Society Aff:* ASAE, SCSA, AGU, ASEE.

Schwab, Richard F
Business: Allied-Signal Inc Engineered Materials Sector, P.O. Box 1139R, Morristown, NJ 07960
Position: Mgr Process Safety & Loss Prevention *Employer:* Allied Signal Inc. *Education:* MSChE/Chem Eng/NY Polytechnic Inst; BChE/Chem Eng/CCNY *Born:* 3/22/32 Native of NYC. Upon grad from CCNY in 1954 became inspector for the Factory Ins Assoc. Assumed position of Supervisor in Chem Dept in 1960. Joined Allied Chem Corp (now Allied Signal Inc) in 1969 as Loss Prevention Engr. Became Mgr, Process Safety & Loss Prevention in 1976. PE in CT, NJ, Certified Safety Profl. Elected Fellow of SFPE in 1981; Fellow of AIChE in 1984. Mbr of various NFPA Technical Cttes such as Dust Explosions, Explosion Protection Systems Comm, Elec Equip in Chem Atmospheres. Past Chrmn of AIChE Loss Prevention Ctte; Past Chrmn of AIChE Safety & Health Div (1984). Lecturer for AIChE Continuing Educ Courses. Enjoys classical music and opera. *Society Aff:* SFPE, AIChE, ACS

Schwan, Anthony V
Business: 2701 North 16th St, Ste 108, Phoenix, AZ 85006
Position: Pres *Employer:* Anthony V Schwan & Assoc Inc *Education:* BS/CE/IL Inst of Tech; Pre-Engrg//No State Teachers Coll *Born:* 8/30/31 Tony Schwan began his professional career in 1954 as a structural engr with Merritt, Chapman & Scott, Chicago; his structural consulting career began in 1960 when he joined Johannessen & Girand, Consulting Engrs, Phoenix, as structural design engr and later advanced to VP and Member of the Bd of Dirs. As Pres of A. V. Schwan & Assocs, Inc, Tony is personally involved to some degree in each of the many and varied projects handled by his organization; high-rise hotels, school complexes, commercial and industrial manufacturing plants, shopping centers, medical clinics, residential facilities, social clubs, and specialty structures. *Society Aff:* SEAA, ASCE, NSPE, EERI, ACEA

Schwan, Herman P
Home: 99 Kynlyn Rd, Radnor, PA 19087
Position: Professor. *Employer:* University of Pennsylvania. *Education:* PhD/-/Univ

Schwan, Herman P (Continued)
of Frankfurt, Germany; Dr. habil/Univ of Frankfurt, Germany; Dr. h.c./Univ. of Pennsylvania *Born:* Aug 1915 Aachen, Germany. PhD 1940, Dr habil 1946 Univ of Frankfurt, Germany. Dr. Sci h.c. 1986 Univ Pennsylvania, Specialties: Biophysics, Biomedical Engrg. Res Asst to Asst Dir Max Planck Inst Biophysics Frankfurt, Germany 1937-47; Res Scientist Philadelphia Naval Base 1947-50; Univ of Penn Medical Sch & Moore Sch of Electrical Engrg 1950- ; Prof & Dept Chmn & Head Electromedical Div 1952-73 & Dept of Bioengg, Sch of Engrg & Appl Sci 1973- . Alfred Filler Moore Prof Biomed. Elec Engrg, Sch Engrg & Appl. Sci. Leading contributor to dev of biomedical engrg in USA; 200 scientific papers & book chapters. Cons NIH. Foreign Mbr Max Planck Soc. IEEE Morlock Award, Rajewsky Prize for biophysics, Philadelphia IEEE Achievement Award, natl Acad of Engrg. Fellow IEEE & AAAS. A.V. Humboldt US Senior Scientist Award. Edison Medal IEEE 1983. Am Soc Engrg Ed Award 1983. Centennial Award, IEEE, 1984. *Society Aff:* IEEE, AAAS, Biophys Soc; BMES, German Biophys Soc.

Schwanhausser, Edwin J
Home: 11 Euclid Ave 5A, Summit, NJ 07901
Position: Retired. *Born:* 1894 Jersey City NJ. Jersey City High Sch; ME 1915 Stevens Inst of Technology; Honorary Degree- Dr of Engrg 1959 Stevens Inst. Retired-P Pres & Vice Chmn Worthington Pump & Machinery Corp. P Pres: Diesel Engine Manufacturers Assn, Compressed Air & Gas Inst, Inter-allied Foundries of NY State, Chamber of Commerce Buffalo NY. Trustee Emeritus: Stevens Inst of Technology, Asst Chmn Finance Ctte; P Pres Alumni Assn. Fellow ASME. P Mbr Council- Chmn Finance Comm. Reg PE 7563 NJ. Extensive hunter & fisherman.

Schwartz, Daniel M
Home: 2190 Washington St, Apt. 1204, San Francisco, CA 94109
Position: President. *Employer:* Foothill Engg, Inc. *Education:* BA/Sch of Engg/Stanford. *Born:* March 1913. BA Stanford Univ Sch of Engrg 1933. Bus Admin Certificate Prog 1955-62 Univ of Utah. Mech Engr Pacific Gear & Tool Works San Francisco 1933-36; Falk Corp Milwaukee 1936-40; Dev Engr Dravo Corp Pittsburgh 1940-46; Senior Vice Pres & Genl Mgr Tractor div EIMCO Corp Salt Lake City 1946-66; Prog Mgr Lockheed Missiles & Space Co Inc Sunnyvale Calif 1966-78. Consulting Engr, Foothill Engr Inc 1978. Mbr Utah Bd of Engrg Examiners 1955-58, Utah Apprenticeship Council 1959-63, Utah Coordinating Council on Higher Education 1966. Dir Community Servs Council of Salt Lake 1963-65, Utah Symphony 1963-66. Fellow ASME; Mbr SAE, American Defense Preparedness Assn. More than 50 US patents on transmissions, loaders, tractors, marine railways, & vacuum filters. *Society Aff:* ASME, SAE, ADPA.

Schwartz, Henry G, Jr
Business: 801 North Eleventh, St Louis, MO 63101
Position: VP/Corp Prin Environ. *Employer:* Sverdrup Corp. *Education:* PhD/Environmental Health Engg/CA Inst of Tech; MS/Sanitary Engg/WA Univ; BS/Civil Engg/WA Univ. *Born:* 1938. Post-doctoral Fellow 1965-66 Caltech. Joined Sverdrup & Parcel 1966. Proj Mgr for many studies, state-of-the-art evaluations, & designs for wastewater treatment, air pollution control & solid waste disposal systems. As Vice Pres & Corp Principal is responsible for all study, design, planning, construction mgmt & operations aspects of environ engrg. Mbr ASCE, NSPE, Water Pollution Control Fed, Air Pollution Control Assn; Diplomate American Acad Environ Engrs. Authored numerous publs. Bartow Award (American Chem Soc); Sigma Xi. President (1985-86) Water Pollution Control Federation; EPA Management Advisory Group-Chairman 1987. *Society Aff:* AAEE, WPCF, ASCE, NSPE, APCA.

Schwartz, Mischa
Business: 1206 SW Mudd Bldg, Columbia University, New York, NY 10027
Position: Professor. *Employer:* Columbia University. *Education:* PhD/Applied Physics/Harvard; MEE/EE/Poly Inst of Brooklyn; BEE/EE/Cooper Union. *Born:* 9/21/26. US Army 1964-46. Proj Engr Sperry Gyroscope Co 1947-51, basic studies in radar sys, sys design & engrg; Prof of Electrical Engrg Poly Inst of Brooklyn 1952-74, Hd of Dept 1961-65; Prof of Electrical Engrg & Computer Sci Columbia Univ 1974- Author of 7 books on books on communications, Signal Proc, & Comp Communication Networks, Fellow IEEE Dir, 1977-79; Mbr Bd of Governors, IEEE Communication Soc 1972-1978; VP 1982-3; Pres 1984-1985; Chmn IEEE Information Theory Group, 1964-1965; Chmn Comm C, US Natl Comm of URSI. Fellow AAAS, Mbr ACM, AAUP, Tau Beta Pi, Eta Kappa Nu, Sigma Xi. Awarded NSF Sci Faculty Fellowship 1965-66. Distinguished Visitor Award, Australian-Amer Educational Foundation July 1975. Mbr of Editoral Bds: Networks, Computer Networks, Performance Evaluation. Ed Medal IEEE 1983; Columbia Univ Great Teachers Awd 1983; included Top Ten Elec Educators, IEEE Spectrum Survey, "Outstanding Elec Engrs of the Century." Gano Dunn Award, Cooper Union, 1986. *Society Aff:* IEEE, AAAS, ACM, AAUP, ΣΞ

Schwartz, Morton D
Business: School of Engrg, Long Beach, CA 90840
Position: Prof. Computer Science and Engineering *Employer:* CA State Univ. *Education:* PhD/Engr/UCLA; MS/Engr/UCLA; BS/Engr/UCLA. *Born:* 10/11/36. B.S. 1958, M.S. 1960, Ph.D. 1964 in Engrg from UCLA. work experience at Aerospace Corp., Hughes Aircraft Corp., Rockwell Corp. and TRW Space Systems Group. Previously was Assoc Dean for Instruction in the School of Engrg and Chair of Electrical Engineering at CSULB. Primary interest in biomedical engrg, elec engrg and computer science. Past editor of Journal of Clinical Engrg. Past Pres of IEEE/Engrg in medicine & biology soc. Res interest include the dev of data base management systems and programming languages. *Society Aff:* IEEE, ASEE, AAMI

Schwartz, Murray A
Home: 30 Orchard Way North, Potomac, MD 20854
Position: Manager for Materials Research (Physical Scientist) *Employer:* Bureau of Mines US Dept/Interior. *Education:* Bs/Ceramic Engg/Alfred Univ. *Born:* 11/13/20. Native of Yonkers, NY. Served with US Navy 1944-46. Conducted res in materials science & tech, recycling, nuclear, etc. Supervised Ceramics/Cermets Res of AF Aerospace Res Lab. Was Asst Dir for Ceramics Res at IITRI. Joined Bureau of Mines in 1972 as Res Supervisor Ceramics, became Staff Ceramics Engr in Washington Headquarters in 1974 and is now Program Mnger for Materials Res. Served as exec secy on FCCST's Ctte on Matls from 1975 to 1984 for interagency coordination of Federal matls R&D prog. Served in White House Science Policy Office (OSTP) from 1982 to 84 for materials. Became ACerS Fellow in 1968; was former ACerS VP and Chrmn of Govt Liason and Honorary Members Selection Cttes. Also active in ASM & APMI. Was Chrmn of FMS CTTEE on Educ and is ACerS trustee to FMS and FMS delegate to AAES Engrg Affairs Council. *Society Aff:* ACerS, FMS, ASM, APMI, AAAS, AAES

Schwartz, Nathan
Home: 314 Baxton St, N Syracuse, NY 13212
Position: Prof. *Employer:* Syracuse Univ. *Education:* PhD/Phys/Cornell Univ; BS/Engg Phys/Univ of IL. *Born:* 11/18/22. Prof of EE, Syracuse Univ since 1965. Current interest in Control, communications, and gen engg education. GE Co Electronics Lab 1953-65; Ferrite mtls & devices, microwaves, & optics. USNR Radar maintenance officer, WWII. *Society Aff:* IEEE, APS, AAAS, ΣΞ.

Schwartz, Perry L
Home: 51 Farms Rd, Freehold, NJ 07728
Position: Principal *Employer:* Intertech Assocs *Education:* MS/Ind Engr-Comp Sci/NYU; BEE/EE/City Coll of NY *Born:* 7/29/39 Native of Brooklyn, NY. Taught Electrical Engrg at City Coll of NY 1962- 1971. Communications Engr AIL & ITT specializing in Satellite Communications. Program Mgr Western Electric Co - Established standards for CATV Industry - 1964-1967. Designed first Two-way CATV System - 1967. Dir of Engrg Warner Cable 1972. Designed implemented first 10 pay TV Systems in the US. Opened present communications consulting firm in 1974. Designed & implemented first Government owned & operated Voice/Data Tele-

Schwartz, Perry L (Continued)
phone System with Microwave By-Pass Facilities; 1984 - Atlantic County, N.J. Senior Member IEEE, Facilities Chrmn IEEE Intl Convention 1964-1969. Panelist American Arbitration Association. Adjunct faculty, engrg dept, Ocean County Coll 1980-1983. Adj Faculty- Telecommunications, Rutgers Univ since 1984. Sr Charter Mbr NARTE. Charter Member Intelligent Buildings Institute. *Society Aff:* IEEE, NSPE, ACEC, NARTE, IBI, NFPA.

Schwartz, Richard F
Business: Elec Engrg Dept, Houghton, MI 49931
Position: Professor. *Employer:* Michigan Technological University. *Education:* BEE/EE/RPI, Troy, NY; MEE/EE/RPI, Troy, NY; PhD/EE/Univ of PA, Phila, PA. *Born:* 1922 Albany NY. Signal Corps 1944-46. Teaching & Res RPI 1946-48. Engr RCA Camden NJ 1948-51. Moore Sch of Electrical Engrg Univ of Penn, res staff 1951-59, faculty 1959-73, Graduate EE Chmn 1967-72. Michigan Technological Univ Houghton Mich Prof & Dept Head Electrical Engrg 1973-79 . Reg PE in Pa & Mich. Specialization: communication, electromagnetics, acoustics. Author or coauthor of 36 publs; 3 patents. Senior Mbr IEEE; Mbr AAAS, ASEE, NSPE/MSPE, ASA, AES, Cat Acoust Soc, Eta Kappa Nu, Sigma Xi, Kiwanis, & numerous community orgs. *Society Aff:* IEEE, AAAS, ASEE, NSPE, ASA, AES

Schwartzman, Leon
Home: 1475 Remsen Ave, Brooklyn, NY 11236
Position: Dir-Prog Dev *Employer:* Unisys-Shipboard & Ground Systems Group *Education:* MS/EE/Polytechnic Inst of Brooklyn; BS/EE/Polytechnic Inst of Brooklyn. *Born:* Feb 6, 1931. Native of Brooklyn NY. Served in US Navy 1952-54. With Sperry Gyroscope since 1957 specializing in microwave & antenna design, assigned Engrg Dept Mgr for Microwave Engrg 1973, assumed additional responsibility as Prog MgrDome Radar Progs 1975. Promoted to Mgr 1982 and to Dir-Pro Dev, 1986. years of serv for the Long Island Chap for Antennas & Propagation holding many offices including Chmn. Reviewer for Transactions on Antennas & Propagation; Fellow IEEE. Enjoy painting in oils, classical music and photography. *Society Aff:* ΣΞ, IEEE, ADPA, ASOC, AFA

Schwarz, Ralph J
Home: 33 Wood Lane, New Rochelle, NY 10804
Position: Vice Dean of Engrg & Applied Science. *Employer:* Columbia University. *Education:* PhD/Elec Engg & Physics/Columbia Univ; MS/Elec Engg/Columbia Univ; BS/Elec Engg/Columbia Univ *Born:* June 13, 1922 Hamburg, Germany. Additional graduate studies at Polytechnic Inst of Brooklyn & NYU. Faculty mbr Columbia Univ 1949- , Prof of Electrical Engrg 1958- . Chmn Dept of Electrical Engrg 1958-65, 1971-72, Assoc Dean 1972-75, Acting Dean 1975-76, 1980-81, Vice Dean & Thayer Lindsley Prof 1976- . Teaching & res in system theory, noise theory, communications, & pattern recognition; cons in same areas. Chmn IEEE Circuit Theory Group 1963-65. Board of Governors, IEEE Communications Society, 1986-88, Co-author of *Differential Equations in Engrg Problems* 1954 & *Linear Systems* 1965. Dir Armstrong Memorial Res Foundation, Trustee, Associated Univ, Inc, Dir, GEM Consortium. Fellow IEEE. Reg PE in NY. *Society Aff:* IEEE, SIAM, ASEE

Schwarz, Steven E
Business: Electrical Engg Dept, Berkeley, CA 94720
Position: Prof. *Employer:* Univ of CA. *Education:* PhD/Elec Engg/CalTech; AM/Physics/Harvard; BS/Physics/CalTech. *Born:* 1/29/39. Los Angeles, CA. Experience with Hughes Res Labs, Bell Labs, IBM Res Labs. Interests include millimeter-wave devices, infrared detection, lasers, elementary engg education. Guggenheim Fellow 1971-72. *Society Aff:* IEEE, APS.

Schwarz, William H
Business: 3400 N Charles St, Dept Chem Engg, Baltimore, MD 21218
Position: Prof. *Employer:* Johns Hopkins Univ. *Education:* Dr Engg/Chem Engg/Johns Hopkins Univ; MS/Chem Engg/Johns Hopkins Univ; BS/Chem Engg/Johns Hopkins Univ. *Born:* 6/26/30. Joined chem engg at Stanford Univ in 1958 as acting asst prof. Appointed to full prof in 1967. Joined faculty of Dept of Mechanics at the Johns Hopkins Univ in 1968. Now prof and chrmn of the Dept of Chem Engg at the Johns Hopkins Univ. Res interests include rheology and non-Newtonian flows, physical acoustics, turbulence and heat and mass transfer. Enjoys fishing, sailing and collecting dwarf conifers. *Society Aff:* AIChE, APS, ASA, ARS.

Schwarzkopf, Florian
Home: 1 Half Penny Circle, Savannah, GA 31411
Position: Consultant *Employer:* Self *Education:* MS/Phys Chem/Inst of Tech; PhD/Phys Chem/Inst of Tech. *Born:* 1/16/21. in Vienna, Austria. Schooling Karlsbad, Czechoslovakia, college-Univ in Prague, Czech. After WWII left Czechoslovakia for Ecuador, SA, became Chief Chemist of La Cemento Nacional. Also taught physics and physical chemistry as Full Prof of Univ in Guayaquil. 1951 transferred to Univ of MN as Res Asst, Inst of Analytical Chemistry. 1952-1953 Res Chemist, Lithium Corp of Am, Minneapolis. 1953-86, Kennedy Van Saun Corp, Danville, PA. Mgr of Test Plant, Dir of Res, VP-Product Dev, VP-Sales & Engg. 1954-1956 leave of absence with UNKRA in Korea as technical advisor to mineral processing industry. 1986 to present self employed mineral processing consultant. Registered P.E. in Pa & Ga. *Society Aff:* ASTM, AIME, AIChE, ACS.

Schwegel, Donald R
Home: 690 Oceola Dr, Algonquin, IL 60102
Position: President *Employer:* Baxter & Woodman, Inc. *Education:* MS/CE/Purdue Univ; BS/CE/Purdue Univ *Born:* 4/28/39 Native of Chicago, IL. Served with US Army Corps of Engrs & Med Service Corps 1962-64. With Baxter & Woodman, Inc since 1964. Assoc 1969, VP 1975. Exec. V.P. 1980, President 1987. Baxter & Woodman, Inc Proj Mgr in charge of environmental engrg design & construction projects. *Society Aff:* AAEE, WPCF, AWWA

Schweikart, Herbert C
Home: Box 228 RD 4, Reading, PA 19606
Position: Retired *Born:* May 1908. Aero E 1935 Univ of Cincinnati. Employed Gilbert Assocs 1935- 73, retiring as Exec Vice Pres. Active in ASME, Mgr local sect, Editor Power Test Codes, Mbr Power Test Code Standing Ctte. Received 'Certificate of Award' & Fellow Mbrship ASME. Reg PE in Pa, NY, Mass, Ohio, & Natl Bureau.

Schweiker, Jerry W
Business: 8420 Delmar Blvd, St Louis, MO 63124
Position: Vice Pres *Employer:* Engr Dynamics Intl *Education:* PhD/T&AM/Univ of IL; MS/T&AM/Univ of IL; BS/Mining Engr/Univ of IL *Born:* 07/01/31 Served in US Army 1953-55. Technical Specialist in Structural Dynamics. McDonnell Aircraft Co 1961-65. Prof of Eng Mech St Louis Univ 1965-71. Since 1971, VP and Principal with EDI. Specializing in acoustics, noise control, vibrations and structural dynamics. Reg PE in IL and MO. Pres of Midwest Noise Council 1978. *Society Aff:* ASEE, SEE, SESA

Schweitzer, Otto R
Business: Dept of Chem Engrg, Lafayette, LA 70504
Position: Prof *Employer:* Univ of SW LA *Education:* PhD/ChE/Wayne State Univ; MS/ChE/Wayne State Univ; BS/ChE/Wayne State Univ *Born:* 5/9/32 Native of Detroit, MI. Held engrg positions in the Res Labs Div of Bendix Corp, Exxon Corp, the Missile Div of Chrysler Corp, and Goodyear Atomic Corp. Taught chem engrg at Wayne State Univ 1957-62. Author of several articles on systematic problem solving. Reg PE in LA. Mbr of Tau Beta Pi, Phi Lambda Upsilon, Sigma Xi, Omega Chi Epsilon. *Society Aff:* AIChE

Schweller, David
Home: 62 Richfield Street, Plainview, NY 11823
Position: Pres *Employer:* DBS Associates, Inc. *Education:* BS/Engg Phys/NYU College of Engg. *Born:* 9/4/33. Began nucl engg career at Combustion Engg 1955-60.

Schweller, David (Continued)
Responsible for planning, performing & analyzing reactor experiments leading to a Geneva Paper at the 2nd Intl Conf on Atomic Energy 1958. Worked for the Martin Nucl Div from 1960-1962 as Hd of Experimental Reactor Group. Participated in the initial start-up of 10 reactor facilities from 1955-1968. Began US Govt career in 1962 with the Atomic Energy Commission at Hdquarters, developing agency reactor safety prog. Served at Brookhaven from 1963-87 (AEC-ERDA-DOE) in increasingly responsible capacities. Became mgr '76 responsible for admin of the contract for the operation of Brookhaven Natl Lab. Established DBS Associates, Inc 1987. Providing consulting services in safety, security and management of nuclear facilities.

Schwenk, Francis C
Home: 6709 Brigadoon Dr, Bethesda, MD 20034
Position: Consultant *Employer:* Self *Education:* MS/ME/Univ of IL; AMP72 Harvard Bus. Sch.; BS/ME/PA State Univ. *Born:* 5/22/28. in Reading, PA; attended schools in West Lawn, PA. Employment Record: 1951- 1958 - Natl Advisory Committee for Aero, Cleveland, OH. 1958-1960 - NASA Hdquarters, Space Systems Planning. 1960-1973 - NASA/AEC Space Nucl Systems, Asst Mgr. 1973-1985 - NASA hdquarters, Dir, Res Div; Mgr, Space Utilization Systems; Mgr, SBIR Program. At present - Consultant in engrg & mgmt.

Schweppe, Joseph L
Home: 4987 Dumfries Drive, Houston, TX 77096
Position: President. *Employer:* Houston Engrg Research Corp. *Education:* PhD/ChE/Univ of MI; MS/ChE/Univ of MO; BS/ChE/MO. *Born:* Jan 1921. PhD 1950 Univ of Mich; BS, MS Univ of Mo. Native of Trenton Mo. Jr Chem Engr at TVA prior to WWII. Served with US Navy 1943-46. Now Cpt USNR-Retired. Plant tech serv with E I duPont; Senior Proj Engr with C F Braun & Co; Prof & Chmn of Mech Engrg at Univ of Houston. Founded Houston Engrg Res Corp, an info & control systems prods & servs co, in 1960. 13 tech papers; 4 patents. Reg PE in Calif & Texas. *Society Aff:* AIChE, ASME, IEEE, NSPE.

Schwinghamer, Robert J
Business: Marshall S F Ctr, Huntsville, AL 35812
Position: Director, Materials & Processes Lab. *Employer:* National Aeronautics & Space Admin. *Education:* MSM/Mgmt/MIT; BSEE/EE/Purdue. *Born:* 3/12/28. Senior prod & process engrg group leader Sylvania Electric Prods 1950-57. Supervisory Res Engr Army Ballistic Missile Agency 1957-60. R&D in aerospace with NASA 1960- . Reg P E in IN, OH & AL. Assoc Fellow AIAA; Mbr ASM & SAMPE Fellow. Hold 9 US & 7 foreign patents. More than 50 papers pub. recipient ASTME Res Medal, SAMPE Highest Award, Medal NASA Exceptional Service Medal, NASA Outstanding Leadership Medal. *Society Aff:* AIAA, ASM, SAMPE.

Schwope, Arthur D
Home: 3336 Brendan Dr, Columbus, OH 43220
Position: Consultant. *Employer:* Self. *Born:* Aug 27, 1920. B Chem E Marquette Univ, Milwaukee Wisc; MS Met Ohio State Univ, Columbus Ohio. Mbr Amer Inst of Met Engrs, Amer Soc for Metals; Fellow Amer Soc for Metals; Mbr Inst of Metals (UK).

Sciammarella, Cesar A
Illinois Inst. of Tech, Dept. of Mechanical and Aerospace Engineering, 3110 S. State St, Chicago, IL 60616
Position: Dir, Exp Stress Analysis Lab *Employer:* IL Inst of Tech *Education:* PhD/CE/IL Inst of Tech; Dip/CE/Univ of Buenos Aires *Born:* 08/22/26 Buenos Aires, Arg. Supvr Design & Stress Anal, Hormigon Elastico Inversor, 51-53; Special Assignments Engr, Ducilo, Inc, 53-54; Tech Dir, Zofra, Inc, 55- 56; Sr Res Engr Reactor Engr, Argentine Atomic Energy Comm, 56-57; Assoc Res Engr, IL Inst of Tech, 58-59; Prof Eng Sci & Mech Univ of FL, 61-67; Prof Appl Mech & Aerospace Eng, Polytech Inst of Brooklyn, 67-72; Prof, Univ of Buenos Aires, 56-57; Arg Army Eng Sch, 52-57; Lectr, Brit Sci Res Coun, 66; Visiting Prof PolytechInst of Milano, Univ of Cagliari, Italy 79; Polytech Inst of Lausanne, Switzerland, 79; Univ of Poitiers, Poiters, France 81; Consultant to USA Govt, UN & private Indus. Publications, more than 90 articles to prof journals; Patentee in the field. Honors: Sigma Xi Soc. 1966 Faculty Award; Award for Dist Service to Appl Mech Reviews, 72 ASME 72; Soc of Exp Stress Anal. Frocht Award, 80; Hetenyi Award 1981; Fellow of SESA 1981; Fellow of ASME 1982. *Society Aff:* ASME, ASTM, SESA, OSA

Scipio, L Albert, II
Home: 12511 Montclair Dr, Silver Spring, MD 20904
Position: Univ Prof of Space Sci. *Employer:* Howard Univ. *Education:* PhD/Struct Mechanics/Univ of MN; MS/Struct/Univ of MN; BCE/Civ Engg/Univ of MN; BS/Arch/Tuskegee Inst. *Born:* 8/22/22. Profl career extends over 41 yrs as engr, researcher, consultant, & teacher, specializing in applied mechanics. Academic appointments include those at Univ of MN, Cairo Univ, Tuskegee Inst, Univ of Puerto Rico, Univ of Pittsburgh, and Howard Univ. With Howard Univ since 1967. Appointed to Univ Prof Chair in 1970. Mbr of Army Sci Bd, Bd of Visitors of AF Inst, in addition to consultants including the Natl Sci Fdn, Dept of the Army, & Dept of Housing and Urban Dev. Academic mbrships include several engg & scientific honor socs. D B Steinman Award 1958. Publications include eight books. *Society Aff:* AIAA, CMH, APS.

Scisson, Sidney E
Business: 1401 S Boulder, Tulsa, OK 74119
Position: Chrmn of the Bd. *Employer:* Fenix & Scisson, Inc. *Education:* BS/Gen Engg/OK State Univ. *Born:* 2/4/17. Native of Danville, AR. 1948 - Founded Fenix & Scisson, Inc; presently Chrmn of the Bd. Designed and supervised the construction of first mined LPG storage cavern. Under his direction the firm has designed and constructed over 90% of all mined underground storage in US; has overseen dev of engg and construction services for both the US Govt and private industry. 1945-1948 with Pate Engg Tulsa, OK; 1942-1945 US Naval Reserve, Annapolis, MD; 1939-1943 Served with US Corps of Engrs, Tulsa, OK. Mbr-Natl Academy of Engg; OK State Univ Engg Hall of Fame; Distinguished Alumni Award from OK State Univ and AR Tech Univ. *Society Aff:* AIME, ASCE, OSCE, NSPE.

Scoble, Edward J
Home: 6020 Oakwood Drive, Apt. 2B, Lisle, IL 60532
Position: Director Pharmaceutical Technology *Employer:* Brown & Root Inc, Chicago, ILL 60148 *Education:* BSChE/Chem Engg/NCE; MAdE/Admin Engg/NY Univ. *Born:* July 16, 1918. US Navy 1941-47 - to Lt Cdr. Proj Engr Allied Chem Co 1947-51. Amer Cyanamid Co 1951-59 , incl Lederle Labs, J H Breck Inc; Consumer Prods Div; Shulton Inc Internatl Div - Engr to Mgr of Manufacture, Div V P. Reg P E NY, Iowa. Fellow AIC. *Society Aff:* AIChE, AIC.

Scofield, Gerald A
Home: PO Box 48, RD 4, 17 Lakeside Dr, Clarks Summit, PA 18411
Position: Pres. *Employer:* Sauquoit Industries Inc. *Education:* BS/CE/Univ of TX. *Born:* 1/27/31. Native Los Angeles CA. Served 3 yrs in US Marine Corps. Technical Service Engr, Houston Plant Rohm & Haas Co. Production Mgr, Warren RI Plant Rohm & Haas. Marketing Product Mgr, Rohm & Haas. VP Sauquoit Fibers Co, Pres & principal Sauquoit Industries Inc since 1977. *Society Aff:* AIChE.

Scofield, Gordon L
Home: P.O. Box 1085, Rapid City, SD 57701
Position: Pres *Employer:* SD School of Mines & Tech Foundation *Education:* PhD/Mech Eng/Unif of OK; MS/Mech Eng/Univ of MO-Rolla; BS/Mech Eng/Purdue Univ. *Born:* 9/29/25. Teaching pos at S Dak State Univ., Univ of Mo-Rolla, Mich Tech Univ as Prof & Hd of the Mech Engrg-Engrg Mech Dept, SD Sch of Mines & Tech as Dist Prof, ME & VP and Dean of Engrg, & currently SD Sch of Mines & Tech as Pres. Author tech papers in energy conversion & thermal radiation. Engrg cons in R&D for indus & gov. Pres SAE 1977, Bd of Dir SAE 1966-68, Chrmn St Louis Sect SAE 1962-63; Bd of Dir ECPD 1973-75; Outstanding Alumni

Scofield, Gordon L (Continued)
Award Univ of Mo-Rolla 1975. Tau Beta Pi, Pi Tau Sigma, Phi Kappa Phi, Sigma Xi. *Society Aff:* SAE, ASME, ASEE.

Scordelis, Alexander C
Home: 724 Gelston Pl, El Cerrito, CA 94530
Position: Prof of Civil Engrg. *Employer:* Univ of Calif - Berkeley. *Education:* SM/Civil Engg/MA Inst of Tech; BS/Civil Engg/Univ of CA, Berkeley. *Born:* Sept 1923. Native of San Fran, Calif. Army Corps Egrs 1943-46. Awarded Bronze Star, Purple Heart for service in ETO. On fac at U C Berkeley since 1949. Asst Dean 1962-65; V Chmn 1969-73. Teaching & res in fields of struct theory, shell structs, bridge structs, reinforced & prestressed concrete. Cons on long span shell roofs, bridges, spec structs & computer applications. Fellow ASCE, ACI; Mbr IASS, SEAONC. Mbr of cttes on Concrete Shells, Concrete Bridges & Deflections. ASCE Moisseiff Award 1976 and 1981, ASEE Western Elec Award 1978, Mbr Natl Acad of Engg 1978. *Society Aff:* ASCE, ACI, NAE.

Scorziello, Louis J
Business: 380 Green St, Cambridge, MA 02139
Position: President. *Employer:* Scorziello Assocs Inc. *Education:* Assoc/ME/Franklin Tech /Inst. *Born:* Feb 6, 1925. Franklin Tech Inst. Native of Boston Mass. USN, Motor Torpedo Boat Squadron 1942-46. 1950-59 as Sr Engr for City of Boston School Dept. 1959- , involved in design of HVAC & Fire Protection sys. Mbr ASHRAE & NSPE. Assumed current respon as Pres of Scorziello Assocs Inc in Apr of 1970. Guest lectr Elec Council of New England Energy Conservation & Energy Recovery. *Society Aff:* ASHRAE.

Scott, Alexander R
Home: 420 Commonwealth Drive, Warrendale, PA 15080
Position: Exec, Dir. *Employer:* TMS *Education:* MA/Personnel Psch/Rugers Univ; BS/History/VMI. *Born:* 6/15/41. *Society Aff:* TMS, ASM.

Scott, Charley
Home: PO Box 3834, Mississippi State, MS 39762
Position: Associate Vice President for Academic Affairs *Employer:* MIssissippi St Univ *Education:* PhD/Heat Transfer/Purdue Univ; MSME/Mech Engg/GA Inst of Tech; BSME/Mech Engg/MS State Univ. *Born:* 6/10/23. Native of Starkville, Miss. Asst Engr of Manhattan Proj 1944-46. Taught & held res pos with Miss State Univ, Meridian Jr Coll, & W Vir Univ 1946-63. With the Univ of Alabama in admin assignments 1963-84. Fellow ASME 1969. Chmn ASME Sect 1969 MSU Dir Intercollegiate Athletics 1984-1985; Dir Special Projects 1986-Present; V P Acad Affairs 1986; Assoc V Pres Acad Affairs 1987-Present; Chief Administrative Officer of Graduate School 1987-Present. *Society Aff:* ASME.

Scott, Don
Business: 2638 Delta Ln, Elk Grove Village, IL 60007
Position: Pres. *Employer:* Fermco Biochemics Inc. *Education:* PhD/Enzynology/IIT; MBA/Bus Adm/Univ of Chicago; MS/Enzymology/Cornell Univ; BS/Fermentation/ Cornell Univ. *Born:* 7/8/25. *Society Aff:* AIChE, ACS, AACC, ASQC, IFT, ΣΞ, ΦΛΥ.

Scott, Donald S
Business: Dept of Chem Engg, University of Waterloo, Waterloo Ontario, Canada N2L 3G1
Position: Prof. *Employer:* Univ of Waterloo. *Education:* PhD/Chem Engg/IL; MSc/ Chem Engg/Alberta; BSc/Chem Engg/Alberta. *Born:* 12/17/22. Edmonton, Alberta, Canada. Worked with Imperial Oil Ltd. in Northwest Territories and with Natl Res Council of Canada in Ottawa. From 1949-64 was Asst and Assoc Prof at Univ of British Columbia; Chrmn of Dept of Chem Engg at Univ of Waterloo, 1964-69, Acting Dean of Faculty 1969-70. Assoc, Dean of Engineering, 1980-. VP, 1969-70, and Pres, 1970-71, of Canadian Soc for Chemical Engg. About 70 res publications in fields of reactor engg, two-phase flow, hydrometallurgy. Present res interests in energy conversion (pyrolysis processes) and multiphase reactors. Also initiated teaching, and author of text, on technical innovation. Enjoy golf and philately. *Society Aff:* AIChE, CSChE, ACS.

Scott, Frank M
Business: 75 S Sixth Ave, Unit 105, La Grange, IL 60525
Position: Conslt Engr *Employer:* Self employed *Education:* BS/Elec Engg/Univ of NB. *Born:* 12/27/16. 1940-68 Allis-Chalmers in Milwaukee & Chgo as Motor-Generator design Engr, Hd of Special Motor-Generator Groups, Regional Specialist for Heavy Equip & Mgr W Central Area. 1942-46 Lt Col overseas with Corps of Engrs & Genl Staff Corps. 1968 Harza Engrg Co: VP 1976, Sr Assoc 1976, Assoc 1973; Hd, Mgmt. Opns Elect UKL Section 1968-71, Asst to Dir, PW & Energy Mgmt Group 1971-72, Hd, Electric Utilities Section 1972-73, Dir PW & Energy Resources Mgmt Group 1973-79, Opm & area Mgr, USA & Canada and Asst Dir, Business Dev. OPM 1979-80, V Pres, USA & Canada Area Mgmt Group 1980 to 1984. Fellow ASME, V P ASME Region VI 1977-79, Mbr ASME Exec Comm of Council 1977-79. Chrmn ASME Comm on Plan & Org 1980-81, & Mbr 1979-81, Mbr ASME Bd of Governor 1982-83, Pres ASME 1983-84, Sr Mbr IEEE, Mbr ANS, NSPE, & Western Soc of Engr (Pres 1962-63). *Society Aff:* ANS, WSE, ASME, IEEE, NSPE.

Scott, James L
Business: PO Box X, 5500, C-103, Oak Ridge, TN 37830
Position: Prog Mgr. *Employer:* Oak Ridge Natl Lab. *Education:* PhD/Met Engg/Univ of TN; MS/Met Engg/Univ of TN; BS/CE/Univ of TN. *Born:* 5/22/29. Memphis, TN; son, the late Hugh Maury & Annie Katherine (Wadley) Scott. Married Nancy Jane Bolinger, Aug 8, 1953. Children: Marshall Stuart, Dana Katherine. Instr, Chem Engg, the Univ of Tn, Knoxville, TN, 1953-56. Researcher, Nuclear Fuels, Oak Ridge Natl Lab, 1956-58; Group Leader, Fuels Evaluation, 1958-66; Supervisor, Ceramics Tech, 1966-74; Mgr, Fusion Energy Mtls, 1974-. Mbr, US AEC High Temperature Fuels Committee, 1960-70. Chrmn, Oak Ridge Chapter, ASM, 1963-64. Fellow, ASM, 1980. ANS: Chrmn, Mtls Sci & Tech Div, 1976-77; Fellow, 1977; Chrmn, Natl Prog Committee, 1977-79; Exceptional Service Award, 1980. Patentee in field. Contribute numerous articles to profl journals. Episcopalian. Pres, Bearden PTA, Knoxville, 1962-63. *Society Aff:* ANS, ASM, ACS.

Scott, John W
Home: Box 668, Ross, CA 94957
Position: VP (Retired) *Employer:* Chevron Res *Education:* BS/Chem/Univ of CA, Berkeley; MS/Chem Engrg/Univ of CA, Berkeley *Born:* 05/27/19 Native of Berkeley CA. US Army Ordnance Dept, Europe and Asia,1941-1946. Various res and dev position, Chevron Res Co, 1946-67. VP Chevron Res, 1967-84. Retired 1984. Fields: Hydroprocessing, reforming, cracking, catalysis, synfuels, organization and mgmt of res. 62 patents, 20 publ. Chrmn res, Data Info Services, Amer Petroleum Inst 1971-5, 1979-80. Chrmn advisory bd, chem engrg dept, UC Berkeley, 1972-75. Awards ctte AIChE, 1979-84. Elected Natl Academy of Engrg 1982. AIChE Chem Engrg Practice Award, 1978. Fellow AIChE, AAAS. *Society Aff:* AIChE, ACS, AAAS, API, NAE

Scott, Lester F
Business: 777 N Blue Pkwy, Lee's Summit, MO 64063
Position: Sr Staff Qual Engr. *Employer:* Western Elec Co Inc. *Born:* Mar 1926. BS Univ of Arkansas 1951. Joined Hercules Inc as Qual Control Supr. In 1958 accepted pos of Qual Control Engr with Western Elec. Was promoted to Sr Qual Engr 1961 & to present pos in Oct 1974. Have dev num qual control techs & sys for electron device manufacture. Now serving as cons on qual & costs at Kansas City Works & other Western Elec locations. Have dev & taught SQC courses at Kansas Univ & at Western Elec. Have presented papers at QC seminars & tech mtgs. Cert Qual Engr, Reg Qual Engr Calif. P Chmn Kansas City Sect ASQC - a Past ASQC V P. Recently complete second term as Dir-at-Large. Hobbies: golf & fishing.

Scott, Norman L
Business: 1701 E Lake Ave, Glenview, IL 60025
Position: President. *Employer:* Consulting Engrs Group Inc. *Education:* BSCE/Civ Engg/Univ of NE. *Born:* 10/17/31. Reg PE FL, IL, MD, MN, VA, TX. Struct Engr IL. Sales Engr R H Wright & Son, Ft Lauderdale, FL 1956-58; Mgr Wright of Palm Beach, W Palm Beach FL 1958-59; Exec Secy Prestressed Concrete Inst Chicago, 1959-63; Gen Mgr Wiss, Janney Elstner & Assocs, Northbrook, IL 1963-67; Pres Consulting Engrs Group Inc Glenview, IL 1967-. USAF 1954-56. *Society Aff:* ASCE, NSPE, ACI, ACA, ACEC, PCI, PTI

Scott, Norman R
Business: Room 1215, EECS Bldg, 1301 Beal St, Ann Arbor, MI 48109-2122
Position: Prof of Elec Engrg and Computer Science *Employer:* Univ of Michigan. *Education:* PhD/Elec Engrg/Univ of IL; SM/Elec Engrg/Mass Inst of Tech; SB/Elec Engrg/Mass Inst of Tech *Born:* May 15, 1918 N Y. 2nd Lt to Major 1941-46 Army Signal Corps Labs at Wright Field, Ohio. SB and SM, MIT 1941; Ph.D, Univ of Illinois, 1950. Fac mbr Dept of Elec & Computer Engrg, The Univ of Mich 1951- . Editor IEEE Trans on Computers 1961-65. Author: Electronic Computer Tech 1970. Computer Number Sys and Arithmetic, 1984. Fellow IEEE. Special int: electronic computer sys architecture. *Society Aff:* IEEE, ACM

Scott, Robert A, IV
Business: 1530 Polk St, Corinth, MS 38834
Position: Owner *Employer:* Scott Engrg Co *Education:* MS/Sanitary Engr/MS State Univ; BS/Civil Engr/MS State Univ *Born:* 08/22/49 Worked as res asst on dev of mathematical model of approximately 50 river- miles of Tombigbee River for Weyerhaeuser while in grad school. Began employment with Scott Engrg Co full-time in 1975 as junior partner. Became owner of company in 1986. The company employs 8 and offers consulting services in cadastral and topographic surveys, water and sewer facilities design, highway and bridge design, materials testing, and soil investigations. Natl Dir of SAME 1981-83. Elected to grade of Fellow in SAME in 1981. *Society Aff:* NSPE, SAME, WPCF, NFIB

Scott, Roderic M
Home: 240 Chestnut Hill Rd, Stamford, CT 06903 *Employer:* Retired *Education:* PhD/Astronomy/Harvard Univ; MA/Astronomy/Harvard Univ; PS/Physics/Case. *Born:* Apr 9, 1916 Sandusky Ohio. Instr Phys Vaderbilt Univ 1941; Res Assoc Harvard 1942-45; Res Physicist Sharples Corp Phila 1946-49; with Perkin Elmer Corp 1948- : V P 1965-76 , Ch Scientist 1965-69, Tech Dir Optical Tech Div Danbury Ct 1969-1980 retired. Bd Dir Perkin Fund. Fellow Optical Soc Amer (Dir, David Richardson Medal 1973); Mbr US Natl Comm Internatl Commn Optics (V P 1970- 72), Phys Soc, Soc Photogrammetry, Soc PhotoOptical Instrumentation Engrs (Treas 1971-73, Exec V P 1973-75, Fellow 1975-). Presidents Medal 1986. *Society Aff:* OSA, SPIE, APS, AAAS

Scott, Roger M
Business: RFD 2, Box 144, Peterborough, NH 03458
Position: Wire Mill Consultant. *Employer:* Self-employed. *Education:* BS/Engg/Brown Univ. *Born:* Apr 14, 1905 Worcester Mass. With New England Butt Co, Providence R I 1928-48 as Ch Engr & Sales Mgr. With Morgan Const Co 1948-76, Ch Engr & Sales Mgr in Wire Drawing Machinery Dept. Hold sev pats on Wire & Textile machinery. During WW II dev pkging equip for field telephone wire used by the Signal Corps US Army. In 1952 dev first commercial Dead Block for the wire indus in the USA. Mbr ASME 1932, Fellow 1962, Life Fellow 1970. Mbr ASME Natl Nominating Ctte 1958; served as Chmn of the Worcester & Providence Sects of ASME. Mbr Providence Engrg Soc, Worcester Engrg Soc; Reg P E NH, Mass, RI. *Society Aff:* ASME.

Scott, Ronald F
Business: Mail STA 104-44, California Institute of Technology, Pasadena, CA 91125
Position: Professor of Civil Engg. *Employer:* CA Inst of Tech. *Education:* ScD/Soil Mechs/MIT; SM/Civil Engg/MIT; BSc/Civil Engg/Glasgow Univ Scotland. *Born:* Apr 1929. Married; 3 sons. Soils Engr, Arctic Const & Frost Effects Lab, Corps of Engrs 1955-57 Racey, McCallum and Assocs Toronto 1957-58. Caltech 1958- . Books: 'Principles of Soil Mech' Addison Wesley 1963, 'Soil - Mechs & Engrg' McGraw-Hill 1968, 'Geological Hazards' second edition Springer Verlag 1978, "Foundation Analysis–, Prentice-Hall, 1981. Mbr U S Natl Acad of Engrg. Erskine Fellow, 1981, Newcomb Cleveland Award 1976, Guggenheim Fellow, Churchill Fellow 1972 - 1973. Norman Medal ASCE 1972. Walter Huber Res Prize ASCE 1969. Middlebrooks Award ASCE 1982, Terzaghi Lecturer ASCE 1983, Ranking Lecturer, British Geotechnical Society, 1987. *Society Aff:* AGU, ASCE, NAE

Scottron, Victor E
Business: Box U-37, Storrs, CT 06268
Position: Prof of Civ Engrg (Emeritus) *Employer:* Univ of CT *Education:* DrEng/Envir Engrg/Johns Hopkins Univ; MS/Applied Mech/Columbia Univ; BS/Nav Arch/Webb Inst *Born:* 5/6/15 Taught at Cooper Union 1939-41, Columbia Univ 1943-48, Univ of CT 1948-85. Univ of CT; Assoc Dean of Engrg, 1965-69; Dir, Inst of Water Resources, 1974-80; Dir, Seagrant, 1980-85. Asst Nav Arch, Taylor Basin, 1941-43. Visiting Prof, Delft, Metherlands, 1971; Padova, Italy, 1977. Natl Pres, Chi Epsilon, 1966-68. VP and Trustee, Antiquarian and Landmarks Soc CT, 1971- present. Mbr: Sigma Xi, Tau Beta Pi, Phi Kappa Phi. Author: papers on water resources. PE CT & NY. *Society Aff:* ASCE, SNAME, CSCE, IAHR, PIANC.

Scoville, Loren P
Home: 1175 Archer St, San Diego, CA 92109
Position: Retired. *Education:* MS/ChE/CA Inst of Tech; BA/Phys/Univ of Redlands. *Born:* 2/7/06. Sr Engr, Texaco, Inc 1931-44; VP, Jefferson Chem Co 1944-54; Dir of Engg, then Divisional Gen Mgr, then Dir of Dev, Diamond Alkali Co 1954-63; Pres AZ Agrochemical Corp 1963-67. At present a volunteer consultant to small business as a mbr of the Service Corps of Retired Executives. Past mbr of the council of AIChE, also Chrmn of the NY Sec & Natl Prog Chrmn. Past dir of the Natl Agri Chem Assn. *Society Aff:* AIChE.

Scriven, L E
Business: Dept of Chem Engrg & Mtls Sci, 421 Washington Ave SE, Minneapolis, MN 55455
Position: Prof, Consultant, Exxon Fac Fellow *Employer:* Univ of MN *Education:* PhD/ChE/Univ of DE; BS/ChE/Univ of CA, Berkeley *Born:* 4/11/31 Shell Dev 1956, Univ of MN 1959. Teaching: chem engrg principles, lab, mtls sci, science of porous media, res participation; postgrad fluid mech, thermodynamics, interfacial phenomena, appl computational math, res and reporting. Res: capillary hydrodynamics, drop dynamics, coating flows, coating rheology and calendering (with C. W. Macosko), wetting hydrodynamics and drying (with H. T. Davis); computer simulation by molecular dynamics, Monte Carlo, and finite element methods; porous media science, cryomicroscopy, microstructured fluids and mechanisms of enhancing petroleum recovery (with H. T. Davis); surfactant colloids (with W. G. Miller and H. T. Davis). Service: Gen Res Ctte chrmn; res consultant, NAE and NRC committees, technical expert for UN Ind Dev Organization. *Society Aff:* AIChE, APS, SPE, Chem Soc, SIAM, Soc Rheol, TAPPI

Sculley, Jay R
Business: Washington, DC 20310
Position: Asst Secy of Army for Res Devel & Acqu *Employer:* Pentagon *Education:* PhD/Env Engg/Johns Hopkins Univ; MSE/Env Engg/Johns Hopkins Univ; BS/Civil Engg/VMI. *Born:* 8/6/40. Native of Englewood, NJ. Served in USAF, 1962-65. Design engr Ft E I duPont Co. With VA Military Inst since 1970, except for one yr as gen mgr, Corrugated Services, Inc, Dallas, TX. Former Pres, VA Intercollegiate Soccer Assoc. Enjoys stamp collecting and photography and skiing. Currently Asst Sec of the Army for Res Dev and Acquisition. *Society Aff:* ASCE, ORSA, SAME.

Scully, Marlan O
Business: Optical Sci Ctr, Tucson, AZ 85721
Position: Prof of Phys & Optical Sci. *Employer:* Univ of Arizona. *Education:* BS/Engg Physics/Univ of WY; MS/Physics/Yale; PhD/Physics/Yale. *Born:* Aug 1939. MS & PhD Yale, BS Univ of Wyoming. Fac of Yale, MIT & Univ of Arizona. Presently Prof & Dir of Lab for Quantum Optics, Univ of Ariz and Prof of Optics, Italian Natl Inst Optics (Florence). Cons to Sci Applications Inc, Litton Indus, US Army, USAF, Los Alamos Sci Labs, United Technologies. Rice Univ (E E), NTSU (Physics), Univ of Colo (Physics). Adolph E Lomb Medal (Optical Soc of Amer), Guggenheim & Sloan Fellowships. Fellow AAAS & Optical Soc of Amer and Amer Phys Soc. Publ extensively in textbook & prof literature. Other prof honors: Jt Council on Quantum Electronics, Internatl Comm on Optics. Hobbies: ranching & sport judo. *Society Aff:* AAAS, OSA, APS.

Sculthorpe, Howard J
Home: 1707 N Livernois Rd, Troy, MI 48084
Position: Chief Engr; retired *Employer:* Sperry Vickers Div/Sperry Corp. *Education:* MBA/-/MI State Univ; ME/-/Stevens Inst. *Born:* May 5, 1934. 1956 ME Stevens Inst of Tech, Hoboken N J. Worked for Douglas aircraft 1956-59. Joined Vickers 1959 at Torrance Calif, serving in var engrg pos for Marine & Ordnance Dept, Indus Div & N Amer Group. Was respon as Ch Engr for design, dev & engrg support of piston pumps, motors & transmissions for the mobile & indus fields. Mbr Amer Soc for Metals, Soc of Automotive Engrs, Fluid Power Soc & Amer Powder Met Inst. Retired. *Society Aff:* SAE, ASM, FPS.

Scurlock, Arch C
Business: 123 N Pitt St, Alexandria, VA 22314
Position: President. *Employer:* Research Indus Inc. *Education:* ScD/Chem Engg/MIT; MS/Chem Engg/MIT; BS/Chem Engg/Univ of TX; BA/Physics/Univ of TX. *Born:* 1/29/20. in Beaumont, TX. BS Chem Engg Univ of TX 1941, BA Physics 1941; MS MIt 1943, ScD 1948. Asst dir chem Engg Res Assocs 1948-49; Pres Atlantic Res Corp Alexandria VA 1949-62, Chmn Bd 1962-65, Dir 1965-67; Pres, Dir Res Indus Inc Alexandria VA 1968- ; Chrmn, Trans Technology Corp (now trading on the Amer Stock Exchange); Dir, Com-Dev, Inc; Dir, Halifax Engg, Inc. *Society Aff:* AAAS, ACS, AICE, AMS, AIAA.

Scutt, Edwin D
2060 Maplewood Ave, Abington, PA 19001
Position: Retired *Education:* Doc. Mech Engineer/Power Plant Design/Rensselaer Poly. Int.; BS/ME/RPI *Born:* 02/16/11 Ph.D., Mech Engrg, Rensselaer Polytechnic 1935. Joined Leeds & Northrup Co 1935. 1943-56 in charge of Steam Power Section; 1956-59 Head of Steam & Process Industries Section, Market Devel Div; 1959-69 reported to VP responsible for marketing & tech aspects of L&N's Control Systems. 1969-75, reported to VP International Operations, working in marketing & tech aspects of L&N's DEB systems, Japan & Australia. During 40 years of active work at L&N has been concerned with application requirements for automatic control & instrumentation in the power field. Awarded Instrument Society of America Philip T. Sprague Award for technical contributions & pioneering efforts in adapting new principles of automatic control to once-through steam electric units for pressure operation, 1964. Became ASME Fellow for his contributions in Mech Engrg in 1969. Mbr of Tau Beta Pi & Sigma Xi. Reg PE in PA. *Society Aff:* ASME, ISA

Seabloom, Robert W
Home: 9707 41st St Pl NE, Seattle, WA 98115
Position: Prof Civil Engg. *Employer:* University of Washington. *Education:* MSCE/San Engr/Univ of WA; BSCE/Civil Engg/Univ of WA. *Born:* July 18, 1924 Tacoma Wash. BSCE & MSCE Univ of Wash. In US Navy 1943-46. Worked for Seattle Engrg Dept & Wash St Health Dept prior to joining faculty of Univ of Wash 1954. Areas of interest include water supply & treatment, waste water collection & treatment & solid waste mgmt. Ser as Secy, V Pres & Pres Seattle Section ASCE. *Society Aff:* ASCE, AWWA, WPCF, AEEP, APWA, AAEE.

Seaborg, Glenn T
Business: Lawrence Berkeley Lab, Berkeley, CA 94720
Position: Prof & Assoc Dir L Berkeley Lab. *Employer:* University of California. *Education:* PhD/Chemistry/US-Berkeley; AB/Chemistry/UCLA. *Born:* Ishpeming Mich. AB 1934 UCLA. Univ of Calif, Berkeley: PhD 1937; Res Assoc 1937-39; on faculty 1939- ; Chancellor 1958-61; designated Univ Prof of Chem 1971; Assoc Dir Lawrence Berkeley Lab 1954-61 & 1972- , & Head Nuclear Chem Res 1946-58 & 1971-75. Co-discovered 10 transuranium elements plus numerous isotopes used in medicine & industry. Nobel Prize for Chem 1951; AEC Enrico Fermi Award 1959. During WWII headed Plutonium Chem Work Univ of Chicago Met Lab (Manhattan Dist). Chmn US AEC under 3 presidents 1961-71. Over 45 hon degrees including D Eng, Mich Tech Univ; Sci & Engrg Award Drexel Inst of Tech 1962; Wash Award Western Soc of Engrs 1965; Foreign Mbr Swedish Acad of Engrg Sci. *Society Aff:* ACS, NAPA, AAAS, NAS.

Seaburn, Gerald E
Business: P.O. Box 23184, 5510 Gray St, Suite 118, Tampa, FL 33623
Position: Dir, VP, & Branch Manager *Employer:* Law Environmental, Inc *Education:* PhD/CE/GA Inst of Tech; MS/CE/Univ of Cincinnati; BS/CE/Drexel Univ; BS/Engrg/Geneva Coll *Born:* 3/2/40 Gerald E. Seaburn Grad from Geneva Coll and Drexel Univ with BS degrees. He received a MS degree in Civ Engrg from the Univ of Cincinnati in 1965 and a PhD degree from GA Inst of Tech in 1976. His dissertation res was conducted in the area of urban stormwater runoff. Dr. Seaburn is a Reg PE in FL & NY. His profl career has included many studies of urban hydrology, stormwater mgmt, groundwater supply, pollution and effluent disposal on NY, GA & FL. He has taught courses in hydrology, urban hydrology, hydrogeology and engrg, and has a long list of publications on these topics. He is a mbr of Chi Epsilon, Sigma Xi. *Society Aff:* ASCE, FES, ASAE, AGU, AWRA, AAAS, WPCF

Seader, J D
Business: MEB 3290, University of Utah, Salt Lake City, UT 84112
Position: Prof, Chem Engg Dept. *Employer:* University of Utah. *Education:* BS/ChE/Univ of Ca, Berkeley; MS/ChE/Univ of Ca, Berkeley; PhD/ChE/Univ of WI, Madison. *Born:* Aug 16, 1927 San Francisco. Group Supr Chevron Res Corp 1952-59; developed a widely used vapor-liquid equilibrium correlation for hydrocarbons; Principal Scientist for Rocketdyne 1959-66; conducted thermal protection res for all rocket engines of Saturn vehicle; Prof of Chem Engrg Univ of Idaho 1966-67 & at the Univ of Utah 1967- . Disting Teaching Award 1975. AIChE Annual Institute Lecture 1983. Dir of AIChE 1983-85. Fellow of AIChE. Assoc. Editor of IEC Research. Has served as tech cons to 25 companies. Author of 5 books & 80 articles; current res in tar sands, process synthesis and design. *Society Aff:* AIChE, ACS

Seale, Robert L
Business: Nuclear Engrg Dept, Tucson, AZ 85721
Position: Prof of Nuclear Engrg/Head of Dept. *Employer:* Univ of Arizona. *Education:* PhD/Physics/Univ of TX; MA/Physics/Univ of TX; BS/Physics/Univ of Houston *Born:* March 1928. Native of Rosenberg Texas. Employed by Genl Dynamics/Fort Worth 1953-61 in various engrg capacities, including Ch of Nuclear Design & Opers. Prof of Nuclear Engrg at Univ of Arizona 1961- , Head of Dept of Nuclear Engrg 1969- . Tech interests include fission reactor sys analysis & nuclear safety. *Society Aff:* ANS, APS, NSPE

Searl, Edwin N
Home: 932 Kevin Rd, Knoxville, TN 37923
Position: VP (retired). *Employer:* Insurance Services Office. *Education:* BS/Fire Protection Engrg/IL Inst of Tech. *Born:* 12/31/13. Aurora, IL. Grad IIT in 1935. Employed Missouri Inspection Bureau, St Louis, 1935-48. Western Actuarial Bureau, Chicago, 1948-71. Gen Mgr 1963-71. Directed operations of fire insurance rating

Searl, Edwin N (Continued)
bureaus in 20 midwestern states, Insurance Services Office, NY, VP 1971-76. Responsible for commercial property and multi-line insurance operations countrywide, including all ISO engrg activities. Active in tech ctte activities of Soc of Fire Protection Engrs and Natl Fire Protection Assoc. Mbr, Underwriters Labs Fire Council. Sponsor, Fire Dept Instructors Conference. SFPE Executive Ctte, 1961-66. NFPA Standards Council and Bd of Dirs, 1970-76. SFPE Finnegan Award, 1970. Elected Fellow, SFPE, 1979. Enjoys music, gardening, hiking. *Society Aff:* SFPE.

Sears, John T
Business: Dept Chem Engrg, Morgantown, WV 26506
Position: Prof - Chem Engrg *Employer:* West Virginia Univ. *Education:* PhD/ChE/Princeton Univ; Bs/ChE/Univ of WI. *Born:* Nov 1938. Worked at Brookhaven Natl Lab & Esso Res & Engrg; at West Va Univ 1969-present, Chem Engrg; assoc in Regional Res Inst. Specialties: Solid- Gas Reactions, Ultrasonics, Economic Effects of Technology. Mbr ASEE. Dow Outstanding Young Faculty Award; Western Electric Fund Award for Instruction. Chmn-Grp 4-AIChE. *Society Aff:* AIChE, ASEE, ACS.

Sears, Raymond W
Home: 372C New Bedford Lane, Jamesburg, NJ 08831
Position: Consultant. *Employer:* Self. *Education:* BA/Physics/Ohio Weslyan Univ; MSc/Physics/Ohio St Univ *Born:* Jan 1906. Bell Telephone Labs - mbr Tech Staff 1929-70, Surface Physics, Electron Devices, Solid St Devices, Lasers - Dir Univ Relations & Tech Employment 1965-70; Bellcomm Inc 1962-66, Dir Opers Analysis - Communications - Manmachine Interface Division. AIP 1971-76, Dir Manpower Division; IEEE Secy/Bd 1969-73; UET mbr of Bd 1970-84, President 1980; IEEE Foundation Inc Treas 1973-present; Engrg Foundation 1971-1981; Engrg Index Inc Treas 1974-79; APS Cons 1971-76. *Society Aff:* IEEE, APS.

Sears, William R
Business: Aero & Mech Engrg Dept, Tucson, AZ 85721
Position: Prof/Aerospace & Mech Engrg. *Employer:* Univ of Arizona. *Education:* PhD/Aeronautics/CA Inst of Tech; B Aero E/Aero Engrg/Univ of MN. *Born:* 3/1/13. in Minneapolis, MN. Instructor & Asst Prof at CalTech 1937-41; Ch of aerodynamics & flight-test at Northrop Aircraft 1941-46; Cornell Univ 1946-74 founder & Dir of Grad Sch of Aerospace Engrg, Dir of Ctr for Applied Math, JL Given Prof of Engrg. Mbr NAS; NAE; Amer Acad of Arts & Scis; AIAA, Hon Fellow. Bendix Award ASEE; Prandtl Ring; Von Karman Lecturer; Lanchester Lecturer; Pendray Award, Reed Aeronautics Award (of AIAA). Airplane owner & pilot (instrument, multiengine) 7500 hrs. Hobbies: chamber music; desert gardening; racket-ball. Mbr Acad Natl De Ingenieria De Mexido; Mbr Deutsche Gesellschaft fuer Luft & Raumfahrt. *Society Aff:* AIAA.

Seaver, Philip H
Home: 16 Harbor Ave, Marblehead, MA 01945
Position: Pres & Chief Operating Officer *Employer:* The Badger Co, Inc *Education:* B/ChE/Cornell Univ; 46th Advanced Management Prog/Bus Management/Harvard Bus School *Born:* 7/2/20 Ridgewood, NJ; s. John Eliot and Helen (Benson) Seaver; m. Anne Lillian Laskowy, April 12, 1947; children - John Benson, Scott Hall. Process engr, E. B. Badger & Sons Co, Boston 1943-44, 46-51; process engr, sales engr, sales mgr, vp, executive vp, pres, dir Badger Co, Inc, Cambridge, MA 1951--. Served to Lt. (j.g.) USNR, 1944-46. Phi Kappa Sigma. Clubs: Chemists (NYC); Harvard, Algonquin (Boston); Eastern Yacht (Marblehead). Patentee in field. Dir, National Conference of Christians and Jews (Boston). Officer and dir of various Badger subsidiaries U.S.A. and abroad. *Society Aff:* AIChE, SCI

Sebastian, Edmund J
Home: 6563 Shabbona Rd, Indian Head Park, IL 60525
Position: Chief Mechanical Engineer. *Employer:* Swick Associates, Inc *Education:* BS/ME/IL Inst of Tech; MS/ME/IL Inst of Tech. *Born:* Mar 27, 1926; native of Chicago Ill. US Navy 1944-46. Res Engr 1951-57; Principal Engr for GPE Controls Div of Genl Precision Corp, specializing in automatic control of interstate petroleum pipelines; Sr Staff Engr & Supr of Res for Inst of Gas Technology; 196673 V Pres of P&W Engr Inc; Ch Mech Engr of De Leuw Cather & Co 1973-78, involved in design of industrial facilities, sys & pollution abatement. Fellow ACEC; Mbr ASME. Reg P E 1978-80 - Div Hd-Mechanical Engg - Harza Engg Co - environmental engg; 1980- present, Chief Mech Engr - Swick Associates, Inc. *Society Aff:* WSE, ASME, ACEC.

Sebo, Stephen A
Business: 2015 Neil Ave, Columbus, OH 43210
Position: Prof *Employer:* OH State Univ *Education:* PhD/EE/Hungarian Acad of Sci; MS/EE/Budapest Polytech Univ *Born:* 6/10/34 Utility Co Lab Engr 1957-61. Asst Prof 1961-66, Assoc Prof 1967-74, Full Prof from 1974. Ford Fdn Postdoctoral Fellow 1967-68. Since 1968 with OH State Univ. Res interest is dev of elec energy transmission systems and high voltage engrg. Developed several sr, grad, and short course topics in power systems. 1981 recipient of Edison Elec Institute's Power Educator Award. 1981 recipient of IEEE-PES Prize Paper Award. Since 1982 AEP Professor of Power Systems Engineering. *Society Aff:* IEEE

Seborg, Dale E
Business: Dept of Chem and Nuclear Engg, Santa Barbara, CA 93106
Position: Prof *Employer:* Univ of CA, Santa Barbara *Education:* PhD/Chem Engg/Princeton Univ; BS/Chem Engg/Univ of WI, Madison. *Born:* 3/29/41. Native of Madison, WI. Taught at the Univ of Alberta from 1968-1977. Joined the Univ of CA, Santa Barbara as a Prof of Chem Engg in 1977. Served as Dept Chrm from 1978-1981. Also serves as an industrial cons on problems involving process modeling, simulation and control. Is the author or co-author of over 80 papers and two books in the areas of process control and computer control techniques. Has received two awards: Joint Automatic Control Conference Best Paper Award and the Tech Achievement Award from the AIChE Southern Ca Section. Has served as the director of three organizations: the American Automatic Control Council, the AIChE Computing and Sys Tech Div, and the ASEE Ch.E. Div. *Society Aff:* AIChE, ISA, IEEE.

Sebulsky, Raynor T
Home: 501 Glengary Dr, Pittsburgh, PA 15215
Position: Gen Mgr/Products. *Employer:* Gulf Res & Dev Co. *Education:* PhD/ChE/Carnegie-Mellon Univ; MS/ChE/Carnegie-Mellon Univ; BS/ChE/Carnegie- Mellon Univ. *Born:* 12/21/32. Native of Wheeling, WV. Process engr, polyester and polyester film, with duPont (1959-1962). With Gulf Res & Dev Co since 1962. Principal activity has been res & dev in petrol processing and petrol products. Present position is Gen Mgr, Products. Patents and publications in refining & Catalysis. Shared Allan P Colburn award (AIChE) with H L Toor in 1964. *Society Aff:* AIChE, ACS.

Sech, Charles E
Business: 8554 Monticello Dr, West Chester(Cincinnati), OH 45069
Position: Pres *Employer:* Charles E Sech Assoc Inc *Education:* BS/ChE/Univ of MI *Born:* 2/4/22 St. Joseph, MI. Res/production engr, 1944-49, Sharples Chems (now div Pennwalt Corp) Wyandotte, MI; asst to W. L. Badger, Ann Arbor, 1949-57; VP, 1957-58, Pres, 1958-63, W. L. Badger Assoc, Inc.; Pres Charles E Sech Assoc Inc, Cincinnati, 1963-; Dir Port Huron (MI) Paper Co; Reg PE, MI. Author, patentee. Specialities - chem engr, high temperature heat transfer/conservation, fired heaters, Dowtherm systems, synthetic fuels, accident and malperformance diagnoses/expert testimony. Past dir - local MI Soc PE Chapter. *Society Aff:* AIChE, NSPE

Sechrist, Chalmers F, Jr
Business: Assistant Dean, College of Engrg, Univ. of Illinois, 207 Engineering Hall 1308 W. Green St. Urbana, IL 61801
Position: Prof. of Electrical & Computer Engrg and Assistant Dean of Engrg.

Sechrist, Chalmers F, Jr (Continued)
Employer: Univ of IL. at Urbana-Champaign *Education:* PhD/EE/Penn State Univ; MS/EE/Penn State Univ; BE/EE/Johns Hopkins Univ. *Born:* 8/23/30. 1959-1965 staff engr and project director, HRB-Singer, Inc, state college, PA; investigated VLF radio wave propagation via the lower ionosphere; 1965-1984 prof of elec engrg, Univ of IL, Urbana; half-time teaching and half-time res in aeronomy lab; res areas include theoretical and experimental studies of the lower ionosphere, ionospheric D & E regions, and the atmospheric sodium layer; laser radar studies of the upper atmosphere; 1984-85 acting assoc head of the Dept of Elec and Comp Engrg.; 1985-86 assoc head for instructional programs in the Dept of Elec. and Comp Engrg; 1986-present assistant dean in the College of Engrg at UIUC. *Society Aff:* IEEE, AGU, AMS, ASEE, AAAS, ΣΞ.

Seckman, Thomas C
Business: 2135 Blakemore Ave, Nashville, TN 37212
Position: President *Employer:* Smith Seckman Reid Inc. *Education:* BS/ME/W VA Univ. *Born:* 10/16/30. Mbr Tau Beta Pi. Experience includes 2 yrs in USAF R/D Command, 5 yrs with Carrier Air Conditioning Co & 18 yrs in cons engrg. Founder & current Pres Smith Seckman Reid Inc & current Pres Gresham Lindsey Reid Ltd, overseas design firm. Mbr & P Pres local chap ASHRAE & Cons Engrs of Tenn. Mbr & Chmn city of Oak Hill planning comm. Mbr Nashville Rotary Club & current secy to Rotary Foundation Board. *Society Aff:* ASHRAE, ACEC.

Secord, Lloyd C
Business: 4195 Dundas St West, Toronto Ontario, Canada M8X 1Y4
Position: President. *Employer:* DSMA ATCON LTD. *Education:* BSc/Mech Engrg/Queen's Univ. *Born:* 8/28/23. in St Thomas, Ontario. 1945 Joined Turbo Res Ltd; upon merger with A V Roe Canada Ltd in 1946, progressed in Gas Turbine Div to position of Ch Design Engr; joined Paul Dilworth & Co 1953 as an Assoc, this firm became cons engrg firm of Dilworth Secord Meagher & Assocs Ltd 1958. Pres of DSMA ATCON LTD, managing the nuclear, space & equipment (accelerators, telescopes, remote manipulator sys, thermal vacuum chambers, gas turbines) fields. Dir of Canadian Nuclear Assn, Dir of Electric Vehicle Assoc, Pengalta Research & Development Ltd. Pres 1969 & Mbr of Sci Council of Canada 1972-75; Fellow Engrg Inst of Canada & Canadian Aeronautical & Space Inst; Mbr Assn of prof Engrs, Assn of Cons Engrs of Canada, Soc of Automotive Engrs. Space Ctte-Aerospace Indus Assoc of Canada. *Society Aff:* AIAC, EVAC, SCC, CNA, APEO, SAE.

Sedeora, Tejinder S
Business: 2300 NE Adams St, Peoria, IL 61639
Position: Chief Met. *Employer:* Westinghouse Air Brake Co. *Education:* BS/Met Engg/MO Sch of Mines & Met; BS/Chemistry/Punjab Univ. *Born:* 8/27/39. Started as a Met Engr at Gunite Foundry. As a Rand D Coordinator at Barber Colman Co was responsible for all the mtls related projs from inception to product utilization. Various jobs performed at Sundstrand Corp included working on IVHM for Breeder Reactor, mtls selection & performance study for machine tools, met trouble shooting for Aerospace components mfg, setting up heat-treat, plating & lab for a plant in Singapore. Assumed the present position in Jan 1978 with responsibilities of all the met, chem & nondestructive activities for the co. Home improvement, writing, reading, golf and racket ball are leisure time enjoyments. *Society Aff:* ASM, AWS.

Seebass, A Richard
Business: Dean, College of Engr, Boulder, CO 80309
Position: Dean. *Employer:* Univ of CO. *Education:* BSE/Aeronautical Engr/Princeton '58; MSE/Aeronautical Engr/Princeton '61; PhD/Aerospace Engrg/Cornell '62 *Born:* 3/27/36. Dean, College of Engrg and Applied Science, Univ of CO 1981-Prof, Aero and Mech Engg, and Math, Univ of AZ, 1975-1981. National Academy of Engrg, Tech Dir, AIAA, Chrmn NAS/NRC Aero and Space Engg Bd, NAS Committee on the SST Sonic Boom, Air Force Scientific Advisory Bd, Fellow, AIAA, AAAS, Assoc Editor, AIAA Journal *Journal, American Institute of Aeronautics and* Astronuactics Editorial Bd, *Annual Review of Fluid Mechanics* and *Physics of Fluids.* Prof and Assoc Dean College of Engg, Cornell Univ, 1972-1975. *Society Aff:* AIAA, SXI, TBΠ, AAAS.

Seebold, James G
Business: Chevron Corporation, P.O. Box 1627, Richmond, CA 94802-0627
Position: Staff Engr. *Employer:* Chevron Corporation. *Education:* Engr/ME/Stanford Univ; MS/ME/Stanford Univ; BS/ME/Stanford Univ. *Born:* 9/6/33. Native of S CA. Served as Engg Officer, USS Lexington, 1956-1959. An aerospace expatriot, specialized in advanced systems & low-gravity fluid mechanics. Since 1966, with Chevron Corp. Principle consulting responsibilities in combustion, acoustics, fluid mechanics, environmental engrg, and energy mgmt. Teach courses in noise control & combustion for AIChE & Stanford Ind TV Network. Pres (1984) & Treas, Inst of Noise Control Engg. BS (1956), MS (1960), Engr (1965), Mech Engg, Stanford Univ. Reside in Atherton, CA with wife Alicia & daughters Suzy & Katie. *Society Aff:* INCE, AFRC.

Seed, H Bolton
Business: Dept of Cvl Engrg, Berkeley, CA 94720
Position: Professor of Civil Engineering. *Employer:* University of California. *Education:* PhD/Civil Engg/London Univ; SM/Soil Mechs/Harvard Univ; BSc/Civil Engg/London Univ. *Born:* 8/19/22. Has been a mbr of teaching staffs at London Univ & Harvard Univ; 1950- has been on staff of Univ of Calif, Berkeley; is now Prof of Cvl Engrg. Also serves as a cons to numerous major engrg cos & govt agencies. Author of over 180 papers; has been recipient of 14 awards by ASCE for res contribs, the Lamme Award and the Vincent Bendix Award of Amer Soc of Engrg Educ, and the TK Hsieh Award of the British Royal Soc and the Inst of Civil Engrs; elected to Natl Acad of Engrg in 1970 and National Academy of Sciences, 1986; presented many memorial lecturing including Tarzaghi Lecture of ASCE, Rankine Lecture of Inst of Civil Engineers, and Carillo Lecture of Mexican Society of Soil Mechanics and Foundation Engineering. *Society Aff:* ASCE, SSA, USCOLD, SEAONC, EERI.

Seedlock, Robert F
Home: 140-B4 Birch St, Falls Church, VA 22046
Position: Consulting Engineer *Employer:* Self *Education:* MS/CE/MIT; BS/Eng/USMA, West Point. *Born:* 2/6/13. 2nd Lt Corps of Engrs USA 1937 to Maj Genl 1963; commanded reconstruction Burma Rd 1944-45; Div Engr MO River; Sr Mbr UN Armistice Comm, Panmunjom Korea; Dir Military Const; Cdr US Army Engr Ctr, US Bd of Engrs of Rivers & Harbors; Miss River Comm. Mem US Del Far Eastern Co & NATO Min Conf; Pres, YUBA Industries, VP Standard Prudential Corp. Program Director, Ralph M. Parsons Company. US Disting Service Medal, French Legion of Honor, Chinese Order of Yun Hui, Metro-Atlanta Engr of the Yr 1975; Engr of Yr in Govt GSPE 1976. Reg PE. Hon Mem ASCE, Fellow, SAME, PIANC, USCOLD. Hobbies: antique cars, violin, golf. *Society Aff:* ASCE, SAME, USCOLD, PIANC.

Seeley, Gerald R
Business: Angola, IN 46703
Position: Dean School of Engg. *Employer:* Tri-State Univ. *Education:* PhD/Civil Engg/Univ of MN; MS/Civil Engg/Univ of MN; MS/Engg Mechanics/Univ of WI; BS/Mech Engg/Univ of WI. *Born:* 10/9/40. in Wausau, WI. Dev Engr with Union Carbide Corp, Stellite Div 1963-1964 and Honeywell, Inc, Aerospace Div, 1966-70 (Part Time 1970-1972). Assumed responsibilities as Dean of Engg in Sept, 1978. Chrmn, CE Dept 1976-1978. Asst Prof CE Dept 1973-1976. Excellence in Teaching Award CE Dept 1974, 1975. Twenty- three technical papers and articles. Chrmn, IL-IN Sec, ASEE 1975-1976. Mbr ASEE Technical Comm on Aesthetics in Design NASA-ASEE Summer Fellowship 1978, Member ASEE Comm on Teaching Methods, CE Div., PEE Rep. ISPE. *Society Aff:* ASCE, ASEE, NSPE, ΣΞ, ASEM

Seeley, Ralph M
Business: Applied Research Laboratory, PO Box 30, State College, PA 16801
Position: Assoc Prof of Engrg Res. *Employer:* Penn State Univ *Education:* MSEE/EE/PSU; BSEE/EE/Duke. *Born:* 12/22/30 Native of Candor, NY. Editor of 'Duk Engr' at Duke. Mbr of Tech Staff, Bell Telephone Labs, Whippany 1952-55, & completed the post-graduate training program there. With Penn State since late 1955, specializing in information processing & control of underwater missiles. 4 papers, 6 patents. Active in recreational trails and river conservation; also comm mental health and comm day care programs. *Society Aff:* IEEE, ΣΞ

Seely, John H
Home: 1538 Beloit Ave, Claremont, CA 91711
Position: Prof *Employer:* CA State Poly Univ *Education:* MMS/ME/Syracuse Univ; ME/Stevens Inst of Tech *Born:* 9/24/21 Native of Pensacola, FL. Employed by IBM Co in Poughkeepsie, NY in various engrg and engrg mgmt positions. Holder of several patents in electronic packaging & heat transfer and recipient of a fellowship from IBM for one yr of study in Thermosci at Stanford Univ. Since retirement in 1976, prof of grad & undergrad Mech Engrg at CA Poly, Pomona. Author of two textbooks in heat transfer & thermodynamics and 40 papers & magazine articles. Reg PE and Fellow of ASME. *Society Aff:* ASME

Seely, Samuel
Home: 37 Brainard Road, Westbrook, CT 06498
Position: Professor of Electrical Engineering. (retired) *Employer:* Univ of Rhode Island. *Education:* PhD/Physics/Columbia Univ; MS/Physics/Stevens Inst of Tech; EE/Elec/Poly Inst Bklyn. *Born:* 5/7/09. Teaching: Instr & Asst Prof CCNY; Assoc Prof Naval PG School; Prof & Hd Syracuse Univ 1947-56; Prof & Hd, Case Inst 1956-64; Visiting Prof, Johns Hopkins Univ, Oklahoma State Univ, Chalmers Tech Univ (Sweden), Univ of Conn, Univ of Mass; at URI since 1972, retired 1979. WWII, Staff Mbr MIT Radlab, radar dev; Fulbright Lectr Greece 1959-60 (recipient of Silver Cross); Hd Engrg Sect NSF 1963-64. Cons to indus & govt. Author or co-author of 16 published texts on network theory, electronics, eletromagnetic fields, electromechanical energy conversion signal processing. Fellow: IEEE, American Physical Society. *Society Aff:* APS, IEEE, ASEE.

Seemann, Gerald R
Home: 3372 Albedo St, Hacienda Heights, CA 91745
Position: Pres. *Employer:* Developmental Sciences, Inc. *Education:* PhD/Astro Sci/Northwestern Univ; MS/ME/OK State Univ; BS/ME/Tech Univ. *Born:* 3/29/37. After PhD took Post Doc at Von Karman Inst (63-4), then Mgr of Reentry Phys Litton Space Lab (64-68), chief of thermo optics McDonnell Douglas Astro (68- 70). Founded Developmental Sciences, Inc 1970 which is a diverse small business actively engaged in high tech (RPVS) and product mfg. Current CEO. He is Reg PE who has published over 75 papers on aerospace engg subjects and two patents. Holds many academic awards including "Distinguished Engr" designation by TX Tech 1978, Receipent of Gustus L. Larson Memorial Award, ASME, 1979 has lectured here and Europe, listed in Who's Who West, American Men & Women of Sci, Who's Who in Technology. AVVS Outstanding Contributor Award 1985. *Society Aff:* AIAA, ASME, NSPE, AUVS.

Segerlind, Larry J
Business: Agr Engg Dept, E Lansing, MI 48824
Position: Prof. *Employer:* MI State Univ. *Education:* PhD/Agri Engg/Purdue Univ; MS/Agri Engg/MI State Univ; BS/Agri Engg/MI State Univ; BS/Math/MI State Univ; MS/Math/MI State Univ. *Born:* 2/15/37. Faculty positions at OH State Univ (1963-70) and MI State Univ (1970- present). Present responsibility: Teaching engg mechanics related courses to agri engg students. Specific res interests are finite element applications in agri, computer-aided design, and physical properties of fruits and vegetables. Received ASAE Young Educator Award 1976. Div editor of EPP-Div of ASAE, 1976-79. Author of *Applied Finite Element Analysis* (1976), (1984). *Society Aff:* ASAE, ASEE, ASME, AAM.

Seglin, Leonard
Business: 799 Broadway, New York, NY 10003
Position: Pres. *Employer:* Econergy Assoc. *Education:* Am/Math/Harvard Univ; BChE/CE/NYU. *Born:* 4/4/17. Albany, NY. Formerly associated wih Ethyl Corp, Bechtel Corp, & FMC Corp in various technical & mgt positions up through Dir of Process R&D (with FMC). Presently Pres of Econergy Assoc of NY, NY. Also Adjunct Prof of Chem Engg at Manhattan College. Author of one technical book & numerous professional articles, inventor of 20 patents, & recognized authority on synthetic fuels from coal & other fossil resources and on assessment of technical and economic risk of private or industrial investments Dir of NY. Section AIChE, mbr AIChE, Harvard Club of NYC, & NY Acad of Science, & Assoc. Chem. and Chemical Engineers. *Society Aff:* AIChE, NYAS.

Segner, Edmund P, Jr
Business: Memphis State Univ, Assoc VP for Research, TN 38152
Position: Assoc VP for Res *Employer:* Memphis State University. *Education:* PhD/Structural Engr/Texas A&M Univ; MS/CE/Univ of Texas (Austin); BS/CE/Univ of Texas (Austin) *Born:* March 28, 1928. Native of Austin, Texas. Prof exper with United Gas Pipe Line Co, Forrest & Cotton Inc, Cons Engrs, & General Dynamics, all in Texas. Served as Instructor, Asst prof & Assoc Prof of Civil Engrg at Texas A&M Univ; Prof of Civil Engrg at Univ of Oklahoma, Univ of Alabama; also Asst Dean, Assoc Dean of Engrg at Univ of Alabama. Currently Assoc VP for Res & Prof of Civil Engrg at Memphis State Univ. Reg PE in 4 states. Mbr ASCE, NSPE, ASEE & many others. Past state or natl officer in ASCE & ASEE. Author of publs. *Society Aff:* ASCE, ACI, ASEE, NSPE, NCURA, NCAR, ΦΚΦ, TBΠ, SUPA

Sehn, Francis J
Home: 3515 Brookside Dr, Bloomfield Hills, MI 48013
Position: Board Chairman Comau Productivity Systems Inc *Employer:* FIAT *Born:* April 1918. Native of Johnstown, Penn. 11 years night school-Lawrence Inst-Univ of Detroit & Wayne State Univ. Reg Prof Mech Engr. Chief Die Engr, Fisher Body Tank Div, General Motors Corp during WWII; founded The Fran Sehn Co, an Interntl consulting organization as well as Press Automation Systems & Coilfeed Systems Inc. Pres Soc of Mfg Engrs 1965-66. Faculty Wayne State Univ, School of Appl Mgmt Modern Tech for Stamping Plants. Recipient of 1975 Joseph Siegel Memorial Award by Soc of Mfg Engrs, for outstanding contribs. Author of Die Design Handbook (1955); Trustee of Detroit Inst of Tech. Mbr Exec Reserve US Dept of Commerce. Chairman Comau Productivity Systems Inc. Past member Lucas Industries Inc Board & Motor Wheel Corp Thompson Industries Div Sheraton Hotels - present member Fruehauf Corp Board of Directors. *Society Aff:* SME, SAE, IPE, PE, ID.

Seichi F Konzo
Home: Clark-Lindsey Village 6110, 101 West Windsor Road, Urbana, IL 61801
Position: Prof of Mech Engrg Emeritus. *Employer:* Univ of Ill - Urbana (Ret). *Education:* MS/Mech Engg/Univ of IL; BS/Mech Engg/Univ of WA. *Born:* Aug 2, 1905 Tacoma Wash. Started res in fuels & combustion in 1929; res dev of residential warm-air heating sys & residential cooling sys 1930's & 40's. Assisted in formation of Small Homes Council. Started teaching 1947. 1960-71 administrative positions. Publ 5 books & hundreds of tech & semi-tech articles. Awarded E K Campbell Award, F Paul Anderson Medal, Fellow & Life Mbr grade ASHRAE. Coll Alumni Medal from Univ of Ill. Now Cons to Small Homes Council. Print collector & gardener. *Society Aff:* ASHRAE.

Seide, Paul
Business: University of Southern California, Los Angeles, CA 90089-1114
Position: Professor of Civil Engineering. *Employer:* Univ of Southern California. *Education:* PhD/Eng Mechs/Stanford Univ; MAeroE/Aero Engg/Univ of VA; BCE/Civil Engg/City College of NY. *Born:* July 22, 1926. Native of NY City. held res positions with Natl Advisory Ctte for Aeronautics, Northrop Corp, TRW Systems, &

Seide, Paul (Continued)
Aerospace Corp 1946-65. Prof of Civil Engrg Univ of Southern Calif 1965- ; Vice Chmn Dept of Civil Engrg 1971- 73, 1981-83 . NSF Sr Postdoctoral Fellow, Univ of Newcastle, England 1964- 65; Albert Alberman Visiting Prof, Technion-Israel inst of Tech 1975. Visiting Prof, Univ of Sydney, Australia, and Univ of Canterbury, New Zealand 1986. Fellow ASME 1972, Fellow American Academy of Mechanics, 1980. Author of 'Small Elastic Deformations of Thin Shells', Noordhoff Internatl Publishing, 1975. *Society Aff:* ASME, ASCE, AAM, ΤΒΠ, ΣΞ, AAUP, NYAS

Seidel, Edmund O
Home: 9507 Valley View, San Antonio, TX 78217
Position: Owner. *Employer:* Edmund O Seidel & Assoc Consulting *Education:* MS in CE/Structures/Univ of TX; BS in CE/Civ Eng/Univ of TX. *Born:* 12/5/22. Native of New Braunfels, TX. Served as US Naval Reserve Officer 1944-1946. Taught mtls testing lab at UT at Austin 1946-1948. Field Engr for E M Freeman & Assoc in Shreveport, LA 1948-1950, Structural Engr for Austin Co in Freeport, TX 1950-1952, Structural Group Leader for Brown & Root in Lake Charles, LA and Houston, TX 1952-1954, Structural Engr for Frank Drought in San Antonio, TX 1954-1960, Self Employed Consulting Engr 1960-present. Reg in TX and LA. VP, TX Sec ASCE 1979, Pres San Antonio Branch 1970, Pres Bexar Chapter TSPE, 1974., Asst Scout Master 1956-1960, Pres Beethoven Maennerchor 1973-1982, Fiesta San Antonio Comm Sec 1981, VP 1982. *Society Aff:* ASCE, NSPE, ACI, XE.

Seiden, Edward I
Home: 4132 Marian St, La Mesa, CA 92041
Position: Retired *Employer:* Genl Dynamics Corp, Convair Div. *Born:* May 6, 1923 in Manila Philippines. Lt, MacArthur's Philippine guerilla forces 1942-45. BS Indus Tech, Natl Univ, San Diego, Calif. With Convair Weight Group 1951 to 1978 was respon for new space systems mass properties 1971-78. Weight Group Supr 1967-70. Conducted company res on weight & cost factor prediction methods; predicted & controlled mass properties of F-106, Sea Dart, 880, 990 & other Convair aerospace vehicles. Mbr SAWE 28 years; Fellow SAWE 1975; Honorary Fellow SAWE 1981; San Diego SAWE Chap; P Dir, Pres; Tech Chmn 1976 Southwestern Regional SAWE Conference; SAWE Intl Exec VP, 1977. Authored 4 SAWE tech papers. Enjoys handyman projects & traveling.

Seider, Warren D
Business: Dept of ChE, 220 S. 33rd St/D3, University of Pennsylvania, Phila, PA 19104
Position: Prof. *Employer:* Univ of PA. *Education:* PhD/ChE/Univ of MI; MS/ChE/Univ of MI; BS/CE/Poly Inst of Brooklyn. *Born:* 10/20/41. Prof of ChE at Univ of PA, Asst Prof - 1967, Assoc Prof - 1971, Full Prof - 1984. Res in Computer-aided Process Design, Math modeling of chem processes. First Chrmn of CAChE (Computer-aids for Chem Engg Educ) Committee. Past Chrmn of AIChE CAST (Computing and System Tech) Div. Dir of AIChE, 1984-86. Author of two books and over 30 research papers. *Society Aff:* AIChE, ACS.

Seidl, Ludwig H
Home: 2250-A Noah St, Honolulu, HI 96816
Position: Prof. *Employer:* Univ of HI. *Education:* DrSc/Naval Arch/Vienna Univ of Tech; MS/Naval Arch/Vienna Univ of Tech; BS/ME/Vienna Univ of Tech. *Born:* 7/19/39. Vienna, Austria. Started 1964 as naval arch & marine engr at Gothaverken shipyard, Sweden. Came to the US in 1965. Worked wiTH M Rosenblatt and Son, Inc, NY & NESCO, Natl Engg Sci Co, Wash DC, joined the graduate faculty of Dpt. of Ocean Engrg, University of Hawaii, in 1968, formed Ocean Engg Consultants, Inc 1972, providing service & software for analysis of offshore terminals & mooring systems. In 1978 joined PAMESCO, engaged in design of SWATH-type vessels. Res grants from NSF, MARAD, Sea Grant. Author of 8 technical papers at Patent on SWATH-type vessel. *Society Aff:* SNAME, PIANC, ASCE

Seidman, Arthur H
Home: P.O. Box 712, Sheffield, MA 01257
Position: Prof of EE, Sch of Engg. *Employer:* Pratt Institute. *Education:* MA/Physics/Hofstra Univ; BEE/Ele/CCNY. *Born:* Jan 1923. Graduate work at Columbia Univ. Employed in indus from 1950- 59; last 5 years at Sperry Gyroscope Co (Sr Engr); taught in EE Dept at CCNY from 1954-59 part time & from 1959-62 full time. With Pratt Inst since 1963. Promoted to full Prof of EE in 1971; appointed Acting Dean in 1975-78. Author of numerous articles & papers in elctronics; contributor to Encyclopedia of Physics (1974, Van Nostrand Reinhold); co-author of Semiconductor Fundamentals (1963, Wiley), Electronics Circuit Analysis (1972, MacMillan); Co-editor Handbook for Electronics Engrg Technicians (1976, McGraw-Hill); etc. Series Editor "Wiley Electrical and Electronics Technology Handbook Series" and Associate Editor, Solid State Technology - Contributor to Encylopedia Britannica "Yearbook of Science and the Future." Mbr ASEE, AAAS, Tau Beta Pi, Eta Kappa Nu, Sigma Pi Sigma; Sr Mbr IEEE. Enjoy ham radio (K2BUS) & collecting political buttons & items (mbr APIC). *Society Aff:* TBP, HKN, SPS, IEEE, AAAS, ISHM.

Seidman, David N
Business: Northwestern Univ, 2145 Sheridan Rd, Evanston, IL 60201-9990
Position: Prof. Materials Sci *Employer:* Northwestern University *Education:* PhD/Physical Met/Univ of IL; MS/Met/NYU; BS/Met/NYU. *Born:* 7/5/38. Cornell Univ: Post-doctoral Assoc 1964-66; Asst Prof 1966-70; Assoc Prof 1970-76; Prof 1976 to 1985. Robert Lansing Hardy Gold Medal of the AIME, 1967. John Simon Guggenheim Fdn Fellow, 1972-73, 1980-81, Fellow Amer Physical Soc 1984. Prof. Hebrew Univ of Jerusalem 1983-85, Prof. Northwestern Univ. 1985 to present. Visiting Scientist, Centre d' Etudes Nucleaires de Grenoble 1981 and Centre Nationale d' Etude Telecommunications Meylan 1981. Lady Davis Visiting Professorship at Hebrew Univ of Jerusalem 1978, 1980. Visiting Assoc Prof, Tel-Aviv Univ, 1972. Visiting Sr Lectr, Technion-Israel Inst of Tech, 1969. Res interests: interfaces, surfaces, imperfections in metals & simiconductors, radiation damage, field-ion & atom-probe microscopy, transmission electronmicroscopy and analytical microscopy. *Society Aff:* AIME, APS, MRS

Seidorf, Christian E
Home: 3937 Orchard Rd, Cleveland Heights, OH 44121
Position: Dept Coordinator. *Employer:* General Electric Company. *Born:* July 30, 1923 Newark, NJ. Undergraduate degree from Tusculum Coll in Tenn (1947); MS from Case-Western Univ, Cleveland, Oh 1951. Employed by The Cleveland Graphite Bronze Co 1951-56; with GE Co since then as materials engr, process dev engr, wire operation supr, Dept Coordinator for Ecology-Safety- Energy. Have contributed many proprietary process innovations. Served virtually every possible function in the Cleveland Chap of the ASM; have served on two Natl Cttes of ASM; presently serving as the VP of the Cleveland Tech Soc Council (comprised of 50 tech & scientific societies in Northern Ohio).

Seigle, Leslie L
Home: 1 Saywood Ln, Stony Brook, NY 11790
Position: Prof. *Employer:* State Univ of NY. *Education:* DSc/Met/MIT; MS/Met/Univ of PA; BChE/ChE/Cooper Union. *Born:* 6/13/17. Dumbarton, Scotland. Res Met Inco Res Lab 1938-1946; Mgr Met Res Lab Sylvania Elec Prod 1951-1965; Prof Mtls Sci Dept SUNY, Stony Brook, 1965- present. Interests are in Thermodynamics of solid solutions, diffusion in solids, theory of solid state sintering, high temperature protective coatings on metals. *Society Aff:* ASM, AIME, ΣΞ.

Seiler, George R
Home: 30 Burnett St, Glen Ridge, NJ 07028
Position: Partner *Employer:* Profit Planning Associates. *Education:* MS/ChE/MIT; BS/ChE/MIT; BS/Physics/William & Mary. *Born:* 6/27/34. Formed Profit Planning Associates, management consultants, in 1978. Prior to this was VP & Gen Mgr, Precision Gas Products, Inc. Began career in research, moving to planning, mktg, finance & genl bus mgmt. *Society Aff:* AIChE, PF, CDA

Seinfeld, John H
Business: Dept of Chem Engrg, Pasadena, CA 91125
Position: Louis E. Nohl Prof & Exec Officer of Chem Engrg *Employer:* CA Inst of Tech. *Education:* BS/ChE/Univ of Rochester; PhD/ChE/Princeton Univ. *Born:* 8/3/42. Asst Prof 1967-70, Assocc Prof 1970-74, Prof 1974- , Exec Officer for Chem Engrg 1973-Louis E. Nohl Prof 1980-, Calif Inst of Tech. Donald P Eckman Award of Amer Automatic Control Council 1970; Camille & Henry Dreyfus Foundation Teacher Scholar Grant, 1972; Curtis W McGraw Res Award of ASEE, 1976. Allan P Colburn Award of AIChE, 1976 Thiry second Institute Lecturer of AIChE, 1980. Public Service Medal of NASA, 1980. Elected to Natl Academy of Engrg, 1982. Wiliam H Walker Award of AIChe, 1986. George Westinghouse Award of ASEE, 1987. Res interests in air pollution control & systems theory. *Society Aff:* ACS, AIChE, ASEE, APCA, SPE

Seinuk, Ysrael A
Home: 82 Tennis Place, Forest Hills Gardens, New York, NY 11375
Position: Pres *Employer:* Ysrael A Seinuk P.C. *Education:* CE//Univ of Havana *Born:* 12/21/31 Native of Havana, Cuba. Graduate Civil Engr 1954. Structural Engr Saenz- Cancio - Martin - Havana 1957. Own Practice Havana. Head Structural Dept Ministry of Public Works, Cuba 1960. Chief Engr, Hertzberg & Cantor, NYC 1961- 70. Principal Office of I.G. Cantor to date. Pres Ysrael A Seinuk P.C. 1976-date. Prof at Architectural School, The Cooper Union for the Advancement of Science & Art 1968-date. Former Trustee, French-Poly Clinic Hospital. Member American Arbitration Association Licensed PE NY, NJ, FL, MA, AR, DE, PA, DC. *Society Aff:* ACI, CIB, NAEP, NYCCI, PTI

Seiple, Willard Ray
Business: 2929 Glen Albyn Dr, Santa Barbara, CA 93105
Position: Consultant (Self-employed) *Education:* MS/Eng Geology/USC; BS/Geology/USC. *Born:* July 29, 1924 Gibsonburg, Ohio. Proj Engr for B F Goodrich Co & Firestone Tire & Rubber Co, specializing in design dev & manufacture of Fuel Celles, reservoir liners & inflatible dams. With Dames & Moore 1965-1980; respon for mgmt of multi-disciplinary teams for site selection & environmental assessment on power plant & liquefied natural gas projects. Responsible for geotechnical evaluation of earth fill dams, prior to construction. Provided construction surveillance and geotechnical evaluation of earthfill dams during construction. Evaluated proposed modifications to existing earthfill dams. Managing Principal-inCharge, Santa Barbara office of Dames & Moore 1974-77. Resident Geotechnical Engr, Wolf Creek Nuclear Power Plant Proj near New Strawn KS 1977-1980. Mbr So. CA section So Cal Section So Cal section AEG, ASCE, GSA, AIME. Reg Geologist Calif & Maine; Certified Engrg Geologist Calif. Vice Chmn AEG 1974. *Society Aff:* AEG, ASCE, GSA, AIME.

Seireg, Ali A
Business: 1513 University Ave, Madison, WI 53706
Position: Professor of Mechanical Engineering. *Employer:* The University of Wisconsin. *Education:* PhD/ME/Univ of WI; BSc/ME/Univ of Cairo. *Born:* Oct 1927. Taught at Cairo, Marquette; currently Prof Mech Engrg Univ of Wisc. Advisory Engr & Cons Falk Corp 1956- ; Biomechanics Cons V A Center 1964- . Author of Mech Sys Analysis. Chmn Design Engrg Div 1978. Chmn US Council of Int Fed of Theory of Mechanics. Chmn Computer Tech Ctte, Res Ctte on Lubrication Policy Bd Communications, Policy Bd Genl Engg & Automation Res Council. 1970 George Westinghouse Award ASEE, 1973 Richards Mem Award ASME, 1974 E P Cornell Award AGMa. 1978 Machine Design Award ASME Chmn Int Conf Design Aut. Advisor World Conf on Tribology. Co-Chmn Fifth World Congress of Theory of Machines. Chrmn Century II Int Comp Tech Conf. Ebaugh Prof, Univ Florida, Gainesville 1986-. Editor-in-chief, SOMA "Engineering for the Human Body" 1986- . Honorary Mbr, Chinese Mech Engr Soc. *Society Aff:* SESA, AGMA, ASEE.

Seki, Hideo
Home: 5-11-3 Kamiuma, Setagaya Tokyo, Japan 154
Position: Laboratory Director *Employer:* Unified Science Laboratory *Education:* DR/Communication/Tokyo Inst of Techn; BS/Elec Engg/Tokyo Inst of Techn. *Born:* October 13, 1905. MS Tokyo Inst of Technology; Dr Engrg TIT 1943 for works on internal noise in telecommunications radio receivers. Internatl Telecommunications Co 1932-47; Railway Tech Labs, Japanese Govt Railways 1947- 49; Radio Res Labs, Ministry of Postal & Telecommunications 1949-53; Tech Dir, Iwatsu Electric Co Ltd 1953-67; Prof Govt Univ of Electro-Communications Tokyo 1967-71; Visiting Prof Univ of Hawaii 1971-72; 1979-83 Tokai Univ Takanawa Campus. Award by Minister of Postal & Telecommunications 1973. 1984- Unified Science Lab Dir, engaged in a research on special method of accelerated learning or suggestopedia. *Society Aff:* IEEE, SALT.

Sekimoto, Tadahiro
Home: B14-9, Nishikata 2-Chome, Bunkyo-ku, Tokyo 113, Japan
Position: President *Employer:* NEC Corp *Education:* Doctor/Engng/Tokyo Univ *Born:* 11/14/26. Native of Kobe City Japan. 1948-present NEC Corp, assumed current responsibility as Pres 1980. Contributed to IEEE on ctte of Tokyo Sect. Pub works on Satellite Time Div Multiple-Access Experiment & many others. Awarded Japanese Governmental prize for dev of SPADE sys 1976. IEEE Fellow Award 'for contrib & leadership in the field of digital communications' 1976. Recipient of the Prize from the Minister of State for Sci and Tech Agency. Prime Minister's office 1976, of the Purple Ribbon Medal from His Majesty the Emperor of Japan 1982, of the Edwin Haward Armstrong Achievement Award from the IEEE 1982. *Society Aff:* JEIDA, OAS, EIAJ, CIAJ

Sekine, Yasuji
Business: 7-3-1, Hongo, Bunkyo-Ku, Tokyo, Japan
Position: Prof *Employer:* Univ of Tokyo *Education:* Dr of Engg/Elec Engg/Univ of Tokyo; MS/Elec Engg/Univ of Tokyo; BS/Elec Engg/Univ of Tokyo *Born:* 12/07/31 Main interest lies in power system engrg. He received many awards from Inst of Electrical Engrs of Japan and also the grade of Fellow of IEEE for his contribution to the methodology of power sys analysis, planning, control and operation. He is currently the chairman of Japanese Natl Ctte of CIGRE. VP, IEEJ, mbr of Organizing Ctte of PSCC, a co-editor of Intl Journal of Elec Power and Energy Sys, and the chrmn of State Examination Bd for Licensed Energy Conservation Engrs of Japan. Visiting professors of Federal Inst. of Technology of Zurich, Univ Texas at Arlington, Xian Jiaotong Univ and several other universities. *Society Aff:* IEEJ, CIGRE, JAE

Selander, Carl E
Business: USBR - DFC PO Box 25007, Code D-1512, Denver, CO 80225
Position: Head Polymer Concrete & Struct Sect. *Employer:* Bureau of Reclamation E & R Center. *Education:* BS/ChE/Univ of Denver. *Born:* October 1926. Trainee at USBR 1948, advanced to Materials Engr 1949. Responsible for many innovative uses of materials in construction, maintenance & repair progs. Active in ASTM D20.23 RTRP Sys Ctte work as well as num in-house teams. 1972- , responsible for R&D of concrete-polymer materials for USBR & for structural res & testing. Has assisted on almost all major USBR field projs. Author of num tech articles & reports. ASCE Thomas Fitch Rowland Prize 1974. Reg P E Colo. Mbr: ASCE, DFC, PE Group, Soc of Sigma Xi. *Society Aff:* ASCE, ΣΞ.

Selbach, Lawrence G
Home: 213 Can-Dota Avenue, Mt Prospect, IL 60056
Position: Vice-President. *Employer:* Bodine Electric Company. *Born:* Feb 1919. MBA-University of Chicago. BSEE Ill Inst of Technology; BS Univ of Ill. Reg P E. Mbr IEEE. Has been in Fractional Horsepower Motor field 1946-present with Bodine Electric Co; Design Engrg & Ch Engr 1963-73, V P Engrg 1973present. Hobbies: Toastmasters Internatl, sailing, golf.

Selby, Joseph C
Home: RR 2, 215 - 28th Ave NW, Cranbrook B C, Canada VIC 4H3
Position: Instructor & Consultant. *Employer:* East Kootenay Community Coll & self. *Education:* MA/Economics/Simon Fraser Univ; MBA/Operations Research/Simon Fraser Univ; BASc/Mechanical Eng/Univ of BC. *Born:* June 1936. Native of British Columbia. Design Engr Dupont of Canada 1959-60; Dist Mgr B C Hydro 1960-62; Lubrication, Sales & Construction Engr Gulf Oil Canada ltd 1962-69; Warehouse Opers Engr Weldwood of Canada ltd 1969-75. Presently Co-ordinator: Bus Admin East Kootenay Community Coll as well as practicing as independent Construction Cons. Reg P E in province of B C. Former Chmn, Transportation, Secy & Treas B C sect SAE. 1972 Russell Springer Award for SAE paper. Hobbies: camping, antique automobiles & canoeing. *Society Aff:* APEBC.

Self, Norman L
Business: PO Box 12830, Pensacola, FL 32575
Position: Supt Manufacturing Services *Employer:* Monsanto Textiles Co. *Education:* BS/Ind Engg/Univ of AR. *Born:* 12/26/33. Native of AR. Worked as chem engr for Alcoa 1956-62. Work as cost evaluation engr for DuPont 1962-63. Joined Monsanto Co in Textiles Fibers Div as ind engr in 1963. Held position of Gen Superintendent of Engg & Services at Monsanto's largest plant. Responsibilities included ind, elec, & mech engg depts, all maintenance, mtls handling, & service operations. Also held position of Supt. Fiber Mfg including Yarn, Polymer, Non Wovens, and Resins. Present position of Supv. Manufacturing Services. Mbr of Engg Advisory Council for Univ of FL, and Acadamy of Industrial Engg for Univ. of Ark.

Self, Sidney A
Business: Mech Engg Dept, Stanford, CA 94305
Position: Res Prof. *Employer:* Stanford Univ. *Education:* PhD/Phys/London Univ; MA/Engg/Cambridge Univ; BSc/Phys/London Univ. *Born:* 4/16/28. Native of London, England. Employed 1949-64 British Admiralty at Services Electronics Res Lab in res on Storage Tube Displays, High Power Microwave Tubes, Microwave Discharges & Gas Lasers. Lectr in Engg, Cambridge Univ 1964-5. Since 1965, at Stanford as Sr Res Assoc in Plasma Res Inst working in gas discharges, plasma waves and instabilities, and since 1971 as Res Prof in Mech Engg Dept. Current res activities include dev of diagnostic techniques for combustion and plasma flows, electrostatic particulate cleanup, plasma and gas-discharge physics, radiative heat-transfer and optical properties of materials. Author/co-author of over 150 publications. *Society Aff:* MRS, APS, OSA, ΣΞ.

Selig, Ernest T
Business: Marston Hall, Rm 28, Dept of Civ Engrg, Amherst, MA 01003
Position: Prof *Employer:* Univ of MA *Education:* PhD/CE/IIT; MS/Applied Mech/IIT; BME/ME/Cornell Univ *Born:* 11/25/33 BME with distinction, Cornell 1957; MS in Mechanics, Ill Inst of Tech, 1960; PhD in CE, Ill Inst of Tech, 1964. Served on res staff of IIT Res Inst 1957-68. Mbr of Civ Engrg Faculty of State Univ of NY at Buffalo 1968-78. Joined Univ of MA Faculty beginning 1978. Previously natl officer of Geotechnical Engrg Div of ASCE and Soil and Rock Ctte of ASTM. Editor of ASTM Geotechnical Testing Journal. Served as chrmn of cttes of ASCE, ASME, ASTM, TRB, SAE. ASME Pi Tau Sigma Gold Medal Award 1961. ASTM D-18 Outstanding Service Award 1979; ASTM Distinguished Service Award in Publications 1983, ASTM Award of Merit and Fellow 1984. Mbr of Tau Beta Pi, Pi Tau Sigma, Phi Kappa Phi, Sigma Xi, Chi Epsilon. Prof Engr MA and NY. *Society Aff:* ASCE, ASTM, AREA, TRB

Selk, John E
Business: 2500 West Sixth St, Lawrence, KS 66044
Position: Exec VP *Employer:* Landplan Engrg, P.A. *Education:* MS/CE/Univ of KS; BS/CE/Univ of KS *Born:* 8/3/49 Began professional career with KS State Hgwy Commission, 1972. Design engr on major urban freeway projects, 1972-1975. Responsible for geometrics review of all interstate and urban inter-changes, special applications of traffic safety devices to problem locations, 1975-1978. Joined Peters-Williams-Kubota, P.A., and directed Engrg/Surveying Division, 1978-1979. Inc Landplan Engrg, P. A., December 1979. Responsible for operation and direction of firm whose activities include civil engrg, surveying, landscape architecture, community planning and development, 1980-present. Formed TRAF/TRAN Engrg, January 1981, subsidiary of Landplan, providing comprehensive traffic & transportation engrg services. Registerd Land Surveyor, KS. Reg PE, KS, MO, CO. *Society Aff:* ASCE, ITE, ACSM.

Selle, James E
Home: 4755 W 101 St Place, Westministter, CO 80030
Position: Associate Scientist *Employer:* Rockwell International *Education:* PhD/Met Eng/Univ of Cincinnati; MS/Met Eng/Univ of WI; BS/Met Eng/Univ of WI. *Born:* Sept 1931 Waukesha Wisc. Sr Res Specialist & Group Leader Monsanto Res Corp/Mound Lab 16 yrs; G M Res Staff 1 1/2 yrs; Dayton Malleable Iron Co 1 yr; Oak Ridge Natl Lab 1974-80, Rockwell Intl 1980-present, presently working in area of Plutonium compatibility. Other projs have included lithium corrosion, determination of binary & ternary phase diagrams of plutonium, internal friction of neptunium & plutonium, & materials problems associated with radioisotopic heat sources. Developed 2-dimensional color coded 'Summary of Binary Phase Diagrams,' which summarizes important characteristics of all known binary phase diagrams on one wall chart. ASM Wilson Award 1972. *Society Aff:* ASM.

Sells, Harold R
Home: 12122 Fairhope Rd, San Diego, CA 92128
Position: Senior Engr *Employer:* San Diego Gas & Electric Co *Education:* MBA/Management/Columbia Univ; BS/Petro Engrg/KS Univ *Born:* 6/26/17 Native of Effingham, KS. Field Petroleum Engr for Kerr-McGee Corp. Served with US Navy as Engrg Officer 1942-1946. Taught Petroleum Engrg at KS Univ. With Sohio Petroleum Co as Reservoir Engr and Rockefeller Brothers as Assistant Petroleum Mgr. Pres of Sells Consulting Service 1959-1979. Since 1979 employed by San Diego Gas & Electric Co as Senior Geothermal Engr, responsible for geothermal development and reservoir evaluation. Member of TauBetPi, SigTau and BetGamSig. Distinguished Petroleum Engrg Alumnus, KS Univ. Geothermal Technical Editor for SPE of AIME. Registered PE (TX). Hobbies are golf, tennis, bowling, fishing and hunting. *Society Aff:* SPE of AIME, AAPG, GRC

Selm, Robert P
Business: P.O. Box 1648, Salina, KS 67401
Position: Partner *Employer:* Wilson & Co Engrs & Architects *Education:* BS/ChE/Univ of Cincinnati *Born:* 8/9/23 Prior to joining Wilson & Co in 1954, Mr. Selm was a regular officer for five yrs in the US Army Ordnance Corps. Dir of the Ind Div since 1962, Mr. Selm is the Partner-in-Charge of all mechanical, elec, chem and process engrg design. Mr. Selm has been the Chrmn of the State KS Engrg Soc Environmental Resources Ctte since 1968 and served on the KS State Bd of Health's Solid Waste Adv Council, which drafted the Regulations for Implementation of the State's Solid Waste Act. *Society Aff:* NSPE, AIChE.

Selner, Ronald H
Home: 4060 Sylvan Drive, Enon, OH 45323
Position: Chief, NDE Group *Employer:* Universal Technology Corp. *Education:* BS/Physics/Western MI Univ. *Born:* Jan 1930. Tested nuclear reactor components Argonne Natl Lab 1958-70; Sr NDT Engr & Proj Mgr on Air Force contract at Wright-Patterson AFB Dayton Ohio 1970-80; 1980- ; Chief of NDE Group which provides engrg services in NDT/NDE to government contractual programs and industrial clients. Genl Amer Res Div/GATX 1970-75; joined Universal Technology Corp Jan 1976. Dir ASNT 1972-75; Assoc Tech Editor Materials Evaluation. ASNT Fellow Award Sept 1976. Hobbies: photography & stamp & coin collecting. ASNT Lester Honor Lecture 1986. *Society Aff:* ASNT, ASM, AWS

Seltzer, Leon Z
Home: 1012 Raritan Dr, Apt 102, St. Louis, MO 63119
Position: Dean of College & Ch Exec Officer. *Employer:* St Louis University (Parks College). *Education:* BSEAeE/AerobbEngg/Univ of MI; PhC/Pharmacy/Univ of IL. *Born:* April 17, 1914. Aeronautical Engr Douglas Aircraft Co; Prof & Chmn Aeronautical Engrg Va Polytechnic Inst & St Univ; Prof & Chmn Aerospace Engrg West Va Univ. ASEE Chmn Aeronautical Div & Mbr Genl Council; Mbr Scientific Advisor Group for Aviation Sys of US Army; Assoc Fellow AIAA; Natl Pres Sigma Gamma Tau; Natl Aerospace Honor Soc; Chmn Metro-East Regional Council on Interinstitutional Cooperation; ECPD State Guidance Chmn West Va; Reg P E; Chmn Aviation Ctte St Louis Regional Commerce & Growth Assn. 1972 Civic Award for Outstanding Contribution in Field of Aerospace Education by AIAA; Citation for Outstanding Serv to AFROTC 1972. *Society Aff:* AIAA, SAE, ASEE, AHS.

Selvidge, Harner
Home: Box 1128, Sedona, AZ 86336
Position: Consultant. *Employer:* Self. *Education:* SD/Comm Engr/Harvard; SM/Comm Engr/Harvard; MS/Elec Engr/MIT; BS/Elec Engr/MIT. *Born:* Oct 16, 1910. Instructor Harvard 1935-37; Assoc Prof Kansas State Univ 1938-41; Sr Staff Applied Physics Lab Johns Hopkins Univ 1942-45; Dir Special Prods Dev Bendix Aviation Corp 1945-60; Exec V P Meteorology Res Inc 1960-69; Cons Instrumentation 1969- . Pres Sky Mountain Aviation 1983- . Chmn Coconino County Air Pollution Control Hearing Bd 1973- . Naval Ordnance Dev Award 1945, OSRD Serv Certificate 1945. Fellow IEEE, Assoc Fellow AIAA, Mbr Amer Meteorological Soc. Tr & P Pres Soaring Soc of Amer. *Society Aff:* IEEE, AIAA, AMS

Semchyshen, Marion
Business: 1600 Huron Parkway, Ann Arbor, MI 48106
Position: VP Research. *Employer:* Climax Molybdenum Co of Michigan. *Education:* PhD/Met Eng/Wayne State Univ; MS/MEt Eng/Univ of MI; BS/Met Eng/Wayne State Univ. *Born:* August 25, 1917. Native of Detroit Mich. Climax Molybdenum Co AMAX Corp 1942-present; currently VP, Met Market Dev, spent sev yrs as VP of Res of Ann Arbor Res, just prior to present position. Active in sev prof socs; on a num of natl cttes of AIME, ASTM, ASME, ASM, MPC; currently Chmn of ASTM-ASME-MPC Joint Cttes. Presented ASTM Gillette Mem Lecture 1975. Hobbies: golf & classical music. *Society Aff:* AIME, ASM, ASTM, ASME.

Semmelmayer, Joseph A
Business: 3250 Van Ness Avenue, San Francisco, CA 94109
Position: Regional Coordinator Engineering Services *Employer:* Eastman Kodak Company. *Education:* BS/Gen Studies/Univ of Rochester. *Born:* 7/13/30. Native of Rochester NY. Employed by Eastman Kodak Film Testing Div to 1964 in dev work on densitometers, sensitometers & quality control of film prods. Tech Rep for Business Sys Markets Div to 1967; Prod Planner for Motion Picture Markets to 1970. Sales & Engrg Rep to motion picture labs & television stations to 1979. Married; 4 children. Regional Coordinator engineering Services San Francisco to date. *Society Aff:* SMPTE.

Semon, Warren L
Business: 313 Link Hall, Syracuse, NY 13210
Position: Dean/Sch of Computer & Info Science. *Employer:* Syracuse University. *Education:* PhD/Computers/Harvard Univ; MA/Mathematics/Harvard Univ; SB/Meteorology/Univ of Chicago. *Born:* Jan 1921 Boise Idaho. Married Ruth Swift 1945; children: Warren Jr., Nolan, Jonathan, Sue. Meteorologist USAAc 1943-46. Instructor Hobart Coll 1946- 47; student 1947-49; Res Assoc 1949-54; Asst Dir Harvard Computation Lab 1954- 61; Mgr Applied Math Sperry Rand Res Center 1961-64; Mgr Computation & Analysis Lab Burroughs 1964-67; Prof Computer Sci Syracuse Univ 1962- , Dir Sys & Info Sci Prog 1968-76, Dean Sch of Computers & Info Sci 1976- . Fellow IEEE. *Society Aff:* IEEE, ACM, MAA.

Sen, Amiya K
Business: Mudd Bldg, Columbia University, 120th St. & Amsterdam Ave, New York, N.Y 10027
Position: Prof. *Employer:* Columbia Univ. *Education:* PhD/EE/Columbia Univ; MS/EE/MIT; Dipl IISc/EE/Indian Inst of Sci *Born:* 12/14/30. in Calcutta, India. Worked with GE Co as test engr & design analysis engr from 1953-55. With Columbia Univ since 1958, becoming Prof of EE in 1974. Dir of Energy Conversin Lab since 1963. Had various res grants & contracts from AFOSR, NASA & NSF. Published more than fifty papers in scientific journals on energy conversion, solar wind-geomagnetic field interaction, plasma waves, plasma stability & control & plasma turbulence. Made significant contributions in magnetospheric physics, trapped particle instabilities & control of instabilities in fusion plasmas. *Society Aff:* IEEE, APS

Seng, Thomas G, Jr
Business: One Sci Rd, Glenwood, IL 60425
Position: VP. *Employer:* Mtls Res Lab Inc. *Born:* 10/31/40. Native of Winnetka, IL. Worked for Alco Spring Co on Atlas Missle support system. Taught for Metals Engg Inst. Past Pres of ASM Calumet Chapter. Natl Council mbr of ASM's Profl Dev Council, & Marketing & Mgt Committee. Currently VP in charge of sales for Mtls Res Lab Inc. Responsible for co growth in the mech testing field. Hobbies include: golf & old cars. *Society Aff:* ASTM, ASM, ISS.

Senior, Charles R A
Business: 1000 Hess St, Saginaw, MI 48601
Position: V Pres & Mgr. *Employer:* Baker Perkins Inc. *Born:* July 5, 1921 Vancouver BC. BSc (1st class honours) Queens Univ, Kingston Ont 1948. Prod foreman wth Proctor & Gamble, Hamilton Ont 1948-55. Prod G.M. Drew Chem Co, Boonton N J 1955-60. Genl Mkting Mgr, Podbielniak Div, Dresser Indus 1961-67. Genl Mkting Mgr, Chem Machinery Div, Baker Perkins Inc 1967-70. Group Mkting Mgr, Baker Perkins Holdings Ltd, Vice Pres. & Gen. Mgr.Eng. Presently V P & Gen Mgr of the Chem Machinery Div, Dir Baker Perkins Inc, & Chmn Baker Perkins Chem Machinery Ltd, Stoke-on-Trent England. Canadian Navy 1941- 46. Assn: Prof Engrs Ontario 1949-76, AIChE. Dir, Baker Perkins Guittard, SA, Chelles, Paris, France.

Senior, Thomas BA
Business: Dept Elec Eng & Computer Sci, Ann Arbor, MI 48109-2122
Position: Professor. *Employer:* Univ of Michigan. *Education:* PhD/Appl Math/Cambridge Univ; MSc/Appl Math/Manchester Univ; BSc/Appl Math/Manchester Univ. *Born:* 6/26/28. Yorkshire England. 5 yrs with the Royal Radar Estab (RRE) in Malvern England before joining the Radiation Lab of the Univ of MI. Apptd Prof of Elec & Comp Engrg 1969, Dir of the Lab in 1975, Assoc Chmn Dept Elec Engrg & Comp Sci in 1985 and Acting Chmn in 1987. Primarily involved with radar scattering & propagation studies. Editor of "Radio Sci" (1973-78) Chmn of US Natl Comm of URSI (Intl Sci Radio Union) (1982-84) and now Chmn URSE Commission B. Fellow of IEEE & mbr of Sigma Xi, Tau Beta Pi, Eta Kappa Nu. *Society Aff:* IEEE, ΣΞ, TBΠ, HKN.

Senitzky, Benjamin
Business: Route 110, Farmingdale, NY 11735
Position: Prof *Employer:* Polytechnic Univ *Education:* PhD/Physics/Columbia Univ; BS/Elect Eng/Columbia School of Eng *Born:* 11/15/26 Served in US Army from 1943- to 1945. Mbr of Technical Staff of Bell Telephone Labs from 1956 to 1959. Worked for Technical Res Group Inc from 1958 to 1966. Joined faculty of Polytechnic Inst of NY in 1967 and is presently Prof of Electrical Engrg. Res interest is the interaction of radiation and matter. Enjoy playing in chamber music groups and physical activities such as swimming and hiking. *Society Aff:* ASEE, IEEE

Senning, Herbert C
Business: 135 S LaSalle St, Chicago, IL 60603
Position: VP-Engg. *Employer:* Liquid Carbonic Corp. *Education:* BS/ChE/Univ of IL *Born:* 1/23/26. Active in chem & engg admin & mgt for over 30 yrs in various ca-

Senning, Herbert C (Continued)
pacities. Have served on various committees of the Compressed Gas Assn (CGA), & past chrmn of the Air Separation Equip Committee. Mbr of Am Chem Assn, Will County Historical Society, Am Mgt Assn, Exec Club of Chicago, & Am Inst of Chem Engrs for many yrs; also corporate rep on Engrs Joint Council. Lifelong native of IL, Chicago area. Active in Easter Seal Soc, Lions, Elks, & numerous Masonic organizations. Past Pres of Will County Easter Seal Soc, Joliet Shrine Club, & New Lenox Athletic Club. Precinct Committeeman for 20 yrs, Lockport Township Chrmn, 10 yrs. Enjoy golf, fishing, swimming, and boating. Married, wife, Mary Kay; one daughter, Joyce. *Society Aff:* AIChE, AMA, ACS, CGA.

Senoo, Yasutoshi
Business: 7 Horie, Matsuyama, Japan 799-26
Position: Chief Engineer *Employer:* Miura Co Ltd *Education:* Dr. of Engineering/Mechanical Eng./Kyushu University, Japan; Master of Engineering/Mechanical Eng. /Kyushu University, Japan *Born:* 02/25/24 Graduated from Kyushu Univ Japan & received doctor's degree in 1952. Studied at M.I.T. Gas Turbine Laboratory 1954-57 as a Fulbright scholar. Asst Prof of MIT 1958-60. Sr Engrg Specialist of Airesearch Manufacturing Co, Garrett Corp 1960-64. Prof of Kyushu Univ 1964-87 & dir of research inst. At present Emeritus Professor, Kyushu Univ & Chief Engineer of Miura Co. Conferred Moody Award by ASME Hydraulic Div, 1960. Nominated to ASME Fellow, 1985. Also received many awards from Japanese societies. *Society Aff:* ASME

Sensiper, Samuel
Business: PO Box 3102, Culver City, CA 90231
Position: Consulting Engr. *Employer:* Self. *Education:* ScD/Elec Engg/MA Inst of Tech; EE/Elec Engg/Stanford Univ; BS/Elec Engg/MA Inst of Tech. *Born:* 4/26/19. From 1941, except for Doctor's degree yrs, at Sperry Gyroscope, Hughes Aircraft, Aerojet Gen, TRW in asst proj engr to lab mgr positions and from 1970 except 1973-1975 at Transco Products as self-employed consultant in fields of microwave test equip, components, electron tubes, solid state devices and antennas over range from ELF to millimeter waves. 23 patents, some with assocs, 16 papers or communications in IEEE journals and intl symposium proceedings record some of work. Participation in IEEE committees at local and natl level including chairmanship of two intl symposiums. Fellow IEEE and AAAS. MIT Industrial Electronics Fellow, USN Bureau of Ships Commendation. *Society Aff:* IEEE, AAAS, BEMS, AOC, NSPE, FAS, ΣΞ.

Sentman, Lee H
Business: AAE, 105 T. B, 104 S. Matthews, Urbana, IL 61801
Position: Prof of Aero & Astro Eng *Employer:* Univ of IL *Education:* PhD/Aeronautics & Astronautics/Stanford Univ; BS/Aero Eng/Univ of IL *Born:* 01/27/37 Native of Chicago, IL. Developed analysis procedures for the design of near earth satellites for Lockheed Missile & Space Co 1959-65. Mbr of the faculty of the Univ of IL 1965-present. Active research areas include power spectral performance, mode-media interactions and saturation effects in chem lasers, molecular energy transfer in vibrationally excited molecules, kinetic theory of gas-solid interactions, effects of delayed energy release on blast waves. Conslt to various Aerospace companies. Univ of IL Loyalty Award 1977; W L Everitt Undergrad Teaching Excellence Award 1969, Assoc Fellow AIAA. *Society Aff:* AIAA, OSA, APS

Sepsy, Charles F
Home: 206 W 18th Ave, Columbus, OH 43210
Position: Prof & Research Supervisor. *Employer:* Ohio State Univ. *Education:* MS/Mech Engg/Univ of Rochester; BS/Mech Engg/Univ of TN. *Born:* 5/19/24. Fellow ASHRAE. Dir Charles F Sepsy; Prof of Mech Engrg at Ohio State Univ - recog for his univ res in energy conservation principles for bldgs. Active mbr of ASHRAE - Dir & Regional Chmn of Region V 1971-74, National President af ASHRAE 1980-81 named Tech Man of Yr by the Columbus Chap, Pres of the chap 1967. Named Tech Man of Yr by the Columbus Ohio Tech Council in 1975. Has served on a num of tech cttes & task groups & on genl cttes. In addition to teaching duties, has been a cons on indus & military assignments. Degrees from the Univ of Tenn & the Univ of Rochester. Mbr ASME, Hon Fellow Automated Procedures for Engrg Cons Inc; mbr Tau Beta Phi, Phi Kappa Phi, Sigma Xi. Bd of Gov 1981 AAES. *Society Aff:* ASHRHE, ASME, ASEE.

Sereda, Peter J
Home: 153 Roger Rd, Ottawa, Ont, Canada K1H 5C7
Position: Consultant *Employer:* Self *Education:* DSc/Matr Sci/U of Windsor; MSc/Chem Eng/U of Alberta; BSc/Chem Eng/U of Alberta *Born:* 4/25/19. Native of Saskatchewan Canada. Employed since 1944 by the Natl Res Council of Canada as a res engr in the Atomic Energy Proj at Chalk River Ont to 1949 and since with the Div of Bldg Res as res officer - 1961 to 1979 as Hd of Bldg Matls Sec 1966 to 1982 directing an interdisciplinary team of scientists & engrs on the basic properties & behavior of bldg matls. Retired in 1982, acting as a consultant to various agencies. Author of over 50 sci & tech papers. Served in var capacities on cttes of ASTM, TRB of U S Acad of Sci, RILEM, & Amer Ceramic Soc as Tr of Cements Div. Sam Tour Award from ASTM, Gold Medal from EIC, Fellow Amer Ceramic Soc LE Copeland Award of Amer Ceramic Soc., Silver Jubilee Medal, Queen Elizabeth II, Editor, Int Journal on Durability of Bldg Matls. Honorary Degree of Dr of Sci, U of Windsor. *Society Aff:* ACS, OAPE, ASTM

Serna, Francisco J
Business: Felix Berenguer 126, Mexico, D.F., 11000, Mexico
Position: Pres. *Employer:* Inarco. *Education:* BS/Civ Engg/TX A&M Univ. *Born:* 12/3/25. BS in Civ Engrg, TX A & M 1947. Pres 1956-Constructora Franser SA where has directed complex projs ie: slip form, shotcrete, concrete shells, electro-mech construction, etc. Pres 1966-Inarco, consulting group with 100 plus profls, in charge of design stds for Mexico City 1976; bldg code for Guerrero (Acapulco); consutor to dept of environment, London, 1963-68; foreign bldg office, Wash 1967-79; Ministry of Works, Nicaragua 1973 & others. Pres Deplan, SA 1971-74 where coordinated urbn planning city of Panama (El Maranon) and the reconstruction plan for Managua (after 1973 earthquake). Pres 1973- Inarco- eduvision, an education tech group. Lectr in Mexico, US, Guatemala, Panama, Colombia & Chile in engg & construction mgt. Pres 1981- Franina, SA, Pres 1983- Serfran, SA. Fellow, ASCE; Pres Mexico Sec ASCE 1978. Intl Contact Dir ASCE 1983-86. Author Construction Management system published Mexico 1980. Coordinator Mexico City Design Standards, Public Works Dept, 3 vol, 1976. Commissioner for Bldg code, Mexico City 1979 and in 1986. Originator patents on const improvements numbers 110519 and 102666. His biography appears in the 6th ed of Marquis Who's Who in the World. *Society Aff:* ASCE, ASTD, AIAM, SMIEC, EERI, SMIS, SISF

Serovy, George K
Business: Dept of Mech Engrg, Iowa State Univ, Ames, IA 50011
Position: Anson Marston Distinguished Professor *Employer:* Iowa State Univ. *Education:* PhD/Mech Engg & Theo & Applied Mechs/IA State Univ; MS/ME/IA State Univ; BS/ME/IA State Univ *Born:* Aug 29, 1926 Cedar Rapids Iowa. US Navy 1945-46. Res on compressor aerodynamics. NACA (Cleveland) 1949-53. Mech Engrg fac mbr at Iowa State since 1953. Organized turbomachinery teaching & res progs at ISU, Exec Dir ASME Turbomachinery Inst, ASME Gas Turbine Div citation 1972; NASA Cert for res 1975; Fellow ASME 1976. Anson Marston Distinguished Prof in Engg, ISU 1979, Mbr Sigma Xi, Tau Beta Pi, Pi Tau Sigma. *Society Aff:* ASME, AIAA, IAHR, ΣΞ

Seruto, Joseph G
Business: 5263 N 4th St, Irwindale, CA 91706
Position: President. *Employer:* Specialty Organics Inc (Self). *Education:* BA/Organic Chem/OH State Univ; MS/Organic Chem/NY Univ. *Born:* 1/14/12. Elementary & secondary educ in the public schools of Meriden Ct; BS Ohio State Univ 1939; MS N Y Univ 1942. Res chemist in res labs of Amer Cyanamid, Bound Brook N J.

Seruto, Joseph G (Continued)
Author of a num of pats on dyestuffs & intermediates. Pres, Genl Mgr & Main Stockholder of Specialty Organics Inc Irwindale Calif - R&D lab & pilot plant spec in custom synthesis, manufacture, and process res for the manufacture of indus & agri synthetic organic chems & intermediates 1963-. International Consultant. *Society Aff:* AIChE, ACS, AIC

Servais, Ronald A
Home: 2130 Vienna Parkway, Dayton, OH 45459
Position: Prof of Chem Engr. *Employer:* Univ of Dayton. *Education:* DSc/ChE/WA Univ; MS/Fluid Mechanics/St Louis Univ; BS/Aero Engg/St Louis Univ. *Born:* 4/6/42. Aerodynamics Engr, McDonnell Aircraft Corp, 1963-66. Prof, Univ of MO-Rolla, 1968-72. Res Scientist, Reentry Mtls, USAF Mtls Lab, 1973-74. Prof, Univ of Dayton since 1973; Chrmn of Chem Engg Dept, 1975-82. Received Outstanding Engrs and Scientists Award in Dayton, 1978. Consultant in truck & automobile aerodynamics. Res emphasis in fluid flow & heat transfer at high temperatures and process modeling. *Society Aff:* AIChE, AIAA, ASEE, AEMS.

Servi, Italo S
Home: 3 Angier Rd, Lexington, MA 02173
Position: Consulting Engineer *Employer:* Self Employed *Education:* ScD/Metallurgy/MIT; SM/Metallurgy/MIT; ScD/Industrial Chemistry/Univ Milan, Italy. *Born:* 10/3/22. Union Carbide Metals 1951-59: R&D corrosion, high-temp alloys, refractory metals. Special Metals Inc 1959-62: Tech V P, vacuum melting, high temp alloys, powder met. Kennecott Copper Co 1962-1981: Dir Materials Tech, fabricated product & process dev, indus res mgmt methodologies, organizational communication. Self employed consultant since 1981: materials dev and applications, fabrication tech, project planning and control; tech mgmt, marketing research. Formly cons, Natl Acad of Sci; Bd of Dirs TMS/AIME, Chmn Boston Sect AIME. Fellow Inst of Matallurgists (London) since 1983: lecturer in continuing education, Northeastern Univ (concurrent activity). Reg PE Mass. Fellow ASM. Mbr Boston Chapter ASM Exec Ctte. *Society Aff:* TMS-AIME, ASM, SAMPE, APMI.

Sesonske, Alexander
Home: 16408 Felice Drive, San Diego, CA 92128
Position: Prof Emeritus of Nucl Engg. *Employer:* Purdue Univ. *Education:* PhD/ChE/Univ of DE; MS/Che/Univ of Rochester; BChE/ChE/RPI. *Born:* 6/20/21. Has been active in nuclear energy activities since 1943. Initially, with Manhattan Proj in isotope separation dev, then with Los Alamos Scientific Lab in numerous activites including reactor dev. Purdue faculty since 1954 & Prof 1959-86 while continuing nucl-related activities as consultant. Prof Emeritus and independent consultant since 1986. Widely-known as author (with S Glasstone) of "Nuclear Reactor Engineering" and as author of "Nuclear Power Plant Design Analysis–. ANS Arthur Holly Compton Award, 1987. Contbr of numerous articles to profl journals in nucl reactor engg, heat transfer, & nucl fuel mgt. *Society Aff:* ANS, AIChE, ASEE, ΣΞ.

Sessler, Gerhard M
Home: Fichtestrasse 30 B, 6100 Darmstadt, W-Germany
Position: Prof *Employer:* Technical Univ *Education:* Dr rer nat/Physics/Univ of Goettingen; Diplom/Physics/Univ of Goettingen; Vor-Diplom/Physics/Univ of Munich *Born:* 02/15/31 b. in Rosenfeld, Germany. Mbr of Technical Staff, later Spvr, Acoustics Res Dept, Bell Labs, Murray Hill, NJ, 1959-75. Work in electroacoustics, plasma physics, and solid-state physics. Prof of Electroacoustics and Dir, Inst of Electroacoustics, Tech Univ Darmstadt since 1975. Published about 130 articles in scientific and technical journals, granted 18 US patents. Book: *Electrets* (Springer Verlag, 1980, 2nd Edition 1987). Invention: Electret microphone. Awards: Callinan Award, Electrochemical Soc, 1970; Sr Award, IEEE, 1971; Fellow, IEEE, 1976, Thomas W. Dakin Award, IEEE DEIS 1986. Married with Renate Schulz, 1961. Children: Cornelia, Christine, and Gunther. Chrmn, German Acoustics Association (DAGA). *Society Aff:* IEEE, APS, ASA, DPG, VDE, NTG, DS

Sessler, Stephen M
Business: Newcomb & Boyd, One Northside 75, Atlanta, GA 30318-7714
Position: Partner *Employer:* Newcomb & Boyd *Education:* BS/ME/GA Inst of Tech *Born:* 8/9/48 Joined Newcomb & Boyd in 1970, and became Partner in 1984. Developed Newcomb & Boyd's consulting practice in acoustics and noise control. Projects include US Naval Communication Station, Rota, Spain and the Midfield Terminal Complex at Hartsfield Intl Airport in Atlanta. Also Project Mgr and chief mechanical design engr for mechanical and electrical systems in major commercial and institutional bldgs. Projects include Southern Bell Center, Richard B Russell Federal Bldg and Courthouse, and the Coca-Cola Company Headquarters Complex in Atlanta. Chrmn of ASHRAE TC 2.6 from 1978-1981; ASHRAE Willis H Carrier Award 1977; Past Assoc Editor for INCE Noise/News; Past Secretary, GSPE Ethical Practices Committee; 8 papers published on acoustics, mechanical system design, and management of design firms. Registered in 4 states and by the National Council of Engrg Examiners. *Society Aff:* ASHRAE, ASA, INCE, AES, NSPE

Seton, Waldemar
Business: 317 SW Alder St, Portland, OR 97204
Position: President. *Employer:* Seton, Johnson & Odell Inc. *Education:* BS/ChE/Stanford. *Born:* Nov 1929 Portland Ore. With Monsanto 1952-55 in Santa Clara, Avon, & St Louis. Entered pulp & paper field with Western Kraft from 1955-62, with Georgia Pacific from 1962-72: Tech Dir in Toledo & Samoa - as Ch Engr/Chemicals, was Proj Mgr for a num of capital projs. Formed proj mgmt firm in Portland Ore in 1972 as Waldemar Seton Co & formed Seton, Johnson & Odell Inc as a genl engrg firm in 1975. Projs are in chem, petro, pulp & paper, wood products, metals & food indus. *Society Aff:* TAPPI, AIChE.

Settles, F Stanley
Home: 1627 E La Jolla Drive, Tempe, AZ 85282
Position: Corp. Dir., Industrial Manufacturing Engineering (IME) *Employer:* Garrett Corp *Education:* PhD/Ind Engrg/AZ State Univ; MSE/Ind Engrg/AZ State Univ; BS/Ind Engrg/LeTourneau College; BS/Prod Tech/LeTourneau College. *Born:* 10/3/38. Denver Colo. With Garrett since 1962 in Design Engrg, Proj Engrg, Sys Analysis, Matl Mgmt, Ind. Engr Mgmt & Operations Management prior to present role; Pres IIE 1987-88, Dir - Opers Res Div IIE; Central AZ Chap AIIE Pres 1974-75; Mbr 'IIE Transactions' Edit Bd; Fellow IIE; Mbr TIMS, SME, AMer Mgmt Assn, Alpha Pi Mu, Sigma Xi; Mbr, Clerk & Pres Tempe Elem School Bd; Chmn Tempe Bd of YMCA; Natl Exec Ctte of YMCA Parent/Child Progs; Scoutmaster BSA; Maricopa Cty Health Council 1974-76; Bd Mbr ASU Alumni Assn; listed in Who's Who in America; Natl Chief Y-Indian Guide Longhouse 1978- 79. IIE VP - Sys Engrg & Tech Div 1981-83; Natl Council of YMCA 1979-83; Adjunct Faculty Ind. Engrg. ASU 1975-Present. *Society Aff:* IIE, TIMS, SME

Setzler, William E
Business: 277 Park Ave, New York, NY 10017
Position: Exec VP. *Employer:* Witco Chem Corp. *Education:* BCE/CE/Cooper Union. *Born:* 12/20/26. Native of NY. Held various mfg & engg positions at Argus Chem Corp, Brooklyn NY (1949-1966). Merged with Witco Chem Corp 1966 & named VP Engg 1969; Group VP 1971; Exec VP, Mbr Bd Dirs 1975. Hobbies include sailing. *Society Aff:* AIChE.

Sever, Lester J
Home: 2206 Evergreen Dr, Hendersonvl, NC 28739
Position: Retired *Education:* BEE/Power/OH State *Born:* 7/21/12. Born & raised in NW OH. Worked way thru college. Spent 8 yrs in various utility design, application & construction of power plants & substations; six yrs in oil refinery design & maintenance & construction of elec & mech facilities. Thirty one yrs in ind, commercial & institutional bldg & elec facility design. *Society Aff:* IEEE

Severin, Warren J
Business: 122 S Michigan Ave, Chicago, IL 60603
Position: Supt. *Employer:* Natural Gas Pipeline Co of Amer. *Education:* BSEE/Elec Engg/Univ of NB. *Born:* July 1923. Native of Hallam Nebr. US Navy 194346. With Natural Gas Pipeline Co of Amer since 1951. Assumed current respons as Supt of Mech Sys Dept in 1966. Chmn of Diesel & Gas Engine Power Div of ASME 1975-76 - ASME Diesel & Gas Engine Power Award 1974. Enjoy gardening & fishing. *Society Aff:* ASME.

Severinghaus, Nelson, Jr
Business: 612 Tenth Ave, N, Nashville, TN 37203
Position: Pres. *Employer:* Franklin Limestone Co. *Education:* BS/Mining Eng/Univ of AZ; EM/Mining Eng/Univ of AZ. *Born:* 8/27/29. Pres of Franklin Limestone Co, Chmn Franklin Brick Co and V. P. & a Dir of Franklin Industries, Inc in Nashville, TN. He holds a BS in mining engg and an honorary Engr of Mines degree from the Univ of AZ. Began his mining career in 1954 as a mine engr for TN Coper Co, in Ducktown, TN. Subsequently he joined GA Marble Co where he became pres. He is a past chrmn of the Ind Mineral Div of SME-AIME, 1980; Pres of Soc of Mining Engrs-AIME, 1984; Pres of Amer Inst. of Mining, MET and Petroleum Engrs; Dir, Natl Stone Assn. Recipient, Hardinge Award (AIME) 1984. *Society Aff:* AIME, SME-AIME.

Severud, Fred N
Business: 485 Fifth Ave, New York, NY 10017
Position: Consultant. *Employer:* Severud-Perrone-Szegezdy-Sturm. *Education:* CE/Civil Engg/Norwegian Inst of Tech. *Born:* June 1899. Independent practice as a cons engr since 1928. Recog auth on problems of struct design. Noted for his solutions to brick & block masonry problems in the early 1930's, high-rise hosp const in the 1940's, large span in the 1950's. Has lectured widely & is the author of num articles in tech magazines. Fellow ASCE, Hon Assoc Mbr of AIA. Mbr of many other engrg soc's. Ernest E Howard Gold Medal 1964; Engrg News Record Award 1967, Mbr Natl Acad of Engg. *Society Aff:* ASCE, NAE, AIA.

Sevier, William J
Business: 9461 El Cajon Blvd, Suite A, La Mesa, CA 92041
Position: Pres. *Employer:* Sevier Engg, Inc. *Education:* MS/Mech Engg/San Diego State Univ; BS/Mech Engg/San Diego State Univ. *Born:* 4/4/35. Native of San Diego, CA. Employed by Convair/Gen Dynamics while attending college during '50s. Conducted fluid systems analysis on various Convair designed aircraft. Served as Engg Officer with SAC, 1957-60. Returned to Convair 1960-1968. Participated in conceptual designs of space station and space shuttle progs and the mfg of experimental environmental control systems. Holds patent for oxygen separation system. Author of many papers. Since 1970, Pres of consulting engg firm that provides design of HVAC and plumbing systems for new construction and energy conservation studies for new and existing bldgs. Past Pres of San Diego Chap, NSPE. *Society Aff:* ASHRAE, ASPE, CSI, AEE.

Sevin, Eugene
Home: 7534 Westlake Terr, Bethesda, MD 20817
Position: Asst Deputy Undersecretary of Defense (Offensive & Space Systems) *Employer:* Department of Defense. *Education:* PhD/Appl Mech/IIT; MS/ME/CA Inst of Tech; BS/ME/IIT. *Born:* 1/5/28. Native of Chicago, IL. Married to Ruth Hirschson Sevin (1951), daughters Lori, Lynn, Lisa. With IIT Res Inst 1951-1969, Dir Eng Mech Div, 1964-1969. Adjunct Prof, Applied Mech, IIT 1963-1970. Consultant, 1969-1970. Prof & Chrmn, Mech Engg faculty, Ben Gurion Univ, Israel, 1970-1974. Defense Nuclear Agency 1974-86; with Office of Secretary of Defense since 1986. Specialize in structural dynamics, optimization & comp methods. Founding mbr and first chrmn, Computational Methods Committee, Applied Mechanics Div, ASME. Tech Advisor to DOD Shock & Vibration Info Ctr, Mbr Editorial Bd & reviewer for several tech journals. *Society Aff:* ASME, AIAA, NAE

Sewell, John I
Business: Univ of Tennessee, 103 Morgan Hall, Knoxville, IN 37916
Position: Assoc Dean. *Employer:* Univ of TN. *Education:* BS/Agr Engr/Univ of GA; MS/Agr Engr/NC State Univ; PhD/Agr Engr/NC State Univ. *Born:* 8/28/33. Native of Cave Spring GA. Lt US Army Corps of Engrs 1954-56. With Univ of TN, Knoxville since 1962; Assoc Dean of Experiment Station since 1986. Res, teaching & design in soil & water conservation, animal waste mgmt, infrared remote sensing. Lic P E TN. Chmn TN Sect Amer Soc of Agr Engrs 1972. Chmn SE Region ASAE 1973. Pres Univ of TN Chap of Gamma Sigma Delta 1971. Webster Pendergrass Outstanding Serv to Agr Award 1975. Pres Univ of TN Chapter, Phi Kappa Phi, 1979. Sr mbr of Amer Society of Agr Engineers; Colonel, Armor, USAR (Retired); Regional Dir. ASAE 1981-83. *Society Aff:* ASAE, SCSA.

Sexton, Joseph M
Home: 3135 La Quinta Dr, Missouri City, TX 77459
Position: Corporate Dir of Purchases *Employer:* Raymond Intl Builders *Education:* B/IE/Lehigh Univ *Born:* 09/08/18 Native of Newark, NJ. Started technical career with the M.W. Kellogg Co and advanced through a variety of technical and mgmt positions at their NY Office and Jersey City Mgf Plant. Assigned to London office and thence to Paris, France to become Gen Mgr of Kellogg's French Subsidiary Co. Joined Raymond Intl in 1963 in present position of Corporate Dir of Purchases and traffic responsible for policies and procedures in connection with worldwide procurement operation for engrg and construction contracts. Married Marian M. Meleady in 1944 and have four children. Active in Church and civic affairs and former elected councilman of Missouri City, TX. Avid oil painter and outdoorsman and rancher. *Society Aff:* AACE, ASME, NSPE, SAME

Sforza, Pasquale M
Business: Polytechnic University, Route 110, Farmingdale, NY 11735
Position: Prof *Employer:* Polytechnic Univ *Education:* PhD/Fluid Dynamics/ Polytech Inst of Bklyn; MS/Aero & Astro/Polytech Inst of Bklyn; B/AeE/Polytech Inst of Bklyn *Born:* 03/05/41 Native of NY. Attended Polytechnic Inst of Bklyn and stayed on as faculty mbr of its Aerodynamics Labs. Appointed to present position at Polytechnic Inst of NY in 1977. Hd of the Dept of Mech and Aerospace Engrg. at Polytechnic Inst of NY (1983-1986). Res on vortex flows resulted in several patents including those for a vortex wind turbine. Received AIAA technological achievement award for that work. Founded Flowpower, Inc, a fluid power consltg firm, in 1978. Appointed Assoc Editor of AIAA Journal in 1980, Book Ed 1983. Presently serving on Advisory Ctte to the NY State Legislative Commission on Sci and Tech. *Society Aff:* AIAA, ASME, NYAS

Sforzini, Mario
Business: via Savoldo 5, 20125 Milano, Italy
Position: Director. *Employer:* Ente Nazionale/L'Energia Elettrica. *Education:* PhD/Engg/Milan Polytech Inst; MS/EE/IIT - Chicago. *Born:* Sept 1928. Native of Bressana, Pavia, Italy. In the field of elec power transmission: design engr with CIELI Co since 1954 & res engr with Edisonvolta Co since 1958. With the R&D Dept of ENEL (Natl Elec Energy Agency of Italy): Ch Engr respon for the Transmission Div of the Elec Res Ctr since 1963; Dir of this ctr since 1974. Mbr & Chmn of Italian & Internatl (CIGRE, IEC, IEEE) cttes. Mbr of the Steering Ctte of ENEL '1000kV Project'. IEEE Fellow 1976. *Society Aff:* CIE, AEI, CIGRE.

Shaad, George E
Home: 322 Oakridge Dr, Schenectady, NY 12306
Position: Consultant. *Employer:* Self-employed. *Education:* BS/Chem Engr/Univ of KS. *Born:* Sept 1915. Employed by G E Co 1935, spec in paper indus engrg 1945. Respon for G E hdqtrs paper indus engrg 1955 to retirement 1972. Presently cons to paper indus, spec in operational requirements of paper making & processing equip & assoc drive sys. Fellow Tech Assn Pulp & Paper Indus 1968. Engrg Div Award 1968. Secy Engrg Div 1970-72, V Chmn 1972-74, Div Steering Ctte 1974-76. *Society Aff:* TAPPI.

Shabaik, Aly H
Business: 6532 Boelter Hall, UCLA - Engineering, Los Angeles, CA 90024
Position: Prof of Engg. *Employer:* Univ of CA. *Education:* PhD/Mech Engg/ Univ of CA, Berkeley; MS/Mech Engg/RPI, Troy, N.Y.; BS/Mech Engg/Cairo Univ, Cairo, UAR. *Born:* 3/1/37. Born in Egypt. Has been a mbr of the faculty of the School of Engineering, Univ of CA since 1967. Authored numerous tech papers published in national and international journals. Reg Professional Manufacturing Engg in the State of CA. Listed in Who s Who in the West, Who s Who in America, Dictionary of International Biography, Men of Achievement, Eminent Engrs of Tau Beta Pi. UNDP expert; OAS expert; U.S Scientist NSF intl prog; visiting prof Universite de Technologie de Compiegne, visiting prof Universidad Cat'olioca del Chile, Visiting Sci, Kuwait Inst for Sci Res. Areas of expertise: metal-forming, metal cutting, plastic flow of metals, and friction and wear. Has been consultant to various industries in the USA, Europe,and Middle East. Was awarded "Distinguished Teaching Award," ASEE Resident Fellowship, and Ford Motor Co Fellowship. Assoc Ed, ASME. *Society Aff:* ASME, SME, ASM, ASEE.

Shackelford, James F
Business: Dept of Mech Engr, Davis, CA 95616
Position: Prof & Assoc Dean-Undergrad Studies *Employer:* Univ of CA. *Education:* PhD/Mat Sci & Engr/Univ of CA; MS/Ceramic Engr/Univ of WA; BS/Ceramic Engr/Univ of WA; AA/Engg/Yakima Valley College. *Born:* 9/1/44. With Univ of CA, Davis since 1973. Asst Prof 1973-79. Assoc Prof 1979-84. Prof 1984-present, Assoc Dean-Undergrad Studies 1984-present. ASME (UCD Student Branch) Outstanding Teacher Award 1976/77. Consultant to Lawrence Livermore Lab 1977 to present. ASM (Sacramento Valley Chapter) Secy 1975/76, VChrmn 1976-77, Chrmn 1977/78. Chrmn, Univ of CA Committee on Subj A 1978/79, 1979/80. Author of over 40 publications with emphasis on amorphous solids. Author "Introduction to Materials Science for Engineers–, (Macmillan Publ Co, 1985). *Society Aff:* ASM, ACS, TMS.

Shadley, Frederic C
Home: 567 Abilene Trail, Cincinnati, OH 45215
Position: Mgmt Consultant. *Employer:* Self. *Born:* May 1924. MS Harvard 1949; BS Notre Dame 1948. Native of Meadville Pa. Navy Electronics Technician 1944-46; instr at Naval Res Lab school. Factory Technician, then Field Engr for Emerson Elec, St Louis 1949-51. With Avco Corp 1951-76. Ch Engr 1964-66, V P Mkting 1966-69; Div V P, Genl Mgr 1969-76. Began cons 1976. Mbr NSPE, OSPE, IEEE. Mbr Engrg Adv Councils for Notre Dame & OSPE. Awarded 1st Notre Dame Engrg Honor Award 1974. Married, 5 children. Enjoy family activities, bridge, travel.

Shaffer, Bernard W
Home: 18 Bayside Dr, Great Neck, NY 11023
Position: Prof & Consultant. *Employer:* Poly Inst of NY. *Education:* PhD/Appl Math/Brown Univ; MS/Appl Mechanics/Case Inst of Tech; BME/Mech Engg/ College of City of NY. *Born:* 8/7/24. Aero Res Scientist NACA (Now NASA), Cleveland, 1944-47; Special Lecturer Appl Mechanics, Case Inst of Tech 1946-47; Res Assoc Grad Div of Appl Math, Brown Univ 1947-50; Professorial Faculty and Proj Dir-Res Div, NYU 1950-73; Prof Mech and Aero Engrg, Poly Univ of NY 1973-; Consultant to Private Cos; Hon Editorial Advisory Bd, Intl Journal of Mech Sciences; Res Committee of Natl Research Council; Richards Meml Award ASME 1968; res and publications on Theory of Metal Cutting; Elastic-Plastic Analysis of Wide Curved Bars and Cylindrical Shells; Synthesis of Mechanisms; Stress Analysis of Solid Propellant Rocket Assemblies; Analysis of Filament Reinforced Composites *Society Aff:* ASME, AIAA, ΣΞ, ΤΒΠ, ΠΤΣ.

Shaffer, Harold S
Home: 3817 N. MacArthur Rd, Decatur, IL 62526
Position: Retired *Education:* MS/CE/Univ of IL; BS/CE/Univ of Il *Born:* 08/04/24 Native of IL. Served with US Army Corps of Engrs 1943-46 & 1950-51. Geotechnical Engr, Corps of Engrs 1949-50; Engr, IL Div Highways 1951-54. Geotechnical Engr, , J. E. Greiner Co 1954-57; Owner & Chief Engr, Soil Engrg Services, Decatur, IL, 1957-77. President Shaffer, Krimmel, Silver & Assocs Consltg Engrs, Decatur, IL 1977-86. Currently - Retired. *Society Aff:* ASCE, NSPE

Shaffer, Louis Richard
Business: P O Box 4005, Champaign, IL 61820
Position: Deputy Director. *Employer:* US Army Const Engrg Res Lab. *Education:* PhD/CE/U of IL; MS/CE/U of IL; BSCE/CE/Carnegie Mellon *Born:* 2/7/28. Native of Sharon PA. Fac of the CE Dept of UIUC 1954- . In respon charge of const mgmt option 1961-69. Assumed current repson as Deputy Dir of USA Const Engg Res Lab 1969. ASCE Walter L Huber Res Prize 1966; ASCE Construction Mgmt Award, 1978; USA Commander's Award 1978; US Army Meritorious Civilian Service Award 1973 & 1977; Meritorious Sr Exec, 1981; Disting Sr Exec, 1986; ASCE Peurifoy Res Awd, 1986; Coord of CIB Comm Org on Mgmt of Const; Mbr, Intl Activities Committee. *Society Aff:* ASCE, SAME

Shaffer, Richard F
Business: 2 Elm St, Windsor Locks, CT 06096
Position: Mgr - Environmental Tech. *Employer:* C H Dexter Div, the Dexter Corp. *Education:* MS/ChE/Columbia Univ; BS/ChE/Princeton Univ. *Born:* 2/21/17. Resident of Suffield, CT. Production mgr Schlegel Litho 1938-42. Chem engr, Military Explosives, and Experimental Station, E I duPont de Nemours 1942-45. Prof & Dept Chrmn Chem Engr Pratt Inst 1945-55; private Consultant 1955-60; Res Dir, Std Packaging Corp 1960-64. Adjunct Prof & ChE Res Dir NYU 1964-71; Adjunct Prof Environmental Engr RPI Hartford 1972-77; Mgr Environmental Tech, C H Dexter Div, the Dexter Corp; also consultant to other divs, 1971-present. Hobbies: fox hunting, tennis, platform tennis, beagling and carpentry. PE, NY & CT. *Society Aff:* AIChE, ACS, APCA.

Shah, Haresh C
Home: 909 Cottrell Way, Stanford, CA 94305
Position: Prof and Chrmn, Dept of Civil Engrg. *Employer:* Stanford Univ. *Education:* BE/Civil Engg/Univ of Poona; MS/Structural Engg/Stanford Univ; PhD/ Civil Engg/Standford Univ *Born:* 8/7/37. Married to Joan Dersjant; Two children - Boys, Hemant 20 yrs and Mihir 17 yrs; Univ of PA, 1962-68; At Stanford since 1968. Author of more than 150 tech papers and reports, co-author of the book, *Terra Non-Firma: Understanding and Preparing for Earthquakes.* Editorial Board of "Earthquake Spectra," and of "Structural Safety." Advisor and Consultant to Federal, State, and Local governments; Consultant to UNESCO programs in Earthquake Engineering; Consultant to various consulting engineering offices in the U.S. and abroad. *Society Aff:* ASCE, EERI, ACI, SSA

Shah, Kanti L
Business: T J Smull Coll of Engg, Ada, OH 45810
Position: Prof. *Employer:* OH Northern Univ. *Education:* PhD/Environmental Eng/

Shah, Kanti L (Continued)
Univ of OK; MS/Environmental Eng/Univ of KS; BS/Civil Eng/Aligarh Univ, India *Born:* 1/6/35. in India; naturalized US citizen. Worked as a Govt Irrigation Engr and Municipal Engr in India 1955-1960. Sanitary Engr, KS State Dept of Health, 1963- 1967. US Public Health Service grantee for doctoral work 1967-1969. Consultant to the PA Dept of Health for Water Quality problems 1969-1970. Asst and Assoc Prof of Civil Engg at Ohio Northern Univ 1970-1975. Prof of Civi Engg responsible for the environmental engg progs since 1975. Pres of Sigma Xi, OH Northern Chapter 1979-1980. VP of Rotary Intl Club, Ada, OH 1984-85. President Elect Sigma Xi, O.N.U. Chapter 1987-88. *Society Aff:* ASCE, ASEE, ΣΞ.

Shahani, Ray W
Home: 4700 Burbank Dr, Columbus, OH, OH 43220
Position: Program Mgr, Dev. *Employer:* Jeffrey Mining Mach Div, Dresser Industries. *Education:* MBA/-/OH State Univ; MS/Mech Engg/Akron Univ; BS/Mech Engg/MI Tech Univ. *Born:* 6/27/39. MBA OH State Univ; MS Akron Univ; BS MI Tech Univ. 1960-64 specification engr for Waukegan Works, US Steel Corp spec in design & manufacture of mech springs & fabricated Metal forms: 1964-68 res mgr, app res lab. US Steel Corp in wire products. 1968-69 res engr, Struct Matls Div Battelle Columbus Labs. 1971- 72 Proj engr for Marion Power Shovel Div Dresser Industires spec in design of structures, stress analysis fatigue & fracture computation & analysis. 1972-78 mgr advance engg, of Power Shovel div, Dresser Industries respon for advance engg of surface mining machinery, field testing, desing specification. With Jeffrey Mining Mach Div, Dresser Industries since 1978 respon for dev of mining mach. Cert Mfg Engr by soc of Mfg Engrs, Mbr of Soc of Mining Engrs. Taught at Marion Tech College. Enjoy classical music, racquetball, jogging. *Society Aff:* SME, ASM, SAE.

Shahbender, Rabah
Home: 107 Autumn Hill Rd, Princeton, NJ 08540
Position: Senior Mbr Tech Staff retired *Employer:* RCA Labs. *Education:* BEE/EE/Cairo Univ; MS/EE/Washington Univ; PhD/EE/Univ of IL. *Born:* Damascus Syria. Consultant in Communications April 1987- . From 1958 to March 1987, with RCA Labs, conducted research in Satellite Communications; from 1971-75, Hd, Appl Electronics Res, & from 1961-71 Hd, Digital Device Res; from 1960-67 Chmn, Dept of Electronic Phys, Evening Div, LaSalle Coll, Phila Pa. RCA Labs Outstanding Achieve Award for 1982 1974, 1963, 1960. IR - 100 Award for 1969 & 1964. AFIPS Best Paper Award 1963. Fellow Univ of Ill. Fellow IEEE. *Society Aff:* IEEE, AAAS, ΣΞ, HKN.

Shahrokhi, Firouz
Home: P O Box 674, Tulllahoma, TN 37388
Position: Professor & Director. *Employer:* Univ of Tennessee. *Education:* PhD/ME & AE/Univ of OK; BS/ME/Univ of OK. *Born:* July 29, 1938. Grad of Univ of Oklahoma 1965. Prof & Dir Remote Sensing Div Univ of Tenn. Prior to coming to Univ of Tenn, was a fac mbr at Louisiana State Univ. Noted author & editor of 4 books in the area of remote sensing of earth resources & author of more than 50 tech papers. Heavily involved in res & grad level instruction. Key mbr of num natl & internatl tech cttes on resources & environ as well as the White House ad hoc ctte on space travel & U N Ctte on Peaceful Util of Space. *Society Aff:* AIAA.

Shaikh, A Fattah
Home: 8731 N. 64th St, Milwaukee, WI 53223
Position: Prof *Employer:* Univ of WI-Milwaukee *Education:* PhD/Structures/Univ of IA; MS/Structures/Univ of HI; BE/CE/Univ of Karachi *Born:* 08/13/37 b. in West Pakistan, came to USA in 1962 under the East-West Center Fellowship at the Univ of HI, where he received MS degree. Following completion of the PhD at Univ of IA, joined the Univ of WI-Milwaukee in 1967. Major areas of profl interest include behavior and design of reinforced and prestressed concrete structures and soil-structure interaction problems. Co-recipient of the 1971 Martin P. Korn award by the Prestressed Concrete Inst and recipient of the 1979 Amoco Distinguished Teaching Award by the Univ of WI-Milwaukee. Reg PE in WI. Enjoys racquet sports and classical music. *Society Aff:* ACI, PCI, ASCE, ASEE

Shainin, Dorian
Business: 35 Lakewood Circle S, Manchester, CT 06040
Position: President. *Employer:* Shainin Cons, Inc. *Education:* SB/Aero Engrg/MA Inst of Technology. *Born:* 9/26/14. 16 yrs wth United Aircraft Corp. 23 yrs - mgmt cons Rath & Strong Inc, Sr V P. Since 1975 independent cons in Quality & Reliability Control, Productivity Improvement, in problem solving in mfg, R&D, Product Liability Prevention for 35 yrs; prepared & presented tech seminars for Amer Mgmt Assns. Was apptd Fac Assoc Sch of Bus Admin, Univ of Ct. Fellow Amer Soc for Qual Control, & Academician of the Internatl Acad for Quality. ASQC's Brumbaugh Award 1951; Edwards Medal 1970; Eugene L. Grant Award 1981; Exec Secy & V P ASQC. Co-author 8 books; lectured in 11 engrg films. *Society Aff:* ASQC.

Shairman, Alvin H
Business: 340 Main St, Worcester, MA 01608
Position: Prin. *Employer:* Alvin H Shairman Assoc. *Education:* SB/Mech Engg/MA Inst of Tech. *Born:* 7/31/19. in Boston, MA. Educated in the public schools and at MIT. Served in the USNR during WWII. Held various positions as Design Engr, Production and Plant Engr and Plant Mgr for different cos. Opened my own office as a Consulting Engr in 1961 and continue to serve Architects and Bldg Owners in design and specification of Mech and Elec Systems. *Society Aff:* ASME, ASHRAE, NSPE.

Shaler, Amos J
Home: 705 W Park Ave, State College, PA 16803
Position: Pres & Treas. *Employer:* Amos J Shaler, Inc. *Education:* ScD/Met/MA Inst of Tech; SB/Phys/MA Inst of Tech. *Born:* 7/8/17. US Citizen born in United Kingdom, married, three children. South African industry 1940-6 and army 1942-5. Faculty, MA Inst of Tech 1946-53. Prof and Hd of Dept of Met, the PA State Univ, 1953-60. Office of Naval Res 1950-1. NATO 1969-70. Consultant to industry, govt and intl organizations, inventor, entrepreneur in mtls sci and engg, including powder met and mtl systems, refractory mtls, design and failure investigation of buried pipes, tanks and pressure vessels, mixed fuels, physical oceanography, treatment of pollution in seawaters and fresh water bodies, treatment of liquid wastes. *Society Aff:* ASM, NFPA.

Shamash, Yacov A
Business: Elec Engg Dept, Boca Raton, FL 33431
Position: Asst Prof. *Employer:* FL Atlantic Univ. *Education:* PhD/Control/Imperial College of Sci & Tech; BScEngg/Elec/Imperial College of Sci & Tech. *Born:* 1/12/50. in Iraq, held the position of Lecturer from 1973 until 1976 in the Sch of Engg, Tel-Aviv Univ Israel. I was Visiting Asst Prof in the Systems Engg Dept at the Univ of PA for the 1976-77 yr. Since then I have been Asst Prof in the Elec Engg dept at FL Atlantic Univ. My main area of res is control systems theory. Have had more than fifty technical publications. *Society Aff:* IEEE.

Shamblin, James E
Business: 505 Engrg North, Stillwater, OK 74078
Position: Prof of Indus Engrg, Director. *Employer:* Oklahoma State Univ. *Education:* PhD/Mech Eng/Univ of TX-Austin; MS/ME/Univ of TX-Austin; BS/ME/Univ of TX- Austin. *Born:* Mar 24, 1932. Test Engr Pratt & Whitney Aircraft 1954-55; res Engr Southwest Res Inst; Asst, Assoc & Prof of Indus Engrg Okla State Univ 1964- ; Dir of Center for Local Gov Tech 1975- . Outstanding Teacher, Engrg 1969 & 1973, Western Elec/ ASEE Award, Outstanding Engrg Educator 1973; ASEE/Chester F Carlson Award for Innovation in Engrg Educ 1976. 30 plus publs 'Opers Res' McGraw-Hill. Mbr IIE, ASEE, OSPE, NSPE, APWA, ASCE & ICMA. IIE-H B Maynard Innovative Achieve Award, 1977 Reg P E Okla & Texas. Family: wife Norma Jeanne; son Earle; daughter Katherine. Govt Div Award of AIIE, 1981.

Shamblin, James E (Continued)
Alan Pritsker Teaching Award, 1985-86. *Society Aff:* NACE, ASEE, APWA, NSPE, ICMA, IIE, ASCE

Shamis, Sidney S
Home: 11 Stanley Rd, W Orange, NJ 07052
Position: Dean of Engr *Employer:* Polytechnic Inst of NY *Education:* MS/EE/Stevens Inst of Tech; BS/EE/Cooper Union School of Eng *Born:* 07/19/20 Native of Norwalk, CT. Taught in Army Air Force Officers Electronics School 1942-46. Engr for Allen B Dumont Labs 1946-52; Hd of Adv Dev Group. Prof of Elec Engrg, New York Univ School of Engrg and Sci/Polytechnic Inst of New York, 1952- present. Dir, NYU graduate center at Bell Labs, 1957-59; Assoc Dean of Engrg, 1972-81; acting Dean of Engrg since June, 1981. *Society Aff:* IEEE, ASEE

Shane, Paul
Home: 384 Smithfield Ave, Winnipeg, Manitoba R2V OC9 Canada *Education:* BSc/EE/Univ of Manitoba. *Born:* 1913 Vancouver B C Canada. Employed by Manitoba Power Comm until 1961; last pos Sys Supervisory Engr respon for sys protection & communications schemes, & solutions for special problems. With Manitoba Hydro since 1961: Mgr Distrib Design Dept until 1973; Distrib Cons Engr reporting to AGM - Customer Serv. Retired March 1978. Chmn Distrib Stds, Safety, & Metric Cttes. Mbr of sev cttes Canadian Stds Assn. Former chmn T & D Sect, Canadian Elec Assn. V P Prairie Region, Engrg Inst of Canada 1970-72. Currently V P Canadian Soc for Elec Engrg. Fellow EIC 1974. Author sev papers & articles. *Society Aff:* EIC, CSEE.

Shane, Robert S
Home: 7821 Carrleigh Pkwy, Springfield, VA 22152
Position: Principal. *Employer:* Shane Assoc, Inc. *Education:* PhD/Chemistry/Univ of Chicago; BS/Chemistry/Univ of Chicago. *Born:* 12/8/10. Consultant-Natl Mtls Advisory Bd, Office of Deputy Undersecy of Defense for Res & Engg, Smithsonian Inst Sci Info Exchange, Natl Bureau of Stds,dept of Agri,industrial corps & academic inst. Formerly: Staff Scientist, Natl Mtls Advisory Bd; Mgr, Parts, Mtls, Process Engg (Space Div), Design Review (Reentry Systems Div). GE Co; Nucl Specialist, Bell Aircraft Co, Mgr, Chemistry, Ceramics, Powder Met (Atomic Power Div), Westinghouse Elec Corp; Asst Dir, New Product Dev (Davis & Geck), Am Cyanamid Co; Proj Supervisor, Govt Contract Res Wyandotte Chem Co; Plant Chemist, Bausch & Lomb Optical Co; Superintendent, Amecco Chems Co; Res Chemist Gelatin Products Co, Technical Dir, Western Adhesives Co, Chemist, Universal Oil Products Co; Chemist , Stein-Hall Mfg Co, Chemist (Natl Amiline Div) Allied Chem Co. 5 books, 20 papers, 3 patents. Winner of Gold Key Man Award (GE). Stewart Distinguished Service Award (Am Chem Soc) Dept of Agriculture Citation for Distinguished Service. *Society Aff:* AIChE, ASTM, ACS, ASM, SAMPE.

Shanebrook, J Richard
Business: Dept of Mech Engrg, Schenectady, NY 12308
Position: Prof - Mech Engrg. *Employer:* Union College. *Education:* PhD/Mech Engg/Syracuse Univ; MS/Mech Engg/Syracuse Univ; BS/Mech Engg/Syracuse Univ. *Born:* 7/10/38. With Union Coll since 1965. Promoted to Prof 1975 & was Chmn of the Mech Engg Dept 1974-79. Res activities incl three-dimensional boudary layers, artificial heart valves, energy conservation & effective util of renewable energy resources. Artificial heart valve res incls design of heart assist devices as well as "in vitro" testing techs. Outstanding Educators of Amer Award in 1970 & 1971. Enjoys golfing, biking & water sports. *Society Aff:* ASME, TBΠ, AAPT, UCS, SES

Shank, Maurice E
Home: 68 Juniper Lane, Glastonbury, CT 06033
Position: VP, Pratt & Whitney of China, Inc. *Employer:* Pratt & Whitney, United Tech *Education:* ScD/Metallurgy/Mass Inst of Tech; BS/Mech Engrg/Carnegie Mellon Univ *Born:* 4/22/21 US Army 1942-45, Maj. Assoc Prof of Mech Engrg MIT, 1949-60. Author of 22 publ in tech literature. 1986-VP, Pratt & Whitney of China, Inc. resp for technology exchange and working relationships Peoples Republic of China. 1972-85, Dir Engrg- Tech, respon for mgmt of analytical and component tech activites, advanced engine programs, preliminary engine design, advanced performance and vehicle analysis in the commercial aircraft engine bus. Mbr of NASA Aeronau Advisory Cttee and its subctte on Propulsion (Chrmn 1979-81) and Transp Aircraft. Mbr of various univ visiting ctte and govt advisory ctte. Has appeared before US Senate and House Cttee to testify on aeronau and NASA budget authorizations. Mbr, Natl Academy of Engrg, cited for "Outstanding contributions and tech direction in advancing state-of-the-art in aircraft gas turbine tech." *Society Aff:* AIAA, ASME, TMS/AIME, ASM, SAE

Shankle, Derrill F
Home: 4731 W. Barlind Dr, Pittsburgh, PA 15227
Position: Advisory Engr, Advanced Sys Tech. Retired *Employer:* Westinghouse Elec Corp. *Education:* BS/EE/Univ of Pittsburgh. *Born:* May 6, 1922 Kingsport Tenn. Employed by Westinghouse Elec Corp since 1943 except WW II, Radio Matl Officer USNR 1944-46. Sponsor Engr, Midwest Region, applying elec util power equip, planning power sys 1946-59. Mgr Advanced Dev, Transmission, & Field Res 1959-69, supervising analytical & field investigations of EHV transmission to 750 kV. Advisory Engr, Advanced Sys Tech, conducting power sys res & dev 1970-87. Served on natl, internatl, tech & standardization cttes. Author over 40 tech papers. Mbr CIGRE; Fellow Mbr IEEE; Reg P E Pa. Presently retired, consulting and teaching graduate courses in local universities. *Society Aff:* IEEE, CIGRE.

Shankman, Aaron D
Business: 19711 Little Harbor Dr, Huntington Beach, CA 92648
Position: Tech Director. *Employer:* Shankman Assocs. *Education:* BS/Illinois Inst of Tech 1950 *Born:* Nov 4, 1924 Chgo Ill. Work exper incls 12 yrs as a met engr in the aerospace indus with the USAF, Ford Motor Co Aircraft Engine Div - Chgo, with the El Segundo & Long Beach Divs of the Douglas Aircraft Co & with North Amer Aviation in the Space & Autonetics Divs & Ocean Sys Opers. 1959-62 Ch Metallurgist of E & J Heat Treat Inc, commercial heat treaters, Los Angeles. Since 1969, Tech Dir of Shankman Assocs, met & matls engrg cons, Huntington Beach Calif. Major contribs have been in the fields of off-shore drilling & deep ocean mining sys structures. Reg P E Calif MT-984. Fellow ASM 1975. *Society Aff:* ASM.

Shanmugam, K Sam
Home: 3212 Longhorn Dr, Lawrence, KS 66044
Position: Prof *Employer:* University of Kansas *Education:* PhD/EE/OK State Univ; ME/EE/ID Inst of Sci; BE/EE/Madras Univ. *Born:* 1/6/43. Native of India, Res Assoc-Univ of Kansas 1971-73; Assoc Prof of Electrical Engg-Wichita State Univ 1973-79; Mbr of the Tech Staff-Bell Labs Holmdel NJU 1979-80; Assoc Prof of EE-Univ of KA 1980-83; Prof and Dir of Telecommunications Lab, Univ of KA, 1983-present; Consultant to various industrial organizations; Teaching and res in the areas of signal processing and communication systems; Author of many scientific articles including a textbook (Digital and Analog Communication Systems, John Wiley and sons, 1980). Fellow, IEEE, Mbr SAE, and ASEE. Receipient of the 1979 SAE Outstanding Young Engineering Educator Award, the Amoco Foundation and the Gould Award for Outstanding teaching at the Univ of Kansas. Elected as Fellow of the IEEE 1987. *Society Aff:* IEEE, SAE.

Shannahan, Cornelius E
Home: 1 Hilltop Circle, Chappaqua, NY 10514
Position: President *Employer:* Process Management Enterprises *Education:* MChE/-/MIT; BME-/Brooklyn Polytechnic Inst. *Born:* Apr 1924. Tau Beta Pi, M W Kellogg Scholarship Advanced Degree. LTjg USNR WW II. Joined M W Kellogg 1946, progressing through engrg, admin, mgmt pos to Sales Exec East Coast USA. Contrib to tech in cat cracking, partial oxidation, ethylene cracking, as well as asst in reorganizing engrg opers. Joined Rust Engrg Co 1969 - Mgr N Y Sales - promoted to V P Corp Hdqtrs Pgh Pa. Joined Hoechst-Uhde Corp 1972 - V P Sales/Treas. Contrib to tech and licensing of Mobil MTG, Dupont Canada LLDPE, Texaco coal

Shannahan, Cornelius E (Continued)
gasification and commercialization of first plant to adopt ICI AMV Ammonia Process. Organized new company "Process Mgmt Enterprises" in 1984 to offer consulting services to process plants industry. *Society Aff:* AIChE.

Shannette, Gary W
Business: Dept of Met Eng, Houghton, MI 49931
Position: Assoc Prof *Employer:* Michigan Tech Univ *Education:* PhD/Metallurgy/IA State Univ; MS/Metallurgy/IA State Univ; BS/Met Eng/MI Tech Univ *Born:* 01/24/37 Native of Ferndale, MI. Served in US Army 1960-61, Metallurgist for Alcoa, Davenport, IA 1961-1963. Taught at MI Tech 1968 tp present time. *Society Aff:* ASM

Shannon, Eldon B
Business: PO Box 8405, Kansas City, MO 64114
Position: Partner *Employer:* Black & Veatch *Education:* BS/ME/KS State Univ *Born:* 01/10/28 Native of Miltonvale, Kansas. Design engr for Black & Veatch 1950-1952. Served with the Army Ord Corps 1952-1954. With Black & Veatch since 1954. Served as resident engr, mech design engr and proj engr on a large number of projs. Partner and proj mgr since 1977 with responsibility for design and construction mgmt of projs related to power generation. *Society Aff:* NSPE, MSPE

Shannon, Robert E
Business: Texas A&M Univ, Industrial Engrg, College Station, TX 77843
Position: Prof, Indus Engrg. *Employer:* Texas A&M Univ *Education:* PhD/IE/OK State Univ; MSE/IE/Univ of AL; BS/IE/OK State Univ. *Born:* Sept 6, 1932 Blackwell Okla. MS Univ of Alabama; BS & PhD Oklahoma State Univ. 1955-56 Coleman Co Inc Wichita Ks; 1956-58 US Army; 1958-60 Army Ballistic Missile Agency; 1960-65 G C Marshall Space Flight Ctr NASA; 1965-82, Univ of Alabama Huntsville. 1982- , Texas A&M Univ. Reg P E Ala. Author 'Sys Simulation: The Art & Sci' Prentice-Hall, "Engr Mgmt" John Wiley & Sons & num journal articles. Winner 1975 IIE H B Maynard Book-of-Yr Award. Elected Fellow of AIIE, Mbr Edit Bd: IIE Transactions, IEEE Engr Mgt, Adv in comp sim & modeling. Mbr IIE, ORSA, TIMS, SCS, Sigma Xi. *Society Aff:* AIIE, ORSA, TIMS, SCS

Shannon, Robert R
Business: Optical Sci Ctr, Tucson, AZ 85721
Position: Professor. *Employer:* Univ of Arizona. *Education:* MA/Physics/Univ of Rochester; BS/Optics/Univ of Rochester *Born:* Oct 1932. BS Optics & MA Physics Univ of Rochester. Optical Engr, Mgr of Optical Design Dept & Dir Optics Lab at Itek Corp, Lexington Mass 1959-69. Since 1969, Prof of Optical Sci Ctr, Univ of Ariz. Spec in lens design, image analysis & fabrication of large optical components. Mbr Bd of Dir, Optical Soc of Amer 1971-74. Pres New Eng Sect Optical Soc of Amer 1969. Pres Soc of Photo-Optical Instrumentation Engrs, 1979-80. Pres Optical Society of America, 1985. *Society Aff:* OSA, SPIE.

Shannon, William L
Business: 1105 N 38th St, Seattle, WA 98103
Position: Consultant (Self-employed) *Education:* BS/Civil Engr/Univ of WA; SM/Engg/Harvard Univ. *Born:* Oct 1913 Seattle Wash. 10 yrs with US Army Corps of Engrs. Joined father's cons firm, William D Shannon & Assocs 1950. Left in 1954 to co-found Shannon & Wilson Inc. Active mbr num prof orgs. Mbr Natl Panel of Arbitrators, Amer Arbitration Assn. Fellow Amer Soc of Civil Engrs. 1975 Engr of Yr, Cons Engrs Council of Wash. Instrumental in estab prof liability progs for geotech cons. *Society Aff:* ASCE, ACEC, ASFE.

Shapiro, Ascher H
Business: 77 Massachusetts Ave, Cambridge, MA 02139
Position: Inst Prof. Emeritus *Employer:* MA Inst of Tech. *Education:* ScD/Mech Engr/MIT; SB/Mech Engr/MIT *Born:* 5/20/16. Native of Brooklyn, NY. On staff & faculty at MIT since 1938. Chrmn of the Faculty, 1964-65, Hd of Dept of Mech Engg, 1965-74, currently Inst Prof. Visiting Prof, Cambridge Univ, 1955-56. Fields of interest: fluid mechanics, thermodynamics, power & propulsion, biomedical engg. Profl activities include teaching, res, writing of textbooks & scientific articles, educational films. Natl Acad of Sci, Natl Acad of Engg, Am Acad of Arts & Sci. Richards Meml Award ASME, 1960. Worcester Reed Warner Medal, ASME, 1965. Lamme Award, ASEE, 1977. Townsend Harris Medal, CCNY, 1978. Dr of Sci honoris causa, Univ of Salford, 1978, Israel Institute of Technology, 1985. *Society Aff:* ASME, AIAA, BME

Shapiro, Elliot J
Business: 6315 Mill Lane, Brooklyn, NY 11234
Position: President *Employer:* Charles M. Shapiro & Sons, P. C. *Education:* BS/Marine Engrg/US Merchant Marine Acad *Born:* 12/16/29 Born in Bklyn, NY. Sailed as Merchant Marine Engr Officer 1952. Active duty with US Navy 1953 to 1955. Employed by Charles M. Shapiro & Sons, P. C. and predecessor 1955 to date. Past Pres of Kings County Chapt of NY State Soc of Professional Engrs, Chrmn of Codes Ctte of NY State Soc of PE. Mbr of Technical Advisory Group to Div of Air Resources of NY State Dept of Environmental Conservation, Diplomate American Acad of Environmental Engrs and Licensed PE in NY, NJ and CA. *Society Aff:* NSPE, AIChE, APCA, ASTM, NFPA.

Shapiro, Eugene
Business: 91 Shelton Ave, New Haven, CT 06504
Position: Group Supervisor. *Employer:* Olin Corporation. *Education:* PhD/Met Engg/Drexel Univ; MS/Met Engg/Drexel Univ; BS/Met Engg/Poly Tech Inst NY. *Born:* Dec 1944, New York NY. 1969 joined Olin Corp's Metals Res Labs as res engr working on copper, aluminum & nickel base alloy dev. Then Grp Supervisor Mechanical Metallurgy for Olin's Metals Res, respon for res efforts in alloy dev, sheet metal formability, hot working, stress relaxation, microyielding, metal flow, fracture & metal processing. Currently Dir Process Development Dept responsible for Solidification, Mechanical Metallurgy, Electrochemistry and Engineering Groups. *Society Aff:* AIME, ASM, ASTM, ASME.

Shapiro, Howard I
Business: 6315 Mill Lane, Brooklyn, NY 11234
Position: VP. *Employer:* Charles M Shapiro & Sons, PC. *Education:* BCE/Civ Engg/Poly Inst of Brooklyn. *Born:* 4/16/32. Served with Army Corps of Engrs 1953-56. Reg PE in six states. As Prin Structural Engr of Charles M Shapiro & Sons, PC, has participated in the design of hundreds of crane and derrick installations throughout the US and has served as a consultant to both domestic and foreign crane mfg. Has been directly involved in the recent rapid dev in the mathematical analysis of cranes. Actively serves on crane and derrick technical committees developing consensus stds in progs of SAE, ASME and ISO. Author of *Cranes and Derricks*, McGraw- Hill, 1980. *Society Aff:* ASCE, NSPE, AISC, SAE.

Shapiro, Sidney
Business: Dept of Elec Engg, University of Rochester, Rochester, NY 14627
Position: Prof & Chmn *Employer:* University of Rochester. *Education:* PhD/Applied Physics/Harvard Univ; AM/Applied Physics/Harvard Univ; AB/Physics/Harvard Univ. *Born:* Dec 4 1931. Earned degrees at Harvard with PhD res on solid-state maser. 1959 joined A D Little Co, carried out experiments confirming the ac Josephson effect in tunneling between superconductors; 1964-67 at Bell Labs, Murray Hill NJ; then came to Univ of Rochester, most res has concerned high frequency properties & applications of Josephson effect, Professor Electrical Engrg, 1974- 1979 served as Assoc Dean of Coll Engrg & Applied Sci, administering the Coll's graduate progs while continuing to teach & carry on res. Fellow APS, Fellow IEEE. Chairman, Dept of Electrical Eng, 1980 to present. *Society Aff:* APS, IEEE.

Shapiro, Stanley
Business: P.O. Box 1352, Edison, NJ 08837
Position: President *Employer:* Revere Research, Inc *Education:* PhD/Met & Mtls Sci/Lehigh Univ; MS/Engg Sci/RPI; BChE/CE/CCNY. *Born:* 1/3/37. in Brooklyn, NY. 1960-64 Res Engr, United Aircraft Corp: Evaluated high temperature mech properties of ceramic & metalloid MHD candidate mtls. Studies growth & characteristics of in-situ eutectic composites. Olin Corp Metals Res Labs, 1966 through 1978; supervised alloy & process res & dev in nonferrous sys. Joined Revere Copper and Brass Inc 1979 as Dir of Res; respon for all the co's alloy, process & product res & dev progs. Past Chrmn CT Sec AIME; Mbr ASM Advisory Technical Awareness Council. *Society Aff:* ASM, AIME, ASTM, EMSA, ΣΞ, AAAS.

Shapiro, Stephen D
Business: Dept. of Electrical Engineering, Stony Brook, NY 11794
Position: Prof. *Employer:* SUNY at Stony Brook *Education:* PhD/Digital Systems/Columbia Univ School of Engg and Applied Sci; MS/EE/Columbia Univ School of Engg of Appl Sci; BS/EE/Columbia Univ Sch of Engg and Appl Sci. *Born:* 2/18/41. in NYC, NY. PhD Elec Engrg Columbia Univ. Mbr; Tech Staff Bell Labs 1967-1971 engaged in computer software dev. Consultant on computing to industry and commerce, 1971-. Prof, Dept Elec Engg and Computer Science, Stevens Inst of Tech, 1971-1980. Chrm., Dept of Electrical Engineering, SUNY at Stony Brook, 1980- , Leading Prof and Chrm, 1986- . Consulting dir, grad level computer science education program at Bell Labs, 1976-1986. NSF grantee. VChrmn, software and applications technical interest council IEEE computer society, 1978-80. Mbr, exec comm machine intelligence and pattern analysis tech comm, IEEE, 1967-1981. Tech program co- chrmn, 1979 IEEE conference on pattern recognition and image processing. Mbr, 1979 IEEE delegation to People's Republic of China. Charles J. Hirsch Award, IEEE, 1986. Author over 25 tech papers. Pres, Armstrong Memorial Research Foundation, Inc. 1982-84. 1982-85 Chrm, Technical Interchange Ctte, Chrm 1986 Annual Conference. Long Island Forum for High Tech. Sr Mbr, IEEE; Mbr: ACM, Sigma Xi. *Society Aff:* IEEE, ACM, ΣΞ.

Shariful, Islam Mohammad
Home: Gomostapara, Rangpur Bangladesh
Position: Military (Army). *Employer:* Govt of Peoples Republic of Bangladesh. *Education:* BS/-/Rajshahi Univ; BSc/-/Military College of Engg Risalpur. *Born:* 1/24/38. Bangladesh Natl. Grad in sci from Bangladesh & in civ engg from Pakistan. Served with Army Corps Engrs Pakistan 1960-71 & later on with Bangladesh Army to date. Engaged with Army Engr works in different fields specialized in road construction. Worked in Karakoram Hgwys (KKH) in Pakistan. Engaged with civ constructional works as chief engr. Fellow of Inst of Engrs Bangladesh & chrmn Bangladesh Std Inst (Elec Div). Hobby games & reading. *Society Aff:* BDSS, IE (BD).

Sharp, Claude B
Business: 295 N Maple Ave, Basking Ridge, NJ 07920
Position: Asst VP-Distribution Service & Engg. *Employer:* Amer Telephone & Telegraph Co. *Education:* BS/Elec Engg/Univ of IA. *Born:* 6/30/22. in Pittsburgh, PA. Attended Grad Sch of Indus Admin Carnegie-Mellon Univ. Currently resident of Summit NJ. 1944-46 served with US Army Signal Corps; since 1941 with Bell Sys. Formerly VP-Genl Mgr Phila area, VP Mktg & Asst VP Engg, Bell of PA. Currently responsible for Bell Sys Outside Distribution Plant Construction & Maintenance Methods & for Application&s Engg Methods & Procedures for the Bell Sys Operating Cos. Mbr IEEE, Tau Beta Pi, Eta Kappa Nu. Reg PE in PA. *Society Aff:* IEEE, ΤΒΠ, HKN.

Sharp, Howard R
Business: 192 EDC, Bartlesville, OK 74004
Position: Principal Engineer - Mechanical Design *Employer:* Phillips Petroleum Co. *Education:* BS/ME/KS State Univ. *Born:* Oct 10 1923, Erie Kan. 1943-47 served with Navy Air Corps; 1950 started with Phillips Petroleum Co in design of var plants & facilities, 1967 assumed present position, Design Coordinator on var large proj (Ekofisk Phase III, Kenai Alaska LNG). Elected V Pres ASME for Reg X 1977, Chmn Petroleum Div 1975-76, Chmn Natl Membership Ctte & served var sect & regional offices. Hobbies incl fishing, hunting & traveling; active in United Methodist Church holding var offices incl Trustee & as Lay Witness. Mbr API Refining sub-committees on PV & T and Inspection Codes. *Society Aff:* ASME, NSPE, API.

Sharp, Hugh T
Home: 4 Vanwyck, Montvale, NJ 07645
Position: VP, Product Plan & Dev. *Employer:* Sweet's Div., McGraw-Hill Info Sys Co *Education:* BSChE/Chem Engg/Villanova Univ. *Born:* 9/6/29. After service in the US Navy and engg posts in industry, joined McGraw-Hill in 1953 as an editor on the staff of CHEMICAL ENGINEERING magazine. Except for two yrs in company planning post, spent the next nineteen yrs in editorial, marketing and management posts on CHEMICAL ENGINEERING. Moved to McGraw-Hill Information Systems Co in 1972 as Publisher of Building Cost Services and General Manager, Special Systems, Sweet's Catalog Services. Started several new publications and construction cost services. Formed Cost Information Systems Div in 1979. Became Vice President, Product Planning & Development, Sweet's Catalog Files Division, in 1980 and started Sweet's Intl Catalog File. Held many committee posts in AIChE and AACE over the yrs. *Society Aff:* AIChE, IIA, AACE.

Sharpe, Irene W
Home: 4430 Hardwoods Dr, W Bloomfield, MI 48033
Position: Sr Proj Engr *Employer:* Gen Motors Corp *Education:* PDED/Prof Dev/Univ of MI; BS/EE/Howard Univ. *Born:* 9/5/41. Native of VA. Held various positions with US Govt 1963-77. Ford Motor Co 1977-81. ENGA 1981. Joined GM, 1983. SWE Exec Committee Mbr 1975-77 and 1979-81. SWE-Detroit Sec Pres 1978-79. Chrmn SWE natl committees as follows: Convention 1982, Admissions 1980-81. Bylaws 1979-80, Publications 1977-79, Statistics 1975- 77; Mbr natl Nominating Committee 1977-78 and 1981-82. Received Natl Alliance of Businessmen's Award 1970. Presented paper at Engg Fdn Conf on Career Guidance for Women Entering Egg 1973. Listed "International Register of Profiles" Fourth Edition, "The World Who's Who of Women" 1974, "Who's Who Among Black Americans" 1975-76, "Notable Americans of the Bicentennial Era" 1976. Enjoys playing bridge & bowling. *Society Aff:* SWE, ΤΒΠ, SAE.

Sharpe, Roland L
Business: 480 Calif Ave 301, Palo Alto, CA 94306
Position: Chairman of Board *Employer:* Engineering Decision Analysis Co Inc. *Education:* MSE/Structures/Univ of MI; BSE/Structures/Univ of MI. *Born:* 12/18/23. 1942-46 served with USMC; 1948-50 Instructor Civil Engrg Univ Mich; 1950-66 V Pres John A Blume & Assoc Engrs, 1966-73 EVP & Genl Mgr; in addition to position with Engrg Decision Analysis Co Inc: Sr Consultant Applied Technology Council, Palo Alto, Calif. Bd Dir EERI 1971-74, Struc Engrs Assoc of N Calif 1969-71, SEA of Calif 1971-73, Chmn SEAOC Seismology Ctte 1971-73, Chmn ASCE Task Ctte on Shock & Impulsive Loadings 1970-75. Chrmn ASCE Committee on Dynamic Effects-Structural Div 1978-80. Member ASCE Structural Division Executive Committee (Chrmn 1982-83), Member Joint Coordinating Committee- US/Japan Large Scale Testing Program. Fellow ASCE; Mbr ACI, EERI, SEAONC, BSSC, SSA Chrmn, US Joint Ctte on Earthquake Engineering. *Society Aff:* ASCE, ACI, EERI, BSSC, SEAONC, SSA

Sharpe, William N, Jr
Business: Dept of Mechanical Engrg, Johns Hopkins Univ, Baltimore, MD 21218
Position: Prof & Chrmn. *Employer:* John Hopkins Univ *Education:* BS/ME/NC State Univ; MS/ME/NC State Univ; PhD/Mechanics/Johns Hopkins Univ *Born:* 4/15/38. Native of Chapel Hill, NC. Prof Chrmn Dept of Mech Eng, John Hopkins Univ 1983- . Asst Prof, Assoc Prof, Prof at MI State Univ 1966-1978. Prof & Chrmn, Mech Engg Dept, LA State Univ 1978-83. Prof & Chmn, Mech Engg Dept, Johns Hopkins U, 1983- . Res specialty is Experimental Solid Mechanics with emphasis on Laser Interferometry. Pres SEM 1984-85. Fellow of ASME. *Society Aff:* SEM, ASME, ASTM, ASEE.

Sharpless, William M
Home: 63 Riverlawn Dr, Fair Haven, NJ 07701
Position: Research Engineer, retired. *Employer:* Bell Telephone Laboratories. *Education:* EE/Elect Engg/Univ of MN; BSEE/Elect Engg/Univ of MN. *Born:* Sept 4 1904, Minneapolis Minn. Supervisor Radio Techniques Res Dept, Bell Telephone Labs Holmdel NJ, early work incl studies of short wave radio reception & angle of arrival of microwaves, since 1958 respon for res dealing with design of special semiconductor diodes for use as 1st detectors in microwave, millimeter wave & optical systems. Granted 11 patents, author num tech articles. Fellow IEEE, AAAS; Mbr APS, Internatl Scientific Radio Union, Comm A & Sigma Xi; Chmn IEEE Monmouth Sect 1956-57. Reg PE in NJ. *Society Aff:* IEEE, AAAS, APS, URSI.

Sharrin, Jerome I
Home: 5625 Stratford Dr, W Blooomfield, MI 48033
Position: Genl Mgr. *Employer:* IMC Industry Grp, Inc. *Education:* MBA/Bus Admin/Fairleigh Dickenson Univ; BChE/Chemistry/City College of NY. *Born:* Aug 1935, Brooklyn NY. Dir Bus Dev IMC Chemical Grp, Mgr Commercial Dev for IMC Chemical Group Inc; prior positions incl Mgr Opers IMC Chemical Group Inc, Mfg Mgr Arsynco Chemicals, Production Mgr Tenneco Chemicals, Production Mgr, Supervisory Engr & Process Engr Nepera Chemical Div Warner-Lambert. Mbr AIChE since early 1960's: Genl Mgr, IMC Industry Grp, Oils, Resins & Minerals Div. *Society Aff:* AIChE, CDA.

Shatynski, Stephen R
Business: 110 MRC, Dept of Mat Engr RPI, Troy, NY 12181
Position: Asst Prof *Employer:* RPI *Education:* PhD/Metal Eng/OH St Univ; MS/Mat Sci/Syracuse Univ; BS/Physic/Bowling Green St Univ *Born:* 08/16/49 Dir of the Lab for corrosion cleaning and coating of metals at RPI. Work in this lab includes high temperature corrosion, descaling and cleaning of metal surfaces and coating res. Mbr of many national society cttes and is current chrmn of Hudson-Mohawk section of AIME. Serves as a consultant on problems concerning oxidation and metal processing for corrosion protection. *Society Aff:* ASM, AIME, ECS, AAAS

Shaver, Robert G
Home: 9343 Athens Rd, Fairfax, VA 22030
Position: Vice President & Division Manager. *Employer:* Versar Inc, General Technologies Div. *Education:* SB/Chem Eng/MIT; SM/Chem Eng/MIT; ScD/Chem Eng/MIT. *Born:* Dec 1930. BS, MS & ScD from MIT. 1953-55 served with Army Ordnance Corps; prod dev engr for Dewey & Almy Div, W R Grace; solid propellant grp Ch, Atlantic Res Corp; since 1964 with Genl Technologies Corp, now a div of Versar Inc, previous positions were Div Mgr, V Pres & Pres Genl Technologies, 1972 assumed present Versar respon. Va Chap Pres SAMPE. *Society Aff:* AIChE, ACS, SAMPE, SPE.

Shaw, Carl B
Business: Portland Genl Elec Co, SB-B 121 SW Salmon St, Portland, OR 97204
Position: Staff Engr *Employer:* Portland Genl. Elec Co *Education:* BS/Physics/Univ of TX. *Born:* 9/30/24. Reg PE. During past 35 yrs employed as Quality Engr, radiographic specialist in nuclear field; supv respon for heat exchangers & pressure vessel fabrication; employed as res engr in ultrasonics then Natl Mgr for Automation Indus Inc Labs; respon at Battelle & HEDL included Mgr NDT for advanced power reactors; personal interest in training & certification, dev of neutron radiography (US PAtent 3,914,612) & image enhancement. Corp NDT Level III and respon for inservice inspection at PGE. Chmn Tech Council, Secy, VP, Treas, Pres & Chmn of Bd ASNT (1970- 76). Mbr ASTM Ctte E7, formerly IIW 'Expert' on Comm V. Awarded ASNT Gold Medal 1984. *Society Aff:* ASNT, ASME, ASTM, AWS.

Shaw, George B
Business: 14 West First St, Dayton, OH 45402
Position: President *Employer:* Shaw, Weiss & De Naples PC Consulting Engrs *Education:* MSCE/Environ Eng/Univ of Dayton; B/CE/Univ of Dayton; A/ME/Univ of Dayton *Born:* 02/25/40 Native of Dayton, OH. Assoc Prof of Civil Engrg and Mech Engrg, Univ of Dayton, Dept of Civil Engrg 1967 to present. A founding principal in 1968 of the consulting engrg firm of Shaw, Weiss & De Naples, Professional Corp. Mr Shaw is Pres and Chief Exec Officer of the firm and is responsible for directing a professional staff that is actively engaged in the design of approximately $60 million dollars in constr cost of public works projs and industrial consulting contracts. Mr Shaw is a Diplomate of the American Acad of Environmental Engrs and is a Reg Professional Engr in 8 states. *Society Aff:* NSPE, WPCF, AWWA, ASCE, AAEE, ACEC, SAME.

Shaw, Kathryn E
Home: 10008 Stedwick Rd, Gaithersburg, MD 20760
Position: Sr Cost Engr. *Employer:* Bechtel, Inc. *Education:* Assoc/Bus Admin/Davis Bus College; Credits/Engr & Comp Sci/Wesley College; Credits/Human Relations/Golden Gate Univ. *Born:* 8/22/18. Originally from Toledo, OH. Joined construction industry as field cost engr in 1952, which entailed travel to 20 field projs across the USA, Canada, & Puerto Rico. Responsible for all technical services, mtl, labor & subcontract cost control. Authored papers given at AACE conventions & intl DACE Congress in Utrecht, Holland. Elected Dir of AACE (1977-1979); elected Secy (1979-1981). Presently with the Division Services Bechtel Power Corporation in Gaithersburg Maryland. Coordinated 26 projs since 2060. Employed by Bechtel Groups for 20 yrs. Formerly with M W Kellogg. Travel also permitted working with Mission Lutheran Churches in remote areas. Hobbies: gardening & antiques. *Society Aff:* AACE, PMI.

Shaw, Lawrance N
Home: 8715 N.W. 4th Place, Gainesville, FL 32601
Position: Professor *Employer:* Univ of FL *Education:* PhD/Argic Eng/OH St Univ; MS/Agric Eng/Purdue Univ; BS/Agric Eng/ND St Univ *Born:* 03/15/34 Native of St Vincent, MN. Raised on a grain and livestock farm in Red River Valley and helped operate the farm during teenage yrs. Served as Extension E: Agric Eultural Engr for Univ of ME, specializing in potato handling and sugar beet production, 1959-66. Prof in Agricultural Engrg at Univ of FL, specializing in vegetable production mechanization, forest mechanization, and utilization of biomass as engine fuel, 1969 to present. Holds patents on a tomato harvester and a naval stores chipper and has two planting machinery patents pending. Visiting prof at Natl Inst of Agricultural Engg, Silsoe, England, 1976, and Friederich-Wilhelms-Univ, Bonn, F. R. Germany, 1986. Enjoys antique auto and farm machinery restoration. *Society Aff:* ASAE, ΣΞ, AE, ΓΣΔ

Shaw, Melvin P
Business: Dept of Elec & Comp Engg, Detroit, MI 48202
Position: Chairperson, Dept of Elec & Comp Engg *Employer:* Wayne State Univ. *Education:* PhD/Phys/Case Inst of Tech; MS/Phys/Case Inst of Tech; BS/Phys/Brooklyn Coll. *Born:* 8/16/36. in Brooklyn, NY. Sr Scientist & Scientist-in-charge, Microwave Phys, United Tech Corp, 1964-70. Prof of Elec Engg and Physics (1978) Wayne State, 1970- present. Adjunct - Yale (1969), RPI (Hartford, 1966-69), Trinity (1965). Consultant - Energy Conversion Devices (1970-present, retained), Royal Signals & Radar Establishment. Thomson, CSF, Proctor and Gamble, Ex-Cill-O. Dir, Wayne State Res Inst for Engg Sciences, 1976-77. Recipient, Wayne State Faculty Recognition Award, Bd of Governors, 1975. Listed, Outstanding Educators in Am, 1975. Enjoys exercise, cooking, travel & theater. *Society Aff:* APS, IEEE, TBΠ, ΣΞ.

Shaw, Milton C
Business: ECG 247, Tempe, AZ 85287
Position: Prof of Engrg *Employer:* Arizona State Univ *Education:* BS/ME/Drexel Univ; M/EngSc & ScD/Univ Cincinnati; DhC/-/Louvain Univ- Belgium *Born:* 05/27/15 Ch Materials Branch, Lewis NASA Lab; 1946-61 Mech Engrg Dept MIT; 1961-75 Hd Mech Engrg Dept, Carnegie-Mellon Univ, 1975-77 Univ Prof. 1978- Prof of Engr AZ State Univ. Reg PE PA, MA & OH. Mbr NAE Polish Acad (PAN); Honorary Mbr SME, ASME, ABS, ASLE; former Pres CIRP. Westinghouse Award

Shaw, Milton C (Continued)
ASEE, Gold Medal ASTME, Hersey Award ASME, Natl Award ASLE, Wilson Award ASM. McKenna Award, Gold Medal ASME, Fellow AAAS, ASM; Guest Prof AAchen TH, Univ Birmingham UK, Univ CA Berkeley. AZ State Univ former Mbr Natl Materials Adv Bd, NASA Adv Bd. Mbr Bd Engr Fdn. *Society Aff:* ASME, ASLE, ASM, JSME, ASEE, CIRP, ABS, SME, AAAS.

Shaw, Montgomery T
Business: U-136, Institute of Materials Science, Univ of Conn, Storrs, CT 06268
Position: Professor *Employer:* Univ of CT *Education:* PhD/Chem/Princeton Univ; MA/Chem/Princeton Univ; MS/ChE/Cornell Univ; BChE/ChE/Cornell Univ *Born:* in Ithaca, NY. Received undergrad educ at Cornell Univ and grad training at Princeton Univ under Prof Tobolsky. The next six yrs were spent at Union Carbide Corp, Bound Brook, NJ. In 1977 assumed position of Assoc Prof in the Dept of Chem Engrg at the Univ of CT, promoted to Prof of Chem Engrg in 1982. Secy and Exec Ctte, The Soc of Rheology, 1977-1981. *Society Aff:* SOR, SPE, IEEE, ACS

Shaw, Ralph W
Business: Box 95162, Raleigh, NC 27625
Position: Exec Dir. *Employer:* ElectriCities of NC *Education:* BSc/Elec Engg/Univ of NB; MSc/Physchology/Univ of Omaha. *Born:* 4/12/19. in Nuckolls County, NB. 1943-46 US Army Signal Corps; 1946 joined Omaha Public Power Dist, 1959 named Ch Engr, 1965 promoted to Asst Genl Mgr, Oct 1973 named Genl mgr. Reg PE, belongs to Prof Engrs of NC, IEEE. Honored as a Select Alumnus during "Masters Week" March 1974 at Univ NB in Lincoln. Univ NB Omaha Citation for Alumnus Achievement 1978. 1978 named Exec Dir of ElectriCities of NC & Genl Mgr of NC Municipal Power Agencies Number One, Number Two & Number Three. *Society Aff:* IEEE, NSPE.

Shaw, Robert F
Home: 3980 Cote des Neiges Rd c29, Montreal Quebec, Canada H3H 1W2
Position: Senior Consultant *Employer:* Monenco Consult LTd. *Education:* PhD/Hon/Mac Master U; PhD/Hon/Tech U of NS; PhD/Hon/Univ of New Brunswick; PhD/Hon/McGill; B Eng/Civil Engg/McGill. *Born:* 2/16/10. Companion of the Order of Canada; Honorary Doctor of Engrg Technical University of Nova Scotia, Science Mac Master Univ; Science Univ of New Brunswick. Science McGill Univ; Pres 1975-76, Fellow, Sir John Kennedy Medal 1979 & Julian C Smith Medal 1967, The Engrs Inst of Canada; Mbr & Gold Medal 1967 Assoc of Prof Engrs of Ontairo; Gold Medal, Canadian Council of Prof Engrs 1979; Pres, Monenco Pipeline Consultants Limited 1975-78; Special Advisor to the Minister of Indus Dev Gvrmt of Newfoundland Labrador 1979; Deputy Minister 1971- 75, the Dept of the Environ, Canada; Chmn 1969-71 Fnd Engrg Corp Ltd; Deputy Commissioner Genl, Expo '67 1963-68; Pres 1962-63 The Fnd Co of Canada; Pres 1952-53 The Order of Engrs of Quebec; VP & Ch Engr 1951-52 Defence Const Ltd, Canada; 1933-62 var positions The Fnd Co of Canada; Chrmn, Bd of Governors, Univ of New Brunswick 1978-1980, Mbr 1974-1986. *Society Aff:* EIC, OIQ, APEO, CSCE.

Shay, Felix B
Home: PO Box 117, Malvern, PA 19355
Position: Consultant (Self-employed) *Education:* EM/Mining/Lehigh Univ; BS/Mining/Lehigh Univ. *Born:* July 1910, Andover NB Canada. 1931-35 Kennecot Copper Co, Chile; 1935-39 Cia Minera Agua Fria, Honduras; Foote Mineral Co, Philadelphia 1940-45 Plant Mgr, 1946-48 Production Mgr, 1948-62 V Pres Production, 1962-67 Exec V Pres, 1968-70 Grp V Pres, 1970-72 V Chmn Bd; 1972-79 Consultant. Mbr AIME; Chmn Indus Minerals Div Soc Mining Engrs 1974; Fellow Inst Mining & Metallurgy, London. (Mbr-Mininy & Met Society of Amer) Disting Member Award, Soc Mining Engrs 1976. Mbr Martins Dam Club Philadelphia, Curling Club, Amer Wine Soc. *Society Aff:* SME, IMM, MMSA.

Shea, Joseph F
Business: Raytheon Co, 141 Spring St, Lexington, MA 02173
Position: Senior Vice Pres - Engineering *Employer:* Raytheon Co, Executive Offices. *Education:* PhD/-/Univ Mich; MS/-/Univ Mich; BS/-/Univ Mich. *Born:* Sept 1926, NYC. BS, MS & PhD Univ Mich. Married 1947, 5 children. 1944- 47 served to ensign USNR; 1948-50, 53-55 Instructor Engrg & Mechanics Univ Mich; 1950-53 Res Mathematician Bell Telephone Labs; 1955-59 Mil Dev engr; 1959-61 Dir Advanced System R&D, also Mgr Titan Inertial Guidance Prog AC Spark Plug Div, Genl Motors Corp; 1951-62 Space Prog Dir Space Tech Labs; 1962-63 Deputy Dir Systems, Manned Space Flight NASA; 1963-67 Mgr Apollo spacecraft prog office Manned Spacecraft Center; 1968-69 V Pres & Genl Mgr Equip Div Raytheon Co, 1969- 1975 Sr V Pres, Genl Mgr Equip Div. 1975-1981 Group Executive; 1981- Sr VP Engrg of the Raytheon Co. 1965 Fellow AAS, AIAA, Natl Space Club, NAE. *Society Aff:* NAE, NSC, AAS, AIAA.

Shea, Richard F
Home: 255 Inner Dr East, Venice, FL 33595
Position: retired. *Employer:* retired. *Education:* BS/Communications/MIT. *Born:* Sept 13 1903, Boston Mass. Boston English High Sch; 1925-63 electronics engr with num firms, most recently Cons with Genl Electric Co, company engr 26 years; 1963 retired, became cons engr for num firms, Cons Editor John Wiley & Sons, Editor IEEE Transactions on Nuclear Sci, Editorin-Chief IEEE Nuclear & Plasma Sciences Soc. Edited 1st book on transistor applications 'Principles of Transistor Circuits,' John Wiley & Sons 1953, also 'Amplifier Handbook,' McGraw- Hill 1966; editor or author of 4 other books on transistor applications. Lifr Mbr & Fellow IEEE, awarded IEEE Nuclear & Plasma Sciences Soc Special Award for outstanding contributions in field of nuclear sci 1973. Reg PE MA. *Society Aff:* IEEE.

Shea, Timothy E
Home: 92 Pine Grove Ave, Summit, NJ 07901
Position: Retired (VP) *Education:* SM/Elec Engg/MIT; SB/Elec Engg/MIT; SB/Elec Engg/Harvard. *Born:* Aug 6 1898, Newton Mass. 1921-39 Mbr staff Bell Labs; 1941-45 Dir War Res Columiba Univ, founded USN Underwater Sound Lab; Mbr Div 6 NDRC, later Pres Teletype Corp, V Pres Sandia Labs, V Pres Bell Labs, V Pres Western Electric Co, founded Western Electric Engrg Res Center. Author reference book 'Transmission Networks & Wave Filters.' ScD (Hon) Columbia, EngD (Hon) Case. Presidential Medal of Merit for improvements in submarines; USN Disting Civilian Service Medal for work as Chmn Undersea Warfare Ctte, NAS. Mbr NAE; Chmn var NAS proj; Genl Chmn Engrg Council, Manhattan Coll; Former Goverer & Treasurer SMPE; Fellow Acoustical Soc & IRE.

Shea, Timothy G
Business: Engg-Science, 10521 Rosehaven Street, Fairfax, VA 22030-2899
Position: VP. *Employer:* Engg-Science *Education:* PhD/Sanitary Engg/U CA Berkeley; MS/Sanitary Engg/U CA Berkeley; BS/ Civ Engg/Loyola Marymount U, Los Angeles. *Born:* 8/22/39. in Chicago, IL. BS Loyola Univ, Los Angeles; MS & PhD Univ CA Berkeley. Specialized experience in dev and application of municipal and industrial water pollution control technology and in sludge handling and disposal technologies; VP at Engg-Sce since 1987; Res Award, Water Quality Div AWWA 1971. *Society Aff:* WPCF, AWWA, IAWPRC.

Shearer, J Lowen
Home: 136 Redwood La, State College, PA 16801
Position: Professor of Mechanical Engineering, Emeritus *Employer:* The Pennsylvania State University. *Education:* ScD/Mech Engg/MIT; MS/Mech Engg/MIT; BS/Mech Engg/IL Inst of Technology. *Born:* April 25 1921. 1939-44 Sundstrand Machine Tool Co; 1944-46 US Navy; 1948-63 Mass Inst Tech; 1963- , Penn State Univ. Professional interests: teaching & res system dynamics & control, fluid control systems, conservation & resource recovery, solar energy recovery. Reg PE Mass. ASME: Automatic Control Div Exec Comm 1965-70, Basic Engrg Dept Policy Bd 1972-75 & 76-80, Policy Bd Education 1976-80, Richards Memorial Award 1966; ISA Donald P Eckman Award 1965. Sr Technical Editor of "Journal of Dynamic Sys, Measurement & Control–, Trans ASME July 1979 to July 1983. Rufus Olden-

Shearer, J Lowen (Continued)
burger Award 1983. Member of Systems and Design operating group of ASME, 1984- , Retired as Prof. Emeritus of Mechanical Engineering in 1985. *Society Aff:* ASME

Shearer, Marvin N
Business: Agri Engrg Dept, Corvallis, OR 97331
Position: Irrigation Consultant. *Employer:* Self. *Education:* MS/Water Res Dev/MI State Univ; BS/Agri Engr Tech/OR State Univ. *Born:* Feb 1920. Since 1952 Extension Irrigation Specialist, 1963 Prof, state leader of extension progs in irrigation system design, oper & mgmt, state proj leader of trickle irrigation res, classroom instructor & cons in Mexico, Bangladesh, Tunisia, and with indus on equip dev. Chmn: Ore Sect ASAE 1972, Western States irrigation specialist 1967-71, regional trickle irrigation res ctte 1976, Sprinkler Irrigation Assoc Projs Ctte 1971 & Tech Adv Ctte 1972-73 Irr: Assoc Tech Conf 1982. Author 88 publ. SIA Man of Year 1966; Oregon Ext Sr Award 1980; Sigma Xi. ASAE Pac N.W. Agricultural Engr of the year 1985; Chmn: Bd of Governors of the Irrigation Assoc Education Foundation 1985. *Society Aff:* ASAE, USCID.

Shearer, William A, Jr
Home: 4000 Heather Dr, Wilmington, DE 19807
Position: Conslt *Employer:* Self *Education:* BS/ME/Carnegie-Mellon *Born:* 4/14/18. Parnassus, Pa. Son of William A & Catherine Horner Shearer. 1942-45 Army Corp of Engrs, 1945-53 Reserve LtC; 1940-58 E I duPont de Nemours & Co Inc Tech Engr, Genl Maintenance Supt & Asst Mech Supt Du Pont Co, Belle WVa & Orange Tex, 1958 appointed Dept' mtl Engr Plastics Dept, 1971 Dept Engr newly formed Polymer Intermediates Dept, respon for design & const of processes, plant facilities, environ & energy conservation projs for dept. Pres Del Council Engrg Socs 1968-69; Citation US Government 1967; V Pres ASME 1972-74. Reg PE DE 1962; Mbr Profl Affairs and Ethics Ctte 1975-1986, Chrmn 1977-78, VP ASME Bd Prof Practice & Ethics, mbr of ASME Honors Ctte since 1980, ASME Honors CTTE Vice Chmn and Chrmn 1981-1984, Mbr AAES, Honors Ctte 82-84. Member ASME Board of Governors 1986-present. ASME Centennial Medal 1980. Delaware Engr of Year 1984. Married to Joanne Snyder Shearer, one son, Charles Russell Horner Shearer at Washington & Lee Univ. *Society Aff:* ASME, NSPE, AAAS, ROA

Shebesta, Harvey
Home: 2345 Mayfair Dr, Brookfield, WI 53005
Position: Dist Dir. *Employer:* WI Dept of Trans. *Education:* Certificate/Hgwy Traffic/Yale Bureau of Hgwy Traffic; BS/Civ Engg/Univ of WI. *Born:* 12/16/26. Native of Milwaukee, WI. Served with US Army AF Philippine Islands, 1946. With the WI Dept of Trans since 1950. Assumed current position as Dist Dir July 1969. Pres WI Sec ASCE, 1977; Charter Pres WI Sec ITE, 1966; Hd Dept 3 ITE Technical Council, 1968-1970; Dir Dist IV ITE Bd of Direction 1970-1972; Intl Pres ITE 1980. *Society Aff:* ASCE, ITE.

Shedd, Wilfred G
Business: 3724 Maumee Rd, PO Box 10842, Ft Wayne, IN 46854
Position: Pres & Owner *Employer:* Met Processing, Inc & Metals, Inc. *Education:* BS/ME/MI State Univ. *Born:* 3/27/29. Native of Rockford, IL. 1950-1951 field construction engr Surface Combustion Corp. Served as a communications officer USAF with a yr in Korea 1952-1953. Returned to Surface Combustion working as a sales engr, Chicago dist office, 1955-1959. Involved in several unique heat treat furnace applications & designs, ie sold their first atmosphere draw furnace. Became an incorporator in Metals, Inc & Met Processing, Inc commencing commercial heat treating activity in 1960. Involved in application of nitrogen in furnace atmospheres for carburizing & carbonitriding. Sole Owner of Metallurgical Processing, Inc. since 1985. Instrument pilot. Rotarian. *Society Aff:* ASM, MTI.

Sheeline, Randall D
Home: 6213 Camino Verde, San Jose, CA 95119
Position: Proj Mgr. *Employer:* Conslt GWR Assoc *Education:* MS/ChE/Univ of MI; BS/ChE/Poly Inst of NY. *Born:* 4/10/17. Developed concept for converting radioactive waste metallic sodium to a salt, Quadrex Corporation, Campbell, CA, 1981. Developed patented Inert Carrier Radwaste Process for low level nuclear power plant wastes, Chem Systems Div, United Technologies, Sunnyvale, CA, 1972 to 1981. Developed patented continuous Inert Carrier Process for ordnance mtls, three production plants built, Rocketdyne, Canoga Park, CA 1955-1971. Solid propellant rocket dev, BuOrd, Navy Dept, Wash, DC, 194-1955. Gun propellant dev, Picatinny Arsenal, Dover, NJ, 1941-1944. Treas, Secy, VChrmn, Chrmn N CA Sec AIChE 1975-1979. Elected "Fellow" 1977. Prog Chrmn Annual Meeting, Los Angeles, 1968. Chem Engr of the Yr (S CA) 1964. Chrmn Wash, DC Natl Mtg, 1954. Secy, VChrmn Chrmn Natl Capitol Sec. Four patents. Reg PE, DC. N.Ca Sec Delagate to San Francisco Bay Area Engrs Council. Charter mbr Media Technical Info Ctr of N.CA Prof Engrg and Scientific Soc. *Society Aff:* AIChE.

Sheer, Daniel P
Business: 1055 First St, Rockville, MD 20850
Position: Planning Engg. *Employer:* Interstate Commission on the Potomac River Basin. *Education:* PhD/Envir Engg/John Hopkins Univ; BS/Nat Sci/John Hopkins Univ. *Born:* 5/16/48. Native of Island Park, NY. Joined ICPRB in 1974 as Planning Engg. Respon for coord of all water and related land resource planning in the Potomac Basin. Now Dir ICPRB Cooperative Operations section. Respon for coord of reservoir operations in the Potomac for multiple purposes, including Washington Metropolitan Area water supply. Res interests: Water Resources Systems Management, Multiobjective Programming Techniques. Mbr, Control Group, ASCE Water Resources Systems Committee. ASCE Edmund Friedman Young Engrs Prof Achievement Award 1978. Sailor, USCG Licensed Charter Capt. *Society Aff:* ASCE, AWWA.

Sheets, George H
Home: 60 Harman Terr, Dayton, OH 45419 *Education:* PhD/Chemistry/Inst of Paper Chemistry; MS/Chemistry/Inst of Paper Chemistry; BChE/CE/OH State Univ. *Born:* 4/22/15. Native of WA Court House, OH. Attended OSU 1933-37, Inst of Paper Chemistry 1937-41. Have been in Pulp & Paper Industry with Mead Corp since 1941 as Res and Dev Engr; Managing Dir - Res & Dev; Production Mgt; & Corporate Mgt. Assumed current responsibility as Exec VP of Mead in 1968. Past Pres & Dir of the Technical Assn of the Pulp & Paper Industry, Past Alumni Trustee of the Inst of Paper Chemistry Past Dir-Mead Corp, British Columbia Forest Products, Brunswick Pulp & Paper Co & Northwood Forest Industries (Chrmn of the Bd). Retired May 1, 1980. *Society Aff:* TAPPI, CPPA.

Sheets, Herman E
Home: 87 Neptune Dr, Groton, CT 06340 *Education:* ScD/Applied Mech/Univ of Prague; Dipl Ing/Mech Eng/ZUniv of Dresden. *Born:* Dec 24, 1908 Dresden Ger. m. 1942; c. 6. Dipl Dresden 1934; Dr Tech Sci (Appl Mech) Prague 1936. Dir Res St Paul Engrg & Mfg Co, Minn, 1942-44; proj engr Elliott Co Pa 1944-46; Engrg Mgr Goodyear Aircraft Corp Ohio 1946-53; V P Engrg & Res Elec Boat Div, Genl Dynamics Corp 1953-69; Prof Ocean Engrg & Chmn Dept Univ of R I 1969-79. Dir of Engrg Analysis & Tech, Conn 1979-84. Consulting Engineer 1984- . Mbr Natl Acad Engrg; Mbr N Y Acad of Scis; Mbr Conn Acad of Scis & Engrg; Fellow Amer Soc Mech Engrs; Assn Fellow Amer Inst Aero & Astronautics; Fellow AAAS; Mbr Soc Naval Archs & Marine Engrs; Amer Soc Naval Engrs; Marine Tech Soc. Author of num articles in field. *Society Aff:* NAE, ASME, AIAA, NYAS.

Sheffet, Joseph
Home: 75484 Montecito Dr, Indian Wells, CA 92210
Position: Pres, Joseph Sheffet Inc. *Employer:* Joseph Sheffet Inc, Cons Struct Engr. *Education:* MS/Civil Engg/CA Inst of Technology; BS/Civil Engg/CA Inst of Technology. *Born:* June 1910. Ch Struct Engr for Donald B Parkinson, Architect 1938-45; in private practice since 1946. Respon for $5,000,000,000 bldg const. Mbr 1935 Struct Engrs Assn of So Calif (Pres 1958); also Pres 1959 of Struct Engrs Assn of Calif. Mbr 1953 Cons Engrs Assn of Calif (Pres 1966). Dir Cons Engrs Council 1967-68. Mbr Tau Beta Pi, Pi Kappa Delta, Chi Epsilon (Hon Mbr), Sigma Xi (Assoc Mbr,) Amer Arbitration Assn (on Natl Panel of Arbitrators & P Mbr So Calif Ariz Adv Council). *Society Aff:* ACI, ACEC.

Sheikh, Shamim A
Business: Dept of Civil Engineering, Univ Houston, Houston, TX 77004
Position: Assoc. Professor *Employer:* Univ of Houston *Education:* PhD/Structural Eng/Univ of Toronto; MA Sc/CE/Univ of Toronto; B Sc/CE/Pakistan Univ of Eng & Tech *Born:* 05/15/50 Recipient of 1971 Pakistan Univ and Faculty Gold Medals. Supervisory Field engr for Pakistan Industrial Development Corp, 1972. Teaching and research at Univ of Toronto, 1972-78. Design Engr for SWR Engrg Limited and Ontario Hydro, 1978-80. A principal and VP of Kani Assocs Inc, Toronto since 1980. Joined Univ of Houston in Jan 1981. Author of more than 25 technical papers published in ACI, ASCE Journals and proceedings of various conferences in North Amer, Caribbean and Europe. Authored eight major reports on concrete structures and two reports on engrg educ. United Nations conslt to Engrg Univ, Peshawar, Pakistan. Member ACI-ASCE National Committees on Concrete Columns and Lateral Leads. Interests: skiing, squash, rac and photography. *Society Aff:* ACI, EIC, CSCE, PE, EERI

Sheinbaum, Itzhak
Business: 136 West Walnut Ave, Monrovia, CA 91016
Position: Pres. *Employer:* I Sheinbaum Co, Inc. *Education:* MS/ChE/Univ of S CA; BS/ChE/Univ of CA *Born:* 2/12/38. Pres & founder of own co (1975). Broad experience in the design & construction of processing plants; managed the design of a large number of refining units & followed them through construction & startup activity. Devoted time recently to the geothermal industry, specifically in improving innovative methods of converting available geothermal brine into production of power. Possesses a number of patents on the subj & has presented papers describing these techniques. Fields of expertise include proj mgt, petrol refining, and geothermal power plants. refining & geothermal power plants. *Society Aff:* AIChE.

Sheinberg, Haskell
Business: Los Alamos National Laboratory, P O Box 1663 G770, Los Alamos, NM 87545
Position: Powder Metallurgist. *Employer:* Los Alamos Natl Lab. *Education:* BS/ChE/Rice Inst (Rice Univ). *Born:* Dec 1919 Houston Texas. Progress Engr Consolidated Steel Corp 1941-43. US Army Special Engr Detachment with Manhattan Proj at Oak Ridge Tenn & Los Alamos NM. Employed at Los Alamos Sci Lab 1945- , first as chem engr for plutonium purification & fabrication & since Jan 1950 as powder metallurgist. Presently Alt Section Leader Ceramic-Powder Met Sect. Elected Fellow ASM 1975. Chmn Los Alamos Chap ASM. V Chmn NM State Amer Ceramic Soc. Pats & many papers in powder met & ceramics. Elected Laboratory Fellow, Los Alamos Natl Lab 1982. Duties: Elective Powder Technology Projects, programs. *Society Aff:* ASM, ACS, APMI, FPS, TMS (AIME), SAMPE.

Sheladia, Pravin N
Business: 3833 34th St, Mt Rainier, MD 20822
Position: Pres. *Employer:* Williams & Sheladia Inc. *Education:* MS/Civil Engg/Univ of MO; BE/Civil Engr/Sardar Patel Univ. *Born:* 10/25/40. MSCE Univ of Missouri Columbia 1965; BSCE Birla Engrg Coll 1963. Joined State Hwy Dept of Mo & worked with other private cons in diff capacities until 1973. Became V P of Williams & Sheladia Inc 1973; one of the Sr Principals of the firm. Became Pres of the firm in 1978. *Society Aff:* ACEC, NSPE, ARTBA.

Shelby, Billy Lee
Business: 8735 Hamilton Road, Southaven, MS 38671
Position: Dir Advanced Research *Employer:* American Electric *Education:* MBA/Computer Science/Memphis State University; BS/Math/Memphis State University *Born:* 02/02/36 *Society Aff:* SPE, IES, ASM

Sheldon, John L
Home: 179 Dodge Ave, Corning, NY 14830
Position: Staff Res Consultant. *Employer:* Corning Glass Works. *Education:* PhD/Chemistry/Univ of MI. *Born:* Sept 1904 Battle Creek Mich. Teaching Fellow Univ of Mich 1935-39. Corning Glass Works: 1940-69 Res Chemist, Mgr Color TV R&D, Staff Res Mgr TV Prod Div, 1970- Staff Res Cons TV Prod Div - spec in R&D on glass & ceramics for electron tubes & components & light & UV sources. Fellow IEEE & Chmn Admin Comm 1957-65. Chmn EIA ad hoc comm on x-radiation 1965- . Mbr SMPTE, OSA, IES, Electrochem Soc, SID. EIA Engrg Award of Excell 1973. Avid nature photographer; enjoy camping & hiking.

Shell, Gerald L
Business: Peach Court Office Bldg, Suite 105, Brentwood, TN 37027
Position: Pres. *Employer:* Gerry Shell Environmental Engg, Ins. *Education:* MSCE/Sanitary Engg/Univ of CA, Berkely; BSCE/Civil Engg/MI State Univ. *Born:* 9/8/36. in Ecorse, MI, youngest of six children. Entered MI State in 1954 to study Civil Engg. Graduated in June of 1958. Worked three yrs for Dept of Water Res of CA on the dev of the CA Water Plan. Entered Univ of CA in 1961 to studey Sanitary Engg. Graduated with Masters Degree in 1962. Worked three yrs for Infilco (Tucson, AZ) in Process R&D. Worked nine yrs for Eimco (Salt Lake City, UT) also in Process R&D. Became Dir of Sanitary R&D. Worked two yrs for Aware (Nashville, TN) Consultnats as VP and Prin. Started consulting firm "Gerry Shell Environmental Engineers" in 1975. Am Pres and Owner also started lab fir "Tri- Tech Laboratories" in 1977. Am Pres and 50% owner. *Society Aff:* WPCF, ASCE.

Shell, William O
Home: 4532-70th Place, Urbandale, IA 50322
Position: President *Employer:* Shell Consltg Engrs *Education:* BS/Arch Eng/IA State Univ *Born:* 12/10/32 Degree received 1955; US Army 1955-1957; Consoer Townsend & Assoc, Chicago 1957-1959; Stephens-Adamson Mfg Co, Aurora, IL 1959-1965; Founded Shell Consltg Engrs 1965. Areas of practice included Architectural and Structural design for multi-family, commercial and industrial projs up to $25, 000,000 in size. Also provided specialized design for nationally known equip manufactures for materials handling, people moving and water treatment. Currently registered as Architectural, Structural, Civil and Mech Engr, variously, in thirty-one (31) states. Hold patents and copyrights on concrete panel systems and computer programs. Mbr of NSPE since 1963 serving as Chapter, State and Natl Officer. *Society Aff:* NSPE, ACEC, ACI, AISC

Shelnutt, J William, III
Business: UNCC Station, Charlotte, NC 28223
Position: Chrmn, Dept of Engrg Tech *Employer:* Univ of NC at Charlotte *Education:* MS/Sys Engrg/Air Force Inst of Tech; B/ME/Gen Motors Inst *Born:* 11/05/38 Native of Greenville, SC. Worked for Fisher Body Div of Gen Motors Corp briefly before entering Air Force in 1963. Worked as Project (test) Engr, Res Engr in Energetics, and Deputy Dir of Energy Conversion Lab in Air Force. Resigned commission in 1971 to attend Univ of Cincinnati. Worked as Instructor, Asst Prof, and Chrmn of Dept of Mech Engrg Tech at the Univ Ohio Coll of Applied Sci 1972-1978. Came to the Univ of NC at Charlotte as Assoc Prof in 1978, and assumed present position as Chrmn of Engrg Tech in 1979. Present interests include energy conserrvation in buildings and alternative energy. He is a staff assoc of Stanford & Toomey, Inc, a Charlotte consltg firm. *Society Aff:* ASHRAE, ASME, NSPE, ASEE, SME.

Shelton, J Paul
Business: Suite 407, 11166 Main St, Fairfax, VA 22030
Position: Pres *Employer:* Symmetron, Inc. *Education:* BS/Physics/OH State Univ. *Born:* Dec 1931. Grad study in math, phys, elec engrg. Worked in military R&D for indus, non-profit corp, & gov. Helped to found 3 small co's in 1960, 1968, and

Shelton, J Paul (Continued)
1982. Spec microwave antennas, components, & networks. Original contribs in strip transmission line, wideband components, monopulse antennas, multibeam & multimode antennas & assoc networks, microwave lenses. 50 papers, 9 pats. Currently Pres. Symmetron, Inc., Fairfax, VA. Fellow IEEE 1976. *Society Aff:* IEEE, ΣΞ.

Shelton, Robert D
Home: Rt 2, Box 434A, New Albany, IN 47150
Position: Prof. *Employer:* Univ of Louisville. *Education:* PhD/EE/Univ of Houston; MS/EE/MIT; BSEE/EE/TX Tech Univ. *Born:* 9/14/38. Native of Dublin, TX. Ind experience with Adcom, Inc & TX Instruments. Worked on Apollo Space Communications system at NASA in 1963. Instr & Asst Prof at Univ of Houston (1964-68). Assoc Prof at TX Tech (1968-70). Assoc Prof and Prof at Univ of Louisville, 1970-present. Chrmn of AMACS Dept (1974-77). Proj Dir on 14 grants & contracts from NSF, NASA, et al. Hobbies: cross country skiing, canoeing, backpacking. *Society Aff:* IEEE, ACM, IEE, ASEE.

Shemilt, Leslie W
Home: 17 Hillcrest Ct, Hamilton Ontario, Canada L8P 2X7
Position: Fac of Engg/Prof of ChE. *Employer:* McMaster Univ. *Education:* PhD/Phys Chem/Univ of Toronto; MSc/Phys Chem/Univ of Manitoba; BASc/Chem Eng/Univ of Toronto. *Born:* 12/25/19. Souris, Manitoba. Lab & acid plant supr, Defence Indus Ltd 1941-44. Held teaching pos at Univs of Manitoba & Toronto 1944-47, Brit Columbia 1947-60. Hd, Dept of Chmn Engrg Univ of New Brunswick 1960-69. Dean of Engg McMaster 1969-79. Prof of Chem Engg 1979- . Chmn N B Res & Productivity Council 1962-69; Mbr Natl Res Council of Canada 1966-69. Fellow Royal Society of Canada Chem Inst of Canada, AIChE, Soc for Values in Higher Educ. EIC Mbr, APEO, AAAS, ACS. Pres CIC 1970-71, Ed of Canadian Journal of Chem Engrg 1967-85. Visiting Prof Univ Coll London 1959-60, Ecole Polytechnique Federal de Lausanne 1975, Indian Inst Technology Kanpur and Madras, 1975, Univ of Sydney, 1981. Chrmn Tech Advisory Ctte to Atomic Energy of Canada Ltd on Nuclear Fuel Waste Mgmt Program 1979-. Author of 60 papers on applied thermodynamics, corrosion, electrochemical engrg and mass transfer. Assoc Editor Chem Eng Research and Design 1984-. *Society Aff:* ASEE, AAAS, CIC, ACS, AIChE, APEO, EIC, SVHE, MRS, CNS, CSCHE.

Shemitz, Sylvan R
Business: 145 Orange Ave, West Haven, CT 06516
Position: Principal & Chief of Design. *Employer:* Sylvan R Shemitz & Assocs Inc. *Education:* BS/Economics/Univ of PA *Born:* 04/18/28 Sylvan R Shemitz combines the attributes of designer, scientist, entrepreneur, academic and inventor. He is a lighting designer and conslt and the principal in the firm of Sylvan R Shemitz Assocs, Inc. The inventor and proponent of task/ambient lighting, he holds patents for major innovations in lighting equip and is the founder of Elliptipar, Inc, manufacturer of some of those innovations. He is a grad of the Univ of PA. He has been a Visiting Lecturer on the faculty of Yale , Princeton and Penn's grad schools of architecture. Among his credits are the lighting of the Jefferson Memorial in Washington, DC, Yale Law Library, Baltimore Museum of Fine Art; offices for Arco in Phila, Crowley-Maritime in San Francisco, 3M Austin, TX, and Chrysler in Auburn Hills, Mich, as well as institutional facilities of many kinds. *Society Aff:* IES

Shen, Chih-kang
Business: Dept of Civil Engrg, Davis, CA 95616
Position: Prof. *Employer:* Univ of Calif - Davis. *Education:* BS/Civil Engg/Natl Taiwan Univ; MS/Civil Engr/Univ of NH; PhD/Civil Engr/Univ of CA, Berkeley. *Born:* Sept 1932. PhD Univ of Calif; MS Univ of New Hampshire; BS Natl Taiwan Univ. Native of Chekiang China. Prof Univ of Calif, Dept of CE, Davis. While attending U C Berkeley, worked as teaching asst & engrg specialist. Taught at Loyola Univ of La 1965-67. Also served as asst res engr at UCLA. Assumed present teaching pos 1967. Has served as cons to Calif Dept of Transp, US Army Corps of Engrs & other private geotech cons firms. Enjoy reading, music, hiking & photography. Collingwood Prize 1970 ASCE. *Society Aff:* ASCE.

Shen, Chi-Neng
Home: 2 Village Dr, Troy, NY 12180
Position: Prof. *Employer:* RPI. *Education:* PhD/Engg/Univ of MN; MS/Engg/Univ of MN; BS/Engg/Natl Tsing Hua Univ. *Born:* 7/18/17. C N Shen served as an asst prof at Dartmouth College. He joined RPI as an assoc prof for two yrs and now is a prof in Elec Computer & Systems Engg Dept. During 1967-68 he was invited to MIT as a visiting prof of Mech Engg. His publications include a book & about 100 technical papers. His areas of res cover: state estimation & decision, Robotics, Astrodynamics guidance & control, Nonlinear Stability, Structural Dynamics Optimization, Nuclear reactor optimization & solar energy cooling systems. He is listed in Who's Who in Am & Am Men & Women of Sci. *Society Aff:* AIAA, ASME, ANS, IEEE.

Shen, David W C
Business: 33rd and Walnut Sts, Philadelphia, PA 19104
Position: Professor *Employer:* Univ of PA *Education:* PhD/EE/Univ of London; BSc/EE/Tsing Hua Univ *Born:* 01/04/81 Taught at the Univ of Adelaide (Australia), Univ of IL and MIT before joining the faculty of the Moore School of Elec Engrg, Univ of PA in 1955. Sometime a consultant to Compudyne Corp, Hatboro, PA, Technitrol Inc, Philadelphia, and General Dynamics Electronics Div, Rochester, NY. Participated as a US delegate in the US Japan Seminars on Learning Control, Nagoya, Japan 1970 and Gainesville, FL 1973. Acted as Chrmn for the sessions on Cybernetics and Indus during the First and the Third Congress of Cybernetics and Systems held in London 1969 and Bucharest 1975. Also served as a mbr of the American Advisory Ctte of the World Organization of General Systems and Cybernetics. *Society Aff:* IEEE, AAAS

Shen, Hsieh W
Business: Engrg Res Ctr, Fort Collins, CO 80523
Position: Prof of Civil Engrg *Employer:* Colorado State Univ *Education:* PhD/CE/Univ of CA, Berkeley; MS/CE/Univ of MI Ann Arbor; BS/CE/Univ of MI, Ann Arbor *Born:* 07/13/31 Worked as hydraulic engr with Harza Engrg Co & US Army Corps of Engrs. Asst Prof 1967 & Prof 1967- , and Professor-in-chrg, Hydrology & Water Resources Program, 1979- , CO State Univ. Freeman Fellow 1966 ASCE; Guggenheim Fellow 1972; Sr Fulbright Scholar 1973. Serve on Sedimentation Ctte, Stochastic Hydraulics Ctte, Publs Ctte, Hydraulics Div ASCE. Spec in fluvial hydraulics & rivers. Editor of *River Mechs* Vols I & II, *Environ Impact on Rivers, Sedimentation, Stochastic Approaches to Water Resources* Vols I & II, & *Modeling of Rivers*; 1978 Horton Award, American Geophysical Union. *Society Aff:* ASCE, AGU

Shen, Liang C
Business: Dept of Electricl Engrg, Houston, TX 77004
Position: Prof *Employer:* University of Houston *Education:* Ph.D/Applied Physics/Harvard Univ; SM/Applied Physics/Harvard Univ; BS/Elec Engrg/Natl Taiwan Univ *Born:* 3/17/39 Native of Chekiang, China, became US citizen in 1973. Joined Univ of Houston faculty in 1967 after receiving the doctoral degree. Served as the

Shen, Liang C (Continued)
Chairman of the Electrical Engineering Department in 1977-81. Worked at Gulf Research and Development Co as a full-time consultant in 1981-82. Part-time consultant for Gulf Oil, Exxon Prod Res, and Mobil Co. Specialized in antennas, microwaves, and well-logging. Author of a text book "Applied Electromagnetism" and more than 50 technical papers. *Society Aff:* IEEE, SPE, SPWLA, SEG

Shen, Thomas T
Home: 146 Fernbank Ave, Delmar, NY 12054
Position: Sr Scientist. *Employer:* NY State Dept of Env Conservation *Education:* PhD/Env. Engg/RPI; MS/Sanitary Engg/Northwestern Univ; BS/Civ Engg/St John's Univ. *Born:* 8/14/26. Born & educated in Shanghai, China. Came to the US for grad studies in 1958. With the Boeing Co as an assoc engr for two yrs. Prior to the current position as sr res scientist of NY State Dept of Environmental Conservation, worked with WA State Health Dept & then NY State Health Dept from 1964 to 1970. Dr Shen is a diplomat of Am Acad of Environmental Engrs; chrmn of NY State Council of Am Soc of Civ Engrs; a mbr of technical committees of Am Soc of Testing Mtls & Air Pollution Control Assn, with many technical publications in environmental engg, energy related pollution problems, measurement methods, air pollution control techniques, & hazardous waste mgt. Member of the U.S. EPA's Science Advisory Board and an adjunct professor of Columbia University. *Society Aff:* ASCE, APCA, ASTM, AAEE.

Shen, Tony C Y
Business: 800 N Lindberg, St Louis, MO 63166
Position: Sr Fellow. *Employer:* Monsanto Co. *Education:* PhD/ChEng/Univ of IL; MChE/ChEng/Univ of Lousville; BChE/ChEng/Natl SW Assoc Univ Kumming, China. *Born:* Dec 1921. Naturalized 1955. Int Fellow, Joe E Seagram Co, Louisville Ky 1949; Sr Res Chem Engr Inst of Indus Res, Louisville Ky until 1952; joined Monsanto Co 1954 & assumed current pos 1967. More than 60 U S pats & sev hundred foreign pats. Pats are licensed worldwide. Expert in detergent raw-matl & processing. Books, pubs, awards. Fellow AIChE. *Society Aff:* AIChE, AOCS, ΣΞ.

Sheng, Henry P
Business: Dept of Chemical Engineering, Pomona, CA 91711
Position: Prof *Employer:* Cal Polytechnic Univ *Education:* PhD/ChE/Univ of OK; MS/ChE/Purdue Univ; BS/ChE/Univ of ME *Born:* 08/25/31 Patent holder of seven designs and inventions; Recipient of three awards; one is for scoring the highest grade in Natl PE Examination (OH, 1971); Collaborated with three US Symphony Orchestras as a piano soloist; Taught 90 quarter-units of chem engrg and interdisciplinary courses in three state universities; Fields of interest including energy recovery from solid waste; Served as a conslt to 10 public agencies and private firms. *Society Aff:* AIChE, ACS, ASEE, TBΠ, ΣΞ

Shenian, Popkin
Business: 1350 S 2nd St, Coshocton, OH 43812
Position: Mgr - Res & Dev. *Employer:* Genl Electric Co. *Born:* May 1928. Having grad from Univ of Mass 1954 with doctorate in organic chem, spent 6 yrs with Amer Cyanamid. Joined GE 1960, heading many opers concerned with the dev of polyimides, polyesters, & thermoplastics. After serving as Mgr of R&D with Polymer Prods Oper, became Mgr of Mkting & Bus Dev in the Chem Dev Oper, Mgr of Bus Dev for the Plastics Dept & Mgr of Decorative Prods Sect in the Laminated & Insulating Matls Bus Dept. Apptd Mgr R&D Jan 1974.

Shenoi, B A
Chairman, Elect. Systems Engg, Wright State University, Dayton, OH 45435
Position: Prof. & Chairman. *Employer:* Wright State Univ. *Education:* PhD/EE/Univ of IL; MS/EE/Univ of IL; DIISc/EE/Indian Inst of Sci; BSc/Phys/Univ of Madras. *Born:* 12/23/29. Practical and profl experience includes the following: Res in Bell Labs for 4 summers, consultant to several industries. Pres of IEEE Circuits and Systems Soc, Secy-Treas of IEEE Circuits & Systems Soc, Assoc editor of IEEE Transactions on Circuits and Systems. Fellow, IEEE. Current interest in active, passive & digital filters and digital signal processing. *Society Aff:* IEEE.

Shepard, William M
Home: 2224 S Hoyt Ct, Lakewood, CO 80227
Position: VP Operations/Asst Secy *Employer:* Tellis Gold Mining Co., Inc. *Education:* MS/Mining Geology/MO School of Mines; BS/Mining Geology/MO School of Mines. *Born:* 9/16/28. Native of Upper Montclair, NJ. US Army Corps of Engr 1952-1954. Exploration geologist with Bear Creek Mining Co (Kennecott Copper Corp) western US 1954- 1964. Mine geologist Burgin mine, Utah, Kennecott 1965-1968. Staff geologist Brown and Root, Inc 1969-1970. Climax Molybdenum Co, Asst Chief Geologist and Chief, Geology & Exploration, Western Operations 1970-1974, V.P., US Exploration, AMAX Exploration, Inc with responsibility for US metals exploration programs for AMAX. Pres, American Gold Minerals Corp/Altex Minerals 1984-1985 Currently Vice President, Tellis Gold Mining Co., Inc. Dir SME-AIME 1972- 1975, 1976-1978. Dir AIME 1978-1979, VP and Dir AIME 1980. *Society Aff:* SME of AIME, SEG, AIPG.

Shepherd, Dennis G
Business: Sch of Mech & Aero Engg, Upson Hall, Cornell Univ, Ithaca, NY 14853
Position: Prof of Mech Engrg. *Employer:* Cornell Univ. *Education:* BS Eng/Math/Univ of MI; BS Eng/Physics/Univ of MI. *Born:* London England. US citizen 1951. Plant engg England 1935-39. Engr R&D of turboject engines, Power Jets Ltd England 1940-46. Ch Exper Engr, Gas Turbine Div, A V Roe Canada, Toronto 1946-48. Cornell Univ 1948-, Prof ME 1952, Hd Dept of Thermal Engg 1963-65, Dir Sibley Sch of Mech Engg 1965-72. 7 bks on gas turbines, turbo machinery, fluid mechs, aerospace propulsion. Res ints in turbomachinery, propulsion, wind power. Guggenheim Fellow 1954-55; Cornell Soc Engrs Excell in Teaching Awards 1968, 1975. ASME Worcester Reed Warner Medal for contrib to literature 1976. Life Fellow ASME, 1984. *Society Aff:* ASME, AAAS, ΣΞ, AWEA.

Shepherd, John
Business: P O Box 777, Place Bonaventure, Montreal Quebec H5A 1E3
Position: President. *Employer:* J S Mgmt & Shipbldg Cons Ltd. *Born:* Native of Uddingston Scotland. 1956 Royal Tech Coll Glasgow. MIMechE, P Eng Que & NB Canada. Mbr SNAME. C Eng UK. Served with Royal Engrs, Egypt 1946-48. Apprentice ship with Weir/Drysdale Group, plus 4 yrs as Sr Mech Engr, spec in marine power plants & eventually thermal plants. 6 yrs wth St John Drydock as yard mgr respon for all opers until 1973. 9 1/2 yrs with Davie Shipbldg, 6 1/2 as Asst Ch Engr & 3 as V P & Genl Mgr until sale of co 1976. Terminated as V P & Genl Mgr & formed consultancy in assn with Montreal Engrg.

Shepherd, Mark, Jr
Business: PO Box 225474 MS 236, Dallas, TX 75265
Position: Chmn & Ch Corporate Officer *Employer:* Texas Instruments Incorporated *Education:* MS/Elec Engg/Univ of IL; BS/Elec Engg/So Methodist Univ. *Born:* 1/18/23. Dallas Tex. Test engr G E 1942-43. US Navy 1943-46, radar/electronics maintenance. Proj engr Farnsworth TV/Radio Corp 1947-48. Joined Geophys Service Inc (predecessor of Texas Instruments) as proj engr 1948. TI pos: Asst V P, V P/Mgr of SC Div; Exec V P, Pres. Became Chmn 1976. Mbr Bd of Tr SMU, Internatl Council Morgan Guaranty Trust, NAE. Trustee and Councillor Conf Bd. Tr Amer Enterprise Inst. Dir, First Republic Bank Corp, USX Corp. Member US - Japan Business Council. Fellow IEEE. Prof Engr TX; Mbr The Bus Council; Trustee, Committee for Economic Dev; Mbr, Horatio Alger Assoc of Distinguished Americans, Inc & Advisory Cttee for Trade Negotiations, Office of the US Trade Rep. *Society Aff:* SEG.

Shepherd, R
Home: P.O. Box 410, Fawnskin, CA 92333-0410
Position: Prof of Civil Engr *Employer:* Univ of CA, Irvine *Education:* DSc/Structural Dynamics/Univ of Leeds; PhD/CE/Univ of Canterbury; MS/CE/

Shepherd, R (Continued)
Univ of Leeds; BS/CE/Univ of Leeds. *Born:* 06/28/33 educated in England. Worked in aeronautical and civil structural design offices. Faculty mbr Univ of Canterbury, New Zealand 1959-71 and Univ of Auckland 1972-79. Teaching and research interests in structural analysis as applied to earthquake engrg. Fulbright awardee and Visiting Prof, CA Inst of Tech, 1977. Overseas Visiting Scholar, St. John's Coll, Univ of Cambridge, U.K. 1984. Erskine Visiting Fellow, University of Canterbury, New Zealand, 1987. Specialist conslt on computer applications in seismic resistant design in New Zealand, Australia and CA. Joined Univ of CA, Irvine, in 1980. Outside interests include sailing and Rotary. *Society Aff:* ASCE, ICE, SSAm, SEAOC, EERI

Shepherd, William G
Business: 103 Shepherd Laboratories, University of Minnesota, 100 Union St. SE, Minneapolis, MN 55108
Position: Prof Emeritus EE. *Employer:* Univ of Minnesota. *Education:* PhD/Physics/Univ of MN; BEE/Elec Eng/Univ of MN. *Born:* Aug 1911 Ft William Ont Canada. Mbr Tech Staff Bell Tele Labs 1937-47. Prof EE Univ of Minn 1947-79 , spec in the field of phys electronics. Founded the univ's Phys Electronics Res Labs & served as its director. Assoc Dean Inst of Tech 1954-56; 1956-63 Hd, Elec Engrg; 1963-73 V P, Acad Admin; 1973-79 Dir, Space Sci Ctr. Fellow IEEE (Pres 1965-66), Amer Phys Soc, Natl Acad of Egrg & Amer Soc Engrg Educ. Citation Bureau Ships USN 1947. Fellow IRE 1952. NEC Medal of Honor 1965. Minn Engr of Yr 1966. *Society Aff:* IEEE, APS, NAE, IEEE, APS.

Sheppard, Albert P
Business: VPR, CRB-0370, Atlanta, GA 30332-0370
Position: Assoc VP, Res & Prof, Elec Eng *Employer:* Georgia Inst of Tech *Education:* PhD/EE/Duke Univ; MS/Physics/Emory Univ; BS/Physics/Oglethorpe Univ *Born:* 06/06/36 Started career in electronics at Martin-Marietta, Orlando, Res Div in 1960 working in areas of lasers and microwaves. Joined US Army Res Office, Durham, 1963-65 and administered basic research program in electronics and instrumentation. Joined Georgia Inst of Tech in 1965 and held several electronics materials and instrumentation positions in its Engrg Experiment Station unitl 1972 when I became Assoc Dean of Engrg until 1974 when I moved to present position. Was Acting VP for Res for yr 1980. Have published more than 75 papers and reports in electronics, engrg educ and alternate energy areas. Have been active as conslt to indus in energy, electronics and mgmt. *Society Aff:* ASEE, NSPE, IEEE, USRA, ISES, IMPI, NCURA, ΣΠΣ, ΣΞ

Sheppard, Stanton V
Home: 8947 Elsmere Dr, Parma, OH 44130
Position: V P/Genl Mgr. *Employer:* The Ceilcote Co. *Education:* -/Pre-Engg/KY Univ; BS/Chem Engg/Purdue Univ. *Born:* Apr 1927. BSChE Purdue Univ. Tau Beta Pi. Alpha Chi Omega. Native of Indianapolis Ind. Sales, Plant & Ch Engr for Allied Oil Co, Div of Ashland Inc 1950-59. Mgr of Design Engrg, Const Mgr, V P - Opers, Genl Mgr Air Pollution Control Div, V P, The Ceilcote Co, Unit of Genl Signal Corp 1960- . Author sev tech articles on packed wet scrubbers & electrostatically augmented wet scrubbers for control of gaseous & particulate pollutants. Reg P E Ohio. *Society Aff:* ASME, AIChE, NACE, ASHRAE.

Sheppard, William V
Business: 1301 Gervais St, Suite 416, Columbia, SC 29201
Position: Dir of Transportation Services *Employer:* Post, Buckley, Schuh & Jernigan *Education:* BS/CE/The Citadel *Born:* 4/18/41 Joined PBS&J in 1980 as VP and Dir of Transportation Services. Directs the firm's staff of over 200 transportation engrs and planners in a wide range of multi-mode transportation programs including hgwy and bridges, EIS, transportation studies, transit programs, CBD parking & traffic studies, and airport projects. Graduated-Citadel-1964, joined HNTB in KS City. Served 2 years in US Army 82nd Airborne Division. Joined Wilbur Smith and Assocs in 1967. Served as Office Mgr-Cincinnati-1968; Principal Engr-New Haven-1970; Assoc-in- Charge-Los Angeles-1971; Western Regional VP-1974; and Southern Regional VP- 1977. *Society Aff:* ASCE, ITE, NSPE, TRB, ACEC

Sher, Rudolph
Business: Dept of Mech Engr, Stanford, CA 94305
Position: Prof. Emeritus. *Employer:* Stanford Univ. *Education:* PhD/Phys/Univ of PA; AB/Phys/Cornell Univ. *Born:* 5/28/23. 1951-61 Brookhaven Natl Lab. Worked on experimental nuclear reactor physics prblems and neutron cross-section evaluations. Since 1961 on faculty of Mech Engg Dept, Stanford Univ, teaching nuclear engg, with res in above areas plus nuclear safeguards. 1958-59: Visiting Scientist, CEA, Saclay, France. 1968-69: Visiting Scientist, AB Atomenergi, Studsvik, Sweden. 1978-79: Visiting Expert in nondestructive analysis techniques for nuclear safeguards, Intl Atomic Energy Agency, Vienna. 1978-1980: Editor, *Progress in Nuclear Energy*. Co-author of "The Detection of Fissionable Materials by Non-destructive Means" (Amer-Nucl-Soc. 1980). Presently consultant in nuclear power plant severe accident analysis. *Society Aff:* ANS, INMM.

Sherer, Keith R
Home: 105 Barber Rd, Madras, OR 97741
Position: Principal *Employer:* Sherer & Assoc *Education:* MS/Envir Sci/Univ of OK; BSCE/CE/OR State Univ. *Born:* 10/8/35. Native of OR. USMC 1953-1955, Commissioned Officer in US PHS 1962. OR State Sanitary Authority as Asst Dist Engr 1960-1963. Active duty, USPHS - 1963-1982, assigned to the Indian Health Service. Assignments with IHS included working with Indian Tribes in OR, ID, OK, and AZ as Field and Dist Engr, Construction Branch Chief, Deputy Dir and Chief, SFCB, IHS. Presently engaged in intl conslt in approp tech. Outstanding Young Men of America, 1970; Dipl Amer Acad of Envir Engrs; reg PE in OR and AZ. *Society Aff:* AAEE, NSPE, WPCA.

Sheridan, Vincent G
Business: 305 Main St, Catskill, NY 12414
Position: Prin (Self-employed) *Education:* MBA/-/Farleigh Dickinson Univ; BS/CE/CO Univ; BS/Bus/CO Univ *Born:* 04/15/21 Native New Yorker, WWII US Navy. Varied employment in public serv and private industry in civ, ind and elec engrg; property and bus claims in appropriation of land and bus. Currently engaged in private practice as consltg engr, ACEC, land planner and real estate conslt. Has been a value analyst/engrg conslt, bus and real estate conslt for more than 30 yrs. Employment and private practice spans indust engrg in major mfg; const & const supervision of mfg & processing plants & equip; civ engrg projs; site planning for housing projs; economic studies; title engrg & property surveys; property mgmt; insurance claims adjusting; real estate, bus, machinery & equip appraisals; farming and ranching a lifetime pursuit; teaching & lecturing. Educ, US Navy training schools in electronics and mechanics, univ degrees in engrg & bus, grad studies and an MBA. Work experience in mech, elec and building trades; prof exper spans civ, mech, elec, sanitary and struct engrg, value analysis & appraising knowledge competence resulted from employment as a value engr and appraiser; real property and bus intimacy is the result of rehabilitating, managing and selling income producing properties and counselling bus problems. *Society Aff:* XE, ΣΤ, ΤΒΠ, NSPE, ASCE, ASAE, SAME, ASPO

Sherman, Bennett
Business: 90-59 56th Ave, Elmhurst, NY 11373
Position: President. *Employer:* Sherman Engrg Group. *Education:* MS/Opt Eng/Univ of Rochester; BS/Engg/Purdue Univ. *Born:* 1922 NYC. Sr Engr & Res Scientist Farrand Optical Co 10 yrs; Ch Engr Conn Inst Div of Barnes Engrg Co 5 yrs; Sr Engr for Optics at GTE Labs 5 yrs; Founder & Pres Sherman Engrg Group, Cons in Optical Engrg. Natl Convention Chmn Optical Soc of Amer 1973; Chmn Optical Soc of Amer Tech Ctte on Optical Matls 1971-72; pats in optical instruments, spectrophotometry, & photo-optical sys & devices; sci cons & contrib editor on sci devs for Modern Photography magazine N Y. *Society Aff:* OSA, SPSE, AAAS.

Sherman, Herbert
Business: 677 Huntington Ave, Boston, MA 02115
Position: Exec Comm & Staff Mbr Inst For Health Res *Employer:* Harvard School of Public Health. *Education:* DEE/Elec Engg/PIB; MEE/Elec Engg/PIB; BEE/Elec Engg/CCNY *Born:* Feb 24, 1920. Reg P E Mass. USN Radar Officer USS Portsmouth, SiG C & USAF Rome Air Dev C 1940-52. Staff Mbr & Comm Group Leader MIT Lincoln Lab 1952- 76, respon SAGE Comm & Lincoln Exper Satellites 1 through 6. Currently Fac Mbr Harvard Sch of Pub Health & Harvard Med Sch. Cons: var agencies DOD, Natl Ctr Health Serv Res, Peter Bent Brigham Hosp 1958-72, Beth Israel Hosp Boston 1969- . Over 40 pubs in comm, optical signals, ptrn recog, satellites, cardiovascular models, ambulatory med care, cost & qual of med care. Fellow IEEE, AAAS and Soc Adv Med Sys. Mbr Tau Beta Pi, AAAS, Sigma Xi, Biomed Engrg Soc. *Society Aff:* IEEE, BMES, AAAS, AAMSI, ΣΞ, ΤΒΠ.

Sherman, Ralph A
Home: 3101 Leeds Rd, Columbus, OH 43221
Position: Tech Dir - ret. *Employer:* Battelle Mem Inst. *Education:* AB/Chem/Univ of IA; -/Chem/Univ of Toulouse, France; -/Chem/William Penn College. *Born:* Dec 1896 Oskaloosa Iowa. Signal Corps US Army 1917-19. AB Chem Univ of Iowa 1920. US Bureau of Mines Sept 1916-Nov 1917, June 1920-Mar 1930. Battelle Mem Inst Mar 1930-Dec 1961, successively Supr Fuels Div, Asst Dir, Tech Dir. Mbr ASME 1924- . Fellow 1949- . Life Fellow 1959- . Mbr of Council 1948-52. Percy Nicholls Award ASME-AIMME 1948. Hon Mbr Performance Test Codes Ctte ASME 1972- . Life Mbr ASHRAE, Treas 1955-56. Sigma Xi 1935- . Tau Beta Pi 1970- . Reg P E Ohio 1935- .

Sherman, Roger J
Business: 2 World Trade Center, New York, NY 10048
Position: Chrmn of the Bd. *Employer:* Ebasco Services Inc. *Education:* BME/Mech Engg; Grad/Nucl Engg/Columbia Univ/NY Univ College of Engg. *Born:* 2/17/13. Grad studies in nuclear engrg Columbia Univ, mech engrg NYU. Reg PE 23 states. Mbr: ASME (Fellow), Amer Nuclear Soc, Amer Nuclear Energy Coun (Dir), Atomic Indus Forum (Honorary Dir), NSPE, Newcomen Soc, Union League Club NY, Whitehall Club NY, Achieve Award New York Univ; Disting Engr in Private Practice award NYSSPE. US Navy WW II - Lt Cdr. Ebasco Services Inc 1939- , engrg & officer capacities. Exper in design & const of fossil & nuclear central station power plants. *Society Aff:* ANS, ANEC, AIF, NSPE, MPC, ASME.

Sherman, Russell G
Home: 592 Dryad Road, Santa Monica, CA 90402
Position: Pres. *Employer:* Nevada Engrg & Technology Corp. *Education:* BS/Met Engg/Univ of Penn. *Born:* Jan 1926. BS Met E Univ of Penn 1950. WW II & Korean War 1Lt Ordnance Corp. Plant Metallurgist in magnesium & aluminum plants Howard Foundry Co Chicago; Res Metallurgist Titanium Metals Corp Henderson Nevada - developed heat treat procedures for Ti-6Al-4V, developed 8Al-1Mo-1V alloy & early work on hydrogen in titanium; Ch Metallurgist Allen Mfg Co Hartford Ct; Voi-Shan Mfg of Culver City Ca; Group Head Material & Proc Support Missile Div Hughes Aircraft Culver City; V P, Engr & QC Valley-Todeco Sylmar Ca. Chmn L A Chap ASM 1973-74, L A Chap Chrmn 1979-80 AIME; has been in both socs 37 yrs. *Society Aff:* AIME, ASM.

Sherman, Samuel M
Home: 606 Stratford Drive, Moorestown, NJ 08057
Position: Staff Scientist. *Employer:* RCA Government & Commercial Systems. *Education:* BA/Physics/Univ of PA; MA/Physics/Univ of PA; PhD/Electrical Engrg/Univ of PA. *Born:* Sept 1914 Camden NJ. Taught high school physics before WW II. Radar Design & Dev Officer Army Air Forces 1942-46. Res Assoc Univ of Penn 1946-48; Physicist, later div supt Naval Air Dev Center Johnsville Penn 1948-55; RCA Moorestown NJ 1955- , supervisory & staff positions, currently Staff Scientist engaged in radar sys analysis & dev. Served on government adv panels; pub num papers. Fellow IEEE; mbr: Phi Beta Kappa, Sigma Xi, Pi Mu Epsilon. Hobbies: music (piano), linguistics, home movies. *Society Aff:* IEEE, ΣΞ, ΦΒΚ, ΠΜΕ

Sherman, William F
Home: 19967 Doyle Pl W, Grosse Pointe Woods, MI 48236
Position: Dir, Engr Div. *Employer:* Motor Vehicle Mfg Assn. *Education:* BAeE/Univ of Detroit. *Born:* 1/1/13. Native of Detroit. Technical editor the Iron Age, 1937-42, Engr with Automotive Council for War Production & Aircraft Council for War Production; co- author "Motor Vehicle Production" (technical) for Ind College of the Armed Forces; dir of engrg div & res prog for Motor Vehicle Mfg Assn (Automobile Mfg Assn). Retired, 1978. Current exec dir Sci & Engrg Fair of Metropolitan Detroit. Author "The SAE Story–, for publication at SAE Diamond Jubilee. *Society Aff:* SAE, SME, ESD, NAEM.

Sherrod, Gerald E
Business: 399 Park Avenue, New York, NY 10043
Position: Vice President, Senior Professional *Employer:* Citibank N A. *Education:* BS/Petroleum Engg/Univ of Tulsa. *Born:* Nov 1922. Home Tulsa Oklahoma. Served Army Air Corps 1941-45. Engr Texas Pacific Coal & Oil Co 1950-57; joined Citibank 1957, assumed present position of V P Engrg 1961. Treas AIME 1971; Pres Soc of Petroleum Engrs of AIME 1973. SPE Disting Serv Award 1975. Hobbies: antique cars & skiing. *Society Aff:* SPE/AIME, SPEE.

Sherwin, Douglas S
Business: Phillips Petroleum Company, Bartlesville, OK 74004
Position: Pres. *Employer:* Phillips Products Co Inc. *Education:* MA/Economics & Philosophy/Univ of Oxford, England; BS/Chem Engg/Penn State. *Born:* 1918. Chem Engr Phillips Petroleum Co 1942. 1947-51 on leave from Phillips to attend Oxford & serve Foreign Serv Officer Depts of Army & State Berlin. 1952-55 on loan from Phillips to Office of Synthetic Rubber RFC. 1955 Phillips Asst Coordinator Rubber Chems Div, 1957-64 Mgr Market Res & Commercial Dev, 1965-75 VP & Genl Mgr, 1975-79 Exec VP and subsequently Pres Phillips Products Co Inc (Phillips subsidiary). Patents petroleum fractionation processes; articles on mgmt. Penn State's nominee for Rhodes Scholarship. Financial Mgmt Ctte SPI, ACS, AIChE & Amer Economic Assn. Dir of Phillips Products Co., Inc., Canada Western Cordage Co Ltd, and Canada Cordage Inc. *Society Aff:* SPI, ACS, AIChE, AEA.

Sherwin, Martin B
Home: 11500 Karen Drive, Potomac, MD 20854
Position: Exec VP *Employer:* W R GRACE & Co *Education:* BChE/Chem Eng/CCNY; MChE/Chem Eng/Brooklyn Polytechnic Inst; PhD/Chem Eng/CUNY. *Born:* 7/27/38. Halcon Internatl & Scientific Design 1960-64, engaged in process design & dev; CUNY 1964-66 insturctor Chem E; Chem Sys Inc 1966-1980, 5 yrs in cons covering variety of projs & 7 yrs VP R&D, responsible for all aspects of R&D activities including program admin & sales. 2 yrs Managing Dir of Chem Sys Intl, London a consulting office. W R Grace 1980-present Exec VP in the Corporate Research Div. Mbr: AIChE, ACS, Sigma Xi. Listed in Amer Men of Sci. Authored 23 US Patents & num publs *Society Aff:* AIChE, ACS.

Sherwood, A Wiley
Home: 3411 Chatham Road, Hyattsville, MD 20783
Position: Owner Aerolab Supply Company. *Employer:* Aerolab *Education:* MS/Aerospace Engg/Univ of MD; ME/Mech Engg/RPI. *Born:* Jan 13, 1915 St Louis Mo. Started teaching Univ of Md 1939 as instructor. During WW II Proj Engr David Taylor Model Basin while serving in USN. Returned to Univ of Md as prof, 1st Head of Aero Dept Univ of Md, owner of Aerolab designing & manufacturing wind tunnels for educational institutions & res facils domestic & foreign. Fellow ASME. Author 'Aerodynamics' (McGraw-Hill 1946). P Chmn Washington DC Sect ASME & AIAA. Hobbies: golf & tennis. Retired from Univ of MD in Jan 1977. Still active in Aerolab. *Society Aff:* AIAA, ASME.

Who's Who in Engineering

Sherwood, Richard F
Business: 1530 S West Temple, Salt Lake City, UT 84115
Position: Water Prod Supt. *Employer:* Salt Lake City Public Util Water Dept. *Education:* BA/Humanities Engg/Univ of UT. *Born:* 1/24/24. 6 yrs - Humanities & Sociology Maj. Civ Engg Univ of UT Minor. 4 yrs - B-29 Pilot WWII. 28 yrs - Salt Lake City Water Dept Sanitary Engg Water Treatment. 1 yr - United Nations - Peace Pilot Projs - (Sanitary Engg) Mexico. Native of Salt Lake City, UT. Civ Engg student 1941 Univ of UT. Served with Army Air Corp, WWII as B-29 Pilot, 1942-46. Returned to Univ of UT Civ & Sanitary Engg Minor with Humanities Sociology Maj, required in order to work with UN Peace Pilot Projs. 1949-50, served with SLC Water Dept Engrs as design draftsman - Water Treatment & Sanitary Engg from 1951 to present. Rec'd AWWA Geo W Fuller Award 1978. Intermtn Sec AWWA Chrmn 197. Avid sportman. Canoe, hunt, fish, ski, backpack, etc. *Society Aff:* AWWA, WPCF.

Shetron, John H, Jr
Business: 1920 Chestnut Street, Philadelphia, PA 19103
Position: Partner. *Employer:* Blauvelt Engineering Co. *Education:* BCE/-/Clarkson Univ. *Born:* Jan 1926; grew up in White Plains NY. Served with Army Corps Artillery in ETO during WW II. Tau Beta Pi, Brown & Blauvelt 1951-58; Blauvelt Engrg Co 1958-present spec in Civil, transportation and traffic engineering design & supervision of construction, Tech Mgr in charge of Philadelphia branch office 1965-present, Assoc of firm 13 yrs, Partner 1975-present. Licensed P E in 7 states. Mbr: ASCE (Fellow), ITE (Fellow), SAME, NSPE & ASHE. Hobbies: tennis, sailing & skating. *Society Aff:* ASCE, NSPE, ITE, ASHE, SAME.

Shewan, William
Business: Gellersen Ctr, Rm 101, Valparaiso, IN 46383
Position: Prof Emeritus of Elec Engg *Employer:* Valparaiso Univ. *Education:* PhD/Elec Engg/Purdue Univ; MS/Elec Engg/Univ of Notre Dame; BS/Appl Sci/Valpariso Univ. *Born:* 5/24/14. Native of Chicago, IL. Served in the US Navy 1942-45. Rank, Electronics Technician Chief. Active duty in New Guinea and the Phillipine Islands. Power Distribution Engr, Northern IN Public Service Co, one yr. Served in successive ranks of Instructor, Asst Prof, Assoc Prof and Prof at Valparaiso Univ. Duties include: sixteen yrs as Chrmn, Dept of Elec Engg, Dean of the Coll of Engg, two yrs and many Univ comms. Res: Variable Speed Solid-State Drives; Numerical Methods; Automatic Control; Increment Motion Studies. Consultant: US Navy Solid State Power Supplies, Electro-Hydraulic Hoist, etc. Natl Dir of Res, ISA, two yrs. Retired May 1984. Retired May 1984, with rank of Prof Emeritus of Electrical Engg. *Society Aff:* IEEE, ASEE, ΣΞ, TBΠ.

Shewmon, Paul G
Business: 116 West 19th Avenue, Columbus, OH 43210
Position: Professor. *Employer:* The Ohio State University. *Education:* Phd/Met Eng/Carnegie Inst of Tech; MS/Met Eng/Carnegie Inst of Tech; BS/Met Eng/Univ of IL. *Born:* 4/18/30. Stillman Valley Ill. Res Engr Westinghouse Res Lab 1955-58; Prof Carnegie Inst Tech 1958-67; Rossitor Raymond (AIME) & Alfred Noble Prizes (Founder Soc) 1960; Natl Sci Found Fel MPI Phys Chem Goettingen W Germany 1963-64; Argonne Natl Lab Assoc Dir Metallurgical Div 1967-68, Assoc Dir EBR-II 1969, Dir Matl Sci Div 1970-73; Dir Matl Sci Div Natl Sci Found 1973-75; Prof & Chmn Dept Met E Ohio State Univ 1975-83. Prof 84-. Fellow ASM; Mbr: ASM, AIME, AAAS, ANS, Interests: Physical metallurgy kinetics of reactions in solids & materials in energy sys. Howe Medal (ASM) 1977, Mbr USNRC Advisory Com on Reactor Safeguards 1977- , Fellow ANS, Fellow AAAS, Fellow TMS-AIME, Mbr Natl Acad Engr, Mathewson Medal (981) *Society Aff:* ASM, TMS-AIME, AAAS, ANS, MRS

Shield, Richard T
Business: 216 Talbot Laboratory, 104 S Wright St, Urbana, IL 61801
Position: Professor *Employer:* Univ of Illinois/Urbana-Champaign. *Education:* PhD/Applied Math/Durham Univ; BSc/Math/Durham Univ. *Born:* 7/9/29. Held faculty appointments at Brown Univ & CalTech. 1970- , Professor of Theoretical & Applied Mechanics Univ of IL/U-C. (Dept. Head 1970-84) Simon Guggenheim Mem Fellow 1961-62 at King's Coll Newcastle Eng; Alcoa Visiting Prof at Univ of Pittsburgh 1970-71. Coeditor J Applied Math & Physics (ZAMP) 1965- & Elasticity 1971-83; mbr ed bd of SIAM J Applied Math; Associate Editor, J Applied Mech 1979-85. Fellow Amer Acad of Mechanics, US Dir 1970-73, Secy 1973-77, Pres 1977-78; Fellow Amer Advan Science; Dir, Soc Eng Science 1981-86; Pres Assoc Chrmn of Depts of Mechs 1978- 79. Mbr, US Natl Cttee on Theoretical and Applied Mechanics 1981-86. Res interests in elasticity, plasticity & stability theory. *Society Aff:* AAAS, AAM, ASEE, ASME, SIAM, SES, AMA.

Shields, Arthur M, Jr
Home: Rt 1, Box 197, Bluefield, WV 24701
Position: Owner. *Employer:* Shields Engg Co. *Education:* BSCE/Civ Engg/VPI. *Born:* 11/6/28. Native & resident of Bluefield, WV. Served in USMC Aviation. Proj Engr with Dupont at the Hydrogen Bomb Plant Augusta, GA - Design Engr with Hazlett & Erdall, Louisville, KY - Proj engr WV Dept of Hgwys. Founded & present owner of Shields Engg Co-1965. Baptist minister. *Society Aff:* ASCE, ACSM, NSPE.

Shields, Bruce M
Home: 104 Altadena Drive, Pittsburgh, PA 15228
Position: Management Consultant *Employer:* Retired from United States Steel Corporation *Education:* BS/Met Eng/Carnegie-Mellon Univ; MS/Metallurgy/MIT; Cert/Mgt Program for Execs/Univ of Pittsburgh. *Born:* 9/27/22. BS Met Eng Carnegie-Mellon Univ; MS Metallurgy MIT; Cert from Univ of Pittsburgh's Mgmt Program for Execs. Reg PE in Pa. Served in US ARMY Corps of Engineers during WW II advancing from Pvt to 1st Lt. Spent entire industrial career with US Steel advancing through positions of increasing mgmt responsibility in steel plants and headquarters to Director - Metallurgical Engrg. Currently a metallurgical consultant in quality management, iron & steelmaking, process & product metallurgy. Active mbr AISI, AIME, Fellwo ASM. Elder and Trustee in Presbyterian Church. Mbr Bd of Dirs, Silver Beaver Award, Allegheny Trails Council, Boy Scouts of America. *Society Aff:* AISI, ASM, AIME.

Shields, Donald H
Business: 4052 Graveley St, Burnaby, British Columbia V5C 3T6 Canada
Position: Principal Consultant & Assoc. *Employer:* Hardy Assoc Ltd. *Education:* PhD/Civ Engg/Univ of Manchester; DIC/Civ Engg/Imperial College; BSc/Civ Engg/Univ of Saskatchewan. *Born:* 9/2/34. in the shadow of the Transcona Grain Elevator. Escaped Kamsack, Saskatchewan by way of Montreal & London. Consulting engr since 1956. VP of Toronto firm from 1963 to 1968. Prof in Ottawa 1969 to 1978. Principal Consultant & Assoc in Vancouver with special responsibilities for mining, ocean bottom surveys and in situ testing. Happily married with three children at univ and two in high sch/public sch. Active in technical soc affairs & stds council. *Society Aff:* CGS, EIC, ASCE.

Shields, Edward E
Home: 21106 SE 258th St, Maple Valley, WA 98038
Position: District Mgr *Employer:* Interlake Material Handling Div *Education:* AAS/Mfg Engr Tech/Highline College. *Born:* 8/28/42. Native of Seattle Washington, Prof Certified in Material Handling. Staff Aid Fisher Mills Inc 1961-72 with proj responsibilities in Res & Promotion, Production, & Maintenance & Engrg Depts; Air Mac 1972-78, Branch Mgr, major responsibilities concept design, engrg & implementation of material handling sys. Cook-Newhouse & Assocs, Inc, 1979-1982, Proj Mgr, Maj Respons Mgnt Consultant in Industrial Facilities Planning and Sys Design. Private business owner. Currently employed by Interlake material Handling Division, District Sales Manager. Internatl V P IMMS 1975-76; Mbr Engrg Dept Adv Ctte, Vocational Adv Ctte, Highline Coll. Active in youth work. *Society Aff:* IMMS.

Shields, Robert E
Home: 901 N W 14 Court, Miami, FL 33125
Position: Sr. VP *Employer:* Ferendino, Grafton, Spillis and Candela *Education:* MS/Structures/Stanford Univ;BS/Civil/Fresno St Univ *Born:* 3/8/26 Native of CA. Began career with CA Division of Hgwys in construction and bridge design. Chief Engr for Golden Gate Bridge District during major repairs to structure and during planning and implementation of bus and ferry systems. Led transit planning efforts domestically and overseas for major consultant. Currently Senior VP and Dir of Engrg for Ferendino Grafton Spillis and Candela, largest minority-owned consulting firm in the US. Enjoys flying, gliding and sailing. *Society Aff:* ASCE, IBTTA, ITE, TRB, ARTBA

Shier, John W
Business: 315 Peoples Bank Bldg, P.O. Box 8, Port Huron, MI 48060
Position: Executive Vice President. *Employer:* Acheson Industries Inc. *Education:* BSChE/Chem Engr/Wayne State Univ. *Born:* April 1923. Native of Detroit Michigan. Served in US Army Armored Infantry WW II 1943-45. Joined Production Dept Acheson Colloids Co 1949, appointed Genl Mgr 1956, V P Mfg for Acheson Industries Inc & Mbr Bd/Dir 1971, Exec V P 1975. Directed construction or org of mfg co's in US, Netherlands, Eng, Germany, Japan, Brazil & Korea. Disting Engrg Alumnus Award Wayne State Univ 1958; Charter Mbr - Wayne State Univ. College of Engg Hall of Fame 1983. Member Tau Beta Pi. Chmn Detroit Sect AIChE 1954; Dir Peoples Bank of Port Huron; Mbr: AIChE, Eng Soc of Detroit. Hobbies: travel, golf & travel photography. *Society Aff:* AIChE, ESD.

Shifrin, Walter G
Business: 303 East Wacker Drive, Suite 600, Chicago, IL 60601
Position: Sr Vice President. *Employer:* PRC Consoer Townsend, Inc. *Education:* BS/Civil Engr/MIT; MS/Sanitary Engr/MIT. *Born:* June 29, 1933. Native of St Louis Mo. Asst Sanitary Engr US Public Health Serv 1956-58; various positions with Horner & Shifrin Inc 1958-69; PRC Consoer Townsend, INc 1969- , advanced to Assoc 1980, Partner 1972, VP 1976, Sr VP 1979, in charge of St Louis Mo Branch Design Office 1969-1980. 1980 - in charge of international business development and production. Reg P E in 27 states. Pres St Louis Chap MSPE 1971-72; Pres Mo Water Pollution Control Assn 1970, Natl Dir 1972-75. Bedell Award 1975. Hobbies: golf & ice hockey. *Society Aff:* ASCE, AWWA, NSPE, ACEC, WPCF, APWA, TBΠ, XE.

Shigeoka, Dennis K
Home: 250 Naniakea St, Hilo, HI 96720
Position: VP, Treas *Employer:* Okahara, Shigeoka & Assoc, Inc *Education:* BS/CE/Univ of HI *Born:* 12/07/45 Currently VP, treas for Okahara, Shigeoka & Assocs, Inc. Previously was employed at the county of Hawaii, Planning Dept as Planner/Civil Engr and before that, at the state of Hawaii, Dept of Natural Resources, Div of Water and Land Development as a civil engr. Was past pres of Hawaii Soc of PE, Big Island Chapt, 1980-1981 and previously served as their VP, 1979-1980. Is mbr of Rotary Club of South Hilo and holds position of Dir in yr 1981-1982. Previously was mbr of Kiwanis Club of Hilo holding position of Dir 1978. Was Pres of Church's organization, 1974-76. Married with three sons, Reid born 9/27/75, Neil born 5/14/78, Dean born 5/12/80. *Society Aff:* NSPE, ASCE

Shigley, Joseph E
Home: 125 Timber Trail, Pinecone Beach, Roscommon, MI 48653
Position: Ret. (Prof of Mech Engr, Emeritus) *Education:* MS/Eng Mech/Univ of MI; BSEE/Elec Engg/Purdue Univ; BSME/Mech Engg/Purdue Univ. *Born:* 4/10/09. Prof of Mech Engg 1957-78. Author *Machine Design, Kinematic Analysis of Melchanisms, Dynamic Analysis of Machines, Mech Engg Design, Simulation of Mech Sys, & Applied Mechanics of Materials* and others. Fellow ASME. ASME Mechanisms Ctte Award 1974, ASME Worcester-Reed-WArner Medal, 1977., ASME Machine Design Award, 1985. *Society Aff:* ASME.

Shima, Hideo
Home: Takanawa Paircity-607, 14-1, Takanawa 4-Chome, Shinagawa-ku, Tokyo 108, Japan
Position: Special Advisor. *Employer:* Natl Space Dev Agency of Japan. *Education:* -/Mech Eng/Tokyo Imperial Univ. *Born:* 5/20/01. Graduated Tokyo Imperial Univ 1925. Staff then Ch Motive Power Rolling Stock Sect Japanese Natl Railways 1925-42, Ch Engr Exec V P JNR 1942-51 & 1955- 63, planned & constructed high speed 'New Tokaido Line' (Shinkansen); Ch Mech Officer Sumitomo Metal Indus Corp 1951-55 & 1963-69; Pres Natl Space Dev Agency of Japan 1969-77; Pres Japan Railway Tech Serv Inc 1970-77; Pres Japan Railway Engrs Assn 1964-77. Elmerr & Sperry Award 1967; James Watt Internatl Medal 1969. Fellow ASME, Inst of Mech Engrs/England. P Pres Japan Soc of Mech Engrs 1960-61. *Society Aff:* JSME, ASMe, IME, VDI.

Shima, Shigeo
Home: 7-10-14 Koyamacho, Shinagawa-ku Tokyo, Japan 142
Position: Advisor To The Board. *Employer:* SONY Corporation. *Education:* BSc/EE/Waseda Univ Tokyo Japan. *Born:* Aug 29, 1905 Tokyo Japan; citizen of Japan. NHK (Japan Broadcasting Corp) 1933-61, Mbr Sys Engrg Staff 1933-51, Head Acoustics & Audio Freq Res 1951-56, Deputy Engrg Dir 1956-58, Dir NHK Tech Res Labs 1958-61; SONY Corp 1961- , Managing Dir of Bd 1961-72, Sr Managing Dir of Bd 1972-73 (Head Engrg & Dev Div 1961-65, Dir R&D Labs 1965-73), Advisor To The Bd 1973. Expert Ctte Mbr Radio Regulatory Tech Comm of Government Japan 1956-80; Chmn Color CRT Dev Comm 195861; V P Inst TV-E J 1959-60; Auditor Inst ECE-J 1958-60; Bd of Trustees Inst Acoust-E J 1957-75; Dir IEEE (USA) 1967-68, Chmn IEEE (USA) Tokyo Sect 1973-74. Awards: Assn for Post & Telecomm 1942, Assn for Inventions Promotion 1943, Nippon Radio Wave Assn 1950, NHK 1956, Inst TV-E J 1972, Soc for Telecom 1972, Minister of Telecom & Pst 1974, IAEJ 1983, IEEE 1984. Emperor of Japan Medal of Hon with Blue Ribbon 1967, Imperial Order of The Rising Sun 1975. *Society Aff:* IEEE, SID, IECEJ, IEEJ, ITVEJ, IAEJ, AESJ, SMPTE, ASA.

Shimazu, Satoshi D
Business: 1210 Ward Avenue, Honolulu, HI 96814
Position: President. *Employer:* SSFM Engrs, Inc. *Education:* MS/Sturctural Engr/Univ of Ill; BS/Civil & Str/Rose-Hulman Inst of Tech *Born:* 11/18/23. Served WW II 442nd AJA Combat Team. Reg P E in Hawaii and Guam. Pres of SSFM Engrs, Inc 1962- ; also worked at Park & Yee, HC & S Co & Ill Div of Hwys at Peoria; taught structural design part-time at Univ of Hawaii 1963-64. Hawaii Engr-of-the-Yr 1970, Mbr: ASCE, NSPE (P Pres HSPE 1964-65), SEAOH (P Pres 1974), WSCSEA (P Pres 1974), CECH (P Pres 1975-76) Tau Beta Pi; Natl Dir CECH 1976-77. Active in community Chmn of Bd CILO Hawaii 1976-77, Hobbies: fishing, classical music, & martial arts. *Society Aff:* ASCE, CECH, SEAOH, NSPE, HSPE, ACEC

Shimek, C L
Business: 1300 Adolphus Tower, Dallas, TX 75202
Position: Partner. *Employer:* Shimek, Jacobs & Finklea. *Education:* BS/Civil Engrg/Univ of TX, Austin. *Born:* June 8, 1923. Employed by cons engrg firms of Alfred Tamm 1947-48, H N Roberts & Assocs, & Roberts & Merriman 1948-49, Forest & Cotton Inc 1949-69 (V P 1960-66; Pres 1966-69), Shimek, Jacobs & Finklea 1969-present (Partner). Fellow ASCE, Natl Dir NSPE, P Pres & Engr of the Yr Award Dallas Chap TSPE, Honor Mbr Chi Epsilon. Extensive exper in design of waterworks, sewerage & other public works projs; admin & supervision of personnel in these & other related activities; mgmt of engrg & administrative groups. Hobbies: golf, boating & hunting. *Society Aff:* NSPE, ACEC, ASCE.

Shina, Isaac S
Business: PO Box 371-X, Sewickley, PA 15143
Position: Pres *Employer:* Shina & Associates, Inc *Education:* MSE/Civ Engg/Univ of MI; BS/Civ Engg/Am Univ of Beirut; BA/Engg/Am Univ of Beirut. *Born:* 12/23/25.

Shina, Isaac S (Continued)
Reg PE in more than 20 states & holder of Certificate of Qualification from the Natl Council of Engg Examiners. Started his own consulting firm in 1982. Prior to that was Exec VP and Chief Engr of Green Intl, Inc, where he served for nearly 30 years and in his last position was responsible for all foreign work undertaken by the firm. Taught in the Civ Engg Dept at the Univ of MI in 1950/1951 & 1951/1952. He was also Adjunct Assoc Prof in Civ Engg at the Univ of Pittsburgh, 1976-82, leaving that position in 1982. For several years performed as advisor on World Bank and Asian Development Bank assisted projects in road design and construction as well as contract administration in developing countries. Special expertise in design of industrial structures & heavy fdns, hgwys & bridges including bridge rehabilitation & erection, feasibility studies, construction mgt & contract negotiations. Extensive international travel in numerous countries and capability in several foreign languages. Married to the former Jo Ann Hawk of Johnson City Tennessee: Three sons, Jeffrey, Matthew and Michael. *Society Aff:* ASCE, NSPE, EERI, IABSE.

Shingledecker, Ross B
Home: 12001 Eau Galle Road, Caledonia, WI 53108
Position: Metallurgical Director/Mfg Services. *Employer:* Ladish Co/Cudahy Wisconsin. *Born:* March 15, 1923. Native of Clarion County Pa. Machine apprentice Parker Appliance Co 1941-43. Enlisted USMC, served in Engrg Battalion/Iwo Jima & occupation of Japan. BSME Marquette Univ 1950; grad studies Univ of Wisconsin. Ladish Co 1950- ; positions in process control, furnace design, engrg, Met Dir of Forging & Heating; currently Met Dir of Mfg Servs. Mbr ASM, P Chap Chmn & Bd Mbr, currently on Activities Ctte out of Metals Park; Mbr AIME, P Chmn Wisconsin Sect & Bd Mbr, selected Wisconsin Sect Man of the Yr. Mbr BSA, holds Silver Beaver Award. Father of 7 sons & 3 daughters. Enjoys family & gardening.

Shinnar, Reuel
Business: Convent Ave at 140th St, New York, NY 10031
Position: Distinguished Professor of Chem Engrg/Chmn of Dept. *Employer:* City Coll of The City Univ of NY. *Education:* PhD/ChE/Columbia Univ; Dipl Engg/Chem Eng/Technion Haifa, Israel. *Born:* 1923 Vienna Austria. Worked in Industry 10 yrs before PhD; taught at The Technion & Princeton Univ; joined City Coll 1964 as Prof of Chem E; Cons to Chem & petroleum industries. Pub over 60 papers in process dynamics & control, chem reactor design, crystallization, polymerization, tracer experiments, mixing & rheology. Owns sev patents. Active in research in process economics and design, chemcial reactor design and control, and synthetic fuel tech. Mbr: AIChE, ACS, AIAA, NY Acad of Sci & ORSA. Recipient of the Alpha Chi Sigma Award of the Amer Inst of Chem Eng 1979, elected to National Academy of Engineering 1984. *Society Aff:* AIChE, AIAA, ACS, ORSA, NAE.

Shinners, Stanley M
Home: 28 Sagamore Way North, Jericho, NY 11753
Position: Sr Res Section Hd. *Employer:* Sperry Sys Mgmt (Div of UNISYS). *Education:* MS/EE/Columbia Univ; B/EE/City Coll of New York *Born:* May 1933. With Sperry 1958- , assumed current responsibility of Sr Res Section Hd in 1976; Adjunct Prof NY Inst of Technology and The Cooper Union. Author of books: 'A Guide to Sys Engrg & Mgmt,' Lexington Press Lexington Mass 1976; 'Modern Control Sys Theory & Application, Second Edition,' Addison-Wesley Reading Mass 1978; 'Techniques of Sys Engrg,' McGraw-Hill NY 1967; 'Control Sys Design,' John Wiley NY 1964. Fellow IEEE 1973. Awarded P E license by NY 1970 for eminence achieved in engrg. Awarded the Career Achievement Medal by the Engrg and Architecture Alumni of the City Coll, 1980. *Society Aff:* IEEE, ASEE

Shinozuka, Masanobu
Business: Columbia University, 610 Mudd, New York, NY 10027
Position: Renwick Professor of Civil Engineering. *Employer:* Columbia University. *Education:* PhD/Civil Eng/Columbia Univ; MS/Civil Eng/Kyoto Univ; BS/Civil Eng/Kyoto Univ. *Born:* 12/23/30. Asst Prof Columbia 1961-65, Assoc Prof 1965-69, Prof, 1969-present. Mbr: ASCE, AIAA & Fellow of ASME. Chmn & mbr of a num of tech cttes ASCE. Awarded ASCE W L Huber Res Prize 1972 & ASCE State-of-the-Art of Civil Engrg Award 1973. Awarded A M Freudenthal Medal 1978; Awarded Nathan M. Newmark Medal in 1985; Elected mbr of Natl Acad of Engg, 1978. Recipient of a num of NSF Res Grants. Approx 140 tech papers in area of structural engrg, reliability & random vibrations. Cons with indsu & gvmt agencies. *Society Aff:* ASCE, ASME, AIAA.

Shipinski, John H
Business: 449 Gardner St, S Beloit, IL 61080
Position: Manager *Employer:* Warner Electric *Education:* PhD/ME/Univ of WI, Madison; MS/ME/Univ of WI, Madison; BS/ME/Univ of WI, Madison *Born:* 08/25/32 Award for res in internatl diesel combustion. Proj Engr on most successful outboard engine in history of indust. Led agricultural indus to adoption of turbo-charged diesel engines. Proj mgr of largest, all new engine ever built for Ag andindustrial equipment applications. *Society Aff:* SAE, ASME

Shipley, Larry W
Home: 1816 War Eagle Dr, N Little Rock, AR 72116
Position: Pres. *Employer:* Chematco, Inc. *Education:* BS/CE/Univ of AR. *Born:* 5/11/47. Native of Bentonville, AR. Technical Service engr with Aluminum Co of Am 1970-76, specializing in alumina chems. Proj engr with GE Plastics 1976-77 with responsibility for construction, start-up, & process dev of polybutylene terephthalate plant. Elected Pres of Chematco in 1978. Assumed additional responsibilities of Bd Chrmn & Chief Operating Officer in 1979. Interested in astronomy & intl politics. *Society Aff:* AIChE.

Shipman, Harold R
Home: 7108 Edgevale St, Chevy Chase, MD 20815
Position: Water Supply & Wastes Advisor. *Employer:* World Bank. *Education:* MS/Sanitary Engg/Univ of MN; BS/Chem Engg/Univ of MN. *Born:* Feb 1911. Native of Montana. Reg Engr Minnesota. Minn Dept of Health 1937-50, Farm Security Admin 1940-42; Engr Advisor Gov'ts of Korea 1950-51, Turkey 1952-54, Egypt 1954-58; Ch Engr Pan American Health Org & Regional Engr for the Americas, WHO 1958-62; Water & Wastes Advisor World Bank 1962-1976. 1977-present Engineering Consultant. Diplomate American Academy Environ Engrs; Fellow ASCE, APHA; Mbr AIDIS, IWSA, PSA, UN Interagency Subctte Water Resources (Chmn 1971). President, USA Section Inter American Asso of Sanitary Engineers 1981-85. Citations from governments Korea, Israel, Egypt & Bolivia. Johns Hopkins Univ Centennial Fellow. Friendship Medal British Inst Water Engrs. Author num papers on tech, financial & public health aspects of water & wastes in developing countries. Distiguished Mbr Award - AIDIS. *Society Aff:* ASCE, NSPE, AWWA, WPCF, APHA

Shirk, Charles A
Business: 3650 Mayfield Road, Cleveland, OH 44121
Position: Pres/Genl Manager/Ch Exec Officer. *Employer:* The Austin Company. *Education:* BS/Civil Engg/Univ of Notre Dame. *Born:* May 1920 Gary Ind. Joined The Austin Co 1948 Chicago Dist, named Mgr of R&D Div Cleveland Ohio 1961, Dist Engr Chicago 1962, Asst Mgr Chicago Dist 1965, elected Company V P & Mgr of Austin Process Div 1970, elected Company Dir 1971, Pres, GM, CEO Feb 1973. Deputy Bd Chmn 1979. Fellow ASCE. Reg C E in 49 states & DC. Certified by Natl Bureau of Engrg Regis. Mbr NSPE, Amer Welding Soc, Amer Concrete Inst, Cleveland Engrg Soc, Newcomen Soc. *Society Aff:* ASCE, NSPE, ACI, AWS.

Shivler, James F, Jr
Business: P O Box 4850, Jacksonville, FL 32201
Position: Chairman of the Board *Employer:* Reynolds, Smith & Hills. *Education:* MS/Civil Eng/Univ of FL; BSCE/Civil Engr/Univ of FLA. *Born:* 2/17/18. C E faculty Univ of Fla 1940-41; Design Engr present firm 1941, Partner 1950, Pres 1970, Chairman of the Board 1983. Dir Environ Sci & Engrg Inc. Commissioned Officer US Navy Civil Engr Corps 1943-46. Pres Fla Engr Soc 1960- 61, Pres Fla Bd Engr

Shivler, James F, Jr (Continued)
Examiners 1964-70, Chmn NSPE-PEPP 1963-65, Pres NSPE 1972- 73, Fellow ASCE, ACEC; mbr Tau Beta Pi. Awards: 1972 Univ of Fla Disting Alumnus, 1973 ENR Citation for construction indus serv, 1974 NSPE-PEPP, 1970 NCEE Special Serv. ASCE Florida Section, Engineer of the Year 1973; AAES, Vice Chairman, Engineering Affairs Council 1980-81; NSPE Annual Award 1981; NSPE Award for Exceptional Contributions to the Engrg Profession 1984. *Society Aff:* NSPE, ASCE.

Shoaf, Ross T
Home: 147 Seashore Drive, Daly City, CA 94014
Position: Ret. (Traffic & Transportation Engr) *Education:* BS/CE/Univ of CA; /BSus Admin/Harvard Grad, Sch Bus Admin; Cert/Traffic Engg/Yale Bur High Res. *Born:* June 1908 San Francisco Calif. San Francisco City Traffic Engr 1945-62; SF Asst City Engr 1962-72; retired 1972. Int'l Pres Inst of Trans Engrs 1972. ITE Hon. Mem. 1985. Tech Advisor US Delegation UN Conf Traffic Control Vienna 1969, Exec Advisor to Minister of Communications Republic of China/Taiwan on Traffic Admin 1973, Mbr Natl Joint Ctte on Traffic Control 1962-70, Mbr High Res Bd 1965-72. *Society Aff:* ITE.

Shobert, Erle I, II
Home: P O Box 343, St Marys, PA 15857
Position: VP/Technology (Retired). *Employer:* Stackpole Carbon Co/St Marys Pa. *Education:* AB/Math & Physics/Susquehanna Univ; MA/Physics/Princeton; Cert/Physics/Goettingen, Germany; DSc (Hon)/Sci/Susquehanna Univ. *Born:* Nov 1913 Du Bois Pa. Married Marjorie Sullivan April 1939. 2 daughters: Judith (Mrs Edward Marsden) & Margaret (Mrs William Hayes). Inst Internatl Fellow Goettingen Germany 1935-36. Stackpole Carbon 1934- . Chmn Bd of Dir Chemcut Corp, 1978-80. Chmn of the Bd Susquehanna 1978-86. Pres St Marys Youth Council & Elk County Soc Crippled Child & Adults. P Pres ASTM 1971-72, Holm Award 1974; Bd/Dir EJC 1969-71; Penn State Governor's Materials Adv Panel. Fellow IEEE, AAAS, ASTM. Mbr Sigma Xi. Hobbies: photography & travel. *Society Aff:* IEEE, AAAS, ASTM, ΣΞ.

Shockley, Thomas D
Business: Dept of Elec Engr, Memphis, TN 38152
Position: Prof *Employer:* Memphis State Univ. *Education:* PhD/EE/GA Inst of Tech; MSEE/EE/LA State Univ; BSEE/EE/LA State Univ. *Born:* 11/2/23. Pres of SSC, Inc, 1978 to present. Chrmn, Dept of Elec Engg, 1967-83. Previous teaching and res experience: GA Inst of Tech; Univ of OK; Univ of AL. Previous industrial experience: Gen Dynamics, North American Rockwell. Publications and Technical articles: US and foreign including--Inst of Elec and Electronics Engrs Transactions and Proceedings; Indian Journal of Physics; Physics Letters. NFPA Standard 110 (Chairman - Chapters 1 & 2). Memphis - Shelly County, Tennessee Appeals Board. *Society Aff:* IEEE, NSPE, NFPA, ASEE.

Shoemaker, F Glen
Home: 24924 Riverwood, Franklin, MI 48025 *Employer:* Retired from Gen Motors *Education:* BS/EE/Univ of IL *Born:* 06/04/89 Born in Abingdon, IL. Plant foreman for $80 a month was my first job out of coll. More experience as steam engr, aircraft engine instructor during World War I and experimental engr at McCook Field and Franklin Motorcar Co. Joined Gen Motors Res Lab in 1928. Head of Gen Motors Diesel Section in 1934 and Chief Engr at Detroit Diesel in 1938 until retirement in 1954. In the 1930's fathered the still commercially successful Series 71 diesel engine. A highpoint of my retirement was my election as a Fellow of the Soc of Automotive Engrs in 1979. *Society Aff:* SAE, ESD

Shoemaker, George E
Home: 404 Cheswick Pl Apt 452, Rosemont, PA 19010
Position: Reg PE (Self-employed) *Education:* BS/Physics/Princeton; -/Air Service Sch for Radio Officers/Columbia Univ; BS/Elec Engg/MIT. *Born:* Oct 1895. WW I interplane communication dev 1917. Engr Dept of G E Co Nela Park, Dev Dept Phoebus SA & Geneva Switzerland; Serv Div Philadelphia Electric Co; Prof Engrg Cons. Fellow Illuminating Engrg Soc (now Emeritus); P Chmn of the Bd of Fellows, Gold Medal Award Comm & Church Lighting Comm; Rep on ASA Comm of Natl Elec Code (panels 2, 3 & 10); Indus Comm on Interior Wiring Design, Internatl Electrotechnical Comm; Internatl Comm on Illumination (3 Comms). Hobby: 'Christian Action' & underprivileged Philadelphians. *Society Aff:* IES.

Shoemaker, Leonard W
Business: 9235 Katy Freeway, Suite 333, Houston, TX 77024
Position: Pres & Chrmn of Bd. *Employer:* Leonard W Shoemaker & Assoc, Inc. *Education:* BS/Civ Engg/TX A&M Univ. *Born:* 11/11/37. Native of Harris County, TX. Prior to 1970 VP of two of Houston's well known consulting firms. In 1970 formed Leonard W Shoemaker & Assoc, Inc providing services in planning, design and admin of water, sewage and drainage works, subdivisions and related works. Presbyterian ordained deacon and ruling elder. Pres - Sam Houston Chap TSPE 1970-71. Outstanding young engr 1969. Listed in Who's Who in South and Southwest, 1971 to present. Dir & VP - Houston Hunter & Jumper Charity Horse Show. Mbr Greater Houston Builders Assn, and Harris County Heritage and Conservation Soc. Trustee, Waller County Historical Museum. *Society Aff:* NSPE, TSPE, ASCE, NHBA.

Shoemaker, Robert H
Business: 12890 Westwood Ave, Detroit, MI 48223
Position: Pres & Chf Exec Officer, Chmn Bd. *Employer:* Kolene Corporation. *Education:* BS/Finance/Ohio U *Born:* 9/15/21. U of Michigan Extension post-grad studies ferrous metallurgy & related subjects. Pres & Ch Exec Officer Chmn Bd Kolene Corp, world's leading exponent of metal conditioning salt bath tech - corp mgmt coordination of res, dev & engrg progs. Co-patentee of sev inventions involving molten salt bath cleaning equipment & methods for processing metals. Authored many articles in trade publs & presented num tech papers before prof groups both indus & soc: ASM, SAE, AISE, IHEA, AISI. P Pres, Fellow ASM. Active mbr of above soc. ESD, IHEA, ASM. *Society Aff:* ASM, AISI, SAE, ESD, IHEA

Shoemaker, Robert S
Business: PO Box 2689, Grass Valley, CA 95945
Position: President *Employer:* R S Shoemaker LTD *Education:* MS/Inorganic Chem/OR State Univ; BS/Inorganic Chem/OR State Univ; MS/Metallurgical Engg/Univ of WI. *Born:* July 1925 Roseburg Oregon. Prof Met E (Hon) Montana Coll of Mineral Sci & Tech. US Army Corps Engrs 1943-46. Res Engr Union Carbide Metals Co 1953-57; Met Engr Union Carbide Ore Co 1957-62; Supervising Engr/Ch Met Engr 1962-68; Cons Metallurgist 1968- , Mgr of Div Metallurgy 1977-80, Mining & Metals Div Bechtel Corp San Francisco. Vice president, San Francisco Mining Associates, 1981-84. 1984-Independent Consulting Metallurgist. Dir & Treas SME of AIME, Pres SME 1978, Dir & V P AIME, Pres Mining & Met Soc of Amer 1975 & 1976. AIME Richards Award 1974. Author of num tech articles & books. *Society Aff:* AIME, MMSA, IMM, SAIMM, CIMM

Shohet, J Leon
Business: Dept of Elec & Comp Engg, Univ. of Wisconsin, Madison, WI 53706
Position: Prof. & Chairman *Employer:* Univ of WI. *Education:* PhD/Elec Engrg/Carnegie Mellon Univ; MS/Elec Engrg/Carnegie Mellon Univ; BS/Elec Engrg/Purdue Univ. *Born:* 6/26/37. Native of Chicago. Developed one of the earliest interdisciplinary progs in the area of plasmas & controlled fusion at the Univ of WI. Founder of the IEEE Transactions on Plasma Sci. Winner of the Frederick Emmons Termon Award of the ASEE (1977) & the Merit Award, IEEE Nucl & Plasma Sciences Soc (1978). Currently dir of the Univ of WI's Torsatron/Stellarator Lab and the Center for Plasma Processing & Technology. Author of the textbook, "The Plasma State-, along with numerous other scientific & technical publications. Fellow, IEEE & Am Physical Soc. Chairman of the Department of Electrical & Computer Engrg. *Society Aff:* IEEE, APS, ΣΞ, TBΠ, HKN.

Shook, Garry T
Business: 202 Collett, P.O. Box 366, Morganton, NC 28655
Position: Consulting Engr *Employer:* Thompson-Gordon-Shook Engrs *Education:* BS/CE/NC St Univ *Born:* 1/14/50 Founding partner of Thompson-Gordon-Shook Engrs, Inc; a very successful and prestigious engrg firm specializing in bridges. Very active in engrg societies, Secretary-Treasurer of the NC Society of Engrs. Professional work experience includes land surveying, subdivision design, water and sewer system design, hgwy and bridge construction, materials testing, transportation engrg, hgwy and bridge maintenance, power plant construction, design of buildings and facilities for nuclear and coal fired power plants, bridge inspection and rating, and the design of major hgwy structures. *Society Aff:* ACEC, CSI, NCSE, SCSE

Shook, James F
Home: 3313 Powder Mill Rd, Adelphi, MD 20783
Position: Sr. Vice President *Employer:* ARE Inc *Education:* MASc/Transportation Engr/Univ of Waterloo; BS/Civ Engg/Univ of MD. *Born:* 12/27/23. Conducted res for the concrete & aggregate industries, 1949-1956. Chief, Mtls Branch, $27 million AASHO Rd Test, Hgwy Res Bd (TRB), 1956-1961. Several positions with the Asphalt Inst since 1961. Specializes in structural design of asphalt pavements and highway materials. Recognized by Eng News Record as one of the "men who made their mark in construction–, 1973. Author of over 35 published technical papers. Mbr of Tau Beta Pi, TRB committees, Fellow of ASCE, other societies. Reg PE in IL, MD & VA. Active in NSPE. Pres, MD Soc of PE, 1979-1980. *Society Aff:* NSPE, MSPE, ASCE, AAPT, TRB

Shook, William B
Business: 2041 College Rd, Columbus, OH 43210
Position: Prof Emer/Ceramic Engrg Dept *Employer:* The Ohio State University. *Education:* PhD/Cer Eng/Ohio State Univ; B Cer Eng/Cer Eng/Ohio State Univ *Born:* Raised in Scotland & Eng, completed secondary education in Cleveland Ohio during WW II. PhD, BCE Ohio State. Res admin at Engrg Experiment Station OSU Dir of Bldg Res 1956-58; Dir of Ceramic Res at the Station 1961-63; Chmn Ceramic Engrg Dept 1972-82; Prof Emeritus 1983- . Received OSU Disting Teaching Award 1972. Fellow ACS 1972. Hobbies: camping, hiking & canoeing. *Society Aff:* ACerS

Shoolbred, Augustus W, Jr
Home: 125 Fernbrook Circle, Spartanburg, SC 29302
Position: Senior vice President-Manager of Operations *Employer:* Maquire/Beebe *Education:* BS/CE/Clemson Univ *Born:* Dec 29, 1926 Columbia SC. SC Hwy Dept Inspector 1949-59; Design Engr LBC & W-Harwood Beebe Co Spartanburg SC 1950-60, Ch Engr/Dir/Secy 1960-63, V P 1963- 72, Exec V P 1972-75, Pres 1975. Sr VP & Mgr of Operations Harwood Beebe Div, CE Maguire, Inc. Directs opers of Spartanburg Office & 3 branch offices. Served with Corps of Engrs AUS 1945-47. Mbr: NSPE, SC Soc of P E's (State Pres 1975-76; Past V P, Dir & Chap Pres), AWWA, WPCF. Certificate of meritorious serv SCSPE 1966, 1969, 1973, 1976. Chapter Engr of the Yr 1975, State Engineer of the Year 1977. *Society Aff:* NSPE, AWWA, WPCF.

Shooman, Martin L
Home: 12 Broadfield Pl, Glen Cove, NY 11542
Position: Prof EE & CS. *Employer:* Poly Inst of NY. *Education:* DEE(PhD)/EE/Poly Inst of Brooklyn; Master/EE/MIT; Bachelors/EE/MIT. *Born:* 2/24/34. in Trenton, NJ. Ind experience: GE Co (1953-55), Sperry Gyroscope Co (1956- 58), RCA (1961), Grumman (1967). Teaching: MIT, teaching fellow (1955-56), vis assoc prof of EE (1971); Poly Inst of Bklyn, assoc prof of EE (1958-72); Polytech Inst of NY, prof of EE & CS (1973-86), Polytechnic Univ (1986-present). Dir Div of Computer Sci 1981-83. Consultant and lectr to govt & ind: IBM, Grumman, Bell Labs, NASA, many others. IEEE activities & honors: Comp Soc Inter Confs Software Engg, tech chrmn (1973, 75, 76, 79, 85); awards for best technical paper reliability, software (1967, 71, 77(2), 83); annual reliability award 1977, elected fellow 1979. Author: over 80 papers & reports, Chap in 8 books, "Probabilistic Reliability" (1968), "Software Engineering" (1982), "Probabilistic Reliability, 2nd Edition–, 1987. *Society Aff:* IEEE, ACM.

Shoopman, Thomas A
Business: R R 2, Weeping Willow Ln, Rochester, IL 62563
Position: Owner *Employer:* Resource Consultants *Education:* BS/Agri Engg/Univ of IL. *Born:* March 1946. Design Engr Ill Div of Water Resources 1969-73; Proj Engr Michael Baker Jr Inc 1973-75, assigned to Alaska Pipeline Sys Proj responsible for design of drainage & erosion control features along pipeline route; Proj Mgr M & E/ Alstot March & Guillou Inc 1975-1981, responsible for design & mgmt of projs. Projs completed include comprehensive basin reports & designs for flood control, drainage, water supply, water quality, recreation & flood plain mgmt. Owner, Resource Consultants 1981- , specializing in land and water resources engineering. Pres Central Ill Sect ASAE 1971, Sec Capitol Chap ISPE 1973. Hobbies: camping, fishing & boating. *Society Aff:* ASAE, ASCE, SCSA, ISPE.

Shore, David
Home: 1419 McLean Mews Ct, McLean, VA 22101
Position: Division Vice President. *Employer:* RCA Corporation. *Education:* MS/Physics/OH State Univ; BS/Aero Eng/Univ of MI. *Born:* 8/12/19. 1941-54 USAF at Wright Field, last position Civilian Ch Weapons Sys Analysis. 1954-present RCA: Mgr BMEWS Sys Delineation 1958-59, Assoc Dir Advanced Military Sys 1959, Prog Mgr SAINT Satellite 1959-61, Mgr Sys Engrg 1962-65, Ch Engr Communications Sys Div 1965, Ch Defense Engr Defense Electronics 1966, Div V P Advance Progs Dev 1966-1981, Div VP Business Dev 1981- present. Served to 1Lt US Air Corps to 1946. Reg P E NJ. Chmn IEEE Internatl Conf on Communications, Chmn NSIA COMCAC 1969-75, Pres NARPV 1974-75, Natl Res Council USAF Command & Control Studies 1976-78. Mbr Army AMARC. Conslt to DSB 1982, NRAC 1983-present. Mbr Army Sci Bd 1977-present. Career has been dedicated to conception & initiation of advanced military sys & equipment particularly C3I. *Society Aff:* AIAA, IEEE.

Shore, Melvin
Home: 401 Larch Ln, Sacramento, CA 95864
Position: Port Dir. *Employer:* Port of Sacramento. *Education:* BS/Engr/UC Berkeley *Born:* 7/25/24. Started with the Port of Sacramento as "Engr Aide to the Pt Dir" on July 1949. Advanced through the ranks of Chief Engr, etc, to Pt Dir 1963, retired June 30, 1986. Currently serving as a consultant. Have served in various capacities in the Sacramento Sec of ASCE including Pres in 1978. Served as Chrmn of the Bd of American Assoc of Port Authorities 1983. *Society Aff:* ASCE

Short, Byron E
Home: 502 E 32nd Street, Austin, TX 78705
Position: Professor of Mech Engrg Emeritus. *Employer:* Retired *Education:* PhD/Heat Transfer/Cornell Univ; MME/Mech Engg/cornel Univ; MS/Mech Engg/ Univ of TX; BS/Mech Engg/Univ of TX. *Born:* Dec 29, 1901 Putnam Texas. Power Engrg Dept Texaco Inc summers 1926-30. Instructor to Prof Mech E Dept Univ of Texas 1926-73; Prof Emeritus 1973present, Chmn Dept 1945-47 & 1951-53, charge heat power & fluid mechanics lab 1930-65, acting Dean Engrg 1948-49. Cons in heat transfer for sev co's. Res participant Oak Ridge Natl Lab 1956, 57. Author: 'Flow, Measurement & Pumping of Fluids,' 'Engrg Thermodynamics' (with H L Kent & B F Treat), 'Pressure-Enthalpy Charts'(with H L Kent & H A Walls). Res papers & bulletins on heat exchangers & thermodynamics. Mbr & Chmn sev tech cttes of ASME, ASHRAE, NSPE, ASEE. Life Fellow ASME, Life Mbr ASHRAE, NSPE, ASEE. Reg P E Texas. *Society Aff:* ASME, ASHRAE, ASEE, NSPE.

Short, John Patrick
Home: 3912 Greystone, Austin, TX 78731
Position: Sys Mgr/Transp & Pub Safety Div. *Employer:* Eagle Signal Corp. *Born:* Jan 4, 1937. Native of Midwest. BS Physics St Ambrose Coll Davenport Iowa 1958.

Short, John Patrick (Continued)
Joined Eagle Signal 1958 Application Engr, held positions of Sales Engr, Asst Genl Sales Mgr, Genl Sales Mgr, Prod Mgr, Sys Mgr during 18 yr career. Affiliate mbr ITE; mbr TRB, IMSA, NEMA. P Pres AOD Div ITE 1975, current V Chmn NEMA Traffic Sys Sect Tech Ctte, active on ITE & TRB Tech Cttes in area of transportation control sys. Mbr Davenport Iowa Chamber of Commerce Transportation Ctte for past 5 yrs.

Short, Robert A
Home: 1815 Hawthorne Place, Corvallis, OR 97330
Position: Prof/Elec Engg. *Employer:* Oregon State University. *Education:* PhD/EE/Stanford; MS/EE/Stevens Inst of Tech; BS/EE/OR State; BS/Math/OR State. *Born:* Nov 1927. Served with US Infantry 1946-48. Mbr Tech Staff Bell Labs 1952-56; Sr Res Engr Stanford Res Inst 1956-66; faculty at Oregon State 1966- , Prof of Elec E 1969- ; Chmn of Computer Sci 1971-79. Served in ed capacities for IEEE 1968- , Editor-in-Ch 1971-75, on Governing Bd; 1970-72, 1976-78; organized ACM student chap Oregon State 1970; ACM Ed Ctte 1972- ; presently IEEE-CS liaison to ACM for ed activities. *Society Aff:* IEEE, ACM, AAAS, ASEE.

Shortsleeve, Francis J
Business: Old Ridgebury Rd, Danbury, CT 06817
Position: VP International *Employer:* Metals Div, Union Carbide Corp. *Education:* PhD/Phys Metall/Case Inst; MS/Metall/Notre Dame; BS/Metall/Notre Dame. *Born:* 9/8/21. Res engr, Std Oil Co IN, 48-50; asst res engr, metals res labs, electro Metall Co Div, Union Carbide & Carbon Corp, 53-54, asst group mgr, 54-55, group mgr, 55-57, asst dir res, 57-60, dir res, Union Carbide Metals Co, 60-61, prod mgr silicon, Mining & Metals Div, Union Carbide Corp, 61-65, dir mkt, 66-68, VP 68-70, VP & Prod Gen Mgr Vanadium USAAF, 44-45. Am Soc Metals; Am Inst Min, Metall & Petrol Engg; American Iron & Steel Inst; fabrication of semi-furnished products; physical metallurgy; austenite decomposition in steel; relation of flake formation in steel to hydrogen, microstructure and stress. Address: Old Corner Rd, RFD 3, Bedford Village, NY 10506. *Society Aff:* ASM, AIME, AMA, AISI

Shoults, David R
Home: 5449 Gleneagles Drive, Tucson, AZ 85718
Position: Retired. *Education:* C-D-Hon/-/Univ of ID; BSEE/-/Univ of ID. *Born:* 6/23/03. G E Co 1925-45, pioneer in dev of Jet Propulsion for Aircraft in US; V P Engrg Bell Aircraft Corp 1945-47; V P Engrg Glenn L Martin Co 1947-50; Genl Mgr Aircraft Nuclear Propulsion Dept G E Co 1951-60; Rep-Atomic Dev 1960-64, Chmn Ad Hoc Ctte on Internatl Nuclear Liability of Atomic Indus Forum; Cons Atomic Energy 1964- . Fellow IEEE; Mbr SAE . *Society Aff:* IEEE, SAE

Shoup, Terry E
Business: Assistant Dean of Engineering, College Station, TX 77843
Position: Assistant Dean of Engineering *Employer:* Texas A&M University *Education:* BME/ME/OH State; MS/ME/OH State; PhD/ME/OH State. *Born:* 7/20/44. Asst Prof, Assoc Prof & Exec Office of Mech E Rutgers Univ 1969-75; Prof & Assoc Chrmn Mech E Dept Univ of Houston 1975-80; presently Asst Dean of Engrg & Prof. of Mech Engrg. Texas A&M Univ. Mbr: ASME, ASEE, Sigma Xi, Pi Tau Sigma. Program Chmn, ASEE ME Div, Chmn Medical Devices Comm, ASME Design Div, Editor- in-Chief of *Journal mechanism & machine Thoery.* 1974 recipient ASEE Dow Outstanding Faculty Award. *Society Aff:* ASME, ASEE, ΣΞ, ΠΤΣ.

Shoupp, William E
Home: 343 Maple Avenue, Pittsburgh, PA 15218
Position: Technical Consultant (Self-employed) *Education:* PhD/Physics/Univ of IL; AM/Physics/Univ of IL; AB/Physics/Miami Univ. *Born:* 10/5/08. Westinghouse Electric 1938-73, retired V P Res. Awards: Pittsburgh's 'Man of Yr in Sci,' Westinghouse's 'Order of Merit,' Fredrick Phillips Award IEEE, Univ of Ill 'Alumni Honor Award of Disting Serv in Engrg,' Miami Univ Hon ScD, Indiana Inst of Tech Hon ScD. P Pres ANS & Marine Tech Soc, V P NAE, P Chmn of Marine Bd NRC, Mbr Governing Bd NRC 1973-78. Fellow: ANS, ASME, APS & IEEE. Chmn Fossil Fuel Tech Adv Ctte DOE 1975-. Chmn Res Adv Council Miami U. 1979- . Mbr Visiting Ctte to MIT Dept of Ocean Engrg 1973-76, Nuclear Engr Dept 1970-73. P mbr Res Adv Ctte US Coast Guard. P mbr Comm of Natural Resources. *Society Aff:* APS, IEEE, ASME, MTS, ANS, NAE.

Shouse, Corbin W
Business: P O Box 2159, Dallas, TX 75221
Position: VP & Treasurer. *Employer:* American Petrofina Inc. *Education:* BS/Chem Engg/Univ of OK. *Born:* 10/1/29. in Picher, OK. Advanced Mgmt Prog Univ of TX 1961. Mbr: AIChE, Refining Genl Ctte Amer Petroleum Inst 1965-79. Reg PE TX. Mbr Presbyterian Church in US. Mbr Dallas Petroleum Club. Union Carbide Corp 1951-54; US Army R&D Fitzsimmons Army Hospital 1955-56; various positions in petroleum refining & petrochemical mfg Cosden Petroleum Corp 1957-65; VP Cosden Oil & Chem Co 1963; VP Refining Amer Petrofina Co of TX 1967; VP Amer Petrofina Inc 1972; Dir & Group VP Amer Petrofina Inc 1976. Jan 1979 VP & Treasurer, Dir, Chief Financial Officer. *Society Aff:* AIChE, ACS, API.

Showalter, Robert L
Business: 900 W Valley Forge Rd, King of Prussia, PA 18901
Position: President *Employer:* R. L. Showalter & Assoc. *Education:* Bachelor of Eng/ Civil/Villanov Univ. *Born:* 9/18/47. Native of Chalfont, Pa. Active in Consulting field since 1969. Formed his own firm in July of 1980. Responsibilities involve marketing, business development, overall design and management of firm. Restores early 50's English sports cars in spare time. *Society Aff:* ASCE, NSPE.

Showers, Ralph M
Business: Moore Sch of Elec Engrg, Philadelphia, PA 19104
Position: Professor of Electrical Engineering. *Employer:* University of Pennsylvania. *Education:* PhD/EE/Univ of PA; MS/EE/Univ of PA; BS/EE/Univ of PA. *Born:* July 1918. Prof Elec E engaged in res & teaching, res on problems concerning electromagnetic compatibility for over 30 yrs. During WW II Cons to Office of Scientific R&D on short-range communications. Cons to Dept of Navy. Past Chmn CISPR (Internatl Special Ctte on Radio Interference), V P US Natl Ctte of Internatl Electrotechnical Comm, Chmn Accredited Standards Ctte C63 (Electromagnetic Compatibility). Fellow IEEE, mbr AdCom of IEEE Group on EMC, Chmn IEC, Advis. Ctte on EMC (ACEC). *Society Aff:* IEEE, ASEE, AAUP, ORSA.

Shprintz, Lawrence
Business: The Willard Co, Jenkintown, PA 19046
Position: President & Chairman *Employer:* The Willard Co. *Education:* MSChE/Chem Eng/Drexel Univ; BSChE/Chem Eng/Univ of PA. *Born:* 8/27/27. in Philadelphia, PA. Reg P E. Mbr AIChE. Plant Engr Publicker Industries 1950-55; Process Engr Catalytic Inc 1956-59; Proj Mgr United Engrs & Constructors Inc 1959-73; Exec VP Engg & Construction Div Day & Zimmerman Inc 1973-77, Specialty Services Phila Suburban Corp 1977-81, Pres Willard Inc (Sub SD) 1979-81. Pres & Chmn The Willard Co. 1981- . *Society Aff:* AIChE.

Shrader, William W
Business: 144 Harvard Rd, Stow, MA 01775
Position: Consulting Scientist. *Employer:* Raytheon Co. *Education:* MSEE/Elec Engrg/Northeastern Univ; BSEE/Elec Engrg/Univ of MA. *Born:* 10/17/30. Born in China. Grew up in Montpelier, VT & Newton, MA. Joined Boeing Airplane Co, Seattle, in 1953. With Raytheon Co since 1956. Consulting scientist involved with the majority of Raytheon's radar systems. Best known for work in MTI (Moving Target Indication) radar. Seven US patents; 35 publications including chapters in two books, "Radar Handbook–, McGraw-Hill, 1970 & "Radar Technology–, Artech House, 1977. Elected Fellow of IEEE in 1978. *Society Aff:* IEEE.

Shreve, Gerry D
Business: 15565 Northland Dr, Suite 103W, Southfield, MI 48075
Position: Pres *Employer:* Gerry Shreve Assocs *Education:* BS/ME/Lawrence Inst of Tech *Born:* 08/05/31 Born, resides and practices Consltg Mech, Electrical and Energy Engrg in the Detroit Metropolitan area. Abjunct Prof in the Sch of Archit at the Univ of MI 1975-1976. CEC/M Bd Mbr, CEC/M Energy Committee Chrmn and ACEC Energy Committee Mbr 1978-81. CSI Detroit Chapt Mbr, Treas, VP and Pres 1967-75. Project Engr for Minoru Yamasaki Assocs, Inc 1960-1964. Partner in Hoyem, Basso, Adams, Martin & Shreve Consltg Engrs 1965-1975. Founded Gerry Shreve Assocs, Inc in 1975 to practice Consltg Mech, Electrical and Energy Engrg. Enjoys golf, skiing, boating, swimming. *Society Aff:* ASHRAE, AIA

Shrivastava, S Ram
Business: 44 Saginaw Dr, Rochester, NY 14623
Position: Pres, C.E.O. *Employer:* William C. Larsen, P. E., P. C. *Education:* MS/CE/Environ Engrg/Clarkson Coll of Tech; BE/CE/Jabalpur Univ-India *Born:* 07/19/46 A naturalized US citizen living in NY State since 1968. Started as proj engr at the current firm. Active involvement in planning, designing, construction supervisor, and start-up of pollution abatement projs. Areas of expertise include water, wastewater, solid waste, hazardous wastes, Biomass to energy conversion and industrial waste mgmt. Currently a principal in the firm responsible for special proj supervision, domestic and Intl business development and mgmt. *Society Aff:* AAEE, ASCE, WPCF, AWWA.

Shtrikman, Shmuel
Business: Weizman Inst of Science, Rehovoth, Israel
Position: Samuel Sebba Prof of Pure and Applied Physics *Employer:* Weizmann Institute of Science. *Education:* DSc/Electrical Engn Tech; Diplomaed Eng/Electrical Engn Tech; BSc/Electrical Engn Tech *Born:* 1930. Dept of Electronics Weizmann Inst of Sci 1954- . 1986-present: Head, Dept of Electronics. On leave: 1960-62 Sr Staff Physicist Franklin Inst Philadelphia; 1964-65 NSF Sr Foreign Scientist & Visiting Prof Dept of Physics Univ of Penn; 1971-72 Israel Academy Royal Soc Visiting Res Prof at Imperial Coll of Sci London. 1983-1984. Visiting Prof. Univ of CA, San Diego, Dept Engrg. Weizmann Prize for Sci by 1968, Michael Landau Prize 1975. Rothschild Prize for Innovation/export 1984, Armando Kanminitz Prize 1985. Res contrib in Solid State Physics, Instrumentation & Applied Physics. Cons to various industrial orgs. Fellow, IEEE; Fellow, American Physical Society. *Society Aff:* IPS, APS, IEEE

Shubat, George J
Business: G. J. Shubat Company, 370 W. Poplar St, Zionsville, IN 46077
Position: Manager of Engineering (retired) (retired). *Employer:* Diamond Chain Company. *Education:* BS/Met Engg/MI Tech. *Born:* 2/4/15. Native of Caspian Mich. Served with Army Corps of Engrs 1941-46 (final Lt Col). Diamond Chain Co 1947-1980- , assumed responsibility as Mgr of Engrg 1966, Now retired with own consulting firm - "G J Shubat Corporation" specializing in Metals, Materials, Metallurgey & Failure Analysis-Was responsible for Amer Cahin Assn activity in dev of natl & internatl standards for power tranmission & conveyor chains & sprockets. Reg Engr in Ind. Fellow ASM, Sr Mbr & Quality Engr Amer Soc Quality Control. ASM Award - Conferee 2nd World Met Congress 1957 *Society Aff:* ASM, ASQC.

Shubinski, Robert P
Business: 7620 Little River Turnpike, Annandale, VA 22003
Position: VP *Employer:* Camp Dresser & McKee Inc *Education:* PhD/CE/Univ CA-Berkeley; MS/CE/TX A&M; BS/CE/Rice Univ *Born:* 07/15/35 A native of Dallas, TX. He was an instructor at TX A&M Univ from 1958 to 1960 where he taught undergrad courses. From 1960 to 1977, he taught undergrad and grad courses in structural engrg at the Univ of CA at Berkeley and at Davis. He has been with CDM since 1962, where he has been a VP of the Water Resources Div since 1973. As a VP, he is responsible for overall project direction, proj development, client relations and report preparation. Enjoys boating, reading and classical music. *Society Aff:* ASCE, NSPE, WPCF, AAEE

Shuck, Lowell Zane
Home: 401 Highview Place, Morgantown, WV 26505
Position: Pres *Employer:* Tech Dev Inc (TDI) *Education:* PhD/TAM/WVU; MSME/ME/WVU; BSME/ME/W Va Inst of Tech. *Born:* 10/23/36. Employed as Prof, Mech Engg & Mechanics, and Assoc Dir of WVU Engrg Expn St from 1976-80. President, Technology Development Inc (TDI), 1980-present. Employed as Suvry Mech Engr with DOE (ERDA & USBM) from 1970- 1976. NSF Sci Faculty Fellow and Res Engr at WVU 1967-69. Taught and served as Chrmn, ME Dept of WV Tech 1959-67. Presently conslt to DOE and other energy cogs. Author 61 pubs, 15 patents; producer and script writer 4 tech 16 mm films for DOE; Sci Advisor to WV Governor John D Rockefeller IV. Reg PE WV No 4596. Natl Coun of Engrs Examiners Cert No 4230. Assoc Editor ASME Transactions Journal of Energy Resources Tech, Editorial Bd of In Situ Journal. Gov appted mbr WV Coal and Energy Res Advisory Comm. The ASME "Ralph James" Award for 1980; Chmn ASME Engrg Energy Technology Ctte 1979- ; Listed in Who's Who in the East; Who's Who in South & Southwest; American Men & Women in Sci; Who's Who in Tech Today (81); Internatl Who's Who in Engrg (1982-); Leading Conslts in Tech (82-). Mbr Tau Beta Pi, Sigma Xi, SPE of AIME; ASTM *Society Aff:* SPE OF AIME, ASME, ΣΞ, ΤΒΠ.

Shuck, Terry A
Home: 321 Tonawanda Drive, Des Moines, IA 50312
Position: President. *Employer:* Terry A Shuck, Structural Engrs, Inc. *Education:* MSCE/Struct Engr/Univ of IA; BSCE/Civil Engr/Univ of IA. *Born:* Nov 1934. Native of Des Moines Iowa. B J Lambert Faculty Award for Grad study. Mbr Chi Epsilon. Hon C E Soc. Varsity Football 1953, 1954, 1955. Pres Iowa Chap ASCE 1973. Self-employed Structural Cons 1966- . Bridge, Swimming Pool, and Building design. Mbr: ACEC, NSPE, ASCE, ACI, IES & Rotary Internatl. Hobbies: classical music, reading & outdoor activities. *Society Aff:* ASCE, ASEC, ACI, NSPE.

Shuldiner, Paul W
Business: Dept of Civil Engrg, Amherst, MA 01002
Position: Professor. *Employer:* University of Massachusetts. *Education:* DrEng/Transp Engg/Univ of CA, Berkeley; MSCE/Hwy Engg/Univ of Il, Urbana; BSCE/-/Univ of IL, Urbana. *Born:* June 19, 1930 to Abraham & Helen (Golub) S. Married Margot C (Vignet) June 21, 1951; 9 children. BSCE (highest honors), MSCE Univ of Ill; Dr Engrg Univ of Calif/Berkeley; post-grad London Sch Econ. NSF Sci Faculty Fellow, Brookings Fellow. ASCE Res Prize. Ch ASCE Urban Transp Res Council. Co-author 'Technology of Urban Transportation' & 'Analysis of Urban Travel Demands'; founding Editor 'Transportation.' 1966-69 Sr Engr Ch Transp Sys Planning Div (NE Corridor Proj) OHSGT US DOT; 1970-71 Deputy Dir Natl Transp Study conducted by NAS for Presidential Comm. 1975-77 Visiting Lecturer Univ of Ill. 1971- , Prof of Civil Engrg Univ of Mass. *Society Aff:* ASCE, TRB.

Shulhof, William P
Home: 1805 Veterans Memorial Parkway, Saginaw, MI 48605
Position: Technical Director *Employer:* Central Foundry Div of Genl Motors. *Education:* PhD/Petrology/PA State; MS/Petrology/PA State; BA/Geology/Univ of VA. *Born:* BA with Distinction Univ of Va; MS, Dr Philosophy Penn State Univ. Phi Beta Kappa, Raven & Sigma Xi. Industrial career with G M Corp includes sev tech & tech mgmt positions, has specialized in foundry processes 1965- , presently Technical DIR R&D of Central Foundry Div G M Corp. Mbr ASM; mbr & Past Chmn Tech Council Amer Foundrymen's Assn, presently Chairman of the Research Board of the AFS. Holds Amer Foundrymen's Soc Award of Scientific Merit. Received Industry Citation from the Iron Castings Society in 1982 for his advancement of iron casting technology. Author of numerous technical papers and holds one patent. 1980, international exchange speaker at the BICRA International Conference; 1982 presented the U.S. International Exchange paper at the International Foundry Congress. *Society Aff:* ASM, AFS, ITC

Shulhof, William P (Continued)

Shulman, Herman L
Home: 32 Hillcrest Drive, Potsdam, NY 13676
Position: Provost & Dean of Sch of Engg. *Employer:* Clarkson Univ *Education:* PhD/Chem Engg/Univ of PA; MS/Chem Engg/Univ of PA; BChE/Chem Engg/CCNY. *Born:* 2/24/22. NYC, BChE 1942 CCNY Dupont Fellow U of PA. 1947-48, U of PA, MS 1948, PhD (Chem Engg), 1950 U of PA, PE (NY). Chem engr, Gen Motors Corp 1942-43; Barrett Div Allied Chem & Dye Corp 1943-46; res chem engr Publicker Industries Inc 1946- 47; Asst Prof Chem Engr Clarkson College of Tech 1948-51, Assoc Prof 1951-54, Prof 1954-, Chrmn Chem Engg Dept 1958-64, Dir Div Res 1959-73, Dean Grad School 1964-73, VP and Dean Sch of Engg 1968-77, Provost and Dean Sch of Engg 1977-. Fellow AIChE, ASEE; ACS; NSPE, Sigma Xi. Dir St Lawrence County Chamber of Commerce; Mbr St Lawrence County Planning Bd; Commissioner St Lawrence-Eastern Ontario Commission. *Society Aff:* ASEE, ACS, AIChE, NSPE, ΣΞ

Shultis, J Kenneth
Business: Dept of Nuclear Engrg, Manhattan, KS 66506
Position: Professor *Employer:* KS St Univ *Education:* PhD/Nucl Sci & Engrg/U of MI; MSc/Nucl Sci & Engrg/U of MI; BASc/Engrg Phys/U of Toronto. *Born:* 08/22/41 Born and raised in Toronto, Canada. Worked as Scientific Officer in Dept of Mathematics, Univ of Groningen, Netherlands 1968-69 before joining the Dept of NuclearEngrg at KS St Univ where he now purses res interests in numerical analysis, radiative transfer, radiation shielding, energy analyses and risk assessment. He is a licensed Sr Reactor Operator and consults for numerous firms. Designated Black & Veatch Prof in 1978; recipient of national Glenn Murphy Award in 1981 for notable prof contributions to nuclear engrg by ASEE; was von Humboldt Fellow at Univ of Karlsruhe, W. Germany 1980-81. *Society Aff:* ANS, ASEE, AAAS.

Shumaker, Fred E
Business: Alan C. McClure Associates, Inc, 2600 South Gessner, Suite 504, Houston, TX 77063
Position: Vice President *Employer:* Alan C. McClure Associates, Inc. *Education:* BSE/Naval Arch & Marine Engg/Univ of MI. *Born:* 10/28/42. Native of Mt Clemens, MI. From 1965 to 1973 Naval Arch, Chevron Shipping Co specializing in tanker design & construction. 1973 to 1981 with El Paso Marine Co as Naval Arch, Chief Naval Arch, and Mgr Marine Engg. For a fleet of nine liquefied natural gas vessels. Since April 1981, VP Alan C. McClure Assoc Inc. A Naval Architecture and Engrg consulting firm for the offshore and Marine Transportation Industries.. SNAME Activities: Past officer Northern CA Sec, past Chrmn & mbr Exec Committee Gulf Sec, mbr Steering Committee & Chrmn Technical Prog Committee 1979 Spring Meeting/STAR Symposium, past Chrmn & mbr Exec Committee TX Sec. Past technical program chairman, SNAME, Offshore Technology Conference. Enjoy boating, swimming, scuba and country western music. *Society Aff:* SNAME, RINA, NACE

Shumate, William V
Business: PO Box 3707, Seattle, WA 98124
Position: Coordinator, Air Safety Investigation *Employer:* Boeing Commerical Airplane Co *Education:* BS/EE/CO Univ. *Born:* 2/2/24. in Denver, CO. Served in US Navy during WWII. In V-12 prog & NROTC attended Princeton Univ, John Carrol Univ & Univ of Louisville; Joined Boeing Co 1952 working in Customer support; during Boeing career worked in areas of design, Functional Test & Liaison Engg, Field Serv, Accident Investigation, Tech Writing & Reg Dir Customer Supt; assumed present position July 1980. P Governor Dist 2 Toastmasters Internatl. Loaned Exec to United Way of King County 1974. Hobbies: private flying & skiing, mbr: Int Soc of Air Safety Investigators.

Shupe, John W
Business: PO Box 50168, Honolulu, HI 96850
Position: Director, Pacific Site Office *Employer:* US Dept of Energy *Education:* PhD/CE/Purdue; MS/CE/UC Berkeley; BS/ME/KS State Univ. *Born:* 3/30/24. Served in Army Air Corps 1942-45. Faculty & administrative positions at 4 state univs, including Dean of Engrg at Univ of Hawaii for 17 years. Assumed current position 1983. Indus exper with Convair & Hughes Aircraft. Scientific Advisor to Asst Secr for Energy Technology, US Dept of Energy 1977-78. Dir of Hawaii Geothermal Proj; Chmn State Ctte on Alternate Energy Sources for Hawaii; Dir of Hawaii Natural Energy Inst. NSF Faculty Fellowship, ASTM Templin Award, Purdue and Kansas State Outstanding Alumni Awards, HSPE Award as Engr of the Year for Hawaii - 1979. Author of more than 40 tech publs, most recently on alternate energy sources. *Society Aff:* ASCE, NSPE, ASES, GRC.

Shure, Kalman
Business: PO Box 79/Bettis Lab, West Mifflin, PA 15122
Position: Consultant. *Employer:* Westinghouse Electric Corporation. *Education:* PhD/Physics/MIT; AB/Physics/Brooklyn College. *Born:* March 1925 Brooklyn NY. Joined Betis Atomic Power Lab 1951; initial assignment in experimental aspects of reactor shielding changed to design aspects, current responsibility is dev & qualification of analytical techniques for efficient reliable application to assessment of radiation source strengths, shielding requirements, fission prod decay heating & radiation damage. Appointed Cons in Radiation Analysis 1973. Fellow ANS; Mbr Sigma Xi & Amer Physical Soc. ANS Shielding Div Exec Ctte 1960-63, Prog Chmn 1963-66. Mbr ANS Standard 5.1 Working Group on Fission Prod Decay Heat 1974- . *Society Aff:* ANS, APS, ΣΞ.

Shurtz, Robert F
Home: 1200 California Street, San Francisco, CA 94109
Position: Consultant. *Employer:* Self. *Education:* EM/Mining/OH State Univ; MSc/Mineralogy/OH State Univ; BEM/Mining/OH State Univ. *Born:* March 1916. Disting Alumnus Award 1963. Varied exper mining, mineral & chem processing to 1947; Engrg Coll Staff Univ of Texas 1947-51; Ch Engr to Dir Res companies predecessor to Kerr-McGee 1951-56; Basic Inc to V P 1961-64; Cyprus Mines cons 1964-66; Bechtel Corp to V P 1973-77. Consultant in private practice 1977- . 3 patents, 11 publs. Reg in 10 states. Listed in 'Who's Who in the West' 1974. Hobbies: hunting, fishing, photography, travel, skeet & trap shooting, opera, mathematics. *Society Aff:* AIME-SME, MMSA, AAAS, MSA, CIMM.

Shuster, Joseph M
Business: 407 7th St NW, New Prague, MN 56071
Position: President. *Employer:* Minnesota Valley Engineering. *Born:* August 1932 Hibbing Minn. BSChE Univ of Minn 1955 (Evans Scholar). Freeport Sulphur Co 1955-56; Proj Engr Arthur McKee & Co 1956-62; Minn Valley Engrg 1963- , lead MVE as Ch Exec Officer from zero 1963 to one of world's leading cryogenic firms with pants in Minn, Ind & France. Wrote energy report for Congress 1973. Serves on sev bds both civil & corporate.

Siau, John F
Business: College of Environmental Sci & Forestry, Syracuse, NY 13210
Position: Prof. *Employer:* SUNY. *Education:* PhD/Wood Sci/SUNY-Syracuse; MS/Wood Sci/SUNY-Syracuse; BS/CE/MI State Univ. *Born:* 3/30/21. 1943-46 Engrg officer, USNR; 1946-48 Audio engr UT Radio Prod, Huntington, IN; 1948-58 Design engr, vacuum tubes, Gen Elec Co, Schenectady. 1958-66 Prof physics, Paul Smiths College, Paul Smiths, NY; 1966- Prof wood prod engrg SUNY College of Environ Sci and Forestry, Syracuse, NY. Wood Award, Second place, 1968, for PhD thesis work, described in Publication No 9 (see list). This was presented by Wood & Wood Products magazine. Invited to attend NATO Sci Ctte Conf on the Properties of Wood in Relation to Its Structure, Les Ares, France, Nov 17-21, 1975. Grant from NSF to investigate nonisothermal diffusion of moisture in wood. Author of

Siau, John F (Continued)
Transport Processes in Wood, Springer (Heidelberg), 1984. Lic PE in state of NY. *Society Aff:* FPRS, SWST, AWPA, TBΠ.

Sibert, George W
Home: 3536 North 36th Rd, Arlington, VA 22207-4817
Position: Director of Engrg, Army Systems *Employer:* American Electronics, Inc (AMELEX) *Education:* BS/Military Engrg/USMA West Point; MSE/Aeronautical Engrg/Princeton Univ; MBA/Mgmt/Northeastern Univ *Born:* 03/05/37 US Army 1958-85: commanded Co D, 12th Engr Bn(61-62); Troop F, 8th Cavalry (RVN 70-71), 1st Battalion, 8th Infantry(74-76); Army Materials and Mech Res Ctr(79-83). Served in Office of Deputy Chief of Staff for Res, Dev, and Acquisition as DASC for Aircraft Survivability Equipment and Advanced Scout Helicopter(76-79), Dep Dir Sys Review and Analysis Ofc(83), Executive to DCSRDA(84) and Deputy Dir of Army Res and Tech(84-85). Retired as colonel in March 1985. Dir of Finance & Admin of AAES (85-86). Dir of Engrg, American Electronics (1987). *Society Aff:* AHS, AIAA, AAAA, AUSA, TROA, ΦKPHI

Sibley, Earl A
Business: 1105 N 38th St, Seattle, WA 98103
Position: Pres. *Employer:* Shannon & Wilson Inc. *Education:* MS/CE/KS State; BS/CE/ND State. *Born:* 9/13/26. BS ND State, MS KS State. US Marine Corps 1944-46. Instr Civil Engg KS State 1951-52. Asst Prof CE Univ of ID 1952-55. Assoc Prof (Hwy Soils) WA State Univ 1955-59. Proj Engr with Shannon & Wilson Inc 1960-63. Elected VP & mbr of firm 1963. Elected SR VP & Mgr of Seattle office 1968. Elected Pres 1976 & also Chrmn of the Bd in 1979. Enjoys sports & outdoor activities *Society Aff:* ASCE, ACEC, ASFE.

Sibulkin, Merwin
Business: Brown University, Div of Engrg, Box D, Providence, RI 02912
Position: Prof of Engrg *Employer:* Brown Univ *Education:* AE/Aeronautics/CA Inst of Tech; MS/ME/Univ of MD; BS/ME/NYU *Born:* 08/20/26 Born in New York, NY. Served in US Navy 1943-44. Engaged in res work in fluid mechanics heat transfer for the Natl Adv Ctte on Aeronautics, Naval Ordnance Lab, Jet Propulsion Lab and Convair Scientific Res Lab. Joined Brown Univ in 1963. Worked in the area of radiative gas dynamics with application to high speed heat transfer. Current field of interest is combustion with application to fire tech. *Society Aff:* CI, ASME

Siddell, Derreck
Business: Engineering Laboratory, 107 Park St N, Peterborough Ontario K9J 7B5 Canada
Position: Chief Metallurgist *Employer:* G.E. Canada. *Education:* PhD/Met/Univ of Surrey; College Dipl/Met/Lanchester College Advanced Tech; LIM/Met/Rutherford College. *Born:* 8/11/42. Native of Consett, County Durham, England. Technical trainee Consett Iron Co while obtaining assoc mbrship of Inst of Met by part time study. Joined Atlas Steels Co of Welland Ontario after PhD. Res into powder met of tool steels and formability of stainless steels. Joined Canadian GE Nuclear Div in 1973, Res into transient failures & design requirements for nuclear fuel handling systems. Supervised quality assurance group for nuclear fuel handling systems. Now chief metallurgist of GE Canada. Past chrmn Ontario chapt ASM. Fellow Inst of Metallurgists, PE Ont, C.Engr (UK). Enjoy classical and folk music & sailing. Mbr Ontario Socceer Referees Assn. *Society Aff:* APEO, ASM, IM, MS.

Sidebottom, Omar M
Business: 226 Talbot Lab, Urbana, IL 61801
Position: Prof, Dept of Theo & Appl Mech. *Employer:* Univ of Illinois. *Education:* AS/-/Lincoln Jr College; BS/-/Univ of IL; MS/-/Univ of IL. *Born:* Mar 1919. Native of Easton, Ill. Conducted res in Plasticity & Creep since 1942; co-author 45 papers & 3 books. Taught in Theoretical & Applied Mech Dept Univ of Illinois 1945- . Prof since 1957. Mbr ASME (Fellow), ASEE, SESA, ASTM. Enjoys fishing. *Society Aff:* ASME, SESA, ASTM, ASEE.

Sides, Samuel E
Business: Dept of Agri Engrg, Orono, ME 04469
Position: Assoc Prof (Research). *Employer:* Maine Agri Exper Sta-Univ of Maine. *Education:* BS/Agri Engg/Univ of ME. *Born:* Apr 1930. Alpha Zeta. Reg P E Maine. Mbr Amer Soc of Agri Engrs. Engrg Trainee, 2-yrs, 'Grad Training Prog', Allis-Chalmers Corp Milwaukee, Pgh & Cincinnati 1951-53. Proj Engr forage harvesting & handling machinery at Allis- Chalmers' Harvester Works, LaPorte Ind 1953-56. Asst Engr, then Assoc Prof, engrg res on potato and vegetable harvesting, packing & marketing sys with the Maine Agri Exper Station, Presque Isle & Orono 1956- . Amer Soc of Agri Engrs: Chmn-Elect 1973-74, Chmn 1974-76 Acadia Div (Me, NB, NS & PEI). *Society Aff:* ASAE, AZ

Siebeneicher, Paul Robert 2nd
Home: 143 Canterbury Road, Mount Laurel, New Jersey 08054-1414
Position: Dir of Facilities, Indus & Mfg Engineering *Employer:* Infotron Systems Corp *Education:* BST/Elec/Univ of Houston; ASET/Elec/Univ of Houston *Born:* 09/29/43 Native of Houston, TX, educated in the Houston Public Schools. Served in the USNR-1960-66, USN-1963-65. Attended Univ of Houston from 1966-69. Held positions of Mgr of Indl Engrg, Congoleum, Maintenance Engrg, NL Industries; Plant Engr, Camden Iron & Metal Inc, Flexitallic Gasket Co; Proj Engrg, American Can Co. Currently Director of Facilities, Manufacturing & Industl Engrg, Infotron Systems Corp. VP, Region II, Inst of Industl Engrs from 1980-82. Received Award of Excellence, 1976, 84 and 87 four Service Awards, 1974, 78, 79 and 83. South Jersey Chapter-IIE. VITA Volunteer Consultant since 1968; Editorial Quality Audit Panel Mbr, Plant Engrg, since 1971; received Volunteer Organization of the Year Award from Goodwill Industries of S New Jersey, 1974, VP of Operations, 1974-75; Pres, 1975-76, VP, 1974-75, Dir of Newsletter, Profess Dev, Comm Affairs, PR and Special Projs, 1971-Present, all with S.J. Chapter of IIE. President and Chrmn of the Bd, Southern NJ Tech Consortium, 1987 and 88. VP-President elect of S. Jersey Chapter IIE, 1987-88. CERTIFIED MANUFACTURING ENGINEER. Sr Mbr of IIE & SME *Society Aff:* IIE, SME

Siebert, William M
Business: 77 Mass Ave - Rm 36-825, Cambridge, MA 02139
Position: Ford Prof of Engrg *Employer:* MIT. *Education:* ScD/Elec Eng/MIT; SB/Elec Eng/MIT *Born:* Nov 19, 1925 Pgh Pa. Staff mbr & Group Leader (Radar Tech) Lincoln Lab 1953-55. Mbr of the fac of the Elec Engrg & Computer Sci Dept MIT 1952- . Hd of the Communications Biopsyc Group 1970- . Chmn NIGMS Training Ctte on Engrg in Biology & Medicine 1968-71. Pres Commission on Communication & Control Processes IUPAB 1972-75. Res & teaching in the appli of statistical communication theory to the design of radar & communication sys, & to the analysis of physiological & psychological sys. *Society Aff:* IEEE

Siegel, Howard J
Home: 14147 Forestvale Dr, Chesterfield, MO 63017
Position: Director - Production Engineering & Computer Aided Technology *Employer:* McDonnell Aircraft Co. *Education:* BS/Mining Engg/Univ of KY; MS/Metallurgical/Univ of KY. *Born:* Oct 1929. Native of Elizabeth N J. USAF, assigned to AF Matls Lab 1951- 53. Joined McDonnell Aircraft (McDonnell Douglas Corp) 1955. Appointed Ch Metallurgist 1960 & Mgr - Matl & Process Dev Dept 1966. Respon for aircraft matls application, fabrication, assembly & finishing processes & major dev progs to advance matls & process tech. Appt Chf Prog Engr Adv AV-8 Aug 1977, Managed Studies of AV-8B, AV-8Bplus and TAV-8B Adv Vstol Aircraft Appt Program Mgr AV-8 Kingston, UK June 1979. Appt Dir Production Engrg June 1981 respon for Computer Aided Techn, Materials and Processes, Standard Parts, Producibility, Liaison Engr and Loft. Served on Natl Matls Adv Bd Cttes 1962-70. NASA Matls Adv Ctte mbr 1965-70. NASA Matls & Structs Ctte mbr 1971-76. Fellow ASM 1972. USAF Sci Adv Bd Ctte on Advanced Composite Tech 1976. *Society Aff:* AIAA, ASM, NCGA

Siegel, Robert
Business: 21000 Brookpark Road, Cleveland, OH 44135
Position: Senior Research Scientist *Employer:* NASA Lewis Research Center *Education:* ScD/Mech. Engrg./Massachusetts Inst. of Technology; MS/Mech. Engrg./Case Institute of Technology; BS/Mech. Engrg./Case Institute of Technology *Born:* 07/10/27 Born and raised in Cleveland, OH. Served in the Army 1945-47. Heat transfer specialist for General Electric Co., Schenectady, New York 1953-54, & then at the Knolls Atomic Power Lab. 1954-55. Research Scientist at NASA since 1955 where has held the positions of Section Head and Branch Chief. Heat Transfer Memorial Award, ASME, 1970. Fellow ASME, 1977. Coauthor of textbook "Thermal Radiation Heat Transfer," 2nd ed. 1981. NASA Exceptional Scientific Achievement Medal, 1986. Associate Editor, Jour. of Heat Transfer, 1973-83; Jour. of Thermophysics and Heat Transfer, 1986-present. Married with two sons. *Society Aff:* ASME, AIAA

Siegel, Stanley T
Home: 10 Winthrop Rd, Brookline, MA 02146
Position: Dir of Transportation. *Employer:* Town of Brookline, Mass. *Born:* Dec 25, 1923 Wash D C. BSCE Tufts Univ 1945; Cert in Hwy Traffic Yale Univ 1947. CEC, USNR, WW II. Employed in var of transp & traffic engrg assignments with State of N J, City of Chgo, Natl Safety Council, City of Indianapolis, Boston Redev Auth, Computer Traffic Controls Inc, T T Wiley & Assocs, PADCO - Intl in Brazil, E Lionel Pavlo Engrg Co, Town of Brookline. Owner, Diversified Engrg Services. Guest Lectr Northwestern, Purdue, Tufts, Northeastern & Suffolk Univs; Fellow ASCE, ITE. Reg P E Ill, Ind, N Y, Mass. P Pres Ind & N E Sects ITE, P Chmn Dist One ITE. Disting Serv Award N E Sect ITE 1976. Editor N E Engrg Journal 197075; Natl Dir NSPE 1975-76; Pres Mass Soc of Prof Engrs 1976-77. Wood Badge, BSA; True Craftsman Lodge, F&AM; Brookline Kiwanis Club; Winthrop Yacht Club. Wife Hilda; 2 married children, Ruth & Michael.

Siegele, Harold L
Business: Haakon VII's, Gate 5, Oslo, Norway, 3
Position: Chairman *Employer:* Esso Norway Inc *Education:* BS/Chem Engrg/KS St Univ *Born:* 6/3/24 Native of Emporia KS. Served in US Navy during WW II. Joined Creole Petroleum Corp. (Exxon Affiliate in Venezuela) in 1947 and served in several engrg and managerial positions. Named Deputy Mgr Exxon Corp Producing Dept in NY 1969. Returned to Venezuela as Dir of Creole 1973. Pres Esso Libya 1976 - 1979. Sr VP Exxon USA 1980-1981. Assumed present position as Chairman and CEO of Esso Norway Inc. in 1981. Received Francisco de Miranda medal from Venezuelan Gov 1975. Received Distinguished service in Field of Engrg award from KS St Univ 1978. Retired August 1984. *Society Aff:* AIME

Sieger, Ronald B
Home: 3220 Van Tassel Dr, Indianapolis, IN 46240
Position: Supervising Engr. *Employer:* Brown & Caldwell. *Education:* BSCE/Civ Engg/IA State Univ. *Born:* 6/29/45. Design engr in civ-sanitary dept for Black & Veatch. Joined Brown & Caldwell in 1971 & was involved in the dev of advanced wastewater treatment (AWT) concepts. ASCE Samual A Greeley Award in 1977 for technical paper describing the design of a 30 mgd AWT plant. Supervised res & dev on co- combustion of sewage sludge & municipal solid waste. Authored several publications & coordinated seminars on sewage sludge combustion & co-combustion techniques. Presently involved as proj mgr of a solids mgr study for the city of Indianapolis. PE reg in CA & IN. *Society Aff:* ASCE, WPCF, NSWMA, NSMA.

Siegfried, Robert E
Business: One Broadway, Cambridge, MA 02142
Position: Chrmn of the Bd & Ch Exec Office. *Employer:* The Badger Co Inc. *Education:* SM/Chem Engr/MIT; BS/Chem Engr/Lehigh Univ. *Born:* Jan 1922. AUS 1943-46. Process Engr Supr for Stone & Webster Engrg Corp London Eng. 1951-52; Process Eng E. B. Badger & Sons Co., Boston 1947-51 Process Engr Supr Badger Co., Cambridge, Mass. 1952-54; Project Mgr. 1954-56; Project & Engrg. Coordinator 1956-59; Asst. Engrg. Mgr. 1959-60; Engrg . Mgr. 1960-65; Director 1963; Vice President & Engrg. Mgr. 1965-68; President 1968; Chief Exec. Oficer 1971; Chairman of the Board 1977; Director State Street Boston Corp., State Street Bank and Trust Co; Trustee Commonwealth Energy System; Memb. Corp., Trustee, Vice President Boston Museum of Science Member Corp. Northeastern University Reg P E Mass, La, Ky; Mbr AIChE, Fla Engrg Soc, Natl Soc of Prof Engrs, Sigma Xi, Tau Beta Pi; patentee desalination of sea water *Society Aff:* ΣΞ, TBΠ

Siegman, Anthony E
Business: Edward L Ginzton Lab, Stanford, CA 94305
Position: Prof of Elec Engrg. *Employer:* Stanford Univ. *Education:* PhD/Elec Engrg/Stanford U; MS/Applied Physics/UCLA; AB/Harvard Coll *Born:* 11/23/31. Author *Microwave Solid-State Masers* 1964; *An Intro to lasers & Masers* 1971; *Lasers*, 1986. John Simon Guggenheim Fellow 1969, W R G Baker Award IEEE for the best paper publ in any IEEE Transactions during 1971. In 1965 was a Visiting Prof of Apply Phys at Harvard Univ. Fellow of IEEE, Optical Soc of Amer, Amer Phys Soc. Mbr of Natl Acad of Engrg. R. W. Wood Price of Optical Soc of Amer for invention of the Unstable Optical Resanator (1980). Awarded Sr US Scientist Awd by Humboldt Fdn for 1984; elected Fellow of the Amer Acad of Arts & Sci, 1984. Awarded Frederic Ives Medal of Optical Soc. of Amer. for overall distinction in optics (1987). *Society Aff:* IEEE, OSA, APS, AAAS, NAE, AAAS.

Siegmund, Walter P
Home: Cassidy Rd, Pomfret Center, CT 06259
Position: Dir R & D *Education:* PhD/OPTICS/Physics/Univ of Rochester; AB/Gen Sci/Univ of Rochester *Born:* Aug 26, 1925 Bremen Germany. USNR 1944-45 Post-grad res in physiological optics (U of R) 1951-53. Amer Optical 1953-83; Sci Aide to V P for Res 1953-56, working on radically new motion picture process, manufacture of aspheric ophthalmic lenses. Asst Dir of Res 1957-64. Worked in field of fiber optics & lasers. Managed gov res & dev contracts. Technical Mgr, Reichert Fiber Optics, 83-86, Director of R&D, Schott Fiber Opticss 1987- . Respon for tech dev of fiber optics products & processes as well as applications engrg. Fellow Optical Soc of Amer; Fellow SPIE; Karl Fairbanks Award for Indus 1970 David Richardson Medal (OSA) 1977. *Society Aff:* OSA, SPIE, ACS.

Sielbach, Fred A, Jr
Business: 2020 Grand Ave, Billings, MT 59102
Position: Vice Pres. *Employer:* Christian, Spring, Sielbach & Assocs. *Born:* Jan 1930. BSCE Montana State Coll 1952. US Army Corps of Engrs 1952-54, with cons firm from 1954-57 & then with Montana Hwy Dept 3 yrs in Traffic Engr's Office doing traffic engrg interchange design. Was City Engr for City of Helena for 9 yrs. With Christian, Spring, Sielbach & Assocs since 1968. Presently V P as principal in charge of engrg. Also Pres of CSS&A of Wyoming, a wholly owned subsidiary firm in Wyoming. Activities incl flying, photography, backpacking & fishing. Reg P E & L S in Montana; Reg P E Colo & Wyoming.

Siemens, John C
Business: Agri Engg Dept, Urbana, IL 61801
Position: Prof. *Employer:* Univ of IL. *Education:* BS/Agri Engg/Univ of CA; MS/Agri Engg/Univ of IL; PhD/Civ Engg/Univ of IL. *Born:* 2/22/34. From 1963 to 1968 was on Agricultural Engg staff at Cornell Univ. Since 1968 has been on Agri Engg staff, Univ of IL at Urbana-Champaign with responsibilities in farm power and machiney extension and res. Primary interests have been the evaluation of tillage-planting systems and optimization of agricultural machinery for corn and soybean production. Spent one yr with Deere & Co and has served as consultant in several foreign countries. *Society Aff:* ASAE, NIFS.

Sierakowski, Robert L
Business: Dept of Civil Engr, 470 Hitchcock Hall, 2070 Neil Ave, Ohio State Univ. Columbus, Ohio 43210
Position: Prof, Chrmn *Employer:* OH State Univ *Education:* BSc/Engr/Brown Univ;

Sierakowski, Robert L (Continued)
MS/Engr Mech/Yale Univ; PhD/Engr Mech/Yale Univ *Born:* 4/11/37. Experience: 1957-1958 Student Asst, Brown Univ. 1958-1960 Dynamics Engr, Sikorsky Aircraft. 1960-1963 Res Asst, Yale Univ. 1963-1967 Sr Res Scientist, United Aircraft Corp, Res Lab. 1965-1967 Asst Prof, RPI, Hartford Grad Ctr. 1967-1968 Visiting Asst Prof, Univ of FL. 1968-1973 Assoc Prof, Univ of FL. Summers 1972 & 1973 NRC Sr Res Fellow, AF Mtls Lab. 1973-1983 Prof, Univ of FL. Summer 1974 Consultant, AF Mtls Lab, WPAFB. Summer 1976 Consultant, AF Armament Lab, Eglin AFB, Visiting Sci. AFOSR 1982, Chrmn, Dept of Civ Engr OH State Univ, Columbus, OH. *Society Aff:* ASME, AIAA, ASCE, SEM, SAMPE

Siess, Chester P
Home: 805 Hamilton Dr, Champaign, IL 61820
Position: Professor Emeritus of Civil Eng. *Employer:* U of Illinois-Urbana-Champaign. *Education:* PhD/Struct Eng/Univ of IL; MS/Civil Eng/Univ of IL; BS/Civil Eng/LA State Univ. *Born:* 7/28/16. Fac mbr Univ of IL 1941- ; Hd, CE Dept 1973-78; Prof Emeritus 1978- . Hon Mbr Amer Concrete Inst; P Pres, recipien Turner Medal, Wason Medal. Hon Mbr ASCE, recipient Res Prize, Howard Medal, Reese Award. Mbr NAE. Chmn Reinforced Concrete Res Council 1968-80,recipient Boase Award. Mbr Nuclear Regulatory Comm, Adv Comm on Ractor Safeguards 1968- , Chmn 1972; NRC Distinguished Service Award, 1987. Elected Charter Mbr La State Univ Engg Hall of Distinction 1979. Univ. of Ill. College of Engng Alumni Honor Award for Distinguished Service in Engineering, 1985. *Society Aff:* ASCE, ACI, ASEE.

Sievenpiper, Ward
Home: 3440 Home Rd, Alden, NY 14004
Position: V Pres - R&D. *Employer:* Ward Hydraulics Div - ATO Inc. *Born:* July 2, 1912. Alden H S 1928; 4 yrs apprenticeship at machinist trade; 6 yrs supplemental at trade school - tool making & machine design. Supervisory pos with Std Sanitary & Curtiss Wright Corp. In 1943 founded Ward Tool & Machine Co, which later became Ward Hydraulics; served as pres of that co until sale to A T O Inc 1964; assumed pos of V P Engrg R&D at that time - retained in that pos to present. Mbr Automotive Engrg Soc (SAE), Buffalo Engrg Soc, Natl Fluid Power Assn - Corp Rep since 1964, Fluid Power Soc - P Natl Pres plus local chap, Natl Conf of Fluid Power - Dir, Alden Chamber of Commerce - Charter Mbr & P Pres with long involvement in civic dev. Enjoys hunting & fishing.

Siewiorek, Daniel P
Business: Carnegie-Mellon University, Dept. of Computer Science, Pittsburgh, PA 15213
Position: Prof *Employer:* Carnegie-Mellon Univ *Education:* PhD/EE/Stanford Univ; MS/EE/Stanford Univ; BS/EE/Univ of MI *Born:* 06/02/46 Native of Cleveland, OH. Received Honorable Mention Award as Outstanding Young Electrical Engr for 1977 by Eta Kappa Nu and elected Fellow of IEEE in 1981 for contributions to the design of modular computing systems. Serve as conslt to several commercial and government organizations. Helped to initiate and guide the Cm proj for a 50 processor multi-micro-processor sys at Carnegie-Mellon Univ. and major contibutor to nine multiprocessor systems. Participated in the Army/Navy Military Computer Family proj to select a standard instruction set. Served as Assoc Editor of the Computer Sys Dept of the Communications of the Association for Computing Machinery and am currently serving as director of Carnegie-Mellon's center for dependable systems. *Society Aff:* IEEE, ACM.

Siforov, Vladimir I
Business: Kuznetsky Most 20, Moscow Centre, USSR 103897
Position: President. *Employer:* Popov Society. *Education:* Elec Engr/Corresponding mbr of the USSR Acad of Scis Dir/Inst for Problems of Info Transmission of the USSR Acad of Scis *Born* May 1904. Native of Moscow. Grad from V I Lenin Electrotech Inst, Leningrad. Elec Engr from 1929. DSc Tech from 1937, Prof from 1938, corresponding mbr of the USSR Acad of Scis from 1953. Specialty: Radio Tech, Electrocommunication, Electronic, Cybernetic, Sci of Info. Dir Inst for Problems of Info Transmission of the USSR Acad of Scis. Pres of the Popov Soc 1954- . Hd of the Chair of the Moscow Power Inst 1957- . Mbr of URSI 1957- . Fellow IEEE 1967- . Hon Mbr Hungarian Acad of Scis. Hon Dr of the Budapest Polytech Univ. Awarded 7 Orders of the Soviet Union. Laureate of the Internatl TV Symposiums, Montreux 1967- . Awarded 8 Orders of the Soviet Union. *Society Aff:* AS Popov Soc

Sigel, Robert J
Business: Sabine & Essex Aves, Narberth, PA 19072
Position: Chrmn of Bd *Employer:* Robert J Sigel, Inc *Education:* BS/ME/PA St Univ *Born:* 6/24/11 in Pittsburgh, PA, raised in Philadelphia suburbs. Was active in soccer, basketball and track through Coll. Upon graduation, worked on general Mechanical Engrg projects with Philadelphia based consultants before founding the firm of Robert J. Sigel, Inc. Consulting Engrs, in 1950 handling mostly industrial accounts both in the USA and overseas. Present volume $6 million per year. Holds patents on laminar flow air distribution and energy control for large buildings. Enjoys playing golf and watching son Jay play in major amateur tournaments. *Society Aff:* NSPE, ASHRAE, PSPE, PEPP

Signell, Warren I
Home: 12 Kingston Dr, Livingston, NJ 07039
Position: Ch Marine Engr. *Employer:* H M Tiedemann & Co. *Education:* BS/Naval Architecutre & Marine Engg/MIT. *Born:* July 1923. Native of West N Y, NJ. BS MIT. US Navy after grad from MIT - 1944-46. Joined Marine Dept of Foster Wheeler Corp as Proposal Engr 1948, became Engrg Mgr 1960 - respon for design of naval & merchant boilers. Joined J J Henry Co 1969 as VP, Engg respon for the design of naval & merchant main propulsion & auxiliary machinery sys. Joined H M Tiedemann & Co 1974 as VP, Engg, respon for similar fields plus field services, with special emphasis on vibration analysis. Author sev papers to SNAME. Linnard Prize from SNAME 1969 for best paper presented at 1968 Annual Mtg. *Society Aff:* SNAME, ASNE, ASME, IME.

Signes, Emil G
Business: Homer Res Lab, Bethlehem, PA 18018
Position: Engr, Res. *Employer:* Bethlehem Steel. *Education:* SM/Met/MIT; MA/Span Literature/Lehigh; SB/Met/MIT. *Born:* 6/18/40. in Paterson, NJ. After attending MIT, worked as Production Met at Watertown (MA) Arsenal. Was Res Fellow at British Welding Res Assoc 1964-65. Came to Bethlehem Steel in 1966. Publications: *British Welding Journal* 1967, *Welding Journal* (1968, 1972, 1979, 1981), Metals Progress (1969). PhD candidate in Spanish Literature. Long-distance runner. Rugby coach & referee. *Society Aff:* ASM, AWS.

Signs, Cheryl L
Business: 655 Broadway, Suite 700, Denver, CO 80203
Position: President *Employer:* Cheryl Signs Engineering *Education:* BS/Eng Sci/CO State Univ *Born:* 11/14/49 Sole proprietor of Cheryl Signs Engrg established in 1979 as consltg firm specializing in water resource analysis, planning and related computer applications. Served five yrs as sole water resource engr for Westminster, CO. Primary responsibility was water supply and operations. Instrumental in several innovative water mgmt porjs. Water rights engr for Leonard Rice Consltg Water Engrs four yrs. Served as chrmn of NWRA's municipal committee, CCEC's public relations ctte, CO; WPCA's convention committee, CO, ASCE's Secretary, CO and on AWWA's joint government affairs committee, serves on CO Water Congress federal & state water quality committee, CO SWE Career guidance; Chrmn of CO ASCE Water Resources Group. Received Distinguished Service Award from Denver Regional Council of Governments, 1980. *Society Aff:* ASCE, SWE, AWWA, ACEC, NSPE

Sikarskie, David L
Business: Rm 330 Engg Bldg, E Lansing, MI 48824
Position: Chrmn-Prof. *Employer:* MI State Univ. *Education:* ScD/Engg Mechanics/Columbia Univ; MS/Engg Mechanics/Columbia Univ; BS/Civ Engg/Univ of PA. *Born:* 8/3/37. Mbr of the Technical Staff, Ingersoll Rand Res Ctr 1963-66. Visiting Lecturer, Princeton Univ 1965-66. Aerospace Engg Dept, Univ of MI 1966-79; Asst Prof 1966-69, Assoc Prof 1969-75, Prof 1975-79. MI State Univ, Prof and Chrmn, Dept of Metallurgy, Mechanics and Materials Sci, 1979-. Guggenheim Fellow 1959-60, NDEA Fellow 1960-62, Tau Beta Pi outstanding teacher award 1973. *Society Aff:* ASME, SME, SES, ASM.

Sikes, John M
Home: 304 Magnolia Oak Dr, Longwood, FL 32750
Position: Senior VP *Employer:* Dawkins & Assocs, Inc *Education:* B/CE/Auburn Univ *Born* 1/20/36 Native of Douglas, GA. Graduate of Auburn Univ, 1960 - Bachelor of Civil Engrg. Senior VP and Dir of Project Management Dawkins & Assocs, Inc, Consulting Civil/Environmental Engrs, Orlando since 1976. Responsible for managing planning, design and construction phases of multi-discipline engrg services involving civil, environmental and industrial projects including water resource developement and utility systems, wastewater treatment plants for municipalities, electric power utilities and industries, chem plant facilities, ports and harbors, railroads and civil works. Registered PE in FL, MS, and GA. *Society Aff:* NSPE, ASCE, ACEC, WPCF

Silady, Fred A
Business: PO Box 85608, San Diego, CA 92138
Position: Mgr *Employer:* GA Tech Inc *Education:* PhD/Nucl Engr/Univ of IL; MS/Nucl Engr/Univ of IL; BS/Nucl Engr/KS State Univ; -/Philosophy/Immaculate Conception Seminary *Born:* 11/07/47 Native of Shawnee Mission, KS. One of five engr in immediate family. With GA Tech (formerly Gen Atomic) since grad. Performed safety analysis of High Temp Gas Cooled Reactor (HTGR), second generation nucl pwr plant. Author of over two dozen tech papers in the areas of HTGR safety risk assessment. Assumed respons as Branch Mgr of Safety Design in 1980. Then Mgr of Safety and Reliability in Engrg Dev Div in 1984. Currently Mgr of Reactor Systems Engr Dept with respons of Systems Engr, Control and Operations, and Safety & Reliability branches. Past president of local ANS sections at U of I and San Diego. Enjoy family and church activities and basketball. Avid KS State, KS City, and San Diego sports fan. *Society Aff:* ANS

Silberman, Edward
Business: St Anthony Falls Hydraulic Lab, MS River at 3rd Ave SE, Minneapolis, MN 55414
Position: Prof Emeritus, Civ Engr *Employer:* Univ of MN. *Education:* MS/CE/Univ of MN; BS/CE/Univ of MN. *Born:* 2/8/14. Early employment in water resources with MN State Planning Bd, Corps of Engrs, Minneapolis Dredging Co, TN Valley Authority, Civ Aero Authority. Served in Corps of Engrs, 1941-45; discharged as Maj. In 1946 joined St Anthony Falls Hydraulic Lab, Univ of MN. Dir, St Anthony Falls Hydraulic Lab, 1963 to 1974. Retired 1982. Natl pres, AWRA, 1968-69. Mbr, Exec Ctte, Water Resources Planning & Mgt Div, ASCE; Chairman 1981. Commissioner and Treas, Bassett Creek Water Management Commission, Hennepin County, MN. Currently active in Univ teaching and research in water resources, though retired. *Society Aff:* ASCE, AWRA, IAHR, AAAS, SAME.

Silcock, Frank A, PE
Business: PO Box 35557, Houston, TX 77235
Position: Project Mgr *Employer:* Lockwood, Andrews & Newman, Inc. *Education:* BS/Civil Engg/TX A&M Univ. *Born:* 9/24/39. Native of San Antonio, TX. Professional practice of engg in Houston since 1964, in consltg, construction and petrochem E&C. Private practice 1977-85. Reg PE in TX, LA & AL. Active in Boy Scouts (Troop Chairman), Chruch (Bishop's Committee) and community (BD Mgr YMCA and N. Houston Assn). Former chrmn NSPE Chapter Activities & Young Engrs Comms. Founding chrmn TX SPE PE's in Construction Div. Chrmn TX SPE Prof Empl Comm & former chrmn several other Comms. Past-Pres of San Jacinto Chapter TSPE/NSPE, Engrs Council of Houston & Sci/Engg Fair of Houston. Fellow in ASCE. Photography, mechanics, carpentry and music are interests. *Society Aff:* NSPE, ACI, ASCE.

Silkiss, Emanuel M
Business: 50 Hudson St, New York, NY 10013
Position: Pres. *Employer:* Lucius Pitkin INc. *Education:* MS MET E/Metallurical Engrg/Univ of Michigan; BSChemE/Chem Engrg/Univ of Michigan; BSMeT E/Metallurical Engrg/Univ of Michigan; LLB/Law/NY Law School *Born:* 5/24/27. Mbr N Y Bar. Reg P E Penna & Calif. Accredited NACE Corrosion Specialist. US Army WW II - 8th Army Ordnance 1946-47. Asst Res Metallurgist Inland Steel Co 1950-51; Sr Metallurgist Curtiss-Wright Corp 1951-54, high-temp alloy materials; Staff Metallurgist M W KEllogg Co 1954-56, materials, welding & heat treatment problems related to pressure vessel, piping & solid-fuel missile fabrication; Ch Metallurgist Air Prods Inc 1956-57, cryogenic met. Since 1957, Pres & Ch Metallurgist Lucius Pitkin Inc, indus metal cons & testing labs specializing in failure analysis of struct and machinery in all major indus. Mbr AWS, NACE, ASNT, ASME, ASTM. Chmn N Y Chap ASM 1966-67, Chap Achieve Award 1975. *Society Aff:* ASM, AWS, AIME, NACE, ASNT, ASME, ASTM

Silla, Harry
Business: Castle Point, Hoboken, NJ 07030
Position: Prof. *Employer:* Stevens Inst of Tech. *Education:* PhD/ChE/Stevens Inst of Tech; MS/ChE/Stevens Inst of Tech; BChE/ChE/CCNY. *Born:* 12/7/29. Columbia Univ Engr Ctr, NY, (1954-59); Proj Engr; Pilot plant design, construction, operation: AeroChem Res Labs, Princeton, NJ, (1959-64); Proj Leader; designed bench scale facility for propellants, conducted solid propellant combustion studies: Stevens Inst of Tech, Hoboken, NJ, (1964 to present); Prof: North Jersey Sec AIChE; Chrmn (1979): specializing in process engr-teaching, res, consulting: orginated the Chem Engr Design Lab at Stevens which emphasizes mechanical design; process experience; solid wastes, food, explosive, and fluid bed heat treating, energy fuels evaluation for stratecic reserves; future marine fuels; hazardous wastes; coal liquefaction; biomass conversion; mbr, Solid Waste Task Force of the NJ Energy Res Inst. *Society Aff:* AIChE, ACS, ΣΞ.

Sillcox, Lewis K
Home: 241 Clinton St 4, Watertown, NY 13601
Position: Retired. *Born:* Apr 30, 1886 Germantown Pa. DSc Clarkson Coll 1932; D Eng Cumberland Univ 1941, Purdue Univ 1951; LLD Syracuse Univ 1948, Queen's Univ 1955; DHL Norwich Univ 1963. Asst to Pres, Exec V P, V Chmn Bd, N Y Air Brake Co NYC 1927- 59; first Dir Office of Transp, Exec Dept, State of N Y 195960; Prof, Lectr Engrg Purdue Univ 1946- , & var colleges & univs. Gold medals ASME, Inst Mech Engrs; Geo Washington Honor Medal 1954; Freedoms Found Award 1955. Fellow IME, IEEE, AAAS. Hon Mbr ASME, AREA, EIC; Life Mbr ASCE, AIME. Mbr Sigma Xi, Pi Tau Sigma, Tau Beta Pi. Author 'Mastering Momentum' 1941.

Sills, J Thomas
Home: 2941 Edgemont Ct, Allentown, PA 18103
Position: Mgr, Govt & Technical Systems Sales *Employer:* Air Products & Chemicals *Education:* MBA/Marketing/Univ of Buffalo; BCE/Nuclear Power/Cornell Univ *Born:* 09/29/34 Miami, FL. Attended Cornell Univ on John McMullen Scholarship. Elected to membership in, and held offices in, Chi Epsilon and Rod & Bob Civil Engrg Hon Societies. Held several Ctte memberships in both AIChE and ASHRAE, holding chairmanship in the latter. Was Dept Mgr in the dev and marketing of the world's first (and leading) commercially successful liquid nitrogen food freezing process. Married to Patricia Gay Sills, four children. Holder of two patents in cryogenic

Sills, J Thomas (Continued)
refrigeration, and author of numerous articles and presentations to tech societies. Held teaching positions at Cornell and Lehigh Univ. *Society Aff:* AIChE, ASHRAE

Silton, Ronald H
Home: 714 Jackson Drive, Port Clinton, OH 43452
Position: Quality Engr. *Employer:* Gilbert/Commonwealth *Education:* BSEE/EE/Fairleigh Dickinson Univ *Born:* 05/11/51 Native of Butler, NJ. Served as Exec Dir of Student Branch of IEEE in 1972 & 1973. Worked as a field engr for Florida Power & Light until 1977. With Gilbert/Commonwealth since 1978. Assumed current responsbility as Quality Engr in 1986 for the TVA Sequoyah Nuclear Plant. VP, Jackson Ski Club, 1980. Pres, 1981. Enjoy skiing, music, and sports. *Society Aff:* NSPE, IEEE

Silva, Armand J
Business: Dept of Ocean Engrg, Kingston, RI 02881-0814
Position: Prof of Ocean & Civl Engrg *Employer:* Univ of Rhode Island *Education:* PhD/CE-Geotech/Univ of CT; MS/CE/Univ of CT; BS/CE/Univ of CT *Born:* 06/01/31 Chrmn of Ocean Engrg Dept and hold joint appointment in Ocean and Civil Engrg at URI. Also Dir of Marine Geomechanics Res Group with major involvement in research on ocean sediments and marine geomechanics. Previously taught at Univ of CT, then on faculty at Worcester Polytechnic Inst 1958 through 1976 with final position of Prof and Hd, Dept of Civil Engrg. Private conslt on numerous soils, foundations, and marine geotechnical projs. Soils-foundation engr for Thompson and Lichtner Co, Inc 1956-58. *Society Aff:* ASCE, AAAS, MTS, AGU, ASTM

Silva, Ignacio
Business: 525 Lancaster-PO 1498, Reading, PA 19603
Position: VP - Intl. *Employer:* Gilbert Assocs, Inc. *Education:* BS/CE/Univ of Havana. *Born:* 1922 Cuba. Schooling U S (Michigan) & Cuba. Started in Cuba as Field Engr with Frederick Snare Corp, Engrs & Contractors, N Y-Havana-Lima. Genl Supt 1954. Transferred to N Y 1960. Branch Mgr Puerto Rico 1965. Elected Pres 1967. Dir orderly closing of co (resulting from major loss on Bahamian Gov contract started in 1966). Cons to U S Sureties in England 1969. Joined Gilbert Assocs' Mgmt Cons Div 1970. Elected V P 1975 of Gilbert/CommonWealth Internatl Inc, Elected VP of Parent Co Gilbert Assoc in 1977.

Silva, LeRoy F
Business: Engr. Admin. Bldg, Purdue University, West Lafayette, IN 47907
Position: Prof of Elec Engr. *Employer:* Purdue Univ. *Education:* PhD/Elec Engr/Purdue Univ; SM/Elec Engr/MIT; BSEE/Elec Engr/Purdue Univ. *Born:* 11/17/30. Prog Leader, Measurements at Purdue's Lab for Applications of Remote Sensing. He received the BSEE and PhD degrees from Purdue Univ and the SM degree from MIT. He has been with Lincoln Labs, Lexington, MA, Ballistic Res Labs, Aberdeen Proving Ground, MD and CP Electronics, Columbus, IN where he was VP, Engg. Dr Silva is an active consultant and publishes in electronics, magnetics, and remote sensing. His res in remote sensing centers on field instrumentation and the physical basis of spectral band selection. He is a Reg PE - State of IN. His principal hobby is automotive mechanics and he is a certified Gen Automotive Mechanic in the Natl Inst of Automotive Service Excellence (NIASE). *Society Aff:* IEEE, AAAS, OSA, NSPE, ΣΞ.

Silver, Edward A
Business: Faculty of Management, 2500 University Drive N.W, Calgary, Alberta T2N 1N4 Canada,
Position: Prof of Management *Employer:* Univ of Calgary *Education:* BEng/Civil Engg/McGill Univ; ScD/Operations Res/MIT. *Born:* 6/13/37. P Eng Ontario 1971, Alberta 1981. Prof staff mbr of OR Group of Arthur D Little Inc 1963-67. Assoc Prof in Coll of Bus Admin Boston Univ 1967-69. Assoc Prof 1969-72, Prof 1972-81, Assoc Chmn 1974-81, Dept of Mgmt Scis, Fac of Engrg, Univ of Waterloo. Prof of Mgmt 1981- , Univ of Calgary. Author over 70 articles in prof journals, co-author *Decision Sys for Inventory Mgmt & Production Planning*. Cons in operations management, production planning & inventory control (Pres, Maxed & Assocs Inc, consulting firm). Past-Pres of CORS; Mbr IIE, ISIR, TIMS, APEGGA, APICS, ORS (UK), PMI. Chairman of Industrial Engrg Grant Selection Ctte of Natural Sciences and Engrg Research Council (NSERC) of Canada (1980-81), Mbr of NSERC Advisory Ctte on Engrg. (1984-86). Recipient of the 1986 Operations Research Division Award of the Institute of Industrial Engineering. *Society Aff:* CORS, TIMS, APICS, IIE, ISIR, ORS (UK), PMI

Silver, Francis, 5th
Business: P. O. Box 1032, Washington, D. C 20013-1032
Position: Cons Envir Engr & Surveyor of Lands (Self-emp) *Employer:* Francis Silver Environmental Engineering *Education:* BE/Gas Engg/Johns Hopkins Univ; grad work in aerodynamics//Univ of MD. *Born:* 1/4/16. 1937-1942 heavy inorganic chem mfg. 1942-1946 Army Ord, WWII. 1947-1950 flour, feed mfg. 1951-1957 aircraft weight engr, Fairchild & Boeing. Environmental engg, eco-relativity & measurement of life research. Called as expert in gas toxicology on first bill curbing auto exhaust 1958 US Congress & 1963. Pres WV Assoc Land Surveyors 1973-1975; Pres Bd of Trustee, City Hospital 1977-1978. Diplomate Am Acad Environmental Engineers. Author 70 scientific papers, book chapters, etc. Seminars. Consultant to allergists with chemically sensitized patients and specialize in indoor air pollution & environmental house calls for homes, office bldgs & projects since 1962. Recipient Jonathan Forman Award, AAEM, 1982. *Society Aff:* NSPE, ASME, AAAS, ACS, AAEM, ΣΞ, AIChE, AAEE, APCA, APHA, RSH, NYAS

Silver, John E
Business: 5456 McConnell Ave 270, Los Angeles, CA 90066
Position: Owner - Elec Engr. *Employer:* John E Silver & Assocs. *Education:* BS/Elec Engg/LA State Univ. *Born:* Aug 20, 1922 Flint Mich. Served with Special Engrs Detachment in Manhattan Proj at Oak Ridge Tenn during WW II. Elec Engr with E R Little Co in Detroit Mich for 4 yrs & with Ralph Phillips in Los Angeles Calif for 4 yrs. Started John E Silver & Assocs 1954. Sr Partner of Silver, Meckler & Assocs for 5 yrs. Pres Hellman Silver Lober Inc for 5 yrs. Re-estab John E Silver & Assocs 1974, a cons elec engrg firm. Enjoy bowling, golf, sailing. *Society Aff:* ACEC, IES.

Silver, Marshall L
Business: Dept of Civil Engrg, Box 4348, ms 130, Chicago, IL 60680
Position: Prof *Employer:* Univ of Illinois *Education:* PhD/CE/Univ of CA-Berkeley; MS/CE/Univ of CA-Berkeley; BS/CE/Univ of CO *Born:* 11/26/42 M. L. Silver joined the faculty of the Univ of IL in Sept 1969 after completing the work for his Doctorate in Civil Engr from the Univ of CA at Berkeley. He was promoted to the rank of Prof in 1978. His area of specialization is applied soil mech and foundation engrg, and he has conducted lab and field studies for soil dynamics, embankment dam, earthquake engrg, noise, vibration and transportation related research. He is the author of over 50 papers in these areas. He is the Chrmn of the USCOLD Ctte on Ed and Training, Chrmn of the ASTM Ctte on Soil Dynamics and Chrmn of the ASCE Geotech Engrg Ctte on Definitions and Standards. *Society Aff:* ASCE, ASTM, USCOLD

Silvera, Americo
Home: One Rosewood Dr, Stony Poinet, NY 10980
Position: Pres & Owner *Employer:* Amersil Overseas Inc. *Education:* BSc/Architecture/RPI (Rensselaer Polytechnic Institute); BSc/Sciences/Inst Nacional Panama. *Born:* Oct 1912 Republic of Panama. Naturalized USA 1950. With Carrier Internatl Corp Syracuse N Y 1937-77. Has been Ch Engr, Appl Sys Sales Mgr, Export Sales Mgr & V P for Africa-Middle East Zone Hdqtrs in Athens Greece. Fellow ASHRAE & recipient ASHRAE Disting Serv Award - served as Chmn of Internatl Relations Ctte 1963-66. Mbr USA Indus Sector Adv Ctte on Multinatl Trade Negotiations. Author tech articles on 'Concrete Cooling for Dam Const', 'Brewery Refrig' & 'Low Temp Blast Freezing'. Pat on 'Apparatus for Continuous Blast

Silvera, Americo (Continued)
Freezing'. Formerly V Chmn of Bd Carrier Thermofrig Corp - Tehran Iran Mbr of Bd of Dirs Airco Inc - Johannesburg S Africa. Now Pres of Amersil Overseas an Intl Mkting & cons Firm assit USA Firms in establishing mkting operation overseas, with special emphasis on Latin America, Middle East and Africa. Married to former Emma Morales Santiago. Two sons Roger A and Edwin Silvera and 3 by Former wife Hildegard Seelig: Robert K, Ronald E and Ricardo R Silvera. *Society Aff:* ASHRAE.

Silverman, Bernard
Business: ECE Dept, Link Hall, Syracuse, NY 13210
Position: Assoc Prof. *Employer:* Syracuse Univ. *Education:* PhD/EE/Univ of IL; MS/EE/Univ of IL; BS/EE/VPI. *Born:* 8/15/22. in Richmond, VA. Served in Signal Corps, US Army in WWII; currently, Col, US Army Reserves (Ret). Instr in EE at Univ of IL (1947-1953). Engr in Electronics Lab, GE Co, (1954-1958); R&D of applications of magnetic & dielectric devices, primarily ferrites & barium titanate. Since 1958, assoc prof, Elec & Comp Engg Dept, Syracuse Univ. Consultant, NASA, summer, 1967. Principal interests are nonlinear & discrete systems. Current res activities are in the areas of communications & digital signal processig. *Society Aff:* IEEE.

Silverstein, Abe
Home: 21160 Seabury Ave, Fairview Pk, OH 44126
Position: Consultant (Self-employed) *Born:* Amer aeronautical engr. Educ Rose Polytech Inst. Jr Engr, Langley Lab, Natl Adv Ctte for Aeronautics (now Natl Aeronautics & Space Admin) 1929-40; successively Hd Full-Scale Wind Tunnel, Ch Engrg Installation Div, Ch Wind Tunnel & Flight Div, Ch of Res; Assoc Dir Lewis Flight Propulsion Lab 1940-58; Dir of Space Flight Progs NASA 1958-61; Dir Lewis Res Ctr 1961-Nov 1969; involved in environ planning, Republic Steel Corp 1970- . Mbr Internatl Acad of Astronautics, Natl Acad of Engrg; Fellow Amer Inst of Astronautics, Amer Astronautical Soc, Royal Aeronautical Soc. Num awards. 5 hon degrees. Author some 50 sci papers.

Silvestri, George J, Jr
Home: 511 Weir Rd, Aston, PA 19014
Position: Fellow Engr. *Employer:* Westinghouse Elec Corp. *Education:* MSME/ME/Drexel Univ; BSME/ME/Drexel Univ. *Born:* 8/3/27. US Navy, 1945-1946. Westinghouse Steam Turbine Div, 1953 to present. EPRI (on loan), 1973-74 Fossil Fuel and Advanced Systems Div. Performance and evaluation of advanced power generating concepts. Dev of comp progs for calculating steam turbine performance-heat balances and steam properties (1936 K & K and 1967 ASME Steam Tables). Author of 15 published reports and papers. Co- author of *1967 ASME Steam Tables*. Awarded 7 patents. Mbr of ASME Res Committee on Properties of Steam and Power Div Gen Committee. Fellow ASME. *Society Aff:* ASME, ANS.

Sim, Richard G
Business: PO Box 2188, Hickory, NC 28601
Position: General Manager *Employer:* GE. *Education:* BSc/Mech/Univ of Glasgow; PhD/MEch/Univ of Cambridge. *Born:* 9/9/44. Lectr & res fellow Univ of Liverpool 1968/69. Worked for Westinghouse Advanced Reactors Div 1970/71. Joined GE 1972 in the advanced reactors systems dept. Progressed through a variety of positions to mgr of reactor engg in 1977. Played maj role in coordinating & leading liquid metal past breeder reactor dev natlly & intlly. Mgr of engg of distribution transformers for GE in 1978-80. Assumed current position of general manager of Protective Equipment Products in 1980. Published over 20 papers in areas of high temperature mtls & structure behavior & core/reactor tech. *Society Aff:* IEEE.

Simaan, Marwan
Business: Dept. of Electrical Engineering, University of Pittsburgh, Pittsburgh, PA 15261
Position: Prof *Employer:* Univ of Pittsburgh *Education:* PhD/Elect Engrg/Univ of IL; MS/Elect Engrg/Univ of Pittsburgh; BEE/Elect Engrg American Univ of Beirut *Born:* 07/23/46 Taught Electrical Engrg and was affiliated with the Coordinated Sci Lab, Univ of Il 1972-1974. Res Engr with Shell Development Co, 1974-1975. In 1976 he joined the Univ. of Pittsburgh where he is presently Prof of Elect Engrg. Conslt for Gulf R & D from 1979 to 1985, and for Alcoa from 1985 to present. Res interests are mainly in the areas of optimization, control and signal processing. Published over 110 articles, and edited two books in geophysical data processing. He is Assoc Ed for the Journal of Optimization Theory and Applications, and the IEEE Transactions on Geosciences and Remote Sensing. Received numerous teaching awards from students at the Univ. of Pittsburgh. Received the 1985 IEEE Geoscience and Remote Sensing Prize Paper Award. Reg PE in PA. *Society Aff:* IEEE, SEG, ΣΞ, ΦΗΣ, HKN, NY ACAD SCI, AAAI

Simard, Gerald L
Business: Jenness Hall, Orono, ME 04469
Position: Assoc Prof Engg. Retired. *Employer:* Univ of ME. *Education:* PhD/Phys Chem/MIT; BS/Chemistry/Bates College. *Born:* 5/11/12. Native of ME. Res Engr with Battelle Memorial Inst 1938-43. Res Chemist ad then Group Leader Amer Cyanamid Co, 1943-53, Section Hd and Later Mgr Schlumberger Well Surveying Co 1953-67. Assoc Prof Chem Engg Univ of ME 1967-77. Retired. Enjoy sailing. *Society Aff:* AIChE, ACS, ΣΞ, ΦΒΚ.

Simcox, Craig D
Home: 4640 132nd Ave SE, Bellevue, WA 98006
Position: Technology Chief - 767 *Employer:* Boeing Commercial Airplane Co. *Education:* BS/Aero & Astro/IA State Univ; MS/Aero & Astro/Stanford; PhD/Engg Sci/Purdue. *Born:* 9/18/39. Res Scientist at NASA Ames Res Ctr 1962-65. Conducted res in jet exhaust noise, was Contract Mgr for SST Follow-on-Noise Reduction Prof, & has held sev mgmt pos in Noise Tech at Boeing since 1969. Currently Technology Chief for 767 Aircraft. Author 30 tech papers. Mbr Boeing Mgmt Assn. Fellow AIAA. Past Chmn of Pacific Northwest Sect AIAA. Chmn AIAA Aeroacoustic Tech Ctte for 1977-78. Genl Chmn AIAA 4th Aeroacoustic Mtg, Admin Chmn AIAA 12th Fluid Dynamics Conf, AIAA Tech Spec Group Coord, AIAA Deputy Dir Region VI 1978-80, Sci Ctte/Co-Chrmn IUTA/AIAA/ICA Symposium on Mechanisms of Noise Generation, Gettingen, Germany, Aug 1979. Gen. Chairman AIAA 1982 Aerospace Scis Meeting, Orlando, Fla. Arrange Chmn 15th Intersociety Energy Conversion Engrg Conference, Seattle, 1980. Chmn 1984 AIAA Assoc Fellow Grade Cttee. First VP BMA Completed Columbia Univ Exec Program in Bus Administration. Enjoys hiking, skiing, photography, golf & sport cars. *Society Aff:* AIAA, BMA.

Simison, Charles B
Home: 1632 N. Vermilion, Danville, IL 61832
Position: Gen Mgr *Employer:* Gulf & Western Mgf Co *Education:* BS/EE/Purdue Univ; BS/NS/Purdue Univ *Born:* 03/07/28 Indiana native. Served 3 yrs officer US Navy during Korean conflict aboard destroyer then minesweeper. Qualified for command small vessel. Spent 3 yrs Appl Engr Minn Honeywell and 25 yrs Bohn Coil Div G&W Mgf Co. Serving A/C Regrig Indus with Heat Exchangers & Unitary Products. Experience as Application Engr followed by sales mgr, sale & engr dept mgr, dir then VP engr. Currently Div Gen Mgr with sale, engr mfg responsibility. Interest golf and travel. *Society Aff:* ASHRAE, HTRI, ARI

Simitses, George J
Business: 225 North Ave, NW, Atlanta, GA 30332
Position: Prof *Employer:* Georgia Inst of Tech *Education:* PhD/Aero Structures/Stanford Univ; MS/AE/GA Tech; BS/AE/GA Tech *Born:* 07/31/32 Born in Greece; Married 1960 with three children. Author of a text *Elastic Stability of Structures*, Prentice-Hall, 1976, and of over 50 technical papers in Refereed Journals (AIAAJ, J of Applied Mech, Computers & Structures, J of Aircraft, etc); Also has presented 30 papers at natl and intl meetings area of specialty in structural mech with emphasis in stability, structural optimization and composite materials. Has served as major

Simitses, George J (Continued)
prof to 12 PhD students. Has served as conslt to several engrg companies. *Society Aff:* AIAA, ASME, SSRC, ΣΞ, AAM, HSTAM

Simkovich, Alex
Home: R.D. 1, Box 330, Ligonier, PA 15658
Position: Pres *Employer:* Spang Specialty Metals *Education:* BS/Metal/Penn State Univ; MS/Metal/Penn State Univ; PhD/Met Eng/Carnegie Tech. *Born:* Dec 25, 1933 Jacobs Creek Pa. US Naval Officer 1956-59 at Naval Res Lab wth Proj Vanguard. PhD & joined Latrobe Steel Co 1963 - Sr Dev Metallurgist, Mgr of Process Res, Mgr of Primary & Finishing Opers. Tech Dir, Pres of Teledyne Ohiocast; 1971-72 Chmn of the Pgh ASM Chap. Author var papers on steelmaking. Mbr ASM, AIME, AISE, AISI. Recipient of 1965 Elec Furnace Conf Best Paper Award. Presently, Pres Spang Specialty Metals. *Society Aff:* AIME, ASM.

Simmons, Alan J
Business: PO Box 73, Rm D-422, Lexington, MA 02173
Position: Assoc Leader. *Employer:* MIT/Lincoln Lab. *Education:* PhD/EE/Univ of MD; MS/EE/MIT; BS/Phys & Chem/Harvard Univ. *Born:* 10/14/24. Served as Radar Mtl Officer WWII. 1946-48 Res Asst at MIT working on early guided-missile prog. 1948-57 Microwave & Antenna Res at Naval Res Lab, Wash, DC, where received a patent on a widely-used microwave polarizer. Co-founder, Dir of Res & Gen Mgr of Boston Div of TRG, Inc, which developed into a leading US supplier of millimeter-wave components & antennas. Since 1971 at MIT/Lincoln Lab with responsibility for dev of microwave & antenna portions of advanced communication systems based on communication satellites. Elected Fellow of IEEE in 1979. *Society Aff:* IEEE, AAAS.

Simmons, Malcolm E
Business: POBox 1630, Alexandria, LA 71301
Position: Asst State Conservationist/Water Resources. *Employer:* Soil Conserv Serv - US Dept of Agri. *Education:* BS/Agri Engg/LA State Univ. *Born:* 9/22/36. With Soil Conservation Service since 1958. Served as field office engr in Lake Charles La & area engg in Tallulah LA. Received cert of merit for outstanding service as area engr 1972. Served as engr on Watershed Planning- River Basin Studies Staff in Alexandria LA; Served as Staff Leader (Nov 1975- Sept 1979); currently Asst State Conservaionist/Water Resources-Receives cert of merit for outstanding ser in 1977. Received cert of merit for outstanding serv in 1975. Mbr Amer Soc of Agri Engrs; Chmn La Sect 1974. Mbr La Engg Soc, Soil Conservation Soc of Amer, Natl Assn of Conservations Dists. *Society Aff:* ASAE, LES, SCSA, NACD.

Simmons, William W
Business: R1/1196, One Space Park, Redondo Beach, CA 90278
Position: Director, Group Res *Employer:* TRW S&D *Education:* Ph.D./Physics/Univ. of Ill.; M.S./Physics/Univ. of Ill.; B.A./Physics/Carleton College *Born:* 04/24/32 Married (Barbara), 1954; three daughters. Dir Group Research, TRW, 1985-. Principal Scientist/Engr, Lawrence Livermore Natl Lab, 1972-84. Designed, constructed, operated series of solid state laser systems for intertial fusion studies (Argus, Skiva, Nova). Professor, Engrg, UCLA, 1968-72. Sr Scientist, TRW, 1960-68. Awarded Distinguished Prof, UCLA, 1971. Awarded IEEE Simon Ramo Medal for system engrg; 1987. *Society Aff:* IEEE, AIAA, LIA, APS

Simnad, Massoud T
Home: 9342 La Jolla Farms Rd, La Jolla, CA 92037
Position: Adjunct Professor and Consultant *Employer:* University of California in San Diego *Education:* PhD/Matls Sci/Cambridge Univ; BS/Metallurgical Engg/ Imperial College of Science & Technology London Univ. *Born:* 3/11/20. 1945-48 Res Assoc Cambridge Univ; 1949-56 Guest Fellow, Sr Res Staff, Fac Mbr Carnegie-Mellon Univ; 1956-81, Genl Atomic Co (1956-69 Asst Chmn Met & Fuels Dept - 120 staff, 1969 to 1981 Sr Tech Advisor Advanced Energy Sys Div; 1981- , Adj Prof Univ California in San Diego; Consultant; 1962-63 Visiting Prof MIT, Nuclear Engrg & Met Depts. Activities: Matls & Fuels Dev for nuclear energy, fusion & TRIGA reactor; Matls SC and Engrg. Fellow Amer Nuclear Soc, Amer Soc for Metals, AAAS; Cert of Merit ANS 1965. Over 100 papers & Monograph 'Fuel Element Exper in Nuclear Power Reactors' 1971 ANS. *Society Aff:* ANS, ASM, AAAS, EC, ΣΞ

Simon, Albert
Business: Dept. of Mech Engin, Univ. of Rochester, Rochester, N.Y 14627
Position: Prof *Employer:* Univ of Rochester. *Education:* PhD/Phys/Univ of Rochester; BS/Phys/CCNY. *Born:* 12/27/24. NYC. USN 1944-46. Oak Ridge Natl Lab 1950-61, Asst Dir, Neutron Phys Div, worked on fission reactor, shielding & controlled fusion. Gen Atomic Co, 1961- 66, Hd, Plasma Phys Div, worked on fusion reactor concepts. Prof, Dept of Mech Engineering and Prof, Dept. of Physics, Univ of Rochester, 1966-, Chrmn, ME Dept, 1977-84, Sr Scientist, Laboratory for Laser Energetics, 1983-, res on fusion rectors. Author of *An Introduction to Controlled Fusion Research* (Pergamon, 1959) & encyclopedia Americana article on *Controlled* Fusion. Guggenheim Fellow, 1964. Hold three patents on mirror fusion reactor concepts. Former Chrmn, APS Plasma Phys Div (1963-64). Lic PE, NY (since 1981). Chmn, Nuclear Engrg Div, ASEE. 1985-86. *Society Aff:* ASME, APS, ASEE, AAAS.

Simon, Andrew L
Home: 550 Northwood, Akron, OH 44313
Position: Professor & Head. *Employer:* Univ of Akron. *Education:* PhD/Civil Eng/ Purdue Univ; Dipl Eng/Civil Eng/Tech Univ Budapest. *Born:* Dec 1, 1930 Hungary. Worked as hydraulic engr for Hungarian gov design office 'Melyepterv'. In 1956 emigrated to the U S. Until 1958 stress analyst at Babcock & Wilcox Co. PhD Civil Engrg Purdue Univ Jan 1962. Between 1961 & 1965, Prof & Hd of Civil Engrg West Vir Inst of Tech. Since 1965 in same pos at Univ of Akron. Also since 1975, Exec Dir of Inst for Tech Assistance. Extensive cons engrg exper with major corps. World Bank Consultant. Recent books: Energy Resources (Pergamon 1975), Practical Hydraulics (Wiley 1976) 3rd Ed. 1985, Basic Hydraulics (Wiley, 1981). Fire Hydraulics (Wiley 1983), Prin of Statics and Strength of Mtls (W.C. Brown 1983), Reg P E Ohio, FL, W Va, Indiana & N J. *Society Aff:* ASCE, GES

Simon, Herbert A
Business: Schenley Park, Pittsburgh, PA 15213
Position: Prof *Employer:* Carnegie Mellon Univ *Education:* Ph.D./Polit. Sci./U. of Chicago; B.A./Polit. Sci./U. of Chicago *Born:* 06/15/16 Professor of Computer Science and Psychology, Carnegie-Mellon Univ. Research on decision processes, suing artificial intelligence, operations research, and cognitive science approaches. Co-inventor of computer list-processing languages, computer problem solving by heuristic search. Faculty, UC (Berkeley), 1939-42; Ill Inst of Tech, 1942-49; Carnegie-Mellon Univ since 1949. President's Science Advisory Cttee, 1968-72; Council, National Academy of Sciences, 1978-81, 1983-86. Distinguished Scientific Contribution Award, American Psychological Assoc; Turing Award, Assoc for Computing Machinery; Nobel Prize in Economics; Procter Prize, Sigma Xi; National Medal of Science. Author of professional papers and books, including Human Problem Solving, Sciences of the Artificial. *Society Aff:* IEEE, AAAS, NAS, ΣΞ, ACM

Simon, Martin
Home: 105 Lake Cliff Dr, Erie, PA 16511
Position: Ret - Management Consultant *Employer:* Genl Electric Co. *Education:* BS/EE/Penn State Univ. *Born:* Feb 19, 1921. BSEE Penn State Univ 1948. US Army 1941-46. Taught engrg evening courses at Gannon Univ. Design Engr for G E Co, spec in propulsion & control sys for transp vehicles 1948-62. Held var mgmt pos in engrg including Mgr of Engrg Sect, Transp Sys Bus Div in the G E Co. IEEE Natl Tech Paper Award 1955. Sev pats. Currently Sr Assoc of Norman W. Seip and Assoc, Management Consultants.

Simon, Marvin K
Business: 4800 Oak Grove Dr, Pasadena, CA 91109
Position: Sr Res Engr. *Employer:* Jet propulsion Lab. *Education:* BEE/Elec Engr/ CCNY; MSEE/EE/Princeton Univ; PhD/EE/NYU *Born:* 9/10/39. Bell Tel Labs 1961-1963, 1965-1966 Instr of EE, NYU. 1966-1968; Jet Propulsion Lab. 1968-present communications res in synchronization & modulation techniques. Visiting Prof of EE, CIT, 1979; Author of "Telecommunications Systems Engineering–, Prentice-Hall, "Phase-Locked Loops and Their Application–, IEEE Press; Fellow IEEE; Mbr Sigma Xi, Tau Beta Pi, Eta Kappa Nu; Editor of Comm Theory for IEEE Trans on Comm, 1974-1976 and "Spread Spectrum Communications" Computer Service Press, 1984. Chrmn Comm Theory Committee of IEEE ComSoc, 1977- 1980; Tech Prog Chrmn, NTC '77. Consultnt to private Ind & govt in space & satellite communications. NASA Exceptional Service Medal 1979. *Society Aff:* IEEE

Simon, Ralph E
Business: RCA-Solid State Div, Somerville, NJ 08876
Position: Div Vice President. *Employer:* RCA. *Education:* PhD/Physics/Cornell Univ; BA/Physics/Princeton Univ. *Born:* Oct 20, 1930. PhD Cornell Univ; BA Princeton Univ. Mbr Tech Staff RCA Labs 1958-67, working primarily on surface physics, electron emission, TV camera tubes. Held var mgmt pos in RCA Electro-Optics, moving to RCA Lancaster plant in 1968. Div V P, RCA Solid State Div, Electro-Optics & Devices 1975. Div VP RCA Solid State Div Electro Optics and Power Devices 1979. RCA David Sarnoff Team Award for Sci; IEEE V K Zworykin Award. Mbr DOD Advisory Group on Electron Devices. Mbr APS, Sigma Xi, AAAS, Fellow IEEE. *Society Aff:* ΣΞ, IEEE, APS, AAAS.

Simons, Daryl B
Home: 3535 Terryridge Rd, Fort Collins, CO 80524
Position: Pres *Employer:* Simons, Li & Assoc, Inc *Education:* PhD/CE/CO State Univ; MS/CE/UT State Univ; BS/CE/UT State Univ *Born:* 02/12/18 Several yrs experience in teaching, consltg and res. Written over 400 papers; coauthor of two text books; chapt for three texts. Prof, Univ of Wyoming, 1948-1957; Proj Chief, US Geological Survey, Ft Collins, CO 1957-1962. Recipient of ASCE J. C. Stevens Award, 1960; ASCE Croes Award, 1964; Outstanding PE, CO. 1973; Co-recipient, ASCE 1979 Karl Emil Hilgard Hydraulic Prize. Served as hydraulic conslt on several important engrg projs for United Nations; Corps of Engrs dealing with river mech, fluid control, navigation problems, mathematical modeling; also, natl and intl panels delegated to outline res needs in hydraulics, irrigation and drainage. Pres of Simons, Li & Assoc. Assoc Dean and Dir of Engrg Res and Prof of Civil Engrg at CO State Univ. Retired Summer, 1983. *Society Aff:* ASCE, ICID, IAHR, AGU

Simons, Eugene M
Home: 662 Overlook Dr, Columbus, OH 43214
Position: Retired *Employer:* Retired *Education:* PhD/Mech Engr/OH State Univ; MS/Mech Engr/VA Polytech Inst; BS/Mech Engr/Carnegie Inst of Technology. *Born:* Feb 1917. Taught var engrg subjects at CIT, Marshall Coll, & VPI 1937-43. Joined Battelle Mem Inst 1943: Principal Mech Engr 1943-50, Asst Supr Nuclear Engrg & Analysis 1950-56, Fellow 1956-72, Staff Manager 1972-82. Retired Mar. 1982. Supervised num nuclear & mech engrg res projs; was editor of tech journals on Reactor Matls & Sci Pol; authored or coauthored 29 sci publs. Reg P E Ohio. Life Fellow ASME; Sigma Xi, Pi Tau Sigma (Mech Engrg), Sigma Pi Sigma (Phys). Married to Ruth L Simons; 4 children. *Society Aff:* ASME, ΣΞ, ΠΤΣ, ΣΠΣ.

Simons, Gale G
Business: Ward Hall, Manhattan, KS 66506
Position: Prof *Employer:* KS St Univ *Education:* PhD/Nucl Engrg/KS St Univ *Born:* 9/25/39 Engr for Argonne Natl Laboratory from 1968 to 1977, specializing in methods development and fast-reactor instrumentation system design. Head of the Experimental Support Group and Mgr of the Argonne Fast Source Reactor. With KS St Univ since 1977, activities oriented toward education and research in nuclear instrumentation and radiation dosimetry. Dir of the Neutron Activation Analysis Laboratory and the Radiation Shielding Facility. Halliburton Faculty Award, 1979; Sigma Xi Research Excellence, 1980; Chairman Mo-Kan ANS, 1982; mbr Phi Kappa Phi, Sigma Xi, Pi Mu Epsilon, Sigma Tau and Alpha Nu Sigma. *Society Aff:* ANS, HP.

Simonsen, John M
Home: 430 E 1980 N, Provo, UT 84601
Position: V Pres - Engrg. *Employer:* Valtek Inc. *Education:* BSME/Mech Engr/Univ of UT; MSME/Mech Engr/Purdue; PhD/Mech Engg/Purdue. *Born:* May 1927. Taught Mech Engrg Brigham Young Univ 1954-75. Currently V P Engrg Valtek Inc. Formerly Mgr Thermodynamic Aerospace Corp. V P Region VIII ASME 197375. Outstanding Engrs of Utah Award 1957. *Society Aff:* ISA, ASME.

Simpkins, Ronald B
Home: PO Box 1082, Ross, CA 94957
Position: Pres. *Employer:* Simpkins & Assoc, Engineers *Education:* BS/Engg/San Jose State College; AA/Engg/Placer Jr College. *Born:* 10/28/29. Civ Engr, Land Surveyor & Planner involved in large land dev projects - mostly in USA & dealing with heavy earth moving or water ways. Recognized specialist in environmentally pleasing sculptured hill grading & shoreline, lake & bay improvements for regional, municipal, commercial & private projs. *Society Aff:* NSPE.

Simpkinson, Scott Holmes
Home: 9333 Tallyho St 9, Houston, TX 77017
Position: Retired *Education:* BS/ME/Univ of Cincinnati. *Born:* Aug 1919. Native of Piqua Ohio. Hd, Flight Measurements Lewis Res Ctr NASA 1943-58; Ch Launch Opers, Mgr Capsule Opers & Launch Capsule Test Conductor, Space Task Group 1958-60; Mercury Sys Cons 1960-61; Spec Asst to Dir Atlas Vehicle 1961-62; Mgr Test Opers Gemini Prog, Manned Spacecraft Ctr-Houston 1962-67; Asst Prog Mgr for Flight Safety Apollo Prog 1967-72; Mgr Flight Safety Shuttle Prog, Johnson Space Ctr-Houston 1972-1981. Recipient of num group & achievement awards NASA; also 4 exceptional service medals; Pres Medal of Freedom 1970; Disting Alumnus Award Univ of Cincinnati; Sr Mbr AAAS; Amer Soc Mech Engrs; Sr Mbr Instru Soc of Amer; Assoc Fellow AIAA; Mbr Soc Logistics Engrs; Sr Mbr Amer Astronautic Soc Spacecraft Opers; testing; electronics; instrumentation. Enjoys golf, bridge, music, hi-fi & collects stamps, first day covers and commemorative medals. *Society Aff:* ISA, AIAA.

Simpson, David M
12 De Vitre Place, Grove, Wantage, Oxon., OX12 ODA England
Position: Proj Design Engr. *Employer:* Science and Engineering Research Cou (Serc) *Education:* MSc/-/Cranfield Inst of Technology; HND/-/Hull College of Technology; ONC/- /Hull College of Technology. *Born:* 1/2/48. Yorkshire UK. Joined Hawker Siddeley Aviation Ltd 1964. Spec in data res, computer aided design res with applications to H S Hawk advanced trainer. Apptd as Sr Proj Wt Engr at H S A Brough in 1973. Apptd to Advanced Mech Sys Group, HMS DAEDALUS, Royal Navy 1976. SAWE Revere Cup Awd 1973. Dec 1978 Apptd to Space & Astrophysics Div of SERC. Analysis of Drive Mechanisms on the X-Ray Polychromator Experiment for the Solar Maximum Mission Satellite. In charge of Proj Engg on the Coronal Helium Abundance Spacelab Experiment Instrument (Chase) (This flew successfully on Space Shuttle Challenger in July 1985). Devmnt of Thermal Analysis Software for General Spacecraft use. Thermal design of ROSAT Wide Field Camera and Startrackers. Design reponsibility for along track Scanning Radiometer on European Remote Soundings Satellite ERS-1. Director of UK Chapter of the SAWE 1977/81 President of UK Chapter of the SAWE 1981-82 Ints: astronomy, astrophys & classical music. *Society Aff:* RAeS, SAWE, BIS, BAA.

Simpson, George A
Business: 4424 Montgomery Ave, Bethesda, MD 20814
Position: Associate. *Employer:* John J Allen Assoc. *Education:*

Simpson, George A (Continued)
BSE/Mathematics/George Wash Univ; BA/English/Georgetown Univ. *Born:* 4/6/31. in Washington, DC. Served in topographic co, Army Corps of Engrs, 1953-55. In consulting engg, land planning and surveying in Wash, DD area since 1955. Assoc, John J Allen Assoc since 1973. Chrmn, MD PEPP Sec 1975-1977. Pres, Potomac Chapter, MSPE, 1977-1978. VP, MSPE, 1979-1981, Pres, MSPE, 1981- . *Society Aff:* HSPE/MSPE, NSPE/PEPP, WSE.

Simpson, George D
Business: 1300 E 9th St, 700 Bond Ct Bldg, Cleveland, OH 44114
Position: Pres. *Employer:* Havens & Emerson, Inc. *Education:* BSCE/Sanitary/Duke Univ; BS/Math/Univ SC. *Born:* 12/25/23. Chicago, IL; US Navy 1943-1948. Joined Havens & Emerson Consulting Engrs in 1949; elected its Pres 1975. Specialized in design of large water and wastewater treatment systems and in disposal of solids residues. Managed enviromental engg projs for fed, state & regional agencies. Author of numerous papers in the technical press on the above topics. Honorary mbr of WPCF & recipient of the Bedell award. Expert witness & consultant on eivnornmental matters. Phi Beta Kappa, Tau Beta Pi. Married, one daughter; enjoys golf & tennis. *Society Aff:* ASCE, WPCF, AWWA, AAEE, NSPE.

Simpson, John W
Business: 200 Professional Bldg, Hilton Head Island, SC 29928
Position: President *Employer:* Simpson Business Services, Inc. *Education:* BS/US Naval Academy; MS/Elec Engg/Univ of Pittsburgh *Born:* Sept 1914 S Car. Employed at Westinghouse 1937 to 1977. Successively Supervising Engr, Switchgear Div; Proj Mgr, Shippingport Atomic Plant & Tech Dir, Bettis Atomic Power Lab; V P Atomic Power Divs; V P Engrg & Res; Group V P Elec Util Group; Pres Power Sys Co. 1977 to date Senior Associate Consultant Int. Energy Associates Limited. 1980 to Date President Simpson Business Services, Inc. Fellow IEEE, ANS & ASME, Member Science Advisory Com. Notre Dame U., Member D.O.4 Energy Research Advisory Board. Hon Mbr ASME; Mbr NAE; Edison Medal IEEE; Gold Medal for Advancement of Res ASM; Newcomen Gold Medal for Steam Franklin Inst; Former Pres ANS, Former Chmn Atomic Indus Forum. Former Chmn Internatl Energy Assocs Ltd George Washington Gold Medal ASME. Mbr Sci Advisory Com Notre Dame U, Mbr DO4 Energy Research Advisory Bd. *Society Aff:* IEEE, ASME, ANS, NAE.

Simpson, Ted L
Business: Coll of Engineering, Columbia, SC 29208
Position: Prof *Employer:* Univ of SC *Education:* PhD/Engr & Appl Phys/Harvard; MSEE/Elect Engr/Univ of TN; BSEE/Elect Engr/Univ of TN *Born:* 12/13/35 Native of Knoxville, TN. Joined Deco Electronics after graduating Univ of TN in 1956. Returned to UT for MS degree in 1958-59. Specialized in high power VLF transmitting antenna design. Continued res in antenna design at Harvard, received PhD in Engr & Applied Physics in 1970. Joined faculty in Coll of Engr at Univ of Sc in 1968. Appointed Assoc Dean in 1979, promoted to Prof same yr. *Society Aff:* IEEE, ASEE

Simpson, Thomas A
Business: Dept of Mineral Engg, PO Box 1468, University, AL 35486
Position: Assoc Prof, Emeritus *Employer:* Univ of AL. (Ret) *Education:* EM/Mining Engr/Univ of MO; MS/Geology/Univ of AL; BS/Mining Engr/MO School of Mines. *Born:* 10/23/25. Adams, MA. Served with USMC WWII, Pacific, 1943-46, and Korea 1951-52, retired Col USMCR, worked NJ Zinc Co, 1950 as Shoveler and Miner. Worked for USGS as mining engr and geologist, Tuscaloosa, Bessemer, AL, 1953-59; Hydrologist USGS Anthracite region, 1959-61; Chief Economic Geologist AL Geol Survey, 1961-63, Asst State Geologist, Programs and Plans, 1963-75; Lecturer in Geology, Univ of AL, 1966-74; Assoc Prof Dept Civ & Mineral Engg, The Univ of AL, 1975-78; Assoc Prof Mineral Engg, and Acting Hd, Dept Mineral Engr, 1978-79; Assoc Prof Dept Mineral Engg 1975-86 Retired, September 1, 1986 as Assoc Professor Emeritus; Mining Consultant Private Co, Surinam, SA, 1965; Hydrogeology Consultant, FAO-UN, Venezuela, SA, 1968; Mining and Geological Consultant to private individuals, cos 1975-present. Layreader Episcopal Church. Enjoy reading, hunting, travel (foreign & domestic). *Society Aff:* SME OF AIME, GSA, SEE, AIPG, AEG, AAPG, ORDER OF ENGINEERS, ΣΓΕ, ΣΞ.

Simrall, Harry C F
Home: 107 White Dr W, Starkville, MS 39759
Position: Dean Emeritus of the Coll of Engg. *Employer:* Mississippi State Univ. *Education:* MS/Elec Engg/Univ of IL; BS/Mech Engg/MS State Univ; BS/Elec Engg/MS State Univ; Grad Sty/Elec Engg/U of Pittsburgh. *Born:* Oct 16, 1912 Memphis Tenn. Employed by Miss State Univ for 44 yrs, incl Prof & Hd of EE for 10 yrs & Dean of the Coll of Engrg for 21 yrs. Pres-Elect 1969-70 & Pres 1970-71 of Natl Soc of Prof Engrs. V P 1972-74 of Natl Council of Engrg Examiners. Fellow & Life Mbr IEEE. Awarded Disting Serv Cert of Natl Council of Engrg Examiners 1971. Bd of Dir, Engrs' Council for Prof Dev 1971-74. *Society Aff:* IEEE, NSPE, IES, ASEE, ASAE.

Sims, Anker V
Business: 201 S Lake Ave, Pasadena, CA 91101
Position: V P, Consulting Services. *Employer:* The Ben Holt Co. *Education:* MS/Chem Eng/USC; BS/Chem/UC Berkeley. *Born:* 1924 Alameda Calif. MS from USC. BS Univ of Calif Berkeley. Work in R&D, design, econ & feasibility studies with The Ben Holt Co since 1962, V P since 1966. Reg Chem Engr. Mbr AIChE. *Society Aff:* AIChE.

Sims, Chester T
Dept. of Materials Engineering, Rensselaer Polytechnic Inst, Troy, N.Y 12180
Position: Prof, Materials Engrg. *Employer:* Rensselaer Polytechnic Inst. *Education:* MS/Met Engg/OH State Univ; BS/CE/Northeastern Univ. *Born:* 12/14/23. Winchester, MA. Served with Army Chem Corps 1942-45 & 1950-51. Met, then Asst Div Chief, Non-ferrous Phys Met, Battelle Inst Columbus, OH, 1947-49, 1951- 58. Joined Knolls Atomic Power Lab, Schenectady 1958-60 and then GE Mtls and Processes Lab, 1960-1974, as Mgr, Alloy & Joining Met. Then at GE Gas Turbine Div as Mgr, Advanced Mtls Progs, to 1981, when appointed Manager, Materials Information Systems, in the GE Research and Development Center. In 1984, appointed Mgr, Materials Info Syst, GE Corp Engr and Mftg. Now adjunct Prof, Materials Engrg, Rensselaer Polytechnic Inst., Troy, N.Y. Specialize in refractory, nuclear, and superalloy met, and materials databases. Author 70 papers and patents, and editor two books. Natl honors include; ASM Eisenman Award for application of fundamentals to practice in met, Fellow, ASM, Dedicatee, US-Brazil Conference on Role of Refractory Metals in Superalloys. Commodore, Lake George Corinthian Yacht Club. *Society Aff:* AIME, ASM, ASTM.

Sims, Craig S
Business: Dept of Elec Engrg, Morgantown, WV 26506
Position: Full Professor *Employer:* W VA Univ *Education:* PhD/EE/So Methodist Univ; MS/EE/Univ of AZ; BS/EE/Santa Clara Univ *Born:* 01/16/43 Dr Sims received his PhD in Elec Engrg from S Methodist Univ in 1970. He was an Asst Prof of Elec Engrg at OK State Univ from 1970 to 1974. From 1974 to 1981he was Assoc Prof at OK State Univ. He is currently a Prof of Elec Engrg at W VAUniv. His principal areas of interest are control systems, estimation theory, and the application of digital signal processing methods to seismic exploration problems. Dr Sims has more than thirty-five res publications. *Society Aff:* IEEE

Sims, E Ralph, Jr
114 Luther Lane, Lancaster, OH 43130
Position: Associate Professor Industrial & Systems Engrg & Chrmn: The Sims Consulting Group, Inc. *Employer:* Ohio Univ - Athens, OH. *Education:* BAdmE/Indust Engg/NY Univ; MBA/Mktg/OH Univ. *Born:* 10/12/20. Assoc Prof-Industrial Systems Enging, Ohio Univ, Prof background incls pos with cons, mfg, distrib, & res orgs. Has written 5 books, num articles & parts of the 'Production Handbook',

Sims, E Ralph, Jr (Continued)
'Matls Handling Handbook', 'Handbook of Bus Admin'. Encyclopedia of Prof Mgmt. Handbook of Indus Engrg, Warehousing Handbook. Mbr IIE (Fellow) ASME, NSPE, BIME, Fellow, AMC (PPres), IMC Director, AMA, NCPDM. *Society Aff:* IIE, ASME, NSPE, BIME, IMC.

Sims, James R
Home: 2532 Bluebonnet Blvd, Houston, TX 77030
Position: Herman & Geo R Brown Prof of CE. Emeritus *Employer:* Rice Univ. *Education:* PhD/Civil Engg/Univ of IL; MS/Civil Engg/Univ of IL; BSCE/Civil Engg/Rice Institute *Born:* 7/2/18. On fac of Rice Univ 1942-87: Chmn Dept of CE 1958-62, V P & Bus Mgr 1962-74, Herman & Geo R Brown Chair of C E 1974-87. Amer Soc of Civil Engrs, Dir 1966-68, V P 1970-71. Pres 1981-82. United Engineering Trustee 1983-87 Specialty Structural Engineering. *Society Aff:* ASCE, NSPE, ASTM.

Sims, Robert R
Business: 6017 Bristol Pkwy, Fox Hills Business Park, Culver City, CA 90230
Position: Vice Pres *Employer:* Engrg Service Corp *Education:* BS/CE/Univ of WI *Born:* 08/22/39 Native of Milwaukee, WI. Worked for Los Angeles County Road Dept after graduation followed by 5 yrs with a major Subdivision Consltg Engrg firm in Los Angeles. Exec VP and Gen Mgr of Guam office for ESCO Intl A/E in charge of several major hotel designs, Assoc Architect of 300 bed hospital, Subdivisions and Urban renewals for 3 yrs followed by taking present position as VP of Parent Corp, Engrg Service Corp, in Los Angeles. In charge of Land Planning Dept and Mktg. State Dir for CSPE. Reg Civil Engr in Guam, HI, and CA. *Society Aff:* ASCE, NSPE, CSPE, SMPS, PMI, APA

Sinclair, A Richard
Business: c/o Santrol Products, Inc, 11757 Katy Freeway, Suite 1260, Houston, TX 77079
Position: President-Santrol Products, Inc. *Employer:* Santrol Products, Inc *Education:* BS/Mech Engr/Univ of OK; ME/Mech Engr/Univ of OK; PE/Prof Engr TX/- *Born:* 2/5/40. Native of OK City, OK. Res Scientist for NASA - Ames 1964-67. Sr Res Specialist for Exxon Production Res 1967-76, working in Offshore Engrg, Nondestructive Testing, Acid and Hydraulic Stimulation, Instrumentation, fluid flow & heat transfer. Won 1971 SPE Cedric Ferguson Medal for best tech paper & in 1973 the AIME Rossiter Raymond Award for best tech paper by an author under 33, becoming the first person ever to win both awards. Joined Maurer Engrg Inc as Sr Engrg Assos June 1976 to 1981. Later became special projects manager for Maurer & in 1978 became President of Santrol Products Inc, Makers of Super Sand for oil & gas well sand control and hydraulic fracturing. Conslt and President of Well Stimulation, Inc 1981 to present. *Society Aff:* SPE, AAPG.

Sinclair, Charles S
Home: 5003 Bristol Rd, San Diego, CA 92116
Position: Naval Architect/Genl Mgr. *Employer:* SW Marine Architects & Engrs Inc. *Education:* BS/Naval Arch/Univ of MI. *Born:* Aug 1923. Sgt US Army WW II. BS Univ of Michigan 1950. Mbr Tau Beta Pi, Natl Engrg Honor Soc. Reg P E Calif. Mbr Soc of Naval Architects & Engrs and Chmn San Diego Sect 1970-71. Spec in the design & conversion of commercial & fishing vessels. Speak Spanish & have supervised ship const projs in Latin Amer. Founded Southwest Marine Archs & Engrs Inc 1969 to pursue area of specialization. *Society Aff:* SNAME.

Sinclair, Douglas C
Home: 20 Burncoat Way, Pittsford, NY 14534
Position: Professor. *Employer:* Univ of Rochester. *Education:* PhD/Optics/Univ of Rochester; BS/Physics/MIT. *Born:* July 13, 1938 Cambridge Mass. Married, 2 children. BS MIT 1960; PhD Univ of Rochester 1963 (Optics). US Army, 1st Lt 1963-65. Asst Prof Univ of Rochester 1965-67; Tech Dir Spectra-Physics 1967-69; Assc Prof Univ of Rochester 1969-75, Prof 1975- . Optical Soc Lomb Medalist 1968; Editor Journal of the Optical Soc of Amer 1976-78 . *Society Aff:* OSA.

Sinclair, George
Home: 25 Medalist Rd, North York City, Ont, Canada M2P 1Y3
Position: Chmn of the Bd. *Employer:* Sinclair Radio Labs Ltd. *Education:* PhD/EE/OH State; MSc/EE/Univ of Alberta; BSc/EE/Univ of Alberta. *Born:* 11/2/12. Organizer & first Dir of the Ohio State Univ Antenna Lab (now ElectroSci Lab). Dev antenna model techs now universally used by antenna designers. Dev CW tech for measuring radar reflectivity of targets. Prof of EE Univ of Toronto 1947-78. Org first PhD prog in EE. Org Sinclair Radio Labs Ltd 1951 & created the modern multicoupling indus. Dir IRE 1960-62; Dir IEEE 1964-66, 1969-70, 1974-75. Fellow IRE; Engrg Inst of Canada; Royal Soc of Canada, AAAS. Awarded McNaughton Medal, Region 7 IEEE 1975. DSc (Hon) OH State, 1973, Julian C. Smith Medal, EIC, 1980. *Society Aff:* IEEE, CSEE, ASEE, EIC, AAAS, NSPE.

Sinclair, George M
Business: Dept of Theo, Appl Mech, Urbana, IL 61801
Position: Prof of Theoretical & Appl Mech. *Employer:* Univ of Illinois. *Education:* MS/Met Engg/Univ of IL; BS/Met Engg/Univ of IL. *Born:* Mar 9, 1922. 1st Lt 75th Inf Div 1943-46 ETO. Mbr of the staff of the Dept of Theoretical & Applied Mech, Univ of Ill since 1949 with exception of 1 yr period 1951-52 when worked as res metallurgist for Westinghouse Res Labs. Cons to num indus & gov orgs on problems involving mech testing, fracture of metals, esp by fatigue. Joined ASME 1956, mbr & Chmn Metals Engrg Div 1964-68, Div Semi Centennial Award 1971, Fellow 1974, A Nadai Award 1975. Active Mbr ASM & ASTM since 1950; served as chmn of the latter's Res Subctte on Fatigue 1964- 68. Mbr Sigma Xi & Alpha Sigma Mu. Author or co-author approx 70 tech papers dealing with fatigue & failure of metals. Elected Fellow ASM 1981. *Society Aff:* ASME, ASM, ASTM.

Sinclair, Thomas L, Jr
Home: 1443 Ahuawa Loop, Honolulu, HI 96816
Position: Sr Partner (Self-employed) *Education:* MA/FE Studies/Harvard Univ; BA/History/Trinity Coll *Born:* 02/24/14 Born in Yanchow, China (Amer Missionary parents). USNR 1942-68, EnsCdr; PT Boats, Destroyers, Submarines; Mediterranean-So Pacific. 1949-53 Pres and Gen Mgr, Philippine Fleet Industries, Inc; Manila. 1954-55 Advisor, Chinese Navy, Taipeh, Taiwan. 1956-62 Sr Mbr, SEABEE Assocs, Manila. 1963-66 USN active duty, small craft combat operations, Washington, Pearl Harbor, Vietnam. 1967-77 Partner, Holbrook and Sinclair-Marine Surveys and Designs, Honolulu. 1978 present-Pres Sinclair and Assocs, Inc., Naval Architects, Honolulu, Hawaii. *Society Aff:* SNAME, SSCD, ABYC.

Sines, George H
Business: 6532-Boelter Hall, Sch of Engg, Los Angeles, CA 90024
Position: Prof. *Employer:* Univ of CA. *Education:* PhD/Engg/UCLA; MS/Engg/UCLA; BME/Engg/OH State Univ. *Born:* 7/12/23. Asst Prof of Met, Metals Inst, Univ of Chicago, 1953-56; Fulbright Res Prof 1958-59, Tokyo Inst of Tech, Sr NSF Postdoctoral Fellow 1965-66, Belgian Nucl Ctr; ASTM Templin Award; Co-editor, "Metal Fatigue–, McGraw-Hill, 1959, Author, "Elasticity and Strength–, Allyn-Bacon 1969; Co-editor Fourteenth National Symposium on Fracture Mechanics 1981; interests in the several aspects of the mech properties of matls - testing of ceramics and fatigue of metals under complex stress, micro-mech theories, Carbon-carbon composites, crystal imperfections, design to effectively utilize their characteristic properties, & failure analysis. *Society Aff:* ASME, ASM, ASTM, ACerS.

Sinfelt, John H
Business: Exxon Research & Engineering Co, Annandale, NJ 08801
Position: Sr Scientific Advisor. *Employer:* Exxon Res & Engrg Co. *Education:* PhD/Chem Engg/Univ of IL; MS/Chem Engg/Univ of IL; BS/Chem Engg/Penn State Univ. *Born:* 2/18/31. Honorary Dr of Sci, Univ of IL, 1981. Native of Munson PA. With Exxon Research & Engineering Co. Mbr Ntl Acad of Sciences (elected 1979). Mbr Natl Acad of Engrg (elected 1975). Fellow, Amer Acad of Arts

Sinfelt, John H (Continued)
& Sciences (Elected 1980). Recipient of many scientific awards, including: Emmett Award in Catalysis 1973. ACS Petro Chem Award 1976. Dickson Prize in Sci and Engrg, Carnegie- Mellon Univ 1977. Amer Physical Soc Intl Prize for New Matls 1978. President's Natl Medal of Sci, 1979. Perkin Medal Society of Chemical Industry, American Section, 1984. Gold Medal, American Inst of Chemists, 1984. *Society Aff:* ACS, AIChE, CATALYSIS SOCIETY.

Singer, Leslie J
Home: 404 Spring Creek Dr, Richardson, TX 75081
Position: Work Measurement Sec Mgr. *Employer:* Republic Natl Bank of Dallas. *Education:* Bachelor's/Engg Rt/Univ of TX; Assoc/Engg Option/Wharton Co Jr College. *Born:* 8/20/39. Serving as Dir of the IIE Banking & Financial Services Div through Mar 1980. He coordinated establishment of the div. He has also actively participated in IIE Dallas Chapter for eight yrs as a Sr Mbr. Mr Singer was involved in the Methods-Time-Measurement consortium effort which developed MTM-C (MTM Clerical Std Data). Prior to working in the banking industry, Mr Singer worked in the Ind Engg functions at the Boeing Co and the Gen Dynamics Corp. *Society Aff:* IIE.

Singer, Philip C
Business: Dept of Env Scis and Engrg, Univ of NC 201H, Chapel Hill, NC 27514
Position: Professor & Dir, Water Resources Engg Prog *Employer:* Univ of NC *Education:* PhD/Environ Sci and Engrg/Harvard Univ; SM/Environ Sci and Engrg/Harvard Univ; MS/Sanitary Eng/Northwestern Univ; B/CE/Cooper Union *Born:* 09/06/42 Born in Brooklyn, NY. Served as Asst Prof, Dept Civil Engrg, Univ of Notre Dame, 1969-73. Appointed to faculty of Dept of Environmental Sciences and Engrg, Univ of NC, 1973; Assoc Prof, 1973-78; Professor, 1978 to present; Dir of Water Resources Engrg Prog since 1979. Teaching and res interests in water and wastewater treatment, aquatic chemistry. Recipient of McGavran and Newton Underwood Awards for Excellence in Teaching, UNC; NALCO-AEEP Award as Adv of Best PhD Dissertation in Industrial Waste Treatment, 1980; Research Div Award for best Research Paper published by AWWA 1985. Numerous industrial and governmental consulting assignments. Mmbr, Bd of Trustees of Research Div AWWA; Mmbr Water Resources Research Cttee of the Water Sci and Tech Bd of the Natnl Research Council. Author of more than seventy res publications. *Society Aff:* ASCE, WPCF, AWWA, AEEP, ACS, IOA

Singer, S Fred
Business: U.S. Dept. of Transportation, 400 7th St., S.W. (DRP-2), Washington, DC 20590
Position: Chief Scientist *Employer:* U.S. Dept. Transportation *Education:* PhD/Physics/Princeton Univ; BEE/Electr. Engg/Ohio State Univ. *Born:* 9/27/24. US Navy 1944-46. Applied Physics Lab Johns Hopkins Univ 1946-50; Scientific Liaison Officer Office of Naval Res, London 1950-53; Prof of Physics Univ of MD 1953-61; Jet Propulsion Lab of CA Tech 1961-62; Dir of Natl Weather Satellite Ctr Dept of Commerce 1962-64; Dean of Environ & Planetary Scis Univ of Miami 1964-67; Deputy Assistant Secretary US Dept of Interior 1967-70; Deputy Assistant Admin Environ Protection Agency, and Brookings Inst 1971. Gold Medal of Dept of Commerce; Outstanding Young Men Award, Jr Chamber of Commerce. Intl Academy of Astronautics. Currently involved in analysis of energy policy, environ decision making & effects of population growth. Prof Environ Sciences, Univ VA, 1971-87. Chief Scientist, Dept. of Transportation, 1987- . *Society Aff:* AAAS, AGU.

Singh, Amarjit Dr
Home: 4311 Rowalt Drive, Apartment 203, College Park, MD 20740
Position: Visiting Scientist *Employer:* Univ MD. *Education:* PhD/Electron Physics/Harvard Univ; MEngg Sc/Electronics/Harvard Univ; MSc/Physics/Punjab Univ *Born:* 11/19/24. MSc 1945 Punjab Univ; MES 1947, PhD 1949 Harvard Univ; DSc (hon) Punjabi Univ. Publs: 60 papers, 1 bk. Taught electronics & history of sci at Univ of Delhi 1949-53; Proj Leader Microwave Tubes Natl Physical Lab, New Delhi 1953-57; Grp Leader Vacuum Tubes CEERI 1957-59; Heading CEERI 1959- . Visiting Scientist Univ of MI & Bell Tele Labs 1962-63. Director CEERI 1963-84; NCPC/(National Chief Project Coordinator) from 1984-87 UNDP Project CEERI; 1987 onward Visiting Scientist at University of Maryland, USA, Mbr, Res Council; Fellow IEEE (NY). Dist Fellow & SK Mitra Awardee Institution of Electronic & Telecommunication Engrs (New Delhi); Fellow Indian Acad of Scis. *Society Aff:* IETE, IAS

Singh, R Paul
Business: Dept of Agricultural Engrg, Univ of California, Davis, CA 95616
Position: Prof *Employer:* Univ of California *Education:* PhD/Agric Engr/MI State Univ; MS/Agric Engr/Univ of WI; BS/Agric Engr/Punjab Agric Univ *Born:* 06/08/49 Mbr of the Depts of Agric Engrg and Food Sci and Tech, Univ of CA, Davis. Coauthor of the textbooks *Food Process Engrg* and Intro to Food Engrg. Mbr of the bd of editors of Journal of Food Sci; co-editor of Journal of Food Process Engrg. Author of a patent and over 130 technical papers on food engrg topics. Awarded Samuel Cate Prescott Award for Research 1982 from IFT; and AW. Farrall Young Educator Award 1986, ASAE. Participated in several intl profl activities including an appointment by The Natl Acad of Scis to a panel to review *Post-Harvest Food Losses in Sri-Lanka.* Conslt to FAO and UNESCO of United Nations, NASA, NATO and private industry. *Society Aff:* ASAE, IFT

Singh, Rameshwar
Business: Dept of civ Engg & Appl Mechanics, San Jose, CA 95192
Position: Prof. *Employer:* San Jose State Univ. *Education:* PhD/Civil Eng/Stanford Univ; MS/Civil Eng/Auburn Univ; BCE/Civil Eng/Auburn Univ *Born:* 7/2/37. Born in Bihar, India & had my high sch & initial engg education there. I worked for the state govt for about 4 1/2 yrs as an engr-in-charge on various irrigation & river valley projs. I came to the US on Dec 30, 1960 as a student in Auburn Univ in AL. Later I entered the grad sch in Stanford Univ where I received the degree of PhD in Oct, 1965. I taught for two yrs as an asst prof at the Univ of British Columbia in Canada before joining San Jose State Univ in Sept, 1967. I am a prof of Civ Engg & Appl Mechanics. I have authored over a dozen scholarly papers. I am a reg PE in CA & California State President of Natl Soc of PE. I actively consult with local engg firms. I have attended & set up many seminars & continuing education progs. *Society Aff:* NSPE, ASCE

Singhal, Ashok K
Business: 3983 Res Park Dr, Ann Arbor, MI 48104
Position: VP/Dir of Engrg *Employer:* Ayres, Lewis, Norris & May, Inc. *Education:* MBA/Finance/EMU, Ypsilanti, MI; MS/Civil/Univ of CA at Berkeley; BS/Civil/IIT, Kanpur *Born:* 4/8/46. Completed a bachelor's in Civ Engg from IIT Kanpur, India in 1967, & a MS from Univ of CA at Berkeley in Dec, 1968. Started working with Ayres, Lewis, Norris & May, Inc, in Feb, 1969. Obtained profl reg in 1972. Became an Assoc and Hd of the Systems & Studies Dept in July, 1974. Became a Principal in the firm in November, 1980. Have published several articles in AWWA, & 1975 & 1979 Excellence in Engg Design issues of Water & Wastes Engg. Presented a paper at 52nd annual WPCF Conf. Listed in Who's Who in MI, Who's Who n Midwest, & outstanding American. Married to Veena Singhal in July, 1975. Daughter Neha Singhal (born Sept 78) and son Nitras Singhal (March 80). *Society Aff:* AWWA, WPCF

Singpurwalla, Nozer D
Business: The George Washington Univ, Dept of Operations Res, Washington, DC 20052
Position: Prof. *Employer:* George Washington Univ. *Education:* PhD/Statistics/NY Univ; MS/Industr Engr/Rutgers Univ; BS/Mech Engr/BVB College, India. *Born:* 4/8/39. Native of Hubli, India. Parents: Darabsha B Singpurwalla and Goolan N Engr. Prof of operations Res & Prof of Statistics, Director, The Institute of Reliability and Risk Analysis, George Washington Univ. Visiting Prof of Statistics, Stanford

Singpurwalla, Nozer D (Continued)
Univ (1978-1979). Visiting Prof Bhabha Atomic Res Centre, India (1975). Published Several Basic Methodological Papers in reliability, and a standard reference book in reliability.

Sinner, Robert D
Business: P O Box 341, Baton Rouge, LA 70821
Position: General Manager-Engrg, R&D Dept *Employer:* Ethyl Corporation. *Education:* MS/Chem Eng/IA State Univ; BS/Chem Eng/IA State Univ. *Born:* Dec 1, 1928. Joined Ethyl Corp 1951; became responsible for plant Tech Servs 1964, Process Design 1966. Became Tech Dir-Engrg in 1974; Assumed present position in 1982; responsibilities include direction of tech activities for process design, process dev, process eval, and plant tech servs. *Society Aff:* AIChE, AXΣ.

Sinnott, Walter B
Business: 35 Danner Ave, Harrison, NY 10528
Position: Sr VP. *Employer:* Hazen & Sawyer, PC. *Education:* MCE/Sanitary Engg/NYU; BCE/Civ Engg/Cooper Union. *Born:* 12/6/30. Officer in USAF in 1952-53. Joined Hazen & Sawyer in 1953, became partner in 1964 & Sr VP upon inc of firm in 1978. Has directed planning & design of maj municipal water supply & treatment & wastewater collection & treatment projs. Also directed financial & rate studies for established & new municipal agencies. Author of numerous technical papers. Received Fuller Award from NYS Section of AWWA in 1977. Mbr of Town of Harrison Zoning Bd of Appeals for 11 yrs, including chrmn for two yrs. Past Mbr and Chairman of Engg & Construction Committee of AWWA. *Society Aff:* ASCE, AWWA, WPCF, NSPE, AAEE

Sipress, Jack M
Business: AT&T Bell Laboratories, Holmdel, NJ 07733
Position: Director *Employer:* AT&T Bell Laboratories. *Education:* DEE/Elec Engr/Poly Inst of Brooklyn; MEE/Elec Engr/Poly Inst of Brooklyn; BEE/Elec Engg/Poly Inst of Brooklyn. *Born:* 4/9/35. in Brooklyn, NY. Joined AT&T Bell Labs 1958. Currently Dir respon for dev of undersea communications sys. Previous activities include dept hd & supervisory responsibilities for theoretical studies, sys planning & dev for digital transmission sys containing lightguide, coaxial cable, paired cable & waveguide media; satellite communication sys; & exploratory dev of active networks. 9 patents & sev publs in digital transmission & active networks. Fellow IEEE; P Mbr Bd of Govenors of IEEE Communications Soc, P Mbr IEEE Publications Bd, Mbr IEEE Awards Bd, Tau Beta Pi, Eta Kappa Nu, Sigma Xi. P Chmn of Holmdel, NJ Planning Bd. *Society Aff:* IEEE

Sirignano, William A
Business: Schl of Engrg, Irvine, CA 92717
Position: Dean, School of Engrg *Employer:* Univ of CA-Irvine *Education:* PhD/Aerosp Sciences/Princeton Univ; MA/Aeronau Engr/Princeton Univ; BAeroEngr/Aeronau Engr/RPI *Born:* 04/14/38 Native of NY. After completing PhD requirements at Princeton, was invited to remain as Res Staff Mbr and Lectr, appointed to Asst Prof in 1967 then promoted to rank of Assoc Prof in 1969; promoted to Prof in 1973. Transfd to Carnegie-Mellon Univ and received apptmnts as George Tallman Ladd Prof (a distinguished endowed prof) and as Head of the Dept of Mech Engrg. Performed res on a wide variety of physical phenomena in the gen area of combustion, nonsteady fluid mech, transient heat transf and aerosp propulsion. The res has been motivated by the practical issues of space flight, air flight, noise pollution, air pollution, fire safety, and energy conservation and has resulted in about 180 publ and 145 sem on three continents. Has been invited to serve as advisor and conslt to indus, govt agencies, and intl organizations. Served as Treas and Dir of Combustion Inst. *Society Aff:* CI, AIAA, ASME, SIAM, ASEE

Sirjord, Arthur B
Business: 215 Columbia, Seattle, WA 98104
Position: Partner/Dir of Engg. *Employer:* TRA-Arch-Engrs-Planners. *Education:* BS/ME/Univ of Wa. *Born:* 9/20/22 Native of Seattle. US Army Air Corps 1943-46, USAF 1950-51, Carrier Corp 1951-1954. Proj Engr, John Graham & Co, 1954-59; Assoc, J B Notkin & Assoc 1959- 1967, Sr Assoc & Partner TRA, 1967 to present. Assumed partnership in 1968 and served as Partner/Dir of Engg since 1975. Founded Engg Consultants Inc in 1975, and has served as its pres. ECI offers mech, elec, struct & civ engg services to clients outside of TRA. Engineer of the year (1980) Seattle Chapter WSPE. *Society Aff:* ASHRAE, ASME, NSPE, AEE, NFPA, AFEA, BOMA, CECW, ISA, WSPE.

Sirkin, Alan N
Business: One Lincoln Rd Bldg, Miami Beach, FL 33139
Position: Principal. *Employer:* Alan Sirkin, PE. *Education:* MS/Civ Engg/GA Inst of Tech; BS/Civ Engg/Univ of Miami. *Born:* 6/3/44. Pres, Sirkin Bldg Corp/Ex-Dir, Builders Assn of S FL/Dade County Environmental Quality Control Bd/Reg PE/Certified Contractor/Who's Who in Finance & Industry/Dir, VP Univ of Miami Coll of Engg/Various publications, Panel of Arbitrators of the Americon Arbitration Assoc., Consilator for the Home Owners Warranty Proram, Manuscript Reviewer ASCE, Advisor to the Univ. of Miami Civil Engineering Dept., Various Advisory positions, Dir Sierra Club Miami Grp. *Society Aff:* ASCE, AAAS.

Sissom, Leighton E
Business: Box 5005, T.T.U, Cookeville, TN 38505
Position: Dean of Engg. *Employer:* TN Tech Univ. *Education:* PhD/Mech Engg/GA Inst of Tech; MSME/Mech Engg/GA Inst of Tech; BSME/Mech Engg/TN Tech Univ; BS/Industrial Arts/Middle TN State Univ. *Born:* 8/26/34. Native of TN. Educator, author, consultant. Before joining academy, served Chrysler Corp, Westinghouse Electric, and ARO, Inc. Prior to becoming Dean of Engg, was Prof and Chrmn of Mech Engg at TN Tech Univ for 14 yrs. Cons to over 300 industries, insurance cos. law firms and govt agencies on fluid handling, energy utilization, products liability and accident reconstruction. Co-authored three books, translated into several languages, and written over 60 papers. Engineering evaluator for two accrediting bodies and reviewer for several technical journals. Board of Directors, Amer Soc for Engrg Ed. Board of Governor, Amer Soc of Mech Engrs., JETS and Order of the Engineer. Chrm, Engrg Deans Council. Vice Pres., Tau Beta Pi. Nontechnical interests: biorhythm and hypnosis (theory and practice). *Society Aff:* ASME, NSPE, ASEE, SAE, SSS, ASSE, NAFE.

Sisto, Fernando
Business: Mech Engg Dept, Stevens Institute of Technology, Hoboken, NJ 07030
Position: George M. Bond Professor of Mech Engg *Employer:* Stevens Inst of Tech. *Education:* ScD/Aeronautical Engg/MA Inst of Tech; BS/Engg/US Naval Academy *Born:* 8/2/24. Graduated US Naval Academy, a 1946 classmate of Jimmy Carter. Attended MIT, held duPont fellowship and earned 1952 Doctorate in Aeronautical Engg. Six yrs at Curtiss-Wright Res culminated in Propulsion Div Chief position. Stevens Inst of Tech academic appointments began in 1958, was Mech Engg Dept

Sisto, Fernando (Continued)
Hd from 1966 to 1979 and holds George Meade Bond Chair in Mech Engg. Organized Navy/AF/NASA Symposium on Aeroelasticity of Turbine Engines in 1980. Invited lecturer in PRC in 1981. Fellow ASME, Mbr Sigma Xi and Assoc Fellow AIAA. Cons to Westinghouse, United Technologies, Gen Elec, etc and UNESCO/UNDP cons to Natl Aeronautical Lab, India. *Society Aff:* ASME, AIAA, ΣΞ, ASEE, APS.

Sitterly, Charlotte M
Home: 3711 Brandywine St, NW, Washington, DC 20016
Position: Ret. (Res Physicist) *Education:* PhD/Astronomy/Univ CA Berkeley; BA/Math and Astronomy/Swarthmore College *Born:* 9/24/98. BA 1920 Swarthmore; PhD 1931 Univ of CA Berkeley. m. Bancroft Walker Sitterly 1937. Hon doctorates in sci Swarthmore, Univ of Kiel, MI. foreign assoc, Royal Astronomical Soc. Phi Beta Kappa, Sigma Xi. Federal Womens Award, Meggers Award Optical Soc America. P VP American Astronomical Soc, AAAS Sect D; PPres IAU Common "Fundamental Spectroscopic Data–. Specialities: Critical evaluation of data on Selected Atomic Spectra in the visible & ultraviolet regions. Respon for prog on refernce data available on indivudual atomic spectra; this prog is of internatl scope & involves coordination or res in various labs as well as consultation service on the interpretation of spectra of astrophysical interest. In solar physics, the interpretation of various types of solar spectra. *Society Aff:* OSA, APS, AAS, AAAS.

Sittig, Erhard K
Home: In Johannistal 41, 5600 Wuppertal 1, W Germany
Position: Professor. *Employer:* Gesamthochschule Wuppertal. *Education:* Dr rer nat/Physics/Tech Univ Stuttgart; Dipl/Physics/Univ Tubingen; Dipl Imp Coll/Univ of London. *Born:* June 3, 1928 Konigsberg. Diploma in Physics Univ Tubingen; Diploma Imper Coll London; PhD in Physics 1959 Stuttgart. Asst Res Physicist UCLA 1959-61; Senior Scientist Durabond Bearing Co, Palo Alto Calif 1961-63; Mbr of Staff & Supervisor Bell Labs, Murray Hill NJ 1963-76; Prof of Electronics, GH Wuppertal 1976- . Mbr Optical Soc of America; Fellow IEEE. Chmn Group Sonics & Ultrason 1969. *Society Aff:* IEEE, OSA.

Sivazlian, Boghos D
Business: Ind & Systems Engg, Gainesville, FL 32611
Position: Prof. *Employer:* Univ of FL. *Education:* PhD/Operations Res/Case Western Reserve Univ; MS/Operations Res/Case Western Reserve Univ; BS/ME/Cairo Univ. *Born:* 2/11/36. 1962-1965 Mgt Scientist at the B F Goodrich Co Akron, OH. 1966-present Dept of Ind & Systems Engg the Univ of FL. Consultant US Army. 1973-74 Natl Acad of Sciences exchange scholar to Poland. 1980 Fulbright scholar to USSR. *Society Aff:* ORSA, IIE, SIAM, ASEE.

Siver, Dougal H
Home: 212 Orchard Dr W, North Syracuse, NY 13212
Position: Retired. *Born:* March 1916. Native of Watertown NY. Syracuse Univ, Cornell Univ. US Army 1944-46. Several years in field of high pressure hydraulics design & production engrg. 25 years standardization related exper in defense & consumer prod indus including tech papers on standardization for Natl Fluid Power Assn, SME, SAVE, Brigham Young Univ, AFNOR (France). Served on various ANSI cttes, Bd of Counselors ASTM Central NY, Tech Vice Pres & Fellow SES. Participated in organization of Internatl Federation of Standards Engrs, Paris 1974.

Sivetz, Michael
Business: 349 SW 4th ST, Corvallis, OR 97333
Position: Pres. *Employer:* Sivetz Coffee Ent Inc. *Education:* MCE/ChE/Northwestern Tech Inst; BS/CE/Brooklyn Ploy Inst *Born:* 8/17/21. Authority & expert in instant coffee mfg. Author of three books on coffee tech, 1963, 1984 & 1979. Pres & founder of Sivetz Coffee Ent, Inc a coffee consulting & engr firm & mfg of a patented US 3,964,175 fluid bed roasting machine and After-burners suitable for drying too & sold primarily for coffee beans. Formerly proj engr or technocal dir at OR Freeze-dry foods, Am Magnesium, Kaiser Al and Chem, Cia Cacique -Brazil, Cafe Soluble, Nicaragua; Folgers Coffee-Houston, General Foods Corp, the Coca Cola Export Corp & Argonne Natl Lab with extensive travel through & living in South & Central Am, & world wide consulting on coffee subjs. *Society Aff:* AIChE, IFT.

Sivier, Kenneth R
Business: University of Illinois, 102 Transportation Bldg, 104 S. Mathews Ave, Urbana, IL 61801
Position: Assoc Prof *Employer:* Univ of IL *Education:* PhD/Aerospace Engrg/Univ of MI; MS/Aeronautical/Princeton Univ; BSE/Aeronautical/Univ of MI; BSE/Math/Univ of MI *Born:* 12/10/28 Native of Standish, MI. Aeronautical engr with ARO, Inc, AEDC, Tullahoma, TN, 1951-1953. With McDonnell Aircraft, St Louis, 1955-62; Engr in Aerodynamics Dept of Missile Engrg Div 1955-58; Senior and Senior Group Engr in Engrg Labs, 1958- 62, specializing in the development of advanced aerodynamic test facilities. Assoc Research Engr with Propulsion Lab, Univ of MI, 1962-67. With Aeronautical and Astronautical Engrg Dept, Univ of IL since 1967. Teach applied and experimental aerodynamics, flight mechanics, and aircraft design. Additional interest areas include computer aided design (applied to engrg education) and wind energy. Pres, IL Section of AIAA, 1980-81. *Society Aff:* AIAA.

Sivley, Willard E
Home: 937 Frazier Dr, Walla Walla, WA 99362
Position: Chief, Engrg Div, Retired *Employer:* US Army Corps of Engrs, Walla Walla Dist *Education:* MS/Conservation/Univ of MI; BS/CE/Univ of WI *Born:* 1/10/23 Thirty years experience with Corps of Engrs in the planning, design, and construction of large multiple purpose hydro-navigation and flood control projects on the Columbia and Snake Rivers and tributaries. Chief of Engrg Division for eight years in charge of Walla Walla District, Corps of Engrs planning and design programs. Proposed and implemented many environmental programs concerning anadromous fishery. Recipient of the Dept of Army's Meritorious Civilian Service Award. Reg P E, WA state. Past Pres Columbia Section, ASCE. Past Pres Walla Walla Chapter, WSPE. Enjoy golf and fishing. *Society Aff:* ASCE

Sizemore, Earl W
Business: 1387 South Fourth St, Louisville, KY 40208
Position: Managing Dir Louisville Office *Employer:* Miller/Wihry/Lee, Inc *Education:* MS/CE/Univ of KY; BS/CE/Univ of KY *Born:* 10/7/43 Born in Clay County, KY. Son of Jimmy and Lucy White Sizemore. Served with US Public Health Service Commissioned Corps 1968-1970. Employed by KY Dept of Hgwys, Division of Planning 1970-71. Joined Miller/Wihry/Lee, Inc, in 1971. Currently VP and Mgr of Louisville office; responsible for Civil Engrg, Land Surveying and Landscape Architecture Divisions. Specialized in land use planning and design including residential, commercial, industrial and recreational developments. Served as Pres, Louisville Chapter of KY Society of PE 1980-81; Pres, Louisville Engrs and Surveyors Assoc 1979. Married Phyllis Moore. Children: Karen, Peggy, Jill. Enjoy fishing and bicycling. *Society Aff:* NSPE, ASCE

Sizer, Phillip S
Home: 14127 Tanglewood Dr, Dallas, TX 75234
Position: Senior Vice Pres - Tech Dir. *Employer:* Otis Engineering Corp. *Education:* BS/ME/South Method Univ. *Born:* 4/11/26. Tau Beta Pi. Native of Whittier, Calif. Employed by Otis June 1948 as Engr Trainee. Held sev positions to present Sr VP - Tech Dir & mbr Bd/Dir Otis Engrg. Author & co-author of 6 papers; holder of over 55 patents, oil field related. Reg PE in Texas, OK. Licensed in Alberta. Active in ASME Petroleum Div since 1957. Chmn of the Exec Ctte 1974-75. Mbr OTC Exec Ctte 1976- 79. Mbr: ASME codes & STDS Comm 1978-83, Chrmn ANSI/ASME-SPPE (1982-). Mbr-at- Large Indus Dept Policy Bd 1976-79. Chmn, Bd on Accreditation 1981-83 (ASME). Engr of the Year' ASME North Texas Sect 1971. ASME Fellow 1972. ASME Centennial Medal. ASME Dedicated Service Award 1985. Petroleum Div Award 1982, U of TX, Arlington - Hall of Achievement 1983. Active in SPE of AIME, Dir of Dallas Sect 1974-77. Pres of Assoc of Well Hd Equipment

Sizer, Phillip S (Continued)
Mfgs 1976, Mbr Nomads, Kappa Sigma, Who's Who in America (1979-). Who's Who in World (1980-). *Society Aff:* ASME, NOMADS, SPE/AIME, TBΠ, KME

Sjogren, Charles Norman
Home: 504 N Gerona, San Gabriel, CA 91775
Position: Consulting Engineer. *Employer:* Self-employed. *Education:* PhD/Physical Chem/MIT; BS/Chemistry/Univ of MA. *Born:* April 1914. Reg Prof Chem Engr Calif. Teach Fellow & Instructor MIT 1936-40; Union Oil of Calif 1940-48; C F Braun & Co, Staff Cons & Dir of Res 1948-52, Head of Cons Engrg & Res Div 1952-57; Chemet Engrs Inc 1957-72, Pres & Owner; Cons Engr 1972- . Mbr AIChE, ASME, ACS, ASTM. Honorary Soc: Sigma Xi, Phi Kappa Phi. Engaged in wide variety of cons including R&D progs, process design, unit operations, machine & equipment design, & patent litigation. *Society Aff:* AIChE, ASME, ASTM, ACS.

Skalak, Richard
Business: Seeley W. Mudd Bldg, Rm. 610, Columbia Univ, New York, NY 07605
Position: James Kip Finch Prof of Engrg Mechanics *Employer:* Columbia Univ *Education:* PhD/Eng Mech/Columbia Univ; CE/-/Columbia Univ; BS/CE/Columbia Univ *Born:* 02/05/23 in New York City. Radar & Sonar Instructor, US Navy, 1944-46. Married two sons; two daughters. NSF Postdoctoral Fellow, 1960-61, Cambridge Univ, Eng, Dept of Theoretical Physics & Applied Math. NSF Sr Postdoctoral Fellow, 1967-68, Univ of Gothenburg, Sweden, Dept of Anatomy. Visiting Prof, Brown Univ, Div of Engrg 6mos, 1979. Res includes water-hammer theory, non-linear surface waves, and mechanics of blood flow. Over 100 publications on fluid mech and blood flow. *Society Aff:* AAAS, ASCE, ASME, ASEE, TBΠ, ΣΞ, APS, AAM, SES.

Skalnik, John G
Home: 567 Ronda Drive, Santa Barbara, CA 93111
Position: Prof of Electrical Engg. *Employer:* University of California. *Education:* PhD/EE/Yale Univ; MS/EE/Yale Univ; BS/EE/OK State Univ. *Born:* May 30, 1923. Asst & Assoc Prof Yale Univ 1944-65; Prof of Electrical Engrg Univ of Calif, Santa Barbara 1965- : Chmn Electrical Engrg 1968-71, Dean College of Enngrg 1971- . Coauthor of 'Microwave Theory & Techniques,'Microwave Principles,' & 'Theory & Applications of Active Devices.' Mbr IEEE, Electrochemical Soc, American Soc for engrg Education, & Sigma Xi. *Society Aff:* IEEE, ECS, ASEE, ΣΞ.

Skaperdas, George T
Home: 14 Wychview Dr, Westfield, NJ 07090
Position: President *Employer:* George T. Skapaerdas, P.C. *Education:* ScD/Chem Engg/MIT; SM/Chem Engg/MIT; BEngg/Chem Engg/McGill Univ. *Born:* Jan 25, 1914 New York, NY. With George T. Skaperdas, P.C. President 1980-1987. With M. W. Kellogg 1940-79. Senior Cons at Kellogg (1976-1979), Dir of Coal Dev (1975-76) & Mgr of Process Dept (1973-74). Respon for evaluation & dev of novel tech in energy, chem & petroleum fields including coal combustion & gasification, hydrocarbon synthesis (Synthol) methanol fuel synthesis & manufacture of hydrogen, oxygen, acetylene, & herbicide. Adjunct Prof of Chem Engrg NYU (Evening Div) 1947-53. Fellow AIChE (Admissions, Awards, Prog, Res Cttes). Mbr ACS, Sigma Xi. Reg PE in New York and New Jersey. 14 patents; 26 publs. *Society Aff:* AIChE, ACS

Skarda, James J
Home: 303 Robin Dr, Berea, OH 44017
Position: Service Metallurgist. *Employer:* Rolled Alloys Inc. *Education:* BS/Metallurical Engg/Univ of IL. *Born:* July 1926. Native of Chicago Ill; resided in Cleveland Ohio 1951- . Specialize in heat resisting alloys in 600 to 1250 degrees Centigrade. Teach metallurgy courses for ASM affiliate, Met Engrg Inst. Chmn Cleveland Chap ASM 1972, presently active on Natl cttes. Vice Chmn Cleveland Engrg Soc, sales marketing div. Enjoy bridge, chess, & golf. Chmn Bd of Education St Mary's Elementary Sch; Genl Chmn Natl & Internatl Convention 1976. *Society Aff:* ASM, CES.

Skeat, William O
Home: 32 Russell Rd, London, England W14 8HU
Position: Honorary Advisor. *Employer:* Cons Panel Pres British Trspt Relics. *Born:* Aug 9, 1904. BSc (Eng.) King's College London. Native of St Albans Hertfordshire. Practical training in locomotive workshops of Great Northern Railway. Engrg studies at London Univ 1929-33. Asst Editor, later Editor, Institution of Mech Engrs 1933-47; Editor 'British Sci News' (British Council) 1947-50; Secy & Editor Institution of Water Engrs 1950-69. Barry Prize King's Coll London 1932; Graduates Prize Mech Engrg 1940. Author of papers & historical memoranda on water supply; Editor 'Manual of British Water Supply Practice'; Hon Mbr American Water Works Assoc 1965, Inst Water Engrg 1969. Author 'George Stephenson: the Engr & His Letters' 1973. Vice Pres Stephenson Locomotive Soc. Fellow of King's Coll, London. *Society Aff:* American Water Works Assoc.

Skelly, Michael J
Business: One Blue Hill Plaza, Pearl River, NY 10965
Position: Partner *Employer:* Lawler, Matusky & Skelly Engr *Education:* PhD/Environ Engrg/Cornell Univ; ME/Sanitary Engr/Manhattan Coll; BCE/CE/Manhattan Coll *Born:* 4/5/41 Native of New York City. Joined present firm in 1968 and served as project engr, project mgr and Mgr of Environmental Group from 1968 to 1973. Admitted to partnership in 1973 and currently serves as Managing Partner. Served as consultant in solid wastes to USPHS and EPA from 1968-72. Chrmn of Met Section ASCE Energy Group, 1978-81; Dir of Met Section of ASCE, 1981-87: Treas of Met Soc, 1983-85; VP of Met Soc, 1985-86; Pres-Elect of Met Soc, 1986-87; Currently Pres, Met Soc.; Dir of Met Soc of NY Water Pollution Control Federation, 1983-86; Certified by Natl Council of Engrg Examiners., Mbr. of the Exec. Com. of the Energy Div ASCE and current chairman, 1984-88; Mbr. of the ASCE Cttee. on National Energy Policy 1987-88. *Society Aff:* ASCE, WPCF, NSPE, SETAC, IAGLR

Skerkoski, Eugene C
Business: 3605 Warrensvlle Ctr Rd, Cleveland, OH 44122
Position: Principal. *Employer:* Dalton-Dalton-Newport. *Education:* BME/Thermo-Dynamics/Cleveland Status Univ. *Born:* March 15, 1927. Tau Beta Pi. Mech Engr for Austin Co, Osborn Engrg, Avery Engrg, & Dalton-Dalton-Little-Newport. Specializing in design of mech systems inherent in building design- commercial, institutional, insustrial. Twice recipient of top award in natl competition for ingenuity in design. Mbr ASHRAE, ACEC, CECO, CCEA, NSPE, OSPE, CSPE, APTA, CIAC. Served on local &/or natl level as officer or Ctte for ACEC, CCEA, CSPE, APTA. Adv Council: Booth Memorial Hospital (Salvation Army). *Society Aff:* TBTT, ACEC, NSPE, ASHRDE.

Skidmore, Duane R
Home: 960 Lynbrook Rd, Worthington, OH 43085
Position: Prof of Chem Engrg *Employer:* OH State Univ *Education:* PhD/Physical Chem/Fordham Univ; MS/ChE/Univ of IL; PhL/Philosphy/St Louis Univ; BS/ChE/Univ of ND *Born:* 3/5/27 in Seattle, WA and reared in East Grand Forks, MN. Duane R. Skidmore became interested in learning at Sacred Heart School. He graduated in 1944 and after a year in coll as a member of the Enlisted Reserves Corps, was called to active Army service in 1945. He served until November 1946 and re-entered coll at the Univ of ND. His work experience includes industrial experience with E.I. Dupont for 3 1/2 years and faculty service at the Univ of ND for 8 years and WV Univ for 6 years. He works in coal research, publishes, consults, and travels extensively. *Society Aff:* AIChE, ACS, AIME, AAAS

Skilling, Hugh H
Business: Electrical Engrg Dept, Stanford, CA 94305
Position: Professor of Electrical Engineering. *Employer:* Stanford University. *Education:* PhD/EE-Mech/Stanford; SM/EE/MT; Eng/EE/Stanford; BS/EE/Stanford.

Skilling, Hugh H (Continued)
Born: 1905 San Diego. Fellow IEEE, AAAS; Mbr ASEE, Sigma Xi, Phi Beta Kappa. Received IEEE Award & Medal for Education 1965. Mbr King's Coll, Cambridge. Taught electrical engrg Stanford 1929- . Head of Dept 1940-64; visiting at MIT, Cambridge Univ England. Author, books: 'Transient Electric Currents,' 'Fundamentals of Electric Waves,' 'Exploring Electricity,' &'Electric Transmission Lines,' 'Electrical Engrg Circuits,' 'Electromechanics,' 'Do You Teach,' 'Electric Networks,' & 'Teaching.' (Sev books translated into various foreign languages) Author numerous papers, & articles for encyclopedias. Cons in Spain, Alaska, Chile, Hawaii, Philippines, etc. Scientific observer Bikini tests 1946.

Skilling, James K
Home: RD 1, Box 690, Harvard, MA 01451
Position: Staff Scientist. *Employer:* Genrad Inc. *Education:* BSEE/Engg/Univ of CA; MSEE/Engg/Johns Hopkins Univ. *Born:* 5/21/31. Taught electronics at the US Naval Acad while serving in the Navy. Mbr of engg staff of GenRad Inc since 1959, currently as staff scientist in the Board Test Division Designed electronic instruments, including pulse generators & frequency meters; patents on circuits, instruments, and test systems. Principal in dev of comp controlled systems for test of digital circuits. Activities ctr on the applications strategy & design of test systems. Active in IEEE: Stds group, local sections, & the Soc of Instrumentation & Measurement (pres 1979). *Society Aff:* AAAS, ΣΞ, IEEE.

Skilling, John B
Business: 1215 4th Ave, Seattle, WA 98161
Position: Chairman/CEO *Employer:* Skilling Ward Magnusson-Barkshire, Inc. *Education:* BSCE/Civil Engr/Univ of Washington *Born:* 10/8/21. BSCE Univ of Wash. Mbr Natl Acad of Engrg, IASS, SAME, SEAW, IABSE, ACI; Fellow ASCE. Principal Skilling Ward Magnusson Barkshire, Inc, cons structural & civil engrs. In 38 years with firm, has been personally responsible for the structural design of many of the most significant structures in the US including the 110-story twin tower World Trade Center in New York. Elected in the first election of the Natl Acad of Engrg 1965. Received Allied Professions Medal by AIA. Honorary membership in Seattle Chapter American Institute of Architects-1984. Reg PE in 21 states, NEC certified. *Society Aff:* IASS, SAME, SEAW, IABSE, ACI, ASCE.

Skinner, Bruce C
Home: 6 Brookfield Dr, E Lyme, CT 06333
Position: Head, Dept. of Engrg *Employer:* US Coast Guard Acad. *Education:* Naval Engr/-/MIT; MS/ME/MIT; BS/Gen Engr/US Coast Guard Acad. *Born:* 1/23/38. Coast Guard Officer in deck & engg billets aboard ship 1959-1962. Commanding Officer of isolated 30 man Coast Guard Loran Statin in AK 62-63. Staff Naval Arch - Technical Div - Merchant Marine Safety Div of US Coast Guard Hdquarters, Wash DC, 1965-1967. Prof in Dept of Appl Engg, USCG Acad, 1967-present. Currently Head, Department of Engrg VP-Soc of Naval Arch & Marine Engrs. *Society Aff:* SNAME, ASEE, TBΠ, ASNE.

Skinner, Willis D
Business: 45 School St, Boston, MA 02108
Position: Partner *Employer:* Black & Veatch *Education:* BS/CE/KS St Univ *Born:* 02/09/32 Native of Council Grove, KS. With Black & Veatch since 1955, specializing in water supply, treatment and distribution. Assigned to Black & Veatch Internatl from 1972 to 1980, serving as a vp latter three yrs, resp for business dev and mgmt of overseas civil/environmental projs. Admitted to partnership January 1981, and assigned to Northeast Office in Boston, MA as office manager. *Society Aff:* AWWA, AIDIS, NSPE

Skjei, Roger E
Business: 130 Jessie St, San Francisco, CA 94105
Position: Vice President. *Employer:* URS/John A Blume & Assocs, Engrs. *Born:* Jan 1, 1922. MS Engrg Sci, BS Geological Engrg Univ of Calif Berkeley. Vice Pres URS/Blume 1966- , Geological Engr 1959-66. Supervise a staff of research engrs, earth scis specialists, & a computer group. Also Asst Proj Mgr for long-range structural response studies being conducted for the US Energy Res & Dev Admin at its Nevada Operations Office. Have participated in coastal engrg & waterfront site investigation studies in Calif & South America, including environ studies for marinas & commercial waterfront devs, meteorological, oceanographic, & geologic studies, & wind, wave, & swell forecasting.g. Fellow Earthquake Engrg Res Inst. Seismological Soc of America. Assoc of Engrg Geologists. Tau Beta Pi. Sigma Xi.

Sklansky, Jack
Business: Dept of EE, Irvine, CA 92717
Position: Prof. *Employer:* Univ of CA. *Education:* Eng ScD/EE/Columbia Univ; MSEE/EE/Purdue Univ; BEE/EE/CCNY *Born:* 11/15/28. Prof of EE, Info & Comp Sci, & Radiological Sciences at the Univ of CA, Irvine (UCI). Reg PE in the State of CA. At UCI he directs the Focused Research Program on Image Engrg, which is devoted primarily to res on computer systems for the analysis & generation of digital images & the modeling of three-dimensional scenes from images. He has published over 100 papers & 3 books in his fields of interest. The books are *Pattern Recognition: Introduction and Foundations* (published by Dowden, Hutchinson & Ross), *Pattern Classifiers and Trainable Machines* (published by Springer Verlag), and *Biomedical Images and Computers* (published by Springer-Verlag). *Society Aff:* IEEE, ACM, TBΠ, HKN, ΣΞ

Skogerboe, Gaylord V
Business: Dept of Agri & Chem Engg, Ft Collins, CO 80523
Position: Prof. *Employer:* CO State Univ. *Education:* MS/Civ Engg/Univ of UT; BS/Civ Engg/Univ of UT. *Born:* 4/1/35. in Cresco, IA. Resident Engr on construction of Porcupine Dam in northern UT, 1961-63. Developed Cutthroat flow measuring flume while at UT State Univ 1963-68. While at CO State Univ, responsible for dev of Best Mgt Practices for Salinity Control in Grand Valley (western CO) 1968-77, & campus leadership for Water Mgt Res Proj (Pakistan & Vietnam) 1974-1980. Pres of small consulting firm, Aqua-Man, Inc. Presently VP, ISEM. *Society Aff:* ASAE, ASCE, AWRA, IA, ISEM.

Skolnick, Alfred
Home: 5432 N Carlin Springs Rd, Arlington, VA 22203
Position: V Pres, Adv Tech *Employer:* ORI, Inc. *Education:* PhD/Systems Science-Elec Engrg-Applied Math/Polytech Univ, NY; MS/EE/US Naval Postgrad School; MA/Math/Columbia Univ; BS/Math/Queens Coll *Born:* 8/15/30 Aerophysicist, Chance Vought Aircraft (1952). Joined US Navy (1953). Assigned to tactical missiles, strategic missiles, deep submergence, advanced ship design, combat systems integration, weapons development. Navy commendation medal for technical and scientific contributions to surface effect ship development (1971). Navy League's Parsons award for scientific and technical progress, outstanding contributions and exceptional leadership in directed energy weapons (1979). Gold Medal Award, American Society of Naval Engrs (ASNE) for outstanding professionalism and leadership in directed energy weapons system development (1981). Legion of Merit, US Navy, for Exemplary Leadership in personally directing extraordinary laser weapon feasibility demonstrations which "Form the core of all directed energy weapon development for the United States" (1983). National council member, and elected VP (1982-1985) ASNE. Elected ASNE Nat Pres (1985-87); Reelected (1987-89). Engr of the Year (1986), D.C. Council of Engg and Architectural Socs. Adjunct faculty member, Univ of VA and Northern VA Community Coll. Guest lecturer, Defense Systems Management Coll. Soc of Sigma Xi, Cosmos Club, Washington, DC. *Society Aff:* ASNE, IEEE, SNAME, ADPA, USNI.

Skolnik, Merrill I
Home: 8123 McDonogh Rd, Pikesville, MD 21208
Position: Superintendent, Radar Div. *Employer:* Naval Research Labroatory. *Education:* DrEng/EE/Johns Hopkins Univ; MSE/EE/Johns Hopkins UNiv; BS/EES/Johns Hopkins Univ. *Born:* 11/6/27. Superintendent Radar Div of Naval Res Lab

Skolnik, Merrill I (Continued)
Washington DC 1965- . Author of "Introduction to Radar Sys," & Editor of "Radar handbook." Fellow IEEE. Soc of Scholars, Johns Hopkins Univ; Distinguished Alumnus Award, Johns Hopkins Univ; Heinrich Hertz Premium Institution of Electronic & Radio Engrs 1964. Prior employment with Inst for Defense Analyses, Electronic Communications Inc, MIT Lincoln Lab, Sylvania & Johns Hopkins Radiation Lab. Visiting Prof Johns Hopkins 1973-74. Interests include radar, electronic sys, antennas & gaseous electronics. Navy Distinguished Civilian Service Award - 1982. IEEE Harry Diamond Award - 1983. IEEE Centennial Medal - 1984. Member, National Academy of Engineering. *Society Aff:* IEEE.

Skomal, Edward N
Home: 1831 Valle Vista Dr, Redlands, CA 92373
Position: Retired *Employer:* Aerospace Corp *Education:* BA/Phys/Rice Univ '47; MA/Phys/Rice Univ '49 *Born:* 04/15/26 Native KS City, MO. US Navy WW-2, '44-46. Mrrd 1951. Wife Elizabeth B, children Susan B., Catherine A., Margaret E. Prof Affltns: res Engr with Mobile Oil Field Res Labs, Dallas TX, Natl Bur Stds and Harry Diamond Labs, Wash, DC, Sylvania Microwave Phys Lab, Palo Alto, CA, Motorola Solid State Sys Div, Phoenix, AZ. Aerospace Corp, CA. Author technical books and res articles. Elected Fellow IEEE 1980, Rcpnt IEEE, EMC Soc Stoddart Award 1980. Rcpnt IEEE, EMC Soc. Certificate of Achievement 1971, Rcpnt IEEE Vehicular Soc Tech paper award 1970. Aptd mem Presidential Joint Tech Adv Cncl on EMC 1964, reaptd 1971. Chrmn 1978-83 IEEE Tech Cmmt Electro Mag Environments, Chrmn IEEE/EMC Tech Cttes 1984- , Asst Editor IEEE/EMC Trans 1983-86. *Society Aff:* APS, ΣΞ, URSI, IEEE

Skov, Arlie M
Business: Two Lincoln Centre, 5420 LBJ Freeway, Suite 800 LB25, Dallas, TX 75240
Position: Dir, Prod Tech *Employer:* Sohio Petroleum Co. *Education:* BS/Pet Engg/Univ of OK. *Born:* 9/21/28. Native of Noble Co, OK. Served US Army Corps of Engrs 1951-52. Joined Sohio 1956. Mgr Special Projs, 1966, Asst Mgr Engg 1975, and Mgr Production Planning, 1977. Tech Rep, ANAST (Alaska Natural Gas Transportation System), 1980, Mgr New Tech Dev 1981, Dir Prod Tech 1984, with responsibility for res, dev, and tech services in prod tech for Sohio Petro Co. Mbr of SPE-AIME Bd of Dir, 1972-74, Mbr of AIME Bd of Dirs, 1977-1979, and AIME VP, 1979. Past Natl Pres of Pi Epsilon Tau (Honor Soc of Petroleum Engg). Received 1975 Distinguished Service Award from the OK Petroleum Council. *Society Aff:* SPE.

Skov, Ebbe R
Home: 18215 Cape Bahamas La, Houston, TX 77058
Position: VP. *Employer:* Haldor Topsoe Inc. *Education:* MSChE/CE/Royal Tech Univ, Copenhagen, Denmark. *Born:* 7/31/36. and graduated in Denmark and worked since 1960 in Brazil and USA as Production Mngr 1963, Sen Process Engg with C F Baun 1964, Proj Mngr and Tech Dir 1966-76 with W R Grace & Co, and VP Proj and Marketing for So Am for Haldor Topsoe Inc since 1977. Mbr of AIChE, ACS, DIF, FEANI, and Reg PE. Was granted several US patents. Speaks several foreign languages (Port, Ger, FR, SP, Dan, Eng). Special interest in the area of Tech Dev of Chem Processes where res, piloting and economic evaluation come together in new process design. *Society Aff:* AIChE, ACS, DIF, FEANI, PE.

Skovholt, Joseph W
Home: 1430 Raymond St, St Paul, MN 55108
Position: Consultant. *Employer:* Honeywell Inc. *Education:* BEE/EE/Univ of MN. *Born:* 7/21/08. After grad, I served as an Illuminating Engr for Northern States Power for seven yrs. In 1943, I joined Honeywell, Inc, retiring in 1973 after 30 yrs service. I continue to serve as a part time consultant. For a number of yrs, I served as an Asst Chief Engr, supervising a design group of more than 50 engrs. For the last 19 yrs, I was a Mgr of Engg Planning of a Honeywell Div. In 1939- 40, and again the past eight yrs, I have served as a Natl Officer of Theta Tau, a Natl PE Fraternity. *Society Aff:* MSPE, NSPE, ESStP.

Skromme, Arnold B PE
Business: 1100 13th Ave, East Moline, IL 61244
Position: Proj Mgr, Future Product Programs. *Employer:* John Deere Harvester Works. *Education:* BSAE/Agr Engr/Iowa State Coll *Born:* 04/01/17 Reg PI Ill 1966- . Inventor & co-inventor - over 40 pats. Mbr Tau Beta Pi, Alpha Zeta, Sigma Tau Delta, Alpha Epsilon. 1947-50 Res Engr Pineapple Res Inst Honolulu, Hawaii; 1950-55 Asst Ch Engr John Deere Ottumwa Works, Ottumwa Iowa; 1955-70 Mgr Product. Engrg John Deere Spreader Works, E Moline Ill. 1955-70, Mgr Value Engrg John Deere Harvester Works 1970-78. ASAE Natl VP 1965-68, Fellow Dec 1971; Mbr of ANSI Organ Mbr Counc 1970- ; Mbr SAE 1958- ; Mbr SAVE 1970-78; VChrmn Central IL Chap SAVE; Natl VP SAVE 1977-78; received Profl Achievement Citation from IA State Univ in 1977; elected to Certified Value Engr by SAVE 1976; Proj Mgr Future Programs. J Deere Harv 1978- . Author of numerous technical papers; Outstanding Engr of the Year, Quad City Section, ASAE, 1984. *Society Aff:* ASAE, SAE

Skromme, Robert B
Business: Box 322, Detroit, MI 48232
Position: Chief Engineer. *Employer:* Massey-Ferguson Inc. *Born:* BS Agri Engrg Iowa State Univ 1951. Employed by J I Case Co in respon design engrg pos until 1962. Group leader for Sunbeam Corp until joining Massey- Ferguson 1964. Assumed current respon as Ch Engr of Design & Dev of Worldwide Agri Implements & Hay & Forage Machines in 1972. Reg P E Wisc. Mbr ASAE 1952- . Served as V Chmn of Quad City Sect, V Chmn & Chmn of Wisc Sect, & currently Chmn of ASAE PM-03 Stds Ctte. Enjoy bridge, camping & foreign travel.

Slack, Archie V
Home: Wilson Lake Shores, Sheffield, AL 35660
Position: President. *Employer:* SAS Corp. *Education:* MS/Chemistry/Univ of TN; BS/Chemistry/E TN State Univ. *Born:* Apr 1912. Chemist & engr for Tenn Valley Auth 1941-73. Spec in fertilizer process dev & gaseous pollutant emission control. Last pos at TVA, Ch Chemical Engr. Pres of SAS Corp 1973- . Advisory services for some 40 utils & related orgs in the area of emission control. Chmn 1970 ACS Fertilizer Div. Author 8 tech books & over 50 papers. *Society Aff:* ACS, AIChE, APCA.

Slack, Lyle H
Home: 702 Crestwood Dr, Blacksburg, VA 24060
Position: Assoc Prof of Matls Engrg. *Employer:* Vir Polytech Inst & St Univ. *Born:* Jan 1937. PhD Ceramic Sci 1965 Alfred Univ, Alfred N Y; BS Alfred. Native of Whitesville N Y. Served as a Ceramic Engr for the O Hommel Co & Lexington Labs. After PhD, studied dielectric thin films at Bell Labs, Murray Hill. Recent res on switching effects & crystallization in amorphous semiconductors, thin films on glass. Synthesis of solid semiconductors & films for solar photovothasis for solar energy control. Electronic transport in & optical characteristics of semiconducting oxide thin films on glass. Synthesis of solid semiconductors and films for solar photovoltages. *Society Aff:* AmCerSoc, ECS, ASM.

Slamecka, Vladimir
Home: 1661 Doncaster Dr NE, Atlanta, GA 30309
Position: Prof, Sch of Info & Computer Sci. *Employer:* Ga Inst of Tech. *Education:* DLS/Library Sci/Columbia Univ; MS/Library Sci/Columbia Univ; BS/Chem Engg/Brno Univ of Technology. *Born:* 5/8/28. in Brno, Czechoslovakia. Educ in chem engg, info & computer sci R&D in food engg (Australia 1951-54), seawater desalination (W Germany 1955-56, US 1956-59), info processing res (Documentation Inc 1962-64). Prof Sch of Info & Computer Sci, GA Inst of Tech 1964- ; Dir, 1964-78; Clinical Prof of Medicine, Emory Univ, 1979- . Res in automated language processing, biomedical computing, natl info sys. Mbr NY Acad of Sci, Sigma Xi, NAS/NRC Ctte on Intl Sci & Tech Info Progs, Dept of Commerce Technical Advisory

Slamecka, Vladimir (Continued)
Bd, US Atl Ctte for UNESCO/PGI, AFIPS Educ Ctte, AAAS, ACM, ASIS. Cons in Egypt, Ecuador, Venezuela, China, Indonesia. *Society Aff:* ACM, AAAS

Slana, Matthew F
Home: 6S330 Greenwich Ct, Naperville, IL 60540
Position: Mbr of Tech Staff/Dept Hd *Employer:* AT&T Data Systems Group Naperville Ill *Education:* MS/EE/Univ of WI; BS/EE/Univ of Notre Dame. *Born:* 10/11/35. Born, Waukegan, Ill. BSEE, Univ of Notre Dame, 5/57; MSEE, Univ of Wisconsin, 6/58. Further graduate work at Polytechnic Inst Brooklyn. Mbr of Technical Staff, Bell Tele Labs, 1958- . Military Computer Research in logic and devices, cryogenics, and radiation effects, Minuteman Missile safety studies, 1958-1964. Electronic Switching System Development, including logic, device and network studies, transmission, crosspoint device research and development, time division analog and digital switching, military seare voice communications system development, home communications development, 1964-1979. Computer systems, including software development, hardware development, and computer engrg and architecture of micro-and mini-computers, 1979- . Promoted to present position, 1979. Midwest College of Engrg, Prof, 1969-78, Bd of Trustees, 1971- 1978, elected Vice-Chrmn, 1974-1975, Chrmn 1975-1976, Hd of EE Dept, 1974-1978. Lecturer, Digital Switching Short Course, UCLA, 1978-. Mbr, Etta Kappa Nu; Sr Mbr IEEE. *Society Aff:* IEEE.

Slate, Floyd O
Business: Hollister Hall, Cornell Univ, Ithaca, NY 14853
Position: Prof of Engrg Matls. *Employer:* Cornell Univ. *Education:* PhD/Chemistry/Purdue Univ; MS/Chemistry/Purdue Univ; BS/Chemistry/Purdue Univ. *Born:* 7/26/20. Indiana. BS 1941, MS 1942, PhD 1944, Chemistry, Purdue Univ. Manhattan Proj 1944-46. Asst Prof CE Purdue 1946-49. Prof Engrg Matls Cornell Univ 1949- . Indus cons in concrete. Teaching & res in concrete, engrg matls sci & low-cost housing. Fellow or Mbr ACI, ASTM, ACS, AIC, ASCE. About 80 pubs, mostly concrete. Travel, lecturing and study in about 100 countries in low-cost housing and concrete. ACI Wason Res Medal 1956, 1963, 1972. ACI Anderson Medal 1983. ACI Wason Medal for Most Meritorious Paper 1986. Cornell's Excell in (Engrg) Teaching Award 1976. Sr Fellow (low-cost housing in Orient) East-West Ctr 1976. Enjoy travel, stamps & music *Society Aff:* ACI, ASTM, ACS, ASCE, AIC.

Slater, Arthur C
Home: 1323 Central Ave, San Carlos, CA 94070
Position: Principal Cost Engineer *Employer:* Bechtel Group, Inc. *Education:* BSEE/Ele-Pwr/IL Inst of Technology; Cert/Bus Mgt/Univ of CA-Berkeley; Masters/Bus Admin/Golden Gate Univ. *Born:* 11/11/23. Air Force 1942-45. Field Eng for Genl Elec Co - installation & start-up of large Turbine Generators 1949-52. With Bechtel Corp since 1952. Mgr Tech Estimating, Proj Mgr, Mgr of Services. Princ. Cost Eng. Mbr IEEE since 1952 & AACE since 1962. Pres AACE 1973-74. Pres of AACE San Fran Sect 1969-70. Cert of Recognition AACE 1974. Award of Merit AACE-1980. Cert Cost Engg July 1976. Mbr AACE Cert Bd. Commissioner Boy Scouts of Amer 1964-1980. Enjoy camping, fishing & woodworking. Retired 9/86. *Society Aff:* IEEE, AACE.

Slattery, John P
Business: Appliance Pk-Bldg 6, Louisville, KY 40225
Position: Mgr Engrg, Rm Air Cond Dept. *Employer:* Genl Electric Co. *Born:* July 1927. BEE Catholic Univ of Amer. Joined the G E Co 1951, working in var training assignments until 1954. Began work in Air Cond Design in 1954. Assumed present pos of Mgr - Engrg Rm Air Conditioner Dept in 1969. Mbr ASHRAE & Reg P E Penna. Papers on air cond design & efficiency considerations publ by ASHRAE & Assn of Home Appliance Manufacturers. Spare time avocation - flying.

Slaughter, Gerald M
Business: PO Box X, Oak Ridge, TN 37830
Position: Section Mgr *Employer:* Martin Marietta Energy Systems - Oak Ridge Natl Lab. *Education:* M Met Eng/Metallurgy/Renss Poly Inst; B Met Eng/Metallurgy/Renss Poly Inst. *Born:* 6/8/28. Native of Ilion NY. Employed at ORNL since 1951. Supvd Welding and Brazing Dev Grp from 1963-76. From 1976-present has supvd Eng Matls Sect involving respon for Mech Properties, Corrosion, Enrichment Materials Dev, Pressure Vessel Tech, Nondestructive Testing Dev, Materials Joining and Metal Processing. Author Joint AEC/ASM Monograph 'Welding & Brazing of Nuclear Reactor Components'. Elected Chmn of AWS Brazing & Soldering Ctte 1974. P Chmn of both Reactive & Refractory Metals Ctte & Brazing Res Ctte of Welding Res Coun. ASM Awd of Merit in 1970; elected fellow ASM 1976. Adams Lecturer of AWS in 1979. 8 pats in the field of metal joining. Elected Chmn of ASME-ASTM-MPC Joint High-temp. Properties Ctte in 1981. Elected Natl Trustee of ASM in 1983. Tech Advisory Ctte of Metal Properties Coun since 1977. Chmn of Dissimilar Metal Weld Task Grp of Steam Power Panel Ctte since formation in 1976. *Society Aff:* AWS, ASM, ASNT, ASTM.

Slaughter, John B
Business: Main Admin Bldg, College Park, MD 20742
Position: Chancellor, Univ of MD at Coll Park *Employer:* State of MD *Education:* PhD/Engrg Physics/Univ CA, San Diego; MS/Engrg/UCLA; BS/Engrg/KS State *Born:* 3/16/34. Attended public schools in Topeka, KS. Worked as Elec Engr for Gen Dymanics Convair, 1956-60, and for Navy Elec Lab Ctr, 1960-1975, in San Diego. Served as Dir, Appl Phys Lab, Univ of WA, 1975-77, and received Pres appointment to be Asst dir of the Natl Sci Fdn in 1977. Academic VP and Provost of WA State Univ. 1979-1980. In Sept 1980, received Pres appointment as Dir of the Natl Science Foundation. In Nov, 1982 became Chancellor of the Univ of MD at College Park. Fellow IEEE, Natl Acad of Engrg, Bd of Dirs AAAS, Tau Beta Pi, Eta Kappa Nu, Bd of Dirs Monsanto, Comm Credit, Baltimore Gas and Elec, Editor, Computers and Elec Engrg. *Society Aff:* IEEE, AAAS

Slavens, Wayne E
Home: 2708 30th St, Moline, IL 61265
Position: Consultant, Agri Engg. *Employer:* Personal John Deere Harvester Works (Retired). *Education:* BSME/Design Mech Engg/IA State Univ. *Born:* May 7, 1921 Des Moines Iowa. BSME, ISU. Boeing Aircraft 1943. Ch Engr McGrath Mfg, Omaha 1946. Joined John Deere Des Moines Works 1947, became Sr Div Engr respon for corn harvesting equip & crop dryers. To John Deere Harvester Works 1963 as Asst Ch Engr. Mgr, Prod Engrg since 1966. Mbr ASAE 1948, Fellow 1975. 'Engr of Yr' award by Quint-Cities Engrg Council 1969. Assisted with formation & charter mbr Iowa Sect ASAE 1957. Helped form Crop Dryer Mfg Council of FIEI, Chmn 1961-62. Mbr SAE. Author sev papers. Holder of 15 pats on agri machinery. Currently consultant, agricultural engineering & expert in product liability litigation. *Society Aff:* ASAE, SAE.

Slavin, Joseph W
Home: 8203 Excalibur Ct, Annandale, VA 22003
Position: President *Employer:* Joseph Slavin and Associates *Education:* BS/Marine Engg/US Merchant Marine Acad. *Born:* Feb 8, 1927 Boston Mass. s. Ambrose & Evelyn (Tuttle) S; m. Arlene Harris June 4, 1949; children - Elaine (Mrs Kevin Stiles), JoAnne, Patricia. Engr U S Lines 1949-50; mech engr Stone & Webster Engrg Corp Boston 1954; with Interior Dept 1954-70, Asst Dir Utilization & Engrg Wash D C; 1969-70 Natl Marine Fisheries Serv, NOAA, Commerce Dept; Asst Dir Opers 1970-71; Assoc Dir Resource Utilization 1971- . Mbr U S Natl Comm Internatl Inst Refrig; Sci Advisor Regrig Res Inst; Pres Camelot Civic Assn 1969-71, Camelot PTA 1973; Fellow Amer Soc Heating, Refrig & Air Cond Engrs; Mbr Marine Tech Soc, AAAS. Contrib prof journals. Internatl expert in fisheries util & represent U S Gov at natl & internatl mtgs. chief U S fisheries rep to Codex Alimentarius Comm; has num of pubs on fisheries & fisheries util. Enjoy boating. *Society Aff:* ASHRAE, TRRF, IIR.

Slaybaugh, Gregory S
Business: 82 Ionia, N.W, Grand Rapids, MI 49503
Position: VP *Employer:* Daverman Assocs, Inc *Education:* BS/CE/Univ of MI *Born:* 7/10/42 Born in Jackson, a MI native and graduated from Univ of MI in 1965 majoring in Civil Engrg. Have been in the private sector of engrg since graduation including design and administrative positions specializing in structures and airports. Received professional registration in states of MI and FL. Currently hold position of VP Production at Daverman Assocs, Inc. Activities include former Chapter Pres in MI Society of PE and co representative to American Consulting Engrg Council and state and local Chamber of Commerce. Past recipient of Chapter award Young Engr of Year from MSPE. Married and have two children. *Society Aff:* NSPE, ASCE, AAAE, AWWA, ACEC

Slaymaker, Frank H
Home: 134 Glen Haven Rd, Rochester, NY 14609
Position: Retired *Employer:* Retired *Education:* EE/Elec Engg/Univ of NB; BS/Elec Engg/Univ of NB. *Born:* Apr 1914 Lincoln Nebr. Stromberg-Carlson Co & Genl Dynamics Corp 1941- 70: Acoust Engr; Ch Engr Sound Equip Div; Assoc Dir Res; Dir ASW Progs. Res incl electron carillons, sonar, speech analysis-synthesis, optical signal processing. Univ of Rochester: taught Musical Acoustics 1970-75, Biomed Acoustics cons 1975- 78: Tropel, Inc: Scientist 1979-1985. Fellow IEEE, Acoust Soc Amer. Mbr Sigma Xi. Lic P E NY (inactive). Natl Chmn Prof Group on Audio, IRE 1958-59; Bd of Dirs Rochester Engrg Soc 1959-66. Play french horn in Penfield Symphony. *Society Aff:* IEEE, ASA, ΣΞ.

Sleicher, Charles A
Business: Dept of Chem Engg BF-10, Univ. of Washington, Seattle, WA 98195
Position: Chair, Chem Engg. *Employer:* Univ of WA. *Education:* PhD/ChE/Univ of MI; MS/ChE/MIT; BS/Chmistry/Brown Univ. *Born:* 8/15/24. After undergrad degree at Brown in chemistry in 1946, entered US Navy, then MIT in chem engg. Worked one yr and entered PhD prog at MI in 1952. Joined Shell Dev Co in 1956. In 1959-60, held NSF post-doctoral Fellowship for studies in turbulence at Univ of Cambridge. Joined Dept of Chem Engg at Univ of WA in 1960 as assoc prof and became chrmn of dept in 1977. Res interests include heat transfer, fluid mechanics and extraction with particular interest in turbulence and turbulent transport.. *Society Aff:* AAAS, AIChE, ACS.

Slemon, Gordon R
Business: Univ of Toronto, Toronto, Ontario, Canada M5S 1A4
Position: Prof of Elect Engrg *Employer:* Univ of Toronto *Education:* DSc/EE/Univ of London; PhD/EE/Univ of London; DIC/EE/Imperial Coll; MASc/EE/Univ of Toronto *Born:* 08/15/24 Dean, Faculty of Applied Sci and Engrg, Univ of Toronto: 1979-86. Chrmn, Dept of Elec Engrg, U of T: 1966-1976. Pres, Elec Engrg Consociates: 1976-1979. Chrmn, Univ of Toronto Innovation Fdn: 1980-1985. Chrmn, Exec Cttee: 1985- . Chrmn, Microelectronics Dev Centre: 1983- . Chrmn, Canadian Accreditation Bd: 1984-5. *Society Aff:* IEEE, IEE, EIC, CSEE, ASEE

Slepian, David
Home: 212 Summit Ave, Summit, NJ 07901
Position: Res Mathematician; Prof EE. Retired *Employer:* AT&T Bell Labs *Education:* PhD/Phyiscs/Harvard; MA/Phyiscs/Harvard. *Born:* June 30, 1923 Pittsburgh Pa. Univ of Mich 1941-43. US Army Signal Corps 1943-46. MA 1947, PhD 1949 Physics Harvard Univ. Parker Fellow in physics from Harvard at Univ of Cambridge Eng & Sorbonne Paris 1949-50. AT&T Bell Tele Labs 1950-1982, Univ of Hawaii 1970-81. Have worked in a var of areas of appl math with major ints in probability, info theory & communication theory. Mbr Natl Acad of Scis, Natl Acad of Engrg. Fellow Inst of Math Statistics, AAAS, IEEE. *Society Aff:* NAS, NAE, IEEE, AAAS.

Sletten, Carlyle J
Home: 106 Nagog Hill Rd, Acton, MA 01720
Position: President, Solar Energy Tech. Inc *Employer:* Set. Inc. *Education:* BS/Physics/Univ of WI (Madison); AM/Applied Physics/Harvard Univ (Cambridge). *Born:* 1/13/22. Native of Wisconsin. WW II, Radar Weather Officer. Air Force Cambridge Res Labs since 1948, Dir Microwave Phys Lab 1958-75. Formerly Fulbright Prof Madrid 1963-64; Expert Cons for U N Dev Prog Brazil 1975-78. Fellow IEEE; Pres Antennas & Propagation Soc of IEEE 1973; Chmn IEEE Antenna Stds Ctte. Phi Beta Kappa. Pres & Chmn of Bd of Solar Energy Tech Inc, Senior Member Technical STAFF, GTE-Comm. Syst Div, Needham, MA 1983-to date. *Society Aff:* IEEE-APS, URSI.

Slichter, William P
Business: 600 Mountain Ave, Murray Hill, NJ 07974
Position: Exec Dir, Res-Matls Sci & Engrg. *Employer:* Bell Labs. *Born:* Mar 31, 1922 Ithaca N Y. m. 1950; 4 children. AB Phys Chem Harvard 1944, AM 1949, PhD Chem Phys 1950. Mbr tech staff Bell Tele Labs Inc 1950-58, Hd Chem Phys Res Dept 1958-67; Chem Dir 1967-73; Exec Dir Res - Matls Sci & Engrg 1973- . US Army 1943-46, Reserve 1946-51, Capt. Natl Acad of Engrg; Fellow Amer Phys Soc (high polymer phys award 1970); ACS; AAAS. Phys chem of high polymers; nuclear magnetic resonance spectroscopy.

Slider, H C
Business: Ohio State Univ, 140 W. 19th Ave, Columbus, OH 43210
Position: Prof Emer & Conslt *Employer:* OH State Univ *Education:* MS/Petrol Engrg/OH State Univ; BEM/Petro Engrg/OH State Univ *Born:* 6/26/24 Author of *Practical Petroleum Reservoir Engrg Methods*, Pennwell 1976, and *Worldwide Practical Petroleum Reservoir Engineering Methods*, Pennwell 1983. Employed by Shell Oil Co. as a Field Engr & later promoted to Div Reservoir Engineer. Joined the chem faculty at Ohio State U. in 1956 where he taught a series of petro engrg courses. Served many major oil companies as consultant. Has taught more in industry than at Ohio State where he was a full time Prof for 30 years. Has taught and/or consulted on all continents except Australia in 13 countries. Distinguished lecturer for Society of Petroleum Engrs 1978-79. *Society Aff:* SPE, AIChE, ASEE.

Sliepcevich, Cedomir M
Business: 1215 Westheimer Dr, Norman, OK 73069
Position: Geo Lynn Cross Res Prof of Engrg. *Employer:* Univ of Oklahoma. *Education:* PhD/Chem Engr/Univ of MI; MS/Chem Engr/Univ of MI; BS/Chem Engr/Univ of MI. *Born:* 10/4/20. Born and raised in Anaconda, MT. Bachelors, Masters and Ph.D. degrees (chemical engineering) U. Mich. Faculty U. Mich 1943-55. Since 1955, faculty U. Oklahoma; George Lynn Cross Research Professor of Engineering. Consulting engineer since 1941. Numerous awards, honors and citations from technical societies, professional associations and academia, including Curtis McGraw Award, International Ipatieff Research Prize, George Westinghouse Award, National Engineer of the Year, National Academy of Engineering, William M. Walker Award, Oklahoma Hall of Fame, and American Gas Association Gas Industry Research Award. *Society Aff:* ΣΞ, AIChE, ACS, ASEE, NSPE & OSPE, NAE, AAAS, OAS.

Slifka, Richard J
Home: 3205 Tallyho Dr, Kokomo, IN 46902
Position: Staff Engg, Product Assur *Employer:* Delco Electronics Corp. *Education:* BSEE/Power/Univ of WI. *Born:* Sept 1927 Milwaukee Wisc. Field engr with AC Spark Plug Div of GMC on gun & bomb sights. Joined engrg in 1956 & worked on the guidance for missile sys. Became Prog Mgr of Advanced Missile Sys 1969. Transferred to Automotive Electronics 1970. Proj leader of Crash Detector, which is used in GM's Air Cushion Restraint Sys. Became Mgr of Monitor & Control Sys for Security & Energy Conservation Sys 1976. Assumed current respon as Staff Engrs All Electronics Product Assurance Dept hd in 1978. H S Isbrandt Award 1973. Mbr SAE. Private pilot. *Society Aff:* SAE, ΘT.

Slivinsky, Charles R
Home: 508 Simmons Ct, Columbia, MO 65203
Position: Prof. *Employer:* Univ of MO. *Education:* BSE/EE/Princeton; MSEE/EE/AZ; PhD/EE/AZ *Born:* 5/20/41. Have been teaching at MO-Columbia since 1968, except for leaves at the Lockheed Missiles and Space Co and at Wright-Patterson AFB. Have taught and developed many courses in elec engg ranging from automatic control theory to computer software methodology to digital filtering. Conduct res of appl nature under sponsorship by the USAF and Navy and NSF. Currently the Chrmn of Electrical Computer Engrg at the Univ. Enjoy running, reading widely, and gadjets. *Society Aff:* IEEE, ASEE, Sigma Xi

Sloan, Martha E
Business: Dept of Elec Engrg, Houghton, MI 49931
Position: Prof *Employer:* Michigan Tech Univ. *Education:* PhD/Stanford; MS/EE/Stanford; BS/EE/Stanford *Born:* 7/18/39. Native of Aurora Ill. Res engr for Lockheed Missiles & Space Co, spec in communications 1961-63. Taught at Frankfurt Internatl School, Germany. With Dept of Elec Engrg Mich Tech Univ, Houghton, since 1969. Assoc Prof since 1976. Author of 'Computer Hardware & Org' SRA 1976. ASEE Dow Outstanding Young Fac Award 1975. ASEE Frederick Emmons Terman Award 1979. IEEE Computer Soc Governing Bd 1977-82, Pres, 1984-85. *Society Aff:* IEEE, ACM, SWE, AAAS

Sloane, Neil J A
Business: Rm 2C-376, Murray Hill, NJ 07974 *Employer:* Bell Labs. *Education:* PhD/Cornell; MS/Cornell; BA/Univ of Melbourne; BEE/Univ of Australia *Born:* 10/10/39. Author of 4 books: *A Handbook of Integer Sequences, A Short Course on Error-Correcting Codes, The Theory of Error-Correcting Codes* (with F.J. MacWilliams) and *Hadamard Transform Optics* (with M. Harwit); and of 100 papers on communications, combinatorics, sphere-packing, group theory, etc. *Society Aff:* IEEE, AMS, MPA

Slocum, Ernest F, PE
Business: Vice-President, Stetson-Harza, 185 Genesee St, Utica, NY 13501
Position: VP *Employer:* Stetson-Harza *Education:* BME/ME/Cornell Univ *Born:* 5/11/20 Native of Dalton, PA. Served in US Army in France and Germany, 1942-1945. Have practiced as a consulting engr 1951 through present in the fields of development of engrg systems, facility fire protection and sanitary systems. Projects accomplished include industrial power plants and extensive manufacturing facilities, institutional facilities including hospitals numerous public schools and higher education facilities. Served in France in 1973 as consultant to a large tire manufacture for the design of two large industrial plants built in SC. As Sr Mechanical Engr and VP of Stetson-Harza am responsible for project development and energy related systems concept and design. Registered Engr-NY, PA, SC. *Society Aff:* ASHRAE, ACEC, CEC

Slonaker, Robert E, Jr
Business: Dept of Chem Engrg, Lewisburg, PA 17837
Position: Prof, Dept of Chem E. *Employer:* Bucknell Univ. *Education:* PhD/ChE/IA State Univ; MS/ChE/Bucknell Univ; BS/ChE/PA State Univ. *Born:* Dec 21, 1923 Pottstown Pa. Grad from The Hill School 1943. Served with Army Ground Forces WW II Europe. Teaching at Bucknell since 1951; Past Chmn Dept of Chem Engrg 1964-1986. Appointed Full Prof 1967. Visiting Prof: Univ of Pittsburgh (Spring, 1977) and Southern IL Univ (summers, 1981 and 1984). Lindback Award - Disting Teaching 1968. Elected Fellow of the AIChE (1980). Have served as met cons Piper Aircraft Corp 1951-56; Chem Engr E I duPont 1955; Advisor to Fac of Engrg Catholic Univ at Cordoba Argentina 1966. Admissions Ctte, AIChE; P Chmn Central Penna Sect AIChE; Chmn Chem Engrg Div ASEE 1976-77. Res ints: mass transfer opers, single crystal production & properties. Hobbies: music (piano) & outdoor activities with family. *Society Aff:* AIChE, ASEE, ASM, ΣΞ.

Slonneger, Robert D
Business: College of Engr, P.O. Box 6101, West Virginia Univ, Morgantown, WV 26506
Position: Prof., Asst. Dean of Engr. *Employer:* WV Univ. *Education:* MSME/ME/Univ of TEXAS (Austin); BSME/ME/Univ of OK. *Born:* 4/27/23. in IL raised in WI, OK and TX. Ordnance Officer, WWII. Instructor and Asst Prof ME at Univ of TX (Austin). With WV Univ since 1954. Specialties: thermodynamics, heat transfer, HVAC. At present Asst. Dean for undergraduate programs. Hobby: travel. *Society Aff:* ASME, SAE, ASEE.

Slowter, Edward E
Business: 505 King Ave, Columbus, OH 43201
Position: Retired. *Education:* ChE/ChE/OH State Univ; MSc/ChE/OH State Univ; BChE/ChE/OH State Univ. *Born:* 11/1/12. Retired V P, Secy & Treas of Battelle Mem Inst. Prof Engr with extensive exper in sci investigations in chem engrg & met fields. P Pres of NSPE. Recd OSPE Citation Award 1970; Cert of Award for Outstanding Services NSPE 1970; Meritorious Serv Award 1976 Coll of Engrg OSU. NSPE V Chmn, Central Region PEI 1967-68; NSPE Chmn, Central Region PEI 1969-71; NSPE V P, Central Region 1971- 73; NSPE Pres-Elect, 1976-77; ASM Treas 1975-76. NSPE Pres 1977-78; ASM Treas 1979-81. *Society Aff:* NSPE, ASM, ASEE

Small, Mitchell J
Home: 1151 McIntyre, Ann Arbor, MI 48105
Position: Engr *Employer:* Univ of MI *Education:* MS/Env and Wtr Res Engrg/Univ of MI; BS/Civil Engrg-Public Pol/Carnegie Mellon Univ *Born:* 6/11/53 Born in Pittsburgh, PA. Graduated from Carnegie-Mellon Univ with Univ Honors and worked for Allegheny County Health Dept evaluating water quality and flooding problems on small streams in Pittsburgh area. Employed as Engr for Hydroscience, Inc, 1975-1978, where developed methodologies for stormwater evaluation and control. Number of papers published, one of which, *Stormwater Interception and Storage*, received 1980 Wesley W. Horner Award from ASCE. Returned to graduate school in 1978 where research involves long-range transport of air pollution and acid rain. Enjoys camping and all kinds of music. *Society Aff:* ASCE

Smalley, Arthur L, Jr
Home: 438 Hunterwood Drive, Houston, Texas 77024
Position: Consultant *Employer:* Davy McKee Corp. *Education:* BSChE/ChE/Univ of TX. *Born:* 1/25/21. Born in Houston, TX. With Celanese Chem Co 1942-1970 in plant operations, engg & construction. Served as Dir of Engg from 1968 responsible for all maj capital plant expenditures. Sales Exec with Fish Engg Co 1970-1972. With Matthew Hall Inc 1972-87 serving as Pres & Chief Exec officer of this intl engg & construction co for the oil, chem, food and pharmaceutical process industry. Mbr AIChE. Reg TX PE. Active in Boy Scouting & awarded Silver Beaver. On Intl Advisory Bd of "Encyclopedia of Chemical Processing & Design–. Named Distinguished Graduate from Univ of Texas, Department of Engineering in 1987. *Society Aff:* AIChE.

Smalley, Harold E, Sr
Business: School of Ind & Sys Engrg, Atlanta, GA 30332
Position: Regents' Professor *Employer:* GA Inst of Tech. *Education:* PhD/Health Systems Engrg/Univ of Pgh; MSIE/Ind Engrg/Purdue Univ; BSIE/Ind Engrg/Univ of AL *Born:* 04/09/21 Mbr Alpha Pi Mu. Indus engr Stockham Valves Inc 1941-43. US Navy 1943-44. Asst Prof Univ of AL 1947-50; Assoc Prof Univ of Ct 1950-55; Asst to V Chancellor Univ of Pittsburgh 1955-58; Prof of Indus Engg GA Tech 1958-68; Regents' Prof GA Tech since 1968; Dir Health Sys Res Ctr, 1969-84; Prof engr & cons. Founder of Hosp Mgmt Sys Soc; Fellow & Life Mbr, Inst of Indus Engrs. Book-of-the-Yr Award IIE 1968; Disting Res Award IIE 1974. Disting Engg Alumnus Award Purdue Univ 1976. Frank & Lillian Gilbreth Award IIE, 1979. Book-of-the-Yr Awd HMSS 1983. *Society Aff:* IIE, HIMSS.

Smallwood, Charles, Jr
Business: Box 7908, Raleigh, NC 27695-7908
Position: Prof of Civil Engrg. *Employer:* N Carolina State Univ. *Education:* BS/San Eng/Case; SM/San Eng/Harvard. *Born:* May 20, 1920 Phila Pa. Prof of CE - Sanitary Engrg N C State Univ, Raleigh. Fellow ASCE, Mbr AWWA - P Pres N C Sect; Mbr APWA, N C WPCA, Bd of Control WPCF 1973-76. Fuller Award AWWA. Special ints in indus waste control & abatement & in stream sanitation & water qual. *Society Aff:* ASCE, AWWA, NCWPCA, APWA.

Smally, Donald J
Business: 133 S McIntosh Rd, Sarasota, FL 33582
Position: President. *Employer:* Smally, Wellford & Nalven Inc. *Education:* BSME/-/Univ of Cincinnati. *Born:* Aug 1922 Cleveland Ohio. US Army 1942-45, Master Sgt Ordnance Dept, European Theatre. BSME Univ of Cincinnati 1949. Fla resident since 1950. Ch Engr Mosby Engrg Assocs 1952-55. Pres Smally, Wellford & Nalven Inc 1956. Reg P E & Land Surveyor Fla. Pres Cons Engrs Council of Fla 1968. Dir Amer Cons Engrs Council 1965-66. Participated & served as chmn of natl cttes ACEC. Presented ACEC testimony to Congress as Chmn of Natl Water Resources Ctte. And worked on ACEC Financial Mgmt for conslt Engrs programs and manuals. And worked on ACEC Financial Mgmt for conslt Engrs programs and manuals. for consl Pilot. Active in num civic & cultural orgs. Treasurer ACEC 1980-1982. *Society Aff:* ACEC, NSPE, AWRA.

Smaltz, Jacob J
Business: Ind Engg Dept, Manhattan, KS 66506
Position: Prof. *Employer:* KS State Univ. *Education:* MS/Ind/KS State Univ; BS/Gen Engg/Bradley Univ. *Born:* 6/14/17. Native of Fulton, IL. Faculty mbr of KS State Univ since 1940. Reg engr in KS. Certified Safety Profl. Mgr, Remote Computing Lab in Engg, 1972-present at KS State Univ. Prof, KS State Univ, 1952-present. Assoc Prof, KS State Univ, 1949-1952. Asst Prof, KS State Univ, 1946-1949. Instr, KS State Univ, 1940-1946. *Society Aff:* IIE, ASEE, ASSE, ACGIH.

Smathers, James B
Business: Radiation Oncology Dept, Los Angeles, CA 90024
Position: Prof and Director of Medical Physics; Dept of Radiation Oncology *Employer:* UCLA *Education:* PhD/Nuclear Engr/Univ of MD; MS/Nuclear Engr/NC State College; BNE/Nuclear Engr/NC State College. *Born:* 8/26/35. Career to date has concentrated on the application of nuclear sci to medicine. Previous positions were held at Atomics Intl Inc, Walter Reed Army Inst of Res, Dept of Nuclear Engg, TX A&M Univ. and Bioengrg Div, Texas A&M Univ. Married to Sylvia Lee Rath in 1957, 4 children. *Society Aff:* ANS, HPS, AAPM, Rad Res, NSPE, CSPE.

Smedes, Harry W
Business: Office of Waste Isolation, U.S. Dept of Energy NE-330 GTN, Washington, DC 20545
Position: Senior Technical Advisor *Employer:* US Dept of Energy *Education:* PhD/Geology/Univ of WA; BS/Geology/Univ of WA. *Born:* 1926. USNR 1944-46. Instr Kansas State Univ 1951-53. Res Geologist USGS since 1953, northern Rocky Mtns. Field mapping, studies of mineral deposits, struct, stratigraphy, volcanic & plutonic rocks, environ geol, computer-aided land-use planning/resource mgmt. Currently studying hardrock sites for nuclear waste disposal. Mbr U S Treaty team to monitor peaceful nuclear explosions in Russia. Fellow Geol Soc of Amer, Mineralogical Soc of Amer, Explorers Club; Mbr Assn Engrg Geol, Amer Soc Photogrammetry, Colo Sci Soc, Geolical Society of Washington DC - P Pres's Award, Sigma Xi. AIL Award Amer Soc Photogrammetry. Num sci publs, maps. Hobbies: hiking, skiing, photography, choir, music. 4 sons. *Society Aff:* GSA, MSA, AEG, ASP, AAAS.

Smelcer, Glen E
Home: 1235 Madera St, Dubuque, IA 52001
Position: Chief Design Analyst. *Employer:* MEDAC. *Education:* BSME/Machine Design/Univ of WI; MS/Math and Mech/Univ of IA. *Born:* 5/3/31. Began working for Mech Design Analysis Cons in Moline IL in 1957. Opened Chicago area branch 1959. Became Chief Operating Officer 1967. Was employed by Intl Harvester Farm Equip Res & Engg Ctr as a design analyst 1959-1970. Hold US Patent 346700, 3269464, 3279831, 3302959 and 3417891. Reg PE in IL 1967, WI 1971 and IA 1971. Certified for Planning and Design of Protective Construction by Dept of Defense 1970. Moved Mech Design Analysis Cons to Dubuque IA 1970. Reg Acronym MEDAC 1974. Cons firm listed by Hazan Intl, Paris, France 1979. Reliability Analysis 1984, Engineering Analysis 1986: John Deere Dubuque Works. *Society Aff:* NSPE, SAE, ISPE.

Smeltzer, Walter W
Business: Dept of Mtls Sci. & Eng, Hamilton, Ontario, Canada L8S 4M1
Position: Prof. *Employer:* McMaster Univ. *Education:* BSc/Engg Chemistry/Queen's Univ; PhD/Physical Chemistry/Univ of Toronto. *Born:* 12/4/24. Prof of Mtls Sci & Eng. with res interests in the fields of corrosion, reactiviy & surface properties of materials. 1977: elected Fellow, Am Soc for Metals. 1979: elected Fellow, Royal Soc of Canada. 1981: elected Doctor Honoris Causa of The University of Dijon 1968: Centennial Medal for Service to the Nation (Canada). 1986: ASM Sauveur Achievement Award. Res Engr: Natl Res Council of Canada (1948- 50), Aluminum Labs (1953-55). Carnegie Inst of Tech (1955-59). British Res Council Sr Res Fellow (1976); France-Canada Exchange Prof (1979); NATO Sr Res Fellow (1979). *Society Aff:* ASM, ECS, NACE, CIMM.

Smetana, Frederick O
Business: Dept of Mech & Aerospace Engr, Raleigh, NC 27695-7910
Position: Prof. *Employer:* NC State Univ. *Education:* PhD/Engrg/USC; MSME/Mech Engr/NC State Univ; BME/Mech Engr (Aero Opt)/NC State Univ *Born:* 11/29/28. Flight test analyst, Douglas Aircraft; flight test engr & 1st Lt USAF, 1953-1955, Res Scientist, Univ of S CA, 1955-1962. Faculty, NC State, 1962- present. Author of over 100 scientific articles & reports on solar energy, cryogenics, plasma physics, flight test techniques, aerodynamic analysis, control system design, rarefied gas dynamics, info retrieval, parachute aerodynamics, wind energy systems; two books *Fortran Codes for Classical Methods in Linear Dynamics*, McGraw-Hill, 1982, and *Computer-Assisted Analysis of Aircraft Performance, Stability, and Control*, McGraw-Hill, 1984. *Society Aff:* AIAA, AVS.

Smidt, Dieter
Business: Postfach 3640, D-7500 Karlsruhe 1, Germany 07247/822550
Position: Dr Physics. *Employer:* Kernforschungszentrum - IRE *Born:* Mar 1927. Univ of Karlsruhe 1946-56. Dr rer nat 1954. Deutsche Babcock u Wilcox 1956-60. Nuclear Res Ctr Karlsruhe 1960. Since 1965 Full Prof Nuclear Engrg, Univ of Karlsruhe; Dir of Inst for Reactor Dev, Nuclear Res Ctr Karlsruhe. Reactor Safety Comm. Fellow ANS. Nuclear ints: nuclear safety.

Smidt, Fred A
Business: Code 4670, Naval Research Lab, Washington, DC 20375-5000
Position: Branch Hd *Employer:* Naval Res Lab. *Education:* PhD/Phys Chem/IA State Univ; BSc/Technical Sci/Univ of NB. *Born:* 7/19/32. born in Sioux City, IA, and spent childhood in NB. Served in USAF 1954-56. Career included Ames Lab, 1956-62; Sr Scientist in Hanford Lab, Richland, WA, 1962-69 (GE and Battelle); Naval Res Lab, 1969-present except for 15 mo assignment as coordinator of matls dev prog for breeder reactor at DOE, 1977-78. Named Hd Materials Modification and Analysis Branch, NRL in 1982. Have published approx 120 papers in area of radiation effects on matls, alloy dev and ion implantation of materials. Hold 4 pats, and have served on task groups & committees for breeder and fusion reactor matls dev, ion implantation. and coatings technology. Chrmn of Washington, DC chapter ASM 1978-79 & recipient of ASTM Dudley Award 1979, Fellow ASM 1983. Federal Laboratories Technology Transfer Award 1987. *Society Aff:* ASM, AIME, MRS, ΣΞ.

Smit, Raymond J
Home: 1312 Ardmoor, Ann Arbor, MI 48103
Position: Partner *Employer:* McNamee, Porter & Seeley. *Education:* MS/Civil Engg/Univ of MI; BS/Civil Engg/Univ of MI. *Born:* 9/21/28. in Detroit, MI. Reg P E MI & NY. Elected to the MI House of Reps for 4 terms 1967-74. Partner in Ayres, Lewis, Norris & May, Cons Engrs 1961-70. Since Jan 1977, Partner McNamee, Porter & Seeley, Cons Engrs. 1969 Fuller Award AWWA; 1972 Conservation Legislator, MI United Conservation Club; 1974 Hon Mbr, MI Soc of Archs; 1974 Environ Qual Award, US EPA; 1975 Civil Gov Award ASCE; 1976 & 1978. Mbr MI Bd of Regis, Prof Engrs 1975-present, twice elected Chrmn, Dir, Natl twice elected chmn, Dir Natl Soc P Es 1979-present, 1980-81 chmn, MI Sec AWWA. *Society Aff:* ASCE, NSPE, WPCF, AWWA, APWA.

Smith, A Leonard
Home: 11359 W 59th Pl, Arvada, CO 80004
Position: Consultant (Self-employed) *Education:* Engr of Mines/Mining/CO School of Mines. *Born:* 3/30/15. Native of Denver, CO. Served with Army Corps Engrs 1941 to 46. Mtls Engr with Bureau of Reclamation 1946-54. Served as Mtls and construction engg advisor to the govts of Jordan, Ceylon, Sudan, and Brazil for the US AID missions 1954- 71 under auspices of the Bureau of Reclamation and Bureau of Public Rds. Retained by private consultants as geotechnical engr 1972-77. Established own geotechnical consulting practice 1977-to present. Retained as City Engr for City of Wheat Ridge, 1977 to 1981. CO. Hobbies: golf, hunting, travel. *Society Aff:* NSPE, PEPP, AIME.

Smith, Alexander F
Home: Cricket Springs, Box 308, Geigertown, PA 19523
Position: Pres & CEO *Employer:* Gilbert Associates, Inc. *Education:* MS/ME/Lehigh Univ; BS/ME/Lehigh Univ *Born:* 2/7/29 Native of Reading, PA. Employed by Gilbert Assoc, Inc since 1955. Extensive engrg and managerial background in the electric utility industry as well as experience as a mechanical engr with assignments for industrial facilities, municipalities, and state and federal governments. Experience has included responsibility for design and engrg of power generating stations, management of engrg and construction services and corp functions. Elected VP in 1968; Member of the Bd of Dirs since 1970, and Pres since 1979. Holder of two patents: Condenser Steam Space Divider, 1965, and co-holder of Thermometer Well for Pipes, 1963. Listed in "Who's Who in America–, "Who's Who in the East–, and "Who's Who in Finance and Industry–. Enjoy tennis and gardening. *Society Aff:* ASME, NSPE.

Smith, Allen N
Home: 2545 Cottle Ave, San Jose, CA 95125
Position: Prof. *Employer:* San Jose State Univ. *Education:* PhD/ChE/OR State Univ; MS/ChE/GA Inst of Tech; BS/ChE/Tulane Univ. *Born:* 10/24/21. Chem Engg education; born New Orleans, LA Oct 24, 1921. Chem Engg Shell Oil Co, Wilmington, CA, 1943-46; Assoc Prof Chem Engg Univ of Louisville (KY) 1948- 52. Founder and first chrmn of the chem engg dept at San Jose State Univ. Elected Fellow of Am Inst of Chem Engrs in 1977; founder & first dir of the South Bay subsection of the No CA section of AIChE; Mbr Am Inst of Chem Engrs, Am Soc Engg Education, Tau Beta Pi, Sigma Xi, Phi Lambda Upsilon. *Society Aff:* AIChE, ASEE.

Smith, Allie M
Business: 101 Carrier Hall, Univ of MS, MS 38677
Position: Dean of Engrg *Employer:* Univ of MS *Education:* B/ME/NC St Univ; MS/Mech & Aero Engrg/NC St Univ; PhD/Mech & Aero Engrg/NC St Univ. *Born:* 6/9/34 Native of Tabor City, NC. Employed in aerospace industry and with academic institutions since 1956 specializing generally in the thermal sciences and particularly heat transfer. With Univ of MS since 1979 assuming responsibility as Dean of the School of Engrg and Prof of Mechanical Engrg. Recipient of National AIAA Thermophysics Award for 1978. Fellow, AIAA. Assoc Editor, AIAA Journal and Journal of Thermophysics and Heat Transfer. Chrmn, AIAA Thermophysics Technical Committee, 1975-1977. Chrmn, AIAA Terrestrial Energy Systems Technical Committee, 1979-1981. General Chrmn, AIAA Thermophysics Conference, 1975. General Chrmn, AIAA Aerospace Sciences Meeting, 1979. Recip of AIAA Space Shuttle Flag Plaque Award, 1984, & AIAA Hermann Oberth Award, 1985. *Society Aff:* ASME, AIAA

Smith, Arnold R
Business: 136 Summit Ave, Montvale, NJ 07645
Position: VP *Employer:* URS Co, Inc *Education:* BE/CE/Yale Univ *Born:* 3/26/18 Worked with various consulting engrg firms both in the US and abroad. Headed own construction firm. Present position since 1977. Formerly pres of McPhee, Smith, Rosenstein Engrs, Consulting Civil and Sanitary Engrs. Specialize in all aspects of water-water supply, wastewater collection and treatment, drainage, etc. Hold patent on "Novel Sewage Treatment Apparatus–. "Engr of the Year–, Bergen Co (NJ) NSPE, 1977. Awarded Brigham Medal, NYWPCA *Society Aff:* ASCE, WPCF, AWWA, NSPE, APWA

Smith, Bob L
Business: Dept of Civil Engrg, Seaton Hall, KSU, Manhattan, KS 66506
Position: Prof of Civil Engrg *Employer:* KS State Univ *Education:* PhD/Transp Eng/Purdue Univ; MS/Structures/KS St Univ; BS/CE/KS St Univ *Born:* 01/26/26 Native of Emporia, KS. Reg professional engr and land surveyor, KS. 1948- date: Mbr of KSU and Civil Engrg Faculty; since 1960 has specialized in transportation engrg; special interest: highway safety, traffic engrg, highway esthetics. 1971-1975: lecturer in series of nationwide seminars on highway safety design. Editor, *Dynamic Design for Highway Safety* (FHWA, 1975). 1972- 79: Chrmn, ASCE Ctte on Geometric Highway Design. *Practical Highway Esthetics* was prepared by the ctte and published in book form by ASCE, 1977. 1980-83: Proj dir for *Handbook of Operating Practices for Low Volume Rural Roads.* 1976-87, co-director, State of Kansas "Traffic Assistance Services for Kansas" (TASK Project). *Society Aff:* ASCE, ITE, ΣΞ, ΦΚΦ, XE

Smith, Brice R, Jr
Business: 801 North 11th St, St Louis, MO 63101
Position: President *Employer:* Sverdrup Corp. *Education:* MS/Str Engg/MA Inst of Tech; Bs/Civil Engg/Univ of MO Columbia. *Born:* 3/6/29. Joined Sverdrup & Parcel 1952 as struct engr. Was involved in the admin of intl projs for Sverdrup & Parcel 1959-64. Since 1964 has been respon for the financial affairs of Sverdrup org. Currently Pres and CEO of Sverdrup Corp., Pres of Sverdrup Investments and Partner in Sverdrup & Parcel. Also Chmn and CEO of Convention Plaza Redevelopment Corporation. *Society Aff:* ASCE, NSPE

Smith, Cameron M
Home: P.O. Box 3571, 5803-A 13th Ave S.E, Lacey, WA 98503
Position: Hd, Civil Engrg *Employer:* Saint Martin's College *Education:* PhD Engrg/Struct CE/Yale Univ; MA Engrg/Struct CE/Yale Univ; MS CE/Struct CE/Univ of WA; BS CE/Civil Engrg/Univ of WA *Born:* 01/02/18 Native of Seattle, WA. Taught at Yale, Univ or CO, Univ of CA, Naval Post Grad School, Univ of Toledo, Cleveland State Univ 1947-55, 1965-72. USNR. Worked with TAMS, Aerojet Genl, J F White Contracting Co, & TRW Inc. Exec Dir of Lake Erie Regional Transp Auth (LERTA), in charge of major feasibility study for major hub airport to serve NE OH, a Lake Erie site considered. Assoc Fellow AIAA; Mbr ASCE, APICS, (Secy); Mbr Cleveland Engrg Soc, OSPE. Reg PE CO, MA. *Society Aff:* ASCE, AIAA, TROA, WSPE

Smith, Carl E
Home: 8704 Snowville Rd, Cleveland, OH 44141
Position: President. *Employer:* Smith Electronics Inc. *Education:* Prof EE/EE/OH State Univ; MSEE/EE/OH State Univ; BSEE/EE/IA State Univ. *Born:* Nov 18, 1906 Eldon Iowa. Pres Smith Electronics Inc, & Past Pres Assn of Fed Communications Cons Engrs. Founder Carl E Smith Cons Radio Engrs. Founder Cleveland Inst of

Smith, Carl E (Continued)
Elec. V P Engrg United Broadcasting Co of Cleveland Plain Dealer until 1953, leave of absence to War Dept, Asst Ch Oper Res Staff - commendation award for except civilian serv 1946. Distinguished Alumnus Award OH State Univ, 1974. Distinguished Achievement Citation, Iowa State U 1980, Honor Leadership Award, Cleveland Institute of Elec 1981. Dist Service Award, Natl Religious Broadcasters, 1984. Engg Achievement Award, Natl Assoc of Broadcasters, 1985. Reg P E Ohio & D C. Achievements: Dev antennas to increase AM broadcast coverage, researched & introduced circular polarization to improve FM & TV broadcast serv, conducted Natl Assn of Broadcasters Seminars on Directional Antennas & Digital Techs, authored num tech publs. *Society Aff:* RCA, AFCCE, IEEE, CES, ASEE, NAB.

Smith, Charles E
Business: Electrical Engrg Dept, Anderson Hall, Univ, MS 38677
Position: Chrmn & Prof *Employer:* Univ of MS *Education:* PhD/EE/Auburn Univ; MS/EE/Auburn Univ; B/EE/Auburn Univ *Born:* 6/8/34 Native of Clayton, AL. While pursuing advanced degrees from 1959 to 1968, was employed as a Research Assistant with the Auburn Univ Research Foundation. In 1968, he accepted position of Assistant Prof of Electrical Engrg with the Univ of MS. Advanced to the rank of Assoc Prof in 1969. Appointed Chrmn of the Electrical Engrg Dept in 1975. Currently, Prof and Chrmn. Main areas of interest are the application of electromagnetic theory to microwave circuits and antennas. Recent research has been on the application of numerical techniques to microstrip transmission lines and low-frequency antennas, antennas in lossy media, and automated microwave measurements. *Society Aff:* IEEE, ASEE, ΦΚΦ, ΣΞ.

Smith, Charles H
Business: 8390 Delmar Blvd, St Louis, MO 63124
Position: V P, Production - Distribution. *Employer:* St Louis Cty Water Co. *Education:* MS/Sanitary Engg/Univ of IL; BS/Civil Engg/Univ of IL. *Born:* Oct 25, 1933 Danville Ill. Married. BSCE Univ of ILlinois 1955; MS Sanitary Engrg Univ of Illinois 1963. Reg P E Ill & Mo. Reg Public Water Supply Operator Ill. Cert Water & Waste Water Operator Mo. Lt Col USAF Reserve. Mbr NSPE, MSPE, AWWA, NPCF, ASCE, Natl Assn of Water Co's, Mo Water & Sewerage Conf. Mbr AWWA Cttes on Disinfection & Water Qual Goals. *Society Aff:* NSPE, MSPE, AWWA, ASCE, WPCF, NAWC, MWSC.

Smith, Charles O
Home: 1920 College Ave, Terre Haute, IN 47803
Position: Consultant *Employer:* Self *Education:* ScD/Metallurgy/MIT; SM/Mech Engg/MIT; BS/Mech Engg/WPI. *Born:* 5/28/20. US Navy 1943-46. 28 yrs teaching engrg at WPI, MIT, Univ of Detroit, Univ of Nebr. Rose-Hulman Inst of Tech. 14 yrs indus exper at Alcoa Res Labs & Oak Ridge Natl Lab. Author 4 books & 150 papers dealing with design, matls, Products Liability & Engineering Education. Cons to attorneys & indus orgs in failure analysis & safety eval. Active on natl cttes of the ASEE, ASME. Tau Beta Pi, Sigma Xi, Pi Tau Sigma, Merryfield Award, ASEE. *Society Aff:* ASME, AIME, ASEE, ASM International.

Smith, Clarence R
Home: 1807 Beryl St, San Diego, CA 92109
Position: Struct Fatigue Strength Specialist (Self-employed) *Born:* Feb 1908. Attended Stanford Univ 1927-31. Genl Dynamics - Convair 1941- 70. Dir struct fatigue res 1953-70. Dev Linear Strain Theory & Smith Method for predicting fatigue life. Authored 'Tips on Fatigue' & co-authored 'Analysis & Design of Flight Vehicle Structs'. Also publ more than 50 magazine articles, tech govt reports & papers presented at var engrg soc's. Fellow SESA 1976, Tatnall Award 1969, San Diego Engr of Yr Award (Soc of Prof Engrs) 1968. Reg P E Calif. *Society Aff:* SESA, ASTM.

Smith, Cyril Stanley
Home: 31 Madison St, Cambridge, MA 02138
Position: Institute Prof Emeritus. *Employer:* MIT. *Education:* DSc/Metallurgy/MIT; BSc/Metallurgy/Univ of Birmingham. *Born:* Oct 4, 1903 Birmingham England. Came to U S 1924. Res metallurgist, Amer Brass Co 1927-42; in charge of met at Los Alamos Sci Lab 1943-46; Founder first dir of the Inst for the Study of Metals Univ of Chgo 1945-61; Inst Prof MIT 1961-69, Prof Emer 1969- . Mbr Genl Adv Ctte U S Atomic Energy Ctte 1946-52; Pres's Sci Adv Ctte 1959; Smithsonian Council 1966-76. Recipient of many awards from tech, and learned soc's in US & abroad. Res & publ in phys met, on genl principles of struct, & on the hist of tech, and relations between art science and technology. *Society Aff:* AIME, ASM, APS, NAS, SHOT, HSS.

Smith, Daniel J, Jr
Home: 919 University Ave, Burbank, CA 91504
Position: Sr Engr. *Employer:* Metro Water Dist of So Calif. *Education:* BS/Civil Engg/Univ of So CA. *Born:* Sept 1914 Los Angeles Ca. Reg C E Calif. US Army 1942-46. Engr with Koebig & Koebig Inc 1954-67, spec in earthquake engineering and struct design of sewage & water treatment plants, storage tanks, marine structs & pipelines. With Metro Water Dist of So Calif since 1967. Projs incl water treatment plants, aqueducts, distrib facils & dams. Since 1971 spec in earthquake engrg for water storage, treatment & distrib sys. Mbr ASCE; Struct Engrs Assn of So Calif. Earthquake Engrg Res Inst. ASCE Thomas Fitch Rowland Prize 1976. Enjoy semi-classical music, golf & photography. *Society Aff:* ASCE, SEAOSC, EERI.

Smith, Darrell W
Business: Dept. of Metallurgical Engrg, Michigan Technological Univ, Houghton, MI 49931
Position: Prof of Metall Engrg *Employer:* MI Tech Univ *Education:* PhD/Metallurgy/Case Western Res Univ; MS/Metallurgy/Case Western Res Univ; BS/Metall. Engrg/MI Tech Univ *Born:* 7/31/37 Native of So CA. Process metallurgist with Babcock & Wilcox Co (1959-1962). Research metallurgist with Lamp Division of General Electric Co (1962-1970). Joined faculty of MI Tech Univ in 1970. Am currently Prof of Metallurgical Engrg. Research and teaching interests are in physical metallurgy and process metallurgy (particularly powder met). Am currently Vice Chrmn of ASM's Powder Met Committee and teach courses in powder metallurgy and physical metallurgy for ASM. Is an active consultant to both indus and govt on metallurgical problems. Enjoys fishing and flying single engine aircraft. *Society Aff:* ASM, APMI

Smith, David B
Home: 6 Pastern Lane, Blue Bell, PA 19422
Position: Prof Emeritus. *Employer:* Drexel Univ. *Education:* MS/EE/MIT; BS/EE/MIT. *Born:* Dec 3, 1911. Philco Corp 1934-61; V P Res & Dir 1945 Philco; Philco-Ford V P Res & Engrg 1961-64; Prof Sys Engrg & Chmn Grad Group Sys Engrg, The Moore School, Univ of Penna 1964-67; Lecturer- Drexel Univ 1979- Pres H R B Singer Inc 1967-69; Lectr The Moore School 1969-72; Prof Engrg Mgmt & Dir Engrg Mgmt Prog, Drexel Univ 1972-78; Dir Narco Scientific Mbr Natl Television Sys Ctte, Radio Tech Planning Bd, Charter Mbr Joint Tech Adv Ctte; V Chmn 2nd Natl Television Sys Ctte; advisor/cons to Dept of the Army, Dept of State, Dept of Commerce; Fellow IEEE, Mbr Amer Assn for Engrg Educ. Author & patentee. *Society Aff:* IEEE, ASEE, AAAS.

Smith, David B
Business: Agri & Bio Engrg Dept, PO Box 5465, MS State, MS 39762
Position: Agric Engr *Employer:* MS State Univ *Education:* PhD/Agric Engrg/Univ of MO; MS/Agric Engrg/MS State Univ; BS/Agric Engrg/MS State Univ *Born:* 4/6/42 Dr Smith's involvement in pesticide application research for 22 yrs has resulted in 95 technical publications including 2 book chapters. He is a member of Gamma Sigma Delta, Tau Beta Pi, Sigma Xi and Alpha Epsilon; is a registered PE in MO and MS; and was selected by ASAE to receive the 1980 FIEI Young Researcher Award. He has made numerous technical talks, of which 21 were by invitation. He was invited to serve on 2 natl task forces. These were sponsored by CAST and EPA-Purdue Univ respectively. He has served as an officer or member of several techni-

Smith, David B (Continued)
cal ASAE Committees, ESA-ASAE-(1977-80) and ASAE-ESA(1978- present) Liasion Representative. *Society Aff:* ASAE, ESA

Smith, Donald D
Home: RD 2, Box 138A, Tunkhannock, PA 18657
Position: Pres *Employer:* Smith, Miller & Assoc Inc *Education:* BS/Engrg/PA State Univ *Born:* 1/23/26 Wilkes-Barre, PA. US Army Engr Corps 1944-46: employed by Firm Architects & Engrs 1948-53: started private practice 1953 as Roushey & Smith. Successor firm Smith, Miller & Assocs, Inc organized 1974, providing architecture, engrg and related construction design with staff of 35. Pres, treasurer, & chief executive officer of SMA, and of the Kingston Development Corp Real Estate Firm, all located at 189 Market St Kingston, Luzerne Co, PA 18704. Past Pres of PA Society of PE, Wilkes-Barre Rotary Club, PA State Engrg Society, Penn State Alumni Club, and Susquehanna River Basin Assoc. Active in many community, civil, social, & fraternal affairs. *Society Aff:* NSPE, ASHRAE, IEEE, IES

Smith, Donald R
Business: University of Missouri-Kansas City, 600 W. Mechanic, Independence, MO 64050
Position: Dir-UMC/UMKC Coord Engrg Progs. *Employer:* Univ of Missouri. *Education:* PhD/Physics/Univ of CO; MA/Physics/Univ of CO; MS/Nuclear Engr/Air Force Inst of Tech; BS/Mech Engr/KS St Univ. *Born:* 9/5/31. Retired Lt Col AF (22 yrs active duty). Military service: 3 yrs Aircraft Nuclear Propulsion Proj; 8 yrs fac AF Acad as Assoc Prof of Physics; 3 yrs as Proj Dir, Defense Nuclear Agency; 3 yrs as Ch of Plans, AF Office of Sci Res. Currently Resident Dir of UMC/UMKC Coord Engrg Progs in Kansas City. *Society Aff:* ASEE, ANS, NSPE, ASME

Smith, Donald S
Business: 1st St & Normal Ave, Chico, CA 95926
Position: Full Prof. *Employer:* CA State Univ. *Education:* PhD/Mech Engr/Univ of CA; MS/Mech Engr/Univ of CA; BS/Mech Engr/Univ of CA. *Born:* 12/23/26. 10 yrs as prof of Mech Engg including chairmanship of the Mech Engg Dept. 15 yrs of diversified engg and mgt consulting activities. Recipient of several Natl Sci Fdn and Natl Aeronautics and Space Admin fellowships. Author of publications in technical journals on combustion and air pollution. *Society Aff:* ASME, ASEE.

Smith, Donald W
Business: 305 East 63rd St, New York, NY 10021 *Employer:* Andrews & Clark, Inc *Education:* M/CE/Polytech Inst of Brooklyn; BS/CE/Swarthmore Coll *Born:* 6/26/26 Native of Long Island, NY. Field engr and land surveyor on urban housing programs for Levitt & Sons, Inc 1947 to 1952. Joined Andrews & Clark as a hgwy design engr in 1952, became chief engr in 1959 and pres of the consulting engrg firm in 1975. Technically responsible for the planning and design of several major parkway and freeway programs and award winning parks. Active in education as trustee of Friends World Coll and Friends Academy, both on Long Island. *Society Aff:* ASCE, NSPE, AISC

Smith, Dwight D
Home: 4407 Beechwood Rd, Hyattsville, MD 20782
Position: Consultant. *Employer:* Self. *Education:* BS/Agri Engr/KS State Univ; MA/Agri Engr/Univ of MO. *Born:* Aug 1905. Westinghouse Elec 1928-29; Instr Univ of Mo 1929-33; Field & Res Engr & Proj Supr USDA Soil Conservation Serv 1933-52; Agri Res Serv USDA 1952-70, as Res Supr Mo 1952-56, Erosion Res Leader Beltsville 1956-61, & Asst Dir for Water Mgmt Res 1961 until retirement 1970. Unit Award for Superior Serv USDA 1959; Mo Honor Award for Disting Serv in Engrg, Univ of Mo 1961; John Deere Gold Medal Award 1968, Amer Soc of Agri Engrs; Life Fellow ASAE, Life Mbr ASA. *Society Aff:* ASAE, SCSA, ASA.

Smith, Easley S
Home: 304 Overlook Dr SW, Blacksburg, VA 24060
Position: Exten Specialist, Agri Engrg. *Employer:* Vir Polytech Inst & State Univ. *Education:* MS/Agri Engg/VA Polytech Inst & State Univ; BS/Agri Engg/VA Polytech Inst & State Univ. *Born:* May 1924. Native of Nottaway Cty, Vir. Army Air Force 1943-45 as 1st Lt & bomber pilot. Territory Mgr for Oliver Corp 1948-56. Extension Specialist, Agri Engrg at VPI & SU since 1956. Currently respon for Extension educational progs involving distrib of bulk fertilizers & lime; dev & promotion of reduced & notillage corn & soybean production practices; state 4-H petro power prog; farm machinery costs, & safe & proper use. Mbr ASAE since 1948; Chmn Vir Sect ASAE 1971; Mbr Vir 4-H All Star Chap since 1940; Epsilon Sigma Phi (Extension Hon Frat), Gamma Sigma Delta (Honor Society of Agriculture). *Society Aff:* ASAE.

Smith, Edgar H
Home: 1410 Adrian Blvd, Fort Atkinson, WI 53538
Position: Pres *Employer:* Shawnee Engrg Inc. *Education:* BS/Argi Eng/Penn State Univ. *Born:* 11/18/29. Raised on a dairy farm in PA. Served 4 yrs in US Navy Sub Service. PE. Have 13 yrs experience at Butler Manufacturing Co managing a design engg dept. As part of top management at a div, helped establish: div sales goals, product lines, and financial budgets. Had 4 yrs experience as one of three organizers and owners of a design engineering and metal fabrication corp. Have had experience in sales, bus management and special equipment design engg. Have also had 7 yrs experience at Sperry New Holland in farm equipment design, res and dev, and testing of new machinery. Presently doing consulting engineering work on mechanical design safety analysis & accident investigation. *Society Aff:* ASAE.

Smith, Edward J
Home: 614 Third St, Brooklyn, NY 11215
Position: Emeritus Prof of Elect Engrg & Dir Comp Sci Div *Employer:* Polytechnic Inst of NY *Education:* PhD/EE/Polytech Inst of Bklyn; M/EE/Polytech Inst of Bklyn; B/EE/Cooper Union *Born:* 12/12/20 Born in NY City. Research engr, Remington Rand Co, 45-47; instructor in electrical engrg, NY Univ, 47-48. Joined the Polytechnic Inst of Bklyn as research fellow, 48; held various ranks to prof, 59; Head, Dept of Electrical Engrg 67-71 and 79-81; Dir, Computer Science Division, 75- 81, and 83-85. Visiting Prof, Technische Hogeschool Te Eindhoven, Netherlands, 63-64. Earlier work in nonlinear magnetics, magnetic amplifiers and logic devices. Recent activities focus on education and research in computer science, including computer organization and architecture, logic design and switching theory. Emeritus Professor, Polytechnic Univ. 1986. *Society Aff:* IEEE, ACM, ASEE, AAAS, NYAC

Smith, Edward M
Home: RD 2-20341 Wineland Rd, Butler, OH 44822
Position: Ret (VP) *Education:* ME/ME/OH State Univ; BME/ME/OH State Univ; BA/Engg/Denison Univ. *Born:* 12/15/13. Native of Mansfield, OH. Res Engr Bailey Meter Co 1937-1946. Chief Engr Barnes Mfg Co 1946-1950. Founded Gorman Rupp Industries Div 1950. Designed and developed most of present line of small chem handling & metering pumps a well as hypohyperthermia machines for medical control of total body temperature. 24 patents issued. VP & dir of engg Gorman Rupp Industries Div 1958-1978. Dir Gorman Rupp Co 1968 to present. Continue as engg consultant to div. Enjoy farming.

Smith, Edward P
Home: 1301 Hawthorn Rd, Schenectady, NY 12309
Position: retired *Education:* BS/EE/KS State Univ. *Born:* Dec 1917. Grad of G E's Advanced Engrg Prog, Modern Engrg course, Advanced Mgmt course. Design engr for G E, spec in large AC & DC motors & generators. Mgr of engrg of var M&G products for 25 yrs. Before assuming present assignment as Mgr Internatl Gas Turbine Progs - Asia, Eastern Europe & N Africa - was Mgr of Engrg G E of Brazil, respon for power transformers, locomotives, hydro generators, hydro turbines & large AC & DC motors. Fellow IEEE 1964. Natl Dir NSPE 1957-60. Dir NYSSPE 1953-61. Charter Mbr CIGRE Ctte of Brazil 1971. Enjoy gardening & golf. Retired

Smith, Edward P (Continued)
1983 - Active in professional and civic organizations. *Society Aff:* IEEE, AISE, AMA, NSPE.

Smith, Edward S
Home: 3181 Medinah Circle, Lake Worth, FL 33467
Position: Sr Mtls Engr. *Employer:* Pratt & Whitney Aircraft. *Education:* MS/Met Engr/Univ of Pittsburgh; BS/Met Engr/VPI. *Born:* 12/4/24. Native of NYC. Served in Army Air Corps WWII & retired from US Naval Reserve Civ Engr Corps. Advanced art of vacuum induction & consumable electrode melting of high temperature metals mostly for turbine engine applications in 1950's. As Pres of Electralloy Corp, pioneered the new argon/oxygen process & developed the process technically & commercially into a tool for the production of most grades of steels. Mbr Tau Beta Pi, Phi Kappa Phi, Sigma Gamma Epsilon; ASM Fellow; recipient of AIME Clarence Sims Award. Dir AIME Elec Furnace Div, 1975-78; Dir Midwest Chapter PA Engr Soc 1972-75, Registered Prof Engr, PA Ohio, FL. *Society Aff:* ASM, AIME, AISE, ADPA.

Smith, Edwin E
Business: 140 West 19th Ave, Columbus, OH 43210
Position: Prof *Employer:* OH State Univ *Education:* PhD/ChE/OH State Univ; MS/ChE/OH State Univ; B/ChE/OH State Univ *Born:* 1/18/23 Research Prof with Engrg Experiment Station, and Prof, Dept of Chem Engrg, The OH State Univ since 1949. Research interests include industrial pollution control and application of chem engrg principles to analysis of fire systems. Responsible for the development of the OSU Release Rate Apparatus, a fire test system for evaluating fire performance of materials and products. *Society Aff:* AIChE, ACS, ASTM, NFPA

Smith, Frank M, Jr
Business: 21 W. Church St, Jacksonville, FL 32202
Position: Sr VP *Employer:* Charter Oil Co. *Education:* BS/ChE A&M Univ *Born:* 11/14/47. Native of Houston, TX. Began career as a Process Control Engr at the Houston Refinery. Assumed position of Process Supervisor (Powerhouse facilities) in Sept, 1974. Promoted to Technical Operations Supervisor (Chief Process Engr) in early 1975. Promoted to Mgr, Engg at Houston Refinery in mid 1977. Co's energy Conservation Coordinator, responsible for submitting DOE required Energy Consumption reports to the API through Dec, 1978. Promoted to VP, Info Systems Dev in Jan, 1979. Responsible for coordinating the dev of info systems (from a business & mgmt viewpoint) throughout Charter Oil Co. In Mar, 1980, transferred to Charter's Alaska Project, as VP, Operations. Promoted to current position as Sr VP, Planning and Development for Charter Oil Co in Sept, 1981. *Society Aff:* AIChE.

Smith Frederick A Jr
Business: Dean of the Academic Bd, US Military Acad, West Point, NY 10996
Position: Formerly Dean of the Academic Bd *Employer:* US Army *Education:* PhD/Theoretical & Applied Mech/Univ of IL; MBA/Bus/Geo Wash Univ; MS/Mech Engrg/Johns Hopkins Univ; BS/Gen Studies/US Military Acad *Born:* 12/08/21 Brigadier Gen Frederick A Smith Jr, was named Dean of the Academic Bd, US Military Acad, in Aug 1974. Prior to his appointment, he was Prof and Hd, Dept of Mechanics. He graduated from West Point in 1944 and was commissioned in the Infantry. He served in the European Theater of Operations in the 5th Infantry Regiment, 71st Div; in the Korean War with the 5th Regimental Combat Team; in Germany as Commander of the 2d Battalion (Mechanized), 87th Infantry, and at 7th Army Hdqtrs in Stuttgart; in the Vietnam War at Macv hq; and as Special Asst to the Chief of Army Res and Dev. He attended the Army's Command and Gen Staff Coll (where he later served as an instructor), the Armed Forces Staff College, and the Indust Coll of the Armed Forces. A native of Dallas, TX, he is married to the former Katherine B. Egerton of Baltimore, MD. The have two daughters, both serving as majors in the US Army. *Society Aff:* ΦΚΦ.

Smith, George E
Business: 600 Mountain Ave, Murray Hill, NJ 07974
Position: Hd, MOS Device Dept. *Employer:* Bell Tele Labs Inc. *Education:* PhD/Physics/Univ of Chicago; MS/Physics/Univ of Chicago; BA/Physics/Univ of PA. *Born:* May 1930. Joined Bell Labs 1959; studied the elec properties & band structs of semimetals. Made Dept Hd of the Device Concepts Dept 1964. Primary ints are new semiconductor devices for logic & memory applications, the device physics of semiconductor-insulator interfaces & chargecoupled devices. Mbr Pi Mu Epsilon, Phi Beta Kappa & Sigma Xi; Fellow IEEE; Fellow APS. Stuart Ballantine Medal 1973; Morris N Liebmann Award 1974. 14 pats, 11 pats pending & approx 40 publ articles. *Society Aff:* IEEE, APS.

Smith, George F
Business: 3011 Malibu Canyon Rd, Malibu, CA 90265
Position: Sr V Pres & Res Labs Dir. *Employer:* Hughes Aircraft Co. *Education:* PhD/Physics/Caltech; MS/Physics/Caltech; Bs/Physics/Caltech. *Born:* May 9 1922 Franklin Ind. Joined Hughes Aircraft Co 1952. Personal res on thermionics & secondary electron emission, direct view storage tubes, laser applications. In 1960-61, conducted first laser radar experiments. Dir of Hughes Res Labs since 1962. V P of Hughes Aircraft Co 1965-81, Sr VP since 1981. Prior to joining Hughes, was a founding engr of Engrg Res Assocs, now part of Univac Div, Sperry Rand. Taught at Caltech & USC. Mbr Army Scientific Advisory Panel 1975-78. Fellow APS & IEEE. Mbr Sigma Xi, Tau Beta Pi, AAAS. *Society Aff:* IEEE, APS, AAAS, ΣΞ, ΤΒΠ.

Smith, George H
Home: 7430 Miami Lakes Dr, Miami Lakes, FL 33014
Position: Partner & Principal Transit Engr. *Employer:* Gannett Fleming Engineers *Education:* BS/Structural/Drexel Univ. *Born:* 7/6/11. Native of Phila Pa. Taught in Radio Technician Schools, US Navy 1943-45. Reg PE PA, NJ, NY, MS, LA, FL & CA. Lifetime in all phases oper & maintenance of urban transit sys, incl rapid rail trasit, bus & streetcar. Ch Transit Engr in design & const of Phila Lindenwold Rapid Transit Line. Respon charge design & const of num bus garages & other transit facils. P Chmn Plant & Engrg Div of Amer Pub Transit Assn. Abr ASCE, SAME, ASHE, NSPE, APTA. *Society Aff:* APTA, ASCE, NSPE, ASHE, SAME.

Smith, George S
Home: 1715 N E Naomi Pl, Seattle, WA 98115
Position: Prof Emeritus. *Employer:* Univ of Washington. *Education:* BS/EE-Professional Degree/Univ of WA. *Born:* Oct 1889. Native of Nebr. Genl Elec Test 1916-18. Elec Engr for Internatl Coal Prods Corp 191921. Teaching EE Dept Univ of Washington 1921-61. Genl Elec Lightning Arrester Dept Pittsfield Mass 1927-28 - inspection of var high voltage labs & attended 6 wks Prof Conf at G E 1937. Mbr AIEE 1918; Fellow 1951 Amer Soc for Engrg Educ 1950. Chmn AIEE Pacific Coast Elec Space Heating & Heat Pumps 1957-59. Res in magnetic measuring, elec transicuts, & elec heating & heat pumps, resulting in 15 publ papers in AIEE, ASHVE, London Electrician & others, as well as 4 Univ of Wash Exp Station bulletins. Sev cons assignments before & after retirement. Hobbies: woodwork with rare woods - rock cutting & polishing - garden work. Awarded title "Engineer of the Year" 1957 by Seattle Section of Society of Professional Engineers. *Society Aff:* IEEE.

Smith, George V
Home: 104 Berkshire Rd, Ithaca, NY 14850
Position: Consulting Engr (Self-emp) *Employer:* Retired *Education:* ScD/Metalluray/Carnegie Mellon U; BS/Metallurgical Eng/Carnegie Mellon U *Born:* 4/7/16. Native of Clarksburg, WV. Employed by US Steel Corp in Fundamental Res Lab 1941-1955. From 1955 to 1970, was Francis Norwood Bard Prof of met engg. Cornell Univ. From 1957-1962, also asst dir of Sch of Chem & Met Engg & Prof of Engg Physics, Cornell Univ. Full-time consulting, 1970 to 1982, now retired. Honorary fraternities: Sigma Xi, Tau Beta Pi, Phil Kappa Phi, Alpha Sigma Mu. Authored more than 100 technical publications including two textbooks. *Society Aff:* ASM, AIME

Smith, Halfred C
Business: Bethlehem Plant Office, Bethlehem, PA 18016
Position: Asst General Manager *Employer:* Bethlehem Steel Corp - Beth Plant. *Born:* Apr 25, 1929. BS Met Engrg from Michigan Tech Univ in 1951. Mbr Alpha Sigma Mu - Met Hon Soc. Native of Battle Creek, Mich. Employed 1951 by Bethlehem Steel Corp, assuming present pos of Asst Gen. Mgr. 1977. Respon for Bethlehem Plant operations in the production of struct shapes, tool steels, castings & forgings. Have had a major involvement in the manufacture of very large forgings, primarily for the elec util indus. Mbr ASM, serving as Chmn of Lehigh Valley Chap in 196465. Received the Chap's Bradley Stoughton Award 1975.

Smith, Herman E
Home: 2612 Warwick Dr, Oklahoma City, OK 73116
Position: Owner. *Employer:* Herman E Smith & Assoc. *Education:* BS/Mech Engr/ OK State Univ. *Born:* 11/26/18. Native Oklahoman. Educ-BSME, OSU, 1939. Educ-BSME, OSU, 1939. Served in Army Air Force 1942-1946 and 1951-1953, discharged rank of Maj. With HTB, Inc, Architectural-Engg-Planning firm 1948 to 1976 as Sr VP and Secy-Treas. Established own firm in 1976, providing consulting services to the profession. Reg PE in 22 states. Mbr State Bd of Reg for PE and LS 1972-. Natl Dir ACEC 1975-81. Fellow of ASCE and ACEC. Bd of Visitors, Sch of Civ Engg, OK State Univ 1972-1975. *Society Aff:* ASCE, ASME, NSPE, SAME, ACEC, NCEE, APWA.

Smith, Howard W
Home: 1612 Crescent Rd, Lawrence, KS 66044
Position: Prof. *Employer:* University of Kansas. *Education:* PhD/Aero Engr/Okla State Univ; MS/Aero Engr/Wichita State Univ; BS/Aero Engr/Wichita State Univ *Born:* 11/24/29. The Boeing Co 20 years; Assoc Dean of Engrg 1974-76, Univ of Kansas to date. Reg PE 4591 in Kansas. Assoc Fellow AIAA, Outstanding Faculty Adv of the Year 1973. Mbr ASEE Aero Div Exec Bd, Outstanding Campus Activity Coordination 1973-74; Mbr SAME, Tasker Howard Bliss Medal Winner 1974; Mbr SESA, SAMPE, AHF, MAEGC, AIAA Shuttle Plaque Recipient (1984); conslt to Cessna Aircraft 1983-1984. AT&T Award (1987 - $1,500). *Society Aff:* AIAA, SEM, ASEE, SAME, AHF, SAMPE, MAEGC, ΩΚΩ, TBPI.

Smith, Hubert C
Business: 233 Hammond Bldg, Dept of Aerospace Engg, Univ Park, PA 16802
Position: Asst Prof. *Employer:* PA State Univ. *Education:* PhD/Systems Engg/Univ of VA; MS/Aerospace Engg/PA State Univ; BS/Aero Engg/PA State Univ; BA/ Liberal Arts/Gettysburg College. *Born:* 6/20/30. Served as USAF Aircraft Maintenance Officer, 1952-54. Served, as civilian, progressively, as Aero Engr, Proj Engr, and Chief, Engg Operations at Olmsted AFB, PA, 1959-65. Joined the PA State Univ as Res Asst in 1965, promoted to Res Assoc in 1968, Asst Prof in 1971. Author of books: *Introduction to Aircraft Flight Test Engineering, Performance Flight Testing, The Illustrated Guide To Aerodynamics*; numerous papers on aircraft performance and aerospace education. Instr of first aviation education workshop in PA in 1970.FAA licensed commercial pilot, flight instr, and accident prevention counselor. Assoc. Fellow of AIAA. *Society Aff:* AIAA, ASEE, ASAE, SAE.

Smith, Isaac D, Jr
Home: 133 Sampa Rd, Mt Pleasant, SC 29464
Position: V P of Laboratories and Special Projects. *Employer:* Soil Consultants Inc. *Education:* BS/CE/Citadel. *Born:* 9/29/30. USN 1951-55. Reg PE in SC & GA. Mbr: ASCE. State Pres 1975; Charleston Civil Engrs Club; Const Specification Inst, Charleston Chap Dir & P Pres; SC Soc of Prof Engrs; ASCE Contact Mbr; The Citadel Student Chap (1963-1986), US Natl Soc of the Intl Soc of Soil Mechs & Foundation Engrs. Appointed to panel of arbitrators, Amer Arbitration Assn. Natl Railway Historical Soc. Listed Who's Who Southeast 1975-76. Hobby-Model Railroading. *Society Aff:* ASCE, CSI.

Smith, J F Downie
Home: 900 Holly Lane, Boca Raton, FL 33432
Position: Retired. *Education:* DS/Heat Transfer/Harvard Univ; MS/Heat Transfer/ Harvard Univ; ME/Mech Engg/VPI; SM/Mech Eng/GA Tech. *Born:* Oct 1902. Taught at Georgia Tech, Va Poly Inst, & Harvard Graduate Sch of Engrg. Dean of Engrg, Dir of Engrg Exper Stat & Dir of Engrg Extension Ser for 11 years Iowa State Univ. Chmn of Adv Council on Engrg Manhattan Coll. Advisor in Engrg to sev Colls. Exec Engr for R&D at Budd Co; Ch Engr United Shoe Machy Co; Vice Pres Carrier Corp & Pres of Carrier R&D Co. Honorary Mbr Iowa Engrg Soc. Served 6 years as advisor to Natl Bureau of Standards. Served in many capacities in ASEE including council. Pub 50 papers on Heat transferaNd Education etc. 8 patents. Many activities in ASME, Fellow 1947- . Organized Machine Design & Rubber & Plastics Divns. Served on Res Council Power Test Codes, Bd on Technology, Res Planning, & Mbr of Council From 1969-Dir, Boca Raton Natl Bank. *Society Aff:* ASME, ASEE.

Smith, J Herbert
Business: Box 107 Commerce Court Post O, Toronto, Toronto Ontario M5L 1E2. Canada
Position: Consulting Engineer. *Employer:* Retired. *Education:* DSc/Hon/Univ of New Brunswich; DSc/Hon/Assumption Univ; MSc/Elec Engg/Univ of New Brunswick; BSc/Elec Engg/Univ of New Brunswick. *Born:* 11/21/09. Cons Engr; Retired Chmn, De Havilland Aircraft of Canada Ltd. Dir: Acres Cons Servs Ltd; Banister Continental Ltd; Heitman Canadian Realty Investors; Rio Algom Ltd; Rio Tinto Holdings Ltd; Canadian Imperial Bank of Commerce; Sun Life Assurance Co of Canada; Lornex Mining Corp Ltd. Retired as Chmn & Ch Exec of Canadian Genl Electric Co Ltd 1972 after 40 years of service. Pres Assn of Prof Engrs of Ontario 1953; Mbr Engrg Inst of Canada; Fellow Inst of Electrical & Electronics Engrs. *Society Aff:* IEEE, EIC.

Smith, Jack
Business: College of Applied Sci & Engg, Arizona State Univ - West Campus, Phoenix, AZ 85023
Position: Prof of Engg *Employer:* Univ of TX at El Paso *Education:* PhD/EE/Univ of AZ; MS/EE/Univ of AZ; BS/EE/Univ of AZ *Born:* 11/28/27 Attended Stanford Univ under an NSF Faculty Fellowship. Employment history is as follows: GE Co 1952-56; Prof during the yrs 1956-64. Univ of TX at El Paso: Assoc Prof and Prof of Elec Engrg; Asst Graduate Dean for 4 yrs; Dean College of Engrg sin 23699325oftheJof thece 1976-82. Professor Arizona State University - West Campus 1986-present. Electrical and Electronics Engrs; American Geophysical Union; American Society of Engrg Educ. Major interests: Electromagnetics - optical and high frequency radio wave propagation characteristics. *Society Aff:* IEEE, ASEE, AGU, ΣΞ.

Smith, Jack L
Home: 12 Tulane Ave, Pocatello, ID 83201
Position: Div VP. *Employer:* J R Simplot Co. *Education:* MBA/Bus Adm/ID State Univ; BS/Bus Adm/ID State Univ; BS/ChE/Univ of UT. *Born:* 10/29/31. Since graduating from the Univ of UT, I have worked as a utility corrosion engr, a refinery instrumentation engr, a chem proj engr, & in several capacities with my present employer: process engr, Chief Process Engr, Dir of Process Engg & Environmental Control, VP for Dev & Planning. I am responsible for: Long-Range Planning, Res & Dev & Environmental Affairs. I am a PE (ID) & have been Chrmn, ID Sec, AIChE & Chrmn, Mfg Environmental Committee, The Fertilizer Inst. I have published several papers & hold two US patents (both joint with other inventors). I recently served on the Governor's (ID) Task Force on Envirosafe (hazardous waste disposal site). *Society Aff:* AIChE, ACS, APCA, AMA.

Smith, Jack R
Business: Elec Engrg Dept, Gainesville, FL 32601
Position: Professor of Elec Engrg & Psychiatry. *Employer:* University of Florida. *Education:* BS/EE/USC; MS/EE/USC; PhD/EE/USC. *Born:* 1935 ND. Elec Engr: Hughes Aircraft Co, Jet Propulsion Lab, Aerospace Corp. Joined Univ of Fla in

Smith, Jack R (Continued)
1964, Prof of Electrical Engrg & Psychiatry. ASEE Southeastern Sect Western Electric Award for excellence in Engrg Education 1975. Mbr Inst of Electrical & Electronic Engrs, sleep Research Society, & Audio Engrg Soc. *Society Aff:* SRS, IEEE, AES.

Smith, James A
Home: 737 S. Bristol St, Lakewood, CO 80228
Position: Chief Structural Engr *Employer:* The Engrs Collaborative, P.S.C. *Education:* Bachelor of Sci/Civil Eng/Univ of FL. *Born:* 4/6/39. Native of Pensacola FL. Served with USNR 1958-60. Structural Engr for Phillip R Jones Cons Engr; Dept Hd, stockholder Phillip R Jones & Assoc Inc; Pres of James A Smith & Assocs Inc; Proj Engr, Mickey & Fox, Inc. Young Engr of the Yr NWF chap FES 1975-76. Mbr ACEC, NSPE, ASCE, FES, & FICE. Mbr of the Nazarene Church. 1973 Awards Vol of Outstanding Young Men of America; 1974 Vol of Who's Who in FL. Enjoy fishing, boating & family activities. *Society Aff:* NSPE, ASCE

Smith, James G
Business: College of Engg & Tech, Carbondale, IL 62901
Position: Professor *Employer:* Southern Illinois University. *Education:* PhD/Engg Physics/Univ of MO at Rolla; MS/Elec Engg/Univ of MO at Rolla; BS/Elelc Engg/ Univ of MO at Rolla. *Born:* May 1930. Native of Benton, Ill. Served with UN forces in Korea 1951-53. Taught at Univ of Mo at Rolla 1957-66. With Southern Ill Univ at Carbondale 1966- . Chmn of Electrical Sciences & Systems Engrg Dept 1971-80. Served as ASEE Ill-Ind Sect Chmn 1970-71 & mbr of Council of Sects East. Mbr of Exec Council of Zone II 1971-72. Indus exper includes the Boeing Co in antennas & electromagnetic fields. *Society Aff:* IEEE, ASEE, AAAS, ΤΒΠ, ΦΚΦ, ΣΠΣ.

Smith, James L
Business: Dept of Agri Engr, Laramie, WY 82071
Position: Prof & Hd *Employer:* Univ of WY *Education:* PhD/Agri Engg/Univ of MN; Masters/Civ Engg/Univ of IL; BS/Agri Engg/Univ of IL. *Born:* 2/16/37. Native of Geneseo, IL. Served on faculty of Univ of UT & IL Inst of Tech and CO State Univ. Proj Engr with Intl Harvester Co, 1967-1969 specializing in soil dynamics. Appointed Head of Agricultural Engrg Univ of WY, 1981. Fulbright Lecturer and visiting prof. Univ of Nairobi, Kenya, 1987-88. Current teaching and res activities include machine design and dev related to land application of wastes, strip mine reclamation, surface irrigation efficiency and erosion control. Foreign Univ service in Mexico and Kenya. Reg PE in CO, WY, & AZ. *Society Aff:* ASAE, WPCF, ASEE

Smith, James R
Business: Box 5011, Cookeville, TN 38505
Position: Prof. *Employer:* TN Tech Univ. *Education:* PhD/IEOR/VA Tech; MS/ Statistics/VA Tech; BS/IE/VA Tech. *Born:* 11/15/43. Native of Gate City, VA. Worked as Industrial Engr at Holston Defense Corp, Kingsport, TN from 10-66 to 6-69. Grad Student and Instructor at VA Tech's IEOR Dept 6-69 to 8-71. On Faculty of Dept of Industrial Engr at TN Tech Univ from 9- 71 to present. *Society Aff:* AIIE, ASQC.

Smith, Jerome A
Business: 800 N Quincy, Arlington, VA 22217
Position: Tech Dir. *Employer:* Office of Naval Research. *Education:* PhD/Aero/CA Inst of Tech; MS/Aero/CA Inst of Tech; BSE/Aero Engg/Univ of MI. *Born:* 4/17/40. Faculty mbr Princeton Univ Dept of Mech and Aero Engg; 1967-72 Asst Prof, 1972-78 Assoc Prof 1978 Prof. Research emphasis on experimental gas dynamics, particulary gas physics. Contributed to understanding of high temp gasses using shock tubes; studied low density, nearly free molecular flows and high density turbulent hypersonic flows; examined characteristies of candidate high energy laser systems. Experience with flow diagnostics utilizing lasers, electron beam fluorescence and automated data acquisition. 1979 Tech Dir of ONR, determining tech scope and objectives of the Navy's basic research program conducted in univs and inhouse labs. *Society Aff:* APS, ΣΞ.

Smith, Joe M
Home: 760 Elmwood Dr, Davis, CA 95616
Position: Prof. *Employer:* Univ of CA. *Education:* ScD/Chem Eng/MIT; BS/Applied Chemistry/Cal. Tech *Born:* 2/14/16. Text books in chem engg, thermodynamics & kinetics. Prof at Purdue 1945-57, Northwestern 1957-61, & Univ of CA Dav 1961-. Fulbright & Guggenheim Awards, Netherlands, Spain, Chile, Argentina, Ecuador, Mbr ACS, AIChE, Nat Acad of Engg. W.K. Lewis, R.H. Wilhelm and W.H. Walker Awds of American Inst of Chem Engrs, and Outstanding Educator, ASEE, Chemical Engrg Section. *Society Aff:* AIChE, ACS.

Smith, John W
Business: Civ Engg Dept, Memphis, TN 38152
Position: Prof. *Employer:* Memphis State Univ. *Education:* PhD/Civ Engg/Univ of MO; MS/Civ Engg/Univ of MO; BS/Civ Engg/Univ of MO. *Born:* 11/18/43. A native of DeSoto, MO. Worked for Esso Res, an engg co, prior to joining Memphis State Univ in 1970. Have served as asst associate and presently as full prof of Civ Engg at MSU. Presently involved in res and teaching in the Civ Engg dept. Res efforts have lead to being selected as Distinguished Researcher in 1979 at Memphis State Univ. *Society Aff:* AWWA, WPCF.

Smith, Joseph Seton
Home: 4912 Countryside Ln, Lyndhurst, OH 44124
Position: President. *Employer:* Seton-Scherr. *Education:* DScEng/EE/NY Univ; MEE/EE/NY Univ; BS/EE/IA State Univ. *Born:* July 1925. Fellow IEEE. Res Assoc with NYU 1946-55. Originated courses in Digital Computer Design. System Design for original transistorized computers for Burroughs Corp. Prog Mgr for Range Ships at Sperry Rand. Div Pres for Communivations & later for Guidance & Control Systems of Litton Indus. Sr VP of A-T-O for Electrical & Electronics, & Const & Mining Machinery Operations. Chmn NY Sect IRE 1957; Sigma Xi, Eta Kappa Nu, Pi Mu Epsilon. *Society Aff:* IEEE.

Smith, Julian C
Home: 711 The Parkway, Ithaca, NY 14850
Position: Prof *Employer:* Cornell University. *Education:* CHEME/CHE/Cornell U; BCHEM/Chemistry/Cornell U *Born:* March 1919. Native of Montreal Canada. (US Citizen). Res Engr, Group Head duPont Co 1942-46; Asst Prof Chem Engrg Cornell 1946, Assoc Prof 1949, Prof 1953-86; Emeritus Prof 1986- Dir of continuing education 1965-71; Assoc Dir Chem Engrg 1973, Dir 1975-83 ; Cons to duPont Co 1955- , Rockwell Intl 1977- ; other companies & government agencies. Co-author, 'Unit Operations of Chem Engrg'; about 50 tech articles; contributor ,sect editor Perry's 'Chem Engrs Handbook.' Fellow AIChE; Mbr ACS, Soc Chem Ind. Enjoy piano playing, singing, golf, home computers. *Society Aff:* AIChE, ACS

Smith, Kenneth A
Business: 555-17th St, Rm 1820, Denver, CO 80217
Position: VP Technology. *Employer:* Anaconda Industries. *Education:* PhD/Chem Engg/Univ of WI; MS/Chem Engg/Univ of WI; BS/Chem Engg/Univ of WI. *Born:* Dec 30, 1921 LaCrosse Wisc. Joined Sinclair Oil Corp Harvey Ill res org in mid-1940's; became Mgr of Sinclair's Corporate Computer Servs in New York, NY 1960; elected Vice Pres of Sinclair Petrochemicals 1964; following Sinclair/Atlantic Richfield merger transfered to Philadelphia becoming Vice Pres of Res & Engrg for ARCO Chem Co 1973. 1977-date - VP, Technology, Anaconda Industries, Div of The Anaconda Co, Subsidiary of Atlantic Richfield Co. Mbr AIChE, ACS, API, IRI, Tau Beta Pi, Sigma Xi, Phi Lambda Upsilon, & the Petroleum Club. *Society Aff:* AIChE, ACS, IRI, SCI-NA, CSMRI.

Smith, Kenneth R
Business: 4405 Talmadge Rd, Toledo, OH 43623
Position: Chrmn. *Employer:* Finkbeiner, Pettis & Strout Ltd. *Education:* BSCE/Civil

Smith, Kenneth R (Continued)

Engg/Oh Univ. *Born:* Sept 1930. USAF 1952-54. Finkbeiner, Pettis & Strout 1954- . Past Chrmn Ohio Sect AWWA,; P Pres Toledo Sect ASCE; P Vice Chmn Cons Engrs of Ohio; P Trustee Engrs Foundation of Ohio. Received 'Young Engrs Award' Ohio Soc of Prof Engrs 1964. Reg PE in Ohio, Mich, NY, Ind, Ky, N. Carolina, S. Carolina & Natl Council of State Bds of Engrg Examiners. Specialized exper in Water Supply, Treatment & Distribution, Wastewater Collection Systems & Treatment, Water Utility Mgmt, & marine work associated with the water & wastewater field. Interested in duck & pheasant hunting, fishing, & early american antiques. *Society Aff:* NSPE, ASCE, AWWA, WPCF, SAME.

Smith, Lawrence H

Home: 71 Evergreen Lane, Glastonbury, CT 06033
Position: President. *Employer:* ORCOMATIC, INC - Div of Eagle-Picher Industries. *Education:* BChE/Chem Engg/Rens Poly Inst. *Born:* 11/16/34. Native of Cornwall NY. Graduated from Storm King Sch (Coll Prep); received Holloway US Navy Scholarship to attend Rensselaer Polytechnic Inst attaining BSChE. Commissioned & served 3 years as line officer in Atlantic Distroyer Fleet. Chemist & Senior Engr for E I duPont 1959-67; held positions as Prod Sales Mgr, Marketing Mgr, Mfg Mgr, & Genl Mgr for Ohio Rubber Co 1967-76. Pres Orcomatic Inc 1978-Present. Mbr Conn Rubber Group. Hobbies include tennis & officiating scholastic soccer. R.I. Rubber Group, Amer Chem Soc, Amer Inst of Chem Engrs. *Society Aff:* ACS, AIChE, CRG, RIRG.

Smith, Lawrence L

Home: 1123 NW 107 Terrace, Gainesville, FL 32601
Position: Deputy State Matls & Rsch Engr *Employer:* FL Dept of Transportation *Education:* BS/CE/Univ of FL; MS/CE/Univ of FL *Born:* 7/10/32 in Orlando, FL. Serve with heavy construction battalion, US Army 1956-58. Research engr with FL State Rd Dept 1958-62. Construction engr 1962-64. With FL Dept of Transportation since 1964. Held current position since 1969. Responsible for planning and conducting pavement research. Authored or co-authored over 30 technical and research reports. Publications accepted by TRB, ASCE, ACI and AAPT. Member of three TRB committees. Pres FL Section ASCE 1973; pres FL Council of Engrg Societies 1975. Engr of year FL Section ASCE 1975. Former member two Professional Activities Committees ASCE. Currently visitor of Accreditation Bd for Engrg and Technology (ABET). *Society Aff:* ASCE, NSPE

Smith, Leo A

Business: 207 Dunstan Hall, Auburn, AL 36830
Position: Prof. *Employer:* Auburn Univ. *Education:* PhD/Ind Engg/Purdue Univ; MSIE/Ind Engg/GA Tech; BIE/Ind Engg/GA Tech. *Born:* 5/22/40. Native of Waycross, GA. Married in 1962 to Mary Ruth Boggs of Russell, KY. Three sons. Military service as Lt assigned to US Army Human Engg Labs, 1964- 1965. Joined Ind engg faculty of Auburn Univ, Auburn, AL in Jan, 1969. Academic specialization in ind applications of human factors engg (ergonomics). AIIE Ergonomics Div Dir 1978-1979. Hobbies include fly fishing. *Society Aff:* IIE, HFS, ASSE, AIHA.

Smith, Leo C

Business: 9 Greenway Plaza, Houston, TX 77046
Position: Sr VP. *Employer:* The Coastal Corp *Education:* BS/ChE/Univ of UT. *Born:* 6/15/32. Responsible for the gen mgt of the Coal & Chem Div of the Coastal Corp. Has led the Div through a series of expansion projs leading to the creation of the largest single underground coal mine West of the MS. Has served as Pres of the Wycon Chem Subsidiary since 1972. Was responsible for the startup & operations of this grass roots ammonia based fertilizer complex in 1964 as Production Supt. Prior employment with US Steel Corp, Shell Chem Co and GE Co. *Society Aff:* AIChE.

Smith, Leroy H

Business: 1 Neumann Way, PO Box 156301, Cincinnati, OH 45215-6301
Position: Manager, Turbomachinery Aero Technology *Employer:* GE Aircraft Engines *Education:* Dr. of Engineering/Mech. Engrg./The Johns Hopkins University; MS/Mech. Engrg./The Johns Hopkins University; BE/Mech. Engrg./The Johns Hopkins University *Born:* 11/03/28 Dr. Smith joined GE Aircraft Engines in 1954. He has held a succession of individual contributor and managerial positions concerned with aerodynamic research, design, and development of turbomachinery, particularly axial flow compressors and fans. His work has resulted in significant advances in compressor aerodynamic design technology. He is currently Manager of Turbomachinery Aero Technology. *Society Aff:* ASME

Smith, Leslie G

Home: 2305 Southmoor Dr, Champaign, IL 61821
Position: Prof. *Employer:* Univ of IL. *Education:* PhD/Phys/Univ Cambridge; MA/Phys/Univ Cambridge; BA/Phys/Univ Cambridge. *Born:* 11/14/27. Native of England. Came to USA 1951, naturalized, 1959. Scientist at AF Cambridge Res Ctr, 1953-1958; at GCA Corp, 1958-1972. Assoc Dir, Aeronomy Lab, Dept of Elec and Comp Engg, UIUC, 1972-85. Prof of Elec and Comp Engg, UIUC, 1975-present. Darton Prize, Royal Meteorological Soc, 1954. Profl interests: Ionospheric res using instrumented sounding rockets; atmospheric electricity. Over 50 publications in technical journals. *Society Aff:* IEEE, AGU.

Smith, Leslie S

Home: 1230 Del Webb Blvd, Sun City Center, FL 33570
Position: Retired *Employer:* Consultant *Education:* BS/Mech Engg/NJ Inst of Tech. *Born:* 8/23/17. Eagle Scout, B-29 flight engr, WWII. Stone & Webster Engg Corp, Boston, '45-'53, electric power design projs. Metcalf & Eddy, Boston & Greenland, '53- '57. Boston Edison Co '57-'82, Sr Mech Engr. Scout leader, 4 yrs. Town Planning Bd, 4 yrs. Lay leader of church, 7 yrs. Chrmn of Library Trustees '71-'81. Chmn, Boston Section ASME '68. Pres, Engg Soc of New England, '73. VP, New England Region, ASME '78-'80. Wife, Doris. Children, Douglas & Patricia. 6 grand children. Speed Walking 10 miles/wk. Enjoys classical music, art, and travel. *Society Aff:* ASME.

Smith, Lester C

Home: 1840 Asylum Ave, West Hartford, CT 06117
Position: Retired. *Employer:* Retired. *Education:* BS/Chem Engg/MIT. *Born:* Aug 1904. Junior Engr with Spencer Turbine Co specializing in air movement 1925; Asst Ch Engr concerned with air mocvement invloving Turbo Compressors, Gas Boosters, Indus Vacuum Cleaners 1932; became Ch Engr 1944, Vice Pres 1947, Dir 1952, assumed Presidency 1957. Responsibility that of Ch Exec Officer directing all facets of Co operation. Hold 14 patents on Indus Vacuum Cleaning, Dust Collecting, & Air Flow Devices. Special interest in engrg aspects of air movement as it pertains to Air Pollution. Retired 1971. Reg PE in Conn. Life Fellow ASME, Vice Pres 1966-67, Dir 1963-66. *Society Aff:* ASME.

Smith, Lester H, Jr

Business: 2135 Blakemore Ave, Nashville, TN 37212
Position: Partner. *Employer:* Smith Seckman Reid Inc. *Education:* BE/Elec Engg/Vanderbilt Univ. *Born:* 5/24/30. Served USAF 1951-52. Electrical Engr Monsanto Co 1954-56, engaged in design & field supervision const projs; Engr Vanderbilt Univ 1956-62, responsible for plant maintenance, remodeling, long-range planning, contract admin & inspection; Administrator Vanderbilt Sch of Medicine 1962-67, in business areas of personnel, finance, budgeting, planning; founding partner & Ch Electrical Engr 1968-75 Smith Seckman Reid Inc, Pres 1976. Mbr IEEE & its I&CPS Group. Reg PE in 13 states. Private pilot. Pres, Consulting Engrs of Tennessee, 1982-83. Treasurer, Amer Conslting Engrs Council, 1984-86. Author, Chapter 2, IEEE White Book, *Electrical Design for Hospitals*. *Society Aff:* IEEE, ACEC, NSPE.

Smith, Levering

Home: 1462 Waggamanciy, Mclean, VA 22101
Position: Consultant. *Employer:* Self. *Education:* BS/-/US Naval Acad. *Born:* March

Smith, Levering (Continued)

1910. Commissioned Ensign US Navy 1932 var assignments 1932-47; Dept Head Dep Tech Dir US Naval Ordnance Test Station 1947-54; commanding officer US Naval Missile Test Facility White Sands NM 1954-56; Tec Dir Strategic Systems Proj Dept Navy Washington 1956-65; Dir 1965-1977. Deocrated: 3DSM; C N Hickman Award ARS 1957; Parsons Award US Nave League 1961; Conrad Award US Navy 1966; Hon Knight Commander Order of British Empire 1972. Forrestal Award-Natl Security Ind Assn - 1977. Fellow AIAA; Mbr Natl Acad of Engrg, ASNE (Gold Medal 1961), ADPA (Medalist 1968), Sigma Xi. *Society Aff:* AIAA, ASNE, ADPA, $\Sigma\Xi$.

Smith, Marion L

Business: 2070 Neil Ave, Columbus, OH 43210
Position: Associate Dean, Engineering, Emeritus. *Employer:* Ohio State University. *Education:* Master of Sci/Mech Engr/OH State Univ; Bachelor of Sci/Mech Engr/ Louisiana State Univ. *Born:* June 1923. As prog officer of Coll 1958-84 was responsible for curriculum, evening, co-op, part-time & continuing education, student servs, placement, affirmative action programs. Chr of Ohio Engrg Regis Bd. Chr, NCEE Uniform Exam Committee. Taught thermal systems courses in mech engrg at Ohio State Univ & directed res on combustion & miniature engine generators 1947-58. Co-author of textbook 'Fuels & Combustion.' Engrg exper: Cooper Bessemer Corp, E I duPont engrg dept, & Co Commander, Army Ordnance 1945-46. Fellow ASME. Also active in ASEE, NSPE, SAE, NCEE, community service. Hobbies: golf, music. *Society Aff:* ASME, ASEE, SAE, NSPE.

Smith, Marion P

Business: 13350 US Hwy 19, St Petersburg, Clearwater, FL 33516
Position: Principal Reliability Engineer *Employer:* Honeywell Inc. *Education:* BS/EE/LA State Univ. *Born:* Dec 1925. Additional work at Okla A&M, Yale & Univ of S Fla. 'Engr of the Year' Fla 1970. ASQC Outstanding Serv Award 1968-69. Fellow IEEE 1973; P Chmn IEEE Reliability Group. Appointed Quality & Reliability Advisor, US Army Mgmt Training Agency 1975; Tech Advisor, US Natl Ctte of Internatl Electrotechnical Comm TC-56 - Reliability & Maintainability 1975; VP, US Natl Comte of Intl Electrotech Comm 1979- ; USA Delegate 1st NATO Conference on Quality 1973; Engrg supervision Bell Telephone, Armed Forces Security Agency, Vitro & Honeywell. Currently Principal Engineer - Reliability Honeywell Avionics Div. Pres 1973-75 Natl Assn Retarded Citizens. Mbr Natl Adv Council Dev Disabilities. Annual Reliability Award 1978-IEEE Reliability Group. Honeywell Community Service Award 1979. White House Conference on Inflation 1975. White House Conference on Handicapped 1977 National Chairman for Governmental Affairs - National Association for Retarded Citizens, 1977 to present. *Society Aff:* IEEE, ASQC.

Smith, Mark K

Home: Main St, PO Box 189, Norwich, VT 05055
Position: Management Consultant. *Education:* PhD/Geophysics/MIT; BS/ Geophysics/MIT. *Born:* Feb 1928. Joined Geophysical Serv Inc as res seismologist & later made Dir of Res Engrg & Dev; 1963 promoted to Mgr RD&E for TI's Sci Serv Div & named V Pres of GSI; later named Exec V Pres of GSI & Pres 1966; in 1967 promoted to V Pres of Texas Instruments & Mgr of Sci Serv Div; in 1969 became Mgr of newly formed Information & Automation Serv Div of TI; appointed V Pres of Corp Dev 1972. *Society Aff:* NAE

Smith, Milton L

Business: P.O. Box 4130, Dept of Industrial Engrg, Lubbock, TX 79409
Position: Prof *Employer:* TX Tech Univ *Education:* PhD/IE/TX Tech Univ; MS/IE/ TX Tech Univ; BS/IE/TX Tech Univ. *Born:* 5/30/39 Spent early years in Childress, TX. Served with US Air Force from 1962 to 1965 as a Management Engrg Officer in the Azores and Delaware. Became faculty member at TX Tech Univ in 1968. Teaching interests in applied quantitative methods and computer simulation; research interests in agricultural systems, hailstone hazards to materials, and scheduling and control of production systems. *Society Aff:* IIE, ASEE, ORSA

Smith, Morton C

Home: 163 El Gancho, Los Alamos, NM 87544
Position: Consultant, Geothermal Energy *Employer:* Los Alamos Nat'l Lab, Univ of Calif. *Education:* MS/Metallurgy/Lehigh Univ; Met Eng/Metallurgy/SD School of Mines; BS/Met Eng/SD Sch of Mines. *Born:* Sept 15, 1917 Selby, SD. Coll teacher, Metallurgy 1939-45, 1946-54; Indus Metallurgist 1945-46; Staff Mbr, Alternate Group Leader, Group Leader Los Alamos Scientific Lab 1954-1985 (alloy dev, carbon & graphite R&D, fuel elements, melting penetrators, geothermal energy). Independent consultant, 1985- . Reg PE in Colo. Tech author. Fellow ASM. Disting Scientist of the Year NM Acad of Sci 1974. Lead Participant, NATO-CCMS Geothermal Energy Pilot Study. Proj Leader US-Italy Cooperative Geothermal Prog. Organizing Ctte, Second UN Geothermal Smyposium. Geothermal panels, cttes, advisory bds. *Society Aff:* ASM, NMAS.

Smith, Neal A

Business: 2015 Neil Ave, Columbus, OH 43210
Position: Prof, Emeritus *Employer:* The OH State Univ *Education:* MS/EE/OH State Univ; B/EE/OH State Univ. *Born:* 2/10/19 Born in Norwich, OH, son of Miner and Bessie Smith. Graduate Zanesville, OH high school. Married Faye Schlupe 1942. Electrical Engr, Federal Machine and Welder Co Warren, OH 1941-44. Lt jg USNR electronics officer 1944-46. OH State Univ, Elec Engrg Dept: Instructor 1947-54, asst prof 1954-60, assoc prof 1960- 65, prof 1965-84, prof emeritus 1984- . Teaching areas: circuits, machines, energy conversion, power systems, high voltage. Consultant: electric power systems, electric shock, fires of electrical origin. NSF grantee 1961-62, member of Eta Kappa Nu, Tau Beta Pi. Senior Mbr of PES Society, Ind App Society, and Education Group of IEEE. Mbr ASEE. *Society Aff:* ASEE, IEEE

Smith, Nelson S, Jr

Business: Dept of Elec Engg, Morgantown, WV 26506
Position: Prof. *Employer:* WV Univ. *Education:* DSc/Elec Engg/Univ of Pittsburgh; MS/Elec Engg/WV Univ; BS/Elec Engg/WV Univ. *Born:* 8/9/29. Served in USAF 1948-1952. With WV Univ since 1956, Prof since 1970. Consultant to US Bureau of Mines since 1957, US Dept of Energy since 1979. Res interests include dielectric absorption currents, electrets, electrostatic precipitation, electrogasdynamics, semiconductor gas sensors. NSF Sci Faculty Fellow 1960, 61. *Society Aff:* IEEE, ASEE.

Smith, Norman B

Business: Stanley Building, Muscatine, IA 52761
Position: Chief Chem Engr *Employer:* Stanley Consultants Inc. *Education:* BS/Chem/Univ of IL; B S/Chem Eng/Univ of IL. *Born:* May 1927 Lincoln Ill. Fermentation Plant Supt 1952-58, responsible for production, recovery & shipping indus fermentation prods; R&D Chem Engr 1958-66, res included process dev to manufacturer corn starch & dextrose. US & foreign patents granted. With Stanley Cons 1966- , present. Chief Chem Engr. Respon for chem engrg & discipline activities including quality assurance programs, new capabilities, continuing education and chem engrg skills dev. Special expertise in process energy conservation, atmospheric fluidized bed combustion, flue gas desulfurization tech, fuel alcohol processes, corn wet milling industry, and utility plant corrosion control. Also respon for design of major projs involving chem engrg. Reg PE. Sr Mbr ACS, AAAS, SIM & NSPE. Dir Iowa Engrg Soc. Mbr IL Chap Triangle Fraternity 1949- . *Society Aff:* ACS, AAAS, SIM, NSPE, IES, AIChE, ASTM.

Smith, Paul D

Business: 1440 Broadway, Oakland, CA 94612
Position: Pres. *Employer:* Gage-Babcock & Assocs. *Education:* BCE/Civ Engg/Univ of Santa Clara. *Born:* 12/12/23. Served with Army Air Corps in WWII. Fire Protection Engr and Chief Engr for Johnson & Higgins (CA) 1950-57. With GBA since 1957 in Sr, Supervisory and Proj Engg and Mgt positions on projs in North Americ and Europe. Pres since 1964. Was pres of No CA-NV Chap (1966) and So CA-AZ

Smith, Paul D (Continued)
Chap of SFPE (1970-71). Chair several NFPA technical committees and SFPE Committee on Engg Reg. *Society Aff:* SFPE, ASCE, CEAC, ACEC.

Smith, Paul R,
Home: 1675 York Ave, New York, NY 10028
Position: Pres *Employer:* PRS Energy Industries, Inc *Education:* MS/ME/Worcester Polytech Inst; BS/Aero Engrg/Parks Coll-St Louis Univ; AS/ME/Worcester Jr Coll *Born:* 7/4/44 Native of Shrewsbury, MA, having 17 yrs of mgmt experience in engrg activities for the power generation, aerospace, and defense indus. He has written a text book for McGraw-Hill on piping sys in power plants, indus and commercial facilities. Proj Mgr experience has been gained at NPS and EPM, as well as PRS Energy Indus, which he founded and was CEO. The corp performed engrg conslt, engrg training, and exec search natly. He has served as Lead Engr for Stone and Webster and Principle Engr for EBASCO Services. He has also served as intl mech expert on design, fabrication and manpower requirements. Mr. Smith is a former mbr of an IEEE Working Grp, co-authored an ASME paper and was filmed making a video tape on nucl power plants supports. *Society Aff:* NSPE, NYPE, ASME

Smith, Peter
Home: 1792 Truscott Drive, Mississauga Ontario, Canada L5J 1Z9
Position: Dir Res & Dev Branch. *Employer:* Ministry of Transpt & Communic. *Education:* BSc/Civil Engg/Univ of Leeds England. *Born:* 7/20/26. Airfield const Royal Air Force Germany & England 1947-54; concrete projs for aeronautical & nuclear res establishments 1954-57; Sr Matls Engr (concrete) Ontario Dept of Hwys prior to leading matls res 1965-71; Dir Engrg R&D Branch 1971-78. Dir Sys R&D Branch 1978-80. Reg PE & Const Matls Specialist in Ontario. Pres 1982-83 and Fellow ACI; Mbr ICE, CSCE, ASTM, RTAC. Prof recognitions: RTAC President's Medal 1966, 70; TRB Award 1971; Stanton Walker Lectureship Univ of MD 1971; ACI Distinguished Serv Award 1972; APEO Silver Engrg Medal 1973; R E Davis Lecturer ACI 1975, ASTM Charles B Dudley Award 1983. Personal Honors: Queen's Silver Jubilee Medal, 1978 *Society Aff:* ACI, CSCE, ICE.

Smith, Peter W
Business: Bell Communications Research, 331 Newman Spgs Rd, Red Bank, NJ 07701
Position: District Res Mgr, Guided Wave and Opto Electronics Research *Employer:* Bell Communications Res Inc *Education:* PhD/Physics/McGill Univ; MSc/Physics/McGill Univ; BSc/Math & Physics/McGill Univ *Born:* 11/03/37 Worked at Canadian Marconi Co, Montreal Canada 1958-59. From 1963-83 mbr of technical staff at Bell Labs, Holmdel, NJ. USA working on quantum electronics res. In 1970 was on leave of absence from Bell Labs as visiting MacKay Lectr in the Dept of EE, Univ of CA, Berekely. In 1978 was on leave of absence from Bell Labs as visiting res scientist, Lab D'Optique Quantique, Ecole Poly, Palaiseau, France. Since 1984 District Res Mgr, Bell Communications Res Inc, Holmdel NJ. Fellow of IEEE & Optical Soc of Am. Assoc Editor IEEE Journal of Quantum Electronics 1976-1979. Assoc Editor Optics Letters 1979-1982 Treas, IEEE Quantum Electronics and Applications Society, 1981, 1982. VP IEEE Quantum Electronics and Applications Soc 1983, and Pres 1984. Recipient IEEE Quantum Electronics Award 1986. *Society Aff:* IEEE, OSA, APS, AAAS, SPIE, CAP.

Smith, Pierson K
Home: Bean Rd RD 1, Norristown, PA 19401
Position: Vice President, Engineering. *Employer:* Baeuerle & Morris Inc. *Born:* April 1914. Graduated from Cornell Univ 1938. Native of Philadelphia Pa. Cost Engrg & Equipment Design Philadelphia Coppersmithing Co Inc 193841; Vice Presi Engrg Vulcan Corp, Manufacturers of Heat Treating Furnaces 1941-47; Philadelphia Coppersmithing Co Engrg & then Pres 1947-58; Vice Pres Engrg Baeuerle & Morris Inc, Designers & Manufacturers of Process Equipment 1958-1981. *Society Aff:* AIChE.

Smith, Raymond L
Home: P.O. Box 498, Chassell, MI 49916 *Employer:* R. L. Smith, Inc. Houghton, *Education:* PhD/Met Engg/Univ of PA; MS/Met Engg/Univ of PA; BS/Min Engg/Univ of AK. *Born:* 1/25/17. Vanceboro Maine. Asst Prof Metallurgy Univ of Alaska prior to WWII. Army Ordinance Corps 1943-46. Senior Res Metallurgist Franklin Inst Philadelphia 1953, then Tech Dir until 1959. Prof & Head Met Engrg & Coordinator of Res, Mich Technological Univ beginning 1959, Pres of Univ 1965-79. Pres. Houghton Daily Mining Gazette 1980-81; Pres. RLS Inc 1981-present. Areas of expertise: engineering education, minerals and metals supplies, research management low temperature mech properties, ultra high purity metals, diffusion. Pres 1979-80 ASM; Dir Lake Shore Inc, Lake Superior & Ishpeming Railroad. Dir. Cupper Range Co. D-engrg-Michigan Technological University, D-engr-South Dakota School of Mines, DSc Western Mich Univ, LLD Northern MI, Univ. Fellow TMS, ASM. D Robert Yarnell Award Univ of Penn; Honorary Mbr ASM Distinguished Alumnus Award Univ of Alaska; President's Award for Distinguished Citizenship Northern MI Univ Distinguished Life Mbr Alpha Sigma Mu. Outstanding Serv Award, Air Force ROTC. Contributor of numerous articles to metallurgical sci journals. *Society Aff:* ASM, AIME.

Smith, Raymond V
Business: 2021 N Old Manor, Wichita, KS 67208
Position: Rehab Engr *Employer:* Rehabilitation Engrg Ctr *Education:* PhD/Engrg Science/Oxford Univ; MS/Fuel Tech/Univ of UT; MS/ME/Univ of CO; BS/ME/Univ of CO *Born:* 11/17/19 Employment History: Design Engr at Boeing Aircraft Co, 1941-44 - 1948-49, Instructor, CO School of Mines, 1949-52, Research Engr at Sandia Corp 1952-53, NM St Univ, Asst Prof 1953-54, Univ of UT, Assoc Prof, 1954-57, CO St Univ, Assoc Prof 1957-61, Natl Bureau of Standards, Engr & Sec Chief 1961-71, Wichita State Univ, Prof & Chrmn, ME Dept 1971-78, Wichita State Univ, Prof 1978-1984. Self employed Rehab. Engr 1984-present. Attached UK, AERE, Harwell, Eng 1966-68, 1978, 1981. India Inst of Science, 1981, 1982, 1984. Honors and Awards: Cryogenic Engr Conf, Best Paper, 1963, NBS Distinguished Authorship Award, 1964, NASA Technical Utilization Award 1972. *Society Aff:* ASME, ΠΤΣ, RSA

Smith, Richard A
Business: PO Box 911, Baker St, Wheeling, WV 26003
Position: Pres. *Employer:* OH Valley Steel Co. *Education:* MS/Civ Engg/Univ of Pittsburgh; MPW/Public Works/Univ of Pittsburgh; BS/Civ Engg/WV Univ. *Born:* 9/22/47. Served as Dir of Public Works, City Engr, Planning Dir and City Mgr in Wheeling, WV. Founder and Pres of R A Smith & Co, consulting engg mgt firm. In addition to being Pres of OH Valley Steel Co, also occupy position of Pres of Terracom, Inc, real estate dev firm, and Capital Properties, Inc, real estate holding co. Mbr of bd of dir - Wheeling Society for Crippled Children, Wheeling Dollar Bank, Municipal Auditorium Bd, Mount de Chantal Visitation Academy, and OH Valley Industrial and Business Dev Corp. Received first Rockwell Fellowship to Univ of Pittsburgh Public Works Ctr. *Society Aff:* ASCE, NSPE.

Smith, Richard G
Home: RDI Box 277, Center Valley, PA 18034
Position: Dir *Employer:* AT&T Bell Labs *Education:* PhD/EE/Stanford Univ; MS/EE/Stanford Univ; BS/EE/Stanford Univ *Born:* 1/19/37 Received Natl Science Foundation and General Electric Fellowships for Graduate Study. Joined Bell Labs, in 1963 and has worked on solid state lasers, nonlinear optics and devices for fiber optics. Currently dir of Lightwave Devices Lab. Fellow of IEEE (1981); member Phi Beta Kappa, Tau Beta Pi. Co- chrmn conference on Laser and Electro-Optical Systems (CLEOS) 1980; Program Chrmn, 1978. Pres, IEEE Quantum Electronics and Application Society (1981). Married, three children. Enjoy sports. *Society Aff:* IEEE, OSA, APS

Smith, Robert E
Business: 124 W. Church St, Dillsburg, PA 17019
Position: Chrmn of the Bd *Employer:* Capitol Engrg Corp *Education:* BS/CE/MIT

Smith, Robert E (Continued)
Born: 12/12/19 Native of York, PA. Served in US Army Air Force 1941-1946. With Capitol Engrg Corp since 1946 beginning as a Jr principal responsible for sanitary design. Served as Pres 1959-1986 with supvry responsibilities for all of the civil disciplines practiced by the firm with strong emphasis on hgwy and bridge engrg. Both domestic and overseas projects. Fellow, ASCE; Fellow, ACEC; and Pres, CEC of PA 1971-73. Currently serves the firm as consultant and chrmn of the bd. *Society Aff:* ASCE, ACEC, NSPE, SAME.

Smith, Robert L
Home: 2915 Harvard Rd, Lawrence, KS 66044
Position: Deane Ackers Prof of Civil Engrg. *Employer:* University of Kansas. *Education:* MS/Mech & Hydraulics/Univ of IA; BS/Civil Engg/Univ of IA. *Born:* 10/31/23. In Schaller, IA. Asst Prof Univ of2; Exec Dir IA Natl Res Council 1952-55; Ch Engr KS Wat Res Bd 1956-62; Glen Parker Prof of Water Resources Univ of KS 1962-66; Special Asst Office of Sci & Technology & Chmn Comm on Water Resource Res Federal Council for Sci & Technology, Exec Offic of the Pres, 1966-67; Chmn Dept of Civil Engg Univ of KS 1966-72; Ackers Prof of Civil Engg 1972- . Consultant, Black & Veatch Conslts Engrs 1968-, Conslt Honeywell Corp 1984-85. Mbr NAE. KS Engg Soc Outstanding Engr of the Yr Award. Pres KS Sect ASCE KS 1948-5. Recipient USGS Centennial Plaque, 1980. Mbr Water Science & Technology Bd, Natl Res Council, 1982-86; Mbr Comm on Engrg and Tech Sys, Natl Res Council, 1983-86. *Society Aff:* ASCE, NSPE, AWWA, AGU, AAAS, ASEE, AIH, NAE

Smith, Robert M
Business: 245 West 55 St, New York, NY 10019
Position: Executive Vice President. *Employer:* Du Art Film Labs Inc. *Born:* Been at Du Art for the past 27 yrs. Prior to that a Photpgraphic Reconnaissance Officer USAF. Attended OH State & Pace College. Elected three times to public office in the State of NJ. Fellow of SMPTE & presently Pres of SMPTE & Dir of the Assoc of Cinema & Video Labs.

Smith, Rodney W
Business: 100 South Main St, P.O. Box 46, Bridgewater, VA 22812
Position: VP *Employer:* Patton, Harris, Rust & Assoc, P.C. *Education:* BS/CE/VA Polytech Inst & State Univ *Born:* 7/29/44 1967-72 Hercules, Inc, Radford VA; 1972-72 VA State Water Control Bd, Richmond, VA; 1972-76 Central Shenandoah Planning District Commission, Staunton, VA. With Patton, Harris, Rust & Assocs, Bridgewater, VA since 1976. Assumed responsibility as VP in 1980. Office Mgr and Principal-in-Charge of office in Shenandoah Valley, VA. Principal-in-Charge of WV coal environmental office. PE- VA, WV & KY. Publications: Effect of Land Use Management on Flood Predictions, Bulletin 103, VA Water Resources Research Center (1977) and Rock Type and Minimum 7-Day/10-Year Flow in VA Streams, VA. Water Resources Research Center (1981). *Society Aff:* NSPE, VSPE, WPCF, AWWA, NWWA

Smith, Ronald E
Business: 1300 Piccard Dr, Suite 110, Rockville, MD 20850
Position: Vice President. *Employer:* Woodward-Clyde Consultants. *Education:* PhD/Geotechnical Engg/NC State Univ; MS/Geotechnical Engg/NC State Univ; BS/Civil Engg/NC State Univ. *Born:* Nov 1938. Native of Rocky Mount NC. Taught at the Dept of Civil Engrg Univ of Fla 1966-70; joined Woodward-Clyde Cons in 1970 as Proj Mgr. Became an Assoc in 1973; named Vice Pres in 1975. Became Mgr of Woodward-Clyde Cons Washington DC area office in 1976. Reg PE in 5 states. Active in both civic & prof activities. Author of a number of tech papers. *Society Aff:* ASCE, SME, NSPE, AEG.

Smith, Ross W
Home: 1730 O'Farrell, Reno, NV 89503
Position: Prof Dept of Chem & Met Engrg & Chmn *Employer:* University of Nevada. *Education:* PhD/Mineral Engrg/Stanford Univ; SM/Metallurgy/Mass. Inst. of Technology; BS/Mining Engrg/Univ of Nevada, Reno. *Born:* Dec 11, 1927. PhD Stanford Univ; SM MIT; BS Univ of Nevada. Native of Calif. 1 year mining engr for Consol Copper Mines Corp 1950-53; 2 years US Army; Res Asst MIT 1953-55; Process Engr Portland Cement Assoc 1955-58; Process Engr Colo Sch of Mines Res Inst 1958-60; Prof of Metallurgy SD Sch of Mines & Tech 1960-66; Acting Inst Stanford & Chem Engr US Geol Survey 1966-68; Prof Univ of Nevada 1968- . Author of over 40 tech papers. Mbr AIME, ACS, AIChE, Sierra Club, Wilderness Soc; have been chmn mineral proc fund ctte, SAE/AIME; mbr various met education cttes; have lead wilderness study group for Sierra Club; active athlete- 1972 Natl Open. AAU 50 mile race champion (AAU all-American); AAU Masters Marathon Champion 1975 & 1978 ; hold US age US group running records, all PAAAU Cross Country Team 1972-75. *Society Aff:* AIME, AIChE, ACS, ΣΞ.

Smith, Russell L, Jr
Business: 677 Ala Moana, Suite 1000, Honolulu, HI 96813
Position: President. *Employer:* Smith, Young & Assoc, Inc. *Education:* BA 1948/Math & Physics /Univ of Hawaii; 1938-1941/Poli Sci & Econ/Stanford Univ *Born:* 12/25/19. Petaluma California. Army Air Forces 1942-45; Hawaii Air Natl Guard 1946- 67. AF Command Pilot, retired Col. Reg Civil-Sanitary Engr, Hawaii 711, Guam 23. P VP Austin, Smith & Assoc Inc 1959-75. Master planned sewerage & water systems for Trust Territory of Pacific Islands; master planned water sys Amer Samoa & Guam. Water Cons to Guam 1966-74. Designed first wastewater plants on Islands of Kauai, Maui, & Hawaii; first tertiary plant State of Hawaii. Pres ASCE, Hawaii 1961; WPCA, Hawaii 1967; CEC, Hawaii 1970. VP ACEC 1976-78. Pres-Elect, ACEC 1981-82. Pres ACEC 1982-83; Chrmn ICED 1982-1983; Hawaii Engr of the Year 1976. Pres. The Russ Smith Corp, 1975-85; Chief Engr & Dir of Public Works, City & County of Honolulu, 1985 & 1986; Pres. APWA, Hawaii 1985; Pres. Smith, Young & Assoc, Inc. 1987-. *Society Aff:* ACEC, ASCE, NSPE, SAME, WPCF, AWWA, APWA.

Smith, Seymour D
Home: 12 Hollywood Dr, Plainview, NY 11803
Position: Assoc Prof. *Employer:* Long Island Univ. *Education:* Engr/Ind Engg/Poly Inst of NY; MBA/Quant Systems/Hofstra Univ; BEE/Electronics/Cooper Union. *Born:* 10/19/22. Native of NYC. Practised engg for 25 yrs, particularly in hydraulic and electrical applications engg, and configuration and prog mgt. Was with Greer Hydraulics, Inc from 1951-1959 and Bendix Corp from 1964-1968. Assumed current faculty position in 1968. Am responsible, in part, for grad Mgt Engg and undergrad Computer Science prog dev. Have done extensive consulting work in computer programming, including scientific, business and engg applications. Was director of Garsite Products, Inc. of Deer Park, NY from 1980 to 1982. Was Computer Science Dept Chrmn from Sept. 1981 - August 1984. *Society Aff:* TBΠ

Smith, Verne E
Business: PO Box 3395 Univ Sta, Laramie, WY 82071
Position: Industrial Engineer. *Employer:* Laramie Energy Technology Center. *Education:* MS/Civil Engg/Univ of WY; BS/Civil Engg/Univ of WY. *Born:* May 1940. BS, MS Univ of Wyoming. Civil Engr with US Forest Service 1963-66 in Wyoming & Montana. Cons Engr in public utilities & land use surveying in Montana 1967-69. With Resources Res Inst at Univ of Wyoming 1969-78, specializing in surface hydrology. Planning of hydrocarbon energy recovery experimental projects 1978- . Reg PE & Land Surveyor in Montana & Wyoming. Pres of Wyoming Sect ASCE 1974-75. Enjoy basketball & hunting. *Society Aff:* ASCE.

Smith, W Tilford
Home: 129 James River Dr, Newport News, VA 23601
Position: Ret. (Sr VP) *Education:* BS/Civil Engr/NC State Univ. *Born:* 10/9/07. Harvard Advanced Mgmt Prog 1955. With Newport News Shipbuilding Engrg Div 1929, Vice Pres Production 1963, Exec Vice Pres 1964, Senior Vice Pres 1966, Shipbuilding Cons 1973. Shipbuilders Council of America Exec Ctte, Recipient of

Smith, W Tilford (Continued)
First Distinguished Engrg Alumnus Award at NC State Univ 1966. Mbr Soc Naval Arch Mar Engrs; Naval Engrs, Navy League Served as Shipbuilding Cons in Singapore 1974 with Internatl Exec Service Corp. *Society Aff:* SNAME.

Smith, Waldo E
Home: 1330 Massachusetts Ave NW Apt #1016, Washington, DC 20005
Position: Consultant *Employer:* Self *Education:* BE/Civil Engr/Univ of IA; MS/Civil Engr (Hydraulic)/Univ of IA; Some Work towards PhD/Univ of IL *Born:* 8/20/00. Hydraulic engr with cons engrs 1924-27; Instructor Theoretical & Applied Mechs Ill 1927-28; Assoc Prof & acting Head Civil Engrg Robert Coll, Istanbul 1928-31; Asst Prof Civil Engrg ND State 1931-35; Hydraulic Engr Muskingum Conservatory Ohio 1935-39; Hydraulic Engr Federal Government 1939-44; Exec Secy/Dir American Geophysical Union 1944-70, retired as Exec Dir Emeritus; Delegate to 8th to 18th Genl Assemblies of the Intl Union of Geodesy & Geophysics, and to various other international meetings on aspects of Geophysics. Professorial Lecturer Civil Engrg, George Washington Univ 1946-61. Staff Natl Graduate Univ on sci curriculum dev; Consultant on water, geophysical energy and resource problems. Fellow, Natl Dir ASCE; Fellow, Council Mbr AAAS; Fellow AGU, new Waldo E Smith Medal. Cosmos Club. In 1982 the American Geophysical Union established the Waldo E Smith medal for outstanding service to Geophysics In 1983 the first award was made to WES. 2 children. *Society Aff:* ASCE, AGU, AAAS, CC.

Smith, Walter J
Home: 37 Dix St, Winchester, MA 01890
Position: PE (Self-employed) *Education:* SM/Chem Eng/MIT; SB/Chem Eng/MIT. *Born:* Sept 17, 1904 Boston Mass. Reg PE in Commonwealth of Mass. With Arthur D Little Inc Cambridge Mass 1924-69; chem engr & group leader R&D Div; prod & process dev with emphasis during past 25 years on air cleaning & air pollution control. Patents in fields of hygrometry & air cleaning. ASTM: Chmn Standing Ctte on Res 1960-65, charter mbr Ctte D-22 on Methods of Sampling & Analysis of Atmospheres, Chmn D-22 1964-66, Award of Merit & Fellow 1968, Honorary Mbr 1974. Honorary Mbr APCA 1976. Fellow AIChE 1979. Other interests: home workshop, gardening, music, & photography. *Society Aff:* AIChE, ACS, ASTM, APCA.

Smith, Warren J
Home: 2588-D El Camino Real, Suite 248, Carlsbad, CA 92008
Position: Chief Scientist *Employer:* Kaiser Electro-Optics *Education:* BS/Optics/Inst of Optics; Univ of Rochester *Born:* 8/17/22. Born in Rochester, NY. Physicist Manhattan District Oak Ridge TN 1944-45; Ch Optical Engr Simpson Optical Mfg Co Chicago 1946-59; Mgr Optics Sect Raytheon Santa Barbara CA 1959-61. Dir R&D with Infrared Indus Inc Santa Barbara 1961-83; 1983-87 VP R&D with Santa Barbara Appl Optics (mfgr of precision optics), respon for optical design & engg. Currently Chief Scientist Kaiser Electro-Optics, Carlsbad CA. Author *Modern Optical* Engrg, McGraw Hill 1966 & chapters in *Handbook of Military Infrared Technology*, ONR 1966, *The Infrared Handbook*, ONR 1978, *Handbook of Optics*, McGraw Hill 1978. Mechanical Design & System Handbk, McGraw Hill 1985. Teach short courses *The Design of Optical Systems & Practical Optics* at Univ of WI. also courses in Lens Design in association with Genesee Computer Center. Fellow Soc Photo Optical Instrumentation Engrs Governor 1980-1982 Pres 1983 awarded Gold Medal of the society 1985. Fellow Optical Soc America, Dir 1972-74. VP 1978; Pres Elect 1979; Pres 1980. Jr. Past Pres. 1981. Sailing & Tennis. *Society Aff:* OSA, SPIE, ISOE.

Smith, Warren L
Business: RM 2C-285, AT&T-Bell Laboratories, 555 Union Blvd, Allentown, PA 18103
Position: Supervisor. *Employer:* AT&T Bell Laboratories *Education:* BS/EE/Univ of WI. *Born:* July 1924. Served with US Navy 1943-45, 1951-54. Joined Bell Labs in 1954, Supervisor 1962- . Responsible for dev & design of precision crystalcontrolled oscillators & associated equipment as well as dev of Monolithic Crystal Filters. Fellow IEEE, Mbr of Tech Ctte on Time & Frequency of GIM 1973- . Mbr US Delegation of IEC Tech Ctte 49. Received C B Sawyer Memorial Award 1976. Hobbies include photography, hunting, fishing, & boating.

Smith, Wilbur S
Business: 1330 Lady Street, Suite 609, Columbia, SC 29201
Position: Owner. *Employer:* Wilbur S. Smith Management. *Education:* MS/Elec Engg/Univ of SC; BS/Elec Engg/Univ of SC. *Born:* Sept 6, 1911 Columbia SC. Grad Transportation Harvard 1937, Hon LLD 1963 Univ of SC. Hon LHD 1975 Lander Coll. Reg PE in 50 states, District of Columbia, Puerto Rico, United Kingdom, Queensland Australia, Hong Kong & New Zealand. Chmn of Bd Eno Foundation for Transportation Inc. Mbr of BD/Dir Koger Properties Inc, The Koger Company, Rank Development, Inc. Former Pres American Road & Transportation Builders' Assn. Former Mbr Disting Practitioner-Lecturer Prog Coll of Business Admin, Univ of Georgia. Trustee Univ of SC: College of Business Administration; Educational Foundation, Chm Planning Committee. mbr USC President's National Advisory Council. Mbr Phi Beta Kappa, Tau Beta Pi & Phi Sigma Kappa. Prof Mbrships: AICE, ASCE, TRB, ITE, NSPE, NAE, IRF. Honorary Member, Chi Epsilon. *Society Aff:* AICE, ASCE, IEEE, ITE, NSPE, NAE.

Smith, William A, Jr
Home: 729 Tan Tara Sq, Raleigh, NC 27615
Position: Prof; Coordinator, Advanced Program Development *Employer:* North Carolina State Univ. *Education:* D EngSc/IE & OR/NY Univ; MS/IE/Lehigh Univ; BS/-/Naval Acad. *Born:* 7/13/29. Native of Parkersburg WV. Pres IIE 1975-76; VP Tech & Div Affairs 1972-74. Conslt (full time), Advanced Automation Engrg, Northern Telecom Inc., 1984-5. Prof, IE 1973-present; Hd of Industrial Engrg Dept, NC State 1973-82; Founder & Dir Productivity Res Ext Prog, NC State, 1975-1984; Chrmn Southern Section (IE) of ASEE 1979-80; Captain Naval Reserve retired. Prof/Assoc Prof Industrial Engg 1966-73; Dir Computing Lab 1957-67 Lehigh Univ. Res studies in quality improvement, productivity, human resource dev, mfg systems and matl flow 1977-present; infor sys effectiveness, 1967-74. Pioneer in computer use in engineering education 1957-1970. Cons in organization effectiveness, mgmt sys design, computer-based sys, mgmt planning & infor processing for indus, hospital, banking & educational insts. VChrmn, NC Governor's Comm on Governmental Productivity 1977-1983. Chrmn, Assoc for Cooperation in Engg 1979. Chrmn, AAES Coordinating Ctte on Productivity & Innovation 1980-82; Mbr, Exec Ctte, AAES Public Affairs Ctte, 1980-83; Chrmn, AAES Public Affairs Ctte 1984-85; Chrmn, Natl Productivity Network 1984-5. *Society Aff:* IIE, ΣΞ, AΠM, ASEE, TIMS, CASA OF SME

Smith, William J
Business: American Lane, Greenwich, CT 06830
Position: Exec VP/Pres Paper Sector *Employer:* American Can Co. *Education:* BME/Mech Engg/Syracuse Univ. *Born:* 6/27/26. Advanced Mgmt 1967 Harvard Business Sch; MA Bus 1976 Dartmouth Inst. Business career American Can Co: Plant Mgr 1960-62, Tech Dir Paperboard 1967-68, Vice Pres Plastics & Paperboard Mfg 1969-71, Vice Pres & Genl Mgr Genl Packaging 1972, Vice Pres & Genl Mgr Operations Technology 1973-74, Senior Vice Pres Technology & Dev 1975- . Responsible for R&D Engrg & Mfg Technology at American Co. Native of Cranford NJ. Chrmn, Corp Oper Ctte 1979-80/Exec VP-Pres, Paper Sector operating responsibility for $1.4 billion paper sector in US, Canada & Intl. US Navy 3 years. m. Dolores Masson 1953, 5 children.

Smith, William P
Home: 1107 W Campus Dr, Lawrence, KS 66004
Position: Prof of Elec Engr Emeritus. *Employer:* University of Kansas. *Education:* PhD/Elec Engr/Univ of TX; MSEE/Elec Engr/Univ of MI; BEE/Elec Engr/Univ of MN. *Born:* Jan 1915; native of Superior Wis. Taught at the Univ of Minn & Texas before coming to the Univ of Kansas 1950; Chmn of Elec Engrg Dept 1955-65 &

Smith, William P (Continued)
Dean, Sch of Engrg 1965-78. Employed as engr by Commonwealth Edison Co & several cons firms. Active duty US Naval Reserve 1941-45; present rank Capt USNR Ret. Served as cons on engrg educ in Colombia & Venequela. Spec int in direct elec energy conversion. Pres 1968-69 Eta Kappa Nu. *Society Aff:* HKN, TBΠ, ΣΞ, ΔTΔ, IEEE ASEE.

Smith, William R, Sr
Home: 522 Miner Rd, Orinda, CA 94563
Position: Mgr Materials & Quality Serv Dept. *Employer:* Bechtel Group Inc. *Education:* BS/MET/Univ of CA, Berkeley. *Born:* April 1917. US Navy 1938-45. GE 1952-69 specializing in materials & welding for nuclear plant equipment, Mgr Material Engrg 1958-69; Bechtel Corp 1969- , Mgr Materials & Quality Servs Dept responsible for providing servs in the fields of materials & welding for engrg & const of electric power & indus plants. Active in dev of Natl codes & standards for 20 years. Mbr ASME Ctte on Nuclear Codes & Standards, Boiler & Pressure Vessel Ctte, & Vice Chmn Subctte on Nuclear Power. Fellow ASME. Enjoy fishing & hunting. *Society Aff:* ASMe, AWS, ASTM, NSPE, ASNT.

Smith, Yancey E
Home: 3350 Bluett Dr, Ann Arbor, MI 48105
Position: Res Mgr. *Employer:* AMAX Materials Res Ctr *Education:* PhD/Met E/Univ of MI; MBA/Finance, Marketing/Univ of MI; MSE/Met E/Univ of MI; BSE/Met E/Univ of MI. *Born:* 6/21/37. Dr Smith spent his first two yrs as a met on nucl fuel mtls R&D with GE Co. He then returned to the Univ as a Res Asst to study the notch sensitivity of superalloy sheet on mtl for supersonic transport applications. Upon joining Climax Molybdenum in 1967, he directed his efforts toward an evaluation of transformation characteristics of high strength low alloy steels. Over the past ten yrs, he has supervised significant alloy devs in high strength low alloy steels, carburizing steels, rail steels, normalized steels & oil industry steels. He presently has mgt responsibility for res on stainless steels, elevated temperature steels, oil industry steels, tool steels, molybdenum metal and superalloys. New respon as Mgr Contract Res involves seeking external funding for res ctr. Also, recently became Fellow of ASM. He has about 25 publications. *Society Aff:* AIME, ASM, ASTM, APMI.

Smithberg, Eugene H
Business: College of Engrg Boston Univ, 110 Cummington St, Boston, MA 02215
Position: Assoc Dean *Employer:* College of Engrg Boston Univ *Education:* BME/Mech Engg/City College of NY; MME/Mech Engg/Brooklyn Poly Inst; D Eng Sc/Melch Engg/NY Univ. *Born:* June 1922. Native of New York City. Served in US Navy 1944-46. Varied parttime indus & cons exper while teaching mech engrg 1950-67. Graduate Dean Newark Coll of Engrg (now NJIT) 1967-80. Active in design & operation of minority engrg prog 1968-80. Acting Vice Pres for Academic Affairs, NJIT. 1976- 80 Responsible for graduate & undergraduate progs in engrg & progs in engrg technology, architecture & indus administration. Mbr of NJ Student Assistance Bd 1978-81. Prof of Aerospace and Mechanical Engrg; Assoc Dean, College of Engrg., Boston Univ 1981-1987 Exec Ctte Mbr, Masspep 1981-1987. Registered Professional Engr. *Society Aff:* ASME, ASEE.

Smolinski, Adam K
Home: Polna 54 m48, Warszawa Poland 00-644
Position: Prof (retired) *Employer:* Technical Univ of Warszawa Poland. *Education:* Dr Engr/-/Tech Univ Warszawa *Born:* Oct 1910. Dipl Engrg & Dr Engrg Tech Univ of Warszawa. Radio Engr in State Radio & Telephone Works in Warszawa 1933-44. Lecturer 1945 & Prof 1946-81 (retired) in Electrical & Electronics Faculty of Tech Univ of Warszawa. Head of Dept 1949-81 & Dean 1950-52. Mbr of the Polish Acad of Scis 1962. Fellow IEEE New York 1974. Vand Life Fellow 1984. Fellow IEE London 1975. Pres of the Polish Natl Ctte of URSI 1972- . Chmn of URSI Comm D 1975-78. VP of URSI 1978-84. Polish State Prize 1964. Cdr's Cross of the Polonia Restituta Order 1973. Order of the Banner of Labot (second class) 1980. Honorary Teacher of the Polish Peoples Republic 1981. IEEE Centennial Medal 1984. Medal of N. Copernicus of the Polish Acad of Scis 1985. *Society Aff:* SEP, PTETIS, TNW.

Smoot, L Douglas
Home: 1811 N 1500 East, Provo, UT 84601
Position: Prof of Chem Engg & Dean, Engg & Tech. *Employer:* Brigham Young Univ. *Education:* Phd/Chem Eng/Univ Wa; MS/Chem Eng/Univ Wa; BES/Chem Eng/Brigham Young Univ; Bs/Chemistry/Brigham Young Univ. *Born:* 7/26/34. 18 yrs at Brigham Young Univ; 4 yrs with Lockheed Propulsion Co., 1 yr at CA Inst of Tech; summers with Hercules, Phillips Petroleum & Boeing. Cons to 30 companies & agencies in energy, combustion & propulsion areas in US & Europe. Principal investigator for over $10 million in res from indus & government. Active res in coal mine fires, combustion in pwr generators generators & kinetics of coal gasification. Mbr of AIChE, ASEE, AIAA & Combustion Inst. has pub of presented over 100 articles, and two books on Coal Combustion. Racketball, tennis & horses. *Society Aff:* ASEE, AICHE, AIAA, Comb Inst

Smyers, Daniel J
Home: 8657 Wittmer Rd, Pittsburgh, PA 15237
Position: Chief Engineer *Employer:* Hall Industries Inc. *Education:* BS/Civil Engg/Carnegie Mellon Univ. *Born:* 5/21/20. in Pittsburgh, PA. Resides in Pittsburgh PA. Served in 1st Detachment Navy Const Battalion, 4 yrs WWII in South Pacific. Employed by Gulf Res, Swindell- Dressler, Rust Engg, Dravo, Comstock, Envisco & Hall Industries as Administrater & Structural Engr in charge of design of facilities & equipment for steel, paper, petroleum & transit inds-PPres-Wittmer Park Water Assn; PPres Pittsburgh Sect ASCE, CMU Forum, CMU North Hills Clan & VP CMU Natl Alumni Assn. Received Evening Student Achievement Award 1959, CMU Alumni Serv Award 1969, & Award of Appreciation from Pittsburgh Sect ASCE 1976. Reg PE in PA, TN & CA. Enjoys boating, skiing & music *Society Aff:* ASCE, SAME, SME, RI.

Smylie, Robert E
Business: Washington, DC 20546
Position: Assoc Adm, Space Tracking and Data Syst *Employer:* NASA, Headquarters *Education:* MSME/ME/MS State Univ; BSME/ME/MS State Univ; MS/Mgt/MIT. *Born:* 12/25/29. Native of Brookhaven, MS. USN 1948. Taught ME at MS State Univ. Started career in 1956 as design engr with Douglas Aircraft Co. Joined NASA-Manned Spacecraft Ctr in 1962 where for the next decade was responsible for the technical mgt & supervision of the Crew Systems Div activity in support of the Apollo Prog. Deputy Assoc Administrator for Aero & Space Tech NASA Hdquarters 1973-1976. Deputy Dir, Goddard Space Flight Ctr, 1976-1980. Appointed Associate Administrator, space Tracking and Data Systems, April 1980. Assoc Fellow AIAA, Fellow AAS, Victor Prather Award, NASA Exceptional Service Medal. Tau Beta Pi, Pi Tau Sigma, Kappa Mu Epsilon. *Society Aff:* AAS, AIAA.

Snavely, Cloyd A
Business: c/o Marvalaud, Inc, P.O. Box 331, Westminster, MD 21157
Position: President *Employer:* Technovation Advisory Services Inc. *Education:* PhD/Met/OH State Univ; MS/Met/Columbia Univ; BS/Met/Columbia Univ; AB/Pre- engg/Columbia Univ. *Born:* 5/8/17. at Massilon, OH. Active duty, US Naval Reserve, 1941-1945, rank of Ensign to Lt, Sr Gr, Battelle Fellow at the OH State Univ, 1945-1947. Doctoral dissertation on theory for the character of chromium electrodeposits. Battelle Meml Inst, Columbus, OH. Res Engr, 1946, Asst Div Chief, 1949, Asst Dept Mgr, 1954. Steel Improvement & Forge Co, Cleveland, OH Mgr Electrochem Div, 1960- 1962. Battelle Meml Inst, Columbus, OH, Asst Dept Mgr 1962-1963. Battelle Dev Corp, Mgr Dev Dept, 1963-1969. Dev Engg, Inc, Denver, CO Pres 1969-1970. Founded Technovation Mgt, Inc, Denver, CO, Pres, 1969-1970. Natl-Std Co, Niles, MI, Technical Consultant, 1970-1972, Mgr Res Lab 1972-1979, Mgr Contract Res & Dev 1979-. Technovation Advisory Services, Inc., Co-

Snavely, Cloyd A (Continued)
lumbus, Ohio, President, 1980-. Married 1941, three daughters. Main career interests, electrochemistry, physical & extractive met, energy related tech & invention development. *Society Aff:* ASM, ECS, AIME.

Sneck, Henry J
Home: 21 Bolivar Ave, Troy, NY 12180
Position: Prof *Employer:* RPI *Education:* PhD/Engrg/RPI; M/Engrg/Yale Univ; B/ME/RPI *Born:* 11/9/26 Member of RPI ME Dept since 1953. Taught fluid mechanics, heat transfer, mechanics, machine dynamics, and machine design. Publications in the field of hydrodynamic lubrication and fluid mechanics problems. Assoc editor of ASLE and ASME journals. Active consultant to industry on a wide range of multidisiplinary mechanical engrg problems. Hobbies include tennis, auto-repair, and classical music. *Society Aff:* ASME, ASLE

Sneckenberger, John E
Home: 521 Woodhaven Dr, Morgantown, WV 26505
Position: Prof *Employer:* WV Univ *Education:* PhD/Engrg/WV Univ; MS/ME/WV Univ; BS/ME/WV Univ; AA/Pre-Engrg/Hagerstown Jr Coll *Born:* 8/17/37 My professional experiences include 17 years of undergraduate and graduate instruction and research at WV Univ, summer employments with several industrial companies and with NASA, EPA, and NIOSH government groups, and participations on natl executive committees such as ASME Sound and Vibrations Committee. Research results on synchronous active vibration control, nozzle noise reduction, and community transportation noise impact have been published. Sponsored work has also pursued engine-muffler acoustics. Recent involvements have been related to coal feeder system efficiency, instrument measurement of impact noise, and safety analysis of robotics in automated manufacturing. *Society Aff:* ASME, SAE, ASEE, EX

Snedeker, Robert A
Home: 17 Smithshire Estates, Andover, MA 01810
Position: Vice President. *Employer:* Merrimac Paper Co Inc. *Education:* PhD/Chem Eng/Princeton Univ; MS/Chem Eng/MIT; BS/Chem Eng/MIT. *Born:* Aug 3, 1928 New York, NY. m. 1952; c. 3. Standard Oil Co Ind fel Princeton 1953-54, Am Cyanamid Co fel 1954-55, PhD 1956. Engr photo prod dept E I du Pont de Nemours & Co 1955-58, res supv 1958-67; Tech Dir Scott Paper Co 1967-70; Vice Pres Mfg Merrimac Paper Co 1970- . USAF 1951-53, 1st Lt Am Chem Soc; Am Inst Chem Engr; Soc Photography Sci & Engrg. Polymer fabrication; photopolymerization; coating & drying; papermaking techniques; pollution control. *Society Aff:* ACS, AIChE, SPSE, ΣΞ.

Sneed, Ronald E
Business: Box 7625, Raleigh, NC 27695-7625
Position: Prof. *Employer:* NC State Univ. *Education:* PhD/Bio & Agr Engg/NC State Univ; BS/Agr Engg/NC State Univ. *Born:* 11/23/36. Native of Oxford, NC. Served in US Army in 1960, mbr of USAR since that time, present rank of COL. Extension Water Mgt Specialist at NCSU since 1960 with exception of one yr spent with US Army Corps of Engrs, Wilmington, NC, in their grad fellowship prog. Served on various committees of ASAE Soil and Water Div, including chrmn of the Div. Grad of US Army War College, Consultant to several US and foreign companies. Served on various committees of IA, Received 1981 Man of the Year Award from IA. *Society Aff:* ASAE, SWCS, IA, ΣΞ.

Snell, Absalom W
Business: 104 Barre Hall, Clemson, SC 29631
Position: Assoc Director-SC Agri Expt Station *Employer:* Clemson University. *Education:* PhD/Agri Engg/NC State Univ; MS/Agri Engg/IA State Univ; BS/Agri Engg/Clemson Univ. *Born:* April 1924. Air Corps pilot during WWII. Asst Prof, Assoc Prof, Prof & Dept Head in Agricultural Engrg Dept Clemson Univ 1949-75. Chmn of Directorate of the Water Resources Res Inst 1964-75. Current position Assoc Dir of the Agricultural Experiment Station Clemson Univ. Additional offices & responsibilities include mbr of Exec Bd, Universities Council on Water Resources 1974-77; Chmn of Governor's Adv Ctte on Water Resources 1966-69; cons to Water Resources Comm 1968-76; Secy & Vice Chmn Southeast Region ASAE 1958-60; Regional Dir ASAE 1971-73. Mbr of 9 prof & honorary orgs. *Society Aff:* ASAE, ΦΚΦ, ΤΒΠ, ΓΣΔ, ΣΞ.

Snell, Robert R
Business: Dept of Civil Engrg, Seaton Hall, Manhattan, KS 66506
Position: Prof & Head Dept of Civil Engg. *Employer:* KS State Univ. *Education:* PhD/Civil Engg/Purdue Univ; MS/Civil Engg/KS State Univ; BS/Civil Engg/KS State Univ. *Born:* 4/17/32. Native of St John, KS; Served with US Army 1955-57; Ford Fdn Fellow, Purdue Univ 1960-63; Asst and Assoc Prof of Civil Engg, KS State Univ 1963-67; Ford Fdn Resident in Engg Practice, The Rust Engg Co, Pittsburgh, PA 1967-68; Prof of Civil Engg, KS State Univ 1968-present; Head, Dept of Civil Engg, KS State Univ 1972-present; Pres KS Sec of ASCE 1972; Natl Dir of ASCE 1984-1987; Pres KS Engg Soc 1978; Enjoy camping and reading. *Society Aff:* ASCE, NSPE, ASEE.

Snell, Virgil H
Business: P.O. Box 8405, Kansas City, MO 64114
Position: Gen Ptnr/Manager of Engineering, Power Div *Employer:* Black & Veatch, Engrs-Archts *Education:* BS/Arch Engrg/KS State Univ *Born:* 11/5/31 Native of Winfield, KS. Served in US Air Force, 1954-1956, graduated with honor from Installations Engr School, USAFIT, OH, served as engr in AK. With Black & Veatch since 1956 as engr for power generating stations. Dept Head of Civil-Structural Dept, Power Division, 1976. General Partner of Black & Veatch, 1981. Mgr of Engrg, Power Division, 1986. Selected as honor member of Phi Alpha Epsilon by ICSU Chapter, 1985. Selected as honor member of Chi Epsilon by KSU Chapter, 1979. ABET Ad Hoc Accreditation visitor for engrg tech 1978-1983. Pres of Engrs Toastmaster Club of KS City, 1966-1967. Registered PE in KS and MO. Enjoy water skiing, scouting, Americana. *Society Aff:* ASCE, NSPE, NSAE

Snider, David G
Home: 3142 S. Oak Ave, Springfield, MO 65804
Position: Director of Public Works. *Employer:* City of Springfield, Missouri. *Education:* BS/Civil Engg/Univ of MO-Columbia. *Born:* 10/10/37. Native of Kansas City, Mo. Held various positions with the Mo State Hwy Dept in the Maintenance & Traffic Divs 1959-66. Asst Dir of Traffic & Trans City of Des Moines 1966-72. Traffic Engr City of Springfield Mo 1972-73. Dir of Public Works 1973- . Mbr Trans Res Bd, Inst of Trans Engr, NSPE, MSPE, APWA, & ASTM. Reg PE in Mo & Iowa. Hobbies are golfing & fishing. *Society Aff:* MSPE, NSPE, ITE, ASTM, APWA.

Snitzer, Elias
Business: 38 Henry St, Cambridge, MA 02139
Position: Res Fellow-Fiber Optics *Employer:* Polaroid Corp *Education:* PhD/Physics/Univ of Chicago; MA/Physics/Univ of Chicago; BS/EE/Tufts Univ *Born:* 2/27/25 Polaroid Corp, 1984 to present; Mgr, Fiber Optics program; UTRC, 1977-84, work in fiber optics, plasma display panels, glass tech; American Optical Corp, 1959-1976, fiber optics, glass lasers, glass tech. 1971 Geo W. Morey Award of American Ceramics Society. 1979 Quantum Electronics Award of IEEE. Elected to Natl Academy of Engrg in 1979 and CT Academy of Science and Engrg in 1980. 28 publications, 53 patents. Married to Shirley Ann Wood 1950, five children. *Society Aff:* APS, OSA, IEEE, ACS

Snoddy, Jack T
Business: 2601 NW Expressway Suite 903 E, Oklahoma City, OK 73112
Position: Owner. *Employer:* JTS Engg Serv. *Education:* BS/Elec Engg/OK State Univ. *Born:* 12/24/39. Native of Oklahoma City, OK. Began profl career as design engr for OK Gas & Elec Co. Served with US Army Security Agency & Strategic Communications Command, Europe, 1963-1966. Served with another consulting

Snoddy, Jack T (Continued)
engg firm and OK City Public Schools prior to entering private practice in 1978. Currently designing solar oriented geodesic residential structures with water to air heat pump. At present, serves as Vice President (Marketing) and Secretary-Treasurer of Standard Drawworks, Inc., a designer and manufacturer of drawworks for drilling rigs; is responsible for purchasing the raw materials for the drawworks. *Society Aff:* NSPE.

Snoeyink, Vernon L
Business: 208 North Romine, Dept of Civil Engrg, Urbana, IL 61801
Position: Prof *Employer:* Univ of IL *Education:* PhD/Water Resources Engrg/Univ of MI; MSE/Sanitary Engrg/Univ of MI; BSE/CE/Univ of MI *Born:* 10/10/40 Native of Dutton, MI. Worked for Metcalf and Eddy from 1968 to 1969 as a Sanitary Engr. With Univ of IL, Dept of Civil Engrg since 1969 as an assistant prof (1969-73), and assoc prof (1973-77), and have been Prof of Environmental Engrg since 1977. Have taught courses in water and wastewater treatment, and in water chem, and have specialized in research on water purification using activated carbon. Authored *Water Chem*, John Wiley & Sons, 1980, with D. Jenkins. Served as a consultant to the World Health Organization, the US Environmental Protection Agency and several private companies. *Society Aff:* ASCE, WPCF, AWWA, AEEP

Snoeys, Raymond AJ
Business: Dept. of Mech. Engrg, Celestijnenlaan 300B, Leuven (Heverlee), Belgium B-3030
Position: Professor *Employer:* Kath. Univ. Leuven *Education:* doctor/mech. engr./K.U. Leuven; ingenieur (master)/mech. engr./K.U. Leuven *Born:* 10/04/36 After graduation at K.U. Leuven I joined the research group of the Belgian Metalworking Industries 1962; I became chief engineer and got a doctoral degree on the analysis of grinding stability 1966; Was engaged in a research team of the Univ of Cincinnati dealing with grinding problems of high strength thermal resistant alloys 1967; Associate prof. at K.U. Leuven 1968; Prof. at K.U. Leuven 1978 (head of lab for production engrg. research); Dean of faculty of engrg 1981; President elect for CIRP 1987. *Society Aff:* ASME, SME, CIRP, KVIV

Snow, Phillip D
Business: Civil Engrg Dept, Union College, Schenectady, NY 12308
Position: Assoc Prof *Employer:* Union Coll *Education:* PhD/Environ Engrg/Univ of MA; MS/Environ Engrg/Univ of MA; MS/Geol/Syracuse Univ; BS/Geol/Marietta Coll *Born:* 3/23/43 Academic work in geochem and environmental engrg. Research has been in lake restoration with emphasis on water and sediment chem as well as lake systems modeling. Also, principal assoc in consulting firm, Enviromed Assocs, specializing in lake restoration projects, sediment and water sampling, analysis, and interpretation. *Society Aff:* ASCE, WPCF

Snow, Richard H
Business: PO Box 2128, Bartlesville, OK 74005
Position: Dir *Employer:* Natl Inst for Petroleum and Energy Res (NIPER), a Div of ITT Res Int *Education:* PhD/ChE/IIT; MS/ChE/VPI; AB/Chemistry/Harvard Univ. *Born:* 4/26/28. Dr. Snow is in charge of a former govt lab, the Bartlesville Energy Tech Ctr, now a contract res organization in the fields of oil and gas prod, processing and thermodynamics of hydrocarbons, and fuel/engine res. He has been with ITT Res Inst since 1956, was an Engrg Adv from 1973 to 1977, and was Mgr of Chem engrg res from 1977 to 1983. His main technical interest has been math modelling of processes by computer, including chem equilibria and kinetics, particle behavior, and size reduction of ores. He has been active in the development of the IITRI RF process for the in situ recovery of oil from shale. He has sixty publications and four patents. *Society Aff:* AAAS, AIChE, AIME.

Snowden, Wesley D
Business: 1805 136th Pl. NE Suite 104, Bellevue, WA 98005
Position: Pres. *Employer:* Alsid Snowden & Assoc. *Education:* MBA/Bus (Mgt)/Seattle Univ; BSCE/Environmental/Univ of WA. *Born:* 1/20/44. Married Elizabeth Marie Diers 9/11/64, 2 children; Jeffrey Glenn & John Wesley. Course Director EPA Air Pollution training Inst 1967-69. Mgr of Environmental Services at Valentine Fisher & Tomlinson, 1969-76. Founder & Pres Alsid, Snowden & Assocs, 1976 to present. Served as 1st Lt USPHS, 1967-69. Reg PE WA State & certificated in Am Acad of Environmental Engrs. Published EPA reports on Emissions from Asphaltic Concrete Plants & Fireplaces. Editor of monthly "Stack Sampling News–, Delegate to 1980 White House Conference on Small Business. Mbr of Rainier Yacht Club, Seattle Municipal League, Newport Yacht Club. *Society Aff:* NSPE, ASCE, APCA, SES, AMA.

Snowman, Alfred
Home: 121 Huguenot Ave, Englewood, NJ 07631
Position: Dir R&D. *Employer:* Semi-Alloys *Education:* MBA/Indust Mgmt/Fairleigh Dickinson Univ; ARSM/Met/Royal Sch of Mines; BSc(Eng)/Met Engrg/London Univ. *Born:* 7/11/36. Born and educated in London, England. Worked in US since 1958 in a number of R&D positions with Philco Corp. Kawecki Chem Co & Assoc Metals & Minerals Corp. From 1967 to 1979 with Gen Cable Corp as mgr met res & mgr ind products. Potters Ind Inc as Dir, R&D 1979-83. Currently a Technical Dir at Semi-Alloys Inc. Chmn NY Chapter ASM 1972/73. Published papers on creep characteristics of aluminum, shot peening and semiconductor packaging mtrls, and contributed articles to ASM Metals Handbook Vol 7 (1972) and Vol 2 (1979). *Society Aff:* ASM, AIME, IOM, ARD.

Snyder, Bernard J
Business: Washington, DC 20555
Position: Dir,Three Mile Island Prog Off *Employer:* US Nuclear Regulatory Comm *Education:* PhD/Nuclear Engrg/Univ of MI; MME/Nuclear Engrg/Cornell Univ; BME/Mech Engrg/Cornell Univ *Born:* 9/21/35 Currently Dir of NRC office responsible for regulatory oversight of the cleanup from TMI-2 accident. Previously Assistant Dir for Policy Review, NRC; AEC/ERDA Assistant Project Mgr, Project Engr and Executive Assistant to Project Mgr Clinch River Liquid Metal Fast Breeder Project; Assistant Project Mgr, Fast Flux Test Facility; Project Mgr - Hallam Nuclear Power Facility Decommissioning; Sr Reactor Engr - AEC; Sr Engr - Westinghouse Astronuclear Lab; Engr - Naval Reactors, AEC, Washington, DC. *Society Aff:* ΣΞ, ΤΒΠ, ΠΤΣ

Snyder, Charles T
Business: Mail Stop 200-3, Moffett Field, CA 94035
Position: Dir of Aeronautics and Flight Systems *Employer:* NASA-Ames Res Ctr. *Education:* Engr/Aero & Astro/Stanford Univ; MS/Aero & Astro/Stanford Univ; BS/Aero Engg/Univ of Wichita. *Born:* 7/2/38. Native of Belle Plaine, KS. Worked for Beech Aircraft Corp 1956-62. Joined NASA-Ames Res Ctr in 1962. Conducted aircraft flight dynamics res on direct lift control, wind shear effects, ground effects, simulation techniques, SST flight characteristics & certification criteria, & two-segment approach. Led cooperative interagency (NASA/FAA) & intl (US, UK, France) res investigations, 1969-72. Received Dryden Meml Fellowship from Natl Space Club, 1972. Since 1974, responsible for direction of broad res & tech prog with emphasis on STOL and VSTOL aircraft, rotorcraft, & aviation safety. Mbr AHS Bd of Dirs & NASA Aeronautics Advisory Subcommittee. *Society Aff:* AIAA, AHS, ΤΒΠ, ΣΓΤ.

Snyder, David A
Home: 3108 Huntmaster Way, Owings Mills, MD 21117
Position: Chief Technical Advisor *Employer:* US Dept of Defense *Education:* MS/EE/Johns Hopkins Univ; MS/Computer Sci/Johns Hopkins Univ; BS/EE/Johns Hopkins Univ *Born:* 12/25/50 Graduate of the "A-Course" of the Baltimore Poly Inst, an engrg-oriented high school. Won the natl Westinghouse-Johns Hopkins Scholarship Award in engrg. Worked in transducer design group at Westinghouse and also worked for MTD Res & Dev, AAI Corp and Western Elec. Served in many positions within Tau Beta Pi - elected to Exec Counc (Bd of Dirs), 1982-1986. Lec-

Snyder, David A (Continued)
turer in the Computer Sci Dept at Towson State Univ since 1979. Currently chief tech advisor of a computer support div at an intelligence agency within the Dept of Defense. Enjoy genealogical res with intent to publish family history. *Society Aff:* IEEE, TBΠ, HKN

Snyder, Harold J
Home: 833 Jewel St, New Orleans, LA 70124
Position: Dir. *Employer:* Snyder Technical Lab. *Education:* PhD/Met Engg/Univ of Pittsburgh; MS/Met Engr/Univ of Pittsburgh; BS/Met Engg/Univ of Pittsburgh. *Born:* 5/22/24. Native of Pittsburgh, PA. Reg profl met engr in states of LA & PA. Began 35 yr engg career while serving with US Naval Reserve aboard LST'S in Pacific Theatre during WWII. Most noteworthy achievements from a standpoint of patents, devs, & publications were made during employ with Westinghouse Electric Corp and Mellon Inst in the 1950 to 1970 era. Co-inventor of Westinghouse Elec Corp's silver base control rod mtls presently employed in the commercial power plants having pressurized water reactors. Named Fellow ASM 1977. Presently serving second term as Chrm New Orleans Chapter ASM. Assumed present duties in 1971. *Society Aff:* ASM, AWS, AIME.

Snyder, Herman C
Home: 533 29th Ave, East Moline, IL 61244
Position: Chief Product Dev Engr. *Employer:* International Harvester. *Education:* BS/Agri Engg/Purdue. *Born:* March 1930. BS Agri Engrg (Power & Machinery) 1952 Purdue Univ. Served with USAF 1952-56. With Internatl Harvester 1956- . Traveled extensively with Internatl Harvester testing & developing combines & allied headers used for grain harvesting. Now managerially responsible for lab & field testing & developing combines. Active in ASAE having held sev officer position both locally & regionally & served on Natl Cttes. Mbr of Church of Peace Rock Island Ill having held office of Pres & Secy. Married with 3 children. Enjoy fishing & home workshop. *Society Aff:* ASAE.

Snyder, James J
Business: 18901 Euclid Ave, Cleveland, OH 44117
Position: Sr. Met Engr *Employer:* Gould Ocean Systems. *Education:* BMetE/Met/Cleveland State Univ; MS/Metallurgy & Materials Science/Case Western Reserve Univ *Born:* 9/4/31. Served in US Navy as an electronics technician. Met with US Steel 1960-69. Covered carbon steel rods, wire and cold roll carbon and stainless strip areas. With TOCCO Div 1969-79. As Chief Met holds two process patents in induction thermal treatment applications. Presently a Sr. metengr with Gould Ocean Systems responsible for material selection, failure analysis, NDE of composites, electron microscopy and dev advanced materials. Active on ASM committees. Authored "P/M Porous Parts–, Vol 7, 9th Ed., and "Failures of Iron Castings–, Vol 11, 9th Ed. for ASM Metals Handbook. Enjoy classical music, jogging and handball. *Society Aff:* ASM, APMI, SAMPE

Snyder, James N
Business: Dept of Comp Sci, 1304 W. Springfield Ave, Urbana, IL 61801
Position: Hd of Dept. *Employer:* Univ of IL. *Education:* PhD/Phys/Harvard Univ; MA/Phys/Harvard Univ; BS/Phys/Harvard College. *Born:* 2/17/23. Mbr faculty Phys Dept, Univ of IL at Urbana-Champaign, 1949-present; Assoc Prof, Digital Computer Lab, 1957-1958, Prof, 1958-70, Prof Computer Sci and Physics, Hd, Dept of Computer Sci, 1970-present; Hd, Comp Div, Midwestern Univ Res Assn, Madison, WI, 1956-57. Mbr of Governing Bd, IEEE Computer Soc, 1976-80; Chrmn of AFIPS Education Committee, 1975-76. *Society Aff:* AAAS, IEEE, ACM, APS.

Snyder, Julian
Business: 1965 Sheridan Dr, Buffalo, NY 14223
Position: VP. *Employer:* VSSR Architects/Consulting Engrs. *Education:* PhD/Civil Engg/Northwewstern Univ; MS/Civil Engg/IL Inst of Tech; BS/Civil Engg/IL Unst of Tech. *Born:* 1/28/28. Engg for Office of J Fruchtbaum Assoc, Buffalo, NY 1951-52, served in Army Chem Corps 1952-54. Structural engg for Alfred Benesch & Co and Nelson, Ostrom, Baskin, Berman, 1955-57 in Chicago, Taught Civil Engg at IIT 1957-70 and hold adjunct professorship at SUNYAB presently; became a partner at Office of J Fruchtbaum Assoc, 1970. Am presently a prin in the sucessor firm, Van Wert, Snyder, Sklarsky, Rowley (VSSR) Arch/Consulting Engrs; and responsible for structures and computing for VSSR. PPres of Buffalo Section, ASCE. Active in magic, running, racquetball and badminton. *Society Aff:* ASCE, ACI, ASEE, SAME, NSPE.

Snyder, Lynn E
Home: 1641 Creekwood Dr, Brownsburg, IN 46112
Position: Project Engineer *Employer:* Allison Gas Turbine Div., General Motors *Education:* PhD/ME/Purdue; MSME/ME/Purdue; BSME/ME/Purdue. *Born:* 1/20/44. Raised near Rossville Ind. Graduate Instructor in Res Purdue Univ 1967-70; Res Engr with Pratt & Whitney Aircraft responsible for developing aeroelasticity design systems 1970-74; Development Engr with Detroit Diesel Allison specializing in unsteady aerodynamic analysis in turbomachinery 1974-77. ASME Gas Turbine Award for the most outstanding tech paper 1974; Vice Chmn Central Ind AIAA Sect 1976. Graduated with highest honors, Tau Beta Pi, Pi Tau Sigma Supervisor of Dynamics Analysis with Detroit Diesel Allison 1977-80 Chief, Structural Mechanics with Allison Gas Turbines 1980-85. Chief Project Engineer, Advanced Engines 1986-. *Society Aff:* ASME, AIAA, AHS, AAAA.

Snyder, M Jack
Business: 505 King Ave, Columbus, OH 43201
Position: Assist Dept Manager *Employer:* Battelle-Columbus Division. *Education:* BSc/Chem/OH State Univ *Born:* Oct 12, 1921. Held number of res & mgn positions since joining Battelle in 1943. Successively Res Engr, Assoc Div Ch, Dept Mgr, Asst to Dir, Prog Office Mgr, Assist Dept. Mgr. Major R&D exper in fields of cement & concrete technology, glass technology, structural ceramics, surface chemistry & physics, chem engrg. Wason Medal ACI. Fellow American Ceramic Soc. Active in community & church. Pres of Library Bd, Trustee of Church, Drama Group. *Society Aff:* ACS, ACS, ACI, NICE, ΣΞ.

Snyder, Richard L
Home: 4625 Van Kleek Dr, New Smyra Beach, FL 32069
Position: Ret. (Self-employed) *Education:* BSEE/-/Lehigh Univ. *Born:* Nov 1, 1911 Pittsburgh Pa. Res Engr for Farnsworth TV, RCA, Inst for Advanced Study; Asst Prof Univ of Pa; Ch Engr EDVAC proj; Branch Ch Aberdeen Proving Ground; Dept Mgr Hughes Aircraft. Private res & engrg cons. Developed sev types of photomultipliers, scanning microscope, storage tubes, computer memories, tape handlers, & disc stores. *Society Aff:* IEEE, AAAS, ΣΞ.

Snyder, Robert D
Business: UNCC Station, Charlotte, NC 28223
Position: Dean of Engrg *Employer:* Univ of NC-Charlotte *Education:* PhD/Mech/WV Univ; MS/ME/Clemson Univ; BS/ME/IN Inst of Tech *Born:* 4/15/34 Native of Lancaster, PA. Graduated Lancaster High 1952. After period as apprentice machinist, obtained BSME IN Inst of Tech 1955. Instructor Clemson Univ 1957-59. Assistant Prof of Engrg Mechanics Clemson 1959-60. NSF Faculty Fellow Brown Univ 1960-62. Assistant Prof of Theoretical and Applied Mechanics 1964-65, Assoc Prof 1965-68, Prof 1968, Assoc Chrmn for Graduate Studies in Mechanical Engrg and Mechanics at WV Univ. Co-authored two successful textbooks on engrg mechanics. Moved to the Univ of NC at Charlotte as Prof and Chrmn of Engrg Science, Mechanics and Materials 1975 and became Dean of Engrg in January 1977. *Society Aff:* ASME, ASEE, NSPE, ΣΞ

Snyder, Thoma M
702 Meadow Lane, Los Altos, CA 94022
Position: Consulting Engineer. Retired *Employer:* General Electric Co. *Education:* PhD/Physics/Johns Hopkins Univ. *Born:* 1916. Married Charlotta Untiedt; 3 children. Instructor Physics, Reactor Dev Princeton 1940-43; Bomb Dev Los Alamos 1943-45; Mgr Physics, Sodium Cooled Breeder, Naval Reactors, Knolls Atomic Power Lab 1946-56; Mgr Res KAPL 1956-57; Mgr Physics General Electric Vallecitos Atomic Lab 1958-64; Cons General Electric Nuclear Energy Systems Div: Res & Engrg VAL 1964-68, Reactor Fuels & Reprocessing 1968-72, Prod & Quality Assurance 1972. Reg Nuclear Engr State of Calif. Phi Beta Kappa, Sigma Xi; Fellow American Physical Soc; Fellow American Nuclear Soc; P Chmn, Cttes - P Pres Northern Calif Sect. AEC Tech Advisor: Physics Progs, Internatl Confs, Fellowship Bd. Citations: Secy of Navy, Secy of War 1945. Steinmetz Award 1979. GPU Nuclear Corp General Office Review Bd. Cincinnati Gas & Elec Operations Review Ctte, Shippingport Decommissioning Readiness Review Ctte. *Society Aff:* ANS, APS.

Snyder, William R
Home: 6 Oriole Dr, Wyomissing, PA 19610
Position: Retired. *Education:* DSC/Hon/Albright College; BS/EE/Penn State. *Born:* 1903. Native of Pa, lived in Philippines for 30 years. BS Penn State; Honorary DSc Albright. Distribution Engr for Met Ed Co 1924-30. Manila Electric Co as Elec Engr, Asst Elec Mgr, Vice Pres Operations, Pres successively 1930-60. Pres Met Ed Co 1961-68. Responsible for rebuilding of completely destroyed Manila Electric Co property after WWII. Responsible for Dev of Met Ed Co power system in 1960's including nuclear power station on 3 mile island in Susquehanna River near Harrisburg Pa. Fellow IEEE. Retired 1968. *Society Aff:* IEEE, NSPE.

Sobel, Matthew J
Business: 312A Harriman Hall, Suny at Stony Brook, Stony Brook, NY 11794-3775
Position: Leading Prof. *Employer:* SUNY at Stony Brook. *Education:* PhD/Operations Res/Stanford Univ; AM/Math Statis/Columbia Univ; BS/Indus Engg/Columbia Univ; AB/Pre-engg/Columbia College. *Born:* 12/24/38. Have applied, taught, & dev new techniques in opers res. Worked for Western Elec Co, US Publ Health Serv, and Stanford Res Inst. Yale Univ faculty 1967-77, GA Inst of Tech faculty 1977-86, SUNY at Stony Brook 1986 (Inst. Dec. Sci., Harriman School, Appl. Math. & Stat.) Specialties include probabilistic models, production, congestion & competition. Consultation on problems in bus, indus, & govt. *Society Aff:* ORSA, TIMS, ΣΞ, ALPHAΠM.

Sobol, Harold
Business: 1200 N Alma Rd, Richardson, TX 75081
Position: Vice President Engineering and Advanced Technology *Employer:* Rockwell International, Telecommunications Group *Education:* PhD/EE/Univ of MI; MSE/EE/Univ of MI; BSEE/EE/CCNY. *Born:* June 1930. Native of Brooklyn NY. Res Assoc Univ of Mich 1952-59, worked on radar, missle guidance & travelingwave tubes; IBM Res Lab 1960-62, worked on superconductors; RCA Labs 1962-73 staff engr & Head Communication Tech, worked on plasmas, klystron, microwave solid state devices & circuits; Rockwell Internatl, Collins Radio 1973- . Presently Vice President Engineering and Advanced Technology, responsible for dev of microwave LOS radio, multiplex & ancillary equipment, lightwave transmission, switching, network termination. IEEE. Fellow, Microwave Natl Lecturer 1970, Dallas Sect Outstanding Engr Award 1975, Centennial Medal 1984. Administrative Ctte Microwave Soc 1972- . Chmn Electronic Components Conf 1974, Internatl Solid-State Circuits Conf 1975, Pres IEEE Microwave Soc 1978. *Society Aff:* IEEE, APS.

Socie, Darrell F
Business: 1206 W Green St, Urbana, IL 61801
Position: Assoc Prof *Employer:* Univ of IL *Education:* PhD/Theo & Appl Mech/Univ of IL; MS/Metallurgical Engrg/Univ of Cinn; BS/Metallurgical Engrg/Univ of Cinn *Born:* 10/29/48 Project Engr for Structural Dynamics Research Corp, specializing in failure analysis from 1972-74. Assistant Prof of Mechanical Engrg, 1977-1981. Assoc Prof since 1981 at the Univ of IL at Urbana-Champaign. Research Activities have included developing design methodology for preventing fatigue and fracture in service. SAE Teetor Award 1981. *Society Aff:* ASTM, ASM, SAE, NSPE

Socolow, Robert H
Business: Engrg Quadrangle, H-104, Princeton, NJ 08544
Position: Prof *Employer:* Princeton Univ *Education:* PhD/Physics/Harvard Univ; MA/Physics/Harvard Univ; BA/Physics/Harvard Univ *Born:* 12/27/37 As Dir of Princeton Univ's Center for Energy and Environmental Studies, I am responsible for a research enterprise which seeks to identify and to clarify complex, value-laden issues in energy and environmental policy several years before they become front-line concerns. Our research focuses on the energy conservation, indoor air quality (including radon), energy and developing countries, and arms control. Many of my own contributions are summarized in four books: *Patient Earth* (Holt, Rinehart 1971) with John Harte; *Efficient Use of Energy* with K. Ford, G. Rochlin, M. Ross; *Boundaries of Analysis: An Inquiry into the Tocks Island Dam Controversy* with H. A. Feiveson, F. W. Sinden; and *Saving Energy in the Home: Princeton's Experiments at Twin Rivers. Society Aff:* APS

Soderstrand, Michael A
Business: Dept of Electrical and Computer Engg, Davis, CA 95616
Position: Prof *Employer:* Univ of CA. *Education:* PhD/EE/Univ of CA; MS/EE/Univ of CA; BS/EE/Univ of CA; AA/Engg/Sacramento City College. *Born:* 7/31/46. Dr Soderstrand worked at various radio and tv stations, and for 11 yrs at Sandia Labs, Livermore, CA before joining the Univ of CA, Davis in 1972. He is author of 4 text books, 2 IEEE press research books, and over 80 technical papers in various journals. He is currently on the editorial board for the IEEE press. *Society Aff:* CAS, CSPE, ASEE, NSPE, IEEE.

Soedel, Werner
Business: Sch of Mech Engg., Purdue Univ, W Lafayette, IN 47907
Position: Prof. *Employer:* Purdue Univ. *Education:* PhD/ME/Purdue Univ; MSME/ME/Purdue Univ; 4th yr ctf/Automotive engr/GM Inst; Ingenieur/ME/Staatliche Ingenieurschule Frankfurt. *Born:* 4/24/36. Prague, Czechoslovakia (Sudetengerman). Parents: Hermann Soedel and Gertrud Fritsche. In W Germany after 1945. Schooling in Bavaria and Hesse. Jr/Sr engr in automotive design with Gen Motors (Opel, W Germany, and GM Tech Ctr, MI), 1957- 63. Teaching/res asst 1963-67, asst/assoc prof 1967-75 at Purdue, presently full prof. US citizen since 1968. Res and teaching in Vibration (solids, liquids and gases). 135 refereed papers, 152 res reports, 5 monographs, one textbook on shell and plate vibrations, Won Solberg Teaching Award twice, Ruth and Joel Spiro Award. Ralph Coates Roe Award (ASEE), 1986. Hobby: Engg history. Married to Ann S Greiber. Children: Sven (1962), Fritz (1967), Dirk (1969), and Leni (1973). *Society Aff:* ASME, ASA, AAM, ΣΞ, ASEE.

Soehrens, John E
Home: 395 S Los Robles Apt. 32, Pasadena, CA 91101
Position: Engineering Consultant. *Employer:* Self-employed. *Education:* MS/Civil Eng/Univ of CO; BS/Civil Eng/Yale Univ. *Born:* July 23, 1909. Native of Rhode Island. 15 years in civil serv with Bureau of Reclamation & The Panama Canal; staff cons for C F Braun & Co 1948-73. Mbr of subcom of ASME Boiler & Pressure Vessel Code. Principal activities: structural design and analysis, design of storage tanks, pressure vessels & refinery piping, engrg mathematics, and computer procedures. Fellow ASME & ASCE. Retired 1973. Enjoy golf, camping, & fishing. *Society Aff:* ASME, ASCE, Sigma Xi

Sogin, Harold H
Business: 6823 St Charles Ave, New Orleans, LA 70118
Position: Prof & Head *Employer:* Tulane Univ *Education:* PhD/ME/IL Inst of Tech; MSc/ME/IL Inst of Tech; BSc/ME/IL Inst of Tech *Born:* 12/14/20 Native of Chicago, IL. Asst Prof Mech Engrg IL Tech 1953-55. Asst Prof Engrg, Brown Univ, 1955-57. Assoc Prof Engrg, Brown Univ, 1957-60. Prof Mech Engrg, Tulane Univ 1960-

Sogin, Harold H (Continued)
present. Head of dept since September 1978. Principal investigator on NSF grants 1962-64, 69-71, 72-75. *Society Aff:* ASME, ΣΞ

Sohn, Hong Yong
Business: Dept Metallurgical Engrg, University of Utah, Salt Lake City, UT 84112-1183
Position: Prof *Employer:* Univ of UT *Education:* PhD/Chem Engrg/Univ of CA, Berkeley; MSc/Chem Engrg/Univ of New Brunswick; BS/Chem Engrg/Seoul Natl Univ *Born:* 8/21/41 Born in Korea. US citizen. PhD in Chem Engrg, Univ of CA, Berkeley, 1970; Research Assoc at SUNY, Buffalo; Research Engr for DuPont Co (Engrg Tech Lab, Wilmington, DE); with Univ of UT since 1974. Currently, Prof of Metallurgy and Metallurgical Engrg and Adj Prof of Chemical Engrg. Research activities in extractive metallurgy, combustion of solids and liquids, and oil shale processing. Coauthored and coedited 7 books and published some 130 papers. 1 patent. Camille and Henry Dreyfus Foundation Teacher-Scholar Award, 1977. US Dept of Energy of Energy Fossil Energy Lecturer. Fulbright Distinguished Lecturer, 1983. Enjoy various sports and travel. *Society Aff:* TMS-AIME, AIChE, KSEA

Sohr, Robert T
Home: 140 LeMoyne Pkwy, Oak Park, IL 60302
Position: Technical Dir *Employer:* R. T. Sohr; PChE *Education:* BS/Chem/Valparaiso Univ; MBA/Econ/Univ Chicago; PChE/AIC/-; PCh/AIC/- *Born:* 4/4/37. Developer of more than 20 "State of the art" solutions for air-odor abatement all over the world. 25 plus technical publications-world wide. Include: printing, welding, foundry core rooms-pouring-cooling-shakeout, insulation, paper mills, solvents, rubber, grain processing, corn milling, fish, snack foods & many others. Developer of TV-sonar fish locating unit. Intl contract bridge authority. *Society Aff:* AIChE, AIC, ACS, APCA, AFS

Sokolowski, Edward H
Business: 8927 Internatl Dr, San Antonio, TX 78216
Position: President. *Employer:* International Aerial Mapping Co. *Born:* March 1927. BCE New York Univ. Civil Engr & Assoc with Vogt, Ivers & Assocs of Cincinnati Ohio; Ch Engr San Antonio Office of Vogt, Ivers & Assoc; Vice Pres, Genl Mgr of Internatl Aerial Mapping Co of San Antonio Texas, assumed responsibility as Pres 1970. Survey & mapping projs throughout the US, Bahamas, Mexico, Nicaragua, Panama & Okinawa. Served as: Pres of Texas Sect ASCE; Pres of Texas Inst of Traffic Engrs; Chmn Cartographic Surveying Ctte, ASCE. Serving as: Natl Dir ASCE, dist 15 Texas, New Mexico & Mexico; Pres Alamo Heights Rotary Club San Antonio Texas; Vice Chmn Exec Ctte, Surveying & Mapping Div ASCE. Reg PE in Ohio, Texas, Washington, West Va, Iowa, & New Mexico; Reg Photogrammetric Engr in West Va; Reg Public Surveyor in Texas. Mbr American Soc of Photogrammetry.

Soland, Richard M
Business: Sch of Engg & Appl Sci, George Washington Univ, Washington, DC 20052
Position: Prof *Employer:* George Washington Univ. *Education:* PhD/Math /MIT; BEE/Elec Engg/RPI. *Born:* 7/27/40. Mbr of the Technical Staff, Res Analysis Corp, 1964-71. Assoc Prof, College of Business, The Univ of TX at Austin, 1971-76. Assoc Prof, Ecole Polytechnique de Montreal, 1976-78. Visiting faculty positions in Denmark, Finland, France and Venezuela. Consultant on missile allocation problems. Author or co-author of 40 articles dealing with Bayesian statistics, discrete and nonconvex optimization, location theory, missile defense, and multiple criteria decision making. Currently Prof of Operations Res, The George Washington Univ, and responsible for the undergrad program in Systems Analysis and Engrg. *Society Aff:* IEEE, ORSA, IIE, CORS, MPS, TIMS

Solberg, James J
Business: Sch of Ind Engrg, Purdue Univ, W Lafayette, IN 47907
Position: Prof. Director, Engrg Research Center *Employer:* Purdue Univ. *Education:* PhD/Ind Engg/Univ of MI; MS/Ind Engg/Univ of MI; MA/Math/Univ of MI; BA/Math/Harvard. *Born:* Joined Purdue in 1971. Co-authored widely used textbook in Operations Res; won IIE Book of the Yr Award in 1977, IIE Distinguished Award in 1982. Director of Engineering Research Center for Intelligent Manufacturing Systems, one of the national ERC's created by the National Science Foundation in 1985. Wife: Elizabeth. Children: Kirsten & Margaret. *Society Aff:* IIE, ORSA, SME.

Solberg, Ruell F, Jr
Business: P. O. Drawer 28510, San Antonio, TX 78284
Position: Principal Eng *Employer:* Southwest Research Inst *Education:* MS/ME/Univ of TX-Austin; MBA/Bus Adm/Trinity Univ; BS/ME/Univ of TX-Austin *Born:* 7/27/39 Research engr with Applied Research Labs (formerly Defense Research Lab), Austin, TX, specializing in research, development, design, and fabrication of electromechanical equipment for underwater acoustics use. At Southwest Research Inst since 1967, specializing in research, development, design, fabrication, and testing of electromechanical equipment. Projects have ranged from equipment used on NASA's Apollo, Skylab, and Space Shuttle to shipboard, shorebased, and airborne radio direction finding antenna systems. Contributor to technical, professional, and historical publications and conferences. Prominent in field. Reg PE in the state of TX. *Society Aff:* ASME, SAWE, NSPE, HFS, ΤΒΠ, ΠΤΣ, ΣΞ, ΣΙΕ, ISA.

Solien, Vernon L
Business: 2601 No Univ Dr, Fargo, ND 58102
Position: Consulting Structural Engineer. *Employer:* Solien Cons Structural Engineer. *Education:* BS/Civil Eng/ND State Univ. *Born:* 5/18/25. Served US Marine Corps 1944-46. Engr US Bureau of Reclamation 1950-53; Design Engr Cerny Assocs, Architects & Engrs Minneapolis Minn 1953-64; Design - Structural design of commercial buildings in steel & reinforced concrete; Cons Structural Engrg firm Solien CSE in Fargo North Dakota 1964-. Reg PE in North Dakota, Minn, South Dakota, Iowa, Wisc, Mont, Wyoming. Mbr North Dakota Soc of Prof Engrs. Mbr ASCE. *Society Aff:* ASCE, NSPE.

Soliman, Ahmed M
Business: Dept of Electrical Engrg, Boca Raton, FL 33431
Position: Assoc Prof *Employer:* FL Atlantic Univ *Education:* PhD/EE/Univ of Pittsburgh; MS/EE/Univ of Pittsburgh; BS/EE/Cairo Univ *Born:* 11/22/43 Native of Egypt. Taught at Cairo Univ, the American Univ at Cairo, the Univ of Pittsburgh, Coll of Steubenville, San Francisco State Univ, FL Atlantic Univ, Univ of Petroleum and Minerals, Dhahran, Saudi Arabia. Has published over 100 papers in the area of multivariable network synthesis, active RC filter design, active R filters and compensation of operational amplifier circuits. Received the State Engrg Science Prize from the Scientific Research Academy, Egypt. Received the first class science medal from the Pres of Egypt for services to the field of engrg in 1977. *Society Aff:* IEEE

Solomon, Alvin A
Business: Nucl Engg, W Lafayette, IN 47907
Position: Prof *Employer:* Purdue Univ. *Education:* BS/Mech Eng/U of IL; MS/Theor & App Mech/U of IL; PhD/Mat'ls Sci/Stanford Univ *Born:* 8/17/37. Conducted post-doctoral res at the Ctr d'Etudes Nucleaires at Saclay, France, 1968/69 & res on nucl mtls at Argonne Natl Lab from 1969 to 1974. Accepted position as Assoc Prof in the Sch of Nucl Engg of Purdue Univ in 1974 specializing in basic res & teaching in the area of nucl mtls. Res interests include mech behavior of metals & ceramics, radiation effects, sintering, ceramic fabrication. Currently has rank of Prof. *Society Aff:* ACS, ANS, AAUP

Solomon, George E
Business: 1 Space Park, Redondo Beach, CA 90278
Position: Executive Vice President *Employer:* Electronics & Defense Sector, TRW Inc. *Education:* BS U. of Wash. 1949; MS (1950), PhD (1953) Calif. Inst. of Tech-

Solomon, George E (Continued)
nology *Born:* 07/14/25. Seattle, WA. Joined TRW Inc 1954, Dir of Sys Res & Analysis Div 1958-65, VP in 1962. Dir of Sys Labs 1965-8; Dir of Marketing & Reqts Analysis 1968-71. Gen Mgr TRW Sys Group 1971-81. Exec V Pres & Gen Mgr, TRW Electronics & Defense Sector 1981-. Mbr Natl Acad of Engrg, Fellow AIAA and AAS, Mbr Bd of Gov Aerospace Industries Assn, Mbr VFW, AUSA, AFA, AFCEA, Phi Beta Kappa, Tau Beta Pi, & Sigma Xi. *Society Aff:* NAE, AIAA, AAS, NSIA, ADPA, EIA

Solomon, James E
Business: 2900 Semiconductor Dr, Santa Clara, CA 95051
Position: Mgr, Linear R&D. *Employer:* Natl Semiconductor. *Education:* MS/EE/Univ of CA; BS/EE/Univ of CA. *Born:* 7/20/36. in Boise, ID. Spent 3 yrs at Motorola Systems Res Lab in Riverside, CA working on Radar & missile control. In 1963, started up Linear IC R&D at Motorola in Phoenix. Activities included design of IC op amps, regs, video, audio, TV. Joined Natl Semiconductor in 1971 as hd of linear IC dev. There he developed a broad line of linear IC's including BI-FET op amps, Telecom IC's, MOS A/O's & DAC's. Present activities include MOS data acquisition, speech synthesis & recognition. Spare time activities include skiing, diving, a Hobie cat, & home computing. *Society Aff:* IEEE.

Solomon, Lee H
Business: 4775 Indian School Rd, Albuquerque, NM 87110
Position: Pres. *Employer:* Plateau, Inc. *Education:* BS/Chem Engg/Univ of MI. *Born:* 1/4/35. Following Grad from college employed by Exxon USA in various engg, economic and mgmnt positions for 10 yrs. Next three yrs spent with Esso Interamerica in mgmnt of South American Refineries. In 1971 a founding partner of the consulting firm Turner, Mason & Solomon - Dallas, Tx. During the next 5 yrs I was active in energy policy consulting for major companies and the Federal Government. Joined Plateau, an independent refiner and marketer in the mountain states as Pres and Chief Exec officer in 1978. *Society Aff:* AIChE, NSPE.

Solomon, Richard H
Business: 356 South Loomis St, Naperville, IL 60540
Position: owner *Employer:* Self employed *Education:* BS/Fire Protection/Illinois Institute of Technology *Born:* 06/02/33 Native of Bloomington, IL. Worked in municipal engrg dept, Illinois Inspection & Rating Bureau 1955-66. Served in research & devel lab - USN, 1956-7. Established own consulting engrg practice 1966-current. Engrg Practice serves local, county and state governmental units in the field of fire protection engrg; dealing with development of codes and standards for construction, analysis of fire apparatus design, fire station distribution, communications, water system design for fire protection. Bd mbr, SFPE 1985-current. Member various cttees SFPE, NFPA. *Society Aff:* SFPE, NSPE, NFPA

Somasundaran, Ponisseril
Business: 911 S.W. Mudd Building, School of Engineering & App. Sc, Columbia University, New York, NY 10027
Position: La Von Duddleson Krumb Prof of Mineral Engg *Employer:* Columbia Univ. *Education:* PhD/Miner Tech/U CA, Berkeley; MS/Miner Tech/U CA, Berkeley; BE/Metal/Indian Inst of Sci; BSc/Chem/Kerala Univ. *Born:* 6/28/39. Presently La Von Duddleson Krumb Prof at Columbia Univ. 20 yrs of profl experience and 15 yrs of managerial experience in industry and academia. Holds 3 US patents. Author/editor of 9 bks and 200 publications. Chrmn of Mineral Processing Div Soc Min Engrs-AIME - 1982/83. Mbr of Bd of Dirs and Exec Ctte (84) of Soc. Mining Engrs/AIME - 1982/85. Chrmn of intl and natl conferences like Natl Sci Fdn Workshop on Minerals, Intl Symposium in Res: Engrg Mbr of Natl Acad of Scis Cttee on Accessory Element-Phospate Panel Editor-in-Chief of *Colloids and Surfaces* journal and assoc editor of other journals. Conslt to Exxon, IBM Englehard Colgate-Palmolive, Goodrich, NIH, NSF, Interests in mineral processing, surface and colloid chem and enhanced oil recovery. Received most distinguished achievement in Engrg Awd from AINA, 1980; Pub Bd Awd, of the Soc of Mining Engrs AIME, 1980. Antoine M Gaudin Awd of Soc Min Engrs-AIME 1982 Millman of Distinction. Fellow of inst of mining & metallurgy (London) 1984. Keynote speaker at a number of natl and intl/symposiums. Distinguished Mbr Soc Min Engrs, AIME, 1983. Member, National Academy of Engrg, 1985 Robert H. Richards Award of AIME, 1987 Arthur F. Taggart Award of SME, 1987. Chmn, Engrg Foundation Conference Committee, 1985-. *Society Aff:* SME/AIME, ACS, AIChE, SPE, IMM, FPS, SPE/AIME.

Somers, Richard M
Home: 8428A Charles Val Ct, Baltimore, MD 21204
Position: Exec Asst To VP & Genl Mgr - Retired. *Employer:* The Bendix Corp. *Education:* EE/-/Rensselaer Poly Inst. *Born:* Dec 22, 1904. Mathematics Instructor RPI; Engr RCA Communications, LI; many years with T A Edison, West Orange NJ - Dir of Engrg. Bendix Corp Teterboro, then as Dir of Engrg Kansas City Div (Atomic Energy/Weapons), then various managerial & staff positions with Bendix, Towson Md. Retired June 1970. Life Fellow IEEE, Reg PE in NJ, Kansas, & Mo. 58 US Patents. Tau Beta Pi. Boy Scouts of America Silver Beaver. Hobby: 'fixing' electro-mechanical things with the tools & materials commonly available at home.

Somerville, John E
Business: 1301 Vista Ave, Boise, ID 83705
Position: Senior Vice President. *Employer:* James M Montgomery, Cons Engrs Inc. *Education:* MS/Sanitary Engg/Stanford; BS/Civil Engg/Univ of WI. *Born:* April 1940 Hayward Wisc. Field Engr with Alaska Hwy Dept 1963-64; Sanitary Engr with Metcalf & Eddy 1964-65; Principle with Kingman Engrs on water & wastewater projs 1966-70; with JMM 1970-. Senior Vice President & Regl Mgr of Boise, ID; Portland, OR; Ogden, UT; Salt Lake, UT; Denver, CO; Anchorage, AL Offices. In charge of environmental studies, water & wastewater, & civil & municipal hydropower & mining projs in Northwest. Dir of S F Bay Sect of Calif Water Pollution Control Assn 1973-74; Chmn Prof Dev Comm of ISPE 1974-75; Mbr ASCE Nonpoint Source Ctte 1975-76; Chmn of Advisory Ctte to State of Idaho on Wastewater Design Criteria 1976. Enjoy hunting & fishing. *Society Aff:* AWWA, ASCF, WPCF, NSPE.

Somerville, Mason H
Home: 4402 85th St Apt 281, Lubbock, TX 79424
Position: Dean, Coll of Engrg *Employer:* Texas Tech Univ *Education:* PhD/ME/PA State Univ; MS/ME/Northeastern Univ; BS/ME/Worcester Poly Inst. *Born:* 12/21/41. Worcester, MA 1941; Received BSME 1964 Worcester Poly Inst, MSME Northeastern 1966, PhD (ME) PA State Univ 1972. Worked for: Worcester Gas Light Co, 1964-65; Norton Co, 1965-6; Instr of Mech Engg at PA State, 1967-71, Sr Engr, Westinghouse Bettis Atomic Power Lab, 1971-73; Asst Prof ME Univ of ND, 1973-4, Mgr, Engg Experiment Station, UND 1974-7, Assoc Prof of ME, UND, 1976- 1980, Dir (1977-1978), Engrg Experiment Station, Univ of ND, Grand Forks, ND, Solar Energy advisor to the State energy Office 1978-1980, 1980-1984 Prof & Dept Hd, Mech Engrg Univ of AK, 1984-present, Dean and Prof of ME Coll of Engrg, TX Tech Univ. Dir on bd of Mid-American Solar Energy Complex Corp. Conducted $1.3 million in res since 1973, has published over 40 reports and 30 papers all in energy. Reg PE in ND No. 1989, AR No. 5425. *Society Aff:* ASME, ASHRAE, ASEE.

Sommer, Alfred H
Home: 16 Wildwood Circle, Wellesley, MA 02181
Position: Consultant (Self-employed) *Education:* PhD/Phyiscal Chem/Berlin Univ. *Born:* Nov 1909 Frankfurt, Germany. Baird- (later Cinema-) Television Co London 1936-46; EMI-Res Labs Hayes England 1946-53; RCA-Res Labs Princeton NJ (Fellow Tech Staff) 1953-74; Cons for Thermo-Electron Co Waltham Mass 1974-. R&D concerned with new photoemissive & secondary electron emitting materials & with devices using these materials, such as Television Camera Tubes, Image Intensifiers, & Photomultipliers. Books: 'Photoelectric Tubes' (Methuen 1946, 2nd ed

Sommer, Alfred H (Continued)
1951); 'Photoemissive Materials' (John Wiley 1968. Reprinted 1978 with Update References.). Fellow IEEE. Mbr American Physical Soc. *Society Aff:* IEEE, APS.

Sommer, Richard S
Business: 46 Main St, New Canaan, CT 06840
Position: President *Employer:* Sommer Industries, Inc. *Education:* BChE/Chem Engg/Cornell Univ. *Born:* July 19, 1941. Mgr Projs & Planning Esso Chem Corp, 1963-69, responsibilities included design & const of chemical plants. Corporate Vice Pres, Secy for Combustion Equipment Assocs Inc 1969-79 Exper has included administration, engrg & market dev of (1) resource recovery projs- fuel & by-prods recovered from municipal refuse; & (2) sulfur dioxide projsSO2 & particulate removed from boiler flue gases. Former Mbr Bd/Dir of CEA Internatl, CEA Internatl GmBh, CEA-Canada, Carter-Day, Hart-Carter, Columbia Univ. Grad. School of Bus. Admin-Executive Committee on Organization. Currently Pres and owner of Sommer Indust Inc an Energy/Environ Serv and Supply. Also US Representative to Coal Industry Advisory Board of International Energy Agency, (Paris, France) Member of Bd/Dir. Industrial Energy Inc (Wash DC), and Kone Wood Inc (Atlanta, GA) , Arbitrator for American Arbitration Association. *Society Aff:* AIChE, AAA

Sommerfeld, Jude T
Home: 2924 Clairmont Rd, Apt I-7, Atlanta, GA 30329
Position: Prof *Employer:* GA Inst of Tech *Education:* PhD/ChE/Univ of MI; MSE/ChE/Univ of MI; B/ChE/Univ of Detroit *Born:* 2/4/36 in Elmwood Place, OH, raised in Detroit, MI. Sr systems engr with Monsanto Co (1963-66), participated in development of the FLOWTRAN system. Chief process systems engr (1966-68) and then dir of process engrg (1968-70) with BASF- Wyandotte Corp, responsible for corp-wide scientific computer applications, process improvement and systems engrg. With GA Inst of Tech since 1970, as assoc prof (1970-75), prof (1975-present) and assoc dir (1981-present) of chem engrg. Visiting prof in England (1973, 1980) and Chile (1979, 1984). Registered PE in GA. Tech interests include computer applications, process design and energy conservation. Enjoys travelling, sports and classical music. *Society Aff:* AIChE, ACS, ISA, NSPE

Sommers, Otto W
Home: 7514 Shadylane, San Antonio, TX 78209
Position: Ret. (City Public Service Bd) *Education:* BS/EE/TX A&M Univ. *Born:* Nov 20, 1906 San Antonio Texas. Tau Beta Pi. Fellow AIEE. Reg PE in Texas. San Antonio Public Service Co 1929-42; City Public Service Bd 1942-71, Genl Mgr 1958-71. *Society Aff:* AIEE.

Sonderegger, Paul E
Business: 825 J St, Box 80358, Lincoln, NE 68501
Position: President. *Employer:* Hoskins-Western-Sonderegger Inc. *Education:* BS/Chem Eng/Univ of NB. *Born:* 5/29/22. Served in US Navy-Seabees 1943-45. Mgr & later Pres of Western Labs, material engrg cons to 1969; Partner 1958-66, & Pres 1966-71 of Miller-Warden- Western, a cons engrg firm specializing in foreign hwy engrg. As Pres of Hoiskings-Western-Sonderegger Inc, a multi-disciplined cons firm since 1972, directs foreign & domestic opers. Recipient of Japanese Government Ministry of Const Award for engrg consultation on Japan Expressways. *Society Aff:* ASCE, AAPT, NAPA.

Sonenshein, Nathan
Home: 1884 Joseph Dr, Moraga, CA 94556
Position: Pres. *Employer:* Sonenshein and Assoc. Inc. *Education:* BS/Elec Eng/US Naval Acad; MS/Naval Const & Mar Eng/MIT. *Born:* 1915 Lodi NJ. Graduated from US Naval Acad 1938; MS in Naval Const MIT 1944; Advanced Mgmt Prog Harvard 1965. As an Engrg Duty Officer, served in various assignments in ship design, const & maintenance. Promoted to Rear Admiral in 1965. Was Cdr Naval Ship Systems Command 1969-72 & Dir for Energy for DOD 1972-74. Upon retirement from naval service in 1974 and until 1984 was Asst to the President Global Marine Dev Inc, an org devoted to developing advanced marine systems primarily in energy field offshore. Now Pres of consulting firm in these fields. P Pres American Soc of Naval Engrs 1970-71, currently a Vice Pres for Life of American Soc of Naval Architects & Marine Engrs. Mbr Sigma Xi. Listed in Who's Who in America. Recipient of ASNE's Harold E. Saunders Award for 1982. Appointed by Pres. Reagan to 2-year term, 1984-86, on National Advisory Committee for Oceans & Atmosphere. *Society Aff:* ASNE, SNAME, ΣΞ.

Song, Charles CS
Business: Dept of Civil & Mineral Engr, Minneapolis, MN 55455
Position: Prof *Employer:* Univ of MN *Education:* PhD/CE/Univ of MN; MS/Mech & Hydraulics/St Univ of IA; BS/CE/Nat Taiwan Univ *Born:* 1/12/31 Research experience in cavitation and hydroelasticity prior to 1970. Currently specialized in mathematical modeling of unsteady flows in pipes and open channels, variational principles, and sediment transport. Teaching courses on fluid mechanics, open channel flows, hydraulic transients, and mathematical modeling. Consltg experience in sewer and pipe network modeling, seismic analysis of submerged bodies, etc. Received 1980 J C Steven award. *Society Aff:* ASCE, IAHR, ΣΞ, AWRA.

Sonnemann, Harry
Home: The Rotonda 907, 8360 Greensboro Dr, McLean, VA 22102
Position: Consultant *Education:* BSEE/Electronics/Polytechnic Inst of Brooklyn. *Born:* Sept 1924. Medical Electronics 1943-47; Nuclear Instrumentation 1947-50; Hudson Labs Columbia Univ 1951-64; Dept Head to 1962, Asst Dir; Res in ASW, acoustics oceanography & ocean engrg; Asst Dir Field Engrg then Acting Deputy Dir Advanceed Res Projs Agency 1964-68, engaged in Nuclear Test Detection Res; Special Asst (Electronics) 1968-76, Special Asst (ASW & Ocean Control) to the Asst Secy of the Navy (R&D) 1976-77; Asst to the Chief Engr 1977-78, Deputy chief Engr Natl Aeron & Space Admin 1978-82. NASA Deputy Chief Engr System Engg & Tech 1982-85; NASA Asst Chief Engr 1985-86. P Pres Ridgewood Pistol Club, Willow Brook Figure Skating Club, Ice Club of Washington. Treasurer Art League of Northern Va. Pres, Rotonda Condominium Unit Owners Assoc 1982-1984, McLean, VA. *Society Aff:* IEEE.

Sonnenfeldt, Richard W
Home: 4 Secor Drive, Port Washington, NY 11050
Position: Professional Director *Employer:* Self *Education:* BSEE/-/John Hopkins Univ. *Born:* July 3, 1923. Graduate studies 1951-54 Univ of Pa. Mgr of Opers RCA Corp 1949-61; Genl Mgr Digital Systems Div Foxboro Co 1962-64; CEO, Pres Digitronics Corp 1965-70; Vice Pres RCA Corp 1970-1979. CEO Electronic Indus Engrg Corp 1971-73. Executive VP National Broadcasting Corp Inc 1979-82. Dean, Graduate School of mgmt, Polytechnic Inst of New York, 1982-1984, Prof of Mgmt, Polytechnic Inst of New York 1982- ; Dir, The Foxboro Co, NYSE, 1981- Dir Lee Enterprises, NYSE, 1982- , Dir, Decision Data Computer Corp, 1984- , Dir, Steinbrecher Corp 1984- , Dir Compuflight Inc, 1984-. 35 patents; over 20 pub papers. Honorary Fraternities: Omicron Delta Kappa, Tau Beta Pi, Member Council on Foreign Relations, Fellow IEEE, mbr AMA, mbr Adv Council, the Johns Hopkins Univ. *Society Aff:* IEEE, AMA, CFR, ISA, NACD

Sonnino, Carlo B
Home: 7206 Kingsbury Blvd, St Louis, MO 63130
Position: Professor of Met Engrg Consultant. *Employer:* Universtiy of Missouri - Rolla. *Education:* PhD/Metallurgy/Univ of Milano (Italy); LLD/-/Univ of Milano (Italy). *Born:* May 12 1904 Torino Italy. PhD Univ of Milano Italy. Director of Res Italian Aluminum Co; Pres of LAESA Cons Firm Milano; Tech Advisor of Boxal, Fribourg Switzerland, Beaurepaire France; Tech Advisor of Thompson Brand France; Materials Engrg Mgr of Emerson Electric, St Louis; Cons for Monsanto, & other major firms in USA & abroad; Prof of Met Engrg Washington Univ, St Louis 1960- 68, Univ of Mo-Rolla 1968- . Inventor of Synthetic Criolite Process, Introduced first aluminum can in the world. Ents & papers in the field of metallurgy, surface treatment of aluminum, corrosion. Honorary Mbr Alpha Sigma Mu; Fellow

Sonnino, Carlo B (Continued)
ASM. Klixon Award 1960; Knight Commander of the Italian Republic in recognition of outstanding technological achievements. ASTM; Chair, G107 on galvanic corrosion. NACE, etc. Art collector. *Society Aff:* ASM, ASTM, AAAS, AIA, SME, AFP, APMI, ΣΞ ΑΣΜ, NACE, SAMPE

Sonquist, John A
Home: 4463 Nueces Dr, Santa Barbara, CA 93110
Position: Professor. *Employer:* University of California. *Education:* PhD/Sociology/Univ of Chicago. *Born:* January 22 1931. Past Chairman Association for Computing Machinery Special Interest Group on Social Science Computing; Past Editor A C M - S I G S O C Bulletin. Director Sociology Computing Facility Univ of Calif Santa Barbara. P Res Assoc & Head Computer Servs Facility Inst for Social Res, Univ of Mich. One of the developers of OSIRIS statistical analysis system (Mich). Author of several books on computer-based statistical analysis in social sci. Teach computer sci, info systems design, computer applications in social sci, res methodology. *Society Aff:* ACM, ASA, ASA.

Soo, Shao L
Home: 2020 Cureton Dr, Urbana, IL 61801
Position: Professor of Mechanical Engrg. *Employer:* Univ of Illinois at Urbana-Champaign. *Education:* SD/Mech Engrg/Harvard Univ; MS/TE/GA Inst of Tech; BS/ME/Natl Chiaotung Univ *Born:* 03/01/22 Native of Beijing China; naturalized US citizen. Ranks through Assoc Prof Princeton Univ 1951-59; Prof Univ of IL 1959- in charge of Multiphase Flow Res Lab. Fellow ASME; Hon Mbr Pi Tau Sigma. Appl Mech Rev Awd. Intl Freight Pipeline Awd, 1980. Alcoa Res. Awd 1985. Con to NASA 1971-75, NIH 1972- , FEA 1975-6, NATO 1973, ANL 1975- , & Dir S L Soo Assoc Ltd; Kumar Consulting, Inc. Mbr USRA-NASA 1971-75, Sci Adv Bd EPA 1976-78, Review Ctte for Energy Conversion JPL, 1984. AGARD Lecturer, von Karman Inst of Fluid Dynamics, Brussel Belgium; AVA, Goettingen Ger 1973; Fulbright-Hays Distinguished Lecturer Natl Univ, Buenos Aires Argentina 1974; Invited lecturer, 1980, Invited Mbr, 1986, Academia Sinica, Invited lecturer, Brit Inst Chem Engrs, 1981. Mbr, Advisory Bd, Intl Powder Inst, London, 1977- ; Editorial Bd Int J Multiphase Flow 1973- , Int J Pipeline, 1980- , Intern J Particulate Sci & Tech 1982- ; Intern J Sci & Engrg, (Liberia) 1983; VP Fine Particle Soc 1983-5; Advisor World Bank Korean Energy Trans Study, 1979. Special field: Multiphase Flow (Introduced the field via paper: Soo, I/EC Fund *4* 1965, 426). *Society Aff:* ASME, ASEE

Soohoo, Ronald F
Home: 568 Reed Dr, Davis, CA 95616
Position: Professor of Elec Engrg & Comp Sci. *Employer:* University of California. *Education:* PhD/EE/Stanford Univ; MS/EE/Stanford Univ; SB/EE/MIT. *Born:* Sept 1, 1928 Canton China. SB 1948 MIT; MS 1952, PhD 1957 Stanford. Exper: Asst Engr Pacific Gas & Electric Co 1948-51; Dir of Res Analysis Cascade Res Corp 1954-57; Res Physicist & Staff Mbr MIT Lincoln Lab 1957-61; Assoc Prof Engrg & Applied Sci Caltech 1961-64; Prof 1964, Chmn 1964-70 Dept of Electrical Engrg & Computing Sci Univ of Calif at Davis; ECPD engrg curricula accreditation. Honors & Awards: Fellow IEEE 1970- ; Fellow NATO, CNRS France 1970; NSF Faculty Res Participation Fellow IBM 1974. Author of numerous physics & engrg articles & 3 books. Cons to many labs & indus. *Society Aff:* ΣΞ, APS, IEEE.

Sorber, Charles A
Business: Dean, School of Engineering, University of Pittsburgh, 240 Benedum Hall, Pittsburgh, PA 15261
Position: Dean *Employer:* School of Engrg, Univ. of Pittsburgh *Education:* PhD/CE/Univ of TX-Austin; MS/Sanitary Engrg/PA State Univ; BS/Sanitary Engrg/PA State Univ *Born:* 9/12/39 Native of Wilkes-Barre, PA. Served with US Army 1961-65 and 1966-75. Dir, Environ Quality Div, USA Med Bioengrg R & D Lab (1971-75). Dir, Ctr for Applied Res and Tech, Univ of TX at San Antonio (1975-1980). Asst Dean, Coll of Sci and Math, Univ of TX at San Antonio (1976-77). Acting Dir, Div of Earth and Physical Sci, Univ of TX at San Antonio (1977-80). Assoc Dean, Coll of Engrg, Univ of TX at Austin (1980-1986), Currently, Dean, School of Engineering and Prof. of Civil Engrg, Univ. of Pittsburgh. Reg P E: PA & TX. Res in health effects of land application of wastewater, disinfection, and water reuse have resulted in more than 120 papers, chapters, and reports. *Society Aff:* ASCE, APHA, AWWA, WPCF, IAWPRC, ASEE, AAEE, NSPE, ΣΞ, ΤΒΠ, ΧΕ, ΦΚΦ

Sorensen, Kenneth E
Business: 150 S Wacker Dr, Chicago, IL 60606
Position: VP *Employer:* Harza Engrg Co *Education:* MS/CE/Univ of MN; B/CE/Univ of MN *Born:* 1/16/19 Native of Minneapolis, MN with Kimberly Clark Corp 1939-40. Lt USNE, with the Panama Canal 1940-42. US Naval Construction Battalion 1942-45. With Harza Engrg Co since 1946. VP since 1957. Mainly engaged in river basin and project planning, includling large dams and hydroelectric plants on such rivers as the Congo (Zaire), Columbia (USA), Tigris (Iraq), Indus (Pakistan), Caroni (Venezuela), and Parana (Argentina-Paraguay. Presently, Chief Technical Advisor to the Pres at Harza Engrg. *Society Aff:* ASCE, USCOLD, NSPE

Sorenson, Harold W
Home: 2511 Loring St, San Diego, CA 92109
Position: Professor of Engrg Sciences *Employer:* University of California, San Diego *Education:* PhD/Control Theory/UCLA; MS/Control Theory/UCLA; BS/Aero Engg/IA State. *Born:* 8/28/36. Native of Omaha, NB. Worked at Gen Dynamics/Astronautics in San Diego, CA from 1957-1962 and at AC Electronics Div of GM in El Segundo, CA from 1963-1966. Was a Guest Scientist 1966-1967, at the Inst for Guidance and Control in Munich, Germany. Joined the faculty of Univ of CA, San Diego in 1968 and currently is Prof of Engg Sciences. Cofounded Orincon Corp in 1973 and served as Pres from its founding until 1980. Serves as CEO at present. Elected Fellow of IEEE for res contributions in 1977. Elected President of IEEE Control Systems Soc in 1980. *Society Aff:* IEEE, ORSA.

Sorgenti, Harold A
Business: 1500 Market St Rm 3600, Philadelphia, PA 19101
Position: Pres. *Employer:* ARCO Chem Co (Div of Atlantic Richfield Co). *Education:* MS/ChE/OH State Univ; BS/ChE/City College of NY. *Born:* May 28, 1934 Brooklyn NY. Prin Engr Battelle Mem Inst 1956-59; Development Engr Atlantic Richfield Co 1959-62, Group Leader chem process development 1962-69, Dir Process Development 1969-71, Mgr Process Development & Analysis 1971-72, Mgr Planning & Evaluation 1972-74, Tech Dir R&E 1974-75, Vice Pres Products Div R&E 1975-76. He was elected SR VP of Atlantic Richfield Co in Nov, 1978, and apptd Pres of ARCO Chem Co in Jan, 1979. Bd of Dir Coord Res Council 1974-76; Mbr ACS & AIChE. A native of New York City. *Society Aff:* AIChE, ACS.

Sorkin, George
Home: 7315 Rebecca Dr, Alexandria, VA 22307
Position: Consultant *Employer:* Institute for Defense Analyses. *Education:* BS/Chem. Met./Univ of AZ. *Born:* 8/03/17. Over 45 years of increasingly greater responsibilities in R&D, Prod & Mgmt in Metals & Chem Indus & Navy Dept. Chemist & Metallurgist American Smelting & Refining Co 1938-41; Plant Supt Gulton Ind 1941-42; US Navy Dept 1942-1980. Retired as Dir of Ship Systems R&T Office. Managed a large diverse program in ship materials; structures; hull, machinery and electrical systems and hardware; ship and personnel survivability and protection; energy conservation; pollution abatement; corrosion; ocean engineering; and manufacturing tech. Participate in many tech, honorary, prof adv socs & cttes. Authored many papers & lecture in field of interest. Prof Achievement Award ASE 1963; Solberg Award for R&D ASNE 1970; Navy Dept Distinguished Civilian Serv Award 1974. Prof Metallurgical Engr in DC. Currently Consultant to Government and Industry in field of materials, structures, fracture, military critical technology, export

Sorkin, George (Continued)
control of production technology, critical and strategic materials utilization and recycling R&D. *Society Aff:* ASM, ASNE, ΦΛΥ, WRC, ISSC, ASE, ADPA.

Soroka, Walter W
Business: Mech Engrg Dept, Berkeley, CA 94720
Position: Prof Emeritus. *Employer:* University of California. *Education:* ScD/Mech Engg/MIT; SM/Mech Engg/MIT; SB/Mech Engg/MIT. *Born:* 9/18/08. Taught at MIT 1930-37; R&D in refrigeration & military aircraft indus 1937- 47; taught at Univ of Calif Berkeley 1947-74. Fellow ASME 1968, ASA 1968. Natl Pres SESA 1962-63; Univ of Calif Berkeley Citation 1974. At various times 1956- 74 served as Chmn Divs of Engrg Design, Mechanics & Design, Applied Mechanics, Dept of Continuing Education, Asst Dean of Engrg for Internatl Cooperative Progs. Consultant to indus on architectural acoustics & noise control. Prof Emeritus 1974- . *Society Aff:* ASME, SESA.

Sorokach, Robert J
Home: 110 Mill Run Rd, Youngstown, OH 44505
Position: Chrmn Industrial Engr. *Employer:* Youngstown State Univ. *Education:* MSE/Mech Engr/Univ of Akron; BE/Mech Engr/Youngstown State Univ. *Born:* 5/10/30. in Sharon, PA. Grad from Youngstown State Univ with the BE degree in 1961, and from the Univ of Akron with the MSE in 1963. Completed additional grad work in Ind and Systems Engg at the Univ of Pittsburgh. Mbr of IIE. Prior to appointment as Chrmn of the Ind Engg Dept in 1967, was a mbr of the faculty of the Mech Engg Dept at Youngstown Univ. Formerly associated with the Tranformer Div of Westinghouse Elec Corp. Served in various capacities, including design engg and cost control. Consulted for several local cos and the Youngstown Hospital Association. Reg PE in the State of OH. Hobby is golf. *Society Aff:* IIE.

Sorrell, Furman Yates
Home: Box 7910, Raleigh, NC 27695-7910
Position: Prof. *Employer:* NC State Univ. *Education:* PhD/Aero/CA Inst of Tech; MS/Aero/CA Inst of Tech; BS/ME/NC State Univ. *Born:* 7/14/38. Res Engr, Pratt & Whitney Aircraft 1961-62. Dept of Aerospace Engg Sci, Unv fo CO, Boulder CO 1966-1968. NC State Univ Raleigh, NC 1968-present. Dir of Grad Prog 1970-1974. Prof 1976-present. Assoc, D Y Perry Assoc Mech Engrs 1974-1975 (On leave from NC State Univ). Technical Director, N.C. Alternative Energy Corp. 1981-83 (on Leave From N.C. State Univ). Grants and/or Contracts from NASA, NSF, Army Res Office, NIH, NOAA. Consultant, unsteady flow, shock physics. *Society Aff:* APS, AGU, ASME.

Sorrells, Robert R
Home: 824 S Ridgeway Dr, Cleburne, TX 76031
Position: Senior Engineering Specialist. *Employer:* Vought Corporation. *Education:* BS/Physics/TX CHristian Univ; MBA/Mgmt/TX Christian Univ. *Born:* 12/4/36. Native of Cleburne TX. MA properties engr for Vought Corp 1966- , respon for mass properties optimization of aerospace sys. Internatl Vendor Coordinator Soc of Allied Weight Engrs 1974-78, Int's Mbrship Chrmn SAWE 1979; Revere Cup Award for outstanding tech presentation at 1976 SAWE conf. Hobby is pocket knife collecting. *Society Aff:* SAWE.

Soteriades, Michael C
Home: 3380 Stephenson Pl NW, Washington, DC 20015
Position: Prof, Dept of Civ Engg. *Employer:* Catholic Univ of America. *Education:* ScD/Soil Dynamics/ MIT; Dr Engr/Structures/Nat Tech Univ Athens; Dipl Eng/-/Nat Tech Univ Athens *Born:* 3/25/23. Res Asst: MA Inst of Tech (53-54). Consultant: A Woolf & Assts (52-54). Hd, Dept of Struct Design & Spec Office of the Dist Engr Volos, Greece (1954-56). Assoc Engr: A Woolf & Assts Boston, MA (56-57). VP & Treas Doxiadis Assts Inc Wash DC (58-61). Ordinary Prof, Catholic Univ of America, Wash DC (61-date). Chrmn, Dept of Civ Engr Catholic Univ of America, Wash DC (75-81). Areas of specialization: structures structural dynamics, earthquake engr, soil mechs & fdns, comp software programming, steel stacks, guyed towers. *Society Aff:* ΣΞ, ASCE, ACI

Soto, Marcello H
Business: PO Box 1963, Harrisburg, PA 17105
Position: VP. *Employer:* Gannett Fleming Transportation Engrs, Inc *Education:* BS/Civ Engg/Univ of CO. *Born:* 4/19/25. With Gannett Fleming since 1954. Chief of Structural Sec since 1966 & Asst Transportation Dir. Specializing in hgwy & railway bridges; transportation structures such as tunnels, stations, maintenance facilities; & pollution control & hydraulic structures. Reg in 14 states. Pres of PA Engg Fdn. Scoutmaster (BSA). Thomas Fitch Rowland Prize, 1979. *Society Aff:* ASCE, NSPE, AREA, AWS.

Soudack, Avrum C
Business: Electrical Engrg Dept, Vancouver British Columbia, Canada V6T 1W5
Position: Professor of Electrical Engineering. *Employer:* University of British Columbia. *Education:* PhD/EE/Stanford; MS/EE/Stanford; BScEE/EE/Univ of Manitoba. *Born:* July 1934 Winnipeg Canada. Faculty of UBC 1961- . Assoc Editor of Simulation 1966-69. Sci Marv Emerson Award for Outstanding Teacher 1972. Major outside interests are photography & music.

Soukup, David J
Business: 25 Technology Park, Norcross, GA 30092
Position: Associate Director *Employer:* Institute of Industrial Engineers *Education:* M.S./Ind. Engr./Univ. of Tennessee; B.S./Systems Engr./Univ. of Arizona *Born:* 12/25/55 Native of Cleveland, Ohio. Attended high school and college in Tucson, Arizona. Was Assistant Secretary-Treasurer for the Tau Beta Pi Association from 1976-84. Joined the staff of the Institute of Industrial Engineers in February 1984. Received Professional Engineer's license in 1987. I serve as IIE's liaison to other engineering societies in the US and abroad. I am the staff support for the Education Policy Bd and Research Advisory Council. My outside interests include tennis, jazz, ethnic folk music, and travel. *Society Aff:* IIE, TBΠ, NSPE, ASEE, CESSE

Soutas-Little, Robert
Business: Department of Biomechanics, Michigan State University, E. Lansing, MI 48824
Position: Prof & Chrmn. *Employer:* MI State Univ. *Education:* PhD/Mechanics/Univ of WI; MS/Mechanics/Univ of WI; BS/Mech Engg/Duke Univ. *Born:* 2/25/33. Faculty mbr, Univ of WI, 1962-63; OK State Univ, 1963-65; MI State Univ 1965-. Chrmn, Dept of Mech Engg, 1972-77. Western Elec Teaching Award 1970-71. Author, *Elasticity* Prentice-Hall, 1973. Consultant, AC Spark Plug, CBS Res Lab, BF Goodrich Co, Ford Motor Co, Lawrence Livermore Lab. Chaired Dept of Biomechanics since 1977, res unit in muscular-skeletal studies. Summer progs: USAID India, 1965; Glaciological and Arctic Sciences Inst, 1968-73. Goldberg Chair, Bioengineering, Technion, Israel, 1982. *Society Aff:* SES, ASB, ASME

Sowa, Walter A
Home: 113 Park St, Nanticoke, PA 18634
Position: Professor. *Employer:* The Pennsylvania State University. *Born:* Jan 5, 1925. BSEE 1950 Pa State Univ; MS Physics 1965 Wilkes Coll. Technology instructor for 20 years. Co-authored text with James Toole, 'Special Semiconductor Devices.' This text also selected & translated to Chinese. Authored, 'Active Devices for Electronics.' Received ASEE MidAtlantic Western Electric Outstanding Teacher Award 1975; received Outstanding Teacher Award from Pa State Univ Engrg Soc 1973. Active in local Planning Comm, Recreation Bd, & Chamber of Commerce.

Sowell, Robert S
Business: Box 7625, North Carolina State University, Raleigh, NC 27695-7625
Position: Prof. *Employer:* NC State Univ. *Education:* PhD/Biological & Agri Engg/NC State Univ; MS/Agri Engg/Kan. State U; BS/Agri Engg/Miss State Univ. *Born:* Aug 1939. Served in US Air Force as Opers Res Analyst, Hq Military Airlift Command 1967-70. Res & teaching Biological & Agricultural Engrg Dept, NC State

Sowell, Robert S (Continued)
Univ 1970- . Graduate Administrator 1987- . Res area systems analysis in agriculture. Applications of microcomputers in agriculture. Teaching microcomputers in agriculture. Mbr of the American Soc of Agricultural Engrs. Chmn NC Sect of American Soc of Agri Engrs 1974, Chmn T-6 committee (Operations Research), ASAE, 1978-79 and T-6 Committee & Agricultural Systems Analysis 1986-87. Officer in USAFR. Recipient of 1979 ASAE Paper Award. *Society Aff:* ASAE

Sowers, George F
Business: 2749 Delk Rd, Marietta, GA 30067
Position: Sr Consultant. *Employer:* Law Engg, GA Tech. *Education:* MS/Civ Engg/Harvard; BS/Civ Engg/Case Inst of Tech. *Born:* 9/23/21. Mr Sowers specializes in geotechnical work: the fusing of geology, engg design & construction. He has been on Consulting Bds for Power Cos, US Bureau of Reclamation, Architect of the Capitol, Wash, DC, & the World Bank. He has been a consultant on dams, fdns & earth & rock construction throughout the world. He has written more than 100 technical papers & is author or co-author of three books. "Introductory Soil Mechanics and Foundations, Earth and Rockfill Dam Engineering–, & "Foundation Engineering" (Edited by Leonards). He was selected as the 1979 Terzaghi lecturer of the ASCE, he received the 1975 Clemens Herschel Award by the Boston Soc of Civ Engrs, & the 1977 ASCE Middlebrooks Award. *Society Aff:* ASCE, NSPE, AEG, GSA.

Sowers, William A
Business: PO Box 4038, Roanoke, VA 24015
Position: Pres. *Employer:* Sowers & Assoc., P.C. *Education:* BS/Civil Engr/VA Tech; BS/Architectural Engr/VA Tech. *Born:* April 1923; native Virginian. US Army 1944-46, Counter Intelligence Corps. Continuously engaged in the practice of Cons Engrg in Roanoke Va 1948- . Principal in present firm specializing in building systems since 1953. Fellow of American Cons Engrg Council & served the Natl Council as Secy-Treasurer 1965-67 & as Pres 1971-72. Served on Bd of Roanoke Chamber of Commerce, Roanoke Kiwanis Club & Christ Lutheran Church. Bd of Examiners, Roanoke City Electrical Code & Bd of Appeals, Roanoke City Building Code. Chrmn Roanoke City Planning Commission *Society Aff:* ACEC, IES.

Space, Charles C
Business: PO Box 4850, Jacksonville, FL 32201
Position: Vice President. *Employer:* Reynolds Smith & Hills. *Education:* BME/Mech Engg/GA Tech. *Born:* 10/27/35. BME 1957 GA Tech. Native Rochester, NY. 10 yrs exper in Design & Applications Petrochemical Chem Food & Process Indus, 15 yrs in AIE Consulting. Currently VP Reynolds Smith & Hills, respon for design of atmos--heric fluidized bed steam generator facility for DOE. Facility will burn high sulfur coal without scrubbers. Formerly respon for RS&H Environ & Process Div, Tampa FL Engg Div & Miami FL Regional Office. Currently VP Prof & Pub Affairs, ASME. *Society Aff:* NSPE, FES, ASME.

Spachner, Sheldon A
Home: 17 Country Vlg Way, Media, PA 19063
Position: Pres *Employer:* Metal Systems Development, Inc. *Education:* PhD/Metallurical Eng/IL Inst of Tech; MS/Physics/Northwestern Univ; BS/Physics/Northwestern Univ. *Born:* 3/28/24. Native of Chicago IL. US Air Force & Airborne Artillery 1943-45. IIT Res Inst 1953-65. Conducted R&D work in physics & mech metallurgy, & metalforming process dev. Sr Scientist 1958-65. Joined E W Bliss Engrg Res & Dev Div in 1967, which became the G&W AD&E Ctr. Managed activity in Coal-Base Fuel Dev and Production, metal handling, and metalforming process and machine system dev. Mbr ASM. Sigma Xi. Sigma Pi Sigma. Current Position: 1984-Present. Pres, Metal Systems Development, Inc. Titanium Recovery Process Development; extrusion & forging system development. Past Position: 1980-1982 VP of Technology, Gulf & Western Advanced Fuels Tech Div. Fields of Specialization: Titanium extractive metallurgy, heavy metal forming. *Society Aff:* ASM, ΣΞ, ΣΠΣ

Spacil, H Stephen
Business: GE Corp R&D, Kowa 35 Bldg, 14-14, Akasaka 1-chome, Minato-Ku, Tokyo 107 Japan
Position: Scientific Rep. *Employer:* GE R&D Ctr. *Education:* PhD/Phys Met/MIT; MS/Phys Met/MIT; BS/Phys/MIT. *Born:* 12/14/30. Asst Prof, MIT, 1956-59. Staff mbr, GE R&D Ctr, Schenectady, 1959-79. Specialized in application of thermodynamics & kinetics to mtls processing and stability. Transfer to GE Corp R&D Far East Liaison, Tokyo, 1979. Divisional Chrmn, ECS, 1978; Fellow, ASM, 1978. *Society Aff:* ASM, ECS.

Spaeth, Edmund E
Home: 4856 Rochelle Ave, Irvine, CA 92714
Position: Pres, Amer Tech & Ventures Div *Employer:* American Hospital Supply Corp. *Education:* PhD/Applied Mechanics/Calif Inst of Tech; MS/Chem Engr/Calif Inst of Tech; BS/Chem Engr/Stanford Univ *Born:* 3/11/41. Major fields: convective diffusion, interfacial phenomena. Postdoctoral Fellow Caltech 1967. Sr Engr Biosatellite Proj, Jet Propulsion Lab 1967-69, while Capt US Army. Joint faculty appointment Dept of Chem Engrg & Dept Surgery, Washington Univ St Louis 1969-73; also Dir Biological Transport Lab. With Amer Hospital Supply Corp 1973- . Assumed current position 1983, responsible for dev of selected new business opportunities, including venture start-up. Natl Speakers Bureau AIChE 1971; Chmn Prod Dev Comm HIMA 1975-77; Chmn Postdoctoral Fellowship Comm ASAIO 1973-75; Prog Comm ASAIO 1976-79; NIH Consensus Panel on Biomaterials, 1982. *Society Aff:* AIChE, ASAIO, AHA, BMES

Spaght, Monroe E
Home: 70 Arlington House, London SE1 7NA, England Engla, England
Position: Director, Retired. *Employer:* Royal Dutch/Shell Co. *Education:* PhD/Chemistry/Stanford Univ; MA/Chemistry/Stanford Univ. *Born:* Dec 9, 1909. Joined Shell Oil Co 1933 as res chem, Vice Pres Shell Dev Co 1946, Pres 1949; Exec Vice Pres Shell Oil Co 1953, Pres 1961, Chmn 1965-70. A Managing Dir Royal Dutch/Shell Group of Co 1965-70. Dir Royal Dutch Petroleum Co & Shell Oil Co. 1970-80.Trustee Stanford Univ 1955-65; Dir Stanford Res Inst 1953-70. Currently Dir of American Petroleum Inst; Inst of Internatl Education (Chmn 1971-74). 7 honorary degrees, various awards & citations in USA & abroad. Johnson Lecturer & Medallist; Royal Swedish Academy of Engrg Scis, Stockholm 1966; Chem Indus Medal 1966; Honorary Mbrship Chemists' Club 1967. Mbr NAE. *Society Aff:* NAE, AICE, ACS, AAAS, SCI.

Spalding, Henry A
Business: 1 Morgan St, 608 Broadway, Hazard, KY 41701
Position: Metallurgical Engr - Civil & Mining Architect. *Employer:* H A Spalding Inc *Born:* March 1899. General & widely diversified engrg practice; ed private tutors; inventor metallurgical processes. Pres H A Spalding Inc. Outstanding Citizen Award, Hazard Civic Clubs 1958. Mbr AIME, ASCE Ky, Soc of Prof Engrs, CEC, NSPE, etc. Co-Author Engrg Vest Pocket Book. Contributor articles in field. Mbr Ky Geological Survey Adv Bd. Married Gertrude Petrey. Fellow ASCE, ARA, ACSM, Legion of Honor, AIME. *Society Aff:* AIME, ASCE, KSPE, ARA, ACSM

Spalding, Vernon B
Business: 655 W 13 Mile Rd, Madison Heights, MI 48071
Position: President. *Employer:* Spalding, DeDecker & Assocs Inc. *Education:* BCE/-/Univ of MN. *Born:* Jan 1922. BCE 1949 Univ of Minn. 3 years Ch Engr D L Briegal Assocs, specializing in expressway utilities; current Mich Natl CEC Dir, P Pres CEC/Mich. Fellow ACEC. F Dir Design Profs Insurance Co. 20 years Pres of Spalding, DeDecker & Assocs Inc, Cons Engrs. Expert in municipal engrg, municipal contracts & const insurance. 8 years Chmn City Building Code Bd of Appeals. Mbr American Arbitration Assoc. *Society Aff:* ASCE, ACEC.

Spamer, James S
Home: 512 Holden Rd, Towson, MD 21204
Position: Pres. *Employer:* James S Spamer & Assoc. *Education:* BS/Civ Engg/Univ of MD; MS/Civil & Sanit Engg/Univ of MD; Cert/Sanitary/Univ of IL. *Born:* 3/14/25. Native of Towson, Baltimore County, MD. Attended Univ of IL during WWII service. Grad: Univ of MD BSCE 1946. Grad work 46/48 Instr/CE. In Private Practice of Civ Engg & Surveying since 1953. Reg in MD, DC, DE, & GA. Past Pres Chesapeake Chap and the MD Soc of PE (1976/77). Secretary and VP Engg Soc of Baltimore, Inc (1979), Pres (1981), Fellow ASCE, MAE, MD Soc Surveyors. Hobby: E African travel and correspondence. Charter Mbr: H L Mencken Soc of Baltimore. 32- Mason, Shriner. *Society Aff:* ASCE, NSPE.

Spandorfer, Lester M
Business: PO Box 500, Blue Bell, PA 19424
Position: Dir, Office Information Systems *Employer:* Sperry-Univac. *Education:* PhD/EE/Univ of PA; MS/EE/Univ of MI; BS/EE/Univ of MI. *Born:* 10/16/25. Bell Telephone Labs 1948-50; taught undergraduate courses, Mgr Computation Center Univ of Pa; Sperry Univac 1957- , Mgr LARC CPU proj, contributed to circuitry of Univac III and 1100 Series, to system dev of 1700, 1800, 1900 prods. Dir Technology Planning, Corporate Staff 1967-72, active in LSI evolution; Tech Dir Peripherals Div, Currently active in office Automation Field, 14 US Patents, 17 Publs. Fellow IEEE; Sigma Xi; Tech Prog Ctte, Internatl Solid State Circuits Conf 1967-72; Computer Elements Ctte, IEEE Computer Group; 1968 US Tech Prog Ctte, IFIP; papers referee, IEEE & AFIPS journals. *Society Aff:* IEEE.

Spangler, Carl D, Jr
Home: 2400 Mockingbird, Ponca City, OK 74601
Position: Director NAR Div Process Engrg Dept. *Employer:* Continental Oil Co. *Born:* March 15, 1926 Chickasha Okla. BSChE 1950 Okla Univ. 18 months US Air Force WWII. 5 years with Black Sivalls & Bryson Inc, in gas process equipment design. 7 years with C W Nofsinger Co in petrochem & petrochems process engrg. Continental Oil Co, process engr 1962- in petrochems & petroleum refining. Dir of North American refinery process engr 1969- . Mbr AIChE, ASME, API. Chmn of Fuels & Petrochems Div of AIChE 1976.

Spanner, Jack C
Home: 2042 G Wash Way, Richland, WA 99352
Position: Manager Materials Applications. *Employer:* Westinghouse-Hanford Co. *Education:* MS/Met Engg/WA State Univ; BS/EE/CO State Univ; MS/Engg/Univ of Southern CO. *Born:* 5/5/32. AS degree Univ of Southern CO; BSEE CO State Univ; MSMetE Washington State Univ. Liscensed Prof Engr. Specialized in nondestructive testing for the past 20 yrs. Author of numerous publs on NDT including a book entitled, *Acoustic Emmission: Techniques & Applications.* Editor of ASTM STP-571, Monitoring Structral Integrity by Acoustic Emmission. Mgr of NDT & Matls Applications at Westinghouse-Hanford Co, Richland WA. Previous employers include the GE Co & the Battelle Memorial Inst. Instrumental in the formation of the Acoustic Emission Working Group (AEWG) in 1967, & Chmn of the ASTM Subctte on Acoustic Emission since its inception in 1972. Charter Chmn of the Columbia River Sect ASNT; P Chmn of the Richland Sect IEEE; Mbr ASM & ASME. Elected to 3-yr term in ASNT Bd of Dirs in 1978. Elected ASNT Fellow 1974. *Society Aff:* ASNT, ASM, ASME.

Spanovich, Milan
Business: 4636 Campbells Run Rd, Pittsburgh, PA 15205
Position: Pres & Principal Engr *Employer:* Engrg Mechanics, Inc *Education:* BS/Civ Engrg/Carnegie-Mellon Univ; MS/Civ Engrg/Carnegie-Mellon Univ *Born:* 02/19/29 Native of Steubenville, OH. Served with USAF 1952-1954. Charter member of E. D'Appolonia Assocs, Consulting Engrs, of Pittsburgh, PA 1957-61. Conducted research on soil dynamics at Univ of NM 1961-63. Founded Engrg Mechanics, Inc (EMI), Consulting Geotechnical Engrs, in Pittsburgh, PA, in 1963. Presently Pres and Principal Engr of EMI. Enjoys skiing, fishing and golf. Author of numerous papers on Soil Mech. *Society Aff:* ASCE, NSPE, ASTM, ACEC, SEE

Sparks, Morgan
Business: Sandia National Labs, Albuquerque, NM 87185
Position: President. *Employer:* Sandia/National Laboratories *Education:* PhD/-/Univ of IL; MA/-/Rice Univ; BA/-/Rice Univ. *Born:* 7/6/16. Bell Telephone Labs 1943-72, semiconductor res, transistor dev, tech mgmt; Exec Dir Semiconductor Component Dev 1959- ; VP Tech Info Personnel 1969; VP Electronic Technology 1971. In 1972 assumed present position. Sandia's primary mission is res, dev & engg in ordnance phases of nuclear weapons. Also substantial programs in energy R&D--especially in areas of solar energy, electron beam & lightion fusion, & nuclear reactor safety. Fellow: APS, IEEE, AIC. Mbr: Natl Acad of Engg & AMer Chem Soc. *Society Aff:* APS, ACS, IEEE, AIC.

Sparling, Rebecca H
650 West Harrison Ave, Claremont, CA 91711
Position: Retired. *Education:* MS/Physical Chem/Vanderbilt Univ; BA/Chemistry/Vanderbilt Univ. *Born:* 6/7/10. Metallurgist, foundries in South and Midwest, 1931-44. Chief Matls Engr, Turbodyne Corp 1944-50. Design Specialist, Mstls, Genl Dynamics/Pomona, 1951-68. Main interests were matls for high temperature service, and nondestructive testing. Since retiring, work on problems on energy production and air pollution, serving with public agencies as well as civic groups and engg societies. Frequent speaker explaining technical subjs. Reg Mechanical Engr in CA. Awarded Gold Medal by Soc of Women Engrs 1957; given Outstanding Engr Merit Awards by Los Angeles Engrs' Week 1965, by Inst for Advancement of Engg in 1978, and by Orange County Engineering Council 1978. *Society Aff:* ASM, ASNT, SWE, AWIS.

Sparrow, Ephraim M
Home: 2105 West Hoyt Avenue, St. Paul, MN 55108
Position: Division Director *Employer:* National Science Foundation *Education:* Ph.D./Mech. Engrg./Harvard University; M.A./Mech. Engrg./Harvard University; M.S./Mech. Engrg./M.I.T.; B.S./Mech. Engrg./M.I.T. *Born:* 05/27/28 Worked at Raytheon Co & the NACA Lewis Research Center after completion of Ph.D. at Harvard. Professor of Mechl Engrg at Univ Minnesota, 1959-85. Guided research for approximately 70 Ph.D. theses and 100 Master's theses. Author of more than 550 papers on heat transfer and related subjects. Beginning in 1986, took leave of absence from Minnesota to work at the National Science Foundation, first as a Program Dir and later as a Div Dir. *Society Aff:* ASME

Spatz, D Dean
Business: 5951 Clearwater Drive, Minnetonka, MN 55343
Position: President & Chmn of the Board. *Employer:* Osmonics Inc. *Education:* ME/Engrg/Dartmouth Coll; BS/Engrg/Dartmouth Coll; BA/Sci/Dartmouth Coll *Born:* 3/20/44. Native of Albuquerque NM. Degrees from Dartmouth College. Pres & founder of Osmonics Inc, Minnetonka Minn. Osmonics, Inc. designs, manufactures and markets machines, systems and components used in the separation of fluids. The Company manufactures disposable semi-permeable membranes and other filter materials for use in fluid separation; reverse osmosis (RO) & ultrafiltration (UF) membrane; membrane elements; RO/UF water purification & waste water treatment equipment; & multistage centrifugal pump which Osmonics manufactures and sells under name of Tonkaflo. Reg PE. Mbr: AIChE, AWWA, ASTM, AES, WSIA. Active in reverse osmosis membrane technology since 1964 with initial contracts from the Office of Saline Water, for membrane dev 1964- 68. Mgr of Engrg for Aqua Technology Inc, an ion exchange & filter equipment manufacturer 1968-69. Awarded 4 Patents in membrane field. Author of numerous papers on metal recalmation, indus waste water treatment & food processing using membranes as well as papers on water purification. Received Putnam Food Award for Sugar Reclamation System using Reverse Osmosis 1975, Chemical Processing Award for Energy

Spatz, D Dean (Continued)
Conservation 1982, Industrial Equip. News Award for Innovation in Material Reclamation 1982. *Society Aff:* AIChE, AWWA, ASTM, AES, WSIA, ACS.

Spaulding, Philip F
Business: 911 Western Ave, Ste 400, Seattle, WA 98104
Position: Pres *Employer:* Nickum & Spaulding Assoc Inc *Education:* MS/Naval Arch & Marine Engrg/Univ of MI; BS/Commercial Engrg/Univ of WA *Born:* 11/27/12 Mr Spaulding has spent his lifetime in the marine field. Employed 6 months by Bethlehem Steel Co, then 11 years by Todd Shipyards Corp as Asst Chief Engr; Naval Architect; Supt of steel construction; Principal Engrg Estimator; and Planning Supt. Founded Philip F. Spaulding & Assoc in 1952 and for 19 years was its pres. In 1971 he merged with W. C. Nickum & Sons to form Nickum & Spaulding Assoc, and since 1976 has been its pres. Spaulding is licensed in States of WA and AK; honored as Engr of the Year, Puget Sound Engrg Council, 1971; Puget Sound Maritime Man of the Year, 1973; and awarded the David Taylor Gold Medal by SNAME, 1979 for notable achievement in the field of naval architecture and marine engrg. *Society Aff:* SNAME, NSPE

Spaulding, Westbrook H
Business: 436 Main St, Lewiston, ME 04240
Position: VP *Employer:* Alberti: Larochelle & Hudson *Education:* BS/CE/Univ of ME *Born:* 07/14/45 Currently VP and Structural Div Mgr of Aliberti, LaRochelle and Hodson Engrg Corp, Inc, Lewiston, ME. Prior experience includes four yrs in the structural steel fabrication industry in ME and one yr with a major paper industry. P Pres ofthe Maine Section American Society of Civil Engrs. *Society Aff:* ASCE, AISC

Spauschus, Hans O
Business: Georgia Tech Research Institute, Georgia Institute of Technology, Atlanta, GA 30332
Position: Director-Energy and Materials Sciences Laboratory *Employer:* Georgia Tech Research Institute *Education:* AB/Chem & Math/IL College; MS/Phys Chem/Tulane Univ; PhD/Phys Chem/Tulane Univ. *Born:* June 1923. GE Co Transformer Lab 1950-53; Res Assoc Major Appliance Labs 1956-68, specializing in physical chemistry related to refrigeration & air- conditioning; Mgr Physical Scis Lab 1968-80, Director- Energy & Materials Sciences Laboratory Engineering. Experiment station Georgia Institute of Technology. Fellow ASHRAE. Senior Mbr IEEE, AAAS, IIR, Klixon & Wolverine Awards, ASHRAE; Steinmetz Award GE Co. Advanced Materials and Composites, Thermoelectric Power Generation, systems applications of solar energy ; halocarbon chemistry; heat pumps; technology appraisal; technology transfer; R&D strategic planning. *Society Aff:* ASHRAE, AAAS, IIR, IEEE.

Speakman, Edwin A
Home: 9018 Charles Augus Dr, Alexandria, VA 22308
Position: Staff Adv Scientific & Cryptologic. *Employer:* US Army Intelligence & Security Command. *Education:* BS/Phyiscs/Haverford College. *Born:* 8/14/09. Instructor in Physics Haverford Coll 1931-34; Radio Engr Philco Corp 1934- 40; Asst Supt Naval Res Lab 1940-49; Vice Chmn R&D Bd Dept of Defense 1949-52; Vice Pres & Genl Mgr Guided Missle Div Fairchild Airplane Corp 1952-57; Vice Pres Missile Range Opers RCA Serv Co 1958-65, Dir Office of R&D US Dept of Transportation 1967-68, Scientific Adv US Army Security Agency 1968- . Fellow IEEE; Assoc Fellow AIAA. Navy Meritorious Civil Serv Award; Navy Commendation for work in EW. Natl Chmn Prof Group Mil Electronics IEEE 1958; Chmn Internatl Space Electronics Symposium IEEE 1965. Life mbr, Assn of Oldcrows, Natl Sec 1976-81; US Army Inscom Meritorius Civil Serv Award 1984; Radio Amateur W3AUR 1928-date. *Society Aff:* IEEE, AIAA, AOC, AFCEA

Spear, Clyde C
Business: 555 E Walnut St, Pasadena, CA 91101
Position: Supervising Engineer. *Employer:* James M Montgomery. *Born:* Jan 4, 1922. Attended Ind Univ, Business Administration; attended AWWA Mgmt Seminars, & Water & Sewage Seminars Purdue & Ill Univs. Navy Pilot WWII 1942-45. Water Utilities Mgr Delphi Water Works 1954-60. Water & Sewage Utilities Mgr, Columbus Ind 1960- . Comprehensicve Master Planning & Dev Water & Sewage Utilities. Starting with James Montgomery, Cons Engr Inc, moving to the Philippines, cons with the Philippine Government as Supervising Engr in the dev of their water resources & facilities. P Chmn Ind Sect AWWA 1972, recipient of the Fuller Award Ind Sect AWWA 1976. Enjoy all sports. Golf & fishing.

Spear, Jo-Walter
Business: John Sexton Contractors, 1815 South Wolf Road, Hillside, IL 60162
Position: Dir, Coporate Development Division *Employer:* John Sexton Contractors *Education:* MSc/Phy Chem/Univ of Pa; AB/Chem Bio/Rutgers State Univ. *Born:* 1942. Miscellaneous graduate courses in physiology & chemistry at Glassboro State Coll. Taught in Elsinboro & Vineland Public Schools until 1973. With PQA 1973-1978. Responsible for dev of laboratory capability & the design & oper of a modern environ lab. Fellow American Inst of Chemist. Mbr WPCF & ACS. Accredited Prof Chemist. Enjor fishing, tennis, & music. 1983 John Sexton Contractor. Dir, Corporate Development Div - Corporate responsibility for development of new disposal facilities; expansion of existing facilities; acquision of related businesses. 1978-1979 Delaware River Basin Commission Sanitary Engineer Water Quality Branch Resp. for evaluation of Eng. reports and technical studies relating to water pollution control implementation as part of the comprehenive water resources program in the Delaware River Basin. 1979- John Sexton Contractors, Director, Chemical Process Division - Corporate responcibility for implementation and operation of Chemical waste disposal facilities. Fellow American Inst. of Chemist. Mbr WPCF, AICHE, NSWMA, ASTM, accredited Prof Chemist Certified Chemical Eng. Faculty Midwest College of Eng- environmental Eng. Enj fishing, tennis & music. 1983 John Sexton Contractor; Dir, Corp Dev Div. Corp respon for dev of new disposal facilities; expansion of existing facilities; acquisition of related businesses. *Society Aff:* WPCF, NSWMA, AIChE, AIC, APWA.

Spearman, Rupert B
Home: P.O. Box 18857, Spokane, WA 99208-0857
Position: Retired *Education:* BS/CE/Univ of CO *Born:* 6/25/08 Native Crawford, NB. With US Bureau of Reclamation 1933-71, rose from rodman to Chief Civil Engrg Div, Boulder Canyon Project, Boulder City, NV 1933- 56; Chief Office Engrg Div Durango, CO. Projects office 1956-63; Area Engr, Upper Columbia Development Office, Spokane, WA, 1963-71, responsible for water resource planning Upper Columbia area. Principal Engr, Intl Engrg Co, San Francisco 1971-72. Study dir, responsible for prefeasibility studies 12000 MW hydroelectric "Itaipu Project" for Brazil and Paraguay. Received Superior Accomplishment Award 1961. Regional Dir's Commendable Service and Interior's Meritorius Service Awards 1971. Registered, CO and NV. *Society Aff:* ASCE, USCOLD

Spechler, Jay W
Home: 2 Spinning Wheel Lane, Tamarac, FL 33319
Position: Director, Engineering *Employer:* American Express Co. *Education:* B.S./Industrial Management/New York University; M.S./Industrial Engineering/Stevens Institute of Technology; Ph.D./Economics/Columbia Pacific University *Born:* 02/06/34 Serves as an adjunct professor of Indust Engrg, Univ of Miami and as adjunct professor of Economics at the Embry-Riddle Aeronautical Univ. Currently holds the rank of Colonel, USAFR. Previously Dir of Management Services, Florida Power & Light Co.; Internal Consulting Mgr, Gulf & Western Industries. Authored "Administering The Company Warehouse & Inventory Function-; "3-SIGMA Communication Strategies" and has published fifty articles in the professional literature. Lectures internationally on engrg & mgmt topics. Mbr, I.E. Council on Engrg; Tau Beta Pi; Omicron Delta Epsilon. *Society Aff:* IIE

Specht, Theodore R
Business: 469 Sharpsville Ave, Sharon, PA 16146
Position: Fellow Engineer. *Employer:* Westinghouse Electric Corp. *Education:* BEE/Elect Eng/Univ of MN; MS/Elect Eng/Univ of Pittsburgh. *Born:* July 8, 1918. Native of Minneapolis Minn. Design & dev engr with Westinghouse 1941- . Assignments included instrument, power, & rectifier transformer & saturable core reactor design; transformer magnetic core dev; control of transformer audible sound; & application problems of transformers. Also participated in the design & testing of the Government power line carrier proj for civilian emergency alerting, & Westinghouse HVDC proj. Fellow Mbr IEEE. *Society Aff:* IEEE.

Specter, Marvin M
Business: 174 Brady Ave, Hawthorne, NY 10532
Position: Managing Partner (Self-employed) *Employer:* M.M. Specter. PE-LS (Partnership) *Education:* MSc/Civil/Columbia Univ; BE/Civil/Johns Hopkins Univ. *Born:* Nov 9, 1927. Reg PE & LS in NY, NJ, Conn, & Vt. PE in Mass. Prof Planner in NJ. NCEE Certificate as PE. Practice in Cons Engrg Planning, Land Surveying, & forensic engrg. Pres Westch Chap Pres of NYS Soc. PE's 1968, 'Engr of the Year' 1972. State Pres, NYS Soc of Prof Engrs 1972. Natl Dir NSPE 1966-76. Founded NYS Prof Engrs in Private Practice Sect (PEPP). NSPE PEPP Vice Chmn (Northeast) 1974-76. NSPE Vice Pres (Northeast) 1976-78. Chrmn NSPE Const & By Laws Comm 1978-80. Chrmn NSPE Forensic Engr Comm 1978-80; Mbr NSPE Bd of Ethical Review 1979-82. NSPE Pres 1982-83. Fellow ASCE. Charter Mbr NYS Assoc of Professions. Chmn Town of Eastchester Planning Bd & Town Conservation Comm; 1970- Chrmn Westchester Co NY Environ Mgmt 1976-78; Pres Westchester Municipal Planning Federation 1976-78; Mbr Westchester Co Indust Dev Agency 1985- . Founding Pres NAFE 1983, continues as NAFE Secy. Rec'd NSPE Award (highest honor) 1986. *Society Aff:* ASCE, NSPE, SAME, NAFE.

Spector, Leo F
Home: 436 Dorset Place, Glen Ellyn, IL 60137
Position: Editor, 'Plant Engrg' Magazine. *Employer:* Technical Publishing Co. *Education:* BS/Mech Engg/Univ of KS. *Born:* 11/10/23. BSME Univ of Kansas. Army Air Force pilot 1943-46. Reg PE in Mo. Formerly, regional editor, 'Machine Design,' & Editor, 'Assembly Engrg.' Joined 'Plant Engrg' in 1970, named Editor in 1974. Served with several natl standardization cttes (ASTM, USAI). Mbr of ASME, SPJ, Tau Beta Pi, Sigma Tau & Pi Tau Sigma. Standing genl chmn of Natl Plant Engrg & Maintenance Conf and related regional conferences. Author of numerous tech articles and speaker at tech meetings and tech journalism seminars. Enjoy photography, classical music, & radio-control aircraft modeling *Society Aff:* ASME, SPJ, ΤΒΠ, ΣΤ, ΠΤΣ.

Spedden, H Rush
Home: 4131 Cumorah Dr, Salt Lake City, UT 84124
Position: Mineral Processing Engrs Cons (Self-employed) *Employer:* Self *Education:* MS/Min Process/MT Sch Mines; BS/Mining Engr/U of WA *Born:* May 1916. Mineral Dressing Engr (honorary) 1964 Mont Sch of Mines. Instructor 1940-42 & Asst Prof 1946-52 mineral engrg MIT. US Government mission to Bolivia 1942-44. Res group Mgr to Dir of Res, Union Carbide Ore Co 1952-64. Dir of Res, Kennecott Res Center 1964-74. Adjunct Prof of of Metallurgy, Univ of Utah 1970- . Honors: Robert H Richards Award, AIME; Distinguished Mbr SME of AIME. President Soc of Mining Engrs 1970; Vice Pres AIME 1971; Chmn 1964 & Co- Pres 1968 Internatl Mineral Processing Congress. *Society Aff:* SME, MMSA

Speece, Prof RE
Business: Drexel University, 3200 Chestnut, Phila, PA 19104
Position: Prof *Employer:* Drexel Univ *Education:* PhD/Sanitary Engr/MIT; ME/CE/Yale Univ; BE/CE/Fenn Coll *Born:* 8/23/33 Native of Galion, OH. Tenured prof at: Univ of IL Urbana 1961-1966. NM State Univ 1966-1970, Univ of TX-Austin 1970-1974. Presently Betz Chair Prof of Environmental Engrg, Drexel Univ Phila, PA; WPCF Eddy Medal 1966; AEEP Distinguished Faculty Award 1970; AEEP Distinguished Lecturer (North America) 1978; American Fisheries Society - Honorable Mention for Best Paper 1973; ASCE - Chaired Organizational Committee for Natl Symposium on Reaeration Research 1975; Consultant to 26 industries and governmental Agencies; Hold 10 US and Foreign Patents; 15 Refereed Publications from Intl Conferences; 34 Refereed Publications in US Journals: 1 book; JJ Croes Medal 1983 - Amer Soc Civil Engrg; Engrg Sci Award of Assoc Env Engrg Prof for Outstanding PhD Dissertation-Advisor 1982. *Society Aff:* ASCE, WPCF, AWWA, AFS, ASM

Speich, Gilbert R
Business: 125 Jamison Lane, Monroeville, PA 15146
Position: Res Consultant. *Employer:* US Steel Corp. *Education:* BS/Met Engg/IL Inst of Tech; MS/Met Eng/Univ of WI; ScD/Metallurgy/MIT. *Born:* 12/17/28. in Chicago, IL. Served in Army Ordnance Corp 1953-55.Res Consultant, US Steel Corp 1958- , specializing in structure-property relations in steels. Chmn of Ferrous Metallurgy & Mathewson Gold Medal Committees of AIMe. Fellow ASM. Received Hitchett Medal of Metals Society (England) in 1978 Enjoys tennis & chess. *Society Aff:* ASM, AIME, Metals Society.

Speiser, Ambros P
Business: BBC Brown Boveri Res Center, CH 5401 Baden, Switzerland
Position: Director of Research. *Employer:* BBC Brown, Boveri & Co Ltd. *Education:* PhD/EE/Swiss Fedl Inst of Technology; Dipl/EE/Swiss Fedl Inst of Technology. *Born:* Nov 1922 in Switzerland. Early computer design work at ETH. Dir, IBM Res Lab Zurich 1955-66. Dir, of Res of Brown Boveri 1966- , responsible for all res activities on the corporate level. Textbooks: Digitale Rechenanlagen 1961, Impulsschaltungen 1963. Pres IFIP 1969-72. Mbr Bd of ETH. Mbr Swiss Government Task Force on Energy 1974-78. Honorary Mbr Zurich Physical Soc. Fellow IEEE. IFIP Silver Core. Honorary Member Aargau Society for Sciences. Honorary Doctor's degree, ETH. President, National Academy of Engineering. *Society Aff:* IEEE.

Speitel, Gerald E
Business: 306 Evesham Commens, Route 73 and Evesham Rd, Marlton, NJ 08053
Position: President. *Employer:* Speitel Associates/BCM Engineers *Education:* MS/Engg Mgmt/Drexel Univ; BS/Civil Engg/Drexel Even College. *Born:* 2/4/30. SrVP, BCM Engineers, Inc. and Pres. of Speitel Asso. Previously associated with John G. Reutter Asso. as Cief of Engineers and VP. Current ASCE Zone 1 VP, Past ASCE Nat. Dir., Past Pres. ASCE/NJ, Past Pres Consulting Engrs. Council of NJ. Registered Prof. Eng., Land Surveyor, Professional Planner or Landscape Architect in various states in the USA and in Puerto Rico. *Society Aff:* ACEC, ASCE, NSPE, WPCF.

Spence, Robert
Business: Dept of EE., Imperial College, Exhibition Rd, London SW7 2BT England
Position: Prof of Information Engrg. *Employer:* Imperial College. *Education:* PhD/Electronic Engg/London Univ; Dipl/EE/London Univ; BSc/EE/London Univ; D.Sc(Eng)/EE/London Univ. *Born:* 7/11/33. in Hull, England. After his doctorate worked in Res Div of Gen Dynamics/Electronics, USA. As a Lectr, Reader then Prof at Imperial College since 1962, his res included circuit theory, the use of computers in teaching, computer-aided design & human-comp interaction. His res group was responsible for the Minnie interactive-graphic circuit design facility. Author of "Linear Active Networks" (1970) & "Resistive Circuit Theory" (1974); co-author of "Tellegen's Theorem and Electrical Networks" (1970) "Sensitivity and Optimization" (1980) & "Circuit Analysis by Computer" (1986). Co-edited IEE Journal "Electronic Circuits and Systems–. A founder, & Chmn, of Interactive Solutions Ltd. Enjoys photography and walking. *Society Aff:* IEEE, IEE.

Spencer, Domina Eberle
Home: 75 S Alphonsus St 2101, Boston, MA 02120
Position: Prof of Mathematics. *Employer:* Univ of CT. *Education:* SB/Physics/MIT; SM/Math/MIT; PhD/Math/MIT. *Born:* 9/26/20 in New Castle, PA. Friends Select

Spencer, Domina Eberle (Continued)
School PA. SB 1939, SM 1940, PhD 1942 MIT. Asst Prof Physics Am Univ 1942-43, Tufts Coll 1943-47, Brown Univ 1947-50; Assoc Prof Univ of CT 1950-60; Prof of Mathematics 1960- . Co-author with Prof Parry Moon of 9 bks. Author of over 200 papers in applied mathematics, lighting, design, vision & color. Cons in illuminating engr. Design of luminous ceilings, aperture lamps, macrofocal conics IES Gold Metal 1973 (with Prof Moon). Disting Alumna Award, Friends' Select School, 1987. *Society Aff:* IES, AMS, MAA.

Spencer, George L
Business: 116 East Sola St, Santa Barbara, CA 92101
Position: President. *Employer:* Archer-Spencer Engrg Assocs Inc. *Born:* March 20, 1921 Omaha Neb; Calif resident for over 45 years. BS MechE Univ of Calif, Berkeley. Reg PE in Calif M-5291. Vice Pres & Production Mgr of Archer-Spencer Engrg Assocs Inc 1963-70, Pres 1970- . Asst & Ch Engr US Navy Attack Transport WWII. Mbr of ASHRAE, ACEC. Mbr of Bd of Appeals & Engrg Representative, Code Ctte for the County of Santa Barbara. Involved in Mech Cons Engrg 1949- . Hobbies are tennis, water skiing, (a sports addict in general). *Society Aff:* ACEC, ASHRAE, CEAC.

Spencer, Max R
Business: 277 Park Ave, New York, NY 10172
Position: Vice President-Operations, World Grain Division *Employer:* Continental Grain Co. *Education:* BS/ChE/Purdue Univ. *Born:* Sept 26, 1921. Served with US Army Artillery in ETO 1943-46. With Continental Grain Co 1962- , Vice Pres 1965- . Mbr AIChE, Ctte Chmn for North American Export Grain Assn; Mbr of Natl Grain & Feed Assn; Author of sev tech papers on grain handling. Reg PE in 3 states. Committee Chmn NFPA Committee on Agri dusts; Member Federal Grain Inspection Service (USDA) Advisory Ctte. *Society Aff:* AIchE, GEAPS, NFPA, NGFA, NAEGA.

Spencer, Ned A
Business: 1820 D Madison Blvd, McLean, VA 22101
Position: Associate Department Head. *Employer:* The MITRE Corp. *Education:* BS/EE/MIT. *Born:* Aug 28, 1925. SBEE MIT. Bell Aircraft Corp in 1947; Joined Wheeler Labs in 1948 & contributed to many devs in the field of microwaves & antennas, became Ch Engr 1967; later transfered to the parent Corp Hazeltine in 1969 where became Dir of Tech Planning. In 1971 joined the MITRE Corp in the Air Transportation Systems Div, presently an Assoc Dept Head in Advanced Dev. Principal activities relate to planning & dev of improved surveillance of air traffic, collision avoidance sys, & new landing guidance systems. Fellow IEEE. *Society Aff:* IEEE, ION.

Spencer, Robert C
Business: 723 North Ave, Schenectady, NY 12345
Position: Manager Engineering. *Employer:* General Electric Co. *Education:* ScB/ME/Brown Univ; MS/ME/Columbia Univ. *Born:* 8/10/26. Native of Emglewood NJ. Upon graduation employed in Genl Electric's Large Steam Turbine Dept as Steam Design Engr & later as Advance Design Engr, became Mgr of Steam Design Engg responible for thermodynamic design & performance of large steam turbines 1961, Mgr of Mech Design Engg 1968 respon for engg & drafting design of main rotating & stationary turbine parts. Mgr of Design Engg 1977 respon for overall design and performance of large steam turbines. Active in ASME at local, natl & internatl levels. Co-author of definitive papers on steam turbine performance and of 1967 ASME Steam Tables. Enjoy outdoor activities both summer & winter. George Westinghouse Gold Medal ASME 1970. Tau Beta Pi, Sigma Xi. Reg PE in NY. *Society Aff:* ASME.

Spencer, William J
Business: Livermore, Albuquerque, CA 94550
Position: Director. *Employer:* Sandia Labs. *Education:* PhD/Physics/KS State Univ; MS/Math/KS State Univ; AB/Physics Ed/William Jewell College. *Born:* 9/25/30. Joined Bell Labs in 1959. Supervisor, Piezoelectric Devices Group, 1960-68. Mgr of Piezoelectric Devices Dept, 1968. Dir of Univ Relations and Technical Employment, 1973-73. Dir of Microelectronics, Sandia Labs, 1973-78. Respon for R&D in silicon integrated circuits, computer-aided design, hybrid microcircuits, discrete components, electronics subsys for nuclear weapons and silicon solar cells for solar energy programs. Presently Dir of sys Dev, Sandia Labs. Respon for nuclear weapon sys including R&D, design, production engr, stockpile maintenance, and applied mechanics activity. Jointly a Res Prof of Medicine, The Univ of NM Sch of Medicine, involved in dev of insulin delivery sys for diabetics. *Society Aff:* IEEE, AAAS.

Spendlove, Max J
Home: 13121 Clifton Rd, Silver Spring, MD 20904
Position: Cons. *Employer:* Self-employed. *Education:* BS/EE/Univ of UT. *Born:* 12/16/12. UT, Engr for UT elec utility to 1940; US Bu Mines, UT, MI, and MD to 1950. US Bu Mines Resource Recovery Res MD, Asst Dir Metallurgy Res, Wash DC, Res Dir Bu Mines College Park MD. Metallurgy Res Ctr 1970 to 1973. Technical spokesman for Fed Govt at conf in many US and European cities, Bolivia and Japan. Mining and Metallurgy People to People Delagate to South East Asia, 1987. Author or coauthored 26 technical publications and presented over 29 technical papers. Received 3 Dept of Int superior performance awards, and AIME 1977 Distinguished Service Award. Has 4 patents; is a licensed PE, and mbr of AAAS, AIME, and honor societies Tau Beta Pi and the Sigma Xi. Cons, resource recovery, since 1973. *Society Aff:* AIME, AAAS, ΤΒΠ, ΣΞ.

Sperber, Philip
Home: 30 Normandy Heights Rd, Convent Station, NJ 07961
Position: President *Employer:* REFAC International, Ltd. *Education:* Juris Dr/Law/Univ of Maryland; BS/EE&Chem Eng/NJ Inst of Tech. *Born:* 2/29/44. Philip Sperber has held mgmt, legal, engg and sales positions, most recently as Pres of REFAC, Pres of APRO Scientific, Group Exec for ITT, VP for Cavitron, and partner in Blair, Olcutt & Sperber. Posts held include Trustee, Licensing Exec Soc; VP, Ultrasonic Indus Assoc; Judge, Amer Arbitration Assoc; Section Co-Chrmn, Assoc for Advancement of Medical Instrumentation; Chrmn, Food, Drug and Cosmetic Section, NJ Bar Assoc; Chrmn, NJ Inventors Hall of Fame; and Section Chrmn, Amer Soc for Testing and Materials. Named Eminent Engineer by Tau Beta Pi, author 100 papers and thirteen books, and biographee in Marquis' Who's Who in the World. *Society Aff:* IEEE, AIChE, LES, IEA, IPA, ABA.

Sperry, Philip R
Business: PO Box 14448, St. Louis, MO 63178
Position: Chief Metallurgist, Corp *Employer:* Consolidated Aluminum Corp. *Education:* BS/Met/Univ of Notre Dame. *Born:* 7/6/20. Employment by Prof Paul A Beck as metallographer while pursuing undergrad work at Notre Dame led to co-recipient of AIME Mathewson Gold Medal award for 1953. In 32 yrs with aluminum industry (Kaiser, Olin & Consolidated), achieved intl recognition for interpretation of aluminum alloy microstructure. Principal author of ASM Handbook (Vol 8) article on Met of Aluminum. Holds 30 patents for alloys & met processes, plus many publications. Contributor to 1984 book, ALUMINUM: Properties and Physical Metallurgy, published by ASM. Mbr of Aluminum Assoc Sr Technical Cttee. *Society Aff:* TIMS-AIME, ASM.

Spetnagel, Theodore J
Home: 855 Kipling Drive, NW, Atlanta, GA 30318
Position: Deputy Engr *Employer:* HQ, Second US Army *Education:* MS/CE/GA Inst of Tech; BS/CE/Clemson Univ *Born:* 05/26/48 Born Chillicothe OH, raised Kingsport TN. Currently Deputy Engr, HQ Second US Army. 1979-84 Chief Minor Construction Sect, HQ US Army Forces Command. 1978- 79 Civ Engrg, HQ Ft McPherson. 1971-78 Chief Design Engr (Civ/Struct), Atlantic Building Sys, Inc. Reg PE in seven states. Recent Awards - 1984 Who's Who in the World. 1984 Young Engr of the Year, 30 Greater Atlanta Engr Organizations. 1982 Engr of the Year in

Spetnagel, Theodore J (Continued)
Government, GA Soc of PE. Dir GA Sect ASCE, 1984. NSPE-PEG Bd of Governors, 1981. Chrmn GA PE in Government, 1980-82. Dir Atlanta Post SAME, 1982-83. *Society Aff:* ASCE, NSPE, SAME, XE, TBΠ

Spicer, Clifford W
Business: 600 Grant St, Pittsburgh, PA 15230
Position: Area Mgr Met Service - Eastern Operations *Employer:* US Steel Corp. *Education:* BS/ChE/Univ of Cincinnati. *Born:* March 9, 1919. Certificate in the field of Exec Dev 1971 Temply Univ, Sch of Business Administration. Mbr ASM, appointed Fellow 1975, received 25 year Mbrship Award 1975; AIME Natl Open Hearth Ctte 3rd place award Chicago Sect off the record meeting 1968. Licensed Prof Engr in the Commonwealth of Pa by examination. *Society Aff:* ASM, AIME.

Spiegel, Walter F
Business: 321 York Rd, Jenkintown, PA 19046
Position: Pres. *Employer:* Walter F Spiegel, Inc Consulting Engrs. *Education:* MS/Mech Engr/Univ of PA; BS/Mech Engr/Univ of PA. *Born:* 11/7/22. Presidential Mbr (1972-73) and Fellow of American Society of Heating, Refrigerating and Air Conditioning Engrs (ASHRAE). Received ASHRAE F. Paul Anderson Medal in 1986, ASHRAE's highest technical award. (1975-79) Pres of Phila Chapter of Consulting Engrs Council (1979-81) and VP of Commission E-1 of Int'l Inst of Refrigeration. Registered in 7 states, and Pres of Walter F Spiegel, Inc, Consulting Engrs, founded in 1963. Has done research and written papers on special air conditioning systems and energy mgt, as well as book and Chapters in ASHRAE Handbook. Mbr Sigma Tau and Tau Beta Pi. *Society Aff:* ASHRAE, NSPE, ACEC.

Spiegler, Kurt S
Business: Dept of Mechanical Engrg, Univ of CA, Berkeley, CA 94720
Position: Prof of Mech Engrg *Employer:* Univ of CA-Berkeley *Education:* PhD/Chem/Hebrew Univ, Jerusalem; MS/Chem/Hebrew Univ, Jerusalem *Born:* 5/31/20 Vienna. Academic experience: Prof in Residence of Mechanical Engrg, Univ of CA-Berkeley, 1964-78, now Emeritus. Prof of Chem Engrg, MI Tech Univ, 1977- 81. Industrial experience: Gulf Research and Development Co, British Petroleum, Dead-Sea Salt Works, Apollo fuel cell. Research on water desalination, applications of non-equilibrium thermodynamics, transport processes in membranes, solar energy. 4 books, 60 technical articles, 7 patents. Editor-in-Chief *Water* Pollution; on Editorial Bd of 2 intl journals. 1980 Research Award, Japan Society Promotion of Science; Chrmn of 2 Gordon Research Conferences; Co-dir of a NATO Adv Study Inst. *Society Aff:* AEE, ACS, ΣΞ

Spiel, Albert
Home: 98 DeHaven Dr, Yonkers, NY 10703
Position: Vice President. *Employer:* Nabisco Inc, Protein Foods Div. *Born:* Aug 9, 1921 New York, NY. MChE 1949 Polytechnic Inst of NY; BChE 1944 CCNY. Metallyrgist Manhattan Proj, Houdaille-Hershey Corp Decatur Ill 1944-46. Dev Engr Kraftco Corp 1949-54; Sect Head Genl Foods Corp, Tarrytown NY 1954-62; Mgr Engrg Servs DCA Food Indus Inc 1962-68; with Nabisco Inc 1968- , Vice Pres Process Systems 1975- . Also directed Process Dev for R&D for 2 years. Currently in Mfg & D for protein prods. Mbr IFT, ACS, AIChE, Licensing Execs Soc. Mbr Standards Ctte AIChE. M. to Jacqueline Grundstein; one son, Douglas.

Spielvogel, Lawrence G
Business: Wyncote House, Wyncote, PA 19095
Position: Consulting Engineer. *Employer:* Lawrence G Spielvogel Inc. *Education:* BS/MechE/Drexel Univ *Born:* June 2, 1938 Newark NJ. Pres Lawrence G Spielvogel Inc, Wyncote Pa 1970- . Lecturer Univ of Pa Graduate Sch 1971-78. Visiting Lecturer Yale Univ 1975-81. Served to 1Lt US Army 1963-65. Reg PE in Pa & 48 states. Certified Plant Engineer (AIPE) Mbr: ASHRAE, Chmn Water Heating Ctte 1968-70, Chmn Computer Ctte 1973-75, Chmn Std 90 Panel 5 1974-83, Distinguished Service Award 1975, Award of Merit 1976 Crosby Field Award 1981; American Soc of Plumbing Engrs; American Cons Engrs Council (Energy Ctte); Illuminating Engrg Soc, Dir 1976-79 . *Society Aff:* ASHRAE, ASPE, ACEC, IES, ASME, ISES, NCEE.

Spigai, Daniel J
Business: 1500 N Beauregard St, Alexandria, VA 22311
Position: Partner. *Employer:* Howard Needles Tammen & Bergendoff. *Education:* BS/Civ Engg/Manhattan College. *Born:* 3/4/33. Born in NYC, NY. Joined HNTB in 1957, elected to partnership in 1974. Chairman of the HNTB Partnership and Executive Committee 1986-7. Partner-in-Charge of offices in Alexandria, VA, Atlanta, GA, Raleigh, NC and Charleston, SC. Also has responsibility for a broad spectrum of airport planning and design projects throughout the US. Reg PE in 23 states. Active member of several professional organizations, including ARTBA, ACEC, IBTTA, AAAE, AOCI, ACC, TRB. *Society Aff:* ASCE, NSPE, ARTBA, AAAE.

Spillane, Leo J
Business: 601 Jefferson, Suite 535, Houston, TX 77002
Position: Pres. *Employer:* Gulf States Asphalt Co, Inc. *Education:* PhD/Organic Chemistry/Univ of MN; MS/Organic Chemistry/Univ of MN; BS/Chemistry/College of St Thomas. *Born:* 8/23/16. Pres of Gulf States Asphalt Co since 1970; Mgr of Res, Monsanto Co & Lion Oil Co 1951-1969; Chief Chemist, Reaction Motors, Inc, 1949-1950; Group Leader, Allied Chem Corp, 1942-1948. Carried on res & directed process dev in petrochems, petrol products, synthetic fibers & rocket fuels. Commercially significant contributions were achieved in biodegradable surfactants, aromatic chems, asphalt products & in hydrocarbon cracking & recovery tech. *Society Aff:* AIChE, ACS, API.

Spiller, W R
Home: Boca Raton Hotel, Villa 1515, 501 E. Camino Real, Boca Raton, FL 33432
Position: Retired. *Education:* BS/ME/Univ of PA. *Born:* July 1899 Philadelphia Pa. White Motor Co, Cleveland Ohio 1922-39 to Ch Engr; Harris Corp 1939-69, Dir Ch Engr to Vice Pres, Res & Engrg & Senior Vice Pres Engrg of the Corp responsible for dev & advisor in acquisition of printing presses & other graphic arts machinery. Fellow & Life Mbr ASME. Vice Pres of Cleveland Engrg Soc & of Res & Engrg Council of the Graphic Arts, Graphic Arts Tech Soc. 30 patents in automotive & graphic arts machinery. *Society Aff:* ASME.

Spillers, William R
Business: Dept of Civil Engrg, Troy, NY 12181
Position: Professor of Civil Engineering. *Employer:* Rensselaer Polytechnic Inst. *Education:* PhD/Eng Mechs/Columbia Univ; MS/Civil Engg/Univ of CA (Berkeley); BS/Civil Engg/Univ of CA (Berkley). *Born:* Aug 1934. Native of Fresno Calif. Struct Engr for John Blume Assoc, San Francisco 1956-57. Associated with Columbia Univ as student them Prof 1957-76. Joined RPI in 1976. Guggenheim Fellow 1968. NSF Fellow 1975. Specialty: computers & struct mechs. Cons on sev major structs. Pub 90 tech papers & 4 books; contributed pieces to 3 other books. Books: Automated Struct Analysis, Pergamon 1972; Basic Questions of Design Theory, N Holland 1975; Iterative Structural Desingn, N Holland 1976, Introduction to Structures, Horwood, 1984. *Society Aff:* ASCE

Spillman, Charles K
Home: 3209 Windgate Cir, Manhattan, KS 66502
Position: Professor Hd. Dept of Agric Engg *Employer:* Kansas State University. *Education:* PhD/Agri Engrg/Purdue Univ; MS/Agri Engrg/Univ of IL; BS/Agri Engrg/Univ of IL; AS/PreEngrg/Vincennes Univ *Born:* 2/26/34 US Marine Corps 1953-56. Farm Bldgs Extension Specialist 1962-66 MI State Univ. Joined Kansas State Univ 1969 to tech & conduct res on environmental control in farm bldgs & for dev of feeding facilities for farm animals. Recent res has been dev of solar heat collection units for animal shelters, bldg mgmt & solar energy for greenhouses. Hobbies: fishing & hunting. ASAE offices: KS Sect Chrmn 1970-71; Structs & Environ Div Steering Ctte 1975-81, V Chrmn 1978-79, Chrmn 1979-80; Natl Nominating

Spillman, Charles K (Continued)
Ctte 1976-77; and Dir (Midcentral Region), 1982- 84. *Society Aff:* ASAE, ASEE, ASHRAE.

Spina, Frank J
Business: 1020 Seventh North St, Liverpool, NY 13088
Position: Partner *Employer:* Calocerinos & Spina *Education:* B/CE/Syracuse Univ *Born:* 7/9/30 Native of Syracuse and graduate of Syracuse Univ. Did graduate work in sanitary engrg at same univ until employment with Central NY consulting firm. Worked with same firm for 12 1/2 years (1955-1968) until forming my own firm with E. J. Calocerielos, (Calocerielos & Spina). I am currently a member of the City of Syracuse, City Planning Commission. *Society Aff:* WPCF, APWA, ACEC, NYSSPE, AWWA

Spinks, John L
Home: 26856 Eastvale Rd, Rolling Hills, CA 90274-4007
Position: Pres & Gen Mgr *Employer:* Environmental Emissions Engrg Co. *Education:* BS/ME/Univ of KY; PhD/Eng (Hon)/World Univ. *Born:* 6/19/24 Deputy Dir, Engrg Div, USAFR, for 190 prof engrs, scientists, and chemists. Directed their air and water pollution control programs. Was responsible for air quality in L.A. County. Pioneered development of engrg principles for air pollution control techniques. Directed development of "Total Carbon Analysis" procedure for determining organics. Co-author "Air Pollution Engrg Manual–. Natl and Intl conslt for Gov't Air Quality Agencies. Registered PE in CA, LA, TX, DE, WI, NH,, OK, KY, and MS. Commissioned a "KY Colonel" by Gov John Y. Brown. Diplomate, AAEE; Fellow, Inst Adv Engrg; recipient USAF Commendation Medal for meritorious service as consulting engr. Mountaineering instructor; marathon long-distance running lecturer; Little League Baseball mgr; golf league dir; certified reserve police officer. Post grad work in ME at USC and UCLA. *Society Aff:* AAEE, APCA, ASME, NSPE.

Spinrad, Bernard I
Business: 261 Sweeny Hall, ISU, Ames, IA 50011
Position: Prof and Chrmn *Employer:* Dept of Nuclear Engrg, IA State Univ *Education:* PhD/Phys Chemistry/Yale Univ; MS/Phys Chemistry/Yale Univ; BS/Chemistry/Yale Univ *Born:* 04/16/24 Born and raised in NY City. 40 years in nuclear energy: ORNL, 1946-48; ANL, 1949-67 and 1970-72 (Dir, Reactor Engrg Div, 1957-63); IAEA, Dir, Div of Nuclear Power and Reactors, 1967-70; OR State Univ, 1972-82; Chrmn, Dept of Nuclear Engrg, IA State Univ, 1983-present. Conceived Savannah River reactor design, patents on other reactor concepts. Fields of interest: reactor phys, nuclear fuel cycle and economics, advanced nuclear applications; nuclear safeguards, energy futures. Major contributor to: *–Energy in Transition,* 1985-2010– (Freeman, 1980); *–Energy in a Finite World–* (Ballinger, 1981); and *–Nuclear Energy: A Sensible Alternative–* (Plenum, 1985). *Society Aff:* ANS, APS, ACS, AAAS, ΣΞ, ASEE

Spiro, Irving J
Business: 4924 Mammoth Ave, Sherman Oaks, CA 91423
Position: ‡Senior Engineer *Employer:* The Aerospace Corporation. *Education:* MS/Math & Engg/UCLA; BS/Mech Engg/IL Inst of Tech. *Born:* Sept 1913 Chicago Ill. Reg PE. Taught at UCLA 1942-45; Genl Mgr Roxmar Engrg Co, Hollywood, Cameras 1946-50; Ch Engr Borman Engrg Inc, North Hollywood 1950-55, developed smog measuring instruments used by LA County; Ch Mech-Optical Engr Aerophysics Corp, Santa Barbara, Optics for Missile Guidance 1955-58; Mgr Space Technology Labs & Aerospace Corp 1958- , specializing in Infrared Photo- Optical Engrg. Mbr Tau Beta Pi, Sigma Xi, Sigma Lambda Tau, Optical Soc of America, Soc Photo-Optical Engrs (Pres L A Chap), Natl Bd of Governors, Assoc Editor, Optical Engrg. Elected Fellow, SPIE, 1978; Elected Fellow OSA, 1979. Enjoy classical music & golf. *Society Aff:* TBΠ, ΣΞ, SPIE, OSA, CORM

Spitler, E Eugene
Home: Chevron U.S.A, 575 Market Street, San Francisco, CA 94105
Position: General Manager, Product Engineering *Employer:* Chevron U.S.A. *Education:* PhD/Mech Engrg/Purdue Univ; MSME/Mech Engrg/Purdue Univ; BSME/Mech Engrg/Purdue Univ *Born:* Aug 1931. Early life near Wabash Ind. Brief engrg exper General Motors Res 1953,54. Served in US Army 1954-56. Res Engr Chevron Res starting in 1959, specializing in internal combustion engine fuels & lubricants; Mgr Fuels Div 1969-77, responsible for all fuels R&D, Standard Oil (Calif). Mgr, Environmental Planning, Chevron USA 1977-78, Asst to Pres, Chevron Res, 1978-1980. Asst. Secretary, Standard Oil (Calif), 1980, Vice President, Chevron Research (1981-1985) Gen Mgr, Prod Engrg, 1985-present. Served SAE Bd/Dir 1976-78. Enjoy tennis, hiking, music, sport cars, & studying philosophy. Married, 3 children. *Society Aff:* SAE, ASEE

Spitz, Albert W
Home: 437 N Sterling Rd, Elkins Park, PA 19117
Position: President. *Employer:* Self-employed. *Education:* BS/ChE/Univ of PA. *Born:* Nov 28, 1912. Served as Area Engr Frankford Plant, Allied Chem Corp, & Proj Mgr Central Engrg Dept, American Cyanmid Co New York, NY. Started Albert W Spitz & Assocs in 1949. Work includes chem & met plant processes & equipment design, pollution control systems. Also, adjunct faculty mbr Environmental Engrg Dept, Drexel Univ Philadelphia Pa. *Society Aff:* ACS, AICHE, APCA, AAAS.

Spitz, Peter H
Home: 145 Edgemond Rd, Scarsdale, NY 10583
Position: Pres. *Employer:* Chem Systems Inc. *Education:* MS/CE/MIT; BS/CE/MIT. *Born:* 5/31/26. Process Design Engr at Std Oil Dev Co (1949-1956). Proj Mgr & Proj Dev Sec Hd at Scientific Design Co. VP & Asst to Pres at Halcon Intl (1956-1964). Founded Chem Sys Inc in 1964 & has headed co since that time. Chem Sys is a consltg firm specializing in long range and strategic planning for companies in the chem and petroleum industries. Spitz also spearheaded the establishment of Chem Sys Res, Inc an affiliated firm engaged in power dev and tech assessment. As pres of both firms, which have a broad intl clientele, he has presented a number of important lectures to industry and government groups and assocs, dealing with long range chemical industry and energy trends. *Society Aff:* AIChE, TBΠ, SCI.

Spitzer, Robert R
Business: Box 644, 1025 N Milwaukee St, Milwaukee, WI 53201
Position: Pres. *Employer:* Milwaukee School of Engg. *Education:* PhD/Animal Nutrition & Medical Physiology/Univ of WI; MS/Animal Nutrition and Biochemistry/Univ of WI; BS/Agri/Univ of WI. *Born:* 5/4/22. Native of Burlington, WI. Elected Pres of Milwaukee School of Engg in 1977. Coordinator, Food for Peace Prog, US State Dept, 1975-1976. Currently, Education Comm Chrmn, Nutrition Fdn, NY; Sr Advisor, Allis-Chalmers Corp; Dir, Mirro, Inc, Tracy & Son, and the Larsen Co, Inc. Past Pres, WI Mfg Assn and Univ of WI Intl Alumni Assn. Univ of WI Distinguished Service Award, 1972. Iowa State Univ World Food Lecturer, 1977. MI State Univ Distinguished Lecturer, 1976. *Society Aff:* ASEE.

Spitzig, William A
Business: 125 Jamison Ln, Monroeville, PA 15146
Position: Associate Research Consultant. *Employer:* US Steel Corp. *Education:* PhD/Matls Sci/Case-Western Reserve; MS/Matls Sci/Case-Western Reserve; BS/Metallurgy/Cleveland State. *Born:* Sept 1931. Metallurgist Thompson-Ramo-Wooldridge 1960-62; Matls Engr Lewis Res Center, NASA 1962-66; Res Metallurgist US Steel 1966-68, Senior Scientist 1968. Associate Research Consultant 1975. USCG 1950-54. Special interests in fracture & strengthening mechanisms in steels, titanium alloys, aluminum alloys & polymers. ASM scholarship at Cleveland State. Mbr ASM, Tau Beta Pi, & Sigma Xi. *Society Aff:* ASM, TBΠ, ΣΞ.

Splinter, William E
Business: 223 LW Chase Hall, Lincoln, NE 68583-0726
Position: George Holmes Distinguished Prof and Head *Employer:* University of Ne-

Splinter, William E (Continued)
braska. *Education:* PhD/Agri Engg/MI State Univ; MSc/Agri Engg/MI State Univ; BSc/Agri Engg/Univ of NB. *Born:* Nov 24, 1925. Instructor Mich State Univ 1952-54; Assoc Prof & Prof Biological Agri Engrg Dept, NC State Univ 1954-68; Prof & Hd Agri Engrg Dept, Univ of Nebraska 1968-84. George Holmes Distinguished Prof and Head, 1984-. Teaching Exper includes courses in Instrumentation, Human Factors Engrg & Mathematical Modelling of Plant Growth. Res includes electrostatic deposition of pesticides, plant growth dynamics, human factors engrg, physical properties of biological materials. 91 tech publs; 5 patents. Fellow ASAE, Dir of Pubs 1972-74, Regional Vice Pres 1976-78, Pres 1978-79, Fellow AAAS, Mbr SAE, ASEE, NSPE, AAAS, Flying Engrs Mbr. Ex. Bd AAES, 1980. Mbr, Natl Academy of Engr, 1984. *Society Aff:* ASAE, SAE, ASEE, NSPE, AAAS, NAE.

Spooner, Stephen
Business: Sch of Chemical Engrg, Atlanta, GA 30342
Position: Prof *Employer:* Georgia Tech *Education:* ScD/Metallurgy/MIT; BS/Metallurgy/MIT *Born:* 04/02/37 Native of Lincoln, MA. Worked with Raxtheon, Inc and Lab for Electronics while attending MIT. Joined GA Tech in 1965 where I have taught metallurgy and have done crystal structure-related res in metals, alloys, minerals and biological materials using x-ray and neutron diffraction. During leave of absence in 1975 at Oakridge Natl Lab, participated in development of small angle neutron scattering experiments on irradiated materials, superconductors and magnetic alloys. Chrmn, structures activity of materials Sci Div of ASM. *Society Aff:* AIME, APS, ASM, ACA

Sporseen, Stanley E
Home: 3184 Oakcrest Dr NW, Salem, OR 97304
Position: Consultant *Employer:* Haner, Ross & Sporseen *Education:* BS/Civil Engr/Univ of WA; Pr Eng/Civil Engr/Univ of WA *Born:* 4/30/04 Twenty four years with US Army Engrs & 25 years in private practice. One year with Republic of Venezuela, 6 month Montreal Engr Co & 4 assignments with United Nations. 75% of above time was on Hydro Development. At present time am on FERC Engrg Review Bd also Hydro Consultant with Haner, Ross & Sporseen. Prof Engrg in WA, OR, MT, AK. *Society Aff:* ASCE, Con EngLo, MICE, USCOLD

Spotts, Merhyle F
Business: Northwestern University, Tech Inst, Evanston, IL 60201
Position: Professor Emeritus. *Employer:* Northwestern University. *Education:* PhD/Applied Mechs/Univ of MI; MA/Mathematics/OH State Univ; BSME/Mech Eng/OH Northern Univ. *Born:* Dec 1895, Iowa. 1927-35 engr & designer Brown Steel Co & Jeffrey Mfg Co, Columbus Oh; 1938-41 taught Johns Hopkins Univ, 1941-65 Northwestern Univ. Author 4 textbooks on mech design, many tech articles. Received ASME Worcester Reed Warner Medal 1968 'for an outstanding contribution to the permanent literature of engineering.'; ASME Century II Medallion, 1980; Machine Design Award, 1980. *Society Aff:* ASME (Life Fellow).

Spradley, J Calvin
Business: 1500 E. Bannister Rd, Kansas City, MO 64131
Position: Dir, Design & Const Div PBS *Employer:* Gen Services Admin *Education:* BS/Arch Engr/Univ of KS *Born:* 05/10/30 Responsible for alteration, maintenance and repair, design and construction of federal structures in KS, NB, IA and MO. Programs include Energy Conservation/Mgmt, active/passive Solar Design, Handicapped Accessibility, Art- in-Arch, Life Cycle Cost-based Design and the creation of Efficient, Readily- Altered Work Environments. Serves as Chrmn, Planning Commission, mbr, arts council and mbr, Mayor's Capital Improvements Task Force, city of Leawood, KS. Chrmn Bd of Elders, Cherokee Christian Church, Prairie Uiusge, KS. Mbr, KS Uty Engrg Extension Advisory Council to the Univ of MO Sch of Engrg. MS/CE/Univ of MO, 80% complete. *Society Aff:* AIAA, APWA, NCMA

Spreiter, John R
Home: 1250 Sandalwood Ln, Los Altos, CA 94022
Position: Prof of Appl Mechanics & of Aero & Astro. *Employer:* Stanford Univ. *Education:* PhD/Engg Mechanics/Stanford Univ; MS/Engg Sci/Stanford Univ; BAeroE/Aero Engg/Univ of MN. *Born:* 10/23/21. Raised and attended public schools in Staples, MN. Res engr and scientist NACA & NASA Ames Res Ctr, Moffett Field, CA 1943-69. Flight Res 1943-46. Theoretical Aerodynamics 1947-62. Chief Theoretical Studies Branch (Space Sciences) 1962-69. Lectr (1951-68) and prof (1968-). Stanford Univ. Consultant, Nielsen Engg & Res, Inc, 1969-85, RMA Aerospace 1985- . Mbr & sometimes chrmn of numerous committees of NASA, Aeronomy, Interunion Commission on Solar-Terrestrial Phys & the Intl Scientific Radio Union. Enjoys tennis, swimming, skiing, travel, reading, and conversation. *Society Aff:* AIAA, AAAS, APS, AGU, RAS, ΣΞ, ΤΒΠ.

Spreitzer, William M
Home: 22663 N Nottingham Dr, Birmingham, MI 48010
Position: Manager-Planning *Employer:* Research Labs General Motors Corp. *Education:* PAeE/Aero Engr/Univ of Detroit; BAeE/Aero Engr/Univ of Detroit *Born:* Aug 1929. 1951 joined GM Res Labs, 1951-61 automotive gas turbine engine R&D & applications; 1961-66 Staff Asst to Mgr; 1966-87 Hd Transportation Res, 1987 assumed present respon. Chmn AIAA Mich Sect 1960-61; Mbr SAE Res Council 1968-72, Mbr SAE Total Transportation Ctte 1971-73, Chmn SAE Transportation Systems Activity 1973-74; Mbr Civic Affairs Council Engrg Soc of Detroit 1969- 72; Mbr Natl Res Council Assembly of Engrg Ctte on Transportation 1970- , Bay Area Rapid Transit (BART) Impact Program Adv Ctte 1972-79; Mbr Trans Res Bd Group 1 Council 1984-86; Chairman Transportation Research Board Group 5 Council 1986- . Mbr Trans Res Bd Ctte on Educ and Training 1981-86; Mbr Visiting Ctte Univ of Pittsburgh 1973-75; Mbr Indus Adv Univ of Detroit 1974- . *Society Aff:* AIAA, SAE, ORSA, TRF, NDTA, ESD.

Sprenkle, Raymond E
Home: 2149 Rayburn Rd, East Cleveland, OH 44112
Position: Dir of Education until April 1961. *Employer:* Bailey Meter Company. *Education:* BME/ME/Bucknell Univ. *Born:* Jan 21 1895, Waynesboro Pa. 1918-19 served with Meteorological Div US Signal Corps; 1919-22 Service-sales engr Bailey Meter Co, 1922-46 in charge meter engrg, 1946-61 Dir Education until retirement. ASME Fellow 1960, Mbr Fluid Meter Res Ctte 1927-63, V Chmn 1950-60, Emeritus Mbr 1964- , Standards Ctte on Measurement of Fluid Flow 1975- , presented num ASME papers on Fluid Flow Measurement. Sev patents pertaining to this technique. Twice delegate to France & Germany 1959-60, representing US at Internatl Standardization Organization meetings. *Society Aff:* ASME.

Spriggs, Richard M
Business: Natl Res Council, Natl Acad of Sci JH817, 2101 Constitution Ave, NW, Washington, DC 20418
Position: Senior Staff Officer/Staff Director *Employer:* National Research Council *Education:* PhD/Ceramic Engrg/Univ of IL; MS/Ceramic Engrg/Univ of IL; BS/Ceramics/PA State Univ *Born:* May 1931. BS Penn State ; MS & PhD Univ Ill. 1952-54 Lt USNR; 1958-59 Sr Res Engr Ferro Corp; 1959-64, Sr Staff Scientist & Ceramic Res Grp Leader Avco Corp, 1964-80 Lehigh Univ, 1964-67 Assoc Professor of Met Engr, 1964-70 Assoc Dir Materials Res Center, 1967-80 Professor of Met & Materials Eng, V Pres for Admin. Special interests include high technology ceramics. Fellow Inst of Ceramics; Internatl Inst for Sci of Sintering; Fellow, Pres 1984-85 Amer Ceramic Soc. Ross Coffin Purdy Award 1967; Hobart M. Kramer Award 1980. Trustee, Federation of Matls Socs (1978-84). 1979-80 Visiting Sr Staff Assoc, Natl Matls Advis Bd (NMAB), natl Res Counc, Since 1980, Sr Staff officer, NMAB 1980-date; Staff Dir, Bd on Assess of NBS Progs, 1984-date. *Society Aff:* ACerS, NICE, ΣΞ, AAAS, ASTM PLUS OTHERS

Springer, Charles
Business: Dept of Chem Engg, Fayetteville, AR 72701
Position: Prof. *Employer:* Univ of AR. *Education:* PhD/CE/Univ of IA; MS/CE/Univ of IA; BS/CE/Univ of IA. *Born:* 2/22/29. Native of IA. Asst Prof, Assoc Prof,

Springer, Charles (Continued)
Prof at Univ of AR since 1965. Interested in Corrosion & mass transfer, especially diffusional processes relating to release of chems to the environment. Previously with Dow Chem Co and Trane Co. *Society Aff:* AIChE, ASEE, NACE, ACS.

Springer, E Kent
Business: University Park-BHE315, Los Angeles, CA 90089-0231
Position: Emeritus Professor & Director of FCCCHR *Employer:* University of Southern California. *Education:* BS/Mech Engr/Univ of So CA; MS/Mech Engr/Univ of WI *Born:* 9/17/12. in Bellingham, WA. 1935-39 Plant Engg, Fluid Packed Pump Co; 1939-41 Mech Engg PRECO; 1941 Mech Engr Cons Western Steel Corp; 1941-46 Instructor & Asst professor Univ Wisc; 1946-1977, Asst, Assoc & Prof of Mech Engr Univ of So CA, 1977-Emeritus Professor of Mech Engr. 1964- , Dir Fnd for Cross-Connection Control & Hydraulic Res at USC. Natl Pres Pi Tau Sigma 1962-68; VP Region IX ASME 1970-72, Life Fellow ASME. Intl authority on protection of potable water sys from backflow of pollutants or contaminants Emeritus Prof of Mech Engg-1977 Emer. Dir of FCCHR at USC. *Society Aff:* ASME, ASEE, AWWA.

Springer, George S
Business: Dept Aero/Astro, Stanford, CA 94305
Position: Prof *Employer:* Stanford Univ *Education:* Ph.D/M.E./Yale; M.S./M.E./Yale; M.Eng/M.E./Yale; B.Eng/M.E./U of Sydney *Born:* 12/12/33 *Society Aff:* ASME, AIAA, SAE, APS

Springer, Karl J
Home: 111 Shalimar Dr, San Antonio, TX 78213
Position: VP *Employer:* Southwest Research Institute *Education:* M.S./Physics/Trinity University; B.S./Mech Engrg/Texas A&M University *Born:* 4/14/35 Engineer, Engines, Fuels and Lubricants (SwRI) 1957. Active duty as Lt. USAF Propulsion Lab, Wright Air Dev Center, 1958-1960. Field Engineer, DuPont Co., 1960-1962. Internationally known for pioneering research on odor, smoke and a host of other exhaust emissions from diesel and gasoline powered trucks, buses and cars. Dir of Emissions Research Dept., SwRI, 1966-1986. VP of Automotive Products and Emissions Research Div 1986. ASME Centennial Award Medallion. Author of over 50 technical papers and reports, Fellow ASME 1984, Diplomate AAEE 1986, Reg PE TX. Native of San Antonio, Texas. *Society Aff:* ASME, SAE, ASTM, ASEE, RESA

Springer, Melvin D
Business: Ind Engg Dept, Fayetteville, AR 72701
Position: Prof. *Employer:* Univ of AR. *Education:* PhD/Math Stat/Univ of IL; MS/Math Stat/Univ of IL; BS/Math/Univ of IL. *Born:* 9/12/18. M D Springer received his BS, MS, & PhD degrees from the Univ of IL. He has served as Dir of Reliability Res (A C Electronics Div) & Operations Analyst (Defense Systems Div) for GM Corp, and as Sr Operations Analyst for Technical Operations, Inc. He has taught in the Math Depts at the Univ of IL & MI State Univ, and presently on the faculty of the Ind Engg Dept at the Univ of AR. He has published papers in numerous engg, math, & statistical journals, and holds mbrship in several engg & statistical socs. He is also the author of a recent book entitled *The Algebra of Random Variables* (Wiley-Interscie, 1979). *Society Aff:* IIE, ASEE, ASQC, IMS, ASA, ORSA.

Sproat, Robert L
Business: Benson-East, Jenkintown, PA 19046
Position: Vice President Manufacturing & Engrg. *Employer:* SPS Technologies, Inc. *Education:* MS/Metallurgy/Penn State Univ; BS/Metallurgy/Penn State Univ. *Born:* Oct 5 1921, State College Pa. 1943-44 employed by Wright Aero Corp; 1944-46 served in US Navy; 1948 Metallurgy Facility Penn State Univ; since 1948 with Standard Pressed Steel Co, 1968 elected Pres Engrg & Dev, 1974 assumed current respon as V Pres Mfg & Engrg. ASM Fellow 1970. Penn State ASM David Ford McFarland Award 1975; Philadelphia Chapter ASM Del Valley Metal Man of Year 1975. *Society Aff:* AIME, ASM.

Sproles, Max R
Business: 1133 15th St. NW, Washington, DC 20005-2701
Position: Sr VP Business Dev *Employer:* De Leuw, Cather & Co *Education:* MS/CE/GA Inst of Tech; BS/CE/VA Polytech Inst *Born:* 02/26/34 With Bureau of Public Roads as Highway Engr from 1958 to 1965, NC State Highway Commission from 1965 to 1970, promoted in 1967 to State Planning & Res Engr. From 1970 to 1973 served as Exec Dir, Highway-Rail Progs with the Association of American Railroads. Served as VP of Harland Bartholomew & Assocs from 1973 to 1981. Joined De Leuw, Cather & Co in 1981 as VP of Marketing and Sales. Promoted to Senior V.P. Business Development in 1984. Has been principal advisor to the State Rail Progs Office of the Assoc of American Railroads since 1973. Active mbr of numerous profl transp engrg organizations. *Society Aff:* AREA, ASCE, ITE, NSPE, TRB, ARTBA

Sproul, Otis J
Business: Univ. of New Hampshire, Coll of Engrg & Physical Sci, Durham, NH 03824
Position: Dean & Wheelabrator-Frye Prof *Employer:* Univ of NH *Education:* ScD/Sanitary Engrg/WA Univ-St Louis; MS/CE/Univ of ME-Orono; BS/CE/Univ of ME-Orono *Born:* 7/9/30 Dover Foxcroft, ME; married Dorothy R. Estabrook June 8, 1952; children, Bryce J. and Dana C. (deceased); registered PE, ME; received Rudolph Hering Medal, ASCE 1972; co-founder IOA, 1973; editor, Journal Environmental Engrg Division, ASCE, 1980-1982; Diplomate, AAEE; environmental engrg researcher and educator; research speciality: virus, cyst and other organisms in water, their inactivation and removal by treatment processes and survival in the natural environment; contributor to professional journals. *Society Aff:* ASCE, NSPE, WPCF, ASEE, AAUP, AAEE, AWWA, IOA, IAWPR, SWE

Sproule, Robert S
Home: 4731 Connaught Ave, Montreal, Quebec, Canada H4B1X5
Position: President *Employer:* Sproule Hydro Inc. *Education:* S.M./Gas turbine/M.I.T.; B. Eng./Mechanical/McGill *Born:* 09/04/15 Consultant on design, manufacture and modernization of hydraulic turbines. Managed the hydraulic turbine business of Dominion Engrg, 1954-71, when such world record installations were made as the 190,000 h.p. fixed blade propeller turbines for Rocky Reach and Francis turbines for Churchill Falls and Grand Coulee *3 Power House. These Grand Coulee turbines were, by a wide margin, the world's most powerful hydraulic turbines, up to the end of 1977. Was Manager, Hydro Generation Systems Development, for G.E. Canada, 1973-80, where the activity was modernisation of hydro plant. Society Aff:* ASME, EIC

Spruiell, Joseph E
Business: Dept of Materials Science & Engineering, University of Tennessee, Knoxville, TN 37996-2200
Position: Prof. *Employer:* Unv of TN. *Education:* PhD/Met Engr/Univ of TN; MS/Met Engr/Univ of TN; BS/ChE/Univ of TN. *Born:* 10/13/35. Taught met & polymer sci at Univ of TN since 1963. Maj interest are in structure-property-processing interrelationships for mtls. Full Prof 1970. Assoc Hd, Chem, Metallurgial & Polymer Engrg 1982. Hd, Materials Sci and Engrg 1984. *Society Aff:* ASM, SPE, IPPS, SR.

Spunt, Leonard
Home: 8629 Keokuk Ave, Canoga Park, CA 91306
Position: Prof *Employer:* CA State Univ at Northridge *Education:* MS/ME/USC; BS/Struct Engr/UCLA *Born:* 02/07/40 Native of Boston, MA. Raised and educated in Los Angeles. Res Engr for North American Aviation 1964-66. With CA State Univ at Northridge, CA since 1965. Author of *Optimum Structural Design*, Prentice Hall 1971. Holder of patent 3954437 *Modular Dome Structure*. Currently Prof of Engrg.

Spunt, Leonard (Continued)
Res interests: Structural Optimization, Modular Structures, Solid Mech, Mbr AIAA. *Society Aff:* AIAA.

Squier, L Radley
Business: P.O. Box 1317, 4255 Oak Ridge Rd, Lake Oswego, OR 97035
Position: President. *Employer:* L R Squier Assocs, Inc. *Education:* PhD/Civil Engg/Univ of IL; MS/Civil Engg/Univ of IL; BS/Civil Engg/Rutgers Univ. *Born:* May 6 1933, Alton Ill. 1961-65 Sr Engr Shannon & Wilson Inc, geotechnical cons Seattle, 1967-71 Principal Engr; 1971 founded L R Squier Assocs Inc, cons geotechnical engrs, Lake Oswego Ore. Assoc Reporter Session on Earth & Rock Fill Dams, 7th Internatl Conference on Soil Mechanics & Foundation Engrg 1969. Served to Ltjg USNR 1955-59. Mbr US Comm on Large Dams, ASCE, ASTM, Internatl Soc Soil Mechanics & Foundation Engrg, Sigma Xi, Phi Kappa Phi, Pres Geotechnical Engrg Div, Ore Metropolitan Sect ASCE, 1975. Dir, Consulting Engineers Council of OR (1979). Secretary (1980) Vice President (1981), Pres (1982) 1981 founded Geo-Tech, Explorations, Inc, Portland, Oregon. Contribute articles to prof journals. *Society Aff:* ASCE, CECO, ASTM, ACEC, USCOLD, ΣΞ, ASFE.

Squillace, Stephen S
Business: 455 W Fort St, Detroit, MI 48226
Position: Associate, Director Electrical Engrg. *Employer:* Smith, Hinchman & Grylls Assocs, Inc. *Born:* April 30 1925. BSEE Univ Mich 1947. 1947-50 Jr Engr, City of Detroit; 1950-52 Sr Engr for H E Beyster Corp; 1952-57 Ch Electrical Engr, Victor Gruen & Assocs; 1957-73 Ch Electrical Engr, Hyde & Bobbio Inc; currently Corporate Dir of Electrical Engrg & part time Professor, Univ Detroit. Fellow IES; Engrg Soc Detroit; US Natl Ctte of CIE. Pub: 'Visual Comfort Probability I, II & III' in Detroit Edison's publ Professional Speaking for Architects & Engrs; 'Comprehensive Predetermination of Contract Losses' presented at Annual Tech Conference IES, Chicago; 'Analysis of an Automotive Drafting Task for Visual Performance' to be presented at Annual Natl Conference IES, Cleveland 1976.

Squire, Alexander
Business: 3000 George Washington Way, Richland, WA 99352
Position: Deputy Managing Dir *Employer:* WA Public Power Supply System *Education:* SB/Electrochem Engrg/MIT *Born:* 9/29/17 Dornock, Dumfrieshire, Scotland; emigrated to US 1928; Boston Latin School 1935; MIT 1939; Executive Program, Columbia Univ 1958. Research Metallurgist Handy and Harman, Fairfield CT, 1939-1941; Production Metallurgist Sullivan Machinery Co, Michigan City Ind 1941-1942; Chief, Powder Metallurgy Section, Watertown Arsenal Lab 1942-1945. Mgr Metallurgical Development, Westinghouse Elec 1946-1950; Mgr Materials Engr, Westinghouse Bettis Lab 1950-1954. Project Mgr Bettis Submarine Fleet Reactor 1954-1958; Mgr Materials Dept; Mgr Destroyer Project. Assistant Division Mgr, then Division Mgr Westinghouse Plant Apparatus Division 1962-1969. Dir Purchases and Traffic Westinghouse Elec Corp 1969-1971. Pres Westinghouse Hanford Co 1971-1979. Deputy Managing Dir WA Public Power Supply System 1980 to present. *Society Aff:* ANS, AIME, ASM, ADPA, ASA

Squire, William
Business: Dept of Mech & Aerospace Engg, Morgantown, WV 26506
Position: Prof. *Employer:* WV Univ. *Education:* MA/Mathematics/Univ of Buffalo; BS/Chemistry/CCNY. *Born:* 9/22/20. Native of NY, NY. Held positions at Natl Bureau of Stds, Cornell Aeronautical Lab, Bell Aircraft and Southwest Res Inst. At WV Univ since 1961, technical interests Fluid Mechanics (Turbulent flow, combustion shock waves), Applied Mathematics and Numerical Methods. Author of "Integration for Engrs and Scientists" and "Hits and Misses–. *Society Aff:* AIAA, APS, SIAM, ACM.

Squires, Arthur M
Home: 2710 Quincy Ct, Blacksburg, VA 24060
Position: Univ Distinguished Prof (Emeritus) (VPI & SU) *Employer:* Self; writer & consultant *Education:* PhD/Physical Chem/Cornell Univ; AB/Chemistry/Univ of MO. *Born:* 3/21/16. Native of Neodesha, KS. Manhattan Proj, 1942-46. Hydrocarbon Res, Inc, 1946-59. Consultant, 1959-67. The CCNY, 1967-76. Frank C Vilbrandt Prof of ChE, VA Tech, 1976-82; Univ Distinguished Prof, 1978-86. Univ Distinguished Prof (Emeritus), 1986- . Res in iron ore reduction; coal gasification, combustion, flash hydrogenation, & liquefaction; "fast" and "turbulent" fluidization; fluidization of sticky particles; ash agglomeration in fluidized-bed coal gasification; hot gas cleaning; panel bed contacting; vibrated beds. Henry H Storch Award, ACS, 1973. Inst Lectr, AIChE, 1977. Mbr, Natl Acad of Engg. Fellow, Am Acad of Arts & Sciences. Active as profl musician, NY Pro Musica, 1952-60. *Society Aff:* AAAS, AIChE, AIME, ASME, ACS.

Staats, Gustav W
Home: 6124 N Lake Drive, Whitefish Bay, WI 53217
Position: Emeritus Prof. *Employer:* Retired. *Education:* PhD/EE/IIT; MS/EE/IIT; BS/EE/IIT. *Born:* 11/30/19. Forest Park, IL. Engr (1942-1956) & engg mgr (1957-1964) at Allis-Chalmers Corp, responsible for design & dev of large conductor cooled steam turbine generators. With the Univ of WI-Milwaukee since 1965, currently emeritus prof of elec engg. Eta Kappa Nu recognition of young elec engrs (1953), E W Seeger Meml Award of Milwaukee IEEE (1975). Mbr Eta Kappa Nu, Tau Beta Pi, Sigma Xi. Life Fellow, IEEE. Seven patents. *Society Aff:* IEEE.

Stables, Benjamin J, Jr
Business: 1500 Meadow Lake Pkwy, P.O. Box 8405, Kansas City, MO 64114
Position: Partner *Employer:* Black & Veatch Consltg Engrs *Education:* BS/EE/Univ of MN *Born:* 07/20/24 Born in Philadelphia, PA. Served in US Navy 1943-46. With Black & Veatch, Consltg Engrs continuously since 1951. Design, field, and project engrg on elec utility projs through 1965; Electrical Engrg Dept Hd 1966-76; Control Engrg Dept Hd 1977-80. Assumed current position as Partner-in-Charge of transmission, substation, distribution, and energy mgmt sys projs in 1981. Held elective offices through Chrmn in Kansas City Sec IEEE 1966-70. Mbr of several ctte and working groups responsible for developing and reviewing engrg standards in IEEE Power Engrg Soc 1966-80. Enjoy opera and golf. *Society Aff:* CIGRE, IEEE, NSPE

Stacey, Weston M, Jr
Business: School Nuclear Engg, Atlanta, GA 30332
Position: Callaway Prof Nuc Engr. *Employer:* GA Inst of Technology. *Education:* PhD/Nuclear Engg/MIT; MS/Nuclear Sci/GA Tech; BS/Physics/GA Tech. *Born:* 7/23/37. Nuclear Engr & Mgr at Knolls Atomic Power Lab 1962-64 & 1966-69; Head of reactor theory & code dev 1969-73; Dir Controlled Thermonuclear Fusion Res Prog 1973-77, Argonne Natl Lab. Callaway Prof Nuc Engr of GA Tech 1977- ; Snr US participant Int Tokamak Reactor Workshop (IAEA) 1979-81. Pub more than 100 papers & 4 books on nuclear reactor theory & fusion reactor physics & tech. Fellow, Bd/Dir 1974-77, Chmn Math & Comp Div 1973-74 ANS; Mbr APS, AAAS Tech Publ award of Reactor Physics Div ANS 1971-72. Outstanding Achievement in Fusion Award, ANS, 1981. *Society Aff:* ANS, APS.

Stach, Joseph
Home: 418 Hubler Rd, State College, PA 16801
Position: Consultant. *Employer:* Self. *Education:* PhD/EE/Penn State; MS/EE/Penn State; BS/EE/NJIT. *Born:* 8/21/38. Dev of hydrogen injection process for defect free boron junctions used in PMOS, CMOS, TTI, I2L, ECL, linear IC's and high efficiency PplusN solar cells. Developed the model for chlorine inc in oxides used for MOS & bipolar processes which explains, based on the reversal of oxygen & chlorine activities at the oxide-silicon interface, why chlorine resides at this interface. Mbr, IEEE, ECS & Advisory Editor to Solid State Tech. Worked at Bell Tel Labs, 1966-1967, where he co-invented the low threshold duel dielec A12O3 - SiO2 gate structure for MOS devices. At PA State he founded & directs the Solid State Device Lab as well as consults on the area of IC process tech. Joined MA Tech Park Corp in Feb 1984 as Exec Dir. The first project of MTPC is the MA Microelectronic Center a 48 million dollar joint venture with state government, public and private

Stach, Joseph (Continued)
universities as well as industry. Raised in excess of $21 million for equipment from industrial concerns. Retired to private consultanting on Feb 1987. Mbr Bd of Directors of SEMI. Chairman of Board, Diamond Materials. Institute, Inc. State College, PA. *Society Aff:* IEEE, ECS.

Stacha, John H
Business: Dallas City Hall, Suite 4A North, Dallas, TX 75277
Position: Assistant Director. *Employer:* Dallas Water Utilities. *Education:* BS/Geol Engg/TX A&M Univ. *Born:* Feb 16, 1934. Native of Clifton Texas. Mgmt Relations Univ of Michigan; Urban Mgmt MIT. Reg P E in Texas. Geologist Atlantic Richfield 1957; began with Dallas Water Utilities Dept 1958 as Jr Engr, then Construction Engr, Deputy Dir of Construction & Maintenance, now Asst Dir of Opers. Mbr Amer Water Works Assn 1967- , AWWA Intl VP 1980-81, AWWA Intl Pres Elec 1981-82, AWWA Intl Pres 1982- 83, AWWA Genl Policy Council 1979- , AWWA Standards Council 1973-81, AWWA C-700 Ctte on Meters 1970-81, Dallas Plumbing Appeals & Adv Bd 1973-81, Texas Sect AWWA Bd Chmn 1973-74, Bd Mbr 1971-75, Texas Water Utilities Assn Bd Mbr 1974-75, Texas Water Conservation Assn Bd Mbr 1974-76, City of Dallas Employees' Retirement Fund Bd Chmn 1975, Bd Mbr 1972-74, City Employees' Credit Union Bd V Chmn 1973-74. *Society Aff:* AWWA, WPCF.

Stack, John R
Business: PO Box 8405, Kansas City, MO 64114
Position: Partner *Employer:* Black & Veatch Consltg Engrs *Education:* BS/Arch Engrg/KS State Univ *Born:* 05/15/33 Has been continuously associated with Black & Veatch since 1955. He has served in various capacities on numerous elec power generation projs, with current responsibilities as Partner-in-Charge of major projs. Mr. Stack has served as Field Engrg Proj Mgr on several power related construction projs. He is reg PE in KS, MO, OK, OH, CO, TX, MI, MN, AZ, NM, NV and WI. He served as pilot in the USAF. *Society Aff:* ASCE, NSPE, NSAE

Stack, Vernon T
Home: 2625 Hedrick Rd, Harleysville, PA 19438
Position: Mgr Process Engineering *Employer:* Woodward-Clyde Consultants *Education:* BChE/-/NC State Univ *Born:* 9/20/24. Entered environmental engg field with NC State Bd of Health in 1949. With Union Carbide 1951-1959 as Group Leader of environmental projs. In Dev Dept. With Roy F Weston, Inc 1959-73 as VP of R&D & Principal Consultant. Pres of Weston & Stack, Inc (an instrument co) 1963-73. Obtained five instrumentation patents. Returned to Union Carbide, Environmental Systems Dept, 1973-75, as Mgr of Municipal Projs. With Betz-Converse-Murdoch, Inc from 1975-1981 as VP of Concept Design & Operations Dept. Pres Smokey Stack, Inc (consulting service to consultants) 1982-1984. Woodward-Clyde Consultants 1984-86 as Mgr, Process Design. Vortech Treatment Systems 1986-87 as Chief Applications Engineer. In 1987 reestablished Vernon Smokey Stack, Environmental Consultant. *Society Aff:* AIChE, ASTM, WPCF, AAEE

Stacy, T Don
Business: 200 E. Randolph, Chicago, IL 60601
Position: Mgr, Production *Employer:* Amoco Production Co (Intl) *Education:* PhD/Engrg/MS State Univ; MS/Petrol Engr/LA Tech Univ; BS/Petrol Eng/LA Tech Univ *Born:* 01/13/34 Native of Houston, TX. Taught petroleum engrg at MS State Univ. Served in the USAF in 1958-61. Reached rank of Capt USAF. Served as staff evaluation engr and property acquisitions for Amoco 1970. Chief engr, Amoco Canada, and served four yrs as mgr of res for Amoco Production Co. Treas of SPE of AIME 1977-80. Pres elect SPE-1981. *Society Aff:* SPE, API

Stadelmaier, Hans H
Home: 906 Brooks Ave, Raleigh, NC 27607
Position: Prof. *Employer:* NC State Univ. *Education:* Diplom-Physiker/Engg Phys/Univ of Stuttgart; ScD/Engg Phys/Univ of Stuttgart. *Born:* 11/14/22. Native of Stuttgart, Germany, naturalized US citizen. With NC State Univ since 1952, Prof of Materials Engg. since 1959. Grad instruction and systematic res on alloys containing boron, carbon, nitrogen, permanent magnet materials, electronic mtls. Fellow, ASM. *Society Aff:* TMS-AIME, ASM, DGM.

Staebler, Lloyd A
Home: 2543 E 32nd St, Erie, PA 16510
Position: Consulitng Engr (Self-employed) *Education:* BS/EE/Univ of MI. *Born:* March 1909. Native of Ann Arbor Michigan. BSE Univ of Michigan; Grad work Temple Univ. 8 yrs refrigeration dev Kelvinator prior to WW II. Defense effort at Ford Motor Co. 1944-49 appliance (domestic) dev engrg with AVCO & Universal Cooler; Philco appliance div Ford Motor Co 1949-73, Dept Head for more than 20 yrs responsible for R&D activities related to household appliances (particularly refrigeration and air conditioning). 11 patents, 17 publs. ASHRAE Fellow, recipient of Disting Serv Award 1967 & Wolverine Award for outstanding publ 1948. Charter mbr AIMC 1958. Personal interests: conservation & environ concerns church activities, music & ham radio. *Society Aff:* AIMC.

Staehle, Roger W
Business: 107 Lind Hall, 207 Church St SE, Minneapolis, MN 55455
Position: Dean, Inst of Tech *Employer:* Univ of MN *Education:* PhD/Met Eng/OH St Univ; MS/Met Eng/OH St Univ; BS/Met Eng/OH St Univ *Born:* 2/4/34 Roger W. Staehle is dean of the Inst of Tech where he started as of Feb 1, 1979. Formerly he was prof metallurgical engrg at OH St Univ and dir of the Fontana Corrosion Ctr. He is a mbr of the Natl Acad of Engrg, a Fellow of the Amer Soc for Metals, formerly the Intl Nickel Prof of Corrosion Sci and Tech. In addition, he has been chrmn of major cooperative activities between the US and Japan and between the US and the Soviet Union. He presently serves on the Bd of Dirs of Data Card Corp, Donaldson Company, Inc, MN Cooperation Office, and Midwest Research Inst. *Society Aff:* ASM, NACE, AIME, ANS

Staff, Charles E
Business: Box 759, Upper Montclair, NJ 07043
Position: Pres *Employer:* Staff Industries, Inc. *Education:* Dokton Ingineer/Chem/Tech Hochschule-Germany; BS/ChE/Univ of MI *Born:* 12/28/08 Born in Detroit, MI. Employed in Res & Dev by Carbide and Carbon Chemicals, S. Charleston, WV 1935-44, transferred to Bakelite Div 1944-52 in Bloomfield, NJ, transferred to Union Carbide Plastics in NY 1952-62 at which time started Staff Industries as fabricators of large linings for preventing seepage from lagoons, reservoirs, etc. *Society Aff:* ACS, ASAE

Stafford, Jack D
Home: Rt. 3, Box 34A, Princeton, WV 24740
Position: V Pres *Employer:* Gates Engrg Co *Education:* BS/Math & Physics/Concord Coll; Chem Engrg/NC St U *Born:* 4/19/36 Native of Princeton, WV. Worked as Development Engrg for 3 yrs for Celanese Corp. Received 3 patents. 7 yrs Chief of Quality Assurance, Rockwell Intl Corp (North Amer Aviation). Served 5 yrs as Assistant St Highway Commissioner for St of WV. Street in Princeton, WV named in my Honor "Stafford Drive." 8 yrs V Pres- Civil, Gates Engrg Co. P Chrmn Proffessional Engrs in Private Practive, WV Chapter, P Pres Princeton Jaycees, P Pres Concord Coll Fnd, P Chrmn Concord Coll Bd of Advisors; Mason, Shriner, Methodist. Married to former Carol Blelins, Three Children. *Society Aff:* NSPE, WPCF, AWWA

Stafford, James P, Jr
Home: 326 McAuley Drive, Vicksburg, MS 39180
Position: Retired *Education:* MS/Agri Engrg (Structures)/IA State; BS/Agri Engrg (Structures)/MS State. *Born:* Oct 1918. Native of Vicksburg Miss. Reg P E in Miss. Served with Army Corps of Engrs 1941-45; U S Army Reserve thru 1973; Graduate C&GS Assoc Crse 1963; Extension Course Ind Coll Armed Forces 1972; Meritorious Serv Medal 1973. Engr for construction of DeGray Dam & Power Plant 1963-

Stafford, James P, Jr (Continued)
1973. Chief Construction Division Vicksburg Dist C.E. 1973-1986 supervising placement of over 600 million dollars of flood control and navigation works in Miss, Arkansas & La. prior to retirement May 1986. SAME Goethals Medal 1976. Mbr: ASCE, SAME, USCOLD, ROA. *Society Aff:* ASCE, SAME, USCOLD, ROA

Stafford, Joseph D
Business: Occidental Eng. Co, 2100 SE Main St, Irvine, CA 92714
Position: Vice President & Chief Engineer. *Employer:* Occidental Internatl Engineering Co. *Born:* Feb 3 1924. BSChE U/Michigan 1949; Oklahoma St U Grad Sch 1951. 195161 Fertilizer Mfg Phillips Petrol Corp; 1962-72 Mfg Supt & Mgr R&D Farmers Chem Assn; Occidental Petrol Corp & it's subs Garrett R&D 1972- : Plant Mgr, Ch Engr, VP & Ch Engr. Mbr AIChE 1950- ,& 4-d Ctte. Safety in Ammonia Plants & Related Facilities. Vice President-Manager of Engineering Occidental Engineering Co Irvine CA.

Stafford, Stephen W
Business: Dept of Metallurgical Engr, El Paso, TX 79968
Position: Assoc Prof; Assistant Dean College of Engr. *Employer:* Univ of TX. *Education:* PhD/Mech Engr & Mat Sc/Rice Univ; BS/Met Engr/UTEP. *Born:* 7/21/48. Assoc process metallurgist with Armco Steel 1970-1972. Postdoctoral Fellow Rice Univ 1975, Visiting Asst Prof Mech Engr - Univ of Houston. Robert A Welch Fdn Fellow, Rice Fellow, ASARCO Scholarship recipient. *Society Aff:* AIME, ASM, ASNT, ΤΒΠ, ΑΣΝ, ΣΞ.

Staggs, Harlon V
Home: 429 Bristol Rd, Lexington, KY 40502
Position: Retired *Education:* BS/EE/Univ of KY *Born:* 9/19/16 Resident of Lexington, KY. Served 4 1/2 yrs in US Navy in 1941. Retired as Commander USNR. Worked four yrs with consulting engrg firm before Co-founding own firm in 1952. Retired. Registered as Mechanical and Elec Engr in KY. Now serving on Advisory Council of Univ of KY Coll of Engrg. P Pres of Consulting Engrs Council of KY. *Society Aff:* NSPE, ACEC

Stahl, Frank L
Home: 209-11 28th Road, Bayside, NY 11360
Position: Chief Engr, Transportation Div *Employer:* Ammann & Whitney. *Education:* BS/Struct Eng/Zurich (Switzerland). *Born:* 1/29/20. Ammann & Whitney Cons Engrs 1946- ; served as Proj Engr for Design & Const on Throgs Neck & Verrazano-Narrows Bridges. Proj Mgr on improvements to Golden Gate, Williamsburg, and Royal Gorge Bridges. Lecturer and author of papers on suspension bridge design, const, performance & quality control. Fellow: ASCE, ASTM (Former V Chrmn Ctte A1 and Chmn Subctte on Steel Reinforcement), Mbr: Amer Inst of Steel Const, Res Council on Structural Connections of the Engrg Foundation, Intl Assn for Bridge & Structural Engrg, Intl Bridge, Tunnel & Turnpike Assoc. ASCE Thomas Fitch Rowland Prize 1967, ASCE Innovation in Civil Engrg Award of Merit 1983, ASTM Award of Merit 1982. 1986 Gold Award, The James F. Lincoln Arc Welding Foundation; 1987 ASCE Metropolitan Civil Engineer of the Year. *Society Aff:* ASCE, ASTM, AISC, IABSE, IBTTA, RCSC.

Stahlheber, William H
Home: Wyndham Dr, York, PA 17403
Position: Pres. *Employer:* Baeuerle & Morris, Inc. *Education:* BSME/Mech Engg/Washington Univ. *Born:* 7/17/30. Belleville, IL. s Herbert Adolph and Josephine Ella (Knewitz)S; BS in Mech Engg, Washington Univ, St Louis, 1952; married Shirley Ann Boefer, Sept 8, 1951; children - Sharon Ann, Linda Christine. Dev eng Union Carbide Corp, Oak Ridge, 1952-55; mgr marketing Trane Co La Crosse, WI, 1955-69 exec VP Alco Control Co St Louis, 1969-71; VP York Div Borg Warner Corp, York, PA 1971-76; Chmn. Baeuerle & Morris Inc, King of Prussia, PA, 1976- ; President Stahl, Inc, York, PA 1986- ; Director of Penn Lane Industries, Inc. Director of Suburban Enterprises, Inc. Mbr ASME, Am Inst Chem Engrs, Am Soc Heating, Refrigeration and Air Conditioning Engrs. Presbyn Club; Country of York. *Society Aff:* ASME, AIChE, ASHRAE.

Stair, William K
Business: Perkins Hall 101 UTK Campus, Knoxville, TN 37916
Position: Assoc Dean & Dir. Engr. Exp. Sta. Retired. *Employer:* Univ of TN. *Education:* MS/ME/Univ of TN; BS/ME/Univ of TN. *Born:* 10/1/20. Wm Kenneth Stair, a native of Clinton, TN, is Assoc Dean for Res, College of Engg, the Univ of TN, Knoxville, TN. He received BS & MS degree in Mech Engg from the Univ of TN & joined the Univ faculty as Prof of Mech Engg until becoming Asst Dean in 1970 & Assoc Dean in 1972. He has an active res interest in fluid sealing & lubrication & has directed a number of res projs at UTK. He is a mbr of Tau Beta Pi, Sigma Xi, Pi Tau Sigma, Phi Kappa Phi. Retired. *Society Aff:* ASME, ASLE.

Staker, Rodd D
Home: 646 W 70th St, Kansas City, MO 64113
Position: Mgr of Admin Power Div *Employer:* Burns & McDonnell Eng Co *Education:* MBA/Business Admin/Univ of MO; BS/CE/Univ of KS *Born:* 6/27/43 Served to lieutenant US Navy 1966-69. Joined Burns & McDonnell Eng Co 1969 as assistant structural engr in Environmental-Civil Div; resident const engr 1970-72; project structural engr, Power Div 1973-77; assistant project mgr, power plant projects 1977-80; Registered Professional Engineer; Mgr of Admin, Power Div 1980-present. MSPE Western chapter Young Engr of the Month 1977. Beta Gamma Sigma (honorary business society). Sigma Phi Epsilon, outstanding alumni award 1981, Editor alumni newsletter 1974-79. Director and pres, Burns & McDonnell Credit Union 1975-81. Who's Who in Finance and Industry. Who's Who in the Midwest. Reg PE. *Society Aff:* NSPE, ASCE

Stallings, William W
Business: 7777 Leesburg Pike, Falls Church, VA 22043
Position: Dept Mgr. *Employer:* CTEC, Inc. *Education:* PhD/Comp Sci/MIT; MS/EE/MIT; BS/EE/Univ of Notre Dame. *Born:* 7/27/45. Currently, mgr of systems design and analysis for CTEC, Inc, a computer systems design and dev firm. Formerly VP of CSM Corp, a consulting firm for the health care field. Also, formerly sr analyst for the Ctr for Naval Analyses and principal engr for Honeywell Info Systems. Author of numerous profl papers. *Society Aff:* IEEE, ACM, SGSR.

Stallmeyer, James E
Business: 208 N Romine St, Urbana, IL 61801
Position: Prof of Civil Engrg *Employer:* Univ of IL *Education:* PhD/CE/Univ of IL; MS/CE/Univ of IL; BS/CE/Univ of Il *Born:* 8/11/26 Covington, KY. Married Mary K Davenport, on April 11, 1953. Six children Cynthia M, James D, Michael J, Catherine A, John C, Gregory E. Univ of Il Dept of Civil Engrg, Asst Prof 1952-57, Assoc Prof 1957-60, Prof 1960-pres. Recipient Adams Memorial Mbr Award of Amer Welding Soc, 1964. Consultant to industry and government agencies on failures, special design problems, fatigue and welded structures. *Society Aff:* ASCE, AREA, AWS, ACI, ASTM, ASM, ASME, SEM

Stamm, Robert C
Business: P O Box 3, Houston, TX 77001
Position: Vice President. *Employer:* Brown & Root. *Education:* BS/Mech engg/Columbia Univ Sch of Engg. *Born:* 7/2/25. in Philadelphia, PA. Reg PE. Tau Beta Pi. Lt USAAF 1944-46. Proj Engr Westvaco Corp, NY City, 1950-56, Mgr Design and Const 1956-59, Engrg Group Mgr 1959-69, Chief Engr 1969-75, Corp Mgr/Energy and Property Conservation 1975-76; VP Brown & Root Inc, Houston 1976- . TAPPI Fellow; Num association posts past 30 years, including TAPPI Dir, Exec Ctte Mbr & Chmn Technical Operations Council, Finance Ctte Mbr, Chmn Engg Div, Chmn Corrosion Ctte and member NCASI & API Cttes. Mbr AAAS, ACI, ASME, CPPA, ROA. Pres & Dir Cooperative Aptment Corp. *Society Aff:* TAPPI, AAAS, ACI, ASME, CPPA, ROA.

Stamper, Eugene
Business: 73 Cranford Pl, Teaneck, NJ 07666
Position: Prof of Mech Engg. *Employer:* New Jersey Institute of Technology. *Education:* BME/Mech Engg/CCNY; MME/Mech Engg/NYU. *Born:* March 1928. Native of New York, NY. Mbr Mech Engrg Dept NJ Inst Technology 1952- ; assumed current position 1972. Mbr: ASME, ASHRAE, ASEE, Tau Beta Pi, Pi Tau Sigma. Chair ASHRAE Tech Ctte on Bldg Energy Calculations. Editor 3rd edition handbook 'Heating & Air Conditioning' (Indus Press). Author 'Mech Engrg for Prof Engrs Exam' (Hayden Book Co). DOE delegate to Internatl Energy Agency, working party on Bldg Energy Conservation. Fellow ASHRAE. *Society Aff:* ASHRAE, ASME, ASEE, ΣΞ.

Stampfl, Rudolf A
Home: 210 Clover Lane, Ambler, PA 19002
Position: Dir Sys Dept. *Employer:* Naval Air Development Center. *Education:* PPhD/EE & Communications/Inst of Tech, Vienna; MS/EE/Inst of Tech, Vienna; BS/Engr/Inst of Tech, Vienna. *Born:* Jan 1926 Vienna Austria. Came to US 1953. Electronic Res Lab Army Dept Ft Monmouth NJ 1953-59; NASA Goddard Space Flight Center Greenbelt Md 1959-73; Visiting Lecturer UCLA 1964-65; Naval Air Dev Ctr, Dir AETD 1973- . Communications Navig 1977-Sys 1979. Recipient Proj SCORE Superior Performance Award (Army Dept), TIROS Group Achievement Award (NASA), Harry Diamond Award 1967. Assoc Fellow AIAA. IEEE Fellow; held offices of AES Soc, Chmn Wash Chap. Genl Chmn EASCON 1970. Membership Dev Ctte AES Bd of Governors. Pres, AES 79 & 80. Contributed articles to prof journals & chaps to books. Patentee in field. Main achievement: design leader of NIMBUS spacecraft. *Society Aff:* IEEE, AIAA.

Stancell, Arnold F
Home: 15 Woodside Dr, Greenwich, CT 06830
Position: VP, US Producing-Oil/Nat'l Gas *Employer:* Mobil Corp. *Education:* PhD/Chem Eng/MIT; BChE/Chem Eng/CUNY *Born:* PhD MIT 1962; BChE CUNY 1958. Reg Engr NY. Visiting Prof Chem Engrg MIT 1970-71. Mobil 1962- ; moved through various jobs in res mgt in petrochemicals and plastics (1962- 70) to chemical business planning (72-76) to VP & Gen Mgr of plastics packaging film business (77-79) to Mgr Planning of Mobil Corp (80-81); then VP (82-84) of Mobil Europe Mktg & Refining business to VP planning of worldwide mktg & refining (84-86), and now VP for US Producing (oil and natural gas) business. Mbr Visiting Ctte MIT. Mbr Advisory Ctte Univ. of Delaware. Author of *Diffusion Through* Polymers in Mark & Tobolsky's Polymer Sci & Materials. Numerous articles and patents. *Society Aff:* ΤΒΠ, ΦΛΥ, ΣΞ.

Stander, Richard R
Business: Rr 153 Orange St, PO Box 1321, Mansfield, OH 44901
Position: Chmn of the Bd *Employer:* Mansfield Asphalt Paving Co. *Education:* BCE/CE/OH State Univ. *Born:* 1/3/19. in Mansfield, OH. Inspector, OH Dept of Hgwys Testing Lab, 1940; US Army Corp of Engrs, 1941-1946, (Includes 3 1/2 yrs with Engr Bd on bridge & vehicle res & dev); Mansfield Asphalt Paving Co, 1946 to date, Pres since 1959, Now Chmn. Following offices held related to engg: Chrmn, The Road Info Prog, 1984-86, Exec Ctte, Natl Construction Industry Council, Chrmn, Am Road & Transportation Builders Assn, 1978. Pres, Natl Asphalt Pavement Assn, 1963-1965. Pres, OH Contractors Assn, 1962. Dir, Assn of Asphalt Paving Technologists, 1969-1970. Chrmn, Transportation Res Bd, Committee, A2F06. Exec Committee, OH State Univ Engrs Committee of 100, 1968 to date. Mbr, Bd of Consultants, Eno Fdn for Transp 1978- 80. Mbr, Transp Res Bd of OH, 1973 to date. Charter Mbr & first Pres, Richland County Regional Planning Commission Awards received: OH State Univ Centennial Alumnus Award, 1970. OH State Univ College of Engg. Distinguished Alumnus Award, 1970. Natl Asphalt Pavement Assoc Industry Recognition Award, 1979. Oh Contractors Assoc, Hall of Fame, 1979. Life Mbr, ASCE. CIT Rebuilding Amer Award, 1986. *Society Aff:* ASCE, NSPE, SAME, AAPT.

Standley, David
Business: Room 2000/100 Cambridge, Boston, MA 02111
Position: Commissr Dept Environ Quality Engrg. *Employer:* Commonwealth of Massachusetts. *Born:* Sept 1929. Native of Beverly Mass. SM Harvard Univ 1957; BS Tufts Coll 1951. Naval Officer (destroyer) 1951-54. Active in public health & environ engrg field in Mass 1955- ; spec in air quality control, radiological health & noise control. Also involved in energy & transportation issues. Employed by State & City of Boston & private cons firms. Taught at Northeastern, Tufts & Harvard. P Mbr ALA Natl Air Conservation Comm, FEA Environ Adv Ctte. Genl Prog Chmn 1975 APCA Annual Meeting. APCA 1st V P 1975-76, Pres 1976-77.

Standley, Richard A, Jr
Home: 788 Brighton Avenue, Portland, ME 04102-1016
Position: Chief Engineer. *Employer:* Self *Education:* /Electrical/Univ of ME *Born:* May 1926 Beverly Mass. Educated Portland Me Deering High 1944. USAAF 1944-45 UMO - Feb 46 - Feb 49. ICS 1951-54 Civil Engrg. Reg P E 1958 Land Surveyor 1968. 1949-66 Asst Engr Maine State Hwy Comm, resident on hwy const & paving projs Spec in bituminous paving 1960-66; Materials Engr on Interstate Hwy Const in No Maine 1966 for L M Pike & Sons of Laconia NH; returned to MSHC 1967 as Asst to Engr of Bituminous Materials, assumed position of Bituminous Materials Engr when MSHC became Maine DOT 1973. Retired from MDOT in 1986. Formed Private Practice "Bimenco--as Pres & Chief Engr. of firm, for general consulting, specifications and related projects. Mbr: Asphalt Paving Technologists, NSPE, Maine Assn of Engrs. Hobbies: photography, S.C.C.A. rallying, gardening. *Society Aff:* NSPE, AAPT, MAE, AAA, AARP.

Stanek, Eldon K
Business: Houghton, MI 49931
Position: Dept Head *Employer:* Michigan Technological Univ *Education:* PhD/EE/IL Inst of Tech; MS/EE/IL Inst of Tech; BS/EE/IL Inst of Tech *Born:* 12/12/41 Has taught at 3 major universities: IL Inst of Tech, WV Univ, MI Tech Univ. Currently Prof and Head of elec engrg at MI Tech. Studied under NSF and NASA fellowships. Conducted substantial research supported by agencies such as the US Bureau of Mines, NSF, Allegheny Power System, Appalachian Power Co, Westinghouse Education Fnd and the Defense Elec Power Admin. Authored over 65 journal articles and papers. Currently a mbr of the NSPE Educational Fnd Bd of Trustees, Chrmn of the IEEE/IAS Mining Industry Ctte and Chrmn of the Douglas Houghton Chapter of the MI Soc of P E's. *Society Aff:* IEEE, NSPE, ASEE

Stanislao, Joseph
Business: Coll/ Engrg & Architect, Fargo, ND 58105
Position: Dean of Engineering & Architecture. *Employer:* North Dakota State University. *Education:* EngScD/Ind Engg/Columbia Univ; MS/Ind Engg/Penn State Univ; BS/Ind Engg/TX Tech Univ. *Born:* Nov 1928 Manchester Conn. US Marine Corps 1948-51. Grad Asst Indus Engrg Dept Penn State & Asst Engr Naval Ordanance Res Lab 1958-59; Asst Prof NC State Univ 1959-61; Dir R&D Darlington Fabrics Corp 1961-62; Asst & later Assoc Prof of IE & Ocean Engrg Univ of Rhode Island 1963-71; Chmn & Prof Indus Engrg & Assoc Dean Engrg Cleveland State Univ 1971-75; Dean of Engrg & Architecture & Prof of IE ND State Univ 1975- . Mbr of num honor socs. Special assignments for Asian Productivity Org 1972-75. Pres & CEO of XOX Corp., 1985- . *Society Aff:* IIE, ASME, ASEE, ASAS, NSPE.

Staniszewski, Bogumil E
Home: Mokotowska 17 M.13, Warsaw, Poland 00-640
Position: Professor & Dir of the Institute *Employer:* Warsaw Technical Univ *Education:* Dr. Habil/Mech. Eng./Warsaw Tech. University; Ph.D./Mech. Eng./Warsaw Tech, University; M.Sc/Mech. Eng./Lodz Tech. University *Born:* 04/18/24 Native of Warsaw, Poland. Taught in Warsaw public schools. Field engineer for an internal combustion engines company 1946-49. Since 1949 with Warsaw Technical Univ. Currently Full Professor and Dir of the Institute of Heat Engrg. 1969-73 VP

Staniszewski, Bogumil E (Continued)
of the Univ. Since 1969 mbr of the Polish Academy of Sciences. Area of activity: heat transfer, thermodynamics and energy conversion. Author and co-author of several books and scientific papers. Served many times as visiting prof at foreign institutions, among them at three U.S. Univ.. Enjoy classical and modern music. *Society Aff:* ASME, PAS, WSS, PSME

Stanitz, John D
Home: 14475 E Carroll Blvd, Univ Heights, OH 44118
Position: Consulting Engr. *Employer:* Self-employed. *Education:* ScD/ME/MIT; MS/ME/MIT; BS/ME/MIT. *Born:* 5/13/20. Chief of Centrifugal & Supersonic Compressor Res Branch at NACA (now NASA) Lewis Res Ctr, 1943-1953; Asst Proj Engr for Advanced Gas Turbine Dev at AiRes Mfg Co, 1953-1955; Mgr of Mech Product Dev at TRW, Inc, 1955-1966; & self- employed consultant on the fluid-dynamic design of compressors, turbines and pumps since 1966. Author of numerous papers & reports on the fluid mechanics of turbomachinery. NACA reports on two- and three-dimensional, compressble flow in centrifugal compressors & axial-flow trubines are the basis for the classical theory of these machines in many modern textbooks. Elected Fellow of ASME in 1979. *Society Aff:* ASME, ΣΞ.

Stanley, Bill L
Business: P O Box 4261, Pasadena, TX 77502
Position: Vice President, Secretary/Treasurer. *Employer:* Ventech Engineers Inc. *Born:* Nov 1938. BSChE Univ of Texas. Worked with Diamond Shamrock & Houston Res Inst before founding Ventech Engrs 1966. Ventech uses an engineered approach to recycling process equipment. Dir of Ventech Engrs, Trans Plant Constructors, Houston Res Inc Aquasol. Has 5 children & one boat.

Stanley, Marland L
1734 Rainier, Idaho Falls, ID 83402
Position: Div Mbr/Electrical Engineering *Employer:* EG&G Idaho. *Education:* MS/Physics/Western IL Univ; BS/Physics/Western IL Univ. *Born:* 10/8/34. Native of Augusta IL. Res Physicist Borg Warner Res Center spec in thermoelectric matls res; Idaho Natl Engrg Lab 1967- , developed transducers for nuclear applications. Managed operational activity on Semiscale Prog; currently manages Electrical Engrg Div supporting INEL Water Reactor Safety Res Progs including LOFT, PBF & Semiscale. Hobbies: experimental electronics, radio control, fishing & flying. *Society Aff:* ISA.

Stanley, Richard H
Business: Stanley Bldg, Muscatine, IA 52761
Position: Chairman *Employer:* Stanley Consultants Inc. *Education:* MS/Sanitary Engg/State Univ of IA; BS/Mech Engg/IA State Univ; BS/Elec Engg/IA State Univ. *Born:* 10/20/32. BS in Mech Engrg, Iowa State Univ 1955; BS in Elec Engrg, Iowa State Univ 1955; MS in Sanitary Engrg, Univ of Iowa 1963. Married Mary Jo Kennedy Dec 20, 1953; children: Lynne Elizabeth, Sarah Catherine, and Joseph Holt. With Stanley Cons since 1955 except for 2 yrs active duty as US Army Corps of Engrs officer. Chairman and director, Stanley Consultants, Inc.; chairman and director Stanley/Mettee-McGill-Murphy, Inc; chairman and director Stanley/Wantman Inc. Reg P E. Fellow Amer Cons Engrs Council (Pres 1976-77). Chrmn, Committee on Fedl Procurement of Architectural and Engg Services 1979. Chrmn Natl Construction Industry Council, 1978-79. Mbr Construction Action Council of the US Chamber of Commerce, 1976- . Author of numerous technical and professional articles and papers. Mbr Iowa Engrg Soc (Pres 1973-74, Distinguished Serv Award, 1980.); Cons Engrs Council of Iowa (Pres 1967). Received Prof Achievement Citation for Coll of Engg, IA State Univ, 1977, Mbr Engrg Coll Adv Council, Iowa State Univ 1969- (Chrmn 1979-81); Vice Chairman and Dir HON Indus Inc; President & Dir Stanley Found; Muscatine Chamber of Commerce (Pres 1972-73). Presbyterian
e (Elder). *Society Aff:* NSPE, ACEC, ASCE, ASME, IEEE, ASEE.

Stanners, James E
Home: 200 Ridley Blvd. Apt. 106, Toronto, Ontario M5M 3M2 Canada
Position: President. *Employer:* Stanners Consultants Ltd *Education:* BASc/-/Univ of Toronto. *Born:* July 1920. Native of Toronto. Worked as a struct designer in sev well known cons struct engrg firms before joining Shore & Moffat as Ch Struct Engr 1955. Held this pos until 1962 when became Partner in Charge of Struct Engrg in Shore & Moffat & Partners. In 1971 formed the partnership of Stanners & Aylon Cons Engrs, spec in concrete. Dir 1970-73 ACI. Elected Fellow ACI 1973. In 1976 went into private practice as concrete specialist with Stanners Consultant Ltd. *Society Aff:* ACI, APEO.

Stanton, Curtis H
Business: 500 S Orange Ave, Orlando, FL 32801
Position: Exec V P & Genl Mgr. *Employer:* Orlando Utils Comm. *Education:* BME/Mech Eng/Univ of FL. *Born:* June 1918 Key West Fla. Genl Elec Co Schenectady from June 1940 to Apr 1947. Steam turbine engrg & dir of sales engrg training prog. Orlanto Utils Asst Genl Mgr Apr-Dec 1947. Exec V P & Genl Mgr Jan 1948- . Pres Orlando Chamber of Commerce 1965; Pres Fla Municipal Utils Assn 1960; Pres Central Fla Fair 1962; Chmn Fla Elec Power Coord Group 1974. Hobby: big game fishing. Pres Amer Water Works Assoc 1978-79. *Society Aff:* AWWA, ASME, IEEE.

Stanton, George B, Jr
Business: P.O. Box 490, Lodi, NJ 07644
Position: Pres. *Employer:* Amer Hazard Control Consultants Inc. *Education:* MChE/Chem Engg/Polytechnic Inst of Brooklyn; MA/Safety & Health/NYU - The Ctr For Safety; MBA/Mgmt/NYU - The Grad School of Business; BChE/Chem Engg/Polytechnic Inst of Brooklyn. *Born:* 11/3/26. Founder, Amer Hazard Control Consultants, Inc, international consultants to chemical manufacturers and users, local and State governments, and attorneys on safety, health and environmental hazards of chemicals. Practice includes explosions, fires, spills, and toxic effects - both prevention and mitigation - of materials of all kinds. Chemical Safety Engineer for Lodi, NJ, and Lyndhurst, NJ. Teach grad safety engg and industrial hygiene as Adjunct Prof, NJ Inst of Technology and at New York Univ - The Center for Safety. Former Chf, Occupational Health, State of NJ. In industry, served as Product (Business) Mgr, Paisley Products; Div Engr, PPG Industries; Plant Mgr, Technical Mgr, Intl Mgr, Rubber & Asbestos Corp; Chf Chemist, Arvey Corp. Reg PE-NJ; Certified Industrial Hygienist; Certified Safety Professional; Fellow, Royal Society of Health; Chrmn, Safety Div, ASME, 1980; Pres, NJ Section, AIHA 1976-77. Centennial Medal Award ASME, 1980; First Prize, Paper Award, ASSE, 1980. Thirteen publications. Bd of Education, youth, church leader. *Society Aff:* AIChE, ASME, ASTM, NSPE, ASSE, AAIH, AIHA, SSS, NFPA, RSH.

Stapinski, Stephen E
Home: 49 Fifth Ave, Haverhill, MA 01813
Position: Pres *Employer:* Merrimack Engrg Services *Education:* BS/Civ Eng/Merrimack Coll *Born:* 1/20/53 Instructor Civil Engrg Merrimack College Engr w/ Piantidosi Assoc; Survey Chief Stevens Land Survey Co; Survey Chief A H Lehman Co, NY; Asst District Traffic Engr, MDPW, Reg Land Surveyor, MA; Project Construction Engr, W K Rust Co; Attended Catholic Univ & Georgia Inst of Tech. *Society Aff:* ASCE, ACSM, NSPE, AAPW

Staples, Basil G
Home: 275 Colwick Rd, Rochester, NY 14624
Position: Retired. *Education:* MS/Biochemistry/Univ of ME; BS/Biochemistry/Univ of ME. *Born:* Aug 1914 Eliot Me. College apprentice to foreman Genl Chem Co 1936-42. Foreman to Supt U S Rubber Co Inst, W Va 1942-45, Research Chemist to Senior Engineer, The Pfaudler Co., Rochester, N.Y. 1945 to 1979. Mbr ACS. Fellow Amer Inst of Chemists. Mbr Natl Assn of Corrosion Engrs. Fellow Amer Ceramic Soc (V P 1974). Mbr ASTM (Chmn Comm F-3 1976). Mbr Natl Inst of Ceramic Engrs. Mbr Alpha Gamma Rho & Alpha Zeta. Patentee in field. Home: 275

Staples, Basil G (Continued)
Colwick Rd, Rochester N Y 14624. Currently retired but consulting in the fields of gaskets and corrosion. *Society Aff:* ACS, ACerS, ASTM, NICE, NACE, AIC.

Staples, Carlton W
Home: Box 731, 53 Bridle Path, Marstons Mills, MA 02648
Position: Prof Emeritus of ME. *Employer:* Worcester Poly Inst. *Education:* MSME/ME/WPI; MEd/Education/MA State; BSME/ME/Tufts. *Born:* 7/9/24. A retired full Prof of ME at WPI formerly teaching with maj emphasis on machine design & machine dynamics. Also worked extensively on student-ind proj advising. Ind consulting with many cos in the areas of machine dynamics and design. Works extensively in the design of cam & non circular gear systems and has developed several original design methods in these areas. Presently very active in product liability litigation involving expert evaluation of various designs. *Society Aff:* ΣΞ, ASME, ASEE

Staples, Elton E
Home: 1671 Legion Dr, Winter Park, FL 32709
Position: Retired. *Education:* BS/Met/MIT. *Born:* 10/25/03. in Dighton, MA. With Dorr Oliver, Inc, 1926-1931, Met process equip dev. Gulf Oil Corp until 1939, ind dept sales. Hevi Duty Elec Co (now Lindberg Div Gen Signal Corp) 1940-1968, sales engg & mgt, finally Div Pres; Pres, Koyo Lindberg Hevi Duty, Ltd, Japanese partner, 1966-68. From 1942-1946 in US Army. Dir Submaine Mine Sch, & Chief, Gen Supplies. Quartermaster Corps, ETO. Pres, Ind Heating Equip Assn 1956 & 1964. Mbr, Natl Defense Exec Reserve & Metric Study Advisory Panel, both ad hoc groups of US Dept of Commerce. *Society Aff:* ASM.

Staples, Lloyd W
Home: P.O. Box 3133, Eugene, OR 97403
Position: Consultant (Self-employed) *Education:* PhD/Mineralogy, Geol/Stanford Univ; MS/Econ Geol/Univ of MI; AB/Math Geol Metallurgical Eng/Columbia Univ *Born:* 7/08/08 Native of Jersey City, NJ Research and teaching at MI Coll of Mines (1930- 33), OR St Coll (1936-37), Univ of OR (1939-present). Chief geologist and Mining engr, Horse Heaven Mines (1937-44), Cordero Mining Co (1941-45). Consultant to Eugene Water & Electric Bd (Carmen Smith Hydro Project), St. Joe Minerals Corp; Bd of Consultants, Portland General Elec (since 1971). Sci consultant at UNESCO, Paris and Iraq (1968-69). Chmn, OR Sect AIME (1941-42). Gugghenheim Fellow, Mexico (1960-61). Citation, OR Acad Sci (1973). Honors Award, Amer Fed Min Soc (1972). Distinguished Mbr Citation AIME (1976). OR Reg Engr, OR Reg Engrg Geologist. Interested in Mexican and Canadian travel and exploration, landscaping. *Society Aff:* AIME, MSA, GSA, ΣΞ, ΦBK.

Stapleton, Herbert N
Home: 6666 E Cooper St, Tucson, AZ 85710
Position: Professor - ret. *Employer:* Univ of Arizona - Tucson. *Education:* MS/Agri Engg/KS State Univ; BS/Agri Engg/KS State Univ. *Born:* July 1907. Native of Kansas. Teaching & res Penn State; extension at UVM; rural electrification with a Vermont utility; prior to WW II. Served with Army AAA & Corps of Engrs 1942-46. Dept Hd Univ of Mass 1947-55. Agri engr with a Vermont grassland dairy 1955-63. Prof Univ of Ariz 1963-73. R&D time about equal between educ & indus from testing engines on alcohol-gasoline blends 1933- 35 to cotton production optimization through computer simulation of the cotton plant 1965-73. ASAE Engrg Concept of Yr 1974. Enjoy innovation experimentation & travel. *Society Aff:* ASAE, NSPE, ASEE, IEEE, AAAS.

Stark, Henry
Home: 7 Orchard Grove, Loudonville, NY 12211
Position: Prof *Employer:* RPI. *Education:* BEE/EE/CCNY; MS/EE/Columbia Univ; Dr Eng Sc/EE/Columbia Univ *Born:* 5/25/38. Antwerp, Belgium. Married - 2 children Lawrence J & Richard D. Author of over 50 technical papers in signal processing, pattern recognition, statistical optics, communication theory, systems theory. Author, with F B Tuteur of Yale Univ of *Modern Electrical Communications: Theory and Systems*, Prentice-Hall (1979). Editor and co-author of *Applications of Optical Fourier Transforms* Academic Press (1981). Sr res engr at Riverside Res Inst (1962-1969). Visiting Scientist at Israel Inst Tech (1969-1970). Assoc Prof at Yale (1970-1977). Visiting Prof at U Penn (fall 1983), Visiting Prof at Univ CA San Diego. Presently Prof at RPI, Consultant to GE, Air Force, Natl Acad of Sci. Fellow Optical Society of America. Receipient of grants and contracts from NSF, RADC, others. *Society Aff:* IEEE, OSA

Stark, Louis
Business: 1901 Malvern, Fullerton, CA 92634
Position: Group Sr Scientist. *Employer:* Hughes Aircraft Co. *Education:* MS/EE/MIT; BS/EE/MIT. *Born:* Apr 5, 1926. BS & MS EE MIT. Presently Group Sr Scientist at Hughes Aircraft Co, Ground Sys Group - has respon for the Group IR&D Program. Major field of individual contrib has been R&D of phased array radar antennas, in tech & mgmt pos. Elected Fellow IEEE 1971. *Society Aff:* HKN, ΣΞ.

Stark, Robert A
Home: 359 Beverly Rd, Camp Hill, PA 17011
Position: Project Engineer. *Employer:* Gannett Fleming Corddry & Carpenter. *Born:* June 1939. BSCE Michigan Tech Univ 1962. Native of Menominee Mich. With Gannett Fleming Corddry & Carpenter Inc 1962- . Served as Field Engr 1962-64, Design Engr 1964-66 & Proj Engr 1966-76. Respon for planning, genl supervision & design of water supply projs incl dev of water sources, water treatment, water storage, transmission & distrib of water. Active mbr AWWA. Reg P E Pa. Enjoy golf & gardening.

Stark, Robert E
Home: 64 Pine Way, New Providence, NJ 07974
Position: Asst. Vice President *Employer:* Gibbs & Cox Inc. *Education:* MS/NA&ME/MIT; BS/Eng/USNA. *Born:* May 1921. Engrg Duty Officer US Navy for 26 yrs, spec in ship design. Prof of Naval Arch MIT 1965-68. V Pres M Rosenblatt & Son 1970-72. With Gibbs & Cox Inc 1972- ; currently FFG-7 Proj Mgr. Mbr Soc of Naval Architects & Marine Engrs; V P 1975-77; Chmn New England Sect 1967-68. Mbr Tau Beta Pi, Sigma Xi, Amer Soc of Naval Engrs. Lic P E NY. *Society Aff:* SNAME, ASNE.

Stark, Robert M
Business: Dept of Civil Engrg, Newark, DE 19716
Position: Professor. *Employer:* Univ of Delaware. *Education:* PhD/Applied Sci Civil Engg/Univ of DE; MA/Physics/Univ of MI; BA/Math- Physics/Johns Hopkins. *Born:* Feb 6, 1930 N Y. Asst Dean Engrg & Asst Prof of Math Cleveland (Ohio) State Univ 1957-62; Visiting Assoc Prof MIT 1972-73; Instr, Asst Prof, Assoc Prof, Prof of CE & Mathematical Scis Univ of Delaware 1962- . Mbrships, offices & apptments: AAAS, ASCE, ASEE, Operations Res Soc of Amer, Natl Res Council. *Society Aff:* ORSA, TIMS, ASCE.

Starke, Edgar A, Jr
Business: Dean, Sch of Engrg and Applied Sci, Thornton Hall, Charlottesville, VA 22901
Position: Earnest Oglesby Prof and Dean *Employer:* Univ of VA *Education:* PhD/Met Eng/Univ of FL; MS/Met Eng/Univ of IL; BS/Met Eng/VA Polytech Inst *Born:* 5/10/36 Native of Richmond, VA. Served US Army 1954-56. Research Engr, Savannah River Lab 1961-62, GIT Asst Prof 1964-68, Assoc Prof 1968-72, Prof 1975, Dir of Fracture and Fatigue Res Lab 1978, Earnest Oglesby Prof of Materials Sci at Univ of VA 1982. Mmbr, Center for Advanced Studies 1982-83. Visiting Sci Oak Ridge Natl Lab, Summer 1967, Visiting Sci Max- Planck-Instit Stuttgart, Germany, 1971. Assumed current respon as Dean, Sch of Engrg and Applied Sci, at Univ of VA, in 1984. Author of numerous publications in sci journals. Pres GA Chapter AIME 1967-68, Bd of Dirs TMS-AIME 1978-79. Bd of Review, Metallurgi-

Starke, Edgar A, Jr (Continued)
cal Transactions A, Mbr of the DARPA Res Council, Academic Advisory Ctte of the Aluminum Assoc. Consultant to Lockheed Corp, Reynolds Metals, Northrop Corp. Married Donna Frazier 1961, Two Children, John A, and Karen L. *Society Aff:* AIME, ΣΞ

Starkey, Neal E
Home: 106 Bonaventure Dr, Greenville, SC 29615
Position: Ret (Consultant) *Education:* BSEE/Power/Univ of NB. *Born:* May 1917. Native of St Paul Nebr. With G E Co since 1940. Designed Control Sys, then managed var engrg components for gas turbines. Mgr of Engrg for gas turbines 1962-68. Mgr Large Gas Turbine Engrg 1969-73. Mgr & Cons of gas turbines at present. Fellow ASME. Reg P E NY & SC. Hobbies: amateur radio, water sports & music. *Society Aff:* ASME.

Starkey, Walter L
Home: 7000 Coffman Rd, Dublin, OH 43017
Position: Professor Emeritus. *Employer:* Ohio State Univ. *Education:* PhD/Mech Engg/OH State Univ; MSc/Mech Engg/OH State Univ; BME/Mech Engg/Univ of Louisville. *Born:* Oct 1920. Prof career has been in teaching, res & indus cons, esp in fields of machine design, mech failure analysis & stress analysis. 27 pubs & sev pats. Served on num high level cttes in the univ & for prof soc's. Machine Design Award 1971 ASME. Fellow 1972 ASME. Dir of Co-op Prog 1975-77. Acting Chmn, Dept of ME OSU 1976-77. Consultant to attorneys. *Society Aff:* ASME.

Starling, Kenneth E
Business: Energy Center F318, Norman, OK 73019
Position: Professor. *Employer:* Univ of Oklahoma. *Education:* PhD/Gas Tech/IL Inst of Tech; MS/Gas Engr/IL Inst of Tech; BS/Chem Engr/TX A & I; Bs/Chem/TX A & I. *Born:* Mar 9, 1935. Res, Inst of Gas Tech 1958-62. Postdoc - transport theory - Rice Univ 1963. Sr Res Engr Esso Prod Res 1963-66. Prof Univ of Okla 1966- . Visiting Prof Univ of Leuven Belgium 1972. Dir Sch of Chem Engrg & Matls Sci Univ of Okla 1974-75. Res areas: correlation of thermophys properties, simulation, flow measurement. Chmn AIChE Cryogenics Ctte 1972-75. Chmn Engrg Sciences, Oklahoma Acad Sci 1974. Chmn Intl Sch of Hydrocarbon Measurement 1981-present. *Society Aff:* AIChE, SPE, ACS, ASEE.

Starr, Charles E, Jr
Home: 75 A New England Ave, Summit, NJ 07901
Position: Cons. *Employer:* Am Assn of Engg Societies. *Education:* BSChE/Chem Engg/MIT. *Born:* July 1909. Native of Philadelphia. BS Chem E MIT Grad courses in Chem E Louisiana State Univ. Instructor Univ of Miami; Chem Engr 3 yrs du Pont; Exxon Res & Engrg 33 yrs, started as Chem Engr, Deputy V P 4 yrs, V P & Dir 6 yrs. 10 patents, 11 tech articles & 1 book. Proj Mgr during WW II for 100 octane aviation gasoline, dev of butadiene proj & special material used in aviation gasoline to increase power for glider lifts for Normandy Invasion. Mbr AIChE & ACS. Cons to AAES. Enjoys contemporary & classical music, ballet. *Society Aff:* AIChE, ACS.

Starr, Chauncey
Business: Electric Power Research Inst, 3412 Hillview Avenue, Palo Alto, CA 94304
Position: President Emeritus *Employer:* Electric Power Research Institute. *Education:* PhD/Physics/ Rennselaer Inst; EE/Electric Power/Rennselaer Inst. *Born:* April 14,1912. Dean UCLA Sch of Engrg & Applied Sci 1967 following 20 yr indus career during which he served as V P of N Amer Rockwell & Pres of its Atomics Internatl Div. Became associated with Manhattan Dist in its early days at Univ of Calif/Berkeley Radiation Lab & then at Oak Ridge. Following WW II pioneered in dev of nuclear propulsion for rockets & ramjets, miniaturizing nuclear reactors for space & developing atomic power electricity plants. Founder & P Pres ANS; Mbr Bd/Dir & Past V P Atomic Indus Forum. Mbr and P Director of AAAS. Foreign Mbr Royal Swedish Academy of Engrg Sciences. Mbr and P VPres NAE. Nom. "OFficer" in French Legion of Honor, Feb 1978; recd 1979 Walter H Zinn Award for Outstanding Contributions to Advancement of Nuclear Power given by ANS. Founders Award for application of his scientific planning and mgmt talents to establishment of industrywide energy technology R&D organization (EPRI), presented in 1980 by 7th Energy Technology Conference; Honorary Doctorate of Engr., 1980, Swiss Federal Inst of Tech (ETH). *Society Aff:* NAE, AAAS, ANS, APS, SRA.

Starr, Dean C
Home: 28 Fieldstone Drive, Whippany, NJ 07981
Position: Vic President Engineering & Res. *Employer:* Amax Specialty Metals. *Education:* PhD/Metallurgy/Univ of UT; MS/Metallurgy/Univ of UT; BS/ Metallurgy/So Dak Sch of Mines. *Born:* 4/24/21. in Tulare, SD. In navy in S Pacific WWII. Metallurgist with G M; Asst Res Prof 3 yrs Univ of CA; Joined Wilbur B Driver 1954, BBP engg & Res 1971- . Joined Amax Specialty Metals, Alloy Div 1979 as VP Eng & Res. Visiting Sr Lecturer Grad Sch Stevens Ist Tech. ASTM Sam Tour Medal 1955. ASM Fellow & 25 yr mbr; ACS Fellow. Underwriters Lab advisor. Pub over 70 papers, holds 10 patents. Recognized for work in electrical resistance alloys, thermocouples & heat resistant alloys. OO Pres Metal Sciences Club. active in community affairs - Pres Fieldstone Civil Assn, Pres Whippany Chorus & Ban Assn, Church Elder. Hobbies: golf, bowling & people. *Society Aff:* ASM, AIME, ASTM, AICE, SME.

Starr, Eugene C
Home: 7830 SW Canyon Drive, Portland, OR 97225
Position: Consulting Engineer. *Employer:* Bonneville Power Administration. *Education:* EE/Professional/OR State Univ; BS/Elec Engg/OR State Univ. *Born:* Aug 1901 Falls City Ore. High Voltage Res Engr G E; Prof Elec E OSU; US OSRD 1942-45; Ch Engr Bonneville Power Admin 1954-61; Cons Engr BPA 1939-54 & 1961- ; Misc cons servs--high voltage engrg including EPRI, BNW Labs, Electrobas Brazil SA, etc. IEEE Life Fellow, various cttes & P Dir; CIGRE US Rep SC No 14 a-c/d-c Plants; various prof socs. Disting Serv Award Oregon State Univ 1976; Engr of the Yr, Prof Engrs of Ore 1965; Habirshaw Award IEEE 1968; Lamme Medal IEEE 1980; Disting Serv Award & Gold Medal USDI 1958. BPA Administrator's Award for Distinguished Service, 1982; IEEE Centennial Medal, 1984. Author of more than 40 tech papers. *Society Aff:* IEEE, ANS, NSPE, AAAS, CIGRE.

Starr, James E
Business: Measurements Group, PO Box 27777, Raleigh, NC 27611
Position: President Measurements Group. *Employer:* Vishay Intertechnology Inc. *Born:* April 1927. US Naval Electronics & Advanced Radar Schs; Electronics Engrg RCA Inst. Previous positions: Ch Engr Control Devices Inc; Ch Electronics Engr Budd Instruments; Res Dir Allegany Instruments Co; Genl Mgr & Ch Engr Micro-Measurements prior to election as Pres 1962, appointed Pres Measurements Group Vishay Intertechnology Inc 1973. Recognized as outstanding authority in gage design, gage carrier resin sys, gage attachment methods, sensor instrumentation, transducer design & construction & strain gage test instrumentation. Has lectured widely. 14 patents relating to sensor & transducer field. Albert F Sperry Medal Award ISA 1975. Murray Lecturer SESA 1976. Sr Mbr ISA; Mbr SESA. Lazan Award, SESA, 1980. *Society Aff:* ISA, SESA.

Startzenbach, Herman R
Business: Capitol City Center, Harrisburg, PA 17105
Position: Asst Vice Pres & Dir/Genl Engrg Div. *Employer:* Gannett Fleming Corddry & Carpenter. *Born:* Nov 1926. MS Regional Planning, BS Penn State Univ. Reg P E in sev states. Mbr ULI & Amer Inst of Planners; P Pres Engrs Soc of Penn. Extensive exper in planning & engrg of new towns & coml, indus, institutional & military complexes. As Dir of Genl Engrg Div is responsible for design of projs of a special or unusual nature, airports & bldg servs including design of mech/electrical/ HVAC sys, Energy Conservation & Mgmt.

Staszesky, Francis M
Business: 800 Boylston Street, Boston, MA 02199
Position: Pres *Employer:* Boston Edison Co *Education:* MS/ME/MIT; BS/ME/MIT *Born:* 4/16/18 Native of Wilmington, DE and grad from Wilmington HS. Prof Career: Union Oil Co of Calif, E I du Pont de Nemours & Co, and Boston Edison Company since 1948. At Boston Edison Co, Superintendent of Engrg & Construction in 1957, elected V Pres in 1964, Exec V Pres in 1967, and Pres since 1979. Dir of Boston Edison Co, Shawmut Corp and Shawmut Bank of Boston, NA, Fellow ASME, currently Chmn of The Atomic Industrial Forum, Inc, and Reg PE in MA. *Society Aff:* ASME, IEEE

Staton, Marshall
Home: 4321 Galax Drive, Raleigh, NC 27612
Position: Chief Sanitary Engineer. *Employer:* North Carolina Div of Health Servs. *Education:* MSSE/Sanitary/Univ of NC; BCE/Civil/Clemson Univ. *Born:* Sept 1921 Wadesboro NC. BCE Clemson Univ; MSSE Univ of NC. Worked with municipal water & sewerage sys as Dist Engr for State Bd of Health 195465; Asst Ch Engr responsible for planning & field opers 1965-69; assumed current position 1970 as Ch Engr in charge of administering statewide sanitary engrg progs including planning, training, supervision & surveillance of public water supplies, solid waste mgmt, vector control, food & lodging, shellfish sanitation & other significant environ health activities. Chmn NC Sects AWWA- WPCA 1973. George Warren Fuller Award 1976. Dir AWWA 1976-79. *Society Aff:* AWWA, WPCF, APHA.

Statton, Charles D
Home: 796 Buena Vista Ave, Montecito, CA 93108
Position: Dir & VP. *Employer:* Bechtel Group of COS. *Education:* BS/Mech Engg/IA State Univ. *Born:* 12/26/23. Native of Boone, IA. Served in the Air Force during WWII after which returned to IA State Univ to earn BS degree in Mech Engg. Upon grad joined WI Power & Light Co. In 1957 joined the Bechtel Group of Cos. Progressed through various levels of responsibilities and in 1972 was named VP of Bechtel Power Corp. In 1976 became a Dir of the Bechtel Group. In 1977 awarded Profl Achievement Citation in Engg from IA State Univ. A reg PE in 11 states. Enjoys golf & fishing. *Society Aff:* ASME, NSPE.

Statz, Hermann
Business: Raytheon Research Div, 28 Seyon Street, Waltham, MA 02254
Position: Tech Dir and Assit Div Mgr-Res *Employer:* Raytheon Co *Education:* Dr rer nat/Physics/Univ of Stuttgart, Germany; Diplom Physiker (MS)/Physics/Univ of Stuttgart, Germany *Born:* 1/9/28 Doctorate in Physics (1951) from the Technische Hochschule of Stuttgart. Post doctoral work at Technische Hochschule Stuttgart and later at MIT, Cambridge, MA, with Prof John C Slater in Solid State and Molecular Theory Group. Joined Raytheon in 1953 as Group Leader. Was named Asst Div Gen Mgr, Res Div in 1958, and Tech Dir and Asst. Div Gen Mgr, Research Div in 1969. Currently active in the field of GaAs microwave and digital devices, Si submicron CMOS and lasers. Fellow of IEEE and APS. Mbr, Bd of Editors, Journal of Applied Physics and Applied Physics Letters 1969-71. Past mbr and chairman of Sperry Awards Ctte. *Society Aff:* IEEE, APS

Staub, Fred W
Home: 1186 Godfrey Lane, Schenectady, NY 12309
Position: Consulting Engr, Heat Transfer *Employer:* General Electric Corp Research & Dev. *Education:* MME/Mech Eng/Rensselaer Poly Inst; BME/Mech Eng/ Rensselaer Poly Inst. *Born:* 4/5/28. Responsible for R&D in heat transfer at Corp R&D Ctr 1968- , special field 2 phase flow & heat transfer with over 40 tech publs. USAEC-EURATOM Assignment with French CEA Grenoble 1966. AIChE Exec Comm Heat Transfer Div, elected 1971- 74; Chmn ASHRAE & AIChE Heat Transfer Cttes 1969-75. Editor Heat Transfer Chap ASHRAE Handbook of Fundamentals 1972. Sigma Xi. Fellow ASME. Coolidge Fellow Gen Elect Co. Visiting Fellow Cambridge Univ 1979, ASME Melville Medal 1980. *Society Aff:* ASME, AIChE.

Staunton, Jack L
Business: 2 Park Ave, New York, NY 10016
Position: Partner. *Employer:* Staunton & Freeman Consulting Engrs. *Education:* BS/Sanitary Engg/MIT. *Born:* 12/25/12. in NY. Dipl Amer Acad of Environ Engrs, Life Mbr ASCE, Mbr AWWA & WPCF. Reg NY, NJ, PA, CO, CT, Wash, DC, RI, MN, VT & VA. Assoc Seelye, Stevenson, Value & Knecht 1947-57; initiated own cons engg practice 1958. Designed 1st continuously operating water supply sys in permafrost regions approx 1950. Contributed to Vols 1 & 2 *Data Book for Civil Engrs* by Elwyn Seelye, *Architectural Engg* & *Architectural Graphic Standards*, 7th Edition, Published by F W Dodge Corp. Resident of Westport, CT. *Society Aff:* AAEE, ASCE, AWWA, WPCF.

Staunton, John J J
Home: 310 Wesley Avenue, Oak Park, IL 60302
Position: Engineering Consultant. *Employer:* Self-Employed. *Education:* DE/-/Midwest College of Engg; EE/Thesis Degree EE/Univ of Notre Dame; MS/ Physics/Univ of Notre Dame; BS/Elec Engg/Univ of Notre Dame. *Born:* July 4, 1911 Binghamton NY. Head Physics Dept DePaul Univ 1936-38; Coleman Instruments, later div of The Perkin - Elmer Corp, 1938-78, Dir of Res 1944-66, Retired 1978. Currently Engineering Consultant in the field of Spectrophotometry & Temperature Controlled Instrumentation. 38 US patents. Chicago Tech Socs Council Award of Merit 1970. Fellow IEEE 1970. Mbr: OSA, Sigma Xi, K of C. Reg P E in Ill. Hobbies: organ design, scientific innovation. *Society Aff:* IEEE, OSA, ΣΞ.

Stavenger, Paul L
Home: 16 Broadview Road, Westport, CT 06880
Position: Director of Technology US. *Employer:* Dorr-Oliver Incorporated. *Education:* BS/Chem Eng/Univ of IL; MS/Chem Eng/Univ of MI. *Born:* March 1924. Native of La Grange Ill. US Navy 1943-45. R&D Dorr Co Westport Lab 1948-54; Mgr Centrifugal Technology Div Dorr-Oliver 1954-62, Mgr Food & Pharmaceutical Technology Div 1962-67, Mgr Membrane Technology 1967-72, Dir Separation Prod Group & Dir Technology US 1972- . Field of specialization: liquid-solid separations. Fellow & Chmn Admission Comm AIChE, Mbr Ed Review Bd CEP, Chmn Fairfield County Sect AIChE. Chmn NY Sect AIChE, Chmn Publications Comm AIChE. *Society Aff:* ACS, IFT, AADCC.

Stavis, Gus
Home: 19 Huff Road, Wayne, NJ 07470
Position: Department Manager; Retired. *Employer:* Kearfott Division/The Singer Co. *Education:* BEE/Elec Eng/CCNY; BA/Accounting/Wm Paterson College; MBA/-/Fairleigh Dickinson U. *Born:* June 5, 1921 NYC. ITT Federal Telecommunication Labs 11 yrs. USN WW II 2 yrs. Res Assoc Harvard Radio Res Lab 1 yr; Kearfott 1952-86, primarily active in air navigation electronics for entire career. Currently retired. 50 patents issued plus sev pending. NJ Res Council Award for Outstanding Invention (TACAN) 1967. Fellow IEEE. Licensed P E NJ. Tau Beta Pi, Eta Kappa Nu. *Society Aff:* IEEE, ION.

Steadman, H Douglas
Business: 7073 San Pedro, San Antonio, TX 78216
Position: Pres. *Employer:* W E Simpson Co, Inc. *Education:* MS/Civ Engg/Univ of TX; BS/Civ Engg/Univ of TX. *Born:* 10/12/26. Early life and public school education in Ft Worth, TX. USNR from 1944 to 1946 attaining the rank of Lt (jg). Joined W E Simpson Co in 1948 as engr-in- training. Became Secy and Dir in 1968 and assumed current position as Pres in 1986. Reg PE in four states. have served TX Sec ASCE as VP - Technical and Bexar Chapter TSPE as Pres. Selected as Engr of the Yr Bexar Chapter in 1978. Deacon Church of Christ, Pres Castle Hills PTA 2 yrs, Zoning Bd City of Castle Hills. Enjoy handiman projs and fishing. *Society Aff:* ASCE, ACEC, NSPE.

Steakley, Joe E
Business: 1504 Main St, Greensboro, AL 36744
Position: President. *Employer:* Southern Resources Mapping Corp. *Education:* BA/Geography/Boston Univ; MS/Geography/Boston Univ; MBA/Bus Admin/ Auburn Univ; Dipl/Military Mgmt/Air War College. *Born:* 1923 Cleburn Texas. Served 2nd Infantry Div throughout WW II. Served Defense Mapping Agency as geodisist, photogrammetrist, Ch of Topography Div & Sr Staff Exec. Graduate USAF Air War Coll. Organized, trained technicians & developed aerial mapping co to serve civil engrg & woodlands indus in Mid-South; introduced advanced techniques to do topographic mapping in heavily vegetated areas & coastal regions. Pres ASP, Pres Mid-South Region ASP. Mbr Amer Congress on Surveying & Mapping. Assoc Fellow AIAA. Affiliate ASCE & Alabama Soc of Prof Land Surveyors. *Society Aff:* ASP, ACSM, ASCE

Stear, Edwin B
Home: 9122 Home Guard Drive, Burke, VA 22015
Position: Chief Scientist USAF *Employer:* US Air Force *Education:* PhD/EE/UCLA; MS/ME/USC; BS/ME/Bradley Univ *Born:* 12/08/32 Mbr Tech Staff Hughes Aircraft Co, 1954-59. Project Officer, USAF Flight Control Lab, 1961-63. Mgr, Control and Communication Research Lab, Lear Siegler, 1963-64. Asst/Assoc Prof of Engrg, UCLA, 1964-69; Prof, Dept of Elec Engrg and Computer Sci, Univ of CA, Santa Barbara, 1969-79 and Dept Chmn, 1975-79. Principal tech advisor to the Air Force Chief of Staff, 1979- . Reg PE, St of CA. Mbr Air Force Scientific Advisory Bd, 1971-79. Professional consultant to industry and government, 1964-79. Author/Co-Author numerous engrg articles, research papers, and book chapters. One patent. Assoc Editor, *Amer Inst of Aeronautics and Astronautics Journal of Aircraft*, 1974-77. Fellow, IEEE. *Society Aff:* IEEE, AIAA, AAAS

Stearns, H Myrl
Home: 246 La Cuesta Drive, Menlo Park, CA 94025
Position: Director & Consultant. *Employer:* Varian Associates. *Education:* Doctor of Sci(Hon.)/Communications/Univ of ID; EE/Comm/Stanford U; BS/EE/Univ of ID. *Born:* 4/24/16. BSEE Univ of Idaho 1937; EE Stanford 1939. 1939-41 Asst to Chief Engr Gilfillan Bros; 1941-48 Asst Proj Engr to Res Engr (Dept Head) Sperry Gyroscope Co; 1948-57 Co-founder Varian Assocs, Exec V P & Genl Mgr; 1957-64 Pres Varian, 1964-present Dir & Cons. Honorary D Sc Univ of Idaho 1960. Chrmn, Crocker Capital (venture capital Co.) 1971-1975. *Society Aff:* ΣΞ, ΤΒΠ, IEEE, AAAS.

Stearns, S Russell
Business: Thayer Sch of Engg, Hanover, NH 03755
Position: Prof Ret.; Consulting Engineer *Employer:* Darmouth Coll *Education:* MSCE/CE/Purdue Univ; CE/CE/Thayer Sch, Dartmouth Coll; AB/CE/Dartmouth College. *Born:* 2/28/15. NH native; Dartmouth AB, CE; Purdue MSCE; engr with private consulting firms. US Army, Corps of Engrs, and industry; Prof of CE since 1953, planning, design, and operation of transportation systems, structural fdns; res and planning on Cold Regions facilites and logistics-AK, Canada, Greenland, UN consultant to Poland; Dir, Pres, 1983-84 ASCE; Mbr NH Bd of Reg for PE 1955-83; mbr and past chrmn of Sch Bd, Planning Bd, Airport Authority, State and Natl committees; love gardening and the outdoors. *Society Aff:* ASCE, ASEE, TRB, NSPE.

Steckl, Andrew J
Business: Ctr for Integrated Electronics, Troy, NY 12181
Position: Director *Employer:* Rensselaer Polytech Institute *Education:* PhD/EE/Univ of Rochester; MS/EE/Univ of Rochester; BS/EE/Princeton Univ *Born:* 5/22/46 Dr Steckl's initial professional specialty was in the field of infrared, working in industrial labs at Honeywell and Rockwell. Since 1976, he has been in academia, as a faculty mbr in the Elec, Computer and Sys Engrg Dept at Rensselaer Polytech Inst in Troy, NY. At RPI Dr Steckl initiated and developed a program of research in integrated circuits with government and industrial support. This activity led to the establishment in 1981 of a multi-disciplinary Ctr for Intgegrated Electronics. Dr Steckl, as Dir of the Ctr, directs the activites of faculty and students from various disciplines (cf Materials, EE, Physics, etc.). Dr Steckl's non-scientific interests include hiking, kayaking and literature. *Society Aff:* APS, IEEE

Steele, C Jay
Business: PO Box 277, Roswell, GA 30077 *Employer:* Steele & Sons. *Education:* BS/-/IA State Univ. *Born:* 1/-/28. at Anamosa, IA. After brief employment as chem engr by Esso Std Oil Co, in 1951 formed own corrosion control firm, nationwide bus & merged it with conglomerate in 1968. Since has formed Mainstay Corp & Steele & Sons, Inc, which firms produce and apply spec corrosion barrier for concrete sewer pipes & structures. Mbr NACE since 1952, Dir 1972-5. Centrl Com mbr & teacher at Appalachian Underground Corrosion Short Chourse at Univ of WV.

Steele, Earl L
Business: 435 Anderson Hall, Lexington, KY 40506
Position: Professor *Employer:* Dept of Elec Engrg/Univ of Kentucky. *Education:* PhD/Physics/Cornell Univ; BS/Physics/Univ of UTAH. *Born:* Sept 1923. Transistor R&D G E Co Motorola Inc, & Hughes Aircraft Co. Worked at N Amer Aviation Autonetics Div on Laser Dev & Solid State Circuit Simulation. Presently Prof of Elec E Univ of Kentucky with major interest in Solid State Device Electronics. Past Adjunct Prof Ariz State Univ & Univ of Calif/ Irvine. Fellow IEEE; mbr Eta Kappa Nu & Tau Beta Pi & Sigma Xi; Mbr APS and ISHM. *Society Aff:* IEEE, ΣΞ, APS, ISHM, ASEE, ΤΒΠ.

Steele, Lendell E
Home: 7624 Highland St, Springfield, VA 22150
Position: Consultant. *Employer:* Self *Education:* BS/Chemistry/Geo Washington Univ; MA/Economic/Amer Univ. *Born:* 5/5/28. R&D Officer USAF 1951-53; Met Engr US Atomic Energy Comm 1966-67; otherwide employed at Naval Res Lab in nuclear matls res. Currently private technical consult. Fellow ASM & RESA; ANS award & prize for contrib in radiation effect on metals 1972, ASTM Dudley Medal 1973, Wash Acad of Sci Engrg Award 1962. Author, Editor of 8 books on nuclear radiation effects in engrg matls for high temparature & nuclear applications. Award of Merit & Fellow ASTM 1978, Navy Superior Civilian Service Award 1976, named Profl Engr CA, 1976. ASTM Bd of Dirs for term 1980-82, Vice Chrmn ASTM Bd of Dirs, elected Chrmn ASTM for 1985. Pres Federation of Materials Soc - 1984. Technical consultant 1986- . *Society Aff:* ASTM, ASM, ANS, ASME, FMS.

Stefan, Heinz G
Business: Dept of Civ & Min Engg, St Anthony Falls Hydr Lab, Minneapolis, MN 55414
Position: Prof & Assoc Dir. *Employer:* Univ of MN. *Education:* DrIng/Hydromechanics/Univ of Toulouse; IngHydraulicien/Hydraul Engrg/Univ of Toulouse; DiplIng/CE/Tech Univ Munich. *Born:* 6/12/36. Educated at Tech. Univ of Munich, Stuttgart and Univ. Toulouse, France. Positions at Toulouse (1960-63), Berlin (1965-67), MN (1963-65 and since 1967). Experience in water resources and environmental engg. Researcher and educator. Over 50 articles in referreed profl. journals & 60 reports on cooling water, pollutant transport, mixing zones, lake and reservoir hydromechanics, water intake and discharge structures, water quality models, hydraulic structures. Consultant to industry and govt agencies. Chrmn & mbr of NAS, ASCE, AGU, ANSI committees. Editor "Surface Water Impoundments–, 1660 pp. publ. by ASCE, 1981, and "Advancements in Fluid Mechanics & Hydraulics–, 1200pp. publ. by ASCE, 1986. *Society Aff:* ASCE, AGU, AWRA.

Steffen, Alfred J
Home: 2863 Ashland St, West Lafayette, IN 47906
Position: Associate Professor/Environ Engrg Retired *Employer:* Purdue University. *Education:* BS/CE/Univ of WI; Grad Studies/Univ. of WI & Univ. of Mn.. *Born:* 1911. Native of Milwaukee Wisc. Grad work Univ of Minn & Univ of Wisc. Pre WW II Dist Sanitary Engr Wisc State Bd of Health. Army Corps of Engrs WW II in charge of water & sewage, 6th Serv Command stationed principally in Chicago 1942-46. Environ Engrg Wilson & Co Inc 18 yrs. 1965-70 V P of R B Carter Co & Genl Mgr Water & Wastewater Equipment Div; 1970-77 Assoc Prof Environ Engrg Purdue Univ & Co-Chmn Purdue Indus Waste Conf. Pres 1965 Water Pollution Control Federation & Life Mbr; Diplomate Amer Academy of Environ Engrs. Authored 40 pub papers. Lectured in Eng, Germany, Sweden & Japan. 1977 Assoc. Prof. Emeritus Purdue Univ. *Society Aff:* WPCF, AAEE.

Steffen, John F
Business: 2333 Grissom Drive, St Louis, MO 63141
Position: President. *Employer:* John F Steffen Assocs Inc. *Born:* 1936. Cons Engr reg 30 states. 23 yrs practice. Designed mech/electrical sys for hospitals, office bldgs & educational facils. Projs include US Pavilion Montreal Expo '67. Formerly on faculty Univ of Wisc/Milwaukee, presently faculty mbr Wash Univ St Louis Mo where engrg education was received, instructs on computer sci/engrg. Public speaker on 'Life Cycle Analysis,' 'Cost Estimating,' 'Const Mgmt' & 'Value Engrg'; has writtenn computer progs for these activities. Mbr Missouri & Amer Cons Engrs Coucil. Serves on St Louis Electrical Bd of Trade. P Pres Natl Assn Flying Engrs. Has multi-engine pilot license & real estate broker's license.

Steffen, John R
Business: Gellerson Bldg, Valparaiso, IN 46383
Position: Assoc Prof. *Employer:* Valparaiso Univ. *Education:* PhD/Mech Engg/ Rutgers, State Univ of NJ; MS/Mech Engg/Univ of Notre Dame; BS/Mech Engg/ Valparaiso Univ. *Born:* 1/29/43. Taught at Mercer County Community College before joining faculty of Valparaiso Univ in 1974. Author of several papers dealing with Mechanism Design. Active mbr of ASME serving as Chrmn of Mbrship Dev of Design Div and Campus Faculty Advisor of Student Sec. Recipient of 1979 Ralpah R Teetor Award presented annually by Natl SAE. Areas of instruction include Machine Design, Mechanisms, Vibrations, and Stress Analysis. Mech Design Consultant and Reg PE for state of IN. *Society Aff:* ASME, ASEE.

Steffen, Juerg
Business: ASULAB SA, 6 Passage Max Meuron, 2001 Neuchatel Switzerland
Position: Hd Laser Optics Group *Employer:* ASULAB S.A. *Education:* PhD/Appl Physics/Univ of Berne; MS/Math & Physics/Swis Fed Inst Techn *Born:* 12/11/42 Born in Zurich, Switzerland, married, 4 children. Public and high schools (1949-61) in Zurich; physics studies at Swiss Fed Inst of Tech (1961-66) and at Univ of Berne (1966-72). 1972-74 Research Mgr. Research Inst. Pierres Holding in Thun, Switzerland, working on development of lasers and laser material processing systems and optical quality inspection equipment. 1974-81 research/development mgr, V pres, LASAG AG, laser manufacturing company in Thun, active in lasers and applications in materials processing, in measurement and control technics and in medicine. Since 1981 Hd of laser-optics group of ASULAB SA, central R/D labs of the SMH Group, Neuchatel, investigating laser application problems in materials processing and optical metrology. Since 1978 teaching at Techn Hochschule Darmstadt (West Germany). He holds a fellowship of the Optical Soc of Amer. *Society Aff:* OSA, SPIE, EPG, SPG, SGOEM

Steidel, Robert F, Jr
Business: 5138 Etcheverry Hall, Berkeley, CA 94720
Position: Prof of Mechanical Engrg, The Assoc Dean of Engrg *Employer:* University of California/Berkeley. *Education:* D Eng/ME/UC Berkeley; MS/ME/COlumbia Univ; BS/ME/Columbia Univ. *Born:* July 6, 1926. Asst Prof Mech E 1949-52; Asst Prof to Prof Mech E Univ of Cal 1954- ; Chmn of Dept 1969-74; The Associate Dean of Engineering 1981-86, Cons Bonneville Power Admin 1949-54; Jet Propulsion Lab 1962-68; Lawrence Livermore Lab 1955- . Fellow, I Mech E.; Fellow, Chmn, San Francisco Sect 1967, ASME; mbr: ASEE, Sigma Xi, Tau Beta Pi, Pi Tau Sigma. ASEE Western Electric Fund Award 1973 & Chester F Carlson Award 1974. Univ of Cal/Berkeley Rep to Pacific 10 Conf, Pres 1973-74. *Society Aff:* IMechE, ASME, ASEE.

Steier, William H
Business: MC0438, Los Angeles, CA 90007
Position: Prof Elec Engr *Employer:* University of Southern California. *Education:* PhD/EE/U of IL; MS/EE U of IL; BS/EE/U of Evansville *Born:* 5/25/33. Asst Prof Univ of Ill 1960-62; Mbr Tech Staff Bell Telephone Labs Holmdel NJ 1962-68; Assoc Prof & Prof USC 1968- , Co-Chmn Elec Engrg Dept 1970-84, with responsibility for electro-physics area, Univ res centers on laser devices, high energy lasers & optical materials; Cons on high energy lasers & optics Northrop Corporate Labs. Mbr: Eta Kappa Nu, Sigma Xi, AAAS, ASEE; Fellow IEEE. *Society Aff:* HKN, ΣΞ, AAAS, IEEE, APS, OS

Steige, Walter E
Business: 3812 Spenard Rd. Suite 100, Anchorage, AK 99503
Position: President *Employer:* Crews, Macnnes Hoffman/Vitro *Education:* BS/Civil Eng/CO State Univ. *Born:* 7/15/36. 22 yrs cons exper. Former Genl Mgr Crews, Maclnnes & Hoffman & a prin of the firm; and corp Pres. Now mgr of the branch office noted for Vitro Design Assoc and a director of Vitro Engrg. Exper in design of lighting, communications, controls & pwr distribution in coml, indus & institutional bldgs. Also exper in design of power distribution facils & their related sub stations & preparation of sys analysis, mapping, cost & feasability studies. Mgmt respons for branch office with total personnel of fifty. Mblr of NSPE, ASPE, ACEC, PEPP. Past Alaska Chrmn of PEPP. Currently Pres of CEC Alaska. Is also a private pilot and enjoys outdoor activities and participates in competitive tennis. *Society Aff:* NSPE, ACEC.

Steigerwald, Bernard J
Business: US EPA, Res Triangle Park, NC 27711
Position: Dir/Office of Air Quality. *Employer:* US Environmental Protection Agency. *Education:* Bs/Civil Eng/Cast Inst Tech; MS/Sanitary Eng/MA Inst of Tech; PhD/Environ Eng/Case Inst Tech. *Born:* Oct 1931 Cleveland Ohio. PhD, BS Case Inst of Technology; MS MIT. US Public Health Serv 1954-59; Asst Prof Case Inst 1959-62; US Dept HEW 1962-69; V P Zurn Environ Engrs 1969-70; Dir Office of Air Quality Planning & Standards US EPA 1970-76; Dir, Off of Regional Programs, EPA 1976- . *Society Aff:* APCA.

Steigerwald, Edward A
Business: 23555 Euclid Ave, Cleveland, OH 44117
Position: General Manager. *Employer:* TRW Turbine Components Div Metals. *Education:* BS/Metallurgy/Case Inst of Tech; MS/Metallurgy/Case Inst of Tech; PhD/Metallurgy/Case Inst of Tech. *Born:* April 1930. Native of Cleveland Ohio. BS, MS PhD Case Inst of Technology. Asst Mgr Met Res Cleveland Twist Drill Co 1953-57; US Army Chem Center 2 yrs; TRW Inc Met Engr & Mgr Met Res; Mgr Engrg TRW Metals, worked in investment casting of nickel base superalloys for airfoils, Genl Mgr TRW Metals 1973-1981. VP and Gen Mgr, Castings Div 1981-1983, VP Productivity 1983. Pub in areas of fracture mechanics, stress corrosion 1959-60. Chmn Cleveland Chap ASM & AIME. Sev natl cttes, Fellow ASM. Outstanding Young Engr 1967. Cleveland engrg socs. *Society Aff:* AFS, ASM, AIME.

Steigerwald, Robert F
Home: 3530 Charter Pl, Ann Arbor, MI 48105
Position: Area Manager *Employer:* Bechtel Group, Inc *Education:* PhD/Met/RPI; BMetE/Met/RPI. *Born:* 12/31/35. in Detroit. Res Met at E I DuPont. Dev Mgr Climax Moly. Res Mgr Climax Molybdenum. Chief operating officer Harbor House. Area Office Mgr at Bechtel. Specialist in met of corrosion resistant alloys. *Society Aff:* ASM, AIME, NACE, ASTM.

Steiglitz, Kenneth
Business: Dept. of Computer Sci, Princeton, NJ 08544
Position: Professor *Employer:* Princeton Univ *Education:* Eng ScD/EE/NYU; M/EE/

Steiglitz, Kenneth (Continued)
NYU; B/EE/NYU *Born:* 1/30/39 Born in Weehawken, NJ and joined Princeton Univ in 1963, where he is now professor. Consults for government and industry. Author of *Introduction to Discrete Systems*, Wiley, 1974; and co-author of *Combinatorial Optimization: Algorithms and Complexity*, Prentice-Hall, 1982. Author or co-author of more than 100 papers. Fellow of the IEEE (1981), for "contributions to the theory and application of digital signal processing." Received Technical Achievement Award of the IEEE Acoustics, Speech, and Signal Processing in 1981, the Society Award of that society in 1986, the IEEE Centennial Medal in 1984. *Society Aff:* IEEE, ΣΞ, TBΠ, HKN, ACM

Steimle, Stephen E
Business: Steimle & Assoc, Inc, PO Box 865, Metairie, LA 70004
Position: Pres. *Employer:* Self. *Education:* PhD/Environmental Engr/Tulane Univ; MS/Environmental Engr/Tulane Univ; BSCE/Civ Engr/Tulane Univ. *Born:* 12/12/40. Dr Steimle was born in New Orleans, LA, on 12-12-40 & is the son of E N Steimle a machinist. He received the degrees of BS, MS, & PhD from Tulane Univ, and completed his formal education in 1969. He has served as a civ engr for Modjeski & Masters, public health engr for the City of New Orleans, & asst prof of environmental engg at Tulane Univ. In 1973 he founded the consulting firm of Steimle, Schroeder, Smalley & Assoc, Inc. He served two (2) yrs as Pres at the LA Water Pollution Control Assn (mbr Water Pollution Control Federation). He is the husband of the former Joyce T Brown & is the father of four children. *Society Aff:* ASCE, WPCF, AWWA, AIChE, LES, IAWR.

Stein, Bland A
Home: 732 Jouett Dr, Newport News, VA 23602
Position: Branch Hd. *Employer:* NASA-Langley Res Ctr. *Education:* MS/Met Engg/VA Tech; BS/ME/CCNY; Post Grad/Matl Eng/NC State; Post Grad/Pub Admin/Univ of VA. *Born:* 2/27/34. Native of NYC. Res Engr with NACA/NASA-Langley Res Ctr since 1956. Asst Hd, Mtls Res Branch, 1972. Head, Advanced Materials Br, 1981. Hd, Polymetic Materials Br, 1983. More than 50 papers & natl presentations on mech properties & environmental effects on a broad range of metallic mtls & advanced composites for aerospace applications. 8 group achievement awards for planning & directing res progs on aerospace mtls & failure analysis. Awarded US Civ Service Commission fellowship-Education for Public Mgt-1978. Serves on natl technical committees for AIAA & ASTM. Assoc Fellow AIAA. ASM Hampton Rds Chapter Chrmn, 1973. Temple Sinai Trustee. Currently developing new concepts for rapid adhesive bonding of metals and composites (patents applied for). Temple Sinai "Man of the Year" 1984. Who's Who in Govt, Who's Who in the South/Southwest. Enjoys family, camping, classical music, sports. *Society Aff:* AIAA, ASM, ASTM, ACS, SAMPE.

Stein, Dale F
Business: Michigan Tech Univ, Houghton, MI 49931
Position: President *Employer:* Michigan Technological University. *Education:* PhD/Metallurgy/Rensselaer Polytechnic Inst; BS/Metallurgy/Univ of MN. *Born:* 12/24/35. G E Co 1958-67; Univ of MN 1967-71; MI Technological Univ 1971- . AIME Hardy Gold Medal 1965 & ASM Giesler Award 1967. Primary res area deformation & fracture of matls. Mbr adv cttes or bds NSF, Dept of Energy, & The Met Soc AIME. Past Pres TMS/AIME. Fellow, TMS/AIME, Fellow, ASM; Fellow, AAAS, Member National Academy of Engineering 1986- , Bd of Dirs, Michigan Biotechnology Inst, Mbr High Tech Task Force, State of Michigan. *Society Aff:* AIME, ASM, TBΠ, ΠΚΦ, ΣΞ, AAAS, NAE

Stein, Edward E
Business: AFRPL/LK, Edwards AFB, CA 93523
Position: Deputy Ch Liquid Rocket Division. *Employer:* USAF Air Force Rocket Propulsion Lab. *Born:* Nov 1926. Native of Gary Ind. BSME Purdue Univ 1950. USAF Pilot 195155; Fuel Sys Engr AF Aeropropulsion Lab 1956-59; AF Rocket Propulsion Lab Liquid Rocket Div 1960-70, Solid Rocket Div 1971-74, present position 1974- . AIAA Sect Awards Chmn.

Stein, Peter K
Business: 5602 East Monte Rosa, Phoenix, AZ 85018
Position: President. *Employer:* Stein Engineering Services Inc. *Education:* MS/Engg/MIT; BS/Mech Engg/MIT; BS/Bus Admin/MIT. *Born:* Dec 19, 1928 Vienna Austria. Educated Shanghai China & MIT; Instructor MIT 1950-55; Instrumentation Engr AiRes Mfg Co Phoenix Ariz 1955-59; Prof of Engrg Ariz State Univ 1959-77; Pres SES Inc 1950- . Fellow, 20-yr serv award Western Regional Strain Gage Ctte, Frocht Education Award SESA; IEEE Fellow; mbr Ad-Com IEEE-G-IM; NAS Evaluation Panels for 2 NBS Divs; ISA Fellow; NCSL Bd/Dir; ASU Faculty Achievement Award 1965; Outstanding Educators of Amer 1970; mbr: Tau Beta Pi, Pi Tau Sigma, ASME; Full Mbr Sigma Xi. AGARD/NATO lecturer. US Delegate to 3 Tech Cttes Internatl Measurement Confederation. Developed Unified Approach to Engrg of Measuring Sys adopted widely in US, Canada & Europe. Over 180 papers, books, handbook chaps & other publs. Over 320 papers & lecture series presented. Principal res interest concept res in measurement sys. *Society Aff:* IEEE, ISA, SESA, ASME, WRSGC.

Stein, Seymour
Home: 56 Great Meadow Road, Newton Centre, MA 02159
Position: President. *Employer:* SCPE, Inc. *Education:* PhD/Applied Physics/Harvard; MS/Applied Physics/Harvard; BEE/EE/CCNY. *Born:* April 1928. SCPE, Inc 1979-pres, consulting to government & industry. Founder & Pres Stein Assocs Inc 1969-79 (subsidiary Adams- Russell Co 1970). Co-author 'Communication Sys & Techniques' McGraw-Hill 1966 & 'Modern Communication Principles' McGraw-Hill 1967. IEEE Info Theory Group Annual Paper Award 1964. Fellow IEEE. Continues active tech work. Specialist in Communication Sci & Digital Signal Processing applications. *Society Aff:* Tau Bet Pi, Eta Kap Nu, Sigma Xi, IEEE(Fellow)

Stein, Theodore
Home: 30 Branford Rd, Hastings-on-Hudson, NY 10706
Position: Vice President. *Employer:* Halcon International Inc. *Born:* Oct 1927. ScD, SM, SB from MIT. Served USN 1946-48. Dev Engr Standard Oil Co Indiana 1955-60, detailed study of kinetics of catalytic reforming & correlations are basis for computer control sys for Standard Oil Catalytic Reformer; joined Scientific Design Co Inc 1960 Process Engr, appointed Dir of Engrg Dev 1964. 1970 elected a V P of Halcon, 1971 appointed Pres Halcon Computer Technologies, computer software co spec in engrg progs. 1976 Dir of Dev; 1978 Dir of Research..

Stein, Theodore W
Business: 2 Park Ave, NY, NY 10016
Position: VP. *Employer:* Halcon Res & Dev. *Education:* ScD/CE/MIT; SM/CE/MIT; SB/CE/MIT. *Born:* 10/1/27. Native of Evansville, IN. Served with USN 1945-1948. Chem Engr with Std Oil (IN) 1955-1960. With Halcon since 1960. Sr VP in charge of res. Previously had been Vp of Dev & Pres of Halcon Comp Tech, Inc. *Society Aff:* AIChE

Steinberg, Bernard D
Business: Moore School of Electrical Engrg/D2, Philadelphia, PA 19104
Position: Professor. *Employer:* University of Pennsylvania. *Education:* PhD/EE/Univ PA; MSEE/EE/MIT; BSEE/EE/MIT *Born:* 1924 Brooklyn. BSEE, MSEE from MIT 1949; PhD Univ of Penn 1971. Res Div Philco through mid 50's in radar backscatter & radar signal processing. A founder Genl Atronics Corp Phila 1956, V P & Tech Dir 15 yrs. A founder and chrmn Interspec Inc, Phila 1979. Worked in signal processing for radar, HF Communication, hydroacoustics, seismology. Recent work in self-adaptive signal processors, particularly in large antenna arrays. Currently Prof Univ of Penn & Dir its Valley Forge Res Center, developing large, self-adaptive microwave imaging sys called Radio Camera. Fellow IEEE, mbr US Comms B & C of Internatl Scientific Radio Union. *Society Aff:* IEEE, ΣΞ, HKN, URSI, AAAS

Steinberg, Meyer
Business: Building 526, Upton, NY 11973
Position: Head Process Scis Div. *Employer:* Brookhaven National Laboratory. *Education:* BChE/-/Cooper Union; MChE/-/Polytechnic Inst Brooklyn; PE/Prof. Engr. Lic. /NYS. *Born:* 7/10/24. Holds PE License. Sev yrs on Manhattan Proj & heavy chem indus; joined Dept of Applied Sci Brookhaven Natl Lab, now Hd of Process Sci Div, Dept of Applied Sci. Major fields of interest: dev of use of nuclear energy and fossil energy for the synthesis & production of synthetic fuels and indus chem, matls scis, dev of novel energy & fuel conservation processes. Mbr AIChE, ACS, AAAS, AIC, NYAS, Long Island Sect AIChE, Sigma Xi, & Ans, Lic Prof Engr State of NY. *Society Aff:* AIChE, ACS, AAAS, ANS, AIC, NYAS, ΣΞ.

Steinberg, Morris A
Home: 348 Homewood Road, Los Angeles, CA 90049
Position: VP-Sci Retired *Employer:* Lockheed Corp *Education:* DSc/Met/MIT; BS/Met/MIT; MS/Met/MIT *Born:* Sept 1920. Native of Hartford Conn. Joined Lockheed Aug 1958 holding positions of Mgr of Material Sci Lab, Deputy Ch Scientist & Dir of Technology Applications. Prior to that date Horizons Inc Head of Metallurgy Dept; MIT Instructor in metal processing, foundry, mech metallurgy & powder metallurgy. 9 patents, 9 pending, has pub extensively. Fellow ASM & AIC. Mbr Natl Materials Adv Bd & NAS. Cons to USAF Scientific Adv Bd. Reg P E Calif, Consultant to NASA. Fellow AIAA and AAAS and IAE. Mbr, ASEB - Air Force Studies Bd. Retired 1986. *Society Aff:* NAE, ASM, AIC, AIAA, AAAS

Steinbrugge, John M
Business: 280 Newport Center Dr, Newport Beach, CA 92660
Position: Pres *Employer:* STB Structural Engrs Inc. *Education:* BS/Civil Engg/OR State Univ. *Born:* Nov 1923 Tucson Ariz. Served to LCDR USNR in WW II. Structural Engr Moffatt & Nichol Long Beach Cal 1948-52; C F Braun & Co Alhambra Cal 1952-54; Supervising Structural Engr State of Cal 1954-62; Partner in cons structural engrg firm Johnson-Nielsen-Steinbrugge Riverside Cal 1962-64; Partner Steinbrugge & Moon Structural Engrs Newport Beach Cal 1964-72; Principal Steinbrugge & Thomas Inc Cons Structural Engrs Newport Beach Cal 1972-80; Principal Steinbrugge, Thomas & Bloom Inc. Cons. Structural Engrs, Newport Beach, Cal. 1980-85. President STB Structural Engs Inc, Newport Beach, Ca 1985- . *Society Aff:* ASCE, ACI, CEAC, SEAC, PCI, AAA.

Steiner, Bruce W
Business: Natl Bureau Standards, MAT A163, Gaithersburg, MD 20899
Position: Scientist *Employer:* National Bureau of Standards. *Education:* PhD/Phys Chem/Princeton Univ; AB/Chem/Obertin College *Born:* 5/14/31. Postdoctoral res Univ of Chicago 1958-61. Res Scientist Natl Bureau of Stds. Program Offic for Admin & Molecular Physics, Natl Sci Fnd (Dept of Commerce Sci & Technology Fellow) 1976-77. Co-founder Council for Optical Radiation Measurement: Chmn Tech Council Optical Soc of Amer 1978-79. Chmn U.S. Natl Comm, Intnatl Commission for Optics, 1985-87. *Society Aff:* OSA, AIC, APS, ACS, TMS.

Steiner, George G
Business: 2049 Bartow Ave, New York, NY 10475
Position: Exec General Mgr *Employer:* Riverbay Corp *Education:* M/Plant Engr/Polytech Inst TNS; BS/ME/School of Electro Mech Engr Clu *Born:* 11/28/25 After graduation, I became an Engrg Supvr in the Bldg Maintenance and Plant Engrg organization. After receiving a Masters Degree in Plant engrg I was promoted to Chief Engr for a co with approximately 1,200 employees with our own property construction engrg dept. After arriving in the US I was the Chief Plant Engr of a plastic extrusion co and later became Mgr of Plant Engrg for a real estate co with 2,000 units for 15 years. At the present time I am the Executive General Mgr for the biggest real estate housing project in the US which consists of 15,300 units - 3 shopping centers - power plant operation and a construction management dept totalling 875 white collar personnel plus 58 engrg supvrs. *Society Aff:* ASME, AIPE, NSPE

Steiner, Henry M
Business: Sch of Eng & Appl Sci, Geo Wash Univ, Washington, DC 20052
Position: Prof. *Employer:* Geo Wash Univ. *Education:* PhD/CE/Stanford; MS/CE/Stanford; BS/ME/Stanford. *Born:* 8/20/23. in San Francisco, CA. Practicing engr in US & overseas from 1944-1959. Joined faculty of Mexico City College in 1959, & taught at Stanford, Univ of TX at Austin, Visiting Prof. Univ. of Calif., Berkeley, & presently at GWU. Author of four books on logistics, urban transport, rural road planning, and public investment. Founder & Past Pres International Transport & Economic Development Chapter, TRF. *Society Aff:* ASCE, TRF.

Steiner, John E
Home: 3425 Evergreen Point Rd, Bellevue, WA 98004
Position: V P/Corporate Product Evaluation. *Employer:* The Boeing Airplane Company. *Born:* BS Univ of Wash; Masters Aeronautical Engrg MIT. 1964 selected Man Of The Yr by Aviation Week & City of Seattle for dominant role in 727 dev. 1965 Prod Engrg Master Design Award, later Elmer A Sperry Award, 1974 Thulin Medal in Sweden. Fellow AIAA & Royal Aeronautical Soc. Participated in every Boeing coml proj since joining co; including 314, 307, 377, 707, 727, 737, 747. Hobby: sailing.

Steiner, Ronald S
Home: 1061 NE 181 St, N Miami Beach, FL 33162
Position: VP - Branch Mgr *Employer:* Fogel & Assoc, Inc. *Education:* MSCE/Civ Engrg/Villanova Univ; BSE/Civ Engrg/Cooper Union. *Born:* 11/23/44. Joined Fogel & Assocs, Inc in 1981 to head their Fort Lauderdale office. Responsible for construction scheduling, estimating, proj mgmt and construction claims support on various projects in the United States and other countries. Previously employed as Mgr of Scheduling for the Kaiser Transit Group on dev of the Miami Metrorail Transit Program. Before that, employed for 11 years by PA D. O.T., Philadelphia Dist. Initially, design coordinator of roadway projects including one involving first use of a two-way reinforced earth slab in U.S. Received ASCE Collingwood Prize, 1976, for work and paper on proj. Later, on construction mgmt team supervising completion of I-95 through Philadelphia, including relocation of transit line into interstate median. Elected to Who's Who in Sci and Tech, 1981. Registered Professional Engr. *Society Aff:* ASCE, PMI, NSPE, CSI.

Steinhorst, Gerald L
Home: 2359 Mattos Dr, Milpitas, CA 95035
Position: Facilities Engg Mgr. *Employer:* Zilog, Inc. *Education:* BSEE/Elec Engg/Roosevelt Univ; -/Engg/Univ of WI. *Born:* 4/2/32. Native of Janesville, WI. Air Force veteran, Korean war. Worked 9-1/2 yrs with Motorola Semiconductor, Phoenix, AZ as Facilities Engr and Equip Design Engr. Worked 9-1/2 yrs for Fairchild Semiconductor, Mt View, CA, as Facilities Engg Mgr. Have been responsible for elec design and construction of semiconductor plants totaling over 1,000,000 sq ft. Locations include: Mexico City; Seoul, Korea; Hong Kong; Jakarta; Singapore; Healdsburg, CA; Mtn View, CA; San Jose, CA; Cupertino, CA; and Nampa, ID. Assumed current responsibility of Facilities Engg Mgr with Zilog, Inc in 1979. *Society Aff:* NSPE, CSPE, AIPE.

Steinmeyer, Daniel E
Business: 800 N Lindbergh, St Louis, MO 63166
Position: Engineering Fellow *Employer:* Monsanto. *Born:* April 1936. MS Michigan; MA Washington Univ; BS Northwestern. Native of St Louis. 1959-present Monsanto as process designer, heat transfer & energy conservation specialist; Process Design Mgr on olefins venture 1976; Dir of Engg in Brasil 1978; Manager Process Energy Technology 1980; assumed current responsibility in 1985. Past Chmn Heat Transfer & Energy Conversion Div AIChE. Instructor in AIChE continuing education series on 'Process Design for Energy Conservation. Author of sev papers & chapters in Perry's Chem Engrg Handbook and Kirk Othmer's Ency Chem Technology.

Stell, George
Business: Dept. of Chemistry, SUNY, Stony Brook, NY 11794-3400
Position: Professor of Chem & Engrg *Employer:* State Univ of NY *Education:* PhD/Math/NYU; BS/Physics/Antioch Coll *Born:* 1/2/33 Was Assoc Research Scientist, Inst of Math Sciences, NYU, 1961-64 and Belfer Grad School, Yeshiva, 1964-65. Asst Prof, PIB, 1965-66; Assoc Prof, PIB 1966-68. Assoc Prof, Dept of Mechanics, SUSB, 1968-70; Prof, Dept of Mechanical Engrg, 1970-present; Prof, Dept of Chemistry, SUSB (joint appointment with Mech Eng), 1979-present. Have consulted for Lawrence Livermore Lab, IBM, NBS; Principal Investigator since 1970 for various NSF and Petroleum Research Fund Research Grants; Principal Investigator since 1979 for DOE Contract; Principal Investigator since 1982 for NATO Research Grant. John Simon Guggenheim Memorial Fellow, 1984. *Society Aff:* APS, AMS, ΣΞ

Stelle, William W
Home: 60 Wood Road, Bedford Hills, NY 10507
Position: Retired *Employer:* Consltg Engr in Civil Engrg *Education:* B/CE/NYU & Princeton Univ *Born:* 10/27/17 in Ossining, NY. Tau Beta Pi Tuttle Award. Served in US Army 1941-45. Discharged as Capt. Employed by American Electric Power Service Corp 1948-55 and 1956 to date. Presently Div Mgr, Civ Engrg Div, supervising more than 50 Engrs, Geologists and Engrg Technicians. *Society Aff:* ASCE

Stellwagen, Robert H
Business: 22255 Greenfield 236, Southfield, MI 48076
Position: Executive Vice President. *Employer:* DiClemente-Siegel Engrg, Inc *Education:* BS/Mech Engg/Univ of MI. *Born:* 4/23/27. Design Engr Hyde & Bobbio Inc 1950-59; Pres Stellwagen & Mouw Inc 1959-73 spec in design of HVAC sys, process piping & electrical sys for institutional & indus bldgs. Assumed current responsiblities 1973. Hobbies: flying, skiing & tennis. *Society Aff:* NFPA, NSPE.

Stengel, Robert F
Business: D-202, Engrg Quadrangle, Princeton, NJ 08544
Position: Prof *Employer:* Princeton Univ *Education:* PhD/Aero & Mech Sci/Princeton Univ; MSE/Aero & Mech Sci/Princeton Univ; MA/Aero & Mech Sci/Princeton Univ; SB/Aero & Astro/MIT *Born:* 9/1/38 Robert Stengel is Prof of Mech and Aerospace Engrg at Princeton Univ. He is currently conducting res in artificial intelligence, nonlinear and optimal control theory, sys identification, automation, man-machine sys, and microprocessor applications. He teaches graduate courses on optimal control and estimation and on aircraft dynamics, and undergraduate courses on space flight engrg and flight guidance. Prior to his Princeton appt, he was a section leader at The Analytic Sciences Corp, a group leader at the MIT Draper lab, a USAF officer, and an aerospace technologist with NASA. He recently completed the book *Stochastic Optimal Control: Theory and Application* (J. Wiley & Sons, Inc., 1986) and has authored or co-authored more than 100 technical papers and reports. *Society Aff:* AIAA, IEEE

Stenger, Edwin A
Business: 2980 Spring Grove Ave, Cincinnati, OH 45225
Position: Pres. *Employer:* Queen City Steel Treating Co. *Education:* MetE/Met Engg/Univ of Cincinnati; BBA/Accounting/Univ of Cincinnati. *Born:* 1/31/26. Since a WWII Navy tour, has served in various capacities at Queen City Steel Treating prior to becoming Pres in 1968. Activities include Past President of the Metal Treating Inst, & service on the bds of the Cincinnati Ind Inst, OH Info Committee, & mbrships in the Engrs and Scientists of Cincinnati & Rotary. Worked toward the dev & establishment of the Stenger Heat Treating Lab at the new Engg Complex at the Univ of Cincinnati, signifying three generations of family participation here. Hobbies include bowling, golf, & antique auto restoration. 1980 recipient of the ASM Eisenman Award. *Society Aff:* ASM.

Stenger, Harvey G
Business: PO Box 852, Auburn, NY 13021
Position: Pres. *Employer:* Stenger Engg, PC. *Education:* BS/ME/Clarkson College of Tech. *Born:* 9/12/34. Pres & Chief Engr for Stenger Engg, PC, a product dev, res & consulting firm in the fields of filtration, heat pumping, compression equip, & energy mgt. Holder of patents on compression equip & control systems. Base experience includes Engg & technical mgt assignments with Carrier Corp, Bomac Inc, & Syracuse Univ. These involved mgt of quality control, reliability, product dev, & product evaluation for air conditioners, heat pumps, furnaces, humidifiers, dehumidifiers, control systems, & related testing & mfg facilities. Developer of advanced, cost effective, & reliable residential coolness storage air conditioner. Mbr of Tau Beta Pi, Pi Tau Sigma, & Alpha Pi Mu. *Society Aff:* ASHRAE.

Stenuf, Theodore J
Home: 4861 Coventry Rd, Syracuse, NY 13215
Position: Dist Teaching Prof. *Employer:* SUNY College of Environ Sci & Forestry. *Education:* PhD/CE/Syracuse Univ; MChE/CE/Syracuse Univ; BChE/CE/Syracuse Univ. *Born:* 2/27/24. in Vienna, Austria. Immigrated to US in 1937. Served in US Navy 1943-46. 1949-53 Res Asst/Assoc Syracuse Univ, Inst of Ind Res; 1953-59 Service and Res Engr, E I duPont de Nemours; 1959-60 Corrosion Engr, ESSO Res and Engg Co; 1960- date Prof at State Univ of NY College of Environmental Sci and Forestry teaching chem engg. Promoted to "Distinguished Teaching Prof" 1977 for "unique teaching accomplishment, expertise and scholarship–. Served on and chaired many natl committees. Very active in scouting. *Society Aff:* AIChE, TAPPI, ACS, ΣΞ, ΑΧΣ.

Stephan, David G
Home: 6435 Stirrup Road, Cincinnati, OH 45244
Position: Sr. Engineering Advisor, Hazardous Waste Engineering Research Laboratory. *Employer:* US Environmental Protection Agency. *Education:* PhD/Chem Engg/OH State Univ; MSc/Chem Engg/OH State Univ; BChE/Chem Engg/OH State Univ. *Born:* Feb 1930 Columbus Ohio. Battelle Mem Inst, AEC Feed Materials Production Center & US Public Health Service Taft Center 1952-65. Dir Res & Asst Commissioner R&D Federal Water Quality Admin & Dir Res Prog Mgmt for EPA in Wash DC 1965-75. Dir Industrial Environmental Research Lab & Hazardous Waste Engineering Research Lab for EPA in Cincinnati, OH 1976-85. Specialties include fabric air filtration, advanced waste treatment & wastewater renovation. Bd/Dir MTS 1970-73; Governing Bd of Internatl Assn on Water Pollution Res & Control 1976-85; Meritorious Achievement & Superior Serv Awards DHEW; Bronze Medal & Gold Medal EPA; 1981 Environmental Div Award of AIChE; 1983 Distinguished Engr of the Year Award of the Engrg Soc of Cincinnati. Disting Engrg Alumnus Ohio State Univ; Fellow AIChE. *Society Aff:* AIChE.

Stephens, Donald L
Business: 790 Garden Ave North, Renton, WA 98055
Position: Dir-Tech Ctr *Employer:* PACCAR Inc *Education:* BS/ME/Univ of WA *Born:* 12/28/28 Native of Seattle. Served in Army Signal Corps 1946-47. Joined PACCAR, Inc in 1952. Designed propellers, valves, components of military vehicles, railroad cars, and heavy duty trucks. Served as chief engr Peterbil Motors (division) 1970-77. Assumed present position as dir of corporate tech ctr and designer of new tech ctr in 1978. Director SAE 1980-82. *Society Aff:* SAE, ΣΞ

Stephens, Jack E
Business: Box U-37, Storrs, CT 06268
Position: Prof of Civil Engg. *Employer:* Univ of CT. *Education:* PhD/Hgwys/Purdue; MS/Civil/Purdue; BS/Civil/Univ of CT. *Born:* 8/17/23. A native of western OH. Three yrs service in US Army with active service in European Theater, WWII. BS 1947, Univ of CT, MS 55 and PhD 59 Purdue. Instructor to Prof 1947--Civil Engg, Univ of CT. Dept head 65 to 72. Res on hgwy topics with CT Dept of Transportation, maj area materials. Past pres CT Sec ASCE, CT Soc of Civil Engrs, and Education Div of ARTBA. Comm Activities in ASTM, TRB, and AAPT. Current chrmn Transportation Comm of the CT Academy of Sci and Engg. *Society Aff:* ASCE, TRB, AAPT, ASP, ACSM, ASEE, ARTBA, AAUP, CASE.

Stephens, John C
Home: 1111 NE 2nd St, Ft Lauderdale, FL 33301
Position: Consultant (Self-employed) *Education:* BS/MI Engr/Univ of AL; MS/Geology/Stanford Univ. *Born:* 9/22/10. Worked as field Engr & Geologist in Baja CA & AL, 1932-34. Employed by USDA in 1934 (SCS-ARS) & served in Southeastern & Northwestern USA as field engr, res engr, res admin in fields of water res & soil conservation until 1976; including Supervisor, Everglades Res Proj, Dir SE Watershed Res Ctr; Chief, NW Br, Soil & Water Conservation Res Div, & MS Valley Res Area Dir. (Employed as Chief Water Control Engr, Dade Co FL 1945-48). Consultant to US-AID RE: Forested watershed Res for ASEAN countries 1980. Self-employed since 1976 as Consultant in Water Resources & Agri Engr, Ft Lauderdale, FL. Certificates of Appreciation for Meritorious Service, Dade Co, FL 1948; as Chrmn Irr & Dr Div, 1971-72, as Chrmn Mgt Group D (-Wet Divs-) 1972-73, ASCE; John Deere Award, 1976, ASAE. 1986- . Mmbr: Panel on Land Subsidence, Ctte on Ground Failure Hazards, Natnl Res Coun. *Society Aff:* ASCE, ASAE, AGU, FCSSS

Stephens, Warren C
Business: P O Box 734, Sioux Falls, SD 57101
Position: Principal. *Employer:* Sayre/Stephens Consulting Engrs, Inc. *Education:* DS/Civil Eng/SD State Univ. *Born:* Feb 1943. BSCE South Dakota State Univ. Howard Needles Tammer & Bergendoff 6 months in 1969 as Bridge Designer; 1969-present Sayre/Stephens Inc as Proj Mgr on sanitary, sewer water treatment & distribution, hwys & subdivisions. Stockholder & Mbr Bd/Dir Sayre/Stephens Inc. Reg P E & L S in S Dakota. 2nd V P S Dakota Sect ASCE 1974, Pres 1975. Mbr: AWWA, ASCF, NSPE & WPCF. *Society Aff:* SDSU.

Stephenson, Edward T
Business: Bethlehem Steel Corp, Research Dept, Bethlehem, PA 18016
Position: Senior Scientist *Employer:* Bethlehem Steel Corporation. *Education:* PhD/Met E/Lehigh Univ; MS/Met E/MIT; BS/Met E/Lehigh Univ. *Born:* Nov 7, 1929 Atlantic City NJ. Engrg Duty Officer USN 1951-54. Res Engr, Sr Res Engr & Sr Sci Bethlehem Steel Corp 1956- , spec in physical metallurgy. Mbr 1965-68 & Chmn 1968-69 ASM Transactions Ctte; Chmn 1969-70 ASM Publ Council, Chmn 1971 Exec Ctte of Met Transactions, Mbr 1974-75 & Chmn 1976 Joint Comm for Met Transactions, Mbr 1977-84 and chrmn 1984-87 ASM comm for Intnl Met Rev. Grossman Award 1965 ASM. *Society Aff:* ASM, AIME, IEEE, ΣΞ, ΤΒΠ.

Stephenson, Junius W
Home: 38 Sunset Road, Demarest, NJ 07627
Position: Prin. *Employer:* Havens & Emerson Inc. *Education:* BSCE/CE/Union College. *Born:* 2/4/22. B, 1922 in NY City. BS in CE Union College 1944. Instructor in Engg Union Coll briefly after graduation. Havens & Emerson Inc Cons Environ Engrs 1944- , Prin 1970- , Saddle Brook NJ. Specializes in solid waste mgmt, resource & energy recovery. Chmn ASME Solid Waste Processing Div 1965-66 & 1979-80. Received Div award for outstanding serv 1976. Centennial Medal 1980. Chmn 1968 Natl Incinerator Conf. Author of num papers on solid wast mgmt. Mbr: ASCE, ASME, APWA, ISW, ISWA, APCA, ACEC, WPCF. Reg PE in NC, TN, ME, PA, NY, NJ, CT, MI, VA, WI, MA, Oklahoma. Diplomate AAEE. Trained Value Engineer. *Society Aff:* ASCE, ASME, APCA, ACEC, WPCF

Stephenson, Robert E
Business: College of Engineering, Salt Lake City, UT 84112
Position: Associate Dean. *Employer:* University of Utah. *Born:* Aug 7, 1919 Nephi Utah. BSEE 1941 Univ of Utah; MSEE Caltech 1946; PhD Purdue Univ 1952. Mbr: Sigma Xi, Eta Kappa Nu, Theta Tau, Tau Beta Pi, IEEE. Reg P E Utah 1458. Faculty Univ of Utah 1946- ; faculty Univ of Ky 1960-62. Visiting Prof Bandung Indonesia 1960-62. Assoc Dean Coll Engrg Univ of Utah 1971-1986. AIEE Utah Sect Secy 1955-56, V Chmn 1956-57, Chmn 1957-58; IEEE Utah Sect Secy 1964-65, V Chmn 1965-66, Chmn 1966-67. Textbook 'Computer Simulation for Engrs' Harcourt Brace 1971.

Stephenson, Robert L
Home: 1309 Shady Ave, Pittsburgh, PA 15217
Position: Consultant. *Employer:* Retired. *Education:* AB/Chemistry/Princeton Univ. *Born:* 2/11/13. Native of Pittsburgh, PA. Began career in 1935 at Duquesne Steel Works of US Steel. Developed a "Hardenability Slide Rule" for accurately predicting the hardenability of steel from its chem composition. Chief Met of Duquesne Works 1949 to 1951. Res Engr & Chief Res Engr, ore reduction and hot-metal production 1951 to 1978. Published many technical articles on ironmaking and coedited a book on ironmaking and one on direct reduction. Fellow of ASM in 1976. Recipiant of the AIME T L Joseph award for significant contributions to the advancement of ironmaking in 1979. Dist Mbr of the Iron & Steel Soc of AIME in 1983. *Society Aff:* ASM, AIME, AISE, AISI, MS.

Sterbenz, Frederick H
Home: 9 Irving Lane, New Hyde Park, NY 11040
Position: Associate *Employer:* Howard Needles Tammen & Bergendoff *Education:* BCE/Structures/NYU *Born:* 7/1/26 Native of metropolitan New York area. With Howard Needles Tammen & Bergendoff, Consulting Engrs, since 1948 and since 1974 has been an Associate of the firm. Experience has centered on bridges of all types and toll roads. Served as project engineer of the Second Delaware Memorial Bridge (suspension), Cooper River Bridge (cantilever) and has expertise in the design and repair of many movable bridges including the Woodrow Wilson Memorial Bridge (bascule) and the Pennsylvania Railroad Bridge over the Chesapeake & Delaware Canal (lift). He is author of an article on the Cooper River Bridge published in Civil Engrg. *Society Aff:* ASCE, IABSE, NSPE, NYSSPE, AWS, IBTTA

Sterling, Charles F
5 Parkinson Street, Needham, MA 02192
Position: Asst. State Traffic Engineer for Traffic Oper. *Employer:* Mass Dept of Public Works. *Education:* ME/Civil Engg/Penn State Univ; Cert/Traffic & Transp Engg/Bureau of Highway Traffic; BS/Civil Engg/Lowell Tech Inst; MPA/Public Mgmt/Harvard Univ. *Born:* May 21, 1943 Boston Mass; M.P.A. Harvard Univ; Certificate in Traffic & Transportation Engrg Bureau of Hwy Traffic; ME Penn State; BS Lowell Tech; Inst of Transportation Engrs P Pres Award 1974; Asst. State Traffic Engr-In charge of Traffic Opns.; Major-Corps of Engrs (Reserve), Active in Soc of Military Engrs, ITE, Chrmn of Tri State Transportation Planning Ctte. *Society Aff:* ITE, MAUDEP, BTG, SAME.

Stern, Arthur C
Business: Dept of Environ Sci/Eng, Chapel Hill, NC 27514
Position: Emeritus Prof of Air Hygiene. *Employer:* Univ of North Carolina/Chapel Hill. *Education:* ED/Engg/Stevens Inst of Technology; MS/Engg/Stevens Inst of Technology; ME/Engg/Stevens Inst of Technology. *Born:* March 14, 1909 Petersburg Va. 1968-78 Prof Air Hygiene Dept Environ Sciences & Engrg, Sch of Public Health Univ of NC/Chapel Hill. Life Fellow ASME; Diplomate Amer Academy of Environ Engrs & Amer Bd of Indus Hygiene; P Pres, recipient Richard Beatty Mellon Award Air Pollution Control Assn. Editor 'Air Pollution' 3rd edition, 8 volumes (Academic Pres 1976-86); Co-author 'Fundamentals of Air Pollution' (2nd Ed) Academic Press 1984. Mbr NAE. Recipient: Gordon Maskew Fair Award, American Academy of Environmental Engineers 1983. Christopher Barthel Award, International Union of Air Pollution Prevention Associations 1983. *Society Aff:* ASME, APCA, AIHA.

Stern, Arthur P
Home: 606 North Oakhurst Dr, Beverly Hills, CA 90210
Position: President *Employer:* Magnavox Advanced Prods and Sys Co *Education:* Dipl Ing/Elect Engrg/Swiss Fed Inst of Tech; MS/Elec Engrg/Syracuse Univ *Born:* 7/20/25. Budapest Hungary. Married Edith Marguerite Samuel; ch: Daniel, Claude, Jacqueline. Res Engr Jaeger Inc Basel 1948-50; Instructor Swiss Fed Inst of Tech 1950-51; Engr, Engrg Mgr & Mgr Electronic Devices and Applications Lab G E Co 1951-61; Dir Engrg Electronics Div Martin Marietta Corp 1961-64; Dir Opers De-

Stern, Arthur P (Continued)
fense Sys Div Bunker-Ramo Corp 1964-66; V P & Genl Mgr Advanced Prods Div The Magnavox Co 1966-79. Pres, Magnavox Adv Prods and Sys Company 1980 -. Author of 20 tech articles, co-author 2 tech books. 12 US & foreign patents. Genl Chmn Intl Solid State Circuits Conf 1960. Dir IEEE 1970-77, Secy 1972, Treas 1973, V P 1974, Pres 1975, Fellow 1962, Centennial Medal Award 1984; Fellow, Amer Assoc for the Advancement of Sci 1982. *Society Aff:* IEEE, AAAS, ΣΞ, AAS

Stern, Ferdi B, Jr
Home: 22 Davelin Road, Wayland, MA 01778
Position: Consultant *Education:* MS/Applied Mech/MIT; BE/Mech Eng/Tulane Univ. *Born:* June 1, 1919. Native of New Orleans La. Magnaflux Corp 42 yrs; Res Dept Chicago, Field Engr Houston Texas, then New England. Has lectured & participated in educational progs in field of Nondestructive Testing & Experimental Stress Analysis. P E in Texas & Mass. Mbr: ASME, SEM, ASNT, ASM, Sigma Xi. Tatnall Award SESA 1970. Fellow ASNT 1974 & SEM 1976. Retired from Magnaflux Corp. April 1983. Dec. 1983-, NDT Instructor, Training & Certification Office, Army Materials Technology Lab. Watertown, Mass. *Society Aff:* ASNT, SEM, ASME, ASM, ΣΞ.

Stern, Louis I
Business: 6 Commerce Drive, Cranford, NJ 07016
Position: Partner *Employer:* Dames & Moore *Education:* MS/CE/Soil Mech/Harvard Univ; BS/CE/Soil Mech/ MIT *Born:* 6/7/29 Born in New York City. Served in US Army - 1952-54. Soils Engr for Capitol Engrg Corp, Dillsburg, PA. Joined Dames & Moore in 1957. Became Principal-in- Charge, 1964; Partner, 1968. Served as firm-wide Technical Mgr of Soils and Foundation Engrg. Presently principal on varied soil mechanics and foundation engrg projects in US and overseas. Principal on aquifer thermal energy storage project; Long Island, NY. *Society Aff:* ASCE

Stern, Richard M
Business: 1520 Eastlake E, Ste 201, Seattle, WA 98102 *Employer:* TML/STERN ENGINEERS *Education:* BS/CE/Univ of ND *Born:* 3/21/13 Initial training in HVAC as student engr with Carrier Corp followed by air conditioning and heating work for contractors in Spokane. Mechanical engr for Austin Co on Navy shore station work in Washington and Oregon in WWII. Established professional office in 1947. Early work in development of floor heating and ventilating systems for slab-on-ground applications. Active in use of unitary and modular equipment for energy economy in semi-central systems. Various service on Tech Cttes for ASHRAE, and on Seattle code Cttes. *Society Aff:* NSPE, ASHRAE, ASPE

Sternberg, Eli
Business: Div Engrg Appl Sci, Pasadena, CA 91125
Position: Professor of Mechanics. *Employer:* Caltech. *Education:* PhD/Mechanics/IL Inst Tech; MS/Civil Eng/IL Inst Tech; BCE/Civil Eng/NC State Univ *Born:* Nov 1917. Native of Vienna Austria; in U S since 1939. Taught at IIT 1945-56 (Prof of Mech since 1951); Div Appl Math Brown Univ 1957-64; Div Engrg Appli Sci Caltech since 1964. Visiting professorships in Netherlannds, Japan, Chile. Res in continuum mech, elasticity theory. Fulbright Fellow 1956-57 & 1970-71; Guggenheim Fellow 1963-64; Sr Res Fellow, British Sci Res Council 1968. Hon DSc N C State 1963, Hon DSc Technion 1984. Mbr Natl Acad of Engrg since 1975, Mbr Natl Acad of Sci since 1979.

Sternberg, Ernest R
Home: 38 Court Sq SW, Graham, NC 27253
Position: Engrg Consultant. *Employer:* Ernest R. Sternberg, Inc. *Education:* MS/ME/Univ of WI; BS/ME/Univ of Southern CA. *Born:* 7/3/14. Served as Sales Engr, dir of Engg, VP, Sterling Motor Truck Co Inc, Milwaukee, WI. Asst VP Production White Motor Co; Dir of Engg Autocar Div; ch Engr & then VP Engg White Truckk Div; Corp Engg Staff White Motor Corp; retired Feb 1975 as Dir Adv Truc Engg, White Truck Grp. Served on SAE Genl Stds Council and Tech Bd, 1976 SAE Buckendale lectr. *Society Aff:* SAE.

Sternlicht, Beno
Home: 2520 Whamer Lane, Schenectady, NY 12309
Position: Tech Director & Bd Chmn. *Employer:* Mechanical Technology Inc. *Born:* Mar 12, 1928. BSEE Union Coll; MS in Engrg & PhD Columbia Univ; Hon DSc Union Coll. 30 yrs of indus exper as Dev Engr, Staff Engr, Tech Dir & Cons Engr with G E Co. Co - founder of MIT. Presently Tech Dir, & Chmn/Bd of Mech Tech Inc. Co is engaged in products & services of electro-mech tech, energy conversion machinery, conservation equipment, qual assurance & metrology equipment. Auth many articles & books. Patents in field. Pres & Chmn of Bd of VITA 1967-72. Chmn NASA Ctte on Energy Tech & Space Propulsion 1970-1973. Recipient of ASME Machine Design Award 1966. Member National Academy of Engineering. Member of National Energy Task Force. *Society Aff:* ASME, ASLE, AIAA.

Sternstein, Sanford S
Business: Materials Engrg Dept, Troy, NY 12181
Position: Professor *Employer:* Rensselaer Poly Inst *Education:* PhD/ChE/RPI; BS/ChE/Univ of MD *Born:* 6/19/36 Professor Sternstein is the W W Walker Professor of Polymer Engrg at Rensselaer Polytechnic Institute. His research specialties include the viscoelastic, fracture, crazing and plastic deformation of engrg plastics; the mechanical properties of high performance composites; solid polymer-solvent interactions and inhomogeneous swelling; rubber elasticity and network theory; and instrumentation for rheological measurements. He is a fellow of the Amer Physical Soc and mbr of the executive ctte of the Division of High Polymer Physics. Also, he is the Pres of the Dynastatics Instruments Corp. *Society Aff:* AICHE, ACS, SR, APS

Stetson, John B
Business: 185 Genesee St, Utica, NY 13501
Position: Principal-Chief Exec Officer *Employer:* Stetson-Dale, Architects and Engrs *Education:* MS/Eng/Princeton Univ; BS/CE/RPI; BS/Marine Eng/US Naval Academy *Born:* 11/27/28 Before assuming the presidency of Dale Engrg Company in 1963, John B Stetson enjoyed a 13-yr career as an officer in the US Navy Civil Engrg Corps. Highlights of his experience include assignments in Madrid, Spain as project manager for construction of the US Naval Bases in that country; design officer and project engr in the Southwest Div of the Bureau of Yards and Docks, and; civil engrg consultant ot the Latin American Navies. Published articles include *Navy's New Carrier Berthing Facilities at San Diego,* and *Geology Study for Navy Pier Construction.* *Society Aff:* ASCE, SAME

Stetson, Laverne E
Business: 253 Chase Hall, Lincoln, NE 68583-0729
Position: Agri Engr. *Employer:* Agri Res USDA. *Education:* MS/Agri Engg/Univ of NE. *Born:* Aug 1933. Native of Crawford Nebr. Res on radiofrequency elec fields for control of insects in stored grain & to improve seed germination 1962-71. Presently in res to coord irrigation scheduling with elec demands to reduce power demands & to improve safety & reliability of elec equip used in irrigation. Mbr ASAE, IEEE, Tau Beta Pi, Sigma Tau, Sigma Xi, Gamma Sigma Delta. Reg P E. Chmn Special Tech Subctte which dev new articles 675, on irrigation machines for 1975 Natl Elec Code. *Society Aff:* ASAE, IEEE.

Stetson, Robert F
Home: 6918 Tanglewood Rd, San Diego, CA 92111
Position: Matls Technologist. *Employer:* Self *Education:* CE/Ferrous Metallurgy/PA State Univ. *Born:* 10/20/28. Discharged from Army Air Corps 1947. Supervisor Insp, Babcock & Wilcox through 1958. Retired 1986 from GA Technologies after 25 years service. Currently self employed as a materials technologist. Served as Chmn San Diego Chap ASM, was instrumental in estab cert of met technicians & technologists. Was awarded first ICET-ASM cert in this field. Awarded ASM Chap Assoc Engrg Award. Served on ASM Exec Ctte of San Diego Chap of ASM 10 yrs. Past Mbr of the Bd of Trustee for ICET, A by lawbody of NSPE. Recipient of the

Stetson, Robert F (Continued)
ASM Natl 1979 Engg Associates Achievement Award, Patentee in field. Recipient of the 1980 Lincoln ARC Welding Foundation Award "For One of the Most Outstanding Achievements of 1979 with Welded Design, Engineering and Fabrication–. Fellow of ASM International Class of 83. *Society Aff:* ASM Int, ICET

Stetson, Robert L
Home: 215 Angelita Ave, Pacifica, CA 94044
Position: Vice President. *Employer:* Pacific Nuclear Fuels Inc. *Education:* BS/Chem Engg/WA State Univ. *Born:* June 1918. Native of Port Orchard Wash. Worked in pulp & paper indus prior to WW II. Served as naval officer in Bureau of Aeronautics 1943-46. Res engr in naval labs spec in radioactive contamination-decontamination field testing. With Amer-Std & Tracerlab as reactor applications engr & sales mgr, Lawrence Livermore Lab in transuranium processing plant design, & US Navy in ship reactor overhaul & refueling. Estab Pacific Nuclear Fuels Inc 1973 spec in nuclear fuels patent work and related consulting. Reg P E State of Calif 1969- . Enjoy hiking & participating in performing arts.

Stevens, Frederick
Business: 1800 Century Pk E, Los Angeles, CA 90067
Position: Sr VP & Group Exec Construction Group. *Employer:* Northrop Corp. *Education:* MSEE/Engg/CA Inst Tech; AB/Math & Physics/Whitman College. *Born:* 6/13/22. AB Whitman College Walla Walla Washington 1944; MSEE Caltech 1947; Advanced Mgmt Prog Harvard Bus Sch 1969. Served 10 yrs as mbr of Air Force Sci Adv Bd; presently "Meritorious Civilian Serv" Award by USAF 1968. Genl Chmn IEEE WINCON Convention on Aerospace 6 Electronic Sys 1968, & apptd Fellow IEEE Dec 1969. Joining Northrop in 1947, he held var position & assumed present position a Sr VP & Group Executive-Construction Group in 1977. *Society Aff:* IEEE.

Stevens, Gladstone T, Jr
Business: Indus Engrg Dept, Arlington, TX 76019
Position: Prof & Chmn of Indus Engrg. *Employer:* Univ of Texas at Arlington. *Education:* PhD/Industrial Engg/OK State Univ; MS/Mech Engg/Cas Inst of Tech; BS/Mech Engg/Univ of OK. *Born:* Dec 16, 1930. Taught at Lamar Univ 1962-64 & Okla State 1966-75; Proj Engr E I duPont 1956-59; Res Engr Thompson-Ramo-Wooldridge 1960-62; assumed current pos Sept 1975. Sr Mbr & Past Natl Dir Qual Control & Reliability Div 1974 IIE; Natl Pres & Mbr of Exec Council Alpha Pi Mu 1976-present; E L Grant Award ASEE 1973; Reg P E Okla & Texas; Mbr Omicron Delta Kappa, Tau Beta Pi, Sigma Tau, Sigma Xi. *Society Aff:* IIE, ASEE

Stevens, Karl K
Home: 3533 NE 6th Dr, Boca Raton, FL 33431
Position: Prof *Employer:* FL Atlantic Univ. *Education:* PhD/Theo & Appl Mech/Univ of IL; MS/Theo & Appl Mech/Univ of IL; BS/Mech Engg/KS State Univ. *Born:* 1/24/39. Topeka, KS, Employed with Sandia Corp, Bell Telephone Labs, and US Army Ballistic Res Labs; private conslt. Prof of Engg Mech, OH State Univ, 1965-1978. With FL Atlantic Univ since 1978, specializing in finite element methods vibrations, and stress analysis. Reg PE in FL & OH. Contributor to profl journals and author of "Statics and Strength of Materials–, Prentice-Hall, 1979, 2nd ed. 1987 Past Chrmn Prof Interest Council III and mbr Bd of Dir, ASEE, 1981-83. Chrmn Palm Beach Section ASME. FAU-Rotary Outstanding Researcher Award, 1983. Alumni Award for Distinguished Teaching - OH State Univ, 1975. *Society Aff:* ASME, ASEE, SEM, SNAME.

Stevens, Philip A
560 Sycamore St, Tiffin, OH 44883
Position: Director *Employer:* Stevens Consulting Group *Education:* SM/Ind Mgt/MIT; BIE/Ind Engg/OH State Univ. *Born:* 7/13/34. Native of Toledo, Ohio. Dir, Stevens Consulting Group since 1983. Prior to that, VP Admin Natl Machinery Co, Tiffin, OH, since 1970. Served as VP European Operations 1962-96 in West German subsidiary for Natl Machinery. Pres Tiffin - Seneca public library, Mbr DIIE, Rotary Intl. Trustee of Terra Technical College, Fremont, Ohio and Natl Machinery Foundation, Tiffin, Ohio. He is a Certified Financial Planner (CFP) Pres of Tiffin Rotary Club registered Prof Eng in OH. *Society Aff:* IIE, IAFP.

Stevenson, Albert H
Home: 2 Huckleberry La, Greenwich, CT 06830
Position: Conslt Engr *Employer:* Independent *Education:* MS/Sanitary Engrg/Harvard Univ; BS/CE/Union Coll *Born:* 5/18/14 Independent Conslt Engr (1982 to date). Formerly VP for intl operations, Malcolm Pirnie, Inc, consulting engrs (1971 to 1982). Career officer in US Public Health Service (1941-1971) retiring as assistant surgeon general (rear admiral) and chief engr. Served as chief sanitary engr, Indian and Alaskan Native Health Service (1956-1966), assigned as chief sanitary engr, Federal Civil Defense Administration (1954-1956), served as deputy dir, Taft Sanitary Engrg Center during its development (1947-1954), and as sr assistant engr in USPHS Regional Office (1941-47). Early experience as assistant engr with Malcolm Pirnie Engrs and with NY State Health Dept (1937-1941). Pres, AAEE (1968-1969). Honorary Chief Blackfeet Indian Tribe. Author of numerous papers on environmental engrg and health subjects. Travelled extensively in more than 100 countries. *Society Aff:* AAEE, ASCE, AWWA, AIDIS, NAS

Stevenson, Gerald L
Business: PO Box 2008, Lakeland, FL 33806
Position: Group VP *Employer:* Jacobs Engineering Group Inc *Education:* BS/Chem Engg/Univ of MO; MS/Chem Engg/Univ of MO; AMP/Mgmt/Harvard Business School; Honorary Professional Degree Univ of Mo. *Born:* 1/16/37. MBA program St Louis Univ. Native of Salem, IL. Served in US Army. Refinery Technologist for Shell Oil Co. Instru Dept of Chem Engg & Chem, Univ of MO. Held pos of Prod Engr, Proj Mgr, Expansion Engr & Dev Engr with Intl Minerals & Chems. Served as VP with D J Stark Group of cos. Served as Sr VP and General Manager with Davy McKee Corp With Jacobs Engrg Group, Inc since April 1984 as Group VP. Reg PE Ontario. Mbr AIChE, AIME, ACS & The Fertilizer Soc (London). Exec Bd of Engrg & Construction Contracting Division of AIChE. *Society Aff:* AIChE, ACS, TFS, SME/AIME.

Stevenson, James R
Home: 1140 Fernwood Dr, Schenectady, NY 12309
Position: Mgr - Planning & Operations Retired *Employer:* Genl Electric Co. *Education:* BSEE/Elec Engr/Univ of TN. *Born:* 1/13/21. Attended the Univ of TN & was assigned to the Manhattan Proj Los Alamos while in the service. Joined the Analytical Sect of Central Station Engg in Schenectady in 1948 & became an Application Engr in th SE in 1954. From 1957-64 was Mgr - Application Engg for the southeastern part of the country. In 1964 returned to Schenectady as mgr Power Plant Automation Operation. In 1969 was apptd Mgr - AC Transmission Engg. In 1977 became Mgr Planning & Operations of the Electric Utility Sys Engg Dept. Retired. *Society Aff:* IEEE.

Stevenson, John D
Home: 21099 Claythorne Rd, Cleveland, OH 44122
Position: Vice President. *Employer:* J D Stevenson Cons. *Born:* May 1933. PhD Case Inst of Tech; BSCE Vir Military Inst. 1st Lt Army Corps of Engrs 1954-56. Field Engr McDowell Const Co 1956-57. Taught Civil Engrg Vir Military Inst 1957-62. Cons Engr & Mgr of Struct Engrg Westinghouse Pressurized Water Reactor, Nuclear Service Dept 1966-74. Mgr Eastern Operations EDAC Inc 1974-76. V Pres & Genl Mgr J D Stevenson Cons, Div of Arthur G McKee & Co 1976-79. VP Woodward Clyde Consultants 1979-80. Pres Stevenson & Assocs 1981- . Mbr & Chmn of a num of nuclear power stds & code cttes of ASCE, ASME, ANS, ACI, & ANSI. Winner Mossieff Award ASCE 1971. *Society Aff:* ASCE, ASME, ACI

Stevenson, Lawrence G
Business: 4625 Roanoke Pkwy, Kansas City, MO 64112
Position: V Pres & Genl Mgr - Sales. *Employer:* J F Pritchard & Co. *Born:* Apr 1923. BSChE Kansas State Univ; MBA Univ of Missouri. Native of Kansas. Served in the Army Air Corps 1943-46; trained at Caltech as Weather Officer. First employed as R&D Engr with Linde Div of UCC. With Spencer Chem Co/Gulf Oil Co 1951-69, as Process Dev, Commercial Dev & last pos as Mgr Chems of Korea Gulf Oil Co. With Lummus Co 1969-74, first as Deputy Managing Dir of Lummus Far East & later as V P of Lummus-Iran respon for Middle East commercial activity. Joined J F Pritchard & Co 1974 as V Pres - Mkting & currently V Pres & Genl Mgr - Sales.

Stevenson, Warren H
Business: Purdue University, Sch. of Mech. Engr, W. Lafayette, IN 47907
Position: Prof. *Employer:* Purdue Univ. *Education:* PhD/ME/Purdue Univ; MSME/ME/Purdue Univ; BSME/ME/Purdue Univ. *Born:* 11/18/38. Prof Stevenson initiated the Appl Optics prog in the Sch of Mech Engg at Purdue in 1965 upon joining the faculty. He has conducted extensive res in this area with particular emphasis on laser measurement techniques. His current res interests include laser velocimetry, optical particle sizing, and infrared holography. He is the author of over fifty technical papers. In 1973 he received the US Sr Scientist Award of the Alexander Von Humboldt Fdn. He is active on various profl committees and has served as a consultant to several cos & govt agencies. *Society Aff:* OSA, ASME, ASEE, LIA.

Stevenson, William D, Jr
Home: 2706 White Oak Rd, Raleigh, NC 27609
Position: Prof Emeritus (Self-employed) *Education:* MS/EE/Univ of MI; BSEE/EE/Carnegie Inst of Tech; BSE/ME/Princeton Univ. *Born:* 7/21/12. Pittsburgh, PA. Author "Elements of Power System Analysis," McGraw-Hill Book Co 4th ed 1982. Former faculty mbr Clemson & Princeton Univs. Former Assoc Dept Hd & Grad Administrator for EE, NC State Univ. Received Naval Ordnance Award 1946 for work in guided missiles, Special Citation from Edison Electric Int, 1977. Fellow of Inst of Elec & Electronics Engrs. *Society Aff:* IEEE, ASEE, ΦΒΚ, ΣΞ, ΦΚΦ, HKN.

Stever, H Guyford
Business: Foreign Secretary, National Academy of Engineering, 2101 Constitution Ave, NW, Washington, DC 20418
Position: Foreign Secretary *Employer:* Natl Acad of Engineering *Education:* PhD/Physics/CA Inst of Tech; AB/Physics/Colgate Univ. *Born:* Oct 1916 Corning N Y. After receiving PhD 1941, served as a staff mbr of the Radiation Lab MIT for a yr before joining the London Mission of the Office of Sci R&D. Following WW II, returned to MIT as Prof of Aeronautical engrg. Remained there until 1965, leaving the post of Hd of the Depts of Mech Engrg, Naval Arch & Marine Engrg to become Pres of Carnegie-Mellon Univ. Active in acad & civic orgs in Pgh. Assumed pos as Dir of the Natl Sci Found 1972-76. Science Advisor to Pres of US 1976-77. Dist Pub Serv Medal 1968 Dept of Def; Chmn Chrmn, Assembly of Engg, Natl Res Council 1979-83; Corporate Dir, Trustee and Consultant 1977-; Pres, University Research Assoc 1982- ; Foreign Secretary, Natl Academy of Engrg 1984-. *Society Aff:* NAS, NAE, AIAA, APS.

Stewart, Bill R
Business: Agri Engrg Bldg-Rm 303, College Station, TX 77843
Position: Extension Agri Engr. *Employer:* Texas Agri Extension Serv. *Education:* BS/Agri Engg/TX A&M Univ; MS/Agri Engg/TX A&M Univ; PhD/CE/TX A&M Univ. *Born:* Feb 1930. Engr with USDA Soil Conservation Service 1951-53; Shell Oil Co 1953-55. Taught Agri Engrg 1955-72 at Texas Tech Univ & Texas A&M Univ. Assumed present pos Jan 1972 with statewide respon for educ progs in agri bldgs, rural housing, environ control. Chmn SW Region ASAE 1973-74. Bd of Dir 1974-76. *Society Aff:* ASAE.

Stewart, David C
Business: 84 South St, PO Box K, Carlisle, MA 01741
Position: President. *Employer:* Carlisle Engrg Mgmt Inc. *Education:* BS/IE/Northeastern Univ. *Born:* Aug 1922. Established industrial engineering consulting firm in 1975, currently 8 professionals, operating nationally, Cons in Matls Handling Sys Engrg, Materials Management Systems & Facil Design Mgmt for Indus Plants & Distrib Ctrs; formerly V P Indus Div, V P Clt Office, V P Dist Offices (Clt, Denver, Portland Ore, Jakarta) wth Chas T Main Inc Engrs; PE, PCMM, Intl Pres IMMS 1974-75, Adv Bd 1976-80; Life Mbr Appalachian Mtn Club. 000h. *Society Aff:* IMMS

Stewart, Hugh B
Home: La Crescenta-Bx 738, Rancho Santa Fe, CA 92067
Position: Preisdent *Employer:* Nutevco *Education:* PhD/Physics/OH State Univ; Ms/Physics/OH State Univ; BS/Phyiscs-Chem/Kent State Univ. *Born:* Dec 1916. Work exper in vibration engrg, exper reactor phys & nuclear engrg. Res engr for Allison Div of G M 1941-45; with Knolls Atomic Power Lab 1946-59 - critical experiments, nuclear analysis & nuclear engrg mgmt; with Genl Atomic 1959- , Chmn Nuclear Analysis & Reactor Phys Dept, Asst Dir of he Lab, V P HTGR Fuel & Matls Div Engrg. Fellow Amer Nuclear Soc; Mbr APS, Sigma Xi; author & contributor of articles for books, prof journals & tech soc's. Presently, Dr Stewart is the owner & Pres of NUTEVCO, a nuclear technology evaluation comp located in San Diego, CA NUTEVCO does energy technology consulting for both industry & gov. *Society Aff:* APS, ANS.

Stewart, Jack C
Business: P O Box 3166, Tulsa, OK 74101
Position: Project Manager *Employer:* Agrico Chem Co. *Education:* BChE/Chem Engr/OH State *Born:* 12/8/22. 3 yrs in ETO US Army. 7 yrs exper molded & extruded rubber compounder. 30 yrs exper in nitrogen fertilizer chems incl amonia, urea, nitric acid, ammonium nitrate & fertilizer solutions. Initial 2 yrs of nitrogen chems exper as process engr. Next 19 yrs as Process Supt, Urea & Nitric Acid, 3 yrs as Tech Supt, 6 yrs overseas Design and Operation. Indus exper with Goodyear, Vistron & Agrico. Have major knowledge & exper in urea production by Inventa, Toyo-Koatsu, Stamicarbon & Snamprogetti progress incl tech aspects. Currently on assignment in Trinidad and Tobago, operating and de bottlenecking a 1620 MT/D Snamprogetti Urea Unit with NSM Fluidized Bed Granulation. *Society Aff:* AIChE

Stewart, James P
Home: PO Box 411, Princeton, NJ 08540
Position: Mgmt Consultant (Self-employed) *Education:* M Eng/ME/Cornell Univ. *Born:* 9/25/06. in Rochester, PA. LLD Rider College. Elliott Co Jeannette PA 1928-43, Sales Engr to Dir Commercial Res; Borg Warner Supcharger Div Milwaukee 1943-46, Asst Genl Mgr; DeLaval Turbine Trenton NJ, Spec Sales Rep1946 to Pres & Dir 1951- Ret 1966; independent mgmt cons 1966- ; Former Dir Burns & Roe & sev affiliates. ASME Life Fellow 1961. Hon. Tr Pace Univ. Life Dir NJ State C of C (Pres 1959- 61). Former Tr Cornell Univ, Rider & Brainclift (Pres 1968-69) Colls. Former Pres Hydraulic Inst, Compressed Air & Gas Imst. Former Dir NJ Bell Tel, NAM, MAPI, NJ Natl Bank. *Society Aff:* ASME.

Stewart, JM (Jim)
Business: PO Box 2200, Streetsville Postal Station, Mississauga, Ontario L5M 2H3 Canada
Position: Sr VP *Employer:* Du Pont Canada Inc *Education:* BASc/CE/U of Toronto; PhD/CE/U of London. *Born:* 04/04/34 Native of Toronto area. Worked for Dow Chem for three yrs following graduation. With Du Pont Canada since doctorate degree in 1962. Held positions in Res including Res Centre mgr and is patent holder in polymerization tech. Seconded to natl Defense Coll 1973-74. Following various marketing positions was mgr of Corp Planning 1978-80, VP, Corp Projs & VP Corp Dev., VP mfg. At present is Senior Vice President. Active in Chem Inst of Canada (past director and Fellow) and Canadian Soc for Chem Engrg. Pres of CSChE 1980-81. Vice Chairman, Ontario Division, Canadian Manufacturers Ass.

Stewart, JM (Jim) (Continued)
Married to teacher and has two daughters. Hobbies include canoeing, fly fishing, duck hunting and cross country skiing. *Society Aff:* CIC, CSChE.

Stewart, John D
Business: 3198 Chestnut St 6000, Philadelphia, PA 19101
Position: Consultant *Employer:* Genl Elec Co. *Born:* Sept 1924. Canadian Army 1942-45. PhD Aero Engrg Univ of Toronto 1952. Performed aerodynamic tests at the Natl Res Council of Canada & then joined Convair, Fort Worth, to do aerodynamic analyses on the F111. In Jan 1957 joined the Missile & Space Div of G E to conduct aerothermodynamic studies on reentry missiles & space vehicles. In 1959, was apptd Mgr of Aerophys Engrg. Became Mgr of Sys Tech 1961 & in 1962 named Mgr of the Aeromech & Matls Lab. Assumed respon as Genl Mgr of the Res & Engrg Dept Jan 1968. Assumed current responsibility 1978, 1968-70: Mbr of the AIAA Tech Ctte on Missile Sys. June 1972-78, serving on the Bd of Mgrs of Spring Garden Coll. 1976-79 serving on the Adv Panel on 'Experiential Partnership for the Reorientation of Teaching' at Drexel Univ from a NSF grant.

Stewart, Larry E
Business: Dept Agri Engrg Univ MD, College Park, MD 20742
Position: Chrmn, Dept of Agri Engg. *Employer:* Univ of MD. *Education:* PhD/Agri Engg/Univ of MD; MSAE/Agri Engg/WV Univ; BSAE/Agri Engg/WV Univ. *Born:* 5/31/37. Native of Salt Rock, WV. Joined faculty of Univ of MD as an Agri Engg Specialist with the Cooperative Ext Service in August 1961. Maj prog responsibilties in areas of farmstead mechanization & farm safety progs. Became Chrmn of Dept of Agri Engg at the Univ of MD in Jan 1978. Active in ASAE on local & natl level; received 6 blue ribbon awards for Ext education progs from ASAE; immediate past-pres, Wash, DC-MD Section ASAE (1978-79). Enjoy music, hunting, fishing, beach-combing, & most sports. *Society Aff:* ASAE, ASEE, NIFS, CAST.

Stewart, Warren E
Home: 734 Huron Hill, Madison, WI 53711
Position: Prof of Chem Engrg. *Employer:* Univ of Wisconsin-Madison. *Education:* ScD/Chem Engg/MA Inst of Tech; MS/Chem Engg/Univ of WI; Bs/Chem Engg/Univ of WI. *Born:* July 1924. Married May 1947; 6 children. Proj engr with Sinclair Res 1950-56, spec in catalytic process kinetics & proj eval. Asst Prof of Chem Engrg, Univ of Wisc 1956-58; Assoc Prof 1968-61; Prof 1961- ; Dept Chmn 1973-78. Visiting Prof Universidad Nacional de la Plata Argentina 1962. Assoc Editor, Computers & Chem Engg. Cons to sev indus & to Univ of Wisc Fusion Tech Prog. Mbr, Mathematics Research Ctr, Univ of Wisconsin - Madison. Co-author textbook 'Transport Phenomena' & about 90 res papers on transport phenomena, numerical methods, reactor modelling & design. Congregationalist; Fellow AIChE; Mbr ACS. Hobby: music. *Society Aff:* AIChE, ACS, ASEE, ΤΒΠ, ΦΒΚ, ΣΞ, ΑΧΣ.

Stice, James E
Business: Univ Texas, b2202 Main Bldg, Austin, TX 78712
Position: Dir, Ctr for Teaching Effectiveness & T. Brockett Hudson Prof of Chem Engrg *Employer:* Univ of Texas at Austin. *Education:* PhD/Chem Engg/IL Inst of Tech; MSChE/Chem Engg/IL Inst of Tech; BSChE/Chem Engg/Univ of Ark. *Born:* 9/19/28. Armour Res Found Fellow, Ill Inst of Tech 1949-51, master's degree. Process engr Visking Corp 1951-53. Chem engr with div of W R Grace 1953-54. Asst Prof of Chem Engrg Univ of Arkansas 1954-57. Instr Ill Inst of Tech 1957-60; Res Assoc 1960-62; PhD 1963. Assoc Prof & Prof of Chem Engrg Univ of Arkansas 1962-68. Assoc Prof & Dir Bureau of Engrg Teaching, Univ of Texas 1968-73. Prof of Chem Engrg & Dir Univ of Texas Ctr for Teaching Effectiveness 1973-85. T. Brockett Hudson Prof of Chem Engrg, 1985-present. Ints: improvement of teaching, process control & thermodynamics. ISA Instrumentation Tech Award 1965. Visiting Prof, Universidad//Iberoamericana, Mexico City Summer 1977. Received Gen Dynamics Award 1980 and Western Elec Fund Award 1981 for excellence in engrg teaching. Received Chester F Carlson Award (ASEE) for innovation of engrg teaching in 1984. Named Fellow of American Society for Engineering Education, 1987. Reg PE in AK & TX. *Society Aff:* AIChE, ASEE

Stickels, Charles A
Home: 2410 Newport Rd, Ann Arbor, MI 48103
Position: Principal Res Engr. *Employer:* Ford Motor Co - Engineering & Manufacturing Staff *Education:* PhD/Met Engg/Univ of MI; MSE/Met Engg/Univ of MI; MA/Math/Univ of MI; BSE/Chem & Met Engg/Univ of MI. *Born:* 4/6/33. in Detroit, MI. US Navy as Destroyer Engr Officer 1956-59. Joined Ford Motor Co Res Lab 1963. Genl area of res: ferrous phys met. Special expertise in a) metal processing as practiced in the auto indus, incl metal forming & steel heat treatment & b) failure analysis of automotive parts. VP ADDRG 1975-76; Chmn ASM Sheet Metal Forming Activity 1971-75. Chmn ASM Acad for Metals Committee 1979; Editor, Journal of Heat Treating 1982-Present; Fellow, Amer Soc for Metals. *Society Aff:* TMS-AIME, ASM, ΣΞ.

Stickler, David C
Home: 110 Bleecker St, Aps 23C, New York, NY 10012
Position: Res Professor *Employer:* NY Univ. *Education:* PhD/EE/OH State Univ; MSC/EE/OH State Univ; BSc/Math/OH State Univ. *Born:* Apr 1933. Res Assoc The Antenna Lab, Ohio State Univ; Mbr Tech Staff Bell Tele Labs; Sr Engr Sys Control Inc; Sr Res Assoc, Appl Res Lab Penn State Univ. Active in res progs concerned with electromagnetic diffraction & the design of electromagnetic radomes, thermoelastic effects & ocean acoustic propagation. Publ articles in prof journals, taught grad level courses in electromagnetic theory, acoustics - seminars on ocean acoustics. Mbr Sigma Xi.

Stickley, C Martin
Home: 8108 Horseshoe La, Potomac, MD 20854
Position: Vice President and General Manager for Advanced Technology *Employer:* The BDM Corp *Education:* PhD/EE/Northeastern Univ; MS/EE/MIT; BS/EE/Univ of Cinn *Born:* 10/30/33 Fellow, IEEE; DoD Outstanding Meritorious Civilian Service Award; Distinguished Engineering Alumnus Award, U. of Cinn. Tau Beta Pi; Eta Kappa Nu; Omicron Delta Kappa; Sigma Xi. Outstanding Senior Engr at UC. Officer in USAF, 1958-1962. Built first laser in USAF at AFCRL; chief of Laser Physics Branch, 1965-1971. Joined Defense Advanced Research Projects Agency; became Dir for Materials Sciences; initiated first programs in GaAs IC's & RSR technology. Joined Energy R&D Admin. in 1976 as first Dir of Laser Fusion; obtained authorization for NOVA, and Particle Beam Fusion Accelerator. Joined BDM in 1979; responsible for programs in optical computing AI, expert system shells for materials processing, sensor hardening against lasers and many others. Married to Dorothy Riggs; three children. *Society Aff:* IEEE, SPIE, ΤΒΠ, ΣΞ

Stief, Robert D
Business: P.O. Box 8361, S Charleston, WV 25303
Position: Dir of Engg. *Employer:* Union Carbide Corp, Central Engr Dept *Education:* MChE/Syracuse Univ; BSChE/Univ of NM. *Born:* 4/18/26. Native of Cleveland, OH. With Union Carbide Corp since 1952. Previously (1951-1952) with Los Alamos Scientific Lab. Dir of Engg for Union Carbide since 1972. Presently responsible for Engr Tech and Design for process support, safety and environmental systems, energy systems, process computer control, and other support areas. Also responsible for international project management. Responsible (1975-1980)for Process Engg, Chem Engg Tech, Environmental Systems Engg, Energy Systems Engg, Process Comp Control, Special Instrument Dev, & Process Safety & Fire Protection Engg. Responsible 1972-1974 for intl design & construction. Fellow of AIChE. *Society Aff:* AIChE.

Stier, Howard L
Business: 1271 Ave. Of The Americas, New York, NY 10020
Position: Consultant *Employer:* United Brands Co. *Education:* PhD/PlantPhysiol-Hort/Univ of MD; MS/Horticulture/Univ of MD; BS/Agri Educ/Univ of MD. *Born:* Nov 28, 1910. Consultant United Brands Co. 1980- ; Corp VP

Stier, Howard L (Continued)
United Brands 1978-79; 1973-78, V P Res, Dev & Qual Control, United Brands Co; joined United Fruit Co 1961 as Corp Dir of Qual Control. Active in Amer Soc for Qual Control 25 yrs, serving as Dir, V P, Pres & Chmn of Bd; was elected Fellow ASQC 1961. Fellow AAAS. Mbr Inst of Food Technologists, Amer Statistical Soc, Biometrics Soc. Author 'Qual Control in the Food Indus', Sect 31, 3rd Edition of Qual Control Handbook, McGraw-Hill. *Society Aff:* AAAS, IFT, ASA, ASHS.

Stiffler, Jack Justin
Home: 34 Lake Shore Dr, Hopkinton, MA 01748
Position: Exec VP *Employer:* Sequoia Sys, Inc *Education:* AB/Physics/Harvard; MS/Elec Engr/ Caltech; PhD/Elec Engr/Caltech *Born:* May 22, 1934 Mitchellville Iowa. Mbr tech staff Jet Propulsion Lab, Pasadena Calif 1959-67, spec in communication & synchronization sys design & analysis. Raytheon 1967-81, Sequoia 1981- ; current ints incl fault-tolerant computer arch design & reliability modeling. Author 'Theory of Synchronous Communications' Prentice-Hall 1971; co-author 'Digital Communications' Prentice- Hall 1964; contributing author 'Fault-Tolerant Computing' Prentice-Hall 1986; num tech pubs; 7 pats. IEEE Fellow; mbr Phi Beta Kappa, Sigma Xi. *Society Aff:* IEEE, ΣΞ, ΦBK.

Stiles, Alvin B
Home: 1301 Grayson Rd, Welshire, Wilmington, DE 19803
Position: Res Prof *Employer:* Univ of DE *Education:* BChE/-/OH State Univ; MS/ChE/OH State Univ *Born:* 07/16/09 Born in Springfield, OH. Education: OH State Univ, BChE 1931, MSc 1933. Employed by duPont 1931 to 1974, Res Fellow. 63 Patents, many publs and chapters in books. Two books *Catalyst Manufacture & Catalyst Supports & Sappo-Lod Catalysts.* Principle discipline: catalyst res, characterization and industrial application. Co-founder Konawha Valley Sect of AIChE. Employed by the Univ of DE, Dept of Chem Engrg 1974 to present. Co- founder of Center for Catalytic Sci and Tech, Assoc Dir. Distinguished Alumnus OH State Univ Coll of Engrg, Eminent Engr, Tau Beta Pi. Recognized for Outstanding Contributions to Catalysis, Phila Catalysis Soc 1980. *Society Aff:* AIChE, ACS, ΣΞ, TBΠ, AM CHEMIST

Stiles, Charles E
Home: 264 Charles St, Meadville, PA 16335
Position: Consulting Engr. *Employer:* Self. *Education:* BS/Math/Edinboro State College. *Born:* 7/16/29. Native of Meadville, PA. Field Engr for Mullen Engg Co-1947-51 Taught in Army Corps of Engg-1951-53. Field Engr and Corp Secy of Hammond-Britton & Assocs-1953-1960. Formed C E Stiles and Assocs a Consulting Engg Firm in Municipal and Civ Engg-1960 to present. *Society Aff:* ASCE, ACSM, NSPE.

Stillman, Gregory E
Business: 1406 W Green St, Urbana, IL 61801
Position: Prof. *Employer:* Univ of IL. *Education:* PhD/EE/Univ of IL; MS/EE/Univ of IL; BSEE/EE/Univ of NE *Born* 2/15/36. in Scotia, NB. Grad from Univ of NB in 1958. Served as officer and pilot in USAF (Strategic Air Command) 1958-1963. Attended grad school in elec engg at Univ of IL 1963-67. From 1967 until 1975 worked on characterization of high- purity GaAs and compound semiconductor devices and avalanche photodiodes at MIT Lincoln Lab; since 1975 has been a Prof of Elec Engg at the Univ of IL, conducting res on compound semiconductor alloys and devices. Fellow in IEEE in 1977, Pres of Electron Devices Soc of IEEE 1984-85. *Society Aff:* EDS, IEEE, APS, ECS

Stinchcombe, James D
Business: Scott Plaza II, Phila, PA 19113
Position: Div-VP. *Employer:* Scott Paper Co. *Education:* BSME/ME/Univ of MI. *Born:* 5/3/22. Native of Flint, MI. Served with the USAF, 1942-1946. Was grad from the Univ of MI, where I was elected to Tau Beta Pi & Pi Tau Sigma. In 1949, joined Scott Paper as an engg trainee. Attended the Inst of Paper Chemistry for special studies in 1954. Assumed current responsibilities as VP of Div Engg. In this capacity am responsible for maj capital facilities projs for the Packaged Products Div. Mbr of Engg Advisory Bd of Widener Univ - Enjoy golf and gardening. *Society Aff:* TAPPI.

Stinson, Don L
Home: 1213 Garfield, Laramie, WY 82070
Position: Owner *Employer:* Don Stinson (Consulting Engineer) *Education:* PhD/Chem Engg/Univ of MI; Ms/Chem Engg/Univ of MI; BS/Chem Engg/Univ of OK. *Born:* Oct 1930 Hominy Okla. Worked for Phillips Petro Co, Gulf Res & Dev Corp & Univ of Wyoming & Arnjac Corporation before establishing own practice as a consulting engineer. Areas in int & res incl natural freezing for desalting water, energy recovery, secondary recovery of crude oil & natural gas, hydrocarbon fluid properties, coal recovery techs, underground waste disposal & reservoir inhomogeneities. *Society Aff:* ACS, AIChE, SPE, NSPE.

Stitch, Malcolm L
Business: P.O. Box 101, Florham Park, NJ 07932
Position: Engrg Associate *Employer:* Exxon Research & Engr Co *Education:* PhD/Physics/Columbia Univ; BS/Physics/So Meth Univ; BA/French/So Math Univ. *Born:* Apr 1923. Army 1943-46. Microwave device dev 1953-56 Varian Assocs. Maser dev 1956-60, laser dev & engrg 1960-68 at Hughes Aircraft as Group Hd, Sect Hd, Dept Mgr, Ch Scientist R&D Div. Laser dev, engrg 1968-73 at Korad Dept. Union Carbide Corp as Asst Genl Mgr. Cons Naval Res Lab, Univ So Calif, others 1973-74. Laser isotope enrichment at Exxon Nuclear as Section Mgr. 1974-81; Advanced instrumentation dev of Exxon Res. & Engrg Co since 1981. Fellow, SPIE 1981; Dev first application of lasers: rangefinding 1961. Fellow IEEE 1971; Fellow, Dir (So Calif) Inst Adv of Eng 1971-73; Fellow SPIE 1981; Chmn Electronics Indus Assoc, Laser Sect 1968-71; Adj Prof EE Univ of So Calif 1974- 76; Prog Chrmn and/or session organizer num IEEE, SPIE or DOD sponsored symposia 1963- . Organized with Dept of Commerce Intl Laser Colloquium in Paris 1969 to inaugurate opening of U S trade center. Many prof publs. Enjoy skiing, classical music. *Society Aff:* IEEE, OSA, APS, AAAS, SPIE.

Stivers, Theodore E
Business: T.E. Stivers, Chairman, T.E. Stivers Associates Inc, P.O. Box 550, Decatur, GA 30031
Position: Chairman *Employer:* T E Stivers Assocs, Inc. *Education:* BS/Milling Tech/KS State Univ. *Born:* Jan 26, 1920 Cleveland Tenn. Darlington School, Rome Ga 1931-35; Penn State Univ, State Coll Pa 1942, post grad work - Mech Engrg. Engrg Officer US Navy. 1953 estab cons firm; registration in over 25 states. Pres Cons Engrs Council Ga 1970; 'Cons Engr of Yr' 1972. GSPE 'Engr in Private Practice for Metro Atlanta' 1972. Pres Amer Soc of Agri Cons 1972-73; 'Disting Serv Award' 1973. State Bd of Regis for Prof Engrs & Land Surveyors 1970-80. Pres Natl Council of Engrg Examiners 1976-77. Dir Natl Council of Prof Serv Firms 1974-79. NCEE 'Distinguished Service Award' 1980. ASAE 'Cyrus H McCormick Award' 1982. *Society Aff:* NCEE, ASAE, ASAC, NSPE, ACEC.

Stockel, Ivar H
Home: 3 Joan Dr, New York, NY 10956
Position: Dir of R&D. *Employer:* St Regis Paper Co. *Education:* ScD/Mech Eng/MIT; MS/Mech Eng/MIT; BS/Mech Eng/MIT. *Born:* 4/30/27. Taught at US Naval Postgrad Sch 1950-54 & at US Naval Acad 1954-56. With St Regis Paper Co since 1956. Assumed current respon of Dir of R&D 1968. Corp Representative Indus Res Inst; Mbr of NSF Indus Panel on Sci & Tech. LCDR USNR Ret. *Society Aff:* CPPA, TAPPISoc of Rheology.

Stocker, Gerard C
Home: 76 Stockmar Dr, Basking Ridge, NJ 07920
Position: V P of Engrg & Mkting. *Employer:* Thomas Assocs Inc. *Education:* MS/Mgt Eng/Newark College of Engg; BE/ChE/Stevens Inst of Tech. *Born:* 12/13/44. Licensed as a Professional Engr in 19 states. Member of AIChE and NSPE. Employed in various engineering capacities for Chevron Oil Company, Witco Chemical Company, and Mesco Tectonics, Inc. Since 1973 VP of Engrg and Marketing for Thomas Assocs, Inc, engrg and construction firm specializing in liquefied petroleum gas systems. Author of several publications dealing with safety aspects of liquefied petroleum gas plants. Technical conslt and technical witness to LP-gas firms. Listed in current edition of "Who's Who in the East-. *Society Aff:* NSPE, AIChE.

Stockwell, Joe
Home: 6550 Branford Ct, Cincinnati, OH 45236
Position: President *Employer:* Pedco E&A Services Inc. *Education:* CE/Civ Engg/Univ of Cincinnati. *Born:* 10/6/11. in Cincinnati, OH. Proj Supt John Bryon State Park, Natl Park Serv 1934-40; Asst Ch Engr The Drackett Co 1940-45, designed & operated soy bean processing plant; Assoc Harold M Hermann & Assocs 1945-52, designed major distilleries & breweries; Partner Hall McAllister Stockwell 1952-75, designed field facilities Greater Cincinnati Airport, mech branches of Cincinnati Covention Ctr, Rockdale Temple, the Jewish Hosp, Northern Kentucky State Univ, Kings Dominion Amusement Park. Mbr ASHRAE, ASTM, Engrg Soc of Cincinnati, Bankers Club, Cincinnati & Kenwood County Clubs. 1975-1981 Partner A-E Design Assoc Designed Bldg Utility Sys for major pharmaceutral plants & chem plants. 1981 Pres Pedco E&A Services Inc. Designers of Industrial Plants. *Society Aff:* ASHRAE, ESC.

Stockwin, Herbert G
Business: 5950 Fairview Rd, Charlotte, NC 28210
Position: Vice President. *Employer:* Chas T Main Inc. *Born:* Jan 12, 1924. Bach of EE Clarkson Coll of Tech, Potsdam N Y June 1949. Employed by Chas T Main Inc 1955- . Presently in charge of Charlotte Dist Office. Respon for admin, production & sales effort of the firm in the SE region of the U S. Career activities incl Elec Engr, Proj Mgr & Const Mgr. Projs both domestic & foreign incl Panama, S Amer, Far East. Mbr IEEE & Natl Soc of PE's.

Stoddard, Stephen D
Business: PO Box 11, Los Alamos, NM 87545
Position: Ldr- Ceramics-Powder Met Sect. *Employer:* Los Alamos Sci Lab of U of Calif. *Education:* BS/Cer Engg/Univ of IL; -/Mathematics/CT College; -/Basic Engg/College of Puget Sound. *Born:* 2/8/25. US Army combat infantryman 1943-46 - ETO-BSM, CIB, PHM, USAR Capt CE. BS Ceramic Engg Univ of IL 1950. Coors Porcelain Co 1950-52. At Los Alamos since Oct 1952 in the Mtls Tech Groups. Amer Inst of Chem, Sigma Xi & Amer Ceramic Soc. Fellow of Amer Ceramic Soc & AIC. PACE Award of NICE 1964. Reg P E NM. P Chmn ACS Nuclear & Refractory Divs, P Tr, VP & Treas ACS, Pres of ACS 1976-77. Kiwanis, Elks, Masons, Goodwill Industries-Bd of Govs, Cty Commissioner, Justice of the Peace & Municipal Judge, Materials Tech Assoc, Inc-Pres & Treasurer 1978-present. Bd Mbr - Bank of Los Alamos 1983-present. Bd Mbr - Southrail Indian Arts Assoc 1987- . Bd Mbr - State of New Mexico - Nature Conservatory 1987- . *Society Aff:* ACS, NICE.

Stodola, Edwin King
Business: PO Box 360872, Melbourne, FL 32936
Position: Engrg Consultant-Electronics, Electronic Warfare, Systems *Employer:* Self Employed *Education:* EE/Elec Engg/Cooper Union; BS/Elec Engg/Cooper Union. *Born:* 10/31/14. Independent Consultant: Radar, communications, electronic warfare, and aviation. Born 10/31/14. Education EE, BS in EE; Cooper Union. 17 years with U.S. Army at DOD and ERADCOM, Ft. Monmouth; also with Reeves Inst Corp, SURC, etc. Papers and special reports on radar, electronics, aviation, etc in Proc IEEE, Aeronautics & Astronautics, AFCEA Signal, AOPA Pilot, and others. Over 20 pats in radio, radar, electronics. Fellow IEEE (citation: For contributions to dev of extended-range radar systems.), Fellow Radio Club/Amer, Assoc Fellow AIAA, Mbr NSPE, Assoc of Old Crows (Electronic Warfare Assn), others. Rated commercial pilot. Registered Prof Eng, FL & NJ. *Society Aff:* IEEE, AIAA, NSPE, AFCEA.

Stoecker, Wilbert F
Business: Dept of Mech & Ind Engg, Urbana, IL 61801
Position: Prof. *Employer:* Univ of IL. *Education:* PhD/Mech Engg/Purdue Univ; MS/Mech Engg/Univ of IL; BS/Mech Engg/Univ of MO at Rolla. *Born:* 12/15/25. Prof of Mech Engrg at the Univ of IL at Urbana-Champaign; author of three textbooks *Refrigeration and Air Conditioning, Principles for Air Conditioning Practice,* and *Design of Thermal Systems*; ASEE Western Electric Award; four Effective Teaching Awards from alumni; ASHRAE F Paul Anderson Award; Prof Degree in Mech Engrg from the Univ of MO at Rolla; Halliburton Education Leadership Award; Ralph Coats Roe Award, Am Soc. for Engr Educ.; Industrial consult; worked twelve summers in industry; play piano and organ. *Society Aff:* ASHRAE, ASME, IIR, ASEE.

Stoke, George W
Business: State Univ of N Y, Stony Brook, NY 11794
Position: Prof & Dir, Electro-Optical Sci Lab. *Employer:* State Univ of N Y Stony Brook. *Education:* BSc/Physics/Univ of Montpellier (France); Dipl Ing/Optics/Inst of Optics (Univ of Paris); DrEsSc/Physics/Sorbonne (Univ of Paris). *Born:* 1924. PhD Physics Sorbonne. Internatl auth on optical diffraction gratings, coherent optics & holography. His influence upon the creation of the fields of optical computing, digital & optical image processing has been most profound, extending his classic expers (some performed with Dennis Gabor, Nobel- prize winning originator of the field) & the writing of the world's first book on the subject (1966) that many consider the foundation of modern holography. Has been serving as cons to many of the world's leading institutions & govs, & was Visiting Prof at Harvard Univ 1970-73. Fellow APS, IEEE, OSA. Allan Gordon Mem Award 1971 SPIE. *Society Aff:* APS, IEEE, SPIE, OSA, SPSE.

Stoker, Warren C
Home: 4 Neptune Drive, Groton, CT 06340
Position: Pres Emeritus. *Employer:* The Hartford Grad Ctr. *Education:* EE/EE/RPI; MEE/EE/RPI; PhD/Physics/RPI. *Born:* Jan 1912. EE, MEE, PhD RPI. Joined Rensselaer fac in 1934, apptd Prof of EE 1951, Dir of Computer Lab 1952; Dir Hartford Grad Ctr 1955, Dean of the Ctr & Assoc Dean of Grad Sch in Troy 1957, V P & Tr RPI of Conn 1961, Pres & Tr 1974- 76. Pres Emeritus 1976. Mbr ASEE, Newcommen Soc, Sigma Xi, RESA, Eta Kappa Nu, Tau Beta Pi. Fellow, Life Member IEEE, Mbr Hartford Club. P Pres Conn Council on Higher Educ, Corporator of Mechanics Savings Bank Hartford. *Society Aff:* IEEE, ASEE, ΣΞ, HKN, TBΠ.

Stokes, Charles A
Home: 2355 Kingfish Rd, Naples, FL 33962
Position: Chrmn. *Employer:* Stokes Consulting Group. *Education:* ScD/ChE/MA Inst of Tech; ChE/ChE/Univ of FL; BSChE/ChE/Univ of FL. *Born:* 10/15/28. Floridian. Asst Prof ChE, MIT Chief Technical Officer, Cabot TX Butadiene, and Columbian Carbon (Cities Service). VP & Dir Petrocarb, Columbian Carbon, PCI, Columbian Intl, Plastics Mtls. At Cities responsible for tech & planning for chem & polymer operations. Consultant petrochemicals, fossil energy, and synthetic fuels since 1969. Expert in resource recovery waste-to-energy processes. War Production Bd 1944-45. Chrmn of Advisory Bd and annual meetings of Pittsburgh Coal Conversion Conf. Served as *Director Industrial Research Institute and Director AIChE, and Chairman Boston Section.* Engr of the Yr, Central Jersey 1976. Outstanding Technical Achievement of the Yr, FL Engg Soc 1979. Engineer of the Year, Florida Engrg Society 1984-85. Fellow AIChE Member of National Society of Professional Engineers. *Society Aff:* AIChE, ACS, NSPE, FES.

Stokes, Vijay K
Business: P.O. Box 8; K-1, 3A23, Schenectady, NY 12301
Position: Mechanical Engineer *Employer:* General Electric Company *Education:*

Stokes, Vijay K (Continued)
Ph.D./Mech. Eng./Princeton University; M.S.E./Mech. Eng./Princeton University; B. Sc. Eng. (HONS)/Mech. Eng./Banaras Hindu University, India *Born:* 08/26/39 Native of India. Professor at the Indian Inst of Tech, Kanpur, and Chairman of the Mech Engrg Dept 1974-77. Visiting Unidel Assoc Professor in the Dept of Chem Engrg, Univ of DE 1970-71. Senior Staff Engr, Foster-Miller Assoc, Inc., Waltham, MA 1971-1972. Joined General Electric's R & D Center in 1978 and worked on a variety of problems including the analysis of a novel washing machine, and the analysis of a process for making amorphous metal ribbons. Since 1980 worked on developing a comprehensive mechanical technology for load-bearing applications of plastics, including analysis procedures for rigid plastic foam structures, thermoplastic sheet forming, and vibration welding of plastics. Forty papers, fourteen patents and one book. *Society Aff:* ASME, SPE

Stoll, Robert D
Business: 611 SW Mudd Bldg, New York, NY 10027
Position: Prof of Civ Engg. *Employer:* Columbia Univ. *Education:* EngScD/CE/Columbia Univ; MS/CE/Columbia Univ; BS/CE/Univ of IL. *Born:* 8/12/31. Native of Lincoln, IL. Served in US Navy, Corp of Civ Engrs, 1953-55. Taught courses in soil mechanics and fdn design since 1957. Involved in res on the mechanics of particulate mtls such a soil and ocean sediment. Became associated with Lamont-Doherty Geological Observatory in 1968 and since leave carried out a prog of res on sediment acoustics and gas hydrates. Consultant to various firms on problems involving soil mechanics and fdn vibrations. *Society Aff:* ASCE.

Stoll, Ulrich W
Home: 2121 Hall, Ann Arbor, MI 48104
Position: Pres. *Employer:* Stoll, Evans & Assoc. *Education:* MSE/Structral/Univ of MI; BSE/Civ/Univ of MI. *Born:* 12/22/24. in Detroit, MI. Served as 2nd Lt Pilot USAF, WWII. Soils & mtls engr, S Bureau of Reclamation 1950 to 1956. Instr at Univ of MI, & staff engr MI Dept of Hgwys 1957 to 1963. Principal in geotechnical consulting firm 1963 to present. Past pres MI Sec, ASCE. Mbr Ann Arbor Bldg Appeals Bd 1970-80; Mbr Washtenaw County BPW 1978-82, Outstanding Civ Engr Award, 1978, Ann Arbor Branch, ASCE; Bd of Dirs, Glacier Hills Retirement Home. Mbr Washington Cnty Rd Commission 1984. Enjoy sailing & travel. *Society Aff:* ASCE, ASTM, ASFE.

Stoller, Sidney M
Home: 19 The Serpentine, Roslyn Estates, LI, NY 11576
Position: Crmn & CEO-Retired *Employer:* Self *Education:* BCE/CE/CCNY. *Born:* 3/24/24. Founder (1959), and retired (1/1/85) Chief Exec of the S M Stoller Corp which, since 1973, has been a subsidiary of Arthur D Little, Inc. The co has provided consulting services on technical, economic & policy considerations covering about half of the nuclear power projs committed in the US, & has provided such services to elec utilities, US govt agencies & foreign utilities. Direct contributions to the co's profl assignments include witnessing in numerous administrative and legal proceedings. Author of numeous technical articles & co-editor of the Nuclear Reactor Handbook, Volume 4 - Reprocessing; before forming co was VP Engg for Vitro Engg Co (1948-1959) & was on staff of Foster Wheeler Corp. *Society Aff:* NSPE, AIChE, ANS, AAA.

Stoloff, Norman S
Business: Mtls Engg Dept, Troy, NY 12180-3590
Position: Prof. *Employer:* RPI. *Education:* PhD/Met/Columbia Univ; MS/Met/Columbia Univ; BMetEng/Met Engg/NYU. *Born:* 10/16/34. Native of Brooklyn, NY. Jr Engr with Pratt & Whitney Aircraft Corp 1956-58. Res Scientist with Ford Scientific Lab 1961-65. Joined faculty of RPI in 1965, tenured 1967, promoted to Prof in 1970. Teach courses in mech properties, fracture of metals & alloys. Conduct res on deformation & fracture of intermetallic compounds, environmental effects on fatigue & fracture. Mbr, Exec Committee of Inst of Metals Div - AIME 1974-76; Chrmn, Hudson-Mohawk Chapter of AIME 1970-71. Author of one hundred and thirty technical articles, co-editor of five books; married, four children. Ed "Reviews on High Temperature Materials, Freund Publ Co Tel Aviv, Israel, 1978-1982 Assoc Ed High Temperature Materials, Freund Publ Co 1982-present, Vice Chairman, Editorial Ctte Interntl Materials Reviews 1982-present. *Society Aff:* ASM, ASTM, AIME, MRS

Stone, H Nathan
Home: 38 Sun Valley Dr, Worcester, MA 01609
Position: Vice Pres & Director Res & Engrg. *Employer:* Dresser Industries, Inc. *Education:* PhD/Chem Eng/Univ of IL; MS/Chem Eng/Univ of IL; BS/Chem Eng/Univ of NH. *Born:* May 1920, Claremont NH. 1943-46 active service WWII in Navy, final rank Ltjg USNR; 1943 Chem Engr US Bureau of Mines; 1950-52 Res Engr Norton Co; 1952 started similar position with present employer, sev promotions to present position of V Pres & Dir Res & Engrg 1965. Reg PE; Mbr & Chmn Mass Bd of Registration of Prof Engrs & Mbr Natl Council of Engrg Examiners for 5 years. *Society Aff:* ACS, ACS, NSPE.

Stone, Harold S
Business: ECE Dept, Amherst, MA 01003
Position: Prof. *Employer:* Univ of MA. *Education:* PhD/EE/Univ of CA; MSEE/EE/Univ of CA; BSE/EE/Princeton Univ. *Born:* 8/10/38. At Univ of MA since 1974. With SRI Intl, 1963-1968. At Stanford Univ 1968- 1974. Res in comp arch, distributed computers, microprocessors, & parallel computers. IEEE CS Governing Bd 1972-3, 1979-80. Author, co-author, & editor of six text books on computer arch, computer programming, & discrete math. Consulting editor, McGraw-Hill. Visiting prof, Univ of CA at Berkeley in 1977- 1978. *Society Aff:* ACM, IEEE CS.

Stone, Harvey H
Business: Rd 1, P.O. Box Q, Tidioute, PA 16351
Position: President *Employer:* Northwest Engrg, Inc *Education:* BCE/Civ Eng/RPI *Born:* 3/24/42 Native of Oceanside, New York. Employed by US Army Corps of Engrs, NY District, 1964-66. Employed by A W Cross Constuction Co, Inc as project engr in Portugal, Newfoundland and Thule, Greenland 1966-68. With Northwest Engrg, Inc, Tidioute, PA since 1968. Joined firm as project engr, in present position as Pres of Corp since 1980. Past Pres of Bucktails Chapter, PA Soc of PEs. Received Excellence Award for bridge design in 1980 from PA Consulting Engrs Council and in 1982 for inovative storm drainage design. Reg PE in PA and VT. Licensed as Surveyor, Sewage Enforcement Officer, Sewage Treatment Plant Operator and Water treatment plant operator in PA. *Society Aff:* NSPE, ACEC, ASHE, PEPP

Stone, Henry E
Home: 6805 Castlerook Drive, San Jose, CA 95120
Position: Consultant *Employer:* Self *Education:* MS/Engrg/Union College; BS Eng/Mech Eng/Univ of Buffalo *Born:* 02/10/22 Native of Germany. Served with Army Engrs in World War II. Worked over 38 years at GE, mainly in nuclear power related activities. 24 years Knolls Atomic Power Lab in design construction operation. testing of naval power plants including training of personnel. 14 years with commercial activities in engrg. Elected VP in GE in 1978. Presently consultant with electric utilities. *Society Aff:* ASME, ANS, NAE, Tau Bet Pi

Stone, Joseph K
Home: 3180 Lake Shore Dr 11F, Chicago, IL 60657
Position: Steelmaking Consultant (Self-employed) *Education:* MBA/Bus/Univ of Chicago; MetE/Metallurgy/Lehigh Univ; BS/Metallurgy/Lehigh Univ. *Born:* 10/3/14. Career, including portion of Army service, was in steel indus. With Kaiser Engrs for 20 yrs as Mgr Steel Plant & Process Dev, Author of L-D Process Newsletter, num tech papers, 6 patents; now in retirement, writes quarterly column for Steel Times International USA Update. Early exper include oper of pilot plants investigating pneumatic steelmaking for US Steel, plant metallurgical work for Jones & Laughlin & Weirton Steel. Mbr AIME, AISE & VDEh. Reg PE IL & CA. Interests include sailing, travel & oenology. *Society Aff:* AIME, AISE, VDEh.

Stone, Lawrence H
Business: 231 Johnson Ave, Newark, NJ 07108
Position: Pres. *Employer:* Kason Corp. *Education:* MBA/Marketing/NY Univ; MChE/Chem Engg/NY Univ; BChe/Chem Engg/Polytechnic Inst of Bklyn. *Born:* 8/19/31. Native of NY. Began his career as a res engg for General Foods Corp. Switched to sales engg first for Ecopumps and then Sharples Centrifuges. Became General Manager for Teknika Division, Lodding Corp in 1965. In 1967 formed Kason Corp a company specializing in particulate separation machinery. In 1977 became Chmn on the Bd and Chief Exec Officer for both Kason Corp and its Canadian Parent. Pres of the Process Equipment Manufacturers Assoc. Authored a tech article in the field of part separation in 1979. Guest Lecturer Clarkson Coll "Screening Machinery." *Society Aff:* AIChE.

Stone, Morris D
Home: 1308 Macon Ave, Pittsburgh, PA 15218
Position: Consulting Engr (Self-employed) *Employer:* Self *Education:* PhD/App Mathematics/Univ of Pittsburgh; MS/Mech Engg/Harvard Univ; BS/Mech Engg/Harvard Univ. *Born:* Dec 1902, Cambridge Mass. 1923-24 ASME Steam Table Res; 1925-34 Supervisory Res Engr, Westinghouse Electric Corp; 1934-71 Mgr, V Pres R&D, Dir & Cons, United Engrg & Foundry Co; presently Cons Engr. War-time cons to Materials Adv Bd, AEC & NATO; Chmn ASME-AOA shell & bomb mfg cttes. Hd Westinghouse Advanced Mech Design Sch & Adjunct Lecturer at Univ Pittsburgh 1925-34; Lamme Fellow 1930; Fellow ASME 1968. Enjoys classical music, reading internatl history & history of technology. *Society Aff:* ASME, AISE, AIME, ADPA.

Stone, Raymond V, Jr
Business: PO Box 28100, San Diego, CA 92128
Position: Executive Vice President. *Employer:* Neste, Brudin & Stone Inc. *Born:* June 1921, Berkeley Cal. BSCE-sanitary 1947 Univ Cal Berkeley; MS Sanitary Engrg 1948 Harvard, received Clemens Hershel Award. Worked as Res Chemist; Asst Dir Sanitary Engrg Res Proj; Assoc & Sr Sanitary Engr, State Public Health Dept; Exec Officer, Santa Ana Regional Water Quality Control Bd; 1963 joined Neste, Brudin & Stone 1963. Currently Pres Calif Water Pollution Control Assn, 1st V Pres Amer Public Works Assn, San Diego Chap. Mbr APWA, ASCE, AWWA, CCCE & LS, CEAC, SAME, CWPCA & WPCF, Sigma Xi & Delta Omega. Enjoy sailing, photography, restoring 1958 Thunderbird & avacado grove.

Stone, Richard G
Home: 1875 Grand View Dr, Oakland, CA 94618 *Born:* 1918. BSME Univ Minn 1940; MSME Univ Colo 1951. 1940-44 Indus Engr E I duPont; 1944-45 Engrg Officer US Merchant Marine; 1946-52 taught Mech Engrg at Univ of Denver; 1952-55 Res Engr, nuclear weapons dev, ACF Indus Albuquerque NM; since 1955, UC Lawrence Livermore Lab, Proj Engr, Div Hd & since 1971 Assoc Dept Head of ME Dept in design, R&D, mostly on nuclear weapons dev. Retired from the Lawrence Livermore Lab in Jan 1979. Now acting as a consultant. Enjoy sailing.

Stoops, Charles E
Home: 2346 Lynn Park Dr, Toledo, OH 43615
Position: Professor Emeritus of Chemical Engineering. *Employer:* University of Toledo. *Education:* PhD/Chem Engg/Purdue; BChE/Chem Engg/OH State Univ. *Born:* 12/17/14. Taught at Lehigh & Clemson Univ, served as Dept Chmn at Clemson & Univ of Toledo; 1944-67 with Phillips Petroleum Co at various res mgmt levels in Res & Atomic Energy Divs; other indus exper at Oldbury Electrochemical Co & Publicker Alcohol Co. Hold several patents, author tech papers. Fellow AIChE, received Herbert C Thober Award from Toledo Sect as outstanding chem engr. *Society Aff:* AIChE, ACS, AAAS.

Storb, John W
Business: 410 N. Easton Road, Willow Grove, PA 19090
Position: President *Employer:* Storb Inc *Education:* BS/Mil Art & Eng/West Point; MS/CE/Ga Inst of Tech *Born:* 4/26/24 1945-55 US Air Force, Installations Officer; 1955-64 Gulf Oil Corp - Regional Engr; 1964-present - Consulting Engrg Co - Owner Consultant to Petroleum Industry *Society Aff:* NSPE, SAME, PEPP

Storm, David R
Business: C/O Storm Engineering, 15 Main St, Winters, CA 95694
Position: Owner/Principal. *Employer:* Storm Engg. *Education:* BS/Civ Engr Hydraulics/Univ of NV, Reno; MS/Water Quality/Univ of CA, Davis; PhD/Environ Engrg/Univ of CA, Davis. *Born:* 5/7/31. David Storm has accumulated over 30 yrs of varied engg experience, the last 10 as hd of the environmental engg firm, Storm Engg. He received a BS in Civ Engg from the Univ of NV, Reno and advanced degrees, MS & PhD, from UC Davis. His res interests include land application tech for wastewater & wastewater solids and sanitary microbiology for both of which he has lectured and published. He is licensed in four (4) states as a Civ Engr and in CA as an Agri Engr and maintains mbrships in a number of profl socs. A recent JWPCF article co-authored by Storm and J Theis of the Sch of Medicine, UC David, reported on nearly four yrs of res on the presence and persistence of intestinal parasite ova in sewage sludge from US 12 urban communities. Storm has recently been involved in the planning and design of wastewater treatment facilities for both large and small table wineries in CA, WA, and the eastern Mediterranean. Storm is on the editorial staff of the *Practical Winery*, a tech journal for the wine industry and writes a bi- monthly article on winery water and waste. *Society Aff:* ASCE, WPCF, ASEV.

Storm, Herbert F
Home: 136 Fernbank Ave, Delmar, NY 12054
Position: Consultant (Self-employed) *Education:* DScEng/Elec Engg/Tech Univ, Vienna, Austria. *Born:* 6/28/09. in Vienna Austria. 1946 joined Genl Elec Co, served in GE's Corporate R&D in Schenectady NY until 1974. R&D work in magnetic amplifiers, semiconductor devices & circuits, solid-state pwr electronics. Taught at graduate sch of Rensselaer Polytechnic Inst, Troy NY. Over 20 patents, publs include book *Magnetic Amplifiers* translated into French, Japanese and Russian. IEEE Fellow, and positions of leadership including Founder & Pres of IEEE Magnetics Society, founder & 1st Chmn IEEE Intl Magnetics Conference (INTERMAG), Magnetics Group Award 'In recognition of outstanding contributions...,' 1968; Honorary chmn IEEE INTERMAG Conference, 1966; Honorary Life Mbr, IEEE, 1975; IEEE Magnetics Society Achievement Award, 1982; IEEE Centennial Medal, 1984. *Society Aff:* IEEE, ΣΞ.

Stormoe, Donovan K
Business: 662 Cromwell Ave, St Paul, MN 55114
Position: Mgr of Engrg *Employer:* Soil Exploration Co. *Education:* B/CE/Univ of MN *Born:* 12/29/40 Native of Hatton, ND. Served with Air National Guard 1963-69. Employed by Twin City Testing and Engrg Lab as technician, inspector and engr. Joined Soil Exploration Company as Project Engr in 1972 and named Mgr of Engrg in 1979. Responsible for Soil Explorations Company policy and consulting/testing service for environmental engrg. Served as Consulting Engrs Council Advisor to MN Pollution Control Agency. Treasurer-MSPE; MN Young Engr of the Yr 1974. Hobbies include reading, fishing, hunting, golf and skiing *Society Aff:* ASCE, NSPE, PEPP, ASTM, ACEC

Storms, James G
Business: 303 California Street, Garfield, WA 99130
Position: Chief Engr *Employer:* J E Love Company *Education:* BS/AgE/Univ of ID *Born:* 8/26/39 Native of Eastern WA and Northern ID. Grad from Univ of ID in 1961. Development Engr for Green Giant Co 1961-68. With J E Love Co since 1968. Named Chief Engr in 1979. Secretary-Treasurer Inland Empire Section ASAE 1979 and 1980. Regional Dir Pacific Northwest Region ASAE 1981-83. Dir Indland Empire Christmas Tree Association 1981-84. Scoutmaster Troop 577. Enjoy fishing and raising Christmas trees and nursery stock. *Society Aff:* ASAE

Storry, Junis O
Home: 105 Sunnyview, RR 3, Box 26, Brookings, SD 57006-9407
Position: Dean of Engineering Emeritus, and Professor Emeritus, Electrical Engineering Department *Employer:* South Dakota State University. *Education:* PhD/Elec Engg/IA State Univ; MS/Elec Engg/SD State Univ; BSEE/Elec Engg/SD State Univ. *Born:* 3/16/20. Astoria SD. 1942-46 Engr in Electric Sect Bureau of Ships, Navy Dept; 1946 design engrg with Reliance Electric; 1946 joined SDSU, 1st coordinator of SDSU Center for Power System Studies, 1971 assumed position of Acting Dean of Engrg, 1972-82 Dean, also Dir of Engrg Res & Engg Extension, 1982 Amdahl Distinguished Prof of Engrg. Summer exper with architects, Commonwealth Assocs & Commonwealth Edison. Special interest in electric power. Enjoy the great outdoors. *Society Aff:* IEEE, ASEE, NSPE.

Story, Vern R
Home: 2908 Mobley St, San Diego, CA 92123
Position: Weapons Handling Systems Engineer - Navseacen Pac *Employer:* DOD-Navy, Navseacen Pac, San Diego *Education:* ASEE/Electronics/Valporaiso Tech Inst; BSEE/Engg/Univ of MI. *Born:* 6/27/36. Self supporting as technician through univ. 10 years as Design & Sr Engr; since 1969 with SDCE; currently studying for business degree. Hobbies: organic gardening, dirt bikes & camping. *Society Aff:* IEEE.

Stott, Alfred F
Business: 57-W Grand Av, Chicago, IL 60610
Position: President. *Employer:* Aero-space Institute. *Education:* BS Aero Engg/Airplane Desing/Hancock College. *Born:* Oct 4 1911. Hon DSc Ind Northern Univ. 1932-35 Milwaukee Parts Corp, Milwaukee Wis, in charge of production, testing & service of Tank Aircraft Engines; 1935-59 V Pres & Dean of Engrg Aeronautical Univ, Chicago Ill, during this period also Asst Dir Training for 5 year US Army Air Corps Prime Contract to train Air Corps Mechanics; 1959- , Pres Aero-Space Inst. Aero Engr cons with indus; Reg PE Ill, FAA certificates as A&P Mechanic, Ground Instructor, Parachute Rigger. Assoc Fellow AIAA & Soc Licensed Aircraft & Engine Technologists; Prof Racing Pilots Assn; OX-5 Aviation Pioneers; Silver Wings Frat; P Master & Secy of Masonic Lodge; Boy Scouts; P Asst Chmn Aiviation Div ASEE. *Society Aff:* AIAA, SLAET.

Stott, Charles B
Business: Tucson, AZ 85744
Position: Advisory Engr *Employer:* IBM Corp *Education:* MS/EE/Cornell Univ; BS/EE/Tufts Univ *Born:* 6/27/24 Currently a mbr of the AdCom and Pres of the 8000 mbr IEEE Engrg Mgmt Soc. P Chmn of the Denver Chapter of Engrg Mgmt Soc, P VChmn of the Denver Sect of IEEE, Junior P Chmn of the Tucson Sect of IEEE, and VChmn of the Tucson Region, AZ Council of Engrg and Scientific Assoc (ACESA). Most of 32 yrs with IBM have been spent in mgmt positions in the product development labs in Endicott, NY, Boulder CO, and Tucson, AZ. He holds 2 US Patents, and is listed in Who's Who in the West. In the avocational area, received the Silver Beaver Award from the Boy Scouts of Amer. With wife, Barbara-Ann, serves as Southern AZ Regional Coordinators for United Methodist Marriage Encounter. *Society Aff:* IEEE, ACM SIGPLAN/APL, ΣΞ, ΤΒΠ

Stott, Martha A
Business: 160 E Grand Ave, Chicago, IL 60611
Position: Corp Secy-Treas. *Employer:* Aero-Space Institute. *Education:* BS/Aerodynamics/Aeronautical Univ. *Born:* Oct 8 1917, Western Springs Ill. Graduate of Lyons Township High Sch, Lagrange Ill, Natl Honor Soc student; BS Aero Engrg Aeronautical Univ, Chicago Ill 1937. 1937-59 Prof tutoring Math, Sci & Aircraft subjects; 1945-59 Tech Training Cons to aircraft indus; 1959- , Sr Aeronautical Engrg Instructor & Corporate Secy-Treasurer, Aero-Space Inst Chicago Ill.

Stott, Thomas E
Home: Swallow Hill Dr, Cummaquid, MA 02637
Position: Pres *Employer:* Thomas Stott & Assoc *Education:* BS/ME/Tufts University *Born:* 05/14/23 Married & 5 children. 1984 to present - Consultant, Thomas Stott & Assoc. Treas & mbr bd of dir ASME International Gas Turbine Inst Ch, Gas Turbine Cttee on Standards, International Standards Organization, Geneva. 1964-84 Pres ASEA-Star Inc. - International (Swedish base) Co designing manufacturing & installing marine steam & gas turbines systems in large merchant & Naval ships. 1956-64 Engrg & mgt positions - shipbuilding div, Bethlehem Steel Co - specialized in gas & steam turbine marine proposcoal. Honors - Fellow ASME, ASME Centenial Medalist, 1981 ASME Sawyer Award (major contributions to industry - gas turbines marine) U.S. Naval Officer - WW II. Publications in field. *Society Aff:* ASME, SNAME, ISO

Stotz, John K, Jr
Home: 3 Settlers Way, Setauket, NY 11733
Position: Asst to the Vice President Product Integrity and Environmental Protection. *Employer:* Grumman Corporation *Education:* BSME/Mech Eng/Cornell Univ. *Born:* 7/14/23. 1943-46 Army Signal Corp, 1st Lt; 1948 joined Grumman Aerospace Corp as Instrumentation Engr, respon for flight test instrumentation & data sys for experimental aircraft & ground tests, 1963 assigned Lunar Module (LM) Prog as Mgr Test & Support, respon for test & launch operations planning, support of field sites, training & logistics: Mgr Space Progs Proposals for 3 yrs joined Grumman Energy Sys Inc in 1977, as Manager, Wind Programs. Grumman Corporation, April 1981.Treasurer Instrument Soc Amer 1975-76. Commercial Pilot. *Society Aff:* ISA.

Stoudinger, Alan R
Business: Darling St, Angola, IN 46703
Position: Chrmn, EE Dept. *Employer:* Tri-State Univ. *Education:* PhD/EE/UT State Univ; MSEE/EE/CO State Univ; BSEE/EE/Tri-State Univ. *Born:* 4/16/38. Dr Stoudinger is chrmn of the elec engg dept at Tri-State Univ; primarily an engg univ located in the northeast corner of IN. Currently, Dr Stoudinger served as the Intl Pres of Eta Kappa Nu, the Elec Engg Honor Soc. Consulting interest is directed toward instrumentation & computer control applications. Dr Stoudinger is married, has four daughters & resides in Angola, IN. Other interests and activities include cabinet-making, hiking, and racquetball. *Society Aff:* HKN, IEEE.

Stout, Arthur W, Jr
Business: 1 State St Plaza, New York City, NY 10004
Position: President. *Employer:* Todd Shipyards Corporation. *Education:* BChE/ChE/Tulane Univ. *Born:* June 19 1920, Mobile Ala. BChE Tulane Univ 1941. 1943-46 US Navy, Cdr USNR retired; 1946-48 Todd Shipyards Hoboken Div, 1948-55 New Orleans Div, Genl Supt; 1955-60 Avondale Shipyards, Genl Supt; 1960-67 NY Shipbuilding Corp, Exec V Pres & Works Mgr; 1967-74 Todd Shipyards Houston Div, Genl Mgr, 1974-75 Corporate V Pres, 1975 elected Pres & Dir. Chmn Bd Designers & Planners, Inc; Mbr Amer Bureau of Shipping, SNAME; Dir Natl Ocean Industries Assn. *Society Aff:* ABS, SNAME, NOIA, ASNE.

Stout, Bill A
Business: Agri Engrg Dept, Texas A & M Univ, College Station, TX 77843
Position: Professor of Agricultural Engrg. *Employer:* Texas A&M University *Education:* BSc/Agr Engr/U of NE; MSc/Agr Engr/MI State U; PhD/Agr Engr/MI State U *Born:* 7/9/32. Neb. 1953 Res Engr with John Deere Waterloo Tractor Works; 1955-81, Instructor to Prof of Agri Engrg at MSU, 1970-75 Dept Chmn, on leave: farm power & machinery specialist FAO/Rome 1963-64, Visiting Prof Univ of Cal at Davis 1969-70, 1981- Prof of Agri Engrg at Texas A&M. Res & teaching related to agri machinery dev. Specialities: fruit & vegetable harvest & handling, physical properties. Extensive exper with mechanization in developing countries. Since 1975, work has dealt with energy education & focusses on conservation. Current research on computers and electronics applied to agriculture. Numerous pubs on energy use & conservation in the food sys, solar energy, biomass fuels & wind power. Fellow in ASAE and AAAS. Registered Profession Engr. *Society Aff:* NSPE, ASAE, ASEE, AAAS

Stout, George S
Home: 510 Hawaii Ave, PO Box F2430, Freeport Grand Bahama Island, The Bahamas
Position: Gen Mgr. *Employer:* Syntex Corp, Bahamas Chem Div. *Education:* BSCE/CE/Univ of MI. *Born:* 7/18/29. - Buffalo, NY. Work history - Merck & Co, Inc - 1951-1966. Syntex Corp - 1966-present. *Society Aff:* AIChE, ACS.

Stout, Glenn E
Business: Univ of IL, 208 North Romine, Urbana, IL 61801
Position: Director *Employer:* Water Resources Ctr *Education:* BS/Math/Findlay Coll; MS/Meteorology/Univ of Chicago; DSc/Sci/Findlay Coll *Born:* 03/23/20 Developed atmospheric sciences research program of IL St Water Survey between 1949-69 involving use of radar to measure precipitation and to detect tornadoes, research on variability of precipitation and raindrop sizes. Served as the NSF Scientific Coordinator for the National Ctr for Atmospheric Research between 1969-71, Assistant Chief, IL St Water Survey between 1971-73, and Dir of the Water Resources Ctr and Professor of Environmental Sciences, Univ of IL from 1973 to present. Research Coordinator for Illinois-Indiana Sea Grant Program 1987. Research on restoration of lakes and reuse of sediment in agriculture. In July 1984, I assumed the Exec Directorship of the Intl Water Resources Assoc, Sec General 1985-91. Author and co-author of 40 tech publs. *Society Aff:* AMS, AGU, AWRA, ΣΞ, IWRA.

Stout, James J
Home: 2831 N Van Buren St, Arlington, VA 22213
Position: Retired *Education:* BSCE/Civil Engg/KS State Univ. *Born:* 9/26/16. in NJ. Employed by Federal Power Comm, & its successor, Federal Energy Regulatory Comm, Wash DC, 1939-80, 1962-71 Ch Div of River Basins, 1971-80 Dir Div of Licensed Projs, in charge of all tech phases of FPC's hydroelectric licensing program, pub several tech papers on hydroelectric power. Independent Cons 1980-83. Fellow ASCE, Pres Natl Capital Sect 1965-66; Mbr Amer Water Resources Assn. *Society Aff:* ASCE, AWRA.

Stout, Koehler S
Home: 1327 W Granite st, Butte, MT 59701
Position: Assoicate Dean, Engg Div (Retired). *Employer:* Montana Coll of Mineral Sci & Tech. *Education:* MS/Geol Eng/MT Coll of Mineral Sci & Tech; BS/Min Eng/MT Coll of Mineral Sci & Tech; LLB/Amer Law Proc/LaSalle Ext Univ; Doctor Engrs/(Hon)/Mt Univ System. *Born:* Sept 1922, Helmville Mont. 1943-46 served in Army Ordnance European theatre of operation; Mine Engr to Mine Capt at Mt Hope Iron Mine, Dover NJ; since 1952 with Montana Tech in Engrg Div, Engrg Sci & Mining Depts, Acting Dean of Academic Affairs. P Pres Mont Mining Assn, World Museum of Mining; P Chmn Mont Prof Engrs & Land Surveying Bds. Admitted to practice law in Mont Courts & Mont Federal District Courts. Enjoy fishing & prospecting, commercial pilot. *Society Aff:* SME, NSPE, MBA

Stout, Melville B
Home: 1417 Morton Ave, Ann Arbor, MI 48104
Position: Professor Emeritus, Electrical Engrg. *Employer:* University of Michigan. *Education:* MS/Elec Engg/Univ of MI; BSEE/Elec Engg/Univ of MI. *Born:* Oct 17 1895, Pittsburgh Pa. Univ of Pittsburgh 1914-17; 1917-19 served in France, 15th US Engrs. 1920-22 Westinghouse Airbrake Co; 1922-64 Instructor/Professor Electrical Engrg Univ of Mich, since 1964 Professor Emeritus. Chmn Mich Sect AIEE 1942-43. Several tech articles. Speciality is electrical measurements & instrumentation. Author 'Basic Electrical Measurements.' Mbr Tau Beta Pi, Eta Kappa Nu, Sigma Xi; Fellow IEEE, ISA, AAAS. *Society Aff:* IEEE, ISA, ΤΒΠ, HKN, ΣΞ

Stout, Robert D
Business: Whitaker Lab 5, Lehigh University, Bethlehem, PA 18015
Position: University Professor *Employer:* Lehigh University. *Education:* BS/Metallurgy/Penn State; PhD/Metallurgy/Lehigh Univ. *Born:* Jan 1915, Reading Pa. 1935-39 Carpenter Steel Co; 1939- , faculty mbr at Lehigh Univ, 1960 -1980, Dean of graduate sch. Speciality is welding metallurgy. Author of 135 papers & a book. Pres AWS 1972-73, AWS awards: Lincoln Gold Medal 1943, Adams Lecturer 1960, Spraragen Award 1963, Jennings Award 1974, Thomas Award 1973;IIW Houdremont Lecturer 1970; ASM Stoughton Award 1952, A E White Award 1972, Fellow, Cons in welding & mech metallurgy. *Society Aff:* AWS, ASM.

Stout, Thomas M
Home: 9927 Hallack Avenue, Northridge, CA 91324 *Employer:* Retired *Education:* PhD/EE/Univ of MI; MSE/EE/Univ of MI; BS/EE/IA State. *Born:* Nov 1925. 1947-48 Dev Engr Emerson Electric; 1948-54 Instructor & Asst Professor Univ of Wash; 1954-56 Res Engr Schlumberger Instrument; 1956-58 various positions with Ramo-Wooldridge, 1958-64 TRW & 1964-65 Bunker-Ramo Corp concerned with early installations of digital computers for indus process control; 1965-85, Founder & Pres Profimatics Inc, engrg cons on computer applications for process indus; 1985- , retired. Fellow ISA; Sr Mbr IEEE; Mbr AIChE, TAPPI, ASEE & other prof & honorary socs. Reg PE; num publs; 4 patents. *Society Aff:* ISA, IEEE, AIChE, TAPPI, ASEE, SCS, ORSA, TIMS, AISE, NSPE, CSPE

Stoyak, Michael G
Business: 5085 Reed Rd, Columbus, OH 43220
Position: Member. *Employer:* Burgess & Niple, Limited. *Education:* BS/Civil Engg/Penn State. *Born:* May 26 1922 Green County PA. BSCE Penn State Coll. Reg P E Ohio, PA, Class III Operator- Wastewater & Water Supply Works. 1943-49 with Curtis Wright Corp, preparation of aircraft oper, maintenance & repair manuals for military aircraft; 1949- , with Burgess & Niple Ltd, Mbr of firm in charge of engrg services during const (acting as agent of Owner in bidding, contract award, contract negotiations, approval of materials, federal aid, financing & final acceptance of const projs), & resident/ engrs/inspectors on water, wastewater & related projs. Reports on water & wastewater facilities. Mbr NSPE, OSPE, ASCE, WPCF & AWWA; AWWA Dir Ohio Sect, Utility Man-of-Year Award 1974, Ohio Sect chmn 1973-74. Enjoy golf, photography & music. *Society Aff:* nSPE, WPCF, ASCE, AWWA.

Strahle, Warren C
Business: School of Aerospace Engrg, Atlanta, GA 30332
Position: Regents' Prof *Employer:* GA Inst of Tech *Education:* PhD/Aero Eng/Princeton Univ; MA/Princeton Univ; MS/Aero Eng/Stanford Univ; BS/ME/Stanford Univ *Born:* 12/29/38 Native of Whittier, CA. Mbr of Tech Staff with Aerospace Corp, 1964-67. Mbr of Professional Staff with Institute for Defense Analyses, 1967-68. With GA Tech since 1967. Numerous refereed publications and several research awards. Consulting in areas of aerospace propulsion, noise and fires. *Society Aff:* AIAA, CI, ASEE, ΣΞ

Strain, Douglas C
Business: 13900 NW Science Pk Dr, Portland, OR 97229
Position: Chairman of Board 1980-85 Vice Chairman 85- *Employer:* Electro Sci Indus Inc. *Education:* BS/EE/Caltech. *Born:* 1919 Spokane Wash. Office of Sci R&D 1941-46; Design Engr Beckman Instruments 1947-49; V P Res & Engrg Brown Electro Measurements 1949-53; Pres Electro Sci Indus 1953-80. Chmn of Bd 1980-85 NAS/NAE Adv Panel to Elec Div NBS 1967-75, Chmn 1971- 73. NBS Metric Adv Panel 1969-70. Natl Pres ISA 1970-71; ISA Fellow; Life Mem IEEE 1985; former SAMA Bd Mbr; P E, Bd of Dir Volunteers in Tech Assistance. Chmn Pacific Univ & Internatl Coll of Caymans. Dist Alumni Award, Cal Tech 1986 Reed Coll Howard Vollum Award for Sci & Tech 1975. Honorary Doctor of Humanities, Int. College of Caymans 1979. Married, 2 children. Trustee OR Grad Ctr 1979- *Society Aff:* ISA, SAMA, PMA, IEEE, PE.

Straiton, Archie W
Home: 4212 Farwest Blvd, Austin, TX 78731
Position: Ashbel Smith Prof of EE. *Employer:* Univ of Texas at Austin. *Education:* PhD/Physics/Univ of TX at Austin; MA/Physics/Univ of TX at Austin; BSEE/Elec Engg/Univ of TX at Austin. *Born:* Aug 1907 Arlington Texas. Married Esther Mc-

Straiton, Archie W (Continued)
Donnald 1932. Children: Janelle (Mrs Henry Holman), Carolyn (Mrs John Erlinger). Engr Bell Tele Labs, NYC 1929-30, Instr to Prof & Hd. Engr Texas Coll of Arts & Indus 1931-43. Assoc Prof to Ashbel Smith Prof, Univ of Texas at Austin 1943- . Chmn EE Dept 1966-71. Acting V P & Dean Grad Sch 1972-73. Fellow IEEE 1952. Mbr Acad of Engrg 1976. Res in Radio Wave Propagation incl Atmos Efforts on Millimeter Radio Waves. *Society Aff:* ASEE, IEEE.

Strand, Donald R
Business: 1660 W 3rd St, Los Angeles, CA 90017
Position: Assoc. *Employer:* Brandow & Johnston Assocs. *Education:* MS/Arch Engg/Univ of TX; BS/CE/Univ of CA. *Born:* Oct 1931 Chgo Ill. Struct Engr for bldg design with Brandow & Johnston Assoc. since 1957 & became an Assoc 1962. Reg Struct Engr Calif & Oregon. Served as Treas & V P Los Angeles Sect ASCE 1968-73 and Pres 1986. Dir SEAOSC & SEAOC 1974-77 & Chmn Seismology Ctte 1975. Pres SEAOSC 1981 and SEAOC 1983. Active on Natl ASCE & ACI cttes & has presented papers to these groups. *Society Aff:* ASCE, ACI.

Strandberg, Malcom W P
Business: 26-353, Cambridge, MA 02139
Position: Professor. *Employer:* MIT. *Education:* PhD/Physics/MA Inst of Tech; SB/Engg Sci/Harvard College. *Born:* Mar 9, 1919 Box Elder Montana. Joined MIT June 1941 as a Res Assoc in the Radiation Lab, & has remained there. Present pos is that of Prof of Physics. An expert in microwave physics, solid state physics, & of late in the physics of biological matter. Fellow IEEE, APS, Amer Acad of Arts & Scis, & Amer Assoc for Adv of Sci. *Society Aff:* APS, IEEE, APT.

Stratman, Frank H
Business: 5725 E River Rd, Chicago, IL 60631
Position: President. *Employer:* Reclasource Corp. *Education:* BSChE/Chem Engg/Univ of Louisville. *Born:* Oct 1920. Native of Louisville Ky. US Navy 1942-46. Genl Mgr Genl Amer Process Equip to 1961. Pres Infilco Inc to 1966. Genl Mgr Chgo Bridge Horton Process Div to 1970. Pres K-G Indus Inc (Berwind) to 1975. Pres Reclasource Corp (Berwind) to present. *Society Aff:* AIChE, AIME.

Straton, John W
Business: 201 N Kanawha, Beckley, WV 25801
Position: President & Chief Operating Officer. *Employer:* Gates Engrg Co. *Education:* BSEM/Coal/W VA Univ; MBA/Indus Mgmt/Univ of PA. *Born:* 1922 W Vir. P E 14 states. Worked as indus engr, mining engr & coal co mgr before beginning cons in 1961. Dir all mining cons of Gates until 1971, then became respon for all mining, civil & arch work of all offices. Respon for design of 8 million ton/yr underground coal mine. Presented or publ 15 tech papers, 1 reprinted in German mining magazine with internatl circulation. Reclamation Bd of Review (surface mining) 1971- . Chmn WVSPE-PEPP 1968-69; Pres WVSPE 1970-72; Natl Dir 1972-76; Chmn ACE's Coord Ctte on Energy 1976. Chrmn, Coal Div of SME 1979; SME dir 1980- . *Society Aff:* WVSPE, NSPE, AIME, UPADI, ACE.

Strattan, Robert D
Business: 600 S College Ave, Tulsa, OK 74104
Position: Prof of Elec Engr. *Employer:* Univ of Tulsa. *Education:* PhD/EE/Carnegie-Mellon Univ; MS/EE/Carnegie-Mellon Univ; BS/EE/Wichita State Univ. *Born:* 12/7/36. Specialist in Radar Cross Section Analysis and Radar Absorber Material Design and Application with Tulsa Div of Rockwell Intl. Joined faculty of Elec Engg at the Univ of Tulsa as Chrmn from 1968 to 1975. Acting Chrmn of Elect Engrg from 1983 to 1985. Consultant in Microwave Electromagnetics. Pres of OK Engg and Tech Guidance Council 74-75, Pres Tulsa Chapter of OK Soc of Professional Engrs 77-78, "Outstanding Engr" Award from Tulsa Chapter OSPE-77, Teetor Award from SAE, 1982, Chrmn of Tulsa Section IEEE 82-83, Vice Chrmn of SW Region PEE, NSPE 81-83. Chrmn of 1987 IEEE Region 5 Conference, 1987. *Society Aff:* IEEE, NSPE, SAE, ASEE.

Stratton, Frank E
Business: Civil Engrg Dept, San Diego State University, San Diego, CA 92182
Position: Professor *Employer:* San Diego St Univ *Education:* PhD/CE/Stanford Univ; MS/CE/Stanford Univ; BS/CE/San Diego St *Born:* 12/20/37 Born in San Diego County, CA. Assistant Professor of Civil Engrg, San Diego St Univ 1966. Assoc Professor, 1969; Professor, 1972-86. Chmn of Civil Engrg Dept, 1972-76. Asst Dean, 1983-86. Editor of Newsletter, Environmental Engrg Div of ASCE, 1973-76. Mbr Exec Ctte of ASCE's Environmental Engrg Div, 1978-82. Chmn of Exec Ctte, 1982. Dir of San Diego Regional MESA Program (to bring minority high school students into science and engrg careers) 1979-86. Engineering Educator. Specialist in water and wastewater treatment. Diplomat, American Academy of Environmental Engineers. *Society Aff:* ASCE, WPCF, AWWA, AAEE

Stratton, John M
Home: 2605 Clarendon Rd, Sedalia, MO 65301
Position: Corp Pkging Engr. *Employer:* Rival Mfg Co, Kansas City Mo. *Born:* Mar 22, 1941. BS Univ of Wisc - Stout. Native of Menomonie Wisc. USAF 1963-67. Pkging Engr for 3M Co 3 yrs for Reflective & Decorative Prods Divs 2 yrs. Pkging Engr for McQuay-Perfex Inc Mpls Minn; last 3 yrs Pkging Engr for Rival Mfg Co Kansas City Mo. Pres Kansas City Chap Soc of Pkging & Handling Engrs 1975-76, 1976-77. Enjoy photography & wood working.

Stratton, Julius A
Business: Rm 14N-112, Mass Inst Tech, Cambridge, MA 02138
Position: President Emeritus. *Employer:* MIT. *Education:* ScD/Math Phys/ETH, Zurich; SM/Elec Engrg/MIT; SB/Elec Engrg/MIT *Born:* 5/18/01. Joined MIT fac 1928. Subsequently Prof Physics, Dir Res Lab of Electronics, Provost, VP, Chancellor, Pres. Retired 1966. Pres Emeritus, 1966. Chmn Ford Found 1966-71. Chmn Comm on Marine Sci, Engrg & Resources 1967-69; Mbr Natl Acad of Sci (VP 1961-65), Amer Philosophical Soc, Founding Mbr Natl Acad of Engrg, Fellow Amer Acad of Arts & Scis, IEEE (Life Fellow), APS. Medal for Merit 1946; Distinguished Public Service Award, USN 1957; Medal of Honor IRE 1957; Faraday Medal, Brit Inst of Elec Engrs 1961. Num hon degrees. Life Mbr MIT Corp. Life Tr Boston Museum of Sci, Mbr Emeritus Corp, Charles Stark Draper Lab. Fellow, Amer Assn for the Advancement of Sci; Mbr, Coun on Foreign Relations, Sigma Xi, Tau Beta Pi; Eminent Mbr, Eta Kappa Nu *Society Aff:* NAS, NAE, APS, AAAS, IEEE, AAAS, American Physical Society

Straub, Harold E
Home: 3117 Ripplewood, Garland, TX 75042
Position: Chief Engr *Employer:* . *Education:* -/Hon of Mech Engr/Univ of MO, Rolla; MS/ME/Univ of IL; BS/ME/Univ of MO, Rolla. *Born:* 10/3/19. US Army 1941-45. Taught Heat Power 1949-51 Univ of Ill. Conducted res on room air distrib at Univ of Ill 1951-55. Res Engr for Titus Mfg 1955. Mgr Tech Dev for Environ Elements Corp, subsidiary of Koppers Inc Oct 1975. Past (Mbr) Chmn Air Diffusion Council Engrg Ctte; Reg P E Ill & TX. ASHRAE Fellow, Life Mbr. Mbr ASHRAE TC5.2 on Room Air Distribution. Enjoy sailing. Chief Engr Titus Products Div Philips Ind 1982. *Society Aff:* ASHRAE.

Straubhar, John J
Business: 250 S Beechwood Ave, Boise, ID 83709
Position: Pres *Employer:* J-U-B ENGRS Inc. *Education:* BS/Civ Engrg/UT State Univ; Assoc/Surveying/OR Tech; -/Gen/BYU *Born:* 8/8/35. Attended BYU & OTI. Native of Idaho. Started engrg interest in 1953 by working on survey crew. Served in USAF. Civ engr Idaho Dept of Hwys (EIT); Proj engr/proj mgr for Hood Corp, const of large irrigation proj; design engr for Johnson, Underkolfer & Briggs, Cons Engrs, spec in public works engrg; Proj Supt A & J Cons - earthfill dam, irrigation & pub works projs; Partner Riedesel & Straubhar Cons Engr, merged with J-U-B ENGRS 1971. Assumed respon for bus dev in 1976; As Pres respon for admin of co and all area offices of JUB, which include Boise, Couer d'Alene, Nampa, Twin

Straubhar, John J (Continued)
Falls, ID, plus Seattle and Kennewick, WA plus marketing admin of co, on Bd of Dirs of Aerial Mapping Co. of Boise, Also Exc VP of AMC, Pres of Cogeneration, Inc. A Hydro-Electric Power Generating Firm, in Idaho, Dir ACEC for CEI 2 yrs. Presently VP of CEI - 1974-75. Active in civic affairs, politics & prof soc's. Enjoy flying airplanes, skiing, sailing, golf, outdoor sports, hunting, fishing, *Society Aff:* ACEC, ASEE. AWWA, NSPE

Strauss, Alvin M
Home: 6520 Crest Ridge Cir, Cincinnati, OH 45213
Position: Dept Hd. *Employer:* Univ of Cincinnati. *Education:* PhD/Mechanics/WV Univ; AB/Phys/Hunter College. *Born:* 10/24/43. *Society Aff:* AAAS, SAE, SAME, AGU, AAM.

Straw, Henry F
Business: P.O. Box 2100, Denver, CO 80201
Position: Mgr-Project Dev *Employer:* Texaco Inc *Education:* BS/Petro Eng/Univ of TX *Born:* 7/9/28 Native of Gatesville, TX. Joined Texaco in 1951 and held several positions in South TX Producing Operations until 1964 at which time was named District Petro Engr at Billings, Montana. Named Assistant Div Petro Engr at Denver, CO in 1968, and in 1974 was appointed Staff Engr of the Alternate Energy and Resources Dept. Named Coord-Uranium Projects in 1981 and assumed current position as Mgr- Project Dev in 1983. Responsible for coordination and supervision of tech and economic evaluations, planning, engrg, design, construction, and operations of mines and facilities required for commercial exploitation of Texaco's coal and mineral reserves. Chmn, Denver Sect - SPE, 1974-75; Chmn, SPE Membership, 1977- 78; chmn, SPE Tech Coverage, 1982-83. *Society Aff:* SPE, SME, PE, GRC.

Strawbridge, Randall A
Home: 15 S Auburn Ave, Richmond, VA 23221
Position: President/Engr. *Employer:* Randall A Strawbridge Inc. *Education:* BCE/Structures/Cleveland State Univ; MSCE/Structures/Northeastern Univ. *Born:* Dec 29, 1935. MSCE Northeastern Univ; BCE Cleveland State Univ. Native of Townville Pa. Served in Army as CWO-Surveyor Tech 1960-63. Design Engr for Camp, Dresser & McKee Boston Mass in environ facils. One yr of study toward PhD at Lehigh Univ, Pa. Struct Engr with Torrence, Dreelinn, Farthing & Buford, Cons Engrs, Richmond Va. Started own cons engrg firm 1970; incorp 1974. Secy-Treas Richmond Branch ASCE; won engrg excell award from CEC/Va 1976. *Society Aff:* ACEC, ASCE.

Streebin, Leale E
Business: 202 West Boyd, Room 334, Norman, OK 73019
Position: Dir, CEES *Employer:* Univ of OK *Education:* PhD/CE/OR St Univ; MS/CE/OR St Univ; BS/CE/IA St Univ *Born:* 6/21/34 in Blockton, IA June 21, 1934. Surveyor for Henningson, Durham and Richardson, 1954 to 1957. Surveyor US Army, 1957 to 1959. Project Engr for Powers, Willis and Assocs, 1961 to 1964. Joined the faculty of Civil Engrg and Environmental Science at the Univ of OK in 1966 as Assoc Prof, Prof, 1971, Dir since 1979. Consultant to numerous industries and governmental agencies. Author of more than 30 articles and papers on industrial waste treatment, resource recovery and water reuse. Hobbies include jogging and skiing. Member of SigXi and PhiKapPhi. *Society Aff:* ASCE, NSPE, APWA.

Street, Robert L
Business: Dept of Civil Engrg, Stanford, CA 94305
Position: Prof Fluid Mechs & Applied Math, VP for Information Resources & Dir, Environmental Fluid Mechanics L *Employer:* Stanford Univ. *Education:* PhD/Fluid Mech/Stanford; MS/Hydraulics/Stanford Univ. *Born:* 12/18/34. Native of Honolulu Hawaii. Served in USN Civil Engr Corps 1957-60 (Proj Mgr Pacific Missile Range 1958-60). With Stanford Univ since 1962; Prof of Fluid Mech & Appl Math 1970- ; Assoc Dean for Res, Sch of Engrg 1971-83; Chmn Dept of Civil Engrg 1972-80; Vice Provost 1983-87; V Pres for Information Resources 1987- ; Sr Postdoctoral Fellow, Natl Center for Atmospheric Res 1978, Senior Queen's Fellow in Marine Science (Australia) 1985, Author: *Partial Differential Equations*, 1973; co-author: *Elementary Fluid Mechanics*, 5th Ed (1975-76), 6th Ed (1982); Mbr Rep to Univ Corp Atmos Res 1974- . (Mbr Bd of Trustees 1983- ; Chrmn Bd of Trustees 1987-) Mbr AAAS, ASME (Knapp Award 1985), AGU, ASCE (Huber Res Prize 1972), ASEE, Tau Beta Pi, Phi Beta Kappa, Sigma Xi. Teaching & res ints incl theoretical & exper hydrodynamics & numerical simulation of turbulent flow sys. Enjoys body surfing, volleyball, tennis as recreation. *Society Aff:* AAAS, ASCE, ASEE, ASME, AGU, ΣΞ, TBPI, ΦBK.

Streetman, Ben G
Business: Elec Engg, Urbana, IL 61801
Position: Prof. *Employer:* Univ of IL. *Education:* PhD/EE/Univ of TX; MS/EE/Univ of TX; BS/EE/Univ of TX. *Born:* 6/24/39. Prof of Elec Engg and Res Prof (Coordinated Sci Lab), Univ of IL at Urbana. He received the PhD in EE from the Univ of TX, Austin, in 1966. From 1964-1966 he was an Oak Ridge Inst of Nuclear Studies Fellow at the Oak Ridge Natl Lab. He joined the faculty of the Univ of IL in 1966, where his teaching and res is in the area of semiconductors and solid state devices. He is the author of "Solid State Electronic Devices" (Prentice-Hall, 1972, 2nd Edition, 1980). He has been elected to Tau Beta Pi, Eta Kappa Nu, and Sigma Xi. He is a Fellow of the IEEE and serves on the Exec Committee of the Electronics Div, Electrochemical Soc. Received Frederick Emmons Terman Award (ASEE) 1981. *Society Aff:* IEEE, ECS.

Strehlow, Roger A
Home: 505 S Pine St, Champaign, IL 61820
Position: Prof. Emeritus *Employer:* Univ of IL. *Education:* PhD/Phys Chemistry/Univ WI; BSc/Chemistry/Univ WI. *Born:* 11/11/25. I have worked on flame propagation, both laminar and turbulent, and the initiation and structure of detonation waves. My current interests are flammability limits and extinction mechanisms, blast waves from nonideal sources, accidental explosions and fire investigation. *Society Aff:* AAAS, ACS, APS, CI, AIAA, AIChE, NFPA, AΞΣ, ΣΞ.

Streicher, Michael A
Home: 1409 Jan Dr, Wilmington, DE 19803
Position: Research Professor. *Employer:* Univ of DE. *Education:* PhD/Metallurgy/Lehigh Univ; MChE/Chem Eng/Syracuse Univ; BChE/Chem Eng/Rensselaer Polytech Inst. *Born:* 9/6/21. Employed from 1949-79 at duPont Exper Station in res on engrg matls. Since 1979 Research Professor, Dept of Mech Engg, The Univ of DE., Newark, DE. Author 45 papers on sci of corrosion, new corrosion test methods, new stainless steels; 12 US pats. Cons on corrosion problems. Chmn Corrosion Div of Electrochem Soc 1958-60, The Gordon Conf on Corrosion 1962, Corrosion Subcomm WRC 1966- . Mbr NRC Comm on Chromium Utilization 1975-76, NSF Delegation on Corrosion to Russia 1975. Young Author & Turner awards of Electrochem Soc 1949, W R Whitney Award NACE 1972, Fellow ASM 1970, Sam Tour Award, ASTM, 1979. *Society Aff:* ASM, ASTM, AAAS, ΣΞ

Streifer, William
Business: Spectra Diode Labs, 80 Rose Orchard Way, San Jose, CA 95134
Position: Research Manager *Employer:* Spectra Diode Labs *Education:* PhD/Eng/Brown Univ; MSEE/EE/Columbia Univ; BEE/EE/CCNY. *Born:* 9/13/36. Lectr, Elec Engg, CCNY 1957-59. Res Engr, Heat & Mass Flow Analyzer Lab, Columbia Univ 1958-59. Sucessively, Asst, Assoc, & Prof of EE, Univ of Rochester, 1962-72. Visiting Assoc Prof, EE, Stanford Univ 1969-70. Principal Scientist, Xerox Palo Alto Res Ctr 1972-77, Res Fellow 1977-80. Sr. Res. Fellow 1980-85. Lectr, EE Dept, Stanford Univ 1976-79. Director, Ctr. for High Tech. Materials, Univ of New Mexico, Research Manager, Spectra Diode Labs 1986-present. Editor IEEE Journal of Quantum Electronics 1982-88. Fellow IEEE & Optical Soc of Am. Traveling lecturer IEEE Quantum Electronics & Applications Soc 1984-85. IEEE Centennial Service Medal 1984. IEEE Jack Morton Medal 1985. *Society Aff:* IEEE, OSA.

Streight, H R Lyle
Home: 17-174 Dufferin Rd, Ottawa, Ont K1M 2A6 Canada
Position: Intl Consulting. *Employer:* Retired *Education:* DSc BA/Honorary/Univ of Waterloo; PhD/Chem/Univ of Birmingham; MA/Chem & Math/Univ of British Columbia; BA/Chem & Math/Univ of British Columbia *Born:* 5/31/07. in New Westminster BC Canada. 1851 Exhibitioner 1929; Plummer Medal, Engg Inst Can 1958; Queen Elizabeth Medal, 1977. Postdoctorate Oxford Univ 1932-33. Managerial positions with Imperial Chem Industries, Canadian Industries & Dupont of Canada. Later with Canadian Enterprise Dev Corp & Canadian Exec Service Overseas. Engaged in res, dev, design of modern process equip, modernization of plants. Author of technical papers published in scientific journals in Canada & Europe. Patents in CD and other countries. Has been actively engaged in promotion of chem engg in Canada & foreign countries. Past pres of The Chem Inst of Canada. *Society Aff:* FCIC, ENG, AIChE

Streitmatter, Vern O
Home: 56 Winthrop Dr, Riverside, CT 06878
Position: Pres. *Employer:* USM Assoc, Inc. *Education:* BS/ChE/Univ of IL. *Born:* May 1921. Native of Central Ill. BSChE Univ of Ill. Grad Sch of Bus Admin Carnegie Tech. USNR 1944-46. Tech, production & admin pos with Victor Chem Works - Stauffer Chem Co. R T Vanderbilt in the indus minerals & specialty organic chems business 1969-85. Present position, Pres of USM Assoc, Inc., Manufacturers' Representatives for Industrial Packaging Machinery Suppliers. Mbr AIChE, AIME. Part-time farmer. *Society Aff:* AIME.

Stremler, Ferrel G
Business: 1425 Johnson Dr, Madison, WI 53706-1691
Position: Prof. *Employer:* Univ of Wisc - Madison. *Education:* PhD/Elec Engg/Univ of MI; SM/Elec Engg/M Inst of Technology; BS/Elec Engg/IL Inst of Technology; AB/Math/Calvin College. *Born:* 3/10/33. in Lynden, WA. Radar technician US Army 1953-55. Res engr in radar & optics at Willow Run Labs, Univ of MI 1960-67. With Univ of WI Madison 1967- , Senior Technical Staff Member, Environmental Research Inst of MI, 1986-87. Res ints in communication sys, radar sys, meteorological instrumentation & data processing. Western Elec Award ASEE 1975; Chancellor's Award Univ of WI 1976; Officer of Madison Sect IEEE 1973-77. Author: Intro to Communication Sys, 1977, 1982. *Society Aff:* IEEE, ASEE, ΣΞ

Strickland, James M, Jr
Business: 1315 Franklin Rd SW, Roanoke, VA 24016
Position: Partner. *Employer:* Hayes, Seay, Mattern & Mattern. *Education:* BS/Civ Engg/VA Military Inst. *Born:* 3/17/29. Engg officer with US Army Corps of engrs 1951-53. Civ Design Engr with consultanting firm and municipal waterworks 1953-56. With Hayes, Seay, Mattern & Mattern, Architects/Engrs/Planners since 1956. Currently a partner in the firm responsible for the practice of environmental engg. Projs include water, wastewater, flood control and solid waste facilities for industry, municipal, state and fed agencies. Reg PE in ten states. Enjoy sailing and fishing. *Society Aff:* AAEE, AWWA, ACEC, NSPE, WPCF.

Strickland, Jerry R
Home: 11710 Winchire Circle, Houston, TX 77024
Position: President. *Employer:* Wilson-Strickland Inc. *Education:* BS/ChE/Ark Univ. *Born:* BSChE Univ of Arkansas, with advanced degree work in bus admin at Babson Inst. Attended mgmt progs at Harvard & Texas A&M. Career work in oil field process equip, reactive metals, indus photography, & steel plate fabrication of heavy wall pressure vessels. Respons have ranged from Dist Engr to V P of Mkting for 50 million dollar sales oper. Currently pres & founder of service co providing expertise in refinery & petrochem plant turnarounds. Publ articles on applications for matls of const for process equip, use of instant photography in indus, & appli for special matls in refineries. Directorships in 3 corps; Mbr AIChE, API, Houston Chamber of Commerce. *Society Aff:* AIChE.

Striegel, Richard R
Business: 1243 Alpine Rd, Suite 210, Walnut Creek, CA 94596
Position: VP/Gen'l Mgr *Employer:* PRC Toups *Education:* BS/CE/US Military Acad *Born:* 2/8/44 Grad from West Point in 1966. Spent 5 yrs in Army including 1 yr in Vietnam. Worked as Chief of Party on Survey Crew for the City of Vista, CA, for one yr before joining PRC Toups in 1973 as Draftsman/Chmn. Became Project Mgr on Flood Plain Studies in 1976. Assigned as Genl Mgr of Norhtern CA Branch Office in 1978. Became V Pres in 1979. *Society Aff:* ASCE, NSPE, APWA

Stringer, Loren F
Business: 9015 Cliffside Drive, Clarence, SPE Corporation, Collins, NY 14031
Position: Chief Engr./Pres *Employer:* Stringer Power Electronics Corp. *Education:* PhD/Math/Univ of Pittsburgh; MS/EE/CalTech; BS/EE/Univ TX. *Born:* 9/28/25. Dr. Stringer joined the Westinghouse Elec Corp in 1947. In 1956, he became Mgr, Dev Engrg, Sys Control Div, Buffalo. In 1972, Dr. Stringer became Div Engrg Mgr, Ind Equip Div, Buffalo. In this capacity, he was responsible for planning and implementing strategic engrg power electronic & industrial automatic control progs for both the Ind Equip and Ind Sys Div. With the formation of the Power Electronics & Drive Systems Div in 1979, Dr. Stringer became the Div Engrg Mgr. Dr. Stringer became Chief Engr of that Div. in 1981. Dr. Stringer is the recipient of a Westinghouse Special Patent Award, the Westinghouse B G Lamme Scholarship, the Westinghouse Order of Merit, the 1984 IEEE Newell Award for outstanding contrib to power electronics and the 1985 Lamme Medal of the IEEE. Dr. Stringer holds 25 patents. Past Chrmn of the IEEE Static Power Converter Ctte, P Chrmn of the IEEE Power Semiconductor Ctte, P Chrmn of the NEMA Static Power Converter Section, P Chrmn of the Tech Support Subctte of ANSI Ctte C34, P Tech Advisor, USNC for TC22, SC22B, SC22D, and SC22E of the Intl Electrotech Commission. Dr. Stringer is P Secretary of SC22G on Static Power Converters for Motor Drive Systems and P Chairman of the IEEE/IAS Industrial Static Power Conversion Systems Department. Dr. Stringer was elected a fellow of the IEEE in 1980. In 1985, Dr. Stringer retired from Westinghouse & founded the Stringer Power Electronics Corp., Specialists in power conversion & automatic control systems technology, in Clarence, NY of which he is Pres & Chief Eng. *Society Aff:* IEEE.

Stringfellow, Gerald B
Business: Dept of Electrical Engrg, Salt Lake City, UT 84112
Position: Professor *Employer:* Univ of UT *Education:* PhD/Mat Sci/Stanford; MS/Mat Sci/Stanford; BS/Mat Sci/Univ of UT *Born:* 4/26/42 Received PhD from Stanford Univ-1968. Spent 13 yrs at Hewlett-Packard Labs, Palo Alto CA - Headed group of 18 scientists, engrs and technicians working on new semiconductor materials and devices. In Sept 1980 became professor of Electrical Engrg and Materials Sci and Engrg at Univ of UT. Published over 110 papers in scientific and engrg journals. Bd of Editors: 1) *Journal of Electronic Materials,* 2) *Journal of Crystal Growth.* Chairman - Electronics Materials Ctte of AIME. *Society Aff:* APS, ECS, EMC/AIME, IEEE

Stringham, Glen E
Business: Dept of Agri and Irrigation Engg, Logan, UT 84322
Position: Prof. *Employer:* UT State Univ. *Education:* PhD/Civ Engg/CO State Univ; BS/Agri Engg/UT State Univ. *Born:* 8/30/29. Born in Lethbridge, Canada. Taught in AE dept of CA State Poly Univ, 1955- 1957. Extension work at USU, 1957-1960. Res Scientist at UT Water Res Lab 1965- 1967. Dept of Agri Engg at USU 1967 to the present 1984. Maj responsibilities are in the area of surface irrigation. Have had foreign consulting assignments, Colombia, Saudi Arabia, Kenya, Honduras, Phillipines, and El Salvador. Served 2 yr term (1977-79) on the Bd of Dirs of ASAE. Active in community service and lay leadership positions in the Latter Day Saint Church. *Society Aff:* ASAE, ASCE, ASEE, SCSA.

Stripeika, Alexander J
Business: P O Box 808, Livermore, CA 94550
Position: Proj Manager. *Employer:* Lawrence Livermore Lab. *Education:*

Stripeika, Alexander J (Continued)
BS/EE/Cleveland State Univ. *Born:* 9/9/18. US Navy Electronics Schs Bowdoin Coll & MIT. US Electronics Officer 1944- 46. Electronics Engrg Lawrence Berkeley Lab 1946-52, working on nuclear counting instrumentation. Lawrence Livermore Lab 1952- , Div Leader 1958-79, Energy Mgmt 1980- , Sr Mbr IEEE, Mbr & VP Nuclear & Plasma Scis Soc, Mbr Assoc of Energy Engrs. *Society Aff:* IEEE, NPSS, AEE.

Strohbehn, John W
Business: Thayer School of Engrg, Hanover, NH 03755
Position: Sherman Fairchild Prof of Engrg *Employer:* Dartmouth College. *Education:* PhD/Elec Engg/Stanford Univ; MS/Elec Engg/Stanford Univ; BS/Elec Engg/Stanford Univ. *Born:* 11/21/36. San Diego, CA. Educated at Stanford in Elec Engg. Joined the faculty of the Thayer Sch of Engg, Dartmouth College, Hanover, NH in 1963 where he presently holds positions as Sherman Fairchild Prof of Engg and Adjunct Prof of Medicine, (Radiation Therapy) Dartmouth Medical Sch. His res efforts have been in the field of radiophysics including microwave and optical propagation through the atmosphere, and in biomedical engg including image processing, tomography, and hyperthermia in cancer therapy. *Society Aff:* AAAS, IEEE, OSA, URSI, AAPM, NAHG.

Strohman, Rollin D
Business: Agr Eng Dept, Cal Poly, San Luis Obispo, CA 93407
Position: Professor *Employer:* Cal Poly *Education:* PhD/Agr Eng/Purdue U; MS/Agr Eng/Univ of IL; BS/Agr Eng/Univ of IL; BS/Agr Sci/Univ of IL *Born:* 10/29/39 Grew up on farm in IL. While at the Univ I spent three summers at John Deere. Since 1969, I have taught at Cal Poly in the Agricultural Engrg Dept in the areas of Physical Properties of Agricultural Materials, Agricultural Process Engrg, Photogrammetry, and Remote Sensing. Research interest - microprocessors and computer aided mapping. Present Rank - Full Professor. 1976 - Sabbatical at U. C. Berkeley in Photogrammetry and Remote Sensing. 1981 - Chmn Southern CA Sect ASAE. Summer 1981 Recipient of NSF Science Faculty Professional Development Award which was spent at Deere and Co working on CAD/CAM 1982-sabbatical at Deere and Co working with Finite Element Analysis and Cast Iron Property Prediction. *Society Aff:* ASAE, ASPRS

Stroman, Robert M
Home: 9035 Cliffside Dr, Clarence, NY 14031
Position: Consultant *Employer:* Self employed *Education:* BS/Chem Engg/Univ of Rochester; BS/Bus Admin/Univ of Buffalo; AMP/Mgt/Harvard. *Born:* Apr 1924. BSChE Univ of Rochester; BS Bus Admin Univ of Buffalo; Advanced Mgmt Prog Harvard Bus Sch. Phi Beta Kappa, Tau Beta Pi, Beta Gamma Sigma. Ltjg US Naval Res WW II. Res, Production, Process Engrg & Sales pos at Durez Plastics 1945-67, Managing Dir Hooker Chemicals in Belgium 1967-70, Genl Mgr Hooker in Ft Erie Canada 1971, Bus Mgr Molding Compounds at Durez Div of Hooker 1971-73, Genl Mgr Puerto Rico Chem 1974-76. Formerly Natl Dir SPE, Mkting Mgr - Industrial Chems 1976-77; Dir Engg Procurement 1977-82. Bus. Mgr. Energy From Waste 1983-85, consultant in plastics, waste, and real estate dev. 1986-87. *Society Aff:* AIChE, SPE.

Strombotne, Richard L
Home: 24401 Peach Tree Rd, Clarksburg, MD 20871
Position: Dir, Office of Automotive Fuel Economy Stds. *Employer:* Nat'l Hwy Traffic Safety Admin. *Education:* PhD/Physics/Univ of CA, Berkeley; MA/Physics/Univ of CA, Berkeley; BA/Physics/Pomona College. *Born:* 5/6/33. Native of Watertown, SD. Thesis in nuclear magnetic resonance. Measured heplus & fine structure at NBS, 1962-68. With US Dept of Transportation since 1968. Planned and initiated program to assess effects on environment of supersonic aircraft fleet in 1971. Planned and directed Automotive Energy Program 1972-76. Chief, Energy & Environment Div, 1973-78. Director, Office of Automotive Fuel Economy Stds since 1978. Led or participated in many studies involving interactions among transportation, energy, and air quality. Recevied Departmental awards for Superior Achievement & Meritorious Achievement. *Society Aff:* APS, SAE, AAAS.

Strong, Benjamin R, Jr
Business: 1615 Broadway, Oakland, CA 94612
Position: Sr Tech Consultant *Employer:* ECHO Energy Consultants, Inc. *Education:* MS/Thermoscience/Stanford Univ; BS/Thermo-ME/Univ of IL - Chicago; *Born:* 12/25/46 Raised in Chicago, IL. With Sargent & Lundy Engrs to 1973 specializing in fluids and safety analysis. Joined EDS Nuclear, Inc. in 1974 consulting in nuclear power; specializing in thermal-hydraulics, heat transfer and safety analysis. Formed the Thermal-Hydraulics Section and led the group until assuming respons as a Senior Tech Specialist. Early participant in the (EPRI) Nuclear Safety Analysis Center investigation TMI-2. Mbr of the ANS-56.10 Standards Writing Group. Ten Publications. Principle in the recently formed ECHO Energy Consultants, Inc. responsible for tech and business dev in nuclear and solar energy. Art and music avocations. *Society Aff:* AIAA, ANS, ASME, NSPE

Strong, James D
Home: 3008 Dartmouth Road, Alexandria, VA 22314-4824
Position: Staff Consultant *Employer:* American Society for Engineering Education *Education:* B.S./Engr-Gen/Yale University; U.S. Army Engineer School; U.S. Army Command & General Staff College *Born:* 05/09/09 Native of Minnesota. Served in Army Corps of Engineers in World War II & Korea. Executive Officer, Ordnance Training Center. Director of Education, European Command Intelligence School. General Staff duty in U.S., Europe, Far East. Engineer Group Commander, Korea. Department Director, U.S. Army Engineer School. Professor of Military Science & Dept. Chairman, also Adjunct Prof, College of Engrg, Univ of CA, Berkeley. Dir, Educational Services Div, National Soc of Prof Engs. Sr Commissioner, Engrg Manpower Commission, AAES. Awards: Tasker H. Bliss Medal, SAME; Distinguished Service, AAES. Reg PE. Avocation: professional tennis official & player. Married, four children. *Society Aff:* ASEE, ASCE, SAME

Stroud, Roger G
Business: 600 Lynnhaven Parkway, Virginia Beach, VA 23452
Position: President *Employer:* Stroud, Pence and Assoc, Ltd *Education:* ME/Struct Eng/Old Dominion Univ; BS/CE/NC St Univ *Born:* 11/3/47 Design Engr for the Newport News Shipbuilding & Dry Dock Co 1970-73, responsible for the structural design of several major portions of Nuclear Powered Submarines. Joined a private consulting firm in Norfolk, VA in 1973 and two yrs later helped found the consulting firm of Craig and Stroud, Ltd, the parent firm of Stroud, Pence and Associates, Ltd, where he currently serves as Pres. Responsible for the structural design of many major building structures in the Southeastern region of the US. Adjunct lecturer at Old Dominion Univ. Bd Mbr and P Chmn of the Volunteers of Amer, a national voluntary service organization. President of the Tidewater Chapter of the Virginia Soc of Prof Engrg 1987-88. *Society Aff:* ASCE, NSPE, ACI, PEPP, ACEC

Stroup, Larry D
Business: Box 3325, Albuquerque, NM 87190
Position: President *Employer:* Dale Bellamah Land Co, Inc *Education:* BS/CE/Univ of IL *Born:* 12/27/42 Born in IL - Career developed through employment with consulting engrg firms in highway and subdivision engrg. Licensed as a reg surveyor and/or engr in 12 states throughout the Midwest, South, and Southwest. Currently pres of one of the largest land development firms in the Southwest. *Society Aff:* NSPE, ASCE

Strull, Gene
Business: Mail Stop 3519, Bx 1521, Baltimore, MD 21203
Position: Genl Mgr, Advanced Tech Div *Employer:* Westinghouse Elec Corp. *Education:* BS/Elec Engg/Purdue Univ; MS/Elec Engg/Northwestern Univ; PhD/Elec Engg/Northwestern Univ. *Born:* 5/15/29. Prog for Mgmt Dev, Harvard Univ. 1967, Predoctoral NSF Fellowship 1952-54. Native of Chgo Ill. With Westinghouse Elec Corp since 1954, with Matls Engrg, Semiconductor Div, Air Arm & now Sys Dev

Strull, Gene (Continued)
Div. Currently Genl Mgr Advanced Tech Div. Lectr Univ of Pittsburgh 1954-58. 21 pats, 40 pubs & 100 talks & seminars. Army Sci Bd, S NASA Adv Ctte pos since 1967. Fellow IEEE. Reg PE Md. Assoc Mbr Defense Sci Bd. *Society Aff:* IEEE.

Strum, Robert D
Home: PO Box 4461, Carmel, CA 93921
Position: Assoc Prof. *Employer:* Naval Postgrad Sch. *Education:* MS/EE/Santa Clara Univ; BS/EE/Rose Poly Inst. *Born:* 9/23/24. Born in LaCrosse, WI, on Sept 23, 1924. He served in the USN during WWII and in 1946 received the BSEE, degree from Rose Poly Inst. Strum worked for GM, then returned to Rose to teach in the mech & elec engg depts from 1948 to 1958. In 1958, Strum joined the faculty of the Naval Postgrad Sch where he currently serves as Prof of EE. He was awarded the MSEE degree from Santa Clara Univ in 1964. Strum's main areas of interest are control theory, digital signal processing & educational tech as applied to engg educ; he has co-authored several texts in circuit & linear system theory. He was a sr Fulbright Res Scholar in England during the academic yr 1970-71, working with British Univs in applying educational tech to engg educ. In 1971, he received an ASEE Western Elec Award for outstanding contribution to engg educ. In addition to his other academic assignments, strum currenlty prepares self-instructional modules for US Naval Officers on and off campus. *Society Aff:* AAAS, IEEE.

Struth, Bert W
Home: 30 Blair Rd, Armonk, NY 10504
Position: Exec V Pres. *Employer:* Chem Systems Inc. *Education:* BChE/ChE/NY Poly Inst; MChE/ChE/NY Poly Inst; DChE/ChE/NY Poly Inst. *Born:* Aug 11, 1927. Doctorate Chem Engrg New York Poly Inst 1955; MS 1952 & BS 1948. Employed by Allied Chem 1948-50 in chem plant operations. Joined Esso Res & Engrg 1954. Filled pos in Process Planning of Refinery & Petrochem facils, Process Dev & Process Design. Major contrib in dev of high temp short residence time steam cracker for olefins. In 1965, joined Chem Systems Inc, a cons & engrg co in NYC. Advanced to Mgr of Process Engrg, V P. Current pos Exec V P. Respon for Chem Sys U S cons & engrg opers. V P & Bd Mbr 1969, Exec V P 1975. *Society Aff:* AIChE.

Strutt, Max J O
Business: Kraehbuehlstr 79, Zurich, Switzerland 8044
Position: Prof & Director. *Employer:* Swiss Federal Gov. *Education:* DSc/Elec Engg/Univ Delft Holland; MSc/Elec Engg/Univ of Delft Holland. *Born:* Oct 2, 1903. Dutch citizen. Res Engr Philips Co Eindhoven Holland 1927- 48. Prof & Dir Dept Advanced Elec Engrg Swiss Fedl Inst of Tech 1948-74. Industrial cons. 1974- . 9 books on Electronics & Transistors, 7 in German, 2 in American. More than 70 U S patents. More than 400 sci papers. D Eng Hon degree Univ of Karlsruhe 1950. Gauss Medal 1954. Fellow IEEE 1956. McKay Prof Elec Engrg U C Berkeley 1960 through 1963. Sr Sci Adv Ctte NASA Proj 1962-63. Foreign Sr Scientist Fellowship NSF Wash 1966-67. Hon Mbr Electronics Assn of Japan 1965. Hon Mbr Institute Electronics & Communications Engrs Japan 1967. Life Fellow IEEE. *Society Aff:* IEEE.

Stryker, Clinton E
Home: 7032 N Barnett Ln, Milwaukee, WI 53217
Position: President. *Employer:* Maypro Inc. *Education:* EE/Electrical/Armour Inst of Technology. *Born:* Feb 27, 1897 Chgo Ill. Asst Prof EE Armour Inst of Tech. Ch Engr Fansteel Met Corp 1923-35. Partner Mc Kinsey Kearney Cons 1936-40; Exec V P Nordberg Mfg Co 1940-46; Pres Adel Precision Prods Corp 1946-48; Pres Maysteel Prods Corp 1948-74; Pres Maypro Inc 1964- . Reg P E Wisc. Fellow IEEE, Mbr SAE. Pres United Community Services Milwaukee 1954-57; Bd of Dir Goodwill Indus Amer 1963-69; Bd of Dir Milwaukee YMCA since 1951. Adv Bd St Mary's Hosp Milwaukee since 1953; Pres Amer Metal Stamping Assn 1957-58; Tr Ill Inst Tech 1962-65; Ill Tech Medal of Honor 1968. *Society Aff:* IEEE, SAE.

Stuart, Thomas A
Business: Elec Engg Dept, Toledo, OH 43606
Position: Assoc Prof. *Employer:* Univ of Toledo. *Education:* PhD/EE/IA State Univ; ME/Engg/IA State Univ; BSEE/EE/Univ of IL. *Born:* 2/6/41. Circuit design and test engr during 1963-69. Employed by Martin, Honeywell, and Collins Radio. Grad res asst at IA State during 1969-72. Asst Prof in EE at Clarkson College from 1972-75 (maj interest: computer applications in elec power engg). Assoc Prof in EE at Univ of Toledo from 1975 to present (maj interests: comp modeling of elec power systems, machines, and power electronics). *Society Aff:* IEEE.

Stubberud, Allen R
Business: Department of Electrical Engineering, University of California, Irvine, CA 92717
Position: Prof EE (on leave) & Division Director NSF *Employer:* Univ of CA, National Science Foundation *Education:* PhD/Engg/Univ of CA; MS/Engg/Univ of CA; BS/EE/Univ of ID;. *Born:* 8/14/34. Native of Glendive, MT. On faculty of UCLA from 1962 to 1969 as Assoc Prof. Since 1969, has been with Univ of CA, Irvine. Formerly Dean & Currently Prof of Elec Engrg (on leave). From 1983-85 was with US Air Force as Chief Scientist. Since 1985 has been with NSF as Director of Electrical, Communications, and Systems Engineering Division. Has authored or co-authored over 70 articles & books on estimation & control. Fellow of the IEEE. Past Chair of the Orange County Sec. Past Chair of Los Angeles Council. Vice Chair of Region VI. Board of Directors of ECM/ECI. Is Assoc Fellow of AIAA. Appointed Chair of the Guidance & Control Technical Committee 1/1/78-1/1/80. Chair of the 1980 WESCON Profl Prog Committee. Reg PE in the State of CA. *Society Aff:* IEEE, AIAA, ΣΞ, ΤΒΠ, ΣΤ, ORSA.

Stubbles, John R
Business: Res Dept, Independence, OH 44131
Position: Staff Engr *Employer:* LTV Corp *Education:* PhD/Phys Chem/Univ of London; BSc/Metallurgy/Univ of Manchester *Born:* May 16, 1934 near London England. BSc (1st class honors) in met 1954 from Univ of Manchester. PhD Phys Chem 1957 from London Univ. 2 yrs of postdoc res at Princeton Univ were followed by 4 yrs of lecturing in high temp met at Manchester. From 1963 to 1979, held var tech pos with the Youngstown Sheet & Tube Co (Lykes Corp). Enjoys studying the hist of met, playing golf, & listening to classical music. Former Chmn of PCSC (AIME). Former Chmn of Long Range Planning Ctte (ASM). Former Chmn of Continuing Education Ctte (ISS AIME). Certified Mgr (NMA), Certified Quality Engr (ASQC), Mbr of AISI Steel Fellows Program. Presently with LTV Steel Corp as Staff Engr, Res Dept. Married, 3 sons. *Society Aff:* ISS-AIME, AISE, ASM, ASQC

Stubbs, William K
Home: 1614 Andover Rd, SE, Grand Rapids, MI 49506
Position: Principal *Employer:* Stubbs Consultancy *Education:* BS/EE/Univ of Cincinnati. *Born:* 5/26/25. Covington, KY. USAF Officer & Aerial Navigator 1943-1945. Employed 1950 to 1962 by Lear Inc specializing in Dev, Design & Proj Mgt of electromech & electronic equip & systems for aerospace applications. Asst Engg Mgr of Electromech Div 1959 to 1962. Joined Rapistan Inc in 1962 as VP Dev. Became VP Engg, responsible for both Product Dev & Engg activities in 1965. In 1970, assumed additional responsibilities for Overseas Operations, Controls Operations, & consulting services. From 1978 until 1980, responsible for Intl & Systems Engg Groups, Controls Operations, & Consulting Services. Joined Mannesmann Demag Corporation in 1980 as Vice President for Corporate Development. In 1982 became Sr VP responsible for Material Handling Sys Activities in N Amer since 1982, Pres Barrett Electronics Corp. Established Stubbs Consultancy in 1986 serving as Consultant in Business Strategy and Planning, Acquisition, Technology Assessment & Transfer. *Society Aff:* IEEE, IMMS.

Stude, Robert A
Business: 800 W 47th St, Ste 600, Kansas City, MO 64112
Position: VP *Employer:* Boyd, Brown, Stude & Cambern *Education:* MS/CE/Univ of KS; BS/CE/Univ of TX-Arlington *Born:* 9/23/38 in Topeka, KS. Design Engr, Howard Needles, Tammen & Bergendoff. Partner and VP, Boyd, Brown, Stude & Cambern, Consulting Engrs since 1968. Past Pres, American Concrete Inst MO Chapter. Past Pres, Kansas City Post, Society of American Military Engrs. Served in US Marine Corps Reserves 1956 to 1963. Reg PE, States of MO, KS, IA, MI, OK, LA, and CO. *Society Aff:* SAME, NSPE, ASCE, ACI

Stuetzer, Otmar M
Home: 708 Lamp Post Circle SE, Albuquerque, NM 87123
Position: Dept Mgr. *Employer:* Sandia Labs. *Born:* 1912 Nuernberg Germany. PhD 1938 Univ of Munich. 1936-45 Dept Hd, Aircraft Radio Res Inst (FFO): radar. 1946-55 Brannch Chief, Wright Patterson AFB Ohio: vacuum tubes, transistors. 1955-63 Dir of Dev, Genl Mills Inc Mpls Minn: Electronics Div. 1963 to present, Dept Mgr, Sandia Labs, Albuquerque N M: electronic components. Lilienthal Soc Award 1942. Fellow APS, IEEE. Enjoys birdwatching & arguments.

Stuever, Joseph H
Business: 5815 Melton Drive, Oklahoma City, OK 73132
Position: Pres *Employer:* Stuever & Associates *Education:* BS/ME/Univ of Notre Dame *Born:* 3/19/32 Native of OK, worked in nuclear research in Albuquerque/Los Alamos area 1958-68; returned to OK in 1968. Initiated Stuever & Associates Consulting Engrs in 1969 specializing in industrial and pollution control processes. Continued to assist industrial clients in design of manufacturing and pollution control processes designed to meet requirements of governmental regulations. Pres and principal engr of Stuever & Associates. *Society Aff:* NSPE, AAEE, APCA, WPCF, SES, ASM

Stufflebean, John H
Home: 4117 East Kilmer St, Tucson, AZ 85711
Position: Retired *Education:* BS/Civil/Univ of MO. *Born:* Oct 1918 Brookfield Mo. Const inspector Mo Hwy Dept 1940, Jr Engr CB&QRR Co 1941-42. Served in Corps of Engrs 1942-44 in Design & Dev Sec Engr Amphibian Command. Released 1st Lt. Sr Engr Southern Pacific Co 1945-46 in respon charge of main line design & relocation. Joined Blanton & Co 1946 as office engr progressing to Ch Engr in 1948 & to Exec V P 1966 & to Pres 1969. Retired 1986. Pres Ariz SPE 1955, V P NSPE 1960-62, Pres NSPE 1963-64. P Pres Award 1964 NSPE. Mbr 1974-79 ECPE EE&A Ctte. "Honor Award for Distingusihed Service in Engr–, 1976, Univ of MO-Columbia. "Distinguished Citizen Award" 1976, Univ of AZ. President, Old Pueblo Club 1980-81. *Society Aff:* NSPE, ASCE, ACEC, AWWA, SAME.

Stukel, James J
Home: 2650 Lakeview #1610, Chicago, IL 60614
Position: Executive Vice Chancellor & Vice Chancellor for Academic Affairs *Employer:* Univ of IL at Chicago *Education:* PhD/ME/Univ of IL Urbana-Champaign; MS/ME/Univ of IL Urbana-Champaign; BS/ME/Purdue Univ *Born:* 3/30/37 Native of Joliet IL. Research Engr WV Pulp and Paper Co, Covington, VA 1959-61, Asst Prof Univ IL Urbana-Champaign 1968-71; Assoc Prof 1971-75; Prof 1975; Dir office of Coal Research and Utilization 1974-76; Dir office of Energy Research 1976-80; Dir Public Policy Program Coll of Engrg 1980-; Assoc Dean and Dir Engrg Exp. Station Coll of Engrg 1984-85. Vice Chancellor for Research & Dean Graduate College Univ of Illinois at Chicago 1985-86, Exec Vice Chancellor & Vice Chancellor for Academic Affairs, Univ of Illinois at Chicago 1986-87. Exec Sec Midwest Consortium on Air Pollution 1972-73, Chrmn Bd of Dirs, 1973-75; Member, Coal Study Panel, Energy Resources Commission, State of IL 1976; Chrmn, Panel on Dispersed Electric Generating Technologies, Office of Technology Assessment, US Congress; received State-of-the-Art of Civil Engrg Award (ASCE) 1975; Research interests are Aerosol Science and Environmental Public Policy; Member PiTauSig and PhiKapPhi.. *Society Aff:* ASME, ASCE, APCA, ASEE.

Stults, Frederick C
Business: 1007 Market St, Wilmington, DE 19898
Position: Consultant. *Employer:* E I duPont de Nemours & Co. *Born:* Oct 1922. PhD Univ of Washington 1949; electronics Princeton & MIT; BS Univ of Kansas 1944. US Navy 1944-46. With duPont 1949- ; Res Engr 1949-53; Res Supr 1953-55; Tech Supt (cellulose sponge plant) 1955-59; Tech Supt (cellophane plant) 1959-62; business analysis group mgr 196265; internatl bus cons 1966- . Fellow Amer Inst of Chem Engrs. Specialties: planning/forecasting, info sys, bus/financial analysis.

Stump, Wayne D
Business: P O Box 464, Golden, CO 80401
Position: NDT Manager. *Employer:* Rockwell Intl, Rocky Flats Plant. *Education:* BS/Physics/Univ of Denver. *Born:* Jan 1923. Served in 5th Marine Div during WW II. Started in nondestructive testing at Rocky Flats Plant AEC mfg facil now operated by Rockwell Internatl Energy Systems Group, in 1952. Set up & supervised NDT dev group for 18 yrs. Estab a plant NDT training prog in 1954 which is still active. Has publ sev papers & authored or co-authored num classified documents. Served on ASNT Natl Educ Ctte for sev yrs. Elected an ASNT Fellow 1975. Wife: Medora E. *Society Aff:* ASM, ASNT.

Stumpe, Warren R
Business: 5101 W. Beloit Road, Milwaukee, WI 53214
Position: VP & Chief Tech Officer *Employer:* Rexnord Inc. *Education:* PhD (Honorary)/-/Milwaukee Sch of Engrg; MIE/Ind Mgmt/New York Univ; MS/Mech Engrg/Cornell Univ; BS/-/U.S. Military Academy *Born:* 07/15/25 Native of NY. Served with Army Corps Engrs. 1945-54--combat engr, resident engr & Asst Prof--Mech, USMA. With AMF Inc 1954-63 to Deputy Gen Mgr responsible for design, development, manufacturing & installation of ground support equipment for missile progs. With Rexnord since 1969. Unitl 1971 with Material Handling Div--Dir in charge of Sys Mgmt office in Stamford, CT. From 1971 to present at corporate headquarters in Milwaukee. Currently VP-Research & Technology, & Chief Technical Officer, responsible for Corporate Research and Innovation Group which fosters internal growth through produce development, product test & evaluation. Also functionally responsible for all company-wide tech activity. Registered PE in States of NY, WI, and FL. Pres, Industrial Research Inst 1985-86. *Society Aff:* ΤΒΠ, ΦΚΦ, IRI

Stumpers, F Louis H M
Home: Elzentlaan 11, Eindhoven, Netherlands 5611 LG
Position: Scientific Counsilor - ret. *Employer:* Philips Res Lbs, Univ/Nijmegen & Utrecht. *Education:* Prof Signal Processing/Phys/Utrecht, 1977-81; Prof/Elec/Nijmegen, 1969-82; Prof/Inf and Circuit Theory/Bochum, 1974-75; Doctor/Tech Sci/Delft, 1946; Doct/Exam/Utrecht, 1937 *Born:* Aug 30, 1911. Over 100 papers, 4 pats. Philips res 1928-35, 1938-73: noise, modulation, info theory, electromagnetic compatibility. Chmn Comm VI URSI (info & circuit theory, electromagnetic waves) 1963-69. V P URSI 1975-81. Dir Region 8 IEEE 1975, 1976. Chmn Benelux Sect 1965-67. Prog Chmn Eurocon 1974 & Telecom 1975. Chmn Info Theory Symposia Brussels 1962, Noordwijk 1970. Prog Chmn EMC Symposia Montreux, Zurich 1975, 77, 79, 81, 83, 85, 87; Wroclaw 1976, 78, 80, 82, 84, 86, 88. Chmn CISPR 1967-73. Mbr Royal Acad of the Netherlands 1969-, Life Fellow IEEE. IEEE EMC Group Awards 1973 & 1975, Veder Prize 1956, 1972. Mbr Netherlands delegation ITU-CCIR 1956-78. Mbr Bd Radio Astronomy Netherlands 1948-82. Prof Electronics Nijmegen 1969-1982. Prof Digital Signal Processing in Physics, Utrecht, since 1977. Prog Chmn Telecom 75, 79 (ITU). V Chrmn Prog Comm Eurocon Venice 1977, Stuttgart 1980. Columbus Gold Medal City Genoa 1976, IEEE Award in Intl Comm 1978, CCIR Hon Award 1978, Hon Mbr Hungarian Acad of Scis 1978. Hon Mbr Dutch Radio & Electronics Soc 1980, (NERG), VP. EMC Symp Zurich 1983-85. Hon Chrmn Telecom progr comm ITU World Forum 1983. Chrmn URSI Ctte E (Electromagnetic noise, interference) 1983-87- . Mbr Progr Comm Eurocon 1984, Brighton. Mbr Progr Comm Eurocon 1986 Paris. Golden Honor Award of Distinction SEP 1984 (Polish Eng Soc.); Friedrich List Plaquette, Dresden, DDR 1985;

Stumpers, F Louis H M (Continued)
Mbr Progr. Comm EMC Symposium Limoges 1987; Mbr Organizing Comm Budapest Coll. Microwave Comm. 1974, 78, 82, 86; Res. Ass. MIT, Cambridge Mass 1952-53. Professor Information & Control Theory, Ruhr University, Bochum, FRG 1974-75, Emer. professor Nijmegen since 1982. *Society Aff:* NERS, IEEE, KNAW, HAS

Stuntz, John W
Home: 110 Green Spring Dr, Annapolis, MD 21403
Position: Exec VP *Employer:* Westinghouse Elec Corp *Education:* MS/EE/Univ of MD; BS/EE/Univ of MD *Born:* 2/25/23 From Washington, D.C. lecturing while obtaining MSEE in 1950. Joined Westinghouse, 1952, as senior engr working on airborne radar systems studies. 25 Yrs spent in key mgmt roles in evolution of many significant defense and space systems: development of side looking radar, low light level TV and electro- optical systems (Apollo), advanced sensors for space (Gemini), airborne radars for AWACS anf F16, and implementation of air space mgmt system for Morocco. Current repson - Exec VP and Deputy to Pres, Defense - 1980. Chrmn, Aerospace Tech Council (AIA), Westinghouse Order of Merit, 1973. Enjoys sports and music. *Society Aff:* IEEE, AIA

Sturdevant, Bruce L
Business: R. W. Beck & Assoc, 660 Bannock, Denver, CO 80204
Position: Partner & Mgr *Employer:* Central Design Division R. W. Beck & Assoc *Education:* BS/ME/Univ of IA *Born:* 12/5/22 in What Cheer, IA. Fighter Pilot, US Air Corp in World War II. BSME, Univ of IA in 1948. With Stanley Consultants from 1948 to 1969 rising to VP, Head of Power Group and Member of Bd of Dirs. With R. W. Beck and Assocs since 1969. Currently Partner, Mgr of Central Design Division and member of Executive Committee. Participates in design of power generation, central heating plants and dist sys, and industrial projects. *Society Aff:* ASME, NSPE, ACEC, AMA.

Sturgeon, Charles E
Business: P O Box 7497, Birmingham, AL 35253
Position: Sr. Vice Pres *Employer:* Vulcan Matls Co. *Education:* AMP/Bus/Harvard; MBA/Bus/Tulsa Univ; BSChE/Chem Engr/KS Univ. *Born:* May 1928 Cherryvale Ks. Grad work Univ of Tulsa, mgmt. Harvard AMP 77 Army Chem Corps 1951-53. Res Engr Amoco Chem 1953-56. Vulcan Matls Co 1956- . 20 yrs operations & marketing. V P Mkting 1973; V P Mfg 1974. Pres - Chem Div 1977-87, Sr. V.P., VMC 1987. Who's Who in Amer, SE Ed 1975. Who's Who in Am. 41st Ed-1981. *Society Aff:* AIChE.

Sturgeon, Franklin E
Business: 610 Madison Ave, Grove City, PA 16127
Position: President. *Employer:* Sturgeon Engrg Inc. *Education:* BS/Ind Mgt/Carnegie-Mellon Univ *Born:* Mar 1918. BS Carnegie-Mellon Univ 1940. Native of Pgh Pa area. Plant Engrg Dept Cooper-Bessemer Corp 1940-43. Plant Engr Cooper-Bessemer Corp 1943- 47. Entered private practice as cons engr Apr 1947, offering cons mech & elec services to architects & indus. Named Pres of Sturgeon Engrg Inc July 1, 1975 on initial incorporation. Reg P E Pa, N Y, Ohio, W Vir. Mbr NSPE, ACEC, ASHRAE. Enjoy golf & bowling. *Society Aff:* PSPE, CECPA, NSPE, ASHRAE, ACEC

Sturgul, John R
Business: PO Box 1, Ingle Farm, 5098 Australia
Position: Prof Mining Engr *Employer:* S Aust Inst of Tech *Education:* BS/Mining/MI Tech; MS/Math/AZ; PhD/Mining/IL *Born:* 1/3/40 Native of Hurley, WI. Fulbright Fellow (1972) to Univ of Queensland (Australia); previous affiliation with Univ of AZ (8 yrs); New Mexico Tech (6 yrs); faculty appointments at Melbourne Univ and New South Wales (Australia); Mbr TAC of ABET; Mbr, Bd of Dirs of ABET; elected to membership in Mining & Metallurgical Soc of Amer (1979); Reg PE (AZ); Chmn, Minerals Div of ASEE; Tau Beta Pi, Phi Kappa Phi, Pi Mu Epsilon, Sigma Epsilon (V Pres-Western Region). *Society Aff:* AIME, AMC, Aus IMM, ASEE, MMSA.

Sturman, Gerald M
Business: 250 W 34th St, New York, NY 10001
Position: Sr Vice Pres. *Employer:* Parsons, Brinckerhoff etal. *Education:* PhD/Structures/Cornell; MS/Applied Mechs/Columbia; BS/Civil Engg/Columbia; BA/Math/Columbia. *Born:* Mar 11, 1936 NYC. AB 1956, BS 1957, MS 1959 Columbia; PhD 1963 Cornell. Mgr Struct Engrg Lab Cornell 1959-63, Asst Prof & Ford Postdoc Fellow MIT 1963- 67. Presently Sr V P & Partner Parsons, Brinckerhoff, Quade & Douglas Inc. Specializes in proj dev & mgmt of large multidisciplinary planning & engrg projs in the constructed facils field incl transp, water resources, urban & regional planning, ports & harbors, major structs, & environ & energy services. Wason Medal for res, Amer Concrete Inst 1963. Mbr Tau Beta Pi. *Society Aff:* AIP.

Stutzman, Leroy F
Home: 26 Lynwood Rd, Storrs, CT 06268
Position: Prof & Engr Consultant. *Employer:* Univ of CT. *Education:* PhD/ChE/Univ of Pittsburgh; MS/CE/KS State Univ; BS/ChE/Purdue Univ. *Born:* 9/5/17. Prof & Chrmn of CE, Northwestern Univ, 1943-57; Dir of Res Remington Raud Univac, 1957-59; Headed team which set-up first doctorate prog in S Am, 1959-61; Univ Prof & Hd of CE, Univ of CT 1963-; Guest Prof of Twente Univ (Netherland); Padava (Italy); Hacettepe (Turkey); Univ of Vienna (Austria): Mbr Bd of Dirs, Control Data Corp 1975-. Presented over 250 technical talks in 31 countries. Over 75 published papers & books. *Society Aff:* AIChE, ACS, AAAS, ASEE, NYAS.

Stynes, Stanley K
Home: 20161 Stamford, Livonia, MI 48152
Position: Professor of Engg. *Employer:* Wayne State Univ. *Education:* PhD/Chem Engg/Purdue Univ; MS/Chem Engg/Wayne State Univ; BS/Chem Engg/Wayne State Univ *Born:* 1/18/32. Primary career has been in teaching & res with interests in multi-phase transport & air pollution control methods. Dean of WSU College of Engg since 1972-85. Mbr of the Bd of Dirs of Energy Conversion Devices, Inc; Engg Soc of Detroit, Sci Court & Res Inst, Inc., Science Center of Metropolitan Detroit and currently President of the Mich Soc Prof Engrs. *Society Aff:* AIChE, ACS, NSPE, ASEE, AAAS

Stys Z Stanley
Home: 200 Jefferson Rd, Princeton, NJ 08540
Position: Vice President. *Employer:* Brown Boveri Corp. *Education:* MS/Mech Engg/FIT Zurieh, Switz. *Born:* Feb 1920. MS Mech Engrg Fedl Inst of Tech, Zurich Switz. 1945-48 Brown Boveri, Baden Switz, Steam & Gas Turbine Dept. 1949 CADE Power Plant, Buenos Aires Argentina. 1950 Brown Boveri Corp NYC, Proj Engr. 1954 Hd, Mech Dept. 1956 V P. Presented over 30 papers on the subject of compressors, steam turbines & gas turbines to ASME, IEEE, Amer Power Conf & other orgs, as well as delivered many talks to ASME local sects, etc. 1961-66 Mbr Exec Ctte Gas Turbine Div ASME. 1965 Chmn of this ctte. Organized & was Chmn of the First Internatl Gas Turbine Conf in Zurich Switz 1966. April 1975 Gas Turbine Magazine Profile Award. 1976 EEI/ ASME Prime Mover's Award. Enjoys skiing, jogging, dancing. *Society Aff:* ASME, ANS.

Su, Kendall L
Business: 225 N Ave, NW, Atlanta, GA 30332
Position: Regents' Prof. *Employer:* GA Inst of Tech. *Education:* PhD/EE/GA Inst of Tech; MS/EE/GA Inst of Tech; BS/Mech & EE/Amoy Univ. *Born:* 7/10/26. Native of Fujian Province, China. Has been a mbr of faculty at GA Tech since 1954, Regents' Prof since 1970. IEEE Fellow. Fields of interest include network theory, active filters, electromagnetic compatibiity, & engg aspects of Chinese language. Author of three textbooks & approximately 50 technical papers. *Society Aff:* IEEE.

Sublett, James E
Business: 15708 Minnesota Ave, Paramount, CA 90723
Position: President. *Employer:* Process Equipment Co. *Education:* BE/Chem Engg/Univ of So CA. *Born:* Apr 1925. BE Chem Engrg USC 1951. Process & Unit Operations Engr Union Oil Co 1951-57; Sales Mgr Jacobs Engrg Co Equipment Div 1957-62; established Process Equipment Co 1962, specializing in design & sales of mass transfer & fluid handling systems. *Society Aff:* AIChE.

Subramanian, R Shankar
Business: Dept of Chem Engg, Clarkson University, Potsdam, NY 13676
Position: Prof & Chrmn *Employer:* Clarkson Univ *Education:* PhD/Chem Engg/Clarkson Univ; MS/Chem Engg/Clarkson Univ; BTech/Chem Engg/Univ of Madras. *Born:* 8/10/47. Experience in res and teaching for fourteen yrs. Current res interests in transport phenomena, colloidal systems, and mtls processing in space Principal Investigator of planned Space Shuttle experiments on physical phenomena in containerless processing sponsored by NASA; John Graham Jr. Award 1978, Distinguished Teaching Award, 1981, Clarkson, Dow Outstanding Young Faculty Award 1980, ASEE, Editorial Advisory Bd, Journal of Colloid and Interface Science, 1984-86. *Society Aff:* AIChE, AAAS, ACerS, ASEE, ΣΞ, APS

Suciu, Spiridon N
Business: PO Box 11508, St Petersburg, FL 33733
Position: Gen Mgr. *Employer:* GE Co. *Education:* PhD/ME/Purdue Univ; MS/ME/Purdue Univ; BS/ME/Purdue Univ. *Born:* 12/11/21. Native of Flint, MI. Served as torpedo officer in the USN during WWII. Joined Aircraft Gas Turbine Engine Div of GE Co Jan 1951. Performed & managed applied res in combustion, rocket engines, & space propulsion. Managed design and dev of components for commercial & military aircraft engines. Managed engg design dev, production & servicing of heavy duty gas turbines for power generation, mech drives, & ship propulsion. Later, responsible for assessing energy systems tech & developing Corp plans. Now, Gen Mgr responsible for production of components for nucl weapons progs. *Society Aff:* ASME, AIAA, AMA.

Sucov, Eugene W
Home: 4 Bayard Rd, Pittsburgh, PA 15213
Position: Manager, Inertial Confinement Programs *Employer:* Westinghouse Fusion Power Sys Dept *Education:* PhD/Physics/NY Univ; MS/Physics/NY Univ; BS/Physics/Brooklyn College. *Born:* 10/27/22. Studied strength of glass & diffusion processes at Pittsburgh Plate Glass Res Ctr; with Westinghouse Res Labs 1963-78. Superviced experimental res on optical radiation & gas discharge physics applied to lasers, plasmas, illumination, lamps; with Westinghouse Fusion Power Sys Dept 1978- . Mgr programs in Inertial Confinement Fusion; prin investigator, classified DOE contract to study design of ICF power plants. Publ over 40 papers; received 4 pats; elected Fellow IES, Intl Comm on Illumination; enjoy tennis, canoeing, folk dancing *Society Aff:* APS, IEEE, ΣΞ, ANS

Sudan, Ravindra N
Business: Cornell University, 369 Upson Hall, Ithaca, NY 14853
Position: IBM Prof of Engg & Deputy Dir. *Employer:* Cornell Univ. & Cornell Theory Center *Education:* PhD/EE/Univ of London; DIC/EE/Imperial College; DIISc/EE/Indian Inst of Sci; BA/English Lit/Univ of Panjab. *Born:* 6/8/31. From 1952 to 1961 Sudan worked in elec machine tech, switching & power transmission. From 1962 onwards he switched to the fields of plasma & space physics, controlled thermonuclear fusion and in the dev of high power electron and ion beam physics. He has held visiting appointments at Stanford, Lawrence Berkeley Lab, Inst for Advanced Study, Princeton, Culham Laboratory, England & the Intl Ctr for Theoretical Phys, Trieste, Italy. During 1974-1975 he was the Scientific Advisor to Dir of Res, Naval Res Lab, WA. He was Dir of Cornell's Lab of Plasma Studies 1975-85. He is consultant to natl labs, served on the editorial board of Phys. Fluids, "Nuclear Fusion–, & "Comments in Plasma Physics" and has organized a number of conferences in his field of res. He is a fellow of IEEE and of the APS and holds U.S. patents in intense ion beam sources and inertial fusion. *Society Aff:* IEEE, APS, AAAS

Sudduth, Charles Grayson
Home: 1851 Boca Ave, Los Angeles, CA 90032
Position: Super. Civil Engineer III *Employer:* Los Angeles County Dept. of Public Works *Education:* MPA/Public Admin/USC; MS/Civil Eng/USC; BS/Civil Eng/USC. *Born:* Sept 1935; native of Los Angeles Cty. At Long Beach Water Dept 1962-63 as Design Engr; at L A Cty Flood Control Dist 1963-70 working successively on water conservation facilities, groundwater studies & runoff studies, supervised soils engrg report reviews & specifications recommendations & concrete res; 1970-80 Head, Soils Engrg Section, L A Cty Engr with respon for soil report review, soil report preparation & proj consultation. 1980-84 Head, Engr Const Sec with respon for construction inspection of underground Utilities and the Soils Engr Sect Operations. Daniel W Meade Prize 1971 ASCE. 1984-1987 Head, Soils Engineering Section and Later Geotechnical Engineering Section with same responsibilities as previously described. 1987-Special Administrative Assignment to consolidate Division Files. *Society Aff:* ASCE, CSPE, NSPE, AEG, AWWA, APWA, ISSMFE

Sudheimer, Richard H
Home: 4358 Carlo Dr, Kettering, OH 45429
Position: Operations Research Analyst. *Employer:* Aeronautical Systems Div USAF. *Education:* B/Aero Engrg/U of Minn; -/Sloan Fellow/Stanford Univ. *Born:* Sept 1933; native of Waconia Minn. Sloan Fellow at Stanford Univ; BS Univ of Minn. Associate dev engr with Honeywell 1956-57. Served with Air Force 1957-60; civilian aerospace engr with Aeronautical Sys Div (ASD) specializing in sys analysis 1960-66; Supr Aerospace Engr ASD 1966-70; assumed current position 1970, with respon as leader of major studies which provide viability of alternatives for decision makers. Sigma Xi & AIAA Mbr. Author of numerous tech reports. Natl Interfraternity Conference Dir 1972-77; Natl Council Triangle Fraternity 1972-75; Natl Pres Triangle Fraternity 1974-75. Natl Pres Natl Interfraternity Conference 1977-78; Treas Natl Interfraternity Fnd 1978-1982; Dir Triangle Education Fnd 1976-present. Enjoys traveling & gardening. VP Natl Interfraternity Fdn 1982-86; Pres Interfraternity Res & Advisory Counc 1984-1986. Chairman of the Board Natl Interfraternity Fdn 1986-present. *Society Aff:* ΣΞ, AIAA, AFA, AOPA, AFDPA

Suehrstedt, Richard H
Home: 398 Race St, Berea, OH 44017
Position: President. *Employer:* Marine Consultants & Designers Inc. *Education:* BSE/Naval Arch/Univ of MI. *Born:* Served US Army Transportation Corps 1954-56. Joined Marine Cons & Designers Inc 1951; presently respon for all engrg of firm which consists of bulk cargo ship designs, surveys, marine feasibility studies, vessel conversions & new construction, dry bulk & self-unloading vessels. P Chmn Great Lakes & Great Rivers Section SNAME 1971-72; Mbr: SNAME, Propeller Club of US. Reg P E Ohio. *Society Aff:* SNAME.

Suffield, Frederick G
Home: 817 West Fir, Sequim, Washington 98382
Position: Consultant *Employer:* FG Suffield & Associates *Education:* Privately Educated/MBS & BS Physics *Born:* 10/22/20 1940-46 Westinghouse, contributed to radar that detected aircraft at Pearl Harbor, first night fighter radar for Navy. 1946-49 led team that developed first radar for commercial aircraft, and radars for Presidential aircraft Independance and Columbine. Extensive consulting work in fields of management and planning. Have had issued approximately 14 patents, and presented approximately 50 papers. Dir of Region 6 of the IEEE (12 Western Sts) and Mbr of the Bd of Dirs of the IEEE, Assoc Fellow AIAA, Reg PE. *Society Aff:* IEEE, AIAA, AOC, AFA

Suggs, Charles W
Business: Box 7625, Raleigh, NC 27695
Position: Prof. *Employer:* NC State Univ. *Education:* PhD/Bio & Agri Engg/NC State Univ; MS/Bio and Agri Engr/NC State Univ; BS/Agri Engr/NC State. *Born:*

Suggs, Charles W (Continued)
5/30/28. Invented, developed and introduced successful automatic harvesting and handling system for tobacco. Presently working on automation of other agricultural tasks including seedling transplanting and machine control. Also engaged in res in agriculturally related man-machine systems. This work involves noise, vibration, thermal stress, mental loading, vehicle control and other factors which are adversely modified by machines. Registered as a Professional Engineer in NC. *Society Aff:* ASAE, HFS.

Suits, Chauncey G
Home: Crosswinds, Pilot Knob, NY 12844
Position: Vice Pres & Dir of Research-retired. *Employer:* General Electric. *Education:* DSc/Physics/Eid Tech Hochschule-Zurich; BA/Physics/Univ of WI. *Born:* 3/12/05. Joined G E Res Labs as Physicist 1930; Res on non-linear circuits, elec arcs in the 1000 atmospheric pressure range; 79 pats issued; established G E Knolls Lab 1950; established Knolls At Po Lab 1953; Ch Div 15 OSRD Countermeasures 1942-46; Naval Res Adv Comm 1956-64 Chmn 1958-61; Ex Comm Sigma Xi 1962-65; Dirs of Ind Res 1945-65 Chmn 1962-63; NYS Sci & Tech Found 1967-82; V Chmn 1969-82; AMA VPres & Chmn Res Div 1956-58; Mbr NAS; Founding Mbr NAE; mbr Amer Acad Arts & Sci; mbr Amer PhilSoc.; US Pres Medal for Merit; King's Medal for Ser in the Cause of Freedom (UK); Eta Kappa Nu Awards; Proctor Prize AAAS; Medal, Indus Res inst; Medal for Res ASM; Disting Serv Award AMA; Frederik Philips Award IEEE; Hon Degrees: Union Coll, Hamilton Coll, Drexel Inst; Marquette Un; Renselear Poly, Inst. *Society Aff:* NAS, NAE, APS, IEEE, AAAS, APS.

Sullins, John K
Home: 1242 Linville St, Kingsport, TN 37660
Position: President. *Employer:* Sullins Ecology Engineering. *Education:* BS/Chem Engg/Univ of TN. *Born:* Nov 1923 Ogden Center Mich. Raised in Western N C & East Tenn. Married with 3 children. BSChE Univ of Tenn 1948; P E Apr 1969. US Naval Aviator & LSO 1942-46; served in the Pacific. The Mead Corp 1949-73; prod, res, tech, environ & managerial positions. Formed cons firm 1973. Presently affiliated with Environ Protection Specialists, a Canton Grp Co. Author of tech papers in field of waste treatment; pats issued & pending in fields of of Chem Engrg, Mech Engrg, Electro-Chem & Waste Treatment. Enjoy gardening, photography, golf, restoring antiques, working with plastics & wood. *Society Aff:* AIChE, NSPE, TSPE, APCA, WPCA, TAPPI.

Sullivan, A Eugene
Business: Unite 302 Union Wharf, Boston, MA 02109
Position: Pres *Employer:* Robert W Sullivan Inc-Cons Engrs. *Education:* MS/Sanitary Engg/Harvard Univ; BS/Mathematics/Boston College. *Born:* Aug 1931. Reserve Officer USPHS. Pres Robert W Sullivan Inc, Cons Engrs; Cvl Serv Examiner for Mass. Author of several tech articles & course manual; Prof of San Engrg Tech of Franklin Inst of Boston. Holder of patent relative to waste water disposal. Cons Engrs Council of New England: Dir 1971-73, V Pres 1973-75, Pres 1975-76, Natl Dir 1976-77. Enjoys civic & community involvement. Former City Councillor, Deputy Mayor & Chmn of Housing Authority. Formerly VP of local hospital & bank. Formerly Trustee & Mbr of Bd of Investment of Winchester Savings Bank. Design Engrg Award 1982-ASPE. *Society Aff:* NSPE, ASCE, ASPE, ACEC, SFPE.

Sullivan, Cornelius P
Home: 74 Coach Rd, Glastonbury, CT 06033
Position: Senior Project Matls Engineer *Employer:* United Technologies Corp. *Education:* BS/Metallurgy/Metallurgy/Univ of Notre Dame; MS/Metallurgy/Metallurgy/MIT; ScD/Metallurgy/MIT. *Born:* Sept 28, 1929; native of Schenectady N Y. ScD & SM from MIT; BS from Univ of Notre Dame. Holder of Climax Molybdenum Fellowship at MIT. Serin US Navy 1952-54. Joined United Technologies 1962 as res metallurgist; assumed present position as Sen Proj Matl Eng 1977; respon for Minestructural Analysis. Mbr: ASM & Sigma Xi. AWS Spraragen Award 1967 *Society Aff:* ASM Intnatl, ASTM.

Sullivan, James T
Business: John Fitch Indus Pk, Warminster, PA 18974
Position: Product Manager. *Employer:* Penna Pacific Corp. *Born:* July 1926; native of Phila Pa. BS LaSalle Coll; short course in Packaging Engrg, Purdue. With US Marine Corps 1944-46. Respon for prod design & marketing for PennPac Div, Penna Pacific Corp - specializing in heavy duty packaging for indus. Mbr Soc of Packaging & Handling Engrs; Pres Phila Chap SPHE 1967-69; Dir Natl SPHE 1967-69; Dir Phila Chap SPHE 1964-75. Also mbr of Packaging Inst USA & Sigma Pi Epsilon (hon packaging fraternity).

Sullivan, John M
Business: 227 E. Palace Ave, P.O. Box 283, Santa Fe, NM 87504-0283
Position: Pres *Employer:* Sullivan Design Group, Inc. *Education:* MS/CE/Stanford; B/CE/GA Tech *Born:* 11/03/42 Registered Professionl Engr in 7 states. Registered Land Surveyor in 2 states. Born Nov 3, 1942. Professional experience: Harza Engrg Co, Chicago IL; Captain, US Army Corps of Engrs; ID Water Resource BD, Boise, ID; Regional Engr, NY State Urban Dev Corp, Albany, NY. VP Dev, Coastland Corp, VA Beach, VA; Gen Mgr Wittaker Community Dev Corp, Santa Fe, NM. Recipient of Ford Fdn Fellowship for Graduate Studies at Stanford Univ. Army Commendation Medal and Citation, Army Airborne Medal, NMSPE Outstanding Service Awards (4), NMSPE Legislative Roundhouse Award, Eagle Scout Award. Mbr Chi Epsilon, Nat'l Civil Engrg Honorary Soc; Kappa Kappa Psi; Nat'l Music Honorary Soc; NSPE, NMSPE, ASCE, ACEC, CECNM; Licensed commercial pilot (instrument rating). Pres, NMSPE. MSCE Stanford Univ 1965; BCE Georgia Tech 1964 *Society Aff:* NSPE, CECNM, NMSPE, ASCE, ACEC.

Sullivan, T Darcy
Home: 6956 Riverwood Dr, Knoxville, TN 37920
Position: Principal Transportation Engr *Employer:* MCI Consulting Engineers Inc *Education:* MSCE/Transp Engr/Northwestern Unv; BSCE/Civil Engr/MI State Univ. *Born:* Sept 8, 1936. Staff Engr for Barton-Aschman Assoc 1961-62; Traffic Engr for the Ill Dept of Transportation 1962-65 & 1967-72; Asst Dir Engrg Div Northwestern Univ Traffic Inst 1965-67; Director, Dept of Traffic Engr 1972-80 Sr Assoc Envirodyne Engrs Inc 1980-1984; Self-employment 1984-86. Mbr Inst of Transportaton Engrs (Pres Ill Sect 1971 & Tenn Div 1976). Sr Transp Engr, MCI Consulting Engr Inc. 1986- . *Society Aff:* ITE.

Summer, Virgil C
Business: 1426 Main Street, Columbia, SC 29201
Position: Chairman Emeritus *Employer:* SC Electric & Gas Co. *Education:* M.S./Engineering/University of S.C.; Graduate/Internatl. Correspondence Schools *Born:* 08/21/20 Utility company Chairman Emeritus; b. Spartanburg, S. C., 1920; grad. Internat. Corr. Schs.; M.S., U.S.C.; married. With S. C. Electric & Gas Co., Columbia, 1937--, v.p. electric ops. and engring., 1966-67, sr. v.p., 1967-77, pres., chief operating officer, 1977-79, pres., chief exec. officer, 1979-82, chmn. bd., chief exec. officer, 1982-85; chmn. bd. 1985-86; also Chmn. Midlands Tech. Edn. Found; mem. Riverbanks Park Commn., Lexington County Health Edn. Found., National Urban League Commerce and Industry Council, Palmetto Bus. Forum; trustee Lowman Home. *Society Aff:* ASME, IEEE, NSPE

Summer, Virgil C, Jr
Business: Chairman Emeritus, S.C. Electric & Gas Co, Columbia, SC 29218
Position: Chmn of the Bd Emeritus *Employer:* South Carolina Electric & Gas Co. *Education:* MS/Mech Engrg/USC. *Born:* 8/21/20. Named Engr of Year 1968 by the Columbia Chapter of the SC Soc of Prof Engrs, Columbia SC. Fellow ASME; Mbr NSPE. With SC Elec & Gas Co 1937- ; current position Chmn of the Bd Emeritus & mbr of Bd of Dirs. Past Ch Richland-Lexington Counties Comm for Tech Educ. 1978-Received Hon Doctor of Laws Degree from Newberry College; 1978-Was named "Natl Man of the Year" by ICS; 1978-Presented "Distinguished Alumnus" Award by ICS. Married Vera Boland; 3 children - Brenda (Mrs J A Nunamaker), Mike, Kenneth. Enjoy hunting, fishing & gardening. *Society Aff:* ASME, NSPE, TBΠ.

Summerlin, James C
Business: 1609 Broadway, Little Rock, AR 72206
Position: Pres. *Employer:* Summerlin Assoc, Inc. *Education:* MS/Civ Engg/LA Tech; BS/Civ Engg/LA Tech. *Born:* 10/9/39. Attended LA Tech on football scholarship. Obtained Half-Time Instructorship for Grad School. Stress Analysts for Gen Dynamica/Ft Worth, '63 & '64. Structural Engr for Rust Engg Co and J E Sirrine Co, '64, '65 & '66. Civ/Structural/Planning Engr for Garver & Garver, Inc, Engrs, '67-'74. pres of Summerlin Assoc, Inc, Engrs, Planners, Surveyors, since beginning in '74, a 21-person firm of 3 Reg PES engineers and 2 Reg Land Surveyors working on gen civ and structural engg projs. State Pres of ASPE in 1974. Received the ASPE "Most Outstanding Young Engr of the Yr" award for the state in 1972. The firm provides general civil and structural engineering services to industry; municipal, county, state and federal government agencies; general contractors and architectural firms. *Society Aff:* NSPE, ASCE, APA, SAME, AWWA.

Summers, John Leo
Home: 10797 W 61st Ave, Arvada, CO 80004
Position: NDT Consultant *Employer:* self *Education:* Dipl/bus/Muscatine Jr College; Cert/Reg Medical Technician/Univ Hospital of IA. *Born:* Sept 1918; native of Ardon Iowa. Muscatine Jr Coll Iowa: 2 yr assoc degree, 1 yr business; Univ of Iowa: Reg Medical X-Ray Tech; Univ of Colo: 4 yrs 1960-64. Medical X-Ray Tech supervised by V C Flower MD, Oak Park Ill. NDT Area Mgr 24 yrs Rocky Flats Plant, Dow Chem Co & Rockwell Internatl (Prime Contractor for ERDA, formerly AEC) 1952- . Supervise NDT pers performing prod quality tests incl radiaton gages, radiography, ultrasonics, eddy current, dye penetrant & leak testing. Prepared over 50 procedure manuals pertaining to the operating & testing requirements. P Chmn of ASNT Colo Section; Mbr ASM; received ASNT Fellow Award 1974; Mbr of Select AD Hoc Ctte ASNT (Natl) for Level III Certification. Reg P E Calif 1025. Mbr of ASNT Bd of Dirs. Chrmn, ASNT Education and Qualification Council 1984. Served on a task group for DOT - AL pipeline - Publication (Landolt, Stump, Summers) "Visual Comparative Melthod for Radiographic Determination of Defect Thickness–, ASNT Matl Evaulation - Oct 1978. Hon recipient of the ASNT Tutorial Award, 1984. Promoted to NDT Sr Principal Engr, June, 1983. Retired - November 1, 1985. *Society Aff:* ASNT, ASM

Summers, Joseph B
Business: 887 N Irwin St, PO Box 1122, Hanford, CA 93232
Position: Pres. *Employer:* Summers Engg, Inc. *Education:* MS/Civil Engrg/CO Univ; BS/Civil Engrg/Univ of IA *Born:* 3/5/23. Served in Army Air Corps 1942-45, European theater. Experience includes 4 yrs in hydraulic lab, irrigation Engr with public dist and prin Engr with cons firm. Entered private practice 22 yrs ago. Specialize in water supply, irrigation systems, drainage and municipal facilities. Cons to numerous water and irrigation districts, county and the UN. Presented papers before the Intl Congress on Irrigation and Drainage, New Delhi, India 1966, and Athens, Greece 1978. Served on Panel of Experts on 11th Intl Congress on Irrigation and Drainage, France 1981. Appointed to Intl Committee on Irrigation and Drainage Working Group on Draninage at Melbourne, Australia in 1983. Elected to Exec Committee of the U.S. Committee on Irrigation and Draninage in 1983. Also operate a producing walnut orchard. *Society Aff:* ASCE, ASAE, ΣΞ, CEAC, USCID.

Sumner, Billy T
Business: 404 James Robertson Parkway, Nashville, TN 37219
Position: Exec V Pres *Employer:* Barge, Waggoner, Sumner & Cannon *Education:* BE/Civ/Vanderbilt Univ *Born:* 7/13/23 Principal - Barge, Waggoner, Sumner and Cannon since 1955. George P Rice, Consulting Engr, New Orleans - one yr. Dixie Poultry Processors, Franklin, TN - one yr. Universal Concrete Pipe Co, Columbus, OH - two yrs. Pol, Powell & Hendon, Consulting Engrs, Birmingham, AL - five yrs. ENR Award 1976. ACEC's Edmund Friedman Award 1974. NSPE/PEPP Award 1973. Pres ACEC - 1975-76. V Pres - Amer Inst of Consulting Engrs - 1973. Mbr - Bd and Exec Ctte NSPE - PEPP Chmn - 1969-70. Trustee - AAEE - since 1981. Tau Beta Pi. Army, Asiatic Pacific, WW II, Major, Corps of Engrs. Sigma Nu Social Fraternity. *Society Aff:* AAEE, ACEC, APWA, ASCE, AWWA, NSPE, SAME, WPCF

Sumner, Eric E
Business: AT&T Bell Laboratories, 480 Red Hill Rd, Middleton, NJ 07748
Position: V Pres *Employer:* Bell Telephone Labs Inc. *Education:* Prof/EE/Columbia; MA/Physics/Columbia; BME/Mech Eng/The Cooper Union. *Born:* 12/17/24. With BTL 1948- ; dev of switching apparatus, electronic switching sys, pulse code modulation transmission, ASW Sys, Loop Transmission, Customer Network Operations. Assumed current respon as V Pres in 1981, now respon for Operations Systems and Network Planning; Fellow IEEE: Mbr of National Academy of Engg since 1985; Mbr of Honor Soc's: Tau Beta Pi, Pi Tau Sigma. Mbr of Adv Bd to Georgia's Tech Inst Sch of EE, Chrmn of the Adv Council to Cooper Union School of Engrg. Member of Bd of Dir of IEEE, 1986-87. Member of National Research Council's Bd on Telecommunications-Computer Applications. Awarded: Alexander Graham Bell Medal of IEEE in 1978; Gano Dunn Medal for Engg Achievement from the Cooper Union in 1974; the Cooper Union Disting Alumni Citation in 1985. *Society Aff:* IEEE.

Sun, Hun H
Business: Drexel University, Department of Electrical and Computing Eng, 32nd & Chestnut Sts, Philadelphia, PA 19104
Position: Prof Dept of Elec Engg. *Employer:* Drexel University. *Education:* PhD/EE/Cornell; MS/EE/Univ of WA; BS/EE/Chiao-Tung Univ, China *Born:* Mar 1925. PhD Cornell Univ 1955; Postdoc res & NIH Special Fellow MIT 1963-64. Prof of Elec Engrg & Biomedical Engrg 1959- , Dir of Biomedical Engrg & Sci Prog 1963-74, Chmn of Dept of Elec Engrg 1973-78, E O Large Prof of Elec Engr 1978- , Drexel Univ. Ed of IEEE Transactions on Biomedical Engrg 1972-78, Adjunct Prof of Physiology, Temple Med Sch; Res Prof of Surgery, Hahneman Med Coll; Chmn 27th Anual Conference in Engrg in Media & Biology, Phila Pa 1974. Mbr Adv Ctte IEEE-EMB 1969-73. Editor-in-chief, Annals of Biomedical Engrg, 1984- . Mbr of Study Section on Surgery and Bioengineering, Natl Inst of Health, 1981-85. *Society Aff:* IEEE, ASEE, BES

Sundaram, Meenakshi R
Business: Dept of Industrial Engineering, Box 5011, TTU, Cookeville, TN 38505
Position: Professor *Employer:* Tennessee Technological Univ *Education:* PhD/IE/Texas Tech University; MSc/Prod Engrg/Madras Univ India; BE/ME/Madras Univ India *Born:* 06/20/42 A naturalized U.S. citizen born in India. Worked in industry and taught engrg at Madras Univ in India before migration to USA. Was an asst prof in the school of engrg at Old Dominion Univ, Norfolk, VA, for 3 years from 1976-1978. Awarded tenure at Old Dominion. Moved to SUNY Coll of Tech at Utica, NY, to develop Industrial and Manufacturing baccalaureate degree progs. Assumed assoc prof position in Industrial Engrg at Tennessee Tech Univ at Cookeville, TN, effective Jan 1980. Registered PE in VA. Promoted to full Professor in 1986. Teaching, consltg, and research interest are in the areas of manufacturing sys design, group tech, scheduling, and process planning. Dev a micro computer and computer graphics lab facility in the IE dept at Tech and is developing a machining cell and a metal cutting research laboratory. *Society Aff:* IIE, SME, ASEE, ASME

Sundaresan, Sonny G
Business: Buncombe Rd, Greer, SC 29651
Position: Mgr, QC & Met. *Employer:* Homelite-Textron. *Education:* MBA/Bus Admin/Clemson-Furman Univs; MS/Met Engr/Univ of Cincinnati; BTech/Met

Sundaresan, Sonny G (Continued)
Engr/IIT. *Born:* 4/22/47. Native of Madras, India. Sr Met with Firestone-Elec Wheel Co, Quincy, IL 1971-73. With Homelite-Textron, since 1973. Presently mgr, quality control and met; implemented mtls specification & process specification systems & developed technical services lab, towards achieving value-effective QC system. chrmn ASM Old-South Chapter 1976-77, Mbr of Natl Handbook Committee & contributor to metal progress tech forecasts. Reg PE, SC. Certified quality and reliability engr (ASQC). Evening collee instr at Greenville Technical College. Enjoys travel, photography, golf & tennis. *Society Aff:* ASM, ASQC, ASTM.

Sunderland, J Edward
Business: Dept of Mech Engrg, Univ of Mass, Amherst, MA 01003
Position: Professor. *Employer:* University of Massachusetts. *Education:* PhD/Mech Eng/Purdue Univ; MS/Mech Eng/Purdue Univ; SB/Mech Eng/MIT. *Born:* Oct 26, 1932. Previously held faculty positions at Northwestern Univ, Ga Tech & N C St Univ; was Reynolds Prof of Mech Engrg at N C St Univ 1970-72. Received res award from Soc of Sigma Xi, a papers award from ASAE & is Fellow ASME. Presented testimony before Senate Commerce Ctte on pending energy legislation June 1975. Received Massachusetts Centers of Excellence Award, August, 1987. *Society Aff:* ASME, ASHRAE, IFT.

Sundstrom, Donald W
Business: U-139, Storrs, CT 06268
Position: Professor *Employer:* Univ of CT *Education:* PhD/ChE/Univ of MI; MS/ChE/Worcester Poly Inst; BS/ChE/Worcester Poly Inst *Born:* 11/18/31 Native of Worcester, MA. Industrial experience as group leader at Linde Div of Union Carbide and as research supervisor at Allied Chemical. Previous teaching experience at Univ of MI and Univ of Cincinnati. Professor of Chem Engrg at Univ of CT since 1970. Author of over 50 technical articles and textbook on *Wastewater Treatment.* Current research interests are environmental engrg, biochemical engrg and rheology of polymers. *Society Aff:* AIChE, ASEE, AIP

Sununu, John H
Home: 24 Samoset Dr, Salem, NH 03079
Position: Governor. *Employer:* State of New Hampshire *Education:* PhD/ME/MIT; MS/ME/MIT; BS/ME/MIT. *Born:* July 2, 1939. Ch Engr Astro Dynamics Inc 1960-65; Assoc Prof Tufts Univ 1966- ; Assoc Dean Coll of Engrg Tufts Univ 1968-73; Pres Thermal Res Inc 1966-82, Pres JHS Engrg Co 1966-82, Chmn of Bd of Dirs SCORE Inc 1970-82. Elected N H House of Representatives 1973-74. Elected Gov State of NH 1982, Reelected Gov 1984 and 1986. SAE Ralph R Teetor Award 1972; Outstanding Educator of Amer 1971; elected Eminent Engr to Tau Beta Pi 1971. NAS Visiting Panel on Ghanian Sci Policy & Priorities; NAE Ctte on Public Engrg Policy; Pres's Council on Environ Quality Adv Ctte in Advanced Power Sys; Dept of Transportation Subctte on Transportation Energy. *Society Aff:* ASME, AAAS, ΣΞ, ΤΒΠ.

Suozzi, Joseph J
Business: Bell Telephone Labs, Whippany, NJ 07981
Position: Department Head. *Employer:* Bell Telephone Labs. *Education:* PhD/EE/Carnegie-Mellon; MEE/EE/Catholic Univ; BEE/EE/Catholic Univ. *Born:* March 1926. PhD EE Carnegie Mellon; MEE, BEE Catholic Univ. Native of Glen Cove NY. Instructor, Catholic Univ 1949, 50. Worked in Magnetics Div, US Naval Ordnance Lab until 1955. With Bell Telephone Labs since 1955. Currently Head of Energy Systems Engrg Dept. Responsible for all planning, economic studies & systems engrg for all power systems in Bell System. Alumni Outstanding Achievement Award from Catholic Univ 1973. Fellow IEEE 1973; Pres IEEE Magnetics Soc 1969, 70.

Supple, Richard G
Business: 213 Truman Ave, Albuquerque, NM 87108
Position: President *Employer:* Bridgers & Paxton *Education:* BS/ME/Purdue Univ *Born:* 3/7/32 Grad from Purdue Univ in 1954. Received Commendation Medal from Secretary of the Army for design of a projectile recovery systems. Branch Mgr for Amer Blower from 1960-65 with responsibility for sale and application of heating, ventilation, and air conditioning equipment. Elected V Pres of Bridgers and Paxton Consulting Engrs, Inc in 1969 and Pres in 1979. Responsible for many innovative cooling, heat recovery, and solar heating systems. Active in ASHRAE having served as Regional VChmn for Energy Mgmt, and presently serving ad Dir and Regional Chmn. *Society Aff:* ASHRAE

Suran, Jerome J
Business: Bldg 30 EE, 1285 Boston Ave, Bridgeport, CT 06602
Position: Staff Exec. *Employer:* General Electric Co. *Born:* Jan 11, 1926. BSEE 1949 Columbia Univ; graduate studies Columbia Univ, Ill Inst of Technology; Dr. Engrg honors causa Syracuse Univ. Held engrg positions at J W Meaker Co & Motorola Inc prior to joining GE in 1952. Mgr Electronics Lab, active in dev of solid-state devices & circuits, & in the mgmt of the lab opers. Co-author of 2 books on transistor circuits; pub 50 papers in prof journals; hold 18 patents. Non-resident instructor, MIT 1959-63; adjunct Prof, Syracuse Univ. Vice Pres Publications and Educ Activities, IEEE; IEEE Pres, 1979 was active on sev cttes, NAS; Mbr Dudley Observatory Ctte. Mbr accreditation ctte, ECPD. Reg PE in NY and CT. *Society Aff:* IEEE

Susskind, Charles
Business: UC Coll of Eng, Berkeley, CA 94720
Position: Professor. *Employer:* University of California. *Education:* PhD/EE/Yale; MEng/EE/Yale; BS/EE/Caltech. *Born:* 8/19/21. Prague, Czechoslovakia; educated there & in England. USAAF 1942-45. BS 1948 Caltech; MEng 1949, PhD 1951 Yale. Res Assoc, Lecturer, Stanford Univ 1951-55; Univ of CA, Berkeley faculty 1955- , Prof 1964-; Asst Dean Engrg 1964-68, in UC systemwide admin 1969-74. Fellow IEEE; Mbr Instn Electronic & Radio Engrs, London; Bioengg Soc, AAAS, History of Sci Soc, Soc for History of Tech, Tau Beta Pi, Sigma Xi (Pres Berkeley Chap 1972-73). Author, 'Fundamentals of Microwave Electronics,' W M Chodorow 1964; 'Exporting Tech Education,' W L Schell 1968; 'Understanding Technology,' 1973, 'Twenty-five Engrs and Inventors,' 1976; 'Ferdinand Braun,' W F Kurylo, 1981; 'Electricity and Medicine: A History of Their Interaction,' W M Rowbottom. Editor, 'Encyclopedia of Electronics,' 1962; 'Heinrich Hertz: Memoirs, Letters, Diaries,' W M Hertz, 1977. Cons Government, indus, publishing; Dir, San Francisco Press Inc. Tech speciality: electron physics; bioengrg; societal effects of tech. *Society Aff:* IEEE, IERE, BES, HSS, SHOT, AAAS

Sussman, Martin V
Business: Chem Engrg Dept, Medford, MA 02155
Position: Professor of Chemical Engineering. *Employer:* Tufts Univeristy. *Education:* PhD/Chem Eng/Columbia; BChE/Chem Eng/CCNY. *Born:* PhD 1958, MS Columbia Univ; BS CUNY. With Tufts Univ since 1961, Chmn Dept 1961-71. Initiated graduate prog. Res: chromatography, nucleation, thermodynamics. Visiting Prof, MIT spring 1976. Engrg Education Coordinator, NSF prog N Delhi 1967. Started Chem Engrg Dept, Robert Coll, Istanbul 1958-61. Senior Engr, DuPont Textile Fibers 1953-58. Educational Cons, AID, Ford Foundation in India, W Africa, & S America. Garland Publ; NIH Special Fellowship Award, Weizmann Inst 1968; Tau Beta Pi, 'Eminent Engr' 1975; Fulbright-Hays Lectureship Award 1977; Books: 'Availability (Energy) Analysis' Mulliken House 1980. 'Elementary General Thermodynamics,' Addison-Wesley 1972. Fellow AIC. Patents, papers. ERSKINE Prof Univ of Canterbury, NE 1983. Meerhoff Fellow, Weizmann Inst, 1984-85. Married, 3 children. *Society Aff:* AIChE, ACS, TBP, AAUP.

Sussman, Victor H
Home: 300 S Vernon, Dearborn, MI 48124
Position: Dir Stationary Source Envir Control. *Employer:* Ford Motor Co. *Education:* BChE/Chem Eng/City College of NY. *Born:* 7/26/30. Indus Hygiene Engr, NY State Dept of Labor 1953-55; Public Health Engr, Public Health Service,

Sussman, Victor H (Continued)
US Dept of Health, Education & Welfare 1955-58, during this period conducted a study of petroleum refinery emissions in Los Angeles; Dir, Bureau of Air Pollution Control, Pa Dept of Health & Pa Dept of Environmental Resources 1958-72. Mbr Temple Univ Faculty. Guest Lecturer on Air Pollution, Pa State Univ & Federal Training Prog. Mbr Adv Cttes to US EPA & ASME. Cons to UNESSO for air pollution study in Iran & Pan American Health Org for government cons in Brazil. P Chmn of State & Territorial Air Pollution Prog Administrators. APCA: Pres 1969-70, AAEE Bd Mbr 1972-73 Profl Engrs Licence, PA. *Society Aff:* APCA, AAEE.

Sutcliffe, Harry
Business: 50 Beale St, San Francisco, CA 94119
Position: Mgr, Boston Office. *Employer:* Bechtel Incorp. *Education:* MS/Civil Engr/CA Inst of Technology; BSC/Civil Engr/Manchester England; Bus Man Cert/-/Univ of CA, Berkeley. *Born:* March 1924 England. MSCE 1951 Caltech; BSc Honors 1950 Manchester. Soil Mechanics Asst, LM&S Railway 1941-53. (Time out for war & coll). Lt, Royal Engrs 1942-46. Bechtel Inc- . Field Engr to Proj Mgr on dam design & const 1953-63; Proj Mgr tunnel projs in Calif, Colo, Mass, & Austrian Alps 1963-74; Proj Mgr on Northeast Corridor High Speed Passenger Rail Studies 1974-75. Fellow ASCE, Fellow Inst CE (London). Mbr NAS/NAE & ASCE working cttes on tunneling, and contract mgt. RETC Exec Ctte 1980-81. Proj Mgr and MBTA Redline Ext NW, Cambridge MA & Mgr Boston Office 1975-to date. Visiting Lectr MIT 1979-80. Enjoy opera & early cartography. *Society Aff:* ICE, ASCE, BTS, USCOLD, SAME.

Suter, Vane E
Business: Unocal Oil & Gas Div, Unocal Corp, 1201 W. 5th St, Los Angeles, CA 90051
Position: V Pres, Operations *Employer:* Unocal Oil & Gas Div - Unocal Corp *Education:* BE/Petro Eng/Univ of So CA *Born:* 2/14/29 With Unocal since 1954, 20 yrs with Oil and Gas Div as Reservoir Engr, Secondary Recovery Engr, Div Engr and District Operations Mgr. 10 yrs with Geothermal Div since 1974 as District Mgr and as V Pres Operations. Since 1984, with Oil & Gas Div as V Pres, Operations. Have served as a mbr of the CA St Bd of Registration for PEs, and am a reg PE in the Sts of CA and TX. *Society Aff:* SPE, API

Sutera, Salvatore P
Business: Washington Univ, Box 1185, St Louis, MO 63130
Position: Professor. *Employer:* Washington University. *Education:* PhD/Mech Engg/CA Inst of Techn; MSc/Mech Engg/CA Inst of Techn; BMe/Mech Engg/Johns Hopkins Univ. *Born:* Jan 1933. Fulbright Fellow, Paris France 1955-56. Faculty Mbr, Brown Univ 1960-68; Exec Officer of Div Engrg 1966-68. Moved to Washington Univ 1968 to become Chmn, Mech Engrg. Chmn until 6/82. Chmn until 6/82. Resumed Chairmanship 9/85. Visiting Prof, Univ of Paris VI 1973. Visiting professor Monastir University, Tunisia 1986. Primary res interests: biomechanics, technology of artificial organs, flow of suspensions. Teach dynamics, fluid mechanics, heat transfer & mechanics thermodynamics. Mbr ASME, ASHRAE, ASEE, ASAIO, Internatl Soc Biorheology, AAAS, Sigma Xi. Licensed PE in Mo. *Society Aff:* ASME, ASHRAE, ASEE, ASAIO, AAAS, NASB

Sutherland, Donald C
Business: 2140 S Ivanhoe St, Suite 210, Denver, CO 80222
Position: Chrmn of the Bd *Employer:* Sutherland Engrs, Inc *Education:* BS(CE)/Civ Engrg/Univ of CO *Born:* 11/2/09 Native of Glengarry, Pictou County, Nova Scotia, Canada. Employed by CO Div of Highways prior to WW II. Served with 10th Army Engrs in South Pacific 1944- 46. Engaged in Civil Engrg and Land Surveying practice 1946-Present. Specialized in Highway and railroad location and design in US and South America. Instrument rated pilot. P Pres and Life Mbr CSE. P Pres El Jebel Air Patrol, Fellow-Life Mbr ASCE, City Engr, City of Federal Heights, CO 1976-Present, Town Engr, Town of La Veta, CO 1984. Distinguished Alumnus Awd, Private Practice Category, Univ of CO, 1983. *Society Aff:* ASCE

Sutherland, Earl C
Home: 2565 Dexter Ave N, Suite 401, Seattle, WA 98109
Position: Principal. *Employer:* Earl C Sutherland & Assoc. *Education:* MS/Metallurgical Engr/MI Tech Univ; BS/Metallurgical Engr/MI Tech Univ; MBA/Business/Portland State Univ. *Born:* 7/23/23. Native of Detroit, MI, US Army 1943-46, Captain AUS, US Army Reserve 1946- 75; Retired Col, Logistician. Ten years engr IBM Corp. Five yrs engg refractory metals, Fansteel Corp. Four yrs operating foundry in fabricating plant, Sao Paulo, Brazil, as tech dir mfg mgr. Precision Castparts Corp and Dir of Res and Engg, Omark Industries. Res Scientist on nuclear rocket materials, NASA. Prin and VP, MEI-Charlton Inc before establishing own practice in 1975. *Society Aff:* NSPE, AIChE, AIME, ASM, NACE, ACEC, CECW, ACEC.

Sutherland, Ivan E
Business: 1035 Devon Rd, Pittsburgh, PA 15213 *Employer:* Sutherland, Sproull, & Associates, Inc. *Education:* PhD/EE/MIT; MS/EE/Caltech; BS/EE/Carnegie Tech. *Born:* 5/16/38. While serving with US Army in 1963 was stationed at the Natl Security Agency. Was Dir for the Inf Processing Techniques Office of ARPA in the Defense Dept 1964-67. While Assoc Prof of EE at Harvard 1966-68, his res resulted in the const of equipment for the Head-mounted, 3-Dimensional Display. Left Harvard Univ 1968 to become VP of R&D of Evans & Sutherland Computer Corp. Fletcher Jones Prof of computer Sci at CA Inst of Technology 1976-80. Mbr NAE, NAS, Defense Sci Bd, Former Mbr Naval Res Adv Comm. Awarded the 1st Zworykin Award by NAE 1972. 1980-present at Sutherland, Sproull & Assoc, Inc. *Society Aff:* NAS, NAE, IEEE, ACM.

Sutherland, Robert L
Business: P O Box 54, Laramie, WY 82070
Position: Conslt Engr *Employer:* Self *Education:* BS/Mech Engg/Univ of IL; MS/Theoretical & Applied Mechs/Univ of IL. *Born:* May 1916. Res & dev pos with Firestone Tire & Rubber, Borg-Warner (Borg & Beck Div), G M (Buick), & Aeronca Aircraft Corp. 1948-80 taught in Mech Engrg depts, State Univ of Iowa & Univ of Wyoming (Hd of Dept 1960-70 now Prof Emeritus of Mech Engg). Consulting engineer since 1980. Pres Skyline Engrg Inc since 1972. VP for Region VIII ASME 1965-67. Fellow ASME 1972. Reg P E Ill, Iowa, Wyoming. Mbr Pi Tau Sigma, Sigma Xi, Tau Beta Pi. Author 'Engrg Sys Analysis' Addison-Wesley Publ Co 1958. Lic private pilot since 1940. *Society Aff:* ASME, SAE, ΠΤΣ, ΣΞ, ΤΒΠ

Sutphen, Russell L
Home: 225 Seneca St, Chittenango, NY 13037
Position: V President. *Employer:* O'Brien & Gere Engrs Inc. *Born:* Aug 8, 1931 Watertown N Y. Grad Oneida High School 1949; Syracuse Univ BCE 1953. Pitometer Assocs Inc 1955-57. Joined O'Brien & Gere 1957, became a principal engr in 1961, a partner in 1966, & a vice pres in 1971. Dir O'Brien & Gere Engrs Inc. Respon for studies, investigations & design of water supply & sewage facils & in charge of the firm's Sanitary, Water, Design & Const Engrg Divs. *Society Aff:* ACEC, AWWA, NSPE.

Sutro, Frederick C, Jr
Business: 600 Grant St, Pittsburgh, PA 15230
Position: Asst to VP. *Employer:* USS Chemicals Div, US Steel Corp. *Education:* BS/Ind Engg/Yale Univ. *Born:* 6/21/20. Native of Basking Ridge & Elizabeth, NJ. 2nd Lt (Aircraft Maintenance Engg), US Army Corp, 1943-44. Tech Sales Rep, Union Carbide Corp, 1944-50, selling and Tech Service, plastics resins. Spencer Chem, 1954-60, various tech & marketing positions. Cabot Corp, 1960-66. Since 1966, USS Chemicals as Gen Mgr, Plastics 1966-1978; 1978-Ass't to V.P. Plastics & Olefins; Also, 1979- V.P. & Genl Mgr - TEX/USS Polyolfins Co, a joint venture of Tex & USS chemicals. SPE - National Councilman, 1955-61, VP, 1958; Pres, 1959; Exec Comm Hee, 1957-61. SPI - Dir, 1973-78; Chmn, Public Affairs Comm Hee

Sutro, Frederick C, Jr (Continued)
(respon for all public issues affecting the plastics industry), 1974-76; Treasurer - 1976-78. Enjoy tennis, golf, skiing, photography. *Society Aff:* SPE, SPI, CDA, AMA.

Sutton, C Roger
Home: 1129 Central Ave, Downers Grove, IL 60515
Position: Sr Assoc. *Employer:* Roger Sutton Assocs. *Education:* MS/ChE/Univ of Detroit; BChE/ChE/Univ of Detroit. *Born:* Oct 20, 1605. m. Catherine Donnelly 1927. Teaching Fellowship Univ of Detroit 1927-28. 15 yrs in automotive indus 1932-41. Last 6 yrs as Sr Met Engr, Chrysler Corp. From 1941-50 engaged in heat & corrosion-resisting alloy casting as Dir; Engrg & met Genl Alloys Co & as independent cons. Entered nuclear energy field 1950 as Sr Engr & Sect Mgr, Reactor Engrg Div, Argonne Natl Lab. Joined INCO Inc 1955 as, ultimately, Supr; Stainless Steel Dev. Returned to Argonne Natl Lab 1965 as Asst Dir; Reactor Engrg Div & Mgr 1000 MW LMFBR Follow-on Studies. Since 1970, Roger Sutton Assocs. Reg P E Mass & Calif. Fellow ASM 1974. Since 1974, Chmn NMAB ctte on Atmos Purification & Control in Closed Sys. (Natl Acad of Sci). *Society Aff:* ASM, ASTM.

Sutton, Carl
Home: 7349 Sleepy Hollow Dr, Tulsa, OK 74136
Position: Retired *Education:* BS/ChE/Univ of KS. *Born:* Jan 1922. Native Kansan. R&D Engr Phillips Petroleum; Staff Engr Hudson Gas & Oil; Secy Gas Processors Assn, an internatl trade assn for LP-gas indus 1963-87. Basic responsibilities: assn's res prog with approx 10 investigators, ctte liaison for dev of indus test procedures & specifications, overall control for 8 regional & 1 natl meetings. Officer through Chmn Ark-La-Tex Sect ACS, Officer through Chmn Tulsa Sect AIChE, Dir & thru Chrmn Fuels & Petrochemicals Div AIChE, Editor, AIChE Publication, Energy Progress, sev natl ctte assignments AIChE. Reg PE in OK, TX and LA. Col military intelligence US Army Reserve. Retired June 1987. *Society Aff:* AIChE, ACS.

Sutton, George E
Business: Youngstown State Univ, Youngstown, OH 44555
Position: Dean, Engrg *Employer:* Youngstown St Univ *Education:* PhD/ME/MI St Univ; MSE/ME/Univ of FL; BS/ME/WV Univ *Born:* 6/3/23 Born at Blandville, WV, Served RCAF and US Navy during WW II and US Navy during Korean conflict as Aviator; Retired Civil Engineer Corps Reserve Lieutenant Commander. Engr with Industrial Engrg and Construction Co (Design/construct coal handling installations), Fairmont, WV. Research Engr and Instructor, Univ of FL, taught at MI State, AZ, AZ State, Univ of Nevada-Reno. Diverse consulting assignments during academic yrs. Dir of Professional Services, Natl Council of Engrg Examiners two yrs. Dean, William Rayen School of Engrg, YSU, 1976-date. Reg PE in FL, AZ, NV and OH. *Society Aff:* ASEE, NSPE, ASME, ASHRAE, AAHE

Sutton, George W
Business: 2385 Revere Beach Pkwy, Everett, MA 02149
Position: VP *Employer:* Avco Everett Resch Labs, Inc *Education:* PhD/Engg/CA Inst of Tech; MSME/Phys/CA Inst of Tech; ME/Physics/CA Inst of Tech; BME//Cornell Univ *Born:* At Avco Everett Resch Lab, Inc. Dr Sutton is Staff VP. His respons include high energy laser propagation damage, and systems studies; and the AERL Independent Resch nd Dev Tech Plan. He previously directed the Gov't Laser Dev Program Office which was respon for the engrg dev of complete laser sys for military applications. His respons include the supersion of programs for the design, production and delivery of high energy lasers to the 3 services, and several application studies. He is Editor-In-Chief of the AIAA Journal, a Fellow of the American Inst of Aeronautics and Astronautics; recipient of the Arthur D. Fleming Award in 1965 for outstanding gov't service; and the 1980 AIAA Thermophysics Award. He has taught graduate courses in MHD and ionized gases at MIT, Stanford, and the Univ of PA., He was the Chrmn of the Critical Tech Export Group for export controls on High Energy Lasers for DOD/AIA, and has authored 60 papers. *Society Aff:* AIAA

Sutton, Robert L, Jr
Home: 194 Evelyn Street NW, Marietta, GA 30060
Position: County Engineer. *Employer:* Cobb County. *Born:* Nov. 1931. MBA Ga State; BSCE Ga Tech. Served with Army Corps Engrs 1955-57. Various positions as ch design engr, production engr, office mgr, Asst to Public Works; County Engr Cobb County Ga June 1970; VP & Gen Mgr Arthur Pew Construction Co July 1977; named to current position as Director of Public Works for Cobb County Ga Jan 1979 with responsibility for roads, water & sewer, airport, drainage, bond constr projs Member of Water Pollution Control Federation; AWWA, ASCE, Ga Society Professional Eng., Ga. Water & Pollution Control Association. Cobb Commissioners Award; WSB Eager Beaver Award; WPCF Arthur Sidney Bedell Award; GSPE Eng. of Year in Construction Award.

Suzuki, Keiji
Home: 3-23-27 Soshigaya, Setagayaku Tokyo, Japan
Position: Honorary Professor. *Employer:* Technological Univ of Nagaoka *Education:* BE/-/Hamamatsu Tech Coll. *Born:* June 1911. Graduate Hamamatsu Tech Coll; PhD in Elec E Univ of Tokyo 1950. Joined NHK, Japan Natl Broadcasting Corp, 1932; as Res Mbr of Tech Res Labs worked on microwave relay, color TV, TV signal transmission & TV recording techniques; Sr Res Engr 1965. Nov 1968 TEAC Managing Dir & Ch of R&D Dept on audio equipments & memory devices for computer. 1975- , Prof Aichi Inst of Technology. 1980- , Prof The Technological Univ of Nagaoka (Governmental). Received letter appreciation from COMSAT regarding achievements in TV relay transmission via satellite during Tokyo Olympics. Elected Fellow IEEE Jan 1969. Natl Purple Ribbon Award & Medal of Honor 1965. Vladimir K Zworykin Prize Award IEEE 1967. May 1977, Honorary mbr of the Inst of Television Engrs of Japan. April 1984, Honorary Prof of the Tech Univ of Nagaoka. July 1984, Prof of Hiroshima Inst of Tech. May 1986, Honorary Mbr of the Inst of Electronics and Communication Engrs of Japan. *Society Aff:* IECEJ, IEEJ, ITEJ, MSJ.

Svendsen, Sven B
Business: 3250 Wilshire Blvd, Los Angeles, CA 90010
Position: Sr VP *Employer:* Daniel, Mann, Johnson & Mendenhall *Education:* MS/CE/Norwegian Inst of Tech *Born:* Pilot WWII, RNAF England and Canada; Consulting Engr, dams, hydropower, Oslo, Norway; Chief Engr for airbase construction, Greenland; Partner, Severud and Assoc., Consultants, N.Y.C.; Tech Dir, Spanish Airbase Program, Madrid, Spain; Gen Mgr. Scientific Inst, Caracas Venezuela. Joined Daniel, Mann, Johnson, & Mendenhall in 1961 as Mgr of Far East Activities, Tokyo, Japan, VP Int't Operations; VP Domestic Opearations; Sr VP Operations; Pres & CEO Tech Mgmt Services, Inc. Mbr Venezuelan Delegation to United Nations Conference, 1958; Mbr, Atomic Energy Commission, Venezuela, 1960. *Society Aff:* ASCE, CEAC, SAME, NIF, IAE

Svore, Jerome, H
Home: 250 Oak Hill Dr, Trophy Club, Roanoke, TX 76262
Position: Retired *Education:* MS/San Engr/Harvard; BA/Mech Engr/Univ of N Dak; PE/Mech Engrg/Honorary. *Born:* 12/29/11 Field engr with Fed Gov't 1934-38, Staff Engr ND State Health Dept 1938-45, Chief Engr ND State Health Dept 1945-53, State Health Officer 1953-57, In charge of Fed Water Pollution in South West 1958-66, as a commiussioned officer of the US Public Health Service. Directed Public Health Service Environmental Programs in Washington DC and Cincinnati OH 1967-68. Assigned to EPA as Regional Admin of KS City Regional Office with rank of Asst Surgeon Gen. 1969. Retired in 1976. Presently completing term on the BD of AAEE. Also serve on Bd of the Nat'l Sanitation Foundation, Ann Arbor. President of the Municipal Utility District, Town of Trophy Club, TX. *Society Aff:* ΣΞ, AWWA, AAEE.

Swabb, Lawrence E, Jr
Home: 92 Addison Dr, Short Hills, NJ 07078
Position: Retired *Education:* PhD/Chem Eng/Univ of Cincinnati; MS/Chem Eng/Univ of Cincinnati; BS/Chem Eng/Univ of Cincinnati. *Born:* Oct 1922. 1Lt US

Swabb, Lawrence E, Jr (Continued)
Army Ordnance Dept 1943-46. Joined Esso Res Labs Baton Rouge La 1951 in refinery process R&D, V P Exxon Res & Engrg Co 1968 petroleum res 1974 synthetic fuels res. 1980 Corporate Services. 1982 retired. UC Disting Alumnus Award 1969. Mbr AIChE & AAAS. Elected to NAE 1977. *Society Aff:* AIChE, AAAS, NAE.

Swalin, Richard A
Business: Allied Corp, P.O. Box 3000R Plaza, Morristown, NJ 07960
Position: VP-Research & Development *Employer:* Allied Corp *Education:* PhD/Metallurgy/Univ of MN; BS/Metallurgy/Univ of MN. *Born:* 3/18/29. in Minneapolis, MN. Res Assoc G E Res Lab Schenectady NY 1954-56; 1956-60 Asst & Assoc Prof Univ MN, 1960 apptd Prof, 1963-68 Hd Sch of Mineral & met Engrg, 1968-71 Assoc Dean of Univ of MN Inst of Tech, 1971-77 Dean. 1963 guest at Max Planck Inst Phys Chemie Gottingen Germany. 1977-1980 VP, Eltra Corp; 1980- VP, Research & Dev, Allied Corp; 1973-77 & 80, Dir, Medtronic Co; 1976-77, Dir, Donaldson Co; 1973-77, Dir, Buckbee-Mears Corp: 1971-75, Dir, Sheldahl Co. 1971, Recipient, NATO Fellowship in Science. Cons, Honeywell, G E Co. *Society Aff:* APS, AAAS.

Swan, David
Home: Point Rd, Wilson Point, South Norwalk, CT 06854
Position: Retired *Education:* BENG/Met Engrg/RPI *Born:* 05/02/20 Born in Arlington, Nj. US Navy (Pacific Ocean-Destroyers) WW II. Joined Union Carbide in 1946 as Res Engr, During career with UCC served as VP-Res, Linde Div, VP-Tech-Metals Div, Mgr-Planning & Gen. Mgr-Defense & Space Sys Dept. Joined Kennecott Copper Co as VP-Tech in 1966. Retired from Kennecott in 1982. Honors-Fellow, AIME; ASM; Poly Inst of NY; Demers Gold Medal-RPI; Krumb Lecturer-AIME; Orton Lecture-Am Cer Soc Offices-Pres-Met Soc (AIME); Smelter Control Res Assoc; Smelter Env Res Assoc; Dir Indust Res; Dir-AIME; CO Sch Mines Res Fdtn; Eng Fdnt; Carborundum Co; Quebec Iron & Titanium Co Adv Comm Nat Mat Adv Bd (NSF); Comm on Scope & Conduct of Mat Res (NAS); Met Adv Comm-Univ Penna; Carnegie-Mellon Univ; Eng Adv Comm-RPI; Poly Inst NY. *Society Aff:* AIME, ASM

Swan, Rens H
Business: P32-76/PO Box 16858, Philadelphia, PA 19142
Position: Manager/Weight Technology *Employer:* Boeing Vertol Company. *Education:* BS/ME/Drexel Univ. *Born:* 1926. Electronic Tech USN 1945. 1954 Licensed P E Penn. 1950 Piasecki Helicopter Corp (now Boeing Vertol Co); increasing responsibility as Weight Analyst, Weight Engr & R&D Proj Weight Engr for Helicopter & V/STOL A/C. 1958 Ch Weight Engr responsible for all mass properties prediction, analysis & control. 1968 Technology Mgr 107/CH-46 responsible for coordination & application of all Technology efforts in support of H-46 & 107 Helicopter progs. 1970- , Mgr Weight Technology. 1960 Natl Exec V P, 1974 elected Fellow SAWE, 1978 Licensed PE MD. 1985, Elected Honorary Fellow SAWE. *Society Aff:* SAWE, AHS.

Swanson, A Einar
Business: P.O. Box 8405, Kansas City, MO 64114
Position: Partner-in-Charge *Employer:* Black & Veatch Consulting Engrs *Education:* B/ME/Univ of MN *Born:* 10/8/26 A partner in the firm and serves as project mgr for various Power Div major fossil projects; provides consulting services for other study and generating plant projects, and periodically provides expert witness testimony at official public hearings. A native of MN joined Black & Veatch in 1966 as Dir of Nuclear Activities, directing the firm's work in various areas of nuclear work from personnel development to managing nuclear projects. Prior to joining Black & Veatch was with Northern States Power Company for 19 yrs involved in nuclear engrg responsibilities - mainly Pathfinder Atomic Power Plant. Enjoys the outdoors - hunting, fishing. *Society Aff:* ASME, ANS, NSPE, NCEE, PEPP

Swanson, Bernet S
Home: 2N151 Swift Road, Lombard, IL 60148
Position: Professor of Chemical Engineering. *Employer:* Illinois Institute of Technology. *Education:* PhD/ChE/IL Inst of Techn; MS/ChE/IL Inst of Tech; BS/ChE/IL Inst of Techn. *Born:* Nov 1921 Chicago Ill. Mbr faculty Ill Inst of Technology 1944-86, Dept Chmn 1967-72. Mbr: AIChE, ISA & ASEE. Winner of Journal Award & Donald Eckman Award ISA, Western Electric Award ASEE. P E Ill. Cons in control & gas explosions. Author 'Electronic Analog Computer' 1965. Hobbies: fishing & woodwork. *Society Aff:* ASEE, ISA, AIChE.

Swanson, Norris P
Home: 1311 North 43 Street, Lincoln, NE 68503
Position: Agricultural Engineer. *Employer:* USDA/Agricultural Research Service. *Education:* BS/Agri Engg/IA State Univ. *Born:* Oct 1916. Native of Clarinda Iowa. P E Texas. USDA Soil Conservation Serv, Texas 1939-41 & 1945-48. US Army & USN 1941-45; Engr Officer, destroyers. Transferred to SCS Res Div 1948 & USDA Agricultural Res Serv 1954 Texas. Cooperative res with Dept Agricultural Engrg Univ of Nebr & Nebr Agricultural Experiment Station, Lincoln 1955-80. Collaborator 1980- Journal publs include irrigation, soil erosion control, control & utilization of beef feedlot runoff. USDA Certificate of Merit 1968. Chmn Mid-Centrol Region 1971-72, Dir 197476, Fellow ASAE, Fellow SCSA, SCSA Councilman 1981-84. *Society Aff:* ASHE, ASA, SSSA, SCSA.

Swarner, Lawrence R
Home: 2548 South Dover Way, Lakewood, CO 80227
Position: Chief Land Resources Mgmt Branch. *Employer:* US Bureau of Reclamation/USDI. *Education:* MS/Soils Engg/OR State Univ; BS/Agri Engg/OR State Univ. *Born:* June 1914. Irrigated farm background in E Oregon. Irrigation Engr USDA Medford Ore conducting res on crop water requirements & irrigation practices 1941-45; Water resource dev & mgmt Bureau of Reclamatinon/Div of Water Oper & Maintenance USDI 1945- . Specialty in land dev, irrigation practices, irrigation sys oper & maintenance. Author or co-author of over 20 tech papers or bulletins. Reg P E. P Chmn Pacific-NW, P Chmn Natl ASAE; Soil & Water Steering & Exec Cttes; ASAE Fellow. Pacific-NW ASAE Engr of the Yr 1968. Water Resource Dev Cons to Japan, Thailand, Egypt, Tunisia, Spain & Tanzania. *Society Aff:* ASAE, USICID, NSPE.

Swartwout, Robert E
Business: West Virginia Univ, Engineering Sciences Bldg, Morgantown, WV 26506
Position: Prof *Employer:* Coll of Engrg *Education:* PhD/EE/Univ of IL Urbana/Champaign; MS/EE/IL Inst of Tech; BS/EE/IL Inst of Tech *Born:* 12/06/21 Began career as electrical draftsman. In WW II was Signal Corps radio ofcr in US and overseas. Electronic instrumentation and closed circuit TV for ITT Federal Labs in Ft Wayne, Indiana. At UIUC assisted in designing ILLIAC II computer. Since 1962 at WV Univ as Prof of Electrical Engrg and Graduate Advisor. Was Chrmn of WVU faculty Senate for one year. Was first chrmn of Steering (exec) committee for Multiple-Valued Logic group which in now a technical committee of IEEE Computer Society. In 1981 awarded Computer Society Certificate of Appreciation for contributions to Multiple-Valued Logic Intl Symposia. *Society Aff:* IEEE, IEEE, CS, ΣΞ

Swartz, Stuart E
Business: KS State Univ, Manhattan, KS 66506
Position: Prof *Employer:* Dept of Civil Engg *Education:* PhD/CE/IL Inst of Tech; MS/CE/IL Inst of Tech; BS/CE/IL Inst of Tech *Born:* 10/17/38 Native of Chicago, IL. Resch since 1961 in design of concrete structures, models of shell structures, stability of concrete structures, cracking and fracture of concrete. Nat'l Pres, Society for Experimental Stress Analysis 1982-83. Founder and first Pres of KS Chpt, ACI. Registered Prof Eng in KS. Chrmn, ASCE-Engrg Mechanics Div (EMD), Committee on Experimental Analysis and Instrumentation 1973-75; Chrmn, ASCE-EMD Programs Committee 1978-82; Chrmn SESA Committee on Monographs 1978-80; Chrmn, SESA Nat'l Meetings Committee 1980-82; Nat'l Treasurer, SEM 1986-87. Other Committees: ASCE-ACI Joint Committee on Shell Design and Construction,

Swartz, Stuart E (Continued)
ACI Models of Concrete Structures, ACI Fracture Mechanics. *Society Aff:* ACI, ASCE, SEM, BSSM

Swearingen, John E
Business: 200 E Randolph Drive, Chicago, IL 60601
Position: Chairman of the Board. *Employer:* Continental Illinois Corp *Education:* MS/Chem Eng/Carnegie-Mellon Univ; BS/Chem Eng/Univ of SC. *Born:* Sept 1918. Joined Standard Oil 1939 as Chem Eng Whiting Ind Res Dept. Elected Dir Standard Oil 1952; V P Production 1954; Pres 1958; Ch Exec Officer 1960, Chm & CEO 1965-1983 CHM CEO Continental Corp 1984-. Mbr NAE, Dir Continental Illinois Corp., Lockheed Corp., ADN Corp., Sara Lee Corp., TR Carnegie-Mellon Univ., Fellow AIChE. *Society Aff:* ΦΒΚ, ΤΒΠ, ΣΞ, ΟΔΚ, NAE, ACS, AIChE.

Swearingen, Judson S
Business: 2235 Carmelina Ave, Los Angeles, CA 90064
Position: Pres. *Employer:* Rotoflow Corp. *Education:* PhD/ChE/Univ of TX; MS/ChE/Univ of TX; BS/ChE/Univ of TX. *Born:* 1/11/07. San Antonio, TX, Partner, San Antonio Ref Co, 33-38; asst prof chem engrg, TX, 39-40, assoc prof, 40-41, prof 41-42; turbine designer, Elliott Co, 42-43; div engr, Kellex Corp, 43-45; Pres, Rotoflow Corp, 46-, Rotoflow AG, Switz, 71-; Swearingen Bros, Inc, 64-82; Statham-Swearingen, Inc, 59-62. US Natl Acad Engrg, Mexican Natl Acad Engrg, Natl Soc PE. Am Chem Soc; Fellow, AIChE; Am Inst ME; Am Soc for Testing Mtls. Low temp gas separations; turboexpanders; petrol & gas; seals, pumps & centrifugal compressors; absorption refrig. *Society Aff:* AIChE, ASME, ACS, ASTM.

Swearingen, Thomas B
Business: 1835 Terminal Dr, Suite 230, Richland, WA 99352
Position: VP *Employer:* McDaniel & Assoc *Education:* PhD/ME/Univ of AZ; MS/ME/WA State Univ; BS/ME/KS State Univ *Born:* 10/31/31 Thomas B. Swearingen is VP McDaniel & Assoc, with 25 yrs of experience in Design Engrg, Resch and Dev, and Project Mgmt for Coal-Fired and Nuclear Power Plants, Nuclear Materials Facilities, Military and Aerospace Projects, and Industrial Facility Projects where he managed consulting enginering activities and supervised the activities of design and project mgmt personnel. *Society Aff:* ASME, ASHRAE, NSPE, ΣΞ

Sweeney, James L
Business: Dept of Eng - Economic Systems, Stanford, CA 94305
Position: Prof. *Employer:* Stanford Univ. *Education:* PhD/EES/Stanford Univ; SB/EE/MIT. *Born:* 3/22/44. Prof James L. Sweeney has focused his activities on the application of economic methods and mathematical modeling to natural resources issues, particularly energy planning and policy analysis. At Stanford Univ, he serves as a Prof in the Dept of Engrg-Economic Systems and has served as the Dir of the Energy Modeling Forum, and the Chrmn of the Inst for Energy Studies, and until recently, the Dir of the Ctr for Economic Policy Research. Outside of Stanford, Prof Sweeney serves as a conslt or advisor to industry, research organizations, consulting firms, law firms, and U.S. and foreign governments. *Society Aff:* AEA, IAEE, ES.

Sweeney, Joseph P
Home: 108 South Hills Dr, Hershey, PA 17033
Position: Vice President, Technology *Employer:* AMP Inc. *Education:* M. S./Mathematics/Eastern New Mexico University; B. S./Mathematics/Eastern New Mexico University *Born:* 8/14/23 Native of Wichita Falls, TX. Served with the US Navy in the Atlantic and Pacific Theaters, 1942-45. Taught Mathematics and Chemistry at Eastern NM Univ prior to joining North American Aviation as an Electrical Design Supervisor in 1952. In 1955 joined AMP Inc and presently serves as Corp VP, Technology. In addition to corporate duties serves as a reviewer for the National Science Foundation programs and as an Industrial and Curriculum Advisory Member with a number of universities. *Society Aff:* AAAS, IEEE

Sweeney, Neal J
Business: 17 Amelia Pl, Stamford, CT 06904
Position: VP. *Employer:* Crawford & Russell. *Education:* MSChE/CE/NYU; MBA/Marketing/NYU; BSChE/CE/Univ of Notre Dame. *Born:* 10/27/36. Native of Long Island NY. In 1958 joined W R Grace (Clifton NJ) in engg and tech service. Moved to Baton Rouge for maj position in semi-works & commercial polyolefin Proj. Joined M W Kellogg in 1961 & had a maj role in the dev, design & worldwide startup of new generation of all centrifugal ammonia plants. Moved into proj mgr in 1968 & managed maj ethylene and methanol projs. Joined Crawford & Russell in 1970 in Stamford as proj mgr. Moved into sales in 1972 & assumed increasing responsibility including estimating & cost control. Appointed VP sales in 1975 and VP Stamford Operations April 1980. Enjoy golf & fishing. *Society Aff:* AIChE.

Sweeney, Thomas B
Business: 3650 Mayfield Road, Cleveland, OH 44121
Position: Senior Vice President/Design & Research. *Employer:* The Austin Company. *Education:* BS/CE/Lafayette College. *Born:* May 1927. Structural Engr & Estimator The Austin Co 1950-57; Ch Engr Austin Brazilian subsidiary 1957-63; 1964 appointed Pres Austin Europe; 1972 assumed present position with corporate responsibility for all engrg performed by The Austin Company as well as direction of its res activities. President Cleveland Engrg Soc Mbr: ASCE, ACI, AWS. *Society Aff:* ASCE, ACI, AWS, CES, NSPE.

Sweers, John F
Business: 1220 Mound Ave, Racine, WI 53404
Position: Dir-Distribution & GAME Mfg. *Employer:* Western Publishing Co Inc. *Education:* BS/Ind Engg/VPI. *Born:* Nov 1925 Chicago Ill. BSIE VPI 1946. P E Mass. Genl Mgr Physical Distribution Div & Plant Mgr Fayetteville Western Publishing Co. P Pres Wayne Warehousing Corp. Fellow IIE. Various regional & natl offices IIE 1960-76, past Pres & mbr Exec Ctte. Adv Bd IEOR, VPI. Ohio State Physical Distribution & Logistics Adv Bd. Engrg Curriculum Adv Bd Univ of Wisc at Parkside. *Society Aff:* IIE.

Sweet, John W
Home: 459 Newton Street, Seattle, WA 98109
Position: Retired from Boeing Co. *Employer:* Cons. *Education:* BS/Engr/Univ of WA. *Born:* May 1910. Native of Seattle. Engrg The Boeing Co Jan 1935; appointed Ch Metallurgist 1946; supervised Materials Engrs in selection of materials & their processing for propeller aircraft, jumbo jets, spacecraft & supersonic aircraft; retired 1973. Mbr: Alpha Sigma Mu (Hon Met Engrg Fraternity), Natl Adv Ctte for Aeronautics, Subctte on Aircraft Structural Materials 1957, Bd of Trs ASM 1961- 63, Fellow ASM 1976. Licensed Met Engr Wash 1947- . Hobbies: cons, fishing, gardening, travel. *Society Aff:* ASM.

Sweet, Larry M
Business: GE Corp. Res. & Dev, PO Box 8, Rm K1-SC17, Schenectady, NY 12301
Position: Mgr, Artificial Intelligence Branch *Employer:* General Elec *Education:* PhD/ME/MIT; ME/ME/MIT; MS/ME/MIT; BS/ME/Univ of CA. *Born:* 1/2/48. Larry directs res & dev in Knowledge-based Systems, Image Understanding, and Parallel Computing at CRD, helping establish GE as a technical leader in these fields. Previously, he directed dev of factory automation tech as Mgr of the Control Theory and Systems Program. From 1974-82 he was an Assoc Prof in the Dept of Mech and Aerospace Engg at Princeton Univ, a Guggenheim Foundation Fellow, and worked at the Japanese Natl Railways Tech Inst in Tokyo. A native of CA, Larry completed his undergrad studies at the Univ of CA, Berkeley in 1969, and received his PhD from MIT in 1974. *Society Aff:* ASME, AAAI, IEEE, ΤΒΠ, ΣΞ.

Sweetland, David L
Business: PO Box 504, State College, PA 16801
Position: President. *Employer:* Sweetland Engg & Assocs. *Education:* BS/Civil Engg/PA State Univ; MS/Civil Engg/PA State Univ. *Born:* 12/9/42. Native of LaJunta, CO. Served at lt, Jr, Grade, with Coast and Geodetic Survey supervising hydrographic survey crew and surveying of obstruction charts of airports, 1965-1967. con-

Sweetland, David L (Continued)
struction engr for Glenn O Hawbaker, Inc, 1968-70. pres of Sweetland Engg & Assoc, Inc since its founding in 1970. Consulting Engr practicing civil engg, land planning, site design, land surveying, soils investigation, feasibility studies and reports. Cert Sewage Enforcement Office of PA, 1975. Pres of Mid-State Chapter PSLS. Enjoy antiques and sailing. *Society Aff:* NSPE, PEP, ACSM, PSLS.

Swenson, George W, Jr
Business: Elec Engrg Dept, U of Ill, 1406 W Green St, Urbana, IL 61801
Position: Prof Electrical Engrg & Astronomy. *Employer:* University of Ill/Urbana-Champaign. *Education:* PhD/EE/Univ of WI; EE/EE/MI Tech; SM/EE/MIT; BS/EE/MI Tech. *Born:* Sept 22, 1922. Signal Corps Officer US Army WW II. Faculty positions Wash Univ/St Louis, Univ of Alaska, Mich State Univ 1951-56; present position 1956- ; Hd, EE Dept, 1979-85. Visiting Scientist Natl Radio Astronomy Observatory 1964-68 & Caltech 1972-73. Dir Vermilion River Observatory Univ of Ill 1968-81. Fellow IEEE. Acting Head Dept of Astronomy Univ of Ill 1970-72. Mbr Federal Govrenment & Natl Res Council Adv Panels. Natl Acad of Engg (Mbr)-1978. Fellow AAAS. Guggenheim Fellow, 1984. *Society Aff:* IEEE, AAAS, OSA

Swensson, Gerald C
Home: 422 Oak Valley Rd, Media, PA 19063
Position: Consultant (Self-employed) *Education:* MS/Marine Engg/MIT; BS/Naval Arch & Marine Engg/MIT. *Born:* 8/26/23. Served as Navy Battleship officer in WWII. After grad sch (1949), worked at Bethlehem Steel Co-Quincy Shipyard in ship machinery design area. Became chief turbine engr in 1957. Shipyard mfg steam propulsion turbines of 7,000-27, 000 hp. Became asst chief engr in 1963. Promoted gas turbine propulsion for naval and merchant vessels. Joined Sun Ship (1964) as hd machinery design group. Became engg mgr (1973) for LNG carrier design group. Joined R&D Dept (1978) for dev & advanced ship propulsion projects. Chrmn of Marine committee, gas turbine div of ASME, 1971-72. Chrmn of Phila Sec of SNAME, 1978-79. PE in PA & MA. *Society Aff:* SNAME, ASME, IME.

Swerdfeger, John H
Home: 928 Keith Rd, West Vancouver BC, Canada V7T 1M3
Position: Consulting Engineer. *Employer:* Self. *Education:* BASc/Civil Engg/Univ of BC. *Born:* Nov 30, 1922. BSCE Univ of British Columbia. 1944-49 various engrg positions, 1949-75 St Partner large Architect/Engr firm, 1967-75 Pres Unecon Engrg Cons Ltd & Chmn of Bd Mech Cons Ltd. Sev yrs mbr, then Chmn Civil Bd of Examiners, Assoc Prof Engrs of BC. Num local & natl offices, Engrg Inst of Canada, 1967-68 Pres Engrg Inst of Canada. Reg P E in BC & Yukon Territory. *Society Aff:* FEIC, MCSCE.

Swerdlow, Nathan
Home: 1919 Chestnut St, Phila, PA 19103
Position: Consultant (Self-employed) *Education:* BS/EE/Drexel Univ. *Born:* 6/5/07. Native Philadelphian. Was Sr Advance Engr for GE, responsible for design and application of isolated phase bus. Retired in 1972 & continued as Consultant on bus for several utilities & consulting engrs. Recipient of 17 patents on bus & author of 10 technical papers published in IEEE transactions. Mbr of committee to revise ANSI std C37.20. Mbr of IEEE working group 70.1. Named IEEE fellow in 1977. Reg PE in PA. *Society Aff:* IEEE.

Swerling, Peter
Home: 1136 Corsica Dr, Pacific Palisades, CA 90272
Position: Bd Chrmn *Employer:* Tech Service Corp *Education:* PhD/Math/UCLA; MA/Math/UCLA; BA/Econ/Cornell Univ; BS/Math/Caltech *Born:* 03/04/29 Born in New York City. Lived in Southern CA since 1930. Brought VP in Beverly Hills & attended elementary & high school there. Worked at Rand Corp (Engrg Staff) 1948-61; Conductron Corp (Dept Hd), 1961-64; Radar Consultant, 1964-66; Founder and Chrmn, Tech Service Corp, 1966-Present. Also visiting Asst Prof of EE, Univ of IL, 1956-57; Adjunct Prof of EE, Univ of Southern CA, 1965- 67. Also founding mbr, Bd of Trustees, Crossroads School of Arts and Scis, Santa Monica (1970-Present), and founding mbr, Bd of Dirs, Centurion Savings and Loan Assn, Los Angeles (1979-Present). Married 23 yrs to Judith Ann Butler Swerling, three children: Liz (20), Carole (17), Steven (15). Hobbies: reading, music, skiing, swimming, tennis, bridge. *Society Aff:* NAE, IEEE

Swern, Leonard
Business: 1290 Ave of Americas, New York, NY 10104
Position: VP Tech Programs. *Employer:* Sperry Corp. *Education:* AB/Physics/Columbia Coll; MA/Physics/Columbia Univ. *Born:* Feb 1925. Educ: AB/Physics/Columbia; MA/Phys/Columbia. Addl work toward PhD in microwave spectroscopy. Worked at Columbia Radiation Lab on millimeter wave components. Involved in microwave physics res & dev at Sperry Gyroscope 1949-60. Publ 10 papers in this field. Engaged in early work on microwave radiometry. Administered Sperry's internal R&D Prog 1960-69. Assumed present respon at Sperry Rand corp level in 1969. Chmn of Corp's Tech Advisory Ctte Fellow IEEE, Mbr Natl Sci Found Indus Panel on Sci & Tech. Mbr Tech Advisory Ctte, US Postal Service. Mbr NY Acad of Sci. *Society Aff:* APS, IEEE, NYAS.

Swift, Calvin T
Home: 55 Cherry Ln, Amherst, MA 01002
Position: Professor *Employer:* Univ of MA *Education:* PhD/Physics/William and Mary; MS/Physics/VA Polytech Inst; BS/Physics/MIT *Born:* 02/06/37 Res Engr, North American Aviation 1959-62. Aerospace Engr, NASA Langley ResCenter, specializing in Microwave Remote Sensing and Reentry Communications, 1962-81. Prof of Elec and Computer Engrg, Univ of MA 1981- . Secretary-Treas IEEE Antennas and Propagation Society 1973-76. Editor, IEEE transactions on Geoscience and Remote Sensing, 1980-84. Recipient of 1977 Distinguished Service Award, IEEE Council on Oceanic Engrg. Fellow, IEEE. *Society Aff:* IEEE, URSI, AGU.

Swift, Clinton E
Home: 4 Tyler Ave, Branford, CT 06405
Position: Chief Engineer. *Employer:* Custom & Precision Prods Inc. *Education:* BSME/Mech Eng/Univ of IL. *Born:* May 1905 Hinsdale Ill. Field engr in automatic welding for Westinghouse Elec. R&D in welding of copper alloys for Anaconda Amer Brass Co & Ampco Metal Co, granted 4 pats in this field. Pres of Connecticut Welders Inc for 11 yrs & Pres of Swift Welding Corp for 11 yrs. Life Mbr AWS & ASM. Hobbies: reading & classical music. *Society Aff:* AWS, ASM.

Swift, Fred W
Business: 201 Engg Bldg, Memphis, TN 38152
Position: Assoc Dean. *Employer:* Memphis State Univ. *Education:* PhD/Industrial Engr/OK State; MS/Mech Engr/Univ of AL; BS/Aeronautical/Notre Dame Univ. *Born:* 7/4/36. Native of Springfield, MO. Engr for Douglas Aircraft, 1958-61. Designer for NASA on Saturn vehicles including Moon program, 1961-70. Specialty areas were dynamics and design of control system. Taught Systems analysis, operations research, Statistics and other engg courses at Univ of AL, Univ of MO and Memphis State Univ. Currently serve as Assoc Dean of Engg and Dir of Industrial Systems program. I have been a consultant to natl and intl cos on various projects. Past Pres of local IIE chapter and currently serve as regional VP and mbr of Bd of Trustees of IIE. *Society Aff:* IIE, SME, ASEE, ΤΒΠ.

Swift, Roy E
Home: 321 Melbourne Way, Lexington, KY 40502
Position: Prof Emeritus. *Employer:* Univ of Kentucky. *Education:* DEng/Metallurgy/Yale Univ; MEng/Univ of WA; MEng/Metallurgy/Univ of UT; BS/Mining/MO Sch of Mines (Univ of MO); NMD/Nutrition/John F. Kennedy Univ Medical Arts & Sciences. *Born:* 11/4/11. Mbr AIME, ASM, ASEE, KSPE, Alpha Chi Sigma. Also Alpha Sigma Mu, Tau Beta Pi, Sigma Gamma Epsilon & Sigma Xi. Lexington Exchange Club (Pres), Lexington Toastmasters Club (P Pres). Prof registrations: Met Engrg, Chem Engrg, Mining Engrg in KY; Met Engrg &

Swift, Roy E (Continued)
Mining Engrg In IN. Assoc Prof of Mining Engrg 1951-56. Prof of Met Engrg 1956 to present, Univ of KY (Dir of Foundry Educ Found Prof 1957-63). 1940-51: var acad assignments at LSU, Univ of AL, Purdue Univ & Univ of NV. 1934-40: with the mineral indus in the fields of met geol & mining engrg. Now Prof "Emeritus" of the Univ of KY (1977-present). Since retiring have earned the degrees: Doctor of Naturopathic medicine and Dr of Osteopathy from the Anglo-Amer Inst of Drugless Therapy in Refrew, Scotland, United Kingdom. Now practicing as a consultant: Holistic Healthful Living. Also, continuing on the PhD in Nutritional Sci with the Texas Healthology Inst, located at New Braunfels, Texas. Licensed in the state of NC to practice Naturopathy. Granted the degree of Doctor of Nutritional Medicine by the John F. Kennedy University of Medical Arts and Sciences in 1986. Certified Nutrimedicist by the American Nutritional Medical Assoc, 1987. Registered Consult of Holistic Healthful Living by the American Holistic Health Sciences Assoc. *Society Aff:* AIME, ASM, ASEE, AHHSA, ANMA, USMA

Swim, William B
Business: Box 5014. T.T.U, Cookeville, TN 38505
Position: Prof. *Employer:* TN Tech Univ. *Education:* PhD/Mech Engr/GA Inst of Tech *Born:* 11/18/31. Specialization in fluid dynamics and fluid machinery lead to work in flow noise and then into acoustics and noise control. Res group leader in fluid dynamics in air conditioning industry for 10 yrs. Currently Prof of Mech Engg and Director of the Noise Control Facility at TN Tech Univ. Active in teaching, research in fan performance and noise and consulting in fluid dynamics and noise control. ASME. Past Chrmn of Fluid Mechanics Comm & Current Chmn Rotating Machinery Noise Comm of ASME. Mbr, ASA, ASHRAE, Chrm Std Committee on Lab Meas of Flow Losses in Duct & Fitting, Reg PE. Elder United Presbyterian Church. Pastimes: hiking, swimming, cycling, woodworking, and automobile restoration. *Society Aff:* ASME, ASHRAE, ASA, ASEE.

Swink, Earl T
Home: 910 Preston Ave, Blacksburg, VA 24060
Position: Prof Emer - retired. *Employer:* Va Poly Inst & State Univ. *Education:* MS/Agr Engr/VPI & SU; BS/Agr Engr/VPI & SU. *Born:* Sept 7, 1907. Grad trainee Westinghouse Elec & Mfg Co 1930-31. Agri Engr Va Elec & Power Co 1932-35; VPI & SU fac since 1935 as Asst Prof Agri Engrg, Assoc Prof Agri Engrg, Prof. and Hd Agri Engrg Dept 1954-67 & Ext Leader Spec Progs 1967-70. Prof Emeritus 1970- . Mbr ASAE 1930- ; Chmn Elec Power & Processing Div 1953-54, Chmn Va Sect 1956, Chmn Educ & Res Div 1959-60, Natl Bd of Dir 1961-63, V P 1963-66. Chmn Agri Engrg Div ASEE 1959-60. Chmn SW Va Chap Va Soc PE's 1962-63. Co-Founder Va Farm Electrification Council 1945, Bd of Dir 1945-53 & Chmn 1953-57. Pres So Assn Agri Engrs & Voc Agri 1958-59. Apptd to Agri Bd, Natl Res Council, Natl Acad of Sci 1968. Author or co-author some 45 bulletins, circulars & leaflets on agri engrg subjects. Fellow ASAE 1959. Mbr Tau Beta Pi, Epsilon Sigma Pi, Omicron Delta Kappa. *Society Aff:* ASAE, ΤΒΦ, ΟΔΚ, ΕΣΦ.

Swisher, George M
Business: Box 5015, Tennessee Tech Univ, Cookeville, TN 38501
Position: Assoc Dean of Engg and Prof of Mech Engr. *Employer:* Tenn Tech Univ. *Education:* PhD/Mech Engr/OH State; MS/Mech Engr/OH State; BSME/Mech Engr/Univ of Cl. *Born:* July 1943. Native of Lancaster Ohio. Taught sys engrg at Wright State Univ (Dayton Ohio) 1969-73. Teaching mech & sys engrg at Tenn Tech 1973-79. Associate Dean of Engineering 1979- . Summer NASA Res Fellow 1972 & 1975. Prof Engr in Ohio. Author 'Intro to Linear Sys Analysis' 1976. *Society Aff:* ASME, ASEE, AAUP, ΤΒΠ, ΠΤΣ, ΦΗΣ.

Sylvester, Nicholas D
Business: 600 S College, Tulsa, OK 74104
Position: Dean of Engrg *Employer:* The Univ of Tulsa *Education:* PhD/ChE/Carnegie-Mellon Univ; BS/ChE/OH Univ *Born:* 04/16/42 Native of Cleveland, OH. Teaching experience at Carnegie-Mellon Univ and the Univ of Notre Dame before going to The Univ of Tulsa in 1972. Assumed current position as Dean of Engrg and Applied Scis in 1978 after serving one year as Chrmn of the Div of Petroleum and Chem Engrg. Indust experience with Shell Oil Co and Amoco Production Co. Is a reg PE in OK and a conslt to numerous energy related companies. Has directed more than 30 funded res projs totaling nearly two million dollars and has authored more than one hundred tech pubs. *Society Aff:* AIChE, AIME, NSPE, ACS, APS, ASEE

Symington, Kenneth A
Home: 2025 Cherry St, Philadelphia, PA 19103
Position: Pres-S S White Div. *Employer:* Pennwalt Corp. *Born:* 1/17/32. Chem Engg Rensselaer Polytechnic Inst. Proj & Process Engr assoc with Proctor & Gamble & Std Brands Inc. Also work in ceramic engg. Extensive prof service in Latin Amer, Europe & the Middle East. Fluent in 4 languages. With Penwalt Corp (S S White Dental Div) since 1965. Have been genl mgr of sev domestic & intl divs of the co. Also corp positions in corp & profit planning include Asst to Pres/Chrmn of Bd now Pres of SS White Dental Div. Mbr AIChE.

Symonds, Paul S
Business: Div of Engrg, Providence, RI 02912
Position: Prof of Engrg (Research). *Employer:* Brown Univ. *Education:* PhD/Engg Mechs/Cornell Univ; MS/Engg Mechs/Cornell Univ; BS/Physics/Rensselaer Polytechnic Inst. *Born:* 1916. Instr in engrg mech Cornell 1941-43; physicist US Naval Res Lab 1943-47; Asst Prof Brown Univ 1947, Assoc Prof 1951, Prof 1954-1983, Prof (Research) and Prof Emeritus 1983- , Chmn Div of Engrg 1959-62. Fulbright res scholar 1949-50 Cambridge Univ & 1957-58 Swansea Wales; ICI Fellow 1950-51 Cambridge Univ; Guggenheim Fellow 1957-58 Swansea; NSF Sr Postdoc Fellow 1964-65 Oxford. Fellow ASCE, ASME & Amer Acad of Mech. Res pubs in struct mech, esp inelastic effects due to dynamic loading. *Society Aff:* ASCE, ASME, AAM, IABSE.

Symons, George E
Home: 86 Edgewood Ave, Larchmont, NY 10538
Position: Engineering Editor/Consultant. *Employer:* Self. *Education:* PhD/San Eng Chem/Univ of IL; MS/Chem & Bact/Univ of IL; BS/Chemistry/Univ of IL. *Born:* 4/20/03. Experience in water & wastewater plant design & oper; first chief chemist of Buffalo NY Sewer Auth. Lectr, Editor & Cons. P E Lic NY. Fellow ASCE, APHA; Hon Mbr, Diven Medalist & Fuller Awardee AWWA; Hon Mbr, Bedell Awardee & Emerson Medalist WPCF; Diplomate AAEE; Mbr ACS, AIChE, NSPE, AIDIS, Dir AWWA 1968-71, VP 1971-72, Pres Elect 1972-73, Pres 1973-74; Trustee AAEE 1965-71, Secy 1970-71; Hon Mbr AWWA 1968, Diven Medalist, 1976; Pres NYWPCA 1944-45; Dir WPCF 1948-51, Emerson Medalist 1969, Hon Mbr 1962. Mgr Spec Projs, Malcolm Pirnie Inc 1970-78; Engg Editor/Consultant since 1978. *Society Aff:* ACS, AIChE, AIDIS, AWWA, APHA, ASCE, NSPE, WPCF, AAEE.

Symons, James M
Home: 7510 Lawyer Rd, Cincinnati, OH 45244
Position: Sanitary Engr. *Employer:* US Environmental Prot Agency. *Education:* ScD/San Engr/MIT; SM/San Engr/MIT; BCE/Civ Engr/Cornell Univ. *Born:* 11/24/31. Native of Champaign, IL. Asst Prof of Sanitary Engg at MIT from 1957-1962, Joined the Fed Govt in 1962. Early responsibility for res in the area of water quality control in reservoirs. Assumed current responsibility in 1970 as Chief of Drinking Water Treatment Res. Am responsible for developing drinking water unit processes to meet the Natl Drinking Water. Regulations as issued by the US Environmental Protection Agency. Enjoy golf & photography. *Society Aff:* AWWA, WPCF, ASCE.

Symons, Robert S
Home: 290 Surrey Pl, Los Altos, CA 94022
Position: Product Line Manager, Electron Devices Div *Employer:* Litton Systems Inc. *Education:* MS/EE/Stanford; BS/EE/Stanford. *Born:* July 1925. Resident of Los Altos Calif. Engr Eitel-McCullough Inc 1946- 47; Engr Heinz & Kaufman 1948; Engr Pacific Electronics 1949; with Varian Assocs Inc 1950-83. Klystron Dev Engr 1950-58; Mgr Radar & Accelerator Tube Dev 1958-66; Mgr Radar & Accelerator Tube Dev & Mfg 1966-73; Mgr Energy Progs, Corp Res 1974-77. Sr Scientist Palo Alto Microwave Tube Div 1978-83. Product Line Mgr. Litton Electron Devices Div, 1983- . Associate Editor, IEEE Transactions on Electron Devices 1981-. Fellow IEEE. Mbr Phi Beta Kappa, Tau Beta Pi, Commonwealth Club of Calif. Served to 1st Lt AUS 1950-53. *Society Aff:* IEEE.

Synck, Louis, J
Business: P.O. Box 447, Amarillo, TX 79178
Position: Div Chrmn *Employer:* Amarillo College *Education:* MS/Math/Carnegie Tech; BME/Mech Engrg/Univ of Dayton. *Born:* 03/08/22 Born in Coldwater, OH. Served as radar officer, US Naval Reserve, World War II. Carnegie Tech graduate school, 1947-49. Employed Mutual Benefit Life home office, Newark, 1949-57, in actuarial study and agency res. With Amarillo College as math instructor since 1961; Chrmn Dept of Math & Engr since 1973; Chrmn Div of Scis & Engrg since 1980. Widower with family of 4 grown children. Interests: classical music, piano, philately and air covers. *Society Aff:* ASEE, AMATYC.

Syverson, Aldrich
Home: 222 St Antoine, Worthington, OH 43085
Position: Prof Emeritus. *Employer:* Retired. *Education:* PhD/Chem Engg/Univ of MN; BChE/Chem Engg/Univ of MN. *Born:* Mar 1914. Native of Minn. Indus exper: Tech Mgr, Organic Chems Dev; Plant Mgr Avon Lake Exper Sta, B F Goodrich Chem Co. Res Dir O M Scott. Cons to indus in reaction kinetics & process design. Acad exper: Syracuse Univ, Assoc Prof 1948-50; Ohio State Univ fac in Chem Engrg 1950-77, serving as Dept Chmn 1967-76, Prof Emeritus 1977- . Major res int in adsorption & heterogeneous catalysis. Hormel Res Found Fellow 1942; Columbus Tech Council Man of Yr 1975; Fellow AIChE 1974. *Society Aff:* AIChE, ACS, ASEE.

Szablya, John F
Business: EBASCO Services Inc, 10900 NE 8th St. #500, Bellevue, WA 98004-4405
Position: Electrical Consult Engr *Employer:* Ebasco Services Inc. *Education:* Doctorate/Economics/Jozsef Nador Univ of Budapest; Dipl/ME/EE/Jozsef Nador Univ of Budapest; Dipl/Economics/Jozsef Nador Univ of Budapest; Dipl/Educ/Jozsef Nador Univ of Budapest. *Born:* 6/25/24. Born Budapest, Hungary. 1947 to 1956, design engr at Ganz Works, faculty at his alma mater, & consultant to Hungarian Central Elec Res Inst & Hungarian Stds Inst. 1957 to 1963, faculty at Univ of British Columbia. 1963-82, faculty at WA State Univ. Was Visiting Prof at Univ of the West Indies & at Technische Universitat Braunschweig. Since 1981 consulting engr. with Ebasco Services Inc. Published more than 100 papers & reports. Fellow IEEE & IEE. Active with IEEE Committees. Reg: States of AK, CO, ID, MT, OR, WA & WY. Province of British Columbia, Great Britain, Register of Higher Technical Professions. *Society Aff:* IEEE, IEE, TSGB, OVE, ΣΞ.

Szabo, August J
Business: P O Box 52115, Lafayette, LA 70505
Position: Secy-Treas (Principal). *Employer:* Domingue, Szabo & Assocs Inc. *Education:* MS/Sanitary Engr/Harvard Univ; BS/Civil Engr/LA State Univ. *Born:* Sept 27, 1921. Engr officer US Army 1943-46; Engr Corps of Engrs (3 mos); Public Health Engr La (9 1/2 yrs); Assoc Prof CE USL (8 yrs); Principal, cons engrg firm Domingue, Szabo & Assocs Inc since Apr 1, 1957. Reg P E La, Miss; Reg Land Surveyor La. Mbr ASCE; Diplomate Amer ACAD of Environ Engrs; Mbr ASEE; Pres La Engrg Soc 1970-71; Dir NSPE 1973-77. Arthur Sidney Bedell Award 1968 (LPCA & WPCF); 25 yr mbr La Conf on Water Supply & Sewerage; Prof Servs Award (Leo Odom Award) 1976 LES; Fac Advisor & Contact Mbr USL Student Chap ASCE; Mbr Chi Epsilon CE Honor Soc. *Society Aff:* ASCE, AAEE, ASEE, NSPE, AWWA, WPCF.

Szabo, Barna A
Business: Campus Box 1129, Washington University, St. Louis, MO 63130
Position: Professor *Employer:* Washington Univ *Education:* PhD/CE/State U of NY at Buffalo; MS/CE/State U of NY at Buffalo; BASc/Mining Engrg/Univ of Toronto. *Born:* 09/21/35 Native of Hungary, naturalized U S citizen. Mining Engr, Internatl Nickel Co of Canada (1960-62); Engr Acres LTD, Niagara Falls, Canada (1962-68); on the faculty of the Sch of Engrg of Washington Univ since 1968. Albert P. and Blanche Y. Greensfelder Prof of Mechanics since 1975. Dir, Center for Computational Mechanics since 1977. Main res interest: finite element tech. Reg PE in MO. *Society Aff:* AAM, SES, IACM, ASME

Szawranskyj, William M
Home: 798 Lauren Ct, Webster, NY 14580
Position: Commissioner of Public Works & Town Engr. *Employer:* Town of Webster. *Education:* BS/Civ Engg/TN Tech Univ; MS/Mech Engg/RIT. *Born:* 9/27/46. Native of Gunsberg Germany. Proj Engr with Woodward Assoc Consulting Engg specializing in water and waste water treatment. Proj Mgr from 1971 till 1973 with Sear-Brown Assoc Engg specializing in Commercial and residential land planning and engg. Sr Assoc from 1973 till 1978 with Passero-Scardetta Assoc Engg Specializing in land planning and in charge of civ engg sec. Assumed current position as Public Works Commissioner in 1978. Responsible for Engg, Hgwy, Sewer, Water and Bldg Depts. Reg PE and Land Surveyor. Mbr of the Bd of Dir of the Rochester Engg Soc. Officer in the local sec of ASCE. Received honors for use of innovative methods in drainage control and in Zoning and Planning proceedures. Enjoy photography and sports. Mbr of the West Webster Fire Dept and the Ukrainian Catholic Church of the Epiphany. *Society Aff:* ASCE, NSPE, WPCF, ACSM, APWA.

Szczesny, Walter A
Business: City Building, Springfield, OH 45502
Position: Director. *Employer:* Clark Cty/Springfld Trans Coord Ctte. *Education:* BCE/Civil Engg/OH State Univ; -/Transportation Eng/Yale Univ *Born:* Sept 15, 1936 Cleveland Ohio. Ohio State Univ. 1960 B. of Civil Engr. Yale Univ, Bureau of Hwy Transp 1964. Prof Engr Ohio. Transportation Engr Dayton Ohio 1960-65. Dir of Transp Study for Clark Cty & Springfield Ohio - TCC. Respon for dir staff & coop with local govs & citizens, the Ohio Dept of Transp, FHWA, & UMTA in estab local transp policies, progs & plans. Pres Ohio Inst of Transp Engrs 1972. Secy Ohio Planning Conf 1975- . Mbr Inst of Transportation Engrs, Amer Soc of Civil Engrs & Am Planning Assoc. *Society Aff:* ITE, ASCE, APA.

Sze, T W
Business: Dept of Elec Engrg, Pittsburgh, PA 15261
Position: Professor *Employer:* Univ of Pittsburgh *Education:* PhD/EE/Northwestern Univ; MS/EE/Purdue Univ; BS/EE/Univ of MO *Born:* 09/13/22 Educator; born Shanghai, China; (came to US 1945, naturalized 1962); W. Frances Tung, August 29, 1952; children - David, Deborah, Daniel. Served to major in Chinese Nationalist Army, 1941-45. NATO Sr Fellow NSF, 1967; AIEE Prize Paper Award 1962. Mbr, Sigma Xi, Tau Beta Pi, Eta Kappa Nu. BS Univ of MO 1948; MSEE Purdue Univ 1950; PhD Northwestern Univ 1954. Univ of Pittsburgh, Asst Prof 1954-56; Fessenden Prof 1957-62, Westinghouse Prof 1962-65, Assoc Dean 1970-77, Acting Dean 1973-74. Currently Prof of Elec Engrg and Dir of Pattern Recognition Lab, and Adjunct Prof of Shanghai and Xian Jiaotong Universities in China. Author of books and technical paper in digital computer, system optimization and image processing. *Society Aff:* IEEE, ASEE

Szego, George C
Home: Box 340, Warrenton, VA 22186
Position: Pres. *Education:* PhD/CE/Univ of WA; MS/CE/Univ of WA; BS/CE/CCNY. *Born:* 8/10/19. Prof & Hd, Chem Engg Dept Seattle Univ, 1948-56; Mgr, Space Power & Space Propulsion, GE Co, 1956-59; Sr Engr, Space Tech Labs,

Szego, George C (Continued)
TRW, 1959-61; Sr Staff, Inst for Defense Analyses, 1961-1970. Pres, Chrmn InterTech/Solar Corp, 1970- 1981; Pres, Dr George C. Szego, PE & Assocs, Inc. Chrmn, Natl Prog Committee, AIChE; Chrmn, Elec Power Systems Technical Committee, ARS & AIAA; Founding Chrmn, IECEC, elected "Man of the Year;" VP, Solar Energy Industries Assoc; Chrmn, Bds of ITC/Solar Italia, ITC/Solar do Brasil, Ltda, ITC/Solar Korea, ITC/Solar Iberica, The Gasohol Corp. Reg PE, 19 states. *Society Aff:* Fellow, AIChE, NSPE, AAAS, VSPE, ISES.

Szekely, Julian
Business: MIT Room 4-117, Dept Matls Sci, Engrg, Cambridge, MA 02139
Position: Prof of Matls Engrg. *Employer:* MIT Room 4-117 *Education:* DSc/Engrg/Univ London; PhD/Engrg/Univ London; BSc/Engrg/Univ London. *Born:* Nov 23, 1934. P E New York. Taught at Imperial Coll, SUNY/Buffalo; at present, Prof of Matls Engrg at MIT. Also cons to gov & indus. Co-author texts: 'Rate Phenomena in Process Met', 'Process Optimization', 'Gas-Solid Reaction', & 'Fluid Flow Aspects of Metals Processing'. Res ints: transport phenomena in materials processing, optimization & modelling, environ & energy problems. Mathewson Gold Medal AIME, Sir George Beilby Gold Medal RIC, Prof Progress Award AIChE & Curtis McGraw Res Award ASEE; Howe Mem Lecture, AIME; Extractive Metallurgy Lecturer, AIME; Mbr, Nat Acad of Engrg; Guggenheim Fellow. Enjoy classical music, squash, skiing & sailing. *Society Aff:* AIME, IChemE (British).

Szentirmai, George
1353 Sarita Way, Santa Ana, CA 95051
Position: Pres. *Employer:* DGS Assocs *Education:* PhD/EE/Polytech Inst Brooklyn; PhD/EE/Tech Univ Budapest; MS/EE/Tech Univ Budapest. *Born:* Oct 1928 Budapest Hungary. Taught in Hungary & at Polytech Inst of Brooklyn. Spent 3 yrs in England, mainly at Std Telephones & Cables. Bell Labs: mbr of tech staff 1959-68, cons 1968-75. Assoc & Full Prof Cornell Univ 1968-75. Staff scientist Electronics Res Div of Rockwell Internatl 1975-80. VP Advanced Dev, Compact Div of CGIS 1980-82. Pres DGS Assocs, 1982- . Publ about 35 papers & edited a book. Spec in circuit & sys analysis & design & computer aided design. Fellow IEEE, Sr Mbr IEE (UK). Enjoys tennis & classical music. *Society Aff:* IEEE, IEE (UK).

Szeri, Andras Z
Business: 641 Benedum Hall, Pittsburgh, PA 15261
Position: Prof. and Chairman, ME Prof. Mathematics *Employer:* Univ of Pittsburgh *Education:* PhD/Eng/Univ of Leeds-England; BSc/Eng/Univ of Leeds-England *Born:* 6/6/34 Present primary appointment: Prof & Chmn of Mech Eng. Secondary Appointment: Prof of Math Univ Pittsburgh; Consultant, Westinghouse Elec Co., Assoc Editor, Journal of Tribology. Hungary. Married, '62, Mary J. Parkinson. Children: Andrew, '64, Elizabeth, '67, Maria, '70. Education: Deak Gimnazium, Gyor 49-53; University of Sopron, 53-56; Hungary. University of Leeds, UK, B.Sc. (Hon) 1st class 1959, Ph.D. 1962. Employment: Res Engr. The English Electric Co. 62-64; British Council Prof. of Fluid Mech., U. Santa Maria, Chile; Asst. Prof. 67-70, Assoc Prof 70-75, Prof 75- , Dept Mech Engrg, 78- Dept Math, 84- Chmn, Dept Mech Engg, Univ of Pittsburgh. Publications: 50 Journal papers, mainly Fluid Mech, Lubrication; TRIBOLOGY, Hemisphere, 1980. Res interests: Viscous Flow, Flow stability, Non-Newtonian Fluids, Heat Transfer, Tribology. Soc Mbrsp: Fellow ASME, AAM, SIAM, SES, Soc Nat Phil. *Society Aff:* AAM, ASME, SIAM, SES, Society for Natural Philosophy

Szewalski, Robert
Home: 19 Batorego, Gdansk, 80-251 Poland
Position: Prof/Mbr Polish Acad of Sci. *Employer:* Tech Univ Gdansk/Polish Acad of Sci. *Education:* Drhabil/Engg/Tech Univ Lwow; Dr/Engg/Tech Univ Lwow; MS/Engg/Tech Univ Lwow. *Born:* Aug 16, 1903 Nisko Poland. Habilitation 1938. Asst Tech U Lwow 1927. Assoc Prof 1938, Prof 1944, Prof Tech U Gdansk 1945. Rector, Pres Tech U Gdansk 1951-54. Full Mbr Polish Acad of Sci 1960. Visiting Prof Univ of Stuttgart Germany 1975, Visiting Prof Brown Univ, Providence R I 1976. Doctor honoris causa Tech Univ Gdonsk, 1978. Doctor h.c. Tech Univ. Poznan, 1987. Author 150 papers on theory & design of steam & gas turbines; high efficiency steam & gas turbine cycles; turbine blading of extreme length. Life Fellow ASME; Fellow IME London; Gdansk Sci Soc, Pres 1970-72. Comdr's Cross Ord Polonia Restituta; Gold Medal Tchechoslovak Acad Sci 1966; Officer Polish Air Force 1939. *Society Aff:* GTN, TMTS, SIMP.

Sziklai, George C
Home: 26900 St Francis Rd, Los Altos Hills, CA 94022
Position: Consulting Scientist (Self-employed) *Education:* ABS/Physics/Univ of Budapest; Dipl Eng/Chemistry/Tech Inst of Munich; Baccl/Phys Math Chem/Univ of Budapest. *Born:* 7/9/09. Native of Budapest. m. Violet Jambor. Ch Engr Polymet Corp 1933-35. Supervisory & Res Engr RCA in NY, Bloomington Ind & Princeton NJ 1939-56. Asst to VP Res & Engrg, Dir of Res Westinghouse Elec Corp 1956-67. Sr Mbr Res Lab Lockheed Missile & Space Co 1967-75. Cons 1976- . Lectured at Columbia, Princeton, Indiana, RPI. UCLA & Stanford Univs. Life Fellow IEEE, former Chrmn of Princeton Sect, Editor Bd 1950-65. Mbr NY Acad of Sci, Optical Soc of Amer, Sigma Xi. Received the 1947, 1951 & 1952 RCA Awards for 'outstanding res'. *Society Aff:* IEEE, OSA, ΣΞ, NYAS.

Szirmay, Leslie v
Home: 948 Winnona Dr, Youngstown, OH 44511
Position: Prof *Employer:* Youngstown State Univ *Education:* PhD/ChE/Denver Univ; ME/Nuclear Eng/IA State Univ; MS/ChE/Univ of Detroit; Dipl/Chem/Eotvos Univ Budapest *Born:* 11/13/23 In 1969, Dr. Szirmay joined the Chem Engrg Dept of Youngstown State Univ where he is Prof and Chrmn of the Chem and Metallurgical Engrg Dept at the present time. He has over twenty years of experience in various fields of the petrochemical, metal processing and nuclear industries, such as research, development and plant engrg, while working in Eurpope and North America, and he is registered as a Profl Engr in the US and Canada. His primary interests are: gas separation, air and water pollution and radiation from nuclear fallout. He was also involved in various types of consultations on the industrial and governmental level. He was conslt to the Dept of Energy on coal liquefaction. His publications and inventions cover mainly the theory and practice of adsorptive separation. *Society Aff:* AIChE, ANS, ACS, PE OH, PE ONT.

Taback, Harry
Business: Marsh & McLennan, Inc, 1221 Ave of the Americas, New York, NY 10020
Position: Managing Dir *Employer:* M&M Protection Consultants. *Education:* B of Eng In ChE/Chem Eng/NY Univ; M Science/Business/Columbia Univ *Born:* 3/21/45. Mr. Taback is currently a Managing Director of Marsh & McLennan, Inc. and Assistant Manager for the worldwide operations of M&M Protection Consultants, a hazard control consulting firm. In addition to his duties as Assistant Manager, he is also responsible for the direct marketing and sales activities for the entire organization and the consulting activities of two worldwide groups that provide consultation to the Hydrocarbon and Chemical Industries and the Gas and Electric Utility Industries. Chemical engrg graduate of New York Univ; has a Master of Science degree in Business Policy from Columbia Univ. Member of NFPA and AIChE. *Society Aff:* AIChE, NFPA.

Tabar, William J
Business: 4801 N 63rd St, Boulder, CO 80301
Position: President. *Employer:* AMF Head Division. *Born:* Feb 24, 1931. MSChE W Va Univ; BSChE Penn St Univ. Served as 1st Lt USMC 1953-55. With Union Carbide 1953-69 as process dev engr - polymers, business analyst & mfg mgr; with Rochester Button Div Duplan Corp 196974 as V Pres Mfg, administering 7-plant complex; joined AMF Head as Dir of Opers 1974, Pres since Sept 1975 of div mfg & marketing snow skis & tennis equipment. Enjoy skiing & classical music.

Tabisz, Edward F
Home: 14 Hillholme Rd, Toronto Ontario, Canada M5P 1M2
Position: Retired (General Manager) *Education:* -/-/IL Inst of Tech; -/-/Northwestern Univ. *Born:* Oct 1906 North Chicago Ill. With Underwriters Lab Inc 1925-50, engaged in life, fire & accident projects; Genl Mgr of Underwriters' Lab of Canada, Scarborough Ontario 1951-71 (retirement); organized staff & established testing facility in Canada. Mbr ULC Fire Council 1952-71; Chmn 1963-71. Chmn NFPA Ctte on Liquid Fuel Burning Equipment 1952-72; mbr other NFPA cttes. Mbr ASTM 1950- 70. NFPA Tech Ctte Serv Award 1966 & 1971; Chmn Qualifications Bd 1961-64, Mbr Exec Ctte 1964-72, V Pres 1965-69, Pres 1969-71 SFPE. Mbr Committee on the Natl Fire Code of Canada 1976-1982. Elected Fellow SFPE 1984. *Society Aff:* SFPE, CFSA, IAEI.

Tabler, L Earl, Jr
Home: 115 Sandralayne Dr, Corapolis, PA 15108
Position: Manager Project Controls. *Employer:* Dravo Engineers and Constructors Pittsburgh *Education:* BS/Civil Engr/Univ of IA. *Born:* Feb 1927; native of Council Bluffs Iowa. Served in US Navy 1945-46; started as a design engr for Dravo Corp 1952; advanced through field engr, planning & scheduling engr, Proj Engr to preset position in 1978. P Chmn ASCE's Const Div Exec Ctte; P Chmn ASCE Ctte on Project Planning; Thomas Fitch Rowland Prize 1966. Mbr AACE. Reg P E in Pa. Enjoy woodworking & gardening. *Society Aff:* ASCE, AASE.

Tachmindji, Alexander J
Home: 5314 Falmouth Road, Bethesda, MD 20816
Position: Sr VP *Employer:* The MITRE Corp *Education:* SM/Marine Eng & Naval Arch/MIT; BSc/Marine Engrg/King's Coll, Univ of Durham, England; BSc/Marine Engrg/King's Coll, Univ of Durham, England *Born:* 02/16/28 in Athens. Came to the US in 1950, naturalized in 1958. Involved in the formulation, direction and control of advanced tech progs for the Defense Advanced Research Projects Agency (Deputy Dir, 1973-75) and the Inst for Defense Analyses. With the MITRE Corp since 1975 as Chief Scientist. Named VP in 1976. VP and General Mgr, Washington C3I Operations (1979-1984). Current position Sr VP. Holds Supercavitating Propeller and Ventialated Propeller Patents. Editor of the Journal of Defense Research. Fellow, Royal Inst of Naval Architects; Fellow, AAAS. Recipient of US Naval Meritorious Civilian Service Award, 1956, and Secretary of Defense Civilian Meritorious Medal, 1975. *Society Aff:* AAAS, AIAA, SNAME, INA, NECIES, ΣΞ.

Tadeusz K Slawecki
Business: Chem Engrng Dept, Youngstown State U, Youngstown, OH 44555
Position: Prof *Employer:* Youngstown State Univ. *Education:* PhD/ChE/Univ of PA; MS/ChE/Univ of PA; BS/ChE/Univ of IL. *Born:* 11/4/19. Engr, educator; b E Chicago, IN, BS, Univ IL, 1943; MS Univ PA, 1948, PhD 1952; m Louise O Elder, Mar 18, 1945. Res Assoc Univ PA 1950-53; Res Engr. Frankford Arsenal, Phila 1953-56, cons 1957-58; cons engr, mgr, Gen Elec Co Phila, 1956-68; dir chem engrg and chem div, chmn. Chem engrg dept Ind Inst Tech, Ft Wayne, 1968-71 Essex Intl prof in chem engrg, 1968-71; Chmn, chem engrg and met engrg dept Youngstown (OH) State Univ 1971-85; Professor 1971- ; cons Villanova Univ 1956- 57, Franklin Elec Co 1968- ; tech adviser Grid Craft Inc 1969-85. Served with AUS, 1944-46. Mem Amer Inst Chem Engrs, Amer Chem Soc, Amer Soc Engg. Edn, Sigma Xi, Unitarian. Tech. Advisor HTP Corp. 1980-. *Society Aff:* AIChE, ACS, ASEE, ΣΞ

Tadjbakhsh, Iradj G
Business: Dept of Civ Engr, Troy, NY 12181
Position: Prof. *Employer:* RPI. *Education:* PhD/Mechanics/RPI; MS/Civil Engr/Univ of MI; BS/Civ Engr/Univ of MD. *Born:* 1/4/28. in Iran. Obtained BS, MS, PhD from Univ of MD, Univ of MI & RPI respectively 1954, 56 & 59. A post doctoral fellow at Courant Inst Math Sciences 59-60. Joined IBM as res staff mbr Watson Res Ctr Yorktown Hts 1961-69. Taught at RPI, NYU yr 61-62, 73-74 & 67 respectively. Assumed adm duties & teaching in Iranian Univs yrs 70-79. Areas of interest include structural engg, appl mechanics & math. *Society Aff:* SES, ASCE, ASME.

Taflin, Charles O
Business: Progressive Consulting Engrs, Inc, 6040 Earle Brown Dr, Minneapolis, MN 55430
Position: Supt Plant Operations. *Employer:* Minneapolis Water Works. *Education:* BS/Civ Engg/Univ of MN. *Born:* 7/4/34. MN native. Following BS degree did grad work in sanitary engg & hydraulics. Served as US Army Officer, Medical Service Corps, 1958 to 1963, stationed in W Germany. Worked as design engr for Mpls Water Works & Schoell & Madsen, Hopkins, MN. Taught grad level course in water plant design Univ of MN 1975-76. Since 1975 in charge of water treatment, pumping, storage & lab facilities for Mpls Water Works. Chrmn N Central Sec AWWA 1974; AWWA Fuller Award 1976. Married Marlys Melberg 1958; children Daniel, David, & Mary. Mbr Evangelical Covenant Church. *Society Aff:* AWWA, IWSA, AWRA.

Taggart, Robert
Business: 9411 Lee Highway, Suite P, Fairfax, VA 22031
Position: President. *Employer:* Robert Taggart Inc. *Education:* BS/Naval Arch/Webb Inst of Naval Arch. *Born:* Oct 1920. Naval officer WWII, employed by Maritime Comm, Naval Res Lab, Army Transportation Corps, Navy Bureau of Ships & Reed Res before forming Robert Taggart Inc 1958. Conceived & designed 1st ship dynamic positioning sys 1961. Has lectured at several univs, published numerous papers on acoustics, ship propulsion, ship maneuvering & ocean engrg; authored 3 books; Editor of two books; holds 3 US pats. Licensed P E in Va & D C. Mbr SNAME (Vice Pres 1981-83), ASNE (Pres 1974-75), ASA, MTS, AOO, NCIES & PIANC. *Society Aff:* SNAME, ASNE, ASA, MTS.

Tahiliani, Jamnu H
Business: 825 Ridge Lake Blvd, Suite 195, Memphis, TN 38119
Position: Pres *Employer:* Jamnu H Tahiliani Consulting Engrs *Education:* MS/Engg/Memphis St Univ; BS/CE/MSU Baroda, India *Born:* 2/20/36 in India. Worked for the Government of India in Design of Bridges on National Highways. Served as an executive Engineer during 1964-1969. Came to US to do MS in Structural Engrg in 1969. Worked for 10 yrs with consulting firm of Pickering-Wooten-Smith and Weiss. On 3-1-79 started own consulting firm. Structural design by firm include Steel and Reinforced and Prestressed Concrete Buildings and Bridges. *Society Aff:* NSPE, ACI

Taiganides, E Paul
Business: c/o Agricultural Engineering, Ohio State Univ, Columbus, OH 43210
Position: Project Manager/Professor. *Employer:* U N Food & Agri Orgn. *Education:* PhD/Environ Eng/IA State Univ; MS/AE/IA State Univ; BA/Agri Engr/Univ of ME. *Born:* Oct 1934 Macedonia Greece. Attended HS in Veroia; received scholarship to study in USA. Joined faculty at Ohio St Univ 1965 as Assoc Prof & was promoted to Prof 1969; served as cons in India for Ford Foundation 1966, in Europe for World Health Orng, for Food & Agri Orgn as Proj Mgr of Animal Waste Mgmt & Utilization in Singapore, since 1975. Lectrd on environmental issues throughout N Amer & Europe (East & West) & in Asia. Authored papers pulb in US, Europe; some were translated & pub in Japanese; book on "Animal Wastes–. Editor of Intl Journal, "Agricultural Wastes.–. Served as chmn of following natl cttes: Career Guidance, Animal Wastes, Prof Dev Dept, Internatl Cooperation Year in ASAE. Chmn of Environ Engrg & Agri Engrg Div ASEE. Recipient A W Farall Award (1974, ASAE). Reg PE. *Society Aff:* ASAE, ASEE.

Takagi, Joji
Business: 220 Mineola Blvd, Mineola, NY 11501
Position: Exec VP. *Employer:* Charles R Velzy Assocs Inc. *Education:* BCE/Civil Engg/NY Univ. *Born:* 1/31/22. Lb. Jan 1922. BCE NYU. Served as Instr in drafting, descriptive geometry & surveying, NYU. After serving in US Army & as Hydraulic Engr on Surface Water Studies with USGS, became Proj Engr for Charles R

Takagi, Joji (Continued)
Velzy Assocs. Later was given charge of firm's L I offices, incl mgmt of corp functions. Fellow ASCE, Mbr NSPE, APWA, AWWA & WPCF. Recipient of William F Knack Memorial Award NYU. 2nc V Pres & Dir 1969-72 NYS SPE. *Society Aff:* ASCE, WPCF, NSPE, APWA, AWWA.

Takagi, Noboru
Home: 3-6-20, Kitashinagawa, Shinagawa, Tokyo, Japan 140
Position: Prof, Pres *Employer:* Tokyo Engineering Univ *Education:* DR/Elect/Univ of Tokyo; BE/Elec/Univ of Tokyo *Born:* 6/19/08 His professional career has included research and teaching at Physio- Chemical Institute (1931), Tokyo Institute of Technology (1931-34), Nihon Univ (1934-42), Univ of Tokyo (1942-69), Nihon Univ (1969-1980) and Dean, Nippon Electronics Engrg College (1980-86). Pres, Tokyo Engineering Univ (1986-present). He was the director of the Institute of Space and Aeronautical Science, Univ of Tokyo, and the Space Development Center, Agency of Science and Technology. He is now active in the field of standards as President and then Immediate Past President of the International Electrotechnical Commission in Geneve since 1977. IEEE Activities: Fellow '80, TOKYO Section Chairman (1977-78), Executive Committee (1977-present). *Society Aff:* IECE, IEE, ITE.

Takahashi, Yasundo
Home: 135 York Avenue, Berkeley, CA 94708
Position: Professor of Mechanical Engineering, Emeritus. *Employer:* University of California, Berkeley. *Education:* Dr Eng/Mech Engg/Univ of Tokyo, Japan. *Born:* June 1912 Nagoya Japan. Design for Natl Railway of Japan 1935-37; 1937- 40 Asst Prof Yokohama Tech Coll Japan; 1940-44 Assoc Prof Nagoya Univ Japan; 1944-58 Prof at Univ of Tokyo Japan; 1958-79 Prof of Univ of Calif. Sr technical advisor, Mikuni Berkeley R. & D. Corp., 1981-present. Bd of Dirs Japan Soc of Mech Engrs 1950-54 (Honor Mbr 1982). ASME: Chmn of Automatic Control Div 1967-68, Sr Tech Ed ASME-Trans Ser G (Dynamic Sys, Measurement & Control) 1974-75, Fellow 1972. Life Fellow 1978. Oldenburger Medal 1978. Hon Dr Control Univ of Grenoble, France. Education Award, 1981, Am. Automatic Control Council. Author 'Control' & several other books. Hobbies: travel, pictures, personal computing & classical music. *Society Aff:* ASME, ISA

Takeda, Yukimatsu
h33-1, Shiba 5-Chome/Minato-Ku, Tokyo 108 Japan
Position: Advisor. *Employer:* NEC Corporation *Education:* PhD/EE/Ryojun Inst of Tech; BS/EE/Ryojun Inst of Tech. *Born:* 4/24/11 Native of Yamagata Japan. Dr Engrg Ryojun Inst of Tech. Prof, Ryojun Inst of Tech until 1945; Dev Mgr microwave electron tubes 1946-54 & semiconductor devices 1954-56 in Elec Communication Labs, Nippon Telegraph & Telephone Public Corp (NTT), Japan; with Nippon Elec Co Ltd (NEC) 1957- ; Genl Mgr, Semi- conductor Div NEC 1958-64; V Pres Engrg NEC 1965-71, Pres NEC Sys Lab Inc, Lexington Mass 1972-79. Gen Mgr, NEC Inst of Tech Education in Tokyo Japan 1979- . IEEE Fellow Mbr 1972. Mbr Inst Electronics & Telecommunication Engrs Japan, Inst Television Engrs Japan and Inst Electrical Engrs Japan. Enjoy classical music. *Society Aff:* IEEE, IETE, ITE, IEE.

Talaat, Mostafa E
Business: Dept of Mech Engg, College Park, MD 20742
Position: Prof. *Employer:* Univ of MD. *Education:* PhD/EE/Univ of PA; MS/EE/Univ of PA; BS/EE/Univ of Cairo. *Born:* 5/16/24. Cairo, Egypt; Natl US, EE, BSc (Hon), Cairo, 46, MSc, 47, PhD (EE), 51, PA. Asst, MIT 47-49; instr, PA, 51; proj engr Westinghouse Elec Corp 51-52; spec mach designer, Star-Kimble Motor Div, Miehle Press, 53; sr, res proj engr, Elliott Co Div, Carrier Corp, 53-59; mgr energy conversion & asst, dir, engg and res, nucl div, Martin-Marietta Corp, 59-64; Prof Energy Conversion, Dept ME, Col Engg, Univ MD, College Park, MD, 64-present. Energy conversion including magnetoplasma dynamic, thermionic, fuel cell & thermoelectric energy conversion as well as bilogical power sources & solar energy storage & conversion. Several patents, awards & numerous publications. IEEE Power Div Natl First Prie 56-57, Martin-Marietta Corp awards for outstanding creative contributions, 60-63. Mbr Sigma Xi, 47; Honorry Pi Tau Sigma, 64; Eminent Engr Tau Beta Pi 77; Enjoys ballet & opera. *Society Aff:* IEEE.

Talbert, John J, III
Home: Rd 638 PO Box 6, Josephine, TX 75064
Position: Manager Plant Engineering. *Employer:* E-Sys Inc Garland Div. *Education:* BS/ME/Univ of TX, Austin. *Born:* Apr 2, 1938 Waco Texas. With US Army 1961-63 at Army Electronic Proving Grounds Arizona. Joined E-Sys predecessor Co 1963 as Design Engr in Facilities Dept. Assumed Manager Plant Engr position 1978. V Pres AIPE 1973-78. Pres AIPE 1978-79; Chrmn-AIPE Fnd 1979-80; Cert Plant Engr Cert Fallout Protection Analyst Prof Engr-Texas. Hobbies: Horticulture & hunting. Chmn AIPE Certification Board 1984. *Society Aff:* AIPE.

Talbert, John T Jr
Business: P O Box 3333, 916 South 17th Street, Wilmington, NC 28406
Position: Pres *Employer:* Talbert, Cox & Assoc. Inc. *Education:* BS/CE/Citadel *Born:* 10/23/22 Mr. Talbert founded Talbert, Cox & Associates, Inc. in 1954 and as president of the firm has directed its growth and diversification since that time. He has more than thirty five years of private and public sector engrg and planning experience working with private businesses, industries, municipalities, county, state and federal government agencies, authorities, commissions and foreign governments. Mr. Talbert is a licensed pilot. He enjoys flying and duck hunting. *Society Aff:* ASCE, NSPE

Talbot, Douglas W
Business: 303 Lynnhaven Parkway, Suite 202, VA Beach, VA 23452
Position: Pres *Employer:* Talbot Mgmt Enterprise *Education:* BS/CE/VA Military Inst *Born:* 2/20/35 Served as Assistant City Engr for Richmond, Va, City Eng for Independence, MO and Director of Public Works in Hopewell, VA. from 1957 to 1968. Began private practice of Talbot and Associates, Ltd in 1968-an engrg, architectural, planning and surveying firm with a staff now in excess of 100. Awarded Young Eng of the Year in MO, served as Chairman of the Urban Planning Division ASCE, past Chairman of the Urban Planning Div ASCE, current Chairman of ASCE National Land Use Policy Committee. Served as founder and Pres of Hopewell Chapter of VSPE, lectured at Univ of VA, Old Dominion Univ, Tidewater Community College and several national conferences on various technical subjects. Enjoys golf and skiing. *Society Aff:* NSPE, ASCE, APWA.

Talbot, Eugene L
Business: PO Box 98, Magna, UT 84044
Position: Staff Materials Technical Specialist. *Employer:* Hercules Inc. *Education:* PhD/Metallurgy/Univ of UT; MS/Met Engg/Univ of UT; BS/Chem Engg/Univ of UT; BS/Mech Engg/LSU. *Born:* Jan 18, 1921. Adjunct Prof of Materials Univ of Utah; Head of Materials Dev & Materials Specialist for Hercules Inc 1963- . Inorganic Applications Leader 3M Co 1950-63. Areas of expertise: adhesives, surface chem corrosion, resins, composites, insulation. Natl Pres Soc for Advancement of Material & Process Engrg 1975-76; V Pres 1974-75. *Society Aff:* SAMPE, ASM.

Talbot, Thomas F
Business: UAB Sch of Engrg, Birmingham, AL 35294
Position: Professor and Chrmn, Mechanical Engineering Dept *Employer:* University of Alabama at Birmingham *Education:* PhD/Mech Engg/GA Inst of Tech; MS/Mech Engg/CA Inst of Tech; BME/Mech Engg/Auburn Univ. *Born:* July 1930. Native of Birmingham Ala. Served USAF 1952-56; Brigadier Gen USAFR Ret. Sr Practice man Tenn Coal & Iron Div US Steel Corp 1956-58; Asst Prof Mech Engrg Georgia Tech 1958-65; Assoc Prof Materials Sci & Mech Design Vanderbilt Univ 1965-67; Assoc Prof & Prof Univ of Ala, Birmingham 1967- & Dir of Continuing Engrg Educ 1972-78, & Chrmn, Mech Engrg 1983- . Recipient NSF Sci Faculty Fellowship 1962. Reg P E Ala, Ga, Tenn. Member, State of Alabama Bd of Regist for Prof

Talbot, Thomas F (Continued)
Engrs and Land Surveyors 1985-Present. Mbr ASM, ASME, ASEE, AWS, Amer Assn Univ Profs, National Council of Engrg Examiners, Natnl Soc of Prof Engrs, Soc of Automative Engrs, Newcomen Soc, Reserve Off Asso, Air Force Asso, Alabama Soc of Prof Engs. Sigma Xi, Phi Kappa Phi, Tau Beta Pi, Pi Tau Sigma, Omicron Delta Kappa, Lamda Chi Alpha. Chrmn, Bd of Dirs, Amer Alloy Product, Inc, Cullman, AL Amer Alloy Prods is small Stainless Steel Foundry with about 20 employees, Jan 1977-present, Enjoy private flying, travel & outdoor activities. *Society Aff:* ASM, ASME, ASEE, AWS, AAUP, SAE, ΠΤΣ, ΣΞ, ΦΚΦ, ΤΒΠ, ΛΧΑ, NCEE NSPE, ASPE.

Talbott, John A
Business: 7 SE 97th Ave, Portland, OR 97216
Position: Pres. *Employer:* Talbott Engrs, Inc *Education:* BS/Mech Engr/OR State Univ. *Born:* 12/5/25. City Engr Albany, OR 1949-50; Mech Engr Stevens & Thompson, Cons Engr Portland, OR 1952-54; Asst Ch Engr OR Steel Mills, Portland, OR 1954-60; Pres Talbott Engrs, Inc Portland, OR 1960- ; Lectr Engr Univ of Portland, 1964; Instr Engr OR Sys Higher Educ 1947-49; Mbr Air Quality Control Adv Comm City of Portland 1964-68; Chrmn, Mayor Portland Advisory Comm 1969-70; Served to 1st Lt US Army Engrs 1944-47, 1950-52. PE; ASCE A Greensfelder Const Prize 1974; CEC US Engg Excellence Honor Award 1970. ASME Centennial Medallion 1980. P E of OR. Engr of the Year 1983. ASME VP 1972-74, Tau Beta Pi, Pi Tau Sigma. Contrib to Prof Journals. Inventor of mechanisms, cons on bridges, cableways, steel mfg, indus plants, equip, accident analysis. Hobbies: wilderness backpacking, skiing. *Society Aff:* ASME, ASCE, SAE, ASHRAE, NSPE, ACEC, CSI, ICBO, ASTM, ANSI, AAFS.

Talbott, Laurence F
Home: 66 Benton Way, San Luis Obispo, CA 93401
Position: Department Head, Industrial Technology *Employer:* CA Poly State Univ. *Education:* EdD/Industrial Education/UT State Univ; MEngr/Civil Engg/CA Poly State Univ; MBA/Industrial Mgt/Univ of S CA; AB/Gen Engg/San Diego State Univ; Certificate/Intl. Graduate School/University of Oxford (England); BA/History/University of State of New York. *Born:* 11/17/20. San Diego, CA. Parents, George and Mary Lanz Talbott. Married Patsy Anne Davis, 1963; children, Michael Laurence, Mary Anne, Susan Alice. Elec Engr, Convair San Diego 1950-51, C F Braun & Co 1951-52, Mgr of Facilities Engg, North American Aviation 1952-62, Mgr Test Quality Control Rocketdyne 1962-66. Prof and Assoc Dean, School of Engg and Tech 1966-1979. Dept. Head Industrial Tech 1979- present. American Inst of Plant Engrs, VP Professional Dev, recipient of 1976 Intl Service Citation Award. CA Assoc of Industrial Tech, Pres 19 75-76. AAES, Educational Affairs Council. National Association of Industrial Technology, Pres. 1984-85, Authored articles in journals of AIPE, AIEE and Air Pollution Control. Reg PE, CA. Certified Plant Engr. *Society Aff:* NAIT, O.E., ΕΠΤ

Talerico, John P
Home: 85 Burkhardt Lane, Harrington Park, NJ 07640
Position: Sr. Vice Pres *Employer:* Frederic R. Harris, Inc. *Education:* BS/Civ Eng/Rutgers Univ *Born:* 6/24/24 Native of Long Branch NJ. Served with Army of US 1942-45. Civil Eng US ArmySignal Corps and Monmouth County Engrg Dept with PRC Engrg, formerly Frederic R Harris, Inc. Consulting Engs Since 1951. Assumed Current Responsibility as Sr Vice President and Office Mgr in 1973. Manager N.J. office since 1979 Since 1987 name change back to Frederic R. Harris, Inc. In charge of Major Civil Engrg Projects Worldwide including Harbors and Waterways, Highways, Bridges, Airports, Mass Transit Systems, Environmental Studies, Author *Harbors, Breakwaters and Ports* and *Levees* for Encyclopedia Americana Member Consulting Engineers Council of NJ, National Society ofPE, American Society Civil Engineers, Enjoy Gardening and golf. *Society Aff:* ASCE, NSPE, CEC

Talian, Stephen F,
Business: Senior Engineer, Gannett Fleming, PO Box 1963, Harrisburg, PA 17105
Position: Partner *Employer:* Gannett Fleming Water Resources Engineers *Education:* MSCE/Hydraulics/Hydrology/Lehigh Univ; BS/CE/Lehigh Univ *Born:* 12/9/40 Native of Bethlehem, Pa. Taught at Lehigh during MSCE program, and EIT and PE review courses for Penn State. Performed studies for Ground Water Branch of US Geological Survey, 1964-66. With Water Supply Section, of Gannett Fleming Consulting Engineers since 1966. Made partner of firm in 1979. Conducts water system and ground-water investigations, authors water system planning reports, consults on water system management and finances, provides expert testimony and other services. Enjoys golf, basketball, running, and other sports. *Society Aff:* ASCE, AWWA, ACEC

Tallamy, Bertram D
Business: 2500 Q St NW, Washington, DC 20007
Position: Proprietor. *Employer:* Bertram D Tallamy, C E. *Education:* CE/Civil Engr/Rensselaer Polytechnic Inst; Dr Engg (hon)/Civil Engr/Rensselaer Polytechnc Inst. *Born:* Dec 1, 1901. Ptnr Fretts Tallamy & Sr 1929-44; Successively Dept Supt N Y St DPW, 1945-47 Ch Engr. 1948 Supt.1948-56; Chmn Ch Ex Officer N Y St. Thruway Authority 1950-57; Fedl Hwy Administrator 1957-61; Ptnr Tallamy Byrd Tallamy & MacDonald 1961-70; Former Chmn US Delegation, Inter-American Hwy. Conferences in Panama, Bogota, Colombia S A, & Pio de Janeiro Brazil; Mbr Moles; Natl Honor Mbr Chi Epsilon, Mbr Tau Beta Pi, Received Roy Crumb Award & many Natl Citations in Engrg; Reg P E: N Y, N J, Wash DC. P Pres Amer Assoc State Hwy Officials, Life Mbr ASCE and NSPE and NYSSPE. *Society Aff:* ASCE, NSPE, NYSSPE.

Talley, Wilson K
Business: Dept of Applied Science, Davis, CA 95616
Position: Prof *Employer:* U of Calif *Education:* PhD/Nuclear Engrg/Univ of CA-Berkeley; SM/Physics/Univ of Chicago; BS/Physics/Univ of CA-Berkeley *Born:* 1/27/35. pres & Dir, Fannie and John Hertz Foundation (1972-). Served as Assistant Administrator for Research and Development US EPA (1974-77); Study Director of the Commission on Critical Choices for Americans (1974), Asst VP U of CA (1971-74); Head Div of Theoretical Physics Lawrence Livermore National Lab (1971). Consultant to Lawrence Livermore National Lab, John Controls, Inc., Penn Central Federal Systems, Inc., Biophotonics Technology, Inc., Assistant Secretary of the Army (RDA), Chrmn of Army Sci Bd (1983-86), Mbr of Defense Sci Bd (1983-86), Dir of Helionetics, Inc (1982-86), Marinco, Inc (1983-86). Author of papers on applications of nuclear energy, environmental policy, coauthor of "Constructive Uses of Nuclear Explosives" McGraw Hill (1983). *Society Aff:* ANS, APS.

Tallian, Tibor E
Home: 36 Dunminning Road, Newtown Square, PA 19073
Position: consultant tribologist *Employer:* Tallian Consulting Corp *Education:* Master/Mech Eng/Budapest Institute of Technology Hungary *Born:* 10/18/20 Born in Budapest, Hungary. In the U.S.A. since 1956. Rolling Bearing R & D since 1950, practicing tribologist consultant since 1985. Employed by SKF Industries Inc. (rolling bearings) from 1957 to 85, last as VP and chief technical officer. Published 2 books, 40 papers on contact tribology. Principal technical interests: failure diagnosis (expert systems), fatigue prediction, vibration in rolling contact. Fellow, ASME, Fellow and life member STLE, winner National Award STLE. *Society Aff:* ASME, STLE

Tallmadge, John A, Jr
Business: Chemical Engrg Dept, Drexel University, Philadelphia, PA 19104
Position: Prof of Chem Engr *Employer:* Drexel Univ *Education:* PhD/ChE/Carnegie Mellon Univ; MS/ChE/Carnegie Mellon Univ; BS/ChE/Lehigh Univ *Born:* 2/19/28 Research engr with duPont 1953-56 in Textile Fibers dept. Assistant and associate professor at Yale Univ 1956-65. Sr visitor at Imperial College (London) 1965. With Drexel since 1966. Acting chrmn-1967 and 1974. Fulbright Prof at Univ New South Wales (Australia) 1974. Visiting Prof at Univ of CA (Berkeley) 1982 and Univ of

Tallmadge, John A, Jr (Continued)
WA (Seattle) 1983. Fellow of AICHE. Over seventy technical publications, chapters in two books. Research on liquid coatings, rheology, fluid dynamics, water treatment, process metallurgy, separations and transport phenomena. *Society Aff:* AIChE, ASEE, ACS.

Talman, Woods G
Home: 2624 Thorntree Dr, Pittsburgh, PA 15241
Position: Mining Consult (Self-employed) *Education:* BS/CE/VA Military Inst; Command & Special US Army; General Staff/Assoc Course *Born:* 05/08/10 Native Virginian. Careers, one mining coal, ore, limestone, other military. 38 years US Steel, worked in and managed all aspects. Lastly at Pittsburgh headquarters, chief inspector and conslt to 5 districts. Military - CMTC 1925, high sch and coll, CCC, WW II, retired colonel. Recognitions include: Howard N. Eavenson Award, Natl Safety Council Citation, The Donald S. Kingery Award and Prof Award for Coal Mining Health, Safety and Research. Ninth Annual Inst Blacksburg, Va. First recipient. Commanded Aviation Battalion, ETO, received Bronze Star and Croix de Guerre. *Society Aff:* AIME, SME

Talton, Raymond S
Business: PO Box 1551, Raleigh, NC 27602
Position: Consultant. *Employer:* Carolina Power & Light Co. *Education:* BS/Mech Engg/NC State Univ. *Born:* 12/12/14. Native of Smithfield, NC. Employed with CP & L since July 1939; retired in 1977 from position of VP-Sys Engg & Const. Reemployed as Consultant on the sale of interests in generating plants. Previous positions held: VP-Engg June 1968- Feb 1972; Engg Cons system planning Jan 1963-June 1968; engg positions in power, plant, production, & maintenance. *Society Aff:* ASME, NSPE.

Tamarelli, Alan W
Home: 49 Wexford Way, Basking Ridge, NJ 07920
Position: Chmn and Chief Exec *Employer:* Dock Resins Corp, Linden, NJ. *Education:* PhD/Chem Engrg/Carnegie-Mellon Univ; MBA/Internatl Bus/NY Univ; MS/Chem Engrg/Carnegie-Mellon Univ; BS/Chem Engrg/Carnegie-Mellon Univ. *Born:* 8/13/41. Born in Aug 1941. Taught as Asst Prof at Carnegie-Mellon Univ. Served as Captain, US Army. Employed by Exxon Corp from 1966 to 1970 as Engr and Proj Leader in new ventures and petrochem processing. With Englehard Corp from 1970 to 1983 in various capacities, including Gen Mgr (New Ventures), Asst to Exec VP, Gen Mgr (Petroleum Processing), VP (Chem Oper), Group VP (Chem Group) and Sr VP (Energy and Environ Sys Group) respon for bus segments encompassing proprietary chems, catalysts, licensed processes, electrochem, chem raw matls, engr sys, and pollution control equipment and also respon for strategic planning diversification and R&D. Chmn and Chief Exec since 1983 of Dock Resins Corp, Linden, NJ, a manufacturer of specialty chem resins for adhesives, indust coatings, sealants, printing inks, paper coatings, and custom polymer synthesis. *Society Aff:* AIChE, ACS, CDA, NASCP, TBΠ, ΣΞ, NYAS

Tamblyn, Robert T
Business: Suite 200, 2 Sheppard Ave E, Willowdale, Ontario, Canada M2N 5Y7
Position: Chrmn *Employer:* Engineering Interface Limited. *Education:* BASc/-/Univ of Toronto. *Born:* March 22, 1921. PE in Ontario & Alberta. 4 years Lt, Royal Canadian Engrs Overseas; 12 years Mgr, Air Cond Div, Canadian Ice Machine Co; Chmn, York Central Dist High School Bd 1958; 15 years Chmn, Tamblyn, Mitchell & Partners Ltd, large Canadian cons in design of mech bldg services; since 1973, Chmn Engrg Interface Ltd, cons to other consultants in concept design; Cons Editor, Canadian Consulting Engr Magazine.; 8 articles in ASHRAE Journal; 2 patents in air conditioning design. *Society Aff:* ASHRAE.

Tamir, Theodor
Business: 333 Jay St, Brooklyn, NY 11201
Position: Prof. *Employer:* Poly Inst of NY. *Education:* PhD/Electrophys/Poly Inst of Brooklyn; MS/EE/Technion, Israel Inst Tech; Ingenieur/EE/Technion, Israel Inst Tech; BS/EE/Technion, Israel Inst Tech. *Born:* 9/17/27. Bucharest, Romania. PhD from Bklyn Poly, NY; BS, Dipl Ing and MS from Technion, Israel. Wife: Hadassah; children: Jonathan & Yael. With Poly Univ., NY since 1958; Prof since 1969; Hd, Dept of Elec Engg 1974-79. Consultant to industrial & govt electronics labs. Fellow, IEEE, IEE (England) and Opt Soc Am; Mbr URSI; PE, NY State. Awarded IEE Inst Premium (1964) & Electronics Sec Premium (1967); also, IEEE Special Recognition (1968). Published over 90 articles in profl journals & books, mostly on wave propagation in electromagnetics, optics & acoustics. Enjoys classical music, photography & travel. *Society Aff:* IEEE, OSA, IEE, URSI.

Tan, Chor Weng
Business: 51 Astor Place, New York, NY 10003
Position: Dean, School of Engineering. *Employer:* The Cooper Union. *Education:* BS/Mech Engrg/Univ of Evansville; MS/Mech Engrg/Univ of IL; PhD/Mech Engrg/Univ of IL. *Born:* PE in NY. Taught at Cooper Union 1963- ; Asst Prof of Mech Engrg 1963- 67; Assoc Prof of Mech Engrg 1967-72; Prof of Mech Engrg 1972- ; Acting Chmn of Mech Engrg Dept 1971-73; Head, Div of Engrg 1973-75; Dean, School of Engg 1975 ; Exec Officer & Dir of the Cooper Union Res Fnd . *Society Aff:* ASME, ASEE.

Tanaka, Richard I
Home: 10321 Shadyridge Dr, Santa Ana, CA 92705
Position: Pres *Employer:* Lundy Electronics & Systems Inc. *Education:* PhD/EE & Physics/CA Inst of Techn; MS/EE/Univ of CA, Berkeley; BS/EE/Univ of CA, Berkeley. *Born:* 12/17/28. Prof: Lundy Electronics & Systems Inc, Glen Head, NY 1987-present, Pres; Systonetics Inc, Anaheim, CA 1980-1986, Pres. & CEO; Internal Tech Resources Co, Tustin, CA 1977-80, Pres: Calif Computer Prods, Inc, Anaheim, CA 1966-77, Sr VP; Lockheed Res Lab, Palo Alto, CA 1957-65, Sr Mbr for Computer Res, Mgr Computer Res Dept; Hughes Aircraft Co 1953-57, Culver City, CA, Res Scientist; North Amer Aviation, Inc 1951-53, Downey, CA, Res Engr. Societies: Intl Fed for Info Processing (IFIP): Pres 1974-77, Trustee 1970-74, US Delegate 1970-78; Hon Mbr 1979; Amer Fed of Info Processing Societies (AFIPS): Pres 1969-71, VP 1967-68, Bd/Dir 1965-69, Distinguished Service Award 1983, Genl Chrmn 1964 Fall Joint Computer Conference; IEEE Computer Soc: Chmn 1965-66, Bd/Dir 1964-68; Fellow IEEE. Phi Beta Kappa, Tau Beta Pi, Eta Kappa Nu. *Society Aff:* IEEE, ACM.

Tanenbaum, Morris
Business: 540 Broad St, Newark, NJ 07101
Position: Pres. *Employer:* NJ Bell Tel Co. *Education:* PhD/Phsyics Chem/Princeton; MA/Chem/Princeton; BA/Chem/Johns Hopkins. *Born:* Nov 10, 1928 Huntington, West Va. BA Chem, Johns Hopkins Univ 1949; MA, PhD Physical Chem, Princeton Univ 1950, 1952. AT&T VP, Engrg & Network Services 1976- ; Bell Telephone Labs Bd/Dir 1976- , Exec VP 1975-76, Dir Solid State Dev Lab 1962-64, Res Div 1952-64; Western Elec Co VP, Transmission Equip 1972-75, VP Engrg 1971-72, Genl Mgr Engrg 1968-71, Dir R&D 1964-68; Mich Bell Telephone Bd/Dir 1976- ; Shawmut Corp Bd/Dir 1975- . Natl Res Council Governing Bd 1974- ; NAE (Council 1974-); Fellow IEEE, APS; Mbr ACS, AIME, Phi Beta Kappa, Sigma Xi; ASM Campbell Lecturer 1975. Dev of first successful silicon diffused transistors, & pnpn diodes, ldr of group that discovered practical materials for superconducting magnets, 7 patents, numerous tech articles.

Tang, Wilson
Business: 3122 Newmark, 208 N Romine, Urbana, IL 61801
Position: Prof *Employer:* U of IL *Education:* PhD/CE/Stanford Univ; MS/CE/MIT; BS/CE/MIT; CE/-/Stanford Univ. *Born:* 8/16/43 Native of Hong Kong. Came to USA for college education in 1962. Eng for Fay, Spofford and Thorndike and Sargent and Lundy. Taught at Univ of ILL since 1969. Visiting Prof at Norwegian Geotechnical Institute, Imperial College, London in 1976-77, and National Univ of Singapore in 1983. Consultant to industry and governmental agencies. Faculty advisor to Student Chapter of ASCE. Officers and chairman of national committees. Coauthor of series of books entitled Probabilistic Concepts in Engrg Planning and Design. Outstanding Teacher Award, 1980. Guggenheim Fellow, 1976. Enjoy music, tennis and reading. *Society Aff:* ASCE, ISSMFE, ASEE

Tankus, Harry
Business: 6400 W Oakton St, Morton Grove, IL 60053
Position: President. *Employer:* Crane Packing Co. *Education:* Dip/ME/IIT. *Born:* Aug 23, 1921 Bialystok Poland. Diploma mech engrg Armour Tech 1942; student Univ of Ill 1946-47; Graduate Mgmt course Univ of Chicago 1966. Insp dept head Buick div General Motors Melrose Park Ill 1942-44; Crane Packing Co Chicago 1947- , Specification Engr 1947-53, Ch Engr 1953-62, Asst Vice Pres Engrg 1962-64, Asst Vice Pres Sales & Engr Cons 1964-71, Vice Pres Prod Sales 1971-76, Pres 1976- . Served with US Army 1944-46. Decorated Purple Heart. ASLE Pres 1975-76, mbr Bd/Dir 1970- . Mbr Bd of Governors 1970- . Mbr ASM, Certificate of Recognition for Design 1965. ASTM Chmn C-5 Subctte Carbon- Graphite 1965, Exec Comm 1962-66. Mbr ASME, SAE, WSE. 13 patents, foreign & domestic. Mason (Shriner): Moose. Chrmn 1970 of Ill Inst of Tech. Inst for Indus Innovation through Tribology: (IIIT) Bd of Dir-1980. Alumni Bd of Dir: Ill. Inst of Tech (IIT) 1980-82. Pres Oakton College Educational Fdn: 1979-82. Mbr Adv Bd: Niles Township Sheltered Workshop. Bd of Dirs: Junior Engrg Tech Soc (JETS) in Ill. (Sponsored by Univ of Ill). Bd of Trustees: Skokie Valley Community Hospital 1981-83. Holder of 15 patents (Foreign and Domestic). Pres Crane Packing 1976-1982; Chrmn 1982- ; 1981 Bd. Dirs United Way of Skokie Valley; 1981 General Campaign Chrmn United Way of Skokie Valley; 1982, V.P. United Way of Skokie Valley; 1983, Pres, United Way of Skokie Valley; 1983, Dir United Way of Suburgan Chicago, Pres Council Lutheran Social Services of Illinois. *Society Aff:* ASME, SAE, ASCE, ASM, WSE.

Tannen, Leonard P
Business: 500 Ann Page Rd, Horseheads, NY 14845
Position: VP Manufacturing, Bakery/Grocery Div, Mfg Group *Employer:* Great Atlantic and Pacific Tea Co. *Education:* MS/Biochemical Eng/MIT; BChE/Chem Engg/CCNY. *Born:* Jan 1937. Native of New York, NY. Process Engr E R Squibb & Sons 1959- 62; Assoc Dir R&D Genl Mgr Operations Accent Internatl, Div of IMC 1962-72 ; Mfg Dir & Corporate Dir of Engrg & Dev Wm Underwood Co 1972-74; Vice Pres & Genl Mgr of Fermentation Design Inc, Subsidiary of New Brunswick Scientific Co Inc 1975- 78. VP, Manufacturing Bakery/Grocery Div A&P, 1978. Basic background in general management of businesses in fermantation and food operations involving antibiotics, enzymes, biochemicals, microbial & viral insecticides, food additives, and a broad line of grocery, detergent, frozen food, baked goods and printing. *Society Aff:* IFT, ACS, AIChE.

Tanner, Robert H
Business: P O Box 533, Naples, FL 33939
Position: Consulting Eng (Self-employed) *Education:* MSc/Acoustics/Imperial Coll Univ of London; BSc/Comm/Imperial Coll, Univ of London *Born:* 7/22/15 Television pioneer with British Broadcasting Corporation's first high- definition station 1938-39. Served in Royal Artillery and Royal Signals (Major) during WWII. BBC Research Dept. 1939-46. Emigrated to Canada in 1946 to work for Northern Electric and later Bell-Northern Research. Headed various development depts. 1947-68. Director of info 1968-72; founded company magazine *Telesis*. Loaned to Canadian Dept. of Communications 1972-75 as Director of Industrial Research. Also part-time acoustical consultant 1955-75; acoustical design of Stratford Shakespearean Festival Theatre, Ontario. Now full time accoustical consultant covering SE United States. McNaughton Gold Medal (IEEE Canadian Region) 1974. President IEEE 1972, IEEE Haraden Pratt Award 1981. *Society Aff:* IEEE, ASA, EIC, NSPE, IEE (UK)

Tanonaka, Clarence K
Business: 190 South King St, Suite 2085, Honolulu, HI 96813
Position: VP/Treasurer *Employer:* Park Engrg Inc. *Education:* BS/CE/Univ of HI *Born:* 10/25/28 Has been consulting eng for over 26 years. For the past 14 years has been responsible for planning and administering civil engrg projects in public works, engrg reports and land development. Has been responsible for coordinating the design and construction of highways and roadways, sewage systems, water systems, drainage facilities and flood control projects, commercial and industrial developments, marina development, redevelopment projects, airport improvement projects and schools, parks and other educational and recreational projects. *Society Aff:* CEC, ASCE, APWA

Tantala, Albert M
Business: 4903 Frankford Ave, Phila, PA 19124 *Employer:* Private practice. *Education:* MS/Civ/Villanova Univ; BS/Civ/Univ of PA; BA/-/Univ of PA. *Born:* 10/13/38. *Society Aff:* ASCE, NSPE, PSPE, PEPP, SAME.

Tao, William K Y
Business: 2357 59th St, St Louis, MO 63110
Position: President. *Employer:* William Tao & Assocs Inc, Cons Engrs. *Education:* MS/ME/Washington Univ; BS/ME/Chekiang Univ China; BS/EE/Southwestern Univ China. *Born:* 10/27/17. Pres William Tao & Assocs Inc, Cons Engrs; Affiliated Prof Schs of Architecture & Engrg & Trustee of Washington Univ. Bd Mbr IES 1975-78, Chrmn Energy Ctte, IES 1974-75; Chmn Lighting & Thermal Environment Ctte, IES: panel Mmbr, ASHRAE Standard 90-75; ASHRAE Standard 100.3 Chrmn of Building Tech Adv Ctte of Mo Energy Agency 1976-77; Bd Mbr NAAB, 1980-81; Bd Mbr IERI 1979-82. Actual Apecifying Engr Award 1965; St Louis Electrical Man of Year Award 1975; Washington Univ achievement award 1978. *Society Aff:* ASHRAE, IES, NSPE

Tapley, Byron D
Business: Dept of Aerospace Engrg, Austin, TX 78712
Position: W R Woolrich Professor *Employer:* The University of Texas at Austin. *Born:* Native of Charleston Miss. BSME 1956, MSEM 1958, PhDEM 1960 Univ of TX at Austin. Fellow AIAA, AAAS, IEEE, Mbr AGU, AAS, Amer Astronautical Soc, Intl Astronomical Union, COSPAR, Soc of Engrg Sci. Mbr Pi Tau Sigma, Sigma Xi, Tau Beta Pi, Phi Kappa Phi, Natl Acad of Mechanics. Chrmn AAAS Sect M(Engrg) 1971- 72. ECPD EE&A Ctte, Chmn Region IV 1972-76. NASA Adv Ctte: Guidance & Control, Universities Space Res Assoc; NASA Experiment Teams for GEOS-C & SEA-SAT; NASA TOPEX Sci Steering Gp. Editor *Celestial Mechanics* Journal. Assoc Ed, AIAA Jr. Guidance and Control and AGU Geophysical Reviews. Chmn AIAA Astrodynamics Ctte 1975-77. Chmn, NRC Ctte on Goodery. Mbr, NRC Geophysical Research Board. Pub 3 books, 73 reports & tech articles. Listed in 8 reference directories & Who's Who.

Taracido, Manuel E
Business: 9400 South Dadeland Blvd, Suite 100, Miami, FL 33156
Position: Principal *Employer:* Wolfberg/Alvarez/Taracido & Assoc *Education:* M/Solid Mech/So IL Univ; B/Eng/So IL Univ *Born:* 11/25/45 Graduated in 1969. Worked as a Structural Engr for local Miami Consltg Firms until opening his own A/E Firm in 1976. The firm grew from 10 employees in 1977 to present 92 employees located in offices at Miami and Fort Lauderdale. Taught Physics and Mathematics at Miami Dade Community Coll from 1974 to 1980. Under his own direct supervision and guidance, firm won the 1979 Prestress Concrete Inst Award and the Award of Excellence of the American Concrete Inst for the Outstanding Concrete Structure in the State of FL. *Society Aff:* ASCE, NSPE, ACEC, ACI

Taradash, Samuel
Home: 13026 Paseo Del Verano, San Diego, CA 92128
Position: Consulting Engr (Self-employed) *Education:* BS/Civil/IL Inst of Technology. *Born:* 12/10/15. Graduate work Ill Inst of Technology; additional graduate work Youngstown State Univ. Reg PE in Ill. Fellow ASCE; Mbr Institution of Civil Engr

Taradash, Samuel (Continued)
(England), British Geotechnical Soc (England), Honors: Chi Epsilon, Tau Beta Pi, Sigma Tau. Authored numerous papers on Underground Construction. Lectures given at universities throughout US to Graduates & Undergraduates. Mbr numerous Natl Cttes on Use of Underground. Travels extensively throughout US, Canada, Europe, & South America on most major underground projs transportation, water, diversion & sewer projs. Retired from Commerical Shearing Inc, January 1, 1979. *Society Aff:* ASCE, ICE, BGS.

Taraman, Khalil S
Business: Assoc. Dean of Engg, Lawrence Tech, 21000 W. Ten Mile, Southfield, MI 48075
Position: Prof & Chrmn. *Employer:* Univ of Detroit. *Education:* PhD/Ind Engg/TX Tech Univ; MSc/Mfg/Univ of WI; MSc/Mfg/Ain Shams Univ; BSc/Mfg/Ain Shams Univ. *Born:* 7/10/39. Married Sanaa Roushdy 1968 m. 3 children. Instr, Ain Shams (Cairo, Egypt) 1964-67, Res Asst, WI Univ 1967-69, Res. Fellow, TX Tech 1969-70, Detroit Univ: 1970 Asst Prof, 1973 Assoc Prof, 1977-1986 Prof. & Chrmn, Mech Engrg., obtained 1.3 million dollars in grants. Lawrence Institute of Technology, Associate Dean of Engineering & D.I.T. Professor of Manufacuturing Engineering 1986. Coordinated R & D grants of $434,000 during this past year. Conslt: Ford, GE, United Nations and Bendix. Mbr. SME, ASM, ASEE, ASME, Egyptian Amer Scholars, Pi Tau Sigma, Alpha Pi Mu and Tau Beta Pi. At SME: Education C 1975-77, Chairman 1987-88, Honor Awards C 1979-81 and 1987-88. Machining Tech since 1971, Chrmn, Material Removal Council 1979-84. SME International Posture and SME Fellow C, Intl Dir 1981-83 and 1984-86. Published 46 papers, Edited two books. Listed in Who's Who in Midwest, Outstanding Educators of America, Intl Who is Who in the Arab World, Who's Who in Frontier Science and Technology and Who's Who in the World. *Society Aff:* ASM, ASME, EAS, SME, ASEE

Tarantine, Frank J
Home: 1221 Cherokee Dr, Youngstown, OH 44511
Position: Prof *Employer:* Youngstown State Univ *Education:* PhD/ME/Carnegie-Mellon Univ; MS/ME/Univ of Akron; BE/ME/Youngstown St. Univ *Born:* 5/27/35 Native of Youngstown, OH. Previous engineering employment with Commercial Shearing, Inc. Design Eng, and Automatic Sprinker Corporation of America, Hydraulic Eng. Supervisor of the Graduate Program in Mechanical Engrg at Youngstown State Univ. Engrg Consulting work specializing in Vibrations, Testing, Combustion and Mechanical Design Analysis. *Society Aff:* ASME, ASEE, ΣΞ, ΦΚΦ.

Tarbox, Robert M
Business: P.O. Box 4157, Burlingame, CA 94010
Position: Design Manager, Carrejon Coal Project *Employer:* Morrison-Knudsen Co., Inc. *Education:* MSCE/-/Univ of CA, Berkeley; BS/-/US Military Acad. *Born:* Jan 1918. Serv in various Army Engr Troop Units & in Staff positions 1941-67. Div Engr North Central Div 1967-69; Commanding Genl US Army Engr Const Agency Vietnam 1969. Dir of Const US Military Asst Command Vietnam 1970. Retired from US Army As BGen 1971. Joined Intl Engrg Co Inc 1971, Vice Pres. Reg PE in DC & CO. Fellow SAME & ASCE. SAME: Dir Natl Soc 1981-84, Dir West Point Post 1964-65, Pres Frankfort Post 1966-67 Regional V Pres 1968-69, Award of Merit 1966. *Society Aff:* ASCE, SAME.

Tardos, Gabriel I
Home: 232 Fycke Ln, Teaneck, NJ 07666
Position: Asst Prof. *Employer:* CCNY. *Education:* DSc/MEngg/Technion; MSc/MEngg/Technion; DiplEng/Energy Engg/Polytechnick. *Born:* 5/14/46. A native of Romania, obtained his Dipl-Eng degree in Energy Engg in Bucharest in 1969 and his MSc and DSc degree in Mech Engg at the Technion-Israel Inst of Tech the last in 1977. Was a Grad Fellow at Carnegie-Mellon Univ in Pittsburgh in 1976 and a Res Assoc and Post Doctoral Fellow at the CCNY in 1978. Assumed the current position as Asst Prof in the Dept of Chem Engg, the City College of the City Univ of NY in 1979. Is responsible for coordinating res on fluid-particle systems, dust cleaning devices and coal powder tech. *Society Aff:* AIChE, APCA.

Tarika, Elio E
Business: Old Ridgebury Rd, Danbury, CO 06810
Position: Corp Exec VP *Employer:* Union Carbide Corp. *Education:* MBA/-/Univ of Chicago; MS/CE/Newark Coll; BS/CE/Univ of IL. *Born:* 7/20/26. born in Cairo, Egypt 1926. Grad from Univ IL, BS in chem engrg 1949, MS in chem engrg, Newark Coll of Engrg 1951, MBA Univ Chicago, Exec Prog 1960. Assoc with Union Carbide when The Visking Corp (now Films-Packaging Div) joined the corp in 1957. Operations mgr, food casings in 1960, exec vice-pres div in 1964 and pres 1967. Asst to the pres of Union Carbide 1971, and a corp VP in 1972. Elected sr VP of the corp 1977. In 1982 elected exec VP of the corp. As of Sept 1982, his respons include Union Carbide Europe SA, Union Carbide Agri Prods Co, Inc and the following domestic divs: Specialty Polymers and Composites; Films Packaging, Home & Automotive Prods; Silicones & Urethane Intermediates; Specialty Chems; Medical and Indust Prods Dept. Mbr AIChE & Tau Beta Pi. Lives in New Canaan, CT, married, three children.

Tarman, C William
Business: 3250 Wilshire Blvd, Los Angeles, CA 90010
Position: VP & Dir *Employer:* Daniel, Mann, Johnson, & Mendenhall *Education:* BS/CE/IL Inst of Tech *Born:* 11/6/23 Project mgr with DeLeuw Cather and Co for 9 yrs on large transportation and civil engrg projects. Over a period of 12 yrs was Chief Civil Engr, Gen Mgr, and Pres of Meiscon Corp engaged in civil engrg in computer aided design and drafting. Joined DMJM in 1971 as VP in charge of Civil Sys Div. Assumed position of Group Mgr for Regional Group in 1975. Appointed to the Bd of Dirs in 1978. Registered Engr in 17 states. Author of several papers, active in civic affairs. *Society Aff:* ASCE, NSPE, TRB

Tarn, Tzyh-Jong
Business: PO Box 1040, St Louis, MO 63130
Position: Prof. *Employer:* Washington Univ. *Education:* DSc/Control Systems Engg/Washington Univ; MS/Chem Engg/Stevens Inst of Tech; BS/Chem Engg/Cheng Kung Univ. *Born:* 11/16/37. Presently associated with: Washington Univ, St Louis, MO. Started in 1968. Assumed current responsibility as Prof in the Dept of Systems Sci and Mathematics in 1977. Hold visiting positions at Imperical College in London, the Univ of Rome, and Nagoya Univ in Japan. Principle expertise involves: Mathematical Systems Theory, Robotics and Automation. *Society Aff:* IEEE, SIAM.

Tarrants, William E
Home: 12134 Long Ridge Ln, Bowie, MD 20715
Position: Program Analyst, Off of Occupant Protection *Employer:* Nat Hwy Traffic Safety Admin US Dept of Transportation *Education:* PhD/Indus Safety/NY Univ; MSc/Indus Engg/OH State Univ; BIE/Indus Engg/OH State Univ. *Born:* 12/9/27. Served with USAF 1951-57, attaining rank of Cpt. Served as indus engrg instructor & res assoc OR Group at Ohio State Univ. Asst Prof & Res Assoc Center for Safety NYU. Ch Div of Accident Res Bureau of Labor Statistics 1964-67. Dir Manpower Dev Natl Hwy Traffic Safety Admin, US Dept of Transportation 1967-1980. Scientist, Office of Program and Demonstration Evaluation, Natl Hwy Traffic Safety Admin, 1980-1983. Chief Program Analyst, Office of Occupant Protection, Natl Hywy Traffic Safety Admin, 1983- . Part-time faculty member, Johns Hopkins University, 1984 to present. Pub over 100 prof articles. Serves on editorial bd of 2 internatl scientific journals. Serves as editor-in-chief, Prof. Journal "Traffic Safety Evaluation Research Review." Holds licenses as Reg PE & CSP. Mbr Bd/Dir & Officer ASSE: 1st Vice Pres 1975-76, Pres-Elect 1976-77, Pres 1977-78. Chairman, Accreditation Council, ASSE, 1979 to present. Elected ASSE Fellow in 1972, 1st Place Award Natl Tech Paper Contest 1960, 1962, & 1966. Founders Day Award, NYU, 1963. Author of 3 books: A selected bibliography of reference materials in Safety

Tarrants, William E (Continued)
Engineering and related fields, 1967; Dictionary of terms used in the safety profession, 1971; and measurement of safety performance, 1980. Outstanding Performance Award, US Dept of Transportation 1973 and 1986. *Society Aff:* IIE, ASSE, HFS, ERS, AAAS, SRA.

Tartaglia, Paul E
Business: Engrg & Tech, Norwich Univ, Northfield, VT 05663
Position: Head, Div of Engrg & Tech *Employer:* Norwich Univ *Education:* Dr of Eng/Design/Univ of Detroit; MS/Controls-ME/Northwestern Univ; BS/ME/Univ of Detroit *Born:* 9/30/44 Asst prof of ME at Univ of Mass 1970-75. Instituted Program for Master Thesis in Industry. Recipient of SAE Ralph R Teetor Award. After leaving U Mass worked as project Engr for Rodney Hunt Co, Orange, MA and was Chief Eng for Computerized Biomechanical Analysis, Inc, Amherst, MA. Presently Head of Engrg and Technology Division at Norwich Univ, Northfield, VT. Have consulted extensively in Design and Product Liability Areas; Received Patent on Flow ControlValve with Texas Instruments, Inc. *Society Aff:* ASME, ASEE

Tartar, John
Business: Dept of Computing Sci, Edmonton Alberta, Canada T6G 2E1
Position: Prof. *Employer:* University of Alberta. *Education:* PhD/EE/AZ State Univ; MS/EE/OK State Univ; BS/EE/OK State Univ. *Born:* Jan 1931. USAF 1950-54, instructor in radio communication. North American Rockwell Corp 1961-69, dir of hybrid computer facility, responsible for missilt & aircraft simulation studies & man-in-the-loop studies for Apollo. Avionic systems principal investigator, responsible for assessment of microelectronics & computing systems architectures for application to airborne systems. Prof Dept of Computing Sci, Univ of Alberta 1969- . Res activity in microprogramming, multiprocessor systems, & minicomputer networks. Mbr Eta Kappa Nu, Tau Beta Pi, Sigma Xi, IEEE, ACM. *Society Aff:* ACM, IEEE, ΤΒΠ, ΗΚΞ, ΣΞ.

Tasi, James
Business: Dept of Mechanical Engrg, Stonybrook, NY 11794
Position: Prof *Employer:* State Univ of NY *Education:* PhD/Eng. Mech/Columbia Univ; MS/CE/Univ of IL; B/CE/NYU *Born:* 12/6/33 in NYC. Associate Research Scientist with Martin Marietta Corporation 1961- 1965. Post-Doctoral Fellow at the Johns Hopkins University 1965-1966. Taught at State Univ of NY at Stony Brook from 1966 to present. Prof since 1972. Research fields include: thermoelastic dissipation in crystals, stability and vibration of composite shells, nonlinear shock wave analysis of crystal lattices. *Society Aff:* ASME, AAM, ΣΞ.

Tassios, Dimitrios
Business: 323 High St, Newark, NJ 07102
Position: Prof *Employer:* NJ Inst of Technology *Education:* PhD/ChE/Univ of TX; MS/ChE/Univ of TX; Dipl/ChE/Nat Tech Univ, Greece *Born:* 08/27/37 Native of Greece. Served in the Greek Army and worked for the Nuclear Research Center, "Democritos," Athens, Greece. Associated with NJ Inst of Tech since 1966. Served as officer and Chrmn (1978) of the North Jersey AIChE Section and as a mbr of Natl AIChE committees. Fullbright fellow (1962), mbr of Tau Beta Pi and Xe. Received from AIChE the "Grateful Appreciation Award" (1971); Outstanding Counselor Award (1973); As Chapter Advisor, the "Outstanding Chapter Award" (1971-1980); Certificate of Recognition (1978). Author of over fifty presentations and papers. Symposia Chrmn (AIChE, ACS). Dir and Lecturer, "Applied Distillation," Center for Profl Advancement. Industrial Conslt. Listed in: Outstanding Educators of Amer; American Men and Women of Sci; Who's Who in Amer. *Society Aff:* AIChE, ASEE, ACS

Tassoney, Joseph P
Home: 1314 Desert Willow Lane, Diamond Bar, CA 91765
Position: Prof *Employer:* CA State Polytechnic Univ, Pomona *Education:* MS/ChE/Newark Coll of Engrg; BS/ChE/Univ of Pittsburgh *Born:* 04/04/26 in McKeesport, Pa. Taught at the Univ of Pittsburgh and PA State Univ- McKeesport Campus for a period of 10 yrs. (1954-59) & (1959-1964) coupled with consltg work in Industrial & Sanitary Wastes. Over 20 yrs Industrial experience with Union Carbide, Oakridge Tenn, American Cyanamid Corp. Bituminous Coal Research, Texaco, Inc and Occidental Petroleum Corp with 20 US Patents in Energy related areas tempered with profl registration in NJ, NY, PA, CT, and CA. Enjoy golfing, bridge, and flying (private pilot). *Society Aff:* AIChE

Tassos, Thomas H, Jr
Business: PO Box 18725, Charlotte, NC 28218
Position: Sr. V P and Director of Engineering *Employer:* J N Pease Associates. *Education:* BCE/Sanitary Engg/NC State Univ. *Born:* Oct 29, 1925 San Antonio Texas. C E NC State Univ. WW II US Army Infantry 2 1/2 yrs. Sr. V. Pres/Director of Engineering J N Pease Assocs. P Pres NC Sect ASCE, P Pres Rotary Club of E Charlotte, P Pres S Piedmont Chap of Prof Engrs of NC. Reg P E in NC, Fla, Ga, Nev, SC, Tenn, Texas, & Va. Holds 'Certificate of Qualification' from Natl Council of Engrg Examiners. Mbr Charlotte Engrs Club & Mem United Methodist Church. *Society Aff:* ASCE, NSPE.

Tatinclaux, Jean-Claude P
Business: Hydraulics Lab, Iowa City, IA 52242
Position: Assoc Prof *Employer:* Univ of IA *Education:* PhD/Mech & Hyd/Univ of IA; MS/Mech & Hyd/Univ of IA; Eng. Dip/Mining Eng/Ecole des Mines, France *Born:* 9/20/42 Native of France. Educated in France and the US - Res Engr at the Bassin des Carenes, Paris, from 1969-73 - Joined the Faculty of the Univ of IA Coll of Engrg and Inst of Hydraulic Res in Fall 1973. In charge, there, of the Ice Engrg Res Program since 1975. On leave of absence at the Cold Regions Resch and Eng. Lab of the US Army Corps of Eng. in 81-82. Sectry of the Committee on Ice Problems of the Int'l Assoc for Hydraulic Resch 1981-pr. Fulbright grant 164. *Society Aff:* ASCE, ΣΞ

Tatlow, Richard H, III
Business: 10 E 40th St Suite 3400, New York, NY 10016
Position: Chairman of the Board. *Employer:* Abbott Merkt & Co Inc. *Education:* CE/CE/Univ of CO; BS/CE/Univ of CO. *Born:* May 27, 1906. BS 1927, CE 1933 Univ of Colorado. Reg P E, L S & Planner. Partner Harrington & Cortelyou 1929-40; US Corps of Engrs Col 1941-46, Legion of Merit; Pres, Chmn Abbott Merkt & Co Inc 1946-1984; Chmn Abbott Merkt Architects Inc 1972- . Chrmn, Abbott Merkt International Inc 1976- . Dir NUS Corp 1960- 1980; Chmn Bldg Res Adv Bd, Natl Academy of Sci 1962-64; V P EJC 1963-67; Mbr SST-Sonic Boom Ctte 1964-71; Bd/Dir, V P Animal Medical Center 1958-1970; Fellow ASME, ASCE (Pres 1968, Dir); mbr Natl Academy of Engrg; AICE. P Mbr UET, Chrmn. Awards: Iron Ring Canada 1961, ENR Citation 1968, Univ of Colorado Disting Engrg 1968, ASCE Met Sect C E of Yr 1968, NSPE/PEPP 1969, Univ of Colorado Associated Alumni Citation 1970, Univ of Colorado George Norlin 1970. Mbr: Newcomen Soc North Amer Cosmos (DC), Chevy Chase (DC), Union League (NYC). Registered PE in 20 states. *Society Aff:* ASCE, ASME, ACEC.

Tatman, D Russell
Business: 2005 Concord Pike, Wilmington, DE 19803
Position: President. *Employer:* Tatman & Lee Associates Inc. *Education:* Bach/Civil Engg/Univ of DE; Mast/Civil Engg/Univ of DE. *Born:* Oct 1936. Served with Army Corps Engrs 1961-63. Engr Hercules Inc & E I duPont de Nemours & Co; V P & Dept Mgr Environ Engrg Dept E H Richardson Assocs 1970-75; Pres Tatman & Lee Assocs Inc Nov 1975- , spec in Environ Engrg. Past Dir Dist V ASCE; P Pres, V P, Dir & Treas Delaware Sect ASCE. P Tres of CEC of DE. DE Tr to Chesapeake Water Pollution Control Assn 1973-75. Mbr NSPE. Hobby: tennis. *Society Aff:* ASCE, WPCF, NSPE, CEC.

Tatum, Finley W
Business: Dept. of Elect. Engr Box 19016, Arlington, TX 76019
Position: Professor of Elect. Engr. *Employer:* Univ. of Texas at Arlington *Education:* PhD/EE/TX A&M Univ; MSEE/EE/Columbia Univ; BSEE/EE/Columbia Univ. *Born:* July 1914, Kent Tex. 1935-47 various positions then Engrg Supervisor, Amer Dist Teleg Co NY; 1947-80 , SMU, 1951-73 Chmn Dept Electric Engrg, 1973-80 , Asst Dean Sch of Engrg & Applied Sci. 1952-54 Mbr Tech Staff Bell Telephone Labs, NY NY; 1960-61 Visiting Lecturer Tex A&M Univ. Fellow IEEE; Chmn EE Div ASEE 1968-69; Chmn Assn of EE Dept Heads 1971-72; 1971-78, Engg Cons Baylor Univ Medical Center; 1980- Univ. of Texas at Arlington Professor of Elect. Engr.. *Society Aff:* IEEE, ASEE.

Tatum, Joe F
Business: P.O. Box 1649, Hattiesburg, MS 39401
Position: Vice Pres *Employer:* Willmut Gas & Oil Co *Education:* BE/EE/Vanderbilt Univ *Born:* 04/06/25 in Hattiesburg, MS. Attended Riverside Military Acad in Gainesville, GA. In 1945 was employed by Westinghouse Electric Corp in Pittsburgh, PA. Was transferred to Sharon, PA to the transformer div and left employment of Westinghouse to return to Hattiesburg for employment by Willmut Gas & Oil Co. Hae received numerous patents in the pressure control, heating application and corrosion fields. Has been active in the formation of savings and loan institutions and other corps. Is married to the former Lady Jean Barker of Nashville, TN. They have three children who are married and reside at Hattiesburg, MS. They have seven grandchildren. *Society Aff:* NACE, NSPE, ASTM

Taub, James M
Home: 5686 N. Camino De La Noche, Tucson, AZ 85718
Position: Materials Conslt *Employer:* Retired *Education:* MS/Met Engrg/Case Inst of Tech; BS/Met Engrg/Case Inst of Tech. *Born:* July 26 1918. 1940-42 Metallurgist Republic Steel Corp; 1942-44 Res Asst for steel cartridge case dev, Case Inst Tech; 1944-74 Grp Leader Materials Tech Grp, Los Alamos Scientific Lab, respon for dev of casting & fabricaton tech for uranium metal, established integrated materials grp to work with metals, ceramics & plastics, dev fabrication tech for complex uranium-loaded graphite fuel elements for high temperature nuclear rocket engine, 1974-75 Materials Cons LASL; 1975- , Cons ERDA & DOE, 1975-1979. Recipient Ernest O Lawrence Award, AEC 1963; Fellow ASM 1970; Mbr ASM & ANS. 1980-81, On Staff of Asst Secy for Nuclear Energy, Dept of Energy. *Society Aff:* ASM, ANS.

Taub, Jesse J
Home: 29 Winthrop Rd, Plainview, NY 11803
Position: Chief Scientist. *Employer:* AIL Division of Eaton Corp. *Education:* MEE/EE/Polytechnic Inst of NY; BEE/EE/City College of NY. *Born:* April 1927, New York NY. 1949-55 Naval Applied Sciences Lab, Microwave Engr; 1955- , AIL serving as Sr Engr, Sect Hd, Div Cons & now Ch Scientist, made many contributions to microwave & millimeter wave circuit tech including over 45 publ & currently respon for AIL's Central Res activity. Fellow IEEE 1967, active in many IEE Microwave Theory & Techniques Soc Cttes. Enjoy classical music & archaeology. *Society Aff:* ΣΞ, IEEE.

Taubenblat, Pierre W
Home: 358 No. 4th Ave, Highland Park, NJ 08904
Position: VP and Dir *Employer:* AMAX Pacific Resources *Education:* MS/Ind Mgmt/Poly Tech Inst of Brooklyn; Engr/Electrometal/Univ of Grenoble *Born:* 09/29/30 Began working at St. Gogain, France in the corporate R&D labs. Have since been involved in R&D, dealing with new processes related to casting, melting and powder metallurgy as well as alloy dev. Mbr of tech Bd of Metal Powder Indus Fed, Mbr of the Editorial Ctte of Intl Journal of Powder Metallurgy. Mbr of the nonferrous ctte AIME. Mbr NJ Bd ASM. Author of many tech articles related to copper base mtls in cast, wrought and powder form. Edited several books dealing with nonferrous metals and powder metallurgy. Received patents in process metallurgy as well as copper and zinc base mtls. Received 1985 Distinguished Service Award from the Metal Powders Industries Federation. Enjoy classical music, skiing and surfing. *Society Aff:* AIME, ASM, APMI

Tauc, Jan
Business: Div of Engrg, Brown Univ, Providence, RI 02912
Position: L Herbert Ballou Professor of Engineering & Physics *Employer:* Brown University. *Education:* Ing Dr/EE/Technical Univ Prague; RNDr/Physics/Charles Univ, Program; DrSc/Physics/Czechoslovak Acad of Sci. *Born:* April 15 1922, Czechoslovakia, came to US in 1969. Ing Dr in EE Technical Univ, Prague 1949; RNDr Charles Univ, Prague 1956; DrSc in Physics Czechoslovak Acad of Sciences 1956. 1949-69 Hd Semiconductor Dept, Inst of Solid State Physics of Czechosl. Acad Sciences; 1964-69 Professor experimental physics, Charles Univ, 1968-69 Dir Inst of Physics; 1969-70 Mbr Tech Staff Bell Labs, Murray Hill NJ, 1970-78, Cons; 1970- , Professor of Engrg & Physics, Brown Univ 1983-1988, Director, materials Research Lab, Recipient National Prize from Czechoslovak government 1955 & 1969, Senior U.S. Scientists Award from the Humboldt Foundation 1981. Frank Isakson Prize of the American Physical Society 1982. *Society Aff:* APS, EPS, AAAS

Tauchert, Theodore R
Business: Dept of Engg Mechanics, Anderson Hall, Lexington, KY 40506
Position: Prof. *Employer:* Univ of KY. *Education:* DEngg/Solid Mechanics/Yale Univ; MEngg/Solid Mechanics/Yale Univ; BSE/Civ Engg/Princeton Univ. *Born:* 9/3/35. Mbr of the faculty of Princeton Univ from 1964 to 1970, and presently a Prof in the Dept of Engg Mechanics at the Univ of KY. Has had experience as a structural engr with Sikorsky Aircraft, and has been a consultant to the Bishop Engg Co and the Gen Tech Corp. Published a book and over 70 papers in technical journals. *Society Aff:* ASCE, ASME, ASEE, SES.

Taylor, Allan D
Business: 45 Fremont St, San Francisco, CA 94105
Position: Manager of Metallurgy *Employer:* Bechtel Civil & Minerals, Inc. *Education:* BMetEngr/Met Engr/Univ of MN. *Born:* 11/9/26. Native of Albany, OR. Active in mining and mineral testing for 15 yrs prior to engg degree in 1960. Bechtel Corp (met plant design) since 1960 as process, proj, and chief met engr prior to present position as engg mgr in 1979. Special profl interest in comminution, including autogenous grinding. Enjoys met & technical history & golf. *Society Aff:* AIME, MMSA.

Taylor, Arthur C, Jr
Business: Dept of Mech Engg, Lexington, VA 24450
Position: Prof of Mech Engr *Employer:* VMI. *Education:* PhD/Structures/OH State Univ; MS/Structures/OH State Univ; BS/Civil Engg/VMI. *Born:* 8/8/23. Taught various engr subjects at VMI since 1949, hd of mech engr dept at VMI from 1957-80, currently prof of Mechanical Engrg. Summer work and consltg work with VA Dept of HGwys, VEPCO, US Weather Bureau, Harry Diamond Labs, Office of Noise Abatment and Control in the EPA. *Society Aff:* ASME, ASEE.

Taylor, Barry N
Business: Natl Bureau of Standards, Room B258, Bldg 220, Gaithersburg, MD 20899
Position: Chief, Electricity Div *Employer:* US Govt, Dept of Commerce, Natl Bur Stds. *Education:* PhD/Physics/Univ of PA; MS/Physics/Univ of PA; AB/Physics/Temple Univ. *Born:* 3/27/36. Native of Phila, PA. Served as an Instructor and Asst Prof of Physics, Univ of PA, 1963-66. Mbr Technical Staff, RCA Labs, Princeton, NJ, 1966-70. Joined natl Bureau of Stds in 1970 as Chf of the Absolute Elec Measuremnns Sect. Served as Chief of the Electricity Div 1974-78, and Chf of The Elc Measurements & Stds Div 1978-1983, Electricity Division 1983-present. Res interests include electron tunneling, superconductivity, Josephson effect, precise electrical measurements and stds, absolute realization of electrical units, and fundamental constant determinations and data analysis. Published about 50 papers on these subjects. Received RCA Outstanding Achievement Award in Sci in 1969, Franklin Inst

Taylor, Barry N (Continued)
Wetherrill Medal in 1975, and Dept of Commerce Silver Medal in 1975. Coauthor of The Fundamental constants and Quantum Electrodynamics, Academic Press, 1969. Fellow of APS, IEEE, and Washington Acad of Sci. Enjoys jogging and biking. *Society Aff:* IEEE, APS, ΣΞ.

Taylor, Byron L
Business: 32251 N Avis Dr, Madison Heights, MI 48071
Position: President. *Employer:* IPE Cheston. *Education:* MBA/Marketing/MI State; MSEE/Engg/UCLA; BS/Physics/Wayne State. *Born:* May 27 1930. 1951-55 served with US Air Force in radar systems; Sr Res Engr Autonetics Div of Rockwell, respon for design & analysis of radar & computer systems; Induction Process Equipment, manufacturers of induction heating & forging equip: Dir Res, V Pres Engrg & currently Pres, designed full line of solid state power supplies ranging from 20KW to 3000KW, at 1KHz, 3KHz & 10KHz. Mbr IEEE.

Taylor, C Fayette
Home: 24 Monmouth Court, Brookline, MA 02146
Position: Consultant (Self-employed) *Education:* ME/ME/Yale Univ; PhB/ME/Sheffield Sci School, Yale Univ. *Born:* 9/24/94. in NYC, 1894. Secy of Class at Yale. Lt USN in charge of aircraft engine testing during WWI. Then engg in charge aircraft engine lav, US Army Air Service, Dayton, OH. 1923-26 Asst Chief Engg and Chief Engg, Wright Aeronautical Corp, Paterson, NJ. 1926-60, Prof of Automotive Engg, MA Inst of Tech. 1960- present, consultant on internal combustion engine problems. Mbr NASA Power Plant Committee. 1918-1936, Author of "The Internal Combustion Engine in Theory and Practice–, MIT Press, Cambridge, MA. Fellow, AAAS and SAE. *Society Aff:* SAE, ΣΞ, INST AERO SCI.

Taylor, Charles E
Business: 231 Aerospace Engineering Building, Gainesville, FL 32611
Position: Professor of Engineering Sciences *Employer:* University of Florida *Education:* PhD/Theo & Appl Mechs/Univ of IL; MS/Engg Mech/Purdue Univ; BS/ME/Purdue Univ. *Born:* 3/24/24. West Lafayette Ind. Married & father of 2 sons. Served with Army Infantry during WWII; taught at Purdue, Univ of Ill and Univ of Fla; Visiting Prof in India, Univ of Colo, Univ Calif at Berkeley, Technische Hogeschool in Delft and at US Military Acad; 1952-54 Struc Res Engr at David Taylor Model Basin. Honorary Mbr SESA (Pres 1966-67, M M Frocht Award 1969, M Hetenyi Award 1970 & 73, W Murray Lectureship 1974, FG Tatnall Award 1983), Fellow Soc of Engrg Sci (Pres 1978), Fellow in ASME (Assoc Editor of Journal of Applied Mech 1971-74); ASEE; Fellow in Amer Acad of Mechanics; Fellow in AAAS, (Mbr Natl Acad of Eng). Specialize in optical methods of stress analysis, particularly photoelasticity & holography. *Society Aff:* SEM, SES, ASME, ASEE, AAAS, AAM.

Taylor, Edgar R, Jr
Home: 1070 Old Gate Rd, Pittsburgh, PA 15235
Position: Advisory Engineer. *Employer:* Westinghouse Electric Corp. *Education:* BEE/Power/Cornell Univ. *Born:* March 1929, Jacksonville Fl. Taught in advanced degree courses for Westinghouse & Univ Pittsburgh; 1951-53 served with Army Signal Corps, Bronze Star for work in planning rebuilding of electric utilities of S Korea; 1952 Advanced Dev & Transmission Systems Engr for Westinghouse, 1961-62 Proj Engr for Apple Grove 775 kV Proj; 1963-66 Sponsor Engr for Southeast; 1967-71 Mgr Transmission Lines & Terminals; 1971-81 Advisory Engr and 1982-85 Consulting Engineer, T&D Systems Engrg. Presently Advisory Engr. in Advanced Systems Technology, Westinghouse. Fellow IEEE, Life Mbr Eta Kappa Nu & Tau Beta Pi. Reg PE Pa. Amateur genealogist. *Society Aff:* IEEE, HKN, TBΠ, NGS, CIGRE.

Taylor, Edward S
Home: 16 Tabor Hill Rd, Lincoln, MA 01773
Position: Prof Emer (MIT) *Employer:* Ret. *Education:* BS/ME/MIT *Born:* 01/26/03 After graduation worked for public service of NJ and Wright Aeronautical. In 1927 went to MIT as Instructor in Aircraft Power Plants. Forty-one years at MIT becoming Prof of Aeronautics in charge of the Gas Turbine Laboratory. Ret 1968. Books: The Internal Combustion Engine (with C.F. Taylor) int. textbook 1938, 1948, 1961; Dimensional Analysis for Engineers Oxford Press 1974; motion picture "Secondary Flow–, Educational Services 1965; Sylvanus Albert Reed Award 1937; Lester Gardner Lecturer 1970; Robert H. Goddard Award 1973; very successful patent 2175999 vibration isolation of aircraft power plants. *Society Aff:* AAAS, AIAA, ASME, SAE

Taylor, George B
Home: PO Box 299, Oneida, TN 37841
Position: Vice President of Research & Dev. *Employer:* Tibbals Flooring Company. *Education:* MS/Nuclear Engg/W VA Univ; BS/Chem Engg/W VA Univ. *Born:* Sept 1936, Fairmont WVa, 1960 Chemical Engr Koppers Co, dev of wood- impregnation processes & products; 1962 Chemical Engr US Bureau of Mines, cost estimation & evaluation of conceptual chem processes utilizing coal; 1964 Nuclear Res Engr W Va Univ, res in woodpolymer composite materials; 1967 Mgr Radiation Process Center, Arco Chem Co, commercialization of radiation processes particularly impregnated wood-polymer flooring material; 1968 Asst V Pres Radiation Internatl Inc, in charge of Wood Products Div; 1972 V Pres Tibbals Flooring Co, respon for R & D Dept, Chem Div & Purch Dept. Reg PE NJ, Tenn & WVa. *Society Aff:* AIChE, ΩΧΕ, FPRS, SME

Taylor, Howard P J
Business: Tallington, Stamford Lines, England
Position: Chief Engr. *Employer:* DOW MAC Concrete Ltd. *Born:* April 1940, Sale UK. BSc Univ Manchester; PhD City Univ, London. Exper: cons, res, contracting. Res interests in struct concrete, cracking, shear, detailing, computer aided design & marine uses; Tech Mgr of UK res programme into use of concrete for offshore strucs. Dir Dow Mac Concrete Ltd. Mbr ICE, IStructE UK, ACI. Assoc CEB Commission V; IStructE Oscar Faber Medal Winner; ACI-ASCE Raymond C Reese & 1974 State of the Art Ctte Award Winner; IStructE Oscar Faber Award Winner; IStructE Henry Adams Medal Winner.

Taylor, John J
Home: 15 Oliver Court, Menlo Park, CA 94025
Position: V P *Employer:* Electric Power Research Institute *Education:* DSc (Hon)/DSc/St. John's Univ; MA/Math/Univ of Notre Dame; BA/Math/St. John's Univ. *Born:* Feb 1922. AB St John's (NY) 1942; MS Notre Dame 1946; DSc (Hon) St John's 1974. Engaged in nuclear energy dev for past 34 years, 1950-1981 with Westinghouse, respon for radiation shielding for 1st nuclear powered submarine & for nuclear core for 1st nuclear powered naval surface ships; 1967-70 Engrg Mgr commercial pressurized water reactor design & dev, 1970-76 respon for fast breeder reactor dev, 1976-81 V Pres, overall respon for pressurized water reactors. Presently, V P, Nuclear Power, Electric Power Research Inst. Westinghouse Order of Merit 1957. Fellow ANS, AAAS; Mbr APS, NAE. *Society Aff:* ANS, APS, AAAS, NAE.

Taylor, Kenneth G
Home: 959 Liawen Ct NE, Atlanta, GA 30329
Position: Consultant *Employer:* Simons-Eastern Co. *Education:* BS/Civil Engrg/Univ of Rhode Island. *Born:* 3/9/21. 1943-46 Capt Army Corps Engrs; 1946-58 Const Mgmt with J E Sirrine Co; since 1958 founding Dir & Exec V Pres Eastern Engrg Co, through 1980 now Simons- Eastern Co. Retired Pres of Company. Currently Consultant. Charter Mbr & Pres DeKalb Chap GSPE, Pres GSPE 1973-74 Natl Dir 1977-79; Pres Atlanta Optimist Club 1972; Charter Director and Pres Ga Engr Foundation 1976. Engineer of Year, Atlanta Metro Area 1974; Engr of Year Ga 1976; AIPE Professionalism Award 1974. Chrmn Pulp and Paper Industry Advisory Ctte for Multidisciplinary Engrg Program Ga Inst of Technology. 4 terms as vestryman & Sr Warden, St Bartholomew's Episcopal Church. Wife & 3 adult children, graduates of UNC RISD & GSC and 4 grandchildren. Dir on Bd Trustees for Ga.

Taylor, Kenneth G (Continued)
Tech Research Corp, Inducted in URI Hall of Fame 1984. *Society Aff:* GSPE, NSPE, ASCE, TAPPI.

Taylor, Leonard S
Business: Elec Engg Dept, College Park, MD 20742
Position: Prof. *Employer:* Univ of MD. *Education:* PhD/Phys/NM State Univ; MS/Phys/NM State Univ; AB/Phys/Harvard College. *Born:* 12/28/28. From 1960 to 1964 employed by GE Co; during 1962-3 a Fulbright Lecturer at the Univ of Madrid. From 1964 to 1967 on the faculty of Case-Western Reserve Univ; during 1964-65 a visiting prof at the Natl Poly Inst of Mexico. Presently holding a joint appointment in the Depts of Elec engg and Radiation Oncology, Univ of MD. Mbr of the US Natl Committee of URSI, Fellow of the IEEE, Fellow of the American Society for Laser Surgery and Medicine. Distinguished Alumnus, NMSU. *Society Aff:* IEEE, APS, OSA, BEMS.

Taylor, Lot F
Business: 509 N 6th, Garden City, KS 67846
Position: Owner *Employer:* Taylor and Assoc, Inc *Education:* BS/CE/KS St Univ; BS/Animal Husbandry/KS State Univ *Born:* 06/30/39 Graduate-Manhattan High Sch 1957, 3 years Marine Corps, Degrees-Animal Husbandry 1961; Cvl Engrg 1972. Registered in State of KS-PE 1976; Licensed Land Surveyor 1977. Practiced with Foster-VanGundy two years. Practice and part owner of Evans-Bierly-Hutchison for 6 year period. I then bought the branch office of EBH in Garden City forming Taylor and Assocs January 1980. I employ a Registered PE and Registered Landscape Architect. We average 16-18 fulltime people. We do basic engrg work; cvl engrg municipal work for Southwest KS. I recently received licenses for Oklahoma and Colorado. The firm has recently opened a Soils, asphalt, and concrete testing lab. The firm has recently open a Soils, asphalt, and concrete testing lab. *Society Aff:* ASCE, KES, ASTM, NSCE, AWWA, ACI.

Taylor, Mervin Francis
Home: 429 Shady Lane, Pasadena, MD 21122
Position: Weight Control Engineer. *Employer:* Retired. *Education:* -/Mech Engr/Univ of NC; -/Aero Engr/Johns Hopkins Univ. *Born:* Dec 1913, Sea Level NC. Mech Engrg Univ N C, 3 years, Aeronautical courses at Johns Hopkins Univ. Weight Control Engr & Ch Weight Control at Martin Marietta Corp, Baltimore Div for 32 years until retirement in 1969; Cons for Weight Control, then Weight Control Engr at Westinghouse Electric's Oceanic Div until 1976. Honorary Fellow & P Natl Pres Soc of Allied Weight Engrs, Chmn Natl Awards Ctte 5 years. Hobby- sailing. *Society Aff:* SAWE.

Taylor, Michael A
Business: Civil Eng Dept, Davis, CA 95616
Position: Assoc Prof/Principal *Employer:* Univ of CA/Self *Education:* PhD/Structural Eng/Univ of CA-Berkeley; MS/Structural Eng/Cornell Univ; BSc/CE/Univ of Manchester-England *Born:* 11/28/38 1961-66 Engr in England, USA and Barbados, W. Indies. 1969-present Prof of Civil Engrg Univ of CA at Davis. 1977-79 Dir N. CA Chapter ACI. 1979-80 Pres N. CA Chapter ACI. Mbr ASTM Committee cq. West Coast representative for ISE. Self- employed conslt since 1963. Phi Kappa Phi, Joe W. Kelly Award Berkeley 1968. *Society Aff:* ACI, ISE

Taylor, R William
Business: 1575 I St. N.W, ASAE, Washington, DC 20005
Position: Pres *Employer:* Amer Soc of Assoc Executives *Education:* MS/Journalism/OH Univ; BS/Chemistry/Murray State Univ. *Born:* 7/28/29. Publications Mgr for Soc of Petrol Engrs (1953-63); Exec Dir and Secy of Am Inst of Mining, Met & Petrol Engrs (1963-68); Exec VP & Gen Mgr of Soc of Mfg Engrs (1968-81). Pres of Am Soc of Assn Executives since 1981. Past Pres of Council of Engg & Scientific Soc Exec (1976-77). Chairman of President Reagan's Citation Program for Private Sector Initiatives; Pres of ASAE Foundation; Board mbr of Washington, D.C. Board of Trade. Former Bd mbr of ASEE, EJC, JETS and AAES. Assoc Editor of *Petroleum Production Handbook. Society Aff:* SME, ASAE.

Taylor, Robert G
Business: 40 Grove St, PO Box B, Wellesley, MA 02181
Position: Partner and Mgr Wellesley Office. *Employer:* R. W. Beck and Assocs *Education:* BS/EE/Univ of WA *Born:* 09/04/27 Grad from Univ of WA in 1958 and joined the Seattle home office of R. W. Beck and Assocs, Consltg Engrs, as an Elec Engr. In 1967, relocated to MA and was elected a Partner in the firm. Since then have been Partner and Mgr of the Firm's Wellesley, MA office serving the Northeast region of the country. Specialize in contractual arrangements for power supply and interchange of power between utilities and electrical energy users - evaluation of power cost alternatives - rate negotiations - financing construction programs - utility mgmt reviews - utility property valuation. Reg PE in 18 states and the District of Columbia. *Society Aff:* IEEE, NSPE, AEE

Taylor, Snowden
Business: Stevens Inst. of Tech, Hoboken, NJ 07030
Position: Prof of Physics and Engg Physics. *Employer:* Stevens Inst of Tech. *Education:* PhD/Physics/Columbia Univ; AM/Physics/Columbia Univ; ME/Engg/Stevens Inst of Tech. *Born:* 6/25/24. Faculty mbr at Stevens Inst of Tech since 1958. I have taught gen physics to engg students for many yrs. Have also taught "Introduction to Nuclear Physics and Nuclear Reactors" to physics and engg physics students. I have worked very closely for 19 yrs with the Stevens Technical Enrichment Prog (STEP) which seeks to facilitate the entry of minority and disadvantaged students into the engg and technical professions. I am faculty advisor to STEP and STEP students. I was Chrmn of the Committee on Undegrad Curriculums for many yrs. I have also conducted res and supervised grad students in Particle Physics. *Society Aff:* APS, ΣΞ, AAUP, FAS, UCS.

Taylor, William E
Business: 12500 Gladstone Ave, Sylmar, CA 91342
Position: Mgr Process Dev. *Employer:* Spectrolab, Inc. *Education:* PhD/Met Engr/Purdue Univ; MS/Economics/AZ State Univ; BS/Met Engr/Purdue Univ. *Born:* 11/15/21. Worland, WY, married, one son. First Lt, US Field Artillery, 1942-46. Met, Oak Ridge Natl Lab, 1950-52; Sr Engr & Dept Mgr, Motorola, Inc, 1952-60; Prof Met Engr, MI State Univ, 1960-62; Operations Mgr Semiconductor Mtls, Motorola, Inc, 1962-69; VP, Finch & Taylor, Inc, 1969-75; Pres & Dir, Camelhd Dev Corp, 1970-80; Sr Engr & Dept Mgr, Spectrolab, Inc, 1975-. Consultant, 1960-62, Control Data Corp, Motorola, Inc, Oak Ridge Natl Lab. Officer, AZ Chapter ASM, 1953-57, Chrmn 1957; Bd of Dirs. Maricopa County Health Planning Council, 1974- 75; Bd of Dirs, Timeplex, Inc, 1977-. Distinguished Engg Alumnus, Purdue Univ 1967. *Society Aff:* ASM, AIME, AAAS, NYAS.

Taylor, William H
Business: 18 Washington Ave, Haddonfield, NJ 08033
Position: Owner (Self-employed) *Education:* MS/CE/Univ of PA; BS/CE/Princeton Univ *Born:* 10/30/28 Following service as line ofcr in US Navy, joined firm now known as Taylor, Wiseman & Taylor, Consit Cvl Engrgs, Mount Laurel, NJ. As principal owner and Pres, had responsibility for direction of wide variety of cvl engrg projects in design and construction services. Sold interest in that company to assocs and entered individual practice in Aug 1981. Provides leadership to several prof and civic organizations; currently serving in six directorships including Bd of Direction, ASCE. Licensed engr, surveyor and planner in three states. Became twelfth surveyor general of West Jersey in 1963. *Society Aff:* ASCE, NSPE, SAME

Teague, Jo M, Jr
Home: The Regency Palace Hotel, P.O. Box 927000, Amman, Jordan
Position: VP, Dir Glass Technology (Retired). *Employer:* Intl Exec Service Corp *Education:* BS/Cer Engrg/GA Tech; MS/Cer Engrg/OH State U; PhD/Glass Tech/OH State U. *Born:* Nov 1916, Atlanta Ga. 1940 joined Owens-Illinois, Engr Clarion Pa; 1941- 46 US Army Engrs, retired LtCol; 1946 rejoined Owens-Illinois Qual/Spec Supervisor, Clarion, at Toledo Oh Headquarters served successively as Ch Phys Sect, Asst Dir Res; Dir Inorganic Res; V Pres Tech Dir Gl Container Div, V Pres Corp Staff & Assoc Tech Dir, currently Retired (6/1/78) VP Dir Glass Tech International Operations, respon for administering multimillion dollar overseas Tech Assistance Programs. Fellow Amer Ceramic Soc & British Soc Glass Tech. Distinguished Alumnus Award, Ohio State Univ & Ga Tech. July '84 Appt Dir of Opers - JORDAN by Intl Exec Service Corps, Stanford CT. Avid golfer & part-time fisherman.

Teal, Gordon K
Home: 5222 Park Lane, Dallas, TX 75220
Position: Consultant. *Employer:* self. *Education:* PhD/Physical & Inorganic Chemistry/Brown Univ; ScM/Physical Inorganic Chemistry/Brown Univ; AB/Math & Chemistry/Baylor Univ; ScD (Hon)/-/Brown Univ; LLD (Hon)/-/Baylor Univ. *Born:* 1/10/07. 1930-53 Mbr res staff Bell Labs, initiated pioneering res that produced 1st germanium & silicon single crystals. Co-author 1st paper on achievement of junction transistor, 1st microwatt amplifier. Inventor grown junction single crystal technique, key technology in latter achievements. Founded & led TX Instruments res labs in developing 1st commercially feasible silicon transistor, 1954, progressed Asst VP & Dir Central Res Labs to VP Ch Scientist 1953-72; hdquartrd in Europe 1962-65; first Dir NBS Inst for Matls Res 1965-67. 64 patents, key publens. Hons include 1966 Inventor of Yr, The Patent, Trade-mark and Copyright Research Institute, George Washington Univ; 1968 IEEE Medal of Honor; 1970 ACS Award for Creative Invention; Mbr NAE 1969-., Bd/Dir IRE 1959 and 62; VP AAAS 1969; Hon Mbr Sigma Pi Sigma Physics Hon Soc, 1959; Distuished Alumnus, Baylor Univ 1965; Brown Univ Grad Sch 50th Anniv Commemoration Citation, 1978; Omicron Delta Kappa Outstanding Alumnus Award, Baylor Univ, 1978; Wilfred T. Doherty Award, 1974; Amer Inst Chem 50th Anniv Meeting Honor Scroll, Fellows' Lecturer, 1973; Citation, TX Instruments, 1980; IEEE Centennial Medal, 1984; Semmy Award, Semiconductor Equipment and Materials Inst, 1984. *Society Aff:* IEEE, APS, ACS, IEE, AIC, ECS, NAE, ΣΞ, ΣΠΣ, DIR, IRI, AAAS

Teale, James M
Home: 48 Pinecrest Dr, Woodcliff Lake, NJ 07675
Position: Principal. *Employer:* J M Teale Associates. *Education:* MS/Ind Engg/Stevens Inst of Technology; BChE/Chem Engg/Clarkson College of Technology. *Born:* 10/24/28. Reg PE NJ. 1950-53 Officer in Army Corps Engrs; Lederle Labs Process Engr, scale-up, design, economic studies; 1959-64 Beecham Products Inc V Pres Mfg & Dir Corporate policies re purchasing, quality control, packaging, engrg, production, distribution; 1964-69 Dir Technical Services, Clairol Div Bristol Myers Co, Process/Proj Engr, quality control, mfg control of sub-contractors; 1970- , Pres & Dir Pedco Products Inc, plastic thermoforming, & Principal J M Teale Assocs, cons Chem/Indus Engr. Num articles pub in engrg journals & trade publ. Sr Mbr AIChE & IIE; Mbr Chem Engrg Prod Res Panel 1974 & Adv Panel 1975. *Society Aff:* AIChE, IIE.

Teare, B Richard, Jr
Home: Wellspring Hse Wash Ave Ext Apt G71, Albany, NY 12203
Position: Univ Professor Engineering Emeritus. *Employer:* Carnegie-Mellon University. *Education:* DEng/Elec Engg/Yale; MS/Elec Engg/Univ of WI; BS/EE/Univ of WI. *Born:* Jan 1907, Menomonie Wisc. Hon DSc TriState Coll; Hon DSci Cleveland State Univ. 1929-33 Genl Electric Co; 19331939 Instructor, Asst Professor, Electrical Engr Yale Univ; Hon Dr of Engg, CMU; since 1939 at Carnegie Inst Tech (now Carnegie-Mellon Univ) as Buhl Professor, Dean Graduate Studies, Dean Coll Engrg & Sci, Univ Professor of Engrg. Res: hysteresis motors, copper covered steel conductors Eng, educ. Major teaching interest: courses in which student learns to deal with real prof engrg problems. Pres ASEE 1959-60; Pres AIEE 1962; active in merger with IRE; V Pres IEE 1963; Educational Cons, World Bank & other organizations. ASEE George Westinghouse Medal 1947, Lamme Medal 1963; IEEE Education Medal 1963. SAME Tasker H Bliss Medal 1977. Hobby interests: travel, music, photography, Doctor of Engrg, Yale Univ 1937.

Tebo, Julian D
Home: 30 Sutton Place, Verona, NJ 07044
Position: Consulting Engineer. *Employer:* Bell Telephone Labs (retired). *Education:* BEE/Elec Engg/Johns Hopkins Univ; Dr Eng/Elec Engg/Johns Hopkins Univ. *Born:* July 1903; native of Harpers Ferry W Va. With Bell Telephone Labs 1928- 68; Head Tech Relations Dept 1961-68; Asst Project Engr on Radar & Sonar during WWII; Ed Bell Sys Tech Journal & Bell Labs Record 1949-57. Mbr Engrg Foundation Bd 1959-68; Chmn Res Ctte 1965-67 & mbr Exec Ctte 1960-63 & 1964-68; Mbr AIEE Bd of Dirs 1955-59, Chmn AIEE Ctte on Tech Opers 1955-56, Chmn AIEE N Y Section 1950-51; Life Fellow IEEE; Fellow N Y Acad of Sciences; Mbr APS; mbr Adv Council, Soc for the History of Tech 1957-67; Pres Montclair Soc of Engrs 1957- 58; mbr Sigma Xi & Eta Kappa Nu. Author & Lectr; Special Lectr in Elec Engrg N J Inst of Tech 1968-69; Patentee protection devices between power & telephone lines. Verona N J Bd of Educ 1954-57; mbr Verona Borough Council 1957-70, Pres 1962 & 65. Mbr Ntl Inventors Hall of Fame Selection Ctte 1973-81. Mbr Advisory Council, Verona, NJ Bd of Education 1977-. *Society Aff:* IEEE, e SHOT, APS, NYAS.

Tedeschi, Frank P
Home: 8440 Royalview Dr, Cleveland, OH 44129
Position: Dean of BEET Program *Employer:* Electronic Tech Inst *Education:* MS/Elec Engr/SUNY at Buffalo; BS/Elec Engr/Youngstown State Univ. *Born:* 10/9/38. He previously taught at Cleveland State Univ, Cleveland OH, Cuyahoga Community College in OH, and was Chrmn of Elec Tech at Miami Dade Jr College in FL. He helped design the launch system for the Minuteman Missile at Sylvania Electronic Products. Mr Tedeschi has authored and co-authored eight books. His most recent publication is titled, *101 Microprocessor Projects, Hardware and Software*, published by Tab Books. *Society Aff:* TBΠ.

Tedesko, Anton
Home: 26 Brookside Cir, Bronxville, NY 10708
Position: Consulting Engr *Education:* CE/-/Inst of Tech, Vienna; Dipl Engr/-/Univ of Berlin; DSc/-/Tech Univ. Vienna. *Born:* Reared & educated in Austria. Came to U.S. in 1927. CE Degree Inst of Tech, Vienna, Austria 1926; Dipl Engr Univ of Berlin; DSc Inst. of Tech Vienna; Asst Prof Inst Tech, Vienna; DEngr (hon) Lehigh Univ 1966; DSc (hon) Univ Vienna 1978. Thompson-Starrett Dir 1960-61. Cons USAF HQ 1955-70, for 35 yrs with Roberts & Schaefer Co, as VP 1955-67; designer & supvis const: arenas, air terminals, bridges, hangars, prestressed concrete, shell structures; evaluated struc failures, rehab failed structures, arbitrator of engrg & const. disputes. Prin of URSAM joint Vent & struc engr respon design NASA's manned Lunar Landing program, assembly & launch facil incl. VAB, largest build. on record, which received Outstanding Civil Engrg Achievement Awd 1966. Established own cons engrg office 1967. Mbr of numerous consult bds; Active ASCE, ACI (PDir), AREA, SAME, RCRC (exec comm), PSRC, IASS, IABSE (US rep Perm Comm). Lectr at numerous Univ; author of articles. Lindau Award 1961; ENR Citation 1966; Boase Awd 1978; Natl Acad of Eng 1967; Intl Awd of Merit in Struct Eng 1978; Hon Mbr ACI 1973; Hon. Mbr ASCE 1978; Hon. Mbr IASS 1979. *Society Aff:* ASCE, ACI, AREA, SAME, RCRC, PSRC, IASS, IABSE

Tedmon, Craig S, Jr
Business: Power Systems Sector, W2K, Fairfield, CT 06431
Position: Staff Executive - Technology Oper. *Employer:* GE Co Power Systems Sector *Education:* BS/Metallurgy/MA Inst of Tech; MS/Metallurgy/MIT; ScD/Metallurgy/MIT. *Born:* 1/19/39. in Pueblo, CO. Native of Seattle Wash. Employed at G E R/D Center 1964-80; from 1964-70, Staff Scientist; 1970-74 Mgr Surfaces & Reactions Branch; 1974-77 Mgr Power Sys Lab; 1977-78 Mgr R&D Applications

Tedmon, Craig S, Jr (Continued)
Operation; 1978-80 R&D Mgr Energy Sci & Engg; 1980- Staff Executive, Power Systems Technology Operation, Faifield Ct. Present mgment responsiblity long and short term technical directions of GE's power system businesses. *Society Aff:* ASM, ECS, AAAS.

Teer, Kees
Business: Philips Research Labs, N V Philips Gloeilampenfabrieken, Netherlands
Position: Executive Director. *Employer:* N V Philips' Gloeilampenfabrieken. *Education:* PhD/Tech/Univ Delft; MSc/Tech/Univ Delft. *Born:* June 6, 1925. Educ: Tech Univ; Master of Tech Sci 1949, Dr of Tech 1959. Employee: Philips Res Labs 1950; Dept Head Acoustics & Recording 1959; Deputy Dir 1966; Exec Dir Electronic Sys, Philips Res Labs 1968. Natl Award for work in Colour TV Transmission Sys 1955. Fellow IEEE 1973. Mbr Royal Dutch Acad of Sciences, 1977. Chmn of the Lab Mgmt, 1982. *Society Aff:* NERG, KIVI, KNAW.

Tefft, Phillip W
Business: PO Box 866, Columbus, OH 43216
Position: Chmn of the Board - Pres. *Employer:* The Claycraft Co. *Born:* Apr 1917 Darlington Pa. Bachelor Ceramic Engrg June 1940, Ohio St. Univ. Ceramic Engr for The Claycraft Co June 1940; progressed to Plant Supr, Genl Supr, V Pres, becoming Pres in Apr 1952; Amer Ceramic Soc Fellow Apr 1956; Chmn Alfred Univ Bd of Trustees 1968-75; Disting Alumnus Ohio St Univ Coll of Engrg 1974; Hon Dr of Laws Alfred Univ June 1975.

Tegtmeyer, Rene D
Home: 1332 Timberly Ln, McLean, VA 22101
Position: Asst Commissioner for Patents. *Employer:* US Patent & Trademark Office. *Education:* BS/ME/WA Univ; JD/Law/Geo Wash Univ. *Born:* 1/5/34. Born St Louis, MO. US Air Force 1956-59 (Pilot); US Air Force Reserve 1960- 1968 (Pilot); N Amer Aviation (Structures Engr) - 1956 Asst; US Patent and Trademark Office 1959 - date (Asst Commissioner 1971-date); Amer Political Sci Assoc Congressional Fellowship Program 1967-1968; Dept of Commerce Gold Medal Award 1980; NJ Patent Law Association Jefferson Medal 1985; Mbr VA State Bar and District of Columbia Bar *Society Aff:* ABA, AIPLA, FBA, ТВП, ПТΣ.

Teich, Malvin C
Business: Dept of Electrical Engineering, New York, NY 10027
Position: Prof *Employer:* Columbia Univ *Education:* PhD/Quantum elect/Cornell; MS/Elec engrg/Stanford; SB/Physics/MIT *Born:* 05/04/39 Native of NYC. Joined MIT Lincoln Lab. Lexington, MA in 1966. With Columbia Univ as Prof since 1967. In 1969 received IEEE Browder J. Thompson Award for paper "Infrared Heterodyne Detection–. In 1973 appointed Fellow of the John Simon Guggenheim Memorial Foundation. From 1977 to 1979 served as a mbr of Editorial Advisory Panel for Optics Letters. In 1981 received Citation Classic Award of Current Contents (Inst for Scientific Info). In 1983 elected Fellow of Optical Soc of America. *Society Aff:* IEEE, OSA, APS, ASA, ΣΞ

Teller, A J
Home: 3140 S. Ocean Blvd, Palm Beach, FL 33480
Position: Consultant *Employer:* Research Coltrell *Education:* PhD/Chem Eng/Case-Western Reserve; MChE/Chem Eng/Brooklyn Poly; BChE/Chem Eng/Cooper Union. *Born:* June 1921. Practice is dev design & const of environ control sys. Received Business Week Award for Environ 1970, AIChE Annual Inst Lect 1972. Vaaler Award, 1978 Author Sci Articles & patents in separation sys. Ed Sections Perrys Handbook & Fertilizer Sci & Tech. President, Teller Env Systems 1970-1986, Consultant Research Coltrell 1986- . Dean Engrg & Sci The Cooper Union 1963-70. Served as Chmn NECORE, mbr Natl Tech Adv Council EPA, Chmn Air Comm AIChE. Married Sherry R Apr 1946. Son Richard. *Society Aff:* AIChE, ACS, APCA, TAPPI

Tellier, Roger A
Home: 18 rue Gallieni, Saint Leu La Foret, France 95320
Position: Controleur General Adjoint *Employer:* Electricite de France. *Education:* -/Elec Engg/Ecole Superieure d'Electricitie. *Born:* 2/3/26. near Paris. Ingenieur ESE (Ecole Superieure d'Electricite, Paris, France). Military Service in French Navy. Joined Electricite de France (EDF) in 1947, specializing in cables in the Res Organization. In 1967 seconded to CIGRE (Internatl Conference on H V Sys); since 1968 back with EDF Distrib Dept. Successively: Tech Asst to the Paris Area Regional Dir, Head of plant & equipment div (Distrb HQ); 1974-1983 Deputy Head of Engrg, covering operations, investment, plant & equipment (from 20 kV & below); now Adviser to the Executive (International Affairs). Laureate Mbr SEE (France), Fellow IEEE, Fellow IEE (England). Chairman of CIRED (1985-1987) (International Conference on Electricity Distribution) Chevalier de L'ordre National du Merite. *Society Aff:* SEE, IEE, IEEE.

Temes, Gabor C
Home: 2015 Stradella Rd, Los Angeles, CA 90024
Position: Prof *Employer:* Dept of Elec Engrg UCLA *Education:* PhD/Elect Engr/Univ of Ottawa; Dipl Phy/Physics/Eotvos Univ; Dip; Ing/Elec Engr/Tech Univ of Budapest. *Born:* 10/14/29. Asst Prof at Techn Univ of Budapest 1952-56; proj engr Measurement Engrg Ltd 1957-59; Dept Head Bell-Northern Res 1959-64; Res Group Leader Stanford Univ 1964-66; Corp Cons Ampex Corp 1966-69; with UCLA 1969- , currently as Prof Assoc Ed Journal of Franklin Inst; former Ed IEEE Trans on Circuit Theory. Fellow IEEE. Co-winner of 1968 and 1981 IEEE-CAS Darlington Award. Co-author of several books. Other honors: Outstanding Engineer Merit Award, Inst. for the Advancement of Engineering, 1981, Western Electric Fund Award, ASEE, 1982; IEEE Centennial Medal, 1984; Andrew Chi Prize Paper Award, IEEE-IMS, 1985; Education Award, IEEE-CAS, 1987. *Society Aff:* IEEE.

Temple, Derek A
Home: 18 Heathfield, Royston, S Cambs SG85BW UK
Position: Dir Gen (Ret) *Employer:* RTZ Group *Education:* PhD/Metallurgy/Univ of Cambridge; BS/Metaullurgy/Univ of London-Imperial Coll *Born:* 06/08/23 Harrow, UK. Technical Education Royal Sch of Mines, London and Trinity Hall, Cambridge. Works Metallurgist Airspeed Ltd, Portsmouth (1945/6) DSIR Research Student, Cambridge (1946/49) Nuffield Foundation Travelling Scholar (1949/50) section leader, Research Dept, ISC Ltd Avonmouth (1950-57) Chief Metallurgist ISP Ltd, Avonmouth (1957-70) Dir & Chief Exec ISP Ltd; Avonmouth (1970-1982), Dir, RTZ (Bristol) Ltd (1981-1982). Responsible for development and exploitation of Zinc-Lead Blast Furnace process. Pres, Inst of Mining and Metallurgy (1979-80). AIME Extractive Metallurgy Lecturer (1980), Dir-General, Zinc-Lead Development Assoc, London (1982-86), Consultant 1986-Date, Fellow, Fellowship of Engrg (1981), Freeman, City of London (1984), Liveryman, Worshipful Co of Engrs (1984). *Society Aff:* TMS(AIME), IMM, IOM

Templeton, Elmer E, III
Business: 200 Union Sq Bldg, Suite 6, Marietta, OH 45750
Position: Pres Elmer E Templeton & Assocs. *Employer:* Elmer E Templeton. *Education:* MS/Petro & Natural Gas/Penn State; BS/Petroleum/Maritta College. *Born:* Oct 18, 1934. Taught many short courses for Petroleum Engrs in Ohio & West Va & Fedl Power Comm, Wash D C; cons for various major & independent producers. V Chmn St of Ohio Oil & Gas Bd of Review. Indus Fellowship Marathon Oil Co summer 1972. Phillips Petro Co 1961-. Fac Petroleum Engg Dept, Marietta College 1967-78, Chrmn of Dept 1973-78. Reg P E 7573 Okla; E-44089 OH Soc of Prof Well Log Analysts; Mbr Soc of Petro Engrs AIME; Outstanding Educ Educator of Amer 1975; Pi Epsilon Tau. *Society Aff:* SPE, SPWLA.

Templeton, Robert E
Home: 12718 Old Oaks Dr, Houston, TX 77024
Position: Project Control Manager *Employer:* M.W. Kellogg *Education:* MBA/Mgt/NY Univ; MSCE/Civil Engg/Carnegie Mellon; BSCE/Civil Engg/Carnegie

Templeton, Robert E (Continued)
Mellon. *Born:* 6/21/31. Joined M.W. Kellogg June 1954; worked as Design Engr 1954-58, contract Status Analyst 1958-60, Asst Mgr contract Status 1960-63; Proj Engr 1963-66, mgr contract Status 1966-68, Mgr Sales Forecasting 1968-71, Mgr Sales & Market Forecasting 1971-73, Veture Analysis Mgr 1973-74, Mgr Analysis & Methods 1974- 76, Mgr Proj Cost Servs Apr 1976- May 1981, Mgr Cost Mgmt Services May 1981- May 1986, Project Control Manager, May 1986- . American Assn of Cost Engr: "Chairman Association Standards and Recommended Practices Committee." Amer Assn of Cost Engrs: Fellow Mbr 1977: PPres 1973, Pres 1972, Admin VP 1971, Tech VP 1970, Dir 1968-69, Chmn 1st Certification Bd 1976-79. AACE Award of Merit (1977), Award of Recognition 1980. Sigma Xi: VP, Kellogg Br 1976-77. Engrg Manpower Comm 1968-. Reg PE: NY. CCE-Certified Cost Engr AACE. Fonn Villas Civic Assn Pres 1976. *Society Aff:* AACE, PMI, AMA, AHSM

Tendick, John Phillip
Business: 17260 W North Ave, Brookfield, WI 53005
Position: President. *Employer:* J P Tendick & Assocs Inc. *Education:* BS/ME/Univ of MO. *Born:* 8/29/22. Mbr Pi Tau Sigma. Joined G E on Student Course; transferred to AEC until end of war; then Westinghouse for 13 yrs, becoming Branch Mgr at Madison & Milwaukee; started J P Tendick & Assocs 1957 & engage in Electro-Mech Engrg under that name. Both firm & J P Tendick are registered with the St of Wis. Presently Mbr CEC. Married to Ruth M; 5 children. Enjoy hunting & fishing. *Society Aff:* ACEC.

Tenenbaum, Michael
Home: 1644 Cambridge, Flossmoor, IL 60422
Position: Dir. *Employer:* Inland Steel Co. *Education:* PhD/Met/Univ of MN. *Born:* 7/23/13. Was Pres Inland Steel Co; 1971-1978 VP Res 1966-68 & VP Steel Mfg 1968-71. Chrmn & Dir of several Inland subsidiaries. Dir of Continental-IL Bank. Former VP & Dir AIME; P Pres TMS-AIME; currently mbr Iron & Steel Inst (Intl & Am); the Metals Soc; AIME-ISS and TMS; AISE; SAE; Western Soc of Engrs. Recipient of Univ of MN Outstanding Achievement Award 1967; AIME Howe Meml Lectureship 1969 ASM/TMS-AIME Disting Lectr in mtls 1976; Fellow AIME-TMS 1970; Fellow 1970 & Disting Life Mbr 1974 ASM; Mbr NAE. Received DSc Northwestern Univ 1974; AIME Fairless Award 1974; Disting Mbr ISS-AIME 1975; WSE Washington Award 1977: ASM Medal Advancement of Research; AIME Honorary Member 1979; Metals Society (British) Bessemer Medal 1980.. *Society Aff:* ASM, AIME.

Tennant, Jeffrey S
Business: Dept of Ocean Engg, Boca Raton, FL 33431
Position: Prof *Employer:* FL Atlantic Univ. *Education:* PhD/ME/Clemson Univ; MS/ME/Clemson Univ; BS/ME/Clemson Univ. *Born:* 7/10/41 Began profl career as res engr for Pratt & Whitney. Entered teaching in the field of Ocean Engg at FL Atlantic Univ. Subsequently taught mech & aerospace engg at the Univ of TN, Knoxville. Returned to FAU in 1975 to teach Ocean Engg Dept. Chairman 77-81. Interior Dean of College 82-83. Active consltg practice in Marine engrg and gen mech engrg. *Society Aff:* ASME, ASEE, NSPE.

Tennant, Otto A
Home: 6176 Terrace Dr, Johnston, IA 50131
Position: Pres *Employer:* Tennant Engrg. *Education:* MA/Eco/Drake Univ; BS/Gen Engrg/IA St Univ *Born:* 09/12/18 Before World War II, worked as test engr for Gen Elec. Served with Army Air Corps 1942-1945 as Meteorological and Navigator Ofcr. Was Application Engr for Westinghouse until 1967. Until 1982 has held various Marketing positions with Iowa Power. At present, is pres of a consulting firm. Have been heavily involved in Company's and also trade association's conservation efforts. Registered - State of IA. Regional VP of Illuminating Engrg Society (1970-1972); Pres-IA Engrg Society (1972-1973); NSPE-PEI Vice Chrman (1974-1976); NSPE Vice Pres (1977-1979); Pres-Elect (1980-1981); NSPE Pres (1981-1982); IA Engrg Society Ansen Master Award 1978, Dist Service Award 1983, IA State Bd of Engrg and Land Surveying Examiners (1983-1986) Chrmn 1984-1985. *Society Aff:* NSPE, IEEE, ASHRAE, IES

Tenney, Mark W
Business: 744 W Washington St, South Bend, IN 46601
Position: President. *Employer:* TenEch Engrg, Inc. *Education:* ScD/Civil Engr/MA Inst of Tech; SM/Sanitary Engr/MIT; SB/Civil Engr/MIT. *Born:* 12/10/36. Native of Chicago, IL. Design engr with Greeley and Hansen, 1960-61. Asst Prof, 1965-67, and Assoc Prof, 1967-73, Univ of Notre Dame. Co-founder, TenEch Engrg, Inc, 1969; full-time with TenEch in 1973. Currently serve as President of firm. Diplomate Amer Acad of Environmental engrs. WPCF Harrison Prescott Eddy Medal, 1973. Mbr Sigma Xi and Chi Epsilon. Brigadier Gen, Corps of Engrs (USAR). *Society Aff:* ASCE, ACEC, WPCF, AWWA, SAME.

Terman, Frederick E
Home: 445 El Escarpado, Stanford, CA 94305
Position: Vice President & Provost Emeritus. *Employer:* Stanford University. *Education:* AB/EE/MIT; Engr/EE/Stanford; BA/Chem Engg Stanford *Born:* June 7, 1900. Hon ScD Harvard 1945, British Columbia 1950, Syracuse 1955, So Methodist Univ. 1977 Instr to Prof 1925-37; Exec Head Elec Engrg 1937- 45; Dean of Engrg 1945-58; Provost 1955-65; V Pres 1959-65, Emeritus 1965- , all at Stanford Univ. Pres SMU Fnd for Sci & Engrg 1965-74; Org & Dir Harvard Radio Res Lab 1942-45. Author 'Electronic Measurements' (with J M Pettit) Radio Engrs Handbooks 1943; Electronic & Radio Engg 1932, 1937, 1947, 1955. Spec interest in electronics & educ. ASEE Hall of Fame 1968; Hon Mbr ASEE 1966; ASEE Lamme Award 1964. Mbr NAE & NAS. Pres IRE 1941. IEEE Educ Medal 1956; IEEE Medal of Honor 1950; IEEE Founders Award 1962. Decorated by British Govt 1946; US Medal for Merit 1948; Sigma Xi Pres 1975; Republic of Korea Order of Cvl Merit 1975; Natl Medal of Sci 1976. *Society Aff:* APS. ΦBK, ΣΞ, APS, ASEE, IEEE.

Terman, Lewis M
Business: IBM Research Center, P.O. Box 218, Yorktown Heights, NY 10598
Position: Research Staff Member. *Employer:* IBM Research Center. *Education:* PhD/EE/Stanford; MS/EE/Stanford; BS/Physics/Stanford. *Born:* 8/26/35. Native of Palo Alto, CA. Employed at IBM Watson Res Center 1961- , as Res Staff Mbr & Mgr; worked on sys logic design, magnetic & semiconductor memory & MOS technology & circuits. Recipient IBM Outstandinging Invention Awards 1971, 1974 & 1976 & Outstanding Contrib Award 1968. Active IEEE; elected Fellow 1975; Former Mbr Solid State Circuits Council, chrmn Solid State Circuits Ctte 1976-79 & Tech Prog Ctte ISSCC; Ed IEEE Journal of Solid-State Circuits 1974-77. Vice Chmn 1982 Int Solid State Circuits Conf. *Society Aff:* IEEE, ΣΞ, AAAS.

Terry, Sydney L
Home: 58 Moross Rd, Grosse Pointe Farms, MI 48236 *Education:* MAE/Automotive Engg/Chrysler Inst; BS/Mechanical Engg/Stanford Univ. *Born:* Apr 1920. With Chrysler Corp 1941- ; V Pres Engrg 1968, V Pres Safety & Emissions 1970-72, V Pres Public Respon & Consumer Affairs 1973-80. Was respon for contact with fedl govt on product matters in regulated areas with emphasis on engrg aspects of such regulations; also respon for overseeing energy conservation & consumer affairs matters. Holder of several pats in automotive field. Dir of Amer Natl Standards Inst & Natl Safety Council. P Dir of Soc of Auto Engrs. Rr of Detroit-Macomb Hospitals Assn. Fellow Engrs of Detroit. Mbr Tau Beta Pi & Phi Beta Kappa. Enjoy golf, tennis & music. *Society Aff:* SAE, NSC.

Tesmer, Dale L
Home: 732 Marseilles Ave, Upper Sandusky, OH 43351
Position: Dir of Field Services. *Employer:* Floyd Browne Assoc, Limited. *Education:* BS/Mining/MO School of Miners. *Born:* 9/13/24. Native of Wisconsin. Served with US Army 1943-45. Formerly employed as Asst Plant Engg for US Gypsum Co. Assoc as Job Engg with F H McGraw & Co for the construction of power plant,

Tesmer, Dale L (Continued)
Cassville, VI; AEC plant, Paducah, KY; Reynolds Aluminum sheet mill, McCook, IL; rod mill for American Steel & Wire Co at Cleveland and Pittsburgh Steel Co billet mill; trubine installation for Westinghouse in Philadelphia. Presently supervise construction of municipal, state, federal, commercial and industrial projects for Floyd Browne Assoc, Limited. Hold Prof Engrs License for states of OH, MI, IN, and MO. *Society Aff:* NSPE.

Tessier, Gaston A
Business: 116 S Cove Rd, Burlington, VT 05401
Position: President. *Employer:* Ventess Inc. *Born:* Nov 1922. Postgrad Stevens Inst Tech; BS St Michael's Coll. 194365 Dev Engr Mgr; Div Prod Mgr; Operations Mgr, Genl Mgr for Genl Foods Corp; initially spec in div & prod of field rations, dehydrated foods, instant coffee, as well as new applications of dehydration, spray drying & cryogenic tech; subsequently spec on plant const, marketing, operations & mgmt. 1965-70 V Pres Corp Marketing Internatl & Prod Dev for Basic Vegetable Prods Inc. 1971-72 Chmn Ch Exec Officer Healthco Inc. 1972-75 Exec Dir & developer of health care delivery orgn, Univ Health Center Inc. Co-founder & Pres Ventess Inc 1973 - venture capital, mgmt & real estate dev. Active in educ, community & civic dev. Enjoys sailing. *Society Aff:* AIChE.

Testa, Rene B
Business: Dept of Civil Engg, New York, NY 10027
Position: Prof. *Employer:* Columbia Univ. *Education:* EngScD/Civil/Columbia Univ; MS/Civil/Columbia Univ; BEng/Civil/McGill Univ. *Born:* 5/30/37. A faculty mbr at Columbia Univ since 1963; Dir of the Carleton Lab since 1965, a facility for testing of materials, structural elements, and structures. Lic PE in NY and NJ. Consultant to govt and industry on problems in structural design, structural failures, and field instrumentation of bridges and other large structures. President of Tegja Inc. a consulting engineering firm. *Society Aff:* ASCE, ASME, ASTM, SSRC.

Tewinkel, G Carper
Home: Rt 1, Box 1504, La Grande, OR 97850
Position: Ret. (Civil Engr) *Education:* MCE/Civil Engg/Syracuse Univ; Bs/Mech Engg/WA State Univ. *Born:* Jan 1909; native of Spokane Wash. Served US Dept Agriculture prior to 1941. Served Coast & Geodetic Survey (now NOAA Natl Ocean Survey) Wash D C 1941- 72; specialty photogrammetric charting. Led dev of C&GS sys of analytic Aerotriangulation. Pres Amer Soc for Photogrammetry 1960; Ed 'Photogrammetric Engrg' 1965-75; Fairchild Award 1967; Dept Commerce Gold Medal Award 1967. pres Comm on Aerotriangulation of Internatl Soc for Photogrammetry (ISP) 1964-68; Genl Secy ISP 1968-72; V Pres 1972-76. Fellow & Life Mbr ASCE; Life Mbr Amer Congress on Surveying & Mapping & ASP. Retired in 1972. *Society Aff:* ASP, ASCE, ACSM

Tewksbury, Dennis L
Home: 11 Harvard St, Concord, NH 03301
Position: Tech Dir of Structural Services *Employer:* Kimball Chase Company, Inc. *Education:* BSCE/Civil Engg/New England College. *Born:* 10/26/43. Attended Univ of NH Grad Sch. Former Part-time Instr in Struct Engrg at New England College. Tech Dir of Structural Services, Kimball Chase Company, Inc. Reg P E: NH, MA, FL, ME, & VT. P Vice Chmn City Conservation Comm; Church Deacon. Mbr ASCE & NSPE. Positions held in NH Sect ASCE: Treas, Secy, V P, Pres-Elect, Pres, PPres, various ctte chmnships & newsletter ed. P Secy ASCE New England Council; P Chmn Natl ASCE Ctte on Younger Mbrs and Ctte on Student Services. NH Young Engrg of Yr-1978. First recipient of the NH Young Engr of Year Award, PPres of the Greater Concord Chamber of Commerce, Mbr ASCE Natl Education Division Executive Ctte. Pastimes: golf, fishing, boating, skiing, and hiking. *Society Aff:* ASCE, NSPE.

Thackray, Arnold W
Business: Univ of Pa, 215 South 34th St, Philadelphia, PA 19104
Position: Prof; Hist & Soc of Sciences. *Employer:* University of Pennsylvania. *Education:* PhD/natural Sciences/Cambridge Univ; BS/Natural Sciences/Bristol Univ. *Born:* 7/30/39. Gas Plant Manufacs Dev Council Scholarship to Bristol Univ: Hd of R/D for Robert Dempster & Sons gas & chem engrs, Elland Yorks. Fellow of Churchhill College Cambridge: MA & PhD. Visiting Lectr Harvard Univ. With Univ of PA 1968- , currently director of the Beckman Center for History of Chemistry, Editor of Osiris, a Research Review of the Hist of Sci and its Cultural influences; and Joseph Priestly Professor of Hist & Soc of Sci. Fellowship - John Simon Guggenheim Mem Found 1971, Ctr for Advanced Study in Behavioral Sci 1974-1983 & All Souls College Oxford 1977. Natl Lectr Sigma XI 1976; 1978 & 1979; Council, Hist of Sci Soc, 1976-85; Soc for Hist of Tech 1974-77; 78-Council, Soc for Social Studies of Sci 1975-80-81;Publs incl: Atoms & owers 1970, John Dalton 1972, Sci & Values 1974, & Toward a Metric of Sci 1978, Gentlemen of Sci 1981, Chemistry in America 1984. *Society Aff:* AAAS, HSS, AHA, AIChE

Thackston, Edward L
Business: Box 133 - Station B, Nashville, TN 37235
Position: Prof & Chrmn, Dept of Civil & Environ Engrg *Employer:* Vanderbilt Univ *Education:* PhD/Env & Water Resources Eng/Vanderbilt Univ; MS/Sanitary Eng/Univ of IL; BE/CE/Vanderbilt Univ *Born:* 04/29/37 Native of Lebanon, TN, and was Lebanon city engr and a design engr for the City of Nashville. He has taught at Vanderbilt since 1965, except for 1972-1973, when he served as Staff Asst for Environmental Affairs to TN Governor Winfield Dunn and has been department chairman since 1980. In 1974, he was named "Conservationist of the Yr" by the TN Conservation League, for promoting environmental policy within TN state government. He has done consltg and research in environ policy, industrial waste treatment, thermal pollution, aeration and mixing in streams, and effects of impoundments on water quality. *Society Aff:* ASCE, AEEP, AWWA, WPCF

Thaemert, Ronald L
Home: 3637 N. County Rd 27E, Bellvue, CO 80512
Position: Sr Water Res Engrs *Employer:* Self *Education:* BCE/Civil Engr/Colo. St Univ; MSCE/Civil Engr/Colo St Univ; Ph.D/Civil Engr/Colo St Univ *Born:* 2/13/41 Native of Colorado. Served in US Navy Civil Engineer Corps 1964-1968. President Water & Environment Consultants Inc 1972-1977. Water Resource Consultant to Indonesia, 1977-1979. Deputy Project Manager for German Engineering Consortium for Water Resource development in Saudi Arabia 1979-1980. Associated with IECO and Agricultural Water Management Practice 1980-1983. Currently conslt to agricultural water users and municipalities for water mgmt and dev, and Pres of Applied Computer Processes, Inc. *Society Aff:* ASCE, USCID, USNI, NRA

Thakur, Ganesh C
Business: Gulf R & D, P. O. Box 36506, Houston, TX 77036
Position: Director *Employer:* Gulf Oil Co *Education:* PhD/Petrol Engrg/PA St Univ; MA/Math/PA St Univ; MS/Petrol Engrg/PA St Univ; MBA/Bus Admin/Houston Baptist Univ; BS/Petrol Engrg/Indian Sch of Mines *Born:* 03/08/47 Native of village Dumaria, Godda, S.P., Bihar, India. Taught in Dumka Zila Sch. Received Gold Medal in BS Engrg from Indian Sch of Mines in 1970. Came to USA in 1970. Worked as a cons with Scientific Software Corp in Denver from 1973- 76. Worked as a sr res engr with Amoco Production Co from 1976-77. Working with Gulf Oil Co from Aug 1977. Assumed current responsibility as Dir of Enhanced Recovery Operations with Gulf Res & Dev Co (subsidiary of Gulf Oil Co). Responsibilities include managing the technical srvcs activities in the Enhanced Recovery area for Gulf Oil Co. Has published several technical papers and is a technical editor of Society of Petrol Engrs. Reg PE in OK & CO. Enjoys travelling and social activities. *Society Aff:* AIME

Thau, Frederick E
Business: 140 St at Convent Ave, NY, NY 10031
Position: Assoc Dean. *Employer:* City College of NY. *Education:* DEnggSc/Control Theory/NYU; MEE/Communication and Control/NYU; BEE/Elec Engg/NYU.

Thau, Frederick E (Continued)
Born: 12/2/38. Born in the Bronx, NY on Dec 2, 1938. Received the BEE, MEE and Engg ScD degrees in 1959, 1961, and 1964, respectively, from NYU, NY. From 1959 to 1965, was a mbr of the technical staff at Bell Labs, and from 1965 to 1969, was a Sr Staff Scientist at Singer-Gen Precision. Since 1969, has been with the City Univ of NY; now Assoc Prof of Elec Engg and Assoc Dean for Grad Studies in Engg. He has been the recipient of a number of res grants from NASA, NIH, and NSF involved with applications of modern feedback control theory to attitude control, microorganism growth, and process control problems. Also completed a number of summer res projs involving applications of modern control theory to power processing circuit design at Bell Labs. Current res interests are in the areas of implementation of suboptimum filters and controllers. *Society Aff:* IEEE, ΣΞ, ΤΒΠ, HKN.

Thaxton, Guy W
Home: 2800 West Main Street, Tupelo, MS 38801
Position: Retired. *Employer:* Retired. *Education:* BSEE/Elec Engg/MS State Univ. *Born:* Sept 1894 Lake, Miss. Taught as Asst Prof of EE at Miss State & at Georgia Tech. Distribution Apparatus Engr with Westinghouse in Atlanta. Pres & Genl Mgr Mid South Public Serv Co. Div Engr Miss Div, Tenn Valley Authority. Regional Engr, Ch Engr & Ch Engrg Div 1935-45 Rural Electrification Administration Washington DC & St Louis Mo. Rural Dist Mgr, General Cable Corp., New York, N.Y. and Washington, DC. Regional Power Supply Coordinator Defense Electric Power Administration, Interior Dept 1951. Power Engr & Ch Electric Power Branch, Atomic Energy Comm, Washington DC 1951-1960 from which retired July 31, 1960. Fellow IEEE; Mbr Tau Beta Pi & Phi Kappa Phi. Reg PE in Mo. *Society Aff:* IEEE.

Thayer, Richard N
Home: 250 Chatham Way 525, Mayfield Heights, OH 44124 *Education:* BS/Physics/Univ of Pittsburgh. *Born:* June 1907 Pittsburgh Pa. Electric lamp design work, principally fluorescent, at Nela Park Labs GE Co 1928-69 (retired). Dev Engr 1928-42; Group Leader 1942-45; Supervisor design & quality 1945-55; Mgr discharge lamp design 1955-63; Mgr fluorescent lamp design 1963-69. Author sev tech papers on fluorescent lamps. Fellow IES. P Chmn Ohio Sect APS; Trustee Cleveland Citizens League 1958-64. *Society Aff:* IES.

Thayer, Royal C
Business: Municipal Construction div (WH-547), 401 M St, SW, Washington, DC 20460
Position: Southern Program Manager. *Employer:* US Environmental Protection Agency. *Education:* BS/Mech Engr/VA Poly Inst. *Born:* Jan 1920. Native of West Va & Va. Served as Artillery Cpt 3rd Army Europe WWII. Designed sight Bofor's 40mm gun. 22 years with Hampton Roads Sanitation District, 10 years responsible charge of all planning, design, const & opers. Regional Engr VaSHD. With US EPA 1969- . Currently responsible for portion of $18 billion const grant prog in 13 Southern states through 2 regional offices. Principal engr coordinator of 1973 & 74 Natl Needs Surveys. Mbr Bd of Control WPCF, & Quarter Century Operators Club. EPA Special Achievement Award 1974 & 1980,& Silver Medal 1975. *Society Aff:* WPCF.

Thayer, Stuart W
Business: Lykes Ctr 300 Poydras, New Orleans, LA 70130
Position: Vice President Engineering. *Employer:* Lykes Bros Steamship Co Inc. *Education:* BS/Mar Transp/MIT. *Born:* 4/27/24. BS from MIT. Native of Oxford, MA. Joined Lykes in 1948, in charge of Ship Replacement Prog 1954, VP Engg 1970. Respon for design & const of 40 cargo ships 1958-87, including 3 revolutionary barge carrying vessels with 2000-ton elevator on stern. Have been active on tech cttes of Council of American Ship Operators, American Inst of Merchant Shipping, Amer Merchant marine Inst & American Natl Metric Council. Active in SNAME and have served on numerous ctte. Presently a VP of The Soc. Past mbr Maritime Transportation Res Bd of Natl Acad of Sci. Author of 20 tech papers. 1973 LA Maritime Man of the Yr. Enjoys fishing. *Society Aff:* SNAME.

Theakston, Franklyn H
Business: School of Engrg, Guelph, Ontario N1G 2W1
Position: Professor & Vice President. *Employer:* University of Guelph. *Education:* BSc/Math & Physics/Acadia Univ; BE/Civil Engg/NS Tech College; ME/Mech/NS Tech College. *Born:* Sept 1919 Halifax Nova Scotia. 4 years as officer in Royal Canadian Engrs WWII. Prof in Agri & Civil Engrg Univ of Guelph 1950- , now full Prof. Since 1956 specializing in Farm Buildings, Surveying, Struct Design. Awards: Metal Building Manufacturers Award, ASAE 1963; Fellow ASAE, EIC & CSAE, Acting Dir Sch of Engrg, Univ of Guelph 1965-67; Vice Pres Cons Engrs, Morrison, Hershfield, Theakston & Rowan Ltd. Main area of res: snow, wind & ice, & agricultural engrg. P Chmn North Atlantic Region ASAE; P Chmn Wellington- Waterloo Br EIC; P Pres Canadian Soc of Agricultural Engrs. *Society Aff:* CSAE, ASAE, EIC, ADEO, CAUT.

Theiss, Ernest S
Home: 7419 Route 43, Kent, OH 44240
Position: Engineering Consultant. *Employer:* Engg Scientific Translation Assocs. *Education:* MS/Mech Engg/Case Inst of Tech; BS/Mech Engg/Case Inst of Tech. *Born:* May 16, 1910 Detroit. Goodyear Engr 1932-37; Case Graduate Asst 1937-39; Duke Univ Asst Prof ME 1939-46; Davey Compressor Co Ch Engr 1946-56; NRM Corp Mgr Rubber Machinery, Extruder Design 1956-66, Engrg Cons Extruder Div 1966-75; Engg Scientific Translation Assocs Pres 1975- . Author: USAF Air Compressor Manuals; Goodyear Aircraft Manuals; NRM Manuals Plastic Extruders, Rubber Extruders; Twin Screw Extruders. ASME: Vice Pres Region V 1951-53; Secy Region IV, Secy, Region V, 1949-51; 1945-46; Life Fellow. Akron Univ Adv Bd, Engrg 1955-76. Mbr NSPE & SPE ASME-100 Anniversary Medal, 1980. Wife's Name-Irmgard. *Society Aff:* ASME, NSPE, SPE.

Themelis, Nickolas J
Business: Henry Krumb School of Mines, Columbia Univ, New York, NY 10027
Position: Professor of Mineral Engineering *Employer:* Columbia University, Themelis Assoc. *Education:* PhD/Metall/McGill Univ; BEng/Chem Engg/McGill Univ. *Born:* 4/25/33. Res with Strategic Matls Corp, Mgr Engg Div of Noranda Mines Ltd 1962-72. Joined Kennecotte in 1972, as VP, Res and Engg, Kennecott Minerals Co. Was apptd Corp VP of Technology in 1979. Apptd. Prof of Mineral Eng., Columbia Univ 1980. Co-author of textbook, *Rate Phenomena in Process Metallurgy*, 80 tech papers, pats in extractive metallurgy including Noranda Continuous Smelting process. AIME Extractive Metallurgy gold medals 1968, 1973, Extractive metallurgy Lect, 1984 McConnell Award, 1987 I.M.M. Lecturer. *Society Aff:* NAE, TMS, MMSA, CIM

Theodore, George J
Business: 1011 Nicollet Mall, Minneapolis, MN 55403
Position: Sr VP *Employer:* Setter, Leach & Lindstrom, Inc *Education:* BS/EE/Cleveland State Univ *Born:* 9/29/36 Served as Assistant Chief Electrical Engr for Osborn Engrg in Cleveland, OH before taking the position of Dir Electrical Engrg with Setter, Leach & Lindstrom, Inc. Architects, Engrs and Planners, Minneapolis, MN. Presently serves as Sr VP and Treasurer of that firm as well as mbr of the Bd of Dir. Served as Treasurer of the Minneapolis Chapter of the National Society of Professional Engrs in 1976, currently a mbr; as well as mbr of the Illuminating Engrg Socity, the Inst of Electrical and Electronic Engrs, Tau Beta Pi and Professional Services Management Assoc. Resigtered Professional Engr in over 20 states. *Society Aff:* NSPE, IEEE, IES, ΤΒΠ

Theodore, Louis
Business: Manhattan College Pkwy, Riverdale, NY 10471
Position: Prof. *Employer:* Manhattan Coll. *Education:* EngScD/ChE/NYU; MChE/ChE/NYU; BChE/ChE/Cooper Union. *Born:* 4/19/34. Dr Theodore joined the facul-

Theodore, Louis (Continued)
ty at Manhattan Coll with the rank of Instructor in the fall of 1960. He has taught courses in Transport Phenomena, Kinetics, Statistics, Mathematics for Chem Engrs, Air Pollution and Its Control and Hazardous Waste Incineration; has served as a lecturer in Transport Phenomena for industry and education here and abroad; and has regularly presented invited lectures and seminars on air pollution control equipment and hazardous waste management. He has consulted for several industrial cos in the field of computer applications and environmental control, participated as res consultant on numerous EPA sponsored proj, served as a reviewer for publishing cos, and supervised NSF environmental projs. *Society Aff:* ΦΛΥ, ΣΞ, ΤΒΠ, ROYAL HELLENIC SOC, AMERICAN CHEM SOC, AMERICAN SOC FOR ENGG EDUCATION, AIR POLLUTION CONTROL ASSOC.

Theofanous, Theofanis G
Business: Chemical and Nuclear Engrg, Santa Barbara, CA 93106
Position: Prof and Dir, Cntr for Risk Studies & Safety. *Employer:* Univ CA, Santa Barbara *Education:* BS/Che/Natl Tech Univ Athens, Greece; PhD/ChE/Univ of MN *Born:* 05/21/42 1969-1985, Purdue Univ. Prin res interests in multiphase processing and application to chem and nuclear reactor safety. Extensive indust and govern consltg. 1985-present, Univ of CA, Santa Barbara. *Society Aff:* AIChE, ANS, ΣΞ

Theros, William J
Home: 232 E Roseville Rd, Lancaster, PA 17601
Position: VP Engg. *Employer:* Culbro Corp. *Education:* BS/Mech Engr/Univ of Miami. *Born:* 8/30/25. Native of Lancaster, PA. Navigation Instr - WWII and Korean Conflict - 1st Lt. Accounting & Bus Courses - Franklin & Marshall College - Springfield State College - Trinity Univ. Completed courses in Toastmasters Intl and Mgt Services Inc. Licensed Machinist - Armstrong Inc (1941-43). Plant Engr - RCA (1952-54 specializing in power plant facilities) Culbro Corp - 1954 to present. 1954-1972 Asst Chief Engr; 1972-1976 Chief Engr. 1976 to present - VP - Engg. Job responsibility: Engg all plant expansions, renovations and maj maintenance for all nine div. Holds PE Licenses in six states. *Society Aff:* ASME, NSPE, AIPE.

Thieblot, Armand J
Home: 1675 Fountainhead Rd, Hagerstown, MD 21740
Position: President. *Employer:* A J Thieblot & Assocs. *Born:* Started Aircraft Engrg in Paris, France 1923 joined Breguet as Junior Engr. Then Bernard. Came to US. Co Designer Columbia Air Liners 1928. Fokker Aircraft 1929 Stress Engr Genl Aviation, Asst Dir of Res. Fairchild 1923, Ch of Stress 1937, Ch Engr to 1950. Pres Thieblot Aircraft, Dir Vitro Corp of America. 1960 Fairchild, Dir of Engrg 1962 Pres Vulcana Corp. Now Engrg Cons. Reg PE in state of Md. Was also PE in DC. Honorary Fellow SAWE.

Thiele, Ernest W
Home: 1625 Hinman Ave, Evanston, IL 60201
Position: Retired. *Education:* ScD/Chem Engg/MIT; LLD/Hon/Notre Dame; MS/Chem Engg/MIT; BS/Chem Engg/Univ of IL; AB/Humanities/Loyola (Chicago). *Born:* 12/8/95. Student Engr Peoples Gas Light & Coke Co 1920-23. Chem Engr Standard Oil Co, Ind 1925; Asst Dir Res 1935; Assoc Dir 1950. Visiting Prof, Chem Engrg Notre Dame 1960-70. Engr, Trail, BC Heavy Water Plant Manhattan District 1942-43. Mbr US Air Force Proj, Nuclear Propulsion of Aircraft 1948. Con Joint Congressional Ctte on Atomic Energy 1949. Fellow American Inst Chem Engrs, Founders Award 1966; Mbr American Chem Soc. Mbr, Natl Acad of Engrg, 1980. Various tech papers. 28 patents. *Society Aff:* AIChE, ACS, AIC, ΣΞ.

Thielman, David J
Home: 4281 Fireside Dr, Williamsville, NY 14221
Position: Consulting Civil Engineer *Employer:* Self *Education:* BSCE/Civil Engr/Univ of Detroit. *Born:* 5/6/40. Native of Buffalo, NY. PE in 4 states. June, 1960-June, 1971 - NY State Dept of Transportation, including positions of Reg Computer Liaison Engr, Project Engr for Design and Construction and Reg Traffic Design Engr. June, 1971 to Present - Consulting Engineer - Project Manager for such projects as highways, sewage and water works, dams, marinas, and land development. Has served as Pres Western NY ABCD and Buffalo ASCE, and member of ASCE Surface Drainage Group and Highway Geometric Design Cttes. Active in Infrastracture funding program. Also member of Rotary International. *Society Aff:* ASCE, ITE, NSPE, ABCD.

Thielsch, Helmut
Home: 140 Shaw Ave, Cranston, RI 02905
Position: President *Employer:* Thielsch Engrg Assoc, Inc. *Born:* 11/16/22. Auburn Univ 1943; Graduate studies at Univ of Mich & Lehigh Univ, Pres Thielsch Engrg Assoc, Inc since 1984; Vice Pres ITT Grinnell; with ITT 1954-1983. Prior engrg responsibilities with Allis Chalmers, Black, Sivalls & Bryson, Welding Res Council, Lukens Steel, Eutectic Welding Alloys. Has served as Cons to large number of utilities, paper mills, chem plants & others covering failure analyses, materials evaluations & quality assurance. Fellow ASME, ASM, ASNT, Mbr AWS, ASTM, ASQC, ANS, NACE, ASPE, ACS. AWS Adams Lecture Award; RI Engrg Freeman Award; Author of book, ' Defects & Failures in Pressure Vessels & Piping.' (Reinhold 1956) & of 150 engrg articles covering various aspects of failure analysis, piping, pressure vessel materials, quality assurance, etc, & of chapters in Piping Handbook & handbooks pub by ASME, ASM & AWS. Active on various cttes of ASME, ASTM, ANSI, etc. Reg PE in RI, Mass, Me, NJ, Ga, Calif. *Society Aff:* ASME, ASM, ASNT, AWS, ASPE, ANS, ASTM, ACS, NACE, ASQC

Thierfelder, Charles W
Home: 1535 Country Club Dr, Lancaster, PA 17601
Position: Div VP, Prod Assurance *Employer:* RCA Corp. *Born:* Feb 3, 1925 Slater Mo. BSEE Okla Univ; MS Physics Franklin & Marshall; attended Northwestern Univ Business Sch. Served in US Navy during WWII & again 1951-53 attaining the rank of Lt. Joined RCA in 1946 as Prod Dev Engr, became Mgr of Design & Standardizing 1954, promoted to Mgr Black & White Picture Tube Engrg 1956, became Mgr Picture Tube Engrg 1961, appointed Mgr Engrg for the Television Picture Tube Div 1965, appointed Div Vice Pres of Mfg in 1973, became Div Vice Pres of Tech Progs in 1975. Appointed Div VP of Prod Assurance in 1977. Hold sev patents. Mbr IEEE.

Thigpen, Joseph J
Home: 1116 Carey Ave, Ruston, CA 71270
Position: Retired *Employer:* Louisiana Tech University. *Education:* PhD/Mech Engr/Univ of TX; MS/Mech Engr/Univ of TX; BS/-/US Military Acad; BS/Mech Engr/Louisiana Tech Univ. *Born:* 2/4/17. Native of Ruston La. Served in US Army Corps Engrs 1941-47, 27 months SW Pacific Theater, 3 yrs Dept of MA&E USMA. Joined faculty of Dept of Mech Engrg La Tech 1947. Military leave 1951-53, Dept MA&E USMA. Head Dept of Mech Engrg, La Tech 1953-75; Dean Coll of Engrg 1976-1982. Fellow ASME 1971. Pres La Engrg Soc 1973-74. Retired June 1982. *Society Aff:* ASME, ASEE, NSPE.

Thoem, Robert L
Home: 1712 Huntingford Dr, Marietta, GA 30068
Position: Manager, Hazardous Waste Engrg *Employer:* Engrg-Sci *Education:* MS/Sanitary Engrg/Rutgers Univ; BS/Civ Engrg/IA State Univ *Born:* 08/26/40 Native Davenport, IA. Served with USPHS in NYC 1962-65 in water pollution control. With Stanley Conslts from 1966 to 1983. Responsibilities for environ projs including water resources, water pollution control and solid waste mgmt. Hd, Environ Engrg in Muscatine IA hdqtrs 1972-76. Hd, Resource Mgmt 1976-1982 and Hd, Operations 1982-83 in Atlanta, GA branch office. Assoc Chief Environ Engr for Corp 1980-83. With Engrg Sci from 1983-Present in Atlanta, GA branch office (1983-86). Sr Proj Mgr with responsibilities for developing and managing projs in solid and hazardous waste mgmt, water pollution control and other environ areas. Mgr, Hazardous Waste Engrg (1986-present). Pres & Treas IA Engrg Soc Chapter. VChrmn & Treas GA PEPP. Outstanding Service Award GA PEPP 1980. Reg PE

Thoem, Robert L (Continued)
in five states. Non engrg activities: Chrmn Cub Pack, Pres and External Affairs Chrmn Willow Point Homeowners Assoc, Admin Bd Methodist Church, Treas Sunday School Class, Deacon and Elder Presbyterian Church, VP and Pres. PTA. *Society Aff:* NSPE, WPCF, ASCE, AAEE

Thoma, Charles W
Business: Box 50, Arkansas City, KS 67005
Position: Pres. *Employer:* Charles W Thoma & Assoc *Education:* BS/Civ Engg/TX A&M. *Born:* 10/14/27. Reyno AR - Grad Biggers AR High School 1944. 33 mo US Army Corps of Engrs 1st Lt Korea 1951-1952 - TX Hgwy Dept-Bridge Engr 1953-1954. VP Scott Engg Co 1955 1960 Ft Worth TX & Watertown, SD 1960 to 1972 Partner Stitzel & Thoma Architects & Engrs Arkansas City, KS 1973-date. Owner CWT & Assoc. *Society Aff:* ASCE, NSPE, KES.

Thoma, Frederic A
Business: PO Box 8788, Trenton, NJ 08650
Position: Manager of Gear Engineering. *Employer:* Transamerica Delaval Inc. *Education:* BME/Machine Design/GA Inst of Tech. *Born:* March 24, 1925 Oak Park Ill. BME 1949 Ga Tech. US Marine Corps 194346. Gear Designer: Giddings & Lewis, Fond du Lac Wis 1949-51; Warner & Swasey, Cleveland Ohio 1966-67. Gear Engr for De Laval Turbine Inc, Trenton NJ 1951-66, 1967- , positions of Asst Ch Engr, Ch Engr & Mgr Gear Engrg. Liscensed PE in NJ. Mbr: American Bureau of Shipping Gear Adv Panel, Navy/Gear Indus Ctte, ASME Gear Rating Standards Ctte, SNAME. Chmn: American Gear Manufacturers Assn High Speed Gear Ctte, AGMA Marine Gear Ctte, American Petroleum Inst Manufacturers Subctte on High Speed Gears. Received SNAME 'Vice Adm E L Cochrane Award' 1972 & AGMA 'Outstanding Contribution Award' 1972. Guest Lecturer at Pa State Univ 1962, Univ of Wisc 1974, & Texas A & M Univ 1975. Contributed articles to prof journals & author of numerous tech papers. *Society Aff:* AGMA, SNAME, ASNE, NGIC, API.

Thomann, Gary C
Business: Wichita, KS 67208
Position: Assoc Prof *Employer:* Wichita State Univ *Education:* PhD/EE/Univ of KS; MS/EE/Univ of KS; BS/EE/Univ of KS *Born:* 07/23/42 Dr Thomann is an Assoc Prof of Elec Engrg at Wichita State Univ (WSU). He teaches courses in Elec Engrg and is Dir of the Energy Studies Lab at WSU, which conducts wind and elec utility res. He is a Tech Ctte mbr of the KS Elec Utilities Res Prog, which funds utility related research in KS. He is also chrmn of the PE in Education section of the KS Engrg Soc. Dr Thomann has published energy res papers in a variety of tech journals. He received his PhD from the Univ of KS, and before coming to WSU was a principal investigator for NASA. *Society Aff:* IEEE, ASEE, HKN, NSPE, ΦΚΦ, AWEA

Thomann, Robert V
Business: Environmental Engg & Sci, Riverdale, NY 10471
Position: Prof. *Employer:* Manhattan College. *Education:* PhD/Oceanography/NYU; MCE/Sanitary Engg/NYU; BCE/Civ Engg/Manhattan College. *Born:* 9/1/34. NY, NY. Married, seven children. Public Health Service 1956-1966. Technical Dir of DE Estuary Comp Study, 1962-1966. With Manhattan College since 1966. Current res is in fate of toxic substances in Great Lakes. Responsible for the Direction of Environmental Engg & Sci Grad Prog. Ordained Deacon, Roman Catholic Church. Enjoy classical music and camping. *Society Aff:* ASCE, WPF, ASL&O.

Thomas, Arthur N, Jr
Business: 16555 Saticoy St, Van Nuys, CA 91409
Position: VP, Special Programs. *Employer:* ISC/Marquardt Co. *Education:* BS/Aero Engg/Univ of MN; MS/Aero Engg/Univ of MN. *Born:* 9/27/22. Dive bomber pilot USMC, WWII. Chief Engr, Marquardt Corp 1950-1969. Branch Mgr, Propulsion, McDonnell Douglas Corp St Louis l919-1976,Deputy for Air & Missile Defense. Office Asst Secy of the Army 1976-1978. VP, Special Programs ISC/Marquardt. 1978- present,Hobbies - soaring & fishing. *Society Aff:* AIAA.

Thomas, Benjamin E
Home: 418 Sherwood Dr, Webster Grove, MO 63119
Position: Retired. *Education:* MSc/Organic Chem/Rutgers Univ; BSc/Organic Chem/Rutgers Univ. *Born:* 9/3/96. Native of VT. Later lived in NJ. Joined Monsanto Co in June 1917. Worked until 1935 supervisor & production of organic chem at the co's St Louis Plant - its parent plant. From 1935-1943 gen supt of mfg of all chem produced at this plant, 1943-1952 gen mgr of all production at 5 plant locations: St Louis, Boston, Norfolk, Seattle & Nitro, WV. 1952 until ret in 1964. Dir of foreign mfg of all the co's organic chem. Enjoy golf, fishing, gardening & bridge. *Society Aff:* AIChE, ACS.

Thomas, Carcaterra
Home: 4210 Isbell St, Silver Spring, MD 20906
Position: Pres *Employer:* E/A Design Group, Chartered *Education:* MCE/Structural Engrg/CCNY; BCE/CE/CCNY *Born:* 11/11/22 NYC. Served with the Army Corps of Engrs, 1944-1946. Structural engr for several consltg engrg firms prior to entering private practice. Partner of Smislova and Carcaterra, 1961-1966. Principal of Carcaterra and Assocs, Consltg Engrs, 1966-1976. Partner, Chatelain, Samperton & Carcaterra, 1976-1978. Pres, E/A Design Group, Architects & Engrs, 1978-1984. Established individual conslt practice in 1984 as Thomas Carcaterra, P.E. consltg engr. Named Man of the Year in 1968 and 1975 by Consltg Engrs Council of Metropolitan Washington. *Society Aff:* ASCE, ACI, PCI.

Thomas, Carl H
Home: 5845 Hyacinth Ave, Baton Rouge, LA 70808
Position: Prof. *Employer:* LA State Univ. *Education:* PhD/Agri Engr/MI State Univ; MS/Agri Engr/LA State Univ; BS/Agri Engr/Clemson Agri College. *Born:* 4/4/25. Native of Holly Hill, SC. Served Dec 1944-Nov 1946 USM Corp, farmed 1942- 44. IHC Tractor Salesman Summer 1950. Res Assoc in Agri Engr at LSU 1951-53. Asst Prof LSU 1953-58. Assoc Prof 1958-70. Hd of Dept 1969-76. Prof 1970-1983, Prof. Emeritus 1984. NSF Faculty Fellow 1965-66. Sabbatic leave 1964-66. Did Res Cotton Mechanization 1953-69; Horticultural Crops Mechanization Res 1977-1984. Served in all offices BR chapter LES, Pres 1973-74. Served in all offices State LES. Pres 1978-79. Mbr Gamma Sigma Delta, Alpha Zeta, Omicron Delta Kappa and Phi Kappa Phi Honorary Soc's. Served USDA-CSRS, Wash, 1976-77. *Society Aff:* ASAE, NSPE, ASEE.

Thomas, Cecil A
Business: Room 341, 550 W Ford St, Boise, ID 83705
Position: Hydrologist. *Employer:* US Geological Survey. *Education:* BS/Civil Engg/Utah State Univ. *Born:* 5/5/15. Native of ID. Engg with US Bureau of Public Roads 1937-39. Hydrologist and engg with US Geological Survey 1939-79. Author or coauthor of US Geological Survey Publications including Water Supply Papers, Bulletins, Water Resources Investigations, and other reports relating to magnitude and frequency of floods, flow durations and other characteristics of stream flow in ID for use by engg, planners, and managers. Officer, Southern ID Section ASCE, 1973-76. Licensed PE and Land Surveyor, ID. Bishop LDS (Mormon) church 1959-68. Six children and 27 grandchildren. *Society Aff:* ASCE.

Thomas, Charles A
Business: 7701 Forsyth Blvd, St Louis, MO 63105 *Born:* Feb 1900. AB, DSc Transylvania Univ; MS MIT. GM Res, one of the developers of tetraethyl lead. Formed Thomas & Hochwalt Labs. Dir Res Monsanto; Coordinator, Chemistry Manhattan Proj 1943-46; Head, Clinton Labs Oskridge 1946- 48; Exec Vice Pres Monsanto 1947-51; Pres & CEO Monsanto 1951-60; Chmn of Bd Monsanto 1960-65; Chmn Tech Ctte 1965-70; P Pres & Chmn ACS; NAS; founding Mbr Natl Acad of Engrg. 13 Honorary Degrees. Merit Medal from Pres Truman 1946; Indus Res Inst Medal 1947; American Inst of Chemists Gold Medal 1948; Mo Award, Distinguished Serv

Thomas, Charles A (Continued)
in Engrg 1952; Perkin Medal 1953; Priestly Medal 1955; Order of Leopold 1962; Societe de Chimie Industrialle, Palladium Medal 1963. Retired Chmn of Bd, Washington Univ.

Thomas, David G
Home: 113 Morningside Dr, Oak Ridge, TN 37830
Position: Tech Mrg, SDI Fast Track Shield Program *Employer:* Oak Ridge Natl Lab. *Education:* PhD/ChE/OH State Univ; MSc/ChE/OH State Univ; BS/ChE/OH State Univ. *Born:* 7/21/26. Native of St Clairsville, OH, Res Engg for Battelle Meml Inst 1948-50. With Oak Ridge Natl Lab since PhD in 1952. Assumed current responsibility as Tech Mgr for SDI Fast Track Shield Program in 1986. Natl Sci Fdn Sr Post Doctoral Fellow at Cambridge Univ 1960-61. Ford Fdn Prof of Fluid Mechanics (Part-time) at Univ of TN since 1964. Consultant to Aerojet Nucleonics (1962) & TVA (1967- 70). Fellow of AIChE. *Society Aff:* AIChE, ACS, AAAS, Soc. Rheology

Thomas, David R, Jr
Business: 1500 S 50th St, Phila, PA 19143
Position: Chrmn of Bd. *Employer:* Arcos Corp. *Education:* Masters/CE/Cornell Univ; Bachelor/Chemistry/Cornell Univ. *Born:* 1/10/15. Spent entire profl career with Arcos Corp, Phila, PA, serving as Dir of Res and Engg 1938-1955, VP 1946-1955, Pres 1956-1976, 1978-1979, currently Chrmn of the Bd. Active in Am Welding Soc, serving as its Pres 1960-1961 and Treas 1966- 1969. Currently active in technical committees of the Am Welding Soc, Am Soc for Metals, Metal Properties Council, Welding Res Council, Intl Inst of Welding. A licensed PE in PA. Mbr, Franklin Inst. *Society Aff:* AWS, ASTM, ASM.

Thomas, Donald E
Business: Beulah Road, Pittsburgh, PA 15235
Position: Consulting Scientist. *Employer:* Westinghouse Res & Dev Center. *Education:* DSc/Metallurgical Engg/Carnegie Mellon Univ; MS/Metallurgical Engg/ Carnegie Mellon Univ; BS/Metallurgical Engg/Carnegie Mellon Univ. *Born:* Oct 1918. Native of Pittsburgh Pa. Served as officer in USNR during WWII. With Westinghouse Electric Corp 1949- ; Bettis Atomic Power Lab 1949-59; Astronuclear Lab 1959-71, in various capacities relative to the dev of materials for nuclear power systems. Fellow ASM; Mbr AIME, Fellow ANS, TMS (UK). Nuclear Metallurgy Ctte of AIMEChrmn Exec Ctte, Materials Sci & Technology Div of ANS. Co-recipient of ASM Engrg Materials Achievement Award for the dev of Zircaloy alloy systems & their contribution to nuclear energy 1972. *Society Aff:* AIME, ASM, ANS, TMS.

Thomas, Donald H
Business: Philadelphia, PA 19104
Position: Assoc Dean of Engrg *Employer:* Drexel Univ *Education:* PhD/Eng Design/ Case Inst of Tech; MS/ME/Drexel Univ; BS/ME/Drexel Univ *Born:* 12/01/33 in Phoenixville, PA, entered the Drexel Inst of Tech Co-op BSME program in 1951. Began teaching in ME Dept at Drexel in 1956. Attended Case Inst of Tech, Engrg Design Center, 1961 to 1965. Awarded PhD in 1965. NSF Sci Faculty Fellow 1962-1964. Mbr of Sigma Xi, Tau Beta Pi and Pi Tau Sigma Honorary Fraternities. Teaching and research areas include dynamic systems and controls, biomechanics, myo-electric control systems, prosthetics, orthotics, engrg design and innovative teaching methods. Conslt on vehicle dynamics, fluid power control systems, machine design, product liability and safety. Who's Who in American Education, 1968 and Who's Who in the East, 1970. *Society Aff:* ASME, SAE, ASEE

Thomas, Eddie F
Home: 4629 Owendale Dr, Fort Worth, TX 76116
Position: Chief, Quality Prog/Proj Mgmt *Employer:* General Dynamics Corp. *Education:* MS/Quality Sys/Univ-Dallas; MS/Engrg Admin/So Methodist Univ; BS/ EE/NC State *Born:* Native of Charlotte NC. Chief Prod Assurance for General Dynamics Corp's Fort Worth Div, responsibilities include Quality Prog and Proj Mgmt. Author. Lecturer. Part-time instructor at local coll. ASQC VP, Past Mbr of Annual Reliability & Maintainability Symposium Mgmt Ctte. Mbr of Natl Mgmt Assn. Reg PE in TX 1961-76. Sr Mbr IEEE 1968-76.

Thomas, Frank A, Jr
Home: 6617 Myrtle, Orange, TX 77630-8741 *Employer:* Retired *Education:* PhD/Mech Engg/GA Inst Tech; MS/Mech Engg/GA Inst Tech; BS/Mech Engg/ Purdue Univ. *Born:* Jan 1924 Atlanta Ga. Served with Marine Corps 1943-46. Indus Hygienist with Ga Dept of Public Health & US Atomic Energy Comm. Asst Prof & Res Assoc at Ga Inst of Technology. Prof, Head Mech Engrg Dept, Dean of Engrg, Vice Pres Academic Affairs, Pres, Regents Prof at Lamar Univ. Dean & Prof. Engrg Technology. Community & Technical College, University of Toledo, 1974-81. Vice Chancellor Academic Affairs, Alamo Community College District 1981-85. Retired 1985. Reg PE in Ga & Texas. Mbr ASME, ASEE, Pi Tau Sigma, Sigma Xi, Tau Beta Pi, Phi Kappa Phi. Vice Pres ASME 1971-73. Baptist Deacon. Enjoy Spanish, photography, outdoors, & automotive. *Society Aff:* ASME, ASEE, ΣΞ, ΦΚΦ, ΤΒΠ, ΠΤΣ

Thomas, Gareth
Business: Dept of Mtls Sci & Mineral Engg, Hearst Bldg, Berkeley, CA 94720
Position: Prof of Met. *Employer:* Univ of CA. *Education:* ScD/-/Cambridge Univ; PhD/-/Cambridge Univ; BSc/Met/Univ of Wales. *Born:* 8/9/32. Came to Berkeley 1960; Full Prof 1966, Assoc Dean, Grad Div 1968-69; Asst to the Chancellor 1969-72 Acting VChancellor, Academic Affairs 1971-72, Sr Faculty Scientist, Mtls & Molecular Res Div LBL. Numerous awards, honors & prizes, among which are: EMSA Natl Award Physical Sciences First Prize, 1965, 1975, 1976. ASM Metallographic Prizes. E.O. Lawrence Award, The Rosenhain Medal, Curtis McGraw Res Award (ASEE), Bradley Stoughton Award (ASM). Fellow ASM, AIME and Royal Mic Soc (London). IMS Metallographic Exhibit Award, Montreal, 1978. 1st Prize Metallographic Exhibit, 1979. Patents 1964 - Patent for Process for Improving Strength & Corrosion Resistance of Aluminum Alloys (US Patent 3,133, 839). 1978 - (DOE) Duplex Steels for strength: weight improvement (US Patent 4, 067,756). *Society Aff:* AIME, ASM, RMS, EMSA, IFSEM.

Thomas, James R, Jr
Home: 48 Deer Harbour Dr, Salisbury, MD 21801
Position: Engg Partner. *Employer:* George, Miles & Buhr. *Education:* BS/Civ Engr/ Univ of MD. *Born:* 5/26/44. Born and raised in Cambridge, MD. Worked with Army Corps of Engrs in Water Supply Branch, followed by position with David Volkert & Assoc of Wash, DC as civ engr working on design of Interstate Hgwy System and Water & Sewer Systems in Wash Area. Worked with present consulting firm since 1969. Promoted to Associate at age 31. Admitted to partnership in Jan, 1978. Currently responsible for managing all of firm's work within state of DE and is partner in charge of engr business dev activities. Active in local church, school and Chamber of Commerce activities. *Society Aff:* ASCE, NSPE, WPCF, ACEC.

Thomas, Jerry K
Business: 2222 Jackson Blvd, Rapid City, SD 57702
Position: President. *Employer:* Thomas & Associates Cons Engrs *Education:* Bach/Civil Engg/Univ of MN. *Born:* Nov 7, 1925 Turton SD. Served WWII with US Navy in 7th Fleet as electronics technician. BSCE with Distinction 1948 Univ of Minn. Entered private practice of engrg in 1948 at Rapid City with Staven Engrg culminating as Vice Pres 1961-65. Formed own practice 1965 as a partnership serving all levels of government & the private sector in civil, hwy, airport & struct disciplines with primary emphasis in transportation field. Inc in 1973. Pres of CECSD 1968 & Natl Dir CECUS 1970 & 71. Registered Prof Engr in SD, ND, NE and WY. Registered Land Surveyor in SD, ND, NE and WY. *Society Aff:* NSPE, ACEC, ASCE, SDES, APWA, ACSM, AAAE, SDPLS

Thomas, John B
Home: 900 Kingston Rd, Princeton, NJ 08540
Position: Professor of Electrical Engineering. *Employer:* Princeton University. *Education:* PhD/EE/Stanford; MSE/EE/Stanford; BSEE/EE/Johns Hopkins; AB/ Physics/Gettysburg. *Born:* July 1925. Served with US Army in WWII. Engr Koppers Co, Baltimore 1945- 52; heavy indus plant design, high voltage engrg, & administration. Faculty mbr, Dept of Electrical Engrg, Princeton Univ 1955- . NSF Senior Postdoctoral Fellow 1967-68 at Berkeley & Stanford. Author of 3 textbooks & over 175 tech papers, principally in statistical communication theory & related topics. *Society Aff:* IEEE, IMS.

Thomas, John H
Business: 436 Lattimore Hall, University of Rochester, Rochester, NY 14627
Position: Univ Dean of Grad Studies and Prof of Mech and Aerospace Sc and of Astronomy *Employer:* Univ of Rochester *Education:* PhD/Eng Sci/Purdue Univ; MS/Eng Sci/Purdue Univ; BS/Eng Sci/Purdue Univ *Born:* 04/09/41 Native of Chicago, IL. Purdue Univ, BS (Engrg Sci) 1962, MS 1964, PhD 1966. Grad Instructor and NSF Summer Fellow at Purdue Univ, 1962-1966. NATO Postdoctoral Fellow at Dept of Applied Math and Theoretical Physics, Unv of Cambridge, England, 1966-67. At Univ of Rochester since 1967. Visiting Scientist at Max-Planck-Inst for Physics and Astrophysics, Munich, Germany, 1973-74. Visiting Fellow at Worcester College, Univ of Oxford, England, 1987-88. Frequent Visiting Astronomer at National Solar Observatory. Research interests include geophysical and astrophysical fluid dynamics and solar magnetohydrodynamics. Author of over fifty scientific and technical publications. Recipient of several research grants from NSF, NASA, Air Force, etc. Mbr of ASME, ASEE, Am Phys Soc, Am Astron Soc, Am Geophys Union, Internatl Astron Union, AAAS. Reg PE in NY State. *Society Aff:* ASME, APS, AAS, AGU, IAU, AAAS, ASEE

Thomas, Joseph M, Jr
Business: 1003 City Hall, Atlanta, GA 30303
Position: Chief, Div of Traffic Engg. *Employer:* Bureau of Traffic & Transportation. *Education:* IET/Indus/GA Tech Eng Ext Div; Cert/Traf Eng/Northwestern Univ; Cert/Traf Eng/GA Tech; Cert/Traf Mgt/Penn State. *Born:* 1/29/39. Degree, IET from GA Tech Engg Extn Div & SIT. Cert Traffic Engr Northwestern Univ, Cert Trans Mgt, Penn State Univ, Traffic Engr with City of Atlanta 1959- . Chief, div of Traffic Engg 1979- with primary respon for traffic sys mgmt & computerized control sys. Part time instructor in Traffic Engg courses at GA Tech 1964- . Pres GA Div ITE 1971; Pres Southern Sect ITE 1973. Chairman, Dist. 5 ITE Bd. of Directors, 1980. Pres, Urban Traffic Engrg Council, 1984. Recipient of Hoose Distinguished Service Award (SSITE) 1980. Recipient of Hensley Award (SSITE) 1976; Registered prof Engr. Enjoy camping & coaching baseball. *Society Aff:* ITE, NSPE, UTEC.

Thomas, Julian B
Home: 2222 Huntington Ln, Fort Worth, TX 76110
Position: Consulting Engr (Self-employed) *Education:* ME/Mech Engg/TX A&M; BS/ Mech Engg/TX A&M; Hon/Disting Alumnus/TX A&M; LLD/Doctor of Laws/TX Christian Univ. *Born:* 7/19/91. Engr Texas Power & Light Co 1912-17; Vice Pres, Pres, Chmn of the Bd Texas Electric Serv Co, Fort Worth 1930-63; Vice Pres of Texas Utilities Co, Dallas 1945-63; Dir Fed Res Bank of Dallas 1955-58; Pres Old Ocean Fuel Co; Pres Texas Atomic Res Foundation 1958-59; Coast Artillery USA 1917-19, discharged as Cpt. Fellow IEEE, ASME; State Bd Registration Prof Engrs 1941-51. Cons US AID Taiwan, Philippines, Korea 1963-65. Cons Engr 1963- . Honorary: Distinguished Alumnus Texas A&M Univ, Honorary LLD Texas Christian Univ. *Society Aff:* IEEE, ASME, TSPE.

Thomas, Leonard W, Sr
Home: 1604 Buchanan St, NE, Washington, DC 20017
Position: Electronics Engr (Self-emp) *Education:* BS/Elec Engg/Auburn Univ *Born:* 5/11/09. Engr Navy Dept Wash DC 1942-1961. Responsible for Navy position on interference reduction. Staff Electronics Engr DoD Electromagnetic Compatibility Analysis Ctr, Annapolis, MD 1961-1970. Responsible for Plans and Progs. Mbr ANSI Committee C63, 1944 to date. Secy ANSI C63 1977-85. Fellow IEEE 1978 cited for "Leadership in Electromagnetic Compatibility and development of interference measurement instrumentation & Standards–. Reg PE, DC. Mbr USA Delegations to CISPR meetings, 1946 to 1960. Secy CISPR Subcommittee A 1974-1977. Author many technical papers published and delivered before profl and governmental organizations. Enjoys classical pipe organ music. Secretary, IEEE Electromagnetic Compatibility Society Board of Directors 1965 to 1981. Recipient, Laurence G. Cumming Award (1980) of IEEE EMC Society. *Society Aff:* IEEE, NSPE, IEE, ASNE, SAE, WSE, DCSPE.

Thomas, Marlin U
Business: Dept of Industrial Engg, 113 Elec Engg Bldg, Columbia, MO 65211
Position: Prof. *Employer:* Univ of MO. *Education:* PhD/Industrial Operations Engg/ Univ of MI; MSE/Ind Engg/Univ of MI; BSE/Ind Engg/Univ of MI. *Born:* 6/28/42. Prior to joining the Univ of MO, was a technical planning and analysis exec for Chrysler Corp. Has held academic appointment at the Naval Postgrad School, Monterey, CA and the Univ of WI-Milwaukee, Chrmn of the Industrial & Mfg Engg Option Program. Also has prior industrial experience with Owens-IL, Inc, Technical Ctr in Toledo, OH. An active researcher and consultant. Numerous publications in the operations res and industrial engg literature. Reg PE and in addition to participation in several technical and profl socieities, served on several natl committees and review bds. *Society Aff:* ORSA, IIE, ASA, SIAM, ΣΞ.

Thomas, Michael E
Business: Sch of Industrial & Sys Engg, Atlanta, GA 30332
Position: Prof & Dir. *Employer:* GA Inst of Technology. *Education:* PhD/Op Research/Johns Hopkins Univ; MS/Chem Engg/Univ of TX, Austin; BS/Chem Engg/ Univ of TX, Austin. *Born:* 5/10/37. Worked for Union Carbide as production engr in vinyl chloride unit at TX City Plant; Res Engr ECI designing thinned arrays. Faculty Mbr at FL 1965-78. Served as Cons to GE Co, US Army Corps of Engr, Container Coorp of Amer, MAPS Inc. and Hewlett Packard Worked for one yr for Natl Bureau of Stds. Currently serving as Dir. Interested in sports. Past Pres of Operations Res Soc of Amer. *Society Aff:* ORSA, TIMS, IIE.

Thomas, Richard E
Home: Rt 1 Box 1175, Bryan, TX 77803
Position: Professor & Director. *Employer:* Texas A&M Univ, College of Engrg. *Education:* B Aero E/Aero Eng/OH State; BA/Math/OH State; MS/Aero Eng/OH State; PhD/Aero Eng/OH State. *Born:* 12/29/25. Native of Logan OH. Aero Engr Air Tech Intelligence Ctr, Wright-Patterson AFB 1951; Res Assoc Aerodynamic Lab, OH State Univ 1952-56; Assoc Prof then Prof Dept of Aerospace Engrg, TX A&M Univ 1964-69; Prof & Dept Hd Aerospace Engrg, Univ of MD 1969-71; Prof of Engrg & Dir of the Ctr for Urban Progs, TX A&M Univ 1971-72. Associate Dean & Acting Dean 1972-74. Now, Professor of Aero Engg & Director Ctr for Strategic Technology. TX A&M Univ. Assoc Fellow AIAA. Received Standard Oil Foundation Award for Distinguished Achievement in Teaching 1968. *Society Aff:* AIAA, ASEE, ΤΒΠ

Thomas, Roland E
Home: 7226 Bell Dr, Colorado Springs, CO 80918
Position: Res Scientist. *Employer:* Kaman Sci Corp. *Education:* PhD/Elec Engg/Univ of IL; MS/Elec Engg/Stanford; BS/Elec Engg/NM State Univ; BS/Agri Engg/NM State Univ. *Born:* April 12, 1930 Austin Texas. Son of Melvin Aubrey & Myrtle (Nelson) T; m. Juanita Rhodes June 15, 1951, c. Lynnertte, Christine, Lee Everett. Cdr 2Lt USAF 1952, advanced through grades to Col 1966; Cons Engr Wright-Patterson AFB Ohio 1953-57; Assoc Prof USAF Acad 1959-63, permanent Prof 1963-79, Prof & Head Dept of Astronautics 1965-66, Prof & Head Dept of Electrical Engrg 1966-79. Retired in Grade of Brig Gen, Mar 1979. Res Scientist in Nucle-

Thomas, Roland E (Continued)
ar effects with Kaman Sci Corp, Colorado Springs, 1979- ., IEEE, Res Soc of America, Sigma Xi, Sigma Tau, Tau Beta Pi. *Society Aff:* IEEE, ASEE.

Thomas, Seth R
Business: Delaware Ave, NW, Warren, OH 44485
Position: Dir of Plating and Finishing *Employer:* Thomas Steel Strip Corp. *Education:* MS/Met Eng/Lehigh Univ; BS/Met Eng/Lehigh Univ; BA/Arts & Sci/Lehigh Univ. *Born:* May 5, 1941. Secondary education in Thomaston, CT public schools. BA Arts & Scis 1963 Lehigh Univ, Bethlehem Pa; BSMetE 1964; MSMetE 1966. Employment history includes Materials Div, Texas Instruments Attleboro Mass, Teledyne Rodney Metals, New Bedford, MA, & current employer. Mbr ASM, ASTM, Sigma Xi, AESF. Chmn RI Chap ASM 197273. Secy 1975 & Chmn 1976 ASM Adv Tech Awareness Council, Chmn surface Treating & Costing Tech. Div. 1982-84. Chmn Warren, OH chap ASM 1984-85. *Society Aff:* ASM, ASTM, ΣΞ, AESF

Thomason, Michael G
Business: 107 Ayres Hall, Univ. Tennessee, Knoxville, TN 37996
Position: Prof of Computer Science *Employer:* Univ of TN. *Education:* PhD/EE/Duke; MS/EE/Johns Hopkins; BS/EE/Clemson. *Born:* 1/4/44. After graduating with BS in 1965, worked at Westinghouse Defense & Space Ctr, Baltimore, until 1970. Received MS in 1970 & PhD in 1973. Prof at UT- Knoxville since 1973 asociated with Comp Sci Dept. Teaching and res in digital systems formal languages & automata, scene analysis, image understanding, applicatons of discrete math. Author of several papers & a textbook in these reas. Reviewer for IEEE transactins & other journals. *Society Aff:* IEEE, ACM, ΣΞ, PRS.

Thompson, A Ralph
Business: Dept of Chemical Engrg, Kingston, RI 02881
Position: Prof Emeritus Chem Engrg *Employer:* University of Rhode Island. *Education:* PhD/Chem Engg/Univ of PA; BASc/Chem Engg/Univ of Toronto. *Born:* Sept 1914. Chem Engr Canadian Indus Ltd 1936-40; Faculty Univ of Pa 1940-52; Prof & Chmn Chem Engrg, Univ of RI 1952-72; Dir Water Resources Center 1966-80 . Cons Sun Oil Co, Monsanto Chem Co & others. Reg PE in Pa & RI. Fellow AIChE, Chmn RI Chap; AAAS. Mbr ACS, ASEE, AAUP, American Water Resources Assn, Sigma Xi (Pres RI Chap), Tau Beta Pi, Phi Kappa Phi (Pres RI Chap), Phi Lambda Upsilon. NAWID (Natl Assoc of Water Inst Dirs) Chrmn. *Society Aff:* AIChE, ACS, AAAS, ASEE, ΣΞ, ΤΒΠ, ΦΚΦ, ΦΛΕ.

Thompson, Anthony W
Business: Dept of Met & Mtl Sci., CMU, Pittsburgh, PA 15213
Position: Prof. *Employer:* Carnegie-Mellon Univ. *Education:* PhD/Metallurgy/MIT; MS/Metallurgy/Univ of WA; BS/Mater Sci/Stanford *Born:* 3/6/40. CA native. Was with the Jet Propulsion Lab (Caltech) for one yr after BS. After receiving PhD in 1970 spent 3 yrs at Sandia Labs, Livermore, and 4 yrs at the Sci Ctr of Rockwell Intl before joining Carnegie-Mellon. Interests encompass the relation of microstructure to a wide variety of mech properties, particularly deformation and fracture. Editor, *Metallurgical Transactions*. Fellow, ASM. *Society Aff:* AAAS, ASM, TMS-AIME.

Thompson, Bill M
Business: Phillips Petroleum Company, 17 Phillips Bldg, Bartlesville, OK 74004
Position: Mgr, Geoscience & Tech Div *Employer:* Phillips Petroleum Co *Education:* BS/Petrol Engrg/Univ of TX *Born:* 09/12/32 Native of Leona, TX. Joined Phillips Petroleum Co in 1954. Held various engrg positions in production, reservoir, drilling, and offshore construction. Named Engrg Mgr-Norwegian Operations in 1973, Director of Drilling and Production-Europe-Africa in 1975, Operations Mgr-Eastern Div (USA) in 1977 and Mgr-Geoscience and Tech Div of Research and Development in 1978. Responsible for developing new tech for upstream operations in petroleum and minerals. Treasurer and mbr of Bd of Dirs of SPE. Vice Chrmn of Visiting Committee for Petroleum Engrg Dept, Univ of TX at Austin. Enjoys golf and hunting. *Society Aff:* SPE

Thompson, Brian J
Business: University of Rochester, 200 Admin Bldg, Rochester, NY 14627
Position: Provost. *Employer:* University of Rochester. *Education:* PhD/Applied Physics/Univ of Manchester; BSc/Applied Physics/Univ of Manchester. *Born:* June 10, 1932; came to US 1963. Married Joyce Emily Cheshire; c. Karen, Andrew. Demonstrator in Physics Univ of Manchester 1955-56, Asst Lecturer in Physics 1957-59; Lecturer in Physics Univ of Leeds, England 1959-62; Senior Physicist, Technical Opers, Burlington Mass 1963-65, Mgr Phys Optics Dept 1965- 66, Dir Optics Dept 1966-67; Mgr Tech Opers West & Tech Dir, Beckman & Whitley, Calif 1967-68; Prof & Dir, The Inst of Optics, Univ of Rochester 1968-75; Prof & Dean, Univ of Rochester 1975-1984. William F May Prof of Engrg 1982-86. Prof and Provost, 1984-present. Army of Great Britain 1950-52. Fellow Inst of Physics & Physical Soc Great Britain, Optical Soc of America, Soc of Photo-Optical Instrumentation Engrs (President's Award 1967) (Kingslake Medal & Prize 1978); (Pezuto Award 1978); (Gold Medal, 1986); Mbvr APS, AAAS. Assoc Editor Optical Engrg 1973-76; American Editor Optica Acta 1981-85; Assoc Editor JOSA 1966-77; Editorial Bd Laser Focus, Optics Communications 1978-86. *Society Aff:* OSA, SPIE, APS, AAAS, ASEE.

Thompson, Charles B
Home: 16044 Ocotillo Dr, Fountain Hills, AZ 85268
Position: Consultant (Self-employed) *Education:* MS/Hydraulic Engr/State Univ of IA; BS/CE/Univ of NM *Born:* 10/22/17 1940-43, ordnance Dept, US Army, St Louis. Headed inspection staff of 400 employees. 1943-46, Asst Prof, Cvl Engr, Univ NM, hydraulics instruction. 1948- 57, Chief Tech Div, State Engr Office, Santa Fe, supervised 44 personnel, water resources projects. 1958-59, Cvl engr-planner, Litchfield-Whiting- Bowne, Bangkok, Thailand, city planning project. 1959-61, Hydraulic Engr, US Study Commission-TX State-water resources plan. 1961-76, Water Resources Engr, US State Dept, AID, 2 yrs Tunisia, 4 yrs Ghana, 9 yrs USA, planning and supervision construction of overseas projects. 1976-78, Water Resources Engr, Harza Engr Co, Chicago, Eastern Hemisphere Project Mgmt. 1976-87, Consltg Engr, self-employed. Registered PE-11 states. *Society Aff:* IAHR, IWSA, ASCE, AWWA, NSPE, IWRA, ICLD, IAID, PIANC

Thompson, David R
Business: Head, Agric. Engg Dept, 109 Ag Hall, OSU, Stillwater, OK 74078
Position: Prof & Head *Employer:* Oklahoma State *Education:* PhD/Agri Engg/MI State Univ; MS/Agri Engg/Purdue Univ; BS/Agri Engg/Purdue Univ. *Born:* 4/4/44. Native of IN. Faculty mbr at the Univ of MN 1970-85. Prof & Head Agricultural Engrg Dept. Oklahoma State Univ. 1985- . Operations engr for Green Giant Co, 1978-1979 while on sabbatical leave from the Univ of MN. Responsibility at Green Giant included definition & evaluation of potential and proposed capital projs. Chrmn, Food Engg Div, ASAE, 1977-1978; Dir, ASAE 1982- 1984. *Society Aff:* ASAE, ASEE, IFT, ASHRAE

Thompson, Donald E
Home: 2113 Briarcliff, Springfield, IL 62704
Position: President *Employer:* Crawford et al Inc, Cons Engrs. *Education:* BS/Civil Engg/Univ of IL. *Born:* Sept 21, 1930 Decatur Ill. US Army Corps 1953-55. Crawford, Murphy & Tilly Inc Cons Engrs 1955- . Reg PE in Ill and Missouri; Reg Structural Engr in Ill. Pres, Ill Cons Engrs Council 1973-74; Pres ASCE, Central Ill Sect 1967. Mbr ASCE, NSPE, Ill Soc of Prof Engrs, APWA, American Cons Engrs Council. *Society Aff:* NSPE, ACEC, APWA, ASCE, SAME

Thompson, Dudley
Home: 5 McFarland Dr, Rolla, MO 65401
Position: Vice Chancellor Emeritus. *Employer:* University of Missouri-Rolla. *Education:* PhD/Chem Engg/VPI & SU; MS/Chem Engg/VPI & SU; BS/CHem

Thompson, Dudley (Continued)
Engg/VPI & SU. *Born:* 1/9/13. Personnel Asst 1935-38; Control, Production Chemist 1938-39; US Army 1941- 46, Col AUS (Retired); Intructor to Assoc Prof Chem Engg, VPI & SU 1946-55; prof Chem Engg 1956-78; Chmn Chem Engg & Chem 1956-63; Dir Sch of Engg 1964-65; Dir IRC 1964-65, Dean of the Fauclty 1964-65; Dean of Faculties 1964-73; Acting Chancellor 1973-74; Vice Chancellor 1974-78; UMR. Res Fellow ASEE at TVA 1947- 48. Cons & Dev Engr, GE. Sr Cons Engr, Monsato. Ret PE in VA & MO. Chmn 3 AIChE Symposia on Ultrasonics. Author of articles on ultrasonics. Vice Chancellor Emeritus & professor Emeritus of Chem Engg 1978- , UMR. *Society Aff:* AIChE, ACS, AIP, IEEE, ASEE, AIME, ΣΞ.

Thompson, Earl R
Home: 23 Harvest Lane, Glastonbury, CT 06108
Position: Assistant Director of Research for Materials Technology *Employer:* United Technologies Res Center. *Education:* DSc/Mat Sci/Univ of VA; MSC/Met Engr/NC State Univ; BSC/Met Engr/NC State Univ. *Born:* Jan 9, 1939 Lenoir NC. Res Scientist, Reynolds Metal Co 1961-62; Instructor in Materials Sci, Univ of Va 1962-64; with United Technologies 1965- , assumed current position as Assistant Director in 1986. Visiting Prof, Federal Polytechnic Sch, Lausane Switzerland summer 1973. Tech publs. ASM Grossman's Authors Award 1970. Co-chmn of 1972 Conf on In Situ Composites. Chmn of 1974 AGARD Meeting of Experts. Mbr Solid State Scis Panel, NRC. Chmn Mat Tech Committee, AIAA 1978-79. Mbr AIME, ASM, AIAA, ACS, Sigma Xi, Tau Beta Pi, Assoc Fellow AIAA & Fellow ASM. Enjoy church related youth work, Church Pres 1983- 1985. *Society Aff:* AIME, ASM, ACS, ASTM

Thompson, Everett S
Business: 611 Cascade W Pky SE, Grand Rapids, MI 49506
Position: President. *Employer:* Williams & Works Inc. *Education:* BA/Pub Adm/MI State Univ; BS/CIvil Eng/MI State Univ. *Born:* 5/19/24. Detroit MI; lived there until 1943 at which time entered the Military Serv & served until 1946 in the So Pacific & Japan with the signal corps. BSCE, BA Public Administration From Mich. State Univ. Asst & City Mgr in several communities in MI upon graduation from Mich. State Univ in 1951. Joined Williams and Works in 1956, main area of work in the fields of civil engg & community planning. Reg PE in MI & 8 other states. Natl Dir of NSPE 1969-73, VP & Chmn of NSPE/PEPP 1973-74. Pres of ACEC, 1980-81. *Society Aff:* NSPE, ASCE, ACEC.

Thompson, Herbert
Home: 3020 Mockingbird Ct, Clearwater, FL 33520
Position: Chrmn, Pres *Employer:* Sunkey *Education:* MChE/ChE/NYU; BCE/CE/NYU. *Born:* 12/14/30. Native of Pleasantville, NJ. Employed by Pfizer, Inc 1954-63 developed patented process for direct compression of sorbitol for pharmaceutical products. Worked for R P Scherer Corp 1964-1969 as Asst to Exec VP. Founded Thompson Labs 1969. Pres of Natl Pharmaceutical Alliance 1978-79. Vice President of Key Energy Enterprises and President of Key Medical Products Division. Currently Chrmn & Pres, of Suncoast Medical. Pres Gateway (St Petersburg) Rotary Club. *Society Aff:* ACS, AIChE.

Thompson, Homer F
Home: 4101 Lochridge Rd, No Little Rock, AR 72116
Position: Retired. *Education:* MSCE/Structures/MO Univ at Rolla; BSCE/Sanitary/MO Univ at Rolla; BSChE/Indus/MO Univ at Rolla. *Born:* Feb 21, 1907 St Louis Mo. Tau Beta Pi, Phi Kappa Phi, Chi Epsilon. Employed as City Engr Rolla Mo; Engr Russell & Axon, St Louis; Asst Prof Missouri Univ at Rolla. Chmn Engrg Div, Ark Tech; and Little Rock District, Corps of Engrs. Fellow ASCE. Reg PE in Mo & Ark. Coordinated planning of lock & dams 4 & Toad Suck Ferry Lock and Dam, Ark River; WWII served as S-3 340th Engrs on Alaska Hwy; P Pres Little Rock Branch & Mid-South Sect ASCE, & Ark Post SAME. Retired June 1974. *Society Aff:* ASCE, SAME, AARP, NARFE.

Thompson, Howard Doyle
Business: Sch of Mech Engg, Purdue Univ, W Lafayette, IN 47907
Position: Prof of ME. *Employer:* Purdue Univ. *Education:* PhD/ME/Purdue; MS/Engg Sci/Purdue; BS/ChE/U of UT. *Born:* 4/17/34. US Citizen. Born Cedar City, UT. Served with USN 1957-60. Faculty, Purdue Univ, since 1962. Full Prof since 1974. Author of twenty Journal articles and many reports. Co-editor of three sets of Proceedings of Laser Velocimetry Workshops. Pratt & Whitney Aircraft, E Hartford, CT 1969-70. Assoc Res Scientist. Arnold Engineering Development Center, Tullahoma, TN 1980-81. Senior Mechanical Engineer. Reg PE (IN). Married to Patricia Ann Frei; five daughters; Bishop, Church of Jesus Christ of Latter-Day Saints 1975-80. *Society Aff:* ASME, AIAA.

Thompson, Hugh A
Business: Sch of Engg, New Orleans, LA 70118
Position: Dean of Engg. *Employer:* Tulane Univ. *Education:* PhD/ME/Tulane Univ; MS/ME/Tulane Univ; BS/ME/Auburn Univ. *Born:* 3/24/35. Currently serves Tulane Univ as Dean of Engg. Prior to appointment to that position in 1976, was Prof of Mech Engg at Tulane, having responsibility for instruction of courses in heat transfer, math, and nuclear reactor theory. Res interests were largely centered around problems of vibration and dynamic response of systems of the electric utility industry. Industrial experience includes four yrs as a process engr in petrol refining. *Society Aff:* IEEE, ASME, ASEE.

Thompson, Jack W
Business: P.O. Box 160039, Irving, TX 75016
Position: Conslt *Employer:* Enercon, Inc. *Education:* BSME/-/SMU. *Born:* 8/12/25. in Dallas, TX. Attended TX A&M Univ, BSME-SMU AUS-WWII Parachute Field Artillery, 11th Airborne Div. Currently Forensic Conslt to the surety indus & legal profession, mech contractors, mech engrs, gen contractors, owners and developers. Construction or design problems, operating problems, claims, claim avoidance, schedules, recovery programs, contract defaults, insurance claims, construction mgmt surety/construction conslt. President, Sam P. Wallace Co, Inc. Previously complete charge Mech Contracting Dallas Div-SAM P. Wallace Co. Respon for bidding, procuring, planning, execution, commissioning, oper & testing of mech & refrig sys for multi million dollar instlations, for commercial, institutional, power & treatment plants. Dir & officer Natl, State & Local Mech. Contractors Assn, Dist Serv Awd (MCAA 1980); Dir Natl Environ Balancing Bureau, Dir & Founding Mbr NTX Contractors Assn; Mech appeals Bd-City of DA. Fellow ASHRAE, Dist Serv Awd 1971, Engr of the Yr (Dallas Chap 1975), Bd Mbr 1963-69, 71-72. Reg PE & NCSBEE, Master Plumber and licensed HVAC contractor. *Society Aff:* ASME, NSPE, TSPE, ASHRAE.

Thompson, James B
Business: 445 S Figueroa - 3500, Los Angeles, CA 90071
Position: Senior Advisor-Operations (Partner). *Employer:* Dames & Moore. *Education:* MS/CE/Harvard Grad Sch of Eng; BS/ME/Univ of CA. *Born:* 1922. Joined Dames & Moore 1953, became partner 1954. Primary responsibility for performance of engrg assignments are soil mechanics & foundation aspects of projs which have ranged in size from small structures to multi-million-dollar facilities throughout the free world - steel mills, chem plants, refineries, power plants, etc. Authored a number of papers & articles which have appeared in publs of ASTM & HRB & in Public Works Magazine. Mbr of faculty of the 8th Annual Conf on Soil Mechanics & Foundation Engrg sponsored by Center for Continuation Study, Univ of Minn. Visiting Prof Univ of Ill 1972-75. Bd/Dir 1971-74 ARBA. Fellow ASCE, ACEC; Mbr ARBA, AAPA, NSPE. *Society Aff:* ASCE, ACEC, NSPE, AAPA, ARTBA.

Thompson, James E
Business: Department of Electrical Engineering, Arlington, TX 76019
Position: Prof and Chmn *Employer:* Univ of Texas *Education:* PhD/EE/TX Tech Univ; MS/EE/TX Tech Univ; BS/EE/TX Tech Univ *Born:* 01/24/46 Native of Lubbock, TX. Mbr of the Technical Staff of Northrop Corp 1974- 1976, specializing in

Thompson, James E (Continued)
high power gas laser research and development. With the Univ of SC from 1976 to 1983. Assumed rank of Prof in 1981. Appointed Tamper Chair Prof in 1980. Prof and Chrmn of Electrical Engrg, Univ of TX at Arlington, 1983. Current interests and research in the fields of pulsed power, high voltage tech, electrical breakdown phenomena, lasers, and electro-optics. *Society Aff:* IEEE, OSA, APS.

Thompson, John W
Home: 2 Yarra St, Kew Victoria Australia 3101
Position: Consultant Metallurgical Engr *Education:* Dipl/Met Engg/New South Wales Inst of Tech C. Eng (Chartered). *Born:* 10/5/14. Native of Melbourne Australia. Res Met 1935 for Australian Iron and Steel Co. Supt of Blast Furnaces 1943-52. Works technical supt 1952-59. Mgr technical services res & adm 1960-68. Gen mgr quality & technical services 1968-78 at Corporate Hq of Broken Hill Proprietary Co. Chrmn RR Res Committee on heavy haul iron ore RR of NW Australia. (Hamersley & Mt Newman Cos) Special interest rail & track devs. Exec mbr of Aust Welding Res Assn. VP (1974-77) of Intl Inst of Welding. Chrmn Structural Steel committee, mbr of council, mbr of metal and bldg stds bd. Stds Assn of Australia. Mbr of ISO stds committees TC17-Steel. Retired from BHP Co 1978. Awarded Silver Medal of Australian Inst of Metals 1979 & Fellow of Am Soc of Metals for Services to Met Engg especially in dev of high tensile rails, & oil & gas pipeline, & offshore platform steels. Quality Control and Metallurgical Engrg. *Society Aff:* IM, ASM, AusIMM, AIME, IMMA, AWI, IE (Aust).

Thompson, Joseph C
Home: 6993 Easton Ct, Sarasota, Florida 34238
Position: Consultant *Employer:* Self *Education:* BS/ME/University of Illinois *Born:* 11/22/20 Born in Chicago. Graduated Lane Technical High School. Started with GATX November 1945. Became Chief Tank Designer 1947. In 1978 named Manager of Engrg and Construction for GATX Terminals Corporation, the largest owner/operator of liquid storage terminals in the US. Retired in 1983 to become consultant in large scale liquid storage and handling. ASME VP and Council Member 1968. Elected ASME Fellow in 1984. Licensed PE in IL, LA, TX. Section author "Liquid Storage and Handling" Perry's Handbook of Chemical Engrg, Fourth Ed. Author of many papers on Fire Safety for Large Petroleum Storage Tanks. *Society Aff:* ASME, WSE, ILTA, API

Thompson, Joseph G
Business: Dept of Mech Engrg, Manhattan, KS 66506
Position: Prof *Employer:* KS State Univ *Education:* BES/Mech Engr/Brigham Young Univ; MSME/Mech Engr/Purdue Univ; PhD/Mech Engr/Purdue Univ *Born:* 8/15/35 Reared in Warsaw, IN. Engr at TRW Corp 1961-62. Instructor at Purdue Univ 1962-66. Developed Automatic Controls program as Asst Prof of Mech Engrg at Univ of TX, Austin 1966-71. Presently, Prof of Mech Engrg specializing in the application of microcomputers to automatic control systems. Author of numerous tech papers and reports. Recipient of numerous awards for excellence in teaching and res Dir of many res contracts and grants for advancing the practice of Automatic Control Sys Design. Cooperative Indus/Univ Dir of Ctr for Res in Computer Controlled Automation. Leader in Church and Community. *Society Aff:* ASME, SAE, SME, ASEE, ASHRAE

Thompson, Ken
Business: 600 Mountain Ave, Room 2C519, Murray Hill, NJ 07974
Position: Mbr of Staff *Employer:* AT&T Bell Labs *Education:* MS/EE/U CA Berkeley *Born:* 02/04/43 Mbr of Technical Staff at Bell Laboratories 1966-present. Co-author of UNIX Time-sharing System. Co-author of BELLE former world champion computer chess program. *Society Aff:* NAE, AAAS, NAS

Thompson, Kermit K
Home: 506 West 3rd St, Ellinwood, KS 67526
Position: President. *Employer:* Thompson Construction Co Inc. *Education:* BSME/Mech/KS State Univ. *Born:* Oct 11, 1916. Wichita Kansas Elementary through 2 years Wichita State Univ; BSME 1940 Kanasa State Univ. Various positions in Engrg Dept, Boeing Airplane Co Wichita 1940-45, highest, Mgr Tech Serv for Engrg. Taught classes in War Training Support WWII. Licensed Engr in Kansas, 2482. All offices of Kansas Engrg Soc, Pres 1971-72. Awarded 'Engr of Distinction' 1973, formerly Natl Dir. Serving on NSPE 1977-78 Nominating Ctte & Education Ctte. Owner, Thompson Const Co Inc, a hwy & heavy const firm. Staff includes 2 Reg PEs. *Society Aff:* NSPE, KES

Thompson, Lewis D
Business: Rt 4 Box 390, DeRidder, LA 70634
Position: President. *Employer:* Thompson Polymer Specialists. *Education:* BS/Chem Engrg/LA Tech. *Born:* 07/25/43. BS La Tech Univ; MS studies Bradley Univ. Native of Rosepine La. Res Engr, Caterpillar Tractor Co specializing in the polymer field 1965-69; Proj Engr with Boise Southern Co 1969-71; with Thompson Polymer Specialists since 1971, assumed responsibilities as Pres & am founder of the same. Am responsible for manufacture & sales of spceialized polymer prods for company. Listed in Who's Who in La, Who's Who in the South, Who's Who in Engrg; past master Masonic lodge, Ducks unlimited Mbr, NRA mbr. Enjoy hunting, fishing & sports. *Society Aff:* AIChE.

Thompson, Lyle I
Home: Box 55, Rigby, ID 83442
Position: Owner. *Employer:* Thompson Engineering. *Education:* BS/Civil/Univ of ID. *Born:* 11/3/40. Licensed PE & Land Surveyor in Idaho & Wyoming. Owner of cons engrg firm 1973- ; Asst County Engr 1972-73; 4 years in engrg with Utah State Road Comm 1963-67; 3 years in engrg with Idaho Dept of Hwys 1960-63. PPres of ID Soc of Prof Engrs. VP of ID Soc of Prof Engrs; P Secy ISPE both in Southeast Chap Soc. Mbr American Cons Engrs Council & State Soc (CEI); ASCE & State Soc; NSPE; ISPE; NACE. Enjoy outdoor sports & recreation. *Society Aff:* NSPE, ASCE, ACEC, NACE.

Thompson, Marshall R
Business: 208 N Romine, Urbana, IL 61801
Position: Professor of Civil Engineering. *Employer:* University of Illinois. *Education:* PhD/Civil Engg/Univ of IL; MS/Civil Engg/Univ of IL; BS/Civil Engg/Univ of IL. *Born:* July 1938. Native of Astoria ill. Various staff positions at the Univ of Ill since 1960 in res & teaching. Achieved rank of Prof of Civil Engrg in 1970. Cons activities in pavements, railroads, soils, & materials for government agencies, indus, & contractors. Special interest in materials, soil stabilization, pavements, & railroad systems. ASCE Huber Prize 1970; AW Johnson Award TRB 1970. Enjoy atheltics & traveling. *Society Aff:* ASCE.

Thompson, Philip A
Business: Dept. Mech. Engrg, RPI, Troy, N.Y 12180-3590
Position: Professor *Employer:* Rensselaer *Education:* ScD/M.E./MIT; M.S./M.E./RPI; B.S./M.E./RPI *Born:* 09/10/28 Native of Galesburg, Il. Served with the US Army Field Artillary 1946-48. Plant Engineer for WV Pulp and Paper Co. & Albany International. Current responsibility is Full Prof, Dept Mech Engrg, Rensselaer Polytechnic Inst. Have served as Program Mgr for the US Dept of Energy. Visiting Scientist at the Max-Planck-Institut fur Stromungsforschung for several years. Humboldt Award 1975. Fellow, ASME. Enjoy classical music and woodworking. *Society Aff:* ASME, AAAS, APS, ACS

Thompson, Robert C
Business: HQ USAF-PRE, Pentagon, Washington, DC 20330
Position: Major General, Dir AF Engrg Servs. *Employer:* United States Air Force. *Born:* July 1920. MS, BS Univ of Md. Native of Altoona Penn. US Army Air Corp/US Air Force Engr 1944- . Assumed current responsibility as Dir of Air Force Engrg & Servs May 1975. Responsible for direction of US Air Force Engrg & Servs personnel & activities throughout the world. Presently serving as Pres SAME; Chmn

Thompson, Robert C (Continued)
Bd/Dir of the Air Force Commissary Serv; Mbr Bd/Dir of the Army-Air Force Exchange Serv.

Thompson, Truet B
Home: 230 E. Garfield St, Tempe, AZ 85281
Position: Conslt Engr *Employer:* Truet B Thompson, Electrical Engr, Inc *Education:* PhD/EE/Northwestern Univ; MS/EE/OK St Univ; BS/EE/LA Poly Univ; BS/Math/LA Poly Univ *Born:* 03/24/17 Native of Shreveport, LA. Service in Air Forces (1942-45) aerial navigator and radar ofcr. Mbr of electrical engrg faculties at OK State Univ (1948-59) and AZ State Univ (1959 to 1983). Chrmn, Dept of Electrical Engrg at AZ State Univ (1960-1967). Field engr, elec utilities section of C.H. Guernsey and Co (1950- 1954). Extensive consltg in areas of elec power, elec safety, elec standards and codes, hospital elec safety, and clinical engrg for clients in twenty states including several natl industrial firms. Cabell Fellow at Northwestern Univ. Mbrship Tau Beta Pi, Eta Kappa Nu, Phi Kappa Phi honor societies. *Society Aff:* IEEE, NSPE, ASSE, ANSI, NFPA, ASEE, NAFE, CIGRE.

Thompson, Walter Dale
Business: 4822 Albemarle Rd, Suite 201, Charlotte, NC 28205
Position: Principal *Employer:* Wilber, Kendrick, Workman & Warren, Inc *Education:* BSE/Engg/Univ of NC-Charlotte *Born:* 11/08/46 Joined Wilber, Kendrick, Workman & Warren Architects-Engrs-Planners in 1969, a partner of firm since 1978. Primary responsibilities are for structural designs and coordination of project engrg with architectural. Past Pres of PE of North Carolina, Past Pres of South Piedmont Chapter of PE NC. Received Chapter's Young Engr of the Year Award in 1978. *Society Aff:* NSPE, ASCE

Thompson, William, Jr
Home: 601 Glenn Rd, State College, PA 16803
Position: Prof. *Employer:* Penn State Univ. *Education:* PhD/Eng Acoustics/Penn State Univ; MS/EE/Northeastern Univ; SB/EE/MIT. *Born:* 12/04/36. Native of Cape Cod, MA. Engr with Raytheon Co, Submarine Signal Div, Wayland MA, 1958-1960; design and dev of underwater electroacoustic transducers. Sr Engr, Cambridge Acoustical Assoc, Inc, Cambridge, MA, 1960-1966; Analytic studies of problems in underwater acoustics such as radiation and scattering by submerged elastic structures. Res Asst, Appl Res Lab (ARL), Penn State Univ, State College, PA, 1966-1972; joint academic appointment (asst prof) with ARL and Dept of Eng Sci and Mechanics, 1971; assoc prof, 1978; Prof., 1985; Coll of Engrg Outstanding Teaching Award, 1984; hd of Transducer Group at ARL 1971-1981; design and dev of electroacoustic transducers for advanced guidance systems for acoustic homing torpedoes. *Society Aff:* IEEE, ASA, SES.

Thompson, Willis F
Home: 128 Litchfield Turnpike, Bethany, CT 06525
Position: Retired *Born:* 12/11/95 My work has involved the location and design of power plants in the US and Europe. *Society Aff:* ASME

Thompson, Wm A, Jr
Home: 128 Orchard Lane, Wise, VA 24293
Position: Chairman *Employer:* Thompson & Litton Inc. *Education:* BS/Civil Engg/VA Military Inst. *Born:* 6/9/30. Native of Bluefield, West Virginia with consulting practice devoted to coalfield regions of Southwest Virginia. Served with US Air Force 1952-55. Founded Thompson & Litton in 1956, specializing in civil and mining engrg for small municipalities and coal industry. Projects include Regional Utility systems, new Community Development, coal mines and related operations. Pres CEC/VA 1970. Past Chmn of Eads Trustees (ACEC's Penstion Trust for member firms). Visiting Instructor in Mining Mgmt at Clinch Valley College of the University of Virginia. Dominion Bank Director. Elder and teacher in Presbyterian Church of the US. *Society Aff:* ASCE, NSPE, ACEC.

Thomsen, Charles L
Business: Charles L. Thomsen & Associates, 2403 San Mateo N.E. Suite W 22, Albuquerque, NM 87110
Position: Consulting Engineer *Employer:* Charles L Thomsen & Associates *Education:* BS/Mech Engg/Univ of NE. *Born:* 1/10/27. Mech Engr, Leo A Daly Co Architects-Engrs 1952-61; Partner Olsson, Burroughs & Thomsen Cons Engrs 1961-63; Prin Mech and Elec Engg, 1963-1980 The Clark Enersen Partners, Architechts-Engrs. Present position, Consulting Engr Charles L Thomsen & Assoc: Pres NE Chap ASHRAE 1961-62. Mbr ASHRAE, IES, NSPE & ISES. Reg PE in NE, IA, WYO, NM, MO, CO, AZ & TX. NCEE Cert. *Society Aff:* ASHRAE, NSPE, IES, ISES.

Thomson, C Ross
Home: 68 Westpark Dr, Ottawa Ontario K1B 3E5, Canada
Position: VP, Fire Res *Employer:* Canadian Wood Council. *Education:* BSc/Geol-Math/Carleton Univ. *Born:* 10/27/29. Born in Montreal, Quebec; Assoc of Insurance Inst of Canada; Inspector, Special Risks Div Canadian Underwriters' Assn, 1953-56; Supervisor Insurance and Fire Protection (Prairie & Pacific Regions) Canadian Pacific Railway 1957-65; CWC 1965-; Pres Montreal-Ottawa Chapter Soc of Fire Protection Engr 1973-75; SFPE Bd of Dirs 1975-80. Chrmn ASTM Cttee on Standards, 3rd Vice Chrmn ASTM Cttee E-5 (Fire Standards) & mbr ASTM Ctte D-7 (Wood); ISO TC 92 on Fire Tests; ULC Comm on Fire Tests; Standing Comm on Fire Performance Ratings (Natl Bldg Code of Canada); Chrmn, Standing Comm on Hazardous Materials. Processes and Operations (Natl Fire Code of Canada). *Society Aff:* SFPE, ASTM.

Thomson, Quentin R
Home: 4730 Camino Luz, Tucson, AZ 85718
Position: Prof. *Employer:* Univ of AZ. *Education:* PhD/Engg Economics/CA Coast Univ; MS/ME/Univ of AZ; BS/ME/GA Tech. *Born:* 11/14/18. Born in Lake Charles, LA, Nov 14, 1918. Grad from Landry High Sch as valedictorian in 1936. Spent four yrs at GA Tech employed as experimental test engr by Pratt & Whitney Aircraft 1940-41; US Navy Submarine Serv for 10 yrs. Received two Bronze Star medals; made 11 war patrols. Taught at Univ of AZ from 1950 to 1981 with four yrs out, 1963-1967, as chief engr for Krueger Mfg Co. Principal Q R Thomson & Assoc, 1981-present. Obtained MS and PhD while teaching part-time and in summer. *Society Aff:* ASME, ASHRAE.

Thomson, Robert F
Home: 4645 Chamberlain Dr, St Clair, MI 48079
Position: Retired. *Education:* PhD/Met Engg/Univ of MI; MS/Met Engg/Univ of MI; BS/Met Engg/Univ of MI. *Born:* Feb 1914. Foundry Metalurgist, Chrysler Corp, 3 years; Internatl Nickel Corp, 5 years; joined GE Res Labs in 1950 as Asst Dept Head Met Engrg, named Dept Head 1952, Tech Dir 1962, Exec Dir 1969. Holder or co-holder of 27 patents. Fellow ASM 1970; delivered ASM Edward DeMille Campbell Memorial Lecture 1963. Presented the American Foundrymen's Soc John A Penton Gold Metal 1955. Was Natl Trustee of ASM. Mbr ASM, AFS, Sigma Xi, Tau Beta Pi, & Phi Kappa Phi. *Society Aff:* ASM, AFS, ΤΒΠ, ΣΞ, ΦΚΦ

Thomson, William J
Business: Dept of Chemical Engrg, Pullman, WA 99164
Position: Prof and Chair *Employer:* WA State Univ *Education:* PhD/ChE/Univ of ID; MS/ChE/Stanford Univ; BChE/ChE/Pratt Inst *Born:* 05/15/39 Upon receipt of MS degree, worked in aerospace related projects at AVCO RAD and NAVAL ORD LABS. Served in US Navy 1963-1966, with final rank of Lt. Prof of Chem Engrg at Univ of ID, 1969-1980. Research engr at Union Oil Research, 1972- 73, at SRI Int, 1980. Conslt to Oil Shale Industry since 1974. Prof and Chair of Chem Engrg at WA State Univ as of Jan, 1981. *Society Aff:* AIChE, ACS

Thomson, William T
Home: 980 Andante Rd, Santa Barbara, CA 93105
Position: Emeritus Professor of Engineering. *Employer:* Univ of California, Santa

Thomson, William T (Continued)
Barbara. *Education:* PhD/Elec Engg/Univ of CA, Berkeley; MS/Elec Engg/Univ of CA, Berkeley; BS/Elec Engg/Univ of CA, Berkeley. *Born:* March 24, 1909. Taught at U.C. Davis, Kansas State Univ, Cornell Univ, Univ of Wisc, UCLA, & UCSB. Life Fellow ASME, Assoc Fellow AIAA, Guggenheim Fellow to Germany 1962. Fulbright Res Prof to Japan 1957. Cons to aerospace indus. Author of 5 textbooks. *Society Aff:* ASME.

Thorbecke, Willem H
Business: P.O. Box 840, Suite 609, 600 Commerce Drive, Coraopolis, PA 15108
Position: Chrmn and CEO *Employer:* Energy Support Services, Inc. *Education:* BSc/Engg & Bus Adm/MIT. *Born:* 7/4/24. Born in Paris, France. During WWII served in RAF as bomber pilot. After grad, joined Royal Dutch Shell group & stationed in US, Europe & Japan before becoming Hdquarters Mgr of Plastics & Resins Dept. Joined Mobil Chem Co in 1960 as Regional Exec-Far East before becoming Gen Mgr-Intl Operations. Elected Pres of Dravo International in 1975. Assumed current responsibility as Chrmn of the Bd of Dir of Energy Support Services, Inc. in 1982. Established joint ventures in the Philippines, Taiwan, Thailand, Indonesia, Saudi Arabia, UAE, Italy and the Netherlands. *Society Aff:* NACD, WAC, AMA.

Thorbjornsen, Arthur R
Business: Department of Electrical Engineering, The University of Toledo, Toledo, OH 43606
Position: Prof *Employer:* Univ of Toledo. *Education:* PhD/Elec Engg/Univ of FL; MSE/Elec Engg/Univ of AL; BSEE/Elec Engg/Univ of WI. *Born:* 9/27/36. in Winter, WI. Served as Electronics Technician in US Navy from 1954 to 1957. Worked as Elec Engr with the Boeing Co from 1962 to 1967 and with Northrop Corp from 1967 to 1968. Joined the Univ of Toledo as Asst Prof of Elec Engg in 1972, promoted to Assoc Prof in 1977, and to Prof in 1983. Teaches courses in circuit theory, electronics, integrated circuits, and computer-aided design Principal Investigator for two Natl Sci Fdn res grants between 1974 and 1978 and for one USAF res grant in 1982 and 1983. *Society Aff:* IEEE, ΣΞ, ΦΚΦ, ASEE.

Thorfinnson, Stanley T
Business: 2909 West 7th Ave, Denver, CO 80204
Position: Exec VP. *Employer:* Woodward-Clyde Consultants. *Education:* BE/Civil Engg/Yale Univ. *Born:* Nov 1925. Graduate Study in Soil Mechanics MIT. US Navy CEC WWII. Omaha District Corps of Engrs 1946-60; Woodward-Clyde Cons, Cons Engrs, Geologists & Environmental Scientists 1960- , Principal, Dir & Vice Pres 1964- , Managing Principal - Rocky Mountain Region. Fellow ASCE; Mbr USCOLD, SAME, NSPE, ACEC, SME of AIME. Reg PE in 13 states. Enjoy hunting, fishing, skiing, golf, & photography. *Society Aff:* ASCE, ACEC, NSPE, SAME, USCOLD.

Thorn, Donald C
Home: 579 San Moritz Dr, Akron, OH 44313
Position: Prof Emeritus Elec Engg. *Employer:* University of Akron. *Education:* BS/EE/TX A&M; MS/EE/Univ of TX, Austin; PhD/EE/Univ of TX, Austin. *Born:* June 1929 Fort Worth Texas. Reg PE in NM, OH & TX. US Army Signal Corps 1951-53. The Univ of NM 1958-67; The Univ of Akron 1967-87; Consulting Engineer 1960- . Married Margaret Landers 1953; 2 daughters & 1 son. Accident Investigation, Electromagnetics. *Society Aff:* IEEE, NSPE, NAFE.

Thorn, Robert B
Business: 603 Topeka Ave, Topeka, KS 66603
Position: Owner *Employer:* Finney & Turnispeed *Education:* BS/CE/KS State Univ *Born:* 6/7/26 Native Kansan and practiced engr primarily in KS. High interest in ASCE Student Chpt Activities. Contact Mbr to KS State Student Chpt since 1958. Recently chrmn ASCE Committee on Student Services. Chrmn ASCE Committee on Registration of Engrs in 1972 and ASCE. Liason mbr to NCEE Honor mbr of KS State Chpt Chi Epsilon. Active in KSU alumni activities, church, boy scouts, and PTA. Served as chrmn of Enviro Health Committee of the State of KS. Currently on Advisory Bd to the Topeka-Shawnee Co. Board of Health. Enjoy people and golf. Part time teaching KS Univ and K.S.U. Civil Engrg Depts. *Society Aff:* ASCE

Thorne, Henry C
Home: 271 Scenic Drive, Tower Lakes, IL 60010
Position: Pres *Employer:* Petrachem International *Education:* MBA/Finance/Northwestern Univ; BChE/Chem Engg/Cornell Univ. *Born:* Nov 1929. With Amoco Chems Corp & its affiliates, in various technical & managerial positions, 1952 to 1986. Formed own company, Petrochem International in late 1986; currently serving as its president. Certified Cost Engr; Natl Pres AACE 1967-68; Delegate to Internatl Cost Engrg Council 1975- ; Elected Chrmn of the Council in 1984. AACE Award of Merit 1976; elected Fellow AACE 1977; Life Mbr in 1984. Mbr Exec Ctte, AAES Int'l Affairs Council, 1982- . Received award for outstanding serv, Ill Soc Prof Engrs 1964. Mbr AIChE, IIE, ACS, CDA. Recognized authority & lecturer in engrg economy. Numerous articles. Co-authoring book, 'Managing Capital Expenditures - Techniques & Applications.' Present continuing education seminars on Venture & Risk Analysis for sev engrg socs. *Society Aff:* AACE, AIChE, CDA, ACS.

Thornley, George C
Business: 615 Wesley Dr, Suite 212, Charleston, SC 29407
Position: Principal *Employer:* Life Cycle Engrg, Inc. *Education:* BS/EE/Univ of SC *Born:* 04/22/45 Native of Holly Hill, SC. Accepted position as electrical engr with Charleston Naval Shipyard upon graduation from USC. Developed expertise in generator, motor generator and rotating machinery problems. Was senior project engr responsible for technical development of vibration analysis and other non- intrusive monitoring and maintenance programs for specific propulsion and auxiliary ship systems at the Naval Sea Systems Command (NAVSEA). Assumed current position as Pres of LCE in 1976.

Thornton, Clarence G
Home: 28 Glenwood Rd, Colts Neck, NJ 07722
Position: Dir US Army Elec Tech & Devices Lab. *Employer:* US Army Laboratory Command. *Education:* PhD/Phys Chem/U of MI; MS/Phys Chem/U of MI; BS/Phys Chem/U of MI. *Born:* Aug 3, 1925. Teaching fellowship physical chemistry Univ of Mich 1948- 51; sect Head, Sylvania Electric Co Ipswich Mass; 1951-54; Mgr semiconductor dev, Philco Corp 1955-60; Dir Res & Engrg, Philco Corp, microelectronics div 1960-71; present Dir, Electronics Technology & Devices Lab, US Army Electronics 1953-74. Fellow IEEE; Mbr Electrochem Soc, AAAS, ACS, American Crystal Assn, Sigma Xi, Phi Lambda Upsilon, Phi Kappa Phi, Alpha Chi Sigma, US Army R&D A Achievement Award 1976. Sr. Exec. Service Award, 1980, 82, 83, 84, 85; US Army Laboratory of the Year Award, 1980, 83. Presdl. Rank Award Sr Exec Service, 1986. Except Civ Serv. Medal, Dept of Army, 1985. Numerous chmnships IEEE 1959-76; 35 publs, 6 patents in electron devices.

Thornton, Keith C
Home: 6 Foxhill Road, Montvale, NJ 07645 *Education:* MS/Civil Engg/Univ of CA Berkeley; BS/Civil Engg/Univ of CA Berkely. *Born:* 4/6/31. VP of T Y Lin & Assocs a CA firm 1963-70. Mgr & prin partner 1970-78. Mbr ASCE, ACI, & PCI. Respon for the follwoign noteworthy projs: 32-story FDR Post Office-NYC; the Garden State Art Center (hung roof cable), Holmdel NJ; the Watergate Office/Aptmt complex, Washington DC; the Broome Cty Meml Arena (space frame), Binghamton NY. Avid bridge & tennis player. Bought out partnership (T Y Lin & Associates-NY) and changed name to Keith Thornton & Associates commencing Jan 1, 1979. *Society Aff:* ASCE, ACI, PCI, NPA, PTI.

Thornton, Stafford E
Business: West Va Inst Tech, Montgomery, WV 25136
Position: Director, Center for Applied Business, Engineering & Technology *Employer:* W Va Inst of Tech *Education:* MCE/Struct. Engr/U of VA; BCE/Civil Engr/U of VA *Born:* July 29, 1934. Native of Roanoke Va. Res Scientist & Instructor, Univ of Va 1959-64; joined faculty at W Va Inst of Technology 1964, appointed Chmn Civil Engrg Dept 1965. Fellow ASCE, Pres Charleston Branch 1969, Pres W Va Sect 1975; Chmn District 6 Council 1976 Natl Dir 1978-81 (ASCE); Natl VP 1982-84 (ASCE); Trustee, United Engrg Ctr, 1984; Chmn W Va Tech Faculty Assembly 1971-73; Chmn Adv Council of Faculty to the W Va Bd of Regents 1972-73. Appointed to W Va Registration Bd for PEs, 1980; Elected Pres W Va Assoc of Licensing Bds, 1981. Enjoy golf. *Society Aff:* ASCE, WVSPE, ASEE, NCEE.

Thornton, William A
Business: Prime-Color, Inc, 27 Harvard Rd, Cranford, NJ 07016
Position: President *Employer:* Prime-Color, Inc. *Education:* PhD/Physics/Yale Univ; MS/Physics/Yale Univ; BS/Univ of Buffalo. *Born:* June 1923. MS, PhD Yale Univ; BS Univ of Buffalo. Native of western NY. Served with Army Air Force weather service 1942-46. Res Assoc, GE Res Lab 1951- 56; Westinghouse Lamp Divs 1956-83. Worked with electroluminescence, luminescent materials. Identified 'prime colors' of human vision in 1966. Developed commercial lamps based on the prime-colors to yield greater clarity, brightness, & preferred coloration. Developed a simple picture of metamerism, & a commercial lamp to assess degree of metamerism. Developed a more accurate correlate to perceived brightness than the footcandle, & a brightness meter to measure it. Mbr APS, OSA; Fellow IES; ISCC. Founded Prime-Color, Inc. in 1983; research on design of spectral composition of lamp emission. *Society Aff:* APS, OSA, IES, ISCC.

Thorpe, Robert L
Business: 292 Court St, Auburn, ME 04210
Position: Pres *Employer:* Harriman Assoc *Education:* BS/EE/Univ of ME. *Born:* 4/19/29 Chief Exec Officer for Architecural-Engrg firm of 70 employees with all architectural and engrg disciplines designing commercial, office, institutional, educational, and Governmental facilities. Holder of Cert from Natl Council of Engr Examiners, Value Engrg Cert *Society Aff:* IES, ASHRAE, IEEE, NSPE.

Thorsen, Richard S
Business: 333 Jay St, Brooklyn, NY 10201
Position: Dean of Res and Grad Studies *Employer:* Poly Inst of NY. *Education:* PhD/ME/NYU; MS/ME/NYU; BS/ME/CCNY. *Born:* 10/6/40. Resident of Syosset, NY. Married and has children. He is Prof of Mech and Nuclear Engg at the Poly Inst of NY where he is also Dean of Res and Grad Studies. He is widely published in heat transfer and solar energy literature and has led major govt and industry sponsored res projects in these areas. Working closely with Aerospace and Arch and Engg Design companies he has led the University's dev of Interactive Computer Graphics and Computer Aided Design (CAD) labs and curricula. He also has responsibility for the Univ-wide CAD applications and integration of computers into the various curricula. Dr. Thorsen is a member of ASEE, ASME, AIAA, ISES and ICGS, and has consulted for major corporate and government organizations. *Society Aff:* ASME, AIAA, ASEE.

Thorson, John M, Jr
Business: Control Data Corp, 2300 Berkshire Lane, Plymouth, MN 55441
Position: Sr Conslt *Employer:* Control Data Corp *Education:* BS/EE/IA State Univ *Born:* 12/16/29 I worked at Northern States Power and Control Data Corp as an electrical engr and mgr at various power system operation, protective relay, communication, interference, control system and training positions. I served on the Bd of Dir of IEEE, Regional Activities Bd, Educational Activities Bd, US Activities Bd and am Dir of Region 4. I was Twin Cities Section Chrmn (3 states) and General Chrmn of the 1980 Summer Power Meeting. I serve as US Advisor/expert to CIGRE on communications, on interference, on power system operation and on telecontrol. I am a PE in MN and ND. *Society Aff:* IEEE, IEEE-PES, IEEE-Comm.

Thorstensen, Thomas C
Business: 66 Littleton Rd, Westford, MA 01886
Position: President. *Employer:* Thorstensen Laboratory. *Education:* PhD/Chem/Lehigh; MS/Chem/Lehigh; BS/Chem/Univ of MN. *Born:* Nov 29, 1919. Engrg Officer WWII. Held sev indus jobs, & taught at Lowell Technilogical Inst, & as an offshoot of the res work done at the inst, Thorstensen Lab was established in 1959. Lab is for independent res, cons, & testing. Pub book, 'Practical Leather Technology' 1969. Received the Alsop Award 1971, American Leather Chemists Assn for excellence in leather sci. Pres AICA 1974-76.

Thouret, Wolfgang E
Business: 2321 Kennedy Blvd, North Bergen, NJ 07047
Position: Director Engineering & Research. *Employer:* Duro-Test Corp. *Education:* Doctor/Elec Engg/Techn Univ of Karlsruhe, West Germany; MS/Physics/Techn Univ Berlin, Germany; BS/Physics/Techn Univ Berlin, Germany. *Born:* 8/27/14. Dr *Society Aff:* IEEE, SMPTE, IESNA, APS, OSA.

Throdahl, Monte C
Business: 7811 Carondelet Ave, Clayton, MO 63105
Position: Retired Senior Vice President, Monsanto Co. *Employer:* Conslt *Education:* BS/Chem E/IA State. *Born:* March 1919 Minneapolis Minn. BS Iowa State Univ. Sr. Vice Pres, Mbr Bd/Dir, Monsanto Co. Began career with Organic Chemical Div as a Res Chemist; subseguently became Dir of Dev, Dir of Res, & Dir of Marketing; appointed Genl Mgr Internatl Div 1965, elected Vice Pres; elected to Monsanto's Bd/Dir & Exec Comm 1966. Mbr AAAS, ACS, SCI, CDA, & National Academy of Engineering P Dir IRI & Natl Assn of Mental Health. Fellow American Inst of Chemists. Fellow of American Institute of Chemical Engineers. Authored many tech articles & hold patents from US & 5 foreign countries. *Society Aff:* SCI, ACS, AIChE, AAAS, AIC, NAE, CDA.

Throop, James W
Business: 1700 W. 3rd Ave, Flint, MI 48502
Position: Prof of Process Eng *Employer:* General Motors Inst *Education:* MS/Eng/MI State Univ; BME/Mech Electrical Eng/Gen Motors Inst *Born:* 08/25/31 Joined GMI in 1962. Named Prof in 1968, Rodes Prof in 1976. Currently Chrmn of Metal Working Fluids Div of SME, VChrmn of Standards Committee of Computer Aided Mfg Intl, (CAM-1), a non-profit research orginization, Co-Chmn of GM Corporate Robot performance Committee. Active in computer aided MFG, CAD, Robotics, Computer Graphics. *Society Aff:* ASME, SME

Thuente, Sylvan T
Business: 150 College St, P. O. Box 4246, Macon, GA 31208
Position: Exec Vice Pres *Employer:* Harrison, Thuente & Assocs, P.C. *Education:* MSA/Mgmt/GA Coll; B/EE/GA Tech *Born:* 07/01/36 Exec VP of an electrical and mech consltg engrg firm doing design work throughout the Southeastern United States. Presently also serves as a dir for the Consltg Engrs Council of Georgia, a mbr of the American Consltg Engrs Council. Formerly an instructor of electronics for the State of GA Univ Sys after a stint in the US Air Force as an aviator and electronics ofcr. Having been born and raised in the Midwest, he has moved to the South to stay. He enjoys singing, fishing, and playing softball. *Society Aff:* IES, AEE, CSI, ASPE

Thuesen, Gerald J
Business: Sch of Ind and Systems Engr, Georgia Tech, Atlanta, GA 30332
Position: Prof. *Employer:* GA Inst of Tech. *Education:* PhD/Ind Engg/Stanford Univ; MS/Ind Engg/Stanford Univ; BS/Ind Engg/Stanford Univ. *Born:* 7/20/38. Mbr of ind engg faculty at GA Tech since 1968 having previously taught at the Univ of TX at Arlington and Stanford Univ. Co-author of two engineering economy texts, *Engineering Economy* and *Economic Decision Analysis* while teaching undergrad

Thuesen, Gerald J (Continued)
and grad courses in engg economics, statistical decision theory, and replacement analysis. Res and consulting on the methods of capital budgeting and the economics of elec power generation. Vice-President Publications (1979-1981); Member, Board of Trustees (1979-1981) for IIE. Member, Board of Directors, ASEE (1977-1979), Presently Editor for *The Engineering Economist*. Presented the Eugene L. Grant Award by ASEE for 1977. Assoc Editor for *The Engineering Economist* 1975-1980 and Departmental Editor for *IIE Transactions* 1976-1980. Recipient Eugene L. Grant Award, 1977 and Fellow IIE, 1987. *Society Aff:* IIE, ASEE, ORSA, TIMS, ΣΞ.

Thullen, Edward J
Business: 1407-116th Ave, NE, Bellevue, WA 98004
Position: Regional VP *Employer:* Protection Mutual Ins, Co *Education:* MBA/Econ/Univ of Chicago; BS/Fire Portection/IL Inst of Tech *Born:* 06/08/33 Regional Mgr for a mbr of the Factory Mutual System. Sub-offices in Los Angeles and San Francisco. We are involved on a daily basis with the fire protection engrg requirements of on insurances most of whom belong to Fortune's 500. Reg PE. Former officer of Wisconsin SFPE Chapter. *Society Aff:* SFPE

Thumann, Albert
Business: 4025 Pleasantdale Rd, Suite 420, Atlanta, GA 30340
Position: Exec Dir *Employer:* Association of Energy Engrs *Education:* BS/Elect Eng/CCNY; MS/Indus Engr/NYU; MS/Computer Sci/NYU *Born:* 03/12/42 Albert Thumann, PE, CEM, is a Cert State Auditor in GA, a Cert Energy Mgr and a licensed Prof Engr in NY, KY, and GA. A thirteen yr veteran of the Bechtel Corp, he is presently Exec Dir of the Assoc of Energy Engrs. He holds a BSEE from City Univ and a MSEE and MSIE from NYU. He is a noted lecturer and author of eight books. *Society Aff:* ASAE

Thun, Rudolf E
Home: 228 Heald Rd, Carlisle, MA 01741
Position: Dir Device Technology *Employer:* Raytheon Co. *Education:* PhD/Physics/Univ of Frankfurt, Germany; Diploma/Physics/Univ of Frankfurt, Germany. *Born:* Jan 30, 1921. Native of Berlin Germany; naturalized 1961. Res Physicist for DEGUSSA, Hanau Germany 1950-55, & for US Army Corps of Engrs 1955-59. Various res & managerial positions in microelectronics with IBM 1959-67. With Raytheon Co since 1967 in the following positions: Mgr of Microelectronics, Tech Dir of Semiconductor Div, Mgr of Elec Prod Design Lavb, Mgr of Advanced Electronics Lab Missile Systems Div, Dir Device Tech. Fellow IEEE & APS. Award for technical achievement, US Corps of Engrs 1959. IEEE Group Award 1975. *Society Aff:* APS, IEEE, ISHM, AVS.

Thune, Arne
Business: 47 Elm St, New Canaan, CT 06840
Position: Consultant. *Employer:* Thune Assoc PC Consult Engrs *Education:* BS/Civ Engg/Univ of Idaho *Born:* 5/22/31. Native of Norway. Worked for Wash State Bridge Dept and private consultants as Structural Engr. Opened own Consulting Engg Office in 1967. Married Ensaf Elmasry in 1956. Two daughters, Miriam and Britt. President of S.W. Chapter of CSPE 1975-76, President of CSPE 1979-80. National director of NSPE 1980, 1981. *Society Aff:* NSPE, ACI, PCI, AISC.

Thurlimann, Bruno
Business: Swiss Federal Institute of Technology, CH-8093 Zurich Switzerland
Position: Professor of Structural Engineering. *Employer:* Swiss Federal Inst of Technology Zurich *Education:* PhD/Civ Engrg/Lehigh Univ, 1951; Diploma/Civ Engrg/Swiss Fed Inst of Tech, 1946. *Born:* Feb 6, 1923 Switzerland. Res Assoc Brown Univ 1952; Res Prof, Lehigh Univ 1953-60; Prof of Structural Engg, Swiss Federal Inst of Technology, Zurich. Fellow ASCE, Honorary Mbr ACI; Honorary Pres Int Assoc for Bridge & Struct Eng (IABSE), Mbr Natl Acad of Engr, Mbr Steering Grp Council Tall Bldgs and Urban Habitat. Comite Europeen du Beton, Swiss Engrs & Architects Soc. Moisseff Award 1964, Norman Medal 1963, Res Prize 1961 ASCE. Honorary Doctor's Degree, Univ. of Stuttgart, Fed. Rep. Germany, 1983; Ernest E. Howard Award, 1986 ASCE. *Society Aff:* NAE, ASCE, ACI.

Thurman, Herman R, Jr
Home: 47 Wedgewood Rd, Newark, DE 19711
Position: Retired *Employer:* Eng Dept DuPont Co *Education:* MSCE/Civil Engg/Lehigh Univ; CE/Civil Engg/Univ of Cincinnati. *Born:* 4/7/26. Grew up in Cincinnati OH. Was co-op student and worked at Wolf Creek Dam. After graduate studies at Lehigh, went to the DuPont Engg Dept where employed for 36 yrs. In early years was structural designer of reactor and disassembly buildings and waste tanks at Savannah River Plant. Also, did structural design of power houses and manufacturing buildings for new plants of the DuPont Co. In later yrs helped redesign waste storage tanks at SRP. Also, was a stress analyst for tall stacks, pressure vessels, chlorinators and explosion accident investigations. Married, 3 children. Enjoy golf and tennis. Retired from DuPont Engg. Dept. in 1986, now a part-time consultant for First Contact Design Inc. *Society Aff:* ASCE, TBΠ.

Thurston, Marlin O
Business: 2015 Neil Ave, Columbus, OH 43210
Position: Critchfield Prof *Employer:* The Ohio State University. *Education:* BA/Physics/Univ of CO; MS/Physics/Univ of CO; PhD/Elec Engg/OH State Univ. *Born:* Sept 20, 1918. Served with the Army Signal Corps & Air Corps 1942-46. Acting Head, Dept of Physics, USAF Inst of Technology 1946-49, & of Dept of Electrical Engrg 1946-52. Joined faculty of the Ohio State Univ as Assoc Prof 1955; Prof 1959-82; Chmn 1965-77. Res & cons in the area of solid-state electronics. Fellow IEEE; Mbr APS, ASEE, Sigma Xi, Tau Beta Pi, Eta Kappa Nu, Aircraft Owners & Pilots Assn. Critchfield Prof and Dir, Electron Device Lab 1977-82; Prof Emeritus 1982- ; Pres, Thurston-Bell Assoc, Inc, 1982- ; Pres, Neoprobe Corp 1983- . *Society Aff:* IEEE, APS, ΣΞ, TBΠ, HKN.

Ticer, Philip F
Home: 1864 Swan Circle, Costa Mesa, CA 92626
Position: Vice President - Marketing. *Employer:* Edgington Oil Co, Inc. *Education:* MS/Chem Engrg/Univ of WA; BS/Chemistry/Stanford Univ. *Born:* Nov 1922. MSChE from Univ of Wash; BS from Stanford Univ. Native of Portland Ore. Served in Army Air Force 1943-46. With Union Oil Co of Calif in various refinery & crude/product supply assignments 1949-74. Assumed current responsibility for Supply & Distribution for Powerine 1974. Hobbies: golf & tennis Joined Edgington Oil Co Inc on Jan 1, 1978. *Society Aff:* AIChE.

Tiedemann, Herman H
Business: Box 28, Texas City, TX 77592-0028
Position: Conslt *Education:* BChE/-/CCNY; MChE/Physic Chem, Engg Law/CCNY; Grad/-/Univ of PA. *Born:* Doctorate studies at Univ of PA Engrg Instr; Rohm & Haas engrg; Engrg, Plant & Div Mgr GAF 23 yrs; Mgr & Tech Dir Linden Chlorine Prods 2 yrs; V Pres PCF Machinery Corp 2 yrs; Prof Engr 1946. AIChE Active Mbr 1949. Ten US patents. Papers given in Australia, Canada, France, Germany, Sweden. Interests: chlorine, nitrations, & petrochem engrg & production in Europe, Intl Engg Consultant, 1974 to present. Clients in Australia, Brazil, Canada, South Africa, United States, Yugoslavia. Expert witness for International Arbitration, Paris. *Society Aff:* AIChE.

Tien, Chang L
Business: Mech Engrg Dept, Berkeley, CA 94720
Position: Prof of Mech Engrg *Employer:* University of California - Berkeley. *Education:* PhD/Mech Engrg/Princeton Univ; MA/Mech Engrg/Princeton Univ; MM/Mech Engrg/Univ of Louisville; BS/Mech Engrg/Natl Taiwan Univ *Born:* July 24, 1935 China. Naturalized US citizen. Faculty mbr at Univ of Calif, Berkeley 1959- ; Prof of Mech Engrg 1968- , Chmn Mech Engrg Dept 1974-81, V Chanc-Res 1983-85. Reg P E Calif; Tech Cons Lockheed Missiles & Space Co 1962- 80, Genl Elec Co 1972-80. Recipient of several disting teaching & res awards at Univ of

Tien, Chang L (Continued)
Calif, Berkeley; Guggenheim Fellow 1965; ASME Heat Transfer Memorial Award 1974; ASME Fellow 1974; ASME Gutus Larson Memorial Award 1975; Mbr NAE 1976; AIAA Thermophysics Award 1977; Sr US Scientist (Humboldt) Award, Federal Republic of Germany 1979; Sr US Scientist Fellowship, Japan Soc Promotion of Sci 1980; Honorary Res Prof, Chinese Acad of Sci, 1981; AIChE-ASME Max Jakob Memorial Award, 1981; Honorary Prof, Huazhong Inst of Tech, 1981, Qinghua Univ, 1982, and Shanghai Jiatong Univ, 1983; Visiting Viola D Hank Chair Prof, Univ of Notre Dame, 1983; Prince Distinguished Lecturer, AZ State Univ, 1983; AIAA Fellow 1985; ASME Distinguished Lecturer, 1987; Hawkins Memorial Lecturer, Purdue Univ, 1987. *Society Aff:* ASME, AIAA, ASEE, AAAS

Tien, Chi
Business: Chem Engrg & Mat Sci, Syracuse, NY 13210
Position: Prof Chem Engrg & Mat Sci Dept. *Employer:* Syracuse University. *Education:* BS/Chem Eng/Natl Taiwan Univ; MS/Chem Engg/KS State Univ; PhD/Chem Engg/Northwestern Univ. *Born:* Oct 8, 1930. PhD Northwestern Univ. Taught at Univ of Tulsa 1957-59; Univ of Windsor 1959-63; Syracuse 1963- . Cons for govt & lil indus. Reg Engr in Ontario Canada. Mbr AIChE, CIC. Speciality in heat transfer, fluid particle spparation, advanced waste treatment. *Society Aff:* AIChE, AAUP, FPS.

Tien, James M
Business: ECSE Department, Rensselaer Polytechnic Institute, Troy, NY 12180-3590
Position: Acting Depart Head *Employer:* RPI. *Education:* PhD/Operations Res/MIT; EE/Systems Engg/MIT; SM/Systems Engg/MIT; BEE/Elec Engg/RPI. *Born:* 3/27/45. Dr Tien was an Area Res Dir at Urban Systems Res and Engg, Cambridge, MA (1973-1975); a Proj Dir at the Rand Corp, New York, NY (1970-1975); and a Mbr of the Technical Staff at Bell Telephone Labs, Holmdel, NJ (1966-1969). He joined the Department of Elec, Computer, and Systems Engg at Rensselaer Poly Inst in 1977, and has been the Acting Department head since 1986. He has also been VP at Public Systems Evaluation, Inc, and EnForth Corp in Cambridge, MA (since 1975). His focus is on the development & application of computer and systems analysis techniques to large scale systems, including public sector issues, computer networks, data base mgmt systems, and expert systems. *Society Aff:* ORSA, TIMS, IEEE, URISA, SMIS

Tien, John K
Business: Krumb Sch of Mines, New York, NY 10027
Position: Howe Chair Professor of Met & Materials Sci. *Employer:* Columbia University. *Education:* BS/Mech Engr/Worchester Poly Tech; MEng/App. Sci & Engr/Yale University; MS/Met & Mat'ls Sci/Carnegie-Mellon Univ; Ph.D./Met & Mat'ls Sci/Carnegie-Mellon Univ *Born:* 6/4/40. PhD Carnegie-Mellon Univ 1968; BS WPR 1962, MEng Yale 1964. Res Supr Chase Brass & Copper Co 1964-65; Res Sci Pratt & Whitney Aircraft 1968-72; with Columbia Univ 1972- , supervising res & instructions in physical-mech met. Specializes in high temp matl behavior. Author over 200 papers & patents & cons to many major corps. ASM Bradley Stoughton Best Young Professor Award 1975; Young Man of Amer 1974 US Jr Chamber of Commerce. *Society Aff:* ASM, TMS, MRS

Tien, Ping K
Business: Room 4F-307. Crawsfords Corner Road, Holmdel, NJ 07733
Position: Department Head. *Employer:* AT&T Bell Laboratories. *Education:* PhD/Elec Engg/Stanford Univ; MS/Elec Engg/Stanford Univ; BS/Elec Engg/Natl Central Univ, China. *Born:* 8/2/19. Checkiang, China. Joined Bell Labs 1952 & has been active in various fields incl microwave electronics, electromagnetic field theory, ferromagnetism, ultrasonics, superoconductivity, lasers, integrated optics & thin-film technology. Mbr Sigma Xi, Fellow of IEEE & Fellow of Optical Soc of Amer; Mbr NAE; Mbr NAS. *Society Aff:* IEEE, OSA, NAS, NAE, ΣΞ.

Tierney, John W
Business: 1230 Benedum, Pittsburgh, PA 15261
Position: Prof *Employer:* Univ of Pittsburgh *Education:* PhD/ChE/Northwestern Univ; MS/ChE/Univ of MI; BS/ChE/Purdue Univ *Born:* 12/29/23 Sr. Research Engr, Pure Oil Co, 1948-1953. Asst Prof Purdue Univ, 1953- 1956. Mgr, Digital Control Dept, Univac Div, Sperry Rand, 1956-1960. Associate Prof, Univ of Pittsburgh, and Visiting Prof, Universidad Tecnica Federico Santa Maria, Valparaiso, Chile, 1960-1962. Prof, Univ of Pittsburgh, 1962-present. Fulbright Lecturer, Universidad de Barcelona, Barcelona, Spain, 1968-1969. Research interests and publications in process modelling, digital control, calculation methods for staged separation systems, and reaction engrg. *Society Aff:* AIChE, ACS, ASEE

Tierney, Martin J
Home: 1025 E Lakeview Dr, Baton Rouge, LA 70810
Position: VP. *Employer:* Brown Eagle Corp. *Education:* Grad Study/Elec Chem/Stuttgart Tech; BS/Chemistry/Middlebury College. *Born:* 7/31/15. Born in Waterbury CT - Uniroyal Inc after grad. After two yrs studied electrochemistry in Germany. Took part in the synthetic rubber prog during the war with Uniroyal & after the war for two yrs on leave to US Strategic Bombing Survey & to the Office & Rubber Reserve of RFC - Rubber Reserve Production Mgr for GR-S plants. Returned to Uniroyal in market res-commercial dev until 1964 when became Pres of Rubicon Chem Inc a 50-50 joint venture of Uniroyal & Imperial Chem Ind Ltd. Retired as Pres of Rubicon in 1977 & joined Brown Eagle Corp a contract personel services firm. *Society Aff:* AIChE, ACS, CDA, CMA.

Tiernon, Carlos H
Business: 901 Glasgow Ave, PO Box 1, Fort Wayne, IN 46801
Position: President *Employer:* Deister Concentrator Co, Inc *Education:* BS/Met Eng/Univ of MO-Rolla *Born:* 12/15/30 Served with Army Corps of Engrs 1954-1956. Company Commander. Field Engr for the Deister Concentrator Co, Inc 1956-1969. VP 1970-1978. Pres 1978- . Developed and introduced suspended multiple deck Coal Washing and Mineral Concentrating Tables in North Amer, South Amer, and Europe. Internationally recognized Coal Preparation Engr and Mineral Dressing Engr. Developed and introduced unique hydraulic Flotation Cell 1980. Mbr of Exec Bd of Coal Div of SME of AIME. Ft Wayne Mayor's Citation 1970 for rescue of mother and child from fire. Recreational acitivities, golf, fishing, and hunting. *Society Aff:* AIME

Tietz, Thomas E
Home: 12640 Corte Madera Ln, Los Altos Hills, CA 94022
Position: Consultant-Materials Sciences *Employer:* Self Employed *Education:* PhD/Phy Metallurgy/Univ of CA, Berkeley; MS/Phy Metallurgy/Univ of CA, Berkeley; BS/Mech Engg/Univ of CA, Berkeley. *Born:* 7/22/20. Design Engr Lane-Wells Co, Los Angeles 1944-47; Res Engr Univ of Calif, Berkeley 1947-50; Sr Metallurgist Stanford Res Inst 1954-59; Mgr Met & Composites Lab Lockheed, Palo Alto Res Lab 1959-63 & 1965-81; Senior Member Lockheed Palo Alto Research Lab 1982-83, Consultant-Materials Sciences 1983-Present. Visiting Lectr in Met Stanford 1957-58. Reg Mech Engr Calif. Mgr AIME, Chmn San Francisco IMD Section 1963-64, high temp alloys ctte 1963-67, struct materials ctte 1965-72; Mbr ASM, Fellow 1976, seminar ctte 1971-76; Mbr AIAA, Materials Tech Ctte 1977- 80; Sigma Xi. Author (with J W Wilson): Behavior & Properties of Refractory Metals, Stanford Univ Press 1965, Russian translation 1969, Spanish translation 1971. Mbr Natl Acad of Scis Natl Matls Advisory Bd Ctte on Amorphous and Metastable Matls 1978-79. Guest Sci at the Max-Planck-Institut for Metallforschung, Stuttgart 1979. Visiting Sr Staff Assoc, Natl Acad of Scis, Natl Materials Advisory Bd 1981-82. *Society Aff:* ASM, AIME

Tiller, Frank M
Business: Chem. Eng. Dept, University of Houston, Houston, TX 77004
Position: MD Anderson Prof of Chem Engg. *Employer:* Univ of Houston. *Education:* PhD/ChE/Univ of Cincinnati; MS/ChE/Univ of Cincinnati; BChE/ChE/

Tiller, Frank M (Continued)
Univ of Louisville; D Honoris Causa/-/Univ of Brazil; D Honoris Causa/-/State University of Rio De Janeiro *Born:* 2/26/17. MD Anderson Prof of Chem Engg. Formerly Dean of Engg and Dir of Intl Affairs at Univ of Houston. Consultant and lecturer in solid-liquid separation. Res primarily in field of compressible cakes including basic flow theory, deliquoring, delayed cake filtration, thickening, centrifugation, & washing. Founder of Intl Consortium of Filtration Res Groups. Directed AID progs establishing grad engg and nation-wide univ modernization in Brazil. Also progs in Costa Rica, Ecuador, and Mexico. AIChE Colburn Award 1950, Filtration Soc Gold Award 1970, D Honoris Causa Univ of Brazil 1962, D Honoris Causa State Univ of Rio de Janeiro 1966. *Society Aff:* AIChE, ASEE, FS, FPS.

Tilles, Abe
Business: 2663 Pillsbury Ct, Livermore, CA 94550
Position: Consulting Engr (Self-employed) *Education:* PhD/Elec Engg/Univ CA, Berkeley; MS/Elec Engg/Univ CA, Berkeley; BS/Elec & Mech Engg/Univ CA, Berkeley. *Born:* 3/9/07. Specialties Electric circuit safety. Power transmission, distribution. Forensic Engrg, Nuclear sci applications. Engg education. Employment Univ CA. Israeli Inst Tech. Lawrence Lab. Los Angeles Power Light. Pacific Gas and Elec. Southwest Engg. Mare Island Navy Yard. Kaiser Richmond Shipyard. Societies Sigma Xi, Berkeley, Treas, Bd. ASEE Div Continuing Studies, Regional chrmn, natl bd. Relations with Industry, Regional Chrmn. IEEE Fellow, 5 natl committees, Chrmn Regional Prize Papers and Regional Awards, Chrmn San Francisco Sec. Founder, editor, the San Francisco Engr. Founder, pres, Proj Southwest. Chrmn, San Francisco Engg Council. Recipient, Founder Societies Alfred Noble Award. Reg PE, CA EE-2. *Society Aff:* IEEE, ASEE, ΣΞ, HKN

Tillinghast, John A
Business: Liberty Lane, Hampton, NH 03842
Position: Sr VP-Technology *Employer:* Science Applications International Corp *Education:* BS/Mechanical Eng/Columbia Univ; MS/Mechanical Eng/Columbia Univ *Born:* 04/-/27 BS in Mech Engrg, Columbia U, 1948, MS, 1949. With American Elec Power Service Corp, NYC 1949-79, Exec VP Engrg and Constrn, 1967-72, Sr Exec VP, VChrmn Engrg and Constrn 1972-79. Sr VP-Tech, Wheelabrator-Frye Inc, Hampton, NH; 1979-83 Science Applications International Corp. La Jolla, CA. Sr VP-Tech, The Signal Co. La Jolla, CA, 1983-86. Since 1986 Sr VP ASME Fellow; mbr Natl Acad of Engrg, Chrmn, Energy Engineering Board, NRC. IEEE. Patented generating unit control system. *Society Aff:* NAE, ASME, IEEE, TBΠ

Tillitt, James C
Business: 2101 Hennepin Ave, Minneapolis, MN 55405
Position: Pres & Owner. *Employer:* Tillitt & Assoc, Inc. *Education:* MS/Civ Engg/Univ of MN; BS/Arch Engg/Univ of IL. *Born:* 6/6/26. Pres and chief engr of Tillitt & Assoc Inc since 1968, having joined the firm in 1960. Prior to that time was on the staff of Schuett - Meier Co, a Minneapolis engg firm. Past pres of the Consulting Engrs Council of MN a mbr of the Minneapolis Chap of Rotary Intl, the American Society of Civ Engrs and a mbr of the MN Soc of PE-served as pres of the Minneapolis chap. In responsible charge of bridge design for hgwy, railroad and pedestrian bridges. These bridges include the Intl Bridge between Intl Falls, MN and Ft Frances, Ontario across the Rainy River. *Society Aff:* NSPE, ACEC, SAME, NAEP, ASCE.

Tillman, Erland A
Home: 300 St. Ives Dr, Severna Park, MD 21146
Position: Assoc VP/Project Dir *Employer:* Daniel, Mann, Johnson, & Mendenhall *Education:* MS/Math & Educ/Univ of RI; BS/CE/Univ of RI *Born:* 02/28/12 Retired from Corps of Engrs, US Army as Colonel after 28 yrs of widely varied assignments covering all phases of engrg and construction mgmt, logistics planning and support admininstration. Following this service for eight yrs as Dir of Engrg and Construction for San Francisco Bay Area Rapid Transit District during construction of that 75-mile sys. Since 1974 has been with DMJM and has been Project Dir for design of initial section of Baltimore Region Rapid Transit Sys. Is closely involved in rail transit committee activities of American Public Transit Association. *Society Aff:* ASCE, SAME, PIANC, APTA

Tillman, Frank A
Business: Dept of Ind Engg, Manhattan, KS 66506
Position: Dept Hd. *Employer:* KS State Univ. *Education:* PhD/Ind Engg/Univ of IA; MS/Ind Engg/Univ of MO; BS/Ind Engg/Univ of MO. *Born:* 7/22/37. Native of Linn, MO. Reg PE in CA. 1969-present, Prof and Hd, Dept of Ind Engg, KS State Univ. 1972-present, Dir Prog Control, Exec Office of the Pres, Price Commission, Wash, DC. Fellow Inst of Ind Engg 1986; 1966-69, Assoc Prof and Hd, Dept of IE, KS State Univ. 1965-66, Asst Prof, KS State Univ. 1963-65, Instr, Univ of IA. 1961-63, Operations Res Analyst, Std Oil Co. 1960-61, Instr, Univ of MO. *Society Aff:* IIE, IEEE, ORSA, TIMS.

Tillquist, George E
Business: 515 S. Flower St, Los Angeles, CA 90071
Position: Dir Strategic Tech and Planning *Employer:* Atlantic Richfield Co. *Education:* BS/ChE/Northwestern. *Born:* Mar 22, 1930 Evanston Ill. Held various engrg & mgmt positions with Sinclair Res & Sinclair Petrochems 1952-69, incl acting as Corp Secy for Sinclair Petrochems 1963-68; with Atlantic Richfield Co in various assignments since Atlantic-Sinclair merger in 1969; assumed position of V Pres ARCO Chem Co Div Atlantic Richfield Co in Jan 1976 with respon for coordination of chem joint ventures & mgmt of chem specialty businesses. Officer and Dir of various Arco Chem Co subsidiaries and joint ventures including Oxirane Chemical Co., Centennial Hydrocarbons Co., Nihon Atlantic, Atlantic Richfield Hanford Co. and Arco Nuclear Co. Became VP Arco Chem Co respon for Finance & Planning in 1977. Became Mgr Strategic Planning for Atlantic Richfield Co in 1979 and Dir of Strategic Tech and Planning in 1984. *Society Aff:* ACS, AIChE, API, PEI.

Tilsworth, Timothy
Home: 1900 Raven Drive, Fairbanks, AK 99709-8358
Position: Professor *Employer:* University of Alaska. *Education:* PhD/Environ Engg/Univ of KS; MSCE/Sanitary/Univ of NB; BSCE/Civil/Univ of NB *Born:* 4/6/39. Pres AK Sect ASCE 1972-73. Chi Epsilon 1968. Listed Who's Who & Who's Who in the West 1975-87, Amer Men & Women of Sci 1974-87, Community Leaders & Noteworthy Amers 1975, Chmn Univ Assembly, Univ of AK 1975-76. Mbr ASCE, AWWA, APWA, WPCF, AEEP. Instructor Civil Engg, Univ of NB 1966; Proprietor, Alaska Arctic Environmental Services, Environ Cons 1972- . USPHS Trainee 1967-70. Asst to the Pres for Acad Affairs, Univ of AK, 1976-78. Hobbies include fishing, canoeing, rafting & shetland sheepdogs. *Society Aff:* ASCE, AWWA, AEEP, WPCF.

Timascheff, Andrew S
Home: 242-West-22nd Ave, Vancouver, BC, Canada V5Y 2G4
Position: Hon Dir. *Employer:* Research Institute of Hydro-Quebec. *Education:* Dr Ing/-/Techn Univ Munich; Dr Ing/-/Techn Univ of Munich; Dipl Ing/-/Techn Univ of Karlsruhe. *Born:* 1899 St Petersburg Russia. 1929-51 Scientific Div Siemens-Schuckert, Berlin, Erlangen (trans behariour app & sys); 1951-63 Genl Engrg Dept ALCAN, Montreal (transmission lines, smelter operation); 1963-66 Visiting Prof Univ Laval, Quebec; 1966-69 Prof 'Titulaire' Univ Laval, Quebec (el machinery, behaviour of sys); 1969-79 Cons I R E Q, Varennes. Publ about 40 papers (machinery, stability, transm lines, forces, skin effect, bundle conductors, icing, dielectr properties of atmosphere); authored 1st German monogr on stability 1940. IEEE Fellow 1967; EIC Fellow 1970; Ross Medal 1963. *Society Aff:* IEEE, EIC.

Timby, Elmer K
Home: Apt. H-102 Pennswood Village, Newtown, PA 18940
Position: Member, Advisory Board. *Employer:* Howard Needles Tammen & Bergendoff. *Education:* CE/Civil Engg/OH State Univ; BCE/Planning & design/OH

Timby, Elmer K (Continued)
State Univ. *Born:* Dec 19, 1905. Cvl Engrg Faculty Princeton Univ 1928-49 (resignation from Chmnship). On leave WWII USNR, design, outfitting, const, oper, maintenace floating dry docks. 1949- ptnr HNTB, Cons Planners, Engrs & Architects; specialized transportation facilities, emphasis on feasibility, design, constr, oper, maintenance projs financed by revenue bonds. Active in natl & internatl prof learned socs. Hon Awards from OSU, ASCE (2), NSPE, ARBA (2), IABSE. Biographical listings incl: Who's Who in the World, Blue Book (London), Dictionary of Internatl Biography (London), Internatl Businessmen's Who's Who. In retirement active in public works legislation and procedures; also in transfer of technology to developing nations. *Society Aff:* ASCE, NSPE, IABSE, ARTBA.

Timm, Lyle A
Home: 1815 Maple St, Bethlehem, PA 18017
Position: President. *Employer:* Timm Materials of Construction Inc. *Education:* MS/ChEng/Univ WI-Madison; BChE/Ch Eng/Marquette Univ. *Born:* 1922, Milwaukee Wisc. Res Engr Allied Chem Corp 1945-50; dev & corrosion engr for Sharples Chems & Penwalt Corp 1950-58. Mgr of Sales in Mid-Atlantic area for Saran Lined Pipe Co (Dow Chem) 1958-72; Pres Timm Materials of Const, Inc. 1971- , a sales firm representing manufacs of corrosion & process equipment. P Chmn Lehigh Valley Section AIChE; Mbr NACE. Reg P E Mich. Enjoy out-of-doors, sports, furniture crafting, photography. *Society Aff:* AIChE, NACE.

Timmerhaus, Klaus D
Business: Engrg. Ctr, University of Colorado 424, Boulder, CO 80309
Position: Associate Dean of Engineering. *Employer:* University of Colorado. *Education:* BS/Chem Engrg/Univ of IL, 1948; MS/Chem Engrg/Univ of IL, 1949; PhD/Chem Engrg/Univ of IL, 1951. *Born:* Sept 10, 1924. Process Design Engr, Calif Res Corp 1952-53; Chem Engr Natl Bureau of Standards, Boulder Colo Cryogenic Lab, summers 1955, 57, 58 & 61; Cons 1955 to present; Univ of Colo Prof 1953- : Assoc Dean of Engrg & Dir of Engrg Res Center 1963- . Served at Natl Sci Foundation, Wash D C, Div of Engrg Sept 1972-Dec 1973. Recipient of 2nd Samuel C Collins Award for Cryogenic Tech 1967; recipient of 23rd Geo Westinghouse Award for Outstanding Teaching 1968; recipient of Alpha Chi Sigma Awd for Chem Engrg Res 1968; recipient AIChE Founders Awd, 1978; recipient of AIChE Eminent Chem Engrg Awd 1983; recipient of NSF Distinguished Public Service Awd 1984. Mbr AIChE 1958- ; Pres 1976. Elected Mbr nAE 1975; Mbr Sigma Xi 1951- , Pres 1986; Elected Mbr Austrian Acad of Sci 1976; Elected Mbr VDI 1983; Mbr AIChE, ASEE, AAAS, IIR, AAS, AAEE, Sigma Xi, Sigma Tau, Tau Beta Pi, Alpha Phi Omega, Phi Eta Sigma, Phi Lambda Upsilon. Reg P E Colo 1966- . *Society Aff:* AIChE, ASEE, AAAS, AAS, AAEE, IIR.

Timo, Dominic P
Home: 872 Hereford Way, Schenectady, NY 12309
Position: Manager - Struct Dev Engrg. *Employer:* General Eelctric Co. *Education:* BS/EE/Univ of WA; -/GE Adv Engg/Program, A, B, C, Courses. *Born:* Aug 1923. BSEE Univ of Washington. G E Advanced Engrg Program. Held various individual contributor & mgmt positions in G E 1047- . Respon for struct dev of large steam turbines for past 25 yrs. Author of over 35 tech publs. Elected Fellow ASME 1975.

Timoshenko, Gregory S
Home: 4805 Gulf of Mexico Dr, Longboat Key, FL 33548
Position: Professor Emeritus of Elec Engrg. *Employer:* University of Connecticut, Storrs. *Education:* PhD/Elect Engr/Univ of MI; Dipl Ing/Appl Physics/Tech Univ of Berlin. *Born:* Nov 1, 1904 St Petersburg Russia; son of Stephen P & Alexandra Timoshenko. 1934-39 Instr in Elec Engrg MIT; 1939- Univ of Conn (1946-68 Hd Elec Engrg Dept). Reg P E in Conn (aircraft ignition sys, submarine sys). Fellow APS 1941- ; Life Fellow IEEE 1971- ; Mbr N Y Acad of Sci 1963- ; Life Mbr ASEE 1971- ; mbr Sigma Xi, Iota alpha, Tau Beta Pi, Sigma Pi Sigma, Phi Kappa Phi. *Society Aff:* APS, IEEE, ASEE.

Tiner, Nathan A
Home: 1017 Skyline Dr, Laguna Beach, CA 92651
Position: Mgr. *Employer:* Microprocess Lab Inc. *Education:* PhD/Met Engg/Stanford Univ; MA/Met/Stanford Univ; BA/Engg/Stanford Univ. *Born:* 11/3/17. Dev engr for 5 yrs for US Steel Corp, Res specialist for 5 yrs for Rockwell Intl Corp, Chief scientist for 13 yrs for McDonnell Douglas Astro Co. Asst dir for Engg Res Ctr for 5 yrs and presently Consultant Mgr for Microprocess Lab Inc since 1971. Author of over 40 technical papers and 2 books in failure analysis and metallurgical res. *Society Aff:* ASM, AIME.

Ting, Thomas CT
Business: M/C 246, Box 4348, Chicago, IL 60680
Position: Professor *Employer:* Univ. of Illinois at Chicago *Education:* Ph.D./Applied Mathematics/Brown University; B. Sc./Civil Engineering/National Taiwan University *Born:* 02/09/33 He was a Visiting Professor at Stanford University in 1972-73, an Associate Editor of Journal of Applied Mechanics in 1975-82. He is a Fellow of the ASME, an honorary member of the Society of Theoretical and Applied Mechanics of the Republic of China. He is listed in many Who's Who, such as American Men and Women of Science, World Directory of Mathematicians, Dictionary of International Biography, etc. *Society Aff:* ASME

Tinker, E Brian
Business: Saskatoon Saskatchewan, Canada
Position: Vice President (Administration) *Employer:* University of Saskatchewan *Education:* BE/Chem Engg/Univ of Saskatchewan; MSc/Chem engg/Univ of Saskatchewan; PhD/Chem Engg/Univ of Saskatchewan. *Born:* Jan 1932 Yorkton Saskatchewan. Production Engr Shell Oil Co 1954-57; faculty mbr Univ of Sask, Saskatoon 1957-68; Univ of Sask, Regina 1968-72; Acting Dean of Grad Studies Univ of Saskatcewan, Regina 1972; V Princ 1972-74; V Pres Univ of Regina 1974-1981. V. Pres (Admin) Univ of Sask. 1981- .Dir of Sci Affairs, Chem Inst of Canada 1973-75; V Pres 1975, Pres 1976, Canadian Soc of Chem Engrg; Mbr ASPE Saskatchewan, Natl Assn of Corrosion Engrs, ASEE. *Society Aff:* APES, NACE, ASEE, CSChE.

Tinkler, Jack D
Home: 4460 Lower River Road, Lewiston, NY 14092
Position: V.P. of Engineering *Employer:* Hooker Chemical Corporation *Education:* PhD/ChE/Univ of DE; MCE/ChE/Univ of DE; BS/ChE/Univ of IL. *Born:* 4/12/36. Native of Lansing, IL. Joined Sun Oil Co in 1963 as Process Dev Engr. Left Sun Oil in 1977 with title of Res Prog Mgr. Joined Occidental Res Corp (ORC) in 1977 as VP of Engg. Promoted to Sr VP of ORC in Sept 1979. Enjoy golf. Promoted to VP of Engineering for Hooker Chemical Corporation in Jan., 1981.. *Society Aff:* AIChE, ACS.

Tinney, William F
Home: 4145 SW 36th Pl, Portland, OR 97221
Position: Head Analysis Section. *Employer:* Bonneville Power Admin-D/Int. *Born:* May 5, 1921; native of Portland Ore. BS & MS 1948 & 1949 Stanford. Radar Officer US Army Air Force WWII. With Bonneville Power Admin since 1950. 1st assignments in substation design & hydro resources. Since 1955 have been engaged in application & dev of methods for computer solution of power sys problems. Credited with pioneering sparse matrix methods in computation. Gold Medal Dept of Interior 1976. Prize papers in Power Engrg Soc 1970 & 1975.

Tippett, G Ross
Home: 1323 Beechwood, Abilene, TX 79603
Position: Owner. *Employer:* Ross Tippett & Assoc, Consulting Engr. *Education:* MS/Mech Engr/TX A&M Univ; BS/Mech Engr/TX A&M Univ. *Born:* 12/29/24. 1943-1945 US Navel Reserve Officer; 1946-1948 Grad Student and Instr in Mech Engg Dept at TX A&M Univ. 1948-1954 Mech Engr for W TX Utilities Co, Abilene, TX except for 19 1/2 months of active duty in USNR. 1954-1970 Partner in

Tippett, G Ross (Continued)
Tippett & Gee Consulting Engrs. 1970-1972 Pres of Tippett & Gee, Inc, Consulting Engrs. 1972-1977 Pres & Chrmn of Bd of Tippett & Gee Engg Services Corp. 1977-present Owner of Ross Tippett & Assoc, Consulting Engrs, Abilene, TX. *Society Aff:* ACEC, NSPE, ASHRAE, CEC-T

Tisdale, Reginald L
Home: 2801 Pound Dr, Tallahassee, FL 32312
Position: Director *Employer:* Barrett-Daffin-Carlan *Education:* BS/Civil Eng/Univ of FL. *Born:* 2/1/41. Native of Winter Garden FL. Grad with honors in 1965 from Univ of FL. Served as Hydrologist in Orange County, FL, 1965-66. 1967-76 Design Engg and Project Mgr for Howard, Needles, Tammen and Bergendoff. Specializing in highway design. Presently VP in charge of Engg for Barrett Daffin and Carlan. Respon include overall management of engg projects in home office. Pres FL Section ASCE 1977-78. Avid duck hunter. *Society Aff:* ASCE. NSPE, TBΠ, ΣT.

Tisdel, Joseph E
Business: Box 400, 113 Main St, Canton, NY 13617 - 0400
Position: Principal *Employer:* Tisdel Associates *Education:* BCE/Transportation/RPI. *Born:* 04/06/30 Native of Buffalo, NY. Served with US Army 1953-55. Design and field engr with Sargent-Webster-Crenshaw and Folley, and later with Lewis-Dickerson Assocs, during construction of St Lawrence Seaway and Power Project. Partner in Akins- Tisdel Assocs between 1963-69. Principal of Tisdel Assocs, consltg engrg, since 1969. Analysis and design responsibility for roads, airports, railroads, waste and water treatment facilities, dams, bridges, industrial parks and bldgs, and municipal and recreational faciltles. Registered PE in NY, VT and PA and registered LS in NY. Certified by Natl Council of Engrg Examiners. *Society Aff:* NSPE, ACEC, WPCF, AWWA

Tittel, Frank K
Business: Rice University, ECE Dept, Houston, TX 77251-1892
Position: Prof *Employer:* Rice Univ *Education:* PhD/Physics/Oxford Univ, England; MA/Physics/Oxford Univ, England; BS/Physics/Oxford Univ, England *Born:* 11/14/33 Born in Berlin, Germany in 1933. Received the MA and PhD degrees in Physics from Oxford Univ, Oxford, England in 1959. Research Physicist at the Gen Elec Research and Development Center, Schenectady, NY 1959-67. Since 1967 at Rice Univ Houston, TX as a Prof in Electrical Engrg Research. Interests include Laser Devices, Nonlinear Optics and Laser Spectroscopy. Author of over 100 papers in Quantum Electronics. Fellow of the Inst of Electrical Engrs and mbr of Amer Physical Soc and Optical Soc of Amer. *Society Aff:* IEEE, OSA, APS

Tizian, Sylvan L
Business: 99 Prospect St, Stamford, CT 06902
Position: President. *Employer:* Tizian Engineering Associates Inc. *Education:* EE/Elec Engg/College of the City of NY; BSE/Engg/College of the City of NY. *Born:* 12/13/11. Native of NY City. Ch Elec Engg USN Blimp Base Const, Brunswick GA 1941-42; Genl Mgr United Ship Repair Yards, Brooklyn 1943-45; Ptnr cons engg firm, Muzzillo, Kreitner & Tizian 1945-48; Ptnr Muzzillo & Tizian 1948-60; Pres Tizian Assoc 1960-68; Pres Tizian-Ibarra Assoc Puerto Rico 1962-69; Pres Tizian Engg Assoc, Inc 1968- , Assoc Kassner-Tizan Corp 1979- . Hobbies: modelmaking, collecting ceramic bottles. *Society Aff:* IEEE, NSPE, AEE, ASME.

Tobias, George S
Home: Backbone Rd, Sewickley, PA 15143
Position: President. *Employer:* Envirotrol, Inc. *Education:* MSChE/Chem Engr/OH State Univ; BChE/Chem Engr/OH State Univ. *Born:* April 1916. 1939-41 Chem Engr Eastman Kodak; 1941-46 served to Maj AUS; 1946-61 Sr Engr to Asst Dir, Esso Res & Engrg Co; 1961-65 Mkt-Res-Tech, Serv Coordinator Esso Internatl; 1965-70 Tech Dir Pittsburgh Activated Carbon Co, Calgon Corp; 1970 assumed present respon as President, Envirotrol Inc, for manufacture & reactivation of granular activated carbons. Expertise in adsorption, activated carbon, liquid & gas phase; treatment of indus & municipal waste & water; air pollution control & gas purification; motor & aviation fuels, lubricants & additives. Patentee in field. *Society Aff:* AIChE, ACS, AIC, SAE.

Tobin, Henry G
Business: 10 W 35th St, Chicago, IL 60616
Position: Assistant Director. *Employer:* IIT Research Institute. *Education:* MS/EE/IL Inst of Technology; BS/EE/Univ of IL. *Born:* June 1934, Chicago. BS Univ of Ill; MS Ill Inst of Tech. With IITRI since 1957, current respon include mgmt of Electronic Systems Sect comprising instrumentation, recording, field operations, biomedical engrg & instrument services, also respon for United Kingdom operations & marketing to indus clients, previously respon for medical device test & evaluation facility. V Pres ISA 1974-76; Treasurer Natl Engrg Consortium 1973-79. *Society Aff:* ISA, IEEE, AAAS.

Todd, David B
Home: 51 E Hannum Blvd, Saginaw, MI 48602
Position: Tech Dir. *Employer:* Baker Perkins Inc. *Education:* PhD/Chem Engg/Princeton; MS/Chem Engr/Northwestern; BS/Chem Engg/Northwestern. *Born:* 12/21/25. in Chester, PA, grew up in ME. US Navy 1943-47, USNR to 1972. Fulbright fellow, Technical Univ Delft, Netherlands 1950. Engr and supervisor at Shell Dev Co 1952-63, specializing in fluidization, extraction, and polymer processing, Mgr of engg, Podbielniak div, Dresser Industries 1963-67. since 1967, technical dir, Baker Perkins Inc with worldwide process engineering respon for polymer process equipment. PE license in IL and MI. chrmn, Mid-MI Section AIChE 1974-75, and selected Chemical Engr of the Year, 1979. Have served on United Way, symphony, church, and sch district boards. *Society Aff:* AIChE, AAAS, SPE, ASSE, ACS.

Todd, Frederick H
Home: –Sea Lodge–, 10 B Harbour Rd - Beadnell, Chathill Northumberland, England NE67-5BB
Position: Retired (Self-employed) *Education:* BSc/Naval Arch/Durham Univ, England; PhD/Naval Arch/Durham Univ, England *Born:* 01/06/03 Armstrong, Whitworth & Co Shipbldrs, 1919-1926. Newcastle, England. Research scholar, 1926-1928, Durham Univ, England. BSc 1925, PhD, 1931. Scientific Staff, Ship Div, Natl Physical Lab, England, 1928-1940, 1942-1948. Prof of Naval Arch, 1940-42, Durham Univ, England. 1948-1957 Head, Hydrodynamics Lab, US Navy Dept, Washington, DC. Md, Head, Ship Hydrodynamics Lab, Natl Physical Lab 1959-62 England. 1962-1969 Scientific Adviser to Dir-Naval Ship Research & Dev Lab, US Navy, Washington, DC. 1969-1972 Conslt, Gibbs & Cox, Naval Architects, NY. Medals: For services to Naval Arch, 1931 Gold Medal, North East Coast Inst, Newcastle, England; 1965 Gibbs Gold Medal, Natl Acad of Scis, USA; 1967 Davidson Gold Medal, Soc of Naval Architects, NY. *Society Aff:* RINA, SNAME

Toebes, Gerrit H
Business: Sch of Civ Engg, W Lafayette, IN 47907
Position: Prof. *Employer:* Purdue Univ. *Education:* Civ Engr/Hydraulic Engg/Delft Tech Univ; PhD/Fluid Mechanics/MIT. *Born:* 1/30/27. Engg experience in the Netherlands, Norway, MA. Taught fluid mechanics, hydraulics, water resource engg, systems analysis at Purdue Univ & abroad. CE Mech Div ASCE; Chrmn Div I, IAHR, & other offices in ASCE, IAHR, AWRA, ASEE. Hd Hydraulic & Systems Engg, Sch of Civ Engg, Purdue Univ. Consulting: flow induced vibrations; reservoir systems operations; engg-economic analysis with engg consultants, TVA, UNESCO, & public agencies in the Netherlands. Personal: classical music, graphic arts, literature. *Society Aff:* ASCE, ASEE, AWRA, AGU.

Tognoni, Hale C
Home: 1525 West Northern, Phoenix, AZ 85021
Position: Pres *Employer:* Mineral Economics Corp *Education:* BS/Geol Engrg & Mining/Univ of NV; LIB J.D./Law & Mineral Econ/Univ of ID & Amer Univ; -/Geo Engr/Colo Sch of Mines; -/Civil Engr/VA Military Inst. *Born:* 04/16/21 Served

Tognoni, Hale C (Continued)
to 1st Lt, CE, Aug 1942-46; married, George-Ann Neudeck, Mar 13, 1947; children, Becky Lou, Brian Hale, David Quenton, Sandra-Ann, Jeffrey R.; Foreman, Molina sect. New Idria Quicksilver Mining Co, 1940-41; geo asst USGS, Steamboat Springs, NV, 1946-48; geo engr Anaconda Mining Co, Butte, Mont, also Darwin, CA 1948-50; staff asst hgwy test engr, Corp Engrs Ft Belvoir, 1951; geo engr Magma Mine, Superior, AZ 1953; reg as PE 1953; Asst Dir res AZ Legislative Cncl 1953-54; state mineral examiner AZ Land Dept, 1954-55; in private practice as cons engr 1954 and lawyer 1955; ptnr law firm Tognoni, Parsons & Gooding (later Tognoni, Parsons, Birchett and Gooding) then Tognoni & Pugh, 1956-71; lecture engr-law AZ State Univ 1962-74; Pres Multiple Use, Inc, Mineral Economics Corp, Minerals Trust Corp, Minerals Services Corp. *Society Aff:* AIMME, NSPE, ABA

Tointon, Robert G
Home: 6305 26th St, Greeley, CO 80634
Position: Pres *Employer:* Phelps, Inc *Education:* BSCE/-/KS State Univ *Born:* 5/19/33 Native of KS. Pilot in Air Force, 1956-59. Engr and superintendent for EB Construction Co. on military and missile gen construction work. Joined Hensel Phelps Construction co in 1963 as project engr and promoted to pres in 1975. Phelps is a general contractor with $250 million annual revenues working in most western states. Served on Bd of dirs of Mountain Bell, Affiliated Bank Shares of CO, Greeley National Bank, Greeley Chamber. Past pres of UNC Foundation and Assoc General Contractors, CO Chptr, Dir of CO State Univ Fdn, KS State Univ Fdn, CO Alliance for Bus, and Mbr Amer Bus Conference.

Toler, James C
Business: Georgia Tech Research Institute, Georgia Tech, CRB, Rm 329, Atlanta, GA 30332
Position: Mgr *Employer:* Georgia Tech *Education:* MS/EE/GA Inst of Tech; BS/EE/Univ of AR *Born:* 01/31/36 Electronics Test Engr with LTV Electro-Sys, Garland, TX Jun 1958 to Oct 1960. Unit mgr for Saturn/Apollo Electromagnetic Compatibility (EMC) at NASA/Marshall Space Flight Ctr, Huntsville, AL from Oct 1960 to Jun 1966. With Georgia Tech Engrg Experiment Station, Altanta, GA since Jun 1966. Initially mgr of EMC research progs. Currently, Mgr of Biomedical Research Div. 76 publications on electromagnetics applied to EMC and biomedical areas. Pres of IEEE EMC Society 1977-1978. IEEE Fellow 1980. *Society Aff:* IEEE, AAMI, BEMS, IMPI, BRAGS

Tollison, Gordon T
Business: 404 South Eleventh St, P.O. Box 516, Oxford, MS 38655
Position: Principal Consltg Engr *Employer:* Precision Engrg Corp *Education:* BS/ChE/Univ of MS *Born:* 07/27/47 Process and production engrg experience with B. F. Goodrich Chem Co, Union Carbide Corp, and Keystone Mtl Moulding Co., specializing in chem process start- ups and mgmt. With Univ of MS Physical Plant Dept, 1972-80, as Sanitary Engr, Environmental Engr, and Supervising Engr. Respon engrg and mgmt of water and wastewater utilities; coordination, inspection, and administrative reporting on all construction projects. Dept won 1977 NSPE Government Profl Development Award. Pres and Chrmn of Bd of Environmental Utilities Services, Inc, a private utility co, 1977-present. Principal in Precision Engrg Corp, consltg engrs and land surveyors, since 1977. Became conslt and corporate officer in 1980, specializing in chem, environmental, and civil engrg. Founded Southern Energy Consultants in 1980, Private Business specializing in Turnkey installation of Electronic energy mgmt and automated control systems in commercial, industrial & institutional markets. Presently co-owner and mgr of Field Engrg for Southern Energy Consults. Selected as winner of "Young Engr of the Year" Award, 1979, by Miss. Engrg. Society. *Society Aff:* NSPE, AIChE, WPCF, AWWA

Tomalin, Peter M
Business: Alcoa Labs, Alcoa Center, PA 15069
Position: Secton Head. *Employer:* Aluminum Co of America. *Born:* Dec 1940, Alexandria Va. BSMetE Lehigh Univ. 10 years exper in mfg of aluminum extrusion & drawn tube products with respon in process engrg & QA; since 1973 in R&D mgmt with major respon developing improved fabricating techniques & process & mech lubricants for prod lines including extrusion, drawn tube, wire, rolled shaped prods & forging; exper in government contract work for metal forming. Mbr ASM & active in Extrusion & Drawing Tech Activity. Enjoy sports & civic work.

Tomasino, Paul
Business: PO Box 16488, Temple Terrace, FL 33687
Position: President. *Employer:* Tomasino & Associates, Inc. *Education:* BCE/Civil Engg/Univ of FL. *Born:* July 1934, Tampa Fl. Reg PE in Fla & Ohio; Certified Fall-out Analyst. US Navy Const Battalion; Designer & Proj Engr for Watson & Co, Tampa; V Pres & Secy Amer Engrg, New Port Richey; Dir Public Works & City Engr (since 1965) for City of Temple Terrace, Fl; Town Engr, Sneads Fla; Founder & Pres Tomasino & Assocs Inc, Cons Engrs since 1971, respon for admin, business dev & proj mgmt. Fla Sect ASCE Dir 1966, V Pres 1970 & 71, Pres 1972. Pres Temple Terrace Chamber of Commerce 1976-77. Pres Tampa No Rotary 1981-82. *Society Aff:* ASCE, NAHB, AWWA, ASTM, NFIB.

Tomiyasu, Kiyo
Business: General Electric Co, Philadelphia, PA 19101
Position: Consulting Engineer. *Employer:* General Electric Company. *Education:* PhD/Eng Sci & Appl Phys/Harvard Univ; MES/Eng Sci/Harvard Univ; MS/Communication Eng/Columbia Univ; BS/Electrical Eng/CA Inst of Tech. *Born:* Sept 25 1919, Las Vegas Nev. 1948-49 Electronics Instructor at Harvard Univ, Cambridge Mass; 1949-55 Engrg Sect Hd for microwave res at Sperry Gyroscope Co, Great Neck NY; since 1955 microwave cons engr for Genl Electric Co in Palo Alto Calif, Schenectady NY & now at Valley Forge Space Ctr, Philadelphia Pa. Fellow IEEE, Hon Life Mbr IEEE Microwave Theory & Techniques Soc, IEEE Microwave Theory and Tech Soc 1980 Microwave Career Award, Microwave Dist Service Award 1987, IEEE Awards Bd Mbr, IEEE Board of Directors 1985-1986, Gen Elec-tric'c Charles Proteus Steinmetz Award for sustained achievements, 1977; IEEE Centennial Medal recipient, 1984; Mbr APS. Author *The Laser Literature, an Annotated Guide.* Hold patents, publ many papers on microwaves, antennas, scattering & remote sensing. *Society Aff:* IEEE, MTT-S, GRS-S, AES-S, APS.

Tomkowit, Thaddeus W
Business: 6226 Brandywine Bldg, Wilmington, DE 19898
Position: Mgr Logistics & Works Supplies Section. *Employer:* E I duPont de Nemours & Co. *Education:* BS/ChE/Columbia; ChE/ChE/Columbia. *Born:* 9/10/18. Since 1942 employed by E I de Pont de Nemours & Co Chemicals & Pigments Dept, Wilmington exper in res, dev, design, maintenance, const & operations, presently Mgr Logistics & Works Supplies Section, Production Div areas of inters in process dev, works engg, organization, personnel dev, and physical distribution. AIChE: Mbr of Council 1970, VP 1971, Pres 1972, recipient Founders Award 1975. Mbr Sigma Xi. *Society Aff:* AIChE, ASEE.

Tomlinson, Henry D
Business: C. Vargas & Associates, Ltd, 8596 Arlington Express-way, Jacksonville, FL 32211
Position: Senior Vice President *Employer:* C. Vargas & Associates, Ltd. *Education:* PhD/Water Resources & Environ Engr/Vanderbilt Univ; MS/Sanitary Engr/MIT; BS/CE/Univ of TN *Born:* 06/14/26 Experience includes employment with Tennessee Health Dept, US Public Health Service, KY Stream Pollution Control Commission, Assoc Prof at Washington Univ, St. Louis, MO, and consltg in environmental engrg with RETA, Inc., Envirodyne Engrs & Reynolds, Smith and Hills. Inc. Presently Sr. Vice President, C. Vargas & Assoc., Ltd. Jacksonville, FL. Listed in Who's Who in FL, Who's Who in Engineering, Who's Who in South & Southwest & 2000 Notable Amer. Awarded NSF Engrg Fellowship; Honorary Mbr, Chi Epsilon. Diplomate, American Academy of Environmental Engrs. Registered PE in 15 states. Received "Engineer of the Year" Award in 1987 from NE FL Chapter of the Florida Engi-

Tomlinson, Henry D (Continued)
neering Society. Married to former Mary Anne Frost of Shelbyville, TN and have 3 children: Hank, Kenneth and Diane. *Society Aff:* ASCE, APCA, AWWA, NSPE, SAME, WPCF, NYAS, FES/NSPE, ΣΞ

Tomlinson, John L
Business: Chemical and Materials Engrg Dept, Pomona, CA 91768
Position: Prof *Employer:* CA State Polytechnic Univ, Pomona *Education:* PhD/Metallurgy/Univ of WA; MA/Physics/Univ of OR; BA/Physics/La Sierra Coll *Born:* 09/15/35 Physicist, Naval Weapons Center, Corona, 1956-63 and 67-69. Research Engr, Boeing Co, Seattle, 1963-64. Ford Foundation Pre-Doctoral Fellow, Metallurgical Engrg Dept, Univ of WA, Seattle, 1964-67. Prof, Chem and Materials Engrg Dept, CA State Polytechnic Univ, Pomona, 1969-present. Received Charles Babbage Award, Institution of Electronic and Radio Engrs (London), 1976. Conslt and author of papers on properties of solid and liquid alloys, thermodynamics and phase equilibria, heat treatment of dental alloys, and failure analysis. *Society Aff:* ASM, APS, AACG, TMS-AIME

Tompkins, Curtis J
Business: Coll of Engrg, Morgantown, WV 26506
Position: Dean *Employer:* West Virginia Univ *Education:* PhD/Indus & Sys Engrg/GA Inst of Tech; MS/IE/VA Polytech Inst & State Univ; BS/IE/VA Polytech Inst & State Univ *Born:* 07/14/42 Dean of the Coll of Engrg at WV Univ since July, 1980, was Chrmn of the Industrial Engrg Dept at WV Univ from 1977 to 1980. He was a mbr of the faculties at the Univ of Virginia from 1971 to 1977 and Georgia Inst of Tech from 1968 to 1971. Dr. Tompkins served as an industrial engr with E.I. Dupont during 1965 to 1967. He was a co-op student with Appalachian Power Co from 1961 to 1963. Pres 1988-89 and mbr of the Bd of Trustees of the Inst of Industrial Engrs during 1983-89. Dr. Tompkins was First VP & mbr of Board of Directors of the Amer Society for Engrg Education during 1985-87. He was a mbr of the Engrg Accreditation Commission of the Accreditation Bd for Engrg and Tech 1981-86. Mbr of Board of Governors of the AAES 1987-90. He is a mbr of the Morgantown Water Comm and the Public Land Corp of WV. He received his BS and MS degrees from VA Polytechnic Inst and his PhD degree from the GA Inst of Tech. *Society Aff:* IIE, ASEE, AAES.

Tomshaw, John
Business: 200 Boyden Ave, Maplewood, NJ 07040
Position: Mgr-Chem Div of Res & Testing Lab. *Employer:* Public Service Electric & Gas Co. *Education:* MS/Metallurgy/Stevens Inst of Technology; BS/Chem Engg/NCE (now NJIT). *Born:* Sept 17 1918, North Jersey. 1943-45 served in US Air Force & in R&D Air Reserve, retired as LtC; advanced from Chemist to Mgr Chem Div, respon for mgmt labs involved in boiler water chemistry, air & water sampling & analysis, radiation monitoring, paint, fuels & oils for Public Serv Electric & Gas Research Corp, Research & Testing Lab. Served on ASTM D-2, D-22, & presently D- 19 Cttes, & on ASTM 'Project Threshold.' Reg PE NJ 16500. Mbr ACS, SAS, AFA, ASC. Authored 5 pub tech articles. *Society Aff:* ACS, SAS, AFA, ASTM.

Tong, Peter P
Home: 328 W Mariposa, Redlands, CA 92373
Position: Exec VP *Education:* MS/EE/Univ of WI; BS/EE/KS State Univ *Born:* 4/2/41 in China. BSEE Magha Cain Laude 1963 KS State Univ. IBM Rochester summer student engr program. MSEE U of WI and taught undergraduate level engrg courses at WI Campus. Assoc Engr at Rochester IBM. Founder of T & T Tech, Inc. Small businessman of the year, State of WI, 1973. Div Pres of Technicon until 1978. All commercial experiences are in Data Acquisition and Medical Data handling field, Presently, CEO of Burdick Corp and Pres of Kone Instruments, Inc Design & Mfg of cardiology measurement and blood analysis equip. *Society Aff:* IEEE

Tonias, Elias C
Business: Tonias Engineers, 65 West Broad Street, Rochester, NY 14614
Position: Principal *Employer:* Tonias Engineers *Education:* MSCE/Structures/OH State Univ; BSCE/Structures/Rensselaer Poly Inst. *Born:* 1/30/34. in Thessalonki, Greece. Arrived in the USA on Dec 10, 1950. Joined Erdmon, Anthony, Assoc 1957 as a Structural Engr; Comp Joined Erdmon, Anthony Assoc 1957 as a Structural Engr; Comp Ctr Dir - 1962; Assoc of firm in charge of Electronic Computing, Land Dev and Hydraulic Projs - 1967; Partner 1978. Established present firm 1980. Projs include bridge, dam, tunnel, sewer and PUD design and construction inspection, flood control and waste water mgt and planning, Profl Reg in the States of OH, NY & PA. Pres, Rochester Sec ASCE; Chrmn, Comp Practices Res Council and Tech Council of Comp Practices ASCE; VChrmn NY State Council ASCE; Pres, Monroe Soc PE. Chrmn, Pub Comm TCCP-ASCE. Adjunct Prof, RIT since 1966. (Roc Inst Techn) Author of tech and profl papers. *Society Aff:* ASCE, NSPE, ASTM, DPMA

Tonnies, Frederick E
Home: 65 S Bay Ave, Massapequa, NY 11758
Position: Manager, Systems Operations. *Employer:* Sperry Div, Sperry Rand Corp. *Born:* March 1926. BSEE Columbia Univ. 1944-47 served in USN; Field Engr for Sperry Gyroscope Co specializing in flight instruments, radar & bombing navigation systems, 1956 appointed Proj Engr for dev of Ship's Inertial Navigation System, 1968 assumed current position respon for Quality Control/Reliability, Field Engrg, Test & Operatins Engrg for Fleet Ballistic Missle Submarine Navigation Subsystem. Enjoy model railroading & boating.

Toole James M
Home: 1000 Pike Road, Bear Creek TWP, Wilkes Barre, PA 18702
Position: Chief Satellite Communications Engrg Section *Employer:* Dept. of. Defense/US Army *Education:* PhD/Matls Engg/PA State Univ; MS/Physics/Wilkes College; BS/Physics & Engrg/PA State Univ. *Born:* Feb 1 1936, Wilkes Barre Pa. BS Physics & Engrg Pa State Univ; MS Physics Wilkes Coll; PhD Materials Engrg Pa State Univ. 1961-63 Design Engr with Eastman Kodak Co, Rochester NY; 1963-65 Radio Corp of Amer, solar arrays for satellite systems, respon for lunar orbiter series; 1965 became Instructor of Physics at Wilkes Coll, 1969 became Chmn of Dept of Engrg. Chmn NE Penn Chap ASM, Mbr ASM Natl Education Ctte; Chmn Architect/ Engr Selection Ctte, Commonwealth of Penn; received ASM President's Award 1971; 1978-1981 VP of Luzerne County Community College; 1981 Dept, of Defense/US Army, Chief, Satellite Communications Engrg Section. *Society Aff:* IEEE, ASM, ARRL.

Tor, Abba A
Home: 48 Cochrane Av, Hastings-on-Hudson, NY 10706
Position: Principal *Employer:* Tor and Partners *Education:* MS/CE/Columbia Univ; CE/Civil Eng/Israel Inst of Tech, HAIFA *Born:* 11/01/23 Practiced and taught Engrg in Israel, 1946-1954; In service training Grant from the US Bureau of Reclamation-specialized in Soils and Structures engrg; Sr Designer, project engr, and project mgr with Ammann & Whitney 1957-1964; In private practice as Conslt for Industrialized Housing Methods 1964-1966; Partner in Pfisterer, Tor, & Assocs, and successor firm: Tor, Shapiro & Assoc & Tor & Partners. Past Mbr of Bd of Dirs, NY Assoc of Consltg Engrgs; currently officer Conn. Engineers in Private Practice. Current responsibilities include project mgmt and overall office mgmt. *Society Aff:* ASCE, NSPE, CEC, ACI, PCI

Torby, Bruce J
Business: 1250 Bellflower Blvd, Long Beach, CA 90840
Position: Prof. *Employer:* CA State Univ. *Education:* PhD/ME/Univ S CA; MS/Engg/Univ CA; BME/ME/CUNY. *Born:* 3/12/39. Began career as res engr with Autonetics, Inc. Went to CA State Univ, Long Beach afterwards & have been there since (1961). Have had various consultancies including several yrs as an expert examiner for the CA Bd of Registration for PE. One yr guest Prof at Chalmers Univ, Gothenburg, Sweden (1977). Areas of interest are: numerical methods, continuous system simulation, dynamics. Current res: Locomotive wheel/track loading, the whirling of rotors. Mbr ASME. Reg in CA. Tau Beta Pi, Who's Who in the West, Dictionary of Intl Biography. NSF Sci Faculty Fellowship. Personal interests: sailing, travel, intl education. *Society Aff:* ASME, ASEE.

Torda, Paul T
Home: 2127 California St, N.W, Washington, D.C 20008
Position: Consultant *Employer:* Self *Education:* Ph.D./Appl. Mech./Brooklyn Polytechnic Institute; Dipl. Ing./Aeron. Eng./Technical University, Berlin. *Born:* 9/22/11 Born in Budapest, Hungary. Has 51 years of industrial and consulting, and 41 years of teaching and research experience (at Polytechnic Inst of Brooklyn, Univ of Il Urbana, and Il Inst Technology Chicago). Developed and implemented the Education and Experience in Engineering (E3) Program of which he is emeritus director. Published widely in the fields of aerothermochemistry, fluid mechanics, turbomachine theory, combustion, and engrg education. Founded the Central States Section of The Combustion Institute, and co-founded, with Dr. Victor Paschkis, the Technology and Society Div of the ASME. *Society Aff:* AAAS, AIAA, ASME

Torgersen, Paul E
Business: Coll of Engrg, Blacksburg, VA 24061
Position: Dean, College of Engineering. *Employer:* Va Polytechnic Inst & State Univ. *Education:* BS/IE/Lehigh Univ; MS/IE/OH State Univ; PhD/IE/OH State Univ. *Born:* Oct 1931, 1953-55 served in USAF; 1955-67 taught at Okla State Univ, recipient of Outstanding Teacher Award 1963; 1967-70 Hd Dept of Indus Engrg & Operations Res, VPI & SU, since 1970 Dean, Coll of Engrg. Received Distinguished Alumnus Award, Coll of Engrg Ohio State Univ 1971; with W J Fabrycky & P M Gare, AIEE H B Maynard Book-of-the-Year Award for 'Industrial Operations Research' 1972. Elected Fellow of Institute of Industrial Engrs 1976; Elected to the National Academy of Engrg 1986. Interest in tennis. *Society Aff:* ASEE, IIE, AAA, SOLE.

Torgow, Eugene N
Home: 9531 Donna Ave, Northridge, CA 91324
Position: Consult & Prof (part time) Cal State Univ, Northridge *Employer:* Eugene N. Torgow, Consultant *Education:* BSEE/Electronic/Cooper Union; MSEE/Electromagnetic Fields/Polytechnic Inst of Brooklyn; Engineer/Elec Engrg/Polytechnic Inst of NY. *Born:* Nov 1925, New York NY. 1947-59 res in microwave component tech at Microwave Res Inst, PIB, 1959-68 supervised engrg progs for aerospace applications; Ch Engr at Dorne & Margolin, & later at Rantec Div of Emerson Electric Co; 1968-1985 at Hughes Aircraft Co managing engrg efforts on sev major missile progs. 1985-present, consultant on engineering mgmt & training, Prof. of Elec. Eng at Cal State Univ, Northridge (Part time). Served on num IEEE cttes including Chmn of Microwave Theory & Techniques Grp, Fellow IEEE 1968; Fellow Inst for Advancement of Engrg 1972. Awarded Distinguished Service Award by Microwave Theory and Techniques Soc in 1978. *Society Aff:* IEEE, ΣΞ, IAE

Torma, Arpad E
Home: 122 Mustang Dr, Socorro, NM 87801
Position: Dean of Engrg *Employer:* NM Inst of Min & Tech *Education:* PhD/ChE/Univ of British Columbia; MSc/Chem/Laval Univ; Dipl/ChE/Swiss Fed Inst of Tech *Born:* 10/30/32 Prof of Metallurgy; Dir of the State Mining and Mineral Resources Research Inst; and Dir of the Sullivan Ctr for In-Situ Mining Research at NM Inst of Mining and Tech in Socorro, NM. Formerly, he was associated with the Quebec Dept of Natural Resources as head of Metallurgy Div at the Mineral Research Ctr. His principal research interest and courses taught are related to electro-hydro-, and pyrometallurgy, solid and liquid ion exchangers, bacterial leaching and solid-gas reactions. He is author and coauthor of 3 books, 12 patents and over 100 scientific and technical journal articles. *Society Aff:* AIME, CMMI, NMES, AAES.

Torng, Hwa C
Home: 418 Winthrop Dr, Ithaca, NY 14850
Position: Prof. *Employer:* Cornell Univ. *Education:* PhD/EE/Cornell Univ; MS/EE/Cornell Univ; BS/EE/Natl Taiwan Univ. *Born:* 8/12/32. Dr. Torng is an educator, a researcher & a conslt in the fields of telecomm, computer networks and computer engrg. His current interests lie in the areas of telecomm, computer network design and processor design and computer aided design. He is a conslt to several indust orgs. He joined the Cornell faculty in 1960. He has held visiting positions with Bell Lab, Natl Taiwan Univ. The Univ of Rochester & Lawrence Livermore Lab. He is the author of two books & 40 technical papers. He is an editor of IEEE Trans. on Computers. *Society Aff:* IEEE, ACM, ΣΞ, ΦKN.

Toro, Jorge A
Business: PO Box 9684, Santurce, PR 00908
Position: Senior Engineer. *Employer:* Capacete, Martin & Associates. *Education:* MS/Structural Engg/Univ of IL; BS/Civil Engg/Univ of Pr. *Born:* July 5 1947, Arroyo Puerto Rico. BSCE Univ of Puerto Rico June 1970; MS Univ of Ill Feb 1971. 1971-72 Structural Engr for Puerto Rico Water Resources Authority; 1972- , Structural Engineer for Capacete, Martin & Assocs. Mbr 'Colegio de Ingenieros, Arquitectos y Agrimensores,' Reg PE in PR. Mbr Sociedad de Ingenieros Estructurales de Puerto Rico, ACI & ASCEI; P Pres, V Pres & Secy Puerto Rico Sect ASCE. Mbr Tau Beta Pi & Phi Kappa Phi. Enjoy classical music & chess. *Society Aff:* ASCE, ACI.

Toro, Richard F
Home: 589 Foothill Rd, Bridgewater, NJ 08807
Position: Exec V.P. & Principal. *Employer:* Recon Systems Inc. *Education:* MChE/ChE/Univ of DE; BS/ChE/Lafayette College. *Born:* 11/11/39. Consulting Environmental Engr with Chem Engr background. Exec. VP & Principal in ind consulting firm of Recon Systems since 1973. Previously VP of Princeton Chem Res. Chrmn of Intersociety Committee on methods of air sampling & analysis. Active on committees of Air Pollution Control Assn. Mbr of Phi Beta Kappa, Sigma Xi: Honorary Socs. Patents & publications, including book on energy & resource recovery from municipal solid waste. *Society Aff:* AIChE, ACS, APCA, ISC.

Torpey, Paul J
Business: 4 Irving Pl, New York, NY 10003
Position: Director Energy Conversion & Fuels R *Employer:* Consolidated Edison Co of NY, Inc. *Education:* MS/Mech Engg/Columbia; BS/Mech Engg/Columbia. *Born:* 9/9/38. 1960-70 Mbr Tech Staff Bell Telephone Labs; Aug 1970 assumed present position, participating in formation of new res dept. Served as Acting Admins of Empire State Electric Energy Res Corp (ESEERCO), a statewide electric utility res consortium 1974-75. Publ in energy conversion related subjs. Chrmn of ESEERCO Fossil Fuel & Advanced Generation Committee through 1979 & Chrmn of Advanced Fossil Power Sys Task Force of the Electric Power Research Inst thru 1978. Elected ASME Fellow in May 1976. Elected Vice President, Research, ASME, 1984, 1986. *Society Aff:* ASME.

Torrance, Kenneth E
Business: Upson Hall, Ithaca, NY 14853
Position: Professor *Employer:* Cornell University *Education:* PhD/Mech. Engrg./Univ. of Minnesota-Minneapolis; MSME/Mech. Engrg./Univ. of Minnesota-Minneapolis; BS/Mech. Engrg./Univ. of Minnesota-Minneapolis *Born:* 08/23/40 Native of Minneapolis, MN. Engrg educator & researcher in heat transfer, combustion and numerical methods in engrg. Research assoc at the U.S. National Bureau of Standards 1966-68 and at the National Center for Atmospheric Research 1974-75. With Cornell Univ since 1968. Currently Prof of Mech and Aerospace Engrg and mbr of the Program of Computer Graphics. Served as Assoc Dean of Engrg 1983-86. Author of one book and numerous articles in professional journals. Fellow, ASME 1986. Best paper award from ASME Heat Transfer Div 1982. *Society Aff:* ASME, APS, Comb. Inst., AGU, AAAS

Torrey, Paul D
Business: 5000 East Bee Cave Road, P. O. Box 3318, Austin, TX 78764
Position: Oil Producer Petroleum Engr (Self-employed) *Education:* BS/Petroleum Geology/Univ of Pittsburgh; Pet Engg/Petroleum Engg/Unibb of Pittsburgh; ScD/ Petroleum Engg/Marietta College. *Born:* June 28 1903, W Feliciana Parish La. Univ of Pittsburgh Petroleum Geology 1925, Petroleum Engrg 1927; Hon ScD Marietta Coll 1952. US Geological Survey Rocky Mountain States, Ark & Penn, pioneered secondary methods in Penn fields, installed 1st systematic water-floods E Okla & W Texas. Chmn num API Cttes, Div of Production; 1st Chmn IOCC Secondary Recovery & Pressure Maintenance Ctte. Work in Canada, Alaska, Zaire, Germany, Columbia, Brazil & Indonesia; teacher at Univ of Tex. Legion of Merit 1952 from Pres Harry S Truman for exploration & dev of radioactive minerals in AZ, UT, Canada & Zaire. Carll Award SPE 1973; Hon Mbr AIME 1974; Outstanding American in the South 1975; Bicentennial Award NPCA 1976. *Society Aff:* SPE, AIME, AAPG

Torvik, Peter J
Home: 2000 Harvard Blvd, Dayton, OH 45406
Position: Prof and Head *Employer:* Air Force Inst of Tech. *Education:* PhD/Engr Mech/Univ of MN; MS/Engr Mech/Univ of MN; BS/Aero Engr/ Univ of MN; BA/ Music/Wright State Univ *Born:* 12/6/38. On the faculty of the Air Force Inst of Tech since 1964 as Asst Prof, Assoc Prof, and Prof (1973). Head, Department of Aeronautics and Astronautics (1980) Offered grad courses in Engg Mechanics and directed the res of more than fifty grad students. Authored more than seventy technical papers and reports primarily concerning structural dynamics and material response. Serves professional societies as active comm mbr and organizer of conferences. Presently serving on committees of ASME and AIAA and as a reviewer for numerous journals. Consultant to various Air Force organizations. On sabbatical during 1979 at The OH State Univ. Fellow, ASME. *Society Aff:* ASME, AAM, AIAA, ASA, ASEE, SR, SES.

Touhill, Charles J, Jr
Home: 2206 Almanack Ct, Pittsburgh, PA 15237
Position: Pres *Employer:* Baker/TSA Inc *Education:* PhD/Environ Engg/RPI; SM/ Sanitary Eng/MIT; BCE/Environ Engg/RPI. *Born:* 8/27/38. As Pres of Baker/TSA is respon for projs in toxic and hazardous matls mgmt, industrial water & wastes mgmt, ecosystems and environmental planning; previous positions include Chmn & Ch Exec Officer, Morris Knowles Inc; VP Engrg-Sci, Inc; Mgr Water & Land Resources Dept Battelle-NW. Dipl & Trustee 1971-77 and 1983-1986 Amer Acad Environ Engrs; Chmn AIChE Environ Div 1977; Editorial BD 1975- 77 "Environmental Sci & Technology." Editorial Advisory Board, "Environmental Progress-. *Society Aff:* AIChE, ASCE, ACS, AAEE, WPCF, AWWA

Toups, John M
Business: 1500 Planning Research Dr, McLean, VA 22101
Position: Pres & CEO. *Employer:* Planning Res Corp. *Education:* BSCE/Civ Engg/ Univ of CA. *Born:* 1/10/26. I received my BSCE degree from the Univ of CA at Berkeley, after which I held several positions as a civ engr in CA. In 1956 I founded Toups Corp, a civ engg firm, which became a Planning Res Corp (PRC) co in 1970. In 1973 I became a mbr of the PRC Bd of Dirs & in 1974 I relinquished the presidency of Toups Corp to become a Sr Vp of PRC. In 1977 I was elected Pres of PRC & named Chief Exec Officer in 1978. I am a reg engr in CA, FL, IL, MD, MI, NY & TN. *Society Aff:* ACEC, ASCE, NSPE, AWWA.

Towers, John R
Home: 39 Cambridge Dr, Oak Brook, IL 60521
Position: Manager *Employer:* Chemical Waste Management, Inc. *Education:* MS/Sanitary Engg/Univ of IL; BS/Civil Engg/Univ of IL. *Born:* April 21 1930. Served in Air Force as Lt during Korean War; 1957-59 Cons engr with Pfeifer & Shultz & Assocs; 1959-1986 with Armco as Sales Engr, Sr Sales Engr & Dist Engr & Account Manager, Metal Products Div. Reg PE in Ill. Chap Pres, State Pres & Natl Dir Ill Soc of Prof Engrs; Mbr NSPE Ethics & Practices Ctte & Dir NSPE Education Foundation. 3 years as Trustee Midwest Coll of Engrg; 7 yrs Bd Mbr of Oak Brook Sch District 53; Mbr Western Soc of Engrs Washington Award Comm & active in Boy Scouts. Formerly Pres, Sch District 53 Bd of Education. Elder- Christ Church of Oak Brook; Board Member - Christ Church of O. B. Pre-School Board; Recipicnt 1981- ISPE Illinois Award; Listed in Who's Who in the Mid-West. Vice President Univ of Illinois Civil Engrg Alumni Assoc, 1986 to present, with Chemical Waste Management, Inc as Program Manager, Waste Reduction Services. *Society Aff:* NSPE, WPCF, AWWA.

Towne, Allen N
Home: 113 Madbury Rd, Durham, NH 03824
Position: Consultant, Insulation Products *Employer:* Self *Education:* MS/Ind Adm/ Union college; BS/ChE/Northeastern Univ. *Born:* 8/20/20. in North Andover, MA. Served in US Army in WWII (1940-45). Worked for the GE Co for 19 yrs (1947-1965) as Mfg Engr, Product Engr & as a Product Mgr specializing in elec insulation. Was Pres & Gen Mgr of New England Mica Co from 1966-1973. Assumed position as Gen Mgr of MWI Eastern Insulation Plants (Essex Group of United Technologies) in 1974. Holds three US Patents. In 1980 was appointed Director, Insulation Products. In 1982 became a consultant and established his own company. *Society Aff:* AIChE.

Towne, Harmon L
Business: PO Box 778, Morris, Manitoba R0G 1K0 Canada
Position: Dir of Engrg *Employer:* Brock Valley Ltd *Education:* BS/Agric Eng/Purdue Univ *Born:* 09/26/39 Grad from Purdue Jan, 1963. Two yrs with Public Service Co of IN as power use advisor. Four yrs with Doane Agricultural Service of St. Louis, MO. Designed over 75 bldgs for agricultural use. With Aerovent Fan and Equipment 1968-1974, Lansing, MI. Designed crop drying and livestock ventilation equipment. Surperior Equipment Mfg 1974-1977 as Product Engr-Crop Drying Equipment and later Dir of Engrg. Joined Brock Mfg in 1977 and assumed position of Dir of Engrg in Dec, 1977. Reg PE in MO, IN and OH. Promoted to Asst Gen Mgr for Brock Mfg. 1981. Promoted to VP/Gen Mgr for Brock Valley Ltd., June 1983. *Society Aff:* ASAE

Towner, Orrin Wilson
Home: 11208 Beech Rd, Louisville, KY 40223
Position: Consulting Engr (Self-employed) *Education:* EE/Elec Engg/Univ of KS; BS/ Elec Engg/Univ of KS. *Born:* 1903 Peterson Iowa. Fellow IEEE, 3 time Chmn Louisville Section; Mbr SMPTE. Prof Engr Kentucky. Recipient of Pres Certificate of Merit 1948. Co-Chmn NIAC Subctte Emergency Alerting Sys Test & Evaluation Ad-Hoc Ctte 1966-68; Chmn Louisville Indus Adv Ctte 1964-68; Chmn AMST Engrg Ctte 1963-68; Chmn TASO Ctte 1.4 1957-58; Chmn NAB Engrg Exec Ctte 1946-48. Engrg Cons 1968- ; Dir of Engrg WHAS Inc 1938-68; Assoc Dir Airborne Instruments Lab Columbia Univ, Div of War Res 1942-45; Dir Alhambra Branch 1943-44; Bell Telephone Labs 1927-38. *Society Aff:* IEEE, SMPTE, AFCCE.

Townes, Charles H
Business: Dept of Physics, Berkeley, CA 94720
Position: University Professor of Physics. *Employer:* University of California-Berkeley. *Education:* PhD/Physics/CA Inst of Tech; MA/Physics/Duke Univ; BA/ Modern Languages/Furman Univ BS/Physics & Modern Languages/Furman Univ. *Born:* 7/28/15. Staff Mbr Bell Telephone Labs 1939-47; prof of Physics Columbia Univ 1948- 61; V Pres & Dir of Res Inst for Defense Analyses 1959-61; Provost & Prof of Physics MIT, 61-66 & Inst Prof 66-67; Univ Prof Univ of CA 1967- ; Microwave spectroscopy, quantum electronics, radio & infrared astronomy. Dir Genl Motors & The Perkin-Elmer Corp. Nobel Prize for Phys 1964; Morris Liebman & David Sarnoff Awards & Medal of Honor, IEEE; Natl Inventors Hall of Fame 1976; Niels Bohr International Gold Medal, 1979 Mbr NAS; Fellow APS, Pres 1967; Foreign Mbr Royal Soc of London. *Society Aff:* NAS, APS, IEEE, OSP.

Townsend, Herbert E
Business: Bethlehem Steel Corp, Homer Res Labs, Bethlehem, PA 18016
Position: Sr Research Fellow *Employer:* Bethlehem Steel Corp. *Education:* PhD/Matls Sci & Eng/U of PA; BS/Met Eng/Drexel Univ *Born:* 7/1/38. Native of Phila, PA. Joined the Res Dept of Bethlehem Steel Corp in 1967 and presently Senior Research Fellow. Work resulted in original contributions to several fields including transport & thermodynamics of molten salt mixtures, thermodynamics of elevated-temperature metal-water systems, stress corrosion cracking and hydrogen embrittlement of steels, and the dev of corrosion- resistant steel products. Current responsibilities include res in coatings & corrosion. Enjoys competitive distance running. *Society Aff:* ASM, NACE, SAE.

Townsend, James Skeoch
Business: Agricultural Engineering, Univ of Manitoba, Winnipeg, Manitoba R3T 2N2 Canada
Position: Prof. *Employer:* University of Manitoba. *Education:* PhD/Agri Engg/ Cornell Univ; Ms/Agri Engg/Cornell; BASc/Mech Engg/Univ of Toronto; BSA/Agri/ Univ of Toronto. *Born:* Apr 1934. PhD & MSc Cornell Univ; BASc & BSA Univ of Toronto. Native of Belwood Ontario. Taught in Ontario secondary schs prior to grad studies at Cornell. With Univ of Manitoba 1968- . Leave of absence to serve with Canadian Internatl Dev Agency in Thailand at Khon Kaen Univ for 20 mos 1973-74. Respon for res & teaching Agri Engrg Univ of Manitoba. Chmn N Central Region ASAE 1971; Prog Chmn for Power & Machinery section CSAE 1971 & 1975; Chmn & Prog Chmn EPP- 41 Ctte ASAE 1975-76; Certificate of Service ASAE 1972. Regional Dir, North Central Region, ASAE, 1985-87; Private Consultant to Canadian Internatl Dev Agency in Thailand and Burma, 1982- , and to CUSO in Burma, 1987; Prog Chmn for A-612 Intl Div of ASAE, 1983-85. Enjoy jogging & fishing. On leave 1979-82 to work with the Intl Rice Res Inst in Burma, on small-scale mechanization proj. *Society Aff:* ASAE, ASEE, CSAE, ΣΞ, ΦΚΦ.

Townsend, John W, Jr
Business: 6010 Executive Blvd, Rockville, MD 20852
Position: Associate Administrator. *Employer:* NOAA - US Dept of Commerce. *Born:* Mar 1924 Washington D C. ScD, MA & BA Williams Coll in physics. Army Air Force 1943-46. Res phys with Naval Res Lab 1949-58; pioneer in upper atmospheric res using sounding rockets; with NASA 1958-68 becoming Deputy Dir of Goddard Space Flight Ctr; instrumental in planning & dir majority of sci & application satellite projs in 1960's; appointed Deputy Admin of Environ Sci Serv Admin 1968 & Assoc Admin of Natl Oceanic & Atmospheric Admin 1970 (both component parts of Dept of Commerce). Mbr NAE.

Townsend, Marjorie R
Home: 3529 Tilden St, NW, Washington, DC 20008-3194
Position: Consultant *Employer:* Self *Education:* BEE/EE/Geo Wash Univ *Born:* 03/12/30 Native of Washington, DC. Natl Bureau of Standards, 1948-1951; radon testing; Naval Research Lab, 1951-1959, sonar research - developed frequency multiplication system, analog logic computer and new submarine detection and classification technique; Goddard Space Flight Center, 1959-1980, developed space hardware from DC to microwaves, patented digital telemetry system, was Project Mgr for three Small Astronomy Satellites and, from 1976-1980, responsible for all advanced mission planning at GSFC; consltg in aerospace and electronic systems; Space Amer, 1983-1984, VP for Space Sys Dev. Chrmn, Washington Section, IEEE, 1974-1975; Pres, Wash Acad of Sci, 1980-1981; Vice- Chrmn, Natl Capital Sec, AIAA, 1984-1985 (chairman 1985). Nineteen awards include NASA's Exceptional Service and Outstanding Leadership Medals, Federal Women's Award, GWU Alumni Achievement Award. Enjoy travelling. *Society Aff:* IEEE, AIAA, SWE, AAAS, AGU

Townsend, Miles A
Business: Dept of Mech & Aerospace Engrg, Charlottesville, VA 22901
Position: Wilson Professor & Chairman *Employer:* University of VA *Education:* PhD/ME/Univ of WI; MSc/Th & Appl Mech/Univ of IL; BSc/ME/Univ of MI *Born:* 4/16/35 Buffalo NY. In industry 1958-68: 1958-59 with GE Co, Evendale OH; 1959-63 with Sundstrand Corp, Rockford IL-res on hydrostatic transmission and controls; 1963-64 and 1966-68 with Twin Disc Inc, Rockford, IL-res and design of vehicle controls, and Project Engr in charge of variable-speed drives and controls; 1964-66 with Westinghouse Electric Corp., Sunnyvale CA-Senior Engr in charge of controls. 1968-69, Lecturer at Univ of WI; 1969-71 Nat'l Sci Foundation Fellow at WI. 1971-74, Associate Prof of Mech Engrg Univ of Toronto, Canada and Assoc of the Cockburn Centre for Engrg Design. 1974-81 Prof of Mech Engrg, Vanderbilt Univ, Nashville, TN, with appointment in Biomedical Engrg and Orthopaedics (Medical Sch). Assumed current position Jan 1, 1982. Over 80 published papers and 150 resch and engrg reports 7 patents and many disclosures. Mbr of several journal editorial bds and nat'l professional committees. *Society Aff:* ASME, AAAS, ΣΞ, ΦΚΦ

Toy, Wing N
Business: AT&T Bell Labs, 200 Park Plaza Rmn 1Z306, Naperville, IL 60566-7050
Position: Supervisor *Employer:* AT&T Bell Labs *Education:* PhD/EE/Univ of PA; MS/EE/Univ of IL; BS/EE/Univ of IL *Born:* 02/03/26 Served with US Navy 1944-46. At Bell Labs, contributed to design and development of three generations of fault-tolerant computers for electronic telephone switching sys and telecommunication sys. Taught computer courses at the Univ of CA at Berkeley, Northwestern Univ and Univ of Wisconsin Extension. Conslt for Dept of Air Force on Fault-Tolerant Full Authority Electronic Engine Control prog. Member of IEEE Computer Society Board of Governors. Editor of IEEE TC on Computers. Mbr of Editorial Bd of Computer Magazine. Mbr of IEEE Ad Hoc Accreditation Ctte for Electrical/ Electronic Tech Curricula. Co-author of three computer engrg text books. Published 16 technical papers on logic circuits, error detection schemes, fault-tolerant computer design and electronic switching sys. Holder of 25 US patents. IEEE Fellow (1981) and Bell Lab Fellow (1983). *Society Aff:* IEEE Computer Society

Tozer, George K
Home: 22 Oakhurst Rd, Beverly, MA 01915
Position: Senior Vice Pres *Employer:* Metcalf & Eddy Inc *Education:* BS/Sanitary Eng/Northeastern Univ *Born:* 01/22/26 As a Sr Vice Pres and Regional Mgr have been involved in all aspects of water pollution control engrg with Metcalf & Eddy and specialized in the mgmt of wastewater projects of all sizes and degrees of complexity. A mbr of the firm since 1954, having more than 10 yrs of mgmt level experience at Metcalf & Eddy in both wastewater and water engrg. *Society Aff:* WPCF, AWWA, CEC

Trabert, George R
Business: 575 Market St 1638, San Francisco, CA 94105-2856
Position: Product Quality Mgr. *Employer:* Chevron USA Inc. *Education:* BS/Mech Engr/UC Berkeley. *Born:* 12/17/32. Native of San Francisco, CA. Torional Vibration Analysis with Enterprise Engine Div, DeLaval 1955-57; with Std Oil Co of CA 1957-77, worked as Field Lubrication Engr until 1966, respon for new automotive support. 1977-79 Chevron Chem Co. Oronite Additive Div Supervised technical service support for marketing of lubricating oil additives, product planning and marketing strategy for additive packages 1979- . Chevron USA Inc product engg dev of new lubricants. Author NCGI Spokesman Articles. Hobbies include backpacking and golf. *Society Aff:* ASLE, SAE.

Tracey, James H
Business: Seaton Hall, Manhattan, KS 66506
Position: Head E.E. Dept *Employer:* KS State Univ *Education:* PhD/EE/IA State Univ; MS/EE/IA State Univ; BS/EE/IA State Univ *Born:* 10/6/34 He is currently engaged in teaching and res in the area of digital system design. Previously, he was a prof at the Univ of MO-Rolla. Dr. Tracey received the BS, MS and PhD degrees at

Tracey, James H (Continued)
Iowa State Univ and served as asst prof there for one yr. Dr. Tracey is a mbr of Phi Kappa Phi, Tau Beta Pi, Eta Kappa Nu, Pi Mu Epsilon, IEEE, ACM and ASEE

Trainor, Eugene F
Business: 286 Congress St, Boston, MA 02210
Position: Sr VP *Employer:* Cygna Energy Services *Education:* MSc/Mgmt/RPI; BSc/Gen Engrg/US Coast Guard Acad *Born:* 09/12/28 A native of MA, joined Cygna Energy Services in 1980 as VP Quality Assurance and Mgmt Services. Respon for Quality and Mgmt consltg. Prior assoc with Stone & Webster Engrg Corp as Mgr Quality Assurance and Chief Engr Engrg Assurance. Dev the first quality assurance program approved by the Atomic Energy Commission for an engr constructor. Assoc with Bethlehem Steel Corp and Gen Dynamics Corp in the Naval nuclear program in various capacities in production and test engrg, as well as Proj and Process Engrg Mgr. Enjoys sailing and wood carving. *Society Aff:* ASME, ASQC

Trandel, Richard S
Business: Mech Engineering Dept, Lexington, VA 24450
Position: Dept Head *Employer:* VA Military Inst *Education:* PhD/ME/Univ of VA; MS/ME/VA Polytech Inst & State Univ; BS/CE/VA Military Inst *Born:* 04/19/37 Native of Chicago, IL. Served in Army Corps of Engrs from 1960-62. Worked as consltg engr, taught at Univ of VA and currently serves as dept chrmn of the newly established Mech Engr Dept at the VA Military Inst. PE in the State of VA. For the past 5 yrs have served as an appointed mbr of the VA Fuel Conversion Authority. Major interest is thermodynamics, solar energy, and energy conservation. *Society Aff:* ASME, LYNCHNOS SOC, ΦΚΦ, ΣΞ, VSPE.

Traub, Joseph F
Business: Computer Sci Dept, 450 Computer Science Bldg, New York, NY 10027
Position: Edwin Howard Armstrong Prof Comp Sci *Employer:* Columbia Univ. *Education:* PhD/Appl Math/Columbia Univ; MS/Appl Math/Columbia Univ; BS/Phys-Math/CCNY. *Born:* 1932. Supvr. Res Group Bell Labs, Murray Hill, NJ 1959-70; Prof & Hd Computer Sci Dept Carnegie Mellon Univ 1971-79. Edwin Howard Armstrong Prof of Comp Sci, Prof of Mathematics & Chrmn, Comp Sci Dept, Columbia Univ; Dir, NY State Ctr for Computer and Info Sys. Chmn Pres's Adv Ctte for Computing, Stanford Univ 1975; Mbr Central Steering Ctte Computer Sci & Engrg Res Study, Natl Sci Fdn 1974-80. Fellow AAAS, Counc 1971-74; Mbr Sci Counc IRIA Rocquencourt France 1976-1979. Ed Journal Assoc Computing Machinery 1970-76. Cons Hewlett-Packard Corp 1981, IBM 1984. Pres, Consortium Scientific Computing, Princeton, 1986-87. Founding Editor, Journal of Complexity, Academic Press, 1985- . Mbr, Natl Acad of Engg, 1985- . Chmn, Comp Sci & Tech Bd, Natl Res Council 1986- . Bd of Govs, NY Acad of Sci, 1986- . *Society Aff:* ACM, AMS, IEEE, AAAS, SIAM.

Traudt, Joseph G
Business: PO Box 5328, Northville, MI 48167
Position: Pres. *Employer:* J G Traudt Co. *Education:* BSChE/CE/Univ of Toledo. *Born:* 12/2/37. Native of Toledo, OH. Part time instructor in math at Univ of Toledo 1961. Technical Service Engr & Production Dept Hd at BASF-Wyandotte Corp, J B Ford Div, 1960-1965. Sales Engr, Sales Mgr, & Operations Mgr for Mooney Process Equip Co, a mfg & distributor of corrosion resistant fluid handling equip, 1965-1978. Currently involved in sales & marketing of majr equip items to the chem process industries, specializing in the area of heat & mass transfer. *Society Aff:* AIChE.

Trauger, Donald B
Business: Oak Ridge Lab PO Bx 4500 N, Oak Ridge, TN 37831-6254
Position: Senior Staff Assistant to the Director *Employer:* Martin Marietta Energy Systems, Inc. *Education:* -/Physics & Engr/Columbia Univ; AB/Physics/NB Wesleyan Univ. *Born:* 6/29/20. in Exeter NB. Employed by Columbia Univ Manhattan Dist Proj 1942-44; Union Carbide 1944- . Spec in dev of gaseous diffusion membranes for uranium isotopes separation & distillation separation of lithium isotopes. Hd ORNL Irradiation Engg Dept for nuclear fuel testing 1954-64; dir Gas-Cooled Reactor Prog 1964-70; Assoc Dir for Nuclear and Engg Technology 1970-84; Now Sr Staff Asst to the Dir with responsibility for several Nuc Prog. studies. Fellow ANS. Hon Dr of Sci NWU 1973. Hon Doctor of Sci, TN Wesleyan College, 1977. *Society Aff:* APS, AAAS, ANS, RSA, ΤΒΠ, ΣΞ

Traugott, Fritz A
Business: 5895 Enterprise Parkway, East Syracuse, NY 13057
Position: Chairman of the Board/Treasurer *Employer:* Robson & Woese Inc. *Education:* BS/Mechanical Engineering/Fed Gov Eng Inst. *Born:* Mar 1928. ME Federal Government Engineering Institute, Austria; graduate studies Syracuse Univ. Received US Achievement Award 1952. Joined Robson & Woese, Inc., Consulting Engineers 1956, presently Chairman of the Board and Treasurer. ASHRAE - Past Director and Chairman, Region I, Member and Chairman various Chapter and Society Committees; elevated to Fellow Grade 1980. Received ASHRAE Distinguished Service Award 1979; Guest lecturer Syracuse University. Author of several published technical articles. Avid Environmentalist. Member Technology Club of Syracuse. Member Syracuse Rotary, President 1986-87. Paul Harris Fellow 1982. Delegate to China as part of People to People Citizen Ambassador Program 1984. Enjoys boating, classical, music, photography, reading, skiing, tennis. *Society Aff:* ASHRAE, ACEC, RI.

Trautman, DeForest L (Woody)
Home: 2625 Middlesex Dr, Toledo, OH 43606
Position: Mgt & Data Processing Consultant. *Employer:* Self. *Education:* PhD/Elec Engg/Stanford; MS/Carnegie Inst of Tech; BS/Elec Engg/Carnegie Inst of Tech. *Born:* June 14, 1920. PhD Stanford (Elec Engr 1949); BS & MS Carnegie. Mbr engrg faculty Carnegie, Stanford, UCLA, SUNY at Stony Brook. Pulse Communications, radio propagation & sys Hughes R&D Labs 1953-58; Sci & Engrg educ projs Ford Foundation, OECD, UNESCO 1958-68; inst res & planning SUNY & Univ Toledo: educ objectives & outcomes, mgmt & budgeting processes in postsecondary educ, data structs, computerized planning & resource allocation processes. Decision-Oriented Mgmt & data processing consultant for bus, industry & education. Tech in regional economic develop. Mbr Tau Beta Pi, Sigma Xi, ASEE; Fellow IEEE 1972; Toledo Engr of the Year 1978, World Future Soc. *Society Aff:* IEEE.

Trautman, William R
Home: 760 Chestnut Hill Rd, East Aurora, NY 14052
Position: Consltg Engr *Employer:* Self Employed *Education:* BS(Eng)/Mech Engrg/Univ of Buffalo. *Born:* 2/20/21. Native of Buffalo, NY, Engr Curtiss-Wright Lab 1943-45; Tech Admin 1946-52; Plant Engr Cornell Aero lab 1952-56. Ptnr Trautman Assocs (Buffalo NY) 1956-82; Pres Trautman King Marluart Assoc PC. Reg PE: NY, MI, PA, OH, NJ, FL. NYSSPE; State Pres 1968-69; Natl Dir NSPE 1965-77. Metitorious Serv Award 1973 & 78. Engr of the Yr Award 1978. Chmn, NSPE/PEPP Professional Libility Insurance Ctte 1979-81, Also Mbr ASHRAE & ASME. *Society Aff:* NSPE, ASME, ASHRAE.

Tree, David R
Business: Sch of Mech Engg, W Lafayette, IN 47907
Position: Prof & Asst Hd *Employer:* Purdue Univ. *Education:* PhD/ME/Purdue Univ; MS/ME/Brigham Young Univ; BSE/ME/Brigham Young Univ. *Born:* 8/18/36. Wanship, UT, taught & conducted res since Sept, 1963. Res Areas: HVAC/R Equip & systems, energy usage in residential dwellings, thermal systems modeling. Became Asst Hd Sch of Mech Engg Sept, 1979. Enjoy home gardening. Editor, Internat'n'l Journal of Refrigeration. *Society Aff:* ASHRAE, IIR, ASEE.

Treffs, David A
17 Deering Ave, Lexington, MA 02173
Position: Manager - Product Quality Assurance. *Employer:* Polaroid Corporation. *Education:* MS/Engg Mgmt/Northeastern Univ; BS/Marine Engg/SUNY Maritime Col. *Born:* Mar 21, 1933. B Marine Engrg SUNY Maritime Coll; MS Engrg Mgmt Northeastern Univ; Certificate of Advanced Educational Study (CAES) in Educational Research at Boston College. Lic Marine Engr. Captain US Coast Guard Reserve (Retired). Genl Elec 1955-60: pulse-jet engine res; airborne opto-mech design. Perkin-Elmer 1960-63: Proj Engr for baloonborne & orbiting light-weight optics. Itek Corp 1963-69: prog Engr for high-acuity photo sys for aerial reconnaissance sys, prog mgr for NASA test installation; Bus Mgr for Advanced Dev Div. Polaroid Corp 1969- : Mgr Hardware Evaluation Lab; Mgr Prod Quality Assurance, corp respon for photo sys incl cameras, film & flash sources. Mbr, Corporate Faculty, Process and Product Optimization (Taguchi Methods). SPIE Governor 1972-79; SPIE Serv Awards (2) 1976. (1) 1978. Elected Fellow 1977 SPIE. *Society Aff:* SPIE, ASQC, ROA.

Trelease, Frank J, III
Home: 3228 Locust Dr, Cheyenne, WY 82001
Position: Assistant State Engineer *Employer:* WY State Engineer's Office *Education:* BS/CE/Univ of WY; MS/CE/CO State Univ *Born:* 05/24/37 Assistant State Engineer (State of WY). Experience includes: Mgr Wright Water Engrs Cheyenne Office, 1979-84. BRW/Noblitt, 1978. Dir WY Water Planning Program, 1968-1978, WY Natural Resources Bd, 1966-1968. Colorado Water Conservation Bd, 1965-1966. Denver Water Bd, 1964. Wright Water Engrs, 1963-1964. US Bureau of Reclamation & CO State Univ, summers 1956-1961. *Society Aff:* ASCE, AWWA, USCD, ΣΧΙ

Tremper, Randolph T
Business: 24400 Hoghland Rd, Richmond Hts, OH 44143
Position: Mgr-Engg. *Employer:* General Electric Co. *Education:* BS/CerEng/Univ of CA, Berkeley; MS/MatSci/Univ of CA, Berkeley; PhD/MatSci/Univ of UT. *Born:* 6/6/44. *Society Aff:* ACS.

Tretter, Steven A
Home: 601 Hawkesbury Terr, Silver Spring, MD 20904
Position: Assoc Prof. *Employer:* Univ of MD. *Education:* PhD/EE/Princeton Univ; MA/EE/Princeton Univ; BSEE/EE/Univ of MD. *Born:* 5/28/40. in Greenbelt, MD. With Hughes Aircraft Co, Culver City, CA from Jan to Sept 1966. Since 1966 has been with the Elec Engg Dept at the Univ of MD, College Park, MD and is presently an Assoc Prof. Has consulted for several govt agencies and private cos including Naval res Lab and RIXON, Inc. Specialtie include communications, error correcting codes, high speed wire-line modems, and digital signal processing. Author of the book *Introduction to Discrete-Time Signal Processing*, Wiley, 1976. *Society Aff:* IEEE.

Treves, David
Business: Dept of Electronics, Weizmann Inst of Science, Rehovot Israel
Position: Professor. *Employer:* Weizmann Institute of Science. *Education:* DSc/Elec Engrg/Technion; MSc/Elec Engrg/Technion; Ing/Elec Engrg/Technion; BSc (c.1.)/Elec Engrg/Technion. *Born:* June 1930 Italy. Since 1953 he has been associated with Weizmann Inst & has been involved in study of magnetization process; magnetic materials; magnetooptic memories & laser sys; coherent optics and microwave antennas. He spent several yrs in the USA as a Res Assoc at Pomona Coll, Claremont, CA. Mbr of the Tech Staff at Bell Labs, Murray Hill, NJ, at the Ampex Res Lab at Redwood City, CA, as World Trade Exchange Fellow at IBM, San Jose, CA, and at Xerox Palo Alto Res Ctr. 1982-86: Hd of Elec Dept at the Weizmann Inst of Sci. 1986-present: On sabbatical at Xerox-Optimem, Sunnyvale, CA. *Society Aff:* IIEEE, OSA, APS, SPOIE, IPS.

Trexler, Jay E
Home: 44 W Franklin St, Topton, PA 19562
Position: Consulting Engr. *Employer:* Self Employed. *Education:* BE/Chem Engg/Johns Hopkins Univ. *Born:* 8/13/20. My diverse background includes 16 yrs experience in engg in the refinery and petrochemical industry. This was followed by shorter tenures in the weighing and materials handling industry, glass industry and the field of mfg of valves and controls. I have been self employed as a consulting engr for the past 10 yrs. I rose through the chairs to pres of Toledo Tech Council in 1959-1960. This is a Council of some 30 professional societies in Northwestern OH. Hobbies include photography and golf. *Society Aff:* AIChE, NSPE, IEEE.

Treybig, Harvey J
Home: 2504 Timberline Dr, Austin, TX 78746
Position: Pres *Employer:* ARE Inc *Education:* MS/CE/Univ of TX-Austin; BS/CE/Univ of TX-Austin *Born:* 02/10/42 Born, reared, and educated in TX. In 1965 upon graduation from the Univ of TX he began his profl career with the TX Highway Dept where he served as an Assoc Design Engr till 1969. After leaving the TX Highway Dept he took a Research Engr position with the Univ of TX, Ctr for Highway Research. In 1970 he joined the firm ARE Inc as a Research Engr and served in that capacity through 1973. In 1973 he was name Pres and Chief Engr of ARE Inc and continues to serve in that capacity. He is a Registered PE in TX. *Society Aff:* ASCE, TRB, APWA, NSPE

Triassi, Michael A
Business: 85 Metro Park, Rochester, NY 14623
Position: President *Employer:* Sear-Brown Associates P C. *Education:* BSCE/Civil Engg/Marquette Univ. *Born:* 8/2/42. Rochester N Y. Mbr Chi Epsilon. Capt in US Army Corps of Engrs. Joined Sear-Brown 1968 becoming Prin 1978. V.P. in charge of operations. Principal-in- charge on many of large private & public construction projects. P E New York State. Pres Rochester Section ASCE. Pres Monroe Chap PE. Chmn N Y State Council ASCE. Chairman Professional Engineers in Private Practice NYSSPE. *Society Aff:* ASCE, NSPE, APWA, SAME, RES.

Tribus, Myron
Business: 77 Mass Ave, Cambridge, MA 02139
Position: Dir - Ctr for Advanced Engrg Study. *Employer:* Mass Inst of Tech. *Education:* PhD/Engg/UCLA; BSc/Chemistry/UC Berkeley. *Born:* Oct 30, 1921 San Francisco. Capt USAF 1942-46. Instr to Prof UCLA 1946- 61; Dean Thayer Sch of Engrg, Dartmouth 1961-69; Asst Secy for Sci & Tech, Dept of Commerce 1969-70; Sr V Pres R&E Xerox Corp 1971-74; Prof & Dir Center for Advanced Engrg Study, MIT 1975; Bd of Dirs Spaulding Fibre Co 1962-65; Carpenter Tech Corp 1965-69; Bd of Governors, Israel Inst of Tech Haifa 1972- . Mbr IEEE NSPE, ASEE; Fellow ASME, NAE; Awards: Thurman Bane IAS 1945; Wright Bros Medal SAE 1946; Heat Transfer Div Memorial Award 1968; Alfred Noble Prize 1952; Hon Doctorates Oakland Univ, Rockford Coll. Books, publs in thermodynamics, heat transfer & decision theory. *Society Aff:* ASME, NSPE, IEEE, ASEE, NAE.

Trick, Timothy N
Business: Dept. of Electrical & Computer Engr, 1406 W. Green St, Urbana, IL 61801
Position: Prof. and Dept. Head *Employer:* Univ of IL. *Education:* PhD/EE/Purdue Univ; MS/EE/Purdue Univ; BS/EE/Univ of Dayton. *Born:* 7/14/39. Native of Dayton, OH. Employed by Univ of IL, Urbana-Champaign, since 1965. Currently Prof and Head of Elec and Computer Engg and Res Prof of Coordinated Sci Lab. Visiting Assoc Prof, Univ of CA, Berkeley, 1973-1974. ASEE-NASA Summer Faculty Fellow 1970, 1971. Teach courses and conduct res in integrated circuits, computer-aided circuit design, and digital signal processing. IEEE Fellow, IEEE Bd of Dir (1986-87) Pres (1979), Pres-Elect (1978), Secy/Treas (1976-1978), mbr of Admin Committee (1973-1976), Assoc Editor (1971-1973), IEEE Circuits and Systems Soc. Received Guilleman-Cauer Award for best paper published in IEEE Transactions on Circuits and Systems, 1976. *Society Aff:* IEEE, ASEE, NSPE

Triebwasser, Sol
Business: J Watson Res Center, Box 218, Yorktown Hts, NY 10598
Position: Mbr Research Review Bd. *Employer:* IBM Corporation. *Education:*

Triebwasser, Sol (Continued)
PhD/Physics/Columbia Univ; MS/Physics/Columbia Univ; BA/Math/City Univ, NY. *Born:* Aug 16, 1921. US Army Signal Corps 1942-46 Capt; Phys Instr Brooklyn Coll 194751; employed by IBM 1952- . Publs in solid state physics & devices in ferroelectricity, photoconductivity, injection lasers, MOS devices. Respon for Applied Res in Large Scale Intergation & Semiconductor Device Physics. Was Asst Dir Applied Res & Prog Dir Advanced Technology, Corporate Staff. Fellow IEEE & APS. Presently FET Program Mgr. *Society Aff:* IEEE, APS, AAAS.

Triffet, Terry
Business: Office of the Dean, Coll of Engrg, Tucson, AZ 85721
Position: Actg Dean *Employer:* Univ of AZ *Education:* PhD/Struc Mech/Stanford Univ; MS/CE/Univ of CO; BS/Arch Engrg/Univ of Co; BA/Liberal Arts/Univ of OK *Born:* 6/10/22 1948-50 Instructor (Civil and Architectural Engrg), Univ of CO. 1950-55 General Engr (Rocket and guided missile research and development, Section Head 1953-55), US Naval Ordnance Test Station, China Lake, CA. 1955-59 Research Engr (Radiological effects research, Project Mgr 1955-56, Branch Head 1956-59), US Naval Radiological Defense Laboratory, San Francisco, CA. 1959-76 Professor of Mechanics and Materials Science (associated with Center for Applied Mathematics from 1968), MI State Univ. 1976-1984 Associate Dean (for Research), Coll of Engrg, and Dir of Engrg Experiment Station at the Univ of AZ., 1984-, Acting Dean of Engrg. Consultant: 1959-65 US Department of Defense. 1965-68 Battelle Memorial Inst. 1980- US Dept. of Energy. 1984- National Aeronautics and Space Adm. 1966-69 Lear-Siegler Instrument Div. 1966-67 and 1972-73 Sr Research Fellow 1985-86 Distinguished Visiting Scholar in Mathematical Physics at the Univ of Adelaide, South Australia. *Society Aff:* SES, IEEE, APS, AMS, ACM

Trigger, Kenneth J
Business: 222 Mech Engrg Bldg, Urbana, IL 61801
Position: Professor of Mechanical Engineering Emeritus. *Employer:* University of Illinois-Urbana. *Education:* MS/Mech Eng/MI State; BS/Mech Eng/MI State. *Born:* 9/6/10. 1933 & 1935 taught Mech Engrg at Mich St, Swarthmore Coll & Lehigh Univ. On Mech Engrg Staff at Univ of Ill 1938-77; Head of Indus-Prod Div 1951-76. Res in metal cutting & assoc material removal problems. Author or co-author of some 30 papers on metal cutting & temps in machining; contrib to ASM & SME handbooks. Cons on metal removal & allied probs to numerous companies. Co-recipient of ASME Blackall Award 1957; recipient of ASTME Res Medal 1959; Life Fellow ASME; Life Fellow ASM, Fellow Soc Mfg Engrs & Sigma Xi. *Society Aff:* ASME, ASM, SME, ΣΞ.

Triggs, James Frederick
Business: 1000 Grandview Ave, Pittsburgh, PA 15211
Position: Prin. *Employer:* J Fred Triggs & Assocs. *Education:* Doctor of Sci/Engg Sci/St Vincent College; /Arts/Carnegie Inst of Tech; /Liberal Arts/St Vincent College. *Born:* 1/9/07. in New Haven, CT, attended St Vincent College and other colleges. Taught in SVC Engg Courses, 1953-1957. Reg PE in PA, OH, WV, MD. Partner in Arch/Engg firms: Triggs and Pierce, Triggs, Myers and Assoc. Conducted Professional Practice in Forensic, Civil and Environmental Engg 1945 to Present. Served as Bd Mbr, Advisory Bd for Bldg Construction 1957 to Present for PA Dept of Labor & Industry. *Society Aff:* AAAS, ASCE, NSPE, SAME, AAEE, PSPE.

Trinidad, Adolph A, Jr
Home: 175 Violet Dr, Pearl River, NY 10965
Position: Exec V Pres. *Employer:* Andrews & Clark Inc. *Education:* MS/CE/Columbia Univ Sch of Engrg; BS/CE/Columbia Univ Sch of Engrg. *Born:* 3/31/25. With Andrews & Clark Inc Cons Engrs in transportation 1950- , as designer, proj mgr, ch struct engr, Ch Engr, V Pres 1965- . Exec VP 1981; Fellow ASCE; Fellow ACEC; Mbr Natl Soc & N Y St Soc of Prof Engrs, Amer Concrete Inst, Amer Inst of Steel Constr, Tau Beta Pi. Prof Planner in N J; Prof Engr in N Y & 12 other states. Past Pres & Dir of N Y Assn of Cons Engrs. *Society Aff:* ASCE, ACEC, NSPE, SAME, AWS, ACI.

Trinkle, Arthur H
Business: 131 N. San Gabriel Blvd, Pasadena, CA 91107
Position: President. *Employer:* Metrex Sys Corp *Education:* -/Civ Engrg/Purdue Univ. *Born:* Mar 1926; Native of Chicago Ill. With Field Artillary forward observation Europe WWII. Office Engr in field offices Stone & Webster, Foster Wheeler Corp. V Pres Baker Smith, N Y. Dist Mgr with Jack Ammann Photogrammetric Engrs. Originated & managed EHV Transmission Line Dept for Fairchild Aerial Surveys. Founded Metrex 1966, which acquired Brewster Pacific Maps 1969 & installed interactive computer graphics, geographic database mgmt sys 1974. Fellow ACSM Natl Bd of Dir Amer Soc of Photogrammetry 1974-77, Pres SW Region 1973; Member IEEE. *Society Aff:* IEEE, ICMA, ASP, APWA, ACSM.

Trivelpiece, Alvin W
Business: 1000 Independence Ave, Washington, DC 20585
Position: Dir, Ofs of Energy Research *Employer:* US Dept of Energy *Education:* PhD/EE & Phys/CA Tech; MS/EE/CA Tech; BS/EE/CA Poly State Coll. *Born:* 03/15/31 in Stockton, CA. Taught at UC Berkely 1958-66 (EE, Asst Assoc Assoc Prof); Prof Physics, Univ of MD, Coll Park 1966-76 1966-76; Asst Dir for Res, CTR Div, USAEC 1973-75; VP Eng & Res Maxwell Labs, San Diego, 1976-78; Corp VP, Sci Application, Inc, La Jolla, CA 1978-1981; Dir, Office of Energy Research, US Doe, 1981- ; Fulbright Scholar 1958, Guggenheim Fellow 1966; Distinguished Alumnus, Cal Poly 1978; Author, "Principles Plasma Physics" McGraw Hill 1973; 100 papers; "Slow Wave Progation in plasma Waveguider" 1967. Patents; Conslt to Government and Industry. *Society Aff:* IEEE, APS, AAAS.

Trocino, Frank S
Home: 3962 Spring Blvd, Eugene, OR 97405
Position: Pres. *Employer:* F S Trocino, Inc. *Education:* BS/Chem Engg/Univ of IA. *Born:* 8/26/28. Native of Oelwein, IA. 1951 worked as process engg in ordnance plant. Served in Army 1953-55 in thermodynamic res project. Superindendant of grain processing plant involved in soybean extraction, alcohol and vitamin manufacture. Joined Borden Chem 1958, designing formaldehyde plants. Transferred to Oregon plant as plant engg. 1965 tech dir of Bohemia, Inc. Conducted res in forest products. Developed a procerss for extracting wax from bark, making total chem utilization of bark. This process won three awards including 1974 "Chemical Engineering" magazine merit award for personal achievement in chem engg. 1979 formed own corp as chem consultant. *Society Aff:* AICHE, ACS, FPRS.

Trojan, Paul K
Home: 3939 Vorhies Rd, Ann Arbor, MI 48105
Position: Prof *Employer:* Univ of MI. *Education:* PhD/Met Engg/Univ of MI; MS/Met Engg/Univ of MI; BS/Met Engg/Univ of MI. *Born:* 9/8/31. Co-op student (1952-53) & Sr Engr (1964) with Ford Motor Co. Since 1956 have taught 15 different courses in Mtls & Met at Univ of MI-Dearborn & Ann Arbor campuses. Interim Engrg dean-Dearborn (1980-82). Have published over 60 technical articles; primarily in process met. Also co-author of two textbooks on engg mtls & their processing. Past experience includes profl consulting for over 50 industrial clients. Recipient of ASM Howe Medal (1963) & AFS Pangborn Gold Medal (1978). Profl interest is in res, dev, microstructure control & defect analysis; primarily in cast metals. Extra-curricular interest is in outdoor activities. *Society Aff:* ASM, AFS, ASEE.

Trolio, Andrew E
Business: Dept of Continuing Education, Broomall, PA 19008
Position: Pres. *Employer:* Broomall Industries Inc. *Born:* 8/26/29. USAF Schools; 1951 Grad Advanced Curriculum Electronics, Teaching & Measurements & Evaluation; attended Villanova Univ 1954. Inventor & holder of several pats & many pats still pending or applied for. Founder of Adtrol Inc & held position of Pres & Dir 9 yrs; Pres of Lin-Lea Corp; P V Pres & Dir of The Ottawa Corp; P Dir of Instrumetation Marketing Corp; Founder, Pres & Dir of Broomall Indus. Mbr & P Governor SPIE; AOPA. Mbr Lincoln Natl Bank Adv Ctte. *Society Aff:* SPIE, ASP.

Trombly, Arthur J
Business: 1137 Suffield, Birmingham, MI 48009
Position: Retired Independent process consultant *Education:* BS/Chem E/Univ of Detroit *Born:* 2/11/15. Native of Detroit. In US Army, WWII; served on Manhattan Proj, awarded Legion of Merit. Headed high-temperature fluidization dev group for Am Metal Co 1946-50. With Du Pont engg dept (1950-68) in various dev & design positions on titanium dioxide, polyester, synthetic leather projs. In independent practice 1968-71. Headed process design group for Kling-Leopold Inc, Phila, 1971-73. With Cunningham Engrs since 1973 as Chief Process Engr, responsible for process & instrumentation design. VP since 1976. In 1979, appointed Dir of Engg responsible for all engg design functions. Appointed Pres 1981. Retired 1983. *Society Aff:* AIChE.

Trotter, Ide P, Jr
Business: Feedstocks and Energy Dept, B-1920 Machelen, Belguim
Position: Supply Mgr *Employer:* Essochem Europe, Inc. *Education:* BS/Chem Engg/TX A&M Univ; MA/Chem Engg/Princeton Univ; PhD/Chem Engg/Princeton Univ. *Born:* 10/27/32. Colombia, MO, 10/27/32. Raised in TX. Served 1st Lt Chem Corps. Research Engg Humble Oil and Refining,1958-64. Patents, publ in chem reactor design, computer control. Head, Catalytic Cracking Research, Esso Research & Eng, 1965- 67. Refining Headquarters, Corp Planning, Humble Oil & Refining, 1967-70. Tech Mgr, 1970-72. Process Superintendent, 1972-74, Billings MT, Refinery, Exxon Co, USA. Sr Advisor, Logistics Dept, Exxon Corp, 1974-78. Gen Mgr, Logistics, Esso Standard Sekiyu KK, Tokyo, Japan, 1978-81. Supply Manager, Feedstocks and Energy Dept, Essochem Europe, Inc, 1981. AIChE, Program Committee, Chmn, Engg Sci & Fundamentals Group, 1964-67. *Society Aff:* AIChE, ACS, ΣΞ

Trowbridge, Roy P
Home: 17377 Westmoreland, Detroit, MI 48219
Position: Director - GM Engrg Standards. (Retired) *Education:* BAME/Mech Engg/Leland Stanford, Jr Univ; Assoc of Arts/Mathematics/Sacramento Jr College. *Born:* Dec 16, 1918 Sacramento Calif. V Pres Codes & Standards, ASME 1966-68; ANSI VP 1969-70; Pres ANSI 1971-74; V Pres Codes & Standards ASME 1976-78 and 1978-80. Mbr AMSE Bd of Governors 1981-82. Bd of Dirs ANSI, 1966-1980; Bd of Dir ANMC 1973 to 1982. Fellow ASME, SAE, SES; and Engrg Soc of Detroit; Mbr Tau Beta Pi. Received SES Leo B Moore Medal 1969; SME Introprofessional Award 1975; 1976 Howard Coonley Medal from ANSI, 1981 SAE Arch T Colwell Medal, 1981 AMSE Codes Standards Medal; 7 certificates of award from ASME & 2 from SAE. Numerous contribs to dev & mgmt of natl & internatl standards *Society Aff:* ASME, ANMC, SES, SAE, TBΠ.

Troy, John J
Home: 1312 Rimer Dr, Moraga, CA 94556
Position: Principal *Employer:* Gage-Babcock & Assoc, Inc. *Education:* BE/Mech Engr/Youngstown Univ. *Born:* 9/3/34. Native of Sharon, PA. Joined Automatic Sprinkler Corp of America before grad, advanced to Asst Mgr Special Hazards Dept - dev and application of unique fire protection systems throughout US; Technical Expert - Fire Protection, The Dow Chem Co; Corporate Loss Prevention Engr, BASF Wyandotte Corp; presently consultant in fire protection, security, safety. Maj projs - Program Mgr Fire/Life Safety, major rapid transit systems - Atlanta & Dade Co, Proj Engr - Fire Protection for Doe Memorial Lib, UC Berkeley; SFPE - Natl Dir 1979-1984, No Cal Chapter Pres 1977-78; NSPE - Diablo Chapter Pres 1978-1980; 1985-88 NFPA Comm mbr Rail Rapid Transit Std; PE - CA, GA, OH, WA, MI. *Society Aff:* NSPE, SFPE, SSS.

True, Newton F
Business: 55 E Monroe St, Chicago, IL 60603
Position: Hd, Estimating Div *Employer:* Sargent & Lundy *Education:* MBS/Bus/CA Coast Univ; BS/Engrg/CA Western Univ *Born:* 03/28/26 Native of IL. Served with the US Navy from 1944 to 1946. IL Sch Dist 201 Chrmn of Building Ctte for Bd of Education 11 years and Chrmn of Church Building Ctte from 1974 to 1975. Has been with Sargent & Lundy for 36 years. Assumed current position of Hd, Estimating Div in 1974. Became Assoc of the firm in 1978. Mbr of AACE since 1974, holding various positions on the regional and natl levels. Enjoys photography, gardening, and golfing. *Society Aff:* AACE

Trueblood, Walter D
Business: PO Box 8405, Kansas City, MO 64114
Position: Mgr of Project Services. *Employer:* Black & Veatch Engineers-Architects. *Education:* BS/ME/Univ of MO. *Born:* 1/15/31. Joined Black & Veatch 1957. Directs the Pwr Div's admin services with prime respons in the coord of div policy formulation and implementation, performance analysis, cost control planning, and personnel performance analysis, procurement control & planning, and personnel administration. His respons include the control of microfilm, technical publications, library, document control, and word processing services for the div. Mr Trueblood served 7 yrs as the Mech Engg Dept Hd where he was respon for dept competence of engg design and production. These respons included control of technical acceptability oif mechanical engg design through continuing education of dept personnel, dev of departmental stds and policies, and maintaining a dept attitude of thoroughness and accuracy. Mbr ANSI, ASME, ASTM; Regl VP Reg VII ASME 1974-75 and 1975-76, has held past offices in every category through local chapter ASME. Mbr Bd/Dir Univ of MO Engg Alumni Organization, Chairman Mech. Eng. Dept Advisory Board Univ of Mo. *Society Aff:* ASME, ASTM, ANSI.

Truesdell, Clifford A
Business: Il Palazzetto, 4007 Greenway, Baltimore, MD 21218
Position: Prof of Rational Mech. *Employer:* Johns Hopkins Univ. *Education:* PhD/Math/Princeton; Dotting/Mech Engrg/Milan Poly, Italy; DSc/Engrg/Tulane; FilD/Phys/Upsala, Sweden; Dr Phil/Sci/Basel, Switzerland. *Born:* 2/18/19. I have attempted to study rational mechs and its history at the clearest conceptual level and with the best mathematics available. Prof math Ind Univ, Bloomington, 1950-61; prof rational mechs Johns Hopkins Univ, Baltimore, 1961-. Recipient Bingham medal Soc of Rheology, 1963, Gold medal and internat prize Modesto Panetti, Accademia di Scienze di Torino, 1967, Birkhoff prize Am Math Soc, 1978. Mem Soc for Natural Philosophy (founding mbr dir 1963), Socio Onorario dell Accademia Nazionale di Scienze, Lettere ed Arti (Modena, Italy), Academie Internationale d'Histoire des Sciences (Paris), Istituto Lombardo Accademia di Scienze e Lettere (Milano, Italy), Istituto Veneto di Scienze, Lettere ed Arti (Venice, Italy), Accademia delle Science dell Istituto di Bolegna, (Italy), Accademia Nazionale dei Lincei (Rome), Academie Inter de phil de Sci (Bruxelles, Belgium), Author 20 books, 300 papers. Accademia delle Scienze (Turin, Italy), Academia Brasileira de Sciencias (Brazil). *Society Aff:* SNP.

Truesdell, Dan A
Home: 69 N Park Blvd, Glen Ellyn, IL 60137
Position: Vice Pres/Mgr Mid West Operations. *Employer:* Peter Cooper Corporation. *Education:* BSChE/Chem Engg/OH State Univ. *Born:* 2/15/13. Cleveland Ohio. Pilot plant oper, refinery oper & tech serv U O P Inc & Pure Oil Co 1936-52; plant mgmt & bus dev Processors Inc 1952-59; plant mgmt, proj mgmt & corporate mfg Velsicol Chem Corp 1959-69; established grass roots chem mfg oper Mexico for Velsicol; plant mgmt, Mfg Exec & Company Officer Peter Cooper Corps 1969-79. Conslt, Plant Mgmt 1980 to date. *Society Aff:* AIChE.

Truitt, Marcus M
Home: 504 E Ailsie Ave, Kingsville, TX 78363
Position: Prof of Civil Engr. *Employer:* TX A & I Univ. *Education:* PhD/Envr Engr/Johns Hopkins Univ; Engr/Engr Geol/Stanford Univ; SM/Soil Mechanics/Harvard Univ; BSCE/CE/OK State Univ. *Born:* 1921 Enid, OK. Pilot, Army Air Corps, WWII. Soil Mechanic in Central American 1949-1951. Consulting Practice in addition to academic position 1952- 1965, Soil Mechanics and Municipal Engr. Asst Prof, Assoc Prof, Prof, Dept Chairman in Civil engr., and now Prof. 1951-pres. Tx

Truitt, Marcus M (Continued)
A & I Univ. Author, Soil Mechanics Technology, Prentice Hall Publsh., 1981. Registered Prof Engr Registered Public Surveyor. *Society Aff:* ASCE.

Trull, Olin C
Business: PO Drawer 707, Bennettsville, SC 29512
Position: VP. *Employer:* Powell Mfg Co, Inc. *Education:* BS/Agri Engr/NC State Univ. *Born:* 6/28/34. Began employment with Powell Mfg Co, Inc as an Engr in 1956. After time out for six months in the US Army as a commissioned officer, worked as a Design Engr until 1962, when became Supervisor of Engg. In 1965, became respon for all mfg as Works Mgr. In 1971, assumed respon as VP-Operations, in charge of Engg with additional administrative duties. Given responsibility as VP-Operations in 1985, responsible for planning purchasing and manufacturing. Active in ASAE. Chrmn of NC Section ASAE in 1977. Active on ASAE Cttes. *Society Aff:* ASAE.

Trump, John G
Business: Bldg N-10, Cambridge, MA 02139
Position: Prof Emeritus & Sr Lecturer. *Employer:* MIT Dept Elec Engrg & Computer Sci. *Education:* DSc/Elec Engrg/MA Inst of Tech; MS/Phys/Columbia Univ; BS/Elec Engrg/Polytech Inst of Brooklyn. *Born:* Aug 21, 1907 NYC. Prof Emeritus & Sr Lecturer EE & CS Dept & Dir of High Volt Res Lab MIT; Consultant, High Volt Engr Corp. Chmn, Bd of Trs Lahey Clinic Found. Awards: The King's Medal for Freedom 1947, Presidential Citation 1948, Lamme Med AIEE 1960, Engrg Soc of N E Award for 1967, Power Life Award-IEEE 1973. Gold Medal, Amer Coll of Radiology. IEEE Fellow-Life Mbr. Worked on dev of high voltage Van de Graaff accelerators for sci & medicine, application of megavolt X-rays & electrons to cancer therapy, dielectric strength of compressed gases & vacuum, application of compressed gas insulation to underground power cables, disinfection of municipal wastewater sludges by high energy electron irradiation. Hon Mbr, Amer College of Radiology 1975; Mbr, Natl Acad of Engg, 1977..

Trumpler, Paul R
Business: 1442 Phoenixville Pike, West Chester, PA 19380
Position: President. *Employer:* Trumpler Associates Inc. *Education:* PhD/Mech Engrg/Yale Univ; BS/Mech Engrg/Lafayette Coll. *Born:* Nov 1914. Ill Inst of Technology Instructor 1939-41, Prof in charge of machine design 1949-57; Univ of Penn Prof of Mech E 1957-69; Heat transfer Dev Engr M W Kellogg Co 1942-47; Centrifugal Compressor Engr Clark Bros Co 1947-49; Founder, Pres, Chmn of the Bd Turbo Res Inc Cons Mech Engrs 1961-74. Founder, Trumpler Assocs Inc. Cons Mech Engrs 1974. Pres Turbotherm Corp & Turbo Res Foundation. Patentee, author of num tech articles & 'Design of Film Bearings.' Mbr, Chmn Heat Transfer Div 1953, Life Fellow ASME. Fellow, ASEE. Specialties: film bearings & turborotor dynamics. *Society Aff:* ASME, ASEE, ΤΒΠ, ΠΤΣ, ΦΒΚ, ΣΞ.

Trumpler, William E
Home: 728 Pine Ridge Road, Media, PA 19063
Position: Consulting Engineer. *Employer:* Trumpler Associates Inc. *Education:* BSME/Mechnical/Lehigh Univ. *Born:* Aug 1916. Westinghouse Electric Corp Lester Penn Design, Res & Dev Engrg of steam turbines & related machinery prior to 1951; Mgr of mech design for large steam turbine engrg 1951-70, special work in vibratory analysis of steam turbine blades & discs, application of plastic flow properties of metals to design of turbine parts at high temperatures; Adv Engr blade vibration problems 1971-74. Cons Engr turbo machinery, materials engrg & pressure vessels 1975 Trumpler Assocs Inc. P E state of Pa. Life Fellow member, P Chmn Materials Div, mbr of cttes, Boiler & Pressure Vessel Code 1960-76 ASME. Council Award 1971. Mbr ASTM, active Joint Com on Effect of Temp on Properties of Metals. Hobbies: music, yachting, tennis & skiing. *Society Aff:* ASME, ASTM.

Trutt, Frederick C
Business: 453 Anderson Hall, U. of Kentucky, Lexington, KY 40506-0046
Position: Prof & Chairman of Electrical Engrg. *Employer:* U. of Kentucky. *Education:* PhD/EE/Univ of DE; MS/EE/Univ of DE; BS/EE/Univ of DE. *Born:* 6/28/39. Native of NY, NY. Capt in Army Artillery Corps 1965-67. Res Elec Engr with Army Corps Engrs 1967-69. Team Chief, Army Mtl Command 1969-72. Assoc Prof 1972-81, Prof 1981-85 of EE at the Pennsylvania State Univ involved in teaching & research in elec power systems & machinery. Currently Prof. & Chairman of EE at the Univ of Kentucky. IEEE Outstanding Paper Award 1963, 1982, 1983. *Society Aff:* IEEE, ASEE.

Truxal, John G
Business: Coll Engrg/Applied Sci, Stony Brook, NY 11794
Position: Professor. *Employer:* SUNY at Stony Brook. *Education:* ScD/EE/MIT; BA/-/Dartmouth. *Born:* 2/19/24. Taught Elec Engg Purdue 1950-54; Polytechnic Inst of Brooklyn 1954-72 (last 10 yrs as VP); Dean Engg Stony Brook 1972-76. Mbr NAE, Fellow IEEE & AAAS. Cons or Adv Ctte Mc-Graw Hill Book Co, NSF, Ford Motor Company ISA Education Award 1963, ASEE Westinghouse AWard 1965, ORSA Lanchester Prize 1972, IEEE Education Medal 1974. *Society Aff:* IEEE, ISA, AAAS, ASEE.

Tsai, Keui-wu
Business: Dept of Civ Engrg, San Jose, CA 95192
Position: Prof & Chrmn of Civ Engrg *Employer:* San Jose State Univ *Education:* PhD/Geotech Engrg/Princeton Univ; MS/CE/Princeton Univ; MA/-/Princeton Univ; BS/CE/Natl Taiwan Univ *Born:* 01/22/41 Born in Taiwan, naturalized in 1973. Active conslt in the field of geotech engrg, specialing in settlement and stability of struct on soft and compressible soils and related construction problems. Reg CE in CA and Taiwan. Pres of ASCE San Jose Branch, 1978-79. Pres of CAETDA, 1985-87. Teaching, res and consltg since 1967. Assumed current respon as Chrmn of Civ Engrg at San Jose State Univ since 1981. Civ Engrg Prof of the Year 1976 and 1984. Sigma Xi and Chi Epsilon. *Society Aff:* ASCE, ISSMFE, CAETDA

Tsai, Steve Yuh-Hua
Home: 7130 Summit Rd, Darien, IL 60559
Position: Proj Leader *Employer:* Argonne Natl Laboratory *Education:* PhD/CE/Univ of IL; MS/Hydr Eng/SD State Univ; BS/CE/Natl Taiwan Univ *Born:* 10/23/46 Sr Engr, Sargent and Lundy Engrs in Chicago, 1973-1975. Involved in advanced analysis of hydraulic and hydrologic designs of nuclear and fossil-fuel power plants. Joined Argonne Natl Lab in 1979 as an Environmental Scientist and Project Leader. Profl and research interest in pollutant transport in surface and groundwater systems, thermal discharges, and hydrologic impacts associated with radioactive and hazardous wastes disposal. Reg PE in WI. Awarded 1980 Karl Emil Hilgard Hydraulic Prize of ASCE. *Society Aff:* ASCE, AGU

Tsao, Ching H
Home: 3930 Wiateria, Seal Beach, CA 90740
Position: Prof. *Employer:* CA State Univ. *Education:* PhD/Engg/Mechanics/IIT; MS/Civ Engg/MI State Univ; BS/Civ Engg/Chiao-Tung Univ. *Born:* 11/16/20. Native of China. Taught at USC 1953-55. Did research in stress analysis at Hughes Aircraft Co 1956-61. Aerospace Corp 1961-65. Taught at CSULB since 1965. *Society Aff:* ASEE, SESA.

Tsareff, Thomas C, Jr
Business: Code T27/P O Box 894, Indianapolis, IN 46206
Position: Sr Experimental Metallurgist. *Employer:* Detroit Diesel Allision, GMC. *Education:* BS/Chemistry/Butler Univ. *Born:* Aug 1926. Native of Indianapolis Ind. BS Butler Univ. Army Intelligence 1945-46; Marine Corps Artillery Opers 1950-51. Detroit Diesel Allison 1951- , Gas Turbine Failure Analyst, advanced joining dev; AEC Indus Loanee to Bettis Atomic Power Labs 1959-60; Materials Specialist Military Compact Reactor & Nuclear Rocket Study Prog 3 yrs; Researcher Hot Corrosion Resistant Alloy Dev, Boron Fiber-Epoxy Composite Structures. Current T56 Gas Turbine Aircraft & Indus Engine Materials/Processes Specialist & Engrg Mate-

Tsareff, Thomas C, Jr (Continued)
rials/Processes/Inspection Specification Specialist. DDA rep to AMS Div SAE (Commodity Ctte B Chmn) T Ams Editorial Consultant, IN Vocational Teacher License.. Received ASM Award of Merit 1969. *Society Aff:* ASM.

Tschebotarioff, Gregory P
Home: 26 George St, Apt 5, Lawrenceville, NJ 08648
Position: Retired (Self-employed) *Education:* DrIng/Soil Mech/Tech Hochschule Aachen; Dipl Eng/Civil Ing/Tech Hochschule, Charlottenburg. *Born:* 2/15/99. Native of Pavlovsk, Russia. 1916-20 Artillery officer Imperial & White Armies. Engg with contractors & consultants in Europe, 1926-28. Design & Research Engg with Egyptian Govt, 1929-36. 1937-64 Civil Engg Faculty, Princeton Univ; in charge soil engg res projects for two US Govt Agencies.1956-70 Associate King & Gavaris, Consulting Engg. Author (McGraw-Hill) Book on Soil Mechanics & Foundations, 1-st Ed 1951, translated Spanish, Japanese, Russian; 2nd Ed, 1973, translated Portuguese. Autobiography, 1964, "*Russia, My Native Land*–. Prof articles & discussions 1977, Hon M ASCE. Karl Terzaghi Award, Geotech Div ASCE, 1979. *Society Aff:* ASCE, NSPE.

Tsontakis, Stephen G
Business: Box 1180, Riverhead, NY 11901 *Employer:* Steve G Tsontakis Assoc, PC *Education:* MSME/-/Poly Inst of Brooklyn; MSME/Mech Des/Poly Inst of Brooklyn. *Born:* 10/15/30. in Springfield, MA, served in the US Army 1951-1953. Design, systems integration & engg mgt on several Aerospace Projs, including the Apollo/Lem prog for Grumman Aerospace Corp. Consulting & Design Engg in the areas of Bldgs, Rds, Drainage, Water Supply Systems & Errossion Control. Pres of Engrg Firm & Construction Co, operations mgr for a domestic water supply co. Presently working primarily in the area of complete turn key design & construction mgmt of buildings & other structures. Licensed in FL & NY, NCEE Certified. *Society Aff:* NSPE, NFPA

Tsoulfanidis, Nicholas
Business: Nuclear Engrg, Univ of Missouri-Rolla, Rolla, MO 65401
Position: Prof *Employer:* Univ of MO-Rolla *Education:* PhD/Nuclear Eng/Univ of IL; MS/Nuclear Eng/Univ of IL; BS/Phys/Univ of Athens, Greece *Born:* 05/06/38 Born in Ioannina, Greece. Worked at Nuclear Res Ctr Democrits before coming to the USA in 1963. Faculty mbr of the Univ Of MO-Rolla since 1968. Worked for GenAtomic, San Diego, CA, in 1974-75; for AK Power and Light Co, during the summers of 76 - 79. Hd of the Nuclear Engrg Program at the Univ of MO-Rolla 1981-85. Vice Chancellor Acad Affairs 1985-86. On leave at Nuclear Research Center Cadarache, France, 1986-87. Author and co-author of many tech papers. Author of the text "Measurement and Detection of Radiation–, McGraw-Hill, 1984. Specializing in radiation transport (neutrons and Gammas), Nuclear fuel cycle, and Health Phys. *Society Aff:* ANS, HPS, NSPE

Tsui, Yuet
Business: Dept of Civil Engr, Pokulam Rd, Hong Kong
Position: Lecturer of Civil Engg. *Employer:* The Univ of Hong Kong. *Education:* PhD/CE (Geotechnical)/Duke Univ; BSC (Eng)/CE/Univ of Hong Kong. *Born:* Sept 1941. Native of Mainland China. PE (NC). Joined faculty Hong Kong Tech Coll 1968-69; Geotech Engr Stone & Webster Engrg Corp 1974-75 involved in nuclear & hydroelectric power projs, specialized in numerical methods in Geomechanics; Asst Prof Civil Engrg Duke 1975-79. Lecturer, Civil Engg Univ of Hong Kong 1979- . Conslt to H K Soil Mech Ltd, Hong Kong. Awarded 1975 Collingwood Prize of ASCE. *Society Aff:* ASCE, ΣΞ, XE.

Tsujiuchi, Jumpei
Business: 4259 Nagatsuta, Midori-ku, Yokohama, Japan 227
Position: Prof *Employer:* Tokyo Inst of Tech *Education:* Dr of Eng/Applied Physics/Univ of Tokyo; BS/Astronomy/Univ of Tokyo *Born:* 08/18/27 Native of Wakayama, Japan. Research Scientist specializing in optical instruments at Government Mech Lab, Tokyo, 1951-67. Visiting Scientist at Inst d'Optique, Paris, France 1958-60 on leave of absence from the Lab. Since 1967, Prof of applied optics at Tokyo Inst of Tech. Dir of Imaging Sci and Engrg Lab, Tokyo Inst Tech, 1972-73, 1978-79, 1984-85. VP of Intl Commission for Optics 1975-78, 1978-81. JSAP: Best Paper in Optics Award 1962. SPSTJ: Technical Award 1980. Pres of Intl Comm for Optice 1981-84. *Society Aff:* OSA,SPIE

Tu, King-Ning
Business: T J Watson Res Ctr, Yorktown Heights, NY 10598
Position: Sr Mgr. *Employer:* IBM. *Education:* PhD/Appl Phys/Harvard Univ; MS/Mtls Sci/Brown Univ; BS/ME/Natl Taiwan Univ. *Born:* 12/30/37. Born on mainland China. Joined IBM Res Ctr in 1968 and assumed current position as sr mgr of Thin Film Sci in the Physical Sci Dept since 1978. Spent a sabbatical yr from 1975 to 1976 at the Cavendish lab, Physics Dept, Cambridge Univ, England as a Sci Res Council Sr Visiting Fellow. Fellow of APS and President of MRS in 1981. Co-edited two books - "Low Temperature Diffusion and Applications to Thin Films," Elsevier, 1975 and "Thin FIlms-Interdiffusion and Reactions–, Wiley, 1978. Has 90 publications in the areas of device met, thin films and interfaces. Other interests include Chinese calligraphy and photography. *Society Aff:* APS, ASM, AVS, MRS.

Tuck, Edward F
Business: Boundary, 1900 W. Garvey Ave So., Suite 200, West Covina, CA 91790
Position: General Partner *Employer:* Boundary *Education:* BS/Communications/Univ of MO; Elec Engr (Hon)/EE/Univ of MO. *Born:* 7/5/31. GM Corp 1953-4; US Army Electronic Warfare 1954-6; Lenkurt Elec Co (Subs GT&E), Mgr Microwave Engg, Mgr, Microwave Mfg, 1957-64; Co-founder, Kebby Microwave Corp 1964-5; Asst Dir, Transmission Dev, ITT Corp, 1965-8; VP, Technical Dir, ITT N Am Telecommunications, 1968-72; GM, Pres Tel-Tone Corp, 1972-4; Pres, Edward Tuck & Co Inc, 1974-1985 (included VP Marketing & Engg) Am Telecommunications Corp (now a subsidiary of Condial), 1976-9). General Partner. The Boundary Fund (a venture capital fund) 1986-present. Bd mbr of: Dytel Corp, Schaumburg, IL; Dest Corp, Wilpitas, CA; Joslyn Corporation, Chicago. First Prize, AIEE Communications Div, 1963. *Society Aff:* IEEE, AAAS, IREE.

Tucker, Irwin W
Home: 1810 Crossgate Ln, Louisville, KY 40222
Position: Prof Environmental Engg. Emeritus; Pres. NCEB *Employer:* Retired. *Education:* PhD/Chemistry/Univ of MD; BS in Engg/Chemistry/Geo Wash Univ. *Born:* 10/30/14. Native of NY. Prior to WWII was engaged in agri res at USDA in WA and CA. Early experience led to advances in frozen food production tech and innovative grain alcohol tech. After services with US Natl Security Resources Bd and with maj intl tobacco cos as Dir of Res and bd mbr, joined faculty of Univ of Louisville focusing his attention on environmental activities. Directed solid waste and air pollution studies and initiated and participated in educational series of TV presentations. Was a founder and currently Pres of Natl Council for Environmental Balance (NCEB) and serves as editor of extensive series of environmental books and pamphlets. *Society Aff:* ACS, ΣΞ, ΑΧΣ.

Tucker, L Scott
Home: 615 S Eldridge, Lakewood, CO 80228
Position: Exec Dir. *Employer:* Urban Drainage & Flood Control Dist. *Education:* MSCE/Water Res/U of AZ; BSCF/Civil Engr/U of NE *Born:* 8/25/39. Native Omaha, NB. Served in Army Corps Engrs 1962-64. Proj Engr, Dep Prog Dir ASCE Urban Water Resources Res Prog 1966-70. Res Assoc CO State Univ 1970- 72. Exec Dir Urban Drainage & Flood Control Dist, Denver, CO 1972-present. Sec, VP, Pres CO Sect ASCE. Sec, Treas, VP, Pres CO Chap APWA. 1979 APWA Top Ten Public Works Leaders of the Yr. Sec, Chrmn ASCE Urban Water Resources Res Council, Treas, VP, Pres organizing mbr Natl Assn of Urban Flood Mgt Agencies, Chmn APWA/IWR Exec Coun, Unesco consultant 1977 for establishing urban hydrology res in Uruguay, Brazil, & Argentina, mbr NSF Working Group on Flood Hazard Mitigation Study. *Society Aff:* ASCE, APWA, WPCF, NAUFMA

Tucker, R C Jr
Home: 61 Ridgeway Drive, Brownsburg, IN 46112
Position: Associate Director - Technology *Employer:* Union Carbide Corp. *Education:* PhD/Met/IA State Univ; MS/Met/IA State Univ; BS/Chemistry/ND State Univ. *Born:* 3/1/35. US Army Ordnance officer; worked on guided missile dev (now Lt Col USAR, Ret.).As a chemist at Ames Lab, AEC, developed high purity metals, properties of Y & Cr. With Union Carbide since 1967, currently Associate Director - Technology, Coatings Service Dept. Responsible for R&D of metallic, cermet, & ceramic plasma, detonation gun, PVD & CVD coatings, primarily for wear & corrosion resistance. Nine patents & over 50 publications on composite mtls, high & low temperature wear, aqueous corrosion, oxidation, hot corrosion, thermal barriers, & plasma & detonation gun deposition. Past chrmn, etc of ASM Chapter, chrmn of ASTM, NACE, & WRC subcommittees/task forces, symposium & session chairman program comm. International Metallurgical Coatings Conf. Fellow of ASM International. *Society Aff:* ASM, AIME, ASTM, ACS, NACE, WRC, ΣΞ.

Tucker, Randolph W
Business: 5252 Hollister Rd., Suite 500, Houston, TX 77040
Position: VP - Engrg Mgr *Employer:* Rolf Jensen & Assoc, Inc *Education:* MM/Mgmt Policy Marketing, Finance/Northwestern Univ; BS/FPE/IIT *Born:* 12/03/49 Author of numerous articles on fire safety and bldg codes. Lectured on fire safety topics at numerous profl seminars and at univs in the US and abroad. Joined Rolf Jensen & Associates in 1974. Established Houston office in 1981. Active in Model Building Code groups. Registered P.E. in IL, FL, GA, LA and TX. President Houston Chapter SFPE. Past President Houston Chapter SMPS. Listed in 1981 edition of US JC's Outstanding Young Men in America. *Society Aff:* SFPE, NFPA, SMPS, PSMA

Tucker, Richard C
Business: Dames & Moore, 7101 Wisconsin Ave, Bethesda, MD 20814
Position: manager, Washington Office *Employer:* Dames & Moore *Education:* MS/CE/GA Inst of Tech; B/CE/GA Inst of Tech *Born:* 09/21/41 Native of Baltimore, MD. Served with several consltg firms, the US Army Corps of Engrs, and as Staff Water Resources Engr with the Natl Water Commission before joining Dames & Moore in 1974. Now Manag vp and partner (Ltd.) of the firm's Wash office and Dir of the firm's Water Resources practice. Pres of the Amer Water Resources Assoc in 1980. Pres of US Geog. Comm of Int'l Water Resources Assoc 1985-88. Officer and Ctte Chrmn of numerous professional societies. Active in Presby Church and Pres Bd of Trustees. Enjoys sports, travel and reading. *Society Aff:* ASCE, AWRA, SAME, WPCF, IWRA, AWWA, ADPA

Tucker, Richard L
Home: 1362 Lost Creek Blvd, Austin, TX 78746
Position: Professor of Civil Engineering. *Employer:* The University of Texas/Austin. *Education:* PhD/CE/Univ of TX; MS/CE/Univ of TX; BS/CE/Univ of TX. *Born:* July 1935. Native of Hereford Texas. 12 yrs faculty mbr Univ of Texas/Arlington, last 6 yrs Assoc Dean Engrg; V P Res Luther Hill & Assoc Inc Dallas 1974-76; joined Univ of Texas faculty 1976 as Leader of Const Prog in Civil Engrg. Hobbies: hunting, golfing & other sports. C.T. Wells Professor of Project Management, Civil Engineering, University of Texas at Austin. Director of Construction Industry Institute. CEO, Tucker & Tucker Consultants. *Society Aff:* ASCE, NSPE, ASEE, ACI, PMI, ASTM, AACE, SAME, Moles.

Tucker, Tracy C
Business: Southern Div, Nav Fac Eng Com, PO Box 10068, Charleston, SC 29411
Position: Commanding Officer *Employer:* Southern Div, Naval Facilities Engrg Command *Education:* PhD/Civil Engg/Univ of IL; MS/Civil Engg/Univ of IL; BCE/Civil Engg/RPI; BS/US Naval Acad. *Born:* April 21, 1939. Reg P E Penn. Native of Sheridan NY. 1960- , C O on active duty as Capt Navy Civil Engr Corps. SEABEE Battalion Opers Officer in Vietnam, Engrg Instructor CEC Officers School Port Hueneme Calif, Officer in Charge of Const for military const projs. Regional Finalist White House Fellows Competition 1973. SAME Moreell Medal 1976. *Society Aff:* SAME.

Tucker, W Henry
Business: Dept of Chem Engrg, Angola, IN 46703
Position: Prof & Chrmn *Employer:* Tri-State University. *Education:* ScD/ChE/MIT; SM/ChE/MIT; BS/ChE/Univ of VA. *Born:* 7/7/20. Manhattan Proj. Sevl Inc 6 yrs Res Supr Absorption & Compression Refrigeration. Whirlpool Cons. Purdue 16 yrs Advisor, Chem Engg Cheng Kung Univ Taiwan. Sabbatical Swiss Fedl Inst Zurich. Proj Sch Purdue-Eli Lilly 1967. Supr Chem Engg Cooperative Education Purdue. AIChE Winston Churchill traveling Fellowship Britair 1969. Lilly Endowment Mid-Career Fellow 1978. Prof & Chmn Chem Engg Tri-State Univ 1969- . Pres. Campus Ministries, Expertise in loss prevention, air environ control, cost & optimization engg, future energy technology. Hobbies: sailing, singing, selfhelp projs. *Society Aff:* ASEE, AIChE, ΣΞ.

Tucker, William
Business: 1515 Broad Street, Bloomfield, NJ 07003
Position: Exec. VP - Technology *Employer:* C-E Lummus. *Education:* MS/ChE/CCNY; BS/ChE/CCNY. *Born:* 5/22/19. Native of Brooklyn NY. US Eng Dept & US Navy Dept 1940-42; C-E Lumnus 1942- , named Ch Process Engr 1961, VP & Mgr Lumnus Tech Enter 1972, VP Engrg R&D & Ch Tech officer 1975, Sr Vp Tech & Planning 1979 Executive Vice President - Technology & Chief Tech. Officer - 1981. *Society Aff:* AIChE.

Tulenko, James S
Business: 202 Nuclear Science Center, University of Florida, Gainesville, FL 32611
Position: Prof and Chmn Nuclear Engg Sci Dept *Employer:* Univ of Florida *Education:* MS/Nuc Eng/MIT; MEA/Eng Adm/Geo Wash Univ; MA/App Eng/Harvard Univ; BA/App Eng/Harvard Coll *Born:* 06/01/36 Native of Holyoke, MA, has specialized in engrg operations. Hd of Nuclear Fuel Engrg Operations for Babcock and Wilcox from 1974 to 1980. Has since specialized in computer aided engrg. Headed Babcock and Wilcock Nuclear Computer Operations from 1980-1983. Has since been made Mbr of Chief Exec's Automation Task Force specializing in engrg automation and Mgr; Engg Automation for Babcock and Wilcox. In 1986, was named Prof and Chmn of the Nuclear Engg Scis Dept at the Univ of Florida. Was granted tenure in 1987. Has taught at both Harvard Univ and at Geo Wash Univ. A P Chrmn of the Fuel Cycle Div and the VA Section of the Amer Nuclear Soc, was named a fellow of the Amer Nuclear Soc in 1978, and awarded the 25th Silver Anniversity award for outstanding contribution to the nuclear indust. Has publ over 25 papers at ASME and ANS. Currently is chmn ABET Nuclear Engg and is a Univ Accreditor for EAC. *Society Aff:* ANS, ASME, SME, ASEE.

Tulkoff, Joseph
Home: 246 Sequoia Dr, Marietta, GA 30060
Position: Dir of Mfg Tech *Employer:* Lockheed-GA Co *Education:* BS/IE/GA Inst of Tech *Born:* 05/28/29 Native of Ashland, KY. With Lockheed-GA since 1951, beginning as process planner. Named Dir of Mfg Tech in 1980. Respon for Air Force/Lockheed Industrial Modernization Incentives Program (IMIP). Integrated Computer-Aided Mfg Program (ICAM), and Lockheed's Coordinator of Generative Process Planning (Genplan) Tech. Fellow, IIE; IIE Aerospace Div Dir (83-84); CAM-I P Chrmn Advanced Technical Planning Ctte; SME/CAM-I computer-aided process planning instructor; SME/CASA certified Mfg Engr; Chrmn, GA Tech Indust Advisory Bd's Computer Integrated Mfg Program. IIE Excellence in Productivity Award, 1982; IIE Aerospace Div Award, 1982. Conslt, factory of the future, CAD/CAM, and mfg productivity. *Society Aff:* IIE, CAM-I, SME, CASA (SME)

Tullis, James P
Business: UT State Univ, Logan, UT 84322
Position: Prof. *Employer:* UT State Univ. *Education:* PhD/Civil Engg/UT State Univ; BS/Civil Engg/UT State Univ. *Born:* 7/24/38. Currently a prof of Civil Engg at UT State Univ. Teaching experience in the areas of fluid mech and hydraulics. Research has been primarily in experimental hydraulics, including cavitation, drag reduction, hydraulic transients, modeling of hydraulic structures, and testing of valves and pumps. Was recipient of the Huber Research Prize from ASCE in 1977 and recipient of the Researcher of the Yr award from the College of Engrg at USU in 1979. Is a registered PE and has served as consultant to municipalities, consulting engg firms, and industry. Was a contractor in commerical construction prior to academic career. *Society Aff:* ASCE, IAHR

Tully, William P
Home: 109 Emann Drive, Camillus, NY 13031
Position: Professor & Dean of Engineering *Employer:* State Univ of NY/Coll of Environ Sci. *Education:* PhD/Civil Engg/Syracuse Univ; MS/Civil Engg/Northeastern Univ; BS/Civil Engg/Northeastern Univ. *Born:* 12/13/40. Native of Braintree Mass. Const Engr E H Porter Co Peabody Mass; Structural Engr LeMessurier Assocs Boston Mass; Academic appointment SUNY-CESF 1966- , Dean Sch of Environ & Resource Engrg 1980- . Instructional & res interests in water resources, hydrology & structures. Reg P E NY. Mbr ASAE, ASEE, COFE & TAPPI. Mbr & Pres Central NY Chap ASCE. Hobbies: wide range of athletic & cultural activities. *Society Aff:* ASCE, ASAE, ASEE, COFE, TAPPI.

Tummala, Lal R
Business: Dept of Elec Engg & Systems Sci, E Lansing, MI 48824
Position: Assoc Prof. *Employer:* MI State Univ. *Education:* PhD/EE/MI State Univ; MS/Ind Electronics/Indian Inst of Tech; DMIT/Electronics/Madras Inst of Tech; BSc/Math, Phys, Chemistry/Andara Univ. *Born:* 5/22/44. As an electronics engr, participated in the dev of instrumentation systems for human learning res experiments in 1967. Res assoc on Environmental Systems prog 1970-72 at MI State Univ. Joined the faculty of elec engg & systems sci dept at MI State Univ in 1972. Published several papers in systems area & recently edited a book on Systems Modeling. Systems consultant to US/IBP prog at Univ of CA, Berkeley, 1972-74. Served in various capacities for IEEE. Current interests are in the areas of control theory, Environmental Systems, and microcomputer-controlled instrumentation, robotics. *Society Aff:* IEEE, SMC, CSS.

Tung, Chi C
Home: 7908 North Carolina State Univ, Raleigh, NC 27695-7908
Position: Professor. *Employer:* Dept Civil Engrg/N C State Univ. *Education:* PhD/Structures/Univ of CA, Berkeley. *Born:* March 24, 1932 Shanghai China. Asst Prof Civil Engrg Univ of Ill Urbana Ill 1964-69; Assoc Prof Civil Engrg N C State Univ Raleigh N C 1969-76, Prof N C State Univ Raleigh N C 1976. Res in probabilistic methods in Civil Engrg, fatigue life of hwy bridges, water waves. Mbr: Amer. Geophys. Union. *Society Aff:* ASCE, AAM.

Tung, David H H
Business: Cooper Sq, New York, NY 10003
Position: Prof. *Employer:* Cooper Union. *Education:* PhD/Struc Engg/Univ of MI; MSCE/Struc Engg/Cornell Univ; BSCE/Civ Engg/Chiao Tung Univ. *Born:* 9/2/23. in Hong Kong. Came to US in 1948. Naturalized in 1958. Held various positions related to design & construction of structures, in Giffels & Vallet, Hardesty & Hanover, Severud-Elstad-Krueger, & Praeger-Kavanagh-Waterbury. Was proj mgr in charge of design & construction of Arecibo Radio Observatory, a 1, 000-ft world's largest spherical fixed-dish, steerable-feed radar-radio telescope in Puerto Rico. With Cooper Union since 1965. Full prof since 1968. C E Curriculum Dept Chrmn, 74-76. Contribute articles to profl journals & active in consulting practice. Believe in educating engrs with broad views. *Society Aff:* ASCE.

Turchick, Albert W
Business: 345 East 47th Street, New York, NY 10017
Position: Director of Technical Services. *Employer:* American Society of Civil Engineers. *Education:* BS/Civil Engg/Penn State; BS/Mining Engg/Penn State. *Born:* 1932. Native of Coalport Pa. Served with Army Corps of Engrs 1954-56. Urban Planning Engr Michael Baker Jr Inc 1957-60; Reports Engr F G Browne & Assocs 1960-62; ASCE Staff 1962- , assumed current responsibility as Dir of Tech Servs 1968. Responsible for organization, coordination & mgmt of technical activities conducted by ASCE. Has served as Pres of Parish Council, VP of Home- Sch Assn & VP Parish Sch Bd. *Society Aff:* ASCE.

Turin, George L
Business: Dept Elec Engg & Comp Sci, Berkeley, CA 94720
Position: Prof. *Employer:* Univ of CA. *Education:* SB/EE/MIT; SM/EE/MIT; ScD/EE/MIT *Born:* 1/27/30. Native of NYC. Prior to joining faculty at Univ of CA (Berkeley) in 1960, was Staff Mbr of Philco Corp; MIT Lincoln Lab; Edgerton, Germeshausen & Grier; Hughes Aircraft Co. Since 1960, has also been a consultant to numerous industrial and governmental organizations. Honors: John Simon Guggenheim Fellowship; British Sci & Engrg Res Council Visiting Fellow; Fellow, IEEE; Member, National Academy of Engineering. Specializes in communication theory and systems; author, books and numerous papers in this field. Past Chairman, Dept Elec Engg & Comp Sci, UC, Berkeley; Past Dean of Engrg and Appl Sci, UCLA. *Society Aff:* IEEE.

Turinsky, Paul J
Business: Dept of Nuclear Engineering, NCSU, PO Box 7909, Raleigh, NC 27695-7909
Position: Head-Dept of Nucl Engr *Employer:* North Carolina State Univ *Education:* PhD/Nucl Eng/Univ of MI; MSE/Nucl Eng/Univ of MI; M/BA/Univ of Pittsburgh; BS/ChE/Univ of RI *Born:* 10/20/44 As Asst Prof of Nuclear Engrg at RPI (1970-73) completed research on neutron cross section and reactor physics theories. Held senior engrg and mgmt positions in Water Reactor Div of Westinghouse working on nuclear fuel mgmt and core operational/safety performance. Presently Prof and Dept Head of Nuclear Engrg at NCSU with ongoing research in nuclear fuel mgmt, and numerical algorithms for parallel architecture computers with nuclear engineering applications. Present Natl Activities-ANS: Chrmn, Reactor Physics Div; Mmbr, Professional Development Comm.; Advisory Editor "Nuclear News" and "Nuclear Science and Engineering–. ASEE: Mmbr, Honors and Awards Cttee. *Society Aff:* ANS, ASEE, AAAS.

Turk, Julius
Business: 5711 W Park Avenue, St Louis, MO 63110
Position: Metallurgist-Dir of Quality Control. *Employer:* Paulo Products Company. *Education:* BS/Chemistry/Univ of IL. *Born:* June 1916. Employed Carnegie Ill Steel, Gary Works, Met Dept. Instructor Army AF Tech Training Sch-Chanute Field. Plant Metallurgist-Process Control Supr Emerson Electric Armament Div; now Metallurgist-Dir of Quality Control Paulo Prods Coml Heat Treating-4 plants, responsible for related Met problems, cons, quality control of heat treat procedures & techniques. P Chmn St Louis Chap ASM. Mbr ASM Handbook cttes on Austempering, Metallography of Carbon & Alloy Steels, Mbrship. Chmn Quality Control Activity Ctte, Heat Treat Div ASM. MTI Standing Ctte Quality Assurance. *Society Aff:* ASM.

Turkdogan, Ethem T
Business: US Steel Corp Res Lab, Monroeville, PA 15146
Position: Senior Research Consultant. *Employer:* US Steel Corporation. *Education:* PhD/Metallurgy/Sheffield Univ, England; MMet/Metallurgy/Sheffield Univ, England; BMet/Metallurgy/Sheffield Univ, England. *Born:* 9/12/23. Hd Physical Chemistry Sec British Iron & Steel Res Assn 1951-59; with US Steel Res Lab 1959- , presently Sr Res Cons. Brunton Medal of Univ of Sheffield 1951, Andrew Carnegie Silver Medal of The Iron & Steel Inst 1953, The Hunt Medal 1967, The Mathewson Gold Medal 1975 of the Met Soc of AIME, I-R 100 Award 1971, Iron & Steel Soc Howe meml Lecturer 1978, John Chipman Award of ISS 1978, The Kroll Medal of

Turkdogan, Ethem T (Continued)
Metals Soc 1978. In the field of chem metallurgy, process thermodynamics & related sujs, pub about 150 papers, sev patents & chaps of books. Fellow Inst of Mining and Metallurgy; Mbr: Metals Soc AIME, Canadian Inst of Mining & Metallurgy. *Society Aff:* MS, IMM, AIME, CIM.

Turksen, Ismail B
Business: Dept of Indus Engrg, University of Toronto, Toronto Ontario M5S 1A4 Canada
Position: Professor *Employer:* University of Toronto. *Education:* BSc/IE/Univ of Pittsburgh, MSc/IE/Univ of Pittsburgh; PhD/IE-OR/Univ of Pittsburgh. *Born:* 12/20/37. Turgutlu Turkey to Huseyin & Aslye ([illegible]). Asst Prof 1970-74, Assoc Prof 1974-83, Prof 1983- , Univ of Toronto. Mbr Bd/Advisorys Ency Computer Sci & Tech, Amer Inst of Indust Engrs. Dir Canadian Affairs IIE. Regional VP - Canada - IIE. Founding Pres CSIE - Canadian Soc for Industrial Engg. Director-NATO-ASI-CIM, Associate Editor IJAR. Mbr: CORS, IIE, APEO. APET, TORS. Contributor of articles to international journals, ency, books and conferences. *Society Aff:* CORS, IIE, APEO, NAFIP, TORS, IFSA, APET.

Turley, John C
Business: 2245 Parktowne Circle, Sacramento, CA 95825
Position: President. *Employer:* John Turley & Assoc Inc *Education:* BS/Refrigeration & Air Condition/CA State San Luis Obispo CA Poly *Born:* 4/25/30. Mech Design Engr, Proj Engr, Ch Engr, Principal Sanford-Alessi-Turley until 1975; spec in heating, air conditioning, ventilation, refrigeration, plumbing sys design for all types of bldgs. Cogeneration system design. Energy conservation retrofit design. Self-employed 1975-78. Turley-Gribben & Assoc formed 1978. Turley-Gribben-Shinneman Inc formed 1984. Mech Cons. Part-time teaching Sacramento City Coll Mech-Electrical Technology Dept. P Pres Sacramento Valley Chap ASHRAE. Mbr CEC. Hobbies: hunting, fishing, boating. Firm name changed January 1, 1986 to John Turley & Assoc., Inc. *Society Aff:* CEC, CEAC, ACEC, ASHRAE.

Turn, Rein
Business: Dept Computer Sci, Calif State Univ, 18111 Nordhoff St, Northridge, CA 91330
Position: Prof of Computer Science *Employer:* CA State Univ Northridge *Education:* PhD/Engrg (Computers)/UCLA; MS/Engrg/UCLA; BS/Engrg/UCLA *Born:* 04/09/31 in Tartu, Estonia. Info scientist with The Rand Corp, Santa Monica, CA, 1963-1976. Staff engr with TRW Defense and Space Systems Group, Redondo Beach, CA, 1977-78. Professor at the Comp Sci Dept of the CA State Univ, Northridge since 1978. Research and teaching in computer sci, especially computer architecture, security and privacy, man-computer applications, and societal impacts of computers. Author of "Computers in the 1980s" (1974), and editor of "Advances in Computer System Security–, (1981, Volume 2 in 1984). Former Chrmn, IEEE Technical Committee on Security and Privacy and of AFIPS Working Groups on Transborder Data Flows and Resiliency of computerized soc. Speaker at numerous natl and intl conferences. Elected Fellow of IEEE in 1986 for contributions for providing privacy protection and security in computer systems. *Society Aff:* IEEE/CS, ACM

Turnbell, Robert C
Business: 519 Tennessee Ave, Ft Wayne, IN 46805
Position: Chmn of Bd *Employer:* Turnbell Engg Co, Inc. *Education:* BS/Civ Engg/Purdue Univ. *Born:* 9/23/15. Native of Tipton, IN. Employed by Norfolk & Western Railway and predcessors 1937-69, including positions of architectural engr, structural designer, div engr, asst regional engr, asst to general mgr. Broad exposure to design for bldgs, bridges, tracks, electronic yards, also economics of operation and maintenance. Employed by Westenhoff & Norvick 1969-72, specializing in railroad projs. In his own private practice 1972 to date, specializing in land dev and storm water control projs, also railroad projs. In 1977-78, consultant to State of IN for Federally Funded Railroad Rehabilitation Prog. Enjoys water color painting and boating. *Society Aff:* ASCE, NSPE, AREA.

Turnbull, Dale W
Home: 14732 Hillbrook N, Chagrin Falls, OH 44022
Position: Vice Pres *Employer:* Caterpillar Inc. *Education:* BS/Agri Engrg/KS State Univ; Advanced Prog for Exec/-/Carnegie Mellon Univ *Born:* 02/15/27 Educated Eskridge, KS. Served in USAF. Joined Caterpillar Tractor Co upon graduation from KS State Univ in 1950. Held many different positions in Sales & Mktg in the US, Canada & in Europe. Have traveled extensively for Caterpillar worldwide. Elected Corporate VP 1978. Current position is Pres of Caterpillar Industrial Inc., a wholly owned Caterpillar subsidiary with responsibility for dev, mfg & mktg the Caterpillar line of industrial lift trucks & automated guided vehicles.

Turnbull, Robert J
Business: Dept of EE, 1406 W. Green St, Urbana, IL 61801
Position: Prof. *Employer:* Univ of IL. *Education:* PhD/EE/MIT; SM/EE/MIT; SB/EE/MIT. *Born:* 7/26/41. Robert J Turnbull was born in Wash, DC on July 26, 1941. Since 1967 he has been with the Univ of IL at Urbana where he is now Prof of Elec and Nucl Engg. His current res interests are in the fields of electrohydrodynamics & controlled thermonuclear fusion. *Society Aff:* IEEE, APS.

Turner, Daniel S
Business: P O Box 1468, University of Alabama, Tuscaloosa, AL 35487
Position: Hd of CE Dept and Exec Dir of the TSM Assoc. *Employer:* University of Alabama *Education:* PhD/CE/TX A&M Univ; MS/CE/Univ of AL; BS/CE/Univ of AL *Born:* 12/09/45 Native of Tuscaloosa, AL. Civil Engrg Off in Air Force. Asst Prof Engrg Tech Prog at GA Southern Coll 1973-76. Owned and Operated Turner-Meadows Land Surveying Co. Univ of AL 1976, Assoc Prof, Dir of Engrg Tech Prog 1981-83, Hd of Civil Engrg 1984-present, Exec Dir of res/serv organ (TSM), principal investigator on 20 res proj last 6 yrs, 75 tech pubs in last 6 years, Outstanding Teacher in Coll of Engrg 1983-84. Specialty areas: traffic operations, geometric design, pavement analysis and tort liability. PE and Prof Land Surveyor. Chi Epsilon, Tau Beta Pi and Phi Kappa Phi hon frats. Outstanding Young Men of Amer 1976, GA; 1981, AL. Who's Who in Technology Today. Gov's Task Force on Drunk Driving, Tech Conslt to AL Legislature. Capstone Engrg Soc, St Dir of JETS (Jr Engrg Tech Soc). *Society Aff:* ASCE, ITE, NSC, TRB

Turner, Howard E
Home: 3425 Hillock Lane, Wilmington, DE 19808
Position: President *Employer:* Turner & Turner Consultants, Inc. *Education:* MSChe/Chem Engg/Univ of MN. *Born:* 4/16/16. Native of Evanston IL. Teaching Asst Chem E Univ of MN. E I du Pont de Nemours & Co Inc 1940-81. Field Engr on dev studies then Field supervision & mgmt to 1956. Then Mgr Chem Engrg Specialist Cons serving entire co. Major interest include fine particle tech and industry-academic interaction such as Design Partnerships, Intern progs. Fellow AIChE, Sigma Xi, Tau Beta Pi, Alpha Chi Sigma. Reg P E DE. Pres and Business Mgr Intl Fine Particle Research Inst, Inc. Amer representative Specialist Exhibitors, Ltd. Croyden, England. Visiting Scholar NY City College ChE Dept. *Society Aff:* AIChE.

Turner, James Howard
Business: 120 Engineering Center, Athens, GA 30602
Position: Editor. *Employer:* Amer Assn Vocatl Instruc Materials. *Education:* BSAE/Agri Engr/Univ of GA; MSAE/Agri Engr/Univ of GA. *Born:* May 24, 1920. Taught Vocational Agri prior to WW II. Served USAF WW II & Korean War as Aircraft Mechanic & B-29 Flight Engr. Irrigation Sales Engr, Proj Engr SCS. P Pres Ga ASAE & Ga Assoc Irrigation Distributors. Author & Editor of many books & other instructional materials in Engrg Technology. Certified auto mechanic. EducationV P Athens Toastmasters Club 1779. Faculty Advisor Alpha Gamma Rho fraternity. *Society Aff:* ASAE, AVA.

Turner, John H
Business: PO Box 1306, Sausalito, CA 94966
Position: Pres. *Employer:* Hutchinson & Co. *Education:* MBA/Mgt/FL State Univ; BE/ChE/Vanderbilt Univ. *Born:* 11/14/40. Native of Pensacola, FL. Served in Army Intelligence & Security Agency 1964-66. Gen Foreman for Raychem Corp Menlo Park, CA 1966-69. Northwest Regional Rep Norton Co-Chem Process Products Div. Marketing & Bus Dev Mgr - Ind Products Div - Fibreboard Corp 1972. Western Regional Mgr - Environmental Equip Div - FMC Corp 1972-74. Pres Hutchinson & Co 1974. Specialists in sales and engg of air pollution control systems, bulk mtl handling systems, chem process equip. *Society Aff:* AIChE.

Turner, Ralph E
Home: 28 Chelmsford Road, Rochester, NY 14618
Position: Dir Industrial Trade Relations (Retired). *Employer:* Eastman Kodak Company. *Education:* BS/Engg Physics/Univ of IL. *Born:* [illegible]/11/12. Native of Chicago. Trainee WI Steel 1936-40; Steam Engg Dept Youngstown Sheet & Tube Co, Sales Engr G E Co X-Ray Dept 1941-56; Eastman Kodak Co 1956- , assumed current mktg respons as advisor to Radiography Markets Sales Mgr 1965. Pres 1971, Chmn of Bd 1972, Fellow 1974 ASNT; Chmn E-7 Ctte 1970-72, Fellow & Award of Merit 1976 ASTM. Mbr: AFS, ANS, ASME. Retired from Eastmand Kodak on Aug 1, 1977. Received Hon Mbr Award from Amer Soc for Testing & Mtls (June 1979). Hobbies: wood & metal working, piano, organ, cello. Received Hon Mbr Award from ASNT (Oct 1980). *Society Aff:* ASTM, ASNT.

Turner, Richard L
Business: Seattle, WA 98122
Position: Prof. *Employer:* Seattle Univ. *Education:* PhD/Elec Engg/Univ of WA; MS in EE/Elec Engg/Drexel Univ; BS in EE/Elec Engg/Drexel Univ. *Born:* 10/8/23. From Phila, PA; worked at naval aircraft unit and for Thomas A Edison in NJ, then taught electrical engg at Drexel Univ to 1955, with one yr in Caracas, Venezuela. Instructor at Univ of WA 1955 to 1962 where thesis res was in the area of traveling-wave tubes. Currently Prof of Elec Engg at Seattle Univ. Has been area dir for IEEE student activities and presently serves on the natl Council of Dirs of Tau Beta Pi. Present interest is in circuit synthesis and digital circuitry. *Society Aff:* ASEE, IEEE, ΤΒΠ, ΣΞ, HKN.

Turner, Warren H
Home: 10 Old Glen Rd, Convent Station, NJ 07961
Position: Dir/Network Business Relations. Retired. *Employer:* American Telephone & Telegraph Co. *Education:* BS/EE/MIT. *Born:* May 1926. Native of NJ. Ens 1946, Lt 1951-53 USNR. Joined NJ Bell Telephone Co 1947 & served in variety of operational & engrg assignments; shifted to AT&T Genl Depts 1963; appointed Engrg Dir/ Sys Planning 1967; 1978 directed an org providing tech & administrative guidance to equipment & maintenance engrs in Bell operating co's. Appointed Director-Network Business Relations in 1984. Retired 1986. Mbr: Bldg Res Adv Ed. 1976-81. Reg P E in NJ. Retired. *Society Aff:* IEEE.

Turnquist, Ralph O
Business: Mech Engg Dept, Kansas State Univ, Durland Hall, Manhattan, KS 66506
Position: Prof of Mech Engg *Employer:* KS State Univ *Education:* PhD/ME/Case Inst of Tech; MS/ME/KS State Univ; BS/ME/KS State Univ *Born:* 8/10/28 Native of Lindsborg, KS. Served as Lieutenant in US Air Force at Edwards AFB, CA 1952-54. Design and Dev Engr for Westinghouse Aviation Gas Turbino Div specializing in jet engine fuel systems, 1954-59. In Mech Engrg Dept at KS State Univ since 1965. Full Prof since 1975. Respons for teaching and resch in fluid power, automatic controls and engrg design. Enjoy workworking and sports. *Society Aff:* ASME, ASEE, ISA, FPS

Tuteur, Franz B
Business: 2197 Yale Station, New Haven, CT 06520
Position: Prof. *Employer:* Yale Univ. *Education:* PhD/EE/Yale Univ; Masters/EE/Yale Univ; BS/EE/Univ of CO *Born:* 3/6/23. Born in Frankfurt-am-Main, Germany. Served in US Army Signal Corps, 1944 to 1946. Engr with Raytheon Mfg Corp 1946-47. instructor Yale Univ 1950-51, Asst Prof, 1951-57, Assoc Prof 1957-65, Visiting Assoc Prof, Univ of CA Berkeley, 1965-66, Prof of Engg and Applied Sci, Yale, 1966-81, Prof of Electrical Engrg since 1981, Visiting Marine Physicist, Scripps Inst of Oceanography, San Diego, CA 1978. Res and Teaching in control systems, communications, statistical detection theory, underwater acoustics, electroencephalography. Coauthor "Control System Components–, McGraw-Hill, 1960, and "Modern Electrical Communications–, Prentice Hall, 1979. Dir of Grad Studies in Engr & Appl Sci, at Yale, since 1978. *Society Aff:* IEEE

Tutt, Charles L, Jr
Home: 3401 Westwood Pkwy, Flint, MI 48503
Position: Dean Emeritus. *Employer:* Gen Motors Inst. *Education:* Doctor of Engg// Norwich Univ; Mech Engr/ME/Princeton Univ; BSE/Engg/Princeton Univ. *Born:* 1/26/11. A native of Coronado, CA and Colorado Springs, CO. Started his career with Buick Motor Div, GM Corp 1934 as a Special Assignment Engr 1940-46. Asst Prof. ME Princeton Univ. 1940-44 Asst ASME. 1944-46 Assoc Editor. Product Engg, McGraw-Hill Publishing Co 1946-50 Asst to Pres, Gen Motors Inst; Admin Chrmn 1950-60; Dean of Engg, 1960-69; Dean of Academic Affairs, 1969-75; Retired, Dean Emeritus Jan 1, 1975. Served as Trustee Norwich Univ 1965-77. Mbr Engg Fdn Bd 1963-75, Chrmn 1967-74. VP Region V, ASME, 1964-66. 94th Pres, ASME 1975-76. Mbr Bd of Dirs SME 1972-78. Mbr Bd of Dirs ECPD 1974-80. Enjoys fishing, photography and woodworking. *Society Aff:* ASME, ASM, ASEE, SAE, SME, NSPE.

Tutt, Richard D
Home: 2347 Yew Lane, Friday Harbor, WA 98250
Position: Consultant/Air Distrib & Acoustics. *Employer:* Self. *Education:* Grad/Mech/IN Inst; Grad/Academic/Lapel, IN HS. *Born:* Feb 1916 Cincinnati Ohio. Ch air-conditioning engr Allison Div GMC 1940- 46; Ch Engr Tuttle & Bailey Div Allied Thermal Corp 1946-63; V P Engrg Indus Acoustics Co 1964-67; V P R&D Kodaras Acoustical Lab 1966-67; V P Krueger & Rink Divs Lear Siegler Corp 1967-72. Author of text 'Principles of Air Distribution.' Bd/Dir Sheet Metal & Air Conditioning Contractors Natl Assoc 1963-66; Founder & 1st Pres Air Diffusion Council; Mbr 1942- , Fellow 1972, Disting Serv Award 1973 ASHRAE. Awarded 9 patents for air distribution & acoustic devices. Private Cons Air Conditioning & Acoustics. *Society Aff:* ASHRAE.

Tuttle, Albert D
Home: 617 Lacey Dr, Endwell, NY 13760 *Education:* BS/Elec Engg/RPI. *Born:* Jan 11, 1917. WW II served with Navy in radiation lab of MIT 1943-46. Joined NYSEG 1938 & has served as Operating Supt-Liberty Dist, Engr-Sys Planning Sect Elec Engrg Dept, Operating Supt Western Area, Genl Operating Supt with company-wide responsibilities, Asst Ch Elec Eng, Ch Engr, Asst V P-Engrg Planning, V P Engrg Planning, V P & Ch Engr & V P Generation. Mbr NY State Soc of Prof Engrs & Fellow IEEE. *Society Aff:* IEEE, NSPE.

Tuttle, Raymond E
Home: 65 Oyster Reef Dr, Hilton Head, SC 02298
Position: Retired. *Education:* BChemE/Chem E/Cornell Univ; -/Post Grad Courses/Carnegie Tech. *Born:* 4/27/23. Recently retired after six years in executive capacity with epoxy resin mfgr. Previous 25 yrs with Intl Min & Chem Corp & subsidiaries as Managing Dir-NYMA, BV (Netherlands); Exec-VP Sobin Chem; Pres-IMC Chlor-Alkali; Genl Mgr - Ind Chem Div & Biochem Div; Dir of Corp Dev; Asst Plant Mgr & Engg Mgr of Phosphate Chem Plant. Earlier positions as Sr Technologist at Natl Petrochem & Pitt-Consol Coal Co (Synthetic fuels). Tau Beta Pi, Phi Kappa Phi Honoraries. Veteran of US Army European and Pacific Theaters WWII. *Society Aff:* AIChE, ΑΔΦ, ACS.

Tuttle, W Norris
Home: 1048 Lowell Road, Concord, MA 01742
Position: Consulting Engineer. *Employer:* Self. *Education:* PhD/Physics/Harvard; SM/Elect Comm Engg/Harvard; AB/Physics/Harvard. *Born:* March 1902. 1930-42 & 1945-68 Engrg Cons General Radio Co R&D of electronic measuring & testing equipment; 1942 Cons to Chmn Office of Scientific R&D. 1942-44 Head Radar Subsect Operations Analysis Sect Hq 8th AF England; 1945 Operations Analyst Hq 20th AF; 1945-64 Cons in Operations Analysis Hq USAF Washington DC. Fellow IEEE & AAAS. Mbr APS & Operations Res Soc of Amer. Awarded Medal of Freedom 1946. *Society Aff:* IEEE, IEEE-C&S, APS, ASA, ORSA.

Tweedy, Robert H
Business: Deutz-Allis Corp, PO Box 933, Milwaukee, WI 53201
Position: Manager Strategic Business Planning. *Employer:* Deutz-Allis Corp. *Education:* BS/Agri Engrg/IA State Univ. *Born:* March 24 1928, Mt Pleasant Iowa. Parents R James & Olatha (Miller), married Genavieve F Strauss, sons Bruce & Mark, stepdaughter Maryellen Francis. 2 years US Navy. Tau Beta Pi, Phi Eta Sigma, Pi Mu Epsilon, Alpha Zeta, St Patrick, ASAE Honor Award, Phi Kappa Phi, University Engineering Honor Award 4 years. 1953-64 Sr Engr John Deere Waterloo Tractor Works; 1964-69 Marketing Representative US Steel; 1969- , Mgr of Prod Planning & recently Mgr of Strategic Business Planning, Agri Equip Grp Deutz-Allis Corp. ASAE: Pres. 1981-82, V Pres 1974-78, Genl Chmn Headquarters Bldg Proj 1968-70, Mbr since 1952, elected Fellow 1976; Mbr SAE; Farm & Industrial Equip Inst: Chmn Agri Res Ctte, Mbr Safety Policy Adv Ctte; Mbr Natl Safety Council Farm Conference. Num tech papers, several patents. Who's Who in Midwest, Wisc Men of Achievement, Who's Who in Finance and Industry, Intl Who's Who in Engrg, Wisc Agri E. of year 1980, Professional Achievement Citation in Engrg-Iowa State University 1983, Professional Engineer Registration 3239 in Iowa 1954 to date. Mason. *Society Aff:* ASAE, SAE.

Tyer, Arnold J
Home: 1500 Mtn Dr, Little Rock, AR 72207
Position: Pres. *Employer:* Crist Engrs, Inc. *Education:* BSCE/Civ Engg/Univ of AR. *Born:* 9/23/23. Born in McCrory, AR 9/23/23. BS in CE from Univ of AR 1951. Wife, Johnnie Sandage Tyer; Children - David, Lee Ann, John, Paul. Employed by Crist Engrs, Inc 1952-present Pres since 1969. US Army from 1943 to 1946. Mbr American Soc of Consulting Engrs; Pres of Consulting Engrs Council of AR 1966-1967. Mbr of ACEC natl Bd of Dir from 1968-1970; American Water Works Assn; Water Pollution Control Fed; American Soc of Civ Engrs; Natl Soc of PE; Mbr of Panel of Arbitrators of American Arbitration Assn; Engr of Yr in AR 1977. *Society Aff:* ACEC, WPCF, AWWA, ASCE, NSPE.

Tynes, Rex A
Home: 808 Karen, Fort Morgan, CO 80701
Position: Consulting Engineer. *Employer:* self. *Education:* BS/EE/Texas Tech Univ. *Born:* April 7 1915, Hereford Tex. Graduate studies at Rutgers & Univ of NM. 1940-44 Signal Corps Labs; 1944-46 Rural Electrification Admin; 1946-58 cons practice; 1958-63 Public Serv Co of NM; 1963-74 Nev Power Co; 1974- , cons practice. Mbr NSPE, Fellow ASCE. Chmn NM BD of Engrg Examiners 1962-63, Mbr Nev State Bd of Engrg Examiners 1966-76, Mbr, CO State Bd of Engg Examiners, 1978-. Mbr Natl Council of Engrg Examiners. Reg PE in Colo, Utah, Nev, Nebraska, Ariz & Calif. Nev 'Engineer of the Year' 1972. Amateur radio operator, enjoy photography & music. *Society Aff:* NSPE, ASCE.

Tyson, Christopher G
Business: 1000 16th St NW, Washington, DC 20036
Position: Genl Mgr, Cades & Stds Activities. *Employer:* American Iron & Steel Institute. *Born:* April 1929, Philadelphia. BE, ME Yale Univ. 1952-54 engr with Raymond Internatl; 1957-62 with Connell & Assoc; 1963-68 Ch Engr Dade County Bldg & Zoning; 1969-70 Assoc Crain & Crouse Inc; 1971-76 Ch Civil Engr, Dade County Public Works; 1975-76 V Chmn Fla Bd of Bldg Codes & Standards; currently Genl Mgr Codes & Standards Activities, Amer Iron & Steel Inst. Contributor to Fla Hurricane Survey Report 1965. Married Christiane Muller 1953, 2 daughters & 5 sons. Dir Dist 10, 1972-75 ASCE.

Udell, Debra M
Home: 3214 Kingsbridge Ave, Bronx, NY 10463
Position: Mechanical Engr *Education:* BS/ME/Univ of Lowell; Cert/Private Intl Law/Hague Acad of Intl Law. *Born:* 05/31/56 NYC, NY. Humanitarian Mechanical Engr active in independent theoretical research, design, and development and author of resulting papers in: Gamma Particles, Genetic Engrg, Medicine, Computers, NDT, Economics, EW Systems, Meteorology, Law, Space Sciences, and Intl Policy/Strategy. inventor of the AMS- Atmosphere Modification Satellite. Summer 1984 - Cert in Priv Intl Law - The Hague - Peace Palace, The Netherlands. Latest Publications: Nov 1979- Introduction to Design in the Automation of NDT Techniques-9th World Conference on NDT-Melbourne, Australia, Mar 1982-Grumman Flexible 870 Bus: Case in Point for NDT-ASNT Spring Conference. 1986-Silver Medal Award, Geneva Switzerland 14th Int'l Exhibition of Inventions; 1987-Gold Medal Award, NY, U.S.A. 11th Annual Worldwide Inventors Expo '87 U.S.A. Enjoys astronomy, photography, chess, sports, music. *Society Aff:* NSPE, NYAS, ASNT, SWE, AWIS, AST

Udo, Tatsuo
Home: 13-27, 2-Chome, Honkomagome, Bunkyo-Ku, Tokyo, Japan 113
Position: Vice President *Employer:* CRIEPI *Education:* Dr/EE/Tokyo Univ; B/EE/Tokyo Univ *Born:* 09/15/25 Native of Tokyo, Japan. With CRIEPI since 1949, specializing insulation of High Voltage Elec Power Systems. From 1966 to 1968, with BPA, USDI, temporally. Since 1983, General Dir of Abiko Research Center, CRIEPI. The current largest research project is UHV transmission. *Society Aff:* IEEJ, IEE

Ueng, Charles ES
Business: School of Civil Engineering, Georgia Inst. of Technology, Atlanta, GA 30332
Position: Prof *Employer:* Georgia Inst of Tech *Education:* PhD/Applied Mech/KS State Univ; MS/Applied Mech/KS State Univ; BS/CE/Natl Cheng-Kung Univ *Born:* 09/08/30 Native of China. Naturalized US citizen. Asst Structural Engr, Taiwan Power Co, 1954 and 1958. On faculty at Georgia Tech since 1964. Published in the following areas: Static and Dynamic Analysis of Sandwich and Composite Structures; Inflatable Structures; Superplastic Metal Forming and Mfg; Seismic Analysis of Instrumention Cabinets and Panels for Nuclear Power Plants. Recipient of research grants from NSF, NASA and SME. Conslt for Reliance Elec, Westinghouse, Combustion Engrg, Bechtel Power Corp, etc. Chrmn, Task Committee on Composite Structures, Aerospace Div, ASCE; Chrmn, Structures & Materials Aerospace Div, ASCE. Chrmn, Publi Comm, Aerospace Div, ASCE. *Society Aff:* ASCE, ASEE, SES, ΣΞ, ΦΚΦ

Uenohara, Michiyuki
Home: 22-14 Minami-Yukigaya 4, Ota-Ku Tokyo, Japan 145
Position: Dir, Exec Vice Pres *Employer:* NEC Corporation. *Education:* Dr of Engrg/Elec Engrg/Tohoku Univ; PhD/Elec Engrg/OH State Univ; MS/Elec Engrg/OH State Univ; BE/Elec Engrg/Nihon Univ. *Born:* Sept 1925 Kagoshima Japan. Dr Engrg Tohoku Univ, Japan; PhD, MS Ohio State Univ; BE Nihon Univ, Japan. Mbr Tech Staff, Bell Labs 1957-67. Supervised a group working on paramps & microwave devices. Joined NEC Corp 1967, served as Mgr of Electron Device Lab, Quantum Device Lab, Memory Lab of Central Res Labs. Genl Mgr of Central Res Labs 1972. Dir & Vice Pres 1976. Dir & Sr VP 1980; Dir & Exec VP 1982. Inada Memorial Prize from IECEJ; Distinguished Paper Award from Natl Electronics Conf; Distinguished Alumnus Award from OSU; Fellow of IEEE; Foreign Associate Member of National Academy of Engineering; Member of Engineering Academy of Japan. *Society Aff:* IEEE, IECEJ, JSAP.

Ufford, Page S
Home: 12 Lenape Ln, Newark, DE 19713
Position: Retired *Education:* SB/Chemical Engr/MA Inst of Technology; AB/Chemistry/Middlebury College. *Born:* 7/19/21. Native of Middlebury, VT. Joined DePont's Industrial Engg Div in 1944. Left after 3 months to serve as 2nd Lt with US Marines. Rejoined Du Pont in 1946 and worked as Field and Technical Supervisor for sevl plants. Transferred to Engg Dept, Design Div in 1956. Respon for major plant design and proj administration for Du Pont's Polymer Products Dept through 1984. Natl AIChE Chrmn Equipment Testing Procedures Committee with objective to formulate and promulgate suggested testing performance characteristics of Chemical Engg equipment 1979- 80. Past Chairman-1981 Profl Engr - DE. Elected Fellow AIChE 1984. Retired 12/31/84. *Society Aff:* AIChE.

Ugelow, Albert J
Business: 1111 S Federal Hwy, Stuart, FL 33497
Position: Supervising Mech Engrg *Employer:* Ebasco Services *Education:* ME/Mech Engrg/Stevens Inst of Tech; BSc/Nuclear Engrg/Columbia Univ *Born:* 03/16/50 Has been employed by Ebasco Services since graduating from coll. During that time has been involved in many aspects of the mech and nuclear design of both nuclear and fossil fueled power plants. In current position supervises and directs the activities of over 30 engrs and designers. Respon for all mech engrg performed in the Stuart, FL office. A reg PE in NY and FL. Was recipient of the NY State Soc of PE's Young Engr of the Year award in 1982. An active runner who has organized co running clubs in various branch offices. Also involved in community activities such as public school system, religious organizations, etc. *Society Aff:* ANS, ASME, NSPE

Ugiansky, Gilbert M
Business: 15200 Shady Grove Rd, Suite 350, Rockville, MD 20850
Position: Pres *Employer:* Cortest Engrg Services, INC. *Education:* PhD/Metallurgical Engr./Ohio State University; MS/Engr. Materials/Univ of Maryland; BS/Metalurgical and Chemical Engr./U of Maryland *Born:* 05/27/42 Pres, Cortest Engrg Services, Inc. (CES), Rockville, MD, and Houston, TX. CES addresses technical needs in materials performance, engrg, and technical computer software dev and marketing. Twenty-four years of metallurgy, corrosion research, & research mgmt at the National Bureau of Standards (NBS). Dir, NACE-NBS Corrosion Data Center (1982-87); Dept of Commerce Bronze Metal recipient for leadership in establishment & dev of this collaborative program (1984). Pres, Federation of Materials Societies (1987 and 1988). Twice Director of NACE. US Delegate to International Standards Organization Cttee on Corrosion. Chaired ASTM Subcttee on Stress Corrosion Cracking and Corrosion Fatigue. NACE Accredited Corrosion Specialist *(812). Society Aff:* ASM-Intl, ASTM, NACE, FMS

Ugural, Ansel C
Business: 1000 River Road, Teaneck, NJ 07666
Position: Professor Mech Engrg. *Employer:* Fairleigh Dickinson University. *Education:* PhD/Engg Mechs/Univ of WI; MS/Mech Engg/Univ of WI; BS/Mech Engg/Ankara Tech College. *Born:* Sept 18, 1934 Aydin Turkey. USA citizen. Proj Engr, Kiekhaefer Corp 1961-63; Instructor, Univ of Wis 1963-65; Assoc Prof of mech engrg 1966-73, Prof & Chmn of mech engrg 1973-1982, Prof of Mech Engrg 1982-present time. Fairleigh Dickinson Univ. Cons to various companies 1966- . Mbr ASME, ASEE. Contributed articles to prof journals; pub textbooks, 'Advanced Strength & Applied Elasticity' 2nd SI Ed 1987. 'Stresses in Plates & Shells' 1981.. *Society Aff:* ASME, ASEE.

Uhl, James B
Home: RD 8 Box 80, York, PA 17403
Position: Assoc Prof. *Employer:* Penn State Univ. *Education:* MS/Agri Engr/OH State Univ; BS/Agri Engr/Penn State Univ. *Born:* 9/30/31. Prog chrmn of Mech Engg Tech for Commonwealth Campus system of the Penn State Univ. Faculty fellowship NASA Goddard Space Flight Ctr and USDA Beltsville Res Ctr to evaluating thermal sensors for earth resource satellites. Industrial Consultant evaluating thermal efficiency of heating, air conditioning and refrigeration systems. *Society Aff:* ASEE.

Uhl, Robert H
Business: Watkins & Associates, Inc, 446 East High Street, Lexington, KY 40508
Position: Vice President. *Employer:* Watkins & Associates Inc. *Education:* BSCE/Civil Engineering/Purdue Univ. *Born:* Oct 1922 Evansville Ind. BS 1948 Purdue Univ. Served in Army Corps of Engrs 1942-46, 1951-52. 5 years with W G Duncan Coal Co. Joined Watkins & Assocs Inc, Engrs-Architects-Planners in 1955. Assumed present responsibility as Vice Pres, Transportation Div in charge of surveys, planning, design & const inspection for highways, streets, airports, thoroughbred horse racetracks in 1969. Fellow ASCE; Pres Ky Sect ASCE 1972. Reg PE in Ky, Ind, Mich, West Va, Va, NC, Md, & NJ. *Society Aff:* ASCE, SAME.

Uhl, Robert J
Business: 213 Truman NE, Albuquerque, NM 87108
Position: President. *Employer:* Uhl & Lopez Engrs, Inc. *Education:* BSEE/Elec Engg/KS State Univ; BS/Bus Admin/KS State Univ; Assoc-Sci/Pre- Engg/Wentworth Military Acad. *Born:* 2/18/23. Engr, Chant Elec 1950-51; Elec Engg, Corps of Engrs Military Design 1951- 54; Cons Engr, private practice in Albuquerque 1954-56; Partner, Uhl & Lopez Elec Engrs 1956-60, Uhl & Lopez engrs, Inc, Pres 1960- . Mbr Univ of NM Prof Adv Ctte 1969, UNM Medical Seminalr Faculty 1971, UNM EE Seminar Instructor 1972-77. NM Adv Council of Amer Arbitration Assn, NM Elec Trade Bd 1967-78. Chmn, NM Elec Code Ctte 1967-78; Pres, Cons Engrs Council of NM for 2 terms. Mbr Sigma Tau, Eta Kappa Nu, IEEE, CSI, IES, NMSPE, Instructor, UNM Sch of Arch 1977-79, CECNM Charter Mbr, IAEI, NFPA, & AAA. Dir, ACEC 1967-68. Fellow Mbr ACEC; Mbr, AIA- AGC-CECNM Liaison Ctte 1962-71, 1978- . *Society Aff:* ACEC, NMSPE, CECNM, AAA, IEEE, IES, CSI.

Uhler, James W
Home: 419 W. 30th, Kennewick, WA 99337
Position: Construction prog Mgr. *Employer:* UNC Nuclear Industries, Inc. *Education:* BS/Civil Engg/Univ of AZ. *Born:* 1/30/31. Served with Army Corp of Engrs 1952-54. Construction engineer for Western Pipe Line Inc. Proj Mgr for Howard S Wright and Assocs for construction of nuclear test facilities at the Natl Reactor Testing Station,, Proj Field Engr for Bechtel Corp for construction of the Fast Flux Test Facility at Hanford. President, Uhler and Assocs, Inc, Consulting Engrs offering profl engg and construction mgmt services for major irrigation projs, municipal facilities and nuclear facilities. Construction Prog Mgt, UNC Nuclear Industries, responsible for design and construction of nuclear facilities. Enjoy hunting, fishing and skiing. Mbr and PChrmn - Mid Columbia Bldg Appeals Commission; Past VP & Dir United Way; Past Pres Tri City County Club. Lic PE - WA, OR, ID; Chapter Pres, WPSE. *Society Aff:* NSPE, PEPP.

Uhlig, Herbert H
Home: PO Box 444, Duncan Road, Hancock, NH 03449
Position: Consultant (retired) *Employer:* MA Inst of Tech. *Education:* PhD/Physical Chem/MA Inst of Tech; ScB/Chemistry/Brown Univ. *Born:* 3/3/07. Res Assoc MA Inst Tech 1937-40; assoc prof 1946-53; prof 1953-72; emeritus 1972-; Res assoc GE Co, 1940-46; Electrochem Soc: Palladium Medal 1961; Pres 1955-56, honorary mbr; Natl Assoc Corrosion Engrs; Whitney Award 1951; Intl Corrosion Council: 1963-81, Chairman 1975-78. Institution of Corrosion Science and Technology (Great Britain), Honorary Fellow, U.R. Evans Award, 1980. Editor Corrosion Handbook, 1948; author Corrosion and Corrosion Control, 1963, 1971, 1985; also various scientific and technical articles to profl journals. *Society Aff:* ACS, ECS, ASM, NACE.

Uhlir, Arthur, Jr
Business: Coll of Engrg, Medford, MA 02155
Position: Prof Electrical Engrg *Employer:* Tufts University. *Education:* PhD/Physics/Univ of Chicago; SM/Physics/Univ of Chicago; MS/Chem Engrg/IL Inst of Tech; BS/Chem Engr/IL Inst of Tech. *Born:* 2/2/26. Employment: Douglas

Uhlir, Arthur, Jr (Continued)
Aircraft 1945; Armour Res Foundation 1945-48; Bell Telephone Lab 1951-58; Microwave Assocs Inc 1958-69; Computer Metrics 1969-70; Prof of Electrical Engrg Tufts Univ 1970- , Dean Coll of Engrg 1973-80. Fellow IEEE, for theory, dev & application of varactor diodes in parametric amplifiers. Mbr APS, AAAS, Sigma Xi, Gamma Alpha. *Society Aff:* IEEE, APS, AAAS, ΣΞ.

Uhrig, Robert E
Business: PO Box 14000, Juno Beach, FL 33408
Position: Vice Pres Advanced Systems and Technology *Employer:* Florida Power & Light Co. *Education:* BS/Mech Engr/Univ of IL; MS/Engr Mechs/IA State Univ; PhD/Engr Mechs/IA State Univ. *Born:* 8/6/28. Employment: IA State Univ 1948-60; Military Leave (Instructor US Military Acad, West Point) 1954-56; Univ of FL1960-73; Chairman, Dept. of Nuclear Engineering 1960-68 Leave of Absence (Deputy Asst Dir of Res, Dept of Defense) 1967-68; Dean, Coll of Engrg 1968-73; FL Power & Light Co 1973- , Dir of Nuclear Affairs 1973-74, VP of Nuclear Affairs 1974-75, VP Nuclear & Genl Engrg 1975-78, VP, Advanced Sys & Technology 1978- , Mbr DOE Fossil Energy Advisory Committee 1978-81. Received Richard Memorial Award, ASME-Pi Tau Sigma 1969; Secy of Defense Civilian Serv Award 1968 among others. Chmn Engrg Adv Ctte, NSF 1972-73; P Bd Mbr, ECPD, ANS. Fellow ANS 1970. Authors, "Random Noise Techniques in Nuclear Reactor Sys," 1970. Fellow ASME 1980, Fellow AAAS 1980. Distinguished Achievement Citation, Iowa State Univ Alumni Assoc; 1980. Conslt to U.S. Congress Office of Tech Assessment, 1978- . Chrmn NASINRC Ctte on Nuclear Manpower Requirements 1981-82. Mbr of NASINRC Ctte on Intl Cooperation in Magnetic Fusion Research 1983- . Mbr of Research Advisory Ctte 1981-83 and chmn of Advanced Power Systems Div 1983 of the Electric Power Research Inst. Chrmn of Utility Coal Gasification Assoc 1982-83. Chrmn of the Three Mile Island Operation Accelerated Retraining Program Review Ctte 1979-80, 1984-. *Society Aff:* ANS, ASME, NSPE, AAAS

Uicker, Joseph B
Home: 14585 Lamphere, Detroit, MI 48223
Position: Sr Vice Pres *Employer:* Smith, Hinchman & Grylls *Education:* MS/ME/Univ of Detroit; BS/ME/Univ of Detroit *Born:* 03/29/40 Native of Detroit, MI. Joined Smith, Hinchman & Grylls in 1964, specializing in air conditioning sys design. Became an Assoc of the firm in 1971; Asst Dir, Health Facilities Div 1973-75; Corporate Dir of Mech Engrg 1975; VP 1978. Director of Professional Staff 1983-present. Served with US Army Corps of Engrs 1966-67. Registered PE in MI and NC. *Society Aff:* NSPE, ASHRAE, ASME, SAME

Ulaby, Fawwaz T
Business: Radiation Lab, Dept of Elec Eng and Computer Sci, Ann Arbor, MI 48109
Position: Prof. *Employer:* Univ of MI *Education:* PhD/Electrical Eng/Univ of TX; MSEE/Electrical Eng/Univ of TX; BS/Physics/American Univ, Beirut *Born:* 2/4/43. Born in Damascus, Syria on Feb 4, 1943. Joined Dept of Elec Engg of the Univ of KS in 1968. Formerly J.L. Constant Distinguished Prof of Elec Engg & Dir of the Remote Sensing Lab. Field of interest is microwave remote sensing. Received the Univ of KS 1973 Gould Award for "distinguished service to higher education," the 1975 Eta Kappa Nu MacDonald Award as "an outstanding Elec Engg Prof in the US of Am." Pres, Geosci Electronics Soc of IEEE, 1979-1981, the IEEE Centennial Medal in 1984, and numerous other awards. Editor, Transactions IEEE Geos & Remote Sensing, 1985-88. Enjoy bldg & flying kites. *Society Aff:* IEEE, ISP, URSI

Ulam, Frederick A
Home: 1867 Miramar Dr, Ventura, CA 93001
Position: Manager, Special Projects *Employer:* McLaughlin Research Corporation, Camarillo, CA *Education:* MS/Engg Sys/Univ of Ca, Los Angeles; MS/Sys Mgmt/ Univ of Southern CA; BS/Engg/US Naval Acad. *Born:* July 1922 Mt Lebanon Pa. Served in Navy 1944-48, 1950-56. Plant Engr, US Gypsum Co Warren Ohio; Electrical Engr, Pa Transformer Co & Combustion Serv & Equipment Co; Production Engr, Tech Support Dept, Pacific Missile Range. Head Quality Assurance, Mgr Missile Production Support, Phoenix Missile In-service Engrg Mgr Naval Missile Center. Special quality assurance assignment to Tech Dir; Hd, Electronic Warfare Div; Hd, Missile Sys Div, Chief Engr Electronic Warfare Projects, Pacific Missile Test Center. Mgr Special Proj McLaughlin Research Corp Camarillo, CA. P. Chmn Ventura Pacific Sect, AIAA, P Pres, Pt Mugu Crows Club, AOC, Mbr, American Defense Preparedness Assoc, Vice Pres. Missile Technology Historical Assoc. *Society Aff:* AIAA, AOC, ADPA, MTHA.

Ulbrecht, Jaromir J
Business: National Bureau of Standards, Bldg. 221, Room B250, Gaithersburg, MD 20899
Position: Chief *Employer:* Chemical Process Metrology Div, Natl Bureau of Standards *Education:* PhD/CE/Inst of Chem Tech; Ing/CE/Czech Inst of Tech. *Born:* 12/16/28 1984 till present: Chief, Div of Chem Process Metrology, National Bureau of Standards. 1978-1983: Prof & Chrmn, Dept Chem Engg, SUNY at Buffalo. 1968-1978: Prof, Dept of Chem Engg, Univ of Salford, England. 1963-1968: Hd. Labor Engg Rheology, Czechoslovak Acad of Sciences, Praque, Czechoslovakia. 1957-1963: Deputy Hd, Div of Res, Dept Chem Engg, the Synthetic Rubber Co, Zlin, Czechoslovakia. 1952-1954: Process Engr, Coal Tar Processing Plant, Ostrava, Czechoslovakia. Honors/Awards: Alexander von Humboldt Fellow, 1967. CSAV award for Outstanding Scholarship, Editor: 1976 Chem Engrg Communications; and 1985 Chemical Engineering Concepts & Reviews; Patents/Publications: 6 patents, 2 books, "Mixing of Liquids by Mechanical Agitation" & "Non-Newtonian Liquids–, 100 publications. *Society Aff:* AIChE, ICHEME, SOC RHEOL, ΣΞ RESEARCH SOCIETY.

Ulp, Richard B
Home: 3811 Copper Kettle Rd, Camp Hill, PA 17011
Position: VP *Employer:* Gannett Fleming Inc. *Education:* MCE/Water Resources/ Cornell Univ; BS/CE/Bucknell Univ *Born:* 12/6/34 Native of Central PA Employed by Erdman, Anthony, Assoc. Consulting Engrs 1958 to 1982. Assoc of firm 1963 to 1982; Exec VP 1979 to 1982. Pres of Consulting Engrs Council of PA. 1981-82. Pres Bucknell Engrg Alumni Assoc 1982 to 1984. Registered Prof Engr in PA, NY, OH, WV, VA, MD, DE, NJ. Married with 4 children. Active in Chamber of Commerce. Now employed by Gannett Fleming, Inc., Engineers and Planners. Feb 1982 to date. Employed as VP for Proj. Dev. *Society Aff:* ASCE, NSPE, AWWA, ACEC.

Ulrich, Werner
Business: AT&T Bell Laboratories, Naperville, IL 60566
Position: Attorney *Employer:* AT&T Bell Labs. *Education:* DrEngSci/EE/Columbia Univ; MS/EE/Columbia Univ; BS/EE/Columbia Univ; MBA/Bus Adm/Univ of Chicago; JD/Law/Loyola Univ of Chicago. *Born:* 3/12/31. With AT&T Bell Labs since 1953. Specialized in system design of electronic telephone switching systems (ESS). Contributed substantially in the areas of design of reliable systems & real time control of telephone systems. Hd, Exploratory Dev of ESS, Software Dev for UNICOM, No 2 ESS, 1964-8; Dir, Advanced Switching Tech, 1968-77; Hd, Maintenance Systems Design Architecture, 1977-81; Member, Legal and Patent Staff, 1981-. Fellow, IEEE. Member of Illinois Bar. *Society Aff:* IEEE, ABA, ISBA, AIPLA.

Un, Chong K
Business: P.O. Box 150 Chongyangni, Seoul, Korea
Position: Prof. of Electrical Engineering *Employer:* Korea Adv. Inst. of Science & Tech. *Education:* Ph.D/Elec. Engrg./University of Delaware; M.S./Elec. Engrg./ University of Delaware; B.S./Elec. Engrg./University of Delaware. *Born:* 08/25/40 Born in Seoul, Korea. Asst Prof of Elec Engrg at Univ of Maine, Portland 1969-73. Research Engineer for Stanford Research Inst, specializing in digital speech coding and processing. Currently Prof of Electr Engrg at Korea Advanced Inst of Science and Tech since 1977, teaching and doing research in digital communications and

Un, Chong K (Continued)
signal processing--served Dept Chairman (1979-81) and Dean of Engrg (1982-83). Published over 120 papers and 50 reports, and holds 7 patents. Received 1976 L.G Abraham Prize Paper Award from IEEE Communications Society, several achievement awards, and National Order of Merits from the Government of Korea. Fellow of IEEE. *Society Aff:* IEEE.

Underwood, Ervin E
Home: 613 Lorell Terr NE, Atlanta, GA 30328
Position: Professor Sch of Chemical Engrg. *Employer:* Georgia Institute of Technology. *Education:* ScD/Metallurgy/MIT; SM/Metallurgy/MIT; BS/Met Engg/Purdue. *Born:* 1/30/18. in Gary, IN. Employment: US Steel Corp Gary Works 1936,41: Mayor, Artillery, 66th Inf Div, Europe 1941-46. Battelle Meml Inst, Columbus OH 1954- 62; Lockheed Missiles & Space Co, Palo Alto CA, as Sr Scientist 1962-65, Lockheed-GA Co Marietta GA, as Assoc Dir Matls Res 1965-71; Alcoa Prof of mech Engg, GA Tech 1972-74; Guest Scientist, Max-Planck-Inst for Metal Res, Stuttgart Germany 1974-75; Prof Metallurgy Sch of Chem Engr 1975- . Book: Quantitative Stereology 1970, Addison Wesley. Pres, Intl Soc for Stereology 1972-75. Fellow ASM, RMS. Numerous publs on matls res & stereology. *Society Aff:* ASM, ISS, ASTM.

Underwood, Wyatt R
Home: 915 Sierra Pl SE, Albuquerque, NM 87108
Position: Professional Engr & Consultant (Self-employed) *Education:* BS/Arch Engg/ TX Tech College. *Born:* Sept 1912 Abilene Texas. Served US Navy 1942-45, retired Cpt USNR. Reg PE in NM 1937, Texas 1966. Opers Partner Genl Contracting 10 years. Pres NM AGC 1958. Cons Engrg 1963- . Mbr NSPE 1946, one of first in state; chaired State Chap Constitution Ctte; first pres Albuquerque NSPE Chap; chaired State Legis Ctte obtaining first updating of Engrg Act. Served on 5 municipal bds for combined total of 25 yrs. Mbr ACEC 1967. Active in church. Ints: golf, music. *Society Aff:* NSPE, ACEC.

Ungar, Edward W
Business: 505 King Ave, Columbus, OH 43201
Position: Director *Employer:* Battelle Columbus Laboratories *Education:* PhD/ME/OH State Univ; MS/ME/OH State Univ; BS/ME/CCNY *Born:* 02/06/36 Joined Battelle in 1957 as a researcher. Served in various mgmt positions and assumed current position of Dir of BCL in 1978. Also, currently VP of Battelle Memorial Inst (1978). Native of NYC and enjoy tennis and golf. *Society Aff:* ASME, AAAS

Ungar, Eric E
Business: 10 Moulton St, Cambridge, MA 02238
Position: Chief Conslt Engr *Employer:* Bolt Beranek and Newman, Inc *Education:* Eng Sc D/ME/NYU; MS/ME/Univ of NM; BS/ME/WA Univ *Born:* 11/12/26 in Vienna, Austria. Emigrated to US in 1939. Active duty in US Army 1945- 48; reserve duty 1948-51. Worked on atomic weapons development at Sandia Corp, NM, 1951-53. On faculty of NY Univ 1953-58, taught vibrations, machine design and dynamics, kinematics, engines. With present employer since 1958, engaged in research and consltg primarily in vibrations, noise, structural dynamics; pioneered in viscoelastic damping applications, vibration analysis techniques for complex structures; conceived and managed major projects. Authored over 100 technical articles. *Society Aff:* ASME, AIAA, ASA, INCE

Ungar, J Stephen
Home: 25 Thornbury Pl, Vincentown, NJ 08088
Position: Ret. (Chief Perf Engr & Genl Supt) *Education:* ME/-/Poly Inst of Brooklyn. *Born:* 9/22/02. Prior to WWII, Engr in Performance Engrg Bureau of Con Edison of NY Inc. Served US Navy as Engrg Officer in Const & Opers 1942-46. Exec Dev Prog, Con Edison 1946-51. Recalled by US Navy 1951-53; served as Inspector of Naval Matl, Bridgeport CT; administrative control of Navy procurement, R&D contracts & inspection of all Navy matl in the State of CT; received commendations for Navy servs; retired as Cpt USN. Ch Performance Engr, Con Edison 1960-66, duties incl performances of all stations in system, including Indian Point, the nuclear station. Genl Superintendent 1966-68, duties expanded to develop a training sch for operating personnel. Developed a procedure for selecting condenser tube matl for long service & slow fouling rate. ASME granted award certificate for services as Mbr & Chmn of Res Comm on Condenser Tubes 1963-67. Elected Fellow ASME 1973. Mbr of Hon Engr Soc, Tau Beta Pi & Pi Tau Sigma. Lic Prof Engr-NY *Society Aff:* ΤΒΠ, ΠΤΣ.

Unger, Hans-Georg
Woehlerstr 10, D-3300 Braunschweig, Germany
Position: Professor of Electrical Engineering. *Employer:* Technische Universitat Braunschweig. *Education:* Dr-Ing/Elec Engrg/Tech Univ; Dipl-Ing/Elec Engrg/ Braunschweig. *Born:* Sept 14, 1926. Dipl Ing 1951, Dr Ing 1954 Tech Univ of Braunschweig, Dr Ing Eh (hc) Tech Univ Munich Dev Engr & Head of Microwave Res, Siemens AG Munich/Germany 1951-55. Mbr Tech Staff & Dept Head in Res on Guided Waves & Transmission Technique, Bell Labs Holmdel & Murray Hill NJ 1956-60. Prof of Electrical Engrg, Univ of Braunschweig 1960- . Dept Head of Electrical Engrg in Braunschweig 1961-64. Mbr Bd/Dir German Communication Engrg Soc 1967-71, 1975-79. Fellow of the IEEE. Patents, publs, & textbooks in electronics, microwaves & optical communications. *Society Aff:* ITG, BWG, IEEE.

Unger, Stephen H
Home: 229 Cambridge Ave, Englewood, NJ 07631
Position: Prof. *Employer:* Columbia Univ. *Education:* BEE/EE/Poly Inst of Brooklyn; SM/EE/MIT; ScD/EE/MIT. *Born:* 7/7/31. With Bell Labs about 5 yrs. Res in switching theory, digital systems, pattern recognition. Supervised group developing software for ESS No 1. At Columbia Univ since 1961. Res on logic circuits (authored book), comp langauges. Courses taught: digital systems, logic circuits, com programming, tech & soc. Visiting prof Univ CA at Berkeley, Danish Technical Univ. Guggenheim Fellow 1967. Helped found IEEE committee on social implications of tech - currently acting chrmn involved in IEEE activities related to engg ethics. IEEE Fellow. NSF grants in switching theory, applications of tech to facilitate the democratic process, ethical responsibilities of engg socs. *Society Aff:* IEEE, ACM, AAAS.

Unger, Vernon E
Business: 207 Dunstan Hall, Auburn, AL 36847
Position: Professor & Head *Employer:* Auburn Univ *Education:* Ph.D./Ind. Eng./ Johns Hopkins University; M.S./Mgt. Sc./Johns Hopkins University; B.E.S./Ind. Eng./Johns Hopkins University *Born:* 12/14/35 Native of Federalsburg, Md. Worked for AAI Corp., a defense contractor, as an engineer and administrator for 9 years. Taught on the faculty of Georgia Inst of Technology for 11 years, serving as assoc dir of research for the school of Industrial and Systems Engrg for 4 years. Currently Head of Industrial Engrg at Auburn Univ. Member of Publication Policy Bd and Research Advisory Council of the IIE. *Society Aff:* IIE

Ungrodt, Richard J
Home: 11737 No. Granville Rd, Mequon, WI 53092
Position: VP Emeritus. *Employer:* Milwaukee Sch of Engg. *Education:* MS/-/EE-IIT; BS/EE/MSOE; DRE/Honorary/MSOE; DRET/Honorary/Wentworth Institute. *Born:* 2/20/19. PE, VP-Emeritus, Regent (1974) Regent Emeritus and Corp mbr at Milwaukee Sch of Engg since 1966. Involved for 40 yrs as Prof in Elec Engg in teaching and admin providing leadership in developing new engg & tech curricula. Active leadership roles in natl and local technical, profl, religious, and civic organizations. Serves as consultant in Engg & engg tech education in the US and overseas. Six yrs of ind experience included the Manhattan Proj at Allis Chalmers. Honors include the ASEE James H McGraw Award for Outstanding Service and Leadership in the Dev of Engg Tech Education, and the IEEE Milwaukee Sec Meml Award, and the WSPE PE in Education Award, ASEE Fellow, & Honorary Membership. IEEE Cen-

Ungrodt, Richard J (Continued)
tennial Awd Dr of Engrg Tech (Hon) Pres, ASEE (1984-85). Dr. Engrg (Hon.) *Society Aff:* ASEE, IEEE, SME, ASEM, NSPE

Untermyer, Samuel
Home: 201 Escobar Rd, Portola Valley, CA 94025
Position: Founder, Formerly Chrmn *Employer:* National Nuclear Corp. *Education:* BS/-/MIT. *Born:* Nov 25, 1912. While working at Oak Ridge, Argonne, & General Electric developed basic designs used in light water & heavy water power reactors, holding the patent on the boiling water reactor. Developed nondestructive nuclear fuel assay equipment manufactured by Natl Nuclear. Gold Medal, Newcomen Society 1980 Mbr ASME 1932, Fellow 1973. Charter Mbr ANS 1954, Dir 1964-67, Fellow 1963. Reg PE in Calif 1976. *Society Aff:* ASME, ANS.

Unterweiser, Paul M
Business: Amer Soc for Metals, Metals Park, OH 44073
Position: Asst. Director, Reference Publications *Employer:* American Society for Metals. *Education:* MS/Metallurgy/NY Univ; BS/Chem Engr/NY Univ. *Born:* 11/18/16. Married Gloria M (Tucci); c. Paul M Jr, Carl H. Ch Met Indus Elec Div, Fed Tele & Radio Corp 1941-46; Met, Alloy Cast Inst 1946-51; Met Supvr, Wright Aero Div, Curtiss-Wright Corp, 1951-55; Met Editor, The Iron Age 1955-59; Senior Editor, ASM "Metals Handbook" 1959-76; Mgr, Publs Dev 1976-79. Asst Dir Ref Publs 1979-. 12 US patents. Fellow AAAS; Fellow Inst of Metallurgists, Mbr AIME, ASM, SAE Definitions Comm, Overview Panel, Natl Sci Foundation. Editor, "Machining Difficult Alloys' 1962, "Forging Design handbook" 1972, "Source Book in Failure Analysis" 1974, "Case Histories in Failure Analysis" 1979. "Worldwide Guide to Equivalent Irons and Steels" 1979; "Worldwide Guides to Equivalent Nonferrous Metals and Alloys" 1980; "Failure Analysis: The British Engine Technical Reports" 1981. *Society Aff:* AAAS, AIME, ASM, MS, ISS.

Unwin, Gordon D
Business: 2515 A St, Anchorage, AK 99503
Position: Pres. *Employer:* Unwin Scheben Korynta Huettl Inc. *Education:* BSCE/Civ Engg/Lehigh Univ . *Born:* in Anchorage, AK. Airport Engr with the State of AK, Div of Aviation 1962-1969. Design Engr with DOWL Engrs specializing in utilities & airports 1969-1972. In 1972 formed Unwin Scheben Korynta Huettl Inc, which now has a staff of 40 providing complete Civ and Structural Engg, Land Surveying and Architectural services. Currently specializing in airport planning and design. VP, Bd of Reg Architects, Engrs and Land Surveyors, State of AK 1974-1981. Mbr GSA Public Advisory Bd, Region 10 1978-1979. Mbr Anchorage Geotechnical Commission 1977-1980. Mbr American Arbitration Assoc 1976-1980. *Society Aff:* ASPE, NSPE, PEPP, ASPLS.

Upadhyay, Rajendra N
Business: PO Box 76, Brea, CA 92621
Position: Supvr Reservoir Eng *Employer:* Union Oil Co of CA *Education:* MS/Petro Engrg/Univ of CA-Berkeley; BS/Petro Engrg/Indian Sch of Mines, India *Born:* 11/18/41 Born in India. Migrated to USA in 1964. Worked as a petroleum engr for Oil & Natural Gas Commission of India (1961-63), and for Columbia Gas System, Columbus, Ohio (1965-68). Employed by Union Oil Co of CA since 1968, currently as Supervisor of Reservoir Engrg. Involved in the development of oil, gas, and geothermal reservoirs worldwide. Chrmn of Reservoir Engrg Committee, Soc of Petroleum Engrs (1981-82). Technical Editor, SPE (1980-84). Mbr of the Advisory Bd of Tulsa Univ Fluid Flow Projects. Mbr of The Bd of Directors, Computer Modelling Group. Enjoy tennis and photography. *Society Aff:* SPE.

Upadhyaya, Belle R
Business: Dept. of Nuclear Engineering, The University of Tennessee, Knoxville, TN 37996-2300
Position: Assoc Prof *Employer:* Univ of TN *Education:* PhD/Sys Engrg/Univ of CA, San Diego; MASc/Applied Mechs/Univ of Toronto, Canada; BEng/Mech Engrg/ Univ of Mysore, India; Pre-Univ/Sci/St Aloysius Coll, Mangalore, India *Born:* 05/04/43 Belle R Upadhyaya received the PhD degree in sys engrg from the Univ of CA, San Diego in 1975. In 1972 he worked as a summer engr at Gen Atomic Co, San Diego. Since 1975 he has been with the Dept of Nuclear Engrg, the Univ of TN, Knoxville, where he is presently an associate professor. His current work includes research and dev in data processing and sys analysis applied to nuclear power reactors, and process industry. He is a conslt to the Oak Ridge Natl Lab. Dr Upadhyaya is a reg PE in the State of TN. He is currently a mbr of the Book Publishing Ctte of the Amer Nuclear Soc, and a liaison representative of the IEEE Control Sys Soc tech activities bd on energy. Dr Upadhyaya has publ more than fifty articles in the areas of sys analysis, data processing, and applications to nuclear power sys and process control systems. *Society Aff:* IEEE, ANS, ISA, ASEE

Upatnieks, Juris
Business: E.R.I.M, PO Box 8618, Ann Arbor, MI 48107
Position: Research Engin *Employer:* Envirom Res Inst of MI *Education:* MSE/EE/Univ of MI; BSE/EE/Univ of Akron *Born:* 05/07/36 Born in Riga, Latvia, came to US in 1951. Joined ERIM (Formely Williow Run Laboratories, Univ of MI) in 1960. Worked on Development of Off-Axis Holography and Coherent Optics. Lecturer at Univ of MI Electr & Comp Engin Dept since 1971, Adjunct Assoc Prof since 1974. Published over 40 papers, received 14 patents. Awarded Robert Gordon Memorial Award from SPIE in 1965, R. W. Wood Prize from OSA in 1975, Holly Medal from Amer Soc of Mech Engin in 1976. Fellow mbr of OSA and SPIE. *Society Aff:* OSA, IEEE, SPIE

Upp, Donald L
Business: Diesel Equipment Div GMC, 2100 Burlingame SW, Grand Rapids, MI 49501
Position: Staff Assistant to Purchasing Dept *Employer:* General Motors Corp. *Born:* Sept 1931. MS Univ of Dayton; MetE Univ of Cincinnati. With GM since 1949. Current position of Staff Asst (Purchasing) since 1972, with present div since 1979. Prod design patents. Other activities included participation in organizing the Montgomery County Joint Vocational Sch, the Engrg & Sci Center, Dayton, and the Affilliate Socs Org for the Engrg & Sci Center. Former Mbr ASM Natl Mbrship Ctte. Active in Amateur Radio & local civic cttes & orgs. *Society Aff:* ASM

Upthegrove, Wm R
Business: Aerospace, Mechanical & Nuclear Eng'r'g, University of Oklahoma, Norman, OK 73019
Position: Regents' Prof of Engrg *Employer:* Univ of OK, Sch of Aerospace, Mech & Nucl Engrg *Education:* PhD/Metall Engrg/Univ of MI; MSE/Metall Engrg/Univ of MI; BS/Metall Engrg/Univ of MI. *Born:* 11/10/28 Served with USN Pacific Fleet 1950-53. Fac mbr & head of new School of Met Engrg, Univ of OK 1956-62. Supr of new Powder Met Res prog at Int'l Nickel Co Res Labs 1962-64. Hd of Dept of Mech Engrg Univ of TX at Austin 1964-70. Dean Coll of Engrg, Univ of OK 1970-81. Regents' Prof of Engrg, Sch of Aerospace, Mech, and Nucl Engrg, 1981- . Past Chrmn of Engrg Coll Council; Reg PE in OK and TX. Published papers in diffusion in metals, powder metallurgy, engrg educ, and manpower. Current prof ints incl use of matls in design applications, engrg educ & intl coop tech progs. *Society Aff:* ASM, ASME, AAAS, TMS of AIME.

Urbach, John C
Home: 142 Crescent Ave, Portola Valley, CA 94025
Position: Consultant *Employer:* Self *Education:* PhD/Optics/Univ of Rochester; MS/ Physics/MIT; BS/Physics/Univ of Rochester. *Born:* Feb 18, 1934. NATO Postdoc Fellowship in Sci, Royal Inst of Tech, Stockholm Sweden 1962. Joined Xerox Corp Jan 1963. Worked on photographic & optical image eval, as well as on unconventional imaging processes, esp thermoplastic image recording & its appli to holography. Assumed mgmt respon for this work 1967. 1970-85 initiated and managed progs in lasar scanning, optical data, storage, optical communication and display, systems technology activities for Xerox Corporate Research Group. Currently working as an independent consultant in electro-optical technology and technical management. *Society Aff:* OSA, SPSE, SPIE

Urban, John S
Business: Edwards and Kelcly Inc, 70 South Orange Ave, Livingston, NJ 07039
Position: Sr. VP *Employer:* Edwards & Kelcey, Inc. *Education:* BS/CE/NJ Inst of Tech *Born:* 7/1/33 Joined Edwards and Kelcey in 1956, made a VP in 1973 and Sr VP 1984 is chief of the hwys and Public Works Div. Registered engr in the states of NJ, DE, and GA. Past Pres of the Consulting Engrs Council of NJ. Mbr of ASCE, NJSPE, and IBTTA (Robert Ridgeway Award Winner). Lives in Basking Ridge NJ. *Society Aff:* ASCE, CEC/NJ, IBTTA, NJSPE

Urban, Robert A
Business: 300 Keith Bldg, Cleveland, OH 44115
Position: Pres *Employer:* Byers Engrg Co *Education:* BS/Mech Engrg/Case Western Reserve Univ. *Born:* June 15, 1924. Mbr Cleveland cons engrg firm since 1945, partner since 1954. Spec in design of mech sys for bldgs. Presented paper on early work on variable volume air cond sys. Mbr ACEC, NSPE, Fellow ASHRAE, Tau Beta Pi. PPres Cons Engrs Council of Ohio. P Natl Dir ACEC. Mbr Bd of Dir & P Pres Grace Hosp Cleveland. *Society Aff:* ACEC, NSPE, ASHRAE.

Uribe-Angulo, Gustavo
Business: Calle 37 8-43-Piso 12, Bogota, Colombia
Position: Gerente. *Employer:* Restrepo y Uribe Ltda. *Education:* MS/Highways/OH State Univ; CE/Civil Eng/Universidad Nacional. *Born:* May 1925 Bogota. Civil Engr from Universidad Nacional de Colombia 1950. MS Ohio State Univ 1952. Maintenance Planning Engr & Ch Engr Equip Dept, Ministry of Public Works of Colombia. From Aug 1955 to date Partner of 'Restrepo y Uribe Ltda', cons engrs firm with 40 engrs & a personnel of 250 people. Respon for the hwy design & hwy supervision depts, publs 'Hwy Design', 'Pavement Failures at Bus Stops'. Pres ASCE Rep of Colombia Sect 1974-76, Dir 'Sociedad Colombiana de Ingenieros' 1960, 1964, 1968. *Society Aff:* ASCE, SCI, ITE, TRB.

Uriegas-Torres, Carlos
Home: Chichen Itza 301, Mexico City, D.F. 03600 (Mexico)
Position: Director *Employer:* UTYA (Own) *Education:* -/CE/Natl Univ of Mexico; BS/CE/Univ of TX *Born:* 09/20/21 A registered Cvl Engr and a Certified Cost Engr in Mexico. Over 30 yrs experience in the Mexican oil industry, with specialization in Construction Mgmt and Cost Engrg. In 1981 he founded UTYA, a profl services firm for the construction industry. A founder mbr of SMIEC (Mexican Society of Cost and Economic Engrg), appointed Honorary Mbr. He has a long teaching career and is a mbr of the Consltg Bd of the Cost Engrg MS Prog en la Salle Univ. Author of numerous papers and a book on Economic Analysis of Engrg Systems. *Society Aff:* ASCE, AACE, SMIEC, CICM, PMI, ASEE.

Urkowitz, Harry
Home: 9242 Darlington Rd, Philadelphia, PA 19115
Position: Staff Scientist. *Employer:* RCA, Electronic Systems Dept. *Education:* PhD/EE/Univ of PA; MS/EE/Univ of PA; BS/EE/Drexel Univ. *Born:* 10/1/21. Native of Phila, PA. US Army AF 1942-45. Worked for Philco corp 1948-64, Genl Atronics Corp 1964-70, RCA Corp 1970- . Elected Fellow IEEE 1973. Chmn signal Theory Subctte, Std Coord Ctte, IEEE Grp on Info Theory. P Chmn & organizer Phila Chap. IEEE Grp on Info Theory. P Chmn Phila chap IEEE Gp on Curcuit Theory. Adj Prof of EE Drexel Univ. *Society Aff:* IEEE.

Urquhart, Andrew W
Business: Corporate R&D, PO Box 8, Schenectady, NY 12301
Position: Branch Mgr. *Employer:* GE Co. *Education:* PhD/Met/Dartmouth College; MS/Engg Sci/Dartmouth College; AB/Engg Sci/Dartmouth College. *Born:* 8/24/39. Native of VT. Served as Naval Officer, 1962-1967, assigned to Div of Naval Reactors, USAEC, responsible for technical admin of reactor mtls dev progs. Worked as a consulting engr for Creare, Inc, Hanover, NH, during 1967-1968. Joined GE Corporate Res & Dev in 1971 as a met in corrosion res. Assumed present position as Mgr, Inorganic Mtls & Reactions Branch in 1975. Responsible for supervising res on high temperature chemistry & corrosion, high pressure phenomena, hard mts dev, & metal & rock cutting. Chrmn, Eastern NY Chapter, ASM, 1978-1979. *Society Aff:* ASM, Electrochemical Soc., NACE, AAAS

Ussery, Harry B
Business: PO Box 16509, Greenville, SC 29606
Position: VP. *Employer:* Southeastern - Kusan. *Education:* BS/Chem Engg/LA Tech Univ. *Born:* 11/23/41. Native of Shreveport, LA. Worked for Monsanto Co 11 yrs, 1963-74. Held variety of Engg, Manufacturing, and Marketing positions in Petro Chem and Plastic business. Headed start-up of new ABS Polymer facility in Antwerp, Belguim. Have been with Kusan, Inc. 5 yrs, 1974-present. First position was VP and Gen Mgr of Allastics, a div of Kusan in Atlanta, GA. Currently, hold pos of VP for Southeastern-Kusan, Inc. Reg PE in LA and OH. *Society Aff:* AIChE, SPE, SPI, NSPE.

Utela, David R
Home: 18118 60th Ave NE, Seattle, WA 98155
Position: Engrg Instructor. *Employer:* Everett Community Coll. *Education:* MS/Mech Engr/Univ of WA; BS/Mech Engr/Univ of WA; AA/Pre-Engr/Centralia, CC. *Born:* Oct 26, 1943. BSME & MSME Univ of Washington; AA Centralia CC. Native of Wash State. Res Engr for the Boeing Co 1966-70. Taught engrg & math at Bellevue & Shoreline Community Colleges. Currently teaching at Everett CC. Mbr Tau Beta Pi, Amer Soc for Engrg Educ. ASEE Outstanding Young Fac Dow Chem USA 1975. Enjoy hiking, biking, sailing & skiing. *Society Aff:* ASME.

Uthlaut, Herbert T, Jr
Business: S Central Bell-PO 771, Birmingham, AL 35201
Position: Div Staff Manager-Network Planning. *Employer:* South Central Bell. *Education:* BSEE/EE/Auburn Univ. *Born:* 1/5/27. BSEE Auburn Univ 1950. US Army Signal Corps 1944-48, 1951-53. Joined Bell Sys 1953, serving with Southern Bell, AT&T, & presently with South Central Bell. Pos incl transmission, outside plant & data engg plus staff respon for South Central Bell's long range engg, planning & const budge mgmt. Presently respon for long range network planning staff functions for S Central Bell. Sr Mbr IEEE; Mbr Bd of Governors IEEE Communication Soc 1971-73; Chmn IEEE Natl Telecommunications conf 1978, V Chmn IEEE Comm Soc conf Bd. *Society Aff:* IEEE, TBΠ, HKN, ΦΚΦ.

Utic, Phillip J
Business: 631 S Adams St, Green Bay, WI 54301
Position: Gen Mgr. *Employer:* Green Bay Water Dept. *Education:* BS/Civ Engg/ Univ of WI. *Born:* 6/24/29. Native of Oshkosh, WI. Served with Army Corps of Engrs 1951-53. Assumed current responsibility of Gen Mgr of Green Bay Water Utility in 1967. Chrmn Am Water Works Assn 1972. Fuller Award AWWA 1974. Enjoy hunting, fishing, skiing. *Society Aff:* AWWA.

Utku, Senol
Business: 122 Engg., Duke University, Durham, NC 27706
Position: Prof of Civil Engg & Prof of Computer Sci. Director of Graduate Studies (by 7/1/87) Director of Unde *Employer:* Duke Univ. *Education:* ScD '60/Structural Mech/MIT; MS '59/Structural Mech/MIT; Dipl Ing '54/Structural Mech/Istanbul Tech *Born:* Suruc, Turkey, 1931; married, 64; 2 children. Res engr, IBM Mathematics and Applications Dept, NY, 59-60; asst prof, MIT, 60-62; assoc prof, METU, Ankara, 62-63; exec chief, ITU Computation Ctr, Istanbul, 63-65; tech staff mbr, JPL, CalTech, 65-70; assoc prof, Duke Univ, 70-72; prof of civil engg, Duke Univ, 72-79; prof of civil engg and prof of computer sci, 79- ; director of undergraduate studies, 80-87; director of graduate studies, 87- . Univ Prof of Numerical Analysis, ITU, Istanbul, 1977; co-author of the *Elementary Structural Analysis*, 3rd ed, McGraw-Hill, 1976; co-author of the *Dynamics of Offshore Structures*, John Wiley &

Utku, Senol (Continued)
Sons, 1984; mbr of NATO Computer Sci Study Group, Brussels, 1968-70; Fulbright student at MIT, 1957-58. *Society Aff:* ASCE, ΣΞ, XE, ASEE, SIAM, AAM.

Utlaut, William F
Business: US Dept of Commerce, Boulder, CO 80303
Position: Dir, Inst for Telecommun Sci. *Employer:* Natl Telecommunications & Info Admin. *Born:* 1922 Sterling Co. PhD EE Univ of Colorado. US Navy 1943-46. Design Engr G E Schnectady N Y. Joined U S Dept of Commerce 1954 - mgmt respons: lead res projs involving radio wave transmission & telecommunication studies. Received Dept of Commerce Gold Medal Award; Disting Engrg Alumnus Award Univ of Colorado. IEEE Fellow; Mbr IEEE Communications Soc; Member IEEE Com Society Policy Board and Nominating Board; Chmn of. USSG-1 (CCIR); Comm C (USRI). Reg P E Colo. Mbr Engrg Dev Council Univ of Colo. Chmn Denver Sect IEEE 1966-67; Chmn IEEE Internatl Conf on Communications Boulder 1969.

Uyehara, Otto A
Home: 544 So Bond St, Anaheim, CA 92805
Position: Prof. Emeritus *Employer:* Univ of WI. *Education:* PhD/ChE/Univ of Wi; MS/ChE/Univ of WI; BS/ChE/Univ of WI. *Born:* 9/9/16. Birth place-Hanford CA. Parents - Rikichi and Umi Uyehara. Children - Kenneth Otto, Susan Joy, Emi Ryu. Wife - Chisako (Suda) Uyehara, - married Aug 12, 1945. Education BS-1942, MS-1943, PhD 1945. Univ of WI Res Assoc 1945, Asst Prof 1949, Assoc Prof-1952, Prof 1957. ASME Oil & Gas Div Award 1951, (with El Wakil & PS Myers) Benjamin Smith Teaching Award, Fellow SAE, Prof Emeritus 1982 (Univ of WI), SAE Horning Award. 2 time recipient SAE Colwell Award, Inst of Mech Engrs Dugald Clerk Prize (1970). Made an honorary member of Japan Society of Mechanical Engineering (April 1987). *Society Aff:* SAE, ASME, JSME

Uys, Johannes M
Home: 3524 Parthenon Way, Olympia Fields, IL 60461
Position: Dir Quality Control. *Employer:* Jones & Laughlin Steel Corp. *Education:* ScD/Met Engrg/MIT; MSc/Chem/PTA Univ; BSc/Chem/PTA Univ. *Born:* 10/17/25. Born in Republic of S Africa, Naturalized US Citizen. After graduating from Pretoria Univ worked as Researcher in Coal and Steel. Upon grad from MIT joined Allegheny Ludlum Steel Corp, then Youngstown Sheet and Tube Co was VP of Tech Serv of this Co until its merger into Jones and Laughlin Steel Corp, now Dir Quality Control Western Div of J & L, presently Dir Tech Servs, Melouth Steel Prod Corp. *Society Aff:* AIME, ASM, ASTM, AISE.

Vachon, Reginald I
Business: 6855 Jimmy Carter Blvd, Atlanta, GA 30071
Position: Pres. *Employer:* Vachon, Nix & Assoc. *Education:* PhD/ME/OK State; MS/Nuclear Sci/Auburn Univ; LLB/Law/Jones Law Sch; BME/Mech Engg/Auburn Univ. *Born:* 1/29/37. Over twenty yrs experience in Industry, Govt and Education. He holds a Bachelor of Mech Engg Degree, a MS in Nuclear Sci and received the PhD in Mech Engg from OK State Univ in 1963. He is a Reg PE in GA, AL & MS. He earned the LLB, is admitted to practice law in AL and is a mbr of the American and AL Bar Assns. His activities are centered around the areas of systems engg, energy systems, energy policy, total environmental planning and proj mgt. He is the author of over 100 publications. *Society Aff:* ASME, AIAA, ASEE, ABA, NSPE.

Vacroux, Andre G
Home: 1689 Lake Ave, Highland Park, IL 60035
Position: Dean of Engrg *Employer:* Il Inst of Tech *Education:* PhD/ME/Purdue Univ; MS/EE/Univ of Notre Dame; Diplome d'Ingenieur/EE/Ecole Superieure d'Electricite (France) *Born:* 05/22/36 Native of Paris, France - Diplome d'Ingenieur, Ecole Superieure d'Electricite, MSEE Notre Dame, PhD ME Purdue. Asst and Assoc Prof of Electrical Engrg, IIT 65-72; Teaching and Research in systems theory and digital systems. Mbr of the Technical Staff, Bell Telephone Laboratories, Holmdel, NJ, 72-75; Work on M13 digital multiplex. Prof and Chrmn, dept of Electrical Engrg IIT, 75- 80. Since 1980, Dean, Armour Coll of Engrg, IIT. Director, Integrated Information & Telecommunications Systems Center (IITSC). Prof interests include small computers and their applications, telecommunications. Conslt to major corps. Reg PE in CA. *Society Aff:* IEEE, ASEE, AAAS, WSE

Vadasz, Leslie L
Business: 2625 Walsh Ave, Santa Clara, CA 95051
Position: Sr VP. *Employer:* Intel Corp. *Education:* BEE/EE/McGill Univ. *Born:* 9/12/36. Sr VP in charge of the Corp Strategic Staff for Intel Corp. He joined the firm in 1968 & was responsible for MOS IC dev. In 1972, he became Dir of Engg responsible for all tech & component dev. In 1975, he was elected VP. In 1976, he became Asst Gen Mgr of the Microcomputer Div & in 1977 Gen Mgr of the Microcomputer Component Div. He was elected a Sr VP in 1979. Previously, he was with the semiconductor R&D lab of Fairchild Camera & Instruments from 1964 to 1968. His work dealt with circuit & process dev for both MOS & bipolar integrated circuits. In 1961 he joined Transitron Corp. His responsibilities included silicon process & device dev. In 1977 he was elected a Fellow of the IEEE for "leadership in the dev of semiconductor memories & microcomputer components.–. *Society Aff:* IEEE.

Vairavan, K
Business: College of Engg and Appl Sci, Milwaukee, WI 53201
Position: Prof. *Employer:* Univ of WI. *Education:* PhD/Elec Engg/Univ of Notre Dame; MS/Appl Sci/George Washington Univ; BE/Elec Engg/Univ of Madras. *Born:* 7/9/39. Native of India. Naturalized American citizen. Jr Engr, Madras State Electricity Bd, 1962-63. Served as mbr of technical staff at the Bell Telephone Labs in 1970 and united the Univ of Hiroshima, Japan in 1980. With the Univ of WI-Milwaukee since 1968. Current res interests are in the area of computer architecture and parallel processing. Directed several res projs. Other interests include computer applications in developing countries, and tech educ and training in Japan. Has served as a visitor for ABET for the accreditation of Computer Engrg progs. *Society Aff:* IEEE.

Vaishnav, Ramesh N
Business: Dept. of Civil Engrg, The Catholic Univ. of America, Washington, DC 20064
Position: Professor of Civil Engineering. *Employer:* The Catholic University of America. *Education:* PhD/Theo & Applied Mech/Univ of IL; MSE/Struct Engg/Univ of MI; BE/Civil Engg/Univ of Gujarat; BSc/Chemistry/Univ of Gujarat. *Born:* 4/2/34. Divisional Overseer, Public Works Dept Junagadh India 1955-56; Structural Engr, Vogt, Ivers & Seaman Assocs Cincinnati Ohio 1958-59; Prof of Civil Engrg, Catholic Univ Washington DC 1961- ; Dir, Center for Artificial Intelligence and Robotics, Catholic Univ 1985- . Courses taught: elasticity, nonlinear elasticity, plates & shells, continuum mechanics, steel design, reinforced concrete design, computer-aided C.E. design. Res: nonlinear elasticity & viscoelasticity, biomechanics, creep of concrete, artificial intelligence and robotics. Varied cons involving finite element & other methods of stress analysis. Fellow ASME, Wash Acad of Sciences; Sof of Rheology, Natl Inst of Health Bethesda Md 1967-1980. Reg Prof Engr, MD. *Society Aff:* ASME (FELLOW), SOC RHEO, AAM, AAAS, CVSDS, SOC NAT PHIL, ΣΞ, INT SOC BIORHEO, N.Y. A.SC, IAMM, SES, TBΠ ASCE, AAAI, NSPE, WSE, WAS, DCSPE

Valasek, Joseph
Home: 300 Seymour Place SE, Minneapolis, MN 55414
Position: Emeritus Prof *Education:* PhD/Physics/Univ of MN; MA/Physics/Univ of MN; BS/Physics/Case Inst of Tech *Born:* 04/27/97 in Cleveland, OH of Czech parents. After employment at the Natl Bureau of Standards for two yrs, spent the next forty-six yrs teaching Physics and conducting research at the Univ of MN. Natl Research Fellow in 1921-22. Discovered Ferroelectricity, a dielectric analog of Ferromagnetism, in 1920. Author of two textbooks on Optics, twenty articles in Physics

Valasek, Joseph (Continued)
journals, and four articles in the Encyclopedia Britannica. Emeritus Prof of Physics since 1965. Married Leila E. Munson in 1924; have two daughters, Frances and Marion. *Society Aff:* APS, OSA, AAAS

Valencia, Samuel
Business: Army Mat & Mech Ctr, Watertown, MA 02172
Position: Materials Engineer. *Employer:* Army Materials & Mechanics Res Ctr. *Education:* BS/Metallurgy/TX Univ. *Born:* 8/16/15. BS TX Univ at El Paso. Served as Civilian engr in gvmt since May 1951. Prior to joining gvmt, employed as Senior Metallurgist for Amer Smelting & Refining Co. With gvmt have served in variety of positions. Current resps incl tech mgmt at several Dept of Defense Info Analysis Ctrs, tech coordination of 2 Dept of Defense Technology Coordinating Papers & mgmt of Independent Res & Dev Prog. P Chmn Boston Chap, ASM. Appointed mbr Mass Bd of Registration of Prof Engrs & Land Surveyors for 3 5-yr terms. Elected V Chrmn of MA. Bd of Registration of Prof. Engrs & Land Surveyors for 1978 and 1979. *Society Aff:* ASM, NCEE, NSPE, AIME.

Valentin, Richard A
Business: 9700 S. Cass Ave, Bldg 208, Argonne, IL 60439
Position: Dir, NRC Programs *Employer:* Argonne Natl Laboratory *Education:* PhD/Applied Math/Brown Univ; ScM/Applied Math/Brown Univ; MBA/Bus Admin/Univ of Chicago; BS/ME/IL Inst of Tech *Born:* 03/16/38 Native of Chicago, IL. Joined staff of Argonne Natl Lab in 1964 where his research interests have focused on applied mechanics problems related to structural design of advanced nuclear reactors. Published extensively in areas of numerical analysis, elasticity, dynamic plasticity, elevated temperature structural analysis, fluid-structure interactions, and extreme loading of reactor components. Several yrs as mgr of Structural Systems Analysis Section and various specialized projects. Named Sr Mechanical Engr in 1979. Current position: Special Asst to Associate Lab Dir and Dir of Argonne's Nuclear Regulatory Commission research efforts. *Society Aff:* ASME, SNP, IASMiRT

Valentine, William M
Home: 3901 97th SE, Mercer Island, WA 98040
Position: Retired Consulting Engr *Employer:* Pres & Maj Stockholder of Video Stories, Inc *Education:* -/ME/Univ of WA; -/Philosophy/Seattle Univ. *Born:* 10/16/15. Reg PE in WA, CA, MT and AK. Attended Univ of WA. Founded Valentine, Fisher & Tomlinson, Consulting Engrs, Seattle, WA 1956. (Now VFT) Holder of Patent 3, 272,472, Air Handling Device. Responsible engr in charge of Mech design projs since 1949, specializing in boiler plants, steam distribution, air conditioning, heating, air pollution, and energy conservation. Contributor of technical articles to engg publications. Civic efforts: Knights of the Round Table Pres, Seattle Table, 1972-74; Intl VP 1974-77; Mercer Island Bd of Appeals 1952-84; Moose, Elks, Engrs' Clubs. Now a Producer of Video Documentaries for television and general commercial use with staff of 5. Mbr of Intl Television Assoc NY Acad of Sci; AAAS, ASHRAE. *Society Aff:* AAAS, ASHRAE, ITVA, NYAS.

Vallette, William J
Business: 19 Mansion Dr, Topsfield, MA 01983
Position: VP *Employer:* New England Training Inst *Education:* SB/Bus & Engg/MIT. *Born:* 1/11/20. Ind eng Stone & Webster; dir of Ind Eng, electronics Div of CBS; Mgr of Mfg, Itek Corp; Mgr of Mfg Eng and Mgr Mfg LEM, RCA Aerospace Div; VP Value Mgmt Consultants; Corp Dir of I E and Pkg Dev, Polaroid Corp, VP Mfg Northeast Electronics, VP Mfg Chloride Pyrotector. Reg PE MA, NH & CA. Elected to many IIE offices including Natl Pres, Exec Comm 1961-65, Bd of Trustees 1958-66. Served on MSPE Bd of Dir, Eng Manpwr. Comm, and EJC. Past Chmn NE Ind Eng Council, Mbr Natl Council on Ind Eng. Authored technical papers on work meas, incentives, prod control, employee training, and oper res. Mbr Delta Upsilon, Alpha Pi Mu, NSPE, MSPE, IIE (Fellow, Life Mbr). Member, New Hampshire State Board of Registration for Engineers, Architects, and land Surveyors, Member Governor of NH's Advisory Roundtable. *Society Aff:* IIE, NSPE, MSPE, AΠM.

Valpey, Robert G
Business: Calif Poly State Univ, San Luis Obispo, CA 93407
Position: Dean, Sch. of Eng'g & Tech. *Employer:* California Polytechnic State Univ. *Education:* PhD/Engg/US Mech/Univ of IL; MS/ME/Univ of CO; BME/ME/Cornell Univ; BS/Military Sci & Engg. *Born:* Feb 1923. 20 years US Air Force 1945-65. During that time taught at US Military Acad, West Point, & US Air Force Acad, Colo. Also engr & an Engrg Mgr, Equipment Lab, Wright-Patterson Air Force Base & Special Projs Office, Secy of the Air Force. Army & Air Force medals for engrg servs. 1st Dean of Engrg, Calif State Univ Fullerton. Some time cons to Rockwell Internatl. P Chmn, Calif State Univ & Colls Engrg Deans; P Chmn statewide Engrg Liason Ctte. Hobby: 'classic' cars, sailing. *Society Aff:* NSPE, AIAA, ASEE, ASME.

Van Antwerpen, Franklin
Business: 16 Sun Rd, Basking Ridge, NJ 07920
Position: Emeritus Secy/Exec Dir, AIChE (Retired). *Education:* AE/Chem Engg/Newark Inst Tech; BS/Chem Engg/Newark College Eng; Dr Eng/-/NJ Inst Tech. *Born:* Nov 28, 1912 Patterson NJ; son of William & Frances (Wambach) Van A. graduate study 1938-40 Columbia; Honorary Degree (Dr Eng) 1975 NJ Inst of Tech. Married Dorothy Hoedemaker, June 3, 1939; c. Franklin Stuart, Virginia Evelyn (Mrs Walter Rosenberger). Mgr, Munitex Corp 1935-38; Managing Editor, Industrial & Engrg Chemistry 1938-46; founder, Pub Editor Chem Engrg Progress, Business Mgr 1946-54, Founder, Pub Editor Monograph & Symposium Series 1950-78 AIChE; Founder pub AIChE Journal & Internatl Chem Engrg 1954- ; Secy, Exec Dir AIChE 1955-78, Publisher 1955-78. Fellow AIChE, AAAS, AIC, Hon Fellow Inst Chem Eng (Great Britian);; Mbr ACS, CESSE (P Pres 1975-76, Pres 1974-75), Interamerican Confederation of Chem Engrg (Treasurer 1969-73, Pres 1975-). Club: Chemists (trustee, P Pres) (NYC). Author articles & editorials chem engrg magazines. Co- author: '50 Years of Amer Inst Chem Engrs Award: Founders (AIChE); 1st Recrp of F. S. Van Antwerpen Award for service to AIChE. *Society Aff:* AIChE, AAMS, AIC, IChE, IMIQ.

Van Bladel, Jean G
Business: St Pietersnieuwstraat 41, Ghent B-9000, Belgium
Position: Professor of Electrical Engineering. *Employer:* University of Ghent. *Education:* PhD/EE/Univ of WI; -/Elec Engg/Brussels Univ. *Born:* July 24, 1922 Antwerp Belgium. Radio Engr 1948 Brussels Univ; PhD EE 1950 Univ of Wisc. Head, Radar Dept, MBLE Brussels 1951-54. Prof appointments at Washington Univ St Louis, & the Univ of Wisc 1954-64. Dir of Lab for Electromagnetism, Univ of Ghent 1964- , Dean Faculty of Engg 1976-78. Fellow IEEE, IEE. Guggenheim Fellow 1962-63. Montefiore Prize 1965. Visiting Professorships at various univs. Secy genl URSI Oct 1979- . Honorary doctor, Univ of Liegh, 1987. *Society Aff:* IEEE.

Van Bladeren, John
Business: 123 NW Flanders St, Portland, OR 97209
Position: Vice Pres, Engrg & Gas Supply Div. *Employer:* Northwest Natural Gas Co. *Born:* Aug 27, 1920 Amstelveen, The Netherlands. Educated in Amsterdam, The Netherlands: Grade Sch, High Sch, Coll in Electrical Engrg. Reg PE in Oregon & Wash. Received the Beal Medal, AGA's Natl Award for the best tech paper presented in the year 1965, 'The Effects of High Voltage Direct Current Transmission on Buried Metallic Pipelines.' Mbr: Prof Engrs of Oregon, Pres 1975-76; NSPE, Natl Dir; ASME; ASM; Pacific Coast Gas Assn; Portland Chamber of Commerce; American Gas Assn.

Van Buren, Maarten
Home: Hampton Ln, New Canaan, CT 06840
Position: VP Manufacturing & Engrg *Employer:* Lever Brothers Co. *Education:* MS/Chem Engr/Univ of Dordrecht, Netherlands; MBA/Finance/Univ of Chicago. *Born:* 2/27/41. Started career as design engr, Swenson Div, Whiting Corp, which in-

Van Buren, Maarten (Continued)
cluded many start-ups across USA and Canada (5 yrs). Co-publisher of article on "Urea Crystallization in CEP.–. Held positions in Corp Engg and Chemical Div of Abbott labs to manage engg and mfg depts (9 yrs). Assumed respon as Genl Mgr Engg Div of Lever Brothers Co in 1978. Apptd Dir of Mfg effective January, 1980. Apptd VP, Mfg & Engrg 1982. Enjoys Sports. *Society Aff:* AIChE, AMBA, JDF.

Van Buskirk, William C
Home: 1137 9th St, New Orleans, LA 70115
Position: Prof & Hd Biomedical Engg. *Employer:* Tulane University. *Education:* BS/Eng Sci/USMA, West Point, NY; MS/Aero & Astro/Stanford Univ; PhD/Aero & Astro/Stanford Univ. *Born:* March 1942 Utah. Served with US Air Force 1964-67. With Tulane Univ since 1970, assumed current responsibility as Chmn of Biomedical Engrg in 1974. Mbr ASME, ASEE, BMES, & other Socs. ASEE - Dow Outstanding Young Faculty Mbr 1974. ASME - Fellow, 1985, ASEE - Outstanding Biomedical Engineering Education Award, 1986. *Society Aff:* ASME, ASEE, BMES, ORS.

Vance, Carl B
Home: 7520 Hoover Rd, Indianapolis, IN 46260 *Employer:* Retired 5/1/80 *Education:* BSME/Mech Engg/Univ of OK. *Born:* April 1915 Bagnell Mo. Sales-Service Engr, Bailey Meter Co; Senior Engr, Cleveland Elec Illum Co; Supt, Power & Distr Jayhawk Ordnance; with Indpls P&L company since 1945. Mgmt training course, Univ of Mich & Columbia Uni. VP, Power Production 1963-75. Sr VP-oper 1975-78. Exec VP-Oper 1978-80. Retired 5/1/80. Director & Consultant to 5/1/85. Coordinate & supervise all operating functions of company. Elected Dir Jan 1979 & Mbr Exec Comm. PE in Ohio & Ind. Mbr Ind Engrs Registration Bd 1959-62. Fellow ASME 1968. Mbr EEI Prime Moves Comm 1948-1966 (Chmn, 1963-1965). Commissioner- EJC Natl Engrs Comm on Air Resources 1970-71. Chmn Indianapolis APC Bd 1968-70, Mbr 1968-76. Mbr Indiana AP Bd 1969-70, Tech Cons 1968-1980. Mbr Bd/Dir Indianapolis Scientific & Engrg Foundation 1971-74. Deacon, Elder, & Trustee Fairview Presbyterian Church. Hobby, fishing. *Society Aff:* ASME, NSPE.

Vance, John M
Home: 2908 Chaparral Circle, Bryan, TX 77802
Position: Prof *Employer:* Texas A&M Univ *Education:* PhD/ME/Univ of TX; MS/ME/Univ of TX; BS/ME/Univ of TX *Born:* 10/5/37 Native of Houston TX. Ph.O. from Univ of Texas at Austin. Plant Engr for ARMCO Steel, 1960-62. Resch Mech Engr for Texaco, 1963-64. Group Leader for Engrg Analysis at TRACOR, Military Products Div, 1967-69. Asst and Assoc Prof at Univ of FL, 1969-78. Assoc and Full Prof at TX A&M Univ, 1978-present; Assoc Head of Dept of Mech Engrg 1980-81. Battelle Scientific Advisor to US Army 1971-78; Eleven summer appointments at Pratt & Whitney, Southwest Research Institute, Shell Research, U.S. Army. 40 published papers. Book pub by John Wiley on "Rotordynamics of Turbomachinery–. *Society Aff:* ASME, ASEE

Van Cott, Whitfield R
Home: 3047-121st, Toledo, OH 43611
Position: Division Head *Employer:* City of Toledo *Education:* MS/Structures/Univ of Toledo; BS/CE/Univ of Toledo *Born:* 11/09/46 I began working for the City of Toledo in 1971. In 1973, I obtained my PE License for OH in Civil Engrg. During this time, I was promoted to Div Head for the Div of Solid Waste which maintained all residental collection and disposal services. In 1978, I was again promoted to the Div Head of Water Reclamation, which is responsible for the wastewater treatment operation, sewer maintenance and ditch maintenance for the Toledo area. In 1977, I achieved a diplomate status in the AAEE for solid waste, and in 1979, I received my Class III OH Wastewater License. *Society Aff:* ASCE, WPCF, APWA, AAEE

Van de Riet, James L
Home: 1608 Arnold Circle, Virginia Beach, VA 23454
Position: Sr. Vice Pres/Office Mgr *Employer:* Buck, Seifert and Jost, Inc *Education:* MS/Santary Eng/VPI & SU; BS/CE/VPI & SU *Born:* 09/20/42 Native of Norfolk, VA. Pres of Student Chapter ASCE and elected to Chi Epsilon during senior yr of coll. Employed by Buck, Seifert & Jost, Consltg Engrs in 1965, became mgr of VA Office in 1970, named partner and VP in 1973. Appointed to Virginia Sewerage Regulations Advisory Committee in 1984 by Gov. Robb. Pres of VA Water Pollution Control Association, 1981. Deacon, Treas, Music Dir, Glad Tidings Church (A/G). *Society Aff:* NSPE, ASCE, AWWA, WPCF, AAEE

Vandegrift, Erskine, Jr
Business: P.O. Box 2727, Birmingham, AL 35202
Position: Project Engineer. *Employer:* American Cast Iron Pipe Co. *Education:* BS/ME/Auburn Univ. *Born:* Aug 22, 1921. Served in US Navy 1943-46, responsible for dev of machinery for production of pipe & related prods. Birmingham's Engr of the Year 1971. Vice Pres ASME Region XI 1970-72. Recipient of the Region XI Meritorious Serv Citation 1975. Recipient of Silver Beaver Award, Boy Scouts of America. Moxderator of Birmingham Presbytery 1976. Mbr of Tau Beta Pi. Enjoy tennis & good music. *Society Aff:* ASME, NMA, AFS.

VandeLinde, V David
Business: Charles and 34th Sts, Baltimore, MD 21218
Position: Dean of Sch of Engrg *Employer:* Johns Hopkins Univ *Education:* PhD/Sys & Comm Sci/Carnegie-Mellon Univ; MS/EE/Carnegie Inst of Tech; BS/EE/Carnegie Inst of Tech *Born:* 08/09/42 Native of St Albans, WV. Joined the faculty of Dept of Elec Engrg at Johns Hopkins in 1967. Named Prof in 1977 and Dean of the GWC Whiting Sch of Engrg in 1978. Areas of interest include: robust estimation and detection and digital signal processing. *Society Aff:* ASEE, IEEE, SIAM

Vanden Berg, Glen E
Home: 1003 Broadmore Cir, Silver Spring, MD 20904
Position: Assoc Regional Admin. *Employer:* US Dept of Agriculture. *Education:* PhD/Agri Engr/MI State Univ; MS/Agri Engr/MI State Univ; BS/Agri Engr/SD State Univ. *Born:* 8/12/30. Grew up on farm in SD. US Army 1952-1955. Conducted res emphasizing soil- machine mechanics as related to tillage tools and traction devices at the Natl Tillage Machinery Lab, Auburn, AL (1958-1965). Managed agri engg res (1965-1972) and across the bd agri res (1972-present) for the US Dept of Agri. Actively involved in promotion/evaluation systems for res scientists and methods for planning res. Co-author of book "Soil Dynamics in Tillage and Traction–; ASAE Cyrus Hall McCormick Medal 1979. *Society Aff:* ASAE.

Vanden Boogart, Rick C
Home: 908 Michigan Lane, Kaukauna, WI 54130
Position: Risk Control Rep. *Employer:* MSI Insurance *Education:* B.S./Fire Protection & Safety Eng./Illinois Institute of Technology *Born:* 05/15/53 Born and raised in Little Chute, WI. Graduated from St. John High School in 1971. Graduated from Illinois Inst of Tech with B.S. in Fire Protection and Safety Engrg 1975. Insurance agent with Aetna Life & Casualty, 1975-78. Loss Control Mgr. at Integrity Mutual Insurance in Appleton, WI, 1978-86. Sr. Loss Control Rep. with CNA Insurance, 1986-87. Risk Control Rep. with MSI Insurance. Responsible for Co-Operative risk control for accounts in WI & Upper Mich. Society of Fire Protection Engineers - WIS. Chapter Pres.; A.S.S.E. Mbr; Married; Two Daughters; Enjoy golf, fishing, hunting, football. *Society Aff:* SFPE, ASSE

Vander Horst, Leo
Home: 7448 Eastern SE, Grand Rapids, MI 49508
Position: Sr Prof Manager. *Employer:* Williams & Works Inc. *Education:* BSCE/Civil Engg/MI State Univ. *Born:* July 20, 1933 The Hague Netherlands. Education: High Sch, The Netherlands; completed 1 year in Aeronautical Engrg; BSCE Mich State Univ. P Pres Western Mich Branch of ASCE; P Secy, Treasurer, Vice Pres & Pres of Mich Sect ASCE; Panel Mbr, American Arbitration Assn 1974-date; Mbr Bd/Dir, Williams & Works (total employment 320, total engrs 75); Mbr & P Chmn, City of

Vander Horst, Leo (Continued)
Kentwood Mich Planning Comm 1967-76. Mbr NSPE, MSPE, CEC, ASCE. Mbr Ad-Hoc Visiting Ctte of Engrg Council for Prof Dev 1968-72. *Society Aff:* ASCE, NSPE, MSPE.

Vander Linden, Carl R
Business: Ken-Caryl Ranch R&D, PO Box 5108, Denver, CO 80217
Position: VP Dir Contract Research and Dev. *Employer:* Manville Service Corp *Education:* PhD/Chem Engg/IA State Univ; BS/Chem Engg/Univ of WA. *Born:* Sept 1923. Native of Pella Iowa. US Navy 1943-46. Instructor in Chem Engrg, Iowa State Univ 1946-50; with Johns-Manville since 1950 with specialization in mineral processing diatomite & perlite, synthetic silicates, building & const materials. Dir R&D since 1969; Dir Contract Res and Applied Technology since 1973. Chmn NJ Chap AIChE 1963-64, Dir AIChE 1979-81. Pres Perlite Inst 1974-76. Fellow AIChE. *Society Aff:* ACS, NIBS, BFC.

Vander Sande, John B
Business: Room 13-5025, Cambridge, MA 02139
Position: Professor of Materials Sci *Employer:* Massachusetts Inst of Technology. *Education:* PhD/Mat Sci/Northwestern Univ; BS/Mech Engg/Stevens Inst of Tech. *Born:* March 1944. Fulbright Scholar 1970-71, Univ of Oxford. At MIT since 1971, promoted to Assoc Prof 1975 promoted to Prof 1981. Have teaching responsibilities at the undergraduate & graduate level. Major res interests in the areas of microstructures, second phase precipitation, defect structures, & transmission electron microscopy of materials. Mbr ASM, Electron Microscopy Soc of America, & Bd of Review of 'Met Transactions'. Prog Ctte Mbr EMSA. Internatl Metallographic Exhibit Award 1975. Outstanding Teacher Award, MIT 1976. *Society Aff:* ASM, EMSA.

Van Der Sluys, William
Home: 2426 Lakeshore Court, Crown Point, IN 46307
Position: Ret *Employer:* Self *Education:* ME/Mechanical Eng./Stevens Institute of Technology *Born:* 08/15/12 Native of Paterson, NJ. Developmental work at Chrysler Eng 1934-51. With Pullman, Inc. 1951-77 (Pullman-Standard Div) as Mgr Engrg Design, Dir. R&D, General Mgr Passenger Eng, Dir of Marketing, V.P. & General Mgr of Passenger Operations. Originated push-pull operation of commuter trains. Presently Chairman General Awards Cttee ASME. Formerly V.P. Region VI. *Society Aff:* ASME

Vander Velde, Wallace E
Home: 50 High St, Winchester, MA 01890
Position: Prof of Aeronautics & Astronautics. *Employer:* Massachusetts Inst of Technology. *Education:* ScD/Instrumentation/MIT; BSAE/Aero Eng/Purdue Univ. *Born:* 6/4/29. Mbr of the faculty of the Dept of Aeros & Astronautics, MA Inst of Technology 1957- . Teaching & res in the areas of stochastic estimation, control sys, and inertial navigation sys. Author of books and papers on flight control systems, nonlinear control, and fault tolerant control. Member of the IEEE and Fellow of the AIAA. *Society Aff:* AIAA, IEEE.

Vander Voort, George F
Home: 3740 Brandeis Ave, Bethlehem, PA 18017
Position: Engineer. *Employer:* Bethlehem Steel Corp, Res Dept. *Education:* BS/Met Eng/Drexel Univ; MS/Met Eng/Lehigh Univ. *Born:* 9/1/44. Joined Bethlehem Steel in 1967. worked in Bethelehem Plant Met Dept until 1972 & in R&D since 1972. Respon for failure analysis & metallographic evaluations. Res conducted primarily on tool & die matls & heavy forgings. officer in Lehigh Valley Chap, ASM; named Outstanding Young Mbr in 1974. Instructor-Metal Engg Inst. Adjunct faculty member, Lehigh University elected P (1981-83) of the Internatl Metallographic Soc & Genl Chrmn of 1979 Annual Convention. Mbr of Committee E4 on Metallography, ASTM. Secretary of E4.14 on quantitative metallography. Two pat granted, sevl pending. Publs on failure anlaysis and metallography. Hobbies include photography & painting. *Society Aff:* ASTM, ASM, IMS, ISS.

Vander Woude, Jack C
Home: 7 Bramlette Place, Longview, TX 75601
Position: Retired *Employer:* Texas Eastman Co. *Education:* BS/Chem Engg/IL Inst of Tech; AB/Chemistry/Calvin College. *Born:* 10/27/18. Mbr Tau Beta Pi & Phi Lambda Upsilon. Native of Chicago Ill. Joined Tenn Eastman Co, Div of Eastman Kodak Co in June 1941 as a Chem Engr. Served in US Navy May 1944-Aug 1946 as Deck Officer & Navigator. Transferred to Texas Eastman Co Aug 1950, appointed Vice Pres & Works Mgr Aug 1970 & Pres July 1975. Retired November, 1983. Enjoy symphonies, semi-classical music, travel, & golf. *Society Aff:* ACS, AIChE, AXΣ, TBΠ.

van der Ziel, Aldert
Home: 4405 Grimes Ave So, Minneapolis, MN 55424
Position: Professor of Electrical Engineering. *Employer:* University of Minnesota. *Education:* PhD/Physics/Univ of Groningen; MA/Physics/Univ of Groningen; BA/Physics/Univ of Groningen. *Born:* Dec 1910 The Netherlands. Staff mbr Philips Res Lab, Eindhoven Netherlands 1934-47; Assoc Prof Physics, Univ of British Columbia Vancouver Canada 1947-50; Prof of Electrical Engrg, Univ of Minnesota 1950-1981; 1981-Professor Emeritus. Graduate Res Prof, Univ of Fla Gainesville Fla 1968- . Res Field: noise in electronic devices & materials. ASEE Western Electric Award 1967. ASEE Vincent Bendix Award1975- 1980 IEEE Education Medal; Honorary degree Eindhoven University of Technology Eindhoven The Netherlands 1981. Honorary Degree, Univ Paul Sabatier Toulouse France 1975. 1980 IEEE Education Medal; Fellow IEEE 1956. Member Nat. Academy Engineering, 1976. Interested in botany & theology. *Society Aff:* APS, IEEE, NAE.

Vanderbilt, M Daniel
Business: Civil Engrg Dept, Ft Collins, CO 80523
Position: Professor of Civil Engineering. *Employer:* Colorado State University. *Education:* PhD/Civil Engr/Univ of IL; MS/Civil Engr/Univ of IL; BSSE/Civil Engr/Univ of OH. *Born:* 2/9/34. Army Corps of Engrs 1957-59. Joined CSU Staff in 1963. Ctte serv: ACI 104, Chmn 1965-73; ACI 421, joint with ASCE (CCMS), Chair 83-85; ACI 442. ASCE Methods of Structural Analysis, Chair 83-86. Fellow ACI, Delmar L Bloem Distinguished Serv Award 1973. ASEE Resident in Engrg Practice 1974, Northern Colorado Construction Industries Award, 1986. Author of 1 book & numerous papers. *Society Aff:* ASCE, ACI.

Vanderboom, Steve A
Business: 3121 Nicollet Ave, Minneapolis, MN 55408
Position: Pres. *Employer:* PACE Labs, Inc. *Education:* MS/Civil Engg/Univ of MN; BS/Civil Engg/SD School of Mines & Tech. *Born:* 10/25/51. in Platte, SD. Have resided in Minneapolis since grad school. Primary interest is application of engg skills to solve environmental problems. After four yrs with consulting engg-lab firms, opened PACE Labs, Inc. As Pres and Dir of Engg at PACE (Professional Analyticl Chemistry & Engg), activities include mgt, promotion, and engg. Am 1979 Chrmn of MN Professional Engrs in Private Practice and Chrmn of MSPE Continuing Education committee. Was named Young Engr of the Yr in 1979 by MSPE. With wife Julie (RN), enjoy golf, swimming, professional activities, and travel whenever possible. *Society Aff:* NSPE, ASE, WPCF.

Van Domelen, John F
Business: Norwich Univ, Northfield, VT 05663
Position: VA Acad Aff & Faculty Dean *Employer:* Norwich Univ *Education:* PhD/CE/Univ of WI-Madison; MS/Water Resources Mgmt/Univ of WI-Madison; BS/Applied Physics/MI Tech Univ *Born:* 10/19/42 Raised in Falls Church, VA. Spent five yrs as a Regular USAF Officer in various electroptical research assignments and service in Viet-Nam. After separation spent one yr with Charmin Paper Products Co as an Engr-Mgr. Entered Grad Sch in 1970 specializing in the Quantification of Water Pollution Parameters using photographic remote-sensing tech-

Van Domelen, John F (Continued)
niques. Have been with Norwich Univ since 1974/Assoc Prof of Civil Engrg, 1980. Head of Engrg & Tech, 1979. VP Acad Aff & Faculty Dean, 1985. Pres, VT Section ASCE 1980-81. Recvd IEEE Centennial Medal - 1984. Research includes hydrological investigations and wastewater treatment & disposal. Mbr VT Air Natl Guard. Avid cross-country skier and racquetball player. *Society Aff:* ASCE, ASEE

VanDriesen, Roger P
Business: 1515 Broad St, Bloomfield, NJ 07003
Position: Dir, Refining Mktg *Employer:* Lummus Crest Inc. *Education:* BChE/Chem Engrg/Cooper Union. *Born:* Sept 1931. BChE Cooper Union. Native of Rockford, Ill. Proj Engr, Gas Equipment Engrg Co 1955-57; field assignment with US Air Force setting up trailer mounted hydrogen plants. With Cities Service R&D Co 1957-76. Responsible positions in the pilot plant & commercial dev of residual oil hydrogenation & hydrocracking process. Vice Pres, Commercial Dev for hydrogenation & synthetic fuels. Present position as Dir, Refining Marketing for Lummus Crest Inc. Enjoy canoeing, fishing & skiing. Holder of 24 patents; author of several tech papers. *Society Aff:* AIChE, LES.

Van Dusen, Harold A
Home: 1629 Drexel Blvd, South Milwaukee, WI 53172
Position: Senior Engineer. *Employer:* McGraw-Edison Co, Halo Lighting Div *Education:* BS/EE/Purdue. *Born:* Aug 1922. Bellingham, Wa. Served USAAF, Pacific Theater WWII. 37 years experience with McGraw-Edison Co. in research & prod dev outdoor lighting. Now consultant & have conducted res & written many papers on subjects dealing with interaction of various environmental effects on outdoor lighting equipment; ie wind induced pole vibration, thermal effects, dirt depreciation, breakage & degradation of glassware & plasticware, tce. Dev computer progs to design lighting installations. Married, 3 children. 17 patents. Reg PE. Fellow IES. Instructor in several lighting seminars. *Society Aff:* IES

Van Dyke, Milton D
Business: Applied Mechanics, Stanford, CA 94305
Position: Professor of Applied Mechanics. *Employer:* Stanford University. *Education:* PhD/Aeronautics/Caltech; BS/Engg Sci/Harvard Coll. *Born:* Aug 1, 1922 Chicago; son of James Richard & Ruth (Barr) Van Dyke. Married Sylvia Jean Agard Adams, June 16 1962; children, Russell B, Eric J, Nina A, Brooke A., Byron J., Christopher M.(triplets). Res Engr NACA (later NASA) 1943-46, 50-54, 55-58; Visiting Prof, Univ of Paris 1958-59; Prof of Aeronautics, Stanford 1959-75, Applied Mechanics 1975- . Guggenheim & Fulbright Fellow 1954-55. Mbr APS, Phi Beta Kappa, Natl Acad of Engrg, American Acad of Arts & Scis. Author: Perturbation Methods in Fluid Mechanics 1964; An Album of Fluid Motion 1982 Parabolic Press; Editor, Annual Review of Fluid Mechanics 1969- . *Society Aff:* APS, NAE, AAAS.

Van Dyke, Robert P
Home: 1019 Melissa Drive, San Antonio, TX 78213
Position: General Manager. *Employer:* City Water Board. *Education:* BS/Civil Engg/State Univ of IA; MS/Sanitary Engg/State Univ of IA. *Born:* 9/14/28. Reg PE in TX, IA, & CO. Col, USAFR (Ret); Hon V Cons of Bolivia in San Antonio 1968-1985. Amer Water Works Assn activities incl: Mgmt Div Trustee, Chmn, & P Chmn 1970-76; Mgt Div Representative to the Tech and Professional Council 1978-1982; V Chrmn of the Technical and Professional Council 1979-81; Bd/Dir 1973-75 and 1980-1981; Mbr Exec Ctte 1974-1975; 1972 TX Sect Utility Man of the Yr (Fuller Award); Chmn of Southwest Sect AWWA 1969-70; Elected Hon Mbr June 1977; Mbr General Policy Council 1982-1986; Chmn General Policy Council's AWWA Association Awards Committee 1982-1986; Mbr Water Utility Council 1986-present; Abel Wolman Fellow 1987; Distinguished Public Service Awardee 1985. Life mbr of the Reserve Offices Assoc of the US. Elected to ROA's Brigade of Volunteer for Outstanding Service to the Reserve Offices Assoc 1971. Mbr The Retired Officers Assoc 1982-present. *Society Aff:* AWWA, ASCE, ROA, ΘT, ΣΞ, TROA.

Van Echo, Andrew
Home: 8211 Jeb Stuart Rd, Rockville, MD 20854
Position: Sr Scientist & Metallurgical Engr, Manager Metallurgy, Absorbers and Standards *Employer:* US Dept of Energy. *Education:* BSc/Metallurical Engg/OH State Univ; Grad Course/Bus/Lewis Inst; Grad Course/Foreign Language/Montgomery Coll. *Born:* Jan 1918 Barton Ohio. R&D on metal vapor deposition for BMI 1941-42; on steel processing for US Steel Corp 1942-43; on uranuim & aluminum for Univ of Chicago, 1943-45 on Manhattan Proj., Joslyn Mfg & Supply Co 1945-63, bearing & foundry div & Tech Dir of stainless steel div; USAEC, now USERDA, now USDOE, Senior Scientist & Met Engr 1963- . Mbr Joint US-USSR Working Group on Electrometallurgy 1973-78 , Chmn ASTM Ctte A-10 1970-72; Deputy Chmn ASTM Ctte A-1 1972-73 & Chmn ASTM Ctte A-1 1973-80. Fellow ASTM; ASTM Award of Merit 1973; ASTM Honorary Award 1975; Tau Beta Pi. Pres Sigma Gamma Epsilon 1941-42. Contributing author: *Metallurgy of Uranium & Its Alloy & ASM Metals Handbook*, Vol II. Mbr ASM, ASM Chap Chmn 1957-58, ASM Dev Comm 1964-65. AEC-ERDA-DOE Liaison Rep on Natl Materials Adv Bd, *Refractory Metals Sheet Rolling panel, Panel on Vanadium*, 'Ctte on Electroslag Remelting & Plasma Arc Remelting," ctte on Fatigue Crack Initiation; & ctte on steel research." Mbr for U.S. to Intl Working Group on Fast Reactors (IWGFR) Specialists' Meeting on Reactor Absorber Materials, June 7-10, 1983, Obninsk, USSR. *Society Aff:* ASM, ASTM, ADPA.

VanGilst, Carl W
Business: SR 15, Milford, IN 46542
Position: Mgr of Engr *Employer:* Chore-Time Equip. *Education:* MSAE/Agri Engr/Purdue; BSAE/Agri Engr/IA State. *Born:* 11/9/40. I'm orginally from SE IA where my family operates a grain & hog farm. Employments include New Holland Mach Co, Jamesway, Star Agri-Products, & AG Best Corp. I was chief engr of Star Agri-Prod for six yrs & VP of AG Best Corp for two yrs. I currently manage the Engr Dept at Chore-Time. *Society Aff:* ASAE.

Van Gulik, Joe L
Business: 543 Third St, Lake Oswego, OR 97034
Position: President *Employer:* Van Gulik & Assoc *Education:* BE/Sci/EE/Netherlands-Endhoven *Born:* 04/28/26 Founder of Van Gulik & Assocs in 1970. His responsibility besides overseeing corporate operations, encompasses the very rapidly growing concern for the efficient use and control of electrical energy from a variety of sources. These aspects of electrical power utilization include generation, switching, control & distribution as well as efficient end use of the delivered product. Brings to this responsiblity over 25 yrs research and production efforts as both project & chief engr for a number of firms in the northwest. Prof engr reg in OR, WA, AK, ID, NV, and British Columbia. *Society Aff:* IEEE, AWS, ASM, NSPE, SAE

Van Gundy, Richard G
Business: P O Box 4, Ellsworth, KS 67439
Position: Owner. *Employer:* Van Gundy & Associates. *Born:* Feb 14, 1926 Capron Okla. Grad Moline H S Moline Kansas 1944; attended Washburn Municipal Univ Topeka Kansas 1944-46; BSCE 1949 Univ of Kansas Lawrence Kansas, Hwy Materials Sch I & II; Communications Personal & Group (Kansas Univ extension) 16 hrs. Licensed P E Kansas & Oklahoma, licensed Land Surveyor Kansas. Mbr APWA, Mbr NSPE, KES, Pres 1976-1977 Kansas Soc. Kans Natl Dir in NSPE 1949-51 Party Chief State Hwy Comm; 1951-66 Resident Engr State Hwy Comm Pratt Kansas; 1966-67 Design Engr Foster & Company Engrs Ellsworth Kansas, 1968- 69 Design Engr & 1/2 yr as Mgr; 1968-75, Owner & Mgr Van Foster-Van Gundy & Assoc Ellsworth Kansas; Owner & Mgr Van Gundy & Assoc 1976- .

VanHaaren, Frans I
Business: Enka, NC 28728
Position: Tech VP. *Employer:* American Enka Co. *Education:* Masters/Process Dev/Tech Univ Delfl, Netherlands. *Born:* 11/16/35. Product and process dev experience was obtained during a six-year res period in Akzo's Central Res Inst in Arhem, The Netherlands. Subsequently I have obtained experience in chem engg, control engg, computer process control, and facilities engg as asstdir of Akzo's Central Engg Group. In the current function as Tech VP of American Enka Co, my responsibilities are - mgt of the Res, Central Engg and the Marketing Dev Depts. *Society Aff:* AIChE, ACS, KIVI.

Van Hare, George F, Jr
Business: Chemical Group Eng'g Dept, Wayne, NJ 07470
Position: Chief Materials Handling Engineer. *Employer:* American Cyanamid Company. *Education:* BChE/Chem Engg/Pratt Inst. *Born:* 10/08/23 Aircraft Armorer USAF 1942-45. proj Engr 1948-56 Chem Const Corp Sr Proj Engr 1956-67 Amer Cyanamid Co respon for chem process plant design, 1967-present has respon for design of matls handling equipment for all new & expanded co plants. PPres NY Chap Internal Matl Mgmt Soc. P Internatl VP Mbrship IMMS, now Internatl Dir IMMS. Mbr: Tau Beta Pi, PCMH. P E in NY and NJ. Assoc Mbr AIChE, Mbr Fairfield County Sect AIChE. Hobbies: golf, fishing & hunting. *Society Aff:* IMMS, AIChE.

Van Horn, Kent R
Home: 373 Fox Chapel Road, Pittsburgh, PA 15238
Position: Vice President Res & Dev (retired). *Employer:* Aluminum Company of America. *Education:* BS/Metallurgy/Case Western Univ; MS/Metallurghy/Yale Univ; PhD/Metallurgy/Yale Univ. *Born:* July 20, 1905. Honorary DSc Case Western Reserve Univ; Has held offices of Dir of Res & V P R&D. ASM Honorary Mbr, Fellow, Gold Medal, Pres, Cambell Lecture; SNDT Honorary Mbr, Fellow, Pres, Mehl Honor Lecture; AIME Fellow, Inst of Metals Lecture. The Metals Soc London The Platinum Medal. St Cair Medal Metallurgie de la Francaise. *Society Aff:* ASM, NDT, AIMME.

Van Horn, Lloyd D
Business: 950 Threadneedle, Suite 200, Houston, TX 77079
Position: VP. *Employer:* Setpoint, Inc. *Education:* PhD/ChE/Rice Univ; BA/ChE/Rice Univ *Born:* 3/25/38. Currently VP & Dir for Setpoint, Inc, an engg consltg co in the fields of process automation & process comp control. Was a founder of Setpoint, Inc in Feb, 1977. Have consulted in process comp control since 1972, primarily in the petrl refining & petrochems mfg areas. From 1966-72, worked for Shell Oil Co in the Houston Refinery & in Hd Office in process dev & mfg operations. *Society Aff:* AIChE, ISA

Van Houten, Leonard E
Home: 314 North Maple Ave, Greenwich, CT 06830
Position: Proprietor *Employer:* L.E. VanHouten and Associates *Education:* M/CE/RPI; B/CE/RPI. *Born:* 12/20/24 Born in state of NJ. Served in USNR in World War II. Married Marie Regan 1947. Two children - Elizabeth M and Peter J. Started career with Frederic R. Harris, Inc, consltg engrs in 1947. Left as Sr. VP in 1965 to form Van Houten Assocs, Inc. Merged with Dravo Corp in 1976 and served in corporate headquarters in Pittsburgh, PA as a VP and Chrmn and CEO of Dravo Van Houten, Inc. Left Dravo in 1982 to join Parsons Brinckerhoff organization as VP. Formed Parsons Brinckerhoff Van Houten Div. Left Parsons Brinkerhoff in 1986 and formed L.E. VanHouten and Associates, mgmnt and engrg consultants. Enjoy sailing, skiing and the arts. *Society Aff:* NSPE, ASCE, SNAME, USCOLD, ACI, AWS, SAME, PIANC, NWF, SPE of AIME.

VanKirk, Craig W
Home: 15080 W 77th Dr, Golden, CO 80401
Position: Assoc Prof. *Employer:* CO School of Mines. *Education:* PhD/Pet Engr/CO School of Mines; MS/Pet Engr/Univ of S CA; BS/Pet Engr/Univ of S CA. *Born:* 8/4/45. Native of S CA. Resident of Denver, CO since 1969. All education in petroleum engg, receiving degrees from the Univ of S CA and CO School of Mones. Early industry experience with Exxon and Shell Oil Co in field operations, production engg, formation education, and reservoir engg. Intl petroleum consultant since 1974 specializing in reservoir studies, petroleum property economic evaluations, and teaching short courses. Joined staff of Petroleum Engg Dept at the CO School of Mines in 1978. Author of technical articles and papers on petroleum engg. Active in SPE of AIME. Enjoys backpacking and fishing. *Society Aff:* AIME, NSPE, PEC.

vanLangendonck, Telemaco
Business: Rua Avare 497, Sao Paulo 01243, Brazil
Position: Director. *Employer:* THEMAG Engenharia. *Education:* Prof/Univ Sao Paulo; Doutor em Arguitetura/Faculdade de Arquitetura USP; Doutor em Ciencias Fis Matem/Escola Politecnica de S Parlo Engeheiro Civil/Escola Politecnica de S Parlo. *Born:* April 2, 1909 Bage Brazil. Civil Engr 1931, Dr Physical Sciences 1942 Polytechnical Sch of Sao Paulo. Prof Sao Paulo Univ 1942- ; Res Engr ABCP (Brazilian PCA) 1936- ; Dir THEMAG cons engrs 1961- . Awards: Academia Nacional Ciencias EFN Argentina, Academie Royale Sciences OM Belgium 1969, Investigador Honorario LNEC Portugal 1972, Engr of the Yr Brazil 1973, Honorary Mbr ACI USA 1974. Has written many books on Structural Engrg & Reinforced Concrete. Presently V P ABPE (Brazilian Assn Bridges & Structural Engrg) & Dir IBRACON (Brazilian Concrete Inst) & Brazilian Delegate to CEB (Comite Eurointernatioanl du Beton) & Mbr of Permanent Committee IABSE (Internatl Assoc for Bridge & Structural Engg).. *Society Aff:* IESP, ABPE, ABNT, IBRACON, ABMS

Van Lint, Victor A J
Home: 1032 Skylark Dr, La Jolla, CA 92037
Position: Manager-Experimental Physics Division *Employer:* Mission Res Corp. *Education:* PhD/Phisics/CA Inst of Tech; BS/Physics/CA Inst of Tech *Born:* 5/10/28. in Indonesia. Lived in CA, Netherlands, NM prior to grad. Instr, Princeton Univ 1954-55. Enlisted man US Army 1955-57, served at Sandia Base. Technical Staff Gen Atomic & Gulf Gen Atomic 1957-1973, becoming VP of Gulf Energy & Environmental Systems and mgr of Rad Tech Div in 1970. Pres Intelcom Rad Tech 1973-74. Consultant 1974-75. Mbr of Technical Staff at Mission Res Corp 1975-present. Mbr of MRC Bd of Dirs 1976-78. Leave of absence to serve as Special Asst, to Deputy Director Science & Technology, Defense Nuclear Agency 1982-83. *Society Aff:* IEEE, APS, ΣΞ.

Van Manen, J D
Business: Haagsteeg 2-POBox 28, Wageningen, The Netherlands
Position: President. *Employer:* Netherlands Ship Model Basin. *Born:* Feb 27, 1923. MSc Naval Architecture 1949; PhD 1951 Univ of Technology Delft. Pres 1972- ; Netherlands Ship Model Basin, a scientific indus serv lab for shipbuilding, shipping & ocean engrg in Wageningen 1948- . Extraordinary Prof in resistance & propulsion of ships Univ of Technology Delft 1962- . SNAME President's Award 1952, Gold Medal of Royal Institution of Engrs in Netherlands 1972. Fellow Royal Inst of Navigation 1975, Honorary Mbr SNAME 1976. Mbr of the Royal Netherlands Academy of Scis, 1978. Ridder in de Orde van de Nederlandse Leeuw, 1979, David W. Taylor Medal 1984 (SNAME).

Vanmarcke, Erik H
Home: 27 Forty Acres Drive, Wayland, MA 01778
Position: Prof of Civil Engg. *Employer:* Massachusetts Inst of Technology. *Education:* BS/Civil Engg/Univ of Louvain; Ms/Civil Engg/Univ of DE; PhD/Civil Engg/MIT. *Born:* Aug 1941. Native of Belgium. Joined MIT faculty 1969; 1974-77, holder of Gilbert W Winslow Career Dev Professorship MIT. Res, teaching & cons in earthquake engrg, dam safety & reliability analysis applied to soils & structures. ASCE Raymond C Reese Res Prize 1975. Editor-in-chief. Journal of Structural Safety *Society Aff:* ASCE, EDRI, USCOLD.

Van Ness, Hendrick C
Business: Ricketts Building, Troy, NY 12180-3590
Position: Institute Professor of Chemical Engg. *Employer:* Renssalaer Polytechnic Institute. *Education:* DEngr/ChE/Yale Univ; MS/ChE/Univ of Rochester; BS/ChE/ Univ. of Rochester. *Born:* 1/18/24. Earlier Education public sch Greenwich NY. Instructor Univ of Rochester 1945-47; Asst Prof Purdue Univ 1952-56; with RPI 1956-. 1958-59 Fulbright Lecturer King's College Newcastle upon tyne Eng, 1966 Visiting Prof Univ of CA/Berkeley, Institutet for Kemiteknik, 1977 Danmarks Tekniske Hojskole, Lyngby, Denmark. Fellow AIChE. Mbr ACS, Sigma Xi, Tau Beta Pi. Co-author:(with J.M. Smith) *Introduction to Chemical Engineering* Thermodynamics, 1959, 1975, 1987; (with M.W. Zemansky and M.M. Abbott) *Basic Engineering Thermodynamics,*1966, 1975; (with M.M. Abbott) *Schaum's Outline of Theory & Problems of Thermodynamics,* 1972; (with M.M. Abbott) *Classical Thermodynamics of Nonelectrolyte Solutions: With Applications to Phase Equilibria,* 1982. *Society Aff:* AIChE, ACS, ΣΞ, ΤΒΠ.

Vanoni, Vito A
Business: Caltech, Pasadena, CA 91125
Position: Ret. (Prof of Hydraulics) *Education:* PhD/Civil Engg/CA Inst of Tech; MS/ Civil Engg/CA Inst of Tech; BS/Civil Engg/CA Inst of Tech. *Born:* Aug 1904. Res Proj Mgr US Soil Conservation Serv 1938-47; faculty Caltech 1940-74, Emeritus 1974; Res & cons in sedimentation, hydraulic structures & coastal engrg; Cons to US Army Corps of Engrs, US Soil Conservation Serv & others. Taught special courses in Chile 1962 & Venezuela 1967, 1971-75. Honorary Mbr: ASCE, Sigma Xi, Internatl Assn for Hydraulic Res, Tau Beta Pi. Natl Acad of Engg. Karl Emil Hilgard Prize 1976, Naval Ordinance Dev Award 1945, Outstanding Civilian Serv Medal US Army 1974. Hobbies: travel, rivers & gardening. *Society Aff:* ASCE, IAHR, ΣΞ.

Van Pelt, Richard H
Home: 915 Birchwood Drive, Washington, IL 61571
Position: Div Mgr/Res Metallurgy. *Employer:* Caterpillar Tractor Co. *Education:* BS/Met Engg/Univ of IL. *Born:* April 11, 1922. Served in USN as Engrg Officer on USS Wasp 1943-46. 1946- , Caterpillar Tractor Co, currently Div Mgr of Res Metallurgy Div. Fellow ASM, served on ASM Bd of Trustees 1973-76. Hobbies: stamp collecting & lapidary work. *Society Aff:* ASM, AWS.

Van Raalte, John A
Business: CN 5300, Princeton, NJ 08543-5300
Position: Director *Employer:* David Sarnoff Res Ctr. *Education:* Ph.D/EE/Sol.St.Phys./MIT; Engineer's degree/EE/MIT; MS/EE/MIT; BS/EE/MIT *Born:* 04/10/38 Past director Display Systems Research, RCA. Labs; Currently Director Materials & Process Technology - David Sarnoff Research Center. Currently Pres Soc. for Information Display-Past VP, Treasurer, Secr. SID; Past general Chairman & Program Chairman SID International Symposia; Past Chairman MIT Educational Council of Northern NJ; Past Director MIT Club of Princeton. Fellow SID. Fellow IEEE. Past Program Chairman Electro '85. Recipient RCA Outstanding Achievement Award 1971. *Society Aff:* IEEE, SID, ΣΞ, HKN, ΤΒΠ

van Reenen, Willem J
Business: 105 Technology Pkwy, Norcross, GA 30092
Position: Vice President. *Employer:* Atlanta Testing & Engrg Company. *Born:* July 17, 1941. BSCE Lamar Univ; MSCE Ga Tech. Staff Engr & Dept Mgr Law Engrg Testing Co Atlanta Ga 1965-69; Design Engr H J Nederhorst Const Co oGouda Netherlands 1969-70; V P in charge of engrg Atlanta Testing & Engrg Co 1970- . Mbr: Chi Epsilon,, ASCE, Amer Cons Engrs Council; Treas & Mbr Exec Ctte Assn of Soil & Foundation Engrs. Hobbies: tennis, fishing & gardening.

Van Riper, Peter K
Home: 1311 Holly Leaf, Diamond Bar, CA 91765
Position: President *Employer:* West Coast Design, Inc. *Education:* BSCE/Civil Engg/ Univ of CT. *Born:* 9/3/31 Lt USAF 1954-56 Military Const. Experienced in aeronautical/aerospace engrg & cons civil engrg. Pres. West Coast Design, Inc. spec in public works planning & indus land dev progs. Professional background in development of major industrial/commercial office park developments in U.S. and Asian countries. Pres SAME 1972, Mbr Bd/Dir; ASCE legislative delegate 1974-84; Mbr: NCEE, APWA, ASCE, CEAC. Reg Civil & Mech Engr Calif, Reg P E Hawaii, Fla, Utah & Colorado. Hobbies: flying & tennis, Golf. *Society Aff:* ASCE, NCEE, SAME, APWA.

Vansant, Robert E
Home: 435 East 55 Street, Kansas City, MO 64110
Position: Partner and Proj Mgr/Civil-Environ Div. *Employer:* Black & Veatch, Engrs - Architects *Education:* JD/Law/U of MO - Kansas City; CE/Civ Engrg/U of MO - Rolla; BSCE/Civ Engrg/U of MO - Rolla. *Born:* 2/10/30. in Clinton, Active duty US Army Corps of Engrs 1951-53, Reserve Officer 1951-1981, Black & Veatch 1953-present as Design Engr, Specif Engr, Engr- Attorney, Hd Specifications & Standards Sect, Asst Engrg Mgr, Proj Mgr & Partner, Young Engr of the Yr 1965 Western Chap of MO Soc of Prof Engrs. Natl Pres 1973-74 Const specifications Inst. Author of *Vansant's Law* formerly regular column. Chmn AWWA Steel Tank Ctte. Mbr Engrs Joint Contract Documents Ctte representing CSI. Particular interest in const contract law, const contract mgmt, engr's prof liability. *Society Aff:* NSPE, ASCE, CSI, AWWA, ACI, WPCF, SAME, ABA, NACE, SAVE.

van Schilfgaarde, Jan
Home: 2625 Redwing Rd, Suite 350, Ft Collins, CO 80526
Position: Director/Mountain States Area. *Employer:* Agricultural Research Service/ USDA. *Education:* PhD/Agri Eng & Soil Physics/IA State; MS/Agri Eng/IA State; BS/Agri Eng/IA State. *Born:* 2/7/29. Naturalized US citizen. Asst, Assoc, then Full Prof Dept Agri Engrg NC State Univ 1954-64; Water Mgmt Specialist, Assoc Dir, then Dir Soil & Water Conservation Res Div Agri Res Serv Beltsville MD 1964-72; Dir US Salinity Lab USDA Riverside CA 1972-84; Dir Mountain States Area, Agri Res Serv Ft Collins CO 1984- , primary interest in drainage engrg & physics of water flow through soils. Fellow, P Chmn Soil & Water Div ASAE; mbr, P Chmn Res Ctte Irrig & Drainage Div ASCE; Fellow ASA. Mbr Soil Conservation Soc of Amer. Editor ASA monography "Drainage for Agri." Walter Huber Civil Engrg Prize ASCE 1970. Paper Awards ASAE 1964 & 1965, John Deere Gold Medal ASAE 1977. Royce Tipton Award ASCE, 1986. *Society Aff:* ASAE, ASCE, ASA, SCSA, SSSA, USCID, ISSS

Van Sciver, Steven W
Business: 921 ERB, 1500 Johnson Dr, Madison, WI 53706
Position: Prof *Employer:* Nuclear Engrg Dept, Univ of WI *Education:* PhD/Physics/Univ of WA; MS/Physics/Univ of WA; BS/Eng Physics/Lehigh Univ *Born:* 03/13/48 Dr. Van Sciver is a Professor in the Nuclear Engrg and Engrg Physics Dept, Univ Wisconsin-Madison. He is also Associate Director of the Applied Superconductivity Center. He received his B.S. degree (summa cum laude) in engrg physics, Lehigh Univ, 1970. His M.S. and Ph.D. degrees are in low temperature condensed matter physics from the Univ of Washington. Since 1976, Dr. Van Sciver has been at the Univ of Wisconsin in cryogenic research associated with superconducting magnet technology and space cryogenic systems. His current interests include: experimental studies of the heat transfer and thermal stability of composite superconductors in He II, and investigations of forced flow liquid helium for space cryogenic systems and large scale accelerator technology. He is author of one book *Helium Cryogenics,* Plenum Publishing 1986, and 90 articles on low temperature physics and engineering. *Society Aff:* IEEE, APS, AAAS

van Straten, Gerrit S
Business: Consulting Engineers, 1460 W. Lane Ave, Columbus, OH 43221
Position: President. *Employer:* Van Straten Engg. *Education:* BSME/Mech Engrg/OH State U. *Born:* July 1932 Baltimore Md. Served in Army 1954-56. Proj Engr N Amer Aviation 1961-66, spec in production engrg. Developed new titanium forming

van Straten, Gerrit S (Continued)
process. Practiced P E Van Straten Engrs 1966-70; Partner Van Straten & Edwards Engrs 1970-78. Pres Gerrit S. Van Straten, Inc., Consulting Engrs. Developed new mech sys for bldg energy conservation & environ control. Reg P E nationally & specifically in 22 states. Mbr: ASHRAE (Regional Energy Chmn 1975, Chap Pres 1980). Mbr Ohio Governor's Task Force for Energy Conservation, helped author Ohio Energy Code. Mbr: ASPE (Chap. Vice pres. 1986-87). *Society Aff:* ASHRAE, ASPE, AEE, ACEC.

Van Sweringen, Raymond A, Jr
Business: 1600 E. Linden Ave, P. O. Box 51, Linden, NJ 07036
Position: Engr Advisor *Employer:* Exxon Research & Engrg Co *Education:* B/ME/Cornell Univ; BS/ME/Cornell Univ *Born:* 04/01/22 Native Cleveland, OH. Served with Army Ordnance Dept 1943-46 as an artillery Ordnance Sch instructor and in 1951-52 as guided missile liaison ofcr to Signal Corp. With Exxon Res & Engrg Co since 1947. Assignments included research in combustion for turbo and ram jets, burner fuels and supervisor of instrumentation and research equipment design sections. Currently Engrg Advisor responsible for pilot plant tech improvements related to process design, equipment evaluation and computer sys applications. Outside interests: white water canoeing, sailing and camping. *Society Aff:* ISA, AIChE

Van Thyne, Ray J
Home: 1070 Valley Lake Dr, Inverness, IL 60067
Position: Dir of Tech Dev *Employer:* Packer Engrg & Surface Tech Corp *Education:* MS/Met Eng/Ill Inst of Tech; BS/Met Eng/Ill Inst of Tech. *Born:* 05/31/27. Served with Naval Air Corp. BS & MS from IIT. Asst Dir Metals Res Div at IIT Res Inst; 1950-66, Pres Surface Tech Corp; 1973- , Pres Surfalloy Corp. Former Chmn AEC Metallographic Grp & representative to Geneva 'Atoms for Peace' Conference. Pub 35 tech papers, 35 US patents & 75 foreign patents. 1973 ASM Coatings Award. 1977 IR 100 Award. Fellow, Amer Soc for Metals. Enjoy water sports *Society Aff:* ASM, AIME, SME, LES.

Van Tine, Charles L
Home: 13133 N. Caroline, Chillicothe, IL 61523
Position: Plant Engineering Manager *Employer:* Caterpillar, Inc. *Education:* BSC/Civil Engineer/Iowa State University *Born:* 12/14/34 Civil Engrg graduate from Iowa State Univ. Thirty-one (31) years with Caterpillar, Inc. in various Plant and Facility Engineering positions. Active in pollution control, energy conservation, and development of computerized maintenance programs. Principal outside activity with the Amer Inst of Plant Engrs. Former Chapter Presidents of the Peoria, IL and San Francisco, CA Chapters - also has held positions as Group Dir of Regional VP, and is currently National Pres of AIPE (1988). *Society Aff:* AIPE

Van Valkenburg, Mac E
Business: College of Engineering, 1308 West Green Street, Urbana, IL 61801-2900
Position: W.W. Grainger Professor of Electrical Engineering; Dean, Coll Engrg *Employer:* University of Illinois. *Education:* PhD/Elec Engg/Stanford; MS/Elec Engg/MA Inst Tech; BS/Elec Engg/Univ of UT. *Born:* Oct 5 1921, Union Utah. Mbr of faculty at Univ of Utah, Stanford Univ, Princeton Univ, Univ of Ill. Visiting appointments at Univ of Colo, Univ of Calif at Berkeley. Univ of HI at Manoa Univ of Arizona. ASEE George Westinghouse Award 1963, IEEE Education Medal 1972. ASEE Lamme Award 1978 Guillemin Prize, Guillemin Foundation, 1978 Univ of IL Halliburton Engg Education Leadership Award 1979. 1987 IEEE Circuits and Systems Award, V Pres & Bd/Dir IEEE, formerly Editor of 'Proceedings of the IEEE' & 'IEEE Transactions on Circuit Theory', formerly Editor in Chief, IEEE Press. *Society Aff:* IEEE, ASEE.

VanValkenburgh, Kent
Business: Box 470346, Tulsa, OK 74147
Position: Pres *Employer:* Vanco Engg Co. *Education:* BA/Chemistry/Westminster College. *Born:* 1/14/42. Sales engr & owner-partner of Mfg's Rep firm, engaged in design & sales of mech process equip to many different industries including CPI, food processing, waste treatment, mineral processing, rubber & plastics, & petrol. Active on bds of church, YMCA, hospital, & local profl organizations. *Society Aff:* AIChE, WPCF.

van Vessem, Jan C
Home: 19 Lissevenlaan, Waalre, Holland 5582 KB
Position: Industrial Consultant *Education:* PhD/Chemistry & Physics/Utrecht Univ; MSc/-/Utrecht Univ; BSc/Chemistry & Physics/Utrecht Univ. *Born:* Nov 1920, Den Helder Holland. Since 1965 Fellow IEEE. Taught in college during WWII; 1947 joined the Philips Co, after work in Radio Tube labs became Ch Engr for design of solid state products & was respon for 23 years for all product design work of diodes, transistors & integrated curcuits, since 1974 Tech Dir Electronic Components & Materials Div N V Philips Gloeilampenfabrieken Eindhoven, Holland, respon for Strategic Prod Dev & Dev Coordination. After mandatory retirement from Philips Co since 1981 Industrial Consultant in the field of R&D Management and micro electronics. Hobbies: painting, graphical arts & jazz. Since 1982 also mbr Supervisory Bd of M.E.C.O. BV., 's Hertogenbosch, a company specialising in metalplating components for the electronic industry and supplying plating equipment, and member Advisory Board of HYMEC, B.V., in SITTARD, a company specializing in hybrid integrated circuits. *Society Aff:* IEEE.

Van Vlack, Lawrence H
Business: Univ of Mich, Dept Materials Science & Engineering, Ann Arbor, MI 48109-2136
Position: Professor. *Employer:* The University of Michigan. *Education:* PhD/Geology/Univ of Chicago; BS/Ceramic Engrg/Iowa State Univ. *Born:* July 1920. Served in Navy 1945-46; 1942-43 with US Steel Corp as a Ceramist, 1943-52 Petrographer, 1952-53 process Metallurgist; 1953 joined Univ of Mich staff as Assoc Prof of Engrg & advanced to Professor in 1958, 1967-73 Chmn of Dept; serves as cons for various companies. Author of 11 books & over 150 tech papers. Specific interest in material sci & engrg education. Fellow 1964 A Cer S, Fellow 1971 ASM; Fellow 1980 AAAS Gold Medal 1984 ASM: A.E. White Distinguished Award 1985 ASM. *Society Aff:* ACS, ASM, AIME, ASEE, NYAS, ASTM, NICE

Van Vliet, Pieter
Business: President's Office, University of Regina, Regina, Sask, Canada S4S0A2
Position: Special Assistant to the President *Employer:* University of Regina *Education:* BSc/ME/Univ of Saskatchewan-Canada; Eng Dip/ME/HTS-Holland *Born:* 12/03/30 Born and educated in Holland. Served with Dutch Armed Forces 1953. Emigrated to Canada 1953. Mech Engr with White Motor Co 1953-1955. With SaskTel since 1955. Held various engrg and mgmt positions, including Chief Engr 1975-1979. From 1980-1981, VP responsible for design, provisioning and operation of Network Serv, including Oper and Directory Serv. Also responsible for implementing SaskTel's Fibre Optic Broadband Network Prog. 1981-1986, VP Cust Serv, responsible for Planning and Dev, Mktg and Cust Serv. Chrmn Res - 1969, Pres, APES, 1978, Pres, Canadian Council of PE, 1980-81. Chrmn Saskatchewan Engrg Adv Council to Univ of Saskatchewan and Univ of Regina. Fellow of Engrg Inst of Canada, 1987, Sr. V.P. Eng. Inst. of C'DA 1981 Dist Award, Assoc of P. Eng of Saskatchewan. Enjoys filming, skiing, gliding and soaring. Current: Special Asst to the Pres of the Univ. of Regina, Respons for R&D Inst for fibreoptics and intelligent buildings network technologies. *Society Aff:* APES, EIC, CSME

Van Vorst, William D
Home: 751 Enchanted Way, Pacific Palisades, CA 90272
Position: Prof. Emeritus *Employer:* UCLA. *Education:* PhD/Engg/UCLA; SM/ChE/ MIT; ChE/ChE/Rice; BS/ChE/Rice. *Born:* 8/20/19. Early experience in WWII Chem Warfare res followed by indust R&D in Combustion directed toward the design of jet propulsion and gas turbine engines. Remained with UCLA after obtaining PhD, where he has developed several of the courses in chem engg and applied thermody-

Van Vorst, William D (Continued)
namics. While at UCLA, he has also served as a conslt to several industrial firms, largely in the areas of combusion and engg economy evaluation. He has also served on several intl projects, largely devoted to engrg curricula dev. Countries in which he has had considerable experience are Indonesia, the Philippines, Thailand and Turkey. Research interests include: alternate energy sources and fuels, energy generation from waste, and hazardous waste. *Society Aff:* AAAS, AIChE, ASEE, IAHE, NSPE, TBΠ

Van Weele, David P
Business: 3 Lazare Ln, Islip, NY 11751
Position: Owner. *Employer:* Peter J VanWeele & Co. *Education:* BCE/Civ Engg/Rensselaer Poly Inst. *Born:* 7/30/41. Native of Long Island, NY. Worked in his fathers surveying office prior to attending Rensselaer Poly Inst. Grad from RPI in 1963. Entered Commissioned Corp of the US Coast and Geodetic Survey for two yrs. In 1965 rejoined firm of Peter J VanWeele & Co specialists in reality subdiv design. Became a partner of firm in 1972 and in 1976 became sole owner. Has testified as an expert witness on Land Surveyig in Fed, State and local courts as well as conducted seminars on the use of desktop computers. Interests include travel, restoration of old automobiles. *Society Aff:* ASCE, ASP, ACSM, NSPE, NYSPLAS.

Van Winkle, Stephen N
Home: 6205 Post Oak Rd, Peoria, IL 61615
Position: District Planning Eng *Employer:* Il Dept of Transportation *Education:* M/ME/Hwy & Traffic Eng/TX A&M Univ; BS/CE/TX A&M Univ *Born:* 04/24/39 Native of Peoria, IL. Conducted transportation related research (TX Transportation Inst) 1961-62. Officer, Air Defense Artillary USAR 1962-64 (ground to air guided missiles) with IDOT since November 1964. Served as District Traffic Engr 1974-76. Assumed District Planning Engr position December 1976. Responsible for the Programming, Design, Environmental impact analysis, and Public Relations for all roadway contracts constructed within District 4 (50,000,000 average annually). Named Engr of the Yr by IAHE (1974), Young Engr of the Yr by Peoria Area Chapter ISPE (1975). Received IL Award from Peoria Area Chapter ISPE (1980). Pres, IL Section ITE (1981). Fellow-Natl ITE. *Society Aff:* ITE

Van Wylen, Gordon J
Home: 92 E 10th St, Holland, MI 49423
Position: President. *Employer:* Hope College. *Education:* ScD/Mech Engr/MIT; MSE/Mech Engr/Univ of MI; BSE/Mech Engr/Univ of MI; AB/Lib Arts/Calvin College. *Born:* Feb 1920. 1942 worked with duPont; WWII Naval officer in submarine service; 1946-48 taught at Penn State; 1951-72 Univ of Mich, 1958-65 Chmn ME Dept, 1965-72 Dean Coll of Engrg; 1972- , Pres Hope Coll, Holland Mich. Author 'Thermodynamics' 1958, co-author with R E Sontag of 3 other thermodynamics books. Fellow ASME & AAAS. Active in church & community. *Society Aff:* ASME, AAAS.

Vanyo, James P
Home: 717 Sea Ranch Dr, Santa Barbara, CA 93109
Position: Prof. *Employer:* Univ of CA. *Education:* PhD/Engg (Appl Mechanics)/UCLA; JD/Law/Chase Law College; MA/Astronomy/UCLA; BSME/Mech Engg/WV Univ. *Born:* 1/29/28. Industrial experience in automation (7 yrs, Dayton, OH), aerospace engg (7 yrs, Remanco, Marquardt, and Litton, Los Angeles). NASA Doctoral Scholarship at UCLA. Since 1970 on faculty at UCSB. Maj interest is rotational mechanics especially rotating fluids applied to physics and engg. NASA grants to study rotating liquids in Space Shuttle experiments. NSF grant to study ancient Earth-Sun-Moon dynamics using data presented in fossil stromatolites. *Society Aff:* APS, AGU.

Varello, Paul J
Business: 6565 West Loop South, Bellaire, TX 77401
Position: VP - Div Mgr. *Employer:* Daniel International Corporation. *Education:* BE/Civil Eng/Villanova Univ. *Born:* Oct 9 1943, Rockaway Beach NY. Civil Engr for State of Calif, Dept of Water Resources on Aqueduct Proj; Office Engr & Proj Engr for Dravo-Atkinson- Groves on Carley V Porter Tunnel Proj, Gorman Calif; Proj Mgr Beartrap Access Structure Works; Engr for Dravo Corp, Bellevue Wash; Const Mgr for Daniel Internatl on Bath County Pumped Storage Proj for Va Elec & Power Co, Richmond Va, Proj Mgr on Merck Proj, Albany Ga, & Mgr of Projs Daniel Corporate Headquarters SC, & Div Mgr Houston Office of Daniel Int'l. Winner of AAP Greensfelder Const Prize 1971 ASCE. *Society Aff:* ASCE.

Varnerin, Lawrence J, Jr
Business: 600 Mountain Ave, Murray Hill, NJ 07974
Position: Department Head. *Employer:* AT&T Bell Laboratories. *Education:* SB/Physics/MIT; PhD/Physics/MIT. *Born:* 7/10/23. 1943-46 served with Army Ordnance; 1949-52 Sylvania Electric Products Electronics Div, Microwave TR and ATR Tubes; 1952-57 Westinghouse Res Labs, research in ultrahigh vacuum & surface physics, 1957-present Bell Tele Labs, dev of solid state matls & devices, presently Hd, Microwave Device Dept, responsible for dev of microwave low noise & high pwr GaAs FETs, reliability Ga AsICs, Served terms 1971-74, 1975-78, & 1980-83 on Administrative Committee of Magnetics Society; Fellow IEEE, APS. I am presently Assoc Editor, Journal Magnetism & Magnetic Materials. *Society Aff:* FMS.

Varrese, Francis R
Home: 258 Mt Pleasant Ave, Ambler, PA 19002
Position: Sr Principal Materials Engr *Employer:* Honeywell Process Control Div. *Education:* BS/Met Engg/Lehigh Univ. *Born:* 9/1/39. Ambler Penn. Previously employed by Pratt & Whitney Aircraft Co, Haynes Stellite, Div of Union Carbide, & Robert Wooler Co; Standard Pressed Steel Co, served in various met positions before assuming respon as Mgr of Materials Res. Active Mbr ASM for 28 years holding num chap offices, including Chairman, Philadelphia ASM & Mbr natl ASM Adv Tech Awareness Council & National Chapter & Membership Council. Alpha Sigma Mu & Allen S Quier Awards 1961, Young ASM Member Award 1974. Fellow of the American Soc for Metals. Presently employed at Honeywell. *Society Aff:* ASTM, ASM Int.

Vasatka, Richard
Business: 1011 Nicollet Mall, Minneapolis, MN 55403
Position: Pres. *Employer:* Setter, Leach & Lindstrom, Inc. *Education:* BCE/CE/Univ of MN; BS/CE/Univ of MN. *Born:* 8/11/30. and educated in Minneapolis-St Paul, grad Univ of MN, then served in US Army Corps of Engrs 1952-54. Struct Engr and VP, Magney, Tusler and Setter, 1955-1963. Pres, Vanguard Engrs until 1967 when this consulting engg firm merged with Setter, Leach & Lindstrom, Architects and Engrs. Currently Pres and Chief Exec Officer of that firm. Pres NW Sec ASCE, 1963, Chrmn PEPP, Sec MN Soc of PE (NSPE) 1965, VChrmn, ASCE Underground Construction Res Council, 1968-1976. Currently Pres American Underground Space Assn, Chrmn Civil Engineering Dept. Advisory Council, Univ of MN. Reg PE in over 30 states. Listed in *Who's Who in Midwest*. *Society Aff:* ASCE, NSPE.

Vasquez, Numan H
Business: Panama Canal Comm, APO, Miami, FL 34011
Position: Chief Engineering Division *Employer:* Panama Canal Commission *Education:* MPA/Public Administration/Indiana University; BSEE/Electrical Engineer/ Purdue University *Born:* April 1934, Panama City Panama. BSEE Purdue Univ; MPA Ind Univ. With Panama Canal Co since 1954, assumed current respon in 1983. Sr. Mbr IEEE, Sect Chmn 1976; Mbr Canal Zone Soc of Prof Engrs, V Pres 1973-74, Pres 1975, Natl Dir 1976; Mbr Bd of Registration for Engrs & Architects of Panama 1975; Mbr Panama Soc of Engrs & Architects. Enjoy classical guitar music. *Society Aff:* IEEE, NSPE.

Vassell, Gregory S
Business: 180 East Broad St, Columbus, OH 43215
Position: Sr VP. *Employer:* Am Electric Power Service Corp. *Education:* DiplIng/EE/Technical Univ; MBA/Corp Finance/NYU. *Born:* 12/24/21. in Moscow, Russia. Joined Am Elec Power Service Corp in 1951 as Asst Engr. Became Hd of the High-Voltate Planning Sec in 1962, Chief Planning Engr - Bulk Power Supply in 1967, Asst VP - Bulk Power Supply Planning in 1968. Served as Mbr of the Fed Power Commission's Technical Advisory Committee on transmission in connection with the 1970 Natl Power Survey. Was elected VP - System Planning & a Dir in 1973 & Sr VP - System Planning in 1976. Elected Fellow of IEEE in 1978 "for contributions to the planning of reliable & economic electric power systems–. Elected Mbr of NAE in 1980 for contributions in the field of electric energy supply, both as to its technical factors and its societal implications. *Society Aff:* IEEE.

Vaudrey, Calvin
Business: 409 22nd Ave So/PO 298, Brookings, SD 57006
Position: President. *Employer:* Banner Assocs Inc. *Education:* PE/Civil Engg/Univ of WY; MS/Civil Engg/Univ of WY; BS/Civil Engg/Univ of WY. *Born:* April 1924, Wyoming. 1943-45 served in US Army Air Corps. Res PE in SD, Wyo & Minn. Taught Civil Engrg at SD State Coll 9 years; associated with present company 31 years, Pres since 1969, offices located in Brookings & Rapid City SD, Laramie & Cheyenne Wyo & Grand Junc CO. Chrmn of Bd of Horizons, Inc; Bd of Dir NWPS. Mbr ACEC, ASCE, NSPE, ACI, AWWA, WPCF; serve on sev adv bds; P Pres Greater SD Assn. Outstanding Engineer Award 1976. Principle hobby- golf. *Society Aff:* ACEC, ASCE, NSPE.

Vaughan, David H
Business: 311 Seitz Hall, Agricultural Engineering Dept, Blacksburg, VA 24061
Position: Assoc Prof, Agric Engr *Employer:* VA Polytechnic Inst & S. U. *Education:* PhD/Bio & Ag Eng/NCSU; MS/Bio & Ag Eng/NCSU; BS/Bio & Ag Eng/NCSU *Born:* 04/19/45 NDEA Fellow, Grad/Research Teaching Asst (1967-73). Major, US Army Reserve. Faculty responsibilities in teaching and research. Dept leader in energy research including conservation and alternate resources (solar, biomass, etc) for agriculture. Author of over 60 publications (18 refereed). Over 50 scientific presentations to natl audiences. Chrmn, Dept Grad Prog. Served as officer and on several state, regional, and natl committees of profl groups. Chrmn-SE Region ASAE, Natl ASAE Solar Energy and Student Paper Awards Committees. Assoc Editor, Transactions of ASAE. *Society Aff:* ASAE, ΣΞ, ΓΣΔ, AE, ROA

Vaughan, John M
Business: 151 N Delaware, Indianapolis, IN 46206
Position: Vice President Engineering. *Employer:* Inland Container Corporation. *Born:* BS Tulane; MS & PhD Inst of Paper Chemistry. 1955-58 Naval Aviator; since 1963 with Inland Container Corp, 1968 elected VP Engr & Environ Affairs. Active in civic, community & educational affairs. Dir Miami Univ Pulp & Paper Foundation; Mbr Ind State Environ Mgmt Bd, Ind Scientific Adv Bd, Indus Adv Bd to schools of Science & Engr at IUPUI, Bd/Dir Indianapolis Symphony, Mbr Tau Beta Pi, Alpha Chi Sigma, Chamber of Commerce, Indianapolis, Economic Club of Indianapolis. Current nominee for National Dir TAPPI. Elder 2nd Presbyterian Church.

Vaughan, Richard G
Business: 3700 Coors Rd NW, Albuquerque, NM 87120
Position: Owner *Employer:* Richard G. Vanghan & Assoc. *Education:* PhD/Civil Engg/Univ of IL; MS/Civil Engg/Univ of NM; BS/Civil Engg/Univ of MN. *Born:* Dec 1932. NSF Fellowship 1962-63. 1958-68 Asst Professor & Res Assoc UNM. Pres NM Sect ASCE 1970-71; Pres NMSPE 1975-76; ASCE Dir, District 15 1977-80. Active Reg in NM, Tex, Colo & Ariz. Chmn Environ Planning Comm 1966-72; Chmn Albuquerque City Comm 1973-74. Mbr ASCE & NSPE AD Hoc Visiting Ctte for ECPD 1976-81. Albuquerque Engineer of the Year 1976. Mbr GSA Region 7 Public Adv Panel on Architectural & Engrg Services 1975-76. Mbr Dean's Adv Council UNM 1975-80. Mbr Exec Comm Greater Albuquerque Chamber of Commerce. Married, 2 children. *Society Aff:* ACEC, ASCE, NSPE, ACI.

Vaughan, William W
Business: Univ of Alabama in Huntsville, Research Institute, Huntsville, AL 35899
Position: Associate Director, Research Institute *Employer:* Univ of Alabama in Huntsville *Education:* PhD/EngSci/Univ of TN; BS/MathPhySci/Univ of FL. *Born:* 9/7/30. Native of Auburndale, FL. Served with US Air Force 1951-55. R&D Meteorologist for Air Force Armament Ctr, Spec in test support instrumentation. Physicist with Army Ballistic Missile Agency working on engg design and mission analysis problems. With NASA's Marshall Space Flight Center since 1960. Assumed current resp for mgnt of space flight experiment dev and sci-engg aero-space vehicle dev interface proj. Chief atmospheric scis div 1976-86. Joined staff Univ of Alabama in Huntsville as Associate Director, Research Institute, in 1986- . Contr articles to profl jours. Recipient Exceptional Service medal NASA, 1971. Inst of Environmental Sciences, Simpson Tech Editors Award (1984). Mem Am Meteorol Soc AIAA (Losey Atmospheric Scis award 1980), Am Geophys Union, AAAS, Sigma Xi. Atmospheric Sci; Aerospace engrg and tech (3002). Current work: Dev and coordination of res initiatives; mgmt of multi-discipline res programs; mgmt of Res Inst. *Society Aff:* AIAA, AAAS, AMS, AGU, ΣΞ.

Vaughn, Kenneth J
Business: 9233 Ward Pkwy Suite 300, Kansas City, MO 64114
Position: President *Employer:* Larkin & Assoc. *Education:* MS/CE/Univ of KS; BS/CE/Univ of KS. *Born:* 3/20/36. Structural designer Wilson & Co Engrs, Salina KS 1957-60. Grad work in structures & instructor of CE, Univ of KS 1960-63. Assoc Engr Larkin & Assoc Consulting Engrs 1964-1966. Partner 1966, Pres. 1985. Primary responsibility for all structural design of bridges, treatment plants & fdns. State Chrmn Ethics & Practice Committee KS Engg Soc 1973. Young Engr of the Yr Eastern Chapter KES 1971. Pres Kansas City Engineers Club 1984. *Society Aff:* ASCE, NSPE, ASTM, APWA, ACI.

Vaughn, Robert L
Business: PO Box 3504. 0/10-04 B/101, Sunnyvale, CA 94086
Position: Director of Productivity *Employer:* Lockheed Missiles & Space Co., Inc. *Education:* MS/Admin Mgmt/CA State Univ-Northridge; BS/Mech Engg/IA State Univ-Ames. *Born:* 9/19/22. Currently Corporate Director, Productivity, Lockhead Missiles & Space Company, Inc, Sunnyvale, CA. with Lockhead-Corp since Jan 1953, Previous Chief Mfg Engr LMSC Space Systems Div, Prog Mgr-Advanced Structures-Composites Programs, Lockheed Calif Co. Lectured internatlly & at LCC, UCLA, & SWU. Authored 75 articles, books & tech papers; 5 US Letters Patents. MSBA Cal State Univ, BSME Iowa State Univ. Fellow ASME, IAE, SME, & Honorary Fellow, Institution of Production Engineers, MBR, ASM, NSPE & NMA. P Pres San Fernando Valley Engrs Council & Exec Ctte ASME/PE Div. Reg PE 2 disciplines in Calif. Certified Mfg Engr SME. 1981-82 Pres of SME; Dir, 1972-1983 of SME. Past Intl VP, Sect & Treas. of SME. Intl Gold Medal Award SME 1969; Seven awards from SFVEC, NSPE & SME for engrg achievements, Calif State Legislative Resolution Award, Calif State Poly Univ President's Recognition Award, Honorary Mbr, College & Univ Mfg Education Council, Iowa State Univ, College of Engrg. Professional Achievement Citation in Engrg, The 1983 AM Award "For Leadership in Developing Advanced Aerospace Manufacturing Techniques that reflect a commitment to both producibliity and efficiency–. NASA Space Shuttle Technology Award. *Society Aff:* SME, ASME, NSPE/CSPF, NMA, IAE, IPRODE

Vawter, R Glenn
Business: 1600 Broadway, Suite 1150, Denver, CO 80202
Position: VP. *Employer:* Tosco Corp. *Education:* Engr/Petrol Engr/CO Sch of Mines. *Born:* 8/10/38. Native of CO. After grad from college first worked in training & engg at Marathon Oil Co in oil production, exploration & gas processing assignments. In 1964 joined Tosco Corp & served in various proj engg & mgt positions

Vawter, R Glenn (Continued)
culminating in current position, VP, Oil Shale Operations. Engg work at Tosco was principally devoted to the res, dev, & commercial design of an oil shale retorting process-Tosco II. Reg engr in UT & CO. *Society Aff:* AIME, AIChE.

Veatch, Ralph, W, Jr
Home: 1231 East 21st Place, Tulsa, OK 74114
Position: Resch Supervisor *Employer:* Amoco Production Co *Education:* PhD/Eng Sci/Univ of Tulsa; MSc/Petro Eng/Univ of Tulsa; BSc/Petro Eng/Univ of Tulsa *Born:* 7/11/36 A registered Professional Engr in LA, OK, and TX. A Supervisor of Hydraulic Fracturing Research with Amoco Production Co 1976-prsnt; an Assoc Prof of Petro Engrg 1974-75; a Research Engr in Amoco's Drilling and Production Practices Section, 4 yrs; a Petro Engr with Amoco in drilling, producing, and enhanced recovery operations, 7 yrs; mbr of a 7-man Amoco tech team to China 1980. Industry service: Engrg Coordinator, Tight Gas Reservoirs Task Group, Nat'l Petro Council; mbr Industry Advisory Panel, DOE Western Tight Gas Sands, Multi-well Experiments 1980-prsnt; mbr/chrmn Gas Research Institute, Research Coordination Council, Gas Supply Panel 1984-prsnt; API Subcmtte on Evaluation of Well Completion Materials; chrmn API Tech Adv Cmtte on Rheology Research. For SPE: Distinguished Lecturer, "Massive Hydraulic Fracturing–; Education and Professionalism Cmtte; Well Completions Tech Cmtte; Steering Cmtte, SPE Forum on Earth Stresses; DeGolyer Distinguished Service Award Cmtte, Tech Coverage Cmtte, Distinguished Author; chrmn 1984 SPE/DOE Symposium on Enhanced Oil Recovery; Speakers Bureau; chrmn Hydraulic Fracturing Reprint Cmtte; Hydraulic Fracturing Monograph Cmtte; Local Section Office/Director 1981-prsnt. Professional Publications/Patents - 15. Pi Epsilon Tau. *Society Aff:* SPE, ΠΕΤ

Veazey, Sidney E
Business: Naval Systems Engineering Dept, Annapolis, MD 21402
Position: Dept Chairman *Employer:* US Naval Academy *Education:* PhD/Physics/Duke Univ; BS/EE/US Naval Acad *Born:* 09/18/37 Born Wilmington, NC 1937. Navy Nuclear Power & Submarine Schs. Duty on Destroyer and 3 Nuclear Submarines. Designated Engrg Duty Officer in 1971. Last assignment-Officer-in-Charge White Oak Laboratory and Deputy Commander for the Naval Surface Weapons Center at Silver Spring, MD. Currently Chrmn Naval Systems Engrg Dept, USNA, Annapolis, MD. Natl Council Mbr of ASNE & Natl Trustee of the USNA Alumni Assocation. *Society Aff:* ASNE, ASEE

Veenstra, H Robert
Home: 801 29th St, West Des Moines, IA 50265
Position: Chrmn *Employer:* Veenstra & Kimm, Inc *Education:* BS/CE/IA St Univ *Born:* 10/21/21 Born and educated in IA. Served in US Army from 1942-1946. Graduated IA State Univ in March, 1947. Employed by Stanley Engrg Co, Muscatine, IA, April, 1947 to June, 1961. Co-founded Veenstra & Kimm June, 1961. Assumed present position as Chrmn, April, 1980. Authored numerous articles. Awarded Anson Marston Award for outstanding service by IA Engrg Society, 1962. Distinguished Service Award by IA Engrg Society, 1984-85. Past Dir to NSPE, Past Pres IA Consltg Engrs Council, Fellow ACEC, ASCE, Mbr AWWA, WPCF, NSPE, APWA, ASCM & ASTM. *Society Aff:* ASCE, NSPE, ACSM, ASTM, ACEC, AWWA, WPCF, APWA

Veinott, Arthur F, Jr
Business: Dept Oper Res, Stanford, CA 94305
Position: Professor & Chairman. *Employer:* Stanford University. *Education:* Eng ScD/IE/Columbia Univ; BS/IE/Lehigh Univ; BA/Arts & Sci/Lehigh Univ. *Born:* 10/12/34. BA & BS Lehigh Univ 1936, Eng ScD Columbia Univ 1960. Stanford Univ 1962- 1967- , Prof of Operations Res, 1975- , Chmn Dept of Operations Res, Phi Beta Kappa, Tau Beta Pi, Fellow Inst of Mathematical Statistics, Guggenheim Fellow, VP Publ 1973-76 & Council Mbr 1971-73. Inst of Mgmt Scis. Founding Editor 1974- , *Mathematics of Operations Research*; Dept Editor 1969-73 & Assoc Editor 1963- 69 *Management Science*; Assoc Editor 1972-75 *Annals of Statistics*; editor & co-editor 3 books; author & co-author 30 tech papers. Res specialisties: inventory theory, dynamic programming. lattice programming & mathematical programming. *Society Aff:* TIMS, ORSA, IMS, AMS, ES, SIAM, ASA, MPS.

Veinott, Cyril G
Home: 4540 Gulf Mexico Dr 306, Longboat Key, FL 33548
Position: Consultant (Self-employed) *Education:* DEng/Elec Engg/Univ of VT; EE/Elec Engg/Univ of BT; BS/Elec Engg/Univ of VT. *Born:* 2/15/05. in Somerville, 27 yrs with Westinghouse in dev & design of elec machinery; 17 yrs with Reliance Elec Co. as Chf. Ac Engr. and Chf. Engr. Analyst. Pioneered use of built-in thermal overload protection in motors, also use of computers in electric machinery design. Taught courses in computer-aided design of induction machines; 2 yrs Guest Professor at Laval Univ in Quebec; Volunteer Exec for Intl Exec Service Corps, 2 terms in Mexico & 2 in Brazil; Guest Lecturer, Univ of MO at Rolla, Extension Div, since 1972. Citations: Westinghouse Silver W; Rotating Machinery Comm of IEEE Power Englg Soc; IEEE Nikola Tesla Award, 1977; Medal & Golden Plaque from Yugoslav Soc for Promotion of Scientific Knowledge; 14 patents; 4 books & 50 papers pub. Hobbies: travel & photography. Fellow IEEE. *Society Aff:* IEEE, PES, SI.

Veith, Ferdinand S
Business: 235 E 42nd St, New York, NY 10017
Position: Vice President Manufacturing. *Employer:* Pfizer International Inc. *Education:* SBChE/Chem Engg/MA Inst of Tech. *Born:* Nov 8 1922, Brooklyn NY. 1943-46 served in US Naval Reserve; 1947-52 Chem Engr Chas Pfizer Inc, 1952-53 Proj Engr; 1953-58 Proj Engr Pfizer Internatl Inc, NY, UK & France, 1958-61 Plant Mgr Pfizer do Brasil, Sao Paulo Brazil; 1961-67 Dir Production Pfizer Internatl, 1967- , V Pres Mfg & Dir, respon through related positions above dealing with mfg & plant facilities dev in the area of pharmaceuticals & fine chemicals including fermentation & synthesis, presently respon for internatl mfg operations in 40 countries abroad. Mbr AIChE; Mbr & Fellow AIC. Reg PE in NY. Enjoy sailing, skiing & tennis. *Society Aff:* AIChE, AIC, ISPE.

Vela, M A
Business: 8707 Katy Fwy #103, Houston, TX 77024
Position: President. *Employer:* VELCO Engineering Inc. *Education:* BS/Chem Eng/Univ of TX; MS/Chem Engg/Univ of TX. *Born:* June 6 1925, Edinburg Tex. 1942-46 served with US Armed Forces in Europe; successively associated with Phillips Petroleum Co, Monsanto Chem Co, Pace Co & Fluor Corp; 1970-75 Mgr Process & Computer Engrg Crawford & Russell, Houston; President VELCO Engrg Inc, Cons Engrs, Houston. Patents & publ in separations field. Mbr AIChE, TSPE, NSPE & Internatl Adv Bd Encyclopedia of Chem Processing & Design, formerly Mbr-Dir Bd/Dir Fractionation Res Inc. *Society Aff:* AIChE, NSPE, TSPE, ACM.

Veletsos, Anestis S
Home: 5211 Paisley, Houston, TX 77096
Position: Brown & Root Prof. *Employer:* Rice Univ. *Education:* PhD/Struct Engr/Univ of IL; MS/Civ Engg/Univ of IL; BS/Civ Engg/Robert College. *Born:* 4/28/27. Mbr of Civ Engg faculty, Univ of IL at Champaign-Urbana, 1953-64, advancing to full Prof in 1959. Mbr of Rice faculty since 1964, where he served as Chrmn of Civ Engg, 1964-1971. Has held visiting appointments at the Univ of CA, Berkeley & at univs in Brazil & India. Author of over 100 publications. Recipient of the Norman Medal, Huber Res Prize & Newmark Medal of ASCE, & of Distinguished Alumnus Award, Civ Engg Alumni Assn, Univ of IL. Elected to mbrship of the Natl Acad of Engg in 1979. His service to profl societies includes the Chrmnship of the Engg Mechanics Div of ASCE, the vp of EERI and mbrshp in the Natl Sci Fnd Ctte for Engrg, Civil & Environ Engrg and Earthquake Hazards Mitigation Program. He is an active lectr & consultant for industry. *Society Aff:* ASCE, EERI.

Velkoff, Henry R
Business: Dept of Mech Engr, Columbus, OH 43210
Position: Prof *Employer:* Ohio State Univ *Education:* PhD/Mech Engr/OH State Univ; MS/Mech Engr/OH State Univ; BS/Aero Engr/Purdue Univ *Born:* 05/14/81 Native of Ft Wayne, IN. Flight test analyst at Lockheed CA 1942-43. Research Engr at Aerotor Assocs-jet helicopter design 1944. Research Engr with US Air Force helicopter rutors at Wright Field 1944-1950. Supervisor helicopter rutor research Wright Field 1950-1957. Supervisor helicopter and small gas turbine engine development 1957-1960 at Wright Field. Conslt in advanced propulsion Wright Field, 1960-63. Prof, Mech Engrg OH State Univ 1963-present. Research in electrofluid mechanics, heat transfer, helicopter rotor wakes and boundary layers. Editor, Technical Dir and VP American Helicopter Soc. Conslt to Sikorsky, Boeing on helicopters. Chief Scientist Army Aviation 1972-1974, Staff Scientist 1985-87. Served with US Army 1944-1947, and US Air Force 1951-1953. *Society Aff:* AHS, ASME, AIAA, ESA

Veltman, Preston L
Home: 212 Old County Rd, Severna Park, MD 21146
Position: VP Special Projs - Retired. *Employer:* W R Grace & Co Research Div. *Education:* PhD/Physics Chem/Univ of WI; MS/ChE/MI College of Mines; BS/Min & Tech/MI College of Mines. *Born:* July 18 1912, Grand Rapids Mich. 1938-43 Asst Dir Fuels Res, The Texas Co; 1943-45 Plant Engr Manhattan Dist, Los Alamos Sci Lab; 1945 joined Davison Chem Corp (now Grace), 1950-54 Mgr R&D Div, 1955-59 Res Dir W R Grace & Co, 1960 assumed current respon as V Pres Res Div, Special Projs. Mbr AIChE, ACS, Sigma Xi, Maryland Bio-Medical Engrg Sic, Maryland Governor's Sci Resources Adv Bd. Retired activity as consultant in biomedical fields. Now technical consultant to Niro Intl., National Medical Care, and Carbo Med Canada. *Society Aff:* ACS, AIChE.

Velzy, Charles O
Business: 355 Main St, Armonk, NY 10504
Position: Pres. *Employer:* Charles R Velzy Assoc, Inc. *Education:* MS/San Engr/Univ of IL; BS/Civil Engr/Univ of IL; BS/Mech Engr/Univ of IL *Born:* 3/17/30. Design engr and proj engr with Nussbaumer, Clarke and Velzy. Experience in design of sewer systems, water supply and distribution systems, municipal and industrial wastewater treatment facilities, solid waste mgt, energy from refuse, and air pollution control projs. Bd of Dirs since 1966 and Pres since 1976 of Charles R Velzy Assoc, Inc. Reg PE in 15 states. Authored 30plus technical papers. Past Chrmn ANSI Z228 on Incineration; US Technical Advisory Group to ISO TC146 on Air Quality; Mbr Bd of Govt ASME 1983-84, Chrmn Solid Waste Processing Div ASME 1973-74, active other ASME Committees. Fellow ASME. *Society Aff:* ASME, ASCE, NSPE, APCA, ASTM, APWA, WPCF, AWWA, AAEE.

Venable, Emerson
Business: 6111 Fifth Ave, Pittsburgh, PA 15232
Position: Consulting Chemist & Engineer. *Employer:* Hedenburg & Venable. *Education:* BS/Chemistry/Univ of Pittsburgh. *Born:* Dec 3, 1911. Chemist Agfa-Ansco 1933-35; Res Chemist MSA Co 1935-37; 44- 46; Res Engr Westinghouse Co 1937-44; Dir of Res Freedom-Valvoline Oil Co 1946- 51; Cons with Hedenburg & Venable 1951- . Reg P E in Pa; Cert by Amer Bd of Indus Hygiene; Accredited Chemist by Amer Inst of Chemists; Lectr in Chem & Engrg Univ of Pgh 1937-46. Pres Amer Inst / Chemists 1971-72; Natl Dir for Region III Amer Chem Soc 1971-74; Natl Secy Assn Const Chemists & Chem Engrs 1960-64; Fellow AAAS, APHA; Hon Mbr Amer Inst of Chemists. Mbr Allegheny Cty Air Pollution Advisory Bd (1970-to date; 1987 appointment). Pats in gas masks, gas detectors, elec insulation, fire & explosion control devices. *Society Aff:* AAAS, APHA, AIChE, NSPE, AIC, ACS.

Venable, Joseph W
Business: 1 Lone Palm Pl, Lakeland, FL 33801
Position: Vice Pesident. *Employer:* Pine Lake Chemicals Inc. *Education:* BS/Chemistry/Univ of AL; MS/Physical Chem/Univ of AL. *Born:* 1/23/24. Dotham Ala. ONR grant on methyl hydrazine res; group leader: res, production, internatl sales - Virginia-Carolina Chem Corp; V Pres - Gulf Design Corp (Engr-Constr); V Pres Conserv Inc (Chem Manufac). Mbr AIChE, ACS, ISES. Dir Pine Lake Chems Inc & Investment Corp. Patents: fluorine chems from silicon tetra fluoride. Current intr solar energy applications. *Society Aff:* AIChE, ACs, ISES

Venable, Wallace
Business: Mechanical & Aero Engineering, West Virginia University, Morgantown, WV 26506-6101
Position: Associate Professor. *Employer:* West Virginia University. *Education:* EdD/Engg Education/WV Univ; MSES/Engg Sci/Univ of Toledo; BA/Physics/Cornell Univ. *Born:* Apr 1940; native of Pittsburgh Pa. Reg P E in W VA. Joined faculty of W VA Univ 1966; involved in dev & evaluation of programmed instruction in engrg courses. Mbr ASME, AIAA, ASEE, NSPE; Chmn Prof Interest Council IV, Mbr, Bd of Dir, ASEE, 1980-82. ASEE Dow Young Teacher Award 1974. *Society Aff:* ASEE, ASME, NSPE.

Vendola, Arthur N
Business: 43 Cedar St, New Britain, CT 06052
Position: Principal / President. *Employer:* A N Vendola & Assocs/A N Vendola Inc. *Education:* BE/Civil/Yale Univ; MS/Engrg/Univ of CT *Born:* 12/22/34. Native of New Haven, CT. Served with Army Corps Engrs 1957-58. 1958-68 Prin Engr with various Engrg firms involving mgmt of large civil, sanitary, structural and environ projs. 1968 Prin A N Vendola and Assocs, Consulting Engrs and Pres A N Vendola, Inc. Construction Mgrs, providing all phases of engrg services throughout the eastern seaboard. Firms have been in continual growth and expanding pattern. Currently, past Pres of CT Soc Civil Engrs Sect of Amer Soc of Civil Engrs; Chmn, St of CT, Bd of Matl Review, CT. Building Code. Fellow ASCE. Married to former Donna DiMugno; Three (3) children. Cert NCEE Reg CT, NY, RI, NH, MA, NJ, DE, PA & VA. *Society Aff:* NSPE, ASCE, ACEC.

Ventrice, Marie B
Business: Dept of Mech Engg, Cookeville, TN 38505
Position: Prof.& Interim Dir, Center for Electric Power *Employer:* TN Tech Univ. *Education:* PhD/Mech Engr/TN Tech Univ; MS/Mech Engr/Auburn Univ; BSES/Engr Sci/TN Tech Univ. *Born:* 10/17/40. MS res was concerned with the fracture toughness of high-strength adhesive- bonded joints. PhD res investigated the effect of acoustic oscillations on Reynolds number dependent processes, using an experimental analog technique, with the maj area of application being the understanding of combustion instability in liquid propellant rocket engines. Continued this work after receiving PhD. Carried out res in passive solar energy and congeneration. Since March 1974, have been faculty mbr at TN Tech; teaching both undergrad and grad courses in the thermal/fluid sciences. Since Sept 1985, have been the interim director of the center for Electric Power at TN Tech. *Society Aff:* AIAA, ASEE, ASME, ISES/ASES, NSPE, ΣΞ, SWE

Ver, Istvan L
Home: 10 Glen Rd, Lexington, MA 02173
Position: Principal Cons. *Employer:* BBN Laboratories Inc. *Education:* BS/EE/Tech Univ Budapest; MS/EE/Tech Univ Aachen, W Ger; PhD/Acoustics/Tech Univ Munich, W Ger. *Born:* 12/22/34. R&D Engr. Rhode & Schwarz GmbH Munich, spec in design of acoustical measurement instruments; joined BBN 1965, active in cons, res & prod dev. Publs & pats in field; lectr at MIT; Hd Experimental Acoustics & Facility Design Group of BBN; Elected Principal Cons BBN 1976. Fellow Acoustics Soc of Amer; Mbr Bd of Dir Inst of Noise Control Engrg. Recipient of the 1978 Sen US Scientist Award of the Alexander von Humboldt Foundation, Germany. Recieved the 1979 Crosby Field Award of the Amer Soc of heating, Refrig & Air-Conditioning Engrs (ASHRAE) for the "Best Technical Paper in 1978–. Enjoys literature, travel & tennis. Resides in Lexington MA. *Society Aff:* ASA, INCE.

Ver Planck, Dennistoun W
Home: 7885 Torrey Ln, La Jolla, CA 92037
Position: Consultant (Self-employed) *Education:* DEng/Elec Eng/Yale; MS/Elec Eng/MIT; BS/Elec Eng/MIT. *Born:* Jan 29, 1906. Life Fellow IEEE, Fellow ASME, Fellow ANS. Elec Engr Genl Elec Co Schenectady & Phila 1929-36. Asst Prof EE Yale Univ 1936-40. Engr Naval Ordnance Lab 1940-42 (degaussing res); Cdr USNR Bur of Ordnance 1942-46 (in charge degaussing). Navy Disting Service Award, Order of Brit Empire (Military Div) 1946. Prof of ME & ME Dept Head Carnegie Inst of Tech 1946-60. Mgr gas- cooled fast reactor prog Genl Atomic Co 1960-71. Nuclear reactor cons 1971- . Author: 'Engineering Analysis' with B R Teare, Jr 1954; contrib of articles to prof pubs. Chmn ASME Pgh Section 1954-55; Chmn ASME Bd on Educ 1958-59. *Society Aff:* ASME, IEEE, ANS.

Verbeke, Frank G, Jr
Home: 8050 Armour, San Diego, CA 92111
Position: President. *Employer:* Alturdyne. *Education:* BSE/Mech/Univ of MI. *Born:* Dec 11, 1934 Detroit Mich. Mech-Dev Engr wth Continental, Curtiss-Wright & Solar; conceived Alturdyne 1970 with aim toward use of gas turbine engine throughout indus; also sole owner of Verbeke Assocs, Alturdyne Internatl, Swift Alturair & Alturservice. Recent activities relate to industrialization of the rotary engine. Mbr SAE, ASME, EAA; Reg P E for Calif with NSPE, IEEE, CAF, ASHRE, IEE, CAF. *Society Aff:* NSPE, SAE, ASME, EAA, IEEE, ASHRE, CAF.

Verduzco, Miguel A
Business: Leibnitz 1-8o Floor, Mexico D F, Mexico 5
Position: Director General. *Employer:* Asesoria Tecnica Indus-S A de C V. *Born:* Mar 16, 1940 Monterrey Mexico. BSChE Inst Tecn de Monterrey Mexico; PhD Met Engrg IIT 1966. Tech Mgr in R&D Dept La Salle Steel Co Hammond Ind; Adjunct Prof & lectr IIT Chicago; Cons to Mexican Govt & at this time Dir Genl Asesoria Tecnica Industrial, S A de C V in Mexico City. Dir of 2 Org's of Amer States progs related to tech transfer & dev. Pub over 10 tech papers & 2 pats. Presented tech seminars in USA & So Amer. P Pres Mexico ASM Chap 1974.

Verheyden, Virgil E
Business: 1501 Roakoke Blvd, Salem, VA 24153
Position: Mgr - Engrg/Drive Sys Dept. *Employer:* General Electric Co. *Education:* BS/EE/Univ of TX. *Born:* Sept 1921. US Air Force 1943-46. 7 mos Radar Sch Harvard & MIT. With G E Co since grad BSEE Univ of Texas 1947. 19 yrs control sys engrg design & 1st level mgmt for steel mill processes & gas turbine applications. 3 yrs mgr of section with respon for Marketing, Engrg & Product Accounting with 65 tech grads & support. Since Jan 1973 Mgr-Engrg Drive Sys Dept, providing Dev, Prod Engrg & Prod Service necessary to supply DC drive sys for heavy indus processes worldwide. Mbr IEEE since coll; Assoc Mbr AISE for several yrs. *Society Aff:* IEEE, AISE.

Verhines, Jack G
Business: 510 W Texas, Artesia, NM 88210
Position: President. *Employer:* Scanlon & Associates Inc. *Born:* May 30, 1932. Graduate of Univ of New Mexico 1958. Civil Engr. Reg P E in New Mexico June 15, 1962.

Verhoeven, John D
Business: 104 Metallurgy Bldg, Ames, IA 50011
Position: Professor of Metallurgy. *Employer:* Iowa State University. *Education:* BS/Ch Engr/Univ of MI; MS/Met Engr/Univ of MI; PhD/Met Engr/Univ of MI. *Born:* 8/26/34. Faculty Dept Materials Sci & Engrg at Iowa St Univ 1963- . Area of res: solidification of metals. Area of teaching: Physical Met Publs: Textbook, *Fundamentals of Physical Metallurgy*; approximately 70 papers in area of solidification, and physical metallurgy. Outstanding Teacher Award Iowa St Univ 1976. *Society Aff:* ASM, AIME, Metals Soc, AACG.

Verhoff, Frank H
Home: Rt 7 Box 307, Morgantown, WV 26505
Position: Prof *Employer:* West Virginia Univ *Education:* PhD/ChE/Univ of MI; MSE/ChE/Univ of MI; B/ChE/Univ of Dayton *Born:* 10/19/42 Spent first eighteen yrs in Putnam County OH. Have worked for PPG Industries and Miles Laboratories. Have taught previously at Univ of Notre Dame. Presently pursue research on mathematical modeling of bioengrg and separation problems. Have consulted for a variety of companies including Johnson and Johnson, Exxon, Army Corps of Engrs, etc. Translate technical Russian articles. *Society Aff:* AIChE, ACS, AGU, IAGLR

Verink, Ellis D, Jr
Home: 4401 NW 18th Pl, Gainesville, FL 32605
Position: Disting Service Prof *Employer:* Univ of FL. *Education:* PhD/Met Engg/OH State; MS/Met Engg/OH State; BS/Met Engg/Purdue. *Born:* 2/9/20. Born in Peking, China. Active duty US Navy 1941-46. Presently Commander USNR (Ret). Engr and Sec Hd Dev Div, Aluminum Co of Am 1946-59 Developing new products of aluminum for chem & petrol industries. Mgr, chem & petrol ind sales, Aluminum Co of Am 1959-62. Joined faculty Dept Mtls Sci & Engg Univ of FL 1965. Chrmn & Prof Dept of Mtls Sci & Engg Univ of FL, 1973-86. Specialties Corrosion & Mtls Selection. Bd of Dirs, TMS/AIME, Pres 1984. Past Mbr, NACE. SamTour Award of ASTM, 1978. Teacher-Scholar of Yr, Univ of FL, 1979. Willis Rodney Whitney Award in Corrosion Sci, NACE 1982; Dist Alumnus Ohio State Univ, 1983; FL Blue Key Dist Fac Award, 1983; Dist Serv Prof 1984- . *Society Aff:* TMS/AIME, NACE, ECS, AWS, ASM.

Verlo, Clayton C
Business: 2701 Lake St, Melrose Park, IL 60160
Position: Director of Engineering Services. *Employer:* The Richardson Co. *Education:* MBA/Behavioral Sci/Univ of Chicago; BSME/Mech Eng/Univ of WI. *Born:* 8/4/21. Served in Army Air Force 1940-45. Student Engr York Copr 1948; Field Engr Westerlin & Campbell Co 1949-51; joined Richardson 1951; assumed present position 1968. Pres AIPE 1975-76. Hobbies incl reading & fishing. *Society Aff:* AIPE, APCA, WPCF.

Vermeulen, Theodore
Business: Dept of Chemical Engineering, Berkely, CA 94720
Position: Prof of Chem Engrg *Employer:* Univ of California *Education:* PhD/Physical Chem/UCLA; MS/ChE/CA Inst of Tech; BS/ChE/CA Inst of Tech *Born:* 05/07/16 With Union Oil 1937-39 and Shell Dev Co 1941-47. Founding Chrmn of Chem Engrg at Univ of CA (Berkeley), 1947-53. Miller Research Prof 1959-60. Dir, Water Tech Ctr, since 1980. Fulbright Prof in Belgium and France, 1953-54. Guggeheim Fellow, Cambridge Univ, 1963-64. Visiting prof Natl Univ of Mexico 1967; S China Inst of Tech 1981. Section editor, Chem Engrs Handbook. Co-editor, Advances in Chem Engrg, 1961-81. Dir Memorex Corp, 1963-81. Research and teaching in chem kinetmcs and reactor design, separation processes (ion exchange, liquid exgraction), mixing and dispersion, coal liquefaction, desalination. Walker Award of AIChE 1971; Chem Eng Div Lecturer Award, ASEE, 1979. *Society Aff:* AIChE, ACS, AWWA, ASEE, ANS

Verneuil, Vincent S
Business: 1400 N Harbor Blvd, Fullerton, CA 92635
Position: Vice President. *Employer:* Simulation Sciences Inc. *Education:* BS/ChE/LA State Univ; MChE/ChE/Univ of DE; PhD/ChE/Univ of DE. *Born* Aug 1939. MS & PhD Univ of Delaware; BS 1961 LSU. Native of Louisiana. 1965-67 Asst Branch Ch, Proj Leader, US Army Nuclear Defense Lab, Edgewood Md 1967-69; E I duPont de Nemours & Co, Cons Engr; with Simulation Sci Inc 1969- , V Pres of Tech Opers 1972- , respon for use of "Process" Simulator by petroleum, petrochem, gas & chem indus throughout world. *Society Aff:* AIChE.

Veronda, Carol M
Home: 3707 Camelot Dr, Annandale, VA 22003
Position: Hd, Indentification Sys Branch. *Employer:* US Naval Res Lab. *Education:*

Veronda, Carol M (Continued)
BS/EE/Caltech. *Born:* Aug 1920. Field & Test Engr Genl Elec Co 1942-43. Navy Radar Schs Bowdoin Coll, MIT & Bell Tel Labs. Submarine Force Pacific Fleet WW II. Microwave electron device design & dev with Philips Labs, Sperry Electronic Tube Div (Great Neck N Y & Gainesville Fla) & Varian Assocs, Bomac Div; Ch Circuits & Antennas Branch NASA Electronics Res Ctr 1965-70 respon for engrg dev of space communications & telemetry sys; 1970-78 Ch Electromagnetic Tech Div, Transportation Sys Center Cambridge Mass, dev & applying electronic tech to transportation sys. Sr Mbr IEEE; Fellow AAAS. 1979-Hd Identification Sys Branch US Naval Res Lab Washington, DC. *Society Aff:* IEEE, AAAS.

Versnyder, F L
Home: 6A Edgewater, Limewood Avenue, Branford, CT 06405
Position: Asst Dir of Res for Matls Tech. Retired. *Employer:* United Technologies Res Center. *Education:* BS/Metallurgy/Univ of Notre Dame *Born:* May 1925; native of Watertown N Y. Served AUS 1943-45; 1950 joined Genl Elec Co, field of high temp alloys. Awarded Henry Marion Howe Gold Medal ASM 1954. 1955-61 G E Res Lab field of creep fracture also solidification. Joined Pratt & Whitney Aircraft (United Tecnologies Corp) 1961, Asst Dir Advanced Materials Lab; named Assoc Dir 1964. In 1966 awarded George Mead Gold Medal, United Aircraft Corp for engrg achievement. 1971 Mgr Materials Engrg & Res Lab. Arch T Coldwell Merit Award 1971 SAE. 1972 awarded Dickson Prize, Carnegie- Mellon Univ. 1973 awarded Francis J Clamer Medal, Franklin Inst, also Life Fellow. Fellow ASM 1973. 1975 Coll of Engrg Honor Award of Univ of Notre Dame 1975. Engr Materials Achievement Award ASM. Hon Mbr Alpha Sigma Mu, Member, National Academy of Engineering 1981. PE-CT-Reg. No.10701, 1978. Fellow TMS-AIME 1983. Eli Whitney Award 1986 CT Patent Law Assoc. National Medal of Technology 1986 President of the U.S. *Society Aff:* ASM, TMS-AIME

Vertes, Michael A
Business: PO Box 868, Greenwich, CT 06836
Position: Pres. *Employer:* Metachem Inc. *Education:* MCE/ChE/NYU; BCE/ChE/CUNY *Born:* 5/14/36. 1961-65 Res & Proj Engr, Electro-chem Energy Conversion for space prog. 1965-1978 AMAX Inc. Dir of Product Planning & Projs - Molybdenum Div in charge of construction projs in $50 million range. Advanced Mgt Prog, Harvard Bus Sch 1976. 1978 Formed Metachem Inc, intl consulting, engg & metals & chems trading. Projected 1979 sales $2.2 million. *Society Aff:* AIChE, ACS.

Verzuh, James M
Business: PO Box 2553, Billings, MT 59103
Position: Regional Engr, UM Region *Employer:* US Bureau of Reclamation. *Education:* BS/CE/Univ of CO. *Born:* 6/12/35. Native of Crested Butte CO. Military Service US Army (Signal Corps). Civ Engr with Denver Water Bd (Dillon Dam, Dillon CO) 1960. Began career with Bureau of Reclamation 1961 as designer of irrigation works at Bureau's Engg and Res Ctr, Denver, CO. Transferred to Regional Office, Billings, MT, 1969, assumed Chief, Design Branch position 1973, appointed Regional Engineer 1980. Wife, Julie; Sons, Rudy, Eric, John, Jim. NSPE Bur of Reclamation Engr of the Year 1984. Advisory Ctte Montana State Univ Dept of Civ Engg. *Society Aff:* ASCE, USCID.

Vesic, Aleksandar S
Business: Sch of Engrg, Durham, NC 27706
Position: Jones Distinguished Prof & Dean. *Employer:* Duke University. *Education:* DSc/Engrg/Univ of Belgrade; Dipl Ind/Struct Engrg/Univ of Belgrade; BSc/Civil Engrg/Univ of Belgrade. *Born:* 1924 Yugoslavia. Postdoc study at Ghent, Belgium, Manchester England & MIT. Struct Engr, cons & univ lectr in Yugoslavia 1949-56. Res Engr Natl Geotechn Inst of Belgium 195658. Asst & Assoc Prof Ga Tech 1958-64. With Duke Univ as Prof 1964- ; Disting Prof 1971- ; Chmn Cvl Engrg Dept 1968-74; Dean of Engrg 1974- . On leave from Duke 1971-72 as Fellow of Churchill Coll, Cambridge England. Chmn ASCE Ctte on Deep Foundations 1968-74; Chmn HRB Ctte on Theory of Pavement Design 1965-70; Chmn ASTM Subctte on In-Situ Bearing Tests 1965-70. Honorary doctorate in Sci, Univ of Ghent, 1981. Thomas Middlebrooks Award ASCE 1974; Hwy Res Bd Award Natl Res Council 1969; Outstanding Cvl Engr Prof Award Chi Epsilon 1967; Prize for Sava Bridge Design Yugoslavia 1953; Prize for Scholastic Achievement Natl Comm Univ Yugoslavia 1949. *Society Aff:* ASCE, ASTM, ASEE, TRB, APWA.

Vesilind, P Aarne
Business: Dept of Cvl Engrg, Durham, NC 27706
Position: Professor and Chairman of Cvl Engrg. *Employer:* Duke University. *Education:* BSEC/CE/Lehigh; MSCE/CE/Lehigh; MSSE/San Eng/UNC; PhD/Environ Eng/UNC. *Born:* June 13 1939 Tallinn Estonia. Undergraduate degree in Cvl Engrg Lehigh Univ; MSCE Lehigh; MSSE Univ of N C Chapel Hill; PhD Environ Engrg UNC. Work experience incl 1 yr as Res Fellow with Norwegian Inst for Water Research Oslo & R&D Engr with Bird Machine Co Walpole Mass. Presently Prof and Chairman Civil Engrg Duke Univ , engaged in teaching environ engrg. Awards: Geo C Bunker Award UNC & Collingwood Prize ASCE; Fullgright-Hayes Lectureship, New Zealand; Outstanding Teacher Award Duke Univ; Earl I Brown Teaching Award, Duke Univ; Student Service Award, Duke Univ. *Society Aff:* AEEP, ASCE, WPCF, ASEE.

Vest, Gary D
Business: Osaf/Riq Pentagon, Washington, DC 20030-1000
Position: Dir for Environment, Safety and Occupational Health. *Employer:* Department of Air Force *Education:* MUP/Urban Planning/Univ of WA; BA/Pol Sci/Univ of ID. *Born:* Mar 1946 Boise ID. Assoc City Planner Redmond Wash 1969-71; Cvl Engrg Officer/Land Use & Airport Planner US Air Force 1971-74; Community Planner HQ US Air Force 1974-82: in current capacity 1983- . Full Mbr Amer Inst of Cert Planners, Amer Planning Assoc & Soc of Amer Military Engrs. *Society Aff:* APA, AICP, SAME.

Vest, Robert W
Business: Potter Engr Center, West Lafayette, IN 47907
Position: Turner Prof of Engg. *Employer:* Purdue University. *Education:* PhD/P Chem/IA State Univ; BS/Chem/Purdue Univ. *Born:* Oct 1930. Mbr NICE, ASM, ISHM, Fellow ACerS & awarded Ross Coffin Purdy Award ACerS. Currently holds Basil Turner Disting Prof of Engrg at Purdue Univ. Res interests incl: transport & dielectric properties of ceramic materials; solid state chem; electroceramic devices; has dev Turner Lab of Electroceramics at Purdue. Has held position of Chemist with Natl Lead Co & Monsanto Corp; came to Purdue from Sys Res Labs 1966 as Prof of Engrg in Schs of Materials Sci & Met Engrg & Elec Engrg. Served as Hd, Sch of Matls Engg 1973-78. Has authored over 40 papers & 2 book chapters. *Society Aff:* ACerS, NICE, ASM, ISHM.

Vetelino, John F
Business: EE Dept, Orono, ME 04469
Position: Prof. *Employer:* Univ of ME. *Education:* PhD/EE/Univ of RI; MSEE/EE/Univ of RI; BSEE/EE/Univ of RI *Born:* 10/17/42. Native of Westerly, RI. Prof, EE at the Univ of ME, since 1969. Maj areas of res interest include solid state, electromagnetics & acoustics. Served on Natl Sci Fdn review panels for res & sci education. Actively involved in res in microwave acoustics & sci educ. Consultant with Naval Underwater Systems Ctr, Newport, RI. Allied Corp, Morristown, NJ. Hobbies include antique automobiles & fishing. *Society Aff:* HKN, ΦΚΦ, ΤΒΦ, IEEE

Veziroglu, T Nejat
Business: College of Engrg, Coral Gables, FL 33124
Position: Dir - Clean Energy Res Inst. *Employer:* University of Miami. *Education:* PhD/Heat Transfer/Univ of London; DIC/Advanced Studies/Imperial College, London; BSc/Mech Engg/Univ of London; ACGI/Mech Engg/City & Guilds College, London. *Born:* 1/24/24. Tech Cons & Dir Steel Silos Div, Office of Soils Prods Ankara Turkey 1952- 56. Pres & Chmn of Bd MKV Const Co Istanbul Turkey 1957-62. With Univ of Miami Coral Gables 1962- ; currently Prof of Mech Engg,

Veziroglu, T Nejat (Continued)
College of Engg; Heat Transfer, Solar energy, nuclear energy & hydrogen energy sys. Elected Fellow ASME, Inst of Mech Engrs & American Assoc for Advancement of Science. Founding Pres of Intl Assn for Hydrogen Energy. Pub over 120 papers, ed 60 book & is ed of Intl Jorunal of Hydrogen Energy. *Society Aff:* ANS, AAUP, ASEE, SES, ΣΞ.

Viaclovsky, Syl A
Business: 12301 Kurland Dr, P.O. Box 1700, TX 77001
Position: Supervisor. *Employer:* Houston Lighting & Power Co. *Education:* BS/Engrg Sci/Univ of TX. *Born:* 2/27/42. Native of Wallis, TX. Reg PE in TX. Employed with SW Res Inslt June 1965- Nov 1973; primary respon incl matl evalutation studies, non-destructive testing of nuclear power plant components & tubular steel transmission towers & quality assurance audits. Employed by Houston Lightning & Pwr Co Dec 1973; promoted to Site Quality Assurance Supr for S TX Nuclear Proj Oct 1975, respon for all quality assurance & quality control activities at S TX Nuclear Proj site; apptd Supervisor of Quality Assurance Support Div respon for all auditing, document reviews, training & vendor surveillance functions in support of nuclear & fossil projects; Promoted to operations QA. Supervisor responsible for nuclear plant preservice and inservice inspections June 1980. Mbr ASM, ASNT & ASME. *Society Aff:* ASM, ASNT, ASME.

Viani, Ronald E
Business: 570 Lexington Ave, New York, NY 10022
Position: Mgr Power Systems Engg. *Employer:* GE. *Education:* MS/Aero Engg/NYU; BS/Aero Engg/NYU. *Born:* 10/31/45. A native of NYC, received a BS in engg from NYU in 1967 & was employed by Grumman Aircraft as a gas turbine performance engr from 1967 thru 1972, while studying for an MS degree, attained in 1971. Joined the Export Sales & Services Div of GE Co in 1972 & was appointed Mgr, Power Systems Engg in 1979. Currently responsible for the conceptual design of all power generation facilities for the Intl Market. *Society Aff:* ASME.

Viano, David C
Business: Biomedical Science Department, Warren, MI 48090-9055
Position: Principal Scientist *Employer:* General Motors Research Labs *Education:* PhD/Applied Mech./CA Institute of Technology (Caltech); MS/Applied Mech./CA Institute of Technology (Caltech); BS/Elect. Engrg./University of Santa Clara *Born:* 05/07/46 Biomedical Science Post-doctorate, Univ of Zurich 1972-74. Researcher 1974-79, and Manager 1979-, crash injury program, Biomedical Science Dept and Leader, Safety Research Prog, 1987-, GM Research Labs specializing in automotive safety engrg, impact injury biomechanics & occupant protection systems. Contributes to automotive safety policy & technology in General Motors. Member, NAS Cttee on Federal Trauma Research 1984-85, and serves on CDC/HHS Injury Research Review Cttee 1986-. Chairman, SAE Automotive Body Activity 1985-, Safety Advisory Cttee 1985-, and Passenger Protection Cttee 1980. 1981 and 1985 SAE Isbrandt Medal. AAAM Executive Cttee 1986- and Bd of Dir 1984-. ASME Fellow 1986 and Assoc Editor, Journal of Biomechanical Engrg 1982-. Adjunct Professor Engrg, Wayne State Univ 1981-. Bd of Dir, National Head Injury Foundation, 1986-. *Society Aff:* SAE, ASME, AAAM

Vicens, Guillermo J
Home: 42 Bannan Drive, North Andover, MA 01845
Position: VP *Employer:* Camp Dresser & McKee Inc *Education:* PhD/Water Resources/MIT; SB/CE/MIT; SM/CE/MIT; *Born:* 11/27/48 Native of Ponce, PR. Asst Prof at MIT after completing grad studies there. Joined Resource Analysis, Inc in 1974 and was VP of the firm when it merged with Camp Dresser & McKee Inc in 1978. Responsible for broad range of activities in consulting services in the areas of protection and management of water resources for federal, state, and municipal clients. Currently Mgr of Water Resources Groups in Boston Office of CDM. Author of numerous papers. Chrmn of North Andover, MA Conservation Commission. *Society Aff:* AGU, ASCE, AWRA, AIH

Vick, Columbus E Jr
Business: PO Box 3306B, Raleigh, NC 27606
Position: Pres & Chrmn of Bd *Employer:* Kimley-Horn & Assocs, Inc *Education:* MSCE/Transp Eng/NC State Univ; BS/CE/NC State Univ *Born:* 11/08/34 For more than 25 yrs Ed Vick has served the field of transportation as consltg engr on a wide variety of planning and design projects, as Pres and CEO of Kimley-Horn, and as a leader in profl and technical soc activities. Vick's broad perspective as a transportation planner, and his considerable expertise in public participation and presentation, are frequently tapped for the firm's work on planning, transit, parking and private development projects. He is a past board mbr and Fellow of the Inst of Transportation Engrs, a Fellow in ASCE, and has served on various working committees of NSPE, ACEC, and ARTBA. Reg PE in fifteen states. *Society Aff:* ITE, ASCE, NSPE, APA, ACEC, AICP, AREA, ARTBA

Vick, Harold D
Business: 4431 Embarcadero Dr, W Palm Beach, FL 33407
Position: Executive Vice President. *Employer:* Kimley-Horn & Assocs Inc. *Education:* BSCE/CE/Duke Univ; MS/Transp Eng/NC State Univ; MBA/Business/ Nova Univ. *Born:* Dec 8, 1939 Jacksonville total. Served as officer US Navy 1961-65; Res Asst Hwy Res Prog N C St 1965-67; Transportation Planning Engr Mel Connor & Assocs Inc, Tallahassee Fla 1967-68; with Kimley-Horn & Assocs Inc 1968- , V Pres & Principal 1970-79, Ex. VP in charge of firm's total operations. Secy- Treas, V Pres & Pres Fla Section ITE 1973-75. P Pres's Award Fla Section ITE 1968. Cvl Engr of Yr Palm Beach Branch ASCE 1974. Mbr ITE, ASCE, NSPE, FES. APA, AICP, PSMA Reg FL, NC & KY, Pres FL Sec ITE 1975. Representative to Dist 5 Bd of Dirs 1979-80. *Society Aff:* ITE, ASCE, FES, NSPE, APA, AICP, PSMA, FSITE, PBCPC.

Vicory, Freeman M
Home: 2925 Sunnymede Ct, Topeka, KS 66611
Position: Squad Chief, Bridge Design Sect. *Employer:* Kansas Dept of Transportation. *Education:* BS/Civil Engg/KS State Univ. *Born:* Dec 22, 1923 Greenleaf Kansas. Grad Greenleaf HS 1941. WW II, Ensign USNR Pacific Theater 1943-46. BS Cvl Engrg Kans St Univ 1948. Began career Kans Hwy Comm (now Dept of-Transportation) Sept 1948; current position Squad Ch Bridge Design Section. Reg P E Kans 1731. Kans Section ASCE Secy 1967-70, V Pres 1975, Pres 1976; Secy & Chmn Dist 16 Council ASCE 1968-71. Mbr 1st Lutheran Church. Hobbies: music, photography, gardening, fishing & hunting. *Society Aff:* ASCE.

Vidal, Jacques J
Business: 3531 Boelter Hall, Los Angeles, CA 90024
Position: Prof. *Employer:* UCLA. *Education:* PhD/Nucl Engg/Univ of Paris; MSEE/-/Uni of Liege; Dipl/Nucl Engg/Natl Inst of Nucl Sciences. *Born:* 4/18/30. Dr Jacques Vidal left an academic position in Belguim in 1963 to become a Prof of Comp Sci at UCLA. His scientific res activity has ranged from hybrid computing, sensitivity analysis, on line computing, and, more recently, the bioengg of man-machine communication. His current work is at the frontier of engg with the neurosciences, with an eye toward the modelling & analysis of some aspects of brain function. He is the author of some forty scientific articles & communications, as well as several chapters to hardcover books. *Society Aff:* IEEE, SNNAS

Vidal, James H, Jr
Business: 3565 Piedmont Rd, NE, PO Box 4659, Atlanta, GA 30302
Position: Mgr - SE Region, Elec Utility Sales Div. *Employer:* GE Co. *Education:* BIE/-/Univ of FL. *Born:* 1/13/23. Native of Gainesville, FL. Served in WWII in Europe & a holder of the Purple Heart. With GE since 1947. Assumed present position as Mgr SE Region, Elec Utility Sales Div, July 1968. Have 22 yrs of managerial experience in broad scope of ind & utility products. Mbr of FL Council of 100 & serve on Univ of FL Engg Advisory Council. Hobbies: golf, tennis, fish, hunt, sail. *Society Aff:* IEEE.

Videon, Fred F
Business: Dept. of Civil & Ag Engrg, Bozeman, MT 59717
Position: Professor Civil Engineering. *Employer:* Montana State University. *Education:* PhD/CE/Univ of IL; MS/CE/CO State Univ; BS/CE/CO State Univ. *Born:* 10/4/34. in Hayden, CO. On faculty of MT State Univ 1965- . Mbr ASCE, PPres MT Sect. Mbr. Ed. Div. Exec. Comm. & Common St. Serv.; Mbr Amer Soc for Engg Educ. Received Western Electric Fund Award ASEE for Pacific NW Sect 1975. Mbr Natl Soc of Prof Engg. Hon Socs: Phi Kappa Phi, Sigma Xi & Chi Epsilon. Reg P E in MT. *Society Aff:* ASCE, ASEE, NSPE.

Vidosic, Joseph Paul
Home: 7 Walnut Street, Cooperstown, NY 13326
Position: Professor & Dean Emeritus - Retired. *Employer:* Georgia Tech & Middle Ga Coll. *Education:* PhD/Design Engg/Purdue Univ; MS/Phys-Math/Stevens Inst of Tech; ME/Mech Engg/Stevens Inst of Tech. *Born:* June 1909. Life Fellow ASME; Life Mbr ASEE. Regent's Prof of Mech Engrg, Ga Tech - Emeritus; Dean of Admin, Middle Ga Coll - Emeritus. Authored 3 textbooks in design engrg & cutting tech areas; also mechanics section in Standard HDK (Marks') for Mech Engrg as well as 3 articles in Encyclopedia Britannica. Ser in WW II; now retired Col AF Reserve. Awarded ASME Cert of Appreciation for preparing Engrg Merit Badge for BSA. Active in res & indus cons. Wrote numerous tech papers. Held chmnships in ASME & ASEE sections, divs & cttes; Tau Beta Pi, Pi Tau Sigma, Sigma Xi & other hons. Enjoy wrld travel & presenting travelogs. *Society Aff:* ASME, ASEE.

Vierling, Bernard J
Business: Locust Grove, Flint Hill, VA 22627
Position: Retired *Education:* BA/Mech Engg/Stanford Univ. *Born:* Dec 1914 Bakersfield Calif. BSME Stanford Univ 1936. Engr Douglas Aircraft Co 1936-39, worked on DC2, DC3, DC4 & DC5; Supr Engr, Supt of Maintenance, Dir Engrg & Maintenance Pa-Central Airlines 1939-47; participated in dev of hot air heating sys adopted by C47 in WW II & in refinement of braking sys later used in B29. Pres Aircraft Adv Inc 1947-48; Pres Aircraft Supply Corp 1948-51; Pres Helidusters Inc 1951-52; Ptnr Aircraft Supply Co 1954-62; Dir FAA Maintenance Serv 1962-65. Received Natl Award for Econ Achievement from Pres Johnson 1964. Deputy Dir US Supersonic Transport Dev, FAA & Dept of Transportation 1965-71; Acting Dir 1969-70. Dir Morgantown PRT Sys Dev, Urban Mass Transportation Admin 1971-73. Received Dept of Transportation Secy's Award 1973. Special Counselor to Assoc Admin of R&D 1973-74; UTMA. Dir Bus/Paratransit Tech UMTA R&D 1975-79. Farmer & Consultant 1979- ; Farmer, Consultant in Transp 1979-. *Society Aff:* ASME, SAE, IAS.

Viest, Ivan M
Business: PO Box 1428, Bethlehem, PA 18016
Position: Consulting Structural Engineer *Employer:* Self Employed *Education:* PhD/Engg/Univ of IL; MS/Civil Engg/GA Inst of Tech; Ing/Civil Engg/Slovak Tech Univ. *Born:* 10/10/22. native of Bratislava Czecholsovakia. Employed on various const sites, in struct design offices & at Slovak Tech Univ WW II & immediately thereafter. Conducted structures research at Univ of Ill 1948-57; headed Bridge Res Branch AASHO Road Test 1957-61; with Bethlehem Steel Corp 1961-82; managed and conducted technical market development activities in steel consuming industries; self employed 1983- ; offering consulting services in the field of civil engrg structures. V Pres ASCE 1973-75. Const Award ENR 1972. Honary membership in ASCE 1981). Elected to National Academy of Engineering in 1978. *Society Aff:* NAE, ASCE, TRB, ACI, AISC, IABSE.

Vieth, Wolf R
Home: RD 2 - Box 253, Belle Mead, NJ 08502
Position: Professor. *Employer:* Rutgers University. *Education:* ScD/Chem Engg/MIT; MSc/Chem Engg/OH State; SB/Chem Engg/MIT. *Born:* May 1934. Educator. Married Peggy Schira; 4 children. From Asst Prof to Assoc Prof Chem Engrg MIT 1961-68; Dir Sch of Chem Engrg Practice MIT 1965-68; Prof & Chmn chem & biochem engrg Rutgers Univ 1968-78; cons to govt & indus. Invention Award duPont Co 1960; St Albert the Great Medal for Sci, Aquinas Coll 1952. MIT/Tech Univ Berlin exchange Fellow 1967. Disting Foreign Sci Award, Japan Soc for the Promotion of Sci 1975. Mbr Sigma xi, Phi Lambda Upsilon, AIChE, Fellow NY Acad of Sci; ASEE, ACS, Fellow AIC. Author 100 publs & 10 pats in field of specialization. Book: *Membrane Systems: Analysis and Design*, Hanser Press, 1988. Editor J Mol Catalysis. *Society Aff:* AIChE, ASEE, ACS.

Viets, Hermann
Business: University of Rhode Island, College of Engineering, Kingston, RI 02881
Position: Dean of Engrg *Employer:* Univ of RI *Education:* PhD/Astronautics/Bklyn Polytech Inst; MS/Astronautics/Brklyn Polytech Inst; BS/Aerospace Eng/Bklyn Polytech Inst *Born:* 01/28/43 Native of Quedlinburg, Germany; raised in NY. Served as Research Assoc, Brooklyn Polytechnic. Awarded NATO Post-Doctoral Fellowship at von Karman Inst, Brussels, Belgium 1969-70. Research Group Leader and Engr at the USAF Aerospace Research Labs 1970-76. Prof at Wright State Univ, Dayton, OH 1976-1981. Assoc Dean for Research, Coll of Engrg, WV Univ 1981-83. Moved to current position as Dean of the Coll of Engrg at the Univ of RI in 1983. Respon for the admin of the Coll prog with 120 employees including 75 faculty and an annual budget in excess of $6.5 million including $2.5 million in external res funding. Bd Chrmn, Precision Stampings, Inc, Beaumont, CA. Author of 50 publications and 7 patents. Married to former Pamela Deane and father of Danielle, Deane, Hans, Hillary. *Society Aff:* AIAA, ASEE, SME, DGLR

Viggers, Robert F
Business: 900 Broadway, Seattle, WA 98122
Position: Prof of Mech Engr *Employer:* Seattle Univ. *Education:* MS/ME/OR State Univ; BS/ME/Univ of WA. *Born:* 1/18/23. Teaching: Seattle Univ-mech engg since Fall 1949. OR State Univ-mech engg, 1948-1949. Univ of WA-gen engg 1946-1947. PE: Licensed in OR 1949, WA 1950. Consulting since 1949 - machine design, fluid mechanics, thermodynamics, and invention dev. Res: With Reconstructive Cardiovascular Res Ctr of Providence Medical Ctr since 1959 - heart valve prosthesis dev and dev of apparatus for cardiovascular system experimentation. Military Service: Navy radar technician, 1944-1946. *Society Aff:* ASME, ASEE.

Vigilante, Frank S
Business: 295 North Maple Ave, Basking Ridge, NJ 07920
Position: A VP *Employer:* American Tel & Tel Co *Education:* MS/EE/NY Univ; BS/EE/Univ of CA. *Born:* 3/15/30. Native of Brooklyn NY. Began Bell sys career 1957 as mbr of tech staff at Bell Labs. Dev Electronic Swiching Sys 1957-70; AT&T Dir of Switching 1970-73; Exec Dir of Switching sys Engrg 1973; VP, Corporate Engg, Western Elec Co in Nov of 1977. Joined American Telephone and Telegraph Co on Sept 1, 1981 as Asst VP, Business Mktg. Fellow IEEE; V Chmn Interantl Switching Symposium 1972; Chm Internatl program Ctte, Internatl Switching Symposium. Selected on Outstanding Young Amer Engr 1964 Eta Kappa Nu. Awarded 15 pats. Enjoy sailing & tennis. *Society Aff:* IEEE, ТВП, HKN.

Villalon, Leonardo A
Business: 1200 St. Charles Ave, New Orleans, LA 70130
Position: Sr VP, Operations *Employer:* Waldemar S. Nelson & Co, Inc *Education:* EE/Power/Univ de la Habana *Born:* 11/07/32 Native of Havana, Cuba. Electrical Engrg degree in 1955. Sales and Applications Engr, Gen Elec Cubana SA 1955-1960. Joined the Waldemar S. Nelson Co in 1963 as design engr. Named asst VP in 1966; Mgr of Electrical Engrg in 1970; VP in 1974; Sr VP, Operations in 1980. Reg PE in LA, TX, and CA. *Society Aff:* NSPE, IEEE, ISA, API, LES

Villani, Vincent F
Home: 1317 Mt Vernon Pl, Charleston, WV 25304
Position: Assoc Director *Employer:* Union Carbide. *Education:* MS/Management Engr/NJ Inst of Tech; BS/Chem Engr/Lafayette College. *Born:* 12/01/40 in NYC. US Army Commander of Explosive Disposal Detachment 1962-65. Mfg Engr to 1966

Villani, Vincent F (Continued)
and Dept Head of Production and Maintenance Activities at Union Carbide Chem and Plastics Plant in Bound Brook, NJ to 1974. Transferred to UCC Engg Dept in Charleston, WV in 1974 as Proj Mgr/Process Group Leader. Promoted to Energy Systems Engg Mgr in 1976 and to Assoc Dir in 1979. Responsible for Hydrocarbons, Energy Systems, Syn Gas, and Coke Process Engg. Active in church activities (Lay Minister and youth education). Enjoy jogging and tennis. *Society Aff:* AIChE.

Villard, Oswald G, Jr
Business: 333 Ravenswood Ave, Menlo Park, CA 94025
Position: Senior Scientific Advisor/Professor *Employer:* SRI International and Stanford University *Education:* PhD/Elec Engg/Stanford Univ; EE/Elec Engg/Stanford Univ; BA/English/Yale Univ. *Born:* Dobbs Ferry N Y. 1916; Special Res Assoc, Mbr Sr Staff Harvard Univ Radio Res Lab 1942-46; Acting Asst Prof, Asst Prof & Assoc Prof 1946-55, Prof Elec Engrg 1955- , Dir Radio Sci Lab 1958-73 Stanford Univ; Dir Ionospheric Dynamics Lab Stanford Res Inst 1970-72; Sr Sci Adv 1972- . Mbr USAF SAB; 1961-75; Chmn, Div Adv Grp, Electronic Systems Div; Mbr Naval Research Advisory Ctte, 1967-75, Chmn 1972-75. Member, advisory board, National Security Agency 1979-86; Author of 60 publs in ionospheric radio propagation & radio communication techniques. NAS 1961; NAE 1966. IRE Morris Liebmann Memorial Prize 1957. Phi Beta Kappa; Sigma Xi. Meritorious Civilian Service Medal, Dept of the Air Force; Secretary of Defense Medal for outstanding Public Service; IEEE Centennial Medal; Dir: California Microwave, Inc. *Society Aff:* IEEE, URSI, AAAS, NAS, NAE, AAAS, AGU

Vinal, Edgar B, Jr
Home: 29 Skyline Dr, Plainville, CT 06062
Position: Exec VP *Employer:* Greiner, Inc. *Education:* BSE/Civ/Univ of CT. *Born:* 3/6/29. 1951-1953 - Served in US Army Topographic Battalion in Philippine Islands. 1953-1962 - Proj Mgr with consulting firm of P W Genovese & Assoc, New Haven, CT in charge of Hgwy & Civ Engg Depts. 1962-1968 - Asst City Engr, New Haven, CT. 1968-1970 - Cit Engr, New Haven, CT - in charge of maj urban renewal and water pollution abatement progs. 1970-present Exec VP - Greiner, Inc Chief Executive Office - In charge of Northeast Division. Reg PE in CT, NY, NJ, PA, MD. Land Surveyor in CT, PA. *Society Aff:* ASCE, NSPE, TRB.

Vines, Darrell L
Business: Office Dean Engr, Box 4200, Lubbock, TX 79409
Position: Prof. *Employer:* TX Tech Univ. *Education:* BS/Sci/McMurry College; BS/EE/TX Tech Univ; MS/EE/TX Tech Univ; PhD/EE/TX A&M Univ. *Born:* 3/7/36. Native Texan. Worked at TX Instruments, Inc 1960-1962. Taught part-time at TX Tech & TX A&M where he received PhD in EE in 1967. Undergrad lab dir, advisor, grad faculty, faculty senate, etc since 1966. Received three outstanding teaching awards - Western Electric, Std Oil of IN., & ABELL Award. IEEE Education Society Pres 1987, Bd of Dirs 1978 & 1979. Region 5 Dir 1978-1979. Consultant & summer employee in petrol, govt electronics, transportation, & communications industries. Enjoys old car restoration. Appointed distinquished visiting Prof at USAF Academy 1981-1982, Associate Dean Undergraduate Programs in Engrg, TTU. *Society Aff:* IEEE, ASEE.

Vines, George W
Home: 333 Sunset Dr, Wilmette, IL 60091
Position: President. *Employer:* The Cooper Corporation. *Education:* BS/ChE/Univ of Pittsburgh. *Born:* Feb 1918; native of Pittsburgh Pa. Aircraft Maintenance Engrg Officer, S Pacific Theatre 4 yrs WW II. Contracting Engr 12 yrs; Treadwell Const Co & Heyl- Patterson Pgh Penna. Bulk Material Handling Chem Process Equipment, Atomic Energy & Missile Components. V Pres Sales 14 yrs Phillips Mfg Co Chicago Ill, metal cleaning, pollution control equipment - speaker, lectr metal cleaning AES. Acquired Cooper Corp 1976 Degreasing & Process Tanks Manufac. Enjoy golf & bridge. *Society Aff:* AIChE.

Vinokur, Herman R
Home: 620 Galahad Rd, Plymouth Meeting, PA 19462
Position: Pres. *Employer:* Vinokur-Pace Engg Services, Inc. *Education:* BSME/ME/PA State Univ. *Born:* 2/21/29. Honorable discharge US Army 1948. Air Conditioning Dev Engr, Philco Corp 1952-55; Service Engr, DE Refrig Co 1955-56; Mech Engr, Cronheim-Weger Arch 1956-61. Formed Vinokur-Pace Engg Services 1961-Design/energy conservation for bldgs. Phila Bicentennial NPS projs include Franklin Court Museum & Liberty Bell Pavilion. 1970 First Prize Yale Math Bldg Competition with Arch Venturi-Rauch and 1984 First Prize Newport News, VA Cultural Ctr Competition with Architect Dagit-Saylor. Visiting lectr at Univ of PA Grad Sch of Arch since 1975. Reg Engr - Twelve states and DC; NCEE Certificate of Qualification. Special Examiner for Phila Civ Service Bd, May 1966. Mbr Plymouth Township Municipal Authority since 1977. *Society Aff:* ASHRAE, IES, NSPE, PSPE, AEE.

Viscomi, B Vincent
Business: Dept of Civil Engg, Easton, PA 18042
Position: Prof *Employer:* Lafayette Coll. *Education:* PhD/Civil Engg/Univ of CO; MS/Mech Engg/Lehigh Univ; BS/Mech Engg/Drexel Univ. *Born:* 9/21/33. PE - PA, NJ. Res Engr, Phila Elec, 1957-61. Nuclear Reactor Engr, Phila Elec, 1961-64. Asst to Assoc Prof of ME, Lafayette Coll, 1964-10. Prof and Hd of CE, Lafayette Coll, 1972-present. Conlt Naval Sea Systems Command, Wash, DC; Consumer's Res, Inc., Wash, NJ; Resource Recovery Services, Trenton, NJ; various engrg, legal and insurance firms. Recipient, Jones Faculty Award for Teaching and Res Excellence, 1969; Lindback Award for Distinguished Teaching, 1976; Student Council Superior Teaching Award - 1973, 1977. Listed in Who's Who In The East; Who's Who in Technology Today; International Who's Who In Engineering. Mbr Lehigh Valley Sec ASCE, Sigma Xi, Tau Beta Pi, Easton Area Joint Sewer Authority, Energy Advisory Ctte to Congressman Ritter, Lehigh-Delaware Devel Council, Bd of Health - City of Easton. *Society Aff:* ASEE, ASCE, ASTM.

Visich, Marian Jr
Business: Coll of Engineering & Applied Sciences, Stony Brook, NY 11794
Position: Assoc Dean of Engrg *Employer:* State Univ of NY *Education:* PhD/Applied Mech/Polytech Inst of Bklyn; MAeE/Aeronautical Eng/Polytech Inst of Bklyn; BAeE/Aeronautical Eng/Polytech Inst of Bklyn *Born:* 01/08/30 Faculty of the Polytechnic Inst of Bklyn 1956-74 attaining the rank of Prof of Aerospace Engrg. Joined the faculty of the Coll of Engrg and Applied Scis of the State Univ of NY as Prof of Engrg in 1974. Presently Assoc Dean of Engrg. Exec Committee AIAA, L.I. Section 1967-1971. PE, State of NY. *Society Aff:* AAAS, AIAA, ASEE, AAUP, ΣΞ, ΤΒΠ

Viskanta, Raymond
Business: School of Mech Eng, Purdue University, West Lafayette, IN 47907
Position: W.F.M. Goss Distinguished Professor of Engineering *Employer:* Purdue University. *Education:* PhD/Heat Transfer/Purdue Univ; MSME/Heat Transfer/Purdue Univ; BSME/-/Univ of IL. *Born:* 7/16/31. in Lithuania. Asst & Assoc Mech Engr, Reactor Engg Dir Argonne Natl Lab 1956-62; Assoc Prof of mech engg Purdue Univ 1962-66, Prof 1966- ; Visiting & Springer Prof of mech engg Univ of CA Berkeley 1968-69; Guest professor at the Tech Univ of Munich 1976-77. Visiting Prof. of mech engg Tokyo Institute of Technology 1983. Author of over 300 tech papers & other pubs in heat, mass & radiation transfer. Sr US Scientist Award, Alexander von Humboldt Foundation 1975; ASME Heat Transfer Memorial Award 1976; AIAA Thermophysics Award 1979; ASEE Senior Research Award 1984; Max Jakob Memorial Award (ASME/AIChE) 1986; U.S. National Academy of Engineering 1987. Fellow ASME; Mbr AIAA, AAUP. Reg P E in IL. *Society Aff:* ASME, AIAA, CI.

Viswanathan, Chand R
Business: 7732 Boelter Hall, Los Angeles, CA 90024
Position: Professor & Chrmn. *Employer:* University of California. *Education:*

Viswanathan, Chand R (Continued)
PhD/Solid State Phyiscs/UCLA; MS/Applied Physics/UCLA; BSc (Hons)/Electronics/Univ of Madras, India. *Born:* Oct 1929. Naturalized US citizen. Current field of interest is solid- state electronics, semiconductor devices & integrated circuits. Taught engrg UCLA for last 17 yrs; currently Prof of Engrg & Chrmn, Elec Engg Dept, Sch of Engrg & Applied Sci. Cons to various semiconductor indus. Received Disting Teaching Awards from Grad Students Assn 1972 & from UCLA Alumni Assn & Academic Senate 1976; received Western Elec Fund Award from ASEE for excellence in instruction of engrg students 1974. Author of several publs in solid-state & semiconductor electronics. Fellow IEEE. *Society Aff:* IEEE, APS, ASEE

Visweswaran, Subramaniam
Home: 119 Independence Ave, Tappan, NY 10983
Position: Mgr, Mfg Standards *Employer:* Am Cyanamid Co. *Education:* MBA/Ind Mgt/Fairleigh Dickinson Univ; BSC (Tech)/ChE/Univ of Madras. *Born:* 1/27/40. Subramaniam Visweswaran, b St Thomas Mt, Madras, India, Jan 27, 40; S Sambamurthy Subramania Iyer & Kamalam; student Univ Madras, 55-61, BSc (Tech) CHE; Univ Bombay 65-68, Grad dip Ind E; Fairleigh Dickinson Univ NJ, 72-75, MBA; Married Vijayalakshmi, son Subramaniam, 20; daughter Shuba, 16; Supt Premier Fertilisers, Madras, India, 61-63; Mgr (Prod), Lever Brothers, Bombay, 63-69; Technical Mgr, Lake Soap Industries, Tanzania, 69-71; Sup, Cooper Labs, Wayne, NJ, 71-73; Sr Mgt Engr, Dept Mgr, Prod Supt, Lederle Labs Div, Am Cyanamid, Pearl River, NY; was Plant Mgr of Amer Cyanamid Plant, at BULSAR, INDIA from 81- 82; Sr Mbr, IIE, Chapter Dev Chrmn 76-77k VP 77-78, Pres 78-79 & 79-80; Meritorous Awards from IIE; Past Pres, Bhrathi Soc of An NY; Mbr, Hindu Temple Soc of NA; Speaker for Pkg Inst, SME, IIE, etc; *Society Aff:* IIE.

Viterbi, Andrew J
Business: 3033 Science Park Rd, San Diego, CA 92121
Position: Sr VP & Chief Scientist *Employer:* M/A-COM, Inc *Education:* PhD/EE/USC; SM/EE/MIT; SB/EE/MIT *Born:* 3/9/35. Native of Bergamo Italy. Communication Res at Jet Propulsion Lab 1957-63; Prof of Engrg & Applied Sci UCLA 1963-75; Co-founded LINKABIT Corp 1968, Exec V Pres LINKABIT (full-time) 1973-1982; Pres, M/A-COM LINKABIT, Inc 1982-1984; Sr VP & Chief Scientist, M/A-COM, Inc 1984- ; also Adjunct Prof Univ of Calif San Diego. Fellow IEEE; Mbr Sigma Xi; Christopher Columbus Internatl Communications Award 1975; Annual Outstanding Info Theory Paper Award IEEE 1968; Natl Electronics Conference Award for Outstanding Original Paper 1962, Natl Acad of Engg, 1978. Co-recipient AIAA Aerospace Communication Award 1980, IEEE Alexander Graham Bell Medal for exceptional contributions to the advancement of telecomm, 1984. *Society Aff:* NAE, IEEE

Vitovec, Franz H
Business: Dept of Mech Engrg, Edmonton, Alberta CD T6G 2G8
Position: Prof Emeritus *Employer:* Univ of Alberta. *Education:* Dr tech Sci/Eng Materials/Tech Univ Vienna; Dipl Ing/Mech Eng/Tech Univ Vienna *Born:* 6/7/21. Taught at Univ of MN 1952/58, Res on creep & fatigue of heat resistant alloys. Prof of Met & Nucl Engg, Univ of WI, Madison, 1959/1966. Res on hydrogen attack of steels & irradiation effects. Prof of Met Engg, Univ of Alberta & Chrmn of Dept of Mineral Engg 1970-1980. Fellow of Am Soc for Metals 1977. Prof. Emeritus 1986. *Society Aff:* ASM.

Vittucci, Rocco Victor
Home: 4534 Warren St N W, Washington, DC 20016
Position: Consultant *Employer:* Self Employed *Education:* MME/Heat Transfer/Cornell Univ; ME/Heat-Power/Cornell Univ; Cert/Adv Mgmt/Univ of Pittsburgh. *Born:* Oct 5 1913. Reg P E in DC, DE. Presently R&D management support. Completed 36 years with Navy Dept, most recently as R&D Prog Admin developing tech for advanced ships, subs & boats for US Navy. Formerly taught engrg Univ of Texas. Author of textbook 'Mechanical Engineering Laboratory Manual'. Designed propulsion plants for most large US Navy ships built during & following WW II. Received a number of awards. Designed machinery and power plants for Amer Gas & Elec Co & E I DuPont de Nemours & Co. Active in ASME (Fellow); ASNE (Council, Pres's Award); ASE (P Pres). Formerly in ASEE, SES & SNAME. Pub papers on both engrg & mgmt. Hobbies: music, oil painting. *Society Aff:* ASME, ASNE, ASE.

Vivian, J Edward
Business: Rm 66-448, Cambridge, MA 02139
Position: Prof of Chem Engg. *Employer:* Mass Inst of Tech. *Education:* ScD/Chem Eng/MA Inst Tech; MS/Chem Eng/MA Inst Tech; BEng/Chem Eng/McGill Univ. *Born:* July 6, 1913 Montreal Que, Canada. Came to US 1936; naturalized 1942. Res Engr Canadian Pulp & Paper Assn 1936; MIT Cambridge Mass Teaching Asst 1937- 38; Asst Dir Bangor Chem. Engrg Practice Sch Station 1938-41; Dir Buffalo Chem Engrg Practice Sch Station 1941-43; Res Engr Govt Proj 1943-46; Assoc Prof Chem Engrg 1946-56; Prof Chem Engrg 1956-80; Exec Officer Dept Chem Engrg 1974-79; Prof Emeritus 1980-; Cons Chem Engr 1946- ; Visiting Prof Birla Inst Tech & Sci Pilani Rafasthan India 1972. Fellow AIChE (Colburn Award 1948) Mbr TAPPI, Amer Chem Soc, AAAS, ASEE, Sigma Xi. *Society Aff:* AIChE, ACS, ASEE, TAPPI, AAAS.

Vizy, Kalman N
Home: 16 Clearview Dr, Spencerport, NY 14559
Position: Physicist, Scientific Advisor *Employer:* Eastman Kodak Co. *Education:* BES/Eng Sci/Cleveland State Univ; MS/Theor Physics/John Carroll Univ. *Born:* 7/7/40. Taught in Parma OH until 1967. Licensed P E in NY & Certified Fallout Shelter Analyst for DOD office of Civil Defense. Joined Eastman Kodak Res Labs 1967 & functioned as Sr Res Physicist to 1980. Has since become scientific advisor to Health Sciences Markets Division Management. Joined Rochester Inst of Tech adjunct faculty 1968 & teaches Modern Phys & Advanced Engg Math since then. Winner of ASP Autometric Award for best paper 1975. Intersociety coordinator SPSE. Licensed Photogrammetrist by Am Soc of Photogrammetry. Excellence in Teaching Award from RIT for 1980. Worldwide lecturer on medical diagnostic physics and engrg. *Society Aff:* NSPE, ASP, SPSE, ASME, AAPT, AAPM, SMPTE, SID

Vlad, Ionel-Valentin I
Home: 10, Stockholm St, Bucharest, Rumania R-71222
Position: Chief Lab *Employer:* Natl Center of Physics *Education:* Dr Eng/Elec/Polytech Inst Bucharest; Dipl Eng/Elec/Polytech Inst Bucharest. *Born:* 09/22/43 Mbr of the research staff of the Inst of Atomic Physics, Bucharest, and from 1975, Chief of Holography and Optical Info Processing Lab in the Natl Center of Physics-Laser Dept, Deputy Chief of Dept Lasers from 1984 Stage at CGE-Marcoussis, France, in 1969. From 1968, associate lecturer at the Polytechnic Inst of Bucharest. Author of two books and over 100 papers reports; "T. Vuia" Award of Rumanian Acad 1978. Mbrships including: Rumanian Natl Comm Physics, Fellow Opt Soc Amer, European Physical Soc. Mbr of editorial advisory panel of the journal "Optics Letters" (U. S.A.). *Society Aff:* RNCP, OSA, AIP, EPS, DGaO.

Vockroth, John H, Jr
Home: 1014 Prescott Ave, Scranton, PA 18510
Position: City Engineer *Employer:* City of Scranton *Born:* 06/12/37 Native of Scranton, PA-US Navy, Mobile Construction Battalion No.6, 1957 to 1959. Field Engr for Albright and Friel, Inc (Now Betz Eng Co) consltg engrs, 1959 to 1962. President, John H. Vockroth Assoc, Consltg Engrs, 1962 to present. Scranton City Engineer, Jan, 1981 to present-see attached condensed experience record. Active in community affairs and profl organizations. Presently a mbr of a court appointed commission to fix the boundary lines of Lehigh County, PA. Associate of two multi-discipline Architect Engrg firms. Enjoy Real Estate speculating, classical music, hunting and fishing. *Society Aff:* NSPE, PSPE

Voelcker, H B
Business: MAE, Upson Hall, Cornell University, Ithaca, NY 14853
Position: Charles Lake Professor of Mechanical Engrg *Employer:* Cornell Univ. *Education:* PhD/Engg/Imperial College of Sci & Technology London; SM/Elec Engg/MIT; SB/Mech Engg/MIT. *Born:* 1930. HERBERT B. VOELCKER holds the Charles Lake Chair in the Sibley School of Mechanical & Aerospace Engineering at Cornell University, and is Director of Cornell's Manufacturing Engineering & Productivity Program (COMEPP). He spent 1985-86 at the National Science Foundation, helping to organize first the Design & Manufacturing Division in the Engineering Directorate and then the new Computer & Information Science & Engineering Directorate. Prior to that he was a member of the electrical engineering faculty at the University of Rochester for 24 years. While at Rochester he founded the Production Automation Project in 1972 and served as its Director until 1985. He served in the U.S. Army for seven years, has worked or consulted for various industrial and governmental organizations, and has won various research prizes, teaching prizes, and fellowships. *Society Aff:* IEEE, ASME, ACM, SME.

Vogel, Chester T
Business: 1271 Ave of Americas, New York, NY 10020
Position: Partner *Employer:* Kallen & Lemelson *Education:* B/ME/RPI *Born:* 01/16/37 Design and sales engr for Johnson Controls Co from 1958 to 1965. Reg PE in NY, NJ, CT and FL. *Society Aff:* ASHRAE, ASME, NSPE, NCCC, ASHE, ASSE, ASPE

Vogel, F Lincoln
Home: Voorhees Rd RD3, Whitehouse Station, NJ 08889
Position: Pres & Gen Mgr *Employer:* Intercal Co *Education:* BS/Met Engrg/Univ of PA; MS/Met Engrg/Univ of PA; PhD/Met Engrg/Univ of PA *Born:* Nov 30, 1922. Mbr Tech Staff Bell Labs 1952-59, pioneering res on dislocations in crystals; Mgr Advanced Materials RCA 1959-63, R&D on silicon & gallium arsenide as electronic materials; Dir semiconductor res, Sprague Electronic Materials Co 1963-68, R&D on mos transistors & ion implantation; Univ of Pa Visiting Prof 1968-69 & Res Prof 1973-83, res on high conductivity graphite intercalation compounds. Chmn Inst of Metals, AIME 1970-71; Engrg Soc's Library Bd 1972-81, Pres. ESL board. 79-81 Editor, Synthetic Metals. 1983 - Pres & Gen Mgr, Intercal Co. *Society Aff:* AIME, APS

Vogel, Fred J
Home: 1806 Beverly Cir, Clearwater, FL 33546
Position: Transformer Consultant (Self-employed) *Education:* BS/Elec Eng/MIT. *Born:* Apr 1893. Testing 1915-16; USN civil employee, submarine const 1917-18; Westinghouse Elec 1919-43; Div Engr Large Power Transformer & 1st model Mark 18 Torpedo, US Navy Prof Elec Engr ITT 1943-51; Cons Allis Chalmers 1952-61 Pratt & Whitney 1962-63; McGraw Edison 1964-69; others to present. Chmn Sharon Section IEEE, CHmn IEEE Trans Ctte & various subcttes. Co-Or Comm 4 IEEE St Comm Fellow IEEE. Recipient Habirshaw Award & Medal. Received many pats & contributed many papers; Certificate of Commendation, Navy Dept-1947. *Society Aff:* IEEE.

Vogel, Herbert D
Home: 3033 Cleveland Ave N W, Washington, DC 20008
Position: Consulting Engr (Self-employed) *Education:* Dr Ing/Civil Engg/Berlin Tech Univ; CE/Hydraulics/Univ of MI; MS/Civil Engg/Univ of CA; BS/-/US Military Acad. *Born:* Aug 1900. Served 30 yrs in Corps of Engrs US Army, incl cvl works assignments as Dist & Div Engr, Engr of Maintenance Panama Canal & Lt Gov C Z. Served during WW II in SW Pacific from Australia to Japan as G-4 USASOS & USASCOMC & Cdr Base M in Philippines. After retirement from Army was Chmn of Bd TVA 1954-63 & Engr Adv World Bank 1963-67. Reg P E in N Y, Tenn, Texas & D C. Hon Mbr ASCE & Internatl Comm PIANC; Hon Mbr Public Works Historical Society Recepicent President's Award ASCE 1979 Hon Mbr SAME, Mbr Nat'l Acad. Engrg, ACEC & Royal Soc London; Mbr ICOLD & NSPE. Decorations incl DSM, Legion of Merit & Knight of the Grand Cross (Thailand). *Society Aff:* ASCE, SAME, ACEC, NAE.

Vogel, Richard C
Business: 3412 Hillview Ave, Palo Alto, CA 94303
Position: Technical Advisor *Employer:* Electric Power Research Inst *Education:* PhD/Physical Chemistry/Harvard Univ; MS/Biophysical Chemistry/PA State Univ; BS/Chemistry/IA State Univ. *Born:* 1/28/18. in Ames, IA. He was married in 1944 & has three children. From 1946 to 1949 Dr Vogel was Asst Prof of Chem at IIT. He was a Sr Chemist at Argonne Natl Lab from 1949-54. Assoc Dir of the Chem Eng Div from 1954-63 & Div Dir from 1963 to 1973. From 1973-83 he has been employed by Exxon Nuclear Co. He is now employed by EPRI. He is a mbr of the Am Chem Soc, a fellow of the Am Nucl Soc & a fellow of the Am Inst of Chem Engrs. He has received the Robert E Wilson Award in Nuclear Chemical Engg from AIChE. He has done res in physical inorganic chem, fluorine chem, pyrochem processes, separations processes, & chem problems in nucl reactor safety. He is currently responsible for EPRI's source term work. *Society Aff:* AIChE, ACS, ANS.

Vogelman, Joseph H
Home: 48 Green Dr, Roslyn, NY 11576
Position: President. *Employer:* Vogelman Development Co. *Education:* DEE/Electronics/Polytechnic Inst of Brooklyn; MEE/Electronics/Polytechnic Inst of Brooklyn; MBA/Acctg/CCNY; BS/Mathematics/City College of NY. *Born:* Aug 18, 1920 New York, N Y. Signal Corps AUS 1942-44; Ch Test Equipment Section 1944-47, Ch Dev Branch 1947-51 Watson Lbs USAF; Ch Scientist Genl Engrg 1951-54, Dir of Communications 1954-59 Rome Air Dev Ctr N Y; V Pres & Dir Capehart Corp 1959-64; Dir of Electronics 1964-67, V Pres 1967-73 Chromalloy Amer Corp; Cons 1964-72, Dir & Ch Scientist 1973- Orentreich Foundation for Advancement of Sci; V Pres & Dir 1968-73 CRO-Med Bionics Corp; V CHmn of Bd & Sr V Pres 1968-73 Laser Link Corp; Dir TheatreVision Inc 1971-73; Pres 1973 Vogelman Dev Co. Dir 1983-85 CompuPix Tech Inc. Over 100 publs & over 100 pats. Mbr IEEE (Fellow), AAAS (Fellow); Mbr Sigma Xi, Eta Kappa Nu, N Y Acad of Sci. Reg P E in N Y & N J. Outstanding Performance Award USAF 1957. *Society Aff:* IEEE, AAAS, NYAS, NSPE.

Vogl, Thomas P
Business: Environmental Research Institute of Michigan, 1501 Wilson Blvd, Suite 1105, Arlington, VA 22209
Position: Research Manager *Employer:* Environmental Research Institute of Michigan (ERIM) *Education:* BA/Phys Sci/Columbia Univ; MS/Physics/Univ of Pittsburgh; PhD/Elec Engr (Sys Sci)/Carnegie-Mellon Univ *Born:* 07/10/29 1952-60, 1961-1974: Mgr, Optics, Westinghouse Res Lab, Pittsburgh, PA. Res on photoelectric effect, thermal imaging devices and detectors, corporate optical consltg service. 1960-61, Head Intraved Sect, Hughes Res Lab, Malibu, CA 1974-1977: NAS-NRC Staff Officer, Div of Medical Scis. Committees: Prosthetic Res in VA, Health Care Resources in VA, others. 1973-79 Adj (Full) Prof of Radiation Biophysics, Depts of Radiology and Pediatrics, Columbia Univ NY 1977- present: Natl Inst of Health, Bethesda, MD '77-'79: Exec Secy, Natl Comm on Digestive Diseases, '79-'86: Expert, Nutrition Coordinating Committee, NIH. '86-present: Research Manager, Environmental Research Institute of Michigan, Arlington, VA. 1957-1975: Instructor, UCLA short courses; 1973-74 Mbr, NAS-NRC Committee on Phototherapy of the Newborn and Chrmn of its subcommittee on Bioengrg Aspects. 50 publications, 5 patents. Fellow, Optical Soc of America *Society Aff:* OSA, ASP, AAAS, AAWH

Vogt, Edward G
Home: 4804 Randolph Dr, Annandale, VA 22003
Position: Oper Res Analyst *Employer:* US Army/Oper, Test & Eval Agency *Education:* PhD/ChE/Univ of MI; MS/ChE/Univ of MI; BS/ChE/Tri-State College. *Born:* 3/2/19. Pilot plant operator, UOP, 1941-43. Res Assoc in high temperature met, Dept Engg Res, Univ of MI, 1943-46. Asst Prof of Chem Engg, WSU 1946-50.

Vogt, Edward G (Continued)
Res Engr in lube oil processing, M W Kellogg Co, 1950-54. Ran vacuum melting operation at Universal-Cyclops Steel Corp, 1954-59. Performed Operations Res studies in naval tactics, strategy & long range planning for Ctr for Naval Analysis, 1959-72 & for Ketron, Inc, 1972-78. Ran own co, Dunhill of Alexandria 1978-81. Conducting evaluation of weapon systems for US Army Operational Test & Evaluation Agency 1982-present. *Society Aff:* AIChE, ACS, ORSA.

Vogt, Howard W
Home: 13259 Aleppo Dr, Sun City West, AZ 85375 *Education:* BS/Chem Engr/Univ of Rochester; MBA/Management/Univ of Rochester. *Born:* Nov 1919. Exec Dev Program. With US Air Force 1942-46; Pilot. Photo Engr, Eastman Kodak 1949- . Color Photo process design for Kodachrome, Ektachrome & Eastman Reveral Intermediate Films. Dir Photo Tech 1973. Fellow SMPTE; Kalmus Gold Medal Award 1970. Mbr ACS, SPSE, Rochester Engrg Soc. Photographic Soc of America (PSA). Retired 1981. Retired 1981. *Society Aff:* SPSE, PSA.

Vogt John E
Home: 3333 Moores River Dr. /206, Lansing, MI 48911
Position: Sanitary Engineer. *Employer:* Retired *Education:* MPH/Pub Health Engg/Univ of MI; BSCE/San Engg/Univ of IA. *Born:* 5/12/16. Native of Iowa City Iowa. WK Kellogg Fndtn. Fellowship in Public Health Engrg 1937; San Engr Mich Dept of Public Health 1940-78; apptd Dir 1960 of what is now Bur of Environ & Occupational Health. Retired from MDPH Jan 1978. Employed by Mc Namee, Porter & Seeley Consulting Engrs, Ann Arbor, MI. July 1978 to July 1984 as San Engr; Retired from active practice July 1984; served with Army Corps of Engrs & San Corps 1941-46. Reg PE in MI. Mbr of Mich & Natl Soc of Prof Engrs. Hon Mbr AWWA 1975; Recipient of AWWA Fuller Award. Chrmn of Conference of State San Engrs 1970-71. Diplomate Environ Engrg Inter - Soc Bd. Recipient 1976 US EPA Environ Quality Award in Mich. Enjoy traveling, reading & photography. *Society Aff:* NSPE, AWWA, AAEE, CSSE.

Vogt, John W
Business: 8660 Astronaut Blvd, Cape Canaveral, FL 32920
Position: VP. *Employer:* Stottler, Stagg & Assoc. *Education:* BS/Civ Engg/Univ of FL. *Born:* 12/28/36. Native of Lake Wales, FL. Attended Lakeland, FL public schs. Design Engr with Wellman-Lord, Inc, & Lakeland FL Brown Engg Co, Cape Canaveral, FL Proj Mgr with MDC Systems Corp, NYC. With Stottler, Stagg since 1969. Elected to FL State Senate in 1972, reelected in 1976 and 1980. Chrmn of Natl Resources Committee. Natl Conf of State Legislatures, Natl Resources Committee. FL Engg Soc Legislative Effectiveness Award 1981. FL Wildlife Federation Conservation Legislator of Yr, 1979 and 1981. FL Engg Soc Local Chapter Engr of the Yr, 1973. Wife Tonie, daughters D'Anna, Suzanne, Leeanne, Lisa, & Vicki. Enjoy tennis & folk guitar. *Society Aff:* NSPE.

Voiculescu, Ion Alexandru
Business: Str 13 Decembrie Nr 94, Brasov Romania 2200
Position: Sr Res Engr. *Employer:* INMT-Brasov Branch. *Education:* PhD/ME/Univ of WI; Diploma Engr/ME/Univ of Brasov-Romania. *Born:* 6/17/37. Native of Craiova, Romania. Testing & field engr for UTB Tractor Factory 1961-66. Assoc Res specializing in automotive diesel engines at the Inst for Res & Dev for Automobiles and Tractors (ICPAT) Brasov. With the Natl Inst for Thermal Engines (INMT) since 1976. Part-time teaching Heat Transfer & Thermodynamics at The Univ of Brasov. Fulbright scholar, under Sr Fulbright - Hays Program, USA-1971. SAE Horning Meml Award 1978. "Cintarea Romaniei" Award 1980 and 1985. Enjoy music, skiing and automobile rally. *Society Aff:* CNIT.

Voigt, Robert C
Home: 2421 N 69th St, Kansas City, KS 66109
Position: Chief Civil/Structural Engineer *Employer:* The C.W. Nofsinger Co. *Education:* MS/Civil Engg/Univ of KS; BS/Civil Engg/MIT. *Born:* Dec 25, 1937; native of Atchison Kans. MS Cvl Engrg Kans Univ 1967; BS MIT 1959. Designer Detailer with Howard Needles, Tammen & Bergendoff 1959-66, Ch Struc Engr Schlup, Becker & Brennan 1966-74, V Pres & Ch Struc Engr 1974-1981, Chief Civil/Structural Engineer-The C.W. Nofsinger Co. 1981-present; responsibilities include all design & construction plans for all civil/structural portions of the company projects. Mbr. ASCE, Enjoy marksmanship, big game hunting, classical music, photography & literature. *Society Aff:* ASCE.

Volcy, Guy C
Home: 26 Allee Du Levrier, Le Vesinet, France 78110
Position: Independent expert, consultant, arbitror. *Employer:* Bureau Veritas - Paris. *Education:* MS/Techn Sci/Polytechnical Univ Gdansk; BS/-/Polytech Univ. Gdansk *Born:* April 1923. Taught same Univ, marine engrg, practical experience new bldg steam & diesel driven ships, shipboard & repairs 1950-57. With Bureau Veritas 1958- in hull & machinery, respon for trouble shooting activity alignment problems of line shaftings crankshafts, main gearings. 1962 discovered unconventional deformations of aft part of hull girder; 1972 introduced integral treatment of static & vibratory phenomena of ships & res of forced vibration resonators. Fellow Inst Marine Engrs, Mbr SNAME, V Adm E L Cochrance Award 1975 (as 1st non-American) SNAME, Fellow RINA. Mbr ATMA Mbr ASME. Mbr NECIES - 52th Memorial Andrew Laing lecture (as 1st Frenchman and 2nd non-Briton)-1983. Author 110 publs & lectures. Enjoy music, sailing, tennis, 7 foreign languages. Visiting Senior Lecturer Univs New Castle upon Tyne, Rio De Janeiro, Hong Kong, Trieste, Madrid. Chrmn of Intl Cooperation and Marine Engrg Systems (DCMES). Since 1986 Visiting Prof of JMU World Maritime Univ. Since 1987 retired from position of scientific Dir of Msf Branch of B.V.; active as independent Expert consultant. In Aug 1987 honored by French Marine Ministry with title CHEVALIER of MARITIME MERITS. Involvement in several Arbitrations and Litigations as technical Expert. *Society Aff:* ATMA, RINA, SNAME, ASME, NECIES, IME.

Volkin, Richard A
Business: 504 Beaver St, Sewickley, PA 15143
Position: Corp VP *Employer:* Green Intl Inc *Education:* MS/Ind Mgt/London Univ; BS/Marine & Elec Engg/MA Maritime Acad. *Born:* 6/29/42. Reg PE in MA & NH. Previous experience since 1963 includes licensed Merchant Marine Engrg Officer, Gen'l Mgr. Proj Mgr & Chief Engr of Large Consultant Engrg firms. Presently Corp VP of Green Intl Inc and Pres of Engrg Co, Inc responsible for all Natl & Intl operations & office mgt. Involved with engrg, design, construction inspection and consulting to all governmental municipal, ind & private clients located in the US and abroad. An officer in the US Naval Reserve, enjoys boating & fishing. *Society Aff:* AIChE, NSPE, CSI, WPCF, APCA.

Vollmar, John R
Home: 225 Oak Hill Dr, Hatboro, PA 19040
Position: Partner. *Employer:* LTK Engineering Services *Education:* BS/Elec Engr/Drexel Univ. *Born:* 11/8/29. Native of Phila. Served with Corps of Engg US Army 1952-54. Sales and application engr Locomotive and Car Equip Dept GE Co 1954-1959. Rail transportation equip engr LTK&A 1959-65. Became partner in charge of rail transportation equip engg LTK&A in 1965. Became Vice President LTK Engineering Services 1985 (Successor corp. to LTK&A). Responsible for engg services resulting in electric commuter cars for service on Long Island RR, New Haven a N Jersey Conrail commuter operations; rapid transit cars for service in S Jersey, Wash, DC and Boston. Chrmn Rapid Transit subcommittee of Land Transportation Committee of IEEE. *Society Aff:* IEEE, NSPE.

Vollmer, James
Home: 609 Centre St, Haddonfield, NJ 08033
Position: President *Employer:* James Vollmer Assoc, Inc. *Education:* PhD/Physics/Temple Univ; MA/Physics/Temple Univ; BS/Gen Sci/Union College; AMP/Mgmnt/Harvard Bus Sch. *Born:* 4/19/24. Native of Phila, PA; Taught at Temple Univ; was Res Leader with Honeywell; was Sr V.P. RCA - is currently pres

Vollmer, James (Continued)
of James Vollmer Assoc, Inc, a consulting and advisory group for high technology business. Has been broadly involved in educ, res & bus. Fellow IEEE, AAAS; Phi Beta Kappa, Sigma Xi, Sigam Pi Epsilon & Eta Kappa Nu, IEEE Centennial Medal. *Society Aff:* IEEE, AAAS, APS, NYAS, ΣΞ, ΦΒΚ, ΣΠΕ, HKN.

Vollmer, Reuben P
Home: 29840 Knoll View Dr, Rancho Palos Verdes, CA 90274
Position: Partner. *Employer:* Vollmer-Gray Engg Labs *Education:* BS/ME/Milwaukee School of Engg; AAS/Metallurgy/Milwaukee School of Engg. *Born:* 9/24/35. Mr Vollmer was educated in metallurgy and mech engg and has practiced in the field of accident analysis for approximately 20 yrs. He is a Reg PMechE and Safety Engr in CA. He has testified in many municipal, county, state, and fed cts across the country as an expert in his field and has lectured at various univs and professional organizations relating to accident analysis. Mr Vollmer has been prin in an engg consulting firm and engg lab for ten yrs. He enjoys yacht racing, scuba diving, photography and jogging. *Society Aff:* SAE, ASM, ASTM.

Volpe, John A
Business: American Embassy, APO, NY 09794
Position: Ambassador (Retired). *Employer:* US Governmetnt. *Education:* Assoc Degree/Architectural Eng/Wentworth Inst. *Born:* Mass. Grad Wentworth Inst. Entered genl contracting field 1933; formed own const firm 1941 - the Volpe Const Co. In Navy Cvl Engrg Corps WWII, reached grade of Lt Cdr. Was served as Mass Commissioner of Public Works 1953; Governor of Mass 1961-63 & 1965-69; Secy of Transportation 1969-72; appointed Ambassador to Itlay 1973-77. P Soc Dir, V Pres & Pres; Fellow & Hon Mbr SAME, 31 Hon Degrees. *Society Aff:* SCE, SAME.

Von Aulock, Wilhelm H
Home: Box 84 Fairview Dr, Bedminster, NJ 07921
Position: Retired *Employer:* Retired *Born:* Jan 24, 1915 Pirna Germany. Dipl Ing Tech Univ Berlin 1937; Dr Ing EE Tech Univ Stuttgart 1953. Specialist US Navy Dept Wash D C, torpedo countermeasures, EM fields in seawater 1947-53; with Bell Labs 1954- ; microwave ferrite materials & devices, phased array radar antennas; Hd Nuclear Effects Dept 1962-70; sys studies of telephone loop plant 1971-77;, Dir, Technical Support, Amer Bell, International 1977-78; Fellow IEEE 1974. 2 pats, several publs & 2 books on ferrite materials & applications. Retired 1985. *Society Aff:* IEEE.

Von Bargen, Kenneth L
Business: Dept of Agricultural Engrg, Lincoln, NB 68583-0726
Position: Prof *Employer:* Univ of Nebraska-Lincoln *Education:* PhD/Agr Eng/Purdue Univ; MS/Agr Eng/Univ of NB; BS/Agr Eng/Univ of NB *Born:* 04/06/31 Specialize in agricultural equipment systems utilizing economic models and simulation models of crop, soil, climate and equipment. Research has involved fuel use in agricultural field operations, hay and crop residue harvesting and handling systems, energy required for particle size reduction of hay and crop residues and field performance of agricultural tractors. Chrmn of the Bd of NB Tractor Test Engrs, 1981, and Bd mbr since 1976. Chrmn of the NB Section of ASAE, 1981-82. Mbr of the Bd of Dirs of the American Forage and Grassland Council, 1977-80. *Society Aff:* ASAE, AFGC, SAE

von Behren, Fred W
Business: Village of Cross Keys, Baltimore, MD 21210
Position: President. *Employer:* Fred W von Behren Inc. *Education:* BSME/Mech Engr/Johns Hopkins. *Born:* Nov 1920 Balto Md. Balto Polytech Inst 1939; Engr Condair Corp 1945-48; Engr-Designer Henry Adams Inc 1949-52; Ch Engr Riggs Distler & Co Inc 1952-56 & 1959-63; Div Hd Bernard Johnson & Associates 1956-59; Pres Fred W von Behren Inc 1963-81; Assoc Principal RTKL Assoc, Inc 1981. USNR 1943-45; presently active as officer, dir or trustee of various religious, prof, civic & serv orgs; contrib articles to prof publs. Prof practice strongly oriented to design & replacement of utilities & associated modernization of large existing bldgs. *Society Aff:* NSPE, ASHRAE, AEE.

Vonderschmitt, Bernard V
Home: Box 147, New Hope, PA 18938
Position: Vice President & Genl Manager. *Employer:* RCA Corp Solid State Division. *Born:* Oct 14 1923. BSEE Rose Polytechnic Inst; MSEE Univ of Penn. 1944 RCA Engr, Component Design Projects, Camden NJ; US Navy Radar Officer; returned to RCA in Deflection System Dev for monochrome & color TV receivers, 1959 Solid State Div, Mgr Applications Microelectronics Dept, engrg mgmt positions of increasing responsibility, 1959 Mgr Integrated Circuits, became Div V Pres Solid State Integrated Circuits, Jan 1973 appointed V Pres & Genl Mgr Solid State Div, respon for engrg, mfg & marketing of integrated circuits, power transistors, thyristors, rectirectifiers, power hybrid modules & electro-optics devices for commercial, consumer, indus & aerospace markets. 13 patents; authored num tech papers. 1971 co-recipient RCA's David Sarnoff Outstanding Achievement Award in Engineering, citing efforts & leadership in timely dev of superior integrated circuits f.

Vondrick, Art F
Business: 5312 N 12th St, Phoenix, AZ 85014
Position: Exec VP. *Employer:* Authur Beard Engs Inc. *Education:* BSCE/Sanitary/Univ of WI. *Born:* Nov 1923. Reg PE Ill, Az, Texas, California, & New Mexico; Certified Grade IV Operator Water & Wastewater. Bd/Dir Ariz Water Resources Ctte; Bd/Dir Central Ariz Proj Assn; Pres Water Pollution Control Federation 1970-71; Pres Elect 1969-70, V Pres 1968-69, Bedell Award 1960; AWWA Water Utility Man of the Year 1972; Distinguished Service Award for Environmental Control 1972; Hon Citizen of Tex; Chmn Governor's Task Force on Emergency Planning for Water 1966; Diplomate AAEE; Select Soc of Sanitary Sludge Shovelers; Life Member -AWWA. *Society Aff:* ASCE, AWWA, WPCF, AAEE, SAVE.

Von Eschen, Garvin L
Home: 3758 Chevington Rd, Columbus, OH 43220
Position: Prof Emeritus *Employer:* Ohio State University. *Education:* MS/Aero Engg/Univ of MN; BAeroE/Aero Engg/Univ of MN. *Born:* July 1913, Morristown Minn. 1936-46 taught engrg, Univ of Minn; 1939-40 Engr Lockheed Aircraft; 1944-46 Cons Aerodynamicist, Aeronautical Div MPLS Honeywell; 1946-79, Professor & Chmn Aeronautical Engrg Dept Ohio State Univ, Res Supervisor, OSU Res Found 1947-79, Prof, Dept of Aero-Astro, Engg 1979-80. Cons to USAF & Indus. Res specialist N Amer Aviation 1955, 57. Mbr & Chmn var faculty comm & councils at Ohio State; Mbr of Engrg Educ & Accreditation Comm of ECPD 1965-70; Mbr ASEE NASA Space Engrg Comm 1971-75. Active pilot since 1939. Dir 1971 ECPD. *Society Aff:* AIAA, APS, ASEE.

Von Flue, Frank W
Home: 921 E Walnut Ave, El Segundo, CA 90245
Position: Program Mgmt. *Employer:* Rockwell International Corp. *Education:* MBA/Industrial Mgmt/Univ of So CA; BS/Engg/Univ of So CA. *Born:* Nov 1925, El Segundo Calif. After serving in Army Air Corp during WWII, graduated USCBS 1950, MBA 1967. 12 years at AiResearch-Garrett in proj, systems & mfg engrg, specialized in aerospace heat transfer & turbo-machinery equip design, involved in qualification tests, equip modifications & support operations; 2 years with Teledyne Corp as Applications Engr; joined Rockwell Internatl Space Div, Advanced Programs over manned & unmanned space flight activities, managed elements of major proposal for Space Shuttle Orbiter definition, Shuttle System Integration & Global Positioning Satellite, respon for proposal policy on advanced space systems. Program Dev Mgr over Satellite Power Sys Contract & Analysis of Space Construction Sys. Currently Business Mgr of Space Trans Sys Oper. ASME: Mbr, recipient Church Award 1976 for contributions to field of mech engrg education, P Chmn Los Angeles Sect; Fellow Inst for Advancement of Engrg (FIAE); Reg PE Calif. Mbr

Von Flue, Frank W (Continued)
Ntl Mgmt Assoc. Enjoys building houses & high desert living. *Society Aff:* ASME, NMA, FIAE.

von Gierke, Henning E
Home: 1325 Meadow Lane, Yellow Springs, OH 45387
Position: Dir Biodynamics & Bioeng Div *Employer:* Aerospace Med Res USAF *Education:* Dr Eng/Communication/Tech Univ; Dipl Ing/Engrg/Karlsruhe, Germany *Born:* May 1917, Karlsruhe Germany. 1944-47 Res acoustics, applied physics & lecturing; since 1947 with Aerospace Medical Res Lab of US Air Force, res on effects of noise, vibration, impact, blast on man 's safety & performance, since 1961 Dir Biodynamics & Bioengrg Div. 0ver 140 journal/book publ, acoustic patents. Served on many natl adv cttes on aviation noise, environ safety & space program; active in natl & internatl standardization in bioacoustics, ride quality, effects of shock & vibration on man. Clin Prof Wright State Univ. Assoc Professor Ohio State Univ; Mbr NAE, Internatl Acad of Aviation & Space Medicine; Hon Fellow IES; Fellow ASA, Aerospace Medical Assn; Mbr Inst of Noise Control Engrg. Dept of Defense Distinguished Civilian Service Award 1963; Liljenkrantz Award 1966 & Tuttle Award 1974 of Aerospace Medical Assn. Lissner Award Amer Soc of Mech Engrs 1983. *Society Aff:* ASA, BMES, ASMA, IES, NAE.

VonGonten, William D
Business: Petroleum Engg Dept, College Station, TX 77843
Position: Dept Hd. *Employer:* TX A&M Univ. *Education:* BS/Petroleum Engr/TX A&M Univ; BS/Geology/TX A&M Univ; MS/Petroleum Engr/TX A&M Univ; PhD/Petroleum Engr/TX A&M Univ. *Born:* 2/15/34. After completing BS degrees, worked for Mobil Oil Co as a production engr and the US Air Force before entering Grad School. Completed grad work in 1966 and joined the staff of the Petroleum Engr Dept at TX A&M Univ as an Asst Prof. Assumed current position of Dept Hd and Prof of Petroleum Engg in 1976. Past Mbr of the Bd of Dir and E&A Committee of ABET, lecturer and past chrmn of the Education and Accreditation Comm for the Soc of Petroleum Engrs. Past Member of the Bd of Directors of the Soc of Petroleum Engineers. *Society Aff:* SPE, AAPG, ASEE, SPWLA.

Von Ohain, Hans
Home: 5598 Folkestone Dr, Dayton, OH 45459
Position: Sr Res. Engr. *Employer:* Univ. of Dayton Res Inst. *Education:* PhD/Phys/Univ of Goettingen. *Born:* Dec 1911. Ch Jet Engine Dev Div Heinkel-Hirth Co, Stuttgart Germany; patented & designed turbojet engine which achieved world's 1st turbojet flight, Aug 1939 Rostock Germany; 1945-47 Cons to US Navy, Kornthal Wurtemberg, Germany; performed theoretical work on advanced air-breathing propulsion systems for USAF at WPAFB Ohio, Supervisory Aeronautical Res Engr, Thermomechanics Res Branch Aeronautical Res Labs, 1963-75 Ch Scientist Aerospace Res Labs, presently Ch Scientist AF Aero Propulsion Lab. Fellow AIAA from 1970, received AIAA Goddard Award 1966; Distinguished Mbr Jet Pioneers' Assn of USA 1974. Retired 15 Jan 1979, is currently gas physics consultant to USDOE; Univ of Dayton. Res Inst, Department of Defense Disting Civilian Service Award 1979; Mbr Ntl Academy of Engrg 1980. *Society Aff:* AIAA, NAE.

von Riesemann, Walter A
Home: 7928 Woodhaven Dr NE, Albuquerque, NM 87109
Position: Supervisor *Employer:* Sandia National Laboratories *Education:* PhD/Civil Engg/Stanford Univ; MS/Civil Engg/Univ of IL; BS/Civil Engg/Polytechnic Inst of Brooklyn. *Born:* Feb 1930, Brooklyn NY. 1948-52 served with US Army Corps of Engrs; 1959- 60 res engr with Alcoa Res Labs, New Kensington Pa; since 1960 with Sandia Labs, Albuquerque NM, respon for structural safety analyses of light water reactor nuclear power plants. ASCE Robert Ridgway Student Award 1958; NSF Fellowship 1965-68; ASME Pressure Vessels & Piping Literature Award 1975. Pres NM Sect ASCE 1975-76; Pres NM Sect ASME 1980-81; Fellow ASCE; Mbr ASME & ANS. Active in backpacking & camping. *Society Aff:* ASCE, ASME, ANS.

von Roeschlaub, Frank
Home: 1 Pleasant View Dr, Gloversville, NY 12078
Position: Consulting Engineer. *Employer:* self *Education:* D Eng/Elec Engg/Yale Univ; BS/Elec Engg/Yale Univ. *Born:* May 11 1912, Denver Colo. 1936-46 design & dev engr for Genl Electric Co in synchronous machinery & power system protection; 1946-77 with Ebasco Services Inc as Cons Engr, respon for formulating power plant electrical engrg practices, power system planning, protection & automatic oper, extensive work in trouble analysis & special tech problems. Since 1977, active as consulting engineer in private practice. Fellow IEEE; Mbr Sigma Xi. Reg PE. US Representative to CIGRE Ctte on Protection 1968-76; Membership in IEEE natl tech cttes has included Standards 1961-69, Rotating Machinery, Power System Relaying (Chmn 1954-56). *Society Aff:* IEEE, ΣΞ, CIGRE.

Von Tersch, Lawrence W
Business: Mich State Univ, East Lansing, MI 48824
Position: Dean, College of Engineering. *Employer:* Michigan State University. *Education:* PhD/Elec Engg/IA State Univ; MS/Elec Engg/IA State Univ; BS/Elec Engg/IA State Univ. *Born:* March 1923. 1946-56 Instructor to Professor Electrical Engrg, Iowa State Univ; 1956- , Dir Computer Lab Mich State Univ, 1958-61 Professor Electrical Engrg Chmn Dept, 1965-68 Assoc Dean Engrg, 1968- , Dean. *Society Aff:* IEEE, NSPE.

von Turkovich, Branimir
Business: 201 Votey Bldg, Burlington, VT 05401
Position: Professor & Department Chairman. *Employer:* The University of Vermont. *Education:* PhD/Mech Engg-Physics/Univ of IL; D Naval Engg/Naval Engg/Univ of Madrid, Spain; MS/Naval Engg/Univ of Madrid. *Born:* Dec 23 1924, Zagreb Croatia. MS & D Naval Eng Univ of Madrid, Spain; PhD Univ of Ill. 1952-57 Sr Res Engr Kearney & Trecker Corp, Milwaukee WI; 1957-70 Professor Mech & Indus Engrg Univ of Ill at Urbana-Champaign; since 1971 Professor & Chairman Mech Engrg Dept Univ of Vt; 1967-68 Visiting Professor Polytechnic Institute, Turin Italy; 1976 NATO Visiting Professor Italy. Chmn 1975-76 Production Engrg Div ASME; since 1972 Dir Vt Amer Corp, Louisville Ky. Silver Medal Italian Soc for Mechanics 1968; Engineer of the Year 1974 Vt; SME Research Medal 1976; Fellow ASME; Mbr CIRP, APS, Sigma Xi, SME. Res areas: mech metallurgy, metal physics, production engrg. *Society Aff:* ASME, CIRP, SME, APS.

von Wettberg, Eduard, F Jr
Home: 210 North Rd, Wilmington, DE 19809
Position: Retired. *Education:* PhD/ChE/Yale Univ; BS/ChE/Yale Univ. *Born:* 11/24/05. 1930-70 DuPont Co in plant & equip design, process dev, justification & siting of plants, negotiation of contracts & particularly supervision & admn of engg R&D, for 10 yrs managed group of 30-50 prof level res engrs. Pres Del Valley Sect AIChE 1955-56; Pres Del Engg Assn 1954-55; Mbr Engg Foundation Bd 1965-77; Chmn EF Projs Ctte 1970-73, Mbr EF Proj Ctte 1968-date, Chmn EF Conferences Ctte 1974-76 Mbr EF Conferences Ctte 1970-date. *Society Aff:* AIChE, AAAS, EF.

Von Winkle, William A
Home: 105 Gardner Ave, New London, CT 06320
Position: Tech Dir for Res *Employer:* Navy Underwater Sys Ctr *Education:* PhD/Fluid Mech/Univ of CA, Berkeley; MS/EE/Yale; BS/EE/Yale. *Born:* 11/29/28 Tech Dir for Res at the Naval Underwater Sys Ctr. Prof. of Math. & Eng. at UCONN and U New Haven, Member Univ. Rhode Island Engineering Advisory Council. US Representative to the Scientific Ctte of Natl Representatives of NATO's SACLANT ASW Res Centre in La Spezia, Italy. Founding Mbr of the CT Acad of Sci and Engrg. Fellow, IEEE and Acoustical Society of America; Member, Cosmos Club. *Society Aff:* IEEE, ASA, ΣΞ, COSMOS.

Vook, Richard W
Business: 209 Physics Bldg, Syracuse, NY 13244-1130
Position: Prof. *Employer:* Syracuse Univ. *Education:* PhD/Phys/Univ of IL; MS/

Vook, Richard W (Continued)
Phys/Univ of IL; BA/Phys/Carleton College. *Born:* 8/2/29. Born in Milwaukee, WI. Staff scientist, IBM, Yorktown Hts, NY in Cryogenics Res Dept, 1957-1961. Sr Res Physicist, Franklin Inst Res Labs, Phila, PA, in Physical Met Lab, 1961-1965. Assoc Prof Mtls Sci, Syracuse Univ, 1965-1970, Prof 1970-84, Prof of Physics 1984-present. Over 120 res publications in the following principal areas: thin films, surface science, epitaxial growth, crystal imperfections using x-ray and electron diffraction, transmission and scanning electron microscopy, and Auger electron spectroscopy. Present res in thin films and surface science. Elected to mbrship in Phi Beta Kappa, Sigma Xi, Pi Mu Epsilon, LB Pfeil Medal and prize for 1983 from the Metals Soc of Great Britain. *Society Aff:* APS, AVS, EMSA, MRS

Voorhees, Alan M
Business: 7798 Old Springhouse Rd, McLean, VA 22101
Position: President. *Employer:* Alan M Voorhees & Assocs, Inc. *Born:* Dec 1922, New Brunswick NJ. BSCE from RPI; MCP from MIT; Certificate from Yale Univ Bureau of Highway Traffic. City Planning Engr, City of Colorado Springs; Traffic Planning Engr Automotive Safety Foundation, Washington DC; 1961 founded own firm, providing transportation & urban planning services worldwide. Chmn Exec Ctte 1972, Transportation Res Bd; Pres 1968-69 & Mbr of Bd 1961-64 Amer Inst of Planners; Mbr num other prof soc. TRB Award; ITE Past President's Award; 1st ASCE Harland Bartholomew Award.

Voorhees, Roland
Home: 72 Peach Hill Rd, Darien, CT 06820
Position: Retired. *Education:* BS in Engrg/Chem Engrg/Princeton Univ; Chemical Engrg/Rensselaer Polytechnic Inst *Born:* 4/22/13. 6. Philadelphia PA. Attended RPI 1930-31; Graduated Princeton Univ 1935 summa cum laude, BS in Engrg. Employed continuously by Union Carbide 1935-1978; Assoc Dir Engr Chemicals Div, 1953-55; var mgmt posts, 1955-64; Asst Sec & Treas UCC 1964-78, respon for corporate secy's dept, investor relations, pension fund mgmt; Retired 1978. AIChE: Fellow, Dir 1959-61, also var cttes & posts; Dir ECPD 1961-64; Mbr ACS, Chemists' Club. Married Sheila H Connolly June 29 1935; deceased Dec 31, 1983. Married Betty G. Anderson 1985. Principal recreational interests swimming, power yachting. P Cdr Darien, Conn Power Squadron. *Society Aff:* ACS, AIChE

Voskuil, Walter H
Home: 2173 Vale St, Reno, NE 89509
Position: Retired *Education:* PhD/Geology/Univ of WI. *Born:* 8/22/92. 1921-24 Asst in Geology Univ Wisc; 1924-30 Asst Professor Wharton Sch, Univ Penn Philadelphia; 1931 Conference Bd NYC; 1931-60 Ch Mineral Economist Ill State Geological Survey, Urbana Ill; 1960-70 Disting Visiting Professor Mineral Economics, Univ Nev Mackay Sch Mines, Reno Nev. Served with USNR 1917-18. Mbr AIME, Mineral Economics Award 1974; Assn Amer Geographers, Geological Soc Amer; Sigma Xi: author *Minerals in Modern Industry* 1930; *Minerals in World Industry* 1955; *A Geography of Minerals* 1969. *Society Aff:* AIME, GSA, AAG, ΣΞ.

Voytko, John D
Business: Box 334, Madison, PA 15663
Position: Project Mgr *Employer:* Westinghouse Elec, Synthetic Fuels Dept *Education:* BS/CE/Univ of Pittsburgh. *Born:* 9/11/33. Westinghouse Electric, Mgr, Keystone Project. Exec performance of Keystone Coal-to-Methanol feasibility study sponsored by DOE. The proposed project to be located in Western Pennsylvania, would convert 18,000 TPD of coal to 100,000 BPD of methanol. Mr Voytko maintains responsiblity for the preliminary facility design, economic analysis, market analysis, environ, health and safety analysis and the overall mgmt plan to bring the facility into commerical reality. *Society Aff:* AIChE, ESA, STC

Voznick, Henry P
Business: 1574 Batavia St, Orange, CA
Position: VP & Gen Mgr. *Employer:* Burkert Contromatic Corp. *Education:* Masters/Chem Engg/Lehigh Univ; BS/Chem Engg/Univ of VA. *Born:* 1/30/30. Currently Exec VP & Genl Mgr of Burkert Contromatic Corp., designers and manufacturers of solenoid control valves and flow control systems. Previously with Lockheed in program mgmt and Rohm & Haas Co in chem plant engg. Pres of Precision Meas Assoc 1980-81 and 1983-84. Exec VP of Precision Meas Assoc, 1979-80, Secy, 1978 and Pres of the Los Angeles Chap, 1979. Mbr of the Advisory Bd of Meas Sce Conf and a Mbr of ISA & AIChE. Prof Reg: CA, Control sys Engg, Publications: (1) Nuclear Reactor Handbook, Sec 4.2, 1964, on Heat Exchange Equipment for Higher Temperature; (2) "Hight Temperature Heat Transfer System" *Chemical Engineering*, 1964; (3) "Mixing Viscous Materials" AIChE Journal, 1958; (4) "The Electronic Thermometer for Human Use" Medical Electronics, 1979. Pats: (1) Intervolameter and Method USP 3,700, 971, 1972; (2)Advanced Electronic Device for timing Rocket firing, USP 4,050,309, 1973: (3) Single Wire Multiplex Control for Aircraft Weapons, USP 3,803,974, 1974; Non Contact temperature measuring Dev, USP 4,456,390, 1984; Multiple Probe Temperature measuring sys, USP 4,436, 438, 1984 Thermocouple Surface Probe, USP- 4,454,370, 1984. *Society Aff:* PMA, AIChE, ISA, SME.

Vrablik, Edward R
Home: 1404 N Walnut Ave, Arlington Hgts, IL 60004
Position: Pres. *Employer:* Estech Inc. *Education:* BS/CE/Northwestern Univ; MBA/Marketing/Univ of Chicago. *Born:* 6/8/32. Dir of Ind Marketing, Eimco (Now Envirotech) 1956-1961. Dir Ind Marketing and Planning, Swift & Co 1961-68. VP & Gen Mgr Swift Chem Co (Div of Esmark, Inc) 1968-73. Pres, Estech Inc (Div of Esmark, Inc) 1973-pres.

Vreeland, Robert P
Home: 31 Faculty Rd, Durham, NH 03824
Position: Retired *Born:* Aug 6 1911, New York City. BS Yale 1932; MS Columbia 1933; MEng Yale 1941. 1933-36 Surveyor & const supervisor, Natl Park Serv; 1936-39 Structural Engr for const co; 1942-45 Maj Army Corps of Engrs in Pacific area; 1946-66 taught civil engrg at Yale; 1966-76 Univ of NH, 1973-76 Chmn. Author sev books & articles on surveying. Dir & V Pres ASCE. *Society Aff:* ASCE

Waag, William R
Business: 509 Madison Ave, Ste 1406, New York, NY 10022
Position: Managing Dir *Employer:* Hydrocarbon Energy Inc *Education:* BS/PET Eng/PA State Univ *Born:* March 1925. BS Penn State Univ. Native of Pittsburgh, PA. 1st Lt Army Air Corp 1943-46. Petroleum Reservoir Engr for Sohio Oil Co. spec in property acquisitions to 1954. With Empire Trust Co and The Bank of New York as VP and Head of Oil and Natural Gas Dept., extending loans and services to the petroleum industry through 1969. Since 1970 with Drexel Burnham Lambert Inc, investment bankers, resp for mergers and acquisitions, private placements and underwritings relative to the petroleum & coal mining industries. In 1980 established my own co with two other engrs - Hydrocarbon Energy Inc. HEI makes private investments in oil and gas drilling investing about $10 million each yr. Dir SPE 1959-62; VP & Trustee UET 1973-76. Trustee UEF 1978-81. Enjoy sailing and golf. *Society Aff:* SPE.

Waber, James T
Business: 2145 Sheridan Blvd, Evanston, IL 60201
Position: Prof. *Employer:* Northwestern Univ. & Michigan Tech. Univ. *Education:* PhD/Met Engg/IIT; MS/CE/IIT; BS/CE/IIT. *Born:* 4/8/20. Native of Chicago. Obtained BS, MhS & PhD from IL Inst of Tech. Was Res Asst Prof at IIT in 1947; left to become Staff mbr at Los Alamos Scientific Lab. Sr Post-doctoral Fellow to Univ of Birmingham (England, 1960). Became Prof of Mtls Sci at Northwestern Univ 1967 to date. Fellow Am Physical Soc, Fellow of (British) Inst of Metallurgists: Chartered Engr. PE. Author of over 135 papers. Editor of several books. Holder of several patents. Senior Awardee, Alexander von Humboldt Found, also currently Prof. Physics, Michigan Tech. Univ. Prof. Eng. (Illinois). *Society Aff:* ASM, AVS, TMS, MRS.

Wachter, Robert V
Home: 2813 NE 153 St, Vancouver, WA 98686
Position: Consulting Engr (Self-emp) *Employer:* Self *Education:* BS/EE/WA State Univ. *Born:* 3/17/20. Native of Vancouver, WA. With the exception of two years service in the US Naval Radar Program During WW 2, pursued a 34 year career of Alcoa engineering assignments involving the design, construction, and operation of world-wide electrical facilities for aluminum production. Since 1976, and until recent semi-retirement, practiced privately in the industrial power conversion field. Active participant in both National and IEC Standards for large static power conversion equipment. Member, IEEE Standards Board 1971-1973; President, IEEE Industry Applications Society 1977; IEEE Fellow; 1973 recipient, IAS Outstanding Achievement Award. Outside interests include Shriner's Hospital, Masonry, Church, Amateur Radio, and workshop. Society Affil: IEEE. *Society Aff:* IEEE.

Wachtman, John B, Jr
Home: 20 Independence Court, Piscataway, NJ 08854
Position: Director, Center for Ceramics Research *Employer:* Rutgers University *Education:* PhD/Physics/Univ of MD; MS/Physics/Carnegie-Mellon Univ; BS/Physics/Carnegie-Mellon Univ. *Born:* 2/6/28. Born in Conway, SC. Married Edith Virginia Matheny Aug 27, 1955. With Natl Bur of Stds 1951-1983: Dir, Center for Matls Sci 1978-1983. Spec assignments; Vis Prof, Northwestern Univ 1971; Prog Mgr, Congressional Office of Technology Assessment 1974-75. Fellow ACERS, Fellow APS, Pres 1975 Fed of matls Soc, Pres 1978 American Ceramic Society, Orton Fndtn., Mbr NAE & Natl Matls Adv Bd., Gold Medal, Commerce Dept 1971; Stratton Award 1975; Hobart Craner Award 1978. Dorn Lecturer 1981, Orton Lecturer 1981. Lecturer (Part-time) in Ceramics at John Hopkins Univ 1981, Appointed Dir, Ctr for Ceramics Research and Sernior Prof, Rutgers University in March 1983. *Society Aff:* ACerS, ACS, APS.

Waddington, Harold T
Business: P O Box 1963, Harrisburg, PA 17105
Position: Sr. VP, Gannett Fleming Affiliates & Pres., Gannett Fleming Valuation and Rate Consultants *Employer:* Gannett Fleming Valuation and Rate Consultants, Inc. *Education:* BS/Commerce & Engr/Drexel Inst of Technology. *Born:* 5/10/30. Reg PE in PA 1962. Joined Gannett Fleming Corddry & Carpenter Inc in 1953. Served two yrs with Army Corps of Engrs 1953-55. Attained current position in 1981. Named VP of firm in 1976. Directs studies & provides expert testimony related to regulation of public utilities. Resp for work concerning valuation & depreciation of utility plant, cost of service, and rate design. Studies are also conducted for purposes of sales & acquisitions, condemnations, & insurance. *Society Aff:* NSPE, IIE, AGA, NAWC, AWWA.

Wade, Glen
Business: Dept EE & Comp Sci, Santa Barbara, CA 93106
Position: Professor. *Employer:* Univ of California. *Education:* PhD/EE/Stanford; MS/EE/Univ of UT; BS/EE/Univ of UT *Born:* 3/19/21. Born in Ogden, UT. Studied at Univ of UT & obtained PhD at Stanford Univ. At Stanford appt Assoc Prof of Elec Engg. Joined Raytheon co 1960 & was Asst Gen Mgr of Res Div. Went to Cornell Univ in 1963 as Dir of Sch of Elec Engg & holder of endowed J Preston Levis Chair in Engg. Served the IEEE in various capacities (Bd/Dir & Exec Comm 1970-72, Editor of the Proceedings of the IEEE 1977-1980). Published over 180 papers & awarded sevl patents. Recd Eta Kappa Nu Award 1955. Recd Natl Electronics Conf Annual Award 1959. Vis Prof at Tokyo Univ 1971. Fulbright-Hays Lectureship at Univ of Madrid in 1972-3. Distinguished Teaching Award for 1977 from UCSB Acad Senate. IEEE Fellow. "Special Chair" Prof at National Taiwan University in 1980-81. Received IEEE Centennial Award, 1984. United Nations Consultant at Nanjing (China) Institute of Technology, 1986. *Society Aff:* IEEE, APS, TBΠ, HKN

Wade, L Preston
Business: 2310 Langhorne Rd, P.O. Box 877, Lynchburg, VA 24501
Position: Chief Exec Officer. *Employer:* Wiley & Wilson *Education:* BS/Civil Engr/VA Polytechnic Inst & State Univ (VPI&SU) *Born:* June 30, 1933 Lynchburg, VA. Ist Lt USAF; joined Wiley & Wilson 1958, Pres 1973-80. Named Chmn of Bd, Jan 1980. Named Chief Executive Officer Jan. 1986. Reg PE VA, NC, DC, GA, IN, KY, MD, OH, AL, FL, MI, NV, TX, UT, SC, TN, WVA, CA. Mbr: NSPE, (Natl Dir 1975-79), VA Soc Prof Engrs (Pres 1970- 71), VPI & SU Alumni Assn, ASCE, VA Assn of Professions. Recipient Societas Cincinnatorium Institutda Award, 1955; VSPE Dist Serv Award 1971. Central VA "Engineer of the Yr" Award, 1977. Advisory Bd, Salvation Army, Lynchburg, Va. Lynchburg Fine Arts Center Newcomen Soc. Biographical Listings: Who's Who in the South & Southwest; Who's Who in Finance & Industry. Active in church and civic affairs. President, Greater Lynchburg Chamber of Commerce, 1976-77. VSPE Engr of the Yr - 1982. *Society Aff:* NSPE, VSPE, ASCE, NSNA.

Wade, Larry R
Business: 4010 Stone Way N, Seattle, WA 98103
Position: Principal Engr & Managing Partner. *Employer:* Hammond, Collier & Wade-Livingstone *Education:* MS/Civil Engr/Univ of WA; BS/CE/Univ of WA. *Born:* Nov 16,1940 Spokane, WA. Mbr ACEC,ASCE, SAME, AWWA, ASTM. Treasurer of Seattle ASCE 1976-77 and Assoc Rep 1974-75. Principal Engr, Secy-Treas and Managing Partner of Hammond, Collier, & Wade-Livingstone Associates. Spec in Municipal Engrg. Has taught Civil Engrg courses at local university. Recd several design excellence awards from HUD, CECW, ASCE and Water & Wastes Engg. VP Seattle ASCE 1979-80, Pres Seattle ASCE 1980-81, Secy-Treas Pacific Norwest Council ASCE 1979-80 *Society Aff:* ACEC, AWWA, ASCE, SAME.

Wade, Robert G
Business: Structural Engineering Associates, 101 W. 11th, Suite 1010, Kansas City, MO 64105
Position: Chairman/CEO *Employer:* Structural Engrg Assocs, Inc *Education:* BS/CE/Univ of MO Columbia *Born:* 11/21/33 Officer US Air Force 1956-58. With Carter-Waters Corp 1958-62 designing and detailing prestressed concrete structural components. With present firm since 1962. Owner and principal since 1976. Primary structural design and administrative responsibilities on major building projects. Also, short-to- medium span bridge design. Registered states of MO and KS. President, Consulting Engrs Council of MO. Past Pres, Midwest Concrete Industry Board. Past President, KS City Sec, Amer Soc of Civil Engrs. Former member, Educational Needs Assessment Committee, North Kansas City School District. Presented tech lectures. *Society Aff:* ACEC, ASCE, PCI, ACI, NSPE

Wadell, Robert P
Business: 1860 El Camino Real, Ste 300, Burlingame, CA 94010
Position: President/Chief Executive Officer *Employer:* Wadell Engineering Corp. *Education:* MSCE/Civil Eng/Univ of CA; BSCE/Civil Eng/Univ of CA, Berkeley. *Born:* 7/19/44. Reg Civil Engr, Licensed Multi-Engine Pilot, Pres of nationally recognized airport planning-eng consulting firm. Mbr of ASCE; Edmund Friedman Young Engr Award for Prof Achievement 1974; Natl ASCE Chrmn of Engrg Mgt Cttee 1974. Outstanding in ASCE Activities in CA 1971. West Coast mbr of ASCE Engg Mgmt Div Exec Committee. ASCE Air Transport Div Committees on Planning & Terminal Facilities. *Society Aff:* ASCE, AAAE, AOCI.

Wadlin, George K
Home: RFD #1, Box 10, Stockton Springs, ME 04981
Position: President *Employer:* George K. Wadlin Consulting, P.A. *Education:* PhD/CE/Carnegie Inst of Tech; MS/CE/Univ of ME; BS/CE/Penn State Univ. *Born:* 9/20/23 in New Haven, CT and raised in West Orange, NJ. Served in US Army 1942-46 in European Campaign. Civil engrg univ prof for over 31 yrs including serving as Head, Civil Eng. Dept at the Univ of Maine 11 yrs. and Michigan Tech Univ for 9 yrs. Director, Education Services, ASCE 1978-86. Registered professional engr since 1956 and involved in numerous consulting positions relative to structural engrg design and investigations of building distress. Author of numerous publi-

Wadlin, George K (Continued)
cations related mainly to engrg education. Retired October 1, 1986 and formed a personal corporation in Maine. *Society Aff:* ASCE, ASEE, TBΠ, XE, ΣΞ

Wadsworth, Gayle B
Business: PO Box 16858 M/S:P38-21, Philadelphia, PA 19142
Position: Mgr, Materials Engg. *Employer:* Boeing Vertol Co *Education:* BS/Physics, Solid State/Birmingham So Coll; -/Materials Math/Univ of TN; -/Engg/Wichita State Univ. *Born:* 9/22/36. in Birmingham, Alabama. BS Birmingham Southern College. Grad studies in metallurgy and mathematics. Metallurgical Res Asst at Oak Ridge Natl Lab. Metallurgist at Combustion Eng. With the Boeing Co since 1958. Assumed present position as Manager, Materials Engineering in 1980. Manager res, dev, and selection of materials for the Boeing Helicopter Company. Own and operate materials consulting firm specializing in product reliability. Intl V Pres of SAMPE, P Chmn Wichita Chapter ASM, Mbr Electron Microscope Soc of Amer, Mbr of Amer Helicopter Soc. *Society Aff:* ASM, SAMPE, EMSA, AHS.

Wadsworth, Milton E
Business: 209 Brwng Bldg, Salt Lake City, UT 84112
Position: Distinguished Prof, Metallurgy & Dean *Employer:* Univ UT Coll Mines & Min Ind. *Education:* PhD/Metallurgy/Univ of UT; BS/Metal Engr/Univ of UT. *Born:* 2/9/22. Univ of UT faculty 1948 present. Academic rank: Prof, Chrmn Dept of Metallurgy 1957-66. Ford Fnd Consultant Univ of Philippines 1958-70. Chrmn Dept of Mining, Metallurgical and Fuels Engrg 1974-76. Assoc Dean, Coll of Engrg 1973-74, Assoc Dean College of Mines & Mineral Ind 1976-1983, Dean, 1983- present. Res interests: Surface chemistry of mineral sys, mineral dressing, hydrometallurgy, solution mining and extractive metallurgy processes. Mbr ASM, AIEM, Sigma Xi, Phi Kappa Phi, Tau Beta Pi, Dist Res Award, Univ of UT 1972-73, Best Paper Awards 1957, 1967 AIME, Taggart Awards AIEM 1975 and 1977, James Douglas Gold Medal, AIME 1978, Distinguished Teaching Award, Univ of UT 1978, Grade of Fellow AIME 1976, Distinguished Life mbr ASM, 1984, Distinguished Mbr AIME, 1978, Mbr Natl Acad of Engrg 1979, Honorary degree, U of Liege Belgium 1979- , U of U Rosenblatt Prize 1986. Who's Who in America and American Men of Science, Henry Krumb Lecturer, AIME 1979. Holds 4 Patents; Author or co-author 121 Tech publs. Distinguished University of Utah Professor, 1983; Antoine M Goudin Award 1984; Mineral Industry Education Award, 1981. *Society Aff:* ASME, AIME, ASM,MPF, ΣΞ, ΦΚΦ, TBΠ, ACS, ES, CIMM.

Waegemann, August E
Home: 274 Union Blvd, Ste 444, Lakewood, CO 80228
Position: Consultant *Education:* BS/CE/Univ of CA; AA/CE/SF Jr. Coll *Born:* 7/18/19 Served as Navy Dept Physicist during WWII. Employed by city of SF for 4 yrs as airport and structures designer and went into private practice in 1950. Registered in CA 1949 as C.E. and in 28 other states. Currently practicing in CO with consultation in SF and Reno. Specializing in the design of structures of all kinds. *Society Aff:* ASCE, ASME, ASPE, ACI

Waggoner, Eugene B
Home: 336 Seawind Drive, Vallejo, CA 94590
Position: Conslt Engrg Geologist *Employer:* Self-employed *Education:* MA/Geology/UCLA; BA/Geology/UCLA *Born:* Jan 7, 1913. Staff Geologist to Chief Engr, US Bur/Rec 1945-54; Consulting Engrg Geologist 1954-60; VP and co-mgr, Woodward-Clyde-Sherard, 1960- 67; Pres 1967-73. Retired 1973. Currently private consultant. Pres CEC/CO 1961- 62. Natl Pres CEC/US 1966-67. Fellow ASCE, ACEC, ASFE, Mbr Assn of Engrg Geologists. Reg in 9 states. Served as Consultant on major civil projects in many countries. Mbr School Bd and many other civic groups in CO and CA. Authored numerous tech papers. Interested lapidary; enjoys music. Honorary Mbr Assoc Engrg Geologists. *Society Aff:* ACEC, ASCE, ASFE, AEG, NAE.

Wagner, Aubrey J
Home: 201 Whittington Dr, Knoxville, TN 37923
Position: Private Consultant. *Employer:* Self-employed *Education:* BS/CE/U of Wisc. *Born:* Jan 1912. BS Univ of Wisc; hon PhD Newberry College(SC)and Lenoir Rhyne (NC). Employed TVA 1934 as enrg aide, progressing to Gen Mgr, 1954, to Bd of Directors (Presidential appt), 1961, Served as Chrmn, 1962 - 1978 Dist Serv Citation, Univ of Wisc 14th Annual Engrs Day 1962 and Tenn Tech Univ 1981; N.W. Dougherty Award, Univ of Tenn, Coll of Engrg, 1962; NAE 1973, Amer Nuc Soc Walter H. Zinn Award, 1978; licensed PE, Tenn. Served on various Federal committees. Consulted and advised overseas on resource mgt problems. Boy Scout work since 1935; Exec Council, Lutheran Church in America, 1962-70. Enjoy woodworking, photography, fishing and sports. *Society Aff:* NAE, TSPE, TSK, TBP, NSPE, AAAS.

Wagner, Charles L
4933 Simmons Drive, Export, PA 15632
Position: Consulting Engr. *Employer:* Retired; private practice *Education:* MS/EE/Univ of Pittsburgh; BS/EE/Bucknell Univ. *Born:* 11/23/25. Served in US Navy 1943-46. Joined Westinghouse 1946. Sponsor Engineer, Elec Utility Engrg 1950-67. Mgr Transmission Systems Engrg 1967-76. Consulting Engr 1976-85. Retired-1985. Fellow IEEE, Mbr IEEE Pwr System Relaying & Switchgear Committees. Chrmn Switchgear Committee 1973-74. IEEE Standards Board 1975-76. Chrmn ANSI C37 Switchgear Committee 1971-86. Pres PES 1984-85. VP 1982-83. Chrmn PES TOD 1977-81. PES TOD Committee 1973-81. PES Prize Paper Award 1973. IEEE Standards Medallion 1980. Switchgear Cttee Distinguished Service Award 1982, IEEE Centennial Medal 1984. IEEE CP Steinmetz Award 1985, Relay Cttee Distinguished Service Award 1985. Instructor in Graduate Schools of University of Pittsburgh & Carnegie Mellon University. Mbr CIGRE. Reg PE in Pennsylvania. *Society Aff:* IEEE, PES.

Wagner, Harvey A
Home: 12900 E Outer Drive, Detroit, MI 48224
Position: Vice President. *Employer:* Overseas Advisory Assoc., Inc. *Education:* Dr Engrg/Engg/Lawrence Inst of Technology; BSME/Power/Univ of MI. *Born:* Jan 2, 1905. With Detroit Edison 1928-70, Cons Engr 1970--. Mbr Detroit Bd of Water Commissioners, 1952-60 (Pres 1955-56) Trustee Natl Sanitation Fdtn 1965--. Recipient Dist Alumnus Award U of Mich 1953 and Sesquicentennial award as outstanding exec and nuclear power consultant, U of Mich 1967. Certificate of pub service, FPC 1964. Fellow ASME, ANS and Engrg Soc of Detroit (Pres 1968-69). Mbr NAE, Tau Beta Pi, Phi Kappa Phi. Author over 60 papers in field. Consulting Engineer. *Society Aff:* ASME, ANS, NAE.

Wagner, J Robert
Home: 2996 Runnymede Dr, Norristown, PA 19401
Position: Prof & Consit *Employer:* Phila Coll of Textiles & Science *Education:* MS/Textile Tech/NC St Univ-Raleigh; M Phil/Textile Industries/Univ of Leeds-England; BS/Textile Engrg/Phila Coll of Textiles & Sci *Born:* 01/08/32 Presently Prof and Past Dir of Textile and Apparel Research at the Phila Coll of Textiles and Sci and Conslt to the Textile and Paper Industries. Published numerous papers and have obtained more than a dozen patents. Past Research Mgr and Chief Product Engr of Formex Co, a div of Huyck Corp. Designed the world's first synthetic Fourdrinier wire, inside press fabric and open mesh drier fabric and also developed fireproof nonwovens used in the Space Shuttle Astronaut space suits. Charter mbr of Lafayette Hill, PA and Greeneville, TN. Junior Chamber of Commerce. Secy of the latter. Secy of the Dryer Fabric Subcommittee and Chrmn of Testing Section, Mbr at Large, Secy, Vice Chrmn and Chrmn of Nonwovens Div of TAPPI. Present Chrmn of TAPPI Publications Cttee, Nonwovens News Letter editor and TAPPI Journal Nonwovens editor. Received TAPPI Paper Synthetics Div award in 1981. *Society Aff:* AATCC, TAPPI, INDA, NRA, ΦΨ

Wagner, John, Jr
Business: 110 Essex Ave, Narberth, PA 19072
Position: Pres *Employer:* PSG Corrosion Engrg, Inc. *Education:* BS/CE/Univ of PA. *Born:* 5/27/24. Reg PE, PA & other states. Mr Wagner has undertaken many special studies of corrosion and corrosion related problems. Mr Wagner has participated regularly as a mbr of the instructional staff for the Univ of WV - "Appalachian Underground Corrosion Short Course–. Mr Wagner's lectures have been on the theoretical background of polarization effects in studies of corrosion and practical approaches to the solution of corrosion problems. In addition to instructional lectures, Mr Wagner has authored technical papers on corrosion for presentation at various engg and technical soc meetings. *Society Aff:* AIChE, NACE.

Wagner, John L
Business: 615 Washington Rd, Pittsburgh, PA 15228
Position: Pres. *Employer:* Benfield Corp. *Education:* MS/CE/State Univ of NY; MBA/Bus Adm/State Univ of NY; BS/CE/Univ of WI. *Born:* 2/3/35. Native of NE WI. Joined Linde Div of Union Carbide Corp in 1957 as Dev Engr. Specialized in adsorption processes & developed PSA (Pressure Swing Adsorption) processes for hydrogen purification & air separation. Other positions included engg mgr - process design, product mgr - engg products, and two yrs in Japan as mgr engg products for Union Carbide Japan. With Benfield Corp (Subsidiary of Union Carbide) since 1977. Appointed pres of Benfield in 1979. *Society Aff:* AIChE, LES.

Wagner, Victor G
Business: 3333 Founders Lane, Indianapolis, IN 46268
Position: Principal Engr. *Employer:* Howard Needles Tammen & Bergendoff. *Education:* BS/CE/Univ of WI. *Born:* Oct 1925. Vet WWII. Native Waterloo, Wisc. Former employment as Chf Water Pollution Control Sec, Indiana State Bd of Health and Dir, Sanitary Engrg, Can- Tex Industries. President, WPCF 1976. Chrmn, Govt Affairs Comm, WPCF 1968-73. Arthur Sidney Bedell Award, WPCF 1965. Married. 6 children and 8 grandchildren. *Society Aff:* WPCF, AAEE, ASCE, AWWA, NSPE.

Wagoner, Mike N
Business: South Elm-Eugene St, Greensboro, NC 27420
Position: Chief Engr. *Employer:* Carolina Steel Corp. *Education:* MBA/Marketing/UNC-G; BS/Civ Engg/NC State Univ. *Born:* 12/4/41. & raised in Greensboro, NC. Employed with Carolina Steel Corp since grad from NCSU. Charter mbr of the Engg Surveying Joint Action Committee, a state wide committee of five socs that were responsible for the rewrite and revision to the state engg law over the yrs. Chrmn of ESJAC 1976, 1978. Pres of State Sec of ASCE 1977. Dir of NC Soc of Engrs 1978, 1979. Active with Local Technical Inst as a mbr & as chrmn of advisory committee (GTI) 1974-1979. Recipient of Rotary Intl Group Study Exchange Award 1975 to Australia. Sports include fishing, water skiing, & handball. *Society Aff:* ASCE, NSPE.

Wagoner, Robert H
Business: Dept Metallurgical Engrg, 116 West 19th Ave, Columbus, OH 43210
Position: Professor *Employer:* OH State Univ *Education:* PhD/Met Engrg/OH State U; MS/Met Engrg/OH Sate U; BS/Met Engrg/OH State U. *Born:* 1/8/52 NSF Postdoctoral Fellow and mbr of Brasenose Coll at Univ of Oxford, 1976- 77. Phys Dept, Gen Motors Res Labs 1977-1983 as Staff Res Scientist. Current position: Prof of Metallurgical Engrg, OH State Univ. Chnmn of TMS-AIME Shaping and Forming Ctte. 1981 and 1983 Rossiter W Raymond Memorial Award (AIME) for paper entitled "Measurement and Analysis of Plane-Strain Work Hardening–, 1981 Hardy Gold Medal, 1984 Presidential Young Investigator Award, Professional interests: metal plasticity theory, applied elasticity. Enjoy tennis and duplicate bridge. *Society Aff:* AIME, ASM, AΣM.

Waguespack, Otis J
Business: PO Box 7931, Metairie, LA 70010
Position: Pres. *Employer:* Otis J Waguespack & Assoc, Inc. *Education:* BS/Chem Engg/LA State Univ. *Born:* 12/12/25. Native of Vacherie, LA. Grad from LSU, in 1949, after having served three yrs with the Navy, in the Pacific. My entire profl career has been working for oil cos or oilfield related equip mfg. I have worked in design, applications, mfg, operations, maintenance, & training. Served on two engg doctorate comittees, at Tulane Univ. The co-author of two US patents & a reg engr in Chem & Mech Engg. Pres & sr consultant of a firm engaged by local and foreign oil cos, performing consulting engg services on oil & gas production facilities in design & operator training. *Society Aff:* AIChE, ASME, AIME.

Wahl, Arthur M
Home: 3300 NE 36 St, Ft Lauderdale, FL 33308
Position: Retired. *Education:* PhD/Math & Mech Engg/Univ of Pittsburgh; MS/Mech Engg/Univ of Pittsburgh; BS/Mech Engg/IA State Univ. *Born:* Oct 1901. Churdan, Iowa. Res Engr, Adv Engr and Consultant, Westinghouse Res Labs 1926-73, spec in applied mech, stress analysis & machine design fields. Pubs include 65 tech articles, mainly in ASME Transactions; bk, Mechanical Spring 1963; sections in Marks', Kent's and ASME Handbooks. Exec Comm, Applied Mech Div ASME 1957-62; Pressure Vessel Res Comm; SAE Tech Bd Sprng Comm. ASME Richards Memorial Award 1949, and Machine Design Award 1965. Westinghouse Order of Merit 1950; Marston Medal, Iowa State Univ 1966; Hon Mbr ASME 1973. *Society Aff:* ASME, SEM.

Wahler, William A
Home: 1420 Byron St, Palo Alto, CA 94301
Position: Principal *Employer:* W A Wahler Consulting Engrg & Geology. *Education:* BS/Civil Engg/Univ of CO; MS/Civil Minor Geology Engg/Univ of CO. *Born:* 11/5/25. Colorado. Postgraduate studies Harvard, MIT, Geo Washington Univ US Dept Interior Mgt Trng Prog. Five yrs, 1948-52, US Bur/Rec. Founder/partner Denver soils engrg firm, 1953-55. Proj Engr Bechtel Corp 1955-60. Founded W A Wahler and Assoc 1960, & W A Wahler, Consulting Engineers & Geologists, 1979. Reg Prof/Civil Engr in 24 states; Reg Eng Geologist, CA. Chrmn: USCOLD Tailings Dam Comm; or Fellow: AIME, ASCE, AEG, CEAC, ACEC, CEAC, USCOLD, ICOLD, GSA, ISSM&FE, GSA & ASFE. Specialist in geological hazards engrg, and dam and reservoir engrg. *Society Aff:* AEG, ISSM&FE, ASCE, ACEC, USCOLD, ICOLD, CEAC, GSA, ASFE.

Wahls, Harvey E
Business: Civ Engrg Dept, Box 7908, Raleigh, NC 27695-7908
Position: Prof & Assoc Dept Hd. *Employer:* NC State Univ. *Education:* PhD/CE/Northwestern Univ; MS/CE/Northwestern Univ; BS/CE/Northwestern Univ. *Born:* 8/8/31. in Evanston, IL. Served as instr at Northwestern and Worcester Poly Inst. Since 1960, faculty mbr at NC State Univ where currently Prof & Assoc Dept Hd in Civ Engrg. Outstanding Teacher Award, 1973. Specialist in geotechnical engg with res publications in soil compaction, consolidation, and soil response to dynamic loads. Reg PE and consultant in geotechnical engg. Fellow, ASCE. Active in profl societies including Chrmn, ASCE Soil Properties Committee, 1969-74; ASCE 1975 Geotechnical Specialty Conf; and TRB Committee on Mechanics of Earth Masses, 1975-81. Member, ASCE Geotechnical Division Executive Committee, 1980-84. Secretary, U.S. National Society for Soil Mechanics and Foundation Engineering, 1985-date. *Society Aff:* ASCE, ASTM, ASEE, TRB, USNS/ISSMFE.

Wahren, Douglas
Business: P.O. Box 1039, Appleton, WI 54912
Position: VP *Employer:* Inst. of Paper Chemistry *Education:* Docent/Che/Royal Inst of Tech, Sweden; Dr of Sci/Paper Tech/Royal Inst of Tech, Sweden; PHD/Paper Tech/Royal Inst of Tech; MCE/Paper Tech & Tool Machines/Royal Inst of Tech *Born:* 3/12/34 Joined Swedish Forest Products Lab in Stockholm, Sweden and passed up through the ranks; Dir of Res there 1969-73. Sabbatical 1 yr with Beloit Corp., Beloit WI at their res lab. Concurrent with STFI job also full prof at the Royal Inst 1970-73, Dep of Paper Tech. Received honorary title of Prof (from the

Wahren, Douglas (Continued)
King of Sweden) 1973. VP for Res with KMW Corp, a Swedish machinery builder, from 1974-1978. Joined the Inst of Paper Chem as VP for Res in 1979. TAPPI Fellow 1981. *Society Aff:* TAPPI, CPPA, SPCI, BPBMA

Waid, Donald E
Home: 2801 W. Yorkshire Dr, Muncie, IN 47304
Position: Cons Energy & Pollution Control (Self-employed) *Employer:* Self *Education:* -/Meteorology/NYU; BS/ChE/Purdue Univ. *Born:* 3/24/17. Reg PE, IN, Diplomate Am Acad Environmental Engrs. Engr. of the Year, 1984, Ind. Soc. Prof Engrs. Holds patents on air heating burner design. Author of 40plus papers & parts of five books in air pollution control, Ind Hygiene, Waste Fuel Utilization, Energy Conservation and Burner Design. Consultant in design of Ind Burner Equip, Special Applications & Air Pollution Control Equip. Design of Air Heating Burners, Hydrogen & Waste Gas Utilization. Control of Carcinogenic Nitrosamines in Spray Drying & Malt Drying of Food Products through Low NO2 production from the drying burner. *Society Aff:* AIChE, ASM, APCA, AIHA, NSPE, AAEE.

Waidelich, Donald L
Business: University of Missouri, Dept. of Elect. & Comp. Engr, Columbia, MO 65211
Position: Professor Emeritus of Electrical & Computer Engrg *Employer:* Univ of Missouri-Columbia. *Education:* BS/Elec Engg/Lehigh Univ; MS/Elec Engg & Math/Lehigh Univ; PhD/Elec Engg/IA State Univ. *Born:* May 3,1915, Allentown, PA. Assoc Dir of Engrg Experiment Sta 1954-58 and Chrmn, Dept Elec Engrg 1961-62, Univ of Missouri. Fulbright Grants: Cairo Univ, Egypt 1951-52 and Univ of New South Wales, Australia 1961-62. Consultant, USNavy Electronics Lab 1948-52; US Atomic Energy Comm 1953-57 and 1966-70; Natl Aeronautics and Space Admin 1962-66; McDonnell Douglas Corp. 1982-87, Hughes Aircraft Co 1972-87. AIEE Fellow 1956, IRE Fellow 1960, IEEE Fellow 1964. Sigma Xi Research Award 1976 Halliburton Teaching Award 1983. Univ of Missouri Honor Award for Distinguished Service in Engrg, 1986. *Society Aff:* IEEE, NSPE, ASEE, ASNT, ASTM.

Wainscott, Lotcher A
Business: 22400 Barton Rd, Suite 200, Grand Terrace, CA 92324
Position: Pres *Employer:* L A Wainscott & Assocs, Inc *Education:* BS/CE/Univ of IL *Born:* 10/16/31 in Columbia, MO. Served in USAF 1955-57. Served as Asst to Field Coordinator on addition to Union Oil Co refinery in Wilmington, CA prior to Oct, 1960. Until Jan, 1963, served as field coordinator on Experimental Organic Cooled Reactor Project near Idaho Falls with C. F. Braun & Co. Served as Project Engr for Land Development Div of A. A. Webb & Assocs from 1963-76. Responsibilities included land planning and drainage. Developed computer progs for Clary and IBM 1130 computers. Am currently Pres of L A Wainscott & Assocs, Inc, formed in 1976. Reg Civil Engr in AZ and CA. Enjoy reading and history. *Society Aff:* ASCE, NSPE

Wainwright, Sam H
Business: 2306 Ind Rd, Dothan, AL 36302
Position: Pres. *Employer:* Wainwright Engg Co Ing. *Education:* BS/Civ Auburn Univ. *Born:* 2/23/32. Organized Lee, Wainwright & Ham, 1960; Reorganized into Wainwright Engg Co, Inc, 1965; Specializes in Sanitary & Airport Engg; Reg PE, 13 states; Reg Land Surveyor, 7 states; Offices; AL Bd of Registration, 1973 to present - Secy, VChrmn, Chrmn; NCEE Pres, VP, Dir, Chrmn Land Surveying Committee; Consultant Uniform Exams Qualifications Committee; NSPE State Dir; ASCE Fellow & State Dir; ACEC State Pres & Dir. *Society Aff:* NSPE, ASCE, ACEC, NCEE, SAME.

Waisman, Joseph L
Business: 5301 Bolsa Ave, Huntington Beach, CA 92647
Position: Program Mgr. *Employer:* McDonnell Douglas Astron Co. *Education:* PhD/Engg/UCLA Sch of Engg & Appl Sci; BS/Metallurgical Engg/Univ of IL. *Born:* 3/10/19. Positions held: Chf Metallurgist, Chf Materials & Process Engr, Asst Chf Aerospace Syst Engr, Dir R&D, Dir, Energy Progs. Reg PE-Mech Engrg, Metallurgical Engrg-Calif. Mbr: AIME, Fellow ASM, ASTM, Assoc Fellow AIAA, Sigma Xi, Alpha Sigma Mu, Tau Beta Pi, NAS Materials Adv Bd - Consultant 1969-74. Invented preload indicating device for fasteners. Taught extension course, metallurgy, tool engrg, fatigue, titanium. Coauthor book *Metal Fatigue*. *Society Aff:* AIME, ASM, ASTM, AIAA, ΣΞ, ΑΣΜ, ΤΒΠ.

Wait, James R
Business: ECE Dept, University of Arizona, Tucson, AZ 85721
Position: Prof of Electrical Engineering and Geosciences *Employer:* University of Arizona *Education:* BASc/Physics/Univ of Toronto; MASc/Physics/Univ of Toronto; PhD/Elec Engg/Univ of Toronto. *Born:* Jan 1924. Canadian Army Radar Technician 1942-45. Res engr with Newmont Exploration Ltd. Arizona 1950-52; Radio Physics Lab Ottawa 1952-56; US Dept of Commerce Boulder Co 1955-. Prof. of Elect. Eng. Univ of Ariz 1980- . Prof Adjoint of Elec Engrg, Univ of Colo 1962-; Consultant, Inst for Telecommunication Science, Boulder Co 1971-; Fellow of the Cooperative Inst for Res in Environmental Sciences (CIRES), Univ of Colo, Boulder CO 1968-1980; Visiting Prof, Harvard 1966-67. Secy of the USNC of the Intl Union of Radio Science 1975-78. AAAS Boulder Scientist Award 1957; US Dept of Commerce Gold Medal 1959; IEEE Fellow Award 1962; IEEE Harry Diamond Award 1964; US Chamber of Commerce Flemming Award 1964; NOAA's annual Res & Dev Award 1973. Vander Pol Gold Medal (URSI), Helsinki Aug 1978. IEEE Centennial Medal (1984). The IEEE Geoscience & Remote Sensing Achievement Award for 1985. *Society Aff:* IEEE, URSI, NAE.

Wait, John V
Home: Dept. of Elec. & Computer Engineering, Tucson, AZ 85721
Position: Prof. *Employer:* Univ of AZ. *Education:* PhD/EE/Univ of AZ; MS/EE/Univ of NM; BS/EE/Univ of IA. *Born:* 10/1/32. in Chicago. Taught at Univ of IA, NM, CA (Santa Barbara), FL (Cape Kennedy), and AZ. 1st Lt in USAF. Currently Prof of Elec Engr at Univ of AZ. Co- author of two texts on Operational Amp. & Simulation. NSF Fellow, 1960. Sr Mbr, IEEE and SCS. Sigma Xi, Tau Beta Pi, Eta Kappa Nu, Omicron Delta Kappa. *Society Aff:* IEEE, SCS.

Waite, William P
Business: Dept of Elec Engg, Univ. of Arkansas, Fayetteville, AR 72701
Position: Prof. *Employer:* Univ of AR. *Education:* BS/Elec Engr/Univ of MO; MS/Elec Engr/Univ of KS; PhD/Elec Engr/Univ of KS. *Born:* 8/27/33. Native of Osceola, MO. Worked as proj engr, radar systems for Bendix Corp Kansas City Div 1956-1965. Sr Res Engr, Ctr for Res in Engr Sci, Univ of Kansas, 1965-1970. Prof, Dept of Elec Engr, Univ of AR since 1970. Served as principal investigator on over a dozen research grants and contracts in the areas of radar systems, radar cross section determination and microwave remote sensing. *Society Aff:* IEEE, ASPRS, ΣΞ, URSI, HKN.

Waitkus, Joseph
Home: 25 Chestnut St, Wellsville, NY 14895
Position: Consultant. *Employer:* Private Practice. *Education:* BE/ME/Johns Hopkins Univ; Masters/ME/Brooklyn Poly Inst; -/Nuclear/Univ of MI; -/Nuclear/Stevens Inst of Technology; -/Bus. Admin/Harvard Bus Sch. *Born:* Sept 2, 1906. Adv Mgt Pro Harvard Sch Bus Admin; Nuclear Enrg Univ of Mich, Stevens Inst of Tech. Retired 42 yrs service C-E Air Preheater Div Combustion Engrg Co. Positions: Mgr. Erection & Service, Industrial Sales Dev, Asst to Tech Mgr, Tech Advisor to Pres. Elected two terms Trustee of Village of Wellsville, NY, Chrmn Wellsville Water & Light Comm, Bd Chrmn and Administrator Jones Memorial Municipal Hospital. Mbr several professional engrg societies. ASME Fellow. PPres Steuben Chapter NYSSPE and Rotary Club. Contributed generously to tech press on heat exchange, nuclear energy, environmental control. NY State licensed. Presently in private consulting practice. *Society Aff:* ASME, NYSPE, AISE, ACS, NSPE.

Wakeland, Howard L
Business: College of Engg, 207 Engg Hall, Urbana, IL 61801
Position: Assoc Dean. *Employer:* Univ of IL at Urbana-Champaign. *Education:* BS/Agr Engrg/Univ of IL; MS/Agr Engrg/Univ of IL. *Born:* 7/22/27. Assoc Dean of Engg for Undergrad Programs, 1968-present; Prof Agr Engg, 1968-present, Univ of IL, Urbana: Mbr BOD ABET, 1974-1986; Exec Comm Engr Accred Coun 1981-83; VChmn ECPD Counc, 1979; Mbr, EdAC of AAES 1980-1986; ASEE Vincent Bendix Awd 1980; CEC/IL Dist Serv Awd 1981. Pres JETS, 1976-78; BOD JETS, 1972-1988: BOD Mara Edu Foundation, Kuala Lumpur, Malaysia 1986-present; BOD IAESTE, 1971-present: Chmn IAESTE:US 1985-present: Mbr, 1972-present and Chmn 1973-75, 1984-85 Natl Exec Comm on Guidance: Chmn, Task Force Minority Engg Educ Effort (ME3), 1972-75: USAID Conslt, J Nehru Agri Univ, Jabulpur, India 1971: Chmn ASEE Pub Pol comm 1970-71: Mbr Amer Delegation UPADI, Mexico 1966: BOD Illidata, Inc. 1960-72: BOD Intl Farms Sys 1970-72: Advisory Bd, Intl Tech Educ Conslt (ITEC) 1978-1984: Partner, Wakeland construction 1978-present; Hd US/IAESTE delegation to Intl Coun, Montpelier, France 1984; Instanbul, Turkey 1985; Lisbon, Portugal 1986; Budapest, Hungary 1987: ASEE US Engineering Deans Delegation to China, 1984; MUCIA Edu Conslt, Indonesia, 1984-present. *Society Aff:* ASAE, ASEE, NSPE.

Wald, Henry J
Business: 85 Willow St, New Haven, CT 06511
Position: Consulting Engg. *Employer:* Self-employed. *Education:* MS/Arch/Columbia Univ; BME/Mech Engg/Cooper Union. *Born:* Nov 1922. BME Cooper Union, MS Architecture Columbia Univ. US Army Ordnance Dept 1943-46. Reg PE. Engrg Exp: Westinghse; Kurt Versen Co; Voorhees, Walker, Foley & Smith; Guy B Panero; Firm of Edward E Ashley. Gen Partner (1956- 70),Wald & Zigas Consulting Engrs. Dir, Health Sciences Facilities Plng and Construction, Columbia Univ (1972-75). Dir, Program & Facilites Planning, Yale- New Haven Medical Center, Inc (1975-79). Principal, Jansen & Rogan, Cons Engrs (1980-83). Lecturer; Lighting theory & design, Pratt Institute. Adjunct Assoc Prof of Architecture, Columbia Grad School of Architecture and Plng. Tech Adv Panel-Architectural and Engineering News. Fellow IES; Chrmn NY Sec, Natl Chrmn, School and College Lighting Committee. Mbr, Energy Mgmt Sys Committee. Commission Internationale d'Eclairage. Expert Committees on Interior Ltg & Actinic Effects. Pi Tau Sigma. Cooper Union, Alumni Council, Trustee (1983-86). Design Awards: AIA, HUD, IES Applied Ltg Competition. Bd of Dirs, Hospice, Inc (1972-78), Yale Psychiatric Inst (1977-79). *Society Aff:* IES.

Waldhauer, Fred D
Home: 25 Hance Rd, Fair Haven, NJ 07701
Position: Director *Employer:* Resound Corp *Education:* MSEE/Engg/Columbia Univ; BEE/Engg/Cornell Univ. *Born:* 12/6/27. 1927, Brooklyn, NY Married to Ruth I. Waina; 5 children. RCA 1948-55; Bell Labs 1956-87; Director, Resound Corporation, 1987- . Work has centered on feedback processes and digital communications. Author of papers on circuit theory, quantized feedback, high speed digital transmission, and feedback amplifiers. Coauthor of early transistor text and author of book on *Feedback*. Currently engaged in hearing health care. Fourteen patents. Cofounder and secretary of Experiments in Art and Technology in 1966. Fellow, IEEE; mbr, AES; Ewg PE, NJ. *Society Aff:* IEEE, AES.

Waldner, John W
Business: 307 E McCarty St, Indianapolis, IN 46206
Position: Statistician. *Employer:* Eli Lilly and Co. *Education:* BS/Genl Bus/Butler Univ. *Born:* June 23,1922. Native of Indianapolis, IN. Served USMC in World War II. With Eli Lilly and Co since 1942, wide variety of assignments, currently attached to research functions working as Quality Control Consultant for the Pharmaceutical Mfg operations. Recd Edward J.Oakley Award, Midwest Conference Bd, 1967. Fellow ASQC & P Dir at Large ASQC P Vice President ASQC, Chrmn Midwest Conf Bd, ASQC and a wide variety of other ASQC positions. Enjoy golf and tennis. *Society Aff:* ASQC.

Waldron, Kenneth J
Business: Dept of Mech Engrg, Columbus, OH 43210
Position: Nordholt Prof *Employer:* OH State Univ *Education:* PhD/ME/Stanford Univ; MEngSc/ME/Univ of Sydney; BE/ME/Univ of Sydney *Born:* 2/11/43 Sydney, Australia. Engr, Australian Iron and Steel Co Ltd. 1963-64. Acting assistant prof, Stanford Univ 1968-69. Lecturer/Senior Lecturer Univ of New South Wales (Australia) 1969-74. Assoc Prof, Univ of Houston 1974-79. Assoc Prof/Prof/Nordholt Prof OH State Univ 1979. Author of over 60 technical articles and conference proceedings. Researcher and conslt in mech sys design, computer aided design and robotics. Registered professional engr in TX. *Society Aff:* ASME, SAE, ASEE, SME.

Waligora, John M
Home: 3716 Prairie Dunes Drive, Sarasota, FL 34238
Position: Consultant. *Employer:* Merck & Co, Inc. *Education:* MSinEE/EE/Univ of MN; BEE/EE Power/Univ of MN *Born:* 8/01/10 BEE/EE, Univ MN. MSEE, EE Univ of MN. Born and Educated MN. Worked in EE at Northern States Power Co, Phila Elec Co, Newport News Shipbldg Born & Dry Dock Co; Marine Engr Supervisor of Shipbldg USN, Newport News VA. Dir of Engrg, Merck Sharp & Dohme Div of Merck & Co Inc, W Point PA, Chrmn, N PA Water Authority for 10 yrs. Pres of N Penn Hospital; Intl Pres of Amer Inst of Plant Engrs 1957. Fellow, AIPE, 1968. AIPE Richard H Morris Medal, Plant Engr of the Yr 1964. Distinguished Serv Awd N Penn Chamber of Commerce 1957. *Society Aff:* AIPE

Walitt, Leonard
Business: 6269 Variel Ave, Suite 200, Woodland Hils, CA 91367
Position: Pres. *Employer:* Numerical Continuum Mechanics, Inc. *Education:* PhD/Fluid Mechanics/Univ of CA; MSE/Fluid Mechanics/Univ of CA; BSME/ME/CCNY. *Born:* 5/5/35. Native of NYC. Engr with N Am Aviation, 1958 to 1962. Responsibilities included aerodynamic design of inlet & inlet control systems for B-70 bomber. Sr aerodynamics engr for Ram jet engines at the Marquardt Corp 1962 to 1966. Developed numerical methods calculating internal & external viscous flow fields from 1966 to 1975 for Applied Theory, Inc & Thermo-Mech Systems Co, respectively. Pres of Numerical Continuum Mechanics, Inc since 1975. Developed software to analyse 3-D viscous flow fields about missiles, fuselages & within turbo machines; author of 11 engr soc papers & journal articles; SAE Arch T Collwell Merit Award 1978. *Society Aff:* AIAA.

Walk, Frank H
Business: 600 Carondelet St, New Orleans, LA 70130
Position: Chrmn of the Bd *Employer:* Walk, Haydel & Assocs, Inc *Education:* BS/ME/LA St Univ; BS/CE/LA St Univ *Born:* 08/30/20 Decatur, AL, resides New Orleans, LA. Served as ofcr with Corps of Engr, US Army, 1942-47; VP, chief engr Dunham Wilson Mfg Co, Baton Rouge, LA, 1946-52; managing assoc Wayne & Assoc New Orleans, 1953-58; Pres Walk, Haydel & Assocs, Inc New Orleans, LA, 1959-present. Louisiana Engr Soc (Lockett Award for public service 1977), Soc Am Military Engrs (Outstanding Service Award 1972), Am Consltg Engrs Council (Fellow), Natl Soc of Prof Engrs, Am Soc Mech Engrs. (Chrmn N.O. Section, 1965, Chrmn Petroleum Div 1977-78, Awarded Silver Medallion 1979), Am Soc of Cvl Engrs (Fellow). Am Petroleum Inst, Consltg Engrs Council, LA (Pres, 1974). Registered PE LA, MS. Pres, New Orleans Chamber of Commerce, 1980; Mbr State Bd of Commerce & Industry, 198q. Hobbies: Hunting, fishing, hiking. *Society Aff:* ASME, ASCE, NSPE, ACEC

Walker, Bruce H
Business: 400 Wakara Way, Salt Lake City, UT 84108
Position: President *Employer:* Drilling Research Laboratory, Inc *Education:* BS/Math Method & OR/Columbia Univ; BS/Math & Physics/Whitman College. *Born:* 9/22/46. Began with Christensen Inc (formerly Christensen Diamond Products) as a consultant in 1969, later became Operations Res Engr followed by Group

Walker, Bruce H (Continued)
Leader, Applied Res, then Manager of Research in 1978, and finally Product Manager in 1982. In 1983 became Pres of Drilling Research Laboratory, a wholly owned subsidiary of Terra Tek where full-scale testing of drilling, production and completion equipment and techniques under simulated deep-well conditions are done for industry and government. Has made many tech presentations, has published sevl papers and holds sevl patents. Awarded 1974 SPE-AIME Cedric K. Ferguson Medal. Mbr ASME and Distinguished Mbr of SPE-AIME. *Society Aff:* SPE-AIME, ASME.

Walker, Charles Allen
Business: 307A Mason Laboratory, Yale Univ, New Haven, CT 06520
Position: Prof Emeritus of Chem Engg *Employer:* Yale Univ. *Education:* DEng/Chem Engg/Yale Univ; MS/Chem Engg/Univ of TX at Austin; BS/Chem Engg/Univ of TX at Austin. *Born:* 6/18/14. in Wise County, Texas. Yale faculty mbr since 1942 (Master of Berkeley Coll. 1959-69; Dir of Undergraduate Studies 1970-74; Dept Chrmn 1974-76, 1981- 84; Staff Mbr, Institution for Social and Policy Studies 1970-84). Active in ACS (Ch. Petroleum Res Fund Adv Bd 1972-81) and Sigma Xi; (Northeast Reg Dir 1976- 78). Res int in water pollution control, photochemical engrg, catalysis, social sciences in energy and environmental problems. *Society Aff:* AIChE, ACS, AAAS, ASEE

Walker, Eric A
Business: Rm 222A Hammond Bldg, University Park, PA 16802
Position: President Emeritus. *Employer:* Penn State University. *Education:* ScD/Elec Engg/Harvard; MS/Bus/Harvard; BS/Elec Engg/Harvard. *Born:* April 29,1910 Long Eaton England. Came to U.S. 1923. Married Josephine Schmeiser 1937 - two children, Brian, Gail. Taught at Tufts College and Univ of Connecticut. Was successively professor, dean, vp and president of The Pennsylvania State Univ retiring 1970. Vice Pres and Dir of Alcoa 1970-75. Awarded a number of hon. degrees. PPres: EJC, ASEE, NAE. Fellow: Acoustical Soc, IEEE, ASEE, Ben Franklin Fellow of the Royal Soc of Arts. Chrmn of Natl Science Bd 1966-68. Member, Institute for Defense Analyses Board. Now director of several industrial corporations and consultant on engineering mgt and technology. *Society Aff:* IEEE, ASA, ASHRAE, ASEE.

Walker, Gary Wayne
Business: 1445 E. Los Angeles Ave, Suite 213, Simi Valley, CA 93065
Position: Owner *Employer:* Gary Walker & Assoc *Born:* 02/11/48 Employed as Mech Engr for consltg firms for 8 yrs. Established Energy Analyst Prog for the Sacramento Municipal Utility District, Sacramento CA, held the position of Principal Energy Analyst. Supervised a section of engrs engaged in auditing energy and power consumption for the utilities' large commercial/industrial customers. These engrs under my general direction reviewed and evaluated energy usage; promoted energy efficient equipment and lighting; prepared written energy conservation plans and other recommendations, and reports; estimated plan costs, forecasted cost benefit results, and produced life cycle studies. Established Gary Walker & Assocs Consltg Engrs in 1980, providing mech/elec engrg. *Society Aff:* ASHRAE, NSPE

Walker, Gene B
Home: 2408 Cypress, Norman, OK 73072
Position: Assoc Dean. *Employer:* College of Engr, OU. *Education:* BS/EE/Univ of TX; MS/EE/Univ of TX; PhD/EE/Univ of TX. *Born:* 2/24/32. in TX. Worked in area of electromagnetic wave propagation & radar meteorology at the Univ of TX Elec Engg Res Ctr from 1959-64. Did applied res at Southwest Res Ctr in San Antonio, TX, in radio direction finding, 1964-67. Have been in teaching at the Univ of OK in Elec Engg; tenured in 1970.

Walker, H Carl
Home: 2429 Highpointe Drive, Kalamazoo, MI 49008
Position: Majority Owner & President *Employer:* Carl Walker Engineers, Inc. *Education:* MSE/CE/Univ of MI; BSE/CE/Univ of MI *Born:* 12/22/35 1960-1962 Chief Engr, Precast/Schokbeton Inc, Kalamazoo, MI. 1962-1963 Design Engr Miller-Davis Engrs Inc, Kalamazoo, MI. 1963/1965 Design Engr/Assoc T.Y. Lim & Assoc, Chicago, IL: 1965-1982 Pres Carl Walker & Assoc Inc, Kalamazoo, MI. - 1980 - 1981 Adjunct Prof of CE , Univ of MI, Ann Arbor, MI, 1980 - Volunteer in Mission Bush Pilot. Bd of Global Ministries. United Methodist Church, Liberia, W Africa. 1983-1985 Managing Partner, Walker & Cagley, Consulting Engineers, Kalamazoo, Michigan; 1985-present, President, Carl Walker Engineers, Inc., Kalamazoo, Michigan. *Society Aff:* ASCE, NSPE, ACI, NPA/PCC, PCI, IMPC

Walker, Hugh S
Home: 2828 Nevada, Manhattan, KS 66502
Position: Prof *Employer:* KS State Univ *Education:* PhD/ME/KS State Univ; MS/Eng Mech/LA State Univ; BS/ME/LA State Univ *Born:* 7/31/35 Native of Mooringsport, LA. Enrolled in 1953 at LS State Univ on athletic scholarship which was given up as a sophomore to concentrate on engrg studies. Taughts Engrg Mech at LSU 1957-60. With KS State Univ since 1960. Teach courses and direct graduate students in the areas of internal combustion engines and controls, dynamics, machine vibrations, stress analysis, acoustics and engrg analysis. Reg PE in KA and LA. Serve frequently as consultant and expert witness for lawyers. Enjoy pastimes of auto repair and the watching of high school and college sports. Recent efforts involve extensive use of microcomputers. *Society Aff:* ASME, ASEE, SESA, ΣΞ

Walker, John N
Business: S-129 Agr'l. Science Bldg. North, Univ of Kentucky, Lexington, KY 40506
Position: Assoc Dean, College of Agriculture *Employer:* University of Kentucky. *Education:* BS/Agri Engg/PA State Univ; MS/Agri Engg/PA State Univ; PhD/Agri Engg/Purdue Univ. *Born:* Feb 1930. Native of Erie, PA. Served with US Navy 1951-54 on active duty. Retired with rank Commander. Ext. Engr for Penn State Univ 1954- 58. Joined faculty at Univ of Kentucky 1960. Apptd Chmn of Agr Eng Dept 1974-81. Acting Dir Kentucky Inst for Minerals & Mines Res, Title III Programs 1981-82. Appt'd Assoc Dean College of Agriculture 1982. Served as Chrmn of the Structures and Environment Div of ASAE 1972, Dir for Prof Dept 1975-1977, Dir for Educ and Res 1981-83, V Pres for Administrative Council 1984-87. In 1974 recd 'Great Teacher Award' from Univ of Kentucky Alumni Assn and the MBMA Award for work in advancing the knowledge and science of farm buildings from ASAE, elected Fellow American Society of Agricultural Enginers 1979. *Society Aff:* ASAE

Walker, John R
Home: 9315 E Center Ave 5C, Denver, CO 80231
Position: Retired. *Education:* EE/-/RPI. *Born:* 1901 Erie, PA. Employed by US Bur/Rec:1926-40, design of hydro-electric features for dams; 1945-60, Reg Supervisor of Power for Missouri River Basin Proj. Army, 1940-45, Lt.Col. Chief EE for State Dept. Karachi, Pakistan 1960-62. Chief EE for International Engrg Co, San Francisco 1962-64. Power Consultant for power projects in Cuba, Iran, Korea, Bangladesh, etc. Fellow and Life Member of IEEE, Vice Pres AIEE, 1954-56. Reg. PE Colorado.

Walker, John S
Business: Dept Theor & Appl Mech, Urbana, IL 61801
Position: Professor. *Employer:* University of Illinois. *Education:* PhD/Theo Fluid Mechs/Cornell Univ; BS/Naval Archit-Marine Eng/Webb Inst. *Born:* 5/25/44. Mbr of the faculty of the Dept of Theoretical and Applied Mech at the Univ of IL at Urbana-Champaign holding positions of Asst Prof from 1971-75, of Assoc Prof from 1975 to 1978 Asst Dean from 1980 to 1981 and Prof from 1978 to present. Research on crystal growth, composite materials, fusion technology and homopolar generators. 1976 recipient of the Pi Tau Sigma Gold Medal from ASME. *Society Aff:* ASME, ANS, ASEE, AAM, SIAM, ΣΞ, SES.

Walker, Joseph H
Home: 661 Los Diamantes, Green Valley, AZ 85614
Position: Consultant (Self-employed) *Education:* BS/Chem Engrg/Rose Polytechnic Inst; ChE/Chem Engrg/Rose Polytechnic Inst. *Born:* Aug 23,1914. Native of Terre Haute, Indiana. Plant Engr & Res Chemist, Louisville Cement Co 1936-40. Svd with Ordnance Dept in World War II from 1940- 46; released from active duty as Lt Col. VP for R&D, Portland Cement Assn 1960- 72. Special travel and studies regarding research and development in cement, cement manufacturing process, and concrete in many foreign countries. Active in several committees in American Society of Testing and Materials Committee C-1 on cement, and in ACI. President, ACI 1975. Hobbies: photography and golf. *Society Aff:* ASTM, ACI.

Walker, Leland J
Business: 528 Smelter PO Box 951, Great Falls, MT 59403
Position: Chairman of the Board. *Employer:* Northern Engrg & Testing Inc *Education:* BS/CE/IA State Univ.; Hon. Dr./Engineering/MT. State Univ. *Born:* April 1923. Native of Fairfield, Montana. Served in Civil Engr Corps, USNR, 1944-46, 1951-53. Field engr for US Bur/ Rec 1946-51,1953-55. Vice Pres, Wenzel and Co, Consulting Engrs 1955-58. Co-founded Northern Testing Laboratories in 1958; Pres 1958-75, became Chr 1975. Pres ASCE 1976-77; Vice Pres 1965-67; Dir 1961-64. Pres of CEC/Mont 1970-71. Secy of American Council of Independent Labs 1973-75. Pres of Montana State Univ Endowment and Res Fndtn 1973-75. Vice Pres, Montana Energy Res Inst 1974-76. Director, Treasurer, Vice Chairman, Great Falls Federal Savings & Loan. Fellow, ASCE, CECUS, AAAS; member ASEE; Honor Mbr, Chi Epsilon, Tau Beta Pi; Pres ABET 1980-83 VP 1978-80, dir 1975-78; Pres Independent Labs Assurance, Ltd 1977-79; Trustee, Rocky Mountain College 1976- 81; Trustee, Dufresne Foundation; Trustee, President, Mt. Deaconess Medical Foundation; Recipient, Distinguished Engineering Alumnus, IA.ST.U. 1979; Receipient, Distinguished Alumnus Award, Mont. State U. 1981; Dir Montana Power Co; Dir Slettem Construction Co; Dir, American Association of Laboratory Accreditation; Mbr, Natl Academy of Engrg; VP, Dir. - Serv Corp of MT; VP, Dir. Danforth Instruments; Dir, MT Sci & Tech Bd; Dir Natnl Center for Policy Analysis for Acid Rain; Dir, Entech, Inl; Dir. TLL, Inc; VP & Trustee, MT Sch for Deaf & Blind. *Society Aff:* ASCE, ACEC, ASEE, ACIL, ASFE, USCOLD

Walker, M Lucius, Jr
Business: 2300 Sixth St, NW, Washington, DC 20059
Position: Dean, Sch of Engrg *Employer:* Howard Univ *Education:* PhD/Mech Engr/Carnegie Inst of Tech; MSME/Mech Engr/Carnegie Inst of Tech; BSME/Mech Engr/Howard Univ. *Born:* 12/16/36 Howard Univ since 1966: dean, Engr (1978-); Acting Dean; Asst Dean; Assoc Dean; Chrmn, Mech Engr; Acting Hd, Mech engr, Council of Univ Senate. Visiting Prof, Univ of MA; Visiting Sr Staff Mbr, Intl Res & Tech Corp. Consultant: Harold Sanders & Assoc, Howard Univ Hospital Cardiovascular Renal Res Team, A. K. Nellums & Assoc. Mbr: Center for Manufact Engrg, Natnl Bureau of Standards; Lister Hill Center of the Natnl Lib of Medicine; Engrg Manpower Commission; Bd of Trustees, Carnegie-Mellon Univ, Ad Hoc Visitor for ABET. Author of technical publications, reports, articles on minorities in engrg. Honor: Ralph R. Teetor Award. Reg PE. Fraternity: Kappa Alpha Xi. Wife Oswaldene (nee Cocking) Walker, DDS. *Society Aff:* ASEE, ASME, ΤΒΠ, ΣΞ, ΠΜΕ.

Walker, Paul N
Business: Dept of Agricultural Engrg, Urbana, IL 61801
Position: Assoc Prof *Employer:* Univ of IL *Education:* PhD/Engr/Univ of MA; MS/Agr Engr/Univ of HI; BS/Agr Engr/Univ of AR *Born:* 10/08/48 Raised on a farm near Dover, AR. Worked as an engineering trainee with Intl Harvester in Hinsdale, IL during summers of undergraduate program. With Univ of IL since 1974. My responsibilities are half teaching and half research.My main research interest is the study of rate processes related to agriculture. Projects have included work on computerized crop management, aquaculture, irrigation, drainage, and heating greenhouses with reject warm water. Recreational interests include wookworking, photography and camping. *Society Aff:* ASAE

Walker, Richard O, Jr
Business: Civ Engg Bldg, W Lafayette, IN 47907
Position: Assoc Prof. *Employer:* Purdue Univ. *Education:* MSE/Civ/Princeton Univ; BSE/Civ/Princeton Univ. *Born:* 9/26/23. From 1948 until 1974 employed by consulting firm, Abbott, Merkt & Co, Inc NY, NY, in various capacities: Designer, Principal, Secy, VP, Mbr of Bd of Dirs. After short period with TAMS, was consulting engr in private practice. 1978 appointed Assoc Prof, Div of Construction Engg & Mgt & The Sch of Civ Engg, Purdue Univ, Past Pres, Am Inst of Consulting Engrs; Past Pres, AA Potter Chapter, ISPE; P Pres, Princeton Club of NY; P Sr Warden, Christ Church, Bronxville, NY; Past Mbr, Planning Commission, Bronxville, NY; Mbr, Dev Systems and Services Council, Fellow ULI; reg in 10 states. Past Vestryman, St. Johns Episcopal Church, Lafayette, IN.; Bd of Directors, Tippecanoe: County Building Authority. *Society Aff:* ASCE, ASTM, ULI, ASEE, IES.

Walker, Robert B
Home: 4816 Southwest 310 St, Federal Way, WA 98003
Position: Vice Pres Operations. *Employer:* Reichhold Chemicals Inc. *Born:* Nov 1926. BSc from U of Alberta. U.S.citizen. Employed by Reichhold in R&D, Sales and production since 1948. Assumed current position in 1973. Resp for 9 West Coast plants. Products include synthetic resins, industr. chemicals and glass fibers. Interests are reading, music, horticulture.

Walker, Samuel R, Jr
Business: 8950 Kemphood, Houston, TX 77080
Position: Owner. *Employer:* Sam Walker & Assoc. *Education:* BSCE/Structural/Univ of Houston. *Born:* 6/28/13. 1935-41, field engr TX Hgwy Dept & Humble Oil Co. 1941-1945 field engr Todd Houston Shipyard (US Maritime Commission). 1945-1955 Div Supervisor Engg & Construction Gulf Oil Corp, 1956 Div Engg Mgr for Lift Slab, Inc, 1957-1961 VP of T-Stress, Inc. 1961-1970 sr structural engr for the Lummus Co, Foster-Wheeler Corp, & Bechtel Corp. 1970-79 consulting engr for Engg Sci, Inc, C B Southern, Inc, Colt Engg Co, Signal Chem Co, Stauffer Chem Co, Anchor-Wate, Inc. Gulf Coast Waste Disposal Authority, Hutchison-Hayes Intl, Lawrence-Allison & Assoc, Cooper Industries, 1979-82 Bufete Engr & Const Int'l, Process Projects Int'l. *Society Aff:* NSPE, PEPP, SAME.

Walker, Stanley E
Business: E.C. Jordan Co, 261 Commercial St, P.O. Box 7050, Portland, ME 04112
Position: VP *Employer:* E.C. Jordan Co. *Education:* MS/Civil Engg/Univ of ME; BS/Civil Engg/Univ of ME. *Born:* May 8,1940, Minot, Maine. Reg PE ME, NH, VT NY, NJ, Miss., Minn., Ill., Fla, CN, RI, Mass. Tau Beta Pi; Phi Kappa Phi; Army Corps Engrs 1964-66, Vietnam. Woodward, Clyde, Sherard & Assoc., Clifton, NJ, Asst. Proj. Engr 1966-68. Jordan Gorrill Assoc, Portland, Maine, Proj Engr 1968-71, Chf Soils Engr 1971-74, President 1974-85; E.C. Jordan Co. Portland, Maine 1968-- , V.P 1984- . ASCE, Maine Sec, Director 1970-72, President 1974-75. NSPE, Eastern Maine Chap, Pres 1972-73; Maine Soc Dir 1976- 78, Young Engineer of Year 1974. Supporting mbr Trans Res Bd. Assoc Mbr, Geological Soc of Maine. Mbr, Amer Soc for Testing and Materials; Mbr Project Mgmt Inst. M. Meredith Osgood, 1962; c: Scott S, Andrew O, James E. *Society Aff:* ASCE, NSPE, ASTM, TRB, PMI.

Walker, U Owen
Home: Box 1224, Albuquerque, NM 87103
Position: Eng *Employer:* US Dept of Interior *Education:* BE/Civil Engg/Yale Univ; /Basic/Millsaps College; Certificate/Plant Layout & Facilities Planning/Univ of KS. *Born:* 10/9/25. Native of AR. Served US Navy, World War II. Engr involved in Construction, Plant & Facilities Engg & Energy Conserv., 1946-Present. Professional Engr AR (1796) & MS (2488). Certified Plant Engr (391), AIPE. Pres, Natl Dir, Chapter 77, AIPE, 1965-70. TN State Chrmn, Professional Engrs in Govt Practice Sec, NSPE, 1965-67. "Plant Engr of Yr" 1968, Morris Medal, AIPE. "Mr. Industry

Walker, U Owen (Continued)
Award–, July 1969, Editors, Industrial Maintenance & Plant Operation Magazine. Listed: "Who's Who in South/Southwest–, 11th Edition, 1969; "Engrs of Distinction–, 1st Edition, 1970, EJC; "Dictionary of Intl Biography–, 7th Edition, 1971; 'WHO'S WHO IN ENGIEERING' 4th Edition, 1980; 'WHO'S WHO IN ENGINEERING' 5th Eidition, 1982; 'WHO'S WHO IN TECHNOLOGY TODAY', 2nd edition, 1981, "WHO'S WHO IN TECHNOLOGY TODAY', 3rd Edition, 1982, Volume 4 - Civil and Earth Sciences. "WHO'S WHO IN THE WEST–, 19th Edition, 1984-85; "INTERNATIONAL WHO'S WHO IN ENGINEERING–, 1st Edition, 1984. Certified Energy Mgr, AEE; Certified Records Mgr, ICRM; Accredited Personnel Mgr, PAI. *Society Aff:* ASCE, AIPE, AMS, AEE.

Walker, Walter W
Home: 5643 E 7th St, Tucson, AZ 85711
Position: Senior Scientist *Employer:* Hughes Aircraft Co. *Education:* BS/Met Engr/Univ of AZ; MSc/Met/Univ of AZ; PhD/Met/Univ of AZ. *Born:* 1/14/24. Native of AZ. Served four yrs apprenticeship in foundry prior to service in WWII. Served with US Navy 1943-45. Res Engr CA Res & Dev Co, Livermore, CA & ORNL 1952-53. Supv of Met at Hughes Aircraft Co 1953-59. Lectr in Met Univ of AZ 1959-1967. Assoc Prof of Met Engr Univ of AZ 1967-73. Sr Staff Engr add Met Group Leader 1973-84, Senior Scientist - 1984-present. Responsible for HAC long range efforts in Mtl Shortage Planning. Hobbies include mountaineering, caving, backpacking, bagpipes & Ufology. Principal, ANDECO (AZ nondestructive Testing Co) 1980-present. Level III NDT Examiner. *Society Aff:* ASM, AIME, ASTM, IMS, ASNT, ABANA, NYAS

Walker, William F
Business: 1300 Main St, Houston, TX 77001
Position: Prof & Chairman *Employer:* Rice Univ *Education:* PhD/ME/OK St Univ; MS/Aero Engr/Univ of TX; BS/Aero Engr/Univ of TX *Born:* 12/1/37 Native of Sherman, TX. Industrial experience with Ling-Temco-Vought Corp and Sandia Laboratories. VP-The Technical Corp. Served as consultant for numerous industrial and governmental organizations. *Society Aff:* ASME, ASAIO, TBΠ

Walker, William Hamilton
Business: 1106 Newmark Lab, 208 N Romine St, Urbana, IL 61801
Position: Assoc Prof *Employer:* Univ of IL at Urbana-Champaign *Education:* PhD/CE/Univ of Il-Urbana-Champaign; MS/CE/Univ of Il-Urbana Campaign; BS/CE/Univ of Mass-Amherst. *Born:* 12/28/34 in Brookline, MA, son of William A and Ingeborg Thorkilsen Walker; married Shirley Ackerman, Nov 3, 1962; Children--William Franklin and John Hamilton. Has served on the faculty of the Univ of IL at Urbana-Champaign since 1961: instructor 1961-63, asst prof 1963-68, assoc prof 1968-, asst head dept 1973-76; Assoc. head dept 1985- . Teaching and research in structural mechanics and dynamics, field testing of highway bridges, and fatigue life expectancy. Consulting in structural dynamics to various govt agencies and industry. Pres of Central Ill Section, ASCE. *Society Aff:* ASCE, ASEE, ΣΞ, TBΠ, XE

Wall, Alfred S
Home: 115 Wagush Trail, Medford Lakes, NJ 08055
Position: Mgmt Consultant (Self-employed) *Employer:* Retired *Education:* BS/EE/Lehigh Univ *Born:* 1/21/12. Native of NJ. Joined Western Electric in 1935 and rose to level of Inspection and Test Dept Head. Principal subsequent employment was with Weston, ITT and from 1961 to 1974 with RCA in Govt & Commercial Systems Staff Reliability and Maintainability Mgr and Product Safety Administrator. Have special interest in motivation, writing, training programs, public speaking and the application of value engrg to such disciplines as reliability, maintainability and quality assurance. Since 1974 retired. Testimonial Award for Leadership 1968 ASQC. VP Relations, 1971-73 ASQC. *Society Aff:* ASQC, IEEE

Wall, Evern R
Business: Box 982, El Paso, TX 79960
Position: President, Chairman & CEO *Employer:* El Paso Electric Co. *Education:* BS/EE/NM State *Born:* 11/20/32. Borger, Texas. Reared in Hobbs, NM. Began with El Paso Elec Co as Part Time Night Service Clerk, 1954, while attending college. Joined full time June 1957 as Jr Eng Progressed to Engr, Industrial Relations Mgr, VP, Exec VP to Pres & CEO May 1976. Reg PE TX; Bd of Dir EEI. Past Pres TX Atomic Energy Research Fdn. Served as Pres & Dir of El Paso Boys Club, Yucca Council Boy Scouts, El Paso Rehabilitation Council, United Way. Past Pres WEST, Chrmn of Bd, Franklin Land and Resources; Dir of State Natl Bank; Dir NMSU Fdn; Past Pres Rocky Mtn Elec League; Chrmn of Renaissane 400; Pres El Paso Chamber of Commerce; 1976 Distinguished Alumnus Award, 1983 Reach for Excellence Award; 1983 Ballet El Paso Man of the Year Award, Honorary Doctorate of Laws, NMSU; 1984 Humanitarian Award by Natl Jewish Hospital and Asthma Res Ctr, Governors' Task Force on Private Sector Initiatives; Governors Emergency Task Force on Jobs and Unemployment Fund; Governors Outstanding Volunteer

Wall, Fred J
Business: Box 9175 Attn M-101, Lester, PA 19113
Position: Mgr Materials Eng Lab. *Employer:* Westinghouse Electric Corp. *Education:* MS/Metallurgical Engg/Univ of PA; BS/Math/Penn State Univ. *Born:* 1/8/29. Active in the ASTM Committe A-1 on nickel base castings, mbr of the Joint ASTM-ASME Committee on Effect of Temperature on the Properties of Metals, Chrmn of the High Temp Alloys Committee of the AIME, Chrmn of the Power Activity of ASM. Dir of Matls Sys & Design Div, ASM. Ten yrs experience in superalloy dev for both aviation & industrial gas turbines turbines & for past twelve yrs, Mgr of the Metallurgical, Chem & Mech Labs at the Westinghouse plant located in Lester, PA. *Society Aff:* ASM.

Wall, Nelson C
Business: 2620 San Mateo, Albuquerque, NM 87110
Position: Vice President *Employer:* Solar America Inc. *Education:* Dr/Politcal Sc/Univ of Havana; BS/mech Engg/GA Tech. *Born:* 10/18/28. Havana, Cuba. Naturalized US citizen. Universidad de la Habana, Doctor Ciencia Politica, 1954; Univ of Okla, Industrial Dev Institute, 1969. United Railways of Havana, 1948-50; Barrenos y Equipos de Cuba 1950-52; Asst. to President, Ferrocarriles Occidentales de Cuba 1952-59; Ma's Bottling Co 1960-63; Principal Res. Eng & Dir. Intl Prog. Div, GA Inst of Tech 1964-1980; Vice President Intl, Solar America Inc. 1980-present. Resp for all international projects conducted by Solar Am Inc. in areas of research, evaluation, and implemention of energy alternatives. Active in ASME, AIDC, P Reg Advisor SBA, sev offices in IIE. Latest two, Natl Exec VP and VP Intl Operations. Order of the Engrs 1975; Past Natl President Cuban Soc Mech Engrs. *Society Aff:* IIE, ASME, AIDC, NSPE.

Wallace, Eugene E
Business: Wagner Div., McGraw Edison Co, St Louis, MO 63141
Position: Vice President-Brake Engrg *Employer:* Wagner Div., McGraw Edison Co. *Born:* Nov 1920. BSME, Washington Univ (St.Louis). Native of St.Louis, MO. Naval aviator 1942-45. Rank of Commander USNR. With Wagner since 1947. Successive positions of development engr, supervisor-brake development, chf engr - automotive products, and vice pres -engrg (1966). Active in SAE as Chrmn-Brake Committee, Chrmn-Automotive Council, Mbr-Tech Bd and Mbr Bd/ Dir. Hobbies include hunting/fishing and outdoor activities. *Society Aff:* SAE.

Wallace, Floyd C
Business: 1500 Meadowlake Pkwy, Kansas City, MO 64114
Position: Retired *Education:* BSc/Gen Engrg/Univ of Cincinnati. *Born:* 6/11/18. Served with 8th USAF 1943-45. Design Engr, Project Engr and Proj Mgr with Black & Veatch since 1946. Responsible for design of fossil power plants, electric transmission and distribution facilities. Partner in Black & Veatch since 1956. Enjoy scuba diving, underwater photography and travel. *Society Aff:* ASME, IEEE, NSPE, AWS.

Wallace, James D
Home: 3501 Bayshore Blvd. No. 803, Tampa, FL 33629-8933
Position: Ret. (Radio Engr) *Education:* MA/Physics/Univ of MS; BA/Physics/Univ of MS. *Born:* March 1904, Gloster, Miss. Was employed 44 yrs at Naval Res. Lab. During latter years served as Head, Communication System Sec and thereafter as Consultant, spec in Military Communications. Have servedd on Advisory Group on Reliability of Electronic Equipment and as Navy rep of CCIR. Awarded several patents on electronic devices and was an occasional contributor to technical literature. Fellow IEEE. Active in IRE (an IEEE predecessor) serving on various committees, arranging tech mtgs, drafting standards and recommending awards. *Society Aff:* IEEE.

Wallace, James M
Business: P.O. Box 450, Clifton Park, NY 12065
Position: Senior Vice President *Employer:* Greenman Pederson Assoc. *Education:* BS/CE/Univ of MI.; MPA/Pub Adm Urban Plng/SUNY. *Born:* 3/8/24. BSCE Univ of Mich June 1948, Master of Pub Admin, State Univ of NY at Albany (Urban Plng) 1967, Army Corps of Engrs 1942-45. Employed by NY State Dept of Transportation in Traffic and Transportation Engrg 1948-67. Private practice with James M Wallace Assoc 1967-69. Priv practice with Wallace, Champagne Assoc, 1969-76. Private practice, Ptnr Greenman Pedersen Assoc 1976- . Mbr NSPE, ITE and ASPA. Licensed PE in nine states and D.C. *Society Aff:* ITE, ASPA, NSPE, AAFS, TRB

Wallace, John M,
Home: Orchard Hl Rd Mason Is, Mystic, CT 06355
Position: Dir of Engrg. *Employer:* Naval Submarine Base New London. *Education:* BE/CE/Yale Univ. *Born:* Feb 1925. Native of Janesville, Wisc. Served with Marine Corps 1943-46. Asst City Engr, New London, Connecticut. With US Navy since 1956 as Dir of Engrg, Rota, Spain; Chf Engr for Resident Officer in Chg of Construction, Subase New London. Assumed current position of Dir of Engrg, Subase New London 1971. Served as Vice Chrmn of Plng Bd, City of New London. Pres of Eastern CT Yale Alumni Club. State Pres, CT Soc of Prof. Engrs, Natl Chrmn, PEG/NSPE. Trustee of Mitchell College, New London Reg PE and LS in Connecticut. *Society Aff:* NSPE.

Wallace, Norval D
Home: 16 Club Grounds So, Florissant, MO 63033
Position: Dean, School of Engg *Employer:* Southern IL Univ. *Education:* PhD/Mathematics/St Louis Univ; MS/EE/Univ of MO-Rolla; BS/EE/Univ of MO-Rolla. *Born:* 7/27/32. Eight yrs with Emerson Elec Co as a radar sys analyst. One year with McDonnell-Douglas as a Sr Grp Engr with radar sys design as specialty. Two yrs with the Univ of MO-Rolla as an Assoc Prof in the Dept of Mathematics. Eighteen yrs with Southern IL Univ-Edwardsville. Currently Dean of the School of Engg. Special interests are probability and statistics, operations research with applications to engg problems. *Society Aff:* IEEE, ORSA, IIE.

Wallace, Paul B
Home: 6134 Hemlock, Shawnee Mission, KS 66202
Position: Chief Metallurgist. *Employer:* Superior Metal Trtg & Equip Co. *Born:* Sept 10,1923. BS U of Notre Dame 1947. Served US Marines WWII. Fifteen yrs Chf Metallurgist, Metallurgical, Inc. Mpls., Minn. Proj Engr Missile Dev Programs ABMA and Thiokol Chem Huntsville, Ala; Founder & President Advanced Materials Technology, Inc. Roseville, Minn. Instructor St. Paul Vocational. Chapter Chrmn and Mbr Technician and Handbook Committee, ASM. Reg PE Minn. Hobbies: Tennis, Bridge, Reading and Dancing.

Wallace, Ralph H
Home: 718 E State St, Mason City, IA 50401
Position: Retired *Employer:* Wallace Holland Kastler Schmitz & Co. *Education:* BS/Civil Engg/IA State Univ. *Born:* Feb 25,1916, Williamburg, Iowa. Field Artillery 1941-46; 76th Inf. Div. Captain. Field Engr, L.W.Mahone, 1946-48. Wallace Engrg Co 1948-56. Partner, Wallace & Holland 1956-64. Partner Wallace Holland Kastler & Schmitz 1964-69. Pres 1969-'81. Sch Bd Mbr. Mason City Iowa 1952-64, Pres 1964. Mbr Iowa Bd of Engrg Examiners 1958-67, Chrmn 1962,1963,1967. Iowa Bd of Regents 1967-73. Iowa Pub Broadcasting Netwks Bd 1971-76. Bd/Dir Iowa State Mem Union 1967-73. Bd/Dir First Natl Bank, Mason City Iowa 1969-81. NCEE 1958-67, Dir 1970-72. Fellow ASCE, Dir 71-72, VP 1976-77. Fellow, ACEC. Mbr AAAS, ASEE, NSPE. Awarded Dist Serv Cert, NCEE 1968. Dir and bd mbr, North Iowa Medical Center, 1979-present. Tau Beta Pi-1981. *Society Aff:* ACEC, ASCE, AAAS, ASEE, NSPE.

Wallace, Robert T
Business: 3873 Sulphur Spring Rd, Toledo, OH 43606
Position: Consultant. *Employer:* Self. *Education:* MSIM/Indust Mgt/MIT; BSChE/Chem Eng/Univ of MI. *Born:* 1920, A consultant in the management of new opportunities and energy. Associated with Union Carbide, Libbey-Owens-Ford and for 21 yrs Owens-Illinois in New Products, the last 9 as Corp Vice Pres. Became consultant 1975. Fellow AIChE, reg PE in Ohio . 'Toledo Engr of the Year' in 1965 and Chrmn of Polymeric Materials Council at Princeton Univ 12 yrs. Married to Virginia Frey, daughter Marjorie Stearns. *Society Aff:* AIChE.

Wallace, Ronald Stephen
Home: 1131 8th Ave West, Birmingham, AL 35204
Position: Analyst/Corp Planning. *Employer:* Cameron Iron Works, Inc. *Education:* JD/Law/Univ of AL; MBA/Bus/Univ of AL; BS/Metallurgical Engg/Univ of AL. *Born:* 6/30/53. Native of Birmingham, AL. Holds BS Metallurgical Engg *cum laude*, MBA, and Law degrees from Univ of AL. Mbr of numerous organizations among which are ASM, AIME, ABA & Theta Tau Natl Pro Engr Fraternity, Tau Beta Pi & Omicron Delta Kappa. Presently employed by Cameron Iron Works, Inc of Houston, TX as an analyst in the Corporate Planning Div. *Society Aff:* TBΠ, ABA, ASM, AIME.

Wallenstein, Gerd D
Home: 298 Austin Ave, Atherton, CA 94025
Position: Lecturer *Employer:* Stanford University *Education:* MS/Cybern Sys/San Jose State Univ; PhD/Intern Planning/Stanford Univ. *Born:* Jan 1913, Berlin, Germany. Studied EE at Berlin 1931-33. Radio engr in Berlin. from 1939-47, in Shanghai and Tientsin, China. With GTE Lenkurt since 1948 as Applications Engr; 1956 VP Prod Plng. Concerned with dev of multichannel telecommunications worldwide. VP Plng from 1963-70. Active in International Telecommunication Union. Chrmn of CCITT/CCIR Spec Study Group GAS/3 'Economic & Technical Comparisons of Transmission Systems 1973-1980. Fellow IEEE. Bekesy medal of Hungarian Post Office Research Institute. Teaches interdisciplinary course on international telecommunication standardization at Stanford University. *Society Aff:* IEEE.

Wallenstein, Howard M
Home: 84 Spring Valley Rd, Paramus, NJ 07652
Position: VP. *Employer:* Southland Frozen Foods. *Education:* BChE/ChE/Poly Inst of NY. *Born:* 5/8/37. Communications officer USS Cambria (APA-36) 1958-1962. Process mgr, refined syrups, CPC Intl 1968-1972, achieved comp control of processing, environmental compliance of Yonkers Sugar Refiney. Bus mgr of corn syrup business, CPC Intl 1972-1975. Plant mgr & dir of plant operations, Modern Maid Food Products, elected vp-production Southland Frozen Foods 1978. Responsible for all field, production, & engg functions. Active in community affairs. Enjoy flying & holds class 3 private pilots license. *Society Aff:* AIChE.

Waller, Robert
Business: NUS Corporation, 910 Clopper Rd, Gaithersburg, MD 20828
Position: Manager, Regulatory Compliance Programs. *Employer:* NUS Corporation *Education:* PhD/Envir Engg Sci/Johns Hopkins Univ; MS/Envir Engg/RPI; BS/Chem Eng/RPI. *Born:* 9/2/37. in Newburgh, NY. Joined the NY State Dept of

Waller, Robert (Continued)
Health as a specialist in the Water Supply Section, 1958-1962, after receipt of PhD in 1966, employed with the Engg Dept of the DuPont Co, Newark, DE, as an ind waste engr. Hd of Environmental Engg Sec of Hittman Assoc, Inc, 1969-1972. VP of Environmental Quality Systems Inc (EQSI) ,1972-1980, responsible for proj mgt & technical performance. Specialist in advanced waste treatment, hazardous waste management, residue mgt, & water quality mgt. Special UNESCO consultant 1977. Mbr of NSF Inspection team for Miyagi-Oki (Japan) earthquake, 1978. Director of Public Works Programs, NUS Corporation, 1980-83. Assistant General Manager & Manager, Regulatory Compliance Programs, NUS Waste Management Services Group, 1983-present. *Society Aff:* AIChE, AWWA, WPCF.

Wallis, Bernard J
25315 Kean Ave, Dearborn, MI 48124
Position: Chmn of the Bd *Employer:* Livernois Engineering Co. *Education:* BS/ME/Detroit Inst of Technology. *Born:* Dec 1910. Pres of Livernois Engrg Co since 1949 and Pres of Livernois Automation Co since 1965. Cos are engaged in the engrg and mfg of tooling and equipment for the metal stamping industry. Reg PE Michigan in 1949 in the field of Mech Engrg. Chrmn Detroit Chap1 of SME 1959-60. Dir Natl Soc Mfg Engrs 1974- 76. Recd Merit Award SME 1971. Life mbr of SME 1976. Received in 1979 from the SME the Joseph Siegel Meml Award. Elected Fellow, SME 1986. *Society Aff:* SME, SAE, ESD, NSPE.

Wallis, Clifford M
Home: RFD 1, Moretown, VT 05660
Position: Professor Emeritus. *Employer:* Retired. *Education:* DSc/-/Harvard Univ; BS/EE/Univ of VT. *Born:* March 7,1904; At Univ of MO-Instructor in EE 1928-34; Asst Prof 1934-40; Assoc Prof 1940-44; Prof of EE 1944-70; Chrmn Dept of Elec Eng 1947-67. Visiting Prof of Comm. Eng Harvard Univ, Summer 1941; Spec Res Assoc Harvard Underwater Sound Lab 3/44-10/45; Res Assoc Edward Street Lab, Yale Univ Summer 1953. Consultant for US Navy UnderWater Sound Lab, Summers 1954-70. Dir Prof Dev US Navy Sound Lab New London CT 1970-72. Fulbright Lectureship, Univ of Ankara, Turkey 1960-61; Fulbright Lectureship Nat U of Taiwan, Taipei, Taiwan 1967-68. Mbr Research Comm of AIEE several yrs in 1950's. Vice Chrmn of Comm one year. Recd Award for Dist Serv in Engrg from Univ of Missouri April 1973. Fellow-Amer Inst of Elec & Electronic Engrs. Mbr-Amer Soc of Engg Education. *Society Aff:* ASEE, IEEE.

Wallis, Graham B
Business: Thayer Sch of Engg, Hanover, NH 03755
Position: Prof. *Employer:* Dartmouth College. *Education:* PhD/Engg/Cambridge Univ; SM/ME/MIT; BA/Engg/Cambridge Univ. *Born:* 4/1/36. Rugby England. '59-62, UK Atomic Energy Authority. '61, Fellow of Trinity College Cambridge. '62-present, Dartmouth. Ludwig Mond Prize, Inst Mech Engg '62. Moody Award ASME '71. ASME Centennial Medal 1980, Interests: two-phase flow, heat transfer, nucl safety. *Society Aff:* ASME

Wallis, William E
Business: 8031 Broadway, San Antonio, TX 78209
Position: Pres *Employer:* Wallis & Assocs Consltg Engrs, Inc *Education:* BS/ME/TX A&M Univ *Born:* 12/25/25 Native of TX. Engrg Ofcr US Navy 1942-1945, 1950-1953. Field engr for HVAC manufacturers, three yrs. Engrg contracting, three yrs. Project engr with consltg engrg firm three yrs. Principal engr of own consltg engrg office for twenty-five-yrs. Extensive experience in the analysis, design, & construction contract administration of major HVAC & Electrical Sys. Large tonnage central chilled water plants & .60 KV to 15.0 KV electrical distribution & utilization. Author of statewide energy conservation plans for multibuilding owners. Active in solar energy & sys application. Enjoy hunting & home shop work. *Society Aff:* ACEC, NSPE, PEPP

Wallmark, J Torkel
Business: Gothenburg, Sweden, 41296
Position: Professor. *Employer:* Chalmers Univ of Technology. *Education:* Doctor/Electronics/Royal Inst Technology, Stockholm Sweden. *Born:* 1919. Teknologie doktor, Royal Inst Technology, Stockholm Sweden 1953. R Inst Technology, electron tube research, 1943-53; AB Standard Radiofabrik, electron tube dev 1944-45; Natl Bd of Tech Res, research grants 1950-51; Elektrovarmeinstitutet, solid state res 1952-53; RCA Labs, Princeton NJ electron tube and solid state res 1947-48, 53-64,66-68; Chalmers Univ of Tech 1964-66, 1968- . Prof of solid state electronics. 1983 Prof of innovations. Wallmark Award, Royal Swed. Acad. Science 1954. RCA Laboratories Achievement Awards 1948,57, 62. David Sarnoff Outstanding Team Award, RCA Labs 1964. 10 books, 20 patents, 70 scientific publications. *Society Aff:* IVA, VVS, IEEE, AAAS, VA.

Walsh, Edward K
Business: Dept. Engineering Sciences, University of Florida, Gainesville, FL 32611
Position: Prof. *Employer:* Univ of FL. *Education:* PhD/Applied Mathematics/Brown Univ; BE/Mech Engg/Union Coll. *Born:* 2/19/31. Native of Phila, PA. Employed by Genl Elec Co, Schenectady, NY until 1962; attended Union Coll extensional Div, Schenectady, NY and received BE, Melchanical Engg, 1963. Employed at Melchanical Technology, Inc, Latham, NY 1962-63. Attended Brown Univ, Providence, RI 1963-66, receving a PhD in Applied Mathematics in 1967. Was associated with the Mellon Inst of Sci as a Postgraduate Fellow from 1966-68, and joined the Dept of Civil Engg, Carnegie- Mellon Univ in 1969. Left Carnegie-Mellon Univ in 1970 to join the faculty of the Coll of Engg, Univ of FL, Gainesville, FL. Consultant for the Sandia Labs, Albuquerque, NM. Enjoy scuba diving, bicycling, and racquetball. *Society Aff:* SNP, ΣΞ.

Walsh, John M, III
Business: 785 Central Ave, Murray Hill, NJ 07974
Position: President. *Employer:* Bull & Roberts, Inc. *Education:* BChE/Cornell Univ. *Born:* Dec 1936. With Bull & Roberts, Inc since 1959 as engineer, asst to president, vice president and president. PE N Y and New Jersey. Mbr AIChE, ACS, SNAME, AOCS. *Society Aff:* AIChE, ACS, SNAME, AOCS.

Walsh, Myles A
Home: 60 Barnabas Rd, Falmouth, MA 02540
Position: Professor. *Employer:* Massachusetts Maritime Academy. *Education:* PhD/Aero Eng/Caltech; MBA/Business/Univ of CT *Born:* March 1944. PhD from Caltech; MS from Caltech; MBA from U.Conn; BA from Harvard. Asst Proj Engr for Pratt & Whitney Aircraft 1967-71 specializing in high-temperature fuel cells and gas turbines. Chairman, Engrg Dept at Massachusetts Maritime Academy 1971-82. PE in Connecticut and Massachusetts. Consultant in electro-chemistry to Cape Cod Research. Enjoy bridge and chess. *Society Aff:* ES

Walsh, Thomas J
Business: Case WRU, 10900 Euclid Ave, Cleveland, OH 44106
Position: Adj. Prof. Chemical Eng./Sr Research Engineer *Employer:* Case Western Reserve University *Education:* PhD/CE/Case Western Reserve Univ; MChE/CE/RPI; BChE/CE/RPI. *Born:* 7/17/17. Native of Troy, NY. Taught Chem Engg at RPI, Case WRU (Prof at CWRU), & Cleveland State Univ. Worked with Sohio, NASA, TRW, & SCM Corp. Fields of specialty: unit operations, distillation heat transfer, environment control, fuels, thermodynamics, energy conversion. Awards from ACS, CTSC (Cleveland Technical Socs Council) Engineer of Year 1985-86. Joined SCM Corp in 1966. Retired 1980 Adj. Prof CWRU 1980 to present, Adj. Prof. at Cleveland State Univ; retired 1986; consultant. *Society Aff:* AIChE, ACS, CES, ΣΞ, TBΠ, ΑΧΣ, ΦΛΥ.

Walston, William H, Jr
Business: Mech Engg Dept, College Park, MD 20742
Position: Assoc Prof. *Employer:* Univ of MD. *Education:* PhD/ME/U of DE; MSME/ME/U of DE; BSME/ME/U of DE *Born:* 4/13/37. Native of Salisbury, MD.

Walston, William H, Jr (Continued)
Married with two children. Served in US Army Ordnance 1963-65. Have been at Univ of MD since 1965 in Mech Engg Dept. Areas of interest include vibrations, dynamics, applied math and design. Acoustics Noise. Active in ASME, holding several local offices including section chrmn. Recently res activity in the area of ground vehicle aerodynamics, the area of aerial release mechanisms for insects and in the area of industrial and community noise control. *Society Aff:* ASME

Walter, C Richard
Home: 53 Highland Ave, Eastchester, NY 10707
Position: Pres. *Employer:* Hazen & Sawyer, PC. *Education:* MSCE/Civil Engg/Northwestern Univ; BCE/Civil Engg/Manhattan College. *Born:* 4/17/29. Native of Yonkers, NY. Post grad courses at MIT & Columbia. Reg PE in 7 states. Industrial waste engr for Lederle Labs. With Hazen & Sawyer since 1952, spec in reports & designs for water and wastewater treatment programs. Became partner 1962, Pres when firm incorporated in 1978. Lectured at Manhattan College & the Univ of WI. VP 1972-76, and Pres 1976-77 NY Assn of Consulting Eng. Natl Dir of Water Pollution Control Federation 1977-80. Pres and Trustee of Consulting Engrs Life & Health Insurance Plan 1979-87. Active in church & public sch affairs, Eastchester, NY. *Society Aff:* ASCE, AWWA, NYWPCA, ACEC, NYACE

Walter, Francis J, Jr
Business: P O Box 889, Savannah, GA 31402
Position: District Engineer. *Employer:* Corps of Engineers. *Born:* Sept 1932. Master Industrial Engrg NYU; Master Military Arts & Science, USAC&GSC; BSCE, Notre Dame. Reg PE DC 1961. Native of Staten Island, NY. Entered Army in 1954 as Dist Military Grad from Notre Dame. Held numerous troop command and staff assignments (9 yrs). Served in staff assignments in Office of Chief of Engrs, Off of Personnel Operations and Office of Secretary of the Army (9 yrs). As Savannah District Engr since May 1976, resp for Corps of Engrs Civil Works Program in much of Georgia and Military Construction in the Carolinas and Georgia. SAME Natl Dir 1971-76; SAME Gold Medal 1976; SAME Fellow. Enjoys upland game hunting.

Walter, Gordon H
Business: 7 S 600 County Line Rd, Hinsdale, IL 60521
Position: Mgr, Materials Engrg *Employer:* Intl Harvester Co. *Education:* BS/Met Eng/IL Inst of Tech *Born:* 12/27/34. Mr Walter grad from IL Inst of Tech in 1958 with a BS degree in Met Engg. His ind experience began in 1953 as a Cooperative Engg Student. Past responsibilities at IH include mtls specifications, various met res functions, Mtls Engr for the Agri/Ind Engg Dept, & mgr of Metals Res. Presently, he is Mgr of Materials Engrg & Specification for Agri Engrg. Mr Walter is a mbr of ASM and SAE & serves on ISTC Div 33, Gear Met, ISTC Div 8, Carbon & Alloy Steel Hardenability, SAE Fatigue Design & Evaluation Committee. *Society Aff:* SAE, ASM, AIME

Walters, James V
Home: PO Box AB, University, AL 35486
Position: Prof of CE *Employer:* Univ of AL *Education:* PhD/CE/Univ of FL; MS/CE/GA Inst of Tech; B/CE/GA Inst of Tech *Born:* 5/13/33 Native of Dublin, GA. USPHS commissioned officer 1956-59. Assistant Prof of CE, Univ of AL, 1959-63. Assoc Prof of CE, Univ of AL, 1963-70. Prof of CE, Univ of AL, 1970-87. Director WPCF, 1955-69. Mbr Exec Ctte, WPCF Bd of Dir, 1967. WPCF Arthur Sidney Bedell Award, 1967. Ford Foundation Engrg Resident, Intl Paper Co, Moss Point, MS, Mill, 1966-67. Founding Dir, Tuscaloosa Testing Laboratory, Inc, 1966-84. Chrmn of the Bd, TTL, 1982-87. Engr Officer, Dir, USPHS, Inactive. First Pres, AL Assoc for Water Pollution Control, 1977-78. Mbr of the Univ of AL Environ Inst for Waste Mgmt Studies, 1983-87. *Society Aff:* ASCE, WPCF, AWWA, ACS, APCA, TAPPI

Walton, David G
Business: 2828 E 45th St, Indianapolis, IN 46205
Position: Associate. *Employer:* James Associates Arch & Engrs. *Education:* BSME/HUAC/Case. *Born:* Oct 1924. BSME Case Inst of Technology. Native of Pittsburgh,PA. Served with US Army combat engrs 1942-46 - European Theatre. Design engr and proj mgr with J M Rotz Eng Co 1953-71. Assumed current resp of mgt of mech and elec engrg dept 1971. Reg PE in Indiana, Ohio and Illinois. Mbr ASHRAE, ASPE, NSPE and ISPE. National Director for ACEC. Mbr of Bd/Dir of Consulting Engrs of Indiana. Enjoy golf, sailing and walking. *Society Aff:* ASHRAE, ASPE, NSPE, ISPE.

Walton, Edward H
Home: 144 Knoll Dr, Hamden, CT 06518
Position: Retired. *Education:* BS/Engg/Swarthmore College. *Born:* Jan 1913. Held engrg positions with Gibbs & Hill, Amer Radiator, Babcock & Wilcox 1934-42. With United Illuminating Co 1942-69 in various engrg and operating positions to VP, Engrg 1959. Resp included construction of new generating, transmission and distribution facilities; participation in regional plng & inter-company pooling arrangements; oversight of hiring and indoctrinating grad engrs; and in later yrs, assessment of environmental impact. With ASME 1969-1978 as Dir of Planning & later as Deputy Exec Dir. Fellow ASME (Vice President 1962-66) Mbr IEEE, NSPE, Sigma Tau, Pi Tau Sigma, Tau Beta Pi. Civil activities included: Pres, United Fund of Greater New Haven VP, Quinnipiac Council, BSA; Chrmn, Town Plan & Zoning Comm. Hamden, CT. *Society Aff:* ASME, IEEE, NSPE.

Walton, Harold V
Home: 291 E McCormick Ave, State College, PA 16801
Position: Professor Emeritus *Employer:* Retired *Education:* BS/Agri Engg/Penn State Univ; MS/Agri Engg/Penn State Univ; PhD/Agri Engg/Purdue Univ. *Born:* June 17, 1921. m.Velma P Braun. c.H.Richard Walton, Marilyn J. Friedersdorf, Carol A Coates. General El. Co Engr 1943-45. Teaching & research: Penn State U 1947-61, Univ of MO. 1962-76, Penn State U 1976-1985, Retired Aug 1, 1985. Chrmn Dept of Agr Engrg Univ of MO 1962-69. Chf of Party, Univ of MO-USAID contract for Univ Dev, India 1969-71. Hd, Dept of Agr Engrg, Penn State Univ 1976-1985. Mbr ASEE. Fellow ASAE and Director 1967-69 & 1985-87. *Society Aff:* ASAE.

Walton, John W
Home: 8116 Millview Dr, Brentwood, TN 37027
Position: Deputy Director *Employer:* Tenn Div of Air Pollution Control. *Education:* MS Engr/Air Pollution/Univ of Cincinnati; BS/Civil/Univ of MO at Rolla. *Born:* July 11,1937 St Louis Missouri. Air Pollution Control Assn; Organizer and Sec Chrmn, So Sec 1969-70; Chrmn, S-12 Comm, Natl, 1971-73; Dir Natl Bd/Dir 1973-75; VP Natl 1975-76; KY Soc Prof Engrs, Capitol Chap Secy 1967-68; Georgia Soc Prof Engrs, NW Chap Dir 1970-71; Reg PE in Tennessee, Kentucky and Missouri. *Society Aff:* APCA, NSPE.

Walton, Ray D, Jr
Home: 19205 Germantown Rd, Germantown, MD 20874
Position: Engrg Program Mgr *Employer:* Argonne National Laboratory *Education:* MS/Chem Engg/OR State Univ; Bs/Chem Engg/OR State Univ. *Born:* Jan 1921, US Army 1943-46 Captain FA; Chem Engr, Gen Elec Co, Chem Proc irradiated Nuclear Fuel 1947-56; Chem Engr, Idaho Operations Office, US AEC 1956-60; Operations Analysis, US AEC Washington DC 1960-64, 1966-68; Sr Officer, Intl Atomic Energy Agency, Vienna, Austria 1964-66; Operational Safety, US AEC 1968-70; Radioactive Waste Mgt 1970-75; Nuclear Fuel Cycle & Production, US ERDA 1976-77. High-Level Waste Program Engr, Office of Waste Operation and Technology, US Dept of Energy 1977-86. Nuclear Fuel Cycle and radioactive waste mgt expert. Resp for US DOE Dev and planning for long-term mgt Defense High-Level Radioactive Waste; Currently Nuclear Engrg Program Coordinator/Mgr for Argonne Natnl Laboratory's office of International Energy Dev Programs in Germantown, MD 1986-Date. Past Chrmn Nuclear Engrg Div, AIChE. Resp for process design Waste Calcination Facility Idaho Nuclear Test Sta 1958-60. Led IAEA Nuclear Power Mission to Turkey 1965. *Society Aff:* AIChE, ASF.

Walton, Robert Edward
Business: 7346 S Alton Way, Denver, CO 80210
Position: President. *Employer:* Walton Associates, Inc. *Born:* July 11,1923 Chicago, Illinois. Grad Univ of Texas, Coll of Engrg 1945. Bliss Electrical School, Takoma Park, MD 1943. Grad work APL Johns Hopkins Univ Silver Spring, MD 1946-47. m.Sandra C Walton Aug 31, 1968. c.Robert E, Jr., Martha C., James B., Nancy K. Present position 1955- . Reg PE 34 states and Canada. Mbr NSPE, ACEC, CEC, PE of Colo. Sr. Mbr IEEE. Mbr Denver Athletic Club, Pinery Country Club.

Waltz, Gerald D
Business: 25 Monument Circle, P. O. Box 1595B, Indianapolis, IN 46206
Position: Sr. Vice President *Employer:* Indianapolis Power & Light Company *Education:* MBA/Bus. Adm./Butler University;BS/Elec. Engrg./Rose-Hulman Institute of Technology *Born:* 02/05/39 Native of Terre Haute, IN. Employed since June, 1960, at Indianapolis Power & Light Co. Progressed through various Engrg Organizational responsibilities to become a company officer December 1, 1978. Promoted to Sr VP - Engrg & Operations May 1, 1986. Active in IEEE, served as Central Indiana Section Chairman 1978-1979. Serves on Indiana State Bd of Registration for Prof Engrs & Land Surveyors; former ch of the bd. Presently serving on EPRI Research Advisory Cttee and Chairman of EPRI's Coal Combustion Systems Div Advisory Cttee. ISEF Community Service Award, 1984. Enjoys golf and fishing. Reg PE. *Society Aff:* IEEE, NSPE, ASME, ISEF, APCA

Walvekar, Arun G
Business: Box 4230 Rm205, Las Cruces, NM 88003
Position: Professor *Employer:* New Mexico State University. *Born:* May 1942. PhD Illinois Inst of Tech 1967. MS from IIT 1966. BSME Bombay Univ 1964. BSEE Bombay Univ 1963. Native of Bombay India. Asst. Prof of Math at Northeastern Illinois State Coll 1967-68. Was Assoc Prof of I.E. at Texas Tech Univ prior move to New Mexico State Univ in 1976. Mbr Alpha Pi Mu, Sigma Xi, Sigma Iota Epsilon, Tau Beta Pi. Gave mgt seminars in South America. Enjoy golf and classical music.

Wamack, Norman S
Business: 911 W Main St, Chattanooga, TN 37402
Position: Supervisor Welding Eng Dept. *Employer:* Combustion Engineering Inc. *Education:* BS/Metallurgical Engg/Univ of TN-Knoxville. *Born:* Oct 1941 Chattanooga. BSME Univ Tenn 1964 as co-op student. Grad stud at UT. Joined C-E in 1964 in Corp R&D Lab on welding and materials projects. Transferred to mfg dept 1968 and became supvr in 1970, supervising engrs and technicians in both lab and admin functions. Mbr AWS Bd/Dir 1974-79, AWS 1st VC of Welding Qualification Corm, Chmn-elect for AWS Technical Activities Comm (TAC) for 1980-83, AWS Comm C5b Chrmn Shielded Metal Arc Wldg, AWS Chap Chrmn for re-write of Welding Handbook on Submerged Arc Welding and mbr for Surfacing Chap, Past Mbr ASTM A1.06, mbr ASME serving as alt on 2 Code subcommittees, Chrmn of Advisory Comm to Welding Dept of State Area Voc Tech School. Mbr on advisory comm to the County Dept of Education. Enjoy 35 mm photography, music, most sports. *Society Aff:* AWS, ASME, ASTM.

Wambold, James C
Business: 301 Mech Engr, Univ Park, PA 16802
Position: Prof. *Employer:* Penn State Univ. *Education:* PhD/ME/Univ of NM; MS/ME/Carnegie Tech; BS/ME/Penn State. *Born:* 11/24/32. Engg Educator, born Emmaus, PA, Nov 24, 1932. Systems designer, Sandia Corp 1959-63, Instr ME, UNM 1963-67, Prof of ME, PA State Univ 1967-present. Currently active in tire-road interface studies including road roughness measurements & its effect on vehicle dynamics, skid resistance & hydroplaning. Also active in vehicle dynamics, random vibration, interactive graphics, dynamic modeling (including finite element analysis), digital vision systems, and robotics. *Society Aff:* ASME, ASTM, TRB, NAS.

Wambold, John N
Home: 394 Glenside Rd RD 3, Mountain Top, PA 18707
Position: Conslt *Employer:* Self-employed *Education:* BS/Ceramic Tech/PA State Univ. *Born:* 07/13/22. BS Penn State Univ in Ceramic Technology and Stevens Inst in Powder Metallurgy. Reg Engr in PA. Mbr: ACerS, ASM, AMPI, NSPE. Has held various offices in local chapters of these societies, served on natl Pub Serv Ctte of ASM, as chrmn of Prof Engrs in Industry for N.E.PA and as Bd/Mbr of PA. Soc. of Prof Engrs in Industry. Industrial experience has been related to the high temperature processing of powered materials using the newer techniques of consolidation by mechanical and isostatic pressure as related to high quality aircraft and military spec materials. *Society Aff:* ASM, AMPI, NSPE.

Wandmacher, Cornelius
Business: 1722 Larch Ave. Apt. 421, Cincinnati, OH 45224
Position: Prof of Engrg Education and Dean Emeritus *Employer:* University of Cincinnati. *Education:* PhD D Engrg/Hon/Rose Hulman Inst; PhD D r/Hon/Poly Inst of NY; Master/Civil Engg/Poly Inst of NY; Bachelor/Civil Engg/Poly Inst of NY. *Born:* Sept 1,1911 Bklyn NY. Poly Inst. NY, D Engrg(hon) 1969; Rose-Hulman, D Engrg (hon) 1976; plant engrg. W.Va. Pulp & Paper Co., 1936-38; Engrg Examiner, Municipal Civil Serv Comm NY 1939-51; Bridge des., NYCRR 1939; Struc des, Phelps Dodge Corp 1940; faculty mbr & admin officer Poly Inst NY 1938-51; hd Dept CE, Assoc Dean, Dean Coll Engrg, Univ Cincinnati 1951-74; prof engrg. educa, UC 74- 81. Mbr Ohio Registra Bd Prof Engrs 1962-72; Dist Serv Award NCEE 1970; various offices in ASEE, President 1974-75; ECPD Dir 1966-73, mbr educa & accred. comm 1968-72; chr prof train comm 1952-54,57-59; EJC dir 1975-78; Chr ACE, 1976-78. Author:'Metric Units in Engineering-Going SI'; various articles on engrg edu & prof devel. Mbr Cosmos Club Washington, DC. Dir & Chr Metric Practice Comm, Amer Natl Metric Council 1976-81; VC ASTM E-43 Metric Practice Comm; Hon Mbr ASCE 1977 & ASEE 1979; Freund Award, Co-Op Educa, 1980. *Society Aff:* ASEE, ASCE, ANMC, NCEE, NSPE, ASTM.

Wane, Malcolm T
Home: 38 Mile Rd, Suffern, NY 10901
Position: Prof. *Employer:* Columbia Univ. *Education:* EM/Mining/Lehigh Univ; MS/Mining/Columbia Univ; BS/Mining/Lehigh Univ. *Born:* 1/2/21. Research Technician & Mineral Engr. US Steel '50-'53 Mining Engr & Technician - American Metals '56-'57 Asst Prof Columbia '56-'64 Assoc Prof Columbia '64-'69 Prof Columbia '69 to date. Assoc of Behre Dolbeer Mining & Mineral Consultants '76 to date, Geol Conslt to NYC Bd of Dirs, NYC Transit Authority. *Society Aff:* AIME.

Waner, John M
Home: 1249 Holly Rd, Webster, NY 14580
Position: Dir, Sensitized & Chem Prod Prog. *Employer:* Eastman Kodak Co. *Education:* BS/Chem/Union College. *Born:* Feb 25,1923. BS Union College, Schenectady, NY; native of Fort Plain, NY. Served in the US Army, European Theater of Operations during WW II. With Eastman Kodak since 1947, starting in Res Labs; Field Engrg Service Rep; Chf Engr; Dist Sales Mgr in Hollywood, CA. serving in Motion Picture and Television Industry. Instructor in color photography in night school UCLA, USC. Project Consultant in Manufacturing, then to current position. An SMPTE Fellow 1963; SMPTE Bd/Governors 1965-66; Natl Conf Prog Chrmn 1966; Kalmus Gold Medal in 1967. *Society Aff:* SMPTE.

Wang, Albert S
Business: 32nd and Chestnut Sts, Phila, PA 19104
Position: Prof *Employer:* Drexel Univ *Education:* PhD/Aero Engrg/Univ of DE; MS/Struct Mech/U of NV; BS/CE/Nat Taiwan Univ. *Born:* 7/13/37 Native of Chefoo, China. Taught high school Math/Science following graduation from the National Taiwan Univ in 1959. Came to the US in 1961 and became a naturalized citizen in 1974. Have been on the Faculty of Engrg, Drexel Univ since 1967. Engaged in teaching and applied research in the broad area of structural mechanics. Am a consultant to several federal and industrail institutions, including Air Force Materials

Wang, Albert S (Continued)
Lab, Lawrence National Lab, Lockheed Space and Missiles Co, General Motors Corp, etc. *Society Aff:* ASME, SAMPE, ASTM.

Wang, Chiu-sen
Business: 221 Hinds Hall, Syracuse, NY 13210
Position: Prof. *Employer:* Syracuse Univ. *Education:* PhD/CE/CA Inst of Tech; MS/CE/KS State Univ; BS/CE/Taiwan Univ. *Born:* 12/3/37. Native of Taichung, Taiwan. With NYU Medical Ctr 1968-69. With Syracuse Univ since 1969. Res interests in gas cleaning, aerosol sci, and inhaled particles. Spent sabbatical leave at Kyoto Univ, Japan, 1978. *Society Aff:* AIChE, AAAS, ACS, AIHA, ΣΞ

Wang, Franklin F Y
2 Forsythe Meadow Lane, Stony Brook, NY 11790-1841
Position: Prof of Engrg *Employer:* State Univ of NY at Stony Brook *Education:* PhD/Ceramics/Univ of IL-Urbana; MS/Glass Tech/Univ of Toledo; BA/Chem/Pomona Coll *Born:* 09/19/28 I have been teaching in the Coll of Engrg at Stony Brook since 1966. Prior to that, I worked in industries, such as A. O. Smith Corp, Sperry Rand Research Ctr, since 1956. I enjoyed both aspects of engrg, and am very interested in the manufacturing processes and the means to reduce the energy consumptions in these processes. My specialties are in electronic materials and silicon processing. *Society Aff:* A Cer Soc, A Chem Soc, IEEE, MRS, APS, AACG.

Wang, George C
Business: Box 1468 Dept of Mineral Engrg, University, AL 35486
Position: Prof *Employer:* Univ of AL *Education:* PhD/Petro Eng/Univ of TX; MS/Petro Eng/Univ of Tulsa; BS/Mining Eng/Yunnan Univ *Born:* 6/18/23 Is prof of Petroleum Engrg in the Department of Mineral Engrg of the Univ of AL. Where he has taught since 1975. Previously, he taught two yrs at Montana Coll of Mineral Science and Technology and served in various engrg positions for Chines Petroleum Corp and Mobil Oil Co. Research activities include miscible gas oil recovery processes, heavy oil recovery and geothermal energy. He has written many papers and technical reports and served as consultant to a number of oil and gas companies and government organizations. *Society Aff:* SPE, AIME

Wang, Jaw-Kai
Business: Ag Eng Dept. Univ of Hawaii, 3050 Maile Way, Honolulu, HI 96822
Position: Professor of Agr Engrg. *Employer:* University of Hawaii. *Education:* PhD/Agr Eng/MI State Univ; MS/Agri Eng/MI State Univ; BS/Agri Eng/Natl Taiwan Univ *Born:* 3/4/32. Nanjing China. Naturalized US citizen. Faculty, Univ of Hawaii since 1959. Chrmn Agr Engrg Dept 1964-75. Current interests: Aquacultural Engrg (Shrimp, Oyster, Prawn, Nori) Agr & Acquaculture Systems Analysis. Consultantship: World Bank, FAO/UN, IFAD, Rockefeller Foundation, Universal Tankship (Delaware), AID, Intl Rice Res Inst, Pacific Concrete and Rock Co Ltd, US Army Civilian Admin, Taiwan Sugar Cooperation. Sr Invitational Fellow East-West Center Food Institute 1973- 74. Reg PE Hawaii. Mbr: Sigma Xi, Pi Mu Epsilon, World Mariculture Soc, Gamma Sigma Delta, ASAE, Engineer of the Year, Pacific Reg, ASAE 1975. Chrmn, Pacific Region, ASAE 1975-76. Co-Dir, Intl Sci & Educ Council, 1979. Reg PE HI. Fellow, Amer Soc of Agri Engrs (elected 1980). *Society Aff:* ASAE, WAS

Wang, Leon R L
Business: Department of Civil Engineering, Old Dominion University, Norfolk, VA 23508
Position: Prof and Chrmn of Civil Engineering *Employer:* Old Dominion Univ *Education:* ScD/Structures/MIT; MS/Structures/Univ of IL; BSCE/Civ Engg/Natl Cheng Kung Univ. *Born:* 6/15/32. Dr. Leon Ru-Liang Wang, born June 15, 1932 in Canton, China, obtained his BS degree from Natl Cheng-Kung Univ, in Taiwan 1957, MS from Univ of IL in 1961 & ScD from MIT in 1965; joned RPI in 1965, from Assist Prof to Assoc Prof of Civ Engg; and as Professor of Civil Engineering at the University of Oklahoma from September 1980-84; interested in Earthquake Engg; Lifeline Earthquake Engrg; Structural Mechanics, Model Analysis, Steel & reinforced concrete designs and computer applications and currently as Prof and Chrmn of Civil Engrg at Old Dominion Univ from Sept 1984. *Society Aff:* ASCE, ACI, ASEE, EERI.

Wang, Ping-chun
Business: 333 Jay St, Brooklyn, NY 11201
Position: Prof *Employer:* Polytechnic Inst of NY *Education:* PhD/Engrg/Univ of IL; MS/CE/Univ of IL; BS/CE/Nat Central Univ of China *Born:* 2/10/20 in China. Supervising engr with Seelye Stevenson Value & Knecht from 1951 to 1960. Assoc prof, Stevens Inst of Tech from 1960 to 1963. Prof Polytechnic Inst of NY, 1963 to present. Author of textbook *Numerical and Matrix Methods in Structural* Mechanics, John Wileyd & Sons publisher. Consultant in structural engrg and structural dynamics. *Society Aff:* ASCE, EERI

Wang, Shien T
Business: Dept of Civ Engg, Lexington, KY 40506
Position: Prof *Employer:* Univ of KY. *Education:* PhD/Struct Engrg/Cornell Univ; MS/Struct Engrg/MI State Univ; BS/Civ Engrg/Natl Taiwan Univ *Born:* 8/24/38. in Changsha, Hunan, China. Obtained BS degree from Natl Taiwan Univ and came to the US in 1962, naturalized in 1975; Married to Lung Chu Sun in 1969, two children; Res interests: Structural stability; Behavioral Design of thin-walled steel structures; computer modeling and applications; finite element methods; computer aided design and computer graphics; CAD applications in construction and underground mining structures; 40 technical articles in profl journals and conference proceedings; Mbr and chrmn, ASCE and SSRC Committees. *Society Aff:* ASCE, ASEE, SSRC, XE, TBΠ, ΣΞ.

Wang, Shih-Ho
Business: Dept of Elec and Comp Engrg, Davis, CA 95616
Position: Prof *Employer:* Univ of CA, Davis *Education:* PhD/EECS/Univ of CA; MS/EECS/Univ of CA; BS/EE/Natl Taiwan Univ. *Born:* 6/29/44. Native of Kiangsu, China. Did res at Univ of Toronto and NASA Langley Res Ctr. Taught at Univ of CO for four yrs and Univ of MD at College Park for six years. Spent one year at the Office of Naval Res, Arlington, VA. Interested in Control System and Robotics. Currently Prof of Electrical and Computer Engrg, Univ of CA, Davis. *Society Aff:* IEEE.

Wang, Shu-Yi
Business: Ctr for Computational Hydrosci & Engrg, University, MS 38677
Position: Prof and Dir *Employer:* Univ of MS *Education:* PhD/Fluid Mech/Univ of Rochester; MS/Aero Sci/Univ of Rochester; BS/ME/Cheng Kung Univ *Born:* 9/21/36 Native of Chung-King, China. Has taught at Univ of MS since 1967 specializing in Hydrodynamics, Numerical Modeling and Simulation of Fluid Flows. Directed research projects supported by NSF, AFOSR, ARO, USDA, etc. Received 1975 Ralph R. Teetor Award from SAE, Special Award from AIAA, 1979. Elected as one of the directors of AIAA AL Section, 1979, Assoc Fellow of AIAA, 1981 and, Exec Council Mbr of ISCME, 1980. Principal Editor of Finite Elements in Water Resources, Vol. III, 1980, Applied Numerical Modeling Vol. IV, 1984, and River Sedimentation Vol. III 1986. Nat Tech Ctte Mbrs of professional societies including AIAA, ASME, ASCE, IAHR & ISCME. *Society Aff:* AIAA, ASME, ASCE, IAHR, NSPE.

Wang, Shyh
Home: 8636 Thors Bay Rd, El Cerrito, CA 94530
Position: Prof. *Employer:* Univ of CA. *Education:* PhD/App Phys/Harvard Univ; MA/App Phys/Harvard Univ; BS/EE/Chiaotung Univ. *Born:* 6/15/25. in China, & naturalized in 1962. Awarded BS, Chiaotung Univ, 1945; MA, 1949 & PhD, 1951 in Applied Phys, Harvard Univ. Res Fellow, Harvard Univ 1951-1953; Engg Specialist, Semiconductor Div, Sylvania Electric Products, Gen Tel & Electronics, 1953-1958, Assoc Prof 1958-64, & Prof 1964-present, Dept of Elec Engg & Comp

Wang, Shyh (Continued)
Sci, Univ of CA, Berkeley. Guggenheim Meml Fdn Fellow 1965-66; Fellow IEEE; Mbr APS; Fellow OSA. Known for his work in magnetic resonance and relaxation phenomena, non-reciprocal integrated-optics devices, distributed Bragg reflector lasers, & semiconductor injection lasers. His recent interests are integrated optics, microstructure engg of electronic & optical devices, interface & surface studies of semiconductors, & elec & optical properties of amorphous semiconductors. He has over 100 publications in scientific journals and conference proceedings, and six US patents. *Society Aff:* IEEE, APS, OSA.

Wang, Su-Su
Business: 216 Talbot Lab, Dept of Theoretical & Appl Mechanics, Urbana, IL 61801
Position: Prof & Director *Employer:* Univ of IL. *Education:* ScD/Mech Eng/MIT; SM/Eng Mech/Natl Taiwan Univ; BSE/Eng Mech/Natl Cheng- Kung Univ *Born:* 5/10/48. in Nanking, China. Lecturer of mechanics in Chinese Army Inst of Tech, 1970-71. Received Doctor of Sci Degree from MIT, 1974. Res staff and Faculty of Engg, MIT, 1974-77. Faculty of Engg, Dept of Theoretical and Appl Mechanics, Univ of IL, since 1977. Prof & Director, National Center for Composite Materials Research, 1986. Elected to mbr of American Academy of Mechanics, 1976. Currently, mbr of Structure and Mtls Committee, Aerospace Div, ASME; mbr of Composite Materials Mech Ctte, Applied Mech Div, ASME; mbr of E-9 (fatigue), E- 24 (fracture) and D-30 (high modulus fibers and composite) Committees of ASTM; mbr of SPRC, ASCE. Engaged in res and teaching on fracture, fatigue, impact, mechanics and mech behavior of polymer-matrix, metal-matrix and ceramic-matrix, composites, and adhesively bonded joints. Author and coauthor of 90 technical papers and reports. *Society Aff:* AIAA, ASME, AAM, ASTM.

Wang, Yui L
Business: 1400 N Harbor Blvd, Fullerton, CA 92635
Position: President. *Employer:* Simulation Sciences Inc. *Education:* PhD/Chem Engg/Caltech; MS/Chem Engg/Caltech; BS/Chem Engg/Cheng Kung Univ. *Born:* May 1935. MS and PhD from Caltech. BS from Cheng Kung University, Taiwan China. Native of Shanghai, China. Senior Engr at Burroughs Corp and C F Braun & Co, specializing in process simulation. Co-founder of Simulation Sciences Inc in 1966. Assumed Presidency in 1971. *Society Aff:* AIChE, ACS.

Wangsgard, Lew A
Business: 624 N 300 W, Salt Lake City, UT 84103
Position: President *Employer:* Nielsen, Maxwell & Wangsgard. *Education:* BS/Civil Engg/UT State Univ. *Born:* 1/29/33. in Ogden, UT. Attended Weber State Coll; Served with Army Corps Engrs 1953- 55. Design Engr for Nielsen & Maxwell 1958-71. Became prin stockholder in 1971, firm name changed to Nielsen, Maxwell & Wangsgard. Co-recipient Eric Ryberb Scholarship US Univ 1958. P Pres Ogden Exec Assn; P Pres ASCE UT Sec; P Pres PTA; P Pres CECU; P Chrmn Weber Cty Solid Waste Bd. City Engr So Ogden City, No Ogden City, UT. VP Power Mountain Ski Resort. Active mbr Church of Jesus Christ of Latter-Day Saints. Mbr-State of UT Hazardous Waste Committee. Enjoys skiing, boating, hunting & fishing. *Society Aff:* ACEC, ASCE, AWWA, WPCF.

Wanielista, Martin P
Business: College of Engrg, Orlando, FL 32816
Position: Prof of Engg & Chmn, Civ Engg & Envir Sci. *Employer:* Univ of Central FL. *Education:* PhD/Environ Engr/Cornell; MS/Sanitary/Manhattan; BS/Civil Engg/ Univ of Detroit. *Born:* 12/7/41. Secy-Treas of AEEP, Pres of the E Central Br, FL Sec, ASCE in 1975. Reg PE in FL. Spent 2 yrs in the Army Medical Service Corps with rank of Captain. Now Chrmn of Civil Engs at UCF. Published 35 tech articles & editor of 5 proceedings in Environ Engg. Recd 15 awards related to Environ Engg in the last 5 yrs, wrote 2 books. *Society Aff:* AWWA, NSPE, AEEP, FES, WPCF, ASCE.

Wanjura, Donald F
Business: Cropping Systems Research Lab, Route 3, Box 215, Lubbock, TX 79401
Position: Agri Engr. *Employer:* US Dept of Agri. *Education:* PhD/Agronomy/Univ of AZ; MS/Agri Engr/Clemson Univ; BS/Agri Engr/TX A&M Univ. *Born:* 7/30/38. and raised in Weimar, TX. Employed by USDA since 1962. Res responsibility in plant stress detection and evaluation including the use of remote sensing technology, and mathematical modeling of cotton growth and dev. Authorship of 57 res articles; Chrmn, TX Sec ASAE 1978; CDR in US Navy Reserve. (Retired). *Society Aff:* ASAE, ΣΞ.

Wankat, Phillip C
Business: Sch of Chem Engrg, W Lafayette, IN 47907
Position: Prof of Chem Engg. and Head Freshman Engg. *Employer:* Purdue University. *Education:* PhD/ChE/Princeton; MSED/Education/Purdue; BSChE/ChE/Purdue *Born:* July 1944. Asst and Assoc Prof & Prof at Purdue. Head Freshman Engg, 1987. Departmental teaching award 1974 & 1979. All engrg Teaching award (Potter Award) at Purdue 1979. Dow Outstanding Young Faculty Award (ASEE) 1980, Research on separation processes with emphasis on chromatography, cyclic techniques and cascades. Student Chapters Cttee of AIChE. Chrmn 1976, Vice Chrmn 1974-75. Sabbatical at Univ of Calif, Berkeley 1976. Sabbatical at LSGC, ENSIC, Nancy, France 1983-84. Mbr of ACS, ASEE, AIChE. ChE Div ASEE: Prog Chrmn, 1980; Secry/Treas 1980-1982; Chrmn 1986-87. Western Electric and George Westinghouse awards of ASEE, 1984. *Society Aff:* AIChE, ACS, ASEE, AACD.

Wanket, Achiel E
Business: 630 Sansome St, Rm 1249, San Francisco, CA 94111
Position: Chief, Engg Div. *Employer:* US Army Engr Div, S Pacific. *Education:* MBA/Bus Adm/Univ of MI; BS/Civ Engg/Univ of Detroit. *Born:* 10/5/28. Since 1975, Chief of the Engg Div of the US Army Engr Div, S Pacific, San Francisco, responsible for staff supervision of all engg activities carried on by three subordinate Dist Offices relating to planning & design of civil works projs authorized by the Congress, & military construction progs of the Army & AF. Served in Detroit Dist of the Corps of Engrs in positions of progressive responsibility from 1958 to 1972. Four-yr tour in AF 1952-56. Mbr ASCE, ICID, APWA, SAME (Mbrship Comm), USCOLD (Chrmn, Publications Comm), Commonwealth Club of Calif, Sr Exec Assoc. *Society Aff:* ASCE, USCOLD, SAME.

Wantman, Joel N
2000 Lombard St, West Palm Beach, FL 33407
Position: Pres. *Employer:* Stanley/Wantman, Inc. *Education:* BS/Civil Eng/Univ of AZ; MS/Civil Eng/NY Univ. *Born:* June 27,1941 Bklyn NY. Jr Engrg for Amman & Whitney, Consulting Engrs and Hazen & Sawyer, Consulting Engrs in NYC 1963-68. Engr with Buchart-Horn, Consulting Engrs in York, PA. 1968-69. Proj Engrg with Adair & Brady, Consulting Engrs in West Palm Beach, FL 1969-73. Formed Wantman & Assoc Consulting Engrs spec in Land Development and Environmental Design offices in West Palm Beach, Orlando & Ft. Myers Fla. Started Meridian Surveying and Mapping Inc. 1978, Pres. Started Realty Designs Inc. 1981-Sec. Tres. MBR Businessmen's Assn of the Palm Beaches 1974-1980; Merged Wantman & Assoc, Inc. and Meridian Surveying and Mapping, Inc. with Stanley Consultants, Inc. of Iowa in 1986 and formed SCI Companies. Pres of Stanley/Wantman, Inc. and Stanley/Meridian, Inc. Mbr of SCI Bd of Dir. Secy 1975, Pres 1976. Mbr AWWA, FICE, FES, ASCE (Secy 1976) member Palm Beach Citizens Task Force. Leisure time spent: Golf, Tennis, Polo. *Society Aff:* AWWA, FICE, FES, ASCE.

Ward, Allen W
Business: 8177 South Harvard, Suite 335, Tulsa, OK 74137
Position: Pres & Owner. *Employer:* Ward Assoc Engg. *Education:* MS/Metal Engr & Sci/Univ of AL; BS/Mech Engr/Clemson Univ. *Born:* 8/31/33. Native of Birmingham, AL. Served in Marine Corps as pilot, 1956-1960, proj dev in helicopter recovery procedures through NASA Space vehicle progs. Engr and exec mgt for small through large foundries in new product dev casting design, machinability, and quality control requirements since 1955. Since 1971 operates independent engg consulting firm, owner. Specializes in metallurgical engg services to small through multinational corps in natl, foreign, and intl projs. Developed practical application infrared thermography for energy progs, productivity, and maintenance. Pres, Ward Engg Co, Reg PE, Canada and USA. President OSPE National Board NSPE. Service on airport and water advisory trust appointed by city. *Society Aff:* NSPE, OSPE, SME, AFS.

Ward, George D
Home: Lower Church Hill Rd, Washington Depot, CT 06794
Position: President. *Employer:* Ward Douglas & Co. *Education:* BChE/Chem Engg/ Cornell Univ. *Born:* Dec 30,1922 Rochester NY. Grad Phillips Exeter Academy 1940, Cornell U 1947. BChE. Employed Exxon Res & Engrg 1947-66. Mgr European Engrg Office 1959- 62. Asst Gen Mgr Exxon Engrg 1962-64. Deputy VP Res 1964-66. VP Exxon Chemical 1966-71. Pres Ward Assoc & Ward Douglas & Co 1971--. Dir LCP Chemicals & Plastics 1972-82. Dir. First Nat. Bank Litchfield 1986- . Dir & Pres Westken Petroleum Corp 1980- . Mbr AIChE, SPE, AAAS, Cornell Soc Engrs, Tau Beta Pi. Trustee & Vice Chairman Wyneham Rise School, Trustee & Pres. Steep Rock Assoc, Directed major refinery planning, design & construction Europe and Africa 1951-64. Directed res iron ore reduction, synthetic proteins, alternative energy sources 1964-66. Directed plastics, fibers, rubber, blg materials businesses 1966-71. Chief Exec Chemical and Coppermining Co 1973-78. Chief Exec oil and tar sand development Co 1980-. Mgmt & Engg Consultant 1971- . *Society Aff:* AIChE, AAAS, SPE.

Ward, Harold R
Home: 23 Hilltop Dr, Bedford, MA 01730
Position: Consulting Scientist. *Employer:* Raytheon Co. *Education:* MS/EE/Univ of S CA; BEE/Electronics/Clarkson College. *Born:* July 1931. Native of Alden, NY. Electronic development engr for Cornell Aeronautical Lab and Hughes Aircraft Co. Served with the Army Chemical Corp 1954-56. Res. engr with Sylvania for 7 yrs doing radar system analysis. With Raytheon's Equipment Div since 1964. Contributed to the design, analysis and testing of over a dozen radar systems. Currently a Consulting Scientist. Has over 20 publications including the 'Handbook of Radar Measurement' with D K Barton. Fellow IEEE and Reg PE in Massachusetts. Enjoys photography, hunting, and amateur radio. *Society Aff:* IEEE.

Ward, Howard O
Business: 30 Eldredge St, Binghamton, NY 13901
Position: Pres *Employer:* Northeastern Test & Bal, Inc *Education:* BS/ME/Syracuse Univ *Born:* 4/3/18 Native and resident of Candor, NY. Joined Pratt & Whitney Aircraft Div in 1940 and served as field installation engr until 1953. Entered consulting engr field and was part owner of firm from 1965 specializing in HVAC and Plumbing design. Won Grand Conceptor award from CEC-NYS in 1975 for well water heat pump systems design in Binghamton Governmental Complex. Retired Sept 30, 1983 & started present job Mar 1984. Past ASHRAE Region 1 Vice Chrmn for Energy Mgmt. Elected "Fellow" of ASHRAE Jan, 1986. *Society Aff:* NSPE, ASHRAE, NEBB

Ward, Joseph S
Home: 61 Norwood Ave, Upper Montclaire, NJ 07043
Position: President and Chief Executive Officer *Employer:* Converse Ward Davis Dixon, Inc. *Education:* MS/Civil Engg/Rutgers Univ; BCE/Civil Engg/Manhattan College. *Born:* Jan 23,1925 in New York City where I received Elementary and High School education. MS from Rutgers University (1948). BCE from Manhattan College (1946). Doctorate study at Columbia University (1948-51). Asst Prof at the Cooper Union (1948-57); Sr Visiting Lecturer at Stevens Inst of Tech (1968-70). Pres of Joseph S Ward Inc & Joseph S Ward Intl Inc & partner of Joseph S Ward and Assoc Consulting Geotechnical Engrs in Caldwell NJ (1950-78). Pres of Converse Ward Davis Dixon, Inc (1978-present) and Chief Executive Officer (1981-present). VP ASCE 1975-77. Pres ASCE 1979-80. Resides in Upper Montclair, NJ and Normandy Beach, NJ with family. *Society Aff:* ASCE, NSPE, ACEC.

Ward, Julian R
Business: PO Box 3, Houston, TX 77001
Position: VP. *Employer:* Brown & Root, Inc. *Education:* BS/ChE/Rice Univ. *Born:* 8/3/35. More than 21 yrs of engg experience in the chem, petrochem, & petrol industries. This background includes 8 yrs of engg mgt experience preceded by 13 yrs in process engg. Is an expert in all technical aspects of petrochem projs & has specialized in corrosion control & other types of problem solving. As vp of tech for the Petrol & Chem Engg Div, is currently responsible for mgt of all engg discipline skills, dev of technological methods, & control of discipline performance quality. Is reg PE in TX & LA, MS. *Society Aff:* AIChE, NACE, GPSA.

Ward, Lew O
Business: 502 S Filmore, Enid, OK 73701
Position: Owner. *Employer:* Ward Petroleum Corp *Education:* BS/Petroleum Engr/ Univ of OK *Born:* 7/24/30. Apprentice Engr, 1952, Delhi Oil Corp, Alice, TX; Field Engr, 1953, Delhi Oil Corp, Casper, WY; Pipeline Engr, Army US, 1953-1955; Dist Engr, Delhi-Taylor Oil Corp, Tulsa, 1955-1956; Partner, Ward & Gungoll Oil Investments, 1956- present; Owner, Ward Oil Co, 1965-present, Originate, Direct & Supervise Exploration, Completion and Production of over 200 Oil & Gas Wells; Mbr Governors Advisory Council on Energy 1978; Dir, Community Bank; Dir, Bass Hospital; VP, Independent Petrol Assn of America - 1978; Pres, OK Independent Petrol Assn, 1978-80. Trustee, Phillips University. State Chairman, US Olympic Cttee. Bd of Visitors, Coll Engerg, Univ. OK. Governors Rep. Interstate Oil Compact Comm. Univ Bd, Pepperdine Univ. Private Pilot, Scuba Diving, Skiing. *Society Aff:* SPE

Ward, Richard E
Business: Dept of Industrial Engrg, Morgantown, WV 26506
Position: Assoc Prof *Employer:* W VA Univ *Education:* PhD/IE/W VA Univ; MS/ IE/W VA Univ; BS/IE/PA St Univ *Born:* 1/19/42 Native of Harrisburg, PA. Project Engr for American Can Co after receiving his BS in 1964, and later MS and IE for NYC Transit Authority following his MS in 1968. Project Engr on US Dot sponsored Personal Rapid Transit R&D project in Morgantown WV 1970-75. WVSPE Young Engr of the Year 1974. Visiting Research Fellow at Univ of Leeds, England, 1975-76. With WVU since 1976. Pres local chapter WVSPE 1973. National Dir of Transportation and Distribution Div of IIE 1980-81. Enjoys tennis, skiing, theatre and music. Lectures on Wine & Wine Tasting for major US wineries. *Society Aff:* IIE, APICS, TIMS, NSPE/WVSPE.

Ward, Robert C
Business: Agri & Chem Engr Dept, Ft Collins, CO 80523
Position: Prof *Employer:* CO State Univ. *Education:* PhD/Agri Engr/NC State Univ; MS/Agri Engr/NC State Univ; BS/Agri Engr/MS State Univ. *Born:* 7/4/44. in Swanson, Wales, Great Britain. Raised in MS Res Assoc with NC State Univ 1969-70. With CO State Univ since 1970 (1970-75 Asst Prof, 1975-1980 Assoc Prof, 1980-present Prof). Sabbatical leave 1976-77 with Water Quality Inst, Horsholm, Denmark and US Envr Protection Agency, Las Vegas, NV. Res and teaching areas are water quality monitoring and mgt, on-site home sewage systems and Operations Res. VP, Rocky Mtn Region, ASAE, 1974 and 1978, SAME Gunlogson Countryside Engr Award 1976. CO State Univ Durrell Award for creativity and excellence in teaching 1974. *Society Aff:* ASAE, NWWA, AGU, AWRA, WPCF.

Ward, Sol A
Home: 185 W End Ave, New York, NY 10023
Position: Consultant, Previously Principal Engineer *Employer:* Ebasco Services Inc. *Education:* PhD/City Planning/Union Grad Sch; BA/Urban Planning/Goddard College. *Born:* 8/16/24. 30 yrs experience in construction engg in diversified responsible positions. Consultant Ebasco Services Inc, previously Principal Engr, on visiting fac-

Ward, Sol A (Continued)
ulty of Pratt Inst - taught construction mgt, construction cost analysis, ecological & geological problems in construction. Adjunct prof of ind tech at Kean College of NJ. Visiting prof at CA State Poly Univ and Univ of MD, Construction Engg. Author of *Cost Control in Design and Construction*, McGraw- Hill 1980 & *Urban Planning and Architecture*, Peter Owen Ltd London 1971. Held posts at AACE as Prog Chrmn, Technical VP, Administrative VP, Dir & Chrmn of Technical Committee. Seminar lectr at IEEE and SAM. Certified Cost Engr. Dir of seminars and Sr Mbr of IIE. Mbr of Authors Guild. *Society Aff:* AACE, PMI, IIE

Ward, Thomas L
Business: Dept of Indus Engrg, Louisville, KY 40292
Position: Prof of Engrg Mgmt and Industrial Engrg *Employer:* University of Louisville *Education:* PhD/Ind & Sys Engg/Univ of S CA; MS/Ind & Sys Engg/Univ of S CA; MS/Sys Engg/W Coast Univ; BS/Phys/Univ of TX. *Born:* 5/12/30. Native of Norfolk, VA. Educated in the primary and secondary schools of Austin, TX. Engr for maj airframe manufacturers from 1953-59, engr for scientific instrument mfg from 1959-69, consulting engr from 1970-75, Asst Prof of Industrial and Systems Engg at Univ of Southern CA from 1975-78, Prof and Chair of Industrial and Mfg Engg at CA State Poly Univ from 1978-80, Member Technical Staff of Jet Propulsion Laboratory at California Institute of Technology from 1980-81, Chair of Industrial Engineering at University of Louisville 1981-86, and Prof of Engrg Management and Industrial Engrg 1981-present. Mbr of the College-Industry Council on Material Handling Education from 1976-80. Alpha Pi Mu, Omega Rho, Sigma Xi, and Tau Beta Pi. *Society Aff:* IEEE, IIE, SME, ASME

Ward, William J, III
Business: General Electric, CRD, CEB 473, K-1, P.O. Box 8, Schenectady, NY 12301
Position: Res Engr. *Employer:* General Elec Corp R & D. *Education:* PhD/Chem Eng/Univ of IL; MS/Chem Eng/Univ of IL; BS/Chem Eng/PA State *Born:* 10/4/39. BSChE Penn State; MS, PhD, Chem Engrg, Univ of IL. Joined General Electric Res & Dev Ctr in 1965. Recd Alan P Colburn Award of AIChE in 1974. Areas of work include membrane gas separation processes, barrier jilius for packaging, and catalysis. *Society Aff:* AIChE, AAAS.

Wardell, Roland D
Home: 5917 Bernard Pl, Edina, MN 55436
Position: Ret *Education:* Bach/Met Eng/Univ of MN *Born:* 2/26/22. Native Minnesotan. Naval Aviator WWII, served with fighter bomber Squadron VBF 152. After degree in Met Engg, was Res Fellow at Univ of MN 1949-50. With Honeywell Inc from 1950-1981, progressed from Assoc Met to Supervisor of Mtls Engg. Responsible for mtls selection and lab testing of wide range of mtls used in electronic & electromech control systems. Reg Profl Met Engr and also Reg Corrosion Specialist. Presently Chrmn of ASTM Ctte B4 on electrical contacts. Family oriented but also enjoy archeology & hobby farming. Retired June 30, 1981. *Society Aff:* ASM, NACE

Warder, Richard C, Jr
Business: Dept of Mech and Aero Engr, Columbia, MO 65211
Position: Prof. *Employer:* Univ of MO - Columbia. *Education:* PhD/Mech Engr/ Northwestern Univ; MS/Mech Engr/Northwestern Univ; BS/Mech Engr/SD School of Mines. *Born:* 9/30/36. Asst Prof Mech Engg and Astronautical Sciences - Northwestern Univ (1963- 65). Sr tech Staff and Mgr - Energy Processes Res, Space Sciences Lab - Litton Industries, Beverly Hills, CA (1965-68). Mech Engg Dept - Univ of MO-Columbia (1968 to date). Prog Mgr and Head - Resources Sec, Res Applications Dir, Natl Sci Fdn (1974-76). *Society Aff:* AAAS, AIAA, APS, ASME, ASEE.

Wardwell, Richard E
Home: RFD 2, Box 920, Carmel, ME 04419
Position: Faculty. *Employer:* Univ of ME. *Education:* PhD Candidate/Geotechnical Engg/CO State Univ; MS/Soil Mechanics/Univ of ME; BS/Civil Engg/Univ of VT. *Born:* 3/1/47. in upstate NY, now reside in Carmel, ME; Worked 4 summers for Rist Frost Assoc, cons engrs prior to serving 4 yrs as a Naval Flight Officer in the US Navy: Upon discharge, became commanding officer of a civil engg Seebee Reserve Unit in Bangor, ME; Worked 3 yrs for Jordan Gorrill Assoc as a proj mgr specializing in geotechnical engg proj; Assumed current position in 1976; res and cons interests include the behavior of organic soils and saturated sludges from municipal and industry waste water treatment facilities with organic breakdown; As a charter mbr of the ME Professional Dev Council, instrumental in developing a continuing education prog for engrs in the State of ME; Published over 10 professional papers, most of them dealing with the behavior of soils with fiber breakdown, and the continuing education of civil eng. *Society Aff:* ASCE.

Ware, Charles H, Jr
Business: 13902 N Dale Mabry Hwy Suite 117, Tampa, FL 33618
Position: Consultant. *Employer:* Commercialization Insights (sole propietor) *Education:* PhD/Chem Engg/Univ of PA; MS/Chem Engg/Univ of PA; BSE/Chem Engg/Princeton Univ. *Born:* 7/8/27. Native of New York City; served in USA Counter Intelligence Corps 1952-54. Equipment Sales engr for manufacturer's rep; Jr Process Engr Atlantic Refining. With Texaco Inc 1959-74, devoting the last 10 yrs to development and application of new methods in research/development/ design. Ad; Assoc Prof, Columbia Univ 1968-69; Adj Prof, U of PA 1975. Adj Prof, Manhattan College 1978-80. Own consulting business 1974-, specializing in the mgmt of R&D for commercialization. For AIChE: Mid-Hudson Sec, Treas 1961-62, V Chrmn/Chrmn 1963- 64; Govt Interaction Cttee Chrmn 1977-80; Natl Speakers Bureau 1963-; Machine Computation Ctte Chrmn 1973-76; computers & systems Tech Div. Formation Comm 1976-77, Exec Comm 1978; Fellow 1980 Lecturer, Today Series, 1975-. Tennis, Squash. *Society Aff:* ACS, ASA, AIChE, AMA.

Ware, Lawrence A
Home: 1265 Melrose Ave, Iowa City, IA 52240
Position: Prof Emeritus. *Employer:* University of Iowa. *Education:* PhD/Physics/SUI; MS/Physics/SUI; EE/Elect Engr/SUI; BE/Elect Engr/SUI. *Born:* May 1901. All university work in Univ of Iowa. Taught High School one year. 3 1/2 years in BTL, NY. Taught 2 yrs in Physics, Montana State College, Bozeman. Since 1937 in EE, Univ of Iowa One summer 1945 on Proximity Fuse work, U of Iowa. Two summers, Stromberg Carlson Co on Radar.-(42 & 43). Officer in Cedar Rapids Section at its beginning. One summer Univ of AL, Geophysical work, 1956. *Society Aff:* IEEE.

Ware, Willis H
Business: The RAND Corporation, 1700 Main St, Santa Monica, CA 90406
Position: Sr Computer Scientist. *Employer:* The RAND Corp. *Education:* PhD/EE/Princeton; MS/EE/MIT; BS/EE/Univ of PA. *Born:* 8/31/20. Native of Atlantic City, NJ. A pioneer in the computing field, he joined The RAND Corp in 1952 & subsequently became Hd of the Computer Sciences Dept. He is currently the Sr Computer Scientist of the Corp Res Staff. He was the first & double term chrmn of AFIPS & has held numerous natl offices in the IEEE of which he is a Fellow. In 1972-73 he was the chrmn of DHEW Secy's Advisory Cttee on Automated Personal Data Sys. In 1975 he was appt by Pres Ford to be a mbr of the Privacy Protection Study Commission and was V Chrmn of it. He also chairs or is a mbr of numerous govt advisory cttees, and in 1975, was a DPMA Man-of-the-Year. In 1979 He was awarded the Exceptional Civilian Service Medal by the U. S. Air Force. In 1984 awarded the IEEE Centennial Medal. He is a member of the National Academy of Engineering. *Society Aff:* IEEE, ACM, AAAS.

Warfield, John N
Home: 2517 Huntingdon Rd, Charlottesville, VA 22901
Position: Forsyth Prof. *Employer:* Univ of VA. *Education:* PhD/EE/Purdue Univ; MSEE/EE/Univ of MO; BSEE/EE/Univ of MO; AB/Math/Univ of MO. *Born:* 11/21/25. in Sullivan, MO. Served in US Army 1944-46. Faculty mbr in elec engg at seven univs. Sr Res Leader at Battelle Meml Inst. Dept Chrmn at Univ of VA for 3

Warfield, John N (Continued)
1/2 yrs, with chaired profship. Inventor of solid-state goniometer. Books on computers, societal systems. Developed interpretive structural modeling process. Developed natl plan for environmental education. Past Pres IEEE SMC Soc. Fellow of IEEE. Western Elec Award for Excellence in Teaching. Outstanding Contribution Award, IEEE SMC Soc. *Society Aff:* IEEE, SMC.

Warmann, Robert A
Home: 43 Karmel Ct, Defiance, MO 63341
Position: Pres *Employer:* Alternatives In Engrg *Education:* BS/ME/Univ of MO *Born:* 11/21/48 Born and raised in St. Louis, MO. Field Engr for Factory Mutual Sys, 1971- 74. Joined Protection Mutual as Dist Engr 1974. Founded Alternatives In Engrg April 1984 to provide Loss Prevention Engrg Consulting. Secy-Treas Greater St Louis Chapter SFPE, 1979-80; Pres, 1980-82. Licensed PE, MO, since Feb 1978. Presented paper to NFPA Convention in 1979 on "Home Fire Protection and the Handicapped Child–. Capt, US Army Reserves, Engr Corp. Hobbies including hunting, black powder shooting and computers. *Society Aff:* SFPE, NFPA, NSPE, MSPE

Warnecke, Edward M
Business: 1775 Commerce Dr, Atlanta, GA 30302
Position: Regional Sales Mgr. *Employer:* Eastman Kodak Co. *Born:* Nov 4,1922 - BS Fordham Univ, native of New York City. Served with US Army Coast Artillery & Infantry 1943-46. Recd Bronze Star Medal - Valor Chemist and Service Engr to Motion Picture and Television Industries in NYC 1946-69. Assumed present position 1969 directing sales and engrg staff Mbr SMPTE 1946- present, Fellow mbrship grade 1965 - held position as NYC Sec Chrmn, Natl Mbrship Chrmn and elected to governorship of natl org serving two terms 1970-73.

Warner, Harry B
Home: 485 St Andrews Dr, Akron, OH 44303
Position: Consultant. *Employer:* F Goodrich Co. *Education:* BSChE/OH-Engr/OH State Univ; MS/OH-Engr/OH State Univ; PhD/Engr/Akron, Univ. *Born:* July 7,1916 Columbus Ohio. Joined B F Goodrich 1939 - Tech Control Lab - Res Engr Pilot Plant PVC-SBR Rubber 1940-41-Mfg PVC 194142 - 1942-46 Plant Mgr - Construction and Engr 1947-49. Gr Britain Plant Mgr Dev Ctr-VP Tech 1953, Sales 1957-60 Chemicals-1960-63 Pres BFG Chemical Co. Grp VP B F Goodrich 1963-67 - Pres 1967-74, 1974-present Consultant. Pres Cleveland Engr Sci Ctr 1963-64. PDir Natl AIChE, PDir and VP Mfg and Chem Assn. Texnikoi Outstanding Alumnus Award 1957, Gold Lamme Medal OSU 1962, OSU Centennial Achievement Award 1970. Fellow AIChE. Reg PE Ohio and NY. Hobbies - woodworking, electronics, photography, old cars. *Society Aff:* AIChE, ACS, OPE, NPE, PE.

Warner, Lawrence D
Business: Storch Engineers, Two Charlesgate West, Boston, MA 02215
Position: Supervising Structural Engineer *Employer:* Storch Engineers *Education:* MSCE/Structures/Northeastern Univ; BSCE/Structures/Tufts Univ. *Born:* 9/14/37. Native of Medford MA. Served with Army Corps of Engr, 1961-1963. Structural Engr with Fay, Spofford & Thorndike (1963 to 1973), specializing in bridges and sanitary water treatment plant designs. With Parsons, Brinckerhoff from 1973 to 1980. Duties included being structural dept & proj mgr, responsible for bridges, transit structures, parking garages & bldgs. With Storch Engrs since 1981. Duties include resp for bridge and building designs. Past Pres, 1979 to 1980 and Dir Metropolitan Chapter of MA Soc of PE. General Chrmn, 1978 and Arrangements Chrmn, 1979 to 81; Engr week-Boston Prog (Met Chapter, MSPE & Engr Soc of New England). Enjoy photography, bowling, sailing & swimming. Reg PE in MA. *Society Aff:* ASCE, NSPE, AISC/MSPE.

Warnock, J Gavin
Business: 480 University Ave, Ontario, Canada Toronto M5G 1V2
Position: VP Corp Dev *Employer:* Acres Inc *Education:* BSc/Mech Engg/Univ of Glasgow; DIC/HydroPower/Univ of London. *Born:* 5/28/25. Native of Scotland. Has specialized in hydroelectric power engg, initially involved in design dev and mfg of hydraulic turbines and generators as Mgr, Hydroelectric Div of the English Elec Co with responsibilities both in Canada and UK. Joined Acres Inc in 1964 and in 1977 became VP & Gen Mgr of Power & Heavy Civ Engg Group of Acres American Inc. Actively involved in hydroelectric developments including tidal power and large scale energy storage using underground reservoirs. In 1980 became VP Corp Dev, Acres Consltg Serv Ltd. Past Pres American Underground-Space Assn 1977-1979; Exec Bd Mbr, Intl Water Resources Assn 1973-1979. *Society Aff:* ANS, NSPE, IWRA, AUA, APEO.

Warren, Clifford A
Home: 69 Tuttle Rd, Watchung, NJ 07060
Position: Retired. *Education:* BS/EE/Cooper Union; MS/EE/Stevens Inst of Tech. *Born:* 11/6/13. Formal education was obtained from Cooper Union (BSEE) and MSEE from Stevens Inst of Tech. Employed by Bell Labs in 1931. Primary effort has been on R&D for radars and guided missile systems. Part of five man group to develop first AA fire control radar for Navy for which recd Naval Development Award in 1945. Primary effort last twenty-five yrs has been air defense. Proj Engr for NIKE Hercules 1953-55; NIKE X 1962-67, Sentinel 196769, Safeguard 1969-76. Awarded Army Outstanding Civilian Award for Sentinel and Safeguard. Fellow IEEE 1970. Mbr of Bd - Plantronics, Inc, San Jose, CA. *Society Aff:* IEEE

Warren, Dana White
Home: 108 Sea Cove Rd, Northport, NY 11768
Position: Independent Consultant *Employer:* Self *Education:* BS/Math/Diuty College; BS/EE/MO School of Mines *Born:* July 7,1915 Springfield MO. 1939-40 Consulting Radio Engr. 1940-42 Columbia Broadcasting Corp, Engr in Chg, Radio Frequency Div. 1942-45 Radio Res Lab, Harvard Univ Res Assoc. 1946-65 AIL Div Cutler Hammer Inc, Engrg Consultant. 1965-67 Inst for Defense Analyses (on leave from AIL) Staff Mbr. 1965-80 AIL Div Cutler Hammer, Tech Asst to VP for Operations. 1980--Independent Electronic Consultant Main fields of interest have included Air Navigation and Traffic Control, Antenna Theory, Information theory, Filter theory, Signal and Data processing, and the theory of Adaptive Processes. Fellow IEEE 1957. *Society Aff:* IEEE.

Warren, George E
Home: 333 W CA Blvd, Apt 202, Pasadena, CA 91105
Position: Retired (Bd Chrmn) *Born:* 11//90. IL. Grad Univ of IL, BSCE 1912, CE 1917. Asst Engr Balt; Sewerage Comm one yr. Engr of Construction Contracting Firm 1913-15. Asst Engr Universal Portland Cement Co 1915-21. Asst Gen Mgr Portland Cement Assn and especially charged with dev of res prog 1921-33. VP Southwestern Portland Co Fairborn OH 1933-49. Pres Southwestern Portland Co, LA 1949-61 and Chrmn of Bd 1961-70. Life Mbr ASCE, Joined as Assoc Mbr Chrmn Bd of Dir, Portland Cement Assn 1957-58. Hon Mbr Portland Cement Assn, ASTM. Long time mbr ACI. Reed Turner Medal from ACI 1971 for notable achievement in concrete industry.

Warren, S Reid, Jr
Home: 45 Server Lane, Springfield, PA 19064
Position: Emeritus Professor. *Employer:* University of Pennsylvania. *Education:* ScD/EE/Univ of PA; MS/EE/Univ of PA; BS/EE/Univ of PA. *Born:* Jan 1908. Univ of PA. Teaching, Research, and Administration, School of Engrg and Applied Science, and Dept of Radiology, Medical School, Univ of PA 1933-76. Vice Dean, Moore Sch of Elec Engrg, 1951-54; Asst VP, Engrg Schools 1954-73. Chrmn, AIEE and IEEE Comm; Electrical Techniques in Medicine and Biology 1951-53; Science and Electronics Div Cttee 1954-56; Student Branches Cttee 1963-65; Recognition Awards 1959-61; Chrmn Philadelphia Sec 1950-51. Fellow IEEE 1953. Fellow Physics, Amer Coll of Radiology 1948. Eminent Member Eta Kappa Nu Association 1984. Trustee Episcopal Academy 1958-. Cttee on Science and the Arts, Franklin

Warren, S Reid, Jr (Continued)
Inst 1958- Chrmn 1974. Articles and Books in Electrical Circuits and Fields, and Radiologic Physics. *Society Aff:* IEEE, RRS, HSS, ACR, HPS, AAPM, SHOT.

Warren, William J
Business: P O Box 928, South Pasadena, CA 91030
Position: President. *Employer:* Paul L Armstrong Co, Inc. *Education:* BS/Chem/UC, Berkeley. *Born:* 9/24/29. Sales Engr 1952-64. Owner of Paul L Armstrong Co since 1964. Co is a manufacturers rep fo engineered products used in refinery and chemical plant operations. VP ASME Reg IX 1974-76. Exec Cttee of Council 1974-76, Chrmn Cttee on Reg Affairs 1975-76. Reg Engr State of Calif since 1956. Chmn Meetings Comm, Policy Bd Communication 1979-80. Mbr, 1st Bd of Governors, 1981-82. *Society Aff:* ASME.

Warshaw, Stanley I
Business: National Bureau of Standards, Bldg 101, A603, Gaithersburg, MD 20899
Position: Assoc Director *Employer:* National Bureau of Standards. *Education:* ScD/Ceramics/MIT; BS/Ceramic Eng/GA Inst Tech *Born:* 11/5/31. DCerE GA Inst Tech 57; ScD ceramics MIT 61. Res Asst GA Inst Tech 56-57; MIT 57-61; Harvard Bus Sch, Advanced Mgmt Program, 1978; sr res scientist Raytheon Mfg Co 61-64; res supvr ceramics & metall sect American Standard Inc, 64-68, mgr ceramic tech res & dev ctr, 68-69, mgr mat & chem dept, 1972, gen mgr prod dev & engr lab, 72-75; Dir Ctr for Consumer Prod Tech, Natl Bur of Standards and office of Engrg Standards, Dept of Commerce 1975-80. Dir Off Prod Stds Policy 1981-1986, Assoc Director, 1986-present. Sigma Xi, Phi Kappa Phi, Tau Beta Pi, Keramos. *Society Aff:* ASTM, ANSI, UL.

Warters, William D
Home: 514 Sunnyside Rd, Lincroft, NJ 07738
Position: Asst. VP, Network Technology Research *Employer:* Bell Communications Research (Bellcore) *Education:* PhD/Physics/CA Inst of Technology; MS/Physics/CA Inst of Technology; AB/Physics/Harvard College. *Born:* March 1928, Des Moines Iowa. Joined Bell Labs 1953. Did res in multi- mode waveguide transmission and millimeter-wave repeaters, later was dir of labs resp for transmission performance standards and objectives, dev of long-haul millimeter waveguide, coaxial cable, and satellite transmission systems. Assumed present position in 1983; resp for applied research in optics, high-speed digital electronics and systems, other future network technology. Mbr APS, Phi Beta Kappa, Sigma Xi. Fellow IEEE. Active in educational and cultural affairs; was first president of Monmouth County (NJ) Arts Council, past chrmn of Bd of Trustees of Monmouth College. *Society Aff:* APS, IEEE, ΦΒΚ, ΣΞ.

Wasan, Darsh T
Business: 10 W 33rd St, Chicago, IL 60616
Position: Chrmn and Prof. Chem Engr Dept. *Employer:* Illinois Institute of Technology. *Education:* PhD/Chem Eng/Univ of CA at Berkeley; BS/Chem Eng/Univ of IL at Champaign-Urbana. *Born:* July 1938. Came to US in 1957; naturalized US citizen 1974. With IIT since 1964. Attained current position in 1971. Specialist in mass transfer, interfacial phenomena, separation processes and particle science and technology. Have published over 150 tech papers and ten books and res hon graphs. Consultant to various industries including Exxon; Stauffer Chemical Co. Serving on Tech Adv Cttee of Nelson Industries; serving on Adv Bd of Intl Jrnl of Powder Technology; Editorial Advisory Board of "Surfaces & Colloids" Journal, and Publications Bd of Particulate Sci. & Tech. Journal. P President of the Fine Particle Society, Inc. Recipient Western Elec Fund Award of the ASEE in 1972; Excellence in Teaching Award IIT in 1967, Hausner Award of the Fine Particle Society in 1982, Special Creativity Award of the Natl Sci Foundation. *Society Aff:* AIChE, ACS, AIP, FPS, ΣΞ, SOC. OF RHEOLOGY.

Washington, Donald R
Home: PO Box 1102, St Louis, MO 63188
Position: Sr Project Mgr. *Employer:* Sverdrup & Parcel and Assoc. *Education:* PhD/Sanitary Eng/MIT; MS/Sanitary Eng/Univ of CA, Berkeley; BS/Civil Eng/Univ of CA, Berkeley. *Born:* 10/16/29. Educator 1959-1971, prof, Rensselaer Polytechnic Inst and OH State Univ; Dir, Water Resources Inst, state of OH 1969-71; Dir, Caribbean Field Office, USEPA (GS-15) 1971-73; Owner of Consultint firm, Donald Washington and Assoc, San Juan, PR 1974-78; Sr Proj Mgr, Sverdrup & Parcel and Assoc 1978-79. Reg PE in CA, OH, NY and PR. Pres, Mohawk Hudson Sec, ASCE 1965; Harrison Prescott Eddy Medal, WPCF, 1963; Guest Prof, ETH, Zurich, Switzerland 1967-68. *Society Aff:* ASCE, AEEP, WPCF

Wasil, Benjamin A
Business: Civ Engg Dept, Rm 127 Hinds Hall, Syracuse, NY 13210
Position: Assoc Prof. *Employer:* Syracuse Univ. *Education:* MCE/Civ Engr/NYU; BCE/Civ Engr/NYU. *Born:* 2/23/20. Native of NYC. Started career in civ engg with the TN Valley Authority in 1942. Following WWII, design engr specializing in structure with several consulting engg firms in NYC. Instr & Asst Prof IL Inst of Tech 1949-58. Since 1958 Assoc Prof of civ engg Syracuse Univ. Reg Structural Engr IL, PE NY & IL. Pres Chi Epsilon 1958-1960. Consultant & design specialist to consulting engrs, insurance cos, attornies, industries & governmental agencies. Res & publications in the fields of mtls, structures & mechanics. *Society Aff:* ASCE, ACI.

Wasilewski, Roman J
Business: 1800 G St NW, Washington, DC 20550
Position: Sec Head, MRL Sec. *Employer:* National Science Foundation. *Education:* PhD/Metallurgy/Columbia; BA/Nat Sci/Cambridge, Eng. *Born:* Oct 6,1919, Warsaw, Poland. Served Polish Forces 1939-45. Taught Liverpool University prior to arrival in US in 1956. Research supvr with Engrg Res Dept E I du Pont & Co 1956-67; Sr Tech Advisor, Battelle Columbus Labs 1967- 71; NSF 1971--. Resp for the Materials Res Lab program involving interdisciplinary research at a number of U.S. universities. Mbr of the Bd, TMS- AIME 1976-78 ; Fellow, Institution of Metallurgists (London); Mbr, ASM and The Metals Society (London). Primary interest: the interface between science and technology. *Society Aff:* ASM, TMS-AIME, MET-SOC, INST-MET.

Wasley, Richard J
Business: PO Box 808, Livermore, CA 94550
Position: Research Engineer *Employer:* Lawrence Livermore Lab. *Education:* PhD/Civ-Mech Engr/Stanford Univ; MS/Civ Engr/Stanford Univ; BS/Civ Engr/Univ of CA, Berkeley. *Born:* 6/24/31. Native of San Leandro, CA. Served with US Army 1953-55. Sanitary Engr with CA Dept of Public Health 1955-56. With Univ of CA, Lawrence Livermore Lab since 1961. Assumed current responsibility as Research Engineer in 1985. Prior assignment, 1978-85, was Leader of Nuclear Test Engg. Div. Recipient of Alfred Noble Prize in Engg 1962. Author: Stress Wave Propagation in Solids, 1973; also over 30 articles. *Society Aff:* ASCE, ASME, NSPE, ΣΞ.

Wasserman, Reuben
Home: 5 Cooke Rd, Lexington, MA 02173
Position: President. *Employer:* Datametrics Inc. *Born:* June 1929 N.Y.C. MSEE Univ of Michigan; BSEE City College of NY, and graduate studies from MIT. Grad Res Assoc at Willow Run Res Ctr & Univ of Mich 1953-56; Assoc Dir of Digital Products at Hermes Electronics 1956-61; Exec VP and Founder of Hyperion Industries 1961-67 and President and Founder of Datametrics, Inc 1967--. Also, VP of Solid State Product Plng for ITE-Imperial Corp. Author of many tech papers and mbr of Sigma Xi, Tau Beta Pi, Et Kappa Nu, and ISA President's Advisory Cttee for Industrial Controls.

Wasson, James A
Business: PO Box 109, Morgantown, WV 26507
Position: Prof *Employer:* West VA Univ *Education:* MS/Pet & Nat Gas Eng/Pa St Univ; BS/Pet & Nat Gas Eng/PA St Univ; BS/Meteorology/PA St Univ. *Born:* 7/5/26 Native of Tyrone, PA. Served in USAAF 1945-47, USAF 1951-55. Petroleum Engr for Humble Oil & Refining Co (Exxon) 1956-58. Assistant Prof, Petroleum Engrg, LA Tech Univ 1958-60. Prof, Petroleum Engrg, W VA Univ, 1960-85. Petroleum Engr, US Dept of Energy (formerly US Bureau of Mines and ERDA) 1960-85. Registered Professional Engr, W VA & PA. *Society Aff:* SPE OF AIME, ASEE, ΤΒΠ, ΠΕΤ, ΣΓΕ

Watanabe, Ichiro
Home: 2-21-7 Numabukuro Nakano-Ku, Tokyo 165, Japan
Position: Prof. Emeritus *Employer:* Kanto Gakuin Univ. *Education:* Doctor of Engrg./Mech. Engrg./Tokyo Imperial University; MS/Mech. Engrg./Tokyo Imperial University; BS/Mech. Engrg./Tokyo Imperial University *Born:* 04/13/08 Graduated from Dept of Mech Engrg, Tokyo Imperial Univ 1931. Assoc Prof of Tokyo Imperial Univ and Mbr of Aeronautical Inst of the Univ 1936-1946. Prof of Keio Univ 1946-1974. Awarded "Gijuku Prize" 1955 and "Fukuzawa Prize" 1968 by Keio Univ. Pres of GTSJ 1972. Awarded "Ranju Prize" 1973. Awarded 3rd Class Rising Sun Medal Prize 1983. Prof of Aoyama Gakuin Univ 1974-1977. Prof of Kanto Gakuin Univ. 1977-1982. Consultant, Inst of Technology, Kanto Gakuin Univ 1984-present. Hobbies: Lawn Tennis. *Society Aff:* ASME, JSME, GTSJ, TMSJ

Watchorn, Carl W
Business: Fairhaven C115, 7200 Third Avenue, Sykesville, MD 21784
Position: Retired *Education:* JD/Law/Univ of MD§ LLB/Law/Univ of MD; MS/EE/Lehigh Univ; EE/EE/WPI; BS/EE/WPI. *Born:* April 25, 1900, Waltham Mass. A pioneer in the dev of and the application of differential equations to the economical operation of electric power systems, particularly hydrothermal generating systems, and also in the development and application of probability methods and engrg economics to the economical planning of such systems. Is the author of numerous published tech papers and discussions in the above flds Fellow Award 1962 'For contributions to the economic solutions of planning and operating problems in the electric utility industry.' Other awards: Tau Beta Pi 1921; Sigma Xi 1923. *Society Aff:* IEEE, ΣΞ, ΤΒΠ

Waterhouse, William T
Home: 1337 Syracuse St, Denver, CO 80220
Position: Consulting Engineer. *Employer:* Self. *Education:* BA/Math/Univ of IL; MA/Math/Univ of IL; BS/Civil Engg/Univ of CO. *Born:* April 1911 Burlington Iowa. Construction Engr, Army Corps of Engrs Rock Island, Vicksburg and Galveston Districts 1934-42. Officer, US Army Corps of Engrs 1942-46. Gen Engr Bur/Rec, Chief Engrs Office, Denver as Cost Estimator 1946-64 and Chief, Tech Evaluation Branch 1964-72. Consulting Engr 1972-84. Reg PE and Land Surveyor Colorado. Mbr Tau Beta Pi, Chi Epsilon, Sigma Tau and Pi Mu Epsilon. Fellow, ASCE. Mbr SAME and Colo Engrg Council. President Colo Sec ASCE 1972. Compiled and edited 'Thesaurus of Water Resources Terms'. Awarded Dept of Interior Meritorious Serv Medal. *Society Aff:* ASCE, SAME.

Waterman, Alan T, Jr
Home: 562 Gerona Rd, Stanford, CA 94305
Position: Professor Emeritus Elec Engrg. *Employer:* Stanford University. *Education:* AB/Physics/Princeton Univ; BS/Meteo/CA Inst of Tech; MA/Engr Sci & Applied Physics/Harvard Univ; PhD/Engr Sci & Applied Physics/Harvard Univ. *Born:* 1918; Meteorologist, Ameican Airlines 1940-41; Instructor, Minnesota 1941-42; Res Scientist, Calif Inst Tech 1942-45; Res Scientist, Columbia 1945; Chief Meteorologist, Texas 1945-46; Asst, Harvard 1946-52; Res Assoc, Stanford 1942-57; Assoc Prof of Elec Engrg 1958-63; Prof of Elec Engrg 1964-83. Prof. Emeritus Elec Engr 1983-present. Chrmn, Comm II, Intl Science Radio Union 1961-64; Chrmn, USNC, Intl Union of Radio Science 1970-73; Vice Chrmn, Intl Comm F, URSI 1975-78 Chrmn, Intl Com F Intl Union of Radio Science 1978-81; Chrmn, Antennas & Propogation Group, IEEE 1965- 66; American Meteorological Soc, Amer Geophysical Union; Fellow IEEE 1965. Field of Spec: Radio wave propagation; remote probing of atmosphere. *Society Aff:* IEEE, AGU, AMS.

Waters, Alfred E
Home: 4603 Hillard Ave, La Canada-Flintridge, CA 91011
Position: Consulting Engineer *Employer:* Self *Education:* ME/Gen Engg/UCLA; MS/CE/CA Inst of Tech; BS/CE/CA Inst of Tech *Born:* 03/21/24 Originally from NY. Served in the Army Corps of Engrs from 1942-1946. Sucessively Structural Engr, Project Mgr, Corporate Secy and VP with Quinton Engrs Ltd of Los Angeles from 1949 to 1971. VP of Consolidated Pacific Engrg 1971. Senior Structural Engr with Donald R. Warren Co from 1972 to 1981, and from 1981 to 1985 with Jacobs Engrg Group. Consulting Engineer since 1985. Registered Cvl Engr in CA & AK, and Structural Engr in CA. With ASCE holds Daniel W. Meade Prize and served as Pres of Los Angeles Section and Chrmn of CA State Council 1979-1980. *Society Aff:* ASCE, NSPE, ΤΒΠ, IAE.

Waters, James P, Jr
Business: 1800 14th Ct. S, Birmingham, AL 35205
Position: Consulting Engr. *Employer:* Waters Engg. *Education:* BSEE/Elect Eng/Univ of AL. *Born:* Dec 27, 1927. BS Univ of Alabama 1953. Native Fairfield Alabama. Elec Engr with US Steel Corp 1953-64. US Steel Exec Trng 1954. Electrical Design complete wire Mill Atlantic Steel Co 1965-66. Prof engr private prac 1966--. Design of electrical facil for commercial and industrial bldgs. Spec: schools, banks, churches, auto dealerships and nursing homes, Forensic Application of Elec Engrg with Attorneys and Insurance companies. Mbr American Arbitration Assoc. *Society Aff:* IEEE, AAA.

Waters, Robert C
Business: Engineering Administration Dept, Washington, DC 20052
Position: Prof. *Employer:* George Washington Univ. *Education:* DBA/Bus Econ/Univ of S CA; MBA/Oper Analysis/UCLA; BS/Mech Engr/UCLA. *Born:* 4/27/30. Native of Long Beach CA. Gen Elec Mfg Training Prog, 1956-58; Supervisor, Instrumentation Shop, Jet Engine Dept, 1958-59. With TRW Systems (Space Technology Labs), 1959-68, as Cost Effectiveness Engr and later Market Planning Mgr, Systems Labs. VP and Dir, Transportation Mgt Div, EMSCO Engg and Mgt Sciences Corp, 1969-72. Assoc Prof, Engr Mgt Dept, U of MO - Rolla, 1972-79. Chrmn Engr Economy Div, ASEE, 1977-78. Since 1979 Prof of Engg Admin, George Washington Univ; Chrmn, 1984-1989. Treas, Amer. Soc. of Engr. Mgmt 1980-1984. Visiting Prof., Grad. Sch. of Mgmt., UCLA, 1988. *Society Aff:* AAAS, ASEE, ASEM, ΣΞ.

Watkins, Charles B
Home: 173 Chadwick Rd, Teaneck, NJ 07666
Position: Dean of Engineering *Employer:* City College of New York *Education:* PhD/ME/Univ of NM; MS/ME/Univ of NM; BS/ME/Howard Univ. *Born:* 11/20/42 Mbr of Tech Staff-Sandia Natl Labs 1964-71. Howard Univ 1971-86. Dept Chmn 1973-86. ASME Natl Governors Assoc Fellow 1984-85. Dean School of Engrg City College 1986-. Res activities include computatutional fluid dynamics, gas bearing analysis, experiment and numerical heat transfer. Conslt activities have included work for NSF, US Army, and US Navy. Chmn Region III ASME, Dept Hd Ctte, Mbr K-12 Heat Transfer Ctte ASME. Reg PE, District of Columbia. Mbr Tau Beta Pi, Sigma Xi. *Society Aff:* ASME, SAE, AIAA, NSPE, ASEE.

Watkins, Dean A
Business: 3333 Hillview Ave, Palo Alto, CA 94304
Position: Chrmn of the Board. *Employer:* Watkins-Johnson Co. *Education:* PhD/Engg/Stanford Univ; MS/Engg/CA Inst of Technology; BS/Engg/IA State College. *Born:* Oct 1922. Served as engr unit cmdr in US Army 1943-46. Discharged rank of lst Lt. While a student at Stanford, was co-inventor of the low-noise traveling wave tube in 1952. Author of numerous tech pubs & book 'Topics of Electromagnetic Theory'. Prof 1956-63. Was a design engr for Collins Radio Co and on staff of Los Alamos Scientific Lab 1948-49. Fellow IRE 1958--. Served as consultant

Watkins, Dean A (Continued)
of electron devices to DOD, R&E in Wash, DC 1956-66. Co-founder of Watkins-Johnson Co 1957. Currently Chrmn of Bd & Dir. Trustee of Stanford U 1967-69. Mbr Bd of Regents of U of Calif & Bd of Overseers of Hoover Inst of War Revolution & Peace. Dir & VP WEMA 1967. Mbr NAE 1968. *Society Aff:* IEEE, NAE.

Watkins, Henry H
Business: PO Box 8851, Greenville, SC 29604
Position: Pres. *Employer:* Hunter Watkins & Assoc, Inc. *Education:* BSCE/Civ Engg/Citadel. *Born:* 5/15/25. in Greenwood, SC, son of the late Wm Paul & Genevieve Hunter Watkins. Educated in the Greenwood Public Schools & The Citadel. Married Alice Marie Bryson, 1946. Was with J E Sirrine Co from 1945 until 1963, in the Pulp & Paper and Structural Depts. Founded Hunter Watkins & Assoc in 1963. Mbr Fourth Presbyterian Church, railroad & history buff & enjoys travel & photography. Mbr Natl Model Railroad Assn, Natl Railway Historical Soc. Life Mbr, Assn of Citadel Men. Past Pres, SC Soc of PE. SC Soc of PE "Engineer of the Year," 1981. *Society Aff:* NSPE, ACEC, ASCE, NAFE

Watkins, Peter H
Home: 142 Stoneridge Rd, New Providence, NJ 07974
Position: Retired *Education:* BS/Chem/Univ of AZ; BS/ChE/VPISU; MS/ChE/VPISU; PhD/ChE/VPISU. *Born:* 11/17/19. Served Army Corps Engrs 1942-46, Staff mbr, Los Alamos Labs 1944-46. Mbr Chem Engr Faculty, VPISU 1948-56. Exxon Res & Dev Labs Baton Rogue LA 1956-64, process dev operations. Exxon Res & Eng Co NJ 1964-81. Managed recruitment, employment & placement of all professionals & oversee non-profl employment. Currently retired. *Society Aff:* AIChE, ACS, ASEE.

Watkins, Susan J
Home: 1110 Rudgear Rd, Walnut Creek, CA 94596
Position: Metallurgical Engineer *Employer:* Bechtel Group Inc. *Education:* -/Mtl Sci & Eng/Univ of CA; BE/Mtl Sci & Met Engg/Vanderbilt Univ. *Born:* 8/11/52. Native of Nashville TN. Process engr for Airco Temescal, Berkeley, CA 1974; Met Engr for Airco Central Res, Murray Hill, NJ 1975, specializing in electron beam vapor deposition & sputtering. Proj dev engr for Cutter Labs, Berkeley, CA 1976-1981 specializing in design & dev of injection molded plastic componants for medical equip. Also responsible for dev of new line of electro-mech equip. Metallurgical Engineer for Bechtel Group Inc, Since 1981 consulting on supplier welding and metallurgical problems. Exec committee of ASM Golden Gate Chapter, recipient of Pres's Award 1979. ASM Natl Career Guidance Committee. *Society Aff:* ASM, SWE.

Watkins, William S
Business: 34711 Chardon Rd, Willoughby Hills, OH 44094
Position: Pres & Ch Engr. *Employer:* Wm S Watkins & Assoc, Inc. *Education:* BEE/Elec Engg/Fenn College. *Born:* 4/23/16. Born and educated in the Cleveland Area. Worked as electrician, electrical foreman, plant electrical engr, eventually as Corporate Elec Engr for OH Rubber Co, Willoughby, OH. After 23 yrs in industry opened consulting engg office, providing design for power, lighting and control systems, presently 15 employees. My office developed early progs in elec safety and energy mgt. We teach classes in code compliance and safety for industries, inspectors, contractors and engrs. I am retained by OSHA to instruct Compliance Officers. I lecture on a natl basis and write articles for natl magazines. Active in IEEE and ASSE. *Society Aff:* IEEE, NSPE, IES, AIPE.

Watler, Kenneth G
Home: 14927 Bramblewood Dr, Houston, TX 77079
Position: President. *Employer:* The Walter Company *Education:* BSChE/Ch Engg/Lamar Univ. *Born:* 12/15/37. Worked for 2 yrs as Mgr of Construction for Hess Terminals Div, New Orleans, LA then 4 yrs as Proj Mgr for Delta Engrg in Houston, 3 yrs with Hutchinson-Hayes, as Process & Mech Div Mgr, Pres of Weatherby Engrg for 7 yrs and Pres of Wilcrest Engrg Corp, for 2 yrs, and now Pres of The Walter Company, design and engrg of gas production facilities, gas processing plants, gas compressor stations, gas treating plants, elec generating stations, pipeline stations, gas permeation membrane systems for both onshore and offshore installations, domestically and internationally. *Society Aff:* AIChE, NSPE.

Watson, Albert T
Business: Borg-Warner Chemicals B.V, Cyprusweg 2, 1044 AA Amsterdam, The Netherlands
Position: Pres *Employer:* Borg-Warner Chemicals (Europe) B.V. *Education:* PhD/Physical Chemistry/Vanderbilt Univ; MS/Physical Chemistry/Vanderbilt Univ; BS/Chemistry/Wofford College. *Born:* 10/14/23. Native of Spartanburg, SC Served in US Army in 1943-1946 1st Lt Infantry. Employed by Exxon Res & Exxon Chem Tech, Baytown, TX 1949-1975. Positions included Res Chemist, Polypropylene Tech Mgr, Mgr Plastics Res Labs. With Chemstrand Corp, Decatur, AL 1958-1959 as Group Leader, Fiber Formation. Joined Borg-Warner Chemicals in Wash, WV as Plastics Technical Dir 1975; VP Technical 1978-1979; Holder of 18 US patents; Sr VP 1979-1984; Exec VP 1984-85; Pres, Borg. Warner Chemicals Europe, 1985-present. *Society Aff:* ACS, SPE, AIChE

Watson, Ian C,
Home: 1380 Plumosa Dr, Fort Myers, FL 33901
Position: Pres *Employer:* Rostek Services, Inc *Education:* BS/Chem Eng/Neath College of Tech *Born:* May 1942 London England. Edu at Berkhamsted School and Neath Coll of Technology. Degree in Chem Engrg 1963 and addtl courses in data processing and business admin. Nuclear fuel fouling research 1964 Atomic Energy of Canada Ltd. Design, cost & scheduling, & construction, Bechtel Corp 1965-69. Mgr, Resource Studies, CSR Inc 1969-72. David Volkert & Assoc 1972-77, VP 1973, resp for Environmental Plng and Engrg Div activities. Assoc Mbr AIChE 1975; Assoc Amer Inst Planners; Charter Mbr NWSIA, Dir since 1976; Parliamentarian 1976-78. Enjoy all sports, motor racing and gardening. 1977-81, Island Water Assoc, Inc GM & CE 1979, respon for all operations & maintenance, design & construction of new plant & facilities. 1981-83, Mgr of Intl Sales for Hydranautic Water Systems, Manufactures of Reverse osmosis systems. Since August 1983, Pres of Rostek Service Inc, specialized consltg in membrane separations. *Society Aff:* NSPE, AWWA, NWSIA, FES, WSE, IDA

Watson, James F
Business: PO Box 81608, San Diego, CA 92138
Position: Dir, Advanced Technology Div *Employer:* General Atomic Co. *Education:* PhD/Metl Engr/Univ of MI; MS/Metl Engr/Univ of MI; BS/Metl Engr/Univ of MI. *Born:* 8/26/31. At General Atomic (1962-date), supervised both mtls and process dev progs related to high temperature gas-cooled reactors (HTGR). Supervised work including radiation effects on both core mtls and structural mtls typically used in power-generating equip. Work also included many aspects of welding, corrosion, fatigue, friction and wear, creep and stress-rupture, brittle fracture, and failure analysis for such power plant components, and mtls problems unique to prestressed concrete pressure vessels. Also have been concerned with preliminary studies of mtls problems related to the use of helium gas turbines in gas-cooled reactors. Other profl experience at General Dynamics Corp and the Dow Chem Co. Mbr and Fellow - ASM, Mbr ANS. Reg Prof Engg (Metallurgical), CA. *Society Aff:* ANS, ASM.

Watson, Kenneth S
Home: 1665 Winnetka Rd, Glenview, IL 60025
Position: Editor Indust Wastes Magazine *Employer:* Scranton Gillette Communications *Education:* MS/ChE/W V Univ; BS/ChE/W V Univ *Born:* 12/24/11 As a major in the combat engrs during World War II, commanded a combat engrg Battalion and served in military govt as trade and indus officer for the city of Frankfort, Germany. Originated the initial environmental control prog for the GE Co in 1950 and headed it for 18 years. Directed Kraft, Inc prog of environmental control for 9 yrs. Was elected pres of the Water Pollution Control Federation in1957 and awarded

Watson, Kenneth S (Continued)
the federation's Willem Rudolfs Medal in 1955 and 1978 and the Charles Alvin Emerson medal in 1976. Was elected to honorary membership in AWWA in 1968. *Society Aff:* WPCF, AICHE, AWWA, ACS

Watson, Kenneth W
Business: 1798 South West Temple, Salt Lake City, UT 84115
Position: Principal *Employer:* Eckhoff, Watson & Preator Engrg *Education:* ME/Environ/Univ of UT; BS/CE/Univ of UT *Born:* 01/17/48 and raised in the small community of Parowan, UT. Graduated from Parowan High Sch, Univ of UT (BS '71; MS '74). Eagle Scout. Hydraulics and Sanitary Engr of UT Dept of Transportation 1971-1975. Hydraulics Engr for Salt Lake County Flood Contol - 1975-1977. Author of many publications on water quality, hydraulic and hydrology systems. Co-owner (Principal) Eckhoff, Watson and Preator Engrg; EWP/Systems; EWP/Construction Services Inc; and Rocky Mountain Engrg. All above companies I am co-owner as part of EWP/GROUP, INCORPORATED. Registered PE and Land Surveyor in UT, NV, WY, ID and CO. Love my family, hunting and fishing. *Society Aff:* ASCE, ACSM, AWWA, ACEC

Watson, Velvin R
Home: 21366 Amulet Dr, Cupertino, CA 95014
Position: Chief, Workstation Application Office. *Employer:* NASA - Ames Research Ctr. *Education:* PhD/Aero & Astro/Stanford; MS/Aero/UC Berkeley; Bs/Mech Engg/UC Berkeley. *Born:* June 1932. Res in Comp Graphics, Fluid Dynamics, Plasmadynamics, and Computational Physics. Currently conducting res in computational fluid dynamics & computer technology. Instructing part time at Santa Clara Univ and Stanford Univ in Computer-aided engg. Previously taught numerical method for scientists at San Jose State U. Previously Chrmn of the San Francisco Sec of AIAA, Vice Chrmn, Secy and Treas. NASA Invention Award and Patent for a major contribution in plasma generation theory & technology. *Society Aff:* AIAA, ACM, IEEE, ASEE

Watt, Brian J
Business: 16650 Greenbriar Plaza Dr 260, Houston, TX 77060
Position: President *Employer:* Brian Watt Associates, Inc. *Education:* ScD/CE/MIT; MSc/CE/Univ of Witwatersrand, S. Africa; BSc/CE/Univ of Witwatersrand, S. Africa *Born:* 7/24/40 Born in Johannesburg, S Africa. From 1966 to 1969 studied at MIT and received ScD for research in theoretical soil mechanics. Joined Ove Arup and Partners in 1970 and worked in S Africa, France and Britain on a variety of onshore and offshore projs. Immigrated to Houston, TX in 1977 where he founded Brian Watt Associates, Inc. Specializes in high technology engrg for offshore frontiers. Mbr of sev prof cttes of NRC, ASCE, API. Guest lecturer at more than a dozen institutions in USA and overseas. Enjoys classical music and golf. *Society Aff:* ASCE, SPE, ASFE, ACEC, NSPE

Watt, Joseph T, Jr
Business: P.O. Box 10029, Beaumont, TX 77710
Position: Prof Dir, Engrg Cooperative Education *Employer:* Lamar Univ *Education:* BA/-/Rice Univ; BSEE/EE/Rice Univ; MSEE/EE/Univ of TX, Austin; PhD/EE/Univ of TX, Austin *Born:* 7/16/33 Born in Honolulu, HI; reared in Laredo, TX. Completed Advanced Engrg Prog, GE Co. Prof of Elec Engrg, Lamar Univ, since 1965. Overhauled lab courses. Developed new courses in automatic control, digital sys, and small computers. Dir, Engrg Co-operative Education, since 1984. Mbr, US Army Reserve. Publications in automatic control and small computers. *Society Aff:* IEEE, ISA, ASEE

Watters, Gary Z
Home: 1425 Winkle Drive, Chico, CA 95926
Position: Dean, School of Engrg, Computer Science, and Tech *Employer:* California State Univ., Chico *Education:* PhD/CE/Stanford Univ; MS/CE/Stanford Univ; BS/CE/Chico State College. *Born:* Oct 1935. Taught at Chico State College 1958-61; Taught at Utah State Univ 1963-1980. Served as Asst Dean of the College of Engrg at Utah State Univ 1972-74 and Assoc Dean 1974-76. Prof engrg experience with Stone & Webster Eng Corp, Boston MA and served as a consultant for Hansen & Assoc Bringham City UT; CH2M Hill, Corvallis OR; Valley Engrg Logan UT; Keller Engrg Logan UT; Winzler and Kelley Eureka CA and Johns Mannille Corp, Denver Co. Spec in hydraulic transients, mathematical modeling of fluid flow, design of irrigation systems. Honors include being President Utah Sec ASCE 1975-76; ASEE Ford Foundation Resident in Engrg Practice 1972 and ASEE Outstanding Young Engrg Teacher, Rocky Mtn Sec 1969 Mbr ASCE, NSPE and ASEE. Enjoys flying, golf, skiing and backpacking. Present position is Dean of School of Engineering, Computer Science, and Technology, 1980--Author of one book, coauthor of two others. *Society Aff:* ASCE, ASEE, NSPE.

Watts, Oliver E
Home: 7195 Dark Horse Pl, Colorado Springs, CO 80919
Position: Owner (Self-employed) *Education:* BS/CE/CO State Univ. *Born:* 9/22/39. Native of Craig, CO. Served with Army Corps of Engrs 1962-64; CA Dept of Water Resources 1964-70; CF&I Steel Corp 1970-; United Western Engrs 1970-76; United Planning and Engr 1976-79; Now self employed: Pres Prof Engrs of CO & CO Springs branch ASCE. Enjoys hunting and fishing. *Society Aff:* NSPE, ASCE, PLSC, ACEC, CEC, CECC, PEC, CWC, ICID

Watwood, Vernon B
Business: Dept of Civ Engrg, Houghton, MI 49931
Position: Prof and Chrmn *Employer:* Mich Tech Univ *Education:* PhD/CE/Univ of WA; MS/Eng Mech/Cornell Univ; B/CE/Auburn Univ *Born:* 9/24/35 Born in Opelika, Ala. USN Civil Engr Corps, 1957-60; Asst Prof of Civil Engrg, MS State U., 1960-61; Engr, The Boeing Co., 1962-64; Sr Engr, Esso Production Res Co., 1966-67; Engr Associate, Battelle Northwest Laboratories, 1967-70; Mgr, Applied Mechanics Lab, Franklin Inst Res Laboratories, 1970-73; MI Technological U., 1973- , Prof of Civil Engrg 1979- , Chrmn of Dept 1980- ; Chrmn of ASCE Ctte on Automated Design and Analysis 1979-81; Chrmn of ASCE Ctte on Analysis 1987- Editor of Large Mining Machinery Directory for COSMET; Consulting assignments with manufacturing, engrg, mining and const companies; Reg MI, WA and VA. *Society Aff:* ASCE, AISC, ASEE

Waugh, David J
Business: College of Engg, Columbia, SC 29208
Position: Dean of Engg. *Employer:* Univ of SC. *Education:* M Engg/Solid Mech/Yale; BS/Civil Engg/Univ of SC. *Born:* 9/20/32. in Charleston, WV. US Navy 1954-57. Stress analysis and design, Bendix Corp. 1957-59. NSF Sce Faculty Fellow 1962-64. Univ of SC faculty 1959 - present; Assoc. Dean of Engg 1968-77, Dean 1977 - present. Founding mbr, Southeastern Conference on Theoretical and Applied Mechanics (SECTAM). Founding mbr and board of dir, Assoc for Media-Based Continuing Engg Education (AMCEE). Consultant in structural analysis and mech. Various offices ASCE, ASEE; Bd of Dirs, Engrg Deans Council; Engr of the year-SC; Order of the Palmetto; Fellow of ASEE. *Society Aff:* ASCE, ASEE, NSPE.

Waugh, Richard J
Business: 20 Eglinton Ave W, Toronto Ontario Canada M4R 2E5
Position: VP & Gen Mgr. *Employer:* Assoc Pullman Kellogg Ltd. *Education:* BSc/ME/Kings College, Univ of London. *Born:* 3/2/23. Native of London, England. Served in RAFVR 1943-1947. Emigrated to Canada in 1956. Canadian citizenship 1961. With Lummus Canada, Beckiel Canada, Stone & Webster Canada from 1959-1974. With Assoc Pullman Kellogg (Previously Pullman Kellogg Canada) since 1977. Currently VP & Gen Mgr respon for Toronto operations of APKL. Chrmn, CSChE annual conf 1976. VP CSChE 1977-1978. Pres CSChE 1978- 1979. *Society Aff:* CSChE, CIC, IChE.

Way, A R (Jack)
Business: 6300 Cornhusker Hgwy, Lincoln, NB 68529
Position: Pres *Employer:* Concrete Industries, Inc *Education:* BS/Arch Engrg/KS State Univ *Born:* 10/16/26 1954-56: Design Engrg, Lincoln Steel Corp. 1956-57: Struct Design Engr, Davis, Fenton, Stange & Darling, Architects. 1957-58: Field Engr, Portland Cement Co. 1958-present: Worked from position of struct design engr in the Prestressed Concrete Co managed by Concrete Ind through mgmt of the Prestressed Concrete Co, the Ready Mixed Concrete Co, the contrete block companies on to the present position as Pres of Concrete Industries, Inc which is the parent company to seven concrete products companies.

Way, Stewart
Business: Box 505, Whitehall, MT 59759
Position: Ret *Employer:* Self-Employed Consultant *Education:* SCD/Eng Mechanics/U. of Michigan; MSC/Eng. Mechanics/U. of Michigan; AB/Math/Stanford Univ *Born:* 7/22/08 Born Sewickley PA. 7/22/08. Lived in Calif. 1916-29. Exchange student to Germany 1929/30. Worked Douglas Aircraft, weight and Bal. Eng. 1930/31. Westinghouse Research Labs Pittsburgh, Mechanics Dept, 1933-73, specializing solid mechanics, fluid mechanics and combustion. Contributed dev first American jet engine (turbojet). Did early experimental work on ramjet engine, (electrically heated model). Directed early experimental work on magnetohydrodynamic energy conversion at Westinghouse 1959-70. Directed MHD generation experiments at Westinghouse 1960-73. MHD consultant - 1973-present. *Society Aff:* ASME, AIAA

Way, William R
Home: 1320 Lombard Cresent Apt 49, Montreal Canada H3R 3GI
Position: Retired *Education:* BSc/Elec Engg/McGill Univ. *Born:* May 29, 1896. Elected Fellow AIEE 1953. Pres Dist 10 AIEE Canada 1953. In Industry: Sr Vice Pres, Chief Engr and Director, The Shawinigan Water and Power Co. Director of S.W.& P. Co. and of four associated companies. Retired 1963. *Society Aff:* OEQ, EIC, IEEE.

Wayland, Russell G
Home: 4660 N 35th St, Arlington, VA 22207
Position: Energy Mineral Consultant Member, Virginia Oil and Gas Conservation Board Washington Representative, *Employer:* Retired from USGS *Education:* PhD/Econ Geology/Univ MN; AM/Economic Geology/Harvard Univ; MS/Economic Geology/Univ MN; BS/Mining Engg/Univ WA. *Born:* 1/23/13. Native of AK & Black Hills, SD. Engr or aid with Homestake Mining Co summers '30-'39. 2nd Lt to Lt Col, Corps of Engrs '42-'46. Strategic mineral supply for Army, '42-'45. German mine rehabilitation for War & State Depts, '45- '52. US Geological Survey '39-'42 & '52-'80, including Chief, Conservation Div, '66-'78. As Chief, directed the work of 270 petrol & mining engrs, 44 other engrs, 186 engg technicians, 590 other scientists or aids, & 700 support personnel in supervising & evaluating industry operations involving Fed & Indian mineral leases, onshore & offshore. Energy minerals consultant since '80. Wash Rep, AIPG since '82; Mbr, VA Commonwealth Oil & Gas Conserv Board since '82. Dept. of the Interior Distinguished Service Award. '68. Who's Who in America. Colonel, Army Reserves (retired). Tau Beta Pi. Sigma Xi, Phi Gamma Delta, Cosmos Club. *Society Aff:* AIME, SEG, AIPG, GSA, AEG, MSA, AAAS, AAPG.

Wayman, Clarence Marvin
Home: 2204 Pond St, Urbana, IL 61801
Position: Professor of Metallurgy. *Employer:* Univ of Ill at Urbana-Champaign. *Education:* BS/Metallurgy/Purdue; MS/Metallurgy/Purdue; PhD/Metallurgy/Lehigh. *Born:* 8/12/30. Air Force Matls Lab 1952-54, titanium alloy dev. Univ of IL 1957- . Major res int is in martensitic phase transformations; book, "Introduction to the Crystallography of Martensitic Transformations,–. Mbr ASM, AIME, Japan Inst of Met and Inst of Mets. Bd of Review and Jt Comm, Met Transactions. Guggenheim Fellow, 1969, NATO Lecturer, 1969. Fellow of Churchill College, 1969. Fellow of the Inst of Met 1970. Fellow Japan Soc for the Promotion of Sci 1976. Zay Jeffries Award, ASM 1977. Fellow ASM, 1977. Mathewson Gold Medal, AIME, 1978. Editorial Committee, "Metallography" 1979. *Society Aff:* ASM, AIME, IOM, JIM.

Wayman, Morris
Home: 17 Noel Ave, Toronto Ontario, Canada M4G 1B2
Position: Professor. *Employer:* University of Toronto. *Education:* PhD/Organic Chem/Univ of Toronto. *Born:* 3/19/15. in Toronto, Canada. Tech Dir Columbia Cellulose Co 1952-58; Dir of Res, Sandwell Intl Ltd Consultants 1958-63; Prof Dept of Chem Engg & Applied Chem, Univ of Toronto 1963- . Prof Fac of Forestry, Unit of Toronto 1973- . Pres Morris Wayman Ltd, Consultants, 1965- . Prof Engr Ontario Book: "Guide for Planning Pulp & Paper Enterprises–, Food & Agri Organization, UN 1973; "Wealth and Welfare–, Griffin House, 1979. FAAAS, FCIC, FRSC. *Society Aff:* TAPPI, TSCPPA, CIC, AAAS, ACS, CSChE, RSC.

Waynant, Ronald W
Home: 13101 Claxton Drive, Laurel, MD 20708
Position: Sr. Optical Engineer *Employer:* Food and Drug Administration *Education:* Ph.D./Electrical Engrg./Catholic University; M.S.E.E./Electrical Engrg./Catholic University; B.E.S./Electrical Engrg./Johns Hopkins University *Born:* 10/04/40 Westinghouse Electric Corp., Baltimore, MD. Developed solid state lasers, studied laser breakdown and laser damage to solids, 1962-66. Continued part-time in active imaging while in graduate school. Naval Research Laboratory, Plasma Physics Div., developed first vacuum ultraviolet laser in molecular hydrogen, extended to shorter wavelengths, discovered VUV ion laser, applied traveling-wave excitation 1969-75. Optical Sciences Div., measured excimer laser kinetics, developed far ir gas dynamic lasers, waveguide and rf excited excimer lasers, deep uv lithography 1976-86. Food and Drug Admin, research in laser surgery and medicine. Editor in Chief, Circuits and Devices Magazine; Fellow IEEE, 1987. Serious amateur photographer, tennis player and avid do-it-yourselfer. *Society Aff:* IEEE, OSA, APS, SPIE

Wayne, Burton H
Business: Engrg Analysis & Design, Charlotte, NC 28223
Position: Chrmn, Dept of Eng Anal & Des. *Employer:* University of North Carolina. *Education:* PhD/Elect Eng/MI State Univ; MS/Elect Engr/MI State Univ; BS/Elect Engr/MI State College. *Born:* 11/18/24. Served in US Army Signal Corps 1943-47. Prof MI State Univ East Lansing 1954-63; Prof Univ of NC at Charlotte 1964-70; Chrmn, Eng Analysis & Design Dept, UNCC 1970-. Sen Mbr IEEE, Mbr ASEE, Charlotte Engrs Club, Eta Kappa Nu, Sigma Sigma Phi. NC Professional Engrs. VChrmn Charlotte Sec IEEE 1974-75. Chrmn Sec IEEE 1975-76, Sec, NC Council IEEE 1979. Pres Perimeter Swim League 1966-. Pres Shannon Park Swim Club 1968-69. *Society Aff:* IEEE, ASEE.

Wayner, Peter C Jr
Business: Dept of Chemical Engrg, Rensselaer Polytechnic Institute, Troy, NY 12180-3590
Position: Professor *Employer:* RPI *Education:* PhD/ChE/Northwestern Univ; SM/ChE/MIT; BS/ChE/RPI *Born:* 08/18/34 Born in Taunton, MA. Naval Officer, USN 6/56-9/58. Res Scientist for United Aircraft Res Laboratories 3/63-8/65. Prof of Chem Engrg at RPI 9/65 - present. Chem Engr for Lawrence Livermore Lab (6/76 - 9/76), Contract Researcher for NASA, Ames and Visting Prof at Stanford Univ (9/76 - 7/77). Consultant in fields of Heat Transfer and Interfacial Pheonomena. Elected Fellow of AIChE (1986). Consulting Editor of *Chemical Engineering Progress.* 1985 Chairman of the Heat Transfer and Energy Conversion Division of AIChE. *Society Aff:* AIChE, ACS, ASME

Waynick, Arthur H
Business: Ionosphere Res Lab, University Park, PA 16802
Position: Professor Emeritus. *Employer:* Pennsylvania State University. *Education:* ScD/Communications/Harvard; MS/Physics/Wayne State; BS/Physics/Wayne State. *Born:* Nov 1905. ScD Harvard, Radio Engr with Reno Radio 1922-35. Inst in Physics, Wayne U, Demonstrator in Physics, Cambridge U, Asst Prof in Physics Wayne

Waynick, Arthur H (Continued)
U 1935-40. Electronics Sys Harvard Underwater Sound Lab, Ordnance Res Lab and Ionosphere Res Lab 1940-71. Prog Dir in Engrg NSF 1968-69. Emeritus A Robert Noll Prof of EE. Hd of EE Dept and Dir of Ionosphere Res Lab. Liaison Scientist ONR London, England. PChrmn USA Nat Cttee URSI; PChrmn IEEE, PGAP· PDir IEEE; Fellow IEEE, AGU, IEE. Hon Mbr USNC/URSI. Guggenheim Fellow. Mbr National Acad. of Engrg. *Society Aff:* IEE, IEEE, AGU, URSI.

Wear, Robert J
Home: 71 Hillcrest Drive, Denville, New Jersey 07834
Position: Admin Dean; Retired. *Employer:* Acad of Aero. *Education:* MA/Admin Higher Ed/NYU; BSIE/Ind Engg/Fairleigh Dickinson Univ. *Born:* 12/25/17. in Hazleton, PA. Taught at the Casey Jones Sch of Aero in Newark. Taught and served as Acad Dean & Admin Dean at the Acad of Aero. Was chrmn of the Engg Tech Div and the Tech College Council of the Am Soc for Engg Education. Was an exchange awardee of the English speaking Union to England. Received the 30th Annual James McGraw award for Engg Tech from the Am Soc for Engg Education. Hobbies-photography & sound recordings. *Society Aff:* ASEE, SAE, AIAA

Weart, Harry W
Home: Rt 4, Box 108, Rolla, MO 65401
Position: Chrmn & Prof. *Employer:* Univ of MO. *Education:* PhD/Met Engr/Univ of WI; MS/Met Engr/Univ of WI; BS/Met Engr/Rensselaer Poly. *Born:* 7/10/27. Physical met with special interest in phase transformations & associated fundamental phenomena, such as diffusion & interfaces. Performed metal phys res at Westinghouse Res Lab (1956-60); faculty mbr at Univ of WI-Madison (1953-56), Cornell Univ (1960-64) & Univ of MO-Rolla (1964-present); chrmn, Dept of Met and Nuclear Engg, 1964-1983, Chrmn, Dept of Met Engr and Nuclear Engr; 1983-present, Chrmn, Dept. of Met Engr. Frequent service to AIME, mostly on committees concerned with education & technical publications. Natl pres Alpha Sigma Mu (met honor soc), 1980. Civic service to Community concert assoc & Lutheran Church MO Synod, primarily in fund-raising, budgeting and long-range planning. Principal hobbies are whitewater kayaking and canoeing. *Society Aff:* ASM, AIME, ASEE, ΣΞ.

Weatherly, David M
Business: 1800 Peachtree Rd NW, Atlanta, GA 30309
Position: President. *Employer:* The D M Weatherly Co. *Education:* BS/ChE/High Point College. *Born:* June 1921. Native of North Carolina. Engaged in plant operations, engrg and construction of various types of chem plants from 1942-54 by Hercules Powder Co, Kellex Corp, J A Jones Construction Co & H K Ferguson Co. In charge of engrg & construction of chem plants for John J Harte Co Atlanta GA from 1947-53. Founded The D M Weatherly Co in 1954 and continuously engage since that time as Pres of the company in the engrg and construction of various chem plants both in the USA and other countries. Mbr ACS, AIChE. *Society Aff:* AICHE, ACS

Weaver, Charles Hadley
Business: 402 Communications Bldg, Knoxville, TN 37916
Position: VP Cont Edu & Dean UT Space Inst. *Employer:* University of Tennessee. *Education:* PhD/EE/Univ of WI; MS/EE/Univ of TN; BS/EE/Univ of TN. *Born:* Jan 27, 1920. Engrg Co-op student. Engrg Supvr, Tennessee Eastman Co, Oak Ridge 1943-46. Instructor to full professor, EE Dept, UT, 1946-59. Westinghouse prof and EE Dept Head, Auburn Univ 1959-65. Dean, Coll of Engrg UT 1965; First Chancellor, UT, Knoxville 1968. VP, Cont Edu 1971; Also assumed pos of Dean, UT Space Inst, Tullahoma 1975. Reg Engr Tenn & Ala. Mbr Eta Kappa Nu, Omicron Delta Kappa, Sigma Xi. Hon Mbr Tau Beta Pi,Phi Kappa Phi. Fellow IEEE. Mbr ASEE, Tenn Soc Prof Engrs, Engr of the Yr 1969 East Tenn Chap TSPE. Recipient Air Force ROTC Outstanding Serv Awd 1969 Tasker H Bliss Medal, SAME 1970 and N W Dougherty Eng Award 1976. Extensive consulting work and numerous publications in field. Married, four children. *Society Aff:* HKN, ΟΔΚ, ΣΞ, ΤΒΠ, ΦΚΦ.

Weaver, David R
Business: 2960 E State St, Salem, OR 97310
Position: Regional Traffic Engr. *Employer:* OR DOT. *Education:* -/Engr/OR State College. *Born:* 12/30/35. Regional Traffic Engr for the OR DOT. Pres & owner of Arrow-K, Inc, of Eugene, OR. Pres of GWG & Assoc of Salem, OR. Structural Engr for G-G, Inc, Gervais, OR. Past Pres OR Sec ITE. *Society Aff:* ITE.

Weaver, Graeme D
Home: 1200 Bristol St, Bryan, TX 77801
Position: Res Engr. *Employer:* TX Trans Inst. *Education:* PhD/Civ Engg/TX A&M Univ; MS/Civ Engg/TX A&M Univ; BS/Civ Engg/Univ of TX. *Born:* 2/6/38. Dr Weaver is a res engr & mgr of the Design & Implementation Prog with the TX Trans Inst. His field of specialization is roadway design & operational safety. He has been principal investigator of more than a dozen fed & state res projs concerning hgwy safety. For an ASCE paper concerning safe roadside slope design, he was co-recipient of the 1977 Arthur M Wellington Prize. Dr Weaver, a reg PE is listed in Outstanding Americans in the South & Who's Who in TX & in the South & Southwest. He has authored more than fifty technical papers & reports on hgwy safety. *Society Aff:* TRB, TXITE, XE, ΣΞ.

Weaver, Hobart A
Business: 3300 Lexington Rd, Winston-Salem, NC 27102
Position: Dir of Engr *Employer:* Western Electric Co. *Born:* BSME from VA Tech 1950. Native of Roanoke, VA. Joined Western Elec in Winston-Salem, NC; advanced through ranks from Mfg Engr to Dev Engrg Mgr. Resp for dev of flexible circuits; creator of new organization in WE - Electronic Sys Interconnecting Products. Planned and brought into successful operation new company location for these products in Richmond, VA. Attended Williams Coll 'American Studies for Executives' 1970. ASME Edwin F Church Medal 1974. Mbr ASME, Mbr of Advisory Council, Sch of Engr; NC State Univ; Ctte of 100 VA Tech.

Weaver, James B
Home: 4773 Green Croft Rd, Sarasota, FL 33580-8230
Position: Venture Consultant, Process Industries *Employer:* Venture Services (Part Time-Sole Proprietorship) *Education:* MS/ChE/MA Inst Tech; BS/ChE/MA Inst Tech. *Born:* 2/28/23. After holding positions in the res depts of Godfrey L Cabot Inc and Olin Industries Inc joined Atlas Powder Co in 1954. Co acquired by Imperial Chem Industries 1971 and was named VP for Corp Plng and Appraisal of ICI Americas Inc. in 1973. Also assumed resp for corp res 1975-77 and served as dir of ICI US 1974-77. Retired Sept. 1980 as VP, Venture and Capital Appraisal, to enter consulting as Venture Services. Authored numerous articles on chem cost and profitability estimation including Section 25, Chemical Engr's Handbook (5th Edition). Also authored articles on use of computers in establishing equitable legislative districts & authored a book *Fair & Equal Districts.* Hobbies include jazz clarinet, classical music, tennis, and natl and world affairs. *Society Aff:* ACS, AIChE, NSPE, NML, IIE, AACE, ASEE.

Weaver, Leo
Business: Environmental Engineering Consultant, 6978 Presidio Ct, Cincinnati, OH 45244
Position: Exec Dir & Chief Engr *Employer:* OH River Valley Water Sanitation Commision *Education:* B/Civil-Sanitary/NYU *Born:* 01/04/25 Elem and HS, Clifton, NJ. Served in Combat Engr Battalion WWII. Commission US Public Health Service Engr Officer 1948. Specialized in solid waste field investigations, prog dev. Dir, Natl Water Quality Network, Rob't A. Taft Sanitary Engrg Ctr, Cincinnati 1960-66. Chief, USPHS Solid Waste Prog, 1967. Retired USPHS, 1968. Dir, Washington office APWA 1968-70. Asst Sec WPCF 1970-74. Exec Dir OH River Valley eight state compact for Water Pollution Control 1974-1987. Env. Eng'ring Consult. 1987-present. Dir at large WPCF 1979-82, Pres AAEE 1985-86. Life Mbr APWA, VP Rotary Club of Cincinnati 1981, Private Pilot Certificate, Tech Editor *Refuse Col-*

Weaver, Leo (Continued)
lection Practice 2nd Ed. APWA. Author 30 published papers. *Society Aff:* ASCE, AAEE, WPCF, APWA, AWWA, NSPE.

Weaver, Leslie F
Home: 6034 Franklin Park Road, McLean, VA 22101
Position: Partner (Retired). *Employer:* LBC&W Associates. *Education:* BS/Elect Engr/NY Univ. *Born:* Jan 1907. Native of White Plains, NY. Worked for Westinghouse & Pub Serv Elec & Gas of NJ prior to WWII. Served with Corps of Engrs and AF 1940-60, retiring as Col. Since then, partner in Lyles, Bissett, Carlisle & Wolff, Architects-Engrs-Planners. Appointed adv to Dept of Housing & Urban Dev's 'Operation Breakthrough' 1970, by Secy George Romney; also to Pres Nixon's Comm on Employment of the Handicapped 1970. Pres, Dist of Columbia SPE 1967; Nat Dir 1969-71 then VP, NSPE 1972-74. *Society Aff:* NSPE, NRA, DAV, TROA

Weaver, Lynn E
Business: Florida Inst. of Tech, 150 W. Univ Blvd, Melbourne, FL 32901
Position: Pres *Employer:* FL Inst of Tech *Education:* PhD/EE/Purdue Univ; MS/EE/So Methodist Univ; BS/EE/Univ of MO *Born:* 01/12/30 Native of St. Louis, Mo. Served in US Air Force, 1951-53. Dev. engr, McDonnell Aircraft, 1952-53. Aerophysics Engr, Corvair Corp., 1953-55. Instructor in Elec Engrg at Purdue at Univ, 1955-58. Prof and Hd of the Dept of Nuclear Engrg, Univ of AZ, 1958-1969. Associate Dean, College of Engrg, Univ of OK, 1969-1970. Dir of Office of Environmental Studies, Argonne Associated Univs, 1970-72. Dir of Sch of Nuclear Engrg and Health Physics, GA Inst of Tech, 1972-82. Dean of Engrg & Dist. Prof., Auburn Univ., 1982-87. Pres., FL Inst. of Tech., 1987- . Chrmn, Coordinating Ctte on Energy of the AAES, 1981- . *Society Aff:* ANS, ASEE, IEEE, AAES

Weaver, Robert E C
Business: Box 26518, New Orleans, LA 70186
Position: Assoc Dir *Employer:* Gulf South Res Inst *Education:* PhD/Chem Engr/Princeton; MA/Chem, Economics/Princeton; MS/Chem Engr/Tulane; BSChE/Chem Engr/Tulane *Born:* 3/30/33. Active in process simulation and control, econometric forecastng, resource mgt, real time computing, environ engrg and appl physiology. Charter Trustee of the CACHE Corp (Computer and Curriculum Aids in Chem Engg). Dean of Engrg, Univ TN 1980-83; Head Dept Chem & Mgmt Engrg, Tulane 1977-80; Pres Intl Tech Mgmt Services. Dir of the Tulane Chem Engg Practice School at twelve indus sites and govt labs. Pres of Delta Engg Assoc and exec conslt in energy and environ mgt to the Sec of the LA State Dept of Natural Resources. Co-author of the rules and regulations of the state hazardous waste mgt prog. Author of a series of Economic Atlases on LA (Elec Industry, Refining and Petrochemicls and Renewable Resources) along with over 60 engrg papers. AIChE delegate to AmericanAuto Control Council. Native of LA, married to Karen Coci, one son Robert Christian. Reader in Christian Sci Church. Charter Trustee Leelanau Schools. Glen Arbor, MI. *Society Aff:* AIChE, SPE/AIME

Weaver, Theodore
Business: 3333 Michelson Dr, Irvine, CA 92730
Position: Director of Licensing. *Employer:* Fluor Corp. *Education:* MS/Chem Engg/CA Inst of Technology; BS/Appl Chem/CA Inst of Technology. *Born:* April 1919. From 1941 through 1944 employed as Refinery Engr by Tidewater Associated Oil Co. Joined Fluor Corp in 1944. Filled various assignments as Process Engr and Proj Mgr. Became Mgr of Development in 1953. Resp for market res, engrg dev, process improvement and process licensing, primarily in petroleum and petrochemical field. Assumed present position in 1972. Have spec interest in legislation pertaining to registration of professional engrs. AIChE Alan P. Colburn Award 1956, Founders Award 1975, President 1973, Vice President 1972, Director 1968-70. *Society Aff:* LES, AIChE.

Weaver, Wayne L
Business: 137 Newbury St, Boston, MA 02116
Position: President. *Employer:* Wayne L Weaver & Assoc, Inc. *Education:* BS/Civil Engg/Univ of CO; MS/Civil Engg/Univ of IL. *Born:* June 6,1925 Boulder Colo. MS Univ of Ill, BS Univ of Colo. Structural Engr with US Bur/Rec, Denver CO 1948-53. VP Bull & Co Architect/Engr, Denver Colo, engaged in Design of commercial, industrial & institutional projects. Directed all tech & business activities for co. Co was also engaged in Real Estate Development. 1968 established and became President of Theodore/Weaver/Assoc Inc which name has subsequently been exchanged to Wayne L Weaver & Assocs, Inc (1977) as Structural Consultants and have been engaged in design of numerous building projects throughout New England. *Society Aff:* ASCE, ACI, IABSE, PCI.

Webb, Byron K
Business: 103 Bavre Hall, Clemson Univ, Clemson, SC 29631
Position: Dean & Dir, S.C. Cooperative Extension Service *Employer:* Clemson University. *Education:* PhD/Agri Engg/NC State Univ; MS/Agri Engg/Clemson Univ; BS/Agri Engg/Clemson Univ. *Born:* 2/2/34. Native of Cross Anchor, SC. Since 1955 involved in teaching & res at Clemson Univ. Recd two ASAE paper awards four patents. PPres, SC Sec of ASAE. Past res efforts primarily in fruit & vegetable mechanization & aquacultural engg. Former hd of the Agri Engg Dept at Clemson Univ. Currently Dean & Dir, Cooperative Extension Service, Clemson Univ. *Society Aff:* ASAE, ASEE.

Webb, G Arthur
Home: 4822 Rolling Hills Road, Pittsburgh, PA 15236
Position: President Emeritus. *Employer:* Carnegie Mellon Institute. *Education:* PhD/Chem Engg/Univ of Pittsburgh. *Born:* July 1910. Sr Fellow, Dir of Engrg, Dir of Adm. Assoc. Dir, President, Carnegie-Mellon Inst 1937-75, President Emeritus 1975--. Chrmn PA Air Pollution Commission 1967-71; Chrmn, Allegheny Co Air Pollution Control Advisory Com 1967- 76; Allegheny Conf Community Development 1967-; Plng Comm Whitehall Borough 1960-; Civil Serv Comm Whitehall Borough 1954-; PBd Chrmn, Industrial Health Fndtn; Engrs Soc W. PA; Univ Club, MPC Corp, Oakland Rotary Club, Bd Mbr RIDC Scientific and Res Advisory; Res Appliance Co, II-VI, Inc; APCA; Universities Research Inc; Diplomat American Association for Environmental Engineers; Fellow American Institute of Chemical Engrs; Distinguished Service Award PSPE; Honor Scroll AIC; Disting Serv Award & Award of Merit U of Pittsburgh. *Society Aff:* AIChE, ASME, APCA, ESWP, ACS, AAEE.

Webb, Jervis C
Business: Webb Dr, Farmington Hills, MI 48018
Position: President & Chairman of the Board *Employer:* Jervis B Webb Co. *Education:* BS/ME/MIT. *Born:* 05/22/15. After graduating, Mr. Webb joined his father's corporation, the Jervis B. Webb Company, as an engineer. He worked in every department of the company and was elected president in 1952. Mr. Webb is also president and chairman of the board of Spider Installations, Limited, Hamilton, Ontario, and chairman of the board of Jervis B. Webb Limited, London, England. He also is president and director of Control Engineering Company; Webb Electric Company; Webb Forging Company; Jervis B. Webb International Company; Jervis B. Webb Continental Company; Campbell, Henry and Calvin, Incorporated; Ann Arbor Computer Division; Jervis B. Webb Company of California; Jervis B. Webb Company of Georgia; and Jervis B. Webb Company of Canada, Limited. Mr. Webb also serves as a director of First Federal Savings and Loan Association, Huron forge and machine Company, and the Amer Automobile Assoc (MI). *Society Aff:* ASME, AMHS, CEMA, MHI.

Webb, John R
Business: 464 GPM South, San Antonio, TX 78216
Position: President. *Employer:* W J Enterprises Inc. *Education:* BS/ChE/MS State Univ. *Born:* July 1940. Native of Leland Miss. 1963-64 Process Engr Uniroyal Inc 1964-67 Pilot Plant Engr Copolymer Rubber. 1967,68 Asst Production Supt Baxter

Webb, John R (Continued)
Labs. 1968-70 Res Engr, Celanese. 1970-- President W J Enterprises Inc. Resp for the establishment, development and operations of five offices in Texas involved in Engrg Recruitment manpower planning and personnel development for major chem and petrochemical cos world-wide. Have chaired and served on numerous employment and personnel dev committees on a local, state, and natl level. *Society Aff:* AIChE.

Webb, Ralph L
Business: 307 Mech Engg, Univ Park, PA 16802
Position: Prof *Employer:* PA State Univ. *Education:* PhD/ME/Univ of MN; MSME/ME/Rensselaer Poly Inst; BSME/ME/KS State Univ. *Born:* 2/22/34. Dr Ralph L Webb is Prof of Mech Engrg at PA State Univ. From 1963-77 he was Mgr of heat Transfer Res for the Trane Co, La Crosse, WI. Dr Webb joined Trane after spending two yrs as an experimental engr at the Knolls Atomic Power Lab. He received his PhD from the Univ of MN and has published in the area of heat transfer augmentation, including three US patents on enhanced heat transfer surface. Dr Webb Fellow of the ASME, former Tech Editor of the ASME *Journal of Heat Transfer* and is past chrmn of the Heat Transfer Div Exec Ctte. He currently holds technical editor positions with "Heat Transfer Eng" and the J. Heat Recovery Systems. He teaches academic courses in "Heat Exchanger Design" and "Enhancement of Heat Transfer" at Penn State and is currently performing res in the areas of enhanced heat transfer, condensation and boiling. He teaches professional courses in enhanced heat transfer and heat exchanger design. *Society Aff:* AIChE, ASHRAE, ASME.

Webb, Richard C
Home: 4906 Club House Court, Boulder, CO 80301
Position: Chrmn *Employer:* Webb Engr Co & DATA RAY Corp. *Education:* PhD/EE/Purdue Univ; MS/EE/Purdue Univ; BS/EE/Univ of Denver *Born:* Sept 1915. Instructor/ Prof/ lecturer in various colleges 1939 to date - Purdue U, Iowa State Coll, U of Denver, U of Colorado. Traffic Engr Mtn Sts Tel & Tel 1937-39. Res Engr RCA Labs Princeton NJ (in Color TV) 1945-53. President, Founder & Tech Dir of: Colo Res Corp, Broomfield CO (merged ITT)1956-61, Colo Instruments Inc Broomfield (merged MDS) 1961-71. President/Gen Mgr Colo Instr/ MDS Corp 1971-73. President Webb Engr Co Boulder Colo 1973 to date. President DATA RAY Corp, Broomfield, Colo 1975 to date. Holds 28 patents and has produced 36 products and devices from 1939-75. Numerous publications. Mbr several professional associations and recipient of many awards. *Society Aff:* IEEE, SMPTE

Webb, Theodore S
Business: PO Box 748 (Mail Zone 2810), Ft Worth, TX 76101
Position: VP F-16 Programs *Employer:* Gen Dynamics. *Education:* PhD/Phys/CA Inst of Tech; BS/Phys/OK Univ. *Born:* 3/4/30. Gen Dynamics, Ft Worth Div, 1955-present. Assignments have included: Nuclear Shield Engg; Proj Engg for Missiles, Space Systems & Aircraft; Engg Dir in charge of Aero, Propulsion & Wind Tunnel Testing VP, Res & Engg., and from 1980, VP - F-16 Programs Mbr of Technical Bd & Aerospace Council, SAE; Bd, Engg Fdn, Univ of TX at Austin; Bd, SMU Fdn for Sci & Engg; Bd, Metal Properties Council. Phi Beta Kappa; Sigma Xi; Sigma Pi Sigma; Phi Eta Sigma. Married, Cuba Evans, 1952; two children, Theodore S, III (1956), Kelly Elizabeth (1959). *Society Aff:* APS, AAAS.

Webb, Thomas H
Business: 3092 Broadway, Cleveland, OH 44115
Position: Research Associate. *Employer:* Standard Oil Co (Ohio). *Education:* PhD/Organic Chemistry/Duke Univ; MS/Organic Chemistry/Duke Univ; BA/Chemistry/Univ of VA. *Born:* July 16, 1935. PhD and MA from Duke Univ, BA from Univ of Virginia. Army Chem Corps 1961-63. Proj Chemist for Sinclair Oil Corp spec in lubricant additive synthesis, motor oil formulation, & industrial lubricatn 196168. Lubricants Coordinator for BP Oil Corp 1968-70 resp for tech aspects of marketing and manufacturing lubricants. Res Assoc with The Standard Oil Co (Ohio) 1970-present resp for long range res programs in lubricants and related areas, especially metalworking fluids. Chairman of ASLE's Fluids for Metalworking Cttee 1976-77; chrmn of R&D Div IX on Oxidation of ASTM Committee D-2, mbr of ANSI Committee Z-11 on petroleum products, US delegate to ISO Tech Comm 28 on petroleum products. Mbr City Council of Solon Ohio 1976-79. Hobbies include Boy Scouts and fishing. *Society Aff:* ASLE, ASTM, ANSI, ACS.

Webb, Watt W
Business: Clark Hall, Ithaca, NY 14853
Position: Prof. *Employer:* Cornell Univ. *Education:* ScD/-/MIT. *Born:* 8/27/27. Early yrs Kansas City, MO. Union Carbide Metals Co, Niagara Falls, NY 1947- 1952, 1955-1961 met engr, res scientist, asst dir of res. To Cornell Univ in 1961 now Prof of App Phys and Dir of the Sch of Applied and Engrg Phys. Diverse res interests in condensed matter phys, critical phenomena, fluids & biophys. Consulting, advisory committees, various editorial bds, Guggenheim Fellow. Persistent interest in philosophy & techniques of applying sci. Yachtsman. *Society Aff:* APS, ASCB, AAAS.

Weber, Clifford E
Business: Div Reactor Res and Technology, International Nuclear Programs Div, Washington, DC 20545
Position: International Technical Specialist *Employer:* US Dept of Energy. *Education:* PhD/Phys Chem/Johns Hopkins Univ; BS/Phys Chem/Univ of CA at Berkeley. *Born:* 5/20/18. Native of Fresno, CA. R&D at Johns Hopkins for Manhattan Dist 1942-46; with Knolls Atomic Pwr Lab, Genl Elec 1946-61 spec in nuclear fuel dev. Program mgt at Atomics Intl 1961-68. Joined Atomic Energy Comm 1968. Responsible for planning and analysis of foreign nuclear activities and coordination of international cooperative activities. Fellow ASM 1972. Enjoy golf and skiing. *Society Aff:* ASM, NYAcad of Sci.

Weber, Erich C
Home: 617 Creekmore Court, Walnut Creek, CA 94598
Position: Engr Mgr *Employer:* Bechtel Petroleum Inc *Education:* BME/Mech Engrg/Cornell Univ *Born:* 7/16/27. Grad of St Johns-Ravenscourt Sch, Winnipeg, Canada, and Kemper Military Sch, Boonville, MO. Served in Corps of Engrs, US Army, Alaska. With Bechtel since 1960, and has been resp for proj mgmt and engrg design for refinery facilities, tarsands and oilshale facilities petrochemical & ammonia plants, steam generation facilities including combined cycle gas turbine pwr recovery sys, for plants built in US, Canada, UK & the Far East. Past VP of ASME, Pacific Reg IX 1976 and mbr of Natl Nominating Ctte. Reg PE in CA, Wash and province of Alberta, Canada. Author of an ASME paper on cold weather plant design. Past Mbr of ASME Bd of Governors. Fellow, ASME. *Society Aff:* ASME

Weber, Ernst
Home: P O Box 1619, Tryon, NC 28782
Position: Pres Emeritus - Retired. *Employer:* Polytechnic Univ *Education:* ScD/Elec Engg/Techn Univ, Vienna; PhD/Physics/Univ of Vienna; Dipl Ing/Elec Engg/Techn Univ Vienna. *Born:* Sept 6,1901. Technical Univ. Vienna; Austrian Siemens Co 1924-29, Siemens-Schuckert Co Berlin 1929-30, Polytechnic Inst of Bklyn successively 1930-58, Visiting Prof, Prof, Grad El. Eng, Dept Hd; Pres 1958-69. Chrmn Div of Engrg NRC 1969-74; Mbr NAS 1965 Founding Mbr NAE 1964; Defense Science Bd 1963- 66; Army Scientific Adv Panel 1957-69; Pres IEEE 1963; Pres ECPD 1968-70; Presidential Certificate of Merit 1948; Education Medal AIEE 1960; H Coonley Medal ASA 1966, Founders Award, IEEE 1970; Hon Mbr IEE of Japan 1963; Hon D. Eng U of Mich 1964; Poly Inst of Bklyn 1969. *Society Aff:* NAS, NAE, IEEE, ASEE, APS.

Weber, Homer S
Home: 2788 Peachtree Rd NW, A-4, Atlanta, GA 30305
Position: Prof & Dir Emeritus. *Employer:* Georgia Inst of Tech. *Education:* PhD/Engg Mech/Stanford; MS/GA Tech; BSME/GA Tech. *Born:* Feb 24,1898. Instructor to Prof in the Dept of Engrg Drawing and Mechanics 1924-46; Prof and

Weber, Homer S (Continued)
Dir, Sch of Mech Engrg 1946-62; Prof and Dir Emeritus since 1962. Mbr ASME since 1924; Chrmn of the Atlanta Sec, 1958-59; Fellow 1966. Hon fraternities: Phi Kappi Phi, Tau Beta Pi, Pi Tau Sigma. PE in Georgia. *Society Aff:* ASME, NSPE, ASEE.

Weber, Howard H, Sr
Home: 323 Leeward Lane, Fort Pierce, FL 33450
Position: Pres - Retired. *Employer:* Howard Weber Enterprises. *Born:* March 1896 Allentown PA. York Highschool, Lehigh Univ BS, Msc. American Expeditionary Force, US Army WWI 1917. Chief Inspector Jamestown NY 1919. Asst EE, Utica G & E Co 1920. Chf Engr Wire & Cable Dept Rome Wire Co 1924. Chf Engr Low Tension Cable Dept, General Cable Corp 1927. Chf Engr & Gen Sales Mgr US Rubber Co Wire/Cable Dept 1940. Mktg Mgr, Electrical Products Div, Kaiser Aluminum & Chemical Corp 1957. President, Howard Weber Enterprises (Consultants & Mgt) 1961. Fellow & Life Mbr IEEE. Served on many Industry Cttees, such as NEMA, NFPA, ASTM, IPCEA, AIEE, IAEI, IMSA. Served WPB WWII. Hobbies: golf, fishing, yachting. Retired.

Weber, James H
Business: Dept Chem Engr, Lincoln, NE 68588
Position: Regents Prof of Chem Eng. *Employer:* University of Nebraska-Lincoln. *Education:* BS/Chem Eng/Univ of Pittsburgh; MS/Chem Eng/Univ of Pittsburgh; PhD/Chem Eng/Univ of Pittsburgh. *Born:* Nov 21,1919 Pittsburgh PA. US Army 1941-46 & 1952, 1st Lt. Univ of Neb 1948--. Instructor to Regents Prof, including Departmental Chrmn 1958-71. Consultant to Phillips Petroleum Co 1952-79. Past Consultant to Natl Gas Processors Assn and C F Braun & Co. Author and co-author of 70 scientific and tech articles which have appeared in domestic and foreign journals. Mbr AIChE (Fellow), ACS, SHOT. *Society Aff:* AIChE, ACS, SHOT.

Weber, Joseph L
Business: P O Box 1160, Trenton, NJ 08610
Position: Exec VP/Sec. *Employer:* Warren-Balderston Co. *Born:* March 1923. Ind. Eng Degree Drexal University, Philadelphia, PA. Born in Trenton, New Jersey. National Director of SME - third two-year term.

Weber, Marvin A
Business: 715 E Mission Drive, San Gabriel, CA 91776
Position: Vice Pres, Health Facil. *Employer:* Store, Matakovich & Wolfberg. *Born:* April 1925. BSME, Tri-State University, Angola, Indiana. Native of St. Louis, MO. USN, Naval Aviation Sec 1943-46. 7 years experience as Air Conditioning Application Engr, Refrigeration Div, Curtis Mfg Co St Louis MO. With Store, Matakovich & Wolfberg since 1961. Became Mgr Health Facil Design 1967 and V.P. 1975. Overall mgt resp for all Health Facility Projects. Mbr ASHRAE and NSPE. Specific emphasis on Environmental Designs for critical areas of hospital. Presented tech paper on 'Laminar Flow in the Hospital Surgical Suite' at ASHRAE Convention, Dallas TX Feb 1976. Enjoy travel for pleasure.

Weber, Norman
Home: 1041 E Porter Ave, Naperville, IL 60540
Position: Assoc *Employer:* Sargent & Lundy *Education:* PhD/Mech Engrg/Montana State Univ; MS/Mech Engrg/Univ of So CA; BS/Mech Engrg/CA State Poly Univ (SLO) *Born:* 11/25/34 Native of San Luis Obispo, CA. Worked for the Rocketdyne Div of Rockwell Intl 1957-1968. specializing in applied heat transfer and associated res. With Westinghouse Hanford Co 1971-1974. Joined Sargent & Lundy in 1974 in its thermal-hydraulics analysis section. Appointed Asst Hd of the Nuclear Safeguards and Licensing Div in 1981. Appointed Assoc in 1985. Chrmn, Chicago Sect, Amer Nuclear Soc 1984-1985. Enjoys golf, skiing and woodworking. *Society Aff:* ANS, ASME, WSE, ASES, NYAS

Weber, Richard G
Home: 166 Dannell Dr, Stamford, CT 06905
Position: Manager of Process Engineering *Employer:* Pitney-Bowes Inc. *Education:* DSc/Physical Met/Inst Natl Poly de Lorraine; MS/Met/Univ of CT; BS/ME/Univ of CT. *Born:* 9/4/43. Native of Stamford, CT. After college worked as a metallurgist at Sikorsky Aircraft until receiving MS in 1969. Then spent one yr at the Univ of CT investigating high temperature metal fatigue. Travelled to France in 1970 to complete doctoral res at Ecole des Mines de Nancy and Centre des Materiaux de l'Ecole des Mines de Paris. After completing studies returned to the US in 1973 and joined Machlett Labs as the co sr metallurgist. In 1979 joined Pitney-Bowes and works as manager of process Engineering, served from 1976-1980 on the Stamford Bd of Education Actively involved with Bridgeport Engg Inst as Chrmn of the ME Dept. In 1974 received first prize in Color Class at the Intl Metallographic Exhibit sponsored by IMS and ASM. Enjoy carpentry, fishing and tennis. *Society Aff:* ASM, SME, APMI.

Weber, Walter J, Jr
Business: Water Res Eng Bldg 1-A, Ann Arbor, MI 48109
Position: The Earnest Boyce Distinguished Professor of Engineering *Employer:* The University of Michigan. *Education:* PhD/Water Resources Engg/Harvard Univ; AM/Environ Chem/Harvard Univ; MSE/Civil Engg/Rutgers Univ; ScB/Engg/Brown Univ. *Born:* 6/16/34. Began career with Caterpillar Tractor Co 1956-57. Taught at Rutgers Univ & served with Soil Conservation Serv 1957-59. Joined the Univ of MI as Asst Prof in 1963; advanced to Assoc Prof in 1965, to Prof in 1968 named Distinguished Professor in 1978; and named the Earnest Boyce Distinguished Professor of Engineering in 1987; Dir of the Univs Prog in Water Resources since 1969. Reg PE and active consultant to ind & govt. Pioneered dev & application of physicochemical treatment processes. Author of one text book and co-author of two other ref books; author of over 300 tech pub in the environ & water res field. Tau Beta Pi, Chi Epsilon, Delta Omega, Sigma Xi; Diplomate AAEE; Cert of Merit 1962 ACS; Dist Faculty Award 1967 Univ of MI; Dist Faculty Award 1968 AEEP, VP 1970-71, Dir 1970-72 AEEP; John R Rumsey Meml Award 1975 WPCF; Dis. Prof 1978 Univ of MI; AEEP-Nalco Award for Chem Res 1979; Rudolph Hering Medal 1980 ASCE; Research Excellence Award 1980 Univ. of Michigan; Willard F. Shephard Award 1980 WPCF; Academic Achievement Award 1981 AWWA; Thomas R. Camp Award 1982 ASCE; F.J. Zimmerman Award 1982 ACS; Simon W. Freese Award 1984 ASCE; AEEP-ESI Research Award, 1984; G. Brooks Earnest Award, ASCE, 1985; Elected to the U.S. National Academy of Engrg, 1985. *Society Aff:* ACS, AIChE, ASCE, AWWA, WPCF, UCWR, AEEP, AAEE, ASEE.

Webster, A Russell
Business: P O Box 4867 Rumford Br, Providence, RI 02916
Position: President. *Employer:* Webster Associates Inc. *Education:* BSc/Civil Engrg/Brown Univ; -/Ind Eng/URI; -/Aer Eng/UMA *Born:* 11/01/23 Graduate Brown Univ. Attended URI & UMA. Providence RI native. Laborer & timekeeper in heavy construction prior to WWII. Combat pilot 1943-46 US Army Air Corps. Field Eng Thompson & Lichtner 1949-58. VP Materials Div, M A Gammino Const Co 1958-72. Chf Eng & Prod Mgr Simeone Corp 1972-76. President Webster Associates Inc Consulting Engrs 1976. Specializing in design and production of mineral aggregate construction materials and design of related manufacturing facilities and environmental controls. Participated in research and design of skid resistant pavement, stabilization of bases. President RISPE 1975-76, Outstanding Service Awd RISPE/NSPE 1970. Hobbies - skiing, swimming, photography. Reg PE, Reg Land Surveyor. Engr-of-the-Year Award, RISPE 1983. Dir- NCSA. *Society Aff:* NSPS, RISPE, SME-AIME, NSA, ACI, ACI-NE, NSTM.

Webster, John G
Business: Dept Elec/Cptr Engrg, Madison, WI 53706
Position: Prof of Elec & Computer Engrg. *Employer:* University of Wisconsin-Madison. *Education:* PhD/Elec Engg/Univ of Rochester; MSEE/Elec Engg/Univ of Rochester; BSEE/Elec Engg/Cornell Univ. *Born:* May 1932. Prof of Elec & Computer Engrg at the Univ of Wisc-Madison. Teaching and research in medical instrumentation, specializing in tactile sensors and bioimpedance techniques. Fulbright Fellow 1953; NIH Predoctoral Fellow 1963; ASEE-NASA Summer Faculty Fellow 1970; NIH Res Career Dev Awd 1971; ISA Donald P Eckman Edu Awd 1974. ASEE Western Elec Fund Award 1978. ISA Fellow 1979. IEEE Fellow 1986. Coauthor of *Medicine and Clinical Engineering.* Editor of *Medical Instrumentation: Application and Design.* Co-editor of *Electronic Devices for Rehabilitation.* Co-editor of *Clinical Engrg: Princs & Practices.* Co-editor of *Design of Microcomputer Based Medical Instrumentation.* Co-editor of *Therapeutic Medical Devices: Application & Design.* Co-editor of *Interfacing Sensors to the IBM PC.* *Society Aff:* IEEE, ISA, ASEE, AAMI, BES.

Webster, Karl S
Business: Rm 218, Boardman Hall, Orono, ME 04469
Position: Prof Mech Engr Tech *Employer:* Univ of ME *Education:* MS/ME/Penn State Univ; BS/ME/VT *Born:* 08/18/24 Native of Orleans, Vt. Served in U S Army 1942-46. GE Co. Application engr 1949-52, Fellows Gear Shaper Co 1953. Instructor of NH 1958-62. Associate Prof NH Tech Inst 1962-65. Assoc Prof and Prof Univ of ME 1965. Asst to the VP for Academic Affairs 1970-72. Reg Prof Engr. Publications primarily in pollution control.Principal interest in mechanical design, production, technical consulting. Hobbies: sailing and general aviation. *Society Aff:* ASEE, SME

Webster, Neil W
Business: 555 Plymouth Ave NE, Grand Rapids, MI 49508
Position: Mgr Development Engr *Employer:* LSI Rapistan *Education:* MS/Agr Engg/Univ of MA; BS/Agri Engg/Univ of ME. *Born:* 6/13/38. in Randolph, VT. 4 yrs Gen Elec Engg Trng Prog 1965-70 Sperry-New Holland as a Res Engr. Analyzed future dev areas. 1970-77, Chief Engr of Neapco, resp for product design, testing, engg customer liaison, and product reliability. 1977-1983, VP Engg, Sweet Mfg Co - complete line of grain conveying equipment. 1984-present Mgr of Dev Engg, LSI Rapistan Manuf of unit conveyors. Mbr ASAE, 1963 Chrmn PA sec, Chrmn NAR Prog Ctte. Chrmn Subctte EPP-34. Reg PE, PA 1968, OH 1977. *Society Aff:* ASAE.

Wechsler, Monroe S
Business: 261 Sweeney Hall, Iowa State Univ, Ames, IA 50011
Position: Professor. *Employer:* Iowa State University. *Education:* PhD/Physics/Columbia Univ; AM/Physics/Columbia Univ; BS/Physics/CCNY. *Born:* 5/1/23. Signal Officer, Signal Corps, US Army 1944-47. Elec Engg US Naval Air Magnetics Lab 1947-48. Physicist, Oak Ridge Natl Lab 1954-69 (Hd, Radiation Metallurgy Sec, Solid State Div, 1960-69). Prof, Metallurgy, Univ of TN 1965-69. Chrmn Dept of Metallurgy, and Chief, Metallurgy Div, Ames Lab USAEC, IA State Univ 1970-75. Prof, Dept of Matls Sci & Engg 1970- and Dept of Nuclear Engg 1978- IA State Univ. Fellow APS, ASM, IA Acad Sci. AIME (Chrmn, Nuclear Metallurgy Comm 1975-76). ASM, ANS. Hon mbr, Alpha Sigma Mu. Research interests: Radiation effects on metals, nuclear materials, phase tranformations, shape-memory-alloy heat engines. *Society Aff:* TMS-AIME, ASM, APS, ANS, IAS.

Wedd, Ralph W
Home: 1257 Witham Dr, Dunwoody, GA 30338
Position: Retired *Education:* BS/ChE/KS State Univ *Born:* 12/16/22 Native of Oak Hill, KS. Served with USN 1944-46. With Goodyear Tire and Rubber Co in Compound Dev 1947-49, with Swan Rubber Co as Asst Chemist 1949-51. With Akron Chem Co from 1951. Presently VP and Southern Branch Mgr, also mbr of Bd of Dirs Revlis Corp. Hobby - sailing. *Society Aff:* SRG, ACS, Rubber Div of ACS

Weeks, Harold B
Home: 298 Rood Ave, Windsor, CT 06095
Position: Mgmt Training Consultant *Education:* BS/Bus/Univ of NH. *Born:* 10/17/20. Native of Windsor, CT. PPres of CT Valley Chap ASEE, PReg XI. Affiliated with Boston Chap ASEE - RI Safety Soc - ME Safety Soc, NH Accident Prevention Council. Serves on Safety Cttee for CT Businesses & Industries Assn & the First- Aid & Accident Prevention Committee, Greater Hartford Ch, Amer Rd Cross. Also Safety Ctte Ct Soc for Prevention of Blindness. Reg instructor for NSC DDC in US & Canada, ARC. Multimedia First Aid & ASSE Applied Sys Safety Techniques. At present Private Consultant. *Society Aff:* ASSE, CSP.

Weeks, W F
Business: Geophysical Institute, Univ. of Alaska, Fairbanks, AK 99775-0800
Position: Glaciologist *Employer:* Geophysical Inst/UAF *Education:* PhD/Geochem/Univ of Chicago; MS/Geol/Univ of IL; BS/Geol/Univ of IL *Born:* 01/08/29 Specialist on engrg physics and geophysics of sea and lake ice. Worked on problems in applied glaciology since 1955. Extensive field experience in both Arctic and Antarctic. Current interest is in sea ice properties as they influence offshore operations and design. Glaciologist, Air Force Cambridge Res Center 1955-57; Asst Prof of Earth Sciences, WA Univ (St. Louis) 1957-62; Adjunct Prof. Dartmouth College 1962-85; Res Glaciologist USACRREL 1962-pres; Prof. of Geophysics and Project Scientist, Alaskan Syntheic Radar Facility, Geophysical Institute, Univ of Alaska, 1986-present. Visiting Prof Hokkaido Univ 1973; ONR Prof Chair of Arctic Marine Science 1978-79; Pres Interntl Glaciological Society 1972-75; National Academy of Engrg (elected 1979). Musician (contrabass). *Society Aff:* Int. Glaciol. Soc, Amer Geophys Un, AINA

Weeks, William G
Business: 350 Grove St, PO Box 6835, Bridgewater, NJ 08807
Position: President. *Employer:* H V Weeks Inc. *Education:* BS/EE/NJIT. *Born:* Dec 1920. Reg PE. Native of New Jersey. Army Corps of Engrs 1943-46. Engrg departments of major corporations 1947-56. Taught evening engrg classes NCE (NJIT). With H V Weeks Inc Consulting engrs since 1956; President since 1965. Resp for overall mgt of engrg co. VP of S M Electric Co Inc 1957-75. Mbr NJ Chap NECA; State Pres 1962-63; Natl Governor 1965-66. Commissioner New Jersey Bd of Electrical Examiners 1962-70. Mbr Somerset Raritan Valley Sewerage Authority 1972-1978; 1980-1984; Chairman 1984. Active in community politics. *Society Aff:* NSPE, ACEC, IEEE.

Weertman, Johannes
Business: Materials Sci & Engrg Dept, Northwestern Univ, Evanston, IL 60201
Position: W.P.Murphy Prof of Mater Sci&Eng. *Employer:* Northwestern University. *Education:* DSc/Physics/Carnegie-Mellon; BS/Physics/Carnegie-Mellon. *Born:* May 1925. PostDoc at Ecole Normale Superieure, Paris. Native of Beaver, PA. At Northwestern Univ since 1959 in Materials Science and Engrg Dept and in Geological Sciences Dept. Consultant Los Alamos Scientific Lab since 1967. Visiting Prof at Cal Tech in 1964, at Scott Polar Res Inst of Cambridge Univ 1970-71, and at Swiss Federal Reactor Research Institute in 1986. Formerly Scientific Liaison Officer ONR American Embassy, London and Solid State Physicist at U.S.Naval Res Lab, USMC 1943-46. Fields are creep & fatigue of metals, dislocation theory, and glaciology and geophysics. NAE 1976, Fulbright Fellowship 1950-51, Guggenheim Fellowship 1970-71, Horton Award AGU 1962. Mathewson Gold Medal AIME 1977, Acta Metallurgica Gold Medal 1980. Seligman Crystal of International Glaciological Soc 1983. Weertman Island in Antarctica named after, 1960 by British Antarctic Place Names Committee. Enjoy backpacking and long distance running. *Society Aff:* AIME, ASM, APS, GSA, AGU, IGS, NAE.

Weese, John A
Business: Engineering Technology Dept, Texas A & M University, College Station, TX 77843-3367
Position: Head of Engrg Tech Dept & Prof of Mechanical Engrg *Employer:* Texas A & M University *Education:* PhD/Engr Mechanics/Cornell Univ; MS/Engr Mechanics/Cornell Univ; BS/Mech Engr/KS State Univ. *Born:* 7/24/33. Grad Training Engr-Allis Chalmers 1955. Instructor-Engg Mechanics-Cornell 1955-57. Structural Dynamics Engr Boeing Wichita 1959-60 and 1962-63. Asst Prof of Mechanics and 1st Lt USAF Academy 1960-62. Structural Dynamics Engr Martin Marietta Denver

Weese, John A (Continued)
summer 1963. Denver Univ 1963-74: Res Engr Denver Res Inst 1963- 70; Assoc Prof Mech Engg 1963-67; Prof Mech Engg 1967-74; Grad Chrmn Mech Sciences & Environmental Engg 1968-70; Dean College of Engg 1970-74. 1974-1983 Dean School of Engg and Prof Mech Engg & Mechanics Old Dominion Univ Engg Mechanics Texts Coauthor with Higdon & Stiles. 1983-86 Dir of the Mech Struct & Materials Engr Div of the Natl Sci Fdn. 1986-present, Head of Engr Technology Dept, Head Engr Tech Div Texas Engr Exp. Sta. and Prof of Mech Engr, Texas A & M Univ. Prof Service - serve as a mbr of the engrg Accreditation Comm of the Accredition Bd for Engrg and Tech (ABET) for ASME. *Society Aff:* AIAA, ASME, NSPE, ASEE, SEM.

Wegman, Leonard S
Business: 330 W 42nd St, New York, NY 10036
Position: Chrmn. *Employer:* Leonard S Wegman Co., Inc. *Education:* BS/Civ Engg/CCNY. *Born:* 3/7/11. City of NY 1935-41; Army Engrs 1942-45; Leonard S Wegman Co Inc (Chrmn) 1945 to date. Consulting Engg Services to Fed, State & Local Govt & to industry. Hgwy, Bridges, Water & Waste Water Systems. Innovative Concepts in Transp, Waste Mgt, Resource Recovery & Energy Conservation. Honor Awards: Engg Recovery & Energy Conversion, Engg News Record 1972- Soc of Am Military Engrs, 1979, Mayor of Kings Point, NY 1974- . *Society Aff:* ASCE, NSPE, SAME, WPCF, AAEE.

Wegman, Raymond F
Home: 34 Mt Arlington Rd, Ledgewood, NJ 07852
Position: Consultant *Employer:* T/A Adhesion Associates. *Education:* BS/Chemistry/Seton Hall Univ. *Born:* 9/22/28. Native of NJ. Retired after 30 years with the U.S. Army Reg engr, Calif. Past Mbr of the Bd of Directors Soc for the Advancement of Material and Process Engrg. PChrmn NY Chap SAMPE. P Chrmn NJ Chap SAMPE. Inventor, patented process for treating titanium alloy prior to bonding; Patent for outdoor fatigue tester; Patent for method of surface perparation of aluminum for bonding. Mbr Sigma Xi. Mbr ASTM E6.23, D14. As of Jan 1984 running own co, Adhesion Assocs, a consltg co at 34 Mt Arlington Rd, Ledgewood NJ 07852. *Society Aff:* SAMPE, ASTM

Wehausen, John V
Business: Naval Arch and Offshore Engrg Dept, Berkeley, CA 94720
Position: Prof of Engrg Sci *Employer:* Univ of Calif Berkeley *Education:* PhD/Math/Univ of MI; MS/Physics/Univ of MI; BS/Math/Univ of MI *Born:* 09/23/13 Born in Duluth, Minn. Elem educ in Oak Park, Ill. Univ of Mich 1931-1937. Instructor in Math. Brown Univ 1937-38, Columbia Univ 1938-40, Univ of Missouri 1940-44. Consultant, Operations Evaluation Group, USN 1944-46. Mathematician, David Taylor Model Basin, USN, 1946-49. Acting Head, Mechanics Branch, ONR, 1949-50. Exec Editor, Mathematical Reviews, 1950-56. Dept of Naval Arch and Offshore Engrg, Univ of Calif Berkeley, 1956-date. Visiting Prof, Universities of Hamburg, Nantes, Grenoble, Flinders Univ., the Technion (Haifa), Chalmers Univ of Tech (Gothenburg). Wife: Mary Katherine Wertime. Children: Sarah, Peter, Julia, John. *Society Aff:* SNAME, AAAS, AMS, MAA, NAE

Wehner, Gottfried K
Business: E E Dept, Minneapolis, MN 55455
Position: Professor. *Employer:* University of Minnesota. *Born:* 9/23/10. Dr. Wehner received his doctorate from the Technical University in Munich, Germany in 1939. He became the second recipient of the Welch Award of the American Vacuum Society for outstanding scientific achievements 1971. He serves on the Board of Directors of the AVS, is a Fellow of the APS, IEEE, and a honorary mbr of the American Vacuum Society, and holds memberships in the German and European Physical Societies. His major contributions to science are concerned with plasma oscillations, sputtering by ion bombardment, and surface physics. *Society Aff:* APS, AVS, IEEE.

Wehring, Bernard W
Business: Dept. of Nuclear Engrg, Box 7909, Raleigh, NC 27695
Position: Prof/Director *Employer:* North Carolina State University *Education:* PhD/Nuclear Engg/Univ of IL; MS/Phys/Univ of IL; BSE/Engg Phys/Univ of MI; BSE/Engg Math/Univ of MI. *Born:* 8/3/37. Professor of Nuclear Engrg and Dir of the Nuclear Reactor Program at North Carolina State Univ. Was on the Faculty of the Nuclear Engrg Program at the Univ of Illinois at Urbana-Champaign for 18 years before moving to N.C. State, May of 1984. Also has worked at Babcock and Wilcox Co, Atomic Energy Div (summer), Autonetics, a Div of North Am Aviation (summers), Construction Engg Res Lab (conslt), Oak Ridge Natl Lab (sabbatical), Argonne Natl Lab (conslt), and Los Alamos Scientific Lab (conslt). Expertise involves nuclear instrumentation related to nuclear radiation spectroscopy, and also nuclear data related to nuclear radiation transport and nuclear fission; authored or coauthored over 90 publications. *Society Aff:* ANS, APS, ASEE, IEEE.

Wei, James
Business: M.I.T, Chemical Engineering, Cambridge, MA 02139
Position: Professor of Chem Engrg. *Employer:* MA Inst of Technology. *Education:* BChE/Chem Eng/GA Tech; ScD/Chem Eng/MIT. *Born:* 8/14/30. Advanced Mgt Prof 1959 Harvard Bus Sch. Res chem engrg to sr scientist, Mobil Oil Res Dept 1955-68; Mgr of Long Range Analysis, Mobil Oil Corp 1969-70. Allan P Colburn Prof of Chem Eng, Univ of DE 1971-77. Warren K Lewis, Prof & Hd of Chem Engrg Dept, MIT 1977- . Visiting Prof, Princeton Univ 1962-63; visiting prof CA Inst of Tech 1965. ACS Award in petroleum chemistry 1966. AIChE, Inst Lecturer 1968, Professional Progress Award 1970. William H. Walker Award 1980. AIChE Dir 1970-72; Chrmn of awards cttee 1976- , Coordinator of World Congress on Chem Engrg, Amsterdam 1976. Consulting Editor, McGraw Hill series in Chem Engrg 1964- . Editorial Bd of "Chemical Technology–, "Chemical Engineering Communications–, "Industrial and Engineering Chemistry Reviews–. Chrmn of Catalyst Panel, Cttee on Motor Vehicle Emissions, NAS 1972-74. VChmn of Enviorn. Editor in Chief "Advances in Chemical Engineering–. National Academy of Engineering, member 1978, Membership Committee, Nomination Committee. *Society Aff:* AIChE, ACS.

Weidlinger, Paul
Business: 110 E 59th St, New York, NY 10022
Position: Partner. *Employer:* Weidlinger Associates. *Education:* BS/Eng/Tech Inst Brno; MS/Engg/Polytechnic Inst Zurich. *Born:* Dec 22,1914 Budapest Hungary. Chf Engr, Bur/Rec, Bolivia and Prof of Structural Engrg, Saint Andrew Univ, La Paz, Bolivia to 1943. Chf Engr, Atlas Aircraft Products, NYC to 1946, and later Div Dir, National Housing Agency. Since 1949 Senior Partner, Weidlinger Associates, Consulting Engrs, with offices in NY, Boston & San Francisco. Visiting Lecturer Harvard, MIT & Yale. On Scientific Advisory Bd, USAF, 1957 and Chrmn, NAS Panel for N.B.S. ASCE J James Croes Medal 1963; Engrg News-Record Award 1966; ASCE Moisseiff Award 1975. Designer of many award-winning projects and author of numerous tech papers and articles and of book on aluminum design. Reg PE in 16 states. Fellow ASCE; Mbr ACI, IABSE, NY Academy of Sciences, AIAA. *Society Aff:* ASCE, ACI, NYASC, AIAA, EERI.

Weidmann, Victor H
Business: 201 East County Road 66E, Fort Collins, CO 80524
Position: Principal *Employer:* Weidmann Engineering *Education:* BSCE/Civil Engg/Northeastern Univ. *Born:* March 1926 New York. Served in all theaters of operation in WWII with US Coast Guard. BSCE Northeastern University. Mbr of several prof and tech societies-have held local, state and national offices- Fellow ACEC and ASCE. Left steel industry in 1956 for private practice. Was branch office mgr and principal in consulting firm prior to opening own practice in Indiana. Served on community improvement cttees. Moved to Colorado 1973. Have and continue to practice in all aspects of civil engrg. Dipl, Amer Acad of Environ Engrs. Hobbies include the American Quarter Horse, and farming. *Society Aff:* ACEC, ASCE, AAEE,

Weidmann, Victor H (Continued)
NSPE, AWWA, WPCF, AREA, American Arbitration Association, Institute of Transportation Engineers.

Weigel, Philip
Business: 1835 Dueber Ave, SW, Canton, OH 44706
Position: Dir - Met. *Employer:* Timken Co. *Education:* MS/Met/MA Inst Tech; BS/Met Engg/Lehigh Univ. *Born:* 5/22/33. Native of Knoxville, TN. Served in Army Ordnance Corps. Joined the Timken Co in 1955 & became Dir - Met in 1977. Am concerned with met aspects of mfg tapered roller bearings & alloy steel. Enjoy a variety of outdoor sports. *Society Aff:* ASME, ASM.

Weil, Rolf
Business: Castle Point Station, Hoboken, NJ 07030
Position: Prof. *Employer:* Stevens Inst of Tech. *Education:* PhD/Metallurgy/PA State Univ; MS/Metall Engg/Carnegie-Mellon Univ; BS/Metall Engg/Carnegie-Mellon Univ; M Eng (Hon)/Metall Engg/Stevens Inst of Tech *Born:* 8/5/26. in Germany. Emigrated to US in 1940. Employed as Metallurgist by American Metal Co 1946-1948, Argonne Natl lab 1951-1954. Served US Army 1954-1956. Joined Stevens Inst of Tech as Asst Prof 1956 currently Prof of Materials and Metallurgical Engg since 1967. Consultant to private industry and govt organizations. Prin areas of interest: electrodeposition, corrosion, electron microscopy. Past Chrmn, Electrodeposition Div of Electrochemical Soc; Fellow, Inst of Metal Finishing. *Society Aff:* TMS-AIME, IMF, AESF, ASTM, ES, ΤΒΠ, ΣΞ.

Weiler, Daniel John
Home: 7402 Beverly Manor Dr, Annandale, VA 22003
Position: Vice President. *Employer:* JJH, Inc. *Education:* BSCE/Civil Engg/Marquette Univ. *Born:* July 1930. Native of Milwaukee Wisc. with Bur/Ships 1952-55, served as ILT San Eng with US Army, Plant Eng with A O Smith Corp 1957-58, San Eng with Dist Pub Wks Off US Navy 9N. D. Great Lakes IL. Held these positions, Naval Arch, H.D. of Ship Arrangements, Dir of Naval Architecture Naval Sea Systems Command 1958-1985, Chairman NATO Subgroup "Influence of Human Factors on Ship Design" 1979 to 1985. 1985 to present Vice President, JJH, Inc., Naval Arch & Marine Eng's. Mbr ASNE, SAWE, SNAME, Naval Inst, Navy League of US, ASE. ASE Pres 1967, Treas 1966, Mbrship Chrmn 1965. ASNE Mbr of Council 1976-78 & 1984-86 Cttee Mbr ASNE Day 1971-present, ASNE Exhibit Committee Chair 1974 to present. Chrmn 1969 DC Council of Engrs and Arch Soc. SNAME, Membership Chairman Chesapeake Sect., Member Executive Committee 1985 to present, Vice Chairman elected 1987 to 1988. Elected Fellow 1985. SAWE served on Bd/Dir. Elected Fellow 1987. Author of papers for ASE, ASNE, SNAME, SAWE. Awards ASE Outstanding Serv Awd 1965 and 1969. ASE Prof Achievement Awd 1975, ASE Silver medal Award 1984, Senior Service Awd 1986, Navy's Civilian Superior Serv Awd 1973. Bicentennial E & A Awd from DC Council and ASNE President's Awd 1976, Frank Law Award 1984, National Capital Award 1981 Engineer of Year, NAVSEA Federal Engineer of the Year 1984. Enjoys fishing, music, gardening. *Society Aff:* ASNE, SNAME, SAWE, ASE, Navy League, Naval Institute.

Weill, Jacky
Home: 41, rue du Val d'Or, Saint-Cloud, France 92210
Position: Chief; Dept Indust Devel. *Employer:* French Atomic Energy Commission *Education:* Engr/Elec Engrg/Inst Polytechnique de Grenoble (France); Dr-Engrg/-/Faculte des Sciences de Paris (France) *Born:* 07/27/24 Native of Strasbourg (France). At the French Atomic Energy Commission since 1947. Chief of Electronic and Nuclear Instrumentation Department (300 people) at the Nuclear Res Center of Saclay from January 81 to January 87. Is also involved in standardization. Now, Chrmn of Ctte n 45 "Nuclear Instrumentation" of the Internatl Electrotechnical Commission. Fellow of the IEEE since 1981 "For contributions asa res mgr and leader in the field of electronic control and protection systems for nuclear reactors–. At the present time, Director in the IRDI (Institute de Recherche Technologique et de Developpement Industriel) in charge of Industrial Development. *Society Aff:* IEEE, SEE, SFEN

Weimer, Frank C
Business: Electrical Engineering Dept, 2015 Neil Ave, Columbus, OH 43210
Position: Prof Emeritus *Employer:* Ohio State Univ *Education:* PhD/EE/OH St Univ; MS/EE/OH St Univ; BS/EE/OH Univ *Born:* 07/27/17 Native of Dayton, OH. Test engr with Delco Products Div of Gen Motors Corp 1941. Joined Electrical Engrg faculty of OH State Univ in 1941; Prof and Vice Chrmn of the Electrical Engrg Dept 1968-1983. Research Assoc at MIT, 1946 and at Univ of IL, 1951. Conslt for Battelle Memorial Inst, 1958-1970. Registered PE in OH. Natl Press of EtaKapNu 1972-1973. Enjoys travel and photography. *Society Aff:* IEEE, NSPE, AAAS, ASEE, HKN, ΤΒΠ.

Weimer, Paul K
112 Random Rd, Princeton, NJ 08540
Position: Fellow Tech Staff. Retired. *Employer:* RCA Laboratories. *Education:* PhD/Phsyics/OH State Univ; MS/Physics/Univ of KS; AB/Physics/Manchester College. *Born:* Nov 5,1914. 1942-81 (Retired 1981) research at RCA Labs. Awards include a Television Broadcaster's Award in 1946, the IRE Zworykin Television Prize in 1959, the RCA David Sarnoff Outstanding Achievement Award in Science in 1963, the IEEE Liebmann Prize in 1966, and an hon degree of Doctor of Science by Manchester College in 1968, Member of The National Academy of Engineering since 1981, Fellow of the IEEE. Since 1955. *Society Aff:* IEEE, Fellow

Weimer, Wilbur A
Home: 1407 Bitner Dr, Pittsburg, KS 66762
Position: Consulting Engr *Employer:* Self Employed *Born:* 04/06/05 3 years Elec Engrg KS State Univ. Retired 1970, VP, Peabody Coal Co, with 40 yrs experience in design, engrg and management of 25 mines, railways, river loading docks, deep seaports, mine layouts, valuations and operation analysis. Discovered many million tons coal on three continents. Since 1970, consulting mining engr in USA, Australia, Canada, Philippines and S America. Numerous inventions: "Weimer, Extended Bench Method of Surface Mining–; Pumping Refuse to Disposal; Truck to Railway Car Transfer System and Railway Cable Car Loading and Unloading System. Many convention papers; Chapter: "Planning and Engrg Design of Surface Coal Mines" in Society of Mining Engrs Handbook of 1973. Distinguished mbr award of the SME of AIME in 1980. *Society Aff:* AIME, SME

Weinberg, Alvin M
Business: P O Box 117, Oak Ridge, TN 37830
Position: Director. *Employer:* Institute for Energy Analysis. *Education:* PhD/Physics/Univ of Chicago. *Born:* April 1915. His contributions to nuclear tech cover the systematization of reactor theory;reactor design, include responsibility for the Hanford multiplication factor; reactor systems, where he proposed the use of pressurized water for power reactors, and early urged the importance of breeder development to assure an infinite energy source for mankind; a concern with broad social implication of neclear energy; & his position as Research Dir beginning in 1948 and then as Dir of ORNL since 1955. He has guided the Lab to recognition as one of the world's leading centers of basic and applied research. His view on scientific policy & on the energy crisis have elicited international response. *Society Aff:* NAS, NAE, AAAS, ANS.

Weinberg, Fred
Business: Dept of Metallurgy, Vancouver BC Canada
Position: Professor. *Employer:* University of British Columbia. *Education:* PhD/Metallurgy/Univ of Toronto; MA/Physics/Univ of Toronto; BSc/Eng Physics/Univ of Toronto. *Born:* 4/6/25. Research Scientist in Metal Physics, Mines Branch, Govt of Canada 1951-61; sec head metal physics 1961-67. Prof of Metallurgy, Univ of British Columbia 1967 to present. One of recipients of Charles H Herty Jr Award of AIME Iron and Steel Div. 1974. Recipient of Robert F. Mehl Award and Inst of

Weinberg, Fred (Continued)
Metals Lecturer 1975. Dept Head, Metallurgical Engrg Univ of British Columbia 1980. *Society Aff:* AIME, ASM, CIMME, IM.

Weinberg, Harold N
Home: 11 Stratford Drive, Livingston, N.J 07039
Position: Group V Pres, Process Technology *Employer:* Englehard Corporation *Education:* MSE/Engg Thermo/Univ of FL; BME/Engg/Univ of FL. *Born:* 12/24/29. Miami, FL. Joined Exxon Res & Engg Co in 1951; two yr leave as 1st Lt U.S. Army. Various Exxon assignments in Engineering, R&D, & Business including Petroleum, syn fuels, Chemicals, New Business. Had major role in Flexicoking dev. 1979 became Genl Mgr of Corp Res - respon for commercializing science based innovations. Mbr of NJ Technical Advisory Bd, NJ. Res and Dev Council, 1981/82 Chrm NJ Energy Res Inst., 1983 Mbr, Governor's Task Force on Improving NJ's Econ and Regulatory Climate. Joined Engelhard Corporation 1/87, Group V Pres responsible for Process Technology Group. *Society Aff:* AIChE.

Weinberg, Harold P
Business: 2550 Huntington Ave, Alexandria, VA 22303
Position: Sr Vice President. *Employer:* VSE Corp. *Education:* BS/Metallurgical Engg/VA Polytechnic Inst. *Born:* June 1948. Born in Syracuse NY. Served in US Navy 1944-46. Materials Engr, US Naval Weapons Plant 1949-60. Spec in materials selection, testing and failure analysis. With Value Engineering since 1960. Appt VP in 1961 and Sr VP in 1975.Resp for mgt of Lab Div which is engaged in materials testing, inspection services, failure analysis and product evaluation. Chrmn, Washington Chap ASM 1967. President Washington Area Council of Engrg Labs 1976. Appointed Sr. VP in 1978, Engineering Group. VSE Corp (Formerly Value Engineering Co.) Resp for mgt of Group engaged in Design, Development, Test and Evaluation of Military hardware. *Society Aff:* ASM, AWS, ASTM, NSPE.

Weinberg, J Morris
Business: 19 Blackstone St, Cambridge, MA 02139
Position: Division Manager. *Employer:* Block Engineering Inc. *Born:* May 1939. PhD from Ohio State in molecular spectroscopy. BS from Univ of Maine in engrg physics. With present company since 1965. Resp have included Res Mgr and Mktg Mgr. Presently Div Mgr of Res and Eng Div. General product resp includes: electro-optical devices, process control equipment, clinical instrumentation, and government R&D. Have published extensively in the field of infrared physics and presented papers at numerous natl and intl meetings. Mbr of many professional societies.

Weinberg, Louis
Home: 11 Woodland St, Tenafly, NJ 07670
Position: Professor of Elec Eng & Prof. of Math., Grad. Sch., CUNY. *Employer:* City Coll of the City Univ of NY. *Education:* ScD/Elec Eng/Mass Inst of Tech; SM/Applied Physics/Harvard Univ; BA/English/Brooklyn College. *Born:* July 15,1919 Brooklyn NY. Served with US Air Force as radar officer 1943- 46. Taught at Univ of Mich, Calif Inst of Tech, Univ of So Calif, Mass Inst of Tech Univ. of Tokyo. Was Vice Pres for Information Processing at Conductron Corp 1961-64. Was head of Networks Research and Communications Sec at Hughes Aircraft Co 1951-61. Since 1965 Prof of Elec Eng, Also Prof. of Math., City Col of City Univ of NY. Fellow IEEE and AAAS and New York Academy of Sciences. Alumnus of Year, Bklyn Coll 1964, Thomas Jefferson High School 1971. Winner of Dean Bildersee Scholarship, Brooklyn College, for being first graduating scholar 1941. Received award from Japan Society for Promotion of Science to do research in Japan on Matroids. Enjoys water sports, Racquetball and reading. *Society Aff:* IEEE, AAAS, URSI, NYAS.

Weinberger, Jack
Home: 3000 Valley Forge Circle, Apt 746, King of Prussia, PA 19406
Position: Vice President *Employer:* The Wagner Group *Education:* BS/Pet Eng/PA St Univ; Dip/Bus Admin/Univ of PA; Dip/Naval Sc/US Merchant Marine Academy *Born:* 05/20/23 Reg Prof Engr and Cert Plant Engr. Principal career activities in plant engrg and constr for the process and marine industries. Author and lecturer on environmental and energy conservation and management progs. Current position responsible for Corporate Business Dev. Formerly with Roy F. Weston, Inc and Catalytic, Inc. *Society Aff:* AIChE, NSPE, WPCF, AIPE, AAA

Weinbrandt, Richard M
Home: 909 Canyon View Dr, Laguna Beach, CA 92651
Position: Pres *Employer:* EORCO *Education:* PhD/Petroleum Engrg/Stanford Univ; MS/Mech Engrg/Univ of CA, Berkeley; BS/Mech Engrg/Univ of CA, Berkeley *Born:* 7/20/44 Is Pres of the consulting firm EORCO with offices in Laguna Beach, CA and Jackson, WY. He has over 15 yrs of petroleum industry experience, primarily in the areas of reservoir description and the development and application of enhanced oil recovery processes. He has worked for union oil in the research department and for Aminoil USA as mgr, enhanced oil recovery engrg. His publications include 8 papers on Alkaline Flooding (including operational aspects), 2 papers on Thermal Recovery, 1 paper on CO2, 2 papers on reservoir description and 3 papers on Formation Damage. Mbr of SPE and has served the organization on local, regional and national levels. *Society Aff:* AIME

Weindling, Joachim I
Home: 204 Fairfield Dr, Wallingford, PA 19086
Position: Prof *Employer:* Polytech Univ *Education:* PhD/Operations Res/Columbia Univ; MS/IE/Columbia Univ; B/ME/CCNY *Born:* 2/18/27 Dir, Operations Res Prog. Was head of IE & OR Dept, 1973-74. Reg PE, NY & PA. Also taught mech and ind engrg at City Coll (NY), Drexel Univ, and Widener Univ, and part-time at NYU and Swarthmore Coll. Ten yrs in industry in shock and vibration control as VP of Engrg, Vibration Mountings & Controls, Inc and Ch Engr, Korfund Dynamics Corp. Consultant in operations res, ind and mech engrg to NYC-Rand Corp, Picatinny Arsenal, Chase-Manhattan Bank, Western-Electric and others. Author of several technical papers and co-author of one book. Was NSF Sci Faculty Fellow and ORSA-TIMS visiting lecturer. *Society Aff:* ORSA, ASEE, IIE, ASME, ΣΞ

Weiner, Jerome H
Business: Div of Engg, Providence, RI 02912 *Employer:* Brown Univ. *Education:* PhD/Appl Math/Columbia Univ; AM/Mathematics/Columbia Univ; BS/Mech Engg/Cooper Union. *Born:* 4/5/23. I have always been concerned with the theory of the Mech behavior of solids. Initially, this question was studied on the continuum level, with special emphasis on thermal stesses, and I am co-author with B A Boley of "Theory of Thermal Stresses" (J Wiley, NY 1960). More recently, I have studied these phenomena on the atomistic level with special emphasis on dislocation dynamics and the mechanical behavior of polymers, and have written a book "Statistical Mechanics of Elasticity" (J Wiley, NY 1983) which deals with some of these questions. *Society Aff:* ASME, AMS, APS, ΣΞ.

Weiner, Robert I
Business: 305 W. Chesapeake Ave, Towson, MD 21204
Position: Pres *Education:* MAE/Mech Engg/NYU; BAE/Mech Engg/NYU. *Born:* 10/21/33. Profl safety and mech engr (PE mech-MD, PE Safety-CA, CSP by exam). In private practice as prin of Weiner & Assoc, Consulting Engrs for over 15 yrs. Specializes in solution of unique safety problems and accident reconstruction for govt, industry and the legal profession. Author of many technical articles, reports and safety stds. Officer of the Consultants Div of ASSE and Fellow of SSS. Holds five US patents. Recognized as expert inproduct safety in 10 states and DC. Holds many awards and certificates from US Govt agencies and maj corps for solutions to aerospace, nuclear and product safety problems. *Society Aff:* AIAA, ASSE, NSPE, SSS, SAE, WSO, VSI, NSPI.

Weinert, Donald G
Business: 2029 K St, NW, Wash, DC 20006
Position: Exec Dir. *Employer:* Natl Soc of PE. *Education:* MSE/Engrg/Purdue Univ; BS/Military Science & Engrg/US Military Acad. *Born:* 9/16/30. BS US Military Acad; MSE Purdue Univ; Participant in the Inst of Mgt, Northwestern Univ; 2nd Lt Corps of Engrs 1952; Military engg proj design, construction and mgt in US, Germany and Korea; Water Resource Proj Mgt as Dist Engr, Little Rock, AR; Sr Staff Officer, Hq Dept of Army and Office, Chief of Engrs; Study Dir, Army Real Property Mgt; Specialties in Operations Res and Systems Analysis, Nuclear Weapons Effects; Parachutist; Retired 1978 as Brigadier Gen; Reg PE State of TX; mbr NSPE, SAME, Rotary. Married, 5 children, resides in McLean, VA. *Society Aff:* NSPE, SAME.

Weinert, Peter C
Home: 214 Barberry Rd, Highland Park, IL 60035
Position: Ret. (Consulting) *Education:* BSChE/Chem Engrg/MIT. *Born:* May 1,1914 in Paris, France. Author of technical articles on Catalytic Polymerization, Production of Ethylbenzene, Isomerization, Isomer Separation. *Society Aff:* AIChE.

Weinfurter, Robert W
Business: 8431 Quivira Rd, Lenexa, KS 66215
Position: Exec VP *Employer:* Terracon Consultants, Inc *Education:* BS/Eng Geol/Lawrence Univ *Born:* 05/17/28 Native of Appleton, WI. Served with US Air Force 1946 to 1949. Field engrg geologist for Soil Testing Services, Inc. 1953 through 1955; Bechtel Corp 1957 through 1955; Bechtel Corp 1957 through 1959 for American River Hydroelectric proj. Assumed responsibility for the O'Hare Field proj 1959 through 1961. Founded Soil Testing Services of WI in 1962 serving as Pres until merger of the firm in 1971. Admn VP of Soil Testing Services, coordinating all domestic and overseas offices. Principal and founding of Terracon Consultants, Inc 1980. Resp for the KS/OK Div. Dir of ASFE 1976 through 1981 serving as Treasurer and Pres of the organization. *Society Aff:* ASFE, NSPE, ASTM, AIME, AEG

Weinman, Eugene A
Home: 4649 Norwood Drive, Wilmington, DE 19803
Position: Senior Consultant *Employer:* E I du Pont de Nemours & Co Inc. *Education:* MChE/Chem Eng/Polytech Inst of Brooklyn; BChE/Chem Eng/CCNY. *Born:* Nov 6, 1922 New York, Served in WWII, Instructor in Math at Poly, Development Engr with MW Kellogg. With Du Pont since 1952. Deputy Dir of European Design, assumed present position 1972. Acts as internal consultant for design problems and major programs. Specialist in process development and commercialization. Secy of Delaware Valley Chap 1957-59. Natl Dir of AIChE, 1978-80. AAAS and Sigma Xi. Reg PE in Del. VP of Temple Beth Emeth since 1973, Bd/Dir of Jewish Family Service and Jewish Federation of Del. Natl Treas of AIChE since 1981-82. *Society Aff:* AIChE, ACS, AAAS, ΣΞ.

Weinreb, Sander
Business: 2015 Ivy Rd, Charlottesville, VA 22903
Position: Assistant Director *Employer:* Natl Radio Astronomy Observ. *Education:* PhD/EE/MIT; BS/EE/MIT *Born:* 12/9/36. Res Asst at MIT 1960-1963, engaged in investigations of varactor frequency multipliers & digital autocorrelation techniques. Joined Lincoln Lab 1963, designing radiometric equip for Haystack antenna. Joined Natl Radio Astronomy Observatory 1965. Was Hd of Electronics Div until 1977. During 1977-78 was a visiting Res Assoc at Univ of CA, Berkeley. He is presently Assistant Director at NRAO & is in charge of Central Dev Lab. Have been an advisor to the Netherlands Fdn for Radio Astronomy & has served on the Arecibo Scientific Advisory Committee. *Society Aff:* IEEE, URSI

Weinschel, Bruno O
Business: One Weinschel Lane, Gaithersburg, MD 20877
Position: President; Chief Engr *Employer:* Weinschel Engineering Co Inc. *Education:* DrEngg/-/Technische Hochschule; Hon Dr/Sc/Capitol Inst of Tech. *Born:* 5/26/19. Technische Hochschule. City College, NY 1939. Grad Study Columbia Univ 1939-40. Bklyn Polytechnic Inst 1943. Amer Univ 1957-58. Dr of Engrg, Technische Hochschule, Munich, Germany 1966. Secretary, Cttee SC-46D, Intl Electrotechnical Commission on Coaxial Connectors since 1982. Principal Investigator for NSF Study on Engrg Utilization since 1983. Honorary DR-Science, Capitol Inst of Tech. Chrmn of US Commission I of the Intl Scientific Radio Union 1967-70. IEEE VP for Profl Activities 1978 & 1979 (2 terms); IEEE Secretary 1980. Chrmn of Exec Ctte, Conf for Precision Electromagnetic Measurement, (CPEM) 1976-78. Dir, Precision Measurements Assoc, 1978, 1979. Fellow, IEEE, 1966. Fellow IEE, 1977 Chmn, Engr Affairs Council of AAES 1980-81 (2 Terms). Mbr of NSPE (National Society of Professional Engineers). *Society Aff:* IEEE, IEE, NSPE, URSI.

Weinstein, Alvin S
Business: 123 South St, Concord, NH 03301
Position: Emeritus Prof of Mech Eng & Pub Policy. Adjunct Prof. of Law, Franklin Pierce Law Ctr. *Employer:* Self. Consultant. *Education:* PhD/Mech Eng/Carnegie-Inst of Tech; MS/Mech Eng/Carnegie-Inst of Tech; BS/Mech Eng/Univ of MI; JD/Law/Franklin Pierce Law Ctr. *Born:* 6/12/28. Served in US Army 1945-47. Faculty mbr Carnegie-Mellon Univ since 1955. In addition to research in matls processing, am engaged in studies of the interaction of law & technology in product lability litigation & safety standards for products. ASME Melville Medal, 1972; ASEE Western Elec Award 1973. Chrmn ASME Cttee, Design, Engrg and the Law. Co-author of *Products Liability & the Reasonably Safe Product*, John Wiley & Sons. Emeritus Prof. Engrg. Carnegie-Mellon Univ. 1985- ; Adj. Prof. of Law, Franklin Pierce Law Center, 1987. Now engaged in legal-technical consulting in Products liability for industry, insurers and legal profession. President of Weinstein, Romualdi & Assoc., Consulting Engineers. Member, evaluation panel for Office of Technology Assessment 1985-1987. Lecturer in Products Liability for American Management Association and Soc. of Automotive Engineers. *Society Aff:* ASME, ASEE, AISE, ASTM.

Weinstein, Norman J
Home: 105 Reimer St, Somerville, NJ 08876
Position: Pres. *Employer:* Recon Systems, Inc. *Education:* PhD/CE/OR State Univ; MChE/CE/Syracuse Univ; BChE/CE/Syracuse Univ. *Born:* 12/31/29. Born Rochester NY 1929. Exxon Res & Engg Co 1956-1966 (Engg Assoc). Princeton Chem Res, Inc 1966-1969 (Dir, Engg & Dev). Founder & Pres Recon Systems, Inc, Three Bridges, NJ 1969-present. Consulting experience in process engg, economics, fluidized solids, catalytic processes, & environmental engg. Leader in used oil recovery, energy from solid waste, & hazardous waste mgt. Many publications & patents, including book on "Thermal Processing of Municipal Solid Waste for Resource and Energy Recovery." Formerly mbr AIChE environmental div exec committee & chrmn of air section. Formerly Pres, Somerville Board of Education and Borough Council. Currently Chrmn, Somerset County Democratic Cttee. *Society Aff:* AIChE, ASTM, ACS, NYAS.

Weinstock, Walter W
Business: Borton Landing Rd, Morrestown, NJ 08057
Position: Principal Scientist. *Employer:* RCA. *Education:* PhD/EE/Univ of PA; MS/EE/Univ of PA; BS/EE/Univ of PA. *Born:* 8/18/25. Airborne Radar Design at Philco 1946-49. With RCA since 1949 specializing in systems eng with emphasis on radar & processing for ddefense systems. Maj efforts include Land Based Talos, the Ballistic Missile Early Warning Sys (BMEWS) & AEGIS. As the Principal scientist of Missile & Surface Radar, he is the Chief Technical Advisor to its VP & Gen Mgr. Fellow IEEE; RCA David Sarnoff Medalist 1972. Tau Beta Pi, Pi Mu Epsilon, Sigma Tau and Eta Kappa Nu *Society Aff:* IEEE.

Weir, Charles R, Jr
Business: P. O. Box 371, 233 Race St, Ambler, PA 19002
Position: VP/Treas *Employer:* C. Raymond Weir Assocs, Inc *Education:* Dipl/CE/Drexel Inst Tech *Born:* 04/05/27 Native of the Ambler area. Served in the US Naval Reserve 1945 to 1947. Mbr and Past Elder Supplee Presbyterian Church.

Weir, Charles R, Jr (Continued)

Mbr CBMC. Served on Review Committee to Comm of PA Dept of Environmental Resources preparing Guide Lines for "Storm Water Mgmt." Established Seminars on Storm Water Mgmt and 'Surveying and Legal Aspects of Real Property' for the PA Soc of Land Surveyors, was Chairperson for the first Storm Water Seminar and supervised the rest. Haven been legislative chrmn of PSLS for 6 yrs. Have lectured at PA State Univ surveyors conference and Storm Water Symposium. *Society Aff:* MSPE, PSLS, WPCF, ACSM

Weir, James R, Jr

Business: P O Box X, Oak Ridge, TN 37830-6134
Position: Assoc Director, Metals & Ceram Div. *Employer:* Martin Marietta Energy Systems *Education:* MS/Met Engr/Univ of TN; BS/Met Engr/Univ of Cincinnati *Born:* Dec 29,1932. PE Ohio 27668. 1st Lt Army Corps Engrs 1957-59. With Oak Ridge Natl Laboratory 1959 to date. Currently Assoc Director of Metals and Ceramics Div of ORNL. Atomic Energy Commission E O Lawrence Memorial Award 1973. Dist Alumnus Award, Univ of Cincinnati 1974. Fellow, ASM 1974. Tenn Soc of Prof Engrs Significant Achievement Award 1976. Fellow AAAS 1978. Tau Beta Pi 1979. *Society Aff:* AAAS, ASM.

Weir, John P

Business: 20 N Wacker Dr, Chicago, IL 60606
Position: President *Employer:* Paul Weir Co *Education:* BS/Mining Engr/PA St Univ; BS/ChE/Purdue Univ *Born:* 07/14/23 Pres of Paul Weir Co consulting mining engrs and geologists specializing in coal and minerals of similar occurence. The co was established in 1936 and Jack Weir has been employed by Weir Co since 1949. Experience has included work in all major coal areas in USA as well as assignments in Australia, Germany and S Africa. Considered to be an authority on valuation of coal mines, coal reserves, and coal companies. *Society Aff:* AIME, ACEC, IME (UK), AIMM.

Weis, Alfred E

Business: American Can Co, American Lane, Greenwich, CT 06830
Position: Mgr Quality Assurance. *Employer:* Dodge Mfg Div-Reliance Elec Co. *Born:* Feb 21,1932. Served 4 yrs USAF 1951-55. BS Industrial Mgt from Purdue Univ 1961. Reliability Engr with Bendix Missile Div 1961-64. Employed as Quality Control Engr at Dodge Mfg Corp 1964 and progressed through various management positions to present position in 1972. Chrmn of div Product Safety Control Cttee. Articles on statistics and quality control published in tech magazines. ASQC Activities: Fellow Mbr, Certified Quality Engr, Certified Reliability Engr, PChrmn South Bend-Mishawaka Sec, Mbr Bd/Dir, Chrmn of Inspection Div. Active in local politics. Enjoy golf.

Weisberg, Leonard R

Business: Honeywell Inc, Honeywell Plaza, Minneapolis, MN 55408
Position: Vice President, Res & Engrg *Employer:* Honeywell Inc. (Aerospace and Defense Group) *Education:* MA/Phys/Columbia Univ; BA/Phys/Clark Univ. *Born:* 10/17/29. Native of NYC. With RCA Labs, Princeton, NJ, 1955-71; Dir of Semiconductor Device Lab 1969-71. With Itek Corp, Lexington, MA, 1972-75; VP of Central Res Labs. With Dept of Defense, 1975-79, Dir of Electronics & Physical Sciences. With Honeywell Inc., Aerospace and Defense Group, 1980-present, Vice President of Res and Engrg. Pioneering res in GaAs mtls & devices. In Dept of Defense was responsible for most electronics tech progs including radar, undersea acoustics, night vision, avionics, electronic warfare, C3, computers.In Honeywell, responsible for technology & engrg strategies and the Systems and Research Center. Received the VHSIC Founders Award from the Department of Defense in 1987. Fellow IEEE. *Society Aff:* IEEE, APS, AIA, ADPA.

Weislogel, Stacy

Business: OSU Dept of Aviation, Box 3022, Columbus, OH 43210
Position: Prof and Dept Chrmn *Employer:* The OH State Univ *Education:* Juris Doctor/Law/Capital Univ; MS/Ind Adm/Purdue Univ; B/Aero and Astro Eng/OH St Univ; *Born:* 06/16/39 Sr. faculty mbr and chrmn of the OH State Univ Dept of Aviation. Responsible for the university's academic and res prog in aviation, and oversees the flight operations and training prog, air trans serv, and univ owned and operated airport. Specialist in aviation law, the aviation system, aircraft performance and flight test engrg. He was Chrmn of the AIAA General Aviation Systems Tech Ctte 1979-80. He serves as a consultant to the aviation indus, has authored several papers, and served as principal investigator on several industry, NASA and FAA res projs. His experience in indus includes General Dynamics Corp (1963-65) and Landrum and Brown Aviation Consultants (1965-68). He holds airline transport pilot and flight instructor certificates. *Society Aff:* AIAA, SETP, ABA, NAFI, ISASI, UAA.

Weisman, William I

Home: 3737 E 48th Place, Tulsa, OK 74135
Position: Vice President *Employer:* Chemical Marketing Services *Education:* BChE/Chem Eng/Univ of MN; BBA/Bus Adm/Univ of MN. *Born:* Aug 1918. Univ of Minn. Lt US Navy WWII. Joined Ozark-Mahoning Co 1948. Asst to Pres 1961, President 1964. Author and Co-Author of papers on submerged combustion evaporation, sodium sulfate processing, production & markets, fluorspar production & markets, chapter on sodium sulfate in fourth edition, Industrial Rocks & Minerals. Mbr AIChE, AIME, ACS. Retired from Ozark-Mahoning Co. Aug. 31, 1983. Joined Chemical Marketing Services Sept. 1, 1983. *Society Aff:* AIChE, ACS, SME, AIME.

Weiss, Alfred

Business: 300 Broad St Box 10745, Stamford, CT 06904-1745
Position: Pres. *Employer:* Mineral Sys Inc. *Born:* 4/20/28. in Surabaia, Indonesia. Came to US in 1951. Naturalized in 1958. Recieved secondary education in Netherlands; Pres of Mineral Sys Inc a CT based consulting and technical services firm to the international mineral communit. Previous experience includes Bus Advisor to Exxon Corp, NYC; Asst Dir Bureau of Mines, Wash, DC; Dir, Technical/Operational Sys, Kennecott Copper Corp, NY & Dir - Metal Mining Div Computing Ctr, Kennecott Copper Corp in Salt Lake City. Earlier headed the Sys & Data Processing Div of Exploration Services Grop and Statistical Unit of Bear Creek Mining Co. 1978 Daniel C Jackling Award recipient, Mbr Bd of Dirs AIME and SME of AIME, 1980 Henry Krumb Lecturer, 1981 Pres of Society of Mining Engrs of AIME, Distinguished Mbr Award SME-AIME Class of 1976. PChrmn Intl council APCOM. Mbr Mining Club, NYC.

Weiss, Alvin H

Business: Dept of Chem Engg, Worcester Polytech Inst, Worcester, MA 01609
Position: Prof of Chem Eng. *Employer:* Worcester Polytechic Inst. *Education:* PhD/Phys Chemistry/Univ of PA; MS/ChE/Newark College of Engg; BS/ChE/Univ of PA. *Born:* 4/28/28. Dir of the Catalysis Soc and a Fellow of AIChE. Maj specializations are catalysis and petrochemical processing. Ind experience includes Colgate-Palmolive Corp and Houdry Process Corp. Invented and developed Houdry Litol and Detol processes. UNIDO Chief Technical Advisor to Petrochemical Complex of Bahia Blanca, Argentina. UNIDO Expert in Petrochemicals. *Society Aff:* AIChE, ACS, DECHEMA, AAUP, ISSOL, CS.

Weiss, Gerald

Home: 141-04 Coolidge Ave, Briarwood, NY 11435
Position: Professor of Elect Eng'g *Employer:* Cooper Union *Education:* D/EE/Polytech Inst of Brooklyn; SM/EE/Harvard Univ; B/EE/Cooper Union *Born:* 08/03/22 Born 8/3/22, Cologne, Germany; in USA since 1939. US Army Service 1945-47. Dev engr, 1943-58, for Liquidometer Corp, Arma Corp, and W.L. Maxson Corp, instrumentation and control systems. 1985-86 Electrical Engrg Dept at the Polytechnic University (Brooklyn NY). Professor Emeritus 1986-. Secretary, 1960-71, of the American Automatic Control Council (AACC). *Society Aff:* IEEE, ISA, ΣΞ, ΤΒΠ, ASEE

Weiss, Herbert G

Home: 28 Barberry Rd, Lexington, MA 02173
Position: V.P. Engineering *Employer:* U.S. Windpower, Inc. *Education:* SB/Elect Eng/MA Inst of Techology. *Born:* Oct 1918. Instructor MIT in EE Dept. Mbr Radiation Lab 194145, Physics Div Los Olomos Scientific Lab 1945-48, Computer Dept of Raytheon Mfg Co 1948-51, MIT Lincoln Lab 1951--. Directed Radar Div 1966-70, Air Traffic Control Div 1970-74, Surveillance & Control Div 1974-79. U.S. Windpower, Inc. V.P. Engineering 1979--. Active in the development of large antennas and high-power radars for astronomy and space object detection. Engrg Dir of N.E. Radio Astronomy Corp. Presidential Certificate of Merit 1946, Air Force Outstanding Achievement Award 1964, IEEE Editorial Bd 1965-70, Fellow IEEE, Fellow AAAS, Sigma Xi, Amer Physical Soc, Inst/Navigation. Enjoy sailing, hiking, skiing, music and gardening. *Society Aff:* IEEE, APS, ION, AAAS.

Weiss, Irving

Business: 171 Madison Ave, New York, NY 10016
Position: Vice President Engrg. *Employer:* Kryos Energy Inc. *Education:* BChE/Chem Eng/CCNY; MBA/Indus Mgmt/CUNY. *Born:* July, 1935. Graduated City College of N.Y. Jan 1957 with BChE and from City University of NY June 1966 with MBA with major in Industrial Engineering. Have been working in Cryogenics since 1957 starting with Air Products from 1957- 59 with American Messer from 1959-69, with Chemical Construction Corp from 1969- 71, with Synergistic Services 1971-75 and with Kryos Energy from 1975-present. My supervisory experience was as Mgr of Process Engrg at both American Messer 1967-69 and Chemico. At Synergistic Services I was VP Process Engrg. At the present time I am co-owner of Kryos Energy, a consulting firm spec in Cryogenics, LNG, Air Separation. I am a mbr of AIChE and had been a guest lecturer at the New England Gas Assn Gas Operations School from 1975-1979. *Society Aff:* AIChE, ISA, AGA.

Weiss, Kenneth E

Business: AFRPL/TE, Edwards AFB, CA 93523
Position: Chief, Test & Support Div. *Employer:* Air Force Rocket Prop Lab. *Born:* Oct 1926. MS Aeronautical Engrg, Air Force Inst of Tech; BS Aeronautical Engrg, Univ of Illinois. Native of Quincy, IL. Rated Air Force Senior Pilot. Associated with the various aspects of rocket propulsion technology and systems development since 1956. Was Air Force Proj Engr resp for the dev of the rocket engines for the X-15 airplane and the Titan II missile. Held engrg mgt and supv positions at various line & staff levels throughout Air Force Systems Command, including the Dir of Propulsion at Command Headquarters. Assumed current position in 1972.

Weiss, Martin

Home: 79 Spiers Rd, Newton, MA 02159
Position: VP & General Mgr *Employer:* Metcalf & Eddy Services, Inc *Education:* MS/SE/Northeastern Univ; BS/CE/Northeastern Univ *Born:* 04/30/34 Native of Stoneham, MA. Patent Examiner, US Patent Office, 1957-1958. Engr, Metcalf & Eddy, to 1963. Proj Engr, Camp, Dresser & McKee, 1963-1967. With Metropolitan District Commission (Greater Boston) since 1967. Resp included; Superintendent, Deer Island Sewage Treatment Plant; Dir, Environmental Planning; Chief Engr over Parks, Water and Sewerage and Construction Divisions. Since 1982, VP and General Mgr, Metcalf & Eddy Services, Inc. Since 1982, VP and General Mgr, Metcalf & Eddy Services, Inc. Since 1981, Associate, Metcalf & Eddy, responsible for contract operation and maintenance services. Dir, WPCF, 1981-1984, Hatfield Award, 1974. Pres, New England WPCA, 1980. Dir, Treasurer, AMSA, 1978-81. Dir, AMWA, 1981. Diplomate, AAEE. Reg Prof Engr, Certified Wastewater Treatment Plant Operator, Grade VII, MA. Enjoys golf, hunting and fishing. *Society Aff:* AAEE, AMSA, WPCF, AMWA.

Weiss, Richard R

Home: 40170 N 107th St W, Leona Valley, CA 93550
Position: Chief Scientist. *Employer:* AF Rocket Propulsion Lab. *Education:* PhD/Mech Engg/Purdue Univ; MS/Mech Engg/Univ of Southern CA; BS/Aero Engg/Univ of MI. *Born:* Nov 4,1934 Detroit Michigan. Worked at the Rocket Propulsion Lab since 1959, previously serving in various section, branch and division positions resp for combustion tech and liquid rocket systems. Assoc Fellow AIAA, previously serving 3 yrs on Propellants and Combustion Tech Cttee. Has been chrmn for several national-level tech meetings and has served as session chrmn at tech society mtgs on numerous occasions. Over 20 pubs on various rocket propulsion topics. *Society Aff:* ADPA, AIAA, CI.

Weiss, Stanley

Business: Commerce Center-Suite 580, 744 N 4th St, Milwaukee, WI 53203
Position: President *Employer:* Stanley Weiss Ltd, Conslt Engrs *Education:* ScD/Metallurgy/MIT; SM/Metallurgy/MIT; BMetE/Metallurgy/Poly Inst of Brooklyn. *Born:* April 1929. ScD in Metallurgy from MIT 1964, SM Met from MIT 1955, B. Met.E from Polytechnical Inst of Bklyn 1951. Over ten yrs of industrial experience (mostly with the General Electric Co) in gas turbine, steam turbine, electronics, lighting, motors, switchgear and welded product divisions as a metallurgical and welding engr. Res Assoc directing research at MIT from 1964- 68. Recipient of several GE Managerial Awards and Wasserman Medal from AWS 1976. PChrmn of AWS Handbook, Natl Cttee 1974-76. Consultant to major industrial corporations. Currently Prof Emeritus, Materials Dept College of Engrg and Applied Science, Univ of Wisc-Milwaukee. Pres, Stanley Weiss Ltd., Consulting Engrs, Reg PE in Wisc and Commonwealth of Mass. Primary areas of expertise include welding engrg, materials joining, failure analysis, and physical metallurgy. *Society Aff:* ASME, ASM, AWS, ΣΞ.

Weiss, Vladimir S

Home: 69 Mapplewood Rd, Mississauga, Ont, Canada L5G 2M7 Canada
Position: Pres (of 4 companies) *Employer:* Weiss Systems Inc. *Education:* MSc/Mech Engg/Univ of Vienna, Austria; MSc/Mech Engg/Univ of Zagreb, Yugoslavia; BSc/Mech Engg/Univ of Zagreb, Yugoslavia. *Born:* 8/23/31. From 1957 to 1965 was Thermal Design Engr, Res Engr and then Head of Heat Transfer Lab for MLW Ltd. In 1965 started Weiss & Assocs, in 1966 Con-Des Inc., in 1976 Weiss Engrg Ltd., in 1979 Weiss Cons Inc., in 1983 Heiss Engrg Corp (USA), and in 1987 Weiss Systems Inc. consulting engrg cos specializing in process controls & instrumentation. Was active in ISA, held local section positions up to Montreal Sec President, and then became Dist VP resp for Society activities across Canada. Currently Pres of Instrument Society of America-Worldwide, the very first Canadian to be so honored. Married with two children. *Society Aff:* NSPE, CSME, AIChE, ISA, CSPE

Weiss, Volker

Business: 409 Link Hall, Syracuse, N.Y 13244
Position: Professor of Engineering and Physics, 1986-present. *Employer:* Syracuse University. *Education:* PhD/Solid State/Syracuse Univ; MS/Physics/Syracuse Univ; Dipl Ing/Physics/Vienna, Austria. *Born:* Sept 1930. Dipl. Ing. Physics from Tech Univ of Vienna, Austria. Res Engr at DEMKA Steel factory in Holland; on the faculty of Syracuse University since 1957; Chem. Solid State Science and Tech Prog 1961 to 1977; Dir Syracuse University Institute for Energy Research 1976-1980. VP for Research and Graduate Affairs, 1978-86. Authored over 80 papers on mech behavior of materials and physical metallurgy; editor of Sagamore Conf Series on Materials Science and other monographs. Fulbright Scholarship 1953/54; NASA Minor Award 1968; Adams Award AWS 1972; Fellow ASM 1976. Enjoy classical music, skiing, sailing. *Society Aff:* ASM, ASTM, ΣΞ.

Weissenburger, Jason T

Business: 8420 Delmar Blvd - 303, St Louis, MO 63124
Position: Pres. *Employer:* Engg Dynamics Intl. *Education:* Doctor of Sci/Appl Mechanics/Washington Univ; MS/Appl Mechanics/Washington Univ; BS/Mech Engg/Washington Univ. *Born:* 12/11/32. Founded Engineering Dynamics International in 1970. Formerly Tech Specialist in Structural Dynamics with McDonnel Aircraft Corp and Prin Consultant Engg with McDonnell Automation Co Lecturer and Ad-

Weissenburger, Jason T (Continued)
junct Prof of Applied Mechanics, WA Univ, 1955-1970. Presented numerous seminars, lectures, talks and technical papers on noise and vibration. Reg PE in MO, IL, FL, NC, VT and WI. Bd of Dirs, Engrs' Club of St Louis, 1972-74 and 1978-80, Pres elect Natl Council of Acoustical Conslts, Past Chrmn Engrg Sch Advisory Council, Washington Univ. One of the founders of the Midwest Noise Council. City Council of Univ City, MO, 1970-78. Vice Chm Alumni Board of Governors, Washington University. *Society Aff:* ASHRAE, ASME, NSPE, INCE, ASA, NCAC.

Weissmann, Sigmund
Business: Dept Mechanics & Mtls Sci, College of Engg, Piscataway, NJ 08854
Position: Prof. *Employer:* Rutgers Univ. *Education:* PhD/Phys Chem/Brooklyn Poly Inst; MS/Phys Chem/Brooklyn Poly Inst; BS/CE/CCNY. *Born:* 7/1/17. Native of Vienna, Austria. Arrived USA 1939. Served with US Army Ordnance 1945-46. Joined Rutgers Univ, College of Engg 1949, Dir of Mtls Res Lab College of Engg from 1952; Natl Lecturer Sigma Xi 1962; Reicpient of Howe Gold Medal ASM 1962. Editor of JCPDS-Intl Centre for Diffraction Data Swarthmore, PA (Metals & Alloys). Consultant to: Lawrence Radiation Lab, Livermore, CA (1964-75); US Steel (1958-1968). Lady Davis Visiting Professorship, Hebrew University, Jerusalem, 1980. NSF Research Creativity Award, 1983. 1. Lady Davis Visiting Professorship, Hebrewe University, Jerusalem, 1980 2. NSF Research Creativity Award, 1983 Enjoy classical music, give annual, classical piano recitals. *Society Aff:* ASM, AIME, ACA, MRS.

Weisz, Paul B
Business: Univ of Penn, Dept. Chem. Engineering, Towne Building/D3, Philadelphia, PA 19104
Position: Distinguished Prof (Univ. of Pa.) Consultant, R&D Strategy *Employer:* Univ of PA *Education:* BS/Pysics/Auburn Univ; ScD/Chem Technology/Swiss Fed Inst of Technology. *Born:* July 1919. 1940-46: Bartol Res Foundation, Swarthmore, PA, cosmic rays, radiation physics; MIT radiation Lab, Loran simulators, electronics. 1946 to date: Mobil Oil Corp Res Assoc, Sr Res Assoc, Sr Scientist (1960); Mgr, Process Res Sec; Mgr, Central Res Div, Mobil R & D Corp. Visiting Prof (Chem Engrg) Princeton University 1974-75. Distinguished Professor, Univ of Penn, 1984-, Editor, Advances in Catalysis, Academic Press 1956- . Current interests: interdisciplinary phenomena in basic applied science and engrg, industry, and society. *Society Aff:* ACS, APS, AIChE, NAE.

Weisz, William J
Business: 1303 E Algonquin Rd, Schaumburg, IL 60196
Position: Vice Chairman & Ch Exec Officer. *Employer:* Motorola Inc. *Education:* BS/Elec Eng/MIT. *Born:* Jan 8,1927 Chicago. Tau Beta Pi, Eta Kappa Nu, Sigma Xi. Joined Motorola Inc 1948 Junior Engr. 1961-VP & Dir of Communications Products. 1965-VP & Gen Mgr, Communications Div. 1968-Motorola Inc Bd/Dir. 1969-Exec VP & Asst Chf Oper Officer. 1970-President. 1972-Chf Operating Officer. 1980-Vice Chairman, Past: Chmn IEEE, Vehicular Communications Grp; Chmn EIA Tech Cttee TR8-Land Mobile Services and Land Mobile Comm. Sec, and Communications & Industrial Electronics Div; FCC Land Mobile Adv Cttee. Past: Chairman EIA Bd of Governors, MIT Bd of Trustees, Present: MIT Development Cttee, Visiting Cttees to Sloan School of Mgt and School of Elec Engrg & Computer Sciences Past Chairman; Awards & Honors; Fellow IEEE; Natl Electronics Conf Awd of Merit 1970; Freedom Foundation of Valley Forge Awd 1974; Hon Dr of Bus Admin St. Ambrose College 1976, EIA Medal of Honor 1981. *Society Aff:* IEEE, RCA, ΤΒΠ, HKN, ΣΞ

Weitzenkorn, Lee F
Home: 107 Cresent Bluff, Austin, TX 78734
Position: Sr VP - Retired. *Employer:* Armco, Inc. *Education:* RPI/BSChE *Born:* 10/27/14. Native of Pittsburgh, PA. Apprentice engr Babcock & Wilcox 1937-39 to Rustless Iron & Steel, Baltimore, MD 1939 (became Armco Inc in 1945) Chief Met 1947-57. Asst to Works Mgr 1957-60. Dir Quality & Met, Armco Inc Middletown, OH 1960-67. Asst VP - Res, 1967-70. VP-Res & Tech 1970-74. Sr VP Res & Tech 1974- 76. Now retired, enjoys golf, swimming, and travel. *Society Aff:* ASM, AIME, ΤΒΠ.

Welber, Irwin
Business: 1600 Osgood St, North Andover, MA 01845
Position: Exec Dir, Transmission Sys. *Employer:* Bell Telephone Labs. *Education:* MEE/Elec Engg/Rensselaer; BSEE/Elec Engg/Union. *Born:* March 3,1924. Native of Amsterdam NY. Taught at RPI 1948-50. Served with Army Signal Corps 1943-46. Became a mbr of tech staff at BTL in 1950. After completing BTL's 3-yr Communications Development Training Prog, specialized in work on transmission sys development, particularly long-haul microwave radio relay, submarine cables, TASI, satellite communications and millimeter waveguide. On the Telstar project was resp for ground communication equipment & tech plng with NASA and foreign participants. Currently has responsibility for development of Radio, Wire and Coaxial Systems for use in the Bell System Network IEEE Fellow 1973, mbr of AAAS and IEEE. Authored several tech articles and granted two patents. *Society Aff:* IEEE, AAAS.

Welbourn, John A
Business: 1119 Chas Nat Plaza, Charleston, WV 25301
Position: Consultant. *Employer:* WEX Corporation *Education:* BSE/ChE/Princeton Univ. *Born:* 4/9/19. Petrochem plant engg, 13 yrs. Air conditioning engg, construction, & maintenance, 15 yrs. Since 1975, have concentrated on Energy Conservation & Mgt. Developed simple analytical methods, cost effective conservation procedures, & training progs for plant engrs & mech contractors. Taught seminars at Univs of WI, Auburn, Purdue, TX and VPI. Energy mgt consultant to college which won 1st prize in NACUBO Energy Cost Reduction Prog. Energy consultant to Am Museum of Sci & Energy. Technical advisor to Oak Ridge Associated Univs for energy audits on hospitals. Past pres ASHRAE chapter & VP, Air Conditioning Contractors of Am Registered PE in WV, FL, PA, TX. *Society Aff:* AIChE, AIPE, ASHRAE, AEE, NSPE.

Welch, Albert F
Business: GM Bldg 3044 W Grand Bl, Detroit, MI 48202
Position: Mgmt Coordinator. *Employer:* General Motors Corp. *Education:* MBA/Mfg Mgmt/MI State Univ; BSEE/Communications/Tufts Univ. *Born:* Oct 1922. WWII Army Europe Bronze Star Medal and Three Battle Stars. 1948 to present at GM: Wide range of product engrg & mfg engrg assignments, primarily in applying electronics and computer. Tau Beta Pi, SAE, SME, Benefactor of Paris-based College Intl Res de Production (CIRP), American Assn of CIRP Industrial Sponsors (AACIS). Presently: Bd/Dir SME and Michigan State AMP Club; President AACIS. Reg PE Mich and Calif. Certified Mfg Engr SME. SME 1975 Progress Award for 'outstanding work and leadership in systems engrg...in improving the productivity and efficiency of modern mass production.' Recreations - playing competitive sports and enjoying a large family (13 children). *Society Aff:* SME, SAE, AMP.

Welch, Arthur
Business: 16 W 260 83rd St, Hinsdale, IL 60521
Position: Mgr, Flexible Mfg Sys Dev *Employer:* Intl Harvester *Education:* BS/ME/UCLA *Born:* 09/07/32 Served in USAF, '52-'56. Rockwell Intl, Los Angeles Div, 22 yrs. From journeyman machinist to numerours mgmt positions, including Mgr Fabrication; NC Programming; Tooling. Responsible for developing CL data as control media for NC machines which is currently proposed before ANSI as an Intl Standard. Develop strategy and direction for Northrop Corps. Aircraft Divs use of computers in manufacturing and integrating CAD and CAM. Presently, Mgr, Flexible Mfg Systems Development, Intl Harvester, Hinsdale, IL. Received awards from SME; Society of PE; San Fernando Council of Engineers. Registered PE, CA. Certified Mfg Engr, CASA/SME. Past Chrmn AIA CAMTAG. Intl Dir, SME. *Society Aff:* SME, NCS

Welch, H William
Business: Dept. of Electrical Engineering, Arizona State University, Tempe, AZ 85287
Position: Professor of Engineering *Employer:* Arizona State University. *Education:* PhD/EE/Univ of MI; MS/Physics/Univ of MI; BA/Physics/DePauw Univ. *Born:* Oct 1920. Began career in electronics at Radio Res Lab Harvard U during WWII. Joined U of Mich as research physicist in 1946. Conducted and directed research on microwave & solid state devices, electronic countermeasure and combat surveillance. Joined teaching faculty as assoc prof in 1953. Appt Prof of Elec Eng in 1955. Mbr of various Dept of Defense advisory cttees from 1954-62. Joined Motorola as Director of Research, Military Electronics Div 1957. Started and became Gen Mgr of Control Systems Div. Retd to academic career at Arizona State Univ in 1967. Associate Director then Dir of the Center for Research in Engg & Applied Sciences, ASU 1968-1980. Major interests now microwaves, communication and control, technology & public policy. *Society Aff:* AAAS, IEEE, ΣΞ.

Welch, J David
Home: 10942 Wickshire Way, Rockville, MD 20852
Position: Retired *Education:* MSE/Soil & Mechs/Princeton Univ; BSCE/Civil Engg/ Univ of WI. *Born:* Dec 1924. Ensign WWII USNR. Res Asst and Instructor, Princeton Univ 1947- 49. Chf Soils Engr, Howard, Needles, Tammen & Bergendoff 1949-60; example projects - Delaware Memorial Bridge, New Jersey Turnpike, West Virginia Turnpike, Maine Turnpike, Etc. Partner, Welch and Associates, Consulting Engrs and Geologists, Summit, NJ 1960-69. Merged with Joseph S Ward and Assoc, Consulting Geotechnical Engrs as Partner 1969-76; soil mechanics, engrg geology, rock mechanics, and environmental geology; Partner-in-charge Washington DC office 1973-76. In private practice of geotechnical engrg and image enhancement/ marketing development for design professionals, 1976-78. Dir of Mktng/ Geotechnical Consultant; Johnson, McCordic & Thompson, PA, Silver Spring, MD. Retired 1981. Pres CEC/NJ 1969-70, Natl Dir ACEC from New Jersey 1971-72. *Society Aff:* ASCE, SLPS, ACEC.

Welker, J Reed
Business: 401 West Main, Suite 220, Norman, OK 73069
Position: President *Employer:* Applied Technology Corp. *Education:* BS/Chem Engg/ Univ of ID; MS/Chem Engg/Univ of ID; PhD/Chem Engg/Univ of OK. *Born:* 12/1/36. Res Engr & Proj Dir Univ OK Res Inst 1965-73; Proj Dir, Visiting Assoc Prof Univ of OK 1973-77. Consulting engr, Univ Engrs Inc 1965-77, VP 1972-77. Pres, Applied Technology Corp 1977- . Major int: Fire protection & safety, Liquefied natural gas technology, Fundamental fire res. *Society Aff:* AIChE, ACS, NFPA, AGA, CI.

Welkowitz, Walter
Home: 138 Highland Ave, Metuchen, NJ 08840
Position: Professor. *Employer:* Rutgers University. *Education:* PhD/Elec Engg/Univ of IL; MS/Elec Engg/Univ of IL; BS/Elec Engg/Cooper Union. *Born:* Aug 3,1926 Bklyn NY. Res assoc at Columbia Univ 1954; 1955 to 1964 employed by Gulton Industries Inc in the design and development of medical instruments and medical instrument systems; since 1964 at Rutgers Univ in charge of program in biomedical engrg; research interests are in the analysis of cardiovascular systems and the design of heart assist devices; prof and chrmn of the Biomed. Eng. dept and adjunct prof Dept of Surgery (Bioengineering) of Robert Wood Johnson Medical School, UMDNJ; mbr of Tau Beta Pi Eta Kappa Nu, Pi Mu Epsilon, Phi Kappa Phi and Sigma Xi, Fellow IEEE. *Society Aff:* IEEE, ASAIO, SMB, NYAS.

Wellech, Edmund Heinrich
Home: 75 W Fifth St, Corning, NY 14830
Position: Retired *Education:* M.E. *Born:* 1894 Vienna, Austria. Raised and educated there. Served in WWI 1914-18 Reserve Officer Commander of Anti-Aircraft Battery, 1918-24 Mgr of Chemische Fabrick Stockerau, 1924-75 Corning Glass Wks, Development Engr, Asst Plant Mgr, Asst Chf Engr, Mgr of Mech Res Dept, 1970-75 Consultant to Process Res Ctr. Work: Continuous glass fiber, all-glass Radio Tube Sealbeam Headlights, tube drawing processes, Radomes and other Govt work, glass feeding and forming. Nine patents. ASME Mgr, Fellow.

Weller, Edward F
Home: 3790 Lakecrest, Bloomfield Hills, MI 48013
Position: Dept Hd. *Employer:* GM Res Labs. *Education:* EE/-/Univ of Cincinnati. *Born:* 11/30/19. Mr Weller grad from the Univ of Cincinnati with an Elec Engg Degree. He joined GM Res Labs in 1946. He has been a Dept Hd since 1962. Mr Weller holds 20 patents & is author of technical papers & articles on automotive instrumentation, gas analyzers & non-destructive inspection techniques. Weller is a Fellow of the Inst of Elec & Electronic Engrs, & a mbr of Eta Kappa Nu and Sigma Xi. He received the Distinguished Alumnus Award from the Univ of Cincinnati in 1973. Mr Weller is married to the former Mary Elizabeth Rourke. *Society Aff:* IEEE.

Weller, Lloyd W
Business: 1500 Meadow Lake Pkwy, Kansas City, MO 64114
Position: Partner. *Employer:* Black & Veatch. *Education:* MSEE/Environ Health Engg/Univ of KS; BS/Sanitary Engg/Univ of IL; BS/Civil Engg/KS State Univ. *Born:* 11/27/21. WWII, Army 1944-46. With Black & Veatch since 1946, partner since 1964. Current projs include wastewater facilities for Johnson County, and Wichita KS, KS City, MO & Rochester, NY, water facilities for Wichita, KS & KS City, MO. Mbr of Exec Cttee, WPCF 1974-75, Natl Dir WPCF 1972-75, Chrmn ANSI A21 Ctte, 75-79, AWWA Cttee on Grooved & Shouldered Joints, Chrmn, KS Consulting Engrs 1979/80; Natl Dir ACEC 1980/81 Chairman and Safety Awards subcommittee WRF, 1979/83. Chrmn Facilities Dev Subcommittee TPC, WPCF; Chrmn, Water Pol Mgmt; ASCE 1979/80 Diplomate AAEE, Fellow ASCE, Mbr NSPE, KCE, AWWA, SAME, WPCF, ACEC. Reg KS, IL, MI, MO, NY, SD, NJ. Enjoy tennis, photography, stamp collecting, sailing. *Society Aff:* ASCE, NSPE, AWWA, WPCF, SAME, KCE, MSPE, KES

Wellman, Paul
Business: P O Box 391, Ashland, KY 41101
Position: Mgr Process Economics. *Employer:* Ashland Oil Inc. *Education:* BSChE/Engg/W VA Univ. *Born:* March 1928, Native of West Virginia. Served US Army 1950-52. Infantry squad leader, wounded Sept 1951. Employed for 16 yrs by the US Bureau of Mines, Dept of Interior, worked on synthetic fuels and economic evaluation of processes. Joined Ashland Oil Inc Jan 1975 as Mgr of Process Economics in the Synthetic Oil Dept. Have published about 20 papers in the contemporary literature having to do with synthetic fuels and/or process economics. Mbr of AIChE and AACE. Mbr Bd/ Dir AACE, Elected 1975. Mbr Bd/Dir AACE. Elected 1982. Mbr Bd/Dir AACE. Elected 1982. *Society Aff:* AACE, AIChE.

Wells, Albert John, Jr
Home: 297 Holly Rd, Marshfield, MA 02050
Position: Mktg Manager. *Employer:* BBN Instruments Corp. *Education:* BS. *Born:* 10/4/32. BSChe Univ of Rhode Island 1954. US Naval Reserve 3 1/2 yrs. E I duPont de Nemours & Co 5 yrs, in Engg Dept, Newark Del. 5 yrs with The Hays Corp in various sales and mktg positions in NYC and Mich City, IN. 7 yrs with the Foxboro Co in NY, NJ and Foxboro, MA. as Sales Specialist and Mktg Mgr. Specialized in applications of analytical instruments to processes. Own business distributing electrical material, epoxy resins, solvents, and other material throughtout New England. Panterhnicon, manufacturers representative for a variety of instrumentation and control products. Presently managing the mktg dept of BBN Instruments Corp. BBN Mfg Piezo Electric Accelerometors and Portable Noise Monitors. *Society Aff:* ISA, AIChE, MANA.

Wells, Clarence A
Home: 2071 N Altadena Dr, Altadena, CA 91001
Position: Retired. *Education:* BS/Math & Physics/Occidental College. *Born:* Feb 5,1895 N.H. Conn. 1917 Mbr PBK, EKN, Fellow IEEE 'for tech & edu contribs in

Wells, Clarence A (Continued)
communication transmission engrg' Prof Eng Cal 1917-70 with Pac Tel in L.A. (except USA 1918) 1917-19 Engr outside plt. 1920-36 head group of engrs (voice transmn design), 1937-60 Trans & Prot Engr., dept head of appr 100 engrs & 30 assts resp for trans design & elec prot. including ind. interference of all com. systs in So. Cal. 1961-64 A.V.P. Cent Tel Co in Las Vegas, Nev in chg of engrg & consultant on ops.; during period of rapid expansion held orders filled, service impr., fundamental plan developed, jt use agrmt. made with power co. Served or chaired many sec & natl cttees of IEEE. As chrmn of LA Sec assisted in expanding its tech activities in the 1950's. *Society Aff:* IEEE.

Wells, Frank H
Home: 9 Edward Road, Oxford England OX1 5LH
Position: Electronic Design Consultant (Self-employed) *Education:* MSc/-/Imperial College of Science and Technology, London, England; DIC/- /Imperial College of Science and Technology, London, England. *Born:* 1915 London England. Design of Television Receivers & Parabolic Aerial Arrays with Mazda Valve Co, to receive first TV broadcasts in London. 1939-43: Radar Establishment (TRE) England: design of GEE navigation system being responsible for aircraft equipment and control electronics for ground transmitting stations. 1943-45: Naval Research Lab. Washington, DC: electronic circuit development for joint English/American radar identification project. 1946-78: Atomic Energy Research Establishment, England: variety of nuclear electronic instrumentation including nanosecond techniques. Joint author of first book on *Millimicrosecond Pulse Techniques. Society Aff:* IEEE, IEE.

Wells, Herbert C
Business: 4505 Maryland Pkwy, Las Vegas, NV 89154
Position: Lecturer in Engrg, Las Vegas Dept of Engrg *Employer:* Univ of NV *Education:* MS/Mining Engrg/Univ of CA; BA/Geol Mining/Univ of CA; AS/Sci Geol/Pasadena Comm Coll *Born:* 4/11/27 Attended Pasadena J. C. for two yrs before entry into Army Corps of Engrs. Transferred to Berkeley for BS & MS. Oil exploration with United Geophysical for two yrs then to Climax, CO for 4 yrs. From there to Titanium Metals (1/2 yrs) with part-time teaching at Univ of NV, Las Vegas. The teaching continued through the 4 yrs spent at Blue Diamond Gypsum as plant engr and Mill Supt. This lead to full-time teaching with dev of degrees in Geology and Engrg, I served as Dean (3 1/2 yrs) and dept Chrmn (another 3 1/2 yrs) during this time. For 6 yrs I have been faculty part-time Engr with the Bureau of Mines BCEL. Consltg with small firm until it shuts down in mid 1983. *Society Aff:* AGU, AIME, NSPE, SAME, AAPG, SEG, ASEE.

Wells, Larry G
Business: Rm 109, Dept of Agri Engrg, Lexington, KY 40546
Position: Assoc. Professor *Employer:* Univ of KY *Education:* PhD/Agr Engrg/NC St Univ; MS/Agr Engrg/Univ of KY; BS/Agr Engrg/Univ of KY *Born:* 07/28/47 Faculty, Dept. of Agric. Engrg, 1974. Teaches grad and undergrad courses in design, dev and mgmt of agricultural machinery. Res areas include design and dev of agr machinery, dynamic soil-machine interactions, hydrology and sedimentology of agr and surface mined lands. Authored or co-authored 16 referred journals & articles and 28 prof papers. Mbr of 2 prof societies, 6 hon societies and 2 "Who's Who" prof listings. Licensed as prof engr, 1979. Prof consulting in hydrology, sedimentology, rheology, machinery dev, and machine systems analysis associated with agr, urban dev and surface mine reclamation. *Society Aff:* ASAE, ISTVS

Wells, Martin G H
Business: P.O. Box 88, Crucible Res Center, Pittsburgh, PA 15230
Position: Technical Dir *Employer:* Colt Industries *Education:* PhD/Phys Metallurgy/Royal Sch of Mines, Imperial Coll of Sci and Tech London Univ England; BSc/Metallurgy/Royal Sch of Mines, Imperial Coll of Sci and Tech *Born:* 3/18/35 Native of Redhill, Surrey, England. Resch and Senior Resch Engr for Jones and Laughlin Steel Corp. 'specializing in ferrous phase transformations.' Assumed present position in 1970, with respons for Resch and Dev in alloy, tool and valve steels and titanium. Chrmn, Pittsburgh Chptr AIME, 1974. Chrmn, Metals Handbook Committee, ASM, 1979-present. Enjoy tennis, canoeing and natural history. *Society Aff:* AIME, ASM, ASTM

Wells, Robin D
Home: Rt 2, Filer, ID 83328
Position: Research Associate *Employer:* Univ of ID *Education:* MS/Agri Eng/Univ of ID; BS/CE/Univ of ID; BS/Agri Eng/Univ of ID *Born:* 12/05/48 Mr. Wells was born in Twin Falls, ID. He graduated from the Univ of ID in 1972 having participated in ASAE, ASCE, and student government while living in a fraternity he helped organize, Alpha Kappa Lambda. His education was financed with his own custom harvesting business. In 1972, he became a commissioned officer in the NOAA Corp and was assigned to hydrographic survey operations. In 1976 he resigned as a Lieutenant, returned to the Univ of ID and studied hydraulic effects in oil shale retort. The Univ hired him in 1978 to study water delivery costs in pump-irrigation agr in ID, a position he currently occupies. Mr. Wells is a reg prof engr in the state of ID. *Society Aff:* ASAE

Wells, Thomas W
Business: 1200 St. Charles Ave, New Orleans, LA 70130
Position: Vice President *Employer:* Waldemar S Nelson and Co, Inc *Education:* MS/CE/Univ of IL; MBA/Management/Loyola Univ; BS/CE/Univ of FL *Born:* 09/19/43 Structural engr for large water and wastewater facilities in OH and FL. Structural engr and proj mgr for design and constr of large power plant and quarters facilities for offshore petroleum projs, for design of lock and dam projs, airfield and support facilities and other Corps of Engrs projs, and for design, studies, and reports for service buildings, drainage, wharves, docks, etc, for industrial plants in Gulf Coast. Devised method of analysis and design of massive concrete elements in navigation locks. Registered in eleven states. 1981 Pres LA Chapter ACI. Elder and teacher in Presbyterian Church. *Society Aff:* ASCE, NSPE, ACI, LES, SAME

Wells, William P
Business: 420 Hawthorne Lane, Charlotte, NC 28204
Position: President. *Employer:* Mechanical Engineers Inc. *Education:* BS/Mech Engg/Univ of NM. *Born:* 1912. Before forming the firm of Mechanical Engrs Inc in 1949 was employed as Design Engr with US. Indian Service for 2 yrs; as Manufacturer's Rep for 4 1/2 yrs and as Contractor for 7 yrs Full member of CSI, ASME, NSPE & Life Fellow of ASHRAE. Served as President of the Prof Engrs of No.Carl. and its So Piedmont Chapter; as President of Southern Piedmont Chap of ASHRAE, and Reg Dir of ASHRAE for 3 yrs. Reg in NC,SC,MD. *Society Aff:* NSPE, ASHRAE, CSI, ASME.

Wells, William R
Business: Howard R. Hughes School of Engineering, University of Nevada, Las Vegas, 4505 Maryland Parkway, Las Vegas, NV 89154
Position: Dir, School of Engrg *Employer:* Univ Nevada, Las Vegas *Education:* PhD/Aerospace/VPI; MA/Applied Math/Harvard Univ; MS/Aerospace/VPI; BS/Aerospace/GA Inst of Tech *Born:* 11/28/36 William R Wells received the PhD degree from the VA Poly Inst in 1968. He has worked as an Aerospace Technologist for NASA at the Langley Res Center and has held faculty positions at VPI and the Univ of Cin. In 1977, Dr. Wells took the position as Chrmn of the Engrg Dept at Wright State Univ. In 1984, Dr. Wells became Dean of the College of Engineering and Technology at Bradley Univ. Currently he is Director of the School of Engineering at the Univ. of Nevada, Las Vegas. Dr. Wells is recognized internationally for his research in the areas of system identification and control theory. *Society Aff:* AIAA, ASEE, NSPE

Welmers, Everett T
Home: 1626 Old Oak Rd, Los Angeles, CA 90049 *Employer:* Retired *Education:* AB/Math & Classics/Hope College; AM/Math & Astro/Univ of MI; PhD/Math & Astro/Univ of MI; ScD/Hon/Hope College. *Born:* 10/27/12. in Orange City, IA. Instructor, Math Dept MI State College 1937-44. Flight Test Engr, Chf of Dynamics, & Asst to the Pres, Bell Aircraft Corp 1944-59. Staff, Inst for Defense Analyses, 1959-60. At Aerospace Corp 1960-1980, initially as Group Dir, Satellite Sys; 1968-77 Asst to the Pres, respon included computer planning, corp res, staff dev, & technical relations; 1977-1980, Corp Historian. Dean, College of Engineering, Northrop Univ. 1980-82. Mbr numerous DoD study groups, former Mbr Air Training Command Advisory Bd, Commissioner, Community Redev Agency, LA. ScD. Hope College 1966. *Society Aff:* MAA, AMS, ΣΞ, AIAA, IEEE, ΦBK, ΦKΦ, TIMS, TBΠ.

Welsh, James N
Business: 2451 Stemmons, Dallas, TX 75207
Position: Pres. *Employer:* Shirco Inc. *Education:* MBA/Finance/Harvard Bus Sch; BS/CE/Rice Univ; BA/CE/Rice Univ. *Born:* 7/17/42. Native of Dallas, TX. Sold ind chems for Procter & Gamble 1965-67. MBA Degree from Harvard Bus Sch 1967-69. Investment banking in Dallas 1969-76 Since 1976, partner in Bright & Co & Pres of Shirco, Inc - a mfg of waste disposal equip. *Society Aff:* AIChE, WPCF.

Welty, Robert
Business: 1501 West First St, The Dalles, OR 97058
Position: President and Chief Engr. *Employer:* Robert Welty Engineers Inc. *Education:* BS/EE/OR State Univ. *Born:* 1/13/15. Field Engr REA 1938-42 and 1944-45. Tech Field Advisor Alaska Communic System. US Army Sig Corps 1942-44. Organized Robert Welty Engrs Aug 1945- . Incorporated 1972. Reg PE several states. P Pres OCEC, Mbr ACEC, PEO, NSPE. *Society Aff:* ACEC, NSPE, IEEE, HKN, ΣT.

Wempner, Gerald A
Business: ESM School Georgia Tech, Atlanta, GA 30332
Position: Professor *Employer:* Georgia Inst Tech *Education:* PhD/Mech/Univ of IL; MS/Mech Eng/Univ of WI; BS/Mech Eng/Univ of WI *Born:* 11/11/28 Served on faculites of Univ IL (1953-1959), Univ AZ (1959-1962), Univ CA (1962-63), Univ of AL (1964-1973), Georgia Tech (1973-present). Res Assoc Stanford (1963-1964). Served as consultant to industry and governmental agencies. Published *Mechanics of Solids* Sijihoff-Noordhoff, 1982. Co-author of *Mechanics of Deformable Bodies* (Charles Merrill, 1961). Recipient of Alexander von Humboldt Award of German Fed Republic (1973); Killam Award, Univ of Calgary (1983). Published numerous articles on solid mechanics, variational methods, stability and theories of shells. *Society Aff:* ASCE, ASME, SES, Amer Acad Mech

Wen, Chin Y
Business: Dept Chem Engrg, Morgantown, WV 26506
Position: Prof of Ch Engrg *Employer:* West Virginia University. *Education:* PhD/Chem Engg/W VA Univ; MS/Chem Engg/W VA Univ; BS/Chem Engg/Natl Taiwan Univ. *Born:* 12/5/28. Chrmn of Dept of Chem Engrg WVU from 1969 to 81. Over 25 yrs experience in fossil energy res funded by private & governmental agencies-NSF, OCR, EPA, NIH, HEW, NASA, Bur of Mines, and DOE-FE. Specialized areas - fossil energy research and development, coal conversion processes, coal technology optimization techniques, fossil energy computer simulation. Chaired natl symposia in above, editor of several energy related pubs, cttee mbr on numerous fossil energy panels including Natl Res Council, and Author of over 100 relevant energy related articles. Author co-author of several books & book chapters relating to coal conversion technology, fluidization, chemical reaction engg, pneumatic transport & reaction flow models. Fellow of Amer Inst Chem Engrs, Alcoa Fnd Award, Benedum Prof of Chem Engrg. *Society Aff:* AIChE, ACS, ASEE, SCE.

Wen, Robert K
Business: Dept of Civ Engg, E Lansing, MI 48824
Position: Prof. *Employer:* MI State Univ. *Education:* PhD/Civ Engg/Univ of IL; MS/Civ Engg/Univ of VA; BS/Civ Engg/St John's Univ. *Born:* 7/1/29. Asst Prof at Univ of IL from 1957 to 1959. On faculty of Civil Engrg/Dept of MI St Univ since 1959; Prof since 1967. Res in structural dynamics, stability, and reinforced frozen soil. Served as conslt with Bechtel Power Corp, Sargent Lundy Engrs, and Lawrence Livermore Lab. *Society Aff:* ASCE, ΣΞ.

Wen, Yi-Kwei
Business: 3118 Newmark Civil Engineering Lab, 208 N. Romine, Urbana, IL 61803
Position: Prof of Civil Engrg. *Employer:* Univ of IL *Education:* BS/CE/Nat'l Taiwan Univ; MESc/CE/Univ of Western Ontario, Canada; DESc/CE/Columbia Univ *Born:* 9/14/43 Has been active in teaching and res in the areas of struct reliability, random vibration and applications to Civil Engrg. Also active in ASCE committees, and intenat'l res activity, including lecturing in Europe, Japan and China. Prof of Civil Engrg since 1981. Enjoy tennis and fishing. *Society Aff:* ASCE, AAM, EERI, IASSAR.

Wenck, Norman C
Business: 545 Indian Mound, Wayzata, MN 55391
Position: VP. *Employer:* Hickok & Assoc. *Education:* MS/Sanitary Engg/IA State Univ; BS/Civ Engg/IA State Univ. *Born:* 5/9/40. Served in USN, Facilities Engg Command at bases in the US & Thialand. VP with Hickok & has managed projs of regional & natl scope including waste load allocation, hazardous waste evaluation, lake restoration projs, EPA res projs and a natl runway evaluation proj past Chrmn Minn. Pollution Control Agency is Advisory Committee on Sludge Disposal and Advisory Committee on Hazardous Waste of CEC/MN EPA-MRA Liason Committee & Member. *Society Aff:* AIChE, ASCE, ASAE, APCA, CEC/MN.

Wendel, Peter K
Business: 11 Pinchot Ct, PO Box 1456, Buffalo, NY 14228
Position: CH BD *Employer:* Wendel Engineers, PC. *Education:* BSCE/Structures/Univ of MI. *Born:* 5/24/32. Professional Engineer NY and PA. Wendel Engineers, P.C. Buffalo and Lockport, NY. Design Engineer '57-70, Partner '70-75, President '75-87, now Chairman. President, Upstate Building Corp.; Chairman, Niagara County Joint Chamber Task Force; Chairman Leadership Niagara; Past President, Consulting Engineers Council of NY State, Past Chairman, Western NY Section NYWPCA; Past Co-Chairman, Exploring Niagara's Future, Past President, Lockport Area Chamber of Commerce. Native Lockport, NY, Naval Officer '55-57. Sailor on Great Lakes, Listed in Who's Who in the East. Organizational Consultant. *Society Aff:* ASCE.

Wender, Leonard
Business: Livingston Ave, Dobbs Ferry, NY 10522
Position: Chief Engg-East. *Employer:* Stauffer Chem Co. *Education:* SM/ChE/MIT; BChE/-/CCNY. *Born:* 10/12/25. Native of NY, NY. Instr in Chem Engg at CCNY & later asst prof at Cooper Union. Process Engr with the M W Kellog Co, specializing in process design of new, res-developed processes. Joined Stauffer Chem Co Engg Dept as Mgr of Chem Engg in 1969. Assumed current responsibility as chief engr-East in 1978. Enjoy photography. *Society Aff:* AIChE.

Wendl, Michael J
Home: 24 Zinzer Ct, St Louis, MO 63123
Position: Technical Specialist. *Employer:* McDonnell Aircraft Corp. *Education:* MS/Controls/WA Univ; BSEE/Elec/WA Univ. *Born:* June 1934, Sacalaz, Rumania. BSEE & MS (Control Engrg) Washington U. Mbr Tau Beta Pi, Delta Phi Alpha & AIAA. PE State of Missouri. 1958-59 Instructor Physics & Elec Engrg, Washington U. Most of professional life spent in design & development of advance guidance and flight control systems at McDonnell Aircraft Corp. Conducted feasibility studies related to Flight Propulsion Control Coupling, Aircraft Energy Mgt, and Advanced Fighter Technology Integration. Performed analysis and pilot-in-the-loop simulation studies on Control Configured Vehicles, fly-by-wire flight control technology & active control augmentation concepts. Directed development of the Model

Wendl, Michael J (Continued)
F-15 aircraft fuel system analysis program. Recipient of the 1974 Wright Brother Award SAE. *Society Aff:* AIAA

Wengenroth, Reece H
Home: 65 Pheasant Drive, New Canaan, CT 06840
Position: VP. *Employer:* Parson Brinckerhoff Centec, Inc. *Education:* MCE/Civil Engg/NY Univ; BSCE/Civil Engg/MA Inst of Tech. *Born:* 1/9/20. in Middletown, NY. Captain Corps of Engrs 1942-45 Progressed from Bridge Engr to Chf Engr while working with several onsulting firms 1945-62. Exec VP of Westenhoff and Novick Inc 1962-75. Currently VP of PBC, a railway consulting firm. Engg involvement included over 300 highway & railroad bridges, 200 miles of interstates & urban highways & rapid transit facilities. Licensed PE in 23 states. Fellow ASCE & ACEC; Mbr AREA & NSPE. VP & Secy of CECI 1972-74; Pres of Chicago Chapter ISPE 1971-72. *Society Aff:* ACEC, ASCE, AREA, NSPE.

Wenk, Edward, Jr
Home: 15142 Beach Dr NE, Seattle, WA 98155
Position: Emeritus Prof of Engrg Public Affairs, and Social Management of Technology. *Employer:* University of Washington. *Education:* DR Eng/CE/Johns Hopkins; MSc/Applied Mech/Harvard; BE/CE/Johns Hopkins. *Born:* 1/24/20. Navy specialist in submarine strength and ship dynamics 1941-56. Chrm Dept of Engrg Mechs, Southwest Res Inst, 1956-1959. First science advisor to Congress 1959. Chief, Science Policy, Congr. Res. Service, 1964-66. Technology advisor staff to Presidents Kennedy, Johnson and Nixon 1961-70. Prof of Engrg & Public Affairs, Univ WA 1970-83 Now Emeritus; Affiliate Prof, Seattle Univ 1984. Director, prog in Soc Mgt of Tech, 1973-78. Univ WA. Visiting Scholar Harvard 1976, Univ Sussex 1977, NAE, NRC Assembly of Engrg, Natl Acad of Public Admin, Dr Sc Hon Univ of RI, Tau Beta Pi, Sigma Xi, Chi Epsilon, WA. State Govener's Award. ASME Fellow 1974. SESA Pres 1957. Engineer Educator of Year, Puget Sound Eng. Council, Dist Alummus, John Hopkins Pres Int Assoc Impact Asses 1981. Mbr Congress Technology Assessment Advisory Council 1973-79. President's Natl Advisory Cmtee on Oceans and Atmospher 1971. Chmn, Adv Council, Cousteau Soc. Designer, Designer Aluminaut Submarine. Author, "The Politics of the Ocean." –Margins for Survival." –Tradeoffs: Imperatives of Choice in a High-Tech World." Consultant to US, Sweden, Philippines, Australia, UN Dir URS Corp. *Society Aff:* ASCE, ASME, ASPA, NAE.

Wennerstrom, Arthur J
Home: 4300 Fair Oaks Rd., Apt. 5, Dayton, OH 45405
Position: Chief, Compressor Rsch Grp *Employer:* Aero Propulsion Lab. *Education:* D.Sc./Turbomachinery/Swiss Federal Inst. of Technology; M.Sc./Aeronautical/Massachusetts Inst. of Technology; B.Sc./Mechanical/Duke University *Born:* 01/11/35 Early work experience from 1958-65 included work for a U.S. aerospace manufacturer, a Swiss gas turbine manufacturer, and three years as an officer in the US Air Force. From 1967 to date, he has been employed at Wright-Patterson AFB, first as head of the Fluid Machinery Research Group at the Aerospace Research Lab and, since 1975, in his present job. Noted for significant contributions to compressor design now in use by industry; in particular for fostering the introduction of sweep in compressor blades and vanes. Fellow of ASME and AIAA. 34 publications and 5 patents. *Society Aff:* ASME, AIAA

Wensing, Donald R
Business: 1100 First St, King of Prussia, PA 19406
Position: Mgr Materials Engr. *Employer:* SKF Ind Inc. *Education:* BS/Met/Grove City College. *Born:* 5/5/30. Served two yrs as a commissioned officer in the US Marine Corps & then joined the Aluminum Co of Am as a production met. In 1956, accepted in quality control engr with Marlin Rockwell Corp in Jamestown, NY, later advancing to position of chief met &, upon the acquisition of MRC by TRW, received title of mgr-mtls & processes. Joined SKF in 1969, to be active in area of mtls processing & heat treatment, playing a role in the liaison between engr/tech and the outlaying SKF mfg plants. As a result of work in the metalworking field, he has been listed in Who's Who in the East & Who's Who in Iron & Steel. *Society Aff:* ASM, ASTM, SAE, AFBMA.

Wentz, Charles A, Jr
Business: Phillips Petroleum Company, Bartlesville, OK 74004
Position: Development Manager *Employer:* Oil Shale Division *Education:* PhD/CE/Northwestern Univ; MS/CE/Univ MO; BS/CE/Univ MO. *Born:* 10/12/35. Native of Edwardsville, IL. Employed by Phillips Petrol Co since 1961 in various positions, including res & dev, marketing, financial & proj engg activities. Enjoy hunting, fishing & gardening. *Society Aff:* AIChE, SPE, ACS.

Wentz, Edward C
Home: 342 Dutch Lane, Hermitage, PA 16148
Position: Consultant (Self-employed) *Education:* BS/Elec Eng/Univ of MN; MS/Elec Eng/Univ of Pittsburgh; EE/Elec Eng/Univ of MN. *Born:* 7/3/05. Sec Mgr, Instrument Transformer and Large Power Dev Engg 1942-65, Prof, Elec Eng. Penn State Univ 1965-69, Consultant. Chrmn, Sharon Sec IEEE 1952, Chrmn ASA C57.13 Instrument Transf Cttee 1963-73, presently mbr IEEE Relay Ctte. Author "Transformers for the Elec Power Industry–, McGraw Hill 1959, and of a considerable number of IEEE papers. Fellow Mbr IEEE. Author Transformer Section, Standard Handbook for Elec Enggs (McGraw Hill) 11th Edition. *Society Aff:* IEEE.

Wentz, William H
Business: Aero Engr Dept Box 44, Wichita, KS 67208
Position: Distinguished Prof *Employer:* Wichita St Univ *Education:* PhD/Engr/Univ of KS; MS/Aero Engr/Wichita State Univ; BS/Mech Engr/Wichita State Univ *Born:* 12/18/33 Native of Wichita, KS. Served in USAF 1955-57 as aircraft maintenance officer. Instructor of Mechanical Engrg at Wichita State Univ. 1957-58. Propulsion and Aerodynamics Engr at the Boeing Co, 1958-63. Asst, Assoc, Distinguished Prof, of Aeronautical Engrg at Wichita State Univ, 1963-present. Also Director, Center for Basic and Applied Research at WSU. Teaching and res in aerodynamics, and Principal Investigator for a number of res projs for NASA, and various industrial firms. Res publications in vortex flows on delta wings, airfoils, control surfaces and flaps, automobile and truck drag, and wind turbines. Recipient of 1981 AIAA General Aviation Award for res on airfoils, control surfaces and high-lift devices. *Society Aff:* AIAA, SAE, ΣΓΤ, ΤΒΠ, ASEE

Wenzel, Harry G, Jr
Business: Dept of Civ Engg, Urbana, IL 61801
Position: Prof Civ Engg. & Asst. Dean *Employer:* Univ of IL. *Education:* PhD/Civ Engg/Carnegie-Mellon Univ; MS/Civ Engg/Carnegie-Mellon Univ; BS/Civ Engg/Carnegie-Mellon Univ. *Born:* 9/4/37. Native of Pittsburgh, PA. Joined Dept of Civ Engg, Univ of IL in 1964. Assoc Prof since 1971. Taught courses in hydrology, hydraulic engg, fluid mechanics, and drainage design. Res interests include open channel flow, urban drainage design & simulation, computer models. Acting dir of Univ of IL Water Resources Ctr 1972. Sabbatical leave at Stanford Univ & CO State Univ, 1973-74. ASCE Walter L Huber Res Civ Engg Prize, 1977. Sabbatical leave at Colorado State Univ 1981-82. Assistant Dean, College of Engg, 1986. *Society Aff:* ASCE, AGU, ASEE.

Wenzel, Leonard A
Home: 517 Fifteenth Ave, Bethlehem, PA 18018
Position: Prof Dept Chem Eng *Employer:* Lehigh University. *Education:* PhD/Chem Eng/Univ of MI; MS/Chem Eng/Univ of MI; BS/Chem Eng/PA State Univ. *Born:* Jan 21,1923. Married Mary E. Leathers Oct 21,1943 and has four children. Served with AUS 1944; with USNR 1944-46. MS and PhD Univ of Mich 1948 and 1950. Res chemist, Colgate-Palmolive Co 1949-51. Faculty mbr, Lehigh Univ since 1951 spec in thermodynamics, heat transfer, and cryogenic engineering. Consultant to industry. Since 1962, Chrmn, Dept of Chem Engrg. Unesco Expert and Proj Coordinator, Bucaramanga, Colombia 1969-70. Fulbright Lecturer, Colombia 1971.

Wenzel, Leonard A (Continued)
Unesco Consultant Venezuela 1973-80. Founder, Lehigh Valley Sec AIChE 1955. Chrmn 1960-61. Author of five books and numerous research articles. *Society Aff:* AIChE, ACS, ASEE, ΣΞ

Weppler, H Edward
Home: 171 Main St, Apt 44, Maidson, NJ 07940
Position: Retired Engg Dir - Tech Stds & Regional Planning. *Employer:* American Tele & Tele Co. *Education:* BS/Elec Engg/Purdue. *Born:* 9/4/15. With Bell Telephone System, 1937, with Michigan Bell; went to AT&T Co, New York in 1950. Directed organization responsible for technical aspects of major regulatory and legal matters. Active in radio spectrum mgt circles. Mbr Frequency Mgt Adv Council NTIA. Has participated in US delegations to numerous intl radio conferences. *Society Aff:* HKN, ΤΒΠ.

Werden, Robert G
Business: P O Box 414, Jenkintown, PA 19046
Position: President. *Employer:* Robert G Werden & Assoc., Inc. *Education:* BSIE/Mech/Lehigh Univ/1937. *Born:* Feb 1916. Employed by York Corp from 1937-58. President own consulting engrg firm since 1958. Holds 2 patents. Reg PE in PA, NJ, DE, MD, WV, AL, IA, NH, IL, VA, D.C. Fellow ASHRAE and ACEC. Extensive experience in energy conservation, and heat pumps. Golf and sailing. *Society Aff:* ASHRAE, SAME, ACEC, NSPE.

Werner, Fred E
Business: Westinghouse R&D Ctr, 1310 Beulah Rd, Pittsburgh, PA 15235
Position: Mgr, R&D Patents & Library *Employer:* Westinghouse Elec Corp. *Education:* ScD/Met/MIT; BS/Met/MIT. *Born:* 9/22/27. Instr, MIT, 1950-53. With Westinghouse Elec R&D Ctr since 1955. Held positions as mgr of met (30 profls) & magnetics (9 profls) depts. Have been responsible for mtls (metallic) used in power equip. Mgr of R&D patent system, library, licensing, and tech info retrieval. Structural & magnetic steels. Physical met, metals processing, non-destructive testing, magnetics. Fellow, ASM. Active in past in Natl ASM & AIME committees, presently treas and advisory committee mbr of IEEE/AID magnetics conf. *Society Aff:* ASM, AIME, IEEE.

Werner, James T
Home: 8317 Lupine Circle, Fort Worth, TX 76135
Position: Engrg Chief *Employer:* General Dynamics. *Education:* BS/ME/Ford Engr Sch. *Born:* 01/19/24. BS Ford Engrg School. Native Detroit, MI. Worked FoMoCo during and after WWII in Engrg Dept as Sr Engr. Moved Texas in 1951 to work in the Engrg Dept of General Dynamics specializing in Mass Properties Assumed task of Mass Properties Group Engr upon award of YF-16 contract. Currently in chg of all Mass Properties Group effort, at General Dynamics/Ft Worth. Wife Doris, Engr. Son, Douglas, Engr Mgr IBM Corp. PDir SAWE, Intl Sr VP SAWE 1973 & 78 Exec VP 1979, Pres 1980-82, Fellow, Honorary Fellow, Currently VP/Tech Dir *Society Aff:* SAWE.

Werner, John E
Business: Advanced Technology Center, Penn State University, 120 South Burrowes St, University Park, PA 16801
Position: Dir of Res & Dev Retired from Bethlehem Steel Corp 11/85. *Employer:* Penn State University *Education:* MS/Metallurgy/PA State Univ; BS/Metallurgy/PA State Univ. *Born:* 10/25/32. Appointed dir for res & dev of the Adv Tech Ctr in the Office of Indus Dev at the Penn State Univ. in Jan, 1986. Joined Bethlehem Steel Mgr Training Program 1954. Assigned to Lackawanna Plant, NY as Metallurgical Engr, Steelmaking Quality Control. 1955-56 served with US Army in Germany, Enter Graduate School in 1958. Rejoined Bethlehem Steel in 1960 as a Res Engg, Steelmaking in the Res Dept. Promoted to Sup of Steelmaking Processes Res in 1965 and Asst Sect Mgr of the Steelmaking Section in 1966. In 1969 was appointed Sect Mgr of the Phsical Metallurgy Sect. Appointed Asst Div Mgr, Product Metallurgy in 1974 and Asst Div Mgr, Primary Processes REs in 1976. Appointed Div Mgr, Raw Materials & Chem Processes Res in 1978. Appointed Dir of Res 1982 and Dir Tech Transfer & Ventures 1984. Retired from Bethlehem Steel, Oct, 1984. Served as Chmn of the Process Tech Committee of AIME and the BD of Dir of the Iron & Steel Society. Currently sec ASM Govt & Public Affairs Committee, Chrmn Engrg Advisory Ctte-Lafayette Coll, Bd of Dirs-Metals Property Council. *Society Aff:* ASM, AIME-ISS, ITI.

Werner, Richard L
Business: 3000 Troy Rd, Schenectady, NY 12309
Position: Mgr Computer Services. *Employer:* C T Male Associates. *Education:* MSCE/Hydraulics/Cornell; MBA/Finance/Northeastern; BSCE/Hydraulics/Buchnell. *Born:* Dec 25,1938. MSCE Cornell Univ, MBA Northeastern Univ. BSCE Bucknell Univ. Reg PE in NY and MA. Born in New York City. Worked for Bechtel Corp and Stone and Webster Enrg Corp in field of Hydraulics and Computer Applications to Civil Engrg. Presently working for C.T.Male Assoc, P.C., Schenectady as Mgr of Computer Services. Assumed position as a Director of co in 1972 and Treasurer in 1976. Taught courses at Cornell Univ and Union College. President of ASCE, Mohawk Hudson Sec, 1975; Secy, NY State Soc of Prof Enrs, Mbr CEC. Bd/Dir United Cerebral Palsy Assn of Schenectady, Schenectady County YMCA and Niskayuna Rotary Club. Enjoy sailing, tennis, handball, classical music. *Society Aff:* ASCE, NYSSPE.

Wernick, Jack H
Business: Div. Mgr., Materials Science Res, Bell Communications Research, Red Bank, NJ 07701-7020
Position: Div Mgr, Materials Science *Employer:* Bell Communications Research *Education:* PhD/Phy Met/Penn State Univ; MS/Phy Met/Univ of MN; BMETE/Met Engg/Univ of MN. *Born:* 5/19/23. Native of St Paul, MN. Served with Army Corps Engrs Los Alamos 1944-46. Assumed current responsibility 1984. Over 20 prof level people in dept. Consultant to govt agencies, OST, NBS, NSF. Penn State ASM McFarland Award. Elected to Natl Acad Engg March, 1979. Fellow Met Soc. AIME, Fellow ASM, Fellow APS, AAAS, IEEE, Eletrochemical Soc, Fellow NYAS. Coauthor two books, twenty- four patents, 230 tech papers *Society Aff:* AIME, APS, AAAS, IEEE, ASM, NAE, Electrochemical Soc.

Wert, Charles A
Home: 1708 W Green St, Champaign, IL 61820
Position: Prof of Metallurgy & Hd of Dept. *Employer:* University of Illinois. *Education:* PhD/Physics/State Univ of IA; BS/Physics/Morningside College; MS/Physics/State Univ of IA. *Born:* Dec 31,1919. Staff mbr, Radiation Lab, MIT, 1943-45. Specialty, microwave antennae. Staff mbr, Inst for the Study of Metals, Univ/Chicago 1948-50. Mbr, teaching faculty, Dept of Metallurgy and Mining Engrg 1950–. Head of dept since 1966. 1987: Named to endowed chair: Ivan Racheff Professor of Materials Engineering. Areas of interest: physics of metals, diffusion, internal friction, gases in metals, coal science. Recipient of Alumni Achievement Award, Morningside College 1973. Univ of Illinois College of Engrg: Stanley H Pierce Award for Excellence in Teaching 1970. Recipient of US Senior Scientist Award, Humboldt- Stiftung, Bonn, FRG, 1981 and 1987. Spend leisure time in woodworking and gardening. *Society Aff:* AAAS, APS, ASM, TMS.

Wert, James J
Business: Box 1621 Sta B, Nashville, TN 37235
Position: George A. Sloan Professor of Metallurgy *Employer:* Vanderbilt University. *Education:* PhD/Metallurgical Engg/Univ of WI; MS/Metallurgical Engg/Univ of WI; BS/Metallurgical Engg/Univ of WI. *Born:* 1/9/33. Served with US Army in Korea 1953-55. Assoc Engr with Westinghouse, specializing in the design of nuclear fuels. Res Scientist for the A O Smith Corp studying deformation of brittle materials. With Vanderbilt Univ since 1961, and apptd Chrmn of the Materials Science Dept in 1969. Served as Chrmn of the Materials, Mechanics and Structures Div, and Dir of the Ctr for Coating Sci and Tech. Chrmn of Dept of Mech and Materials

Wert, James J (Continued)
Engrg from 1975-1982. Consultant to TVA, Murray-Ohio, VXTRA, Natl Academy of Science, Avco, Dupont, Westinghouse, Arnold Engrg Ctr, Temco, Gen Elec, Ford, and NASA. Adams Award from AWS 1969, Tau Beta Pi Award 1974, visiting prof at Cambridge Univ 1974. Senior Fulbright Lecturer at Damascus and Alepps Universities in Syria. Res activities include tribology and deformation behavior. Tau Beta Pi Award, 1978 Western Electric Award, 1979; Society Affiliations - ASME, EMSA; ASM, AWS, AIME, AAUP, ASEE. *Society Aff:* AIME, ASM, AWS, ASME, ASEE, EMSA, ASTM.

Wertheim, R H
Business: Lockheed Corporation, 4500 Park Granada Boulevard, Calabasas, CA 91399
Position: Senior Vice Pres.-Science & Engineering *Employer:* Lockheed Corporation *Education:* MS/Physics/MIT; BS/Naval Science/US Navy Acad; -/AMP/Harvard Business School. *Born:* Nov 1922. Native of Carlsbad, New Mexico. Naval Ordnance Engr specializing in major weapon system mgt. From 1977 to 1980 Dir of Navy Strategic Systems Projects resp for all aspects of research, development, test evaluation, production and operational support of submarine ballistic missile systems. Distinguished Service Medal (2 awards) US Navy 1979, 1980; Parsons Awrd, US Navy League 1971; ASNE Gold Medal 1972; Hon Mbr, ASNE; Fellow, AIAA; Mbr Sigma Xi, member NAE, mbr Tau Beta Pi, mbr Cosmos Club. *Society Aff:* ASNE, AIAA, ΣΞ, NAE, TBΠ.

Werts, Robert W
Business: PO Box 1018, Reading, PA 19603
Position: Asst VP System Operations. *Employer:* GPU Service Corp. *Education:* BS/EE/PA State Univ. *Born:* 11/28/16. Native of Reading, PA. Employed by elec utility in distribution engg systems prior to WWII. Served with Signal Corps & Army Air Corps 1942-1946 in radar systems. Since 1946 active in dev of EHV elec networks & mine mouth generating stations. Assumed current responsibility, Asst VP - System Operations, in 1971. Served on numerous local, state, natl & intl committees on power system dev and operations. *Society Aff:* IEEE, NSPE, AREA, CIGRE.

Wesler, Philip
Business: 999 Asylum Ave, Hartford, CT 06105
Position: President. *Employer:* Engg Consultant of Hartford, PC. *Education:* BCE/Civil Engg/Clarkson College. *Born:* 6/17/29. in Stamford, CT. Mbr Tau Beta Pi. Designer-Hardesty & Hanover 1948. Joined Fraioli-Blum-Yesselman Structural Consultants, Norfolk, VA 1950, NY 1954 Assoc. Opened new Hartford, CT office-Prin 1965. Founded the Engg Consultants of Hartford, PC 1978. Performed structural design: Hartford Civic Ctr (Recd Engg Excellence Award-CT Engrs); SNETCO Communications Ctr; Jai-Alai Fronton. Co- authored article, Jan 1974 Civil Engg Raising Civic Ctr Space Frame Roof. Mbr ACEC, NSPE/PEPP, ASCE, ACI. Recd Clarkson College "Golden Knight" Award. Reg Engr 6 states. *Society Aff:* NSPE, ACI, ASCE, ACEC.

Wesley, Fred A
Home: 16 Colesbery Dr, New Castle, DE 19720
Position: Project Engr. Retired *Employer:* E I du Pont de Nemours & Co. *Education:* BSCE/Civil Eng/Univ of KY; AB/Math/Berea College. *Born:* Dec 1,1921. AB Berea College; BSCE Univ of KY. Native of Lynch, Kentucky Served with US Navy 1943-46. Worked with Miami City Engr's Office and American Bridge Co. Joined du Pont in 1952. Retired in 1986 as a Project Engineer in Du Pont Engrg Project Div. Have held offices of Secy, VP, and Pres in Delaware Sec of ASCE. Grade of Fellow in ASCE. Reg Engr in Delaware and Pennsylvania. Enjoy music, camping and hiking. *Society Aff:* ASCE.

Wesner, John W
Home: 63 Glendale Dr, Freehold, NJ 07728-1357
Position: Supervisor, Prod Design *Employer:* AT&T R&D *Education:* PhD/Mech. Eng./Carnegie Mellon University; MS/Mech. Eng. California Institute of Technology; BS/Mech. Eng./Carnegie Institute of Technology *Born:* 07/14/36 George Westinghouse Scholar at Carnegie Tech; NSF Fellow at CalTech. Served as officer, US Army Signal Corps. Designs at Westinghouse (1964-68) include nuclear reactor control rod drives, new fuel shipping container, LMFBR, FFTF. AT&T design projects include Transaction Telephone, first AT&T Personal Computer, STARLAN, MERLIN communications system. ASME: Chair (1983-87), Design Education Cttee; 1986 Decennial Education Conference (Exegetes for Design); Executive Cttee of Design Engrg Div; Currently chair Design for Manufacturability Cttee; Elected Fellow, 1987. NSF Design Theory & Methodology Ad Hoc Advisory Cttee. Registered PE. 37 years active with Boy Scouts (Eagle; Silver Beaver). Hobbies: Model Railroading, War Gaming. *Society Aff:* ASME, ΣΧΙ

Wesselink, Robert D
Business: 2623 E. Pershing Rd, Decatur, IL 62526
Position: Vice President *Employer:* Blank, Wesselink, Cook & Assoc. *Education:* BSCE/Civ Engg/IA State Univ. *Born:* 6/26/32. Native of Sibley, IA. Field Engr for the OH Dept of Hgwys from 1954-1956. Design Engr on Toll Hgwy Construction 1956-1959. Proj Mgr with consulting firm for design of Interstate Hgwys 1959-1965. Established partnership of Blank and Wesselink & Assoc as engg consultants in 1965 to provide mech, structural, elec, civ, sanitary and process engg services currently serve as Mbr. Bd of Dirs, Secy., Treas. and V.P. for Blank, Wesselink, Cook and Assoc, Inc. Pres of Central IL Chapter of IL Soc of PE in 1973. President of Consulting Engineers Council of Illinois 1981-1982. *Society Aff:* ASCE, NSPE, ACEC.

Wessenauer, Gabriel O
Home: 2931 Nurick Dr, Chattanooga, TN 37415
Position: Consultant. *Employer:* -. *Education:* BS/CE/Carnegie Inst of Tech now Carnegie Mellow Univ. *Born:* Oct 1906. Native of Sewickley, PA. 1927-35 investigated hydro-electric possibilities for West PA Elec Co; 1935-38 project plng engr for multipurpose projects of TVA. In 1938 power supply engr in TVA's power depts; 1944-70 TVA's Mgr of Power with full resp for its power program. Since 1970 consultant on electric power, including nuclear & environmental aspects of power generation. *Society Aff:* ASCE.

Wessler, Melvin D
Home: 447 Benton Dr, Indianapolis, IN 46227
Position: Pres. *Employer:* M D Wessler & Assoc. *Education:* MBA/Mgt & Adm/IN Univ; BSEE/Elec Engg/IN Inst of Tech. *Born:* 8/2/32. Native of Arenzville, IL. Proj Engr for Intl Pharmaceutical Firm. Proj Mgr and Regional Mgr for Intl Consulting Firm prior to forming MD Wessler & Assoc, Inc in 1975. Responsible for overall mgt and operations of the co providing Malti-Discipline Services in Elec, Environmental, Sanitary & Civ Engg to Municipal, County & State Govts, private developers & ind firms. Dir IN Soc of PE 1974, Pres Consulting Engrs of IN 1975. Chrmn of Bd of Dirs Consulting Engrs of IN 1976. Dir Amer Conslt Engrs Council 1977. Enjoy boating, fishing, hunting. *Society Aff:* IEEE, NSPE, ACEC, ISPE.

Wesstrom, David B
Route 1, Box 320, Hockessin, DE 19707
Position: Retired *Education:* ME/-/Stevens Tech. *Born:* April 2, 1904 New York City. Employed by M W Kellogg Co, Lummus Co, Travelers Indemnity Co, finally with Du Pont Co Engrg Dept 1940-69, consulting on special problems in stress analysis and pressure vessel and piping design, retiring as Sr Mech Engr. Was mbr until 1969 of ASME Boiler & Pressure Vessel Cttee since 1950, Subctte on Unfired Pressure Vessels since 1931 and various other subcttees, and Pressure Vessel Res Cttee from 1945. Also mbr Joint API- ASME Cttee on Unfired Pressure Vessels. ASME Fellow 1965. J. Hall Taylor Medal 1967. PE in NY, NJ and Delaware. *Society Aff:* ASME.

West, Arnold Sumner
Home: 3896 Sidney Rd, Huntingdon Valley, PA 19006
Position: Sr. Technical Specialist Corp Govt & Reg. Affairs. *Employer:* Rohm and Haas Co. *Education:* MS/ChE/PA State; BS/ChE/Univ of PA. *Born:* 1/12/22. With Rohm & Haas Co 1946- ; Currently, Sr. Technical Specialist Corp Govt & Regulatory Affairs; previous assignments include Mgr of Petroleum Chem Res Dept, Supt of Semi-Works and pilot plants, Supvr of Economic Evaluations, Supvr of Res Computing Lab, Supvr of Corrosion Lab, Process Design Leader, Process Engr. Pres of AIChE 1977; VP 1976; Treas 1973-75; Dir 1964-66. Fellow AIChE. Mbr ACS, SAE, NSPE, WPCF. Reg PE in PA. Secy of Lower Moreland Township Autho 1970- . Dir EJC 1976-1979; Vice-Chairman AAES Public Affairs Council 1981- , Chrmn 1982-1983; Mbr AAES Bd of Governors 1981-1983. Pres, United Engrg Trustees 1986-1987. AIChE Founders Award 1977; Hamilton Award 1982; Van Antwerpen Award 1983. *Society Aff:* AIChE, ACS, SAE, NSPE, WPCF.

West, Clinton, L
Business: 526 Stevens Ave, Yuba City, CA 95991
Position: VP Engrg *Employer:* Yuba City Steel Products Co *Education:* MS/Inst & Controls/Case Inst of Tech; BS/ME/CA Inst. of Tech *Born:* 07/02/35 Res Engr for the Warner & Swasey Co from 1957 to 1964, with Yuba City Steel 1964 to present. Reg in OH and CA. Hour four patents. Past chrmn No. Cal - Nevada Section ASAE. *Society Aff:* ASAE, AWS, FPS

West, Harry H
Business: 212 Sackett Bldg, University Park, PA 16802
Position: Prof of Civil Engrg *Employer:* Pennsylvania State Univ. *Education:* PhD/Structural Engg/Univ of IL; MS/Civil Engg/Penn State Univ; BS/Civil Engg/ Penn State Univ. *Born:* June 1936. Originally from Ringwood, NJ. Faculty member in Civil Engrg at Penn State Univ 1958-61, 1962-63, 1967-present. Served in US Air Force 1961-62. Taught at Univ of Illinois 1966-67. Primary resp at Penn State have been undergraduate and graduate teaching and research in structural analysis and design. Major research in cable-supported structures and bridge engrg. Author of *Analysis of Structures, An Integration of Classical and Modern Methods*, published by John Wiley & Sons, 1980. Mbr ASCE, ASEE. ASCE Moisseiff Award 1970. Penn State Alumni Outstanding Teaching Award 1975. ASEE Western Elec Teaching Award 1977. Penn State Engg Society Society Premier Teacher Award 1986. Aff: ASCE, ASEE. *Society Aff:* ASCE, ASEE.

West, James R
Home: 432 Stratford Rd, So Hempstead, NY 11550
Position: Retired. *Education:* PhD/ChE/Univ of Pittsburgh; MS/ChE/Univ of Pittsburgh; BS/ChE/Univ of Pittsburgh. *Born:* Dec 1914. Native of McKeesport PA. 1937-39 Asst Mellon Inst Pitts., 1939- 46 Fellow, 1946-60 Sr Fellow in Sulphur, 1960-63 Asst Mgr Research, Texasgulf Inc, 1963-69, Mgr Res, 1969-72 Mgr Res and Engrg, 1972-1976, VP, Res & Engrg 1976-1980, VP, Res, Engrg and Construction. Enjoy square dancing. *Society Aff:* AIChE, ACS, AIC.

West, John M
Home: 1608 S.E. 40th Terrace, Cape Coral, FL 33904
Position: retired *Education:* MS/Physics/Univ of IA; BS/Physical Sci and Math/NE OK State Coll *Born:* 01/18/20 1941-43, ballistics work for du Pont. Continuously involved in nuclear activities since 1943. 1943-44, Manhattan Proj, du Pont. 1944-49, Hanford Works, du Pont and GE. 1949-57, Argonne National Lab, directing reactor dev activities, including CP-5 res reactor and EBWR boiling water plant. 1957-65, VP and Exec VP of General Nuclear Engrg Corp. 1965-85, exec positions in nuclear activities at Combustion Engrg. 1974-85, VP and senior VP of Nuclear Power Sys with responsibility for dev, design, manufacturing, proj, and commercial activities. Charles A. Coffin Award from GE in 1949; 1984 Walter H Zinn Award from Power Div of ANS. Charter mbr and Fellow of ANS. Mbr, National Academy of Engrg. Retired Feb 1, 1985. *Society Aff:* ANS, NAE

West, Ozro E
Home: 2329 Laguna St, San Francisco, CA 94115
Position: VP *Employer:* O. E. West Engrs & Assoc, Inc *Education:* MS/Constr Engrg/ Univ of CA; BS/CE/Univ of CO *Born:* 07/16/36 Principal of Proj Mgmt Consulting Co; O.E. Engr & Assoc, Inc; Served on Bd of Dirs- Pacific Engr & Constr Services Co, San Francisco; - Principal in A E firm; GARTNER/WEST Assoc, San Francisco, - Dir of Olympic Proj Services Group, Portland OR; - Guest Lecturer in US and S America on Proj Mgmt Techniques. Over 25 yrs experience in constr management, project planning, and Critical Path Scheduling in Europe, Japan, Australia, and the Middle East. *Society Aff:* ASCE, PMI, ACEC, PE.

West, Robert C
Business: 801 N Eleventh St, St Louis, MO 63101
Position: Chrmn of Bd & Pres. *Employer:* Sverdrup Corp. *Education:* BCE/Civil Engg/GA Inst of Technology. *Born:* 4/26/20. US Army, WWII, supervised heavy construction. With Sverdrup since 1953. Involved with: highway & railroad projects; and design of major and/or award winning bridges, US and abroad. Officer Sverdrup since 1968. As chief operating officer, 1973, undertook organization study to establish new functions and realign executive responsibilities. President and CEO 1975, board chrmn 1976. Reg PE 12 states. Officer/fellow mbr of numerous professional and tech societies, ACEC, ASCE, ASTM, NSPE, British Tunneling Society. Bd mbr/officer of community, educational and business organizations-Goodwill Industries, BSA, United Way, Drury College First Natl Bank (St Louis), Genl Steel Industries, Angelica Corp. *Society Aff:* ASCE, NSPE, ACEC, ASTM, MSPE.

West, Robert F
Business: Wayne, NJ 07470
Position: Group Dir. *Employer:* American Cyanamid Co. *Education:* PhD/Chem Engg/Columbia Univ. *Born:* 12/22/20. Licensed PE in NJ and NY. Accredited by Natl Council of State Bds of Engg Examiners. Expertise in Engg Admin. *Society Aff:* AIChE, ACS, ΣΞ.

Westbrook, Jack H
Business: FNB 120 Erie Blvd, Schenectady, NY 12305
Position: Cons. Matl Info Serv. *Employer:* Genl Electric Co. *Education:* ScD/Metallurgy/MIT; M MetE/Met. Eng/RPI; B MetE/Met Eng/RPI. *Born:* 8/19/24. Native of Troy, NY. Joined GE in 1949 as metallurgist, Res & Dev Ctr, affiliated successively with the General Physics Lab, Metallurgy & Ceramics Lab, Program Planning Operation & Physical Chemistry Lab. Mgr. Matl's Info. Services, 1971-1981. Apptd to present pos 1981. More than 100 pubs on a wide variety of materials, metals, ceramics, glass, minerals, & semiconductors as well as on technical information products and services, several received distinction thru natl & internatl awards. Also edited 11 bks & holds 5 patents. US Navy 1944-46. Adjunct Prof, RPI 1957-59. Consultant to NMAB-NAS, NSF, Army, Air Force, ARPA, NASA, OTA. Spec tech ints: materials information systems, mechanical properties, intermetallic compounds, & grain boundaries. Avocational int in history of science & technology. Trustee 1971-1980 EI. Campbell Lecturer, ASM 1976, Chmn U. S. Nat'l CODATA Comm. 1982-85. *Society Aff:* ASM, AIME, ACerS, EcS, ASTM, SHOT.

Westcott, Ralph M
Home: 3635 Oak Cliff Dr, Fallbrook, CA 92028
Position: Consulting Engr. *Employer:* Ralph M. Westcott, Cons Engr. *Education:* -/Physics/UCLA. *Born:* Jan 1906 Redlands Calif. 1940-48 founded and operated water treating service co. 1949-present consulting engr on water quality, corrosion scale, purity etc. Partner Pomeroy & Westcott, Engrs 1949-51; 1952-64 Pres Holladay & Westcott, Cons Engrs; 1964-67 parttime Exec Dir CEAC; self employed since 1964. ASHRAE F. LM. Dist Serv Awd, Council 13 yrs; CEAC Pres two terms 1957-58; ACEC Fellow, Founder, Pres 1959-60; NACE Lm. Certificate as corrosion Specialist; Chmn TPC-T-TK. Non Chem Water Treating Devices; AWWA LM.; IAE Fellow; Engr of Yr Award from CEC-Washinton 1960. PE & CE Calif; PE Nev. LA

Westcott, Ralph M (Continued)
Chamber of Commerce Director 3 yrs 1960-63. *Society Aff:* ASHRAE, NACE, AWWA, ACEC.

Westendorp, Willem F
Home: 17 Front St, Schenectady, NY 12305 *Education:* EE/EE/Tech Univ of Delft Netherlands. *Born:* May 7,1905 Amsterdam Netherlands. Elec Engr Delft, Tech Univ 1928. Adv Course in Engrg, GE 1928-30. Doctor of Engrg (hon) RPI 1947. Came to US 1928, naturalized 1934. Res Lab GE, gas discharges 1929-35; High Voltage X-rays, Betatrons, Synchrotrons 1935-54; Thermonuclear Research 1954-66; computations on superconducting coils, magnetic fields and electric fields 1966-70. Consultant 1970 to 1980; retired 1970, computations eddy current losses in transformers. Recipient John Price Wetherill Medal Franklin Inst. Fellow and Life Mbr IEEE. Fellow APS. Mbr Science Cttee IEEE 1958. *Society Aff:* IEEE, APS.

Westerback, Arne E
Home: 593 Maureen Ln, Pleasant Hill, CA 94523
Position: Corrosion Engrs. *Employer:* East Bay Municipal Utility Dist (Retired) *Education:* BS/Elect Eng/Lawrence Inst of Tech *Born:* 10/27/20. Reared in upper peninsula of MI, served in Army during WWII, received BS in EE degree in 1948. Elec Distribution estimator & engr with Pacific Gas & Elec Co for 7 yrs & Utilities Engr with CA Public Utilities Commission for 3 yrs. Supervised Corrosion Control Unit of East Bay Municipal Utility Dist for 22 yrs. Retired in 1983. Pub several articles on corrosion in NACE & AWWA publications. Won 1969-70 AWWA Distribution Div Award for article in AWWA Journal. Reg Elec & Corrosion Engr in CA. Since retirement doing part time consulting in corrosion control. *Society Aff:* NACE

Westerhoff, Russell P
Home: 576 Highland Ave, Ridgewood, NJ 07450
Position: Consultant (Self-employed) *Education:* SM/Civil Engg/MIT; SB/Civil Engg/MIT. *Born:* Dec 27,1904 Paterson NJ. With Ford, Bacon & Davis 1928-73 in valuation, construction and engrg depts. Prof activities ranged from engrg and economic reports to design and supvn of construction of industrial, chem and steam electric pwr plants. 1951 elected VP Engrg and Dir, subsequently became chf engr, VP Operations, President in 1965 and Ch/Bd 1966. Also director, president and ch of bd of subsidiary companies, Ford, Bacon & Davis Construction Corp, Ford, Bacon & Davis Canada Ltd., Ford, Bacon & Davis Texas Inc and Ford, Bacon & Davis Utah Inc. Mbr NSPE, NJSPE. Fellow ASCE, ASME. Reg PE 2 states. NJSPE Engr Awd 1970. Natl Dir NSPE 1958-60, 63-65, 5 yrs Trustee NJSPE. Gen editor of 'Resource Recovery Technology for Urban Decision- Makers' 1976. Trustee Paterson NJ Orphan Asyhan Asylum Assn. *Society Aff:* NSPE, ASCE, ASME.

Westermann, Fred E
Business: Dept. of Mat'ls Sc & En, University of Cincinnati, Cincinnati, Ohio 45221
Position: Prof. *Employer:* Univ of Cincinnati. *Education:* ChE/MetE/Univ of Cincinnati; MS/MetE/Univ of Cincinnati; PhD/MetE/Univ of Cincinnati. *Born:* 3/14/21. A native of Cincinnati, OH. Following grad from the undergrad cooperative education prog at the Univ of Cincinnati, was employed at Inland Mfg Div GMC. Served in the US Army 1944-46, then returned to the Univ for grad work. Joined the faculty as instr in 1948 and advanced to full Prof in 1963. Named Peter E Rentschler Prof of Metallurgical Engg in 1966. Assumed a maj role in developing the undergrad Metallurgical Engg Prog. *Society Aff:* ASM, ASEE, AFS

Westfall, Herbert C
Business: Tower Bldg. 7th Avenue At Olive Way, Seattle, WA 98101
Position: Managing Partner. *Employer:* R W Beck and Assoc. *Education:* BS/CE/Univ of CO. *Born:* Oct 1924. Served with US Marine Corps WWII; awarded two Purple Hearts and Silver Star. Two years resident construction engr Oregon State Hgwy Dept. Two yrs design engr Stevens and Thompson, Portland, Oregon. With R W Beck and Assoc since 1952. Elected Partner in 1954, Mng Partner and Chf Engr since 1968. President of R W Beck and Assoc Inc. University of Colorado Distinguished Engrg Alumnus in Private Practice 1976. Reg PE in 17 states. *Society Aff:* ASME, ASCE, NCEE, WSPE, NSPEE.

Westfall, Paul R
Home: 1811 Washington Ave, Parkersburg, WV 26101
Position: Mgr Eng & Maint *Employer:* Borg-Warner Chemicals, Inc *Education:* BS/ChE/W VA Univ *Born:* 07/31/26 Positions with Pittsburgh Coke & Chemicals as Dev and Proj Engr until 9/1/59. With present firm, positions involved site survey, design, and construction of facilities in USA, Canada, Scotland, Holland, Japan, and Australia, and joint venture activities in Mexico. Helped to build and now to operate world's largest ABS plastics plant. Active in community work and in W VA Soc of PE (currently State Pres). Naval Veteran of WW II, mbr of Amer Legion. *Society Aff:* NSPE

Westgate, Charles R
Business: Dept of Elect Eng, 34th and Charles, Baltimore, MD 21218
Position: Prof. & Chmn *Employer:* Johns Hopkins Univ. *Education:* PhD/EE/Princeton; MA/EE/Princeton; BEE/EE/R.P.I. *Born:* 7/19/40. Joined the faculty of Johns Hopkins in 1966, assoc prof in 1970 and prof in 1975. Res interests include semiconductor device physics, microwave instrumentation, integrated circuits and design. Other appts at Johns Hopkins include, Chairman of Electrical and Computer Engineering, Prin Prof Staff of the Applied Physics Lab, and Prof of Mech (joint appt with Electrical Engrg). *Society Aff:* IEEE.

Westin, Harold J
Home: 2504 Manitou Island, White Bear Lake, MN 55110
Position: Pres. *Employer:* Harold J Westin Architects and Engrs, PA. *Education:* BCE/Civil Engg/Univ of MN; BSL/Law/St Paul Coll of Law. *Born:* 5/6/20. Reg PE: AZ, CO, FL, IA, MN, MT, NB, ND, OH, OK, SD, TX, WI. Naval Officer, Pacific Theater, World War II, Amphibious Command. Pres of Harold J Westin, Architects and Engrs, PA. Engaged in design and professional construction mgt of medical, commercial and industrial centers throughout the USA. Lecturer: Contracts and Specifications, Univ of MN, Civil Engg. Author of: Engg text books; "An Engr Looks at the Law" and "An Engr Applies the Law–. MSPE Engr of the Year, 1971. Developed new integrated professional design and construction mgt concepts. Founded USA operations for Nilcon insulated prestressed concrete structures. Research in the design and construction of energy efficient buildings. *Society Aff:* NSPE, ASCE, MSPE.

Westlake, Leighton D, Jr
Home: 3312 Northside Dr. 610, Key West, FL 33040
Position: Dir of Tech Services *Employer:* City of Key West, FL *Education:* MS/CE/Carnegie-Mellon Univ; BS/CE/Carnegie-Mellon Univ *Born:* 7/14/46 Native of St Louis, MO. Served with Army Corps of Engrs 1968-1975. Project engr with Corps on Chatfield and Bear Creek Dams near Denver. Installation Engr in West Germany. Past Dir of Public Works for Berkeley, MO. In private practice as principal in 1977. Eastern Regional Engr for CRSI in 1980. Current responsibility is for the city of Key West's environmental construction program of $38.8 million and construction of port facilities for the port authority. Received Army Commendation Medal for service during Hurricane Agnes, 1972; Army award for Energy Conservation suggestions, 1975. Listed in Who's Who in the Midwest, Community Leaders and Noteworthy Americans. *Society Aff:* NSPE, ASCE

Westmann, Russell A
Business: 7400 Boelter Hall, Los Angeles, CA 90024
Position: Assoc Dean/Prof. *Employer:* Univ of CA, Los Angeles. *Education:* PhD/Civil Engg/Univ of CA, Berkeley; MS/Civil Engg/Univ of CA, Berkeley; BS/Civil Engg/Univ of CA, Berkeley. *Born:* 5/20/36. in Fresno, CA. Upon receiving PhD, taught or conducted res at Univ of CA, Berkeley; Univ of Glasgow, Scotland; CA Inst of Technology; Univ of CA, Los Angeles. Since 1973 served as Prof-Mechs

Westmann, Russell A (Continued)
& Structures Dept and since 1976 Assoc Dean for Sch of Engr & Applied Sci at UCLA. Genl technical interests are in structural mechanics and applied mathematics with res specialty in analytical fracure mechanics. Served as a consultant to indusltry in these fields and currently is a consultant and mbr of the Bd of Dirs of Failure Analysis Assocs. Author of numerous technical papers & reports in field of mechanics. *Society Aff:* ASCE, AAM.

Weston, Donald W
Home: RFD 2 Box 208, Winterport, ME 04496
Position: Consultant (Self-employed) *Education:* BS/ME/Univ of ME. *Born:* May 1918. Native of Madison, Maine. Served in Artillery 1941-46. Plant Eng Gen Foods Corp Caribou, MI 1946-52. Ch Eng Loring AFB Maine 1952-56. Deputy Base Eng Dow AFB Bangor ME 1956-64. Dept Head, Environmental Control Tech Dept, Eastern Maine Voc-Tech Inst 1966-79. PPres ME Soc Prof Eng. Natl Dir NSPE. Past Pres ME Assn Eng. Reg PE Maine, Reg Land Surveyor Maine. *Society Aff:* NSPE, AWWA, WPCF.

Weston, Leonard A
Business: 308 West Basin Rd, New Castle, DE 19720
Position: VP, Gen Mgr. *Employer:* Lehigh Testing Labs, Inc. *Education:* MS/Met & Mtls Sci/Marquette Univ; Registered P.E.(Delaware); BS/Met & Mtls Sci/Marquette Univ. *Born:* 12/20/44. Product Mgr for TX Instruments Met Mtls Div; Patent awarded 1969 for copper-steel multilayer strip mtl for current-carrying springs. Engaged in met failure analysis, consulting & mtls testing at MA Mtls Res 1969-72. Became Ge Mgr of MMR's Delaware Subsidiary, Lehigh Testing Labs, in 1974, and appointed Corporate VP in 1977. Chrmn of ASM Wilmington Chapter, 1978-79. Other interests include scouting, Lions Club, & golf. Also member of ASNT, ASME, and ASTM. *Society Aff:* ASM, ASME, ASTM, ASNT.

Weston, Roy F
Business: Weston Way, West Chester, PA 19380
Position: Chairman of the Board. *Employer:* Roy F Weston Inc. *Education:* MCE/Civil Engg/NY Univ; BS/Civil Engg/Univ of WI. *Born:* June 1911. Doctor of Engineering, Honoris Causa, from Drexel University, Diplomate AAEE. Native of Reedsburg Wisc. Employed by Atlantic Richfield 1939- 55. Initiated Roy F Weston Inc, environmental consultants-designers, in 1955. Mbr NAE; PPres WRA/DRB; PPres AAEE; PDir WPCF; Fellow ASCE; Mbr AIChE, ASTM, PChrmn Industrial Wastes Analysis Sec Cttee D-19; Past Mbr PA Solid Waste Mgt Advisory Cttee; Past Mbr US Chamber of Commerce Cttee on the Environment. Recd NSPE Engr of the Yr Awd 1973; PaSPE. Mbr, Governor's Energy Council (PA); Listed in Who's Who of America Engr of the Yr 1970; Dist Serv Citation, Univ of Wisc 1975; Arthur Sidney Bedell Award 1959; Industrial Wastes Medal WPCF 1950; George Washington Medal, Engrs Club of Phila 1973. PE. *Society Aff:* NAE, AAEE, WPCF, ASCE, AIChE, WRA, ACEC.

Westphal, Warren H
Home: 4398 S Akron, Englewood, CO 80111
Position: President *Employer:* Tellis Gold Mining Co., Inc. *Education:* AB/Geology/Columbia Univ. *Born:* Feb 1925. Born in Easton, Pennsylvania. Was geologistgeophysicist with New Jersey Zinc Co 1947-54. Chf geophysicist with UT Intl 1955-59. Sr geophysicist and subsequently Chrmn Dept of Earth Sciences Stanford Res Inst. 1959-69 resp for world-wide geophysical and geological res operations. VP for Minerals for Intercontinental Energy Corp in charge of exploration and mining of non-petroleum energy and base metal mineral resources from 1969-79. Pres, Spruce Creek Energy Co. 1981-85. Currently Pres, Tellis Gold Mining Co., Inc. Dir 1971-74 SME and 1976-78. Pres Energy Minerals Div AAPG, 1979. *Society Aff:* AAPG, SEG, SEXG, SME, AIPG.

Westwater, J W
Business: Dept. of Chemical Engineering, 213 Adams Laboratory-Box C-3, University of Illinois, 1209 West California Street Urbana, Il 61801
Position: Prof of Chem Engrg *Employer:* University of Illinois. *Education:* PhD/Chem Engg/Univ of DE; MS/Chem Engg/Univ of DE; BS/Chem Engg/Univ of IL, Urbana. *Born:* Nov 1919. Native of Danville, Ill. Joined Chem Engrg Dept Univ of Ill 1948; Head 1962 to 1980. AIChE: Director 1968-70, Institute Lecturer 1964, Walker Award 1966, Fellow 1972. AIChE and ASME: Max Jakob Award 1972. ASEE: Bendix Award 1974. Conf Chrmn, Third Intern. Heat Transfer Conf 1966. Elected to NAE 1974. Research on heat transfer, boiling, condensation, melting, freezing, nucleation, and phase changes. Developed techniques for high-speed motion picture photography with great magnification. *Society Aff:* AIChE, ASME, ACS, ASEE.

Westwater, James S
Business: 1460 Commerce Bldg, Cleveland, OH 44115
Position: Sr VP-Construction & Engg - Retired. *Employer:* Cleveland-Cliffs Iron Co. *Education:* EM/-/MI Tech; BS/Mining Engg/MI Tech; Hon/Dr Engg/MI Tech. *Born:* Dec 6,1912 Aberdeen, Scotland. Hon Dr of Engrg, Mich Tech 1975. US Engrs 1935-37 in channel, levee and soils engrg. Chile Exploration Co Chuquicamata, Chile 1937-40 as blast foreman, explosives testing. Cleveland-Cliffs Iron Co 1940-present: Ishpeming, Mich-Mining Engr to Mgr of Mines. Cleveland, Ohio-VP Mng to Sr VP Construction & Enrg. *Society Aff:* AIME, AISI, AMC.

Westwood, Albert R C
Business: 1450 S Rolling Rd, Baltimore, MD 21227
Position: Corp Dir - Res and Dev *Employer:* Martin Marietta Corp *Education:* DSc/Matls Sci/Univ of Birmingham, UK; PhD/Physical Metallurgy/Univ of Birmingham, UK; BSc (Hons)/Physical Metallurgy/Univ of Birmingham, UK. *Born:* 6/9/32. m.1956. 2 daughters. Tech officer, I.C.I. Ltd., U.K. 1956-58. Joined Martin Marietta Laboratories (then RIAS) in 1958, becoming Assoc Dir 1964, Deputy Dir 1969, and Dir 1974. Promoted to Corp. Dir R&D in 1984. Published over 100 papers, concerned principally with environment-sensitive mechanical behavior of variety of materials. Fellow Inst of Metallurgists 1967, Inst of Physics 1967, ASM 1974, AAAS 1986. Beilby Gold Medal of Royal Inst. Chemistry and Inst Metals 1970. Tewksbury Lecturer, Univ Melbourne 1974, Burgess Lecturer ASM 1984, Campbell Memorial Lecturer ASM 1987, Guest Lecturer Acad. Sciences USSR 1969, 1976, 1978, 1981 and 1986. Elected Mbr Natl Acad of Engrg 1980. Mbr Ntl Materials Advisory Bd, NRC-NAS 1980-85; and Commission on Engineering and Technical Systems, NRC-NAS 1985- ; Mbr Bd of Dirs of Martin Marietta Energy Systems 1984- ; The Metallurgical Society of AIME 1980-3, Financial Planning Officer 1985-8; Indus Res Inst 1984- , Vice President 1987- ; Maryland Humanities Council 1984- . Vice Chairman 1987- ; Mbr Adv Council to Sch of Arts & Sci, The Johns Hopkins Univ 1983- , Sch of Engrg, Univ. of Maryland 1983-; Oak Ridge National Laboratory 1985- ; Mbr Editorial Advisory Bd to Journal Materials Science. Hobbies, music, piano, travel. *Society Aff:* TMS-AIME, ASM-I, NAE, AAAS, IOP.

Westwood, Clarence P
Business: 3276 Commercial St SE, Salem, OR 97302
Position: President. *Employer:* C & H Engg, Inc. *Education:* BS/Civil Engg/OR State Univ. *Born:* April 1933. BSCE Oregon State Univ. Technician: Oregon State Hwy Dept, Fed Hwy Admin. City Engr for Klamath Falls, Ore and Medford, Oregon. VP, Chf Civil Engr for Marquess & Marquess Consulting Engrs. Resp for design and construction of general civil-municipal projects. President, C & G Engrg. Resp for admin and production of a 40 employee organization. Multidiscipline depts specializing in water and wastewater. Mbr NSPE, APWA, AWWA, PEO. *Society Aff:* NSPE, APWA, AWWA, PEO.

Wetherby, David A
Business: 130 Main St, Somersworth, NH 03878
Position: Mgr-Engg. *Employer:* GE Co. *Education:* BSEE/-/Univ of VT. *Born:* 4/4/28. in Timmouth, VT. Began his GE career in 1951 with Aviation & Ordinance

Wetherby, David A (Continued)
Systems. Served in Korea from 1951 to 1953 & rejoined GE Co, holding assignments in Syracuse and Utica, NY. In 1964 he was assigned as Mgr of Engineering at El Co, Ltd, Ireland. In 1969 he was assigned to the TV Bus Dept, Portsmouth, VA, as Mgr-Production Engg & later as Mgr-Component & Subsystem Engg. In Dec, 1978, he was assigned to the Meter Bus Dept in Somersworth, NH as Mgr of Engineering.

Wethern, James D
Business: P O Box 1438, Brunswick, GA 31520
Position: VP *Employer:* Brunswick Pulp & Paper Co. *Education:* BChE/Chem Eng/Univ of WI; MS/Chem Eng/Lawrence Univ; PhD/Chem Eng/Lawrence Univ. *Born:* July 12,1926 Minneapolis, Minn. Held various positions with Central Research Dept of Crown Zellerbach 1951-58 and with the Pulp & Paper Div of Riegel Paper Co 1959-72. Since Feb 1972 have been VP Mfg, Brunswick Pulp & Paper Co. Became Pres June 1980 and transfered to VP Planning in May 1982. Reg PE (ChE) in Louisiana. Was on Res and Tech Cttees of NCASI. TAPPI activities have been numerous including Chrmn of the Coating Cttee and Chrmn of College Relations Cttee. Was on TAPPI Bd/Dir 1973-77 and Chrmn of Local Sec Oper Cttee Enjoy water sports. *Society Aff:* TAPPI, AMA, CPPA, ΑΧΣ

Wetstein, Willis J
Business: Morrison-Maierle/CSSA, Inc, 4621 N. 16th St, Ste. D-401, Phoenix, AZ 85016
Position: Pres *Employer:* Morrison-Maierle/CSSA, Inc. *Education:* BS/Civ Engg/MT State Univ. *Born:* 4/12/35. Grad from MT State Univ in 1957. Joined Morrison-Maierle, Inc as a Sanitary Engr specializing in water & wastewater projs. Became Chief of the Co's Environmental Dept in 1966, responsible for all Environmental Engg projs. In 1978, was named Sr VP, in charge of co operations. Moved to Phoenix 1986, Pres of Morrison-Maierle/CSSA, Inc. Pres of the MT Sec of WPCF in 1974-75. National Director WPCF 1980-83. Received AWWA Geo Warren Fuller Award, MT Sec, 1979. Received WPCF Biedel Award, MT Section, 1986. Favorite past times are golfing & skiing. *Society Aff:* NSPE, WPCF, AWWA, APWA.

Wettach, William
Home: 659 Grove St, Sewickley, PA 15143
Position: VP & Tech Dir. *Employer:* Sterling Div-Reichhold Chemicals. *Born:* March 1910. PhD Univ of Pittsburgh; BS in Engrg and ChE from Princeton Univ, Phi Lambda Upsilon and Sigma Xi. On School Board - Sewickley and Quaker Valley School Dist 1954-66. Bd Mbr of Sewickley YMCA 1954-66. VP of Sterling Varnish Co 1940-69; VP and Tech Dir of Sterling Div-Reichhold Chemicals, Inc 1969- .

Wetzel, Otto K, Jr
Home: 3311 Beverly Dr, Dallas, TX 75205
Position: Pres. *Employer:* Purvin & Gertz, Inc. *Education:* BS/ChE/MA Inst of Technology. *Born:* 10/20/25. Native of OK City, OK. Initial position following graduation from MIT was the Natural Gas Dept of Mobil Oil Corp. In 1950, entered consulting engg with predecessor co of Purvin & Gertz, Inc and performed consulting assignments throughout the world related to economic utilization of hydrocarbons. Have appeared before various regulatory bodies and legal inquiries as a technical witness. *Society Aff:* AIChE.

Wetzel, Richard B
Business: 222 N New Jersey St100, Indianapolis, IN 46204
Position: Owner. *Employer:* Wetzel Engineers, Inc. *Education:* BSCE/Civil Engg/Purdue Univ. *Born:* Sept 1,1928 Indianapolis Indiana. Purdue Univ 1949 BSCE. Indiana reg PE and LS. Fellow ASCE, Vice President, American Consulting Engineers Council, Mbr of the Panel of Arbitrators-American Arbitration Assn, Past-Dir of Consulting Engrs of Indiana, Inc. First Dir of the Indianapolis Dept of Transportation 1970-74 for four and one-half yrs with responsibilities for design, construction, traffic control and maintenance of 3000 miles of roads & highways, approx 600 employees with 20 engrs. Prior to forming Consulting Engr firm in 1967, had been a principal in 2 other consulting engrg organizations; gen mgr and chf engr of a general constuction co, chf engr of a steel fabricating co. US Army Corps of Engrs, The Engineer School, instructor in design and construction of airfields. *Society Aff:* ASCE, AAA, ACEC, ACI, CEI.

Weymueller, Carl R
Business: 614 Superior Ave. W, Cleveland, OH 44113
Position: Senior Editor. *Employer:* Penton/IPC *Born:* Nov 9,1922 in Chicago, IL. Bachelor's Degree in Metallurgical Engrg, U of Illinois Feb 1949. Plant Metallurgist, Republic Steel Corp 1949-57 (Chicago). Assistant, Associate, and Senior Editor; Metal Progress 1957 to 1977.Editor, Metals Engineering Quarterly 1971 to 1977. Asso, then Sr Editor; Welding Design & Fabrication 1977 to present.

Whan, Glenn A
Business: College of Engrg, Albuquerque, NM 87131
Position: Professor *Employer:* University of New Mexico. *Education:* PhD/ChE/Carnegie Inst of Tech; MS/ChE/MT State Univ; BS/ChE/IN Inst of Tech. *Born:* Aug 1930, North Lima, Ohio. Asst Prof, Assoc Prof and Prof chem engrg and nuclear engrg at Univ of New Mexico since 1957. Dir of Los Alamos Grad Ctr 1963- 66. One yr in Lisbon, Portugal, as IAEA tech consultant to Laboratory of Nuclear Physics and Engrg. Chrmn of Nuclear Engrg 1966-71; and Chrmn of Chem and Nuclear Engrg 1971-75 and 1983-85. Asso Dean 1976-79. Consultancies have included Western Interstate Nuclear Bd, USAEC, Sandia Laboratory Los Alamos Ntl Lab, Rockwell Intl, and ACF Industries. Summer research at Los Alamos Sci. Lab, Sandia Lab, Oak Ridge and Argonne Natl Labs. *Society Aff:* ANS, AIChE, ASEE, NSPE.

Wharton, Charles B
Business: Phillips Hall, Ithaca, NY 14853
Position: Prof of EE. *Employer:* Cornell Univ. *Education:* MS/EE/Univ of CA; BS/EE/Univ of CA. *Born:* 3/29/26. From Gold Hill, OR, electronic technician, USN 1944-46. RF & plasma res, Lawrence Labs (Berkeley & Livermore, CA) 1950-62: developed microwave plasma diagnostics. Fundamental plasma wave experiments, Gen Atomic Co 1962-67; experimental confirmation of Landau Damping & Trapped-particle theories; discovery of plasma wave echoes & nonlinear sidebands. Teaching & res, Cornell Univ, 1967-84; plasma heating; intense ion and relativistic electron beams; positrons; plasma diagnostics. Res in wave interactions & ion diagnostics, Max- Planck Inst, Munich Germany 1959-60 & 1973-74. Dir, Intl Sch, Varenna, Italy 1978 and 1982. Visiting prof, advanced energy, Occidental Res Corp Irvine, CA 1979-80. Activities: sailing, skiing, mountaineering. *Society Aff:* APS, IEEE.

Wheaton, Elmer P
Home: 501 Portola Rd Box 8087, Portola Valley, CA 94025
Position: Consultant (Self-employed); Assoc & Dir *Employer:* Marine Development Associates *Education:* BA/Physics/Pomonx College. *Born:* Aug 15,1909. Grad from Pomona College, Claremont, CA 1933. Reg PE. Retired VP & Gen Mgr Lockheed Missiles & Space Co, Sunnyvale, CA R&D Div. 1962- 74: Assoc & Dir, Marine Development Associates 1986- . Resp for broad range of eng and scientific activities and diversification efforts. 1934-61: Corp VP 1960-61 after diverse mgt projects at Douglas Aircraft Co Santa Monica CA. Mbr Natl Advisory Cttee on Oceans & Atmosphere 1973, Fellow Marine Res Cttee, State of Calif 1972, Sea Grant Advisory Panel, State of Calif, AIAA Fellow, Fellow Marine Tech Soc, Fellow of Royal Aeronautical Soc, Fellow American Astronautical Soc. Mbr National Academy of Engineering. *Society Aff:* AIAA, MTS, AAAS.

Wheaton, Fredrick W
Business: Agri Engg Dept, College Park, MD 20742
Position: Prof. *Employer:* Univ of MD. *Education:* BS/Agri Engg/MI State Univ; MS/Agri Engg/MI State Univ; PhD/Agri Engg/IA State Univ. *Born:* 11/17/42. Native of Petoskey, MI. Agri Engr for US Dept Agri Res Service 1965-68. Been with the Agri Engg Dept, Univ of MD since 1968 as res assoc, assoc prof & now prof. Served as Acting Coordinator of MD Water Resources Res Ctr 1979-1981. Responsible for the Aquacultural Engg Res & grad teaching prog with the Agri Engg Dept. Res interests include engg aspects of production, processing & transportation of food & fiber derived from aquatic (fresh, brackish & saltwater) sources. *Society Aff:* ASAE, NSA, AFS, WAS.

Wheby, Frank T
1319 Grant St, Evanston, IL 60201
Position: Consultant *Employer:* Self. *Education:* MS/Civ Engr/Northwestern Univ; BS/Geology/MA Inst of Tech; BS/Civ Engr/WV Univ. *Born:* 9/7/30. Co officer in Corps of Engrs 1952-54. Engaged in dam design and construction for Alcoa 1956-59. Served Harza Engg Co for 15 yrs in a variety of staff capacities and finally as Hd of Geotechnical Div. Founded own firm in 1974, in which he is engaged in individual practice in planning, design, and construction surveillance of geotechnical projs. Worldwide experience (Surinam, Pakistan, Greece, Indonesia, Philippines, Brazil, and Ethiopia) in site evaluation, tunneling, fdns, offshore construction, and slope stability. Enjoys languages, computers, travel. Active in local chapter of profl societies. *Society Aff:* ASCE, AEG, BTS.

Wheeler, Arthur E
Home: 1812 Circle Rd, Towson, MD 21204
Position: Sr VP. *Employer:* Henry Adams Inc. *Education:* BS/ME/Duke Univ. *Born:* 8/28/26. in Phila PA. Served with USNR 1944-46. Instructor Physics RPI Troy NY 1947. Phila Region Engg Mgr, Carrier Corp 1948-59. Sr VP, Dir of Engg, Henry Adams, Inc, Towson, MD 1959- . Mbr Baltimore Fire Safety Cttee 1973. Recipient Hon Conceptor Award CEC/US 1969. Reg PE DE, FL, MD, MA, MI, NY, PA, VA, District of Columbia. Fellow ASHRAE. Pres Baltimore Chap 1968. Chrmn Natl Cttee Large Bldg Air Conditioning Sys 1973-76. Mbr Stds Cttee Energy Conservation in New Bldgs. Mbr Stds Committee. Engr Soc of Baltimore Dir 1970, Mbr American Consulting Engrs Council. Natl Chmn Duke Annual Giving, ASHRAE R&T Ctte mbr. *Society Aff:* ASHRAE, SAME, CEC.

Wheeler, C Herbert, Jr
Home: 638 Franklin St, State College, PA 16803
Position: Prof. *Employer:* Penn State Univ. *Education:* M in Arch/Arch/MA Inst of Tech; B of Arch/Arch/Univ of Penn. *Born:* 6/6/15. Native of Merchantville, NJ. Served arch internship with Austin Co & White Engg Corp in NY prior to WWII. Served with Army Corps of engrs, 1942-45. Disch rank of Major. Returned to White Engg Corp, & designed power plants, warehouses, refineries as chief arch until 1955. Became mgr of engg for Stran Steel Corp in Detroit, MI. In 1958 became Chief Shelter Systems Engg for Curtiss Wright Corp. In 1964 joined the faculty of PA State Univ. Prof of Arch Engg, spec ints in education of srs class & practicing engrs and archs in the field of Prof Practice & mgt. Enjoy travel. Active as a Consultant to Arch/Engg firms, Prof Societies, Govt Agencies & Universities. Chmn of Div 69-70 ASEE/AE. Elected Fellow of the AIA-American Inst of Arch. Now retired as Professor Emeritus; appointed permanent secretary of Professional Development Group of Union Internationale des Architectes & member of its College of Delegates. *Society Aff:* ASEE, AIA

Wheeler, Edward A
Home: 534 Line Road, Hazlet, NJ 07730
Position: Supervisory Training Specialist *Employer:* US Army Communications and Electronics Command *Education:* EdD/Ed Adm/Lehigh Univ; MS/Math/IN State Univ; MA/Phil/Northwestern Univ; BME/Mech Eng/Rensselaer Poly Inst. *Born:* Dec 1924 Amherst, MA. Served with the U.S. Army 1943-1946. Asst-to Assoc. Prof. ME 1950-1964 at Rensselaer, Rose and Virginia Polytechnic. Administrator, continuing education, 1965-1968 for Bell Labs and Hughes Aircraft. Employee Develop Specialist 1969-1978, U.S. Army; Director, Army Product Assurance Industrial Training program 1979 to present. Mbr, Bd/Dir ASEE 1970-1972; Assoc. Fellow AIAA; reg PE in PA and VA. *Society Aff:* NSPE, ASEE, ASTD, PTS

Wheeler, Harold A
Business: Greenlawn, NY 11740
Position: Chf Scientist, Chairman E *Employer:* Hazeltine Corp. *Education:* BS/Physics/George Washington Univ. *Born:* 1903 Minnesota. Engr, Hazeltine Corp, NY area 1924-46,59--. Dir, Chmn 1966. Pres., Wheeler Labs Inc 1957-68, then a subsidiary of Hazeltine Corp. Before WWII the design of broadcast receivers and early TV. Inventor of diode AVC and linear detector for AM receivers. During WWII, the design of IFF equipment for radar. Continued specializing in microwaves and antennas. Over 180 US patents and many foreign patents. IRE & IEEE, Fellow, Medal of Honor 1964. Resident of Smithtown, NY. Stevens Inst of Technology DEngg (Hon) 1978. Author "Hazeltine the Prof" 1978; "The Early Days of Wheeler and Hazeltine Corporation" 1982. George Washington U, D.Sc (Hon) 1972. *Society Aff:* IEEE, AIAA, RCA.

Wheeler, John A
Home: 15 Decatur Dr, Nashua, NH 03062
Position: Gen Mgr Semiconductor Processing Co. *Employer:* Cabot Corp. *Education:* MS/Met/IL Inst of Tech; BS/Engg/IL Inst of Tech. *Born:* 12/12/35. Native of Chicago, IL. Oak Ridge Natl Lab, 1960/65. Genl conference chrmn and editor of proceedings "Diffusion in BCC Metals–" (ASM 1965). While with Union Carbide Corp (1965/1970) and Cabot corporation (1970-present) involved with the development, manufacture, and mktg of high technology matls for aircraft, chemical, electronic and power generation applications. Controller for Steel products and Oil Field Equipment Div 1978-1980. Controller (1980-82) and Gen Mgr US Operations (1982-83) wear tech div. Corporate Staff 1983-86. Gen Mgr Semiconductor Processing Co, (1987-present). Active in community youth activities. Pres. BSA Council (1976-77). *Society Aff:* AAAS, AIME, ASM.

Wheeler, Lewis T
Business: 4800 Calhoun, Houston, TX 77004
Position: Prof. *Employer:* Univ of Houston. *Education:* PhD/ME/Caltech; MS/ME/Univ of Houston; BS/ME/Univ of Houston. *Born:* 9/28/40. Native of Houston, TX. Attended Univ of Houston 1959-1964. Returned from Caltech in 1968 to join faculty. Currently Prof of Math & Mech Engg. *Society Aff:* ASME, AMS, SIAM.

Wheeler, Orville E
Business: Herff College of Engg, Memphis, TN 38152
Position: Dean. *Employer:* Memphis State Univ. *Education:* PhD/CE/TX A&M Univ; MSCE/CE/Univ of MO; BE/CE/Vanderbilt Univ. *Born:* 12/31/32. Memphis, TN; married Mary Bea Rychlik 6-6-56, daughter Lynnette; pilot USN 56-59; Structural Engr Chance Vought 59-60, Hayes Aircraft 60-61, Brown Engg 61- 62; Aerospace Technologist MSFC, NASA 62-66; Proj Str Engg & Design Spec Gen Dynamics 66-72; Chief Structures Engr Bucyrus-Erie Co 72-78; Prof of Civ & Mech Engg & Dean, Herff College of Engg Memphis State Univ 78-present. 1973 Am Iron & Steel Inst Award for Best Engg Ind Product. *Society Aff:* ASCE, ASTM, ASEE, AISC, NSPE.

Wheeler, Robert G
Business: 401 Becton Center, Physics, P.O. Box 2157, New Haven, CT 06520
Position: Prof of Applied Physics *Employer:* Yale Univ *Education:* PhD/Physics/Yale Univ; BS/Eng Physics/Lehigh Univ *Born:* 01/05/29 During the period of 1955-57 he was on active duty with the USAF stationed at the Aeronautical Res Lab, Wright Patterson AFB, OH. In 1957 he returned to Yale Univ as a faculty mbr; he served as Chrmn of the Dept of Engrg and Applied Science from 1971-74; as Director, Div of Physical Sciences from 1974-77; and as Chrmn of Applied Physics from 1981- . He is married with three children. *Society Aff:* IEEE, APS

Wheeler, William C
Home: 105 Hunting Lodge Rd, PO Box 41, Storrs, CT 06268
Position: Prof Emeritus. *Employer:* University of Connecticut. *Education:* BSAE/Agri Engg/Univ of GA; MSAE/Agri Engg/VA Poly Inst; Grad Sty/Agri Engg/ MI State Univ. *Born:* Feb 1914. Grad study MSU, MS, VPI and B.S., Univ of Georgia. Taught Georgia Vocation School and Univ of Tenn prior to Infantry Serv 1942-46. Assoc Prof, Univ of Tenn 1946-53 spec in Rural Electrification. Prof and Dept Hd, Agricultural Engrg Dept, Univ of Connecticut 1953-72. Participated in Teaching, Extension, Research Programs, Active in New England Reg. Res Projects relating to: Poultry Housing, Mechanization of Forage Crops, Mechanization of Fruit and Vegetable Harvest and Improvement of Apple Harvest Efficiency. Chrmn, Connecticut Farm Elec Council Projects, Cttee 1953-68. PChrmn, Tenn and No Atlantic Sections ASAE. Life-Fellow ASAE. Registered Prof Engr in TN. Prof Emeritus since 1973. Enjoy travel and Home Workshop Activities. *Society Aff:* ASAE, NSPE.

Wheeler, William T
Business: 7462 N Figueroa St, Los Angeles, CA 90041
Position: Chairman *Employer:* Wheeler & Gray. *Education:* BS/Civil Engg/CA Inst of Tech. *Born:* Oct 1911. Calif registration and nine additional states. Consulting engr, civil and structural since 1946. Chrmn, Wheeler & Gray. Honorary Mbr: Structural Engrs Assn So Calif, PPres; CEAC, PDir; LA Council Engrg Societies, PPres; ACI, PPres, So Calif Chap; ASCE Fellow; EERI, Fellow, Director; Inst for Advancement of Engrg, Fellow. Awards: Sam Hobbs Service Award, SoCol Chap ACI 1978 Outstanding Engr Merit Awd IAE 1974, Commendation Calif State Legislature 1972, USA Rep UNESCO Intl Cttee 1963 to formulate 'Principles of Earthquake Resistant Design-Developing Countries'. Author: 'Engineers Offer Earthquake Code' ENR 1959, 'Conventional Foundations and Earthquake Problems' 4th World Conf Earthquake Enrg, Santiago, Chile 1969. *Society Aff:* ASCE, ACEC, EERI, ACI

Wheelon, Albert D
Business: 909 N Sepulveda Blvd, El Segundo, CA 90245
Position: Vice President. *Employer:* Hughes Aircraft Co. *Education:* BSc/Eng Sc/ Stanford Univ; PhD/Physics/MIT *Born:* Jan 1929. Sr VP & Group Pres of Space & Communications Group which is responsible for all of the Hughes Space programs and its involvement in advance communications tech and operations. Prior to this assignment, served as VP, Engrg since joining Hughes in 1966 and became broadly involved in the Hughes technical program. Mgr: Prof's Foreign Intelligence Advisory Bd, Fellow IEEE. Mbr NAE. More than 30 scientific papers on missile guidance, radiowave propagation and fluid turbulence, 'Tables of Summable Series & Integrals Involving Bessel Functions.'. *Society Aff:* NAE, IEEE

Wheler, Arthur Gordon
Business: 10 Albany St, Cazenovia, NY 13035
Position: Senior Partner *Employer:* Stearns & Wheler *Education:* MS/Sanitary Engr/ MIT; BS/CE & Physics/Syracuse Univ *Born:* 2/21/21 Married Barbara Jane Carlson, 1952; Children: Bradford Gordon, Georgeann Hobson; Liaison Engr, Glenn L Martin Co, Baltimore, MD, 1941-43; Air Force Intelligence, Pacific, WWII, 1943-46; Research Assistant, MIT, 1950-52; Research engr, Oak Ridge National Lab, 1952; Research Assoc, NY Univ, 1952-53; Sanitary Engr, Nitrogen Division, Allied Chem & Dye, Hopewell, VA, 1953-55, Consulting Sanitary Engr, Stearns & Wheler, Engrs and Scientists, Cazenovia, NY, 1955 to date, designed water, sewage and industrial waste treatment plants; Received Charles Agar Award, 1957 and 1964. Author: "Assaying Techniques for Radioatopic Contaminants in Water Supplies–, NYO No. 4437, published by Atomic Energy Commission, 1952. *Society Aff:* ASCE, ACEC, AWWA, WPCF, AAEE

Wheten, Waldo A
Business: 89 Carlingview Dr, Toronto, Ontario, Canada M9W 5E4
Position: Sr Engg. *Employer:* Underwood McLellan Ltd Consulting Engrs & Planners. *Education:* BSc/Civil Engg/Univ of Saskatchewan. *Born:* July 1914 Saskatoon, 1938-40 Field Engr for Dominion Dept of Agriculture on dam and irrigation works. 1940 seconded to the Dept of Natl Defence on airport construction. 1941-46 commissioned as a Tech Armament Officer with R.C. A.F. 1946-49 Irrigation Engr with Govt of Ceylon on irrigation and land development. 1949-55 with Canadian Westinghouse in various engrg managerial positions. 1955-74 with City of Hamilton in various engrg positions. Appointed City Engr in 1959. 1974 appointed Commissioner of Engrg Regional Municipality of Hamilton-Wentworth. Retired from this position in July 1979 and joined the firm of Underwood McLellan ltd, Consulting Engr & Planners as S Engr. Responsibilities include identifying & securing business assignments in central & southern Ontario & providing technical assistance on specific projs. PChrmn Canadian Sec AWWA. PChrmn Hamilton branch EIC. PChrmn Hamilton Chap APEO. Hobby: Philately. Elected Fellow of the ASCE 1966. Elected Fellow of the EIC 1971. *Society Aff:* EIC, ASCE, AWWA, APWA.

Whicker, Lawrence R
Business: Code 6850 NRL, Wash, DC 20375
Position: Hd Microwave Tech Br. *Employer:* Naval Res Lab. *Education:* PhD/EE/Purdue Univ; MS/EE/Univ of TN; BS/EE/Univ of TN. *Born:* 10/3/34. Employed by Sperry Microwave Elctronics Co from 1958-61. Various Mgmt & Tech Positions at the Westinhouse Defense & Space Ctr 1964-70. Since 1970, Hd of the Microwave Tech Branch, Naval Res Lab. Elected Fellow of the IEEE in 1978, for contributions to non-reciprocal microwave components. Pres Microwave Theory & Techniques Society. Gen Chrmn for 1980 Int'l Mcrowave Symposium. *Society Aff:* IEEE, MTTS, GSU.

Whinnery, John R
Business: Dept EECS, Berkeley, CA 94720
Position: Professor. *Employer:* University of California. *Education:* PhD/EE/Univ of CA, Berkeley; BS/EE/Univ of CA, Berkeley; AA/Engin/Modesto (Calif) J.C. *Born:* 7/26/16. Worked at General Electric Co, Schenectady NY 1937-46; on faculty Univ of Calif Berkeley since 1946 (Dir Electronic Research Lab 1952-56, Chrmn EE Dept 1956-59 and Dean, Coll of Engrg 1959-63); teaching or research leaves at Union College, Hughes Aircraft Co, ETH (Zurich), Bell Labs, Univ of Calif, Santa Cruz, Stanford Univ & Natl Defense Acad Yokosuka, Japan. Honors include Guggenheim Fellowship 1959, IEEE Education Medal 1967; ASEE Lamme Medal 1975; Microwave Career Award IEEE (1976); Distinguished Engrg Alumnus Award UCB 1980; Appointed Univ Prof 1980; Medal of Honor IEEE 1985; Founders Medal NAE 1986; Mbr NAE, NAS and Amer Acad Arts and Science; Bd/Dir IRE 1956-59 and of IEEE 1967-70. Fellow of IEEE and Secy 1970; Fellow OSA, Fellow AAAS. *Society Aff:* IEEE, OSA, APS, ΣΞ, AAAS.

Whipple, Walter L
Home: 1678 Spruce View Dr, Pomona, CA 91766
Position: Design Specialist. *Employer:* Gen Dynamics. *Education:* PhD candidate, Comp Info & Control Engr/Univ of MI; candidate, Univ of MI; BS/Engg Sci/ Harvey Mudd College *Born:* 6/23/40. Field service Rep & Technical writer, Ordinance Div, GE Co, 1962-65. Grad study, math statistics, Univ of CA, Berkeley, 1965-66. Resident Inspector, Welker & Assoc, Marietta, GA, 1966-67. Engr, Apollo Field Service, Raytheon Co 1967-69. Sr Prog/Analyst, Profl Services Div, Control Data Corp, 1969-78. Owner, The Secretariat, Ann Arbor, MI, 1975-78. Design specialist, Systems Dynamics, Gen Dynamics/Pomona, 1978- (lead engr, Phalan dynamics & Divad Support software, project leader, weapon computer systems IR & D project; principal investigator, high order languages IR & D. task; coordinator, technical advisory committee software IR & D J Session Chrmn simulation, 1972 ACM Annual conf. Session Chrmn, financial & econometric modeling, 1975 Winter Comp Simulation Conf. MSE Thesis: Optimal Profit Scheduling. Part-time Instructor, Elec & Comp Engr, CA Polytechnic Univ, Pomona, 1979. Adjunct Instr, Comp Sci, Harvey Mudd Coll, Claremont, 1980-81. This is in addition to my full-time employmemt. *Society Aff:* IEEE, NSPE, ACM, SCS, AIAA.

Whipple, William
Home: 395 Mercer Rd, Princeton, NJ 08540
Position: Assis Dir, Water Supply & Watershed Protec *Employer:* Div of Water Res, NJ Dept of Environ Protec *Education:* CE/Civil Engr/Princeton Univ; MA/ Economics/Oxford Univ; BA/Economics/Oxford Univ; BS/-/U.S. Mil Acad. *Born:* 2/4/09. Louisiana. Grad West Point Service Army Corps Engrs as district and division engr, and Office Chief of Engrs. Retired Brigadier General. Chf Eng NY World's Fair Corp 1960-64. Dir Water Resources Research Inst, Rutgers Univ 1965- 79. Research Prof 1979-81. Author 100 papers and 5 books. Commendation from President of US 1971. Toulmin Award SAME 1975. President, AWRA 1973. Chrmn, Urban Water Resources Res Council ASCE 1973-74. Chrmn, Universities Council on Water Resources 1976-78. ICKO Ibeu Award from AWRA 1978; Julian Hinds Award ASCE 1982. Many consulting engagements. Prime interests water supply, stormwater management, and political economy. *Society Aff:* ASCE, SAME, AWRA.

Whitaker, H Baron
Home: 1905 Wedgewood Dr, Sanford, NC
Position: President - Retired. *Employer:* Underwriters Laboratories Inc. *Education:* BS/Elec Engg/NC State Univ; Hon/LLD/IL Inst of Technology. *Born:* 5/25/13. in Durham, NC. Joined Underwriters Lab 1936. US Army Signal Corps 1941-45. Rejoined Underwriters Lab 1946. In charge of NY operations 1959. VP & Chief Elec Engg 1959. Exec VP 1963. Pres 1964. Chrmn of Natl Elec Code Cttee 1961-67. Vice Chrmn 1968-72. Mbr Bd/Dir Natl Fire Prot Assn & ANSI. Mbr Tau Beta Pi, Eta Kappa Nu. IEEE Fellow. Hon LLD from IL Inst of Tech 1968. Consultant to the Pres; mbr US Natl Committee-Intl Electrotechnical Commission; mbr US Delegation to the Intl Conference on Lab Accreditation - mbr US Delegation to CERTICO of Intl Stds Organization. *Society Aff:* NFPA, IEEE.

Whitaker, Stephen
Business: Dept Chem Engr, Davis, CA 95616
Position: Prof *Employer:* Univ of Calif *Education:* PhD/ChE/Univ of DE; MS/ChE/ Univ of DE; BS/ChE/Univ of CA-Berkeley *Born:* 07/08/32 Prof Whitaker's main res interests are interfacial phenomena, fluid mechanics and transport phenomena in multiphase systems. His teaching interests are identified with three textbooks: *Introduction to Fluid Mechanics*, R.E. Krieger Pub. Co, 1981, *Elementary Heat Transfer Analysis*, Pergamon Press, 1976 and *Fundamental Principles of Heat* Transfer, Pergamon Press, 1977. *Society Aff:* AIChE, ACS.

Whitaker, Thomas B
Business: P.O. Box 5906 College Station, Raleigh, NC 27607
Position: Res Engr *Employer:* USDA - ARS *Education:* PhD/Biol & Agr Eng/OH St Univ; MS/Biol & Agr Eng/NC St Univ; BS/Biol & Agr Eng/NC St Univ *Born:* 05/16/39 Native of Raleigh, NC. Work for USDA, ARS since graduation. Also holds appointment as prof in the Biological and Agri Engrg Dept at NCSU. Res deals with the control, prevention, and detection of aflatoxin in agr commodities. Awarded the National Peanut Council's *Golden Peanut Research Award* in 1980 for developing aflatoxin testing programs for the US peanut industry. *Society Aff:* ASAE, APRES, ΣΞ

White, Alan D
Home: 127 Hillside Ave, Berkeley Heights, NJ 07922
Position: Consultant *Employer:* Self *Education:* MSc/Physics/Syracuse Univ; BA/ Physics/Rutgers Univ. *Born:* 7/6/23. Degrees from Rutgers Univ 1949 and Syracuse U 1952 in physics. Served overseas with armed forces 1943-46. Employed at Bell Tel Labs 1953-1983 as mbr of tech staff working in gas discharge physics, gas lasers, and optical systems. Retired from Bell Labs in 1983. Elected to grade of Fellow IEEE in Jan 1976 for development and subsequent improvements of the visible light helium neon laser. Mbr Phi Beta Kappa, Sigma Xi, OSA, IEEE and SPIE. Recipient David Sarnoff award 1984 for inven of visible light helium neon laser. Presently conslt in lasers and optics. *Society Aff:* IEEE, OSA, SPIE

White, Edwin Lee
Home: 1206, One Beach Drive, St. Petersburg, FL 33701
Position: Retired. *Education:* MS/Comm Eng/George Washington Univ; AB/Physics/ George Washington Univ. *Born:* July 5,1896 Valley City, ND. Mil. Svc.: USNR, WWI; North Atlantic Service, Victory Medal with Clasp WWII; USA Air Corps; Pacific, CBI and European Theaters. Bronze Star. Retired as Colonel. Civil Service: 1922-27 Naval Research Laboratory, D.C. 1927-30 Radio Engr, Signal Corps, Hawaii 1930-55 Federal Radio Commission and Federal Communications Commission. Mbr of numerous U S Delegations to Intl Mtgs on Telecommunications matters. U S Spokesman at ITU Intl Aeronautical Administrative Radio Conference Geneva 1948-49. Fellow IRE and AIEE consequently Fellow IEEE. Consulting Engr 1955-65, 1965 to date. Author. Active in Civic and Academic organizations. *Society Aff:* IEEE, TROA, SAR, QCWA, SAME, AFA, HPA.

White, Francis B
Home: 10 Timber Lane, Randolph, NJ 07869
Position: Mgr Internatl Operations. *Employer:* Foster Wheeler Energy Corp. *Education:* ChE/Chem Engg/Princeton; BSChE/Chem Engg/Princeton. *Born:* Nov 1915 NYC. Sigma Xi. PE Licenses NY and NJ. With Foster Wheeler since 1937. Mng Dir Ishikawajima Foster Wheeler in Japan 1963-65 with responsibility for engrg and sales of process plants. Mgr Intl Operations Process Plants Div Foster Wheeler Energy Corp from 1966, coordinating parent co services to worldwide affiliates and Japanese licensee. Fellow and former mbr Admissions Cttee AIChE and former Chrmn NY Sec. President American Oil Chemists' Soc 1976- 77, General Chairman AOCS World Conference on Edible Oil Processibly, l982 Lecturer Edible Oils and Fatty Acid Processing. Interests: golf, hiking, camping, scouting and gardening. *Society Aff:* AOCS, AIChE, PEA, ACS.

White, Gerald A
Business: Air Products and Chemicals, Inc, PO Box 538, Allentown, PA 18105
Position: VP-Finance & Chief Financial Officer. *Employer:* Air Products & Chems, Inc. *Education:* BCE/-/Villanova Univ; Adv Management Prog/-/Harvard Bus Sch. *Born:* 8/2/34. Native of Long Island, NY. Active duty with USN as naval aviator 1957-62. Joined Air Products in 1962 as staff engr. Subsequently worked in data processing, controllership, Energy Systems, Corporate Planning, and presently Vice President-Finance and Chief Financial Officer. Active in community & civic affairs. *Society Aff:* AIChE, FEI.

White, Gifford E
Business: Box 698, Austin, TX 78767
Position: President. *Employer:* White Instruments Inc. *Education:* MA/Physics/Univ of TX; BA/Physics/Univ of TX. *Born:* Feb 17,1912 San Saba, Texas. Research Geophysicist at Humble Oil & Ref Co 1934-39. Grad School MIT 1939-41. Research eng Sperry Gyroscope Co 1941-47. VP Statham Instruments Inc 1947-52. Pres. White Instruments Inc 1953–. Fellow IEEE. Instrumentation for physical measurements; circuit theory. *Society Aff:* SEG, IEEE.

White, Harold D
Business: Prof. Engineers & Assoc. P.C, 561 University Drive, Athens, GA 30605
Position: President. *Employer:* Professional Engrs and Associates, P C. *Education:* MSAEN/Structures/IA State Univ; BSAEN/Genl/Univ of GA. *Born:* 8/29/10. Native of GA. Established engrg at Abraham Baldwin College. Resident Engr Univ Sys of GA. Special Supvr War Food Admin. Advisor to State War Bd. Life fellow ASAE. Reg PE. P Chrmn of GA & Southeast Region ASAE. State Dir GSPE. P Grand Pres Alpha Gamma Rho Natl Natl Fraternity. Gamma Sigma Delta, Phi Kappa Phi, Aghon, Gridiron, Sigma Xi, Who's Who in S & SE, Who's Who in Amer Ed. Prof Emeritus, Univ of GA, Who's Who in Engg, Who's Who in Education. 39 yrs teaching and Research in Univ sys of GA. Treasurer, Alpha Gamma Rho Bldg Fund Inc. Who's Who in Tech Today. GA ASAE Engr of the Year 1984. Who's Who in Tech Today. GA ASAE Engr of the Year 1984. *Society Aff:* NSPE, ASAE.

White, Harry J
Home: Del Mesa Carmel 69, Carmel, CA 93921
Position: Consultant (Self-employed) *Education:* PhD/Physics/Univ of CA Berkeley; MS/Physics/Univ of CA, Berkeley; BSEE/Elec Engg/Univ of CA, Berkeley. *Born:* 7/29/05. Assoc Group Ldr, MIT Radiation Lab during WWII. Dir R/D Research-Cottrell 1945-60 and Bd Mbr 1958-62. Head, Dept Eng and Applied Science, Portland State Univ 1960-72. Served on Bd of Trustees, Oregon Graduate Ctr 1963-72. Author of 'Industrial Electrostatic Precipitation' Addison-Wesley 1963 and numerous other books and papers, in electrostatic precipitation, air pollution control, gaseous electronics, high-voltage equipment, radar, & elec engg. His most recent book is *Electroslatic Precipitation of Fly Ash*, 1977. Mbr of APCA, Sigma Xi and other societies. Fellow IEEE, APS and AAAS. Recipient of Frank A Chambers Award from APCA 1971, and Larry Faith Tech Achievement Award, APCA, 1979. Award of Merit, Inst of Elec Japan, 1982. International consultant in electrostatic precipitation and industrial air pollution control 1960--. Editorial Bd of J. APCA 1975--. *Society Aff:* IEEE, APS, AAAS, APCA, ΣΞ.

White, J Coleman
Business: 3412 Hillview Ave, Palo Alto, CA 94303
Position: Program Mgr-Plant Electrical Systems and Equipment *Employer:* Elec Power Res Inst. *Education:* BEE/Elec Engg/Cornell Univ. *Born:* Aug 1923 Ft Dodge, Iowa. Served in Army Air Corps 1943-46. Engr for Gen Elec Co from 1947 to 1979, specializing in synchronous, induction, and direct current motor and generator design and development, and in computer applications to the design process. Fellow IEEE. Mbr of IEEE Rotating Machinery Cttee. Reg PE. Received IEEE Tesla Award, 1987. *Society Aff:* IEEE.

White, John A
Home: 5119 Stratham Dr, Dunwoody, GA 30338
Position: Regents Prof of Indus & Sys Eng *Employer:* Georgia Institute of Technology. *Education:* PhD/IE/OH State Univ; MS/IE/VA Poly Inst; BS/IE/Univ of AR. *Born:* 12/5/39. Reg PE VA. Mbr ASEE, IMMS, CLM, NSPE, ORSA, SME, TIMS, WERC. Fellow IIE. Mbr Alpha Pi Mu, Omicron Delta Kappa, Phi Kappa Phi, Sigma Xi. Past Pres IIE, PEVP-Div, Educ & Publications IIE. PVP-Sys Eng & Tech Div IIE. P Dir, Continuing Edu IIE. P Dir Facilities Plng & Design Div IIE. Senior Editor, IIE Transactions. Chairman AAES, 1986, Sec/Treas AAES, 1984. Mbr Bd of Gov, AAES, 1982-83. Past Mbr, Bd of Dirs, Matl Handling Educ Fnd. Co-author 5 text books. Co-recipient 1974 and 1986 IIE Book of the Yr Award. Recipient 1974 Region III IIE Award for Excellence. Recipient 1980 Facilities Planning and Design Div Award, IIE. Recipient 1984 Region IV IIE Award of Excellence. Recipient Distinguished Alumnus Award, College of Engrg, The Ohio State University, 1984. Chmn SysteCon, Inc. Employed as engr with TN Eastman Co, Ethyl Copr, & North Amer Rockwell. Taught indus engg at OH State Univ, VA Poly Inst & GA Tech. Member, National Academy of Engineering, 1987. Honorary Doctorate of Engineering, Catholic University of Leuven, Belgium. Member, Arkansas Academy of Industrial Engineering, 1987. Married, two children. *Society Aff:* CLM, SME, WERC, NSPE, ASEE, IMMS, ORSA, TIMS, IIE.

White, John B
Home: 127 Oak Manor Dr, Natrona Hts, PA 15065
Position: VP, Primary Glass Products. *Employer:* PPG Industries Inc. *Education:* BSME/Mech Engg/OH State Univ. *Born:* Feb 1924. Native of Toledo, Ohio. Married Helen F Collins; four children. USAF 1942-45. With PPG Industries Glass Div since 1952. VP since 1966. Mbr ASME and SAE. Reg PE Ohio and PA. 32nd Degree Scot. Rite Mason. Hobbies include photography, amateur radio and boating. Ohio State Univ Lamme Medal 'Meritorious Achievement' 1974. *Society Aff:* ASME, SAE, NSPE.

White, John K
Business: Beulah Rd, Pittsburgh, PA 15235
Position: Senior Research Scientist *Employer:* Westinghouse Research Lab. *Education:* BS/Chem/Univ of Pittsburgh. *Born:* 1924 Pittsburgh PA. Univ of Pittsburgh. Fellow of ASNT and former chrmn of its Nuclear Components Cttee. Formerly PVRC and ASTM. Taught NDT for Penn State. In charge of Westinghouse Corp. development in NDT. Formerly with ORNL, Curtiss-Wright, and Graham Research Lab. Active in inspectn of nuclear reactors, steam turbines, pipe tubing. *Society Aff:* ASNT.

White, Joseph F
Business: 43 South Ave, Burlington, MA 01803
Position: Technical Dir. *Employer:* Microwave Assoc, Inc. *Education:* PhD/EE/RPI; MS/EE/Northwestern U; BSEE/EE/Case Inst of Tech. *Born:* 6/5/38. Tech Staff at MITRE Corp, Bedford, MA, June 1960 - Dec 1961; Since then with M/A-COM, Inc, Burlington, MA. Currently VP/Tech Dir, Corp. Tech. Invented transmission type diode phase shifter, used in microwave phased array antennas. Received 1975 application award of Microwave Theory & Techniques Group of IEEE for contributions to phased array antenna field. Fellow of IEEE Mbr Eta Kappa Nu & Sigma Xi Honorary Fraternities. Wrote text *Microwave Semiconductor Engineering*, Van Nostrand Reinhold, NY, 1982, 558 pgs. *Society Aff:* IEEE, HKN, ΣΞ.

White, Kenneth N
Business: R R 4, Box 26, Salina, KS 67401
Position: Partner. *Employer:* White, Hunsley & Assoc. *Education:* BS/Civil Engrg/Univ of KS. *Born:* April 14,1932. Elected board mbr of Water Dist No 1, Johnson County, Kansas 1966-67. Young Engr of Yr, MO. 1967. Mbr Kansas Engrg Soc, NSPE, ASCE, AWWA. Reg PE in KS, MO, IA, NE, CO and FL. Chrmn of Kansas Consulting Engrs 1976. Partner of White, Hunsley & Assoc 1969 to present. Proj Engr at Bucher & Willis 1967-69. Bridge Design Engr at Howard, Needles, Tammen & Bergendoff 1955- 67. Mbr Aircraft Owners & Pilots Assn and Natl Pilots Assn, Mbr City of Salina Planning Commission 1981-. *Society Aff:* NSPE, ASCE, ACEC.

White, L Forrest
Home: 434 Avon Rd, Memphis, TN 38117
Position: Mechanical Engineer-Plant Inspection *Employer:* DA-Corps of Engineers *Education:* BS/ME/Univ of TN; Additional work on Masters/CE/Memphis State Univ. *Born:* 7/13/24. Native of Memphis, TN. Served as Pilot in AF 1943-46, 1951-53. Retired Memphis Light, Gas, and Water Div 1974 after 25 yrs in Natural Gas Distribution design, res, and construction. Control (Central) System design for the largest publically owned Utility in the US while Asst Chief Engr-Gas Div. Joint Engg Coordinator 1970-74 responsibility for common utilization of new techniques in house by four Engg Depts while in GE Dept. Chief Engr-Environmental for Div of Public Works, City of Memphis 1974-1980 with responsibility for Res and Dev of special projs including Wastewater Treatment under Public Law 92-500, Contract Document Evaluation, CAMRAS (Computer Assisted mapping and Records Admin Sys). Mechanical Engineer, Plant Inspection, Department of ARMY, Corps of Engineers 1980-present with responsibility for equipment and facilities engineering for River Operations and Navigation in the Memphis District. Contracting Officer Representative. Chrmn, ASME-Mid-South 1964, VP, MSPE 1969, Chrmn, TAC, APWA, CAMRAS 1976-77, Featured Engr (EJC) ASME 1965, MSPE 1970, NJEC 1979. Enjoys horticulture & bowling. Registered PE - TN No. 2883 - MS No. 6329. *Society Aff:* NSPE, ASME, ASP, ACSM.

White, Marvin H
Business: Adv Tech Lab MS3531, Baltimore, MD 21203
Position: Advisory Engr. *Employer:* Westinghouse Def & Elec Sys Ctr. *Born:* Sept 6,1937 Bronx, NY. PhD in EE from Ohio State Univ 1969, MS in Physics 1961 and BSE in Physics & Math 1960 from Univ of Mich, AS from Henry Ford Community College 1957. With Westinghouse since 1961 in the special electron device and integrated circuit areas. R&D on silicon diode camera tubes, electro- optical photodiodes, MOS and CMOS transistors, bipolar-MOS LSI, low/high power microwave bipolar transistors, MNOS memory transistors and CCD's for imaging,

White, Marvin H (Continued)
memory, and signal processing. Fellow IEEE for contributions to electron device R & D especially CCD imaging & MNOS nonvolatile memory transistors.

White, Merit P
Business: Univ of Mass, Amherst, MA 01003
Position: Comm Prof - Civil Eng Dept. *Employer:* University of Massachusetts. *Education:* PhD/Civil Engg/CA Inst Tech; MSCE/Civil Engg/CA Inst Tech; CE/Civil Engg/Dartmouth College; AB/Civil Engg/Dartmouth College. *Born:* Oct 1908. Res Assoc, Harvard and Cal-Tech 1935-39. Asst Prof IIT to 1942. U S Govt to 1948. Univ Mass to present. Consultant to Govt and industry in engrg seismology, structural dynamics. Publications in those areas, also plastic wave phenomena, fluid mechanics. Fellow ASCE, ASME; Institution of Mechanical Engrs; RILEM; Seismological Soc of America. Previously Chrmn of ASCE Engrg Mech Div, Cttee on Engrg Edu, Cttee Honors and Awards. Recd President's Certificate of Merit 1948. Listed in Who's Who in America. *Society Aff:* ASCE, ASME

White, Niles C
Home: 823 Watts Dr SE, Huntsville, AL 35801
Position: Technical Advisor. *Employer:* Atlantic Res Corp. *Education:* BS/ChE/Univ of AL *Born:* 2/14/22. Saragossa, AL, M57, CE. BS AL 50. CE, US Naval Ord Sta 50-51; Redstone Arsenal, 51-56, Supv Aerospace Engr, Army Rocket & Guided Missile Agency 56-64, Chief Solid Propellant Chem Function 64-80, Tech Adv. Atlantic Research Corp. 80-present. Res & Dev Achivement Award Dept Army 61. US Army 43-46, AAAS, ACS, AIChE, AUSA, Rocket Propuslion; Propellants; Combustion; Physical Properties of Elastomers, Chem Safety. *Society Aff:* ACS, AUSA, AIAA.

White, Philip C
Home: 1812 Kalorama Square, Washington, DC 20008
Position: President *Employer:* Owner *Education:* PhD/Organic Chem/Univ of Chicago; BS/Chem/Univ of Chicago. *Born:* May 1913. Native of Chicago. Res Depts of Standard Oil Co (Ind) and Amoco affiliates 1938-75. VP Amoco Oil 1965-69. Gen Mgr, Res, Standard Oil 1969-75. Asst Admin-Fossil Energy ERDA 1975-77. Sr Technical Advisory-Energy Technology DOE 1977-78, President, Energy Consultants, Inc. 1978- . Dir,Coordinating Research Council 1967-75; President 1973-75. Industrial Research Institute, Director, Pres 1971-72. US rep - Exec Bd, World Petroleum Congresses 1968-75. Mbr AIChE, ACS. Hobbies-tennis, golf, skiing. *Society Aff:* ACS, AIChE.

White, Richard N
Business: Hollister Hall, Ithaca, NY 14853
Position: Assoc. Dean for Undergraduate programs. *Employer:* Cornell University. *Education:* BS/Civil Engg/Univ of WI; MS/Structures/Univ of WI; PhD/Structures/Univ of WI. *Born:* Dec 1933. Native of Chetek, Wisc. 2nd Lt. US Army Corps Engrs 1957. Struct Engr for J Strand, Madison Wisc 1958-60. With Cornell Univ since 1961; Prof since 1972; Dir of CEE Sch 1978-1984; Assoc. Dean since 1987. Sr author of 3-vol series, 'Structural Engineering', John Wiley & Sons and co-editor of 'Building Structural Design Handbook', John Wiley & Sons, 1987. Research in concrete structures, structural model analysis, nuclear structures, and timber structures. Fellow ACI and ASCE. Recd ASCE Collingwood Prize 1967. Enjoy photography, woodworking, and outdoor activities. *Society Aff:* ASCE, ACI, ASEE, NSPE.

White, Robert A
Business: Dept of Mech Eng, 1206 W. Green St, Urbana, IL 61801
Position: Prof *Employer:* Univ of IL *Education:* PhD/ME/Gas Dynamics/Univ of IL; MS/ME/Univ of IL; BS/ME/Univ of IL *Born:* 12/16/34 Born in Chicago IL, resident of Urbana, IL. Sr Res Scientist at the Aeronautical Res Inst of Sweden, 1963-1965. Join Univ of IL, 1965. Prof of Mech Engrg, 1972-present. Dir of Automotive System Lab, 1979-present. Fulbright scholar, 1960-61. NATO Fellow, 1968. Thord Gray Fellow, 1968. Field of specialization, rocket and jet engine to flight vehicle integration problems. Secretary of Supersonic Wind Tunnel Assoc (STA), 1980, 1981, Pres 1983. Member Air Forces Studies Board, National Research Council, 1984-1987; Ralph R. Teetor Award, SAE 1986; Associate Fellow, AIAH, 1986; National Treasurer Porsche Club of America, 1981-82. Active in sports and historic car events. *Society Aff:* SAE, ASEE.

White, Robert E
Home: 201 Bryn Mawr Ave, Newtown Square, PA 19073
Position: Consulting Engineer; Prof Emeritus Villanova University. *Employer:* Self. *Education:* DChE/Chem Eng/Brooklyn Poly; MChE/Chem Eng/Brooklyn Poly; BChE/Chem Eng/Brooklyn Poly. *Born:* August 31,1917; wife, Gloria (Ingersoll); 5 children; residence 201 Bryn Mawr Ave, Newtown Square PA; PE Pennsylvania. Res Eng, York Corp 1942-47. Asst Prof Bucknell 1947-48. Chrmn & Prof Villanova 1949-1986; Prof Emeritus. Consulting Engineer-present. various offices in Middle Atlantic ASEE; Delaware Valley AIChE; Natl Dir AIChE 1970-72; Fellow AIChE; Consultant to paper, chemical & petroleum industries. Chrmn, Newtown Township Municipal Authority. Active in civic affairs. *Society Aff:* AIChE.

White, Robert E
Home: 7 Guion Lane, Larchmont, NY 10538
Position: Consulting Engineer *Born:* 1913. BS in CE from Harvard in 1934. Native of New York City. Have been with present firm since 1934 in various positions from field engr to pres. Specialized in heavy construction, cofferdams, caissons, underpinning, moving buildings. Hold patents on rock-anchored tiebacks. Was in charge of firm's work on underpinning and shoring of The White House 1950. A P Greensfelder Construction Prize 1964. Active in ASCE. Founder and Chrmn of Metropolitan Section's Cttee on Soil Mechanics and Foundations. Author of various papers on geotechnics. Winner of Martin S. Kapp Award of the ASCE, 1977.

White, Robert L
Business: Dept of Elec Engg, Stanford, CA 94305
Position: Prof. *Employer:* Stanford Univ. *Education:* PhD/Phys/Columbia; MA/Phys/Columbia; BA/Phys & Math/Columbia. *Born:* 2/14/27. Plainfield, NJ. PhD Columbia, microwave spectroscopy of gasses, 1954. Hughes Res Labs, 1954-1961 res on magnetic mtls, laser mtls. 1961-62 Gen Telephone & Electonic Res Lab, Palo Alto, Hd Magnetics Dept 1963-present. Prof, Dept of EE, Stanford Univ. Chrmn EE Dept, Dir Inst for Electronics in Medicine. Approx 80 publications in solid state physics, esp magnetism, & in neural prostheses especially dev of artifical ear. Consutant to Ampex, Lockheed, IBM, Varian,USN, Dean Witter Co. Initial limited partner, Mayfield Fund & Mayfield II, venture capital. Dir Biostim, Inc.Married, Phyllis Arlt, (1952) four children. *Society Aff:* IEEE, APS, ΣΞ, ΦBK.

White, Robert M
Business: 2101 Constitutn Av NW, Washington, DC 20418
Position: Director, Academy Forum. *Employer:* Natl Academy of Sciences. *Education:* PhD/Chem Eng/Univ of MI; MS/Chem Engg/UNiv of MI; BS/Chem Engg/Cooper Union Inst; -/Sr Mgmt Prog/MIT; -/Law/Depaul Univ. *Born:* March 1916. Asst Prof-Assoc Dean Chem Eng Univ Mich 1942-60. Atlantic Rfg VP-R&D 1952-62, Champion Papers VP-Corp Devel 196266, WR Grace Pres Res Div 1966-67, Case Western Reserve Univ, Dean, Mgt 1967-71, NAS, Dir-Acad Forum 1971--. Managerial and Technical Counselling 1942-60, 1967- ; Innovation, R&D, licensing, mgt. Director, Ferro Corp, Cleveland Ohio, Awards: Professional Progress AIChE 1956; Junior AIChE 1945, George Westinghouse ASEE 1955, Russell U Mich 1945, Sesquicentennial U Mich 1967, Cooper Union Alumni 1974. Cosmos Club, Wash, DC, Currently resp for Forums on Natl issues involving science and technology. *Society Aff:* AIChE, ACS, ORSA, SCI.

White, Robert M
Business: 2600 Virginia Ave, NW 514, Washington, DC 20037
Position: President *Employer:* Univ Corp for Atmospheric Research *Born:* Feb 13, 1923 Boston MA. PhD & MS, MIT; BA, Harvard. Air Force Weather Officer. Re-

White, Robert M (Continued)
search Scientist Air Force Cambridge Res Ctr; Assoc. Dr. Research Travelers Ins Co. Pres Travelers Res Ctr 1959-63. Chf U S Weather Bureau 1963- 65. Administrator Environmental Science Serv Administration 1965-70. Administrator Natl Oceanic & Atmospheric Administration 1970-80. President's Commission; Marine Fisheries Adv Com, World Climate Conf Natl Adv Com Oceans & Atmosphere, Bd Overseers Harvard Univ. on Marine Science, Engrg and Resources. Chrmn, Federal Cttee for Meteorological Services and Supporting Research. Chrmn, Inter-agency Cttee for Marine Science & Engrg. Chrmn, Dept of Commerce Marine Fisheries Advisory Cttee. Chrmn, Domestic Council Sub-Cttee on Climate. US Permanent Rep World Meteorological Org. U S Commissioner, Intl Whaling Commission. Marine Fisheries Adv Com, World Climate Conf Natl Adv Com Oceans and Atmosphere, Bd Overseers Harvard Univ.

White, Stanley A
Home: 1541 Amberwood Lane, Santa Ana, CA 92705
Position: Sr Tech Staff *Employer:* Rockwell International *Education:* PhD/EE/Purdue Univ; MS/EE/Purdue Univ; BS/EE/Purdue Univ *Born:* 09/25/31 Born in Providence, RI. Hon discharge from US Air Force in 1955. Taught at Purdue and UCLA. Engr at Allison Div General Motors and Radiochemistry Corp. Joined N American Aviation (later became Rockwell International) in 1959. Held various technical, line mgmt, and proj mgmt post. 15 patents issued, 70 publications in open literature. Active in IEEE chapters and techical societies. Mbr of International Bd of Dirs of Eta Kappa Nu Assoc. N American Aviation Science Engrg Fellow. Enjoy composing, arranging, performing, and directing choral music. *Society Aff:* AAAS, IEEE, AES, ΣΞ, HKN, TBΠ

White, Thomas L
Home: 721 W Warren Ave, Youngstown, OH 44511
Position: Ret. (Consulting Engr) *Born:* May 30,1903 Youngstown Ohio. Was Chf Engr Commercial Shearing Inc 192451; Chf Consulting Engr 1951-68; retained Consultant since retirement. Coauthor of books 'Rock Tunneling with Steel Supports' and 'Soft Ground Tunneling with Steel Supports'. Also authored numerous papers on Pressure Vessel Stresses and Cold Working of Steel, and lectured on tunnel design & problems. Was Chrmn of Petroleum Div of ASME in 1959. Reg PE, Civil Br, Ohio. Fellow ASME. Mbr AREA. Prior to retirement was mbr of PVRC of Welding Research Council, also ASM and ASTM. For many yrs have been called upon as Consultant for a great many tunnels all over the United States, and also for tunnels in Canada, Europe, South America, Australia, India and others. Life Fellow in ASME. *Society Aff:* ASME, AREA.

White, Trevor
Home: 2421 Branch St, Duluth, MN 55812
Position: VP. *Employer:* Fraser Shipyards Inc. *Education:* BS/Naval Arch/Queen's Univ, Belfast N Ireland. *Born:* Jan 1926. Shipbuilding and design training at Harland & Wolff, Ltd. Hull Engr, Asst Naval Architect, Dir of Engrg with American Ship Building Co 1951-63. With Fraser Shipyards, Inc since 1964 as Dir of Engrg. VP-Engrg since 1973 and VP 1979. Resp for design, proposals, contracts, sales, estimates involving ship construction and reconstruction. Chrmn, SNAME, Great Lakes and Great Riv Sec 1973-74. Author of several papers relating to Great Lakes vessel design, presented at SNAME meetings. Skier, mountain hiker, photographer, runner. *Society Aff:* SNAME, IME.

White, William A
Home: 5505 8th Ave, Sacramento, CA 95820
Position: Exec Dir Emeritus. *Employer:* Calif Council of CE & LS. *Education:* MA/Math/Univ of MT, Missoula; BA/Math/Univ of MT, Missoula. *Born:* March 27,1906 Joliet Montana. Taught high school science two yrs. With Army engrs and Quartermaster Corp 14 yrs as sr computer, principal computer, asst engr, statistician, assoc engr, engr. Secy of Calif bd of reg for engrs 8 yrs. Exec dir with present employer 21 yrs Ed emeritus 10 yrs. Chrmn 1970-71 ASCE Surv & Map Div. ASCE Surveying and Mapping Awd 1975. Wife is the former Virginia Emmett. Son Mack White is civil engr. Daughter Virginia Johnson is teacher. *Society Aff:* ASCE, ACSM, ASAE, CESSE.

White, William F
Business: PO Box 322, Monroe, UT 84754
Position: Conslt *Employer:* Self *Education:* BS/ME/Bucknell Univ. *Born:* Jan 1925. BSME. Engr for Bethlehem Steel Co 1948-52. Sales Engr for filters Dorr-Oliver, from 1952-56. Sales Engr Bird Machine Co Inc from 1956-62, Sales Mgr in 1963, VP Mktg, Process Equipment Div 1971-79 VP Intl 1979 to 1983. 1983 to present a Consultant on Marketing and product development with emphasis on Japan & China. Resp for the sale, application, design and development of liquid-solid separation equipment on which over 10 papers have been presented or published. Active 1974-82 in management in Process Equip Mfg Assoc. 1970 to 1983, a mbr Mass.Bd. of Certification of Operators of Wastewater Treatment Plants. Mbr of AIChE & ASME. *Society Aff:* AIChE, AIMME

White, Zan C
Home: 27 Brittany Woods Rd, Charleston, WV 25314
Position: Gen Mgr-Network Design & Provisioning. *Employer:* Chesapeake & Potomac Tel Co of W.Va *Born:* Native of Elkins, WV. Served in US Army Signal Corp 1943-46. Employed by Bell System since '42. Various positions in Engrg Dept in switching equip, outside plant and transmission engrg. With AT&T as Engrg Dir - Inventory Mgt 1973-75. Gen Mgr C&P Telephone of MD 1975-79. Assumed present responsibility as Gen Mgr-Network in 1979. Pres of WV Sec of IEEE 1964. Mbr Eta Kappa Nu, Tau Beta Pi and Sigma Pi Sigma. Served on Baltimore Consultants Evaluation Bd. Hobbies are golf and fishing. Bd Mbr Kanawha-Ohio Valley Construction Users Council. Mbr Salvation Army Advisory Bd.

Whitehead, Carl F
Home: No. 1 Fiddlers Green Dr, Lloyd Harbor, NY 11743
Position: Group VP *Employer:* Ebasco Services Inc. *Education:* B/Civil Engg/ Rensselaer Polytechnic Inst. *Born:* July 1927. Native of Long Island, New York. Joined Ebasco in 1951 as a Jr Designer and served in increasing levels of responsibility in both the Design and Engrg Depts up to Supvr of Nuclear, Fossil and Hydroelectric projects. Following this was Proj Mgr and later Mgr of Projects on the Ludington Pumped Storage Proj which received the 'Outstanding Civil Engineering Achievement Award' for 1973. Became VP-Engrg in April 1976, Sr VP Proj Operations Feb 1978. Group VP Int'l, Nov 1982. Fellow ASCE, Mbr ICOLD and a reg PE. Hobbies are sailing and golf. *Society Aff:* ASCE, ICOLD.

Whitehead, Daniel L
Business: Roundtop Rd RD 3, Export, PA 15632
Position: Mgr Engrg Labs Dept. *Employer:* Consultant *Education:* MS/Elect Engg/ Cornell Univ; BS/Elect Engg/Univ TN. *Born:* Dec 25,1915 Walland TN. AIEE 1934, Fellow IEEE 1962. Mbr Tau Beta Pi, Phi Kappa Phi, Eta Kappa Nu, Phi Eta Sigma & Scabbard & Blade. Presently Member IEEE-HVTT Subctte, Member ANSI C-68, mbr PSIM, Recognition Working Grp, Safety Practices in High Voltage Measurements. 7 patents in field. Consultant, High Voltage Tech. Married, Lenore Ruth, three children, Barbara, Daniel Jr. and Patricia Gail. Enjoys hunting, fishing and flying. *Society Aff:* IEEE, ANSI.

Whitehead, Ralph L
Business: P O Box 4301, Charlotte, NC 28204
Position: President & Chf Engr. *Employer:* Ralph Whitehead & Assoc. *Education:* BCE, St Univ, 1950 *Born:* 03/01/27 Native North Carolinian, served 2 years with 8th Air Force in Europe. Received broad engineering experience on structural and civil works with city and private organizations before establishing Ralph Whitehead & Assoc in 1959. Have since had responsibility for engineering of projects involving over $100 million in construction costs, including over 165 rail or highway bridges. Reg PE in 10 states. Mbr ACEC, ASCE, AREA, NSPE, ARBA. Served as VP and

Whitehead, Ralph L (Continued)
Pres of So. Branch of NC Sec ASCE; Bd/Dir and President, NC Sec ASCE; Secy-Treas of CEC of N.C. *Society Aff:* ASCE, NSPE

Whitehouse, Gary E
Business: IEMS Dept, UCF, PO Box 25000, Orlando, FL 32816
Position: Prof, IEMS Dept. *Employer:* Univ of Central FL *Education:* PhD/AZ State Univ; MS/IE/Lehigh Univ; BS/IE/Lehigh Univ. *Born:* 8/13/38. Dr Whitehouse has been Prof (Chairman 1978-85) of the Ind Engg & Mgt Systems Dept at the Univ of Central FL since 1978. Previously he served the IE Dept at Lehigh Univ for 13 yrs. He has served as a consultant for Air Products & Chem, Inc, Chrysler Motors, the AF, Alcoa, Martin Marietta, etc. He has written over 120 articles & papers & 3 books. He was a Natl Officer for IIE, ASEE, & Alpha Pi Mu (Ind Engg Honorary). In 1976, he received ASEE's Westinghouse Award for outstanding teaching and in 1978 IIE's H B Maynard Award for Innovative Achivement in Ind Engrg. He is a Fellow of IEE. His latest book "Practical Partners: Computers & Industrial Engineering" won IIE Publication of the Year in 1986. *Society Aff:* ORSA, IIE, TIMS, ASFE, NSPE.

Whitehouse, Gerald D
Business: Dept Mech Engrg, Baton Rouge, LA 70803
Position: Prof Mech Engrg. *Employer:* Louisiana State University. *Education:* PhD/Mech Engg/OK State Univ; MS/Mech Engg/OK State Univ; BS/Mech Engg/ Univ of MO-Rolla. *Born:* 5/17/36. PhD and MS from OK State Univ. BS from Univ of MissouriRolla. Native Sapulpa, OK. Served with US Army Corps of Engrs 1959-62. Taught in Mech Eng Dept at Louisiana State Univ 1966-present. Central VP Pi Tau Sigma 1974-79, Acting Chrmn ME Dept 1976-77, Dir Div of Engg Res, LSU 1978-81, Specializing in Mech Design, Teaching and Research in Mech Design area (Vibrations, Mech Design, Automatic Controls, Mechanisms) *Society Aff:* ASME, AWS, NSPE, SAE.

Whitehurst, Eldridge A
Home: 632B Providence Ave, Columbus, OH 43214
Position: Prof *Employer:* OH St Univ *Education:* MS/CE/Purdue Univ; BS/CE/VA Military Inst *Born:* 05/26/23 Served in US Marine Corps 1943-46; Associate Res Engr Portland Cement Assoc 1947-50; Prof Civil Engrg, Dir TN Highway Res Prog and Assoc Dir Engrg Experiment Station Univ of TN 1952-70; Prof Civil Engrg and Dir TRANSPLEX/OSU The OH State Univ since 1970. Dir ASTM 1966-1969. Current activities primarily in area of pavement skid resistance and vehicle traction. *Society Aff:* ASTM, ARTBA

Whitelock, Leland D
Home: 2320 Brisbane Street, Apartment 4, Clearwater, FL 34623
Position: Chf Engr, Digital Tech. *Employer:* Naval Ship Eng Ctr (Ret). *Education:* BSEE/EE/Carnegie Inst of Technology. *Born:* Sept 1907 Petersburg, Indiana. Attended Purdue Univ and Carnegie Inst of Tech with BSEE degree in 1931 while working at Westinghouse as a Radio Engr. Res Engr at Union Switch & Signal Co 1931-41. Electronics Engr with Navy Dept, Bureau of Ships and Naval Ship Engrg Ctr 1941-74, from 1963 in the position of Chief Engr, Digital Techniques. Fellow IEEE and Wash. Academy of Sciences. Mbr IEEE Bd/Dir IEEE Reg 2 1973-74. Washington Academy of Sciences Bd/Governors 1975-77. Electronics Res & Dev activity incl publications and patents on electronics, microelectronics and systems effectiveness. *Society Aff:* IEEE.

Whitesides, Jack C
Home: Rt 1, box 20B, Tifton, GA 31794
Position: VP Eng. *Employer:* Kelley Mfg Co Div Beatrice Foods Co. *Education:* BS/Agri Engg/Clemson Univ. *Born:* 8/29/31. Native of York SC served with US Army 1950-52, Design Engr with Intl Harvester Co Memphis TN 1956-59, Sr Proj Engr with Lilliston Corp Albany, GA 1959-68. VP, Engg with Kelley Mfg Co, Tifton, GA since 1968. Holds 10 US Pats Chrmn of GA Section ASAE 1977-78, mbr of Agri Engg Advisory committee at Clemson Univ 1976-79, Pres's Honor Club of Beatrice Foods Co 1979, Mbr of Chamber of Commerce, Elks, & Past VP Optimist Club. Ruling Elderly in Presbyterian Church, Chairman Bd of Deacons 1975. *Society Aff:* ASAE.

Whitfield, Jack D
Business: Sverdrup Technology, Inc, 600 Wm. Northern Blvd, Tullahoma, TN 37388
Position: Pres *Employer:* Sverdrup Tech, Inc *Education:* DSC/Gas Dynamics/Royal Inst of Tech, Stockholm, Sweden; MS/Mech Engr/Univ of TN; BS/AE/Univ of OK *Born:* May 16, 1928 Paoli Okla. Sverdrup Technology, Inc (formerly Sverdrup/ ARO, Inc), Mbr Bd of Dirs, Exec VP, Pres; Prof Aerospace Eng, Univ of TN (part-time). Fellow AIAA. Mbr NSPE, TN Soc Prof Engrs, Tau Beta Pi, Tau Omega and Sigma Xi. Employed by ARO Inc in 1954 as Proj Engr, since served as Enrg Supvr, Asst Br Mgr, Br Mgr, Facility Chf, Facility Dir, Exec VP & Pres. Reviewer AIAA Journal. Res Gen H H Arnold Awd 1968 for outstanding contribution toward advancing the state-of-the-art of the aerodynamic and astronautical sciences. 28 AEDC (ARO Inc) Tech Pubs and 56 invited lectures, presentations and publications in the reviewed tech literature *Society Aff:* TBΠ, TΩ, ΣΞ, AIAA

Whitman, Lawrence C
Home: 13 East Lake Drive, Haines City, FL 33844
Position: Retired. *Education:* MS/Physics/Univ of VT; BS/Elect Engr/Univ of VT. *Born:* 1903 Franklin VT. 1932-33. 31 yrs with Gen Elec Co as design, development and research engr in transformers, electrical insulation and heat transfer. Author 25 tech papers and 8 patents. 6 years Full Prof of EE SoDakota State Univ. Conducted insulation research gasses with high voltage impulse, a.c., and d.c. Prize AIEE Papers 1954 and 1962. Consulting work various companies. Mbr Phi Beta Kappa, Eta Kappa Nu. Life Fellow IEEE. Listed Amer Men of Sci. Previously reg PE in Mass and So Dakota. Retired and living in Florida. *Society Aff:* HKN, ΦBK, IEEE.

Whitman, Robert V
Business: MIT. Room 1-342, Cambridge, MA 02139
Position: Prof of Civil Eng. *Employer:* Mass Inst of Tech. *Education:* BS/Civil Eng/ Swarthmore College; SM/Civil Eng/MIT; ScD/Civil Eng/MIT. *Born:* 2/2/28. On Staff and faculty MIT since 1951, except leaves of absence for military service (US Navy Civil Engr Corps), yr at Stanford Res Inst, 1 yr at Cambridge Univ (England) and 1/2 yr at Norwegian Geotech Inst. Researcher in soil mechanics, soil dynamics, earthquake engg and seismic risk. Huber Res Prize ASCE 1963. Fitzgerald Medal (Boston Soc Civil Eng) 1974 Horne Award (Boston Soc Civil Eng) 1977. Terzaghi Lecturer ASCE 1981. Elected Natl Academy of Engg 1975. Past Pres- elect Earthquake Engrg Res Inst. Mbr various governmental advisory committees. Consultant on projects in U.S. and abroad. *Society Aff:* ASCE, EERI, SSA.

Whitney, Eugene C
Home: 249 Cascade Rd, Pittsburgh, PA 15221
Position: Consulting Engr. *Employer:* Independent. *Education:* -/Elec Engg/Univ of Michigan. *Born:* Aug 1913. BSEE Univ of Mich plus numerous courses Univ of Pittsburgh. Experience in design & building variable frequency generators, synchronous condensers, special diesel and steam turbine generators and Mgr of Waterwheel Generator and Synchronous Condenser Engrg (23 yrs). Design and application resp included generators Niagara Falls and Grand Coulee -615 MVA Hydro generators plus many pumped storage Generator/Motors and 4.3 million HP of large synchronous motor. After retirement in 1975 independent part time consulting. IEEE Fellow and mbr of Rotating Machinery, Synchronous subcttee, Pwr Generation Hydraulic Subcttee. PE Penn. *Society Aff:* IEEE.

Whitney, Herbert W
Home: 14224 McPhee Dr, Sun City, AZ 85351
Position: Retired *Education:* BS/Chemistry/IL Wesleyan Univ. *Born:* 12/11/08. Reg PE in IL-Listed in Maquis Fifth Edition Who's Who in the Midwest. Born Chicago,

Whitney, Herbert W (Continued)
IL. Married Kathryn Preisel-2 daughters: Mrs Barbara Edwards, Mt Horeb WI, and Mrs Annette Hidde, Mequon WI. Employed 1929-30 Western elec 1935-37 Int Harv McCormick works 1937-52 Viking Pump Co Cedar Falls, 1956-1974 Chf Metallurgist Kearney & Trecker Milwaukee WI. IA 1952-73 Active Chicago Chapter ASM and Chrmn Milwaukee Chapter. Duties involved heat treating-Lab Supervision- Matl Selection-Failure Analysis-Specification writing Review of all new & revised drawings of present & proposed machines. *Society Aff:* ASM, AFS.

Whitney, James C
Business: 677 Connecticut Ave, Norwalk, CT 06856
Position: VP. *Employer:* Dictaphone Corp. *Education:* PhD/Plasma Phys/Columbia Univ; MS/Engg/Columbia Univ; BS/Phys/Caltech. *Born:* 4/26/42. Yonkers, NY. Grad Fieldston Sch, NYC, 1960; CA Inst of Tech, 1964 (BS); Columbia Univ, 1966 (MS); 1968 (PhD). Thesis on Kinetic Theory of Thermonuclear Reactions. Post-Doctoral Res Fellow, Columbia, 1969. Dictaphone Corp, Rye, NY: Proj Engr, 1970; Group Leader, 1972; Engg Mgr, 1974; Dir of Advanced Dev, 1976; VP of Res and Dev, 1977. Direct Corp's res and product dev efforts, which ctr on electronic office communications and recording products.

Whitney, Lester F
Business: University of Massachusetts, Amherst, MA 01003
Position: Professor. *Employer:* University of Massachusetts. *Education:* PhD/Agri Eng/MI State Univ; MS/Agri Eng/MI State Univ; BS/Agri Eng/Univ of MN. *Born:* March 21,1928 New Bedord Mass. m.Phyllis Burrill, 1 daughter, 6 sons. Consultant, Kelley Onion Farms, Parma Mich 1950-52; Design/project engr, Ariens Co Brillion Wisc 1952-54; Asst Chf Engr, plant supt Wirthmore Feed Div CPC Intl Waltham Mass 1954-59; with Univ of Mass since 1959. NSF Science Faculty Fellow 1963. Sigma Xi, Phi Tau Sigma, Sr mbr ASAE (Chrmn N.A. Region 1972), IFT, MSPE. Reg PE Mass. Consultant Food Processing, Machine Design. Patents: Tillers 1953, Tallow Application in Feeds 1958, Fluidized Bed Cooking(UK) 1975. Current research: Leaf protein processing, byproduct development, quality determination of canned foods, fluidized bed heat transfer. Public Service: Amherst school bldg cttee 1964-67, Amherst plng bd 1965-72 (chrmn 1969-71), Town Meeting Mbr 1964-76. *Society Aff:* ASAE, IFT, NSPE.

Whitney, Roy P
Home: 1709 South Douglas St, Appleton, WI 54914
Position: Retired. *Education:* ScD/Chem Engg/MIT; SM/Chem Engg/MIT; SB/Chem Engg/MIT; MS/-/Lawrence Univ. *Born:* 5/30/13. At MIT, Asst Prof of Chem Engg 1939-45. On loan to US Army Chem Warfare Serv 1942-45. Recd War Dept Cert of Appreciation. At Univ of ME, Dir Dept of Industrial Cooperation 1945-47 and Prof and Acting Hd Dept of Chem Engg 1946-47. At the Inst. of Paper Chemistry, Group Leader in Chem Engg 1947-57, Dean 1956- 76, VP 1958-78. Fellow AIChE (Colburn Award 1948), AIC, TAPPI (Bd/Dir 1969-72, Div Awd 1969, Gold Medal 1980), Mbr ASEE, ACS (Chm Cellulose Div 1978). Pro Bono Labore Award, Finnish Paper Engineers Assn 1978. *Society Aff:* AIChE, TAPPI, AIC, ACS, ASEE.

Whitt, James A
2740 Brigstock Rd, Midlothian, VA 23113
Position: Partner. *Employer:* Austin Brockenbrough & Assoc. *Education:* BS/Civil Engg/VA Military Inst. *Born:* 9/2/32. Native of Farmville VA. Engr with TVA 1954-55; service with US Army Artillery, Germany until 1958. Design Engr with Austin Brockenbrough and Assoc beginning Feb 1959. Elected Partner 1963. Reg Engr in VA and NC. Reg Land Surv in VA. Fallout Shelter Analyst. Qualified Task Team Leader Value Engrg. Mbr ASCE, NSPE, CEC. Reg Engr in W VA, MD, District of Columbia, Reg Land Surveyor in MD. Enjoys fishing. *Society Aff:* ASCE, NSPE, ACEC.

Whittemore, James H
Business: 34 Pontiac Ave, Providence, RI 02907
Position: President. *Employer:* Whittemore Corp. *Education:* BS/Civil Engg/Univ of CT. *Born:* June 1935. 10 yrs with Civil Eng Consultant, Providence RI in positions of draftsman through chief engr. Founded Whittemore & Assoc in 1968 which later became Whittemore Corp in 1972; a small general civil eng org spec in survey layout services for the heavy construction industry. Various offices in RI Sec ASCE; Pres 1975 and later Director. PE in RI, MA, CT. Reg Land Surveyor, RI & MA. *Society Aff:* ASCE, NSPE, ACSM.

Whittemore, Laurens E
Home: Apt. AB-19, Heath Village, Hackettstown, NJ 07840
Position: Retired from AT&T Co. *Education:* MA/Physics/Univ of KS; AB/Math/Washburn Univ. Of Topeka *Born:* 1892 Topeka, Kansas. Sigma Xi. Mbr radio staff Natl Bur of Standards during WWI. Mbr headquarters staff AT&T Co 1925-57 concerned, among other things with Federal regulatory organizations relationship. Participated in several Natl and Intl Radio and Communications Conferences on behalf of the US Govt or AT&T Co. Joined IRE 1916; Fellow 1927; VP 1928. Active in Radio Standardization, Bd/Dir and Bd/Editors. Authored articles on History of Institute of Radio Engrs. Cosmos Club, Wash, DC. *Society Aff:* IEEE, ΣΞ

Whitwell, John C
Home: 318A Stowe Ln, Jamesburg, NJ 08831
Position: Retired. *Education:* ChE/ChE/Princeton Univ; BSE/ChE/Princeton Univ. *Born:* 11/17/09. Faculty, Princeton Univ from 1932; Prof 1946; Emeritus 1974; Maj responsibilities in Planning & Construction of 275,000 sq ft Engg Quadrangle 1955-62; many technical publications, including "Conservation of Mass and Energy" McGraw-Hill, 1969. Consultant & res assoc. Textile Res Inst 1944-. Consultant, number of chem cos, 1953-87. Trustee and treas, Princeton Hospital 1962-71. Dir, Medical Ctr at Princeton Fdn, Inc 1977- ; 1971-2 Presidential appointee to Health Indus Ctte, Cost of Living Council. *Society Aff:* AIChE, (Fellow), ACS, ASA, Fiber Society

Wiars, Dale M
P.O. Box 7, Illbrook, IN 49334
Position: Dir of Facilities *Employer:* SUNY Coll of Tech *Education:* EMBA Candidate/Mgmt/Univ of WI-Milwaukee; BSA/Architecture/Univ of WI- Milwaukee *Born:* 09/27/33 Former Aldermanic Coun Mbr, City of Wauwatosa, 1974-78. Certified Plant Engr (CPE) - Natl. Reg Architect (RA) - Multiple States & NCARB Certificate. Pres of professional consltg firm, Wauwatosa, WI, 1973-79. Recepient of CEFP's "Distinguished Service Award For Leadership In Planning Facilities, 1977–. Written up for work in *The Milwaukee Journal, Milwaukee Sentinel, Wauwatosa* News-Times, *Escanaba Daily News.* Authored articles in *School Shop,* "Heat Recovery System Saves Money–, "Milwaukee's Hang It All Facility–, and in *American School and University,* "Technical College Gets Solar Boost–. Speaker - Bay DuNoc College - "Stressed Membrane Structures–, Escanaba, MI. Presider of "Energy Conscious Design" Seminar - CEFP, Chicago, IL. Designer of UNIDESIC structures. Holder of trade name UNIDESIC. Mgr of architectural/engrg projs, plant engrg, and construction. Dir of Univ Facilities. *Society Aff:* AIPE

Wibberley, Harold E
Business: 580 Northern Ave, Hagerstown, MD 21740
Position: Pres *Employer:* Baker-Wibberley & Assoc, Inc *Education:* BS/CE/Univ of VA *Born:* 7/17/21 Hagerstown, MD, married, June 24, 1946, Wife, Amelia; children–Murray, Bryan, Amelia. Design Engr, City of Hagerstown, 1946-49; founder, Pres, Baker- Wibberley & Assoc, Inc. Consulting Engrs, Hagerstown, MD, 1949 to present. Mbr, MD Game and Inland Fish Commission, 1967-68; Vice Chairman, Bd of Review, MD Department of Natural Resources, 1969-79; Mbr, Exec Bd, Mason-Dixon Boy Scouts of America; World War II USNR; Decorated Purple Heart. Named Distinguished Eagle Scout,National Council Boy Scouts of America 1973. Engr of Distinction 1973. Registered Professional Engr, MD, PA, MA, DC. *Society Aff:* ACEC, ARTBA, AME, MAE

Wiberg, Donald M
Business: 7731 Boelter Hall, Los Angeles, CA 90024
Position: Prof. *Employer:* Elec Engrg Dept, Univ of CA *Education:* PhD/Engrg/Caltech; MS/Engrg/Caltech; BS/Engrg/Caltech. *Born:* 9/20/36. B.S. Caltech 1959, M.S. 1960, Ph.D. 1965. Sr Design Engr, Convair (San Diego), 1964-1965. Asst Prof UCLA 1965-1971, Assoc Prof 1961-1977, Prof 1977- present, in Electrical Engrg. Also Prof Anesthesiology 1979-present. Conslt in Field. Visiting Prof DFVLR, Munich 1969-1970; NATO Sr Fellow, KFZ Karsruhe, 1973; twice Fulbright Fellow, Danish Tech Univ Copenhagen 1976-1977, and Norwegian Inst of Tech, Trondheim, 1983-1984. Author of "State Space and Linear Systems–, McGraw Hill, 1971, and co-editor of "Modelling and Control of Breathing–, Elsevier 1983. Res interests: Sys identification, filtering, modeling, optimal control, non-linear systems. Applications in biomedical, aerospace, nuclear engrg and especially respiratory physiology and anesthesia. *Society Aff:* IEEE, ΣΞ.

Wiberg, Donald Victor
Home: 16520 Lillan Rd, Brookfield, WI 53005
Position: Supvr, Plant & Equip Eng. *Employer:* A O Smith Corp. *Education:* BSME/Mech Eng/Univ of MN. *Born:* Aug 1924. Grad work Univ of Wisc. Native of Minneapolis, Minn. Design Engr American Hoist & Derrick 1949-51. With A O Smith Corp 1951 to present. Currently Supvr of Plant & Equipment Engrg resp for press room equipment and Elec Engrg. Pres of FPS 1970-71. PPres of the Milwaukee Council of Engrg & Scientific Societies. Chrmn of the ANS B11.2 Safety Standard for the Construction, Care & Use of Hydraulic Presses Cte. Mbr various church, civic & professional society cttees. Hon Awd FPS 1971 Presidents Award FPS 1974 and Award of Merit Fluid Pwr Foundation 1972. Chmn FPS Cert Policy Bd. Enjoy travel, music and fishing. *Society Aff:* FPS.

Wichser, Robert F
Business: P O Box 16858, Philadelphia, PA 19142
Position: Program Mgr. *Employer:* Boeing Vertol Co. *Born:* April 1921. BAE & MIE from NYU. Native of Long Island, NY. Served in Army Air Corps 1943-46. Chief Research Engr for Republic Aviation Corp engaged in research and testing of aircraft systems. Joined Boeing Vertol Co in 1964 to work as Project Engr on CH-47. Assumed current position as Program Mgr in 1973 for development of Navy LAMPS helicopter. PE 1954; AHS 1964. Enjoy golf and boating.

Wick, Robert S
Home: 1204 Neal Pickett, College Station, TX 77840
Position: Prof, Nucl Engg. *Employer:* TX A&M Univ. *Education:* PhD/ME/Univ of IL; MS/ME/Stevens Inst of Tech; BS/ME/RPI. *Born:* 12/4/25. Reg PE-TX, USN 1943-1946 (Active), 1946-1956 (Reserve). Babcock & Wilcox, Jet Propulsion Lab Cal Tech (1952-55), Westinghouse Bettis Atomic Power Lab (1955-66). Adjunct Prof UCLA (1954), Univ Pittsburgh (1960). Prof Aerospace and Nucl Engg, TX A&M Univ (1966-present). TX A&M Distinguished Faculty Award 1969. (Std Oil NJ 1946-47). *Society Aff:* ASME, AIAA, ANS, ΠΤΣ, ΣΞ, ΦΚΦ.

Wickizer, Gilbert S
Home: 19 Wister Pl, Aberdeen, NJ 07747
Position: Retired. *Education:* BS/Elect Eng/Penn State Univ. *Born:* 1904 Warren PA. Eta Kappa Nu. Employed by RCA Corp 1926-68 at Riverhead NY and Moorestown NJ. Specialty: research and development in problems of radio reception. RCA Labs Award 1954 'for investigation of wave propagation in the diffraction field leading to a better understanding of the phenomenon.' Fellow IEEE 'for contributions to experimental wave propagation research.' Author or coauthor 12 published tech papers. Active in amateur radio 1921 to present. Tech Asso Soc of Wireless Pioneers. Former semi-pro dance musician. *Society Aff:* IEEE, SOWP, ARRL.

Wicks, Charles E
Business: Chem Eng Dept, Corvallis, OR 97331
Position: Dept Head. *Employer:* Oregon State University. *Education:* PhD/Chem Engg/Carnegie-Mellon Inst; MS/Chem Engg/Carnegie-Mellon Inst; BS/Chem Engg/OR State Univ. *Born:* July 9,1925 Post-grad 1960-61 Univ of Wisc. Professor, Oregon State U 1954-present, tech advisor, US Bur of Mines, Pacific Pwr & Lt, Oregon Metallurgy. E I du Pont 'Year-in-Industry' Awd 1964, AIChE: Asst Prog Chmn Natl Mtg 1957, Prog Chmn Reg Mtg 1963; Prog Vice Chmn Natl Mtg 1969; Symposium Chmn 1969. Chem Engrs of Oregon, President 1973. Author: Thermodynamic Properties of 65 Elements-Their Oxides, Halides, Carbides and Nitrides, 1963; Fundamentals of Momentum, Heat and Mass Transfer 1969, 1976 and 1984. *Society Aff:* ASEE, AIChE, ΤΒΠ.

Wicks, Moye, III
Home: 5823 Yarwell Dr, Houston, TX 77096
Position: Res Assoc *Employer:* Shell Dev Co *Education:* PhD/ChE/Univ of Houston; MS/ChE/Univ of Houston; BS/ChE/Univ of Houston; BS/Math/Univ of Houston. *Born:* 8/18/32. in Houston. Co-Dir AGA-API Proj NX-28 Multiphase Flow in pipes, Univ of Houston, 1960-65. NSF Fellowship 1960-65. Winner 1965 Best Fundamental Paper Award, So TX AIChE. With Shell since 1965. Supervisor & Sr Staff Engr Shell Dev, in charge of fluid mechanics, 1967-75. Currently Res Assoc, Prodn Operns Res Dept, Shell Devel Co. Mgr solids handling & viscous fluids, in Shell's Hd Office, 1976-1981. Corp Secy, Shell Co Fnd, 1982. Chrmn, Technical Committee, Design Inst for Multiphase Processing (AIChE), 1973-78. Author of over 20 technical papers and 20 US patents. Listed in "Who's Who in the South & Southwest–. Elected Fellow, AIChE, 1985. Married, five children. *Society Aff:* AIChE.

Widdoes, Lawrence C
Business: 7505 Fannin, Houston, TX 77054
Position: Vice President. *Employer:* Magna Corp. *Born:* Nov 10,1919. BS Calif Inst Tech 1941. MS Univ of Mich 1947. Served as an officer in US Navy WWII. Res Engr, Kraft Foods. Engrg Mgr, Nuclear Power, Monsanto Chem Co. President, Internuclear Co, a nuclear engrg design and consulting firm. Dir of Research, Petrolite Corp 1960-64. Intl Mktg Mgr, Atomic Power Div,Westinghouse. President, Conresco Corp, a technical venture capital organization. Since 1972 VP, Chem Oper, Magna Corp, marketer and manufacturer of specialty chemicals for crude petroleum production, transportation and refining.

Wideman, Charles J
Business: MT College of Mineral Science & Tech, Butte, MT 59701
Position: Assoc Prof. *Employer:* MT Tech. *Education:* PhD/Geophysics/CO Sch of Mines; MSc/Geophysical Engr/CO Sch of Mines; BSc/Geophysical Engr/CO Sch of Mines. *Born:* 2/7/36. At present I'm directing geophysical research aimed at delineation of possible geothermal sites in MT. 1975-1976 served as Chrmn, Geophysical Engg Dept MT College of Mineral Sci and Tech. Academic interests are earthquake seismology, MT Seismicity, and Uranium occurances in hydrothermal systems. Reg Geophysicist in CA. *Society Aff:* SEG, SSA.

Widera, G E O
Business: Box 4348, M/C 251, Chicago, IL 60680
Position: Prof & Head, Mech. Engr. *Employer:* Univ. Illinois at Chicago *Education:* B.S./Appl. Math. & Eng. Physics/Univ. Wis.-Madison; M. S./Eng. Mech./Univ. Wis.-Madison;Ph.D./Engr. Mech./Univ. Wis.-Madison *Born:* 02/16/38 Born in Dortmund, W. Germany. Attended Milwaukee area elementary and high schools. With the Univ. of Illinois at Chicago since 1965. Visiting appointments at Argonne, Stuttgart, Wisconsin-Milwaukee and Marquette. Assumed present position in 1983. NASA Fellow 1966. Alexander von Humboldt Fellow 1968. Fellow ASME. Technical Editor of ASME Journal of Pressure Vessel Technology. Member of Executive Cttee and Program Chairman, ASME Pressure Vessel and Piping Division. Chairman of ROEL Subcttee of WRC Pressure Vessel Research Cttee. Research interests include stress analysis, plates and shells, composites and deformation processing. *Society Aff:* ASME, ASCE, SME, ASEE, AAM

Widman, Richard G
Business: 6200 Oak Tree Blvd, Cleveland, OH 44131
Position: President & CEO. *Employer:* Consultant. *Education:* BS/CE/Univ of MI. *Born:* 1/28/22. BSCE Univ of MI 1947. Mbr Tau Beta Pi. Served 1943-46 with US Army Corps Engrs. With Arthur G McKee & Co (now Davy McKee Corp) since 1948 as field engr, construction supt, gen mgr, petroleum & chem div, VP petroleum & chem div, Exec VP becoming Pres Chrmn and CEO 1970-79. Retired March 1, 1979. Also dir parent co Davy Corp Ltd of London. Mbr AIChE, AISI, API, Newcomen soc of NA. Interests & involved in various civic affairs, including Rotary, also serving as V Chrmn of OH Motorists Assn. *Society Aff:* TBΠ.

Widmer, Wilbur J
Business: Dept of Civil Engrg, Box U-37, Storrs, CT 06268
Position: Prof *Employer:* Univ of CT *Education:* MS/Sanitary Eng/MIT; B/CE/Cooper Union *Born:* 10/20/18 Born in West New York, NJ. Hull Technician/Scientist for Gibbs & Cox, Naval Architects, New York 1943-47. Teaching at Univ of CT since 1948; currently Pastrof of Civil Engrg. World Health Organization, Sanitary Engr in Lahore Pakistan 1968-70 and in Brazil 1972 organizing univ grad teaching/res progs; also Sanitary Engr for WHO in Egypt, Saudi Arabia, Sudan in 1979 documenting waste stabilization pond practices. WPCF Arthur Sidney Bedell Award, 1972. Elected Fellow in the Royal Society of Health (U.K.), 1971. Married; three children. Enjoy astronomy, philosophy, herpetology, programmable calculators. P Pres, CT Section of ASCE, 1972-73. *Society Aff:* AAEE, ASCE, ASLO, AWWA, WPCF

Widnall, Sheila E
Business: Bldg 37-475, Cambridge, MA 02139
Position: Professor. *Employer:* Mass Inst of Technology. *Education:* BS/Aero/MIT; MS/Aero/MIT; ScD/Aero/MIT. *Born:* 7/13/38. Tacoma, WA. Asst Prof 1964-70; Assoc Prof 1970-74. Appt Prof in 1974. Research in wing theory, unsteady aerodynamics, aerodynamic noise, fluid dynamics, flows with vorticity, wake turbulence. Teaching included aerodynamics, dynamics, acoustics aeroelasticity, aerodynamic noise. In 1974 served as the First Dir of the Office of Univ Research of the US Dept of Transportation. Mbr, Bd/Dir, Fellow, AIAA; Assoc Editor, AIAA Journal of Aircraft, AIAA Aeroacoustics Panel Mbr Advisory Cttee to NSF, DOT, USAF. Mbr Space & Aeronautics Bd NAE. Recd the 1972 AIAA Lawrence Sperry Award and the 1974 Outstanding Achievement Award of the SWE. Bd of Dir, Fellow AAAS; Assoc Ed Journal of App Mech ASME; Fellow, Assoc Ed Physics of Fluids APS; Mbr US Natl Ctte on Theoretical & App Mech; Trustee, Carniegie Corp; Chrmn Faculty MIT 1979-81; Bd of Visitors USAF Acad 1977- 83; Bd of Dir., Pres, Chem-FAB Inc, Aerospace Corp, Draper Labs, Member NAE. *Society Aff:* AIAA, APS, SWE, AAAS, NAE.

Widrig, Francis F
Business: 25200 Telegraph Rd, Southfield, MI 48037
Position: Project Director. *Employer:* Giffels Assoc Inc. *Education:* BSCE/Structural/Univ of MI. *Born:* 7/10/26. Early experience in bridge design and supervision of maintenance work for NY Central RR and the Wayne County (MI) Road Commission. Formerly VP & Chief Engr of Amer Prestress Inc. Formerly partner in consulting firm. Hol/forty Widrig & O'Neill Inc and Pres Widrig O'Neill & Assoc Formerly with Carl Walker & Assoc as VP, in charge of Detroit Office. Joined Giffels Assoc April 1977 as a Project Director. P Pres ACI/MI & ASCE/MI Chap. Fellow ASCE; mbr MI Soc of Prof Engrs, ACI, Prestressed Concrete Inst, Engrg Soc of Detroit. Avid ski and tennis sportsman. Mbr Meadowbrook Country Club, Northville, Mich. *Society Aff:* MSPE, NSPE, ACI, ASCE, PCI.

Wiebe, Donald
Home: 106 Woodland Rd, Sewickley, PA 15143
Position: Consultant *Employer:* A. Stucki Co *Education:* MSEM/Mining Eng/W. Va. Univ; BSEM/Mining Eng/W. Va. Univ. *Born:* 06/30/23 Native of Pittsburgh, Pennsylvania. Research Engr-Mgr, Joy Manufacturing Co. Experimental Station, Pittsburgh, PA, 1949-62. New product dev & engrg relating to mining, construction & industrial machinery & equipment. Mgr Engrg Mechanics, Westinghouse Astronuclear Lab, Pittsburgh, PA, 1962-64. Mgr and V P Research & Engrg, A. Stucki Company, Pittsburgh, PA, 1964-Present. In charge of new product devt & engrg of Stucki railroad products - freight car truck suspension components. Holder of 30 U.S. letters patents relating to freight car trucks, mixed flow fans, particle separators & fluid flow control & measurement. *Society Aff:* ASME, AIME, ISA, SEM

Wiedeman, John H
Business: 1789 Peachtree Rd, NE, Atlanta, GA 30309
Position: Pres. *Employer:* Wiedeman & Singleton Engrs. *Education:* MS/Pub Health Eng/GA Tech; BS/Civil Eng/GA Tech. *Born:* 11/1/35. 2 yrs active service as Lt.j.g. US Navy CEC. Partner, Wiedeman and Singleton, Engrs since 1959. Presently Pres of Wiedaman and Singleton Inc. Pres GA Sec ASCE 1974. Dist 10 ASCE Dir 1975-78, ASCE VP Zone II 1978-80. ASCE National Pres. 1982-83. Mbr WPCF, APHA and AWWA. Specializes in water and wastewater fields and is author of several papers ranging in subjects from pipe line construction to treatment of textile finishing wastes. Mbr Atlant Rotary Club. Alanta Athletic Club. *Society Aff:* ASCE, WPCF, APHA, AWWA, NSPE, APWA, ACI.

Wiedemann, Harold T
Business: PO Box 1658, Vernon, TX 76384
Position: Assoc Prof. *Employer:* TX Agri Expr Sta. *Education:* MS/Agri Engr/TX A&M Univ; BS/Agri Engr/TX A&M Univ. *Born:* 5/10/34. Reared on TX farm. Taught in Agri Engrg Dept of TX A&M Univ from 1958 to 1969. Joined the Univ's TX Agri Experimental Sta in 1969 to conduct res in equip dev for brush control and range improvement. Low-energy grubber designed for brush control has received worldwide attention and concept was displayed at 1977 Worlds Fair for New Tech. A fluffy grass seed metering device developed for grass drill is now manufactured by six companies covering over 85% of market. Developed disk-chain for cost. effective range and seedbed preparation. Authored numerous publications. Have served as pres of TX Sec ASAE and consultant for brush control problems in foreign country. *Society Aff:* ASAE, SRM.

Wiegel, Robert L
Business: 412 O'Brien Hall, Berkeley, CA 94720
Position: Prof of Civil Engrg. *Employer:* University of California. *Education:* MS/Mech Engg/Univ of CA, Bekeley; BS/Mech Engg/Univ of CA, Berkeley. *Born:* Oct 1922. Native of San Francisco, CA. Served to Lt, Ordnance Corps, US Army 1942-46. Res Engr 1946-60; Assoc Prof 1960-63; Prof of Civil Engrg 1963-; Asst Dean, 1963-72; Acting Dean, Coll of Engrg 1972-73; U of CA. Berkeley. Consultant to a number of firms & govt agencies in many aspects of ocean engrg. Author of numerous tech papers & consulting reports, the book 'Oceanographical Engrg' & coordinating editor of the book 'Earthquake Engrg.' Mbr of a number of boards & cttees incl Exec Cttee Waterways, Harbors and Coastal Engrg Div ASCE (1971-76; Chrmn 1974-75) and US Army Coastal Engrg Res Bd. Mbr NAE and mbr of the Marine Bd, NAE; Pres 1972-75 ECOR. Chrmn, ASCE Coastal Engg Res Council, 1978-. *Society Aff:* ASCE, ΣΞ, AAAS, IAHR, PIANC, ASBPA.

Wiegel, Ronald L
Business: PO Box 867, Bartow, FL 33830
Position: Tech Mgr *Employer:* Intl Minerals and Chem Corp. *Education:* PhD/Mineral Processing/Univ Queensland (Australia); MS/Chem Engg/Carnegie - Mellon Univ; BChE/Chem Engg/Univ of Dayton. *Born:* 1/10/36. in Springfield, OH. Previously employed by DuPont, Jones and Laughlin Steel Corp, Univ of MN and currently by Intl Minerals and Chem Corp. Published in the areas: mathematical modeling and computer simulation of mineral processing operations, computer process control and process calculations. Currenty responsible for Process Engg, Process Dev and Analytical Labs of IMC's Fertilizer Group, Phosphate Minerals Div. *Society Aff:* AIME, ΣΞ.

Wiehl, Frederick R
Home: 741 Stevens Ave, Westfield, NJ 07090
Position: Mgr Materials & Reliab.Eng. *Employer:* The Singer Company. *Born:* Nov 4,1923. MS NJ Inst of Tech, BS Lafayette College. Served in US Coast Guard 1942-45. Foundry Metallurgist at Singer 1950-55. Field Metallurgist Asarco 1955-59. Rejoined Singer as Staff Materials Engr 1959. Organized synectics lab at Singer 1962. Assumed current position 1972. Resp for materials and process develoment and selection for consumer products as well as reliability testing and prediction. Won 1st prize Gray Iron Founders Society design contest 1962. Active in Red Cross (First Aid instruction). Enjoy gardening and fishing.

Wiel, Stephen
Business: 505 E King Street, Carson City, NV 89710
Position: Commissioner *Employer:* Public Svc Commission of NV *Education:* PhD/Public Admin/Univ of Pittsburgh; MS/ChE/Stanford Univ; BS/ChE/Stanford Univ. *Born:* 1/9/39. Formed consulting engrg & planning firm in Reno, NV in 1977 in order to promote the concept of living softly. Helped create innovative new solar bldg concept called xen-wall system. Currently serving 4 year term as Public Svc Commissioner in NV. Principal author of regional govt's air pollution control plan & state govt's energy conservation contingency plan. Designer of new solar homes & old homes adapted to solar. Developer of new concepts for solar utilization. Formerly Assoc Prof in mech engrg at Univ of Nevada, Reno. Formerly exec vp of large intl engrg firm in Wash, DC. First eight yrs after college in aerospace industry. Reg PE. *Society Aff:* NSPE, ASHRAE, ASES.

Wieland, Warren R
Business: 911 Locust St 200, St Louis, MO 63101
Position: President. *Employer:* Campbell & Wieland, PC. *Education:* BSCE/Civil Eng/Rolla Sch of Mines & Metallurgy; -/Civil Eng/WI Inst of Technology. *Born:* July 1923 Potosi, Wisc. BSCE Missouri School of Mines & Metallurgy, Rolla, MO. 88th Infantry Div, Fifth Army in Italy. Purple Heart & Cluster Pres of Campbell & Wieland,PC; Engrs/Architects/Planners St.Louis, MO. Resident of Belleville Illinois. PPres Belleville Bd of Edu, Dist 118; PPres of Spec Edu Dist for Handicapped Children; PTreas of CEC/MO - presently on Bd/Dir. Reg PE in 28 states. 1976 recd Prof Degree in Civil Eng from Univ of MO-Rolla for outstanding contribution to engrg profession. *Society Aff:* ASCE, NSPE, ACEC, CSI, ACI.

Wiener, George W
Home: 2348 Marbury Rd, Pittsburgh, PA 15221
Position: Mgr Materials Science Div; Retired March 1, 1986 *Employer:* Westinghouse Elec Corp. *Education:* PhD/Met Eng/Univ of Pittsburgh; MS/Met Eng/Univ of Pittsburgh; BS/Chemistry/Univ of WI. *Born:* Jan 1922 Providence RI. Employed at Res Labs of Westinghouse Elec Corp since 1949. Extensive research and publications on magnetic materials & broad experience in mgt of R & D. Present responsibilities include mgt of R&D on materials technology required for the development of advanced electrical energy systems and improvements in both fossil and nuclear power generation equipment. Taught graduate courses at the Univ of Pittsburgh 1953-68. Fellow ASM 1971. *Society Aff:* ASM, AIME.

Wier, Thomas P Jr
Business: PO Box 58403, Houston, TX 77258
Position: Pres. *Employer:* Wier & Assoc. *Education:* PhD/Chemistry/Rice Univ; MA/Chemistry/Rice Univ; BS/ChE/Rice Univ *Born:* 6/10/19. Native Houston, TX. '42-'43 taught anal chem Rice Univ. '43-'51 Sr Res Chemist & Group Leader (anal res, process res) Shell Oil. '51-'71 Pres, Wood- Protection Co & related cos, produced preserved & fire retardant wood, wood products, lumber, prefabrication, & construction. '71 to present, owner, Wier & Assoc, consultants. Reg PE. Honors: Phi Beta Kappa, Tau Beta Pi, Sigma Xi. Several technical publications & patents. Past pres: Rice Engg Alumni, Houston Tau Beta Pi Alumni, Vacuum Wood Preservers Inst, Gulf Coast Sci Fdn. Past Trustee & VP Space Ctr Meml Hospital Fdn. Chairman and President MultiMark Int'l Inc.; Chairman, Seagate Foundation for the Arts. *Society Aff:* AIChE, ACS, AAAS, WPCF, NFPA

Wiersma, Frank
Business: 501 Ag Sci Bldg. 38, Tucson, AZ 85721
Position: Prof & Agr Engr. *Employer:* University of Arizona. *Education:* PhD/Agr Engg/SD State Univ; MS/Agr Engg/OK State Univ; BS/Agr Engg/SD State Univ. *Born:* 11/30/26. Born and raised on a South Dakota farm. WWII service in US Navy and Korean War service in US Army Corps Engrs. Private contractor for rural water, sewerage and drainage systems. Research in irrigation development 1954-57. With the Univ of Arizona since 1957. Research project leader in the engrg of livestock production systems with emphasis in reducing heat stress in dairy cattle. Developed a new concept in evaporative cooling utilizing horizontal pad systems for agricultural applications. PE in AZ & CA. Active in ASAE, ASHRAE, ASEE and Cast. Bd of Dirs ASAE. Consultant for dairy prod systems in U.S., Mexico and the Middle East. *Society Aff:* ASAE, ASEE, ASHRAE, CAST

Wiese, Warren M
Business: Upper Mountain Rd, Lockport, NY 14094
Position: Chief Engr *Employer:* Harrison Radiator Div GMC *Education:* MSME/ME/Univ of MN; BS/ME/Univ of MN *Born:* 04/14/29 Native of Rochester, MN. Joined General Motors in 1952 at the Res Labs. In 1967, transferred to Dayton, OH as Mgr of A/C & Automotive Products Engrg for the Frigidaire Div until 1972 when promoted to Asst Chief Engr. Assumed responsibility as Dir of Engrg with the formation of the new Delco Air Division in 1975. Currently serving as Chief Engr, Refrigeration Components. Controls, and Materials for Harrison Radiator Div. In 1975-76 Chrmn of the Dayton Section of SAE and in 1981-83 served as mbr of the National SAE Bd of Dirs. In 1979 served on the National Academy of Science Ctte. 1976, listed in the *Ward's Who's Among US Motor Vehicle* Manufacturers, and *American Men and Women of Science* in 1973. Attended Stanford Exec Prog Grad School of Business - Stanford Univ - 1976. *Society Aff:* SAE

Wiesenberg, Jerome D
Home: 81-33 190 St, Jamaica Estates, NY 11423
Position: Principal. *Employer:* Schacher-Greentree Inc. *Education:* MBA/Ind Mgt/CCNY; BBA/Bus Adm/CCNY. *Born:* 12/27/31. Native of NYC. Served in US Army 1953-55. Over 20 yrs of extensive experience in all phases of ind engg & bus mgt in a wide diversity of industries. Reg PE-CA. Lectr & author of many technical articles. Former proj mgr of several multi-million dollar ventures. Former college instr. New prinicpal in mgt consulting firm specializing in purchasing cost reduction & profit improvement progs. Present natl dir of work measurement & methods engg div, & held all positions of responsibility as chapter officer - IIE. *Society Aff:* IIE, APICS, NAA.

Wiesendanger, J U
Home: East Winthrop, ME 04343
Position: Consulting Engr (Self-employed) *Employer:* Self *Education:* CE/-/Cornell Univ; LLB/-/Forham Univ. *Born:* July 1913. With Elwyn E Seelye and Seelye, Stevenson, Value & Knecht, NYC 1936-51; Prepared 1st and 2nd edition of 'Data Books for Civil Engrs' for Mr. Seelye. In charge Civil Engrg Design for Tuttle, Seelye, Place & Raymond WWII projects and after. Private practice 1951 to date; Structural, sanitary, airports. Reg PE NY, ME, NH, VT, MA. Fellow ASCE, PPres, ME Sec. Mbr AAEE, ACI, AISC, ACEC. Rotarian, Fisherman, gardener. *Society Aff:* ACEC, ACI, ASCE, AAEE, AISC.

Wiesenfeld, Joel
Home: 311 Valentine St, Highland Park, NJ 08904
Position: Prof of Civil Engg. *Employer:* Rutgers Univ. *Education:* PhD/Appl Mechanics/Poly Inst of Brooklyn; MS/Civil Engg/MA Inst of Tech; BCE/Civil Engg/City College of NY. *Born:* 4/9/18. Chrmn, Dept of Civil and Environmental Engg, Rutgers Univ, 1970-1980 Asst Dean for Freshman 1980- . Prof of Civil Engg (1958-date). Assoc Prof (1953- 1958); Asst Prof (1947-53). Aircraft Stress Analyst (1941-

Wiesenfeld, Joel (Continued)
46); Superintendent, Highland Park, NJ, Water and Sewer Dept, since 1961. Professional Engrs Registration, NY, NY, FL and WV. Chrmn, State of NJ Uniform Construction Code Advisory Bd. Consultant in structural design, water and sewage works, industrial plants, insurance valuations, and depreciation studies. Conference chrmn, Annual NJ Asphalt Paving Conference (30 yrs). Active participation in continuing education. Mbr of Tau Beta Pi, Chi Epsilon, and Sigma Xi. *Society Aff:* ASCE, ASEE, NSPE, AWWA.

Wiesner, Jerome Bert
Business: 20 Ames Street, E 15-207, Cambridge, MA 02139
Position: President Emeritus and Institute Professor. *Employer:* MIT. *Education:* PhD/EE/Univ of MI; MS/EE/Univ of MI; BS/EE & Math/Univ of MI. *Born:* 5/30/15. Mbr of faculty at MIT for many years. Former dean of MIT Sch of Science from 1964-66, Provost, from 1967-71, pres 1971-1980; President Emeritus and Institute Professor 1980- He was science advisor to Presidents Kennedy and Johnson. Within the scientific community he is recognized as an authority on microwave theory, communications science and engrg and radio and radar propagation phenomena. During WWII was a leader in the development of radar and later one of the principals in the conception of radio transmission by scatter techniques from the earth's ionosphere. Fellow IEEE, AAAS. Mbr NAS, NAE, APS, ASA, AGU, AAUP, Tau Beta Pi and Eta Kappa Nu. *Society Aff:* AAAS, APS, AAUP, NAE, NAS.

Wiggins, Edwin G
Home: 13 Clare Dr, East Northport, NY 11731
Position: Dean *Employer:* Webb Institute of Naval Architecture *Education:* PhD/Mech Engrg/Purdue Univ; MS/Nuclear Eng/Purdue Univ; BS/Chemical Engineering/Purdue Univ *Born:* 06/12/43 Born in Palo Alto, CA. Served in the Navy for 13 years. Tours included various ships, a shipyard and staff duty. Left the Navy to teach at TX A&M Univ at Galveston. After 3 1/2 years there, moved to U.S. Merchant Marine Acad. Assumed position of Dean of Webb Institute of Naval Architecture July 1, 1987. Vice Chairman of the NY Metropolitan Sect of SNAME and Assoc Ed of Computers in Education Journal. Tech interests include thermodynamics, fluid mechanics and heat transfer. *Society Aff:* ASEE, ASME, SNAME

Wiggins, John H
Business: 1650 S Pacific Coast Hy, Redondo Beach, CA 90277
Position: Pres & Mbr Bd of Dir *Employer:* Crisis Management Corp *Education:* PhD/CE/Univ of IL; MS/Geophysics/St Louis Univ; BS/Physical Sci/Stanford Univ. *Born:* 5/12/31. Tulsa, Okla. Worked in nuclear weapons effects, oil well drilling, sonic boom, Earthquake Engineering and other areas. Officer in the US Air Force 1953-58. Employed by Exxon, John A Blume & Assoc and formed J H Wiggins Co in 1966-84. Formed Crisis Management Corp in 1987. Mbr of the BRAB, NRC, Advisor to many governmental agencies. Reg Civil and Safety Engr, reg Geophysicist, Calif. Formed co in new field of Engrg Risk Mgt and has formerly been involved in the following fields: biophysics, underground structure, structural dynamics, drilling technology, rock mechanics, seismology, exploration geophysics, fire technology, shock dynamics, professional liability, socio-technical systems, risk assessment, winner of Moisseiff medal 1965. *Society Aff:* ASCE, AIPG.

Wigglesworth, Robert W
Business: 532 SE 5th Ave, PO Box 1281, Gainesville, FL 32602
Position: V Pres-Surveying Services *Employer:* M K Flowers Inc., A Subsidiary of George F. Young of Florida, Inc. *Born:* 8/27/36. Native of Gainesville, FL; been with the same firm since 1953; partnership 1960-1987; presently assisting Univ of FL surveying prog by teaching surveying drafting and being guest lectr; FSPLS VP 1977, Pres Elect 1979, Pres 1980. FCES VP 1975, Pres 1976. Enjoy traveling & hunting; SAMSOG; ACSM; NSPS; FES; NSPE. *Society Aff:* ACSM, NSPS, NSPE.

Wight, George
Business: 127 S Northwest Hgwy, Barrington, IL 60010
Position: Pres. *Employer:* Wight Consulting Engrs, Inc. *Education:* BSCE/Civil Engrg/IA State Univ; /—/Keller Grad Sch of Mgmt *Born:* 6/7/28. Profl Engg Consultant Middlewest USA. Pres Wight Envir Cons & Wight Cons Engrs Inc Barrington, IL. Dir, Bank of Villa Park, IL. Cons to Municipalities, Utilities, & Developers, Dir Bank of Illinois Bloomington, Normal, IL. *Society Aff:* NSPE, ISPE, ASCE, APWA, APA, AWWA, WPCF, ULI.

Wilbeck, Jerry L
Home: 724 W Santa Fe Trail, Kansas City, MO 64145
Position: VP Electrical Engrg Systems *Employer:* Phoenix Aerospace Inc *Education:* BS/EE/KS State Univ *Born:* 9/10/43 Native of South Hutchinson, KS. Participated in initial test stand checkout and static test firings of S-1C stage of Saturn IV in Huntsville, AL while Co-op student mbr of NASA Engr team (1964-1965). Sales Engr for Ckyogenics manufacturer/distributor. Electrical design engr on major systems for Gates Learjet Corp. Designed and patented several Farm Tillage Tools while Gen Mgr of Farm Equip Manufacturing Co. Traveled North America and Europe extensively while marketing Airborne Electronic Power Equip for present employer.

Wilbourn, Sanford M
Business: 1908 West Eleventh St, P.O. Box C-50, Little Rock, AR 72203
Position: Pres *Employer:* Garver & Garver, Inc *Education:* MS/CE/Stanford Univ; BS/CE/Stanford Univ *Born:* 10/5/26 Lonoke County, AR. Attended public schools and served in US Army, Infantry, 1944-46. Project Engr, LA Dept of Hgwys, Bridge Div, 1948-50. Joined Marion L Crist & Assoc, Little Rock, AR 1950. Served as project engr water and sewage treatment plants until 1954. VP overall responsibility for design 1954-56. Active in re-establishing firm of Garver & Garver, Inc, Little Rock, AR, 1956. VP/Chief Engr, 1956-70. Chrmn/Pres 1970-final responsibility 100-man plus firm. Chrman/CEO - 1986 to present. Past Pres, member AR Bd of Reg. Past VP ASPE. Enjoys music, travel, fishing. *Society Aff:* NSPE, WPCF, FASCE, ACEC

Wilbur, Leslie C
PO Box 97, Berlin, MA 01503
Position: Prof of ME & Dir, Nuclear Reactor Facility. *Employer:* Worcester Poly Inst. *Education:* MS/ME/Stevens Inst of Tech; BS/ME/Univ of RI. *Born:* 5/12/24. Taught mech engr courses at Duke Univ for 8 years prior to coming to WPI in 1957 to teach thermodynamics and power generation courses. Became Dir of the WPI Open Pool Reactor when it was constructed in 1959 and became a full prof in 1961. Have studied nuclear engineering at Oak Ridge Natl Lab, Brookhaven Natl Lab, Argonne Natl Lab, Cornell Univ, N.C. State College and MIT. Reg PE in Commonwealth of MA, WPI rep for Atomic Ind Forum, and listed in AIF speaker's bureau. Served in ETO in WWII, Editor-in-Chief of John WIley & Sons Handbook, Energy Systems, now in preparation with publication scheduled for 1984-85. Fellow, ASME 1983. *Society Aff:* ANS, ASME, ASEE, AAAS, ΤΒΠ, ΣΞ, ΠΤΣ, ΦΚΦ.

Wilbur, Lyman D
Home: 4502 Hillcrest Drive, Boise, ID 83705
Position: Consulting Engr. *Employer:* Self. *Education:* AB/Civil Engg/Stanford Univ. *Born:* 4/27/00. Draftsmn Standard Oil Co of CA 1921. Draftsman, Asst Engr City of San Francisco Hetch Hetchy Proj 1921-24. Designer Merced Irrigation Dist 1924-

Wilbur, Lyman D (Continued)
26. Designing Engr East Bay Municipal Utility Dist 1926-29. Asst to Chf Consulting Engr, Middle Asia Water Economy Serv 1929-31. 1932-70 with Morrison-Knudsen Co Inc from engr to VP Overseas Oper. Dir, also with Intl Eng Co Inc 1956-70 as Exec VP, Pres & Chrmn of Bd and also Resident Partner Atlas Constructors, Moroccan Air Bases and RMK-BRJ Vietnam Military construction. 1971-Consulting Engr. Mbr NAE, NSPE 1975 Award. One of Top Ten Construction Men ASCE 1975. Moles Award 1974. John Fritz Medal 1973. Hon Mbr ASCE 1968, ASCE Presidents Award 1980, Engineering News Record Construction's Man of the Year 1966, Honorary Doctor of Laws College of Idaho 1962, Honorary Doctor of Science University of Idaho 1967. *Society Aff:* ASCE, ISPE, NAE.

Wilbur, Paul J
Business: Dept of Mech Engg, Ft Collins, CO 80523
Position: Prof. *Employer:* CO State Univ. *Education:* BS/Mech Engrg/Univ of UT; PhD/Aerospace & Mech Sci/Princeton Unvi. *Born:* 11/8/37. Prof Wilbur worked for four yrs as a nuclear power engr for the US Atomic Energy Commission before returning to the sch to obtain his PhD. Since completing his formal educ he has worked as a Prof of Mech Engg at CO State Univ. While most of his res has been conducted in the area of elec propulsion of spacecraft and ion implantation of mechanically active surfaces he has also been involved in studies of the application of solar energy sys. He has over 120 pubs on his res. *Society Aff:* ASME, AIAA, ΣΞ.

Wilburn, Darrell G
Business: 1637 Laurel, Lake Oswego, OR 97034
Position: Pres. *Employer:* Assoc Transportation Engr & Planning, Inc. *Education:* AA/Engg/Palomar Jr College; BSE/Civil Engr/CA State Univ at San Diego; MSCE/Transp Engr/Univ of WA. *Born:* 1/21/45. 6/72-11/73 City of San Diego, CA-Civil Engr (Structural and Traffic Depts); 11/73-8/75 City of Eugene, OR-Asst Traffic Engr; 8/75-12/76 Assoc Consultant for Transp Planning & Engg Bellvue, WA; 12/76-4/78 City Traffic Engr, City of Eugene, OK; 4/78-6/79 Transp Engr for OHZM-Hill, Portland, OR; 6/79-date Pres of Assoc Transp Engg & Planning, Inc, Portland, OR. *Society Aff:* ITE, ASCE.

Wilcock, Donald F
Home: 1949 Hexam Rd, Schenectady, NY 12309
Position: President. *Employer:* Tribolock Inc. *Education:* DES/Engg/Univ of Cincinnati; MES/Engg/Univ of Cincinnati; BS/Civil Engg/Harvard Univ. *Born:* 9/24/13. Native of Bklyn NY. Sherwin-Williams 1939-42. Gen Elec 1942-65. Mech Tech Inc, 1965-78, Mgr Tribology Ctr & Dir Res, Tribolock Inc, Pres, 1978- . Fellow ASME, ASLE; ASME Hersey Award 1973, Policy Bd-Research, PChrmn Lubrication Div and Res Comm of Lubrication. Primary concern energy conservation through more efficient lubrication and wear control. Hobbies sailing, music. *Society Aff:* ASME, ASLE.

Wilcox, Lyle C
Business: Pueblo, CO 81005
Position: Pres *Employer:* Univ of S CO *Education:* PhD/Elec Engrg/MI State Univ; MSEE/Elec Engrg/MI State Univ; BSEE/Elec Engrg/Tri-State Coll. *Born:* 8/8/32. Major Res Area: Res adm, ind & univ, educational program dev, specific technical interests in instrumentation, controls, computers. Dean - Coll of Engrg - Clemson Univ 1973-1980. Assoc Dean for Prof Studies of the Coll of Eng Clemson Univ 1970-73; Hd Dept of Elec & Computer Engrg Clemson Univ 1966-72; Chf Engr and Dir of Operations, So Medical Res Support Ctr, Vets Admin Wash, DC. Mgt Conslt local indust and govt, Dir Res contracts for Greenwood Mills Inc, NSF, Dept of Army, NASA. ASEE Prizes and Awds Cttee (Frederick Terman Awd); Self- Study Cttee, So Assn of Colls and Schools IEEE; Chrmn, Admin Unit, SE Sec ASEE. Author of numerous pubs. As of Sept 1, 1984 will be Deputy Asst Secy for Breeder Reactor Programs with the US Dept of Energy in Washington, DC. *Society Aff:* ASEE, IEEE, WICHE.

Wilcox, Marion W
Business: C/ME Dept, Dallas, TX 75275
Position: Prof. C/ME *Employer:* So. Methodist Univ. *Education:* ScD/Engr Mechanics/Notre Dame; MSCE/Civ Engr/IIT; BSCE/Civ Engr/Notre Dame. *Born:* 8/17/22. in Broken Arrow, OK. Public schools, OK, NM, MS. US Navy, 1941-46; V-12, Navy Officers' Training Prog, Notre Dame, 1944-1946. Ind experience in such cos as McDonnell, Bendix Aviation & Gen Dynamics, 1948-1960. Assoc Prof of Mech Engr, Univ of AZ, 1961. Assoc Prof SMU, 1962-1965. Prof SMU, 1966. Prof and Dir, Solid Mechanics Center SMU, 1966-1971, Prof C/ME, SMU, 1966- , Pres, Wilcox Engrg Co, Inc, 1983- . Reg PE TX. *Society Aff:* AIAA, ASEE.

Wilcox, William R
Business: Clarkson Univ, Dept. of Chemical Engrg, Potsdam, NY 13676
Position: Professor, Dir of Center for Advanced Materials Processing, Dir of Center for the Development of Com *Employer:* Clarkson Univ *Education:* PhD/ChE/UC Berkeley; BE/ChE/Univ of So CA. *Born:* 1/14/35. Manhattan Kansas. Mbr Tech Staff TRW Semiconductors 1960-62 spec in solid state diffusion. Hd Crystal Tech at Aerospace Corp 1962-68. Assoc Prof 1968-74 and Prof 1974-75 of Chem Eng and Mat Sci at U So Calif. Prof and Chrmn Chem Eng at Clarkson 1975-86. Director of Center for Advanced Materials Processing and Center for the Development of Commercial Crystal Growth in Space since 1986. Dean of Engineering since 1987. Author of over 150 papers related to crystal growth. Consultant for 25 organizations since 1968. Mbr Exec Cttee 1971-87, Chrmn Western Sec 1974-75, VP 1984-87 Am Assn Crystal Growth. Reg PE NY since 1976. *Society Aff:* AIChE, ASEE, AACG, ECS, ACerS

Wild, Harry E, Jr
Home: P.O. Box 1023, Ormond Beach, FL 32074
Position: President. *Employer:* Briley, Wild & Assoc. *Education:* ME/Bioenviornal Engg/Univ of FL, Gainesville; BCE/Civil Engg/Univ of FL, Gainesville. *Born:* Sept 1940. US Army Staff Sgt. 1971 President Briley, Wild & Assoc. 1972 WPCF Harrison P Eddy Awd. Mbr AMA, APCA, APWA, ASCE, ASTM, AWWA, CEC, CSI, NSPE, PEPP, WPCF, Florida Eng Soc (FES), Florida Inst of Consulting Enrs, (FICE), Daytona Beach Comm Coll Tech Adv Ctte, Flagship Bank Bd/Dir. FES Cttees: Vice Chrmn Legislative, Mbr Ethical Practices. FICE: No Reg Dir 1975-76; Secy Treas 1976-77; Chrmn Leg/Govt Rel; Mbr Fin. Daytona Beach Chap of FES: VP 1974-75; Pres 1975-76. FICE State Pres 1979-80, Flagship Bank: Trust Committee, Loan & Discount Committee, FL Representative from FICE Advisory Committee to EPA's Southeast Region Administrator, FICE VP (state) 1978-1979, Ormond Beach Chamber of Commerce Bd of Governors. Hobbies: sailing, snow-skiing FICE - Vice President, North 1980-81 1980-81 208 Program Advisory Committee Member. *Society Aff:* APCA, ACEC, ACI, AIChE, AMA, APWA, ASCE, ASME, ASPO, ASTM, AWRA, AWWA, CSI, IAPC, NSPE, WPCF.

Wilde, H Dayton
Home: 3013 Avalon Pl, Houston, TX 77019
Position: R/D Coord (Ret). *Employer:* Exxon Co USA. *Education:* ScD/Chem Eng/MIT; MS/Chem Eng/Univ of TX; BS/Chem Eng/Univ of TX. *Born:* Sept 1900 Aguascalientes Mexico. All business career with Humble Oil & Refining Co (now Exxon) in res and dev. Organized and was mgr of Production Res Div. Was mgr of Refining Tech and Res Div during WWII years spec in Toluene for TNT, aviation gasoline and synthetic rubber programs. Subsequently became coordinator for Humble's overall R & D activities in exploration, production and refining. Retired in 1965. Served as Natl Chrmn of Petroleum Br of AIME and as Dir of AIME, AIChE and ASTM. Prin hobbies are electronic projects and color photo processing. *Society Aff:* AIChE, SPE, ACS, ASTM.

Wilde, Richard A
Home: 22482 Alexander Dr, St Clair Shores, MI 48081
Position: Chf Materials Engr. *Employer:* Eaton Corp. *Education:* MS/Met Engr/Univ of IL; BS/Metallurgy/PA State. *Born:* 3/25/17. Native of State College PA. Process

Wilde, Richard A (Continued)
Eng Linde Air Products, Res Eng at Battelle Mem Inst, Raw Materials Eng for Western Elec Co, Plant Metallurgist at Steel Foundry of United Eng & Foundry Co, Asst Dir of Res at Lukens Steel worked on improvement of basic openhearth practice and developed processing of CLAD steel especially A1-Clad steel, with Eaton Corp since 1954 as Chief Materials Eng at the Research Ctr. Responsible for Material & Material Process Dev for Eaton Corp, Field of Expertise - Gear Technology, especially mech of gear scuffing and gear pitting. Chrmn of Detroit Sec of AIME 1962. Hobbies: Floriculture, golf, bird watching, philosophy of life and psychology of parent- child relationship. *Society Aff:* ASM, TMS-AIME, ASTM, SAE.

Wilder, Carl R
Home: 2034 Camino Dr, Escondido, CA 92026
Position: Ret *Education:* BS/CE/Univ of MO - Columbia *Born:* 09/06/13 Native of Cape Girardeau, MO. Reg Prof Engr, CA and CO. Employed by Phillips Petroleum Co in 1936; US Army Corps of Engrs 1936-47; Portland Cement Assoc 1947-retirement in 1978, as Dir its Dept of Energy and Water Resources. Responsible for developing use of soil-cement as riprap substitute for earth dams and embankments and as lining for canals, reservoirs, and water treatment basins. Was secretary of Exec Ctte of ASCE's Irrigation and Drainage Div; also was mbr of Exec Ctte of US-CIDFC. Currently serving six cttes of ASCE, ASTM, ACI and AWWA. Enjoys music, golf, bridge. *Society Aff:* ASCE, ACI, ASTM, AWWA, USCOID, USCIDFC

Wilder, David R
Business: Dept of Materials Sc and Engrg, Ames, IA 50011
Position: Prof & Dept Chrmn *Employer:* IA St Univ *Education:* PhD/Ceramic Eng/IA St Univ; MS/Ceramic Eng/IA St Univ; BS/Ceramic Eng/IA St Univ *Born:* 06/11/29 Member of teaching faculty, Iowa State Univ, 1955 to date. Chmn of Ceramic Engg 1961-76. Chmn of Materials Sci and Engg 1976 to date. *Society Aff:* NICE, ASEE, ACerS

Wilder, Thomas C
Business: 80 Bacon St, Waltham, MA 02154
Position: Consultant (Self-employed) *Education:* ScD/Met & Chemistry/MIT; MS/Met/MIT; AB/Chemistry/Bowdoin College. *Born:* 3/21/35. Native of Boston. With P R Mallory & Co as Electrochemist 1962-63. With Kennecott Copper corp (Ledgemont Lab, Lexington, MA) as Extractive Metallurgist, Group Leader, and Section Head Chem Processes 1963-1978. Specialty was extractive metallurgy of new minerals and refining of product metals. Presently am self-employed Consultant in Chem Metallurgy. Mbr of Electrochemical Soc 1960- present. Mbr of AIME 1960-present. Chrmn of Boston Sec AIME 1979-1980. Rossiter W Raymond Award, AIME, 1968. *Society Aff:* AIME, ECS.

Wilding, John D
Home: 39 Van Brackle Rd, Holmdel, NJ 07733
Position: Consulting Engr. *Employer:* Self employed. *Education:* BS/EE/PA State Univ. *Born:* 6/17/08. Hudson Coal Co Scranton PA, Elec Pwr Dept 1931-37. M W Kellogg NYC, Oil Refinery Design and Construction 1937-41. Officer, US Army Corps of Engrs 1941-45. MW Kellogg NYC, Project Engr Oil Refinery Design, Construction and Spec Projects 1946-54. ASME Dir of Standards and Secy Boiler Code and Pressure Vessel Cttee 1954-70. From 1970 Consulting Engr to ASME Nuclear Pwr Plant Certification Prog, Adv Cttee on Reactor Safeguards Pressure Vessels, Govt of Algeria Natl Standards for their manufactured products, Commonwealth of Pennsylvania Mbr Boiler Advisory Bd. Reg NY and Engr and Land Surveyor PA. Fellow ASME, Mbr ASCE. Avocations: History, Languages, good music, Clocks, Handicrafts. *Society Aff:* ASME, ASCE, AMIME.

Wilding, John V
Business: 210 State St, Hackensack, NJ 07601
Position: Mgr Systems Engg *Employer:* Computran Systems Corp *Education:* MS/EE/Rutgers Univ; BS/EE/Rutgers Univ *Born:* 12/18/29 Officer with US Army Signal Corps 1952-54. 8 yrs with ITT specializing in toll telephone automatic ticketing equip. Asst mgr of applications engrg at Quindar Electronics, Inc. For 10 yrs working with remote control and telemetering equip for utilities. With Computran Systems Corp since 1975 engaged in design and implementation of computerized vehicular traffic control and surveillance systems. Presently mgr of systems engrg. Registered PE. Mbr of Tau Beta Pi, Eta Kappa Nu, Sigma Xi, Mensa and Intertel. *Society Aff:* IEEE

Wildman, Robert F
Business: 111 New Montgomery St, San Fransico, CA 94105
Position: President *Employer:* Wildman & Morris, Inc *Education:* BS/Civil Engrg/Univ of CA, Berkeley *Born:* 3/30/22 Native of Berkeley, CA. Served with US Army Air Corps, 1943-1946. Maintenance and Construction Offices for 20th Air Force. Registered engineer in California, Nevada and Arizona. President of Wildman & Morris, Inc, an engineering and architectural firm since inception 1953. Past president and Director, Structural Engineers Assn. of Calif; Past Director and Secy/Treas. Structural Engineers Assn of California; Past Director, Consulting Engineers Assn of Calif.; Director San Francisco Past Society of American Military Engineers: Fellow ASCE. Mbr Chi Epsilon and Tau Beta Pi. *Society Aff:* ASCE, SAME, ACEC, ADPA

Wiles, Gloyd, M
Home: 68 Barkers Point Rd, Sands Point, NY 11050
Position: Consultant *Education:* BS/Mining Engrg/UC Berkeley CA *Born:* 02/16/98 Hanford, CA. Gen Supt Treadwell Yakon Co; VP Gen Mgr and dir Park City; Consol M Co; Chrmn Mining Co and Gen Mgr Mining Dept National Lead Co; VP & GM Nickel processing Co; Mbr: Canadian Mining and Metallurgical Engrs, American Institute Mining Metallurgical Engrs, Australian Inst Mining and Metallurgical Society, Mining Club Inc, Delta Phi, Theta Tau. *Society Aff:* ΘT, CIM, AIME, MMS

Wiley, Albert L Jr
Home: 3235 Hwy 138, Stoughton, WI 53589
Position: Prof (Tenured) *Employer:* Univ of WI-Madison *Education:* PhD/Radiological Sci & Nuclear Engrg/Univ of WI-Madison; MD/Medicine & Surgery/Univ of Rochester, Rochester, NY; Bach of Engrg/Nuclear Engrg/NC State Univ, Raleigh, NC; 1 year Grad Work/Nuclear Engrg/NC State U *Born:* 06/09/36 Native of Forest City, NC. Engr in indust (paper, nuclear). Lt. Commander and Medical Dir of the US Navy Radiological Defense Lab. Prof of Radiology and Human Oncology at the Univ of WI since 1970. Medical Board Certifications- Nuclear Medicine and Radiation Therapy. Published in Nuclear Medicine, Medical Phys, Biomedical engrg. Listed in Who's Who in Frontier Sci and Tech. Governors Appointee to Wisc Radioactive waste Review Bd (current) and Governor's commission on United Nations. Was the Republican Candidate for US Representative in Congress (WI 2nd Congressional Dist 1984 General Election). Was candidate for Governor of Wisc (1986 primary election). *Society Aff:* IEEE, ANS, ΤΒΠ, HPS, AAPM, ΣΞ, AAAI

Wiley, Carl A
Home: 6627 W 82nd St, Los Angeles, CA 90045
Position: Chief Scientist *Employer:* Hughes Aircraft Co *Education:* BS/Math/Antioch College. *Born:* 12/30/18. Developed first Radar Correlation Guid Sys at Wright Field. Also discovered piezoelectric behavior of barium titanate. At Goodyear Aircraft invented, designed and flew first syn-aperture radar (SAR). Proposed light-pressure space propulsion in May, 1951 Astounding Stories. Pres, Wiley Elect 1962-1963. Designed, flight-tested first microwave mapping radiometer with Dr Allan Love. From 1963-1978, held tech exec pos at Autonetics in radar, sonar, space sci. Turbulence-radiation theories dev, tested. Elected IEEE Fellow for Dev SAR Radar, 1979. Chief Sci at Huges Aircraft doing orbital remote sensing at present. *Society Aff:* IEEE.

Wiley, Richard G
Business: 510 Stewart Drive, N. Syracuse, NY 13212
Position: V.P. Chief Scientist *Employer:* Research Associates of Syracuse, Inc *Education:* PhD/Electrical Eng/Syracuse Univ; MS/Electrical Eng/Carnegie-Mellon Univ; BS/Electrical Eng/Carnegie-Mellon Univ *Born:* 08/25/37 For over twenty-five years, Dr. Wiley has been an important contributor to electronic intelligence and electonic wafare efforts. He has published two books (Artech House 1982 and 1985) on electronic intelligence and presents continuing education courses on the subject. *Society Aff:* IEEE

Wiley, Samuel K
Business: PO Box 249, Bellaire, TX 77401
Position: President. *Employer:* Resource Mgt Consultants. *Born:* March 1944. BSChE from Univ of Texas at Austin; MBA from Univ Houston. Currently President and owner of Resource Management Consultants, an engrg consulting firm specializing in combustion related air pollution control. Perform engrg studies and mfg incinerators, flares and associated equipment for elimination of process vents from refineries and petrochemical plants. Reg PE in Texas and Louisiana.

Wiley, T T
Business: TT Wiley Associates Pc, 1383 Veterans Mem Highway, Hauppauge, NY 11788-3048
Position: Pres. *Employer:* T T Wiley Associates, PC. *Education:* MSCE/CE/Univ of IL; BSCE/CE/Univ of IL. *Born:* 5/17/08. Champaign, IL. Engr 1931-1938 with IL Div of Hgwys. Ten yrs in Detroit as Asst City Traffic Engr. To NYC in 1949 to organize & hd its Dept of Traffic. Consulting Engr from 1962. Pioneer in dev & application of traffic engg, planning, design, & operations techniques, including statewide hgwy use studies, concept & planning for Detroit expressway system, municipal parking systems, traffic actuated signals, neon pedestrian signals, thermoplastic paving markings, actuated control signals, computerized traffic surveillance and guide sign systems. Engg & citizen group awards. Deacon Reformed Church of Am. Choir conducting, singing, & golf. *Society Aff:* ASCE, ITE, NSPE.

Wilhelm, Luther R
Business: University of Tenn, Agri Engg Dept, Knoxville, TN 37996
Position: Professor *Employer:* Univ of TN. *Education:* PhD/ME/Univ of TN; MS/ME/Univ of TN; BS/Agr Engr/Univ of TN. *Born:* 1/25/39. Native of Camden, TN. Served in USAF 1960-67. Assignments included Base & Maintenance Engrg, Joined the Univ of TN Faculty in 1971. Responsibilities include teaching grad and under-grad courses and directing res in food & process engineering & agricultural instrumentation systems; and coordinating computer applications for Dept. Teaching and research interests include: computer applications, agricultural instrumentation, physical properties of agricultural products, handling and processing of food products, and grain drying. Elder in Presbyterian Church. Mbr of Civ Air Patrol. Hobbies include reading (history) and water skiing. *Society Aff:* ASAE, ASME, IFT, ASEE, ASHRAE

Wilhelm, William J
Business: College of Engineering, The Wichita State Univ, Wichita, KS 67208
Position: Dean, College of Engg. *Employer:* The Wichita State Univ. *Education:* PhD/Civil Engr-Structures/NC State Univ; MS/Civil Engr-Structures/Auburn Unv; BME/Mech Engr/Auburn Univ. *Born:* in St Louis, MO and grew up in Mobile, AL. Worked with Palmer and Baker Engrs as a Structural Engr 1958-60. Served as Instructor Engg Graphics, Auburn Univ 1960-64. Joined faculty in Civil Engg at WV Univ 1967 and served as Prof and Chrmn 1976-79. Appointed Dean of Engg, Wichita State Univ 1979. Reg PE in Kansas & WV. Served 13 mos active duty as Officer, Army Corps of Engrs. Mbr of Phi Kappa Phi, Sigma Xi, Tau Beta Pi, Pi Tau Sigma. Chapter Honor Mbr, Chi Epsilon, WV Univ. Recipient 1986 Amer Concrete Inst, Joe W. Kelley Award. *Society Aff:* ASCE, ACI, ASEE, PCI, NSPE

Wilhoit, James C, Jr
Home: Rt 2, McCowan's Ferry Rd, Versailles, KY 40383
Position: Emeritus Prof *Employer:* Rice Univ *Education:* BS/ME/Rice Univ; MS/ME/TX A&M; PhD/Engrg Mech/Stanford Univ. *Born:* 12/22/25 Native Tulsa, OK, Veteran US Nay 1944-1946. Instructor TX A and M 1949- 1951; Sr Aerophysics Engr, Convair, 1953-1954; Assistant Prof of Mech Engrg, Rice Univ 1954-1960; Assoc Prof 1960-1966; Prof 1966-1981; Emeritus Prof 1981. Research and consulting in the field of Mechanics, particularly in strss analysis and vibrations, 1954-1981; part-time prof in the Engrg Mech Dept, The Univ of KY, 1981, 1982, 1984; farmer. *Society Aff:* ASME, ΤΒΠ.

Wilke, Charles R
Business: Dept Chem Eng, Berkeley, CA 94720
Position: Prof of Chem Eng. *Employer:* Univ of Calif,Berkeley. *Education:* BS/Chemical Engg/Univ of Dayton; MS/Physical Chem/State College of Washington; PhD/Chem Engg/Univ of WI. *Born:* 2/14/17. Process Engr, Union Oil Co 1944-45. Instructor Washington State University 1945-46. At University of California at Berkeley 1946 to present: Department Chairman 1953-63; Assistant to Chancellor of Academic Affairs 1967-69. Major Research in Biochemical Engineering: Utilization of Micro-organisms for Chemical Synthesis, Recovery and Utilization of Enzymes, Enzymatic Hydrolysis of Cellulose, Production of Ethanol. President California State Board of Registration for Professional Engineers 1967-69; Advisory Board of the Petroleum Research Fund 1965-70; Annual Lecture Award from the Chemical Engineering Division of the American Society for Engineering Education in 1964; recipient of the Colburn Award from the American Institute of Chemical Engrs in 1951, Inst Lecture 1957, Walker Award 1965; Natl Acad of Engg 1965, Founders Award, AIChE 1986. *Society Aff:* AIChE, ACS, ASM, SAB, ASEE.

Wilkening, Rolland M
Business: 221 Felch Street, P.O. Box 7589, Ann Arbor, MI 48107
Position: Chrmn & Chief Exec Officer *Employer:* CERTECH *Education:* BS/CE/Purdue Univ *Born:* 3/10/26 Native of Deshler, NB. Served with US Navy 1944-46. Graduate Purdue Univ BSCE 1950. 1951-81 Field Engr, Project Engr, VP, Executive VP, Pres, Vice Chrmn of the Bd -- Barton-Malow Co, General Contractors and Construction Mgrs. Chrmn and CEO, Certech Corp. Instrumental in the development and application of the Construction Management Method for both public and private construction. MI Society of PE--Engr of the Year 1974. Purdue Univ--Distinguished Engrg Alumnus 1976. ASCE--Construction Management Award 1980. Enjoys photography, golfing and professional sports. *Society Aff:* ASCE, NSPE, CSI

Wilkenson, Thomas H
Home: P.O. Box BB, Hawi, HI 96719
Position: Dir Safety (Ret). *Employer:* U S Army. *Education:* ML/Ind. Rel. (Eco)/Univ of Pittsburgh; BS/Economics/Bucknell Univ. *Born:* Sept 9,1909. ML from Univ of Pittsburgh; BS from Bucknell. Native of PA. CSP. Served in safety engrg for the US Army in Pittsburgh; Pueblo, Colo; Hawaii; Japan; & Wash DC 1942-73; Army Dir of Safety 1955-73. Lectured in safety at Northwestern and New York Univ 1955-65; Bd of Dir NSC 1959-74. Fed Safety Comm Adv Bd 1956-70. Ch Natl Comm on Films for Safety 1969-71. Emer. Vets of Safety. Fellow Emeritus ASSE. Army Mert & Excep Civ Serv Awds 1954 and 1968. Art. Will. Mem Gold Medal Awd by World Safety Res Inst 1967. Defense Dist Civ Serv Awd 1972. *Society Aff:* ASSE.

Wilkes, J Wray
Business: Univ. of Arkansas, Engr. 309, Fayetteville, AR 72701
Position: Prof. *Employer:* Univ of AR. *Education:* DSc/Ind Engr/WA Univ; MSIE/Ind Engr/WA Univ; BSME/ME/S Methodist Univ. *Born:* 5/10/22. Lectr/Instr Mech Engr, S Meth Univ 1946-49. Lectr/Instr Ind Engr, WA Univ, St Louis '49-54. Asst Prof/Assoc Prof/Prof Ind Engr, Univ of AR '54-present. USNR 1943-46 (Engr officer), Prin Investigator NSF Comp Expansion Grant to Univ of AR 1965-68. Chrmn Comp Sci Grad Studies Univ of AR 1972 to present. Reg PE AR & TX. Assoc

Wilkes, J Wray (Continued)
Supv/Supv Computing Ctr Univ of AR 1960-72. Hd of Instruction/Res Div Computing Sevices, Univ of AR '72-81'. Mbr: Am Inst Ind Engr, Am Soc Engr Educ, Assn for Computing Machinery, Sigma Xi (Pres AR Chapt 1966-67). *Society Aff:* IIE, ASEE, ACM, ΣΞ.

Wilkes, James O
Business: Dept Chem Engrg, Univ. of Michigan, Ann Arbor, MI 48109
Position: Professor. *Employer:* Univ of Michigan. *Education:* PhD/Chem Engg/Univ of MI; MA/Chem Engg/Univ of Cambridge; MSE/Chem Engg/Univ of MI; BA/Chem Engg/Univ of Cambridge. *Born:* 1/24/32. Southampton, Eng. Mbr of chem engrg faculties at Cambridge (1957-60) & Michigan (1960-). Chmn of chem engrg at Michigan (1971-77). Assoc, Trinity Coll of Music, London 1951. Recipient, English-Speaking Union King George VI Memorial Fellowship, 1955. Recipient, Univ of Michigan Distinguished Ser Award, 1966. Recipient, Univ of Michigan Excellence in Teaching Award, 1980 . Coauthor of *Applied Numerical Methods*, Wiley (1969). and of *Digital Computing and Numerical Methods*, Wiley (1973). *Society Aff:* SPE, AIChE, TBΠ, ΣΞ, AGO.

Wilkes, John S, III
Home: 9700 E Iliff Ave, Apt K-133, Denver, CO 80231
Position: Commissioner-Engr *Employer:* State of Colorado, Board of Land Commissioners *Education:* MS/CE/TX A & M Univ; BS/Military Engrg/USMA *Born:* 9/13/37 in Birmingham, AL. Raised in Pamlico County, NC prior to entrance into West Point in 1956. Served with Army Corps of Engrs after graduation from 1960-1980. Served in various command and staff positions in US and Pacific in combat, airborne and construction engrg responsibilities. Last duty with COE was a Deputy District Engr St Louis District. Joined Arch Mineral Corp in July 1980 as Mgr of Government Relations, Western Division 1983-85 was an engineering manager with ITT-FSI in Alaska. Appointed by Governor Lamm of Colorado for a 6 year term as Land Commissioner in January 1985. Pres SAME - St Louis, 1978. Military awards include LOM, BSM, JSCM, MSM, ACM for military engrg achievement and service. *Society Aff:* SAME, USCOLD, PIANC, WSLCA.

Wilkes, Lambert H
Business: Agri Engg Dept, College Station, TX 77843
Position: Prof. *Employer:* TX A&M Univ. *Education:* BS/Agri Engg/Clemson Univ; MS/Agri Engg/TX A&M Univ. *Born:* 12/1/26. Native of SC. Instr-Agri Engg, Univ of AR, Asst Prof, Agri Engg NM State Univ, served as officer, USAF, in aircraft maintenance. Presently-joint appointment with TX A&M Univ (teaching) & TX Agri Experiment Stations (res). Co author on texbook "Farm Machinery and Equipment" McGraw Hill, 43 scientific or journal publications. Sr author on three US patents. *Society Aff:* ASAE, ΣΞ, ΓΣΔ, AZ.

Wilkes, W Jack
Business: 1629 K St, NW, Suite 801, Washington, DC 20006
Position: VP/Mgr. *Employer:* Figg & Muller Engrs, Inc. *Education:* BSCE/Civ Engg/Univ of TX. *Born:* 12/25/18. A native of TX. Spent 37 yrs in progressively responsible positions in govt service. First four yrs in Navy Dept & balance in the Bureau of Public Rds (now FHWA). Assignments in Austin, TX; Manila, PI; Raleigh, NC; & Wash, DC. Retired as Dir, Office of Engg, in 1978, to accept position of VP - Mgr Regional Office, Figg & Muller Engrs, Inc, who specialize in segmental concrete bridges. As BPR's Chief of Bridge Div was responsible for developing ntl stds for bridge design, inspection, specifications & replacement. Enjoys golf, woodwork, & handicrafts. *Society Aff:* ASCE.

Wilkins, J Ernest, Jr
Business: Deputy Genl Mgr, PO Box 1625, Idaho Falls, ID 83415
Position: Deputy Genl Mgr. *Employer:* EG&G Idaho, Inc. *Education:* BS/Mathematics/Univ of Chicago; MS/Mathematics/Univ of Chicago; PhD/Mathematics/Univ of Chicago; BS/Mech Engg/NY *Born:* 11/27/23. Taught Tuskegee Inst 1943-44. Worked at Metallurgical Lab, Univ of Chicago 1944-46; at Scientific Instrument Div, Amer Optical Co 1946-50; at Nuclear Dev Corp of Amer 1950-60; and at Gulf Genl Atomic 1960-70; Dist Prof of Applied Math Physics at Howard Univ 1970-77. Presently serving as Assoc Genl Mgr at EG&G ID, Inc. Mbr Bd/Dir 1967 ANS. Mbr Exec Cttee 1969 ANS. Treasurer 1971 ANS VP 1973-74 ANS. Pres 1974-75 ANS; Chrmn of Army Sci Bd 1978- . *Society Aff:* ANS, AMS, AAAS, ASME, OSA.

Wilkins, Joseph E
Home: 6700 Moss Lake Dr, Chattanooga, TN 37443
Position: Project Manager *Employer:* TN Valley Authority. *Education:* BSME/ME/Univ of KY. *Born:* 7/2/30. Grad BSME Univ of KY 1958. Worked on design & construction of coal-fired power plants 1958-1966, except for one yr in design of navigation lock components. Held responsible positions in design & construction of nucl power plants for public utility 1966-present. Current position includes responsibility for cost, planning, scheduling, security, quality assurance & quality control as well as all onsite engg for constructing two 1125 megawatt elec PWR nucl units. Received award from ASQC in 1978, as a quality achiever. *Society Aff:* NMA

Wilkinson, Bruce W
Home: 143 Clyde James Rd, Pleasant Lake, MI 49272
Position: Prof, Assoc. Dir., Eng. Res. *Employer:* MI State Univ. *Education:* PhD/Chem Engr/OH State Univ; MS/Chem Engr/OH State Univ; BS/Chem Engr/OH State Univ. *Born:* 8/9/28. Born in Shelby, OH. Educated at Fenn College (Cleveland State Univ) and the OH State Univ. Employed in Res, Process, Proj Engg with the Dow Chem Co, Midland, MI 1954-1965. Joined MI State Univ 1965. Designed, built, operated nuclear reactor 1967-1984. Taught chem and nuclear engg courses. Assoc Prof 1968, Prof 1977-present; Assoc Dir, Div of Engrg Res 1979-present. Reg PE OH and MI. *Society Aff:* AIChE, ANS, ASEE, NSPE.

Wilkinson, Dwight A
Home: 290 Haas, Frankenmuth, MI 48734
Position: Chief Metallurgist. *Employer:* Saginaw Steering Gear Div. *Born:* June 1926. Native of Hartford Mich. BS from Mich Tech Univ. Served with Navy Air Corp 1944-46. Metallurgist with Cerro de Pasco Corp Lima Peru 1950-51. With Saginaw Steering Gear since 1952. Chief Metallurgist since 1959. Responsible for material specification, processing, and reliability. Reg PE. 1974-75 chrmn Saginaw Valley Chap ASM. 1968 ASM medal for steel in energy absorbing steering columns. Lost Lake Woods Club mbr.

Wilkinson, Eugene R
Business: 1525 S Sixth St, Springfield, IL 62703
Position: Exec VP *Employer:* Hanson Engrs, Inc. *Education:* BS/CE/Univ of IL *Born:* 10/6/34 Joined Hanson Engrs in 1962 as structural design engr; has designed and supervised design of numerous structures, including bridges, buildings, towers, dams, storage silos; registered in 7 states; named Exec VP 1979; and chief operating officer 1986; elected to Bd of Dir of the Univ of IL CE Alumni Assoc 1978, elected to Bd of Dir of Cons Engrs Council of Illinois 1984, secty/treas 1986-87. *Society Aff:* NSPE, ASCE, AREA, SAVE, ASM, ACEC

Wilkinson, Roger I
Home: Oak Tree Lane, Rumson, NJ 07760
Position: Retired. *Education:* BSEE/Math/IA State Univ. *Born:* March 18,1903 Mason City Iowa. Prof Eng 1950 AT&T, D&R 1924-34, Applied Probability Theory; BTL 1934-68, Head, Traffic Res Dept. Studies: Teletraffic Theory, Delay Theory, Customer Calling Habits. Mbr TBP, Phi Kappa Phi, PME, Eta Kappa Nu (Natl Pres 1933-34). Fellow IEEE. Hon Mbr Intl Teletraffic Congress Comm. 25 tech papers. 10 prof papers. Book 'Nonrandom Traffic Curves & Tables for Engineering & Administrative Purposes.' WWII Oper Anal, So West Pacific, Radar Oper, IFF Studies, Medal for Merit. Hobby: Design and Racing of Sailing Canoes;

Wilkinson, Roger I (Continued)
several natl championships; Commodore American Canoe Assn 1960-61. *Society Aff:* IEEE, ASEE, IAS, HKN, AYRS, ASA.

Will, Herbert
Business: 560 Sylvan Ave, Englewood Cliffs, NJ 07632
Position: VP. *Employer:* HOECHST-UHDE Corp. *Education:* PhD/-/Univ of Munster, Germany. *Born:* 3/7/26. Since 1953 with UHDE GMBH as Process Engr. 1968 Dir of Dept 1. Respon for Oil Refining, Coal Gasification, Ammonia, Methanol and Heavy Water Projs. Since 1979 VP, since 1980 President of HOECHST-UHDE Corp. *Society Aff:* AIChE.

Willard, Beatrice E
Business: Green Ctr, 16th and Arapahoe Sts, Golden, CO 80401
Position: Head, Dept of Environmental Sci and Engrg Ecology *Employer:* CO School of Mines. *Education:* PhD/Botany/Univ of CO; MA/Botany/Univ of CO; BA/Biological Sci/Stanford Univ. *Born:* 12/19/25. Pioneered communication about use of ecological principles in bringing industrial dev into harmony with Earth's ecosystems. Has done extensive work with the mining industry; recently received AIME's Environmental Conservation Distinguished Service Award. Before coming to the CO School of Mines as its first biologist, served four yrs as mbr of the Pres's Council on Environmental Quality. Before that headed Thorne Ecological Inst in Boulder, CO for eight yrs. During that time was a leader in many liaison communication activities between industry and environmentalists. Profl plant ecologist with a specialty in alpine areas. *Society Aff:* ASCE, AIME, ESA, ΣΞ.

Willard, Charles O
Business: Dept 240/020-2, Hgwy 52 & NW 37th St, Rochester, MN 55901
Position: Stds Admin. *Employer:* IBM Corp. *Education:* MS/Education/IA State Univ; BS/Sci Educ/Mankato State Univ. *Born:* 9/8/17. Native of Mankato, MN. Served four yrs in the US Marine Corps in WWII. Taught Physics and Sci in MN high schools. Joined IBM education dept in 1957. Later moved into Mfg and Industrial Engg prior to stds assignment in 1962. Presently is Stds Admin and Metric Coordinator of IBM's Rochester, MN plant. Was elected a Fellow of SES in 1975 and certified in Stds Engg in 1978. VP Admin of Stds Engrs Soc 1978-80. *Society Aff:* SES, ASTM.

Willard Marcy
Home: 142 Lincoln Ave, Suite 781, Santa Fe, NM 87501-2006 *Employer:* Retired. *Education:* PhD/Org Chem/MA Inst Tech; SB/Chem Engr/MA Inst Tech. *Born:* 9/27/16. Military Service: U S Army Chem Corps 1942-45. Academic Exp: Res Asst (Chem) MIT 1946-49. Amstar Corp. Asst Supt Brooklyn Refinery 1937-42; Res & Dev Div, refinery process dev 1949-64, Res Corp Associate 1964-5; VP 1966-1982. Pres ARDUS, Inc. 1983-1985. Evaluation, patenting & licensing of acad inventions. Chmn Bd of Editors "Research Mgmt" 1974-76; Chmn Comm on Pat Matters & Related Legislation, Amer Chem Soc, 1974-1980; Pres, Amer Inst Chem 1984-5; NSF Advisory Com. 1979-1980; CHM NY Chapt Amer Inst Chem 1980-81; The Chem. Club, NY, Trustee, 1981-84; Chm, Library Com. 1983-86. *Society Aff:* ACS, AIC, AAAS, AIChE, SCI, NYAS.

Willardson, Lyman S
Business: Agricultural & Irrigation Engr. Dept, Utah State Univ, Logan, UT 84322-4105
Position: Prof *Employer:* UT State Univ *Education:* PhD/Agr Eng/OH State Univ; MS/CE/UT State Univ; BS/CE/UT State Univ *Born:* 5/10/27 Irrigation Engr with United Fruit Co. in Dominican Republic in Honduras, 1952-54. Univ of Puerto Rico Agric Experiment Station, 1957-57. Res Agric Engr, USDA, Agric Res Service, 1957-74 in UT, OH, and CA. Since 1974, Prof of Agric and Irrigation Engr at UT State Univ. International drainage consultant. Registered Professional Engr in CA. *Society Aff:* ASAE, ASCE, ICID, NSPE, ΣΤ, ΓΣΔ, ΑΕ, ΣΞ, AAAS, ASTM

Willcutt, Frederick W
Home: 6934 33rd St.,NW, Washington, DC 20015 *Education:* SB/Elect Engrg/MIT; SM/Elect Engrg/MIT. *Born:* 1/9/06. With Jackson & Moreland, Boston, Hoboken, NJ 1928-30, PA Pwr & Lt Co,, Allentown 1930-31. WA Ry & Elec Co 1932, Potomac Elec Pwr Co 1932-71. Mgr Syst Plng Dept 1960-71. Lecturer Geo Washington Univ 1941-43. Served with US Marine Corps as Capt 1943-45, Lt Col USMC (Ret) 1966. Reg PE DC. Life Fellow IEEE (Mbr at large council Power Eng Soc 1964-67), AIEE Chrmn Washington Sec, Chrmn Inst Comm on Transfers, Mbr Inst Bd Examiners. Mbr Beta Theta Pi. Republican. Presbyn. Clubs: MIT of WA, Congressional Country (Bethesda, MD). *Society Aff:* IEEE, PES.

Wille, Jerry L
Business: 425 S 2nd St, Ames, IA 50010
Position: Pres *Employer:* Curry-Wille & Assoc, Consulting Engrs, PC *Education:* ME/Agric Engr/IA St Univ; BS/Agric Engr/IA St Univ *Born:* 9/7/51 Native of Williamsburg, IA. Worked with Norval Curry, Consulting Engr from 1973-78, at which time the old firm was terminated, and the new professional corp of Curry-Wille & Associates formed. Currently President of Curry-Wille & Assoc. Active on both the state and national level of ASAE. IA Section P Pres, VP, and secretary-Treasurer. Has served on various committees on the national level. *Society Aff:* ASAE, NSPE

Willems, Jan C
Business: Mathematics Inst, PO Box 800, 9700 AV Groningen, The Netherlands
Position: Prof & Dean of the Department of Mathematics and Computer Science *Employer:* UN of Groningen *Education:* PhD/EE/MIT; MSc/EE/Univ of RI; Electromech Engrg/Engr School/Univ of Ghent- Belgium *Born:* 9/18/39 in Bruges, Belgium. Assistant prof, Dept of Elec Engr, MIT 1968-1973. Since then at present position. Published a book and many papers in the area of control and systems theory. Managing editor systems and control letters; editor Siam J. on control, Rairo Jaune Mathematicae Applicandae, Dynamics Reported, Mathematics of Control, Signals, and Systems, Dynamics & Stability of Systems; past assoc editor IEEE Trans Automatic Control. Married; 2 children. *Society Aff:* IEEE, Dutch Math Soc, SIAM

Willenbrock, F Karl
Business: American Society for Engineering Education, h11 Dupont Circle, Suite 200, Washington, D.C 20036
Position: Executive Director *Employer:* ASEE *Education:* Sc.B/EE/Brown; M.A/EE/Harvard; Phd/EE/Harvard *Born:* N.Y.C., July 19, 1920. Sc.B., Brown U., 1942; M.A., Harvard U., 1947, Ph.D., 1950. Research fellow, lectr. and asso. dean Harvard U., Cambridge, Mass., 1950-67; provost, prof. faculty engring. and applied sci. SUNY, Buffalo, 1967-1970; dir. Inst. Applied Tech., Nat. Bur. Standards, Washington, 1970-76; dean Sch Engring. and Applied Sci., So. Meth. U., Dallas, 1976-81; Cecil H. Green prof. engring. Sch. Engring. and Applied Sci., So. Meth. U., 1976-86; exc. dir. Am. Soc. for Engring. Edn., Washington, 1986--. Contbr. articles to profl. journs. Served with USN, 1943-46. Recipient Disting. Engring. award Brown U., 1962; Gold medal U.S. Dept. Commerce, 1975. Fellow IEEE, AAAS; mem. Nat. Acad. Engring., Am. Phys. Soc., Am. Soc for Engring. Ed., ASTM, Sigma Psi, Tau

Willenbrock, F Karl (Continued)
Beta Phi. *Society Aff:* IEEE, ASEE, NSPE, AAAS, TBΠ, ΣΨ, APS, ASTM, NAE, ABET.

Willey, Benjamin F
Home: 1105 Country Club Rd, Elgin, IL 60120
Position: Water Treatment Consultant *Education:* -/Pre-Medicine/Univ of IL; -/Bio-Chemistry/Univ of IL; -/Bio- Chemistry/North Park College; PE/ChE/Armour Inst of Tech; Certificate/Mgt of Res *Born:* 2/1/13. Forty-five yrs in ind, municipal & power plant water treatment, product and process engg. Ten yrs with Nalco Chem Co as Analyst, Res Engr, Quality Control Mgr, Asst Superintendent. Nearly twenty-three yrs as Chief Chemist and Technical Dir, Elgin Softener Corp. Assumed responsibilities in 1967 as Dir of Chicago's Water Purification Labs. Retired 6/30/80. Numerous Scientific papers published in CEP, & other journals. Assn activities resulted in honors including Distinguished Service Award-Environmental Chemistry-ACS, Univ of WI-Engg Leadership, WCF citation for Technical Leadership, AWWA Water Quality Div-best paper published-1974. Hobbies: Barbershop Quartet Singing, Choral directing & fishing. *Society Aff:* ACS, AIChE, AWWA, CEHA.

Willey, William E
Home: 341 W Wilshire Dr, Phoenix, AZ 85003
Position: Consulting Civil Engr (Self-employed) *Born:* 7/3/09. Prof CE 1953, BSEE 1932 Univ of Ill. Arizona Hwy Dept 1933, Bridge Designer, Planning Survey Engr, Asst Deputy State Engr. 1955-61 State Hwy Engr, Asst Hwy Engr, Transp Plng, Asst Dir Transp Plng. 1974 AZ Dept Transp. 1975 Civil Engr Pub Hearing Admin ADOT. Medallion of Merit U of Ariz 1960. Vet WWII rank of Col in USAF. Noted author and recognized authority on hwys. PPres Western Assn State Hwy Officials, PVP Amer Assn of State Hwy Officials. PPres Phoenix Br and AZ Sec, & Pacific Southwest Conference. Fellow & Life Member ASCE. Developed Sufficiency Rating Sys for Hwys and Priority Programming & uphill passing lanes for trucks on mountain grades. Retired ADOT Sept. 1979. Reg Engr. Past Pres Phoenix Kiwanis Club. Lt. Gov Div IX - Southwest District, Kiwanis, International. Golf and Flower Gardening. Methodist.

Willhelm, A Clyde, Jr
Home: 4361 Ruth Rd, Birmingham, AL 35213
Position: Principal Staff Engineer *Employer:* US Pipe & Foundry Co. *Education:* BS/Met Engg/Case Western Reserve Univ. *Born:* 10/10/21. PE in AL. LCDR in Retired Naval Reserve. Served with Navy 1943-46, Intl Silver Co 1946-58, Hayes Intl 1958-60, Southern Res Inst 1960-68, US Pipe 1968- present. Quality Assurance Mgr, US Pipe Technical Services, 1975-86, Principal Staff Engineer since 1986. Served in Chapter Offices in AIME, ASM, AFS and natl committees in ASM, AFS. Other interests include theology, non-classical music, photography, manual crafts, and TX. *Society Aff:* AIME, ASM, ASQC, AFS.

Williams, Albert J, Jr
Home: 901 Llanfair Rd, Ambler, PA 19002
Position: Consultant. *Employer:* Self. *Education:* AB/EE/Swarthmore College; Dipl/-/Media (PA) High Sch. *Born:* 2/28/03. Bell Tel Labs 1924-27. Leeds & Northrup Co from 1927 as employee in research to 1968, as consultant to 1976. An assignment in 1929 led to deliveries in 1933 of electronic nullbalance recorder. Marriage to Phyllis Zinke in 1937 led to delivery of two sons. There have been challenging worthy problems in measurement, control, and science of golf. Some solutions are recorded in 18 technical papers and 58 US patents. 1968 IEEE Morris E Leeds Award, 1972 ASME Rufus Oldenburger Medal. *Society Aff:* IEEE, ASME, APS, AAAS, FI, ΣΞ.

Williams, Albert T Wadi
Home: 901 Rosegill Rd, Richmond, VA 23235
Position: VP & Dir of Engg. *Employer:* Urban Services Inc *Education:* BSEngg/Structures/Leeds Univ. *Born:* 4/14/22. Native of Sierra Leone W Africa. Studied for degree in Engg in Great Britain and became a chartered civ, structural, hgwy, sanitary & welding engr in the United Kingdom - currently fellow for the last three engg disciplines. I have been in the US since Mar 1968 & am reg as PE in seven states. Currently I am Past Pres of the DC (State) Soc of PE, (the first black pres of a State Soc of PE), a mbr of the NSPE Action 80 Mbrship Committee & NSPE rep on the ASCE Committee on Minorities Prog. Love dancing & interested in space age tech. *Society Aff:* NSPE, ASCE, EERI, ACI.

Williams, Allan L
Business: 6706 Santa Monica Bd, Hollywood, CA 90038
Position: Reg Sales Mgr. *Employer:* Eastman Kodak Co. *Born:* Sept 1922. BS, Physics & Math, from Washington and Jefferson College. Native of Cameron WV. 1943 Tech Supvr with Tennessee Eastman on problems relating to the production of U-235 by the mass-spectrograph technique; beginning 1947 with Eastman Kodak, in Rochester, as Chemist, quality Control Engr, Staff Asst, Cine Film Dev Dept Hd, Dir and Mgr of Product Plng. 1969 to date Regional Sales Mgr, Motion Picture & Audiovisual Markets Div. Mbr of both Motion Picture and Television Academies, American Society of Cinematographers, SMPTE, Soc of Photographic Scientists & Engrs. Enjoy golfing.

Williams, Carl L
Home: Chelsea Farm, PO Box D, St Michaels, MD 21663
Position: Retired *Education:* BS/ChEng/Purdue Univ. *Born:* 12/11/19. Joined Scientific Design Co 1957 as Proj Eval Engr. Later promoted to Asst VP & Asst to Sr VP. Named Sr VP Process, Proj & Opers 1969; Exec VP 1974. Elected Pres & Ch Exec Off of Scientific Design 1974. Mbr Tau Beta Pi, Omega Chi Epsilon, ACS, AIChE. Resident St. Michaels, Md. Elected Chrmn of the Bd 1978. Retired 1982. *Society Aff:* AIChE, ACS.

Williams, Charles D, Jr
Home: 2627 North Highland Ave, Arlington Heights, IL 60004
Position: Mgr., Ind. Eng. *Employer:* US Postal Service *Education:* MS/Bus Adm management/N IL Univ; BS/Ind per super/N IL Univ; MS/Data Systems/De Paul Univ. *Born:* 11/22/41 Ind Eng Ekco Prods Co, 1964-66; Sr Ind Eng Motorola, Inc 1966-68, Corp Mrg Ind Eng Morton Qual Prod 1968; Mgt Cons Price Water & Co 1968-70; Corp Asst Dir Engr 1970-72; Corp Spl Proj O'Bryan Bros 1972-75; Corp Engr Admin Baxter Labs 1975-80; Reg Ind Engr USPO 1980-86; Mgr. Ind. Eng. USPO 1986-pres.; Pres Chaswil Ent 1978-pres speaker intl and nat confs in field. VP of Region VIII 1980-82, IIE & Bd of Dirs (Sr; Pres No Sub IL Chpt 1979-80, Chrmn Mgt Div Reg VIII 78-80) Ind Mgt Soc, Soc Mfg Engr, MTM Assn, ASQC US Jaycees, Sig Iot Eps. Contbr, numerous articles to profl jours Del Ups. *Society Aff:* IIE, ASQC, IMS, MTM

Williams, Charles E
Home: 525 Belmont E 5E, Seattle, WA 98102
Position: Retired. *Born:* March 15,1896 Seattle Washington. 1915-17 Architectural Draftsman, parttime Commercial Radio Operator. 1917-25 Radio Engrg draftsman and Radio Laboratorian Puget Sound Navy Yard. 1925-54 Principal Radio and Electronics Engr 13th Naval Dist. Aug 1921 built and operated BC Radio Receiver in Model T Ford Coupe. Designed and Patented Electrical Musical Instruments 1933. Invented and patented Radio Navigational and Direction Finden Systems 1935 & 1937. 1915-76 mbr Inst Rad Engrs & IEEE. 1952 Fellow IRE & IEEE. Served on Standards and History Cttee IRE & IEEE. Listed in "A Century of Honors" IEEE. *Society Aff:* IEEE, PTG, PSA, SWP, AWS, ASA.

Williams, Clifford D
Home: 1021 Hampton Terrace, North Augusta, SC 29841
Position: Retired. *Education:* CE/Structural/IA State Univ; BS/CE/Cleveland State Univ. *Born:* July 1900. Since 1920 engrg experience has included S.D. State Hwy, NYCRR Bridge Dept. Fisher Aircraft, So Calif Edison, City of Pasadena, CA & 35 yrs private practice University work included head of civil engrg, Fenn College -

Williams, Clifford D (Continued)
five yrs and Univ of Florida - seven yrs. Author and co-author of 3 structural engrg text books and tech articles. ASCE: VP 1967-68, Dir Dist 6 1964-66; Pres SC Sec 1973; Pres Florida Sec 1951; Engr of the Yr, SC Sec 1968; Honorary Member 1980. Distinguished Service Award, Cleveland State Univ. Professional Achievement Award, Iowa State Univ. *Society Aff:* ASCE, CEC, AREA, ACI.

Williams, Clyde E
Business: 150 E Broad St 601, Columbus, OH 43215
Position: President. *Employer:* Clyde Williams & Co. *Born:* Sept 8,1893 Salt Lake City. BS,DSc U Utah, DSc Case Inst Tech,OSU, ED Mich Coll, LLD Marietta Coll. m.Martha Barlow 1919; c.Clyde E, Samuel B, Thomas J. US Smelting Co 1915-16; grad fellow U Utah US Bur Mines 1916; Santa Fe Copper Co 1916-17; US Bur Mines Cornell U Sta 1917-18; Hooker Electro-Chem Co 1918-20; Bur Mines 1920-24; metallurgist Govt Argentina 1924-25; Columbia Steel Corp 1925-29; Battelle Mem Inst 1929-34(dir 35-57 trustee 51-58, pres 53-58). Pres Clyde Williams & Co 1958--, Clyde Williams Inv Mgt Co; chrmn, Clyde Williams Corp 1961--; VP Dir Borne Chem Co 1957-61; dir Rand Corp, Claycraft Co, Capital City Mfg Co 1958-65; F W Bell Inc, Mem adv groups, comns, bds, gvtl, and priv orgns. Recipient Presidential Citation, Medal for Merit. Mbr officer natl, state, local Rep Clubs, Res Dir, Rotary, Engrs, Mining. Scioto, Columbus Clubs.

Williams, David C
Business: P O Box 391, Ashland, KY 41101
Position: Vice President. *Employer:* Ashland Oil Inc. *Education:* BS/Mech Engr/TX Tech. *Born:* 01/-/23 Native of Post, TX. Interrupted college to attend Meteorological Cadet School, Air Force Intelligence Sch. Spent 1 1/2 yrs overseas WWII with 8th and 9th Air Forces. Employed as Eng with United Carbon Co, serving as Design Engr, Process Engr, and Proj Engr on construction of grass-roots plant. Served as Plant Engr, Asst Plant Mgr promoted to Mgr Development & Engrg Services for company. Holding 10 patents in carbon black field, including United Reactor. Co acquired by Ashland Oil, served as Chf Engr for Ashland ChemCo, became Corporate Officer in 1972 in charge of Environmental Affairs, Safety, Security and Product Safety. Nov 1980- President Ashland Development, Inc. *Society Aff:* ASME, AIChE, NSPE.

Williams, David J
Business: 2181 Victory Pkwy, Cincinnati, OH 45206
Position: Partner *Employer:* Burgess & Niple Ltd *Education:* BS/CE/OH N Univ *Born:* 11/19/29 Native of Van Wert, OH; served on the faculty of the Ft Belvoir engrs school 1952 to 1954. Design and field engr on several Dam and Spillway at projects till 1965 then project director On several water management retorts for waters heds, couries, regions and state of OH. Also project Dir on several park and recreation master and design plans. *Society Aff:* SAME, ASCE, NSPE, AWWA, AWRA

Williams, Doyle A
Business: 4800 E 63rd, PO Box 173, Kansas City, MO 64141
Position: Proj Mgr Power Plants *Employer:* Burns & McDonnell Engrg Co *Education:* BS/EE/Finlay Engrg Coll. *Born:* 2/22/33 Native of Ft Smith, AK. Field Engr on Titan Missile Bases in ID and AZ 1961-63. Electrical design engr on power plants (10 units) 1964-73 with 3 yrs. Served in Field. Head of power station electrical design 1972-74. Project Mgr, Coal-Fired power plants 1974-present. Projects managed include a 400 MW plant for S MS EPA (1973-78) and A 257 MW Plant for Plains Electric Coop near Grants, NM (1978-84). These Plants utilize state of the Art Pollution Control and Waste Disposal Systems. Enjoys travel, golf, reading, music, and outdoor activities. Married, 5 sons. *Society Aff:* IEEE, NSPE, MSPE.

Williams, Duane E
Business: Amoco Chemicals Co, 225 N. Michigan MC B402, PO Box 7516, Chicago, IL 60680
Position: Project Engr. *Employer:* AMOCO Chemicals Co. *Education:* BS/Mech Engg/MI Tech Univ; BS/Engg Admin/MI Tech Univ. *Born:* 10/29/30. in Saginaw, MI. Employed 32 yrs with Amoco Corp. primarily in instrument & control sys. Developed the Co's first full-scale implementation of direct computer control to new process facilities at AMOCO refinery, TX City. Currently in AMOCO Chemicals Co as Proj Engr respon for control sys on new facilities worldwide. Participation in ISA include: Mgt of the Computer Control Workshop; Secy & Dir of the Chem & Petr Div; Soc Adv Ctte; VP of the Industries & Sci Dept; Honors & Awards Committee; VP Long Range Planning Dept. *Society Aff:* ISA.

Williams, E Gex, Jr
Business: 2312 Wilton Dr, Fort Lauderdale, FL 33305
Position: President. *Employer:* Williams, Hatfield & Stoner. *Education:* BCE/Civil Engg/Univ of FL. *Born:* 6/10/27. Native of Fort Lauderdale, Florida. US Navy 1945-46. Founding President Williams, Hatfield & Stoner Inc, Consulting Civil Engrs, established 1956. Fellow 1970, Engineer of the Year, an PPres of Florida Engrg Soc. Fellow ASCE. Reg PE in Florida and Alabama. Mbr Tau Beta Pi and Pi Kappa Alpha. Past Commodore Lauderdale Yacht Club. PPres of Rotary Club of Fort Lauderdale North. Director of Glenfed, Inc. *Society Aff:* ASCE, NSPE, AAEE

Williams, Earl C, Jr
Business: Suite 900, James K. Polk Bldg, Nashville, TN 37219
Position: Res Engr. *Employer:* TN Dept of Trans. *Education:* BS/CE/Univ of MD; Certificate/TE/Yale. *Born:* 12/26/20. Corbett, MD; Baltimore Public Schools; Lt, Army Air Corp, WWII; Chief Traffic Engr, Wichita Falls, TX, 1952-59; Chief Traffic Engr, Montgomery County, MD, 1959-62; State Traffic Engr, TN DOT, 1962-77; Res Engr, TN DOT, 1977 to present. Pres, Southern Sec, Inst of Transportation Engrs, 1970; Intl Dir, ITE, 1979-present, Mbr Trans Res Bd, mbr Natl Comm on Uniform Traf Control Devices. *Society Aff:* ASCE, ITE.

Williams, Earle C
Business: President, BDM International, Inc, 7915 Jones Branch Dr, McLean, VA 22102
Position: Pres. *Employer:* BDM International, Inc. *Education:* BEE/Communications/Auburn Univ; -/Grad Studies in Bus Admin/Univ of NM. *Born:* 10/15/29. Army Special Weapons Electronics Officer, 1952-53. Inspection engr and utilities design engr with Std Oil Co (IN), 1954-56. Proj engr and systems engr with Sandia Corp, 1956-62. Joined BDM as Sr Engr in 1962; Pres, Chief Exec Officer, and mbr of Bd of Dirs since 1972. Pres, Prof Services Council 1977- 1979 (Dir 1974-present). Mbr, Bd of Dirs and Intl VP, Armed Forces Communications and Electronics Assn, 1979-82, '84. Chrmn, Naval Research Advisory Cttee, 1986-present (member since 1984). Mbr, Legislative Affairs Committee of NSPE, 1978-80. Mbr, Virginia State Bd for Community Colleges, 1980-1987. Chrmn, Fairfax County (VA) Economic Dev Authority, 1978-80. *Society Aff:* NSPE

Williams, Forman A
Business: MAE Dept, Princeton, NJ 08544
Position: Prof *Employer:* Princeton Univ *Education:* PhD/Engr Sci/CA Inst of Tech; BSE/Aero Engr & Physics/Princeton Univ *Born:* 1/12/34 Previously Visiting Prof at Imperial Coll, London, Sydney Univ, and the Univ of CO; Lecturer and Assistant Prof, Harvard; Prof of Aerospace Engrg, Univ of CA, San Diego Honors include the Daniel and Florence Guggenheim Predoctoral Fellowship, National Science Foundation Pre- and Postdoctoral Fellowship; John Simon Guggenheim Memorial Foundation Fellowship; the Silver Medal of the Combustion Inst; The Humboldt Award. *Society Aff:* AIAA, APS, CI, SIAM

Williams, Frank S G
Home: 700 Island Way 504, Clearwater, FL 33515
Position: President. *Employer:* Frank S G Williams & Assoc. *Education:* BS/Industrial Admin/Yale Univ. *Born:* 11/16/04. Native of Oakland, FL. Employed by Taylor Forge & Pipe Wks from 1925-68 in standards, research and development,

Williams, Frank S G (Continued)
division admin, purchasing, sales & marketing; retired as VP. Chrmn of American Standards Code for Pressure Piping 1948-64, now on Exec Cttee. 30 yrs on ASME Boiler & Pressure Vessel Cttee. J. Hall Taylor Medal and Hon Mbr ASME. ANSI Howard Coonley Medal 1963. President of Frank S G Williams & Assoc Inc 1968 to present, engrg consulting in piping and pressure vessel industries, with emphasis on practical quality assurance in nuclear field. *Society Aff:* ASME.

Williams, Frederick C
Home: 19919 Grandview Dr, Topanga, CA 90290
Position: Chief Scientist *Employer:* Radar Systems Group Hughes Aircraft Co *Education:* MS/EE/UCLA; AB/Math-Physics/UC Berkeley *Born:* 3/9/27 Native of Middleport, NY. Served as ETM 3/C with USNR during WW II. With Raytheon in 1951. With Hughes Aircraft, Radar Systems since 1952. Appointed Chief Scientist in 1979, responsible for tech direction of interceptor radar and synthetic array radar efforts in the Advanced Programs Div. Fellow IEEE (1980). Assoc Fellow and Mbr Space Sciences and Astronomy Ctte AIAA. Mbr Radar Panel and Chmn of Radar Terminology Ctte, AES. Received L A Hyland Award, 1977. Author of numerous publications and holder of 16 patents in fields of radar antennas and radar signal processing. *Society Aff:* AIAA, IEEE, ΣΞ, AESS, APS

Williams, G Bretnell
Business: MK. Ferguson Co, One Erieview Plaza, Cleveland, OH 44114
Position: Vice Chairman *Employer:* MK-Ferguson Co. *Education:* BS/Engrg/Yale Univ. *Born:* 4/3/29. Native of Phila, PA. Commissioned Ensign, US Navy, Bureau of Ordnance, 1951-54. Chief Industrial Engr, Chrysler Corp, 1954-57. Respon for time studies, plant layouts and Methods Dept. Cunningham-Limp Co 1957-76. Started as mechanical Engr-up the ladder through field construction and sales to Pres. With the H K. Ferguson Co Cince 1976, Pres 1978-1985. Ferguson is an engg and construction co operating on a worldwide basis with offices in Cleveland, and San Francisco. Vice Chairman MK Ferguson 1985-present. *Society Aff:* NCEE

Williams, George H
Business: EEICS Dept, Schenectady, NY 12308
Position: Prof. *Employer:* Union Coll *Education:* PhD/Engrg/Yale; MS/Engrg/Yale; BSEE/Engrg/Union; BA/Econ/Union. *Born:* 11/7/42 Dr Williams is Prof of the Electrical Engrg and Computer Science Dept at Union Coll. He is concerned with computer hardware and software designs. His research interest include Computer-Aided Design, Very Large Scale Intergrated Systems, and Programming Methodology. *Society Aff:* IEEE, ACM.

Williams, Gordon C
Home: 1722 Trevilian Way, Louisville, KY 40205
Position: Prof Emeritus. *Employer:* University of Louisville. *Education:* PhD/Chem Engin/U of Wisc; MS/Chem Engin/U of Wisc; BS/Chem Engin/U of Wisc. *Born:* July 1908 Mexico, Maine. PhD, MS, BS in Chem Engr from Univ of Wisc. Instr at Univ of Wisc and Miss State College before coming to Univ of Louisville in 1936. Asst Prof and Assoc Prof of Chem Engr and Prof of Chem Engr 1947-75, Hd of Dept 1947-69. Industrial Experience with Monsanto Chem Co, Union Carbon & Carbide Chem Co and 1957-8 E I DuPont de Nemours Engr Dept. Year-in-Industry Professor. Fellow AIChE. Mbr ACS, ASM, Ky Acad of Sci, AAUP, Phi Eta Sigma, Phi Lambda Upsilon, Tau Beta Pi, Sigma Xi, Phi Kappa Phi. Reg PE KY. Currently Emeritus Prof of Chem Engr U of Louisville and Consultant in Physical Metallurgy and Chemical Engrg.

Williams, Grover C
Business: P O Box 572, Austin, TX 78767
Position: President. *Employer:* Trinity Engrg Testing Corp. *Education:* BS/-/TX A&I Univ. *Born:* June 16,1927 Roswell NM. Active duty Navy Reserve 1945. Reg PE in Texas, New Mexico, Okla. Geotechnical and materials engrg since 1951. Associated with TETCO 1958. Currently Pres, Bd/Dir, Stockholder. Serves as pavement design and construction consultant to numerous private corporations and municipalities. President Texas Council of Engrg Labs 1973-74. President Texas Soc of Prof Engrs 1975-76. Dir NSPE 1974-76 & 1977-81. Mbr United Methodist Church. Mbr Chamber of Commerce and Economic Development Council. Married 1951. Four Children. *Society Aff:* NSPE, CSI, ACIL.

Williams, Harry E
Home: 557 Baughman Ave, Claremont, CA 91711
Position: Prof Engr. *Employer:* Harvey Mudd College. *Education:* PhD/Mech Engr/ Caltech; MS/Mech Engr/Caltech; BME/Mech Engr/Santa Clara Univ. *Born:* 3/11/30. Fulbright Fellow, Univ of Manchester, 1956-57. Engr, Caltech Jet Propulsion Lab, 1957-60. Harvey Mudd College, 1960-. Liaison Scientist, ONR, London, 1966- 67. Visiting Prof, Univ Sussex, 1974-75. Consultant for Aerojet Gen, Azusa; Naval Weapons Ctr, China Lake; Caltech Jet Propulsion Lab. *Society Aff:* ASME.

Williams, James C
Business: Carnegie Institute of Technology, Carnegie Mellon University, Pittsburgh, PA 15213
Position: Dean of Engrg and Prof of Metallurgy *Employer:* Carnegie-Mellon University. *Education:* PhD/Met Engg/Univ of WA; MS/Met Engr/Univ of WA; BS/Met Engg/Univ of WA. *Born:* 12/7/38. Salina, Kansas. Wife: Joanne Rufener, Children: Teresa A. Patrick J. Worked for the Boeing Co 1961-67. Joined tech staff at Science Ctr, Rockwell Intl 1968; appt group leader 1970, then Prog Mgr, Aerospace Group Staff in 1973 and Mgr, Prog Dev in 1974. Joined the faculty of Metallurgy and Materials Science Dept, Carnegie-Mellon Univ in 1975. Appointed President of Mellon Institute in 1981, Dean of Engrg, Carnegie-Mellon Univ, 1983. Mbr National Academy of Engineering, ASM, AIME, Sigma Mu, EMSA, Amer Assoc for Defense Preparedness, DARPA Mat Res Council; Amer Soc for Engrg Ed. Chrmn, US Delegation for Third Intl Ti Conf held in Moscow USSR in May 1976. Ladd Award 1976, Adams Award, Amer Weld Soc 1979, Albert Sauveur Lecturer 1983, ASM Fellow, 1983. Tech interests include Ti metallurgy, fracture, microstructure-property relations and electron microscopy. Recreational interests include photography, and sports. *Society Aff:* ASM, AIME, ASEE, NAE.

Williams, James C, III
Home: 1164 Terrace Acres Dr, Auburn, AL 36830
Position: Prof & Head *Employer:* Auburn Univ *Education:* PhD/Engr/Univ of S CA; MS/Applied Mech/VA Polytech Inst; BS/Aero Engr/VA Polytech Inst. *Born:* 10/11/28 Ocala, FL, Oct 11, 1928, married 1951, two children. Aeronautical Research Intern, NACA, 1951; Lt US Air Force 1951-1953; Teaching Fellow, VPI 1953-1954; Aeronautical Engr, N American Aviation 1954-1957; Research Engr, Engrg Center, Univ of S CA 1957-1962; Assoc Prof 1962-1965, Prof 1965-1972, Prof and Assoc Head 1972-1979, Mech and Aerospace Engrg Dept, NC St Univ; Prof and Head, Dept of Aerospace engrg, Auburn Univ; Assoc Fellow AIAA, Mbr ASME, Mbr ASEE. *Society Aff:* AIAA, ASME, ASEE

Williams, James E, Jr
Home: 5625 Dover Ct, Alexandria, VA 22312
Position: Deputy Asst Sec of the Air Force (Acq & Proc). *Employer:* U S Air Force. *Education:* BS/Aero Eng/IA State Univ. *Born:* 6/23/27. BS IA State Univ 1950. Native of Macon, GA. Employed by USAF at Robins AFB, GA; Hq USAF; and Office of the Secretary of the Air Force since 1950 in capacity of Mechanical Engineer, Aerospace Engineer, and Administrator. Currently involved with new systems acquisition, proceneuent policy and industrial resources.

Williams, Jane
Business: PO Box 936, Richardson, TX 75080
Position: Fluid Tech Engg. *Employer:* Sun Production Co. *Education:* MS/Biomedical Engg/UTA & UTHSCD; BS/ME/UTA. *Born:* 9/19/51. Native of Dallas, TX. Started working for Sun Production Co in 1974. Initially involved in efficient utilization of equip, especially compressors and electric submersible pumps.

Williams, Jane (Continued)
Also involved in energy conservation efforts including overall co equip and educating fellow employees on personal energy conservation measures. Am currently responsible for custody transfer and improving fluid measurement techniques for the co. Pres of TX Sec of Soc of Women Engrs, 1979. Enjoy backpacking & snow skiing. *Society Aff:* SWE.

Williams, John A
Business: Dept of CE, Holmes Hall, 2540 Dole St, Honolulu, HI 96822
Position: Prof CE *Employer:* Univ of HI *Education:* PhD/CE/Univ of CA-Berkeley; M/CE/Univ of CA-Berkeley; BS/CE/Univ of CA- Berkeley *Born:* 8/24/29 Native of San Bernardino, CA. USAF, 1952-56, assigned to Materials Laboratory, Wright Air Development Center, R and D Command. Research Assistant, Waves Project, IER, University of CA, Berkeley, 1957-63. Appointed to faculty of Univ of HI as Assistant Prof of CE, 1963; Assoc Prof, 1967; and Prof, 1972. Conducted model studies for Tsunami Research Program, HI Inst of Geophysics, and held appointment as Associate Researcher, 1969. Additional research involves groundwater tides and wastewater injection. Courses taught include solid and fluid mechanics and engrg geophysics. American Men of Science, 11th, 12th, and 14th ed. Registered CE, HI CA. Amateur geologist and rock collector. *Society Aff:* ASCE, AGU

Williams, Michael M R
Business: Nuclear Engineering Dept, The University of Michigan, Cooley Building, N. Campus, Ann Arbor, MI 48109-2104
Position: Prof of Nuclear Eng. *Employer:* University of Michigan *Education:* DSc/Nuclear Engg/Univ of London; PhD/Nuclear Engg/Univ of London; BSc/ Physics/Univ of London. *Born:* Dec 1935 London. Educated at Ewell Castle School and Croydon Polytechnic Graduate of King's College London 1958 (Physics Hons). Chartered Engr. Engr with CEGB 1961-62. Res Assoc Brookhaven Natl Lab 1962-63. Lecturer in Physics, Univ Birm 1963-65. Reader in Nuclear Engrg, Univ of London 1965-70. Prof of Nuclear Engrg Univ of London since 1970. VP Institution of Nuclear Engs 1970-76. Fellow Inst Physics. Fellow ANS. Pubs in fields of neutron & charged particle transport and reactor noise. Author of three books. Exec Editor Annals of Nuclear Energy & Head of Department of Nuclear Engineering Dept, Queen Mary College, London Univ 1980-1986. Professor of Nuclear Engineering, Nuclear Engineering Dept, Univ of Michigan, 1987- . *Society Aff:* ANS, INE.

Williams, Milton A
Home: 7108 Waterline Rd, Austin, TX 78731
Position: Partner *Employer:* Energy Management Consultants *Education:* BSME/Univ of MN *Born:* 7/20/22. Native of St Paul, MN. Attended Univ of MN for 1 1/2 yrs prior to entering military service in 1943 for 3 yrs. Returned to Univ of MN in 1946 Graduated with a BSME n 1948. Began 30 yr career with Union Carbide Corp in 1948. Held a variety of managerial positions. During last 3 1/2 yrs as Energy Conservation Mgr of the Chems & Plastics Div. Assumed Presidency of Energy Eng Assoc in May, 1978 after retirement from Union Carbide. In 1980, formed Energy Management Consultants specializing in organizing energy conservation programs for industry. Also technical Prog Dir since 1979, for Confs on Ind Energy Technology held in Houston each year. Have acted as industrial consult to the Electric Power Research Institute since 1980. *Society Aff:* AIChE, NSPE.

Williams, Paul C
Home: 3364 E Smith Rd, Medina, OH 44256
Position: Pres *Employer:* Paul Williams & Associates *Education:* BSME/ME/OH State Univ; BS/Marine Engg/US Merchant Marine Acad. *Born:* 1/14/26. Native of Columbus, OH. Lt USNR (Ret). Successive positions with Babcock and Wilcox, Power Generation Group included: test engr, asst to mgr of engg, engg section mgr, mgr gen production control & purchasing agent. Developed the "Sintering Test" an index to superheater slagging, with stock equip since 1971. Received patent for radioactive waste handling equipment. Became VP of nuclear div in 1977. Mbr OSU Engg Dept Committee for Tomorrow. Took early retirement, & started Paul Williams & Assoc in Oct 1986, a mktg & consulting co dealing in the low level radioactive & hazardous waste field. Mbr ASME radioactive systems ctte. *Society Aff:* ANS, ASME.

Williams, Richard A
Business: Dept of Elec Engr, University of Akron, Akron, OH 44325
Position: Prof Engr, Assoc Prof. *Employer:* Univ of Akron. *Education:* PhD/Elec Engr/OH State Univ; MSc/Elec Engr/OH State Univ; BSEE/Elec Engr/OH State Univ. *Born:* 7/21/36. PhD received March 65. Have worked for Sandia Corp (Albuquerque), Ohio Power Co, OH Edison Co, OSU ElectroScience Lab. Have taught at OH State Univ & have been at Univ of Akron for 19 yrs. Lic PE (OH), Commercial Pilot and Commercial Radio Engr. Do consulting work in the Electrical/Electronic fields, specializing in Communication Systems and Scientific Computer Applications. Presently also associated with Loral Systems Div. Married. Four children.

Williams, Richard F
Business: P O Box 4849, Pocatello, ID 83201
Position: President & Gen Mgr. *Employer:* Alpha Engineers Inc. *Education:* BS/Mech Engg/NM State Univ. *Born:* 1921 Nebraska. During WWII served as Engrg Officer on USS Reid- DD369(sunk) and USS Steinaker-DD863. 10 yrs with Pennwalt in Wyandotte, MI; 3 ys with Dow Chemical in Midland, MI, 1 1/2 yrs with American Potato in Blackfoot, Idaho. In private practice since 1961. Organized Alpha Engineers, Inc, consulting engrs in 1966 along with two associates. Presently President of Alpha Engrs Inc offering services in mechanical, structural, electrical, chemical and metallurgical engrg, architecture and land surveying with offices in ID, MT & OR. Enjoys fishing, hunting, backpacking, and skiing in Idaho's great outdoors. *Society Aff:* ASME, NSPE, ACEC.

Williams, Richard L
Business: 4919 Colonial Ave, Roanoke, VA 24018
Position: Pres *Employer:* Richard L Williams Consulting Engr, Inc. *Education:* BS/CE/VPI & SU. *Born:* 2/7/33 in Norflok, VA. Worked for engrg firms in Roanoke and Richmond prior to 1965. In 1965 formed partnership of Shumate & Williams, Consulting Engrs. Added Electrical and Mechanical Partners to form Shumate, Williams, Norfleet & Eddy in 1967. Partner, Smithey & Boynton, Architects & Engrs, 1970-71. Formed Richard L Williams Consulting Engr in 1971. Activities; VP, CEC/VA, Pres-Elect, CEC/VA, 1978; Pres, CEC/VA, 1979; Dir, CEC/VA, 1980. ACEC Private Enterprise Ctte mbr. Charter mbr Roanoke Chapter of CSI. Mbr of Roanoke County Electrical Examining Bd since 1978. Architectural Technilogy Lay Advisory Ctte, VA Western Community Coll since 1970. ACEC, VP 1983-1985. *Society Aff:* ACEC, CEC/VA, ASCE, CSI.

Williams, Robert M
Business: 7037 Canal Blvd, New Orleans, LA 70124
Position: President. *Employer:* Williams Engrg Inc. *Education:* BS/CE/UNiv of MS. *Born:* March 16,1921 New Orleans. s.Williams Horace and Ruby (Mugnier) W.; student Tulane U 1939-42, 46-47; m.Sheila Bosworth Wilkinson 1950; c.Wendelin Dunado, Sheila Wilkinson, Leila Elizabeth. Partner, engr W Horace Williams Co 1949-50; engr W Hor Horace Williams Co div Williams-McWilliams Industries 1957- 63, exec VP in Chg 1963-65; Pres Williams Engrg Inc 1965--. Served to maj., CE AUS 194246;ETO. Decorated Bronze Star. Reg PE, AL, LA, MS, TX. Fellow ASCE. Mbr AISC, ACI, PCI, NSPE, CEC, LES, Beta Theta Pi. Episcopalian. *Society Aff:* LES, ASCE, AISC, ACI, NSPE, CEC, PCI.

Williams, Robin
Business: IBM Research, K52-803, 650 Harry Road, San Jose, CA 95120-6099
Position: Mgr, Dept K52. *Employer:* IBM Res Div. *Education:* PhD/Comp Sci/NY Univ; MSc/Elec Engg/NY Univ; BSc/Elec Engg/Imperial College, London Univ. *Born:* 7/10/41. Joined Mullard Res Labs, Redhill England and worked on optical character recognition simulations and built OCR equipment. 1964 transferred to

Williams, Robin (Continued)
North- American Philips Res Labs and worked on memory sys. 1967 Instructor NYU, 1971 Asst. Prof. Joined IBM Res 1972, worked on Color Display Sys. database for graphics sys. Managed Database & Distribution Sys dept.; currently manager of Office Systems Lab. Published many papers, given numerous talks, chaired ACM, IEEE conferences. 1975-77 was Chrmn of SIGGRAPH (the ACM spec int group for graphics). *Society Aff:* ACM, IEEE, HKN.

Williams, S Hoyt
Business: P.O. Box 1307, Wise, VA 24293
Position: VP *Employer:* Thompson & Litton *Education:* MS/ME/VA Polytech Inst & St Univ; BS/ME/VA Polytech Inst & St Univ *Born:* 12/29/39 Native of Buchanan County VA. Served with US Air Force 1964-1967. Became employed at Thompson & Litton, consulting engrs, Wise, VA in 1967, specializing in water & sewer systems design. Other areas of specialty include dynamite blasting & mechanical failure analysis. Currently serving as Pres of Thompson & Litton, Past Pres of Mountain Empire Chapter, VA Society of PE & Past State Dir. Married to Dawneda Fowler Williams. Two children, Lucretia Gay & Stafford Hoyt III. Member Church of Christ. *Society Aff:* ASHRAE, NSPE, ACEC, SAME.

Williams, Samuel L
Business: Mtl Readiness Command, DRSAR-QAE, Rock Island, IL 61299
Position: Gen Engr. *Employer:* US Army Armament. *Education:* Masters/Engg Admin/Bradley Univ; BS/Met Engg/Univ of IL. *Born:* 4/3/35. Native, Quad-Cities area (Davenport, IA & Rock Island, IL). Worked in heat treating & powder metal res, Burgess-Norton Co, Geneva, IL, following 2 yrs in Army Corps of Engrs. Foil Mill Met one yr - Alcoa. Joined Met Evaluation Sec, lab at Rock Island Arsenal in 1966; headed that section from 1973 to 1977; failure analysis & mech testing. Supervisor of Met Control & Testing, Farmall Plant, Intl Harvester Co until 1979. Currently a Gen Engr for the US Army, Rock Island Arsenal. Past Chrmn, Tri-City Chapter, ASM, Contributor *Metals Handbook*, speaker for ASM, & Instr at local colleges. Beer can collector, enjoy volleyball, water skiing, & boating. *Society Aff:* SM, IMS.

Williams, Steven A
Home: 433A Mona Dr, Hermitage, TN 37076
Position: Sr Engr. *Employer:* E I du Pont de Nemours & Co. *Born:* Sept 16,1951. I completed high school at the age of 17 in Union City, Tenn. I started immediately at Tennessee Tech U in Fall 1969 working towards a BS in ChE. I completed this degree in June 1973. While attending TTU, I recd two scholarships 1)Cities Serv Fndtn Award and 2)Monsanto Award. Mbr NSPE, EJC and AIChE. 1972-73 elected president of AIChE. Mbr Theta Tau (Treas 1972-74).MSChE TTU. Mbr Tau Beta Pi and Sigma Xi. I then was employed by E I du Pont de Nemours & Co. I am presently active in the city AIChE and have participated in Jr Achievement. Presently, I am a Sr Engr with Du Pont dealing most with Process Engrg.

Williams, T Cortlandt
Home: 204 Live Oak Lane, Harbor Bluffs, Largo, FL 33540
Position: Chrmn of Bd (Ret). *Employer:* Stone & Webster Engineering Corp. *Education:* LLD/-/Bucknell Univ; ME/-/Bucknell Univ; BS/ME/Bucknell Univ. *Born:* 6/30/97. Mbr ASME 1939. Fellow ASME about 1963. Field Engr for Stone & Webster Engrg Corp 1923. Later became Resident Engr, Asst Supt of Constructn; Supt of Construction; Proj Mgr; Construction Mgr; Sr Construction Mgr; Vice Pres; Exec Vice Pres; President and retired 1962 as Chrm of Bd. Most outstanding job was that in charge of construction of his company's work during of the original atomic facilities in Oak Ridge,TN. *Society Aff:* TBΠ, ASME, ASME, KΣ.

Williams, Theodore C
Business: Rm 205, 300 Galisteo, Santa Fe, NM 87501
Position: Consultant *Employer:* Self-Employed *Education:* BS/CE/MI St Univ *Born:* 5/6/24 Joined Williams & Works in 1946; elected Pres in 1968; Chmn of the Bd in 1972. P Pres MI Water Pollution Control Assoc past Dir Water Pollution Control Federation; past Dir MI St Chamber of Commerce. Honorary mbr Michigan Engrg Society; honorary mbr Grand Rapids Engrs Club. Particular interest is in the area of innovative approaches to water and waste treatment. Also Chmn of the Bd of Williams & Works, Inc; EDI Engrg & Sci Inc; & W & W Facilities Group, Inc. Retired from The Williams & Works Companies in 1985. Now self-employed as a consultant on water & waste water treatment in Santa Fe, NM. *Society Aff:* AAEE, WPCF, AWWA, IAWPRC

Williams, Theodore J
Business: 334 Potter Center, W Lafayette, IN 47907
Position: Prof of Engrg, Dir Purdue Lab Applied Ind. Ctrl *Employer:* Purdue University. *Education:* BS/ChE/Penn State Univ; MS/ChE/Penn State Univ; MS/EE/OH State Univ; PhD/ChE/PA State Univ. *Born:* 9/2/23. Native of PA. USAF 1942-45, 1951-56 to Captain. Fellow ISA, AIChE, AIC, AAAS, Inst Meas and Control. Hon Mbr SCS. Sr Mbr IEEE. Pres, AACC 1964-66. Pres ISA 1968-69. Pres AFIPS 1976-78. Awarded Sir Harold Hartley Medal, Inst of Measurement and Control, England 1975. mbr Naval Res Adv Cttee, US Navy 1975-78. Chrmn, Automation Res Council 1972-81. *Society Aff:* AIChE, ISA, AIC, AAAS, INST MC, SCS, IEEE, ACS, TAPPI

Williams, Wayne W
Business: Dept of Civil Engrg, Seaton Hall, Manhattan, KS 66506
Position: Prof *Employer:* KS State Univ *Education:* MS/CE/IA State Univ; BS/CE/IA State Univ *Born:* 12/14/22 Born in Powersville, MO. Married Berverly M. Children Marilyn, Wayne, Jr., Charles, Fred, Gary, Sherry, Chris. *Society Aff:* ΦΚΦ, ASTM, TBΦ, XE

Williams, Wendell M, Jr
Business: Sch of Mech Engg, Ga. Tech, Atlanta, GA 30332
Position: Asst Prof. *Employer:* GA Inst of Tech. *Education:* PhD/ME/OH State Univ; MS/ME/OH State Univ; MS/Automtv Engg/Chrysler Inst; BS/ME/GA Tech. *Born:* 5/15/32. Native of Canton, OH. Married to Mary Catharine Linville 1961. Children Debra & Stephen. Coop student with Timken Roller Bearing Co 1950-55. Test & Dev Engr with Chrysler Corp 1955-58. Instr in Mech Engg, OH State Univ 1958-65. Teaching & res in Mech Engg at GA Tech since 1966. Mbr of GA Tech Vehicular Collision Res Team. Consultant on questions of vehicle design. Chrmn SAE Atlanta Sect 1977. Chrmn SAE Natl Student Activities Committee 1978-79. SAE Sections Bd 1978-80. Chrmn Teetor Awards Comm 1987. SAE Bd of Dir 1988-90. Recipient SAE Teetor Award 1979. Reg PE, OH. *Society Aff:* SAE, ASME, ASEE, AAAM, AAUP.

Williams, Wesley B
Home: 1918 Hebron Hills Dr, Tucker, GA 30084
Position: Conslt *Employer:* Hensley-Schmidt, Inc *Education:* B/CE/GA Inst of Tech; Grad/ National Security Management/Industrial Coll of the Armed Forces *Born:* 9/7/22 Born in Conyers, GA. Col. NSAF Ret. Professional eexperience includes 15 yrs. State Govt; 8 yrs industry and 15 yrs consulting engrg; 25 yrs. Active/Res USAF Civ Engrg. FL Office Mgr, Dorr-Oliver, Inc. 1959-'64; Dir, Water Quality Control Services, GA Water Quality Bd, 1965-'72. Since 1975 Principal, Hensley-Schmidt, Inc, responsible for planning/design/const of various water/wastewater facilities including plant capacities up to 120 MGD and a Sewer System for Nigeria's New Capitol City. Registered Engr: GA, FL, SC. WPCF Dir, 1972-'75. WPCF Bedell Award, 1969; WPCF Service Award, 1976; DOD Legion of Merit, 1979. Ga. Water & POllution Contr Assoc Pres 1971, Storey & 5 S Awards, 1986, Honorary Membership, 1987. Presbyterian Elder. Phi Kappa Sigma. *Society Aff:* ASCE, SAME, NSPE, WPCF, AWWA, ROA, AFA.

Williamson, Elmer H, Jr
Home: 1050 Bakersfield Rd, Columbia, SC 29210
Position: Design and Specs Engr. *Employer:* B P Barber & Assoc. *Education:*

Williamson, Elmer H, Jr (Continued)
BS/Civil Engg/Univ of SC. *Born:* 9/20/25. Served in Army Infantry Div 1943-46, European and Pacific Theaters of Operation. m.1947 Margaret Casteen. c.Gloria, Catherine,Sandra, Theresa, Elmer III. Engr City of Greenville SC 1950. Joined firm of B P Barber & Assoc 1951. Principal 1955. VP 197074. Secy 1975 to date. Currently resp for design review and specifications of water treatment and wastewater projects. President SC Soc of Prof Engrs 1966-67. Natl Dir NSPE 1970 to 1977, Mbr ASTM, CSI. Enjoy boating, Treasurer, Dist 26, U S Power Squadrons. *Society Aff:* CSI, NSPE.

Williamson, Merritt A
Business: Box 1576 Sta B, Nashville, TN 37235
Position: Dist Professor Emeritus. *Employer:* Vanderbilt University. *Education:* PhD/Metallurgy/Yale Univ; MS/Metallurgy/Yale Univ; MS/Aeronautics/CA Inst of Tech; MBA/Bus/Univ of Chicago; BE/Metallurgy/Yale Univ. *Born:* 4/1/16. Metallurgist, Scovill Mgr Co 1937-42 and Remington Arms Co 1942-44. Prof Officer, Guided Missiles, US Navy 1944-46. Dir of Tech Res, Solar Aircraft Co 1946-48. Assoc Dir of Dev, Pullman Standard Car Mfg Co 1948-52. Mgr Res Div, Burroughs Corp 1952-56. Dean, Coll of Eng Penn State Univ 1956-66. Ingram Dist Prof of Engrg Mgt, Sch of Engrg, Vanderbilt Univ 1966-present. Pres Yale Engrg Assn 1956. Ch Inst for Certification of Engrg Technicians 1961-63. Chrmn Natl Conf of Admin of Res 1964. Pres ASEE 1959-60. Ch Bd Houseboating Corp of America 1972--. ASEE James H McGraw Award 1974. *Society Aff:* AIME, IIE, ASEE, NSPE, YSEA, IGS, ISGS, ASEM, IEEE.

Williamson, Raymond H
Home: 933 Rothowood Rd, Lynchburg, VA 24503
Position: Retired. *Education:* MS/EE/Union College; BS/EE/IA State Univ. *Born:* April 6,1907 Iowa. Licensed radio amateur 1921. Tau Beta Pi. MSEE Union College 1935. General Elec Electronics engr 1928-71 at various locations incl installation and technical liaison assignments in Canal Zone and Italy. Product design team projects included VLF very-high-power Naval radio telegraph transmitters and WLW 500-kw broadcast. Later yrs in engrg mgt, incl transmitter tech standards chrmnship for Electronic Industries Assn. Mbr IEEE 1931; Fellow 1951. Prof Engr NY 1954. Scoutmaster 1929-44. Fellow American Numismatic Soc 1957. Mbr Lynchburg Electoral Bd 1971-1980; Museum Systems Bd 1976-1981.. *Society Aff:* TBΠ, HKN.

Williamson, Robert B
Business: 773 Davis Hall, Univ of Calif, Berkeley, CA 94720
Position: Professor of Engineering Science *Employer:* Univ of Calif *Education:* Ph.D/Applied Physics/Harvard University; S.B. (cum laude)/Applied Physics; Harvard University; A.B./Physics/Harvard University *Born:* 11/22/33 Was Asst Prof in the Civil Engrg Dept. at MIT 1965-68. In 1968 joined faculty of the Civil Engrg Dept. at the Univ of Calif, Berkeley where he is active in Materials Science & Fire Protection Engrg. He has published papers on many aspects of fire safety, and he has built up one of the principal fire research labs in the USA. This lab is equipped to perform a wide range of large-scale fire research & testing, including the fire growth characteristics of materials & the fire resistance of construction assemblies. He has developed several new full scale fire tests which are more effective in measuring the contribution of materials to fire growth. In Materials Science he has specialized in the "Solidification" of materials and in the investigation of the morphology of materials on both the macroscopic and microscopic scale. He is well known in the materials science community for his article in Progress in Materials Science on the "Solidification of Cement–, (1972). *Society Aff:* ASTM, ACI, CI, SFPE, RILEM, ASCE, ICBO

Williamson, Robert E
Business: Agric Engr Dept, Clemson, SC 29634-0357
Position: Prof Ag Engr *Employer:* Clemson Univ *Education:* Phd/Engr Composite/MS St Univ; MS/Ag Engr/Clemson Univ; BS/Ag Engr/Clemson Univ *Born:* 11/8/37 Native of McConnells, SC. Commissioned service as Cartographer and Construction Officer, USAF, 1959-62. Research Assoc, Clemson Univ 1964-66, and MS St Univ 1966-71. Projects involved research on spray droplet mechanics and development of equipment for ground and aerial application of ag chemicals. From 1971-78, was a mbr of Univ of GA faculty (Coastal Plain Experiment Station) responsible for research and development of machinery and equipment for vegetable harvesting and handling mechanization. Currently, Prof of Agricultural Engrg, Clemson Univ (teaching and research). Registered professional engr. ASAE Research Paper Award, 1975. *Society Aff:* ASAE, ΦΚΦ, ΓΣΔ, ΣΞ, ASEE

Williamson, Thomas G
Business: Dept of Nucl Engr & Engr Physics, Charlottesville, VA 22901
Position: Chmn & Prof *Employer:* Univ of VA *Education:* PhD/Physics/Univ of VA; MS/Physics/RPI; BS/Physics/VA Military Inst *Born:* 1/27/34 Quincey, MA. Married to Kaye Darlan Love, 1960, 3 children. Assistant Prof, Assoc Prof, Prof of Nuclear Engr, Dept of Nuclear Engrg, Univ of VA 1960 to present. Chairman, Dept. of Nuclear Engrg & Engrg Physics, U VA 1977 to present. During Univ leaves have worked for General Atomic, San Diego, CA, Combustion Engrg, Windsor, CT, Los Alamos Scientific Laboratory, Los Alamos, NM. Consultant to the Philippine Atomic Energy Commission, VA Electric and Power Co and Babcock and Wilcox. Academic Leave 1984-85 Natl Bureau of Standards, Gaithersburg, MD. *Society Aff:* ANS, AAAS, ASEE, TBΠ, ΣΧΙ

Willis, Charles A
Business: P.O. Box 36855, 801 E Blvd, Charlotte, NC 28236
Position: Pres. *Employer:* WILLIS/O'Brien & Gere *Education:* BS/Civ Engg/NC State Univ. *Born:* 1/27/37. in Lebanon, PA. Regional Engr for NC Dept of Water Resources, Raleigh, NC from 1959-1962. Opened Charlotte Branch Office of O'Brien & Gere Engrs in Oct 1962. Has served as Managing Engr, Principal Engr, VP, and was elected Pres of O'Brien & Gere, Inc/Engrs in 1976. Effective January 1, 1981, the name of the firm was changedto WILLIS/O'Brien & Gere. Serves on Bd of Dirs of O'Brien & Gere, Inc/Engrs & O'Brien & Gere Engrs, Inc. Responsible for all production, marketing, & admin activities of O'Brien & Gere, Inc/Engrs. Served on Charlotte Mecklenburg Advisory Environmental Quality Council; Advisory Bd of NC Water Resources Inst; Past Chrmn of NC Sections of AWWA & WPCA; VP & Pres Elect of Consulting Engrs Council of NC. Natl Dir of Water Pollution Control Fed. *Society Aff:* ACEC, ASCE, AWWA, NSPE, WPCF.

Willis, Dennis D
Business: 310 A Kentucky Ave, Norton, VA 24273
Position: President *Employer:* Willis, Skeen & Associates P.C. *Education:* BS/CE/VA Polytech Inst. *Born:* 4/29/48 Founder & Pres, Willis, Skeen & Assoc., Inc, 1985. Experience includes service as project engr for Island Creek Coal Co, Buchanan County, Va. Design engr for Wiley and Wilson, Lynchburg VA. Prior to presidency at Thompson & Litton, Inc, Wise, VA. served as VP of Mining Engrg Division. Designed and presented short courses related to mining in five states. Served on ACEC Energy Ctte, Mining Ctte, and served on Govt Affairs Steering Ctte. Presented testimony at Senate & House Subcommittees on Energy and Natural Resources. Awarded Certificate of Recognition by Va. Dept. of Aviation 1985 for outstanding service as a consultant. Married to Elizabeth Short Willis; two daughters, Kristi and Kimberly. Past Mbr Norton City Sch Bd, Christian Church, Past Dir YMCA, Dir VMRA. Hobby: Racquetball, skiing, pilots license, inst. multi-Eng. *Society Aff:* NSPE, ASCE, ACEC, AIME, VMRA

Willis, Edward M
Business: UNCC Station, Dept Engr Tech, Charlotte, NC 28223
Position: Asst Prof. *Employer:* Univ of NC. *Education:* MSCE/Civ Engr/MIT; BS/Engr/USMA. *Born:* 5/7/27. Native of CA. USA Corps Engrs 1951-1976. Retired 1976 as colonel. Former dir of USA Coastal Engr Res Ctr. Former dist engr,

Willis, Edward M (Continued)
USAED Japan. Former Facility Engr Ft Bragg, NC. Pres SAME post 1973-74. Asst Prof Engr Tech, UNCC. Dept Chrmn 1978-79. *Society Aff:* ASCE.

Willis, Homer B
Home: 5812 Wilson Lane, Bethesda, MD 20817
Position: Private Consultant *Employer:* Self employed *Education:* BSCE/Civil Engg/OH Univ. *Born:* 6/17/18. in Lawrence County, OH. Engr consultant in field of water resources dev with emphasis on safety of dams. Prior to Jan 1979, served over 38 yrs on staff of the US Army Corps of Engrs, including five yrs as Chief, Engg Div in the Civil Works Dir of the office of the Chief of Engrs in Washington. Fellow & Life Mbr ASCE. Pres Natl Capital Sec ASCE 1975-76. V Chrmn Exec Cttee, USCOLD 1979. Reg PE in KY & MD Chm ASCE Committee on Dam Safety 1980-81.. *Society Aff:* ASCE, USCOLD, SAME, AWRA.

Willis, Shelby K
Business: Bucher, Willis & Ratliff, 609 W North St, Salina, KS 67401
Position: Founding Partner *Employer:* Bucher, Willis & Ratliff *Education:* Ms/Structural Engg/Univ of IL; BSCE/Structural Engg/Univ of IL. *Born:* 6/9/24. Alton, Illinois. Served 1943-46 US Navy, Pub Wks Officer, NAS, Cecil Field, FL. Between 1947-57 employed by Howard, Needles, Tammen & Bergendoff. In 1957 founded Bucher & Willis. In 1983, firm name changed to Bucher, Willis & Ratliff. Founded Anilas Leasing Inc in 1960 and served as VP & secy to 1979; now Pres. From 1961 to present serve as Pres of Kancen Printing & Adv Inc. KES-NSPE, PTreas, KCE, Nat Dir ACEC, Natl VP ACEC 1977-79, Named outstanding engineer in Kansas by KES in 1981, Natl Pres Elect ACEC 1982-83, Natl Pres ACEC 1983-84. Received Distinguished Alumnus Award from Univ. of IL in 1984. *Society Aff:* ACEC, NSPE, ASCE, ASTM, ACI, ACSM, ASPRS.

Willoughby, Edward
Home: 6455 Anslow Lane, Troy, MI 48084
Position: Director of Env Serv. *Employer:* Giffels Associates. *Education:* BSCE/Civil Eng/Univ of Detroit. *Born:* 1926. BSCE Univ of Detroit 1949. Reg Engr Mich & Fla. Joined Giffels Assoc in 1952, became Dir of Civil Engrg in 1956, Dir of Environmental Services in 1973. Directs pollution abatement & acoustical engrg for this multi- discipline A-E firm. As part owner, assists in business development and establishment of company policies. Has lectured frequently and published over 20 papers relating to industrial waste and noise problems. Mbr WPCF and a Diplomate of AAEE. Avid one design sailor and surf fisherman. *Society Aff:* WPCF, AAEE.

Willson, Alan N, Jr
Business: 6730 Boelter Hall, Los Angeles, CA 90024
Position: Prof & Asst Dean. *Employer:* UCLA. *Education:* PhD/EE/Syracuse Univ; MS/EE/Syracuse Univ; BEE/EE/GA Tech. *Born:* 10/16/39. Native of Baltimore, MD. From 1961 to 1964 he was with IBM Corp. He was Instr in Elec Engg at Syracuse Univ from 1965 to 1967. From 1967 to 1973 he was with Bell Labs, Murray Hill, NJ. Since 1973 has been on UCLA faculty, now Prof of Engg & Applied Sci, & Asst Dean for Grad Studies. Editor of the book "Nonlinear Networks: Theory and Analysis–, IEEE Press, 1974. Former editor of "IEEE Transactions on Circuits and Systems–. Recipient of 1978 Guillemin-Cauer Award of IEEE Circuits & Systems Soc, for coauthoring best paper in *Transactions* during previous yr. *Society Aff:* IEEE, SIAM, ASEE.

Willyoung, David M
Business: 12 Harmon Rd, Scotia, NY 12302
Position: Conltg Engr *Employer:* MacCleggan & CO. *Education:* MS/ME/Union College; BS/ME/Stevens Inst of Tech. *Born:* 5/7/24. Native of northern NJ. Served in USN. Employed by GE Co (1946-87) Retired 1987. Design, dev & engrg mgt responsibilities for large steam turbine generators (1949-1971.) From 1971 to 1980 held strategic planning & business dev responsibilities for GE's Turbine Bus Group (steam & gas turbines & systems). Responsible for ship propulsion studies in 1980 and 1983. During 1981 & 1982 responsible for energy system studies & development. 1983-87 engaged in advanced equip & system design for shipboard application. 37 Patents granted relating to turbine generator elec, mech, excitation & cooling systems, & to combustion power generation cycles. Fellow of ASME & IEEE. Engaged as a private consultant & entrepreneur since retirement from General Electric Co. in 1987. *Society Aff:* IEEE, ASME.

Wilmotte, Raymond M
Home: 2512 Q Street, N.W., 301, Washington, DC 20007
Position: Consultant. *Employer:* FCC. *Education:* ScD/Mech Sci/Cambridge Univ; MA/Mech Sci/Cambridge Univ; BA/Mech Sci/Corpus Christi Col. *Born:* 8/13/01. Paris France. 1921 Natl Physical Lab, England research asst. 1929 came to US to join research staff, Aircraft Radio Corp. 1932, consultant, broadcasting industry. Basic contributions to antenna design. WWII, Wilmotte Laboratory, contributions to proximity fuse, radar, overseas communication installations. 1947 broadcasting and R&D projects. 1959 RCA Prog Mgr, Relay Spacecraft (NASA's experimental active communication satellite). 1962 consultant to major electronics corporations, govt agencies and Academy of Engrg. Currently consultant to FCC. Fellow of IEEE, Bureau of Ordnance Dev Awd. Many papers and articles. 50 patents. 1985 Award FCC "excellence in stimulating development of new technologies for telecommunication–. *Society Aff:* IEEE.

Wilsdorf, Heinz G F
Business: Dept of Mtls Sci, Thornton Hall, Charlottesville, VA 22901
Position: Prof. *Employer:* Univ of VA. *Education:* Doctorate/Met/Gottingen Univ; Dipl-Physiker/Phys/Berlin Univ. *Born:* 6/25/17. Post Doctoral Fellow, Dept of Phys, Gottingen Univ, Germany, 1947-1949. Natl Physical Lab, Pretoria, S Africa. Res Officer, 1950-1954. Principal Res Officer, 1954-1956. Hd, Optics Subdiv, 1955-1956. Franklin Inst Res Labs, Phila, PA. Sr Res Met, 1956-1958. Mgr, Phys of Metals Lab, 1958-1960. Technical Dir, Solid State Sciences Div, 1960-1963. Dept of Mtls Sci, Sch of Engg & Appl Sci, univ of VA, Charlottesville, VA. Prof, 1963-1966 Wills Johnson Prof - 1966-1987. Chrmn, 1963-1976. Dir, Light Metals Center, Univ of VA, 1984- . William G. Reynolds Prof of Mtls Sci, 1987- . *Society Aff:* ASM, EMSA, AIP.

Wilson, Albert E
Business: Box 8060, Pocatello, ID 83209
Position: Prof, Nuclear Engrg *Employer:* Idaho State University. *Education:* PhD/Engg Science/Univ of OK; MS/Nuclear Engg/Univ of NM; BS/Engr Physics/Univ of CO. *Born:* 1/17/27. Formerly with Natl Bur of Stds, Los Alamos National Lab and Univ of OK. With ID State Univ since 1966. Mbr of ANS, ASEE, NSPE. Reg Profl Engr (Nuclear) in ID. Chairman of Eastern Idaho Section ANS, 1981-82. *Society Aff:* ANS, ASEE, NSPE.

Wilson, Andrew H
Home: 580 Wavell Ave, Ottawa Ontario, Canada K2A 3A5
Position: Consultant *Employer:* Self *Education:* MA/Economics/Univ of Glasgow; BSc/ME/Univ of Glasgow. *Born:* 5/30/28. Native of Scotland, in Canada since 1957. Early career in marine engg, hydraulic systems design, and the ball bearing industry. Subsequently with Atomic Energy of Canada Limited at Chalk River, the Economic Council of Canada, the Sci Council of Canada, and the Natl Res Council of Canada. Author of over seventy publications, principally in the field of sci policy. Consultant in research management. Chrmn of the CSME History Committee 1976-1980. Pres of CSME & VP of EIC, 1979-80. Chairman, Canadian Engineering Manpower Council, 1981-83. Pres of Engrg Inst of Canada 1982-83, Chrmn of EIC Fed Govt Liaison Ctte 1982-, Gen Chrmn. Third Intl Conference on Bldg Economics 1984. CSME Certificate of Serv 1984. *Society Aff:* CSME, EIC, ASME, IME, APEO.

Wilson, Arnold
Business: 368 C B, Provo, UT 84602
Position: Prof Cvl Engrg *Employer:* Brigham Young University. *Education:* PhD/Civil Engg/OK State Univ; MS/Civil Engg/Brigham Young Univ; BES/Civil Engg/Brigham Young Univ. *Born:* Feb 1 1933 Payson UT. Lic P E Utah, ID, Alberta Canada, Taught structs cvl engrg at Brigham Young U 1957- . Cons Engr George S Nelson Engrs & Bauman & Christensen Salt Lake City, Rollins, Bwn & Gunnel Engrs Provo. Chmn Mapleton Cty Plan Comm 1968-72; Cty Engr for Mapleton Cty 1972- . Pres Intermountain Chapt ACI 1974-75; Mbr: ACI, Internatl Assn Shell Structs, Sigma Xi. NSF Fellow 1964, Continental Oil Fellow 1965. Named Outstanding Prof, Assoc Students of Brigham Young U 1967-80. Bishop Mapleton 5th Ward LDS Church 1975-80. *Society Aff:* ACI, IASS, ΣΞ.

Wilson, Basil W
Home: 529 Winston Ave, Pasadena, CA 91107
Position: Cons Oceanogr Engr. *Education:* DSc/Eng/Univ of Cape Town; CE/Civil Eng/Univ of IL; MS/Railway Eng/Univ of IL; BSc/Civil Eng/Univ of Cape Town. *Born:* 6/16/09. CapeTown So.Africa. S A Railways & Harbors: Asst Engr, Res 1932-37; Asst Res Engr 1940-52; Commonwealth Fund Serv Fellow, U of Ill 1938-39; Prof Engrg Oceanography Tex A & M U 1953-61; Assoc Dir Eng Ocn, Natl Engrg Sci Co 1961-65; Dir Engrg Ocn, Sci Engrg Assocs 1965-68; Consulting Oceanographic Engr, Pasadena 1968--. Arthur M Wellingon Prize, ASCE 1952. Inst Award, S.A.Inst.CE 1960. Overseas Premium ICE 1968. Norman Medal ASCE 1969. J.G. Moffat-F.E. Nichol Harbor and Coastal Engrg Award, ASCE, 1983. Award for Meritorious Res, SA. Inst. CE, 1984; Emeritus Mbr, NAE, 1984. Author over 120 pubs, journals, books. US delegate PIANC 1957, 1961, US Japan Co-op Coastal Eng Seminar 1964, US Lecturer NATO Advanced Study Inst on Berthing & Mooring Ships, Portugal 1965; England 1973. Alfred E Snape Mem Lecture, SA Inst CE 1975. Chrmn and mbr of numerous ASCE tech cttees. Invited Lecturer 2nd Intl Conf, Structural Safety & Reliability, Munich, W. Germany, 1977. *Society Aff:* ASCE, ICE, SAICE, NAE, AGU, AAAS, ΣΞ.

Wilson, Caldwell J
Home: 7508 Pickard, NE, Albuquerque, NM 87110
Position: VP. *Employer:* Bovay Engrs, Inc. *Education:* BS/Civ/Univ of NM. *Born:* 8/18/08. Born and reared in Paris, TN. Held various engg assignments with Mtn Bell Telephone Co for 37 yrs. Pres of NSPE 1971-72. Received J F Zimmerman award from UNM in 1972. Named outstanding Alumni of College of Engg at UNM in 1979. Joined Bovay Engrs, Inc as consultant in 1973. Elevated to VP and Mgr of Albuquerque Office in 1975. Appointed as mbr of Bd of Registration for PE & Land Surveyors for NM in 1981. *Society Aff:* NSPE.

Wilson, Charles
Business: NJ Institute of Technology, Newark, NJ 07102
Position: Prof of ME. *Employer:* NJ Inst of Tech. *Education:* PhD/ME/Poly Inst of NY; MS/Engg Mech/NYU; MS/ME/NJ Inst of Tech; BS/ME/NJ Inst of Tech. *Born:* 8/2/31. Specialist in Noise Control, Mechanisms & Machine Design. Prof of Mech Engg at NJ Inst of Tech. Taught, conducted res & supervised engg courses at NJIT since 1956. Consultant & expert witness in noise control and machine failure. Licensed PE. Sr Engr in charge of Noise Impact Statement. Past Chrmn, Natural Resources Advisory Committee. Coauthor of textbooks: *Kinematics & Dynamics of* Machinery, *Mechanism-Design Oriented Kinematics & Machine Design-Theory and* Practice. Author of chapters in *HB of the Engineering Sciences*, Industrial Pollution Control, Measurement and Instrumentation and *HB of Industrial Noise Control*. *Society Aff:* ASA, AIP, ΣΞ, ΤΒΠ, ΠΤΣ.

Wilson, Charles W
Business: 34 Woodoak Dr, Westbury, NY 11590
Position: President *Employer:* Charles W Wilson Associates Inc *Education:* MME/Mech Engg/NY Univ; ME/Mech Engg/NY Univ. *Born:* 8/4/21. NYC. Design Draftsman Door Co 1940-42. Served with US Army 1942-46 as Combat Engr Co, Commander Italy. Allied Mission to Greece 1946 disch rank Captain. Employed by Babcock & Wilcox Co 1946 to 1981, Seaworthy Systems 1981-85. Charles W Wilson Associates Inc 1985-present, marine engrg and energy related propulsion applications. Resp for engineering contracts covering marine boilers & nuclear reactors on various commercial & naval vessels. Mbr Bd/Dir Shipbuilders Council of America. Corp rep for Natl Security Industrial Assn & other trade & energy assns & cttees. Chrmn 1971-72 SNAME. Mbr Council 1972-73 & 1986-89 SNAME. Exec Cttee 1971-84 Inst Marine Engrs. Mbr Natl Exec Cttee 1970-84 Propeller Club of the U.S PE NY. Chartered Engr U.K. *Society Aff:* SNAME, PCUS, IME, ASNE, SMC.

Wilson, Claude Glenn
Home: 1685 Country Club Dr, Redlands, CA 92373
Position: President *Employer:* GTS Associates, Inc *Education:* BS/CE/OK Univ; Masters/Pub Adm/Univ of Southern CA. *Born:* 8/13/22. Native of Ardmore, Okla. Served as Capt US Army WWII-3 Battle Stars and Bronze Star Medal. City Engr - Asst City Mgr Ardmore, Okla 1949-55; Design and Field Engr, Freese & Nichols, Ft. Worth, TX 1955-60; Design Engr, Pub Wks Dept City of Riverside Calif 1960-62; Dir of Pub Wks-City Engr and City Mgr, City of Colton CA 1962-72; VP, Neste Brudin & Stone, San Bernardino CA 1972-75; Dir of Pub Wks/City Engr, San Bernardino, CA 1975-1983; Pres, GTS Assocs, Inc., Redlands, CA, 1983. Fellow ASCE. Full Mgr Intl City Mgt Assn. Reg Engr OK, TX, NM, AZ and CA. Active in Calif Soc of Prof Engrs, serving as Reg VP 1972-74, 1st VP 1975-76, State Pres 1976-77 and Natl Dir 1977-78; State Pres, Public Works Officers, League of CA Cities 1979-80; V P - Western Region, Natl Soc of Profl Engrs 1983-85. *Society Aff:* NSPE, ASCE, ICMA, AWWA, APWA.

Wilson, Claude L
Home: PO Box 2848, Prairie View, TX 77446
Position: Consulting Engrg (Self-employed) *Education:* ScD/-/KS State Univ; MS/Mech Engrg/KS State Univ; ME/Mech Engrg/ KS State Univ; BS/ME/KS State Univ *Born:* 11/30/04. Ottawa, KS received elementary and high school education there. Served as Assistant Prof of Mechanic Arts at Prairie View A&M Univ founded School of Engrg 1945, Dean of Engrg 18 yrs, Academic Dean 4 yrs, VP for Physical Plant 4 yrs, is registered Professional Engr in TX, received distinguished service award in Engrg from KS State Univ 1962, Engr of Year for Region IV, TX Society of Professional Engrs 1977, mbr of Tau Beta Pi, Pi Tau Sigma, and Epsilon Pi Tau, Officer in AME Church, married to Lucellustine C Walker, one daughter, Mrs. Rosalind J. Benson. *Society Aff:* ASME, ASTM, ASEE, NSPE, NTA.

Wilson, Delano D
Home: 4 Via Del Zotto, Scotia, NY 12302
Position: President *Employer:* Power Tech, Inc. *Education:* BS/EE/MT State College. *Born:* 4/15/34. Joined the GE Co in 1959. Responsible for numerous studies which defined precedents in electric power system insulation coordination including present techniques of EHV circuit breaker surge suppression. Was responsible for the design and construction of the GE HVDC Simulator. Editor of *Transmission* Magazine. In 1974 joined Power Technologies, Inc. as Principal Engr. He was responsible for development of Compact Transmission Line technology in the U.S. Presently is President and CEO, responsible for all management and operations of the Company. Mr. Wilson is author or co-author of over 58 technical publications and 4 books, is a Fellow of IEEE, and U.S. Representative to CIGRE SC33. *Society Aff:* IEEE, CIGRE, IEC.

Wilson, Donald E
Business: 4301 Dutch Ridge Rd, Beaver, PA 15009
Position: VP & Dir *Employer:* Michael Baker Corp *Education:* Juris Dr/Law/Univ of Pittsburgh; M/CE/Yale Univ; BS/CE/VA Military Inst *Born:* 12/11/27 1950 TO 1953 - Assistant to the District Bridge Engr, PA Dept of Highways. 1953 to 1955 - Engr and VP of Carl J Jacobsen, Inc - design, fabrication and erection of steel rail-

Wilson, Donald E (Continued)
road and highway bridges. 1955 to 1959 - Assistant to Chief Engr, PA Tunnel Commission - responsible for civil engrg design of Fort Pitt Tunnel in Pittsburgh. 1959 to Present - VP, Michael Baker Corp - responsible for contracting, litigation and construction industry claims and arbitration. *Society Aff:* AAAS, ASCE, NSPE, SAME, ABA

Wilson, Drake
Home: 3137 Readsborough Ct, Fairfax, VA 22031
Position: Dir of Military Programs *Employer:* US Army Corps of Engr. *Education:* MS/CE/Princeton Univ; BS/Military Engr/US Military Acad *Born:* 8/17/30 Commissioned into the Corps of Engrs from West Point in 1952. Commanded combat engr troop units in Korea and Vietnam, taught civil engrg at West Point. District Engr at Mobile, AL 1973-76 (Maritime Man of the Year, Port of Mobile, 1975, and Pres, Mobile Post, SAME, 1976). Deputy Dir of Civil Works, 1976-78. Div Engr, Europe 1978-80 (Pres, Frankfurt Post, SAME, 1979). Dir of Military Programs, responsible for military construction and maitenance of facilities for the Army worldwide, 1980 -. Hobbies are sailing, skiing, tennis. *Society Aff:* SAME, ASCE

Wilson, Edward L
Business: 781 Davis Hall, Berkeley, CA 94720
Position: Prof *Employer:* Univ of CA *Education:* D Eng/Struct Eng/Univ of CA, Berkeley; MS/Struct Eng/Univ of CA, Berkeley; BS/Struct Eng/Univ of CA, Berkeley *Born:* 9/5/31 Field Engr for State of CA 1953, 54. Senior Res Engr Aerojet General Corp 1963-65. Prof of Civil Engrg at the Univ of CA since 1965. Served as Chairman of the Div of Structural Engrg and Struct Mechanics 1973-76. At the Univ he teaches courses and conducts resch on numerical methods for the analysis of large structural systems subjected to static and dynamic laods. He has published over 100 reports and papers. The well known computer programs TABS and SAP were developed under his direction. *Society Aff:* ASCE, SEAONC

Wilson, Fred H
Business: 3970 Hendricks Ave, Jacksonville, FL 32207
Position: Sr Partner. *Employer:* Fred Wilson & Assoc. *Education:* BCE/Structures/Univ of FL. *Born:* 9/19/21. Native of Jacksonville, FL. Served with US Coast and Geod Survey 1942-43. Served with USNR 1944-45. Attended Univ of FL 1946-49. Design engr with Reynolds, Smith & Hills, Consulting Engrs Jacksonville, FL 1949-56. Partner, Flood Wilson & Assoc,Consulting Engrs 1956-62. Sr Partner, Fred Wilson & Assoc, Consulting Enrs 1962-present. Reg PE FL, GA, AL, NC. Fellow ASCE. Mbr NSPE, FL E.S., ARBA. Autobiography in Who's Who in the So and SW. *Society Aff:* ASCE, NSPE, FES, ARTBA.

Wilson, Garland
Business: 1576 Sherman St, Denver, CO 80203
Position: VP. *Employer:* RWC, Inc. *Education:* BS/Structures/Univ of CO; BS/Agr Engr/Univ of MO. *Born:* 6/21/40. St Joseph, MO; reg in 5 states; certified consulting engr in CO; principal in RWC, Inc; in-charge of maj bldgs, bridges and civ projs. *Society Aff:* NSPE, IIE, ASCE, ACI.

Wilson, Hugh S
Business: Moore Products Co, Sumneytown Pk, Spring House, PA 19477
Position: VP, International Operations. *Employer:* Moore Products Co. *Education:* BS/Chem Engrg/Pratt Inst. *Born:* 7/1/26. Served AAF 1945. With Moore Products Co since 1951, serving as engr in training, sales engr, Branch Mgr. Mbr ISA since 1954, elected Fellow in 1970. Served as VP, Pub Dept, VP Dist II, Chrmn Edu Cttee, Dir Maintenance Div. Presently Parliamentarian & Asst to Pres for Intl Affairs. Mbr British Inst of Measurement & Control (IMC). Elected Fellow in 1976. Present Assignment with Moore Products Co-VP, Intl Operations ISA-Pres 1977-78. P Pres 1978-79. Enjoys reading, writing & travel *Society Aff:* ISA, IMC.

Wilson, J Caldwell
Home: 7508 Pickard NE, Albuquerque, NM 87110
Position: VP, Mgr. *Employer:* Bovay Engineers Inc. *Education:* BS/Civil Engg/Univ of NM. *Born:* Aug 18,1908. Employed by Mountain Bell Tel Co in various engrg assignments from 1935-73. Married Ruth McCarty in 1937, have one daughter, Rebecca Louise. Served as Pres of NM Soc Prof Engrs 1960-61. Served as Natl Dir NSPE 1961-64, VP Southwest Reg NSPE 1966-67 and 1967-68, Pres of NSPE 1971-72. *Society Aff:* NSPE.

Wilson, Jack P
Home: 1324 Rutland Lane, Wynnewood, PA 19096
Position: Consulting, Expert Witness *Employer:* Retired from P&S 8/31/78 *Education:* BS/Mech Engg/Carnegie Tech; BS/Aero Engg/Carnegie Tech. *Born:* 8/28/13. Taught at Carnegie; taught at Temple Univ WWII. Ind Fce Eng Pennsylvania Industrial Engs Pittsburgh. Mgr Fce Divn to Sect & Bd/Dir. Mbr Phila Drying Machy Co Phila. Chf Eng PDM Div, Chf Eng Chem Warfare & Explosives Div, Chf Eng of Co Dir of R/D,Mgr of Sales Eng, VP of Sales Eng, VP of Proctor Hydroset, VP of Adm Proctor & Schwartz Phila. Consulting Engineering Fields Expert Witness in Technical Fields Engineering Fields Product Liability. Past and present mbr, officer, cttee chrmn ASM, NAM, AIChE, MAPI, Alpha Tau Omega, Phi Mu Alpha, Pi Tau Sigma. Enjoy fishing, gunning, woodworking, serving civic groups. *Society Aff:* ASC, AIChE, SHRS, ATΩ.

Wilson, Joe R
Business: Tech Sta, Ruston, LA 71272
Position: Prof & Hd of Civ Engr. *Employer:* LA Tech Univ. *Education:* PhD/Civ Engr/Univ of TX; MCE/Civ Engr/Rensselaer Poly Inst; BCE/Civ Engr/Rensselaer Poly Inst; BS/-/US Naval Academy. *Born:* 4/5/23. Native of Orange, TX. Served with Pacific Fleet 1944-46 in USS Boston. Served in US Navy Civ Engr Corps until retirement in 1964. Last position was 15th Naval dist Civ Engr (Canal Zone). Joined LA Tech Civ Engg Faculty in 1966 and assumed position of Prof and Hd of Civ Engg in 1976. Enjoy golf and ham radio. Reg PE in TX & LA. *Society Aff:* ASCE, ASEE, LES.

Wilson, Joseph G
Business: 109 Sound Beach Ave, Old Greenwich, CT 06870
Position: President. *Employer:* In-O-Ven Corp. *Education:* BS/ME/Univ of MI. *Born:* 4/11/08. Native of Mexico City. Engaged in petroleum research, engrg, construction. Mgr Shell Oil Manufacturing Eng Dept 1954. VP Newfoundland Refining Co 1968. Pres In-O-Ven Corp 1971. Consultant in energy related fields including design and construction of oil refineries and liquefied natural gas processing facilities; energy recovery from gases containing particulate matter; construction scheduling, planning, and cost control; feasibility studies and economic assessments; engineering contracts and administration. Fellow ASME Mbr API. *Society Aff:* ASME, API.

Wilson, L Edward
Business: Reynolds Road, Franklin, TN 37064
Position: VP. *Employer:* Geologic Assoc, Inc. *Education:* BS/Civil Engg/ TN Tech Univ. *Born:* 5/15/44. Native of Clarksville, TN. Served as engrg officer with Corps of Engrs, 1967-69. Geologic Assoc, Inc, 1969 as a geotechnical and materials engr. Now serves as Pres and Chrmn of the Bd of Geologic Assoc, Inc. Also Bd of Dir MCI/Conslt Engrs, past Pres. Nashville Sect ASCE; Natl Bd of Dir for ASCE from 1981-1984. Serves on Bd of Engrg Advisors for TN Tech Univ. Reg PE in 20 states; Reg Land Surveyor in TN. *Society Aff:* ASCE, ISSMFE, NSPE.

Wilson, Linda H
Home: RD3 Box 387, Fishkill, NY 12524
Position: Staff Engr *Employer:* Internatl Business Machines *Education:* MS/ChE/Univ of Rochester; BS/-/Pratt Institute. *Born:* 3/-/40 Process Design and Evaluation Engr for Texaco Inc. Public Health Engr for Dutchess County (NYS) Health Dept. Founded NANCO Enivornmental Services Inc in 1974 to provide consulting and laboratory services to public and private water and wastewater treat-

Wilson, Linda H (Continued)
ment facilities. Licensed PE in NY 1979 to present - joined IBM to engage in implementation of major pollution control projects. In 1987 changed career emphasis to dev programming. *Society Aff:* AIChE.

Wilson, Myron F
Business: 400 Collins Rd, NE, Cedar Rapids, IA 52498
Position: Group Dir., Operations, Avionics Group *Employer:* Rockwell Intl. *Education:* BS/Mech Engg/Case Inst Tech; BS/Elec Engg/Case Inst Tech; BS/Eng Adm/Case Inst Tech. *Born:* 10/18/24. Native of Cleveland, OH. Served with Navy 1944-46. Employed by Collins Radio in Cedar Rapids, IA as test equip engr 1951. Worked thru various mfg respts 1951-60. 1961 became Chf Engr, Dallas Div of Collins. Dir R&D Div 1961- 64. Became Dir Reliability & Quality Assurance Cedar Rapids 1964. 1971 appt Corp Dir Reliability & Quality Assurance, resp for co-wide reliability and quality assurance policy & direction. In addition, apptd 1972 to Dir of Operations, Avionics Div resp for Collins Cedar Rapids Operations. 1973 apptd VP Mfg, Collins Radio Co. 1976 apptd Dir Quality Assur, Electronics Operations, Rockwell Intl. 1977 Apptd grp dir Product Assurance, avionics and missiles grp, Rockwell Int. 1981 Apptd Grp Dir. Operations, Avionics Grp, Rockwell Internatl. Served on Cedar Rapids Chamber of Commerce Bd/Dir 1972-73. Fellow 1968 ASQC. Region VP IIE 1983. Group VP Chapter Operations IIE 1985. PE, Ohio and Iowa. Dir 1970-71 ASQC. *Society Aff:* ASQC, IIE, IEEE.

Wilson, Nick L
Business: 1120 E 153 St, South Holland, IL 60473
Position: President. *Employer:* Morrison Timing Screw Co. *Education:* MBA/Accounting/Loyola; BS/Chem E/IA State Univ. *Born:* 8/10/44. Research Engr for Amoco Chemicals Corp prior to establishing Change Part Engrg Corp in 1971 and Morrison Timing Screw Co in 1973 where he is President and Chief Financial Officer. Mbr of the Bd/Dir Delta Chi at Iowa State Univ and PDir of AIChE, Chicago Sec. Now resides in Chicago, is an experienced pilot, and is writing a book on guidelines for small businesses. *Society Aff:* AIChE.

Wilson, Percy Suydam
2 Laureldale Ave, Metuchen, NJ 07028
Position: Engineer *Employer:* Self. *Education:* MCE/Hydraulics/Cornell Univ; CE/Hydraulics/Cornell Univ. *Born:* Sept 1896. Tau Beta Pi, Sigma Xi, McGraw Fellow, Asst Engr with James H Fuertes Cons Engr, Water supply & sewerage 1921-25, Engr Spencer White & Prentice, foundations 1926, Engr & Mgr New Rochelle Water Co 1926-27, Supt of Operation, Community Water Service Co 1927-32, Consulting practice 1932-36, Actg Secy & tech asst secy AWWA 1936-39, engr & sales repr for various mfgrs in san eng field 1940 to 1980. Hon M AWWA; Fuller Award 1945; Fellow ASCE; M. WPCF; Diplomate AAEE; Licensed PE NJ & NY; Councilman Boro of Glen Ridge 1955-60; Deacon & Parish Clerk G R Cong Church. Pres Bd of Trustees Glen Ridge Public Library. Personal interest - Genealogy & Local History. *Society Aff:* ΤΒΠ, ΣΞ, ΦΚΤ, ROTARY.

Wilson, Ralph S
Home: 2324 Ada Ct NE, Albuquerque, NM 87106
Position: Dept Mgr. *Employer:* Sandia Labs. *Education:* BS/ME/CO State Univ. *Born:* 6/14/19. Native of Olathe, CO. CO State Univ Honor Engg grad. Engr with Wright Aeronautical Corp Army Air Corp Officer responsible for establishing and operating engine overhaul facilities in Brisbane, Australia from 1942-45. Supervised experimental test at Lycoming and assistant proj engr at Wright Aeronautical. With Sandia Labs since 1950. Supervised atomic weapon dev groups for 18 yrs. Assumed current responsibility for process dev & fabrication of new mtls & components in 1968. Served on ASME Natl Gen Awards Committee & Committee on Honors from 1970-81. A reg PE who enjoys golf, fly fishing, & private flying. *Society Aff:* ASME.

Wilson, Ray W,
Business: 120 E. Market St, Indianapolis, IN 46204
Position: Manager Product Research Dept. *Employer:* Indiana Farm Bureau Coop Assn. *Education:* MS/Agri Eng/Penn State; BS/Agri Eng/Penn State. *Born:* 7/27/42. Hanover PA. Empld by Agricultural Engrg Res Div of USDA in the Virgin Islands 1966. Officer in the U S Army Corps of Enrs 1966-68. Assumed present position in the Product Res Dept in 1969. Present responsibilities include farm building design, coordinator of environmental affairs, and energy conservation officer. Secy, Treas, and VP of local soc NSPE. Membership on various cttees of ASAE. REg PE in Indiana. Now also responsible for quality control laboratory. *Society Aff:* ASAE, NSPE, ASHRAE.

Wilson, Rexford
Business: Postal Bx 81145, Wellesley Hills, MA 02181
Position: President. *Employer:* FIREPRO Incorporated *Education:* BS/Elec Engin/Univ of Mass. *Born:* Dec 1930. BSEE U of Mass. Native of Amherst, MA. Served with Amherst Fire Dept, FIA, NFPA, U of Md Faculty and Fenwal Inc. Since 1970 Pres of FIREPRO Inc, technical consulting ctr for fire protection. Has patented portable automatic fire extinguishing system; authored numerous articles; lectured extensively; investigated major fires; and contributed to NFPA Handbook. Awards - U of Mass Eng Alumni SFPE HATS OFF Founder of Home Fire Mgt, ZeroFire Programs, NFPA Fire Reptg Cttee; co-founder, Evaluating Firesafety Course; Chrmn, SFPE Measurement of Fire Phenomena Cttee; Mbr NFPA Sys Concepts Cttee, NSF-RANN Rev Cttees; Secy Panel 1 GSA Conf on Firesafety of High-Rise Bldgs. Reg PE, PSecy-Treas of SFPE, PPres of New Eng Chap of SFPE, mbr of NFPA, BOCA and NACA. *Society Aff:* SFPA, NEC-SFPA, NFPA, NACA, IMSA.

Wilson, Richard N M I
Business: Dept. of Physics, Lyman Lab. 231, Harvard Univ, Cambridge, MA 02138
Position: Mallinckrodt Professor of Physics *Employer:* Harvard Univ *Education:* D.Phil./Physics/Oxford University; M.A./Physics/Oxford University; B.A./Physics/Oxford University *Born:* 04/29/26 Research Assoc Stanford Univ 1951-2; Rsch Officer Oxford Univ 1952-55; Physics Dept, Harvard 1955-61; Prof Physics, Harvard, 1961-83, Dept Chair 82-85, Mallinkrodt Professor 1983-present. Mbr Adv Cttee, Energy & Environmental Policy Ctr, 1979-present. Has served as consultant to Los Alamos Scien Lab, Oak Ridge Natnl Lab, Lawrence Livermore Lab, et. al. Natnl Acad Sciences Energy Engrg Bd 1982- ; Visiting Cttee, Nuclear Engrg Div, Brookhaven Natnl Lab 1984- ; Chrmn, Visiting Cttee Radiation Medicine, Mass General Hospital, 1984- ; Health Advisory Council, City of Newton, 1984- . *Society Aff:* APS, AAAS, NYAAS, SETAC, SRA

Wilson, Robert C
Business: San Tomas/Central Expry, Santa Clara, CA 95052
Position: Chrmn and Pres. *Employer:* Memorex Corp. *Born:* Jan 1920. BSME Univ of Calif 1941. Native Hazelton, Idaho, US Navy destroyers and submarines 1943-46. Engrg with GE spec electrical pwr distribution. Co-author 'Capacitors for Industry.' VP General Electric 1966. VP Rockwell Intl 1969. President and CEO Collins Radio 1971. Chrmn, President and CEO Memorex Corp 1974. AFCEA, Gold Medal, Natl Academy of Achievement, SME Progress Medal. Enjoy flying, fishing, scuba, golf, jogging.

Wilson, Robert O
Business: 77 Cornell St, Ringston, NY 12401
Position: President. *Employer:* FX Systems Corp. *Born:* April 7,1926. BSME Polytechnic Institute of Brooklyn. Member ISA. Attended Columbia University School of Business Executive Training.

Wilson, Stanley D
Business: 1105 N 38th St, Seattle, WA 98103
Position: Consulting Eng. *Employer:* Self. *Education:* SM/Civil Eng/Harvard Univ. *Born:* 8/12/12. Native of Sacramento, CA. Worked for CA Div of Hwys prior to WWII. Served with Army Corps Engr 1943-46. Disch rank of 1st Lt. Asst Prof of

Wilson, Stanley D (Continued)
Soil Mech & Foundation Engrg, Harvard U 1948-53. Co-founder of Shannon & Wilson, Inc Seattle 1954-78. Has served as consulting engr on major hydroelectric projects throughout the world. Spec interest in instrumentation for field observation of performance of earth structures. *Society Aff:* ASCE, USCOLD, ISSMFE.

Wilson, T Lamont
Home: 1407 Ormsby Ln, Louisville, KY 40222
Position: Consultant (Self-employed) *Education:* MS/Physics/U of Louisville; BS/EE/U of UT *Born:* 6/4/14. Native of Salt Lake City, UT; Broadcast transmitter operator KSL; Chief Engr, Ind Electronics Div, Federal Tel & Radio, (Div of IT&T) 1941-1946; With Votator Div of Chemetron Corp 1946-1977, retired as Mgr of Electronic Engg and Dev; Consultant Dielectric Heating since Sept 1977; Fellow IEEE 1977; Chrmn Louisville Sec IEEE 1970; Chrmn High Frequency Heating Ctte of IAS 1975 to present; Region III Outstanding Engr 1977; Initiated into Tau Beta Pi as eminent engr, 1977; 1979 Achievement Awd, IECI of IEEE: Centenial Medalist, IEEE, 1984: Past mbr Stake Presidency, Louisville KY Stake, Church of Jesus Christ of Latter Day Saints (Mormon). US Member CISPR (IEC) Sub. Com B, Working Group I Member US Delegation CCIR(ITU), IWP1/4, *Society Aff:* IEEE, IECI, IES, TBII, EMC.

Wilson, Thomas B
Home: 6321 Via Colinita, Rancho Palos Verdes, CA 90274
Position: Owner. *Employer:* Thomas B Wilson, Naval Archs & Marine Engrs. *Education:* MS/Naval Arch/Webb Inst; BS/Marine Eng/Webb Inst; BS/Elec Engg/US Naval Acad. *Born:* July 1925. MSplusBS Webb Inst Naval Architecture, BS US Naval Academy. Native of New Orleans, 26 yrs U S Navy with duties involving the design, building, repair and operation of various naval ships. Mgr Design Eng & Sr Mbr Tech Staff, Litton AMTD, Chief Engr, Harco Enrg, VP and Gen Mgr Willard Co, VP Bertel Inc. Consultant on the design of merchant ships, fish and work boats, yachts, hovercraft, and hydrofoils since 1953. Designer M/V Silverado, world's largest commercial fiberglass reinforced plastic vessel. Currently serve on SNAME Hull Structures Cttee and Chrmn Panel HS-10 (Plastics), ABS Spec Cttee on Reinforced Plastic Vessels and on Standards Cttee for the World Dredging Conference. USNA Award Marine Engrg 1948. Enjoy boating, fishing and swimming. *Society Aff:* RINA, SNAME, ASNE, WODA.

Wilson, Thornton Arnold
Business: P O Box 3707, Seattle, WA 98124
Position: Chmn of Bd & CEO. *Employer:* Boeing Co. *Born:* Feb 8,1921 Sikeston Missouri. Graduated from Iowa State Univ 1943 with degree in aeronautical engrg. Joined Boeing 1943, leaving briefly for teaching assignment and advanced study at Iowa State and a yr at Cal Tech for MS in Aeronautical Engrg 1948. Sloan Fellow at MIT in 1952. On Bd/Governors of Iowa State Univ, Member Corporation Development Committee at MIT Fellow AIAA, Business Council, Also Member of Trilateral Commission Stanford Research Institute Council, Bd of Dir of Seattle-First Natl Bank, U.S. Steel, PACCAR & Boeing. Recipient of James Forrestal Award 1975, and Wright Brothers Trophy in 1979. Married to Grace (Miller), three children.

Wilson, William B
Business: Energy Consulting, 1815 E. Westwood Dr, Maryville, TN 37801
Position: Energy Consulting *Employer:* Self *Education:* BSME/Mechanical Engr/Univ of Tennessee. *Born:* Feb 1915 Sparta, TN. BSME U of Tenn 1940. Gen Elec since 1940. Primary experience in conceiving and selecting Industrial Energy Supply Systems. Alternative systems, with controls to optimize reliability and costs, range from process boilers with purchased power and/or large compressor drives-to nuclear plants owned by the Utility or the Industrial. Has authored more than 100 tech papers presented and published in many countries. ASME Fellow, TAPPI Fellow, Sr Mbr IEEE. Chrmn of many cttees and active in leadership of these tech societies. *Society Aff:* ASME, TAPPI, IEEE.

Winchell, Peter G
Business: CMET Bldg, Sch of MSE, W Lafayette, IN 47907
Position: Prof, Materials Engrg. *Employer:* Purdue University. *Education:* PhD/Metallurgy/MIT; BSMet/Metallurgy/MIT; AB/Lib Arts/Univ of Chicago. *Born:* July 1929. BS, PhD MIT 1958. Asst, Assoc, and Full Professor, Purdue U and Visiting Prof at Cornell 1967-68. Also Visiting Prof of Dental Materials at Indiana Univ Dental School. Experimental and theoretical research on structure and properties of hardened steel, dislocation structure in concentration gradients, and experimental studies of dental materials and applications. Industrial metallurgical consultant. Recreation: Tennis. *Society Aff:* AIME, ASM, AAAS, ASEE, AAUP, IADR.

Winchester, Dewey R
Business: 7204 Lakeside Drive, Charlotte, NC 28215
Position: Conslt *Employer:* Self-Employed *Education:* BS/ChE/NC St Univ *Born:* 3/12/19 34 years with J. N. Pease Assocs in design of air conditioning, ventilation, heating, refrigeration, dust collection, process piping, central steam and chilled water plants, chilled water and steam distribution systems. Established concept design, energy and life cycle cost analysis for large projects. Engr and Administrator for Computerized Central Control Monitoring System, three central chilled water plants and chilled water distribution systems and five research-lab facilities at Univ of NC-Chapel Hill. Engr for Westinghouse Nuclear Turbine Components Plant, Winston-Salem, NC, manufacturing facilities for Lance, Inc, Charlotte, NC and Greenville, TX. *Society Aff:* NSPE, ASHRAE

Winchester, Robert L
Home: 935 Westholm Rd, Schenectady, NY 12309
Position: Mgr, Excitation Sys Eng. - Retired *Employer:* General Electric Co. - Retired *Education:* BS/Elec Engg/CA Inst of Technology. *Born:* 2/1/26. Worked for General Elec Co 1948; Generator Enrg, Large Steam Turbine-Generator Dept 1952. Design and advance development engrg. Mgr of Elec Engrg 1962, basic electrical design of large generators rated 100,000 KVA to 1,300,000 KVA. Mgr of Excitation Systems Engrg 1969, design and development of excitation systems used with these large generators. Mgr. Advance Engrg 1983, advance design and dev. Fellow IEEE. Past Chrmn IEEE Rotating Mach Cttee, Power Engrg Society. Retired Dec 1985. *Society Aff:* IEEE.

Windham, Donald R
Business: NOAA, Bay St Louis, MS 39520
Position: Electronics Engr. *Employer:* NOAA Data Buoy Office. *Education:* BS/Mathematics/MS State Univ. *Born:* Dec 7,1932, Bay Springs Mississippi. Taught Military Electronics at Keesler AFB and Redstone Arsenal 5 years. Worked as Res Engr, Boeing Co 4 yrs on Saturn V Test Stand Instrumentation Systems, worked as Data Systems Lead Engr for Gen Elec Co for 4 yrs responsible for total data systems performance from transducers to data analysis. Working as Marine Instrumentation Engr for NOAA Data Buoy Office responsible for systems specification, contract mgt and sensor systems performance. Served in various offices of ISA including Marine Sciences Div Dir 1974-76. *Society Aff:* ISA.

Windisch, Earl C
Home: 5121 W 96th St, Overland Park, KS 66207
Position: Partner *Employer:* Black & Veatch *Education:* BS/EE/Univ of KS *Born:* 6/10/27 Native of Louisburg, KS. Served in US Naval Reserve 1945-46. Employed by Black & Veatch in 1951, as electrical design engr for steam power plants and electric power systems. Entered Partnership in 1981. Presently responsible for all electric power facilities design, analyses, and reports completed by B & V for specific assigned clients. Hobbies are golf & fishing. *Society Aff:* ANS, IEEE, NSPE

Windman, Arnold L
Business: 11 W 42nd St, New York, NY 10036
Position: Vice Chairman *Employer:* Syska & Hennessy Inc. *Education:* BME/ME/CCNY. *Born:* 10/17/26. Design Engr for Frederic E. Sutton Consulting

Windman, Arnold L (Continued)
Engrs 1947-50. Proj Engr for Syska & Hennessy and Principal of the firm since 1950. Became Exec VP and Chf Oper Officer in 1974, Pres 1977. Vice Chairman 1986. Fellow, American Consulting Engineers Council. Pres NY Chap ASHRAE 1967. Pres and Dir NY Assn of Consulting Engrgs 1981-82, and Trustee of Phelps Mem Hosp, Tarrytown, NY 1974-82, Treas Amer Conslt Engrs Council 1982-84, Pres Amer Conslt Engrs Council 1984-85, Pres American Consulting Engineers Council 1985-86. Chmn NY St Bd for Engrg and Land Surveying 1983-84. Mbr of ASHRAE, ASME, Pi Tau Sigma, Tau Beta Pi. Reg PE in NY & 14 other states. *Society Aff:* ASME, ASHRAE, ACEC, Tau Bet Pi

Windolph, George W
Home: 6 Hilldale Dr, Malvern, PA 19355
Position: Dir, Product Assurance. *Employer:* Boeing Vertol Co. *Born:* Nov 1928. BSME from Drexel. Stress analyst with Chase Aircrft Co 194851. With Boeing Vertol Co since 1952, Test Engr 1952-57, Structures Engr 1957-58, Mgr of Engrg Test Labs 1958-62, Chf of Dynamic Systems Design 1962-64, Asst Proj Engr for CH-47 Helicopter 1964-67, Dir of Product Assur 1967 to present. Mbr of AIAA Systems Effectiveness and Safety Cttee for 3 yrs. Mbr of AHS Testing Cttee for 1 yr.

Windt, Ewald C
Business: 1900 Twenty-third St, Cuyahoga Falls, OH 44223
Position: Dir of Plant Operations. *Employer:* Cuyahoga Falls Gen Hospital. *Education:* BME/Mech Engr/Maschinen-Ban-Schule. *Born:* 2/22/28. in Germany. US Citizen, 20 yrs. Reside in Northfield, OH. Res and Dev Engr for Midland Ross specializing in design, consulting and test supervision. Hold two US Patents in Material handling. Proj Engr for Clevite Corp in developing, mfg procedures. With Lear Siegler, Inc as Quality Control Engr. Field Service Engr for Republic Steel, specialty in nationwide installation, service, personnel training, and trouble shooting of machinery. Assumed current position as Dir of Plant Operations for Cuyahoga Falls Gen Hospital in 1974, handling administrative, mangerial, consulting, planning and dev, construction, community relations, technical and financial responsibilities. *Society Aff:* ASME, ASSE, AIPE, NOSHE, OHA.

Winegard, William C
Home: RR 2 Guelph, PO Box 127, Guelph, Ontario N1H 6J6 Canada
Position: Member of Parliament *Employer:* House of Commons *Education:* PhD/Metallurgy/Toronto; MASc/Metallurgy/Toronto; BASc/Met/Toronto. *Born:* Hamilton, Canada Sept 1924. Native of Caledonia, Ontario. Royal Canadian Navy 1942-45. Lecturer, Asst Prof, Assoc Prof, Prof. Dept Metallurgy, Univ of Toronto 1952-67. Asst Dean, Sch of Grad Studies, UofT 1964-67. Pres & Vice Chancellor, U of Guelph 1967-75. Chmn Ontario Council on Univ Affairs 1976-82. Member of Parliament for the riding of Guelph (Ontario) 1984 to present. Fellow ASM. LLD Univ Toronto 1973. ALCAN Awd CIMM 1967. Author of book 'Introduction to the Solidification of Metals' published in 4 languages and of over 100 tech and scientific papers. DSc Meml Univ 1976; Fellow of the Univ of Gulph 1979; LLD Laurentian 1981; Dir-Homewood Sanitarium-Guelph Ontario; Dir- Interntl Dev Res Center-Ottawa; Dir-Ontario Res Fnd. *Society Aff:* ASM, APEC

Winer, Bernard B
Business: 700 Braddock Ave, E Pittsburgh, PA 15112
Position: Consultant *Employer:* Self-Employed *Education:* MS/ME/Carnegie Tech; BS/ME/Carnegie Tech. *Born:* Jan 1920. Native of Pittsburgh, PA. With Westinghouse for 40 years. Specialized in mechanical design and manufacture of large motors and generators. In 1975 assumed responsibility for the Large Motor and Medium Turbine Generator Dept in 1975. Currently a consultant in mechanical engineering area. PChrmn of the ASME Mgt Div and Pittsburgh Chap ASME. Enjoy bridge, golf and discussion groups. *Society Aff:* ASMe, IEEE 365x.

Winer, Loyd E
Business: 3975 Cascade Rd SE, Grand Rapids, MI 49506
Position: Chrmn of the Bd/Sec-Treas. *Employer:* Newhof & Winer Inc. *Education:* MS/Structural Engrg/MI State Univ; BSCE/Civ Engg/MI State Univ. *Born:* 4/23/29. Saginaw, MI; Raised in Grand Rapids, MI; 1952-56 US Navy Carrier Line Officer; 57 Grad Sch; 57 to 67 started as Structural Engr advanced to Chief Structural & Civ Engr to Sr Proj Dir for industrial, Govt & commercial projs for Daverman Assoc; 67 to present Chrmn of the Bd and Sr Dir, Newhof & Winer Inc. Architects-Engrs; Also 67 to 75 Secy-Treas Grand Rapids, Testing Service, Construction Materials testing. Enjoy outdoor activities, hiking & camping and photography, Who's Who in America, Who's Who in Finance and Industry. *Society Aff:* NSPE, MSPE, ACI, AAFC, AEE, ASCE.

Winer, Ward O
Business: Sch of Mech Engrg, Atlanta, GA 30332-0405
Position: Regent's Prof *Employer:* Georgia Inst of Technology. *Education:* PhD/Physics/Cambridge Univ; PhD/ME/Univ of MI; MSE/ME/Univ of MI; BSE/ME/Univ of MI. *Born:* June 27,1936. Demonstrator Physics, Camb Univ 1961-62, Asst Prof-1963-66, Assoc Prof 1966-69, Univ of Mich, Assoc Prof 1969-71, Prof 1971-1984 Regents' Prof 1984-present, at Georgia Tech. Consultnt to several major corporations. 1938 Dist Serv Awd to Outstanding Young Engrg Faculty Univ of Mich 1967, Sigma Xi Awd Georgia Tech Chap 1975. ASME Melville Medal 1975. ASME Mayo D. Hersey Award 1986, Tribology Gold Medal 1986 from Tribology Trust-Great Britain. Author over 190 tech papers, lectured widely in the US and invited lecturer in Denmark, England, France, Sweden, Germany, the Netherlands, Soviet Union, and Japan. Mbr ASME Res Cttee on Lubrication 1966-present Lubrication Div, Exec Cttee 1973-78. Exec Cttee ASME- ASLE Lubrication Conf 1972-80 ASME Committee on Research Needs, Chairman 1979- 1982, ASME Policy Board Research, Member, 1979-1984, Technical Editor, Transactions ASME, Journal of Tribology (1980-1987), Society of Engineering Science, Member Board of Directors (1979-1982), Gordon Res Conf on Lubrication, Friction and Wear - Vice Chrmn 1978, Chairman 1980, Mbr ASLE Bd of Dir (1982- 84), Natl Res Council Steering Cttee on Recommendations for U.S. Army Basic Res (Chrmn 1984-1987), PE in GA. *Society Aff:* ASME, ASLE, SES, AIP-SOR, AAAS, Am. Acad of Mechanics

Wines, Bing G
Business: PO Box 32148, OK City, OK 73123
Position: Consultant *Employer:* Winrock Engr, Inc. *Education:* BS/Petro Engr/Univ of OK *Born:* 6/6/39 Drilling, completion and evaluation of Oil & Gas Wells & Properties in OK, KS, CO, TX, NM and AR. *Society Aff:* SPE-AIME, NSPE

Wing, Arthur K, Jr
Box 294, Grantham, NH 03753
Position: Consultant. *Employer:* Self. *Education:* MS/EE/MIT; BS/EE/Yale. *Born:* 7/1/08. New York City. Concerned chiefly with development & engrg of electron tubes. Federal Telegraph Co 1931-34; RCA 1934-44; Federal Telephone and Radio 1944-45; Federal Telecommunications Lab 1945-62; ITT Electron Tube Div 1962-72 as Vice Pres-Eng. Retired 1972. Subsequently consulting in this field. Life Fellow IEEE. Chrmn IEEE Professional Group on Electron Devices 1961. *Society Aff:* IEEE.

Wing, Omar
Home: White Birch Dr, Pomona, NY 10970
Position: Prof. *Employer:* Columbia Univ. *Education:* Doctor of Engg Sci/EE/Columbia Univ; MS/EE/MIT; BS/EE/Univ of TN. *Born:* 3/2/28. in Detroit. Bell Labs 1952-56. Fulbright Lecturer Chiao-Tung Univ 1961. Ford Engg Resident IBM Res Ctr 1965-66. Guest Prof Tech Univ of Denmark 1973. Fulbright Sr Res Scholar and Guest Prof Eindhoven Univ of Tech, The Netherlands 1979-80. Visiting and Honorary Prof South China Inst of Tech, People's Republic of China 1979. Pres IEEE Circuits and Systems Society 1978. Editor IEEE Transactions on Circuits and Systems 1974-77. Fellow IEEE 1973. Dept Chrmn, Elec Engg and Computer Sci,

Wing, Omar (Continued)
Columbia Univ 1974-78 and 1983-86. Res interests: Computer-aided circuit analysis and design, VLSI system layout and design, parallel computation, circuit theory. *Society Aff:* IEEE.

Wing, Wayman C
Business: 411 7th Ave, New York, NY 10001
Position: Principal. *Employer:* Wayman C Wing, Cons Engrs. *Education:* BSCE/Civil Engg/Univ of WY; MSCE/Civil Engg/Stanford Univ. *Born:* FEb 22,1923 Evanston Wyoming. Capt USAF WWII. Assoc Engr Seelye Stevenson Value & Knecht NYC 1948-60. Principal Wayman C Wing Consulting Engrs NYC 1960--. First Honor Natl for Structural Design of Buildings 1968. Engr of Yr Awd NY 1970. Dist Alumni Awd Univ of WY 1970. Author of numerous articles for professional journals. Fellow ASCE. Former Dir & Officer NY Assn of Consulting Engrs. Tau Beta Pi, NSPE, ACEC. *Society Aff:* ASCE, NSPE, ACEC.

Wingard, M Rex
Business: New Delhi I.D. Dept Sta, Washington, DC 20521
Position: Rep in India. *Employer:* Coop League of USA. *Born:* 1921. BS Rubber Tech Univ of Akron. Native Akron OH. Served with USAF 1942-46 Pilot. Joined Blaw-Knox company experiment station Ann Arbor, became mgr 1948, and was mgr of food processing turnkey serv 1956 when became VP/owner DKA, Chicago, turnkey process engr-contractor. Gen Mgr Austin Co Process Div 1971 when became Exec VP Global Ventures mgt consultants to food processors and suppliers of processing systems. Presently providing tech assitance to Indian cooperatives & advisor to Natl Coop Dev Corp on agricultural product processing mgt.

Wingate, Roger H
Home: Box 93, Winter Harbor Way, Mirror Lake, NH 03853
Position: Retired *Education:* MS/CE/MIT; BS/CE/Tufts Coll *Born:* 11/10/15 Retired in 1980 as Exec VP Liberty Mutual Ins Co. Also served as Dir of that Co. Responsibilities included Liberty's Loss Prev Dept. Composed of over 600 Engs & Technicians as well as Liberty's Research Laboratory. Former mbr Exec Ctte - SFPE, former Dir - CSP. Former Dir - Nat Safety Council and Mass Safety Council. Mbr for six yrs & former Chmn - Nat Advisory Com on Occupational Safety & Health. Reg Prof Engr - State of MA. Current Dir - CT Main Corp - one of the largest Prof Engr Firms in the world. Chmn of the Bd - American Mutual Reinsurance Dir - Shawmut Melrose Wakefied Bank. *Society Aff:* SFPE, ASSE, CSP

Winger, Ray J, Jr
Home: 4050 Ammons St, Wheat Ridge, CO 80033
Position: Chief Drainage & Groundwater Engg (Retired) *Employer:* US Bureau of Reclamation. *Education:* MS/Civ Engg/Univ of CA; BS/Civ Engg/Univ of CO. *Born:* 3/16/19. Born & raised on irrigated farm in Eastern Co. Officer USMC from 1941 to 1945. Chief of survey party & inspector Angostur Dam, SD 1945 to 1947. Chief, Drainage Branch, Huron, SD & Riverton, WY 1949 to 1956. Drainage specialist in Chief Engrs Office USBR 1956 to 1972. Chief, Drainage and Ground Water Engr for USBR 1972-1981, Retired from USBR in 1981. Chrmn Exec Committee, Soil & Water Div ASAE 1976. Chrmn Exec Committee, Div of Irrigation & Drainage, ASCE, 1980. Currently serving as a drainage consultant for World Bank & a number of Foreign Govts. Relax by serving as a Black Belt Instr in Karate and Planning Commissioner, Wheat Ridge, CO. Dist and Meritorious Serv Awards-Dept of Interior, Hancor Award, Amer Soc of Agri Engrs, Dist Drainage Specialist Award, US and Canada Corrugated Plastic Tubing Assoc, Royce J Tinton Award, Amer Soc of Civil Engrs. *Society Aff:* ASCE, ASAE, ICID, NYAS.

Wink, Ralph C
Business: PO Box 2590, S Portland, ME 04106
Position: Dept Hd. *Employer:* Portland Pipe Line Corp. *Education:* MBA/Finance/Southern IL Univ; BSCE/Fluid Mech/Univ of MO. *Born:* 1/3/42. Native of Columbia, IL. Distribution Planning Engr - IL Power Co, 1964-66. Field Engr (Oil Well Logs) - Schlumberger Well Services, 1966-67. Mech Inst and Proj Engr for Shell Oil - Wood River Refinery, 1967-1973. Assumed current position as hd of oil movements dept with Portland Pipe Line in 1973. Responsible for Engg & Mgt functions in transporting on through pipe lines. Pres of Western ME Chapter of PEng - 1976/77. Served as VP State Soc of ME SPE in 1978/79. Currently state Pres ME SPE. *Society Aff:* NSPE.

Winkel, Thomas A
Home: 4685 Cherokee Dr, Brookfield, WI 53005
Position: Dist Chf Plng Engr. *Employer:* Wisc Dept Transportation. *Education:* BSCE/Transp/Univ of WI; BSNS/-/Univ of WI. *Born:* April 1934. BSCE & BSNS from Univ of Wisconsin. Native of Milwaukee, Wisconsin. Served with US Navy 1957-60. Now Commander, CEC, USNR-R. Empl by WIS DOT since 1960 as construction engr and planning engr. Promoted to Dist Chf Plng Engr of the Milwaukee Dist of the Wisc Div of Hwys in 1970. President of the Wisconsin Sec of ITE 1974-75. PE Wisc. Mbr ASCE. Enjoys golf and fishing. *Society Aff:* ITE, ASCE.

Winkelman, Dwight W
Home: 1001 Lewis Cove, Delray Beach, FL 33444
Position: Retired *Born:* 11/5/01. Lohrville Iowa. Hon DSc Morningside Coll Bus Admin, Syracuse Univ. First project as Contractor 1927. Inc 1943 D W Winkelman Co Inc, Syracuse NY. Formerly Pres and Chrmn of its Bd and several affiliated companies. PPres NY State and Natl Chapters Associated Gen Contractors of America. Served on numerous advisory cttees in Washington in connection with engrg, constructon and construction equipment. Fellow ASCE, Hon mbr 1972. Served as Chrmn Exec Cttee Construc Div, Chrmn Constructn Edu and Mgt Cttee ASCE. Trustee Syracuse Univ, former Dir Lincoln First Bank Central.

Winnie, Dayle D
Business: 8500 Culebra Rd, San Antonio, TX 78228
Position: Sr Res Engr. *Employer:* Southwest Res Inst. *Education:* BS/Mech Engr/Univ of WI; Continuing/Mech Engr/Univ of TX. *Born:* 7/20/35. and raised in Brandon WI, Grad from local public schools and Univ of WI in Madison, WI. Served in USAF as Aircraft Maintenance Officer 1958-60. Centralab Div of Globe Union Inc, in New Product Dev Group specializing in thermoset materials post forming res 1960-64. Proj Engr in charge of dev group for automatic dairy products processing equip at Stoelting Brothers Co, 1964-69. Sr Res Engr with Southwest Res Inst specializing in Electromechanical Design, Dept of Electromagnetic Engrg 1969-80, Staff Res Engr 1980- . Awarded several patents. Reg PE WI and TX. Listed in American Men and Women of Sci. *Society Aff:* ASME, ΣΞ.

Winquist, Lauren A
Business: P O Box 75, Ypsilanti, MI 48184
Position: Section Supvr. *Employer:* Ford Motor Co. *Education:* BS/Met Eng/Wayne State Univ. *Born:* July 23,1937. BS Wayne State Univ. Native of Detroit Mich. Majority of industrial experience with Ford wkg in various capacities of process and development engineering, including heat treating, process laboratory, Industrial Processing. Greatest time spent in powder metallurgy. Ten years in supervision. Tech advisory bd Washtenaw Community College. Fourteen tech papers/presentations, Vice Chrmn Powder Metal subcommittee, SME 1969-70. Safety Cttee, MPIF 1974, Powder Metl Oper Cttee ASM 1971-75. Enjoy sailing, music, competitive sports. *Society Aff:* ASM.

Winter, George
Business: Hollister Hall, Ithaca, NY 14853
Position: Professor Emeritus. *Employer:* Cornell University. *Education:* PhD/Struct Engg/Cornell Univ; Dipl Ing/Civil Engg/Tech Univ Munich; Dr Ing/Hon/Tech Univ Munich. *Born:* April 1907 Vienna Austria. Dipl Ing Tech Univ Munich. Design and consulting in Austria and USSR 1930-38. At Cornell Univ since 1938; Chrmn, Dept of Structural Engrg 1948-70. Mbr NAE, F.Am Academy for Arts & Sciences, Hon Mbr ASCE, Hon Mbr ACI. Chrmn Structural Stability Res Council 1974-78. Mbr

Winter, George (Continued)
Permanent Comm Int Assoc Bridge & Structural Engrg. ASCE Moisseiff Award and Croes Medal; ACI Wason Res Medal, Turner Medal and Kelly Award; AISI Tech Mtgs Awd. Abt 100 res publs steel & concrete structures, struct mechs. Co-author with A H Nilson, Design of Concrete Structures, 9th ed 1979. Produced or directed nearly the entire research background for Specification for Design of Cold-Formed Steel Structural Mbrs AISI. Mbr of archaeological expedition to Egypt 1966. Mbr Archaeological Inst of Am, Affiliate, Univ Museum, Univ of PA. Consultant to AISI and others. *Society Aff:* ASCE, ACI, NAE.

Winterkamp, Fred H
Business: DuPont DeNemours (France), S.A, 137 Rue de L'Universite, F-75334 Paris Cedex 07 France
Position: Dir Mfg, Agri Prods (Europe, Mid-East, Africa) *Employer:* Du Pont de Nemours (France), SA *Education:* MSc/ChE/OH State; BS/ChE/OH State. *Born:* 6/30/27. Youngstown Ohio.After graduation, began career with Du Pont Co as Instrument Engr, Sabine Rvr Wks. Transferred to Belle Plant, Belle WV 1957, Houston Plant 1969, Belle 1973, various administrative assignments in production, maintenance and process development. Plant Mgr, Belle Plant 1975. Reg PE, State of OH, Dir ISA Production Processes Div, Standards and Prac Dept 1957-61, ISA Fellow 1965, recipient of ISA Standards and Practices Awd 1967. Married, four children, hobbies classical music, skiing, skiing, sailing. Mbr Tau Beta Pi, Phi Lambda Upsilon. 1980 Dir, Manufacturing Services (Europe), DuPont de Nemours Intl, S.A. Geneva Switerland 1982 - Dir of Manufacturing, Agrichemicals, Europe 1986 - Dir of Manufacturing, Agricultural Products (Europe, Mideast, Africa) DuPont (France) S.A., Paris, France *Society Aff:* ISA.

Winters, Charles E
Home: 8800 Fernwood Rd, Bethesda, MD 20817
Position: Consultant. *Education:* ScD/Chem Engrg/MIT; SM/Chem Engrg/MIT; BS/Chem Engrg/Kansas State. *Born:* July 1916. 1940-43 Mallinckrodt Chem Wks, Proj Engr, Uranium Purification. 1943-47 Manhattan Dist, Uranium Refining and enrichment. Clinton Lab Admin. 1947-61 Union Carbide Corp, Oak Ridge Natl Lab. Engrg Res, Dev, Design. Maj resp for LITR, MTE, HRE, ORR, HFIR. Asst Dir of ORNL. 1962-69 Union Carbide Corp Parma OH, Gen Mgr Fuel Cell Project. 1970-78 Union Carbide Corp, Washington DC Washington Rep, and Asst to the VP. 1963-73 Atomic Safety & Licensing Board Panel, Administrative Law Judge USAEC. Fellow ANS, AIC, Exceptional Civilian Serv Awd, Manhattan Dist, Corps of Engrs. *Society Aff:* ACS, AIChE, ANS.

Winzler, John R
Business: P O Box 1345, Eureka, CA 95501
Position: President. *Employer:* Winzler and Kelly. *Education:* BS/Civil Engg/UC Berkeley. *Born:* May 1930. Native of Eureka Calif. Partner in the formation of the parent firm of Winzler and Kelly in 1951 with most activities devoted to water resource development. Became President of firm in 1964 and is responsible for Administrative duties as well as mgt of Electrical, Mechanical and Industrial Divisions of firm. Served for 10 yrs on Calif Bd of Reg for Prof Engrs, President in 1971. Recd Calif Council of Civil Engrs Public Serv Awd 1972. Named in Who's Who in West. Enjoys skiing, hunting and fishing. *Society Aff:* ASCE, NSPE, ACEC.

Wirges, Manford F
Business: P O Box 300, Tulsa, OK 74102
Position: VP Res. *Employer:* Cities Service Co. *Education:* BS/ChE/Univ of OK; MS/ChE/Univ of OK. *Born:* Jan 18,1925 Beatrice Nebraska. BS, MSChE, Univ of Okla. Married Joan Kelly; one child, Ms. Kelly Wirges. Pres: Cities Service Res & Dev Co. Mbr. AIChE, API, IMS, Twenty-five Yr Club of the Petroleum Industry, Adv Bd College of Engrg & Physical Sci, U of Tulsa; Adv Bd, St. John's Hospital & School of Nursing, Wall Summit Club Tulsa, Bd Mbr Frontier Science Fdtn. Hobbies: Hunting and fishing. Resience: 3208 E 69th St, Tulsa, OK. Mbr on the Fnd for Excellence Bd, Sch of Chem Engg, The Univ of OK;. *Society Aff:* AIChE, API, IMS.

Wirsching, Joseph E
Business: 1200 First City Tower II, Corpus Christi, TX 78478-0601
Position: Opers Mgr *Employer:* HNG Oil Co. *Education:* BS/CE/TX A&M Univ *Born:* 11/15/26 Raised in Tulsa, OK. Employed by Humble Oil and Refining Co (now Exxon Co USA) upon graduation from TX A & M in 1948. Worked in TX and LA for 31 yrs in various engrg and operations management assignments. Served as VP Operations for Canus, Petroleum Inc for 18 months until they sold out. Now employed by HNG Oil Co as Dir Operations Manager. Registered professional engr in TX and LA. Served as Chmn SPE local sections in Midland, TX and New Orleans and as SPE Dir. Served as AIME Dir. *Society Aff:* SPE of AIME

Wirtz, Gerald P
Business: 204 Ceramics Bldg, 105 S. Goodwin Ave, Urbana, IL 61801
Position: Assoc Prof *Employer:* Univ of IL *Education:* PhD/Mat Sci/Northwestern Univ; B/ME/Marquette Univ; BS/Eng/St. Norbert Coll *Born:* 12/22/37 Since 1968. G.P. Wirtz has directed sponsored resch on electronic, catalytic, and electrochemical properties of oxides. From 1966-68 he developed materials for hybrid microelectronics in industry. Internationally, he was an OAS sponsored visiting prof, Instituto Militar de Engenharia, Rio De Janeiro, 1977; Fulbright-Hays Lecturer, Universidade De Aveiro, Portugal 1980; and co- ordinator of NATO sponsored cooperative resch program, Universidade De Aveiro 1981-83. He has taught courses in: Electrical Ceramics, Crystallography and X- Ray Diffraction, Crystal Physics, and Solid Electrolytes. He has consulted on oxide catylysts, high temperature fuel cells, and anodes for cathodic protection of metals. *Society Aff:* ACS

Wirtz, Karl
Business: Kernforschungszentrum Karlsruhe, P O Box 3640, Karlsruhe Germany 7500
Position: Dir, Emeritus Prof. *Employer:* Univ of Karlsruhe. *Education:* PhD/Physics/Univ of Breslau; Habilitation/Physics/Univ of Berlin; Prof/Physics/Univ of Gottingen. *Born:* 4/24/10. Joined Max-Planck Inst for Physics 1937. Participated in early German wk on nuclear energy as mbr of Prof Heisenberg's team since 1940, 1944 Hd Exp Div, MPI for Physics, 1946-57 Prof of Physics Univ of Goettingen, participated 1957 in founding Karlsruhe Nuclear Res Ctr & became full prof. Directed the conceptional design and all physics wk for the construction of the research reactor FR 2. Was instrumental for initiating the German fast breeder project 1960. Author. Fellow ANS. Mbr German Atomic Energy Comm 1956-71. Mbr Bd of German Atomic Forum. *Society Aff:* KTG, ENS, MPG, DAtF, ANS.

Wirtz, Richard A
Business: MIE Dept, Potsdam, NY 13676
Position: Assoc Prof. *Employer:* Clarkson College. *Education:* PhD/ME/Rutgers Univ; MS/ME/Rutgers Univ; BS/ME/Newark College of Engg. *Born:* 8/16/44. A native of NJ, has lived in Potsdam, NY area since joining Clarkson as Asst Prof in 1970; promoted to Assoc Prof and appointed Exec Officer of ME Dept in 1976; as such is responsible for direct supervision of Undergrad ME Prog. Res interests are in free convective heat transfer. Hobbies include auto racing under sanction of Sports Car Club of Am and Can Auto Sports Club and skiing. *Society Aff:* ASME, APS, ΣΞ

Wischmeyer, Carl R
Business: Crawfords Corner Rd, Homdel, NJ 07733
Position: Dir of Education *Employer:* Bell Laboratories *Education:* DSc/Hon/Rose-Hulman Inst of Tech; EE/Rose-Hulman Inst of Tech; M/EE/Yale Univ; BS/EE/Rose-Hulman Inst of Tech *Born:* 10/2/16 Has corporate responsibility for tech, administrative, and management education of Bell Labs employees. He formerly served as professor of electrical engrg, master of Baker Coll, and dir of continuing studies at Rice Univ, mbr of tech staff at Bell Laboratories, research associate at MIT. Radiation Laboratory, resident visitor at Technische Hogeschool te Eindhover, and consultant to industry. He is a licensed professional engr, author of

Wischmeyer, Carl R (Continued)
forty articles on circuits, superconductors, and education, and holds eight paptents in geophysics. He has served on the boards of IEEE, ECPD, ABET, and Rose-Hulman Inst of Tech. He was recently appointed to the NJ St Panel of Science Advisors. *Society Aff:* IEEE, ASEE, NSPE, APS, ASTD

Wise, Carlton T
Home: 200 Aberdeen Dr, Greenville, SC 29605
Position: Exec VP (Ret). *Employer:* J E Sirrine Co. *Education:* BS/CE/Clemson Univ. *Born:* 1906 Newport RI. Structural detailer with Bethlehem Steel Co 1929-32. With J E Sirrine Co since 1933 in structural design and later in environmental and gen civil engrg. VP & Dir since 1959. Retired as Exec VP in 1973. Mbr of SC Bd of Eng Reg since 1967. PPres and Engr of Yr of SC Sec ASCE and SC Sec NSPE. VP NCEE 1976-77. ASCE Greensfelder Prize and NCEE Dist Serv Awd. P Mbr Advisory Bd, Clemson Univ, CE Dept. Enjoy woodworking, drawing and painting, tennis. *Society Aff:* ASCE, NSPE, ACEC, NCEE.

Wise, William F
Business: Route 1, Box 567, Weyers Cave, VA 24486
Position: Owner *Education:* BS/CE/VA Polytech Inst & State Univ *Born:* 12/17/39 Worked in the consulting field for approximately 20 yrs. Was the first engr in VA to obtain approval on a treatment process which produced drinking water from sewerage. Developed and assisted a corp to obtain a patent on a waste treatment device which is thought to be capable of producing a quality effluent from properly aerated open bodies of water. Provided structural designs for numerous economical farm buildings, and cost effective public buildings. Assisted communities in obtaining new utility services through use of innovative and cost effective processes and/or systems. *Society Aff:* ASCE, NSPE, WPCF, PEPP

Wise, William H
Business: 130 W. Adelaide St W, Suite 2900, Toronto, Canada M5H 3P5
Position: VP Planning & Technology. *Employer:* TIMMINCO. *Education:* BS/Metallurgy/Univ of KY. *Born:* Nov 1921. Native of Kentucky. During WWII served in the Army Air Corps from Sgt to Capt in bomber group 1940-45. BS Met Eng from Univ of KY; chosen outstanding senior; elected to Tau Beta Pi 1949. From trainee to proj engr for Union Carbide Corp 1950-55. Production Supt for Pittsburgh Metallurgical Co 1960-61. Plant Mgr at Chromium Mining & Smelting Co 1962-63. From Asst to VP of Oper to VP of Metallurgy and Technology 1964-76. Dir of Metallurgical Soc 1972- 74. President of ISS and Dir of AIME 1974. Hon Mbr ISI of Japan 1975. VP AIME 1976. Pres AIME 1979, VP Planning & Technology, TIMMINCO 1979. Have presented tech papers in the US and France on ferro-alloy operations & furnace design. *Society Aff:* AIME.

Wisepart, Ivor S
Home: 80 McCulloch Dr, Dix Hills, NY 11746
Position: VP & Genl Mgr *Employer:* Computran Systems Corp *Education:* MS/Trans Engr/Polytech Inst of NY; BS/EE/City Coll of City Univ of NY; AB/Math & Lib Arts/Columbia Coll of Columbia Univ *Born:* 4/28/37 in London, England. Emigrated to US in 1947. Employed by Airborne Instuments Laboratory (AIL) Division of Cutler-Hammer Inc in 1960. Performing operations research studies for defense and nondefense applications. Involved in advancing commercial aviation operations and highway research. Joined Computran in 1972. Responsible for management of Systems Studies, Designs, Implementations, and Consulting Activities, related to applications including computers, Telecommunications, Software, and Custom Interfaces. Pres, ITE Met Section, 1981. Chmn, TRB Communications Ctte 1980-1982. Exec Ctte of IEEE Symposium on Alternate Energy Sources. Enjoy tennis and sailing. *Society Aff:* TRB/NRC/NAE, IEEE, ITE, AMA

Wisniewski, John S
Home: 48 War Trophy Ln, Media, PA 19063
Position: Engg Specialist, Mass Properties. *Employer:* Boeing Vertol Co. *Education:* BSME/Mech/SUNY. *Born:* 9/12/29. A 1952 mech engg grad from the State Univ of NY, has over 23 yrs of experience, career specialist employment with three major aircraft mfgs; Boeing Vertol, Bell Buffalo, Douglas-Santa Monica. Experienced in all aspects of mass properties engg. Currently an engg specialist in the preliminary design & res and dev at Boeing Vertol, Phila, PA. Have been an active mbr of the SAWE for 24 yrs. Installed as a Fellow in Munich, Germany in 1977. Held all elected SAWE intl offices: Intl Pres 1979-80; Exec VP 1978-79; Sr VP 1977-78, and was Intl Conf Chrmn in Phila in 1976. *Society Aff:* SAWE, AHS.

Wiss, John F
Business: 330 Pfingsten Rd, Northbrook, IL 60062
Position: Chairman of Board *Employer:* Wiss, Janney, Elstner & Assoc. *Education:* MS/Structures/Northwestern Univ; BS/Civil Engr/Northwestern Univ. *Born:* April 27,1919. Active duty USNR 1940-45. Specializing in structural vibrations, soil dynamics, seismic, and acoustical problems. Research in human perception of transient vibrations, dynamics of structural systems, and strength of masonry walls to explosions (conflagration). Co-founder of present consulting and res engrg firm 1956. ASCE Raymond C Reese Awd 1971 for outstanding contributions to the application of structural engrg research. *Society Aff:* ASCE, SSA, ASA, IES, AGU, SEE

Wissoker, Richard S
Business: 100 Endicott St, Danvers, MA 01923
Position: Mgr-Mkting Services & Planning - Ltg. *Employer:* GTE Intl Inc. *Education:* MBA/Mkting/NY Univ; BS/Elec Engg/Case Inst of Tech. *Born:* Oct 24, 1929 New Rochelle NY. Served in USAF. NY area Sales Engr Day-Brite Lighting 1953-60. Transferred to St Louis 1960 as Sr Applications Engr & subsequently Mgr of Systems Application. Joined GTE Sylvania 1969 as Prod Sales Mgr-Indoor Ltg Equip. Later assigned Prod Mgr-Fluorescent Lamps, Market Application Mgr, and Applications Proj Mgr. Subsequently Prod Mktg Mgr-Ltg for GTE Intl, Mgr-Mkt Planning & currently Mgr-Mkting Services & Planning-Ltg. Reg PE. Fellow IES and VP Operational Activities 1973-75, Dir 1977-80. VP Regional Activities 1980-82. Author and Lecturer on lighting application. *Society Aff:* IES, NSPE.

Witchey, Robert H
Business: 2875 Lincoln St, Muskegon, MI 49441
Position: Mgr Air Melt & Aod Tech *Employer:* Cannon Muskegon *Education:* MBA/Bus Prod & Finance/St Francis College; MS/Bus Adm/St Francis College; BS/Chem Met/Juniata College; -/Met Eng/IL Inst Tech. *Born:* 4/20/29. Native of Altoona, PA. Served in Army Chem Corps 1951-53. Physical Metallurgist U S Steel Corp specializing in stainless and alloy development. With Joslyn since 1958 as Mgr Tech Serv. Became Chf Metallurgist 1974. Spec training in non-destructive testing and quality control. Mbr ASM, ASNT, ASQC, ASTM, ASME, PChap Chrmn ASM. PE Calif. Publications in Iron Age, Metal Progress and Automatic Machining on machining stainless steel. Part time teacher in metals at IVY Tech Coll and Fort Wayne Vocational School. Cannon-Muskegon-Mgr Airmelt and AOD Tech as of 1-15-81. Duties include responsibility for induction melting, AOD Refining and Concasting Tech. *Society Aff:* ASM, ASNT, ASTM, ASME, ASQC.

Witmer, Bill B
Business: Box 1447, Sioux Falls, SD 57101
Position: Livestock Env Spec. *Employer:* G T A Feeds. *Born:* March 1925 Lincoln Nebraska. BS Univ of Nebr 1949. Taught vocational agriculture Palmyra Nebraska Pub Schools 1950-60. Agricultural engr for Eastern Nebraska Pub Pwr Dist 1960-68. GTA Feeds 1968-present. Specialize in design of ventilation systems for livestock confinement bldgs. Hold 2 US patents for equipment for feeding liquid supplement to livestock. Chrmn No Central Reg ASAE 1974-75. Chrmm AFMA Environmental Concerns Ctte 1975-76. Enjoy hunting, fishing and woodcarving.

Witt, John D
Home: 523 Robin Ave, Frankfort, KY 40601
Position: Commissioner, Bur Nat Res. *Employer:* Dept Nat Res & Env Prot. *Born:*

Witt, John D (Continued)
Nov 1928. Graduated from Univ of KY 1950. Air Force 1951-55. Worked for Dept of Transp, serving in Constructn, Right of Way, Design, Program Mgt as Engrg Asst to Commissioner, Asst Dir of Civ of Rural Rds, Dist Engr Flemingsburg, and Asst State Hwy Engr for Rural and Municipal Aid until Jan 1976. After more than 26 yrs, left to accept position of Director, Div of Reclamation, Dept for Natural Resources and Environmental Protection. On March 1,1976 assumed current responsibility as Commissioner of Bureau of Naval Resources. Mbr of Kentucky and Natl Soc Prof Engrs.

Witt, Victor R
Home: 14780 Sky Lane, Los Gatos, CA 95030
Position: IBM Fellow. *Employer:* IBM Corp. *Education:* BSEE/Electronics/NY Univ. *Born:* 5/17/20. 95193/IBM Corp/5600 Cottle Rd/San Jose, CA. *Society Aff:* IEEE.

Witte, Charles R
Business: 900 Liberty Bank Bldg, Main at Ct Sts, Buffalo, NY 14202
Position: Gen Mgr. *Employer:* Acres American Inc. *Education:* BS/Mech/State Univ of NY. *Born:* 1/7/26. Native of Buffalo, NY. Served with Air Force during WWII. Began technical career at August Feine and Sons Co, structural steel fabrication, in 1946 as Draftsman. Became Chief Engr in 1954. Joined Carborundum Co, Niagara Falls, NY in 1964 as Sr Proj Engr. Was proj mgr responsible for design and construction of the first Geodesic Dome (300 ft diameter) used in an industrial complex. With Acres American since 1974. Assumed responsibility as Gen Mgr in 1977. Acres American Buffalo provides engg services to industry and power utilities with a 200 person staff. *Society Aff:* NSPE, NYSPE, ESB.

Witte, Larry C
Business: Dept of Mech Engg, Houston, TX 77004
Position: Prof of Mech Engr. *Employer:* Univ of Houston. *Education:* PhD/ME/OK State Univ; MS/ME/OK State Univ; BS/ME/Univ TX. *Born:* 4/27/39. Native of Hamilton County, TX; Mech Engr for LTV-Astro Div, 1962-63; Res Engr with Argonne Natl Lab, 1965-67, specializing in convection heat transfer and boiling, and indus energy conservation. Joined the Univ of Houston in 1967, Served as Dept Chrmn of Mech Engg at U-Houston, 1972-76; Has researched and published extensively in the fields of explosive vapor formation, boiling and convective heat transfer, and energy related fields; Also has served as consultant in the Thermal Sciences for numerous ind firms. *Society Aff:* ASME, ANS, AIAA, ASHRAE.

Wittes, David R
Business: 121 N 18th St, Philadelphia, PA 19103
Position: Consulting Struc Engr. *Employer:* David R Wittes-Cons Struc Engrs. *Education:* BCE/Civil Engg/Cray College of NY. *Born:* 8/20/32. BCE CCNY. His pioneering experience in concrete flat plate for the Schuykill Housing high-rise proj brought him to Philadelphia as structural consultant to architects, builders & govt agencies. Since 1960 continued to develop structural systems in steel & concrete particularly for multi -story bldgs. Originated a unique threaded space from which recd Lincoln Modern Welding Awd. For Philadelphia Bicentennial, developed the unusual concept of floating cities on water. His hi-rise Casa Fermi Apts, Fire Island Naval Base, and the Australian Griffin Meml are among his projs that have recd ACI & AIA dist bldg commendations. Reg PE. Mbr ASCE, ACI, CEC. *Society Aff:* ASCE, ACI, NSPE.

Wittry, John P
Business: Vice Dean of Faculty, USAF Academy, CO 80840
Position: Colonel-Perm Prof & Vice Dean of Faculty. *Employer:* U S Air Force Academy. *Education:* Prof Engr/Aero & Astro Engg/Univ of MI; MS/Aero Engg/AF Inst of Technology; BS/Aero Engg/St Louis Univ. *Born:* 9/6/29. in Aurora, IL. Regular AF Off 1951- . Prof exper primarily in res & dev proj mgmt - rotary wing & exper fixed wing aircraft, air-breathing, rocket nuclear propulsion, nuclear space power sys. 1962- on faculty USAF Acad-engg mechs, astronaut engg, computer sci. Hd Dept Astronautics & Computer Sci 1972- 78. Apptd to present position as Vice Dean of faculty 1978. Hobbies: sailing, golf. *Society Aff:* ASEE, AIAA.

Witz, Richard L
Home: 1525 N 8th St, Fargo, ND 58102
Position: Professor Emeritus *Employer:* North Dakota State Univ. (Retired) *Education:* BS/Agri Eng/Univ of WI; MS/Agri Eng/Purdue Univ. *Born:* Jan 1916 New Lisbon Wisc. Extension Specialist, Michigan State 1942-45, Research, bldg design, and teaching NDSU since 1945. Sr Mbr and Fellow ASAE. Wood Badge Scoutmaster, Silver Antelope, Council Board, Natl Cttee for Boy Scots of America. Hobbies are scouting, camping, photography and electronics. Reg Prof Engr - North Dakota. *Society Aff:* ASAE, CSAE, ΣΞ, AE, KIWANIS.

Witzel, Homer D
Home: 1700 Gulf Boulevard, Englewood, FL 33533
Position: Consulting Engineer *Employer:* Deere & Co. *Education:* BS/Agri/Univ of Wis; BS/Mech Engg/Univ of Wis. *Born:* Feb 24,1916. Grad Univ of Wisc with BS in Agriculture 1939 and BS in Mech Eng 1940. Has held positions of design, project, div and chief engr and Dir of Implement Engrg, Europe prior to becoming Gen Mgr of a combine harvester factory in Germany in 1964. In 1969, held the corporate staff position of Director, Prod Plng for the largest agricultural machinery mfg co in the world. Mbr of ASAE since 1940 and a Fellow since 1974. Mbr SAE since 1969. Reg PE in Mech Eng since 1948. Awarded 15 US patents. I retired from the position with Deere & Co of Director of Product Planning in Feb 1978. Since then I have created several inventions for special clients while doing Consulting Engineering work for them. *Society Aff:* SAE, ASAE.

Witzel, Stanley A
Home: 213 N 40th St, McAllen, TX 78501
Position: Professor Emeritus. *Employer:* University of Wisconsin. *Education:* BS/CE/IA State Univ; MS/CE/Univ of WI, Madison. *Born:* Nov 1904. Reg PE Wisc. Farm oriented native of Iowa. Early employment - Union Pacific RR, Amer Rolling Mills, Western Electric, and instructor at Texas A&M Univ. In 1930 joined staff of Agricultural Engrg Dept, Univ of Wisc-Madison, retiring in 1973 From 1969-72 established and headed an Agricultural Eng Dept at the new Univ of Ife, Ile-Ife, Nigeria, currently operate a citrus ranch in Texas. Assns: ASAE, ASCE & NSPE. 1972 Cyrus Hall McCormick Gold Medal. 1973 Academy of Achievement Gold Plate Award. 1974 Wisc ASAE Engr of the Yr Awd. Listed in Wisc Men of Achievement. *Society Aff:* ASCE, ASAE, ASPE.

Witzig, Warren F
Home: 1330 Park Hills Ave. E, State College, PA 16803
Position: Prof & Dept Hd. Emeritus *Employer:* Pennsylvania State University. *Education:* PhD/Physics/Univ of Pgh; MS/Elec Engrg/Univ of Pgh; BS/Elec Engrg/RPI *Born:* March 26,1921 Detroit. s.Arthur Judson and Mary (Bender)W. m. Bernadette Sullivan 1942. c.Eric,Leah,Marc, Lisa. Res Engr Westinghouse Res 1942-48. Mgr reactor physics engr Bettis Atomic 1948-60. Co-founder, sr U P Dir NUS Corp 1960-67. Hd, Dept Nuclear Enrg Penn State U 1967-87. Consulting nuclear engr utilities and industry. Chrmn, Gov PA Com. Atomic Energy Dev 1970--. Mbr Bd Mgt YMCA 1955- . Fellow ANS; Consultant to nuclear industry; Reg PE; Presbyn. (elder). Designer S5W submarine reactor 1956-60. Mbr Natl Acad Engg Rad Waste Comm, Serve on Bd of Sevl small nuclear corps, res areas: Nuc Fuel Mgt, utilization of reject heat, reactor safety & siting, tech transfer. Mbr Nuclear Oversight Ctte to Bd of Dirs of PSE&G. Mbr GPUN Bd of Dirs and Nuclear Safety and Compliance Ctte. Mbr Nuclear Operation Ctte, TX Utilities. Mbr Nuc Saf Rev Bd, TVA Watts Bar. *Who's Who; 1972--, Who's Who in Science.* *Society Aff:* ANS, IEEE, APS, AAAS, ΣΠΣ, ΣΞ, HKN

Wlodek, Stanley T
Business: M-87, Bldg 500, General Electric Co, Cincinnati, OH 45215
Position: Engineer, Special Studies *Employer:* AEBG *Education:* BSc/Metall Eng/

Wlodek, Stanley T (Continued)
Queen's Univ; SM/Physical Metall/MA Inst of Tech; ScD/Physical Metall/MA Inst of Tech; PMD/Bus Mgmt/Harvard Bus Sch. *Born:* 9/23/30. in Haiduki, Poland. BSc Queen's Univ, Kingston, Ontario 1952. SM 1954, ScD 1956 MIT. Sec Ldr, Columbium Alloy Sec, Union Carbide Metals Cp Res Labs 1956- 60; Mgr Environ Effects Unit, GE Co, Adv Engg & Tech Dept 1960-65. Research Supervisor, corrosion & coatings Graham Res. Lab J & L Steel Corp (1965-69). Dir R&D Stellite Div Cabot Corp, Cabot Guest Prof at Notre Dame (1969-79). Mgr Process Dev. Matls & Processes Lab LSTG. General Electric Co (1979-85). Currently Engineer Special Studies AEBG General Electric Co., Cincinnati, OH 45215, Main area of experience: physical metallurgy and processing of high performance alloys. Served as Chrmn 1969-71 E & M Div, Electrochemical Soc, and Chrmn, AIME High Temp Alloy Cttee 1975, Genl Chrmn Seven Springs Intl Symposium Committee (1976- 80). Fellow ASM. *Society Aff:* ASM, AIME.

Woelfel, Albert E
Home: 1006 Prince St, Houston, TX 77008
Position: Consultant *Employer:* Self *Education:* MME/ME/Rice University; BSME/ME/Rice Institute *Born:* 12/17/24 1986 - Present Consulting Engineer; 1984 - 86 Koomey, Inc. Director of R & D; 1957 - 84 Cameron Iron Works Research Engr, Chief Engr, Gate Valve Products, Senior Project Development Specialist. 1987 - Present Member, Board of Governors of ASME; 1981 - 83 VP, Region X of ASME. *Society Aff:* ASME

Woerner, Leo G
Home: 1301 Woodshole Rd, Towson, MD 21204
Position: President. *Employer:* CPW Enger Consultants & Designers. *Education:* MS/Elec Engg/Univ of IL; BS/Engg/Swarthmore College. *Born:* 5/21/27. Sigma Xi, Sigma Tau. R&D at Bartol Res Fnd of Franklin Inst, Burroughs Corp Res Lab & Intl Resistance Co. ASTM & MIL Spec Cttes on resistors and wire. Patents on magnetic memory & electronic components. Schick Elec Inc Design Engr 1957; Chf Engr 1961; Dir Engg & Res 1962-65. Patents on elec shaver components and plastic housings. Divisional Mgr of Engg AMP, Inc 1965-67. Dir of Engg Head Ski Co 1967-70 VP Palmer Clark & Woerner Inc 1970-76. Pres of CPW since 1976. Mbr IEEE, SRSNA, ACEC, RIA, AISC, NFPA, SME. Reg PE, MD, MI, AL, & FL. *Society Aff:* IEEE, ACEC, SME, RIA, AAAS, SRSNA, AISC, NFPA

Woglom, James R
Home: Nine Ledgetree Rd, Medfield, MA 02052
Position: Sr VP *Employer:* Camp Dresser & McKee Inc. *Education:* MA/Pol Sci/Comm Plng/Lehigh Univ; AB/Econ/Lafayette Coll; BS/Admin Eng/Lafayette Coll *Born:* 5/17/26 Served with AUS in ETO during WW II. With Morris Knowles Inc - Easton, PA as planner-engr-surveyor 1950-61. Dir-VP Community Planning & Renewal Metcalf & Eddy - Boston, MA 1962-75 with responsibility for creation and development of a dept-div. With Camp Dresser & McKee 1975-81 as VP, Sr VP and Pres of Environmental Planning Div with responsibility for Creation & Development of Div throughout the US. NecplusPE in 46 states. Assoc Mbr Bd of Trustees Lafayette Coll. Mbr several Bds of Ctte of professional societies. *Society Aff:* ASCE, NSPE, SAME, ITE, AICP

Wohlgemuth, Robert E
Business: PO Box 2001 (909 Fannin), Houston, TX 77001
Position: VP - Northern Region. *Employer:* Gulf Oil Co - US. *Education:* BS/CE/Univ of OK. *Born:* 7/30/30. Native of Pampa, TX; grew up in Collinsville, OK. Grad of the Univ of OK. Began career with Gulf Oil Corp in 1952 at Gulf's Port Arthur Refinery. Have held positions at Gulf's offices in Houston, Pittsburgh, London, Atlanta & served as Refinery Mgt at Gulf's Santa fe Srings, CA Refinery. In addition, was on assignment to US Dept of Interior, Office of Oil & Gas, for one yr. Assumed current position in Houston as VP - Northern Region, in Sept 1978. Enjoy hunting, road racing & photography. *Society Aff:* AIChE.

Wojnar, Theodore J
Business: Commander 13th Coast Guard District, 915 Second Ave, Seattle, WA 98174-1067
Position: Commander Thirteenth Coast Guard District *Employer:* U S Coast Guard. *Education:* BSMS/Naval Engrg/USCG Acad; BSCE/Civil Engrg/Rensselaer Polytechnic Inst *Born:* 10/21/30 Civil Engr; US Coast Guard Officer; Tau Beta Pi. Native of Holyoke, MA Served as Engr Officer aboard Icebreaker 'Eastwind' & with Intl Ice Patrol. Field Engr for Coast Guard's Loran-C Navigation System in Pacific and European Theaters. Advanced through grades to Captain 1974. Pub Wks Officer, Training Ctr, Groton, CT 1964-67. Chf Engr Div Europe, London 1967-70. Chf, Civil Engr Br 3rd Dist NYC 1970-73. Chf Civ Engr USCG Wash DC 1975-78. Chief Operations So CA 1978-1980. Chief of Staff 5th Dist 1980-83, Chief Office of Navigation 1983-1986, Commander 13th District 1986- . Decorated CG Achievement Medal, Commendation Medal with 2 Stars. Meritorious Service Medal with 2 stars, Legion of Merit. Mbr SAME, Natl Dir 1973-78. Pres Wash DC Post 1975-76. USCG Acad & RPI Alumni Assn. Propeller Club, Amer Assoc of Port Authorities, Internatl Assoc of Lighthouse Authorities. Chief, Seattle Ch, BSA. *Society Aff:* SAME, IALA, BSA

Wolbarsht, Myron L
Business: Dept of Psychology, Durham, NC 27706
Position: Prof of Ophthalmology & Biomedical Engrg. *Employer:* Duke Univ *Education:* PhD/Biology-Biophysics/Johns Hopkins; AB/Lib Arts/St John's Coll (Annapolis); Chem Engrg/Johns Hopkins; Chem Engrg/Univ of MD *Born:* 9/18/24. Baltimore, MD. Served with the US Army Air Forces (1942-1945) working on bombsights and automatic flight control equipment. Head of Biophysics Dept, National Naval Medical Center until 1968. Conslt to various companies (including: RCA, IBM, NCR, TRW, Xerox) on laser safety and optical hazard analysis. Major interests: laser safety, biomedical engrg applied to ophthalmology in: diabetes & glaucoma, laser surgery, elec tech in diagnosis, and information flow in the visual system. Pres, Laser Inst of America (LIA) 1982-1983, Chrmn, Bioeffects subctte of ANSI Z-136 Ctte on Laser Safety 1967-present. Co-chairman Measurements Sub ctte of ANSI Z-311 Ctte on Lamp Safety, 1980-present. American delegate to Intl Electrotechnical Commission, Tech Ctte 76 Laser Safety (IEC-TC 76). Co-Editor *Lasers in Medicine & Surgery*. Associate editor IEEE, Transactions on Biomedical Engrg, 1962-1969. Exec board of the Vision Ctte, Nat Res Council, 1971-1977. Enjoys fishing and home workshop. *Society Aff:* LIA, IEEE, APS, OSA, ARVO, RSM, SGP, ASLMS.

Wolbrink, Jim F
Business: 25 Technology Park/Altanta, Norcross, GA 30092
Position: Mng Dir Ed. & Pubs. *Employer:* IIE *Born:* Sept 1942. BS Iowa State Univ, grad student ISU. Native Sheldon Iowa. Tech writer/editor for Lawrence Radiation Lab, Livermore CA. Hd of Engrg Pubs and Editor of Engrg Res Inst. Instructor, Iowa State Univ. Joined IIE in 1971 as Editor-in-Chief. Also resp for public relations, Continuing Education. Published in several journals. Frequent speaker. *Society Aff:* IIE.

Wold, Peter B
Business: 915 East Eleventh, Bottineau, ND 58318
Position: Pres *Employer:* Wold Engrg P.C. *Education:* MS/CE/Univ of ND; BS/CE/ND St Univ; AS/Engr/ND St School of Forestry *Born:* 11/19/26 Native of ND and MN. Served in the US Navy in WW 2 and was a Captain in the ND Natl Guard for 12 yrs. Served as field Engr for construction co. Spent seven yrs as a Engrg Coll Instructor before starting Wold Engrg P.C., a firm that specialized in CE work. Have spent many active yrs in community affairs and in politics having spent four yrs as ND Republican Natl Committeeman. Main interests lay in sports, fishing and big game hunting. *Society Aff:* ASCE, NSPE, WPCF

Wolf, Donald J,
P.O. Box 552, Frederick, Maryland 21701-0552
Position: Pres *Employer:* Wood Arts, Inc *Education:* BS/Wood Tech/NC St Univ *Born:* 2/28/25 Registered PE & Wood Technologist. Native of Frederick, MD. Area of specialization restricted to plant layout, product development, product design, process selection, site selection, methods studies and industrial engrg functions for the furniture and wood products industries. Have held positions of chief engr, plant mgr and operations mgr in furniture industry. Past fifteen years have served as consultant to over 100 companies in the US, Canada, and 10 other foreign countries. *Society Aff:* NSPE, PEPP, SWST, FPRS, IIE.

Wolf, Paul R
Home: 4552 McCann Rd, Madison, WI 53714
Position: Professor. *Employer:* University of Wisconsin. *Education:* BS/Civil Engg/Univ of WI, Madison; MS/Civil Engg/Univ of WI, Madison; PhD/Civil Engg/Univ of WI, Madison. *Born:* June 1934. BS, MS and PhD from Univ of Wisc, Madison. Native of Mazomanie Wisc. Highway Engr, Wisc Dept Transp 1960-63. Instructor, Univ of Wisc 1963-67. Asst Prof Univ of Calif Berkeley 1967-70. Assoc Prof and Prof Univ of Wisc Madison 1970 to present. Author of 'Elements of Photogrammetry' and 'Elementary Surveying'. Mbr and Natl Dir of ASP. Mbr of ASCE, ACSM, CIS, Sigma Xi and Chi Epsilon. Reg CE and Reg LS in Wisc. Recd Abrams III Award and 3 presidential citations from ASP, and Fennell Distinguished Teaching Award from ACSM. Hobbies include hunting and fishing. *Society Aff:* ASP, ACSM, ASCE, WSLS.

Wolf, Robert V
Home: 1504 Scenic Dr, Rolla, MO 65401
Position: Prof, Met Engg. *Employer:* Univ of MO. *Education:* MS/Mech Engr/MO Sch of Mines; BS/Mech Engr/MO Sch of Mines. *Born:* 6/5/29. Native of St Louis, MO. On present faculty since 1952. Military leave, 1954-56. Active in committee work and student advisement. Acting dept chrmn, 1970-71. Assistant Dean, School of Mines and Metallurgy, 1981. Teaching and res interests are metal casting, metals processing, nondestructive testing. Partner in Askeland, Kissinger & Wolf, Consultants. Elected natl office in Pi Kappa Alpha fraternity, Alpha Sigma Mu Mtls Honorary, and MSM-UMR Alumni Assn. Outstanding teacher awards, Advisor of the Yr Award, AFS (Natl) Service Citation, Pi Kappa Alpha Outstanding Chapter Advisor. Hobbies are fishing and metal sculpture. *Society Aff:* AFS, AIME, ASM, AWS, ASNT, ASEE.

Wolf, Werner P
Business: Council of Engrg, P.O. Box 2157, New Haven, CT 06520
Position: Prof of Phys & Appl Sci; Chrmn, Council *Employer:* Yale Univ. *Education:* DPhil/Phys/Oxford Univ; MA/Phys/Oxford Univ; MA/-/Yale Univ; BA/Phys/Oxford Univ. *Born:* 4/22/30. Res fellow, Oxford Univ 1954-56; Res fellow, Harvard Univ 1956-57; Fulbright travelling fellow 1956-57; Imperical Chem Ind res fellow, Oxford Univ 1957-59; Univ demonstrator, lecturer 1959-63, lecturer New College 1957-63; Faculty, Yale Univ 1963-present, prof of physics and appl sci 1965-present, dir of grad studies, dept of Engg & Appl Sci 1973-76, Becton Prof & Chrmn, Dept of Engg and Appl Sci 1976-81; Chrmn, Council of Engrg 1981- ; Consultant to ind 1957-present, visiting prof at Technische Hochschule, Munich, Germany 1969; Mbr of prog committee for conf on Magnetism & Magnetic Mtls 1963 & 1965, chrmn 1968; Mbr of advisory committee 1964-65 & 1970-76, steering committee 1970-71; Conf gen chrmn 1971; Mbr of organizing committee for Intl Congress gen chrmn 1971; Mbr of organizing committee for Intl Congress on Magnetism 1967, visiting physicist Brookhaven Nat Lab 1966 & 1968, visiting Sr physical 1970, res collaborator 1972, 1974, 1975 & 1977; Mbr of Ed Ctte (APS), dir of Industrial Graduate Intern Prog 1979. Educator, solid state physics; res in magnetism and microwave resonance, new matls. Hobbies incl travel, skiing, and music. *Society Aff:* APS, ΣΞ, IEEE, AACG.

Wolfberg, Stanley T,
Business: 118-Stoney Creek Rd, Santa Cruz, CA 95060
Position: Principal *Employer:* Stan Wolfberg, P.E. *Education:* MBA/Bus Admin/Stanford Univ; BSME/Mech Engg/Caltech. *Born:* 4/10/17. Reg PE/Ind Engg. Industrial Engr, US Steel thru 1946; Gen Mgr, Retail Clothing Chain thru 1955; Mgt Consultant to present: Prin-Borchardt & Assoc; Prin Cresap, McCormick and Paget; and currently VP Kapner, Wolfberg & Assoc and Principal, Stan Wolfberg, PE. Conslt to Industry, Commerce, Govt & non-prof Insts in areas of Mgmt Operations & Human Resources. Also, 1971-78, Div Administrator at Caltech respon for coordinating and implementing acad, res and administrative functions. Former Natl VP Education & Prof Dev IIE. Recipient IIE Distinguished Service Award and IIE Fellow Award. Fellow, Inst for Adv of Engr Pres--California Legislative Council for Prof Engrs. Mbr teaching faculty, Ind and Sys Engr, Univ of S CA Dir, Productivity Ctr of SW. Mbr Industrial Advisory Council, Ind & Mfg Eng, Cal Poly Univ, Pomona. Active in Safe Boating education as mbr of US Coast Guard Auxiliary. *Society Aff:* IIE, MTM, ASEE, IAE.

Wolfe, Charles M
Business: Washington University, Box 1127, St. Louis, MO 63130
Position: Prof of EE, Dir of Semicond Res Lab *Employer:* WA Univ. *Education:* PhD/Solid State Electronics/Univ of IL; MSEE/EE/WV Univ; BSEE/EE/WV Univ. *Born:* 12/21/35. Native of Fairmont, WV. Served with US Marine Corps, 1955-58. Staff Mbr at MIT Lincoln Lab, 1965-75. Prof of EE at WA Univ in St Louis, 1975-82. Dir of Semiconductor Res lab at WA Univ, 1979-present, Samuel C. Sachs Prof of EE, 1982-present. Res interests in compound semiconductor mtls, devices, & integrated circuits. Elected Fellow of Elec & Electronic Engrs, 1978. Received Electronics Div Award of Electrochem Soc, 1978. *Society Aff:* IEEE, ECS, APS, AAAS.

Wolfe, John K
Business: School of Engineering, Rensselaer Polytechnic Inst, Troy, NY 12181
Position: Prof., School of Engrg *Employer:* Rensselaer Polytechnic Inst *Education:* PhD/Chem/Univ of MD; BS/Chem/Iniv of MD. *Born:* Aug 1914. Native of Washington, D.C. US Naval Res Lab Washington DC 1942- 46. General Electric Research Lab 1946-55. Mgr-Advanced Degree Personnel 1955- 65. Consultant, G E Foundation 1965-68. Mgr University Relations G.E. R & D Center 1968-79. Consultant, OECD (Paris) on Tech Education. Chrmn of Bd Intl Assn for Exchange Students for Tech Exp. Prof of Engrg Science 1979- , Rensselaer Polytechnic Institute. *Society Aff:* ACS, ASEE, AAAS.

Wolfe, John W
Home: 3235 NW Crest Dr, Corvallis, OR 97330 *Education:* PhD/Irrigation Sci/UT State Univ; Dipl Hyd Engr/Theoretical Hydraulics/Delft Tech Inst; MS/Agri Engr/Univ of ID; BS/Agri Engg/SD State Univ. *Born:* 7/14/17. I was a jr engr with the Soil Conservation Service in SD from 1940-42, then spent 4 yrs as an infantry officer. With SCS in OR 1946-47, then joined the faculty of OSU, where I remain. I served 14 mos as irrigation advisor in Iran during 1972-73, 2 yr contract - Egypt - Retired 5/1/81. The irrigation res & teaching in OR has expanded considerably in my 30plus yr tenure. For ASAE, I have served as a technical dir, as a regional dir, & as a mbr of the natl nominating committee. Also, I served as regional chrmn for the PNW region. I'm reg in OR. *Society Aff:* ASAE, IA.

Wolfe, Philip M
Business: Garrett Turbine Engine Co, P.O. Box 5217, Phoenix, AZ 85010
Position: Manager *Employer:* Garrett Turbine Engine Company. *Education:* PhD/Ind Engr/AZ State Univ; MS/Ind Engr/AZ State Univ; BS/Ind Engr/Univ of MO; BS/Bus Admin/Univ of MO. *Born:* July 1941. Native of Neosho, Missouri. Worked for Motorola, Inc from 1967-75 as a Sr Operations Research Analyst and as a Systems Mgr. Served on the Okla State Univ. faculty of Industrial Engrg and Mgt from 1976-1984 as Assoc. Prof. and as Prof. Joined Garrett Engine Company as a Senior Project Leader; currently is Manager of Manufacturing Systems Engineering. During 1974-75 was Dir of Operations Res Div of AIIE, and during 1983-84 was

Wolfe, Philip M (Continued)
Dir of Production and Inv Control Div of IIE. Currently he is Vice-Pres. of Area IV of IIE. Married to Laura Donalson. Have 3 children: Scot, Jaree, and Jennifer. Enjoy reading, jogging and golf. *Society Aff:* IIE, TIMS, SME.

Wolfe, R Kenneth
Home: 4930 Spring Mill Ct, Toledo, OH 43615
Position: Professor *Employer:* Univ of Toledo. *Education:* PhD/ChE/GA Tech; BSCE/ChE/GA Tech *Born:* 9/5/29. Born in Chattanooga, TN & resides in Toledo, OH. PhD in CE at GA Tech, 1956, where he also taught & held the TN Eastman Res Fellowship. Proj Engr specializing in process design at Mallinckrodt Chem, 1955-60. At IBM from 1960- 68 he was successively comp process control engr & chem industry coordinator, recruitment coordinator, sr systems engr, proj mgr, & systems engg mgr. He was Operations Res Mgr at Owens IL during 1968-73. Currently, he holds joint appointments as Prof in Ind Engg & Comp Sci & Engg. During 1976-78 he was Chrmn of PhD prog in Systems Engg. He has contributed many papers to profl journals in chem. Engg, comp sci, control, ind engg, operations res, process design, mgt sci, & computer methods. He is reg engr in OH. *Society Aff:* AIChE, IIE, AAAS, ORSA, TIMS.

Wolfe, William L
Business: Optical Sci Center, Tucson, AZ 85721
Position: Prof *Employer:* Univ of AZ *Education:* MSE/EE/U of MI; MS/Phys/U of MI; BS/Phys/Bucknell. *Born:* 4/5/31 Born in Yonkers, NY. Worked at Johns Manville on specific heat and thermal conductivity of materials, at Sperry on microwave spectroscopy and waveguide couplers, at the Univ of MI on infrared imaging, materials, detectors and other instruments. Then was dept head and chief engr at the Honeywell Radiation Center on infrared and other electro-optical instruments. Presently at the Univ of AZ specializing in scattering, solar energy, lightning, infrared instruments. Edited *The Infrared Handbook*; US Editor of *Infrared Physics*. *Society Aff:* OSA, SPIE.

Wolff, Edward A
Home: 1021 Cresthaven Dr, Silver Spring, MD 20903
Position: Assoc Chief, C&N Div. *Employer:* NASA Goddard Sp Flt Ctr. *Education:* PhD/EE/Univ of MD; MS/EE/Univ of MD; BSEE/EE/Univ of IL. *Born:* Oct 31, 1929 Chicago IL. Electronic Scientist Naval Res Lab 1951-54. Instructor, US Army 1954-56. Proj Engr, Md Electronic Mfg Co/Litton Inds 1956- 59; Proj Engr Electromagnetic Res Corp 1959-61. Engr Mgr Aero Geo Astro/Keltec Incs/ Aiken Inds 1961-67. VP Geotronics Inc 1967-71. Hd Sys Study Offfice 1971- 73, NASA Goddard Space Flight Ctr. IEEE: Fellow, Mbr Bd/Dir 1971-72. Mbr AIAA, NSPE, Sigma Tau, Eta Kappa Nu, Phi Eta Sigma. Author. *Society Aff:* IEEE, NSPE, AIAA.

Wolff, Hanns H
Home: 8624 Caracas Ave, Orlando, FL 32825
Position: Consultant (Self-employed) *Education:* Dozent/El Physics/Technology Univ Berlin; DR Ing/El Eng/Technology Univ Berlin; Dipl Eng/El Commonic/ Technology Univ Berlin; Cand Ing/El Eng/Technology Univ Berlin. *Born:* 12/19/03. Asst to Dir of Res Telephonfabrik AG vorm J Berliner 1928-29. Asst to Patent Attorney 1929-30, Chf Engr Radio AG D S Loewe 1930-44, Lecturer Tech Univ Berlin 1945-47. Chf Res Eng Lavoie Labs 1947-51. Chf Eng Gen Fuse Inc 1951-53. VP Leetronics Inc 1953-56. Staff Eng W L Maxson Corp 1956-60. Chf Elec Eng Republic Aviation Corp 1960-63. Tech Dir Naval Training Equip Ctr 1963-75. Consultant 1975--. Adj Prof Poly Inst of Bklyn 1959-66. Trustee Central FL Museum and Planetarium. Bd/Advisors FL Tech Univ 1970-75. Fellow IEEE. Numerous patents in Electronics and Television. *Society Aff:* IEEE.

Wolfram, Ralph E
Business: 1130 Hampton Ave, St Louis, MO 63139
Position: Dir of Engg. *Employer:* Bank Bldg & Equip Corp. *Education:* BS/Elec Engg/MO Univ of Rolla. *Born:* April 1925. Elec Engr, Sverdrup-Parcel 1950-55. Elec Engr, Leo A Daly Co 1955-56. Bank Bldg Corp since 1956, to date-A national firm with projects from coast to coast, EE specializing in electrical design of commercial bldgs and health care faclities. Presently Dir of Engr. Instructor in Ltg courses for IES and guest speaker at IES mtgs. Reg PE in 24 states and D.C. Reg VP IES 1975-79. Dir 1979-81 Mbr NSPE, VP MSPE, IES, St. Louis Engineers Club. Dir 1979-85 St Louis Electrical Board and Prof Mbr of BOCA, ASHE, NFPA. *Society Aff:* NSPE, IES, ASHE, MSPE.

Wolga, George J
Business: 237 Phillips Hall, Ithaca, NY 14853
Position: Prof EE. *Employer:* Cornell Univ. *Education:* PhD/Phys/MIT; BEngPhys/ Engg Phys/Cornell Univ. *Born:* 4/2/31. Instr of Phys MIT 1957-60; Asst Prof of Phys MIT 1960-61; Asst Prof EE Cornell Univ 1961-1964; Assoc Prof EE Cornell Univ 1964-68; Prof EE & Appl Phys 1968-1980.. On leave US Naval Res Lab 1969-1971 as Branch Hd, Laser Phys Branch. Co-founder, & VP Lansing Res Corp, Ithaca, NY 1964-present. Mbr Bd of Dirs EDMAC Assoc, Inc, Rochester, NY 1978-1980. *Society Aff:* APS, OSA, IEEE, MRS.

Woll, Harry J
PO Box 679, Concord, MA 01742
Position: Div VP & Gen Mgr. *Employer:* RCA Automated Syst Div. *Education:* BS/EE/ND STATE Univ; PhD/EE/Univ of PA. *Born:* Aug 25,1920. PhD Univ of PA. BS ND State Univ. With RCA since 1941. Present position since 1975. Past and current activities include tube and solid state circuitry, electro-optics, lasers, radar & control electronics For Apollo Lunar Module, computers, communications, and automatic test equipment. Fellow, IEEE. Phi Kappa Phi, Reg PE Mass. P Chrmn Aerospace Tech Council, AIA. Chrmn Trustee Moore Sch of Elec Engrg and Mbr, Bd/Overseers, Sch of Engrg, Univ of PA. Moore Sch Dist Alumni Gold Medal. First recipient, David Sarnoff RCA employee Fellowship; IEEE Philadelphia Sec Awd. Enjoy sailing and skiing. *Society Aff:* IEEE, AIAA, AAAS.

Woll, Richard F
Home: 178 Richfield, Williamsville, NY 14221
Position: Fellow Design Engr. Retired *Employer:* Westinghouse Elec Corp. *Education:* BS/EE/Univ of Pittsburgh *Born:* Aug 1915 Pittsburgh PA. Entire working career with Westinghouse Elec in design of AC motors; 1937 to 1946 in East Pittsburgh Works, 1946 to April 1, 1981 in Buffalo Plant. Responsibility for Medium Motor Div in design of new AC motor lines, safety & product liability, securing & maintaining motor apprvls from UL and CSA. Fellow IEEE 1971-present. Retired 1981.

Wollensak, Richard J
Business: 10 Maguire Rd, Lexington, MA 02173
Position: VP, Domestic Program Development *Employer:* Itek Optical Systems Div. of Litton. *Education:* BS/ME/Notre Dame; MS/Mgmt/MIT *Born:* 12/24/30. in Rochester, NY, served as Naval Aviator, 52-56. With Itek Optical Systems Div since 1965. Presently, V.P. for Domestic Program Development. Earlier VP for Human Resources and Public Relations. Earlier Dir of Optics with responsibility for mgt of all optics eng services as well as specific optics studies and hardware progs. From 1956 to 1965, served with Wollensak Optical in various capacities including engr for high speed cameras, proj mgt for opto-mech instruments and production mgr for mfg, assembly and test of all products. SPIE Service includes Pres Boston Chapt SPIE 1971-74; Public Relations VP 1973-74; Governor 1971-73, 1975-76, 1978-80; Fellow of the Soc 1978. President of SPIE 1982. Reg PE, MA, Capt, USNR-R. *Society Aff:* SPIE, AIAA, ADPA, AFA, OSA.

Woller, Myron K
Business: Plant & Structures Div, Newark Intl Airport, Newark, NJ 07114
Position: Mgr *Employer:* The Port Authority of NY & NJ *Education:* BS/Admin Engrg/Syracuse Univ; Cert/Commercial Art/Cooper Union Art School *Born:* 7/11/20 Native of Newark, NJ; Attended public schools and Cooper Union Art School;

Woller, Myron K (Continued)
Army Air Corps, 1942-45, Fighter Pilot, European Theater; Continued military flying, Air Force Reserve and Air National Guard until 1952. Initial Port Authority employment in construction, then 20 years as Aviation Planner for LaGuardia, Teterboro, Kennedy and Newark Intl Airports; Presently, Mgr, Plant & Structure Division, Newark Intl Airport, responsible for major construction, maintenance and planning . ASCE - Pres, NJ Section - 1979-80; Pres, North Jersey Branch - 1976-77; Chrmn, various committees; Editor, National Newsletter, Aerospace Transport division - 1964-68; Licensed PE - NJ. Enjoys tennis, sailing, reading. *Society Aff:* ASCE, APA

Wolman, Abel
Business: 318 Ames Hall, Baltimore, MD 21218
Position: Prof Emeritus, Johns Hopkins Univ. *Employer:* The John Hopkins Univ. *Education:* BA/Johns Hopkins Univ; BS/Engg/Johns Hopkins Univ. *Born:* 6/10/92. Hon Degrees. Dr Engr, Dr Sc, Dr Humane Letters, Dr Laws. Native of Balto Md. Served with US Pub Health Serv in 1913. Chf Engr, Md State Dept Health 1922-39. Prof & Chrmn Dept Sanitary Enrg Johns Hopkins U 1937-62. Consultant federal, state, foreign govts. Mbr NAS and NAE. Recipient Milton S. Eisenhower Medal, Lasker Awd, Natl Medal of Sci, Tyler Ecology Award, Harben Lecturer, The Royal Inst of Pub Health and Hygiene, London; Intl Friendship Medal Award by the Instn of Water Engrs and Sci of England; First Hon Pres of Asoc Interamer de Ingenieria Sanitaria y Ambiental (Ctrl and So Amer), Panama, Aug 1982; Gordon Maskew Fair Medal, Water Poll Ctrl Fedn; Decoration Andres Bello. Presented by the Pres of Venezuela, on behalf of the Acad of Phys and Nat Sci and Math; Prof Eng of the Yr, Balt Chap, Maryland Soc of Prof Engrs; The Balt Pub Works Museum was publicly dedicated on June 4, 1984. It is entitled "The Livable City: Dr. Abel Wolman and the Continuing Work of the Engr." *Society Aff:* NAS, NAE, APWA, ASCE, APHA.

Wolovich, William A
Business: Div of Engg, Providence, RI 02912
Position: Prof. *Employer:* Brown Univ. *Education:* PhD/Elec Sciences/Brown Univ; MSEE/Elec Engr/Worcester Poly Inst; BSEE/Elec Engr/Univ of CT. *Born:* 10/15/37. A Prof of Engg at Brown since 1970. Formally with the NASA Electronic Res Ctr in Cambridge, MA and a ground electronics officer in the USAF. Has published numerous articles on multivariable control systems and robotics, a 1974 text titled "Linear Multivariable Systems" by Springer Verlag, and a 1987 text titled "Robotics: Basic Analysis and Design" by Holt, Rienhart, and Winston. Received a Fulbright Hayes fellowship to lecture at the Univ of Warwick, England during the 1976-77 academic yr. A previous assoc editor of the "IEEE transactions on automatic control" and a current assoc editor of the "IEEE Journal on Robotics and Automation" as well as a fellow of the IEEE. *Society Aff:* IEEE, TBΠ, HKN, SME.

Wolsky, Sumner P
Business: 950 De Soto Rd, Suite 3B, Boca Raton, FL 33432
Position: VP R&D. *Employer:* Ansum Enterprises, Inc. *Education:* PhD/Physics Chem/Boston Univ; MS/Physics Chem/Boston Univ; BS/Chem/Northeastern Univ. *Born:* Aug 21,1926. Employed at Res Div Raytheon Corp 1952-61. Initiated Lab for Physical Science for P R Mallory & Co 1961. Res Dir Laboratory for Physical Science 1961-75. VP, R&D 1975-1981. Resp for R&D in Mallory Co and environmental and energy related matters.Science & Tech of the NSF. Mbr of the Editorial Bds of Intl Journals - Electrocomponent Science and Technology. Author of bk 'Ultra Microweight Determination in Controlled Environments'. Co-Editor new bk series 'Methods and Phenomena Their Applications in Science and Technology'. Specialist in ultra high vacuum,microbalance techniques, sputtering and surface phenomena, batteries, environment, occupational health, and R & D management and organization. January 1, 1981 -- *Society Aff:* ACS,ECS,AVS,IEEE,ΣΞ,APS

Wommack, S Jackson
Home: 3750 N Lake Shore Dr, Chicago, IL 60613
Position: Vice President. *Employer:* Amoco Chemicals Corp. *Education:* BS/ChE/WA Univ; AB/Chem Econ/SW MO State College. *Born:* 9/29/20. m. Edna Brock Wommack, 3 children. Mfg supv and proj engr, Monsanto 1942- 50. Olin Corp 1950-63. Development supt Film Div, Dir-Corp Res, Gen Mgr, Intl Div, Asst Gen Mgr, Energy Div, VP Res & Engr, Organic Div Mobil Oil Corp 1963- 67. VP Corp Dev. President Mobil Chemical Intl Co. Amoco Chemicals Corp 1967- present. VP Plng and Dev, Group VP Intl Div. VP Plastics Products Div, Currently VP-Technology & Development.

Wong, Eugene
Business: Dept EECS, Berkeley, CA 94720
Position: Prof. *Employer:* Univ of CA. *Education:* PhD/EE/Princeton *Born:* 12/24/34. Engr with IBM 1955-1956 and 1960-1962. On the faculty of the Univ of CA since 1962. Have been conslt to Ampex, GenRad, Honeywell, and Computer Corp America. Co-founder of Relational Tech Inc. Prin res interests: stochastic system theory and database mgt. NSF Postdoctoral Fellow 1959-60, Guggenheim Fellow 1968-69, Vinton Hayes Fellow at Harvard 1976-77, Visiting Professor at MIT 1979-80. Fellow, IEEE. Member National Academy of Engineering. *Society Aff:* IEEE, ACM, NAE.

Wong, Kam W
Business: Dept of Civil Engg, Urbana, IL 61801
Position: Prof. *Employer:* Univ of Illinois at Urbana. *Education:* PhD/Photogrammetry/Cornell Univ; MSc/Photogrammetry/Cornell Univ; BSc/ Surveying Engg/Univ of New Brunswick. *Born:* 3/8/40. Has taught graduate and undergraduate courses in photogrammetric and geodetic engrg since 1967. Res interests include analytical photogrammetry, metric vision systems, and satellite positioning systems. Consulting activities include the design of precise geodetic survey systems for measuring ground movement due to underground tunneling, land slide, and solution mining. Received ASP Talbert Abrams Awd 1971, and ASCE Walter L Huber Res Price 1971. Co-authored two books: *Design and Planning of Engineering Systems*, and *Fundamentals of Surveying*. Phi Kappa, Sigma Xi and Chi Epsilon. *Society Aff:* ASPRS, ASCE, ACSM, CISM, SPIE.

Wood, Alan B
Home: Amberley: Killingworth, Newcastle Upon Tyne, England NE12 Ors, Tyne and Wear
Position: Partner *Employer:* Merz and McLellan *Education:* BSc/EE/Durham Univ *Born:* 4/11/20 Partner in the firm of Merz and McLellan, Amberley, Killingworth, Newcastle upon Tyne and Carrier House, Westminster, London. Educated: Gosforth Grammar School and King's Coll, Univ of Durham. Professional career: Apprenticed to CA Parsons (1936) followed by Was service, primarily with the Military Coll of Science. Joined Merz and McLellan in 1950, firstly on distribution design and construction and subsequently on HV transmission networks becoming Head of the Overhead Transmission Lines Dept in 1960. UK Mbr and immediate P Chrmn of CIGRE, Study Ctte 22 dealing with overhead lines. Became Assistant Chief Electrical Engr in 1966, Assoc and Chief Electrical Engr (Transmission) in 1969: taken into Partnership 1970: Chief Electrical Engr 1976. Consultant 1985. *Society Aff:* IEEE, FIEE

Wood, Allen J
Home: 901 Vrooman Ave, Schenectady, NY 12309
Position: Principal Consultant *Employer:* Power Tech Inc. *Education:* PhD/EE/RPI; MS/EE/IL Inst of Tech; BEE/EE/Marquette Univ. *Born:* Oct 1,1925. Allis Chalmers Mfg Co 1949-50; GE Co 1951-59, 1960-69; Hughes Aircraft 1959-60; Power Tech Inc 1969 to date. Founder, secretary and mbr of Bd PTI. Consultant in System and Corporate Planning area. Fellow IEEE, Mbr ANS, AAAS & NSPE. Adjunct Prof in Electric Pwr Engrg at RPI. Hobbies photography & amateur radio. *Society Aff:* IEEE, ANS, NSPE, AAAS.

Wood, Carlos C
Home: 145 Bonniebrook Dr, Napa, CA 94558
Position: Retired *Education:* MS/Aero Eng/CA Tech; MS/Aero Eng/CA Tech; BA/Engg-Sci/College of The Pacific. *Born:* 6/19/13. Native of Turloc, CA. Private pilot 1931. CA Tech 1933-37. Douglas Aircraft 1937-60; Draftsman, research engr, designer, chf preliminary design, chf engr Long Beach Div, dir advanced engrg planning. Sikorsky Aircraft 1960-70; engrg mgr, VP engrg, Past mbr FAA Tech Adv Bd, USAF Scientific Adv Comm, NAE Aerospace Engrg Bd. USA AVSCOM Adv Group, USA TECOM Adv Group, USA-CDC Adv Group, Dir & VP Claude C Wood Co. Presently Regent Univ of Pacific. Enjoys golf, flying, travel, personal computers. Dir-at-Large 1969-72 AIAA. *Society Aff:* AIAA, CdC, NAE.

Wood, Don J
Business: Dept Civil Engrg, Lexington, KY 40506
Position: Prof of Civil Engrg. *Employer:* University of Kentucky. *Education:* PhD/CE/Carnegie-Mellon; MS/CE/Carnegie-Mellon; BS/CE/Carnegie-Mellon. *Born:* 7/28/36. Native of Corry, PA. Taught at Clemson Univ and Duke Univ, consultant for NASA-Lewis Res Ctr. Currently with Univ of KY, Prof of Civil Enrg. Research in the area of transient pipe flow and steady flow in pipe systems and author of 40 technical papers. Developed a computer prog for pipe system analysis which is widely used in USA and other countries. ASCE Huber Research Prize 1975. ASEE Western Electric Fund Teaching Awd 1976. Univ of KY Alumni Great Teacher Awd 1972. *Society Aff:* ASCE, ASME.

Wood, Francis P (SJ)
Business: Seattle Univ, Seattle, WA 98122
Position: Prof Emeritus. *Employer:* Seattle University. *Education:* MSEE/Elec Eng/Stanford Univ; AB/Phil/Gouzaga Univ; STL/Theology/Alma College. *Born:* June 18, 1917 Seattle WA. Entered the Society of Jesus in 1934, AB Gonzaga Univ Spokane WA 1940. S.T.L. Alma College CA 1948. MSEE Stanford Univ 1952. Ordained a Jesuit priest in 1947. Instructor at Gonzaga Univ 1942-44. Asst Prof at Seattle Univ 1952; Assoc Prof 1958; Prof 1965; Visiting Prof - Sogang Univ, Seoul Korea 1974-75; Chrmn of EE Dept 1959 to 1983. NSF Science Faculty Fellow Stanford Univ 1957-58 Power Systems Engr at Bonneville Pwr Admin, summers of 1955,56,57. Res Engr at the Boeing Co, summers of 1958,62,68,69,72, 77, 79,80, 81,82,83. Trustee of Seattle Univ 1958-74, Professor Emeritus 1987. 1983-, Chrmn Seattle Sec IEEE 1968. Mbr ASEE, IEEE, Tau Beta Pi. *Society Aff:* ASEE, IEEE, TBΠ.

Wood, Gerald L
Home: 6602 E Mountain View Rd, Scottsdale, AZ 85345
Position: Conslt Engr *Employer:* W & W Engineering *Education:* BS/Agric Engrg/TX A&M Univ *Born:* 2/24/43 Native of San Saba County, TX. Engr for Allis-Chalmers Corp performing various assignments related to the design and development of Mobile Agricultural Grain Combines. Test engr for Massey-Ferguson Co. Responsible for performance and endurance testing of various agricultural, industrial, and construction prototype mobile machinery. Mech Engr with Natl Pump Co and responsible for the design and development of its line of agricultural and industrial vertical turbine pump. Currently with W & W Engrg, A Consltg Engrg Firm. Chrmn of AZ Section, ASAE 1981-82. Previously held all other offices. Reg PE MO, Mech Engr, AZ, Land Surveyor, AZ. Married 4 sons, 2 daughters. *Society Aff:* ASAE, ASME, SAE, SSS

Wood, Harry G
Home: 849 Mesa Ave, Palo Alto, CA 94306
Position: Packaging Engr. *Employer:* Hewlett-Packard Co. *Education:* BS/English/State Univ of NY at Buffalo. *Born:* Jan 1915. BS ED State U NY at Buffalo. Plng Supvr Morrison Steel Products 1944-51. Sr Methods Engr Schlage Lock Co 1952-6. Packaging Engr Hewlett-Packard Co since 1959. Independent Consultant industrial packaging design. Active industrial engrg edu since 1954. Instructor factory planning Foothill College 1959-60. SPHE Fellow 1971. Packaging Industry Adv Cttee U CA Davis 1969-75. Natl design awards 1963,64,74. President N CA Chap 1964-5. Natl Inst Packaging, Handling & Logistics Engrs: Annual Achievement Award 1974. Several design pats. SPHE Ed Review Bd. 1978-79. 1981-83 originated Hewlett-Packard electrostatic discharge control. 1962 designed & built Ruth Wood Nursery School. Partner, bus. mgr. to date. *Society Aff:* SPHE

Wood, John E, III
Business: One Metroplex, PO Box 7497, Birmingham, AL 35223
Position: Consultant *Employer:* Vulcan Mtls Co. *Education:* PhD/Organic Chemistry/MIT; AB/Chemistry/Lynchburg College. *Born:* 5/20/16. Native of Lynchburg, VA. Employed by various affilites of Std Oil (NJ) 1939-65 in petrochem mfg, oil refining, administrative & exec positions. Pres, Enjay Chem Co 1959-65. Dir, Esso Chem Co 1962-65. Employed by Vulcan Mtls Co 1965-1981 - in admin & exec positions; 1981 - Consultant. *Society Aff:* AIChE, SCI, ACS, AIC.

Wood, Loren E
Home: 905 Coward Creek Dr, Friendswood, TX 77546
Position: Mgr, Technical Planning. *Employer:* TRW Defense Systems Group, Houston *Education:* MS/Math Analysis & Applied Math/Cornell Univ; BS/Mathematics/Brown Univ *Born:* Dec 1927. MS Cornell Univ, BS Brown Univ. Lead Analytical Engr in numerous Army Ordnance, Air Force Armament Projects 1950-56. With TRW Systems since 1956 except 1961-66 at Aerospace Corp. One of few engrs involved throughout ICBM and Manned Spacecraft Programs. Planned, evaluated ICBM tests for ATLAS-THOR-MINUTEMAN, Member. Analysis team selecting mission model, vehicles used for Apollo Lunar Landing Program. Managed Gemini testing. Managed key studies, Apollo, Skylab, Shuttle, currently STS Payloads. Organized Space Prog Applications, Leader, AIAA Pub Affairs activities since 1972. Natl AIAA Outstanding Sec Event Awards, 1972-73, 1973-74. Chrmn, Outstanding AIAA Sec in USA 1975-76. *Society Aff:* ΦBK, ΣΞ, AIAA, AAS.

Wood, Robert A
Home: 59425 Ten Mile Rd, S Lyon, MI 48178
Position: Chief Engr. *Employer:* Star Pak Solar Systems, Inc. *Education:* BSE/Hydraulics-Hydrology/Univ of MI; MS/Public Systems Engg/Univ of MI. *Born:* 4/17/52. Responsible for design and installation of our solar systems. Star Pak has installed about 400 systems for a total of over 100,000 FT2 of collector area in MI and IN. Most astounding and important contribution is a reversible solar assisted heat pump system for heating and cooling. The System uses unblazed plastic panels connected in series with a concrete tank and liquid source heat pump. A bringe solution is the heat transfer fluid. In the heating mode, heat is connected from bota. The air and sun through the unblazed panels. In the cooling mode, heat is retracted through the panels. The tech will revolutionize the solar heating and cooling industry. The system is presently heating and cooling Star Pak's warehouse. *Society Aff:* NPSE, MPSE, ISES, MSEA, MICA, SEIA.

Wood, William G
Business: 12890 Westwood Ave, Detroit, MI 48223
Position: Vice President. *Employer:* Kolene Corp. *Education:* MS/Met Eng/Wayne State Univ; BS/ChE/Wayne State Univ. *Born:* Sept 27, 1919. 1941-1955 Chf Chemist Bower Roller Bearing. Vice Pres Res & Mfg Park Chem Co 1955-60. 1960 Chem and Metallurgical Consultant. Presently VP Technology/Res and Dev Kolene Corp. Has written numerous tech papers and is copatentee of several inventions relating to fused salt systems for metal cleaning and heat treatment. Chrmn ASTM Cttee B10, Secretary, ASTM E43.10, Chrmn Technical Divisions Board ASM, Fellow ASM, Fellow Engineering Society Det. *Society Aff:* ASTM, ESD, ASM,AAAS,SAE, NACE,SAMPE,CAIA.

Wood, William S
Home: 1076 Dunvegan West, West Chester, PA 19382
Position: Consultant. *Employer:* William S Wood and Assoc. *Education:* BSChE/ChE/Purdue Univ. *Born:* June 1913 Indianapolis. Res Engr, Sun Oil Co 1937-52. Safety Engr, Sun Oil Co 1952-72. Chem Safety Consultant, William S

Wood, William S (Continued)
Wood and Assoc 1972--. PE (Pennsylvania) and Certified Safety Profl. Cert in Comprehensive Practice of Industrial Hygiene; Adjunct prof, Temple Univ 1975--. Co-editor, Fawcett & Wood, Safety and Accident Prevention in Chemical Operations. Fellow, AIChE. Fellow ASSE, Reg VP 1972-76. Mbr ACS, AIHA, NFPA. Natl Safety Council 'Distinguished Service to Safety' Award 1972. Past Bd Chrmn, Northeastern Christian Jr College. *Society Aff:* AIChE, ASSE, ACS, AIHA, NFPA.

Wood, Willis B, Jr
Business: 801 S. Grand Avenue, Los Angeles, CA 90017
Position: Exec VP *Employer:* Pacific Lighting Corp *Education:* BS/Petr Engr/Univ of Tulsa; -/Advanced Management Program/ Harvard Business School. *Born:* 9/15/34 Native of KS City, MO. Graduated from Cherryvale, KS Public School System. Petroleum engr with PanAmerican Petroleum Corp (now Amoco) and E B Hall & Co prior to joining Pacific Lighting group in 1960. Previous positions include Exec VP of S CA Gas Co, Pacific Lighting Corp's major subsidiary. Responsible for gas supply, special supply projects and public affairs. Pres & CEO of Pacific Lighting Gas Supply Co. Assumed present position of Exec VP of Pacific Lighting Corporation, in June 1985 with responsibility for land development, financial services & alternative energy. *Society Aff:* AIME, ΠΕΤ, SPE, PCGA, AGA, PEA.

Woodall, David M
Business: Dept of Chem & Nuclear Engrg, Albuquerque, NM 87131
Position: Prof *Employer:* Univ of NM *Education:* PhD/Engrg Phys/Cornell Univ; MS/Nuclear Engrg/Columbia Univ; BA/Phys/Hendrix Coll *Born:* 08/02/45 David M. Woodall is a Prof of Chem and Nuclear Engrg at the Univ of NM and he was Chrmn of the Dept for the period of 1980 through 1983. He is a reg PE in the State of NM. He is a former Chrmn of the Trinity Sect of the Amer Nuclear Soc and has served as the Gen Chrmn of the First and Second Symposia on Space Nuclear Power Sys, held in 1984 and 1985. He has an active res program in plasma phys, pulsed power and nuclear reactor safety, with numerous interactions with the Air Force Space Tech Center, Sandia Natl Labs and Los Alamos Natl Lab. He has been a Univ instructor for more than a decade and has more than twenty-five tech publs to his credit. *Society Aff:* ANS, IEEE, APS, NSPE

Woodburn, Russell
Business: Box 283, University, MS 38677
Position: Professor Emeritus. *Employer:* University of Mississipi. *Education:* BS/Civil Engg/Univ of KY. *Born:* Feb 1907. Native of Central City, KY. Married Mary Frances Miller of Sacramento, Ky 1933. Two daughters, Laura Ann Wall of Oxford, Miss and Mary Ellen Nutt of Franklin, Va. Bridge Engr, Erie RR 1929-30. Field Engr MO Hwy Dept 1930-33. Engr Researcher & Research Administrator, US Dept of Agriculture 1933- 64. This included directorship of the US Sedimentation Laboratory, Oxford, Miss. 1956-61 and Chief, Southern Br, ARS-SWC-USDA with responsibility for research in soil management, water management and watershed engrg, 9 states and Puerto Rico 1961-64. Director, Eng Exp Sta, Coordinator Res & Prof of Civil Eng, Univ of Miss 1964-72. Life mbr ASAE 1970. Fellow ASAE 1975. Author 40 tech papers. Alderman, Oxford, Miss, 1973-1981. Prof Emeritus of Civ Engrg, Univ of Miss 1972 for life. *Society Aff:* ASAE, TBΠ.

Woodbury, Eric J
Home: 18621 Tarzana Dr, Tarzana, CA 91356
Position: Prof Elec Engg & Physics. *Employer:* CA State U, Northridge *Education:* PhD/Physics/CA Inst of Tech; BS/Physics/CA Inst of Tech. *Born:* 1925. U.S. Navy 1944-46. BS 1947, PhD 1951 Calif. Inst. of Tech. Hughes Aircraft Co. 1951-1981 with 9 years in missile guidance electronics and 21 years in laser research and development with emphasis on laser rangefinding and target designation. Co-discover of stimulated Raman scattering in 1962. Retired from Hughes in 1981 as Chief Scientist of the Electro-optical Engineering Division. Professor of Electrical Engineering and Physics at California State University, Northridge 1982-present. Fellow IEEE. *Society Aff:* IEEE, APS, ΣΞ, TBΠ.

Woodbury, Richard C
Business: 459 CB-BYU, Provo, UT 84602
Position: Prof, E Engr *Employer:* Brigham Young Univ. *Education:* PhD/EE/Stanford Univ; MS/EE/Stanford Univ; BS/EE/Univ of UT *Born:* 4/19/31 Native of Salt Lake City, UT. From 1959 to present (1981) teacher of EE at Brigham Young Univ. Oscilloscope circuit designer, Hewlett Packard Co. 1956-59. Current fields of specialization are magnetic devices and monolithic silicon intergrated circuits. Published serveral articles in the fields of magnetic domain wall motion and silicon electron voltaic cells; consulting experience with preferentially etched silicon devices, thermo electric generators, magnetic memory and magnetic force actuators. Active in IEEE. Other interests: hiking, swimming, churchwork. *Society Aff:* ΣΞ, IEEE, TBΠ, ASEE, HKN

Woodman, Richard A
Business: 3664 Grand Ave, Oakland, CA 94610
Position: Pres. *Employer:* Jordan/Casper/Woodman/Dobson. *Education:* MBA/Bus/St Mary's College; BS/Civil Engg/Heald's Eng College. *Born:* March 19, 1940. Specializes in project coordination, contract mgt, specifications and budget estimating. Project engineer on wide variety of projects including containerized shipping facilities, truck terminals and other commercial and industrial buildings.

Woodring, Richard E
Business: 32nd & Chestnut Sts, Philadelphia, PA 19104
Position: Dean of Enrg. *Employer:* Drexel University. *Education:* PhD/Civil Engg/Univ of IL; MS/Civil Engg/Univ of Il; BSCE/Civil Engg/Drexel Univ. *Born:* 11/28/31. Mbr of the faculty of Civil Engrg at Drexel since 1956. Mbr ASCE, & ASEE. Prof Engr Recipient of Ford Foundation Faculty Development Fellowship 1960. Awarded one year Preceptorship in Engineering Practice at Sanders & Thomas Inc Pottstown PA 1969. Proj engr for sev reinforced concrete and structural steel bldgs and bridges. Officer of PRIME (Philadelphia Regional Introduction of Minorities to Engineering) 1974 to present. Awarded George Washington Medal of the Engineer's Club of Phila 1981. Awarded Reginald H. Jones Distinguished Service Award by the Natl Action Council for Minorities in Engrg 1982. Dir Ridgeway Phillips Company, Horsham, PA. Named Engineer of the year (1987) in Delaware Valley. *Society Aff:* ASCE, ASEE, NSPE.

Woodruff, Kenneth L
Business: 2500 Brunswick Pike, Trenton, NJ 08648
Position: Pres. *Employer:* Resource Recovery Services, Inc *Education:* BS/Mineral Processing Engg/PA State Univ. *Born:* 10/10/50. Native of Phoenixville, Chester Co, PA. Served as an Engg Asst at Bethlehem Mines Corp, Grace Mine, Morgantown, PA 1970-71. Res Engr with the Natl Ctr for Resource Recovery, Inc, Wash, DC, 1972-75, responsible for res & dev testing of various unit processes & systems for the recovery of mtls & fuels from municipal waste. Established Resource Recovery Services, Inc in 1975 & have served as Pres since 1976. The firm offers operation & mgt services in the area of solid & liquid wastes as well as ind & post consumer scrap. Have published more than a dozen articles on various aspects of municipal waste recovery & scrap processing. *Society Aff:* AIChE, AIME, ASME, ASTM, NAEP.

Woodruff, Neil P
Home: 2315 Timberlane, Manhattan, KS 66502
Position: Consulting Engineer *Employer:* Self employed *Education:* MS/Agri Engg/KS State Univ; BS/Agri Engg/KS State Univ; -/Engg/IA State Univ. *Born:* July 1919. Native of Clyde Kansas. Farmer 1937-40. Army Air Corps 1941- 45. Agricultural engr & research leader Agic Res Serv USDA and Professor KSU 1949 to retirement 1975; speciality wind erosion mechanics and control. Civil Engr KS Dept Transportation 1977-79, Civil Engr Facilities Planning K-State Univ 1979 to retirement 1984; Consulting Civil Engr Lawn Irrigation and Municipal utilities 1984-present. Fellow ASAE; Tech Dir ASAE 1970-72; Mbr Adv Comm NSF Natural Hazards res

Woodruff, Neil P (Continued)
Univ Colo 1972-75; Mbr Scientific Exch Team to Soviet Union 1974; Hancor Soil & Water Engrg Award 1976. *Society Aff:* ASAE, SCSA, ΣΞ, ΓΣΔ.

Woodruff, Richard S
Home: 4153 Kennesaw Dr, Birmingham, AL 35213
Position: Consultant-Hydro Projs. *Employer:* Self. *Education:* BS/Civil Engg/Univ of AL. *Born:* 3/22/13. Binghamton NY. Alabama Power Co 1937-76 Staff Engr, Generating Plant Tech Services, Power Supply Dept on hydro-electric development, inspection, maintenance, consulting activities and regulatory matters for the company's 14 hydro plants. Southern Co Services Inc 1976-78, Mbr Hydro Plant Plng. Since 1978 Hydro conslt to Corps of Engrs and private engrg firms. ASCE activities: Fellow, Dir Natl Bd 1980-83 Rickey Medal 1972, Sect Ex Comm Pwr Div 1963-68, Chrmn Comm Hydro Pwr Proj Plng Design 1969-72. Pres Birmingham Br 1972 Chrmn Dist 14 Council 1976-77, nominated 'Eng of the Yr' 1967 & 1974. Presentations to technical, service and educational groups. Mbr USCOLD. Enjoy classical music, travel, photography, philately. *Society Aff:* ASCE, USCOLD.

Woods, Donald L
Business: Civ Engg Dept, College Station, TX 77843
Position: Prof/Res Engr. *Employer:* TX A&M Univ. *Education:* PhD/Civ Engg/TX A&M Univ; MS/Civ Engg/OK State Univ; BS/Civ Engg/OK State Univ. *Born:* 10/31/33. Dr Woods-P.E. is a Prof of Civil Engineering & Res Engr in the TX Transportation Inst, TX A&M Univ. He is actively engaged in transportation res. Notable res efforts include hgwy noise reduction, cost effectiveness of roadside safety improvements, criteria for demarking no-passing zones, studies of thermoplastic striping durability, roadway drainage, roadside barriers & pavement markings for low volume roadways. This res has resulted in more than fifty (50) publications. Dr Woods is well known natlly through the more than one-hundred fifty short courses he has conducted over the past fifteen yrs. Dr. Woods is also actively engaged in Consult Eng. *Society Aff:* ITE.

Woods, Donald R
Business: Chem Engg Dept, Hamilton Ontario L8S 4L7 Canada
Position: Prof *Employer:* McMaster Univ. *Education:* PhD/CE/Univ of WI; MS/CE/Univ of WI; BSc/CE/Queen's Univ. *Born:* 4/17/35. Native of Sarnia, Ontario. Worked with a variety of industries: Unilever, Distiller's Co Ltd, British Geon, Imperial Oil, Polysar. Taught at the Univs of WI & Waterloo. Joined McMaster Univ in 1964 with a sabbatical leave at the Rijks Univ, the Netherlands, in 1970. Chrmn 1979-82. Director, Engg & Management Program, 1986- . Author of several books and over 200 articles on separation processes, communication, decision-making, cost estimation, problem solving, colloid & surface phenomena & education. Editor of PS News on problem solving. Received honors for his contributions to education. Chrmn of AIChE Eduction Projs Committee 1978-80. Cofounder of local Heritage Soc, Chrmn of Flamborough Township's Committee of Adjustment and active in church & community affairs. *Society Aff:* AIChE, ASEE, CSChE, ASAC.

Woods, John G
Home: 5928 Devon Pl, Philadelphia, PA 19138
Position: Consulting Engr *Employer:* John G. Woods, P.E. *Education:* BME/Mech Engrg/Cornell Univ; MS/Mech Engrg/Drexel Univ *Born:* 07/21/26 Test Engr, Babcock and Wilcox Co, Alliance, OH 1949-51; Sr Engr, Philco Corp, Phila, PA 1951-54; Sr Dev Engr, IRC Inc 1954-69; Sr Dev Engr Sr Proj Engr and Program Mgr, TRW R & D Labs Elec Components Group, Phila, PA 1969-1982; Chief Engr Fiber Optic Prods, TRW Elec Components Group 1982-1985; Consulting Engr, 1986- . Mbrships: Amer Soc of Mech Engrg; Inst of Electrical and Electronic Engrg; Natl Soc of PE (Natl Dir, 1983-86); PA Soc of PE (Pres, 1984-85); Franklin Inst; Cornell Soc of Engrs; Soc of Photo-Optical Instrumentation Engrs. Registration: Reg PE PA - No. 2627-E. *Society Aff:* NSPE, ASME, IEEE.

Woods, John O, Jr
Business: Suite 715, 4660 Kenmore Ave, Alexandria, VA 22304
Position: VP. *Employer:* FDE Ltd. *Education:* MS/-/Duke Univ; BS/Civ Engg/Citadel. *Born:* 9/20/42. Rock Hill, SC native. Disability retirement from US Army after Vietnam service. Prior to service, coauthored & published paper on my prestressed concrete res in Aug 1968 PCI Journal. Joined FDE Ltd in 1970. Became officer in 1976, assuming current responsibility managing & directing firm in providing consulting structural engg services. In addition to structural design of architectural structures, am experienced in seismic analysis & progressive collapse design. Also responsible for structural evaluation of residential and commercial properties. Experience in historic bldg restoration. Have participated in arbitration cases as expert. Hobbies include politics & sports. *Society Aff:* AAA, ACEC, NSPE, PCI, PTI, VSPE-PEPP, APT.

Woods, John W
Home: 41 Orchard Way So, Potomac, MD 20854
Position: Program Director. *Employer:* NSF, on leave of absence from RPI. *Education:* PhD/Elec Comm/MIT; EE/Elec Comm/MIT; MS/Elec Eng/MIT; BS/Elec Eng/MIT. *Born:* 12/5/43. From 1970 to 1973, he was at the VELA Seismological Ctr, Alexandria, VA working on array processing of digital seismic data. From 1973 to 1976, he was at the Lawrence Livermore Lab, Univ of CA, researching two-dimensional digital signal processing. In 1976 he joined the Elec Computer & Systems Engg Dept at Rensselaer Poly Inst, where he attained the rank of Prof in 1985. He spent the academic year 1985-86 on sabbatical leave supported by a ZWO Fellowship at the Technical Univ of Delft, The Netherlands. Since Jan. 1987 he has been on leave of absence at National Science Foundation, Washington DC, serving as program director of Circuits and Signal Processing. He has co-authored a text in probability & random processes. He is past chairman of The Multidimensional Signal Processing Committee. He is presently serving an elected 3 year term on The Administrative Committee of the IEEE Acoustics, Speech and Signal Processing Society. *Society Aff:* IEEE, AAAS, ΣΞ, ΤΒΠ.

Woods, Raymond C
Business: PO Box 488, Alcoa, TN 37701
Position: Principal *Employer:* J.R. Wauford & Co. *Education:* MS/CE/TN Tech Univ; BS/CE/TN Tech Univ *Born:* 8/27/48 Served with US Army 79th Engrg Battalion, 1968-69. Environmental engr for TWA, 1975-76. Progressed from 1977 as an environmental engr to present position as principal in J.R. Wauford & Co, Consulting Engrs, Inc. Responsible for entire operation of branch office including marketing, planning, design, and construction administration. Active in church work and enjoy golf and hiking. *Society Aff:* NSPE, WPCF, AWWA, ACEC

Woods, W Kelly
Home: 714 Tillman Ave SE, Salem, OR 97302
Position: Professor Nuclear Engg. *Employer:* OR State Univ. *Education:* AB/Chemistry/Stanford Univ; MS/Chem Engrg/MA Inst Technology; DSc/Chem Engrg/MA Inst Technology. *Born:* 12/10/12. Native of Claremore Oklahoma. With duPont, General Electric, and Douglas United Nuclear during 22 yrs at Hanford plutonium production plant in tech support of reactor design and operation. 6 yrs with General Electric in support of commercial boiling water reactors. In Oregon state govt 1971-77. Tech adviser at first Atoms-for-Peace Conference in Geneva, Switzerland 1955. Fellow and former natl Treasurer of ANS. *Society Aff:* ANS, AIChE, ACS.

Woodson, Herbert H
Business: Dean's Office, College of Engineering, ECJ 10.324, Austin, TX 78713
Position: Prof EE *Employer:* Univ of Texas at Austin. *Education:* ScD/EE/MIT; MS/EE/MIT; BS/EE/MIT. *Born:* 4/5/25. Texas native. Two yrs experience as Elec Engr at Naval Ordnance Lab in Maryland. Recd ScD from MIT in 1956. MIT Asst Prof of EE 1956; Assoc Prof 1960; Prof 1964; appt Philip Sporn Prof of Energy Processing, 1967. 1971 became Prof and Chrmn of EE Dept at Univ of Texas, Austin. Alcoa Foundation Professor 1972- 75. Appt Dir of Cntr for Energy Studies UT-

Woodson, Herbert H (Continued)
Austin in 1974. Texas Atomic Energy Research Foundation Prof 1980-83, Ernest H. Cockrell Centennial Chair in Engrg, 1983-present. Associate Dean, College of Engineering, 1986-87. Acting Dean, College of Engineering 1987-present. Associate Director Center for Electro-mechanics, 1978-present. Past Pres of PES IEEE. Mbr NAE. Fellow IEEE. *Society Aff:* IEEE, ASEE, AAAS, NSPE.

Woodson, Thomas T
Business: Engineering Clinic, Claremont, CA 91711
Position: Director, Emeritus *Employer:* Harvey Mudd College. *Education:* BS/Elec Engg/Purdue Univ; MS/EE/OH State Univ. *Born:* 12/13/09. Indiana. Tau Beta Pi, Eta Kappa Nu, Sigma Xi. Dev Engr General Elec Co Res Lab and Maj Appliance Div 1937-45. Mgr of Design and Development Engrg, Maj Appliance Div 1946-56. Developed GE Automatic Washing Machine. 30 patents. Chief Appliance Engr, Waste King Corp 1956-60. Resp for dev, design, prod engrg of Dishwasher and Disposer product line. Taught Engrg Design at UCLA 1961-66. Published textbook "Introduction to Engineering Design" 1966. Co-sponsored USAID Tech Assistance Prog in Northeast Brazil 1962-66. State Dept Officer there 1967- 69. Engrg Fac Harvey Mudd Coll 1969-date. NSF Consultant to Indiana Inst Tech 1971; UNIDO Consultant to Iran 1973; UNESCO Commission on Engg Education 1974, 1976 & 1978. Dir Harvey Mudd Coll Engrg Clinic 1972-77, 1980-81, 1984-85. Freedoms Fnd Award 1979. Dir Minorities Engrg Prog 1978-date. *Society Aff:* IEEE, AAAS.

Woodward, Adin K
Business: 210 Laskin Rd, Virginia Beach, VA 23451
Position: Prin. *Employer:* Woodward & Assoc. *Education:* BS/Naval Architecture & Marine Engg/Webb Inst of Naval Architecture. *Born:* 10/16/20. Native of Norfolk, VA. Active duty as Engr Duty officer in Naval Reserve 1943-46 and 1950-52; present rank Capt, USNR. Estimator and asst superintendent, Norshipco, 1946-50; chief engr, Norfolk Dredging Co 1952-1956; prin in naval architectural firm 1956 to date. Responsible for design and admin; emphasis on inland waterway craft, dredges, very large yachts. Also involved in other business ventures including tug construction, real estate dev. Active in civic league, yachting, restoration of old houses. Former mbr of Technical Review Committee for USS Monitor. *Society Aff:* SNAME.

Woodward, Alan A
Home: 750 S Alton Way 9C, Denver, CO 80231
Position: Retired. *Employer:* Pub Serv Co of Colo. *Education:* -/Applied Electricity/Pratt Inst. *Born:* Sept 1896 Birmingham England. Family moved to USA 1910. Completed course in Applied Electricity Pratt Inst 1917. Signal Corps Serv in France WWI 1918-19. A 45 yr career in the utility industry with Public Service Co of Colorado and associated companies covering many phases of electric power was terminated by retirement in 1964. Served as Supt Prod Steam & Hydro Plants, Actg Chf Mech Engr, Mgr Elec Operations. This comprised responsibility for production, transmission and distribution activities. Fellow ASME. Reg Engr Colo. Enjoy craftwork, golf and photography. *Society Aff:* ASME.

Woodward, Henry E
Business: PO Box 1637, Pinellas Park, FL 33565
Position: President. *Employer:* Woodward Air Balance. *Education:* BCE/Civil Engg/GA Inst Tech. *Born:* 6/27/26. Native of Atlanta, GA. Reg ME in Florida and Georgia. Extensive background air conditioning & refrigeration products applications; mech contracting; air distribution and temperature control product design and application; and engrg mgt. Presently president of consulting engrg firm spec in field testing, balancing and 'trouble-shooting' engineering analysis of HVAC systems. Mbr Florida Eng Soc. 1974 Chap Engr of Yr. 1975 Dist Serv Awd. 1976 Fellow. 1972-77 State Bd/Dir. 1972-75 City of St Petersburg, Florida Plng and Zoning Bd. Mbr ASHRAE. Mbr AABC. 1977 & 1978 State VP FL Engg Society. 1979 State Pres-Elect FL Engg Society. Hobbies: Cooking, bridge and fishing. 1980-81 State Pres FL Engrg Soc. *Society Aff:* ACEC, NSPE, FES, ASHRAE, AABC.

Woodward, James H
Business: 217 Education Bldg, Univ Sta, Birmingham, AL 35294
Position: Sr VP *Employer:* Univ of AL in Birmingham *Education:* PhD/Mechanics/GA Tech; MBA/Bus Admin/UAB; MSAE/Aero Engr/GA Tech; BSAE/Aero Engr/GA Tech *Born:* 11/24/39. Native of Columbus, GA. On faculty of USAF Academy 1965-68 and faculty of NC State Univ 1968-69. Dir of Tech dev with Rust Engg 1969-73. Prof of Civil Engg, Asst VP for Univ Coll, Dean of Engrg at the Univ of AL in Birmingham from 1973-1983. Assumed current responsibility as Sr VP at UAB in 1984. *Society Aff:* NSPE, ASEE, ASCE, AMA

Woodward, John B
Business: Ann Arbor, MI 48109
Position: Prof. *Employer:* Univ of MI. *Education:* PhD/Naval Arch/Univ of MI; MSE/ME/Univ of MI; BSE/ME/VPI & SU. *Born:* 1/23/27. Native of VA. Veteran of WWII. Ten yrs in ship design and construction, followed by grad study at the Univ of MI, leading to faculty position in naval arch & marine engg. Seven children. Author of textbooks and technical papers. Yachtsman, bicyclist, hiker, canoeist. *Society Aff:* ASEE, ASME, SNAME, IME.

Woodward, Richard J
Home: 689 Terra Calif Dr 2, Walnut Creek, CA 94595
Position: Chrmn of Bd Emeritus. *Employer:* Woodward-Clyde Consultants. *Education:* MS/Civil Engg/Univ of CA, Berkeley; BS/Geologic Engg/CO College. *Born:* Dec 1907. Various nontech jobs, until 1941-45 with US Navy Buships at Vallejo Calif. Lecturer CE at UC Berkeley 1946-50. Founding partner, Woodward-Clyde Consultants and predecessor firms; Bd Chrmn 1955-73. Resp for developing firm into one of international recognition in geotechnical, geological and environmental consulting. On Engr'g Criteria Board for the S F Bay Conservation & Development Comm. for several years. FIDIC Standing Comm. on Prof. Liability Insurance; Chmn 1971-76. Director Struc. Engrs. Assoc. of No. Calif., 1958-60. Member, States Tech. Services Comm., US Dept. of Commerce, 1967-70. Bd of Regents Holy Names College, Oakland, CA 1964-1972. Decorated Knight of Magisterial Grace, Sovereign Military Order of Malta. Recipient of Middlebrooks Award, ASCE 1964, The Past-Presidents Award for outstanding contributions to the profession, ACEC 1972. Chmn Geotechnical Comm ASCE, 1968-69. Founding President of Design Professionals Insurance Co. Co-author of Earth and Earth-Rock Dams, Wiley 1963 and Drilled Pier Foundations, McGraw-Hill 1972. Distinguished Engrg. Alumnus Award, UC Berkeley 1983. *Society Aff:* ASCE, ACEC, ΣΞ,XE, ΔE.

Wool, Richard P
Business: Dept of Materials Science and Engrg, Univ Illinois, 1304 W. Green, Urbana, IL 61801
Position: Professor *Employer:* Univ IL *Education:* PhD/Mat Sci & Engr/Univ UT-SLC; MS/Mat Sci & Engr/Univ UT-SLC; BSc/ChE/Univ Coll Cork, Ireland *Born:* 5/26/47 Born and raised in Cork, Ireland. Wife Deborah (Artist), daughters, Sorcha, 12, Meghan, 8 and Bree 5. Major research areas, polymer interfaces, molecular mech of polymers and biodegradable plastics with research group of 3 postdoctoral Fellows, 12 graduate research assistants and 3 undergraduates. Consultant for 10 industrial companies. Presently, 10 yrs at IL, spent 2 yrs on Chem Engrg faculty at CCNY, 1 yr on Mech Engrg faculty in Colorado and 4 yrs in UT. Has 70 publications, book chapters and review articles. Enjoys tennis, sailing and skiing. VP for research, Agri-Tech Industries specializing in corn starch based biodegradable plastics. *Society Aff:* AIChE, ACS, APS, MRS, SPE, AAAS, Soc Rheology

Woolsey, Lawrence S
Business: 5100 W 12th St, Little Rock, AR 72204
Position: Senior Partner. *Employer:* Crist Engineers. *Education:* BS/EE/Univ of Ark. *Born:* 1920. Native of Little Rock, Arkansas. Testing Engineer, Aberdeen Proving Ground, Md. WWII. With Crist Engineers 1945 to date. Organized & direct Tele-

Woolsey, Lawrence S (Continued)
phone Engineering Division. Share management of 50 employee consulting firm with two other Senior Partners. *Society Aff:* ACEC, IEEE, ASCE, TBΠ.

Wooten, Louis E
Business: P O Box 2984, Raleigh, NC 27602 *Employer:* L E Wooten & Co. *Education:* MS/Civil Engg/NC State Univ; BS/Civil Engg/MC State Univ. *Born:* 1/22/94. Instructor & Assoc Prof, NCSU fourteen yrs. Pres L E Wooten & Co 22 yrs. P Pres NC Chap ASCE & NC Chap ACEC. P Natl delegate ACEC. Mbr NC Bd of Registration for Engrs & Land Surveyors 1957-58. fellow mbr ASCE & ACEC. *Who's Who in America. Society Aff:* ACEC, ASCE.

Work, Clyde E
Business: 1215 Wilbraham Rd, Springfield, MA 01119
Position: Dean of Engrg *Employer:* Western New England Col *Education:* PhD/Engrg Mech/Univ of IL; MS/Engrg Mech/Univ of IL; BS/Phys/Univ of IL *Born:* 01/31/24 Born in NE, grew up in KS. Attended Sterling Coll and South East MO State Univ. Teaching and res appointments Univ of IL 1946-53. Assoc Prof RPI 1953-57. Dept hd and Assoc Dean, MI Tech Univ 1957-84. UNESCO expert, Maulana Azad Coll of Tech, Bhopal, India 1968-69. Visiting prof, Univ of Ilorin, Nigeria 1978-79. Assumed current position 1984. Mich Tech distinguished teacher 1948. ASTM Dudley medal 1954. ASEE Western Elec Award 1968. Distinguished educator Mechanics Div ASEE 1984. Pres SESA (now SEM) 1972-73. Author Statics and Dynamics texts and papers on education and structural materials. *Society Aff:* SEM, ASTM, ASEE, AAM, ASME

Worland, Robert B
191 Holly Springs Road, Franklin, NC 28734
Position: Quality Engr, Consultant. *Employer:* Retired. *Born:* Aug 1918 Montgomery Alabama. After Pre-med, Emory Univ, served 4 yrs with US Air Force during WWII. Discharged rank of Captain. An ASQC Fellow and Certified Quality Engr with over 30 yrs of Quality Mgt experience in the Electronics & Missile fields. With Martin Marietta Aerospace for 19 yrs, holding varying positions in Quality Mgt until retirement. Selected for 'Roll of Honor 1969' for outstanding professional contributions in Quality Control education. Chairs 'Advisory Cttee on Education for Quality Control Reliability'; serves on Engrg Technology Advisory Councils of area Community Colleges. Director ASQC 1971-75. Inspecting Engr for BIE, Internatl-Houston and EDG, Inc-Houston; BIE- British Inspecting Engrs; EDG-Engg Design Group. *Society Aff:* ASQC.

Worley, Burton J
Business: 516 W Jackson Blvd, Chicago, IL 60606
Position: Vice President *Employer:* C M St P & P Railroad Co. *Education:* MBA/Mgt/Univ of Chicago; BSCE/Engg/IA State College. *Born:* 11/15/17. Born in Aredale, IA, Nov 15, 1917 - parents Alanzo Sweringen and Cora Elena. Grad from IA State College in July 1939 and then entered service of the Milwaukee RR. Married H Lorraine (Foss) in Dec 1943. Three children Susan, Robert and John. Received MBA from Univ of Chicago in 1966. Appointed to present position of VP of the Milwaukee Road March 1, 1980. Presently resides in Deerfield, IL. *Society Aff:* AREA, AAR.

Worley, Frank L, Jr
Home: 2735 Talbot, W Univ Pl, TX 77005
Position: Prof., Assoc Dean Engrg *Employer:* Univ of Houston. *Education:* PhD/Chem Engr/Univ of Houston; MS/Chem Engr/Univ of Houston; BS/Chem Engr/Univ of Houston. *Born:* 10/9/29. Employed by Stauffer Chem Co 1952-57. With Univ of Houston since 1957. Fulbright Lecturer (Ecuador) 1959-61; Visiting Scientist, EPA at Res Triangle park 1971. ASEE Western Electric Fund Award 1975; Teaching Excellence Award, Univ of Houston 1972. Prog Chrmn 83rd Natl Meeting AIChE; local section officer (3 terms) S TX Sec, AIChE. Dir, Univ of Houston Environmental Wind Tunnel; consultant to govt and private organizations on air pollution problems. Maj res/profl interest: Physical and mathematical modeling of transport, dispersion and reaction of air pollutants. *Society Aff:* AIChE, ASEE.

Worley, Will J
Home: 2106 Zuppke E2, Urbana, IL 61801-6706
Position: Prof of Theoretical & Appl Mechanics. *Employer:* Univ of IL. *Education:* PhD/Engg/Univ of IL; MS/Theo & Appl Mechanics/Univ of IL; BS/Mech Engr/Univ of IL. *Born:* 8/2/19. Principal investigator, USAF contracts and Natl Aero and Space Admin contracts. Contributed papers include creep and fatigue of composite mtls, stress analysis, small and large deflections of plates, optimum design geometry for I-beams with web cutouts and optimum design geometry of shells of revolution. Also papers on analysis of vibration isolation systems with nonlinear characteristics and on direct synthesis of vibration isolation systems. Current interests include active vibration isolation in particular of man in transportation systems. Pres, Worley Systems, Inc, consulting firm engaged in approaches to prevention and analysis of system failure including accident investigation and reconstruction. *Society Aff:* ASME, ΣΞ, ΠΜΕ, ΠΤΣ.

Wormser, Alex F
Home: 24 High St, Marblehead, MA 01945
Position: President. *Employer:* Wormser Engineering Inc. *Education:* BE/ME/Yale Univ. *Born:* July 1932. Journalist with Air Force 1954-56. With General Electric 1957- 64, as advance development engr in gas turbines and instruments. Founder and president of Wingaersheek, Inc 1964-71, specializing in high-performance combustor systems. Formed Wormser Engineering in 1974, Mfg of coal-fired fluidized-bed combusters.Mbr of American Flame Res Cttee, Combustion Institute, AIChE. Has 13 patents. *Society Aff:* AIChE.

Wormser, Eric M
Business: 88 Foxwood Rd, Stamford, CT 06903
Position: President and Dir of Engrg. *Employer:* Wormser Scientific Corp. *Education:* BS/Mech Engr/MIT; Grad/Physics/NYU *Born:* Germany 1921. Graduated MIT 1942 in Mech Engrg. Grad studies physics/ optics, NYU. Research and Development of infrared detectors, instrument and systems 1946-73. 1952-73 with Barnes Engineering Co, Stamford, CT. 1967-73 Exec VP. 1974- formed Wormser Scientific Corp, Consultants on Design of Solar and Alternate Energy Systems. Fellow OSA, CIAA. Mbr ISES. Dir: New England Br ISES and NY Metropolitan Br ISES. Mbr IEEE, ISA, ASHRAE, AAAS, Inst of Physics. *Society Aff:* OSA, ASES

Worobel, Rose
Home: 39 Hampton Ct, Newington, CT 06111
Position: Retired *Education:* MS/Math/RPI; BS/Math/Bates College. *Born:* 7/10/20. Native of Hartford, CT. Hamilton Standard from 1940-77, engineering aide to senior analytical engineer. Responsible for assisting the Aerodynamics Design Analysis Group in solving special problems; propellers, shrouded propellers, wind turbines, Q-Fans, & Prop-Fans. 1977-Retirement in 1983. Recipient of the 1973 SAE Manly Award. *Society Aff:* AIAA,SWE.

Worrell, Wayne L
Business: 3231 Walnut St, Phila, PA 19104
Position: Prof. *Employer:* Univ of PA. *Education:* PhD/Met/MIT; BS/Met/MIT. *Born:* 10/25/37. Assoc Dean, Grad Educ and Res, School of Eng. and Appl. Science, Prof of Mtls Sci & Engrg at the Univ of PA, Phila, PA. At the Univ of PA since 1965. Co-editor of *Progress in Solid State Chemistry.* U.S. Rep on the IUPAC Commission on High Temperatures & Refractory Mtls. Served on several Natl Res Council Committees, including a three yr term as Chrmn of the NRC Committee on High Temperature Sci & Tech. Div Review Committee Mbr for several Natl labs. Consultant to numerous Govt & Ind labs. Published over 80 technical papers, 5 patents, Ed. 2 books. Res interests include electrical and chemical properties of

Worrell, Wayne L (Continued)
ceramics, ceramic sensors and ceramic coatings. *Society Aff:* AIME-TMS, ASM, ECS, ACerS.

Worthington, Donald T
Home: 5516 Flag Run Drive, Springfield, VA 22151
Position: Consultant (Self-employed) *Education:* MSEE/Sys Engg/VPI & SU; BSEE/Comm/Univ of KY. *Born:* Nov 1921. m. Nellie P., two sons, Donald T Jr and David Lee. Ind Coll of Armed Forces 1966. 1948-50 Capehart/Farnsworth mobil communications div & the govt contract sec. 1950 staff engr C&N Labs, Wright Patterson AB. 1951 Air Material Comm staff engr Aircraft Control and Warning Sys. 1952 staff engr spec in problems relating to electronic maintenance of USAF systems. 1953-62 Rome Air Dev Ctr, head system engr on the communications for missile and space sys. 1962- 72 Defense Comm Agency R&D engr for data distribution equipments etc. 1972 Chf of R&D Office, Defense Comm Engrg Ctr. 1979 Consultant Computer and Information Security. Sr mbr IEEE. Author of numerous tech pubs. *Society Aff:* IEEE, AFCEA.

Wortley, Charles Allen
Business: 1513 University Ave, Rm. 266, Madison, WI 53706
Position: Assoc Dean *Employer:* Univ of WI-Madison *Education:* PD/-/Univ of WI; MS/Civ/CA Inst of Tech; BS/-/Antioch Coll. *Born:* May 1934.BS Antioch Coll, MSCE Calif Inst Tech. P.D., Univ of Wisconsin-Madison. Since 1974 with UW-Madison working with adult engrs continuing education. 1957-66 Partner with Lorenzi, Dodds and Gunnill, Cons Engrs and 1966-74 Chf Engr VP Warzyn Engr Co Cons Engrs. Resp for engrg design and mgt. Named Engr of Yr in Private Practice and Engr of the Year in Education by Wisc Soc of Prof Engrs, Pres of of Wisc Sec ASCE and Wisc Prof Engrs in Priv Practice. Hold NCEE Natl Engr Certificate. Licensed Prof Engr and Land Surveyor. Frequent contributor to prof and tech publs. *Society Aff:* NSPE, ASCE, IAHR, ASEE.

Wortman, David B
Business: 1305 Cumberland Ave, PO Box 2413, W Lafayette, IN 47906
Position: Pres *Employer:* Prikske & Assoc, Inc. *Education:* MS/IE/Purdue Univ; BS/IE/Purdue Univ *Born:* 7/30/51 David B. Whortman is President of P&A, Pritsker & Assoc, Inc., a computer software firm specializing in simulation. With P&A since co-founding the firm in 1973, Mr. Wortman has been actively engaged in the development of computer simulation techniques and their application to industrial problems. He has authored numerous papers and presentations on simulation and its applications. He serves as Assoc Ed of SIMULATION, is a Sr. Mmbr of IEE & recip of their 1985 Outstanding Young Indust Engr Award, and holds membership in numerous other professional societies. *Society Aff:* SCS, IIE, ACM, AMA, APICS, SME, PMI

Wozney, Gilbert P
Home: 304 Hunter's Ridge Rd, Concord, MA 01742
Position: Mgr Engrg. *Employer:* General Electric Co. *Education:* BS/ME/Univ of CO. *Born:* 12/11/26. Grad of Genl Elec Advanced Eng Prog A, B, and C Courses. Grad work at MIT and Union College. Joined GE Co in 1949. Held engrg and managerial positions in Matls & Processes Labs, specializing in vibration and mech behavior of matls. Held managerial positions in Gas Turbine Dept. Appt Mgr, Engrg, Mech Dr Turbine Dept in 1971. Assumed current position in 1974 as Manager-Engrg, responsible for design and dev of steam turbines & gears for marine & navy, steam turbines & geneators for power generation & heat recovery steam generators & high performance industrial gearing. Chmn, Hudson-Mohawk Sec ASME 1969; Mbr Advisory Ctte for NY State Comm Colleges 1965-69; Mbr Natl Common College Physics 1967. Mbr ASME. *Society Aff:* ASME.

Wozniak, Joseph J
Business: 200 Ross St 1007, Pittsburgh, PA 15219
Position: Executive Director. *Employer:* Public Pkg Auth of Pittsburgh. *Born:* Jan 4, 1934. Univ of IL BSCE 1958. Was a Traffic Engr with the City of Chicago 1958-60. As the City Traffic Engr of Joliet, IL 1960-66, was resp for the planning, development and execution of all traffic engrg and parking projects. In 1966 joined the Public Parking Authority of Pitsburgh as the Director of Planning. In 1968 was apptd Exec Dir of the Authority and am resp for the administration and execution of the Authority's 52.4 million dollar system. Former Pres of the Mid Atlantic Sec, ITE. Current Pres of the Institutional Municipal Parking Congress. Enjoy water skiing hunting and motorcycle trail riding.

Wray, John M, Jr
Home: 2643 Shoreham Dr, Bon Air, VA 23235
Position: Dir of Operations. *Employer:* VA Dept of Hgwys & Transportation. *Education:* BS/Civ Engg/VA Military Inst. *Born:* 5/8/20. Native of Richmond, VA. Entered engg training prog in VA Dept of Hgwy & Transportation after serving four yrs in WWII with Armored Cavalry in Europe. Carried out engg tasks in VDH&T, including bridge & rd proj inspections, mtl testing, rd & bridge design, & surveying. Have held engg mgt positions at all levels in VDH&T, beginning as Asst Resident Engr through Resident Engr, Asst Dist Engr, Asst Maintenance Engr, Maintenance Engr, and currently Dir of Operations Operations is responsibe for all field operations in this State, & has staff responsibilities of the Dir of maintenance, construction & equip for the Dept. Retired Army Reserve Colonel. Hobbies: tennis & fishing. *Society Aff:* ASCE, AASHTO.

Wray, William R
Home: 4226 Sleepy Hollow Rd, Annandale, VA 22003
Position: Deputy Chf of Engrs. *Employer:* U S Army. *Education:* MS/Civil Engg/TX A&M; BS/Military Engg/USMA. *Born:* Dec 1925. Native of DeQuincy, LA. Served in US Army Corps of Engrs since 1946. Was officer-in-charge of crew which started up and operated Army's first nuclear power plant. One of initial investigators of feasibility of using nuclear explosives for large scale excavation. General officer since 1973, serving in Office, Chief of Engineers as Deputy Director (VP) of Military Construction, as Army's first Director of Facilities Engineering, as Asst Chief of Engrs resp for Army installations planning, facilities programming and environmental matters, as Dir of Military Programs for Corps of Engrs resp for all aspects of Army facilities worldwide-programming, construction, operation and maintence & currently as Deputy Chief of Engineers. Major general. Fellow in SAME. *Society Aff:* SAME, APWA.

Wright, Daniel K, Jr
Home: 2346 Tudor Dr, Cleveland, OH 44106
Position: Armington Prof Emeritus of ME. *Employer:* Case Western Reserve Univ. *Education:* MS/ME/Case School of Applied Science; BS/ME/Case Sch of Appl Sci. *Born:* Sept 6,1913 Belleville NJ. Instructor 1937 Case. Res Engr 1940-41 Plymouth Cordage Co. Returned to Case 1941. Prof 1953, Armington Prof 1973. Chrmn Dept of Solid Mech, Structures, and Mechanical Design 1970-75. Acting Dean of Engrg 1973-74, Chrmn Dept of Mech Design 1975-76. Chrmn Cleveland Sec ASME 1947-48, Chrmn No Ohio Chap SESA 1959-60. Natl Exec Cttee SESA 1959-61, Tech Adv Cttee 1975. Fellow ASME 1974. Max M Frocht Award for Excellence in Teaching SESA 1975. Fellow SESA 1977. Armington Prof Emeritus of ME 1982. *Society Aff:* ASME, SEM.

Wright, Douglas T
Business: University of Waterloo, Needles Hall, Waterloo, Ontario N2L 3G1 Canada
Position: President *Employer:* Univ. of Waterloo *Education:* BASc/Civil Engg/Toronto; MS/Structures/IL; PhD/Engg/Cambridge. *Born:* Oct 1927. Hon degrees: D.Eng Carleton, LL.D Brock, DSc Memorial of Newsfoundland LL.D, Concordia. Athlone Fellow at Trinity College. Duggan Medal EIC. President's Medal CGRA. Centennial Medal for Canada. Fellow Engineering Institute of Canada and American Society for Civil Engineers. Professor and Chairman Dept of Civil Engineering University of Waterloo 1958-63. Dean of Engineering U of Waterloo 1959-66.

Wright, Douglas T (Continued)
Chairman Ontario Ctte on University Affairs 1967-72. Chairman Ontario Comm on Post-Secondary Education 1969-72; Deputy Proj Mgr Secy for Soc Dev 1972-79; Deputy Minister for culture & recreation 1979-1980; present post since July 1, 1981. Tech Ctte service includes numerous EIC, ASCE, Canadian Standards Association, National Research Council of Canada technical cttes incl Chmn of Standing Ctte on Structural Design for National Building Code Cons Engr on spec structs incl Dutch & Mexican Pavilions at Expo 67 & 1968 Olympic Sports Palace. Ontario Place Dome and Forum, 1971; Member, Technical Evaluation Committee, Toronto Domed Stadium, 1984-5; developed comprehensive theories for structural analysis and design of large reticulated shells; author or co-author of numerous papers on structural engineering. Member, Premier's Council on Science and Technology (Ontario); member, Prime Minister's National Advisory Board for Science and Technology; member of Council (representing Canada) for International Institute for Applied Systems Analysis, Laxenburg, Austria. *Society Aff:* ASCE, EIC, IABSE, IASS, AAM.

Wright, Edward S
Business: John Deere Rd, Moline, IL 61265
Position: Dir Gov't Products *Employer:* Deere & Co. *Education:* M.S./Aero mechanical Eng./Princeton U.; B.S./Aero mechanical Eng./Air Force Institute of Technology; B.S./Agriculture/Rutgers U. *Born:* 01/18/30 Directs all activities of John Deere Relative to research, development, design, demonstration, production, and service for new products for governmental applications, including commercial derivations of same. Directs all Deere activities in the Rotary and Stirling engine arena. Serves as Pres of John Deere Technologies International, Inc, a wholly-owned subsidiary of Deere & Co responsible for operation of Deere's Rotary Engine Div and other technology related initiatives. Previously served as Mgr of Component & Engine Planning at Deere (1979-83) and in a variety of energy conversion and engine application projects at United Technologies Research Center (1963-79). *Society Aff:* ASME, ASAE, SAE, ASNE

Wright, Farrin S
Business: Tidewater Research Center, Suffolk, VA 23437
Position: Agric Engr *Employer:* USDA ARS *Education:* PhD/Ag Eng/NC State Univ; MS/Ag Eng/Clemson Univ; BS/Ag Eng/Clemson Univ *Born:* 12/3/36 Native of Cleveland Co., NC. Summer employment with Clemson Univ as Res Asst 1957, 58, and 59. Received BSAE 1959, Clemson Univ. Pursued MSAE 1959-60 with the support of a Nat'l Cotton Council Fellowship. Pursued PhD. AE 1960-66, NC State Univ. Conducted resch in vegetable mechanization, cabbage and sweet potatoes. Improved storage and handling facilities for sweet potatoes. Served USDA ARS 1966-present. Conducted resch to improve harvesting and handling equip of peanuts; developed a direct peanut harvester; studied the effect of tillage production practices on peanut yields and diseases and evaluated irrigation practices and their effects on yield of peanuts, corn, and soybeans. *Society Aff:* ASAE,APRES

Wright, John C Y
Business: 210 Charles St, Sistersville, WV 26175
Position: Pres & Treas and CEO. *Employer:* Wiser Oil Co. *Education:* BA/-/Denison Univ; BEM/Petr Engg/OH State Univ. *Born:* 5/28/25. Masters Petroleum Engr Ohio State Univ 1950. m.Becky Bell Lewis, Lewisburg, WV. Pres, Treas, Dir The Wiser Oil Co. Chairman Director First Tyler Bank and Trust Co. PPres and Trustee Ohio Oil and Gas Assn. VP & Exec Cttee, Kentucky Oil and Gas Assn. Independent Petroleum Assn of Amer Exec Cttee. Sigma Gamma Epsilon, Sigma Chi, Presbyterian Church Elder-Trustee. Texnikoi Award Outstanding Engr Grad Ohio State U 1958. Jt Citizen of Yr Sistersville (with wife) 1973. West Virginia 'Oil and Gas Man of Yr' 1976. *Society Aff:* ΣΓΕ, ΣΧ.

Wright, Kenneth R
Business: 2490 W 26th Ave, Suite 55-A, Denver, CO 80211
Position: President *Employer:* Wright Water Engineers Inc *Education:* BS/Civil Eng/Univ of WI; BBA/Bus/Univ of WI; MS/Civil Eng/Univ of WI. *Born:* 3/10/29. Construction Engr with Arabian American Oil Co 1951-56. US Bur/Rec 1957-58. Partner, Wheeler & Wright water engrs to 1961. President of Wright Water Engrs Inc 1961 to present. Managing partner in Wright-McLaughlin Engrs 1964-1982. Active in planning, design and construction engrg for dams, water and sewer systems, flood control and drainage. Water rights and resources consultant. Member Executive Committee ICID. PChrmn Hydraulics Div, Surface Water Hydrology Cooperation with Local Sections; ASCE. Chrmn AWWA Water Rights Ctte 1982- present. Treas, 12th Congress ICID. Coauthor, Urban Storm Drainage Mgt, Marcel Dekker Inc, 1982. Mbr NSPE, AWWA, APWA, past Dir of ROMCOE, COSF, Pres CMC Foundation. Interior Gold medal for Valor 1958. Past Chrmn, CO State Bd of Registration for Prof Engrs and Land Surveyors. Enjoy mountaineering, scuba, river rafting, photography, and skiing. *Society Aff:* ASCE, NSPE, AWWA, ICID,APWA.

Wright, Marshall S, Jr
Home: 2022 Mock Orange Ct, Reston, VA 22091
Position: Chief Contract Mapping *Employer:* US Geological Survey *Education:* BS/CE/VA Polytech Inst *Born:* 1/22/22 Commissioned officer. Corps of Engrs 1943-46. With US Geological Survey 1946-49 and the Engr Res and Dev Lab 1949-53. Chief Engr of Jack Ammann Photogrammetric Engrs from 1953-61. Was VP and Dir of Surveys and Mapping for Lockwood, Kessler & Bartlett, Inc. Syosset, NY From 1961-75. Rejoined US Geological Survey in 1975. Has served as committee chairman of committees in ASCE and NSPE. Was elected Pres of the American Society of Photogrammetric in 1973. *Society Aff:* ASP, ASCE

Wright, Richard H
Home: 1437 Country Squire Dr, Decatur, GA 30033
Position: Owner. *Employer:* Val/Tec Enterprises. *Education:* BS/ME/GA Tech. *Born:* 9/28/35. BSME (Co-operative) GA Inst of Tech. Native of Monroe, GA. With Ordnance Corps US Army (1st Lt) 1959-61. Associate with Lazenby and Borum for 14 yrs. With Edward and Rosser, Consulting Engrs for 3 1/2 yrs as V P, treas and partner-in-chg of mech engrg. Formed Wright Engg Associates 1973. Mbr of ASHRAE. Reg PE in GA. Enjoys sailing canoeing, music, reading. *Society Aff:* ASHRAE.

Wright, Richard N
Home: 20081 Doolittle St, Gaithersburg, MD 20879
Position: Dir. *Employer:* Ctr for Bldg Tech. *Education:* PhD/Civ Engg/Univ of IL; MSCE/Civ Engg/Syracuse Univ; BSCE/Civ Engg/Syracuse Univ. *Born:* 5/17/32. Syracuse, NY. Employed: PA RR 1953-55; US Army 1955-57; Univ of IL, Urbana 1957-74; Natl Bureau of Stds Ctr for Bldg Tech, Chief Structures Sec 1971-72, Deputy Dir Technical 1972-73, Dir since 1974. Lab profl staff of 150 providing technical bases for performance stds & measurement tech for more useful, safe & economical bldgs. Developed maj progs in computer-integrated const, solar tech & earthquake hazard reduction. Reg engr in NY & structural engr in IL. Res publications in structural analysis, behavior & design, & formation & expression of stds. Pres, Intl Council for Bldg Res, Studies and Documentation, 1983-86. Married Teresa Rios 1957, children: John, Carolyn, Elizabeth & Edward. *Society Aff:* ASCE, ASEE, AAAS, RILEM, EERI.

Wright, Robert E
Home: 28 Kingsbridge, Bristol, TN 37620
Position: Consultant *Employer:* Self employed *Education:* BSME/-/Univ of WI. *Born:* Nov 28,1917 Oconto Wisc. Employed: Monsanto Co 1940-66 as Draftsman, Plant Engr, Design Engr, Mgr of Design,Supt of Design and Construction, Dir of Enrg, Dir of Engrg & Dev, Dir of Process Design, Proj Mgt & Intl Engrg. Wheeling-Pittsburgh Steel Corp 1966-72 as VP Engrg, VP Corp Engrg, VP Corp Dev. Enterprise Fabricators Inc 1972-1981 Chrm of Bd of Dir 1981 to present as Conslt on Coal & Energy. Mbr ASME, AIChE. Reg engr in Missouri, Florida and West Virginia. *Society Aff:* ASME, AIChE.

Wright, Robert R
Business: 6326 Presidential Ct, Fort Myers, FL 33907
Position: Sr Proj Mgr *Employer:* Post, Buckley, Schuh & Jernigan Inc. *Education:* MS/CE/GA Inst of Tech; MS/Environ Engr/WA Univ; BS/CE/Univ of MO-Rolla *Born:* 3/5/37 Professional engr registered in five states with over 20 yrs of environmental engrg experience in both the private and public sector. Currently assistant regional magr and principal associate of a large (700 employees) consulting engrg firm. Functioned as sr project mgr of numerous multi-million dollar environmental engrg projects. Former experience and positions have included director of environmental engrg div for another consulting firm and water dept head of the Water and Wastewater Technical School. *Society Aff:* AAEE, NSPE, ASCE, AWWA, WPCF

Wright, Robert S
Business: 4000 W Chapman Ave, Orange, CA 92668
Position: Vice President. *Employer:* Woodward-Clyde Consultants. *Born:* April 1923. MS 1965 Univ of Calif. BS 1948 Univ of Ill. Native of Minneapolis Minn. Served in US Navy 1944-46. Engaged in structural and foundation engrg for design and construction of railroad facilities, bridges, and heavy industrial structures for EJ & E Ry Co, Texaco Inc and Kaiser Engrs 1948-60. Responsible for field engrg during construc of 10,000,000 cu yd Briones Dam, tunnels and aqueducts for East Bay Municipal Utility Dist 1960-65. Joined Woodward-Clyde Consultants in 1965. Engaged in geotechnical engrg for design and construction of earth dams, bridges, bldgs and underground structures. Assumed current responsibilities as Mgr of So Calif offices in 1975. Mbr ASCE, ISSMFE, ISRM, SEAONC, CEAC, Chi Epsilon, EERI. Reg PE Calif Illinois and Michigan. Enjoy music, flying, sailing and fishing.

Wright, Roger N
Home: 12 Maria Court, Rexford, NY 12148
Position: Prof & Exec Officer *Employer:* Rensselaer Poly Inst. *Education:* ScD/Metallurgy/MA Inst of Tech; BS/Mtls Sci/MA Inst of Tech. *Born:* 11/14/42. Native of Lawrenceville, IL. Sr Res Metallurgist at Allegheny Ludlum Res Ctr 1968-1971. Sr Engr at Westinghouse Res Labs 1971-1974. With Mtls Engg Dept at RPI since 1974, currently Prof & Exec Officer. Interested in large scale metal deformation and fracture, metal working, and lubrication. Active as a consultant specializing in metal working, wire mfg and stainless steel applications. Active in mfg education. *Society Aff:* TMS-AIME, ASME, WA, SME, ASM.

Wright, Theodore O
Home: 18-6842 ALii Drive #2-203, Kailua-Kona, HI 96740
Position: Forensic Engineer, retired *Education:* MS/EE/Univ of IL; BS/EE/Univ of IL *Born:* 1/17/21 Native of Gillette, WY. Junior Appraisal Engr prior to World War II. Served with the Army Air Corps - US Air Force 1941-1965 as a pilot, assistant Lab Chief and Deputy for Engrg, Titan SPO. Was a research specialist and specialist engr with Boeing 1965-81 in a variety of technical and staff assignments. Registered PE 1966. Have published four technical papers - two internationally. NSPE-PEI Vice Chrmn, Western Region 1975-1977. WSPE Distinguished Service Award 1980. WSPE Pres 1981-82. Certified Advanced Metrication Specialist 1981. NSPE Vice Pres Western Region 1985-87. Maintain commerical pilot rating. Enjoy classical music and fishing. *Society Aff:* NSPE, USMA, ASEE, ANMC

Wright, Walter L
Business: 2255 S.W. Canyon Rd, Portland, OR 97201
Position: Sr Vice President. *Employer:* Shannon & Wilson Inc. *Education:* MS/CE/Univ of CA; BC/CE/Univ of WA. *Born:* Feb 1923. Native of Portland Ore. Served in the Army Air Corps 1942-47. Disch rank 1st Lt. Completed formal tech edu 1950. Varied engrg background and experience with govt agencies, aerospace industries, and consultants through 1954. Joined Shannon & Wilson Inc in 1955 and became an Assoc with the firm in 1961. Became a principal in 1964 (VP) in firm's Portland, OR office. Primarily interested in soil mechanics and foundation engrg as well as engrg geology. Still has intense interest in aviation. President CECO 1969-70. *Society Aff:* ASCE, SAME, NSPE, ACEC.

Wright, Wesley F, Jr
Business: 4447 North Central Expressway, Dallas, TX 75205
Position: President. *Employer:* Encon Corp. *Education:* BS in ME/Engg/Univ of TX. *Born:* March 1922. Engrg Officer, destroyers, minesweepers, USNR WWII. Zone Mgr, Servel A C Div 1947-56. Engr Zumwalt & Vinther Consulting Engrs 1956-58. Mgr of Industrial Dept Southern Union Gas Co 1959-68. AGA Award of Excellence for Development of Total Energy Concept. Vice Chrmn, National Engine Use Council. Exec VP, Zumwalt & Vinther 1968-75. Encon Corp 1975. Projects: Central Utility Plant and Tunnel DFW Airport, 25500 Tons. Parque Central, Caracas, 17500 Tons, Centro Empressarial Sao Paulo 6000 Tons. Master Utility System, NTSU. Mbr NSPE, TSPE, ACEC, CEC-T, ASHRAE, PE in Texas. *Society Aff:* ASHRAE, ACEC, CEC-T, NSPE, TSPE Association of Energy Engineers.

Wu, Jain Ming
Business: Univ Tenn, Space Institute, Tullahoma, TN 37388
Position: Bernhard H. Goethert Prof of Aero Eng & Dir, Gas Dynamics Div. *Employer:* University of Tennessee. *Education:* BS/Mech Engg/Natl Taiwan Univ; MS/Aeronauticss/CA Inst of Technology; PhD/Aeronauticss/CA Inst of Techology. *Born:* 8/13/33. Nanking, China. Naturalized US Citizen. Assoc Fellow AIAA 1976. PChrmn AIAA, Tenn Sec. Mbr AIAA Natl Tech Cttee of Missile Systems, Aircraft Design, Sigma Xi Society, Am Physical Society. Visiting Prof, Von Karman Institute for Fluid Dynamics, Rhode- st.-Genese, Belgium, 1970. Awarded Kon-du-Scholar, Natl Taiwan Univ, Teaching Asstship CIT, Institute Scholar CIT, Anthony Scholar CIT, General H H Arnold Awd for Outstanding Contributions Toward Advancing State-of-the-art of Aerodynamic and Astronautical Sciences, AIAA, Tenn Sec 1970. Bernhard H. Goethert Professor, 1985, conferred Honorary Prof., Beijing Institute of Aeronautics & Astro. 1985, Nanjing Inst. of Aeronautics, 1986, Northwestern Polytechnical University, China, 1986. Listed in Amer Men in Sci, Who's Who in Aviation, Men of Achievement, Vol II, etc also Pres, Engg Res and consulting, Inc Consultant to Arnold Engg Dev Center, Boeing Aerospace Co, Seattle and Huntsville, AL. Misserachmitt Bolkow-Blohm Gmbh, Germany. Nielsen Engrg & Res, Inc, MTN View, CA. *Society Aff:* AIAA, ΣΞ, AM. PHY. S.

Wu, James C
Business: Atlanta, GA 30332
Position: Prof *Employer:* GA Inst of Tech *Education:* PhD/ME/Univ of IL; MS/ME/Univ of IL; BS/ME/Gonzaga Univ *Born:* 10/5/31 in Nanking, China. Come to the US in 1953. Naturalized US Citizen, 1964. Married 1957 to Mei-Ying Chang, two children: Alberta Yee-Hwa, 23, Norbert Mao- Hwa, 20. Experience: Assistant Prof, Gonzaga Univ, 1957-59. Research associate, MIT, 1957. Research Specialist - Research Branch Chief, Douglas Aircraft Co, 1959-1965. Professor of Aerospace Engrg, GA Inst of Tech, 1965-present. Contributor to many technical research journals, officers of professional societies, consultant, Chrmn of Bd of Dir, the Chinese American Inst. *Society Aff:* AIAA, APS, ΣΞ

Wu, Keh C
Business: 1 Northrop Ave/3872/62, Hawthorne, CA 90250
Position: Sr Technical Specialist. *Employer:* Northrop Corp. *Education:* MS/Metallurgical Engg/Rensselaer Poly Tech; BS/Mechanical Engg/Natl Chung Cheng Univ. *Born:* Feb 1923. Native of Soochow, Kiangsu, China. Mech Engr for Chinese Petroleum Corp 1946-55. Physical Metallurgist for Watervliet Arsenal, NY 1959- 63. Senior Tech Spec for Northrop Corp, Aircraft Div, 1963 to present. Resp for research programs and contracts in welding metallurgy of titanium alloys and superalloys. Developed weldbonding process for anodized aluminum alloys. Authored and co-authored 28 papers published in national and intl journals in the field of welding. P Dir at Large AWS. PE in Metallurgical Engrg in Calif. Enjoy classical music and photography. *Society Aff:* AWS, ASM.

Wu, S M
Business: Mech Eng and Applied Mechanics Dept, Univ of Michigan, Ann Arbor, MI 48109-2122
Position: Professor. *Employer:* Univ of Wisconsin. *Education:* PhD/ME/Univ of Madison. *Born:* Oct 1924 Chekaing China. Married, two children. Prof of Mech Eng at Univ of Wisc-Madison since 1968. Recipient of AWS's C H Jennings Memorial Awd 1968, ASME's Blackall Machine Tool & Gage Awd 1968, SME's National Education Award 1974, Research Professor in Manufacturing Engineering 1981 and R. & P. Anderson Professorship at the University of Michigan-Ann Arbor 1987. Consultant to a number of industries and govt agencies. Has published extensively in ASME Transactions, IEEE Transactions, Biometrika, Technometrics, JASA, Management Science, Wear, and Internal Journal of Machine Tool Design and Research. *Society Aff:* ASME, SME, ASA, ASEE.

Wu, Tien H
Home: 160 Brookside Oval E, Worthington, OH 43085
Position: Prof *Employer:* OH St Univ *Education:* PhD/CE/Univ of IL; MS/CE/Univ of IL; BS/CE/St John's Univ *Born:* 3/2/23 Worked as civil engr with Deleuw, Cather and Co, Chicago, IL Div of Highways; and Giffels and Valet, Detroit. Taught CE at Michigan St Univ from 1953 to 1965. Has been Prof of CE at OH St Univ since 1965, specializing in geotechnical engrg and soil mechanics. Served as visiting prof at Natl Univ of Mexico, Norwegian Geotechical Inst, Royal Inst of Tech, Stockholm, Punjab Agric Univ, Ludhiana, and Southwest Jiaotong Univ., Chengdu. Practices as a consultant on geotechnical engrg problems. *Society Aff:* ASCE, TRB, ASTM.

Wuensch, Bernhardt J
Home: 190 Southfield Rd, Concord, MA 07142
Position: Prof of Ceramics. *Employer:* MA Inst of Tech. *Education:* PhD/Crystallography/MA Inst of Tech; SM/Phys/MA Inst of Tech; SB/Phys/MA Inst of Tech. *Born:* 9/17/33. Native of Paterson, NJ. Married to Mary Jane Harriman, 1960; children: Stefan R (b 1969), Katrina R (b 1971). Res Fellow, Univ of Bern, Switzerland, 1963-64. Ford Postdoctoral Fellow in Engg (1964-66) & Asst Prof of Ceramics (1964), Dept of Mtls Sci & Engg, MIT; Assoc Prof 1969-74, Prof 1974-present. Acting Dept. Head 1980. TDK Chair of Materials Science and Engineering 1985-present. Visiting Prof, Univ of the Saarland, Saarbrucken, Germany, 1973. Physicist, Max-Planck-Inst Fuer Festkorperforschung, Stuttgart, Germany, 1981. Res interests in Diffusion in Ceramics, Fast Ion Conductors, Crystal Chemistry & Diffraction. Assoc Editor, 1978-1980, *Canadian Mineralogist*, Advisory Editor, 1977-1985, *Physics and Chemistry of Minerals*. Editoral Board, 1981-present, *Zeitschrift fuer Kristallographie*. *Society Aff:* ACS, MSA, ACA, MAC.

Wuerger, William W
Business: College of Engineering, 266 Mech Engrg Bldg, Madison, WI 53706
Position: Professor *Employer:* Univ of WI-Madison *Education:* MS/Mech Engg/Univ of WI; BS/Mech Engg/Univ of WI. *Born:* 4/5/34. Native of Green Bay, WI. Joined the staff of the Univ of WI in 1961 as a coordinator of Engg Inst. Appointed Assoc Dean, Coll of Engrg in 1984. Has instructed courses on engg economy, life cycle costing, CPM, cost accounting, plastics tech, principles of industrial engg, materials handling, plant layout. Experience with Aero Div of Honeywell and Midwest Univ Res Assoc (MURA). Reg PE in WI. Currently Pres of WSPE. Past pres of the Southwest Chapter WSPE and past secy treas of WSPE and past pres Madison-Rock Valley Area Chapter of the IIE. Holds a single-engine commercial pilot license with instrument rating. *Society Aff:* WSPE/NSPE, ASEE.

Wulf, Raymond J
Business: 343 State St, Rochester, NY 14650
Position: Projects Director. *Employer:* Eastman Kodak Co. *Education:* AAS/Photo Technolgoy/RIT; BChE/Chem Eng/NY Univ. *Born:* June 24,1930 New York City. AAS Photographic Tecnology, Rochester Inst of Tech. B Chem Eng, New York Univ. Tau Beta Pi, Fellow and Governor SMPTE. Joined Eastman Kodak Co in 1953. Participated in service engrg to convert motion picture laboratories to use of color film 1953-60. Assisted USAF, NASA establish laboratory facilities at Cape Kennedy 1960-65. Coordinated East Coast efforts to conert TV broadcasters to use of color news film 1964-70. Chief Engr East Coast Div 1967-70. Dir Sales Dev for the Motion Picture and Audiovisual Markets Div 1970-73. Proj Dir 1973, Mgr Corp Dev 1976-. *Society Aff:* SMPTe.

Wulff, Jack G
Business: 1023 Corporation Way, Palo Alto, CA 94303
Position: Pres *Employer:* Wahler Associates. *Education:* BS/CE/Univ of NV. *Born:* 2/20/29. Born in Sacramento, CA. BSCE Univ of NV. 1950, 2nd Lt USAF 1951-53 From 1950-68 employed by State of CA Dept of Water Resources, progressing to Chief of Earth Dam Design in Div of Design & Construction. Since 1968 engaged in consulting Civil & geotechnical eng, attaining position as Presdent of Wahler Associates. Mbr of USCOLD, ASCE, CEAC. Mbr of USCOLD Exec Ctte, Dam Safety Ctte and Ctte on Materials for Embankment Dams. Author of numorous tech publications on dams. Enjoy boating & water skiing. Reg civil eng in 15 states. *Society Aff:* USCOLD, ASCE, CEAC.

Wulpi, Donald J
Home: 3919 Hedwig Dr, Ft Wayne, IN 46815
Position: Metallurgical Consultant (Self-employed) *Education:* BS/Met Engg/Lehigh Univ. *Born:* 5/2/24. Retired from Intl Harvester after 31 yrs, has specialized in failure analysis education & prevention. Taught self-developed course for 16 yrs & wrote book on "How Components Fail–, published by ASM in 1966. Also has done considerable met specification & mtls engg work. *Society Aff:* ASM

Wunderlich, Milton S
Home: 515 Cleveland Ave S, St Paul, MN 55116 *Born:* March 11,1891 Faribault Minn. BS Minnesota 1919, ME 1920. Chf Engr Flaxlinum Co 1920-25. Gen Mgr 1925-29. Dir of Research 1927-32. Research Engr Insulite Co 1932-36. Gen Plant Mgr Int Falls Minn 1936-42. Gen Sales Mgr 1942- 45. Minn Ontario Paper Co Dir of Res 1945-56. Mbr ASME, ASHRAE-Life Mbr. Mbr Tau Beta Pi, Sigma Xi. Fellow ASME. Retired 1958.

Wunderlich, Orville A
Business: P.O. Box 2117, Baltimore, MD 21203
Position: VP & Gen. Mgr. *Employer:* Davison Chemical Division W R Grace *Education:* BS/CHE/Univ of IL. *Born:* 7/7/24 BSChE Univ of IL 1949. Res Engr Pure Oil Co 1949-54. Various technical service-mktg position Davison Div W R Grace 1954-70 in field of catalysis. Genl mgt 1970-73. Assumed position respon for technical activities 1973 as VP-Dev. VP-Marketing 1976. VP & Gen Mgr, Davison Speciality Chemical Co 1978. VP & Gen Mgr, Davison Chemical Div W R Grace 1985. Indus and Emission Control Catalysts and Rare Earths. *Society Aff:* AIChE,ACS

Wundrack, William A
Business: 801 N 11th Blvd, St Louis, MO 63101
Position: VP. *Employer:* Sverdrup & Parcel & Assoc, Inc. *Education:* Profl/EE/Univ MO; BS/EE/Univ MO. *Born:* 6/18/27. Native of St Louis. Entered the engg profession in 1949 with MO Public Service Commission as a field engr. Started with Sverdrup & Parcel in 1951 as elec designer. Responsibilities included proj mgr for design of numerous advanced aero & automotive res & dev facilities including Wind Tunnels & engine test. Experience includes Logic Control Systems for missile launchers. Assumed positions of VP of Sverdrup/Aro & Sverdrup & Parcel in 1977. Mbr of Sverdrup Corp Advisory Council for Sci, Tech & Engg. Reg PE in several states & one foreign country. Mayor City of Sunset Hills, MO. *Society Aff:* AIAA, SAE.

Wyeth, Nathaniel C
406 Ring Road, Chadds Ford, PA 19317
Position: Sr Engrg Fellow. *Employer:* E I du Pont de Nemours & Co. *Education:* BS/ME/Univ of PA. *Born:* Oct 1911. Native of Chadds Ford, PA. Started employement with Du Pont in 1936 after graduation from Univ of PA and assigned to Engrg Dept. After several field assignments as Industrial Engr was apptd Sec Supvr in charge of equipment development. In 1953 became Dev Proj Mgr. In 1960 an Asst Dir of the Mech Dev Lab. In 1963 was named Du Pont's first Engineering Fellow. In 1973, Sr Engrg Fellow. Mbr of SPE, ASME, Delaware Soc of Prof Engrs, NSPE. Hon Mbr of Pi Tau Sigma, named a Fellow by the ASME, appt a mbr of the Museum and Planetarium Cttee of the Franklin Inst. Enjoy building 1/8 scale models of period furniture. Received Achievement Award in Engineering/Technology from SPE May 1981. Retired from Du Pont in 1976. Have been a consultant for Du Pont since retirement. *Society Aff:* ASME, SPE, DSPE, NSPE.

Wylie, Frank B, Jr
Home: 8301 Dinah Way, Louisville, KY 40222
Position: Sr VP-Retired *Employer:* Hazelet & Erdal, Inc. *Education:* MS/Structural Engrg/Univ of TN; BS/CE/MS State Coll *Born:* 4/3/15 in TN, grew up in MS. Worked for American Bridge Co 1937-1942, 1946-1950. Served with Army Antiaircraft Artillery 1942-1944. Seriously wounded in Italy 1944. Retired Major, 1947. Assistant Prof, Civil Engrg, Univ of TN, 1952-1954. With Hazelet & Erdal, Consulting Engrs, since 1954 as senior engr, bridge and structural design and specifications. Assoc 1966, Partner 1976. Sr VP 1982. Registered PE in PA, TN, KY, IL (Structural) and MS. Married, have three children. Member of Methodist Church. Interested in engrg education. Enjoy music, opera and nature. Retired from H & E, 1985. *Society Aff:* ASCE, NSPE, ΤΒΠ, TROA.

Wylie, Robert R
Home: 49 Purchase St, Danvers, MA 01923
Position: Illumination Cons. *Education:* BS/Illumination/MA Inst Tech. *Born:* July 26,1915. Employers, Curtis Lighting, NYC, Morgan Hamel & Engelken, NYWF 1939; Century Lighting; Cincinnati Gas & Electric. Joined Sylvania 1944 as Commercial Engr. Primary concentration applications field with architect and engr and Govt Agency contact, Industrial and Commercial Lighting design. PChrmn Industrial Lighting Cttee IES. PC Chrmn Roadway Lighting Cttee IES. PC Chrmn Papers Cttee IES. Reg PE Mass and NH. PC Chrmn Roadway Lighting Forum Cttee IESNA. Illumination Consultant. *Society Aff:* IESNA.

Wyllie, Loring A, Jr
Business: 350 Sansome St, Suite 500, San Francisco, CA 94104
Position: VP *Employer:* H.J. Degenkolb Assoc, Engrs *Education:* MS/Structural Engrg/Univ of CA-Berkeley; BS/CE/Univ of CA-Berkeley *Born:* 8/21/38 Native of Oakland, CA. After advanced education and two yrs in US Army, joined H.J. Degenkolb Assoc in San Francisco in 1964 and currently VP with that firm. Responsible for structural design of various projects and for investigations of structural distress and failures. Consultant to numerous corporate clients and expert witness in various matters of litigation. Pres, San Francisco Section ASCE, 1980-81. Mbr of ACI Ctte 318, Standard Building Code. Author of numerous papers on structural and earthquake engrg. *Society Aff:* ASCE, EERI, ACI

Wyman, Richard V
Home: 610 Bryant Ct, Boulder City, NV 89005
Position: Prof and Chmn, Dept of Civil and Mech Engg *Employer:* Univ of NV Las Vegas *Education:* BS/Geol Math/Case Western Reserve Univ; MS/Mining Geology/Univ of MI; PhD/Geological Engrg/Univ of AZ *Born:* 2/22/27. and grew up in Painesville, OH, son of Vaughn E and Melinda C Wyman. Served in the US Navy in WWII; studied at Case Inst and Western Reserve Univ, Univ of MI and Univ of AZ. In 1947 married Anne Fenton in Cleveland, OH. Career has been mostly in mining in the Western states and South America as an engr, geologist and corporate officer of various cos. Was asst operations mgr for the Reynolds Elec at NV Atomic Test Site, and is active as a consultant. He joined the faculty at UNLV in 1969. Currently Professor Civil Engg and Chairman of Dept of Civil and Mechanical Engrg at Univ of NV, Las Vegas. Wyman's have one son, William Fenton Wyman. Reg PE (NV) Lic Water Right Surveyor (Nev) Reg Geologist, (CA & AZ) *Society Aff:* AIME, ASCE, AEG, SEG, GSA, ΣΞ,ΦΚΦ.

Wymore, Albert Wayne
Home: 4301 Camino Kino, Tucson, AZ 85718
Position: Principal Systems Engr. *Employer:* SANDS: Systems Analysis and Design Systems. *Education:* PhD/Math/Univ of WI; MS/Math/IA State Univ; BS/Math/IA State Univ. *Born:* 2/1/27. in New Sharon, IA. Married Muriel L Farrell on March 19, 1949. 4 children: Farrell, Darcy, Melanie & Leslie. I was first Dir of the Computing Ctr at the Univ of AZ, first Hd of the Dept of Systems & Ind Engg & first editor of the Journal of Systems Engg. My principal publications are: "A Mathematical Theory of System Engineering: The Elements–, Wiley, 1967, and "Systems Engineering Methodology for Interdisciplinary Teams–, Wiley, 1976. Soon to appear: "A Mathematical Theory of System Design." Retired from the Univ of AZ, June, 87. Now Principal Systems Engr with SANDS: Systems Analysis and Design Systems. *Society Aff:* AAAS, IIE, AMS.

Wynblatt, Paul P
Home: 2303 Clearvue Road, Pittsburgh, PA 15237
Position: Prof of Met Eng & Mtls Sci *Employer:* Carnegie-Mellon Univ *Education:* PhD/Met/Univ of CA, Berkeley; MS/Met/Israel Inst of Tech; BScTech/Met/Univ of Manchester. *Born:* 6/30/35. Israel Atomic Energy Commission, 1958-62: performed res on uranium alloys. Ford Motor Co Scientific Res Lab, 1966-81. Performed res on mass transport phenomena in metals, stability of supported metal catalysts, reactions on metal surfaces, surface composition of alloys, and headed technological assessment/forecasting activity. Carnegie-Mellon Univ, since Aug. 1, 1981: appointed Prof in the Depts of Metallurgical Engrg & Materials Science, and Director of the Center for the Study of Materials since July 1, 1985. Author of more than 70 articles in technical literature. Dir of the Metallurgical Soc (TMS) of AIME, 1977-80. Mbr and officer of various TMS-AIME natl committees. *Society Aff:* TMS-AIME, ASM, AVS, MRS.

Wyndrum, Ralph W, Jr
Business: AT&T-Bell Labs, Holmdel, NJ 07733
Position: Dir, Syst Analysis Center *Employer:* AT&T-Bell Labs *Education:* BS/EE/Columbia Univ; MS/EE/Columbia Univ; Eng ScD/EE/NYU; MS/Bus Adm/Columbia Univ. *Born:* 4/20/37. in NY City. Adj Prof at NY Univ & NCE. Bell Telephone Labs 1963-date. Dir, Syst Analysis Center 1987. Respon for development of product planning, realization, and development tools and process analysis. Author of 40 papers, 6 patents. Editor, Communication Society Transactions. Eta kappa Nu Outstanding Young Electrical Engineer-Hon Ment. Fellow IEEE. Technical Program Chrmn ICC 1976, ISSLS 1978. IEEE Governing Board, 1980-82; Chmn, IEEE Comm Soc Conference Bd 1981-87. Adjunct Prof, 1981-88, Stevens Inst of Tech. *Society Aff:* IEEE.

Wyner, Aaron D
Business: 600 Mountain Ave, Murray Hill, NJ 07974
Position: Hd, Comm Analy Res Dept. *Employer:* Bell Telephone Labs. *Education:* PhD/EE/Columbia Univ; MS/EE/Columbia Univ; BS/EE/Columbia Univ; BS/Math - Physics/Queens Coll. *Born:* 3/17/39. Native of New York City. Has been full and part-time faculty mbr at Columbia and PINY. Since 1963 has been doing res on comm and info theory at Bell Labs. 1969-70 on leave at Weizmann Inst of Sci, Rehovot Israel on Guggenheim Fellowship. Pres of IEEE Info Theory Group 1976 and Fellow of IEEE. Editor, *IEEE Trans in Information Theory*. *Society Aff:* IEEE.

Wyrick, Roger K
Business: 1100 Circle 75 Pkwy, Atlanta, GA 30339
Position: Mgr *Employer:* Inst of Nuclear Power Operations *Education:* MS/Nuclear

Wyrick, Roger K (Continued)
Engrg/Univ of IL; MS/Phys/Univ of CA-Davis; BS/Phys/Indiana Univ *Born:* 12/23/46 Mgr, Operatinal Data Analysis Sect, Inst of Nuclear Power operations (INPO), Atlanta, GA. MS, Nuclear Engrg, U of IL; MS, Phys, U of CA; BS, Phys, Indiana Univ. INPO: respon for equip reliability analysis and transient analysis progs; also involved in INPO evaluation of significant nuclear plant transients and events. FL Power & Light Co: Supervising Engr for Thermal-Hydraulic & Safety Analysis Group; reviewed plant safety analyses and performed thermal-hydraulic and sys transient analyses; mbr of Combustion Engrg Owners Group analysis subctte. Combustion Engrg: Lead Engr in design Transient Safety Analysis group, prepared FSAR safety analyses and developed improved computer models. FERMI Natl Accelerator Lab: staff mbr of accelerator div; participated in dev and operation of Fermi Lab's proton synchrotron. Amer Phys Soc mbr. ANS: Mbr, Session Chrmn at 1978 winter meeting and 1980 summer meeting. Participated in ANS paper reviews. Mbr of Prog Ctte and Session Chrmn for topical meeting on Anticipated & Abnormal Plant Transients in LWRs, Sept 1983. Mbr of NRSD Exec Ctte, 1982-85. *Society Aff:* ANS, APS

Wyskida, Richard M
Home: 225 Spring Valley Ct, Huntsville, AL 35802
Position: Prof *Employer:* The Univ of AL in Huntsville *Education:* PhD/IE/OK St Univ; MS/IE/Univ of AL; BS/EE/Tri-State Coll *Born:* 9/2/35 Native of Perrysburg, NY. Electrical engr for Philco Corp, specializing in missile circuit design. Staff engr with NASA-MSFC, specializing in Technical management. With the Univ of AL in Huntsville since 1968. Research has concentrated upon modeling thermosensitive cushioning materials for application to missile container design. In 1980, published a book entitled *Modeling of Cushioning Systems*, Gordon & Breach Science Publishers. *Society Aff:* IIE, ORSA, ASEE

Wyss, Ralph G
Business: 5930 Beverly, Mission, KS 66202
Position: Dir of Operations. *Employer:* Water Dist 1 of Johnson County. *Education:* BSME, Univ. of Kansas *Born:* Jan 1925 Kansas City Missouri. Army Infantry 1943-46. Project Engr, Sr Engr and Asst Chf Engr,, Kansas City, Missouri Water Dept 1950-59. Research Engr, Smith & Loveless 1959-61. Asst Chf Engr, Chf Engr and Dir of Operations of Water Dist No. 1 of Johnson County, Kansas 1961-present. Reg Engr Kansas, Missouri, Florida. Mbr AWWA 1959,APWA 1980 ASCE 1964. Kansas Sec AWWA Chrmn 1972, Mbrship Chrmn 1972-76. AWWA Diamond Pin Club 1975, AWWA Ambassador Award 1976 and AWWA Natl Dir 1976-79 AWWA George Warren Fuller Award 1977.. BSA-Order of Arrow 1976. Activities: Boy Scouts, bowling, fishing.

Wyszkowski, Paul E
Home: 30 Skyline Dr, Warren, NJ 07060
Position: Consulting Engr. *Employer:* Paul E Wyszkowski, Cons Engr. *Education:* BSCE/CE/Newark College of Engg. *Born:* 3/10/32. Native of NJ. BS in CE 1954 Newark College of Engrg, NJ Inst of Tech. Background incls heavy construction with NY State Dept of Public Works, transportation engg with Louis Berger Associates Consulting Engrs & environmental engrg with CA Dept of Public Health NJ Dept of Health and as VP of Charles J Kupper Inc Consulting Enrs. Operated a private environmental consulting engrg practice 1971-1981. In 1981 joined AT&T Bell Laboratories as Environmental Manager. Reg PE in NJ, NY, PA, OH, AL & CA. Diplomate, AAEE. Mbr ASCE, WPCF, AWWA, NSPE, NJSPE & Chi Epsilon listed in Who's Who in East. *Society Aff:* AAEE, ASCE, WPCF, AWWA, NSPE, NJSPE.

Yacavone, David M
Home: 1888 W Loveland Ave, Loveland, OH 45140
Position: Manufacturing Program Mgr. *Employer:* General Electric Co. - Aircraft Engines *Education:* MBA/Mgnt/Xavier Univ./BE/Met Engg/Youngstown State Univ. *Born:* 10/3/48. Reg PE, State of CA, in Quality Engg. Chrmn of Akron, OH ASM Chapter - 1977, Chrm. of Natl ASM Young Mbrs Advisory Committee 1979. & Chrm. of Miami Valley Sec of ASNT-1981. Formerly with Babcock& Wilcox Co. Naval Nuclear Div & was Sr Quality Engr/Level III Examiner with Goodyear Aerospace Corp. With GE Co, since 1977. Assumed current responsibility as Manufacturing Program Manager in 1986. Other positions held at G.E. Co. were Manufacturing Shop Mgr, Sr. Nondestructive Examination Development Eng, & Program Mgr, Automated Mfr. Cells.

Yachnis, Michael
Home: 4201 Military Rd NW, Washington, DC 20015
Position: Chief Engineer. *Employer:* Naval Facilities Engrg Command. *Education:* DSc/Str Engrg/GW Univ; MEA/Engrg Adm/GW Univ; MSE/Str Engrg/GW Univ; BSCE/Str Engrg/M Tech Coll, Greece; BS/Engrg/Mil Acad, Greece. *Born:* March 22 1922. BS Military Coll, Athens Greece 1943; BSCE M Tech Coll, Athens Greece 1951; MSE Geo Wash Univ 1956, MEA 1962, DSc 1968; 16 Certificates & Diplomas from various univ & sch. Reg PE in DC, Md, Va & Calif. Fellow ASCE; Mbr MTS, WSE, AWS, SAME, NSPE, UMS, Sigma Xi, Xi of Sigma Tau, NAE. Ordained Priest in Greek Orthodox Church. 1943-56 2nd Lt to Maj Corps of Engineers, Greek Army; 1956-63 Specialist & Branch Mgr Structural Engrg Branch Chesapeake Div, Naval Facilities Engrg Command (NAVFAC), 1963-66 Sect Hd Structural Engrg Sect; 1966- 72 cons, Deep Water Structures, since 1972 Ch Engr. Author of 10 maj tech & scientific papers. George W Goethals Medal ASCE, ASCE award for chmnship of Tech Council in Ocean Engrg; Meritorious Civilian Serv Awd; Superior Civ Serv Awd; Achievement - Sr Exec Serv Awd; The George Washington Univ Alumni Achievement Awd; The Dean's of the Sch of Engrg GWU Outstanding Achievement Awd; Engr of the Yr Award by the DC Council of Engrs and Architects; The Sr Exec Serv Achievement Awd; The Presidential Rank Awd for Distinguished Sr Exec Serv. *Society Aff:* ASCE, MTS, WSE, AWS, SAME, NSPE, UMS, NAE.

Yacoub, Kamal
Business: Dept of Elec Engg, Univ. of Miami, Coral Gables, FL 33124
Position: Chrmn. *Employer:* Univ of Miami. *Education:* PhD/EE/Univ of PA; MSEE/EE/Univ of PA; BS/EE/Israel Inst of Tech. *Born:* 11/11/32. Born in Nazareth-Palestine; came to the US in 1956 and was naturalized in 1967. Received PhD in 1961 from the Univ of PA where he taught until he moved to the Univ of Miami in 1966. He is presently Prof and for 15 years served as Chrmn of the Elec and Computer Engrg dept. Areas of specialty are Statistical Communication Theory and Time- Series Analysis. In 1971 he was a recipient of a Res Award from the American Soc for Engg Education. *Society Aff:* IEEE, ΣΞ.

Yaggee, Frank L
Home: 503 Otis St, Downers Grove, IL 60515
Position: Staff Metallurgist. *Employer:* Argonne Natl Lab. *Education:* MS/Met Engg/Univ of MI; BS/Met/Univ of MI; -/Foreign Language/Northwestern Univ. *Born:* 9/4/23. Native of Dearborn, MI. Served with US Navy 1943-1946. Staff metallurgist Mtls Sci Div, Argonne Natl Lab 1949-present. Specialized in all aspects of applied res, dev, & mfg of reactor fuels & cladding 1949-1965. Current resonsibilities include mtls consultant to Reactor Analysis & Safety Div, Safety Res Experiment Facilities, and Reduced Enrichment Res Test Reactor Proj at ANL. Applied mtls res on Light Water Power Reactors leading to predictive models for cumulative damage and cladding breaches. Dev of integrated test methods & systems to simulate environs of nuclear, solar, chem, and conventional power systems. ASM Fellow 1978. *Society Aff:* ASM, ANS, RESA.

Yaggy, Ronald V
Business: 717 3rd Ave. S.E, Rochester, MN 55901
Position: Pres *Employer:* Yaggy Assoc., Inc. *Education:* BS/CE/Univ of IA *Born:* 7/4/36 Native of Mason City, IA. Civil Engr with Wallace, Holland, Kastler & Schmitz, Consltg Engrs 1959-1961 and 1964-1970. Service engr with Northwestern

Yaggy, Ronald V (Continued)
States Portland Cement Co. 1961-1964. Founded Yaggy Assoc, Inc. 1970; currently Pres. Registered PE in MN, IA and WI. *Society Aff:* ASCE, NSPE, AWWA, ACEC.

Yaghoubian, Jack Nejdge F
Business: 1100 Glendon Ave 1000, Los Angeles, CA 90024
Position: Partner. *Employer:* Dames & Moore. *Education:* BS/Civil Engg/Univ of IL; Postgrad Dip/Environ Eng/Northerlands. *Born:* 8/8/34. BSCE Univ of Ill, postgraduate diploma in environ & earth sci from the Netherlands. Partner & Dir of gas market, extensive exper in managing & directing comprehensive worldwide environ studies of power plants, crude oil & natural gas pipelines, Lng liquefication & regasification plants & has given expert testimony in legal proceedings. Mbr ASCE, HRB, SSA, SEASE & CEAC; Reg CE in Calif, Fla, Ill, Penn & Colo. *Society Aff:* ASCE, HRB, SSA, SEASE, CEAC.

Yagoda, Harry Nathan
Business: 210 State St, Hackensack, NJ 07601
Position: Pres *Employer:* Computran Systems Corp *Education:* Ph.D./EE/Polytech of NY; MS/EE/NYU; B/E.E./City Univ. of NY *Born:* 5/19/36 Dr. Yadoda is founder and Pres of Computran Systems Corp. He currently directs a professional staff of about 30 in the application of advanced communications and computer technology to the solution of municipal system engrg problems. Dr. Yagoda designed the nation's first and oldest operational on-line Parking Garage Revenue Control System in Miami, FL; the nation's first and oldest operational traffic signal control computer in West Palm Beach, FL; and the nation's first major municipally owned cable communications system for Paterson, NJ. Dr. Yagoda is active in various professional societies and has served as an Intl Dir of the ITE. *Society Aff:* ITE, IEEE

Yanai, Hisayoshi
Home: 20-14, Amanuma 2-Chome, Suginami-Ku, Tokyo 167 Japan
Position: President & Prof.; Advisor *Employer:* Shibaura Inst of Tech., Toshiba *Education:* Dr of Engg/ EE/Univ of Tokyo; Bach of Engg/EE/Univ of Tokyo. *Born:* 5/19/20. in Japan. BE 1942 and Dr Engg 1953, Univ of Tokyo. Asst Prof 1942-47, Assoc Prof 1947-60 1960-81 Prof 1981- Prof Emer., Univ of Tokyo. Visiting Prof 1966 at Tech Univ Munich, Visiting Prof 1976 at Tech Univ Brunswick, Germany. 1981-Prof at Electrical Engg Dept., 1986-President. Shibaura Inst. of Tech. 1981-Adviser, Toshiba R&D Ctr. REs and educational fields are semi-conductor devices, integrated circuits and optoelectronic devices, Mbr of Japan Sci Council (Vice dir of 5th Div) 1981-85, since 1977 Fellow of IEEE. Distinguished Services Award 1970, two Paper Awards 1957, 1971 from IECE-J, International GaAs Symp. Award 1980 (H. Welker Medal), Das grosse Verdienstkreuz des Verdienstordens from President of FR. Germany 1985 Medal of Honor with Blue Ribbon from Jap. Government. 1986. *Society Aff:* IEICE, IEE-J, IEEE, VDE, OSA.

Yang, Cheng Yi
Home: 6 Welwyn Rd, Newark, DE 19711
Position: Prof *Employer:* Univ of DE *Education:* Doctor of Sci/Civil Engrg/MIT *Born:* 12/17/30 Born in China, American Citizen, undergraduate educ in China, graduate with MS degree at Purdue Univ and DSc at MIT. Taught Civil Engrg at Univ of IL. Now a Prof at Univ of DE. Specialties: Structural Dynamics and Random Vibration, with applications to Earthquake and ocean structural design. A book on "RANDOM VIBRATION OF STRUCTURES" is due to be published in 1986. *Society Aff:* ASCE

Yang, Chih Ted
Business: U.S. Bureau of Reclamation, P.O. Box 25007, Denver, CO 80225
Position: Civil Engr *Employer:* U.S. Bureau of Reclamation *Education:* Ph.D./CE/CO St Univ; M.S./CE/CO St Univ; B.S./CE/Natl Cheng Kung Univ *Born:* 1/23/40 1968-1974, Assoc hydrologist of the IL St Water Survey. 1974-1976, Hydraulic engr of the U.S Army Corps of Engrs North Central Division. 1978, Adjunct Prof of the Univ of MN. 1979 to date, Civil Engr in the Technical Review Staff of the U.S. Bureau of Reclamation, Engrg and Research Center. 1983 to date, Prof Affiliate of CO St Univ and Prof Adj of Univ of CO-Denver. Winner of the 1972 AGU R.E. Horton Award, 1973 ASCE W. L. Huber Research Prize, 1978 U.S. Army Corps of Engrs Certificate of Achievement, and 1980 ASCE J. C. Stevens Award, 1983 US Bureau of Reclamation and NSPE Engr of the Year Award, 1984 US Bureau of Reclamation Special Achievement Award and 1985 Performance Award, and 1986 Pakistan Engineering Congress Gold Medal. Professional experiences include research, teaching, design and technical review in the field of hydraulics, hydrology, sediment transport, river morphology, dams and other hydraulic structures. *Society Aff:* ASCE, AGU, IAHR.

Yang, Edward S
Business: Dept of Electrical Engr, New York, NY 10027
Position: Prof of Elec Engr. *Employer:* Columbia Univ. *Education:* PhD/Engrg & Appl Sci/Yale; MS/Elec Engrg/OK State; BS/Elec Engrg/Taiwan Cheng-Kung. *Born:* 10/16/37. in China and currently Prof of Elec Engg at Columbia Univ, He was associated with IBM Corp first as an engr at the Component Div and later on as a visiting faculty mbr at Watson Research Center. His research interests include p-n junction, heterojunction and Schottky diodes, transistors, SCR, LED, injection laser, solar cells and CRT phosphors. He is the author of a book entitled *Fundamentals of Semiconductor Devices,* McGraw-Hill, 1978. Has published more than 70 articles in technical journals. *Society Aff:* IEEE, APS, MRS.

Yang, Kwang-tzu
Business: Dept Aero/Mech Engrg, Notre Dame, IN 46556
Position: Prof. *Employer:* University of Notre Dame. *Education:* BSME/Mech Eng/IL Inst of Tech; MS/Mech Eng/IL Inst of Tech; PhD/Heat Transfer/IL Inst of Tech. *Born:* Nov 1928, China. 1955 joined Dept of Mech Engrg at Univ Notre Dame as Asst Professor, currently Viola D. Hank Professor of Engineering, Dept Aerospace & Mech Engrg; taught heat transfer & thermodynamics, actively involved in res in boundary layer theory, free convection, unsteady flows, turbulent shear flows with heat transfer, atmospheric diffusion, fire res & hydrodynamic lubrication. Cons to Dodge Mfg Div of Reliance Electric Co in stress analysis, therman analysis, dynamics, tech forecasting & long range planning. Cons to Tyler Refrigeration Corp in refrig sys. Conslt to Nuclear Regulatory Commission. Fellow ASME 1975. ASME Heat Transfer Mem Award 1981. *Society Aff:* ASME, ASEE, AIAA, AAUP, ΣΞ, ΠΤΣ, ΤΒΠ.

Yang, Tah-teh
Business: 315 Riggs Hall Fernow St, Clemson, S.C 29634-0921
Position: Prof *Employer:* Clemson Univ *Education:* Ph.D/Mechanical Engr./Cornell Univ.; M.S./Mechanical Engr./Okla. State Univ.; B.S./Mechanical Engr./Shanghai Institute of Tech *Born:* 08/15/27 Tah-teh Yang was born in Shanghai China. After completing his graduate studies at Cornell Univ he worked at Wright Aeronautical Div of Curtiss Wright Corp, 1960, specialized in internal aerodynamics. He joined mechanical engrg faculty of Clemson Univ in the Fall of 1962 and has held the rank of professor since 1969. He is an ASME Fellow. His early research activities were subsonic diffusers and inverse solutions of internal flows. His recent research activities are applied computational fluid mechanics, regenerative gas turbines, and high efficiency gas turbine designs. *Society Aff:* ASME, AIAA

Yang, Wen-Jei
Home: 3925 Waldenwood Dr, Ann Arbor, MI 48105
Position: Professor *Employer:* University of Michigan *Education:* Ph.D./Mech. Eng./University of Michigan, Ann Arbor; M.S./Mech. Eng./University of Michigan, Ann Arbor; B.S./Mech. Eng./National Taiwan University *Born:* 10/14/31 Born in Taiwan, China. Research Engineer for Ford Scientific Lab, 1957-58 and Mitsui Shipbuilding Company, 1960-61. Fellow Engr for Westinghouse Electric Corp., 1967. Have been with Univ of Mich, Ann Arbor since 1961 and Prof of Mech Engrg since 1970. Invited Visiting Prof at Univ of Tokyo, 1975, Guest Prof at Technical Univ of Berlin, 1975-76 and Visiting Prof at Univ of Bologna, 1986.

Yang, Wen-Jei (Continued)
Consultant to industry and govt, both domestic and foreign. Served in various capacities for prof societies and organizations. Received numerous awards in research, teaching and service from professional societies, university and foreign institutions. Enjoy classical music and swimming. *Society Aff:* ASME, AIAA, AAAS

Yarar, Baki
Home: 13260 Braun Rd, Golden, CO 80401
Position: Prof *Employer:* CO Sch of Mines *Education:* BSc/Chem/METU-Ankara; MSc/Chem/METU-Ankara; PhD/Min Tech/Univ of London; DIC/Min Tech/Imperial Coll-London *Born:* 02/28/41 Obtained PhD and DIC (1969) from Imperial Coll, as a Ford Fdn and Univ of London Scholar. Taught at METU till 1979. Was visiting Prof at UBC, Vancouver BC (1979). Joined CSM in 1980. Active mbr of professional socs, life mbr and VP of Sigma Xi CSM chapter. Was president of Denver Extractive Metallurgy chapter (1986-87) Currently director of CSM Corrosion research center. Holder of 2 certificates of service to Engrg Fdn and one for NATO-ASI on Mineral Processing Design. Speaks Turkish, German, Arabic and English; author of over 70 scholarly publs, co author of one book and co editor of "Interfacial Phenomena in Mineral Processing–. Also Co editor of "Mineral Processing Design–. Worldwide consltg an lecturing experience. *Society Aff:* AIME (SME)

Yarbrough, David W
Business: Box 5013, Cookeville, TN 38505
Position: Prof and chair *Employer:* TN Tech Univ. *Education:* PhD/ChE/GA Inst of Tech; MS/ChE/GA Inst of Tech; BChE/ChE/GA Inst of Tech. *Born:* in CA. Attended public schools in Wash, DC & Charleston, SC. Served two yrs as signal corps officer in US Army. Current active Army Reserve. Mbr of Chem Engg faculty at TN Tech Univ (TTU) since 1968. Served as Assoc Dean of Engr for grad studies & res (TTU) 1976-1979, Chairperson of Chem Engg (TTU) 1987- . Active researcher in energy related areas and physical properties. Prof of Chem Engr since 1976. Adjunct research participant with Oak Ridge Nat'l lab since 1979. Chairman ASTM Subcommittee C 16.21 since 1985. Pres. of TN Acad of Sci-1986. *Society Aff:* AIChE, ASEE, ASTM, ΣΞ, TN ACAD SCI

Yariv, Amnon
Business: 1201 California Ave, Pasadena, CA 91125
Position: Professor. *Employer:* California Institute of Tech. *Education:* PhD/Elec Engg/UC Berkeley; MS/Elec Engg/UC Berkeley; BS/Elec Engg/UC Berkeley. *Born:* 4/13/30. in Tel Aviv Israel. 1948-50 Israeli Army during war of independence. 1959- 64 Mbr of Staff Bell Labs; 1964- , CA Inst of Tech, teach & conduct res in areas of Opt Comm, lasers & electrooptics. Mbr & Fellow IEEE; Fellow OSA & Mbr NAE. 1980 IEEE Quantum Electronics Award. A Founder and Chrmn of the Bd of ORTEL, Inc. *Society Aff:* IEEE, OSA, APS, NAE.

Yates, L Carl
Business: 909 Rolling Hills Drive, Fayetteville, AR 72703
Position: Pres *Employer:* McGoodwin, Williams and Yates, Inc. *Education:* BS/CE/Univ of AR *Born:* 2/15/30 Born, reared, and educated in Northwest Arkansas. Affiliated with McGoodwin, Williams and Yates, Inc., since graduation from the Univ of Arkansas in 1958. Pres since 1966. As such, he directs the activities of the firm and is in overall charge of coordinating & supervising all personnel. He worked in all phases of design, construction & operation in the sanitary engg field. Coordinated the development & construction of two regional water systems, with a total cost in excess of $30 million, from their inception. Is active in several professional organizations with particular involvement in ACEC. *Society Aff:* ACEC, AWWA, ASCE, NSPE, WPCF, AAEE.

Yau, Stephen S
Business: Dept of EE & Computer Sci, Northwestern Univ, Evanston, IL 60201
Position: Walter P. Murphy Prof & Dept Chmn *Employer:* Northwestern Univ, Dept. of Elec Engg & Computer Sci. *Education:* PhD/EE/Univ of IL, Urbana; MS/EE/Univ of IL, Urbana; BS/EE/Natl Taiwan Univ. *Born:* 8/6/35. 1961 joined faculty of Elec Engrg Dept, Northwestern, 1968 became Prof Elec Engg, 1970 became Prof of Elec Engg & Computer Sci Dept when Computer Scis Dept was established, 1972-77 became chmn of Computer Sci Dept, 1977. 1977-87 became Chmn EE and Comp Sci Dppt since the two depts merged in 1977, 1986 became Walter P. Murphy Professor. Pres IEEE Computer Soc 1974-76, Dir IEEE 1976-77; Pres AFIPS 1984-86; dir AFIPS 1972-82; Genl Chmn 1974 Natl Computer Conference and the first IEEE Computer Soc Computer Software Application Conference, 1977; Franklin Inst Louis E Levy Medal 1963; Amer Acad of Achievement Award 1964. Richard E Merwin Award. IEEE Computer Soc, 1981. Centenial Medal, IEEE, 1984. Fellows the Franklin Inst 1972, Fellow IEEE 1973. Fellow AAAS, 1983. Mbr, ASEE, ACM, SIAM, AAAS, Sigma Xi, Tau Beta Pi, Eta Kappa Nu. Res Inst: computer Sys Reliability and Maintainability, Software Engrg, and Distributed Computer Sys. *Society Aff:* IEEE, ACM, SIAM, AAAS, ASEE, ΣΞ, TBΠ, HKN.

Yeargan, Jerry R
Business: Dept. of Electrical Engineering, University of Arkansas, Fayetteville, AR 72701
Position: Dept Hd. *Employer:* Univ of AR. *Education:* PhD/EE/Univ of TX; MSEE/EE/Univ of AR; BSEE/EE/Univ of AR. *Born:* 1/31/40. Native of AR, worked for TX Instruments 1961-1963, 1969. With the Univ of AR since 1967. Res and Publications in MOS devices. Dept Hd of Elec Engg since 1977. Consultant on Concentrator Photovoltaics. . *Society Aff:* IEEE, ASEE.

Yee, Alfred A
Business: 810 1441 Kapiolani Blvd, Honolulu, HI 96814
Position: VP & Technical Director *Employer:* Alfred A Yee Division, Leo A. Daly *Education:* Hon Dr Eng/-/Rose-Hulman Inst; M Eng/Structures/Yale Univ; BS/CE/Rose- Hulman Inst of Technology. *Born:* 8/5/25. In private practice since 1953. Holder of patents on offshore & oceangoing vessels, reinforcing steel bar splices & concrete framing sys for high rise bldgs. Honorary Mbr ACI & ASCE; NAE. Engineer of the Year Award 1969, HI Soc of Prof Engrs; honored in 1971 by McGraw-Hill Publications as 1 of "Men Who Made Marks in 1971–. Mbr PresStressed Concrete Inst 1972-79, currently mbr sev natl htech cttes on precast prestressed concrete construction. Mbr., National Academy of Engineers, 1976. *Society Aff:* PCI, ASCE, ACI, NSPE, IABSE, YEA, SNAME, FES, PTI, ABS, NAE

Yee, Phillip K H
Home: 1885 Paula Dr, Honolulu, HI 96816
Position: Chmn/Board *Employer:* Phillip K. H. Yee & Assoc. Inc. *Education:* B.S./CE/Univ. of HI *Born:* 2/19/16 Consulting Engr - Phillip K. H. Yee & Assoc, Inc. 243 Liliuokalani Ave Honolulu HI 96815 born Feb. 19, 1916 B.S. Univ of HI in Civil Engrg. Practice of professional engrg in Civil Structural & hydraulic branches. Practice encompassed planning, consultation, design of residences apartments, medical clinics commercial bldgs, water & sewage treatment plant facilities & structures, water storage & distribution systems, park & recreational facilities, schools, institutional bldgs. Engr & Asst Superintendent Suburban Water System, City & County of Honolulu, Head of Water Supply Section, U.S. Engrs Office, Preparation of "An Historical Inventory of the Physical, Social, Economic & Industrail Resources of the Territory of HI. Awarded Certificate of Appreciation from U.S. Dept of Commerce for furthering the intl Commerce of the United States as a member of Consulting engrs Councils Trade Mission to Southeast Asia in 1968. Cit Ambass Prog - People to People, Member of Env Water Mgmt Dele to the People's Republic of China, Sep 1983. Pres Ala Moana Investment Corp. Pres. Intercontinental Corp. Who's Who in the West, International Who's Who in Community Service, Two Thousand Notable Americans, World Biographical Hall of Fame, Amer Sec Council, US Congressional Advisory Board-State Advisor. Who's Who in the World (8th Edi. 1987-88 Marquis) Member: Ashtar Command. *Society Aff:* Engineering Associates

Yee, Sinclair S
Business: Seattle, WA 98195
Position: Prof *Employer:* Univ of WA *Education:* Ph.D./EE/Univ of CA-Berkeley; M.S./EE/Univ of CA-Berkeley; B.S./EE/Univ of CA-Berkely *Born:* 1/20/37 in China. Grew up in S.F., CA. Worked for Livermore Lawrence Lab (LLL) prior teaching, specilizing in Semiconductor materials, devices, quantum electronics. Since 1966, served as assistant, assoc and full prof of the electrical engrg dept, Univ of WA. Had been served as consultant to the U.S. Air Force, LLL, Beckman Instrument Co. NIH research fellow two years; IEEE Fellow; Executive VP, S.S. Service Corp. a subsidiary of Savings Association. Enjoys fishing and outdoor activities. *Society Aff:* IEEE

Yegulalp, Tuncel M
Business: 812 SWM, New York, NY 10027
Position: Faculty Mbr. *Employer:* Columbia Univ. *Education:* Dr of Engg Sci/Mining/Columbia Univ; MS/Mining/Technical Univ of Istaubul. *Born:* 11/5/37. Mbr of the faculty of the Henry Krumb Sch of Mines teaching courses in mining engg and decision making methodology appl in mineral industries. His res interests are application of operations res methodology and math statistics in problems of mining and geosciences. He has several publications in the areas of Extreme Value Statistics applications in Earthquake forecasting, simulation, stochastic modelling of mining operations. His most recent res area is underground coal-slurry transportation and related modeling problems, and developing methodology for optimum design of underground coal slurry transport networks. *Society Aff:* AIME, TIMS, ORSA.

Yeh, Hsi-Han
Business: EE Dept, Lexington, KY 40506
Position: Assoc Prof. *Employer:* Univ of KY. *Education:* PhD/EE/OH State Univ; MSc/EE/Univ of New Brunswick; MSc/EE/Natl Chiao-Tung Univ; BSc/EE/Natl Taiwan Univ. *Born:* 11/10/35. in Shanghai, China. Asst Engr for Taiwan Power Co before attending the Univ of New Brunswick, Canada. Joined the faculty of the Univ of KY in 1967. *Society Aff:* IEEE, ΣΞ.

Yeh, Hun C
Business: Cleveland, OH 44115
Position: Prof of Materials Science *Employer:* Cleveland St Univ *Education:* Ph.D./Metallurgy/IL Inst of Tech; M.S./Metallurgy/Brown Univ; B.S./MEI/Natl Taiwan Univ. *Born:* 6/27/35 Professional Employments: 1966-69, Senior Metallurgist, Covnining Glass works, corning, NY 1969-present, Assist. Assoc and full prof (present position), Cleveland St Univ, Cleveland, OH *Society Aff:* AIME-TMS, ACS

Yeh, Kung C
Business: Dept Electrical and Computer Engrg, University of Illinois, 1406 W Green St, Urbana, IL 61801
Position: Prof *Employer:* Univ of IL *Education:* PhD/EE/Stanford Univ; MS/EE/Stanford Univ; BS/EE/Univ of IL. *Born:* Aug 1930. Since 1958 with Dept Elec Engrg, Univ of IL at Urbana- Champaign, Prof since 1967. Fellow IEEE; Mbr AGU, APS & US Comm G and H of URSI. Visiting Prof at NTU 1966 & 1976, Visiting Fellow at the Univ of HI 1967, Assoc in Ctr for Advanced Study UI 1973-74. Received Certificate of Achievement Awd for 1 of many pubs; co-author of book *Theory of Ionospheric Waves.* Assoc Editor 1979-1951, guest editor of a special issue "Radar Returns from the Clear Air" in 1980 and Editor 1983-1981 of all of the journal Radio Sci. A US mbr of the Electromagnetic War Propagation Panel of AGARD, NATO, 1984-1989. *Society Aff:* AGU, APS, AMS, IEEE, URSI.

Yeh, Pai T
Home: 1718 Sheridan Dr, W Lafayette, IN 47906
Position: Prof. Emeritus of Civil Engrg. *Employer:* Purdue Univ. *Education:* PhD/CE/Purdue Univ; MS/CE/Purdue Univ; BS/CE/Natl Checkiang Univ. *Born:* 2/5/20. Native of Canton China. Naturalized in 1961. Engr for Kwok Wha Engg Dev Co, Kuming-Shaughai China 1944-48. Res engr for the Joint Hgwy Res Proj, Purdue Univ since 1953, specializing in air photo interpretation for engg soils and rocks. *Society Aff:* AAAS, ASEE, NSPE, SAME, ASP.

Yeh, William
Business: 4531 C, BH, UCLA, Los Angeles, CA 90024
Position: Prof. & Chairman *Employer:* Univ of CA. *Education:* PhD/Civ Engg/Stanford Univ; MS/Civ Engg/NM State Univ; BS/Civ Engg/Natl Cheng-Kung Univ. *Born:* 12/5/38. With UCLA since 1967. Currently Prof and Chairman, Civil Engineering Department. Areas of teaching and res include Hydrology, Water Resources & Optimization of Large-Scale Water Resources Systems. Has published more than 100 technical papers. Developed optimization models for the CA Central Valley Proj & the Central AZ Proj. Has done consulting work for the State & Federal Water Resources Agencies. Taught intl courses for UNESCO, Brazilian Fnd for Tech Dev of Engrg, & Mexican Govt. Received the 1975 UCLA Engg Alumi Assoc Distinguished Faculty Award. Received the 1981 Engineering Foundation Fellowship Award. Former intercollegiate table tennis champion. Enjoy sports. *Society Aff:* ASCE, AGU, AWRA.

Yelton, Richard L
Business: Old Ridgebury Rd, Danbury, CT 06817
Position: VP, Operations/Technology *Employer:* Union Carbide Corporation. *Education:* BS/Chem Engg/Univ of Cincinnati. *Born:* Oct 1925, Erlanger Ky. 1943-45 aviation cadet in Army Air Force. ChE Univ of Cincinnati in 1949; graduate work at Carnegie-Mellon Univ 1950. With Union Carbide Corp since 1950, involved with petrochemicals mfg in various capacities from Process Engr to Plant Mgr for 23 years, 1974 became V Pres of Distribution, Chemicals & Plastics. 1980 assigned to Solvents & Intermediates Div as VP, Operations/Technology. *Society Aff:* AIChE

Yen, Ben Chie
Business: Dept Civil Engrg, 208 N. Romine St, Urbana, IL 61801
Position: Professor of Civil Engineering. *Employer:* University of Illinois. *Education:* PhD/Fluid Mechs & Hydraulics/Univ of IA; MS/Fluid Mechs and Hydraulics/Univ of IA; BS/ Civil Engg/Natl Taiwan Univ. *Born:* 4/14/35. Reg PE in Ill. 1960-64 Res Assoc at Iowa Inst of Hydraulic Res, 1964-66 at Princeton Univ; since 1966 with Univ of Ill. Visiting Professor Univ of Karlsruhe, Germany 1974, Free Univ of Brussels, Belgium 1983, Natl Taiwan Univ 1983, Univ of Stuttgart, Germany 1983, 1984, Univ. of New South Wales, Australia, 1985; Visiting Res Fellow Hydraulics Res Station, England 1975; Invited Prof Fed Tech Univ at Lausanne, Switzerland, 1982. Specialize in open-channel hydraulics, surface water hydrology, sediment transport, storm drainage & stochastic hydraulics. Chmn ASCE Stochastic Hydraulics Ctte 1976-77, Chmn ASCE Fluids Ctte 1978-80, Chmn Intl Joint Ctte on Urban Storm Drainage, 1982-85. *Society Aff:* ASCE, AGU, IAHR, IWRA, IAWPRC.

Yen, Teh Fu
Business: Dept of CE, Los Angeles, CA 90089
Position: Prof of Environ Engrg and CE *Employer:* Univ of S CA. *Education:* PhD/Chem/VPI; Hon (DS)/-/Pepperdine U; MS/Chem/W VA Univ; BS/Chem/Central China U. *Born:* 1/9/27. Dr Teh Fu Yen, born in Kunming, Southwest China, came to US in 1949, & was naturalized in 1963. In recent yrs, his specialty is in engg related to chem sci & tech. He has been teaching, carrying out res & consulting in the fossil energy sources, particularly the extraction & recovery tech, the conversion (liquefaction & gasification) & the environmental control tech. Also he applies microbiology to natural resources. He has published 10 books, 300 publications, & 10 patents. He also participates in the worldwide scientific & technical activities. *Society Aff:* ACS, AIChE, APS, SPE.

Yeniscavich, William
Business: PO Box 79, West Mifflin, PA 15122
Position: Engineering Manager. *Employer:* Westinghouse Electric Corporation.

Yeniscavich, William (Continued)
Education: PhD/Metallurgical Engg/Carnegie-Mellon Univ; MS/Metallurgical Engg/Carnegie-Mellon Univ; BS/Metallurgical Engg/Drexel Univ. *Born:* June 1934, Penn. 5 years res on effects of irradiation on nuclear reactor fuels, poisons & structural materials; 5 years tech mgmt over manufacture of wrought & cast superalloys for aerospace & chem indus; Visiting Asst Professor in Met Engrg at Purdue Univ 1968; 9 years mgmt of welding dev & application for const of nuclear reactor plants; 4 years Mgr of sub-div for developing & implementing advanced material control concepts for nuclear reactor plants; 3 years Navy representative at Naval Nuclear Core manufacturing facilities, presently Engineering Mgr at Bettis Atomic Power Lab of Westinghouse. MPC Bd/Dir 1970-71; AWS Natl Awards Chmn 1974-75. *Society Aff:* AWS, ASM.

Yenkin, Fred
Business: 1920 Leonard Ave, Columbus, OH 43219
Position: Chrmn of Bd. *Employer:* Yenkin-Majestic Paint Corp & OH Polychem Co. *Education:* BChE/OH State Univ. *Born:* 6/20/11. in Logan, OH, have resided in Columbus since 1920. Taught Coatings Tech & Polymerization Tech as related to Coatings. Served as Chem Engr, Plant Mgr, Tech Dir & Sales Engr. With present employer since 1934. Am serving on Scientific Committee, Environmental Protection Committee, Chem Coatings Committee for the Natl Protective Coatings Assn. Have travelled in Africa, Asia, W Europe, Near East & South America, addressing Technical Coatings Assns & Private Industries. Enjoy travelling, classical music, golf & reading. *Society Aff:* FSCT.

Yenni, Donald M
Home: RR 2 Sanderson Rd, East Jordan, MI 49727
Position: Engineering Associate, retired. *Employer:* Union Carbide, Linde Division. *Education:* BSMetE/Metallurgy/Univ of MI. *Born:* May 19 1917. 1939-40 Foundry Foreman for Spencer Smith Machine Co; 1940- 76 Res Lab Linde Div of UCC, worked extensively in crystal growth & fabrication, welding & Plasma plating. Obtained 22 patents in these 3 areas including basic patents in several Plasma & welding processes. Was active in ASM, elected Fellow 1974; AWS; Amer Rocket Soc, Dir of local chap. Reg PE in NY & Ind. *Society Aff:* ASM.

Yerazunis, Stephen W
Home: 6 Little Bear Rd, Troy, NY 12182
Position: Prof of Chemical Engrg *Employer:* Rensselaer Polytechnic Institute. *Education:* PhD/Chem Engg/RPI; MChE/Chem Eng/RPI; BChE/Chem Eng/RPI. *Born:* 8/21/22. Army Corps Engrs, Manhattan Proj 1943-46; 1947 Genl Electric Co; 1948 Instructor at RPI, 1952 Asst Professor, 1954 Assoc Professor, 1963 Professor, 1967-79 Assoc Dean of Engrg. Cons Knolls Atomic Power Lab 1956-72, NY State Dept Mental Hygiene 1967-71. Mbr 1962-71, & Pres 1971-76, Lansingburgh Central Sch Bd of Education. Fellow AIChE. *Society Aff:* AIChE, ASEE.

Yerlici, Vedat A
Business: Bogazici University, Bebek, Istanbul, Turkey
Position: Prof of Engrg *Employer:* Bosphorus Univ *Education:* DEng/Reinforced Concrete/Tech Univ of Istanbul; MEng/Structures/Yale Univ; BS/CE/Robert College *Born:* 01/26/29 Native of Istanbul, Turkey. Design Engr, Bridge Design Office of the State, Springfield, IL, 1952-53. Reserve Officer, Turkish Army Corps of Engrs, 1954-56. Asst Prof, Lehigh Univ, Bethlehem, PA, 1957-58. Since 1959 Asst, Assoc, and full Prof, Robert Coll, Istanbul, Turkey (became Bosphorus Univ as of 1971). Dean of the Engrg Schs of Robert College and Bosphorus Univ, 1969-82. Visiting Prof, FL Intl Univ, Miami, FL, 1982-84. Author of 2 books on the behavior of reinforced concrete and 38 papers published in various technical journals and intl conference proceedings. *Society Aff:* FERS, IMO, ACI

Yessios, Chris I
Home: 4740 Shire Ridge Road East, Hilliard, OH 43026
Position: Prof *Employer:* The OH State Univ *Education:* PhD/Computer Aided Design & Planning/Carnegie-Mellon Univ; BArch/Arch/Aristotelian Univ; Dip/Law & Econ/Aristotelian Univ *Born:* 8/10/38 Having remained a practicing Architect and Planner since my professional registration in 1967 and an Educator since 1971, for the past 15 years I have also been extensively involved with the use of computers and their graphics capabilites for design. My developmental and research work has been in the areas of spatial synthesis oriented linguistic models, geometric modeling, architectural void modeling and intelligent CAAD systems. It has been reported in over 70 conference presentations, journal publications and book chapters. It has produced such systems as SIPLAN (1973), SID (1978), TEKTON (1980), and ARCHIMODOS (1984), which are currently used by architectural designers and have been commended in the literature. My design projects are primarily single and multi family dwellings and reflect an acceptance for such vernacular principles as contextualism and respect for local forms and culture. Some have appeared in architectural journals. Contrary to the popular belief (which fears that the use of computers in architectural design will lead to repetitions, inflexible and inhumane designs), I accept the computer graphics methods which are rapidly developing to help architecture regain its humanism and functionality. *Society Aff:* ACM, SIGGRAPH, ACADIA, NCGA.

Yetter, John W
Business: PO Box 5406, Denver, CO 80217
Position: Vice President & Manager. *Employer:* Stone & Webster Engineering Corp. *Born:* Sept 1917. 1939 Elec Engr Cornell Univ; 1939-40 Atlantic City Electric Co; 1940-63 Genl Electric Co, involved in generation & transmission engrg & technological studies of EHV transmission; since 1963 with Stone & Webster Engrg Corp, 1st as Ch Electrical Engr, then various engrg mgmt positions, V Pres since 1972, Mgr of Denver Operations Center since May 1974. Elected Fellow IEEE 1968. Reg PE in 10 states.

Yevjevich, Vujica
Home: 309 Yoakum Pkwy, 1401, Alexandria, VA 22304
Position: Research Prof *Employer:* George Washington Univ *Education:* Dr Eng/Hydrology/Serbian Academy of Sci; Master/Hydraulics/Grenoble Sch Hydr Engs; RS/Civil Eng/Belgrade Univ *Born:* 10/12/13 in Yugoslavia. Schooling in Yugoslavia and France. First job as hydraulic engr in Macedonia (Yugoslavia) 1937-1941. In war 1941-1945. From 1944-1958 dir of designing and research organizations, and prof at Univ of Belgrade (Yugoslavia). From Feb 1958 through Sept 1960 senior scientist U.S. Natl Bureau of Standards and U.S. Geological Survey (Washington, D.C., USA). 1960-1979 prof of civil engrg at CO St Univ, Fort Collins, CO. Sept 1979 on, prof and dir, Intl Water Resources Inst, School of Engrg and Applied Science, George Washington Univ, Washington, D.C., and prof of civil engrg, CO St Univ. *Society Aff:* ΣΞ, ASCE, AGU, AAAS, IAHR, IWRA

Yinh, Juan A
Business: Box 6-591 El Dorado, Panama Republic of Panama
Position: Partner. *Employer:* Yinh Y Asociados, SA. *Education:* MS/Struct Engrg/Univ of TX; MA/Mgmt/Univ of OK; BS/Civil Engrg/Univ de Panama *Born:* 11/8/45. Native of Panama City, Panama. With Panama Canal Co in 1967 as field engr. From 1969-70 in the Panama Section of the Pan-American Highway as chf of Technical Dept; also taught at the Univ of Panama and Universidad Santa Maria La Antigua (in Panama) during that time. From 1971-79 in Ingenieria Amado, SA, specialized in steel structures design and construction held: Last position: Technical Mgr & Asst Genl Mgr. Founder partner-1980 of Yinh y Asociados, SA, engaged in consulting, construction and Technologias Ananzadas, S.A., engaged in inspections and appraisals. Pres of ASCE, Panama Sect in 1978. In Bd of Dirs of ACC in 1973. Bd of Dir, Panamanian Construction Chamber, 1983-87. PE license in the Canal Zone, FL and TX. *Society Aff:* ASCE, NSPE, ACI

Yoder, Charles W
1962 Robins Run, Hartford, WI 53027
Position: Partner. *Employer:* Charles W Yoder & Associates. *Education:* BS/Civil

Yoder, Charles W (Continued)
Engg/Penn State. *Born:* June 1911, Pottstown Penna. 1st Chmn Natl Const Indus Council 1975; Dir EJC 1975-77; Pres ASCE 1974, V Pres 1967-68, Dir 1961-64. 1976 'Engineer of the Year,' Engrs & Scientists of Milwaukee; 1975 Alumni Fellow, Penn State Univ; 1975 Community Service Award, Allied Construction Employers Assoc, Milwaukee; 1975 Cited for contributions to construction industry by "Engg News-Record–; 1974 Outstanding Alumnus, Coll of Engrg Penn State; 1974 Distinguished Service Award, Univ of Wisc. Arbitrator Amer Arbitration Assn 1968- ; Mbr Wisc State Bldg Code Review Bd 10 years; Chmn zoning & Planning Comm, Erin Township; active in various civic organizations. Head of engrg-arch firm for 33 years; firm has designed num indus & commercial facilities, mostly in SE Wisc. *Society Aff:* ASCE, ACI, ASTM, CSI.

Yoder, Elmon E
Business: Univ of Ky, Lexington, KY 40546
Position: Agricultural Engineer. *Employer:* US Department of Agriculture. *Education:* MS/Civil Engg/OR State Univ; BS/Civil Engg/OR State Univ; BS/Agri Engg/OR State Univ. *Born:* 1921. Salvage diver with US Army Engrs, WWII; since 1962 Agri Engr with Sci. and Ed. Adm. USDA, respon for R&D of tobacco production mechanization. P Chmn PM-59 & EPP-49 of ASAE; Mbr Tau Beta Pi, Sigma Tau, Sigma Xi & Gamma Sigma Delta; registered prof. engr. (A.E.). *Society Aff:* CAST, ASAE.

Yoder, Richard J
Business: 180 Park Ave, PO Box 101, Florham Park, NJ 07932
Position: Proj Exec. *Employer:* Exxon Res & Engg Co. *Education:* DEng/CE/Yale Univ; MEng/CE/Yale Univ; BS/CE/Univ of IL. *Born:* 12/7/21. in Peru, IL. Joined Exxon Res & Engg Co in 1947, working mainly in the field of process equip design. Transferred to Exxon Corp in 1961 with responsibility in the field of capital investment evaluation & long-range corporate planning. Rejoined ER&E's Petrol Dept in 1969 as Asst Gen Mgr & became Gen Mgr in 1971. Made Gen Mgr of Engg Tech Dept in 1976. Named Proj Exec for New Facilities Prog for ER&E in Dec, 1977. Responsible for planning design & construction for new facilities & space requirements for ER&E. Mbr of AIChE and API. *Society Aff:* AIChE, API.

Yoerger, Roger R
Business: 338 Agri Engrg Sci Bldg, 1304 W Pennsylvania Ave, Urbana, IL 61801
Position: Prof Emeritus *Employer:* University of Illinois UC. *Education:* PhD/Agri Engg-T&AM/IA State Univ; MS/Agri Engg/IA State Univ; BS/Agri Engg/IA State Univ. *Born:* Feb 1929, Merrill Iowa. 1949-56 Agri Engrg staff at Iowa State Univ, 1956-58 Penn State Univ & 1958-1985, Univ of Illinois UC, hd of dept of Agri Engg, res & teaching in farm power & machinery, instrumentation, noise & vibration. Fellow ASEE, Sigma Xi, Gamma Sigma Delta, Alpha Epsilon, Phi Kappa Phi (Dir of Fellowships, Pres 1972-). Patentee of farm machines. Interested in real estate dev. *Society Aff:* ASAE, ASEE, AAAS, CAST.

Yohalem, Martin J
Home: 3201 NE 36th St 7, Fort Lauderdale, FL 33308
Position: President *Employer:* Yohalem Engineering, Inc *Education:* MS/Civil Engg/Purdue Univ; BS/Civil Engg/Case-Western Reserve. *Born:* 9/26/19. Ft Lauderdale Fl. 1942-46 served with civilian & military Army Corps Engrs; 1946-58 Field Engr, Proj Engr, Structural Div Ch for H K Ferguson Co specializing in indus complexes & chem processing plants; 1959 Smith-Yohalem & Assoc; 1960-70 & 1975-77, Martin J Yohalem, PE, respon for structural drawings & specifications for high-rise apartment complexes, banking institutions, schools, churches, port terminal bldgs, etc.; 1970-74 Davis-Yohalem & Assoc; 1974-75 Davis-McAlpine-Yohalem & Assoc, respon for drawings & specifications for structural, mechanical & electrical divisions. 1977 to present-Yohalem Engineering, Inc., consulting civil-structural engineers. Enjoy reading, swimming & attending seminars. *Society Aff:* FES, ASCE, ASTM, ACI, CRSI, CSI, TMS.

Yokell, Stanley
Home: 16 Spruce Rd, North Caldwell, NJ 07006
Position: Consultant. *Employer:* MGT Inc., Consul. Energy's Res *Education:* BChE/Chem Engg/NY Univ. *Born:* May 1922, Brooklyn NY. Served in US Naval Reserve in WWII as Ch Engr & Exec Officer LST; early exper with Kolker Chem in oper & const of agri chem plants; 1948 Process Engr with Indus Process Engrs, 1953 Sales Mgr; left to found Process Engrg & Machine Co, Inc (PEMCO) as V Pres & Ch Engr, 1966 Pres, 1971 Chmn Bd. Pub misc articles, various tech subjects OP & V Journal, Chem Engrg; lecture 7 of 1976 North Jersey AChE Fall lecture series on Practical Aspects of Heat Transfer entitled 'Mechanical Design & Fabrication Considerations.' Mbr Bd of Dir Energy & Resource Consultants, Boulder, CO; Yokell Sales Assoc, Miami Beach, FL; Pres, MGT INC, Upper Montclair, NJ. Hobbies are backpacking, mountaineering, scuba diving, distance swimming, distance running & distance bicycling. Author: "A Working Guide to Shell and Tube Heat Exchangers–. McGraw-Hill, to be published 1982 "Hydraulic Methods in Shell and Tube Heat Exchanger Manufacture & Repair" Houston, Sept 18, 1981 symposium on shell and tube heat exchangers sponsored by AMS, TEMA, ASME, ANS, ASTM. Member special working group on heat transfer equipment, ASME boilers pressure vessel code committee. Licensed PE New Jersey & Colorado.. *Society Aff:* AIChE, ASME, NSPE, AWS, ASNT.

Yoo, Chai Hong
Business: Dept. of Civil Engrg, 1515 W. Wisconsin Ave, Milwaukee, WI 53233
Position: Assoc Prof *Employer:* Marquette Univ *Education:* PhD/Structures/Univ of MD; MS/Structures/Univ of MD; BS/CE/Seoul Natl Univ *Born:* 9/16/39 Taught graduate and undergraduate courses in the area of structural engineering for last ten years. Research grants and publications. Chrmn (81 - 83), Committee on Flexural Members, ASCE, Chrmn, Task Group 14, Horizontally Curved Girders, Structural Stability Research Council. Enjoys classical music and opera. *Society Aff:* ASCE, ASEE, AAEE, ΣΞ,ΧΕ,ΤΒΠ

York, Dennis J
Home: 3 Tennessee Rd, Vicksburg, MS 39180
Position: District Engr *Employer:* US Army *Education:* ME/CE/Univ of FL; BS/EE/SD State Univ *Born:* 02/28/37 Native of Sioux Falls, SD. Serving in US Army Corps of Engrs 1960-Present. Served with combat Engrs in CONUS, Germany and Vietnam. Commanded a construction battalion in Europe. Assumed current position as Commander/District Engr in MS in 1982 respon for a wide variety of water resources, navigation and flood control projs in MS, LA and AR and the Niger River Basin in Africa. Tau Beta Pi. Pres/VP Vicksburg Post SAME 1983-85. *Society Aff:* SAME, ASEM

York, Raymond A
Business: 570 Lexington Avenue, New York, NY 10022
Position: Genl Mgr/Internatl Licensing Dept. *Employer:* General Electric Company. *Education:* BS/EE/Univ of KS. *Born:* May 1917. Military & Indus Electronics RCA. Joined G E 1948 moving through Electronics Assignments to Mgr, Engrg, Power Semiconductors; was responsible for dev of 1st thyristor (SCR); 1967 became Mgr Overseas Bus for Semiconductors; 1969 moved to Internatl Licensing Dept NY, became Genl Mgr 1972; responsible for all G E patent license & tech info agreements overseas. Fellow IEEE, has had major role in semiconductor tech standards work domestically & internationally. *Society Aff:* IEEE.

Yoshida, Susumu
Business: 7-35, Kitashinagawa 6-Chome, Shinagawa-Ku, Tokyo, Japan 141
Position: Senior Managing Director *Employer:* Sony Corp *Education:* PhD/EE/Tohoku Univ; B/EE/Tohoku Univ *Born:* 5/30/23 in Toyama Prefecture, Japan. Joined Sony Corp. in 1953. In 1957, became involved in semi-conductor engrg and developed the black-and white cathode ray tube and receivers. In 1964, was placed in charge of CRT engrg and manufacturing. After final development of the Trinitron receiver in 1968, assumed responsibility for Trinitron CTV produc-

Yoshida, Susumu (Continued)
tion. In 1974, received an Outstanding Paper Award from the IEEE. Received the Medal of Honor with Purple Ribbon from His Majesty The Emperor in 1973 for the invention of the Trinitron. *Society Aff:* IEEE

Yost, Edward F
Business: C W Post Center, Greenvale, NY 11548
Position: Prof. *Employer:* Long Island University. *Education:* DSc/Engg/Columbia Univ; MS/Engg/Columbia Univ; BS/Engg/US Coast Guard. *Born:* Dec 1929. MS, DSc Columbia Univ; BS US Coast Guard Academy. Aviator US Coast Guard; Flight Test Engr Douglas Aircraft in Reconnaissance Aircraft; Principal Engr Fairchild Camera on reconnaissance & intelligence equipment & sys; currently Prof of Engrg Long Island Univ. Principal investigator NASA Apollo 9, Apollo 12, Landsat & Skylab progs. 40 US & foreign patents. Author of 2 books & 40 tech papers. 1971 ASP award for best paper on photo interpretation, 1973 SPSE award for best paper on photographic engrg. *Society Aff:* SPSE, ASP.

Yost, John R
Business: 1725 Warner St, Whitehall, MI 49461
Position: Pres. *Employer:* Muskegon Chem Co. *Education:* MS/CE/Univ of PA; BS/Chemistry/Ursinus College. *Born:* 7/10/23. Process Dev Engr: Sharpes Corp 1949-51, Merck & Co 1951-54. Hd, Chem Pilot Plant: E R Squibb 1954-67. VP, Mfg: Ott Chem 1967-72. Exec VP: Story Chem 1972- 73. Pres: Muskegon Chem 1974-present. Specialities: Fine Chem Mfg. Societies: AIChE, ACS, Past Chrmn, MI Chem Council. *Society Aff:* AIChE, ACS.

Young, C B Fehrler
Business: PO Box 796, Douglasville, GA 30133
Position: President & Chemical Engineer. *Employer:* Young Refining Corp. *Education:* BS/Chemistry/Howard College; MS/Chemical Engg/Columbia Univ; PhD/Electrometallurgy/Columbia Univ. *Born:* 5/13/09. Birmingham Al. In charge of Lecture Demonstration Div, Chem Dept Columbia Univ 1930. Cons in the NE 12 yrs. Established, V P Natl Southern Prods Tuscaloosa Al 1943. Auromet Corp NY Pres 1953-56. Cytho Corp M'Lady Cosmetic & Alabama Southern Warehouses Tuscaloosa Al Pres 1949- . Cracker Asphalt Corp Pres 1955-71. Young Refining Corp Douglasville Ga Pres 1971- . Metals Recycling Alabama 1974- . Honorary Dir Ga Engrg Foundation Inc 1975- . Dir Southern Federal Savings & Loan Assn 1974-1983. Cy Young Corp., Tuscaloosa, Al. - Pres. 1982- Lacymill Corporation, Tuscaloosa, Al., Pres. 1984- *Society Aff:* AAAS, AIChE

Young, Carlos A
Business: 120 Sixth Ave N, Ste 205, Seattle, WA 98109
Position: Sec-Treasurer *Employer:* Systems Architects Engrs Inc *Education:* BS/CE/Univ of Panama *Born:* 1/27/28 After receiving his education at the Univ of Panama, Mr Young was employed by several firms across the nation as an Engr. His work included structural analysis of concrete and steel thin shell structures; design of highrise apt complexes work on nuclear facility, design of wastewater and sewage/stormwater separation sewage pump and lift atations hydraulic and aqueduct projects. As founder of SAE he has been responsible for the development and expansion of the co, staff recruitment training of professional personnel. Mr Young is responsible for all engrg design projects undertaken by this firm. *Society Aff:* ASCE, AWWA, ACEC, ACI, ASCM

Young, Dana
Business: P O Drawer 28510, San Antonio, TX 78284
Position: Consultant/Vice President Emeritus. *Employer:* Southwest Research Institute. *Education:* PhD/Engg Mechs/Univ of MI; MS/ /Yale BS/ /Yale *Born:* Oct 1904 Washington DC. Engr Marland Oil, Shell Petroleum, United Engrs & Constructors 1926-34; Prof Engrg Mechanics Univ of Conn 1934-42; Univ of Texas 1942-50; Univ of Minn 1950-53; Sterling Professorship in C E Yale 1953-62, Dean Engrg 1955-60; Sr V P Southwest Res Inst 1962-70, Cons 1971- ; Cons & Advisor to USAF, Army, NRC, indus. Commissioner Natl Comm on Prod Safety 1968-70. Num publs on elasticity, vibrations, structural mechanics. Honorary Mbr ASME 1967. *Society Aff:* ASME, ASCE, AIAA, SESA, ΣΞ.

Young, Edwin Harold
Home: 609 Dartmoor, Ann Arbor, MI 48103
Position: Professor of Chemical & Met Engrg. *Employer:* The University of Michigan. *Education:* MSE/ChE/Univ of MI; MSE/MetE/Univ of MI; BSE/ChE/Univ of Detroit. *Born:* Nov 1918. Served USN 1943-45; separated as Lt, promoted to LCd 1954, Cdr 1958, Cpt 1964. Chem Engr Wright Air Dev Center Dayton Ohio 1942-43; Instructor Chem E Univ of Detroit 1946-47; Instructor, Asst Prof, Assoc Prof, Prof, Chem & Met Engrg Univ of Michigan 1947- . Teaches process plant design. Res work in heat transfer. Cons to indus & government. Mbr Mich State Bds of Reg for Archs & Prof Engrs 1963-78. Pres 1968-69, V P 1965-67, Dir 196365, NSPE. Fellow: AIChE, Amer Inst of Chemists, ASME, Eng Soc Detroit. Mich SPE 1976 Engr of Yr, Natl SPE 1977 Award, AIChE 1979 Donald Q. Kern. Fellow, ASHRAE. *Society Aff:* AIChE, ACS, ASME, NSPE, ASHRAE, AIC, ASM, ASEE.

Young, Fred M
Business: PO Box 10057, Beaumont, TX 97207
Position: Dean of Engg. *Employer:* Lamar Univ. *Education:* BSME/ME/SMU; MSME/ME/SMU; PhD/ME/SMU. *Born:* Aug 1940. Native of Dallas Texas. Propulsion Engr Genl Dynamics Fort Worth Texas working on aerodynamics of primary air inlet & associated boundary layer bleed sys; Asst/Assoc Prof -- Dir Engrg Grad Studies Lamar Univ Beaumont Texas; Res in heat transfer & unsteady fluid flow; Prof & Head Engrg & Applied Sci Portland State Univ; Cons to Bethlehem Shipbldg for design of ballast, ballast control sys, stability of semisubmersible offshore drilling vessels; Dean of Engg, Lamar Univ Beaumont, TX . *Society Aff:* ASEE, ASME, NSPE.

Young, Hewitt H
Business: Dept of Indus Engrg, Tempe, AZ 85281
Position: Professor of Engineering. *Employer:* Arizona State University. *Education:* PhD/Engg/AZ State Univ; MS/Ind Engr/Case Inst of Technology; BS/Mech Engr/Case Inst of Technology. *Born:* May 16, 1923. P E Scottsdale Ariz. Prof of Engrg Ariz State Univ. Formerly Natl VP for Publications, V P for Education & Prof Dev, Natl V P Region 12, earlier Pres Central Ind & Central Ariz Chaps, elected Fellow 1976, IIE. Cons Editor Production Handbook 3rd Edition. 1960 Ford Foundation Fellow. 1964- 65 Natl Sci Foundation Fellow. NASA/ASEE Summer fellow 1977. USAF/ASEE Summer Fellow 1978. Mbr: Tau Beta Pi, Alpha Pi Mu, Sigma Xi. *Society Aff:* IIE, ASEE, SME

Young, James T
Business: 2330 Victory Pkwy, Cincinnati, OH 45206
Position: Pres. *Employer:* Truman P Young & Assoc. *Education:* BSCE/Civ Engg/OH State Univ. *Born:* 4/24/43. Upon grad from OH State Univ, worked 7 yrs with Armco Steel corp as plant engr in Middletown, OH & technical services, wide flange & construction products in Houston, TX. Joined civ-structural consulting firm of Truman P Young & Assoc in 1973. Past Pres of Cincinnati Sec ASCE, Past Dir of Engg Soc of Cincinnati. ASCE Dean Terrel Award, 1971. Engg Soc of Cincinnati Young Engr of Yr Award, 1978. *Society Aff:* ASCE, ACEC.

Young, Jesse R
Business: One Capitol Mall, 2D, Little Rock, AR 72201
Position: Exec Dir *Employer:* AR Soil & Water Cons Comm *Education:* MSENE/Water Resources/Univ of AR; BSAgE/Soil & Water/Univ of AR *Born:* 12/24/48 Native of Dover, AR. Employed by the State of AR in 1971 as a Water Resources Engr specializing in Water Resources Planning, Dam Safety and Water Rights. Assumed current responsibility as Exec Dir of the Arkansas Soil & Water Conservation Comm in 1985. Am also member of the AR Dept of Pollution Control and Ecology Commission; Chrmn (FY87), Arkansas-Red-White Basins Intera-

Young, Jesse R (Continued)
gency Cttee; mbr, Arkansas-Oklahoma Arkansas River Compact and Red River Compact Commissions. Pres, AR Section of ASCE 1981; Pres, AR Section of AWRA 1980. A Registered PE in the State of AR. Enjoy following the AR Razorbacks. *Society Aff:* ASCE, NSPE, AWRA, ASPE.

Young, John D
Home: 1158 Manzanita Lane, Manhattan Beach, CA 90266
Position: Supervisor/Metallurgy. *Employer:* B-1 Division of Rockwell Internatl. *Born:* Principal responsibilities include Supr over group of engrs & technicians performing (1) failure analysis of B-1, Sabreliner, Genl Aviation Divs test & serv failures & (2) fracture mechanics, mech properties, Met structure studies to improve properties of metallic materials as assistance to efforts of associated staff mbrs. In failure analysis, has interfaced with B-1 mgmt, subcontractors, SPO & AFML reps during process of determining & effecting appropriate corrective actions for prequalification sys test failures so as to not impact B-1 flight schedule (crew module sys, ECS & EBADS, aft fuselage structure, wing carry through structure sys). Has thorough understanding of metallurgy, processing techniques, stress analysis which has proved invaluable in repeatedly & promptly responding to B-1 sys, DVT, 1st flight schedule requirements. Accomplished analyst.

Young, Laurence R
Home: 8 Devon Rd, Newton Ctr, MA 02159
Position: Prof. *Employer:* MIT. *Education:* ScD/Instrumentation/MIT; SM/EE/MIT; SBEE/EE/MIT; AB/Physics/Amherst College *Born:* 12/19/35. In 1957, he was with the Sperry Gyroscope Co, Great Neck, NY, where he worked on the dev of flight control sys. From 1958-62, he was a mbr of the Res Staff at MIT, where he worked on inertial guidance Systems at the Instrumentation Lab and on problems of man-machine interaction at the Electronic Systems Lab. During 1961, he did eye movement res at the School of Medicine, Univ of Puerto Rico. Dr Young is active on several prof and government committees, including the National Research Council's Committee on the Space Station. He chaired the NRC Vestibular Panel of the Summer Study on Life Life Sciences in Space in 1977. He is also a mbr of the National Academy of Engineering, Fellow of the Explorers Club, the Barany Soc. for Vestibular res and a mbr of the editorial bd of Neuroscience. He was a member of the US Airforce Scientific Advisory Bd & the NRC Aeronautics & Space Engg Bd. He was Pres of the Biomedical Engg Soc for the year 1979. His applications to the aerospace medical field have been in instrumentation (eye movement meausrement) and basic appl res in the field of vestibular function. His psychophysical work on semicircular canal and otolith function led to models which are applied in flight simulator motion control. He is Principal Investigator in vestibular experiments scheduled for Flight in Speclab lude visually induced motion effects. *Society Aff:* IEEE, AIAA, BMES, ASTM

Young, Lyle E
Business: W181 Nebraska Hall, Lincoln, NE 68588
Position: Associate Dean, Emeritus *Employer:* University of Nebraska. (ret.) *Education:* BSCE/CE/Univ of Minn; MSCE/CE/Univ of Minn. *Born:* Oct 1919. Cpt Army Air Force WW II. Instructor & grad student Univ of Minn 1945-53; Asst Prof to Prof Engrg Mechanics Univ of Nebr 1953-66, Asst & Assoc Dean Coll of Engrg & Technology 1966-79, Interim Dean 1979-81, Assoc. Dean, 1981-86, Retired 1986. *Society Aff:* ASEE, NSPE.

Young, Maurice I
Business: Dept of Mech Engrg, Newark, DE 19716
Position: Prof *Employer:* Univ of DE *Education:* PhD/Engrg Mech/Univ of PA; MA/Math/Boston Univ; BS/Phys Scis/Univ of Chicago; PhB/Liberal Arts/Univ of Chicago *Born:* 02/10/27 Native of Boston, MA. Joined faculty as Prof in 1968 after eighteen years as Sr Engrg Specialist, Engrg Section Mgr and Advanced Tech Mgr with Philco Comm and Weapons Div, and Boeing-Vertol Co. Conslt for various industries in Dynamics and Vibrations, Flight Dynamics and Aircraft/Helicopter Vibrations, and Sys Dynamics, Control and Optimalization. Enjoys art history, baroque music and travel. *Society Aff:* SES, IASTED, ΣΞ

Young, Ralph M
Home: 3 Wedgewood Ct, Sugarland, TX 77478
Position: Chairman *Employer:* Davy, Inc. *Education:* BS/Chem Engg/Rice Univ. *Born:* 11/27/22. Native of Beaumont, TX. Graduate of Rice University. Served with US Navy 1943-46. Thirty-nine yrs in the engg/construction industry including experience in business dev, operations, and genl mgmt. Assumed successive respon as Pres, DM Intl Inc, and Chairman, Davy, Inc. as a result of merger between Davy Corp Limited and The McKee Corporation. Prior to merger, served as Pres, Arthur G. McKee & Co and mbr Bd of Dirs, The McKee Corporation. Holder of pats covering fractional distillation equipment. Author of publication covering fractional distillation and proj mgmt. Reg Profl Engr in TX, OK, & CA. Mbr of Engg, Construction, Contracting Exec Committee (AIChE) 1975-77. Mbr Texas Society of Professional Engnrs. Enjoys golf, hunting and fishing. *Society Aff:* AIChE, NSPE

Young, Reginald H F
Business: 2540 Dole St, Honolulu, HI 96822
Position: Assoc Dean. *Employer:* Univ of HI. *Education:* DSc/Environmental & Sanitary Engg/WA Univ; MS/CE/Univ of HI; BS/CE/Univ of HI. *Born:* 5/17/37. Native of Honolulu, HI. Served as Planning & Programming Engr in USAF, 1959-62. Asst Proj Engr with Sunn, Low, Tom & Hara, Inc, 1959 & 1962-63. USPHS Res Fellow in Water Supply and Pollution Control 1964-66. Mbr of Univ of HI faculty since 1966 with appointments in the Sch of Public Health (1966-70), Water Resources Res Ctr (1970-79), & Dept of Civ Engg (1970-79), also Asst Dir, Water Resources Res Ctr, 1973-78. Assumed current position of Assoc Dean in 1979. Teaching & res interests in water resources, pollution control, and water quality mgt. *Society Aff:* ASCE, AAEE, AEEP, WPCF, AWWA.

Young, Robert E
Business: 720 F St, Sacramento, CA 95814
Position: Pres *Employer:* The Spink Corp *Education:* BS/CE/Univ of the Pacific *Born:* 10/14/30 I joined The Spink Corporation in 1960 and was made a principal in 1965. Since 1972, I have served the corp as Pres. I hold certificates of registration in civil, structural, and mechanical engrg. I am a specialist in the design of multi-story offices, general purpose bldgs, bridges, marinas, port and harbor facilities, industrial bldgs and process systems and utility systems structures. Extensive experience in the concepts and design of institutional, industrial and commercial structures. Principal-inCharge of the work on the Sacramento/Yolo Port, including both the original design of the bulk handling facilities as well as the recent renovation work resulting from the fire in 1977. Experienced in structual analysis of remodeling projects and intimately familiar with st seismic criteria. *Society Aff:* ACEC, CEAC

Young, Robert T
Home: Route 1, Box 37, Hillsboro, NH 03244
Position: Consultant (Marine) *Employer:* American Bureau of Shipping. *Education:* BSc/Civ Engrg/Tufts Univ. *Born:* Dec 31, 1912 Sri Lanka. June 1935 BSCE Tufts Univ; grad courses in Naval Architecture MIT 1936-37. Bethlehem Steel Corp Shipbldg Div Quincy Mass 1935-37 Design Draftsman, work on design of aircraft carrier 'WASP', cruisers 'QUINCY' & 'VINCENNES' & various merchant vessel designs; Amer Bureau of Shipping 1938 Marine Surveyor stationed at S Portland Maine in charge of Marine Surveyors (approx 400 Liberty ships built here), Tech Staff NY 1938-40, 1949-68 served overseas offices at Buenos Aires & Antwerp as Principal Surveyor & London as Principal Surveyor for W Europe, May 1968 elected V P & returned to NY office, Sept 1968 elected Sr V P, March 1971 elected Chmn of Bd & Pres. Pres ABS Worldwide Tech Servs Inc, Bd Trs of Webb Inst of Naval Architecture, Past Mbr Ship Safety Panel & Ship Adv Ctte Marine Transportation Res Bd. P E License NY. Mbr NY State. Now Retired but still a mbr of the

Young, Robert T (Continued)
Bd of Managed of Amer Bureau of Shiping & Var Tech Cttes connected with the Marine Industry Chmn Emeritus Welding Res Coun. Director, IHI Marine Technology Inc. *Society Aff:* SNAME, RINA, IMA, AWS.

Young, Ronald A
Home: 400 Sunset Blvd, Alpena, MI 49707
Position: Engr-Mgr *Employer:* Alcona County Rd Commission *Education:* BS/CE/MI Tech Univ *Born:* 9/19/50 Native of Alpena, MI. Enjoy hunting, fishing, computer programming and music. Registered PE, MI 1977. Engr Mgr of Alcona County Rd Commission and Dir of Alcona County DPW. *Society Aff:* ASCE, NACE, APWA, CRAM.

Young, Saul
Home: 573 Ridgerest Dr, Yellow Springs, OH 45387
Position: Assoc Prof *Employer:* Univ of Dayton *Education:* Ph.D/IE/Stanford Univ; M.S./Op Res/Univ of WI; B.A./Applied Math./Univ of TX-Austin *Born:* 3/14/41 Accomplishments include publication of numerous articles, two chapters in a textbook, and technical editor on two textbooks on the subject of military applications of modeling. Consulting efforts have included a two-year project with the Aeronautical Systems Div of the U.S. Air Force, working on developing appropriate Management Information Systems for each of more than 20 unique System Program Offices. Other consit work includes design and implementation of a Quality Control Circle Prog for the County govt of Hawaii, in Hilo, Hawaii. Other consulting has been in the area of inventory control with a medical instrumentation mfg corp. Dr Young has also been principal advisor on over 20 graduate master's theses. As sole owner of his own small business for almost 3 years, Dr Young also was able to test the practical application of topics he teaches, including Management Information Systems, Engrg Economics, Prod & Oper Mgt, and Bus Microcomputer Applications. *Society Aff:* IIE, AIDS, AM, ORSA, MORS.

Young, W Rae
Home: 1 Kingfisher Drive, Middletown, NJ 07748
Position: Retired. *Employer:* Former Bell Labs. *Education:* BSEE/Elec Engg/Univ of MI. *Born:* Oct 1915. Native of Lawton Mich. Bell Labs 1937-1979, NY City, Whippany NJ, Holmdel NJ. Sys Engrg in fields of teletypwriter & data, government & military communications, mobile radio, telephone switching. 24 patents. Papers include 'Comparison of Mobile Radio Transmission at 150, 450, 900 & 3700 Mc' 1952. Retired June 1979 as Hd, Radio Studies & Current Sys Dept, respon for long-range planning of mobile radio sys as well as those currently in use, Registered Professional Engineer in NJ. Fellow IEEE 1964. VP-Administration for First Unitarian Church of Monmouth County. Hobbies: saling, duplicate bridge, playing piano & organ, playing string bass in Monmouth Symphony Orchestra. *Society Aff:* IEEE, HKN, ΣΞ, COMM SOC, VTS.

Young, William B
Home: 599 Baker Rd, Hagerstown, IN 47346
Position: Chief Engr/Materials Sci Dept. *Employer:* Engine Products Div/Dana Corp. *Education:* BS/Met Engrg/MI State Univ; BS/Educ, Math/Central MI Univ. *Born:* 1934. Native of Harrison Mich. Taught in Port Austin Mich public schs 1956-59, Staff Metallurgist Flame Plating Dept Linde Div UCC 1961-67; Engine Product Div/Dana Corp 1967- , assumed responsibility of Materials Sci Dept 1969 involving prod dev, specs & serv in area of Materials at Div level. ASM Metals Progress Ed Ctte 1975-77, advisory committee ASM Journal of Applied Metal Forming, 1978-84; ASM Ctte Hwy & Off Hwy & Off Highway Activities Group-Matls Sys & Design Div. 1979 Chmn 1981,1982. ASM Fellow 1978. ASM Material Systems and Design Coun-chrmn 1983-85, ASM Tech Div Bd mbr, 1983-86, ASM prog monitoring ctte 1983-85, Soc of Automotive Engrs, Non Ferrous Casting Ctte mbr 1981-87; Advanced Materials Comm, Corporation of Science & Technology for the State of Indiana 1985-87. *Society Aff:* ASM.

Young, William H
Business: 800 Kinderkamack Rd, Oradell, NJ 07649
Position: VP. *Employer:* Burns & Roe, Inc. *Education:* MS/Gen Engg/Geo Wash Univ; Grad Studies/Bus Admin/Western Reserve Univ; Certif/Reactor Tech/Bettis Sch of Reactor Tech; BS/Naval Arch & Marine Engrg/Webb Inst *Born:* 9/25/34. A leader in the commercial and military application of nuclear power. VP in charge of Project Operations, including all domestic and foreign, nuclear and fassil power plant design and construction. Former Proj Mgr and mbr of the Office of the Pres for Burns & Roe. Also led corporate advance planning and dev efforts. Earlier was Assoc Dir for Submarines in Div of Naval Reactors, US Atomic Energy Commission, for ADM H G Rickover. PE, CA. Wife, Betty & three children, Debbie, Billy, and Beth. Native of Ilion, NY. Enjoys history, theater and opera. *Society Aff:* ASME, ANS, SNAME, AAAS.

Youngdahl, Carl K
Business: Materials & Components Tech Div, Argonne, IL 60439
Position: Senior Mathematician *Employer:* Argonne Nat'l Lab *Education:* PhD/Applied Math/Brown Univ; MS/ME/IL Inst of Tech; BS/ME/IL Inst of Tech; BA/Liberal Arts/Univ of Chicago *Born:* 8/14/34 Native of Chicago, IL. Married with 3 children. Employed by Argonne Nat'l Lab since 1960. Performed resch on thermo-elasticity, dynamic plasticity, and fluid-struct interaction problems for fission breeder reactors and magnetic fusion reactors. *Society Aff:* ASME, AAM, ΣΞ

Youngdahl, Paul F
Home: 501 Forest Ave, 1002, Palo Alto, CA 94301
Position: Consultant *Employer:* Self-employed *Education:* PhD/ME/Univ of MI; MSE/ME/Univ of MI; BSE/ME/Univ of MI *Born:* 10/8/21 Mechanical Engr in private practice. Staff consultant and Dir, Liquid Drive Corp, Holly, Michigan. Consults with companies and attorneys on product liability and has qualified as expert in mechanical design in courts in at least 25 states. Professional experience includes 9 years, Dir of Research, Mechanical Handling Systems, Div. American Chain & Cable, 11 years Prof of Mechanical Engrg, Univ of MI (tenured 9 years). Also worked for duPont, (Industrial and Development Engrg Div). Commissioned officer, USNR 1943-46. Member Tau Beta Pi, Phi Kappa Phi, Pi Tau Sigma. *Society Aff:* ASME, ASEE, NSPE.

Youngs, Jack R
Home: 10832 Lynwood Rd, Pleasant City, OH 43772
Position: Pres. *Employer:* Basic Systems Inc *Education:* BS/CE/Penn State Univ. *Born:* 4/11/27. President of Basic Systems, Inc.; Chief Engineer of Bi-Con Services, Inc.; Principal in Jack R. Youngs & Associates. Born in Albion, PA April 11, 1927. BSCE Penn State University. Active in planning, design and construction of energy, industrial and treatment related project; and registered professional engineer in PA, OH, WV, KY, WA, TN, IN, NY, NV, SC, UT, VA, NB, AZ, CA, accredited by NCEE. Registered professional surveyor in KY & AZ. Fellow ASCE. Director OACE. Hobbies - Antique cars and flying. *Society Aff:* ASCE, NSPE, AACE, NCEE, OSPE, OACE.

Yourzak, Robert J
Business: Robert Yourzak & Associates, 7320 Gallagher Dr 325, Minneapolis, MN 55435
Position: President *Employer:* Robert Yourzak & Associates *Education:* MBA/-/Univ WA; MS/Civ Engrg/Univ WA; BS/Civ Engrg/Univ MN *Born:* 8/27/47. Yourzak, a Prof Engr/Mgt Conslt/Educator, has adv from struct engr/prog rep for Boeing Co, engr/est for Howard S Wright Constrn Co, dir of proj dev & admn for DeLeuw, Cather & Co, sr mgt conslt for Alexander Grant & Co, mgr proj sys dept & proj mgr for Henningson, Durham & Richardson, dir of proj mgt regional offices for Ellerbe Assoc to his present mgt conslt & training firm with five senior professionals in proj/prog/oper/quality mgt. fin & strategic planning, organ dev. & computers. Natl seminar leader & author of "Proj Mgt" & "Mot & Man Proj Team–. Also adj asst prof at Univ of MN Dept of Civ & Mineral Engrg, Mech/Indus Engrg & Sch of

Yourzak, Robert J (Continued)
Mgt teaching "Proj Mgt–. For Seattle Sec ASCE, chrmn of cont edn com giving courses on "Proj Mgt" & "Mot & Man People–. For MN Sec ASCE, Bd Dirs Member since 1981, and President (1986-87). Bd Dir member since 1984 and (President (1985-86) for Twin City Chapter IIE. 1987-88 Vice President-Public Relations on International Bd of Dirs of Proj Mgmt Institute. Founding member and first elected President (1985) of Minnesota chapter PMI. Certified (PMI) "Project Management Professional." Approved "Certified systems professional" by Inst of Computer Profls. Past chrmn of RAGE at Seattle Art Mus, Bd Dirs of FOR at Seattle Repertory Theatre, prog chrmn for Rainier Club, & mbr Mountaineers & BOEALPS. Received "1979 Edmund Friedman Young Engr Awd for Profl Achievement" from Natl ASCE, named "Outstanding Young Man of America" by US Jaycees, & listed in Who's Who in the World/Midwest/West/Finance & Industry/Emerging Leaders in America. *Society Aff:* ASCE, ACEC, ASM, IIE, PMI, ASEM, PSMA, AAA, ASEE, ASTD, XE

Yousef, Yousef A
Home: 105 Long Branch Road, Winter Park, FL 32792
Position: Prof. of Engrg. *Employer:* Univ of Central FL *Education:* Ph.D./Environ Eng/Univ of TX-Austin; M.S/S.E/Univ of TX-Austin; B.S./C.E. /Univ of Alexandria, Egypt. *Born:* 3/27/30 Florida resident, served the Egyptian Ministry of Irrigation for eight years. Studied Environmental Engrg and worked in research for 41/2 years at UT. Taught at Alexandria Univ and headed Radiation Protection Division at atomic energy establishment for 5 years. Returned to U.S. and worked in research for 11/2 years at UT. Joined UCF at Orlando in 1970. Appointed Prof of Engrg, 1976, and Gordon J. Barnett Prof of Environmental Systems Management, 1979. Received 1980 Univ Award for Excellence in Research. Published more than 60 reports and articles on stormwater impacts, natural systems, highway runoff and water resources. A naturalized citizen since 1974. *Society Aff:* NSPE, ASCE, WPCF, AWWA, ASEE.

Yovits, Marshall C
Business: 1125 East 38th St, PO Box 647, Indianapolis, IN 46223
Position: Dean, Purdue Sch of Sci at Indpls *Employer:* Ind Univ-Purdue Univ at Indianapolis *Education:* PhD/Physics/Yale Univ; MS/Physics/Yale Univ; MS/Physics/Union College; BS/Phyiscs/Union College. *Born:* 5/16/23. 1980 - Dean, Purdue Univ Sch of Sci at Indianapolis and Prof of Computer and Info Sci. 1978-79 Prof, Dept of Computer & Info Sci and Prof Dept of Elec Eng, Ohio St Univ. 1966-78 Prof & Chrmn Dept of Computer & Inf Sci and Prof Dept of Elec Eng. 1962-66 Dir Naval Analysis Group of Office of Naval Res. 1962-66 Exec Dir, Military Operations Res Soc. 1956-62 began and headed Info Sys Branch, Off of Naval Res. 1951-56 Physicist, Johns Hopkins Univ, Applied Physics Lab. 1948-50 Instructor of Physics, Yale Univ. 1946-48 Instructor of Physics, Union Coll. 1944-46 Physicist, Natl Advisory Comm for Aeronautics (NACA). 1973-74 Chrmn, Computer Sci Bd, Mbr 1973-present. 1970-Editor of series Advances in Computers, Academic Press. 1970-74 mbr of Biomedical Comm Study Sec of the Natl Inst of Health. Awarded Navy Superior Civilian Service Award 1964. Awarded Navy Outstanding Performance Award 1961. Fellow, Amer Assoc for the Advancement of Sci, 1982. Fellow, Inst of Elec and Electronics Engrs, 1983. E Central Regional Rep, Assoc for Computing Machinery, Elected by mbrship of Region, 1976-79; 1979-1982. Mbr, Nom Comm of EDUCOM, Dec 1978; 1979 and 1980. Council Member-at-Large to Section T (Information, Computing, and Communication) of AAAS 1979-83. Chairperson, Chairperson-Elect, retiring Chairperson of Section T (Information, Computing, and Communication) of AAAS 1985-88. Mbr, Bd of Dir of Institute for Certification of Computer Professionals, 1985-present. Mbr Bd of Directors, Indianapolis Center for Advanced Research, 1980-82. Mbr, Information Processing Committee, Indiana Corporation for Science and Technology, 1984-present. *Society Aff:* IEEE, ACM, IEEECS, AAAS, ΣΞ.

Yow, John R
Business: PO Box 12728, Research Triangle, Park, NC 27709
Position: Pres. *Employer:* Corp Consulting & Dev Co, Ltd. *Education:* PhD/Mech Engg/NC State Univ; BS/Mech Engg/NC State Univ; AA/Pre- Engg/Gardner Webb Jr College. *Born:* 9/2/40 Native of Seagrove, NC. Adjunct Prof at NC State Univ, Raleigh, NC. Founder of Corporate Consulting and Dev Co, Ltd, 1972, serving as Pres of the Co. One of the nation's leading authorities in the field of seismic design and qualification of equipment for nuclear installations. Respon are mgmt and supervision of all consulting engg functions. The company function is: (1) design, analysis and qualification of melchanical and/or electrical equipment for nuclear installations requiring a seismic withstand capability, and (2) design and construction of various industry facilities such as paper mill systems, cast iron foundries, waste handling sys, pre-cast concrete plants, etc. *Society Aff:* ASME, NSPE, IEEE, EERI, ANS, ASHRAE, ASTM

Ysteboe, Howard T
Home: 80 Pleasant View Terr, New Cumberland, PA 17070
Position: Pres *Employer:* PLASTICERT Inc *Education:* BS/IE & M/ND State. *Born:* 11/16/29. 4 yrs in USAF. 1 yr Quality Assurance Engr N Amer Aviation, 9 yrs MICRO SWITCH Div Honeywell in various MFG Engg positions; 1966 began Plastic Molding Oper for Berg Electronics Div of E I Du Pont, 1974-76 respon included all technical training of Personnel in DuPont facilities in Taiwan, ROC. 1976-78 position was Product Engr Mgr. 1978-84 Responsible for vendor selection and design of all vended molds and molded products as well as being a conslt on product design, material selection and processing of molded products. In 1982 PLASTICERT started offering mold design and specialist for small plastic injection molded parts. 1984 retired from DuPont to devote time to PLASTICERT. Charter Mbr Rock Valley & Susquehanna Sections of SPE. Sr Mbr of Both SPE & IIE. Served 3 yrs on Natl Exec Ctte of SPE as Mbr at Large, Treasurer & 2nd VP. Hobbies: Cards & Fishing. *Society Aff:* SPE, IIE

Yu, A Tobey
Home: 222 Gravel Hill Rd, Kinnelon, NJ 07405
Position: Chrmn *Employer:* ORBA Corp. *Education:* PhD/Civ Engg/Lehigh Univ; MS/Mech Engg/MIT; MS/Bus Adm/Columbia Univ. *Born:* 1/6/21. Dr Yu has played an active role as an educator, engr and administrator in his 30-yr dedicated career in mining, processing, transportation and materials handling. As an author of over 140 publications in leading trade journals worldwide, Dr Yu taught at the NYU School of Engg and Cooper Union. He has held positions as Technical Dir of Cia Minera Santa Fe in Chile; VP-Operations, Hewitt-Robins Div of Litton Industries; Pres of A T Yu Consulting Engrs, Inc. He is co-founder and Chrmn of ORBA Corp, recipient of the 1976 and 1978 1980 NSPE and 1977 ASCE Outstanding Engg Achievement Awards. *Society Aff:* NSPE, ASCE, AIME, SAME, SNAME.

Yu, Francis T S
Business: 121 Electrical Engineering, University Park, PA 16802
Position: Professor *Employer:* The Pennsylvania State Univ. *Education:* Ph.D./EE/University of Michigan; M.S./EE/University of Michigan; B.S./EE/Mapua Inst. of Technology (Manila, Philippines) *Born:* 11/12/34 1958-65, he was a teaching fellow, an instructor and a research assoc in the Electrical Engrg Dept at the Univ of Mich. 1966-80 he was on the faculty of the Electrical Engrg Dept at Wayne State Univ. Since 1980, he has been a Prof in the Electrical Engrg Dept at The Penn State Univ. A recipient of the 1983 Faculty Scholar Medal, a recipient of the 1984 Outstanding Research and was named Evan Pugh Prof at Penn State in 1985. He is the author of four books, and published over 250 technical articles. *Society Aff:* IEEE, OSA, SPIE

Yu, Jason C
Business: Dept of Civil Engrg, Salt Lake City, UT 84112
Position: Prof *Employer:* Univ of UT *Education:* Post Doctrual/Transp Eng/Univ of PA; Ph.D/CE/WV Univ; M.S./CE-Transp/GA Inst of Tech; B.S./CE/Nat Taiwan

Yu, Jason C (Continued)
Univ *Born:* 2/5/36 Dr. J. C. Yu is Prof of Civil Engrg, Univ of UT. He was formerly Assoc Prof and Head of Transportation Engrg Division in the Dept of Civil Engrg, VA Polytechnic Inst and St Univ, and Senior Research Staff at the Univ of PA. He is a registered professional engr, and serves several technical committees of professional societies. He is the author of textbook in transportation engineering and of about 100 journal articles and technical reports. He is also actively involved in consulting work in the field of transportation and traffic engrg. *Society Aff:* TRB, ITE, ASCE

Yu, Wei W
Home: Rt 1, Box 125, Rolla, MO 65401
Position: Prof. *Employer:* Univ of MO. *Education:* PhD/Structural Engg/Cornell Univ; MS/Civ Engg/OK State Univ; BS/Civ Engg/Natl Taiwan Univ. *Born:* 7/10/24. Born in China. Taught in Natl Taiwan Univ, 1950-54. Res Engr for Am Iron and Steel Inst, 1960-67, specializing cold-formed steel structural design. Also worked for T H McKaig & Assoc in Buffalo, NY & TRW Systems in Redondo Beach, CA. Joined the faculty of Univ of MO-Rolla in 1968. Prof of Civ Engg since 1972. Teaches structural analysis & design & directs numerous res projs, intl conferences & short courses on cold-formed steel structures. Author of the textbook *Cold-Formed Steel Structures* & published numerous res papers and technical reports. *Society Aff:* ASCE, SSRC, RCRBSJ.

Yu, Yao-nan
Home: 4234 W 14th Ave, Vancouver, BC, Canada V6R 2X8
Position: Prof Emeritus. *Employer:* Univ of British Columbia. *Education:* PhD/Elec Power/Tokyo Inst of Tech; BS/Elec Power/Tokyo Univ of Tech. *Born:* 10/25/09. Born in China, 1909. B.Sc. & D. Eng. Sc., Tokyo Institute of Technology 1936/1962. Elec machine design & mfg, Central Elec Mfg, Kweilin, WWII. Prof of EE & Hd of Natl Hunan Univ, 1943-46. Prof of EE, Natl Taiwan Univ, 1946-1957; hd, 1946-49. Visited U.S. Univs, 1957-60. With the Univ of British Columbia, Canada, since 1960. Prof Emeritus, 1975. IEEE Fellow, 1977. Known for his contribution in tensor analysis of el M/CS & appl of lunear optimal control to power systems. Continuing res in stability & security control. Book on Elec. Power System Dynamics 1983 (AP). *Society Aff:* IEEE.

Yu, Yi-Yuan
Home: 24 Gordon Rd, Essex Falls, NJ 07021
Position: Dean of Engrg *Employer:* NJ Inst of Tech *Education:* PhD/Engrg Mech/Northwestern Univ; MS/CE/Northwestern Univ; BS/CE/Tientsin Univ (China) *Born:* 01/29/23 Born in Tientsin, China, 1923; came to US, 1947; naturalized 1962. Wife: Eileen Wu; children; Yolanda, Lisa. Prof of Mech Engrg, Poly Inst of Brooklyn, 1957-66; Consltg Engr, General Electric Space Div, 1966-70; Distinguished Prof of Aeronautical Engrg, Wichita Stat Univ, 1972-75; Mgr, Components & Analysis, Rocketdyne, Rockwell Intl, 1975-79, Exec engr, Energy Sys, 1979-81; Dean of Engrg, NJ Inst of Tech, 1981-85. Prof of Mech Eng 1985- . Visiting Prof, Cambridge Univ, England, 1960; Advisor, Middle East Tech Univ, Turkey, 1966; Mbr, Ad Hoc Ctte on Dynamic Analysis, USN, 1968-69; Conslt, Intl Advisory Panel, Chinese Univ Dev Proj, 1983. Consultant, David W. Taylor Naval Ships R&D Center, 1987- . Guggenheim Fellow; Air Force & NASA res grantee; Contributor, Handbook of Engrg Mechanics; author of many publs & reports. *Society Aff:* ASME, AIAA, IEEE, ASEE, NY ACAD OF SCI, ΣΞ, ΦΚΦ, ΠΤΣ, ΤΒΠ.

Yue, Alfred S
Business: 405 Hilgard Ave, Los Angeles, CA 90024
Position: Prof of Engg & Appl Sci. *Employer:* Univ of CA. *Education:* PhD/Chem & Met/Purdue Univ; MS/Met/IIT; BS/Met/Chao-tung Univ. *Born:* 11/12/20. Native of China. Taught at Purdue from 1951 to 1956. Res engr at the Dow Chem Co, specializing in fundamental studies of electronic mtls. With Lockheed Palo Alto Res Labs from 1962 to 1969. Assumed current responsibility as Prof of Engg & Appl Sci at UCLA in 1969. Pres, Phi Tau Phi Scholastic Honor Soc of Am from 1978 to 1982. Honorable Professor, Xian Jiao-tong Univ., China 1980. *Society Aff:* AAAS, AIME, ASM, AACG.

Yuen, George A L
Business: P O Box 3378, Honolulu, HI 96801
Position: Director of Health. *Employer:* State of Hawaii. *Born:* Jan 1920. BSCE Univ of Mich 1942. Stress Engr (liaison between USAF & Willow Run Bomber Plant) Ford Motor Co Ypsilanti Mich; Bd of Water Supply City & County of Honolulu Hawaii 1947- ; Mgr & Ch Engr 1967-74, Cons Engr on water dev & sewage & refust disposal to USN in Saipan 1956 & to Trust Territories 1964; Cons to Republic of China on sewer, water & sanitation 1974; Dec 1974- , appointed by Governor of Hawaii as Dir of State Dept of Health. Water Utility Man of the Yr, Chairman, AWWA, Calif Sect 1972. Hobbies: music, art, golf.

Yukawa, Sumio
Home: 4418 Sandpiper Circle, Boulder, CO 80301
Position: Consulting. *Employer:* General Electric Company. Retired *Education:* PhD/Metallurgical Engg/Univ of MI; MSE/Metallurgical Engg/Univ of MI; BSE/Metallurgical Engg/Univ of MI. *Born:* 4/25/25. Native of Seattle Wash. Served in US Army 1944-47. Joined G E Co 1954 as Res Metallurgist, Mgr Metallurgy 1960, Cons Engr 1969; retired from GE Co, 1986; primary specialization in mech properties & design application of materials in power generation equipment including large rotating machinery & nuclear power equipment. Author of 25 tech papers. Fellow mbr ASM 1970, Chmn ASME Materials Div 1975. Fellow mbr ASME 1981. Mbr of sev indus & government tech cttes on materials applications. *Society Aff:* ASME, ASTM, ASM, AIME.

Yurek, Gregory J
Business: Rm 4-136, 77 Massachusetts Ave, Cambridge, MA 02139
Position: Assoc Prof *Employer:* MA Inst of Tech. *Education:* PhD/Met/OH State; MS/Met/Penn State; BS/Met/Penn State. *Born:* 5/28/47. Native of W Wyoming, PA. Ph.D in 1973. Awarded NATO Postdoctoral Fellowship for one yr, at the Technische Universitat of Clausthal, W Germany. Joined Oak Ridge Natl Lab in 1974. Assumed present position as Assoc Prof of Met at MIT in Jan 1976. Current interests are in the areas of high-temperature oxidation, including synthesis of high Tc superconducting oxides, and the development of high temperature alloys through rapid solidification processing. ASM Howe Gold Medal, 1974. *Society Aff:* ECS, Am Cer Soc, ASM, AIME.

Yurick, Andrew N
Home: Paulsboro, NJ 08066
Position: Senior Associate Engineer. *Employer:* Mobil Research & Development Corp. *Education:* BS/Chemistry/PA State Univ. *Born:* Dec 1922. Resides Woodbury NJ. Infantry officer 1944-46. Mobil R&D Corp 1947- ; conducted training sessions to promote on-stream inspection of refinery equipment, helped develop high temperature delay for on-stream ultrasonic metal thickness measurements, Mobil's worldwide cons on nondestructive testing, supervises Corrosion Control Group of Materials Engrg Sect. Active in AEWG, API Subctte on Refinery Inspections, ASNT-Tech Council Offices, BOD, Ad Hoc Cttes To Study Personnel Training & Certification, Review Bd for Level III Certification Without Exam, ASNT Fellow & Level III MT, PT, RT, UT NACE - Philadelphia Sect, NE Region. NACE accredited Corrosion Specialist. Hobbies: travel, sports, gardening. *Society Aff:* AEWG, ASNT, NACE.

Zabel, Dale E
Home: 903 Third Ave, San Manuel, AZ 85631
Position: Met Dir. *Employer:* Magma Copper Co. *Education:* MS/Met/KS State Univ; BS/Ind Tech/KS State Univ. *Born:* 9/26/18. Native of KS. Taught Physical Met at KS State Univ. Served as Lt (jg) with the Navy Amphibious Corps during WWII. Employed by Boeing Airplane Co for 17 yrs as a Res Engr. Natl Mbrship Committee mbr of Am Soc for Metals, & held all local chapter offices in this organization. With Magma Copper Co since 1971, as: Quality Control Dir, Asst Refinery Superintendent, & presently Met Dir. Creative in many types of woods; actively involved in church & community. *Society Aff:* AIME, ASM, ASTM, WA.

Zabel, Hartmut
Business: W. Green, 1110 Dept of Physics, Urbana, IL 61801
Position: Professor of Physics *Employer:* Univ of IL *Education:* PhD/Physics/Univ of Munich; Master/Physics/Tech Univ of Munich; BS/Physics/Univ of Bonn *Born:* 3/21/46 Native to Radolfzell, W. Germany. Education: Tech Univ of Munich, Diplom, 73; Univ of Munich, Ph.D. (Physics) 78; postdoctoral fellow, Univ of Houston, 78-79; since 79 Asst. Prof of Physics, Univ of IL at Urbana-Champaign. Mbr of the Materials Res Lab. Specialization on x-ray and neutron scattering at crystalline and amorphous materials. Resch contributions include investigations of hydrogen phase transitions in metals, lattice dynamical, thermal and struct properties of intercalated graphite, quasicrystalline materials, superlattices and epitaxial layers. Since 1986 Prof of Physics, Univ of IL at Urbana-Champaign. Since 1987 Dir of Materials Research Lab/Dept of Energy Solid State Program. 1985 Guest Scientist, Brookhaven Natl Lab, 1986 Guest Scientist, Riso Nat. Lab, Denmark. *Society Aff:* APS, DPG, AAAS, MRS

Zaborszky, John
Business: Dept of Sys Sci & Math, St Louis, MO 63130
Position: Chairman/Sys Science & Mathematics. *Employer:* Washington University. *Education:* DSc/EE/Technical Univ; Dipl/ME/Univ of Budapest. *Born:* May 13, 1914. Dipl 1937, D.Sc. 1943 with special honors under auspices of Regent of Hungary from Tech Univ of Budapest. Elected to the Natl Acad of Engrg 1984. Received Bellman Heritage Award of American Automatic Control Council 1985. Distinguished Member IEEE Control System Society 1984, IEEE Centennial Award 1983. Indus & academic exper spans engrg, res, teaching & mgmt with contrib to control, filtering estimation, info & power sys. With Washington Univ 1956- , developed Dept Sys Sci & Math of which he is Chmn. Previously Prof Univ of Mo/Rolla, prior to 1949 Ch Engr Budapest Municipal Power Sys. Cons to Westinghouse, McDonnell Douglas, Emerson Electric, Hi-Voltage Equipment, NIH. Pub 2 books & over 60 tech papers. Pres Control Sys Society 1971, Pres Amer Automatic Control Council 1980-82 (VP 1978-80), Director IEEE Division *1 and member IEEE Board 1974/75. Society Aff:* IEEE, ASME, SIAM, ΣΞ.

Zachariah, Gerald L
Business: 1001 McCarty Hall, University of Florida, Gainesville, FL 32611
Position: Dean for Resident Instruction *Employer:* Univ of FL *Education:* Ph.D./Agr. Eng/Purdue Univ; M.S./Agr. Eng/KS St Univ; B.S./Agr. Eng/KS St Univ *Born:* 6/12/33 Native of McLouth, KS. Faculty experience at KS St Univ, Univ of CA--Davis, Purdue Univ, Univ of FL. Chrmn, Agricultural Engrg Dept, Univ of FL, 1975-1980. Dean for Resident Instruction, IFAS, Univ of FL, since 1980. Consultant to industry and Government. Intl Experience. Board of Dirs, ASAE, 1980-1982; 1985-1987. *Society Aff:* ASAE, IFT, NSPE.

Zackay, Victor F
Home: Materials & Methods, Inc, 1014 West Rd, New Canaan, CT 06840
Education: BS/Metallurgy/Univ of CA, Berkeley; MS/Metallurgy/Univ of CA, Berkeley; PhD/Metallurgy/Univ of CA, Berkeley. *Born:* May 1920. Native of San Francisco Calif. Res Metallurgist, Supr & Mgr Ford Motor Co 1954-62; Prof of Metallurgy UCB 1962-79; Assoc. Dean College of Engineering UCB 1969-73; currently President, Materials & Methods, Inc - a consulting organization. Fellow ASM, AIME. Awarded Henry Marion Howe Medal 1961 ASM & Albert Sauveur Award 1971 ASM. Founder & Chmn N Calif Chap TMS-AIME. Principal current prof interest: alloy design, new product development, R & D materials management, Technical Resource Officer for large corporation, Manager - Industry/University programs, technology transfer specialist. Hobbies: 18th & 19th century Amer Art and Furn. *Society Aff:* ASM, AIME.

Zadeh, Lotfi A
Business: Dept of EE & Comp Scis, Berkeley, CA 94720
Position: Prof Elec Engrg & Computer Sciences. *Employer:* University of California/Berkeley. *Born:* Feb 1921. PhD Columbia Univ; MS MIT; studied at Univ of Teheran Iran. Served as mbr of faculty Dept of Elec E Columbia Univ 1950-59; moved to Univ of Calif/Berkeley 1959, served as Dept Chmn 1963-68. Mbr Inst for Advanced Study Princeton 1957. Visiting Prof MIT 1963 & 1968. Editor of Journal of Computer & Sys Sciences. Main fields of interest: sys theory, info processing & decisionmaking under uncertainty. Educ Medal 1973 IEEE. Guggenheim Fellowship 1968. 1973 NAE. Amateur photographer & audiophile.

Zagars, Arvids
Home: 2450 Central Rd, Glenview, IL 60025
Position: Vice Pres & Chief Staff Engr *Employer:* Harza Engrg Co *Education:* Dipl/CE/Riga State Technicum-Latvia *Born:* 08/22/21 Native of Riga, Latvia. Registered Structural Engr-IL, PE Virginia, Alaska. VP Harza Engrg Co since 1977, Past Project Dir of Bath County Pumped Storage Projct, V2.- world's largest. During 35 year service with Harza has been in responsible design charge of several domestic and foreign projects Bath County has received 1987 ACEC Grand Award and Corps of Engrs Distinguished Engrg Achievement Award. Chrm, ASCE Hydro Power Ctte. Recipient of 1986 ASCE Rickey Medal. European hydro experience with Siemens Schuckertwerke AG, Berlin and Linz/Danube on construction of three projects on Enns river in Austria. *Society Aff:* ASCE, NSPE, SAME, ACI.

Zagustin, Elena
Home: 16862 Morse Circle, Huntington Beach, CA 92649
Position: Prof *Employer:* CA State Univ, Long Beach *Education:* PhD/Applied Mech/Stanford Univ; MS/Applied Mech/Stanford Univ; CE/Central Univ Venezuela; MS/Math/Stanford Univ; BS/Math/CA State Univ *Born:* 6/12/37 Published over 100 original articles in professional journals in the fields of solid mechanics, structures, fluid mech, applied math. Asst prof 1964-65, Central Univ, Caracas, Venezuela. Post-doctoral fellow 1965-66 at Northwestern Univ, Evanston, IL. Res Asst 1966-67 at Princeton Univ Assoc prof 1967-72 at CA State Univ, Long Beach. Prof 1972-present at CA State Univ., Long Beach.

Zahn, Markus
Business: Dept of Elec Engg & Computer Sci, High Voltage Research Lab, Building N-10, Cambridge, MA 02139
Position: Prof. *Employer:* Mass Inst of Tech *Education:* ScD/Elec Engrg/MIT; EE/Elec Engrg/MIT; MS/Elec Engrg/MIT; BS/Elec Engrg/MIT. *Born:* 12/3/46. Is presently Assoc Prof of Elec Engrg in the MIT High Voltage Res Lab of the Lab for Electromagnetic and Electronic Sys. Has been Prof with the Dept of Elec Engg at the Univ of FL from 1970-80. Main interest concern teaching & res in electromagnetic fields & forces interacting with media. Has authored the text *Electromagnetic Field Theory: A Problem Solving Approach* (Wiley, 1979) and over 40 tech articles. He is a conslt to Exxon Res & Engrg Co with three patents, Dow Chem Co. Philips Labs, Teleco Oilfield Serv, and Optical Diagnostic Serv. He has received three (3) Excellence in Teaching Awds from Univ of FL Students and was cited for honorable mention for the 1979 C. Holmes MacDonald Outstanding Teaching Awd from Eta Kappa Nu. He was US Organizer of the 1979 US/Japan Seminar on Elec Conduction & Breakdown in Dielectrics, Gen Chmn of the 1980 Second Intl Conf on Magnetic Fluids and Prog Chmn for the 1982 and 1983 Conference on Elec Insulation and Dielectric *Society Aff:* IEEE.

Zahnd, Rene G
Business: 1300 Sooline Bldg, Minneapolis, MN 55402
Position: Pres *Employer:* LanData Assocs Inc. *Education:* M/CE/Inst of Tech Burgdorf, Switzerland; B/Survey/Zurich, Switzerland *Born:* 7/28/25 Citizen of Switzerland. Land survey, geophysical survey and engrg design in Switzerland and Italy. Topographic survey, design and construction of land reclamation, sewerage, rail-

Zahnd, Rene G (Continued)
road, bldgs in East Africa. Supervise survey dept of natl shopping center developer in USA. Now cadastral mapping of corp real estate holdings; civil engineering & land surveying land assemblers for preservation with erection of trail systems and shelters, all owned and operated by private capital. *Society Aff:* ASCE, NSPE, ACSM, LII

Zak, Adam R
Business: 104 S Mathews St, Urbana, IL 61820
Position: Prof *Employer:* Univ of IL *Education:* Ph.D/Aero & Astro Eng/Purdue Univ; M.Sc/Aero & Astro Eng/Purdue Univ; B/ME/Univ of New Zealand. *Born:* 12/24/34 Dr. Zak received his Bachelor of Engrg degree from the Univ of New Zealand in 1956. In 1959 and 1961, respectively, he received the Master of Science and Ph.D. degrees in Aeronautical Engrg from Purdue Univ. From 1961 to 1964 he held a research position with CA Inst of Technology. In 1964 he joined the faculty of the Univ of IL as Assoc Prof in the Aeronautical and Astronautical Engrg Dept, and in 1968 he was promoted to the position of Prof. Dr. Zak's technical interests include structures, viscoelasticity, dynamics, controls, and numerical methods. *Society Aff:* AIAA

Zakin, Jacques L
Business: 140 W 19th Ave, Columbus, OH 43210
Position: Chrmn & Prof Chem Engr Dept. *Employer:* OH State Univ. *Education:* BS/ChE/Cornell Univ; MS/ChE/Columbia Univ; D. Eng Sci/ChE/NYU. *Born:* 1/28/27. Native of Far Rockaway, NY. Served in USNR 1945-46. Worked at Flintkote Co Res Labs (Whippany, NJ) in 1950-51. Socony Mobil Res Labs (Brooklyn, NY) in 1951-56 & 1958-1962. Taught in Dept of Chem Engg at the Univ of MO-Rolla, 1962- 1977. Dir of Minority Engg Prog & Women in Engg Prog at Univ of MO-Rolla. Sabbaticals at Technion, Israel & Naval Res lab, Wash DC. Prof & Chrmn Dept of Chem Engg at OH State Univ 1977 to present. Res on properties of dilute polymer solutions, rheology of concentrated oil-in-water emulsions, turbulent drag reduction & structure of turbulence. *Society Aff:* AIChE, ACS, Soc of Rheology, ASEE.

Zakraysek, Louis
Home: 8432 Brewerton Rd P0B62, Cicero, NY 13039
Position: Mgr/Applied Materials & Processes. *Employer:* General Electric Company. *Education:* BS/Metallurgy/Penn State. *Born:* 12/20/28. USN 1946-48. Instructor Metals Engg Inst & Syracuse Adult Education. Regional chmn & Natl Ctte mbr in prof assns & tech soc. Currently respon for selection, application & use of matls & processes in electronic equipment. Author of 18 pub papers. Who's Who in the East. Fellow Mbr ASM, Pres's Award Syracuse Chap. Listed Amer Men & Women of Science, Pres Award Syracuse Chap. Pres Award Inst of Printed Circuits. Serv Cert Aerospace Council SAE, United Fund Indus Unit Chmn, Federated Fund Unit Chmn, Jr Achievement Fund Drive. *Society Aff:* ASM, AIME/TMS, IMS.

Zambas, Peter G
Business: 515 South Flower, Room 785 WIB, Los Angeles, CA 90071
Position: VP, ARCO Environmental, Inc. *Employer:* Atlantic Richfield Company. *Education:* EM/Mining/Columbia Univ; MS/Mining/Columbia Univ; BS/Mining/Columbia Univ. *Born:* Nov 1933. Native of Athens Greece. Mine Captain Hellenic Mining Co Cyprus 1957-59; Foreman to Asst to Genl Supt Kermac Nuclear Fuels Uranium Mining New Mexico 1959-64; Sr Res Mining Engr Mobil R&D Corp, oil shale dev & other minerals aquisitions 1964-72; Mgr Mining Engrg Atlantic Richfield Co all phases of mining engrg, evaluation, design & related activities of the co 1972-77. VP & Genl Mgr of Subsidiary, ARCO Solar, Inc, all aspects of operation of the co from R&D to Production to Mkting & Sales. Heavy involvement with administration & legislative activities 1977-79; VP ARCO Environ, Inc. In charge of co's new proj on desalination of sea & brackish water 1979- . Issued 2 pats. Coauthored 3 papers pub in AIME transactions. Mbr AIME & Tau Beta Pi, SEIA, ISES, AWWA, NWSIA. Robert Peele Mem Award. Chrmn of Photovoltaic Div of Solar Energy Industries Assoc. *Society Aff:* AIME, SEIA, ISES, NWSIA, AWWA.

Zamborsky, Daniel S
Home: 8853 Falls Lane, Broadview Heights, OH 44147
Position: Vice Pres, Plant Mgr. *Employer:* Aerobraze Corp. *Education:* BS/Chemistry/Case Western Reserve; -/Metallurgical Engg/Cleveland State Univ; Grad Courses/Metallurgical Engg/Cleveland State Univ. *Born:* 2/19/29. Cleveland Ohio. 34 yrs prof exper in iron & steel melting & casting, heat treating, materials & processes selection, control & testing, & failure analysis. Have operational and mgmt responsibilities for div'l. plant of company. SME Sr Mbr, Certified Mfg Engr; ASM: 24 yrs, Chrmn Cleveland Chap 1973-74, natl cttes: Handbook 1971-74, Indus Study 1974, Heat Treat Control 1974-75 (Chrmn 1976-79) Natl Tech Awareness Council 1976, Heat Treating Div Council 1976-87, participant 1976-80-Heat Treat Conferences, MEI Faculty, 1976-present. Natl Long Range Planning Comm 1980-83. Elected to ASM Int'l. Board of Trustees, 1987-90. Hobbies teaching, travel & light classical music. *Society Aff:* ASM, Intl, SME.

Zames, George D
Business: Dept of EE, 3480 Univ St, Montreal Quebec, Canada H3A 2A7
Position: Prof of EE. *Employer:* McGill Univ. *Education:* ScD/EE/MIT; BS/Engg Phys/McGill Univ. *Born:* 1/7/39. Macdonald Prof of Elec Engrg McGill Univ, Montreal. Sr Fellow, Canadian Inst for Advanced Res (1984-). ScD degree Elec Engrg MIT 1960. Sr Scientist at NASA Elec Res Ctr, Cambridge (1965-70), served on faculties of MIT (1960-65), Harvard Univ (1962-63) and the Technion (1972-74). Among awards received are IEEE Field Award for Control Science & Engineering, Killam Fellowship. Guggenheim Fellowship, Natl Acad of Sci RR Assoc (NASA), Athlone Fellowship (Imperial Coll), British Assoc Medal (McGill Univ), AACC Outstanding Paper Prize, five IEEE Trans. Automatic Contrl Outstandg Paper Awards, and a Current Contents Classic Paper Citation. IEEE Fellow, mbr editorial bds of SIAM Journ on Contrl, Sys Contrl Letters, and IMA Journ of Math Contrl and Info, J. Mathematics of Control, Signals, and Communications and author of many papers on Contrl and Non-Linear Sys. *Society Aff:* IEEE, SIAM.

Zanakis, Steve (Stelios) H
Business: Florida International Univ, College of Business Administration, Dept of Decision Sciences & Information Systems, Miami, FL 33199
Position: Prof. *Employer:* Florida International University *Education:* PhD/Mgt Sci/PA State Univ; MA/Statistics/PA State Univ; MBA/Mgt/PA State Univ; MS/Mech-Elec Engg/Natl Tech Univ, Athens, Greece. *Born:* 11/16/40. Currently Prof of Decision Sciences & Information Systems at Florida International Univ (founding chrmn 1982-86). Other experience: Assoc. Prof (1976-80) & Dir (1973-80) of Industrial Engg & Systems Analysis prog at the WV College of Grad Studies (Asst Prof 1972-76). Greek Productivity Ctr (Industrial Engr/Mgt Consultant 1967-68). Consultant to several industrial, hospital and governmental organizations. Guest editor for Management Sciences and European Journal of Operational Research. Author of forty journal articles, two software and four books on "Optimization in Statistics" (North Holland, 1982), "Production Planning and Scheduling" (IIE, 1984), "STAT-EZ: An Easy-to-use Statistical Software for IBM PC & APPLE Microcomputers" (EZ Management Decision Systems, 1986) and "Production/Operations Management Software for the IBM PC" (Wm.C. Brown Publ, 1987). *Society Aff:* TIMS, DSI.

Zande, Richard D
Business: 1237 Dublin Rd, Columbus, OH 43215
Position: Chairman *Employer:* R.D. Zande & Assoc, Ltd. *Education:* BS/CE/MI State Univ *Born:* 4/29/31 Native of Battle Creek, MI Air Force 1951-53. Employed by Shell Chem Corp and Sverdrup & Parcel Engrg Co. 1955-57. Located in Columbus; Engrg work with Michael Baker, Jr. and Rackoff Assocs. 1957-65, before starting consulting engrg firm of R.D. Zande & Assoc, Ltd. in 1965. Firm involved in General Civil Engrg and Surveying. Served on Drainage Task Force with the Mid

Zande, Richard D (Continued)
OH Regional Planning Commission and City of Columbus Planning and Zoning Board. Served as Nat'l Dir of ACEC. Mbr of Columbus Historic Commission. Professional Engrg Registration in 10 states. Honor Award for Engrg Excellence. 1972--CEC; Historic Resource Ctte. *Society Aff:* ASCE, NSPE, ACEC, CEO

Zandman, Felix
Business: 63 Lincoln Hwy, Malvern, PA 19355
Position: President. *Employer:* Vishay Intertechnology, Inc. *Education:* Doctor Engg/Physics/Univ of Paris (Sorbonne) France; ME/Mechs/Univ of Nancy, France. *Born:* May 7 1928, Grodno Poland; citizen USA. 1950-53 Attache of Res, French Natl Center of Scientific REs; 1953-55 Ch Stress Analysis Lab (SNECMA) jet engine factory, France; 1956-62 Dir Basic Res, Instruments Div Budd Co; 1958-66 Lecturer stress analysis at MIT, UCLA & Wayne State; since 1962 Pres of Vishay Intertechnology Inc. 43 patents on photoelasticity, strain gages, moire, resistors; 3 books & num papers on above subject. French Order of Merit for Research & Invention 1960; Edward Longstreth Medal, Franklin Inst 1962 USA; Distinguished Contribution Award SESA 1970 USA.

Zankich, L Paul
Home: 4002 Aikins Ave SW, Seattle, WA 98116
Position: Principal Naval Architect *Employer:* Columbia Sentinel Engineers, Inc *Education:* BSE/Naval Arch & Marine Engg/Univ of MI. *Born:* 9/5/43. Mr Zankich was employed by the Boeing Co as a Naval Architect for 6 yrs designing the presently operating Fleet of hydrofoils. He worked for Todd Shipyards-Seattle Div for 6 yrs as Chief Naval Architect & was a foreman in the Steel Dept. He was Engg Mgr for Marine Power & Equip Co & designer of the "New Evergreen State Class" ferries for the state of WA. He is now a partner and Principal Naval Architect at Columbia Sentinel Engineers. He is chrmn of Pacific Northwest Section of SNAME & a charter mbr of the Puget Sound Section of ASNE. *Society Aff:* SNAME, ASNE, NSPE.

Zapffe, Carl A
Business: 6410 Murray Hill Rd, Baltimore, MD 21212
Position: Professional Engineer. *Employer:* CAZLAB *Education:* BS/Metallurgy/MI Technological Univ; MS/Metallurgical Engg/Lehigh Univ; ScD/Physics & Chem of Metals/Harvard Univ; D Engg/Hon/MI Technological Univ. Univ. *Born:* 7/25/12. 1934-36 employed DuPont Experimental Stat. Postdoctorate res Battelle Mem Inst 1938-43. 1943-45 Rustless Iron & Steel Corp; 1945- , private lab & cons ofc; 1946-55 Civilian Sci Ofc Naval Res. Originator fractography, hydrogen theories for metal defects, stainless uranium steel, submarine vulcanism theory for meteorology and climatology, geohydrothermodynamical theory for Ice Ages and Earth-Noon evolution, magnetospheric theory & M-space for the structuring of space in relativistic physics; author 170 papers, 4 tech movies, 2 books: stainless steels, fracture analysis, X-ray diffraction, steelmaking, foundry practice, welding, metalworking & finishing, electroplating, vitreous enameling, mineralogy, refractories, crystallography, chemistry, physics, geophysics, glaciology, archaeology, relativistic physics, history of sci. Mbr 17 tech socs in fields met, phy, chem, geophys, astronomy. Natl Sci prizes; listed in 9 natl & intl biographies. *Society Aff:* ASM, AWS, ALS, APS, AGU, ISGRG, NACE, ACS, AAAS, AGI, RIP, ASTM, AIC, SHOT, ASP, EX

Zar, Max
Home: 1000 Bob-O-Link Rd, Highland Park, IL 60035
Position: Retired Partner & manager Structural Dept. *Employer:* Sargent & Lundy. *Education:* BS/CE/IL Inst of Tech. *Born:* 1/4/16. in Chicago. 1942 worked in field US Geological Survey prior to coming to Sargent & Lundy in 1942, Jan 1966 became S&L Partner, July 1966 Mgr Structural Dept, respon for all civil & structural engg & architecture performed by firm. Received Civil Engineer of the Year Award, IL Sect ASCE 1968; elected Fellow ACI 1975, Alfred E Lindau Award ACI 1974; Special Citation Award AISC 1976; Structural Engr of the Year Award, Structural Eng Assoc of IL 1979, Distinguished Professional Achievement Award, ILL. Inst. of Tech. 1980. *Society Aff:* ASCE, ACI, NSPE, EERI.

Zarling, John P
Business: School of Engineering, 539 Du Bldg, 306 Tanana Dr, AK 99775
Position: Prof. *Employer:* Univ of AK. *Education:* PhD/Engr Mech/MI Tech Univ; MS/Engr Mech/MI Tech Univ; BS/Mech Engr/MI Tech Univ. *Born:* 3/15/42. Native of WI, Industrial Experience with Shell Oil Co and John Deere. Taught at Univ of WI-Parkside and served as Asst VChancellor. Was hd and Prof of the Mech Engr Dept at the Univ of AK. Presently, Assoc Dean School of Engg and Dir Inst of Northern Engg. Res interests in heat transfer, solar energy, and thermal systems. More than 40 publications in areas of res interest. Past Chrmn of the Milwaukee Sec - ASME. *Society Aff:* ASME, ASHRAE, ASEE.

Zavada, Roland J
Home: 21 Cottonwood La, Pittsford, NY 14534
Position: Corporate Standards Director *Employer:* Eastman Kodak Company. *Education:* MS/Bus-Mktg/Univ of Rochester; BS/Chemistry/Purdue Univ; Dip/Photo Sci/Rochester Inst of Tech. *Born:* March 18 1927. MS Univ of Rochester, BS Prudue, Dipl RIT. 1944-46 Air Corps, USAF 1951-52; since 1952 Prod Engr with Eastman Kodak Co specializing in evaluation & dev of motion picture films, 1975 assumed respon as Tech Assoc & Coordinator for motion picture & television related projects. Became Standards Director for Corporate Standards Administration, 1986. Chmn several SMOTE & ISO committees since 1963; leader of USA delegation to 6 internatl ISO/TC 36 conferences 1972-87; was Engrg V Pres SMPTE for 8 years, elected Fellow 1970; Also Fellow AES and BKTS. Received SMPTE Progess Medal 1985, 1985 Rochester Engineer of the Year Leo East Award and Agfa Gavert Gold Medal for film/video technology 1986. Enjoy flying, woodworking & photography. *Society Aff:* SMPTE, BKSTS, PSA, AES, TAGA, SPSE.

Zdenek, Edna Prizgint
Business: 1121 Prospect St, Westfield, NJ 07090
Position: Principal. *Employer:* Profl Engrs & Planners. *Education:* BSME/ME/Newark Coll of Engrg; Assoc in Sci/Pre-Engr/Staten Island Community College. *Born:* 6/27/42. NEE: Prizgint. Principal and sole proprietor of conslt engrg firm. Licensed PE in NJ, NY, & PA. Licensed Prof Planner in NJ. 1980-1984 Chpt Bd of Dirs, NJ Soc of Architects, Amer Inst of Architects; Prof mbr, NJ State Advisory Bd of Carnival & Amusement Park Safety, 1976 to 1980. Prof mbr, Westfield Bd of Construction Appeals, 1977-1980. Bd of Trustees, NJ Inst of Tech, 1970-1973. NY State Regents Scholarship, 1959-1962. NJIT Outstanding Leadership and Student Activities Scholarship, 1965. Listed in Who's Who in the East, 1979. 1981 Sem on Insul - presented on behalf of The Amer Inst of Architects at Tech Session, NJ Energy Expo, sponsored by Dept of Energy. *Society Aff:* ASHRAE, AAEE, NJSA, AIA.

Zdonik, Stanley B
Home: 174 Lake St, Arlington, MA 02174
Position: VP & Sr Consulting Engineer *Employer:* Stone & Webster Engineering Corp. *Education:* SM/Chem Eng/MIT; SB/Chem Eng/MIT. *Born:* 7/15/16. 1941-42 faculty MIT. Mbr Sigma Xi, AIChE Mbr, Fellow 1976, Natl Nominating Ctte 1976. Reg PE Mass. Mbr, NY Acad of Sciences, (Div Prog Coordinator, Fuels and Petrochemicals Div, AIChE). Author of over 52 books & publ. Mbr Intl Adv Bd Encyclopedia of Chem Processing & Design; Plenary Lecturer at 1st Yugoslav Symposium on the Chemistry & Tech of Petroleum & Petrochemicals. Holder of 2 patents. *Society Aff:* AIChE.

Zebib, Abdelfattah M G
Business: College of Engineering, P.O. Box 909, Piscataway, NJ 08855-0909
Position: Prof *Employer:* Rutgers Univ *Education:* Ph.D./ME/Univ of CO; M.Sc./ME/Univ of CO; B.Sc. /ME/Cairo Univ *Born:* 9/11/46 in Cairo, Egypt. Post doctoral fellow, Univ of CO 1975-76. Prof in Rutgers Univ since 1977. Teaching fluid mech and thermal sci. Research interests include hydrodynamic stability, thermal

Zebib, Abdelfattah M G (Continued)
connection, combustion and boundary layer theory. Play bridge, tennis and also enjoy classical music and fishing. *Society Aff:* APS, SIAM, ASME

Zebrowitz, Stanley
Business: 1914 Lantern Lane, Oreland, PA 19075
Position: President *Employer:* Stelcom Associates *Education:* MS/EE/Univ of PA; MBA/Finance/Temple University; BEE/EE/CCNY. *Born:* Nov 28, 1927 New York City. In US Army Signal Corps. Joined Philco Corp 1949, acquired by Ford Motor Co, became active in design & implementation of large micro-wave & troposcatter sys in locations throughout the world, becoming Engrg Mgr 1965; engrg respon for performance of all ground communication projs in his Div. In 1981 became President, Stelcom Associates, consultants communication system engrg, Lic P E in Pa. Fellow IEEE & a Dir of IEEE Commucications Soc.Enjoys photography & studying history. *Society Aff:* IEEE.

Zebrowksi, Edward R
Home: P.O. 18162, Philadelphia, PA 19116
Position: Principal *Employer:* Pres, T.I.E.S. (Transportation & Industrial Engine Svces) *Education:* BA/Bus Mgmt/Univ of PA. *Born:* 11/30/21. Mgr Sys Dev & Indus Engrg respon for functions of Indus Mgmt Sci; volunteer cons dev of Black Entreprenuer & Frankford Hospital labor improvement through N E Chamber of Commerce Lectr, Indus Engr, Mgmt Sci, Computers, Human Behavior & Motivation at LaSalle Coll, Univ of Pa, Newark Coll of Engrg, Phila Coll of Textiles & Sci, Princeton & Penn St Univ's & Amer Mgmt Assn. Presented & publ papers 'The Use of Computers In Dev Control of Whseing', 'Dev & Acceptance of Formal OR/IE Functions'; computerized admin & control sys analytical & quality control labs, Region II V Pres AIEE. IMMS Prof certification in materials handling (PCMH) & material mgmt (PCMM) public relations dir Engrg Tech Soc's Council, Delaware Valley. Rohm & Hans Co. North America Traffic Mgr, Pres (T.I. E.S.) Transportation & Indus Engrg Services. *Society Aff:* IIE, BCPDM, ATP, AAII.

Zedalis, John P
Home: 8723 Gateshead Rd, Alexandria, VA 22309
Position: International Engineering Consultant *Employer:* Self *Education:* BS/Mining Engg/Univ of MO; Post-grad/Civ/Univ of MO. *Born:* 10/25/28. Collinsville, IL. Served in US Navy 1946-48; US Army Corps of Engrs 1952- 53. Instr, Univ of MO (Rolla) 1954-55. Office Engr, Army Corps of Engrs, St Lawrence Seaway 1955-56; Chas T Main, 1956-57. Seven yrs with DC Hgwy Dept as Supv Hgwy Res Engr. Joined Agency for Intl Dev (AID) in July 1964 as Construction Engr, later became Chief, Transportation Branch, Far East Bureau. Assoc & VP with King & Gavaris Consulting Engrs, 1968-71 & 1973-74. Also was VP Commonwealth Transp Consultants 1971-72, and Chief Transportation Engr, AID Office of Engr 1974-1980. Director of Engineering, 1980-82 responsible for Administration and Management of AID Projects Worldwide. Currently, intl engrg and construction Conslt. US Delegate, Inland Waterways Conferences, United Nations, Bangkok, 1965 & 1966. Special consultant Southeast Asia projs DOT in 1968. Served on various committees Transportation Res Bd. Natl Res Council 1963- 82. *Society Aff:* ASCE.

Zederbaum, Robert B
Business: 84 Park Ave, P.O. Box 3400, Flemington, NJ 08822
Position: Partner *Employer:* Heritage Consulting Engrs *Education:* B.S./CE/Newark Coll of Engrg *Born:* 9/28/49 Professional Engr and Planner and license to operate Class I Water Treatment and Distribution System in NJ. Professional Engr in PA. Project Engr and Chief of Computer Operations with E.T. Killam Assocs until 1978. Editor of co newsletter. Assoc with R.C. Bogart and Assocs heading Water Treatment and Urban Planning Depts. Involved with municipal engrg in 16 communities within Hunterdon and Warren Counties. Currently partner with Heritage Consulting Engrs providing consulting engrg services to tri-state area. Enjoy major home improvement projects and antique car restoration. *Society Aff:* ASCE, CEC

Zeff, Jack D
Home: 4609 Alla RD 1, Marina Del Rey, CA 90291
Position: Executive Vice President *Employer:* Ultrox International *Education:* BS/Chem Engg/Purdue Univ. *Born:* 10/6/23. Born and raised in Evanston, IL. Served with Army Corps of Engrs 1942-45 in WWII. Res Engg for AMF specializing in aerospace life support. Manager of environmental systems with Gatx. Founder and Pres of Westgate Res Corp, which conducts R&D and manufacturing of uv-oxidation water treatment and air pollution control systems which destroy toxic organic pollutants. Now exec VP of successor company, Ultron International, manufacturer of oxidation systems to control hazardous wastes in air and water. Author of numerous papers and articles on water treatment. Holds 10 patents. Enjoys tennis, running, classical music and jazz. *Society Aff:* AIChE, WPCF, ACS.

Zegel, William C
Home: 11011 NW 12 Pl, Gainesville, FL 32606
Position: President and Chief Executive Officer *Employer:* Water and Air Research, Inc. *Education:* ScD/Chem Engg/Stevens Inst of Tech; MS/Chem Engg/Stevens Inst of Tech; ME/Chem Engg/Stevens Inst of Tech. *Born:* 8/4/40. Chem instr Newark Coll of Engrg. Allied Chem Corp Process Res Labs 1964-68; Sr Res Engr, 5 pats on unusual separation processes, water pollution control. Scott Res Labs 1968-72; Sr Engr/Group Ldr, pat on inexpensive CO monitor, contrib to 3-vol 'Environ Engrs Handbk'. RETA 1972-75; Sr Assoc Engr, major proj mgmt respons. Designated tech cons of Regional Air Pollution Study, St Louis. Environ Sci & Engrg Inc; VP/Oper & Prin Engg 1975-79, respon for performance of all prof servs for clients. Water & Air Res; currently Pres & Chf Exec Officer Elected to Tau Beta Pi as Emminent Engineer. Currently Serving on Bd of Dir of Air Pollution Control Assoc. Have Certificate of continued Prof Dev from FL Engrg Soc, Mbr AIChE, PEPP, APCA, NSPE, AAAS, FL. Engrg Soc, Diplomate Amer Acad Environ Engrs, Reg PE FLA, GA & AL. *Society Aff:* APCA, AIChE, TBΠ, NSPE.

Zeien, Charles
Business: 2 World Trade Ctr-9528, New York, NY 10048
Position: President. *Employer:* J J Henry Co Inc. *Education:* MBA/Bus/Harvard Univ; BS/MA & ME/Webb Inst of Naval Arch. *Born:* 8/5/27. With Sun Shipbldg & Dry Dock Co 1952-72 as VP-Engrg & mbr of Bd of Dir. In 1972 joined J J Herny Co as Exec V P & was elected Pres Apr 1975. Mbr Soc of Naval Archs & Marine Engrs; Mbr Soc Naval Engrs, Soc Mil Engrs, Amer Inst of Aeronautics, Royal Inst of Naval Archs. Amer Bureau of Shipping's Tech Ctte, and classification committee. Director of Shipbldrs' Council of Amer's. VChrmn Bd of Trustees Webb Inst of Naval Architecture.

Zelbv, Leon W
Home: 1009 Whispering Pines Dr, Norman, OK 73072
Position: Prof. *Employer:* Univ of OK. *Education:* PhD/Elec Engg/Univ of PA; MS/Elec Engg/CA Inst of Tech; BS/Elec Engg/Univ of PA. *Born:* 3/26/25. in Poland, grad high school in Krakow, Poland. Served with the US Army and PA Natl Guard from 1946 to 1950. Industrial experience with RCA Camden from 1953 through 1967 in various capacities, including consulting. Summer and temporary employments with Hughes R & D Labs, Lincoln Lab, Sandia Corp, Argonne Natl Labs, Inst for Energy Analysis, NASA Lewis Res. Ctr. Taught at Univ of PA from 1961 to 1967; and Univ of OK since 1967. Hobbies include music, reading, and flying (CFI-AGI). Reg in OK and PA; listed in Who's Who in America, American Men and Women of Science. *Society Aff:* IEEE, Franklin Inst, AOPA

Zelman, Allen
Business: Dept of Biomedical Engg, Rensselaer Polytechnic Institute, Troy, NY 12181
Position: Prof. *Employer:* RPI. *Education:* PhD/Biophys/Univ of CA; BA/Biology & Phys/Univ of CA; AA/Biophys/Compton College. *Born:* 2/12/38. Native Los Angeles, CA; married Adala (1971), children Bil (b 1972), Suzette (b 1973); home Troy, NY; 82nd Air Borne Div, Army (1956-59); Post doc Carnegie-Mellon Univ, Biot-

Zelman, Allen (Continued)
ech; Asst Prof of Biophys, Meharry Medical College (1972-75); Asst Prof of Biophys & Biomedical Engg, RPI (1983); Program director for bioengineering and research to aid the handicapped, National Science foundation (1987). Main res interest: biological & synthetic membrane transport phenomenon, rehabilitation engineering. Personal: shooting sports. *Society Aff:* BMES, BS, AAAS, ASEE.

Zeman, Alvin R
Business: 13837 NE 8th, Bellevue, WA 98005
Position: Pres *Employer:* Rittenhouse-Zeman Assoc *Education:* MS/CE/San Jose St Univ; BS/CE/San Jose St Univ *Born:* 1/17/42 Cofounder of Rittenhouse-Zeman & Assoc in 1974, elected pres in 1981. As a Consulting Geotechnical Engr, has gained extensive experience in the analysis of heavy Foundations on soft soils including Seattle's Kingdome. Currently involved in the analysis and monitoring of several state-of-the art projects on AK's North Slope. Past member and chrmn of the Seattle Design Commission. *Society Aff:* ASCE, NSPE, USCOLD, ISSMFE

Zemel, Jay N
Business: Dept. of EE & Center for Sensor Technologies, Univ of Penn, Philadelphia, PA 19104-6390
Position: Prof. *Employer:* Univ of PA. *Education:* PhD/Physics/Syracuse Univ; MS/Physics/Syracuse Univ; MA/Honoris Causa/Univ of PA; BA/Physics/Math/Syracuse Univ *Born:* 6/26/28. My current profl interests lie in the study of microfabricated structures for fundemental research and application to sensors. There is interest in both the detection of chemical species for feedback control purposes & fabrication of electronic devices. The detection study has led to chemically sensitive electronic devices & the evolution of a sci for microfabricated sensors. The gen area can be described as chem electronics. *Society Aff:* IEEE, APS, ISA.

Zeno, Robert S
Home: 1124 Woodview Rd, Hinsdale, IL 60521
Position: Dir, Components Tech Div. Retired *Employer:* Argonne Natl Lab. *Education:* MS/Metallurgy-1946/Penn State Univ; BS/Metallurgical Engg-1941/MI Tech Univ. *Born:* 9/24/19. After earning a BS degree in Metallurgical Engg from MI Tech in 1941, I served five yrs in the Army Corps of Engrs. From 1946-47 I was a Teaching Asst while earning an MS degree in Metallurgy from Penn State Univ. I then joined the GE Co where I accumlated 25 yrs of diversified experience with nuclear and non- nuclear power-generating equip. This experience included various positions of mgt, direction, mtls and processes dev, mfg-engg of nuclear core components, fabrication of large steam-turbine-generator components, gas-turbine dev, quality assurance, and nuclear power plant construction and operation. Previous to my current appointment as Components Tech Div Dir, I served for one yr with Westinghouse as Dir of the Angra Nuclear Proj, the first Brazilian nuclear power plant. Retired from the Argonne National Laboratory 1986. *Society Aff:* ASM.

Zentner, Charles R
Business: PO Box 360, Portsmouth, RI 02871
Position: Principal Engineer *Employer:* Raytheon Co/Submarine Signal Div. *Education:* PhD/Eng Acoustics/PA State Univ; MS/EE/PA State Univ; BS/EE/PA State Univ. *Born:* 2/1/41. Design engr for AC Spark Plug Div of GMC spec in design of airborne radar subsystems. Design engr for Allison Div of GMC respon for design of turbine aircraft engine control sys. With Appl Res Lab of PA St U from 1966-78 as the designer of sonar signal processing sys. Since 1978 with Raytheon Co as principal engineer and Mgr of the Weapons System Section. *Society Aff:* IEEE, ASA.

Zenz, Frederick A
Business: Rt 9-D PO Box 241, Garrison, NY 10524
Position: President. *Employer:* F A Zenz Inc. *Education:* BS/Chemistry/Queens College; MChE/Chem Eng/NY Univ; PhD/Chem Eng/PolyTech Inst of Brooklyn. *Born:* Aug 1, 1922. Specialist in fields of fluidization, conveying & all aspects of FluidParticle tech; 20 yrs indus experience with M W Kellogg Co, Hydrocarbon Res Inc, & Stone & Webster Engrg Co; independent cons & prof since 1962. Dir of PEMM-Corp; Tech Adv to the Bd of Chem Rox Corp, & Pres F A Zenz, Inc; Tech Dir PSRI; Prof Manhattan Coll. Fellow AIChE; author of over 150 papers, books & pats; recipient AIChE Tyler Award 1958; Manhattan Project Gold Key Award 1945; AIChE Chem. Engr'g. Practice Award 1985; McGraw-Hill Personal Achievement Award in Chem. Engr'g. 1986. *Society Aff:* ACS, AIChE, ANS.

Zepfel, William F
Home: 1008-2525 Cavendish Blvd, Montreal Quebec H4B 2Y6 Canada
Position: Dir of Metalurgy. *Employer:* Sidbec-Dosco Ltd. *Education:* BS/Metallurgical Engrg/Univ of Pittsburgh. *Born:* Apr 1925; native of Pittsburgh Pa. BS Met Engrg Univ of Pgh 1949. 3 yrs US Navy during WW II. Joined US Steel Corp in Pgh district 1949, serving in a number of met & mgmt positions; in 1962 went to Venezuela to initiate met & inspection dept at new plant then under const at Matanzas; left Venezuela 1967 to take up current position of Dir of Met at corporate hqrs Sidbec-Dosco in Montreal. Active in natl ctte work with AISI, ASTM, ISO, CSA & ASM. Elected to Board of Trustees of A.S.M. for 3 years 1980-83. Has Extensive Experience in the Tech Aspects of Intl purchasing and selling of finished and semi-finished steel products. *Society Aff:* ASM, AISI, AIME, ASTM, SAE, CSA.

Zerban, Alexander H
Home: 75 Avonwood Rd, Avon, CT 06001
Position: Dean Emeritus of College of Engrg. *Employer:* University of Hartford. *Education:* PhD/Mech Eng/Univ of Michigan; MS/Mech Eng/Penn State; ME/Mech Engg/Polytechnic Univ. *Born:* July 4, 1904. Native of New York N Y. In USN during WW II as Engr Officer, now retired as Capt USNR. Prof of Mech Engrg Penn St 1928-55; Sr Proj Engr United Aircraft 1955-58; Dean of Engrg Univ of Hartford 1958-70, Dean Emeritus 1970- . Reg P E in Penna; named Engr of Year 1970 CSPE. Life Fellow ASME; Life Mbr ASEE. Sigma Xi, Pi Tau Sigma, Phi Kappa Phi, Pi Mu Epsilon, Iota Alpha, Kappa Mu. Co-author texts on Power Plants, Engrg Thermo. *Society Aff:* ASME, ASEE, NSPE, ΣΞ, AARP.

Zersen, Carl W
Home: 436 Colonial Park Dr, Santa Rosa, CA 95401
Position: Ret. (Managing Dir) *Born:* Apr 28, 1903. Concordia Coll Milwaukee Wis; Milwaukee Sch of Engrg; Northwestern Univ Evanston Ill; Hon Degree in EE Milwaukee SOE 1968. Assoc Mbr IES, Full Mbr & Fellow 1953-68, Fellow Emeritus 1968- . Chmn Natl Comm on Lighting Educ 1946-48; produced 1st IES Lighting Fundamentals Course. Chmn Bd of Judges 1947, 1949 & 1952 for Internatl Lighting Expositions & Conferences. Chmn Bd of Judges Lighting Competition for Elec Cont N Y City. Chmn of numerous regional & sect Lighting Compet. Wrote numerous articles on lighting applications which appeared in engrg & trade journals. Spoke: over 6 different radio stations in Chicago; to hundreds of service clubs, PTA's, H S & coll classes, assn's of commerce, church groups. Gave signal from platform at State & Randolph Sts Chicago to Pres Eisenhower to turn on new State St Lighting Sys on evening of Nov 13, 1958. Managing Dir Chicago Lighting Inst 1938-68; Mbr Rotary Club, Chicago 1954-68; Mbr Assoc Exec Forum of Chicago 1949-68; Mbr Chicago Assoc of Commerce & Indus 1945-68; Mbr Milwaukee Sch of Eng-Mbr of Corp 1960-68; Speaker Graduation Class Milwaukee Sch of Engr Jun 1954; Chrmn Speakers Bureau Lights Diamond Jubilee.

Zetasche, James B
Home: 13105 El Camino Real, Atascadero, CA 93422
Position: Associate Professor. *Employer:* Calif Polytechnic State Univ. *Born:* Oct 23, 1938. BS Agri Engrg Texas Tech 1962; MS Cvl Engrg Texas Tech 1967. Farm background. 4 yrs undergrad student work experience USDA, ARS, Cotton Machinery Res; 4 yrs experience Texas A&M - Irrigation Res spec in sub-irrigaton wit plastic pipe. So Plains Res Center Lubbock Texas. 2 US pats in irrigation. Presently Assoc Prof Calif Polytech St Univ, Agri Engrg Dept, San Luis Obispo Calif 93407. P Pres So Calif Sect ASAE. Present Secy-Treas Pacific Region ASAE.

Zevanove, Louis R
Home: 771 Mirador Ct, Pleasanton, CA 94566
Position: Div Ldr/Opers-Electronic Engrg Dept. *Employer:* U C Lawrence Livermore Lab. *Born:* Aug 31, 1933. BSEE June 1956 San Jose St Univ. Worked at NASA Ames in magneto hydrodynamics & high voltage projs until 1960; worked for Texas Instruments as Field Rep for introduction of integrated circuits OEM's; 1962 started at Lawrence Livermore Lab as Reliability Engr Group Leader, dev LLL's microelectronics lab, became Group Leader of nuclear gield support proj; 1968 became Assoc Div Leader of Field Support for Nuclear Electronics; 1969 became Deputy Div Leader; 1971 became Div Leader of Electronics Engrg Dept's Operations Div. 1975 received NSF grant to transfer tech to public sector. Enjoy oceanracing sailboats.

Zia, Paul Z
Home: 2227 Wheeler Rd, Raleigh, NC 27607
Position: Professor of Civil Engrg & Dept Head. *Employer:* North Carolina State University. *Education:* PhD/Structural Eng./Univ of FL; MSCE/Structural Eng/Univ of WA; BSCE/Civil Eng/Natl Chiao Tung Univ, China. *Born:* 5/13/26. Received US citizenship 1962. VP & Ch Struct Engr, Lakeland Engrg Assocs 1953-55; served on faculties of Univ of FL & Univ of CA Berkeley; cons to major engrg firms indus. Elected Fellow Amer Concrete Inst 1974; Fellow ASCE; Mbr PCI, Mbr ASEE, Mbr NSPE, Phi Kappa Phi, Sigma Xi, Chi Epsilon, Tau Beta Pi. PCI Martin P Korn Award 1974; ASCE T Y Lin Award 1975; ASEE Western Elec Fund Award 1976; ASCE Raymond C Reese Award 1976; NCSU Alcoa Fnd Distinguished Research Award 1979-81; Elected Mbr Natl Acad of Engrg 1983; Distinguished Alumnus Award, Univ of FL, 1983; ACI Joe W. Kelly Award 1984; ASEE Benjamin Garver Lamme Award 1986; ASCE NC Section Outstanding Civil Engr Award 1986. *Society Aff:* ACI, ASCE, ASEE, PCI, NAE, NSPE

Zich, John H
Home: 3828 Mystic Vly. Dr, Bloomfield Hills, MI 48013
Position: Retired *Employer:* None *Education:* B.S./Agri. Mach./TX A. & M. Univ *Born:* 4/27/13 Native of Needville, TX. Tractor and Hay Baler Engr for Allis Chalmers Co 1937-1941. Ordanance Test Officer at Abeerdeen Proving Grounds 1942-1946. Service Engr Tractors at Dearborn Motors Corp 1947-1953. Chief Implement Engr for Ford Tractor 1953-1962. Government Regulations Mgr. for Ford Tractor 1962- 1978. Retired 1978. ASAE Gold Metal Award in 1975 for Safety Work. *Society Aff:* ASAE, SAE

Zichterman, Cornelius
Home: 12 Reba Ct, Morton Grove, IL 60053
Position: Vice President. *Employer:* Peterson Enterprises Inc. *Born:* Oct 1922; native of Chicago Ill. V Pres & Dir of Marketing for Peterson Enterprises Inc Glenview Ill; assumed this capacity early 1973. Educ Northwestern Univ & N Park Coll. Has been associated with engrg & sales of motion picture film printing equipment for past 19 yrs; prior 13 yrs were involved in engrg design of radio, TV & papermaking machinery. Very active in SMPTE for many yrs & P Chmn Chicago Secton & P Gov of Natl Midwest Region SMPTE. Enjoys various sports & partakes actively in golf. Woodworking & gardening are good pastimes. *Society Aff:* SMPTE.

Zick, Leonard Paul
Business: 8025 Via Marina, Scottsdale, AZ 85258
Position: Retired. *Education:* MS/Struct Engg/IL Inst of Tech; BS/Civ Engg/CO State Univ. *Born:* 10/1/18. Native of Denver, CO. Joined Chicago Bridge & Iron Co located in Chicago in 1941. Held various positions in res & engg. Was VP & Chief Engr from 1966, through 1979. Retired Dec. 31, 1979 but continue as consultant through 1981. Work has included investigations, tests, designs, consulting & mtl dev for designing and constructing pressure vessels, elevated tanks, low temperature and cryogenic tanks, nuclear containment vessels, offshore structures, and other special plate structures. Have been active in ASME Codes & Stds work since 1956. Served as Chrmn of the ASME Boiler & Pressure Vessel Committee 1971 through 1977 & Chrmn of the Subcommittee on Pressure Vessels before that. Served on ASME Council on Codes & Stds and as ASME VP Pressure Tech Codes & Stds. 1981-85. *Society Aff:* ASME, ASCE.

Zickel, John
Home: 6063 Ranger Way, Carmichael, CA 95608
Position: Professor. *Employer:* California State Univ Sacremento. *Education:* PhD/Appl Math/Brown Univ; MS/Math/Lehigh Univ; BS/Mech Engg/Lehigh Univ. *Born:* Feb 9, 1919. Designer for Amer Safety Table Co 1937-41. Army Corps Engrs 1941-45. Struct Analyst for nuclear reactors & power plants for G E 1952-58; Mgr struct res for rockets & missiles for Aerojet Genl Corp 1958-67; Prof of Mech Engrg CSU Sacramento 1967- , chair of Mech Engrg CSU Sacremento 1981- , spec in machine design & mechanics. Patent on Isotensoid Pressure Vessels. Reg P E Calif, Fellow ASME 1972. Enjoy bridge, Lifemaster ACBL. *Society Aff:* ASME, SEM, ASEE, SAE.

Ziegler, Edward N
Business: 333 Jay St, Brooklyn, NY 11201
Position: Assoc Prof. Chem. Engrg. *Employer:* Polytechnic Univ *Education:* Ph.D./ChE/Northwestern Univ; M.S./ChE/Northwestern Univ; BS/Ch.E./City Coll of NY *Born:* 8/15/38 in Bronx, N.Y.; Industrial Employment-Exxon Research and Engineering 1964-5, Brookhaven National Laboratory 1974-80, and Consolidated Edison Co 1980-present; (Consultant and Summer); Co-Editor of 3 Volumes, "Advances in Environmental Science & Engineering, 1979 and 1980; Co-Editor of 3 Volumes, "Encyclopedia of Environmental Science & Engineering, 2nd Ed 1983; Editorial Advisory Board Member of Water, Air & Soil Pollution (1972-Present). Author of about 50 published articles. Dir of Senior Lab in Chemical Engrg. Dir of Enviromental Studies Program for High School Students. Service Award-OmeChiEps Honor Society. Elected to TauBetPi and SigXi Honor Societies. *Society Aff:* AIChE, APCA

Ziegler, Geza Charles
Home: 1452 Riverbank Rd, Stamford, CT 06903
Position: Dir of Mfg *Employer:* Cognitronics Co *Education:* MSEE/EE/Worcester Polytech Inst; BSEE/EE/Worcester Polytech Inst. *Born:* July 17, 1932 Budapest Hungary. Polytech Univ of Budapest, Worcester Polytech Inst; BSEE 1959, MSEE 1961. Taught: elec engrg, phys & math courses at Bridgeport Engrg Inst, Stamford Ct; Asst Dept Chmn 1966-70, Assoc Dean Stamford 1968-70, Dean of Faculty Stamford 1970-76, Dean of Coll Extension 1976-80 Dean of Danbury 1980-1983. VP of Danbury 1983- . Microwave engr, Microphase Corp Greenwich Ct 195962; Dev Engr Infrared Technology, Barnes Engrg Co Stamford Ct 1962-69; Sr Project Engr TV Technology & EVR, CBS Labs Stamford Ct 1969-75. Chief Engineer - Spanel R&D Stamford, CT 1975-1981. Project Engr - Summagraphics Co. Fairfield, CT 1981-82. 1982-1986 Director of Mfg. Cognitronics Co. Stamford, CT. Mbr ASEE, Conn Soc of Prof Engrs, TELSA of Ct (Hon Mbr). *Society Aff:* ASEE, CSPE, TELSA.

Ziegler, Hans K
Home: 32 E Larchmont Dr, Colts Neck, NJ 07722
Position: Retired. *Education:* PhD/EE/Tech Univ Munich, Germany; MS/EE/Technical Univ Munich, Germany; BS/EE/Technical Univ Munich, Germany. *Born:* Mar 1911 Munich Germany. Teaching at Tech Univ Munich as Asst Prof 1934- 36; with German electronics indus as V P for R&D 1936-47; 1947- at US Army R&D activities at Ft Monmouth N J as Scientific Cons, Asst Dir Res, Dir Astro Electronics Div, Ch Scientist, Deputy for Sci & Dir ETDL until 1977, retired. Respon in early Army Space electronics (SCORE, Vanguard I, Tiros, Courier), overall res mgmt & past 5 yrs emphasis on electronics devices & components. Life Fellow IEEE, Fellow AAS. Exceptional Civil Service Award, 2 Meritorious Service Awards, Antartica Service Medal, AFCEA Meritorious Gold Medal, Distinguished Life Member AFCEA. *Society Aff:* IEEE, AAS, AFCEA.

Ziegler, Waldemar T
Home: 1847 Virginia Ave, College Park, GA 30337
Position: Ret. (Prof of Chem Engrg Emeritus) *Education:* PhD/Physical Chem/Johns Hopkins Univ; MS/Chemistry/Emory Univ; BS/CE/GA Sch of Tech. *Born:* 8/2/10. Manhattan Proj (Columbia Univ) (1944-45), Carbide & Carbon Chems Corp (1945-1946), Teacher (various ranks) chem engg, GA Inst of Tech (1946-1978). Retired as Regents' Prof of Chem Engg (Emeritus), Jun 30, 1978. Consultant to Natl Bureau of Stds Cryogenic Engg Lab as visiting chem engr (summers 1956-58). Res interests: thermodynamics of pure substances & solutions, phase equilibria at low temperatures. Cryogenics. *Society Aff:* AIChE, ACS, AAAS.

Zielinski, Paul B
Business: 310 Lowry Hall, Clemson, SC 29634-2900
Position: Dir, WRRI *Employer:* Clemson Univ *Education:* PhD/CE/Univ of WI; M. S./CE/Univ of WI; B.S. /CE/Marquette Univ *Born:* 9/9/32 Engineering experience with City of Waukesha, WI Engrg Dept as coop student 1953-1956, as Engr 1957 summer. Taught at Marquette Univ 1956-1959, 1964-1967. Taught at Univ of WI 1959-1964 (Part Time). Performed as engr with Reukert and Mielke, Engrs, Waukesha, WI, 1958 and 1959 summers. Assistant Professor of Engrg Mechanics, Clemson Univ 1967 - 1969, Assoc Prof 1969 - 1974. Assoc Prof of Civ Engrg 1974-1978. Assoc Prof of Environmental Systems Engrg 1978; also Dir of S. C. Water Resources Research Inst, 1978 to present. Prof of Environmental Systems Engrg, 1979-1982. Registered PE, SC and WI. Prof, Civil Engrg 1982 to present. Conslt experience. *Society Aff:* ASCE, ASEE, AEEP, ΣΞ.

Ziemer, Rodger E
Business: EE Dept., Univ. of Colorado at Col. Springs, Colorado Springs, CO 80933
Position: Prof of EE. & Chmn. *Employer:* Univ of Colorado at Colorado Springs. *Education:* PhD/EE/Univ of MN; MS/EE/Univ of MN; BS/EE/Univ of MN *Born:* 8/22/37. Native of Amery, WI. After receiving PhD in EE at Univ of MN, served in the USAF from 1965-1968. Taught at the Univ of TN Space Inst & performed res and dev on electronic countermeasures at Wright Patterson AFB. With Univ of MO-Rolla 1968-84. Promoted to Prof in 1972. Joined the Univ of Colorado at Colorado Springs in Jan. 1984 as Professor & Chairman of EE. Co-author of three textbooks in Systems & Communication Theory. *Society Aff:* IEEE, ASEE.

Zierdt, Conrad H, Jr
Home: 4140 Maulfair Dr, Allentown, PA 18103
Position: Retired *Education:* BSEE/Electronics/PA State Univ. *Born:* July 1916 Pittsburgh Penn. Worked in electron-tube manufac, railway- signal, storage battery & business-machines design, test & maintenance; radio & radar-countermeasures design, elec/electronic purchasing through WW II period; transistor design engrg & transistor-integrated circuit reliability engrg & specifications 1952-1981. Fellow IEEE 1965; Reg P E in N Y; Chmn Syracuse N Y Chap IEEE/PGED 1960, Awards Ctte 1965-70; IRPS Mgmt Comm 1971-76; EIA Engr of the Year 1976. Chmn EIA/NEMA JC-13 Ctte on Govt Liaison 1958-1981. Antique-car collector & restorer. *Society Aff:* IEEE.

Zierler, Neal
Business: IDA, Thanet Rd, Princeton, NJ 08540
Position: Tech Staff. *Employer:* Inst for Defense Analys. *Education:* PhD/Math/Harvard Univ; MA/Math/Harvard Univ; BA/Physics/Johns Hopkins Univ. *Born:* 9/17/26. Co-auth of first largescale, effective high-level computer language, 1953- 54. Discovered fundamental idea of algebraic coding theory-the identification of error locations with the roots of certain polynomials-1959. Other basic results in coding for reliability and in the theory and applications of digital sequences, 1958-64. First derivation of the Hilbert space formalism for quantum mech from simple, physically plausible postulates, 1956-59; solution of a form of the hidden observable problem, 1964. Many theoretical results and effective algorithms, with realizations in computer software and hardware, in the field of cryptology, 1965-87. *Society Aff:* IEEE, AMS, MAA, ACM

Zimmerman, Albert
Home: 1919 S Crescent Heights, Los Angeles, CA 90034
Position: President. *Employer:* Albert Zimmerman & Assocs Inc *Education:* BS/ME/IL Inst of Tech. *Born:* 11/13/08. Taught engrg courses at IIT & UCLA; worked at Douglas Aircraft Co's Thermodynamics Dept Dec 1944-48; held position of Ch Mech Engr at Daniel, Mann, Johnson & Mendenhall Architctural firm 1950-56, when established own firm of Albert Zimmerman & Assocs, Cons Mech Engrs; held position as Pres of Mech Engrs Assn during year 1967-68. Enjoy classical music & out-of-doors. *Society Aff:* ΠΤΣ.

Zimmerman, Jerrold
Business: Electro-Optics Operations, 2 Forbes Rd, Lexington, MA 02173
Position: Chief Engr *Employer:* Honeywell, Inc. *Education:* MS/Mech Eng/Univ of Bridgeport; BA/Physics/New York Univ *Born:* 12/24/36 in NY city. Optical Engr at Bausch & Lomb and Perkin-Elmer. Dept Manager at Itek Corp. Joined Honeywell Electro-Optics Div in 1975. Assumed current position as Mgr, GOI New Business Development in 1984. Served as Pres of New England Section of Optical Society of America 1974-1975. Governor SPIE 1980-82. Chrmn Industrial Council SPIE 1982 & elected Fellow SPIE 1982. Active backpacker and bicyclist. *Society Aff:* SPIE, OSA

Zimmerman, Mary Ann
Home: 3801 Connecticut Avenue NW, #215, Washington, DC 20008
Position: Consultant *Employer:* Self *Education:* MSCE/Transpo/Urb Plng/Purdue University; BSCE/Civil Engineering/Purdue University *Born:* 04/22/45 Currently a Washington-based consultant in third world development with recent experience in Southern Africa. 1973-86, directed activities for Cummins Engine Company ranging from new business start-ups and new product introductions to corporate environmental policy. Worked as traffic engineer in Chicago between 1968-73, heading design of State Street Mall. Lifelong commitment to social change, holding leadership positions in national and community-based development organizations. Purdue Civil Engrg Centennial Lecturer 1987, Distinguished Engrg Alumna 1986. Indiana Young Engineer of Year 1979, State Pres 1983-84, NSPE. Fellow 1986, National Board Mbr 1979-83, SWE. Reg Engr, Indiana, Ireland. *Society Aff:* SWE, NSPE, ASCE

Zimmerman, Oswald T
Home: PO Box 606, Durham, NH 03824
Position: President. *Employer:* Industrial Research Service Inc. *Education:* PhD/Chem Engg/Univ of MI; MSE/Chem Engg/Univ of MO; BSE/Chem Engg/Univ of MI. *Born:* Jan 1905. Chem Engr with SimmonsCo 1929-30 & Kerr Mfg Co 1934-35; taught chem engrg at Univ of N D (Instr to Assoc Prof 1935-38) & at Univ of N H (Assoc Prof 1938-41, Prof 1941-70, Dept Chmn 1950-70, Prof Emeritus 1970-); Founder & Pres since 1942 of Indus Res Service Inc. Founder 1956, Hon 1st Pres 1956, Dir 1956-71, Recipient of Award of Merit 1964 of Amer Assn of Cost Engrs. Hon Prof of Chem, Nacional Universidad de San Marcos, Lima Peru 1962. Author & pub of several tech books & numerous articles; Founder & formerly ed & pub of Journal 'Cost Engineering'.

Zimmerman, Richard H
Business: 206 W 18th St, Columbus, OH 43210
Position: Professor - Mechanical Engineering. *Employer:* The Ohio State University. *Education:* MSc/Mech Engg/OH State Univ; BME/Mech Engg/Fenn (Cleve State) Univ. *Born:* Nov 6, 1922 Smithville Ohio; married (Jeanne Yoder); 3 children. BSME Fenn Coll (now Cleveland St Univ) 1943; MSc Ohio St Univ 1946. NACA (now NASA) 1943-45; faculty Mech Engrg Ohio St Univ 1965-70, Exec Asst to Pres & Dir Budget & Resources Planning 1972-76; Prof mech engrg 1976, thermodynamics & energy conversion; res in gas turbines, propulsion, cooling of electronic equipment, life support sys for spacecraft, underground powerplats, accurate equations of state for natural gas mixtures, characteristics of planetary atmospheres. Fellow

Zimmerman, Richard H (Continued)
ASME; mbr Sigma Xi, Pi Tau Sigma, Tau Beta Pi; Reg P E; ASEE Western Elec Award, McQuigg Award, 1977, OSU Alumnis Teaching Award 1959, Retired Sot, 1979, Prof Emeritus, The OH State Univ. *Society Aff:* ASME.

Zimmerman, Robert M
Business: 222 S Riverside Plaza, Chicago, IL 60606
Position: Partner. *Employer:* Greeley and Hansen. *Education:* BS/Civil Eng/Univ of WI. *Born:* 5/14/27. in Madison, WI. Served in US Merchant Marines 1945-47. Joined Greeley and Hansen June 1951. Reg as PE 1955. Named Assoc 1968 & Ptnr 1971; many assignments as design coordinator & const resident engr incl 3 yrs in Columbia S A as Proj Mgr for design & const of sewer sys. Respon as Ptnr incl personnel mgmt. Reg PE IL, VA, MI, WA, SD, WI, MA, AZ, NV. Fellow ASCE, AWWA, WPCF, Amer Acad of Environl Engg, ISPE & NSPE. Enjoy hunting, fishing & golfing. *Society Aff:* ASCE, AWWA, WPCF, NSPE, AAEE.

Zimmerman, Roger M
Home: 2936 Candelita Ct, NE, Albuquerque, NM 87112
Position: Mbr Technical Staff *Employer:* Sandia National Laboratories *Education:* PhD/Struct Mech/Univ of CO; MS/Civil Engg/Univ of CO; BSCE/Civil Engg/Univ of CO. *Born:* 5/15/36. in Rehoboth, NM. Taught at Univ of CO 1959-63 & NM State Univ 1964-79; Asst, Assoc & Acting Dean of Engg 1967-75; Asst Prof, Assoc Prof, Prof & Sr Engr PSL 1964-79, NMSU. Mbr Technical Staff, Rockwell International Science Center 1979-80. Involved in res on autombile occupant safety; sch bus occupant safety, 1975-79; continued res on triaxial strength of concrete 1963-79, res on siesmic sensitivity of bridges 1978-79; res on ultrasonic inspection Space Shuttle Tiles 1979-80. Research on nuclear waste geologic disposal 1980- . Pres NM Sect ASCE 1974-75, received ASEE Dow Chem Co Young Faculty Delegate Award 1969 & NM Ready Mixed Concrete & Sand & Gravel Assn Bicentennial Yr Award 1976 & NMSU Teacher of Yr Award 1979. Major hobbies; sports of handball, & hunting. *Society Aff:* ASCE, NSPE, ISRM.

Zimmermann, Francis J
Business: Dept of Mech Engg, Lafayette College, Easton, PA 18042
Position: Prof of ME, Emeritus *Employer:* Lafayette College. *Education:* ScD/Mech Engr/MA Inst of Tech; ME/Mech Engr/MA Inst of Tech; SM/Mech Engr/MA Inst of Tech; BE/Mech Engr/Yale Univ. *Born:* 4/21/24. in Jersey City, NJ; grew up in New Haven, CT. Married Margaret M Stephens 4/14/50; Children, Stephen R, Marcy J. Education at Yale Univ interrrupted by US Army service (1943-46) in ASTP at Univ of AL and in SWPA. Staff engr in Cryogenics at Arthur D Little, Inc, 1952-55. Asst to Assoc Prof of ME at Yale Univ, 1955-62. Asst Prog Dir for Engg at the natl Sci Fdn 1961-62. Prof of ME at Lafayette College 1962-86; Dept Hd 1978-1982. Consultant in Cryogenics to Arthur D Little, Inc and Air Products and Chemicals, Inc. Papers and patents in the field of Cryogenics. *Society Aff:* ASME, ASEE, CSA, AIAA.

Zimmermann, Henry J
Home: 14 Russet Ln, Lynnfield, MA 01940
Position: Professor Emeritus of Electrical Engineering. *Employer:* Mass Inst of Tech. *Education:* MS/EE/MIT; BS/EE/WA Univ. *Born:* May 11 in St Louis MO. Res interest is signal processing. 1938-40 elec engrg instr Wash Univ. Dev electronic device to present visual indication of pitch for teaching voice inflection to the deaf. 1940 joined MIT Elec Engrg Dept as res asst & appointed Prof 1955. During WW II organized radar training courses which led to establishment of MIT radar sch where he was instr 1941-42, supr of Army courses 1942-44 & Asst Dir 1944-46. 1947 joined Res Lab of Electronics as staff mbr & appointed Dir 1961. Co-authored 'Electronic Circuit Theory' & 'Electronic Circuits, Signals & Systems'. Recipient of Wash Univ 1975 Alumni Achievement Award; IEEE Fellow & Tau Beta Pi, Eta Kappa Nu & Sigma Xi mbr. *Society Aff:* IEEE.

Zimmie, William E
Business: 810 Sharon Dr, Cleveland, OH 44145
Position: Retired *Education:* BSE/Naval Arch/Univ of MI; BSE/Math/Univ of MI; BSE/Mech Engrg/Univ of MI *Born:* 3/18/27 in Detroit, MI. Served in U.S. Merchant Marine 1944-1947. Licensed Chief Engr Steamships of Unlimited Horsepower, Registered Engr, St of OH. Designer and Guarantee Engr, Great Lakes Engrg Works, River Rouge, MI 1952-1953. Fleet Engr, Hutchinson & Co., Managers of The Pioneer Steamship Co. & The Buckeye Steamship Co., Cleveland, OH 1953-1960. President & Chief Executive Officer, Zimmite Corp and subsidiaries, Zimmite Intl and Hyde Products, Inc., Cleveland, OH 1960-1983. Enjoys sailing, golfing, skiing. Mbr Bd of Dirs - ABET (Accreditation Board for Engineering and Technology). *Society Aff:* SNAME

Zink, Carlton L
Home: 1890 24th Ave A, Moline, IL 61265
Position: Product Safety Consultant. *Employer:* Self-employed. *Education:* BSc/Agri Engrg/Univ of NB. *Born:* May 1902; reared on Nebraska farm. Agri Engrg grad Univ of Nebr 1926. Faculty mbr Univ of Ariz 1927-29; Univ of Nebr 1929-43. Engr-in-charge of Nebr Tractor Tests 1930-41. With Firestone Tire, Akron 1943-50; Mgr Farm Tire Dev 1945-50. Deere & Co Moline Ill 1950-68. For 15 yrs corporate representative in dev of product safety standards for agri equipment indus. Reg P E. Recipient ASAE McCormick Medal; Life Fellow ASAE; Mbr Amer Acad of Achievement. 1968- active safety cons primarily in area of farm tractors & equipment. *Society Aff:* ASAE, SAE, NIFS, NSC.

Zinkham, Robert E
Home: 9204 Westmoor Dr, Richmond, VA 23229
Position: Consultant *Employer:* Self *Education:* MS/MMet/Univ Pittsburgh; BS/ME/Carnegie Mellon Univ; BS/Engg/Geneva College. *Born:* 1/19/23. Dev Engr on oil country tubular products with J&L Steel Corp, 1949-1957. Supervisor of elec & mech testing & heat treating sections. Allegheny Ludlum Steel Corp, 1957-1960. Res Engr on drilling methods & tubular products. Gulf Res Corp, 1960-1963. (Granted 3 patents) Since 1963 associated with the Met Laboratory of Reynolds Metals Co in Richmond, VA. As Dir of Mech Met, the chief function involves proj direction concerning mech behavior of aluminum alloys with an emphasis on fatigue & fracture toughness. Technical papers published in these areas. Chrmn of Subcomm on Fracture Toughness of the Metal Properties Council & a mbr of the Editorial Advisory Bd of the Journal of Engg Fracture Mechanics. Retired as Consultant in 1986. *Society Aff:* ASM, ASME, ASTM.

Zinsmeister, Herbert F
Home: 5701 E Glenn-27, Tucson, AZ 85712
Position: VP. *Employer:* Arthur Bd Engg, Inc. *Education:* BSCE/-/Purdue Univ. *Born:* 2/23/23. in Lafayette, IN. BSCE Purdue Univ. Lt (jg) CEC, USNR in WWII. District & Div Mgr Wallace & Tiernan Co 1947-64; VP & Mbr Bd/Dir Clyde E Williams & Assocs 1967-79; assumed additional respon as Dir Environ Engg 1973 & Dir Corp Engg 1975. VP & Mgr Tucson Office, Arthur Beard Engrs, Inc, 1979- . P Chmn IN Water Works Assn; Fuller Awardee 1973. P Pres IN Comprehensive Health Planning Prog. P Pres IN Water Pollution Control Assn. P Pres IN Soc of Prof Engrs & Natl Dir Natl Soc of Prof Engrs. *Society Aff:* NSPE, WFCF, AWWA.

Ziol, Frank J
Home: 3810 Shadowgrove Rd, Pasadena, CA 91107
Position: Professor Emeritus *Employer:* Pasadena City College. *Education:* MS/Tech Ed/Univ of So CA; BS/Ind Eng/Cleveland State Univ. *Born:* Apr 9, 1916 Cleveland Ohio. Prof Emeritus Pasadena City Coll, Pasadena Calif. Prof Engr - Control Sys-St of Calif. Teacher Trainer Calif St Dept of Educ. Cons for Tech & Mgmt Training progs in So Calif area. Fellow, Donald Eckman Educ Award, Distinguished Service Award and Golden Achievement Award of Instrument Soc of Amer. Fellow of Inst for Advancement of Engrg. Phi Delta Kappa, Epsilon Pi Tau. *Society Aff:* ISA, ASEE.

Zippler, William N
Home: 35 Brookdale Rd, Cranford, NJ 07106
Position: Retired - V Pres for Elec Engrg. *Employer:* Gibbs & Cox Inc. *Education:* BS/EE/Univ of PA. *Born:* Mar 19, 1896. Grad Woodbury N J High School; Certificate of Commendation Navy Dept Bureau of Ships Mar 1, 1947 for outstanding services as EE in dev working plans for elec sys for various type vessels of US Navy & contrib to the solution of many complex problems & the improved designs of elec equipment for Navy vessels. Respon for the entire elec plant & commmunication sys on many mercant vessels incl America & SS United States (still the fastest passenger vessel on the high seas). Ch Elec Engr for Gibbs & Cox & Associated Co's for 50 yrs 1920-70. *Society Aff:* ASME, ASNE, ABYC.

Zissis, George J
Business: PO Box 8618, Ann Arbor, MI 48107
Position: Assoc dir - Information & Processing Division & Adj Prof EECS Dept. *Employer:* Environ Res Inst of MI/Univ of MI *Education:* PhD/Physics/Purdue Univ; MS/Physics/Purdue Univ; BS/Physics/Purdue Univ. *Born:* Dec 1922. Army Air Force 1943-46. Sr Scientist Westinghouse Atomic Power Div 1954-55; Univ of Mich in Infrared & Remote Sensing res 1955-73. Mbr and/or Chmn NAS Ctte on Remote Sensing Progs for Earth Res Surveys 1966-76. Public Service Award of Dept of Interior 1972. Ed-in-Ch Remote Sensing of Environ 1971- 78. Fellow OSA, SPIE & AAAS; Mbr EPA Sci Adv Bd, EMA Ctte till 1979. *Society Aff:* OSA, SPIE, AAAS, ΣΞ.

Zitomer, Kenneth J
Business: 1180 MSB 15th & Kennedy Blvd, Phila, PA 19107
Position: Chief Engineer *Employer:* City of Phila. *Education:* BS/CE/Univ of PA. *Born:* 10/24/26. Reading RR, 1948 to 1953 - Asst & Track Supervisor, Maintenance of Way Dept. Phila Water Dept, 1953 to 1958 Engr with Construction Div. LJC Mining & Dev Co, 1958 to 1959 Pres. Phila Water Dept: Div Engr - 1959 to 1964, Chief of Construction Branch - 1964 to 1967, Chief of Design Branch - 1967 to 1972, Deputy Commissioner for Engg - July 1972 to Jan 1981. Chief Engineer Jan 1981 to present Twenty-eight yrs of experience in Design, Construction & Mgt of water & wastewater facilities in City of Phila. Reg Engr in PA-3406E. *Society Aff:* ASCE, NSPE, AWWA.

Zlatin, Norman
Home: 11230 N W 25th St, Coral Springs, FL 33065
Position: V P, Treas, Dir of Machinability Res. *Employer:* Metcut Research Assocs Inc. *Education:* EE/-/Univ of Cincinnati; BS/ME/Univ of Cincinnati. *Born:* 6/24/14. Co-Founder Metcut Res Assocs Inc 1948; retired Apr 1, 1976; retained as cons for Metcut Res Assocs Inc. 1962, 3 mos in Israel under US Operations Mission to Israel studying metal cutting indus; 1965 mbr US-USSR Machine Tool Exchage Group - 3 wks in USSR visiting Machine Tool Plants; 1962 ASTME Gold Medal; Fellow ASME; Author 126 tech papers. *Society Aff:* ASME, SME.

Zmeskal, Otto
Business: 2801 W Bancroft, Toledo, OH 43606
Position: Prof Met Engg, Retired *Employer:* Univ of Toledo. *Education:* ScD/Phys Met/MIT; MS/Chem Met/Armour Inst of Tech; BS/ChE/Armour Inst of Tech. *Born:* 7/16/15. Asst Prof Met - IIT 1941-43: Dir, Dept of Met Engr, IIT, 1946-1954: Res Prof, Mech Engr Univ of FL, 1954-1958. Dean of Engg Univ of Toledo 1958-1970. Prof of Met Engr Univ of Toledo 1970 to date. Dir of Res, Universal Cyclops Steel Corp, Bridgeville Div, 1943-1946. Mbr, Bd of Dir, Knoxville Iron Co 1960- 1968. Prof Emeritus 1985. *Society Aff:* ASM, AIME, MS.

Zmyj, Eugene B
Home: 21 Andover Dr, Short Hills, NJ 07078
Position: Dir of Indus Engg. *Employer:* Warner-Lambert Co. *Education:* MS/Ind Eng/Lehigh Univ; BS/Mech Engg/NJ Inst; BA/Soc Studies/St Basil's College. *Born:* 6/3/30. in Lwiw Ukraine. Reg PE, Cert in MTM. Fellow of IIE, Mbr NSPE, SME, AIMC; active in Amer Inst of Indus Engrs as Chap Pres & Div Dir; contrib articles to proj journals on mgmt controls & work methods Y stds. With Warner-Lambert Co, as Corp Dir of IE respon for IE functions and productivity in 115 mfg facilities, world-wide; previously with the Echlin Mfg Co, the City of NY Amer Can & Western Elec. Adjunct Prof of Mgmt and Industrial Engrg at NJIT. *Society Aff:* IIE, AIMC, SME, NSPE.

Zoino, William S
Business: 320 Needham St, Newton, MA 02164
Position: Prin *Employer:* Goldberg-Zoino & Assoc, Inc *Education:* MS/Civ Engrg/MIT; BS/Civ Engrg/MIT *Born:* 03/12/32 A native of Brockton, MA, Mr Zoino has over 30 yrs of experience in fdn engrg practice in the US and abroad. Upon grad from MIT in 1955, he was employed by EBASCO Services, Inc as fdn engr on maj elec generating facilities. His expertise is in the design and construction of earth, rock, and concrete dams. From 1956-57 he served as a surveyor in the US Navy. He is a cofounder of Goldberg-Zoino & Assoc, Inc, a geotechnical and geohydrological consltg firm in Newton, MA,. Since 1964, he has served as prin geotechnical engr on over 1,000 projects. He is a past pres of the Boston Soc of Civ Engrs Sect of ASCE, past pres of the Assoc of Soil and Fdn Engrs and VP zone 1 ASCE. *Society Aff:* ASCE, ASFE, USCOLD, ASTM

Zoller, David R
Home: 31 Cardinal Road, East Lyme, CT 06333
Position: Principal Engineer *Employer:* General Dynamics/Electric Boat *Education:* M.S./Engr Management/Northeastern Univ; M.S./Naval Arch/M.I.T.; B.S. /ME/Brown Univ *Born:* 12/13/42 Career in shipyard design offices began in 1963 at Newport News Shipbuilding & Dry Dock Co. After earning MS in Naval Architecture, went to work for General Dynamics/Quincy Shipbuilding Division. Transferred to General Dynamics/Electric Boat Division in 1986. Have been involved in all phases of ship design from concept to to detailed working plans. Career has included work in weight control, tank capacities, structural design, vibration analysis, shipyard production liaison and marine coatings. Active in the Society of Naval Architects and Marine Engrs. Various offices in New England Section held including Chrmn, 1979-80. *Society Aff:* SNAME, SAWE, SSPC

Zona, Peter G
Business: P O Box 391, Ashland, KY 41101
Position: Sr Supervisory Cost Engr. *Employer:* Ashland Synthetic Fuels Inc. *Born:* Mar 22, 1925. Polytech Inst of N Y, BCE; N Y U, MBA. Cert Cost Engr, served as: Internatl Pres; V P (Admin & Tech); Dir of Cost Mgmt, Educ & Publicity; Council Bd/Dir; Chmn, Guest Speakers Bureau; Tech Prog Chmn for AACE. 29 yrs of progressively respon supervisory & mgmt experience in engrg & const indus; directly respon for over $8 billion capital project investments. Special qualifications in financial mgmt, estimating & cost engrg, planning/scheduling, contract admin, supervision of const & installations of mgmt cost control sys. Mgr Estimating, Cost Control, Planning/ Scheduling, Financial, Proj Mgr, Supt of Const.

Zonars, Demetrius
Business: 1361 Heritage Rd, Dayton, OH 45459
Position: Acting Deputy Director. *Employer:* Air Force Flight Dynamics Lab. *Born:* Dec 2, 1922; native of Dayton Ohio. BS & MS in Aero Engrg Univ of Mich & PhD Oho State. Respon for tech mgmt of efforts in aerospace vehicle & missll R&D; served as Air Force Flight Dynamics Lab Ch Scientist 1969-75, respon for design of Air Force's 50 million watt arc-heated hypersonic wind tunnel; received Air Force Sys Command Award for Outstanding Achievement in Engrg; joint recipient of Air Force Assn Award in Engrg for 'Redesign of F-111 Aircraft to Achieve Inlet-Engine Capability'. Received Ohio St Univ Disting Alumnus Award. Fellow AIAA; mbr & a natl coordinator of a NATO Adv Group for Aerospace R&D tech ctte.

Zook, Donald G
Business: 100 NE Adams St, Peoria, IL 61629
Position: Asst. Dir. of Mfg. *Employer:* Caterpillar Tractor Co. *Education:* MS/Ag. Eng./Univ of IL; BS/Ag. Eng./Univ of IL *Born:* 3/9/32 Native of Athens, IL. Served with Army 1953-1955. With Caterpillar Tractor Co since 1953. Taught part time at Bradley Univ 1957-1970. Assumed current responsibility as Asst Dir of Mfg in 1983. Responsible for corp decisions for component sourcing and asset mgmt. Currently serving as Pres of the SME. *Society Aff:* ASM, SAE, SME, NSPE

Zorowski, Carl F
Business: IMSE Inst., Box 7915, NC State Univ, Raleigh, NC 27695
Position: R J Reynolds Professor/Dir, Integrated Manufacturing Systems Engineering Institute. *Employer:* N C State University. *Education:* BS/Mech Engg/Carnegie Inst of Tech; MS/Mech Engg/Carnegie Inst of Tech; PhD/Mech Engg/Carnegie Inst of Tech. *Born:* July 14, 1930 Pittsburgh. Instr, Asst Prof, Assoc Prof of mech engrg CIT 1952-62; Assoc Prof, Prof, Assoc Dept Head Mech & Aerospace Engrg Dept N C St Univ 1962-72. Awarded R J Reynolds Professorship in mech engrg 1969; Dept Head Mech & Aero Engr Dept. NCSU 1972-79, Assoc Dean of Engg NCSU, 1979-85, apptd Dir Integrated Mfg Systems Engrg Institute NCSU 1986. Research in mech design, fiber mechanics, & design for automated assembly. OEEC Senior Visit Fellow 1961, ASEE Western Elec Award 1968, Fiber Soc Achievement Award 1970, ASME Pi Tau Sigma Richards Memorial Award 1975. Married, 3 children. Hold SCCA Natl Competition License . *Society Aff:* ASEE, ASME, SME.

Zoss, Abraham O
Business: PO Box 162, Old Bridge, NJ 08857
Position: VP. *Employer:* CPS Chemical Co, Inc. *Education:* PhD/Organic Chemistry/Univ of Notre Dame; MS/Organic Chemistry/Univ of Notre Dame; BS/ChE/Univ of Notre Dame. *Born:* 2/17/17. in South Bend, IN. Entered industry as chemical engr in 1941 specializing in high pressure acetylene chemistry. Various engg, mgmt and exec positions with industrial cos including; General Aniline and Film, MN Mining and Mfg Co, Celanese Corp and Engelhard Minerals & Chemicals Corp. Mbr field info agency Office Technical Services, Commerce Dept, Europe 1946. with CPS Chem Co since 1977 as VP for Corp Dev, Chf Admin Officer & Dir. Centennial of Science Award, Univ of Notre Dame 1965. Enjoy tennis and skiing. *Society Aff:* AAAS, AIChE, ACS, AIC.

Zoss, Leslie M
Business: Valparaiso, IN 46383
Position: Prof. *Employer:* Valparaiso Univ. *Education:* PhD/ME/Purdue Univ; MS/ME/Purdue Univ; BS/ME/Purdue Univ. *Born:* 11/23/26. Native of Lockport, NY. Served in US Army 1945-46, Criminal Investigation Div. Res Engr and Dir of Training for Taylor Instrument Co 1952-58. With Valparaiso Univ from 1958 to present. Served as Dir of Res and Dept Hd of Mech Engg. Reg Mech Engr in IN, Control Systems Engr in CA. Consultant in measurements and control to numerous corps. ISA Education Award 1968; ISA Fellow 1970. Valparaiso Univ Alumni Teaching Award 1977. Numerous publications; author of three books. *Society Aff:* ASME, ISA.

Zowski, Thaddeus
Home: 916 Seminole Rd, Wilmette, IL 60091
Position: VP Principal Mech Engr. *Employer:* Harza Engrg Co - Chicago Ill. *Education:* MS/ME/Univ of MI; BS/ME/Polytech Inst of Warsaw. *Born:* Mar 1914 Ann Arbor Mich. With S Morgan Smith Co York Pa 1941-47 designing hydraulic machinery & special products; with Harza Engrg Co 1974- , becoming Ch Mech Engr 1950 & V P 1968. In Sr Prof Staff capacity respon for basic concepts of mech engrg on Harza projects. Mbr ASME (Life Fellow), ASCE, NSPE (PEPP), ASME Performance Test Code Ctte PTC 18 on Hydraulic Prime Movers & US Natl Ctte of Internatl Electrotech Comm, TC4 on Hydraulic Turbines. Author tech papers & co-author 'Hydraulic Machinery' section, Handbook of Applied Hydraulics. *Society Aff:* ASME, ASCE, NSPE.

Zraket, Charles A
Business: Mitre Corporation, Burlington Rd, Bedford, MA 01730
Position: Exec VP. *Employer:* MITRE Corp *Education:* SMEE/EE/MIT; BSEE/EE/Northeastern Univ. *Born:* 1/9/24. Native of Weston, MA. Served as Group Leader for Advanced Systems Dev at MIT Lincoln Labs, 1952 to 1958. Joined MITRE as Dept Hd for Advanced Systems in 1958. Assumed current position as Exec VP of MITRE in 1978. Responsible for mgt of areas which include defense command & control systems, air traffic control systems, surface transportation, energy systems, environmental control systems & information systems. Fellow, IEEE and AAAS and AAAS. Trustee, MITRE Corp & Hudson Inst. Mbr, Council Foreign Relations. *Society Aff:* IEEE, AAAS, ΣΞ, ΤΒΠ.

Zubick, Gerald M
Business: 840 6th Ave-Box 1207, Huntington, WV 25714
Position: Branch Office Engr-Mgr. *Employer:* L Robert Kimball, Cons Engrs. *Education:* BSCE/Civil Engg/OH Univ. *Born:* Jan 1936 E Liverpool Ohi. Grad E Liverpool H S 1953; Ohio Univ Athens BSCE 1958. Worked as Hull Designer for nuclear submarines, hwy designer for state govt & cons, Distrib Engr for water co & Resident City Planner for Cons. Presently Branch Office Engr-Mgr for Cons doing primarily hwy, municipal, airport, sanitary & drainage projects. *Society Aff:* NSPE.

Zuccaro, Robert M
Business: 10 Palms Plaza, Homestead, FL 33030
Position: Asst Regional Mgr *Employer:* Post, Buckley, Schuh & Jernigan, Inc *Education:* BS/Civil Engrg/Clarkson College of Tech *Born:* 12/3/46 Born and raised in upstate NY. Upon graduation employed by the CT State Hwy Dept as a Civil Engr Intern. Moved to FL and was employed in private practice for 4 yrs as a highway and land development engr for Carr Smith & Assoc. Project Mgr for the last 11 yrs for a number of large dev projects and highways in the South FL area. I am presently the Asst Regional Mgr. - Homestead/FL Keys Region for Post, Buckley, Schuh & Jernigan, Inc., a multi-disciplinary consltg engrg firm of 600plus employees with regional offices in the Southeast US. I enjoy sports, music. *Society Aff:* ASCE, FES, NSPE

Zuck, Alfred Christian
Home: 444 Weymouth Dr, Wyckoff, NJ 07481
Position: Principal. *Employer:* Edwards & Zuck PC. *Education:* ME/ME/Poly Inst of Brooklyn. *Born:* 12/16/24. Consulting Mech Engr. Born in Ridgefield, NJ. Ed: ME/Poly Inst of Brooklyn. Employed by Syska & Hennessy, Inc: 1946 to 1978. Sr VP responsible for Mech & Elec Design of many notable commercial & institutional bldgs including Metropolitan Opera at Lincoln Ctr; LaGuardia Airport Terminal; Nelson A Rockefeller Albany Mall; Time & Life, Exxon, McGraw Hill & Celannese Bldgs at Rockefeller Ctr, NYC. Reg PE in 21 states, Natl Council Engg Examiners. Licensed Profl Planner in NJ. Principal of Edwards & Zuck, PC, Consulting Engrs since 1978. Responsible for Mech. & Elec Design of Meyerson Symphony Center, Dallas, Tex, European Amer Bank Plaza, Hempstead, N.Y. Served US Army 1943-46, USAF 1951-52 attaining rank of Captain. Decorated Bronze Star (2). Fellow, ACEC, past mbr ACEC Natl Ethical Practices Ctte. Mbr NSPE, NYSPE (past Chrmn NYSPEPP), SAME, NYACE (past VP & Dir), NY Bldg Congress (Bd of Govnrs), ASHRAE. Author of numerous articles for prof publs. *Society Aff:* ASHRAE, NSPE, ACEC, SAME, NYSPE, NYACE, NYSPEPP.

Zucker, Gordon L
Business: W Park St, Butte, MT 59701
Position: Prof. *Employer:* MT College Mineral Sci & Tech. *Education:* Dr Engr Sci/Mineral Engg/Columbia Univ; MS/Met Engg/Univ of WI; BS/Metallurgy/MA Inst of Tech; -/Mineral Eng/Penn State Univ. *Born:* 11/24/29. in Providence, RI, secondary education in Quincy, MA specialized in tantalum capacitor R&D and mfg while employed in the electronics industry as sr engr or mgr (1959-1975). Formerly worked for TX Instruments, Union Carbide Corp (Electronics Div), Aerotron-

Zucker, Gordon L (Continued)
Tansitor, Dickson El, and Siemens Corp. Since 1975, Prof of Mineral Processing Engg specializing in coal processing, vermiculite and gold cyanidation. Formerly, Exec Dir, MT Tech Alumni Fdn and formerly its Technical Dir of the Mineral Res Ctr (1976-1979). Presently consulting as President of East West Consultants, Inc. Concurrent with professorship. *Society Aff:* AIME, ACS, ΣΞ.

Zuckerman, Leonard M
Home: 43 Gateway Dr, Great Neck, NY 11021
Position: President. *Employer:* L. M. Zuckerman & Co Inc. *Education:* BChE/Chem Engrg/CCNY, 1944; MChE/Chem Engrg/NYU, 1947. *Born:* 1923. Lic P E in N Y State. 1944 Manhattan Proj for Kellex Corp div of M W Kellogg Co on res tech & high vacuum processing for Oak Ridge Atomic plant. 1947 Standard Brands Inc; participated dev new instant coffee processig tech. 1954 formed & was Pres of Coffee Instants Inc, an instat coffee processor & packer. 1972Pres FDC Sales Inc, a major importer & distrib of processed coffees: soluble, freeze dried & decaffeinated. 1980 Pres, L. M. Zuckerman & Co Inc - Agent, Dealer, Wholesaler, Importer Processed Coffees. Also provided Engrg Services for start up and operation instant and Freeze Dried Coffee Plants. Tech cons to domestic & offshore coffee processing indus. Mbr Bd of Dirs Natl Coffee Assn, Coffee Standards Ctte; Mbr IFT, AIChE, ACS & Green Coffee Assn of N Y. *Society Aff:* PCCA, AIChE, ACS, IFT, NCA, GCNY.

Zummer, Anthony S
Business: 175 W Jackson Blvd, Chicago, IL 60604
Position: Patent Attorney. *Employer:* Self-employed. *Born:* Apr 1, 1929. BSME Purdue Univ 1950; JD Geo Washington Univ 1956; Master of Patent Law John Marshall Law Sch 1959. Native of Kankakee Ill. Designed orthopedic appliances for Pope Foundation 1950-51; designed testing equipment for testing prostheses for Army Prosthetic Res Lab 1951-52; Ordnance Engr USN Gun Factory 1953; Patent Engr GM Corp 1953-56; Assoc in patent law firm Chicago 1956-58; self-employed as Patent Attorney since 1958. Pres Ill Soc of Prof Engrs 1976-77; Natl Dir NSPE 1973-78 .

Zumwalt, Glen W
Business: Dept of Aero Engrg, Wichita, KS 67208
Position: Prof *Employer:* Wichita St Univ *Education:* Ph.D./Mech & Aero. Engr/Univ of IL; MS/ME/Univ of TX; BS/ME/Univ of TX; BS/Naval Sci/Univ of TX. *Born:* 4/21/26 in OK. Served in U.S. Navy, 1944-47; Lt. (j.g). Engr for Sheffield Steel Corp., Houston, TX 1949. Left technical work for two years to serve as traveling staff worker for Inter-Varsity Christian Fellowship. Following M.S. studies, taught at Univ of TX and worked as Research Engr at Defense Research Lab., Austin, TX. From 1959 to 1968, faculty member at OK St Univ., School of Mech. and Aerospace Engrg. Came to Wichita St Univ in 1968 to initiate Ph.D. program in Aeronantical Engrg as Distinguished Prof. Has served as consultant for numerous industrial firms and government agencies. Principal technical area: Gas dynamics and Aircraft De-Icing Systems. *Society Aff:* AIAA, ASEE

Zuniga, Oscar J
Business: 100 E Hillside Rd, PO Box 1788, Laredo, TX 78041
Position: Pres. *Employer:* Zuniga Engg Co. *Education:* BS/ME/Univ of TX. *Born:* 5/30/22. in Laredo, TX. Served in the USN 1943-46. Chief Engr with Air conditioning firm 1949-58. Proj Engr, asst chief engr and civilian coordinator of military construction at Laredo AFB 1958-68. Established Zuniga Engg Co 1968 and still active as Pres. VP of TSPE 1968-70. Pres Gateway Chapter, TSPE 1968-69 & 1977-78 and chosen Engr of the Yr 1968. Principal in Northside Dev Co, Armadillo Construction Co, Zuniga Construction Inc, & Longhorn Air conditioning Co. Dir of City Natl Bank of Laredo. Enjoys hunting, fishing, and camping. *Society Aff:* ASHRAE, NSPE.

Zunkel, Alan D
Home: 55 Red Hill Rd, Warren, NJ 07060
Position: Manager *Employer:* Exxon Minerals Co. *Education:* Doctor of Sci/Met. Engr/CO School of Mines; MS/Met. Engr/CO School of Mines; BS/Met. Engr/MO School of Mines *Born:* 5/3/41 Native of Bethany, MI. Graduated Princeton H.S., Princeton, IL 1959. Aeronautical Engr, U.S. Army, 1967-69. Research Engrg/Mgr St. Joe Minerals Co 1969-1975, specializing in zinc/lead extractive metallurgy. General Superintendent St. Joe Minerals Co 1977-78. Manager, Minerals Technology, Exxon Research and Engrg Co 1978-80. Manager, Minerals Processing Research, Exxon Minerals Co. 1980-1981. Manager, Business Environment and Commodity Analysis, Exxon Minerals Co. 1981-present. Dir, TMS/AIME 1980-1982. Enjoy tennis, travel, stamp collecting, cooking. *Society Aff:* AIME

Zurheide, Charles H
Business: 4333 Clayton Ave, St Louis, MO 63110
Position: President. *Employer:* Zurheide-Herrmann Inc. *Education:* BSEE/Elec Engg/Univ of MO-Columbia. *Born:* 5/9/23. Founder of cons engrg firm 1954; has served as Pres 12 yrs of multi- discipline firm, providing specialized services to indus & govt; also served as Ch Elec Engr of large engrg contractor & prior to that as Indus Proj Engr with elec utility. Chmn state org of both Mo Soc of Prof Engrs & Cons Engrs Council of Mo & Natl Dir of each. Chmn IEEE & IES Chaps in St Louis; also mbr Elec Code Ctte of St Louis & St Louis Cty. Chrmn of MO Engr Reg Bd. Received Honor Award from Univ of Mo 1976. Born in St Louis Mo in 1923. Grad Univ of Mo - Columbia 1944, BSEE. *Society Aff:* IEEE, IES, NSPE, SAME.

Zverev, Anatol Ivan
Home: 914 Hillcrest Rd, Hanover, MD 21076
Position: Consulting Engineer. *Employer:* Westinghouse Elec Corp. *Education:* Dr/-/Acad of Transport. *Born:* Nov 1913; native of Moscow Russia. DrS Acad of Transport Moscow; MS Electrotechnic Inst of Leningrade. Taught in Moscow area tech insts & acad of transport USSR prior to WW II. Ch Communication Engr of Ministry of Transportation 1938-40. With Signal Corps Soviet Army til 1943. Design Engr Siemens & Halske Co Berlin Germany. Post grad work in Techn Hochschule Sharlottenburg. Prof & Dir of Internatl Univ Munich for displaced persons 1945- 47. Stff mbr Atelie de Const Elec de Charleroi (ACEC) Belgium until 1952. 1953 Westinghouse Elec Corp Baltimore; 1958- , Section Mgr, Advisory Engineer, consulting Eng, Dir educ seminar at co; 1967- , Cons Engr of Corp. Author 3 tech books & 52 tech papers in english language. Fellow IEEE & mbr Internatl Reception Soc. Prof Engr in Md & Belgium. *Society Aff:* IEEE.

Zweben, Stuart H
Business: Dept. of Computer & Info Sci, Ohio State Univ, 2036 Neil Ave, Columbus, OH 43210
Position: Assoc Prof *Employer:* OH State Univ *Education:* PhD/Comp Sci/Purdue Univ; MS/Stat & Comp Sci/Purdue Sci; BS/Math/CUNY City Coll *Born:* 4/21/48 Native of Bronx, NY. Worked as systems analyst for IBM Corp 1969-70. Current resch focus is in computer software engrg with particular interest in software quality evaluation. Teaching activities involve software engrg, analysis of algorithms, and systems programming. ACM National Lecturer 1979-81. Chrmn, ACM Central OH Chptr 1977-79. Chrmn, ACM Committee on Chptrs 1980-82. ACM E Central Reg Rep 1982-88. Chrmn, ACM Chapters Bd 1982-85. Acting Chrmn, CIS Dept OH State Univ 1983-84. ACM E Central Reg Rep 1982-88. Chrmn, ACM Chapters Bd 1982-85. Acting Chrmn, CIS Dept OH State Univ 1983-84. Secy/Treas, Computing Sci Accred Bd, 1986-87. *Society Aff:* ACM, IEEE-CS, AAUP.

Zweifel, Paul F
Business: Ctr for Transport Theory & Mathematical Physics, Robeson Hall, VPI & SU, Blacksburg, VA 24061
Position: Univ Dist Prof & Ctr Dir *Employer:* Ctr for Transport Theory & Math Physics *Education:* BS/Physics/Carnegie Inst of Tech; PhD/Physics/ Duke Univ. *Born:* 6/21/29. in New York, NY. G E Co, Knolls Atomic Power Lab 1953-58; Univ of MI Dept of Nuclear Engg Assoc Prof 1958-60, Prof 1960-68. VPI & St Univ

Zweifel, Paul F (Continued)
Phys & Nuclear Engg Prof 1968-71, Univ Prof 1971-75, Univ Disting Prof 1975- . Fellow ANS (Dir 1967-70); Fellow APS; Mbr Amer Math Soc; Mbr Federation Ame Scientists, Secy 1974-75. Mbr AMS, UMI. Visiting Prof Middle East Tech Univ Ankara Turkey, Rockefeller Univ & Univ of Florence, Univ of Ulm, Univ of Milan & Pakistan Inst Nuclear Tech. Ernest O Lawrence Award 1972, John Simon Guggenheim Meml Foundation Fellow, 1974-75. Editor-in-Chief, Journal for Transport Theory & Statistical Physics (1972-82). *Society Aff:* AMS, APS, IAMP, UMI.

Zwicky, Everett E, Jr
Home: 9 Riviera Dr, Latham, NY 12110
Position: Consulting Engineer. *Employer:* General Electric Cp. *Education:* BS/EE/Univ of NM. *Born:* 1925. Ensign USNR 1945-46. Joined G E Co as test engr 1946; with their Steam Turbine Generator Div 1951- , in design & dev engrg. Main emphasis on stress & vibration analysis, numerical analysis & computer applications. Became cons engr in Knolls Atomic Power Laboratory 1987. 2 US pats. Was Adjunct Prof mech engrg Brooklyn Polytech Inst & RPI. Reg P E in N Y. Fellow ASME. *Society Aff:* ASME.

Zwiebel, Imre
Business: Dept of Chem, Bio, and Materials Engineering, Arizona State University, Tempe, AZ 85287
Position: Prof & Chrmn Dept of Chem, Bio, and Mat. Engrg. *Employer:* AZ State Univ. *Education:* BS/Chem Engrg/Univ of MI; MS/Chem Engrg/Yale Univ; PhD/Chem Engrg/Yale Univ. *Born:* June 13, 1932 Budapest Hungary. s. Herman & Bella (Schonberg) Z; came to US 1948, naturalized 1954. m. Barbara E Copeland Dec 23, 1962; c. Karen, Jeffrey H, Kenneth M, Hannah. Design Engr DuPont Co Wilmington DE 1954-57; Res Engr Esso Res & Engrg Co Linden N J 1960-64; Asst Prof Chem Engrg WPI 1964-67, Assoc Prof 1967-71, Prof 1971-79, Dept Chmn 1975-79, Dir WPI Environ Sys Study Program 1970-73; Dir WPI Food Res Program 1976-79. Prof & Dept Chrmn, ASU 1979- . Cons to chem co's; Lady Wilson Fellow Technion Israel 1973-74. Mbr AIChE, ACS, ASEE, Sigma Xi, Phi Lambda Upsilon. *Society Aff:* AIChE, ACS, ASEE, AAAS.

Zwiep, Donald N
Business: 100 Institute Road, Worcester, MA 01609
Position: Prof & Head, Mech E *Employer:* WPI *Education:* D of Eng (Hon)/Eng/Worcester Polytechnic Inst.; M Sci/Mech E/Iowa State U.; B Sci/Mech E/Iowa State U *Born:* 03/18/24 Donald N. Zwiep, engrg educator, B.S. '48 and M.S. '51 in Mechl Engrg, Iowa State Univ and D. of Engrg. '65 (Honorary), Worcester Polytechnic Inst, married Marcia J. Hubers, children, Donna J., Mary N., Joan L., and Helen D. Design engr, Boeing Airplane Co, 1948-50. Ass't/Assoc. Prof. of Mech. Engrg at Colorado State U., 1951-57; Prof. and Head of Mech Engrg at Worcester Polytechnic Inst, 1957 --. Pilot, USAAF in WW II, CBI. Ch, The James F. Lincoln Arc Welding Foundation, 1977 --. Past Pres, 1979-80 and Fellow, Member of AWS, SME, ASEE. Member of Omicron Delta Kappa, Tau Beta Pi, Pi Tau Sigma, Sigma Xi. *Society Aff:* ASME, SME, AWS, ASEE

Zwiep, Donald Nelson
Business: Worcester Poly. Inst, 100 Institute Rd, Worcester, MA 01609
Position: Head of Dept of Mechanical Engrg. *Employer:* Worcester Polytechnic Inst. *Education:* BSci/ME/IA State Univ; MSci/ME/IA State Univ; DEng/Hon/Worcester Poly Inst. *Born:* Mar 18, 1924. In Army Air Force WW II, pilot & then in USAF Reserves (Lt Col). Engr with Boeing Airplane Co 1948-50; mbr of faculty Colo St Univ 1951-57; Prof & Head Mech Engrg Dept WPI 1957- . V P ASME & Chmn Policy Bd, Educ 1972-74; ASME Rep to ECPD Bd/Dir 1965-70. Chm, Bd of Tr The James F Lincoln Arc Welding Foundation of Cleveland Ohio 1976- . Cons mech engr. Fellow ASME 1975; Pres, ASME, 1979-80. *Society Aff:* ASME, ASEE, SME, AWS.

Zwilsky, Klaus M
Home: 11422 Dorchester Lane, Rockville, MD 20852
Position: Executive Director *Employer:* Natl Materials Advisory Bd, Natl Academy of Sciences *Education:* ScD/Metallurgy/MIT; SM/Metallurgy/MIT; SB/Metallurgy/MIT. *Born:* 8/16/32. Several yrs indus experience with New England Materials Lab & Pratt & Whitney Aircraft. 16 yrs of Govt experience with US Navy Ships R&D Center, Atomic Energy Comm, ERDA and the Dept of Energy; 1974-1981 Ch Materials & Radiation Effects Branch, Office of Fusion Energy. 1981 - Executive Director, National Materials Advisory Board, National Academy of Sciences. Pub 30 tech papers in high temp met, nuclear materials, fusion tech, critical and strategic materials. Delegate to several meetings of Internatl Atomic Energy Agency & Internatl Energy Agency. Active in several tech soc's and natl cttes. PChmn Wash D C Chap ASM. Fellow ASM. Trustee ASM 1984-87. Advisory Ctte; Metals and Ceramics Div, Oak Ridge Natl Lab 1981-84. *Society Aff:* ASM, TMS-AIME, MRS, SAMPE.

Zworykin, Vladimir K
Business: David Sarnoff Res Ctr, Princeton, NJ 08540
Position: Hon VP. *Employer:* RCA Corporation. *Education:* EE/-/Petrograd Inst of Tech; PhD/Phil/Univ of Pitts. *Born:* July 30, 1889 Mourom Russia. Came to US 1919, naturalized 1924. Res with Westinghouse Elec & Mfg Co 1920-29; RCA Mfg Co 1929-54. Recipient numerous awards incl Pres Certificate of Merit 1948, Chevalier Cross of French Legion of Honor 1948, Lamme Award 1949, Edison Medal 1952. Recipient Natl Medal of Sci. Served with Signal Corps Russian Army WW I. Fellow IEEE. Hon Mbr SMPTE; Mbr NAE, NAS. Author.

Zwoyer, Eugene M
Business: T. Y. Lin International, 315 Bay St., San Francisco, CA 94133
Position: Executive Director. *Employer:* Amer Society of Civil Engineers. *Education:* PhD/Struct Engg/Univ of IL; MS/Struct Engg/IL Inst of Technology; BS/Civil Engg/Univ of NM. *Born:* 9/8/26. Assumed   position of Exec Dir ASCE 1972. Held admin positions & all faculty ranks at Univ of N M 1948-72. Operated Eugene Zwoyer & Assocs Cons Engrs 1953-72. Elected to Bd/Dir ASCE for term 1968-71. Conducted res & pub reports on behavior of prestressed concrete structs & response of structs to nuclear weapons effects. National Academy of Code Administration (Trustee, 1973-1979). Engineer Joint Council (Director, 1978-1979). American Association of Engineering Societies (Governor, 198C). Engineering Societies Commission on Energy (Director, 1977-). Engineering Information, Inc. (Trustee, 1981-). People to People International (Trustee, 1975-). *Society Aff:* ASCE, NSPE, ACI, ASEE, ASTM, ASAE, AAAS.

Zygmont, Anthony
Business: Villanova, PA 19085
Position: Chairperson EE. *Employer:* Villanova Univ. *Education:* PhD/Elec Engr/Univ of PA; MS/Elec Engr/Drexel Univ; BS/Elec Engr/Villanova Univ. *Born:* 9/16/37. in Phila PA. Worked at the Philco Corp and RCA from 1959 to 1963. During this time activities included the design of communication and control systems. In 1963, I accepted employment with Villanova Univ as an instr. While at Villanova I consulted with GE, the Phila Elec Co, and Goddard Space Ctr. Consulting activities concentrated on res and design of control systems used in the control and tracking of space crafts in addition to personnel training. *Society Aff:* IEEE.

ERRATA

The assembly and error-free presentation of information in a volume as detailed as *Who's Who in Engineering* is a challenge which by its nature can only be imperfectly met. The editorial staff of this book solicits its readership's assistance in learning of any errors which the book contains.

The biographies of several individuals inadvertantly occur out of sequence in the Seventh Edition of *Who's Who in Engineering*. Their names, and the page number on which their biographical information appears, are listed below. At the position where this information should have appeared, the following symbol appears: *e*

Bartholomew, George A.	p. 367
Einspruch, Norman G.	p. 585
Emerson, Howard P.	p. 395
George, Nicholas	p. 582
Halpert, Hugo N.	p. 400
Hartman, Howard L.	p. 395
Johnson, Marion	p. 517
Konzo, Seichi	p. 697
Kozma, Adam	p. 79
Sabersky, Rolf H.	p. 664
Slawecki, Tadeusz	p. 758
Stanley, Stys Z.	p. 752
Tecklenburg, Harry	p. 164
Yokelson, Bernard J.	p. 126

Index of Individuals
by State

Alabama

Aldridge, Melvin Dayne
Anderson, John P
Appleton, Joseph H
Aycock, Kenneth G
Baird, James A
Barfield, Robert F
Barnett, Donald O
Benes, Peter
Berry, Francis C
Black, J Temple
Blythe, Ardven L
Boland, Joseph S, III
Brooks, George H
Brown, Robert A
Brownlee, William R
Bueltman, Charles G
Cain, John L
Carden, Arnold E
Cataldo, Charles E
Chang, Chin-Hao
Chen, Hui-Chuan
Chrencik, Frank
Clark, Charles E
Clark, Melvin E
Conner, David A
Cost, Thomas L
Costes, Nicholas C
Cox, Ernest A , III
Crocker, Malcolm J
Cutchins, Malcolm A
Doughty, Julian O
Downey, Joseph E
Drost, Edward J
Dunnavant, Guy P
Dybczak, Zbigniew W
   (Paul)
Earle, Emily A
Escoffier, Francis F
Evces, Charles R
Frye, John H Jr
Gambrell, Samuel C, Jr
Gibson, Charles A
Goldman, Jay
Green, David G
Greene, Joseph L, Jr
Gungor, Behic R
Hammett, Cecil E
Harrisberger, Lee
Hart, David R
Henry, Harold R
Holder, Sidney G, Jr
Honnell, Martial A
Hood, John L, Jr
Howard L Hartman
Ingram, Troy L
Jackins, George A , Jr
Jones, Edward O, Jr
Jordan, William D
Kallsen, Henry A
Kattus, J Robert
Krishnamurthy, Natarajan
Lane, Charles E III
Lemons, Jack E
Lewis, W David
Lucas, William R
Macintyre, John R
Malone, Frank D
McCarl, Henry N
McCool, Alexander A, Jr
McDonald, Jerry L
McDonough, George F, Jr
McGrath, Philip I, Jr
McKannan, Eugene C
Miller, Edmond T
Mjosund, Arne
Mott, Harold
Myers, Harry E , Jr
Norton, John O
Norwood, Sydney L
Oglesby, Sabert
Perrigin, John G
Planz, Edward J
Posey, Owen S
Puckett, James C

Pumphrey, Fred H
Raney, Donald C
Rey, William K
Reymann, Charles B
Richardson, C D
Robinson, John K
Schwinghamer, Robert J
Simpson, Thomas A
Smith, Leo A
Steakley, Joe E
Sturgeon, Charles E
Talbot, Thomas F
Turner, Daniel S
Unger, Vernon E
Vandegrift, Erskine, Jr
Vaughan, William W
Wainwright, Sam H
Wallace, Ronald Stephen
Walters, James V
Wang, George C
Waters, James P, Jr
Willhelm, A Clyde, Jr
Wood, John E, III
Woodward, James H
Wyskida, Richard M

Alaska

Allen, James
Alter, Amos J
Breeding, Lawrence E
Conover, John S
Crews, Paul B
Eggener, Charles L
Gotschall, Donald J
Hobson, Kenneth H
Hunsucker, Robert D
Johansen, Nils I
Johnson, Ronald A
Kokjer, Kenneth J
Lake, James M
Lee, Harry R
Mack, Edward S
Maneval, David R
Roberts, Thomas D
Sackinger, William M
Steige, Walter E
Tilsworth, Timothy
Weeks, W F
Zarling, John P

Arizona

Anderson, James D
Andrews, Al L
Arbiter, Nathaniel
Bailey, James E
Baird, Avorald L
Baker, Glenn A
Balanis, Constantine A
Baltes, Robert T
Bartlett, Robert W
Beakley, George C
Bedworth, David D
Bell, John F
Bhappu, Roshan B
Bookman, Robert E
Borland, Whitney M
Boston, Orlan W
Bottaccini, Manfred R
Bouwer, Herman
Bowman, E Dexter, Jr
Brinsko, George A
Brooks, Carson L
Brown, Charles T
Butler, Blaine R
Carlile, Robert N
Chen, Chuan F
Cochran, Douglas E
Collins, John S
Conta, Lewis D
Cooperrider, Neil K
Dedrick, Allen R
Demaree, David M
Dorsett, Ronald

Dowdey, Wayne L
Ewing, Ronald L
Fangmeier, Delmar D
Farr, W Morris
Faust, Delbert G
Ferry, David K
Filley, Richard D
Fisher, Gordon H
Frame, John W
Franken, Peter A
Frieden, B Roy
Frohling, Edward S
Gaskill, Jack D
Gomez, Rod J
Greenslade, Wiliam M
Haden, Clovis R
Hamilton, Douglas J
Harris, Michael E
Hart, Lyman H
Hartman, David E
Haynes, Munro K
Henry, Laurence O (Pat)
Herron, William J
Hetrick, David L
Hill, Frederick J
Howard, William G , Jr
Hunt, Bobby R
Hustead, Dennis D
Jimenez, Rudolf A
Johnson, Vern R
Johnston, Bruce G
Jones, Roger C
Jorgensen, Gordon D
Kamel, Hussein A
Kaufman, Irving
Kececioglu, Dimitri B
Kendall, Percy Raymond
Kienow, Paul E
Kiersch, George A
Krapek, Anton
Kuivinen, Thomas O
Kunka, Peter
Larson, Dennis L
Lester, John W
Lim, David P
Lowell, J David
MacCollum, David V
Mattson, Roy H
McFarlane, Maynard D
McGee, George W
McManus, Terrence J
Metz, Donald C
Miller, William E
Moor, William C
Murphy, Daniel J
Neal, Donald K
Nielsen, Michael J
Noodleman, Samuel
Nudelman, Sol
Perkins, Henry C
Perper, Lloyd J
Peterson, Gerald R
Peterson, Harold A
Porcello, Leonard J
Post, Roy G
Prince, John L
Ratermann, Mark J
Renard, Kenneth G
Rogers, Willard L
Rouse, Hunter
Rummel, Robert W
Schwan, Anthony V
Scully, Marlan O
Seale, Robert L
Selvidge, Harner
Settles, F Stanley
Shannon, Robert R
Shaw, Milton C
Shoults, David R
Smith, Jack
Stein, Peter K
Stott, Charles B
Taub, James M
Thompson, Charles B
Thompson, Truet B
Thomson, Quentin R

Tognoni, Hale C
Triffet, Terry
Vogt, Howard W
Vondrick, Art F
Wait, James R
Wait, John V
Walker, Joseph H
Walker, Walter W
Welch, H William
Wetstein, Willis J
Whitney, Herbert W
Wolfe, Philip M
Wolfe, William L
Wood, Gerald L
Wymore, Albert Wayne
Young, Hewitt H
Zabel, Dale E
Zick, Leonard Paul

Arkansas

Ahlen, John W
Asfahl, C R
Babcock, Robert E
Bondurant, Donald C
Bozatli, Ali N
Bryniarski, Americ J
Cole, Jack H
Crisp, Robert M
Crossett, Frederick J
Cummings, Samuel D, Jr
Driggers, William J , Sr
Ellefson, George E
Gattis, Jim L
Gleason, James G
Graham, W William, Jr
Harris, Aubrey L, III
Heiple, Loren R
Imhoff, John L
Jones, Richard A
Killian, Stanley C
Knotts, Burton R
Lucas, Jay
Malone, William J
Matthews, Edwin J
McBryde, Vernon E
McClelland, James E
Miller, Albert H
Miller, Daniel R
Moore, James W
Pearson, James V
Riddick, Edgar K , Jr
Roy, Martin H
Shipley, Larry W
Springer, Melvin D
Summerlin, James C
Thompson, Homer F
Tyer, Arnold J
Waite, William P
Wilkes, J Wray
Yeargan, Jerry R
Young, Jesse R

California

Abdul-Rahman, Yahia A
Acrivos, Andreas
Adam, Stephen F
Adams, William J , Jr
Agbabian, Mihran S
Agnew, Harold M
Aiken, George E
Akin, Lee S
Albert, Edward V
Alford, Jack L
Algazi, Vidal R
Allen, Charles W
Allen, Clarence R
Allen, Gerald B
Allen, Matthew A
Almgren, Louis E
Alpert, Sumner
Altenhofen, Robert E
Amdahl, Gene M
Amelio, Gilbert F

California (Continued)

Marks, Allan F
Marschall, Ekkehard P
Marsh, Alan H
Marshall, Andrew C
Marxheimer, Rene B
Masri, Sami F
Massey, Gail A
Matare, Herbert F
Mathews, Warren E
Maxworthy, Tony
Mayer, Edward H
McCalla, William J
McCarthy, Roger L
McCarty, James E
McCarty, Perry L
McClure, Eldon R
McCoy, George T
McCullough, Charles A
McDonald, Henry C
McDonald, John C
McFee, Raymond H
McGill, Thomas C
McGovern, Sharon A
McGuire, Michael J
McHuron, Clark E
McKillop, Allan A
McKnight, Larry E
McLeod, John H, Jr
McMillan, John Robertson
McMurtry, Burton J
McPherson, Donald J
McSorley, Richard J
McWee, James M
Mead, Carver A
Means, James A
Meckler, Milton
Medbery, H Christopher
Medwadowski, Stefan J
Meecham, Wm C
Mei, Kenneth K
Melese-d'Hospital, G B
Melvin, Howard L
Mendel, Jerry M
Meriam, James L
Merz, James L
Messer, Philip H
Metcalfe, Arthur G
Middlebrook, R David
Miller, Edward
Miller, Robert R
Millet, Richard A
Milligan, Robert T
Milstein, Frederick
Milstein, Laurence B
Minet, Ronald G
Missimer, Dale J
Mitchell, James Curtis
Mitchell, James K
Mitchell, Thomas P
Mitra, Sanjit K
Moffat, Robert J
Mohamed, Farghalli A
Moir, Barton M
Molinder, John I
Moll, John L
Monismith, Carl L
Monson, James E
Moody, Frederick J
Mooney, Malcolm T
Moore, Gordon E
Moore, Raymond P
Moore, Robert J , Jr
Mooz, William E
Moreno, Theodore
Morgan, Donald E
Morgan, Leland R
Moricoli, John C
Morris, Albert J
Morris, Ben F
Morris, Fred W
Morris, Robert L
Morrison, Malcolm C
Mortensen, Richard E
Morton, George A
Moss, Frank H, Jr

Mote, C D, Jr
Mow, Maurice
Muchmore, Robert B
Mudd, Henry T
Mukherjee, Amiya K
Mulder, Leonardus T
Mullen, Joseph
Muller, Richard S
Mulligan, James H, Jr
Munir, Zuhair A
Murashige, James Y
Nack, Donald H
Nadler, Gerald
Nadolski, Leon
Naghdi, Paul M
Neal, Gordon W
Neely, George Leonard
Neff, R Wilson
Nelson, E Raymond
Nelson, Richard B
Nelson, William E
Neou, In-Meei
Newell, Earl D
Newton, Robert E
Nicholson, Richard H
Nickels, Frank J
Nicoletti, Joseph P
Nishkian, Byron L
Nix, William D
Nixon, Alan C
Nobe, Ken
Nolan, Ralph P, Jr
Noonan, Mark E
Noyce, Robert N
Nypan, Lester J
Ohara, George T
Ohlson, John E
Okrent, David
Olander, Donald R
Oldfield, William
Oldham, William G
Olin, John G
Oliver, Bernard M
Olsen, Joseph C
Olsen, Richard A
Olson, Robert R
Olson, Valerie F
Omura, Jimmy K
Ongerth, Henry J
Ono, Kanji
Openshaw, Keith L
Oppenheim, Antoni K
Orchard, Henry J
Ordung, Philip F
Orlob, Gerald T
Oswald, William J
Ott, Dudley E
Packard, David
Paduana, Joseph A
Paik, Young J
Palmer, Roy G
Pantell, Richard H
Paoluccio, Joseph P
Paratore, William G
Parden, Robert J
Parker, Don
Parker, Norman F
Parker, Sydney R
Parks, Robert J
Parzen, Benjamin
Pate-Cornell, M Elisabeth
Paulling, J Randolph, Jr
Pearson, Frank H
Pearson, Gerald L
Pearson, John
Pederson, Donald O
Pefley, Richard K
Pegram, Anthony R
Pehrson, David L
Penner, Stanford S
Penzien, Joseph
Perkins, Roger A
Perrine, Richard L
Perry, William J
Peters, George A
Peters, LeRoy L

Peters, Stanley T
Peterson, Victor L
Pettit, Ray H
Pfister, Henry L
Phillips, Robert V
Phillips, William J, II
Phillips, William R
Pierucci, Mauro
Pigford, Thomas H
Pillsbury, Arthur F
Pinkham, Clarkson W
Pitt, Paul A
Pitzer, Kenneth S
Piziali, Robert L
Platzer, Max F
Plesset, Milton S
Plotkin, Allen
Plumlee, Carl H
Plumtree, William G
Plunkett, Joseph C
Pomeroy, Richard D
Pomraning, Gerald C
Pope, James H
Popov, Egor P
Porter, Nancy J
Posner, Edward C
Potter, Richard C
Prabhakar, Jagdish C
Presecan, Nicholas L
Press, Leo C
Prestele, Joseph A
Pringle, Weston S
Pritchett, Thomas R
Profio, A Edward
Purl, O Thomas
Pursell, Carroll W
Pyke, Robert
Quaney, Robert A
Quate, Calvin F
Raben, Irwin A
Radcliffe, Charles W
Raksit, Sagar K
Ramalingam,
 Panchatcharam
Rambo, William R
Ramey, Henry J, Jr
Randolph, Patrick A
Ratcliffe, Alfonso F
Ravenis, Joseph VJ, II
Raymond, Arthur E
Raymond, Louis
Rechtin, Eberhardt
Recker, Wilfred W
Reed, Willard H
Reichard, Edward H
Reissner, Eric
Remley, Marlin E
Reti, G Andrew
Reynolds, William C
Rhodes, Gilbert L
Rice, Stephen O
Rickard, Corwin L
Ridgway, David W
Riggs, Henry E
Rinne, John E
Rischall, Herman
Robeck, Gordon G
Roberts, George A
Roberts, Sanford B
Roden, Martin S
Rodgers, Alston
Rodgers, Colin
Rodgers, Joseph F
Rolf H Sabersky
Roman, Basil P
Root, L Eugene
Rosen, Harold A
Rosenberg, Richard
Ross, Bernard
Ross, Hugh C
Roth, Bernard
Rothbart, Harold A
Rotolo, Elio R
Rowland, Samuel W, Jr
Rowland, Walter F
Roy, Guy W

Rubinstein, Moshe F
Rubio, Abdon
Rudavsky, Alexander B
Rudee, M Lea
Rumsey, Victor H
Rusch, Willard VT
Russell, John V
Ryan, Roderick T
Ryder, Robert A
Rynn, Nathan
Saffman, Philip G
Salzer, John M
Samulon, Henry A
Sandall, Orville C
Sanders, Karl L
Sanders, Lon L
Sanford, Keith C
Sarkaria, Gurmukh S
Saunders, Robert M
Saunders, Walter D
Sawabini, C T
Saylor, Wilbur A
Schawlow, Arthur L
Scherr, Harvey H
Schiff, Anshel J
Schinzinger, Roland
Schlintz, Harold H
Schmit, Lucien A
Schmitz, Eugene G
Scholtz, Robert A
Schoofs, Richard J
Schott, F W
Schoustra, Jack J
Schrady, David A
Schrock, Virgil E
Schurman, Glenn A
Schwartz, Morton D
Schwarz, Steven E
Scordelis, Alexander C
Scott, John W
Scott, Ronald F
Seaborg, Glenn T
Seed, H Bolton
Seely, John H
Seide, Paul
Seiden, Edward I
Seiple, Willard Ray
Self, Sidney A
Sells, Harold R
Semmelmayer, Joseph A
Sensiper, Samuel
Seruto, Joseph G
Sesonske, Alexander
Sevier, William J
Shabaik, Aly H
Shackelford, James F
Shah, Haresh C
Shankman, Aaron D
Sharpe, Roland L
Sheffet, Joseph
Sheinbaum, Itzhak
Shen, Chih-kang
Shen, Hsieh W
Sheng, Henry P
Shepherd, R
Sher, Rudolph
Sherman, Russell G
Shoaf, Ross T
Shoemaker, Robert S
Shore, Melvin
Shurtz, Robert F
Siegman, Anthony E
Silady, Fred A
Silver, John E
Simmons, William W
Simnad, Massoud T
Simon, Marvin K
Simpkins, Ronald B
Sims, Robert R
Sinclair, Charles S
Sines, George H
Singh, R Paul
Singh, Rameshwar
Skalnik, John G
Skilling, Hugh H
Skjei, Roger E

Connecticut

Colorado (Continued)

Moulder, Leonard D
Mueller, William M
Mumma, George B
Murphy, Vincent G
Nahman, Norris S
Nevin, Andrew E
Newkirk, John B
Nordby, Gene M
Norton, Kenneth A
Olds, Joneil R
Olson, David L
O'Neil, Thomas J
Orth, William A
Osborne, Edward A
Ostwald, Phillip F
Owens, Willard G
Pailthorp, Robert E
Panek, Louis A
Perez, Jean-Yves
Peterson, Gary J
Pforzheimer, Harry Jr
Poettmann, Fred H
Pollard, William S , Jr
Presley, Gordon C
Puls, Louis G
Pyle, Don T
Queneau, Paul B
Ramsay, H J , Jr
Rausch, Donald O
Rautenstraus, Roland C
Reeves, Adam A
Rice, Leonard
Rietman, Noel D
Robertson, James M
Robertson, Lawrence M
Robinson, Charles S
Rockwell, Glen E
Romig, Phillip R
Rossie, John P
Rouse, George C
Rummel, Ward D
Russell, Paul L
Sagar, Bokkapatnam
Tirumala Ananda
Scharf, Louis L
Scherich, Erwin T
Schermerhorn, R Stephen
Schleif, Ferber R
Schuster, Robert L
Seebass, A Richard
Selander, Carl E
Selle, James E
Shen, Hsieh W
Shepard, William M
Signs, Cheryl L
Skogerboe, Gaylord V
Smith, A Leonard
Smith, James A
Smith, Kenneth A
Straw, Henry F
Stump, Wayne D
Sturdevant, Bruce L
Summers, John Leo
Sutherland, Donald C
Thaemert, Ronald L
Thomas, Roland E
Thorfinnson, Stanley T
Tucker, L Scott
Tynes, Rex A
Utlaut, William F
Vanderbilt, M Daniel
VanKirk, Craig W
Vawter, R Glenn
Walker, John R
Walton, Robert Edward
Ward, Robert C
Watts, Oliver E
Webb, Richard C
Westphal, Warren H
Wilbur, Paul J
Wilkes, John S , III
Wilson, Garland
Wittry, John P
Woodward, Alan A
Wright, Kenneth R

Yang, Chih Ted
Yarar, Baki
Yetter, John W
Yukawa, Sumio
Ziemer, Rodger E

Connecticut

Abonyi, Erwin
Allaire, Robert E
Allen, David C
Anderson, Robert M
Baker, Bernard S
Barker, Richard C
Barrow, Robert B
Barton, Cornelius J
Bell, James P
Bell, Thaddeus G
Bennett, William R, Jr
Benson, Walter
Bentley, Lawrence H
Berry, Richard C
Birkimer, Donald L
Bishop, George A, III
Bjorklund, David S Jr
Blizard, John
Boggs, Robert G
Booth, Taylor L
Boutwell, Harvey B
Bradley, Elihu F
Brancato, Leo J
Breinan, Edward M
Bridwell, John D
Brinkmann, Joseph B
Bronzino, Joseph D
Brooks, M Scott
Brown, C Alvie
Bublick, Alexander V
Bueche, Arthur M
Burghoff, Henry L
Buzzell, Donald A
Cadogan, William P
Cagnetta, John P
Campbell, George S
Castorina, Anthony R
Cavaliere, Alfonso M
Chang, Richard K
Charpentier, David L
Cheng, David H S
Cheo, Peter K
Chernock, Warren P
Coogan, Charles H, Jr
Corrigan, Brian
Coughlin, Robert W
Crossley, F R Erskine
Crozier, Ronald D
Cunningham, Walter Jack
Danchak, Michael M
Dappert, George F
Date, Raghunath V
Delgrosso, Eugene J
DeMaria, Anthony J
DeMott, Alfred E
Devereux, Owen F
Dibner, Bern
Dickie, H Ford
Donachie, Matthew J, Jr
Donnalley, James R
D'Ovidio, Gene J
Dresher, William H
Dubin, Fred S
Dunn, Frank W
Dunn, Robert M
Eddy, W Paul
Elijah, Leo M
Ensign, Chester O , Jr
Epstein, Howard I
Fee, Walter F
Fenn, Rutherford H
Fisher, Lawrence E
Fitchen, Franklin C
Fitzgerald, John E
Ford, Curry E
Forger, Robert D
Freeman, Arthur H

Freeman, Mark P
Freeman, William R, Jr
Fried, Benjamin S
Galligan, James M
Gant, Edward V
Garry, Frederick W
Gartner, Joseph R
Gathy, Bruce S
Genovese, Philip W
Giamei, Anthony F
Giavara, Sutiri
Gibble, Kenneth
Goddard, Thomas A
Gryna, Frank M
Gubala, Robert W
Guernsey, Nellie E
Gutzwiller, F William
Haestad, Roald J
Hall, Newman A
Halloran, Joseph M
Halverstadt, Robert D
Hamilton, Stephen B, Jr
Harrison, Otto R
Hauser, H Alan
Heitlinger, Igor
Hellier, Charles J P E
Hill, Richard F
Hirshfield, Jay L
Hix, Charles F, Jr
Hollander, Milton B
Honsinger, Leroy V,
Howard, G Michael
Howell, John T
Hufnagel, Robert E
Hyde, John W
Isaac, Maurice G
Isaacs, Jack L
Jackson, Arthur J
Jackson, Harry A
Jacobson, Nathan L
Jentzen, Carl A
Johnson, Loering M
Jones, Robert A
Kattamis, Theo Z
Kaufman, John E
Kear, Bernard H
Ketchum, Milo S
Klancko, Robert John
Koenig, Herbert A
Kogos, Laurence
Korwek, Alexander D
Kowalonek, John M
Kowalski, Philip L
Kraus, Harry
Kuchinski, Frank Leonard
Lambrakis, Konstantine C
Lander, Horace N
Lanouette, Kenneth H
Lataille, Jane I
Lenard, John F
Levinson, Herbert S
Lien, Jesse R
Long, Richard P
Lowery, Thomas J
Luh, Peter B
Macchi, Anselmo J
Mack, Donald R
Manz, August F
Marsh, B Duane
Marx, Christopher W
McAdams, William A
McCready, Lauren S
McFadden, Peter W
McManus, John A
Mead, William J
Milliken, Frank R
Mills, Robert N
Mindlin, Raymond D
Morral, John E
Morris, Robert L
Murtha-Smith, Erling
Narendra, Kumpati S
Nedom, H A
Neely, H Clifford
Neidhart, John J
Nelson, Burke E

Nelson, Peter R
Nesbitt, Ray B
Ojalvo, Irving U
O'Neil, David A
O'Neill, Patrick H
Owczarski, William A
Pearce, Philip L
Phillips, Aris
Porter, Robert P
Posey, Chesley J
Prober, Daniel E
Proffitt, Richard V
Purdy, Richard B
Radcliffe, Frederick A
Reichl, Eric H
Resen, Frederick Larry
Risoli, Joseph F
Roberts, Robert J E
Ross, Stephen M
Rossi, Boniface E
Rothberg, Henry M
Rudd, Wallace C
Ruhlid, Robert R
Russo, A Sam
Sapega, August E
Schetky, Laurence McD
Schlegel, Walter F
Schreier, Stefan
Schultz, Roger L
Scott, Roderic M
Scottron, Victor E
Seely, Samuel
Shaffer, Richard F
Shainin, Dorian
Shank, Maurice E
Shapiro, Eugene
Shaw, Montgomery T
Sheets, Herman E
Shemitz, Sylvan R
Shortsleeve, Francis J
Siegmund, Walter P
Skinner, Bruce C
Smith, Lawrence H
Smith, Lester C
Smith, William J
Stancell, Arnold F
Stephens, Jack E
Stevenson, Albert H
Stoker, Warren C
Streitmatter, Vern O
Sullivan, Cornelius P
Sundstrom, Donald W
Suran, Jerome J
Swan, David
Swift, Clinton E
Tedmon, Craig S, Jr
Thompson, Earl R
Thompson, Willis F
Thune, Arne
Tizian, Sylvan L
Tuteur, Franz B
Van Buren, Maarten
Vendola, Arthur N
Versnyder, F L
Vertes, Michael A
Vinal, Edgar B, Jr
Von Winkle, William A
Wald, Henry J
Wallace, John M,
Walton, Edward H
Weber, Richard G
Weeks, Harold B
Weis, Alfred E
Weiss, Alfred
Wengenroth, Reece H
Wesler, Philip
Wheeler, Robert G
Whitney, James C
Widmer, Wilbur J
Wolf, Werner P
Wormser, Eric M
Zackay, Victor F
Zerban, Alexander H
Ziegler, Geza Charles
Zoller, David R

Delaware

Barteau, Mark A

Georgia

Georgia (Continued)
Pierce, Allan D
Pierce, G Alvin
Poehlein, Gary W
Porter, Alan L
Powell, H Russell
Prien, John D, Jr
Prince, M David
Ramee, Paul W
Ratliffe, Donald E
Raville, Milton E
Rhodes, William T
Richardson, Brian P
Richardson, Elmo A Jr
Ritter, Guy F
Roberts, Philip J W
Rodrigue, George P
Rouse, William B
Salter, Winfield O
Sanborn, David M
Sangster, William M
Saunders, F Michael
Schafer, Ronald W
Schottman, Robert W
Schwarzkopf, Florian
Sheppard, Albert P
Simitses, George J
Smalley, Harold E, Sr
Soukup, David J
Sowers, George F
Spauschus, Hans O
Spetnagel, Theodore J
Spooner, Stephen
Stacey, Weston M, Jr
Su, Kendall L
Taylor, Kenneth G
Thoem, Robert L
Thomas, Joseph M, Jr
Thomas, Michael E
Thuente, Sylvan T
Thuesen, Gerald J
Thumann, Albert
Toler, James C
Tulkoff, Joseph
Ueng, Charles ES
Underwood, Ervin E
Vachon, Reginald I
van Reenen, Willem J
Vidal, James H, Jr
Walter, Francis J, Jr
Warnecke, Edward M
Weber, Homer S
Wedd, Ralph W
Wempner, Gerald A
Wethern, James D
White, John A
Whitesides, Jack C
Williams, Wendell M, Jr
Williams, Wesley B
Winer, Ward O
Wolbrink, Jim F
Wright, Richard H
Wu, James C
Wyrick, Roger K
Ziegler, Waldemar T

Hawaii

Abramson, Norman
Bathen, Karl H
Blum, Joseph J
Cheng, Ping
Chiu, Arthur N L
Chou, James C S
Chuck, Robert T
Chun, Michael J
Craven, John P
Danzinger, Edward
Fok, Yu-Si
Go, Mateo L P
Hamada, Harold S
Hatch, Henry J
Hirai, Wallace A
Hirata, Edward Y
Hirota, Sam O

Hohns, H Murray
Hughes, Joseph Brian
Inaba, Yoshio
Jakeway, Lee A
Kohloss, Frederick H
Koide, Frank T
Lee, Edgar K M
Lin, Shu
Lo, Donald T
Loughren, Arthur V
Lum, Franklin, Y S
Lum, Walter B
Matsumoto, Michael P
Matsumoto, Yuki
Offner, Walter W
Ogburn, Hugh B
Peterson, Richard E
Pyle, William L
Sarapu, Felix R
Saxena, Narendra K
Seidl, Ludwig H
Shimazu, Satoshi D
Shupe, John W
Sinclair, Thomas L, Jr
Smith, Russell L, Jr
Tanonaka, Clarence K
Wilkenson, Thomas H
Williams, John A
Wright, Theodore O
Yee, Alfred A
Yee, Phillip K H
Young, Reginald H F
Yuen, George A L

Idaho

Anderson, Keith E
Berry, Bill E
Bissell, Roger R
Blake, Wilson
Brockway, Charles E
Edwards, Carl V
Ellsworth, Donald M
Fouladpour, K Danny
Frein, Joseph P
Griebe, Roger W
Hoskins, John R
Humpherys, Allan S
Jacobson, E Paul
Jones, Walter V
Kangas, Ralph A
Karsky, Thomas J
Keys, John W, III
Law, John
Lawroski, Harry
Lineberry, Michael J
Longley, Thomas S
Mayer, Orland C
McFarlane, Harold F
McGee, William D
Miley, Delmar V
Nelson, Sherman A
Pline, James L
Richardson, Lee S
Robinson, Lee
Saul, William E
Somerville, John E
Stanley, Marland L
Straubhar, John J
Thomas, Cecil A
Williams, Richard F
Wilson, Albert E

Illinois

Achenbach, Jan D
Ackermann, William C
Ackley, John W
Ackmann, Lowell E
Adams, Joe J
Adams, William M
Addy, Alva L
Adkins, Howard E
Adler, Robert
Adrian, Ronald J

Albrecht, E Daniel
Alkire, Richard C
Allen, Charles W
Allen, Robert L
Altstetter, Carl J
Amberg, Arthur A
Anderson, Charles J
Anderson, Stanley W
Ang, Alfredo H S
Arzbaecher, Robert C
Axford, Roy A
Aynsley, Eric
Bacci, Guy J, II
Ballotti, Elmer F
Bankoff, S George
Bard, Gerald W
Bardeen, John
Barenberg, Ernest J
Barnett, W John
Barthel, Harold O
Bartlett, Donald L
Bayne, James W
Beck, Paul A
Behnke, Wallace B, Jr
Belytschko, Ted B
Bennett, Richard T
Berg, Eugene P
Berger, Richard L
Bergeron, Clifton G
Bergman Donald J
Berry, Donald S
Bhowmik, Nani G
Birnbaum, Howard K
Blachman, Martin M
Blank, William F
Bloome, Peter D
Boardman, Bruce E
Bode, Loren E
Bodeen, George H
Boesch, Henry J, Jr
Bohl, Robert W
Bolduc, Oliver J
Bono, Jack A
Bonthron, Robert J
Boresi, Arthur P
Bosworth, Douglas L
Bott, Walter T
Boundy, Ray H
Bowen, Carl H
Bowhill, Sidney A
Bowles, Joseph E
Bowman, Louis
Boyd, John H
Boylan, Bernard R
Bradbury, James C
Breen, Dale H
Breyer, Norman N
Bridges, Jack E
Brittain, John O
Broutman, Lawrence J
Brower, Ralph T
Brown, F Leslie
Brown, Richard M
Buchanan, R C
Budenholzer, Roland A
Buehler, Adolph I
Bunker, Frank C
Burtness, Roger W
Burton, Conway C
Butler, Margaret K
Cain, Charles A
Callaway, Samuel R
Campbell, Calvin A, Jr
Camras, Marvin
Carlson, Donald E
Carr, Stephen H
Carreira, Domingo J
Carroll, Paul F
Caywood, Thomas E
Chakrabarti, Subrata K
Chao, Bei T
Chato, John C
Chen, Juh W
Chen, Michael M
Chen, Shoei-Sheng
Chenault, Woodrow C, Jr

Chiswik, Haim H
Chow, Ven Te
Chow, Wen L
Chugh, Yoginder P
Clausing, Arthur M
Clinebell, Paul W
Cloke, Thomas H
Codlin, James B
Cohen, Jerome B
Coleman, Paul D
Collier, Donald W
Condit, Carl W
Conn, Arthur L
Conn, Harry
Conry, Thomas F
Cook, Harry M, Jr
Cooper, Martin
Cordill, John J
Corlew, Philip M
Corley, William G
Cormack, William J
Corten, Herbert T
Costello, George A
Crandall, John L
Crawford, Leonard K
Crist, Robert A
Cross, Robert C
Daehn, Ralph C
Daily, Eugene J
Dalal, Jayesh G
Daniel, Isaac M
Darby, Joseph B, Jr
David, Edward E, Jr
Davis, Philip K
Davisson, Melvin T
DeTemple, Thomas A
Disque, Robert O
Dix, Rollin C
Dobrovolny, Jerry S
Dockendorff, Jay D
Dolan, Thomas J
Dondanville, Laurence A
Dondanville, Leo J, Jr
Donnelly, John I, Jr
Dossett, Jon L
Drickamer, Harry G
Dubin, Eugene A
DuBose, Lawrence A
Duke, Robert E
Duncan, Samuel W
Dundurs, John
Dyer, Harry B
Eck, Bernard J
Eckert, Charles A
Eckmann, Donald E
Ehrenberg, John M
Eilering, John G
Emery, Willis L
Engelbrecht, Richard S
Ernst, Edward W
Eubanks, Robert A
Everhart, Thomas E
Everitt, William L
Ewing, Benjamin B
Eyerman, Thomas J
Farnsworth, George L
Fejer, Andrew A
Ference, George
Figge, Kenneth L
Fine, Morris E
Fintel, Mark
FitzGerald, James E
Flood, Paul E
Foderberg, Dennis L
Footlik, Irving M
Footlik, Robert B
France, Jimmie J
Francis, Philip H
Freedman, Steven I
Freund, Robert J
Frey, Donald N
Friederich, Allan G
Frost, Brian R T
Gaddy, Oscar L
Gage, Elliot H
Gamble, William L

Illinois (Continued)
Widera, G E O
Wight, George
Wilkinson, Eugene R
Willey, Benjamin F
Williams, Charles D , Jr
Williams, Samuel L
Wirtz, Gerald P
Wiss, John F
Wool, Richard P
Worley, Will J
Wright, Edward S
Yaggee, Frank L
Yeh, Kung C
Yen, Ben Chie
Youngdahl, Carl K
Zabel, Hartmut
Zagars, Arvids
Zak, Adam R
Zar, Max
Zeno, Robert S
Zichterman, Cornelius
Zimmerman, Robert M
Zook, Donald G
Zowski, Thaddeus
Zummer, Anthony S

Indiana

Alban, Lester E
Albright, Lyle F
Amber, Wayne L
Amrine, Harold T
Anderson, Philip P, Jr
Arffa, Gerald L
Bailey, Eugene C
Bailey, Herbert R
Barany, James W
Bell, John M
Bennon, Saul
Berrio, Mark M
Betner, Donald R
Bogdanoff, John L
Bonar, Ronald L
Bostwick, Willard D
Bowden, Warren W
Bratkovich, Nick F
Brummett, Forrest D
Carberry, James J
Carlile, V Sam
Caskey, Jerry A
Chao, Kwangchu
Chenoweth, Darrel L
Chiang, Donald C
Clarke, Beresford N
Clarke, Clarence C
Clikeman, Franlyn M
Coates, Clarence L
Cohen, Raymond
Corsiglia, Robert J
Cost, James R
Cotton, Frank E, Jr
Cullinane, Thomas P
Curtis, Kenneth S
Dalphin, John F
Davison, Beaumont
Delgass, W Nicholas
DeWitt, David P
Diamond, Sidney
Dlouhy, John R
Dolch, William L
Donahue, D Joseph
Douglas, Jim, Jr
Dow, John D
Dubois, William D
DuBroff, William
Duket, Steven D
Echelberger, Wayne F, Jr
Erganian, George K
Evans, Robert C
Fassnacht, George G
Fatic, Vuk M
Fehribach, William J
Ferguson, Colin R
Fiore, Nicholas F

Foland, Donald L
Foushi, John A
Fox, Robert W
Fredrich, Augustine J
Friedlaender, Fritz J
Frische, James M
Fritzlen, Glenn A
Fu, King-sun
Fukunaga, Keinosuke
Fulford, P James
Galerman, Raphael
Gaskell, David R
Geddes, Leslie Alexander
Gelopulos, Demos P
Gibson, Harry G
Goldberg, John E
Goldschmidt, Victor W
Goodson, Raymond Eugene
Grace, Richard E
Greene, James H
Gunderson, Morton L
Hall, Allen S, Jr
Hamilton, James F
Hanink, Dean K
Hanley, Thomas R
Hansen, Arthur G
Harder, John E
Harkel, Robert J
Hart, Jack B
Hartsaw, William O
Hayes, John M
Hayt, William H, Jr
Hemphill, Dick W
Herman, Marvin
Hert, Oral H
Hill, William W
Hillberry, Ben M
Hoffman, Joe D
Hopkinson, Philip J
Hughes, Ian F
Hulbert, Samuel F
Incropera, Frank P
Jerger, Edward W
Johnson, Charles L
Johnson, Morris V
Johnson, Orwic A
Johnstone, Edward L
Jones, Howard T
Judd, William R
Kashyap, Rangasami L
Kentzer, Czeslaw P
Kessler, David P
Kinyon, Gerald E
Kivioja, Lassi A
Klein, H Joseph
Kloeker, Delmar L
Knop, Charles M
Krasavage, Kenneth W
Kriesel, William G
Landgrebe, David A
Langhaar, Henry L
LeBold, William K
Lefebvre, Arthur H
Lehmann, Gilbert M
Leipziger, Stuart
Leonards, Gerald A
Levy, Bernard S
Liedl, Gerald L
Liley, Peter E
Lin, Pen-Min
Lin, Ping-Wha
Liu, Ruey-Wen
Lloyd, Fredric R
Lloyd, John R
Lobo, Cecil T
Lucey, John W
Lykoudis, Paul S
Marshall, Francis J
Martin, William R
McCoy, Donald S
McCreery, Robert H
McEntyre, John G
Mehta, Prakash C
Melhorn, Wilton F
Michael, Harold L
Miller, Wally, P E

Monical, R Duane
Monke, Edwin J
Morgan, George H
Mueller, Thomas J
Mueller, Walter H
Murray, Haydn H
Nichols, Edwin Scott
Nolting, Henry F
O'Donnell, Neil B
O'Loughlin, John R
Orr, William H
Owens, C Dale
Pearson, Joseph T
Peppas, Nikolaos A
Peretti, Ettore A
Phillips, Joseph J
Phillips, Winfred M
Plants, Helen L
Potvin, Alfred R
Pritsker, A Alan B
Ragsdell, Kenneth M
Ramkrishna, Doraiswami
Rau, Jim L
Ravindran, Arunachalam
Read, Robert H
Robins, Norman A
Ross, Keith E
Rumbaugh, Max E, Jr
Sain, Michael K
Salisbury, Marvin H
Salvendy, Gavriel
Sandgren, Find
Sargent, Donald J
Sauter, Harry D
Schaefer, Edward J
Schalliol, Willis L
Schoech, William J
Seeley, Gerald R
Shedd, Wilfred G
Shewan, William
Shubat, George J
Sieger, Ronald B
Silva, LeRoy F
Slifka, Richard J
Smith, Charles O
Snyder, Lynn E
Soedel, Werner
Solberg, James J
Solomon, Alvin A
Steffen, Alfred J
Steffen, John R
Stevenson, Warren H
Stoudinger, Alan R
Tenney, Mark W
Thompson, Howard Doyle
Tiernon, Carlos H
Toebes, Gerrit H
Tree, David R
Tsareff, Thomas C, Jr
Tucker, R C Jr
Van Der Sluys, William
Vance, Carl B
Vaughan, John M
Viskanta, Raymond
Wagner, Victor G
Waldner, John W
Walker, Richard O, Jr
Waltz, Gerald D
Wessler, Melvin D
Wiars, Dale M
Winchell, Peter G
Wortman, David B
Wulpi, Donald J
Yang, Kwang-tzu
Young, William B
Zoss, Leslie M

Iowa

Allard, Kermit O
Barnes, Charles Ray
Barrett, Bruce A
Barrett, Ronald E
Baumann, Edward Robert
Berry, William L

Birnbaum, John D
Bowie, Robert M
Brickell, Gerald L
Brock, Harold L
Brown, Ronald D
Buchele, Wesley F
Burger, Christian P
Cagley, Leo W
Carlson, O Norman
Chwang, Allen T
Cleasby, John L
Cooper, Robert H
Coursey, W
Dague, Richard R
Danofsky, Richard A
Davis, Wilbur M
Edwards, Charles E
Ekberg, Carl E, Jr
Evans, James L
Fellinger, Robert C
Finney, William G
Fisher, Garland F
Fosholt, Sanford K
Garber, Dwayne C
Gerlich, James W
Gessner, Gene A
Gschneidner, Karl A, Jr
Hanson, Dudley M
Haug, Edward J, Jr
Holland, Francis E
Husseiny, Abdo A
Jackson, Ralph L
Jain, Subhash C
Jensen, Harold M
Johnson, Gerald N
Johnson, Howard P
Kane, Harrison
Kennedy, John F
Kimm, James W
Klaiber, Wayne F
Knowler, Lloyd A
Krishna, Gopal T K
Kuehl, Neal R
Lance, George M
Landweber, Louis
Larsen, William L
Larson, Maurice A
Liu, Lee
Lohmann, Gary A
Lux, William J
MacDougall, Louis M
Mack, Michael J
Marr, Richard A
McGee, Thomas D
McRoberts, Keith L
Meyer, Vernon M
Miller, Jack C
Mischke, Charles R
Moehrl, Michael F
Olsen, Sydney A
Otto, George
Park, Joon B
Patel, Virendra C
Paulette, Robert G
Peters, Leo C
Pletcher, Richard H
Pulley, Frank L
Read, Jay R
Reusswig, Frederick W
Riley, William F
Rim, Kwan
Sabri, Zeinab A
Sanders, Wallace W, Jr
Scholz, Paul D
Serovy, George K
Shell, William O
Shuck, Terry A
Smelcer, Glen E
Spinrad, Bernard I
Tatinclaux, Jean-Claude P
Tennant, Otto A
Veenstra, H Robert
Verhoeven, John D
Ware, Lawrence A
Wechsler, Monroe S
Wilder, David R

Kansas

Louisiana (Continued)
Wells, Thomas W
Whitehouse, Gerald D
Wilson, Joe R

Maine

Bryant, John H
Christensen, Paul B
Clapp, James L
Doughty, Eric R
Feeley, Frank G, Jr
Fricke, Arthur L
Gates, David W
Gibson, Richard C
Godfrey, Albert L, Sr
Gorrill, William R
Grant, Donald A
Hamilton, Wayne A
Hodsdon, Albert E , III
Holmes, William D
Huff, E Scott
Lachman, Walter L
LaRochelle, Donald R
MacBrayne, John M
McMillan, Brockway
Morgan, Melvin W
Myers, Basil R
Nichols, Clark
Prince, Elbert M
Russell, David O
Simard, Gerald L
Spaulding, Westbrook H
Standley, Richard A, Jr
Thorpe, Robert L
Vetelino, John F
Wadlin, George K
Walker, Stanley E
Wardwell, Richard E
Webster, Karl S
Weston, Donald W
Wink, Ralph C

Maryland

Alexiou, Arthur G
Almenas, Kazys K
Amirikian, Arsham
Anand, Dave K
Armstrong, Ronald W
Astin, Allen V
Bailey, William J
Baldwin, David M
Baldwin, Robert D
Barlow, Jewel B
Beakes, John H
Beard, Arthur H, Jr
Beckmann, Robert B
Belding, John A
Belter, Walter G
Belz, Paul D
Berger, Bruce S
Berger, Harold
Bers, Eric L
Biberman, Lucien M
Billig, Frederick S
Bishop, William H IV
Blake, Lamont V
Blakey, Lewis H
Bochinski, Julius H
Bourdon, Joseph H , III
Boylston, John W
Brancato, Emanuel L
Brillhart, S Edward
Brockway, Daniel J
Brown, Paul J
Burdette, John C , Jr
President
Busey, Harold M
Cacciamani, Jr , Eugene R
Camp, Albert T
Carcaterra, Thomas
Carnes, Wm T
Carroll, Charles F, Jr
Carton, Allen M

Caulton, Martin
Cawley, John H
Chase, Arthur P
Check, Paul S
Cheng, David K
Cheng, Henry M
Chi, Andrew R
Clark, Ezekail L
Clarke, Frank E
Clifford, Eugene J
Cohon, Jared L
Colville, James
Cooper, H Warren, III
Cooper, Norman L
Corotis, Ross B
Costrell, Louis
Cotton, John C
Coulter, James B
Coutinho, John de S
Cunniff, Patrick F
Cushen, Walter E
Dale, John C
Dally, James W
Damiani, A S
Dart, Jack C
Davidson, Frederic M
Dean, Stephen O
Denham, Roy S
Denney, Roger P , Jr
DeRoze, Donald G
Desai, Drupad B
Deutsch, George C
Dickens, Lawrence E
Dieter, George E
Donaldson, Bruce K
Douglas, Charles A
Duffey, Dick
Durelli, August J
Durrani, Sajjad H
Eisenberg, Phillip
Ellingwood, Bruce R
Ellsworth, William M
Emrick, Jonathan E
Engwall, Richard L
Eppler, Richard A
Fairbanks, John W
Felton, Kenneth E
Fernandez, Rodolfo B
Filbert, Howard C, Jr
Fischer, Irene K
Fitzgerald, Edwin R
Fletcher, Stewart G
Fordyce, Samuel W
Foss, James B
Fread, Danny L
Frederick, Carl L, Sr
Frey, Howard A
Frey, Jeffrey
Fuchsluger, John H
Garver, Robert V
Gasser, Eugene R
Gaum, Carl H
Getsinger, William J
Geyer, John C
Gibian, Thomas G
Gillett, John B
Goldman, Alan J
Gompf, Arthur M
Gordon, Daniel I
Gore, Willis C
Graves, Harvey W, Jr
Grayson, Lawrence P
Green, Richard S
Green, Robert E , Jr
Greenberger, Martin
Griffin, Norman E
Guernsey, John B
Gupta, Ashwani K
Haberman, William L
Hahn, Richard D
Hammer, Guy S, II
Hampton, Delon
Hanhart, Ernest H
Hanna, Martin Jay, III,
Harger, Robert O
Harms, John E, Jr

Harris, Benjamin L
Harris, Forest K
Hassani, Jay J
Heil, Dick C
Heins, Conrad P
Hekimian, Norris C
Hermach, Francis L
Hewlett, Richard G
Higbie, Kenneth B
Hirschhorn, Joel S
Hittman, Fred
Hovmand, Svend
Hsu, Yih-Yun
Hubbell, John H
Huddleston, Robert L
Huggins, William H
Hughen, James W
Hyman, David S
Irwin, George R
Jackson, William D
James, Ralph K
James, Robert G
Janes, Henry W
Jaske, Robert T
Jenniches, F Suzanne
Jensen, Arthur S
Johnson, Charles C , Jr
Jolliff, James V
Jones, Robert R
Jones, Wesley N
Jordan, Kenneth L, Jr
Kanal, Laveen N
Kanz, Anthony C
Kappe, Stanley E
Karadimos, Angelo S
Katz, Isadore
Keane, Robert G, Jr
Keiser, Edwin C
Kelly, Keith A
Kim, Hyough (Hugh) S
King, Randolph W
King, T A
Kirby, Ralph C
Kiss, Ronald K
Klinker, Richard L
Klueter, Herschel H
Knight, C Raymond
Knoedler, Elmer L
Knoop, Frederick R, Jr
Koehne, Anthony J
Kramer, Irvin R
Kramer, Samuel
Kravitz, Lawrence C
Kreider, Kenneth G
Kupelian, Vahey S
Kusuda, Tamami
Landsberg, Helmut E
Landsburg, Alexander C
Langston, Joann H
Larson, Clarence E
Lauriente, Michael
Lee, Reuben
Levine, Robert S
Levine, Saul
Levine, William S
Lewis, Bernard T
Lewis, Jack W
Ligon, Claude M
Lin, Hung C
Linaweaver, F Pierce
Lippold, Herbert R , Jr
Lowe, Philip A
Lustgarten, Merrill N
Lyons, John W
MacNeill, Charles E
Manfredi, Robert R
Marcus, Michael J
Marienthal, George
Marks, Colin H
Martin, John J
Mason, Henry Lea
Matthews, Allen R
McAfee, Naomi J
McCallum, Gordon E
McLeod, Robert J
McNeill, Robert E

McVickers, Jack C
Merryman, John, Jr
Meyers, Sheldon
Milano, Vito Rocco
Millard, Charles F
Miller, Richards T
Minneman, Milton J
Mok, Perry K P
Mongan, David G
Morgenthaler, Charles S
Morris, Alan D
Motayed, Asok K
Mudd, Charles B
Mueller, Alfred C
Mundel, Marvin E
Murphy, Charles H
Nachtsheim, John J
Nash, Jonathon M
Naylor, Henry A, Jr
Newcomb, Robert W
Nichols, Kenneth D
North, Harper Q
O'Melia, Charles R
Orrell, George H
Paavola, Ivar R
Page, Chester H
Pai, Shih I
Palmer, C Harvey
Paulus, J Donald
Pell, Jerry
Peppin, Richard J
Perry, Charles W
Pollack, Louis
Prewitt, Judith M S
Pringle, Arthur E
Promisel, Nathan E
Pullen, Keats A
Quinn, James D
Rabinow, Jacob
Ragan, Robert M
Raheja, Dev G
Ramberg, Walter G C
Remer, Bertram R
Resnick, Joel B
ReVelle, Charles S
Rhee, Moon-Jhong
Rice, James K
Richards, John C
Rittner, Edmund S
Ritzmann, Robert W
Robertson, James B, Jr
Robinson, August R
Rockwell, Theodore
Rodger, Walton A
Roethel, David
Rogers, Kenneth A
Roha, Donald M
Roseman, Donald P
Rosenberg, Marvin A
Rosenfeld, Azriel
Ross, David S
Roy, Robert H
Rushing, Frank C
Russell, John E
Sabnis, Gajanan M
Sallet, Dirse W
Saltarelli, Eugene A
Savage, Rudolph P
Sayre, Clifford L, Jr
Scharp, Charles B
Schlimm, Gerard H
Schmerling, Erwin R
Schnabel, James J
Schrader, Henry C
Schulman, James H
Schutz, Harald
Schwartz, Murray A
Schwarz, William H
Schwenk, Francis C
Scipio, L Albert, II
Sevin, Eugene
Sharpe, William N, Jr
Sheer, Daniel P
Sherwin, Martin B
Shipman, Harold R
Shook, James F

Massachusetts

Minnesota

Missouri (Continued)

Batra, Romesh C
Beauchamp, James M
Beumer, Richard E
Bowles, C Quinton
Braisted, Paul W
Brasunas, Anton deSales
Brinkmann, Charles E
Brugger, Robert M
Brumbaugh, Philip S
Bruns, Robert F
Buck, Richard L
Buckingham, Frank E
Cagle, A Wayne
Calkins, Myron D
Callahan, Harry L
Campbell, Bobby D
Carney, William D
Carson, Ralph S
Carter, Robert L
Chen, Ta-Shen
Cheng, Franklin Y
Cheng, William Jen Pu
Clark, John B
Clark, Kenneth M
Coad, John D , Jr
Colteryahn, Henry C
Combs, Robert G
Conner, James L
Crabb, William A
Crawford, George L
Creighton, Donald L
Crocker, Burton B
Crosbie, Alfred L
Cunningham, Floyd M
Davidson, Edwin A
Davis, Robert L
DeCamp, Robert A
Dirscherl, Rudolf
Doll, Paul N
Douty, Richard T
Drosten, Fred W
Dudukovic, Milorad P
Duncan, Donald M
Dutton, Roger W
Eastman, Robert M
Edgerley, Edward, Jr
Elliott, David L
Ellis, Robert M
Emanuel, Jack H
Everhart, James G
Farber, Jack D
Faucett, Thomas R
Feagan, Wilbur S
Feldman, Rubin
Fischer, James R
Foster, Walter E
Fox, Bruce L
Francis, Lyman L
Frisby, James C
Frohmberg, Richard P
Fugate, Charles R
Gaige, C David
Gallen, Donald R
Georgian, John C
German, John G
Gibbs, William R
Gillespie, LaRoux K
Gould, Phillip L
Graham, Malise J
Gray, Glenn C
Hackmann, Robert E
Hahn, James H
Hancock, John C
Haney, Paul D
Hanpeter, Robert W
Harmon, Robert W
Hatheway, Allen W
Hauck, George F W
Hedden, William D
Hedley, William J
Heffelfinger, John B
Hill, John S
Hines, Anthony L
Hirt, George J
Hoag, James D

Hodges, Joseph T
Hollrah, Ronald L
Howard, Robert E
Huang, Ju-Chang (Howard)
Hurlbert, Don D
Johnson, A Franklin
Johnson, J Stuart
Johnson, James W
Jolles, Mitchell I
Kage, Arthur V
Kalin, Thomas E
Kardos, John L
Katz, I Norman
Keller, Walter D
Kimball, Charles N
Kimel, William R
Kintigh, John K
Kirkwood, Thomas C
Kisslinger, Fred
Koopman, Richard J W
Kopetz, Marion J
Kotfila, Ralph J
Krenzer, BK
Lebens, John C
Lee, Quarterman
Lehnhoff, Terry F
Leutzinger, Rudolph L
Lewis, David S
Lillard, David H
Lischer, Vance C
Liu, Henry
Look, Dwight C, Jr
Love, John
Lowery, Anthony J
Loyalka, Sudarshan K
Lyons, Jerry L
Mains, Robert M
Malsbary, James S
Maslan, Leon
Mattei, Peter F
McBean, Robert P
McCall, Thomas F
McCloud, Robert J
McGarraugh, Jay B
McGinnis, Charles I
McKelvey, James M
McKinney, Leon E,
McLaren, Robert W
Meyer, Walter
Meyerand, Russell G
Miles, John B
Miller, Owen W
Miller, Russell L
Mills, George S
Monseth, Ingwald T
Morgan, Robert P
Moshier, Glen W
Motard, Rodolphe L
Moulder, James E
Mueller, Marvin E
Muller, Marcel W
Munger, Paul R
Munroe, Gilbert G
Nau, Robert H
Nelson, Harlan F
Neptune, David B
Nuccitelli, Saul A
O'Connor, John T
Oglesby, David B
O'Keefe, Thomas J
Oliphant, Edgar Jr
Oliver, R C
O'Meara, Robert G
Oresick, Andrew
Patrick, Robert J
Patton, James L
Peters, David A
Peters, Robert C
Peterson, Thorwald R
Phelps, Edwin R
Phillips, Richard E
Pickett, Rayford M
Piest, Robert F
Powell, Michael J
Prelas, Mark A
Prentiss, Louis W, Jr

Pringle, Oran A
Pruitt, Ralph V
Puleo, Peter A
Rau, George J, Jr
Redpath, Richard J
Reece, John D
Reitz, Henry M
Richards, Earl F
Rinard, Sydney L
Roa, William J, Jr
Roberts, J Kent
Robinson, John H
Robinson, M John
Robinson, Thomas B
Rodin, Ervin Y
Roehrs, Robert J
Rosenbaum, Fred J
Ross, Donald K
Ross, James L
Ruf, John A
Salane, Harold J
Satterlee, George L
Sauer, Harry J, Jr
Schaper, Laurence T
Schwartz, Henry G , Jr
Schweiker, Jerry W
Scott, Lester F
Seltzer, Leon Z
Shannon, Eldon B
Siegel, Howard J
Slivinsky, Charles R
Smith, Charles H
Smith, Donald R
Snell, Virgil H
Snider, David G
Sonnino, Carlo B
Sperry, Philip R
Stables, Benjamin J , Jr
Stack, John R
Staker, Rodd D
Stevenson, Lawrence G
Stratton, John M
Stude, Robert A
Swanson, A Einar
Szabo, Barna A
Tao, William K Y
Tarn, Tzyh-Jong
Thomas, Benjamin E
Thomas, Marlin U
Thompson, Dudley
Trueblood, Walter D
Tsoulfanidis, Nicholas
Turk, Julius
Vansant, Robert E
Vaughn, Kenneth J
Wade, Robert G
Waidelich, Donald L
Wallace, Floyd C
Wallace, Norval D
Warder, Richard C, Jr
Warmann, Robert A
Washington, Donald R
Weart, Harry W
Weissenburger, Jason T
Weller, Lloyd W
Williams, Doyle A
Wolf, Robert V
Wolfe, Charles M
Wolfram, Ralph E
Wundrack, William A
Yu, Wei W
Zurheide, Charles H

Montana

Berg, Lloyd
Chittim, Lewis M
Clarkson, Arthur W
Drapes, Alex G
Druyvestein, Terry L
Emerson, C Robert
Griffiths, Vernon
Lacy, Robert R
Nurse, Edward A
Peavy, Howard S

Piper, R Davidson
Sanderson, Robert L, Jr
Sanks, Robert L
Stout, Koehler S
Verzuh, James M
Videon, Fred F
Walker, Leland J
Way, Stewart
Wideman, Charles J
Zucker, Gordon L

Nebraska

Alvine, Raymond G
Backlund, Brandon H
Bender, Frederick G
Bodman, Gerald R
Bousha, Frank N
Brooks, David W
Durham, Charles W
Foster, Edward T, Jr
Freise, Earl J
Gilley, James R
Hawes, Richard D
Hoppel, Susan K
Jenkins, Larry L
Jenkins, Peter E
Johnson, Marvin M
Knorr, David E
Krenzer, Bette A
Krohn, Robert F
Lagerstrom, John E
Large, Richard L
Launer, Milton L
Leuschen, Thomas U
Lieberknecht, Don W
Meedel, Virgil G
Milam, Max
Mills, Robert W
Mullendore, Robert A
Nelson, Russell C
Olsson, John E
Peters, Alexander R
Phelps, George C
Rogers, Gifford E
Schaufelberger, Don E
Schneider, Morris H
Von Bargen, Kenneth L
Voskuil, Walter H

Nevada

Adams, William E
Anderson, James T
Arden, Richard W
Baker, Arthur, III
Beatty, Robert W
Brown, John Webster
Douglas, Bruce M
Gonzales, John G M
Gribben, J Clark
Halacsy, Andrew A
Krenkel, Peter A
Messenger, George C
Montgomery, Max C
Olitt, Arnold
Scheid, Vernon E
Smith, Ross W
Wells, Herbert C
Wells, William R
Wiel, Stephen
Wyman, Richard V

New Hampshire

Allan, Donald S
Ashley, J Robert
Ashton, George D
Baumann, Hans P
Bisplinghoff, Raymond L
Brown, Gordon S
Carroll, Lee F
Chandler, John P
Collins, James J
Crane, Robert K

New Jersey (Continued)
McDonald, Gerald W G
McLellan, Alden, IV
McPherson, John C
Mesner, Max H
Meyer, Andrew U
Meyninger, Rita
Mihalasky, John
Miles, Richard B
Miller, Edward
Miller, Oscar O
Miller, Stewart E
Mills, Robert G
Minter, Jerry Burnett
Mittl, Robert L
Moen, Walter B
Monroe, Paul S
Moore, Robert C
Moritz, Karsten H
Morley, Richard J
Morrow, Darrell R
Mosher, Frederick K
Mosley, Ernest T
Moss, Alan M
Moy, Edward A
Moyer, Kenneth H
Mueller, Charles W
Mueller, William H
Mumma, Albert G
Musa, John D
Nardone, Pio
Nessmith, Josh T
Newhouse, Russell C
Nicholls, Richard W
Noyes, Robert E
O'Brien, Eugene
Ochab, Thomas F
O'Connor, James J
Oishi, Satoshi
Oliver, Billy B
Opie, William R
Orth, H Richard
Otto, Thomas Herbert
Pae, Kook D
Pai, David H
Pandullo, Francis
Pankove, Jacques I
Parker, Jack R
Parks, Lyman L
Pass, Isaac
Patel, C K N
Patton, Willard T
Payne, Matthew A
Pearlstein, Joel P
Pelan, Byron J
Peskin, Richard L
Petkovic-Luton, Ruzica A
Petree, Frank L
Phillips, Wendell E, Jr
Pierce, Harry W
Pinder, George F
Polaner, Jerome L
Powers, Kerns H
Price, Eugene B
Price, Robert I
Purcell, Fenton Peter
Purcell, Leo Thomas, Jr
Purdy, Verl O
Putman, Laurel E
Rabiner, Lawrence R
Ragold, Richard E
Rajchman, Jan A
Raudebaugh, Robert J
Reich, Bernard
Reintjes, Harold
Rennicks, Robert S
Reutter, John G
Richards, Richard T
Ricigliano, Anthony R
Rigassio, James L
Ripa, Louis C
Rivera, Alfredo J
Rivers, Lee W
Rizzo, Edward G
Roche, Edward C Jr
Roe, William P

Rogers, Kenneth C
Rohr, Peter H
Rose, Albert
Ross, Ian M
Royce, Barrie S H
Rubin, Arthur I
Rudkin, Donald A
Russell, Frederick A
Russo, Joseph R
Sadowski, Edward P
Salem, Eli
Salvin, Robert
Samuels, Reuben
Sannuti, Peddapullaiah
Savitsky, Daniel
Sawyer, R Tom
Schapiro, Jerome B
Schenker, Leo
Schlabach, Tom D
Schlink, Frederick J
Schneider, Frederick W
Schneider, Sol
Schoenfeld, Theodore M
Schulke, Herbert A , Jr
Schwab, Richard F
Schwanhausser, Edwin J
Schwartz, Perry L
Sears, Raymond W
Shahbender, Rabah
Shamis, Sidney S
Shapiro, Stanley
Sharpless, William M
Sherman, Samuel M
Siebeneicher, Paul Robert
   2nd
Signell, Warren I
Silla, Harry
Simon, Ralph E
Sinfelt, John H
Sisto, Fernando
Slepian, David
Slichter, William P
Sloane, Neil J A
Smith, Arnold R
Smith, George E
Smith, Peter W
Snowman, Alfred
Socolow, Robert H
Speitel, Gerald E
Sperber, Philip
Staff, Charles E
Stamper, Eugene
Stanton, George B, Jr
Stark, Robert E
Starr, Dean C
Stavis, Gus
Stephenson, Junius W
Stern, Louis I
Stitch, Malcolm L
Stocker, Gerard C
Stys Z Stanley
Swabb, Lawrence E, Jr
Swalin, Richard A
Tassios, Dimitrios
Taubenblat, Pierre W
Taylor, Snowden
Taylor, William H
Thoma, Frederic A
Thornton, Clarence G
Thornton, Keith C
Thornton, William A
Thouret, Wolfgang E
Tien, Ping K
Tomshaw, John
Toro, Richard F
Tschebotarioff, Gregory P
Turner, Warren H
Ugural, Ansel C
Ungar, J Stephen
Van Antwerpen, Franklin
Van Hare, George F, Jr
Van Raalte, John A
Van Sweringen, Raymond A
   , Jr
VanDriesen, Roger P
Varnerin, Lawrence J, Jr

Vieth, Wolf R
Vigilante, Frank S
Vogel, F Lincoln
Vollmer, James
Wachtman, John B, Jr
Waldhauer, Fred D
Wall, Alfred S
Wallenstein, Howard M
Walsh, John M, III
Ward, Joseph S
Warren, Clifford A
Watkins, Charles B
Watkins, Peter H
Wegman, Raymond F
Weil, Rolf
Weimer, Paul K
Weinberg, Harold N
Weinberg, Louis
Weinstein, Norman J
Weinstock, Walter W
Weissmann, Sigmund
Wernick, Jack H
Wesner, John W
Wheeler, Edward A
Whipple, William
White, Alan D
Whitwell, John C
Wiehl, Frederick R
Wiesenfeld, Joel
Wilding, John D
Wilding, John V
Wilkinson, Roger I
Will, Herbert
Wilson, Charles
Wilson, Percy Suydam
Wischmeyer, Carl R
Woller, Myron K
Woodruff, Kenneth L
Wyndrum, Ralph W, Jr
Wyszkowski, Paul E
Yoder, Richard J
Yokell, Stanley
Young, William H
Yu, A Tobey
Yu, Yi-Yuan
Yurick, Andrew N
Zdenek, Edna Prizgint
Zebib, Abdelfattah M G
Zederbaum, Robert B
Ziegler, Hans K
Zierler, Neal
Zippler, William N
Zmyj, Eugene B
Zoss, Abraham O
Zunkel, Alan D
Zworykin, Vladimir K

New Mexico

Abernathy, George H
Albach, Carl R
Alers, George A
Angel, Edward S
Baerwald, John
Ballard, Douglas W
Bardwell, A G
Baum, Carl E
Beck, Edwin L
Bleyl,Robert L
Bradshaw, Martin D
Bridgers, Frank H
Burns, James F
Clark, Arthur J, Jr
Clement, Richard W
DeLapp, Kenneth D
Dickinson, James M
Dorato, Peter
Duncan, Richard H
Durrenberger, J E
Emery, Michial M
Ford, C Quentin
Galt, John K
Garretson, Owen L
Genin, Joseph
Gogan, Harry L

Goodgame, Thomas H
Gregory, Bob Lee
Gross, William A
Gupta, Ajay
Haertling, Gene H
Halligan, James E
Hecker, Siegfried S
Hernandez, John W
Hockett, John E
Holt, Randolph E
Hoover, Mark D
Howard, William J
Humphries, Stanley, Jr
Hurley, George F
Jamshidi, Mohammad
Ju, Frederick D
Kamat, Satish J
Karni, Shlomo
Keepin, G Robert
Keigher, Donald J
Keyes, Conrad G , Jr
Kirchner, Walter L
Kocks, U Fred
Korman, Nathaniel I
Land, Cecil E
Lane, Golden E , Jr
Lawrence, Harold R
Lopez, Arthur M
Lopez, J Joseph
Matuszeski, Richard A
McKiernan, John W
Meyers, Marc A
Miller, Robert E
Molzen, Dayton F
Morgan, J Derald
Mottern, Robert W
Munson, Darrell E
Myers, Gene W
Narud, Jan A
Nunziato, Jace W
Patton, John T
Paxton, Hugh C
Peck, Ralph B
Pettigrew, Richard R
Reinig, L Philip
Rogers, Benjamin T
Sandstrom, Donald J
Savage, Charles F
Schonfeld, Fred W
Sheinberg, Haskell
Smith, Morton C
Solomon, Lee H
Sparks, Morgan
Stuetzer, Otmar M
Sullivan, John M
Supple, Richard G
Torma, Arpad E
Uhl, Robert J
Underwood, Wyatt R
Vaughan, Richard G
von Riesemann, Walter A
Walker, U Owen
Wall, Nelson C
Walvekar, Arun G
Whan, Glenn A
Willard Marcy
Williams, Theodore C
Wilson, Ralph S
Woodall, David M

New York

Abbott, Michael M
Abramson, Paul B
Ackoff, Albert K
Adam Kozma
Adams, Arlon T
Adare, James R
Addoms, Hallett B
Agosta, Vito
Agrawal, Ashok K
Albright, Louis D
Alden, John D
Aldrich, Robert G
Alexander, Donald C

INDEX OF INDIVIDUALS BY STATE

Oklahoma

South Dakota

Texas (Continued)
Brennan, James J
Briley, George C
Broderick, James Richard
Brotzen, Franz R
Brown, Donald C
Brown, James G
Brown, Samuel J
Bruene, Warren B
Bruins, Paul F
Buchsbaum, Norbert N
Bucy, J Fred
Burr, Arthur H
Burr, Richard W
Burrus, C S
Busch, Arthur W
Butler, Jerome K
Buttery, Lewis M
Cadden, James M
Calhoun, John C, Jr
Calhoun, Tom G
Callahan, Frank T
Campbell, Delmer J
Campbell, Stephen J
Canfield, Frank B, Jr
Cardello, Ralph A
Carlile, Robert E
Carroll, Billy D
Carter, Charles L, Jr
Caselli, Albert V
Casey, Leslie A
Caudle, Ben H
Chambers, Fred
Chappee, James H
Chappelear, Patsy S
Cheatham, John B, Jr
Chen, Mo-Shing
Cherrington, B E
Chinners, James E, Jr
Chu, Ting L
Clark, J Donald
Clark, James W
Clarke, John R
Cochran, Robert G
Collier, Samuel L
Collins, Charles E
Collins, Jeffrey H
Collins, Michael A
Collipp, Bruce G
Connally, Harold T
Cooke, James L
Cooke, Robert S
Coppinger, John T
Correa, Jose J
Cory, William E
Cosgrove, Donald G
Cotten, William C, Jr
Covarrubias, Jesse S
Coyle, Harry M
Cragon, Harvey G
Crews, R Nelson
Crocker, John W
Crouse, Philip C
Crow, John H
Crum, Floyd M
Cullinane, Murdock J, Jr
Cummings, George H
Curtis, L B
Dahlberg, E Philip
Dalley, Joseph W
Dalton, Charles
Daniel, David E
Darby, Ronald
Daugherty, Tony F
Davis, Hunt
Day, Roger W
Decker, Howard E
Deevy, William J
deFigueiredo, Rui J P
Demand, Lyman D
Denison, John S
Denman, Eugene D
DeShazo, John J, Jr
deVries, Douwe
DeWitt, John D
Diehl, Douglas S

Dockendorff, Ralph L
Dodge, Franklin T
Dorfman, Myron H
Dotson, Billy J
Dougal, Arwin A
Draper, E Linn
Dromgoole, James C
Dudek, Richard A
Dukler, Abraham E
Duncan, Charles W
Durrett, Joseph B, Jr
Dutton, Granville
Eagan, Constantine J
Eckhardt, Carl J
Edgerton, Robbie H
Eichhorn, Roger
Elliott, Martin A
Erdman, Carl A
Eschenbrenner, Gunther
  Paul
Eschman, Rick D
Everard, Noel J
Fancher, George H
Farmer, Larry E
Farnsworth, David E
Faulkner, C Shults
Fertl, Walter H
Fielding, Raymond E
Finley, J Browning
Finneran, James A
Fischer, Stewart C
Fisher, David L
Fisher, Edwin E
Flawn, Peter T
Fletcher, Leroy S
Focht, John A
Focht, John A, Jr
Ford, Edwin R
Foster, Charles K
Foye, Robert, Jr
Friberg, Emil E
Frick, John P
Frick, Thomas C
Friedkin, Joseph F
Furlong, Richard W
Furr, Howard L
Gaffron, John M C
Galstaun, Lionel S
Gee, Louis S
Geer, Ronald L
Gerard, Roy D
Gerdes, Walter F
Ghosh, Arvind
Giancarlo, Samuel S
Gillette, Roy W
Gilmore, Robert B
Gilruth, Robert R
Glover, Charles J
Gloyna, Earnest F
Goerner, Joseph K
Goland, Martin
Goldmann, Theodore B
Gooden, Charles D
Goodenough, John B
Graeser, Henry J
Graham, Jack M
Gray, Wilburn E
Greene, John H, Jr
Guard, Ray W
Gully, Arnold J
Gunnin, Bill L
Gupton, Paul S
Haisler, Walter E
Halbouty, Michel T
Hale, Bobby L
Hall, Kenneth R
Hamm, William J
Hammack, Charles J
Hann, Roy W, Jr
Hansen, William P
Hardy, William C
Harmon, Kermit S
Harris, William B
Haynes, John J
Hayter, R Reeves
Helland, George A, Jr

Hendricksen, William L
Herman, Robert
Herr, James C
Hester, J Charles
Hightower, Joe W
Hildebrand, Neal J
Hirsch, Robert L
Hixson, Elmer L
Hocott, Claude R
Hoffman, Nathan N
Holley, Edward R
Holliday, George H
Holman, Jack P
Honeycutt, Baxter D
Hopper, Jack R
Howell, John R
Hsu, Thomas T C
Huang, C J
Hudson, Carroll D
Hunsaker, Barry
Hurlburt, Harvey Z
Hussain, A K M F
Hutchinson, Charles A, Jr
Ignizio, James P
Janzen, Jerry L
Jeanes, Joe W
Jenett, Eric
Jenkins, Marie H
Jennings, Nathan C
Jepsen, John C
Johanson, K Arvid, Jr
Johns, Thomas G
Johnson, Horace A
Johnson, Hubert O, Jr
Johnson, R Barry
Johnson, Robert E
Johnson, Robert M
Johnston, Waymon L
Jones, Andrew D
Jones, Marvin R
Jones, Thomas J
Jones, William B, Jr
Joplin, John F
Jorden, James R
Jumper, Billy S
Kaenel, Reg A
Kalteyer, Charles F
Kareem, Ahsan
Kastor, Ross L
Katz, Marvin L
Kennedy, Thomas W
Keplinger, Henry F
Key Joe, W
Kielhorn, William H
Kiesling, Ernst W
Kilby, Jack S
Killebrew, James R
Killough, John E
King, Joe J
Kirk, Ivan W
Kitterman, Layton
Koenig, Charles Louis
Kozik, Thomas J
Kraemer, M Scott
Kraft, Leland M, Jr
Krahl, Nat W
Krause, Henry M, Jr
Krezdorn, Roy R
Kristiansen, Magne
Krizak, Eugene J
Kuers, Marvin M
Kuhlke, William C
LaGrone, Alfred H
Lamb, Jamie Parker
Landes, Spencer H
Langford, Ivan, Jr
Lautzenheiser, Clarence
Lawrence, James H
LeBlanc, Joseph U
Lee III, Thad S
Lee, John A
Lee, William J
Leeds, J Venn, Jr
Leggett, Lloyd W, Jr
Lesso, William G
Levine, Jules D

Lewis, James P
Lewis, Orval L
Lewis, Ronald L
Lienhard, John H
Likwartz, Don J
Lilley, Eric G
Lilly, Edmund D
Lindholm, Ulric S
Lindsey, Kenneth R
Lindsey, Larry M
Little, Doyle
Littmann, Walter E
Loehr, Raymond C
Logan, Horace P
Long, Stuart A
Lou, Jack Y K
Lowe, Wilton J, Jr
Lutes, Loren D
Lutz, Raymond P
MacKenzie, Horace J
Mackey, Thomas S
Magliolo, Joseph J
Malina, Joseph F, Jr
Maloney, Laurence J
Marcus, Harris L
Marion Johnson P E
Marquis, Eugene L
Marshall, A Frank
Marshall, Peter W
Marshek, Kurt M
Martin, Howard W
Martinez, Roberto O
Mason, Raymond C
Matthews, Charles S
Matula, David W
May, Melville M
McBride, Robert R
McClelland, Bramlette
McClure, Alan E
McCormick, William J
McDonald, Dan T
McDonald, Donald
McGrath, Daniel F
McLellan, Rex B
McMorries, Bill R
McNealy, Delbert D
Meier, France A
Meier, Wilbur L, Jr
Mendenhall, Wesley S, Jr
Mercer, Harold E
Mercer, Kenneth K
Meriwether, Ross F
Minor, Joseph E
Mistrot, Gustave A
Mitchell, William L
Moore, John A
Moore, Roland E
Morris, William Page
Morrison, John E, Jr
Morrison, M Edward
Mortada, Mohamed
Motter, Eugene F
Musa, Samuel A
Muster, Douglas
Myrick, H Nugent
Nachlinger, R Ray
Nall, John H, Jr
Nation, Oslin
Noble, Marion D
Nolen, James S
Nordgren, Ronald P
Norton, David J
Noyes, Jack K
Noyes, Jonathan A
Odeh, Aziz S
Ohsol, Ernest O
Olson, Roy E
O'Neill, Michael W
Orr, Charles K
Orr, Robert M
Ostrofsky, Benjamin
Owen, Edwin L
Page, Robert H
Pansini, Anthony J
Parker, Robin M
Paskusz, Gerhard F

Utah

Texas (Continued)
Patterson, David H
Patton, Alton D
Paul, Donald R
Peaceman, Donald W
Pearson, J Boyd, Jr
Peikari, Behrouz
Perkins, Thomas K
Pernoud, Rene B
Perry, Charles R
Phillips, Don T
Phillips, R Curtis Jr
Pickering, C J
Pincus, George
Pinnell, Ray A, Jr
Poage, Scott T
Pollock, Wilfred A
Poole, William S
Powell, Alan
Powell, Allen L
Prehn, W Lawrence, Jr
Preston, Walter B
Prickette, Gerald S
Puckett, Russell E
Purnell, William B
Quance, Robert J
Querio, Charles W
Raba, Carl F, Jr
Rabins, Michael J
Rabson, Thomas A
Rachford, Henry H, Jr
Rady, Joseph J
Raffel, David N
Ragland, John R
Ramsey, Jerry D
Rankin, John P
Rase, Howard F
Rassinier, Edgar A
Reed, Stan C
Reeves, Robert G
Reistle, Carl E, Jr
Rhoades, Warren A
Rhodes, Allen F
Richardson, John H
Richardson, Joseph G
Rider, Bobby G
Rigsbee, H Ken, Jr
Ring, Sandiford
Rizzone, Michael L
Roberts, John M
Robertson, Elgin B
Rodenberger, Charles A
Rogers, Bruce G
Rogers, Cranston, R
Rohlich, Gerard A
Rollins, Albert W
Romani, Fred
Rouse, John W, Jr
Runyan, Edward E
Russell, Fred G
Russell, James E
Sacken, Donald K
Salama, Kamel
Salis, Andrew E
Sandberg, Irwin W
Santry, Israel W, Jr
Scarola, John A
Schmitt, Gilbert E
Schoppe, Conrad J
Schulz, Richard B
Seidel, Edmund O
Sexton, Joseph M
Shannon, Robert E
Sheikh, Shamim A
Shen, Liang C
Shepherd, Mark, Jr
Shoemaker, Leonard W
Short, Byron E
Short, John Patrick
Shoup, Terry E
Shumaker, Fred E
Silcock, Frank A, P E
Sims, James R
Sinclair, A Richard
Singer, Leslie J
Sizer, Phillip S

Skov, Arlie M
Skov, Ebbe R
Smalley, Arthur L, Jr
Smith, Milton L
Sobol, Harold
Sokolowski, Edward H
Solberg, Ruell F, Jr
Somerville, Mason H
Sommers, Otto W
Spillane, Leo J
Springer, Karl J
Stafford, Stephen W
Steadman, H Douglas
Stevens, Gladstone T, Jr
Stice, James E
Straiton, Archie W
Straub, Harold E
Svore, Jerome, H
Synck, Louis, J
Talbert, John J, III
Tapley, Byron D
Tatum, Finley W
Teal, Gordon K
Thakur, Ganesh C
Thomas, Eddie F
Thomas, Frank A, Jr
Thomas, Julian W
Thompson, James E
Tippett, G Ross
Tittel, Frank K
Torrey, Paul D
Treybig, Harvey J
Tucker, Randolph W
Vance, John M
Vela, M A
Veletsos, Anestis S
Viaclovsky, Syl A
Vines, Darrell L
VonGonten, William D
Walker, Samuel R, Jr
Walker, William F
Wall, Evern R
Wallis, William E
Ward, Julian R
Watler, Kenneth G
Watt, Brian J
Watt, Joseph T, Jr
Weaver, Graeme D
Webb, John R
Webb, Theodore S
Weese, John A
Weitzenkorn, Lee F
Welsh, James N
Werner, James T
Wetzel, Otto K, Jr
Wheeler, Lewis T
White, Gifford E
Wick, Robert S
Wicks, Moye, III
Wier, Thomas P Jr
Wiggins, Edwin G
Wilcox, Marion W
Wilde, H Dayton
Wilkes, Lambert H
Willenbrock, F Karl
Williams, Grover C
Williams, Jane
Williams, Milton A
Wilson, Claude L
Winnie, Dayle D
Wirsching, Joseph E
Witte, Larry C
Woelfel, Albert E
Wohlgemuth, Robert E
Woods, Donald L
Woodson, Herbert H
Worley, Frank L, Jr
Wright, Wesley F, Jr
Young, Dana
Young, Fred M
Young, Ralph M
Zuniga, Oscar J

Utah

Allen, Dell K

Allen, Edmund W
Andrade, Joseph D
Armstrong, Ellis L
Atwood, Kenneth W
Barker, Dee H
Baum, Sanford
Bennion, Douglas N
Berrett, Paul O
Boe, Rollin O
Boehm, Robert F
Bosakowski, Paul F
Breitling, Thomas O
Bringhurst, Lynn H
Brophy, James J
Brown, Wayne S
Byrne, J Gerald
Carpenter, Carl H
Case, Edward T
Chabries, Douglas M
Chaston, A Norton
Christiansen, Richard W
Clark, Clayton
Clegg, John C
Coon, Arnold W
Cooper, Harrison R
de Nevers, Noel H
DeVries, Kenneth L
Eckhoff, David W
EerNisse, Errol P
Evans, David C
Firmage, D Allan
Flammer, Gordon H
Fuhriman, Dean K
Gandhi, Om P
Gill, Lowell F
Goodwin, James G
Grow, Richard W
Hanks, Richard W
Hansen, Vaughn E
Haycock, Obed C
Heiner, Clyde M
Herbst, John A
Higginson, R Keith
Hill, George R
Hoeppner, David W
Hogan, Mervin B
Holdredge, Russell M
Huber, Robert J
Hucka, V Joseph
Hunt, Richard N
Irvine, Leland K
Isaacson, LaVar K
Jeppson, Roland W
Johansen, Craig E
Jonsson, Jens J
Karren, Kenneth W
Kelsey, Stephen J
Laramee, Richard C
Larsen, Dale G
Larson, Gale H
Malouf, Emil E
Martin, Bruce A
Marushack, Andrew J
Maxwell, James D
Merritt, LaVere B
Michaelson, Stanley D
Middlebrooks, Eddie Joe
Miller, Gerald, R
Miller, Jan D
Miner, Gayle F
Mostaghel, Naser
Munyan, Leon J
Murdock, Larry T
Nelson, Kenneth W
Nibley, J William
Oblad, Alex G
Olson, Ferron A
Pernichele, Albert D
Peterson, Dean F
Pitt, Charles H
Polve, James H
Probasco, Johnny L
Randolph, Robert E
Reaveley, Ronald J
Rich, Elliot
Richards, Kenneth J

Riesenfeld, Richard F
Rogers, Vern C
Rose, David N
Rosenbaum, Joe B
Sandquist, Gary M
Sherwood, Richard F
Simonsen, John M
Sohn, Hong Yong
Spedden, H Rush
Stephenson, Robert E
Stringfellow, Gerald B
Stringham, Glen E
Talbot, Eugene L
Tullis, James P
Wadsworth, Milton E
Walker, Bruce H
Watson, Kenneth W
Wilson, Arnold
Woodbury, Richard C
Yu, Jason C

Vermont

Agusta, Benjamin J
Anderson, Richard L
Benzing, Robert J
Butler, Donald J
Costello, Frederick J
Dean, Robert C, Jr
Francis, Gerald P
Frost, Paul D
Gratiot, J Peter
Hansen, James C
Hundal, Mahendra S
McCoy, Byron O
McLay, Richard W
Mordell, Donald L
Oppenlander, Joseph C
Outwater, John Ogden
Pope, Malcolm H
Pricer, W David
Pyper, Gordon R
Roth, Wilfred
Smith, Mark K
Tartaglia, Paul E
Van Domelen, John F
von Turkovich, Branimir
Wallis, Clifford M

Virginia

Ackerman, Roy A
Adelberg, Kenneth
Agee, Marvin H
Ailor, William H, Jr
Allen, John L
Ash, Robert L
Attinger, Ernst O
Ball, Richard T
Barnes, Charlie H, Jr
Baum, David M
Beale, D Anthony
Beam, Walter R
Beard, Charles I
Berkness, I Russell
Bidstrup, Wayne W
Bird, George T
Bittel, Lester R
Boge, Walter E
Bogner, Neil F
Boyd, Robert Lee
Boynton, Edgar B
Bradshaw, John P, Jr
Brahms, Thomas W
Bramley, Jenny Rosenthal
Brinson, Halbert F
Broad, Richard
Bronson, Harold R
Buckley, John D
Burbank, Farnum M
Burton, William J
Butler, Gilbert L
Byrd, Lloyd G
Case, James B
Cass, James R, Jr

West Virginia

Index of Individuals
by Specialization

ELECTRIC POWER

EDUCATION (GENERAL)
(Continued)
Thigpen, Joseph J (CA)
Thomann, Gary C (KS)
Thompson, Dudley (MO)
Thorbjornsen, Arthur R (OH)
Tierney, John W (PA)
Tribus, Myron (MA)
Tung, David H H (NY)
Ungrodt, Richard J (WI)
Upthegrove, Wm R (OK)
Utela, David R (WA)
Van Wylen, Gordon J (MI)
VandeLinde, V David (MD)
Venable, Wallace (WV)
Vines, Darrell L (TX)
Wadlin, George K (ME)
Wagner, J Robert (PA)
Wakeland, Howard L (IL)
Waterman, Alan T, Jr (CA)
Waugh, David J (SC)
Weber, Ernst (NC)
Wells, William R (NV)
Wheaton, Fredrick W (MD)
Wheeler, Edward A (NJ)
Wheeler, Orville E (TN)
Wilhelm, William J (KS)
Wilkes, James O (MI)
Willenbrock, F Karl (TX)
Wilson, Joe R (LA)
Winegard, William C
Wittry, John P (CO)
Wolf, Werner P (CT)
Wolovich, William A (RI)
Woodward, James H (AL)
Wuerger, William W (WI)
Yacoub, Kamal (FL)
Yeargan, Jerry R (AR)
Yenkin, Fred (OH)
Yerlici, Vedat A
Yu, Yao-nan
Zachariah, Gerald L (FL)
Zahn, Markus (MA)
Zygmont, Anthony (PA)

ELECTRIC POWER

Acheson, Allen M (MO)
Adomat, Emil Alfred (FL)
Anderson, Arvid E (NJ)
Anderson, John G (MA)
Anderson, Paul M (CA)
Anderson, Stanley W (IL)
Anderson, Walter T (MI)
Andresen, T Richard (MN)
Areghini, David G (CA)
Armstrong, Charles H (CO)
Asbury, Carl E (FL)
Astley, Wayne C (PA)
Avila, Charles F (MA)
Bahder, George (NJ)
Baldwin, Clarence J (PA)
Baldwin, Miles S (PA)
Bandel, John M (NY)
Barkan, Philip (CA)
Barnes, Bruce F, Jr (MO)
Baron, S (NY)
Barrett, Bruce A (IA)
Barthold, Lionel O (NY)
Bartnikas, Ray
Barton, Thomas H
Baum, Willard U (PA)
Beakes, John H (MD)
Beckwith, Sterling (CA)
Behnke, Wallace B, Jr (IL)
Berger, Robert C (PA)
Berry, Bill E (ID)
Bigelow, Robert O (MA)
Billinton, Roy
Bills, Glenn W (CA)
Blankenburg, R Carter (CA)
Boaz, Virgil L (LA)
Boice, Calvin W (WA)
Boyer, Vincent S (PA)

Bradshaw, Martin D (NM)
Brown, C Alvie (CT)
Brown, Gordon S (NH)
Brown, Ronald D (IA)
Brownlee, William R (AL)
Caleca, Vincent (MA)
Caplan, Aubrey G (PA)
Carey, John J (MI)
Carlile, Robert N (AZ)
Carpenter, William L (SC)
Carroll, Lee F (NH)
Cathey, Jimmie J (KY)
Chambers, Fred (TX)
Chartier, Vernon L (OR)
Chaston, A Norton (UT)
Chen, Mo-Shing (TX)
Cisler, Walker L (MI)
Clement, Richard W (NM)
Clements, Kevin A (MA)
Cobean, Warren R (NJ)
Cohn, Nathan (PA)
Colteryahn, Henry C (MO)
Concordia, Charles (FL)
Conner, David A (AL)
Conrad, Albert G (CA)
Cooke, James L (TX)
Corry, Andrew F (MA)
Cottom, Melvin C (KS)
Crippen, Reid P (FL)
Culler, Floyd L, Jr (CA)
Damrell, Charles B (MA)
Dannenberg, Warren B (MA)
Darveniza, Mat
De Mello, F Paul (NY)
DeCamp, Robert A (MO)
DeGuise, Yvon
Denison, John S (TX)
Dickson, George H (WA)
Dillard, Joseph K (SC)
Dolan, John E (OH)
Dopazo, Jorge F (NY)
Dougherty, John J (CA)
Dutton, Roger W (MO)
Dwon, Larry (NC)
Dy Liacco, Tomas E (OH)
Eilering, John G (IL)
El-Abiad, Ahmed H
Ellefson, George E (AR)
Ellenberger, William J (DC)
Ellis, Robert M (MO)
Endahl, Lowell J (DC)
Erickson, Claud R (MI)
Eriqat, Albert K (CA)
Estcourt, Vivian F (CA)
Farber, Jack D (MO)
Fatic, Vuk M (IN)
Fee, Walter F (CT)
Fink, Lester H (VA)
Finney, William G (IA)
Fitzgerald, Joseph P (OH)
Flugum, Robert W (DC)
Fosdick, Ellery R (WA)
Gallup, Robert B (WA)
Gatti, Richard M (PA)
Gee, Louis S (TX)
Gelopulos, Demos P (IN)
Ghandakly, Adel A (OH)
Gibson, Charles A (AL)
Gillette, Robert W (MA)
Gold, S H (CA)
Goldstein, Alexander
Gordon, Walter S (WA)
Gotschall, Donald J (AK)
Gould, Nathan R (CA)
Greenfield, Eugene W (CA)
Greenwood, Allan N (NY)
Grimes, Arthur S (NY)
Gross, Eric T B (NY)
Grumbach, Robert S (OH)
Guernsey, Curtis H, Jr (OK)
Gungor, Behic R (AL)
Gustafson, Robert J (OH)
Halacsy, Andrew A (NV)
Halfmann, Edward S (PA)

Hamilton, Howard B (PA)
Hang, Daniel F (IL)
Hanna, William J (CO)
Hano, Ichiro
Happ, H Heinz (NY)
Haralampu, George S (MA)
Harder, John E (IN)
Harmon, Don M (ND)
Harmon, Robert W (MO)
Hauspurg, Arthur (NY)
Hedman, Dale E (NY)
Herziger, William J (WI)
Hesse, M Harry (NY)
Hileman, Andrew R (PA)
Hingorani, Narain G (CA)
Hiroshi, Sakai M
Hitchcock, Leon W (NH)
Holley, Charles H (NY)
Holway, Donald K (OK)
Horowitz, Stanley H (NY)
House, Hazen E (TN)
Hutchinson, Frank D, III (NY)
Jackson, Ralph L (IA)
Jackson, William D (MD)
Jessel, Joseph J A (VA)
Johnson, Ingolf Birger (NY)
Johnson, Milton R, Jr (LA)
Johnson, William R (CA)
Jones, James A (NC)
Jones, Richard T (PA)
Jorgensen, Gordon D (AZ)
Kaprielian, Elmer F (CA)
Karady, George Gy (NJ)
Keaton, Clyde D (MA)
Kidd, Keith H
Kight, Max H (CO)
Killian, Stanley C (AR)
Kimbark, Edward W (OR)
Kind, Dieter H
Kintigh, John K (MO)
Kirchmayer, Leon K (NY)
Koch, Leonard J (IL)
Koepfinger, Joseph L (PA)
Kohne, Richard E (CA)
Kwiatkowski, Robert W (MA)
Lamm, A Uno (CA)
Lane, James W (MS)
Lauber, Thornton S (NY)
Launer, Milton L (NE)
Laurent, Pierre G
Lawrence, Robert F (PA)
Lee, Thomas H (MA)
Lee, William S (NC)
Leib, Francis E (PA)
Leroy, Gerard L
Leslie, John R
Leung, Paul (CA)
Levi, Enrico (NY)
Lewis, Peter A (NJ)
Lewis, William M, Jr (OH)
Liao, George S (CA)
Light, Frederick H (PA)
Lindsay, Philip J, Sr (WA)
Linke, Simpson (NY)
Liss, Sheldon (CA)
Liu, Lee (IA)
Lockie, Arthur M (PA)
Lofgren, Richard C (WA)
Loftness, Robert L (DC)
Lovejoy, Stanley W (NY)
Lutken, Donald C (MS)
Malik, Om P
Mandil, I Harry (FL)
Marburger, Thomas E (FL)
Marchetti, Robert J (MN)
Marxheimer, Rene B (CA)
McCoy, Byron O (VT)
McNutt, Willliam J (MA)
Meisel, Jerome (MI)
Melvin, Howard L (CA)
Meyerand, Russell G (MO)
Michelson, Ernest L (IL)
Miller, Earle C (MA)
Miller, John G (PA)

Miner, R John (MN)
Moran, John H (NY)
Morgan, J Derald (NM)
Moss, Ralph A (WA)
Moxey, Richard T (PA)
Mozer, Harold M (WA)
Nagel, Theodore J (NY)
Nardone, Pio (NJ)
Nasser, Essam
Nelson, Percy L (MA)
Nichols, Clark (ME)
Nichols, Kenneth D (MD)
Nichols, Richard S (OR)
Nimmer, Fred W (OH)
Noodleman, Samuel (AZ)
Norberg, Hans A (OK)
O'Reilly, Roger P (FL)
Orrok, George A (MA)
Owen, Warren H (NC)
Panoff, Robert (DC)
Pansini, Anthony J (TX)
Paris, Luigi
Park, Gerald L (MI)
Park, Robert H (IL)
Parker, Jack R (NJ)
Patton, Alton D (TX)
Perron, Gilles
Perry, Anthony J (VA)
Peters, Robert C (MO)
Peterson, Harold A (AZ)
Pierce, Ralph (MI)
Poe, Herbert V (SC)
Pollard, Ernest I (PA)
Popham, Richard R (NY)
Qureshi, Abdul Haq (OH)
Ragland, John R (TX)
Randolph, Patrick A (CA)
Rice, James K (MD)
Ringlee, Robert J (NY)
Rivera-Abrams, Carlos (PR)
Robertson, Lawrence M (CO)
Robertson, Richard B (LA)
Roddis, Louis H, Jr (SC)
Roe, Lowell E (OH)
Roettger, Jerome N (MN)
Rossie, John P (CO)
Ruddick, Bernard N (KS)
Ruelle, Gilbert L
Russell, Robert A (KS)
Saba, Shoichi
Sandell, Donald H (IL)
Schaffer, Stanley G (PA)
Schaufelberger, Don E (NE)
Scherer, Harold N, Jr (OH)
Schinzinger, Roland (CA)
Schleif, Ferber R (CO)
Schlueter, Robert A (MI)
Schmidt, Alfred D (FL)
Schmitt, Gilbert E (TX)
Schneider, Frederick W (NJ)
Schott, F W (CA)
Schutz, Harald (MD)
Sebo, Stephen A (OH)
Sekine, Yasuji
Sever, Lester J (NC)
Shane, Paul
Shankle, Derrill F (PA)
Shaw, Ralph W (NC)
Silton, Ronald H (OH)
Slemon, Gordon R
Smith, Leslie S (FL)
Smith, Neal A (OH)
Smith, William P (KS)
Snoddy, Jack T (OK)
Space, Charles C (FL)
Staats, Gustav W (WI)
Stables, Benjamin J, Jr (MO)
Stanek, Eldon K (MI)
Starr, Chauncey (CA)
Staszesky, Francis M (MA)
Stevenson, William D, Jr (NC)
Stoller, Sidney M (NY)

INDEX OF INDIVIDUALS BY SPECIALIZATION

**ELECTRICAL
ENGINEERING**
(Continued)
Galindo-Israel, Victor (CA)
Gallotte, Willard A (WA)
Gallup, Robert B (WA)
Gandhi, Om P (UT)
Gardner, Chester S (IL)
Gaskill, Jack D (AZ)
Gatti, Richard M (PA)
Gaylord, Thomas K (GA)
Geddes, Leslie Alexander (IN)
Geiger, James Edward (TN)
Gerhardt, Lester A (NY)
Gerke, John F (WA)
Gessner, Gene A (IA)
Ghausi, Mohammed S (CA)
Giampaolo, Joseph A (CA)
Gibson, Richard C (ME)
Gibson, Robert C (VA)
Gillette, Jerry M (IL)
Gilliland, Bobby E (SC)
Giordano, Anthony B (NY)
Glasford, Glenn M (NY)
Gonzalez, Rafael C (TN)
Gordon, Walter S (WA)
Gore, Willis C (MD)
Goto, Satoshi
Gowen, Richard J (SD)
Graham, Jack H (OK)
Grant, Ian S (NY)
Gray, Harry J (PA)
Gray, Jon F (OH)
Graybill, Howard W (CA)
Grayson, Leonard C (NJ)
Green, David G (AL)
Gregory, Bob Lee (NM)
Greig, James
Griffiths, Lloyd J (CA)
Grimes, Dale M (PA)
Grogan, William R (MA)
Grove, H Mark (OH)
Grow, Richard W (UT)
Guarrera, John J (CA)
Gungor, Behic R (AL)
Guth, John J, Jr (LA)
Guy, Warren J (PA)
Haag, Kenneth W (IL)
Haber, Fred (PA)
Hacker, Herbert, Jr (NC)
Haddad, George I (MI)
Haddad, Richard A (NY)
Hahn, James H (MO)
Hakimi, S Louis (CA)
Halibey, Roman (NY)
Hall, William M (MA)
Hamid, Michael
Hammer, Kathleen M (MI)
Hammerschmidt, Andrew L (NJ)
Hancock, John C (MO)
Hanna, William J (CO)
Haralampu, George S (MA)
Haralick, R M (CA)
Harbourt, Cyrus O (MI)
Hargens, Charles W, III (PA)
Harger, Robert O (MD)
Harmon, Kermit S (TX)
Harmon, Robert W (MO)
Harrington, Dean B (NY)
Harrington, Roger F (NY)
Harris, Stephen E (CA)
Harrison, Stanley E (VA)
Haus, Hermann A (MA)
Hauser, John R (NC)
Hayes, Thomas B (OR)
Hayt, William H, Jr (IN)
Hazeltine, Barrett (RI)
Heath, James R (KY)
Hedman, Dale E (NY)
Hendricks, Charles D (CA)
Hennessy, John F (NY)

Henning, Rudolf E (FL)
Hermsen, Richard J (CA)
Herskowitz, Gerald J (NJ)
Herz, Eric (NY)
Hewlett, William R (CA)
Heyborne, Robert L (CA)
Hiatt, Ralph E (VA)
Hill, David A (CO)
Hines, Marion E (MA)
Hisao, Kimura
Hix, Charles F, Jr (CT)
Hixson, Elmer L (TX)
Holderby, George D (WV)
Hord, William E (IL)
Horowitz, Stanley H (NY)
Hostetter, Gene H (CA)
Howarth, David S (MI)
Hower, Glen L (WA)
Hu, Sung C (CA)
Huang, Thomas S (IL)
Huber, Robert J (UT)
Hudson, John C, Jr (MS)
Hughes, William L (OK)
Humber, Philip M (FL)
Hunsinger, Bill J (IL)
Hunsucker, Robert D (AK)
Hutchinson, Charles E (NH)
Ibuka, Masaru
Ide, John M (CA)
Ingram, Troy L (AL)
Irby, Raymond F (VA)
Ishimaru, Akira (WA)
Jackson, Ralph L (IA)
Jackson, Stuart P (VA)
Jacobson, David H
Jaron, Dov (PA)
Jenkins, Leo B, Jr (KY)
Jessel, Joseph J A (VA)
Johansen, Bruce E (OH)
Johanson, K Arvid, Jr (TX)
Johnson, Claude W (VA)
Johnson, Gerald N (IA)
Johnson, Ingolf Birger (NY)
Johnson, J Stuart (MO)
Johnson, Loering M (CT)
Johnson, Milton R, Jr (LA)
Johnson, Raymond C (FL)
Johnson, Vern R (AZ)
Johnson, Walter C (NJ)
Johnson, William R (CA)
Jones, Roger C (AZ)
Jones, Thomas J (TX)
Jones, William B, Jr (TX)
Jong, Mark M T (KS)
Jordan, Angel G (PA)
Jordan, Edward C (IL)
Jordan, Kenneth L, Jr (MD)
Jury, Eliahu I (FL)
Kadaba, Prasad K (KY)
Kaenel, Reg A (TX)
Kahng, Seun K (OK)
Kalbach, John F (CA)
Kandoian, Armig G (NJ)
Kaprielian, Elmer F (CA)
Karmel, Paul R (NY)
Karni, Shlomo (NM)
Kashyap, Rangasami L (IN)
Kaufman, Irving (AZ)
Kayton, Myron (CA)
Keller, Arthur C (NY)
Keller, Edward L (CA)
Kim, Kyekyoon (Kevin) (IL)
Kim, Myunghwan (NY)
Kimball, Charles N (MO)
King, Ronold W P (MA)
Kinnen, Edwin (NY)
Kino, Gordon S (CA)
Kirk, Donald E (CA)
Kirkwood, Thomas C (MO)
Klapper, Jacob (NJ)
Klein, Ronald L (WV)
Knorr, Jeffrey B (CA)
Knott, Roger L (OH)
Knotts, Burton R (AR)

Ko, Hsien C (OH)
Kobayashi, Koji
Kohloss, Frederick H (HI)
Koide, Frank T (HI)
Koopman, Richard J W (MO)
Kornhauser, Edward T (RI)
Kramer, Raymond E (OH)
Kreer, John B (MI)
Kresser, Jean V (CA)
Krezdorn, Roy R (TX)
Krieger, Charles H (CA)
Krishna, Gopal T K (IA)
Ksienski, Aharon A (OH)
Ku, Yu H (PA)
Kuh, Ernest S (CA)
Kumar, K S P (MN)
Kurokawa, Kaneyuki
Kushner, Harold J (RI)
Laaspere, Thomas (NH)
Lagerstrom, John E (NE)
LaGrone, Alfred H (TX)
Lake, James M (AK)
Landgrebe, David A (IN)
Lane, James W (MS)
Larrowe, Vernon L (MI)
Lataille, Jane I (CT)
Law, John (ID)
Lawrenson, Peter J
LeBold, William K (IN)
Lechner, Bernard J (NJ)
Leeds, J Venn, Jr (TX)
Leeds, Winthrop M (PA)
Lehmann, Gerard J
Lempert, Joseph (PA)
Leon, Benjamin J (KY)
Leonard, Richard D (CA)
Leroy, Gerard L
Levine, William S (MD)
Lewin, Leonard (CO)
Lewis, Peter A (NJ)
Li, Tingye (NJ)
Liboff, Richard L (NY)
Lin, Pen-Min (IN)
Lin, Shu (HI)
Linder, Clarence H (NY)
Lindquist, Claude S (CA)
Linke, Simpson (NY)
Linvill, John G (CA)
Linville, Thomas M (NY)
Liou, Ming-Lei (MA)
Litchford, George B (NY)
Littrell, John L (CA)
Liu, Ruey-Wen (IN)
Lo, Arthur W (NJ)
Lo, Yuen T (IL)
Lohmann, Carlos A J
Long, Carl J (PA)
Long, Francis M (WY)
Long, Stuart A (TX)
Longerbeam, Gordon T (CA)
Loomis, Herschel H , Jr (CA)
Lopez, J Joseph (NM)
Lord, William (CO)
Lucas, Jay (AR)
Lunnen, Ray J, Jr (PA)
Lunsford, Jesse B (VA)
Macnee, Alan B (MI)
Mailloux, Robert J (MA)
Maler, George J (CO)
Maloney, Laurence J (TX)
Malthaner, William A (FL)
Manfredonia, Savery S (NY)
Mansur, John C (OK)
Marxheimer, Rene B (CA)
Maryssael, Gustave J Ch
Mathes, Kenneth N (NY)
Mathews, Bruce E (FL)
Matsumoto, Yuki (HI)
Mattson, Roy H (AZ)
Mayer, James W (NY)
Mayo, John S (NJ)
McCalla, William J (CA)

McClellan, Dallas L, Jr (GA)
McCollom, Kenneth A (OK)
McConnell, Lorne D (PA)
McDaniel, Willie L (MS)
McDonald, John C (CA)
McDonald, John F (NY)
McIntosh, Robert E, Jr (MA)
McIsaac, Paul R (NY)
McKean, A Laird (NY)
McKnight, Samuel William
McLaren, Robert W (MO)
McLean, True (FL)
McMurray, William (NY)
McNeill, Robert E (MD)
McVey, Eugene S (CA)
McWhirter, James H (PA)
Meditch, James S (WA)
Mei, Kenneth K (CA)
Meindl, James D (NY)
Meisel, H Paul (CO)
Meisel, Jerome (MI)
Melcher, James R (MA)
Melsa, James L (IL)
Mendel, Jerry M (CA)
Mettler, Ruben F (OH)
Meyer, Andrew U (NJ)
Miller, Stewart E (NJ)
Miller, William E (AZ)
Milstein, Laurence B (CA)
Miner, Gayle F (UT)
Miner, R John (MN)
Minnich, Eli B (CA)
Mitchell, J Dixon, Jr (GA)
Mitra, Sanjit K (CA)
Miyairi, Shota
Molinder, John I (CA)
Monseth, Ingwald T (MO)
Monson, James E (CA)
Monteith, Larry K (NC)
Moore, Harry F (PA)
Morgan, J Derald (NM)
Morgenthaler, Frederic R (MA)
Morrison, W Bruce (OR)
Moschytz, George S
Mott, Harold (AL)
Moxey, Richard T (PA)
Moy, Edward A (NJ)
Muckenhirn, O William (OH)
Mukundan, Rangaswamy (NY)
Mulder, Leonardus T (CA)
Mullendore, Robert A (NE)
Mulligan, James H, Jr (CA)
Murata, Tadao (IL)
Murray, John G (OH)
Musiak, Ronald E (MA)
Narendra, Kumpati S (CT)
Nasar, Syed A (KY)
Nau, Robert H (MO)
Navon, David H (MA)
Neuhoff, Charles J (PA)
Newcomb, Robert W (MD)
Nichols, Lee L (VA)
Nickel, Donald L (PA)
Nicolaides, Emmanuel N (FL)
Nietz, Malcolm L (MN)
Nisbet, John S (PA)
Nishino, Osamu
Norris, Roy H (KS)
North, Forrest H (MS)
Novo, A, Jr (FL)
Novotny, Donald W (WI)
Nutter, Roy S , Jr (WV)
Oglesby, Sabert (AL)
Ohlson, John E (CA)
Okress, Ernest (PA)
Omura, Jimmy K (CA)
Orchard, Henry J (CA)
Ordung, Philip F (CA)
Osborne, Merrill J (SC)

ELECTRICAL ENGINEERING
(Continued)

Osepchuk, John M (MA)
Ostrander, Lee E (NY)
Packer, Lewis C (FL)
Paine, Myron D (KS)
Palmer, James D (NY)
Pantazis, John D, PE (IL)
Pantell, Richard H (CA)
Paris, Demetrius T (GA)
Park, Gerald L (MI)
Parker, Don (CA)
Parker, Karr (NY)
Parker, Norman F (CA)
Parker, Sydney R (CA)
Parks, Thomas W (NY)
Paskusz, Gerhard F (TX)
Patel, Purushottam M (NY)
Pearson, J Boyd, Jr (TX)
Peden, Irene C (WA)
Peebles, Peyton Z, Jr (FL)
Pehrson, David L (CA)
Peikari, Behrouz (TX)
Perini, Jose (NY)
Peters, LeRoy L (CA)
Pettit, Joseph M (GA)
Pettit, Ray H (CA)
Phillips, Wendell E, Jr (NJ)
Pierce, Ralph (MI)
Pinkerton, John E (PA)
Piper, Harvey S (PA)
Plunkett, Joseph C (CA)
Polk, Charles (RI)
Pollack, Louis (MD)
Poole, William S (TX)
Porcello, Leonard J (AZ)
Porter, William A (LA)
Posey, Owen S (AL)
Potvin, Alfred R (IN)
Powell, Herschel L (TN)
Prabhakar, Jagdish C (CA)
Press, Leo C (CA)
Prince, John L (AZ)
Pritzker, Paul E (MA)
Prober, Daniel E (CT)
Proffitt, Charles Y (NC)
Pursley, Michael B (IL)
Qureshi, Abdul Haq (OH)
Rabii, Sohrab (PA)
Rabiner, Lawrence R (NJ)
Rabson, Thomas A (TX)
Raemer, Harold R (MA)
Ragland, John R (TX)
Ramakumar, Ramachandra G (OK)
Rankin, John P (TX)
Rappaport, Stephen S (NY)
Rath, Gerald A (IL)
Reece, C Jeff, Jr (NC)
Reggia, Frank (VA)
Revay, Andrew W Jr (FL)
Reynolds, Donald K (WA)
Rhee, Moon-Jhong (MD)
Ricardi, Leon J (MA)
Richards, Earl F (MO)
Riddick, Edgar K , Jr (AR)
Rigas, Harriett B (MI)
Roa, William J, Jr (MO)
Robinson, Denis M (MA)
Robinson, Donald A , Jr (FL)
Roden, Martin S (CA)
Rodrigue, George P (GA)
Rohrer, Ronald A (VA)
Rosenbaum, Fred J (MO)
Rosenberg, Leon T (WI)
Ross, Hugh C (CA)
Ross, Ian M (NJ)
Roth, J Reece (TN)
Rotman, Walter (MA)
Rouse, John W, Jr (TX)
Roush, Robert W (OK)
Rowe, Joseph E (FL)
Roy, Francis C (LA)

Ruddick, Bernard N (KS)
Ruina, Jack P (MA)
Rumsey, Victor H (CA)
Rusch, Willard VT (CA)
Russell, Frederick A (NJ)
Ruston, Henry (NY)
Saba, Shoichi
Salati, Octavio M (PA)
Saletta, Gerald F (IL)
Salis, Andrew E (TX)
Sandberg, Irwin W (TX)
Sander, Duane E (SD)
Sannuti, Peddapullaiah (NJ)
Sato, Risaburo
Saunders, Robert M (CA)
Savage, Charles F (NM)
Schafer, Ronald W (GA)
Schalliol, Willis L (IN)
Scharf, Louis L (CO)
Schell, Allan C (MA)
Scherer, Harold N, Jr (OH)
Scherer, Richard D (KS)
Schott, F W (CA)
Schroeder, Alfred C (PA)
Schulke, Herbert A , Jr (NJ)
Schultz, Sol E (WA)
Schwartz, Morton D (CA)
Schwartz, Nathan (NY)
Schwartz, Richard F (MI)
Schwartzman, Leon (NY)
Schwarz, Ralph J (NY)
Sechrist, Chalmers F, Jr (IL)
Seely, Samuel (CT)
Seidman, Arthur H (MA)
Sekine, Yasuji
Sen, Amiya K (NY)
Senior, Thomas B A (MI)
Senitzky, Benjamin (NY)
Sensiper, Samuel (CA)
Sforzini, Mario
Shaad, George E (NY)
Shamash, Yacov A (FL)
Shamis, Sidney S (NJ)
Sharpe, Irene W (MI)
Shaw, Melvin P (MI)
Shen, Liang C (TX)
Shenoi, B A (OH)
Shewan, William (IN)
Shockley, Thomas D (TN)
Shohet, J Leon (WI)
Showers, Ralph M (PA)
Siewiorek, Daniel P (PA)
Silver, John E (CA)
Silverman, Bernard (NY)
Simmons, Alan J (MA)
Simon, Marvin K (CA)
Simrall, Harry C F (MS)
Skalnik, John G (CA)
Skilling, Hugh H (CA)
Skolnick, Alfred (VA)
Slivinsky, Charles R (MO)
Smith, Charles E (MS)
Smith, Edward J (NY)
Smith, George S (WA)
Smith, Jack R (FL)
Smith, James G (IL)
Smith, Lester H, Jr (TN)
Smith, Neal A (OH)
Smith, Nelson S, Jr (WV)
Smith, Peter W (NJ)
Snoddy, Jack T (OK)
Snyder, William R (PA)
Soderstrand, Michael A (CA)
Soliman, Ahmed M (FL)
Solomon, James E (CA)
Sommers, Otto W (TX)
Soudack, Avrum C
Specht, Theodore R (PA)
Squillace, Stephen S (MI)
Staats, Gustav W (WI)
Stach, Joseph (PA)
Staggs, Harlon V (KY)
Stanek, Eldon K (MI)
Starr, Eugene C (OR)

Steele, Earl L (KY)
Steier, William H (CA)
Steige, Walter E (AK)
Steinhorst, Gerald L (CA)
Stephenson, Robert E (UT)
Stevens, Frederick (CA)
Stevenson, James R (NY)
Stott, Charles B (AZ)
Stoudinger, Alan R (IN)
Stout, Melville B (MI)
Straiton, Archie W (TX)
Strattan, Robert D (OK)
Streetman, Ben G (IL)
Stremler, Ferrel G (WI)
Strohbehn, John W (NH)
Strum, Robert D (CA)
Stuart, Thomas A (OH)
Stubberud, Allen R (CA)
Su, Kendall L (GA)
Summer, Virgil C (SC)
Sun, Hun H (PA)
Swartwout, Robert E (WV)
Swerdlow, Nathan (PA)
Swift, Calvin T (MA)
Tamir, Theodor (NY)
Tao, William K Y (MO)
Tatum, Finley W (TX)
Taylor, Leonard S (MD)
Tellier, Roger A
Temes, Gabor C (CA)
Tennant, Otto A (IA)
Theodore, George J (MN)
Thomann, Gary C (KS)
Thomas, Roland E (CO)
Thompson, James E (TX)
Thompson, Truet B (AZ)
Thompson, William, Jr (PA)
Thorbjornsen, Arthur R (OH)
Thorn, Donald C (OH)
Thorpe, Robert L (ME)
Thuente, Sylvan T (GA)
Thurston, Marlin O (OH)
Tilles, Abe (CA)
Timascheff, Andrew S
Tippett, G Ross (TX)
Toler, James C (GA)
Tracey, James H (KS)
Tretter, Steven A (MD)
Trick, Timothy N (IL)
Trump, John G (MA)
Tuck, Edward F (CA)
Turin, George L (CA)
Turnbull, Robert J (IL)
Turner, Richard L (WA)
Turner, Warren H (NJ)
Tuteur, Franz B (CT)
Uhl, Robert J (NM)
Ulaby, Fawwaz T (MI)
Van Gulik, Joe L (OR)
Van Valkenburg, Mac E (IL)
VandeLinde, V David (MD)
Vasquez, Numan H (FL)
Veinott, Cyril G (FL)
Vetelino, John F (ME)
Vigilante, Frank S (NJ)
Villalon, Leonardo A (LA)
von Roeschlaub, Frank (NY)
Von Tersch, Lawrence W (MI)
Wachter, Robert V (WA)
Wade, Glen (CA)
Waidelich, Donald L (MO)
Wait, John V (AZ)
Walker, Gene B (OK)
Walker, John R (CO)
Wallis, Clifford M (VT)
Walton, Robert Edward (CO)
Waltz, Gerald D (IN)
Ware, Lawrence A (IA)
Warren, S Reid, Jr (PA)
Waters, James P, Jr (AL)

Watkins, William S (OH)
Watt, Joseph T , Jr (TX)
Wayne, Burton H (NC)
Waynick, Arthur H (PA)
Weaver, Charles Hadley (TN)
Webb, Richard C (CO)
Weber, Howard H, Sr (FL)
Weimer, Frank C (OH)
Weinberg, Louis (NJ)
Welty, Robert (OR)
Wentz, Edward C (PA)
Werts, Robert W (PA)
Wessler, Melvin D (IN)
Westerback, Arne E (CA)
Westgate, Charles R (MD)
White, Gifford E (TX)
Whitehead, Daniel L (PA)
Whitman, Lawrence C (FL)
Willenbrock, F Karl (DC)
Willson, Alan N, Jr (CA)
Wilson, Delano D (NY)
Winchester, Robert L (NY)
Wing, Omar (NY)
Wischmeyer, Carl R (NJ)
Wolff, Edward A (MD)
Wolfram, Ralph E (MO)
Woll, Richard F (NY)
Wong, Eugene (CA)
Wood, Alan B
Wood, Francis P (S J) (WA)
Woodson, Herbert H (TX)
Wundrack, William A (MO)
Wyndrum, Ralph W, Jr (NJ)
Yacoub, Kamal (FL)
Yeargan, Jerry R (AR)
Yeh, Hsi-Han (KY)
Yeh, Kung C (IL)
Yu, Francis T S (PA)
Zahn, Markus (MA)
Zelbv, Leon W (OK)
Ziegler, Geza Charles (CT)
Ziemer, Rodger E (CO)
Zurheide, Charles H (MO)
Zygmont, Anthony (PA)

ELECTROCHEMISTRY

Alkire, Richard C (IL)
Giner, Jose D (MN)
Grun, Charles (NJ)
Jalan, Vinod M (MA)
Johnson, James W (MO)
Jorne, Jacob (NY)
McLean, James D (MI)
Mundel, August B (NY)
Nobe, Ken (CA)

ELECTROMAGNETISM

Barker, Richard C (CT)
Baum, Carl E (NM)
Berrett, Paul O (UT)
Bristol, Thomas W (CA)
Carlile, Robert N (AZ)
Chabries, Douglas M (UT)
Collins, Jeffrey H (TX)
Corum, James F (WV)
Della Torre, Edward (MI)
Dienes, Andrew J (CA)
Engl, Walter L
Felsen, Leopold B (NY)
Heller, Gerald S (RI)
Hill, David A (CO)
Joy, Edward B (GA)
Kahn, Walter K (DC)
King, Ronold W P (MA)
Kritikos, Haralambos N (PA)
Kuehl, Hans H (CA)
Muller, Marcel W (MO)
Peters, Leon, Jr (OH)
Porter, Robert P (CT)

ELECTRO-MECHANICAL ENGINEERING

ELECTRONICS
(Continued)

Johnson, Raymond C (FL)
Johnson, Richard C (GA)
Johnson, Walter C (NJ)
Jones, Howard St C, Jr (DC)
Jones, Trevor O (OH)
Jones, Wesley N (MD)
Joslyn, John A (MA)
Kaisel, Stanley F (CA)
Kalbfell, David C (CA)
Kao, Charles K (VA)
Kazan, Benjamin (CA)
Kelleher, John J (VA)
Kendall, John M
Kenji, Kakizaki
Kennedy, David P (FL)
Ketterer, Frederick D (PA)
Kihn, Harry (NJ)
Kilby, Jack S (TX)
King, Donald D (NY)
Kitsuregawa, Takashi
Kodali, V Prasad
Korman, Nathaniel I (NM)
Kosmahl, Henry G (OH)
Kowel, Stephen T (CA)
Kraut, Seymour (MA)
Kravitz, Lawrence C (MD)
Krenzer, BK (MO)
Kressel, Henry (NY)
Kretz, Anna S (MI)
Kretzmer, Ernest R (NJ)
Kreuzer, Barton (NJ)
Kroemer, Herbert (CA)
Kubitz, William J (IL)
Kundert, Warren R (MA)
Kuno, Hiromu John (CA)
Lafferty, Raymond E (NJ)
Lampert, Murray A (NJ)
Lathrop, Jay W (SC)
Law, Harold B (NJ)
Lebenbaum, Matthew T (NY)
Lee, Bansang W (NJ)
Lee, Reuben (MD)
Lehan, Frank W (CA)
Levine, Saul (MD)
Levy, Ralph (CA)
Lewis, Edwin R (CA)
Li, Sheng S (FL)
Lin, Hung C (MD)
Lin, Pen-Min (IN)
Lord, Harold W (CA)
Loughren, Arthur V (HI)
Lucas, Michael S P (KS)
Lustgarten, Merrill N (MD)
Maas, Robert R (CA)
Macnee, Alan B (MI)
MacRae, Alfred Urquhart (NJ)
Marsocci, Velio A (NY)
Martin, William R (IN)
Matare, Herbert F (CA)
Mathews, Warren E (CA)
Matsumoto, Yuki (HI)
Matthews, Allen R (MD)
Mayer, Robert (PA)
McArthur, Elmer D (FL)
McCurdy, Archie K (MA)
McGroddy, James C (NY)
McIntosh, Robert E, Jr (MA)
McIsaac, Paul R (NY)
McLendon, B Derrell (GA)
McMurray, William (NY)
Means, James A (CA)
Mei, Kenneth K (CA)
Meitzler, Allen H (MI)
Mentzer, John R (PA)
Mesner, Max H (NJ)
Middlebrook, R David (CA)
Mihran, Theodore G (NY)
Millman, Jacob (NY)
Minneman, Milton J (MD)

Minozuma, Fumio
Minter, Jerry Burnett (NJ)
Moe, Robert E (KY)
Moll, John L (CA)
Monteith, Larry K (NC)
Moore, Gordon E (CA)
Moreno, Theodore (CA)
Morgenstern, John C (VA)
Morkoc, Hadis (IL)
Morton, George A (CA)
Moses, Kenneth L (IL)
Moss, Ralph A (WA)
Muckenhirn, O William (OH)
Mueller, Charles W (NJ)
Muller, Richard S (CA)
Muroga, Saburo (IL)
Murphy, Gordon J (IL)
Nahman, Norris S (CO)
Nakagawa, Ryoichi
Narud, Jan A (NM)
Nelson, Richard B (CA)
Newhouse, Russell C (NJ)
Nichols, Donald L (MI)
Nishizawa, Jun-ichi
Norman G Einspruch, Dean (FL)
Norris, Roy H (KS)
North, Harper Q (MD)
Noyce, Robert N (CA)
O'Connor, William W (FL)
Okamura, Sogo
Oldham, William G (CA)
Oliner, Arthur A (NY)
Onoe, Morio
Orr, William H (IN)
Packard, David (CA)
Parker, Don (CA)
Parker, Norman W (IL)
Parks, William W (IL)
Parzen, Benjamin (CA)
Patwardhan, Prabhakar K
Pearson, Gerald L (CA)
Pederson, Donald O (CA)
Peikari, Behrouz (TX)
Pence, Ira W (VA)
Peot, Hans G (OH)
Perini, Jose (NY)
Perper, Lloyd J (AZ)
Pestorius, Thomas D (VA)
Petrou, Nicholas V (PA)
Picard, Dennis J (MA)
Piore, Emanuel R (NY)
Pollack, Herbert W (MA)
Powers, John H (NY)
Pricer, W David (VT)
Prince, John L (AZ)
Prober, Daniel E (CT)
Puckett, Russell E (TX)
Pullen, Keats A (MD)
Purl, O Thomas (CA)
Quinn, Will M (VA)
Rajchman, Jan A (NJ)
Rambo, William R (CA)
Ramsay, John F (NY)
Ravenis, Joseph VJ, II (CA)
Richman, Peter (MA)
Riebman, Leon (PA)
Rohrer, Ronald A (VA)
Rose, Albert (NJ)
Rosenbaum, Fred J (MO)
Rosi, Fred D (VA)
Ross, Gerald F (MA)
Ross, Julius (DC)
Rush, Sidney (NY)
Ryder, John D (FL)
Salama, C Andre T
Salzer, John M (CA)
Samulon, Henry A (CA)
Sapega, August E (CT)
Savic, Michael I (MA)
Scanlan, Sean O
Schenker, Leo (NJ)
Schmidt-Tiedemann, K J
Schmitt, Hans J
Schneider, Sol (NJ)

Schulz, Richard B (TX)
Schutz, Harald (MD)
Schwarz, Steven E (CA)
Sears, Raymond W (NJ)
Seeley, Ralph M (PA)
Sensiper, Samuel (CA)
Shahbender, Rabah (NJ)
Shamis, Sidney S (NJ)
Shaw, Melvin P (MI)
Sheldon, John L (NY)
Shelton, J Paul (VA)
Shepherd, Mark, Jr (TX)
Shrader, William W (MA)
Silva, LeRoy F (IN)
Singh, Amarjit Dr (MD)
Sittig, Erhard K
Skolnik, Merrill I (MD)
Sletten, Carlyle J (MA)
Smith, George E (NJ)
Smith, George F (CA)
Smith, Nelson S, Jr (WV)
Smith, Warren L (PA)
Smolinski, Adam K
Snyder, David A (MD)
Soderstrand, Michael A (CA)
Soliman, Ahmed M (FL)
Solomon, James E (CA)
Somers, Richard M (MD)
Sonnemann, Harry (VA)
Speakman, Edwin A (VA)
Spencer, William J (CA)
Stavis, Gus (NJ)
Steckl, Andrew J (NY)
Stein, Seymour (MA)
Stern, Arthur P (CA)
Stillman, Gregory E (IL)
Storm, Herbert F (NY)
Strull, Gene (MD)
Strutt, Max J O
Stuetzer, Otmar M (NM)
Su, Kendall L (GA)
Suran, Jerome J (CT)
Swenson, George W, Jr (IL)
Takagi, Noboru
Takeda, Yukimatsu
Taub, Jesse J (NY)
Taylor, Byron L (MI)
Teal, Gordon K (TX)
Tedeschi, Frank P (OH)
Teer, Kees
Terman, Lewis M (NY)
Thierfelder, Charles W (PA)
Thomas, Leonard W, Sr (DC)
Thornton, Clarence G (NJ)
Thun, Rudolf E (MA)
Tomiyasu, Kiyo (PA)
Toole James M (PA)
Torgow, Eugene N (CA)
Townsend, Marjorie R (DC)
Trick, Timothy N (IL)
Uenohara, Michiyuki
Uhlir, Arthur, Jr (MA)
Unger, Hans-Georg
Urkowitz, Harry (PA)
Vadasz, Leslie L (CA)
van der Ziel, Aldert (MN)
Varnerin, Lawrence J, Jr (NJ)
Veronda, Carol M (VA)
Viswanathan, Chand R (CA)
Vlad, Ionel-Valentin I
Vogelman, Joseph H (NY)
Waldhauer, Fred D (NJ)
Walker, Gene B (OK)
Wallmark, J Torkel
Wang, Shyh (CA)
Ward, Harold R (MA)
Warren, Dana White (NY)
Wasserman, Reuben (MA)
Weill, Jacky
Weinschel, Bruno O (MD)
Weinstock, Walter W (NJ)
Weisberg, Leonard R (MN)

Weiss, Herbert G (MA)
Weller, Edward F (MI)
Wells, Frank H
Westgate, Charles R (MD)
Wheeler, Harold A (NY)
Wheeler, Robert G (CT)
Whicker, Lawrence R (DC)
White, Marvin H (MD)
White, Robert L (CA)
Whitelock, Leland D (FL)
Whitney, James C (CT)
Wiley, Carl A (CA)
Wiley, Richard G (NY)
Williams, Charles E (WA)
Williamson, Raymond H (VA)
Willson, Alan N, Jr (CA)
Wing, Arthur K, Jr (NH)
Wolfe, Charles M (MO)
Wolff, Hanns H (FL)
Woll, Harry J (MA)
Woodbury, Eric J (CA)
Woodbury, Richard C (UT)
Yanai, Hisayoshi
Yang, Edward S (NY)
Yoshida, Susumu
Zemel, Jay N (PA)
Zevanove, Louis R (CA)
Ziegler, Hans K (NJ)
Zimmermann, Henry J (MA)
Zverev, Anatol Ivan (MD)
Zworykin, Vladimir K (NJ)

ELECTRO-OPTICS

Boge, Walter E (VA)
Byer, Robert L (CA)
Cathey, W Thomas, Jr (CO)
Cheo, Peter K (CT)
DeMaria, Anthony J (CT)
Farhat, Nabil H (PA)
Fateley, William G (KS)
Garmire, Elsa (CA)
Harris, Jay H (CA)
Hirschfelder, Joseph D (WI)
Hunsperger, Robert G (DE)
Jacobs, Ira (NJ)
Lakshminarayan, Mysore R (CA)
Massey, Gail A (CA)
Morse, T F (RI)
Rediker, Robert H (MA)
Smith, Richard G (PA)
Statz, Hermann (MA)
Steffen, Juerg
Stickley, C Martin (MD)
Stitch, Malcolm L (NJ)
Upatnieks, Juris (MI)
Wolfe, William L (AZ)
Yu, Francis T S (PA)
Zimmerman, Jerrold (MA)
Zissis, George J (MI)

ELECTROSTATICS

Boll, Harry J (NJ)
Masuda, Senichi
Moore, Arthur D (MI)

EMISSIONS

Dibelius, Norman R (NY)

ENERGY
CONSERVATION

Aldworth, George A
Aspenson, Richard L (MN)
Bigda, Richard J (OK)
Bishop, William H IV (MD)
Buckingham, Frank E (MO)
Busby, Michael R (TN)
Clark, Kenneth M (MO)
Coad, John D, Jr (MO)

ENERGY CONVERSION

ENERGY CONSERVATION
(Continued)
Dale, John C (MD)
Donnelly, Ralph G (OH)
Eilering, John G (IL)
Elovitz, David M (MA)
Feintuch, Howard M (NJ)
Fejer, Andrew A (IL)
Feledy, Charles F (OH)
Finley, James H, Jr (NY)
Foster, Albert L (TN)
Fox, John H
Freedman, Steven I (IL)
Fuller, Robert H (OH)
Gada, Ram (MN)
Gilley, James R (NE)
Gochenour, Donal L , Jr (WV)
Goepfert, Detlef C (OR)
Goss, William P (MA)
Grinnan, James A (FL)
Haberman, William L (MD)
Hansen, Hugh J (OR)
Hausz, Walter (CA)
Heltman, James W (DC)
Holland, John G (CA)
Holman, Jack P (TX)
Jackins, George A , Jr (AL)
Johnston, Walter E (NC)
Jones, Robert R (MD)
Kalin, Thomas E (MO)
Katzen, Raphael (OH)
Kendall, Percy Raymond (AZ)
Kephart, John T, Jr (DE)
Kerr, John R (CA)
Kiesling, Ernst W (TX)
Kimball, William R (CO)
Kreider, Jan F (CO)
Laube, Herbert L (NY)
LeBlanc, William J (LA)
Leon, Harry I (GA)
Newell, John R (FL)
Nichols, Edward J (VA)
Ornes, Edward D (WI)
Perron, Gilles
Perry, Lawrence E , Jr (VA)
Robinson, Donald A , Jr (FL)
Roller, Warren L (OH)
Ross, Donald K (MO)
Rowan, Charles M, Jr (FL)
Salinsky, John L (NY)
Sanborn, David M (GA)
Shelnutt, J William, III (NC)
Shreve, Gerry D (MI)
Socolow, Robert H (NJ)
Spencer, George L (CA)
Spielvogel, Lawrence G (PA)
Steiner, Henry M (DC)
Stelle, William W (NY)
Straub, Harold E (TX)
Tamblyn, Robert T
Thuente, Sylvan T (GA)
Thumann, Albert (GA)
Trandel, Richard S (VA)
Turley, John C (CA)
Vinokur, Herman R (PA)
Walker, Gary Wayne (CA)
Walker, Richard O, Jr (IN)
Wallis, William E (TX)
Weinberger, Jack (PA)
Welbourn, John A (WV)
Wiel, Stephen (NV)
Williams, Milton A (TX)
Wilson, William B (TN)

ENERGY CONVERSION
Addae, Andrews K (DC)
Adomat, Emil Alfred (FL)
Agarwal, Paul D (MI)

Agosta, Vito (NY)
Anderson, J Hilbert (PA)
Arbogast, Ronald G (MI)
Arnas, Ozer A (CA)
Artley, John L (NC)
Aspenson, Richard L (MN)
Atchley, Bill L (SC)
Ault, G Mervin (OH)
Bailey, Eugene C (IN)
Baker, Bernard S (CT)
Baron, S (NY)
Beck, Steven R (TX)
Bement, Arden L, Jr (OH)
Bennion, Douglas N (UT)
Bhada, Ron K (OH)
Biringer, Paul P
Bissell, Roger R (ID)
Bloom, Martin H (NY)
Boice, Calvin W (WA)
Boulet, Lionel
Breeding, Lawrence E (AK)
Brehm, Richard L (CA)
Brown, Harry L (NJ)
Brown, Louis J (DE)
Bruggeman, Warren H (NY)
Budny, Bernard R (WI)
Burton, Robert S, III (CO)
Burton, William J (VA)
Cadoff, Irving B (NY)
Carlson, Walter O (GA)
Carter, Robert L (MO)
Chalker, Ralph G (CA)
Chang, George C (OH)
Chapman, Lloyd E (MA)
Cheng, Shang I (NY)
Chiang, Shiao-Hung (PA)
Chu, Ju Chin (CA)
Chu, Ting L (TX)
Clark, Arthur J, Jr (NM)
Clark, Ezekail L (MD)
Clark, Stanley J (KS)
Cobb, James T, Jr (PA)
Connelly, John R (PA)
Conta, Lewis D (AZ)
Corey, Richard C (VA)
Corry, Andrew F (MA)
Crain, Richard W, Jr (WA)
Creagan, Robert J (PA)
Crynes, Billy L (OK)
Dale, John C (MD)
Dart, Jack C (MD)
Denno, Khalil I (NJ)
Doolittle, Jesse S (NC)
Dorrance, William H (MI)
Dowell, Dr Douglas (CA)
Duffy, Robert E (NY)
Dutta, Subhash (NJ)
Eckert, Ernst R G (MN)
Eggers, Alfred J , Jr (CA)
Eisenberg, Lawrence (PA)
Elliott, Martin A (TX)
Endahl, Lowell J (DC)
Engelberger, John E (PA)
Eschenbrenner, Gunther Paul (TX)
Eskinazi, Salamon (NY)
Esselman, Walter H (CA)
Eustis, Robert H (CA)
Fan, Liang-Shih (OH)
Feledy, Charles F (OH)
Fernandes, John H (RI)
Fischer, James M (MO)
Fonash, Stephen J (PA)
Fosholt, Sanford K (IA)
Frost, George C (NY)
Gaggioli, Richard A (DC)
Gidaspow, Dimitri (IL)
Giner, Jose D (MN)
Glenn, Roland D (NY)
Goldberg, David C (PA)
Goldstein, Kenneth M (WY)
Goodrum, John W (GA)
Goodson, Raymond Eugene (IN)
Grethlein, Hans E (NH)

Griffith, P LeRoy (NJ)
Gross, William A (NM)
Grun, Charles (NJ)
Guthrie, Hugh D (CA)
Halacsy, Andrew A (NV)
Hamilton, DeWitt C , Jr (LA)
Hand, John W (CO)
Hapeman, Martin Jay (PA)
Harman, Charles M (NC)
Harris, William B (TX)
Hasan, A Rashid (ND)
Hausz, Walter (CA)
Hershey, Robert L (DC)
Hesse, M Harry (NY)
Hester, J Charles (TX)
Hill, Richard F (CT)
Hoffman, Myron (CA)
Howard, Robert E (MO)
Huang, C J (TX)
Hurley, George F (NM)
Incropera, Frank P (IN)
Isenberg, Lionel (CA)
Jackins, George A , Jr (AL)
Jackson, William D (MD)
Jacobs, Harold R (PA)
Jahn, Robert G (NJ)
Jenkins, Larry L (NE)
Jenkins, Peter E (NE)
Jennings, Stephen L (KS)
Jimeson, Robert M (VA)
Johnson, James R (WI)
Johnson, Loering M (CT)
Johnston, James P (CA)
Jonke, Albert A (IL)
Karady, George Gy (NJ)
Kaya, Azmi (OH)
Kear, Edward B (NY)
Kendall, Percy Raymond (AZ)
Kiely, John R (CA)
Kim, Rhyn H (NC)
Kindl, Fred H (NY)
Klueter, Herschel H (MD)
Kruger, Charles H , Jr (CA)
Kurzynske, Frank (TN)
Lahey, Robert W (NJ)
Landis, James N (CA)
Lanz, Robert W (WI)
Lathrop, Jay W (SC)
Law, John (ID)
Lee, Harry R (AK)
Levine, Neil M (CA)
Li, Sheng S (FL)
Lindsey, Larry M (TX)
Lo, Robert N (IL)
Lorenzini, Robert A (MA)
Lynch, Edward P (IL)
Mahoney, Patrick F (NY)
Mallard, Stephen A (NJ)
Manvi, Ram (CA)
Mater, Milton H (OR)
McCormick, Robert H (PA)
McFarlane, Harold F (ID)
McHugh, Edward (NY)
Merklin, Joseph F (KS)
Meyerhoff, Robert W (NY)
Middlebrook, R David (CA)
Moeckel, Wolfgang E (OH)
Moran, H Dana (CO)
Murray, Raphael H (PA)
Nasar, Syed A (KY)
Neal, Gordon W (CA)
Nixon, Alan C (CA)
Noodleman, Samuel (AZ)
O'Connor, James J (NJ)
Okamoto, Hideo
Oliver, Paul E (OR)
O'Neil, David A (CT)
Ott, Dudley E (CA)
Palmer, Nigel I (NY)
Pell, Jerry (MD)
Penner, Stanford S (CA)
Perkins, Henry C (AZ)
Pitrolo, Augustine A (WV)
Pope, Michael (NY)

Porter, James H (MA)
Primrose, Russell A (OH)
Ragone, David V (OH)
Reintjes, Harold (NJ)
Reynolds, William C (CA)
Richards, John C (MD)
Roach, Kenneth E (FL)
Roberts, A Sidney, Jr (VA)
Roberts, Irving (VA)
Rockwell, Theodore (MD)
Rogers, Willard L (AZ)
Rotty, Ralph M (LA)
Roush, Maurice D (WA)
Rowley, Louis N, Jr (NY)
Rutherfoord, J Penn (VA)
Ryan, Robert S (OH)
Schiff, Daniel (MA)
Scott, Donald S
Seeley, Ralph M (PA)
Seglin, Leonard (NY)
Self, Sidney A (CA)
Sen, Amiya K (NY)
Shaffer, Richard F (CT)
Shea, Richard F (FL)
Sheets, Herman E (CT)
Sims, Chester T (NY)
Smith, Donald S (CA)
Somerville, Mason H (TX)
Spacil, H Stephen
Spiegler, Kurt S (CA)
Staniszewski, Bogumil E
Stevenson, William D, Jr (NC)
Stokes, Charles A (FL)
Suciu, Spiridon N (FL)
Szewalski, Robert
Talaat, Mostafa E (MD)
Thorsen, Richard S (NY)
Trutt, Frederick C (KY)
Van Vorst, William D (CA)
VanDriesen, Roger P (NJ)
Vertes, Michael A (CT)
Veziroglu, T Nejat (FL)
Walker, M Lucius, Jr (DC)
Walsh, Myles A (MA)
Walsh, Thomas J (OH)
Way, Stewart (MT)
Weaver, Lynn E (FL)
Wilbur, Paul J (CO)
Will, Herbert (NJ)
Willyoung, David M (NY)
Wilson, William B (TN)
Wright, Edward S (IL)
Zdenek, Edna Prizgint (NJ)
Zimmerman, Richard H (OH)

ENVIRONMENTAL ENGINEERING
Adrian, Donald D (LA)
Ahlert, Robert C (NJ)
Alberts, Richard D (FL)
Aldworth, George A
Allen, Merton (NY)
Alley, E Roberts (TN)
Alter, Amos J (AK)
Altwicker, Elmar R (NY)
Anderson, William C (NY)
Andrews, John F (TX)
Annicelli, John P (NY)
Ardis, Colby V (OH)
Arthur, Robert M (WI)
Ashby, C Edward, Jr (PA)
Aufiero, Frederick G (MA)
Aulenbach, Donald B (NY)
Austin, John H (SC)
Aynsley, Eric (IL)
Azer, Naim Z (KS)
Bainbridge, Danny L (WV)
Barnhart, Edwin L (TX)
Barsom, George M (FL)
Bates, Robert L, Jr (FL)
Baum, Richard T (NY)
Belfort, Georges (NY)

ENVIRONMENTAL IMPACT

ENVIRONMENTAL ENGINEERING
(Continued)

Roesner, Larry A (FL)
Rogers, Gifford E (NE)
Rogers, Lewis H (FL)
Rohlich, Gerard A (TX)
Romaguera, Mariano A (PR)
Ross, Edward N (WA)
Rosselli, Charles A (MA)
Roth, John A (TN)
Rowan, Charles M, Jr (FL)
Roy, Guy W (CA)
Ruf, John A (MO)
Russell, David O (ME)
Rust, Thomas D (VA)
Ryder, Robert A (CA)
Saarinen, Arthur W , Jr (FL)
Sacks, Newton N (IL)
Salem, Eli (NJ)
Sanks, Robert L (MT)
Sarriera, Rafael E
Saunders, F Michael (GA)
Schaper, Laurence T (MO)
Schaumburg, Frank D (OR)
Schermerhorn, R Stephen (CO)
Scherr, Richard C (OH)
Schiff, Daniel (MA)
Schmidt, Charles M (OH)
Schwartz, Henry G , Jr (MO)
Scottron, Victor E (CT)
Sculley, Jay R (DC)
Shaffer, Richard F (CT)
Shah, Kanti L (OH)
Shaw, George B (OH)
Shell, Gerald L (TN)
Shen, Thomas T (NY)
Sheppard, Stanton V (OH)
Sherer, Keith R (OR)
Shrivastava, S Ram (NY)
Siau, John F (NY)
Sieger, Ronald B (IN)
Silver, Francis, 5th (DC)
Simpson, George D (OH)
Singhal, Ashok K (MI)
Sinnott, Walter B (NY)
Skelly, Michael J (NY)
Small, Mitchell J (MI)
Smith, David B (MS)
Smith, John W (TN)
Smith, Rodney W (VA)
Snoeyink, Vernon L (IL)
Snow, Phillip D (NY)
Snowden, Wesley D (WA)
Sohr, Robert T (IL)
Somerville, John E (ID)
Sorber, Charles A (PA)
Speece, Prof R E (PA)
Speitel, Gerald E (NJ)
Sproul, Otis J (NH)
Stack, Vernon T (PA)
Standley, David (MA)
Steffen, Alfred J (IN)
Steimle, Stephen E (LA)
Stephan, David G (OH)
Stephenson, Junius W (NJ)
Sternlicht, Beno (NY)
Stormoe, Donovan K (MN)
Streebin, Leale E (OK)
Streight, H R Lyle
Strickland, James M, Jr (VA)
Sussman, Victor H (MI)
Svore, Jerome, H (TX)
Symons, George E (NY)
Taflin, Charles O (MN)
Taiganides, E Paul (OH)
Takagi, Joji (NY)
Tatman, D Russell (DE)
Tenney, Mark W (IN)
Thackston, Edward L (TN)

Thoem, Robert L (GA)
Thomann, Robert V (NY)
Thomson, Quentin R (AZ)
Thumann, Albert (GA)
Tilsworth, Timothy (AK)
Tollison, Gordon T (MS)
Tomlinson, Henry D (FL)
Toro, Richard F (NJ)
Touhill, Charles J, Jr (PA)
Tozer, George K (MA)
Triggs, James Frederick (PA)
Tucker, Irwin W (KY)
Tummala, Lal R (MI)
Ulaby, Fawwaz T (MI)
van Straten, Gerrit S (OH)
Veenstra, H Robert (IA)
Velzy, Charles O (NY)
Vesilind, P Aarne (NC)
Viscomi, B Vincent (PA)
Vogel, Chester T (NY)
Vogt, John W (FL)
Volkin, Richard A (PA)
Voytko, John D (PA)
Waitkus, Joseph (NY)
Waller, Robert (MD)
Walters, James V (AL)
Wang, Chiu-sen (NY)
Wanielista, Martin P (FL)
Washington, Donald R (MO)
Watson, Kenneth S (IL)
Weaver, Robert E C (LA)
Wegman, Leonard S (NY)
Weinstein, Norman J (NJ)
Weiss, Martin (MA)
Weller, Lloyd W (MO)
Wenck, Norman C (MN)
Wessler, Melvin D (IN)
Weston, Roy F (PA)
Wheler, Arthur Gordon (NY)
White, L Forrest (TN)
Wier, Thomas P Jr (TX)
Willey, Benjamin F (IL)
Williams, Theodore C (NM)
Williamson, Elmer H, Jr (SC)
Willoughby, Edward (MI)
Wilson, Linda H (NY)
Woods, Raymond C (TN)
Worley, Frank L, Jr (TX)
Wright, Robert R (FL)
Wyszkowski, Paul E (NJ)
Young, Reginald H F (HI)
Yousef, Yousef A (FL)
Yuen, George A L (HI)
Zegel, William C (FL)
Ziegler, Edward N (NY)

ENVIRONMENTAL IMPACT

Alberts, Richard D (FL)
Angiola, Alfred J (NY)
Bathen, Karl H (HI)
Bjorklund, David S Jr (CT)
Brinsko, George A (AZ)
Hansen, Ethlyn Ann (CA)
Hoover, Mark D (NM)
Kiersch, George A (AZ)
Porter, Alan L (GA)
Rosselli, Charles A (MA)
Socolow, Robert H (NJ)
Talley, Wilson K (CA)
Tsai, Steve Yuh-Hua (IL)
Woglom, James R (MA)

ENZYMES

Scott, Don (IL)

EXPLOSIVES

Grant, Charles H (MN)

Hale, Nathan C (VA)
Pearson, John (CA)
Rollins, Ronald R (WV)
Suter, Vane E (CA)

FAILURE ANALYSIS

Bresler, Boris (CA)
Brinkmann, Joseph B (CT)
Callaway, Samuel R (IL)
Clark, T Henry (TN)
Collins, Jack A (OH)
Felbeck, David K (MI)
George, John A (IL)
Gurfinkel, German R (IL)
James, Arthur M (OR)
Kinser, Donald L (TN)
Larsen, William L (IA)
McCarthy, Roger L (CA)
Murtha-Smith, Erling (CT)
Popelar, Carl H (OH)
Raju, Pal P (MA)
Reinhardt, Gustav (FL)
Spanovich, Milan (PA)
Sweet, John W (WA)
Wulpi, Donald J (IN)

FATIGUE

Bowles, C Quinton (MO)
Findley, William N (RI)
Hoeppner, David W (UT)
James, Lee A (PA)
Jolles, Mitchell I (MO)
Liu, Hao-Wen (NY)
Socie, Darrell F (IL)

FINANCE

Boehringer, Ludwig C , Jr (NY)
Burton, Alan K (CA)
Coonrod, Carl M (KS)
Crabb, William A (MO)
Eyerman, Thomas J (IL)
Goldstein, Edward (NJ)
Korwek, Alexander D (CT)
Mann, Sheldon S (OH)
O'Neill, John J (NC)
Reem, Herbert F (VA)

FIRE

Abderhalden, Ross
Allen, James D (NJ)
Drysdale, David D
Grossmann, Elihu D (PA)
Klinker, Richard L (MD)
Lyons, John W (MD)
Melott, Ronald K (OR)
Smith, Edwin E (OH)

FIRE PROTECTION

Abderhalden, Ross
Allen, James D (NJ)
Almgren, Louis E (CA)
Anderson, Charles R (NJ)
Andrews, Raynal W, Jr (PA)
Berg, Kenneth E (CA)
Boggess, Jerry R (CA)
Bond, Horatio L (MA)
Boulais, Maurice R (RI)
Brown, Martin M (NY)
Bugbee, Percy (MA)
Carlisle, Thomas C , III (NC)
Catanzano, Samuel (PA)
Cobb, Allen L (NY)
Collins, Ralph E (VA)
Crandall, John L (IL)
Dockendorff, Ralph L (TX)
Drysdale, David D
Duke, Robert E (IL)

Feldman, Rubin (MO)
Froyen, Hugo G
Gibble, Kenneth (CT)
Grill, Raymond A (CA)
Hanbury, William L (VA)
Hanna, Martin Jay, III, (MD)
Hartley, Boyd A (IL)
Herbstman, Donald (FL)
Hidzick, George M (IL)
Hill, James E (CA)
Jablonsky, John L (NY)
Jerger, Edward W (IN)
Johnson, Dean E (IL)
Johnson, Joseph E (NJ)
Kampmeyer, John E, (PA)
Keigher, Donald J (NM)
Klinker, Richard L (MD)
Koo, Benjamin (OH)
Kornsand, Norman J (CA)
Lamb, Jamie Parker (TX)
Larson, Vernon M (CA)
Lataille, Jane I (CT)
Lauridsen, David H (NJ)
Levine, Robert S (MD)
Lloyd, John R (IN)
Lucht, David A (MA)
Luhnow, Raymond B , Jr (KS)
Maatman, Gerald L (IL)
McClarran, William H (RI)
McDavid, Frederick R, Jr (VA)
Melott, Ronald K (OR)
Middendorf, William H (KY)
Morley, Richard J (NJ)
Moses, Kenneth L (IL)
Mowrer, David S (TN)
Neal, Donald K (AZ)
Nelson, Harold E (DC)
Nolan, James W (IL)
Osman, Richard R (IL)
Pantazis, John D, PE (IL)
Parks, Lyman L (NJ)
Perry, Robert W (KY)
Rodriquez, Pepe (WI)
Schirmer, Chester W (IL)
Schoneman, John A (WI)
Schwab, Richard F (NJ)
Searl, Edwin N (TN)
Shockley, Thomas D (TN)
Smith, Paul D (CA)
Soedel, Werner (IN)
Solomon, Richard H (IL)
Taback, Harry (NY)
Tabisz, Edward F
Thomson, C Ross
Thullen, Edward J (WA)
Troy, John J (CA)
Tucker, Randolph W (TX)
Vanden Boogart, Rick C (WI)
Warmann, Robert A (MO)
Welker, J Reed (OK)
Williamson, Robert B (CA)
Wilson, Rexford (MA)
Wingate, Roger H (NH)

FLOW CONTROL

Kates, Willard A (IL)

FLUID MECHANICS & HYDRODYNAMICS

Abramson, H Norman (TX)
Acrivos, Andreas (CA)
Addy, Alva L (IL)
Adrian, Ronald J (IL)
Ahrens, Frederick W (MO)
Allen, Harvey G (PA)
Almenas, Kazys K (MD)
Anderson, Harold H
Arndt, Roger E A (MN)

FLUID MECHANICS & HYDRODYNAMICS
(Continued)

Astill, Kenneth N (MA)
Avula, Xavier J R (MO)
Bajura, Richard A (WV)
Banerjee, Sanjoy (CA)
Bankoff, S George (IL)
Barker, Dee H (UT)
Barnett, Donald O (AL)
Barthel, Harold O (IL)
Baumann, Hans D (NH)
Beer, Janos M (MA)
Berger, Bruce S (MD)
Berger, Stanley A (CA)
Berman, Abraham S (MN)
Bethke, Donald G (WI)
Bharathan, Desikan (CO)
Bird, R Byron (WI)
Bloom, Martin H (NY)
Bogue, Donald C (TN)
Bottaccini, Manfred R (AZ)
Bozatli, Ali N (AR)
Brodkey, Robert S (OH)
Brooks, Norman H (CA)
Brown, Frederick R (MS)
Brown, Garry L (CA)
Bruch, John C , Jr (CA)
Burggraf, Odus R (OH)
Campbell, George S (CT)
Carmi, S (PA)
Carrier, George F (MA)
Castro, Walter E (SC)
Cermak, Jack E (CO)
Chakrabarti, Subrata K (IL)
Chan, Shih H (WI)
Chandra, Suresh (NC)
Charles, Michael E
Chen, Chuan F (AZ)
Chen, Rong-Yaw (NJ)
Cheng, Alexander H D (DE)
Chow, Wen L (IL)
Chuang, Hsing (KY)
Chung, Benjamin T (OH)
Chung, Jin S (CO)
Chwang, Allen T (IA)
Clark, Alfred, Jr (NY)
Clarke, Joseph H (RI)
Clauser, Francis H (CA)
Cohen, Ira M (PA)
Cole, Julian D (NY)
Collier, Samuel L (TX)
Comparin, Robert A (VA)
Conly, John F (CA)
Coogan, Charles H, Jr (CT)
Cotton, Kenneth C (NY)
Crane, Roger A (FL)
Cremers, Clifford J (KY)
Cresci, Robert J (NY)
Culick, Fred E C (CA)
Dalton, Charles (TX)
Darby, Ronald (TX)
Davis, E James (WI)
Davis, Philip K (IL)
de Nevers, Noel H (UT)
Deissler, Robert G (OH)
Dobbins, Richard A (RI)
Dodge, Franklin T (TX)
Donaldson, Coleman (VA)
Dowdell, Roger B (RI)
Dukler, Abraham E (TX)
Dussourd, Jules L (NJ)
Eichhorn, Roger (TX)
Eisenberg, Phillip (MD)
Ellsworth, William M (MD)
Emanuel, George (OK)
Epstein, Norman
Eskinazi, Salamon (NY)
Everett, Wilhelm S (CA)
Fejer, Andrew A (IL)
Ferziger, Joel H (CA)
Flammer, Gordon H (UT)
Fleischer, Henry (MI)
Florio, Pasquale J, Jr (NJ)
Forney, Larry J (GA)

Forstall, Walton (PA)
Foss, John F (MI)
Fowler, Jackson E (NY)
Fox, John A (MS)
Fox, Robert W (IN)
Franke, Milton E (OH)
Fried, Erwin (NY)
Giddens, Don P (GA)
Gogarty, W Barney (CO)
Goldschmidt, Victor W (IN)
Goodman, Theodore R (NY)
Gray, Robin B (GA)
Greene, Howard L (OH)
Greif, Ralph (CA)
Griffith, Wayland C (NC)
Gross, Robert A (NY)
Haentjens, Walter D (PA)
Hammack, Charles J (TX)
Hammitt, Frederick G (MI)
Hanks, Richard W (UT)
Hanratty, Thomas J (IL)
Hansen, Arthur G (IN)
Harrawood, Paul (TN)
Hefner, Jerry N (VA)
Henderson, Robert E (PA)
Henry, Harold R (AL)
Hickling, Robert (MI)
Hickman, Roy S (CA)
Hill, William W (IN)
Hoffman, Joe D (IN)
Holl, J William (PA)
Holley, Edward R (TX)
Horlock, John
Horton, Thomas E , Jr (MS)
Hoyt, Jack W (CA)
Hsu, En Y (CA)
Hubbard, Davis W (MI)
Hung, Tin-Kan (PA)
Hussain, A K M F (TX)
Hussain, Nihad A (CA)
Inger, George Roe (CO)
Iotti, Robert C (NY)
Irvine, Thomas F, Jr (NY)
Isaacson, LaVar K (UT)
Jeppson, Roland W (UT)
Johnston, James P (CA)
Jones, Owen C , Jr (NY)
Jordan, Bernard C, Jr (MS)
Kao, Timothy W (DC)
Kapfer, William H (FL)
Kates, Willard A (IL)
Keith, Theo G , Jr (OH)
Kelleher, Matthew D (CA)
Keller, Joseph B (CA)
Kelly, Robert E (CA)
Kennedy, John F (IA)
Kentzer, Czeslaw P (IN)
Kersten, Robert D (FL)
Kessler, David P (IN)
Kline, Stephen J (CA)
Knight, Doyle D (NJ)
Knoebel, David H (LA)
Korpela, Seppo A (OH)
Kovats, Andre (NJ)
Kowalonek, John M (CT)
Krommenhoek, Daniel J, Jr (NY)
Kuhlman, John M (WV)
Kurzweg, Ulrich H (FL)
Lakshminarayana, Budugur (PA)
Landis, Fred (WI)
Landsberg, Helmut E (MD)
Landweber, Louis (IA)
Larock, Bruce E (CA)
Launder, Brian E (CA)
Leadon, Bernard M, Jr (FL)
Leal, L Gary (CA)
Lee, Chang-Sup
Lee, Charles A (TN)
Lee, Richard S L (NY)
Lee, William J (TX)
Leibovich, Sidney (NY)
Lessen, Martin (NY)
Levy, Salomon (CA)

Libby, Paul A (CA)
Lilley, David G (OK)
Lindauer, George C (KY)
List, E John (CA)
Liu, Henry (MO)
Liu, Joseph T C (RI)
Liu, Philip L F (NY)
Loehrke, Richard I (CO)
Marks, Colin H (MD)
Marshall, Francis J (IN)
Martin, C Samuel (GA)
Maxworthy, Tony (CA)
McNown, John S (KS)
McPherson, John C (NJ)
Meecham, Wm C (CA)
Meroney, Robert N (CO)
Michaelides, Efstathios E (DE)
Miller, Joseph S (KS)
Minch, Richard H M (WI)
Moody, Arthur (PA)
Moody, Frederick J (CA)
Morkovin, Mark V (IL)
Morse, T F (RI)
Moulton, Ralph W (WA)
Mueller, Thomas J (IN)
Mujumdar, Arun S
Naudascher, Eduard
Nelkin, Mark S (NY)
Nerem, Robert M (GA)
O'Brien, Edward E (NY)
Ormsbee, Allen I (IL)
Ostrach, Simon (OH)
Pai, Shih I (MD)
Paily, Paily (Poothrikka) P (GA)
Parkin, Blaine R (PA)
Paul, James C (MI)
Peskin, Richard L (NJ)
Peterson, Victor L (CA)
Phillips, Winfred M (IN)
Pierucci, Mauro (CA)
Platzer, Max F (CA)
Plesset, Milton S (CA)
Pletcher, Richard H (IA)
Plotkin, Allen (CA)
Probstein, Ronald F (MA)
Puleo, Peter A (MO)
Purnell, William B (TX)
Ramberg, Steven E (DC)
Raney, Donald C (AL)
Rao, Peter B (OH)
Rawlins, Charles B (NY)
Reed, Joseph R (PA)
Reshotko, Eli (OH)
Reynolds, William C (CA)
Reznik, Alan A (PA)
Richardson, Joseph G (TX)
Roberts, Philip J W (GA)
Robertson, James M (CO)
Rogers, Franklyn C (FL)
Rolf H Sabersky (CA)
Ross, Stephen M (CT)
Rudavsky, Alexander B (CA)
Rudinger, George (NY)
Saffman, Philip G (CA)
Sallet, Dirse W (MD)
Schetz, Joseph A (VA)
Schreier, Stefan (CT)
Schwarz, William H (MD)
Scriven, L E (MN)
Senoo, Yasutoshi
Serovy, George K (IA)
Sforza, Pasquale M (NY)
Shapiro, Ascher H (MA)
Silberman, Edward (MN)
Simcox, Craig D (WA)
Singh, Rameshwar (CA)
Sleicher, Charles A (WA)
Smith, Leroy H (OH)
Smith, Raymond V (KS)
Sogin, Harold H (LA)
Song, Charles C S (MN)
Sorrell, Furman Yates (NC)
Sparrow, Ephraim M (MN)

Spreiter, John R (CA)
Squire, William (WV)
Stanitz, John D (OH)
Street, Robert L (CA)
Strehlow, Roger A (IL)
Strong, Benjamin R , Jr (CA)
Subramanian, R Shankar (NY)
Swim, William B (TN)
Szeri, Andras Z (PA)
Tachmindji, Alexander J (MD)
Thomas, David G (TN)
Thomas, John H (NY)
Thompson, Howard Doyle (IN)
Thompson, Philip A (NY)
Tsai, Steve Yuh-Hua (IL)
Tullis, James P (UT)
Ulbrecht, Jaromir J (MD)
Van Dyke, Milton D (CA)
Vanyo, James P (CA)
Viets, Hermann (RI)
Von Winkle, William A (CT)
Wahren, Douglas (WI)
Walker, John S (IL)
Walker, William F (TX)
Wallis, Graham B (NH)
Wang, Shu-Yi (MS)
Watanabe, Ichiro
Watkins, Charles B (NJ)
Watters, Gary Z (CA)
Whitaker, Stephen (CA)
Wiberg, Donald Victor (WI)
Wicks, Moye, III (TX)
Wiggins, Edwin G (TX)
Williams, Harry E (CA)
Williams, John A (HI)
Wink, Ralph C (ME)
Wu, James C (GA)
Yang, Tah-teh (SC)
Yen, Ben Chie (IL)
Youngdahl, Carl K (IL)
Zakin, Jacques L (OH)
Zebib, Abdelfattah M G (NJ)
Zumwalt, Glen W (KS)

FOOD ENGINEERING

Chung, D S (KS)
Clayton, Joe T (MA)
Harris, Ronald D (IL)
Herum, Floyd L (OH)
Lund, Daryl B (WI)
Mahoney, William P (IL)
McCullough, Charles A (CA)
Murphy, Vincent G (CO)
Olowu, Olayeni
Singh, R Paul (CA)
Whitaker, Thomas B (NC)
Zachariah, Gerald L (FL)

FORENSIC

Abrahamson, Royal T (TX)
Adelberg, Kenneth (VA)
Anderson, Jack W (MN)
Brooks, Frederick M (LA)
Coon, Arnold W (UT)
Cox, William A, Jr (VA)
Dennison, Robert A , Jr (NY)
Gallin, Herbert (NY)
Hohns, H Murray (HI)
Lanz, Robert W (WI)
Merrill, William H , Jr (NY)
Miele, Joel A Sr (NY)
Mullendore, Robert A (NE)
Nawab, Ahmad B (FL)
Peterson, Donn N (MN)

GEOTECHNICAL ENGINEERING, ROCK & SOIL MECHANICS
(Continued)

Johansen, Nils I (AK)
Johnson, Ted D (CO)
Jones, Ronald A (TN)
Jones, Walter V (ID)
Jones, William F (CA)
Judd, William R (IN)
Jumikis, Alfreds R (NJ)
Kane, Harrison (IA)
Keller, George V (CO)
Kennedy, Clyde M (GA)
Keshian, Berg, Jr (CO)
Kevorkian, Hugo (CA)
Kitlinski, Felix T (PA)
Klehn, Henry, Jr (CA)
Klein, Joseph P, III (GA)
Kleiner, David E (IL)
Kleinfelder, James H (CA)
Knight, F James (PA)
Ko, Hon-Yim (CO)
Kraft, Leland M, Jr (TX)
Krizek, Raymond J (IL)
Kruse, Cameron G (MN)
Kulhawy, Fred H (NY)
Ladd, Charles C (MA)
Laguros, Joakim G (OK)
Lake, Thomas D (CA)
Lamont, Joseph, Jr (WA)
LaRochelle, Pierre L
Lawson, Robert T (CA)
Lee, Harry R (AK)
Leonard, Roy J (KS)
Leonards, Gerald A (IN)
Lieberknecht, Don W (NE)
Liu, David C (IL)
Loigman, Harold (PA)
Long, Richard P (CT)
Lum, Walter B (HI)
Lysmer, John (CA)
MacDonald, Donald H
Makdisi, Faiz I
Marr, W Allen (MA)
Marsal, Raul J
Marshall, A Frank (TX)
Mayer, Armand
McBride, Robert R (TX)
McClelland, Bramlette (TX)
McGillivray, Ross T (FL)
McManis, Kenneth L (LA)
McMaster, Howard M (KS)
McRae, John L (MS)
McWee, James M (CA)
Mehta, Prakash C (IN)
Mikochik, Stephen T (NY)
Miller, John H, II (GA)
Millet, Richard A (CA)
Mishu, Louis P (TN)
Mitchell, James K (CA)
Morgan, Melvin W (ME)
Morgenstern, Norbert R
Mosley, Ernest T (NJ)
Munson, Darrell E (NM)
Murdock, Larry T (UT)
Narain, Jagdish
Nethero, Merle F (OH)
Neyer, Jerome C (MI)
Nurse, Edward A (MT)
Olitt, Arnold (NV)
Olson, Roy E (TX)
O'Neill, Michael W (TX)
O'Rourke, Thomas D (NY)
Ortiz, Carlos A (PR)
Otto, Thomas Herbert (NJ)
Paduana, Joseph A (CA)
Panek, Louis A (CO)
Paratore, William G (CA)
Parcher, James V (OK)
Parker, Jack T (GA)
Parsons, James D (NC)
Peck, Ralph B (NM)
Perez, Jean-Yves (CO)
Pernichele, Albert D (UT)

Perry, Edward B (MS)
Phillips, Monte L (ND)
Pickett, Rayford M (MO)
Pierce, Francis C (RI)
Pyke, Robert (CA)
Querio, Charles W (TX)
Ralston, David C (DC)
Reitz, Henry M (MO)
Reti, G Andrew (CA)
Richardson, Gregory N (NC)
Robinson, Lee (ID)
Romig, Phillip R (CO)
Rudnicki, John W (IL)
Russell, James E (TX)
Rutledge, Philip C (PA)
Salisbury, Marvin H (IN)
Samuels, Reuben (NJ)
Sandhu, Ranbir S (OH)
Sangrey, Dwight A (PA)
Sayre, Robert D (VA)
Schaub, James H (FL)
Schmertmann, John H (FL)
Schnabel, James J (MD)
Schoustra, Jack J (CA)
Schroeder, W L (OR)
Scott, Ronald F (CA)
Seed, H Bolton (CA)
Selig, Ernest T (MA)
Shaffer, Harold S (IL)
Shannon, William L (WA)
Shen, Chih-kang (CA)
Shields, Donald H
Sibley, Earl A (WA)
Sikarskie, David L (MI)
Silva, Armand J (RI)
Silver, Marshall L (IL)
Smith, A Leonard (CO)
Smith, Isaac D, Jr (SC)
Smith, Ronald E (MD)
Sowers, George F (GA)
Spanovich, Milan (PA)
Squier, L Radley (OR)
Stern, Louis I (NJ)
Stoll, Robert D (NY)
Stoll, Ulrich W (MI)
Stormoe, Donovan K (MN)
Sudduth, Charles Grayson (CA)
Thorfinnson, Stanley T (CO)
Tsai, Keui-wu (CA)
Tsui, Yuet
van Reenen, Willem J (GA)
Vaughan, Richard G (NM)
Waggoner, Eugene B (CA)
Wahler, William A (CA)
Wahls, Harvey E (NC)
Walker, Leland J (MT)
Walker, Stanley E (ME)
Ward, Joseph S (NJ)
Wardwell, Richard E (ME)
Weinfurter, Robert W (KS)
Wheby, Frank T (IL)
White, Robert E (NY)
Whitman, Robert V (MA)
Williams, Grover C (TX)
Williams, John A (HI)
Wilson, L Edward (TN)
Wilson, Stanley D (WA)
Woodward, Richard J (CA)
Wright, Robert S (CA)
Wright, Walter L (OR)
Wu, Tien H (OH)
Wulff, Jack G (CA)
Yaghoubian, Jack Nejdge F (CA)
Zeman, Alvin R (WA)
Zoino, William S (MA)

GEOTHERMAL

Babson, Edmund C (CA)
Balzhiser, James K (OR)
DiPippo, Ronald (MA)

Dorfman, Myron H (TX)
Jackson, Robert B (WA)
Messer, Philip H (CA)
Nevin, Andrew E (CO)
Sells, Harold R (CA)
Sheinbaum, Itzhak (CA)
Smith, Morton C (NM)

GRAPHICS

Arnette, Harold L (OH)
D'Amato, Salvatore F (NY)
Gattis, Jim L (AR)
Kroner, Klaus E (MA)
Riesenfeld, Richard F (UT)

HARBORS & WATERWAYS

Abonyi, Erwin (CT)
Amirikian, Arsham (MD)
Beeman, Ogden (OR)
Bryant, J Franklin (FL)
Calabretta, Victor V (RI)
Cass, James R, Jr (VA)
Crook, Leonard T (MI)
Galvin, Cyril (VA)
Gaum, Carl H (MD)
Herron, William J (AZ)
Johnson, Joe W (CA)
Johnson, Leland R (TN)
Keith, James M (CA)
Krone, Ray B (CA)
LeMehaute, Bernard (FL)
Liggett, James A (NY)
Neff, R Wilson (CA)
Pierce, Francis C (RI)
Plumlee, Carl H (CA)
Price, Robert I (NJ)
Rein, Harold E (NY)
Stetson, John B (NY)
Thompson, Homer F (AR)
Vanoni, Vito A (CA)
Wilson, Drake (VA)
York, Dennis J (MS)

HEALTH CARE

Clark, Charles E (AL)
Denney, Roger P, Jr (MD)
Durham, Charles W (NE)
Freund, Louis E (CA)
Morris, Ben F (CA)
Pierskalla, William P (PA)
Wiley, Albert L Jr (WI)

HEAT EXCHANGE

Bishop, Eugene H (SC)
Clinedinst, Wendel W (NJ)
Dixon, William T (PA)
Green, Stanley J (CA)
Jacobs, John D (TN)
Lorenzini, Robert A (MA)
Sech, Charles E (OH)
Siegel, Robert (OH)
Simison, Charles B (IL)
Sparrow, Ephraim M (MN)
Supple, Richard G (NM)

HEAT TRANSFER & THERMODYNAMICS

Adams, Ludwig (PA)
Ahrens, Frederick W (MO)
Akins, Richard G (KS)
Alldredge, Glenn E (OK)
Anderson, James T (NV)
Anderson, William M (CO)
Andrews, Robert V (TX)
Ash, Robert L (VA)
Astill, Kenneth N (MA)
Atwood, Theodore (NJ)
Azer, Naim Z (KS)

Bailey, Herbert R (IN)
Banerjee, Sanjoy (CA)
Bankoff, S George (IL)
Bar-Cohen, Avram N (MN)
Barfield, Robert F (AL)
Batch, John M (OH)
Bayazitoglu, Yildiz (TX)
Beck, James V (MI)
Begell, William (NY)
Bell, Kenneth J (OK)
Bergles, Arthur E (NY)
Berkeley, Frederick D (NY)
Berry, Richard C (TX)
Bishop, Eugene H (SC)
Black, William Z (GA)
Bloome, Peter D (IL)
Boehm, Robert F (UT)
Born, Harold J (OK)
Brazinsky, Irving (NJ)
Buffington, Dennis E (FL)
Burghardt, M David (NY)
Callinan, Joseph P (CA)
Carlson, Walter O (GA)
Carmi, S (PA)
Chan, Shih H (WI)
Chao, Bei T (IL)
Chao, Kwangchu (IN)
Chato, John C (IL)
Chen, Chuan F (AZ)
Chen, Ta-Shen (MO)
Cheng, Ping (HI)
Chiang, Donald C (IN)
Chinitz, Wallace (NY)
Chiou, Jiunn P (MI)
Chu, Richard C (NY)
Chuang, Hsing (KY)
Chung, Benjamin T (OH)
Churchill, Stuart W (PA)
Clausing, Arthur M (IL)
Clement, Richard W (NM)
Coit, Roland L (CA)
Cole, Robert (NY)
Comings, Edward W (DE)
Conway, Lawrence (FL)
Coogan, Charles H, Jr (CT)
Cottingham, William B (MI)
Cox, Jim E (DC)
Crane, Roger A (FL)
Cremers, Clifford J (KY)
Crews, Paul B (AK)
Crisp, John N (KS)
Crosbie, Alfred L (MO)
Dale, James D
Danne, Herbert J (OK)
Day, Robert W (MA)
Deissler, Robert G (OH)
deSoto, Simon (CA)
DeWitt, David P (IN)
Drew, Thomas B (NH)
Durbetaki, Pandeli (GA)
Dyer, Harry B (IL)
Eckert, Ernst R G (MN)
Edwards, Donald K (CA)
Eichhorn, Roger (TX)
Erdman, Carl A (TX)
Eustis, Robert H (CA)
Favret, Louis M (LA)
Fenech, Henri J (CA)
Fitzroy, Nancy D (NY)
Fletcher, Edward A (MN)
Florio, Pasquale J, Jr (NJ)
Francis, John E (OK)
Freese, Howard L (NC)
Fried, Erwin (NY)
Gebhart, Benjamin (PA)
Giancarlo, Samuel S (TX)
Giedt, Warren H (CA)
Gilmour, Charles H (WV)
Glicksman, Leon R (MA)
Goff, John A (PA)
Goglia, Mario J (GA)
Goldstein, Richard J (MN)
Gordon, Raymond G (CA)
Graham, Lois (NY)
Graham, Robert W (OH)

HEAVY CONSTRUCTION

HEAT TRANSFER & THERMODYNAMICS
(Continued)
Greif, Ralph (CA)
Griggs, Edwin I (TN)
Haberman, Charles M (CA)
Haberstroh, Robert D (CO)
Halloran, Joseph M (CT)
Hamilton, DeWitt C , Jr (LA)
Hammack, Charles J (TX)
Hammitt, Frederick G (MI)
Herzenberg, Aaron (FL)
Hoffman, Myron (CA)
Holdredge, Russell M (UT)
Holman, Jack P (TX)
Hopper, Jack R (TX)
Hosler, Ramon E (FL)
Howell, John R (TX)
Hrycak, Peter (NJ)
Hsu, Yih-Yun (MD)
Hussain, Nihad A (CA)
Hwang, Charles C (PA)
Imber, Murray (NJ)
Incropera, Frank P (IN)
Irvine, Thomas F, Jr (NY)
Jackson, David B (NJ)
Jacobs, Harold R (PA)
Jefferson, Thomas B (IL)
Jeng, Duen-Ren (OH)
Joffe, Joseph (NJ)
Johnson, Charles E (NH)
Jury, Stanley H (TN)
Kadaba, Prasanna V (GA)
Kelleher, Matthew D (CA)
Kelly, Robert E (CA)
Kestin, Joseph (RI)
Kezios, Stothe P (GA)
Kidnay, Arthur J (CO)
Kim, Rhyn H (NC)
Knoebel, David H (LA)
Korpela, Seppo A (OH)
Kreith, Frank (CO)
Krommenhoek, Daniel J, Jr (NY)
Lamb, Jamie Parker (TX)
Lambrakis, Konstantine C (CT)
Launder, Brian E (CA)
Lawrence, Joseph F (NY)
Leadon, Bernard M, Jr (FL)
Lecureux, Floyd E (CA)
Lee, Richard S L (NY)
Lefebvre, Arthur H (IN)
Lehmann, Gilbert M (IN)
Leipziger, Stuart (IN)
Lenczyk, John P (OH)
Lewis, Clark H (VA)
Lienhard, John H (TX)
Limpe, Anthony T (NY)
Lindauer, George C (KY)
Lloyd, John R (IN)
Loehrke, Richard I (CO)
London, Alexander L (CA)
Look, Dwight C, Jr (MO)
Lottes, Paul A (IL)
Love, John (MO)
Love, Tom J, Jr (OK)
Lucas, John W (CA)
Lumsdaine, Edward (MI)
Lundberg, Lennart A (WA)
Mark, Melvin (MA)
Marks, Colin H (MD)
Marschall, Ekkehard P (CA)
McKillop, Allan A (CA)
McLaughlin, Edward (LA)
Meckler, Milton (CA)
Mewes, Dieter
Miller, Paul L, Jr (KS)
Moffat, Robert J (CA)
Moore, John A (TX)
Morgan, Albert R, Jr (OK)
Moss, Clarence T, Jr (NY)
Mueller, Alfred C (MD)
Mujumdar, Arun S

Mullin, Thomas E (KY)
Nelson, Harlan F (MO)
Norris, Rollin Hosmer (NY)
Oktay, Sevgin (NY)
O'Loughlin, John R (IN)
Otterman, Bernard (NY)
Otto J Nussbaum, P E (PA)
Ozisik, Necati M (NC)
Paolino, Michael A (PA)
Parker, Jerald D (OK)
Pearson, Joseph T (IN)
Peck, Ralph E (IL)
Pfender, Emil (MN)
Pletcher, Richard H (IA)
Potter, Richard C (CA)
Powe, Ralph E (MS)
Revesz, Zsolt
Rich, Donald G (NY)
Richardson, Peter D (RI)
Robertson, Roy C (TN)
Rodin, Ervin Y (MO)
Rolf H Sabersky (CA)
Ruckenstein, Eli (NY)
Safdari, Yahya B (IL)
Saltarelli, Eugene A (MD)
Sandall, Orville C (CA)
Schmidt, Frank W (PA)
Schrock, Virgil E (CA)
Seely, John H (CA)
Servais, Ronald A (OH)
Sesonske, Alexander (CA)
Sibulkin, Merwin (RI)
Siegel, Robert (OH)
Silvestri, George J, Jr (PA)
Smith, Allen N (CA)
Smith, Allie M (MS)
Sogin, Harold H (LA)
Stahlheber, William H (PA)
Staniszewski, Bogumil E
Sunderland, J Edward (MA)
Sununu, John H (NH)
Swearingen, Thomas B (WA)
Taylor, Arthur C, Jr (VA)
Thomas, David G (TN)
Thompson, Hugh A (LA)
Thorsen, Richard S (NY)
Tien, Chang L (CA)
Torrance, Kenneth E (NY)
Uhl, James B (PA)
Van Sciver, Steven W (WI)
Viskanta, Raymond (IN)
Watkins, Charles B (NJ)
Wayner, Peter C Jr (NY)
Webb, Ralph L (PA)
Weber, Norman (IL)
Westwater, J W (IL)
Wirtz, Richard A (NY)
Witte, Larry C (TX)
Wunderlich, Milton S (MN)
Yang, Kwang-tzu (IN)
Yang, Wen-Jei (MI)
Yokell, Stanley (NJ)
Young, Edwin Harold (MI)
Zarling, John P (AK)
Zebib, Abdelfattah M G (NJ)
Ziegler, Waldemar T (GA)
Zimmermann, Francis J (PA)

HEAVY CONSTRUCTION
Beech, Gary D (WI)
Eckstein, Marvin (FL)
Frein, Joseph P (ID)
Gillen, Kenneth F (OH)
Hunt, Charles A (MI)
Janairo, Max R , Jr (PA)
McHuron, Clark E (CA)
Mendel, Otto (NY)
Mooney, Malcolm T (CA)
Ochab, Thomas F (NJ)

HIGHWAYS & BRIDGES
Abonyi, Erwin (CT)
Adams, Joe J (IL)
Baker, James L (TX)
Barber, John W (MI)
Barenberg, Ernest J (IL)
Barros, Sergio T
Bechamps, Eugene N (FL)
Bennett, G Bryce (CA)
Carney, William D (MO)
Chittim, Lewis M (MT)
Coda, Frank M (GA)
Cohen, Louis A (DC)
Cohen, Wallace J (DC)
Comella, William O (SC)
Correa, Jose J (TX)
Cotten, William C , Jr (TX)
Counts, Cecil P (WA)
Covarrubias, Jesse S (TX)
Crawford, George L (MO)
Dajani, Walid Z (PA)
Delyannis, Leonidas T (VA)
Dominicak, Robert H (SD)
Donovan, Robert L (PA)
Flatt, Douglas W (KS)
Frandina, Philip F (NY)
Frein, Joseph P (ID)
Fugate, Charles R (MO)
Gallagher, William L (NY)
Gallardo, Albert J (CA)
Garing, Robert S, Jr (CA)
Genovese, Philip W (CT)
Gibbons, Eugene F (NY)
Gubala, Robert W (CT)
Hammontree, R James (OH)
Harbour, William A (NY)
Hart, Alan S (CA)
Hawley, Frank E (CA)
Hayden, Ralph L (GA)
Haynes, John J (TX)
Head, Julian E (SC)
Herr, Lester A (VA)
Hofmann, Frederick J (NJ)
Hoovestol, Richard A (GA)
Hourigan, Edward V (NY)
Huang, Eugene Y (MI)
Humphreys, Jack B (TN)
James, Robert G (MD)
Jones, Edwin (NY)
Kline, Donald H (NC)
Koshar, Louis D (NY)
Lackey, William M (KS)
Lancaster, Tom R (OR)
Lehman, Lawrence H (NY)
Leonard, Bruce G (NC)
Ludgate, Robert B (PA)
Maas, Roy W (KY)
Mackin, Italo V (PA)
Marquis, Eugene L (TX)
Mendenhall, Wesley S, Jr (TX)
Miller, Robert E (NM)
Miller, William J, Jr (DE)
Moehrl, Michael F (IA)
Mosure, Thomas F (OH)
Mullen, Wesley G (NC)
Neumann, Edward S (WV)
Parsonson, Peter S (GA)
Pettigrew, Richard R (NM)
Ratermann, Mark J (AZ)
Rivera, Alfredo J (NJ)
Robbins, Louis (NY)
Roberts, Bruce E (KS)
Robison, Ralph E (OH)
Satterlee, George L (MO)
Schregel, Peter F (NY)
Schuler, Theodore A (TN)
Shariful, Islam Mohammad
Shebesta, Harvey (WI)
Shook, James F (MD)
Smith, Donald W (NY)
Smith, Robert E (PA)
Snyder, Julian (NY)
Stander, Richard R (OH)

Standley, Richard A, Jr (ME)
Stephens, Jack E (CT)
Sterbenz, Frederick H (NY)
Tallamy, Bertram D (DC)
Tewksbury, Dennis L (NH)
Tisdale, Reginald L (FL)
Warner, Lawrence D (MA)
Weaver, David R (OR)
Webb, Ralph L (PA)
Wengenroth, Reece H (CT)
Wesselink, Robert D (IL)
Wilkes, W Jack (DC)
Wray, John M, Jr (VA)

HOSPITALS
Blakely, Thomas A (NJ)
Denbrock, Frank A (MI)
Nagaprasanna, Bangalore R (FL)

HUMAN FACTORS
Ayoub, M M (TX)
Beakes, John H (MD)
Butts, Bennie J (NC)
Carnino, Annick
Carter, Charles L, Jr (TX)
Danchak, Michael M (CT)
Gilson, Richard D (OH)
Hoag, LaVerne L (OK)
Johnston, Waymon L (TX)
Konz, Stephan A (KS)
Krendel, Ezra S (PA)
Lowen, Walter (NY)
Olsen, Richard A (CA)
Roley, Daniel G (IL)
Salvendy, Gavriel (IN)
Smith, Leo A (AL)
Wang, Albert S (PA)

HYDRAULIC SYSTEMS
Briggs, Edward C (CA)
Chiu, Chao-Lin (PA)
Fernandez, Rodolfo B (MD)
Hanpeter, Robert W (MO)
Lantz, Thomas L (WV)
Maroney, George E (MA)
Schwegel, Donald R (IL)

HYDRAULICS & IRRIGATION
Allen, Keith J (OR)
Alluzzo, Gasper (NY)
Anderson, Keith E (ID)
Arndt, Roger E A (MN)
Ashton, George D (NH)
Ayars, James E (CA)
Barocio, Alberto J
Bhowmik, Nani G (IL)
Bishop, Harold F (CO)
Blaisdell, Fred W (MN)
Bondurant, Donald C (AR)
Borland, Whitney M (AZ)
Boyd, Marden B (MS)
Brill, Edward L (OH)
Bryce, John B
Carey, Donald E (NJ)
Carreker, John R (GA)
Cassidy, John J (CA)
Charley, Robert W (OR)
Christensen, Bent A (FL)
Collins, Michael A (TX)
Cotton, John C (MD)
Cox, Allen L (PR)
Daily, Eugene J (IL)
Dedrick, Allen R (AZ)
Doty, Coy W (SC)
Douma, Jacob H (VA)
Elder, Rex A (CA)
Fangmeier, Delmar D (AZ)
Franques, J Thomas (FL)

HYDRAULICS & IRRIGATION (Continued)

Franzini, Joseph B (CA)
Friedkin, Joseph F (TX)
Giberson, Harry F (CA)
Hansen, Vaughn E (UT)
Hasfurther, Victor R (WY)
Heiser, Will M (PA)
Houston, Clyde E (CA)
Humpherys, Allan S (ID)
Jain, Subhash C (IA)
Jansen, Robert B (WA)
Jednoralski, J Neil (KS)
Johansen, Craig E (UT)
Johnson, Howard P (IA)
Johnson, Joe W (CA)
Justin, Joel B (PA)
Karn, Richard W (CA)
Kennedy, John F (IA)
Keyes, Conrad G , Jr (NM)
Kim, Young C (CA)
Lee, James A (FL)
Li, Ruh-Ming (CO)
Liggett, James A (NY)
Marable, James R, Jr (GA)
Martin, C Samuel (GA)
McNealy, Delbert D (TX)
Mehring, Clinton W (CO)
Munger, Paul R (MO)
Neill, Charles R
O'Brien, Eugene (NJ)
Posey, Chesley J (CT)
Rechard, Paul A (WY)
Ricca, Vincent T (OH)
Rice, Charles E (OK)
Rice, Leonard (CO)
Richards, Richard T (NJ)
Robinson, August R (MD)
Rockwell, Glen E (CO)
Rouse, Hunter (AZ)
Rowland, Walter F (CA)
Sagar, Bokkapatnam
    Tirumala Ananda
    (CO)
Sarkaria, Gurmukh S (CA)
Scherich, Erwin T (CO)
Shearer, Marvin N (OR)
Shen, Hsieh W (CA)
Showalter, Robert L (PA)
Skogerboe, Gaylord V (CO)
Sneed, Ronald E (NC)
Stark, Robert A (PA)
Stefan, Heinz G (MN)
Striegel, Richard R (CA)
Stringham, Glen E (UT)
Summers, Joseph B (CA)
Thaemert, Ronald L (CO)
Thomas, Cecil A (ID)
Thompson, Charles B (AZ)
Trelease, Frank J , III (WY)
Tullis, James P (UT)
Verzuh, James M (MT)
Watters, Gary Z (CA)
Wenzel, Harry G, Jr (IL)
Widmer, Wilbur J (CT)
Wolfe, John W (OR)
Wood, Don J (KY)
Yang, Chih Ted (CO)
Yee, Phillip K H (HI)
Zielinski, Paul B (SC)

HYDRO POWER

Anton, Walter F (CA)
Benziger, Charles P (TN)
Chao, Paul C S (NY)
Ehasz, Joseph L (NY)
Emerson, Warren M (CA)
Gianelli, William R (CA)
Higginson, R Keith (UT)
Hoffman, Phillip R (CA)
Horowitz, George F (CA)
Hougen, Oddvar (CA)
Justin, Joel B (PA)
Kwiatkowski, Robert W
    (MA)

Lacy, Floyd P (TN)
Mace, Louis L (WA)
Mayo, Howard A Jr (PA)
Nixon, David D (MI)
O'Brien, Eugene (NJ)
Pritchard, David F (OH)
Richards, Richard T (NJ)
Rogers, Franklyn C (FL)
Sagar, Bokkapatnam
    Tirumala Ananda
    (CO)
Sanghera, Gurbaksh S (MA)
Satija, Kanwar S (PA)
Sivley, Willard E (WA)
Spearman, Rupert B (WA)
Sporseen, Stanley E (OR)
Sproule, Robert S
Stout, James J (VA)
Zagars, Arvids (IL)

HYDROELECTRIC

Elder, Rex A (CA)
Ghanime, Jean
Gunwaldsen, Ralph W
    (MA)
Hansen, James C (VT)
Hoffman, Phillip R (CA)
Holway, Donald K (OK)
Hunt, Charles A (MI)
Lewis, William M , Jr (OH)
Mayo, Howard A Jr (PA)
McCoy, Byron O (VT)
Miller, John H , II (GA)
Mozer, Harold M (WA)
Puls, Louis G (CO)
Raffel, David N (TX)
Roush, Maurice D (WA)
Sorensen, Kenneth E (IL)
Witte, Charles R (NY)

HYDROLOGY

Aycock, Kenneth G (AL)
Bishop, Floyd A (WY)
Borland, Whitney M (AZ)
Brill, Edward L (OH)
Brutsaert, Wilfried H (NY)
Cannon, Charles E (MA)
Chiu, Chao-Lin (PA)
Clesceri, Nicholas L (NY)
Cragwall, Joseph S, Jr (VA)
Das, Khirod C (VA)
Daubert, Henry (OK)
Fok, Yu-Si (HI)
Fuhriman, Dean K (UT)
Greenslade, Wiliam M (AZ)
Grout, Harold P (CO)
Hagen, Vernon K (DC)
Jennings, Marshall E (MS)
Jones, Everett Bruce (CO)
Keyes, Conrad G , Jr (NM)
Keys, John W , III (ID)
Koelliker, James K (KS)
Lapsins, Valdis (OH)
Linsley, Ray K (CA)
Marino, Miguel A (CA)
Maxwell, James D (UT)
Meredith, Dale D (NY)
Motayed, Asok K (MD)
Murdock, Larry T (UT)
Peters, Robert R (VA)
Pinder, George F (NJ)
Pitt, William A J, Jr (FL)
Pratt, David F (NY)
Probasco, Johnny L (UT)
Quimpo, Rafael G (PA)
Renard, Kenneth G (AZ)
Schuler, Theodore A (TN)
Shen, Hsieh W (CA)
Smedes, Harry W (DC)
Vicens, Guillermo J (MA)
Wainscott, Lotcher A (CA)
Weir, Charles R , Jr (PA)
Williams, David J (OH)

Yang, Chih Ted (CO)
Yen, Ben Chie (IL)
Yevjevich, Vujica (VA)
Zederbaum, Robert B (NJ)

ILLUMINATION & LIGHTING

Allen, Robert L (IL)
Amick, Charles L (MO)
Anderson, Walter T (MI)
Armour, Thomas S, Jr (SC)
Arrigoni, David M (CA)
Bakeman, Charles T (WA)
Baker, Glenn A (AZ)
Bishop, George A, III (CT)
Blackwell, H Richard (OH)
Blackwell, O Mortenson
    (OH)
Bock, John E (NY)
Bonner, Andrew V (CA)
Boylan, Bernard R (IL)
Caplan, Aubrey G (PA)
Carastro, Sam (FL)
Clarkson, Clarence W (NJ)
Coda, Frank M (GA)
Cohen, Norman A (CA)
Demaree, David M (AZ)
Douglas, Charles A (MD)
Eagan, Constantine J (TX)
Enoch, Jay M (IL)
Farnham, Charles P (WA)
Faucett, Robert E (CO)
Fisher, Will S (OH)
Flynn, John E (PA)
Franck, Kurt G (OH)
Fry, Glenn A (OH)
Gianoli, Napoleon H
Goodbar, Isaac (NY)
Guth, Sylvester K (OH)
Helms, Ronald N (CO)
Higgins, Wayne R (WI)
Hill, Ole A, Jr (IL)
Hood, John L, Jr (AL)
Isenberg, Martens H, Jr
    (NY)
Jewell, James Earl (CA)
Kaufman, John E (CT)
Ketvirtis, Antanas
Lemons, Thomas M (MA)
LeVere, Richard C (CO)
Liss, Sheldon (CA)
Maloney, Laurence J (TX)
Marcue, Donald R (CO)
Martinez, Pedro O (FL)
McCully, Robert A (NJ)
Melden, Morley G (NY)
Mitchell, J Dixon, Jr (GA)
Neenan, Charles J (NY)
Neidhart, John J (CT)
Odle, Herbert A (OH)
Patterson, David H (TX)
Projector, Theodore H, P E
    (DE)
Rodgers, Alston (CA)
Roper, Val J (OH)
Roush, Robert W (OK)
Sever, Lester J (NC)
Shemitz, Sylvan R (CT)
Shoemaker, George E (PA)
Spencer, Domina Eberle
    (MA)
Thayer, Richard N (OH)
Thornton, William A (NJ)
Thouret, Wolfgang E (NJ)
Van Dusen, Harold A (WI)
Wald, Henry J (CT)
Wissoker, Richard S (MA)
Wylie, Robert R (MA)
Zersen, Carl W (CA)

IMAGE PROCESSING

Aggarwal, J K (TX)

Angel, Edward S (NM)
Frieden, B Roy (AZ)
Friedlander, Benjamin (CA)
Gattis, Jim L (AR)
Haskell, Barry G (NJ)
Lauterbur, Paul C (IL)
Li, Ching-Chung (PA)
Mersereau, Russell M (GA)
Okochi, Masaharu J A
Prewitt, Judith M S (MD)
Rosenfeld, Azriel (MD)
Schreiber, William F (MA)
Sze, T W (PA)
Vizy, Kalman N (NY)
Woods, John W (MD)

INDUSTRIAL ENGINEERING

Agee, Marvin H (VA)
Alberts, James R (NJ)
Alexander, Suraj M (KY)
Allen, Roy L (TX)
Altenhaus, Julian L R (NJ)
Aly, Adel A (OK)
Amrine, Harold T (IN)
Anderson, Raymond L
    (NC)
Arias, Javier S
Arnwine, William C (CA)
Asfahl, C R (AR)
Avril, Arthur C (OH)
Ayoub, M M (TX)
Babu, Addagatla J (FL)
Bacci, Guy J , II (IL)
Bailey, James E (AZ)
Balestrero, Gregory (GA)
Barany, James W (IN)
Barer, Seymour (NJ)
Barnes, Ralph M (CA)
Barnhill, Walter O (NC)
Baum, David M (VA)
Baum, Sanford (UT)
Beakley, George C (AZ)
Bedworth, David D (AZ)
Belden, David L (NY)
Bennett, G Kemble (TX)
Berry, William L (IA)
Biles, William E (LA)
Birnbaum, John D (IA)
Bishop, Albert B (OH)
Blythe, Ardven L (AL)
Boehringer, Ludwig C , Jr
    (NY)
Bontadelli, James A (TN)
Bowman, Rush A (TN)
Boyer, Robert O (OH)
Braswell, Robert N (FL)
Bringhurst, Lynn H (UT)
Brisley, Chester L (CA)
Brockman, Donald C (MI)
Brooks, George H (AL)
Brown, Robert A (AL)
Brumbaugh, Philip S (MO)
Burnham, Donald C (PA)
Bussey, Lynn E (KS)
Butts, Bennie J (NC)
Byrd, Jack, Jr (WV)
Campbell, Bonita J (CA)
Capehart, Barney L (FL)
Carlile, V Sam (IN)
Carlson, Robert C (CA)
Carson, Gordon B (MI)
Case, Kenneth E (OK)
Cavell, George R (CA)
Chalmet, Luc G (FL)
Chang, Tsong-how (Phil)
    (WI)
Chen, Hui-Chuan (AL)
Cherna, John C
Chisman, James A (SC)
Clark, Charles E (AL)
Comingore, Edward G (CA)
Cook, Leonard C (WV)
Cotton, Frank E, Jr (IN)

INDUSTRIAL ENGINEERING
(Continued)

Crisp, Robert M (AR)
Crotts, Marcus B (NC)
Cullinane, Thomas P (IN)
Daugherty, Tony F (TX)
Deane, Richard H (CA)
Denham, Frederick R
Denny, James O'H III (WV)
DeSpirito, Victor P (NY)
Dickie, H Ford (CT)
Dizer, John T (NY)
D'Ovidio, Gene J (CT)
Draper, Alan B (PA)
Dryden, Robert D (VA)
Dudek, Richard A (TX)
Duket, Steven D (IN)
Dunne, Edward J , Jr (OH)
Eastman, Robert M (MO)
Eckhoff, N Dean (KS)
Elias, Samy E G (DC)
Elsayed, Elsayed A (NJ)
Elzinga, Donald Jack (FL)
Emerson, C Robert (MT)
Enell, John Warren (NJ)
Engwall, Richard L (MD)
Everett, William B (TN)
Fabrycky, Wolter J (VA)
Feorene, Orlando J (NY)
Fey, Willard R (GA)
Filley, Richard D (AZ)
Fleischer, Gerald A (CA)
Flowers, A Dale (OH)
Foote, Bobbie L (OK)
Footlik, Robert B (IL)
Francis, Richard L (FL)
Freund, Louis E (CA)
Fried, Benjamin S (CT)
Frische, James M (IN)
Gambrell, Carroll B (FL)
Geier, John J (NY)
Genheimer, J Edward (MI)
German, John G (MO)
Giffin, Walter C (OH)
Giglio, Richard J (MA)
Gilbreath, Sidney G, III (TN)
Gill, Lowell F (UT)
Gillette, Leroy O (PA)
Givens, Paul E (MS)
Glassey, C Roger (CA)
Glen, Thaddeus M (OH)
Gochenour, Donal L , Jr (WV)
Goldman, Jay (AL)
Goldstein, Irving R (NJ)
Goodell, Paul H (FL)
Gordon, Richard H (PA)
Greene, James H (IN)
Griffin, Norman E (MD)
Groover, Mikell P (PA)
Gross, Donald (DC)
Ham, Inyong (PA)
Hamlin, Jerry L (OK)
Hammond, Ross W (GA)
Hanson, Roland S (OR)
Harer, Kathleen F (FL)
Hayet, Leonard (FL)
Haynes, Frederick L (VA)
Haynes, Patricia Griffith (VA)
Heikes, Russell G (GA)
Heiner, Clyde M (UT)
Henke, Norman W (CA)
Hennessy, Robert L (CA)
Henry, Herman L (LA)
Hess, Daniel E (PA)
Hicks, Philip E (FL)
Hill, Richard L (TN)
Hitomi, Katsundo
Hoag, LaVerne L (OK)
Hodgson, Thom J (FL)
Holzman, Albert G (PA)
Hood, Thomas B (NJ)

Howard P Emerson (LA)
Hulbert, Thomas E (MA)
Hulley, Clair M (OH)
Hunt, Everett C (NY)
Hutchinson, Richard C (FL)
Hwang, Ching-Lai (KS)
Ignizio, James P (TX)
Imhoff, John L (AR)
Ireson, W Grant (CA)
Jacobs, Louis S (IL)
Jaffe, William J (NY)
James, Charles F, Jr (WI)
Janzen, Jerry L (TX)
Jarvis, John J (GA)
Jericho, Jack F (GA)
Jeynes, Paul H (NJ)
Johnson, Dean E (IL)
Johnson, Lynwood A (GA)
Johnson, Marvin M (NE)
Johnston, Burleigh Clay, III (TN)
Kaminsky, Frank C (MA)
Kane, George E (PA)
Kapur, Kailash C (MI)
Karger, Delmar W (FL)
Keenan, Arthur J (CA)
Khalil, Tarek M (FL)
King, Myron D (CA)
Klein, Morton (NY)
Klein, Stanley J (NY)
Kocaoglu, Dundar F (PA)
Konz, Stephan A (KS)
Koss, John P (MA)
Krawczyk, Theodore A (IL)
Kroeze, Henry (NC)
Kroner, Klaus E (MA)
Kuers, Marvin M (TX)
Kvalseth, Tarald O (MN)
Laird, Harry G (MA)
Lamberson, Leonard R (MI)
LaRobardier, Lamont M (NJ)
Lazo, John R (OH)
Leavenworth, Richard S (FL)
Leep, Herman R (KY)
LeMay, Richard P (NY)
Leslie, John H, Jr (KS)
Lesso, William G (TX)
Lewis, Bernard T (MD)
Lindenmeyer, Carl R (SC)
Lockwood, Robert W (CA)
Lorenz, John D (MI)
Louden, J Keith (PA)
Lutz, Raymond P (TX)
MacBrayne, John M (ME)
MacKenzie, Horace J (TX)
Malik, Mazhar Ali Khan
Mallik, Arup K (WV)
Mann, Lawrence, Jr (LA)
Mariotti, John J (MI)
Maxwell, William L (NY)
Mayer, Orland C (ID)
Maynard, Hal B (WA)
McCarthy, Rollin H (NY)
McCullough, Clarence R (OK)
McNichols, Roger J (OH)
McRoberts, Keith L (IA)
McSorley, Richard J (CA)
Meier, France A (TX)
Meier, Wilbur L, Jr (TX)
Mihalasky, John (NJ)
Miller, Owen W (MO)
Mize, Joe H (OK)
Mjosund, Arne (AL)
Mohler, Harold S (PA)
Moir, Barton M (CA)
Montgomery, Douglas C (GA)
Moor, William C (AZ)
Morgan, Donald E (CA)
Morris, Robert G (MI)
Moulder, Leonard D (CO)
Mundel, Marvin E (MD)
Mundt, Barry M (NY)

Nadler, Gerald (CA)
Nagaprasanna, Bangalore R (FL)
Nanda, Ravinder (NY)
Netter, Milton A (OH)
Newton, Edwin H, Jr (VA)
Niebel, Benjamin W (PA)
Olsen, Richard A (CA)
Ostrofsky, Benjamin (TX)
Ostwald, Phillip F (CO)
Otis, Irvin (MI)
Pate-Cornell, M Elisabeth (CA)
Payne, Matthew A (NJ)
Pearlman, Bertrand B (NY)
Perry, Ronald F (MA)
Peterson, Grady F (NC)
Phillips, Don T (TX)
Pickering, C J (TX)
Pigage, Leo C (IL)
Plummer, Ralph W (WV)
Poage, Scott T (TX)
Pollock, Stephen M (MI)
Porras, Octavio R
Power, Richard J (VA)
Pritsker, A Alan B (IN)
Proctor, Charles L
Proffitt, Richard V (CT)
Purswell, Jerry L (OK)
Quaney, Robert A (CA)
Ragold, Richard E (NJ)
Rainey, Frank B (NC)
Ramalingam, Panchatcharam (CA)
Ramsey, Jerry D (TX)
Rathe, Alex W (NY)
Ravindran, Arunachalam (IN)
Reed, Samuel Kyle (TN)
Reynholds, Walter H (NY)
Rigassio, James L (NJ)
Riggs, Henry E (CA)
Rockwell, Willard F, Jr (PA)
Rodgers, Joseph F (CA)
Roettger, Martin (OH)
Rose, David N (UT)
Ross, Gilbert I (NY)
Rothbart, Harold A (CA)
Rotolo, Elio R (CA)
Rouse, William B (GA)
Roy, Robert H (MD)
Rudkin, Donald A (NJ)
Ruhlid, Robert R (CT)
Rutherfoord, J Penn (VA)
Sahney, Vinod K (MI)
Salvendy, Gavriel (IN)
Schaefer, David L (PA)
Schmitz, Eugene G (CA)
Schneider, Morris H (NE)
Schoenfeld, Theodore M (NJ)
Schrader, George F (FL)
Schuermann, Allen C , Jr (OK)
Schultz, Andrew S, Jr (FL)
Self, Norman L (FL)
Settles, F Stanley (AZ)
Sexton, Joseph M (TX)
Shamblin, James E (OK)
Shannon, Robert E (TX)
Siebeneicher, Paul Robert 2nd (NJ)
Sims, E Ralph, Jr (OH)
Singer, Leslie J (TX)
Sivazlian, Boghos D (FL)
Smaltz, Jacob J (KS)
Smith, James R (TN)
Smith, Leo A (AL)
Smith, Milton L (TX)
Smith, William A, Jr (NC)
Solberg, James J (IN)
Sorokach, Robert J (OH)
Soukup, David J (GA)
Spechler, Jay W (FL)
Springer, Melvin D (AR)

Stanislao, Joseph (ND)
Stevens, Gladstone T, Jr (TX)
Stewart, David C (MA)
Sundaram, Meenakshi R (TN)
Sweers, John F (WI)
Swift, Fred W (TN)
Tarrants, William E (MD)
Thomas, Marlin U (MO)
Thuesen, Gerald J (GA)
Tillman, Frank A (KS)
Tompkins, Curtis J (WV)
Torgersen, Paul E (VA)
Tulkoff, Joseph (GA)
Turksen, Ismail B
Unger, Vernon E (AL)
Vallette, William J (MA)
Visweswaran, Subramaniam (NY)
Wallis, Bernard J (MI)
Walvekar, Arun G (NM)
Ward, Richard E (WV)
Ward, Thomas L (KY)
Waugh, Richard J
Weindling, Joachim I (PA)
White, John A (GA)
Whitehouse, Gary E (FL)
Wiesenberg, Jerome D (NY)
Wilkes, J Wray (AR)
Williams, Charles D , Jr (IL)
Williams, G Bretnell (OH)
Wolf, Donald J, (MD)
Wolfberg, Stanley T, (CA)
Wolfe, Philip M (AZ)
Wolfe, R Kenneth (OH)
Wortman, David B (IN)
Wyskida, Richard M (AL)
Young, Hewitt H (AZ)
Ysteboe, Howard T (PA)
Zanakis, Steve (Stelios) H (FL)
Zebrowksi, Edward R (PA)
Zmyj, Eugene B (NJ)

INFORMATION PUBLIC RELATIONS & PUBLISHING

Adamson, Dan K (TX)
Baker, Hugh (OH)
Bolte, Charles C (PA)
Brennan, Peter J (NY)
Butler, Margaret K (IL)
Campbell, John B (NY)
Cowin, Roy B (OH)
Cunny, Robert W (MS)
Dobson, David B (DC)
Filley, Richard D (AZ)
Johnson, Leland R (TN)
Landau, Herbert B (NY)
Marks, Robert H (NY)
Morris, M Dan (NY)
Mueller, Walter H (IN)
Mulvaney, Carol E (IL)
Mumma, George B (CO)
Noyes, Robert E (NJ)
Rossi, Boniface C (CT)
Rossoff, Arthur L (NY)
Willard, Charles O (MN)
Wolbrink, Jim F (GA)

INFORMATION THEORY

Blahut, Richard E (NY)
Carlson, Walter M (CA)
Christiansen, Richard W (UT)
Dickinson, Bradley W (NJ)
Elias, Peter (MA)
Gray, Robert M (CA)
Kadota, T Theodore (NJ)
Scholtz, Robert A (CA)
Slepian, David (NJ)

INFORMATION THEORY
(Continued)
Swerling, Peter (CA)
Tong, Peter P (CA)

INSTRUMENTATION & MEASUREMENT

Alers, George A (NM)
Alexiou, Arthur G (MD)
Ashton, Jackson, A, Jr (LA)
Astin, Allen V (MD)
Attinger, Ernst O (VA)
Babcock, Russell H (MA)
Baker, Raymond G (WA)
Baldwin, Allen J (MN)
Barnett, Donald O (AL)
Baum, Carl E (NM)
Beadle, Charles W (CA)
Beatty, Robert W (NV)
Benedict, Robert P (PA)
Billings, Bruce H (CA)
Blesser, William B (NY)
Bossart, Clayton J (PA)
Bradner, Hugh A (CA)
Brown, Marshall J (FL)
Brown, Roy F (CA)
Buckley, Robert F (MA)
Cancilla, Myron A (CA)
Carlson, Harold C R (NJ)
Cawley, John H (MD)
Chapman, Robert L (CA)
Chase, Robert L (NY)
Chi, Andrew R (MD)
Coles, Donald E (CA)
Cooper, Harrison R (UT)
Costrell, Louis (MD)
Cronvich, James A (LA)
Doty, William D (PA)
Drost, Edward J (AL)
Duffy, Robert A (MA)
Duncan, Richard H (NM)
Dunn, Andrew F
Durrenberger, J E (NM)
Foundos, Albert P (NY)
Freeston, Robert C (NJ)
Fromm, Eli (PA)
Gartner, Joseph R (CT)
Gatti, Emilio
Gerdes, Walter F (TX)
Gilmont, Roger (NY)
Goldberg, Harold S (MA)
Grinnell, Robin R (CA)
Hamilton, James F (IN)
Hammond, Ogden H (MA)
Harada, Tatsuya
Harris, Forest K (MD)
Harting, Darrell R (WA)
Hartz, Nelson W (PA)
Heirman, Donald N (NJ)
Hekimian, Norris C (MD)
Hermach, Francis L (MD)
Herzenberg, Caroline L (IL)
Higinbotham, William A (NY)
Histand, Michael B (CO)
Howland, Ray A (GA)
Hull, Maury L (CA)
Huston, Norman E (WI)
Inman, Byron N (WA)
Jackson, Warren, Jr (OH)
Johnson, Anthony M (FL)
Keepin, G Robert (NM)
Kehoe, Thomas J (CA)
Keithley, Joseph F (OH)
Kind, Dieter H
Knudsen, Dag I (MN)
Kokjer, Kenneth J (AK)
Kopczyk, Ronald J (SC)
Kubick, Raymond A, Jr (CA)
Macintyre, John R (AL)
Maninger, R Carroll (CA)
Marks, Allan F (CA)
Mason, Henry Lea (MD)

McGee, George W (AZ)
McLean, James D (MI)
Miller, Don W (OH)
Moore, Harry F (PA)
Moore, Ralph L (DE)
Moore, William J M
Novak, Al V (NC)
Olin, John G (CA)
Parsonson, Peter S (GA)
Peterson, Robert S (PA)
Povey, Edmund H (MA)
Richardson, Gregory N (NC)
Roemer, Louis E (OH)
Roth, Wilfred (VT)
Ryder, John D (FL)
Sackinger, William M (AK)
Schiff, Anshel J (CA)
Schlink, Frederick J (NJ)
Self, Sidney A (CA)
Selvidge, Harner (AZ)
Sheppard, Albert P (GA)
Simard, Gerald L (ME)
Skilling, James K (MA)
Sprenkle, Raymond E (OH)
Stanley, Marland L (ID)
Staunton, John J J (IL)
Stein, Peter K (AZ)
Stevenson, Warren H (IN)
Strain, Douglas C (OR)
Stripeika, Alexander J (CA)
Taylor, Barry N (MD)
Timoshenko, Gregory S (FL)
Ugiansky, Gilbert M (MD)
Van Sweringen, Raymond A , Jr (NJ)
Voznick, Henry P (CA)
Webster, John G (WI)
Weiss, Gerald (NY)
Wilhelm, Luther R (TN)
Williams, Albert J, Jr (PA)
Windham, Donald R (MS)
Yost, Edward F (NY)
Zemel, Jay N (PA)
Zoss, Leslie M (IN)

INSULATION

Bartnikas, Ray
Middendorf, William H (KY)
Moran, John H (NY)
Sebo, Stephen A (OH)
Thompson, James E (TX)
Udo, Tatsuo

KINETICS & THERMODYNAMICS

Anthony, Rayford G (TX)
Barteau, Mark A (DE)
Calo, Joseph M (RI)
Daubert, Thomas E (PA)
Dorko, Ernest A (OH)
Fisher, Edward R (MI)
Fleming, Paul D III (OH)
Glassman, Irvin (NJ)
McLellan, Rex B (TX)
Palmer, Bruce R (SD)
Rase, Howard F (TX)
Rogers, Benjamin T (NM)
Sheng, Henry P (CA)
Stell, George (NY)
Thomson, William J (WA)

LAW

Anderson, Robert L (GA)
Cozzarelli, Frank, Jr (NJ)
Drobile, James A (PA)
Gallin, Herbert (NY)
Greenberg, Ronald David (MA)
Labosky, John J (MI)

McManus, Terrence J (AZ)
McNamara, Patrick H (DC)
Mingle, John O (KS)
Santilli, Paul T (OH)
Vansant, Robert E (MO)
Weislogel, Stacy (OH)
Wilson, Donald E (PA)

LUBRICATION

Allen, Charles W (CA)
Belfry, William G
Bhushan, Bharat (CA)
Conry, Thomas F (IL)
Devonshire, Grant S
Dow, Thomas A (NC)
Ettles, Christopher M McC (NY)
Fuller, Dudley D (NY)
Kauzlarich, James J (VA)
Lantz, Thomas L (WV)
MacDougall, Louis M (IA)
Neely, George Leonard (CA)
Rabinowicz, Ernest (MA)
Rodkiewicz, Czeslaw M
Safar, Zeinab S (FL)
Sargent, Lowrie B, Jr (PA)
Shaw, Milton C (AZ)
Springer, Karl J (TX)
Stair, William K (TN)
Szeri, Andras Z (PA)
Trabert, George R (CA)
Wilcock, Donald F (NY)

MACHINERY

Abernathy, George H (NM)
Alpert, Sumner (CA)
Anderson, David Melvin (CA)
Bamford, Waldron L
Beltz, Fred W, Jr (NJ)
Blok, Harmen
Borel, Robert J (OH)
Byloff, Robert C (PA)
Campbell, Calvin A, Jr (IL)
Carter, Charles F, Jr (OH)
Coit, Roland L (CA)
Cross, Ralph E (MI)
Daum, Donald R (PA)
Davenport, Granger (NJ)
Davis, Hunt (TX)
Dowdey, Wayne L (AZ)
Ferguson, Arnold D, Jr (LA)
Fondy, Philip L (OH)
Freudenstein, Ferdinand (NY)
Games, Donald W (OH)
Garrett, Roger E (CA)
Gates, Lewis E (OH)
Ginwala, Kymus (MA)
Goering, Carroll E (IL)
Hunt, Donnell R (IL)
Johnson, William H (KS)
Karassik, Igor J (NJ)
Katz, Peter (TN)
King, John E (PA)
Kirk, Ivan W (TX)
Klebanoff, Gregory, Jr (FL)
Kraimer, Frank B (MI)
Larson, George H (KS)
Lithen, Eric E (NY)
Lowen, Gerard G (NJ)
Mack, Michael J (IA)
Matthews, Edwin J (AR)
Mercer, Kenneth K (TX)
Militzer, Robert W (MI)
Moricoli, John C (CA)
Mueller, Jerome J (OH)
Nekola, Robert L (OH)
Nypan, Lester J (CA)
Oldshue, James Y (NY)
Ostermann, Jerry L (KS)

Oxford, Carl J, Jr (MI)
Pickering, George E (MA)
Poe, Herbert V (SC)
Puleo, Peter A (MO)
Pyle, William L (HI)
Raffel, David N (TX)
Ring, Sandiford (TX)
Schneider, Irving (NY)
Schwanhausser, Edwin J (NJ)
Sculthorpe, Howard J (MI)
Shaffer, Bernard W (NY)
Shaw, Lawrance N (FL)
Smith, James L (WY)
Spiller, W R (FL)
Swensson, Gerald C (PA)
Tallian, Tibor E (PA)
Thoma, Frederic A (NJ)
Thomas, Carl H (LA)
Trexler, Jay E (PA)
Wilkes, Lambert H (TX)
Wood, Gerald L (AZ)
Zich, John H (MI)

MANUFACTURING & PRODUCTION

Adare, James R (NY)
Allard, Kermit O (IA)
Allen, Dell K (UT)
Allen, Douglas L (OH)
Anderson, A Eugene (PA)
Andrews, Arlan R (OH)
Armstrong, John C (NY)
Ashwood, Loren F (WI)
Atwood, James D (MO)
Bachman, Henry L (NY)
Bak, Eugene (PA)
Barnes, Clair E (CA)
Bartholomew, Dale C (WA)
Bellows, Guy (OH)
Bennett, Charles H (WI)
Bennett, Ralph W (OR)
Beno, Paul S (WI)
Berg, Eugene P (IL)
Berry, John T (GA)
Berry, William L (IA)
Beu, Eric R (OH)
Bidstrup, Wayne W (VA)
Bjorke, Oyvind
Black, J Temple (AL)
Black, Joseph E (WA)
Bogart, Harold N (MI)
Bonourant, Leo H (TX)
Boothroyd, Geoffrey (MA)
Boston, Orlan W (AZ)
Bowering, Richard E
Brinson, Robert J (SC)
Brock, Harold L (IA)
Brown, Robert A (AL)
Brummett, Forrest D (IN)
Buck, Richard L (MO)
Buehler, Adolph I (IL)
Byer, Thomas G (OH)
Carter, Charles F, Jr (OH)
Casciani, Robert W (OH)
Cavell, George R (CA)
Chastain, William Roy (CA)
Chertow, Bernard (NY)
Chiantella, Nathan A (NC)
Chrencik, Frank (AL)
Churchill, George H
Cimino, Saverio Michael (PA)
Clark, Houston N (CO)
Cline, William E (NJ)
Collins, James J (NH)
Conn, Harry (IL)
Conner, James L (MO)
Cook, Nathan H (MA)
Cooper, Robert H (IA)
Corell, Edwin J (OH)
Cormack, William J (IL)
Cornwall, Harry J (CA)
Coursey, W (IA)

MANUFACTURING & PRODUCTION
(Continued)

Cromwell, Thomas M (CA)
Cross, Ralph E (MI)
Curran, Elton C (FL)
Daley, Joseph F (NJ)
Danne, Herbert J (OK)
Danpour, Henry (MA)
Dappert, George F (CT)
Das, Mihir K (CA)
Davidson, Bruce R (NJ)
De Vries, Marvin F (WI)
Dean, Eugene R (PA)
Demand, Lyman D (TX)
Denning, Anthony J (NJ)
Dickie, H Ford (CT)
Donovan, James (MA)
Dorsett, Ronald (AZ)
Doty, Clarence W (CA)
Dougherty, Robert A (KS)
D'Ovidio, Gene J (CT)
Doyle, James A (TN)
Dwivedi, Surendra Nath (VA)
Earle, Emily A (AL)
Edwin, Edward M (NY)
Emerson, C Robert (MT)
Evans, Gerald W (KY)
Everhart, James G (MO)
Farst, James R (PA)
Ferrigni, George P (NJ)
Ferrise, Louis J (PA)
FitzGerald, James E (IL)
Ford, Frank F (GA)
Francis, Lyman L (MO)
Francis, Philip H (IL)
Frashier, Gary E (MA)
Friar, John II (MA)
Gardiner, Keith M (PA)
Gardner, Leonard B (CA)
Gillespie, LaRoux K (MO)
Gillette, Roy W (TX)
Goddard, William A (CA)
Goering, Gordon D (OK)
Gould, Arthur F (PA)
Groover, Mikell P (PA)
Gurther, H Louis (MI)
Hahn, Robert S (MA)
Hall, Joseph C (TN)
Ham, Inyong (PA)
Hansen, Christian A (NJ)
Hattersley, Robert S (OH)
Hersh, Michael S (CA)
Hibbeln, Raymond J (IL)
Hilbers, Gerard H (MI)
Hitomi, Katsundo
Hladky, Wallace F (PA)
Hobson, Kenneth H (AK)
Hoffman, Russell (OH)
Hollmann, Harold R (PA)
Hora, Michael E (IL)
Horner, John F (IL)
Hornsby, Clarence H, Jr (SC)
Irwin, Kirk R (PA)
James, Charles F, Jr (WI)
Jeffries, Neal P (OH)
Jenniches, F Suzanne (MD)
Jepsen, John C (TX)
Johnson, Walter M (MA)
Johnston, Burleigh Clay, III (TN)
Johnstone, Edward L (IN)
Kalpakjian, Serope (IL)
Karger, Delmar W (FL)
Keefner, Eugene F (CA)
Keenan, Arthur J (CA)
Kegg, Richard L (OH)
Kehlbeck, Joseph H (KY)
Kerpan, Stephen J (CA)
King, Cecil N (GA)
Kittner, Edwin H (KS)
Klein, Philip H (NC)
Kobayashi, Shiro (CA)

Kram, Harvey (NY)
Krieg, Edwin H, Jr (TN)
Kruty, Samuel (IL)
Kurth, Frank R
Lazo, John R (OH)
Lee, Quarterman (MO)
Leep, Herman R (KY)
Levi, Ned S (PA)
Little, Doyle (TX)
Lloyd, Fredric R (IN)
Lorenz, John D (MI)
Ludwig, Ernest E (LA)
Luz, Herbert M (MA)
Mackrell, James J, Jr (PR)
Magarian, Robert J (NJ)
Mainhardt, Robert M (CA)
Marshall, Dale E (MI)
Mason, James A, Jr (NY)
Maynard, Murray Renouf
McBride, David L (OH)
McBryde, Vernon E (AR)
McCoy, George T (CA)
McCoy, Wyn E (OH)
McSorley, Richard J (CA)
McVickers, Jack C (MD)
Mead, William J (CT)
Merchant, M Eugene (OH)
Merryman, John, Jr (MD)
Mills, Robert W (NE)
Mohler, Harold S (PA)
Morrison, Malcolm C (CA)
Moshier, Glen W (MO)
Myers, Robert D (WI)
Nelson, Eric W (PA)
Nemy, Alfred S (NC)
Netter, Milton A (OH)
Nowicki, George L (PA)
Nowlin, I Edward (OK)
Noyes, Jonathan A (TX)
O'Donnell, Neil B (IN)
Oxford, Carl J, Jr (MI)
Patsfall, Ralph E (OH)
Patten, Charles A (PA)
Pearlstein, Joel P (NJ)
Peterson, Gary J (CO)
Petryschuk, Walter F
Phillips, John R (MS)
Poli, Corrado (MA)
Polomsky, John V (MI)
Powers, John H (NY)
Prisuta, Samuel (PA)
Purdy, Verl O (NJ)
Rau, George J, Jr (MO)
Redpath, Richard J (MO)
Rider, Bobby G (TX)
Rikard, Donald A (MI)
Robertson, Elgin B (TX)
Sahney, Vinod K (MI)
Saperstein, Zalman P (WI)
Scher, Robert W (OH)
Schmidt, Alfred O (PA)
Schmitz, Eugene G (CA)
Schoech, William J (IN)
Scott, Roger M (NH)
Shaw, Milton C (AZ)
Sheets, George H (OH)
Shipley, Larry W (AR)
Shulhof, William P (MI)
Siebeneicher, Paul Robert 2nd (NJ)
Sivetz, Michael (OR)
Smith, Richard A (WV)
Smith, W Tilford (VA)
Smith, William J (CT)
Snedeker, Robert A (MA)
Spiel, Albert (NY)
Staples, Elton E (FL)
Stout, George S
Streitmatter, Vern O (CT)
Sturgeon, Charles E (AL)
Sundaram, Meenakshi R (TN)
Swearingen, Judson S (CA)
Taraman, Khalil S (MI)
Taylor, R William (DC)
Thomas, Benjamin E (MO)

Throop, James W (MI)
Tulkoff, Joseph (GA)
Tutt, Charles L, Jr (MI)
Uys, Johannes M (IL)
Van Buren, Maarten (CT)
Van Tine, Charles L (IL)
Veith, Ferdinand S (NY)
Villani, Vincent F (WV)
Walker, Robert B (WA)
Weaver, Hobart A (NC)
Wedd, Ralph W (GA)
Welch, Arthur (IL)
Welsh, James N (TX)
Wensing, Donald R (PA)
Westfall, Paul R (WV)
Wethern, James D (GA)
White, John B (PA)
Wilson, Ralph S (NM)
Wu, S M (MI)
Zook, Donald G (IL)
Zoss, Abraham O (NJ)

MAPPING, PHOTOGRAMMETRY, SURVEYING

Adams, John L (CO)
Altenhofen, Robert E (CA)
Anderson, James D (AZ)
Beazley, Jon S (FL)
Bentley, Lawrence H (CT)
Berry, Carl M (WA)
Blachut, Teodor J
Boge, Walter E (VA)
Brock, Robert H, Jr (NY)
Brown, Robert W (NY)
Burns, Joseph P (MN)
Carr, Grover V (GA)
Cartwright, Vern W (CA)
Case, James B (VA)
Chamberlain, James E (CA)
Clapp, James L (ME)
Cook, George J (OR)
Corlew, Philip M (IL)
Crandall, Clifford J (MS)
Cunningham, Donald J (OK)
Curtis, Kenneth S (IN)
Davidson, Edwin A (MO)
Doyle, Frederick J (VA)
Edwards, Carl V (ID)
Ervin, Fred F (CA)
Evans, Lorn R (WI)
Feldman, Harry R (FL)
Fischer, Irene K (MD)
Gibson, David W (FL)
Goldstein, Lawrence
Greenawalt, Jack O (OK)
Harris, Robert S (FL)
Hayden, Ralph L (GA)
Hirota, Sam O (HI)
Holland, Francis E (IA)
Hopkins, Paul F (NY)
Johnson, Orwic A (IN)
Karcich, Matthew F (CO)
Kivioja, Lassi A (IN)
Kratky, Vladimir J
Lanc, John J (NY)
Landen, David (VA)
Langlinais, Stephen J (LA)
Lippold, Herbert R, Jr (MD)
Lukens, John E
Lyddan, Robert H (VA)
McEntyre, John G (IN)
McEwen, Robert B (VA)
Merchant, Dean C (OH)
Mintzer, Olin W (VA)
Nielsen, John H (WI)
Olson, Charles Elmer, Jr (MI)
Paulson, Donald L (WI)
Reynolds, J David (KY)
Rib, Harold T (VA)
Rossi, A Scott (PA)

Saxena, Narendra K (HI)
Simpkins, Ronald B (CA)
Simpson, George A (MD)
Sokolowski, Edward H (TX)
Steakley, Joe E (AL)
Tewinkel, G Carper (OR)
Trinkle, Arthur H (CA)
Van Weele, David P (NY)
Vockroth, John H, Jr (PA)
Vreeland, Robert P (NH)
Wigglesworth, Robert W (FL)
Wright, Marshall S, Jr (VA)
Zahnd, Rene G (MN)

MARINE ENGINEERING & NAVAL ARCHITECTURE

Achtarides, Theodoros A
Anderson, William M (CO)
Bachman, Walter C (NJ)
Basile, Norman K (NY)
Beltz, Fred W, Jr (NJ)
Benford, Harry B (MI)
Boylston, John W (MD)
Bradford, Ralph J, Jr (CA)
Britner, George F (NY)
Broad, Richard (VA)
Brunner, Richard F (LA)
Calabretta, Victor V (RI)
Chen, Leslie H (VA)
Cheng, Henry M (MD)
Chieri, Pericle A (LA)
Christie, William B
Courtsal, Donald P (PA)
Cowart, Kenneth K (DC)
Cowles, Walter C (NY)
Cox, Alvin E (NY)
Darden, Arthur D (LA)
Davis, Cabell S, Jr (VA)
DiTrapani, Anthony R (VA)
Dotson, Clinton (FL)
Drucker, Jules H (NY)
Eldred, Kenneth McK (MA)
Evans, J Harvey (MA)
Farrell, Keith P
Farrin, James M (FL)
Fei, Fames R (SC)
Femenia, Jose (NY)
Field, A J (CA)
Field, Sheldon B (NY)
Flipse, John E (VA)
Francis, Gerald P (VT)
Freeman, William C (RI)
Fridy, Thomas A, Jr (SC)
Fulton, Robert P (VA)
Ghosh, Arvind (TX)
Goddard, Thomas A (CT)
Goryl, William M (NJ)
Gruber, Jerome M (NJ)
Harden, Charles A (FL)
Harrington, Roy L (VA)
Haskell, Arthur J (CA)
Hassani, Jay J (MD)
Heil, Charles E (CA)
Henry, James J (NY)
Herbert, Robert N (CA)
Hodges, George H (FL)
Hoffmann, Ludwig C (VA)
Holden, Donald A (VA)
Howell, John T (CT)
Hunley, William H (VA)
Hunt, Everett C (NY)
Jack, Robert L (VA)
Jackson, Harry A (CT)
James, Ralph K (MD)
Jan, Hsien Y (NY)
Jasper, Norman H (FL)
Johnston, William N (NY)
Jolliff, James V (MD)
Kakretz, Albert E (NY)
Karadimos, Angelo S (MD)
Keane, Robert G, Jr (MD)
Keil, Alfred A H (MA)

MARINE ENGINEERING & NAVAL ARCHITECTURE
(Continued)

Key Joe, W (TX)
King, Randolph W (MD)
Kiss, Ronald K (MD)
Kossa, Miklos M (CA)
Krepchin, David M (CA)
Kristinsson, Gudmundur E
Landsburg, Alexander C (MD)
Landweber, Louis (IA)
Leopold, Reuven (FL)
Levy, Harry (CA)
Lewis, Edward V (NY)
Lithen, Eric E (NY)
Liu, Donald (NJ)
Long, Chester L (VA)
Maclean, Walter M (NY)
MacMillan, Douglas C (MA)
Macpherson, Monroe D (NJ)
Manganaro, Francis F (VA)
McClure, Alan C (TX)
McCready, Lauren S (CT)
McLaren, T Arthur
Mueller, Jerome J (OH)
Mumma, Albert G (NJ)
Nachtsheim, John J (MD)
Nickels, Frank J (CA)
Nickum, George C (WA)
Nolan, Ralph P, Jr (CA)
Norton, John A (MA)
O'Neil, David A (CT)
Otth, Edward J (VA)
Patrick, Robert J (MO)
Paulling, J Randolph, Jr (CA)
Pergola, Nicola F (NY)
Pierce, Harry W (NJ)
Plude, George H (OH)
Porter, William R (NY)
Powell, Alan (TX)
Price, Robert I (NJ)
Pritchard, David C (FL)
Rein, Robert J (VA)
Ridley, Donald E (MA)
Robertson, James B, Jr (MD)
Robinson, Harold F (FL)
Roseman, Donald P (MD)
Rosenberg, Marvin A (MD)
Rosenblatt, Lester (NY)
Rowen, Alan L (NY)
Savitsky, Daniel (NJ)
Schmid, Walter E (VA)
Shumaker, Fred E (TX)
Signell, Warren I (NJ)
Sinclair, Charles S (CA)
Sinclair, Thomas L , Jr (HI)
Skinner, Bruce C (CT)
Sonenshein, Nathan (CA)
Sorkin, George (VA)
Spaulding, Philip F (WA)
Stark, Robert E (NJ)
Stott, Thomas E (MA)
Suehrstedt, Richard H (OH)
Swensson, Gerald C (PA)
Tachmindji, Alexander J (MD)
Taggart, Robert (VA)
Tatinclaux, Jean-Claude P (IA)
Thayer, Stuart W (LA)
Thornley, George C (SC)
Todd, Frederick H
Van Manen, J D
Veazey, Sidney E (MD)
Vittucci, Rocco Victor (DC)
Volcy, Guy C
Weiler, Daniel John (VA)
White, Trevor (MN)
Wiggins, Edwin G (TX)

Wiggins, Edwin G (NY)
Wilson, Charles W (NY)
Wilson, Thomas B (CA)
Woodward, Adin K (VA)
Woodward, John B (MI)
Young, Robert T (NH)
Zankich, L Paul (WA)
Zeien, Charles (NY)
Zimmie, William E (OH)
Zippler, William N (NJ)

MATERIALS HANDLING & PACKAGING

Agee, Marvin H (VA)
Beck, Paul J, Jr (NY)
Billett, Robert L (CA)
Brockway, George L (GA)
Carstens, Marion R (GA)
Chapman, Robert R (OH)
Falk, Martin Carl (PA)
Farrell, Robert J (OH)
Footlik, Irving M (IL)
Footlik, Robert B (IL)
Frasier, Charles L (VA)
Green, Frank W (CA)
Greene, Irwin R (IL)
Harris, Aubrey L, III (AR)
Himmelman, Gerald L (NH)
Huffman, John R (CA)
Hulm, John K (PA)
Johanson, Jerry R (CA)
Johnson, Charles L (IN)
Laird, Harry G (MA)
Lowery, Thomas J (CT)
Lukenda, James R (NJ)
Mack, Harry R (MI)
Maxted, Wesley R (TN)
McKinlay, Alfred H (NY)
Metzger, Frederick L (OH)
Morrow, Darrell R (NJ)
Reisinger, Robert R (PA)
Russo, Joseph R (NJ)
Schaefer, David L (PA)
Shields, Edward E (WA)
Shore, Melvin (CA)
Stratton, John M (MO)
Stubbs, William K (MI)
Turnbull, Dale W (OH)
Van Hare, George F, Jr (NJ)
Webb, Jervis C (MI)
Wirtz, Gerald P (IL)
Wood, Harry G (CA)
Yu, A Tobey (NJ)

MATERIALS SCIENCE & ENGINEERING

Adamczak, Robert L (OH)
Allen, Charles W (IL)
Altstetter, Carl J (IL)
Amber, Wayne L (IN)
Angus, John C (OH)
Ansell, George S (CO)
Ardell, Alan J (CA)
Argon, Ali S (MA)
Armstrong, Ronald W (MD)
Armstrong, William M
Aseff, George V (GA)
Ault, G Mervin (OH)
Averbach, Benjamin L (MA)
Baker, William O (NJ)
Banicki, John (MI)
Barnes, George C (PA)
Bauer, Jerome L (CA)
Beachem, Cedric D (DC)
Beauchamp, Jeffery O (TX)
Bechtold, James H (PA)
Beck, Paul A (IL)
Becker, Gerhard W M
Beitscher, Stanley (CO)
Bement, Arden L, Jr (OH)

Benzing, Robert J (VT)
Berger, Richard L (IL)
Bergeron, Clifton G (IL)
Beshers, Daniel N (NY)
Bessen, Irwin I (TX)
Best, Cecil H (KS)
Betner, Donald R (IN)
Bever, Michael B (MA)
Bierlein, John C (MI)
Bigelow, Wilbur C (MI)
Bilello, John C (NY)
Blakely, Jack M (NY)
Blight, Geoffrey E
Block, Robert J (OK)
Blum, Michael E
Boardman, Bruce E (IL)
Bornemann, Alfred (MA)
Bourgault, Roy F (MA)
Bowen, H Kent (MA)
Bowles, C Quinton (MO)
Boyd, James (CA)
Bradley, Elihu F (CT)
Brancato, Emanuel L (MD)
Breinan, Edward M (CT)
Briant, Clyde L (NY)
Brittain, John O (IL)
Britzius, Charles W (MN)
Brophy, Jere H (NY)
Brotzen, Franz R (TX)
Broutman, Lawrence J (IL)
Brown, Norman (PA)
Brunski, John B (NY)
Bublick, Alexander V (CT)
Buchanan, R C (IL)
Buckley, John D (VA)
Budinger, Fred C (WA)
Budinski, Kenneth G (NY)
Bunshah, Rointan F (CA)
Burgers, William Gerard
Burns, Stephen J (NY)
Burr, Arthur A (NY)
Burrier, Harold I, Jr (OH)
Burrus, Charles A , Jr (NJ)
Burte, Harris M (OH)
Bush, G Frederick (MI)
Cadoff, Irving B (NY)
Cady, Philip D (PA)
Cagle, A Wayne (MO)
Callaway, Samuel R (IL)
Carbonara, Robert S (OH)
Carden, Arnold E (AL)
Carlson, O Norman (IA)
Carpenter, Joyce E (GA)
Carr, Stephen H (IL)
Carter, Roger V (WA)
Castleman, Louis S (NY)
Cataldo, Charles E (AL)
Chaklader, Asoke C D
Chang, John C (CA)
Chang, Y Austin (WI)
Chin, Gilbert Y (NJ)
Chiswik, Haim H (IL)
Chou, Tsu-Wei (DE)
Christian, Jack L (CA)
Christopher, Phoebus M (KY)
Chung, Deborah D L (NY)
Chynoweth, Alan G (NJ)
Clark, John P, Jr (PA)
Clarke, Beresford N (IN)
Clements, Linda L (CA)
Coffin, Louis F, Jr (NY)
Cohen, Jerome B (IL)
Cohen, Morris (MA)
Cole, Robert (NY)
Conrad, Hans (NC)
Cooper, Thomas D (OH)
Corelli, John C (NY)
Corneliussen, Roger D (PA)
Corten, Herbert T (IL)
Cost, James R (IN)
Cottrell, Alan Howard
Couchman, Peter R (NJ)
Coursey, W (IA)
Courtney, Thomas H (VA)
Cox, Carl M (WA)

Coyne, James E (MA)
Creighton, Donald L (MO)
Cremens, Walter S (GA)
Crilly, Eugene R (CA)
Cross, Leslie Eric (PA)
Culp, Neil J (PA)
Cunningham, John E (TN)
Dadras, Parviz (OH)
Daehn, Ralph C (IL)
Dahlgren, Shelley D (WA)
Daniels, Raymond D (OK)
Darby, Joseph B, Jr (IL)
DeHoff, Robert T (FL)
Dennis, William E (PA)
Deutsch, George C (MD)
DeVries, Kenneth L (UT)
Diamond, Sidney (IN)
Dickinson, James M (NM)
DiDomenico, Mauro, Jr (NJ)
Dill, Frederick H (NY)
Dirscherl, Rudolf (MO)
Disantis, John A (PA)
Dolch, William L (IN)
Donachie, Matthew J, Jr (CT)
Donnelly, Ralph G (OH)
Dougherty, Thomas A (CA)
Drickamer, Harry G (IL)
Drosten, Fred W (MO)
DuBroff, William (IN)
Dudderar, Thomas D (NJ)
Dulis, Edward J (PA)
Dumin, David J (SC)
Dupuis, Russell D (NJ)
Duquette, David J (NY)
Durant, John H (MA)
Duttweiler, Russell E (OH)
Duwez, Pol E (CA)
Eagan, James R (MI)
Edwards, Eugene H (CA)
Egami, Takeshi (PA)
Ellis, Robert W (MI)
Eppler, Richard A (MD)
Ethington, Robert L (OR)
Faust, J W, Jr (SC)
Feldman, David (PA)
Ficalora, Peter J (NY)
Fine, Morris E (IL)
Flaschen, Steward S (NY)
Flemings, Merton C (MA)
Flood, Paul E (IL)
Fonash, Stephen J (PA)
Fonda, LeGrand B (PA)
Forbes, Leonard (CA)
Ford, Curry E (CT)
Foulke, Donald G (NJ)
Frame, John W (AZ)
France, W DeWayne, Jr (MI)
Freche, John C (OH)
French, Robert D (MA)
Frohmberg, Richard P (MO)
Frost, Brian R T (IL)
Fu, Li-Sheng W (OH)
Fuchs, Henry O (CA)
Fuchsluger, John H (MD)
Fullman, Robert L (NY)
Garber, Charles A (PA)
Gardiner, Keith M (PA)
Gaskell, David R (IN)
Gassner, Robert H (CA)
Gavert, Raymond B (VA)
Geller, Seymour (CO)
Gelles, Stanley H (OH)
Gensamer, Maxwell (FL)
Gerberich, William W (MN)
German, Randall M (NY)
Ghandhi, Sorab K (NY)
Giamei, Anthony F (CT)
Gibala, Ronald (MI)
Gillis, Peter P (KY)
Gilman, John J (IL)
Girifalco, Louis A (PA)

MATERIALS SCIENCE & ENGINEERING
(Continued)

Gjostein, Norman A (MI)
Glasser, Julian (TN)
Goehler, Donald D (WA)
Goering, William A (MI)
Goodenough, John B (TX)
Goodrum, John W (GA)
Gott, Jerome E (CA)
Grace, Richard E (IN)
Graham, Charles D Jr (PA)
Grant, Nicholas J (MA)
Gray, Allen G (OH)
Green, Robert E , Jr (MD)
Greene, Joseph E (IL)
Gregg, Henry T, Jr (NC)
Grenga, Helen E (GA)
Grewal, Manohar S (MA)
Gridley, Robert J (NY)
Gschneidner, Karl A, Jr (IA)
Gulbransen, Earl A (PA)
Gurland, Joseph (RI)
Guttmann, Michel J
Hadley, Henry T (WA)
Haisler, Walter E (TX)
Halverstadt, Robert D (CT)
Hanawalt, Joseph D (MI)
Hanneman, Rodney E (NY)
Hanzel, Richard W (IL)
Harms, William O (TN)
Harris, Norman H (CA)
Harwood, Julius J (MI)
Hatch, Marvin D (KS)
Hauser, Consuelo M (CO)
Hauser, H Alan (CT)
Havner, Kerry S (NC)
Hay, D Robert
Heckler, Alan J (OH)
Hellawell, Angus (WI)
Heller, Gerald S (RI)
Henrie, Thomas A (DC)
Herakovich, Carl T (VA)
Herman, Herbert (NY)
Hertzberg, Richard W (PA)
Hibbard, Walter R (VA)
Hibbeln, Raymond J (IL)
Hillert, Mats H
Hockett, John E (NM)
Hoegfeldt, Jan M (MN)
Hoggatt, John T (WA)
Hornbogen, Erhard
Howe, John P (CA)
Hren, John J (FL)
Hubbell, Dean S (FL)
Hucke, Edward E (MI)
Huddleston, Robert L (MD)
Humenik, Michael (MI)
Hurley, George F (NM)
Hurley, Ronald G (MI)
Inouye, Henry (TN)
Jaffee, Robert I (CA)
Jeanes, Joe W (TX)
Johnson, Andrew B (IL)
Johnson, Robert E (TX)
Johnson, Walter R (MA)
Kaplow, Roy (MA)
Kattamis, Theo Z (CT)
Kaufman, Larry (MA)
Kelsey, Ronald A (PA)
Kennedy, Alfred J
Keyes, Robert W (NY)
King, Thomas B (MA)
Kinser, Donald L (TN)
Kittel, J Howard (IL)
Klein, H Joseph (IN)
Klein, Lisa C (NJ)
Klement, William, Jr (WA)
Kocks, U Fred (NM)
Koczak, Michael J (PA)
Kohl, Walter H (MA)
Koo, Jayoung (NJ)
Koppenaal, Theodore J (CA)

Korchynsky, Michael (PA)
Kottcamp, Edward H Jr (PA)
Koul, M Kishen (NJ)
Kraft, R Wayne (PA)
Kramer, Edward J (NY)
Kramer, Irvin R (MD)
Kreider, Kenneth G (MD)
Krenzer, Robert W (CO)
Kulcinski, Gerald L (WI)
Kushner, Morton (WA)
Kyanka, George H (NY)
LaBelle, Jack E (FL)
Lachman, Walter L (ME)
Laird, Campbell (PA)
Lampson, F Keith (CA)
Langdon, Terence G (CA)
Laramee, Richard C (UT)
Laudise, Robert A (NJ)
Lawrence, Leo A (WA)
Leckie, Frederick A (IL)
Lee, Bansang W (NJ)
Lee III, Thad S (TX)
Leidheiser, Henry, Jr (PA)
Lement, Bernard S (MA)
Lemons, Jack E (AL)
Leverenz, Humboldt W (FL)
Lewis, Edward R (NY)
Lewis, Jack R (CA)
Li, Che-Yu (NY)
Li, James C M (NY)
Lichter, Barry D (TN)
Lidman, William G (PA)
Lieberknecht, Don W (NE)
Liedl, Gerald L (IN)
Lindemer, Terrence B (TN)
Lindholm, Ulric S (TX)
Liss, Robert B (IL)
Long, William G (VA)
Lordi, Francis D (NY)
Luborsky, Fred E (NY)
Lucas, Glenn E (CA)
Lucas, William R (AL)
Luce, Walter A (OH)
Luetje, Robert E (MI)
Lurie, Robert M (MA)
MacDonald, Digby D (OH)
Machlin, Eugene S (NY)
MacKenzie, John D (CA)
MacNeill, Charles E (MD)
Maddin, Robert (PA)
Magarian, Robert J (NJ)
Magee, Christopher L (MI)
Magill, Joseph H (PA)
Mal, Kumar M (NY)
Maletta, Anthony G (NY)
Mannheimer, Walter A
Manning, Melvin L (SD)
Marcus, Harris L (TX)
Marshall, Andrew C (CA)
Martin, William R (IN)
Massalski, Thaddeus B (PA)
Masubuchi, Koichi (MA)
Maurer, Robert D (NY)
Mayer, James W (NY)
McCarthy, John J (NY)
McCoy, Herbert E (TN)
McGee, Thomas D (IA)
McGovern, Sharon A (CA)
McHargue, Carl J (TN)
McKannan, Eugene C (AL)
McKnight, Larry E (CA)
McMahon, Charles J, Jr (PA)
Meshii, M (IL)
Metcalfe, Arthur G (CA)
Meyerhoff, Robert W (NY)
Miller, Oscar O (NJ)
Milstein, Frederick (CA)
Mohamed, Farghalli A (CA)
Montgomery, Donald J (MI)
Moore, Peggy B (OH)
Morgenthaler, Frederic R (MA)

Morral, F Rolf (OH)
Morris, Larry R
Morrow, Darrell R (NJ)
Morrow, JoDean (IL)
Moyer, Kenneth H (NJ)
Mukherjee, Amiya K (CA)
Mukherjee, Kalinath (MI)
Munir, Zuhair A (CA)
Munse, William H (IL)
Munson, Darrell E (NM)
Murphy, Daniel J (AZ)
Murr, Lawrence E (OR)
Nadeau, John S
Nathan, Marshall I (NY)
Nelson, Nyal E (VA)
Newkirk, John B (CO)
Nix, William D (CA)
Nowick, Arthur S (NY)
Ohtake, Tadashi
Old, Bruce Scott (MA)
Oldfield, William (CA)
Ono, Kanji (CA)
Orehotsky, John L (PA)
Orr, William H (IN)
Outwater, John Ogden (VT)
Owen, Walter S (MA)
Paprocki, Stan J (OH)
Park, Joon B (IA)
Parthasarathi, Manavasi N (NY)
Patriarca, Peter (TN)
Pearsall, George W (NC)
Penn, William B (PA)
Perepezko, John H (WI)
Perkins, Richard W (NY)
Perkins, Roger A (CA)
Peters, Stanley T (CA)
Peterson, Norman L (IL)
Petkovic-Luton, Ruzica A (NJ)
Petzow, Gunter E
Phelps, George C (NE)
Phillips, Paul J (TN)
Picha, Robert T (MN)
Pinnel, M Robert (OH)
Planz, Edward J (AL)
Plazek, Donald J (PA)
Pollack, Solomon R (PA)
Pollock, Warren I (DE)
Pope, David P (PA)
Prindle, William R (DC)
Pringle, Oran A (MO)
Promisel, Nathan E (MD)
Raba, Carl F, Jr (TX)
Radcliffe, S Victor (DC)
Raymond, Louis (CA)
Reed, William H (KY)
Reucroft, Philip J (KY)
Reymann, Charles B (AL)
Reynolds, Samuel D, Jr (FL)
Rice, James R (MA)
Richard, Terry G (OH)
Richman, Marc H (RI)
Rigdon, Michael A (VA)
Rinaldi, Michael D (MA)
Risbud, Subhash H (IL)
Rischall, Herman (CA)
Roberts, John M (TX)
Rolston, J Albert (OH)
Rose, Robert M (MA)
Rosi, Fred D (VA)
Rossi, Boniface E (CT)
Rossington, David R (NY)
Royce, Barrie S H (NJ)
Rudee, M Lea (CA)
Ruoff, Arthur L (NY)
Russell, Allen S (SC)
Salama, Kamel (TX)
Sandstrom, Donald J (NM)
Santini, William A, Jr (PA)
Sargent, Gordon A (KY)
Saylor, Wilbur A (CA)
Scala, Eraldus (NY)
Schaefer, Adolph O (PA)
Schetky, Laurence McD (CT)

Schlabach, Tom D (NJ)
Schlatter, Rene (NY)
Schloemann, Ernst F (MA)
Schmatz, Duane J (MI)
Schmitt, George F, Jr (OH)
Schonfeld, Fred W (NM)
Schroder, Klaus (NY)
Schulman, James H (MD)
Schwinghamer, Robert J (AL)
Scott, Alexander R (PA)
Scriven, L E (MN)
Seidman, David N (IL)
Seigle, Leslie L (NY)
Sereda, Peter J
Shackelford, James F (CA)
Shaler, Amos J (PA)
Shane, Robert S (VA)
Shank, Maurice E (CT)
Shapiro, Stanley (NJ)
Shatynski, Stephen R (NY)
Shobert, Erle I, II (PA)
Shubat, George J (IN)
Siegel, Howard J (MO)
Sierakowski, Robert L (OH)
Simnad, Massoud T (CA)
Sinclair, George M (IL)
Slack, Lyle H (VA)
Slichter, William P (NJ)
Smeltzer, Walter W
Smith, Charles O (IN)
Smith, Cyril Stanley (MA)
Smith, Yancey E (MI)
Snitzer, Elias (MA)
Solomon, Alvin A (IN)
Spacil, H Stephen
Sparling, Rebecca H (CA)
Sperry, Philip R (MO)
Spitzig, William A (PA)
Spooner, Stephen (GA)
Spruiell, Joseph E (TN)
Squire, Alexander (WA)
Stadelmaier, Hans H (NC)
Stafford, Stephen W (TX)
Stallmeyer, James E (IL)
Starke, Edgar A , Jr (VA)
Stein, Bland A (VA)
Stetson, Robert F (CA)
Stoloff, Norman S (NY)
Stringfellow, Gerald B (UT)
Sundaresan, Sonny G (SC)
Swalin, Richard A (NJ)
Szekely, Julian (MA)
Talbot, Eugene L (UT)
Taub, James M (AZ)
Taylor, William E (CA)
Thomas, Donald E (PA)
Thomas, Gareth (CA)
Thompson, Anthony W (PA)
Thompson, Earl R (CT)
Tietz, Thomas E (CA)
Tomlinson, John L (CA)
Townsend, Herbert E (PA)
Tremper, Randolph T (OH)
Trojan, Paul K (MI)
Tu, King-Ning (NY)
Tucker, R C Jr (IN)
Uhlig, Herbert H (NH)
Underwood, Ervin E (GA)
Upp, Donald L (MI)
Upthegrove, Wm R (OK)
Valencia, Samuel (MA)
Van Echo, Andrew (MD)
Van Raalte, John A (NJ)
Van Vlack, Lawrence H (MI)
Vander Sande, John B (MA)
Varrese, Francis R (PA)
Verink, Ellis D, Jr (FL)
Versnyder, F L (CT)
Vitovec, Franz H
Vogel, F Lincoln (NJ)
Vook, Richard W (NY)
Waber, James T (IL)
Wachtman, John B, Jr (NJ)

MATERIALS SCIENCE & ENGINEERING
(Continued)

Waisman, Joseph L (CA)
Walker, Walter W (AZ)
Wang, Franklin F Y (NY)
Wardell, Roland D (MN)
Wasilewski, Roman J (DC)
Watson, James F (CA)
Weart, Harry W (MO)
Wechsler, Monroe S (IA)
Weertman, Johannes (IL)
Wegman, Raymond F (NJ)
Wehner, Gottfried K (MN)
Weil, Rolf (NJ)
Weinberg, Harold P (VA)
Weir, James R, Jr (TN)
Weiss, Volker (NY)
Weissmann, Sigmund (NJ)
Weitzenkorn, Lee F (TX)
Wells, Martin G H (PA)
Werner, Fred E (PA)
Wernick, Jack H (NJ)
Westbrook, Jack H (NY)
Westermann, Fred E (OH)
Westwood, Albert R C (MD)
Wiehl, Frederick R (NJ)
Wilcox, William R (NY)
Wilde, Richard A (MI)
Wilder, David R (IA)
Williamson, Robert B (CA)
Wilsdorf, Heinz G F (VA)
Wilson, L Edward (TN)
Winchell, Peter G (IN)
Wolfe, Charles M (MO)
Work, Clyde E (MA)
Worrell, Wayne L (PA)
Wynblatt, Paul P (PA)
Yeh, Hun C (OH)
Yue, Alfred S (CA)
Yurick, Andrew N (NJ)
Zabel, Dale E (AZ)
Zackay, Victor F (CT)
Zakraysek, Louis (NY)
Zwilsky, Klaus M (MD)

MATHEMATICS & STATISTICS

Achenbach, Jan D (IL)
Ancker, Clinton J, Jr (CA)
Anderson, Stanley W (IL)
Bacon, David W
Benjamin, Jack R (CA)
Bershad, Neil J (CA)
Borsting, Jack R (DC)
Bose, Nirmal K (PA)
Bruch, John C, Jr (CA)
Calahan, Donald A (MI)
Carlson, Donald E (IL)
Chen, Wayne H (FL)
Cheng, David K (MD)
Clark, Alfred, Jr (NY)
Cover, Thomas M (CA)
Daugherty, Tony F (TX)
Dicker, Daniel (NY)
Douglas, Jim, Jr (IN)
Drenick, Rudolf F (NY)
Elliott, David L (MO)
Finch, Stephen J (NY)
Fine, Terrence L (NY)
Gill, Arthur (CA)
Goel, Amrit L (NY)
Goldman, Alan J (MD)
Goodman, Theodore R (NY)
Greenberger, Martin (MD)
Gross, Jonathan L (NY)
Hanna, Owen T (CA)
Hartmanis, Juris (NY)
Hunter, J Stuart (NJ)
Iwan, Wilfred D (CA)
Katz, I Norman (MO)
Kazemi, Hossein (CO)

Keller, Joseph B (CA)
Kirmser, Philip G (KS)
Knowles, James K (CA)
Lardner, Thomas J (MA)
Lee, David A (OH)
McMillan, Brockway (ME)
Millsaps, Knox (FL)
Moore, Raymond P (CA)
Mortensen, Richard E (CA)
Nagarsenker, Brahmanand N (PA)
Nemhauser, George L (GA)
Nordgren, Ronald P (TX)
Odeh, Aziz S (TX)
Orth, William A (CO)
Peaceman, Donald W (TX)
Perry, William J (CA)
Pomraning, Gerald C (CA)
Porter, William A (LA)
Rachford, Henry H, Jr (TX)
Rajagopalan, Raj (NY)
Ramkrishna, Doraiswami (IN)
Rankin, Andrew W (SC)
Rice, Stephen O (CA)
Richard, Charles W Jr (OH)
Rodin, Ervin Y (MO)
Rohde, Steve M (MI)
Roman, Basil P (CA)
Rosenshine, Matthew (PA)
Saffman, Philip G (CA)
Saibel, Edward A (NC)
Sandberg, Irwin W (TX)
Scharf, Louis L (CO)
Seebass, A Richard (CO)
Segerlind, Larry J (MI)
Singpurwalla, Nozer D (DC)
Sloane, Neil J A (NJ)
Springer, Melvin D (AR)
Stiffler, Jack Justin (MA)
Streifer, William (CA)
Synck, Louis, J (TX)
Tarn, Tzyh-Jong (MO)
Thomas, John H (NY)
Triffet, Terry (AZ)
Veatch, Ralph, W, Jr (OK)
Verhoff, Frank H (WV)
Wallace, Norval D (MO)
Whitwell, John C (NJ)
Zierler, Neal (NJ)

MECHANICAL DESIGN

Adare, James R (NY)
Anderson, Harold H
Barnwell, Joseph H (LA)
Bayne, James W (IL)
Blake, Alexander (CA)
Bonthron, Robert J (IL)
Book, Wayne J (GA)
Brennan, James J (TX)
Brockman, Donald C (MI)
Broward, Hoyt E (FL)
Calder, Clarence A (OR)
Chang, Anthony T (NJ)
Cheatham, John B, Jr (TX)
Collins, Jack A (OH)
Crisp, John N (KS)
Degenhardt, Robert A (DC)
Dilpare, Armand L (FL)
Dix, Rollin C (IL)
Dolan, Thomas J (IL)
Froula, James D (TN)
Gaggstatter, Henry D, Jr (FL)
Gruber, Jerome M (NJ)
Gupta, Krishna Chandra (IL)
Habach, George F (NJ)
Huang, Chi-Lung (Dominic) (KS)
Johnson, Lloyd H (NY)
Katz, Peter (TN)
Kauzlarich, James J (VA)

Larson, Carl S (IL)
Linderoth, L Sigfred, Jr (NC)
Manjoine, Michael J (PA)
Martin, George H (MI)
Meriam, James L (CA)
Moshier, Glen W (MO)
Nelson, Eric W (PA)
Panlilio, Filadelfo (NY)
Pavelic, Vjekoslav (WI)
Ragsdell, Kenneth M (IN)
Ross, Bernard (CA)
Safar, Zeinab S (FL)
Sneck, Henry J (NY)
Tarantine, Frank J (OH)
Tartaglia, Paul E (VT)
Townsend, Miles A (VA)
Vidosie, Joseph Paul (NY)
Waldron, Kenneth J (OH)
Webster, Karl S (ME)
Wells, Larry G (KY)
Wesner, John W (NJ)
West, Clinton, L (CA)
Youngdahl, Paul F (CA)
Zwiep, Donald N (MA)

MECHANICAL ENGINEERING

Abata, Duane L (MI)
Abdel-Khalik, Said I (WI)
Adam, Paul J (MO)
Adams, William J, Jr (CA)
Addy, Alva L (IL)
Adrian, Ronald J (IL)
Advani, Sunder H (OH)
Akin, Lee S (CA)
Alford, Jack L (CA)
Alic, John A (DC)
Allen, Charles W (CA)
Allen, Herbert (TX)
Alvine, Raymond G (NE)
Amberg, Arthur A (IL)
Anand, Dave K (MD)
Anderson, David Melvin (CA)
Anderson, J Hilbert (PA)
Anderson, Raymond L (NC)
Andrews, Raynal W, Jr (PA)
Antrobus, Thomas R (PA)
Appleby, Loran V (NY)
Archang, Homayoun
Arnold, Lynn E (OH)
Aronson, David (NJ)
Arthur, Paul D (CA)
Ash, Robert L (VA)
Ault, Eugene Stanley (MI)
Ayres, James Marx (CA)
Azar, Jamal J (OK)
Bagci, Cemil (TN)
Bahar, Leon Y (PA)
Bahnfleth, Donald R (OH)
Baillif, Ernest A (MI)
Bajura, Richard A (WV)
Baker, Wilfred E (TX)
Barfield, Robert F (AL)
Barkan, Philip (CA)
Barnett, Samuel C (GA)
Barron, Randall F (LA)
Barstow, Robert J (MA)
Batch, John M (OH)
Baxter, Meriwether L Jr (NY)
Bayne, James W (IL)
Beadle, Charles W (CA)
Beakley, George C (AZ)
Beam, Benjamin H (CA)
Beamer, Scott (CA)
Beck, James V (MI)
Bendelius, Arthur G (GA)
Benson, Willard R (NJ)
Bergles, Arthur E (NY)
Bergman Donald J (IL)

Berkeley, Frederick D (NY)
Berlad, Abraham L (NY)
Bernard, James E (MI)
Bertrand, S Peter (LA)
Bethea, Thomas Jesse (SC)
Bezier, Pierre E
Binder, Sol (PA)
Bishop, Thomas R (TX)
Bishop, William H IV (MD)
Bissell, Roger R (ID)
Black, William Z (GA)
Blodgett, Omer W (OH)
Boehm, Robert F (UT)
Bollinger, John G (WI)
Bolt, Jay A (MI)
Bolz, Ray E (MA)
Bowen, Ray M (KY)
Bowman, Thomas E (FL)
Boynton, Edgar B (VA)
Bozatli, Ali N (AR)
Bradford, Ralph J, Jr (CA)
Braisted, Paul W (MO)
Brand, John R (DE)
Bressler, Marcus N (TN)
Bridgers, Frank H (NM)
Briggs, Edward C (CA)
Brighton, John A (GA)
Bringhurst, Lynn H (UT)
Broadman, Gene A (CA)
Brown, Anthony C (NJ)
Brown, F Leslie (IL)
Brown, Harry L (NJ)
Brown, Wayne S (UT)
Bruce, Albert W (CA)
Brummett, Forrest D (IN)
Bruns, Robert F (MO)
Bryers, Richard W (NJ)
Buchele, Wesley F (IA)
Budenholzer, Roland A (IL)
Burkert, John Wallace, Sr (DE)
Burr, Arthur H (TX)
Burr, Richard W (TX)
Burrell, Montrust Q (LA)
Burton, Bruce E (OH)
Burton, Ralph A (NC)
Byars, Edward F (SC)
Cabble, George M (GA)
Cahill, William J, Jr (LA)
Campbell, Henry J, Jr (NY)
Cantieri, William F (OH)
Carden, Arnold E (AL)
Carew, Wm E, Jr (DE)
Carley, Charles T, Jr (MS)
Carlson, Harold C R (NJ)
Carson, Gordon B (MI)
Carson, Larry M (OR)
Carter, Hugh C (CA)
Case, Edward T (UT)
Cashin, Francis J (NY)
Castro, Walter E (SC)
Chaddock, Jack B (NC)
Charley, Philip J (CA)
Cheatham, John B, Jr (TX)
Chelton, Dudley B (CO)
Chen, Michael M (IL)
Chen, Rong-Yaw (NJ)
Chen, Ta-Shen (MO)
Cherna, John C
Cherne, Realto E (NY)
Chewning, Ray C (CA)
Chiang, Donald C (IN)
Chilton, Ernest G (CA)
Chiou, Jiunn P (MI)
Chipouras, Peter A (MA)
Cho, Soung M (NJ)
Chou, James C S (HI)
Chou, Tsu-Wei (DE)
Chu, Richard C (NY)
Chwang, Allen T (IA)
Clark, William R (DE)
Clauser, Francis H (CA)
Cloke, Thomas H (IL)
Cochran, Billy J (LA)
Codola, Frank C (NY)
Cofer, Daniel B (GA)

MECHANICAL ENGINEERING
(Continued)

Leckie, Frederick A (IL)
Lehnhoff, Terry F (MO)
Leidel, Frederick O (WI)
Leigh, Donald C (KY)
Leitmann, George (CA)
LeMay, Iain
Lester, John W (AZ)
Levens, Alexander S (CA)
Levingston, Ernest L (LA)
Levy, Salomon (CA)
Lewis, Edward R (NY)
Lewis, Oliver K (GA)
Libertiny, George Z (MI)
Lichty, William H (MI)
Lightowler, Joseph, Jr (ND)
Liley, Peter E (IN)
Lindholm, John C (KS)
Lindquist-Skelley, Sharon L (VA)
Lindsey, Larry M (TX)
Ling, Frederick F (NY)
Lior, Noam (PA)
Littrell, John L (CA)
Liu, Wing Kam (IL)
Lock, Gerald S H
Lorenzen, Coby (CA)
Loser, Rene F (NY)
Love, John (MO)
Love, Tom J, Jr (OK)
Lowe, Philip A (MD)
Lowen, Gerard G (NJ)
Loyalka, Sudarshan K (MO)
Lux, George R (VA)
Lux, William J (IA)
Mabie, Hamilton H (VA)
MacGregor, Charles W (MA)
Mack, Michael J (IA)
Madeheim, Huxley (NY)
Maeder, Paul F (RI)
Magee, Richard S (NJ)
Manfredi, Robert R (MD)
Manvi, Ram (CA)
Marlowe, Donald E (DC)
Maroney, George E (MA)
Marple, Virgil A (MN)
Marshek, Kurt M (TX)
Martin, George H (MI)
Martin, John J (MD)
Mason, Ted D (IL)
Massart, Keith G (WA)
Mater, Milton H (OR)
Mates, Robert E (NY)
Matsukado, William M (OR)
Matthews, James B (MI)
Maurer, Karl G (ND)
Maxwell, Barry R (PA)
Maxworthy, Tony (CA)
Maybeck, Edward M (NY)
Mayers, Albert J, Jr (LA)
McAleer, William K (PA)
McCarthy, Roger L (CA)
McClure, Eldon R (CA)
McCool, Alexander A, Jr (AL)
McCormick, William J (TX)
McCoy, Wyn E (OH)
McCullough, Hugh (PA)
McDavid, Frederick R (VA)
McFadden, Peter W (CT)
McFeron, Dean E (WA)
McGrath, Philip I, Jr (AL)
McHugh, Edward (NY)
McKillop, Allan A (CA)
McLean, William G (PA)
McLean, William N (IL)
McWilliams, Bayard T (PA)
Meagher, George Vincent
Meckler, Milton (CA)
Meier, Donald R (TN)

Menster, Paul C (OH)
Migliore, Herman J (OR)
Miles, John B (MO)
Miller, Daniel R (AR)
Miller, Edward (NJ)
Miller, Paul L, Jr (KS)
Milligan, Mancil W (TN)
Mischke, Charles R (IA)
Moen, Walter B (NJ)
Moffat, Robert J (CA)
Moffatt, Charles A (WV)
Moltrecht, Karl Hans (OH)
Monroe, Paul S (NJ)
Moore, Roland E (TX)
Moretti, Peter M (OK)
Morgan, George H (IN)
Morin, Herman L (NY)
Morino, Luigi (MA)
Morley, Richard J (NJ)
Mosher, Frederick K (NJ)
Mote, C D, Jr (CA)
Motter, Eugene F (TX)
Mraz, George J (NH)
Mura, Toshio (IL)
Murdock, James W (PA)
Muster, Douglas (TX)
Myers, Phillip S (WI)
Nack, Donald H (CA)
Nardone, Pio (NJ)
Nash, Jonathon M (MD)
Nation, Oslin (TX)
Naylor, Henry A, Jr (MD)
Nee, Raymond M (NY)
Neifert, Harry R (OH)
Nelson, Frederick C (MA)
Nelson, William E (CA)
Neou, In-Meei (CA)
Newman, Malcolm (NY)
Newton, Jeffrey M (VA)
Nichols, Donald E (TN)
Niordson, Frithiof I
Norton, John A (MA)
Nypan, Lester J (CA)
Obert, Edward F (WI)
O'Dea, Thomas J (NY)
O'Donnell, William J (PA)
Ogata, Katsuhiko (MN)
Olds, Joneil R (CO)
Oliver, Paul E (OR)
Olowu, Olayeni
Olsen, Joseph C (CA)
Olsen, Sydney A (IA)
Olson, Donald R (PA)
Olt, Richard G (OH)
Oppenheim, Antoni K (CA)
Otte, Karl H (IL)
Owen, Warren H (NC)
Page, Robert H (TX)
Palmatier, Everett P (FL)
Pan, Coda H T (NY)
Paoluccio, Joseph P (CA)
Park, William H (PA)
Parker, Jerald D (OK)
Parker, Robert B (RI)
Parker, Walter B (FL)
Parks, Lyman L (NJ)
Parks, William W (IL)
Parmley, Robert O (WI)
Pasqua, Pietro F (TN)
Patel, Virendra C (IA)
Paul, Burton (PA)
Paul, Frank W (SC)
Paulus, J Donald (MD)
Pavelic, Vjekoslav (WI)
Payne, Matthew A (NJ)
Pearsall, George W (NC)
Pearson, Allan E (RI)
Pearson, Joseph T (IN)
Pefley, Richard K (CA)
Pei Chi Chou (PA)
Pellett, C Roger (OH)
Perkins, Henry C (AZ)
Perry, Lawrence E, Jr (VA)
Peters, Alexander R (NE)
Peters, Jacques M
Peters, Leo C (IA)

Peters, LeRoy L (CA)
Peterson, Donn N (MN)
Peterson, Gary J (CO)
Pfender, Emil (MN)
Phelan, Richard M (NY)
Phelps, Dudley F (NY)
Phillips, William R (CA)
Phillips, Winfred M (IN)
Picha, Kenneth G (MA)
Polaner, Jerome L (NJ)
Pollock, Wilfred A (TX)
Polve, James H (UT)
Pope, Malcolm H (VT)
Porter, R Clay, (MI)
Potter, J Leith (TN)
Potter, Merle C (MI)
Powell, Herschel L (TN)
Pratt, David T (WA)
Preiss, Kenneth
Preston, Walter B (TX)
Prewitt, Charles E (LA)
Pringle, Oran A (MO)
Proctor, Charles L
Prohl, Melvin A (MA)
Przirembel, Christian E G (SC)
Purcupile, John C (OK)
Rabins, Michael J (TX)
Radcliffe, Charles W (CA)
Randolph, Robert E (UT)
Rastoin, Jean M
Raymond, Louis (CA)
Reaser, Wilbur W (WA)
Redpath, Richard J (MO)
Reece, C Jeff, Jr (NC)
Reeder, Harry C (OR)
Reethof, Gerhard (PA)
Reid, Robert L (WI)
Reinhart, William A (WA)
Rennicks, Robert S (NJ)
Repscha, Albert H (FL)
Rice, Stephen L (FL)
Riddick, Edgar K, Jr (AR)
Rim, Kwan (IA)
Rimrott, Friedrich P J
Roberts, A Sidney, Jr (VA)
Robins, Daniel F (OH)
Rodgers, Oliver E (PA)
Rogers, Willard L (AZ)
Rohr, Peter H (NJ)
Rohsenow, Warren M (MA)
Romaguera, Mariano A (PR)
Roman, Basil P (CA)
Romano, Frank A (MI)
Rosard, Daniel D (PA)
Rosenberg, Richard (CA)
Rosenberg, Ronald C (MI)
Ross, Bernard (CA)
Ross, Donald E (NY)
Roth, Bernard (CA)
Rothermel, U Amel (NC)
Ruisack, Robert G, Jr (KS)
Rusack, John D (NY)
Rush, James R (ND)
Rushing, Frank C (MD)
Russell, Lynn D (MS)
Rybicki, Edmund F (OK)
Sander, Louis F (PA)
Sanders, William T (NY)
Sandgren, Find (IN)
Sandor, George N (FL)
Sandquist, Gary M (UT)
Sanford, Keith C (CA)
Sauer, Harry J, Jr (MO)
Sayre, Clifford L, Jr (MD)
Scavuzzo, Rudolph J (OH)
Schoech, William J (IN)
Scholz, Paul D (IA)
Schurman, Glenn A (CA)
Scofield, Gordon L (SD)
Sebastian, Edmund J (IL)
Seckman, Thomas C (TN)
Secord, George N
Sehn, Francis J (MI)
Seireg, Ali A (WI)

Selby, Joseph C
Selm, Robert P (KS)
Seng, Thomas G, Jr (IL)
Sevier, William J (CA)
Sforza, Pasquale M (NY)
Shabaik, Aly H (CA)
Shairman, Alvin H (MA)
Shanebrook, J Richard (NY)
Sharp, Howard R (OK)
Sharpe, William N, Jr (MD)
Shigley, Joseph E (MI)
Shima, Hideo
Short, Byron E (TX)
Shoup, Terry E (TX)
Shreve, Gerry D (MI)
Shuck, Lowell Zane (WV)
Sigel, Robert J (PA)
Silvestri, George J, Jr (PA)
Simon, Albert (NY)
Simonsen, John M (UT)
Sines, George H (CA)
Sissom, Leighton E (TN)
Sisto, Fernando (NJ)
Skerkoski, Eugene C (OH)
Slonneger, Robert D (WV)
Smith, Donald S (CA)
Smith Frederick A Jr (NY)
Smith, J F Downie (FL)
Smith, Leslie S (FL)
Smith, Lester C (CT)
Smithberg, Eugene H (MA)
Sneck, Henry J (NY)
Snyder, Robert D (NC)
Soehrens, John E (CA)
Solberg, Ruell F, Jr (TX)
Soo, Shao L (IL)
Sorrell, Furman Yates (NC)
Spencer, Robert C (NY)
Spiegel, Walter F (PA)
Spotts, Merhyle F (IL)
Springer, E Kent (CA)
Staggs, Harlon V (KY)
Stair, William K (TN)
Stamper, Eugene (NJ)
Staples, Carlton W (MA)
Statton, Charles D (CA)
Steffen, John R (IN)
Steidel, Robert F, Jr (CA)
Stellwagen, Robert H (MI)
Stern, Richard M (WA)
Stoecker, Wilbert F (IL)
Stokes, Vijay K (NY)
Stukel, James J (IL)
Sturgeon, Franklin E (PA)
Sutherland, Robert L (WY)
Swanson, A Einar (MO)
Talbot, Thomas F (AL)
Talbott, John A (OR)
Tao, William K Y (MO)
Tarantine, Frank J (OH)
Tasi, James (NY)
Taylor, C Fayette (MA)
Tennant, Jeffrey S (FL)
Theiss, Ernest S (OH)
Theros, William J (PA)
Thomas, Donald H (PA)
Thomas, Frank A, Jr (TX)
Thompson, Hugh A (LA)
Thompson, Joseph G (KS)
Thompson, Walter Dale (NC)
Thompson, Willis F (CT)
Tippett, G Ross (TX)
Torrance, Kenneth E (NY)
Tree, David R (IN)
Trueblood, Walter D (MO)
Trull, Olin C (SC)
Tulenko, James S (FL)
Turnquist, Ralph O (KS)
Tutt, Charles L, Jr (MI)
Ugelow, Albert J (FL)
Uyehara, Otto A (CA)
Vachon, Reginald I (GA)
Valentine, William M (WA)
Vance, Carl B (IN)

MECHANICAL ENGINEERING
(Continued)

Vance, John M (TX)
Vandegrift, Erskine, Jr (AL)
Velkoff, Henry R (OH)
Ventrice, Marie B (TN)
Vidosic, Joseph Paul (NY)
Viggers, Robert F (WA)
Voelcker, H B (NY)
Voiculescu, Ion Alexandru
Vollmer, Reuben P (CA)
Von Flue, Frank W (CA)
von Turkovich, Branimir (VT)
Walker, Gary Wayne (CA)
Walker, Hugh S (KS)
Walker, M Lucius, Jr (DC)
Walker, William F (TX)
Wallis, Graham B (NH)
Walston, William H, Jr (MD)
Wambold, James C (PA)
Wang, Shu-Yi (MS)
Wang, Su-Su (IL)
Warder, Richard C, Jr (MO)
Warmann, Robert A (MO)
Wasley, Richard J (CA)
Weber, Homer S (GA)
Webster, Karl S (ME)
Weinstein, Alvin S (NH)
Wells, William P (NC)
Wentz, William H (KS)
Wesstrom, David B (DE)
White, L Forrest (TN)
White, Robert A (IL)
Whitehouse, Gerald D (LA)
Wick, Robert S (TX)
Wiebe, Donald (PA)
Wiggins, Edwin G (NY)
Wilding, John D (NJ)
Williams, G Bretnell (OH)
Williams, Richard F (ID)
Williams, S Hoyt (VA)
Williams, Wendell M, Jr (GA)
Wilson, Charles (NJ)
Wilson, Claude L (TX)
Wilson, Jack P (PA)
Wilson, Ralph S (NM)
Windolph, George W (PA)
Windt, Ewald C (OH)
Winer, Ward O (GA)
Wirtz, Richard A (NY)
Witte, Larry C (TX)
Woelfel, Albert E (TX)
Wollensak, Richard J (MA)
Woods, John G (PA)
Woodward, John B (MI)
Wright, Daniel K, Jr (OH)
Wright, Richard H (GA)
Wright, Wesley F, Jr (TX)
Young, Fred M (TX)
Yow, John R (NC)
Yu, Yi-Yuan (NJ)
Zarling, John P (AK)
Zerban, Alexander H (CT)
Zick, Leonard Paul (AZ)
Zickel, John (CA)
Zimmermann, Francis J (PA)
Zinkham, Robert E (VA)
Zorowski, Carl F (NC)
Zowski, Thaddeus (IL)
Zwiep, Donald N (MA)
Zwiep, Donald Nelson (MA)

MECHANICS, STRESS ANALYSIS, VIBRATION

Abramson, H Norman (TX)
Achenbach, Jan D (IL)

Advani, Sunder H (OH)
Allentuch, Arnold (NJ)
Altiero, Nicholas J (MI)
Anand, Subhash C (SC)
Archer, Robert R (MA)
Argon, Ali S (MA)
Armand, Jean-Louis (CA)
Ault, Eugene Stanley (MI)
Bagci, Cemil (TN)
Bahar, Leon Y (PA)
Bailey, Cecil D (OH)
Bain, James A (PA)
Batra, Romesh C (MO)
Batterman, Steven C (PA)
Belytschko, Ted B (IL)
Berger, Bruce S (MD)
Bernard, James E (MI)
Bert, Charles W (OK)
Beskos, Dimitrios E
Bieniek, Maciej P (NY)
Blake, Alexander (CA)
Bleich, Hans H (NY)
Boardman, Bruce E (IL)
Boettcher, Harold P (WI)
Boley, Bruno A (NY)
Bollard, John R H (WA)
Boresi, Arthur P (IL)
Borg, Sidney F (NJ)
Bowen, Ray M (KY)
Boyd, Landis L (CO)
Bradley, William A (MI)
Brinson, Halbert F (VA)
Brown, Samuel J (TX)
Brunelle, Eugene J, Jr (NY)
Budiansky, Bernard (MA)
Burton, Ralph A (NC)
Butler, Donald J (VT)
Cakmak, Ahmet S (NJ)
Calder, Clarence A (OR)
Carley, Thomas G (TN)
Carlson, Donald E (IL)
Carnes, Walter R (MS)
Chang, Anthony T (NJ)
Chang, Chin-Hao (AL)
Chang, George C (OH)
Chen, Ming M (MA)
Cheng, David H (NY)
Chi, Michael (VA)
Chiang, Fu-Pen (NY)
Clark, Samuel K (MI)
Cliett, Charles B (MS)
Cloud, Gary L (MI)
Codola, Frank C (NY)
Cohen, Raymond (IN)
Conway, Lawrence (FL)
Cooper, William E (MA)
Corten, Herbert T (IL)
Cost, Thomas L (AL)
Cowin, Stephen C (LA)
Cox, J Carroll (SC)
Cranch, Edmund T (NY)
Crandall, Stephen H (MA)
Crossley, F R Erskine (CT)
Cunniff, Patrick F (MD)
Cunningham, Floyd M (MO)
Curreri, John R (NJ)
Cutchins, Malcolm A (AL)
Cutler, Verne C (WI)
Czyzewski, Harry (OR)
Dalley, Joseph W (TX)
Dally, James W (MD)
Daniel, Isaac M (IL)
Datta, Subhendu K (CO)
Davidson, Arthur C
Davis, Evan A (PA)
Davis, Robert L (MO)
Deresiewicz, Herbert (NY)
DiJulio, Roger M, Jr (CA)
Dill, Ellis H (NJ)
Dilpare, Armand L (FL)
DiMaggio, Frank L (NY)
Dixon, Marvin W (SC)
Dohse, Fritz E
Dolan, Thomas J (IL)
Donaldson, Bruce K (MD)

Dong, Stanley B (CA)
Donnell, Lloyd H (CA)
Douglas, Robert A (NC)
Dowell, Dr Douglas (CA)
Drucker, Daniel C (FL)
Dryer, Frederick L (NJ)
Duffy, Jacques (RI)
Dundurs, John (IL)
Durelli, August J (MD)
Dym, Clive L (MA)
Eisenberg, Martin A (FL)
El-Bayoumy, Lotfi E (CA)
Ely, John F (NC)
Engin, Ali E (OH)
Eubanks, Robert A (IL)
Evces, Charles R (AL)
Faucett, Thomas R (MO)
Felske, Armin
Felton, Lewis P (CA)
Fincher, James R (GA)
Findley, William N (RI)
Fisher, Cary A (CO)
Fitzgerald, J Edmund (GA)
Freund, Lambert Ben (RI)
Friederich, Allan G (IL)
Friedmann, Peretz P (CA)
Fu, Li-Sheng W (OH)
Gadd, Charles W (NC)
Gant, Edward V (CT)
Gartner, Joseph R (CT)
Geers, Thomas L (CA)
Genin, Joseph (NM)
Georgian, John C (MO)
Gere, James M (CA)
Gillis, Peter P (KY)
Goldberg, John E (IN)
Goldsmith, Werner (CA)
Graff, Karl F (OH)
Grant, Donald A (ME)
Greif, Robert (MA)
Grunfeld, Michael (CA)
Hamilton, James F (IN)
Hammett, Cecil E (AL)
Hansberry, John W (MA)
Hart, Franklin D (NC)
Hartman, David E (AZ)
Hartsaw, William O (IN)
Harvill, Lawrence R (CA)
Hashin, Zvi
Haug, Edward J, Jr (IA)
Havner, Kerry S (NC)
Haythornthwaite, Robert (PA)
Herakovich, Carl T (VA)
Herrmann, George (CA)
Hetnarski, Richard B (NY)
Hicks, Earl J (OK)
Higdon, Archie (CA)
Hodge, Philip G, Jr (MN)
Hoff, Nicholas J (CA)
Hogan, Mervin B (UT)
Hogg, Allan D
Homewood, Richard H (MA)
Hoppmann, William H, II (SC)
Horvay, Gabriel (MA)
Huang, T C (WI)
Hudson, Donald E (CA)
Hulbert, Lewis E (OH)
Hundal, Mahendra S (VT)
Huston, Ronald L (OH)
Hutchinson, James R (CA)
Irwin, George R (MD)
Iwan, Wilfred D (CA)
Jenike, Andrew W (NH)
Johanson, Jerry R (CA)
Johns, Thomas G (TX)
Johnson, Aldie E, Jr (MA)
Jones, Irving W (DC)
Jordan, William D (AL)
Ju, Frederick D (NM)
Katz, I Norman (MO)
Keaton, Clyde D (MA)
Keer, Leon M (IL)
Kemper, John D (CA)

Kempner, Joseph (NY)
Kenner, Vernal H (OH)
Kerr, Arnold D (DE)
Kim, Thomas J (RI)
Kimel, William R (MO)
Knowles, James K (CA)
Kobayashi, Albert S (WA)
Koenig, Herbert A (CT)
Koh, Severino L (WV)
Koiter, Warner T
Koplik, Bernard (NY)
Korda, Peter E (OH)
Kozik, Thomas J (TX)
Krajcinovic, Dusan (IL)
Kraus, Harry (CT)
Kuivinen, Thomas O (AZ)
Langhaar, Henry L (IN)
Lardner, Thomas J (MA)
Law, E Harry (SC)
Lee, Ju P (MI)
Leigh, Donald C (KY)
Leissa, Arthur W (OH)
Libove, Charles (NY)
Likins, Peter W (NY)
Lin, Tung H (CA)
Lindholm, John C (KS)
Lindholm, Ulric S (TX)
Lutes, Loren D (TX)
Macduff, John N (OR)
MacGregor, Charles W (MA)
Mains, Robert M (MO)
Martin, John B
Masri, Sami F (CA)
McBean, Robert P (MO)
McCormick, Frank J (KS)
McDonald, Patrick H, Jr (NC)
McDonough, George F, Jr (AL)
McElhaney, James H (NC)
McLay, Richard W (VT)
Melosh, Robert J (NC)
Merckx, Kenneth R (WA)
Meriam, James L (CA)
Migliore, Herman J (OR)
Miller, Edward (NJ)
Miller, Robert E (IL)
Mindlin, Raymond D (CT)
Mitchell, Thomas P (CA)
Moffatt, Charles A (WV)
Morrow, JoDean (IL)
Mortimer, Richard W (PA)
Mote, C D, Jr (CA)
Mow, Maurice (CA)
Mow, Van C (NY)
Mura, Toshio (IL)
Murray, William M
Muvdi, Bichara B (IL)
Nachlinger, R Ray (TX)
Nelson, Richard B (CA)
Neou, In-Meei (CA)
Nevill, Gale E, Jr (FL)
Niedenfuhr, Francis W (DC)
Niordson, Frithiof I
Nolan, Robert W (DE)
Noor, Ahmed K (VA)
Nordgren, Ronald P (TX)
Nunziato, Jace W (NM)
Oglesby, David B (MO)
Ojalvo, Irving U (CT)
Ojalvo, Morris (OH)
Orth, William A (CO)
Osborne, Edward A (CO)
Pae, Kook D (NJ)
Pai, David H (NJ)
Palazotto, Anthony N (OH)
Panlilio, Filadelfo (NY)
Papirno, Ralph (MA)
Parker, Paul E (IL)
Parks, Vincent J (DC)
Pelan, Byron J (NJ)
Peppin, Richard J (MD)
Perkins, Richard W (NY)
Peters, David A (MO)

MECHANICS, STRESS ANALYSIS, VIBRATION
(Continued)

Phillips, Aris (CT)
Phillips, James W (IL)
Pian, Theodore H H (MA)
Pih, Hui (TN)
Piziali, Robert L (CA)
Plass, Harold J (FL)
Pletta, Dan H (VA)
Plunkett, Robert (MN)
Pollard, Ernest I (PA)
Popelar, Carl H (OH)
Popov, Egor P (CA)
Raftopoulos, Demetrios D (OH)
Ramberg, Walter G C (MD)
Raville, Milton E (GA)
Rawlins, Charles B (NY)
Reissner, Eric (CA)
Rice, James R (MA)
Riegner, Earl I (PA)
Riley, William F (IA)
Rimrott, Friedrich P J
Robe, Thurlow R (OH)
Robinson, Arthur R (IL)
Roeder, Charles W (WA)
Rogers, Bruce G (TX)
Rogers, Darrell O (MI)
Rollins, John P (NY)
Ross, Arthur L (PA)
Rowlands, Robert E (WI)
Rybicki, Edmund F (OK)
Saibel, Edward A (NC)
Salane, Harold J (MO)
Sanford, Robert J (DC)
Schmidt, Robert (MI)
Schmit, Lucien A (CA)
Schultz, Albert B (MI)
Sciammarella, Cesar A (IL)
Scipio, L Albert, II (MD)
Sevin, Eugene (MD)
Shaffer, Bernard W (NY)
Shahani, Ray W (OH)
Sharpe, William N, Jr (MD)
Shield, Richard T (IL)
Shinozuka, Masanobu (NY)
Sidebottom, Omar M (IL)
Sierakowski, Robert L (OH)
Sikarskie, David L (MI)
Smelcer, Glen E (IA)
Smith Frederick A Jr (NY)
Smith, Paul R, (NY)
Snyder, Robert D (NC)
Sorokach, Robert J (OH)
Soutas-Little, Robert (MI)
Starr, James E (NC)
Sternberg, Eli (CA)
Stevens, Karl K (FL)
Stevenson, John D (OH)
Stone, Morris D (PA)
Strauss, Alvin M (OH)
Symonds, Paul S (RI)
Szabo, Barna A (MO)
Tasi, James (NY)
Tauchert, Theodore R (KY)
Taylor, Arthur C, Jr (VA)
Taylor, Charles E (FL)
Testa, Rene B (NY)
Thomson, William T (CA)
Thurman, Herman R, Jr (DE)
Timo, Dominic P (NY)
Ting, Thomas C T (IL)
Torby, Bruce J (CA)
Torvik, Peter J (OH)
Triffet, Terry (AZ)
Truesdell, Clifford A (MD)
Trumpler, Paul R (PA)
Trumpler, William E (PA)
Tsao, Ching H (CA)
Ueng, Charles ES (GA)
Ugural, Ansel C (NJ)
Ungar, Eric E (MA)
Valentin, Richard A (IL)

Veletsos, Anestis S (TX)
von Riesemann, Walter A (NM)
Wahl, Arthur M (FL)
Walker, Hugh S (KS)
Walsh, Edward K (FL)
Walston, William H, Jr (MD)
Wang, Shien T (KY)
Wang, Su-Su (IL)
Watwood, Vernon B (MI)
Waugh, David J (SC)
Weese, John A (TX)
Weiner, Jerome H (RI)
Wen, Robert K (MI)
Westmann, Russell A (CA)
Wheeler, Lewis T (TX)
Wilcox, Marion W (TX)
Wilhoit, James C , Jr (KY)
Williams, Harry E (CA)
Woodward, James H (AL)
Work, Clyde E (MA)
Worley, Will J (IL)
Young, Dana (TX)
Youngdahl, Carl K (IL)
Yow, John R (NC)
Yu, Yi-Yuan (NJ)
Zwicky, Everett E, Jr (NY)

MEDICAL

Ketterer, Frederick D (PA)
Prewitt, Judith M S (MD)
Waynant, Ronald W (MD)

METAL FORMING, SHAPING & PROCESSING

Allen, Herbert (TX)
Altan, Taylan (OH)
Andrews, Al L (AZ)
Asquith, David J (MA)
Baker, George S (WI)
Batra, Romesh C (MO)
Beck, Paul A (IL)
Becker, Peter E (NJ)
Black, J Temple (AL)
Boulger, Francis W (OH)
Bramfitt, Bruce L (PA)
Brosilow, Rosalie (OH)
Byrer, Thomas G (OH)
Carson, Larry M (OR)
Carter, Joseph R (MA)
Cary, Robert A (PA)
Chase, Thomas L (OH)
Conrad, Hans (NC)
Cox, Elmer J (PA)
Cremisio, Richard S (OH)
Curran, Robert M (NY)
Dadras, Parviz (OH)
Dancy, Terence E
Daykin, Robert P (WI)
Delgrosso, Eugene J (CT)
Dembowski, Peter V (NY)
Dodds, Walter B (WI)
Duncan, Samuel W (IL)
Dwivedi, Surendra Nath (VA)
Fiorentino, Robert J (OH)
Ford, Hugh
Freeman, William R, Jr (CT)
French, Robert D (MA)
Ghosh, Amit K (CA)
Goldman, Manuel (OH)
Gonser, Bruce W (OH)
Goodwin, James G (UT)
Harrington, H Richard (PA)
Hemphill, Dick W (IN)
Hendrickson, Alfred A (MI)
Heydt, Gerald B (PA)
Hoeppner, Steven A (NY)
Hoffmanner, Albert L (OH)
Holfinger, Robert R (OH)

Hook, Rollin E (OH)
Jonas, John J
Kalish, Herbert S (NJ)
Kasper, Arthur S (FL)
Keeler, Stuart P (MI)
Khare, Ashok K (PA)
Kinstler, John R (MI)
Kisslinger, Fred (MO)
Klancko, Robert John (CT)
Kobayashi, Shiro (CA)
Kotfila, Ralph J (MO)
Kuebler, Ronald C (SC)
Kuhn, Howard A (PA)
Kulkarni, Kishor M (OH)
Leach, James L (IL)
Liuzzi, Leonard, Jr (NY)
Loewenstein, Paul (MA)
Lordi, Francis D (NY)
Lovering, Earle W (IL)
Lovgren, Carl A (PA)
Lula, Remus A (PA)
Maciag, Robert J (NY)
Male, Alan T (PA)
Manly, William D (TN)
McCarty, James E (CA)
Melhorn, Wilton F (IN)
Michalak, Ronald S (MI)
Nachtman, Elliot S (IL)
Norwood, Sydney L (AL)
Olling, Gustav (IL)
Phillips, Joseph J (IN)
Poli, Corrado (MA)
Read, Robert H (IN)
Sargent, Lowrie B, Jr (PA)
Shabaik, Aly H (CA)
Smith, Darrell W (MI)
Spachner, Sheldon A (PA)
Stone, Morris D (PA)
Throop, James W (MI)
Tomalin, Peter M (PA)
Trigger, Kenneth J (IL)
Verduzco, Miguel A
Wagoner, Robert H (OH)
Wambold, John N (PA)
Weber, Richard G (CT)
Wright, Roger N (NY)

METALLURGICAL ENGINEERING & METALLURGY

Agarwal, Jagdish C (MA)
Ailor, William H, Jr (VA)
Alban, Lester E (IN)
Albrecht, E Daniel (IL)
Alexander, John A (NJ)
Allen, Charles W (IL)
Altstetter, Carl J (IL)
Amber, Wayne L (IN)
Anderson, Charles J (IL)
Anderson, Eric D (MI)
Anderson, Robert C (TX)
Andrews, Al L (AZ)
Ansel, Gerhard (TX)
Ansell, George S (CO)
Aplan, Frank F (PA)
Arbiter, Nathaniel (AZ)
Ardell, Alan J (CA)
Armstrong, Ronald W (MD)
Arnold, Lynn E (OH)
Asquith, David J (MA)
Austin, James B (PA)
Avedesian, Michael M
Averbach, Benjamin L (MA)
Baker, George S (WI)
Barone, Richard V (RI)
Barrow, Robert B (CT)
Bartell, Howard F (PA)
Bartlett, Robert W (AZ)
Barton, Cornelius J (CT)
Baughman, Richard A (PA)
Beaver, Howard O, Jr (PA)
Beck, Franklin H (OH)
Behal, Victor G

Beitscher, Stanley (CO)
Benjamin, John S (NY)
Bergh, Donald A (MI)
Bernstein, I M (PA)
Bertossa, Donald C (CA)
Beshers, Daniel N (NY)
Betner, Donald R (IN)
Bianchi, Leonard M (PA)
Bigelow, Wilbur C (MI)
Bilello, John C (NY)
Birks, Neil (PA)
Birnbaum, Howard K (IL)
Bishop, Harry L, Jr (NY)
Black, Joseph E (WA)
Block, Robert J (OK)
Bodeen, George H (IL)
Bogart, Harold N (MI)
Bohl, Robert W (IL)
Boily, Michel S
Bomberger, Howard B (OH)
Boorstein, William M (MI)
Bott, Walter T (IL)
Bouman, Robert W (PA)
Bourgault, Roy F (MA)
Bowers, David F (WI)
Boyd, Walter K (TX)
Bramfitt, Bruce L (PA)
Brandenburg, George P (OH)
Brasunas, Anton deSales (MO)
Bratt, Richard W (NY)
Breen, Dale H (IL)
Breinan, Edward M (CT)
Breyer, Norman N (IL)
Briant, Clyde L (NY)
Brick, Robert M (CA)
Brinkmann, Joseph B (CT)
Brooks, Carson L (AZ)
Brower, Ralph T (IL)
Brown, Charles M (NY)
Brown, William F, Jr (OH)
Buchheit, Richard D (OH)
Budinski, Kenneth G (NY)
Bunshah, Rointan F (CA)
Burgess, Price B (MI)
Burghoff, Henry L (CT)
Burrell, Edward R (NY)
Burrier, Harold I, Jr (OH)
Bush, Spencer H (WA)
Byrne, J Gerald (UT)
Cadden, Jerry L (CA)
Cahoon, John R
Cambre, Ronald C (LA)
Cameron, Joseph A (PA)
Canonico, Domenic A (TN)
Capstaff, Arthur E Jr (MA)
Carey, Julian D (NY)
Carlson, O Norman (IA)
Carney, Dennis J (PA)
Carter, Joseph R (MA)
Carter, Roger V (WA)
Cary, Howard B (OH)
Cary, Robert A (PA)
Castleman, Louis S (NY)
Caton, Robert L (PA)
Chakrapani, Durgam G (OR)
Chang, John C (CA)
Chang, Y Austin (WI)
Chase, Thomas L (OH)
Chin, Gilbert Y (NJ)
Chiswik, Haim H (IL)
Clark, Brian M (OH)
Clark, John B (MO)
Clum, James A (NY)
Coffin, Louis F, Jr (NY)
Cohen, Jerome B (IL)
Cohen, Morris (MA)
Coheur, Pierre M E O G
Colandrea, Thomas R (CA)
Cole, Nancy Clift (TN)
Collins, Henry E (OH)
Conybear, James G (OH)
Cooke, Francis W (SC)
Cooper, Thomas D (OH)

METALLURGICAL ENGINEERING & METALLURGY
(Continued)

Nichols, Edwin Scott (IN)
Nicholson, Morris E (MN)
Niedringhaus, Philip W (NY)
Nippes, Ernest F (NY)
Norwood, Sydney L (AL)
Nummela, Walter (OH)
Ogilvie, Robert E (MA)
O'Keefe, Thomas J (MO)
Oliver, Ben F (TN)
Olson, David L (CO)
Olson, Ferron A (UT)
Opie, William R (NJ)
Orava, Raimo Norman (SD)
Orehoski, Michael A (PA)
Otto, George (IA)
Owczarski, William A (CT)
Owen, Edwin L (TX)
Palmer, Bruce R (SD)
Paprocki, Stan J (OH)
Parikh, Niranjan M (IL)
Parthasarathi, Manavasi N (NY)
Paxton, Harold W (PA)
Pearson, Hugh S (GA)
Pearson, William B
Pearson, William H
Pehlke, Robert D (MI)
Pellissier, George E (NY)
Pelloux, Regis M (MA)
Pendleton, Joseph S, Jr (PA)
Pense, Alan W (PA)
Perepezko, John H (WI)
Peretti, Ettore A (IN)
Perkins, Roger A (CA)
Perlmutter, Isaac (OH)
Pesses, Marvin (FL)
Peters, Ernest
Peterson, Warren S (PA)
Petkovic-Luton, Ruzica A (NJ)
Pettit, Frederick, S (PA)
Petzow, Gunter E
Phillips, Albert J (FL)
Phillips, William R (CA)
Picha, Robert T (MN)
Pitler, Richard K (PA)
Pitt, Charles H (UT)
Planz, Edward J (AL)
Pool, Monte J (OH)
Poole, H Gordon (OR)
Pope, David P (PA)
Porter, Lew F (PA)
Powell, Gordon W (OH)
Prager, Martin (NY)
Price, Donald D (WA)
Pridgeon, John W (NY)
Pritchett, Thomas R (CA)
Queneau, Bernard R (PA)
Queneau, Paul B (CO)
Queneau, Paul E (NH)
Quist, William E (WA)
Rack, Henry J (SC)
Radcliffe, S Victor (DC)
Rapp, Robert A (OH)
Rathke, Arlan E (MI)
Raudebaugh, Robert J (NJ)
Ray, Alden E (OH)
Read, Robert H (IN)
Reed, William H (KY)
Reed-Hill, Robert E (FL)
Reinhardt, Gustav (FL)
Restivo, Frank A (MI)
Reynolds, Edward E (MI)
Richards, Earl L (PA)
Richards, Kenneth J (UT)
Richardson, Lee S (ID)
Richmond, Frank M (PA)
Richter, John A (FL)

Rigney, David A (OH)
Rinaldi, Michael D (MA)
Ripling, Edward J (IL)
Ritter, John E, Jr (MA)
Roberts, Earl C (WA)
Roberts, George A (CA)
Roberts, Irving (VA)
Roberts, John M (TX)
Roberts, Philip M (IL)
Robinson, George H (MI)
Roe, William P (NJ)
Rose, Kenneth E (KS)
Rosenbaum, Joe B (UT)
Rosenfield, Alan R (OH)
Rosenqvist, Terkel N
Rosenthal, Philip C (FL)
Ross, Stuart T (MN)
Rostoker, William (IL)
Rowells, Lynn G (IL)
Rozovsky, Eliezer (PA)
Russell, Allen S (SC)
Russell, John V (CA)
Russell, Kenneth C (MA)
Sadowski, Edward P (NJ)
Samuels, Leonard E
Sander, Louis F (PA)
Sandstrom, Donald J (NM)
Santoli, Pat A (PA)
Saperstein, Zalman P (WI)
Sargent, Gordon A (KY)
Sartell, Jack A (MN)
Saxer, Richard K (OH)
Scala, Eraldus (NY)
Schadler, Harvey W (NY)
Scheer, John E (PA)
Schetky, Laurence McD (CT)
Schlabach, Tom D (NJ)
Schlatter, Rene (NY)
Schmatz, Duane J (MI)
Schoppe, Conrad J (TX)
Schroder, Klaus (NY)
Schubert, George H (RI)
Schultz, Jay W (NY)
Schwope, Arthur D (OH)
Scott, Alexander R (PA)
Scott, James L (TN)
Sedeora, Tejinder S (IL)
Seigle, Leslie L (NY)
Selle, James E (CO)
Semchyshen, Marion (MI)
Servi, Italo S (MA)
Shankman, Aaron D (CA)
Shannette, Gary W (MI)
Shapiro, Eugene (CT)
Shapiro, Stanley (NJ)
Shatynski, Stephen R (NY)
Shedd, Wilfred G (IN)
Sherman, Russell G (CA)
Shewmon, Paul G (OH)
Shields, Bruce M (PA)
Shortsleeve, Francis J (CT)
Shubat, George J (IN)
Siddell, Derreck
Signes, Emil G (PA)
Silkiss, Emanuel M (NY)
Simkovich, Alex (PA)
Sims, Chester T (NY)
Skarda, James J (OH)
Smeltzer, Walter W
Smidt, Fred A (DC)
Smith, Darrell W (MI)
Smith, Edward S (FL)
Smith, George V (NY)
Smith, William R, Sr (CA)
Smith, Yancey E (MI)
Snavely, Cloyd A (MD)
Snowman, Alfred (NJ)
Snyder, Harold J (LA)
Snyder, James J (OH)
Sonnino, Carlo B (MO)
Spalding, Henry A (KY)
Speich, Gilbert R (PA)
Spendlove, Max J (MD)
Sperry, Philip R (MO)
Spicer, Clifford W (PA)

Sproat, Robert L (PA)
Spruiell, Joseph E (TN)
St Clair, John M (OH)
St Pierre, George R (OH)
Stadelmaier, Hans H (NC)
Staehle, Roger W (MN)
Stafford, Stephen W (TX)
Staples, Elton E (FL)
Starke, Edgar A , Jr (VA)
Starr, Dean C (NJ)
Steigerwald, Edward A (OH)
Steigerwald, Robert F (MI)
Stein, Dale F (MI)
Steinberg, Morris A (CA)
Stenger, Edwin A (OH)
Stephenson, Edward T (PA)
Stephenson, Robert L (PA)
Stickels, Charles A (MI)
Stoloff, Norman S (NY)
Stone, Joseph K (IL)
Stout, Robert D (PA)
Sullivan, Cornelius P (CT)
Sundaresan, Sonny G (SC)
Sutherland, Earl C (WA)
Sutton, C Roger (IL)
Swan, David (CT)
Sweet, John W (WA)
Swift, Roy E (KY)
Talbot, Thomas F (AL)
Taubenblat, Pierre W (NJ)
Taylor, Allan D (CA)
Temple, Derek A
Tenenbaum, Michael (IL)
Thielsch, Helmut (RI)
Thomas, David R, Jr (PA)
Thomas, Seth R (OH)
Thompson, Anthony W (PA)
Thompson, John W
Tien, John K (NY)
Tiner, Nathan A (CA)
Tomlinson, John L (CA)
Torma, Arpad E (NM)
Trojan, Paul K (MI)
Tsareff, Thomas C, Jr (IN)
Tucker, R C Jr (IN)
Turk, Julius (MO)
Unterweiser, Paul M (OH)
Urquhart, Andrew W (NY)
Uys, Johannes M (IL)
Van Horn, Kent R (PA)
Van Pelt, Richard H (IL)
Van Thyne, Ray J (IL)
Vander Voort, George F (PA)
Varrese, Francis R (PA)
Verhoeven, John D (IA)
Vertes, Michael A (CT)
Vitovec, Franz H
Waber, James T (IL)
Wadsworth, Gayle B (PA)
Wadsworth, Milton E (UT)
Wagoner, Robert H (OH)
Walker, Walter W (AZ)
Wall, Fred J (PA)
Wallace, Paul B (KS)
Wallace, Ronald Stephen (AL)
Walter, Gordon H (IL)
Ward, Allen W (OK)
Wardell, Roland D (MN)
Watson, James F (CA)
Wayman, Clarence Marvin (IL)
Weart, Harry W (MO)
Weber, Richard G (CT)
Wechsler, Monroe S (IA)
Weigel, Philip (OH)
Weinberg, Fred
Weissmann, Sigmund (NJ)
Weitzenkorn, Lee F (TX)
Wells, Martin G H (PA)
Wensing, Donald R (PA)
Werner, Fred E (PA)
Werner, John E (PA)

Wert, James J (TN)
Westermann, Fred E (OH)
Weston, Leonard A (DE)
Weymueller, Carl R (OH)
Wheeler, John A (NH)
Whitney, Herbert W (AZ)
Wilde, Richard A (MI)
Wilder, Thomas C (MA)
Wilkinson, Dwight A (MI)
Willhelm, A Clyde, Jr (AL)
Williams, James C (PA)
Williams, Samuel L (IL)
Wilsdorf, Heinz G F (VA)
Winquist, Lauren A (MI)
Wise, William H
Witchey, Robert H (MI)
Wlodek, Stanley T (OH)
Wolf, Robert V (MO)
Wright, Roger N (NY)
Wu, Keh C (CA)
Wulpi, Donald J (IN)
Wynblatt, Paul P (PA)
Yacavone, David M (OH)
Yaggee, Frank L (IL)
Yarar, Baki (CO)
Yeniscavich, William (PA)
Young, John D (CA)
Young, William B (IN)
Yukawa, Sumio (CO)
Yurek, Gregory J (MA)
Zabel, Dale E (AZ)
Zamborsky, Daniel S (OH)
Zapffe, Carl A (MD)
Zeno, Robert S (IL)
Zepfel, William F
Zinkham, Robert E (VA)
Zmeskal, Otto (OH)
Zucker, Gordon L (MT)
Zunkel, Alan D (NJ)

METEOROLOGY

Crane, Robert K (NH)
Stout, Glenn E (IL)

MICROWAVES

Adam, Stephen F (CA)
Allen, Matthew A (CA)
Ashley, J Robert (NH)
Barlow, Harold M
Bates, Richard H T
Beard, Charles I (VA)
Caulton, Martin (MD)
Cohn, Seymour B (CA)
Cooper, H Warren, III (MD)
Cristal, Edward G (CA)
Cutler, C Chapin (CA)
Garoff, Kenton (NJ)
Garver, Robert V (MD)
Getsinger, William J (MD)
Ishii, Thomas Koryu (WI)
Knerr, Reinhard H (PA)
Lakshminarayan, Mysore R (CA)
Liechti, Charles A (CA)
Love, Allan W (CA)
Mayes, Paul E (IL)
Misugi, Takahiko
Mourier, Georges
Okamura, Sogo
Patton, Willard T (NJ)
Reggia, Frank (VA)
Sard, Eugene W (NY)
Sheppard, Albert P (GA)
Smith, Jack (AZ)
Tomiyasu, Kiyo (PA)
Waite, William P (AR)
Wharton, Charles B (NY)

MILITARY ENGINEERING

Bousha, Frank N (NE)

MINERAL PROCESSING

NONDESTRUCTIVE TESTING (Continued)

Dunn, Frank W (CT)
Eddy, Robert A (MA)
Egle, Davis M (OK)
Erdman, Carl A (TX)
Frandsen, John P (NC)
Graff, Karl F (OH)
Green, Robert E , Jr (MD)
Greene, Arnold H (MA)
Hagemaier, Donald J (CA)
Halsey, George H (PA)
Hellier, Charles J P E (CT)
Henke, George R (CA)
Henry, Edwin B, Jr (PA)
Jenniches, F Suzanne (MD)
Lewis, Heydon Z (CO)
Lewis, William H, Jr (GA)
Lindgren, Arthur R (IL)
Lord, William (CO)
Maddex, Phillip J (CA)
McClung, Robert W (TN)
McEleney, Patrick C (MA)
Mooz, William E (CA)
Morgan, Jack B (PA)
Mottern, Robert W (NM)
Moyer, Richard B (PA)
Nance, Roy A (PA)
Offner, Walter W (HI)
Ostrofsky, Bernard (IL)
Pade, Earl R (PA)
Pherigo, George L (MN)
Phillips, James W (IL)
Rizzo, Edward G (NJ)
Roehrs, Robert J (MO)
Rose, Joseph L (PA)
Rummel, Ward D (CO)
Salama, Kamel (TX)
Selner, Ronald H (OH)
Shaw, Carl B (OR)
Spanner, Jack C (WA)
Stern, Ferdi B, Jr (MA)
Stump, Wayne D (CO)
Udell, Debra M (NY)
Viaclovsky, Syl A (TX)
White, John K (PA)
Wolf, Robert V (MO)
Yacavone, David M (OH)

NUCLEAR ENGINEERING

Abdel-Khalik, Said I (WI)
Abramson, Paul B (NY)
Addae, Andrews K (DC)
Agnew, Harold M (CA)
Agrawal, Ashok K (NY)
Albaugh, Frederic W (WA)
Almenas, Kazys K (MD)
Anghaie, Samim (FL)
Anton, Nicholas G (FL)
Arnold, William Howard (WA)
Axford, Roy A (IL)
Baily, William E (CA)
Balazs, Bill A (OH)
Balent, Ralph (CA)
Bartholomew, Dale C (WA)
Beard, Samuel J (WA)
Beaton, Roy H (CA)
Beckjord, Eric S (DC)
Behnke, Wallace B , Jr (IL)
Bell, Carlos G, Jr (NC)
Binney, Stephen E (OR)
Block, Robert C (NY)
Bohl, Robert W (IL)
Bradish, John P (WI)
Brandon, Robert J (CA)
Brehm, Richard L (CA)
Broad, Richard (VA)
Brugger, Robert M (MO)
Brush, Harvey F (CA)
Buckham, James A (SC)
Busey, Harold M (MD)
Bush, Spencer H (WA)
Cacuci, Dan G (TN)

Cahill, William J, Jr (LA)
Carbon, Max W (WI)
Carelli, Mario D (PA)
Casey, Leslie A (TX)
Check, Paul S (MD)
Chin, Allan (CA)
Chockie, Lawrence J (WA)
Clark, Hugh K (SC)
Cleland, Laurence Lynn (CA)
Clikeman, Franlyn M (IN)
Cobean, Warren R (NJ)
Cochran, Robert G (TX)
Cohen, E Richard (CA)
Cohen, Karl P (CA)
Corelli, John C (NY)
Cornell, Donald H (NY)
Courtney, John C (LA)
Cox, Alvin E (NY)
Cox, Carl M (WA)
Creagan, Robert J (PA)
Cromer, Sylvan (TN)
Cunningham, John E (TN)
Dahl Roy E (WA)
Dalle Donne, Mario
Danofsky, Richard A (IA)
Davis, Milton W, Jr (SC)
Davis, W Kenneth (CA)
Demas, Nicholas G (TN)
Dhir, Vijay K (CA)
Diamond, David J (NY)
Ditmore, Dana C (CA)
Dorning, John J (VA)
Doty, Clarence W (CA)
Duderstadt, James J (MI)
Duffey, Dick (MD)
Durante, Raymond W (DC)
Eaton, Thomas E (KY)
Eckert, Richard M (NJ)
Eckhoff, N Dean (KS)
Edlund, Milton C (VA)
Eggen, Donald T (CA)
Eichholz, Geoffrey G (GA)
Esselman, Walter H (CA)
Evans, Ersel A (WA)
Farr, W Morris (AZ)
Faw, Richard E (KS)
Fee, Walter F (CT)
Feldman, Melvin J (TN)
Fenech, Henri J (CA)
Ferguson, George (DC)
Foster, John Stanton
Fry, Donald W
Fulford, P James (IN)
Furter, William F
Gajewski, Walter M (WA)
Gasser, Eugene R (MD)
Gathy, Bruce S (CT)
Gingrich, James E (CA)
Gomberg, Henry J (MI)
Goodjohn, Albert J (CA)
Grace, J Nelson (GA)
Graham, John (WA)
Graves, Ernest (VA)
Graves, Harvey W, Jr (MD)
Haas, Paul A (TN)
Hahn, Richard D (MD)
Hang, Daniel F (IL)
Hansen, Kent F (MA)
Havens, William W, Jr (NY)
Hendrickson, Tom A (NJ)
Hetrick, David L (AZ)
Hittman, Fred (MD)
Hockenbury, Robert W (NY)
Howe, John P (CA)
Huston, Norman E (WI)
Hutchinson, Frank D, III (NY)
Jacobus, David D (NY)
Jaske, Robert T (MD)
Jens, Wayne H (MI)
Jensen, Roland J (MN)
Jester, William A (PA)
Jones, Barclay G (IL)

Jones, James A (NC)
Kammash, Terry (MI)
Kastenberg, William E (CA)
Kato, Walter Y (NY)
Kerr, William (MI)
Kimel, William R (MO)
Kirchner, Walter L (NM)
Klein, Daniel (PA)
Klema, Ernest D (MA)
Knoll, Glenn F (MI)
Komoriya, Hideko (Heidi) (NY)
Kouts, Herbert J C (NY)
Kramer, Andrew W (IL)
Kulcinski, Gerald L (WI)
Lahey, Richard T, Jr (NY)
Lamarsh, John R (NY)
Lathrop, Kaye D (CA)
Lawroski, Harry (ID)
Leatherman, John E (CA)
Leeds, J Venn, Jr (TX)
Levenson, Milton (CA)
Levine, Saul (MD)
Lewins, Jeffery
Lineberry, Michael J (ID)
Linehan, John H (WI)
Loewenstein, Walter B (CA)
Lowton, Robert B (CA)
Loyalka, Sudarshan K (MO)
Lucas, Glenn E (CA)
Lucey, John W (IN)
Lykoudis, Paul S (IN)
MacPherson, Herbert G (TN)
Mandil, I Harry (FL)
Manson Benedict (FL)
Marcus, Gail H (DC)
Martin, William R (MI)
Martis, Jerome M, Sr (PA)
Mason, Edward A (IL)
Maynard, Charles W (WI)
McCormick, Norman J (WA)
McFarlane, Harold F (ID)
McLain, Milton E (GA)
Meghreblian, Robert V (MA)
Melese-d'Hospital, G B (CA)
Mense, Allan T (VA)
Mesler, Russell B (KS)
Messenger, George C (NV)
Meyer, John E (MA)
Meyer, Walter (MO)
Miley, George H (IL)
Miller, Don W (OH)
Miller, Joseph S (KS)
Mittl, Robert L (NJ)
Moses, Gregory A (WI)
Murray, Peter (PA)
Murray, Raymond L (NC)
Nelson, Peter R (CT)
Nelson, William E (CA)
Nightingale, Richard E (WA)
North, Edward D (TN)
Ohanian, M Jack (FL)
Okrent, David (CA)
Olander, Donald R (CA)
Osborn, Richard K (MI)
Pace, Danny L (MS)
Palladino, Nunzio J (DC)
Pasqua, Pietro F (TN)
Patriarca, Peter (TN)
Patti, Francis J (NY)
Paxton, Hugh C (NM)
Pestorius, Thomas D (VA)
Petree, Frank L (NJ)
Pigford, Thomas H (CA)
Pomraning, Gerald C (CA)
Porter, Nancy J (CA)
Post, Roy G (AZ)
Prestele, Joseph A (CA)
Profio, A Edward (CA)
Raab, Harry F, Jr (VA)
Rasmussen, Norman C (MA)

Rastoin, Jean M
Remick, Forrest J (PA)
Remley, Marlin E (CA)
Reynolds, Roger S (MS)
Rickard, Corwin L (CA)
Rickover, Hyman G (DC)
Ritter, Gerald L (WA)
Ritzmann, Robert W (MD)
Roach, Kenneth E (FL)
Roake, William E (WA)
Roberds, Richard M (TN)
Robinson, M John (MO)
Rodger, Walton A (MD)
Rogers, Vern C (UT)
Rosenberg, Richard (CA)
Ruby, Lawrence (OR)
Sabri, Zeinab A (IA)
Sandquist, Gary M (UT)
Scavuzzo, Rudolph J (OH)
Schrock, Virgil E (CA)
Scott, James L (TN)
Seale, Robert L (AZ)
Sesonske, Alexander (CA)
Sher, Rudolph (CA)
Shohet, J Leon (WI)
Shultis, J Kenneth (KS)
Shure, Kalman (PA)
Silady, Fred A (CA)
Sim, Richard G (NC)
Simon, Albert (NY)
Simons, Gale G (KS)
Simpson, John W (SC)
Smidt, Dieter
Smidt, Fred A (DC)
Snyder, Bernard J (DC)
Snyder, Thoma M (CA)
Spinrad, Bernard I (IA)
Stacey, Weston M, Jr (GA)
Steele, Lendell E (VA)
Stewart, Hugh B (CA)
Stoller, Sidney M (NY)
Stone, Henry E (CA)
Swanson, A Einar (MO)
Szirmay, Leslie v (OH)
Talley, Wilson K (CA)
Taylor, John J (CA)
Theofanous, Theofanis G (CA)
Trauger, Donald B (TN)
Tsoulfanidis, Nicholas (MO)
Turinsky, Paul J (NC)
Turnbull, Robert J (IL)
Uhrig, Robert E (FL)
Untermyer, Samuel (CA)
Upadhyaya, Belle R (TN)
Valentin, Richard A (IL)
Ver Planck, Dennistoun W (CA)
Vogel, Richard C (CA)
Weaver, Lynn E (FL)
Weber, Clifford E (DC)
Wehring, Bernard W (NC)
Weill, Jacky
Whan, Glenn A (NM)
Wick, Robert S (TX)
Wilbur, Leslie C (MA)
Wilkinson, Bruce W (MI)
Williams, Michael M R (MI)
Williams, Paul C (OH)
Williamson, Thomas G (VA)
Wilson, Albert E (ID)
Wirtz, Karl
Witzig, Warren F (PA)
Woodall, David M (NM)
Woods, W Kelly (OR)
Wyrick, Roger K (GA)
Young, William H (NJ)
Zeno, Robert S (IL)
Zweifel, Paul F (VA)

PHYSICS (Continued)

Koller, Earl L (NJ)
Kosmahl, Henry G (OH)
Kramer, Raymond E (OH)
Kravitz, Lawrence C (MD)
Lamarsh, John R (NY)
Levine, Jules D (TX)
Liao, Paul F (NJ)
Liboff, Richard L (NY)
Luborsky, Fred E (NY)
Lunnen, Ray J, Jr (PA)
MacCrone Robert K , (NY)
Macdonald, J Ross (NC)
Mailloux, Robert J (MA)
Mandel, Leonard (NY)
Mason, Edward A (RI)
Massalski, Thaddeus B (PA)
Maurer, Robert D (NY)
McCubbin, T King, Jr (PA)
McGill, Thomas C (CA)
McGroddy, James C (NY)
McLucas, John L (DC)
Mentzer, John R (PA)
Menzel, Erich H
Millman, Sidney (NY)
Mungall, Allan G
Nayfeh, Munir H (IL)
Nelkin, Mark S (NY)
Oertel, Goetz K (DC)
Olson, Valerie F (CA)
Orehotsky, John L (PA)
Owens, C Dale (IN)
Pantell, Richard H (CA)
Patel, C K N (NJ)
Potter, David S (MI)
Pry, Robert H (IL)
Ramberg, Edward G (PA)
Rhee, Moon-Jhong (MD)
Rittner, Edmund S (MD)
Roberds, Richard M (TN)
Rogers, Kenneth C (NJ)
Rosenberg, Paul (NY)
Ruoff, Arthur L (NY)
Rutledge, Wyman C (OH)
Rynn, Nathan (CA)
Schawlow, Arthur L (CA)
Schmitt, Roland W (NY)
Seidman, David N (IL)
Sessler, Gerhard M
Shapiro, Sidney (NY)
Simon, Ralph E (NJ)
Snyder, James N (IL)
Spooner, Stephen (GA)
Strandberg, Malcom W P (MA)
Sucov, Eugene W (PA)
Sudan, Ravindra N (NY)
Symons, Robert S (CA)
Tauc, Jan (RI)
Taylor, Barry N (MD)
Taylor, Snowden (NJ)
Thouret, Wolfgang E (NJ)
Tittel, Frank K (TX)
Townes, Charles H (CA)
Trivelpiece, Alvin W (DC)
Van Lint, Victor A J (CA)
Vanyo, James P (CA)
Vaughan, William W (AL)
Vetelino, John F (ME)
Vizy, Kalman N (NY)
Vollmer, James (NJ)
Vook, Richard W (NY)
Wang, Shyh (CA)
Webb, Theodore S (TX)
Webb, Watt W (NY)
Weeks, W F (AK)
Wehring, Bernard W (NC)
Weiner, Jerome H (RI)
Wheeler, Robert G (CT)
White, Alan D (NJ)
Wolf, Werner P (CT)
Wolga, George J (NY)
Woodbury, Eric J (CA)
Zabel, Hartmut (IL)

PIPELINE

Beggs, H Dale (OK)
Brush, Harvey F (CA)
Culvern, Frederick E (OK)
Kerr, Arthur J (PA)
King, Joe J (TX)
McNamara, Edward J (LA)
Nelson, Ernest O (KS)
Owen, Eugene H (LA)
Ryan, Robert S (OH)
Wood, Willis B , Jr (CA)

PLANT & FACILITIES ENGINEERING

Ackoff, Albert K (NY)
Addoms, Hallett B (NY)
Allaire, Robert E (CT)
Antrobus, Thomas R (PA)
Barrow, Robert B (CT)
Beall, James H (FL)
Beckett, George A (OH)
Bigley, Paul R (OH)
Blakely, Thomas A (NJ)
Blank, William F (IL)
Blazey, Leland W (PA)
Blossom, John S (OH)
Bodenheimer, Vernon B (GA)
Bolduc, Oliver J (IL)
Bombach, Otto F (CA)
Bonnell, William S (PA)
Borel, Robert J (OH)
Bowman, E Dexter, Jr (AZ)
Boyar, Julius W (LA)
Breitling, Thomas O (UT)
Broward, Hoyt E (FL)
Brown, James G (TX)
Byrnes, James J (FL)
Campbell, Charles A (NY)
Campbell, John M (OK)
Carton, Allen M (MD)
Cast, Karl F (LA)
Chapin, William F (CA)
Chappelear, Patsy S (TX)
Conkey, David R (MN)
Creed, Michael W (NC)
Crow, John H (TX)
Davidson, Bruce R (NJ)
Dienhart, Arthur V (MN)
Dlouhy, John R (IN)
Dorsett, Ronald (AZ)
Duchscherer, David C (NY)
Dunnavant, Guy P (AL)
Earl, Christopher B (FL)
Eckhardt, Carl J (TX)
Ellenberger, William J (DC)
Engelberger, John E (PA)
Esgar, Jack B (OH)
Faucher, Richard L
Feldman, Edwin B (GA)
Ferrara, Norman (PA)
Ferrise, Louis J (PA)
Footlik, Irving M (IL)
Foster, Albert L (TN)
Gallen, Donald R (MO)
Gibboney, James A (OH)
Gill, James M (LA)
Goddard, William A (CA)
Gompf, Arthur M (MD)
Gordon, Richard H (PA)
Greenberg, Joseph H (IL)
Grizzard, John L (VA)
Hahn, Jack
Hefter, Harry O (IL)
Heitlinger, Igor (CT)
Hellman, Albert A (CA)
Hira, Gulab G (MA)
Holfinger, Robert R (OH)
Holladay, James F (GA)
Horning, John C (NY)
Howard, J Wendell (TN)
Hulbert, Thomas H (MA)
Hummer, Robert H (VA)
Johnson, Hubert O, Jr (TX)

Johnson, William C (NY)
Johnston, Walter E (NC)
Kittner, Edwin H (KS)
Klion, Daniel E (NY)
Kok, Hans G (NY)
Koss, John P (MA)
Krasavage, Kenneth W (IN)
Kraus, Robert A (OH)
Krizak, Eugene J (TX)
Kulieke, Frederick C, Jr (OH)
Kunkel, Paul F (PA)
Lanker, Karl E (OH)
Lee, Quarterman (MO)
Leonard, Richard D (CA)
Lewis, Bernard T (MD)
Lewis, David W (MA)
Lilly, Gary T (WV)
Lineberry, Michael J (ID)
Mansur, John C (OK)
Marschall, Albert R (DC)
Mason, James A, Jr (NY)
Mason, James M (MA)
Mathias, Robert James
McKnight, Val B (IL)
McLeish, Duncan R (FL)
Meier, Reinhard W
Merims, Robert (NY)
Miller, Robert R (CA)
Moats, Erwin R (OH)
Morris, Ben F (CA)
Morris, Don F (OK)
Morrison, William J (NY)
Nelson, Ernest O (KS)
Neptune, David B (MO)
Pearce, Philip L (CT)
Pflug, Charles E (WI)
Pigage, Leo C (IL)
Purdy, Richard B (CT)
Quinn, James D (MD)
Reich, Herbert H (PA)
Reinig, L Philip (NM)
Reitinger, Robert L (PA)
Roberts, Robert J E (CT)
Ross, Gilbert I (NY)
Ross, James L (MO)
Roudabush, Byron S (DC)
Russell, Fred G (TX)
Saltarelli, Eugene A (MD)
Salvin, Robert (NJ)
Schlegel, Walter F (CT)
Seaver, Philip H (MA)
Shearer, William A, Jr (DE)
Silton, Ronald H (OH)
Spector, Leo F (IL)
Stables, Benjamin J , Jr (MO)
Steiner, George G (NY)
Sweeney, Thomas B (OH)
Talbert, John J, III (TX)
Talbott, Laurence F (CA)
Theros, William J (PA)
Thompson, Robert C (DC)
Van Tine, Charles L (IL)
VanHaaren, Frans I (NC)
Verlo, Clayton C (IL)
Waligora, John M (FL)
Walker, U Owen (NM)
Wallace, John M, (CT)
Watkins, Henry H (SC)
Wesley, Fred A (DE)
Westfall, Paul R (WV)
Weston, Donald W (ME)
Wiars, Dale M (IN)
Wildman, Robert F (CA)
Windt, Ewald C (OH)

PLANT DESIGN

Blythe, Michael E (GA)
Breitling, Thomas O (UT)
Hargis, John C (OK)
Jesser, Roger F (CO)
Lilly, Edmund D (TX)
Pflug, Charles E (WI)

PLASTICS & POLYMERS

Sturdevant, Bruce L (CO)
Walk, Frank H (LA)
Wiars, Dale M (IN)
Windisch, Earl C (KS)
Wolf, Donald J, (MD)

PLASMA

Birdsall, Charles K (CA)
Callen, James D (WI)
Chen, Francis F (CA)
Dorning, John J (VA)
Duderstadt, James J (MI)
Gross, Robert A (NY)
Hickok, Robert L , Jr (NY)
Holland, Leslie A
Humphries, Stanley, Jr (NM)
Kristiansen, Magne (TX)
Kuehl, Hans H (CA)
Levi, Enrico (NY)
Mense, Allan T (VA)
Mourier, Georges
Nakano, Yoshiei
Trivelpiece, Alvin W (DC)
Wharton, Charles B (NY)

PLASTICS & POLYMERS

Andrade, Joseph D (UT)
Armeniades, Constantine D (TX)
Baer, Eric (OH)
Bailey, William J (MD)
Baum, Sidney J (CA)
Belden, Reed H (NJ)
Bell, James P (CT)
Bisio, Attilio L (NY)
Booy, Max L (DE)
Bowman, Mark M (OK)
Brazinsky, Irving (NJ)
Broutman, Lawrence J (IL)
Bruins, Paul F (TX)
Carr, Stephen H (IL)
Christensen, Ronald I (NJ)
Corneliussen, Roger D (PA)
Couchman, Peter R (NJ)
Crilly, Eugene R (CA)
Daniel, Richard A (NJ)
Date, Raghunath V (CT)
DeGuzman, Jose P (CA)
Denn, Morton M (CA)
DuBois, J Harry (NJ)
Duda, John L (PA)
Ely, Berten E (PA)
Ferris, Robert R (PA)
Fitzgerald, Edwin R (MD)
Flanagan, H Russell (TN)
Forger, Robert D (CT)
Forsyth, T Henry (OH)
Fricke, Arthur L (ME)
Fuchsluger, John H (MD)
Funk, William U (OH)
Garber, Charles A (PA)
Gerliczy, George (NJ)
Gigliotti, Michael F X (MA)
Goldmann, Theodore B (TX)
Groves, Stanford E (LA)
Hamielec, Alvin E
Han, Chang D (NY)
Hawkins, W Lincoln (NJ)
Hipchen, Donald E (FL)
Holloway, William J (NH)
Holz, Harold A (NJ)
Isaacs, Jack L (CT)
Jepsen, John C (TX)
Jones, Lee S (CA)
Kaghan, Walter S (FL)
Kardos, John L (MO)
Kenner, Vernal H (OH)
Klasing, Donald E (KS)
Klein, Imrich (NJ)
Koehne, Anthony J (MD)
Kogos, Laurence (CT)

POWER

PLASTICS & POLYMERS
(Continued)

Kolsky, Herbert (RI)
Kowalski, Philip L (CT)
Kramer, Edward J (NY)
Kretzschmar, John R (OH)
Kuhlke, William C (TX)
Kydonieus, Agis (NY)
Lantos, Peter R (PA)
Laurence, Robert L (MA)
Lenz, Robert W (MA)
Li, James C M (NY)
Lichtenwalner, Hart K (NY)
Lindt, Thomas (PA)
Lippe, Robert L (NJ)
Locke, Carl E (KS)
Lopez, Arthur M (NM)
Lynn, R Emerson (OH)
Magill, Joseph H (PA)
Magliolo, Joseph J (TX)
Marion Johnson P E (TX)
Martin, Jay R (PA)
McHugh, Anthony J (IL)
McKelvey, James M (MO)
McLain, William R (TN)
Miller, Edward (CA)
Mueller, Robert Kirk (MA)
Myers, John H (DE)
Nauman, E Bruce (NY)
Notowich, Alvin A (TN)
Pae, Kook D (NJ)
Palmer, Nigel I (NY)
Paul, Donald R (TX)
Peppas, Nikolaos A (IN)
Phillips, Paul J (TN)
Pickering, George E (MA)
Plazek, Donald J (PA)
Prizer, Charles J (PA)
Reucroft, Philip J (KY)
Riccardi, Louis G (NY)
Ricklin, Saul (RI)
Robinson, John K (AL)
Rodriguez, Ferdinand (NY)
Rolston, J Albert (OH)
Rothschild, Paul H (OH)
Saxena, Kanwar B (MI)
Schaffhauser, Robert J (NY)
Schmitt, George F, Jr (OH)
Shaw, Montgomery T (CT)
Shenian, Popkin (OH)
Smith, Lawrence H (CT)
Staff, Charles E (NJ)
Stancell, Arnold F (CT)
Sternstein, Sanford S (NY)
Stokes, Vijay K (NY)
Sundstrom, Donald W (CT)
Sutro, Frederick C, Jr (PA)
Thompson, Lewis D (LA)
Ussery, Harry B (SC)
Vieth, Wolf R (NJ)
Watson, Albert T
Wentz, Charles A, Jr (OK)
Wool, Richard P (IL)
Ysteboe, Howard T (PA)

POWER

Adam, Paul J (MO)
Barnes, Bruce F, Jr (MO)
Bradley, Frank L Jr (NY)
Cagnetta, John P (CT)
Clark, William R (DE)
Crocker, John W (TX)
Daman, Ernest, L (NJ)
Fisher, Garland F (IA)
Fraas, Arthur P (TN)
Freedman, Steven I (IL)
Graybill, Howard W (CA)
Grey, Jerry (DC)
Halibey, Roman (NY)
Heil, Terrence J (WA)
Heitlinger, Igor (CT)
Hollrah, Ronald L (MO)
Jeffries, Ronald F (VA)
Kalin, Thomas E (MO)

Kelly, Alphonsus G
Kephart, John T, Jr (DE)
Kollitides, Ernest A (NY)
Ladd, Conrad M (CO)
Larson, George H (KS)
Leppke, Delbert M (IL)
Loebel, Fred A (FL)
Matthews, S LaMont (OR)
McShane, William R (NY)
Moore, Raymond P (CA)
Nakano, Yoshiei
Ochab, Thomas F (NJ)
Okamoto, Hideo
Olsson, John E (NE)
Phadke, Arun G (NY)
Rector, Alwin H (KS)
Rinard, Sydney L (MO)
Robbins, Joseph E (NY)
Roberson, James E (SC)
Rosard, Daniel D (PA)
Schmidt, Charles M (OH)
Shannon, Eldon B (MO)
Stack, John R (MO)
Sturdevant, Bruce L (CO)
Thompson, Willis F (CT)
True, Newton F (IL)
Weber, Norman (IL)
Wood, Alan B

PRESSURE VESSELS

Fino, Alexander F (PA)
Hansberry, John W (MA)
Mraz, George J (NH)
Pettigrew, Allan
Widera, G E O (IL)
Zick, Leonard Paul (AZ)

PRESTRESSED CONCRETE

Anderson, Arthur R (WA)
Martin, Leslie D (IL)
McCalla, William T (MN)
Naaman, Antoine E (MI)

PRINTING

Dobson, David B (DC)
Geis, A John (PA)
Hughes, Edwin L (FL)

PROCESS ENGINEERING

Adler, Irwin L (NJ)
Adler, Stanley B (TX)
Allen, Dell K (UT)
Allen, Robert S (KS)
Anderson, Frank A (MS)
Arnold, David W (MS)
Ash, Alvin G (WA)
Ban, Thomas E (OH)
Barer, Seymour (NJ)
Barnhart, James H (TX)
Batchelder, Howard R (CA)
Baumann, George P (NJ)
Beavon, David K (CA)
Beckmann, Robert B (MD)
Belden, Reed H (NJ)
Bennett, Marlin J (WA)
Berg, Philip J (PA)
Bertram, Robert L (TX)
Bertran, Enrique (PR)
Berty, Jozsef M (PA)
Blanco, Jorge L (LA)
Blazey, Leland W (PA)
Blue, E Morse (CA)
Bobart, George F (GA)
Bradford, Michael L (LA)
Bramer, Henry (PA)
Brannock, N Fred (CA)
Breitmayer, Theodore (CA)
Bresler, Sidney A (NY)
Brinson, Robert J (SC)
Brown, Robert O (MN)

Brownstein, Arthur M (NJ)
Buchsbaum, Norbert N (TX)
Bushnell, James D (NJ)
Canfield, Frank B, Jr (TX)
Cannon, Charles N (CA)
Caselli, Albert V (TX)
Chamberlin, Paul D (CO)
Cheng, Shang I (NY)
Collins, Charles E (TX)
Conkey, David R (MN)
Connally, Harold T (TX)
Cook, Harry M, Jr (IL)
Cooke, Norman E
Copeland, Norman A (DE)
Cummings, George H (TX)
Danly, Donald E (FL)
Day, Roger W (TX)
De Haven, Eugene S (CA)
Dean, Robert C, Jr (VT)
Delaney, Francis H (NY)
Dickey, Robert, III (PA)
Dockendorff, Jay D (IL)
Dockendorff, Ralph L (TX)
Dodgen, James E (CO)
Dossett, Jon L (IL)
Doumas, Basil C
Draper, Alan B (PA)
Drew, John (FL)
Duker, George H (PA)
Earl, Christopher B (FL)
Eccles, Richard M (NJ)
Falivene, Pasquale J (NJ)
Ferrara, Norman (PA)
Finneran, James A (TX)
Fisher, Edwin E (TX)
Gadomski, Richard T (TN)
Galstaun, Lionel S (TX)
Geddes, Ray L (FL)
Gigliotti, Michael F X (MA)
Gillespie, LaRoux K (MO)
Glover, John W (NC)
Gould, Merle L (LA)
Grimes, William W (OH)
Grossberg, Arnold L (CA)
Groves, Stanford E (LA)
Hart, David R (AL)
Hauck, Charles F (PA)
Hays, George E (OK)
Hirsch, Sylvan R (NJ)
Hoffman, Paul W (PA)
Holmes, Elwyn S (KS)
Holton, John H (NJ)
Hooper, E Dale (NJ)
Hovmand, Svend (MD)
Huff, James E (MI)
Hurlburt, Harvey Z (TX)
Hurt, Kenn (IL)
Inman, Byron N (WA)
Jaco, Charles M, Jr (SC)
James, Alexander (SC)
Jenett, Eric (TX)
Jennings, Nathan C (TX)
Kehde, Howard (MI)
Kelley, Thomas N (NY)
Kelsey, Stephen J (UT)
Kitterman, Layton (TX)
Krase, Norman W (PA)
Kunreuther, Frederick (NY)
Larson, Marvin W (OH)
Leonard, Jackson D (NJ)
Linden, Henry R (IL)
List, Harvey L (NY)
Litzinger, Leo F (MA)
Lofredo, Antony (NJ)
Lowery, Anthony J (MO)
Lundberg, Lennart A (WA)
Maisel, Daniel S (NJ)
Marion Johnson P E (TX)
Martin, Howard W (TX)
Martin, Jay R (PA)
Mason, Donald R (FL)
Mathur, Umesh (OK)
Matsukado, William M (OR)
McBride, J A (CO)

McCall, Thomas F (MO)
McCormick, John E (NJ)
Milligan, Robert T (CA)
Mojica, Juan F
Moore, Peggy B (OH)
Motard, Rodolphe L (MO)
Neil, William N (MA)
Nelson, E Raymond (CA)
Nibley, J William (UT)
Ogburn, Hugh B (HI)
Ohsol, Ernest O (TX)
Parklinson, John R (WA)
Pavelchek, Walter R (PA)
Petrovic, Louis J (MA)
Richardson, Brian P (GA)
Ritter, Gerald L (WA)
Roche, Edward C Jr (NJ)
Rorschach, R I (OK)
Ross, James L (MO)
Savage, Marvin (PA)
Scher, Robert W (OH)
Schlegel, Walter F (CT)
Seglin, Leonard (NY)
Seider, Warren D (PA)
Seidorf, Christian E (OH)
Servais, Ronald A (OH)
Silla, Harry (NJ)
Skov, Ebbe R (TX)
Smith, Frank M, Jr (FL)
Spangler, Carl D, Jr (OK)
Stief, Robert D (WV)
Trombly, Arthur J (MI)
Vela, M A (TX)
Verneuil, Vincent S (CA)
Wagner, John L (PA)
Waguespack, Otis J (LA)
Wallenstein, Howard M (NJ)
Wang, Yui L (CA)
Ward, Julian R (TX)
Wender, Leonard (NY)
Will, Herbert (NJ)
Williams, Steven A (TN)
Zdonik, Stanley B (MA)

PRODUCT DESIGN & ENGINEERING

Allard, Kermit O (IA)
Allen, William S (TX)
Belz, Paul D (MD)
Bennett, Dwight H (CA)
Bennett, Richard T (IL)
Brandinger, Jay J (NJ)
Brown, Delmar L (OR)
Brown, Samuel J (PA)
Tecklenburg, Harry (NY)
Burns, Fredrick B (WI)
Chadwick, George F (NY)
Chilton, Ernest G (CA)
Chipouras, Peter A (MA)
Chou, Robert V (KY)
Cornwall, Harry J (CA)
Davee, James E (GA)
Dembowski, Peter V (NY)
Doret, Michel R (NY)
Dorman, William H (NY)
DuBois, J Harry (NJ)
Dunham, Thomas E (KY)
Earle, Emily A (AL)
Eck, Bernard J (IL)
Ehrich, Fredric F (MA)
Errera, Samuel J (PA)
Eubanks, William H (MS)
Evans, Ralph A (NC)
Ferrante, John A (NY)
Fisher, David L (TX)
Flanagan, H Russell (TN)
Fleischer, Henry (MI)
Forney, Bill E (OK)
France, Jimmie J (IL)
Francis, Lyman L (MO)
Freeman, Arthur H (CT)
Freers, Howard P (MI)
Fried, George (NY)

PRODUCT DESIGN & ENGINEERING
(Continued)

Gerhardt, Ralph A (IL)
Giles, William S (OH)
Greene, Ray P (OH)
Griffin, Daniel M (FL)
Grosser, Christian E (VA)
Hall, Donivan L (OH)
Hamilton, Stephen B, Jr (CT)
Hatheway, Alson E (CA)
Helmich, Melvin J (OH)
Herwald, Seymour W (PA)
Hill, Percy H (MA)
Hodges, Joseph T (MO)
Husby, Donald E (MN)
Huss, Harry O (OH)
Isaacs, Jack L (CT)
Johnson, Andrew B (IL)
Johnson, Ray C (MA)
Jones, Howard T (IN)
Jones, Marvin R (TX)
Kececioglu, Dimitri B (AZ)
Kennemer, Robert E (IL)
Krase, Norman W (PA)
Kreifeldt, John G (MA)
Kress, Ralph H (IL)
Lage, David A (MA)
Laramee, Richard C (UT)
Lee, Low K (CA)
Lehmann, Henry (NY)
Lilley, Eric G (TX)
Linstromberg, William J (MI)
Lottridge, Neil M (MI)
Ludwig, George A (NY)
Lux, William J (IA)
Maneri, Remo R (MI)
McDonald, Robert O (IL)
McGinty, Michael J (OH)
McKinley, Robert W (NH)
Merkert, Clifton S (PA)
Mischke, Charles R (IA)
Morrell, T Herbert (MN)
Morton, Lysle W (NY)
Nielsen, Michael J (AZ)
Noton, Bryan R (OH)
Nutter, Dale E (OK)
Olen, Robert B (OH)
Patton, Willard T (NJ)
Pauly, Bruce H (OH)
Phillips, Benjamin A (MI)
Pickett, Leroy K (IL)
Polaner, Jerome L (NJ)
Powell, T Charles (MI)
Proffitt, Richard V (CT)
Rao, Peter B (OH)
Rau, Jim L (IN)
Rhoades, Warren A (TX)
Ring, Sandiford (TX)
Rizzone, Michael L (TX)
Rumbaugh, Max E, Jr (IN)
Sandgren, Find (IN)
Scheer, John E (PA)
Schneider, Darren B (SC)
Schubert, William L (IL)
Scofield, Gerald A (PA)
Sharpe, Irene W (MI)
Smith, Charles O (IN)
Smith, Edward M (OH)
Sparling, Rebecca H (CA)
Sullivan, James T (PA)
Todd, David B (MI)
van Vessem, Jan C
Vogel, Fred J (FL)
Watkins, Susan J (CA)
Webster, Neil W (MI)
Weinberg, J Morris (MA)
Weiner, Robert I (MD)
Wesner, John W (NJ)
Whitesides, Jack C (GA)
Whitney, Eugene C (PA)
Whitney, James C (CT)
Woerner, Leo G (MD)

Wood, Gerald L (AZ)
Zichterman, Cornelius (IL)

PRODUCT SAFETY

Brennan, James J (TX)
Codlin, James B (IL)
Hirschberg, Erwin E (IL)
Johnston, Waymon L (TX)
Libertiny, George Z (MI)
Mihalasky, John (NJ)
Moll, Richard A (WI)
Morris, Alan D (MD)
Newman, Malcolm (NY)
Ojalvo, Irving R (CT)
Pfundstein, Keith L (IL)
Viano, David C (MI)
Zich, John H (MI)

PRODUCTION

Byrd, Jack, Jr (WV)
Delaney, Francis H (NY)
Goldstein, Irving R (NJ)
Johnson, Lynwood A (GA)
Kraimer, Frank B (MI)
Macklin, Harley R (KS)
Orr, Charles K (TX)
Presley, Gordon C (CO)
Ward, Richard E (WV)

PRODUCTIVITY

Balestrero, Gregory (GA)
Bittel, Lester R (VA)
Chambers, David S (TN)
Engwall, Richard L (MD)
Goldman, Jay (AL)
Gould, Arthur F (PA)
Greene, James H (IN)
LeBlanc, Joseph U (TX)
Matthews, Allen R (MD)
Melden, Morley G (NY)
Rotolo, Elio R (CA)
Scarola, John A (TX)

PUBLIC ENGINEERING ADMINISTRATION

Abbott, Wayne H, Jr (MI)
Adams, William E (NV)
Ahlen, John W (AR)
Almquist, Wallace E (WV)
Alsing, Allen A (OR)
Bellport, Bernard P (CA)
Besson, Pierre A
Bing-Wo, Reginald
Burzell, Linden R (CA)
Calkins, Myron D (MO)
Carney, William D (MO)
Chandler, R L (KS)
Coe, Benjamin P (NY)
Cooper, Norman L (MD)
Despres, David R (MI)
Fremouw, Gerrit D (VA)
Gordon, Ruth V (CA)
Hardin, Clyde D (GA)
Howlett, Myles R (VA)
Kollerbohm, Fred A (MA)
Korf, Victor W (WA)
Kramer, Samuel (MD)
Lang, Martin (NY)
Latham, Charles F (KY)
McBride, Philip R (TN)
McDermott, Joseph M (IL)
McManus, Ralph N
McMillan, Hugh H, (IL)
Medbery, H Christopher (CA)
Morrison, John S , Jr (PA)
Ojalvo, Morris S (DC)
Peake, Harold J (VA)
Power, Richard J (VA)
Schweller, David (NY)
Snider, David G (MO)

Utic, Phillip J (WI)
Wanket, Achiel E (CA)
Woller, Myron K (NJ)

PULP AND PAPER

Atchison, Joseph E (NY)
Baisch, Stephen J (WI)
Barry, John E (OH)
Chang, Nai L (WI)
Doshi, Mahendra R (WI)
Ericsson, Eric O (WA)
Freeland, Manning C (SC)
Grace, Thomas M (WI)
Hershey, Robert V (MN)
Hiett, Louis A (TN)
Nelson, George G, Jr (MS)

QUALITY & RELIABILITY CONTROL

Allen, Robert L (IL)
Anderson, Eric D (MI)
Anderson, Robert T (OH)
Avril, Arthur C (OH)
Babcock, Daniel L (MO)
Barker, Eugene M (CA)
Bartell, Howard F (PA)
Belding, Harry J (PA)
Bird, George T (VA)
Blum, Michael E
Bott, Walter T (IL)
Brooks, M Scott (CT)
Brumbaugh, Philip S (MO)
Burns, Robert G (MA)
Case, Kenneth E (OK)
Chambers, David S (TN)
Charpentier, David L (CT)
Coleman, Edward P (CA)
Cordill, John J (IL)
Cound, Dana M (PA)
Crutcher, Harold L (NC)
Dague, Delmer C (OH)
Dalal, Jayesh G (IL)
Danpour, Henry (MA)
Darden, Catherine H (GA)
Del Re, Robert (FL)
Delve, Frederick D (PA)
Dorsky, Lawrence R (NY)
Early, David D (WA)
Edwards, Ralph P (SC)
Elam, Edward E, III (GA)
Enell, John Warren (NJ)
Evans, Ralph A (NC)
Fiaschetti, Rocco L (CA)
Fonda, LeGrand B (PA)
Friels, David R (TN)
Geoffrion, Louis P (MA)
Gryna, Frank M (CT)
Guernsey, John B (MD)
Haynes, Patricia Griffith (VA)
Heikes, Russell G (GA)
Holder, Sidney G, Jr (AL)
Hornberger, Walter H (MI)
Hultberg, Dwain R (WV)
Hunter, J Stuart (NJ)
Hurd, Walter L Jr (CA)
Husseiny, Abdo A (IA)
Johnson, George W (OH)
Johnson, Marvin M (NE)
Juran, Joseph M (NY)
Kececioglu, Dimitri B (AZ)
Kimberly, A Elliott (MI)
Knowler, Lloyd A (IA)
Kondo, Yoshio
Kydonieus, Agis F (NY)
Lamberson, Leonard R (MI)
Lawrence, Harvey J (CA)
Leavenworth, Richard S (FL)
Lessig, Harry J (IL)
Licht, Kai (OH)
Lingafelter, John W (CA)
Mancini, Gerold A (NY)

QUANTUM ELECTRONICS

Maxton, Robert C (MN)
McAfee, Naomi J (MD)
McBride, J A (CO)
McBryde, Vernon E (AR)
McGrath, Daniel F (TX)
McNichols, Roger J (OH)
Nagarsenker, Brahmanand N (PA)
Naresky, Joseph J (NY)
Nichols, Edward (RI)
Nicholson, Richard H (CA)
Parsons, Donald S (PA)
Peach, Robert W (IL)
Piper, James E (PA)
Plourde, Arthur J Jr (MI)
Raheja, Dev G (MD)
Reich, Bernard (NJ)
Ross, Keith E (IN)
Rubin, Charles H (MA)
Sauter, Harry D (IN)
Scott, Lester F (MO)
Sedeora, Tejinder S (IL)
Shainin, Dorian (CT)
Siddell, Derreck
Singpurwalla, Nozer D (DC)
Smelcer, Glen E (IA)
Smith, James R (TN)
Smith, Marion P (FL)
Stier, Howard L (NY)
Stubbles, John R (OH)
Thomas, Eddie E (TX)
Thompson, John W
Treffs, David A (MA)
Waldner, John W (IN)
Wall, Alfred S (NJ)
Ward, Allen W (OK)
Weis, Alfred E (CT)
Willhelm, A Clyde, Jr (AL)
Williams, Samuel L (IL)
Wilson, Myron F (IA)
Worland, Robert B (NC)
Zierdt, Conrad H, Jr (PA)

QUALITY ASSURANCE

Barker, Eugene M (CA)
Brokaw, Charles H (GA)
Burns, Robert G (MA)
Carter, Charles L, Jr (TX)
Carter, Seymour W (MA)
Charpentier, David L (CT)
Colandrea, Thomas R (CA)
Dague, Delmer C (OH)
Freund, Richard A (NY)
Geoffrion, Louis P (MA)
Grisewood, Norman C (NJ)
Gryna, Frank M (CT)
King, James R (NH)
Kowtna, Christopher C (DE)
Maas, Robert R (CA)
Moriarty, Brian M (DC)
Mundel, August B (NY)
Plourde, Arthur J Jr (MI)
Powell, Michael J (MO)
Shields, Bruce M (PA)
Spechler, Jay W (FL)
Talbott, Laurence F (CA)
Trainor, Eugene F (MA)
Viaclovsky, Syl A (TX)

QUANTUM ELECTRONICS

Ballantyne, Joseph M (NY)
Bass, Michael (CA)
Byer, Robert L (CA)
Casperson, Lee W (OR)
Cheo, Peter K (CT)
Chester, Arthur N (CA)
Crane, Robert K (NH)
Davidson, Frederic M (MD)
Dienes, Andrew J (CA)
Fisher, Edward R (MI)

RADAR

RESEARCH & DEVELOPMENT (GENERAL) (Continued)

Eager, George S, Jr (NJ)
Eccles, Richard M (NJ)
Edwards, James T (CA)
Elliott, Kendall C (WV)
Ellis, Harry M
Ellsworth, William M (MD)
Engesser, Donald G (NJ)
Esgar, Jack B (OH)
Ethington, Robert L (OR)
Everett, Woodrow W, Jr (FL)
Falco, James W (DC)
Fawcett, Sherwood L (OH)
Fear, J V D (PA)
Fearnsides, John J (VA)
Feltner, Charles E (MI)
Fenn, Raymond W, Jr (CA)
Ferguson, Donald M (MA)
Fernandes, John H (RI)
Ferrari, Harry M (PA)
Filbert, Howard C, Jr (MD)
Floyd, Dennis R (CO)
Freche, John C (OH)
Freedman, Jerome (MA)
Freeman, Mark P (CT)
Frei, Ephraim H
Freise, Earl J (NE)
Freitag, Dean R (TN)
Frosch, Robert A (MI)
Fry, Thornton C (CA)
Frye, Alva L (TN)
Gabelman, Irving J (NY)
Galt, John K (NM)
Garoff, Kenton (NJ)
Garrett, Donald E (CA)
Gaumer, Lee S (PA)
Gautreaux, Marcelian F , Jr (LA)
Gaynor, Joseph (CA)
Gerard, Roy D (TX)
Getsinger, William J (MD)
Gianola, Umberto F (NJ)
Gibian, Thomas G (MD)
Giller, Edward B (VA)
Ginwala, Kymus (MA)
Gleason, James G (AR)
Goland, Martin (TX)
Gordon, Eugene I (NJ)
Gottscho, Alfred M (PA)
Gould, Merle L (LA)
Graves, Gilman L, Jr (MS)
Green, Robert P (NH)
Greenberg, Herman D (PA)
Greene, George R (CA)
Greene, John H, Jr (TX)
Greene, Joseph L, Jr (AL)
Greenwood, Allan N (NY)
Gross, John H (PA)
Groth, William R (NY)
Gully, Arnold J (TX)
Gupta, Gopal D (NJ)
Haberman, Eugene G (CA)
Hall, Donivan L (OH)
Halligan, James E (NM)
Halperin, Herman (CA)
Hamilton, William H (PA)
Hammond, Ross W (GA)
Harmon, George W (PA)
Harris, Benjamin L (MD)
Harris, Ronald D (IL)
Harrison, Gordon R (GA)
Hart, David R (AL)
Hartley, Fred L (CA)
Hatcher, Stanley R
Hayashi, Izuo
Hayden, Richard E (MA)
Heer, Ewald (CA)
Hegbar, Howard R (OH)
Heilmeier, George H (VA)
Henry, Richard J (PA)
Henry, Robert J (PA)
Herring, H James (NJ)

Herzog, Gerhard
Heywang, Walter
Hiett, Louis A (TN)
Hillier, James (NJ)
Hipchen, Donald E (FL)
Hirsch, Robert L (TX)
Hirschhorn, Isidor S (NJ)
Hollander, Milton B (CT)
Holloway, Frederic A L (LA)
Holt, Sherwood G (WI)
Hooper, E Dale (NJ)
Howard, William J (NM)
Howarth, Elbert S (PA)
Hsu, Yih-Yun (MD)
Hudelson, George D (NY)
Hughes, Thomas J (CA)
Hulm, John K (PA)
Hulsey, J Leroy (NC)
Hunt, Charles H (PA)
Hurlburt, Harvey Z (TX)
Husby, Donald E (MN)
Husseiny, Abdo A (IA)
Iams, Harley A (CA)
Ireland, Donald R (CA)
Isaac, Maurice G (CT)
Isakoff, Sheldon E (PA)
Jacobaeus, Christian
Jakeway, Lee A (HI)
Jasper, Norman H (FL)
Jatczak, Chester F (OH)
Jenkins, Ivor
Jensen, Arthur S (MD)
Jensen, Harvey M (PA)
Jimenez, Rudolf A (AZ)
Joel, Amos E, Jr (NJ)
Johnson, Charles E (NH)
Johnson, Maurice V, Jr (CA)
Johnson, Richard C (GA)
Johnson, Zane Q (PA)
Jolliff, James V (MD)
Jones, Benjamin A, Jr (IL)
Jones, Charles E (OH)
Jones, Howard St C, Jr (DC)
Jones, Joseph K (NC)
Jones, Marvin R (TX)
Kage, Arthur V (MO)
Kahng, Dawon (NJ)
Kalvinskas, John J (CA)
Kamal, Medhat M (OK)
Karchner, George H (PA)
Karpenko, Victor N (CA)
Katz, Marvin L (TX)
Kaufman, John Gilbert, Jr (PA)
Keane, Barry P (SC)
Kennel, William E (IL)
Kensinger, Robert S (MN)
Kerber, Ronald L (MI)
Kessler, Thomas J , Sr (NY)
Kidder, Allan H (PA)
King, Randolph W (MD)
Kirk, Walter B (PA)
Kleen, Werner J
Kleiner, Fredric (NY)
Koo, Jayoung (NJ)
Korkegi, Robert H (DC)
Kornei, Otto (CA)
Kramer, Daniel E (PA)
Kraus, Wayne P (OK)
Ksienski, Aharon A (OH)
Kuhlke, William C (TX)
Kulkarni, Kishor M (OH)
Kupelian, Vahey S (MD)
Lafferty, James M (NY)
Lafranchi, Edward A (CA)
Lander, Horace N (CT)
Landgrebe, David A (IN)
Lang, Edward W (FL)
Larson, Thurston E (IL)
Lauriente, Michael (MD)
Lawrence, Harold R (NM)
LeBlanc, Joseph U (TX)
Lebow, Irwin L (DC)

Lechner, Bernard J (NJ)
Lederman, Samuel (NY)
Lee, Griff C (LA)
Lenz, Ralph C (OH)
Leopold, Reuven (FL)
Levine, Duane G (NJ)
Lewis, Arthur E (CA)
Limb, John O (NJ)
Ling, Frederick F (NY)
Lipuma, Charles R (NJ)
List, Hans
Littmann, Walter E (TX)
Loftness, Robert L (DC)
Long, Chester L (VA)
Long, Maurice W (GA)
Longini, Richard L (PA)
Lyman, W Stuart (NY)
Lyons, Jerry L (MO)
Lyons, John W (MD)
Mack, Roger G (PA)
Magee, Christopher L (MI)
Mandelbrot, Benoit B (NY)
Maninger, R Carroll (CA)
Manjoine, Michael J (PA)
Marcuse, Dietrich (NJ)
Margiloff, Irwin B (NY)
Marks, Albert B (OK)
Markwardt, L J (WI)
Marschall, Ekkehard P (CA)
Marsh, B Duane (CT)
Martin, J Bruce (OH)
Martino, Joseph P (OH)
Massieon, Charles G (IL)
Mayer, Francis X (LA)
McCawley, Frank X (DC)
McCoy, Donald S (IN)
McDonald, Henry C (CA)
McDonough, George F, Jr (AL)
McFarlane, Maynard D (AZ)
McGrath, William L (FL)
McGuire, Michael J (CA)
McKinley, Donald W R
McKinnell, William P, Jr (CO)
McMulkin, F John
McPherson, Donald J (CA)
McRae, John L (MS)
Mechlin, George F (PA)
Merrill, Roger L (OH)
Mickelson, Cedric G (IL)
Middleton, David (NY)
Millar, G H (FL)
Miller, Raymond E (GA)
Millheim, Keith K (OK)
Milligan, Robert T (CA)
Mills, Robert G (NJ)
Minet, Ronald G (CA)
Mingle, John O (KS)
Money, Lloyd J (DC)
Moody, Arthur (PA)
Moore, W Calvin (PA)
Moreno, Theodore (CA)
Moritz, Karsten H (NJ)
Morris, James G (KY)
Morris, Robert L (CT)
Morrison, John E, Jr (TX)
Morse, Richard S (MA)
Moss, Alan M (NJ)
Mouly, Raymond J (PA)
Murphy, Charles H (MD)
Myron, Thomas L (PA)
Neely, George Leonard (CA)
Nekola, Robert L (OH)
Nelson, Burke L (CT)
Nessmith, Josh T (NJ)
Newell, Earl D (CA)
Newhouse, David L (NY)
Niederhauser, Warren D (PA)
Nolen, James S (TX)
O'Grady, Joseph G (PA)
Ohanian, M Jack (FL)
Ohtake, Tadashi

Olson, Glenn M (IL)
Olt, Theodore F (OH)
Openshaw, Keith L (CA)
Oriani, Richard A (MN)
Ostberg, Orvil S (NY)
Pankove, Jacques I (NJ)
Pao, Yoh-Han (OH)
Paris, Luigi
Pavelchek, Walter R (PA)
Pearson, John W (MN)
Peck, Eugene L (VA)
Peixotto, Ernest D (DC)
Pendleton, Roger L (VA)
Pendleton, Wesley W (MI)
Pershing, Roscoe L (IL)
Persson, Sverker (PA)
Peterson, Donald L (WV)
Peterson, William E (WA)
Piasecki, Frank N (PA)
Plonsey, Robert (NC)
Popovics, Sandor (PA)
Porcelli, Richard V (NY)
Powe, Ralph E (MS)
Powers, Kerns H (NJ)
Pressman, Norman J (DE)
Preston, Frank S (NC)
Prince, M David (GA)
Pry, Robert H (IL)
Pyper, Gordon R (VT)
Queneau, Paul B (CO)
Race, Hubert H (PA)
Ramberg, Steven E (DC)
Reams, James D (OH)
Rechtin, Eberhardt (CA)
Redhead, Paul A
Redmond, Robert F (OH)
Rhodes, John M (MA)
Richardson, C D (AL)
Rivers, Lee W (NJ)
Rizzone, Michael L (TX)
Robertson, Lawrence M (CO)
Robins, Norman A (IN)
Robinson, Arthur S (VA)
Roe, William P (NJ)
Roley, Daniel G (IL)
Rosenberg, Paul (NY)
Rosenberg, Robert B (IL)
Roth, J Reece (TN)
Sabnis, Gajanan M (MD)
Salkovitz, Edward I (VA)
Samaras, Demetrios G (FL)
Sandell, Dewey J, Jr (NY)
Sanks, Robert L (MT)
Santilli, Paul T (OH)
Sargent, Donald J (IN)
Schacht, Paul K (WI)
Schaffhauser, Robert J (NY)
Schenker, Leo (NJ)
Scheuing, Richard A (NY)
Schmerling, Erwin R (MD)
Schmid, David M (PA)
Schmidt-Tiedemann, K J
Schmitt, Hans J
Schnizer, Arthur W (MA)
Schrader, Gustav E (OH)
Schubert, William L (IL)
Schwenk, Francis C (MD)
Scott, John W (CA)
Seiden, Edward I (CA)
Selander, Carl E (CO)
Seruto, Joseph G (CA)
Shaffer, Louis Richard (IL)
Sheinberg, Haskell (NM)
Sherwin, Martin B (MD)
Shima, Shigeo
Shore, David (VA)
Shoults, David R (AZ)
Shoupp, William E (PA)
Shtrikman, Shmuel
Sibert, George W (VA)
Simard, Gerald L (ME)
Simons, Eugene M (OH)
Simons, Gale G (KS)
Sinclair, A Richard (TX)
Sinclair, Thomas L , Jr (HI)

SANITARY ENGINEERING
(Continued)

Coppelman, Daniel P (NY)
Cosens, Kenneth W (OH)
Costello, Charles V (NY)
Costello, Frederick J (VT)
Coulter, James B (MD)
Covert, Paul D (OH)
Crawford, Leonard K (IL)
Cullinane, Murdock J , Jr (TX)
Cullivan, Donald E (MA)
Cywin, Allen (VA)
Daily, Eugene J (IL)
Davis, Jeff W (WA)
Davy, Philip Sheridan (WI)
Dawkins, Mather E (FL)
Deb, Arun K (PA)
Dedyo, John (NY)
DeFraites, Arthur A , Jr (LA)
Dell, Leroy R (MI)
DeLoach, Robert E , II (GA)
Dennis, John F (FL)
Dietz, Jess C (FL)
Dodson, Roy E, Jr (CA)
Dore, Stephen E , Jr (MA)
Dowell, James C (KS)
Druyvestein, Terry L (MT)
Dykes, Glenn M, Jr (FL)
Eckmann, Donald E (IL)
Edgerley, Edward, Jr (MO)
Edzwald, James K (MA)
Eggener, Charles L (AK)
Erganian, George K (IN)
Esvelt, Larry A (WA)
Eunpu, Floyd F (VA)
Ewing, Ronald L (AZ)
Ewing, William C (MI)
Farnsworth, George L (IL)
Fassnacht, George G (IN)
Feagan, Wilbur S (MO)
Fee, John R (CA)
Ferrero, Joseph L (PA)
Fifield, Charles D (MI)
Fisher, Lawrence R, Jr (OK)
Fithian, Theodore A (PA)
Foerster, Richard E (NY)
Force, Richard W (MI)
Foree, Edward G (KY)
Foss, James B (MD)
Furman, Thomas deS (SC)
Gasser, Robert F (NY)
Gassett, Richard B (FL)
Gaynor, Lester (MA)
Genovese, Philip W (CT)
Gerber, H Bruce (PA)
Gibbons, Harry De R (OH)
Giessner, William R (CA)
Gillett, John B (MD)
Gonzales, John G M (NV)
Graham, W William, Jr (AR)
Granstrom, Marvin L (NJ)
Grasso, Salvatore P (NH)
Gray, Glenn C (MO)
Green, Richard S (MD)
Greenleaf, John W, Jr (FL)
Grigoropoulos, Sotirios G
Guy, Louis L, Jr (VA)
Haefeli, Robert J (NJ)
Hartwig, Thomas L (PA)
Hassebroek, Lyle G (WA)
Hawthorne, John W (IL)
Hayduk, Stephen G (NY)
Hayter, R Reeves (TX)
Heil, Dick C (MD)
Henry, Laurence O (Pat) (AZ)
Hernandez, John W (NM)
Hert, Oral H (IN)
Hird, Lyle F (WI)

Hirsch, Lawrence (CA)
Hodges, Raymond D (LA)
Hook, Melvin E (PA)
Hsiung, Andrew K (OR)
Huset, Elmer A (MN)
Ingalls, Larry W (VA)
Jensen, Eugene T (KS)
Jeris, John S (NY)
Johnson, Charles C , Jr (MD)
Johnson, Elliott B (CA)
Johnson, Nolan L, Jr (GA)
Johnson, Walter K (MN)
Jones, John D (OH)
Jones, Raymond M (DC)
Juergens, Robert B (OH)
Kaiser, C Hayden, Jr (MS)
Kaufman, Herbert L (NJ)
Keegan, John J (NY)
Keith, George W (PA)
Kienow, Paul E (AZ)
Kimm, James W (IA)
King, Paul H (MA)
King, Richard L (TN)
Knoop, Frederick R, Jr (MD)
Koehler, Melvin L (OH)
Konrad, William N (WI)
Koptionak, William (CA)
Korbitz, William E (CO)
Kuehl, Neal H (IA)
Kuhn, Paul A (IL)
Kupper, Charles J, Jr (NJ)
LaBoon, John F , Sr (PA)
Lagnese, Joseph F (PA)
Langdon, Paul E , Jr (IL)
Langford, Ivan, Jr (TX)
Latham, Charles F (KY)
Le Van, James H (OH)
Lee, Robert L (OR)
Lenard, John F (CT)
Lennex, Richard B (MI)
Lewis, Willis H (TN)
Liebenow, Wilbur R (MN)
Linstedt, K Daniel (CO)
Lipscomb, Joseph W (PA)
Lischer, Vance C (MO)
Lloyd, William (MN)
Love, William J (NY)
MacNeill, John S (NY)
Malone, William J (AR)
Martenson, Dennis R (MN)
McCallum, Gordon E (MD)
McCloud, Robert J (MO)
McCrate, Thomas A (OH)
McLeod, Robert J (MD)
McMahon, Thomas E (KS)
McManus, John A (CT)
Medbery, H Christopher (CA)
Mercer, Harold E (TX)
Merritt, LaVere B (UT)
Molzen, Dayton F (NM)
Monroe, Edward W (PA)
Montecki, Carl R (FL)
Myers, C Kenneth (PA)
Myers, Harry E , Jr (AL)
Nicoladis, Frank (LA)
Noonan, Mark E (CA)
Novak, John T (VA)
O'Connor, John T (MO)
Ohara, George T (CA)
O'Rourke, James T (MA)
O'Rourke, Robert B (OH)
Oswald, William J (CA)
Owen, Eugene H (LA)
Pailthorp, Robert E (CO)
Parrott, George D (KY)
Pawlowski, Harry M (IL)
Peck, Kenneth M (NY)
Pernoud, Rene B (TX)
Perry, Robert R (VA)
Pfeffer, John T (IL)
Phillips, William J, II (CA)
Pieczonka, Ted J , Jr (NY)
Pirnie, Malcolm, Jr (NY)

Pluntze, James C (WA)
Priestman, John
Purcell, Fenton Peter (NJ)
Purcell, Leo Thomas, Jr (NJ)
Pyper, Gordon R (VT)
Rademacher, John M (IL)
Randall, Clifford W (VA)
Rea, John E Jr (OK)
Reece, John D (MO)
Remus, Gerald J (MI)
Rhett, John T (VA)
Richardson, Elmo A Jr (GA)
Richardson, William H (IL)
Rivas-Mijares, Gustavo
Robeck, Gordon G (CA)
Rodriquez, Pepe (WI)
Roehrig, Clarence S, Jr (KY)
Rollag, Dwayne A (SD)
Roy, Martin H (AR)
Rubin, Alan J (OH)
Ruby, Kenneth W (ND)
Ruiz, Aldelmo (VA)
Ryder, Robert A (CA)
Safhay, Meyer (NY)
Santry, Israel W , Jr (TX)
Schaper, Laurence T (MO)
Schwegel, Donald R (IL)
Scott, Robert A , IV (MS)
Seabloom, Robert W (WA)
Sherer, Keith R (OR)
Shifrin, Walter G (IL)
Shipman, Harold R (MD)
Shoemaker, Leonard W (TX)
Shoolbred, Augustus W, Jr (SC)
Shubinski, Robert P (VA)
Sinnott, Walter B (NY)
Skelly, Michael J (NY)
Smallwood, Charles, Jr (NC)
Smit, Raymond J (MI)
Smith, Arnold R (NJ)
Smith, Charles H (MO)
Smith, Kenneth R (OH)
Snow, Phillip D (NY)
Spamer, James S (MD)
Spina, Frank J (NY)
Staton, Marshall (NC)
Storm, David R (CA)
Sullivan, A Eugene (MA)
Sutphen, Russell L (NY)
Svore, Jerome, H (TX)
Symons, James M (OH)
Szabo, August J (LA)
Tanonaka, Clarence K (HI)
Tassos, Thomas H, Jr (NC)
Taylor, Lot F (KS)
Tenney, Mark W (IN)
Thayer, Royal C (DC)
Thomas, James R, Jr (MD)
Tisdel, Joseph E (NY)
Tozer, George K (MA)
Tyer, Arnold J (AR)
Van de Riet, James L (VA)
Vanderboom, Steve A (MN)
VanValkenburgh, Kent (OK)
Velzy, Charles O (NY)
Vogt John E (MI)
Wagner, Victor G (IN)
Wainwright, Sam H (AL)
Waller, Robert (MD)
Washington, Donald R (MO)
Watson, Kenneth W (UT)
Weaver, Leo (OH)
Weiss, Martin (MA)
Wetstein, Willis J (AZ)
Wild, Harry E, Jr (FL)
Williams, Wesley B (GA)
Willis, Charles A (NC)
Wilson, Percy Suydam (NJ)

Wise, William F (VA)
Wolman, Abel (MD)
Woods, Raymond C (TN)
Zimmerman, Robert M (IL)

SEISMIC

Bolt, Bruce A (CA)
Cole, Eugene E (CA)
Danzinger, Edward (HI)
Frantti, Gordon E (MI)

SEMICONDUCTOR DEVICES

Adam, Stephen F (CA)
Anderson, Richard L (VT)
Ballantyne, Joseph M (NY)
Bloch, Erich (DC)
Brodsky, Marc H (NY)
Bucy, J Fred (TX)
Caulton, Martin (MD)
Dennard, Robert H (NY)
Dow, John D (IN)
Frey, Jeffrey (MD)
Garver, Robert V (MD)
Ghandhi, Sorab K (NY)
Hall, Robert N (NY)
Hess, Karl J (IL)
Hilibrand, Jack (NJ)
Hunsperger, Robert G (DE)
Kahng, Dawon (NJ)
Liechti, Charles A (CA)
Loferski, Joseph J (RI)
Misugi, Takahiko
Ning, Tak H (NY)
Nurmikko, Arto V (RI)
Peck, D Stewart (PA)
Pliskin, William A (NY)
Rabson, Thomas A (TX)
Salama, C Andre T
Senitzky, Benjamin (NY)
Statz, Hermann (MA)
Steckl, Andrew J (NY)
Streetman, Ben G (IL)
Stringfellow, Gerald B (UT)
Wang, Franklin F Y (NY)
Yee, Sinclair S (WA)

SHIPBUILDING

Alden, John D (NY)
Baker, William A (MA)
Cowles, Walter C (NJ)
Davis, Cabell S, Jr (VA)
French, Clarence L (CA)
Hassani, Jay J (MD)
Klinges, David H (PA)
Lee, Chang-Sup
McLaren, T Arthur
Moore, Robert J , Jr (CA)
Newell, John R (FL)
Palermo, Peter M (VA)
Rein, Robert J (VA)
Rosenblatt, Lester (NY)

SIMULATION

Clark, Gordon M (OH)
Duket, Steven D (IN)
Evans, David C (UT)
Gordon, Donald T (CO)
Hetrick, David L (AZ)
Kraus, Wayne P (OK)
Mastascusa, Edward J (PA)
Pritsker, A Alan B (IN)
Raghavan, Rajagopal (OK)
Rubin, Arthur I (NJ)
Rudavsky, Alexander B (CA)
Straubhar, John J (ID)
Wortman, David B (IN)

SOIL & WATER

SOIL & WATER

Aycock, Kenneth G (AL)
Berry, Bill E (ID)
Bondurant, Byron L (OH)
Branscomb, Lewis M (MA)
Bryniarski, Americ J (AR)
Fuhrman, Brian L (WA)
Humpherys, Allan S (ID)
Johnson, Howard P (IA)
Renard, Kenneth G (AZ)
Robinson, August R (MD)
Schottman, Robert W (GA)
Schrader, Gustav E (OH)
Wells, Larry G (KY)
Young, Jesse R (AR)

SOIL MECHANICS

Bauld, Nelson R , Jr (SC)
Brown, Ralph E (GA)
Cheney, James A (CA)
Cornforth, Derek H (OR)
Cox, Ernest A , III (AL)
Dames, Trent R (CA)
Forsyth, Raymond A (CA)
Foye, Robert, Jr (TX)
Gizienski, Stanley F (CA)
Hansen, Kenneth D (CO)
Holliday, Frank J (CO)
Howland, James C (OR)
Jorgenson, James L (ND)
Kalajian, Edward H (FL)
Kim, Jongsol (CA)
Leps, Thomas M (CA)
Lobo, Cecil T (IN)
Marshall, A Frank (TX)
McKittrick, David P (VA)
Mikochik, Stephen T (NY)
Perez, Jean-Yves (CO)
Phillips, Robert V (CA)
Romani, Fred (TX)
Rutledge, Philip C (PA)
Schuster, Robert L (CO)
Selig, Ernest T (MA)
Stern, Louis I (NJ)
Tschebotarioff, Gregory P
(NJ)
Wilkes, John S , III (CO)
Williams, Wayne W (KS)
Wu, Tien H (OH)
Zoino, William S (MA)

SOILS

Arden, Richard W (NV)
Arman, Ara (LA)
Hansen, Kenneth D (CO)
Lundgren, Raymond (CA)

SOLAR ENERGY

Albright, Louis D (NY)
Anderson, Philip P, Jr (IN)
Andrews, Robert V (TX)
Atchley, Bill L (SC)
Bayazitoglu, Yildiz (TX)
Brandt, Kent H (OH)
Bull, Stanley R (CO)
Burdick, Glenn A (FL)
Card, Howard C
Carlson, David E (PA)
Carney, Terrance M (TN)
Chant, Raymond E
Chou, James C S (HI)
Clark, Arthur J, Jr (NM)
Clark, James A (NY)
Clarkson, Clarence W (NJ)
Clausing, Arthur M (IL)
Cohen, Robert M (VA)
Cornog, Robert A (CA)
Crosby, Ralph (LA)
Daly, Charles F (NC)
Decker, Howard E (TX)
Dhanak, Amritlal M (MI)
Doughty, Eric R (ME)

Duffie, John A (WI)
Eibling, James A (OH)
Eldridge, Bernard G (CA)
Eltimsahy, Adel H (OH)
Evans, Robert J (VA)
Faust, J W, Jr (SC)
Felton, Kenneth E (MD)
Gianoli, Napoleon H
Glaser, Peter E (MA)
Gross, William A (NM)
Heinz, Winfield B (CA)
Helfman, Howard N (CA)
Hellickson, Mylo A (SD)
Holdredge, Russell M (UT)
Howe, Everett D (CA)
Hsieh, Jui Sheng (NJ)
Isenberg, Lionel (CA)
Jordan, Richard C (MN)
Kimball, William R (CO)
Kreider, Jan F (CO)
Kreith, Frank (CO)
Lampert, Seymour (CA)
Larson, Dennis L (AZ)
Lavan, Zalman (IL)
Levine, Jules D (TX)
Lior, Noam (PA)
Loferski, Joseph J (RI)
Look, Dwight C, Jr (MO)
Lowery, Gerald W (VA)
McNiel, James S (MA)
Morgan, Leland R (CA)
Navon, David H (MA)
Plunkett, Joseph C (CA)
Ramakumar, Ramachandra
G (OK)
Reed, Willard H (CA)
Reid, Robert L (WI)
Rice, William J (PA)
Richard, Oscar E (IL)
Rockwell, Glen E (CO)
Rogers, Benjamin T (NM)
Russell, T W Fraser (DE)
Safdari, Yahya B (IL)
Sanders, William T (NY)
Smetana, Frederick O (NC)
Szego, George C (VA)
Talaat, Mostafa E (MD)
Taylor, William E (CA)
Wood, Robert A (MI)
Wormser, Eric M (CT)
Yue, Alfred S (CA)

SOLID STATE

Barker, Richard C (CT)
Bergh, Arpad A (NJ)
Birnbaum, Howard K (IL)
Brophy, James J (UT)
Chung, Deborah D L (NY)
Chynoweth, Alan G (NJ)
Conwell, Esther M (NY)
Cotellessa, Robert F (NY)
Crum, Floyd M (TX)
Dow, John D (IN)
EerNisse, Errol P (UT)
Geller, Seymour (CO)
Glicksman, Maurice (RI)
Howard, William G , Jr
(AZ)
Huber, Robert J (UT)
Jordan, Angel G (PA)
Kahng, Seun K (OK)
Ko, Wen H (OH)
Marsocci, Velio A (NY)
McCurdy, Archie K (MA)
Meitzler, Allen H (MI)
Merz, James L (CA)
Miller, Gerald, R (UT)
Muller, Marcel W (MO)
Osgood, Richard M (NY)
Rabii, Sohrab (PA)
Royce, Barrie S H (NJ)
Schmitt, Roland W (NY)
Schulman, James H (MD)
Suran, Jerome J (CT)

Valasek, Joseph (MN)
Woodbury, Richard C (UT)
Yang, Edward S (NY)
Zabel, Hartmut (IL)

SOLID WASTE CONTROL

Alexandridis, George G
(NY)
Alvarez, Ronald J (NY)
Berman, Donald (PA)
Bowen, Carl H (IL)
Dedyo, John (NY)
Hagerty, D Joseph (KY)
Hogan, Brian R (MA)
Kent, James A (MI)
Meyers, Sheldon (MD)
Neubauer, Walter K (NY)
North, James C (TN)
Riddell, Matthew D R (IL)
Sacken, Donald K (TX)
Schloemann, Ernst F (MA)
Sieger, Ronald B (IN)
Smith, John W (TN)
Stratman, Frank H (IL)
Van Cott, Whitfield R (OH)
Woodruff, Kenneth L (NJ)

SPECTROMETRY

Condrate, Robert A (NY)
Miles, Richard B (NJ)

STANDARDS & TESTING

Altschuler, Helmut M (CO)
Astin, Allen V (MD)
Baldwin, Allen J (MN)
Bono, Jack A (IL)
Callahan, Harry L (MO)
Christopher, Phoebus M
(KY)
Clark, Donald E (WY)
Crane, L Stanley (PA)
Cutler, Verne C (WI)
Daniels, Orval R (KS)
Davis, Wilbur M (IA)
Dombeck, Harold A (NY)
Eddy, W Paul (CT)
Ely, Berten E (PA)
Fedorochko, John A (PA)
Fenn, Rutherford H (CT)
Foster, Lowell W (MN)
Gaynes, Chester S (IL)
Harris, Forest K (MD)
Heirman, Donald N (NJ)
Hsu, George C (VA)
Kelly, Thomas R (CA)
Knott, Albert W (CO)
Kornsand, Norman J (CA)
Loudon, Jack M (NY)
Magison, Ernest C (PA)
McAdams, William A (CT)
Mungall, Allan G
O'Grady, Joseph G (PA)
Parsons, Alonzo R (MN)
Pearson, Hugh S (GA)
Pringle, Arthur E (MD)
Rice, Richard L M (DE)
Sanders, Wallace W, Jr (IA)
Schapiro, Jerome B (NJ)
Scharp, Charles B (MD)
Siver, Dougal H (NY)
Snowden, Wesley D (WA)
Staples, Basil G (NY)
Thomas, Leonard W , Sr
(DC)
Trowbridge, Roy P (MI)
Tyson, Christopher G (DC)
Willard, Charles O (MN)

STEAM

Herzenberg, Aaron (FL)
Jeffries, Ronald F (VA)

Precious, Robert W (MI)
Read, Jay R (IA)

STEEL

Birks, Neil (PA)
Catanzano, Samuel (PA)
Emerick, Harold B (PA)
Gillen, Kenneth F (OH)
Kirkendall, Ernest (VA)
Porter, Lew F (PA)
Redline, John G (WV)
Smith, Halfred C (PA)
Woodall, David M (NM)

STRESS ANALYSIS

Austin, Walter J (TX)
Barnwell, Joseph H (LA)
Birnstiel, Charles (NY)
Boresi, Arthur P (IL)
Broome, Taft H (DC)
Campbell, Bobby D (MO)
Chiang, Fu-Pen (NY)
Costello, George A (IL)
Dohse, Fritz E
Gambrell, Samuel C, Jr
(AL)
Herrmann, Leonard R (CA)
Hetnarski, Richard B (NY)
Hudson, Ralph A (OR)
Mall, Shankar (OH)
McMahon, John (NY)
Patterson, J M (GA)
Pipes, R Byron (DE)
Richard, Terry G (OH)
Swartz, Stuart E (KS)
Szabo, Barna A (MO)
Wilhoit, James C , Jr (KY)
Zandman, Felix (PA)

STRUCTURAL

Chen, Ming M (MA)
Cheney, James A (CA)
Conn, Harry (IL)
Conover, John S (AK)
De Serio, James N (NY)
De Silva, Ananda S C (NY)
Eason, Glenn A (NC)
Fleming, John F (PA)
Gennaro, Joseph J (NJ)
Ghosh, Arvind (TX)
Humphrey, James E, Jr
(KY)
Kerkhoff, Harry Piehl van
den (SC)
Knudsen, Clarence V (PA)
Lotta, Joseph G (CA)
McGaughy, John B (VA)
Sandhu, Ranbir S (OH)
Spaulding, Westbrook H
(ME)
Videon, Fred F (MT)
Wang, Albert S (PA)
Weissenburger, Jason T
(MO)
Wilkinson, Eugene R (IL)
Williams, Richard L (VA)
Zak, Adam R (IL)

STRUCTURAL
ENGINEERING

Abel, Carl R (NC)
Adams, Robert R (OR)
Agbabian, Mihran S (CA)
Allen, Edmund W (UT)
Allen, James (AK)
Alpern, Milton (NY)
Amirikian, Arsham (MD)
Anand, Subhash C (SC)
Anderson, Clayton O (MN)
Anderson, John E (MN)
Anderson, Thor L (CA)

STRUCTURAL
ENGINEERING
(Continued)
Ang, Alfredo H S (IL)
Antoni, Charles M (NY)
Appleton, Joseph H (AL)
Armentrout, Daryl R (TN)
Armour, Charles R (GA)
Arnold, Orville E (WI)
Austin, Walter J (TX)
Backlund, Brandon H (NE)
Bacon, Louis A (GA)
Baker, John R (LA)
Baldwin, James W, Jr (MO)
Ball, Richard T (VA)
Barnes, Charles Ray (IA)
Barnes, Stephenson B (CA)
Baron, Melvin L (NY)
Barrett, Michael H (CO)
Bartlett, Donald L (IL)
Bates, Robert T (CO)
Bayer, David M (NC)
Beadling, David R (OR)
Beale, D Anthony (VA)
Beaufait, Fred W (MI)
Becker, Edward P (PA)
Beedle, Lynn S (PA)
Benioff, Ben (CA)
Benjamin, Jack R (CA)
Berrio, Mark M (IN)
Beskos, Dimitrios E
Bezzone, Albert P (CA)
Biddison, Cydnor M, Jr
    (CA)
Bierbach, Edward R (CO)
Biggs, David T (NY)
Billington, David P (NJ)
Billington, Ted F (KY)
Birkeland, Halvard W (WA)
Birnstiel, Charles (NY)
Blank, William F (IL)
Blaylock, Albert J (CA)
Blodgett, Omer W (OH)
Blum, George H (NY)
Blume, John A (CA)
Bobrowski, Jan J
Bodman, Gerald R (NE)
Boggs, Robert G (CT)
Bowles, Joseph E (IL)
Bradley, William A (MI)
Brandow, George E (CA)
Brebbia, Carlos A
Breen, John E (TX)
Bresler, Boris (CA)
Brill, Lawrence F (FL)
Britton, Myron (Ron) G
Brockenbrough, Thomas W
    (DE)
Brockway, George L (GA)
Broderick, James Richard
    (TX)
Brown, Elgar P (OH)
Brown, John Webster (NV)
Browning, David Lee (CA)
Burdette, Edwin G (TN)
Burkart, Matthew J (PA)
Burns, James F (NM)
Busek, Robert H (NJ)
Butler, Donald J (VT)
Calhoun, William D (PA)
Campbell, Bobby D (MO)
Campbell, Regis I (OH)
Campbell, Stephen J (TX)
Carcaterra, Thomas (MD)
Carreira, Domingo J (IL)
Cartelli, Vincent R (NY)
Carter, Nathan B, Jr (MS)
Caseman, Austin B (GA)
Caspe, Marc S (CA)
Cassaro, Michael A (KY)
Cavaliere, Alfonso M (CT)
Chalabi, A Fattah (MA)
Chastain, Theron Z (GA)
Chen, Peter W (NJ)
Chen, Shoei-Sheng (IL)

Chen, Tien Y (OH)
Chenault, Woodrow C, Jr
    (IL)
Cheney, Lloyd T (MI)
Cheng, David H (NY)
Cheng, Franklin Y (MO)
Chiu, Arthur N L (HI)
Chopra, Anil K (CA)
Clark, T Henry (TN)
Clemente, Frank M, Jr
    (NY)
Clough, Ray W (CA)
Cohen, Edward (NY)
Cole, Eugene E (CA)
Colville, James (MD)
Conderman, Charles W
    (NY)
Conway, William B (LA)
Coon, Arnold W (UT)
Cooper, Peter B (KS)
Corley, William G (IL)
Cornell, Holly A (OR)
Cornforth, Robert C (OK)
Corns, Charles F (VA)
Corotis, Ross B (MD)
Cost, Thomas L (AL)
Costello, George A (IL)
Covarrubias, Jesse S (TX)
Cox, J Carroll (SC)
Cravens, Dennis C (KY)
Crawford, Lewis C (KS)
Creed, Michael W (NC)
Criswell, Marvin E (CO)
Crom, Theodore R (FL)
Cross, Edward F (WA)
Cudmore, Russell D (GA)
Cutts, Charles E (MI)
Dalia, Frank J (LA)
D'Angelo, Vincent J (OH)
Daniels, John H (PA)
Darwin, David (KS)
Das, Sankar C (LA)
Davidson, William W (MI)
Dean, Donald L (FL)
Deardorff, Mark E (CA)
DeFalco, Frank D (MA)
Degenkolb, Henry J (CA)
DeLapp, Kenneth D (NM)
Depp, Oren Larry, Jr (KY)
Desai, Drupad B (MD)
Dexheimer, Wallace D (SD)
DiJulio, Roger M, Jr (CA)
Dilger, Walter H
Dill, Ellis H (NJ)
DiMaggio, Frank L (NY)
Disque, Robert O (IL)
Dominguez, Renan G (CA)
Dong, Stanley B (CA)
Donovan, Robert L (PA)
Dotson, Clinton (FL)
Douglas, Bruce M (NV)
Douty, Richard T (MO)
Dubin, Eugene A (IL)
DuBose, Lawrence A (IL)
Duchscherer, David C (NY)
Dumack, Ralph C (PA)
Dyer, Thomas K (MA)
Dym, Clive L (MA)
Ebeling, Dick W (OR)
Efimba, Robert E (DC)
Ekberg, Carl E, Jr (IA)
Eligator, Morton H (NY)
Ellingwood, Bruce R (MD)
Emanuel, Jack H (MO)
Emkin, Leroy Z (GA)
Epstein, Howard I (CT)
Errera, Samuel J (PA)
Ersoy, Ugur
Erzurumlu, H Chik M (OR)
Estes, Edward R Jr (VA)
Eubanks, Robert A (IL)
Evans, Thomas C, Jr (SC)
Everard, Noel J (TX)
Faherty, Keith F (CA)
Farmer, Larry E (TX)
Felton, Lewis P (CA)

Fertig, Marcel M (PA)
Figge, Kenneth L (IL)
Filer, William A
Fincher, James R (GA)
Finkel, Edward B (NJ)
Fintel, Mark (IL)
Firmage, D Allan (UT)
Fisher, John W (PA)
Fitzpatrick, Thomas C (OH)
Fling, Russell S (OH)
Foderberg, Dennis L (IL)
Fok, Thomas D (OH)
Folsom, Jack H (WA)
Foster, Edward T, Jr (NE)
Foster, Edwin P (TN)
Fox, Gerard F (NY)
Franceschi, Bruno J (WI)
Frankfurt, Daniel (NY)
Franklin, J Stuart, Jr (VA)
Fratessa, Paul F (CA)
Frazier, John W (KS)
Furlong, Richard W (TX)
Furr, Howard L (TX)
Gallagher, Richard H (MA)
Gallagher, Ronald P (CA)
Gamble, William L (IL)
Gangarao, Hota V S (WV)
Gant, Edward V (CT)
Gaylord, Edwin H (IL)
Geiger, David H (NY)
Geisser, Russell F (RI)
Genin, Joseph (NM)
Gerber, George C (DC)
Gere, James M (CA)
Gergely, Peter (NY)
Gerstle, Kurt H (CO)
Gesund, Hans (KY)
Ghaboussi, Jamshid (IL)
Gibble, Kenneth (CT)
Gillum, Jack D (CO)
Glenn, Joe Davis, Jr (VA)
Go, Mateo L P (MI)
Gogate, Anand B (OH)
Goldberg, John E (IN)
Goldreich, Joseph D (NY)
Gomez, Rod J (AZ)
Gooden, Charles D (TX)
Goodman, James R (CO)
Goodman, Lawrence E
    (MN)
Goodno, Barry J (GA)
Gordon, Ruth V (CA)
Gosnell, Gary J (CO)
Gould, Phillip L (MO)
Graham, Harry T (OH)
Green, Robert H (GA)
Greene, Leroy F (CA)
Greve, Norman R (CA)
Gribben, J Clark (NV)
Gunnin, Bill L (TX)
Gurfinkel, German R (IL)
Haber, Bernard (NY)
Haddadin, Munther J
Hahn, Ralph C (IL)
Haisler, Walter E (TX)
Hale, Clyde S (MI)
Hall, Francis E (MS)
Hamada, Harold S (HI)
Hamilton, Wayne A (ME)
Hansen, Robert J (MA)
Hanson, John M (IL)
Hanson, Robert D (MI)
Hargis, John C (OK)
Harrington, Burnett W, Jr
    (GA)
Harris, Aubrey L, III (AR)
Harris, Michael E (AZ)
Harstead, Gunnar A (NJ)
Harvey, Francis S (MA)
Hauck, George F W (MO)
Hawkins, Neil M (WA)
Hayes, John M (IN)
Head, Julian E (SC)
Heins, Conrad P (MD)
Hensley, Floyd E (OK)
Herrmann, Leonard R (CA)

Heyer, Edwin F (FL)
Hirsch, Ephraim G (CA)
Hoadley, Peter G (TN)
Hodges, William J (SC)
Hoefle, Ronald A (IL)
Hoffman, Nathan N (TX)
Holland, Eugene P (IL)
Hollrah, Ronald L (MO)
Holt, Randolph E (NM)
Hong, Franklin L (NY)
Hsu, Thomas T C (TX)
Huckelbridge, Arthur A
    (OH)
Hughen, James W (MD)
Hughes, Joseph Brian (HI)
Hulsey, J Leroy (NC)
Hurd, Lewis R (GA)
Hurley, Daniel J (NY)
Hurt, Ronald L (KY)
Hust, James L (MS)
Hutchinson, James R (CA)
Hyman, David S (MD)
Inaba, Yoshio (HI)
Ives, Raymond H, Jr (VA)
Jack, Sanford B, Jr (TN)
Jain, Kris K (FL)
James, Arthur M (OR)
Janney, Jack R (IL)
Jobusch, Wallace E (PA)
Johnson, J Wallace (VA)
Johnston, Bruce G (AZ)
Johnston, Roy G (CA)
Jones, Phillip R (FL)
Jones, Richard C (MI)
Jordan, Frederick E (CA)
Jordan, Michael A (CA)
Jorgensen, Ib Falk (CO)
Jorgenson, James L (ND)
Jumper, Billy S (TX)
Kahn, Lawrence F (GA)
Kain, Charles F (PA)
Kamel, Hussein A (AZ)
Kane, Harrison (IA)
Kangas, Ralph A (ID)
Kapila, Ved P (MI)
Kareem, Ahsan (TX)
Karren, Kenneth W (UT)
Katz, Erich A (NY)
Kealey, T Robert (PA)
Keith, Richard L (CA)
Kelly, Keith A (MD)
Ketchum, Milo S (CT)
Khachaturlan, Narbey (IL)
Khan, Mohammad S (FL)
Kienzle, Gary M (MI)
Killebrew, James R (TX)
Kim, Jai B (PA)
Kinyon, Gerald E (IN)
Klaiber, Wayne F (IA)
Kline, Donald H (NC)
Klippstein, Karl H (PA)
Koo, Benjamin (OH)
Kopetz, Marion J (MO)
Korda, Peter E (OH)
Krahl, Nat W (TX)
Krawinkler, Helmut (CA)
Kriesel, William G (MN)
Krishnamurthy, Natarajan
    (AL)
Kwok, Chin-Fun (VA)
Lacy, Robert R (MT)
Lane, Charles E III (AL)
Lane, Golden E, Jr (NM)
LaRochelle, Donald R (ME)
Lavens, Edmond V (OH)
Lawyer, Donald H (PA)
Leavitt, Sheldon J (VA)
Lee, Edgar K M (HI)
Lee, George C (NY)
Lee, Ju P (MI)
Lehnhoff, Terry F (MO)
LeMessurier, William James
    (MA)
Lepore, John A (PA)
Levy, Matthys P (NY)
Libby, James R (CA)

STRUCTURES

**STRUCTURAL
ENGINEERING**
(Continued)
Libove, Charles (NY)
Lin, Cheng S (CA)
Lin, Tung Yen (CA)
Linders, Howard D (MI)
Lindsay, Philip J, Sr (WA)
Liu, Tony C (VA)
Lo, Donald T (HI)
Lobo, Cecil T (IN)
Lohtia, Rajinder Paul
Long, Adrian E
Lu, Le-Wu (PA)
Luck, Leon D (WA)
Luenzmann, David I (WI)
Lutes, Loren D (TX)
Macchi, Anselmo J (CT)
MacGregor, James G
Mack, Edward S (AK)
Mahan, William E (WA)
Mains, Robert M (MO)
Mandel, Herbert M (PA)
Mandel, James A (NY)
Mangelsdorf, Clark P (PA)
Maniktala, Rajindar K
    (NY)
Mannik, Jaan (OH)
Marchinski, Leonard J (PA)
Mark, Robert (NJ)
Marquis, Eugene L (TX)
Marr, Richard A (IA)
Martin, Ignacio (PR)
Martin, Leslie D (IL)
Marx, Christopher W (CT)
Maslan, Leon (MO)
Matsumoto, Michael P (HI)
Mattock, Alan H (WA)
Maurer, David L (IL)
Maxwell, Clyde V (MS)
McBean, Robert P (MO)
McCalla, William T (MN)
McClellan, Thomas J (OR)
McCormac, Jack C (SC)
McDonald, Donald (TX)
McDonald, Jerry L (AL)
McGarraugh, Jay B (MO)
McGee, William D (ID)
McKim, Herbert P , Jr
    (NC)
McKittrick, David P (VA)
McManus, Ralph N
McNamee, Bernard M (PA)
Medwadowski, Stefan J
    (CA)
Meedel, Virgil G (NE)
Mehta, Prakash C (IN)
Mickey, Forrest R (CO)
Miedtke, Duane R (MN)
Militello, Sam (FL)
Millard, Charles F (MD)
Miller, Jack C (IA)
Minich, Marlin D (OH)
Minor, Joseph E (TX)
Miro, Sami A (CO)
Mok, Perry K P (MD)
Monical, R Duane (IN)
Moore, William E , II (WV)
Morgenthaler, Charles S
    (MD)
Morris, Robert L (CA)
Mostaghel, Naser (UT)
Moustafa, Saad E (LA)
Mudd, Charles B (MD)
Mulcahy, Joseph F (RI)
Munse, William H (IL)
Murashige, James Y (CA)
Murray, Thomas M (VA)
Murtha-Smith, Erling (CT)
Muvdi, Bichara B (IL)
Nadolski, Leon (CA)
Nannis, Walid A (GA)
Nawab, Ahmad B (FL)
Neal, Donald W (OR)
Nelson, George A (GA)

Nelson, Richard B (CA)
Nelson, Sherman A (ID)
Newhof, Paul W (MI)
Nicoletti, Joseph P (CA)
Nilson, Arthur H (NY)
Nishkian, Byron L (CA)
Norman, Joseph H, Jr (VA)
Nuccitelli, Saul A (MO)
Oishi, Satoshi (NJ)
Ojalvo, Morris (OH)
Olson, John W (NY)
Orr, Charles K (TX)
Paavola, Ivar R (MD)
Pai, David H (NJ)
Paik, Young J (CA)
Painter, Jack T (LA)
Palazotto, Anthony N (OH)
Palermo, Peter M (VA)
Palmer, Roy G (CA)
Paulson, James M (MI)
Pegram, Anthony R (CA)
Pekau, Oscar A
Penzien, Joseph (CA)
Perrigin, John G (AL)
Peters, Robert C (MO)
Peterson, Enoch W F (FL)
Peterson, Thorwald R (MO)
Peterson, William E (WA)
Phillips, Aris (CT)
Pierce, Louis F (OR)
Pincus, George (TX)
Pinkham, Clarkson W (CA)
Pinnell, Ray A, Jr (TX)
Pisetzner, Emanuel (NY)
Plewes, W Gordon
Plumtree, William G (CA)
Prickette, Gerald S (TX)
Pridgeon, Hal L , Jr (SC)
Przygoda, Zdzislaw
Pyle, Don T (CO)
Quandel, Charles H (PA)
Radziminski, James B (SC)
Ragold, Richard E (NJ)
Randall, Frank A, Jr (IL)
Rangaswamy, Thangamuthu
    (KY)
Reaveley, Ronald J (UT)
Reed, Stan C (TX)
Reeves, James W (LA)
Reiffman, Norman L (NY)
Reimer, Paul H, Jr (PA)
Revesz, Zsolt
Rey, William K (AL)
Rhodes, Eugene I (NY)
Rich, Elliot (UT)
Rinne, John E (CA)
Ritter, Guy F (GA)
Rivera, Alfredo J (NJ)
Roberts, Sanford B (CA)
Robertson, Leslie E (NY)
Robison, Ralph E (OH)
Roby, Dennis E (IL)
Roeder, Charles W (WA)
Rogers, Bruce G (TX)
Rogers, Darrell O (MI)
Rogers, H Daniel, Jr (NY)
Rolf, Richard L (PA)
Roll, Frederic (PA)
Rosenblueth, Emilio
Rouse, George C (CO)
Roussel, Herbert J, Jr (LA)
Sabnis, Gajanan M (MD)
Saleh, Fikri S (NC)
Salvadori, Mario G (NY)
Samborn, Alfred H (OH)
Sandberg, Harold R (IL)
Sanders, Wallace W, Jr (IA)
Sanghera, Gurbaksh S (MA)
Sarapu, Felix R (HI)
Saul, William E (ID)
Saunders, Walter D (CA)
Saxe, Harry C (KY)
Schlimm, Gerard H (MD)
Schmidt, Albert J (NY)
Schmit, Lucien A (CA)
Schmitt, Frederick C (NY)

Schnobrich, William C (IL)
Schwan, Anthony V (AZ)
Scipio, L Albert, II (MD)
Scordelis, Alexander C (CA)
Scott, Norman L (IL)
Seeley, Gerald R (IN)
Segner, Edmund P, Jr (TN)
Seide, Paul (CA)
Seidel, Edmund O (TX)
Seinuk, Ysrael A (NY)
Sevin, Eugene (MD)
Shaffer, Harold S (IL)
Shah, Haresh C (CA)
Shapiro, Howard I (NY)
Sharpe, Roland L (CA)
Sheffet, Joseph (CA)
Sheikh, Shamim A (TX)
Shell, William O (IA)
Shepherd, R (CA)
Shimazu, Satoshi D (HI)
Shina, Isaac S (PA)
Shuck, Terry A (IA)
Siess, Chester P (IL)
Silcock, Frank A , P E (TX)
Simitses, George J (GA)
Sims, James R (TX)
Smith, Daniel J, Jr (CA)
Smith, James A (CO)
Smyers, Daniel J (PA)
Snell, Robert R (KS)
Snell, Virgil H (MO)
Snyder, Julian (NY)
Solien, Vernon L (ND)
Soteriades, Michael C (DC)
Soto, Marcello H (PA)
Spetnagel, Theodore J (GA)
Spillers, William R (NY)
Spina, Frank J (NY)
Sporseen, Stanley E (OR)
Staker, Rodd D (MO)
Stallmeyer, James E (IL)
Steadman, H Douglas (TX)
Steinbrugge, John M (CA)
Strand, Donald R (CA)
Strawbridge, Randall A
    (VA)
Stroud, Roger G (VA)
Stude, Robert A (MO)
Summerlin, James C (AR)
Sutherland, Donald C (CO)
Swartz, Stuart E (KS)
Tadjbakhsh, Iradj G (NY)
Tahiliani, Jamnu H (TN)
Taracido, Manuel E (FL)
Taylor, Howard P J
Taylor, Michael A (CA)
Tegtmeyer, Rene D (VA)
Thomas, Carcaterra (MD)
Thompson, Walter Dale
    (NC)
Thorn, Robert B (KS)
Thornton, Keith C (NJ)
Thune, Arne (CT)
Thurlimann, Bruno
Thurman, Herman R, Jr
    (DE)
Tillitt, James C (MN)
Tor, Abba A (NY)
Toro, Jorge A (PR)
Tschebotarioff, Gregory P
    (NJ)
Tung, Chi C (NC)
Tung, David H H (NY)
Underwood, Wyatt R (NM)
Utku, Senol (NC)
Vanderbilt, M Daniel (CO)
vanLangendonck, Telemaco
Vasatka, Richard (MN)
Vaughan, Richard G (NM)
Vaughn, Kenneth J (MO)
Veletsos, Anestis S (TX)
Vicory, Freeman M (KS)
Viscomi, B Vincent (PA)
Voigt, Robert C (KS)
von Riesemann, Walter A
    (NM)

Wade, Robert G (MO)
Wagoner, Mike N (NC)
Walker, H Carl (MI)
Walker, Samuel R, Jr (TX)
Walker, William Hamilton
    (IL)
Wang, Leon R L (VA)
Wang, Ping-chun (NY)
Wang, Shien T (KY)
Warner, Lawrence D (MA)
Wasil, Benjamin A (NY)
Waters, Alfred E (CA)
Watkins, Henry H (SC)
Watwood, Vernon B (MI)
Weaver, Wayne L (MA)
Weidlinger, Paul (NY)
Wells, Thomas W (LA)
Wen, Robert K (MI)
Wen, Yi-Kwei (IL)
Wesler, Philip (CT)
West, Harry H (PA)
Westlake, Leighton D , Jr
    (FL)
Westmann, Russell A (CA)
Wheeler, Orville E (TN)
Wheeler, William T (CA)
White, Richard N (NY)
Widrig, Francis F (MI)
Wiesenfeld, Joel (NJ)
Wildman, Robert F (CA)
Wilhelm, William J (KS)
Wilkes, W Jack (DC)
Wilkinson, Eugene R (IL)
Wille, Jerry L (IA)
Williams, Albert T Wadi
    (VA)
Willis, Shelby K (KS)
Wilson, Arnold (UT)
Wilson, Edward L (CA)
Wilson, Garland (CO)
Winer, Loyd E (MI)
Wing, Wayman C (NY)
Winter, George (NY)
Wise, William F (VA)
Wisniewski, John S (PA)
Witte, Charles R (NY)
Wittes, David R (PA)
Woodring, Richard E (PA)
Woods, John O, Jr (VA)
Wright, Douglas T
Wylie, Frank B , Jr (KY)
Wyllie, Loring A , Jr (CA)
Yachnis, Michael (DC)
Yang, Cheng Yi (DE)
Yee, Alfred A (HI)
Yerlici, Vedat A
Yinh, Juan A
Yoder, Charles W (WI)
Yohalem, Martin J (FL)
Yoo, Chai Hong (WI)
Young, James T (OH)
Yu, Wei W (MO)
Zagars, Arvids (IL)
Zar, Max (IL)
Zia, Paul Z (NC)
Zwoyer, Eugene M (CA)

STRUCTURES

Bailey, Cecil D (OH)
Bauld, Nelson R , Jr (SC)
Breyer, Donald E (CA)
Chesson, Eugene (DE)
Colosimo, Joseph L (NY)
Dexheimer, Wallace D (SD)
Driggers, L Bynum (NC)
Gatewood, Buford E (OH)
Goetschel, Daniel B (CA)
Harrington, John M (NC)
Hulbert, Lewis E (OH)
Kingsbury, Herbert B (DE)
Kuzmanovic, Bogdan O
    (KS)
Martin, John B
Milbradt, Kenneth P (IL)

TUNNELING & EXCAVATION

WATER RESOURCES & POLLUTION CONTROL
(Continued)

Eden, Edwin W, Jr (FL)
Edgar, C E , III (MA)
Eisenbach, Robert L (LA)
Engelbrecht, Eugene W (NJ)
Engelbrecht, Richard S (IL)
Erickson, Claud R (MI)
Esvelt, Larry A (WA)
Evans, James L (IA)
Ewing, Benjamin B (IL)
Ewing, William C (MI)
Fagerlund, N David (NJ)
Ferguson, George E (VA)
Field, Richard (NJ)
Fifield, Charles D (MI)
Fischer, Peter A (MN)
Fisher, David L (TX)
Fisher, Lawrence R, Jr (OK)
Fleming, William H , Jr (NJ)
Foerster, Richard E (NY)
Ford, Maurice E, Jr (CA)
Foree, Edward G (KY)
Fouladpour, K Danny (ID)
Franques, J Thomas (IL)
Fredrich, Augustine J (IN)
Fridy, Thomas A , Jr (SC)
Fuhriman, Dean K (UT)
Gaum, Carl H (MD)
Gaynor, Lester (MA)
Germain, James E (PA)
Gill, William H (CA)
Goldstein, Paul (PA)
Golze, Alfred R (CA)
Gonzales, John G M (NV)
Goodman, Alvin S (NY)
Goulding, Randolph (GA)
Graeser, Henry J (TX)
Graham, Malise J (MO)
Granstrom, Marvin L (NJ)
Grout, Harold P (CO)
Haan, Charles T (OK)
Haestad, Roald J (CT)
Hagen, Vernon K (DC)
Haimes, Yacov Y (VA)
Harchar, Joseph J (PA)
Harleman, Donald R F (MA)
Harris, Weldon C (PA)
Harrison, Carter H , Jr (OR)
Harstad, Andrew E (WA)
Hash, Lester J (SD)
Hazen, Richard (NY)
Heiple, Loren R (AR)
Heiser, Will M (PA)
Helweg, Otto J (CA)
Henderson, Angus D (NY)
Higginson, R Keith (UT)
Hildyard, Benjamin G (CA)
Hira, Gulab G (MA)
Hirata, Edward Y (HI)
Hoffmann, Michael R (CA)
Holland, Joe E, Jr (TN)
Holmes, William D (ME)
Hoppel, Susan K (NE)
Horst, Neal A (PA)
Howson, Louis R (IL)
Huang, Chin-Pao (DE)
Huang, Ju-Chang (Howard) (MO)
Hughto, Richard J (MA)
Humenik, Frank J (NC)
Hustead, Dennis D (AZ)
Jacobson, E Paul (ID)
Jaske, Robert T (MD)
Jennings, Marshall E (MS)
Jensen, Eugene T (KS)
Jensen, Marvin E (CO)
Jewell, Thomas K (NY)
Johnson, Stanley E (OH)
Johnson, Walter K (MN)

Jones, Everett Bruce (CO)
Karasiewicz, Walter R (SC)
Katz, William E (MA)
Katz, William J (WI)
Kauffman, Kenneth O (CO)
Kaufman, Herbert L (NJ)
Kelley, Richard B (IL)
Kennedy, David D (CA)
Kennedy, Robert M (CA)
Kimberling, Charles L (OK)
Kipp, Raymond J (WI)
Klingensmith, Russell S (PA)
Knoedler, Elmer L (MD)
Koelliker, James K (KS)
Koelzer, Victor A (CO)
Koenig, Charles Louis (TX)
Korte, George B , Jr (VA)
Krishna, Gopal T K (IA)
Kruse, Cameron G (MN)
Kuranz, Joseph H (WI)
Labadie, John W (CO)
Lager, John A (CA)
Lambert, Jerry R (SC)
Lang, Martin (NY)
Lanouette, Kenneth H (CT)
Lanyon, Richard F (IL)
Larios, Christus J (NY)
Laronge, Thomas M (WA)
Lavens, Edmond V (OH)
Laverty, Finley B (CA)
Lawrance, Charles H (CA)
Layne, Jack E (CO)
Li, Ruh-Ming (CO)
Li, Wen-Hsiung (NY)
Lick, Wilbert J (CA)
Lin, Ping-Wha (IN)
Lindahl, Harry V (TN)
Lischer, Vance C (MO)
Loucks, Daniel P (NY)
Lubetkin, Seymour A (FL)
Lunardini, Robert C (MS)
Madden, Dann M (OR)
Manges, Harry L (KS)
Marks, David H (MA)
Martenson, Dennis R (MN)
Matera, James J (MA)
Mathe, Robert E (CO)
Mathur, Umesh (OK)
Mattei, Peter F (MO)
Mattern, Donald H (TN)
McCallister, Philip (MI)
McClelland, James E (AR)
McClure, Andrew F, Jr (PA)
McCullough, Charles A (CA)
McDonnell, Archie J (PA)
McGinnis, Charles I (MO)
McGuire, John P (NY)
McKinney, Leon E, (MO)
McMahon, Thomas E (KS)
McMasters, Jesse L (OK)
McMillan, Hugh H, (IL)
Meredith, Dale D (NY)
Mielke, William J (WI)
Mitchell, J Kent (IL)
Moak, Charles E (MS)
Moffitt, David C (OR)
Monke, Edwin J (IN)
Moore, Robert C (NJ)
Morris, C Robert (DC)
Motayed, Asok K (MD)
Moyer, Harlan E (OR)
Munger, Paul R (MO)
Munson, James I, Jr (FL)
Myrick, H Nugent (TX)
Neubauer, Walter K (NY)
Neufeld, Ronald D (PA)
North, James C (TN)
O'Meara, Robert G (MO)
O'Rourke, Robert B (OH)
Paavola, Ivar R (MD)
Paily, Paily (Poothrikka) P (GA)
Parker, Robin M (TX)

Parker, William H (VA)
Paulette, Robert G (IA)
Pavia, Edgar H (LA)
Pavia, Richard A (IL)
Peck, Eugene L (VA)
Peterson, Dean F (UT)
Pettigrew, Allan
Piest, Robert F (MO)
Pillsbury, Arthur F (CA)
Pitt, William A J, Jr (FL)
Plate, Erich J
Presecan, Nicholas L (CA)
Price, Bobby E (LA)
Ragan, Robert M (MD)
Reh, Carl W (IL)
Rice, John F (NC)
Ricigliano, Anthony R (NJ)
Riddell, Matthew D R (IL)
Riehl, Arthur M (KY)
Robinson, Lee (ID)
Rowland, Walter F (CA)
Rubin, Alan J (OH)
Russell, Samuel O
Sandza, Joseph G (PR)
Sasman, Robert T (IL)
Saunders, F Michael (GA)
Schad, Theodore M (VA)
Schneider, William J (VA)
Seaburn, Gerald E (FL)
Shah, Kanti L (OH)
Shaler, Amos J (PA)
Shea, Timothy G (VA)
Sheer, Daniel P (MD)
Sherwood, Richard F (UT)
Shoopman, Thomas A (IL)
Shrivastava, S Ram (NY)
Signs, Cheryl L (CO)
Silberman, Edward (MN)
Simpson, George D (OH)
Singer, Philip C (NC)
Singh, Rameshwar (CA)
Singhal, Ashok K (MI)
Skinner, Willis D (MA)
Small, Mitchell J (MI)
Smith, Robert L (KS)
Smith, Verne E (WY)
Sorber, Charles A (PA)
Spatz, D Dean (MN)
Spear, Clyde C (CA)
Spear, Jo-Walter (IL)
Stack, Vernon T (PA)
Staff, Charles E (NJ)
Stefan, Heinz G (MN)
Steimle, Stephen E (LA)
Stephens, John C (FL)
Stevenson, Albert H (CT)
Stout, Glenn E (IL)
Stratton, Frank E (CA)
Strickland, James M, Jr (VA)
Stuever, Joseph H (OK)
Sullins, John K (TN)
Sumner, Billy T (TN)
Taflin, Charles O (MN)
Talian, Stephen F, (PA)
Tallmadge, John A, Jr (PA)
Thackston, Edward L (TN)
Thomas, Cecil A (ID)
Thompson, Charles B (AZ)
Toebes, Gerrit H (IN)
Trelease, Frank J , III (WY)
Tucker, L Scott (CO)
Tucker, Richard C (MD)
Tyer, Arnold J (AR)
Ulp, Richard B (PA)
Van Domelen, John F (VT)
Vanderboom, Steve A (MN)
Vermeulen, Theodore (CA)
Verzuh, James M (MT)
Vicens, Guillermo J (MA)
Vondrick, Art F (AZ)
Walsh, John M, III (NJ)
Walter, C Richard (NY)
Ward, Robert C (CO)
Watson, Kenneth S (IL)
Weaver, Leo (OH)

Weber, Walter J, Jr (MI)
Weinert, Donald G (DC)
Welsh, James N (TX)
Wenck, Norman C (MN)
Wenzel, Harry G, Jr (IL)
Wetstein, Willis J (AZ)
Wheler, Arthur Gordon (NY)
Wibberley, Harold E (MD)
Widmer, Wilbur J (CT)
Wiesenfeld, Joel (NJ)
Wilder, Carl R (CA)
Willey, Benjamin F (IL)
Willis, Charles A (NC)
Winzler, John R (CA)
Wold, Peter B (ND)
Wright, Kenneth R (CO)
Yeh, William (CA)
Yevjevich, Vujica (VA)
Young, Carlos A (WA)
Young, Jesse R (AR)
Yousef, Yousef A (FL)
Zeff, Jack D (CA)
Zielinski, Paul B (SC)
Zitomer, Kenneth J (PA)

WATER TREATMENT

Berger, Bernard B (MA)
Bishop, Stephen L (MA)
Browning, Robert C (NC)
Carpenter, Carl H (UT)
Culver, Everett E (TN)
Harrington, John J (PA)
Karasiewicz, Walter R (SC)
Katz, William E (MA)
Keller, Jack W (CO)
Kennedy, David D (CA)
Rice, James K (MD)
Singer, Philip C (NC)
Snoeyink, Vernon L (IL)
Spiegler, Kurt S (CA)
Stratton, Frank E (CA)
Van Cott, Whitfield R (OH)
Walsh, John M, III (NJ)
Zederbaum, Robert B (NJ)
Zimmie, William E (OH)

WAVE PROPAGATION

Bertoni, Henry L (NY)
Brock, Louis M (KY)
Chu, Ta-Shing (NJ)
Datta, Subhendu K (CO)
Deresiewicz, Herbert (NY)
Douglas, Robert A (NC)
Godleski, Edward S (OH)
Gross, Stanley H (NY)
Horie, Yasuyuki (NC)
Hunsucker, Robert D (AK)
Keiser, Bernhard E (VA)
Knop, Charles M (IN)
Kritikos, Haralambos N (PA)
Levis, Curt A (OH)
Lubinski, Arthur (OK)
Meyers, Marc A (NM)
Riley, William F (IA)

WEAPONS

Badertscher, Robert F (OH)
Jacobs, Donald H
Pei Chi Chou (PA)
Remer, Bertram R (MD)

WEIGHT & CONFIGURATION CONTROL

Anderson, Robert (OH)
Dalton, Robert L (KS)
Damiani, A S (MD)
Duffey, Loren A (CA)
Hollenbeck, Leslie G (MA)